2012 49th ACM/EDAC/IEEE Design Automation Conference

(DAC 2012)

San Francisco, California, USA
3-7 June 2012

Pages 1-677

IEEE Catalog Number:	CFP12DAC-PRT
ISBN:	978-1-4503-1199-1

Copyright © 2012, Association for Computing Machinery, Inc
All Rights Reserved

***This publication is a representation of what appears in the IEEE
Digital Libraries. Some format issues inherent in the e-media version may
also appear in this print version.*

IEEE Catalog Number: CFP12DAC-PRT
ISBN 13:   978-1-4503-1199-1
ISSN: 0738-100X

Additional Copies of This Publication Are Available From:

Curran Associates, Inc
57 Morehouse Lane
Red Hook, NY 12571 USA
Phone: (845) 758-0400
Fax: (845) 758-2633
E-mail: curran@proceedings.com
Web: www.proceedings.com

2012 49th ACM/EDAC/IEEE Design Automation Conference (DAC 2012)

San Francisco, California, USA
3-7 June 2012

IEEE Catalog Number: CFP12DAC-POD
ISBN: 978-1-45031-199-1

2012 Table of Contents

General Chair's Message	i
Proceedings of the 49th Automation Conference®	iii
Committees	iv
Executive Committee	iv
Technical Program Committee	vi
Panel Committees	ix
Industry Liaison Committee	x
Strategy Committee	xi
User Track Committee	xii
PR/Marketing Committee	xiii
Best Paper Award Committee	xiv
Special Session Organizers	xiv
Technical Panel Organizers	xiv
Pavilion Panel Contributors	xv
Tuesday Keynote Address	xvi
Wednesday Keynote Address	xvi
Thursday Keynote	xvii
Perspective Paper Abstracts	xviii
Technical Panel Abstracts	xix
Awards	xxii
Marie R. Pistilli Women in EDA Achievement Award	xxii
The P. O. Pistilli Undergraduate Scholarships for Advancement in Computer Science and Electrical Engineering	xxii
A. Richard Newton Graduate Scholarships	xxii
ACM/IEEE A. Richard Newton Technical Impact Award in Electronic Design Automation	xxii
2011 Phil Kaufman Award for Distinguished Contributions to EDA	xxii
IEEE CEDA Outstanding Service Contribution	xxii
Donald O. Pederson Best Paper Award for the IEEE Transaction on CAD	xxiii
SIGDA Outstanding New Faculty Award	xxiii
ACM/SIGDA Outstanding Ph.D. Dissertation Award	xxiii
IEEE Fellow	xxiii
IEEE Fellow	xxiii
IEEE Fellow	xxiii
49th DAC Best Paper Candidates	xxiii
Reviewers	xxv
Author Index	

2012 Papers

Session 2: E-Health: A Killer Application for Electronic Devices?

Chair: Rajesh Gupta (*Univ. of California at San Diego*)

Biomedical Electronics Serving as Physical Environmental and Emotional Watchdogs

2.1   Rudy Lauwereins (*IMEC*)   1

2.2   **Integrated Biosensors for Personalized Medicine**   6

Giovanni De Micheli, Cristina Boero, Camilla Baj-Rossi, Irene Taurino, Sandro Carrara (*Ecole Polytechnique Fédérale de Lausanne*)

2.3   **Design Challenges for Secure Implantable Medical Devices**   12

Shane Clark, Ben Ransford (*Univ. of Massachusetts, Amherst*); Wayne Burleson, Kevin Fu (*Univ. of Massachusetts, Amherst*)

Session 3: Design Automation for Things Wet, Small, Spooky, and Tamable

Chair: Tsung-Yi Ho (*National Cheng Kung Univ.*)

3.1   **Design of Pin-Constrained General-Purpose Digital Microfluidic Biochips**   18

Yan Luo, Krishnendu Chakrabarty (*Duke Univ.*)

3.2   **Path Scheduling on Digital Microfluidic Biochips**   26

Daniel Grissom, Philip Brisk (*Univ. of California, Riverside*)

3.3   **Realizing Reversible Circuits Using a New Class of Quantum Gates**   36

Robert Wille (*Univ. of Bremen*); D. Michael Miller, Zahra Sasanian (*Univ. of Victoria*)

3.4   **Physical Synthesis onto a Sea-of-Tiles with Double-Gate Silicon Nanowire Transistors**   42

Shashikanth Bobba, Michele De Marchi, Yusuf Leblebici, Giovanni De Micheli (*Ecole Polytechnique Fédérale de Lausanne*)

Session 4: Be Efficient: Low-Power Design Techniques

Chair: Hamid Mahmoodi (*San Francisco State Univ.*)

4.1   **A Semiempirical Model for Wakeup Time Estimation in Power-Gated Logic Clusters**   48

Vivek D. Tovinakere, Olivier Sentieys, Steven Derrien (*Univ. de Rennes 1*)

4.2   **Cost-Effective Power Delivery to Support Per-Core Voltage Domains for Power-Constrained Processors**   56

Michael J. Schulte (*Advanced Micro Devices, Inc.*); Abhishek A. Sinkar, Hamid Reza Ghasemi, Nam Sung Kim (*Univ. of Wisconsin, Madison*)

4.3   **A Hybrid and Adaptive Model for Predicting Register File and SRAM Power Using a Reference Design**   62

Eric Donkoh, Alicia Lowery, Emily Shriver (*Intel Corp.*)

4.4 Coding-Based Energy Minimization for Phase Change Memory   68

Azalia Mirhoseini (*Rice Univ.*); Miodrag Potkonjak (*Univ. of California, Los Angeles*); Farinaz Koushanfar (*Rice Univ.*)

Session 5: Design and Data Security: Is It Even Possible?

Chair: Mohammad Tehranipoor (*Univ. of Connecticut*)

5.1 A Code Morphing Methodology to Automate Power Analysis Countermeasures   77

Giovanni Agosta, Alessandro Barenghi, Gerardo Pelosi (*Politecnico di Milano*)

5.2 Security Analysis of Logic Obfuscation   83

Youngok Pino (*Air Force Research Lab*); Ozgur Sinanoglu (*New York Univ.*); Jeyavijayan Rajendran, Ramesh Karri (*Polytechnic Institute of New York Univ.*)

5.3 Hardware Trojan Horse Benchmark via Optimal Creation and Placement of Malicious Circuitry   90

Sheng Wei (*Univ. of California, Los Angeles*); Kai Li, Farinaz Koushanfar (*Rice Univ.*); Miodrag Potkonjak (*Univ. of California, Los Angeles*)

5.4 On Improving the Uniqueness of Silicon-Based Physically Unclonable Functions via Optical Proximity Correction   96

Domenic Forte, Ankur Srivastava (*Univ. of Maryland*)

Session 6: System Simulation: The Need for Speed!

Chair: Gunar Schirner (*Northeastern Univ.*)

6.1 Transformer: A Functional-Driven Cycle-Accurate Multicore Simulator   106

Qinghao Min, Weihua Zhang, Binyu Zang (*Fudan Univ.*); Jian Li (*IBM Corp.*); Haibo Chen (*Shanghai Jiao Tong Univ.*); Zhenman Fang, Keyong Zhou, Yi Lu, Yibin Hu (*Fudan Univ.*)

6.2 SAGA: SystemC Acceleration on GPU Architectures   115

Sara Vinco (*Univ. of Verona*); Debapriya Chatterjee, Valeria Bertacco (*Univ. of Michigan*); Franco Fummi (*Univ. of Verona*)

6.3 Synchronization for Hybrid MPSoC Full-System Simulation   121

Juan Eusse, Gerd Ascheid, Rainer Leupers, Jovana Jovic, Luis Gabriel Murillo, Sergey Yakoushkin (*RWTH Aachen Univ.*)

6.4 **A Non-Intrusive Timing Synchronization Interface for Hardware-Assisted** 127
HW/SW Co-Simulation

> Yu-Hung Huang, Yi-Shan Lu, Hsin-I Wu, Ren-Song Tsay (*National Tsing Hua Univ.*)

Session 8: Can EDA Combat the Rise of Electronic Counterfeiting?
Chair: Miodrag Potkonjak (*Univ. of California, Los Angeles*)

8.1 **Can EDA Combat the Rise of Electronic Counterfeiting?** 133

> Carl McCants (*Defense Advanced Research Projects Agency*); William Bryson (*Analytical Solutions, Inc.*); Matthew Sale (*U.S. Naval Surface Warfare Center*); Saverio Fazzari (*Booz Allen Hamilton, Inc.*); Farinaz Koushanfar (*Rice Univ.*); Miodrag Potkonjak (*Univ. of California, Los Angeles*); Peilin Song (*IBM Research*)

Session 9: Reliability: From Atoms to 3-D
Chair: Angan Das (*Intel Corp.*)

9.1 **Physics Matters: Statistical Aging Prediction under Trapping/Detrapping** 139

> Jyothi Bhaskarr Velamala, Ketul Sutaria (*Arizona State Univ.*); Takashi Sato (*Kyoto Univ.*); Yu Cao (*Arizona State Univ.*)

9.2 **Library-Aware Resonant Clock Synthesis (LARCS)** 145

> Xuchu Hu (*Cadence Design Systems, Inc., Univ. of California, Santa Cruz*); Walter Condley, Matthew Guthaus (*Univ. of California, Santa Cruz*)

9.3 **Incremental Power Grid Verification** 151

> Farid N. Najm, Abhishek (*Univ. of Toronto*)

9.4 **Analysis of DC Current Crowding in Through-Silicon-Vias and its Impact on** 157
Power Integrity in 3-D ICs

> Sung Kyu Lim, Xin Zhao (*Georgia Institute of Technology*); Michael Scheuermann (*IBM T.J. Watson Research Ctr.*)

Session 10: EDA for Emerging Applications at the Kilometer, Meter, Micron, and Nanometer Scales
Chair: Sai-Wang (Rocco) Tam (*Marvell Semiconductor, Inc.*)

10.1 **Tracking Appliance Usage Information in Residential Settings Using** 163
Off-the-Shelf Low-Frequency Meters

> Deokwoo Jung (*Advanced Digital Sciences Center*); Andreas Savvides (*Yale Univ.*); Athanasios Bamis (*Univ. of Connecticut*)

10.2 **Implementing an FPGA System for Real-Time Intent Recognition for** 169
Prosthetic Legs

Xiaorong Zhang, He Huang, Qing Yang (*Univ. of Rhode Island*)

10.3 **Statistical Design and Optimization for Adaptive Post-silicon Tuning of** 176
MEMS Filters

Fa Wang, Gary Fedder, Larry Pileggi, Tamal Mukherjee, Jonathan Rotner, Xin
Li, Gokce Keskin, Andrew Phelps (*Carnegie Mellon Univ.*)

10.4 **Generic Low-Cost Characterization of VTH and Mobility Variations in LTPS** 182
TFTs for Non-Uniformity Calibration of Active-Matrix OLED Displays

Reza Chaji, Javid Jaffari (*IGNIS Innovations, Inc.*)

Session 11: Facing Dependability: System-Level Solutions and Cybercar Challenges
Chair: Hans-Joachim Wunderlich (*Univ. of Stuttgart*)

11.1 **Towards Fault-Tolerant Embedded Systems with Imperfect Fault Detection** 188

Jia Huang, Kai Huang, Andreas Raabe, Christian Buckl (*fortiss GmbH*); Alois
Knoll (*Technische Univ. München*)

11.2 **Steady-State Dynamic Temperature Analysis and Reliability Optimization** 197
for Embedded Multiprocessor Systems

Ivan Ukhov, Zebo Peng, Min Bao, Petru Eles (*Linköping Univ.*)

11.3 **Considering Diagnosis Functionality during Automatic System-Level** 205
Design of Automotive Networks

Michael Eberl, Michael Glass, Jürgen Teich (*Univ. of Erlangen-Nuremberg*);
Ulrich Abelein (*Audi AG*)

11.4 **Meta-Cure: A Reliability Enhancement Strategy for Metadata in NAND** 214
Flash Memory Storage Systems

Zili Shao, Yi Wang (*The Hong Kong Polytechnic Univ.*); Luis Angel Bathen,
Nikil Dutt (*Univ. of California, Irvine*)

11.5 **EDA for Secure and Dependable Cybercars: Challenges and Opportunities** 220

Hervé Seudié, Ahmad-Reza Sadeghi (*Fraunhofer SIT, and Intel-TU Darmstadt
Security Institute, Germany*); Farinaz Koushanfar (*Rice University*)

Session 12: Volatile or Non-Volatile? That's the Question
Chair: Tei-Wei Kuo (*National Taiwan Univ.*)

12.1 **Software Controlled Cell Bit-Density to Improve NAND Flash Lifetime** 229

Xavier Jimenez, David Novo, Paolo Ienne (*Ecole Polytechnique Fédérale de
Lausanne*)

Observational Wear Leveling: An Efficient Algorithm for Flash Memory Management

| **12.2** | Chundong Wang, Weng-Fai Wong (*National Univ. of Singapore*) | 235 |

12.3 Cache Revive: Architecting Volatile STT-RAM Caches for Enhanced Performance in CMPs 243

Yuan Xie, Chita R. Das, Vijaykrishnan Narayanan, Cong Xu (*Pennsylvania State Univ.*); Asit K. Mishra (*Intel Corp.*); Adwait Jog (*Pennsylvania State Univ.*); Ravishankar Iyer (*Intel Corp.*)

12.4 Point and Discard: A Hard-Error-Tolerant Architecture for Non-Volatile Last Level Caches 253

Jue Wang, Xiangyu Dong, Yuan Xie (*Pennsylvania State Univ.*)

Session 14: Self-Aware and Adaptive Technologies: The Future of Computing Systems?

Chair: Xiaoyun Zhu (*VMware, Inc.*)

14.1 Self-Aware Computing in the Angstrom Processor 259

Martina Maggio (*Lund Univ.*); Anantha Chandrakasan, Anant Agarwal, Yildiz Sinangil, Mahmut Sinangil, Srini Devadas, Eric Lau, George Kurian (*Massachusetts Institute of Technology*); Jim Holt (*Massachusetts Institute of Technology, Freescale Semiconductor, Inc.*); Henry Hoffman, Sabrina Neuman, Jason Miller (*Massachusetts Institute of Technology*)

14.2 The Case for Elastic Operating System Services in fos 265

Charles Gruenwald (*Massachusetts Institute of Technology*); David Wentzlaff (*Princeton Univ.*); Harshad Kasture, Nathan Beckmann (*Massachusetts Institute of Technology*); Lamia Youseff (*Google, Inc.*); Anant Agarwal (*Massachusetts Institute of Technology*)

14.3 A Compiler and Runtime for Heterogeneous Computing 271

Joshua Auerbach, David Bacon, Ioana Burcea, Perry Cheng, Stephen Fink, Rodric Rabbah, Sunil Shukla (*IBM T.J. Watson Research Ctr.*)

14.4 The Helix Project: Overview and Directions 277

Gu-Yeon Wei, David Brooks, Glenn Holloway, Simone Campanoni (*Harvard Univ.*); Timothy Jones (*Univ. of Cambridge*)

Session 15: Why Model? Because Reality is Complicated Enough!

Chair: Ibrahim Elfadel (*Masdar Institute of Science and Technology*)

15.1 Exploring Sub-20nm FinFET Design with Predictive Technology Models 283

Saurabh Sinha, Greg Yeric, Vikas Chandra, Brian Cline (*ARM, Inc.*); Yu Cao (*Arizona State Univ.*)

15.2 **Fast Nonlinear Model Order Reduction via Associated Transforms of** 289
High-Order Volterra Transfer Functions

> Neric Fong, Ngai Wong, Qing Wang, Haotian Liu, Yang Zhang (*The Univ. of Hong Kong*)

15.3 **AMOR: An Efficient Aggregating Based Model Order Reduction Method for** 295
Many-Terminal Interconnect Circuits

> Fan Yang, Xuan Zeng, Yangfeng Su (*Fudan Univ.*)

15.4 **BLAST: Efficient Computation of Nonlinear Delay Sensitivities in Electronic** 301
and Biological Networks using Barycentric Lagrange Enabled Transient Adjoint
Analysis

> Arie Meir, Jaijeet Roychowdhurry (*Univ. of California, Berkeley*)

15.5 **DAE2FSM: Automatic Generation of Accurate Discrete-Time Logical** 311
Abstractions for Continuous-Time Circuit Dynamics

> Karthik Aadithya, Jaijeet Roychowdhury (*Univ. of California, Berkeley*)

15.6 **Chip/Package Co-Analysis of Thermo-Mechanical Stress and Reliability in** 317
TSV-based 3-D ICs

> Moongon Jung (*Georgia Institute of Technology*); David Pan (*Univ. of Texas, Austin*); Sung Kyu Lim (*Georgia Institute of Technology*)

Session 16: Is Formal Verification Ready for the System Level?

Chair: Erik Seligman (*Intel Corp.*)

16.1 **Symbolic Model Checking on SystemC Designs** 327

> Chiao Hsieh, Yen-Sheng Ho, Chun-Nan Chou, Chung-Yang (Ric) Huang (*National Taiwan Univ.*)

16.2 **System Verification of Concurrent RTL Modules by Compositional Path** 334
Predicate Abstraction

> Joakim Urdahl, Dominik Stoffel, Markus Wedler, Wolfgang Kunz (*Univ. of Kaiserslautern*)

16.3 **Equivalence Checking for Behaviorally Synthesized Pipelines** 344

> Kecheng Hao (*Portland State Univ.*); Sandip Ray (*Univ. of Texas, Austin*); Fei Xie (*Portland State Univ.*)

16.4 **Proving Correctness of Regular Expression Accelerators** 350

> Christoph Hagleitner, Mitra Purandare, Kubilay Atasu (*IBM Research - Zurich*)

Sciduction: Combining Induction, Deduction, and Structure for Verification and

| 16.5 | Synthesis | 356 |

Sanjit Seshia (*Univ. of California, Berkeley*)

Session 17: NoCs Next Top Model: From System-Level to Prototype
Chair: Fabien Clermidy (*CEA-LETI*)

| 17.1 | **Cost-Efficient Buffer Sizing in Shared-Memory 3-D MPSoCs using Wide I/O Interfaces** | 366 |

Abbas Sheibanyrad (*TIMA Laboratory/CNRS*); Frédéric Pétrot, Sahar Foroutan (*TIMA Laboratory, Grenoble Institute of Technology*)

| 17.2 | **Attackboard: A Novel Dependency-Aware Traffic Generator for Exploring NoC Design Space** | 376 |

Yoshi Shih-Chieh Huang, Yu-Chi Chang, Tsung-Chan Tsai, Yuan-Ying Chang, Chung-Ta King (*National Tsing Hua Univ.*)

| 17.3 | **Towards Graceful Aging Degradation in NoCs Through an Adaptive Routing Algorithum** | 382 |

Sanghamitra Roy, Kshitij Bhardwaj, Koushik Chakraborty (*Utah State Univ.*)

| 17.4 | **Explicit Modeling of Control and Data for Improved NoC Router Estimation** | 392 |

Siddhartha Nath, Bill Lin, Andrew B. Kahng (*Univ. of California at San Diego*)

| 17.5 | **Approaching the Theoretical Limits of a Mesh NoC with a 16-Node Chip Prototype in 45nm SOI** | 398 |

Sunghyun Park, Tushar Krishna, Chia-Hsin O. Chen, Bhavya Daya, Li-Shiuan Peh, Anantha P. Chandrakasan (*Massachusetts Institute of Technology*)

| 17.6 | **High Radix Self-Arbitrating Switch Fabric with Multiple Arbitration Schemes and Quality of Service** | 406 |

Trevor Mudge, Dennis Sylvester, Ronald Dreslinski, Reetuparna Das, Sudhir Satpathy, David Blaauw (*Univ. of Michigan*)

Session 18: Timing Analysis and Software-Controlled Memory: Are We Safe?
Chair: Frank Slomka (*Univ. of Ulm*)

| 18.1 | **WCET-Centric Partial Instruction Cache Locking** | 412 |

Huping Ding (*National Univ. of Singapore*); Yun Liang (*Advanced Digital Sciences Center*); Tulika Mitra (*National Univ. of Singapore*)

| 18.2 | **Worst-Case Execution Time Analysis for Parallel Run-Time Monitoring** | 421 |

Daniel Lo, G. Edward Suh (*Cornell Univ.*)

18.3 Conforming the Runtime Inputs for Hard Real-Time Embedded Systems   430

Kai Huang, Gang Chen, Christian Buckl (*fortiss GmbH*); Alois Knoll (*Technische Univ. München*)

18.4 STM Concurrency Control for Embedded Real-Time Software with Tighter Time Bounds   437

Mohammed El-Shambakey, Binoy Ravindran (*Virginia Polytechnic Institute and State Univ.*)

18.5 HaVOC: A Hybrid-Memory-Aware Virtualization Layer for On-Chip Distributed ScratchPad and Non-Volatile Memories   447

Nikil Dutt, Luis Angel Bathen (*Univ. of California, Irvine*)

18.6 Age-Based PCM Wear Leveling with Nearly Zero Search Cost   453

Chi-Hao Chen (*National Taiwan Univ.*); Pi-Cheng Hsiu (*Academia Sinica*); Tei-Wei Kuo (*National Taiwan Univ., Academia Sinica*); Chia-Lin Yang (*National Taiwan Univ.*); Cheng-Yuan Michael Wang (*Macronix International Co., Ltd.*)

Session 20: Routing-Driven Design Closure
Chair: Shankar Krishnamoorthy (*Mentor Graphics Corp.*)

20.1 Algorithms and Data Structures for Fast and Good VLSI Routing   459

Christian Schulte, Jens Vygen, Christian Panten, Tim Nieberg, Dirk Mueller, Michael Gester (*Univ. of Bonn*)

20.2 Guiding a Physical Design Closure System to Produce Easier-to-Route Designs with More Predictable Timing   465

Charles Alpert (*IBM Corp.*); Gi-Joon Nam (*IBM Research - Austin*); Natarajan Viswanathan (*IBM Systems and Technology Group*); Cliff Sze (*IBM Research - Austin*); Nancy Zhou (*IBM Systems and Technology Group*); Zhuo Li (*IBM Research - Austin*)

20.3 Rule Agnostic Routing by Using Design Fabrics   471

Gyuszi Suto (*Intel Corp.*)

Session 21: Storing, Computing, and Storing While Computing: The New Face of Non-Volatility in Systems
Chair: Charles Augustine (*Intel Corp.*)

21.1 Making Non-Volatile Nanomagnet Logic Non-Volatile   476

Michael Niemier, Peng Li, Gary Bernstein, Vijay Karthik Sankar, Wolfgang Porod, Xiaobo Sharon Hu, Steve Kurtz, Aaron Dingler, Gyorgy Csaba, Joseph Nahas (*Univ. of Notre Dame*)

21.2 mLogic: Ultra-Low Voltage Non-Volatile Logic Circuits Using STT-MTJ Devices   486

Daniel Morris, David Bromberg, Jian-Gang (Jimmy) Zhu, Larry Pileggi (*Carnegie Mellon Univ.*)

21.3 Future Cache Design using STT MRAMs for Improved Energy Efficiency: Devices, Circuits and Architecture   492

Sumeet Kumar Gupta, Kaushik Roy, Niladri Narayan Mojumder, Sang Phill Park, Anand Raghunathan (*Purdue Univ.*)

21.4 Hardware Realization of BSB Recall Function with Memristor Crossbar Arrays   498

Miao Hu, Hai Li (*Polytechnic Institute of New York Univ.*); Qing Wu, Garrett S. Rose (*Air Force Research Lab*)

Session 22: You Can Count on Me: Why it's OK to be Imprecise or Unreliable
Chair: Qinru Qiu (*Syracuse Univ.*)

22.1 A Methodology for Energy-Quality Tradeoff Using Imprecise Hardware   504

Gabriel Robins, Jiawei Huang, John Lach (*Univ. of Virginia*)

22.2 On the Exploitation of the Inherent Error Resilience of Wireless Systems under Unreliable Silicon   510

Andreas Burg (*Ecole Polytechnique Fédérale de Lausanne*); Christian Benkeser, Christoph Roth (*Eidgenössische Technische Hochschule Zürich*); Georgios Karakonstantis (*Ecole Polytechnique Fédérale de Lausanne*)

22.3 Near-Optimal, Dynamic Module Reconfiguration in a Photovoltaic System to Combat Partial Shading Effects   516

Xue Lin, Yanzhi Wang, Siyu Yue (*Univ. of Southern California*); Donghwa Shin, Naehyuck Chang (*Seoul National Univ.*); Massoud Pedram (*Univ. of Southern California, Los Angeles*)

22.4 Networked Architecture for Hybrid Electrical Energy Storage Systems   522

Qing Xie, Massoud Pedram, Yanzhi Wang (*Univ. of Southern California*); Sangyoung Park, Younghyun Kim, Naehyuck Chang (*Seoul National Univ.*)

Session 23: Optimization to the Rescue of Analog
Chair: Trent McConaghy (*Solido Design Automation, Inc.*)

23.1 A New Uncertainty Budgeting Based Method for Robust Analog/Mixed-Signal Design   529

Jin Sun (*Orora Design Technologies, Inc.*); Priyank Gupta (*Cirrus Logic, Inc.*); Janet Roveda (*Univ. of Arizona*)

23.2 **Variability-Aware, Discrete Optimization for Analog Circuits** 536

Seobin Jung, Yunju Choi, Jaeha Kim (*Seoul National Univ.*)

23.3 **Efficient Multi-Objective Synthesis for Microwave Components Based on Computational Intelligence Techniques** 542

Soheil Radiom, Guy A. E. Vandenbosch, Hadi Aliakbarian, Bo Liu, Georges Gielen (*Katholieke Univ. Leuven*)

23.4 **Non-Uniform Multilevel Analog Routing with Matching Constraints** 549

Hung-Chih Ou, Hsing-Chih Chang Chien, Yao-Wen Chang (*National Taiwan Univ.*)

Session 24: Xterminating Bugs

Chair: Sharad Kumar (*Freescale Semiconductor, Inc.*)

24.1 **X-Tracer: A Reconfigurable X-Tolerant Trace Compressor for Silicon Debug** 555

Feng Yuan, Xiao Liu, Qiang Xu (*The Chinese Univ. of Hong Kong*)

24.2 **Quick Detection of Difficult Bugs for Effective Post-Silicon Validation** 561

Farzan Fallah (*Stanford Univ.*); Nagib Hakim (*Intel Corp.*); Ted Hong, David Lin, Subhasish Mitra (*Stanford Univ.*)

24.3 **Test Data Volume Optimization for Diagnosis** 567

Hongfei Wang, Osei Poku, Xiaochun Yu, Sizhe Liu, Ibrahima Komara, Shawn Blanton (*Carnegie Mellon Univ.*)

24.4 **Invariance-Based Concurrent Error Detection for Advanced Encryption Standard** 573

Xiaofei Guo, Ramesh Karri (*Polytechnic Institute of New York Univ.*)

Session 26: Brain-Inspired Autonomous Computing and Modeling

Chair: Yiran Chen (*Univ. of Pittsburgh*)

26.2 **Accelerating Neuromorphic Vision Algorithms for Recognition** 579

Matthew Cotter (*Pennsylvania State Univ.*); Chaitali Chakrabarti (*Arizona State Univ.*); Vijaykrishnan Narayanan, Michael DeBole, Ahmed Al Maashri, Nandhini Chandramoorthy, Yang Xiao (*Pennsylvania State Univ.*)

26.3 **Statistical Memristor Modeling and Case Study in Neuromorphic Computing** 585

Robinson Pino (*Air Force Research Lab*); Hai Li (*Polytechnic Institute of New York Univ.*); Yiran Chen (*Univ. of Pittsburgh*); Miao Hu (*Polytechnic Institute of New York Univ.*); Beiye Liu (*Univ. of Pittsburgh*)

Session 27: Design, the Next Generation: From Routing to Capturing Design Expertise
Chair: Charles Chiang (*Synopsys, Inc.*)

27.1 **Triple Patterning Aware Routing and its Comparison with Double Patterning Aware Routing in 14nm Technology**   591

Martin D. F. Wong, Qiang Ma, Hongbo Zhang (*Univ. of Illinois at Urbana-Champaign*)

27.2 **GDRouter: Interleaved Global Routing and Detailed Routing for Ultimate Routability**   597

Yanheng Zhang (*Cadence Design Systems, Inc.*); Chris Chu (*Iowa State Univ.*)

27.3 **Standard Cell Routing via Boolean Satisfiability**   603

Nikolai Ryzhenko, Steven Burns (*Intel Corp.*)

27.4 **An Efficient Algorithm for Multi-Layer Obstacle-Avoiding Rectilinear Steiner Tree Construction**   613

Chih-Hung Liu, I-Che Chen (*Academia Sinica*); Der-Tsai Lee (*National Chung-Hsing Univ.*)

27.5 **Avoiding Game Over: Bringing Design to the Next Level**   623

Sameh Galal, Stephen Richardson, Artem Vassilliev, Sabarish Sankaranarayanan, Mark Horowitz, Andrew Danowitz, Megan Wachs, Ofer Shacham, John Brunhaver, Wajahat Qadeer (*Stanford Univ.*)

Session 28: Staying Cool: Modeling Thermal Effects in 3-D and Multicore
Chair: Dhireesha Kudithipudi (*Rochester Institute of Technology*)

28.1 **PowerField: A Transient Temperature-to-Power Technique based on Markov Random Field Theory**   630

Seungwook Paek (*KAIST*); Seok-Hwan Moon (*Electronics and Telecommunications Research Institute*); Wongyu Shin, Jaehyeong Sim, Lee-Sup Kim (*KAIST*)

28.2 **EigenMaps: Algorithms for Optimal Thermal Maps Extraction and Sensor Placement on Multicore Processors**   636

Juri Ranieri, Martin Vetterli, David Atienza, Alessandro Vincenzi, Amina Chebira (*Ecole Polytechnique Fédérale de Lausanne*)

28.3 **An Information-theoretic Framework for Optimal Temperature Sensor Allocation and Full-chip Thermal Monitoring**   642

Huapeng Zhou, Xin Li (*Carnegie Mellon Univ.*); Chen-Yong Cher, Eren Kursun, Haifeng Qian (*IBM T.J. Watson Research Ctr.*); Shi-Chune Yao (*Carnegie*

Mellon Univ.)

28.4 **Optimizing Energy Efficiency of 3-D Multicore Systems with Stacked DRAM** 648
under Power and Thermal Constraints

Ayse Coskun, Jie Meng, Katsutoshi Kawakami (*Boston Univ.*)

Session 29: SOS: Specification, Optimization, and Synthesis in System-Level Design
Chair: Brett Meyer (*McGill Univ.*)

29.1 **Static Dataflow with Access Patterns: Semantics and Analysis** 656

Arkadeb Ghosal, Rhishikesh Limaye, Kaushik Ravindran (*National Instruments Corp.*); Stavros Tripakis (*Univ. of California, Berkeley*); Ankita Prasad, Guoqiang Wang, Trung N. Tran, Hugo A. Andrade (*National Instruments Corp.*)

29.2 **Executing Synchronous Dataflow Graphs on a SPM-Based Multicore** 664
Architecture

Hyunok Oh (*Hanyang Univ.*); Sungchan Kim (*Chonbuk National Univ.*); Junchul Choi, Soonhoi Ha (*Seoul National Univ.*)

29.3 **System-Level Synthesis of Memory Architecture for Stream Processing** 672
Sub-Systems of a MPSoC

Glenn Leary, Weijia Che, Karam S. Chatha (*Arizona State Univ.*)

29.4 **Courteous Cache Sharing: Being Nice to Others in Capacity Management** 678

Akbar Sharifi, Shekhar Srikantaiah, Mahmut Kandemir, Mary Jane Irwin (*Pennsylvania State Univ.*)

Session 30: Future of IC Reliability
Chair: Alesandro Pinto (*United Technologies Research Center*)

30.1 **A Hybrid Approach to Cyber-Physical Systems Verification** 688

Samarjit Chakraborty (*Technische Univ. München*); Anuradha Annaswamy (*Massachusetts Institute of Technology*); Lothar Thiele (*Eidgenössische Technische Hochschule Zürich*); Dip Goswami (*Technische Univ. München*); Pratyush Kumar, Kai Lampka (*Eidgenössische Technische Hochschule Zürich*)

30.2 **Reliable Computing with Ultra-Reduced Instruction Set Co-Processors** 697

Aravindkumar Rajendiran, Sundaram Ananthanarayanan, Hiren Patel, Mahesh Tripunitara, Siddharth Garg (*Univ. of Waterloo*)

30.3 **Identification of Recovered ICs using Fingerprints from a Light-Weight** 703
On-Chip Sensor

Xuehui Zhang, Nicholas Tuzzio, Mohammad Tehranipoor (*Univ. of

Connecticut)

30.4 **Confidentiality Preserving Integer Programming for Global Routing**   709

Hamid Shojaei, Azadeh Davoodi, Parameswaran Ramanathan (*Univ. of Wisconsin*)

Session 32: Breaking out of EDA: How to Apply EDA Techniques to Broader Applications
Chair: Jason Cong (*Univ. of California, Los Angeles*)

32.1 **Design Tools for Artificial Nervous Systems**   717

Louis K. Scheffer (*Howard Hughes Medical Institute*)

32.2 **Dynamic River Network Simulation at Large Scale**   723

Frank Liu (*IBM Research - Austin*); Ben R. Hodges (*Univ. of Texas, Austin*)

32.3 **Humans for EDA and EDA for Humans**   729

Valeria Bertacco (*Univ. of Michigan*)

32.4 **Application of Logic Synthesis to the Understanding and Cure of Genetic Diseases**   734

Pey-Chang Kent Lin, Sunil Khatri (*Texas A&M Univ.*)

Session 33: The Right Placement at the Right Timing
Chair: Saurabh Adya (*Magma Design Automation, Inc.*)

33.1 **Exploiting Die-to-Die Thermal Coupling in 3D IC Placement**   741

Krit Athikulwongse, Mohit Pathak, Sung Kyu Lim (*Georgia Institute of Technology*)

33.2 **ComPLx: A Competitive Primal-Dual Lagrange Optimization for Global Placement**   747

Myung-Chul Kim, Igor Markov (*Univ. of Michigan*)

33.3 **PADE: A High-Performance Placer with Automatic Datapath Extraction and Evaluation through High-Dimensional Data Learning**   756

Samuel Ward, Duo Ding, David Pan (*Univ. of Texas, Austin*)

33.4 **Structure-Aware Placement for Datapath Intensive Circuit Designs**   762

Sheng Chou, Meng-Kai Hsu, Yao-Wen Chang (*National Taiwan Univ.*)

33.5 **GLARE: Global and Local Wiring Aware Routability Evaluation**   768

Charles J. Alpert (*IBM Austin Research Lab*); Sachin S. Sapatnekar (*University of Minnesota*); Douglas Keller, Gustavo E. Tellez, Lakshmi Reddy

(*IBM Systems and Technology Group*); Zhuo Li (*IBM Austin Research Lab*); Natarajan Viswanathan (*IBM Systems and Technology Group*); Cliff Sze (*IBM Austin Research Lab*); Yaoguang Wei (*University of Minnesota*); Andrew D. Huber (*IBM Systems and Technology Group*)

33.6   The DAC 2012 Routability-Driven Placement Contest and Benchmark Suite   774

Natarajan Viswanathan, Charles Alpert, Cliff Sze, Zhuo Li, Yaoguang Wei (*IBM Corp.*)

Session 34: Global Views of Synthesis: Broadening the Scope
Chair: Herman Schmit (*Altera Corp.*)

34.1   Removing Overhead from High-Level Interfaces   783

Megan Wachs, Mark Horowitz, Kyle Kelley, Stephen Richardson, John Stevenson (*Stanford Univ.*)

34.2   On the Asymptotic Costs of Multiplexer-Based Reconfigurability   790

Johnathan York, Derek Chiou (*Univ. of Texas, Austin*)

34.3   SALSA: Systematic Logic Synthesis of Approximate Circuits   796

Swagath Venkataramani, Amit Sabne, Vivek Kozhikkottu, Kaushik Roy, Anand Raghunathan (*Purdue Univ.*)

34.4   Timing ECO Optimization Using Metal-Configurable Gate-Array Spare Cells   802

Iris Hui-Ru Jiang (*National Chiao Tung Univ.*); Yao-Wen Chang, Hua-Yu Chang (*National Taiwan Univ.*)

34.5   Early Prediction of NBTI Effects Using RTL Source Code Analysis   808

Kenneth Butler (*Texas Instruments, Inc.*); Heesoo Kim, Shobha Vasudevan, Jayanand Asok Kumar (*Univ. of Illinois at Urbana-Champaign*)

34.6   Generalized SAT-Sweeping for Post-Mapping Optimization   814

Tobias Welp (*Univ. of California, Berkeley*); Smita Krishnaswamy (*Columbia Univ.*); Andreas Kuehlmann (*Coverity, Inc.*)

Session 35: Adaptive Computing: When, Where, Why, How?
Chair: Philip Brisk (*Univ. of California, Riverside*)

35.1   Accuracy-Configurable Adder for Approximate Arithmetic Designs   820

Andrew B. Kahng, Seokhyeong Kang (*Univ. of California at San Diego*)

35.2 **Recovery-Based Design for Variation-Tolerant SoCs** 826

Anand Raghunathan (*Purdue Univ.*); Sujit Dey (*Univ. of California at San Diego*); Vivek Kozhikkottu (*Purdue Univ.*)

35.3 **A Hybrid NoC Design for Cache Coherence Optimization for Chip Multiprocessors** 834

Ohyoung Jang, Wei Ding, Yuanrui Zhang, Mahmut Kandemir, Mary Jane Irwin, Hui Zhao (*Pennsylvania State Univ.*)

35.4 **Architecture Support for Accelerator-Rich CMPs** 843

Jason Cong, Mohammad Ali Ghodrat, Michael Gill, Beayna Grigorian, Glenn Reinman (*Univ. of California, Los Angeles*)

35.5 **A QoS-Aware Memory Controller for Dynamically Balancing GPU and CPU Bandwidth Use in an MPSoC** 850

Min Kyu Jeong (*Univ. of Texas, Austin*); Nigel Paver (*ARM, Inc.*); Mattan Erez (*Univ. of Texas, Austin*); Chander Sudanthi (*ARM, Inc.*)

35.6 **Metronome: Operating System Level Performance Management via Self-Adaptive Computing** 856

Filippo Sironi, Davide Basilio Bartolini (*Politecnico di Milano*); Simone Campanoni (*Harvard Univ.*); Fabio Cancaré (*Politecnico di Milano*); Henry Hoffmann (*Massachusetts Institute of Technology*); Donatella Sciuto, Marco Santambrogio (*Politecnico di Milano*)

Session 36: Yin and Yang of Memories: The Power-Performance Trade-Off
Chair: Yiran Chen (*Univ. of Pittsburgh*)

36.1 **Adaptive Power Management of On-Chip Video Memory for Multiview Video Coding** 866

Muhammad Shafique, Joerg Henkel (*Karlsruhe Institute of Technology*); Sergio Bampi (*Univ. Federal do Rio Grande do Sul*); Bruno Zatt (*Karlsruhe Institute of Technology*); Fábio Leandro Walter (*Univ. Federal do Rio Grande do Sul*)

36.2 **Heterogeneous Multi-Channel: Fine-Grained DRAM Control for Both System Performance and Power Efficiency** 876

Guangfei Zhang (*Institute of Computing Tech.*); Huandong Wang (*Loongson Technology Corp., Ltd*); Xinke Chen (*Institute of Computing Tech.*); Shuai Huang (*Loongson Technology Corp., Ltd.*); Peng Li (*Institute of Computing Tech.*)

36.3 **Joint Management of RAM and Flash Memory with Access Pattern Considerations** 882

Po-Chun Huang (*National Taiwan Univ.*); Yuan-Hao Chang (*Academia Sinica*); Tei-Wei Kuo (*National Taiwan Univ., Academia Sinica*)

36.4   **Hybrid DRAM/PRAM-Based Main Memory for Single-Chip CPU/GPU**   888

Dongki Kim, Sunggu Lee, Sungjoo Yoo (*Pohang Univ. of Science and Technology*); Dong Hyuk Woo, DaeHyun Kim (*Intel Corp.*); Sungkwang Lee (*Pohang Univ. of Science and Technology*); Jaewoong Chung (*Intel Corp.*)

36.5   **Write Performance Improvement by Hiding R Drift Latency in Phase-Change RAM**   897

Youngsik Kim, Sungjoo Yoo, Sunggu Lee (*Pohang Univ. of Science and Technology*)

36.6   **Constructing Large and Fast Multi-Level Cell STT-MRAM Based Cache for Embedded Processors**   907

Lei Jiang, Jun Yang, Bo Zhao, Youtao Zhang (*Univ. of Pittsburgh*)

Session 38: Probabilistic Embedded Computing
Chair: Vincent Mooney (*Georgia Institute of Technology*)

38.1   **Incorrect Systems: It's not the Problem It's the Solution.**   913

Christoph M. Kirsch, Hannes Payer (*University of Salzburg*)

38.2   **On Software Design for Stochastic Processors**   918

Joseph Sloan, John Sartori, Rakesh Kumar (*Univ. of Illinois at Urbana-Champaign*)

38.3   **What to Do About the End of Moore's Law, Probably!**   924

Krishna Palem (*Nanyang Technological Univ., Rice Univ.*); Avinash Lingamneni (*Rice Univ.*)

38.4   **Obtaining and Reasoning About Good Enough Software**   930

Martin Rinard (*Massachusetts Institute of Technology*)

Session 39: Simulation-Based Verification: New Ways to Harness the Workhorse
Chair: Kerstin Eder (*Univ. of Bristol*)

39.1   **Improving Gate-level Simulation Accuracy when Unknowns Exist**   936

Chris Browy, Kai-Hui Chang (*Avery Design Systems, Inc.*)

39.2 Automated Feature Localization for Hardware Designs Using Coverage Metrics   941

Goerschwin Fey (*German Aerospace Center*); Jan Malburg, Alexander Finder (*Univ. of Bremen*)

39.3 Path Directed Abstraction and Refinement in SAT-Based Design Debugging   947

Brian Keng, Andreas Veneris (*Univ. of Toronto*)

39.4 Checking Architectural Outputs Instruction-By-Instruction on Acceleration Platforms   955

Debapriya Chatterjee (*Univ. of Michigan*); Anatoly Koyfman, Ronny Morad, Avi Ziv (*IBM Haifa Research Lab.*); Valeria Bertacco (*Univ. of Michigan*)

Session 40: Ultra-Low Power Using Subthreshold and Nearthreshold Operation
Chair: Mahadev Nemani (*Intel Corp.*)

40.1 Standard Cell Sizing for Subthreshold Operation   962

Bo Liu, Jose Pineda de Gyvez (*Technische Univ. Eindhoven*); Jos Huisken, Maryam Ashouei (*Holst Centre*)

40.2 Decoupling Capacitor Design Strategy for Minimizing Supply Noise of Ultra-Low Voltage Circuits   968

Mingoo Seok (*Columbia Univ.*)

40.3 Regaining Throughput Using Completion Detection for Error-Resilient Near-Threshold Logic   974

Joseph Crop, Robert Pawlowski, Patrick Chiang (*Oregon State Univ.*)

40.4 Process Variation in Near-Threshold Wide SIMD Architectures   980

Chaitali Chakrabarti (*Arizona State Univ.*); Trevor Mudge, Scott Mahlke, Yongjun Park, Mark Woh, Ronald Dreslinski, Sangwon Seo, David Blaauw (*Univ. of Michigan*)

Session 41: Top Picks of Run-Time Power Management Techniques
Chair: Jian-Jia Chen (*Karlsruhe Institute of Technology*)

41.1 Run-Time Power-Down Strategies for Real-Time SDRAM Memory Controllers   988

Karthik Chandrasekar (*Delft Univ. of Technology*); Benny Akesson, Kees Goossens (*Technische Univ. Eindhoven*)

41.2 Embedding Statistical Tests for On-Chip Dynamic Voltage and Temperature Monitoring   994

Lionel Vincent (*CEA-LETI Minatec*); Philippe Maurine (*Univ. Montpellier 2*); Suzanne Lesecq, Edith Beigne (*CEA-LETI Minatec*)

41.3   **Quality-Retaining OLED Dynamic Voltage Scaling for Video Streaming Applications on Mobile Devices**   1000

Chun Jason Xue (*City Univ. of Hong Kong*); Yiran Chen (*Univ. of Pittsburgh*); Mengying Zhao (*City Univ. of Hong Kong*); Xiang Chen, Jian Zeng (*Univ. of Pittsburgh*)

41.4   **Traffic-Aware Power Optimization for Network Applications on Multicore Servers**   1006

Laxmi Bhuyan, Raymond Klefstad, Jilong Kuang (*Univ. of California, Riverside*)

Session 42: The Dark Side of Test
Chair: Shreyas Sen (*Intel Corp.*)

42.1   **Alternate Hammering Test for Application-Specific DRAMs and an Industrial Case Study**   1012

Rei-Fu Huang (*MediaTek, Inc.*); Hao-Yu Yang, Mango C.-T. Chao (*National Chiao Tung Univ.*); Shih-Chin Lin (*United Microelectronics Corp.*)

42.2   **Goal-Oriented Stimulus Generation for Analog Circuits**   1018

Jayanand Asok Kumar, Shobha Vasudevan, Seyed Nematollah Ahmadyan (*Univ. of Illinois at Urbana-Champaign*)

42.3   **TSV Open Defects in 3D Integrated Circuits: Characterization, Test, and Optimal Spare Allocation**   1024

Krishnendu Chakrabarty, Fangming Ye (*Duke Univ.*)

42.4   **Small Delay Testing for TSVs in 3-D ICs**   1031

Yu-Hsiang Lin, Shi-Yu Huang (*National Tsing Hua Univ.*); Kun-Han Tsai, Wu-Tung Cheng, Stephen Sunter (*Mentor Graphics Corp.*); Yung-Fa Chou, Ding-Ming Kwai (*Industrial Technology Research Institute*)

Session 44: Design Challenges and EDA Solutions for Wireless Sensor Networks
Chair: Roman Hermida (*Complutense Univ.*)

44.1   **Circuit and System Design Guidelines for Ultra-Low Power Processing**   1037

Dongmin Yoon, David Blaauw, Yejoong Kim, Yoonmyung Lee, Dennis Sylvester (*Univ. of Michigan*)

44.2   **Design Exploration of Energy-Performance Trade-Offs for Wireless Sensor Networks**   1043

Ivan Beretta (*Ecole Polytechnique Fédérale de Lausanne*); Francisco Rincon (*Univ. Complutense Madrid*); Nadia Khaled (*Nestlé Research Center*); Paolo Grassi (*Politecnico di Milano*); Vincenzo Rana, David Atienza (*Ecole Polytechnique Fédérale de Lausanne*)

44.3   **Energy Harvesting and Power Management for Autonomous Sensor Nodes**   1049

Jerome Willemin (*CEA-LETI*); Christian Piguet (*Centre Suisse d'Electronique et Microtechnique SA*); Edith Beigné, Jean-Frederic Christmann, Cyril Condemine (*CEA-LETI*)

Session 45: Surviving Timing Challenges in Nanometer Designs
Chair: Florentin Dartu (*Synopsys, Inc.*)

45.1   **Functional Timing Analysis Made Fast and General**   1055

Jie-Hong Roland Jiang, Yi-Ting Chung (*National Taiwan Univ.*)

45.2   **Timing Analysis with Nonseparable Statistical and Deterministic Variations**   1061

Jeffrey Hemmett, Natesan Venkateswaran, Jeremy Leitzen (*IBM Systems and Technology Group*); Jinjun Xiong (*IBM T.J. Watson Research Ctr.*); Eric Foreman (*IBM Corp.*); Debjit Sinha (*IBM Systems and Technology Group*); Vladimir Zolotov (*IBM T.J. Watson Research Ctr.*); Chandu Visweswariah (*IBM Systems and Technology Group*)

45.3   **Reversible Statistical Max/Min Operation: Concept and Applications to Timing**   1067

Debjit Sinha, Natesan Venkateswaran (*IBM Systems and Technology Group*); Vladimir Zolotov (*IBM T.J. Watson Research Ctr.*); Jinjun Xiong (*IBM T.J. Watson Research Ctr.*); Chandu Visweswariah (*IBM Systems and Technology Group*)

45.4   **Predicting Timing Violations Through Instruction-Level Path Sensitization Analysis**   1074

Sanghamitra Roy, Koushik Chakraborty (*Utah State Univ.*)

Session 46: Special Delivery: Challenges in Packaging
Chair: Tan Yan (*Synopsys, Inc.*)

46.1   **A Chip-Package-Board Co-Design Methodology**   1082

Hsu-Chieh Lee, Yao-Wen Chang (*National Taiwan Univ.*)

46.2   **Obstacle-Avoiding Free-assignment Routing for Flip-Chip Designs**   1088

I-Jye Lin, Chin-Fang Shen, Chen-Feng Chang (*Synopsys, Inc.*); Yao-Wen Chang, Yuan-Kai Ho, Hsu-Chieh Lee, Po-Wei Lee (*National Taiwan Univ.*)

46.3   **Clock Tree Synthesis with Methodology of Re-Use in 3-D IC**   1094

TingTing Hwang, Fu-Wei Chen (*National Tsing Hua Univ.*)

46.4   **Can Pin Access Limit the Footprint Scaling?**   1100

Xiang Qiu, Malgorzata Marek-Sadowska (*Univ. of California, Santa Barbara*)

Session 47: Renovate Analog and Mixed-Signal Circuit Simulations
Chair: Chenjie Gu (*Intel Corp.*)

47.1   **Yield Estimation via Multi-Cones**   1107

Rouwaida Kanj (*American Univ. of Beirut*); Rajiv Joshi (*IBM T.J. Watson Research Ctr.*); Zhuo Li, Jerry Hayes (*IBM Research - Austin*); Sani Nassif (*IBM Research - Austin*)

47.2   **Efficient Trimmed-Sample Monte Carlo Methodology and Yield-Aware Design Flow for Analog Circuits**   1113

Wei-Yi Hu, Yi-Kan Cheng, Chin-Cheng Kuo, Yi-Hung Chen, Jui-Feng Kuan (*Taiwan Semiconductor Manufacturing Co., Ltd.*)

47.3   **Towards Efficient SPICE-Accurate Nonlinear Circuit Simulation with On-the-Fly Support-Circuit Preconditioners**   1119

Xueqian Zhao, Zhuo Feng (*Michigan Technological Univ.*)

47.4   **Sparse LU Factorization for Parallel Circuit Simulation on GPU**   1125

Ling Ren, Xiaoming Chen, Yu Wang, Chenxi Zhang, Huazhong Yang (*Tsinghua Univ.*)

Session 48: Heterogenous Platforms: Challenges and Opportunities
Chair: Norbert Wehn (*Univ. of Kaiserslautern*)

48.1   **Is Dark Silicon Useful? Harnessing the Four Horsemen of the Coming Dark Silicon Apocalypse**   1131

Michael Taylor (*Univ. of California at San Diego*)

48.2   **Platform 2012 - A Many-Core Computing Accelerator for Embedded SoCs: Performance Evaluation of Visual Analytics Applications**   1137

Luca Benini (*Univ. di Bologna, STMicroelectronics*); Denis Dutoit, Fabien Clermidy (*STMicroelectronics, CEA-LETI*); Germain Haugou, Thierry Lepley, Bruno Jego, Diego Melpignano, Eric Flamand (*STMicroelectronics*)

Session 50: Hot Chips Running Cool - Energy Efficient Near-Threshold Computing and its Barriers
Chair: David Brooks (*Harvard Univ.*)

50.1   **Assessing the Performance Limits of Parallelized Near-Threshold Computing**   1143

Kory Sewell, Trevor Mudge, David Blaauw, Dennis Sylvester, Nathaniel Pinckney, Ronald Dreslinski, David Fick (*Univ. of Michigan*)

50.2   **Near-Threshold Voltage (NTV) Design - Opportunities and Challenges**   1149

Himanshu Kaul, Mark Anders, Steven Hsu, Amit Agarwal, Ram Krishnamurthy, Shekhar Borkar (*Intel Corp.*)

50.3   **Near-Threshold Operation for Power-Efficient Computing? It Depends**   1155

Leland Chang, Wilfried Haensch (*IBM T.J. Watson Research Ctr.*)

50.4   **Not so Fast my Friend: Is Near-Threshold Computing the Answer for Power Reduction of Wireless Devices?**   1160

Matt Severson, Kendrick Yuen, Yang Du (*Qualcomm, Inc.*)

Session 51: Yielding in an Uncertain World
Chair: Rob Aitken (*ARM, Inc.*)

51.1   **Accurate Process-Hotspot Detection Using Critical Design Rule Extraction**   1163

Yen-Ting Yu (*National Chiao Tung Univ.*); Ya-Chung Chan (*Mstar Semiconductor*); Subarna Sinha (*Stanford Univ.*); Iris Hui-Ru Jiang (*National Chiao Tung Univ.*); Charles Chiang (*Synopsys, Inc.*)

51.2   **Improved Tangent Space-Based Distance Metric for Accurate Lithographic Hotspot Classification**   1169

Xuan Zeng, Jing Guo, Fan Yang (*Fudan Univ.*); Subarna Sinha (*Stanford Univ.*); Charles Chiang (*Synopsys, Inc.*)

51.3   **Simultaneous Flare Level and Flare Variation Minimization with Dummification in EUVL**   1175

Shao-Yun Fang, Yao-Wen Chang (*National Taiwan Univ.*)

51.4   **A Novel Layout Decomposition Algorithm for Triple Patterning Lithography**   1181

Shao-Yun Fang, Yao-Wen Chang, Wei-Yu Chen (*National Taiwan Univ.*)

51.5   **PS3-RAM: A Fast Portable and Scalable Statistical STT-RAM Reliability Analysis Method**   1187

Wujie Wen, YaoJun Zhang, Yiran Chen (*Univ. of Pittsburgh*); Yu Wang (*Tsinghua Univ.*); Yuan Xie (*Pennsylvania State Univ.*)

51.6   **Exploiting Narrow-Width Values for Process Variation-Tolerant 3-D Microprocessors**   1193

Sung Woo Chung, Joonho Kong (*Korea Univ.*)

Session 52: High-Level Synthesis is Not Just About Translation!
Chair: Satnam Singh (*Google, Inc.*)

52.1 Hardware Synthesis of Recursive Functions through Partial Stream Rewriting — 1203

Christian Haubelt, Lars Middendorf (*Univ. of Rostock*); Christophe Bobda (*Univ. of Arkansas*)

52.2 Chisel: Constructing Hardware in a Scala Embedded Language — 1212

Jonathan Bachrach, Huy Vo, Brian Richards, Yunsup Lee, Andrew Waterman, Rimas Avizienis, John Wawrzynek, Krste Asanovic (*Univ. of California, Berkeley*)

52.3 Specification and Synthesis of Hardware Checkpointing and Rollback Mechanisms — 1222

Carven Chan, Sharad Malik, Divjyot Sethi, Daniel Schwartz-Narbonne (*Princeton Univ.*)

52.4 Optimizing Memory Hierarchy Allocation with Loop Transformations for High-Level Synthesis — 1229

Jason Cong, Peng Zhang, Yi Zou (*Univ. of California, Los Angeles*)

52.5 A Metric for Layout-Friendly Microarchitecture Optimization in High-Level Synthesis — 1235

Jason Cong, Bin Liu (*Univ. of California, Los Angeles*)

52.6 Computer Generation of Streaming Sorting Networks — 1241

Marcela Zuluaga (*Eidgenössische Technische Hochschule Zürich*); Peter Milder (*Carnegie Mellon Univ.*); Markus Püschel (*Eidgenössische Technische Hochschule Zürich*)

Session 53: Wild And Crazy Ideas
Chair: Farinaz Koushanfar (*Rice Univ.*)

53.1 CrowdMine: Towards Crowdsourced Human-Assisted Verification — 1250

Wenchao Li, Sanjit A. Seshia (*Univ. of California, Berkeley*); Somesh Jha (*Univ. of Wisconsin, Madison*)

53.2 Extracting Design Information from Natural Language Specifications — 1252

Ian G. Harris (*Univ. of California, Irvine*)

Material Implication in CMOS: A New Kind of Logic

53.3   Elkim Roa (*Purdue Univ.*); Wu-Hsin Chen (*Purdue University*); Byunghoo   1254
Jung (*Purdue Univ.*)

53.4   **Boolean Satisfiability Using Noise-Based Logic**   1256

Pey-Chang Kent Lin, Ayan Mandal, Sunil Khatri (*Texas A&M Univ.*)

53.5   **Cognitive Computing with Spin-Based Neural Networks**   1258

Georgios Panagopoulos, Kaushik Roy, Mrigank Sharad (*Purdue Univ.*);
Charles Augustine (*Intel Corp.*)

53.6   **Capacitance of TSVs in 3-D Stacked Chips a Problem? Not for**   1260
Neuromorphic Systems!

Antoine Joubert (*CEA-LETI Minatec*); Marc Duranton (*CEA-LIST*); Bilel
Belhadj (*CEA-LETI Minatec*); Olivier Temam (*INRIA*); Rodolphe Héliot
(*CEA-LETI Minatec*)

Session 54: Optimizing Embedded Software for High Performance and Reliability
Chair: Rodric Rabbah (*IBM Research*)

54.1   **Communication-Aware Mapping of KPN Applications onto Heterogeneous**   1262
MPSoCs

Jeronimo Castrillon, Andreas Tretter, Rainer Leupers, Gerd Ascheid (*RWTH
Aachen Univ.*)

54.2   **Unrolling and Retiming of Stream Applications onto Embedded Multicore**   1268
Processors

Weijia Che, Karam Chatha (*Arizona State Univ.*)

54.3   **Exploiting Spatiotemporal and Device Contexts for Energy-Efficient**   1274
Mobile Embedded Systems

Chris Ohlsen, Sudeep Pasricha, Charles Anderson, Brad Donohoo (*Colorado
State Univ.*)

54.4   **EPIMap: Using Epimorphism to Map Applications on CGRAs**   1280

Mahdi Hamzeh, Aviral Shrivastava, Sarma Vrudhula (*Univ. of California, Los
Angeles*)

54.5   **Instruction Scheduling for Reliability-Aware Compilation**   1288

Semeen Rehman, Muhammad Shafique, Joerg Henkel (*Karlsruhe Institute of
Technology*)

54.6   **Compiling for Energy Effciency on Timing Speculative Processors**   1297

Rakesh Kumar, John Sartori (*Univ. of Illinois at Urbana-Champaign*)

GENERAL CHAIR'S MESSAGE

Welcome to the new DAC, better than ever!

Dear fellow DAC 49er:

It is a great pleasure to welcome you to the 49th annual edition of the Design Automation Conference, appropriately held in beautiful San Francisco. Since 1964, which is even before the beginning of Moore's law, DAC is the place where electronic systems design meets automation. In his keynote on Thursday, legendary professor Dave Liu will highlight the impact that DAC has had – and continues to have - in shaping our lives and the entire electronics industry. Without exaggeration, DAC is the premier place where thousands of professionals from all over the world converge to exchange ideas, sharpen skills and do business. And we will have a jolly good time doing all that. At the end of each day, join us for a reception to unwind while enjoying a view of the San Francisco skyline.

The 49th DAC in San Francisco is a true 'design rush' with hundreds of presentations, exhibitors, tutorials, panels and much more. The program is carefully designed to maximize personal interaction at all levels. We've reshaped the format of the **technical presentations** into concise 15-minute slots, with a poster session afterwards for in-depth face-to-face discussions. This year we had a significant increase in paper submissions. DAC's meticulous review process ensures that only the very best and most novel work is published at DAC. Domain experts from across the globe worked hard and entered a record 4,000 paper reviews, providing constructive feedback to future generations of EDA professionals.

Continuing on the interactive theme, our new 'Work-In-Progress' session intercepts and reshapes ideas before they are codified in a formal paper. At the session, DAC allows you to interact to shape the advanced research. Similarly our immensely popular 'Wild and Crazy Ideas' session provides food for thought in unexpected directions.

DAC provides a convenient way to widen your engineering skills by attending our concise but information-packed 2-hour **tutorials** on Monday. Whether ESL design, 3-D, analog or physical design, you leave DAC a better engineer.

I'm especially looking forward to attending the lively **panel sessions** on our exhibition floor and as part of the technical program.

DAC is dramatically increasing its focus on **Embedded Systems and Software (ESS)**. In his **Tuesday keynote** address, Mike Muller, the co-founder and Chief Technical Officer of ARM, will shed his light on the future of embedded computing systems. Wednesday will feature the ESS Executive day, to provide managers with timely information to help them make decisions where business and technology in the ESS domain intersect. More than one-third of the entire DAC program is focused on Embedded Systems and Software.

The 49th DAC features a vibrant **exhibition** showcasing nearly 200 companies, including all of the largest EDA vendors and significant foundries. This serves as a convenient one-stop-shop for all of design automation and embedded system design automation.

For the first time our **User Track** presentations will be held on the exhibition floor. The User Track is specifically designed for EDA tool users and feature critical design and methodology challenges as well as case studies of innovative tool use. Wednesday will feature the novel User Track **dual-keynote** by Joshua Friedrich of IBM and Brad Heaney of Intel. This promises to provide a unique view into the kitchen of two dominant IC companies. Since this is where EDA technology meets practice, it's a must-see for every designer.

No less than eight DAC workshops extend the program into several specialized areas of interest, from bio-design automation, high-frequency analog design, heterogeneous computing and much more. Our popular all-inclusive registration package will give you access to all of them.

DAC is also the home for an impressive constellation of eight colocated conferences that complement the DAC program: they include established conferences and symposia such as DFM&Y, HOST, ESLsyn, IWLS, SI2, and SLIP, CELUG and the ACM Student Research Competition.

Last but not least, I'd like to recognize the enormous effort of many hundreds of dedicated volunteers from across all walks of life that make DAC possible by donating their time, expertise and enthusiasm. Without this enthusiastic effort, DAC would not be possible. If you want to join our exciting team in the future, please contact us.

Enjoy your time at DAC and the wonderful city of San Francisco!

Patrick Groeneveld
General Chair, 49th DAC

PROCEEDINGS OF THE 49TH DESIGN AUTOMATION CONFERENCE®

The Association for Computing Machinery
2 Penn Plaza, Ste. 701
New York, NY 10121

Copyright 2012 by the Association for Computing Machinery, Inc. (ACM). Permission to make digital or hard copies of portions of this work for personal or classroom use is granted without fee provided that copies are not made or distributed for profit or commercial advantage and that copies bear this notice and the full citation on the first page. Copyright for components of this work owned by others than ACM must be honored. Abstracting with credit is permitted. To copy otherwise, to republish, to post on servers or to redistribute to lists, requires prior specific permission and/or a fee. Request permission to republish from: Publications Dept., ACM, Inc. Fax: +1-212-869-0481 or permissions@acm.org.

For other copying of articles that carry a code at the bottom of the first or last page, copying is permitted provided that the per-copy fee indicated in the code is paid through the Copyright Clearance Center, 222 Rosewood Drive, Danvers, MA 01923.

Notice to Past Authors of ACM-Published Articles

ACM intends to create a complete electronic archive of all articles and/or other material previously published by ACM. If you have written a work that was previously published by ACM in any journal or conference proceedings prior to 1978, or any SIG Newsletter at any time, and you do NOT want this work to appear in the ACM Digital Library, please inform permissions@acm.org, stating the title of the work, the author(s), and where and when published.

Additional copies may be ordered prepaid from:

ACM Order Department
P.O. Box 11414
New York, NY 10286-1414

email: orders@acm.org
(U.S.A. and Canada)
Fax: +1-800-342-6626

(all other countries)
Fax: +1-212-944-1318

Additional copies of this publication are available from:

IEEE Service Center   +1-800-678-IEEE
445 Hoes Lane   +1-732-981-1393
Piscataway, NJ 08855-1331   +1-732-981-1721 (Fax)
www.ieee.org

EXECUTIVE COMMITTEE

GENERAL CHAIR
Patrick Groeneveld
Magma Design Automation, Inc.
1650 Technology Dr.
San Jose, CA 95110
1-408-565-7654
patrick@magma-da.com

VICE/FINANCE CHAIR
Yervant Zorian
Synopsys, Inc.
700 E. Middlefield Rd.
Mountain View, CA 94043
1-650-584-7120
yervant.zorian@synopsys.com

TECHNICAL PROGRAM CO-CHAIR
Donatella Sciuto
Politecnico di Milano
P.zza L. da Vinci 32
Milano, Italy 2133
+39-02-2399-3662
sciuto@elet.polimi.it

TECHNICAL PROGRAM CO-CHAIR
Soha Hassoun
Tufts Univ.
161 College Ave.
Medford, MA 02155
1-617-627-5177
soha@cs.tufts.edu

PANEL CHAIR
Charles Alpert
IBM Corp.
3120 Castellani Way
Cedar Park, TX 78613
1-512-286-5099
alpert@us.ibm.com

TUTORIAL CHAIR
Michael McNamara
Cadence Design Systems, Inc.
2655 Seely Avenue
San Jose, CA 95134
1-408-914-6808
mcnamara@cadence.com

PAST CHAIR
Leon Stok
IBM Corp.
2070 Rte. 52
Hopewell Jct., NY 12533
1-845-892-5262
leonstok@us.ibm.com

DESIGN COMMUNITY CHAIR
Robert Jones
Intel Corp.
2111 NE 25th Ave., JF4-310
Hillsboro, OR 97124
1-503-712-3555
robert.b.jones@intel.com

EDA INDUSTRY CHAIR
Tiffany Sparks
ARM, Inc.
150 Rose Orchard Way
San Jose, CA 95134
1-408-576-1397
tiffany.sparks@arm.com

PUBLICITY CHAIR
Michelle Clancy
Cayenne Communication
1280 Oakmead Pkwy., Ste. 201
Sunnyvale, CA 94085
1-252-940-0981
michelle.clancy@cayennecom.com

NEW INITIATIVES CHAIR
Nikil Dutt
Univ. of California, Irvine
Dept. of Computer Science
Irvine, CA 92697
1-949-824-7219
dutt@uci.ed

EUROPE/MIDDLE EAST REPRESENTATIVE
Georges Gielen
ESAT-MICAS
Katholieke Univ. Leuven
Kasteelpark Arenberg 10
Leuven, B-3001 Belgium
+32-16-321047
georges.gielen@esat.kuleuven.be

ASIA/SOUTH PACIFIC REPRESENTATIVE
Sri Parameswaran
Univ. of New South Wales
UNSW, Austrailia
+61-9385-4223
sridevan@cse.unsw.edu.au

ACM/SIGDA REPRESENTATIVE
Patrick Madden
SUNY Binghamton
P.O. Box 6000
Binghamton, NY 13902
1-607-777-2943
pmadden@acm.org

IEEE/CEDA REPRESENTATIVE
Al Dunlop
Crossbow Consulting, LLC
PO Box 124
Kattskill Bay, NY 12844
1-518-656-3501
aldunlop@gmail.com

EDA CONSORTIUM REPRESENTATIVE
Anne Cirkel
Mentor Graphics Corp.
8005 SW Boeckman Rd.
Wilsonville, OR 97070
1-503-685-7934
anne_cirkel@mentor.com

CONFERENCE MANAGER
Kevin Lepine
MP Associates, Inc.
1721 Boxelder St., Ste. 107
Louisville, CO 80027
1-303-530-4562
kevin@mpassociates.com

EXHIBITS MANAGER
Lee Wood
MP Associates, Inc.
1721 Boxelder St., Ste. 107
Louisville, CO 80027
1-303-530-4562
lee@mpassociates.com

TECHNICAL PROGRAM COMMITTEE

Soha Hassoun
Technical Program Co-Chair
Tufts Univ.
Medford, MA

David Atienza
Ecole Polytechnique Fédérale de Lausanne
Lausanne, Switzerland

David Bacon
IBM Research
Hawthorne, NY

Iris Bahar
Brown Univ.
Providence, RI

Cristiana Bolchini
Politecnico di Milano
Milan, Italy

Duane Boning
Massachusetts Institute of Technology
Cambridge, MA

John Carulli
Texas Instruments, Inc.
Dallas, TX

Samarjit Chakraborty
Technical Univ. of Munich
Munich, Germany

Naehyuck Chang
Seoul National Univ.
Seoul, Republic of Korea

Karam Chatha
Arizona State Univ.
Phoenix, AZ

Yiran Chen
Univ. of Pittsburgh
Pittsburgh, PA

Kiyoung Choi
Seoul National Univ.
Seoul, Republic of Korea

Pai Chou
Univ. of California, Irvine / NTHU
Irvine, CA

Fabien Clermidy
CEA-LETI
Grenoble, France

Albert Cohen
INRIA
Paris, France

Donatella Sciuto
Technical Program Co-Chair
Politecnico di Milano
Milan, Italy

George Constantinides
Imperial College London
London, Great Britain

Ayse Coskun
Boston Univ.
Boston, MA

Jennifer Dworak
Southern Methodist Univ.
Dallas, TX

Stephen Edwards
Columbia Univ.
New York, NY

Franco Fummi
Univ. of Verona
Verona, Italy

Malay Ganai
NEC Labs America, Inc.
Princeton, NJ

Anne Gattiker
IBM Corp.
Austin, TX

Catherine Gebotys
Univ. of Waterloo
Waterloo, ON, Canada

Andreas Gerstlauer
Univ. of Texas, Austin
Austin, TX

Tony Givargis
Univ. of California, Irvine
Irvine, CA

Soonhoi Ha
Seoul National Univ.
Seoul, Republic of Korea

Ziyad Hanna
Jasper Design Automation, Inc.
Mountain View, CA

Ian Harris
Univ. of California, Irvine
Irvine, CA

Joerg Henkel
Karlsruhe Institute of Technology
Karlsruhe, Germany

Shiyan Hu
Michigan Technological Univ.
Houghton, MI

Xiaobo Sharon Hu
Univ. of Notre Dame
West Bend, IN

Ing-Jer Huang
National Sun Yat-Sen Univ.
Kaohsiung, Taiwan

Mike Hutton
Altera Corp.
San Jose, CA

Paolo Ienne
Ecole Polytechnique Fédérale de Lausanne
Lausanne, Switzerland

Yehea Ismail
Northwestern Univ.
Evanston, IN

Dan Jiao
Purdue Univ.
West Lafayette, IN

Alex Jones
Univ. of Pittsburgh
Pittsburgh, PA

Mahmut Kandemir
Pennsylvania State Univ.
State College, PA

Chandramouli Kashyap
Intel Corp.
Hillsboro, OR

Jaeha Kim
Seoul National Univ.
Seoul, Republic of Korea

Cheng-Kok Koh
Purdue Univ.
West Lafayette, IN

Tei-Wei Kuo
National Taiwan Univ.
Taipei, Taiwan

Fadi Kurdahi
Univ. of California, Irvine
Irvine, CA

Jing Li
IBM T. J. Watson Research Ctr.
Yorktown Heights, NY

Peng Li
Texas A&M Univ.
College Station, TX

Xin Li
Carnegie Mellon Univ.
Pittsburgh, PA

Zhuo Li
IBM Research – Austin
Austin, TX

Sung Kyu Lim
Georgia Institute Of Technology
Atlanta, GA

Bill Lin
Univ. of California at San Diego
La Jolla, CA

Frank Liu
IBM Corp.
Austin, TX

Patrick Madden
SUNY Binghamton
Binghamton, NY

Scott Mahlke
Univ. of Michigan
Ann Arbor, MI

Yiorgos Makris
Univ. of Texas, Dallas
Richardson, TX

Peter Marwedel
Technisch Univ. Dortmund
Dortmund, Germany

Amit Mehrotra
Berkeley Design Automation
Santa Clara, CA

Ting Mei
Sandia National Labs
Albuquerque, NM

Natasa Miskov-Zivanov
Univ. of Pittsburgh
Pittsburgh, PA

Subhasish Mitra
Stanford Univ.
Stanford, CA

Tulika Mitra
National Univ. Of Singapore
Singapore

Kartik Mohanram
Univ. of Pittsburgh
Pittsburgh, PA

Gi-Joon Nam
IBM Corp.
Austin, TX

Michael Niemier
Univ. of Notre Dame
West Bend, IN

Steven Nowick
Columbia Univ.
New York, NY

John O'Leary
Intel Corp.
Hillsboro, OR

Maire O'Neill
Queen's Univ.
Kingston, ON, Canada

Michael Orshansky
Univ. of Texas, Austin
Austin, TX

Ralph Otten
Technische Univ. Eindhoven
Nuenen, The Netherlands

David Z. Pan
Univ. of Texas, Austin
Austin, TX

Preeti Panda
Indian Institute of Technology
New Delhi, India

Dusan Petranovic
Mentor Graphics Corp.
Freemont, CA

Andy Pimentel
Univ. of Amsterdam
Amsterdam, The Netherlands

Arijit Raychowdhury
Intel Corp.
Hillsboro, OR

Sherief Reda
Brown Univ.
Providence, RI

Marc Riedel
Univ. of Minnesota
Minneapolis, MN

Karem Sakallah
Univ. of Michigan
Ann Arbor, MI

Marco Santambrogio
Politecnico di Milano
Milan, Italy

Bing J. Sheu
Taiwan Semiconductor Manufacturing Co., Ltd.
Hsinchu, Taiwan

Thomas Shiple
Synopsys, Inc.
Lexington, MA

Aviral Shrivastava
Arizona State Univ.
Phoenix, AZ

Cristina Silvano
Politecnico di Milano
Milan, Italy

L. Miguel Silveira
INESC - ID/IST - TU Lisbon
Lisbon, Portugal

Vivek Singh
Intel Corp.
Hillsboro, OR

Mircea Stan
Univ. of Virginia
Charlottesville, VA

Takashi Takenaka
NEC Corp.
Kawasaki, Japan

Sheldon Tan
Univ. of California, Riverside
Riveside, CA

Vivek Tiwari
Intel Corp.
Sunnyvale, CA

Rasit Topaloglu
GLOBALFOUNDRIES
Santa Clara, CA

Ting-Chi Wang
National Tsing Hua Univ.
Hsinchu, Taiwan

Martin Wong
Univ. of Illinois at Urbana-Champaign
Urbana, IL

Yuan Xie
Pennsylvania State Univ.
State College, PA

Lamia Youseff
Google, Inc.
Seattle, WA

Qi Zhu
Univ. of California, Riverside
Riverside, CA

Avi Ziv
IBM Haifa Research Lab.
Haifa, Israel

PANEL COMMITTEES

Charles Alpert
Panel Chair
IBM Corp.
Cedar Park, TX

Technical Panel Committee

Dennis Brophy
Mentor Graphics Corp.
Wilsonville, OR

Steve Carlson
Cadence Design Systems, Inc.
San Jose, CA

Olivier Coudert
OC Consulting
Holzkirchen, Germany

Rich Goldman
Synopsys, Inc.
Mountain View, CA

William Joyner
Semiconductor Research Corporation
Research Triangle Park, NC

Andrew B. Kahng
Univ. of California at San Diego
La Jolla, CA

Farinaz Koushanfar
Rice Univ.
Houston, TX

Grant Martin
Tensilica, Inc.
Santa Clara, CA

Noel Menezes
Intel Corp.
Hillsboro, OR

Laura Parker
Mentor Graphics Corp.
Wilsonville, OR

Pavilion Panel Committee

Yatin Trivedi - **Chair**
Synopsys, Inc.
Mountain View, CA

Valery Kugel
Juniper Networks
Sunnyvale, CA

Monica Marmie
Magma Design Automation, Inc.
Santa Clara, CA

Mike Santarini
Xilinx, Inc.
San Jose, CA

Holly Stump
ASTC / VWorks
San Jose, CA

Kathy Werner
Freescale Semiconductor, Inc.
Austin, TX

Jonah McLeod
Kilopass Technology, Inc.
Santa Clara, CA

Troy Wood
Synopsys, Inc.
Mountain View, CA

INDUSTRY LIAISON COMMITTEE

Tiffany Sparks
Industry Liaison Committee Chair
ARM, Inc.
San Jose, CA

Dagmar Berendes
ThinkBold Corporate Communications, Inc.
Campbell, CA

Chuck Byers
Consultant
San Jose, CA

Anne Cirkel
Mentor Graphics Corp.
Wilsonville, OR

Michelle Clancy
Cayenne Communication
Sunnyvale, CA

Brett Cline
Forte Design Systems
San Jose, CA

Diana Dearin
Mentor Graphics Corp.
San Jose, CA

Mike Giafagna
Atrenta, Inc.
San Jose, CA

Patrick Groeneveld
Magma Design Automation
San Jose, CA

Jill Jacobs
MOD Marketing and Events LLC
San Jose, CA

Herta Schreiner
Synopsys, Inc.
Mountain View, CA

Gary Smith
Gary Smith EDA
Santa Clara, CA

David Thon
Cadence Design Systems, Inc.
San Jose, CA

Rob van Blommestein
Jasper Design Automation
San Jose, CA

Cindy Wilson
EVE
San Jose, CA

Lee Wood
MP Associates, Inc.
Louisville, CO

STRATEGY COMMITTEE

Yervant Zorian
Vice Chair/Strategy Committee Chair
Synopsys, Inc.
Mountain View, CA

Magdy Abadir
Freescale Semiconductor, Inc.
Austin, TX

Dennis Brophy
Mentor Graphics Corp.
Wilsonville, OR

Raul Camposano
Xoomsys, Inc.
Cupertino, CA

John Chilton
Synopsys, Inc.
Mountain View, CA

Anne Cirkel
Mentor Graphics Corp.
Wilsonville, OR

Michelle Clancy
Cayenne Communication
Sunnyvale, CA

Nikil Dutt
Univ. of California, Irvine
Irvine, CA

Bob Gardner
EDA Consortium
San Jose, CA

Mike Gianfagna
Atrenta, Inc.
San Jose, CA

Rich Goldman
Synopsys, Inc.
Mountain View, CA

John Goodenough
ARM, Inc.
San Jose, CA

Patrick Groeneveld
Magma Design Automation, Inc.
San Jose, CA

Kathryn Kranen
Jasper Design Automation, Inc.
Mountain View, CA

Vic Kulkarni
Apache Design, Inc. a subsidiary of ANSYS, Inc.
Sunnyvale, CA

Patrick Madden
SUNY Binghamton
Binghamton, NY

Grant Martin
Tensilica, Inc.
Santa Clara, CA

Pankaj Mayor
Cadence Design Systems, Inc.
San Jose, CA

Sani Nassif
IBM Corp.
Austin, TX

Paolo Prinetto
Politecnico de Torino
Torino, Italy

Juan Rey
Mentor Graphics Corp.
San Jose, CA

Gary Smith
Gary Smith EDA
Santa Clara, CA

Tiffany Sparks
ARM, Inc.
San Jose, CA

Tom Spyrou
AMD, Ltd.
San Jose, CA

Lee Wood
MP Associates, Inc.
Louisville, CO

Raj Yavatkar
Intel Corp.
Portland, OR

USER TRACK COMMITTEE

Robert Jones
User Track Chair
Intel Corp.
Hillsboro, OR

EMBEDDED SYSTEMS AND SOFTWARE DESIGN

Tor Jeremiassen
Track Chair
Texas Instruments
Houston, TX

Robert Aitken
ARM, Inc.
San Jose, CA

Sameh Asaad
IBM T.J Watson Research Center
Yorktown Heights, NY

Mike Beunder
Vector Fabrics B.V.
Eindhoven,The Netherlands

Michael Brogioli
Freescale Semiconductor, Inc.
Houston, TX

Pat Brouillette
Roku, Inc.
Scottsdale, AZ

Kaiming Ho
Fraunhofer-Gesellschaft
Erlangen, Denmark

Geert Janssen
IBM Corp.
Yorktown Heights, NY

Ken Knowlson
Intel Corp.
Hillsboro, OR

Laurent Maillet-Contoz
STMicroelectronics
Grenoble, France

Alicia Strang
Marvell Semiconductor, Inc.
Allso Viejo, CA

SILICON DESIGN (Front-End)

Robert Carden
Track Chair
Marvell Semiconductor, Inc.
Aliso Viejo, CA

Srinath Atluri
Cisco Systems, Inc.
San Jose, CA

Benjamin Carrion Schafer
NEC Corp.
Kawasaki, Japan

Jayendra Dwaraka Bhamidipatti
LSI Corp.
Bangalore, India

Amitabh Menon
Qualcomm, Inc.
San Jose, CA

Byeong Min
Samsung
Yongin-City, South Korea

Ambar Sarkar
Paradigm Works, Inc.
Andover, CT

Erik Seligman
Intel Corp.
Hillsboro, OR

Pei Suen
Marvell Semiconductor, Inc.
Aliso Viejo, CA

Rob Sumners
Advanced Micro Devices, Inc.
Austin, TX

Krishnan Sundaresan
Oracle
Santa Clara, CA

SILICON DESIGN (Back-End)

Raj Varada
Track chair
Intel Corp.
Santa Clara, CA

Thomas Brandtner
Infineon Technologies
Villach, Austria

Jarrod Brooks
Cypress Semiconductor Corp.
Lexington, KY

Rajit Chandra
Advanced Micro Devices, Inc.
Santa Clara, CA

Laurent Chaouat
Samsung
Austin, TX

Chihtung (Tony) Chen
Qualcomm, Inc.
San Diego, CA

Gilda Garreton
Oracle
Menlo Park, CA

Ismed Hartanto
Xilinx, Inc.
San Jose, CA

Miguel Miranda
IMEC
Leuven, Belgium

Srinivas Nori
GLOBALFOUNDRIES
Sunnyvale, CA

Nagaraj NS
Texas Instruments
Dallas, TX

Tim Whitfield
ARM, Inc.
Taipei, Taiwan

Matthew Ziegler
IBM T.J. Watson Research Ctr.
Yorktown Heights, NY

PR/MARKETING COMMITTEE

Michelle Clancy
PR/Marketing Committee Chair
Cayenne Communication
Sunnyvale, CA

Brett Cline
Forte Design Systems
San Jose, CA

William Deegan
Bad Dog Consulting
Mountain, View CA

Sonia Harrison
Mentor Graphics Corp.
Wilsonville, OR

Yukari Ohno
Apache Design, Inc. a subsidiary of ANSYS, Inc.
San Francisco, CA

Rob van Blommestein
Jasper Design Automation
San Jose, CA

Lee Wood
MP Associates, Inc.
Louisville, CO

BEST PAPER AWARD COMMITTEE

Luciano Lavagno
Politecnico Di Torino
Torino, Italy

David Atienza
Ecole Polytechnique Fédérale De Lausanne
Lausanne, Switzerland

Sherief Reda
Brown Univ.
Providence, RI

Subhasish Mitra
Stanford Univ.
Stanford, CA

Petru Eles
Linköping Univ.
Linköping, Sweden

Sani Nassif
IBM Research – Austin
Austin, TX

SPECIAL SESSION ORGANIZERS

Giovanni De Micheli
Ecole Polytechnique Fédérale de Lausanne
Lausanne, Switzerland

Robinson E. Pino
Air Force Research Laboratory
Rome, NY

Farinaz Koushanfar
Rice Univ.
Houston, TX

Gi-Joon Nam
IBM Research – Austin
Austin, TX

Saverio Fazzari
Booz Allen Hamilton
Arlington, VA

Christoph Kirsch
Univ. of Salzburg
Salzburg, Austria

Marco Santambrogio
Politecnico di Milano
Milan, Italy

David Atienza
Ecole Polytechnique Fédérale de Lausanne
Lausanne, Switzerland

Jonathan Eastep
Intel Corp.
Hillsboro, OR

Donatella Sciuto
Politecnico di Milano
Milan, Italy

Charles Alpert
IBM Corp.
Cedar Park, TX

Dennis Sylvester
Univ. of Michigan
Ann Arbor, MI

TECHNICAL PANEL ORGANIZERS

Anne Cirkel
Mentor Graphics Corp.
Wilsonville, OR

Frank Schirrmeister
Cadence Design Systems, Inc.
San Jose, CA

Clem Meas
siCAD, Inc.
Cupertino, CA

PAVILION PANEL CONTRIBUTORS

Donald Cramb
EVE-USA, Inc.
San Jose, CA

Leslie Cumming
Skye Marketing Communications
Portland, OR

Gene Forte
Mentor Graphics Corp.
Wilsonville, OR

Kelly Karr
Tanis Communications, Inc.
San Jose, CA

Cindy McDowell
Tanis Communications, Inc.
San Jose, CA

Bill Murray
Bill Murray Consulting, Inc.
Santa Clara, CA

Yukari Ohno
Apache Design, Inc. a subsidiary of ANSYS, Inc.
San Jose, CA

David Park
Synopsys, Inc.
Hillsboro, OR

Daniel Payne
Marketing EDA
Tualatin, OR

Gary Smith
Gary Smith EDA
Santa Clara, CA

Tiffany Sparks
ARM, Inc.
San Jose, CA

Mick Tegethoff
Berkeley Design Automation
Santa Clara, CA

Paul van Besouw
Oasys Design Systems, Inc.
Santa Clara, CA

Jan Willis
Calibra Consulting
Aylesbury, Great Britain

TUESDAY KEYNOTE ADDRESS
June 5, 2012
Room 102/103   8:30am

Mike Muller
ARM, Inc.
Cambridge, United Kingdom

Scaling for 2020 Solutions

Abstract: Comparing the original ARM design of 1985 to those of today's latest microprocessors, Mike will look at how far has design come and what EDA has contributed to enabling these advances in systems, hardware, operating systems, and applications and how business models have evolved over 25 years. He will then speculate on the needs for scaling designs into solutions for 2020 from tiny embedded sensors through to cloud based servers which together enable the internet of things. He will look at what are the major challenges that need to be addressed to design and manufacture these systems and proposes some solutions.

Biography: Mike Muller was one of the founders of ARM. Before joining the Company, he was responsible for hardware strategy and the development of portable products at Acorn Computers and was part of the original ARM design team. He was previously at Orbis Computers who developed network computers. At ARM he was VP, Marketing from 1992 to 1996 and EVP, Business Development until October 2000 when he was appointed Chief Technology Officer. In October 2001, he was appointed to the board of ARM Holdings plc.

xvi-a

WEDNESDAY KEYNOTE ADDRESS
June 6, 2012
Room 102/103   10:45am

Joshua Friedrich
IBM Server and Technology Group
Austin, TX
Brad Heaney
Intel Corp.
Folsom, CA

Designing High Performance Systems-on-Chip

Abstract: Experience state-of-the art design through the eyes of two experts that help shape these advanced chips! In this unique dual-keynote, the design process at two leading companies will be discussed.The speakers will cover key challenges, engineering decisions and design methodologies to achieve top performance and turn-around time. The presentations describe where EDA meets practice under the most advanced nodes, so will be of key interest to both designers and EDA professionals alike.

POWER™ Processor Design and Methodology Directions:
Joshua Friedrich

Processor designs and the EDA tools that support them stand at a key inflection point. The era of Dennard scaling and exponential single thread performance growth is a distant memory. Multi-thread performance continues to grow. However, the gain from simply adding more cores to a die by stepping to the next process node is diminishing due to technology challenges, application bottlenecks, and power/packaging constraints. To continue to deliver the cost-performance gains that drive our industry, designers will need to bring significant innovation to bear by integrating heterogeneous system components and accelerating key portions of the software stack in hardware. This transformation from technology-driven design to innovation-driven design defines new priorities for EDA development compared to prior eras. While timing optimization, power reduction, and support for modular designs remain necessary, differentiation will be achieved by enabling designer productivity through technology simplification, design abstraction, and robust support for heterogeneous IP.

Biography: Joshua Friedrich is a Senior Technical Staff Member and Senior Manager of POWER™ Technology Development in IBM's Server and Technology Group. In his role, Josh leads the physical design, technology direction, and methodology of IBM's future POWER™ processors. Josh has been part of the POWER development team since POWER4™, and on past POWER™ designs, Josh has led multiple design disciplines including power estimation and reduction, hardware characterization, memory subsystem circuit development, and core execution units. Before joining IBM, Josh received his Bachelor of Science in Electrical Engineering from the University of Texas at Austin.

Designing a 22nm Intel® Architecture Multi-CPU and GPU:
Brad Heaney

With each new process technology node and integration of more system components on to a monolithic die, the design methodology challenges must advance to enable validation and implementation of these complex products. With Intel's new 22nm technology, we are designing products with over 1.4 billion transistors and integrating hardware blocks that naturally want different process and design optimizations. The recently launched 3rd Generation Intel Core Process (codename Ivybridge) has an integrated Graphics Processing Unit that has different process and design demands than the CPU Core Processor. With the size and diversity of the product hardware, combined with new advanced process technology features, such as Intel's new tri-gate transistor, more capabilities for silicon debug, coverage, and manufacturing need to be planned and incorporated into the architecture and design implementation. By close collaboration between the process development and design teams at Intel, we are able to develop design methods to ramp these large, complex products into high volume manufacturing at, or ahead of, the schedule on prior products.

Biography: Brad Heaney is an Intel Architecture Group Project Manager and operates out of Intel's Folsom Design Center. Brad is a 25 year veteran at Intel and started his career working on the design of the 80386 family of CPU's and is the holder of four patents for his design work. In the last few years, Brad has been managing the teams that deliver Intel's lead vehicles for ramping new process technologies. Brad's team developed the Penryn CPU, which was a lead vehicle for 45nm process technology. In April of this year, they launched the Ivybridge CPU (3rd Generation Intel Core Processor), which is the lead vehicle for Intel's 22nm process technology. Brad received his Bachelor of Science degree from Drexel University in Philadelphia and his Master of Science in Electrical Engineering degree from Stanford University prior to joining Intel.

xvi-b

THURSDAY KEYNOTE ADDRESS
June 7, 2012
Room 102/103   11:00am

C.L. Liu
National Tsing Hua Univ.
Hinschu, Taiwan

My First Design Automation Conference - 1982

Abstract: It was June 1982 that I had my first technical paper in the EDA area presented at the 19th Design Automation Conference. It was exactly 20 years after I completed my doctoral study and exactly 30 years ago from today. I would like to share with the audience how my prior educational experience prepared me to enter the EDA field and how my EDA experience prepared me for the other aspects of my professional life.

Biography: C. L. Liu received his B. Sc. degree (1956) from the National Cheng Kung University in Taiwan, and his S. M. (1960) and Sc. D. (1962) degrees from the Massachusetts Institute of Technology. He taught at the Massachusetts Institute of Technology, the University of Illinois at Urbana-Champaign, and the National Tsing Hua University in Taiwan. He also served as the President of the National Tsing Hua University from 1998 to 2002.

He is currently the William Mong Honorary Chair Professor of Computer Science at the National Tsing Hua University, an industrial consultant, and the host of a weekly radio show (since 2005). He has published over 180 technical papers, eight technical textbooks and research monographs in the area of EDA, computer –aided instruction, real-time systems, combinatorial optimization, and discrete mathematics, and seven essay collections in the area of science and humanities.

PERSPECTIVE PAPER ABSTRACTS

Session 3: Design Automation for Things Wet, Small, Spooky, and Tamable
TUESDAY June 5 – 11:00am - 11:30am   Room 300
A Microgrid View of Energy Efficient Systems

Speaker: Rajesh Gupta – *Univ. of California at San Diego, La Jolla, CA*
Author(s): Rajesh Gupta – *Univ. of California at San Diego, La Jolla, CA*

Summary: Energy is a precious societal resource, and increasingly rated for its 'quality' or lack thereof as a contributor to greenhouse gases. Modern electrical energy systems operate at the intersection of technological advances in microelectronics, communications, and control. From individual components and systems such as computer systems to their aggregates and enclosures such as data centers and buildings, microelectronic advances in radios, processors, storage and networking are enabling low-cost and effective embedded sensing and its use in operational controls. In the context of energy distribution systems, this trend has led to popular visions of 'smart electrical grids' that dynamically match generation, transmission, and storage for the most efficient and reliable usage of electromagnetic energy. This talk examines how 'microgrids,' which are self-managed grids with local cogeneration capabilities, can be used as testing grounds for the prototyping and testing of smart grid technologies. We examine the emerging computer science problems arising from energy arbitration, alternative energy sourcing and capacity provisioning for computational resources through dynamic deferral of energy loads.

Session 29: SOS: Specification, Optimization, and Synthesis in System-Level Design
WEDNESDAY June 6 – 2:30pm – 3:00pm   Room 308
Embedded Systems - The Neural Backbone of Society

Speaker: Rolf Ernst - *Technische Univ. Braunschweig, Germany*
Author(s): Rolf Ernst - *Technische Univ. Braunschweig, Germany*

Summary: Embedded systems have evolved from single microcontrollers controlling devices to networked systems that jointly control cars, aircraft, buildings, or production lines. Many embedded networks are already reachable via the Internet, but mostly for maintenance and data acquisition purposes. The future will bring a shift towards interoperating networks of embedded systems that use open networks such as the Internet as an integration platform. At the same time, the focus will shift from individual application contexts to large networks of embedded system functions that jointly address societal challenges, such as energy supply, traffic, smart cities and communities, or the aging society. Effectively, networked embedded systems will extend their role as neural backbone of our societies.

This trend towards widely networked embedded systems goes far beyond the "internet of things" concept which makes all "things" accessible through the internet. The Internet of things will rather be a major asset in the future development of embedded systems. In the end, there will be many more embedded systems using the Internet for coordinated embedded systems functions, than people accessing the Internet for information and communication.

In 2006, Europe has established a large funding instrument, the Joint Undertaking (JU) "ARTEMIS" which supports large multinational projects in the area of embedded systems. In 2011, an updated Strategic Research Agenda (SRA 2011) was published which outlines the ARTEMIS vision for the development of networked embedded systems. The presenter is one of the main authors of the SRA. The talk will explain the vision and the SRA approach using scenarios derived from societal challenges to determine research and innovation targets for the upcoming years.

Session 49: Parallelization and Software Development: Hope, Hype, or Horror?
THURSDAY June 7 – 2:45pm – 3:30pm   Room 305
PhD or MD - Who is Better Trained for Building Successful Software Development Tools?

Speaker: Andreas Kuehlmann - *Coverity, Inc. San Francisco, CA*
Author(s): Andreas Kuehlmann - *Coverity, Inc. San Francisco, CA*

Summary: Unlike many other engineering disciplines, software development is often still seen as a magical art: requirements go in on one end, and - at some point - a product comes out on the other end. The larger the project is and the more legacy code it contains, the more unpredictable the process seems. It is astounding to observe the magnitude of diversity of development processes and product quality standards in various software shops. Contemporary buzzwords such as "agile development" or "Application Life-cycle Management" (ALM) only add to the confusion. In contrast, most of the published research in the area of software verification is solely focused on technological aspects, such as pushing model-based design paradigms, improving testing frameworks, or attacking the scaling issue of formal or semi-formal verification approaches. This narrow view completely ignores the fact that success in software development has as much to do with technology as it has with psychology. In this talk, we discuss a number of technological and psychological challenges of software development and argue that development tools must align advanced technologies with sociological and organizational aspects in order to be successful. The tools must not "get in the way of developers", play to their unique psyche, and demonstrate a measurable return of the investment, i.e., time and effort spent. Similarly, their application must fit smoothly into the existing workflow and avoid "off-cycle" processes. The talk will include a variety of concrete examples to illustrate the concepts discussed.We will also provide a comparison of the needs of hardware versus software verification tools and share lessons we have learned during the transition between these two domains.

TECHNICAL PANEL ABSTRACTS

SESSION 1:
TUESDAY June 05, 10:00am - 11:30am | Room 305
Will Reliability be the Death of Moore's Law?
Chair: Ana Hunter - *Samsung, San Jose, CA*
The latest technology roadmap cites "reliability and resilience" as key long-term challenges to continue the Moore's Law scaling. How much will electromigration, aging, and thermal effects limit the benefits of smaller process geometries? Is the cost of reliability (margins, area, power) a showstopper? To what extent will new architectures and design tools mask unreliability? This panel discusses these issues and looks at how reliability challenges could limit product design over the coming decade.
Panelists:
Naresh R. Shanbhag - *Univ. of Illinois at Urbana-Champaign, Urbana, IL*
Jose Maiz - *Intel Corp., Portland, OR*
Sani Nassif - *IBM Research - Austin, TX*
Subhasish Mitra - *Stanford Univ., Stanford, CA*
Goeran Jerke - *Robert Bosch GmbH, Reutlingen, Germany*

SESSION 7:
TUESDAY June 05, 1:30pm - 3:00pm | Room 305
System Models - Does One Size Fit All?
Chair: Brian Bailey - *EETimes EDA Designline, Oregon City, OR*
System-level modeling is a critical part of product design flows. Developing a single model that simultaneously satisfies the needs of software developers, system architects, hardware developers and verification engineers is hard. Time of availability, usage models, accuracy requirements, development effort, and speed vary greatly. Is it possible for one size to fit all? Who will provide the models? Who will pay for them? The panelists will review different aspects of system modeling and discuss which abstraction levels best address specific user requirements.
Panelists:
Stuart Swan - *Cadence Design Systems, Inc., San Jose, CA*
Rick Higgins - *Qualcomm, Inc., San Diego, CA*
John Goodenough - *ARM, Inc., San Jose, CA*
Frederic Risacher - *Research in Motion, Ltd., Waterloo, ON, Canada*
Andrea Kroll - *Tensilica, Inc., San Jose, CA*

SESSION 13:
TUESDAY June 05, 4:00pm - 6:00pm | Room 305
Will Your Next ASIC Ever be an FPGA?
Chair: Kevin Morris – *EE Journal, Portland, OR*
As each technology node increases in cost, the economics for FPGAs become more compelling. FPGAs have gotten larger, faster, and cooler while still maintaining flexibility. Have these factors brought us to a tipping point? Alternatives in process node, shuttles, and 3-D IC offer choices in the cost, performance, and power tradeoffs. What products are on the precipice of a decision in favor of FPGAs? Witness our expert panelists define the tradeoff point between FPGAs and ASICs.
Panelists:
Brent Przybus - *Xilinx, Inc., San Jose, CA*
Misha Burich - *Altera Corp., San Jose, CA*
Bill Lynch - *Huawei Technologies Co., Ltd., Santa Clara, CA*
Dave Ofelt - *Juniper Networks, Inc., Sunnyvale, CA*
Jeanne Trinko Mechler - *IBM Corp., Burlington, VT*
John Frediani - *Advantest America, Inc., Cupertino, CA*

SESSION 19:
WEDNESDAY June 06, 9:00am - 10:30am | Room 305
High-Level Synthesis Production Deployment: Are We Ready?
Chair: Clem Meas - *quickSTART Consulting, Boulder, CO*
High-level synthesis has historically over-promised and under-delivered, but that is all about to change. Or, is it? Are we ready to climb the ladder up to the next level of design abstraction? Watch our panelists debate whether today's technology can handle system validation, IP integration and optimization, power/performance constraints, and design verification challenges. Find out if we are about to connect the world of embedded software development to hardware design.
Panelists:
Eli Singerman - *Intel Corp., Haifa, Israel*
Kazutoshi Wakabayashi - *NEC Corp., Tokyo, Japan*

Mark Johnstone - *Freescale Semiconductor, Inc., Austin, TX*
Mark Warren - *Cadence Design Systems, Inc., San Jose, CA*
Vinod Kathail - *Xilinx, Inc., San Jose, CA*
Andres Takach - *Calypto Design Systems, Inc., Wilsonville, OR*

SESSION 25:
WEDNESDAY June 06, 1:30pm - 3:00pm | Room 305
Is EDA in the Cloud Just Pie in the Sky?
Chair: Nitin Deo - *Concept2Silicon Systems, Cupertino, CA*
Promises of lower costs, seemingly infinite resources, and faster turnaround times make EDA in the cloud an attractive proposition, but skepticism is prevalent. Some object that EDA in the cloud is not new and failed a decade ago. Others worry about security, confidentiality, and data protection. Do traditional time-based licenses fit in this model? Find out whether design in the cloud is ready for primetime.
Panelists:
Michael Buehler-Garcia - *Mentor Graphics Corp., Fremont, CA*
Bruce Jewett - *Synopsys, Inc., Mountain View, CA*
Alex Shubat - *SiCAD, Inc., Dallas, TX*
Anthony Hill - *Texas Instruments, Inc., Dallas, TX*
Pravin Desale - *LSI Corp., Milpitas, CA*

SESSION 31:
WEDNESDAY June 06, 4:00pm - 6:00pm | Room 305
Hot Apps, Cool Phones: Power-Efficient Mobile Design
Chair: Ed Sperling - *Low Power Engineering, San Jose, CA*
Recently, we have focused on techniques for low-power hardware design. But it is not enough. With the advent of app-driven mobile devices, battery life is paramount. We must now consider the impact of software on power consumption, and the EDA industry must look to providing environments that enable modeling, measuring and optimizing the impact of hardware and software interaction on power consumption at the system level. Our panelists explore the technical challenges and potential solutions for designing and verifying these complex power efficient systems.
Panelists:
Jan Rabaey - *Univ. of California, Berkeley, CA*
Emily Shriver - *Intel Corp., Hillsboro, OR*
Alan Gibbons - *Synopsys, Inc., Mountain View, CA*
Narendra Konda - *NVIDIA Corp., Santa Clara, CA*
Barry Pangrle - *Mentor Graphics Corp., Fremont, CA*
David Greenhill - *Texas Instruments, Inc., Dallas, TX*

SESSION 37:
THURSDAY June 07, 9:00am - 10:30am | Room 305
Is 3-D Ready for the Next Level?
Chair: Sachin Sapatnekar - *Univ. of Minnesota, Minneapolis, MN*
Early promises of 3-D IC integration – memory bandwidth and power (wide-IO memory stacks in consumer products), or yield and cost (FPGA die integrated with a silicon interposer) – have now been realized in volume production. What have the design and supply chains learned from the experience of enabling these applications? What will be the next killer applications for 3-D, how will these be enabled across the semiconductor industry, and what key technologies must the EDA industry contribute? Come hear the experts discuss how to "take 3-D to the next level."
Panelists:
Subramanian S. Iyer - *IBM Corp., Fishkill, NY*
Shekhar Borkar - *Intel Corp., Hillsboro, OR*
A.J. Incorvaia - *Cadence Design Systems, Inc., Chelmsford, MA*
Liam Madden - *Xilinx, Inc., San Jose, CA*
Suk Lee - *Taiwan Semiconductor Manufacturing Co., Ltd., San Jose, CA*

SESSION 43:
THURSDAY June 07, 1:30pm - 3:00pm | Room 305
It's the Software, Stupid! Truth or Myth?
Chair: Chris Edwards - *Tech Design Forum, London, United Kingdom*
It's tough to differentiate products with hardware. Everyone uses the same processors, third party IP and foundries; now

it's all about software. But is this true? Since user response, power consumption and support of standards rely on hardware, one camp claims software is only as good as the hardware it sits on. Opponents argue that software differentiates mediocre products from great ones. A third view says only exceptional design of both hardware and software creates great products – and the tradeoffs make great designers. Watch industry experts debate whether it's really all about software.

Panelists:
Serge Leef - *Mentor Graphics Corp., Wilsonville, OR*
Chris Rowen - *Tensilica, Inc., Santa Clara, CA*
Debashis Bhattacharya - *FutureWei Technologies, Inc., Plano, TX*
Kathryn S. McKinley - *Microsoft Research, Univ. of Texas, Austin, TX*
Eli Savransky - *NVIDIA Corp., Santa Clara, CA*

SESSION 49:
THURSDAY June 07, 2:45pm - 5:30pm | Room 305
Parallelization and Software Development: Hope, Hype, or Horror?
Chair: Igor Markov - *Univ. of Michigan, Ann Arbor, MI*
With the fear that the death of scaling is imminent, hope is widespread that parallelism will save us. Many EDA applications are described as "embarrassingly parallel," and parallel approaches have certainly been effectively applied in many areas. Before the panel begins, come hear perspective on software development and the challenges associated with writing good software that are only exacerbated by the growing need to write robust, testable, and efficient parallel applications. Then watch the panelists debate future productive directions and dead ends to developing and deploying parallel algorithms. Find out if claims to super speedups are exaggerated and if the investment in parallel algorithms is worth the high development cost.

Panelists:
Anirudh Devgan - *Magma Design Automation, Inc., Austin, TX*
Kunle Olukotun - *Stanford Univ., Stanford, CA*
Daniel Beece - *IBM Research, Yorktown Heights, NY*
Joao Geada - *CLK Design Automation, Inc., Littleton, MA*
Alan J. Hu - *Univ. of British Columbia, Vancouver, BC, Canada*

AWARDS

Marie R. Pistilli Women in EDA Achievement Award
Dr. Belle W.Y. Wei – *Don Beall Dean of the Charles W. Davidson College of Engineering, San Jose State Univ.*
For her significant contributions in helping women advance in the field of EDA technology.

P.O. Pistilli Undergraduate Scholarships for Advancement in Computer Science and Electrical Engineering
The objective of the P.O. Pistilli Scholarship program is to increase the pool of professionals in Electrical Engineering, Computer Engineering, and Computer Science from under-represented groups (women, African-American, Hispanic, Native American, and physically challenged). In 1989, the ACM Special Interest Group on Design Automation (SIGDA) began providing the program. Beginning in 1993, the Design Automation Conference has provided the funds for the scholarship and a volunteer committee continues to administer the program for DAC. DAC funds a $4000 scholarship, renewable up to five years, to graduating high school seniors.

The 2012 recipient is:
Catherine Agor Mullings

A. Richard Newton Graduate Scholarships
Each year the Design Automation Conference sponsors the A. Richard Newton Graduate Scholarship to support graduate research and study in Design Automation (EDA). Each scholarship is awarded directly to a University for the Faculty Investigator to expend in direct support of the project and students named in the application. The criteria are: the quality and applicability of the proposed research; the impact of the award on the EDA program at the institution; the academic credentials of the student(s); and financial need.

This year's scholarship goes to:

Advisor: **Prof. Yiran Chen** – *Univ. of Pittsburgh*
Student: **Wujie Wen** – *Univ. of Pittsburgh*

Project: NVSim-VX: Variation Aware Emerging Nonvolatile Memory Simulator

ACM/IEEE A. Richard Newton Technical Impact Award in Electronic Design Automation
Altan Odabasioglu – *Gear Design Solutions*
Mustafa Celik – *Synopsys, Inc.*
Larry Pileggi – *Carnegie Mellon Univ.*
For advancing the theory and implementation of model order reduction for efficient circuit analysis via dominant pole/zero methods.

"PRIMA: Passive Reduced-Order Interconnect Macromodeling Algorithm," IEEE Transactions on Computer-Aided Design of Integrated Circuits and Systems, August 1998, Vol. 17, Issue 8, Pages 645-654.

2011 Phil Kaufman Award for Distinguished Contributions to EDA
Sponsored by the EDA Consortium and IEEE Council on EDA
Dr. C. L. David Liu - *William Mong honorary chair professor of Computer Science and former president of the National Tsing Hua University in Hsinchu, Taiwan*
Dave Liu is honored for his work in leading the transformation from ad hoc EDA to algorithmic EDA.

IEEE CEDA Outstanding Service Contribution
Leon Stok – *IBM Corp.*
For significant services as DAC General Chair 2011.

Donald O. Pederson Best Paper Award for the IEEE Transaction on CAD

Umit Ogras – *Intel Corp.*
Paul Bogdan – *Carnegie Mellon Univ.*
Radu Marculescu – *Carnegie Mellon Univ.*
For the paper titled: "An Analytical Approach for Network-on-Chip Performance Analysis," IEEE Transaction on Computer-Aided Design of Integrated Circuits and Systems, December 2010, Vol. 29, No. 12, pp. 2001-2013.

SIGDA Outstanding New Faculty

David Atienza – *Ecole Polytechnique Fédérale de Lausanne*
Outstanding New Faculty for 2012.

ACM/SIGDA Outstanding Ph.D. Dissertation Award

Dr. Tan Yan – *University of Illinois at Urbana-Champaign*
Prof. Martin D. Wong (Dissertation Advisor) – *University of Illinois at Urbana-Champaign*
In recognition of the outstanding dissertation "Algorithmic Studies on PCB Routing."

IEEE Fellow

Luis Miguel Silveira – *Technical Univ. of Lisbon*
For contributions to analysis and modeling of VLSI interconnects.

IEEE Fellow

Steve Trimberger – *Xilinx Inc.*
For contributions to circuits, architectures, and software technology for field-programmable gate arrays.

IEEE Fellow

Naehyuck Chang – *Seoul National Univ.*
For contributions to system-level power characterization, including thermal management

49th DAC Best Paper Candidates

Seven papers were nominated by the Technical Program Committee as DAC Best Paper Candidates. Final decisions will be made after the papers are presented at the Conference.

5.4 On Improving the Uniqueness of Silicon-Based Physically Unclonable Functions via Optical Proximity Correction

Domenic Forte – *Univ. of Maryland, College Park, MD*
Ankur Srivastava – *Univ. of Maryland, College Park, MD*

9.1 Physics Matters: Statistical Aging Prediction under Trapping/Detrapping

Jyothi Bhaskarr Velamala - *Arizona State Univ., Tempe, AZ*
Ketul B. Sutaria - *Arizona State Univ., Tempe, AZ*
Takashi Sato - *Kyoto Univ., Kyoto, Japan*
Yu Cao - *Arizona State Univ., Tempe, AZ*

15.6 Chip/Package Co-Analysis of Thermo-Mechanical Stress and Reliability in TSV-Based 3-D ICs

Moongon Jung - *Georgia Institute of Technology, Atlanta, GA*
David Z. Pan - *Univ. of Texas, Austin, TX*
Sung Kyu Lim - *Georgia Institute of Technology, Atlanta, GA*

17.4 Explicit Modeling of Control and Data for Improved NoC Router Estimation

Andrew B. Kahng - *Univ. of California at San Diego, La Jolla, CA*
Bill Lin - *Univ. of California at San Diego, La Jolla, CA*
Siddhartha Nath - *Univ. of California at San Diego, La Jolla, CA*

18.1 WCET-Centric Partial Instruction Cache Locking

Huping Ding - *National Univ. of Singapore, Singapore*
Yun Liang - *Advanced Digital Sciences Center, Singapore*
Tulika Mitra - *National Univ. of Singapore, Singapore*

27.1 Triple Patterning Aware Routing and its Comparison with Double Patterning Aware Routing in 14nm Technology
Qiang Ma - *Univ. of Illinois at Urbana-Champaign, Urbana, IL*
Hongbo Zhang - *Univ. of Illinois at Urbana-Champaign, Urbana, IL*
Martin D. F. Wong - *Univ. of Illinois at Urbana-Champaign, Urbana, IL*

35.2 Recovery-Based Design for Variation-Tolerant SoCs
Vivek J. Kozhikkottu - *Purdue Univ., West Lafayette, India*
Sujit Dey - *Univ. of California at San Diego, La Jolla, CA*
Anand Raghunathan - *Purdue Univ., West Lafayette, IN*

REVIEWERS

A total of 741 manuscripts were submitted to the 49th DAC. The Technical Program Committee, together with the help of invited expert and external reviewers, selected 168 papers for presentation at the conference. The Conference Executive and Technical Program Committees wish to acknowledge the time and effort spent by the following people who reviewed these manuscripts. Many thanks to all of those who participated and contributed to the success of the conference.

Expert Reviews (Topic experts invited by the TPC Subcommittee Chairs)

Abhijit Davare
Akash Kumar
Alain Girault
Alan Mishchenko
Alastair Donaldson
Alessandro Pinto
Alex Kondratyev
Alexandre Tenca
Alexandros Bartzas
Alper Sen
Ameya Chaudhari
Amith Singhee
An Chen
Anand Raghunathan
Anand Rajaram
Andreas Hansson
Andreas Veneris
Andres Torres
Ann Gordon-Ross
Anna Slobodova
Antonino Tumeo
Antonio Miele
Anup Gangwar
Arjun Rajagopal
Aryabartta Sahu
Ashutosh Chakraborty
Asit Mishra
Azad Naeemi
Bao Liu
Baris Taskin
Bart Vermeulen
Basant Dwivedi
Benjamin Carrion Schafer
Benny Akesson
Bernhard Egger
Bin Li
Bin Wu
Björn Franke
Brady Benware
Brandon Noia
Brett Meyer
Bruce Mcgaughy
Byong Chan Lim
Can Hankendi
Carlo Brandolese
Carlo Galuzzi
Caroline Concatto
Chaoming Zhang
Cheng Zhuo

Chenjie Gu
Chih-Wei Chang
Chirayu Amin
Chokri Mraidha
Chris Chu
Chris Papachristou
Chris Schuermyer
Chris Wilson
Christian Haubelt
Chrysostomos Nicopoulos
Chul-Hong Park
Chung-Ta King
Chunhong Chen
Claire Maiza
Constantin Timm
Corey Goodrich
Danella Zhao
Daniel Kroening
Daniel Tille
David Chinnery
David Hathaway
David Newmark
David Thomas
David Whelihan
Davide Quaglia
Dhireesha Kudithipudi
Dhruva Acharyya
Diana Goehringer
Dimitrios Soudris
Dimitris Gizopoulos
Dong Xiang
Dumitru Potop Butucaru
Duo Ding
Elena Teica
Eli Bozogzadeh
Eli Chiprout
Enamul Amyeen
Eren Kursun
Eric Bracken
Eric Keiter
Erika Cota
Eugenio Villar
Eun Jung Jang
Evan Rosser
Evangeline Young
Fabrizio Lombardi
Florence Maraninchi
Fnu Aatmesh
Francesco Regazzoni
Francisco Cazorla
Francky Catthoor
Frank Vahid
Friedrich Taenzler
Georges Gielen

Georgi Gaydadjiev
Georgi Kuzmanov
Gerard Luk-Pat
Gerardo Pelosi
Gianluca Palermo
Giovanni Beltrame
Giovanni Mariani
Giovanni Squillero
Greg Stitt
Guangyu Sun
Guido Bertoni
Gunar Schirner
Hai Li
Hai Zhou
Haibo Zeng
Hailong Yao
Hans Manhaeve
Hans Vandierendonck
Hans-Joachim Wunderlich
Hao Yu
Harry Foster
Harry Levinson
Haykel Ben Jamaa
Hector Posadas
Henri Fraisse
Hiren Patel
Hiroyuki Tomiyama
Holger Blume
Hua Xiang
Hugo Andrade
Hung-Ming Chen
Ian O'Connor
Ibrahim (Abe) Elfadel
Igor Keller
Igor Markov
Ioannis Koutras
Iuliana Bacivarov
Jacob Kornerup
Jaejin Lee
James C-M Li
Jan Haase
Jarmo Takala
Jason Anderson
Jason Baumgartner
Jay Adams
Jay Bhadra
Jayaram Natarajan
Jean Christophe Madre
Jens Teubner
Jeremy Levitt
Jiajing Wang
Jiang Xu
Jian-Jia Chen
Jilin Tan

Jim O'Reilly
Jing Guo
Joan Figueras
Jogesh Muppala
Johan Lilius
John West
Jon Pimentel
Jonathan Greene
Jongman Kim
Jongwook Kye
Jose Ayala
Jose Ignacio Gomez
Julian Murphy
Kamana Sigdel
Karthik Sankaranarayanan
Ken Albin
Ken Eguro
Kerstin Eder
Kevin Lucas
Koen Bertels
Koji Inoue
Kolin Paul
Krishnaiah Gummidipudi
Krishnendu Chakrabarty
Kun Yuan
Kundan Nepal
Kyle Rupnow
Laleh Behjat
Lars Bauer
Laura Pozzi
Laurent Arditi
Leandro Fiorin
Leandro Indrusiak
Lerong Cheng
Li-Chung Wang
Lijuan Luo
Lin Huang
Luca Carloni
Luciano Lavagno
Luigi Carro
Madhu Mutyam
Mahesh Prabhu
Manuel Prieto Matias
Marc Pouzet
Marcio Juliato
Marco Ottavi
Mariagiovanna Sami
Marisa Lopez-Vallejo
Mark Greenstreet
Masahiro Fujita
Masood Qazi
Mathieu Luisier
Matteo Monchiero
Matteo Sonza Reorda
Matthew Guthaus
Matthew Ziegler
Maurizio Palesi
Mazen Saghir
Meng-Kai Hsu
Michael Engel
Michael Healy
Michael Hsiao

Michael Huebner
Michal Rewienski
Michel Berkelaar
Michel Bourdellès
Michihiro Koibuchi
Mihir Choudhury
Mineo Kaneko
Miroslav Velev
Mladen Berekovic
Mohamed Abu-Rahma
Mohammed Ghiath Khatib
Morteza Biglari-Abhari
Muhammad Shafique
Mustafa Ozdal
Nagib Hakim
Neeraj Goel
Nicholas Callegari
Nick Van Der Meijs
Nicola Nicolici
Nisar Ahmed
Nur Touba
Ozcan Ozturk
Ozgur Sinanoglu
Pao-Ann Hsiung
Paolo Meloni
Patricia Derler
Patrick Longa
Patrick Vuillod
Paul Franzon
Paulo Teixeira
Pete Manolios
Peter Zepter
Petru Eles
Philip Brisk
Philip Leong
Philippe Coussy
Pierre Boulet
Pingqiang Zhou
Prabhat Avasare
Qiang Liu
Qiang Xu
Radu Muresan
Rainer Doemer
Rainer Dorsch
Ramkumar Jayaseelan
Reetuparna Das
Reto Zimmermann
Roberto Suaya
Ruifeng Guo
Russell Tessier
Saibal Mukhopadhyay
Saket Gupta
Saket Srivastava
Salim Chowdhury
Samar Abdi
Sami Yehia
Sander Stuijk
Sandip Ray
Satnam Singh
Savithri Sundareswaran
Scott Davidson
Sebastian Hack

Sergey Gribok
Shangping Ren
Shao-Yun Fang
Shayak Banerjee
Sheng Li
Shengqi Yang
Siddharth Garg
Smail Niar
Smruti Sarangi
Sobeeh Almukhaizim
Solmaz Ghaznavi
Sonali Chouhan
Soontae Kim
Sotirios Xydis
Sourav Roy
Sri Parameswaran
Srinivas Katkoori
Stavros Tripakis
Stefano Quer
Stephan Wong
Steve Kelem
Steve Trimberger
Steven Wilton
Sudarshan Banerjee
Sudeep Pasricha
Sudhakar Reddy
Suhaib Fahmy
Sung Woo Chung
Sungjoo Yoo
Suriyaprakash Natarajan
Syed Alam
Tajana Simunic Rosing
Tan Yan
Taniya Siddiqua
Tarek El-Moselhy
Tiantian Liu
Timothy Fischer
Tiziano Villa
Tm Mak
Todor Stefanov
Tom Vander Aa
Torsten Kempf
Toshinori Sato
Trent Mcconaghy
Tsung-Wei Huang
Tsung-Yi Ho
Umit Ogras
Valeria Bertacco
Vaughn Betz
Victor Kravets
Vikas Chandra
Viresh Paruthi
Vivien Quema
Vladimir Zolotov
Walid Najjar
Warren Hunt, Jr.
William Fornaciari
Xiaochun Yu
Xiaoji Ye
Xiaojian Yang
Xiaoqing Wen
Xiaoxiao Wang

Xijiang Lin
Xuan Zeng
Yanjing Li
Yaping Zhan
Yi Xu
Yinhe Han
Yiran Chen

Yiyu Shi
Yoshi Shih-Chieh Huang
Yoshio Takamine
Yu Cao
Yu Wang
Yun Liang
Yung-Hsiang Lu

Yusuke Matsunaga
Zdenek Kotasek
Zheng Shi
Zhenyu Qi
Zhiru Zhang
Zhiyu Zeng
Zhuo Feng

External Reviews (Technical volunteers affiliated with DAC)

A. C. Rajeev
Abinash Roy
Aditya Mukherjee
Ajay Joshi
Akash Singh
Alberto Ferrante
Alexandro Adário
Ali Ahmadinia
Ali Jafari
Ali Mahdoum
Alok Jain
Altaf Abdul Gaffar
Alvin Jee
Amit Singh
Anastasia Stulova
Andreas Raabe
Andrew Seawright
Angan Das
Anil Pandey
Arda Yurdakul
Arek Zawada
Arif Selcuk Ogrenci
Aritra Hazra
Arshin Rezazadeh
Artjom Grudnitsky
Assem Bsoul
Ateet Bhalla
Aurelia De Colle
Balaji Raman
Balsha Stanisic
Bei Yu
Benfei Wang
Berkin Ozisikyilmaz
Bharath Seshadri
Bill Dougherty
Biswadeep Chatterjee
Bo-Cheng Lai
Bo-Shiun Wu
Brian Mulvaney
Changhwan Shin
Chao Li
Chao Wang
Charles Won
Chen Dong
Cheng-Wu Lin
Chiayi Lin
Chieh Jui Lee
Ching-Yu Chin
Christian Pilato
Christian Zoellin
Chun Zhang

Chung-Yang (Ric) Huang
Chun-Kai Wang
Claudio Brunelli
Colin Ihrig
Cristinel Ababei
Da-Cheng Juan
Dae Hyun Kim
Dan Alexandrescu
Daniel Grosse
Daniel Saab
David Baneres
David Bol
David Papa
Deepak Shankar
Dimitris Nikolos
Dmitry Vasilyev
Dwight Hill
Eduard Cerny
Eduardo Wanderley Netto
Emily Shriver
Ender Yilmaz
Enrico Costenaro
Eric Sherk
Erick Amador
Ethiopia Nigussie
Eyal Bin
Fabrizio Ferrandi
Faisal Mohd-Yasin
Fang Gong
Farhad Mehdipour
Farhad Merchant
Farshad Moradi
Felipe Marques
Francisco Assis Moreira Do Nascimento
Frank Hannig
Frederic Rousseau
Frederik Vermeulen
Fredy Rivera
Gang Qu
Georgios Keramidas
Gjalt De Jong
Goerschwin Fey
Guoqiang Wang
Gustavo Wilke
Hai Wang
Han Wang
Harpreet Gill
Harsh Vardhan
Heiko Falk
Heng Yu
Heon-Mo Koo
Himanshu Thapliyal
Huan Chen

Hwisung Jung
Hyojung Han
Indunil Sikurajapathi
Ing-Chao Lin
Iouliia Skliarova
Iris Hui-Ru Jiang
Jaclyn Dang
Jaeyong Chung
Jagdish Rao
Jai-Ming Lin
Jianchao Lu
Jianyong Xie
Jilong Kuang
Jin Ouyang
Jishen Zhao
Jiwoo Pak
Johann Groszschaedl
John Grout
John Taylor
Joonsoo Kim
Jovana Jovic
Juan Antonio Maestro
Kai-Chiang Wu
Kai-Hui Chang
Kaiming Ho
Karthik Kumar
Karthik Rajagopal
Kazuhiko Iwasaki
Kenneth Francken
Kenneth Mai
Kimihiro Ogawa
Koustav Bhattacharya
Krishnan Sundaresan
Kuei-Chung Chang
Kuntal Nanshi
Lang Lin
Le Jin
Lei Gao
Lluís Ribas Xirgo
Luz Balado
Madhura Purnaprajna
Mahalingam Venkataraman
Mahdi Nikdast
Mahmoud Momtazpour
Manu Jose
Mark Fredrickson
Masanori Kurimoto
Masaru Kakimoto
Massimo Violante
Mathias Silvant
Mathias Soeken
Matthias Gries
Maurizio Martina
Michael Glass

Michael Riepe
Michail Dimopoulos
Michele Petracca
Michiko Inoue
Mihalis Psarakis
Mike Borowczak
Minh Nguyen
Mohamed Abdelhalim
Mohamed Madbouly
Mohit Pathak
Muhammad Adeel Ahmed Pasha
Murthy Palla
Mwaffaq Otoom
Nelson Passos
Nicola Bombieri
Nicola Concer
Nimish Sane
Ognen Nastov
Olivier Gautherot
Omer Khan
P Balasubramanian
Pablo Sanchez
Palkesh Jain
Paul Darga
Paul Schumacher
Paulo Butzen
Peeter Ellervee
Peter Hallschmid
Pierluigi Nuzzo
Pietro Babighian
Piti Piyachon
Po-Cheng Pan
Po-Hsun Wu
Pramod Chandraiah
Prasad Subramaniam
Puneet Sharma
Raimund Ubar
Raj Mitra
Rani Ghaida
Ravikishore Gandikota
Ravishankar Rao
Raviv Gal
Renato Hentschke
Ren-Jie Lee
Robert Walker
Ruijing Shen
Sachin Shrivastava
Sagar Sabade

Said Al-Sarawi
Sambuddha Bhattacharya
Samir Boubezari
Samuel Ward
Sandro Bartolini
Saurabh Hukerikar
Saurabh Kotiyal
Sean Shih Ying Liu
Sebastien Pillement
Seok-Bum Ko
Sergio Bampi
Shalom Bresticker
Shaobo Liu
Shashi Kumar
Sheng-Jhih Jiang
Shreeharsha Balan
Sidharta Andalam
Siong Kiong Teng
Sohaib Majzoub
Somnath Paul
Song Liu
Sreeram Chandrasekar
Sridhar Narayanan
Srobona Mitra
Stefan Wallentowitz
Stephen Blythe
Steven Drager
Subramanyam Sripada
Sujan Pandey
Supriyo Maji
Sushu Zhang
Swapnil Bahl
Swarup Mohalik
Syed Manzoor Qasim
Taemin Kim
Taewhan Kim
Tai-Chen Chen
Teijo Lehtonen
Thomas Brandtner
Thomas Dillinger
Thomas Feller
Thorlindur Thorolfsson
Tong Xiao
Trevor Meyerowitz
Tu-Hsiung Tsai
Tun Li
Ulf Schlichtmann
Valery Sklyarov

Vasily Moshnyaga
Vijay Pasupuleti
Vitali Sokhin
Vittorio Zaccaria
Vivek Joshi
Wai-Kei Mak
Wanderson Roger Azevedo Dias
Wangyang Zhang
Wan-Yu Lee
Wei Wu
Wei_Li Kuo
Weixun Wang
Wolfgang Guenther
Wooyoung Jang
Xiaoke Qin
Xin-Wei Shih
Xiongfei Liao
Xiwei Huang
Xuanxing Xiong
Xuchu Hu
Yang Shang
Yanheng Zhang
Yifan He
Yilin Zhang
Ying Khai Teh
Yoav Katz
Yong Zhang
Youhua Shi
Young-Joon Lee
Young-Su Kwon
Yousra Alkabani
Yu Chen
Yuan-Hao Chang
Yuanzhe Wang
Yu-Chien Kao
Yuchun Ma
Yuhao Wang
Yuko Hara-Azumi
Zhan Chen
Zhangxi Tan
Zhe Feng
Zhen Fang
Zheng Li
Zhonghai Lu
Zijun Yan
Zoran Stamenković

Biomedical Electronics Serving as Physical Environmental and Emotional Watchdogs

Rudy Lauwereins

Imec & KU Leuven
Kapeldreef 75
B-3001 Leuven, Belgium
+32-16-281244

Rudy.Lauwereins@imec.be

ABSTRACT

Over forty years of happy CMOS scaling brought the room-sized super-computer for the nerds into everyone's pocket, literally connecting every-body on earth. In an economy which is based on double digit growth, the obvious next step is to connect everything on earth.

This move redirects the focus from electronics-for-infotainment to electronics helping to solve the mounting societal challenges our earth faces: better and more affordable health care for everyone, safer and more efficient transportation, cleaner and more sustainable environment. Realizing this requires abandoning the traditional keyboard/screen user interface to make the electronic devices autonomous, independent from a human in the loop, and to provide its services hidden in the background.

In this paper, I will first explain why in the background operating electronics recently became feasible in the form of autonomous wireless sensor nodes. Next, I will present a technology roadmap, ranging from sensors measuring physical phenomena, via environmental sensors that combine physical with chemical monitoring, to ultimately emotional sensors that provide instantaneous and objective information about one's emotions. For those worrying about "big brother" possibilities, I will end the presentation with a concrete use case for psychiatric drug approval.

Categories and Subject Descriptors

J3 [**Life and Medical Sciences**]: Health and wellness

General Terms

Design, Experimentation, Security, Human Factors, Verification.

Keywords

Hidden electronics, autonomous sensor nodes, More than Moore, emotion monitoring.

1. FORTY YEARS OF CMOS SCALING

The whole computing history (see e.g. [1] for a nice online overview) started with large, power hungry computing machines accessible to the happy few, typically in research laboratories (see the leftmost point in Figure 1). These machines evolved from hand-operated mechanical calculators, like the ones developed by Blaise Pascal in 1642 and Charles Babbage in 1834, over relay-based electrical machines, like the Z-1 developed in 1935 by Conrad Zuse, to vacuum tube engines, like the Eniac designed in 1943 by Mauchly and Eckert. With the invention of the transistor by Shockley, Bardeen and Brattain in 1947 and especially the invention of the integrated circuit by Jack Kilby in 1958, the path towards miniaturisation lay open. The use of discrete transistors in the fifties reduced the big-hall size of the former vacuum tube computers to room size. The introduction of small integrated circuits in the sixties further reduced the size to occupy a single cubicle and in the seventies even a single desk.

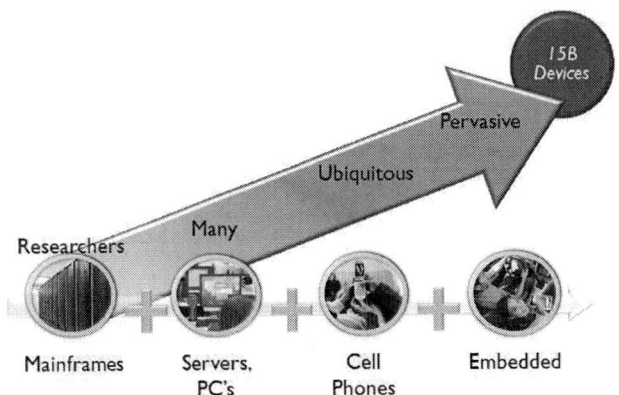

Figure 1. Intel's view on the four major steps in computing history.

In the eighties, the further exponential increase in integration density reduced the size of a computer to fit on top of a desk. This enabled people to install them at home: the desktop personal computer (PC) was born (see the second point from the left in Figure 1). This increased the market size from a few mainframes for the nerds, to hundreds of millions PCs.

In the nineties, transistor density became so high and hence cost-per-function so cheap, that a consumer device could not only combine computing and storage like a PC, but could add communication capabilities: the mobile phone was born (see the third point in Figure 1). Suddenly, literally everyone on earth could become connected at any time and any place: ubiquitous access. The market grew to billions of devices. Not only the computation performance increased by 55% per year, but also the data storage capacity increased exponentially: the 1.5 ton heavy ENIAC computer in 1943 could store 200 bytes; advanced smart phones today can store 32 GByte in a device of 100-200 gram. If this evolution continues, one can video record his/her whole life (24 hours per day during 80 years) in YouTube quality on a single post stamp size memory card in 2024... costing 25 EUR to the end customer! In 2027, the same will be possible in standard TV quality, in 2029 in HDTV quality and in 2030 in 3D HDTV.[1] This example shows how powerful of an exponential evolution becomes when it spans multiple decades. It also indicates that the biggest markets for the electronics industry today are in information processing and entertainment (often called "info-tainment").

What would be the logical next step we anticipate to observe in the second decade of the twenty-first century? Intel (Figure 1) and others predict we will evolve from connecting every-body to literally connecting every-thing, sometimes called the evolution towards an Internet-of-Things [4, 5]. This will grow the potential market to multiple tens of billions of devices. Electronics will become hidden in the background, mostly unnoticed by people. And it will not just provide humans with information processing and entertainment aids, but it will actually help to solve the grand challenges our society faces: better and more affordable health care for everyone, safer and more efficient transportation, cleaner and more sustainable environment. The continuation of exponential density scaling will however not suffice to enable pervasive hidden electronics. Despite the amazing, unparalleled exponential evolution over more than forty years, we still interface our electronic devices in the same way: with keyboard and display screen (Figure 2 and Figure 3). When electronics becomes hidden to people, it should be able to act autonomously [6], without a human in the loop. Autonomous operation hence requires advanced sensors and actuators, next to computation, storage and communication.

2. MORE THAN MOORE ENABLING HIDDEN ELECTRONICS

More than Moore technology [7] allows us to fabricate various structures on top of CMOS chips, using the same equipment and process technologies as used for the fabrication of silicon integrated circuits (Figure 4). The structures that can be built include micro-electromechanical systems (MEMS) like mirrors with controllable tilt angle, physical and chemical sensors like vibrating beams usable to implement an electronic nose, silicon photonic waveguides usable as light valves or light modulators and micro machined coated pillars usable to directly interface to life cells.

---

[1] This is calculated as follows: today's retail price for an 32 GByte SDHC card is around 25 EUR. Over the last 10 years, we observed a doubling of the Flash memory capacity at the same cost every year; the calculation assumes this will continue till at least 2030 (something which is doubtful, but such assumptions are typical for extrapolations...). YouTube quality = 3 Mbytes/minute; standard TV = 2 GByte/hour; HDTV = 8 GByte/hour; 3D HDTV = 16 GByte/hour.

Figure 2. One of the very first personal computers, the PET from Commodore, was interfaced through keyboard and display screen

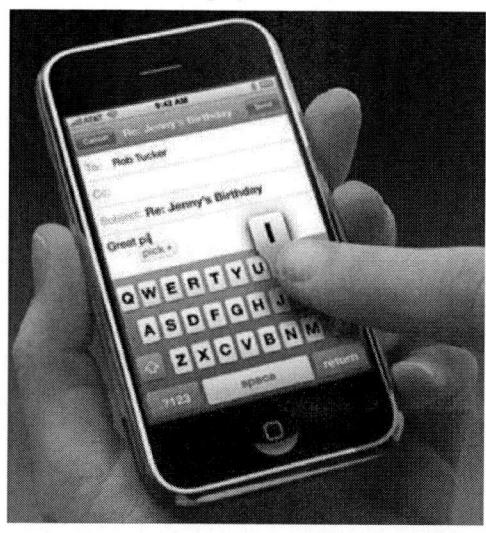

Figure 3. But also today's advanced smart phones, like Apple's iPhone, are still interfaced through keyboard and screen.

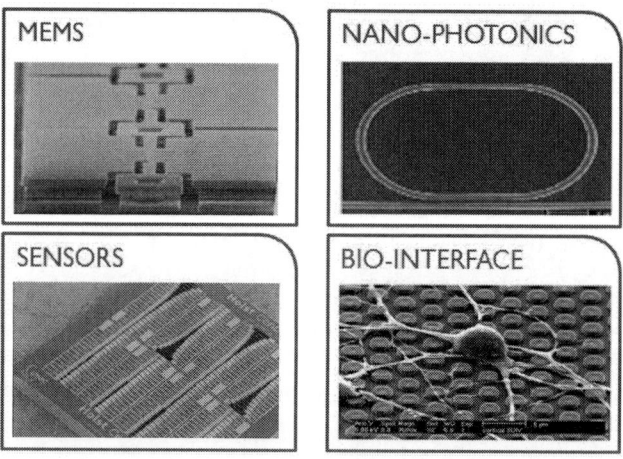

Figure 4. More than Moore structures on top of CMOS chips.

The addition of More than Moore structures to silicon based computation, storage and communication devices offers them the potential to autonomously sense and interact with their environment, thereby eliminating the "human-in-the-loop" [8]. Electronics becomes truly hidden.

3. PHYSICAL, ENVIRONMENTAL AND EMOTIONAL WATCHDOGS

Watchdogs are envisioned as interconnected, smart, autonomous systems enabled by energy-efficient nanotechnology that can operate in an unobtrusive, background manner.

The envisioned applications are based on the concept of a smarter life, e.g. a lifestyle that profits from the instant availability of relevant information, more interconnectivity between devices fitted with all sorts of sensors, and prominently features intentional and intuitive usability: (I) **Physical watchdogs**: These devices will monitor the physical and/or physiological status of individuals in sports, rehabilitation, health and day care, with an awareness of the context of activity of these individuals. With a strong focus on prevention and early diagnosis, these devices will help keep healthcare affordable and accessible to all. (II) **Environmental watchdogs:** These devices will observe ambient conditions for environmental threats, and will be able to communicate with each other to expand their information base. With environmental sensors next to the physical sensors, it will be possible to correlate a person's physical state with the environmental context. (III) **Emotional watchdogs** [9]: These devices will be able to perceive emotional or affective conditions, and will provide helpful functions for the disabled, or could serve to build new generations of smart-driver assistants for automotive and airborne applications. Emotional watchdogs featuring high data security and enhanced personal control will certainly lead to novel societal paradigms that allow improved cognition, communication and prevention.

As the application scenarios above suggest, the impact of deploying such watchdogs on society and the economy is very broad. On an individual level, anyone could benefit from a personal guard, focusing on prevention in health and on personal safety, with the goal of maintaining or improving his or her quality of life in a sustainable society.

Watchdogs are foreseen as interconnected, smart, autonomous systems enabled by energy-efficient nanotechnology; they can be considered as the future of wireless sensors networks [2] (WSN), and by their functionality they can include components of the internet of things [3]. They will be interconnected, not only between themselves, but also through a gateway layer (mobile phones, PDAs, notebooks, tablets) to massive data handling in the cloud (high-performance). By their smartness and complexity they will enable personalized advice and assistance, concerning health and interaction with the environment, far beyond what today's WSN and internet-of-things devices can provide. Watchdog technology will offer unique solutions for new generations of non-invasive biological monitoring, and for future smart apparel with embedded powering and sensing. They will enable unforeseen generations of autonomous robots. The supporting ultra low power technology platform will impact development within other domains such as environmental, building and industrial monitoring, and efficient transportation. It will offer new progress paths for energy-efficient data processing in cloud computing, and change the way mobile computing interacts with humans' needs.

Watchdogs will provide data which will allow us to extract relevant information for a smarter life: making life easier when you are well, helping to efficiently use energy sources, and maintaining or improving mobility and industrial processes without exhausting natural resources. Watchdogs will also play a vital role for those of us who need increasing, or even continuous, support and services due to health problems or reduced mobility or sensory capabilities. Imagine mobile electronic personal assistants being 1000x more energy-efficient than they can be today, so that they could be powered by the energy available in your environment without any power plugs. Imagine part of the energy savings transformed into sensing, communications and interface functions of an invisible system that becomes your day-to-day guard.

4. TOWARDS A ROADMAP FOR WATCHDOGS

Almost all use cases deploy the same set of sensors. From an economical point of view, it hence makes sense to develop a toolbox of standardized sensor devices and platforms, growing over time in coverage, complexity and analysis capabilities, and with decreasing energy consumption needs. This toolbox encompasses physical measurements (motion, temperature, humidity, pressure, conductivity, sound volume and frequency, voltage, ...) as well as chemical and hormone measurements (dehydratation, cortisol, adrenaline, NOx, COx, ...) in any kind of body fluids, breath and air. They should be accompanied by signal conditioning and artifact removal measures. The toolbox should also contain standardized computation, data storage, communication, energy storage and power management components. Implementing a given use case hence boils down to selecting the right components from the toolbox and matching them with use case specific sensor fusion, classification and analysis algorithms.

Since all watchdog use cases are built from the same toolbox, it should be possible to use the same watchdog system differently in different contextual situations. This will be a situational-watchdog that either responds to explicit external signals or coarsely interprets the context to trigger a suitable reconfiguration. As an example, we may think of the situational-watchdog being contacted by the car sensory system to reconfigure the body sensors so that attention in driving is the main goal of detection. Given the increase of voice volume in the car cabin the situational-watchdog may want to switch to a higher priority task, i.e., the monitoring of arousal which is a potentially dangerous situation for the watchdog wearer and for other human beings in the neighborhood. Assume then that, despite whatever action has been taken in response to the detection of a dangerous arousal, the car has an accident. It is not difficult to think that a strong signal in accelerometers or a further exchange with the car sensory system may induce the situational-watchdog to reconfigure the body sensors to perform pain-assessment and analysis of the body vital functions and thus be ready to give updated information to the incoming rescue team…

All use cases put a lot of emphasis on comfort and discreteness. The ultimate goal is to let the sensor systems completely disappear in a person's personal environment: in his/her cloths, watches, spectacles, jewelry. This puts extreme requirements on the packaging: washable, flexible, stretchable, bio-compatible, sealed yet transparent for whatever has to be measured, miniaturized. Recharging secondary batteries or replacing primary batteries is considered cumbersome and a source of discomfort. Ultra high energy efficiency is hence a must. It also puts extreme requirements on the signal capturing and conditioning, especially when the poor body contacts feasible for sensors embedded in cloths or body worn devices like spectacles and jewelry are combined with large motion artifacts in real life situations as opposed to today's use in controlled clinical environments. Countermeasures today typically require high power consumption, increasing the need for ultra low power solutions. System level optimization is clearly a must.

All use cases need a combined progress on multiple axes. Some of the axes, e.g. security and computational performance, need a very fast improvement to enable watchdog acceptance by the general public. Others, e.g. heterogeneous integration and environmental awareness, become important only in later phases of deployment. Figure 5 below lists the various axes of progress and showcases their relative progress needed over time, with the inner circle representation today's state-of-the-art and the outer circle the ultimate functional and non-functional specifications to be reached in next decade.

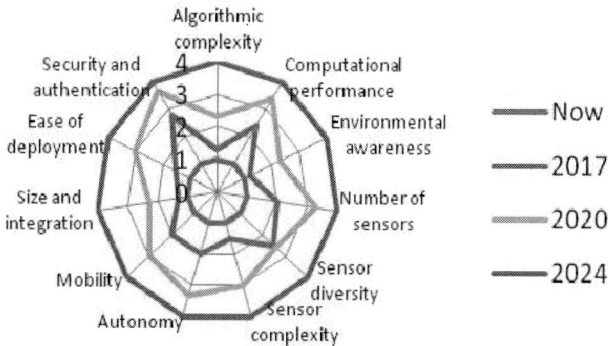

Figure 5. Required technology progress for biomedical watchdogs.

5. USE CASE: OBJECTIVE MEASUREMENT OF EFFICACY OF PSYCHIATRIC DRUGS

Today, medical doctors find it extremely hard to prescribe the right medicine at the right dose for most psychiatric illnesses. Often, symptoms of the illness are vague and subjective and hence the influence of the medication on the symptoms is hard to judge. Current practice is based on the MD interviewing the patient at e.g. the weekly consult and asking him/her questions about how he/she felt every hour during past week. This input is often wrong: people tend to forget how they really felt a week ago, or do not want to admit they feel worse (or better). Improvement or worsening of symptoms cannot only be attributed to the drugs they took: people's feelings are strongly influenced by many external factors including the quality of their sleep, the weather, the discussions or quarrels they had, ... Although it is hard to remember how you felt every hour of the day, an hourly resolution is in fact still too coarse grained, since emotions and feelings change rapidly, with time constants measured in seconds rather than hours.

The objective of deploying physical, environmental and emotional watchdogs in this use case is to complement the traditionally collected subjective data with automatically recorded objective data about the response of the human body to his/her state of mind. It is known that people's feelings have an impact on their autonomous nervous system (ECG, respiration rate and depth, galvanic skin response, temperature, muscle tension), on the ratio of the alpha waves of the left and right part of the brain (EEG), on the chemical composition of their sweat (cortisol, adrenaline), and on their activity level (acceleration of joints, ocular movement).

To rule out the impact of environmental conditions on the state of mind, these body reaction measurements should be correlated with environmental data about the actual weather (temperature, humidity, wind, pressure), the ambient noise (level and frequency) and a general interpretation of the discussions held by the patient (positive, neutral, insulting, demanding, ...).

For privacy reasons, data security is obviously of highest importance. A multi level security scheme with corresponding trusted zones is needed: the patient and the medical doctors he/she appoints, have access to all data; other care takers like psychologists and physiologists can access only part of the data on a need to know bases; if the patient agrees, his/her anonymized data can be uploaded to a central database for epidemiological analysis. Equally important is data authentication, to guarantee that the collected data has not been tampered with, e.g. to prevent data manipulation for polishing up the efficacy of a certain drug.

It is anticipated that the different psychiatric illnesses will all need the collection of the same objective data, but analyzed using different sensor fusion and classification algorithms. It is hence important to develop a toolbox of standardized sensor components that will evolve over time from simple to complex measurements and from bulky, power hungry and uncomfortable boxes to tiny, zero power elements that completely disappear in cloths and jewelry.

Data collection should be unobtrusively captured during normal real life, to avoid that the measurement itself influences the state of mind of the patient. It hence should work reliably and accurately in all environments people live in (from cold to warm, from dry to wet) and under all activity levels (from sleep to extreme sports). Over time, better electrodes, signal conditioning and motion artifact removal should be developed.

Such a system for objective measurement of the efficacy of psychiatric drugs will not only be useful for medical professionals during treatment, but also during the approval process of new drugs.

6. CONCLUSION

Forty years of CMOS scaling enabled us to put a former super-computer for the nerds into every-body's pocket, thereby providing the world population with ubiquitous connectivity. Further miniaturization will make it possible to connect every-thing. However, miniaturization will not suffice: to make electronics hidden in the background, autonomous operation is a must, which implies sensing and actuation. This is enabled by More than Moore technology: the ability to integrate sensors and actuators on top of silicon based computation, communication and storage devices.

Hidden electronics will enable the development of physical, environmental and emotional watchdogs that can continuously overlook a person's well being and safeguard him/her from calamities. The huge computational demands needed for fusing the sensor data combined with the need for autonomous long term operation without replacing or explicit re-charging of batteries, urges for substantial energy efficiency improvements.

The deployment of watchdog systems will enable a wealth of new services, not in the least place in medical treatments, e.g. to determine the efficacy of psychiatric drugs.

7. REFERENCES
[1] The History of Computing Project: www.thocp.net

[2] J. Yick, et al., *Computer Networks*, vol. 52, 2008, pp. 2292-2330.

[3] H. Chaouchi, ed., *"The Internet of Things: Connecting Objects"*, Wiley-ISTE, 2010

[4] K. Ashton, "That 'Internet of Things' Thing", *RFID Journal*, 22 July 2009.

[5] Analyst Geoff Johnson interviewed by Sue Bushell in *Computerworld*, on July 24, 2000

[6] K. M. Razeeb, et al., "A hybrid network of autonomous sensor nodes", *Proceedings of the 2nd European Union symposium on Ambient intelligence EUSAI'04*, ACM New York, NY, USA 2004, pp. 69 – 70

[7] G.Q. Zhang, A. van Roosmalen, *"More than Moore: Creating High Value Micro/Nanoelectronics Systems"*, Springer, 2009, 330 p., ISBN 978-0-387-75592-2

[8] W. Karwowski, *"International encyclopedia of ergonomics and human factors"*, ISBN 041530430X, 9780415304306, CRC Press, 2006

[9] E. A. Kensinger, "The Effects of Emotional Content on Reality-Monitoring Performance in Young and Older Adults", *Psychology and Aging 2007*, Vol. 22, No. 4, 752–764.

Integrated Biosensors for Personalized Medicine

Giovanni De Micheli, Cristina Boero, Camilla Baj-Rossi,
Irene Taurino, Sandro Carrara

EPFL, Lausanne, Switzerland

{giovanni.demicheli, cristina.boero, camilla.baj-rossi, irene.taurino, sandro.carrara}@epfl.ch

ABSTRACT

Biosensors are heterogenous devices, incorporating biological structures combined with electronics, optical or other readout systems. They have been developed for detecting different biomolecules and/or pathogens and represent a key technology for advanced and point-of-care diagnostics as well as patient monitoring. In this paper we present a systematic classification of biosensors described in literature, particularly focusing on nanotechnology-based sensing. Then, we present our approach to develop electrochemical biosensors for measuring metabolites and anticancer drugs, based on a platform for multiple target detection. This platform is modular and achieves a clear separation between the chemical and the electrical components, thus easing design and manufacturing. It shows superior performance thanks to the excellent properties of electron transfer and selectivity showed by enzymes immobilized on carbon nanotubes.

Categories and Subject Descriptors

J.3 [**Life and Medical Sciences**]: Health

General Terms

Measurement

Keywords

biosensors, personalized medicine, point-of-care, integration

1. INTRODUCTION

The integration of biosensors with electrical data acquisition chains and information systems opens new opportunities for health management. In particular integrated biosensors are key elements for *advanced diagnostics*, including portable and disposable devices, and for *monitoring* metabolites and/or drug concentrations, thus enabling (possibly remote) treatment of chronic patients. Drug monitoring in human fluids is important to increase the effectiveness of therapies, and specifically in the case of personalized treatment. Indeed, standard drug therapies are based on randomized clinical trials, and treatments are chosen according to the best mean efficacy,

with improvements in the 20 to 50% patients, while the rest may not completely benefit from the assigned treatments [10]. For all these reasons, the development of an integrated platform to monitor the drug metabolism and the concentration of endogenous compounds in physiological fluids is highly requested. Optimized treatments and follow-up therapies can be easily tuned by using point-of-care devices, which represent a potent and innovative tool for personalized medicine.

Biosensors have been developed for diverse biomolecule and pathogen detection. Disposable electrodes are by far one of the most popular strategy coupled with electronics to develop point-of-care devices. However, system miniaturization becomes highly important and conventional approaches, like disposable electrodes, are a bottleneck for decreasing the size of the system. A potent approach to address this limitation is the integration of the biological layer with the electronic portion of the system. A benefit of integration is better performance with respect to signal-to-noise ratio, especially favorable when dealing with biological signals that are typically weak and noisy. High-density arrays of biosensors and multiple detection can be achieved by reducing the sensor area with microfabrication techniques. Finally, system miniaturization increases also sensor response and requires small samples.

System integration is a key issue for self contained biosensors. Power source, transducer circuitry, control unit, wireless communication are some of the blocks that can be potentially used in biosensing systems. However, the integration of all units may not be a satisfactory solution. Scaling trends for the analog circuit, the digital unit, and the biosensor itself are different, and so heterogeneous technologies may be required [17]. A platform-based design style using heterogeneous components and compositional rules eases the design process and reduces the *non-recurring engineering* (NRE) costs of biosensing systems, thus enabling the introduction of new approaches in the medical arena.

In this paper we present a strategy to develop the sensing block of a biosensor for the detection of endogenous compounds and anticancer drugs, and we compare its performance with the state-of-the-art devices. Before presenting the comparative results, we give an overview of the biosensors used in clinical practice, with particular emphasis on those suitable for integration.

2. CLASSIFICATION

Biosensors are a subgroup belonging to the wide family of chemical sensors. Biosensors may be classified in different ways. IUPAC recommends their classification according to the biological recognition mechanism or the transduction principles [48]. Hereunder, we want to propose an essential classification of biosensors that have been proposed in literature during the last decade.

2.1 Targets

The development of a biosensor is strictly connected to the target we want to detect. In clinical applications there are lots of analytes which are interesting to follow. **DNA** is one of the most important targets that have been studied during the last decades. Applications range from medical diagnosis and genome sequencing to food, pollution, and environmental analysis [6]. There are several methods to detect DNA: the most widely used is typically based on microarray technique, which consists on nucleic acid hybridization and optical readout [35]. Another technique quite popular involves electrical DNA biosensors, based on capacitance measurements [45]. Many researches have been focusing for long time on the detection of **molecules**, too. The most studied metabolite over the last fifty years is by far glucose, which lends to point-of-care device development and self-management for chronic diseases. Glucose biosensors are generally based on electrochemical principles, with a disposable sensing element and a permanent readout device [30]. However, there are other interesting molecules to detect for clinical interest. Over the last two decades there have been proposed biosensors for lactate [31], cholesterol [43], glutamate [38], creatinine[21], etc.

Biomarkers are another large family of biomolecules arising interest, since they are able to point out if a biological process, a disease, or a response to a therapeutic intervention is in progress. Currently, the most popular are the cancer biomarkers, including proteins, peptides, and tumor-related metabolites. One outstanding example is the *prostate specific antigen* (PSA) detection, related to prostate cancer [58], and the carcinoma antigen 125 (CA-125), related to ovarian cancer [47]. Autoimmune diseases present also distinctive biomarkers, typically antibodies or auto-antibodies. *Surface plasmon resonance* (SPR) has showed promising results for the detection of such biomarkers [11]. Biosensors are also promising tools for widespread and cheap screening of infectious diseases by detecting the RNA sequence of virus (for example dengue fever virus) or hepatitis B antigen. Recent works have presented encouraging results for protein detection to achieve the diagnosis of acute myocardial infarction and presence of coronary plaque [11].

Drugs are a further big category of molecules that can be sensed by using biosensors. Their monitoring in patient blood can reveal drug absorption, so that drug supply can be optimized according to the individual, enhancing the therapeutic efficacy. Some examples are detectors for paracetamol (analgesic and antipyretic), theophylline (used as therapy for respiratory diseases), chlorpromazine (antipsychotic), salicylate (antimicrobial agent) [53]. Multi-panel drug biosensors were also proposed for the detection of drugs by using cytochrome P450 in different isoforms. Benzphetamine (used in anti-obesity treatments), cyclophosphamide (used in anti-cancer therapy), Dextromethorphan (cough suppressant), Naproxen and Flurbiprofen (anti-inflammatory compounds) were detected with an electrochemical-based biosensor [9].

2.2 Sensing element

The sensing element is strictly related to the target. Biological systems represent the most selective element and they typically confer specificity to the biosensor. Several biosensors are based on **enzymes**. They are complex macromolecules, largely formed by protein structure, which are able to catalyze a chemical reaction. Generally, the chemical reaction is then transduced in a signal that can be measured and correlated to analyte quantity. The enzymes bind the analyte next to the active site. Enzymes needs a cofactor to work, which is bound to the protein itself. The cofactor is the part of the enzyme typically involved in the oxidation or reduction reaction [44].

Antibodies are another common sensing element. They are able to specifically bind the corresponding antigen, but they do not promote or catalyze any chemical reaction. The antigen, i.e. the target, can be a molecule or a cell (for example a bacteria) [11]. ELISA (Enzyme-Linked ImmunoSorbant Assay) is maybe the most popular analytical assay based on the complex antibody-antigen. An enzyme can be coupled to the antibody as transducer to promote a colorimetric reaction, for example, even if the sensing element remains the antibody [25].

DNA biosensors are primarily based on **nucleic acids** as sensing element. The specificity is conferred by the base-pairing and the strand of nucleic acids can detect genetic diseases, viral infections, and cancer [12]. The strands are often labeled with radioactive or fluorescent compounds, as well as enzymes or electroactive species. The transduction mechanisms, then, is strictly correlated to the used label.

The last category of sensing elements is represented by **receptors**, which are basically cell-membrane proteins. The detected signal is typically electrical, since a charge-flow is measured through an ion-channel [46]. Drugs are mostly the target, even if there are still evident problems to incorporate such receptors onto biosensors [34].

2.3 Transduction mechanism

The transduction mechanism is another big section for a detailed classification. The purpose of the present work is not to give a detailed description of all the mechanisms used for biosensing, but we want to present an exhaustive overview on the main techniques. **Optical biosensors** are typically enzyme-based, since the transduction mechanism is a chemical reaction producing changes in spectroscopic or spectrophotometric properties. Another strategy to perform the detection is by labeling secondary antibodies and DNA strands with fluorescent agents to confer the optical readout [20]. **Surface plasmon resonance**-based biosensors belong to the family of optical sensing and they have increasingly arisen interest in biosensing applications during the last years. This technique consists of the excitation of the interface between a metal and a dielectric by using light waves. If the excitation frequency matches the oscillation frequency of surface charge density, electromagnetic waves propagate along the interface, called surface plasmons. The dielectric medium can be functionalized with biological elements: as soon as the dielectric changes (because the target molecules bind the receptor), there is also a change in the refractive index [56]. Metal layer is mainly functionalized with antibody for the detection of antigens and hormones [11].

Piezoelectric biosensors typically detect mass variation and they are commonly known as *quartz crystal microbalance* (QCM). The quartz disk, or the microcantilever for nanoscale sensors, oscillate because of the application of an alternating electric field. The resonance frequency depends also on system mass. If the surface of the quartz is coated with sensing elements, once the sensing element binds the target, the mass of the system varies and shifts the resonance frequency. Such biosensors have been reported for DNA and pathogens detection and for immunoassays [13].

Surface modification with sensing elements can result in mass variation, as in the case of piezoelectric biosensors, but it can also result in the variation of the **electrical** properties of the surface. If the surface is an electrode, it is possible to quantify such electrical change by measuring the variation of impedance. Two sub-groups belong to the family of impedimetric biosensors. Capacitive biosensors are sensitive to capacitance variation: DNA detectors and immunosensor for tumor biomarkers are often developed according to these principles [50]. The Faradic impedimetric biosensors foresee to couple the antibody with a redox probe: the measured property is the charge transfer resistance [37].

The last category is represented by the **electrochemical** biosensors, which are by far the most reported devices in literature. It is possible to distinguish three sub-groups based on electrochemical sensing. The catalyzed reaction promoted by the enzyme can result in a variation of the electrode potential, while no current flows. Such technique is call *potentiometric*. Ion-selective sensors belong to that family. Potentiometric biosensors have been developed for urea detection in blood, creatinine in biological fluids and immunosensor assays [23]. *Ion charge* or *Field-Effect-Transistors (FET)* are another category of transduction mechanisms where the ion charge variation is monitored. Conventional FET can be modified for biosensing purposes by functionalizing the gate terminal with probes, for example. The binding between probes and targets results in a variation of electric charges at the gate terminal [24]. The functionalization can be applied also to the channel, especially when it is replaced by nanostructures, as nanowires or nanotubes (as discussed in Section 2.4). In this case the binding mechanism is transduced in a conductivity variation of the channel [22]. Finally, there are the *amperometric* biosensors, where current variation is monitored as result of the redox reaction promoted by the enzyme with the target. Amperometric biosensors have had great success in the market, because they can be produced quite easily and inexpensively, and they lend to be integrated in portable devices. Since their development is quite inexpensive, the sensing element (enzyme and electrodes) can be disposable, guaranteing uncontaminated and safe self-measurements. Amperometric biosensors have been developed for many applications: metabolite (especially glucose) and drug monitoring are by far the most common [53].

2.4 Nanotechnology-based biosensors

According to many authors [8], [15], the new frontier of biosensing is nanomaterial employment. Nanomaterials exhibit many interesting properties for biosensing, including dimensions comparable with sensing elements, high electron transfer rate [51], considerable electronic emission [28], and high surface area [2], due to their 3-D structure.

Nanoparticles (NP) are typically metallic, showing interesting electrical and magnetic properties for biosensing applications. NP applied in the biosensing field are often made of gold, because of the numerous ways to modify Au surfaces to obtain high affinity with biomolecules. Silver and platinum are other two reported metals used for NP synthesis with similar behavior. They have presented proper optical properties to be used in biosensing, but also interesting electrical features, as high sensitivity in voltammetry and improved limit of detection in potentiometric techniques [36]. Quantum dots are semiconductor crystals whose size is within 10 nm. Quantum confinement confers different properties to quantum dots with respect to larger particles. In fact, they have remarkable optical properties, suitable to be used as labels for sensing elements [27]. *Core shell* are a subfamily of NP, with a metallic core and an organic or inorganic shell, to improve biocompatibility and reduce particle aggregation [2].

With the improvement of micro and nanofabrication techniques, **nanowires (NW)** have been arising more interest in biomedical applications, since they can interact with biomolecules at the nanoscale. They can be metallic or semiconductor, according to the transduction mechanism. NW are often employed in conductive measurements, when functionalized with proteins, enzymes or antibodies, or field-effect-transistors, as discussed previously in Section 2.3 [39].

Finally, **carbon nanotubes (CNT)** have shown to possess interest-

ing electrochemical properties. The electron current through the nanotube is based on ballistic conductivity, so the measured mean free-path results to be two orders of magnitude higher than the best macroscale conductor [26]. Electron transfer depends on surface conditions: many works have been published regarding emission properties of tips and walls of CNT [7], [29] to explain their high rate. Moreover, they have been largely reported for the absorption of proteins onto their walls, resulting in an excellent immobilization method [4]. Surface modification by using carbon nanotubes can be accomplished in different manner and for several purposes. Directly growing of aligned carbon nanotubes have been proposed on different substrates [15]. Another way to obtain aligned CNT is by self-assembly: surfaces are generally modified with thiols or other functional groups to link nanotubes [40]. Carbon nanotubes can be also randomly dispersed on the electrode surface and many efforts have been addressed to find the right solvent to disperse CNT. Wang *et al.* showed that well-dispersed CNT solution can be achieved by adding Nafion [54]. Carbon nanotubes can be used as a forest of nanomaterials, patterned arrays or as single-sensors. Nanowires and carbon nanotubes can be used to replace the channel in field-effect-transistor for biosensing purposes . As in the case of electrochemical biosensors, biological sensing elements can be adsorbed on NW or CNT surface and modify the conductivity of the channel [52].

2.5 Electrode technology

Disposable biosensors are by far the most common tool sold in the market. They avoid common drawbacks like cleaning process, sterilization procedures, and contamination. On the other hand, fully-implanted monitoring is not possible with disposable electrodes, hampering the development of definitive solutions for the treatment of diabetes. Biosensor integration is definitely needed for such applications. Electrochemical-based sensing is the most suitable approach for the development of integrated biosensors. Amperometric, potentiometric, and impedimetric detection can be easily achieved with CMOS circuits next to the transducer. CMOS technology brings some interesting advantages, especially for electrochemical biosensors, where the signals are weak while the noise is quite high. Signals involved in such measurements are often analog, so the integration of analog-to-digital converters is required as well. CMOS circuits are typically covered with one or more passivation layers, to isolate the chip from the outside and from contaminants. CMOS applied in biology are much more subject to contamination and the wet environment does not guarantee proper working conditions. So, integrated biosensors need hybrid solutions. A really interesting and innovative solution for integrated biosensors was proposed by Guiducci *et al.* [17]: they propose a 3-D integrated system with vertically stacked layers and thru-silicon vias among the different layers. This solution treats each layer with different technologies, particulary suitable for the layer in contact with the biological environment. The authors propose a disposable biolayer, which is not suitable for fully-implanted devices, but can represent a step towards the development of permanent systems. Instead, the other layers designed for the readout, the transmission, the power supply, and the post-processing are permanent.

3. CNT-BASED BIOSENSOR

In the present section we describe one possible strategy to develop a platform of biosensors with the perspective to integrate the electrodes and the electronics in an unique device. Following the classification presented in Section 2, our biosensor can be described as following:

- *Target*: molecules, drugs

- *Sensing element*: enzymes

- *Transduction mechanism*: electrochemical (amperometric)

- *Nanotechnology-based*: carbon nanotubes

- *Electrode type*: disposable, integrated

Afterwards, we compare the performance of our developed biosensors with others found in literature. We will focus on enzyme-based electrochemical biosensors with similar modification of the electrode surface by using CNT and functionalization with the same type of protein.

3.1 Sensor description

Some biosensors are developed by using carbon paste *screen-printed electrodes* (SPE) (Dropsens, Spain) as disposable electrodes. The SPE consist of graphite working and counter electrodes, and Ag reference electrode. Working electrode has an area equal to 13 mm^2. Other molecules are detected by using a microfabricated chip, consisting of five Au microelectrodes, Au counter electrode, and Pt reference electrode. Each working electrode presents an area equal to 0.25 mm^2. Microfabrication details are described in [3]. All the electrode surfaces are modified with *multi-walled carbon nanotubes* (MWCNT - diameter 10 nm, length 1-2 μm - Dropsens, Spain) and functionalized with the enzyme-probe. Two families of enzymes are used for the experiments: oxidases are used for the detection of glucose, lactate, and glutamate, while *cytochrome P450* (CYP) is used for arachidonic acid, ifosfamide, *cyclophosphamide* (CP), and Ftorafur®. Oxidase-based detection is investigated by chronoamperometry with microfabricated Au electrodes and a drop cast solution of MWCNT dispersed in Nafion 0.5%. The working electrode potential is set at +650 mV and the current variation is recorded, since it is proportional to the target concentration. CYP-based sensing, instead, is carried out on screen-printed electrodes modified with MWCNT dispersed in chloroform. A linear-sweep potential is applied forward and backward within a certain potential window, while continuously monitoring the current. The hysteresis plot gives qualitative and quantitative information about the detected target. In particular, the peak hight is proportional to drug concentration and calibration curves can be plotted. Table 1 summarizes the main characteristics of our developed biosensors.

3.2 Results of detection

Here following we will focus on the comparison between our sensors and other similar biosensors found in literature. Table 2 summarizes the main features of the discussed biosensors, like the sensitivity, the linear range, and the limit of detection.

3.2.1 Glucose biosensor

Glucose biosensors have been extensively investigated. Examples regarding diverse surface modification and functionalization are largely reported in literature. Focusing on CNT-based biosensors using *glucose oxidase* (GOD) as sensing element, our biosensor shows the best performance for both sensitivity and limit of detection compared to similar sensors reported in literature. We achieve a sensitivity of 55.5 μA mM^{-1} cm^{-2} in a range from 0 to 1 mM, with a detection limit of 2 μM. Wang *et al.* [55] evaporated a thin Au film onto grown MWCNT and they drop cast GOD on top of the nanotubes, showing a sensitivity of 14.2 μA mM^{-1} cm^{-2}. Another approach can be to mix CNT and GOD in the same solution, with the addition of Nafion to increase the solubility of nanotubes. Tsai *et al.* proposed this approach in [49], where they

Table 1: Features of different metabolite biosensors.

Target	Probe	Technique
GLUCOSE	Glucose oxidase	
LACTATE	Lactate oxidase	Chronoamperometry
GLUTAMATE	Glutamate oxidase	
ARACHIDONIC ACID	custom-CYP	
FTORAFUR®	CYP1A2	Cyclic voltammetry
CYCLOPHOSPHAMIDE	CYP2B6	
IFOSFAMIDE	CYP3A4	

drop cast the mixed solution on glassy carbon electrodes. In a linear range from 0.025 to 2 mM, they got a sensitivity of 4.7 μA mM^{-1} cm^{-2}. Regarding sensitivity, quite similar results were shown by Ryu *et al.*: CNT forme a network on the electrode (defined mat) and GOD is covalently bound to the nanotubes [42]. The highest result in terms of sensitivity was presented in [18], showing 23.5 μA mM^{-1} cm^{-2} for MWCNT-based sensor with *butyric acid* (BA) functionalization and GOD immobilization.

3.2.2 Lactate biosensor

Lactate biosensors have been less studied compared to glucose biosensors, but there have been proposed diverse modification and functionalization, as well. Our sensor shows a sensitivity of 25.0 μA mM^{-1} cm^{-2} within a range from 0 to 1 mmM, and a detection limit of 11 μM. Goran *et al.* [16] drop cast successively N-doped CNT, LOD, and modified-Nafion onto glassy carbon electrode. They obtained higher sensitivity than us, because carbon electrode has better performance than metallic electrodes for the detection of H$_2$O$_2$. In fact, we have already showed similar sensitivity as [16] in our previous work by using carbon paste SPE [5]. However, the linear range is very narrow (from 0.014 to 0.325 mM), which cannot fit with physiological lactate concentration. CNT can be also incorporated with mineral oil to form a paste and used as electrodes [41]. However, the resulting sensitivity is quite low, 0.204 μA mM^{-1} cm^{-2}, in an extended range from 0 to 7 mM. MWCNT have been also incorporated into sol-gel film to form a further matrix for the immobilization of the enzyme. Huang *et al.* deposited the obtained matrix onto glassy carbon electrode, but they still showed a sensitivity ten times lower that our results [19]. In the last example, the authors used titanate instead of carbon nanotubes [57]. The obtained sensitivity is much lower than the previous case, suggesting that carbon gives better performance not only for the nanoscale structure, but also for the material itself.

3.2.3 Glutamate biosensor

Glutamate is a neurotransmitter and its monitoring can be crucial for neurochemical experiments. Lots of microsensors have been proposed to be implanted in the brain and most of them are not based on carbon nanotubes. Pan *et al.*, for example, covered a Pt electrode with Nafion matrix to entrap glutamate oxidase. The detection was carried out in a really narrow range within 1 and 13 μM, obtaining a sensitivity of 16.1 μA mM^{-1} cm^{-2}. Alternatively to Nafion immobilization, Zhang [59] described the entrapment of GlOD in chitosan, with a sensitivity of 85 μA mM^{-1} cm^{-2} within 0 and 200 μM. Similar linear range was also explored by Ammam *et al.* [1]. Differently, they used MWCNT and polyurethane (PU) onto Pt electrodes to increase the sensitivity (384 μA mM^{-1} cm^{-2}) and GlOD was dispersed in polypirrole (PP). All the previously de-

9

Table 2: Comparison of electrochemical enzyme-based biosensors.

	Modification	Sensitivity	Linear range	Limit of detection
	CNT mat + GOD [42]	4.05 μA mM^{-1} cm^{-2}	0.2 - 2.18 mM	–
	MWCNT/Nafion + GOD [49]	4.7 μA mM^{-1} cm^{-2}	0.025 - 2 mM	4 μM
GLUCOSE	MWCNT + GOD [55]	14.2 μA mM^{-1} cm^{-2}	0.05 - 13 mM	10 μM
	MWCNT-BA + GOD [18]	23.5 μA mM^{-1} cm^{-2}	0.01 - 2.5 mM	10 μM
	MWCNT/Nafion + GOD	55.5 μA mM^{-1} cm^{-2}	0 - 1 mM	2 μM
	MWCNT/mineral oil + LOD [41]	0.204 μA mM^{-1} cm^{-2}	0 - 7 mM	300 μM
	Titanate NT + LOD [57]	0.24 μA mM^{-1} cm^{-2}	0.5 - 14 mM	200 μM
LACTATE	MWCNT + sol-gel/LOD [19]	2.1 μA mM^{-1} cm^{-2}	0.3 - 1.5 mM	0.3 μM
	N-doped CNT/Nafion + LOD [16]	40.0 μA mM^{-1} cm^{-2}	0.014 - 0.325 mM	4 μM
	MWCNT/Nafion + LOD	25.0 μA mM^{-1} cm^{-2}	0 - 1 mM	11 μM
	Nafion + GlOD [33]	16.1 μA mM^{-1} cm^{-2}	0.001 - 0.013 mM	0.3 μM
GLUTAMATE	Chit + GlOD [59]	85.0 μA mM^{-1} cm^{-2}	0 - 0.2 mM	0.1 μM
	PU/MWCNT + GlOD/PP [1]	384 μA mM^{-1} cm^{-2}	0 - 0.14 mM	0.3 μM
	MWCNT/Nafion + GlOD	0.9 μA mM^{-1} cm^{-2}	0 - 2 mM	78 μM
ARACHIDONIC ACID	MWCNT + CYP	1140.0 μA mM^{-1} cm^{-2}	0 - 0.04 mM	0.4 μM
CYCLOPHOSPHAMIDE	MWCNT + CYP	102.0 μA mM^{-1} cm^{-2}	0 - 0.07 mM	2 μM
IFOSFAMIDE	MWCNT + CYP	160.0 μA mM^{-1} cm^{-2}	0 - 0.14 mM	2 μM
FTORAFUR®	MWCNT + CYP	883.0 μA mM^{-1} cm^{-2}	0 - 0.008 mM	0.7 μM

scribed sensitivities are higher (up to three orders of magnitude) than the one obtained by our biosensors. In fact, we achieve a sensitivity of 0.9 μA mM^{-1} cm^{-2} and a detection limit of 78 μM. On the other hand, we exploit a wider linear range (from 0 to 2 mM), useful for some particular applications like cell culture monitoring.

3.2.4 CYP-based biosensor

Arachidonic acid is a fatty acid abundant in liver, brain, and muscles. Its detection can be carried out by the isoform CYP102A1, for example. We got a customized CYP isoform from EMPA (St. Gallen, Switzerland) for the detection of fatty acids. Carbon paste SPE are modified as described previously by using MWCNT and the detection is performed by applying cyclic voltammetry. The developed biosensor shows a sensitivity of 1140.0 μA mM^{-1} cm^{-2} within a linear range from 0 to 40 μM, and a detection limit of 0.4 μM. In literature, the detection of such compound is typically optical, as presented by Giovannozzi et al. [14], and noone has described yet electrochemical detection by using CYP450.
Drugs can be detected using different isoforms of cytochrome P450. Here we present the results obtained from the detection of three drugs. Cyclophosphamide and ifosfamide are two alkilating agents, commonly used in anticancer treatments and as immunosuppressant. The third drug is a chemotherapeutic prodrug. Among these three drugs, CP is the only one for which electrochemical biosensors were previously developed. They are typically DNA-based and the signal variation is recorded when the CP interacts with DNA strands under differential pulse voltammetry [32]. We develop an enzyme-based electrochemical biosensor by using three different isoforms of CYP450 for the detection of such compounds. For CP we obtained a sensitivity of 102.0 μA mM^{-1} cm^{-2} in a linear range within 0 and 70 μM and a detection limit of 2 μM. For ifosfamide we got a sensitivity of 160.0 μA mM^{-1} cm^{-2} within a range from 0 and 140 μM, with a detection limit of 2 μM. Finally, Ftorafur® is detected with a sensitivity of 883.0 μA mM^{-1}

cm^{-2} in a linear range within 0 and 8 μM and a detection limit of 0.7 μM. In conclusion, it is the first time that electrochemical biosensors based on MWCNT and CYP are used for the detection of the aforementioned compounds. These results are really promising for the development of integrated biosensors for monitoring of drug mixture in the blood, paving the way to innovative tools for personalized therapy.

4. CONCLUSIONS AND PERSPECTIVES

The aim of the present work is to give an overview on biosensing strategies and developed devices. In particular, our attention is focused on nanotechnology-based biosensors and the possibility of their integration in more complex systems. Then, we proposed one possible strategy to develop biosensors for the detection of some biomolecules and drugs, showing that surface modification of the electrode with nanostructures can enhance the performance in biosensing. The detection of multiple endogenous and exogenous compounds is essential for personalized therapy. All the presented results are really promising in the perspective to develop point-of-care devices. The variety of individual responses to the same treatment requires potent tools for the monitoring of metabolic mechanisms, to optimize therapy management and efficacy. Enhanced sensitivities and lower detection limit can satisfy these demands, as showed by our developed biosensors.

Acknowledgment

The authors would like to thank Dr. L. Thöny-Meyer and the Laboratory of Biomaterials from EMPA for CYP customization and supply. Financial supports are from the i-IronIC project, financed by a grant from the Swiss Nano-Tera.ch initiative and evaluated by the Swiss National Science Foundation, and SNF Sinergia Project (CRSII2 127547/1).

5. REFERENCES

[1] M. Ammam and J. Fransaer. Highly sensitive and selective glutamate microbiosensor based on cast polyurethane/ac-electrophoresis deposited multiwalled carbon nanotubes and then glutamate oxidase/electrosynthesized polypyrrole/pt electrode. *Biosens. Bioelectron.*, 25(7):1597 – 1602, 2010.

[2] T. Asefa, C. Duncan, and K. Sharma. Recent advances in nanostructured chemosensors and biosensors. *Analyst*, 134(10):1980–1990, 2009.

[3] C. Boero, S. Carrara, and G. De Micheli. New technologies for nanobiosensing and their applications to real-time monitoring. In *IEEE BioCAS Conference 2011*, pages 357 –360, November 2011.

[4] C. Boero, S. Carrara, G. Del Vecchio, L. Calzá and, and G. De Micheli. Highly sensitive carbon nanotube-based sensing for lactate and glucose monitoring in cell culture. *IEEE Trans. Nanobiosci.*, 10(1):59 –67, 2011.

[5] C. Boero, S. Carrara, G. Del Vecchio, L. Calzá and, and G. De Micheli. Targeting of multiple metabolites in neural cells monitored by using protein-based carbon nanotubes. *Sensor. Actuat. B-Chem.*, 157(1):216 – 224, 2011.

[6] S. Cagnin et al. Overview of electrochemical dna biosensors: New approaches to detect the expression of life. *Sensors*, 9(4):3122–3148, 2009.

[7] C. Cai and J. Chen. Direct electron transfer of glucose oxidase promoted by carbon nanotubes. *Anal. Biochem.*, 332:75–83, 2004.

[8] S. Carrara, C. Boero, and G. D. Micheli. *Quantum Dots and Wires to Improve Enzymes-based Electrochemical Bio-sensing*. LNICTS 20, Springer, Berlin, 2009.

[9] S. Carrara, A. Cavallini, V. Erokhin, and G. De Micheli. Multi-panel drugs detection in human serum for personalized therapy. *Biosens. Bioelectron.*, 26(9):3914 – 3919, 2011.

[10] S. Carrara et al. Circuits design and nano-structured electrodes for drugs monitoring in personalized therapy. In *IEEE BioCAS Conference 2008*, pages 325 –328, November 2008.

[11] P. D'Orazio. Biosensors in clinical chemistry - 2011 update. *Clin. Chim. Acta*, 412:1749 – 1761, 2011.

[12] B. R. Eggins. *Chemical sensors and biosensors*. John Wiley & sons, 2002. pp. 125-169.

[13] G. Ferreira, A. da Silva, and B. Tomé. Acoustic wave biosensors: physical models and biological applications of quartz crystal microbalance. *Trends. Biotechnol.*, 27(12):689 – 697, 2009.

[14] A. Giovannozzi et al. P450-based porous silicon biosensor for arachidonic acid detection. *Biosens. Bioelectron.*, 28(1):320 – 325, 2011.

[15] J. Gooding. Nanostructuring electrodes with carbon nanotubes: A review on electrochemistry and applications for sensing. *Electrochim. Acta*, 50(15):3049 – 3060, 2005.

[16] J. Goran, J. Lyon, and K. Stevenson. Amperometric detection of l-lactate using nitrogen-doped carbon nanotubes modified with lactate oxidase. *Anal. Chem.*, 83(21):8123–8129, 2011.

[17] C. Guiducci, A. Schmid, F. Gurkaynak, and Y. Leblebici. Novel front-end circuit architectures for integrated bio-electronic interfaces. In *DATE 2008*, pages 1328 –1333, March 2008.

[18] M. Hua, Y. Lin, R. Tsai, and H. Chen. Water dispersible 1-one-butyric acid-functionalised multi-walled carbon nanotubes for enzyme immobilisation and glucose sensing. *J. Mater. Chem.*, 22(6):2566–2574, 2012.

[19] J. Huang et al. A highly-sensitive l-lactate biosensor based on sol-gel film combined with multi-walled carbon nanotube (mwcnts) modified electrode. *Mat. Sci. Eng. C*, 27(1):29 – 34, 2007.

[20] B. Juskowiak. Nucleic acid-based fluorescent probes and their analytical potential. *Anal. Bioanal. Chem.*, 399:3157–3176, 2011.

[21] A. Killard and M. Smyth. Creatinine biosensors: principles and designs. *Trends. Biotechnol.*, 18(10):433 – 437, 2000.

[22] J. Kim, B. Lee, J. Lee, S. Hong, and S. Sim. Enhancement of sensitivity and specificity by surface modification of carbon nanotubes in diagnosis of prostate cancer based on carbon nanotube field effect transistors. *Biosens. Bioelectron.*, 24(11):3372 – 3378, 2009.

[23] R. Koncki. Recent developments in potentiometric biosensors for biomedical analysis. *Anal. Chim. Acta*, 599(1):7 – 15, 2007.

[24] C. Lee, S. Kim, and M. Kim. Ion-sensitive field-effect transistor for biological sensing. *Sensors*, 9(9):7111–7131, 2009.

[25] R. M. Lequin. Enzyme immunoassay (eia)/enzyme-linked immunosorbent assay (elisa). *Clin. Chem.*, 51(12):2415–2418, 2005.

[26] H. Li, W. Lu, J. Li, X. Bai, and C. Gu. Multichannel ballistic transport in multiwall carbon nanotubes. *Phys. Rev. Lett.*, 95:086601, 2005.

[27] Q. Ma and X. Su. Recent advances and applications in qds-based sensors. *Analyst*, 136(23):4883–4893, 2011.

[28] A. Mayer, N. Miskovsky, and P. Cutler. Theoretical comparison between field emission from single-wall and multi-wall carbon nanotubes. *Phys. Rev. B*, 65(15):155420, 2002.

[29] D. McClain et al. Effect of diameter on electron field emission of carbon nanotube bundles. In *Mat. Res. Soc. Symp. Proc.*, volume 901E, 2006.

[30] J. Newman and A. Turner. Home blood glucose biosensors: a commercial perspective. *Biosens. Bioelectron.*, 20(12):2435 – 2453, 2005.

[31] N. Nikolaus and B. Strehlitz. Amperometric lactate biosensors and their application in (sports) medicine, for life quality and wellbeing. *Microchim. Acta*, 160:15–55, 2008.

[32] P. Palaska, E. Aritzoglou, and S. Girousi. Sensitive detection of cyclophosphamide using dna-modified carbon paste, pencil graphite and hanging mercury drop electrodes. *Talanta*, 72(3):1199 – 1206, 2007.

[33] S. Pan and M. Arnold. Selectivity enhancement for glutamate with a nafion/glutamate oxidase biosensor. *Talanta*, 43(7):1157 – 1162, 1996.

[34] J. Pancrazio, J. Whelan, D. Borkholder, W. Ma, and D. Stenger. Development and application of cell-based biosensors. *Ann. Biomed. Eng.*, 27:697–711, 1999.

[35] A. Pease et al. Light-generated oligonucleotide arrays for rapid dna sequence analysis. *P. Natl. Acad. Sci.*, 91(11):5022–5026, 1994.

[36] B. Pérez-López and A. Merkoçi. Nanoparticles for the development of improved (bio)sensing systems. *Anal. Bioanal. Chem.*, 399:1577–1590, 2011.

[37] M. Prodromidis. Impedimetric immunosensors - a review. *Electrochim. Acta*, 55(14):4227 – 4233, 2010.

[38] S. Qin, M. Van der Zeyden, W. Oldenziel, T. Cremers, and B. Westerink. Microsensors for in vivo measurement of glutamate in brain tissue. *Sensors*, 8(11):6860–6884, 2008.

[39] N. Ramgir, Y. Yang, and M. Zacharias. Nanowire-based sensors. *Small*, 6(16):1705–1722, 2010.

[40] G. Rivas et al. Carbon nanotubes for electrochemical biosensing. *Talanta*, 74(3):291 – 307, 2007.

[41] M. Rubianes and G. Rivas. Enzymatic biosensors based on carbon nanotubes paste electrodes. *Electroanal.*, 17(1):73–78, 2005.

[42] J. Ryu, H. Kim, S. Lee, H. Hahn, and D. Lashmore. Carbon nanotube mat as mediator-less glucose sensor electrode. *J.Nanosci. Nanotechnol.*, 10(2):941–947, 2010.

[43] A. Salimi et al. Fabrication of a sensitive cholesterol biosensor based on cobalt-oxide nanostructures electrodeposited onto glassy carbon electrode. *Electroanal.*, 21(24):2693–2700, 2009.

[44] C. Silvestre, P. Pinto, M. Segundo, M. Saraiva, and J. Lima. Enzyme based assays in a sequential injection format: A review. *Anal. Chim. Acta*, 689(2):160 – 177, 2011.

[45] C. Stagni et al. A fully electronic label-free dna sensor chip. *IEEE Sens. J.*, 7(3-4):577–585, 2007.

[46] L. Steller, M. Kreir, and R. Salzer. Natural and artificial ion channels for biosensing platforms. *Anal. Bioanal. Chem.*, 402:209–230, 2012.

[47] D. Tang, R. Yuan, and Y. Chai. Electrochemical immuno-bioanalysis for carcinoma antigen 125 based on thionine and gold nanoparticles-modified carbon paste interface. *Anal. Chim. Acta*, 564(2):158 – 165, 2006.

[48] D. Thevenot et al. Electrochemical biosensors: recommended definitions and classification. Technical report, IUPAC, 1999.

[49] Y. Tsai, S. Li, and J. Chen. Cast thin film biosensor design based on a nafion backbone, a multiwalled carbon nanotube conduit, and a glucose oxidase function. *Langmuir*, 21(8):3653–3658, 2005.

[50] V. Tsouti, C. Boutopoulos, I. Zergioti, and S. Chatzandroulis. Capacitive microsystems for biological sensing. *Biosens. Bioelectron.*, 27(1):1 – 11, 2011.

[51] V. Vojinovic, F. Esteves, J. Cabral, and L. Fonseca. Bienzymatic analytical microreactors for glucose, lactate, ethanol, galactose and l-amino acid monitoring in cell culture media. *Anal. Chim. Acta*, 565(2):240 – 249, 2006.

[52] A. Wanekaya, W. Chen, N. Myung, and A. Mulchandani. Nanowire-based electrochemical biosensors. *Electroanal.*, 18(6):533–550, 2006.

[53] J. Wang. Amperometric biosensors for clinical and therapeutic drug monitoring: a review. *J. Pharmaceut. Biomed.*, 19:47 – 53, 1999.

[54] J. Wang, M. Musameh, and Y. Lin. Solubilization of carbon nanotubes by nafion toward the preparation of amperometric biosensors. *J. Am. Chem. Soc.*, 125(9):2408–2409, 2003.

[55] S. Wang et al. Multi-walled carbon nanotubes for the immobilization of enzyme in glucose biosensors. *Electrochem.Commun.*, 5(9):800 – 803, 2003.

[56] E. Wijaya et al. Surface plasmon resonance-based biosensors: From the development of different spr structures to novel surface functionalization strategies. *Curr. Opin. Solid St. M.*, 15(5):208 – 224, 2011.

[57] M. Yang, J. Wang, H. Li, J. Zheng, and N. Wu. A lactate electrochemical biosensor with a titanate nanotube as direct electron transfer promoter. *Nanotechnology*, 19(7):075502, 2008.

[58] A. Zani, S. Laschi, M. Mascini, and G. Marrazza. A new electrochemical multiplexed assay for psa cancer marker detection. *Electroanal.*, 23(1):91–99, 2011.

[59] M. Zhang, C. Mullens, and W. Gorski. Amperometric glutamate biosensor based on chitosan enzyme film. *Electrochim. Acta*, 51(21):4528 – 4532, 2006.

Design Challenges for Secure Implantable Medical Devices

Wayne Burleson
Department of Electrical and Computer Engineering
University of Massachusetts Amherst
Amherst, MA 01003
burleson@ecs.umass.edu

Shane S. Clark, Benjamin Ransford, Kevin Fu
Department of Computer Science
University of Massachusetts Amherst
Amherst, MA 01003
{ssclark,ransford,kevinfu}@cs.umass.edu

ABSTRACT

Implantable medical devices, or IMDs, are increasingly being used to improve patients' medical outcomes. Designers of IMDs already balance safety, reliability, complexity, power consumption, and cost. However, recent research has demonstrated that designers should also consider *security* and *data privacy* to protect patients from acts of theft or malice, especially as medical technology becomes increasingly connected to other systems via wireless communications or the Internet. This survey paper summarizes recent work on IMD security. It discusses sound security principles to follow and common security pitfalls to avoid. As trends in power efficiency, sensing, wireless systems and bio-interfaces make possible new and improved IMDs, they also underscore the importance of understanding and addressing security and privacy concerns in an increasingly connected world.

Categories and Subject Descriptors

J.3 [**Computer Applications**]: Life and Medical Sciences—*Medical information systems*; C.3 [**Computer Systems Organization**]: Special-Purpose and Application-Based Systems—*Real-time and embedded systems*

General Terms

Security, Design

Keywords

Implantable Medical Devices, IMD Security

1. INTRODUCTION

Implantable medical devices (IMDs) perform a variety of therapeutic or life-saving functions ranging from drug infusion and cardiac pacing to direct neurostimulation. Modern IMDs often contain electronic components that perform increasingly sophisticated sensing, computation, and actuation, in many cases without any patient interaction. IMDs have already improved medical outcomes

for millions of patients; many more will benefit from future IMD technology treating a growing number of ailments.

Because of their crucial roles in patient health, IMDs undergo rigorous evaluation to verify that they meet specific minimum safety and effectiveness requirements. However, *security* is a relatively new concern for regulatory bodies; bug-averse manufacturers have traditionally had little incentive to add security mechanisms that might cause problems or slow down regulatory approval. Perhaps not surprisingly in light of this situation, recent security research has demonstrated that some IMDs fail to meet appropriate expectations of security for critically important systems.

The key classes of IMD vulnerabilities researchers have identified are *control* vulnerabilities, in which an unauthorized person can gain control of an IMD's operation or even disable its therapeutic services, and *privacy* vulnerabilities, in which an IMD exposes patient data to an unauthorized party. Both kinds of vulnerabilities may be harmful to patients' health outcomes, and both kinds are avoidable.

As designers realign themselves with incentives for better security, there are ample opportunities to adapt well-tested security principles to IMD design. This survey paper's goals are to (1) outline design principles for IMD security; (2) highlight the security challenges in designing implantable medical devices, some of which remain open problems; and (3) sketch the defensive measures that researchers have proposed and implemented.

1.1 Security Goals for IMD Design

The term *security* refers to the goal of well-defined, correct system behavior in the presence of adversaries.[1] Security and reliability, both of which define policies and actions under a variety of conditions, form the basis of *trustworthiness* [9]. IMD designers can follow well-founded security practices to avoid pitfalls (§3) and build trustworthy systems. In short, designers should:

- Consider security in early design phases.
- Encrypt sensitive traffic where possible.
- Authenticate third-party devices where possible.
- Use well-studied cryptographic building blocks instead of ad-hoc designs.
- Assume an adversary can discover your source code and designs; do not rely on *security through obscurity*.
- Use industry-standard source-code analysis techniques at design time.
- Develop a realistic *threat model* (§1.2); defend the most attractive targets first.

---

[1] See Bishop's textbook [4] for an introduction to security.

These design principles are not specific to IMDs; they are fundamental security ideas. Applying them to the IMD domain requires special consideration of IMDs' use cases and limitations. For example, the choice of cryptographic system to implement on a tiny biosensor or nonrechargeable heart device can have major implications for device longevity.

Halperin et al. detail some holistic design considerations related to medical-device security [14]. In contrast, this paper focuses on device-level concepts, relating the above principles to three specific classes of IMD (§2).

1.2 Threat Modeling

Threat modeling, which entails anticipating and characterizing potential threats, is a vital aspect of security design. With realistic models of adversaries, designers can assign appropriate priorities to addressing different threats.

The severity of vulnerabilities varies along with the sensitivity of the data or the consequences of actuation; there is no "one size fits all" threat model for IMDs. A non-actuating glucose sensor incurs different risks than a defibrillator that can deliver disruptive electrical shocks to a heart.

Adversaries are typically characterized according to their goals, their capabilities and the resources they possess. Security designers evaluate each threat by considering the value of the target and the amount of effort necessary to access it. Recent work analyzing IMD security and privacy has posited several classes of adversaries, described below.

An *eavesdropper* who listens to an IMD's radio transmissions, but does not interfere with them, can often learn private information with minimal effort. Such a *passive adversary* may have access to an oscilloscope, software radio, directional antennas, and other listening equipment. Several studies have considered this type of adversary and demonstrated that eavesdropping on unencrypted communications could compromise patients' data privacy [15, 20, 23, 26, 28].

An *active adversary* extends the passive adversary's capabilities with the ability to generate radio transmissions addressed to the IMD, or to replay recorded control commands. Halperin et al. demonstrated that an active adversary with a programmable radio could control one model of implantable defibrillator by replaying messages—disabling programmed therapies or even delivering a shock intended to induce a fatal heart rhythm [15]. Jack and Li have demonstrated similar control over an insulin pump, including the ability to stop insulin delivery or inject excessive doses [28, 20].

Another adversarial capability is *binary analysis*, the ability to disassemble a system's software and in some cases completely understand its operation. By inspecting the Java-based configuration program supplied with his own insulin pump, researcher Jerome Radcliffe reverse-engineered the pump's packet structure, revealing that the pump failed to encrypt the medical data it transmitted or to adequately authenticate the components to one another [26]. In contrast to design-time static analysis of source code, a crucial practice that may expose flaws before devices are shipped [18], binary analysis involves inspecting *compiled* code; it can expose flaws in systems that erroneously depend on the supposed difficulty of reverse engineering to conceal private information.

In the context of medical conditions, it may be difficult to comprehend why a malicious person would seek to cause harm to patients receiving therapy, but unfortunately, it has happened in the past. For example, in 2008, malicious hackers defaced a webpage run by the nonprofit Epilepsy Foundation, replacing the page's content with flashing animations that induced migraines or seizures for some unsuspecting visitors [24]. Although we know of no reports

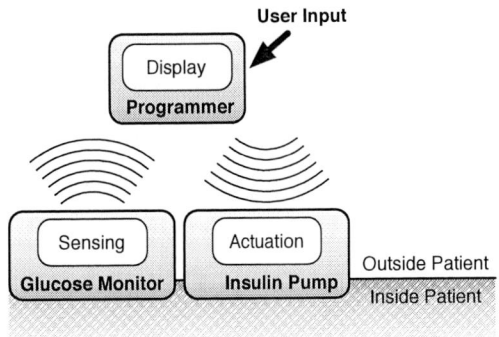

Figure 1: Block diagram of an insulin pump system (IPS), an *open-loop* IMD.

of malicious attacks against IMDs "in the wild," it is important to address vulnerabilities before they become serious threats.

2. DEVICES IN DEPTH

To illustrate the complexity of the design space for IMD security, we offer three examples of IMD systems that pose different security challenges because of their different design and usage. The common thread among all three devices—insulin pump systems, implantable cardioverter defibrillators, and subcutaneous biosensors—is that security is a crucial design concern. Section 2.4 explores commonalities and defensive concepts.

2.1 Insulin Pump: Open-Loop System

Insulin pump systems straddle the boundary between implanted and external systems, including some components that are physically attached to a patient and others that are external. A typical modern insulin pump system (IPS) may include: an insulin infusion pump with wireless interface that subcutaneously delivers insulin, a continuous glucose monitor (CGM) with wireless transmitter and subcutaneous sensor for glucose measurement, and a wireless remote control that the patient can use to alter infusion pump settings or manually trigger insulin injections. The CGM automatically takes frequent glucose readings, presenting the data to the user via a screen or PC, or sending data directly to the pump. The pump automatically provides *basal* doses for insulin maintenance and can also administer larger *bolus* doses to compensate for large insulin spikes that may result from, e.g., a meal. Finally, the remote control provides a convenient interface for the user to adjust pump settings without using the pump controls and screen typically attached at the abdomen. Figure 1 shows a block diagram of an IPS.

Insulin pump systems exemplify *open-loop* IMDs: they require patient interaction to change pump settings. Specifically, the patient's remote control—but not the CGM—directly controls pump actuation. Because the remote-control interface carries crucial information and control signals, initial security studies have focused on finding vulnerabilities at this interface. Li et al. discovered that one IPS's communications were unencrypted, leading to potential disclosure of private patient information (e.g., glucose levels) [20]. They also found that the components failed to check their inputs appropriately, allowing the researchers to inject forged packets reporting incorrect glucose levels to the patient and pump—and more alarmingly, to issue unauthorized pump-control commands. Soon thereafter, two security researchers independently demonstrated full control of IPSes via circumventing authentication mechanisms: Radcliffe compromised the wireless channel of his own (unspecified)

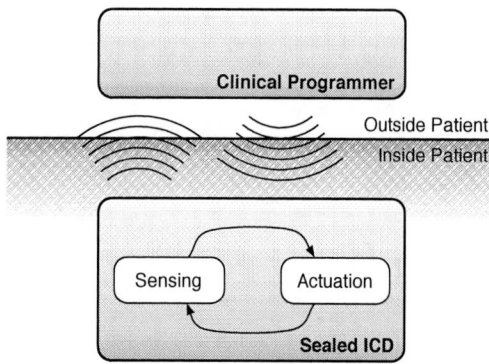

Figure 2: Block diagram of an implantable cardiac defibrillator (ICD), a *closed-loop* IMD.

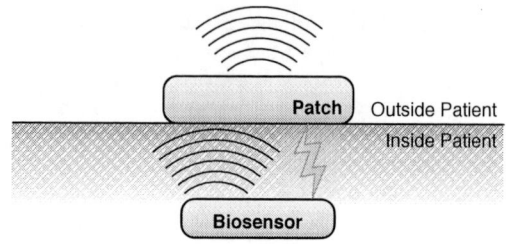

Figure 3: Block diagram of a subcutaneous biosensor. Small biosensors may be injected into the patient and then inductively powered by a patch that relays sensed data to a higher-level wearable device.

IPS [26, 29], and Jack performed a live demonstration in which he remotely controlled and then shut down a volunteer's insulin pump [28]. Jack also demonstrated that certain IPSes responded to anonymous radio scanning with their serial numbers, a privacy vulnerability because of the potential of tracking IPS patients.

2.2 Defibrillator: Closed-Loop System

Like an artificial pacemaker, which continually issues small electrical pulses to heart muscle to maintain a healthy rhythm, an *implantable cardiac defibrillator* (ICD) is implanted under the skin near the clavicle. ICDs extend the capabilities of artificial pacemakers with the ability to issue large (tens of joules) shocks to "reset" an unsustainable heart rhythm (arrhythmia). Figure 2 shows a block diagram of an ICD.

Unlike an insulin pump that accepts patient input via a user interface, a fully implanted device such as an ICD is a *closed-loop* system: under normal circumstances, its sensing function alone dictates its actuation activities. (Closed-loop IMDs typically also have special modes for in-clinic configuration and operation.) Halperin et al. enumerated the security and privacy challenges of closed-loop implanted systems in a 2008 article [14], focusing primarily on the tensions between security and utility.

ICD implantation currently requires invasive surgery with a risk of complications (infection or death) [11], so ICDs are designed to last for at least five years once implanted—resulting in long design and deployment cycles for manufacturers. ICDs draw power from single-use batteries, sealed inside the case, to provide uninterrupted monitoring throughout the device's lifetime and to avoid the heating of tissue that might occur during battery recharging. In conformity with these design choices, ICDs spend most of their time in low-power sensing states. They also include radios for clinical adjustments and at-home status reporting.

A 2008 security analysis of a commercial ICD found vulnerabilities in multiple subsystems [15], including those listed above. Focusing on the ICD's radio link, researchers used open-source software-radio tools to record transmissions between the ICD and a clinical programming console. Offline analysis of these traces revealed patient information in clear text without evidence of encryption. They replayed recorded traces of clinical therapy commands and found that they could control or disable the ICD's therapies with their software radio. Concerning the battery, the study found that a sequence of transmissions from the software radio could keep the ICD's radio in a high-power active mode, indefinitely transmitting packets at a regular rate and dramatically increasing the ICD's power consumption.

2.3 Biosensors for Data Acquisition

Implantable biosensors (Figure 3) are IMDs that measure biological phenomena and send data to a more powerful device for storage or analysis. Biosensors are a broader device category than insulin pump systems or defibrillators, representing a wide range of both signals and signal processing techniques. They are subject to a third set of security and privacy challenges that does not completely overlap with those mentioned above.

Biosensors range from high-data-rate imaging devices for the eye [3] or brain [25] to extremely low-data-rate sensors for glucose [12] or other metabolites in the blood [5]. Actuators that consume biosensor data can control potentially lethal drug-delivery systems [23] or electrical therapies [15]. Keeping biosensor data confidential is important because it can be used in illegal or unethical ways including insurance fraud or discrimination. The provenance (origin) and timestamp information that accompany biosensor readings are also critically important for medical care and must be protected from tampering.

Subcutaneous biosensors present a special set of security and privacy requirements. A subcutaneous sensor [5] involves an implanted biosensor that acts as a *lab on chip*, conducting a small experiment at the molecular or electro-chemical level on the sensor. Current subcutaneous sensors can detect drugs, bio-markers, and antibodies and may eventually examine DNA, and simultaneously log temperature, pH, and other phenomena.

Recent examples of subcutaneous biosensors include *injectable* subcutaneous devices that are remotely powered by a bandage-like patch that also provides a data link to a higher-level wearable device, possibly a body-area-network (BAN) or eventually a higher-level health information system. A related class of devices are low-cost disposable biosensors for detecting infectious disease or critical levels of glucose and lactate in a battlefield or other trauma situation [13]; such devices penetrate the skin for communication and power. These two classes of devices support different threat models because of their different usage parameters.

Biosensors that are fully implanted must communicate wirelessly to transmit through tissue. (Some receive power through tissue as well; recent work has shown that remotely powering biosensors is feasible at gigahertz frequencies that enable millimeter-sized antennas [22].) A key problem with fully implanted sensors is that small, infrequent wireless transmissions may pose a greater privacy risk than large or continuous transmissions. For example, a sensor may take several minutes to complete its task, then deliver only a few bytes of data—giving this information a high value per bit that may make it an attractive target. Short data transmissions necessitate careful use of a cipher, especially if the plaintext sensor data may take only a few different values. The small amount of data also has little inherent redundancy, making error-correction necessary.

When a biosensor includes a patch that is meant to pair with the sensor, additional risks arise. Although eavesdropping on a properly operating tag may be unlikely because of the short (several millimeters for a subcutaneous sensor) nominal transmission range, impersonation of both the clinical reader and the patch are plausible concerns. For example, the patch of an unconscious patient can be removed and replaced by another patch. Similarly, a rogue sensor can upload fraudulent data to a trusted patch. All components involved should authenticate one another using well-studied cryptographic mechanisms, especially during the critically important period when a sensor is first being tested or calibrated.

Biosensors present a diverse set of challenges for security and privacy and a unique combination of constraints. Open problems include: 1) developing more detailed threat models; 2) exploring design alternatives that effectively trade off safety, security, and utility; 3) understanding energy issues, including power depletion and side-channel attacks that exploit the lightweight nature of the biosensor; 4) implementing multiple layers of security to accommodate the multiple stages required to access data from a lightweight sensor (implant to patch to wearable to internet); and 5) understanding the security and privacy implications of future biosensing devices that provide an unprecedented view into the (presumably private) inner workings of the human body. Future devices are likely to include more storage, more complex signal processing, integrated software control, and use of multiple intercommunicating sensors, all of which will complicate security and privacy issues.

2.4 Common Threads

Different classes of IMDs have distinct hardware and usage constraints, but there are important security considerations that apply to many IMDs. Researchers investigating the security and privacy of IMDs have also proposed several domain-specific mechanisms that apply broadly. This section discusses some of these common threads in the context of our example IMD systems.

All of the IMD vulnerabilities disclosed thus far could be mitigated by the use of encryption on radio links. Hosseini-Khayat presents a lightweight wireless protocol for IMDs [17] that leverages well-studied wireless and cryptography technologies and emphasizes low-energy computation. The choice of encryption scheme should consider the nature of the data as well as the device constraints. Fan et al. contribute hardware implementations of the stream cipher Hummingbird [7, 8]. Beck explores the use of block ciphers in IMD security [3].

Unfortunately, encryption is not a panacea for IMD security and privacy vulnerabilities; many questions remain. If the radio link were to use encryption, how would the necessary secret key material be distributed, and by whom? How should an IMD authenticate external entities, and how should it determine whether a particular entity is allowed to communicate with it? Even assuming that each of these questions can be answered, successful implementation of encryption would not completely address known risks. Encryption alone fails to address replay attacks, and previous work has demonstrated that encryption may not sufficiently conceal characteristic traffic patterns [16]. Furthermore, since some IMDs must "fail open" to allow emergency access (e.g., to disable the IMD during emergency procedures), how can it also provide security in non-emergency situations? Should an IMD raise an alarm (perhaps tactile or audible) when a security-sensitive event occurs? These questions are largely open.

Recent research toward addressing these design tensions has proposed new techniques and auxiliary devices to provide fail-open security for IMDs. Rasmussen et al. proposed the use of ultrasonic

distance bounding to enforce programmer proximity [27]. Li et al. proposed body-coupled communications for the same purpose [20], hoping to prevent an adversary from launching a long-range radio-based attack. Both of these distance-bounding techniques require new hardware, but this constraint may not represent a major stumbling block for IPSes or biosensor systems, which are short-lived and non-invasive compared to ICDs.

Researchers have also proposed defenses specifically targeted toward existing ICDs, but which may be useful for other IMDs. Denning et al. proposed that an IMD be paired with a *cloaker* that would provide authentication services whenever it was present, and allow open communication otherwise [6]. Xu et al. proposed the *Guardian*, a device that would pair with an IMD and use radio jamming to defend against eavesdropping and unauthorized commands [30]. Gollakota et al. independently proposed an auxiliary device called the *shield* that would use "friendly" radio jamming to proxy an ICD's communications to an authorized reader [10]. The shield is designed for compatibility with devices that are already implanted, reducing the burden on device designers to address the security vulnerabilities in devices that have not completed their deployment lifecycles.

3. SECURITY PITFALLS

Designing for security has many subtleties. In the context of IMDs, where devices may be physically inaccessible for years, it is particularly important to avoid design errors that lead to failures or recalls later. One common error is believing in *security through obscurity*—relying entirely on proprietary ciphers or protocols for secrecy.

Security through obscurity—relying entirely on the secrecy of proprietary ciphers or protocols—is a common fallacy. Sound security principles dictate that a system's security must not depend on the secrecy of the algorithm or hardware; it is better to use well-studied standard ciphers and spend more design effort protecting cryptographic keys. This principle, commonly known as Kerckhoff's principle,[2] is a fundamental guideline for security design. Following it is essential for resistance against reverse-engineering adversaries.

A recent example that illustrates the hazards of security through obscurity is that of the NXP Mifare Classic smart-card chipset, which is widely used for transit ticketing systems. Nohl et al. reverse-engineered the Mifare Classic hardware and analyzed the underlying cipher and protocol, discovering that it used a flawed implementation of a cipher called Crypto-1 [21]. Crypto-1 supports only a limited key size; the Mifare Classic hardware also implements a predictable random-number generator. These factors combine to allow an adversary to clone a tag in a matter of seconds. The Mifare Classic tag could have addressed these flaws by using established, publicly studied cryptographic primitives rather than ad-hoc proprietary systems.

4. OPEN PROBLEMS IN IMD DESIGN

IMDs are first and foremost intended to improve patients' quality of life. To this end, the primary focus for designers must be device safety and utility. We argue that security and privacy are also important properties that must be part of the design process, but there is the potential for direct conflict between these two sets of properties.

The issue of emergency access highlights some of the tensions

---

[2]First articulated in 1883 by Auguste Kerckhoff in *La Cryptographie Militaire*

that exist among these properties. Requiring users to authenticate to a device before altering its functionality is a boon for security, but it introduces risks in the case of an emergency. A medical professional may need to reprogram or disable a device to effectively treat a patient. As discussed in Section 2.4, encryption or other strong authentication mechanisms could make such emergency measures impossible if the patient is unconscious or the facility does not possess a programming device with a required shared secret.

For some IMDs, including both IPSes and ICDs, designers must carefully weigh the energy costs of encryption against safety and utility. A heavyweight encryption scheme could potentially drain enough energy to require more frequent device replacement—a surgical procedure for ICD patients and a persistent burden for IPS users. Costly encryption could even make the construction and deployment of some subcutaneous biosensors infeasible. It remains to be seen whether ASIC implementations of lightweight algorithms can effectively mitigate this issue because of the lack of public deployments to date [17, 3].

There are no clear-cut methods for resolving these tensions, and there is little publicly available information about whatever steps manufacturers have already taken. While cryptographers and security researchers have long embraced Kerckhoff's principle, device manufacturers employ proprietary systems and generally do not comment (for business reasons) on security measures that they may employ. These closed ecosystems hamper industry-wide progress on shared issues such as security and privacy. Research into whole-system modeling and formal analysis of medical devices [19, 2, 1] offers hope that future IMDs will integrate sound security principles at design time, but the time horizon for industrial adoption may be long.

Recent analyses of implantable medical devices have revealed a number of security and privacy failings, but researchers are developing novel solutions to the problems IMD designers face. By incorporating security and privacy design principles into the development process, IMD designers have the opportunity to address these issues before they become larger threats.

5. ACKNOWLEDGMENTS

This material is based upon work supported by: the Armstrong Fund for Science; the National Science Foundation under Grants No. 831244, 0923313 and 0964641; Cooperative Agreement No. 90TR0003/01 from the Department of Health and Human Services; two NSF Graduate Research Fellowships; and a Sloan Research Fellowship. Its contents are solely the responsibility of the authors and do not necessarily represent the official views of DHHS or NSF.

6. REFERENCES

[1] D. Arney, R. Jetley, P. Jones, I. Lee, and O. Sokolsky. Formal methods based development of a PCA infusion pump reference model: Generic infusion pump (GIP) project. In *Proceedings of the 2007 Joint Workshop on High Confidence Medical Devices, Software, and Systems and Medical Device Plug-and-Play Interoperability*, HCMDSS-MDPNP '07, pages 23–33. IEEE Computer Society, 2007.

[2] D. Arney, M. Pajic, J. M. Goldman, I. Lee, R. Mangharam, and O. Sokolsky. Toward patient safety in closed-loop medical device systems. In *Proceedings of the 1st ACM/IEEE International Conference on Cyber-Physical Systems*, ICCPS '10, pages 139–148. ACM, 2010.

[3] C. Beck, D. Masny, W. Geiselmann, and G. Bretthauer. Block cipher based security for severely

resource-constrained implantable medical devices. In *Proceedings of 4th International Symposium on Applied Sciences in Biomedical and Communication Technologies*, ISABEL '11, pages 62:1–62:5. ACM, October 2011.

[4] M. Bishop. *Computer Security: Art and Science*. Addison-Wesley Professional, 2003.

[5] G. De Micheli, S. Ghoreishizadeh, C. Boero, F. Valgimigli, and S. Carrara. An integrated platform for advanced diagnostics. In *Design, Automation & Test in Europe Conference & Exhibition*, DATE '11. IEEE, March 2011.

[6] T. Denning, K. Fu, and T. Kohno. Absence makes the heart grow fonder: New directions for implantable medical device security. In *Proceedings of USENIX Workshop on Hot Topics in Security (HotSec)*, July 2008.

[7] X. Fan, G. Gong, K. Lauffenburger, and T. Hicks. FPGA implementations of the Hummingbird cryptographic algorithm. In *Proceedings of the IEEE International Symposium on Hardware-Oriented Security and Trust*, HOST '10, pages 48–51, June 2010.

[8] X. Fan, H. Hu, G. Gong, E. Smith, and D. Engels. Lightweight implementation of Hummingbird cryptographic algorithm on 4-bit microcontrollers. In *International Conference for Internet Technology and Secured Transactions*, ICITST '09, pages 1–7, November 2009.

[9] K. Fu. Trustworthy medical device software. In *Public Health Effectiveness of the FDA 510(k) Clearance Process: Measuring Postmarket Performance and Other Select Topics: Workshop Report*, Washington, DC, July 2011. IOM (Institute of Medicine), National Academies Press.

[10] S. Gollakota, H. Hassanieh, B. Ransford, D. Katabi, and K. Fu. They can hear your heartbeats: Non-invasive security for implanted medical devices. In *Proceedings of ACM SIGCOMM*, August 2011.

[11] P. Gould and A. Krahn. Complications associated with implantable cardioverter–defibrillator replacement in response to device advisories. *Journal of the American Medical Association (JAMA)*, 295(16):1907–1911, April 2006.

[12] S. Guan, J. Gu, Z. Shen, J. Wang, Y. Huang, and A. Mason. A wireless powered implantable bio-sensor tag system-on-chip for continuous glucose monitoring. In *Proceedings of the IEEE Biomedical Circuits and Systems Conference*, BioCAS '11, November 2011.

[13] A. Guiseppi-Elie. An implantable biochip to influence patient outcomes following trauma-induced hemorrhage. *Analytical and Bioanalytical Chemistry*, 399(1):403–419, January 2011.

[14] D. Halperin, T. S. Heydt-Benjamin, K. Fu, T. Kohno, and W. H. Maisel. Security and privacy for implantable medical devices. *IEEE Pervasive Computing, Special Issue on Implantable Electronics*, 7(1):30–39, January 2008.

[15] D. Halperin, T. S. Heydt-Benjamin, B. Ransford, S. S. Clark, B. Defend, W. Morgan, K. Fu, T. Kohno, and W. H. Maisel. Pacemakers and implantable cardiac defibrillators: Software radio attacks and zero-power defenses. In *Proceedings of the 29th IEEE Symposium on Security and Privacy*, May 2008.

[16] A. Hintz. Fingerprinting websites using traffic analysis. In R. Dingledine and P. Syverson, editors, *Proceedings of the Privacy Enhancing Technologies workshop*, PET '02. Springer-Verlag, LNCS 2482, April 2002.

[17] S. Hosseini-Khayat. A lightweight security protocol for ultra-low power ASIC implementation for wireless implantable medical devices. In *Proceedings of the 5th*

International Symposium on Medical Information Communication Technology, ISMICT '11, pages 6–9, March 2011.

[18] R. P. Jetley, P. L. Jones, and P. Anderson. Static analysis of medical device software using CodeSonar. In *Proceedings of the 2008 Workshop on Static Analysis*, SAW '08, pages 22–29. ACM, 2008.

[19] I. Lee, G. J. Pappas, R. Cleaveland, J. Hatcliff, and B. H. Krogh. High-confidence medical device software and systems. *IEEE Computer*, 39(4):33–38, 2006.

[20] C. Li, A. Raghunathan, and N. K. Jha. Hijacking an insulin pump: Security attacks and defenses for a diabetes therapy system. In *Proceedings of the 13th IEEE International Conference on e-Health Networking, Applications, and Services*, Healthcom '11, June 2011.

[21] K. Nohl, D. Evans, Starbug, and H. Plötz. Reverse-engineering a cryptographic RFID tag. In *Proceedings of the 17th USENIX Security Symposium*, pages 185–194, July 2008.

[22] S. O'Driscoll, A. Poon, and T. Meng. A mm-sized implantable power receiver with adaptive link compensation. In *Proceedings of the International Solid-State Circuits Conference*, ISSCC '09, pages 294–295,295a. IEEE, February 2009.

[23] N. Paul, T. Kohno, and D. C. Klonoff. A review of the security of insulin pump infusion systems. *Journal of Diabetes Science and Technology*, 5(6):1557–1562, November 2011.

[24] K. Poulsen. Hackers assault epilepsy patients via computer. Wired.com, http://www.wired.com/politics/security/news/2008/03/epilepsy, March 2008.

[25] J. Rabaey, M. Mark, D. Chen, C. Sutardja, C. Tang, S. Gowda, M. Wagner, and D. Werthimer. Powering and communicating with mm-size implants. In *Design, Automation & Test in Europe Conference & Exhibition*, DATE '11. IEEE, 2011.

[26] J. Radcliffe. Hacking medical devices for fun and insulin: Breaking the human SCADA system. Black Hat Conference presentation slides, August 2011.

[27] K. B. Rasmussen, C. Castelluccia, T. S. Heydt-Benjamin, and S. Čapkun. Proximity-based access control for implantable medical devices. In *Proceedings of the 16th ACM Conference on Computer and Communications Security*, pages 410–419, 2009.

[28] P. Roberts. Blind attack on wireless insulin pumps could deliver lethal dose. Threatpost (blog post), http://threatpost.com/en_us/blogs/blind-attack-wireless-insulin-pumps-could-deliver-lethal-dose-102711, October 2011.

[29] D. Takahashi. Excuse me while I turn off your insulin pump. VentureBeat, http://venturebeat.com/2011/08/04/excuse-me-while-i-turn-off-your-insulin-pump/, August 2011.

[30] F. Xu, Z. Qin, C. C. Tan, B. Wang, and Q. Li. IMDGuard: Securing implantable medical devices with the external wearable guardian. In *Proceedings of the 30th IEEE International Conference on Computer Communications*, INFOCOM '11, pages 1862–1870, April 2011.

Design of Pin-Constrained General-Purpose Digital Microfluidic Biochips*

Yan Luo and Krishnendu Chakrabarty
Department of Electrical and Computer Engineering
Duke University, Durham, NC 27708, USA
E-mail: {yan.luo, krish}@duke.edu

ABSTRACT

Digital microfluidic biochips are being increasingly used for biotechnology applications. The number of control pins used to drive electrodes is a major contributor to fabrication cost for disposable biochips in a highly cost-sensitive market. Most prior work on pin-constrained biochip design determines the mapping of a small number of control pins to a larger number of electrodes according to the specific schedule of fluid-handling operations and routing paths of droplets. Such designs are therefore specific to the bioassay application, hence sacrificing some of the flexibility associated with digital microfluidics. We propose a design method to generate an application-independent pin-assignment configuration with a minimum number of control pins. Layouts of a commercial biochip and laboratory prototypes are used as case studies to evaluate the proposed design method for determining a suitable pin-assignment configuration. Compared with previous pin-assignment algorithms, the proposed method can reduce the number of control pins and facilitate the "general-purpose" use of digital microfluidic biochips for a wide range of applications.

Categories and Subject Descriptors

B.2.2 [**Hardware**]: Performance Analysis and Design Aids

General Terms

Algorithms, Performance, Design.

Keywords

Digital microfluidics, electrowetting-on-dielectric, lab-on-chip.

1. INTRODUCTION

Digital (droplet-based) microfluidic biochips are used to handle and analyze picoliter volumes of samples and reagents [1]. By utilizing the principle of electrowetting on dielectric, droplets containing biological samples and reagents can be

---

*This research was supported in part by the National Science Foundation under grant no. CCF-0914895.

manipulated on a chip without any etched microchannels or external pressure sources [1].

Liquid droplets in the digital microfluidic platform are manipulated on an array of discrete unit cells [1]. Each unit cell contains two electrodes that act as a pair of parallel plates [1]. Droplets are actuated by applying appropriate voltages to the electrodes and the sequence of actuation voltages can be stored in a microcontroller or computer memory. By changing the actuation sequence in memory, the chips can be dynamically reprogrammed (i.e., reconfigured). Concurrent manipulation of multiple discrete droplets can therefore be coordinated by control software and voltages applied to the electrodes.

Digital microfluidic biochips are finding extensively applications in biotechnology, e.g., on-chip chemistry for DNA sequencing [1], multiplexed real-time polymerase chain reaction (PCR) [2], and cytotoxicity assays [3]. Design automation methods for digital microfluidics are also receiving attention [4–7]. The complexity of digital microfluidic biochips continues to increase as new applications are targeted by this platform. For example, recently announced commercial products contain up to 5000 electrodes [2,8]. In order to ensure complete reconfigurability and the ability to run any given bioassay on the digital microfluidic platform (i.e., "general-purpose use"), it is desirable that every electrode be controlled by an independent pin. However, a one-to-one mapping between control pins and electrodes (referred to as direct-addressing pin-assignment) is not practical for low-cost disposable chips. A large number of control pins leads to high fabrication cost, large form factors and interconnect routing problems [7].

In order to reduce the number of control pins and to control the digital microfluidic array without significantly affecting concurrent droplet operations, a number of design of optimization techniques have been published in the literature. These techniques can be categorized as being either bioassay-independent [9,10] or bioassay-specific [11–13]. In bioassay-independent techniques such as the use of a bus-phase addressing [9] or cross-referencing [10], the number of control pins required for addressing the electrodes is independent of the target application. For example, cross-referencing requires $m + n$ pins for an $m \times n$ array of electrodes, analogous to row/column-based addressing in memories. Bioassay-specific pin-assignment methods lead to fewer control pins since they utilize knowledge about the operation schedule, module placement, and droplet routing pathways of the target bioassay.

Prior methods on bioassay-specific pin assignment suffer

from three main drawbacks. First, these techniques are not effective for the design of multi-functional biochips, which can be reconfigured post-fabrication for different applications by loading the appropriate control software. General purpose (application-independent) biochips, where software can be used as a differentiator, offer the promise of higher production volume and reduced cost. Second, fluid-handling operations on an application-specific biochip are constrained by the pre-determined pin-assignment, hence post-fabrication tuning of the bioassay protocol, schedule, and droplet routing are not possible. Finally, it is difficult to estimate the number of control pins *a priori* since the number of pins is application-dependent. The cross-referencing technique described above is application-independent; however, it requires a special electrode structure which both top and bottom plates are divided into discrete electrode arrays. This results in increased complexity and higher manufacturing cost [10].

To overcome the above drawbacks, we propose a new method to generate pin-assignment configurations. This method does not depend on actuation sequences of electrodes, or does the scheduling and the routing of droplets. Any target application can be mapped to the array without any restriction on droplet manipulation. The degree of freedom for droplet movement is therefore maximized. The key contributions of the paper are as follows: (1) an analysis of pin-actuation conflicts, and derivation of necessary and sufficient conditions for control-pin sharing to ensure droplet movement (Section 2); (2) an integer linear programming model for designing a pin-assignment with the smallest number of pins (Section 3); (3) a graph-theoretic method to formulate an acceptance test for a pin-assignment configuration and a lower bound on the number of pins (Section 4); (4) a heuristic algorithm that generate a pin-assignment (Section 4); (5) results for experimental prototypes and a commercial biochip (Section 5).

2. ANALYSIS OF PIN-ASSIGNMENT

In this section, we discuss the relationship between droplet movement and voltages applied to the electrodes. Next, the concept of pin-actuation conflict is introduced. Finally, several pin-assignment configurations are analyzed in order to determine the conditions that guarantee conflict-free pin-assignment.

2.1 Pin-actuation Conflicts

To manipulate a droplet that currently resides on an electrode $\mathcal{E}$, appropriate control voltages must be applied to a group of electrodes. This electrode group consists of a central electrode $\mathcal{E}$ and all its non-diagonal adjacent electrodes. Each non-diagonal adjacent electrode is a possible destination for the droplet. Figure 1(a) presents an example. Each square in Figure 1(a) stands for an electrode on the microfluidic biochip; letters such as "A", "B" and "C" stand for the names of control pins that are connected to the corresponding electrodes. The control voltages applied on the control pins are either "High", "Low" or "don't-care". Here we introduce two definitions:

Control electrode group (CEG): An electrode, on which the droplet rests at any given time, and its adjacent, non-diagonal electrodes are defined as the elements of the control electrode group (CEG); see Figure 1(a).

Control pin group (CPG): All the pins that are con-

nected to the electrodes in the CEG are defined as the elements of the control pin group (CPG).

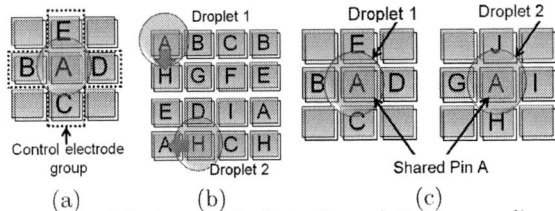

(a)   (b)   (c)

Figure 1: (a) A central electrode and its surrounding, non-diagonal electrodes comprising an electrode group (b) An example of pin-actuation conflicts (c) Groups of electrodes and corresponding control pins. Pin A is the common pin shared by these two groups of electrodes.

When multiple fluid-handling operations are implemented on a biochip with a given pin assignment, pin-actuation conflicts must be considered. An example is shown in Figure 1(b), where droplets D_1 and D_2 are on the array, and they are scheduled to move in the directions of the arrows. The groups of control pins for D_1 and D_2 are {A, B, H} and {A, C, D, H}, respectively. Note that pin H is the common control pin for droplets D_1 and D_2. In order to move D_1 in a designated direction, the voltages applied to A, B, H should be Low, Low, and High, respectively. Similarly, the voltages of A, C, D, H should be High, Low, Low, and Low, respectively, in order to move droplet D_2. Thus, the movement of D_1 requires the application of "High" voltage on pin H, while the movement of D_2 requires the application of "Low" voltage on H. Since it is not possible to apply these voltages to pin H at the same time, the movements of D_1 and D_2 cannot be implemented concurrently.

2.2 Control-pin Sharing and Concurrent Movement of Droplets

We consider the concurrent movements of a pair of droplets to analyze pin-actuation conflicts. Any pin-actuation conflict involving more than two droplets can be studied as a two-droplet problem by examining all possible pairs of droplets.

Assume that there are two droplets and their CEGs are two non-overlapping electrode groups which both contain 5 electrodes. A total of 10 control pins are assigned to the two electrode groups, so that each electrode is connected to an independent pin. It is clear that the two droplets can be moved concurrently to any of the four surrounding, non-diagonal electrodes without pin conflicts. Therefore, the total number of possible concurrent movements for the droplets D_1 and D_2 is $4 \times 4 = 16$.

We next attempt to reduce the number of control pins assigned to the two electrode groups. When nine control pins are assigned to the 10 electrodes, the corresponding pin-assignment configurations can be divided into three categories. Figure 1(c) shows the first category in which the central electrodes of the two groups share the same control pin. The two droplets in Figure 1(c) can be moved concurrently to any surrounding, non-diagonal electrode concurrently without pin conflicts. Thus, there are 16 possible concurrent movements of the pair of droplets.

Figure 2(a) shows the second category, in which two electrodes that are in the same electrode group share one control pin. In this case, droplet 2 can be moved to any surrounding,

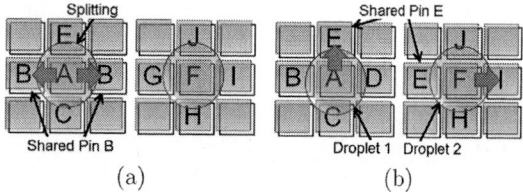

(a)   (b)

Figure 2: (a) Splitting of droplet 1 when two electrodes in the same electrode group sharing one control pin. (b) Droplet 1 and 2 cannot be move concurrently when two electrodes in different electrode groups sharing one control pin.

Table 1: The number of pins assigned to electrodes and corresponding degrees of freedom for droplet movements.

No. pins	Best case	Worst case	No. movement pairs
10	Both droplets can move in all four directions	Both droplets can move in all four directions	16
9	Both droplets can move in all four directions	One droplet can move freely and the other has to stall	8 or 16
8	One droplet can move freely and the other has to stall	Unintentional movement or splitting occurs	Less than 8

non-diagonal electrode. On the other hand, droplet 1 has only two possible directions of movement. This is because the two surrounding, non-diagonal electrodes of droplet 1 are controlled by pin B. If pin B activates the two electrodes at the same time, then droplet 1 will be split. To avoid unintentional splitting of droplet 1, the two electrodes that are connected to pin B must be deactivated. Thus, droplet 1 cannot be moved to these two electrodes, and the number of all possible concurrent movements for the pair of droplets is reduced to $2 \times 4 = 8$.

The third category is shown in Figure 2(b) where two non-central electrodes from the two groups of electrodes share one control pin. Suppose droplets 1 and 2 are scheduled to move in the directions indicated by the arrows. Here pin E is the common control pin for these two droplets. For the movement of droplet 1, the control voltages applied to pin A, B, C, D, and E should be Low, Low, Low, Low, and High, respectively. Similarly, for the movement of droplet 2, the control voltages on pin E, F, J, H, and I should be Low, Low, Low, Low, and High, respectively. Therefore, the status of pin E corresponding to the movements of droplets 1 and 2 are different. To avoid a conflict, droplets 1 and 2 must be moved in different clock cycles. Thus the number of all possible concurrent movements for the droplet pair is $2 \times 4 = 8$.

When eight control pins are assigned to 10 electrodes, we can examine all the possible pin-assignment configurations and their corresponding degrees of freedom for droplet movement. We reach the conclusion that unintentional movement or splitting may occur if the control pin groups of the two droplets share more than one common pin. Table 1 shows the number of pins assigned to two non-overlapping control electrode groups (10 electrodes in total) and the corresponding degrees of freedom for droplet movement.

Based on the above analysis, we obtain the following lemma to ensure the maximum degree of freedom of concurrent droplet movement for any pin-assignment. The lemma provides a necessary and sufficient condition for pin-assignment to avoid unintended movement or splitting of droplets. The proof of the lemma is given in the appendix. Our optimization objective is to minimize the number of control pins while satisfying the conditions of Lemma 1.

LEMMA 1. *For two randomly-selected droplets on the electrode array, the following constraints are necessary and sufficient to ensure that both droplets can move in any direction, or that one droplet can move in any direction while the other droplet does not undergo any unintentional movement or splitting:*

Constraint 1: Any two electrodes in the same control pin group cannot be connected to the same control pin.

Constraint 2: Any two non-overlapping electrode groups cannot share more than one pin.

Note that Constraint 1 and Constraint 2 also ensure that when one droplet is split, the other droplet does not undergo any unintentional movement or splitting. Also, since the movement of a droplet from one electrode to another adjacent electrode is the basic operation for mixing and transportation, Constraint 1 and Constraint 2 can ensure the maximum degree of freedom for any two concurrent fluid-handing operations.

3. ILP MODEL FOR PIN-ASSIGNMENT

In this section, we develop a integer linear programming (ILP) model to optimally solve the pin-assignment problem. On an $M \times N$ electrode array, let $x_{i,m,n}$ be a binary variable defined as below.

$$x_{i,m,n} = \begin{cases} 1 & \text{if pin } i \text{ is connected to Electrode } (m,n) \\ 0 & \text{else} \end{cases} \quad (1)$$

where $1 \leq i \leq L$. The parameter L is the maximum possible index for the number of control pins and the value of L can be set to an easily-determined loose upper bound.

The index of control pin connected to Electrode (m, n) is defined as $P_{m,n}$. It can be expressed as $P_{m,n} = \sum_{i=1}^{L} i \cdot x_{i,m,n}$.

The objective function of the ILP model for a pin-constrained digital microfluidic biochip is defined as:

$$\text{minimize: } C_{max} = Max_{1 \leq i \leq L, 1 \leq m \leq M, 1 \leq n \leq N}\{i \times x_{i,m,n}\} \quad (2)$$

Constraint 1 of Lemma 1 can be expressed as the following set of constraints:

$$x_{i,m,n} + x_{i,m+1,n} + x_{i,m-1,n} + x_{i,m,n+1} + x_{i,m,n-1} \leq 1,$$
$$\forall 1 \leq i \leq L, 1 \leq m \leq M, 1 \leq n \leq N \quad (3)$$

Based on Constraint 2 of Lemma 1, we get the following set of inequalities:

$$x_{i,m,n} + x_{i,m+1,n} + x_{i,m-1,n} + x_{i,m,n+1} + x_{i,m+p,n-1+q}$$
$$+x_{j,m+p,n+q} + x_{j,m+1+p,n+q} + x_{j,m-1+p,n+q} + x_{j,m+p,n+1+q}$$
$$+x_{j,m+p,n-1+q} \leq 3,$$
$$\forall 1 \leq i \neq j \leq L, 1 \leq m \leq M, 1 \leq n \leq N, |p| + |q| \geq 3 \quad (4)$$

The minimax objective function given by (2) can be linearized using standard techniques. This completes the ILP model for optimization. The ILP model is clearly not scalable for large problem instance; nevertheless, we use it later to evaluate the quality of the heuristic solution that we design in the next section. For the $M \times N$ microfluidic array, the number of CEGs is MN. The number of inequalities derived from Constraint 1 and Constraint 2 are O(MNL)

and $O(M^2N^2L^2)$, respectively. Thus for the ILP model introduce above, the number of variables in the ILP model is $O(MNL)$, and the number of constraints is $O(M^2N^2L^2)$. It is important to note that, since L is a loose upper bound for the number of pins, we usually have $L = O(MN)$.

4. HEURISTIC OPTIMIZATION METHOD

The ILP model introduced in Section 3 has high computing complexity and the CPU time is unacceptable for large biochips. In this section, we develop a heuristic algorithm to generate a pin-assignment configuration efficiently; the pseudocode for this algorithm is shown in Part B of the appendix. The algorithm includes two phases; the first phase is to establish a lower bound on the number of pins, and the second phase is to construct the pin-assignment configuration in a greedy fashion. In order to improve the efficiency of the greedy algorithm, we propose a mapping from a pin-assignment configuration to an undirected graph, and then we use the graph to ascertain whether the pin-assignment configuration is a feasible solution. A lower bound on the number of pins from the graph model is also described.

For simplicity, we define two operators. First, the pin-assignment configuration is defined as the operator $\mathcal{C}$, which is a mapping from the set of all electrodes on the chip (E^*) to the set of pins that are assigned to the electrodes (P).

$$\mathcal{C}: E^* \to P \qquad (5)$$

For example, Figure 3(a) shows the coordinate locations for the electrodes and Figure 3(b) shows the corresponding pin-assignment configuration. Here, the electrode at position (i,j) is represented by $E_{(i,j)}$, and $E^* = \bigcup_{i,j} E_{(i,j)}$. When the operator $\mathcal{C}$ is applied to $E_{(1,1)}$, we get $\mathcal{C}(E_{(1,1)}) =$ pin A.

For any electrode $E \in E^*$, we can virtually place a droplet on this electrode and obtain the CEG, referred as E_{group}, for this virtual droplet. By taking any two elements from E_{group}, we can get an unordered pair of electrodes. The set consisting of all such electrode pairs is represented by E_{pair}.

Next, the operator Φ is defined as a mapping from the set of electrodes E^* to the set of electrode pairs E_{pair}.

$$\Phi: E^* \to E_{pair} \qquad (6)$$

For example, as shown in Figure 3(a), when the operator Φ is applied to $E_{(1,3)}$, the following mapping is obtained: $\Phi(E_{(1,3)}) = \{(E_{(1,3)}, E_{(1,2)}), (E_{(1,3)}, E_{(2,3)}), (E_{(1,2)}, E_{(2,3)})\}$.

Based on operators $\mathcal{C}$ and Φ defined above, for any electrode $E_{(i,j)}$ on the layout, a graph $G_{(i,j)}$ can be constructed as follows. First, a droplet is placed virtually on electrode $E_{(i,j)}$ and the corresponding CEG ($E_{group(i,j)}$) is obtained. By applying operator $\mathcal{C}$ to each element of $E_{group(i,j)}$, we get the CPG ($P_{group(i,j)}$) that corresponds to this virtual droplet. Each pin in $P_{group(i,j)}$ is mapped to a node in $G_{(i,j)}$.

By applying operator Φ on E^*, the set of electrode pairs $E_{pair(i,j)}$ can be derived. For each electrode pair (E_x, E_y) in $E_{pair(i,j)}$, we apply operator $\mathcal{C}$ to it and get the corresponding pin pair (P_x, P_y), and add an edge between the two nodes that represent P_x and P_y in the graph. This edge is labeled as $Link_{(E_x, E_y)}$. Since (E_x, E_y) is an unordered pair, we consider the edges with labels $Link_{(E_x, E_y)}$ and $Link_{(E_y, E_x)}$ as the same edge, i.e., we do not distinguish between them.

In this way, a one-to-one mapping from the set $E_{pair(i,j)}$ to the edges in $G_{(i,j)}$ is defined. Based on this definition,

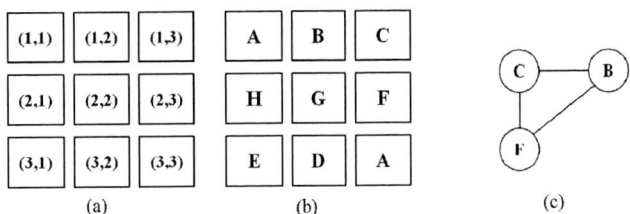

Figure 3: (a) Coordinate locations for the electrodes; (b) pin-assignment for the electrodes; (c) graph corresponding to electrode $E_{(1,3)}$.

we conclude that, if elements in the CEG of $E_{(i,j)}$ are assigned to different pins, then $G_{(i,j)}$ is a complete graph. For example, Figure 3(c) shows graph $G_{(1,3)}$ corresponding to electrode $E_{(1,3)}$ in Figure 3(a).

For a given pin-assignment configuration, let the set of electrodes be defined as E_{layout}. By virtually placing a droplet on electrode $E_x \in E_{layout}$, the graph G_{E_x} for any electrode can be derived. The graph for the pin-assignment configuration, which is written as G_{layout}, is defined as the union of all G_{E_x}, i.e.,

$$G_{layout} = \bigcup_{E_x \in E_{layout}} G_{E_x},$$

where the set of nodes in G_{layout} are obtained by applying the union operation to the set of graphs in G_{E_x} $\forall$ $E_x \in E_{layout}$. The edges with the same label are considered as the same edge in the union graph, i.e., they appear only once [14].

Based on the above definition, any graph G_{E_x} that corresponds to electrode E_x in the layout is a subgraph of G_{layout}. The following lemma results from the structure of G_{layout} and the acceptability of the pin-assignment (Lemma 1). The proof is in Part C of the appendix. Note that depending on the mapping of control pins to electrodes, G_{layout} may be a simple graph (no self-loop and no more than one edge between any two distinct nodes) or a multigraph (either with a self-loop or more than one edge between some pairs of nodes) [14].

LEMMA 2. : *A pin-assignment configuration satisfies Constraint 1 and Constraint 2 of Lemma 1 if and only if the graph G_{layout} is a simple graph.*

The next theorem presents a lower bound on the number of control pins needed to avoid pin-actuation conflicts for any target application. The proof, which is based on Lemma 2, is presented in Part D of the appendices.

THEOREM 1. *Consider an $m \times n$ digital microfluidic array and suppose a pin-assignment configuration with M pins exists such that there are no pin-actuation conflicts. A lower bound on M is given by:*

$$\binom{M}{2} \geq 6mn - 5m - 5n + 2$$

For large values of m and n, the lower bound M_{min} can be approximated as $M_{min} \approx 2\sqrt{3mn}$, and if $m = n = t$, we get $M_{min} \approx 2\sqrt{3}t$. In other words, the number of pins is $\Omega(\sqrt{N})$ when the total number of electrodes in the array is N.

Let us now return to the pseudocode shown in Part B of Appendices, where we determine the lower bound on the number of pins in the first phase. In the second phase, we construct the pin-assignment staring from a single electrode group. First we randomly select a CEG and construct the local pin-assignment as an initial solution. For a single CEG, we can easily satisfy both Constraint 1 and Constraint 2 by assigning a group of independent pins to the CEG. Next we randomly choose an electrode in the neighborhood of the "already-solved" region, and assign an "available control pin" to it. Each time after assigning a pin to a newly-added electrode, we first update G_{layout} by doing the *union* of original G_{layout} with the graph corresponding to the new electrode, and then check by whether the updated G_{layout} is a simple graph. If yes, we successfully expand the feasible local solution. Otherwise, we change the pin assigned to the new electrode until we obtain a feasible solution. In this way, the feasible local solution can be expanded step by one step until it covers the whole layout. When we assign pin to the newly-added electrode, the *complement* of the local graph, referred to as H_{local}, is used to reduce the search scope. All the pins corresponds to isolated nodes in H_{local} are "unavailable pins" and they cannot be assigned to any newly-added electrode. If all the pins are unavailable, we add new pins to the layout.

Assume that the number of electrode on the array is $\mathcal{N}$ and the number of pins assigned to the electrode is $\mathcal{P}$. The computing complexity for checking whether G_{layout} is a simple graph is $O(\mathcal{P}^2)$. Each time when we assign a pin to a newly-added electrode, we have to verify whether G_{layout} is a simple graph. Thus for the worst case, the computational complexity of assigning a pin to a newly-added electrode is $O(\mathcal{P}^3)$. Since all electrodes are added step-by-step to the initial solution, the overall computational complexity of the heuristic procedure is $O(\mathcal{P}^3\mathcal{N})$.

5. EXPERIMENT RESULTS

In this section, we present results for a commercial biochip for an n-plex immunoassay and several university prototype chips. In an n-plex immunoassay, a sample is concurrently analyzed for n different analytes. Sample droplets are mixed with n different reagents, and the mixed product droplets are moved to the detection area which includes an optical detector. The commercial biochip for the n-plex immunoassay consists of more than 1000 electrodes [2, 8] and it is a representative case study for a multiplexed and concurrent bioassay [15]. The layout of this biochip is shown in Figure 4(a).

As seen in Figure 4(a) [8,15], this commercial chip is based on a regular design that consist of two types of unit cells. We design the unit cells in the layout and assign control pins for each unit cell; all other cells can be assigned pins based on the same configuration. By applying the heuristic algorithm proposed in Section 4, we obtain the pin-assignment for the Type I and Type II unit cells, as shown in Figure 4(b) and Figure 4(c). Note here these two unit cells are controlled by independent sets of pins, thus pins 1 to 7 are assigned to Type I unit cell, and pins 8 to 18 are assigned to Type II unit cell.

Table 2 shows the comparison of the number of control pins for the existing design for the fabricated commercial biochip [15], the bioassay-specific algorithm in [15], and the proposed heuristic method, despite being independent of application, leads to comparable or smaller number of control pins than both [15] and the fabricated commercial biochip.

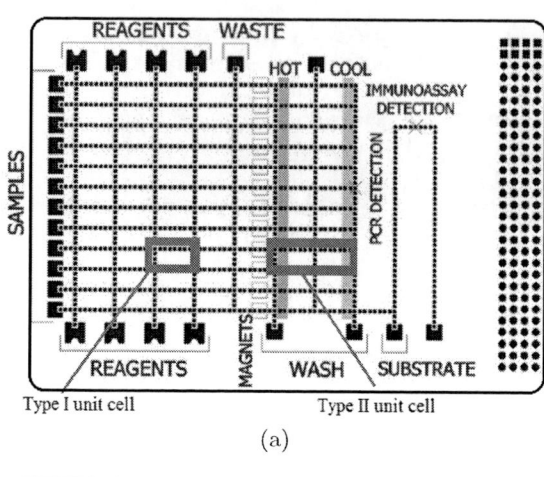

(a)

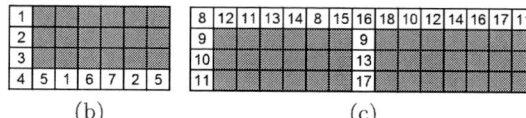

(b) (c)

Figure 4: (a) Layout of the commercial chip [8, 15] and pin-assignment for the unit cells of (b) Type I, and (c) Type II shown in Figure 4(a).

Table 2: Comparison of the number of pins for commercial chip [8,15], using the bioassay-specific method of [15] and the proposed heuristic method.

	No. pins (fabricated chip [15])	No. pins (bioassay-specific method [15])	No. pins (proposed heuristic method)	No. pins (ILP model)	Application-independent lower bound (Theorem 1)
Routing region	7	6	7	7	7
Reaction region	19	13	11	11	10

Since the pin-assignment for the proposed method is bioassay-independent, it also provides the added benefit of being flexible for multiple target applications. Columns 4-6 in Table 2 compare the results derived from the proposed heuristic algorithm, the ILP model and the lower bound given by Theorem 1. It can be seen that the results for the heuristic method are close the lower bound and the optimal result obtained using ILP.

Next we apply the pin-assignment algorithm to the layouts of four university prototypes described in [12], namely a multiplexed assay biochip, a PCR biochip, a protein dilution biochip, and a multifunction biochip. Their pin-assignment configurations are shown in Figure 5. It is important that on the layouts, no electrode is fabricated in these area without pin-assignment. According to the pin-assignment configuration, we also schedule the execution of the corresponding bioassays and derive the assay completion times. Table 3 shows comparisons between the result derived from bioassay-specific method in [12] and the proposed heuristic method. For the multiplexed assay biochip, the protein dilution biochip and the multifunction biochip, the proposed heuristic method can reduce the number of control pins with only a slight increase in the completion times. It is important to note that, for the bioassay-specific designs in [12], it is difficult to subsequently add new operations such as optical detection for droplets or map a new application once the pin-assignment is fixed. However, in the proposed ap-

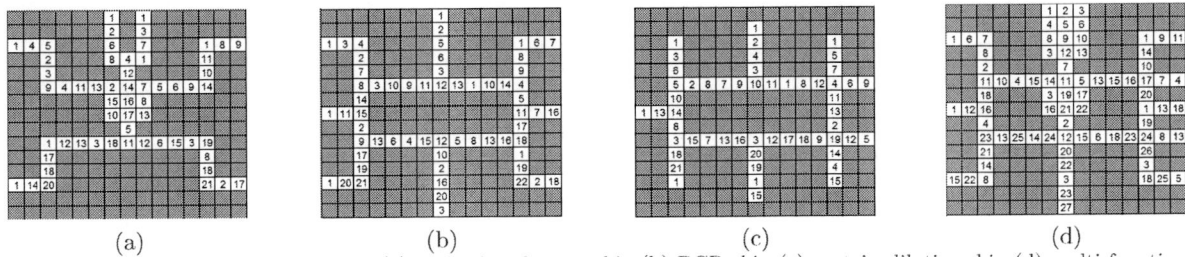

Figure 5: Pin-assignment configurations for (a) multiplexed assay chip (b) PCR chip (c) protein dilution chip (d) multi-functional chip.

Table 3: Comparisons between results of [12] and the proposed method.

Target application	No. pins in [12]	Completion time in [12]	No. pins (proposed heuristic method)	Completion time (proposed heuristic method)
Multiplexed assay chip	25	73 s	21	90 s
PCR chip	14	20 s	22	30 s
Protein dilution chip	26	150 s	20	170 s
Multi-functional chip	37	150 s	27	170 s

Table 4: Comparison between lower bound, ILP model, and heuristic solution.

Array size	No. pins (lower bound)	No. pins (ILP model)	No. pins (heuristic method)
3 × 3	8	8	8
5 × 7	18	21	23
7 × 8	24	29	32
8 × 8	26	32	33
9 × 9	30	Computationally impractical	40

proach, we can adapt fluid-handling operations according to the requirement of bioassay protocols.

Finally, we present a comparison between the lower bound on the number of pins predicted by Theorem 1, the exact optimization results obtained using ILP model, and the results obtained using the heuristic procedure. Without loss of generality, we use various rectangular layouts as test cases. The results are shown in Table 4. The ILP model is solved using Xpress-MP [16], which is a widely used commercial ILP solver. The tool is run on a server with 64 GB of memory and two quad-core 2.53 GHz Intel Xeon processors. The computing time for the 8 × 8 array is about 5 hours while for a 9 × 9 array, the solver runs out of memory. The heuristic method is run on a 2.27 GHz Intel i3 Dual core processor with 2 GB of memory. The CPU times are less than 5 minutes for all cases.

6. CONCLUSION

We have presented a new method for mapping control pins to electrodes in the design of "general purpose" (application-independent) digital microfluidic biochips. We have derived necessary and sufficient conditions for pin-assignment to guarantee conflict-free concurrent manipulation of multiple droplets. We have also presented a lower bound on the number of control pins required for an electrode array. A graph-theoretic "acceptance test" has been developed for a given pin-assignment. An optimization technique based on ILP has been described to automatically derive conflict-free pin-assignments. The ILP model is not scalable to large designs, but it can be used for partitioned designs and to evaluate the quality of heuristic solutions. We have presented an efficient heuristic approach for mapping control pins to electrodes. The heuristic method has been evaluated using a commercial biochip and laboratory prototypes. Compared with previous pin-assignment algorithms, the proposed method can reduce the number of control pins and facilitate the "general-purpose" use of digital microfluidic biochips for a wide range of applications.

7. REFERENCES

[1] R. B. Fair, "Digital microfluidics: Is a true lab-on-a-chip possible?", *Microfluidics and Nanofluidics*, vol. 3, pp. 245-281, 2007.

[2] Z. Hua et al., "Mutiplexed real-time polymerase chain reaction on a digital microfluidic platform", *Anal. Chem.*, vol. 82, pp. 2310-2316, 2010.

[3] I. Nad, H. Yang, P. Park, and A. Wheeler, "Digital microfluidics for cell-based assays", *Lab Chip*, pp. 519-526, 2008.

[4] P-H. Yuh, C.-L. Yang, and Y.-W. Chang, "Placement of digital microfluidic biochips using the T-tree formulation", *Proc. DAC*, pp. 931- 934, July 2006.

[5] M. Cho et al., "A High-Performance Droplet Router for Digital Microfluidic Biochips", *Proc. ISPD*, pp. 1714-1724, April 2008

[6] T.-W. Huang, C.-H. Lin, and T.-Y. Ho, "A Contamination Aware Droplet Routing Algorithm for the Synthesis of Digital Microfluidic Biochips", *IEEE Trans. TCAD*, vol. 29, no. 11, pp. 1682-1695, November 2010

[7] K. Chakrabarty, R. B. Fair and J. Zeng, "Design tools for digital microfluidic biochips: Towards functional diversification and more than Moore" (Keynote Paper), *IEEE Trans. CAD*, vol. 29, pp. 1001-1017, July 2010.

[8] R. Sista et al., "Development of a digital microfluidic platform for point of care testing", *Lab on a Chip*, vol. 8, pp. 2091-2104, 2008.

[9] V. Srinivasan et al., "An integrated digital microfluidiclab-on-a-chip for clinical diagnostics on human physiological fluids", *Lab on a Chip*, vol. 4, pp. 310-315, 2004.

[10] Z. Xiao et al., "CrossRouter: A droplet router for cross-referencing digital microfluidic biochips", *Proc. ASP-DAC*, pp. 269-274, 2010.

[11] C.-Y. Lin and Y.-W. Chang, "Cross-contamination aware design methodology for pin-constrained digital microfluidic biochips", *IEEE Trans. CAD*, vol. 30, pp. 817-828, No. 6, June 2011.

[12] T. Xu and K. Chakrabarty, "Broadcast electrode-addressing for pin-constrained multi-functional digital microfluidic biochips", *Proc. DAC*, pp. 173-178, 2008.

[13] T.-W. Huang and T.-Y. Ho, "A two-stage integer linear programming-based droplet routing algorithm for pin-constrained digital microfluidic biochips", *IEEE Trans. CAD*, vol. 30, pp. 215-228, 2011.

[14] F. Harary, "Graph theory", *Addison-Wesley*, Reading, MA, 1994

[15] Y. Zhao and K. Chakrabarty, "Simultaneous optimization of droplet routing and control-pin mapping to electrodes in digital microfluidic biochips", *IEEE Trans. CAD*, 2011 (accepted).

[16] http://www.fico.com/en/Products/DMTools/xpress-overview/Pages/Xpress-Mosel.aspx

APPENDICES

We present here proofs of Lemma 1-2 and Theorem 1. We also present additional details about our experimental results.

A. PROOF OF LEMMA 1

First, we prove the necessity of Constraint 1 and Constraint 2. In Section 2, the movements of droplets on the layouts of different pin-assignment configurations are discussed. Figure 2(b) shows an example of violating Constraint 1, and the droplet will be split when the voltage on pin B is "High". When Constraint 2 is violated, i.e., two non-overlapping electrode groups share more than one control pins, moving one droplet may result in unintentional movement of the other droplet.

Thus, Constraint 1 and Constraint 2 are the necessary conditions for avoiding unintentional movement and splitting of droplets.

Next, we prove the sufficiency of Constraint 1 and Constraint 2. Assume that we have a pin-assignment configuration that meets the two conditions. Now we randomly select two droplets D_1 and D_2, on the chip. According to Constraint 1, elements in the control electrode group for D_1 must be assigned different pins, and similar assignment is executed for D_2. According to Constraint 2, the control pin groups for these two droplets share either one in common or none at all. As discussed above, the common pin shared by the two control pin groups plays an important role in pin-actuation conflict, so the control pin groups for D_1 and D_2 are assumed to share a common pin (P_{common}).

To implement scheduled operations, corresponding control voltages must be applied to control pins of the droplets. In order to implement operation on droplet D_1, we assume that the control voltage on P_{common} must be S_1. Similarly, in order to implement operation on droplet D_2, the control voltage on P_{common} is assumed to be S_2. S_1 and S_2 are either High or Low signal (which are referred as High and Low). Then we compare S_1 and S_2 for the two cases presented below:

(a) $S_1 = S_2$: This relationship indicates that the actuation voltages required by the concurrently scheduled operations on the common pin are the same, i.e., no pin-actuation conflict occurs. Then, the scheduled operations for the two droplets can be implemented concurrently.

(b) $S_1 \neq S_2$: This relationship indicates that conflict occurs between the actuation voltages for the movements of the droplets. Without loss of generality, we set S_1 to high voltage and S_2 to low voltage. Due to the pin-actuation conflict, D_1 and D_2 cannot move concurrently. Then, a balance-off plan must be scheduled, i.e., one droplet is moved in the scheduled direction, and the other drop maintains its current position. Thus, it is important to consider the actuation signal that should be applied in this plan. According to the physical mechanisms of the biochip, in order to move a droplet from its current position to its destination, low voltage is applied to the electrode under the droplet, and high voltage is applied to the destination electrode. When droplet D_1 moves, high voltage will be applied to P_{common}. If P_{common} is connected to the electrode that is under droplet D_2, then D_2 will stay at its current position without unintentional movement or splitting. If P_{common} is connected to a non-center electrode, high voltage will be applied to the electrode under the droplet. It is important

to note that D_1 and D_2 share only one control pin, and all other control pins for D_1 and D_2 are independent pins. Thus, when high voltage is applied to the electrode under D_2, the movement of D_1 will not be affected. Based on the above discussion, necessity and sufficiency of Constraint 1 and Constraint 2 in Lemma 1 are proved. □

B. PSEUDOCODE FOR GENERATION OF PIN-ASSIGNMENT CONFIGURATIONS

Figure 6 shows the pseudocode for the heuristic algorithm proposed in Section 4.

C. PROOF OF LEMMA 2

Proof: First we assume that Constraint 1 is violated. Thus, an electrode group $E_{group(x)}$ exists in which two elements share the same control pin. According to the definition of edges for G_{E_x}, there will be cyclic edges and multiple edges between two nodes. Thus G_{E_x} is a multigraph. Since G_{E_x} is a subgraph of G_{layout}, G_{layout} is also a multigraph.

We next assume that Constraint 2 is violated. Thus two non-overlapping CEGs E_{group1} and E_{group2} exist, and their corresponding CPGs share more than one common pin. Assume the graph derived from E_{group1} and E_{group2} are G_1 and G_2, respectively. Suppose $E_{x_1} \in E_{group1}$ and $E_{x_2} \in E_{group2}$ are both connected to pin X, and $E_{y_1} \in E_{group1}$ and $E_{y_2} \in E_{group2}$ are both connected to pin Y. According to the definition of the one-to-one mapping from the set of electrode pairs to edges in the graph, electrode pair $\{E_{x_1}, E_{y_1}\}$ is mapped to an edge between the nodes that represent pin X and pin Y. The label of this edge is $Link_{(E_{x_1}, E_{y_1})}$ (or $Link_{(E_{y_1}, E_{x_1})}$). Electrode pair $\{E_{x_2}, E_{y_2}\}$ is mapped to an edge between the nodes that represent pin X and pin Y. The label of this edge is $Link_{(E_{x_2}, E_{y_2})}$ (or $Link_{(E_{y_2}, E_{x_2})}$). Based on the definition of the *union* of two graphs, when G_1 and G_2 are merged to $G_1 \cup G_2$, there will be two edges with different labels between the nodes that represent pin X and pin Y. Thus $G_1 \cup G_2$ is a multigraph. Since $G_1 \cup G_2$ is a subgraph of G_{layout}, we can conclude that G_{layout} also is a multigraph.

According to the definition of G_{layout}, it is easy see that if it is a simple graph, then the pin-assignment configuration satisfies Constraint 1 and Constraint 2. □

D. PROOF OF THEOREM 1

For a $m \times n$ electrode array, and for any electrode $E_{(i,j)}$ in this array, since its corresponding graph $G_{(i,j)}$ is a complete graph, the number of edges can be derived from the number of elements in the CPG. For different positions of the electrodes, the numbers of elements in CPG are different. Thus we calculate the number of edges G_{layout} by classifying electrode into three categories according to their positions, as shown in Figure 7(a).

The first category includes the electrodes located at the corner of the array and the corresponding graph is shown in Figure 7(b). For each graph, there are three nodes and the number of edges is $\binom{3}{2}=3$. Since the number of such electrodes in the $m \times n$ array is 4, we get 4 graphs with 3 edges. The second category includes the electrodes located at four sides but not the corners of the array, and the corresponding graph is shown in Figure 7(c). For each graph, there are four nodes and the number of edges is $\binom{4}{2}=6$. Since the number

1: **Phase 1**: Derive a lower bound on the number of control pins;
2: **Phase 2**: Construct feasible pin-assignment configuration by greedy algorithm;
3: $ElectrodesNeedAssigned$ = {all electrodes on the layout}; // This is the set of electrodes which need to be assigned pins
4: $ElectrodesAlreadyAssigned$ = $\emptyset$; // This is the set of electrodes which have already been assigned pins
5: Randomly selected a control electrode group G;
6: Construct a local feasible solution S_{local} for the selected group;
7: $ElectrodesNeedAssigned$ = {all electrodes on the layout} - G, $ElectrodesAlreadyAssigned$ = G;
8: **while** $ElectrodesNeedAssigned \neq \emptyset$ **do**
9:　　Randomly selected a electrode $E_{neighor}$ in the neighborhood of $ElectrodesAlreadyAssigned$;
10:　　$AssignPin(E_{neighor})$; // Assign a control pin for $E_{neighor}$
11:　　Add $E_{neighor}$ to S_{local}; // Add the new electrode and corresponding control pin to the local solution S_{local}
12:　　$CheckFeasible(S_{local})$; // Check whether the new local solution is feasible solution
13:　　**if** $CheckFeasible(S_{local})$ = True **then**
14:　　　　$ElectrodesNeedAssigned$ = $ElectrodesNeedAssigned$ - $E_{neighor}$;
15:　　　　$ElectrodesAlreadyAssigned$ = $ElectrodesAlreadyAssigned$ + $E_{neighor}$;
16:　　**else**
17:　　　　**while** $CheckFeasible(S_{local})$ = False **do**
18:　　　　　Change the pin assigned to $E_{neighor}$, update the local solution, check whether new local solution is feasible;
19:　　　　**end while**
20:　　**end if**
21: **end while**

Figure 6: Pseudocode for the construction of the pin-assignment configuration using a greedy algorithm.

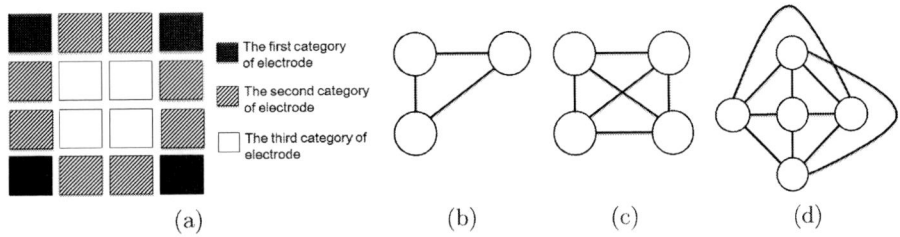

(a)　　　　　　　　　(b)　　　　　(c)　　　　　(d)

Figure 7: (a) Three categories of electrodes and the graphs correspond to electrodes of (b) the first category (c) the second category (c) the third category.

for such electrodes is 2(m+n-4), we have 2(m+n-4) graphs with 6 edges. The third category includes the electrodes located within the array and the corresponding graph is shown in Figure 7(d). For each graph, there are five nodes and the number of edges is $\binom{5}{2}$=10. Since the number of such electrodes is $(m-2) \times (n-2)$, we will get $(m-2) \times (n-2)$ complete graphs with 10 edges.

On the other hand, the number of edges that are contained in two graphs is $(m-1) \times n+(n-1) \times m+2 \times (m-1) \times (n-1)$. Therefore, using the principle of inclusion/exclusion, we set the total number of edge in the graph G_{layout} to be:

$$N_{edge} = 4 \times 3 + 12(m + n - 4) + 10(mn - 2m - 2n + 4)$$
$$- [(m-1)n + (n-1)m] - 2 \times (m-1) \times (n-1)$$
$$= 6mn - 5m - 5n + 2$$

According to Lemma 2, if the pin-assignment configuration satisfies Constraint 1 and Constraint 2, then graph G_{layout} is a simple graph. Assume that there are M control pins in the pin-assignment configuration. The number of nodes in graph G_{layout} is equal to M. The maximum number of edges in the simple graph G_{layout} that has M nodes is given by:

$$N_{edge\_max} = \binom{M}{2} = \frac{M \times (M-1)}{2}$$

Finally, we derive the desired lower bound in the number of control pins. Since N_{edge_max} is an upper limit on the number of edges in a graph, we have the inequality $N_{edge\_max} \geq N_{edge}$. □

For an electrode array with irregular shape, the number of control pins can be lower-bounded in a similar manner.

E. PIN-ASSIGNMENT CONFIGURATIONS FOR SEVERAL BIOCHIPS

The proposed design method is evaluated by applying it to four layouts studied in [12]: a multiplexed assay biochip, a PCR biochip, a protein dilution biochip and a multifunction biochip. Note that for previously published bioassay-specific designs, pin-assignment relies on actuation sequences for each electrode on the array. Once the pin-assignment is fixed, the microfluidic operations process cannot be changed any more. It is difficult to subsequently add new operations such as optical detection for droplets or map a new application. On the other hand, for the approach described here, the pin-assignment configuration is only related to the size and shape of the electrode array. Droplet operations are scheduled after the pin-assignment configuration is determined, thus the scheduling of operations and routing pathways are more flexible. For example, we can add optical detection operations to check the intermediate products in an experiment.

Path Scheduling on Digital Microfluidic Biochips

Daniel Grissom, Philip Brisk
Department of Computer Science and Engineering
University of California, Riverside
{grissomd, philip}@cs.ucr.edu

ABSTRACT

Since the inception of digital microfluidics, the synthesis problems of scheduling, placement and routing have been performed offline (before runtime) due to their algorithmic complexity. However, with the increasing maturity of digital microfluidic research, online synthesis is becoming a realistic possibility that can bring new benefits in the areas of dynamic scheduling, control-flow, fault-tolerance and live-feedback. This paper contributes to the digital microfluidic synthesis process by introducing a fast, novel path-based scheduling algorithm that produces better schedules than list scheduler for assays with high fan-out; *path scheduler* computes schedules in milliseconds, making it suitable for both offline and online synthesis.

Categories and Subject Descriptors

B.7.2 [**Integrated Circuits**]: Design Aids; B.8.2 [**Performance and Reliability**]: Performance Analysis and Design Aids; J.3 [**Life and Medical Sciences**]: Biology and Genetics, Health

General Terms

Algorithms, Design, Performance.

Keywords

Digital Microfluidic Biochip (DMFB), Laboratory-on-Chip (LoC), Electrowetting-on-Dielectric (EWoD), Scheduling.

1. INTRODUCTION

This work presents a scheduling heuristic for digital microfluidic synthesis called *Path-scheduler*. Instead of scheduling each node individually, Path-scheduler schedules sets of connected, dependent operations, called *paths*, to increase utilization and yield better schedules for assays with high fan-out. Path-scheduler computes schedules in milliseconds, making it useful for both offline and online scheduling.

1.1 Background

Microfluidics is a laboratory-on-chip (LoC) technology that manipulates fluids on the micro-liter to nano-liter scale to perform biochemical reactions called assays. In contrast to the first generation of microfluidic devices that transport continuous volumes of fluid through channels by actuating pumps and valves, digital microfluidic biochips (DMFBs) manipulate discrete droplets of fluid to perform assays.

A DMFB is arranged as a 2-dimensional array of electrodes, as seen in **Figure 1(a)**. **Figure 1(b)** shows a droplet sandwiched

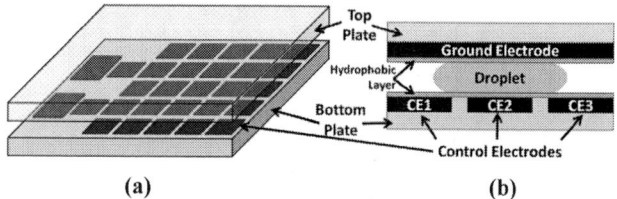

Figure 1. (a) A DMFB 2D array of electrodes; (b) Cross-sectional view of electrode array.

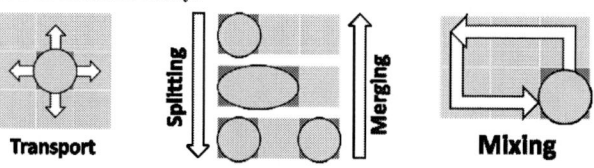

Figure 2. Basic microfluidic operations being performed on 2D array of electrodes.

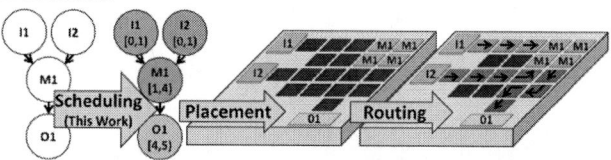

Figure 3. A microfluidic compiler obtains a sequence of electrode activations by *scheduling* a DAG, *placing* DAG operations on the array, and *routing* droplets between operations.

between ground and control electrodes. Although the droplet is centered over CE2, it overlaps neighboring electrodes CE1 and CE3. An activation of CE1 or CE3 will invoke a phenomenon called electrowetting and cause the droplet to flow left or right, respectively, toward the newly-activated electrode [4].

Figure 2 shows how several fundamental microfluidic operations can be performed by activating/deactivating adjacent electrodes in a particular sequence. In addition to droplet transport, splitting, merging and mixing, droplets can be stored on an electrode and input/output from/to reservoirs. Furthermore, individual cells can be equipped with various sensors, cameras and heating elements to perform detection and heating operations [6][11]. These basic operations have been shown adequate to perform an assortment of assays such as in-vitro diagnostics and immunoassays used in clinical pathology [9], DNA polymerase chain reaction (PCR) mixing stages used to amplify DNA [3] and protein crystallizations [12].

A digital microfluidic system consists of two parts: a "wet" array of electrodes, as seen in **Figure 1(a)**, and a "dry" computing device, such as a processor or microcontroller, which sends signals to the microfluidic array to activate electrodes in a pre-determined sequence. This sequence of electrode activations, in turn, causes droplets to perform all the necessary operations (e.g. mixing, merging, transport) to execute an assay.

To obtain the proper sequence of electrode activations, a compiler solves three NP-complete synthesis problems, as seen in **Figure 3**.

26

As input to the compiler, an assay is given as a directed acyclic graph (DAG), which contains the operation dependencies, types and lengths. The compiler first performs resource-constrained scheduling to assign each operation a starting and stopping time-step, ensuring that there are sufficient resources to perform the operations at the scheduled times [3][5][9]. With the newly-scheduled DAG, the compiler then attempts to place/assign each operation to a set of adjacent electrodes at the specified time-steps. Finally, after the operations are placed, droplet routes are computed to transport droplets between dependent operations [2].

1.2 Motivation and Contribution

This paper contributes to the synthesis process by introducing a fast, novel, path-based scheduling algorithm. To date, a list scheduling variant is the fastest scheduling algorithm used for DMFB synthesis [7][9]. When compared to list scheduling, Path-scheduler produces better schedules in competitive times (milliseconds) for assays with high fan-out.

Path-scheduler is also suitable for online synthesis. Currently, assay compilation is performed completely offline due to the complexity of the synthesis process, requiring an assay to be fully-specified before runtime. Ho, Chakrabarty and Pop suggest that "specialized heuristics for the synthesis problems" (scheduling, placement and routing) might enable online synthesis, which would bring new features to DMFBs in the areas of dynamic scheduling, control-flow, fault-tolerance and live-feedback [2]. To date, modified list scheduling (MLS) is the only other scheduler able to generate schedules quick enough to be used in an online manner [7][9]; Path-scheduler can compute better schedules in an online setting for high fan-out assays.

2. RELATED WORK

Our work is closely related to four prior scheduling algorithms that have been proposed in the literature: MLS [7][9], as mentioned above; an optimal scheduler based on integer linear programming (ILP) [9], and two genetic algorithms [5][9]. All of these algorithms solve the resource-constrained scheduling problem, where the DMFB size is known a-priori. This size limits the number of concurrent mixing and storage operations the DMFB can support. The scheduler computes valid start and stop times for each assay operation while ensuring that each DMFB resource is used to process, at most, one operation at each point in time; the objective is to minimize the assay completion time.

Among these four approaches, the genetic algorithms and ILP formulation achieve high quality solutions but with very long runtimes. The genetic algorithms use the iterative improvement paradigm to randomly explore the search space, eventually converging at a local optima; the ILP model employs a commercial solver with an exponential worst-case time complexity to find an optimal solution. Neither of these solutions is appropriate for usage in an online context.

Luo and Akella [3] analyzed a pipelined variant of the PCR assay (**Figure 4(b)**), and developed optimal algorithms for scheduling it (and assays with a similar topology called a "full binary tree") under various resource constraints. In our experiments, both MLS and Path-scheduler found optimal schedules for PCR.

For mixing and dilution operations, larger modules achieve faster runtimes, but consume more area, reducing the availability of resources for other concurrent operations. Su and Chakrabarty [10] developed a genetic algorithm that performs scheduling, module selection, and placement concurrently. The runtime of this

algorithm is too large for use in an online context; however, it illustrates the importance of module selection during synthesis. To enable a fair comparison with MLS, we assume uniform module sizes; incorporating module selection into the path scheduling mechanism, without significantly increasing the runtime, is left open for future work.

3. PATH SCHEDULER

3.1 Definitions and Resource Constraints

Let $G = (V, E)$ be a directed acyclic graph (DAG) in which the vertices (V) are the operations of an assay and the edges (E) describe the dependencies between operations. N_m is the total number of *work modules* (the areas where operations are performed) that can be accommodated on the DMFB. A *general module* is a work module that can perform mixing, merging, splitting, and storage operations. A *special-purpose module* is a general module equipped with a sensor, heater, or other external device; special-purpose modules can use these devices to perform additional operations, such as heating and detection. Let N_{si} be the number of special-purpose modules of type i and N_g be the number of general modules; then:

$$\sum_{\forall i} N_{si} + N_g = N_m \qquad (1)$$

Any time that a work module is not performing an operation, it is free to be used for storage and can be used to store a maximum of s_c droplets at any given time-step. Let sp_{it} be the number of special-purpose operations of type i being processed at time-step t. Let g_t be the number of general operations being processed and s_t be the number of droplets being stored, respectively, at time-step t. Then, the following inequalities must hold:

$$\sum_{\forall i} sp_{it} + g_t + \left\lceil \frac{s_t}{s_c} \right\rceil \leq N_m, \quad \forall t \qquad (2)$$

$$sp_{it} \leq N_i \quad , \forall i, \forall t \qquad (3)$$

In prior work, the feasibility of a given schedule cannot be determined until it is given to a placer and router and successfully mapped to a DMFB. In previous works, a value known as N_a is determined as the number of cells on the DMFB; N_a is then normalized to the number of mixers of a particular size that can fit on the DMFB [9]. N_m, used in this work, is similar to N_a in that it represents the number of work areas of a particular size that have been pre-determined to fit on the DMFB. It is assumed that the modules/mixers can be placed in such a way to allow sufficient room for routing.

Lastly, an I/O port can only process one dispense or output operation at a given time-step and a dispense operation must be bound to a reservoir that contains the appropriate fluid type.

3.2 Approach

A DMFB array contains a fixed number of electrodes on which assay operations can be executed. Unlike a traditional computer where data can be offloaded to a higher-level memory until needed, droplets that are waiting on other dependencies must be stored on the array using the same electrodes that might otherwise be used to perform operations. Consequently, the number of droplets being stored on a DMFB is inversely proportional to the amount of useful work that can be done. Thus, it is important to schedule operations with a goal to minimize the amount of time and number of droplets being stored.

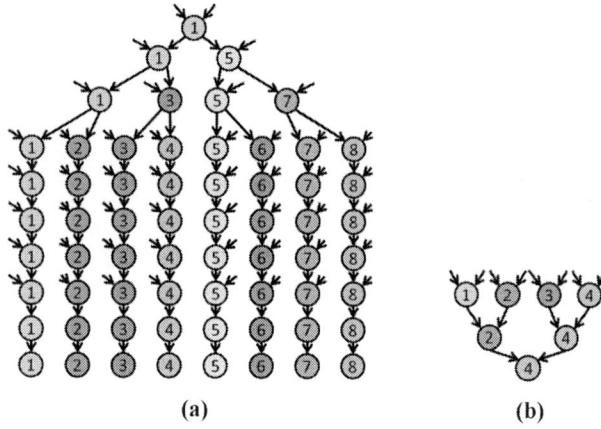

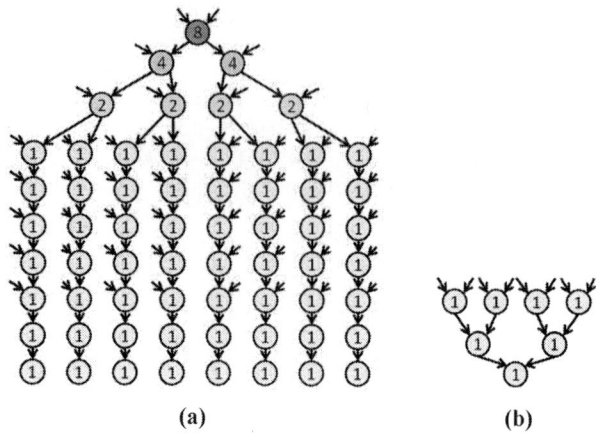

Figure 4. (a) A protein assay with 8 paths; (b) A PCR mixing assay with 4 paths. Input arrows not attached to a node on both sides represent a dispense operation (nodes omitted for clarity).

Figure 6. The independent-path priorities for (a) a protein assay with high fan-out and (b) a PCR assay with high fan-in.

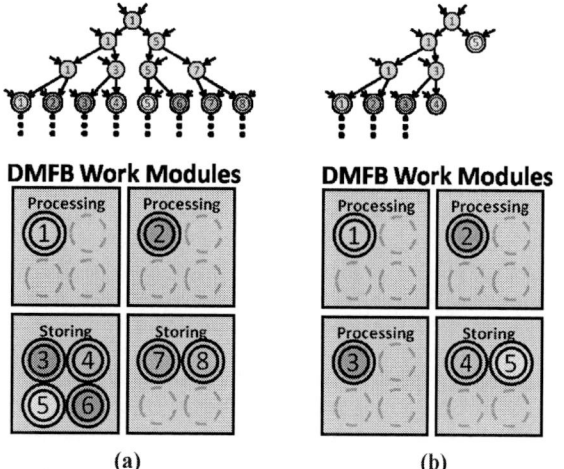

Figure 5. (a) A non-path scheduler may attempt to process all paths simultaneously, forcing two modules to be used as storage; (b) Path scheduler first schedules paths 1, 2 and 3, and uses only one module to store droplets.

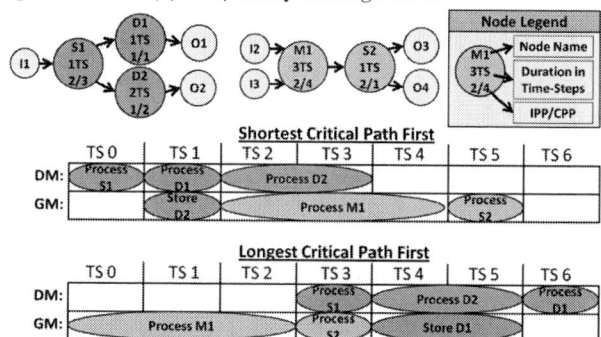

Figure 7. Two schedules for a set of two simple DAGs. The DMFB, in this case, has one general module (GM) and one detect module (DM). Path leaders S1 and M1 both have equal IPPs; if CPP is used as a second, tie-breaking priority, overall runtimes can be reduced if paths with smaller CPPs are chosen first.

5(b)). This approach effectively prevents droplets from being split until at least one of a split operation's children can be processed, reducing the number of droplets being stored in the system and, in turn, increasing work module utilization.

3.3 Priority Function

Path scheduler uses two priorities to efficiently produce schedules that minimize the amount of time a droplet spends in the system. To keep droplets from entering the system (via a dispense or split) till as late as possible, path scheduler sets the first priority of each node to the number of *independent paths*, which is the cumulative number of droplets being output in a node's fan-out.

Figure 6 shows the *independent-path priorities* (IPPs) for assays with high fan-out and fan-in (the protein/PCR assays from **Figure 4**). Recall **Figure 5** with the paths and IPPs from **Figure 4(a)** and **Figure 6(a)**, respectively, in mind. The first node on each path is known as the *path leader*. If the scheduler first schedules path 1, leaders from paths 2, 3 and 5 become candidates for scheduling. According to the IPPs in **Figure 6(a)**, path leaders 2, 3 and 5 have priorities of 1, 2 and 4, respectively. By choosing the lowest priority, path 2 is processed next, preventing additional splits from being made until necessary.

Each node has a second priority called the *critical-path priority* (CPP), which is computed as the length of the longest path (in time-steps) from itself to an output node. As seen in **Figure 7**, if the IPP of two candidate nodes (S1 and M1) are the same (2), the path with the lower CPP (S1, CPP = 3) may result in a shorter overall run-time. Since there is only one detect module, one of the

---

As a simple example, consider the protein assay seen in **Figure 4(a)**, typically used as a benchmark for DMFBs [8][9]. Eight paths can be identified in this assay; path 1 starts with two dispenses, while paths 2-7 originate from one dispense node and one split operation from another path. All paths in **Figure 4(a)** end with an output operation, although this is not the general case, as seen in assays with mix/merge operations (**Figure 4(b)**). It is important to note that the order of paths here is not necessarily unique. Path 1 could be chosen to go down the right side, instead of the left, if all paths from the root split to each output node are the same length. Once the paths are identified, the order in which they are processed is important, as shown in the next paragraph and in Section 3.3.

As seen in **Figure 5(a)**, a scheduler that attempts to schedule single operations at-a-time may attempt to schedule along all eight paths simultaneously (e.g. list-scheduling with a priority function favoring nodes with longer paths to an output) such that there are eight droplets in the system that need to be processed. Assuming the DMFB has enough room for four work modules that can be used to process one operation or store up to four droplets (i.e. $N_m = s_c = 4$), this schedule forces two work modules to be used for storage. However, if operations are first scheduled along paths 1, 2 and 3, only one module is required for storage (see **Figure**

28

droplets from split-node S1 must be stored in the general module until the other detect finishes. Taking the shortest critical path results in less storage time (1 time-step vs. 2 time-steps in **Figure 7**), which provides more opportunity to increase overall system utilization.

3.4 Algorithm

The pseudocode for the path scheduler algorithm is given in **Figure 8**. Before the scheduling process begins, all nodes in the sequencing graph are assigned first and second priorities as described in Section 3.3 and the candidate operations are determined. The initial candidate operations are those whose parents are only dispense operations.

Lines 6-35 describe the main scheduling process that repeats until all candidate nodes have been scheduled. *Lines 7-9* first select the candidate node, or *path leader*, with the lowest priority value (the lowest IPP first, and in the event of an IPP tie, the lowest CPP), reset the scheduling time-step and initialize an empty path.

Lines 10-24 attempt to allocate resources for an entire path of operations starting with the path leader chosen in *Line 7*, and ending with an output or merge operation (see **Figure 4**). *Lines 11-13* attempt to find the earliest gap in time where the current node, S, can fit, given the available resources. If a gap is found, it means there is a resource of type k (general or special-purpose) available from time-step t_i to $t_i + S.duration$, any required input reservoirs are available and that there are sufficient resources to store any incoming droplets from S's parent nodes, if necessary. If a gap is not found in *Lines 11-13*, it means the current path cannot be scheduled at the moment because of some resource conflict (e.g. there is not enough room to store a droplet from one of S's parents (in path P) to S) ; the path is discarded and any resources being temporarily reserved to schedule path P are relinquished. In this case, Path-scheduler will try to schedule the path again later, but will first return to *Line 6* and attempt to schedule another path.

In the event that a gap is found for S, the starting time-step and resource-type are temporarily saved and S is added to the current path P (*Lines 14-16*); however, S is not marked as scheduled. Once S has been added to path P, *Lines 21-22* select the next node to consider adding to the path from S's children. If the new S is an output or unscheduled mixing operation, then path P is a complete path, is ready to be officially scheduled and can break from the path-constructing loop of *Lines 10-24*; otherwise the loop continues and path scheduler attempts to find a gap for the new S.

Finally, in *Lines 26-34*, each operation in the schedulable path P is marked as scheduled and the resources temporarily reserved in *Line 15* to schedule each of path P's operations are officially reserved in *Line 29*. Also, any unscheduled children of the nodes in path P are added to the candidate operations as path leaders.

The edge between the path node and the new candidate operation added in *Line 31* represents a droplet that must be stored indefinitely and accounted for to properly determine resource availability since it has been scheduled to be created, but may not be used for awhile. Thus, when an operation is added to the candidate list in *Line 31*, the corresponding edge is added to a list of droplets being indefinitely stored from their parent's scheduled ending time-step. When path scheduler is finding a gap for operations in *Lines 11-13*, it considers all droplets being stored indefinitely at that point. When an operation is finally scheduled in *Lines 28-29*, any edges/droplets connected to that node that are being indefinitely stored are removed from the indefinite storage list and the finite period that the droplet must be stored for, if any, is accounted for in the system's available resources.

```
1 Given sequencing graph G=(V, E)
2 Given resource constraints N_gm N_dm and s_c
3 Assign priorities for all nodes v ∈ V, ∀v based on IPP and CPP
4 Find candidate operations R = {v_i ∈ V: Type(v_j) == input, ∀j: (v_j, v_i) ∈ E}
5
6 Repeat {
7     Select S ⊆ R : Priority(S) ≤ Priority(r), ∀r : r ∈ R
8     Time-step t = 1
9     Path P = ∅
10    Repeat {
11        Attempt to find earliest time-step t_i and module-type k for S:
12        k ∈ (gc, sp): Avail(ts, k) == true, ∀ts : t_i ≤ ts < t_i + S.duratoin
13        while holding Equations (2) and (3)
14        if (Attempt found a gap for S)
15            Set S.start = t_i , Set S.resType = k
16            Add S to P
17        else // Gap could not be found, S not schedulable now
18            Set P.schedueable = false
19        end if
20
21        Select new
22        S ⊆ S.children: Priority(S) ≤ Priority(S_ch), ∀S_ch: S_ch ∈ S.children
23    } until (  ((Type(S) == mix) AND (S.scheduled == false))
24            OR (Type(S) == output ) OR (P.scheduble == false)  )
25
26    if (P.schedueable == true)
27        for ( ∀p ∈ P ){
28            Set p.scheduled = true
29            Reserve resources for path P
30            for (∀c : c ∈ p.children ∧ c ∉ P )
31                Add c to candidate operations R
32            end for
33        end for
34    end if
35 } until (all candidate operations are scheduled: R = ∅)
```

Figure 8. Pseudocode for the Path-scheduler algorithm.

4. EXPERIMENTAL RESULTS

We implemented our algorithm, as well as two versions of modified-list scheduling in C++; all tests were run on a 64-bit Windows 7 machine with 4GB of RAM, and an Intel Core i7 CPU operating at 2.8GHz.

4.1 Implementation

Our Path-scheduler *(PS)* was implemented as described in **Figure 8**. We re-implemented modified-list scheduling as described in ref. [7] as faithfully as possible. In their work, Su and Chakrabarty describe an *urgency* priority function which sets each node's priority to "the weight of their longest path to the sink" and then "sort[s] them in decreasing order" such that nodes with longer paths are addressed first. We interpret "weight of the longest path" to be calculated in the number of time-steps and compute it similarly to the CPP, described in Section 3.3. Our re-implementation of modified-list scheduling (*MLS_DEC*) processes nodes with higher priorities first.

We also implemented another version of modified-list scheduling with a better-performing priority function. By setting the priorities to our CPP and then sorting them in decreasing order, it causes the scheduler to process a DAG similar to what is seen in **Figure 5(a)**, which is inefficient in terms of utilization. Thus, to be fair to list scheduling, we implemented a new version of modified-list scheduling (*MLS_INC*) which uses CPP for operation priorities, but sorts the nodes in increasing order so nodes with lower priorities are processed first.

4.2 Benchmarks

We used a set of three standard benchmarks: PCR, in-vitro and a protein assay [8]. **Figure 4** displays the protein and PCR DAGs.

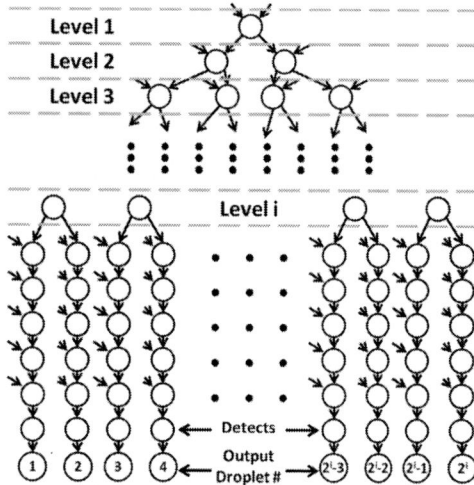

Figure 9. DAG for Experiment 2; i levels results in 2^i output droplets.

4.2.1 Assay Annotations

Each benchmark was converted to a DAG and fed to the three schedulers. The module libraries from ref. [8] were used for operation timings. For the PCR assay, a 2x4 mixer (3s) was used for all mixes. The protein DAG was annotated to use the 2x4 diluter (5s) and 2x4 mixer (3s) for all dilute and mix operations, respectively. The 2-input, 2-output dilute operations used in the protein assay were implemented with consecutive mix and split operation which took a cumulative time of 5s. For the in-vitro benchmark, we used the largest sequencing graph which assays four samples with four reagents for a total of 16 mixes/detects. We used the same mix and detect times as detailed in Table 1, Example 5 of ref. [7].

4.2.2 Experiments

For the first set of experiments, we used the PCR, protein and in-vitro DAGs described in Section 4.2.1 as a base. Then, for each of the three DAGs, we attempted to simultaneously schedule an increasing number of copies of the same DAG, from 1 to 10, to show how each scheduler performs with increasing workloads. As a second experiment, we executed 3 protein DAGs while varying the number of modules (N_m) from 2 to 7. In a third set of experiments, we varied the number of splits performed by the protein assay, which allowed us to evaluate the quality of schedules produced by Path-scheduler as assay fan-out increases. In these experiments, we use the protein assay as the source assay. As seen in **Figure 4**, any path from the root split to an output goes through 3 splits before the final string of 7 operations, resulting in $2^3=8$ output droplets. Thus, we say that the original protein assay has 3 levels of splits. We sweep the number of split levels from 1-8, so that the number of output droplets sweeps from $2^1=2$ to $2^8=256$. As seen in **Figure 9**, the final seven operations are appended to the end of each path after the last level of splits.

4.3 Resource Constraints

For the first and third experiment, we set the number of work modules to four (i.e., $N_m = 4$); the second experiment varies N_m. For PCR, all four modules are general modules. For the in-vitro and protein benchmarks, all four modules are detect modules. Similar to ref. [5], we assume a detect module can be used for any detect operation. The PCR and in-vitro benchmarks have one input for each type of fluid used, while the protein benchmark uses one input for sample fluids, two inputs for buffer fluids and two inputs for reagent fluids.

Table 1. Results of Experiment 1. Lower is better for all metrics.

# DAGS	Protein Assay (118 operations/DAG) # Time-steps (1TS = 1s)			Scheduling Time (ms)			Avg # Storage Modules/TS		
	MLS_DEC	MLS_INC	PS	MLS_DEC	MLS_INC	PS	MLS_DEC	MLS_INC	PS
1	216	197	186	4	2	1	1.35	0.79	0.64
2	599	342	333	12	5	2	2.50	0.80	0.71
3	990	498	480	20	8	4	2.67	0.83	0.74
4	Fail	643	627	Fail	12	5	Fail	0.83	0.76
5	Fail	799	774	Fail	16	8	Fail	0.85	0.77
6	Fail	944	921	Fail	20	13	Fail	0.84	0.77
7	Fail	1100	1068	Fail	25	14	Fail	0.85	0.78
8	Fail	1245	1215	Fail	30	18	Fail	0.85	0.78
9	Fail	1401	1362	Fail	40	22	Fail	0.85	0.79
10	Fail	1546	1509	Fail	50	29	Fail	0.85	0.79

# DAGS	PCR Assay (16 operations/DAG) # Time-steps (1TS = 1s)			Scheduling Time (ms)			Avg # Storage Modules/TS		
	MLS_DEC	MLS_INC	PS	MLS_DEC	MLS_INC	PS	MLS_DEC	MLS_INC	PS
1	9	9	9	0	0	0	0.00	0.00	0.00
2	15	15	15	0	0	0	0.38	0.19	0.19
3	24	21	21	0	0	0	1.20	0.27	0.27
4	36	27	27	1	1	0	1.46	0.32	0.32
5	45	33	33	2	2	0	1.57	0.35	0.35
6	57	39	39	4	2	0	1.66	0.38	0.38
7	66	45	45	5	3	0	1.70	0.39	0.39
8	78	51	51	8	6	0	1.75	0.40	0.40
9	87	57	57	9	6	0	1.77	0.41	0.41
10	99	63	63	12	8	0	1.80	0.42	0.42

# DAGS	in-vitro Assay (80 operations/DAG) # Time-steps (1TS = 1s)			Scheduling Time (ms)			Avg # Storage Modules/TS		
	MLS_DEC	MLS_INC	PS	MLS_DEC	MLS_INC	PS	MLS_DEC	MLS_INC	PS
1	49	44	47	2	1	0	0.68	0.00	0.00
2	119	83	85	9	7	2	1.34	0.00	0.00
3	200	121	123	22	15	3	1.66	0.00	0.00
4	438	159	161	59	27	5	2.58	0.00	0.00
5	580	197	200	101	44	8	2.66	0.00	0.00
6	656	236	238	140	63	10	2.59	0.00	0.00
7	806	273	276	218	88	13	2.66	0.00	0.00
8	915	311	314	265	109	17	2.66	0.00	0.00
9	1065	349	353	364	141	20	2.70	0.00	0.00
10	1174	388	391	442	175	24	2.69	0.00	0.00

4.4 Results and Discussion

Three metrics are presented for evaluation: the completion time in number of time-steps, the time required to compute the schedule and the average number of modules used for storage during each time-step. The last metric represents the scheduler's storage efficiency; the lower this metric, the higher the DMFB utilization.

Table 1 shows the results for Experiment 1. MLS_DEC handles storage poorly, as indicated by its average storage usage; as a consequence, both MLS_INC and PS outperform it significantly. For the protein assay, MLS_DEC attempts to schedule along all paths simultaneously, creating an overwhelming number of storage droplets; for 3 protein DAGs, it already uses 2.67 of 4 modules, on average, for storage. Because of its poor storage handling, it fails to produce feasible schedules for more than 3 protein DAGs because all of the modules are allocated for storage and no assay operations are able to proceed.

The rest of our discussion is limited to MLS_INC and PS. As expected, PS yields great gains with the protein assay, which has high fan-out. In overall schedule quality, PS saves between 9-11 seconds on the first two runs and 37-39 seconds on the last two runs. Furthermore, PS computes schedules 1.94x faster, on average, than MLS_INC. For the interested reader, Supplementary Section S1 shows the scheduled graphs for the protein assay for MLS_DEC, MLS_INC and PS.

Results for the PCR and in-vitro assays are given to demonstrate how PS performs on assays without fan-out since neither of these assays contain a single split. For PCR, PS and MLS_INC yield identical results; PS runs faster than MLS_INC, especially when the number of DAGs increases.

For the in-vitro benchmark, neither PS nor MLS_INC require any storage operations. PS's computed schedules are an average of

Table 2. Results of Experiment 2: scheduling the protein assay using a varying number of modules. Lower is better for all metrics.

	Protein Assay - 3 DAGs					
# Mods	**# TS (1TS = 1s)**		**Sched Time (ms)**		**Avg # Stor Mod/TS**	
	MLS_INC	**PS**	**MLS_INC**	**PS**	**MLS_INC**	**PS**
2	Fail	1177	Fail	3	Fail	0.90
3	775	590	9	2	1.39	0.86
4	480	407	8	1	1.36	0.78
5	345	305	5	1	1.36	0.70
6	295	255	6	1	1.88	0.66
7	250	217	7	1	1.83	0.74
25	60	60	2	1	0.00	0.00

Table 3. Results of Experiment 3: varying the number of splits in the protein assay. Lower is better for all metrics.

	Protein Split Assay					
# Split Levels	**#TS (1TS = 1s)**		**Sched Time (ms)**		**Avg # Stor Mods/TS**	
	MLS_INC	**PS**	**MLS_INC**	**PS**	**MLS_INC**	**PS**
1	71	72	0	0	0.28	0.21
2	106	110	0	0	0.49	0.43
3	197	186	2	1	0.79	0.64
4	389	338	5	2	1.07	0.79
5	757	642	13	4	1.30	0.91
6	1590	1279	29	14	1.54	1.03
7	3456	2644	73	24	1.78	1.15
8	Fail	5570	Fail	49	Fail	1.29

2.7s slower than MLS_INC. The root of this small inefficiency is actually due to input-reservoir conflicts. A conflict resolution step was added to PS and it improved the results for in-vitro to be equivalent with MLS_INC; however, due to its inherent quadratic runtime, this step increased the runtime of PS significantly (e.g. from 24ms to 7439ms for in-vitro run #10) and did not yield improvements for protein or PCR schedules. We describe these issues in greater detail in Supplementary Section S2. Lastly, we note that the runtime of PS is 3.5x to 7.3x faster than MLS_INC in this experiment.

Table 2 shows results for Experiment 2, where we varied the number of work modules while scheduling the protein assay. For 2 modules, PS completed a schedule in 1177 time-steps, while MLS_INC failed to compute a schedule. For $N_m = 3, 4, 5, 6$ and 7, PS's schedules are 185, 73, 40, 40 and 33 time-steps shorter than MLS_INC's schedules, respectively. As the number of modules increase, the schedules tend to converge since there are abundant resources for storage and it becomes algorithmically easier to compute latency-optimal schedules.

Table 3 reports results for Experiment 3, which confirms that PS generates better schedules than MLS_INC when fan-out increases by reducing storage usage. PS loses 1-4 time-steps for the first two levels of splits. However, as the number of splits increase from three to seven levels, PS saves up to 812s (13m 32s). Due to poor memory management, MLS_INC fails to produce schedules beyond seven split levels because it reaches a point where all four modules are used for storage. With the constraints detailed in Section 4.3, PS can theoretically compute schedules up to 12 levels (4096 droplets). We verified this experimentally: PS took 1269ms (1.3s) to compute a schedule of 31h 29m in length.

5. CONCLUSION

We have presented a path-based scheduling heuristic for digital microfluidic synthesis. Instead of scheduling node-by-node, as list scheduler does, Path-scheduler schedules path-by-path, reducing the number of droplets being stored in the system on assays with high fan-out. The increase in storage efficiency leads to an increase in utilization, and in turn, an increase in overall schedule quality. Similar to list scheduling, Path-scheduler produces solutions on the order of milliseconds. As assays grow extremely large, as seen in Experiment 3, the schedules generated by Path-

scheduler will be further appreciated when the compiler attempts to place and route a much smaller schedule.

As synthesis moves to an online setting, short runtimes become increasingly important and assays will likely be scheduled with specialized scheduling heuristics that perform well on a particular assay-class (e.g. multiplexed, high fan-out/in, etc.). Path-scheduler excels on and should be used on assays with high fan-out (easily obtainable info). Even with no a-priori information, Path-scheduler is fast enough that a DMFB could compute schedules with path- and list scheduler and take the best schedule with little penalty. As more fast, high-quality scheduling heuristics emerge, online synthesis will become a growing possibility, bringing a number of new features to DMFBs in the areas of dynamic scheduling, control-flow, fault-tolerance and live-feedback.

6. ACKNOWLEDGMENTS

This work was supported in part by NSF Grant CNS-1035603. Daniel Grissom was supported by an NSF Graduate Research Fellowship.

7. REFERENCES

[1] K. Chakrabarty. Design automation and test solutions for digital microfluidic biochips. IEEE Transactions on Circuits and Systems-I: Regular Papers, 57(1):4-17, January 2010.

[2] T. Ho, K. Chakrabarty, and P. Pop. Digital microfluidic biochips: recent research and emerging challenges. In Proceedings of the Conference on Codesign and System Synthesis, pages 335-343, Taipei, Taiwan, Oct 9-14, 2011.

[3] L. Luo and S. Akella. Optimal scheduling of biochemical analyses on digital microfluidic systems. In Proceedings of the Conference on Intelligent Robots and Systems, pages 3151-3157, San Diego, CA, USA, Oct 29-Nov 2, 2007.

[4] M. G. Pollack, A.D. Shenderov, and R. B. Fair. Electrowetting-based actuation of droplets for integrated microfluidics. Lab on a Chip, 2:96-101,2002.

[5] A. J. Ricketts, K. Irick, N. Vijaykrishnan, and M. J. Irwin. Priority scheduling in digital microfluidics-based biochips. In Proceedings of the Conference on Design Automation and Test in Europe (DATE), pages 329-334, Munich, Germany, March 6-10, 2006.

[6] V. Srinivasan, et al. A digital microfluidic biosensor for multianalyte detection. Proc. IEEE MEMS, pages 327-330, Kyoto, Japan, Jan 19-23, 2003.

[7] F. Su and K. Chakrabarty. Architectural-level synthesis of digital microfluidics-based biochips. In Proceedings of ICCAD, pages 223-228, San Jose, CA, USA, Nov 7-11, 2004.

[8] F. Su and K. Chakrabarty. "Benchmarks" for digital microfluidic biochip design and synthesis. Duke University, Department of Electrical and Computer Engineering, 2006. http://www.ee.duke.edu/~fs/Benchmark.pdf

[9] F. Su and K. Chakrabarty. High-level synthesis of digital microfluidic biochips. ACM Journal on Emerging Technologies in Computing Systems, 3(4): article #16, January, 2008.

[10] F. Su and K. Chakrabarty. Unifed high-level synthesis and module placement for defect-tolerant microfluidic biochips. In Proceedings of Design Automation Conference, pages 825-830, Anaheim, CA, USA, June 13-17, 2005.

[11] K. Ugsornrat, et al. Experimental study of single-plate EWOD device for a droplet based PCR system. In Proceedings of ECTI-CON, pages 6-9, Khon Kaen, Thailand, July 12, 2011.

[12] T. Xu, K. Chakrabarty, and V. K. Pamula. Defect-tolerant design and optimization of a digital microfluidic biochip for protein crystallization. IEEE Transactions on Computer-Aided Design of Integrated Circuits and Systems, 29(4): 552-565, April, 2010.

S1. STORAGE REDUCTION GRAPHS

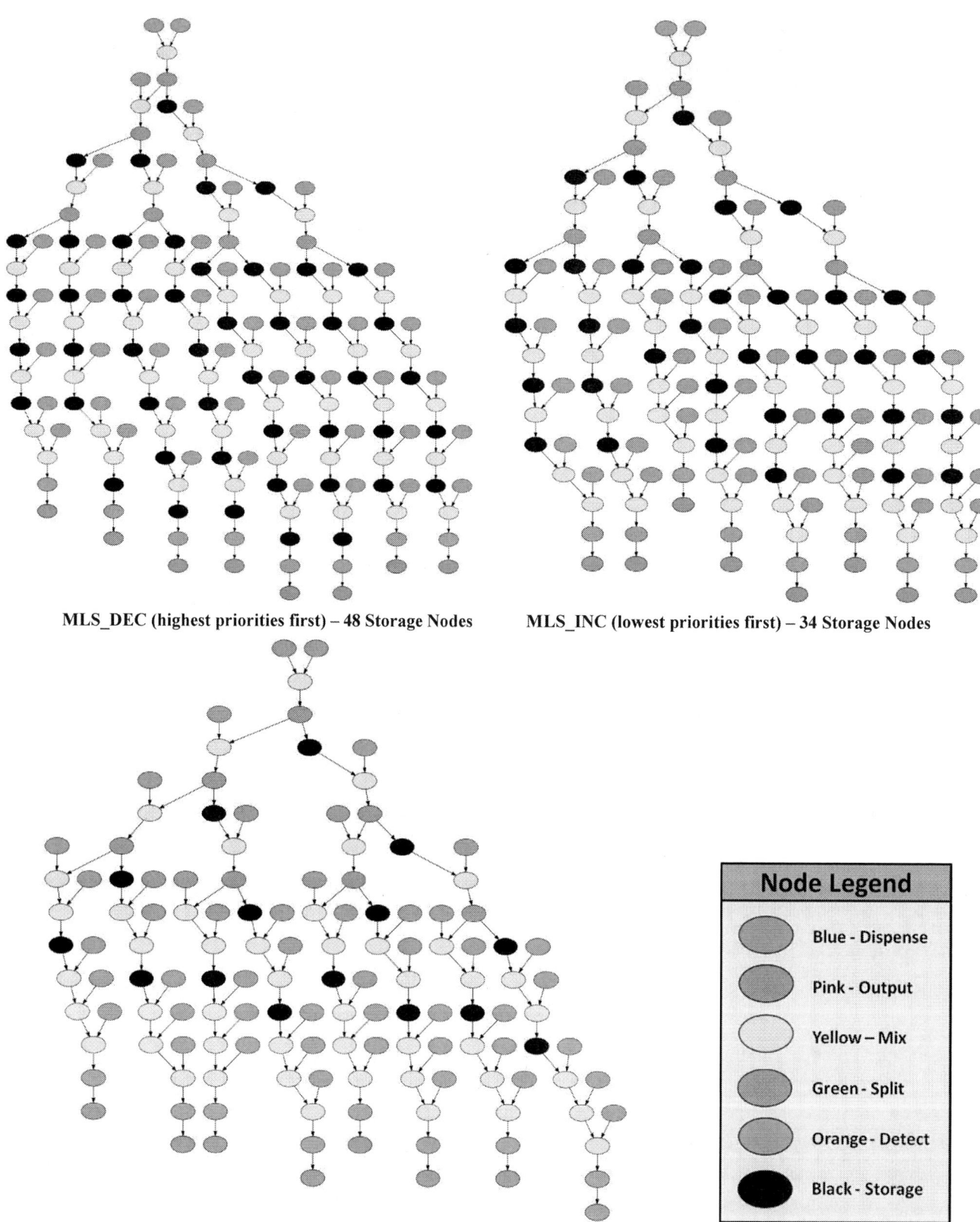

MLS_DEC (highest priorities first) – 48 Storage Nodes

MLS_INC (lowest priorities first) – 34 Storage Nodes

PS (Path Scheduling) – 15 Storage Nodes

Figure S1. Scheduled DAGs for the basic protein assay (timing information was removed for clarity). These graphs have not been bound (placed), and thus, each black storage node represents a droplet being stored for a length of time, possibly in a number of different modules.

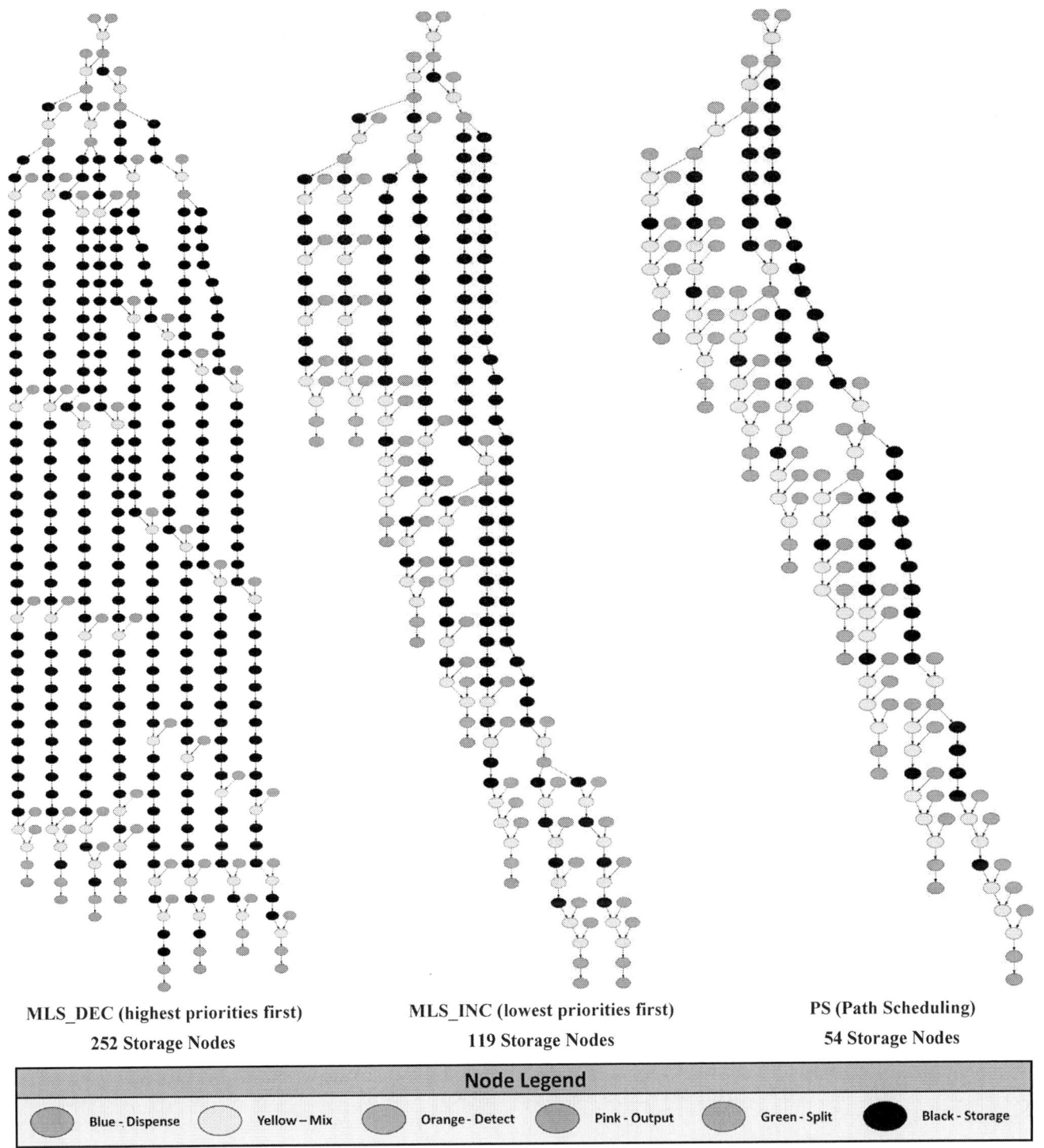

MLS_DEC (highest priorities first)
252 Storage Nodes

MLS_INC (lowest priorities first)
119 Storage Nodes

PS (Path Scheduling)
54 Storage Nodes

Node Legend					
Blue - Dispense	Yellow – Mix	Orange - Detect	Pink - Output	Green - Split	Black - Storage

Figure S2. Scheduled and bound DAGs for the basic protein assay. Each black storage node represents a droplet being stored for an amount of time in a particular module.

Figure S1 shows the scheduled graphs for the basic 118-node protein assay for MLS_DEC, MLS_INC, and PS. The main points of interest are the locations and number of storage nodes. Each storage node represents a droplet being stored for a number of time-steps, possibly in a number of different locations. MLS_DEC inserts storage nodes in-between almost every pair of consecutive non-dispense nodes for a total of 48 storage nodes. MLS_INC is slightly more conservative, while PS inserts only 15 storage nodes, usually after splits occur.

Figure S2 shows the same protein assay that is now scheduled and bound to one of the four specific work modules. Here, the black storage nodes from **Figure S1** have been unrolled and represent a droplet being stored for a number of time-steps in a particular work module. Although placement/binding is beyond

Table S1. Mix times according to in-vitro sample-type.

Sample-Type Mix Times			
S1	S2	S3	S4
5s	3s	4s	6s

Table S2. Detect times according to in-vitro reagent-type.

Reagent-Type Detect Times			
R1	R2	R3	R4
5s	4s	6s	5s

Table S3. Experimental data from experiment 1 with results from PS_IN.

	In-vitro Assay (80 operations/DAG)											
# DAGS	**# Time-steps (1TS = 1s)**				**Scheduling Time (ms)**				**Avg. # Storage Modules/TS**			
	MLS_DEC	MLS_INC	PS	PS_IN	MLS_DEC	MLS_INC	PS	PS_IN	MLS_DEC	MLS_INC	PS	PS_IN
1	49	44	47	45	2	1	0	7	0.68	0.00	0.00	0.00
2	119	83	85	83	9	7	2	66	1.34	0.00	0.00	0.00
3	200	121	123	122	22	15	3	225	1.66	0.00	0.00	0.00
4	438	159	161	159	59	27	5	550	2.58	0.00	0.00	0.00
5	580	197	200	197	101	44	8	1011	2.66	0.00	0.00	0.00
6	656	236	238	236	140	63	10	1700	2.59	0.00	0.00	0.00
7	806	273	276	273	218	88	13	2594	2.66	0.00	0.00	0.00
8	915	311	314	311	265	109	17	3779	2.66	0.00	0.00	0.00
9	1065	349	353	350	364	141	20	5553	2.70	0.00	0.00	0.00
10	1174	388	391	387	442	175	24	7439	2.69	0.00	0.00	0.00

the scope of this paper, we created **Figure S2** from the scheduled graphs seen in **Figure S1** using a simple left-edge binder to better-demonstrate how the various schedules will execute. These results show that MLS_DEC tries to execute all paths concurrently, resulting in much greater demand for storage than MLS_INC or PS. In contrast, PS processes the left side of the assay first, while storing droplets at a few strategic locations on the right side.

S2. IN-VITRO INEFFICIENCY

In this section, we show that the slightly inferior schedules produced by PS for the in-vitro benchmark arise from input reservoir resource conflicts. In the basic 80-node in-vitro benchmark, four samples (S1-S4) are pairwise assayed with four reagents (R1-R4) yielding a DAG with 16 connected components. Each sample-reagent pair is mixed together, sent to a detector for evaluation, and then output. Mix times (**Table S1**) are based on the sample type, while detect times (**Table S2**) are based on the reagent type. All dispense times are 2s.

Figure S3 shows what choices MLS_INC and PS make for the first four assays. Since the IPPs for PS all equal 1, PS and MLS_INC both use only CPPs to help decide which paths/nodes to schedule next. Both choose mix node M6, which mixes S2 and R2 together because the CPP of 7TS (3TS for M6 and 4TS for D6) is the smallest of any sample-reagent pair. From **Table S1** and **Table S2**, the assays with the next three lowest priorities, all tied at 8TS, are S2-R1 (3s+5s), S2-R4 (3s+5s) and S3-R2 (4s+4s). **Figure S3** reveals that PS's first four scheduled DAGs include these 3 assays. However, notice that S2 and R2 were already scheduled and that each of the next three assays contain either S2 or R2. PS first tries to schedule the input operations for S2-R1 at time-step 0, but finds that there is a resource conflict with the input port dispensing S2. Instead of moving to a new connected component, PS sticks with the current one and schedules it as soon as the input port for S2 is available at time-step 2. PS finds that its next component, which assays S3 with R2 (chosen next, at random, from the 3-way tie) has a resource conflict with R2 and must also start at time-step 2. Finally, PS receives the connected component with S2 and R4 and cannot schedule it until time-step 4 because the input port for S2 is busy during the prior time-steps.

If the random order of the ties is the same in MLS_INC and PS, MLS_INC will examine the assays in the same order. However, the behaviors of PS and MLS_INC cause the schedules to differ at this point. PS can go back and add to a partial schedule that is already in place. For example, if PS has already scheduled a path from time-steps 0-10, it can revisit those time-steps and schedule

another path from time-steps 0-10. However, list scheduling (including, but not limited to MLS_INC) takes a constructive approach and does not revisit a time-step once it has moved to the next. For example, if MLS_INC is scheduling a node at time-step 5, it can never go back to time-steps 0-4 to add to the schedule.

When MLS_INC reaches time-step 2, it schedules M6, along with its parent nodes for S2-R2. MLS_INC also examines the three assays with path-lengths (or CPPs) of 8TS, but finds that none can be scheduled at the moment because they all require the input port for S2 or R2. Since MLS_INC cannot come back to this time-step, it greedily examines all assays until it finds three that do not conflict with the inputs of the first assay or with each other. **Figure S3** shows that MLS_INC is able to utilize all input ports in the first two time-steps. Once the first four mixers (and corresponding dispense parents) are scheduled, the DMFB runs out of modules and list scheduler moves to the next time-step.

MLS_INC only releases one module (DM3) before PS. However, even though three of the four modules finish their first assay at the same time for both schedules, MLS_INC has scheduled longer assays, and thus, has less work than PS left to schedule. Both schedulers have nearly identical amounts of resources left due to the fact that PS expended more resources at the beginning of its schedule. This slight skew in scheduling is the cause of the 3-4 time-step deficiencies seen in the PS schedules.

Thus far, we have shown that the inefficiency for the multiplexed in-vitro benchmark is due to resource conflicts with the input reservoirs. The solution is to add some priority to the order in which paths are scheduled based on the input reservoirs required by that path. To test this idea, we modified Path-scheduler so that it no longer considered the CPP as a second priority. If there were a number of assays (mix nodes) with the same IPP priority, the modified Path-scheduler (PS_IN) examines all parents of the tied assays/nodes and schedules the mixers whose dispense parents are available earliest. This solution is quadratic in the number of paths, as it searches through the list of unprocessed (tied) nodes each time PS_IN chooses a new path to schedule.

Table S3 shows the results of PS_IN alongside the original results for the in-vitro diagnostic test from experiment 1. The schedules for PS_IN are now within one time-step (+/-) of the MLS_INC schedules. We attribute the one time-step differences to randomness in the order that ties are ordered by each scheduler. As expected, runtimes for PS_IN are much longer than PS or MLS_INC, which arguably outweighs the benefits of improving the schedule quality by 2 to 3 time-steps.

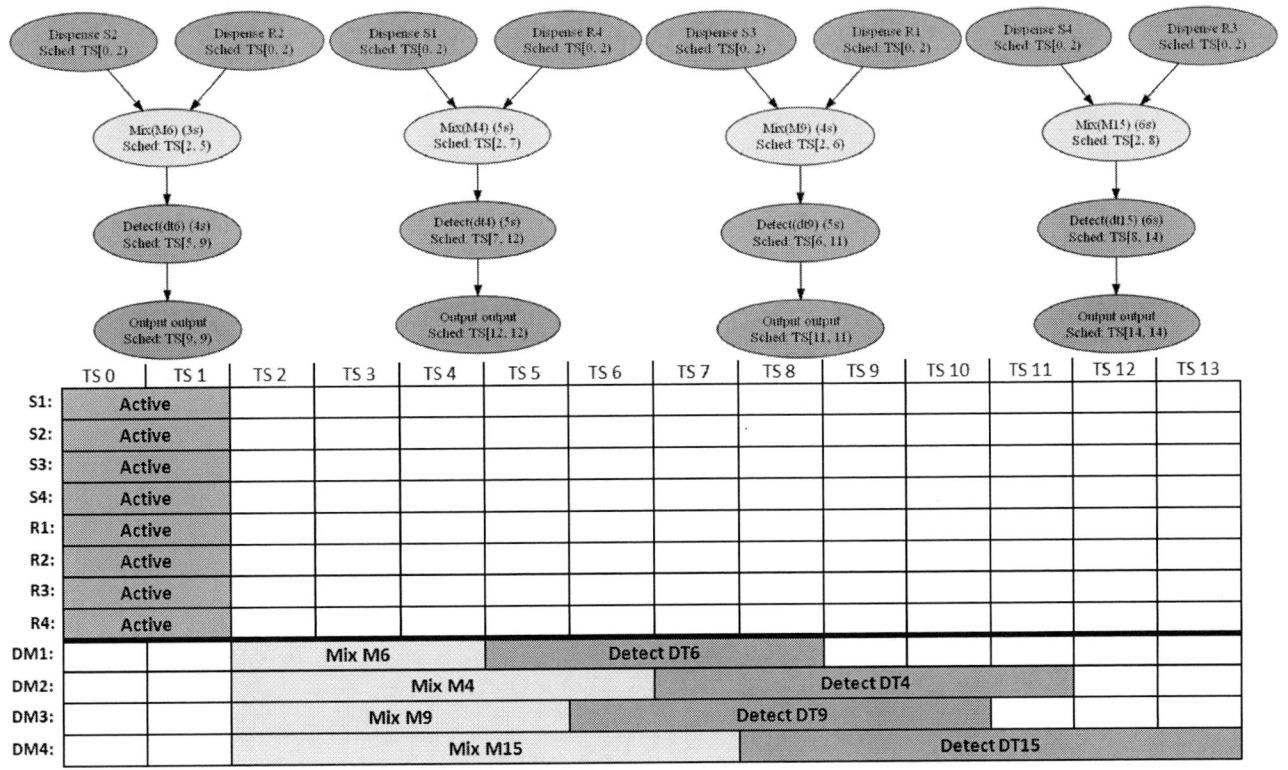

MLS_INC (lowest priorities first)

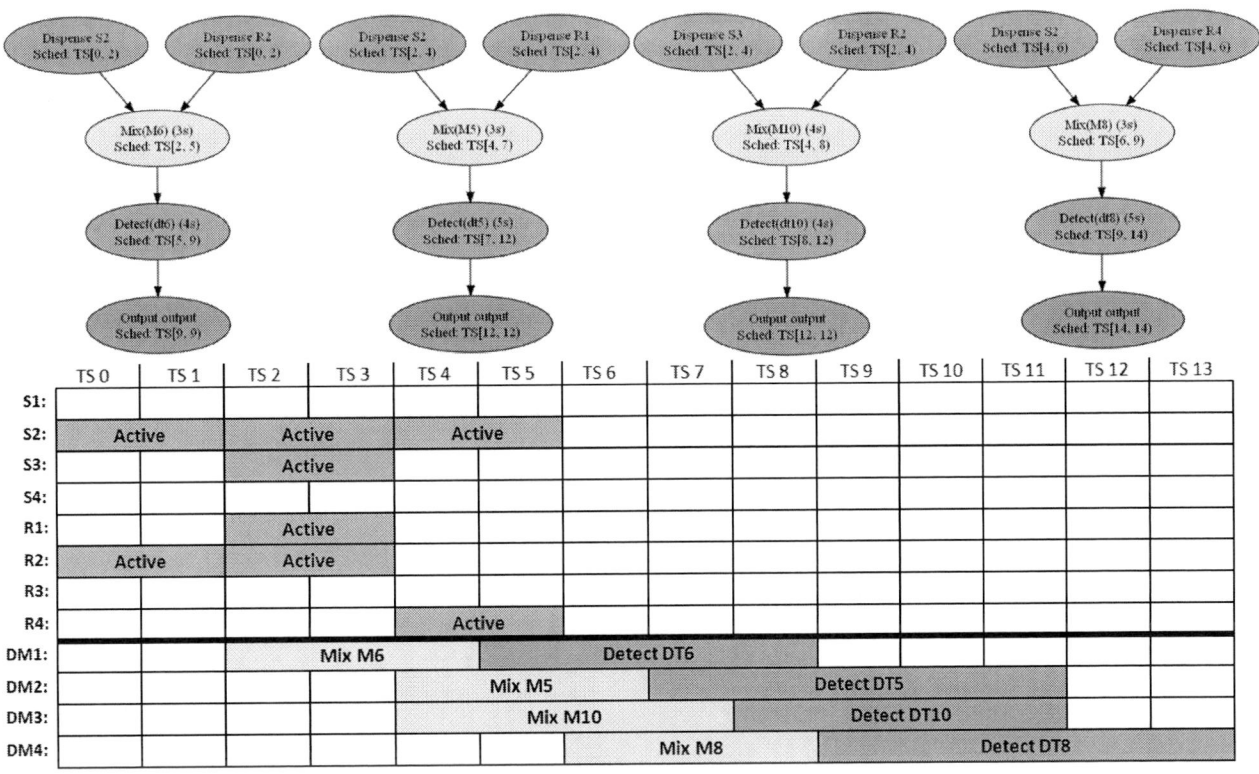

PS (Path-scheduler)

Figure S3. Shows the first four scheduled mixes, along with bindings to work modules and inputs for MLS_INC and PS. S1-S4, R1-R4 and DM1-DM4 represent the four sample input reservoirs, reagent input reservoirs and detect modules, respectively, on the DMFB.

Realizing Reversible Circuits
Using a New Class of Quantum Gates

Zahra Sasanian
University of Victoria
Victoria, BC, Canada
sasanian@uvic.ca

Robert Wille
University of Bremen
Bremen, Germany
rwille@uni-bremen.de

D. Michael Miller
University of Victoria
Victoria, BC, Canada
mmiller@uvic.ca

ABSTRACT

Quantum computing offers a promising alternative to conventional computation due to the theoretical capacity to solve many important problems with exponentially less complexity. Since every quantum operation is inherently reversible, the desired function is often realized in reversible logic and then mapped to quantum gates. We consider the realization of reversible circuits using a new class of quantum gates. Our method uses a mapping that grows at a very low linear rate with respect to the number of controls. Results show that, particularly for medium to large circuits, our method yields substantially smaller quantum gate counts than do prior approaches.

Categories and Subject Descriptors

B.6.3 [**Design Aids**]: Optimization

General Terms

Design

Keywords

Reversible Logic, Quantum Gates, Mapping, Optimization

1. INTRODUCTION

Quantum computation [1] offers the promise of efficient computing for problems that are of exponential difficulty for classical computing paradigms. Here, information is stored in terms of qubits which provide the probabilistic superposition of the Boolean states 0 and 1. This enables solutions for many important problems (*e.g.* database search, factorization, graph problems) significantly faster than with classical approaches (see *e.g.* [2, 3, 4]). The states of the qubits are modified by quantum operations which are inherently reversible and can be represented by unitary matrices.

Considering that many of the established quantum algorithms include a significant Boolean component (*e.g.* the oracle transformation in the Deutsch-Jozsa algorithm, the database in Grover's search algorithm, and the modulo exponentiation in Shor's algorithm), it is crucial to have efficient methods to synthesize quantum gate realizations of Boolean functions. The problem is often approached by a two-stage

procedure: First, a reversible circuit is designed using a reversible gate library. Then, the resulting reversible circuit is mapped into an equivalent quantum circuit.

The synthesis of reversible circuits has been extensively addressed *e.g.* in [5, 6, 7]. In this work, we focus on the mapping of reversible circuits to efficient quantum circuits. So far, the well known NCV quantum gate library (NOT, controlled-NOT and square-root-of-NOT gates) introduced in [8] has been applied to the mapping problem [9, 10]. Different optimization techniques have been introduced *e.g.* in [11]. However, mappings based on the NCV-library become very expensive particularly if large reversible gates are considered which often require so called ancillaries, *i.e.* circuit lines being utilized as temporary work lines only.

In this paper, we consider a modified NCV library motivated by the approach introduced in [12]. We introduce a new methodology for mapping reversible circuits into quantum circuits using this library. We demonstrate that the new library leads to realizations for multiple-control Toffoli gates with far fewer quantum gates than have been found using the NCV library. Our approach uses a structure similar to one recently introduced in [13]. However, while that work addressed a particular application, we here consider the realization of general reversible circuits.

Experiments demonstrate the benefits of the proposed mapping methodology. Compared to the best previously introduced methods, we show that our mapping yields substantially smaller circuits particularly for medium to large scale problems. More precisely, improvements of around 70% can be achieved on average. In the best case, the size of the circuits can even be reduced by approximately 90%. The proposed mapping has other advantages like the direct handling of Toffoli gates with mixed-polarity controls, as well as a better consideration of additional technology-based constraints like the nearest-neighbor constraint.

The remainder of this paper is structured as follows. The next section briefly reviews the basics of reversible and quantum circuits. Section 3 introduces the new gate library motivated by the work in [12] which forms the basis for our mapping methodology. Afterwards, the proposed mappings are presented in Section 4 and evaluated in Section 5. Finally, further benefits of the proposed mappings are discussed in Section 6 and conclusions are drawn in Section 7.

2. PRELIMINARIES

This section presents the background necessary for this paper. Readers interested in more detail should consult the literature, *e.g.* [1].

2.1 Reversible Functions, Gates and Circuits

A Boolean function $f : \mathbb{B}^n \to \mathbb{B}^m$, $\mathbb{B} = \{0, 1\}$ with inputs $X = \{x_1, \ldots, x_n\}$ is *reversible* iff it has the same number of inputs and outputs *i.e.* $n = m$, and it maps each input

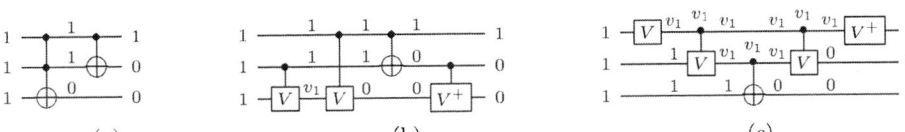

(a)   (b)   (c)

Figure 1: An MCT circuit for a Peres gate (a) and corresponding NCV (b) and NCV-$|v_1\rangle$ (c) circuits.

pattern to a unique output pattern. Otherwise, the function is termed *irreversible*. A reversible function can be realized by a circuit comprised of a cascade of reversible gates with no fan-out and feedback [1].

Several reversible gates have been introduced including the Toffoli gate [14], the Fredkin gate [15], and the Peres gate [16]. A *multiple-control Toffoli (MCT)* gate, a direct generalization of the basic Toffoli gate, has a *target line* x_j and *control lines* $\{x_{i_1}, x_{i_2}, \ldots, x_{i_k}\}$. This gate maps $(x_1 x_2 \ldots x_j \ldots x_n)$ to $(x_1 x_2 \ldots (x_{i_1} x_{i_2} \ldots x_{i_k}) \oplus x_j \ldots x_n)$, *i.e.* the target line is inverted if all the controls have value 1; otherwise the value on the target line is passed through unchanged. The values on the control and unconnected lines always pass through the gate unchanged. An MCT gate with no controls always inverts the target and is a *NOT* gate. An MCT gate with one control line is called a *controlled-NOT (CNOT) gate* (also known as the Feynman gate). The case of two control lines is the original gate defined by Toffoli.

MCT gates are universal in that all reversible functions can be realized using this gate type alone [14]. Fredkin and Peres gates can, for example, be realized using MCT gates. This paper considers reversible circuits composed of MCT gates. An MCT gate is denoted by $T(C; t)$ where $C \subset X$ is the possibly empty set of control lines and $t \in X \setminus C$ is the target line. For drawing circuits, we follow the established convention of using the symbol $\oplus$ to denote the target line and solid black circles to indicate control connections.

EXAMPLE 1. *Figure 1(a) shows an MCT circuit with three circuit lines and two gates that emulates a Peres gate [16]. As shown, this circuit maps the input pattern 111 to the output pattern 100. Note that the gate operations can be applied in either direction, i.e. from the inputs towards the outputs realizing a particular reversible function and from the outputs towards the inputs realizing the inverse of that function. This is because every MCT gate is its own inverse.*

2.2 Quantum Gates & Circuits

The basic unit of quantum information is the qubit whose *state* is written as $|\Psi\rangle = \alpha|0\rangle + \beta|1\rangle$, where α and β are complex numbers such that $|\alpha|^2 + |\beta|^2 = 1$. $|0\rangle$ and $|1\rangle$ are basis states corresponding to the classical 0 and 1 states.

The quantum state of a single qubit can be expressed as a vector $\binom{\alpha}{\beta}$. The state of a quantum system with $n > 1$ qubits can be represented as a normalized (length 1) vector with 2^n elements, called the *state vector*. A quantum circuit is a cascade of quantum gates and the operation of the circuit on the state vector corresponds to the multiplication of appropriate $2^n \times 2^n$ unitary matrices, one for each of the quantum gates [1].

A qubit has a potentially infinite number of values and there is also a potentially infinite number of distinct quantum gates. However, in practice researchers consider circuits composed of a small number of gate types.

The *NCV gate library* was introduced by Barenco *et al.* [8] and contains the following set of quantum gates:

- *NOT* gate $T(\emptyset; t)$: A single qubit t is inverted which is described by the unitary matrix $\left(\begin{smallmatrix} 0 & 1 \\ 1 & 0 \end{smallmatrix}\right)$.

- *Controlled NOT* (CNOT) gate $T(\{c\}; t)$: The target qubit t is inverted if the control qubit c is 1.

- *Controlled V* gate $V(\{c\}; t)$: The operation described by the unitary matrix $\mathbf{V} = \frac{1+i}{2} \left(\begin{smallmatrix} 1 & -i \\ -i & 1 \end{smallmatrix}\right)$ is performed on the target qubit t if the control qubit c is 1.

- *Controlled V^+* gate $V^+(\{c\}; t)$: The operation described by the unitary matrix $\mathbf{V}^+ = \frac{1-i}{2} \left(\begin{smallmatrix} 1 & i \\ i & 1 \end{smallmatrix}\right)$ is performed on the target qubit t if the control qubit c is 1. The V^+ gate performs the inverse operation of the V gate since $\mathbf{V}^+ = \mathbf{V}^{-1}$.

The V and V^+ gates are referred to as *controlled square-root-of-NOT* gates since two adjacent identical V, or V^+, gates are equivalent to a CNOT gate.

EXAMPLE 2. *Figure 1(b) shows an NCV gate circuit which is functionally equivalent to the circuit in Figure 1(a). Note the quantum value output of the first gate. To apply the circuit in reverse, i.e. from output to input, V and V^+ gates must be interchanged as they are the inverse of each other.*

3. A NEW CLASS OF QUANTUM GATES

Although the NCV gate library is universal in the sense that every reversible Boolean function can be realized by a circuit composed of NCV gates [8], other libraries are of interest as they can lead to better circuits, *e.g.* fewer gates. In this work, we focus on a modification to the NCV gate library based on concepts introduced in [12].

If circuits with Boolean inputs use NCV gates only, the value of each qubit at each stage of the circuit is restricted to one of $\{0, v_0, 1, v_1\}$ where $v_0 = \frac{1}{2}\left(\begin{smallmatrix} 1+i \\ 1-i \end{smallmatrix}\right)$ and $v_1 = \frac{1}{2}\left(\begin{smallmatrix} 1-i \\ 1+i \end{smallmatrix}\right)$. The *NOT*, V, and V^+ operations over these four values are:

x	$NOT(x)$	$V(x)$	$V^+(x)$
0	1	v_0	v_1
v_0	v_1	1	0
1	0	v_1	v_0
v_1	v_0	0	1

As shown, *NOT* is a complement operation, V is the cycle $(0 \to v_0 \to 1 \to v_1 \to 0)$, and V^+ is the inverse cycle.

In this work, we adopt a new quantum gate library which we call the *NCV-$|v_1\rangle$ library*. The NCV-$|v_1\rangle$ gate library is composed of (1) the three unitary gates (*i.e.* gates without a control line) performing the *NOT*, V, and V^+ operation as well as (2) single-control versions of these gates. In contrast to the NCV-library, and in keeping with the work in [12], the controlled gates perform the respective operation not when the control line is 1, but rather when the control line is set to the value v_1. We label control connections for NCV-$|v_1\rangle$ gates with v_1 to emphasize this fact.

EXAMPLE 3. *Figure 1(c) shows an NCV-$|v_1\rangle$ circuit functionally equivalent to the circuits in Figures 1(a) and 1(b).*

Besides the benefits in the physical implementation, as discussed in [12], this gate library enables a much more efficient mapping from an MCT gate circuit as we introduce below.

The implementation cost of a quantum gate is heavily technology dependent. Here we assume that all quantum gates, NCV and NCV-$|v_1\rangle$ in particular, have unit cost. Under that assumption the cost of a quantum circuit is the gate count. This is clearly an approximation but a suitable one when considering technology independent optimization.

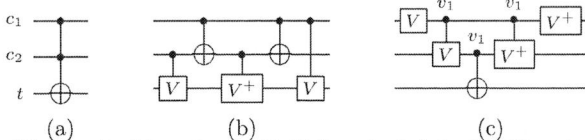

(a)   (b)   (c)

Figure 2: Mapping a Toffoli gate (a) to NCV gates (b) and NCV-$|v_1\rangle$ gates (c).

4. PROPOSED MAPPING METHOD

The common approach to synthesize a quantum circuit implementing a reversible Boolean function has two steps. First, a circuit composed of reversible gates implementing the desired function is synthesized. That circuit is then mapped to a cascade of gates from the target quantum gate library. Optimizations can be applied at various stages of the mapping process (see *e.g.* [11]).

In this section, we briefly review the established mapping methodology based on the NCV library. We then introduce a new mapping approach based on the NCV-$|v_1\rangle$ library. We show that, except for very small circuits, the proposed mapping leads to circuits with significantly fewer gates than the circuits determined using the established NCV mapping.

4.1 Mapping Individual MCT Gates

The well-known [8] optimal mapping of a Toffoli gate $T(\{c_1, c_2\}; t)$ to a cascade of NCV gates is shown in Figure 2(b). As shown, five NCV-gates are required. For MCT gates with more control lines, the number of required NCV-gates increases rapidly.

Table 1 shows the number of NCV gates required to realize MCT gates with up to 15 controls using the approach described in [10]. Besides the number of control lines, the number of ancillary lines available also affects the size of the quantum gate circuit. An ancillary line is a circuit line which is neither used as the target line nor as a control of a Toffoli gate, and is thus available to be used as a temporary work line in the quantum realization. For each number of controls, the rightmost gate count is the lowest possible. Blank entries indicate when the availability of more ancillary lines is not advantageous. The NCV gate counts given in Table 1 are the best known for NCV gate realizations of MCT gates. It is important to note (1) that at least one ancillary line is required for three or more controls and (2) that the gate counts grow quite quickly with the number of controls. In Table 1 for $c \geq 4$, assuming the maximum number of ancillaries required is available, the number of gates is $12c - 28$. If only one ancillary is available, for $c \geq 10$ the number of gates required is $24c - 132$. These formulas have been verified for up to 20 controls and it is believed they will continue to hold for larger numbers of controls.

In contrast, better mappings with much slower linear growth are possible using the NCV-$|v_1\rangle$ library. Therefore, the structure illustrated in Figure 3 for a Toffoli gate with 4 control lines is proposed. Here, the actual operation of the Toffoli gate (*i.e.* the inversion of the target line) is performed by a single CNOT gate controlled by v_1. The V-gates ensure that the control line of the CNOT gate is set to v_1 if, and only if, all control lines c_1 to c_4 are equal to 1. The last V^+ gates are needed to undo the corresponding V operations on the control lines in order to restore their values.

Generalizing this structure, every Toffoli gate with c control lines can be realized using a V gate and a V^+ gate for each control line as well as a single $CNOT$ gate which operates on the target line. This leads to a total of $2c + 1$ NCV-$|v_1\rangle$ gates. While for a Toffoli gate with 1 control line only, this results to a more expensive mapping (3 gates instead of 1), MCT gates with more than 2 controls can be realized with significant reductions in the number of gates

Table 1: Quantum gate counts for MCT gate realizations

| | NCV gates [10] Number of ancillary lines | | | | | | | NCV-$|v_1\rangle$ gates | Ratio |
c	0	1	2	3	4	5	6	$2c+1$	
0	1							1	100%
1	1							3	300%
2	5							5	100%
3		14						7	50.0%
4		20						9	45.0%
5		32						11	34.4%
6		44						13	29.5%
7		64	56					15	23.4-26.8%
8		76	68					17	22.4-25.5%
9		96	88	80				19	19.8-23.8%
10		108	100	92				21	19.4-22.8%
11		132	120	112	104			23	17.4-22.1%
12		156	132	124	116			25	16.0-21.6%
13		180	156	148	136	128		27	15.0-21.1%
14		204	180	172	148	140		29	14.2-20.7%
15		228	204	198	172	160	152	31	13.6-20.4%

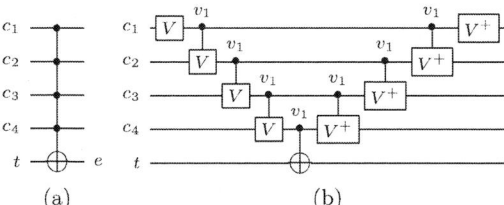

(a)   (b)

Figure 3: Mapping of a 4-control MCT gate (a) to NCV-$|v_1\rangle$ gates (b).

compared to the established NCV mappings. Furthermore, the NCV-$|v_1\rangle$ mappings do not require any ancillary lines.

Table 1 lists the number of NCV-$|v_1\rangle$ gates required for up to 15 controls. The gate counts are significantly lower than the NCV costs. The relative size of the NCV-$|v_1\rangle$ circuits to the NCV circuits drops to about 30% at 6 controls and approaches 20% at 15 controls. We again emphasize the NCV-$|v_1\rangle$ gate circuits require no ancillary lines.

4.2 Mapping a Reversible Circuit

In the last section, the mapping of single MCT gates was considered. We now address how to map circuits composed of MCT gates to a quantum circuit. This can be done by a direct substitution of each single MCT gate by its corresponding quantum gate cascade. Even for this simple approach, the mapping illustrated in Figure 3 leads to significant reductions in comparison to the established NCV mappings. However, even better results can be obtained if further optimizations are applied.

We employ an optimization method based on the ideas presented in [11] extended to handle the new NCV-$|v_1\rangle$ gates. The optimization process involves a *Line Labeling Procedure* (LLP) and a *Gate Reduction Procedure* (GRP). Both are applicable to both MCT and quantum gates.

The LLP is used to assign labels to line segments such that two segments of a line have the same label only if they have the same functionality. The LLP is Procedure 1 of [11] extended to handle NCV-$|v_1\rangle$ gates. The procedure involves a single pass through the circuit from the inputs toward the outputs. A stack of gate operations is kept for each line. As a gate g is processed, the stack for its target line is checked for a sequence of gates realizing the identity operation. When one is found the output line for gate g is assigned the same label that appears on the target line input to the first gate in the sequence. While this procedure only assigns the same label to two segments of the same line that have the same functionality, it is not guaranteed to find all equivalences. We have identified a few such cases, but in general the LLP

38

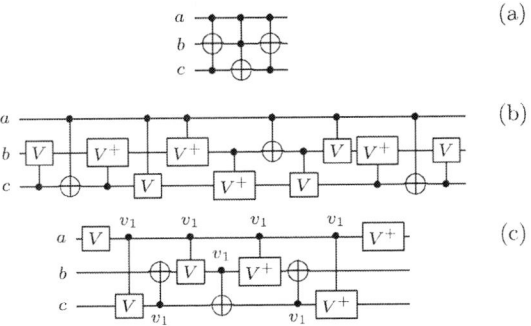

Figure 4: Mapping and optimizing an MCT circuit: (a) MCT (b) NCV (c) NCV-$|v_1\rangle$

finds most equivalences in circuits we have considered.

The GRP is Procedure 2 of [11] extended to handle NCV-$|v_1\rangle$ gates. The procedure starts from the input side of the circuit and processes the gates in order. The LLP is used to label the circuit up to the current gate g of interest. Then, gate g is moved back through the circuit to each of the places that have the same labels on its control lines. Gate g can not be moved past a point where there is a control on the target line of g. As a gate is moved, a list is made that contains gates that can be adjacent to it and have the same target, controls, control labels, and control types. Then, the gates in this list are removed from the circuit and an optimized equivalent sequence, which can be empty when the gates implement the identity function, is inserted in the position of the removed gate closest to the circuit's inputs. The procedure then proceeds for subsequent gates until the end of the circuit is reached.

In our overall approach, the MCT gate circuit is first optimized using the GRP . Then, the appropriate NCV-$|v_1\rangle$ realization is substituted for each gate in the optimized MCT circuit. Finally, the resulting NCV-$|v_1\rangle$ circuit is optimized using the GRP. Because of the regular structure of the new realizations, significant improvements are typical in the NCV-$|v_1\rangle$ optimization step.

EXAMPLE 4. *Consider the circuit depicted in Figure 4(a). No optimization is possible for the MCT gates. Replacing each Toffoli gate with the appropriate version of the realization from Figure 2(b) yields a circuit with 15 NCV gates. Applying the GRP reduces this to the 12 gate circuit in Figure 4(b). In contrast, while using the NCV-$|v_1\rangle$ realization from Figure 2(c) also results in a 15 gate circuit but applying the GRP yields the 9 gate circuit shown in Figure 4(c).*

5. EXPERIMENTAL RESULTS

In this section, we present results obtained by the proposed approach. We first consider the improvements achieved by the new mapping methodology. Then, we also briefly discuss how the resulting circuits have been verified.

5.1 Evaluation

The procedures described in the previous section have been implemented using Python 2.7.1. Our experiments were run on a computer with a Core 2 Duo 2.66 GHz CPU and 4.0 GB RAM. We used a test suite of 138 circuits from RevLib [17]. The results are shown in Table 2. Due to space limitations, the table shows only those circuits for which the improvement was greater than 30% (50 of the 138 circuits).

Each row of the Table gives: (1) The name of the circuit including the RevLib file index number. Note that Fredkin and Peres gates in the RevLib circuits are substituted by MCT gate realizations before applying our techniques. (2) The quantum gate count given on the RevLib site (called

quantum cost in RevLib). (3) The NCV gate count for circuits determined using the mappings from [10] with the optimization techniques described in [11]. (4) The gate count for direct substitution of the $2c+1$ NCV-$|v_1\rangle$ gate realization for each MCT gate in the given circuit. (5) The direct mapping NCV-$|v_1\rangle$ gate count is reported for the circuit found by first applying the GRP to the MCT circuit. (6) The NCV-$|v_1\rangle$ gate count for the circuit from (5) optimized at the NCV-$|v_1\rangle$ gate level using the GRP.

For all 138 circuits, the gate reduction for all circuits in total is 81.8% with respect to the counts from RevLib and 68.7% with respect to the NCV circuits determined using the techniques from [10] and [11].

As the results in Table 2 show, our approach does very well for medium to large circuits since those circuits tend to have more MCT gates with greater than 2 controls than do the small circuits. In addition the smaller circuits often have a high proportion of *CNOT* gates which as noted above require 3 NCV-$|v_1\rangle$ gates. To be precise, our methods yield slightly more gates than the RevLib gate count for 33 of the circuits: 22 with 1-3 extra gates; 4 with 4; 2 for each of 5, 6 and 7; and 1 circuit with 8 additional gates. The great majority of these circuits are small and have a high proportion of CNOT gates, *i.e.* the fact that CNOT is not a primitive in the NCV-$|v_1\rangle$ library has a major effect. In contrast there are 21 circuits where the NCV-$|v_1\rangle$ circuit has more than 1000 fewer gates and a further 6 with from 139 to 984 fewer gates than the circuit reported in RevLib.

The total improvement comes primarily from the new MCT to NCV-$|v_1\rangle$ gate mapping. Besides, the optimizations described in Section 4.2 reduce the gate count by a further 4.5% (MCT gate optimization) and by a further 32.5% (NCV-$|v_1\rangle$ gate optimization) on average. Note that for n-line circuits that have an MCT gate of size n, an ancillary line must be added to map to NCV circuits. Additional lines are never required for NCV-$|v_1\rangle$ circuits. Considering unit delay for all 1-qubit and 2-qubit quantum gates as in [18], NCV-$|v_1\rangle$ circuits are much faster than the NCV circuits as they have lower logic depth.

5.2 Circuit Verification

Verification methods have been applied to confirm that the circuits produced by our methods are functionally equivalent to the original RevLib circuits. This is an interesting problem on its own. The RevLib circuit is binary whereas the circuit our method produces uses 4-valued logic gates. Also our basic MCT gate substitution relies on the fact the inputs are restricted to values 0 and 1.

Our verification procedure uses *Quantum Multiple-valued Decision Diagrams* (QMDD) [19] and is basically the approach described in [20]. The differences are that the gates in the RevLib circuit are treated as 4-valued as are of course the gates in the circuit we produce, and equivalence of the circuits is not just a matter of confirming the equality of the QMDD for the two circuits. Rather, equivalence checking requires a depth-first comparison of the two QMDD that restricts the input line values to 0 and 1, ignoring v_0 and v_1.

The verification procedure is implemented in C. On the computer described above, 115 of the circuit verifications each took only a fraction of a CPU second. However, the larger circuits take significantly longer, *e.g.* just under 3 hours of CPU time for plus127mod8192_162. It is interesting that it is not the largest circuits that take the longest time to verify. More detail can be found in [20].

6. FURTHER BENEFITS

As shown above, applying the NCV-$|v_1\rangle$ gate mapping leads to much smaller quantum circuit realizations in comparison to the established NCV methodology. The NCV-$|v_1\rangle$

Table 2: Experimental Results

Benchmark	Previous Approaches		Proposed Approaches (using NCV-$|v_1\rangle$)			% Improv. wrt [17]	% Improv. wrt [10, 11]
	RevLib [17]	NCV with Opt. [10, 11]	Direct Mapping	MCT Gate Optimization	NCV-$|v_1\rangle$ Gate Optimization		
plus63mod8192_164	45025	19566	6620	5921	2135	95.3	89.1
plus127mod8192_162	73357	35348	12318	10910	3972	94.6	88.8
plus63mod4096_163	32539	14652	5327	4672	1779	94.5	87.9
cycle10_2_110	1202	720	219	219	91	92.4	87.4
hwb9_121	44665	28629	13149	12920	10156	77.3	64.5
hwb9_122	44653	28629	13149	12920	10156	77.3	64.5
hwb9_119	44714	28660	13168	12938	10180	77.2	64.5
hwb9_120	44702	28660	13168	12938	10180	77.2	64.5
hwb8_113	16530	10328	5065	4957	3786	77.1	63.3
hwb8_118	16522	10328	5065	4957	3786	77.1	63.3
hwb8_114	14699	8815	4456	4378	3235	78.0	63.3
hwb8_115	14691	8815	4456	4378	3237	78.0	63.3
ham15_107	1831	1155	836	724	447	75.6	61.3
hwb9_123	22510	14487	9151	9145	5704	74.7	60.6
hwb7_59	5236	3500	2017	1969	1434	72.6	59.0
hwb7_61	3876	2863	1622	1596	1226	68.4	57.2
hwb8_116	7015	4825	3383	3383	2109	69.9	56.3
hwb8_117	7013	4825	3383	3383	2109	69.9	56.3
hwb6_56	1530	1150	766	756	546	64.3	52.5
rd53_130	232	195	112	112	93	59.9	52.3
hwb7_62	2611	1973	1495	1495	957	63.3	51.5
4gt4-v1_74	57	46	31	31	23	59.6	50.0
hwb7_60	4170	2989	2286	2121	1524	63.5	49.0
4gt12-v1_89	45	37	23	23	19	57.8	48.6
4gt4-v0_72	54	34	30	25	18	66.7	47.1
alu-v2_30	114	103	82	79	55	51.8	46.6
mod5adder_128	83	84	59	59	45	45.8	46.4
hwb6_57	1171	872	829	728	473	59.6	45.8
decod24-v3_45	35	35	25	25	19	45.7	45.7
sym9_148	4368	672	1722	616	374	91.4	44.3
sym6_145	777	212	276	187	118	84.8	44.3
ham15_108	453	356	320	321	202	55.4	43.3
mod5adder_127	125	104	75	75	60	52.0	42.3
alu-v2_31	101	83	69	70	48	52.5	42.2
alu-v2_32	39	38	31	28	22	43.6	42.1
hwb5_53	315	282	257	254	166	47.3	41.1
rd53_131	119	90	76	70	55	53.8	38.9
rd53_132	117	90	76	70	55	53.0	38.9
rd53_133	128	72	68	65	45	64.8	37.5
rd53_134	120	72	68	65	45	62.5	37.5
mod5adder_129	77	76	65	65	48	37.7	36.8
4gt4-v0_73	89	49	73	53	31	65.2	36.7
4gt4-v0_80	37	28	21	21	18	51.4	35.7
4gt5_77	28	28	20	20	18	35.7	35.7
one-two-three-v0_97	71	62	57	57	40	43.7	35.5
ham7_104	83	84	91	91	55	33.7	34.5
4gt10-v1_81	34	35	28	28	23	32.4	34.3
decod24-enable_126	86	77	72	72	52	39.5	32.5
decod24-v1_41	22	23	22	22	16	27.3	30.4
4mod7-v1_96	39	33	31	31	23	41.0	30.3

gate mapping has a number of further important benefits.

6.1 MCT Gates with Negative Controls

Thus far, we have assumed that MCT gates have *positive* control lines, *i.e.* that the control lines must all have the value 1 in order to perform the corresponding operation on the target line. But a number of reversible circuit synthesis algorithms, *e.g.* [21, 22], produce circuits with MCT gates that also have *negative* controls, *i.e.* controls that are activated by the value 0. This affects the number of gates in an NCV gate circuit realization.

For example, the circuit in Figure 2(b) has five gates for a Toffoli gate with two positive control lines. If a Toffoli gate has one positive and one negative control line, the number of required gates would remain 5. However, if a Toffoli gate has two negative controls, an additional NOT gate is needed leading to a total of 6 gates [9].

NCV realizations of MCT gates with mixed (positive and negative) controls have been considered in [23]. To give an idea of the results, the NCV realization of a 3-controlled MCT gate has 14 gates for 0 or 1 negative controls, 16 for 2 negative controls, and 18 for 3 negative controls. For 15 controls, the number of NCV gates ranges from 228 to 246 if there is one ancillary line available and from 152 to 168 if 13 ancillary lines are available.

The situation is quite different for our new NCV-$|v_1\rangle$ gate realizations. Consider the structure illustrated in Figure 3. To change a positive control to a negative control, the V gate on that line simply has to be swapped with the corresponding V^+ gate. Hence, our mapping always leads to $2c + 1$ gates for c controls regardless of the mix of positive and negative controls. Thus, in contrast to the NCV situation, the NCV-$|v_1\rangle$ gate circuits handle negative controls for free. And once again, no ancillary lines are required.

6.2 Nearest-Neighbor Constraints

Many quantum technologies, *e.g.* [24, 25], require that a circuit satisfies the *nearest-neighbor constraint*, *i.e.* all controlled gates in a circuit have the control line and the target line adjacent. A circuit can be modified to satisfy this constraint by adding gates which swap the values of lines so that only adjacent lines are used. How best to locate the swap gates has been studied *e.g.* in [26, 27].

As an example, the NCV circuit in Figure 2(b) does not satisfy the nearest-neighbor constraint and can not be made nearest-neighbor by permuting the lines. This also holds for

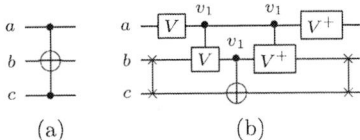

Figure 5: (a) MCT gate $T(\{a,c\};b)$ and (b) NCV-$|v_1\rangle$ gate nearest neighbor circuit.

the general case, *i.e.* existing mappings of MCT gates into NCV quantum gates require additional swap gates to satisfy this constraint. Often, a significant number is required.

In contrast, no intervening lines exist in the proposed structure depicted in Figure 2(c) and Figure 3, *i.e.* the nearest-neighbor condition is already satisfied. Moreover, this mapping remains nearest-neighbor if the target is the top line in which case the structure is inverted. If, however, a Toffoli gate is considered where the target is an inside line, or there are intervening unconnected lines, the structure is no longer nearest-neighbor and swap gates (depicted as two ×'s joined by a line) are required as shown in Figure 5.

As just justified, in many cases the proposed mapping can be directly applied to technologies requiring nearest-neighbor constraints. Developing good heuristics for handling the remaining cases is left for future work. Approaches *e.g.* introduced in [26, 27] may be exploited for this purpose.

7. CONCLUSIONS

The new NCV-$|v_1\rangle$ quantum gate library has been shown to lead to quantum circuit realizations composed of $2c + 1$ gates for MCT gates with c control lines. This is significantly less than the best known quantum realizations based on the NCV library. We have also shown that negative controls for an MCT gate are available at no extra cost. Further, the NCV-$|v_1\rangle$ gate realizations do not require ancillary lines.

Using the NCV-$|v_1\rangle$ library, MCT gate realizations together with extensions to previously introduced optimization techniques lead to very significant gate count reductions especially for medium to large circuits. In fact, on average improvements of around 70% can be achieved.

The MCT to NCV-$|v_1\rangle$ gate mappings have better nearest neighbor properties than do the NCV mappings. Our future work will concentrate on the nearest-neighbor problem and in particular on how to incorporate that constraint into existing optimization procedures. We will also consider the extension to a gate library allowing a range of gate control values. Initial work has shown this can lead to more compact circuits. Such an extension is technology dependent.

Acknowledgment

This work was supported in part by grants from the Natural Sciences and Engineering Research Council of Canada (NSERC), the German Research Foundation (DFG) (DR 287/20-1), and the German Academic Exchange Service (DAAD).

8. REFERENCES

[1] M. Nielsen and I. Chuang. *Quantum Computation and Quantum Information.* Cambridge Univ. Press, 2000.

[2] L. K. Grover. A fast quantum mechanical algorithm for database search. In *Symp. on Theory of Computing*, pages 212–219, 1996.

[3] P. W. Shor. Algorithms for quantum computation: discrete logarithms and factoring. *Foundations of Computer Science*, pages 124–134, 1994.

[4] C. Dürr, M. Heiligman, P. Hoyer, and M. Mhalla. Quantum query complexity of some graph problems. *SIAM Jour. of Comp.*, 35:1310–1328, 2006.

[5] V. V. Shende, A. K. Prasad, I. L. Markov, and J. P. Hayes. Synthesis of reversible logic circuits. *IEEE Trans. on CAD*, 22(6):710–722, 2003.

[6] D. M. Miller, D. Maslov, and G. W. Dueck. A transformation based algorithm for reversible logic synthesis. In *Design Automation Conf.*, pages 318–323, 2003.

[7] R. Wille and R. Drechsler. BDD-based synthesis of reversible logic for large functions. In *Design Automation Conf.*, pages 270–275, 2009.

[8] A. Barenco, C. H. Bennett, R. Cleve, D.P. DiVinchenzo, N. Margolus, P. Shor, T. Sleator, J.A. Smolin, and H. Weinfurter. Elementary gates for quantum computation. *The American Physical Society*, 52:3457–3467, 1995.

[9] D. Maslov, G.W. Dueck, D.M. Miller, and C. Negrevergne. Quantum circuit simplification and level compaction. *IEEE Trans. on CAD*, 27(3):436–444, March 2008.

[10] D. M. Miller, R. Wille, and Z. Sasanian. Elementary quantum gate realizations for multiple-control Toffolli gates. In *Proc. Int'l Symp. on Multiple-valued Logic*, pages 217–222, 2011.

[11] Z. Sasanian and D. M. Miller. Mapping a multiple-control toffoli gate cascade to an elementary quantum gate circuit. In *Proc. Workshop on Reversible Computation*, pages 83–90, 2010.

[12] A. Muthukrishnan and C. R. Stroud. Multivalued logic gates for quantum computation. *Physical Review A*, 62:052309, 2000.

[13] Y. Wang and M. Perkowski. Improved complexity of quantum oracles for ternary grover algorithm for graph coloring. In *Proc. Int'l Symp. on Multiple-valued Logic*, pages 294 – 301, 2011.

[14] T. Toffoli. Reversible computing. In W. de Bakker and J. van Leeuwen, editors, *Automata, Languages and Programming*, page 632. Springer, 1980. Technical Memo MIT/LCS/TM-151, MIT Lab. for Comput. Sci.

[15] E. F. Fredkin and T. Toffoli. Conservative logic. *International Journal of Theoretical Physics*, 21(3/4):219–253, 1982.

[16] A. Peres. Reversible logic and quantum computers. *Phys. Rev. A*, (32):3266–3276, 1985.

[17] R. Wille, D. Große, L. Teuber, G. W. Dueck, and R. Drechsler. RevLib: an online resource for reversible functions and reversible circuits. In *Int'l Symp. on Multi-Valued Logic*, pages 220–225, 2008. RevLib is available at http://www.revlib.org.

[18] H. Thapliyal and N. Ranganathan. Design of efficient reversible logic based binary and bcd adder circuits. *ACM J. on Emerging Technologies in computing Systems*, September 2012.

[19] D.M. Miller and M.A. Thornton. QMDD: A decision diagram structure for reversible and quantum circuits. In *Proc. Int'l Symp. on Multiple-valued Logic CD*, 6 pp., 2006.

[20] R. Wille, D. Große, D. M. Miller, and R. Drechsler. Equivalence checking of reversible circuits. In *Int'l Symp. on Multi-Valued Logic*, pages 324–330, 2009.

[21] K. Fazel, M.A. Thornton, and J.E. Rice. ESOP-based Toffoli gate cascade generation. In *Communications, Computers and Signal Processing, 2007. PacRim 2007. IEEE Pacific Rim Conference on*, pages 206 –209, 2007.

[22] M. Soeken, R. Wille, C. Hilken, N. Przigoda, and R. Drechsler. Synthesis of Reversible Circuits with Minimal Lines for Large Functions. In *Asia and South Pacific Design Automation Conf.*, pages 85–92, 2012.

[23] Z. Sasanian and D. M. Miller. NCV realization of MCT gates with mixed controls. In *Proc. Pacific Rim Conf. on Communications, Computers and Signal Processing*, pages 567–571, 2011.

[24] A. G. Fowler, S. J. Devitt, and L. C. L. Hollenberg. Implementation of Shor's algorithm on a linear nearest neighbour qubit array. *Quant. Info. and Comput.*, 4:237–245, 2004.

[25] S. A. Kutin. Shor's algorithm on a nearest-neighbor machine. In *Asian Conference on Quantum Information Science*, 2006. arXiv:quant-ph/0609001v1.

[26] M. H. A. Khan. Cost reduction in nearest neighbour based synthesis of quantum boolean circuits. *Engineering Letters*, 16:1–5, 2008.

[27] M. Saeedi, R. Wille, and R. Drechsler. Synthesis of quantum circuits for linear nearest neighbor architectures. *Quantum Information Processing*, 10:355–377, 2011.

Physical Synthesis onto a Sea-of-Tiles with Double-Gate Silicon Nanowire Transistors

Shashikanth Bobba[1], Michele De Marchi[1], Yusuf Leblebici[2], Giovanni De Micheli[1]

[1]LSI, EPFL, Lausanne, Switzerland   [2]LSM, EPFL, Lausanne, Switzerland

Abstract

We have designed and fabricated double-gate ambipolar field-effect transistors, which exhibit p-type and n-type characteristics by controlling the polarity of the second gate. In this work, we present an approach for designing an efficient regular layout, called *Sea-of-Tiles* (SoTs). First, we address gate-level routing congestion by proposing compact layout techniques and novel symbolic-layout styles. Second, we design four logic tiles, which form the basic building block of the SoT fabric. We run extensive comparisons of mapping standard benchmarks on the SoT. Our study shows that SoT with *Tile$_{G2}$* and *TileG_{1h2}*, on an average, outperforms the one with *Tile$_{G1}$* and *Tile$_{G3}$* by 16% and 10% in area utilization, respectively.

Categories and Subject Descriptors

B.7.1 [Integrated Circuits]Design Styles: Advanced technologies

General Terms

Design, Layout, Performance, Regular

Keywords

Ambipolar devices, Regular layouts, Silicon Nanowire FET, Tile

1. Introduction

Layout regularity is one of the key features required to increase the yield of ICs at advanced technology nodes [Tejas 07]. Hence, design styles based on regular layout fabrics have the advantage of higher yield as they maximize the layout manufacturability. Various regular fabrics have been proposed throughout the evolution of semiconductor industry, where some recent approaches are discussed in [Lin 09] [Ran 06] [Taylor 07]. In gate arrays fabric style, a sea of prefabricated transistors is customized to obtain a desired logic gate. The flexibility of building generic logic gates comes at a huge cost of area as well as routing overhead, thereby increasing the performance gap between ASICs and gate arrays. With the advent of via programmable gate arrays [Ran 06] and logic-bricks [Taylor 07], the performance gap is minimized. On the other hand, strict design rules, at 22nm technology node and beyond, has led to cell layouts with arrays of gates with a constant gate pitch, which resemble a sea-of-gates layout style. In this work, we define a regular logic *tile* that has an array of prefabricated transistor-pairs grouped together. A desired logic function can be mapped onto an array of logic tiles.

FinFET transistors are successfully replacing planar CMOS transistors beyond 22nm technology node [Hisamoto 00]. Intel has showcased 37% faster chips with low static and dynamic power consumption with their tri-gate transistor technology,

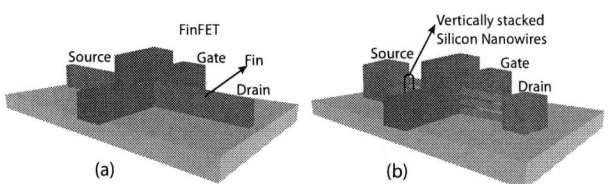

Figure 1. (a) FinFET providing increase in channel area between the source and drain regions (b) Vertically stacked SiNWFET with multiple parallel nanowire channels, each with Gate-All-Around (GAA) control [Saccheto 09].

[Doyle 03], at 22nm node when compared to 32nm planar technology. Following the trend to one-dimensional (1-D) structures, *Vertically Stacked Silicon Nanowire Field Effect Transistors* (SiNWFETs) are a promising extension to the tri-gate FinFETs [Suk 05]. The superior performance of these 1-D channel devices (nanowire FET) comes from a high $Ion/Ioff$ ratio, due to the gate-all-around structure, which improves the electrostatic control of the channel, thereby reducing the leakage current of the device. Figure 1 shows a tri-gate FinFET transistor and a vertically stacked SiNWFET. In addition, SiNWFET exhibit enhanced electrostatics properties, such as polarity control, which are electrically impossible to planar- and Fin- FETs.

Our methodology takes advantage of the electrostatics of these devices, which can be built to be ambipolar, i.e. to exhibit n- and p-type characteristics. By engineering of the source and drain contacts and by constructing independent double-gate structures, the device polarity can be electrostatically forced to either n- or p- type by polarizing one of the two gates. The in-field polarizability of these devices enables the development of new logic architectures, which are intrinsically not implementable in CMOS in a compact form [Jamaa 08]. However, the routing complexity at the device level increases due to the presence of an extra gate, called *polarity gate* (PG).

Typical CMOS layout techniques involve transistors with a single gate. In the traditional approach for CMOS, compact layouts are realized by optimal transistor chaining of p- and n-type transistors [Uethara 81] [Hwang 90]. However, in the case of ambipolar gates, the polarity of the transistor (p-type or n-type) changes with the input signals. Motivated by these observations, we propose compact layout techniques for *Double Gate Silicon Nanowire FET* (DG-SiNWFET). In order to facilitate this, we propose novel symbolic layouts for ambipolar logic with *Dumbell-Stick* diagrams.

As a second contribution, we design an efficient regular layout brick (called as *tile*), which forms the basic building block for *Sea-of-Tiles* (SoT) design methodology. The basic tile for SoT is optimized for area and regularity. Technology mapping, with logic synthesis tools, on various tiles helped us in choosing an efficient tile for realizing SoT. We show that hybrid tile *Tile$_{G1h2}$* and *Tile$_{G2}$*, on an average, outperforms *Tile$_{G1}$* and *Tile$_{G3}$* by 16%, and 10% in total area utilization, respectively.

Finally, we demonstrate $Tile_{G1h2}$ (and $Tile_{G2}$) as a basic building block for the future ambipolar logic circuits.

The remainder of this paper is organized as follows. In Section 2, we present the technology background of ambipolar SiNWFET. In Section 3, we introduce novel symbolic-layouts for ambipolar devices and explain the layout techniques based on dumbbell-stick diagrams. In Section 4, we introduce logic tiles for SiNWFETs and perform technology mapping to find an optimal tile for sea-of-tiles design. We conclude in Section 5.

2. Ambipolar SiNWFET

The advantage of SiNWFETs over other one-dimensional devices such as carbon nanotube transistors, is that SiNWs can be fabricated with a top-down silicon process [Ng 07]. Moreover, SiNWs can be built in vertical stacks, thereby giving highly dense array of nanowire transistors [Sacchetto 09].

Figure 2 show a SiNWFET device structure with SiNWs suspended between source and drain pillars. This SiNW is divided into three sections, which are in turn polarized by two gate-all-around gate regions. The center gate region works as in a conventional MOSFET, switching conduction in the device channel by means of a potential barrier. The side regions are instead polarized by a polarity gate, which controls Schottky barrier thicknesses at the S/D junctions and selects the majority carrier type, thus forcing the device to be either n- or p-type.

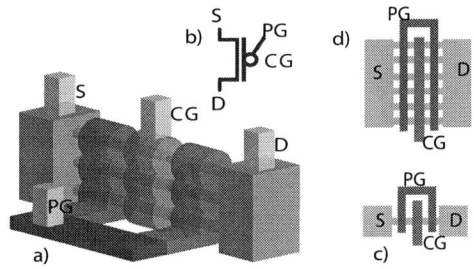

Figure 2. Conceptual structure of the ambipolar DG-SiNWFET: a) 3D view of the device. b) Circuit symbol for the device. c) Top view of the device showing one stack of nanowires forming the channel. d) Large transistor.

A SEM image of an array of vertically-stacked SiNWs, suspended between pillars, before patterning the gates, is shown in Fig. 3a. Figure 3b shows the double-gate SiNWFET after patterning the control and polarity gates. The measured electrical characteristic of the fabricated device is shown in Fig. 4. *Vpg* and *Vcg* correspond to the voltages applied to the polarity gate and control gate respectively. Further device optimization is envisaged for a balanced p- and n- type device.

In order to exploit the unique feature of this device of being polarized electrostatically, a static ambipolar logic family was introduced in [O'Connor 07]. Figure 5a shows a basic logic gate, which can be built with this methodology. An ambipolar transistor constitutes the pull-down network of a Pseudo-CMOS logic gate, having two logic inputs connected to the polarity and the control gate, respectively. In the case of a positive polarity gate input, the transistor behaves as a n-FET, thus producing the output of a classical pseudo-logic inverter. Alternatively, if the polarity gate has a low bias voltage, the transistor behaves as p-type, producing a degraded buffer output characteristic. If we consider this gate as a black box, and see both input signals as logic values, we can see that the gate calculates the XNOR logic function.

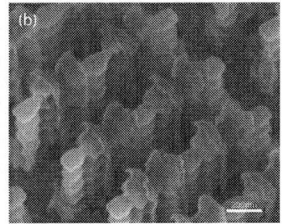

Figure 3. SEM images of a double-gate vertically stacked silicon nanowire FET (a) before the gate patterning; (b) after the gate patterning; Control gate (red); Polarity gate (violet); Active area (green);

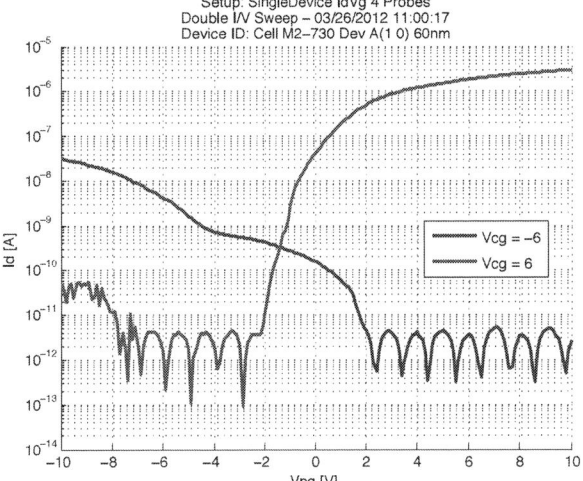

Figure 4. Electrical characteristics of a SiNWFET

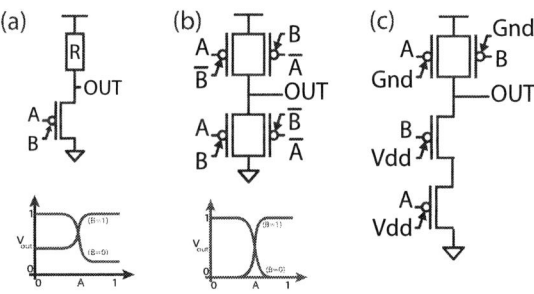

Figure 5. (a) Pseudo-CMOS logic gate with a double gate ambipolar CNTFET in the PDN. (b) Fully complementary XNOR logic gate with opposite polarity transistor pairs in PUN and PDN. (c) NAND logic gate by biasing the polarity gate to either Vdd or Gnd.

In order to obtain a gate featuring full-swing output, a pull-up network substitutes the pull-up resistor, making the logic complementary and each transistor is coupled with another transistor of opposite polarity [Jamaa 09], as in the case of CMOS pass-transistor gates. Figure 5b shows the complete gate, together with its conceptual output characteristic. A two-input XOR function with just 4 transistors portrays the high expressive nature of the double gate ambipolar transistors. CMOS static logic gates (*negative-unate* functions like NAND, NOR, INV, AOI, etc) can be realized by appropriately biasing the polarity gate of the double gate devices. Figure 5c shows a two input NAND gate realized with double gate transistors. The PGs of all the transistors in the pull up network are connected to ground (Gnd), whereas the PGs of all the transistors in the pull down network are connected to supply (Vdd).

3. Layout Technique for Ambipolar Logic Gates

In this section, we first introduce novel symbolic-layouts for ambipolar logic gates, *dumbell-stick diagrams*, based on which we present a layout technique to design complex gates.

3.1 Symbolic Layouts for Ambipolar Logic: Dumbell-Stick Diagrams

Similar to the CMOS stick diagrams, *dumbell-stick diagrams* are proposed for ambipolar devices (in our case DG-SiNWFET) for designing compact layouts by minimizing the cell routing complexity. Figure 6a shows the top view of a DG-SiNWFET (see Fig. 2). The suspended silicon nanowires between the source and drain contacts form the basic dumbell. The control gate and the polarity gate constitute the sticks. Based on this basic building block we present a dumbell-stick diagram of an inverter in Fig. 5d. *A* is the input of the inverter. The nodes *V*, *G*, and *Y* correspond to *Vdd*, *Gnd* and *output* ($\overline{A}$). Transistor pairing, shown in Fig. 5d, is an important transistor placement technique used for layout area reduction. By transistor pairing, two inter-connected pFET and nFET are placed on the same column to minimize the routing complexity as well as to ensure more layout regularity. In Fig. 5e, we show transistor grouping, where the polarity gates of the stacked transistor are connected together. Transistor grouping is unique to ambipolar double-gate devices. In the following section we show the importance of grouping transistors for minimizing the routing overhead introduced by polarity gates.

3.2 Layout Technique for *Unate* and *Binate* Logic Gates

Unate logic functions (e.g. NAND, NOR, AOI, etc) with ambipolar devices are obtained by biasing the polarity gates (PGs) of the *pull-up-network* (PUN) and *pull-down-network* (PDN) to Gnd and Vdd respectively. Hence, all the transistors in the PUN (and PDN) can be grouped together (i.e. PGs of the stacked transistors are connected together), thereby forming one PG for each PUN and PDN. After biasing the PGs, CMOS layout style with transistors aligned according to the Euler paths can be employed [Uethara 81]. The transistors are placed in two parallel rows where all transistors in the PUN are in one row while all the transistors in the PDN are in the other. The main objective is to place transistors in such a way that the gate signals are aligned and drain/source regions of adjacent transistors are abutted. Figure 7a shows an example of a 2-input NAND gate with the PGs biased to either Gnd or Vdd. Figure 7(b,c) shows the final layout of the NAND gate and the dumbell-stick diagram.

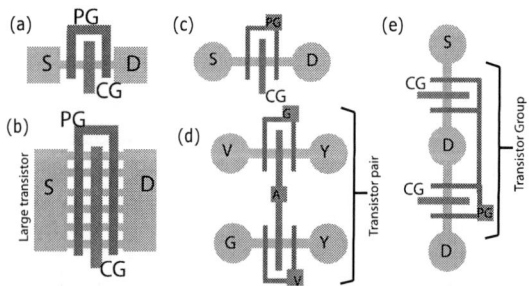

Figure 6 (a) A top view of the DG-SiNWFET shown in Fig. 1. (b) Large transistor. (c) Equivalent dumbell-stick diagram. (d) Dumbell-stick diagram of an Inverter with a transistor pair. (e) Grouping transistor with similar polarity gates.

*3-input XOR gate with 4 transistors is shown in Fig. 11.

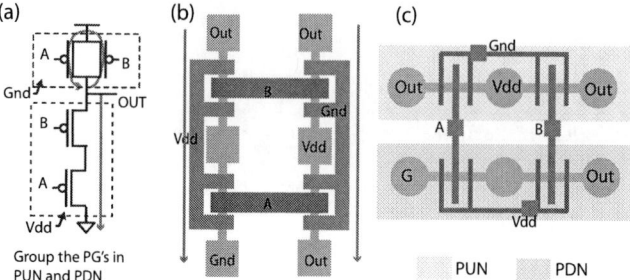

Figure 7. a) Schematic of a static NAND-2 gate by polarizing the ambipolar FET. b) Layout of the NAND gate. c) Dumbell-Stick diagram.

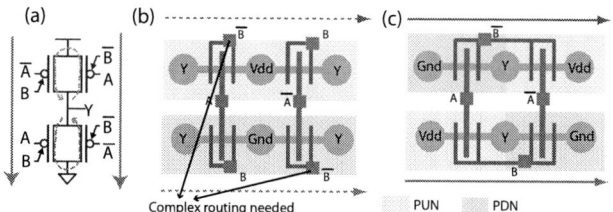

Figure 8. a) Schematic of a 2-input XOR [Jamaa 09] with various Euler paths (b) Dumbell-stick diagram drawn with CMOS style Euler paths (c) Dumbell-stick diagram drawn with Euler paths optimized for transistor grouping.

Efficient implementation of binate logic functions (e.g. XOR and XNOR) is possible by using the polarity gates of the ambipolar FETs as logic inputs. Using the transmission-gate transistor structure of [Jamaa 09], a 2-input (or a 3-input*) XOR gate can be constructed using only 4 transistors. An example of a 2-input XOR gate is shown in Fig. 7a, where all the polarity gates are either connected to logic input B or $\overline{B}$. Unlike for *unate* functions, the polarity gates in the PUN (and PDN) cannot be grouped. In Fig. 7b, we show a dumbell-stick diagram for a CMOS style layout. Since the adjacent transistors cannot be grouped, extra routing effort is needed to connect similar polarity gates together. An efficient implementation is shown in Fig. 7c, where similar polarity gates are grouped together. From the dumbell-stick diagram, we can observe that the PUN and PDN are placed next to each other, which is possible with DG-SiNWFET technology as the transistors are field controlled to make them p-type or n-type.

3.3 Layout Technique for Complex Logic Gates with an embedded XOR/XNOR

Several novel circuit designs and architectures have been proposed which leverage upon embedded XOR functionality of ambipolar logic [Jamaa 09], [De Marchi 10], [Zukoski 11]. In [De Marchi 10], authors have presented the idea of regular logic fabrics and evaluated various complex gates (combination of AND-XOR-OR-INV) based on the number of sub-functions each gate can implement. A key observation is that 2-input XOR/XNOR gates form the main building block of most logic cells. Hence in this work, we focus on layout techniques for complex functions with 2-input embedded XOR function.

Existing CMOS layout techniques have been devised for single-gate transistors and are not applicable to ambipolar transitor network as their polarity (p-type or n-type) changes with the input signals. Hence, modeling a complex gate by two

44

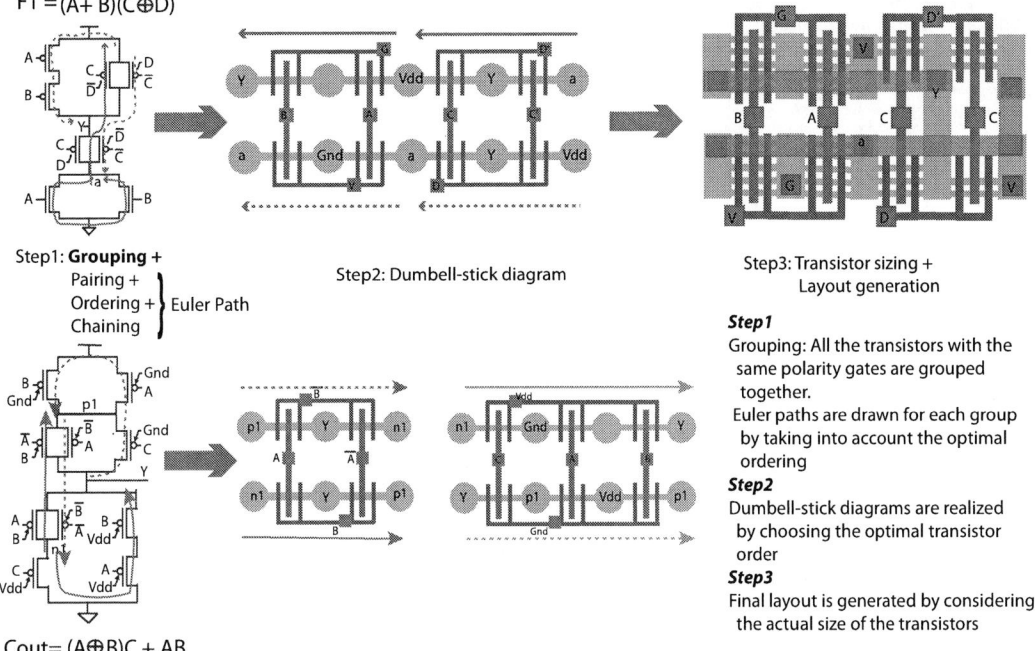

$F1 = \overline{(A+B)(C \oplus D)}$

Step1: **Grouping** +
Pairing +
Ordering +
Chaining
} Euler Path

Step2: Dumbell-stick diagram

Step3: Transistor sizing +
Layout generation

Step1
Grouping: All the transistors with the same polarity gates are grouped together.
Euler paths are drawn for each group by taking into account the optimal ordering

Step2
Dumbbell-stick diagrams are realized by choosing the optimal transistor order

Step3
Final layout is generated by considering the actual size of the transistors

Cout= (A⊕B)C + AB

Figure 9. Layout topology generation of complex gates with embedded XOR operation.

graphs, one for p-type devices and the other for n-type devices, is not feasible anymore.

Figure 9 illustrates the layout topology generation for a *carry-out* logic of a *full-adder* and a configurable regular fabric, *F1* [De Marchi 10]. The procedure is summarized here.

Step1: All the transistors with the same logic input on their polarity gates are grouped together. Dual groups are formed based on the complementary signals on the polarity gates. In the example of *carry-out* logic (Cout in Figure 9), the groups formed by polarity gates B and $\overline{B}$ are dual. Similarly the groups formed by polarity gates Vdd and Gnd are dual.

For each dual group, transistors are chained along the Euler paths. In Figure 9 Euler paths in red are shown for the dual group formed by B and $\overline{B}$, whereas the Euler paths in blue are related to the dual group Vdd and Gnd.

Step2: Dumbell-stick diagrams are derived from the Euler paths, as shown in the Fig. 8. In the case of the regular logic fabric, F1, a dumbbell-stick diagram without any discontinuity in the active area is achieved.

Step3: The final layout of the complex gate is generated from the dumbbell-stick diagram and by extracting the actual size of the transistors from the schematic.

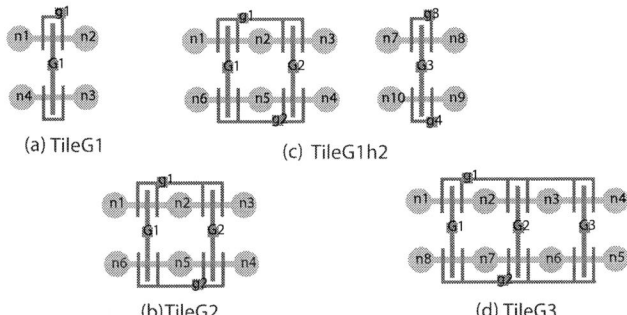

(a) TileG1

(c) TileG1h2

(b)TileG2

(d) TileG3

Figure 10. Dumbell-stick diagrams of various logic tiles considered for SoTs (a) TileG1 (b) TileG2 (c) TileG1h2 (d) TileG3.

4. Sea-of-Tiles (SoTs)

Regular layout fabrics have an advantage of higher yield as they maximize the layout manufacturability. In this work we propose a configurable *sea-of-tiles* (SoTs) architecture, in which an array of logic tiles are uniformly spread across the chip. Four different tiles, shown in Fig. 9, are considered in this work. $Tile_{G1}$, $Tile_{G2}$ and $Tile_{G3}$ are regular logic tiles, where $Tile_{G1h2}$ (hybrid tile) is a combination of $Tile_{G1}$ and $Tile_{G2}$.

4.1 Logic Tiles as Building Blocks

In the previous section, we have discussed on ensuring fine-grain regularity in the layouts by transistor pairing and transistor grouping. Transistor pairing helps in aligning the control gates of the complementary transistors in the PUN and PDN, whereas with transistor grouping polarity gates of adjacent transistors are connected together. By grouping the polarity gates of the adjacent transistors we can reduce the number of *input* pins to the connected fabric, *tile*.

We define a logic tile as an array of transistors, which are paired and grouped together. A $Tile_{Gn}$ is an array of n transistor-pairs grouped together. All the polarity gates of the top/bottom dumbell are connected together. This is the first step towards minimizing the intra-cell routing congestion. In the example of carry-out logic gate of a full-adder (see Fig. 8), $Tile_{G2}$ and $Tile_{G3}$ are employed to realize the gate. Similarly in the case of NAND and XOR (see Fig. 6 and Fig. 7) $Tile_{G2}$ forms the basic building block. Moreover, the technology facilitates in realizing these tiles with a high yield as the silicon nanowires are fabricated in groups.

Table 1. Various logic gates that can be realized by configuring the $Tile_{G2}$

Logic	n1	n2	n3	n4	n5	n6	G1	G2	g1	g2
XOR2	Gnd	Out	Vdd	Gnd	Out	Vdd	A	A'	B'	B
XNOR2	Gnd	Out	Vdd	Gnd	Out	Vdd	A	A'	B	B'
NAND2	Out	Vdd	Out	Out	-	Gnd	A	B	Gnd	Vdd
NOR2	Vdd	-	Out	Out	Gnd	Out	A	B	Gnd	Vdd
INV	Vdd	Out	Vdd	Gnd	Out	Gnd	A	A	Gnd	Vdd
BUF	O1	Vdd	Out	Out	Gnd	O1	A	O1	Gnd	Vdd

45

Table 2. Various logic gates that can be mapped by configuring the contacts and the input signals of the four tiles (#N – Number of tiles, and #UF – Utilization factor).

Gates	Tile$_{G1}$		Tile$_{G2}$		Tile$_{G1h2}$		Tile$_{G3}$	
	#N	#UF	#N	#UF	#N	#UF	#N	#UF
AND2	3	0.6	2	0.6	1	0.75	1	1
AND3	4	0.57	2	0.8	1.38	0.67	2	0.57
AOI21	3	0.6	2	0.6	1	0.75	1	1
AOI221	5	0.56	3	0.625	1.62	0.71	2	0.71
AOI222	6	0.54	3	0.75	2	0.67	2	0.86
AOI22	4	0.57	2	0.8	1.38	0.67	2	0.57
AOI321	6	0.54	3	0.75	2	0.67	2	0.86
BUF	2	0.66	1	1	0.62	1	1	0.67
INV	1	0.66	1	1	0.38	1	1	0.67
NAND2	2	0.66	1	1	0.62	1	1	0.67
NAND3	3	0.6	2	0.6	1	0.75	1	1
NAND4	4	0.57	2	0.8	1.38	0.67	2	0.57
NOR2	2	0.66	1	1	0.62	1	1	0.67
NOR3	3	0.6	2	0.6	1	0.75	1	1
NOR4	4	0.57	2	0.8	1.38	0.67	2	0.57
OAI21	3	0.6	2	0.6	1	0.75	1	1
OAI22	4	0.57	2	0.8	1.38	0.67	2	0.57
OR2	3	0.6	2	0.6	1	0.75	1	1
OR3	4	0.57	2	0.8	1.38	0.67	2	0.57
XNOR2	8	0.57	2	0.8	1.38	0.67	2	0.57
XNOR3	9	0.56	3	0.625	1.62	0.71	2	0.71
XOR2	8	0.57	2	0.8	1.38	0.67	2	0.57
XOR3	9	0.56	3	0.625	1.62	0.71	2	0.71

Figure 10b shows an un-mapped (not configured) $Tile_{G2}$. Various logic functions can be realized by connecting the nodes ($n1$-$n6$) and gates ($g1$, $g2$, $G1$ and $G2$) to appropriate inputs. Table 1 lists various logic functions that can be realized with a single $Tile_{G2}$. However, complex logic functions can be obtained by considering an array of $Tile_{G2}$.

Figure 10 shows four tiles that we consider for the sea-of-tiles architecture. $Tile_{G1}$ is the simplest tile with only one pair of transistors. An array of $Tile_{G1}$ is similar to sea-of-gates. Any Boolean logic function can be mapped on to an array of $Tile_{G1}$. The flexibility of building generic logic gates comes at a cost of area. Moreover, providing access to each and every polarity gate increases the intra-cell routing (Metal1 and Metal2 routing) complexity. $Tile_{G2}$ and $Tile_{G3}$ include two and three transistor pairs, respectively, grouped together. A hybrid tile $Tile_{G1h2}$ is a combination of $Tile_{G1}$ and $Tile_{G2}$, whose polarity gates are not connected. This gives the flexibility of utilizing a part of a tile, when remained un-mapped, by functions with low area utilization. For example, a NAND2 gate when mapped onto a $Tile_{G1h2}$ requires only the segment of a tile with gates G1 and G2. The unmapped part of the tile with gate G3 can be employed either to map an *inverter* or to increase the drive strength of the gate. In Table 2, we report various logic gates that can be configured with the 4 tiles we have considered. The number of tiles required for each gate and their respective area utilization is also presented. It has to be noted that we also consider extra logic needed for generating inverted inputs. For example in the case of XOR2/XNOR2, we have discussed in section 3.2 about realizing with only one $Tile_{G2}$. In the case we use single-rail logic, we take an extra $Tile_{G2}$ for generating the two negated inputs ($\overline{A}$ and $\overline{B}$).

4.2 Optimal Tile: Simulations and Result

In this work we compare four tiles for an efficient implementation of the SoT architecture. Our main objective is to find the best tile, which gives highest area utilization for various benchmarks. Though the techniques presented in this paper are linked to the ambipolar SiNWFETs, the concepts can be extended to all the technologies contending for ambipolar logic.

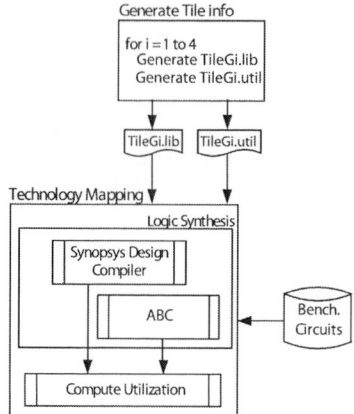

Figure 11. Design flow for finding the best Tile for SoT.

Figure 11 shows our design flow. As a first step, for every tile ($Tile_{Gi}$) we generate a list of logic gates that can be mapped on to it (TileGi.lib) and their respective utilization factor (TileGi.util). Utilization factor takes only the active area into account. For example $NAND2$ when mapped onto a $Tile_{G1}$ has a utilization factor of 0.66, whereas when mapped onto a $Tile_{G2}$ it has a utilization factor of 1. It has to be noted that the number of logic gates that can be mapped to different tiles vary. For technology mapping, we used Synopsys design compiler [DC] and ABC [ABC] synthesis tools to benchmark various circuits.

Table 3 summarizes the results of various benchmark circuits after technology mapping. We report total area utilization for each benchmark when mapped onto four different tiles ($Tile_{G1}$, $Tile_{G2}$, $Tile_{G1h2}$, and $Tile_{G3}$). Technology mapping only uses the cells that are associated with each tile (shown in Table 2). Both the synthesis tools were run with different delay constraints. Area utilization for a benchmark circuit is calculated from the total count of each cell and their respective utilization factors. We did not run simulations to study power and delay, as we assume all the tiles with SiNWFETs. Since we map the same netlist onto four different SoTs, it is reasonable to believe that they have the same delay characteristics.

Examining the results for the four logic tiles, we see that SoT with tiles $Tile_{G1h2}$ (and, $Tile_{G2}$) have a higher area efficiency, 10% (8%) and 16% (14%), when compared to SoT with $Tile_{G1}$ and $Tile_{G3}$, respectively. Though $Tile_{G3}$ and $Tile_{G1h2}$ have the same number of transistors per tile, the hybrid tile outperforms $Tile_{G3}$ with 10% improvement in area efficiency.

Embedded XOR functionality is one of the key features of ambipolar logic gates. With a transmission-gate transistor structure [Jamaa 09], a 2-input and a 3-input XOR/XNOR gate can be constructed using only 4 transistors. In Fig. 11, we show how $Tile_{G2}$ can be the most effective layout possible. In Figure

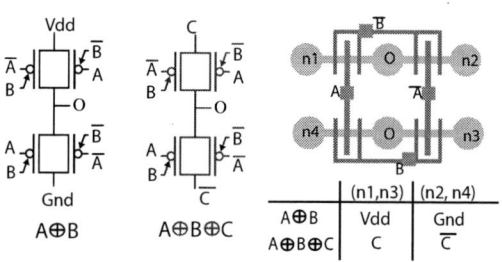

Figure 12. Schematic of a 2-input and 3-input XOR along with the mapping on to a $Tile_{G2}$.

Table 3. Normalized area of various benchmarks when mapped onto a SoT with Tile$_{G1}$, Tile$_{G2}$, Tile$_{G1h2}$, and Tile$_{G3}$ using design compiler [DC] and [ABC].

Bench.	Tile$_{G1}$		Tile$_{G2}$		Tile$_{G1h2}$		Tile$_{G3}$	
	DC	ABC	DC	ABC	DC	ABC	DC	ABC
Dalu	1968	2558	1728	2235	1689	2115	1808	2548
Add64	3946	3004	3693	2664	3483	2483	3560	2740
C5315	4072	5404	3465	4791	3422	4477	3984	5088
C7552	4914	5606	4188	5001	4150	4653	4752	5456
i10	5964	6350	5034	5634	4790	5286	5452	6232
C1908	1132	1778	936	1518	942	1469	1116	1692
C3540	2940	3436	2517	3033	2486	2859	2756	3184
C6288	8462	9336	7227	8253	7373	7744	7580	8000
Des	9392	12482	8142	10623	7910	10323	9016	11912
Average	1	1	0.86	0.87	0.85	0.83	0.94	0.94

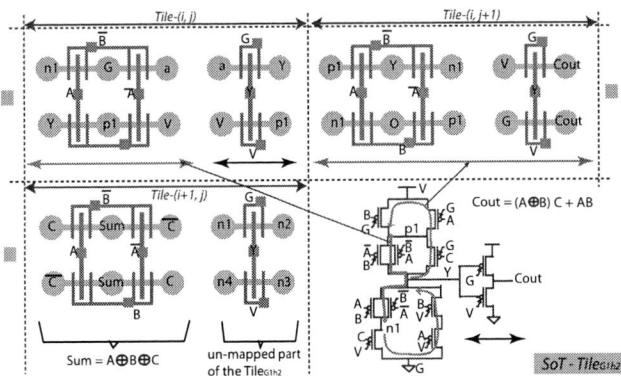

Figure13. A Full-adder mapped on to a Sea-of-Tiles with the hybrid tile *Tile$_{G1h2}$* as the basic building block.

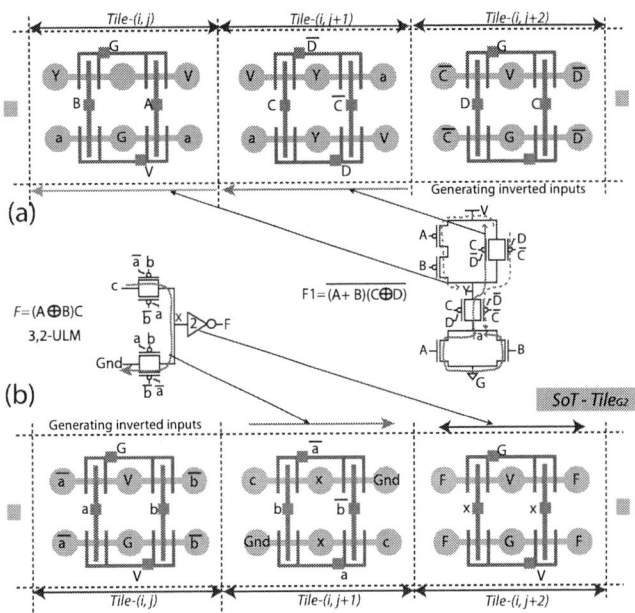

Figure14. Reconfigurable fabrics mapped on to SoT with TileG2(a) Regular computation fabric [Demarchi 10] (b) Universal logic module (3,2-ULM) [Zukoski 11].

13 we show dumbell-stick diagrams of both the *sum* (Sum) and *carry-out* (Cout) logic of a *full-adder*, mapped onto a SoT with *Tile$_{G1h2}$*. The Sum, which is a 3-input XOR of inputs A, B and C, is mapped on to a *Tile-(i+1,j)* of the entire array. The unmapped part of the *Tile-(i+1,j)* can be employed for realizing either an inverter logic gate or can be a part of the neighboring logic gate. Similarly the Cout is mapped on to 2 tiles *Tile-(i,j)* and *Tile-(i,j+1)*.

Several novel reconfigurable blocks have been proposed which leverage upon embedded XOR functionality of ambipolar logic In Figure 14, we demonstrate how a computational fabric (F1) [De Marchi 10] and a universal logic module (3,2-ULM) [Zukoski 11] can be mapped onto a SoT of Tile$_{G2}$. Inverted inputs, for a 2-input XOR functions, are generated with a single tile (*Tile-(i,j)* for 3,2-ULM and *Tile-(i,j+2)* for F1).

With all the three examples, we demonstrate how tiles, *Tile$_{G1h2}$* and *Tile$_{G2}$*, can be the fundamental building blocks for future ambipolar logic circuits.

5. Conclusion

Double-gate SiNWFETs, with an extra polarity gate, are promising contenders for efficient implementation of ambipolar logic [Jamaa 09]. In this work, we present an approach for designing an efficient regular layout fabric, called Sea-of-Tiles. In order to facilitate design, we propose a compact layout technique and novel symbolic-layout styles for ambipolar logic gates. We show that SoT with tiles *TileG$_{1h2}$* and *TileG$_2$*, on an average, outperform the one with *TileG$_1$* and *TileG$_3$* by 16% and 10% in area utilization, respectively. We envisage *TileG$_2$* or *TileG$_{1h2}$* to be the basic building block for the future ambipolar logic circuits.

Acknowledgment

This research was supported by ERC-2009-AdG-246810. The authors would like to thank Davide Sacchetto for his help in fabrication and characterizing the double-gate SiNW transistors.

6. References

[ABC] "Abc system for logic synthesis,"

[DC] "Synopsys design compiler," 2010.

[De Marchi 10] De Marchi, M., *et al.*, "Synthesis of regular computational fabrics with ambipolar CNTFET technology," *IEEE Proc. ICECS*, pp.70-73, Dec. 2010

[Doyle 03] Doyle, B.S., *et al.*, "High performance fully-depleted tri-gate CMOS transistors," *Electron Device Letters, IEEE* , vol.24, no.4, pp. 263- 265, April 2003.

[Hwang 90] Hwang, C.-Y., *et al.*, "A fast transistor-chaining algorithm for CMOS cell layout," *IEEE Trans. CAD*, pp.781-786, Jul 1990.

[Hisamoto 00] Hisamoto, D., *et al.*, "FinFET-a self-aligned double-gate MOSFET scalable to 20 nm," *IEEE Trans. Electron Devices*, pp. 2320- 2325, Dec 2000.

[Jamaa 08] Jamaa, M.H.B., *et al.*, "Programmable logic circuits based on ambipolar cnfet," *ACM/IEEE Proc. DAC*, 2008.

[Jamaa 09] Jamaa, M.H.B., *et al.*, "Novel library of logic gates with ambipolar cntfets: opportunities for multi-level logic synthesis," IEEE *Proc. DATE '09*.

[Lin 05] Y.-M. Lin, *et al.*, "High-performance carbon nanotube field-effect transistor with tunable polarities," *IEEE Trans*, pp. 481–489, September 2005.

[Lin 09] Yi-Wei Lin, Malgorzata Marek-Sadowska, and Wojciech Maly. "Transistor-level layout of high-density regular circuits", *Proc. ISPD, 2009.*

[Ng 07] Ng, R., Wang, T., and Chan, M., "A new approach to fabricate vertically stacked single-crystalline silicon nanowires," *IEEE Proc. EDSSC*, 2007.

[O'Connor 07] O'Connor, I., *et al.*, "Ultra-fine grain reconfigurability using cntfets," *Proc. ICECS*, dec 2007.

[Ran 06] Y. Ran and M. Marek-Sadowska, "Designing via-configurable logic blocks for regular fabric," *IEEE Trans. VLSI*, vol. 14, no. 1, pp. 1 –14, jan. 2006.

[Sacchetto 09] Sacchetto, D., *et al.*, "Fabrication and Characterization of Vertically Stacked Gate-All-Around Si Nanowire FET Arrays," *Proc. ESSDERC*, 2009.

[Suk 05] Suk, S. D., *et al.*, "High performance 5nm radius twin silicon nanowire mosfet (tsnwfet): fabrication on bulk si wafer, characteristics, and reliability," *IEEE Proc.*, IEDM, dec. 2005, pp. 717 –720.

[Taylor 07] B. Taylor and L. Pileggi, "Exact combinatorial optimization methods for physical design of regular logic bricks," *Proc. DAC,* 2007, pp. 344–349.

[Tejas 07] Tejas, J., *et al.*, "Maximization of layout printability/manufacturability by extreme layout regularity," J. Micro/Nanolith. MEMS, 2007.

[Uethara 81] Uehara, T., and Vancleemput, W., "Optimal layout of cmos functional arrays," *IEEE Trans. CAD*, pp. 305 –312, may 1981.

[Zukoski 11] Zukoski, A., Xuebei Yang., Mohanram, K., "Universal logic modules based on double-gate carbon nanotube transistors," *Proc. DAC*, pp.884-889, 2011.

A Semiempirical Model for Wakeup Time Estimation in Power-Gated Logic Clusters

Vivek D. Tovinakere
vivektd@irisa.fr

Olivier Sentieys
sentieys@irisa.fr

Steven Derrien
sderrien@irisa.fr

INRIA/IRISA, University of Rennes 1
22300 Lannion, France

ABSTRACT

Wakeup time is an important overhead that must be determined for effective power gating, particularly in logic clusters that undergo frequent mode transitions for run-time leakage power reduction. In this paper, a semiempirical model for virtual supply voltage in terms of basic parameters of the power-gated circuit is presented. Hence a closed-form expression for estimation of wakeup time of a power-gated logic cluster is derived. Experimental results of application of the model to ISCAS85 benchmark circuits show that wakeup time may be estimated within an average error of 16.3% across 22× variation in sleep transistor sizes and 13× variation in circuit sizes with significant speedup in computation time compared to SPICE level circuit simulations.

Categories and Subject Descriptors

B.7.2 [**Integrated Circuits**]: Design Aids

General Terms

Algorithms, Design, Performance

Keywords

Design automation, leakage current, power gating, wakeup time

1. INTRODUCTION

Power gating has emerged as an important technique to minimize static power and energy consumption in CMOS circuits [3]. As MOSFETs are scaled down to sub-100nm dimensions, an exponential increase in subthreshold leakage current is observed due to the reduction in threshold voltage (V_{th}) to maintain gate overdrive [9]. In ultra-low power circuits with constrained energy budgets, energy consumption due to static currents may dominate its dynamic counterpart for low duty cycle operation. Hence circuit techniques for power gating structures and exploration of power gating

opportunities for automated design of power-gated circuits have received significant attention.

A power gating structure cuts-off bias voltages for MOS devices so that bias-dependent leakage current in the logic circuit reduces significantly in standby state. A simple power-gated circuit shown in Fig. 1 and used in this work consists of a high-V_{th} PMOS sleep transistor connected between power supply rail (V_{dd}) and virtual power supply node, Virtual-Vdd (V_{Vdd}) of the logic cluster. A cluster refers to an ensemble of connected logic gates power-gated by a sleep transistor. The gate terminal of the sleep transistor is connected to a control signal $SLEEP$, to switch the sleep transistor between on and off states. A power-gated circuit operates in three modes in a typical power gating cycle as shown in Fig. 2. When $SLEEP$ is high, power supply to the logic is cutoff; V_{Vdd} decreases and the circuit is said to be in sleep mode. The leakage current decreases exponentially with V_{Vdd} resulting in energy savings. When $SLEEP$ is low, current flows through the sleep transistor to charge circuit capacitances. Due to charging effect, V_{Vdd} increases until it reaches a steady state value less than V_{dd}. We refer to this mode of operation as wakeup mode and the mode of operation after wakeup as active mode.

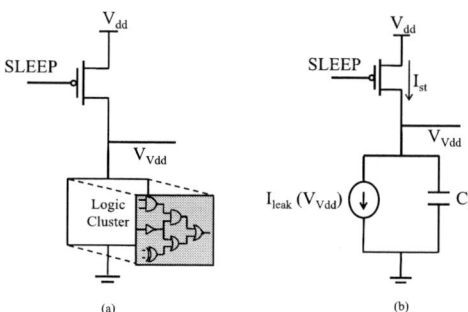

Figure 1: (a) Power-gated logic cluster of header type (b) Equivalent circuit of logic cluster

In this paper, a semiempirical model for V_{Vdd} based on polynomial representation of leakage current in a logic cluster and linear region resistance of sleep transistor is presented. A method to estimate steady-state Virtual-Vdd voltage after wakeup mode using leakage current profiles of constituent logic gates is described. Further, a closed-form expression is derived for estimation of wakeup time of the power-gated circuit. The model for Virtual-Vdd in sleep mode can be used to determine energy savings due to power

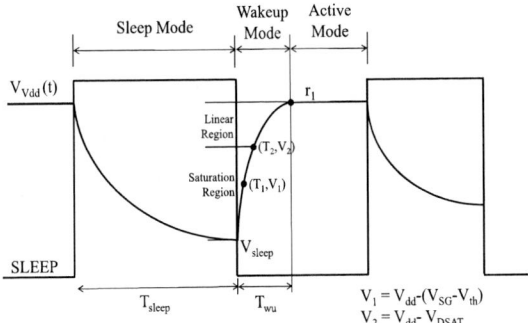

Figure 2: Typical timing instants and modes of operation in a power gating cycle

gating. In other words, some of the key design parameters have been captured in a single model.

The paper is organized as follows. Section 2 gives an overview of related work. Section 3 describes the equivalent circuit of a power-gated logic cluster. In Section 4, the semiempirical approach to estimation of wakeup time is described. The experimental results are presented in Section 5 to validate the model. Section 6 concludes the paper.

2. RELATED WORK

Several works have viewed design of power-gated circuits as an optimization problem of partitioning logic into clusters satisfying constraints of peak current, delay degradation, sleep transistor area, wakeup time and energy savings. Wakeup time estimation is fundamental to logic clustering algorithms proposed in [1], [2] and [6] that require constraints of peak current and wakeup time to be satisfied. In this context, a simple analytical model for estimation of wakeup time is useful especially when it needs to be determined iteratively during an optimization run for a number of candidate clusters. Run-time leakage reduction has been explored in [4] and [10], where only parts of the overall circuit are put to sleep during short periods of inactivity. Along with a high wakeup energy overhead a large wakeup delay in a cluster can result in reduced energy savings. Hence these two parameters have to be carefully considered in the design of power-gated circuits. The problem of wakeup time estimation arises in other scenarios as well. In [5] and [11], a need for wakeup latency estimation arises to quantify the effectiveness of proposed ground-bounce reducing techniques and intermediate strength power gating techniques respectively under a wakeup time constraint. In [14], Xu *et al.* have proposed numerical approaches for estimation of V_{Gnd} as a function of time in sleep mode. To extend the same method to wakeup mode, it is necessary to incorporate size dependent sleep transistor current characteristics. In this case an analytical model for V_{Gnd} would be highly desirable. This analysis is identically applicable to V_{Vdd} in a power-gated cluster with a header type of sleep transistor. Most works have used SPICE simulations or constant current source model [11] for sleep transistors to determine wakeup time. A simple expression for wakeup time based on a restrictive assumption of limited range of transistor operation was proposed in [13]. In this paper, the models are derived taking into account all the regions of sleep transistor operation necessary for applying them to large logic clusters consisting of high leakage cells.

3. POWER-GATED LOGIC CLUSTER MODEL

Models for subthreshold leakage current that capture its exponential behaviour with bias voltages at device level have been described in [12]. In [14], compact models for leakage current have been derived at gate and circuit levels in a hierarchical way. It was shown that the leakage current can be represented by a voltage controlled current source (VCCS) as in Fig. 1(b). In this work, we take a polynomial based approach to derive leakage current profile for the complete circuit. For each type of cell S_i and input pattern j, leakage current is determined at several voltages and the resulting profile is fitted with a polynomial of degree N in V_{Vdd} as given by

$$I_{leak}(S_i, j) = \sum_{k=0}^{N} b_k(S_i, j) V_{Vdd}^k \qquad (1)$$

where $\{b_k(S_i, j)\}$ represents coefficients of the polynomial. We assume a standard-cell based design approach for implementation of the cluster. Therefore, the total static current for $n(S_i, j)$ occurrences of each cell and each input pattern is obtained as

$$I_{leak} = \sum_{i=0}^{P-1} \sum_{j=0}^{R_i-1} n(S_i, j) I_{leak}(S_i, j) \qquad (2)$$

where P and R_i are number of types of cells and number of possible input combinations for cell S_i respectively. As an example, if a logic cluster is composed of a set $S = \{$nand2, inv, nor2, xor2$\}$ of gates, then $P = 4$. For a 2-input NAND gate $R_i = 4$, whereas for an inverter, $R_i = 2$. For notational simplicity, the total leakage current profile of the logic cluster is represented by

$$I_{leak} = \sum_{i=0}^{N} b_i V_{Vdd}^i \qquad (3)$$

in the rest of the paper. Equation (3) has the form of nonlinear resistance. Let C_{ij} denote total input capacitance of S_i with pattern j. The total capacitance of the logic cluster is derived as the sum of capacitances of all the inputs of constituent standard cells as below.

$$C_L = \sum_{i=0}^{P-1} \sum_{j=0}^{R_i-1} n(S_i, j) C_{ij} \qquad (4)$$

4. VIRTUAL-VDD MODEL

4.1 Determination of Steady-State Virtual-Vdd Voltage

Consider the equivalent circuit model in Fig. 1(b). In the wakeup mode, the operating point on the I_{SD} vs. V_{SD} characteristics of sleep transistor moves from saturation region to linear region until V_{Vdd} reaches a steady-state value. The virtual supply node is said to be in steady state when $dV_{Vdd}/dt = 0$, i.e., when there are no changes in V_{Vdd} either on account of short-circuit currents due to changing logic states of internal nodes or due to charging effect. Let the current through the sleep transistor during wakeup and in non-saturation region be denoted by $I_{st,ns}$, the total leakage current at the output of VCCS by I_{leak} and the capacitive load charging current by I_{load}. Then,

$$I_{st,ns} = I_{leak} + I_{load}. \qquad (5)$$

The current through the sleep transistor in non-saturation region is given by the quadratic model

$$I_{st,ns}(t) = \frac{1}{R_{lin}} \left[(V_{dd} - V_{Vdd}(t)) - \frac{(V_{dd} - V_{Vdd}(t))^2}{2(V_{dd} - V_{th})} \right] \quad (6)$$

where R_{lin} is the resistance in linear region. The determination of R_{lin} is described in subsection 4.4. From (3), (6) and $I_{load} = C_L(dV_{Vdd}/dt)$, (5) becomes

$$\frac{dV_{Vdd}}{dt} = -\frac{1}{\tau} \sum_{i=0}^{N} c_i V_{Vdd}^i \qquad (7)$$

where $\tau = R_{lin}C_L$ and $c_i = f_i(V_{dd}, R_{lin}, b_i, V_{th})$ are expressions derived from (3)-(6). To solve for V_{Vdd}, the Nth degree polynomial in (7) is reduced to a quadratic polynomial by least-squares approximation and is expressed in terms of its roots r_1 and r_2 as

$$\frac{dV_{Vdd}}{dt} = -\frac{1}{\tau}(V_{Vdd} - r_1)(V_{Vdd} - r_2). \qquad (8)$$

Both r_1 and r_2 are steady state points of (8). One of the roots r_1 satisfying the interval of validity $V_{sleep} < r_1 < V_{dd}$, is determined to be the steady state Virtual-Vdd voltage. Here V_{sleep} denotes the value of V_{Vdd} at the wakeup transition. In a RC circuit the steady state as defined above is reached at $t = \infty$. However the error in assuming value of V_{Vdd} at onset of active mode to be r_1 is negligible as demonstrated in Section 5.

4.2 Wakeup Mode Virtual-Vdd Model

In order to obtain a model for $V_{Vdd}(t)$ in wakeup mode, the ordinary differential equation in (8) is solved in the non-saturation region and hence, is extended to saturation region by means of approximations. Let at time $t = 0$ the operating point move to non-saturation region so that $V_{Vdd}(0) = V_{initial}$. The solution of (8) satisfying the interval of validity and moving towards r_1 can be written as

$$[V_{Vdd}(t)]_{ns} = \frac{r_1 - r_2 K e^{-at}}{1 - K e^{-at}} \qquad (9)$$

where $K = (V_{initial} - r_1)/(V_{initial} - r_2)$, $a = 1/A\tau$ and $A = 1/(r_1 - r_2)$. From Fig. 2, $V_{initial} = V_{dd} - V_{DSAT}$ where V_{DSAT} is the saturation voltage.

To extend the model to saturation region, the time instant $t = 0$ is moved to sleep-to-wakeup mode transition so that $V_{Vdd}(0) = V_{sleep}$. Let T_{wu} denote the wakeup time defined as the time taken for V_{Vdd} to evolve from V_{sleep} to $0.99r_1$. Further, let V_1 and V_2 be two voltage levels attained by V_{Vdd} at T_1 and T_2 respectively as shown in Fig. 2. The solution (9) does not represent V_{Vdd} in the saturation region, $V_{Vdd} < V_2$ accurately. Therefore corrections are applied to (5) in the first two segments as

$$I_{st}(t) = I_{leak} + I_{load} - \Delta I_0(t) + \Delta I_1(t). \qquad (10)$$

In (10) the time instant $t = 0$ corresponds to sleep-to-wakeup mode transition and $V_{initial} = V_{sleep}$. Let U_T denote the time-shifted unit step function $u(t - T)$. We define

$$\Delta I_0(t) = I_0 \left[U_0 e^{-at} - U_{T_1} e^{-a(t-T_1)} \right] \qquad (11)$$

$$\Delta I_1(t) = I_1 \left[U_{T_1} e^{-a(t-T_1)} - U_{T_2} e^{-a(t-T_2)} \right] \qquad (12)$$

based on heuristics for I_0 and I_1 described in subsection 4.5. In the third interval (9) alone is satisfied and hence no correction is required. Using (9)-(12), the model for Virtual-Vdd in wakeup mode can be derived as

$$V_{Vdd}(t) = \frac{r_1 - r_2 K e^{-at}}{1 - K e^{-at}} + A I_0 R_{lin} \left[U_0 \left(1 - e^{-at} \right) \right] \quad (13)$$
$$\quad - A R_{lin}(I_0 + I_1) \left[U_{T_1} \left(1 - e^{-a(t-T_1)} \right) \right]$$
$$\quad + A R_{lin} I_1 \left[U_{T_2} \left(1 - e^{-a(t-T_2)} \right) \right].$$

In compact form, (13) can be written as $X e^{-2at} + Y e^{-at} + Z = 0$. The solution for t is given by

$$t = \frac{1}{a} \ln \left(\frac{2X}{-Y - \sqrt{Y^2 - 4XZ}} \right). \qquad (14)$$

At $t = T_1$, $V_{Vdd} = V_{dd} - V_{SG} + V_{th}$ corresponding to the criterion $V_{SD} = V_{SG} - V_{th}$. Applying this condition and $V_{SG} = V_{dd}$, X, Y and Z are determined to be

$$\begin{cases} X = A R_{lin} K I_0, \\ Y = K(V_{th} - r_2) - (1 + K)X/K, \\ Z = A I_0 R_{lin} + r_1 - V_{th}. \end{cases} \qquad (15)$$

Similarly for $t = T_2$, $V_{Vdd} = V_{dd} - V_{DSAT}$ corresponding to the condition $V_{SD} = V_{DSAT}$, which gives

$$\begin{cases} X = -A R_{lin} K[I_0(e^{aT_1} - 1) + I_1 e^{aT_1}], \\ Y = K(V_{dd} - V_{DSAT} - r_2) + (X/K) + A R_{lin} I_1 K, \\ Z = -A R_{lin} I_1 + r_1 - V_{dd} + V_{DSAT}. \end{cases} \qquad (16)$$

For wakeup time T_{wu}, $V_{Vdd} = 0.99r_1$, which gives

$$\begin{cases} X = A R_{lin} K[-I_0(e^{aT_1} - 1) + I_1(e^{aT_2} - e^{aT_1})], \\ Y = K(0.99r_1 - r_2) - (X/K), \\ Z = 0.01r_1. \end{cases} \qquad (17)$$

The values of r_1 and r_2 are unaffected due to ΔI_0 and ΔI_1 as (13) satisfies $V_{SD} = V_{DSAT}$ at $t = T_2$ as shown in Fig. 2. For clusters with $0.99r_1 \leq (V_{dd} - V_{DSAT})$, wakeup time $T_{wu} = T_2$ determined from (16) and with the condition $V_{Vdd} = 0.99r_1$.

4.3 Sleep Mode Virtual-Vdd Model

To calculate T_1, T_2 and T_{wu} using (15)-(17) it is necessary to determine V_{sleep}. If the cluster is in sleep state for a time interval T_{sleep}, then $V_{sleep} = V_{Vdd}(T_{sleep})$. It should be noted that for simplicity both mode transitions are assumed to occur at $t = 0$, so that the initial condition for sleep mode can be denoted by $V_{Vdd}(0)$ as for wakeup mode. In sleep mode, the sleep transistor is cut-off so that only a leakage current $I_{st,leak}$ flows through it. I_{load} in (5) is now a discharging current. Hence (5) for sleep mode can be written as

$$-C_L \frac{dV_{Vdd}}{dt} = -(b_0 - I_{st,leak}) - \sum_{i=1}^{N} b_i V_{Vdd}^i. \qquad (18)$$

For each value of V_{Vdd}, the VCCS outputs a current given by (3). Therefore for each value of V_{Vdd} we infer that resistance of the circuit is given by $R_s(V_{Vdd}) = (V_{Vdd}/\sum_{i=0}^{N} b_i V_{Vdd}^i)$.

We refer to R_s as pseudo-resistance in the rest of the paper. Neglecting $I_{st,leak}$ and rewriting (18) similar to (7),

$$\frac{dV_{Vdd}}{dt} = -\frac{1}{R_s(V_{Vdd})C_L}\left[-R_s(V_{Vdd})\sum_{i=0}^{N}b_iV_{Vdd}^i\right]. \quad (19)$$

A numerical solution to (19) is of the form [14]

$$V_{Vdd,j+1} = V_{Vdd,j}e^{-\frac{\Delta t}{R_s(V_{Vdd,j})C_L}} \quad (20)$$

where j denotes a time interval in $[0, T_{sleep}]$ of size Δt. To develop an approximation, we consider a heuristic for choice of R_s as explained in subsection 4.6. Denoting R_{sp} as the pseudo-resistance chosen by applying the heuristic, the model for Virtual-Vdd in sleep mode can be derived as

$$\frac{dV_{Vdd}}{dt} = -\frac{1}{R_{sp}C_L}\prod_{i=1}^{N}(V_{Vdd} - r_i^s) = 0 \quad (21)$$

where r_i^s represents roots of the polynomial in sleep context. Let r_1^s satisfy $r_1 < r_1^s < 0$. Then the approximate solution that moves towards r_1^s from its initial value is given by

$$V_{Vdd}(t) = r_1^s + e^{-\frac{t}{R_{sp}C_L} + K^s}. \quad (22)$$

At the end of active mode, the value of Virtual-Vdd satisfies $r_1 - \Delta V_{Vdd,max} \leq V_{Vdd} \leq r_1$ where $\Delta V_{Vdd,max}$ is the maximum degradation of V_{Vdd} due to dynamically changing inputs of logic cluster. In this work, it is assumed that the power-gated logic cluster remains in active mode for a duration long enough with appropriate input conditions that $V_{Vdd} = r_1$ at the end of active mode. This assumption is mostly true in circuits with adequate positive timing slack. Hence, applying the initial condition that $V_{Vdd}(0) = r_1$, we have

$$V_{Vdd}(t) = r_1^s + (r_1 - r_1^s)e^{-\frac{t}{R_{sp}C_L}}. \quad (23)$$

The value of V_{Vdd} at the end of sleep mode, V_{sleep}, is obtained by substituting $t = T_{sleep}$ in (23).

The energy savings E_s of the power-gated logic cluster in sleep mode with respect to an ungated cluster can be determined by

$$E_s = V_{dd}I_{leak}(V_{dd})T_{sleep} - \int_0^{T_{sleep}} V_{Vdd}I_{leak}(V_{Vdd})dt. \quad (24)$$

4.4 Determination of R_{lin}

To determine the resistance of sleep transistor in linear region, the method proposed in [7] for extraction of series-resistance (R_{sd}) of MOS device is followed. It is described here for completeness. Two operating points $(I_{SD}^{(1)}, V_{SG}^{(1)}, V_{th}^{(1)})$ and $(I_{SD}^{(2)}, V_{SG}^{(2)}, V_{th}^{(2)})$ with $V_{SD} = 0.05V$ are determined from I_{SD} vs. V_{SG} characteristics for a specific width W_{sp} of the transistor. All V_{SG} are chosen such that they satisfy constant mobility condition [7][12] while V_{th} is determined by g_m/I_D method. The drain current $I_{SD}^{(i)}$, for $i = 1, 2$, including the effects of R_{sd} is given by

$$I_{SD}^{(i)} = \mu C_{ox}\frac{W_{eff}}{L_{eff}}\left(V_{SG}^{(i)} - V_{th}^{(i)} - 0.5V_{SD}\right)\left(V_{SD} - R_{sd}I_{SD}^{(i)}\right). \quad (25)$$

Here μ is the constant carrier mobility, $C_{ox} = \epsilon_{ox}/t_{ox}$ is the oxide capacitance, W_{eff} and L_{eff} are effective width and channel length of sleep transistor. From the pair of equations (25), R_{sd} is determined. Further μ is determined from one

of the equations of drain current in (25). Let R_{ch} denote the intrinsic channel resistance. Then $R_{lin} = R_{ch} + R_{sd}$. From [12],

$$R_{lin} = R_{sd} + \left(\frac{L_{eff}}{\mu C_{ox}W_{eff}(V_{SG} - V_{th} - 0.5V_{SD})}\right) \quad (26)$$

Table 1 shows linear region resistances for PMOS sleep transistors of different sizes in an industrial 65nm bulk CMOS technology library with nominal $V_{dd} = 1V$.

Table 1: Linear Region Resistance of PMOS Transistors (at $100°C$, $L = 0.06\mu m$, $V_{SG} = 1V$, $V_{SD} = 0.05V$)

W (μm)	0.54	1.2	2.4	4.8	9.6	12
R_{lin} ($k\Omega$)	2.57	1.203	0.612	0.322	0.167	0.134

4.5 Heuristics for I_0 and I_1

Correction terms in (11) and (12) were applied in (10), to account for saturation region of sleep transistor operation. From (6) the current in saturation region is underestimated by $I_0 = I_{on,sat} - I_{st}(V_{Vdd} = V_{sleep})$ where $I_{on,sat}$ is the saturation drain current. Fig. 3 shows the variation of error in width-normalized estimated drain current $\frac{I_{error}}{W} = \frac{1}{W}(I_{SD} - I_{st})$ with V_{SD} where I_{st} is as determined from (6) for all V_{Vdd}. Similarly for I_1, we choose error in current corresponding to one of the values of V_{Vdd} in the interval $[(V_{dd} - V_{SG} + V_{th}), (V_{dd} - V_{DSAT})]$. From our experiments, we empirically choose $V_{SD} = 0.6V$, at which the error determined from Fig. 3 is $-I_1 = 0.174I_0$.

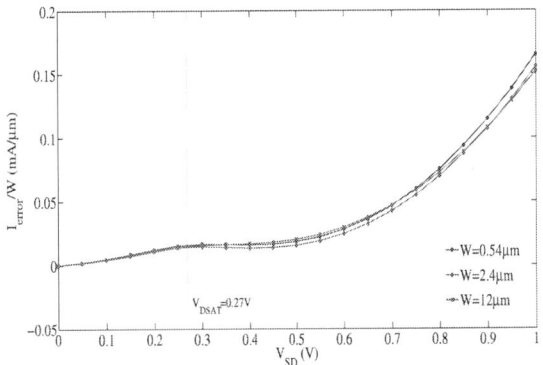

Figure 3: I_{error}/W vs. V_{SD} for 65nm PMOS transistors

4.6 Heuristic for R_{sp}

The voltage dependent pseudo-resistance changes as V_{Vdd} evolves with time according to (22). Hence it can be inferred that the time constant R_sC_L also varies with time. In our experiments, we have observed that in large logic clusters, the values of pseudo-resistance and its dynamic range are less than that for small logic clusters as leakage currents are higher in the former case. A typical variation of pseudo-resistance with V_{Vdd} is shown in Fig. 4 in the next section. The effect of a larger value of pseudo-resistance on V_{Vdd} is that it takes a longer time to change V_{Vdd} levels than with smaller values. Typically, higher values of pseudo-resistance determine V_{Vdd} after about 4 time constants of sleep time. Considering these observations, we choose R_{sp} as the pseudo-resistance at $V_{Vdd} = r_1$.

51

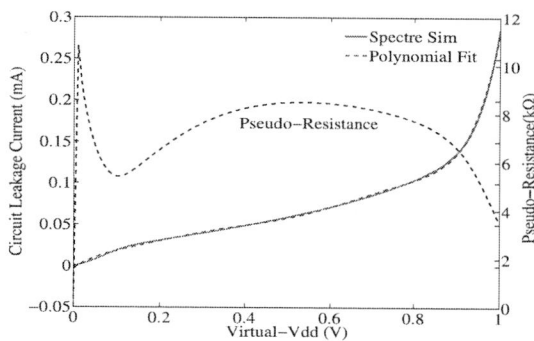

Figure 4: Leakage current and pseudo-resistance profile in c6288

5. EXPERIMENTAL RESULTS

The model was applied to ISCAS85 benchmark circuits [8] listed in Table 4 to validate the approximations proposed. The results were compared with simulations using Spectre circuit simulator of Cadence Virtuoso ICFB. Detailed results are reported for c7552, c6288, c2670 and c432 and a summary of results is provided for all circuits in Table 4.

The circuits were synthesized with two sets of logic gates, {nand2, nor2, xor2, and2, fa, ha, inv} in high-V_{th}(HVT) and {nand2, nor2, xor2, inv} in standard-V_{th}(SVT) process options of an industrial 65nm CMOS technology library. The two sets of circuits present wide variation in leakage current and total circuit capacitance for evaluation. For each logic gate, leakage currents were determined for supply voltage varying between 0 and 1V for all input patterns at an operating temperature of $100°C$ using Spectre. Each of these profiles were then fitted with polynomials of degree 7 using MATLAB. The maximum error between evaluated leakage current and simulated leakage current was less than 3% except near $V_{Vdd} = 0$, where absolute values of leakage current are negligible. Further, the leakage current profile of the complete circuit was determined by weighting the polynomials with number of occurrences in the gate netlist and adding them together to form I_{leak} in (3). A leakage current profile for c6288 is shown in Fig. 4. From this curve, pseudo-resistance is determined at each point in the Virtual-Vdd segment.

One set of Spectre simulations of high-V_{th} PMOS transistor is required for each technology library to determine threshold voltages, constant mobility, saturation voltage and saturation currents. To establish these parameters, I_{SD} vs. V_{SD} characteristics at $V_{SG} = 1V$ and I_{SD} vs. V_{SG} characteristics at $V_{SD} = 0.05V$ and $V_{SD} = 1V$ with $W_{sp} = 0.54\mu m$ were obtained using Spectre.

To compare wakeup time estimation using models with circuit simulations in Spectre, V_{dd} was set to 1V. Without loss of generality, all primary inputs of the circuit were set to logic 0. The evolution of Virtual-Vdd during wakeup and sleep modes in c7552 is shown in Fig. 5 and Fig. 6 respectively. In Table 2 and Table 3, the maximum voltage levels attained by Virtual-Vdd and the wakeup times with sleep transistors of different sizes are given. Table 4 shows average errors(μ_{error}) in estimation of the two quantities for all circuits considered in this work. The wakeup time is esti-

mated by (15)-(17) within an average error margin of 16.3% for $22\times$ variation in sleep transistor sizes. The steady-state Virtual-Vdd is determined within 1.8% on an average from the corresponding results of Spectre simulations. Further, a significant reduction in computation time is achieved for wakeup time estimation using the model compared to Spectre. For example, model calculations in c6288 using MAT-LAB took 21ms compared to 4 minutes in Spectre.

In logic clusters that do not satisfy wakeup dependency [2][6], short-circuit currents are generated due to changing logic states of internal nodes as V_{Vdd} increases towards r_1 in wakeup mode. They create the effect of altering effective resistance of the circuit and hence wakeup time. In other words, the accuracy of wakeup time estimation is reduced when the effects of short-circuit currents are not taken into account as is shown for c499 in Table 2 and Table 3. To address this problem it is necessary to model individual cells for short-circuit currents when both supply voltage and its rise time are varying. This is proposed for future work. The cluster definitions and sleep transistor widths considered in this work are not designed to satisfy wakeup dependency or meet a particular peak current constraint [6] as the problem of logic clustering is not addressed in this work.

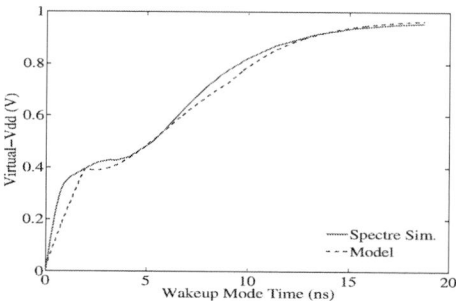

Figure 5: Virtual-Vdd in wakeup mode (W=1.2μm) in c7552

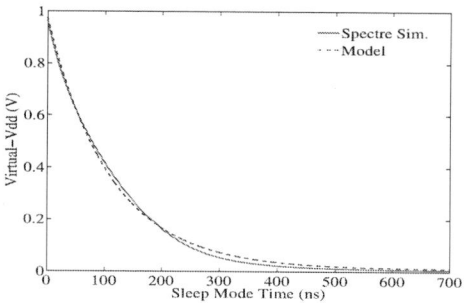

Figure 6: Virtual-Vdd in sleep mode (W=1.2μm) in c7552

6. CONCLUSION

In this paper, a semiempirical approach for estimation of wakeup time of a power-gated logic cluster that relies on only a few basic circuit parameters and one time SPICE level simulations per technology library was presented. A method to determine steady state Virtual-Vdd after wakeup as a function of sleep transistor size and leakage current profile was also described. This was fundamental to development of

Table 2: Maximum Virtual-Vdd after Wakeup and Wakeup Time (HVT Cells)

W (μm)	r_1 (V) Model [Eq. (8)] (Spectre)				Wakeup Time (ns) Model [Eq. (14),(17)] (Spectre)			
	c7552	c6288	c2670	c432	c7552	c6288	c2670	c432
0.54	0.95 (0.93)	0.96 (0.95)	0.97 (0.96)	0.99 (0.99)	40.69 (38.22)	43.24 (46.74)	15.55 (15.04)	4.57 (3.89)
1.2	0.98 (0.96)	0.98 (0.97)	0.99 (0.98)	0.99 (0.99)	18.94 (18.72)	20.27 (22.24)	7.38 (7.25)	2.20 (1.89)
2.4	0.99 (0.98)	0.99 (0.99)	0.99 (0.99)	0.99 (0.99)	9.64 (9.82)	10.35 (11.97)	3.79 (3.75)	1.14 (0.92)
4.8	0.99 (0.99)	0.99 (0.99)	0.99 (0.99)	0.99 (0.99)	5.08 (5.27)	5.46 (6.44)	2.00 (2.02)	0.60 (0.59)
9.6	0.99 (0.99)	0.99 (0.99)	0.99 (0.99)	0.99 (0.99)	2.64 (2.83)	2.84 (3.52)	1.05 (1.11)	0.32 (0.38)
12	0.99 (0.99)	0.99 (0.99)	0.99 (0.99)	0.99 (0.99)	2.13 (2.32)	2.29 (2.95)	0.85 (0.92)	0.26 (0.33)

Table 3: Maximum Virtual-Vdd after Wakeup and Wakeup Time (SVT Cells)

W (μm)	r_1 (V) Model [Eq. (8)] (Spectre)				Wakeup Time (ns) Model [Eq. (14),(17)] (Spectre)			
	c7552	c6288	c2670	c432	c7552	c6288	c2670	c432
0.54	0.73 (0.68)	0.68 (0.60)	0.77 (0.84)	0.97 (0.96)	36.66 (37.10)	62.79 (63.24)	18.06 (13.89)	4.65 (3.49)
1.2	0.85 (0.82)	0.82 (0.78)	0.88 (0.92)	0.99 (0.98)	20.90 (17.16)	34.59 (29.97)	8.17 (6.89)	2.21 (1.74)
2.4	0.91 (0.89)	0.90 (0.87)	0.93 (0.95)	0.99 (0.99)	10.01 (9.07)	16.65 (15.88)	4.01 (3.70)	1.14 (0.96)
4.8	0.95 (0.94)	0.94 (0.92)	0.96 (0.98)	0.99 (0.99)	5.19 (5.02)	8.45 (9.10)	2.07 (2.04)	0.60 (0.57)
9.6	0.97 (0.97)	0.97 (0.96)	0.98 (0.98)	0.99 (0.99)	2.65 (2.77)	4.3 (5.27)	1.06 (1.13)	0.32 (0.37)
12	0.98 (0.97)	0.98 (0.96)	0.98 (0.98)	0.99 (0.99)	2.13 (2.28)	3.45 (4.43)	0.86 (0.94)	0.26 (0.33)

Table 4: Average Relative Errors in Estimation of Maximum V_{Vdd} and Wakeup Time in ISCAS85 Benchmark Circuits

	C_L (pF)		Max. V_{Vdd} μ_{error} (%)		T_{wu} μ_{error} (%)	
Circuit	HVT	SVT	HVT	SVT	HVT	SVT
c7552	2.892	2.993	0.7	2.8	4.7	7.6
c6288	3.171	4.833	0.6	4.3	14.8	11.1
c5315	1.966	2.826	0.6	2.2	12.5	22.3
c3540	1.466	2.037	0.5	1.5	11.2	17.5
c2670	1.148	1.202	0.3	3.0	3.0	7.8
c1908	0.606	0.633	0.2	0.3	21.8	15.4
c499	0.601	0.478	0.2	0.1	24.6	33.7
c432	0.351	0.360	0.1	0.3	16.4	15.1
Mean			0.4	1.8	13.6	16.3

rest of the model. The model in sleep mode can be used to determine leakage energy savings in inactive states of the circuit. In other words, some of the key parameters used as optimization criteria for logic clustering have been captured in closed-form expressions. Our simulations and application of the model to ISCAS85 benchmark circuits with an industrial 65nm CMOS technology library show that on an average wakeup time can be estimated within an error margin of 16.3% over 22× variation in transistor sizes and 13× variation in circuit sizes with significant reduction in computational times compared to SPICE level circuit simulations.

7. REFERENCES

[1] A. Abdollahi, F. Fallah, and M. Pedram. An effective power mode transition technique in MTCMOS circuits. In *Proc. ACM/IEEE Des. Autom. Conf.*, pages 27–32, Anaheim, June 2005.

[2] M. Anis, S. Areibi, and M. Elmasry. Dynamic and leakage power reduction in MTCMOS circuits using an automated gate clustering technique. In *Proc. ACM/IEEE Des. Autom. Conf.*, pages 480–485, New Orleans, June 2002.

[3] S. Henzler. *Power Management of Digital Circuits in Deep Sub-Micron CMOS Technologies*, chapter 5. Springer, 2007.

[4] Z. Hu, A. Buyuktosunoglu, V. Srinivasan, V. Zyuban, H. Jacobson, and P. Bose. Microarchitectural techniques for power gating of execution units. In *Proc. Intl. Sym. Low Power Electronic Des.*, pages 32–37, Newport Beach, USA, August 2004.

[5] S. Kim, C. J. Choi, D. K. Jeong, S. V. Kosonocky, and S. B. Park. Reducing ground-bounce noise and stabilizing data-retention voltage of power gating structures. *IEEE Trans. Electron Devices*, 55(1):197–205, January 2008.

[6] Y. Lee, D.-K. Jeong, and T. Kim. Comprehensive analysis and control of design parameters for power gated circuits. *IEEE Trans. VLSI Syst.*, 19(3):494–498, March 2011.

[7] D.-W. Lin, M.-L. Cheng, S.-W. Wang, C.-C. Wu, and M.-J. Chen. A constant-mobility method to enable MOSFET series-resistance extraction. *IEEE Electron Device Lett.*, 28(12):1132–1134, December 2007.

[8] X. Lu. Layout and parasitic information for ISCAS circuits. http://dropzone.tamu.edu/~xiang/iscas.html.

[9] K. Roy, S. Mukhopadhyay, and H. Mahmoodi-Meimand. Leakage current mechanisms and leakage reduction techniques in deep-submicrometer CMOS circuits. *Proc. IEEE*, 91(2):305–327, June 2003.

[10] S. Roy, N. Ranganathan, and S. Katkoori. A framework for power-gating functional units in embedded microprocessors. *IEEE Trans. VLSI Syst.*, 17(11):1640–1649, November 2009.

[11] H. Singh, K. Agarwal, D. Sylvester, and K. J. Nowka. Enhanced leakage reduction techniques using intermediate strength power gating. *IEEE Trans. VLSI Syst.*, 15(11):1215–1224, November 2007.

[12] Y. Taur and T. Ning. *Fundamentals of Modern VLSI Devices*, chapter 4. Cambridge University Press, 2009.

[13] T.-D. Vivek, O. Senticys, and S. Derrien. Wakeup time and wakeup energy estimation in power-gated logic clusters. In *Proc. 24th Intl. Conf. VLSI Des.*, pages 340–345, Chennai, January 2011.

[14] H. Xu, R. Vemuri, and W.-B. Jone. Dynamic characteristics of power gating during mode transition. *IEEE Trans. VLSI Syst.*, 19(2):237–249, February 2011.

SUPPLEMENTARY PAGES

S1. DETERMINATION OF TOTAL CIRCUIT CAPACITANCE

S1.1. Decoupling Capacitance

In (4), the total circuit capacitance C_L is defined to be the sum of input capacitances of all inputs of all constituient gates of the cluster. In physical implementations with CMOS process technologies, a decoupling capacitance (decap) C_D is generally included between the supply voltage rail (or Virtual-Vdd ring of the power-gated domain) and ground to suppress bounces on supply rails during switching of gate outputs. In the model described in this paper, a decap is not explicitly included. The total circuit capacitance including a decoupling capacitance can be determined to be $C_L + C_D$ since capacitances appear in parallel between Virtual-Vdd and ground.

S1.2. Dependence of Gate Capacitance on Inputs

The gate capacitances of MOS transistors are input dependent. At wakeup, as V_{Vdd} increases some of the logic gates switch to 'Logic 1' while the rest remain at 'Logic 0'. Gate inputs in the fan-out of gate outputs that switch to 'Logic 1' will present a higher output capacitance to the switching gate than to the driving gate remaining at 'Logic 0'. In logic clusters used in the paper, the outputs of each gate is determined from the primary inputs based on the gate function (NAND, XOR etc.) and hence appropriate value of capacitance obtained from SPICE level characterization of standard cell for each of its input is used to determine total capacitance in (4). Further gate terminal capacitances of all MOSFETs in standard cells include parasitic capacitances (fringe and overlap) referred to the gate terminal.

S1.3. Parasitic Capacitance Along Interconnect Lines

Parasitic capacitances along interconnect lines have been neglected in determining total circuit capacitance considering that in cluster based power-gated circuit design, independent clusters have a local distribution of interconnects unlike a distributed sleep transistor network (DSTN) based power gating.

S2. STEADY-STATE VIRTUAL-VDD

The conditions of validity for one of the roots r_1 can be intuitively explained to be $V_{sleep} < r_1 < V_{dd}$ as follows. Let at steady state the pseudo-resistance of VCCS be given by some $R_{ss} = V_{Vdd,ss}/I_{leak}(V_{Vdd,ss})$ where $V_{Vdd,ss}$ denotes Virtual-Vdd in steady state. Then the cicuit at that instant can be represented by Thevenin's equivalent resistance $R_{TH} = \frac{R_{ss}R_{lin}}{(R_{ss}+R_{lin})}$ and Thevenin's equivalent voltage $V_{TH} = \frac{R_{ss}V_{dd}}{(R_{ss}+R_{lin})}$ with total circuit capacitance C_L in series with R_{TH} and V_{TH}. Clearly, $V_{TH} < V_{dd}$. C_L is charged to V_{TH} in steady state which we determine to be r_1 as a solution of (8). For non-zero C_L the inference that $V_{sleep} < r_1$ is trivial.

S3. WAKEUP MODE VIRTUAL-VDD MODEL

In (6), V_{th} corresponds to threshold voltage when the transistor is operating in linear region as determined from Sec-

tion 4.4.

To derive (8), dV_{Vdd}/dt obtained from evaluation of RHS of (7) for a sweep of V_{Vdd} is fitted with a quadratic polynomial in least squares sense using the MATLAB function 'polyfit'. Further r_1 and r_2 are obtained as roots of the quadratic polynomial on the RHS of (8). Equation (9) can be derived from (8) by separation of variables and partial fraction expansion as

$$dV_{Vdd}\left(\frac{A}{V_{Vdd}-r_1} + \frac{B}{V_{Vdd}-r_2}\right) = -\frac{1}{\tau}dt.$$

Equations (10) and (13) denote I_{st} and V_{Vdd} represented by piecewise continuous functions in three intervals: two in saturation region and one in non-saturation region of sleep transistor operation.

The wakeup time T_{wu} is given by (17) under the assumption that $0.99r_1 > (V_{dd}-V_{DSAT})$. For clusters with $0.99r_1 \leq (V_{dd}-V_{DSAT})$, wakeup time $T_{wu} = T_2$ determined from (16) and with the condition $V_{Vdd} = 0.99r_1$.

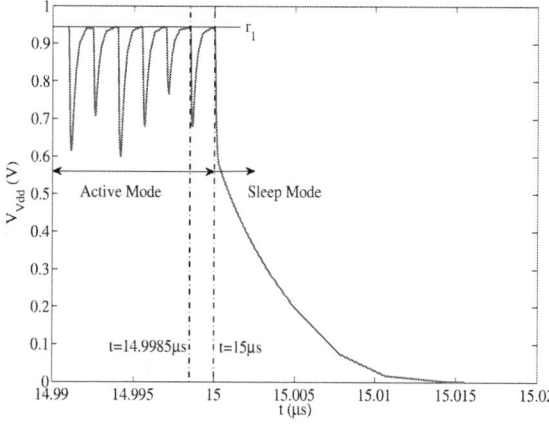

Figure 7: Virtual-Vdd in Active and Sleep Modes

S4. SLEEP MODE VIRTUAL-VDD MODEL

In this section we justify the assumption that at the end of active mode $V_{Vdd} = r_1$ with an experiment. Let $\Delta V_{Vdd,max}$ denote maximum degradation of V_{Vdd} due to dynamically changing inputs of logic cluster. Further, let t_i denote the time instant of end of cycle i in active mode. Clearly, $V_{Vdd}(t_i)$ satisfies $V_{Vdd}(t_{i-1}) - \Delta V_{Vdd,max} \leq V_{Vdd}(t_i) \leq r_1$. We consider the conditions under which $V_{Vdd}(t_{i-1}) = r_1$.

A logic cluster was synthesized for a maximum path delay of 1.0ns and was power-gated by a header type of sleep transistor. In practice, the size of the sleep transistor is chosen for a fixed performance loss or $\Delta V_{Vdd,max}$ in active mode. The circuit was simulated with Spectre circuit simulator with random inputs applied to the circuit at a clock period of 1.5ns, i.e., with a positive slack of 0.5ns in active mode. The variation of Virtual-Vdd voltage in active and sleep modes is shown in Fig. 7. It can be seen that for a maximum duration of path delay, V_{Vdd} degrades by about $\Delta V_{Vdd,max} = 0.3V$ and at the end of active mode time slot of 1.5ns, V_{Vdd} attains a value of r_1. With a sufficient and constant clock cycle period $T = t_i - t_{i-1}$, $V_{Vdd}(t_{i-1}) = r_1$

for all i. The assumption of sufficient positive slack holds for low power, low performance circuits. Therefore $V_{Vdd} = r_1$ can be specified as the initial value of V_{Vdd} in sleep mode in (22).

S5. EXPERIMENTAL RESULTS

To apply the model and to perform simulations on ISCAS85 benchmark circuits all primary inputs were assigned 'Logic 0'. In practice this input combination may not result in minimum leakage current. However the model for estimation of wakeup time described in the paper applies identically to all patterns of primary inputs.

The ISCAS85 benchmark circuits listed in Table 4 were synthesized with both HVT and SVT cells of an industrial 65nm bulk CMOS technology library. Since threshold voltage of MOSFET devices in SVT cells is less than that of HVT cells, the total leakage current in SVT cell implementations of ISCAS85 circuits is higher than that of HVT cell implementations resulting in a lower R_{ss} as defined in Section S2. Hence a lower V_{TH} or r_1 is obtained. This observation is reflected in the results shown in Table 2 and Table 3.

In this work MATLAB was used to evaluate model parameters and hence wakeup time. Alternatively other tools may be used for computations. As an example, the same model, when implemented in a scripting language like Tcl, can be efficiently integrated with standard IC design and analysis flows that use static timing and power analysis tools.

S6. APPLICATIONS

S6.1. Wakeup Energy Estimation

As an extension to estimation of wakeup time presented in the paper, it is possible to determine wakeup energy (E_{wu}).

Wakeup energy is an energy overhead due to sleep-to-wakeup mode transition in a power gating cycle. It is required to determine breakeven energy and hence minimum sleep time for a power-gated logic cluster. Wakeup energy is given by $E_{wu} = \int_0^{T_{wu}} V_{dd} I_{st} dt$ where I_{st} is obtained from (10) and (13) and T_{wu} from (17). It should be noted that, short circuit currents (I_{sc}) that are generated in the internal nodes during wakeup mode are neglected in (5). In this work I_0 and I_1, which are assumed to be constants based on heuristics developed in subsection 4.5, must be replaced with time and V_{Vdd} dependent models. Modeling short circuit currents in logic gates when both supply voltage and its rise time are varying is proposed for future work.

S6.2. Scheduling Power-Gated Clusters

The model for wakeup time estimation presented in the paper may be applied in scheduling of power-gated logic clusters as part of a larger optimization problem. Consider a combinational circuit $\mathbf{C}$. Let $C_i, i = 1, 2, ..., N$ denote N logic clusters obtained by partitioning $\mathbf{C}$ such that they satisfy constraints of minimum sleep transistor area, peak current, maximum delay degradation and minimum wakeup time. The optimization problem referred to wakeup time constraint is stated as follows. Let $T_{wu,i}$ denote the wakeup time of logic cluster C_i and $T_{wu,max}$ the maximum acceptable wakeup time of the overall circuit $\mathbf{C}$. Then,

$$max\left(\sum_{j=1}^{P} T_{wu,j}, \sum_{k=P+1}^{Q} T_{wu,k}, ..., \sum_{l=R+1}^{N} T_{wu,l}\right) \leq T_{wu,max}$$

for some $P, Q, R,...$ such that $P \geq 1$, $Q \geq P + 1,....$ Hence a wakeup schedule for the N logic clusters may be derived. The model presented in the paper may be used to determine each $T_{wu,i}$ during the optimization run.

Cost-effective Power Delivery to Support Per-core Voltage Domains for Power-constrained Processors

Hamid Reza Ghasemi[†], Abhishek A. Sinkar[†], Michael J. Schulte[‡], and Nam Sung Kim[†]

University of Wisconsin-Madison, U.S.A.[†] and Advanced Micro Devices, Inc., U.S.A.[‡]

hamid@cs.wisc.edu, sinkar@wisc.edu, michael.schulte@amd.com, nskim3@wisc.edu

ABSTRACT

Per-core voltage domains can improve performance under a power constraint. Most commercial processors, however, only have one chip-wide voltage domain because splitting the voltage domain into per-core voltage domains and powering them with multiple off-chip voltage regulators (VRs) incurs a high cost for the platform and package designs. Although using on-chip switching VRs can be an alternative solution, integrating high-quality inductors and cores on the same chip has been a technical challenge. In this paper, we propose a cost-effective power delivery technique to support per-core voltage domains. Our technique is based on the observations that (i) core-to-core voltage variations are relatively small for most execution intervals when the voltages/frequencies are optimized to maximize performance under a power constraint and (ii) per-core power-gating devices augmented with small circuits can serve as low-cost VRs that can provide high efficiency in situations like (i). Our experimental results show that processors using our technique can achieve power efficiency as high as those using per-core on-chip switching VRs at much lower cost.

Categories and Subject Descriptors
C.5.4 [VLSI Systems]

General Terms
Design, Management, and Performance

Keywords
Power delivery, voltage regulators, multi-core processors

1. INTRODUCTION

The maximum performance of multi-core processors operating all available cores is typically limited by a power constraint. When a processor runs multiple threads or applications, however, the instructions per cycle (IPC) of each core in the processor can vary notably within each execution interval. Consequently, per-core voltage/frequency (V/F) domains can allow a processor to adjust the V/F of each core dynamically, such that performance is maximized under a power constraint. This approach, however, requires a voltage domain (i.e., VR) per core, while most commercial processors have only one chip-wide voltage domain. This is because splitting the voltage into multiple domains and powering them using multiple off-chip VRs incurs a high cost for the platform and package designs. Alternatively, on-chip switching VRs (i.e., buck converters) integrated with cores can lower the costs associated with the platform and package designs. However, no commercial processors have adopted on-chip switching VRs yet because integrating cores and high-quality inductors for the VRs on the same chip is expensive and technically challenging.

In this paper, we propose a cost-effective power delievery technique to support per-core voltage domains using on-chip low-drop-output (LDO) VRs. The on-chip LDO VRs have not previously been considered for high-performance multi-core processors because they cannot power-efficiently provide a wide voltage range for cores; LDO VRs suffer from high power loss when the voltage drop between input voltage (V_I) and output voltage (V_O) of the LDO VR is large. On the other hand, platform architects must assume that the core voltage can be any value within a given voltage range at which the processor can operate. However, our experiments reveal that the maximum voltage difference between cores at each dynamic V/F scaling (DVFS) interval is small for most intervals when the voltages are optimized to maximize the performance under a power constraint. In other words, the V_O values of multiple LDO VRs can be close to the shared V_I value. Furthermore, many LDO VRs can be implemented at a low cost because they can share their largest component with per-core power-gating (PCPG) devices and, unlike switching VRs, they do not require either large inductors or capacitors.

The key contributions of this paper are as follows:

- We analyze the core voltages when they are optimized to maximize the performance of a power-constrained processor at each DVFS interval. A processor using per-core DVFS can deliver considerably higher performance/power efficiency than a processor using chip-wide DVFS when it exhibits small core-to-core (C2C) voltage variations (Section 2).

- We propose a lost-cost power delivery technique that exploits existing on-chip PCPG devices available in most processors. The cost of an LDO VR can be inexpensive because the VR shares its most expensive component with the PCPG devices. We demonstrate that the power efficiency of LDO VRs can be higher than that of on-chip switching VRs (Section 3).

- We demonstrate that a processor using our proposed technique is as effective as a processor using on-chip per-core switching VRs when running multiple threads or applications (Section 4).

To the best of our knowledge, this is the first study to demonstrate that (i) C2C voltage variations are small at each DVFS interval when the core voltages are optimized to maximize performance under a power constraint, and (ii) on-chip LDO VRs can be implemented with existing PCPG devices cost-effectively.

2. CORE-TO-CORE VOLTAGE VARIATIONS

Most commercial processors support DVFS, but they have only a single chip-wide voltage domain due to the high cost of supporting multiple voltage domains. As more cores share the chip-wide voltage domain, the performance/power benefit of DVFS diminishes. This is because a single voltage domain cannot allow a multi-core processor to effectively exploit runtime performance (i.e., IPC) variations across cores running multiple threads or applications for a given DVFS interval. For example, some cores running threads in memory-intensive phases can operate at lower V/F without impacting the performance while other cores executing threads in compute-intensive phases must operate at higher V/F to maximize performance. Consequently, many researchers have

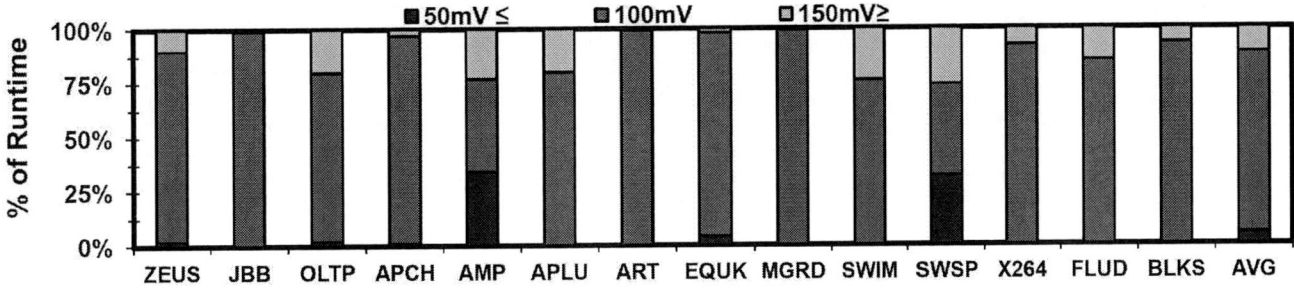

Figure 1: Percentage of intervals exhibiting various maximum voltage differences between cores for a 8-core processor supported by per-core V/F domains. We apply an oracle approach, similar to one used in [1], to determine V/F of each core at each runtime DVFS interval. Each interval is comprised of approximately 10-million executed instructions, which is equivalent to a few hundred microseconds depending on IPC values. A total of 1-billion instructions are executed after 100-million instructions are executed to warm up on-chip caches.

investigated various DVFS algorithms to exploit multiple voltage domains effectively [1,2].

Figure 1 shows that the maximum voltage difference between cores in a processor supported by per-core voltage domains is not large at each DVFS interval. For this experiment, we use four commercial workloads (Apache, JBB, OLTP, and Zeus denoted by APCH, JBB, OLTP, and ZEUS) [3], six SPEC OMP V3.2 benchmarks (Ammp, Applu, Art, Equake, Mgrid, and Swim denoted by AMP, APLU, ART, EQUK, MGRD, and SWIM), and four PARSEC benchmarks (Swaptions, X264, Fluid, and BlackScholes denoted by SWSP, X264, FLUD, and BLKS) [4] running on the GEMS multi-core simulator [5]. An oracle DVFS algorithm [1] is modified to maximize MIPS^3/W for a given maximum power constraint. See Section 4 for the detailed experimental methodology. Because we target server-class multi-core processors under a power constraint, we chose MIPS^3/W to emphasize the performance aspect of processors more than MIPS/W [6]. The maximum voltage difference between cores for the "Per-Core V/F" case is less than or equal to 100mV for at least 90% of the execution intervals in most applications. In the next section, we show that, under such conditions, the power loss of LDO VRs can be lower than the power loss of switching VRs.

3. LDO VRs EXPLOITING PER-CORE POWER-GATING DEVICES

3.1 PCPG-Based LDO VRs

PCPG devices are typically provided for commercial multi-core processors to reduce standby leakage power of idle cores [7]. In active state, a PCPG device incurs a slight voltage drop across it (i.e., between the supply voltage and the actual voltage applied to the core). The voltage drop is inversely proportional to the size (i.e., total transistor width) of the PCPG device for a given amount of total current (dynamic + leakage) drawn by the core. In fact, the voltage applied to the core can be modulated by controlling the effective width (i.e., resistance) of the PCPG device [8]. A PCPG device, which is implemented with many parallel transistors and on/off signal buffers, is similar to the largest component (i.e., the pass device between V_I and V_O) of a typical LDO VR, as illustrated in Figure 2. In other words, an LDO VR can be implemented by augmenting a PCPG device with feedback control circuitry comprised of an error amplifier, an analog-to-digital converter, and a reference voltage generator. In [9], it was reported that the output device and its buffers, both of which can be shared with a PCPG device, accounted for 83% of the total LDO VR area. Since a PCPG device consumes 5%-10% of a core's area [10], we estimate that the extra overhead due to the feedback control circuitry to implement LDO VR is 0.85%-1.7% of the core's area. By contrast, on-chip switching VRs require large inductors and capacitors for

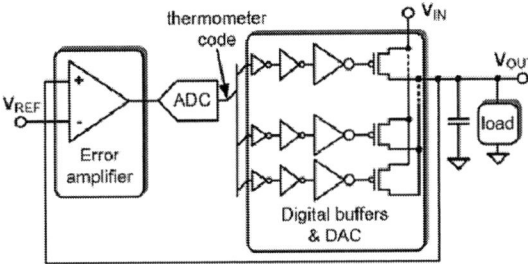

Figure 2: A typical LDO VR architecture; the cartoon is reproduced from [9].

their implementations. As a result, a switching VR has nearly four times larger chip area than a comparable LDO VR [11]. Furthermore, LDO VRs can provide faster transient responses than switching VRs [12] and, unlike switching VRs, they do not inject switching noise in the substrate. This is desirable for the operation of highly sensitive mixed signal circuits.

Figure 3 shows two different approaches to distribute supply voltages to an 8-core processor with per-core V/F domains. Both approaches use a first stage off-chip VR to convert 5V to an intermediate voltage level, V_I, for on-chip per-core VRs; we cannot supply 5V for on-chip switching VRs directly due to the oxide-reliability of nanoscale transistors that implement both VRs and cores. This voltage is further down converted using on-chip per-core VRs to the voltage ($V_O[i]$) required by core i. The arrangement on the left uses LDO VRs (i.e., PCPG devices augmented with the control and reference circuitry to implement LDO VRs). The efficiency of an LDO VR is a function of its V_O/V_I ratio. When the voltages demanded by individual cores are restricted to a limited range (e.g., within 100mV of one another as shown in Figure 1), a

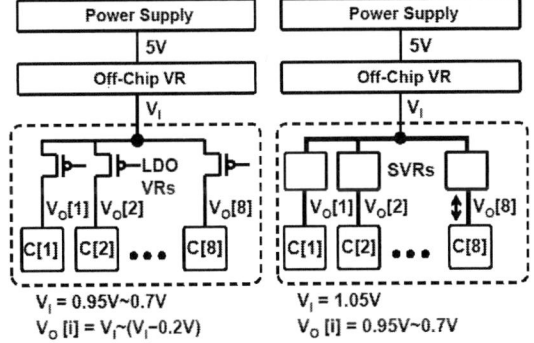

Figure 3: V_I and V_O ranges of LDO VRs (left) and switching VRs denoted by SVRs (right) for per-core voltage domains; we ignore the default voltage drop of the LDO VRs due to the small resistance of the fully turned-on PCPG devices for illustration purpose.

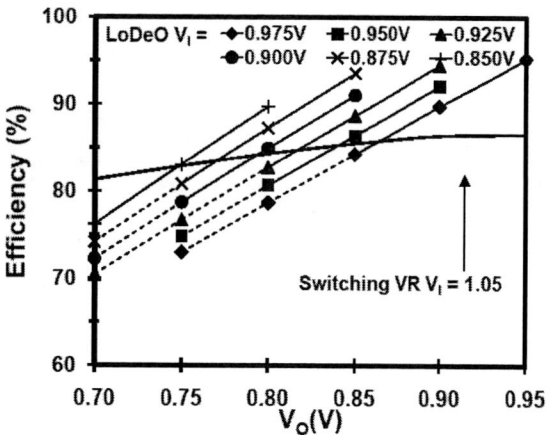

Figure 4: Efficiency comparison between switching and LDO VRs.

high V_O/V_I ratio can be achieved for all the cores by adjusting the V_O of the first stage (i.e., V_I of the second stage) such that it is sufficient to provide the highest V_O demanded by any of cores. Thus, a processor adopting per-core LDO VRs can be tuned to achieve high efficiency by jointly optimizing both their V_I and V_O. The arrangement on the right uses per-core on-chip switching VRs to provide the necessary core voltage. A switching VR uses two active devices, inductors and capacitors to provide high voltage conversion efficiency across a wide range of V_O. This efficiency is primarily determined by the switching losses in the active devices and their conduction losses. The V_I value for switching VRs is fixed to 1.05V in this example. In summary, an LDO VR can be implemented very cost-effectively since (i) it does not require inductors or capacitors [9] and (ii) it can share its largest component (i.e., the output device) with a PCPG device. Furthermore, its efficiency can be very high when cores running multiple threads or applications demand similar voltage values.

3.2 Efficiency Comparison: LDO vs. Swithcing VRs

Figure 4 compares the efficiency of a switching VR with that of an LDO VR. The efficiency of LDO VRs is higher than switching VRs when $V_I - V_O$ (i.e., V_O/V_I) is small, but it becomes lower as $V_I - V_O$ increases. For example, if $V_I - V_O$ is more than 100mV, the efficiency of LDO VRs usually becomes lower than that of switching VRs as shown in Figure 4. We model the efficiencies of both switching and LDO VRs assuming each core consumes the maximum allowed current for each operating voltage. To measure the maximum efficiency of the switching VR at each operating point (i.e., voltage/current), we search for and activate the optimal number of phases out of eight available phases for a given voltage/current; Table 1 summarizes the key design parameters of various VR stages described in this study.

The off-chip switching VR efficiency computation is based on [13] with V_I fixed to 5V. Off-chip switching VR designs built with off-the-shelf components typically have very high efficiencies (> 90%) due to low loss inductors and capacitors. Their efficiency reaches a maximum for a certain load current and then drops with further increase in current due to an increase in conduction losses.

Consequently, as the off-chip regulator V_O for LDO VRs decreases, the efficiency degrades, and thus the overall efficiency of LDO VRs becomes lower than that of switching VRs, as plotted in Figure 4.

Our efficiency analysis assumes a 32nm CMOS process with inductors ($Q = 20$ @ 100MHz), similar to [14], since switching VRs with on-die inductors exhibit poor efficiency; on-package inductors incur packaging design and integration issues, but we do not discuss them in detail in this paper. The design is optimized to achieve a conversion ratio of 1.05V/0.9V at a load current of 16.67A per core (corresponding to a total of 120W for 8 cores at 0.9V) with an efficiency of 88%. An 8-phase topology is used with 63.5nH inductance per phase. As V_O and load current are reduced, the efficiency of switching VRs decreases monotonically. This is because the switching loss constitutes a higher percentage of the output power as the V_O value reduces. The efficiency is strongly dependent on the operating point at which the switching VR design is optimized. For a design optimized for a higher V_O, the efficiency at low output voltage drops more rapidly compared to a design optimized for a lower V_O [15]. In Figure 4, the on-chip switching VRs are optimally designed for $V_O = 0.8V$.

4. EVALUATION

4.1 DVFS Algorithms for Efficiency Comparison

The key objective of this section is to evaluate the effectiveness of the LDO VRs derived from PCPG devices. Thus, we can use various per-core DVFS algorithms optimized for high-performance multi-core processors including the algorithms exploiting C2C frequency and power variations along with thread migrations (TMs) [2]. For the evaluation, we adopt an integer linear programming (ILP) method for the DVFS algorithms. The ILP formulation is similar to one used in [1], which attempts to minimize the power consumption of a multi-core processor for a given performance constraint. We modify the formulation such that we search for the optimal V_O for each core to maximize MIPS^3/W under a power constraint at each DVFS interval as follows:

Objective:

$$maximize\left(\sum_{i=1}^{N} MIPS_i = \sum_{i=1}^{N} \sum_{j=1}^{M} IPC_i \cdot F_{ij} \cdot x_{ij} \right) \quad (1)$$

Constraints:

$$\sum_{i=1}^{N} \sum_{j=1}^{M} P_{ij} \cdot x_{ij} \leq P_{tot\,max} \; and \; \sum_{i=1}^{N} \sum_{j=1}^{M} x_{ij} \leq N \quad (2)$$

where N is the number of cores; M is the number of V_O steps supported by the DVFS algorithm; $MIPS_i$ and IPC_i are the MIPS and IPC of core i; F_{ij} is the frequency of core i at voltage level j; x_{ij} corresponds to one bit of an M-bit binary variable for core i that is guaranteed to assign core i to only one of M possible V/F states (i.e., $\exists i : \sum_{j=1}^{M} x_{ij} = 1$); P_{ij} is the power consumption of core i, which is a function of $V_O[i]$; P_{totmax} is the allowed total power

Table 1: Summary of VR design parameters.

	Off-chip Switching VR	On-Chip Switching VR	On-chip LDO VR
V_I/V_O	5V/1.05V to 5V/0.85V	1.05V/0.95V to 1.05/0.7V	0.95V/0.7V to 0.7V/0.95V
Technology	N/A	32nm	32nm
f_{sw}	300KHz	100MHz	N/A
L/phase	360nH ($r_L = 0.5m\Omega$)	63.5nH (Q= 20 @100MHz)	N/A
No. of Phases	6	8	N/A

Table 2: Summary of DVFS algorithms explored in this study. "Sh," "Se," "PV," and "TM" indicate "Shared," "Separate," "Process Variation," and "Thread Migration," respectively.

Algorithms	Voltage Domain	Frequency Domain	Process Variation Aware	Thread Migration	Off-Chip VR V_O	On-Chip VR	Constraint
ShV/F	Shared	Shared	No	No	Varying	N/A	$V_{O1} = V_{O2} = ... = V_{ON}$
SeV/F	Separate	Separate	No	No	Fixed	SVR	
SeV/F(PV)			Yes	No			
SeV/F(PV/TM)			Yes	Yes			
LDOSeV/F	Virtually Separate	Separate	No	No	Varying	LDO VR	
LDOSeV/F(PV)			Yes	No			
LDOSeV/F(PV/TM)			Yes	Yes			

consumption of the processor; and Eq. (2) is the constraint, respectively. In Eq. (2), the second constraint is to enforce one V_O selection for each core. The V_I for all LDO VRs is determined by taking the maximum value among $V_O[1], V_O[1], ..., $ and $V_O[N]$.

As discussed in [2], this algorithm requires manufacturers to store per-core frequency and power values at each voltage level for DVFS algorithms to exploit C2C frequency and power variations. These values can be characterized by the manufacturer and stored, along with many other processor tuning parameters, in a non-volatile memory of the processor. Like other DVFS algorithms, we also need to predict workload characteristics like the IPC of each thread to assign a proper V/F to each core for the next DVFS interval. Although we can use various methods to predict the IPC of the next interval based on the current IPC, we assume that the IPC value of each thread at every interval is known in advance (as an oracle method). This is to isolate the impact of the IPC prediction from the MIPS^3/W results so that we can fairly compare the efficacy of the two different VR schemes. Finally, we adopt a simple scheme for the TM technique; we assign threads to cores one-to-one in the order of IPC and frequency values. For example, the thread with the highest IPC is assigned to the core with the highest frequency at a given voltage (i.e., the fastest core considering C2C frequency variations). Table 2 summarizes the DVFS algorithms explored in this study and constraints for specific algorithms. Our baseline processor has a single, chip-wide V/F domain using an off-chip VR (i.e., ShV/F).

4.2 Architecture simulation environment

Our processor configuration contains eight cores. Each core is four wide with 32KB private L1 caches and a shared 512KB L2 cache. The cores are connected to each other using crossbar switches. We evaluate different DVFS algorithms using a full-system cycle-accurate simulator, GEMS [5], after we modify GEMS to support per-core frequency domains and TM requiring L1 cache flushing. In addition to four commercial workloads (Apache, JBB, OLTP, and Zeus, six SPEC OMP V3.2 benchmarks (ammp, applu, art, equake, mgrid, and swim), and four PARSEC benchmarks (Swaptions, X264, Fluid, and Black Scholes) [4], we use five mixes of compute- and memory-bound SPEC2006 benchmarks (eight copies of Bzip2, six copies of Bzip2 and two

copies of Libquantum, four copies of Bzip2 and four copies of Libquantum, two copies of Bzip2 and six copies of Libquantum, and eight copies of Libquantum denoted by 8B0L, 6B2L, 4B4L, 2B6L, and 0B8L, respectively) with the processor simulation parameters summarized in Table 3.

4.3 Core frequency and power models

Our objective is to maximize the performance for a given maximum power consumption constraint. To model the maximum power consumption of cores, we assume that (i) the total maximum power consumption of 8 cores is 120W and (ii) 30% of the total power is active leakage [16] at 0.9V. Each core has its own shared L2 cache that shares the V/F domain with the core. Thus, we assume that the L2 power scales with the core power consumption. The power consumption of I/O and other peripheral components including on-chip interconnects, which are tied to other separate fixed voltage/frequency domains, is not included in our analysis since it can be regarded as a fixed power cost for all the cases we explore in this paper; I/O and on-chip interconnects are responsible for ~15% of the total power in Niagara 2 [17].

Due to within-die (WID) C2C frequency and leakage power variations, the power consumption of each core differs. To analyze the impact of WID process variations on the frequency and leakage power consumption of each core, we first generate 100 variation maps for threshold voltage (Vth) and effective channel length (Leff) of transistors in a die and characterize frequency and power consumption by following the methodology presented in [18]: WID correlation distance coefficient $\phi = 0.5$ and WID Vth and Leff variations $\sigma_{sys} = 6.4\%$ and 3.2% of the nominal Vth and Leff values, respectively. We apply the Vth and Leff values of each grid point to a FO4 inverter chain and a dummy circuit, which is comprised of 50% inverters, 30% NAND gates, and 20% NOR gates, to obtain the frequency and leakage power scaling factors of each grid, respectively; NAND and NOR gates in a dummy circuit can have up to 4 inputs and their inputs are assigned randomly with either 1 or 0.

Second, we measure the frequency and leakage scaling factors of each grid point at 0.95V to 0.7V using a 32nm technology model and SPICE. We assume that the frequency of each core is determined by the slowest grid point in the core [18] and the

Table 3: Summary of processor simulation parameters.

Fetch/Issue/Retire	4/4/4	# of Cores	8
IL1	32KB/4-way/64B 3 cycles	Branch Predictor/BTB/RAS	YAGS/1K/32
L2	512KB/8-way/64B 10 cycles	DL1	32KB/4-Way/64B 3cycles
Cache Coherency Protocol	Directory-based MESI	Main Memory (size/block/page/latency)	DDR3-1.6GHz 4GB/64B/4KB/7-7-7-20ns
# of MSHRs	8	Write-buffer entries	16

Table 4: Summary of frequency and power consumption of each core as a funciton of $V_O[i]$. For each core, the frequency (GHz) and power (Watts) are given in the left and right columns, respectively.

$V_O[i]$	Core 1		Core 2		Core 3		Core 4		Core 5		Core 6		Core 7		Core 8	
0.95V	3.6	15.9	3.8	17.9	4.4	18.2	3.9	16.9	3.8	17.2	4.1	19.3	4.4	29.1	4.1	19.7
0.90V	3.2	12.5	3.4	14.0	4.0	14.5	3.5	13.3	3.4	13.5	3.7	15.1	4.0	22.0	3.7	15.3
0.85V	2.8	9.6	3.0	10.8	3.5	11.3	3.0	10.2	3.0	10.4	3.3	11.7	3.5	16.6	3.3	11.8
0.80V	2.4	7.2	2.5	8.1	3.1	8.6	2.6	7.7	2.4	7.8	2.8	8.8	3.1	12.3	2.8	8.9
0.75V	2.0	5.3	2.1	6.0	2.6	6.4	2.2	5.7	2.1	5.7	2.3	6.5	2.6	9.0	2.4	6.6
0.70V	1.6	3.7	1.7	4.2	2.1	4.6	1.7	4.0	1.7	4.1	1.9	4.6	2.1	6.4	1.9	4.7

frequency of the slowest core is 3.2GHz at 0.9V. Then, each core's maximum dynamic power consumption at 0.9V is $\left(F_i / \sum_{j=1}^{N} F_j\right) \cdot 120W \cdot 0.7$ where F_i and F_j are the frequency of core i and j, and N is the number of cores. With the known frequency, voltage, and dynamic power values, we can calculate the maximum core switching capacitance (i.e., C_{dyn}). This allows us to calculate the dynamic power at any given voltage. The leakage power of each grid point is scaled such that the sum of the leakage power from all grid points in a die is equal to 30% of 120W at 0.9V. The sum of the scaled leakage power from all the grid points belonging to a particular core becomes the core's leakage power.

Finally, we allow some cores to run at V/F higher than 0.9V/3.2GHz as long as the total power constraint is satisfied; this is possible when other cores run at V/F lower than 0.9V/3.2GHz. Since all cores in our baseline processor, which uses a per-chip single VR, run at the same frequency, the dynamic power consumption of the processor is lower than when other processors use per-core V/F domains. Thus, we increase the V/F of the processor until 120W is fully used (i.e., 0.9125V and 3.3GHz).

Note that C2C frequency and power variations change across different dies. However, for our analyses, we pick a typical die map from the 100 generated maps because a large amount of simulation time is required to repeat the same experiment for hundreds of die maps. Thus, the MIPS³/W results, which exploit C2C frequency and leakage power variations, represent the value close to the median value of the 100 die maps. Table 4 tabulates the frequency and power consumption of each core a s function of $V_O[i]$. For each core the frequency (GHz) and power (Watts) are given in the left and right columns, respectively.

4.4 MIPS³/W comparison

Impact of VR efficiency on MIPS³/W of Multi-threaded Workloads: Figure 5 compares MIPS³/W of 8-core processors using LDO and switching VRs. Although WID C2C process variations are not exploited and the TM technique is not applied, LDOSeV/F provides 4% higher MIPS³/W than SeV/F on average

(i.e., geometric mean). Note that LDOSeV/F and SeV/F, which do not exploit WID process variations, result in worse MIPS³/W than ShV/F, which uses only an off-chip VR. This is because the power loss by the on-chip VRs (~20% of total processor power consumption) completely negates the benefit of supporting per-core V/F domains for multi-threaded applications.

When both WID C2C process variations and TM are exploited, LDOSeV/F and SeV/F provide 16% and 12% higher MIPS³/W than ShV/F on average; LDOSeV/F leads to 4% higher MIPS³/W than SeV/F whether or not WID C2C process variations and/or TM are exploited. This is because LDO VRs exhibit higher efficiency than switching VRs for most DVFS intervals. This is mainly due to small C2C voltage variations in multi-threaded applications, which allows LDO VRs to provide voltages with higher efficiency than switching VRs, as shown in Figure 4. Note that we include the efficiency of both off- and on-chip regulators to compare LDOSeV/F with SeV/F algorithms. This is because LDOSeV/F requires the off-chip VR to vary its V_O, which results in lower efficiency than the off-chip VR with a fixed V_O (i.e., SeV/F).

Impact of VR efficiency on MIPS³/W of Multi-Program Workloads: A processor executing multiple applications may exhibit more substantial C2C IPC variations than one running multi-threaded applications. This depends on the mix and characteristics of applications. Consequently, supporting a wider range of V_O values using switching VRs may lead to higher MIPS³/W than using LDO VRs under a specified power constraint. Figure 6 shows the MIPS³/W comparison between two processors using switching and LDO VRs when running five mixes of memory- and compute-bound applications; we run the mixes of applications using the multi-core simulator (not in isolation) to accurately model the interaction between applications and memory contentions. When $V_I - V_O$ is larger than 100mV, the power loss by LDO VRs is higher than that of switching VR. However, the power loss by LDO VRs becomes lower than that of switching VRs for the DVFS intervals exhibiting $V_I - V_O$ less than 100mV. Consequently, as long as we have more DVFS intervals with $V_I -$

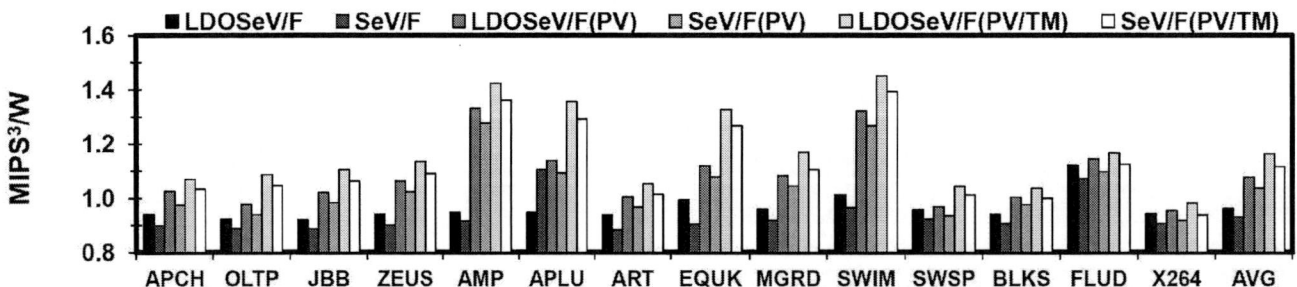

Figure 5: MIPS³/W comparison of 8-core processors supported by LDO (algorithms beginning with the LDOSeV/F prefix) and switching VRs (algorithms beginning with SeV/F) including the power loss from both on- and off-chip VRs. All results are normalized to a processor with ShV/F and include the power loss by the off-chip VR. Each interval is comprised of 10-million executed instructions.

60

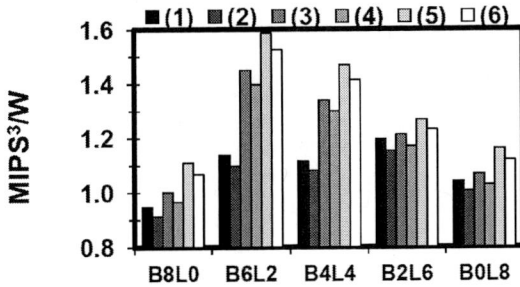

Figure 6: MIPS^3/W comparison of 8-core processors for multi-program workloads. (1), (2), (3), (4), (5), and (6) denote LDOSeV/F, SeV/F, LDOSeV/F(PV), SeV/F(PV), LDOSeV/F(PV/TM), and SeV/F(PV/TM), respectively. All results are normalized to a processor with ShV/F and include the power loss by the off-chip VR.

V_O less than 100mV, the processor using LDO VRs leads to higher MIPS^3/W than the one using switching VRs. Figure 6 shows that LDOSeV/F(PV/TM) has 4% higher MIPS^3/W than SeV/F(PV/TM). This is because, for B4L4, the fraction of DVFS intervals in which LDO VRs have higher efficiency than switching VRs is close to 60% of the total DVFS intervals that are experienced by individual cores.

Impact of DVFS interval on MIPS^3/W: The interval period for applying DVFS algorithms also impacts the benefit of DVFS. In theory, shorter DVFS intervals can capture more C2C IPC variations and thus lead to higher performance/power efficiency. Thus, we reduce the DVFS interval to every 5-million instructions while keeping the TM interval at 10-million instructions. As expected, MIPS^3/W for both LDOSeV/F(PV/TM) and SeV/F(PV/TM) increases, but the relative difference between them remains almost the same. Note that the DVFS interval is often determined by considering both (i) the computational overhead of the DVFS algorithm and (ii) PLL re-locking time for frequency changes. These prohibit a very short interval for a simple threshold-based DVFS algorithm even though the current state-of-the-art off-chip switching VRs can support much faster voltages changes. For example, Microsoft Windows Vista uses 1ms for the most aggressive interval value for DVFS. Finally, in this study, we assume that the on-chip switching VRs use on-package inductors because integrating on-chip high-quality inductors with cores becomes more challenging with technology scaling. On-chip switching VRs that utilize on-package inductors will not be able to keep up with the increasing number of cores due to the physical size of each on-package inductor.

5. CONCLUSION

In this paper, we propose a cost-effective technique to support per-core voltage domains for high-performance server-class processors. We demonstrate that PCPG devices augmented with small circuitry can operate as low-cost LDO VRs. Unlike on-chip switching VRs, LDO VRs do not require on-chip inductors, which are expensive and a major technical challenge for practical use of on-chip switching VRs, but their efficiency becomes poor as their output voltage applied to cores drops (i.e., large difference between input and output voltage of the LDO VRs). Consequently, per-core DVFS using LDO VRs may lead to lower performance/power efficiency than using switching VRs. However, our experiments show that C2C voltages variations are relatively small when the voltages are optimized to maximize performance under a power constraint. After modeling the power efficiency of both LDO and switching VRs using a 32nm technology, we show that the MIPS^3/W of an 8-core processor using LDO VRs is slightly higher than that of a processor using switching VRs. This is because the

efficiency of LDO VRs is higher than that of switching VRs for small C2C voltage variations in each DVFS interval, which was observed through our experiments.

ACKNOWLEDGEMENT

This work is supported in part by a generous gift grant from AMD, a Samsung Advanced Institute of Technology Global Research Outreach Program grant, Wisconsin Alumni Research Foundation, and an National Science Foundation grant (CCF-095360).

REFERENCES

[1] W. Kim, M.S. Gupta, G.-Y. Wei, and D. Brooks, "System Level Analysis of Fast, Per-core DVFS using On-chip Switching Regulators," in *IEEE/ACM Int. Symp. on High-Perf. Comp. Arch. (HPCA)*, 2008, pp. 123-134.

[2] R. Teodorescu and J. Torrellas, "Variation-Aware Application Scheduling and Power Management for Chip Multiprocessors," in *IEEE/ACM Int. Symp. on Comp. Arch. (ISCA)*, 2008, pp. 363-374.

[3] A.R. Alameldeen *et al.*, "Evaluating Non-Deterministic Multi-Threaded Commercial Workloads," in *Comp. Arch. Evaluation using Commercial Workloads (CAECW)*, 2002, pp. 30-38.

[4] Princeton University. PARSEC Benchmark Suite. [Online]. http://parsec.cs.princeton.edu/

[5] M.M.K. Martin *et al.*, "Multifacet's General Execution-driven Multiprocessor Simulator (GEMS) Toolset," *SIGARCH Comput. Archit. News*, vol. 33, no. 4, pp. 92-99, Nov 2005.

[6] D.M. Brooks *et al.*, "Power-aware Microarchitecture: Design and Modeling Challenges for next-generation microprocessors," *IEEE Micro*, vol. 8, no. 6, pp. 26-44, Nov/Dec 2000.

[7] S. Rusu *et al.*, "A 45 nm 8-Core Enterprise Xeon® Processor," *IEEE J. of Solid-State Circuits (JSSC)*, vol. 45, no. 1, pp. 7-14, Jan 2010.

[8] N.S. Kim *et al.*, "Frequency and Yield optimization using Power Gates in Power-Constrained Designs," in *IEEE/ACM Int. Symp. on Low Power Electronics and Design (ISLPED)*, 2009, pp. 121-126.

[9] P. Hazucha et al., "High Voltage Tolerant Linear Regulator with Fast Digital Control for Biasing Integrated DC-DC Converters," *IEEE J. of Solid-State Circuits*, vol. 42, no. 1, pp. 66-73, Jan 2007.

[10] Y. Hoskote, S. Vangal, A. Singh, N. Borkar, and S. Borkar, "A 5-GHz Mesh Interconnect for a Teraflops Processor," *IEEE Micro*, vol. 27, no. 5, pp. 51-61, Sep/Oct 2007.

[11] W. Fu and A. Fayed, "A feasibility study of high-frequency buck regulators in nanometer CMOS technologies," in *IEEE Dallas Circuits and Systems Workshop (DCAS)*, 2009, pp. 1-4.

[12] P Hazucha et al., "Area-Efficient Linear Regulator With Ultra-Fast Load Regulation," *IEEE J. of Solid State Circuits (JSSC)*, vol. 40, no. 4, pp. 933-940, Apr 2005.

[13] Jon Klein. (2006) Fairchild Semiconductors. [Online]. http://www.fairchildsemi.com/an/AN/AN-6005.pdf

[14] P. Hazucha *et al.*, "A 233-MHz 80%-87% Efficient Four-Phase DC-DC Converter utilizing Air-core Inductors on Package," *IEEE J. of Solid-State Circuits (JSSC)*, vol. 40, no. 4, pp. 838-845, Apr 2005.

[15] J. Lee, G. Hatcher, L. Vandenberghe, and C. K. Yang, "Evaluation of Fully-Integrated Switching Regulators for CMOS Process Technologies," *IEEE T. on Very Large Scale Integration (VLSI) Systems*, vol. 15, no. 9, pp. 1017-1027, Sep 2007.

[16] K. Aygun, M.J. Hill, K. Eilert, K. Radhakrishnan, and A. Levin, "Power Delivery for High-performance Microprocessor," *Intel Technology J.*, vol. 9, no. 4, pp. 273-283, Nov 2005.

[17] McPAT: An Integrated Power, Area, and Timing Modeling Framework for Multicore and Manycore Architectures. [Online]. http://www.hpl.hp.com/research/mcpat

[18] S. Sarangi *et al.*, "VARIUS: A Model of Process Variation and Resulting Timing Errors for Microarchitects," *IEEE T. on Semiconductor Manufacturing*, vol. 21, no. 1, pp. 3-13, Feb 2008.

A Hybrid and Adaptive Model for Predicting Register File and SRAM Power Using a Reference Design

Eric Donkoh
Intel Architecture Group
Hillsboro, OR 97124

eric.donkoh@intel.com

Alicia Lowery
Intel Architecture Group
Hillsboro, OR 97124

alicia.p.lowery@intel.com

Emily Shriver
Intel Strategic CAD Labs
Hillsboro, OR 97124

emily.shriver@intel.com

ABSTRACT

This paper presents a predictive SRAM power model that reduces the changes required to adapt existing models to handle new circuit topologies, process corners, and design space exploration. Analytical equations model the impact of varying common characteristics such as bit-width, entries, segmentation, gating, and sizing while topology specific characteristics are captured empirically from a reference design. On distinct topologies of multi-port read, single- and dual-ended writes, this approach demonstrates an error of 5% and 7% for leakage and dynamic power respectively. We show that for a specific topology, any reference configuration can be used for accurate prediction.

Categories and Subject Descriptors

B.8.2 [Performance and Reliability]: Performance Analysis and Design Aid; C.4 [Performance of Systems] - Modeling techniques, Design studies.

General Terms

Algorithms, Performance, Design, Theory

Keywords

Register File, SRAM, Power Model, Leakage Power, Dynamic Power, Reference Design

1. INTRODUCTION

In order to achieve high performance/watt in future deeply-scaled CMOS technologies, accurate prediction of power is critical for early-stage architectural design explorations of performance and power tradeoffs. Register files (RF) consume a significant portion of embedded and high-performance processors [1, 2] power. A large number of studies that explore energy efficiency tradeoffs involve changes to RFs. Hence, accurate power modeling of SRAMs is important for early architectural explorations.

Consider a typical micro-architectural study to explore a range of RF bit/entry sizes for best power/performance tradeoff. A modern processor and SoC could easily have >30 unique and custom RFs [1, 2]. Current parametric approaches for estimating RF power, analytical or empirical, are based on specific topologies and circuit implementations. To model a different RF topology, today's architectural power models and performance simulators [3, 4] either use the existing power model essentially unchanged (inaccurate for the new topology) or modify existing models for

the different topology (time consuming).

Analytical models [5, 6] use device process parameters to calculate the power using analytical equations that model the key capacitances (dynamic) or transistor sizes (leakage) in the RF. To adapt these models to different topologies/technologies require changes to the analytical formulas and parameters. Our proposed approach does not require any changes to the analytical formulas.

Empirical models rely on power simulation on the implemented circuit for the entire SRAM. A major drawback of regression based models [6, 7] is they require the implementation of several RF configurations to curve fit the empirical data for each topology and technology. Applying statistical techniques, such as design of experiments, typically requires at least 5 data points to accurately fit the data. Empirical models are therefore only valid for the specific circuit topology and technology used to generate the model coefficients. Thus, they usually present a method rather than reusable model equations.

Liang [8] presented a hybrid model that empirically captures the power of three array structures (1-bit x 1-entry, 2-bit x 1-entry, and 1-bit x 2-entry) and composes them analytically to obtain the power of an n-bit, m-entry structure. To reuse the model without modification however requires empirical data from the 3 specific array configurations on which the model is based. Moreover, these 1bx1e, 2bx1e, 1bx2e configurations do not exist in real design. As shown in Figure 4 and 5, using very small array configurations as reference to predict power for larger configurations is less accurate due to circuit and layout anomalies that could be magnified in very small arrays.

We present a hybrid model that addresses the aforementioned limitations of adaptability, reusability, and for the first time expands the architect's exploration options to include circuit-level design choices of segmentation, gating, and sizing. Our hybrid model does not calculate the base leakage and dynamic power values which are process technology and circuit topology dependent as in cacti [9]. Instead we rely on a single "reference design" to capture those dependencies and model relative changes from the reference. The empirical "reference design" data, which is an input parameter, captures topology-specific characteristics such as dual-ended/single-ended writes/reads, static/dynamic read, and process technology dependencies. We then analytically model the impact of cross-topology features such as changes in bit-width, entry-count, and common designer choices such as segmentation, gating, and sizing; using the same analytical model for all topologies. To further improve model accuracy and adaptability, we derive an equation for each RF stage independently. This enables the capture of stage-specific characteristics, thereby reducing prediction error and making the model easily adaptable to different SRAM topologies.

The distinct advantages of the modeling approach presented in this paper as compared to previous efforts are:

- A single adaptable model that can accurately predict power for different topologies by only modifying the input parameters of the model.
- Requires only a single reference design empirical data.
- Allows the use of any n-bit, m-entry reference to empirically capture design, topology, and technology specifics.
- Enables design space exploration of circuit implementation choices of gating, segmentation, and device sizing.

We validated our model on fully extracted layout of 3 different topologies, each with ~25 RF configurations. Section 3 presents the model results with an average error range of 5% (leakage) and 7% (dynamic). Section 4 presents scenario application of the proposed model to design space exploration of power sensitivity of two distinct topologies. Section 5 discusses summary and the accuracy of our results, suggesting that the power model presented herein can be used to easily and accurately predict and explore the power of an RF for any topology.

2. POWER MODEL

2.1 General Model Approach

The model is a hybrid of analytical equations and empirical data. We use an analytical approach to model topology independent impacts and empirical data by way of a "reference design" for topology and technology specific characteristics. A "reference design" refers to a circuit implementation of one configuration (bits/entry) of the topology under study from which power and timing data is known or can be obtained. A reference design is required for each distinct circuit topology. We capture the empirical data for each stage of the reference design (Figure 1, 2) and model the relative change in power due to changes in bit-width, number of entries, delay, and common designer choices such as segmentation, gating. We use this approach to make the model adaptable to different design topologies. We model each stage independently. This enables accurate modeling of the unique characteristics of each stage and easy adaptation of the model to different SRAM topologies. The model is of the form:

$$Power_{stage} = f \left\{ \begin{array}{l} RefStagePower, Bit, Entry, \\ Segmentation, Sizing, Gating, AF, SP \end{array} \right\} \quad (1)$$

$$Power_{Total} = \sum Power_{stage} \quad (2)$$

We use both delay and power models to capture the totality of the impact of design choices. In normal design, the increase in array size by additional bits and entries results in the need to increase drive strength for timing. Since leakage and capacitive loading correlates with device size, the impact of bits and entry growth on sizing is modeled by a delay penalty. The model uses a delay threshold number to capture the realistic design scenario where the driver is not upsized for any arbitrary increase in bits or entries but only after a specific threshold.

2.2 Unified Stage Model

We use a single unified model for all stages. Thus, bits and entries are used interchangeably in the model equations depending on the loading seen by the stage driver. The subscript "x" denotes the entity (bits or entry) whose increase (decrease) results in increased (decreased) loading on the stage driver. Subscript "y" denotes the orthogonal entity that does not affect the load on the stage driver. For example, the "x" and "y" entities of a wordline represents "bits" and "entries" respectively since a change in number of bits changes the load on the wordline driver. On the other hand, for the

bitlines "x" and "y" entities represents "entries" and "bits" respectively since bitline driver load depends on the number of entries. The words "entity" ("entities") therefore refers to bit (bits) or entry (entries) depending on stage.

2.3 Register File Topology

Figures 1 and 2 show an illustration of a register file write and read stage definitions. While our model is not specific to this topology, a typical RF/SRAM topology can be broken down to these basic stages. To model a different topology, the reference design is decomposed into component stages. The impact of each stage's distinct characteristic on power and delay is captured by the reference design per stage empirical data.

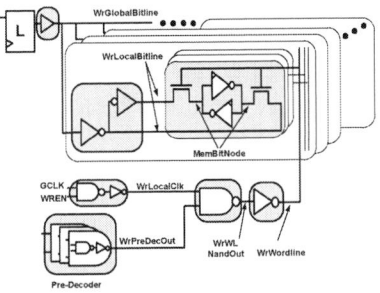

Figure 1: RF Write path showing write stages. Each stage is modeled independently

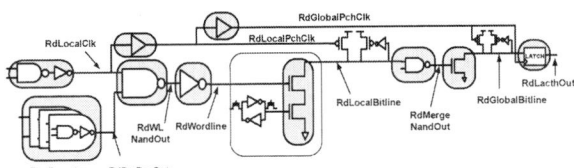

Figure 2: RF read path showing read stages. Each stage is modeled independently

Typically, large array bitlines are segmented into local (primary) and global (secondary) bitlines. A segment is an instance of a physically connected stage node. Thus a stage can have multiple instances of a segment. A global bitline drives (e.g. "WrGlobalBitline" stage) or combines (e.g. "RdGlobalBitline" stage) multiple local bitline segments. Thus the characteristics of a global bitline (number of drivers, gate loading, etc.) depend on the local bitline segments. To capture this in a unified model, two segmentation parameters, $N_{x\_persegment}$ and $N_{x\_perdepsegment}$ are defined. $N_{x\_persegment}$ represents the number of entities per segment while $N_{x\_perdepsegment}$ is the number of entities per dependent segment. The number of entities (bits or entries) for a stage is therefore scaled by its dependent segment as:

$$N_{xest} = Ceil \left[\frac{N_{x\_estimate}}{N_{x\_perdepsegment}} \right] \quad (3)$$

$$N_{xref} = Ceil \left[\frac{N_{x\_reference}}{N_{x\_perdepsegment}} \right] \quad (4)$$

$$N_{xperwireseg} = Ceil \left[\frac{N_{x\_persegment}}{N_{x\_perdepsegment}} \right] \quad (5)$$

$N_{x\_estimate}$: Number of "x" entities to be estimated for the stage.
$N_{x\_reference}$: Number "x" entities of the reference design stage.
$N_{x\_persegment}$: Maximum number of "x" entities per segment.

63

$N_{x\_perdepsegment}$: Number of "x" entities per dependent segment. If the stage has no dependency, $N_{x\_perdepsegment} = 1$

N_{xest} : Number of scaled "x" entities to be estimated for the stage.

N_{xref} : Number of scaled "x" entities of the reference design stage

$N_{xperwireseg}$: Number of scaled "x" entities per physically connected wire segment

2.4 Driver Sizing and Delay Effect (∂_{effect})

Changes in number of bits and entries require resizing of drivers at a specified interval to compensate for the stage slowdown. To account for this in the model, a driver delay threshold D_{Thresh} is defined as the number of entities per interval at which the driver is resized to compensate for any driver delay change. When that specified interval is reached, the driver is resized to drive that number of entities. A driver of that size is used to drive the next D_{Thresh} additional number entities until the threshold is reached before it is resized again. The number of entities that the driver is sized to drive is bounded by $N_{xperwireseg}$.

$$N_{xdriveloadsize} =$$

$$Max\left\{ Min\begin{pmatrix} Floor\left[\frac{N_{xperwireseg}}{D_{Thresh}}\right] \times D_{Thresh}, \\ Floor\left[\frac{N_{xest}}{D_{Thresh}}\right] \times D_{Thresh} \end{pmatrix}, 1 \right\} \quad (6)$$

$$N_{xdriveloadsize\_ref} =$$

$$Max\left\{ Min\begin{pmatrix} Floor\left[\frac{N_{xperwireseg}}{D_{Thresh}}\right] \times D_{Thresh}, \\ Floor\left[\frac{N_{xref}}{D_{Thresh}}\right] \times D_{Thresh} \end{pmatrix}, 1 \right\} \quad (7)$$

$N_{xdriveloadsize}$: Number of entities the driver is sized to drive

$N_{xdriveloadsize\_ref}$: Number of entities the reference design driver is sized to drive

The delay effect which represents the relative change in delay due to change in the number of driven entities relative to the reference design is modeled as:

$$\partial_{effect} = \left(\frac{N_{driveloadsize}}{N_{driveloadsize\_ref}}\right)^{\propto} \quad (8)$$

The $\propto$ value for each stage is derived empirically (via curve fitting) or through simulation.

The estimated stage delay:

$$D_{Stage} = D_{stagedelay\_reference} \times \partial_{effect} \quad (9)$$

$D_{stagedelay\_reference}$: Reference design empirical stage delay.

The relative driver size needed to drive the new number of entities to achieve the same delay as the reference design is:

$$\partial_{effectsize} = \beta \times \partial_{effect} \quad (10)$$

β : Driver delay to sizing ratio. For example $\beta=1$ implies a 1% increase in delay results in 1 % increase in driver size

2.5 Leakage Power Model

2.5.1 Stage Leakage Power ($P_{StageLeakage}$)

The stage leakage is modeled by the driver instance count relative to the reference design.

$$P_{StageLeakage} = P_{stageleakage\_reference} \times SP_f \times$$

$$\left[\lambda + (1-\lambda) \times \partial_{effectsize} \times \left(\frac{N_{xdriver}}{N_{xdriver\_ref}}\right) \times \left(\frac{N_{ydriver}}{N_{ydriver\_ref}}\right) \right] \quad (11)$$

$$N_{xdriver} = Ceil\left[\frac{N_{x\_estimate}}{N_{x\_perdriver}}\right] \qquad N_{xdriver\_ref} = Ceil\left[\frac{N_{x\_reference}}{N_{x\_perdriver}}\right]$$

$$N_{ydriver} = Ceil\left[\frac{N_{y\_estimate}}{N_{y\_perdriver}}\right] \qquad N_{ydriver\_ref} = Ceil\left[\frac{N_{y\_reference}}{N_{y\_perdriver}}\right]$$

$P_{stageleakage\_reference}$: Reference design empirical stage leakage

$N_{y\_estimate}$: Number of "y" entities to be estimated for the stage

$N_{y\_reference}$: Number of "y" entities of the reference design stage

$N_{x\_perdriver}$: Maximum number of "x" entities per driver.

$N_{xdriver}$: Total number of "x" entity drivers

$N_{xdriver\_ref}$: Total number of reference design "x" entity drivers.

$N_{y\_perdriver}$: Maximum number of "y" entities per driver

$N_{ydriver}$: Total number of "y" entity drivers

$N_{ydriver\_ref}$: Total number of reference design "y" entity drivers.

λ : Fraction of reference leakage ($P_{stageleakage\_reference}$) that is fixed (from auxiliary circuits)

SP_f : Leakage signal probability factor of the stage node

2.5.2 Signal Probability Factor (SP_f)

The stage signal probability (SP) represents the probability of a signal being in logic state "1". Since the p-channel and n-channel devices typically have different leakage (due to possibly unequal driver P-N sizing and process technology), the total leakage of a driver is dependent on the SP of a stage output node.

$$Leakage = SP \times L_n + (1 - SP) \times L_p \quad (12)$$

The SP factor accounts for the SP impact of the predicted configuration relative to the reference design.

$$SP_f = \frac{SP_{estimate}+(1-SP_{estimate})\times L_R}{SP_{reference}+(1-SP_{reference})\times L_R} \qquad where\ L_R = \frac{L_p}{L_n} \quad (13)$$

L_R : P-device/N-device leakage ratio; $SP_{estimate}$: SP of the stage to be estimated; $SP_{reference}$: SP of reference design stage

2.6 Dynamic Power Model

2.6.1 Stage Dynamic Power ($P_{StageDynamic}$)

The dynamic power of a stage is a function of the stage capacitance (C), voltage (V), activity (AF) and frequency (F).

$$P_{dynamic} = C \times AF \times V^2 \times f \quad (14)$$

The effects of voltage and frequency on power estimation are captured by the empirical stage power of the reference design. The activity factor and capacitance are the factors that will therefore determine the stage power relative to the reference power. We categorize the capacitance of a stage into three components as shown in Figure 3:

Repeated Capacitance (Φ_r) – This is the fraction of the stage capacitance that is an instantiated multiple of the reference and changes with number of "x" entities.

Non-repeated capacitance (Φ_{nr}) – This is the fraction of the stage capacitance that is not directly dependent on the number of "x" entities but indirectly affected by device resizing as a result of change in the number of "x" entities.

64

Overhead Capacitance (Φ_{ov}) – This is the fraction of the stage capacitance that is not impacted by change in number of entities. This is usually a fixed cap from routing overhead and fixed logic associated with the stage.

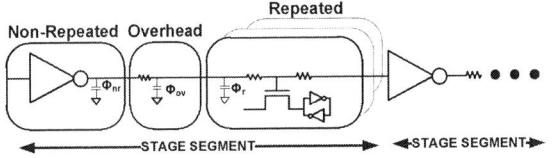

Figure 3: Stage cap definition for dynamic power model

We capture these components individually and analytically model the impact of changes in number of bits, entries, and driver sizing on each of the components.

2.6.1.1 Segment Dynamic Power

The dynamic power of a stage is modeled by the stage segment count relative to the reference design. The stage segment dynamic power is modeled as:

$$P_{segmentdynamic} = \frac{AF_{estimate}}{AF_{reference}} \times P_{dynamicpersegment\_ref} \times$$
$$N_{xstagesegment} \times N_{ystagesegment} \quad (15)$$

$$N_{xstagesegment} = \Phi_r \times N_{rpt} \times \partial_{effectsize\_rpt} + \Phi_{nr} \times$$
$$\partial_{effectsize\_nr} + \Phi_{ov} \quad (16)$$

$$N_{ystagesegment} = \left(\frac{N_{yseg}}{N_{yseg\_ref}}\right)^{(1-G_y)} \quad (17)$$

$$N_{yseg} = Ceil\left[\frac{N_{y\_estimate}}{N_{y\_persegment}}\right] \quad N_{yseg\_ref} = Ceil\left[\frac{N_{y\_reference}}{N_{y\_persegment}}\right]$$

$$P_{dynamicpersegment\_ref} =$$
$$\frac{P_{stagedynamic\_reference}}{\left(Ceil[N_{xsegment\_ref}] \times (\Phi_{ov}+\Phi_{nr})+ N_{xsegment\_ref} \times \Phi_r\right)^{(1-G_x)}} \quad (18)$$

$$where \ N_{xsegment\_ref} = \frac{N_{xref}}{min(N_{xref}, N_{xperwireseg})}$$

$\Phi_r, \Phi_{ov}, \Phi_{nr}$ are the reference design cap components.

N_{rpt} : Fraction of the segment to be estimated.

$P_{stagedynamic\_reference}$: Reference design empirical stage dynamic power

$N_{y\_persegment}$: Maximum number of "y" entities per segment for the stage.

N_{yseg} : Number of "y" entities per stage segment

N_{yseg_ref}: Total number of reference design "y" entity drivers.

$\partial_{effectsize\_rpt}$: Delay effect of repeated cap

$\partial_{effectsize\_nr}$: Delay effect of non-repeated cap

G_x, G_y : Indicates gating of the reference design "x" and "y" entity segments respectively. $G = 1$ (Gated), $G = 0$ (not Gated).

2.6.1.2 Stage Segment Gating

To capture stage segmentation and gating, we define two components of dynamic power for a stage:

Full segment – representing power of a complete segment

Partial Segment – representing power for a fraction of a segment.

We model the gating effect by assuming that all the "Full" and "Partial" segment components are active if the stage is not gated. If the stage is gated, only one full segment is assumed to be active.

$$N_{xfullsegment} = floor\left[\frac{N_{xest}}{N_{xperwireseg}}\right] \quad (19)$$

$$N_{xpartial} = mod\left[\frac{N_{xest}}{N_{xperwireseg}}\right] \quad (20)$$

$N_{xfullsegment}$: Number of instances of a full segment.

$N_{xpartial}$: Number of entities in the partial segment component.

The dynamic power for a stage with gating and segmentation is modeled as:

$$P_{StageDynamic} = Z_{seg} \times N_{seggate} \times P_{stagedyn\_segment} + Z_{part} \times N_{partgate} \times P_{stagedyn\_partial} \quad (21)$$

From equation (15):

$$P_{stagedyn\_segment} =$$
$$P_{segmentdynamic} \mid N_{rpt} = \frac{N_{xperwireseg}}{Min(N_{xref}, N_{xperwireseg})} \quad (22)$$

$$P_{stagedyn\_partial} =$$
$$P_{segmentdynamic} \mid N_{rpt} = \frac{N_{xpartial}}{Min(N_{xref}, N_{xperwireseg})} \quad (23)$$

$$N_{seggate} = \left(Max\{N_{xfullsegment}, 1\}\right)^{(1-G_x)} \quad (24)$$

$$Z_{seg} = Min(N_{xfullsegment}, 1) \quad (25)$$

$$N_{partgate} = 1 - Min(N_{xfullsegment}, G_x) \quad (26)$$

$$Z_{part} = Min(N_{xpartial}, 1) \quad (27)$$

$P_{stagedyn\_segment}$: Full segment dynamic power component

$P_{stagedyn\_partial}$: Partial segment dynamic power component

$N_{seggate}$: Number of gated segments component.

$Z_{part}(0,1)$: Zero out the partial segment component if gated and one or more segments exist.

$Z_{seg}(0,1)$: Zero out the full segment component if total number is less than a segment.

2.7 Total Power

The total power is the sum of all stages. Using the illustrative reference design topology in Figures 1, 2:

$$Power_{ReadTotal} = \sum Power_{Readstages} =$$
$$\sum Power \begin{Bmatrix} RdPredecOut, RdLocalClock, RdWLNandOut, \\ RdWordline, RdLocalPchClk, RdGlobalPchClk, \\ RdLocalBitline, RdMergeNandOut, \\ RdGblobalBitline, RdLatchOut \end{Bmatrix}$$

$$Power_{WriteTotal} = \sum Power_{writestages} =$$
$$\sum Power \begin{Bmatrix} WrPredecOut, WrLocalClock, WrWLNandOut, \\ WrWordline, WrLocalBitline, RdGblobalBitline, \\ MemBitNode \end{Bmatrix}$$

3. RESULTS AND ANALYSIS

3.1 Model Validation

In validating the model, we generated production quality schematic and layout of ~25 configurations (bit x entry) each of 3 different RF topologies using an in-house automation tool. We then extracted the layout parasitics and ran an in-house power analysis tool to obtain the actual dynamic and leakage power for all configurations. The power simulation testbench used were a set of input stimuli that exercised the array read and write operations at a set read/write activity and data toggling rate. To estimate power using the model, for each distinct topology, we picked a single configuration of that topology as the "reference design" and extracted the per stage empirical data for the model empirical parameters. We then used the per stage "reference design" empirical data to estimate the power of the remaining suite of configurations (~24) of that topology using the model. We finally compared our estimated power against the power obtained from the in-house production power analysis tool.

Validation of the model encompasses the following categories: 1) *Reference* validation in which 5 different bit and entry configurations (*4bx4e, 16bx32e, 32bx16e, 64bx48e, and 24bx24e*) of a 1-read 1-write Dual-Ended (1R1W DE) topology (Figure 1) were each used as a reference design to predict the remaining configurations, demonstrating that any reference of the topology can be used; 2) *Topology & Multi-Port* validation in which the model is used to predict other distinct topologies: 1-read 1-write port Single-Ended write (1R1W SE), and 2-read 1-write port dual-ended (2R1W DE) write topologies, using their respective reference design. The DE topology uses a jam-latch memory cell while the SE topology used an interruptible latch memory cell; 3) *Delay effect* validation of the model's ability to predict power in the presence of timing related gate-sizing changes; (4) Multiple benchmarks on a manually designed RF.

3.2 Power Model Results

Figure 4 presents the leakage error distribution for the various RF validations of reference, topology, and delay effect. The average leakage error range is 5% with outliers explained below. The results show that given a representative reference, the model can accurately predict the power of any configuration of that topology.

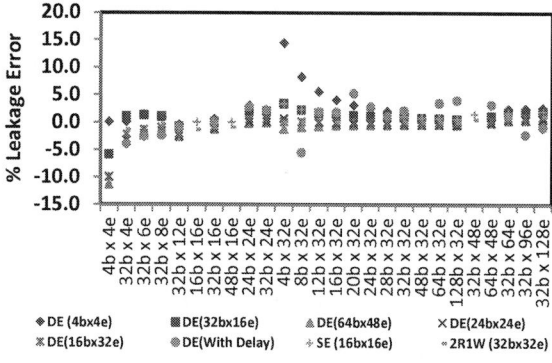

Figure 4. Leakage power error distribution for multiple reference designs, topology, multi-ports, and delay effect.

Leakage outliers are observed when the reference design exhibits circuit characteristics that significantly deviate from the predicted configuration. This is illustrated by the "RdWLNandOut" stage (Figure 2) using a 4bit x 4entry (*4bx4e*) reference. The wordlines are implemented using four 2-input NAND gates. In the inactive mode (leakage state) one of four (25%) NAND input decoded signals from the pre-decoder will always be at logic state "1". This results in leakage through a single N-device of the corresponding NAND gate (input "01"), while the remaining three NAND gates will leak through 2-series N-devices (input "00"). The leakage through a stacked device is considerably lower than a

single device. Since the model will make its prediction based on the reference, this behavior will be replicated, introducing error. Hence, using a *4bx4e* reference to predict a configuration of 32-entries "RdWLNandOut" stage will predict a scenario of 25% (8 of 32) NAND gates leaking though single N-device. However, in actuality only 1 of 32 (3.1%) NAND gates will be leaking through a single device. The impact of this prediction error is dependent on the contribution of the stage to the total power.

Figure 5 presents the dynamic power error distribution for all validation steps across a range of RF configurations. The average dynamic error range is 7% with outliers. Dynamic power is sensitive to layout inconsistencies and the cap ratios of the reference design. The suite of RFs that were generated from automation did not exhibit consistent routing across all configurations. For example, the length of lower level connectivity and routing metals like Metal 1 and Metal 2 varied between configurations. Layout anomalies introduced errors in the actual data against which the model results were compared. This is particularly pronounced with small entry configurations where minor inconsistencies in layout can contribute significantly to a total stage capacitance and hence affect the measured dynamic power. In Figure 6, the model is used to estimate power on different application benchmarks on a manually designed 32bx64e array with good accuracy.

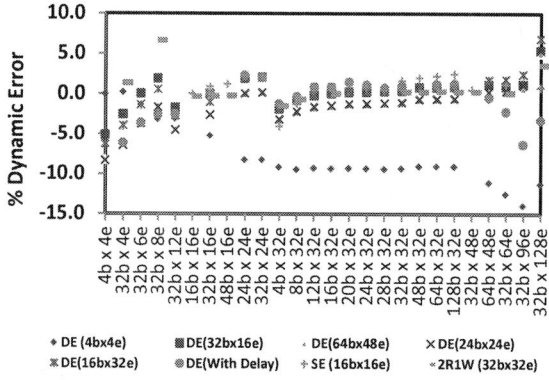

Figure 5. Dynamic power error distribution for multiple reference designs, topology, multi-ports, and delay effect.

4. MODEL APPLICATION

We used our power model to make real world design decisions regarding design topology choices and sensitivity studies. In a typical scenario, an architectural change is proposed to an existing array on the chip. This may be a specialized array. Using our model, the empirical data from the existing array is captured as the reference design to model the power sensitivity to the proposed array size change. Our model can be used to study any number of distinct RF topologies on the chip by using the empirical data from their respective existing reference designs. Figure 7 shows a surface plot of power sensitivity to bits and entries for a topology under study depicting the non-linear power gradient of bit and entry increase for that topology.

In another scenario, there exists multiple circuit implementation options for an RF and the question is the best way to implement a specific array from power, area, timing, and other constraints. For this circuit implementation exploration, we modeled two distinct array topologies, a 1R1W SE and a 1R1W DE array, that have

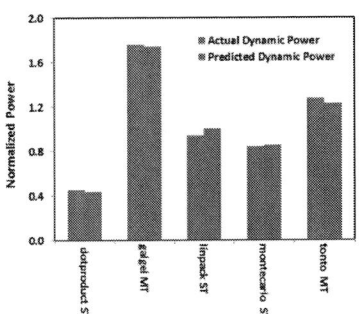
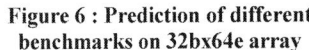

Figure 6 : Prediction of different benchmarks on 32bx64e array

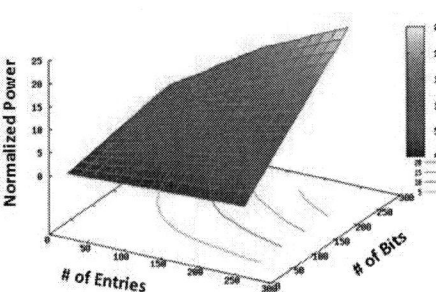

Figure 7: Contour plot of power sensitivity to bits and entries for an array topology under study.

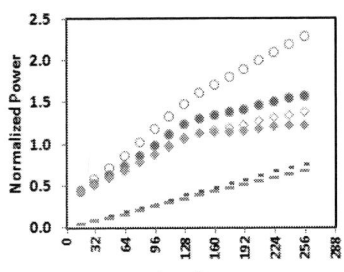

Figure 8: Comparison of two RF topologies as a function of entries, and gating

different SRAM bitcells, physical aspect ratios, and write segmentation. Using a single reference design for each topology, we used the model to predict the impact of bit, entry, and other circuit implementation choices on each topology. Figure 8 shows the relative power comparison of the two topologies as a function of bits and entries. It also shows the impact of write data gating design space exploration on the two topologies. It can be seen from this study that the DE implementation exhibits significantly higher dynamic power cost at high entry configurations relative to the SE configuration, due to the dual-write operation. However, when write data gating is implemented, the dynamic power delta between SE and DE is significantly decreased. The power cost of doubling the entries to an existing DE RF can be mitigated by write data gating. This fact is considered by the architect/designer in the analysis and evaluation of a DE RF entry size increase feature proposal.

5. CONCLUSION

We present a reusable hybrid model in which the analytical formulas remain unchanged for all topologies while incorporating real design choices such as segmentation, gating, and timing impacts; a combination not in previous RF power models. The model is adaptable to other SRAM array structures by using a reference design and decomposing it into component stages, each stage characteristics captured independently by the reference design's empirical data. It allows any representative reference design to be used in estimation and exploration. Changes in topology and/or process require only the empirical data of a single reference to be updated. A typical modern microprocessor has a large number of unique RFs [2], most of which are manually designed and cannot be compiled, making design exploration across various combinations of bits, entries, gating, and segmentation intractable. Using our model, individual unique RFs can be explored using their respective single reference designs without requiring large number of samples as needed for a curve fit approach. Our model is not tied to any specific technology or design style as we do not model the process, device level, or design environment dependent base values but rely on a reference design empirical data to capture these specifics.

We demonstrated how the model enables tradeoffs both within (e.g. gating, segmentation) and across multiple topologies (SE vs. DE) for optimal implementation and provides wider exploration of circuit implementation options. Table 1 summarizes the model's prediction of power consumption across multiple reference configurations, showing good accuracy with the actual RF designs. The proposed reference design approach is also used for delay and area estimations.

Table 1: Maximum leakage and dynamic error for the read and write stages of all validation steps

Topology Used	Ref Design Used	LEAKAGE		DYNAMIC	
		Read	Write	Read	Write
Different References					
DE 1R1W	4B4E	9.2%	7.5%	15.6%	6.7%
DE 1R1W	32B16E	5.4%	7.4%	6.0%	7.5%
DE 1R1W	64B48E	5.5%	2.5%	5.3%	7.9%
DE 1R1W	24B24E	5.5%	2.0%	6.3%	7.5%
DE 1R1W	16B32E	3.3%	2.3%	6.2%	6.9%
Delay Effect Sizing					
DE 1R1W	32B16E	6.0%	6.2%	6.7%	6.6%
Topology/Multi-Port					
SE 1R1W	16B16E	4.4%	2.5%	4.4%	6.3%
DE 2R1W	32B32E	2.2%	2.3%	5.3%	6.0%
Average Max Error		5.2%	4.1%	7.0%	6.9%

6. ACKNOWLEDGEMENT

The authors wish to acknowledge Patrick Chiang (Oregon State University) for his review and valuable feedback, Kurt Kreitzer, Nanda Siddaiah, Patrick Juliano, and Mike Kishinevsky.

7. REFERENCES

[1] N. Kurd et al., "A Family of 32nm IA Processors", *IEEE Journal of Solid-State Circuits, 2011, vol 46, pp 119-130.*

[2] K. Anshumali et al. "Circuit And Process Innovations to Enable High-Performance, and Power and Area Efficiency on the Nehalem and Westmere Family of Intel processors" *Intel Technology Journal, 2010, vol 14, pp 104-127*

[3] N. Vijaykrishnan, et al. "Energy-driven integrated hardware-software optimizations using SimplePower", *ISCA 2000.*

[4] D. Brooks, et al, "Wattch: A Framework for Architectural-Level Power Analysis and Optimization," *ISCA 2000.*

[5] Xuemei Zhao, et al: Design and Realization of a Low Power Register File Using Energy Model, *PATMOS 2002: pp. 268-277.*

[6] Minh Q. Do, et al, "Parameterizable Architecture-Level SRAM Power Model Using Circuit-Simulation Backend for Leakage Calibration," pp.557-563, *ISQED 2006.*

[7] S. L. Coumeri, D. E. Thomas Jr, "Memory Modeling for System Synthesis", *IEEE Transactions on VLSI, June 2000.*

[8] Xiaoyao Liang, Kerem Turgay, David Brooks, "Architectural Power Models for SRAM and CAM Structures Based on Hybrid Analytical/Empirical Techniques", *ICCAD 2007.*

[9] S. Thoziyoor, et al., "CACTI 5.1," *HP Technical Report, 2008,* http://www.hpl.hp.com/techreports/2008/HPL-2008-20.pdf

Coding-based Energy Minimization
for Phase Change Memory

Azalia Mirhoseini
Electrical and Computer
Engineering Department,
Rice University
azalia@rice.edu

Miodrag Potkonjak
Computer Science
Department, University of
California, Los Angeles
miodrag@cs.ucla.edu

Farinaz Koushanfar
Electrical and Computer
Engineering Department,
Rice University
farinaz@rice.edu

ABSTRACT

We devise new coding methods to minimize Phase Change
Memory write energy. Our method minimizes the energy
required for memory rewrites by utilizing the differences between PCM read, set, and reset energies. We develop an
integer linear programming method and employ dynamic
programming to produce codes for uniformly distributed
data. We also introduce data-aware coding schemes to efficiently address the energy minimization problem for stochastic data. Our evaluations show that the proposed methods
result in up to 32% and 44% reduction in memory energy
consumption for uniform and stochastic data respectively.

Categories and Subject Descriptors

B.7.1 [**Integrated Circuits**]: Memory Technologies

General Terms

Algorithms, Performance, Design

Keywords

Phase Change Memory, Energy Efficient Coding

1. INTRODUCTION

The demand for data and information storage has been
upsurging at an unprecedented rate, continually fueling improvements for underlying memory technologies. However,
it has become clear that alternative technologies are necessary to fulfill the requirements of developing devices and applications beyond the near future. In addition to these technology developments, redesigned architectures, tools, and
system-level methodologies are needed to take advantage of
the properties of the latest digital storage media.

One very promising emerging non-volatile storage technology is *Phase-Change Memory (PCM)*. PCM data storage
exploits the large electrical resistance difference between two
states of the phase-change material. In one state, the material is amorphous with a high resistance; in another, the
material is crystalline and highly-conductive. After more
than a decade of dedicated research into new forms of phase
change media, PCM technology is finally available on the
market. Recent announcements indicate advances towards
multi-level phase change memory with improved integration,
retention, endurance, and yield characteristics [19].

This paper aims at minimizing the energy cost of rewriting
to PCM by creating low overhead data encoding methods.
The proposed encoding scheme utilizes PCM bitwise manipulation ability during the word overwrites; only the bits
that are changing for the new word compared to the existing
word in the memory location would require overwriting. Our
new encoding scheme ensures that the energy cost for the
required overwrites is minimized at the expense of adding a
small number of additional bits for encoding. Our formulation and solutions incorporate the fact that the PCM set
and reset energy costs are not equal (see Appendix A). This
asymmetric model captures the inherent physical differences
between the organized crystalline and amorphous state transitions. To the best of our knowledge, this is the first work
that utilizes PCM asymmetric set and reset energy behavior
to minimize energy consumption. Our optimization is easily integrable within the processor architecture and memory
interface with a very low complexity and overhead.

A special case of data encoding for minimizing the unidirectional transitions in the memory is the Rivest and Shamir
Write-Once Memory (WOM) coding which assumed a memory model where the bits could only be set (and could not
be reset) [17]. The goal was to increase the number of effective cycles for memory rewrites. Subsequent work followed, mostly in information theory and coding with the
goal of estimating the capacity and finding more efficient
WOM codes. Applications and extension of this model for
addressing the flash memory device lifetime improvements
were studied [8, 20]. Unfortunately, the assumptions made
in the earlier theoretical work limits their applicability to
PCM because they do not capture PCM bi-directionality
and bit-level access properties.

The large space of possibilities provided by the freedom in both setting and resetting transitions and bit-
programmability motivate the development of new type of
codes that can be applied for improving PCM write energy.
The complicating factors are the new degrees of freedom
and the curse-of-dimensionality resulting from the exponential number of plausible code combinations. To address the
challenges, this paper presents a novel formal handling of the

68

energy minimization that is appropriate for PCM and other storage technologies with bit-level access while simultaneously considering the asymmetric set and reset transition energy costs. The paper's contributions are:

- We introduce a formal treatment and formulation of PCM coding, with the goal of minimizing the energy. We show that the problem is NP-complete.

- A methodology for deriving the optimal bounds for minimum energy data encoding problem is developed.

- We devise an Integer Linear Programming (ILP) formulation that can find the optimal codes. Our ILP framework can integrate both symmetric and asymmetric set/reset costs for different code sizes.

- For runtime and efficiency reasons, we develop an alternative rapid and efficient algorithm for addressing the problem. The method builds upon the smaller optimal codes using a Dynamic Programming (DP) approach.

- We introduce an efficient distribution-aware data encoding method for non-uniformly distributed data.

- Our evaluations on a diverse set of benchmark data show significant gains in PCM energy performance.

The remainder of the manuscript is organized as follows. The relevant literature is surveyed in Section 2. The architecture of the method is presented in Section 3. Section 4 formally defines the energy saving data encoding problem and discusses its complexity. Our method for finding the optimum bounds on the codes is presented in the same section. In Section 5 we introduce the coding algorithms. Evaluations of the methods on several benchmark data sets are presented in Section 6. We conclude in Section 7. Acknowledgements are presented in Section 7.1. We provide complementary methods and discussions in the Appendix.

2. RELATED WORK

The field of resistive memory material has been rapidly growing in recent years, both in research and in terms of industrial prototypes, making PCM the most viable emerging technology for the next generation storage devices [13, 19]. Recent work has shown significant efficiency and improvements in memory structures by integrating the PCM within the storage hierarchy [10, 14, 21].

Previous PCM research has demonstrated that the PCM endurance, reliability, and energy consumption would greatly improve if redundant writes are avoided, i.e., by reading the existing contents of the bits and only programming those bits that must be changed [21]. *Flip-N-Write* is a protocol that adds an indicator bit to each word to determine if the word is inverted or not, [9]. PCM controller can write the data in an inverted form if it requires less number of bit changes. No optimality proof was provided.

Our paper formalizes, provides proofs and generalizes the Flip-N-Write method by devising codes of length $N + K$ for words of length N, where $K \geq 1$. Our approach, for the first time in the literature, considers the asymmetric set and reset energy costs. We will show that significant improvements in energy are achieved at the expense of memory overhead.

Write-Once Memory (WOM) encoding was introduced in [17] to increase the number of writes to uni-directional memories. The NAND flash memory has been modeled as a one-way transitional memory and generalizations of the WOM codes have been applied to it [8, 20]. However, the WOM model and the flash encoding methods do not capture PCM properties including bit-level access and asymmetric energy costs. The bit-level operations for PCM have been used earlier for error correcting codes [18].

The WOM model has also been naturally extended to the family of Write-Efficient Memory [7], with the objective of minimizing the overall number of transitions. However, to the best of our knowledge, the few papers available on WEM have mainly focused on developing loose bounds without providing an optimality guarantee, or they centered on constructing suitable error correcting codes, e.g., [12, 15].

3. ENCODING ARCHITECTURE

Figure 1 presents an abstract view of the placement of the data encoding/decoding module for our method. Our algorithms for devising the codes are run off-line, so their runtime complexity does not affect the realtime chip performance. The energy saving codes resulting from our algorithms are then saved in the memory controller which interfaces to the PCM on one side and to processing units on the other side. The memory controller may also be interfaced to other storage devices in the memory hierarchy. The complexity of runtime encoding and decoding will be discussed in Section 5.3. Our consistent assumption is that the read energy consumption is negligible compared to that of set and reset [21]. More details of the PCM operation and energy characteristics can be found in Appendix A.

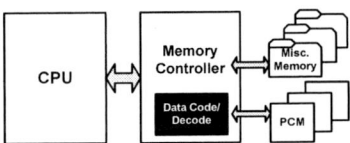

Figure 1: Data encoding/decoding module is a part of memory controller and is interfaced to the PCM.

4. PROBLEM FORMULATION, COMPLEXITY AND BOUNDS

Our goal is to minimize the energy cost associated with writing words to the memory. Each word consists of a fixed number of bits and the energy cost of writing the word is equal to the total cost of the required bit flips (sets/resets).

We provide an optimal encoding scheme that assigns multiple representations (or codes) to each word in the data set. The objective of encoding is to minimize the energy cost for writing the next word of data. The method trades-off the encoding data overhead with resulting energy improvements.

We have shown that assigning the best codes to each word is equivalent to clustering the vertices of a graph where each cluster represents a word (See Appendix B for an example). Clustering should be done such that it yields the minimum distance between the vertices of different clusters. We can formally define our problem as follows:

Problem. Minimize the energy cost of PCM rewrites.

Given. The word and the codeword (symbol) lengths in bits denoted by N and $N + K$ respectively, where $K \geq 1$. Each word is represented by 2^K symbols. The read, set and reset energies are denoted by E_{read} and E_S and E_R respectively.

Objective. Find the best codes for each word so as to minimize the average energy cost of overwrites. We refer to this problem as $\mathcal{P}(N, K)$.

4.1 Problem Formulation

We denote the words by $W_1, W_2, \ldots, W_{2^N}$ and denote the codes corresponding to word W_i by Z_{li}, where $1 \leq l \leq 2^K$.

The cost function C measures the amount of energy consumed to overwrite a symbol by another one. To overwrite Z_{li} with $Z_{l'i'}$, if N_S number of bit sets and N_R number of bit resets are needed, then C would be:

$$C(Z_{li}, Z_{l'i'}) = (N + K).E_{read} + (N_S).E_S + (N_R).E_R. \quad (1)$$

The first term on left shows the energy for reading Z_{li}. This cost is negligible due to PCM high read efficiency. The next two terms show the energy for the overwrite process (setting and resetting) so as to get $Z_{l'i'}$. Similar bits in the two symbols remain untouched. Function Φ gives the energy required to overwrite a currently written symbol Z_{li} by a symbol of the next word $W_{l'}$ that incurs the minimum cost:

$$\phi(Z_{li}, W_{l'}) = \min\{C(Z_{li}, Z_{l'i'}), \quad \forall 1 \leq i' \leq 2^K\}. \quad (2)$$

The Objective Function (OF) can be written as follows:

$$\textbf{OF}: \min\{\mathcal{C}(N, K) = \frac{1}{2^{2N+K}} \sum_{1 \leq l, l' \leq 2^N} \sum_{1 \leq i \leq 2^K} \phi(Z_{li}, W_{l'})\}. \quad (3)$$

The minimization is over all possible partitioning of the symbols to the words. Function $\mathcal{C}(N, K)$ represents the average energy cost of code overwrites for all possible rewrites.

4.2 Problem Complexity

We construct a transformation of the energy minimizing coding problem to a distance-based graph clustering problem where each cluster corresponds to a word (Appendix B). The goal is to minimize the inter-cluster distances. In our problem, the inter-cluster distance is the mean distance between the code symbols in one cluster and the closest code symbol in every other cluster. Extensive prior work on the class of distance-based graph clustering have shown that this problem is NP-complete. The proof was given by a reduction from the set covering problem [11].

4.3 Optimal Bounds on the OF

We develop a method for finding a lower bound for the OF. The average cost of overwriting each symbol Z_{li} with the other words is determined by the following formulation: $\frac{1}{2^N-1} \sum_{l'} \phi(Z_{li}, W_{l'})$ for $l' \neq l$ and $1 \leq l \leq 2^N$. An optimal code assignment is the one that assigns each of the closest $2^N - 1$ symbols to Z_{li} to one of the words $W_{l'} \neq W_l$.

We construct a lower bound for the OF as follows. First, we calculate the distances from each code Z_{li} to all the other $2^{N+K} - 1$ possible codes. Next, the resulting distances are sorted and the average sum of the smallest $2^N - 1$ distances are calculated for each node. The computational complexity of this method is $O(2^{N+K}N + K)$. Based on the fact that the practical values for the memory word and code lengths are chosen relatively small (as we discuss in the evaluation

results) and that the procedure is performed off-line, this method is applicable and gives a lower bound for the OF.

5. ENERGY MINIMIZATION ENCODING

We propose two different approaches for solving the coding problem. Our first solution is based on mapping the problem to an instance of an Integer Linear Programming (ILP). An ILP formulation requires linear objective function and constraints. The variables take integer values. There is a combinatorial complexity associated with assigning values to the variables of our NP-complete problem.

The OF represented in Equation 3 is non-linear since the function ϕ is a distance minimization function. To formulate the OF in a linear form, we define new indicator variables for the distance of each symbol in a cluster to the closest symbol in every other cluster. The OF is equivalent to the average of all these variables. Certain linear constrains are applied to ensure the variables meet the minimum distance criteria. The ILP method finds the optimal solution at the expense of runtimes exponentially increasing with the code size. Due to space limitation, details of the ILP formulation is discussed in Appendix C.

The second solution is based on Dynamic Programming (DP) paradigm for uniformly distributed data. We also develop codings for other data distributions that can further minimize the energy.

5.1 Coding for Uniform Data Distributions

First, we show the optimal coding for solving $\mathcal{P}(N, 1)$. Next, we show how to devise the codes for any $\mathcal{P}(N, K)$ based on the coding solutions for smaller N and K values.

5.1.1 Optimal Coding for $\mathcal{P}(N, 1)$

Claim: Optimal coding of $\mathcal{P}(N, 1)$, for any $N \geq 1$ is achieved by assigning the complement pairs of symbols to the words. The complement of a symbol is derived by flipping all its bits.

Proof: The optimal coding finds $2^K = 2$ symbols, each of size $N + 1$, for each word. For now, we assume that set and reset energies equally. Then the overwrite cost is proportional to the number of bitwise differences for the codes, $E_R = E_S = E$. The average transition cost from each code Z_{li} to all the other words satisfies the following inequality:

$\frac{1}{2^N-1} \sum_{l'} \phi(Z_{li}, W_{l'}) \leq 0.\binom{N+1}{0} + E.\binom{N+1}{1} + \cdots +$
$\frac{N-1}{2} E.\binom{N+1}{\lfloor\frac{N-1}{2}\rfloor} + i_o.\frac{N+1}{2} E.\binom{N+1}{\lfloor\frac{N+1}{2}\rfloor}$, for $1 \leq l' \leq 2^N$.

Where $i_o = 1$ if N is odd and $i_o = 0$ otherwise. The right side of the inequality equals $E.(N+1)2^{N-1}$. The proof of the inequality is as follows. The nearest 2^N codes to Z_{li} should contain all the codes that have zero distance from it (that is Z_{li} itself); the number of such codes is $\binom{N+1}{0}$. It should also include all the codes that are just one bit different from Z_{li}; the number of such codes is $\binom{N+1}{1}$. The next closest set of codes are the ones that are different from Z_{li} in 2 bits and so on. We continue until we reach to the first closest 2^N codes to Z_{li}. In that case, the number of bit differences reach to $\frac{N}{2}$ when N is even and $\frac{N+1}{2}$ when N is odd. This is because the following equation holds:
$\binom{N+1}{0} + \binom{N+1}{1} + \cdots + \binom{N+1}{\lfloor\frac{N-1}{2}\rfloor} + i_o\binom{N+1}{\lfloor\frac{N+1}{2}\rfloor} = 2^N$, where i_o is the same as defined before.

Now, we show that the complement-pair coding assigns all the above 2^N codes to different words. In this case, the average transition cost for each code Z_{li} will be equal to its

	Algorithm 1. DP-based method for energy-aware coding
	Inputs: Word and code lengths: N, N+K; $\mathcal{C}(N,1)$ and optimal coding for $\mathcal{P}(N,1)$ from Section 5.1.1.
⋆	**Finding $\mathcal{C}(n,k)$ and the partitioning index** $index(n,k,1:2)$:
1	for (n=1 to n=N)
2	for (k=1 to k=K)
3	if (k==1)
4	$\mathcal{C}(n,k) = \mathcal{C}(n,1)$;
5	else
6	for (i=1 to i=n-1)
7	for (j=1 to j=k-1)
8	if ($\mathcal{C}(n,k) \geq \mathcal{C}(n-i,k-j)$)
9	$\mathcal{C}(n,k) = \mathcal{C}(n-i,k-j)+\mathcal{C}(ij)$;
10	$index(N,K,1:2)=(i,j)$;
⋆	**Building the codes for $\mathcal{P}(N,K)$:**
11	for (n=1 to n=N)
12	for (k=1 to k=K)
13	if (k==1)
14	$\mathcal{P}(n,k) = \mathcal{P}(n,1)$ from Section 5.1.1;
15	else
16	$\mathcal{P}(n,k) = $ all code combinations from $\mathcal{P}(n-index(n,k,1), k-index(n,k,2))$ and $\mathcal{P}(index(n,k,1), index(n,k,2))$

optimal value and thus the optimal OF is achieved. The sum of bitwise differences of Z_{li} from any complement pair $(Z_{l'1}, Z_{l'2})$, is equal to $N+1$. This is because each bit of Z_{li} is equal to exactly one of the bits of the complement pair. Thus, one symbol of each word has a distance of less than $\frac{N+1}{2}$ bits and the other symbol has a distance of more than $\frac{N+1}{2}$ bits from Z_{li}. This means that all the $2^N - 1$ closest codes to Z_{li} belong to different words. ∎

Note that our complement results for the $K = 1$ case also apply to the asymmetric set/reset costs. The number of sets and resets for traversing from a code to its complement is not symmetric for most of the code words. Recall that our objective is to minimize the average costs over all possible transitions. It can be readily shown that for achieving the mean cost, the average inter-complement distance can replace the two disparate transition costs between the complements. The results of the claim then directly follows.

5.1.2 DP-based approach to $\mathcal{P}(N,K)$

We introduce a DP-based algorithm for solving the general $\mathcal{P}(N,K)$ problem. Our algorithm uses the coding results for $\mathcal{P}(p,q)$ and $\mathcal{P}(r,s)$ to construct the codes for $P(p+r,q+s)$ such that the following bounds can be achieved:

$$\mathcal{C}(p+r, q+s) = \mathcal{C}(p,q) + \mathcal{C}(r,s). \qquad (4)$$

The code construction is as follows. The word W_i of length $p + r$ is partitioned into 2 words, W_i^1 and W_i^2. The first word is the first p bits and the second word is the last r bits of W_i. There are 2^q, $p + q$-bit symbols for W_i^1 and 2^s, $r + s$-bit symbols for W_i^2 that are obtained from solving $\mathcal{P}(p,q)$ and $\mathcal{P}(r,s)$ respectively. We construct the codes for W_i by concatenating all the possible combinations of these two set of symbols which provides a total of $2^q \times 2^s = 2^{q+s}$ codes (of length $p + q + r + s$) for W_i. It can be easily seen that the codes satisfy Equation 4. Based on the above code construction, the DP method breaks N into smaller values

and selects the best partitioning to minimize:

$$\mathcal{C}(N,K) = \min_{i \leq N} \ \{\min_{j \leq i} \mathcal{C}(N-i, K-j) + \mathcal{C}(i,j)\}. \qquad (5)$$

Algorithm 1 provides the details of the DP method. The optimal coding for $\mathcal{P}(N,1)$ is given from the previous part and the algorithm iteratively traverses over all the possible partitions to improve the energy minimization objective (Lines 1-10). The index vector $index(n,k,1:2)$ is used to store the optimal partitioning of (n,k). After finding all the indices, the algorithm builds the codes (Lines 11-16). The complexity of the algorithm is $O(N^2 K^2)$, but recall that this algorithm is run off-line.

5.2 Coding for Stochastic Data

OF 3 minimizes the average energy cost for all the possible word overwrites. Here, we discuss how the inherent stochastic properties for real data scenarios can be exploited for further energy improvements. An important feature is that different words have differing frequencies. To benefit from this fact, instead of weighting all the rewrite energy costs equally, we aggressively optimize our encoding for the rewrites that are more prevalent by assigning different number of codes to the words based on their frequency.

Variable-length and fixed-length coding are two statistical compression techniques. In the variable-length method, shorter codes are assigned to the more frequent words to better improve the compression. However, this adds to decoding complexity and since our main goal is to minimize the energy, decoding efficiency is very important. Thus, we use a fixed-length coding method. We describe our method on text files. The method can be generalized to other data sets with nonuniform frequencies. Our data consists of the lower-case alphabet letters: $W_1 = a$, $W_2 = b$, ..., $W_{26} = z$. Since there are 26 letter, W_i's are 5-bit words.

Let us consider the first 7 most frequent letters of the table, e, t, a, o, i, n and s. The probability that an overwrite occurs on any of these letters (by any other letter) plus the probability that these letters overwrite any other letter accounts for almost 60% of all probable overwrites. Thus, we can benefit a lot by optimizing our coding for these seven letters. To do so, we assign a different prefix to each of these letters such that only the prefixes determine the letter. Since there are 7 letters, the prefixes are 3-bit each and are shown in Figure 2. The prefixes can be interpreted as dictionary indices. The remaining $N + K$ bits of these letters take all the possible 2^{N+K} states. Thus, an overwrite to/by any of these letters requires only adjusting the prefix that is of length 3. The other 19 letters have the prefix (111) as shown in the figure. The remaining $N + K$ bits for the less frequent letters are filled with the codes obtained by solving $\mathcal{P}(N,K)$ as described in Subsection 5. Thus, an overwrite between the letters costs as much as for a regular $\mathcal{P}(N,K)$. By this coding, we assign 2^{N+K} symbols to the highly frequent letters and 2^K codes to the rest of the letters. All the symbols are of length *length of prefix*$+N + K$.

5.3 Runtime Coding/Decoding Complexity

As mentioned in Section 3, our algorithms for developing code words are run off-line. The results of our algorithms are then stored in the memory controller as a look-up table. When writing a new word to the memory, there are 2^K options for the word, where K is a small constant number. In our evaluations we used K in range 1-4. The coding

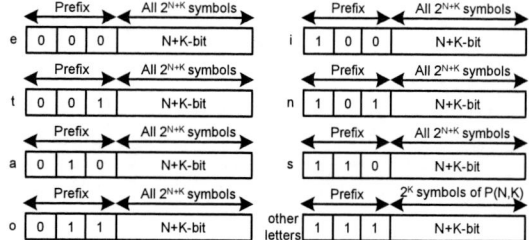

Figure 2: Data-aware alphabet letter codings.

complexity is in the order $\Omega(2^K)$. To do each decoding, the code words can be placed in a binary tree with a depth $K+N$. Searching for a symbol on this tree has an $\Omega(K+N)$ complexity. Thus, both our coding and decoding operations have a very low overhead.

6. EVALUATION RESULTS

We evaluate our energy-aware encoding methods on a variety of benchmark data sets. We perform our evaluations for different relative set and reset energy ratios $\frac{E_R}{E_S}$ and discuss their impact on the energy efficiency. To have a fair comparison, we normalize the costs such that $E_R + E_S = 1$.

6.1 DP-based Algorithm

We analyze DP-based encoding method provided in Algorithm 1 for different word lengths. We compare the average energy costs obtained from this method to that of the no-coding (nc) method and the optimal bound (opt) from Section 4.3. The no-coding method is equivalent to the problem $\mathcal{P}(N,0)$. We denote the average overwrite energy costs for the words for the above three methods as follows; $\mathcal{C}_{nc}(N,0)$, $\mathcal{C}_{dp}(N,K)$ and $\mathcal{C}_{opt}(N,K)$.

Table 6.1, shows the results for the case where $\frac{E_R}{E_S} = 2$ for different word lengths N and number of extra-bits K. Columns six shows the average improvement in the cost obtained from the DP compared to no-coding method. The result shows notable savings. For example, for $N = 8$ and $K = 2$, energy cost is reduced by 72%. Thus, for each word overwrite, we save on average 28% of the energy at the expense of adding 2 bits. The last column shows the DP performance compared to the optimal achievable bounds. Our results show that DP algorithm achieves values very close (in some cases equal) to the optimal bound.

Table 1: DP cost ($\mathcal{C}_{dp}$) comparison against no-coding cost ($\mathcal{C}_{nc}$) and optimal cost ($\mathcal{C}_{opt}$).

N	$\mathcal{C}_{nc}(N,0)$	K	$\mathcal{C}_{opt}$	$\mathcal{C}_{dp}$	$\frac{\mathcal{C}_{dp}(N,K)}{\mathcal{C}_{nc}(N,0)}$	$\frac{\mathcal{C}_{opt}}{\mathcal{C}_{dp}}$
2	.5	1	.37	.37	.75	1
3	.75	1	.55	.55	.73	1
4	1	1	.75	.75	.75	1
4	1	2	.68	.72	.72	.94
8	2	1	1.48	1.48	.74	1
8	2	2	1.42	1.44	.72	.98
8	2	3	1.34	1.39	.69	.96
8	2	4	1.30	1.36	.68	.95

Figure 3 shows the average energy cost $\mathcal{C}(N,K)$ for different $\frac{E_R}{E_S}$ values; N is set to 10 and K is in the range $1,2,\ldots,5$. We see that as the ratio $\frac{E_R}{E_S}$ increases, better energy savings are achieved. This is because our coding scheme aims to optimize the energy consumption by minimizing the number of overwrites. Since resets have a higher energy cost, the minimization impact will be higher for them.

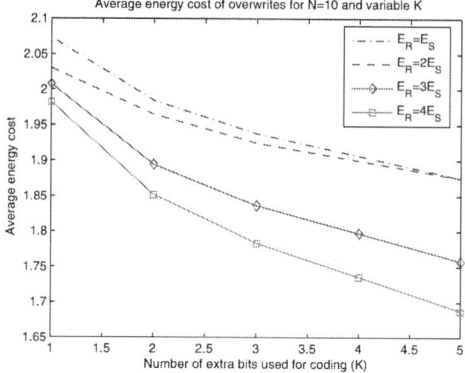

Figure 3: Cost reduction by data-aware coding.

6.2 Performance on Audio and Image Data

We use the encoding method to store audio and image data on PCM. Our benchmark data are from Columbia University audio and Caltech Vision image databases, [1, 6]. The audio data are $msmn1.wav$, $msmv1.wav$, $mssp1.wav$, and $msms1.wav$. The image files are $dcp - 2897.jpg$, $dcp - 2898.jpg$, $dcp - 2899.jpg$ and $dcp - 2830.jpg$. Figure 4 shows the average energy reductions for all file overwrites. More details are outlined in Appendix D.

For audio data, $\mathcal{P}(4,1)$ and $\mathcal{P}(4,2)$ encodings are applied and the results show an energy reduction of 11% and 21% respectively. For image data, $\mathcal{P}(8,1)$ and $\mathcal{P}(8,2)$ encodings are applied and the results demonstrate an average energy reduction by 18% and 28% respectively. The savings are significant and confirm the notable energy improvements of encoding at the expense of adding a few extra bits.

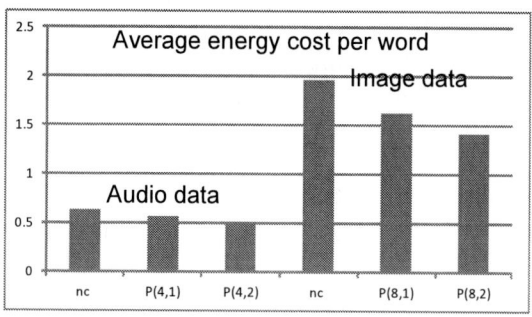

Figure 4: Average energy cost per word, $\frac{E_R}{E_S} = 2$.

6.3 Distribution-Aware Data Coding

We first evaluate English alphabet coding as described in Section 5.2. Then, we provide coding and evaluations for the ASCII characters. We used two text benchmarks, the 31 MB *text8.txt* file from [2], for alphabet evaluations; and the 4.8 MB *KJV.txt* file from [4] for ASCII evaluations.

6.3.1 Alphabet Letters

We encoded the alphabet letters with the distribution-aware encoding. Since there are 26 alphabet letters, $N = 5$; we set $K = 1$, and *Prefix*=3. The codes are of length *Prefix-length*$+N + K = 9$. We evaluated the method on *Text8.txt* data for different test trials. For each trial, we created 100 pairs of vectors by randomly reading the data from the text file. Each vector has 1000 letters. We overwrote the vectors of each pair and computed the average overwrite cost for $\frac{E_R}{E_S} = 2$. The results demonstrate an average 44.1% reduction when compared to the no-coding scheme and 9.3% reduction compared to the uniform coding $\mathcal{P}(5, 2)$.

6.3.2 ASCII Characters

According to the frequencies of ASCII characters from [3], 59% of all the possible rewrites are to/by one of the first 15 most frequent characters. Thus, we optimize our coding for these characters by assigning separate prefixes to them.

The first 15 most frequent characters are: space, e, t, a, o, i, n, s, h, r, d, l, u, m, c. We assigned the following 4-it prefixes to them respectively: (0000), (0001), (0010), (0100), (1000), (1001), (1010), (0110), (0111), (1011), (1101). The prefix for all the other characters is (1111). Since there are 2^7 ASCII characters, $N = 7$ and we set $K = 1$. Thus, the codes will be of length $4 + N + K = 12$. The encoding method is the same as described for alphabet letters.

We evaluated the ASCII coding scheme on the *KJV.txt* file. We created 100 pairs of vectors, each of length 1000 from the file. The first vector in each pair was overwritten by the second vector. We considered $\frac{E_R}{E_S} = 2$. To compare this method with the uniform coding, we encoded the ASCII characters with the codes from $\mathcal{P}(7, 1)$, $\mathcal{P}(7, 2)$ and $\mathcal{P}(7, 3)$ and report the corresponding average costs in the following:

Encoding	Data-aware	$\mathcal{P}(7, 1)$	$\mathcal{P}(7, 2)$	$\mathcal{P}(7, 3)$
Avg cost	1.24	1.42	1.37	1.34

We see that the ASCII data-aware coding, on average, reduces the energy cost to 92% of the best cost achieved from $\mathcal{P}(7, 3)$. Thus, for overwriting each ASCII character, there will be an 8% reduction in the energy cost compared to the results of the uniform encoding. This improvement is at the expense of two extra bits per character.

7. CONCLUSION

We proposed a novel data coding methodology for minimizing PCM write energy. Our approach creates several alternative symbols for each word being written in the memory, trading off energy efficiency with encoding overhead. The new word that is going to be written on the memory is encoded by the symbol with minimum distance to the existing word on that memory location. To address the problem, we developed (i) an ILP-based solution that mostly incurs a high combinational complexity; and (ii) a Dynamic Programming-based approach that combined the smaller optimal codewords. For cases where the distributions of the letters in the alphabet were a priori known, we created a new data-aware algorithm that incorporated those information for further energy reductions. Evaluations on a diverse set of text, image, and audio benchmark data demonstrated the applicability and effectiveness of our new methods.

7.1 Acknowledgments

This research is in part supported by ONR YIP award under grant No. R16480, ARO YIP award under grant No. R17450 and NSF CCF-0926127 award.

8. REFERENCES

[1] http://labrosa.ee.columbia.edu/sounds/.

[2] http://mattmahoney.net/dc/textdata/.

[3] http://millikeys.sourceforge.net/freqanalysis.html.

[4] http://patriot.net/ bmcgin/kjvpage.html.

[5] http://www.gurobi.com/.

[6] http://www.vision.caltech.edu/html-files/archive.

[7] R. Ahlswede and Z. Zhang. Coding for write-efficient memory. *Info and Comp.*, 83(1):80–97, 1989.

[8] J. Anxiao, M. Langberg, M. Schwartz, and J. Bruck. Universal rewriting in constrained memories. In *ISIT*, pages 1219–1223, 2009.

[9] S. Cho and H. Lee. Flip-N-Write: a simple deterministic technique to improve PRAM write performance, energy and endurance. In *MICRO*, pages 347–357, 2009.

[10] G. Dhiman, R. Ayoub, and T. Rosing. PDRAM: A hybrid *PRAM* and *DRAM* main memory system. In *DAC*, pages 664 –669, 2009.

[11] T. F. and Gonzalez. Clustering to minimize the maximum intercluster distance. *Theoretical Computer Science*, 38(0):293–306, 1985.

[12] F. Fu and R. Yeung. On the capacity and error-correcting codes of write-efficient memories. *IEEE Tran. on IT*, 46(7):2299 –2314, 2000.

[13] S. Lai. Current status of the phase change memory and its future. In *IEDM*, pages 10.1.1 – 10.1.4, 2003.

[14] B. Lee, E. Ipek, O. Mutlu, and D. Burger. Architecting phase change memory as a scalable dram alternative. In *ISCA*, pages 2–13, 2009.

[15] T. Mittelholzer, L. Lastras-Montañ Ando, M. Sharma, and M. Franceschini. Rewritable storage channels with limited number of rewrite iterations. In *ISIT*, pages 973 –977, 2010.

[16] M. Qureshi, J. Karidis, M. Franceschini, V. Srinivasan, L. Lastras, and B. Abali. Enhancing lifetime and security of PCM-based main memory with start-gap wear leveling. In *MICRO*, pages 14–23, 2009.

[17] R. Rivest and A. Shamir. How to reuse a write-once memory. In *STOC*, pages 105–113, 1982.

[18] S. Schechter, G. Loh, K. Straus, and D. Burger. Use ECP, not ECC, for hard failures in resistive memories. In *ISCA*, pages 141–152, 2010.

[19] H. Wong, S. Raoux, S. Kim, J. Liang, J. Reifenberg, B. Rajendran, M. Asheghi, and K. Goodson. Phase change memory. *Proceedings of the IEEE*, 98(12):2201–2227, 2010.

[20] Y. Wu and A. Jiang. Position modulation code for rewriting write-once memories. *IEEE Tran. IT*, 57(6):3692–3697, 2011.

[21] P. Zhou, B. Zhao, J. Yang, and Y. Zhang. A durable and energy efficient main memory using phase change memory technology. In *ISCA*, pages 14–23, 2009.

APPENDIX

A. PCM OPERATION AND ENERGY MODEL

A key challenge for non-volatile memory technology, in particular flash, is the high energy cost of writes [19]. The speed of writing and reading from the caches and from the DRAM is often high, and therefore, the number of transitions is higher than the external memories. Therefore, since resistive memory is suggested for replacing and complementing various storage units in the memory hierarchy, saving the energy cost of set and reset transitions is of a high value [19, 16].

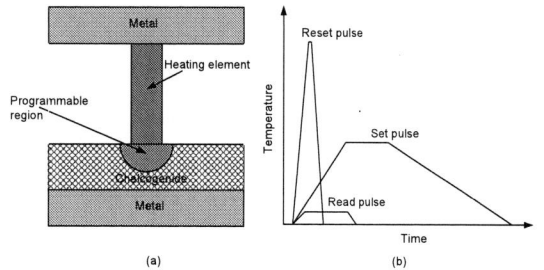

Figure 5: (a) A PCM memory cell; (b) Current pulses for set, reset, and read operations.

As shown in Figure 5(a), the phase change media is placed between an electrode layer and a layer composed of a heater. The current flows through the phase change material from the electrode to the heater. This current is provided as a pulse, and its duration and amplitude controls the temperature needed for the set and reset operations. Heating the phase change material above a crystallization temperature by applying an average current but wide duration pulse results in the set operation. A very high current (melt quenching) pulse with a short duration resets the device to its amorphous state. The read is done by applying a very low amplitude and low power pulse that senses the device resistance. The shape of the three pulses used for set, reset, and read commands is plotted in Figure 5(b), [19]. The energy discrepancy between the PCM set and reset operations has been experimentally shown.

B. A WORD ENCODING/DECODING EXAMPLE

In this example, we describe how one may benefit from coding the PCM data. Here we are solving the problem of finding the optimal coding for 2-bit words with 3-bit codes. We denote the words by $W_1=(00)$, $W_2=(01)$, $W_3=(10)$, $W_4=(11)$ and denote the codes corresponding to the word W_i by Z_{i1} and Z_{i2}, for $1 \leq i \leq 4$; since $K=1$ each word has $2^{K=1}=2$ code representation. The key point is to exploit multiple representations of each word for minimizing the write energy. For instance, if the existing data is Z_{11} and W_2 is to be written on it, among its representations Z_{21} and Z_{22}, the one that incurs the minimum energy cost to overwrite Z_{11} (which in this case is Z_{21}) is selected.

Figure 6 shows a graph representation of the encodings for the 2-bit words shown in separate clusters. The vertices of the graph are the codes and each cluster represents a word.

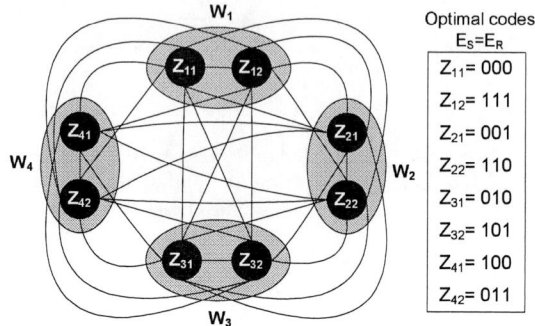

Figure 6: A 3-bit encoding for the 4 words W_1, W_2, W_3, and W_4.

The graph is a directed graph and the weight of each edge shows the cost of overwriting one node with the other. The optimal encoding is derived and provided on the figure. If the code Z_{22} is to be overwritten by a code of W_3, Z_{31} is selected because its energy cost is only equal to E_S. If no coding was used, overwriting W_2 with W_3 would cost the higher value of E_R+E_S. Another example is a cycle of word overwrites (W_1,W_2,W_3,W_4,W_1). Assume that W_1 is coded as Z_{11}. Then, the minimum cost codes would be selected as follows $(Z_{11},Z_{21},Z_{32},Z_{41},Z_{11})$. The cost associated with the code overwrites is $E_S + E_S + E_S + E_R + E_R = 2.E_S + 2.E_R$. Whereas the cost for overwriting the codes without coding is $E_S + (E_S + E_R) + E_S + (2E_R) = 3.E_S + 3.E_R$.

An example of a binary tree for decoding the data with 8 code words ($N + K=3$) for the codes developed in Figure 6 is shown in Figure 7.

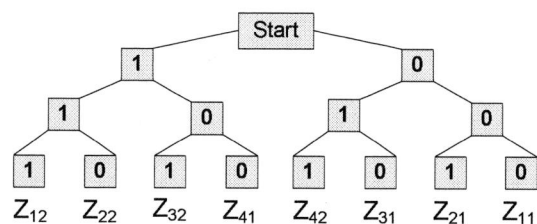

Figure 7: Binary tree for encoding.

C. INTEGER LINEAR PROGRAMMING FORMULATION

To formulate OF in a linear form, we define an index variable that for each symbol, keeps track of the index of the element (in each of the other clusters) with the minimum distance to the symbol. The following set of variables were used in our ILP formulation:

l, l'	Words indices W_l or W'_l for $1 \leq l, l' \leq 2^N$.	
i, i'	Code indices within each cluster, $1 \leq i, i' \leq 2^K$.	
Z_{li}	The i-th code $\in W_l$ for all i.	
$\Phi_{ll'i}$	$\phi(Z_{l'i}, W_l)$ for all l, l', i and i'.	
$w_{ll'ii'}$	$w(Z_{li}, Z_{l'i'})$ for all l, l', i and i'.	
$\Delta_{ll'ii'}$	$w_{ll'ii'} - \Phi_{ll'i}$ for all l, l', i and i'.	
X_{lij}	j-th significant bit of Z_{li} for $1 \leq j \leq (N+K)$.	
$F_{ll'ii'j}$	$w(X_{lij}, X_{l'i'j})$ for all l, l', i, i' and j.	
$Id_{ll'ii'}$	An indicator binary; =0 iff $\Delta_{ll'ii'} = 0$ for all l, l', i and i'.	

The codes representing a word W_l are shown by Z_{li}; $\Phi_{ll'i}$ denotes the cost of overwriting Z_{li} by a code in $W_{l'}$ that requires the minimum overwrite energy; $w_{ll'ii'}$ is the cost of overwriting two codes U_{li} and $U_{l'i'}$. Thus, $\Phi_{ll'i} = \min_{i'} w_{ll'ii'}$. Each code Z_{li} consists of $N+K$ bits and can be written as $(X_{liN+K}, \ldots, X_{li2}, X_{li1})$. The parameter $F_{ll'ii'j}$ is defined to be the cost of overwriting X_{lij} with $X_{l'i'j}$ and its range of values is shown in the table below. Variable $Id_{ll'ii'}$ is an indicator binary variable that indicates if the closest code to Z_{il} in cluster l' is $Z_{i'l'}$ or not.

X_{lij}	$X_{l'i'j}$	$F_{ll'ii'j}$
0	0	0
0	1	E_S
1	0	E_R
1	1	0

Using the above variables, we define our OF and provide constraints to our problem in a way that conforms to the ILP format. Our OF, as written in Equation 3, minimizes the average cost of overwriting the codes for all possible overwrites:

$$OF : \min \frac{1}{2^N . 2^N . 2^K} \sum \Phi_{l'li} \qquad \text{for all } l', l \text{ and } i \text{ variables}$$
(6)

The following constraints define $\Phi_{ll'i}$:
C1. $\Delta_{ll'i} \geq 0$ for all l, l' and i variables,
C2. $\Sigma_{i' \in 1,\ldots,2^k} Id_{ll'ii'} \leq 2^K - 1$,
C3. $Id_{ll'ii'} \leq \Delta_{ll'ii'}$,
C4. $E_R.(N+K).Id_{ll'ii'} \geq \Delta_{ll'ii'}$.

Constraints C1 and C2 set $\Phi_{ll'i}$ not greater than each distance $\Delta_{ll'i}$ and equal to at least one of them respectively; Constraints C3 and C4 define the indicator variable based on the fact that $E_R.(N+K)$ is always grater than $\Delta_{ll'ii'}$.

The following linear constraints set $F_{ll'ii'j}$ as defined in the table above:
C5. $\frac{1}{E_R+E_S} F_{ll'ii'j} + X_{lij} + X_{l'i'j} \leq 2$,
C7. $F_{ll'ii'j} - E_R.X_{lij} - E_S.X_{l'i'j} \leq 0$,
C8. $F_{ll'ii'j} - E_R.X_{lij} - E_R.X_{l'i'j} \geq 0$,
C9. $F_{ll'ii'j} - E_S.X_{lij} - E_S.X_{l'i'j} \geq 0$.

The following constraint defines distance $w_{ll'ii'}$:
C10. $w_{ll'ii'} = \Sigma_{1 \leq j \leq N+K} F_{ll'ii'j}$.

The following constraint is set to ensure that no code is assigned to more than one word; E_S is the minimum cost of overwriting two different codes:
C11. $w_{ll'ii'} \geq E_S$.

The output of the above ILP is the values of X_{lij} that constructs the codes U_{il}. The above constraints are all in linear format and can be readily implemented by any ILP solver. The complexity and runtime for solving the instances

of the ILP for our NP-complete problem exponentially increases with the instance size. In our experiments, we have been able to find the optimal solution by using a limited version of an ILP solver licensed to one user for N and K ($N = 2, 3, 4$, $K = 1, 2$). If one has access to the commercial ILP solvers that run on supercomputers, it is likely possible to find the optimal codes for the practical codes of longer sizes. The longer runtimes can be tolerated since the ILP needs to be used only once and offline (See Section 5.3).

C.1 ILP Results

We used the latest version of Gurobi ILP solver, Gurobi 4.5.2, to solve the ILP defined in Section C, [5]. Gurobi provides free access for academic purposes. The runtime of the solver for solving $\mathcal{P}(4,2)$ is about 30 hours on an Intel Core 2 Duo Processor T9600 computer. Thus, due to the time constraint we were not able to solve the OF for larger problems. However, the authors make the python ILP code available to the interested readers.

D. RESULTS

D.1 Audio and Image Data

We provide more details of the encoding evaluations on audio and image data as described in Section 6.2. The audio data were msmn1.wav, msmv1.wav, mssp1.wav, and msms1.wav and are denoted by $a1$, $a2$, $a3$ and $a4$ in Table D.1. The image files are $dcp-2897.jpg$, $dcp-2898.jpg$, and $dcp-2899.jpg$ and $dcp-2830.jpg$ and are presented by $i1$, $i2$, $i3$ and $i4$ in Table D.1.

Table 2: Average energy costs of DP and no-coding methods for audio data.

Energy costs	nc $\mathcal{P}(4,0)$	dp $\mathcal{P}(4,1)$	dp $\mathcal{P}(4,2)$
$a1 \rightarrow a2$	0.59	0.50	0.43
$a1 \rightarrow a3$	0.66	0.54	0.50
$a1 \rightarrow a4$	0.66	0.62	0.60
$a2 \rightarrow a3$	0.61	0.48	0.39
$a2 \rightarrow a4$	0.59	0.67	0.58
$a3 \rightarrow a4$	0.71	0.60	0.54

Table 3: Average energy costs of DP and no-coding methods for image data.

Energy costs	nc $\mathcal{P}(8,0)$	dp $\mathcal{P}(8,1)$	dp $\mathcal{P}(8,2)$
$i1 \rightarrow i2$	2.20	1.78	1.61
$i1 \rightarrow i3$	2.44	1.66	1.44
$i1 \rightarrow i4$	1.90	1.70	1.58
$i2 \rightarrow i3$	1.66	1.73	1.46
$i2 \rightarrow i4$	1.79	1.38	1.22
$i3 \rightarrow i4$	1.75	1.46	1.19

In both tables, the first column shows the files that are overwritten. For example $a1 \rightarrow a2$ means that $a2$ is overwritten by $a1$. The second and third column show the average overwrite costs for the DP-based algorithm (Data is encoded) and the no-coding method for $\mathcal{P}(4,1)$ (audio data) and $\mathcal{P}(8,1)$ (image) data encodings. For example, the average cost of a word overwrite in $a1 \rightarrow a2$ is 0.50 in DP, while this value is 0.59 in the no-coding method while encodings from $\mathcal{P}(4,1)$ is applied. The forth and fifth columns show the same results for $\mathcal{P}(4,2)$ (audio data) and $\mathcal{P}(8,2)$ (image) data encodings. All results correspond to $\frac{E_R}{E_S} = 2$. Meaningful improvements are achieved by the energy-minimization coding method. The energy cost on average is reduced by our 15.6% and 22.5% for audio and image data respectively.

A Code Morphing Methodology to Automate Power Analysis Countermeasures

Giovanni Agosta
Politecnico di Milano
Piazza Leonardo da Vinci, 32
20133 Milano, Italy
agosta@elet.polimi.it

Alessandro Barenghi
Politecnico di Milano
Piazza Leonardo da Vinci, 32
20133 Milano, Italy
barenghi@elet.polimi.it

Gerardo Pelosi
Politecnico di Milano
Piazza Leonardo da Vinci, 32
20133 Milano, Italy
pelosi@elet.polimi.it

ABSTRACT

We introduce a general framework to automate the application of countermeasures against Differential Power Attacks aimed at software implementations of cryptographic primitives. The approach enables the generation of multiple versions of the code, to prevent an attacker from recognizing the exact point in time where the observed operation is executed and how such operation is performed. The strategy increases the effort needed to retrieve the secret key through hindering the formulation of a correct hypothetical consumption to be correlated with the power measurements. The experimental evaluation shows how a DPA attack against OpenSSL AES implementation on an industrial grade ARM-based SoC is hindered with limited performance overhead.

Categories and Subject Descriptors

C.3 [**Special-Purpose and Application Based Systems**]:
Microprocessor/microcomputer applications;
C.5.3[**Computer System Implementation**]:
Microcomputers[portable devices];

General Terms

Security

Keywords

Power Analysis Attacks, Software Countermeasures, Dynamic Code Transformation, Polymorphic Code

1. INTRODUCTION

The general trend in embedded hardware security shows a large use of cryptographic operations, and an increasing attention towards tamper resistant designs and countermeasures against side-channel attacks like power analysis and fault injection. Indeed, it is effectively proven that the physical access to an embedded device may enable the recovery of sensitive information, which is otherwise supposed to be hidden [1,2,8], through exploiting both the implementation weaknesses of the cryptographic operations and specific features provided by the underlying hardware platform. Differential Power Analysis (DPA) introduced in [6] has been proven a powerful threat that triggered a flourishing research branch with a wide range of improvements and countermeasures both in hardware and software. DPA attacks against an unprotected implementation of a cryptographic algorithm follow a common workflow: first of all they measure the power consumption (*power traces*) of the targeted device for a high number of runs (i.e. considering a high number of input/output values). Subsequently, they select an intermediate operation of the algorithm employing a part of the secret key, and compute an expected consumption for every possible value of the key portion, according to a model of the triggered switching activity (e.g. the Hamming weight of the outputs). Finally, the predicted consumption values are matched against each sample of the recorded power traces to assess which key hypotheses fit better the actual measurements. In this fashion, the secret key can be recovered, one part at time, even if the relevant information is stored within the device in a non accessible way. The principal countermeasures against power analysis are split into two categories [8]: *masking* and *hiding*. Masking aims to invalidate the link between the predicted hypothetical power consumption values, associated to the selected intermediate operation, and the actual values processed by the device. In a masked implementation, each sensitive intermediate value is concealed through splitting it in a number of shares, which are then separately processed. Hence, the target algorithm is modified to correctly process each share and to recombine them at the end of the computation. A masking scheme with only two shares is composed by the values v_m and m, where m is a randomly chosen mask and v_m is a share such that the value v to be protected can be derived as $v=v_m \diamond m$, with $\diamond$ denoting an invertible binary operation. To compensate for this countermeasure, more sophisticated DPA attacks, known as high-order DPAs rely on predicting the consumption of all the operations handling the shares and try to obtain a combination of them independent from the masking values. This value must subsequently be correlated with an analogous combination of the measured consumption values, employing the same techniques of a common (first order) DPA. The technical effort in carrying out an high-order DPA attack quickly grows as the order (i.e. the number of shares) increases just as the time/space resources to be employed in recording a larger number of power traces. It is commonly accepted that a masking scheme with a large number of shares makes DPA attacks either practically unfeasible or inconvenient. Typically, engineering solutions strive to introduce a moderate overhead with respect to the unprotected version of the primitive, resorting to the combination of two-share masking schemes and hiding techniques [9]. Hiding methods aim to conceal the relation between the power consumption and the operations performed by the target algorithm to compute the intermediate values. The protection strategies employed in the open literature, to secure software implementations, are based on execution flow randomization via shuffling the order of some instructions (f.i., permuting the sequence of accesses to lookup tables) and inserting random delays built with dummy operations [9, 10].

To minimize the performance overhead, the execution must be interleaved with delays in multiple places, keeping the individual delays as short as possible. In this way, an attacker faces a cumulative and hardly predictable sum of delays between the start (the end, respectively) of the algorithm and the location of the observed intermediate operation in time [4]. These techniques only affect the time dimension of the power consumption but they do not change the power consumption characteristics of the operations performed by the target device. In spite of the limits of the aforementioned techniques, software countermeasures are well suited for general purpose processors where no dedicated security features are built in at design time. In addition to this broad range of applicability, software-based countermeasures represent also a viable mitigation mean to restore security into hardware-protected systems, where the underlying hardware protections have been compromised, without the need for an expensive part replacement.

1.1 Contributions

The novel approach proposed in this work is a software countermeasure framework based on the combination of a cryptographic algorithm implementation with a *polymorphic engine* which dynamically and automatically transforms the binary code to be protected. Thus, we propose innovative contributions to two common practices in the field: (i) static generation of the protected code; (ii) manual and often application-specific generation of the protected code. Our method moves the code generation at run-time, and enables the generation of many different versions of the protected code at the designer's will, preventing any attacker from both recognizing the exact point in time where the observed operation is executed and understanding how such operation is actually performed. This methodology separates the creative work of identifying replacements for assembly code snippets from the tedious (but amenable to automatization) work of applying such replacements to the entire code. This strategy largely increases the effort needed to predict the value of a sensitive intermediate result of the considered algorithm and hinders the formulation of a correct hypothetical consumption to be correlated with the power measurements. Polymorphic engines are the key component to build a special class of programs: the ones characterized by the ability to modify parts of their own code. In particular, a polymorphic code is composed of two parts: the polymorphic engine, which never changes, and the target code to be modified. Self-modifying code is used in several areas to provide either optimization or obfuscation: dynamic compilers, and especially fragment linking [5], tamper resistant software and protection against reverse engineering of executable code [7]. We adapt self-modification principles to both swap parts of the target algorithm with different, but semantically equivalent, replacements and to implement concepts such as *masking* and *hiding*. In particular, concerning the hiding countermeasure techniques, our approach provides both time-dimension and switching activity hiding, through changing both the type of operation and the time needed to compute the same intermediate value. With respect to state of the art, we provide: (i) a *generalization*, through allowing several types of countermeasures to be applied in a unifying framework; (ii) an *extension*, through providing variants to existing countermeasures; (iii) an increased *variability* of the protected code, through re-generating it as often as needed by means of dynamic code morphing. To this end, we allow the countermeasure designer to specify, for each operation or group of operations, a set of code transformation templates that can be automatically applied to the binary code of the target cryptographic primitive. Sufficient generality is provided to allow the expression of random delays such as those proposed in [4], as well as to replicate other hiding strategies, such as those shown in [3]. The availability of a wide range of transformations in our framework allows

the trade-off between performance overheads and security margin to be finely tuned at design time.

1.2 Case Study

We chose as a case study platform an ARM926-based STMicroelectronics SPEAr SoC, as a representative of a large class of high-end embedded devices where commonly no hardware protection against side channel attacks are employed. The chosen cryptographic primitive is the AES, as implemented in the widely diffused OpenSSL toolkit. The choice of this testbench was driven by the large adoption of this algorithm as a mean to provide data confidentiality. The most common intermediate values employed during an attack to an AES implementation are represented by output of a *load* operation from the S-Box, triggered by the SUBBYTE step and by the output of the post-ADDROUNDKEY state. However, in the selected platform, the consumption model relying on the latter is by far more effective than the one relying on the S-Box, due to the unpredictable power saving on the *load* operations caused by the use of data caches (see Section 4). Even if the number of successful DPA attacks against software implementations on complex SoCs is rather low due to the complexity of such devices, we were able to extract the full AES key from the testbed platform with a sensibly low number of measurements. Once the attack has been proven feasible, we employ the proposed framework to effectively and efficiently counteract the identified vulnerability.

The remainder of the paper is organized as follows. Section 2 provides the definitions necessary to formalize the code morphing operations. Section 3 introduces our code transformation framework, and describes the proposed polymorphic engine. Section 4 presents the experimental evaluation on the target case study. Section 5 provides a brief overview of closely related works. Section 6 draws some conclusions and highlights future directions.

2. SEMANTIC EQUIVALENCE OF CODE FRAGMENTS

Our approach of building a different version of a static binary code at run-time prior to executing it relies on the substitution of each static code fragment with one of its randomly-chosen variants, preserving the black-box behavior of the original static code. The notions of *code fragment* and *semantic-equivalence* between code fragments are formally defined as follows.

DEFINITION 2.1 (CODE FRAGMENT). *A sequence of instructions* $\mathcal{I}=(\mathtt{inst}_1,\dots,\mathtt{inst}_s)$, $s\geq 1$, *is a code fragment when each term* $\mathtt{inst}_j\in\mathcal{I}$ *is executed exactly once,* $\mathtt{inst}_j$ *executes before* $\mathtt{inst}_{j'}$ $\forall j,j'$ *such that* $j<j'\leq s$, *and no other instruction is executed between* $\mathtt{inst}_j$ *and* $\mathtt{inst}_{j+1}$, $1\leq j<s$.

Intuitively, the sequence of terms composing a code fragment must not include any branch or privileged instruction like a supervisor call, an I/O or an I/O MMU-bypass operation.

DEFINITION 2.2 (SEMANTIC EQUIVALENCE). *Let* $\mathcal{I}, \widetilde{\mathcal{I}}$ *be two code fragments, and let* $\mathcal{A}(\mathcal{I})$, $\mathcal{A}(\widetilde{\mathcal{I}})$ *be the sets of* live-out *variables related to* $\mathcal{I}$ *and* $\widetilde{\mathcal{I}}$, *respectively (i.e.: registers and memory locations that from the exit of the code fragment on are read at least once). A semantic-preserving relation between* $\mathcal{I}$ *and* $\widetilde{\mathcal{I}}$ *is an equivalence relation,* $\overset{S}{\sim}$, *where* $\mathcal{A}(\mathcal{I})=\mathcal{A}(\widetilde{\mathcal{I}})$ *and the corresponding written live-out values are the same.*

Given a generic code fragment, the problem of constructing a set of semantically equivalent variants is not practically interesting without a precise characterization of the goals to be achieved (f.i. it is easy to generate an infinite set of equivalent code fragments from a single-instruction loop). Therefore, in the following we formulate sufficient criteria to either map a given code fragment to a semantically-equivalent one or verify if two code fragments have the same semantics. The translation of the original static code is performed through locally scoped substitutions that are easier and more efficient to implement than global ones. Thus, the semantic equivalence of the resulting program is obtained from the composition of semantically equivalent independent code fragments, thus limiting the performance impact of the dynamic translation.

PROPOSITION 2.1 (CODE FRAGMENT EQUIVALENCE).
Let $\mathcal{I}=(\mathtt{inst}_1,\ldots,\mathtt{inst}_s)$, $s\geq 1$, and $\widetilde{\mathcal{I}}=(\mathtt{inst}_1,\ldots,\mathtt{inst}_{\tilde{s}})$, $\tilde{s}\geq 1$, be two code fragments. The semantic equivalence $\widetilde{\mathcal{I}}\overset{S}{\sim}\mathcal{I}$ is preserved if: (1) every register, $\mathtt{reg}_k$, $k\geq 0$, of the CPU is employed in $\mathcal{I}$, and $\widetilde{\mathcal{I}}$ according to the following constraints: (1.a) $\mathtt{reg}_k$ is either not included in any register assignment operations of $\mathcal{I}$ nor $\widetilde{\mathcal{I}}$ or (1.b) $\mathtt{reg}_k$ is used only in $\widetilde{\mathcal{I}}$, where it is spilled to the memory before any assignment and filled back as last action, or (1.c) the collections of possible register values at the end of $\mathcal{I}$ and $\widetilde{\mathcal{I}}$ (computed with the same initial values) must be the same; (2) the memory assignments derived from the sequence of write-operations in $\mathcal{I}$ are preserved in $\widetilde{\mathcal{I}}$, with the same values; (3) any memory assignment performed in $\widetilde{\mathcal{I}}$ but not in $\mathcal{I}$ writes the stack segment in memory locations outside $\mathcal{A}(\widetilde{\mathcal{I}})$, i.e.: in memory locations after the position of the stack pointer at the end of $\mathcal{I}$.

PROOF. The first condition implies that for all CPU registers the corresponding values at the end of $\mathcal{I}$, and $\widetilde{\mathcal{I}}$ are exactly the same. Therefore, the condition required by Definition 2.2 that all *live-out* registers, $\mathcal{A}(\mathcal{I})$ and $\mathcal{A}(\widetilde{\mathcal{I}})$, have the same corresponding values is trivially satisfied. The second condition ensures that memory locations referred in $\mathcal{I}$ and $\widetilde{\mathcal{I}}$ are also written in the same order and with the same values. The last condition allows additional memory assignements in $\widetilde{\mathcal{I}}$ to target a larger portion of the stack segment w.r.t. $\mathcal{I}$, nevertheless restoring the same value of the stack pointer in $\mathcal{I}$ at the end of the fragment $\widetilde{\mathcal{I}}$ as required by the first condition. The memory assignments are guaranteed not to be live-out, which matches the requirement stated in Definition 2.2. □

It is important to note that the sufficient conditions stated in Proposition 2.1 are locally verifiable, while Definition 2.2 employs the liveness property, which must be computed for all memory locations and registers. This requires a global analysis, unfeasible at run-time, and not always feasible at all even at compile-time if the whole address space is considered. By contrast, the conditions in Proposition 2.1 only rely on information which can be retrieved through a linear scan.

3. TRANSFORMATION FRAMEWORK

The proposed framework takes as input a target cryptographic algorithm, and statically compiles it to produce a binary code with proper calls to a run-time library which implements the code morphing engine and is in charge of modifying the target code on the underlying architecture. Figure 1 reports an high level description of the code transformation flow. The static compiler operates within the standard compilation process from source code to machine code. The source program is written as a C code with some specific *annotations*. Code annotations are usually employed to encode more information for the compiler than the explicit source code, such as hints about how to organize or optimize the intermediate representations of the code. In our case, we employ a custom gcc __attribute__ (defined as: __attribute__((secure)) before a function or variable declaration) to specify to the compiler

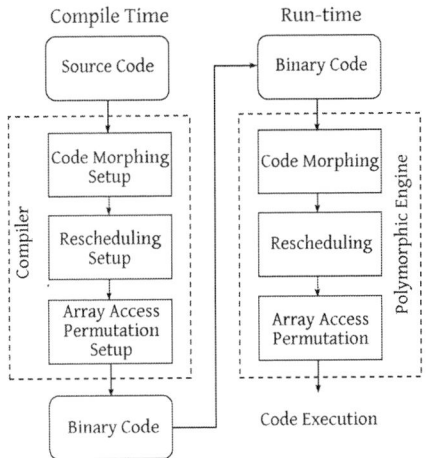

Figure 1: **Compilation and run-time code transformation flow**

which functions and data structures need to be secured. In the case of functions, for each stage of the static compiler (left side of Figure 1), the transformation flow wraps their calls with invocations of the *Code Morphing*, *Rescheduling*, and *Array Access Permutation* run-time routines. The case of array variable declarations is managed by the third stage of the static compiler (*Array Access Permutation Setup*) which makes the access to every array cell as an indirect access through using a further array (with the same length) containing a permutation of indexes. The *Code Morphing Setup* stage also allocates in memory the data structures (*tile set*) containing the knowledge base needed to perform the code morphing, through substituting each code fragment with one of its semantically equivalent variants included in the *tile set*. At run-time, the polymorphic engine, reported on the right-hand side of Figure 1, changes the original binary code through performing three steps: *Code Morphing*, *Rescheduling*, and *Array Access Permutation*. The Code Morphing stage, which is the core of the proposed approach, is described in greater detail in the next section.

The *Rescheduling* step adds a level of obfuscation with respect to the possible recognition (or classification) of the executed code through rearranging the instructions within a finite window, in such a way to preserve the data dependencies. This operation is performed by means of a single scan of the code, from the bottom to the top. At each step, a single instruction $\mathtt{inst}_i$ and a window of k instructions preceding it $\{\mathtt{inst}_{i-1},\ldots,\mathtt{inst}_{i-k}\}$ are considered. The dependencies of $\mathtt{inst}_i$ are computed, and the earliest position $i-h$ ($h\leq k$) which it can take is determined. Then, a random position $i-l$ in the range $[i-h,i]$ is selected, and the instructions are reordered consequently. Finally, the next value of i is set to $i-l$, and the process continues until the start of the code is reached. This technique is based on the well-known code scheduling theory employed in the field of compiler optimization. The code scheduling theory provides a set of semantic preserving schedules for a given code. Where the compiler practices select a schedule to minimize latency and/or power consumption, our goal is to randomly change a given schedule with a different (but equivalent) one to alter the shape and mutual alignment of the power traces.

The *Array Access Permutation* step applies a random permutation to the allocated array indexes, hiding the access patterns to the substitution table of a symmetric cipher. The access pattern hiding technique has been proposed and detailed in [9, 10]. Since the access to substitution tables is not the preferred attack point in our testbed platform, for the sake of brevity, we refer to the aforementioned works for further details.

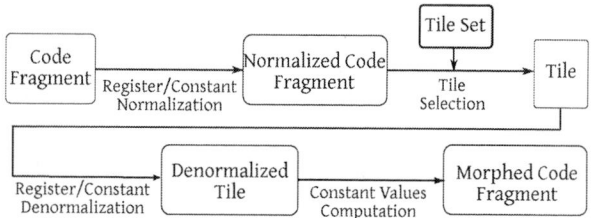

Figure 2: Code morphing engine workflow

3.1 Code Morphing Engine

To apply code morphing to a wide variety of code fragments, it is necessary to represent them in a way that abstracts from the actual registers and immediate values employed (e.g., `add r2,r5,r2` only differs from `add r4,r11,r4` in the registers used, and both can be abstracted to a normal form `add r0,r1,r0`). To clarify the normalization process, we will now formally define the concepts of *register normalization* and *constant normalization*.

DEFINITION 3.1 (REGISTER NORMALIZATION).
Given a code fragment $\mathcal{I}=(\text{inst}_1,\ldots,\text{inst}_s)$, $s\geq1$, where inst_i, with $0\leq i\leq s$, is a data processing assembly instruction specified as `opCode dest,src,operand`*, a register normalization is a map of the register names appearing in the fields {dest, src, operand} of the instruction sequence, into register names starting from* `r0` *for the first register in* inst_1 *on.*

DEFINITION 3.2 (CONSTANT NORMALIZATION).
Given a code fragment $\mathcal{I}=(\text{inst}_1,\ldots,\text{inst}_s)$, $s\geq1$, where inst_i, with $0\leq i\leq s$, is a data processing assembly instruction specified as `opCode dest,src,operand`*, a constant normalization is a map of the immediate values appearing in the fields {src, operand} of each instruction, into immediate values starting from* `#0` *on, to be interpreted as symbolic constants in the resulting instruction.*

Building on the two previous definitions the *normalized code fragment* can be defined as follows:

DEFINITION 3.3 (NORMALIZED CODE FRAGMENT).
Given a code fragment $\mathcal{I}=(\text{inst}_1,\ldots,\text{inst}_s)$, $s\geq1$, a normalized code fragment, $\overline{\mathcal{I}}$, is the sequence of instructions resulting from the application of both register (Definition 3.1) and constant (Definition 3.2) normalization mappings.

In principle, for each normalized code fragment $\overline{\mathcal{I}}_i$, $i\geq1$, a set of $m\geq1$ semantically equivalent fragments $\mathbf{S}_{\overline{\mathcal{I}}_i}=\{\widetilde{\overline{\mathcal{I}}}_{i,0},\ldots,\widetilde{\overline{\mathcal{I}}}_{i,m-1}\}$, $\widetilde{\overline{\mathcal{I}}}_{i,j}\overset{S}{\sim}\overline{\mathcal{I}}_i, \forall j\in\{0,\ldots,m-1\}$, can be written through applying the sufficient conditions specified by Proposition 2.1. Note that any non-trivial $\mathbf{S}_{\overline{\mathcal{I}}_i}$ set must be created manually.

EXAMPLE 3.1. *Consider a code fragment $\mathcal{I}$ composed by a single instruction* inst_1: `eor r5,r0,r4`*, which writes into* `r5` *the bitwise exclusive-or of the values in* `r0` *and* `r4` *($r5\leftarrow r0\oplus r4$). Its normalized form is computed as* `eor r0,r1,r2`*, while a corresponding semantically equivalent fragment is given by $\widetilde{\overline{\mathcal{I}}}=($* `bic r0,r1,r2; bic r3,r2,r1; orr r0,r0,r3`*$)\overset{S}{\sim}\overline{\mathcal{I}}$, where the ARM ISA instruction* `bic r0,r1,r2` *computes $r0\leftarrow r1\wedge\neg r2$. The use of extra registers, as* `r3`*, must be managed through clobbering the temporary registers before their use, and restoring them at the end of the instruction sequence in $\overline{\mathcal{I}}$.*

It is possible to generate large $\mathbf{S}_{\overline{\mathcal{I}}_i}$ sets, which must be stored in memory and used as a knowledge base for the Code Morphing

phase. Therefore, a key issue is to provide a compact representation for them. Note that, for the same $\mathcal{I}$, we can generate several $\overline{\mathcal{I}}_i$ which differ only by the values assumed by some of the constants involved.

Given the normalized fragment $\overline{\mathcal{I}}=($ `and r0 r1 #0`$)$, the designer may want to replace it by applying the following transformation:

$$r0 \leftarrow (r1 \wedge (\#0 \oplus \texttt{const1})) \diamond (r1 \wedge (\#0 \oplus \texttt{const2}))$$

where $\diamond$ is $\wedge$ or $\vee$ and `const1` and `const2` are additional symbolic constants such that `const1`$\oplus$`const2`$=\texttt{0xf...f}$ if $\diamond$ is $\wedge$ and `const1`$\oplus$`const2`$=\texttt{0x0}$ if $\diamond$ is $\vee$. Two semantically equivalent fragments $\widetilde{\overline{\mathcal{I}}}_0, \widetilde{\overline{\mathcal{I}}}_1$, to $\overline{\mathcal{I}}$ are:

$\widetilde{\overline{\mathcal{I}}}_0=($	$\widetilde{\overline{\mathcal{I}}}_1=($
`and r0,r1,#0⊕const1`	`and r0,r1,#0⊕const1`
`and r2,r1,#0⊕¬const1`	`and r2,r1,#0⊕const1`
`and r0,r0,r2`	`orr r0,r0,r2`
`)`	`)`

where `const1` is an additional symbolic constant that can assume any value, whereas the symbolic constants `#0,#1,...` derived from the constant normalization are constrained to their original immediate value, and `r2` is a clobbered register. We formalize the concept of a normalized code fragment augmented with operations on symbolic constants as follows.

DEFINITION 3.4 (TILE). *Given a normalized code fragment $\overline{\mathcal{I}}_i$, a tile t_i is a set of normalized fragments semantically equivalent to $\overline{\mathcal{I}}_i$, distinguished only by the values of additional symbolic constants* `const_j`*. These constants appear as immediate operands in the instructions of the code fragments, either alone or as part of constant expressions containing arithmetic-logic operators and symbolic constants from $\overline{\mathcal{I}}_i$.*

The expressions on symbolic constants are encoded within the bits used to encode the immediate operand fields of each instruction in the tile. As for the code fragments, it is possible to write a set S_{t_i}, which is a collection of tiles for the normalized code fragment $\overline{\mathcal{I}}_i$. For each normalized code fragment $\overline{\mathcal{I}}_0, \overline{\mathcal{I}}_1,\ldots$, the collection of the semantically equivalent tiles $\{\mathbf{S}_{t_0}, \mathbf{S}_{t_1},\ldots\}$, which must be protected, represents the whole *tile set* of the morphing engine. The complete workflow of the code morphing engine, shown in Figure 2, is composed as follows. First, the input code fragment $\mathcal{I}$ is mapped to a normalized code fragment $\overline{\mathcal{I}}$ through the Register/Constant Normalization step. Then, a tile is randomly selected from the Tile Set to replace the normalized code fragment. The Register/Constant Denormalization procedure is then applied to the registers and symbolic constants. Finally, in the Constant Values Computation step, any symbolic constant is replaced by a random value, and the constant expressions encoded in the tile are evaluated to obtain the immediate operands.

4. EXPERIMENTAL EVALUATION

The experimental platform used to provide a validation of our framework is an ARM-based STMicroelectronics SPEAr Head200 development board. The SPEAr SoC is based on a 32-bit ARM926EJ-S processor running at 133 MHz, without any OS, for the sake of more precise analysis of the results. The AES binary, based on OpenSSL 1.0.0d, with 4 T-tables, is run directly from the U-Boot bootloader. The attack exploits the outcome of the *xor* operation in the first ADDROUNDKEY, which is stored in a register.

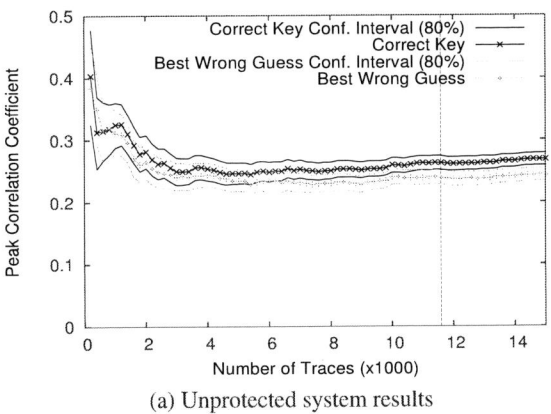

(a) Unprotected system results

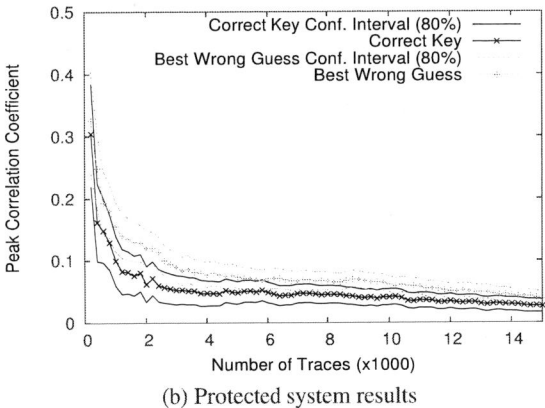

(b) Protected system results

Figure 3: Highest correlation values for correct and best-wrong key guesses as a function of the number of traces, together with their 80% confidence intervals. For the unprotected system, at least 11600 traces are needed to distinguish two separated confidence intervals, i.e. the correct key guess (a). In the case of the protected system (one morph action every 100 traces), all the correlation peaks have statistically negligible values, regardless of the key hypothesis and the increase in the number of traces (b)

4.1 Performance evaluation

Both the protected and the unprotected systems have been evaluated to assess the time overhead introduced by the countermeasure. Through employing a GCC 4.3 based cross-compile toolchain, we observe that the original AES algorithm is composed of 536 instructions, clustered in 20 distinct normalized instructions. We employ three, 4-instruction long, tiles to protect the 64 `eor` code fragments in the whole AES. Thus, instruction replacement adds on average 4 computational instructions for each substituted `eor`, plus a load/store pair to handle *clobbering*. Still, we expect a reduced overhead, as most of the AES execution time is spent in accessing memory for the T-table lookups. The timings have been gathered directly via trace length measurements on the oscilloscope. On average, a run of the unprotected AES algorithm takes 228.8 μs, while a run of the protected one runs for 245 μs, thus resulting in a performance hit of 8.2% of the AES execution time. The overhead due to the code morphing amounts to 90 ms per morphing action. Since the code morphing algorithm is intrinsically memory bound, the majority of this delay is ascribed to accesses to the off-chip memory in our platform. To cope with the additional performance hit, the number of calls to the code morpher over the encryption algorithm runs can be tuned to bring the amortized encryption time within acceptable bounds for the target system. The following section shows how the overhead is reduced to a fraction of the time of one encryption run, and evaluates the security margin provided.

4.2 Differential power analysis

Measured power traces were obtained with an Agilent *InfiniiumDSO80204B* oscilloscope and an active Agilent 1131A differential voltage probe with a 3.5 GHz analog bandwidth. The oscilloscope features 4 independent analog channels, a 2 GHz analog bandwidth, coupled with an 8-bit ADC capable of recording 40 Gsample/s, with a noise floor of 3 mV RMS, and a minimum vertical resolution of 10 mV. The measured power traces have been acquired using a sampling frequency of 500 Msamples/s over an acquisition window of 100 ksamples. The trigger signal is provided via a GPIO pin on the board and collected via a passive probe connected via an Agilent E2697A impedance adapter.

The power measurements have been obtained via measuring the voltage drop at the ends of a 1 Ω SMD resistor inserted on the SPEAr SoC power supply line. To reduce measurement noise, each trace is the result of the average of 32 measurements with the same

plaintext and code. This represents a worst case scenario, as a real world attacker will not be able to choose which code variant is running to get averaged measurements. An attacker might trade off measurements of different plaintexts to gain noise reduction via averaging; however, the maximum number of measurements with a single code variant is bound by a design parameter, i.e. the number of encryption runs before a call to the morphing engine is made. A first order Differential Power Analysis against both unprotected and protected AES implementations has been used as testbench. The employed consumption model is the Hamming weight of one byte of the output of the first ADDROUNDKEY operation. This operation is computed 32 bits at a time, since that is the size of the ARM architecture word length. The analysis computes the sample estimation r of Pearson's correlation coefficient ρ between each sample of the actual power consumption measurements (traces) of the device and the consumption model for each possible hypothetical value of the involved key part. Subsequently, the maximum values of r obtained for each key hypothesis are sorted in decreasing order. For the attack to succeed, the confidence interval I_r of the maximum value should not overlap with any of the others. The correct key hypothesis can be successfully obtained when enough traces are gathered, as the width of I_r decreases when the number of measurements increases. An unbiased estimator for the Pearson correlation coefficient ρ is: $\hat{\rho} = r\left(1 + \frac{1-r^2}{2(n-3)}\right)$, where n is the number of employed samples. To obtain the boundaries of the interval I_r, the probability $\text{Prob}\{\hat{\rho} \in [r_l, r_u]\} \geq \gamma$ is evaluated for a chosen confidence level γ. The theoretical correlation coefficient for the correct key hypothesis ρ_c is 0.250 since the observed operation is performed on 32 bits at a time, while the consumption model takes into account only 8 of them. By contrast, the key hypotheses differing by a single bit from the correct one (the best wrong guessed) has a theoretical correlation coefficient $\rho_w = 0.218$. The values of the estimators $\hat{\rho}_c$, $\hat{\rho}_w$ will converge to ρ_c and ρ_w.

Figure 3a shows that 11600 traces are necessary to distinguish the correct key hypothesis from the best wrong guess with a confidence level $\gamma = 0.8$. Thus the architecture does not provide any embedded protection against side channel attacks, despite the complexity of the SoC.

Two crucial factors for a power analysis success are represented by the knowledge of the implementation strategy of the attacked operation, which allows to infer its consumption model, and the perfect time alignment of each trace. The devised countermeasure

Table 1: Impact of the number of runs among morphing actions on both the security margin of the system (i.e. overlap of the correct key and best wrong guess confidence intervals) and execution time of a single protected AES encryption (optimal tradeoffs in grey). Execution time of a plain AES is 228.8 μs

Code Morphing Interval [no. of runs]	Confidence Intervals Overlap [%]	Average Execution Time
100	79.55	$\times 5.00$
200	79.32	$\times 3.04$
400	79.03	$\times 2.05$
600	78.89	$\times 1.73$
800	78.89	$\times 1.56$
1000	78.98	$\times 1.46$
2000	79.76	$\times 1.27$
3000	79.04	$\times 1.20$
4000	75.16	$\times 1.17$
5000	67.64	$\times 1.15$
6000	56.94	$\times 1.14$
11600	0.00	$\times 1.10$

operates on both factors, thus actively hindering the attack. Consequentially, it is sufficient to perform a code morphing action often enough to avoid the collection of a significant number of traces by the attacker. The maximum number of measurements with the same (albeit unknown) code variant Δn that an attacker is able to collect is thus a design-time-chosen parameter indicating the security margin of the system.

Figure 3b shows the result of the attack performed while our protection methodology was in action with Δn=100. In this case, the confidence interval for the correct key hypothesis and the best wrong guess never separate. In addition to this, the correct key has a sample correlation coefficient lower than the best wrong guess. As a further validation of our approach, we correlated the key hypotheses evaluated through the same consumption model with random values, obtaining a peak sample correlation value higher ($\sim$0.08) than both the previous estimates for $\hat{\rho}_w$ and $\hat{\rho}_c$. Thus, collecting a greater number of traces will not be useful due to the negligible values of the obtained sample correlation estimates.

A practical measure of the security margin is given by the overlap percentage of the confidence intervals of the correct key and the best wrong guess. We note that this measure is a conservative gauge of the actual security margin, as the attacker fails to retrieve the key also when the confidence interval of the best wrong guess is both disjoint and higher than the one of the correct key. In this case, the overlap percentage is zero but the attack does not succeed. However, the opposite scenario does never happen as an overlap of the confidence intervals unquestionably indicates the indistinguishability of the corresponding key hypotheses. The security margin obtained via code morphing for the target platform can be traded off to achieve a lower computational overhead per encryption run. The computational overhead for the encryption is composed of a fixed cost determined by the tile substitution action and an amortized cost over Δn runs due to the call to the morphing engine. Table 1 reports the trends of the security margin (i.e. the confidence intervals overlap) and the average execution time of single AES as a function of the value of Δn. We verified that the security margin of this platform does not report significant hits up to an attack with a half of the traces needed to retrieve the correct key on an unprotected implementation. The optimal trade-off points for this platform are represented by running the code morpher once every 2000–3000 AES runs, as this parameter choice preserves a high security margin, while obtaining an acceptable (around 20%) performance overhead. Raising further the number of AES runs per morphing action drastically reduces the security margin of the system, without a significant performance improvement.

5. RELATED WORK

The distinguishing feature of our solution lies in the code substitution technique, but schemes such as those proposed in [4, 8–10] can be implemented within our framework by means of specific tiles. In these works, the results regarding the described implementations are mostly related to microcontroller platforms and exhibit case-study specific execution times ranging from two to more than fifty times the baseline. When considering attacks based on an a-posteriori model [8], such as *template* attacks, the very high number of code variants provided by our countermeasure would require a prohibitive number of traces to obtain a reliable model, since a significant quantity of information should be gathered for all of them. In [3] the authors propose a framework that employs an information theoretic metric to identify the most sensitive instructions of a software implementation of AES on an 8-bit microcontroller and apply a static local code modification implementing *random precharging*. W.r.t. [3], we propose an automated, dynamic code morphing approach, which can produce a much larger number of different semantically equivalent code versions, and it is also able to apply hiding and masking techniques together in a unified framework.

6. CONCLUDING REMARKS

We presented a framework to automate the application of DPA software countermeasures at run-time and described a code morphing toolchain that proved to be efficient while ensuring protection for any cryptographic primitive. The proposed approach can be applied to either the whole algorithm or to the subset of vulnerable instructions to enhance performances. To our knowledge this is the first work providing this level of protection while being practically viable. The analyzed case study showed how to counter DPA attacks with an acceptable performance overhead. The overhead may be further reduced when protecting a multi-core platform via concurrently executing the encryption routine and the morphing action. Future works will target microcontrollers where the code is in a read-only memory via morphing the execution flow with a series of random jumps along a database of code fragments

7. REFERENCES

[1] A. Barenghi, G. Pelosi, and Y. Teglia. Improving first order differential power attacks through digital signal processing. In *Proc. of the 3rd Int'l Conf. on Security of Information and Networks*, SIN'10, pages 124–133. ACM, 2010.

[2] A. Barenghi, G. Pelosi, and Y. Teglia. Information leakage discovery techniques to enhance secure chip design. In *Proc. of the 5th IFIP WG 11.2 Int'l Conf. on Information Security Theory and Practice: security and privacy of mobile devices in wireless communication*, WISTP'11, pages 128–143. Springer-Verlag, 2011.

[3] A. G. Bayrak, F. Regazzoni, P. Brisk, F.-X. Standaert, and P. Ienne. A first step towards automatic application of power analysis countermeasures. In *Proc. of the 48th Design Automation Conference*, DAC'11, pages 230–235. ACM, 2011.

[4] J.-S. Coron and I. Kizhvatov. Analysis and improvement of the random delay countermeasure of ches 2009. In *Proc. of the 12th Int'l Workshop on Cryptographic Hardware and Embedded Systems*, CHES'10, pages 95–109. Springer-Verlag, 2010.

[5] E. Duesterwald. Dynamic compilation. In *The Compiler Design Handbook*, pages 739–762. CRC Press, Inc., 2002.

[6] P. C. Kocher, J. Jaffe, and B. Jun. Differential power analysis. In *Advances in Cryptology*, CRYPTO'99, pages 388–397. Springer-Verlag, 1999.

[7] C. Linn and S. Debray. Obfuscation of executable code to improve resistance to static disassembly. In *Proc. of the 10th ACM Conf. on Computer and Communications Security*, CCS'03, pages 290–299. ACM, 2003.

[8] S. Mangard, E. Oswald, and T. Popp. *Power Analysis Attacks: Revealing the Secrets of Smart Cards*. Springer-Verlag, 2007.

[9] M. Rivain, E. Prouff, and J. Doget. Higher-order masking and shuffling for software implementations of block ciphers. In *Proc. of the 11th Int'l Workshop on Cryptographic Hardware and Embedded Systems*, CHES'09, pages 171–188. Springer-Verlag, 2009.

[10] S. Tillich and C. Herbst. Attacking state-of-the-art software countermeasures–a case study for aes. In *Proc. of the 10th Int'l Workshop on Cryptographic Hardware and Embedded Systems*, CHES'08, pages 228–243. Springer-Verlag, 2008.

Security Analysis of Logic Obfuscation

Jeyavijayan Rajendran†, Youngok Pino‡, Ozgur Sinanoglu§, and Ramesh Karri†
†Polytechnic Institute of New York University ‡Air Force Research Labs §New York University-Abu Dhabi

ABSTRACT

Due to globalization of Integrated Circuit (IC) design flow, rogue elements in the supply chain can pirate ICs, overbuild ICs, and insert hardware trojans. EPIC [1] obfuscates the design by randomly inserting additional gates; only a correct key makes the design to produce correct outputs. We demonstrate that an attacker can decipher the obfuscated netlist, in a time linear to the number of keys, by sensitizing the key values to the output. We then develop techniques to fix this vulnerability and make obfuscation truly exponential in the number of inserted keys.

Categories and Subject Descriptors

K.6.5 [**Management of Computing and Information Systems**]: [Security and Protection-Physical Security]

General Terms

Security

Keywords

IP protection, Logic obfuscation

1. INTRODUCTION

1.1 Motivation – Preventing IP Piracy

Globalization of Integrated Circuit (IC) design is making IC/Intellectual Property (IP) designers and users re-evaluate their trust in hardware [2]. As the IC design flow is distributed worldwide, hardware is prone to new kinds of attacks such as reverse engineering and IP piracy [1]. An attacker, anywhere in this design flow, can reverse engineer the functionality of an IC/IP. One can then steal and claim ownership of the IP. An untrusted IC foundry may overbuild ICs and sell them illegally. Finally, rogue elements in the foundry may insert malicious circuits (hardware trojans) into the design without the designer's knowledge [3]. Because of these attacks, the semiconductor industry loses $4 billion annually [4].

If a designer can hide the functionality of an IC while it passes through the different, potentially untrustworthy phases of the design flow, these attacks can be thwarted [1].

1.2 Logic obfuscation

Logic obfuscation hides the functionality and the implementation of a design by inserting additional gates into the original design. In order for the design to exhibit its correct functionality (i.e., produces correct outputs), a valid key has to be supplied to the obfuscated design. The gates inserted for obfuscation are the *key-gates*. Upon applying a wrong key, the obfuscated design will exhibit a wrong functionality (i.e., produce wrong outputs).

Consider the circuit shown in Figure 1 which is obfuscated using key-gates *K1* and *K2*. The inputs *I1 – I6* are the functional inputs and *K1* and *K2* are the key inputs connected to the key-gates. On applying the correct key

values (K1=0 and K2=1) the design will produce a correct output; otherwise, it will produce a wrong output.

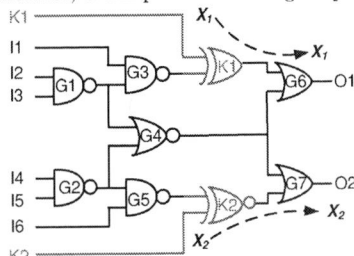

Figure 1: A circuit obfuscated using two key-gates K1 and K2 based on the technique proposed in [1]. By applying the input pattern 100000, an attacker can sensitize key bits K1 and K2 to the outputs O1 and O2, and observe their values.

EPIC [1] incorporates logic obfuscation into the IC design flow, as shown in Figure 2. In the untrusted design phases, the IC is obfuscated and its functionality is not revealed. Post-fabrication, the IP vendor activates the obfuscated design by applying the valid key. The keys are stored in a tamper-evident memory inside the design to prevent access to an attacker, rendering thes key inputs unaccessible by an attacker.

1.3 Attacks against logic obfuscation

The purpose of logic obfuscation is defeated if an attacker can determine the secret keys used for obfuscation. By determining the keys, one can decipher the functional netlist, and make pirated copies and sell them illegally.

We propose an attack where the attacker applies specific input patterns, observes the outputs for these pattern, and deciphers the secret key. To perform this attack, one needs the obfuscated netlist and a functional IC. An attacker can obtain the obfuscated netlist from (1) the IC design, or by reverse engineering the (2) layout, (3) mask, or (4) a manufactured IC as shown in Figure 2. The functional IC, (5) in Figure 2, is bought in the open market.

The value of an unknown key can be determined if it can be sensitized[1] to an output without being masked/corrupted by the other key-bits and inputs. By observing the output, the sensitized key bit can be determined, given that other key-bits (similar to unknown X-sources[2]) do not interfere with the sensitized path.

Once an attacker determines an input pattern that propagates the key-bit value to an output without any interference, it is applied to the functional IC i.e., the IC with the correct keys. Now, this pattern will propagate the correct key value to an output. An attacker can observe this output and resolve the value of that key-bit.

Motivational example 1 (attack): Consider the key input K1 in Figure 1. It will be sensitized to output O1 if the value at the other input of gate G6 is 0 (non-controlling value for an OR gate). This can be achieved by setting I1=1, I2=0 and I3=0. As the attacker has access to the functional IC, one can apply this pattern and determine the value of K1 on O1. For example, if the value of O1 is 0 for that input pattern, then K1 = 0, otherwise K1=1.

---

[1]Sensitization of an internal line *l* to an output *O* refers to the condition (values applied from the primary inputs to justify the side input of gates on the path from *l* to *O* to the non-controllable values of the gates) which bijectively maps *l* to *O* and thus renders any change on *l* observable on *O*.
[2]X-sources: Uninitialized memory units, bus contentions or multi-cycle paths are the source of unknown response bits, i.e., unknown-Xs in testing. They are non-controllable.

83

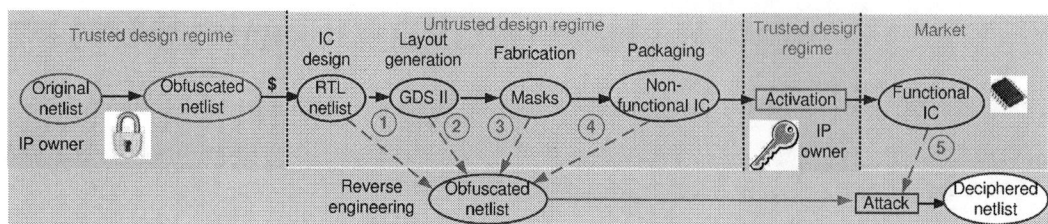

Figure 2: The top blue box represents the EPIC design flow [1]. The design is in the obfuscated form in the untrusted design regime. In the untrusted regime, an attacker can obtain the obfuscated netlist from (1) the IC design, or by reverse engineering the (2) layout, (3) mask, or (4) a fabricated IC, and (5) the functional IC from the market. Using this attack, the attacker can get a deciphered netlist and make pirated copies.

This problem is analogous to the fault sensitization problem in the presence of unknown-X values that can possible block/mask the fault propagation [5]. The key-bits K1 and K2 are equivalent to X-sources X_1 and X_2 in Figure 1.

Similarities and differences between fault detection and key-propagation: Both objectives require an input pattern that sensitizes the fault effect/key bit by

- blocking the effect of some or all of the X-sources/other key bits, and preventing their interference.
- justifying the side input of all the gates on the sensitization path to non-controlling values of the gates

The two problems differ slightly:

- fault detection also involves fault activation by justifying the fault site to the fault value; key propagation requires only sensitization
- fault detection aims at blocking/avoiding unknown X's; key propagation aims identifying unknowns one at a time resulting in an iterative and dynamic process

While an Automatic Test Pattern Generation tool [5] capable of handling X's during test generation can be readily used by the attacker to identify the patterns that decipher key bits, we take a closer look at various interference scenarios for the key bits (unknown sources) with the ultimate goal of building a strong defense, i.e., a smart logic obfuscation technique.

1.4 Smart logic obfuscation

To prevent such attacks, key-sensitization has to be hampered by inserting key-gates in such a way that propagation of a key value will be possible only if certain conditions are forced on other key inputs. As these key inputs are not accessible by the attacker, one cannot force the values necessary to propagate the effect of a key. Thus, brute force has to be employed.

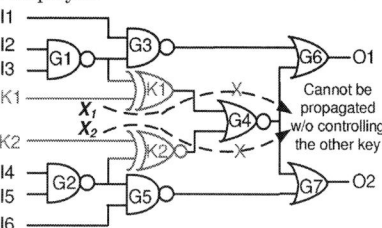

Figure 3: The attacker cannot propagate the effect of key bits K1 and K2 individually to the outputs. Hence, the attacker has to brute force to determine the values of K1 and K2.

Motivational example 2 (defense): Consider the circuit shown in Figure 3 which is the same functional circuit shown in Figure 1 but the two key-gates *K1* and *K2* are at different locations. Here, if the attacker has to propagate the effect of either of the keys, then one has to force a '0' (non-controlling value of NOR gates) on the other input of G4. In order to force this value, one has to control the key inputs, which are inaccessible. Thus one cannot propagate the effect of a key to an output, failing to determine the values of the key.

Depending upon the location of key-gates, different techniques have to employed to propagate the effect of a key. In

Section 2, we describe the different types of key-gates based on their locations and also the strategies that an attacker may follow to decipher the key bits. In section 3, we introduce a graph notation to capture the interference between key bits, enabling algorithmic development for smarter logic obfuscation. Section 4 compares the obfuscation strength and performance results between the random insertion and the proposed logic obfuscation technique.

1.5 Contributions

The contributions of this paper are

- an attack on logic obfuscation based on IC testing concepts
- strategies used by an attacker to decipher keys based on their interference
- a logic obfuscation algorithm based on key-gates interference graph

2. ATTACK STRATEGIES

A logic obfuscation technique can insert the key-gates anywhere in the circuit. Depending upon their location, the attacker develops different strategies to determine the key bits. In this section, we will classify key-gates based on their type of interference with other key-gates and the corresponding strategy used by the attacker.

2.1 Runs of key-gates

A set of key-gates connected in a back-to-back fashion forms a run of key-gates.

Example 3: In Figure 4(a), the key-gates K1 and K2 form a run as they are connected back-to-back.

Figure 4: (a) A run of two key-gates K1 and K2. (b) K3 replaces K1 and K2.

Runs of key-gates reduce the effort of an attacker as they increase the valid key space. If N key-gates form a run, then the valid key space increases from the ideal, 1 valid key, to 2^{N-1} valid keys. In the above example, 01 and 10 are valid keys, one of which suffices for the attacker.

Attack strategy: An attacker can replace a run of key-gates by a single key-gate, thereby reducing the number of key bits. Once the value of that key-gate is determined, one can find the entire valid key space. For example, in Figure 4(b), the attacker replaces K1 and K2 with a key-gate K3. After the value of K3 is resolved as 1, one determines that the valid key space is 01 and 10.

2.2 Isolated key-gates

If there is no path from a key-gate to all the other key-gates and vice-versa, then such a gate is called an isolated key-gate.

Example 4: Consider the gate K1 in Figure 1. As there is no path between K1 and K2, K1 and K2 are isolated gates.

Attack strategy: An attacker prefers isolated key-gates as there is no interference with other key-gates. An attacker

84

identifies a pattern that uniquely propagates the effect of an isolated key-gate's key to an output. One then applies the pattern to the functional IC and determines its value.

As mentioned before, in Figure 1, the pattern 100XXX propagates the value of K1 to output O1. An attacker, upon observing this output, can identify that the value of K1 is 0.

2.3 Dominating key-gates

If there are two key-gates K1 and K2 such that K2 lies on every path between K1 and the outputs, then K2 is called a dominating key-gate.

Example 5: The gate K2 in Figure 5 is a dominating key-gate.

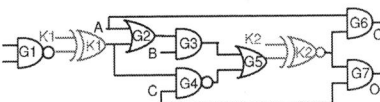

Figure 5: K2 is a dominating key-gate whose key bit value can be determined only after muting the effect of K1. Patterns that make either C = 0 and A=1 or C=0 and B =1 will mute the effect of K1. However, only if A = 1, the effect of K2 can reach O1.

Attack strategy: An attacker can determine the value of K2's key bit only if the effect of K1's key bit is prevented (muted) from reaching key-gate K2 while simultaneously sensitizing K2's key bit to an output. An input pattern that can perform muting as well as sensitization is called the golden pattern. On applying this golden pattern, the attacker can determine the value of K2. If muting of K1 and propagation of K2 cannot be performed simultaneously, then the attacker cannot determine value of K2. In such cases, the golden pattern does not exist, forcing an attacker to employ brute force.

The effect of a key can be muted before it reaches the other key, by using patterns that force controlling values in any of the gates on the path between K1 and K2. If there are multiple paths from key-gates K1 and K2, then the effect of key-input K1 has to be muted on every path.

Example 6: Consider the circuit shown in Figure 5. K2 can be determined only if the effect of K1 is muted. If there is a pattern that justifies the output of G5 to 1, then the effect of K1 will be muted. Patterns that make either C = 0, or A=1 and B =1 will assure this condition, thereby muting the effect of K1. However, the attacker should select the pattern that propagates the effect of K2 to an output. If C =0, G7 blocks the propagation of K2 as its output will always be 0. The condition A = 1, allows K2 to propagate through G6. Hence, an attacker will select the pattern that makes A=1 and C =0, so that one can mute the effect of K1 as well as propagate the effect of K2 to an output.

2.4 Convergent key-gates

Even if there are no paths between two key-gates, the sensitization paths might interfere. Such scenarios happen if these two or more key-gates converge. Depending upon the type of convergence, key-gates can be classified into 1) concurrently mutable, 2) sequentially mutable, and 3) non-mutable key-gates.

2.4.1 Concurrently mutable convergent key-gates

If two key-gates K1 and K2 converge at some other gate, such that K1's key bit can be determined by muting K2, and K2's key bit can be determined by muting K1, then K1 and K2 are called concurrently mutable key-gates.

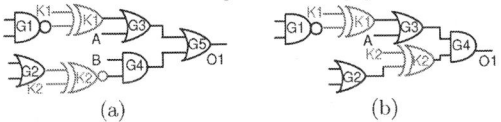

Figure 6: (a) Concurrently mutable key-gates: K1 and K2 converge at G5 and can be muted. (b) Sequentially mutable key-gates: K1 and K2 converge at G4, but only K1 can be muted.

Example 7: Consider the circuit shown in Figure 6(a). The key-gates K1 and K2 converge at the gate G5. The value of K1 can be determined by applying a pattern that mutes K2 (B=0). Similarly, the value of K2 can be determined by applying a pattern that mutes K1 (A=1).

Attack strategy: The attacker determines the golden pattern that mutes one key and simultaneously sensitizes the other key to an output, or vice-versa. If a golden pattern does not exist, then the attacker has to perform brute force only on that set of concurrently mutable key-gates.

2.4.2 Sequentially mutable convergent key-gates

If two gates K1 and K2 converge at some other gate, such that K2's key bit can be determined by muting K1's key while K2's key cannot be muted to determine K1's key, then K1 and K2 are called sequentially mutable convergent key-gates, as they can be deciphered only in a particular order.

Example 8: Consider the circuit shown in Figure 6(b). The value of K2 can be determined by applying a pattern that mutes K1 (A=1), while K2 cannot be muted as it directly feeds the gate where K1 and K2 converge.

Attack strategy: An attacker will first determine K2's value by muting K1 using the golden pattern. One then updates the netlist by replacing K1 with a buffer or an inverter based on the value of K2. Then K1 is targeted. If the golden pattern does not exist, then the attacker has to perform brute force only on that set of sequentially mutable key-gates.

2.4.3 Non-mutable convergent key-gates

If two key-gates K1 and K2 converge at some other gate, such that neither of the key bits can be muted, then K1 and K2 are called non-mutable convergent key-gates.

Example 9: Consider the circuit shown in Figure 3. The key-gates K1 and K2 are connected to the same gate G4.

Attack strategy: To propagate either of the key bits the other one has to be muted. However, as an attacker cannot access key inputs, one cannot force those values. Hence, one is forced to perform brute force attacks.

Input : Obfuscated netlist, Functional IC, Key Inputs
Output: Original netlist
Determine *Runs of Keys*;
Replace them with XOR gates;
Update Netlist;
for *the remaining keys* **do**
    For each *Isolated Key* **do**
        Compute and apply propagation pattern;
        Determine *KeyBits* and update Netlist;
    end
    For each *Consecutive||Concurrent||Sequential key* **do**
        if *there exists a golden pattern* **then**
            Apply the golden pattern;
            Determine *KeyBits*, Update Netlist, Break;
        else
            ApplyBruteForce(), Break;
        end
    end
    For each *Non-mutable Key* **do**
        ApplyBruteForce(), Break;
    end
end

ApplyBruteForce();
For each *possible key combination* **do**
    Generate random input patterns;
    Simulate the patterns and obtain the outputs OP_{sim};
    Apply the patterns on IC and obtain the outputs OP_{exe};
    if $OP_{sim} == OP_{exe}$ **then**
        Valid Key = current key combination;
        Update netlist;
    end
end

Algorithm 1: Attack on logic obfuscation.

2.5 An attacker's action plan

By considering all the different types of interference between key-gates, an attacker uses Algorithm 1 to determine the secret key. The attacker first removes the runs of key-gates and targets the isolated key-gates. Each isolated gate can be removed by one test patterns. After that, one targets consecutively mutable, concurrently mutable, and sequentially mutable key-gates. Only if one is able to generate a golden pattern that simultaneously mutes effects of the other keys and sensitizes the effect of the target key, the value of the target key can be determined. Finally, the non-mutable keys are identified via brute force. As the key bits are identified gradually in every iteration, the corresponding key-gates can be replaced by a buffer or an inverter, possibly changing the type of other key-gates. Thus, in every iteration, the key-gate types need to be re-computed.

3. STRONG LOGIC OBFUSCATION

Strong logic obfuscation hinges on inserting key-gates with complex interferences among them. Next, we relate types of key-gates to the kind of interference they introduce using a graph-based notation.

3.1 Interference graph

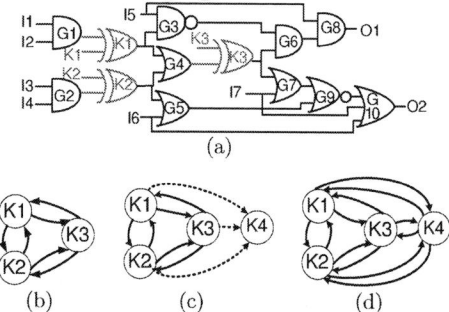

(a)

(b)   (c)   (d)

Figure 7: (a) An example circuit with three key-gates. (b) Interference graph of the key-gates. Non-mutable keys are connected by solid edges. (c) If the new key-gate is inserted at the output G10, it creates mutable edges (dotted lines) with the other key-gates (d) If the new key-gate is inserted at the output G5, it creates non-mutable edges (solid lines) with the other key-gates.

To insert key-gates, we form an interference graph of key-gates. In this graph, each node represents a key-gate and an edge connects two nodes, if two gates interfere. Isolated key-gates are represented with isolated nodes. A run of key-gates is denoted by a single node. Non-mutable key-gates are represented are connected with non-mutable edges, concurrently mutable key-gates are connected with mutable edges. Sequentially mutable key gates are connected by two edges; a non-mutable edge arises from the key-gate that is non-mutable and a mutable edges arises from the key-gate that is mutable.

Example 10: Consider the circuit with three key-gates shown in 7(a). They interfere with each other as follows

- K1 and K2 are non-mutable and so they are connected by non-mutable edges as shown in Figure 7(b).

- The key-gates K1 and K3 converge at the gate G6, hence they are converging key-gates. Specifically, they are sequentially convergent; K3's effect cannot be muted while K1's effect can be muted by applying I5=0. However if I5 is 0, then both key bits are blocked at G8. Hence, K1 and K3 are non-mutable and so they are connected by non-mutable edges as shown in Figure 7(b).

- K2 and K3 converge at the gate G9, through G5 and G7, respectively. However, neither of the key bits can be muted and sensitized individually. For instance, making I6=1, mutes K2 but also blocks the sensitization of K3 at G10. Making I7=1, mutes K3 but also blocks the sensitization of K2 at G10. Hence, K2 and K3 are non-mutable as shown in Figure 7(b).

For a stronger logic obfuscation, the number of non-mutable edges in the interference graph should be maximized, as they force an attacker to perform brute force. On the other hand, if there are more mutable edges, then the attacker can mute the effect of keys and can easily determine their values. Hence, a defender prefers non-mutable edges to mutable edges.

Example 11: Consider the circuit shown in Fig 7(a). If a new key-gate, K4, is inserted at the output of G10, then it creates mutable edges with all the other key-gates. By setting I6=1 or I7=1, the attacker can mute the effects of K1, K2, and K3, and can decipher easily. Hence, G10 is connected with mutable edges with the other key-gates as shown in Figure 7(c).

If the new key-gate, K4, is inserted at the output of G5, then it creates non-mutable edges with the other key-gates as shown in Figure 7(d). Thus, it is better to insert the new key-gates at the output of G5.

Input : Original netlist, KeySize
Output: Obfuscated netlist
KeyGateLocations = {};
Randomly insert 10% key-gates;
Add that location to KeyGateLocations;
Construct KeyGraph;
for $i \leftarrow 2$ **to** *KeySize* **do**
    For each *Gate$_j$* in *Netlist* **do**
        if *Gate$_j$* $\notin$ *KeyGateLocations* **then**
            Cum. Weight = $\sum$ weight of edges in KeyGraph;
            For each *Key-gate$_k$* in *KeyGateLocations* **do**
                Cum. Weight$_j$ += FindMetric(Gate$_j$,
                Key-gate$_k$);
            end
        end
    end
    Select the Gate with the highest Hardness Metric;
    Add the selected gate to KeyGateLocations;
    Insert a key-gate at the output of the selected gate;
    Update KeyGraph;
end

FindMetric(*K1, K2*);
if *K1 and K2 are isolated* **then Return** 0;
if *K1 and K2 are consecutive||concurrent||sequential* **then**
    if a golden pattern exists **then**
    Return weight of mutable edge;
    else Return weight of non-mutable edge;
end
if K1 and K2 are non-mutable **then**
Return weight of non-mutable edge;

Algorithm 2: Insertion of key-gates

3.2 Insertion of key-gates

A defender can use the interference graph to insert key-gates. Algorithm 2 is used to insert key-gates. At every iteration, a key-gate is inserted at a location such that the number of non-mutable edges in the graph is maximized.

Initially, 10% of the total key-gates are inserted at random locations in the circuit. Such random distribution will insert key-gates in different parts of the circuit thereby affecting multiple outputs. Here, we considered 10% for initial distribution (one also can chose a different amount of initial distribution and the impact of this amount on obfuscation is beyond the scope of this paper). Then the graph of key-gates is constructed. Then, the remaining key-gates are introduced iteratively. In every iteration, for each gate in the netlist, we determine the type of edge with the previously inserted key-gate. Depending upon the type of edge, we assign weights; non-mutable edges are given a higher weight than the mutable edges. We then calculate the sum of weights of edges in the graph for that gate. The gate that maximizes the sum of weight of edges in the graph is selected, and a key-gate is inserted at its output. The graph is then updated by including the new key-gate. This procedure is repeated for inserting all the key-gates.

In every iteration, the defender has to check for the presence of golden patterns which might increase the computational complexity of the algorithm. Hence, a defender can assume that there always exists a golden pattern and skip the search for the golden pattern. This is a pessimistic scenario for a defender because some golden patterns might not exist.

4. RESULTS

4.1 Experimental Setup

The proposed technique is analyzed using ISCAS-85 combinational benchmarks. We used the Atalanta testing tool [6] to determine the input patterns for muting and propagation the effects of keys. To obfuscate a circuit with a reasonable performance overhead, we selected the key size as 5% of number of gates in that circuit. While obfuscating a circuit, we assumed that there always exists a golden pattern. While attacking the circuit, we used the techniques proposed in Section 2 where we search for the presence of a golden pattern. For every brute force attempt, we applied 1000 random patterns to determine the value of a key. The area, power, and delay overheads were obtained using the Cadence RTL compiler.

We compared the effectiveness of four types of insertions: random-insertion [1], random insertion with no runs of gates, unweighted insertion where both mutable and non-mutable edges are given the same weight of 1, and weighted insertion where non-mutable edges are given a higher weight (weight = 2) than the mutable edges (weight = 1).

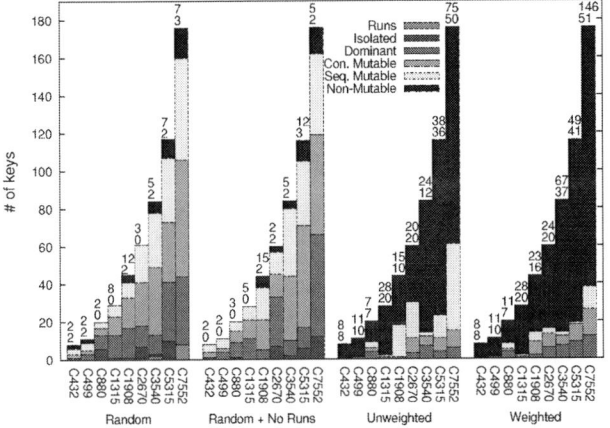

Figure 8: Types of key-gates inserted by different logic obfuscation techniques. Effective key size from an attacker's perspective (top) and from a defender's perspective (bottom) are shown as numbers on top the bars.

4.2 Types of keys and effective key-size

Figure 8 shows the number of types of keys in different benchmarks for different types of insertions. In the random insertion method, most of the keys are concurrently mutable. Some number of keys are inserted in runs benefiting the attacker. Only 30% of keys are non-mutable and sequentially mutable which require brute force approach. In the 'Random + No Runs' method, keys are not inserted in runs thereby increasing the effort of the attacker.

In the unweighted and weighted insertions, around 90% of keys are of non-mutable and sequentially mutable types. Most of the keys in weighted insertion is either non-mutable or sequentially mutable because they are given a higher weight. There are no isolated keys in either of the insertion techniques, as they are not given any weights.

Effective key size: Due to random insertion of the first 10% of key-gates, multiple disconnected graphs might exist within a key-interference graph. The keys in a graph can be either isolated, dominant, or convergent. Since a de-

fender pessimistically assumes that the golden patterns always exist, the effective key size from his perspective is the maximum number of non-mutable keys in a connected key-interference graph. If there are N non-mutable key gates (effective key-size), the number of brute force attempts is 2^{N-1}. However, when an attacker tries to attack, not all the golden patterns will exist. For those keys, he has to try for all possible combinations. Hence, from an attacker's perspective, the effective key size is the largest key size on which brute force is attempted. If the number of brute force attempts is 2^{M}, then the effective key size for an attacker is M.

In Figure 8, the effective key-sizes for a defender and an attacker are shown as numbers on top of the bars. For both the attacker and defender, the effective key sizes of random insertions are less than that of the unweighted and weighted insertions. Therefore, the number of brute force attempts required to decipher the keys inserted using random insertions is exponentially smaller than that of the unweighted and weighted insertions. The attacker's effective key size is always greater than that of the defender's because of the absence of golden patterns which forces the attacker to perform brute force. For example, consider the benchmark C7552, the attacker needs 2^{146} brute force attempts and hence the effective key size is 146. On the other hand, for a defender, the largest number of non-mutable key-gates in a connected graph is 51 and hence the effective key size is 51.

4.3 Number of test patterns

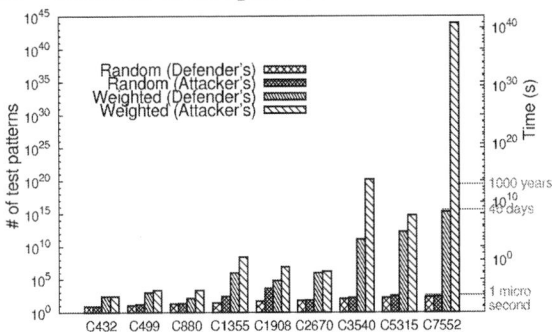

Figure 9: Number of test patterns required by the attacker to determine the keys inserted using random and weighted insertion from an attacker and a defender's perspectives. The time scales are drawn assuming that one billion input patterns can be applied per second.

Figure 9 shows the number of test patterns required to decipher the key from a defender and an attacker's perspectives for the random and weighted insertion method. The time scales are calculated assuming that an attacker can apply a billion patterns per second. On one hand, from a defender's perspective, the number of test patterns are calculated assuming that golden patterns exist. On other hand, from an attacker's perspective, the number of test patterns are more realistic as they are determined using the attack methodology proposed in Section 2. It can be seen that the defender's perspective on timescale is several orders of magnitude smaller than the realistic scenario. For example, in C7552 circuit, a defender thinks that it will take 46 days to decipher the netlist while the attacker will take more than a thousand years. However, from both attacker's and defender's perspectives, a few thousand test patterns are sufficient to figure out the keys when they are inserted randomly. On the other hand, when the weighted key insertion method is used, the number of test patterns required to recover the keys increases by several orders of magnitude. For example, in case of C7552 to about 10^{18} which will take several years to figure out the key bits.

4.4 Effect of the weight of a non-mutable edge

Table 1: No. of non-mutable keys out of the total 176 keys in the benchmark C7552 for different weights of non-mutable edges.

Weight of non-mutable edge	1	2	10	100	1000
# of non-mutable key-gates	115	138	149	156	163

By increasing the weight of the non-mutable edges, the algorithm will create a design that has a large number of non-mutable key-gates. Table 1 shows the number of non-mutable key-gates for different weights of non-mutable edges in one of the ISCAS-85 benchmark circuit, C7552. This circuit was obfuscated with 176 key-gates. While increasing the weight of the non-mutable edges increases the number of non-mutable key-gates in the design, the rate of increase is not the same rate. Increasing the weight from 1 to 2 increases the number of non-mutable key-gates from 115 to 138. But increasing the weight from 2 to 10, increases the number of non-mutable key-gates from 138 to 149.

4.5 Area overhead

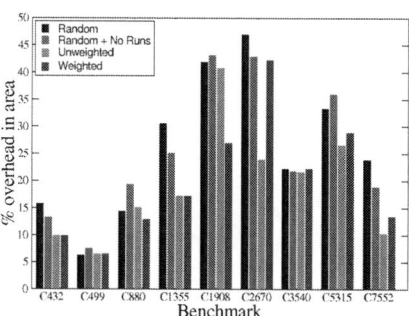

Figure 10: Area overhead for different insertion algorithms.

Figure 10 shows the area overhead for different key-gate obfuscation algorithms. Even though the number of key-gates inserted is 5% of the number of gates in the original design, the area overhead is high as the key-gates are XOR/XNOR gates that consists of a large number of transistors. Unweighted and weighted insertion techniques entail less overhead than random insertion techniques.

4.6 Power-delay product

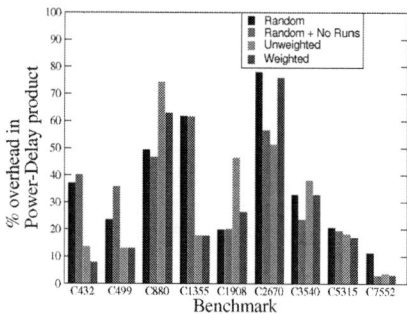

Figure 11: Power delay product overhead for different insertion algorithms.

Figure 11 shows the power-delay product overhead for different insertions. Random insertion yields an average overhead of 25% while weighted and unweighted insertion yields an average overhead of 21%. To minimize this overhead, one can pursue a power and delay constrained obfuscation.

4.7 Logic obfuscation with PUFs

Physical Unclonable Functions (PUFs) are circuits that leverage process variations in IC manufacturing, to produce secret keys. In [1], PUFs are used to give unique keys for each IC even though they are all obfuscated with the same key. The design is first obfuscated with a key and a PUF circuit is attached to it. Upon applying the user key (chal-

lenge) to the PUF, the PUF's response will be the key used for obfuscation. In the proposed attack, the attacker is trying to figure out this response i.e., the key used for obfuscation. On getting this response, the attacker can remove the PUF circuit from the netlist and apply the correct keys directly to the original design. To break the influence of PUFs or any cryptographic algorithms, an attacker can determine the wires that carry these signals and disconnect them.

5. RELATED WORK

Logic obfuscation techniques can be broadly classified into two types—sequential and combinational. In sequential logic obfuscation, additional logic (black) states are introduced in the state transition graph [7]. The state transition graph is modified in such a way that the design reaches a valid state only on applying a correct sequence of key bits. If the key is withdrawn, the design, once again, ends up in a black state. However, the effectiveness of these methods in producing a wrong output has not been demonstrated. In combinational logic obfuscation, as mentioned before, XOR/XNOR gates are introduced to conceal the functionality of a design [1].

Obfuscation is also performed by inserting memory elements [8]. The circuit will function correctly only when these elements are programmed correctly. However, using memory elements will incur significant performance overhead.

6. CONCLUSION

Logic obfuscation is weak when the inserted key-gates are isolated or their effect can be muted. If mutable gates are employed, then the attacker is able to determine the key bits within a second. However, it can be strengthened by inserting key-gates such that their effects are not mutable. In such insertions when the key size is greater than 100, it will take several years for an attacker to determine the key bits. Our analysis reveal that even though a defender pessimistically assumes a smaller effective key size, the actual key size encountered by the attacker is much higher.

IC testing techniques allow designers and testers to peek into the design, by controlling only the inputs and observing the outputs. On one hand, an attacker can use such capability to subvert logic obfuscation. On the other hand, a defender can perform better logic obfuscation by making such process infeasible using the lessons learnt from testing.

7. ACKNOWLEDGEMENT

This material is based upon work funded by AFRL under contract No. FA8750-11-2-0274. Any opinions, findings and conclusions or recommendations expressed in this material are those of the author(s) and do not necessarily reflect the views of AFRL.

8. REFERENCES

[1] J. Roy, F. Koushanfar, and I. Markov, "EPIC: Ending Piracy of Integrated Circuits," *Proc. of Design, Automation and Test in Europe*, pp. 1069–1074, 2008.

[2] "Defense Science Board (DSB) study on High Performance Microchip Supply," http://www.acq.osd.mil/dsb/reports/ADA435563.pdf, 2005.

[3] R. Karri, J. Rajendran, K. Rosenfeld, and M. Tehranipoor, "Trustworthy Hardware: Identifying and Classifying Hardware Trojans," *IEEE Computer*, vol. 43, no. 10, pp. 39–46, 2010.

[4] SEMI, "Innovation is at risk as semiconductor equipment and materials industry loses up to $4 billion annually due to IP infringement," www.semi.org/en/Press/P043775, 2008.

[5] M. L. Bushnell and V. D. Agrawal, "Essentials of Electronic Testing for Digital, Memory, and Mixed-Signal VLSI Circuits," *Kluwer Academic Publishers, Boston*, 2000.

[6] H. Lee and D. Ha, "An efficient forward fault simulation algorithm based on the parallel pattern single fault propagation," *Proc. of IEEE International Test Conference*, pp. 946–955, 1991.

[7] R. Chakraborty and S. Bhunia, "HARPOON: An Obfuscation-Based SoC Design Methodology for Hardware Protection," *IEEE Transactions on Computer-Aided Design*, vol. 28, no. 10, pp. 1493–1502, 2009.

[8] A. Baumgarten, A. Tyagi, and J. Zambreno, "Preventing IC Piracy Using Reconfigurable Logic Barriers," *IEEE Design and Test of Computers*, vol. 27, no. 1, pp. 66–75, 2010.

APPENDIX
(The intuition behind key interference based logic obfuscation)

We will analyze the key-interference graphs of a circuit obfuscated using random and weighted insertion methods both from a defender and an attacker's perspectives. As mentioned in Section 3, the defender always assumes that the golden pattern, that mutes and sensitizes the effect of key-gates simultaneously, does not exist, and constructs his key interference graph accordingly. However, an attacker tries to find the golden patterns to mute key-gates to decipher the value of keys, and constructs his key interference graph accordingly.

A. RANDOM INSERTION

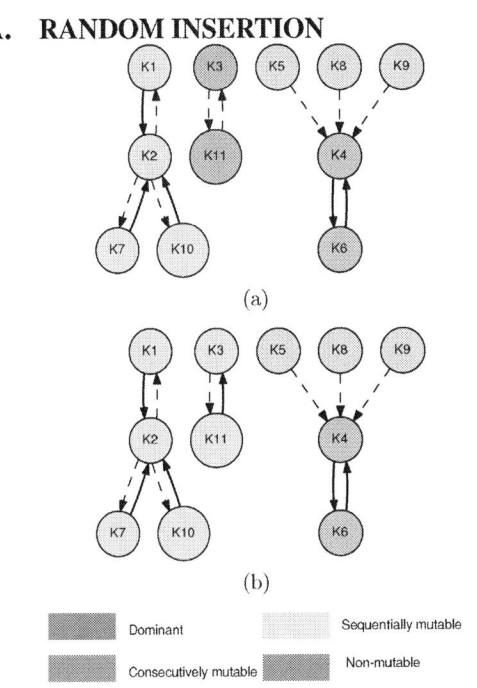

(a)

(b)

Figure 12: Key-interference graphs of C499, an ISCAS-85 benchmark circuit, obfuscated with 11 key-gates. Dotted lines represent mutable edges and solid lines represent non-mutable edges. (a) Key-interference graph from a defender's perspective with an assumption that the edge K11→K3 is mutable. (b) Key-interference graph from an attacker's perspective. The golden pattern to mute the edge K11→K3 does not exist.

Defender's perspective: Figure 12 shows the key interference graph of one of the ISCAS-85 benchmark circuit, C499, which is obfuscated by inserting 11 key-gates. Key-gates K5, K8, and K9 are classified as dominant key-gates[3]. Key-gates K3 and K11 are classified as consecutively mutable key-gates. Key-gates K1, K2, K7, and K10 are classified as sequentially mutable key-gates. Key-gates K4 and K6 are classified as non-mutable key-gates. From a defender's perspective, since there are two non-mutable key-gates, the effective key size is two.

Attacker's perspective: Consider the scenario where an attacker tries to search for the golden pattern for the edge K11→K3 that simultaneously mutes K11 and sensitizes K3. He concludes that such a pattern does not exist. Thus, from an attacker's perspective, the edge from K11→K3 is non-mutable as shown in Figure 12(b). Hence, the key-gates K3 and K11 are classified as sequentially mutable key-gates. As the largest key size on which brute force is attempted is two, the effective key size is two. Notice that even though eleven key-gates are inserted, the effective key size is only two.

B. WEIGHTED INSERTION

---
[3]Please refer Section 2 for the definitions of different types of key-gates.

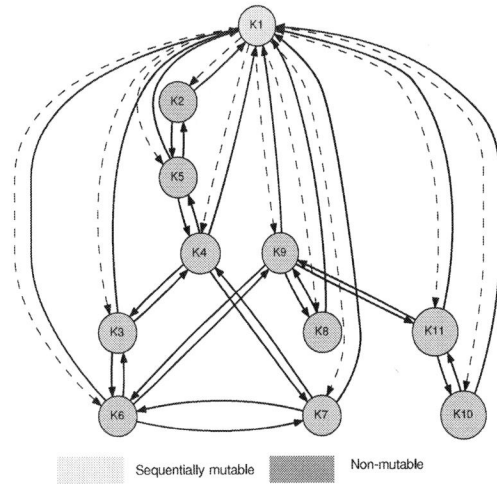

Figure 13: Key-interference graph of C499 from a defender's perspective with an assumption that the edges K1→K2, K1→K5, K1→K10, and K1→K11 are mutable. Dotted lines represent mutable edges and solid lines represent non-mutable edges. Effective key size is 10.

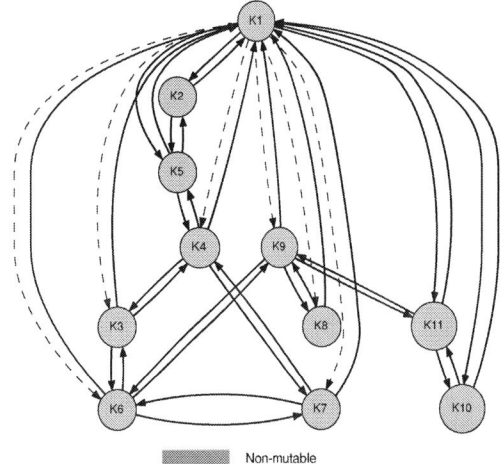

Figure 14: Realistic Key-interference graph of C499 from an attacker's perspective. The golden patterns to mute the edges K1→K2, K1→K5, K1→K10, and K1→K11 do not exist. Hence, K1 is non-mutable increasing the effective key size to 11.

Defender's perspective: As shown in Figure 12(a), the edges K1→K2, K1→K5, K1→K10, and K1→K11 are mutable. Hence, the key-gate K1 is classified as a sequentially mutable key-gate and all the other gates are classified as non-mutable key-gates. From a defender's perspective, since there are ten non-mutable key-gates, the effective key size is ten.

Attacker's perspective: The attacker searches for the golden pattern that mutes the key-gate K1. As such a pattern does not exist, he classifies the edges K1→K2, K1→K5, K1→K10, and K1→K11 as non-mutable. Therefore, the key-gate K1 also becomes non-mutable. As the attacker has to try all combinations of the keys, K1 to K11, the effective key size is eleven. While the effective key size in random insertion is two, the proposed method has an effective key size of eleven.

Hardware Trojan Horse Benchmark via Optimal Creation and Placement of Malicious Circuitry

Sheng Wei[†]   Kai Li[‡]   Farinaz Koushanfar[‡]   Miodrag Potkonjak[†]

[†]Computer Science Department
University of California, Los Angeles
Los Angeles, CA 90095
{shengwei, miodrag}@cs.ucla.edu

[‡]Department of Electrical and Computer Engineering
Rice University
Houston, TX 77005
{kai.li, farinaz}@rice.edu

ABSTRACT

This paper proposes Hardware Trojan (HT) placement techniques that yield challenging HT detection benchmarks. We develop three types of one-gate HT benchmarks based on switching power, leakage power, and delay measurements that are commonly used in HT detection. In particular, we employ an iterative searching algorithm to find rarely switching locations, an aging-based approach to create ultra-low power HT, and a backtracking-based reconvergence identification method to determine the non-observable delay paths. The simulation results indicate that our HT attack benchmarks provide the most challenging representative test cases for the evaluation of side-channel based HT detection techniques.

Categories and Subject Descriptors

K.6.5 [**Management of Computing and Information Systems**]: Security and Protection—*Physical Security*

General Terms

Security

Keywords

Hardware Trojan, benchmark, process variation, gate-level characterization

1. INTRODUCTION

1.1 Motivation

Hardware Trojans (HTs) are malicious components within integrated circuits (ICs) embedded by an untrusted foundry during the manufacturing process. HT detection has recently drawn a great deal of attention because IC outsourcing has become a trend in IC industry; the security and integrity of the manufactured ICs are of great concerns [16].

Since DARPA issued its first call for the study of hardware systems security in 2005 in general and hardware Trojans in particular, over a hundred HT detection techniques have been proposed [15]. Among them, side-channel based HT detection based on power and delay monitoring has become the most focused area [2, 3, 10, 11, 12, 21]. However, until now, there are no standard and publicly accepted test benchmarks for evaluating the various side-channel-based HT detection solutions, making it difficult for IC design companies or researchers to select the most effective HT detection solutions or to compare them.

This paper develops the first systematic way of placing HTs in a target design, in such a way that the most challenging HT test benchmarks can be created. Furthermore, we develop a complete set of HT metrics that can be used to evaluate the difficulty of detecting an arbitrary HT placement. Our idea for HT benchmark creation is to hide the malicious circuitry inside an IC, where the activities of the HTs are either low or unobservable due to the limitations of the side-channel measurements. Our observation is that there are three main conventional test modes that are used for IC testing and HT detection: switching power, leakage power, and delay. Therefore, we create HTs that either do not have an impact or have an exponentially low probability of impacting any of the three test modes.

The impact of HTs can be hidden within the fluctuations due to process variation (PV); its impact can be so small that it is below the sensitivity of modern instruments. There are three key observations behind this claim. The first is that there are many common design structures that have reconvergent paths with high delay discrepancies. Hence, the attacker can easily add malicious circuitry in such a way that the external delay measurements are not impacted. A small example in Figure 1 illustrates this situation. Since path B through gates 1-3-4-5-6-7 is much longer than path A (1-2-H-7), the addition of gate H cannot be detected using delay measurements from inputs $I1$ and $I2$ to output O.

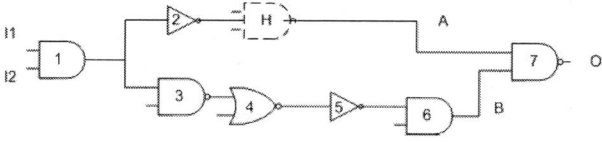

Figure 1: Example of a HT attack with no delay impact.

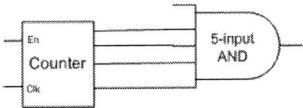

Figure 2: Example of a HT attack with no switching power impact.

Similarly, as shown in Figure 2, it is easy to embed a HT that is activated only on a very rare event (combination of its inputs). The rare event can be triggered only after the IC is active for much longer than any standard testing time. This example shows an AND gate that has many inputs from the most significant bits of a large modulo counter.

Until now, HT detection using leakage current was considered to be the most reliable technique. This is because any gate is subject to several types of leakage energies regardless of its location, inputs, and activation pattern. For example, the subthreshold leakage energy of a gate is given by the following formula [13]:

$$P_{leakage} = 2 \cdot n \cdot \mu \cdot C_{ox} \cdot \frac{W}{L} \cdot (\frac{kT}{q})^2 \cdot D \cdot V_{dd} \cdot e^{\frac{\sigma \cdot V_{dd} - V_{th}}{n \cdot (kT/q)}} \quad (1)$$

where W is gate width, L is gate length, V_{th} is threshold voltage, V_{dd} is supply voltage, n is subthreshold slope, μ is mobility, C_{ox} is oxide capacitance, D is clock period, ϕ_t is thermal voltage $\phi_t = kT/q$, and σ is drain induced barrier lowering (DIBL) factor.

Analysis of the formula indicates that leakage energy depends exponentially on the difference between the supply and threshold voltages. Therefore, an attacker can make the leakage of the HT gates negligibly small, by either resizing W and L or by creating gates that use only high V_{th} transistors. Hence, those gates will have several orders of magnitude lower leakage than regular gates and would be difficult to detect via leakage power measurements.

Our techniques create new types of exceptionally powerful and difficult to detect HTs. The key idea is to use power (or clock) gating in such a way that, in the default mode, the HT is powered off. The HT is activated by a single AND gate whose output is difficult to set to the value '1' (activation condition) by anybody except the attacker. The gate is intentionally aged or implemented to have a very high threshold voltage that corresponds to ultra (exponentially) low leakage energy, which cannot be detected even by state-of-the-art instruments. Therefore, it cannot be detected by any technique that measures switching and/or power. Finally, the HT is placed in such a way that it does not have an impact on delay between any pair of flip-flops. The HT is activated by an input vector sequence that is known by the attacker but otherwise has an exponentially low probability of occurrence.

2. RELATED WORK

2.1 Process Variation

Process variation is the deviation of IC parameter values from nominal specifications after the manufacturing process [4]. There is a wide consensus that PV is the dominant source of challenging problems related to HT detection. For example, it automatically invalidates all approaches based on the existence of a golden model that aim to compare an IC under analysis to its HT-free version. It also easily explains the discrepancy of any global measurement as a consequence of the embedded HT. Recently, a significant progress has been made in the field of HT detection using gate-level characterization techniques where a large set of global measurements is used to calculate variation-dependent characteristics of each gate post-silicon [14, 17, 18, 19, 20, 22]. Therefore, the variability aspects of PV are now addressed well.

However, PV has one more important consequence. It can produce HTs that have such a small impact on the measurements that they are below the sensitivity of modern instruments. Hence, the attacker can easily add malicious circuitry in such a way that external variation caused by the HTs is hidden in the presence of PV.

2.2 Hardware Trojan Detection

Tehranipoor et al. [15] provided a comprehensive survey of HT detection. There are three HT detection techniques that are most relevant to our HT benchmark creation. Jin and Makris proposed using the statistical delay path analysis in wireless cryptographic circuits [8]. However, as our earlier example showed, it would be easy to place HTs that do not impact any delay paths. Also, the attacker can easily resize gates in such a way that HTs are completely hidden. Kim et al. [9] advocate run time detection of HTs after their activation by observing the system bus behavior. Unfortunately, one can easily create HTs that do not alter the system bus characteristics. Agrawal et al. [1] construct IC fingerprints using side-channels (e.g., power and temperature) for a given design and authenticate the IC instances by comparing the fingerprints. However, while their technique does not take PV into consideration, we fully investigate and integrate the PV impact in our benchmark creation process.

3. PRELIMINARIES

There are typically two possible sources of power dissipation on an IC. One is from gate switching (also termed switching power or dynamic power), where the ICs dissipate power by charging the load capacitances of wires and gates. The other source is static power (also termed leakage), where the gates dissipate power due to the leakage current even if they do not switch. The gate-level leakage model is presented in Equation (1). The gate-level switching power model [13] is described by equation (2), where the switching power is dependent on switching probability α, gate width W, gate length L, and supply voltage V_{dd}:

$$P_{switching} = \alpha \cdot C_{ox} \cdot W \cdot L \cdot V_{dd}^2, \quad (2)$$

We use the delay model in [13] that relates the gate delay to its sizing and operating voltages:

$$Delay = \frac{k_{tp} \cdot k_{fit} \cdot L^2}{2 \cdot n \cdot \mu \cdot \phi_t^2} \cdot \frac{V_{dd}}{(ln(e^{\frac{(1+\sigma)V_{dd} - V_{th}}{2 \cdot n \cdot \phi_t}} + 1))^2}$$
$$\cdot \frac{\gamma_i \cdot W_i + W_{i+1}}{W_i}, \quad (3)$$

where subscripts i and $i+1$ represent the the driver and load gates, respectively; γ is the ratio of gate parasitic to input capacitance; and k_{tp} and k_{fit} are fitting parameters.

Furthermore, we employ the aging model proposed in [5] for our IC aging-based HT creation. The time dependence of V_{th} shifts due to negative bias temperature instability

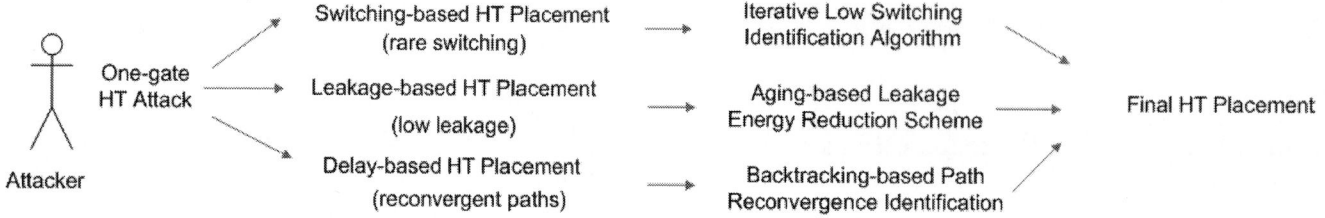

Figure 3: Overall flow for HT creation and placement.

(NBTI) follows fractional power law of the stress time, as shown in the following equation:

$$\Delta V_{th} = A \cdot exp(\beta V_G) \cdot exp(-E_\alpha/kT) \cdot t^{0.25}, \qquad (4)$$

where V_G is the applied gate voltage; A and β are constants; E_α is the measured activation energy of the NBTI process; T is the temperature; and t is the stress time.

4. ONE-GATE HARDWARE TROJAN BENCH-MARK CREATION

4.1 Overall HT Creation Flow

Our idea in creating challenging HT test benchmarks is to embed HTs that induce minimum observable variations into the target design. In order to achieve this goal, we employ a one-gate HT trigger that switches the malicious circuitry on and off during the IC operation. In order to increase the difficulty level for detection, the one-gate HT trigger powers on the malicious circuitry only when a rare event occurs, which is defined and activated by the attacker. In this way, the only observable variation before the activation of malicious circuitry is the single HT gate embedded in the circuit. Therefore, to further complicate the detection attempts, the attacker would hide the single HT gate in the circuit and make it difficult to be detected by the commonly used detection methods.

We consider three possible HT creation models that an attacker may consider to minimize the possibility of detection. The proposed HT models correspond to the three most commonly used side-channels for HT detection, namely leakage power, switching power, and delay. Figure 3 shows the overall flow of creating the three types of one-gate HTs: (1) For the switching-based HT placement, we design an iterative low switching identification algorithm that searches for the most rarely switching locations in the target design; (2) In the leakage-based HT model, we develop an aging-based leakage power reduction scheme to minimize the observable variations in leakage power; and (3) For the consideration of timing-based HT, we employ a backtracking-based algorithm to identify the reconvergent paths in the circuit, where delay variations caused by the HT is not observable. Then, we combine the three sets of locations and find a number of specific locations that minimize the observability of all three properties. Finally, we create challenging HT benchmarks by embedding the one-gate HT trigger at one of these locations.

4.2 Low Leakage HT Benchmark

The leakage power-based HT model corresponds to the HT detection techniques that leverage whole circuit or gate-level leakage power tracing. In this case, the idea for hiding the one-gate HT is to minimize its leakage power consumption. Therefore, the embedded HT gate would cause a limited variation in leakage power and has a high probability of hiding under the measurement errors in the existing leakage power-based detection approaches.

Our implementation of such a low leakage HT gate is based on the observation that the gate-level leakage power decreases exponentially with the increase in the threshold voltage (following Equation (1)), and that the threshold voltage can be increased by IC aging process (following Equation (4)). Therefore, our idea is to intentionally age the embedded HT gate in the post-silicon stage to reduce its leakage power to the greatest extent.

We develop a satisfiability (SAT)-based approach to determine the input vectors that can stress the transistors and age the HT gate. Since the output signal of each gate can be expressed as a boolean expression of the input vectors, SAT can determine the input vectors that generate a specific signal pattern. SAT is one of the first known NP-complete problems. Several very high quality SAT solvers are readily available for delivering fast and accurate SAT solutions [7]. We leverage the SAT solutions to find the aging input vectors that stress the HT gate at the expectant location in the design.

One of the consequences of the aging-based low power HT creation is that it may cause a delay degradation, due to the aging of the one-gate HT as well as a set of other gates in the circuit by applying the selected input vectors. The increased delay may be observable by a timing-based HT detection approach. To address this issue, we compensate for the delay degradation due to aging by employing adaptive body bias (ABB). ABB has been proposed as an effective approach to compensate for the PV impact on performance and power consumption. It provides the ability to manipulate transistor threshold voltage through the body effect and thus enables either a forward or a reverse body effect to change threshold voltage [6]. Here we use ABB to manipulate the threshold voltage of critical gates (e.g., gates that are on the critical path), so that the variation in the circuit delay can be compensated for.

4.3 Rare Switching HT Benchmark

In the switching power-based HT model, the goal is to insert the HT in such a way that it can be switched only by a rare set of input vectors. Consequently, there is a limited probability for the one-gate HT to exhibit any switching activity during the normal IC operation; on the other hand, the attacker can apply the rare input vectors to activate the malicious circuitry at any time. Figure 4 shows our simulation results regarding the switching activities of all gates on ISCAS benchmark C499. Our observation in this example is that all gates can be switched by a certain set of input vectors. Also, there exist gates that switch very often (e.g., more than 50% of the time) and, similarly, there are a small set of gates that have relatively low (but non-zero) switching activities (e.g., less than 5% of the time).

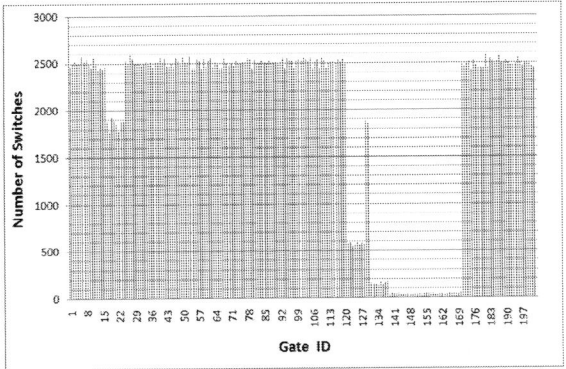

Figure 4: Switching activities of all gates on IS-CAS benchmark C499, under the application of 5000 pairs of input vectors.

We develop a low switching identification algorithm that iteratively searches the locations on the original design and finds out those locations that lead to the most rare switching activities. Then, we connect the obtained input signals to a single HT gate that is expected to have low switching activity. Pseudocode 1 describes the detailed algorithm for finding such a gate set on the target circuit and placing the HT gate.[1] We start with the random simulation results, as shown in Figure 4, and find the most rarely switching gate in the design. Next, we iteratively add one more gate from the design to the candidate group. This gate is the most correlated with the existing gates in the group and has the least switching activity. The algorithm terminates after K iterations and provides us with K locations in the circuit that can drive a rarely switching HT trigger.

4.4 Timing-based HT Benchmark

The delay-based HT model utilizes the limitation in delay measurements that only the delay of one single path is measurable from a specific input to a specific output. Furthermore, in the cases where there are multiple parallel paths between an input/output pair, it is difficult to map the delay measurement to one of the paths that are in parallel. Therefore, the HT gate can be well hidden within one of the parallel paths without being discovered by the existing delay-based characterizations.

Based on this thought, we develop a backtracking-based search algorithm to find out all the possible parallel paths in the target circuit for HT insertion. In particular, we analyze the structure of the netlist and identify the reconvergence points between each pair of input and output. Here we define reconvergence points as the node in the netlist that is the end point of more than one paths. In the case of reconvergence, none of the paths are measurable in terms of delay,

---

[1] $switching(\cdot)$ is the function to find out the switching activity of gate g_i via simulation of random input vectors; and $SAT(\cdot)$ is the procedure to determine whether the specific gates are switchable via SAT problem solving. The details of the SAT approach is introduced in Section 4.5.

Pseudocode 1 Iterative searching algorithm for placing rare switching HT.

Input: Netlist Net;
Onput: A set of locations L that result in rare switching one-gate HT;
1: Find the most rarely switching gate g_0 via simulation of random input vectors;
2: Insert g_0 into L;
3: **for** $i \leftarrow 1; i < K; i++$ **do**
4:     **for all** Gates t that are controlled by the transitive fan-in of L **do**
5:         **if** $switching(g_i) < switching(L)$ && $SAT(L + g_i)$ is solvable **then**
6:             $g_i \leftarrow t$;
7:         **end if**
8:     **end for**
9:     Insert g_i into L;
10: **end for**

---

because it is not clear which path is being measured even though one can measure the end to end delay from a specific input to the reconvergence point. As long as a path is not measurable, it can serve as a difficult case for delay-based HT detection method. The reconvergence identification algorithm converts the netlist of the design to a direct graph. Then, the problem of reconvergence identification converts to a graph theory problem that searches for all the nodes that have an in-degree of at least 2.

4.5 Summary of HT Benchmarks

The three HT models provide us with a systematic way of evaluating an arbitrary HT placement strategy in terms of the difficulty levels for detection. For example, if a single HT gate is embedded at one of the reconvergent paths, where the leakage power consumption is lower than the measurement resolution and the switching probability is small, it would create an ultra challenging case for the HT detection techniques.

Following this idea, we define the first systematic benchmarking strategy for creating and quantifying the HT attacks with various difficulty levels. The difficulty level of a HT attack model can be evaluated using a triplet $< d, l, s >$, where d is a boolean variable indicating whether the HT gate is observable via delay measurement (i.e., whether it is on one of the reconvergent paths); l is a boolean variable representing whether the leakage power of the HT gate is below the resolution of the leakage power characterization, and s is the switching probability of the inserted HT gate at the specific location. We can test and evaluate a HT detection approach using the proposed benchmark, by observing the most difficult level of HT that it can successfully detect.

5. EXPERIMENTAL RESULTS

We evaluate our HT benchmark creation method on a set of ISCAS'85, ISCAS'89, and ITC'99 benchmarks. For each benchmark circuit, we first embed a single HT at the location determined by our approach. Then, we evaluate the leakage power, switching power of the HT gate, as well as its observability under delay measurements. The combination of the three metrics quantifies the difficulty level of detecting such a HT attack.

5.1 Low Leakage-based HT

Table 1 and Table 2 shows the trend of total leakage energy reductions by varying the V_{th} increase during the aging process from 10% to 100%. We observe that the leakage energy can be reduced by up to 28X, which enables the placement of the ultra-low leakage HTs on all circuit locations. Furthermore, we observe that after the delay compensation of the non-HT gates is done using adaptive body biasing, the leakage energy reduction can still be up to 18X. The results indicate that we are able to place the low leakage HT gate without impacting the delay characteristics of the design, which makes the HT difficult to detect using both delay and leakage power-based characterizations. Furthermore, for the larger designs such as C7552 (shown in Table 2), we obtain a larger rate of leakage energy reduction.

Table 1: Leakage energy reduction via aging for HT benchmark creation (Benchmark C6288).

V_{th} Increase	Without Delay Compensation	With Delay Compensation
10%	2.0	1.9
20%	3.7	3.4
30%	6.3	5.6
40%	9.7	8.2
50%	13.4	10.9
60%	17.0	13.2
70%	20.2	15.0
80%	23.1	16.4
90%	25.8	17.5
100%	28.3	18.5

Table 2: Leakage energy reduction via aging for HT benchmark creation (Benchmark C7552).

V_{th} Increase	Without Delay Compensation	With Delay Compensation
10%	2.2	2.0
20%	4.3	4.0
30%	8.7	7.7
40%	16.9	14.4
50%	31.3	25.9
60%	55.2	44.0
70%	91.0	69.9
80%	138.9	102.6
90%	195.9	138.7
100%	256.8	173.7

5.2 Rare Switching-based HT

Figure 5 shows our simulation results for rare switching-based HT creation. The boxplots show the statistical distributions of the switching activities for all gates in the IS-CAS'85 benchmark circuits, obtained from the simulation of 10,000 randomly generated input vectors for each design. For each box in the plot, the lower and upper edges correspond to the 25th and 75th percentiles of the distribution. The line in the middle of each box indicates the median of the distribution. The smallest and largest points are also shown if they happen outside a range from the box. In most cases the switching probability ranges from 20% to 50%, and it is very rare to have gates that can never be switched.

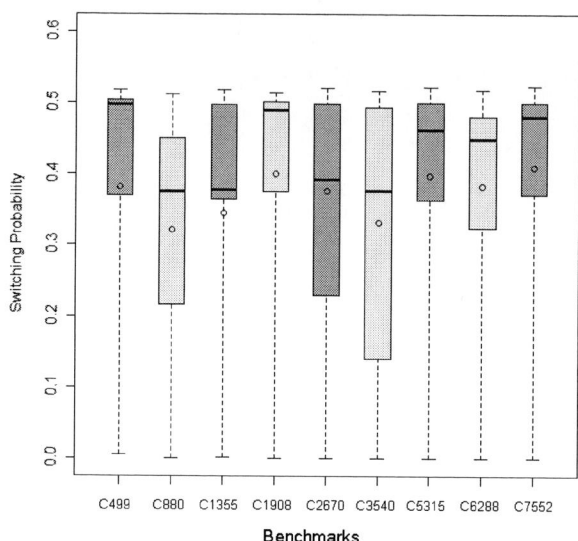

Figure 5: Simulation results of switching activities of all gates on ISCAS'85 benchmarks; 10,000 random input vectors were applied to each design.

However, after applying our iterative low-switching identification algorithm and feeding the obtained input pins to a single AND gate, we obtain a maximum of 0.78% switching probability while simulating 10,000 input vectors. Also, we have used SAT to show that for each AND gate, there is at least one input vector that could activate the malicious circuitry. Therefore, our results indicate that the attacker can use the rare activation condition to trigger the malicious circuitry during the system operation, while the single HT gate with low switching probability is difficult to detect when the malicious circuitry is dormant.

5.3 Timing-based HT

Table 3 summarizes our simulation results regarding delay characterizable gates on a set of ISCAS'85, ISCAS'89, and ITC'99 benchmarks. As discussed in Section 4.4, we cannot characterize the delay of a path if there exist parallel reconvergent path from the input to the output. From the simulation results, we observe that there is no full coverage of all gates in any of the evaluated benchmarks in terms of delay characterization. The highest achieved rate of coverage on the benchmark set is 60%, which still leaves a large portion of the circuit susceptible to HT placement without the risk of being detected.

6. CONCLUSION

We have developed three hardware Trojan attack models for creating HT detection benchmarks. The attack models are based on the consideration of hiding the one-gate HT trigger at low leakage, rare switching, and reconvergent locations on the circuit to bypass the security check by the commonly used side-channel based approaches. We showed that the proposed one-gate HT models can successfully compromise the detection attempts and provide a quantitative metric for evaluating HT detection mechanisms. Simulation

Table 3: Simulation results regarding uncharacterizable gates due to reconvergences. The high percentage of uncharacterizable gates in each design indicates that there is a large number of candidate locations for embedding the non-detectable one-gate HT trigger.

Benchmark	Gates	# Inputs	# Outputs	# Gates Subject to Reconvergence	% Gates Subject to Reconvergence
C499	202	41	32	80	39.6%
C880	383	60	26	208	53.5%
C1355	546	41	32	546	100%
C1908	880	33	25	739	84.0%
C2670	1193	233	140	931	78.0%
C3540	1669	50	22	1542	92.4%
C5315	2307	178	123	1924	83.4%
C7552	3512	207	108	2552	72.7%
S38584	19253	5	304	237	84.3%
B17	32192	37	97	19283	59.9%

results on a set of ISCAS'85, ISCAS'89, and ITC'99 benchmarks verified the effectiveness of the HT benchmarks. The resulting HT benchmark circuits and tools can be found at http://www.cs.ucla.edu/~shengwei/htbench.html.

7. ACKNOWLEDGEMENTS

This work was supported in part by the NSF under Award CNS-0958369, Award CNS-1059435, and Award CCF-0926127 and in part by the DARPA/MTO Grant N66001-11-1-4103.

8. REFERENCES

[1] D. Agrawal et al. Trojan detection using IC fingerprinting. In *IEEE Symposium on Security and Privacy (SP)*, pages 296–310, 2007.

[2] Y. Alkabani and F. Koushanfar. Consistency-based characterization for IC trojan detection. In *ICCAD*, pages 123–127, 2009.

[3] M. Banga and M.S. Hsiao. VITAMIN: Voltage inversion technique to ascertain malicious insertions in ICs. In *HOST*, pages 104–107, 2009.

[4] S. Borkar et al. Parameter variations and impact on circuits and microarchitecture. In *Design Automation Conference (DAC)*, pages 338–342, 2003.

[5] S. Chakravarthi et al. A comprehensive framework for predictive modeling of negative bias temperature instability. In *International Reliability Physics Symposium (IRPS)*, pages 273–282, 2004.

[6] T. Chen and S. Naffziger. Comparison of adaptive body bias (ABB) and adaptive supply voltage (ASV) for improving delay and leakage under the presence of process variation. *IEEE Transactions on Very Large Scale Integration (VLSI) Systems,*, 11(5):888–899, 2003.

[7] N. Een and N. Sorensson. An extensible SAT-solver. In *International Conferences on Theory and Applications of Satisfiability Testing (SAT)*, pages 333–336, 2004.

[8] Y. Jin and Y. Makris. Hardware Trojans in wireless cryptographic ICs. *IEEE Design Test of Computers*, 27(1):26–35, 2010.

[9] L. Kim, J.D. Villasenor, and C.K. Koc. A Trojan-resistant system-on-chip bus architecture. In *IEEE Military Communications Conference (MILCOM)*, pages 1–6, 2009.

[10] F. Koushanfar and A. Mirhoseini. A unified framework for multimodal submodular integrated circuits trojan

detection. *IEEE Transactions on Information Forensics and Security*, 6(1):162–174, 2011.

[11] F. Koushanfar, A. Mirhoseini, and Y. Alkabani. A unified submodular framework for multimodal IC trojan detection. In *Information Hiding*, pages 17–32, 2010.

[12] F. Koushanfar and M. Potkonjak. CAD-based security, cryptography, and digital rights management. In *Design Automation Conference (DAC)*, pages 268–269, 2007.

[13] D. Markovic et al. Ultralow-power design in near-threshold region. *Proceedings of the IEEE*, 98(2):237–252, 2010.

[14] M. Potkonjak, A. Nahapetian, M. Nelson, and T. Massey. Hardware trojan horse detection using gate-level characterization. In *Design Automation Conference (DAC)*, pages 688–693, 2009.

[15] M. Tehranipoor and F. Koushanfar. A survey of hardware Trojan taxonomy and detection. *IEEE Design Test of Computers*, 27(1):10–25, 2010.

[16] M. Tehranipoor et al. Trustworthy hardware: Trojan detection and design-for-trust challenges. *IEEE Computer Magazine*, 44(7):66–74, 2011.

[17] S. Wei, S. Meguerdichian, and M. Potkonjak. Gate-level characterization: Foundations and hardware security applications. In *Design Automation Conference (DAC)*, pages 222–227, 2010.

[18] S. Wei, S. Meguerdichian, and M. Potkonjak. Malicious circuitry detection using thermal conditioning. *IEEE Transactions on Information Forensics and Security*, 6(3):1136–1145, 2011.

[19] S. Wei and M. Potkonjak. Scalable segmentation-based malicious circuitry detection and diagnosis. In *International Conference on Computer-Aided Design (ICCAD)*, pages 483–486, 2010.

[20] S. Wei and M. Potkonjak. Integrated circuit security techniques using variable supply voltage. In *Design Automation Conference (DAC)*, pages 248–253, 2011.

[21] S. Wei and M. Potkonjak. Scalable consistency-based hardware Trojan detection and diagnosis. In *International Conference on Network and System Security (NSS)*, pages 176–183, 2011.

[22] S. Wei and M. Potkonjak. Scalable hardware Trojan diagnosis. *IEEE Transactions on Very Large Scale Integration (VLSI) Systems*, 2011.

On Improving the Uniqueness of Silicon-Based Physically Unclonable Functions Via Optical Proximity Correction

Domenic Forte and Ankur Srivastava
University of Maryland, College Park, MD, USA
{dforte, ankurs}@umd.edu

ABSTRACT

Physically Unclonable Functions (PUFs) are effective for security applications because they generate unique signatures that are resistant to cloning attempts as well as physical tampering. A silicon PUF is a special circuit embedded in an IC that relies on random fabrication process variations to produce a unique signature for its native IC. While current research directions have focused on improving PUF quality at the architectural level, little work has explicitly targeted their fundamental source of randomness, the fabrication process. During IC fabrication, Optical Proximity Correction (OPC) is typically used to suppress manufacturing variations. In this paper, we recognize that this is actually counterintuitive for PUFs. We provide a novel framework which enables OPC to increase the effects of manufacturing variations within PUF circuitry and produce more randomness in PUFs for greater uniqueness and reliability. The proposed OPC techniques are validated using a population of 100 ring oscillator PUFs. Results show that our schemes provide over five times larger variation in ring oscillator delay, improve PUF uniqueness by 5%, and improve PUF reliability by as much as 70% when compared to conventional OPC.

Categories and Subject Descriptors

K.6.5 [**Management of Computing and Information Systems**]: Security and Protection; B.7.2 [**Integrated Circuits**]: Design Aids

General Terms

Performance, Reliability, Security

Keywords

Physically Unclonable Functions, Optical Proximity Correction, Lithography, Process Variation

1. INTRODUCTION

We are becoming more heavily reliant on embedded computing devices in our daily lives and basic infrastructure. For example, embedded devices are used by people for communication, entertainment, and to perform in-store and online purchases. Embedded devices are also critical components in larger systems such as cars, air traffic control systems, etc. Clearly, such systems and devices are responsible for life-critical actions and sensitive/private data. Thus, we must ensure their security by relying on operations such as device/user authentication, protection of confidential information, and secure communication.

Physically Unclonable Functions (PUFs) [9, 7] provide promising solutions to many security issues. Silicon PUFs are novel circuits that can be easily embedded in electronic devices [15, 4]. They exploit random manufacturing variations to generate a unique signature for each fabricated chip. Since the source of their variations are random, the signatures cannot be cloned or predicted even by the manufacturer. Therefore, PUFs can be effectively utilized in device authentication and private encryption/decryption key generation for safe storage and communication [9, 15].

There are three important properties required by PUFs which are discussed in the literature [9, 7]. First, each PUF instance must extract "unique" physical characteristics from its device. This ensures that the signature only belongs to a single device. Second, each PUF output must be "reliable" (generate the same output at all times) in the face of environmental variations. Unreliability in PUF output would mean it cannot be used to accurately authenticate or generate keys. Finally, PUF responses should be "unpredictable" in order to ensure that the secret identifier/key remains safe from outsiders. There has been a great deal of research devoted to new PUF designs for better performance and reliability [15, 2, 4]. However, these works focus on designs at a purely architectural level.

Fundamentally, the source of variations is the manufacturing process itself. The greater the random variation, the better the PUF signature quality (uniqueness, reliability, unpredictability). However, manufacturing variations are extremely detrimental to the other parts of the chip since they increase the probability of yield loss. Therefore, most current chip design methods attempt to generate designs which are immune to such variations. PUFs designed and manufactured by such approaches would fail to possess the level of uniqueness, reliability, etc. desired by practical applications even if PUFs are designed with quality enhancements at architectural level. This is because the design flow followed by architectural level enhancement would suppress the impact of fabrication variations through variability suppressing transistor level and physical design techniques. One could of course generate the PUF layout directly (to be included in the design as a module) thereby avoiding any variability suppression mechanisms. However, variability suppressing Optical Proximity Correction (OPC) generates the mask used for lithography and cannot be avoided [8, 3, 17, 6]. Without OPC, modern lithography and ICs would cease to scale.

The typical objective of OPC is suppression of fabrication variability [17, 6]. In this paper, we reverse the conventional OPC paradigm and propose an OPC-based approach for enhancing the impact of manufacturing variations on PUF quality. There are two other advantages of our approach. First, PUF architectural enhancements could be applied less aggressively, reducing design overheads. Second, the proposed OPC method can be focused on regions of the chip which contain the PUF. The non-PUF circuitry of the design (eg. CPU) could still be manufactured using conventional OPC methods which suppress variability and maintain functional correctness (yield). Since OPC has significant locality in space [3], the proposed and conventional schemes would not compromise each other. Our major contributions are as follows:

- We provide a new objective function for OPC in which the goal is to increase variation in physical characteristics of silicon PUFs. To our knowledge, there has been no prior work which tries to increase fabrication variation.
- We propose an optimization framework and algorithm which improves PUF quality by optimizing our cost function. Results show the proposed OPC algorithms improve PUF inter-distance (uniqueness) by 5% and PUF intra-distance (reliability) by as much as 70% compared to conventional OPC.
- The techniques discussed in this paper are very *flexible* in that they need only be applied to the PUF portion of an IC. They are also *general* in that they can be used to increase variation *in any silicon PUF*.

The rest of the paper is organized as follows. In the next section, we discuss the relevant background and models for PUFs, IC fabrication, and OPC. Section 3 discusses our optimization framework and PUF-aware OPC algorithms which improve PUF quality. Results are discussed in Section 4 and the paper is concluded in the final section.

2. PRELIMINARY

2.1 Physically Unclonable Functions (PUFs)

Silicon PUFs: A silicon PUF is a circuit embedded in an electronic device [9, 15, 4]. The input and output of PUFs are called *challenges* and *responses*. An applied challenge and its measured response is referred to as a *challenge-response pair*. PUFs are designed to exploit manufacturing variations in order to generate a unique signature (set of challenge-response pairs) for every fabricated chip. Since the source of these variations is manufacturing related, this unique signature cannot be cloned or predicted.

There have been many silicon PUF designs proposed in the literature [15, 2, 4]. For the interested reader, a thorough treatment of PUFs is given in [9]. In general, PUF designs can be broken down into two classes [9]: (i) *Delay-based PUFs* exploit the random variations of wire and gate delays. Examples include the arbiter PUF and ring oscillator PUF [15]; (ii) *Memory-based PUFs* exploit the random settling behavior of bistable elements [9]. Examples include SRAM PUFs and butterfly PUFs [9, 4].

Security Applications: Since PUF responses are random, unique, and unclonable, they are promising for many security applications. For example, the unique signatures of the PUF can be used to authenticate the device where the PUF resides [9]. Another interesting application for PUFs is secret key generation for cryptography [15]. Rather than storing the secret key in non-volatile memory (where it can be stolen by side channel and probing attacks), the key is dynamically generated from a PUF response.

Desirable PUF Properties and Metrics: There are three properties that are very important for PUFs:

1. **Uniqueness:** Since a PUF is typically used as a form of identity, then for any particular challenge the difference in responses of any two PUF instances (on separate devices) should be large. A typical measure for uniqueness is called *inter-distance* which is calculated as [7]

$$d_{inter}(C) = \frac{2}{k(k-1)} \sum_{i=1}^{k-1} \sum_{j=i+1}^{k} \frac{\mathrm{HD}(R_i, R_j)}{m} \times 100\% \quad (1)$$

where $\mathrm{HD}(R_i, R_j)$ is the hamming distance between any two responses R_i and R_j from *different PUFs to the same challenge C*, k is the number of chips/devices in the population under test, and m is the number of bits per response. The optimal $d_{inter}(C)$ is 50% [7] which indicates maximum entropy.

2. **Reliability:** The response of a particular PUF instance for the same challenge may vary due to environmental variations (temperature, voltage supply noise) and temporal variations (aging). However, it is desirable for the challenge to be relatively stable so the PUF can be used as an identifier or key. A common measure for reliability is called *intra-distance*. This is calculated by collecting s samples of a response at different operating

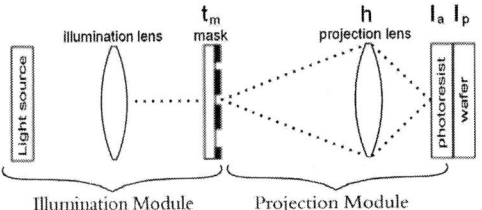

Figure 1: Optical Lithography System

conditions (supply voltage, etc.) and computing [7]

$$d_{intra}(C) = \frac{1}{s} \sum_{j=1}^{s} \frac{\mathrm{HD}(R_i, R'_{i,j})}{m} \times 100\% \quad (2)$$

where R_i is the nominal response of a challenge C to a PUF and $R'_{i,j}$ is the jth sample of R_i for that *same challenge and same PUF instance*. Ideally, $d_{intra}(C) = 0$ which corresponds to no changes in response for challenge C (i.e. 100% reliability).

3. **Unpredictability:** PUF responses should be unpredictable in order to ensure that the secret identifier/key remains safe even from outsiders with privileged information (eg. PUF responses from the same device to different challenges). One measure of unpredictability is the randomness of the responses generated by the PUF (entropy, NIST tests [9, 16]).

2.2 IC Fabrication Process

The fabrication of an integrated circuit (IC) on a silicon substrate involves three basic steps [8]: (1) lithography patterning and etching; (2) semiconductor doping; and (3) film deposition and planarization. With continued semiconductor scaling, these steps have grown more complex and difficult to control. As a result, there are several sources of variability during fabrication such as lens imperfections in the lithography system, optical proximity effects, etching precision (line edge roughness or LER), and the number and position of dopants in the channel [8]. The above variations lead to physical differences in device structures (channel length, oxide thickness, etc.), differences in device electrical parameters (threshold voltage, drain-to-source current, etc.), and also variation in device performance specifications (timing, power). PUF designs attempt to exploit the variabilities imposed by all these parameters.

Manufacturing variations are typically modeled with two components [8]: (i) systematic and (ii) random. Systematic variations result in fabricated chips deviating from the desired specs in a similar and predictable way among the chips. Random variations result in chips that deviate from the specs in truly unpredictable way. In [10], it was shown that random variation is the true source of PUF quality. Since systematic variations can be predicted by models and are the same chip-to-chip, they reduce PUF uniqueness and unpredictability. In this paper, we focus on the lithography step which contains both systematic and random process variations [8].

Lithography Process: Optical lithography is a process by which a photoresist covering a silicon wafer is exposed to optical wavelengths and then developed to form desired patterns/structures on the wafer. An optical lithography system consists of two basic modules [17] which are shown in Figure 1 and discussed below:

1. **Illumination Module:** A mask that represents the target layout is illuminated by a light source through an illumination lens. The mask is typically binary, meaning it consists of only transparent and opaque structures which allow light to reach and prevent light from reaching the photoresist respectively.

2. **Projection Module:** Light is diffracted as it passes through transparent parts of the mask. A projection lens picks up a portion of the diffracted light and projects an "image" (pattern) onto the photoresist. Depending on the type of photoresist, it either hardens or remains soft in presence of light. The portions of the resist that receive enough light are removed by a chemical process. The parts of the wafer that are still covered with resist are protected from etching, doping, and deposition which allow one to form the transistors, interconnects, etc. of an integrated circuit (IC).

Lithography Process Variations: Fundamentally, the optical light pattern falling on the wafer decides what physical structures are fabricated. There are many sources of variation in the optical lithography process which include imperfections in the mask, light source, and lenses as well as variation in the distance between the projection lens and wafer. These sources of variation result in differences between the image expected on the photoresist and the actual image obtained, which in turn result in variation in structures and features (channel length, threshold voltage, interconnect width, etc.) and variation in delay, power, and performance for all manufactured devices.

Lithography Process Modeling: The lithography process is modeled with two basic steps:

1. The mask is projected onto the photoresist creating an "aerial" image I_a. Let $t_m(x, y) \in \{0, 1\}$ denote the pixels at the location x, y for the mask where zero (one) refers to opaque (transparent) pixels. For a coherent imaging system, the aerial image is given by [12, 6]

$$I_a = |t_m * h|^2 \quad (3)$$

where $*$ denotes the convolution operator. h denotes the point spread function (PSF) of the projection lens and is modeled as

$$h(x, y) = \frac{J_1\left(2\pi NA\sqrt{x^2 + y^2}/\lambda\right)}{2\pi NA\sqrt{x^2 + y^2}/\lambda} \quad (4)$$

where J_1 is Bessel function of the first kind, order one [12], NA is the numerical aperture of the projection lens, and λ is the source light wavelength.

2. Once the optical image falls on the photoresist-coated wafer, the photoresist is developed and etched based on image intensity at the corresponding wafer location. If the photoresist material is positive (negative) and the image intensity at a certain location is greater (lesser) than a specific threshold, the resist gets etched out. The post-etch image is called the resist or pattern image I_p and is often calculated as follows [12]

$$I_p(x, y) = \begin{cases} 1, & \text{if } I_a(x, y) \geq I_{\text{th}} \\ 0, & \text{otherwise} \end{cases} \quad (5)$$

where I_{th} is a constant intensity threshold.

Process variations in this optical system are typically modeled by a combination of focus and dose variations [8]:

Focus Variations: Focus variations (defocus) are essentially small changes in the distance between the projection lens and resist/wafer from the ideal setting (zero defocus). This impacts the intensity pattern falling on the wafer. Focus variations are modeled as a change to the PSF of the projection lens as follows [6]

$$h_F(x, y) = \mathscr{F}^{-1}\{(\tilde{H}(f_x, f_y)e^{-j\pi F(f_x^2 + f_y^2)}\} \quad (6)$$

where $\mathscr{F}^{-1}$ denotes the inverse Fourier transform, $\tilde{H} = \mathscr{F}(h)$, f_x and f_y are spatial frequencies, and F denotes the focus error (defocus). Equation (6) is used in place of $h(x, y)$ in Eqn. (3) to produce an aerial image dependent on focal variation $I_a(x, y; F)$ for a given defocus F at the wafer plane. This impacts the pattern image I_p and therefore the final manufactured transistor/wire parameters. Focus variation can model a large class of variabilities including non-planarity of the photoresist surface, imperfect lenses, mask imperfections, etc.

Exposure Dose Variations: Exposure dose variations result from differences in light source intensity, exposure duration, etc. Dose variations are often modeled by replacing the constant I_{th} in Eqn. (5) by a stochastic function [17].

Together, focus and dose variations encapsulate a large class of variabilities associated with the lithography process. *Dose and focal variations are captured by modeling them as random variables with associated probability density functions (PDFs). This randomness is generally detrimental to IC manufacturing but in case of PUFs is exploited to generate device identifying signatures.*

2.3 Optical Proximity Correction (OPC)

As IC features scale downward, it is more difficult to print high resolution patterns [17]. This is because the source light wavelengths are much larger than modern feature sizes which increases their susceptibility to process variations. Optical Proximity Correction (OPC) is a method where the mask structures are optimized to improve the chance of obtaining desired patterns on the wafer. OPC algorithms come in two forms:

Polygon-based OPC: treat the transparent parts of the mask as a set of polygons. Then the polygons are broken down into edge segments which are iteratively moved with the goal of improving a cost function [3]. A typical cost function is the mean square error (MSE) between the desired patterns and patterns generated by the mask. We discuss Polygon-based OPC algorithms in more detail in Section 3.4.

Pixel-based OPC: represent each pixel of the mask as a $\{0,1\}$ decision and model the printed image as a continuous function [12]. Then an optimization problem (typically MSE minimization as well) is solved by using a gradient descent-like algorithm.

Before aggressive scaling of ICs, OPC algorithms assumed that fabrication process variations (focus and dose) had limited effects on printed features. However, more recent work has shown that random dose and exposure variations indeed result in significant spread in the feature dimensions as well as device electrical parameters [17]. In order to address this issue, recent efforts have used polygon-based and pixel-based OPC [17, 6] to build masks that minimize manufacturing variations.

3. PUF-AWARE OPC

3.1 Motivation

As discussed in the previous section, variability-aware OPC algorithms generate masks that are more immune to process variations and make the printed ICs as close to the desired IC as possible. Applying such approaches means that all instances of an IC will have very similar power and delay characteristics. While this is typically highly desirable for most typical applications, such approaches are counterintuitive for PUFs. As discussed earlier, PUFs are an exceptional application which critically depend on random process variation to produce ICs with unique performance and power characteristics.

Existing research in PUFs only improve PUF quality at the architectural level, but since PUF circuitry is subjected to the same variability suppressing OPC methods as the rest of an IC (eg. CPU), PUF circuitry will actually have smaller spread in performance parameters. A more desirable goal for PUFs would be to find masks that are *more susceptible to random process variations* thereby resulting in ICs with better uniqueness, reliability, etc. The key challenge is to design such a mask for the PUF areas of the chip while using conventional variability suppressing schemes for the other parts. Another challenge is to increase random variations while maintaining the same level of or decreasing systematic variation (since they degrade PUF uniqueness). In this section, we investigate this concept which we call *PUF-aware Optical Proximity Correction.*

3.2 Main Concept

PUF-aware OPC is easy to apply and has many advantages for improving PUF quality:

- The proposed PUF-aware OPC is applied after IC design and before fabrication. PUF-aware OPC finds a mask that generates ICs which are more susceptible to fabrication process variations in regions of the IC where a PUF is located.

- PUF circuits with greater random variance in physical characteristics have *more unique and reliable responses*. Furthermore, with these improvements, the circuitry required to measure PUF responses can be made simpler.

- PUF-aware OPC is very flexible. Although non-PUF related applications (CPU, etc.) desire less variability during fabrication than PUFs, our PUF-aware OPC algorithm *need only be applied to the portion of the IC containing the PUF.* The remaining portion of the design can utilize a more conventional OPC

98

objective (eg. minimize mean square error between desired and fabricated features). Since the spatial correlations imposed by OPC decay very quickly with distance [3], variability enhancements will have little impact on non-PUF areas.

- PUF-aware OPC is very general and *can be used to increase variations in any silicon-based PUF* because the masks generated by the algorithm enhance fabrication process variation at the transistor level regardless of PUF design.

- Since the proposed approach is applied during the mask design phase (after the circuit design process) to improve PUF quality, it will not impact area, power, etc., unlike architectural approaches.

For the remainder of the section, we discuss the details of PUF-aware OPC. First, we propose an optimization framework which balances the needs of the PUF with desired feature size. Then, we propose a variant of the standard polygon-based OPC algorithm which is used to solve our optimization problem.

3.3 Optimization Problem

Assume that a desired PUF layout is given. Our goal is to synthesize a mask which produces PUFs which have the following characteristics:

1. The printed features of each PUF should have high random variation. When each PUF has very random features, the challenge-response pairs (signature) produced will be more unique and reliable. Furthermore, systematic variation should be kept low for better PUF uniqueness [10].

2. In order to maintain functional correctness of the PUF architecture, the printed PUF should resemble the shape and dimensions of the target PUF design.

Unfortunately, the above are competing characteristics. First, amplifying both systematic and random variations is counterintuitive for PUFs. Second, a PUF printed with the highest variability will clearly not conform well with the original design. To deal with the first problem, we use compensation methods to the PUF layout/design which counteract (balance out) systematic variation such as those discussed in [10, 5]. *This ensures that PUF responses are effectively determined by only random variation.*

To deal with the second issue, we use an objective function that balances functional correctness with variation enhancement. Let $\hat{I}_p(x,y) \in \{0,1\}$ denote the target resist or pattern image that we wish to generate on the wafer surface (decided by the PUF design/layout). Also, let $I_p^{(t_m)}(x,y;F) \in \{0,1\}$ denote the resist or pattern image obtained using mask t_m under defocus F (see Section 2.2). Because defocus F is a random variable, the associated $I_p(x,y;F)$ is also random and its PDF can be computed using the known PDF of F. *In this paper, we will ignore exposure dose variations since they can be dealt with in a similar manner [17].*

Conventional Objective Function: The mean square error (MSE) between the desired and printed resist patterns ($\hat{I}_p(x,y)$ and $I_p^{(t_m)}(x,y;F)$) can be calculated as follows

$$MSE(F) = \sum_{x,y} (\hat{I}_p(x,y) - I_p^{(t_m)}(x,y;F))^2 \qquad (7)$$

Note that $MSE(F)$ is a random variable since it depends upon defocus F. One can fabricate a chip close to the target design by *minimizing the expected MSE* [6] defined as

$$\mu_{MSE} = \int MSE(F)p(F)dF = \varepsilon_F\{MSE(F)\} \qquad (8)$$

where $\varepsilon_F\{\}$ denotes the expectation operator with respect to F.

Proposed Objective Function: To increase variability and therefore PUF quality, a better objective would be to *maximize the variance of the MSE*

$$\sigma_{MSE}^2 = \varepsilon_F\{MSE(F)^2\} - \mu_{MSE}^2 \qquad (9)$$

Since MSE is a random variable whose PDF depends on the distribution of defocus F, a higher variance in MSE would imply greater

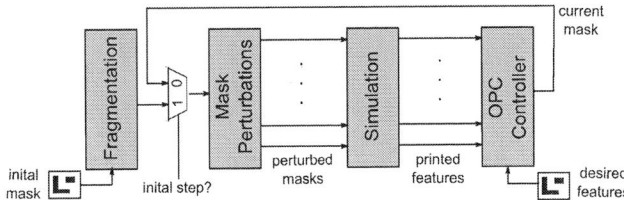

Figure 2: General OPC Algorithm

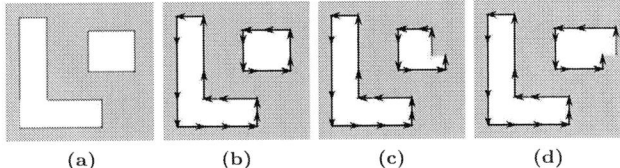

Figure 3: (a) Mask: White (gray) areas correspond to transparent (opaque) pixels; (b) Mask fragmentation; (c) Mask perturbation 1; (d) Mask perturbation 2

variations in the physical parameters of the fabricated features and greater spread in the electrical parameters of the PUF transistors. However, the flaw in this cost function is that it may generate structures that do not resemble the target structure at all. Hence we add a corrective term in this objective that captures the fact that we wish to maximize the MSE variance with minimum impact to the expected MSE. Our objective function combines the above two metrics and we find

$$t_m^* = \operatorname{argmax}\left(\frac{\sigma_{MSE}}{\mu_{MSE}}\right) \quad \text{s.t } t_m(x,y) \in \{0,1\} \; \forall \, x,y \qquad (10)$$

This objective balances the above two characteristics and finds an optimal mask that prefers high variability (large σ_{MSE}) and low error (small μ_{MSE}).

3.4 Optimization Algorithm

As discussed in Section 2.3, there are two basic approaches to solving the mask synthesis problem: polygon-based OPC and pixel-based OPC. In this paper, we use the polygon-based approach for simplicity. In a nutshell, polygon-based OPC iteratively moves "mask edges" to improve an optimization cost function until a stopping criterion is met. The general steps of polygon-based OPC [3, 17] are shown in Figure 2 and described below:

1. **Fragmentation:** An initial mask is used as input to the algorithm. Polygons within the initial mask are fragmented into edges (see Figure 3(a-b)). These fragmented edges are essentially optimization variables. When shorter edges are used, there are more degrees of freedom during the OPC algorithm [3].

2. **Mask Perturbation:** One or more edges are chosen for perturbation (i.e. offset from their current position). For example, a vertical edge can be perturbed by moving one grid unit to the left or right from the current position. Similarly, a horizontal edge can be perturbed by moving it up or down by one grid unit. Edges can be chosen for perturbation in a fixed or random order. Two perturbed masks are shown in Figure 3(c-d).

3. **Simulation:** Perturbed masks will result in a different image intensity and pattern in the resist/wafer. The resulting patterns are computed by a simulation environment.

4. **OPC controller:** The OPC controller compares the simulated patterns with the target pattern and computes objective costs. The perturbed mask is accepted if it results in an improved cost.

5. **Iterate:** The algorithm repeats steps 2-4 until a stopping criteria is met (eg. some minimum or maximum cost).

In this paper, we apply a version of this basic OPC algorithm to solve our optimization problem. We iteratively perturb the mask using a steepest descent-like algorithm. Note that other conventional approaches for OPC use similar iterative greedy optimization methods. Our version of the algorithm is characterized by the following:

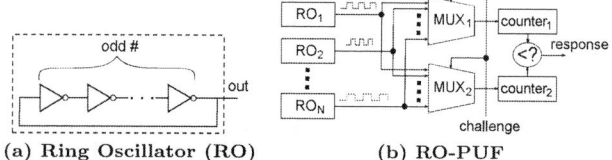

(a) Ring Oscillator (RO) (b) RO-PUF

Figure 4: Ring Oscillator and Ring Oscillator PUF

- For fragmentation, we follow the basic rules discussed in [3].
- In our case, perturbed masks will result in a *random* image intensity on the resist/wafer which is a function of defocus F. $I_p(x, y; F)$ for different random values of F (F's statistical distribution can be characterized by standard methods in Design-For-Manufacturability DFM [11]) are computed by a simulation model that employs Eqns. (3)-(6).
- The objective cost computed by the OPC controller for any mask is given by $\frac{\sigma_{MSE}}{\mu_{MSE}}$ which we attempt to maximize (as discussed in Section 3.3). We assume the defocus F follows a normal distribution (as in [17, 6]) and use Monte Carlo methods to estimate σ_{MSE} and μ_{MSE}.
- A steepest descent based approach is used for mask perturbation. When there is no improvement to the objective cost by perturbing any edge, the algorithm stops.

We investigate two variants of our OPC algorithm:
PUF-aware OPC-1 (P-OPC1): The initial mask used for the optimization process is the desired pattern.
PUF-aware OPC-2 (P-OPC2): This algorithm is the same as P-OPC1, except that the initial mask is the one generated by solving the OPC problem for *minimizing the expected MSE, i.e. the conventional approach.*

Note that the above algorithms rely on the fact that compensation methods have been applied to PUF layout/design which counteract systematic variation (as discussed in Section 3.3).

4. EXPERIMENTAL RESULTS

In this section, we simulate the fabrication of ring oscillator PUF (RO-PUF) circuits using masks generated by conventional and the proposed OPC schemes. Below we describe the RO-PUF architecture, the algorithms used to generate masks, and the lithography parameters used in our experiments. Then we discuss the results.

4.1 Simulation Setup

Ring Oscillator PUF (RO-PUF): A typical ring oscillator (RO) circuit consists of an odd number of inverters as shown in Figure 4(a). The frequency of an RO is determined by the total delay of its inverters and interconnects. A typical RO-PUF is shown in Figure 4(b) and functions as follows. The RO-PUF contains N ROs, which are each expected to have slightly different delay/frequency due to process variation. The challenge (input) to the RO-PUF essentially selects two of the N ROs. The frequencies of the selected ROs are compared by counting the vertical edges of their output. The response is one bit: a zero (one) if the upper (lower) RO has higher frequency than the lower (upper) RO.

In our experiments, each RO consists of three inverters and we use 512 ring oscillators (ROs) for each PUF. Responses are computed using the decoupled neighbor approach [10, 16] where each RO is only used in comparison with its neighbor. Therefore, there are 256 responses total per PUF. *Note that this method counteracts systematic variation because neighboring ROs will have similar systematic delay/frequency variation. When comparing the two ROs, the systematic components cancel each other and responses are a function of random variation only.*

Algorithms: The three algorithms we use to generate masks are denoted as PUF-aware OPC-1 (P-OPC1), PUF-aware OPC-2 (P-OPC2), and Conventional OPC (C-OPC). P-OPC1 and P-OPC2 were discussed in the previous section. C-OPC is an algorithm that minimizes μ_{MSE} and uses an initial mask that is simply the desired pattern.

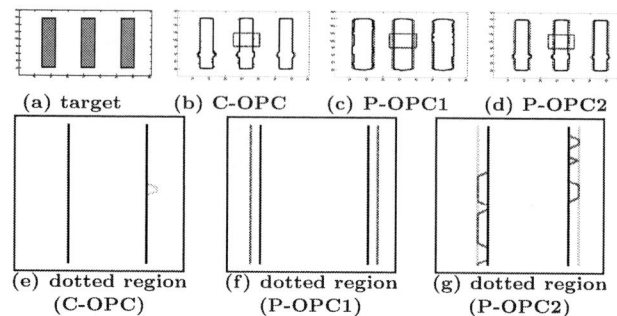

(a) target (b) C-OPC (c) P-OPC1 (d) P-OPC2

(e) dotted region (C-OPC) (f) dotted region (P-OPC1) (g) dotted region (P-OPC2)

Figure 5: Polysilicon patterns for 3 inverter ring oscillator. Entire patterns shown for (a) target, (b) C-OPC, (c) P-OPC1, and (d) P-OPC2. (e-g) show the patterns within the dotted box zoomed in for C-OPC, P-OPC1, and P-OPC2 respectively.

Lithography Parameters: Our target feature size (critical dimension) is 32nm. The wavelength of the optical light is 193nm. We assume the following properties of the lithography process as in [6]. The probability distribution (PDF) for defocus F is a zero mean Gaussian distribution with standard deviation 150nm. For simplicity, the imaging system is assumed to be coherent and dose variations are zero (although our methods could also account for dose variation).

Fabrication Simulation: We experiment with an RO-PUF design with 512 ROs and generate a population of 100 chips. To simulate fabrication of each chip/PUF instance, we choose random focal values F for each RO. To simulate spatial correlations within chips, we determine focal values F by using the quadtree model [1] (see Supplementary Material for details). In short, the area of a chip is divided into four regions, which are each subdivided into four subregions, and so forth. Every region, subregion, etc. is assigned an independent random variable (RV) from an associated PDF. The focal value for an RO is determined by summing the RVs for the region, subregion, etc. where the RO is located. After obtaining a set of samples for F, we apply the three algorithms and obtain three masks corresponding to the different techniques. The 51200 (=512×100) random samples of F are used to simulate fabrication of the polysilicon layer of all ROs in all chips. Standard polysilicon to electrical parameter extraction approaches [14] are used to obtain electrical parameters (effective channel length) of the ROs from the three masks. The electrical parameter distribution is used to calculate the distributions in RO frequencies and estimate PUF uniqueness, reliability, and unpredictability.

4.2 Results and Discussion

In this section, we report and discuss the following for each algorithm: (1) the polysilicon layers produced; (2) the mean and standard deviation of ring oscillator (RO) delay; (3) the average inter-distance (PUF uniqueness measure); (4) the average intra-distance (PUF reliability measure) where voltage supply is subject to variation; and (5) NIST test results (PUF unpredictability).

Polysilicon layers: The target polysilicon pattern is shown in Figure 5(a). Figures 5(b-d) show contours generated for the C-OPC, P-OPC1, and P-OPC2 algorithms respectively for four randomly chosen defocus values. C-OPC obtains patterns closest to the target pattern (low MSE). Comparing the proposed algorithms, the patterns from P-OPC1 are much further from the target pattern (largest MSE) than P-OPC2 (which uses the C-OPC mask as a starting point). Since the different contours are difficult to distinguish in Figures 5(b-d), we also provide higher resolution images in Figures 5(e-g) which correspond to the square regions in the center of Figures 5(b-d) respectively. Visually, the physical contours (and channel lengths) generated by the P-OPC algorithms are more varied per defocus value. Hence, we expect that PUFs generated by P-OPC1 and P-OPC2 will have higher variance in properties such as delay.

Ring Oscillator (RO) Delay: Table 1 shows the mean and standard deviation of ring oscillator delay for the entire RO-PUF

100

Table 1: Comparison of mean and standard deviation for ring oscillator delay

		C-OPC	P-OPC1	P-OPC2
delay	μ	23.08ps	46.05ps	26.91ps
	σ	.658ps	6.88ps	3.85ps

Table 2: Mean (μ) and standard deviation (σ) of inter- and intra-distances for each PUF population.

		C-OPC	P-OPC1	P-OPC2
d_{inter}	μ	47.29	48.96	49.80
	σ	3.12	3.17	3.15
d_{intra}	μ	12.49	4.25	3.32
	σ	2.00	1.25	1.12

population and each algorithm. The smallest variance occurs for the conventional case which is not surprising given that C-OPC minimizes fabrication variation. The proposed P-OPC algorithms have larger variance which should result in improvements to inter- and intra-distances for the PUFs. Note that the mean delay for P-OPC1 is much larger than the other two cases. This can be explained by the polysilicon layers in Figure 5 which show that P-OPC1 tends to increase the polysilicon channel length. With longer channel length, drain-to-source current will be smaller for each transistor and hence delay of each inverter/ring oscillator will be larger.

PUF Uniqueness: For all 100 RO-PUFs, we determined the response to each challenge based on the delay in corresponding ROs. From these responses, we computed the inter-distance of each challenge for the entire RO-PUF population using Eqn. (1). Table 2 shows the inter-distance mean and standard deviation for all three algorithms. C-OPC obtains a mean of 47%. The proposed algorithms obtain means which are closer to the ideal inter-distance (50%). Comparing the two, P-OPC2 is better. The standard deviations in inter-distance for all algorithms are comparable. Overall, there is a 5% improvement in uniqueness compared to the C-OPC generated PUFs.

PUF Reliability (Voltage Supply Variation): We computed the intra-distances for each challenge by comparing responses at one nominal voltage supply ($V_{dd} = .9$) and 100 sample responses from the same RO-PUF with a randomly varying voltage supply (mean .1V, standard deviation .33mV) to emulate environmental variations. Table 2 shows the inter-distance mean and standard deviation for all three algorithms. Ideally, the intra-distance should be 0 for error-free PUF responses. The conventional algorithm clearly obtains the worst intra-distance with largest mean and variance. The closest to the ideal case is P-OPC2 which has the smallest mean and variance. Overall, there is a 70% improvement in reliability compared to the C-OPC generated PUFs. These results are not surprising because P-OPC has been shown above to increase RO delay variance. With larger variation between ROs, a small bit of noise introduced by the voltage supply will have less significant effects on computed response (since two ROs are compared for each response bit).

PUF Unpredictabilty (NIST tests): We tested randomness present in the PUF responses with the NIST test suite [13]. Out of the 15 NIST tests, we use the 7 tests (and 2 variants) which are appropriate for our dataset as discussed in [16]. In contrast to uniqueness which compares the randomness between PUFs, the NIST tests also check the randomness between responses from the same PUF. Due to space limitations, we cannot explicitly show the NIST results (see Supplementary Material). For brevity, we summarize the results as follows. The C-OPC generated PUFs did not have enough randomness and only passed 1 test. The P-OPC1 generated PUFs were an improvement, but still only passed 3 tests. The P-OPC2 generated PUFs passed all the tests.

Summary: The results obtained in this section show that:

- The proposed P-OPC algorithms produce higher variation in the polysilicon layers. This results in over 5 times greater variance in ring oscillator delay than the conventional approach.
- The proposed P-OPC algorithms improve PUF inter-distance (uniqueness) by 5% and PUF intra-distance (reliability) by over 70% compared to the conventional C-OPC algorithm.
- The proposed P-OPC algorithms improved the unpredictability (NIST results) of the PUF response bits and the P-OPC2 generated PUFs pass all the appropriate tests.

5. CONCLUSION AND FUTURE WORK

In this paper, we proposed a novel approach for improving PUF quality. While a large amount of work focuses on improving PUFs at the architectural level, we focused our attention on the PUF fabrication step since manufacturing variations are the source of the most important PUF properties. Compared to conventional OPC, our OPC objective and algorithms generate masks that enhance fabrication variations rather than suppress them. Results clearly show that the PUF quality improves with the proposed techniques. In future work, we plan to investigate similar approaches that target the fabrication process to improve PUF quality.

6. REFERENCES

[1] D. Blaauw, K. Chopra, A. Srivastava, and L. Scheffer. Statistical timing analysis: From basic principles to state of the art. *IEEE Trans. Comput.-Aided Des. Integr. Circuits Syst.*, 27(4):589–607, 2008.

[2] Q. Chen, G. Csaba, P. Lugli, U. Schlichtmann, and U. Ruhrmair. The bistable ring puf: A new architecture for strong physical unclonable functions. In *IEEE HOST 2011*, pages 134–141. IEEE, 2011.

[3] N. Cobb. *Fast Optical and Process Proximity Correction Algorithms for Integrated Circuit Manufacturing*. PhD thesis, University of California, 1998.

[4] J. Guajardo, S. Kumar, G. Schrijen, and P. Tuyls. Fpga intrinsic pufs and their use for ip protection. *CHES 2007*, pages 63–80, 2007.

[5] P. Gupta, A. Kahng, Y. Kim, and D. Sylvester. Self-compensating design for reduction of timing and leakage sensitivity to systematic pattern-dependent variation. *IEEE Trans. Comput.-Aided Design Integr. Circuits Syst.*, 26(9):1614–1624, 2007.

[6] N. Jia and E. Lam. Machine learning for inverse lithography: Using stochastic gradient descent for robust photomask synthesis. *J. Optics*, 12:045601, 2010.

[7] I. Kim, A. Maiti, L. Nazhandali, P. Schaumont, V. Vivekraja, and H. Zhang. From statistics to circuits: Foundations for future physical unclonable functions. *Towards Hardware-Intrinsic Security*, pages 55–78, 2010.

[8] C. Mack. *Fundamental principles of optical lithography: the science of microfabrication*. Wiley-Interscience, 2007.

[9] R. Maes and I. Verbauwhede. Physically unclonable functions: A study on the state of the art and future research directions. *Towards Hardware-Intrinsic Security*, pages 3–37, 2010.

[10] A. Maiti and P. Schaumont. Improved ring oscillator puf: An fpga-friendly secure primitive. *Journal of Cryptology*, pages 1–23, 2011.

[11] D. Pan, P. Yu, M. Cho, A. Ramalingam, K. Kim, A. Rajaram, and S. Shi. Design for manufacturing meets advanced process control : A survey. *J. Process Control*, 18(10):975–984, 2008.

[12] A. Poonawala and P. Milanfar. Opc and psm design using inverse lithography: A non-linear optimization approach. In *Proc. SPIE*, volume 6154, pages 1159–1172, 2006.

[13] A. Rukhin. A statistical test suite for random and pseudorandom number generators for cryptographic applications. Technical report, DTIC, 2001.

[14] S. Shi, P. Yu, and D. Pan. A unified non-rectangular device and circuit simulation model for timing and power. In *ICCAD 2006*, pages 423–428. ACM, 2006.

[15] G. Suh and S. Devadas. Physical unclonable functions for device authentication and secret key generation. In *IEEE DAC 2007*, pages 9–14. ACM, 2007.

[16] C. Yin and G. Qu. A regression-based entropy distiller for ro pufs. 2011.

[17] P. Yu, S. Shi, and D. Pan. True process variation aware optical proximity correction with variational lithography modeling and model calibration. *J. Micro*, 6(3), 2007.

Supplementary Material

S.1. PHYSICALLY UNCLONABLE FUNCTIONS (PUFS)

S.1.1 Main Concept and Terminology

Concept: There is a long history of using random physical features to identify humans. For example, fingerprints are well-known identifiers that are unique to each individual and are difficult to remove/duplicate. A silicon Physically Unclonable Function (PUF) is essentially an extension of this concept applied to chips. Silicon PUFs are special circuits that are added to IC designs. Each silicon PUF possesses random physical characteristics (delay, power, consumption, etc.) which are caused by small differences in transistor and wiring elements which are themselves derived from variations in the manufacturing of ICs. By measuring the characteristics of the PUF, we can uniquely identify an IC.

Terms: The input and output of PUFs are called *challenges* and *responses*. An applied challenge and its measured response is referred to as a *challenge-response pair*.

Types: Generally, silicon PUF designs can be broken down into two classes [9]:

1. **Delay-based PUFs:** exploit the random variations of wire and gate delays. For example, in an arbiter PUF [15], a challenge sets up a race between two paths and the winner of the race determines the response (0 or 1). A ring oscillator PUF compares the frequency generated by two or more ring oscillators to produce a response [15]. The variations in path delays and frequency are attributed to fabrication randomness. Each fabricated chip experiences a unique variability signature.

2. **Memory-based PUFs:** exploit the random settling behavior of bistable elements [9]. Typically, a bistable element is in one of two stable states: 0 or 1. However, if the circuit can be brought into an unstable state, it is unclear how that circuit will behave: it may oscillate between stable states or converge to one stable state. As a result of manufacturing randomness, it has been observed that each bistable element will heavily prefer one stable state over the other [9]. Thus, the settling state is a good candidate for a PUF response. SRAM PUFs and butterfly PUFs are examples of memory-based PUFs [9, 4].

S.1.2 PUF Security Applications

PUFs are convenient for many security applications:

1. **Identification/Authentication:** [9] After manufacturing a device, the manufacturer can record the challenge-response pairs (CRPs) of its PUF in an enrollment phase. After deployment, a device's identity can be verified at any time by the manufacturer by applying any challenge from the enrollment phase to the PUF. Since each PUF provides a unique response and the response can only be measured if one has the physical device, the identity of the device is verified if the response returned is the same as the response recorded during the enrollment phase. To avoid replay attacks, the selected challenge should only be used once to identify the device [9].

2. **Safe Encryption Key Generation:** [15] The safety of cryptographic algorithms critically depends on the protection of encryption keys. Traditionally, keys must be permanently stored in memory of a device and therefore may be susceptible to side channel or invasive probing attacks on memory. However, if a PUF's response (or some derivative of its response) to a unique challenge is used as an encryption key, then the key is physically embedded in the device rather than stored in memory and cannot be obtained by similar attacks.

3. **Tamper Evidence:** Many PUFs have a property that if their physical device is modified, their CRPs also change [9]. This can be used to determine if a device's functionality has been altered in the field.

S.2. OPTICAL LITHOGRAPHY

In order to produce ICs with submicron dimensions, the IC fabrication industry relies on optical lithography systems to print structures on wafers. In the above paper, an optical lithography system is shown in Figure 1. In short, optical wavelengths from a light source are gathered by an illumination lens and passed through portions of a mask. The light diffracts as it is passed through the mask and a portion of it is gathered by a projection lens. The projection lens projects the light onto the wafer surface to essentially create IC structures.

S.2.1 Fundamental Limits

The fundamental limits of an optical lithography system are described in terms of resolution and depth-of-focus.

1. **Resolution:** The resolution of an optical lithography system is the minimum feature size that can be printed and is often written as [17]

$$R = k_1 \frac{\lambda}{NA}$$

where λ is the wavelength of light used and k_1 is a factor that depends on properties of the optical lithography system and the desired IC. NA is the so-called numerical aperture which characterizes the range of angles over which the projection lens can accept light and is given by [8]

$$NA = n \sin \theta$$

where n is the refraction index of the medium in which the lens operates and θ is the maximum angle of light that can enter the lens.

2. **Depth-of-focus (DOF):** The DOF indicates the range of focal variations where the printed patterns remain robust. It is estimated by [17]

$$DOF = k_2 \frac{\lambda}{NA^2}$$

where k_2 is another system and IC pattern dependent parameter.

The goal of the IC industry is to decrease resolution R (feature size) in order to fit more transistors onto an IC. Two ways to accomplish this are by decreasing wavelength λ and increasing numerical aperture NA. Unfortunately, current leading technologies are restricted to $\lambda = 193$nm [8]. Furthermore, increase in NA has an adverse effect on DOF reducing robustness of the optical system to focal variations. Thus, the only way to decrease resolution R and reliably produce desired IC patterns is by decreasing k_1.

S.2.2 Resolution Enhancement Techniques (RET)

There are three major techniques to improve k_1 [8]:

1. **Optical Proximity Correction (OPC):** The mask pattern shape is optimized to produce patterns that are more accurate and robust to optical system variations (defocus and dose).

2. **Phase-Shift Masks (PSM):** Phase shifters are added to certain transparent regions of the mask. The interference of the phase-shifted light with the light coming from unmodified regions of the mask improves the contrast of patterns on the wafer.

3. **Off-Axis Illumination (OAI):** The shape and size of the illumination source is optimized for the specific mask patterns being printed.

In the main paper, we have recognized that the above RET tools can also be used with the opposite goal in mind. Specifically, we want to increase randomness in the patterns for PUF circuitry generated by the optical lithography system while leaving non-PUF portions of the IC unaffected. For this goal, we feel that OPC and PSM are more applicable since they modify portions of the mask which control PUF patterns without affecting other IC components. Although we have only investigated OPC approaches in the main paper, the optimization framework for PSM should be similar. We shall investigate variation enhancing PSM in future work.

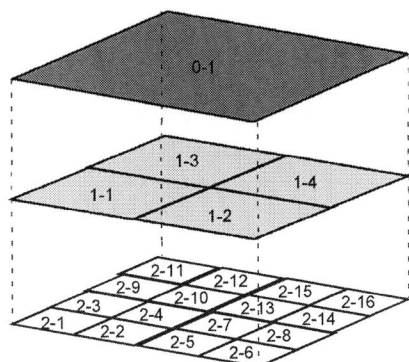

Figure S-2: Quadtree partitioning for a chip. The depth of the tree shown is 3 levels.

S.3. EXPERIMENTAL SETUP

S.3.1 RO-PUF Manufacturing Variations

Manufacturing variations are modeled with systematic and random components. In [10], the authors showed that systematic variations negatively affect an RO-PUF's ability to produce unique and unpredictable chip signatures. They modeled the delay of each ring oscillator (RO) as follows:

$$d_{RO} = d_0 + d_{rand} + d_{sys}$$

d_0 is a constant representing the intended or nominal delay. d_{rand} and d_{sys} are terms denoting the random and systematic variation components of the delay respectively. Environmental noise is ignored for simplicity. In an RO-PUF, the difference in delay of two ROs (Δd_{RO}) essentially determines a response

$$r = \begin{cases} 1 & \text{if } \Delta d_{RO} = \Delta d_{rand} + \Delta d_{sys} < 0 \\ 0 & \text{otherwise} \end{cases}$$

The effects of d_{rand} and d_{sys} on RO delay/frequency and r are summarized as follows. While d_{rand} is random in all ROs and chips, the d_{sys} component is a function of the RO's location on the chip and is the same for each chip. Therefore, in instances where $|\Delta d_{sys}| \gg |\Delta d_{rand}|$, each response r is biased towards one value on all chips. This results in lower uniqueness and higher predictability of PUF responses. In order to counteract systematic variation, the authors in [10] propose that only neighboring ROs be compared. Since neighboring ROs are nearby, they possess similar d_{sys} so $\Delta d_{sys} \approx 0$. As a result, response r depends mostly on Δd_{rand}. In our experiments, we apply this same procedure to ensure that amplified systematic variations do not degrade PUF quality.

S.3.2 Simulation Procedure

The experimental procedure, chip/PUF fabrication, and simulations discussed in Section 4 of the main document are illustrated in Figure S-1 for additional clarity and discussed below. We experiment with an RO-PUF design with 512 ring oscillators (ROs) and generate a population of 100 chips.

Mask Generation: We generate masks using the C-OPC, P-OPC1, and P-OPC2 algorithms. The inputs to each algorithm are (1) an initial mask; (2) a desired wafer pattern; (3) focal distribution for the optical lithography system. For C-OPC and P-OPC1, these inputs are the same and the initial mask/desired pattern are identical. However, P-OPC2 uses the mask generated by C-OPC as its initial mask. Each algorithm is used to generate one mask. The C-OPC algorithm obtains a mask that minimizes the average MSE between the desired and fabricated patterns (μ_{MSE}). The P-OPC algorithms obtain masks that maximize σ_{MSE}/μ_{MSE}. Each mask is then used to generate the polysilicon layer of 512 ROs in 100 chips/PUFs.

Monte Carlo Lithography Simulations: Generating a polysilicon layer for each RO requires two inputs: (1) a focal value and (2) a mask. Random focal values are determined using the

quadtree approach [1] which models inter-chip and intra-chip correlations. In the quadtree modeling approach [1], the area of a chip is recursively partitioned into four equally sized regions. This is illustrated in Figure S-2. The first partition is "0-1" and corresponds to the root of the quadtree. "0-1" is divided into four partitions "1-1", "1-2", "1-3", and "1-4" which form the next level of the tree. The four partitions are subdivided into another four partitions and so forth. Each partition in every level of the tree is assigned a random variable (RV) with its own probability distribution. The spatially correlated variation associated with a gate in the IC is then defined as the sum of the RV at the lowest partition containing the gate and the RVs of the higher partitions that overlap with the gate's position. Correlation exits between gates on a single chip due to the sharing of RVs at higher levels of the quadtree. Correlation between chips exists because the probability distributions associated with each partition are the same for all chips.

In our case, the random variables (RVs) and associated probability distributions correspond to defocus across the chip. The gates in our case are ring oscillators (ROs). We generate RVs for all partitions and levels in each chip/PUF according to some knowns distributions and then compute defocus values as follows. Suppose we want to generate the focal value F for an RO located in partition "2-13" of chip x (see Figure S-2). F is computed by summing the RVs associated with partitions "2-13", "1-4", and "0-1" of chip x. In our experiments, we do this for 512 ROs and 100 chips totaling 51200 random samples. We use these focal values as input to an in-house simulator to generate polysilicon layers for each RO-PUF.

Parameter Extraction and PUF evaluation: We determine physical and electrical parameters from two sources in the manufacturing process: (i) the polysilicon layers generated by the above masks and (ii) independent random sources. In the first case, standard polysilicon to electrical parameter extraction approaches [14] are applied to obtain effective channel lengths for each RO. In the second case, we determine threshold voltage and oxide thickness for each transistor in the chip by taking independent samples from normal distributions (which is the most commonly observed distribution for such parameters [1]). The process parameter distributions are used to calculate the distributions in RO frequencies and estimate PUF uniqueness, reliability, and unpredictability.

S.4. ADDITIONAL RESULTS

Due to space limitations, we could not fit all results obtained from the experiments in the main paper. These results are shown and discussed below.

S.4.1 Uniqueness and Reliability

In Section 4.2, we provided the mean and standard deviation for inter- and intra-distance for the PUF populations. The inter-distance and intra-distance distributions are shown in Figures S-3 and S-4 respectively. Both appear to be gaussian.

S.4.2 NIST test suite (Unpredictability)

The NIST test suite is a standard used for testing the randomness of binary sequences produced by cryptographic random and pseudorandom number generators [13]. The NIST test suite defines an "ideal" random number generator as one that produces a sequence of bits whose statistical properties are similar to that of consecutive flips of a fair coin. It contains 15 statistical tests and each NIST test evaluates the hypothesis that some property of the sequence under test came from the "ideal" random number generator. The sequence "fails" if the statistical properties of its random sequences indicate a clear deviation from the "ideal"; otherwise it "passes".

In this paper, PUF response bits from each PUF population are fed to the NIST test suite. We use the same NIST tests as [16] to check the randomness of PUF response sequence:

1. **Frequency:** calculates the proportion of zeros and ones in the entire sequence. Ideally, this should be close to $\frac{1}{2}$.

2. **Block Frequency:** calculates the proportion of ones within M bit blocks. Ideally, this should be close to $\frac{M}{2}$. For $M = 1$, this test is the same as the frequency test.

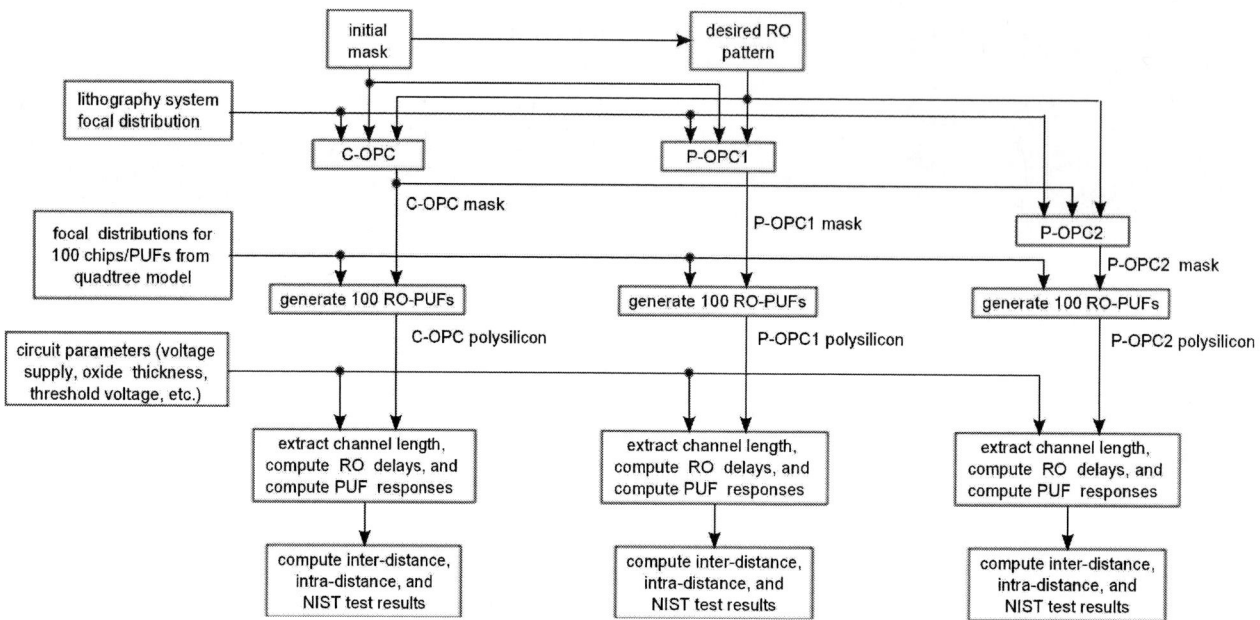

Figure S-1: Experimental Procedure

3. **Cumulative Sums:** calculates the cumulative sum of the partial sequences (ones and zeros) occurring in the tested sequence.

4. **Runs:** compares the total number of runs in the sequence to that expected for a random sequence, where a run is an uninterrupted sequence of identical bits.

5. **Longest Run:** determines whether the length of the longest run of ones within the tested sequence is consistent with the length of the longest run of ones that would be expected in a random sequence.

6. **Approximate Entropy:** compares the frequency of overlapping blocks of two consecutive/adjacent lengths (m and $m + 1$) against the expected result for a random sequence.

7. **Serial:** determines whether the number of occurrences of 2^m m-bit overlapping patterns is approximately the same as would be expected for a random sequence.

Outputs: There are three outputs for the random sequence under test provided by the NIST test suite: (1) *P-value*; (2) the *distribution of P-values*; and (3) *Proportion*. Each sequence is composed of bitstreams. The P-value is the probability that a perfect random number generator would have produced a bitstream less random than the bitstream that was tested. Ideally, bitstream P-values should be uniformly distributed across the [0,1] interval. Proportion is the proportion of bitstreams that pass the hypothesis test (i.e. obtain P-values > threshold).

In our case, the random sequence subject to the NIST test comprises $25600 = 100 \times 256$ bits, where 100 is the total number of PUFs in the population and 256 is the output length for each PUF. One bitstream contains the 256 response bits of one PUF from the population. Suitable test parameters (M, m) are chosen according to the guidelines in [13]. We choose $M = 20$ for the Block Frequency test, $m = 2$ for the Approximate Entropy test, and $m = 5$ for the Serial test.

Results: The output of the NIST tests for C-OPC, P-OPC1, and P-OPC2 generated PUFs are shown in Table S-1. The "C1" through "C10" columns represent the frequency of P-values across the interval [0,1] with each column representing a histogram bin in the range [0,1]. The "P-VALUE" column is a P-value that assesses the uniformity of the P-values for all bitstreams in the PUF sequence under test. The "PROPORTION" column contains the proportion of PUF bitstreams that pass the test (P-value > threshold). Each row corresponds to a different NIST statistical test. A "*" next to the values in the P-VALUE or PROPORTION columns denotes a failure to meet uniformity requirements (P-value ≥ 0.0001) or proportion requirements (PROPORTION ≥ 96) respectively. In the main paper, we report that a population of PUFs passes a statistical test if its row in Table S-1 does not have any "*"s.

As shown in Table S-1, the C-OPC PUFs do not have enough randomness in their responses and only pass the Serial 2 test. P-OPC1 PUFS have a large improvement in uniformity (see C1 to C10 columns) and PROPORTION over C-OPC. P-OPC1 only passes the Runs, Longest Run, and Serial 2 tests. **P-OPC2 passes all tests.**

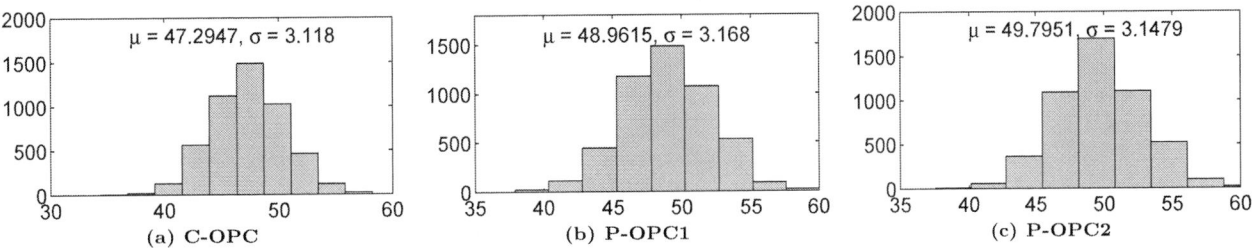

Figure S-3: Uniqueness distribution for each PUF population: x-axis (inter-distance), y-axis (count)

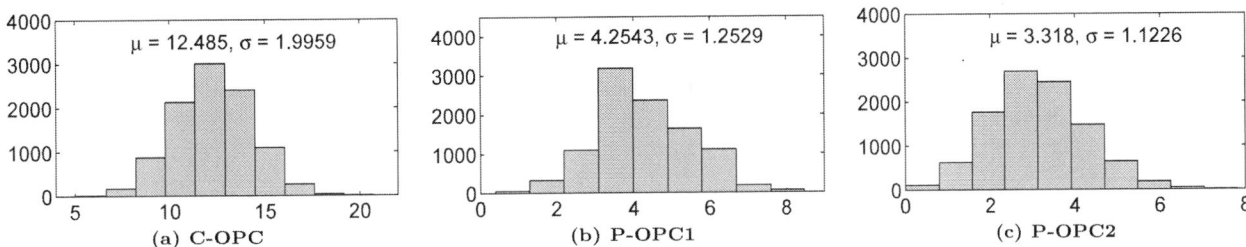

Figure S-4: Reliability distribution for each PUF population: x-axis (intra-distance), y-axis (count)

Table S-1: NIST Results for C-OPC, P-OPC1, and P-OPC2. NIST Parameters $M = 20$ for Block Frequency test, $m = 2$ for Approximate Entropy test, and $m = 5$ for Serial tests. '*' denotes a failure.

C1	C2	C3	C4	C5	C6	C7	C8	C9	C10	P-VALUE	PROPORTION	STATISTICAL TEST
Conventional OPC (C-OPC)												
96	2	2	0	0	0	0	0	0	0	0.000000 *	34/100 *	Frequency
60	6	10	6	7	3	2	4	2	0	0.000000 *	74/100 *	BlockFrequency
95	3	0	2	0	0	0	0	0	0	0.000000 *	35/100 *	CumulativeSums (fwd)
95	1	3	1	0	0	0	0	0	0	0.000000 *	35/100 *	CumulativeSums (bkwd)
26	5	6	10	9	9	7	8	8	12	0.000199	80/100 *	Runs
65	9	15	3	0	3	1	0	4	0	0.000000 *	72/100 *	LongestRun
81	11	2	3	2	0	0	1	0	0	0.000000 *	49/100 *	ApproximateEntropy
62	15	6	5	6	3	0	1	1	1	0.000000 *	62/100 *	Serial 1
20	10	12	8	10	12	9	3	9	7	0.045675	96/100	Serial 2
PUF-aware OPC 1 (P-OPC1)												
43	23	11	6	5	3	3	4	1	1	0.000000 *	80/100 *	Frequency
26	15	10	12	12	5	4	7	5	4	0.000003 *	98/100	BlockFrequency
41	16	15	10	3	3	1	4	2	5	0.000000 *	82/100 *	CumulativeSums (fwd)
46	13	13	7	4	8	0	4	3	2	0.000000 *	80/100 *	CumulativeSums (bkwd)
10	11	5	8	11	9	16	8	12	10	0.574903	100/100	Runs
20	16	13	12	11	10	2	7	4	5	0.000818	97/100	LongestRun
35	9	15	9	8	8	3	3	7	3	0.000000 *	94/100 *	ApproximateEntropy
23	14	9	18	8	3	4	7	7	7	0.000031 *	94/100 *	Serial 1
12	16	6	7	8	14	6	12	11	8	0.275709	97/100	Serial 2
PUF-aware OPC 2 (P-OPC2)												
14	17	8	9	7	10	8	5	8	14	0.171867	97/100	Frequency
13	14	5	12	14	4	9	8	10	11	0.262249	99/100	BlockFrequency
17	4	10	18	10	6	10	5	7	13	0.013569	96/100	CumulativeSums (fwd)
15	14	12	10	4	11	9	5	8	12	0.236810	97/100	CumulativeSums (bkwd)
10	13	10	9	8	7	15	12	9	7	0.719747	100/100	Runs
10	17	8	10	9	8	8	12	4	14	0.224821	99/100	LongestRun
17	9	11	11	7	9	13	4	10	9	0.289667	98/100	ApproximateEntropy
11	12	14	10	6	10	13	4	10	10	0.514124	97/100	Serial 1
10	9	8	10	13	10	8	14	10	8	0.924076	98/100	Serial 2

Transformer: A Functional-Driven Cycle-Accurate Multicore Simulator

Zhenman Fang[1,2], Qinghao Min[2], Keyong Zhou[2], Yi Lu[2], Yibin Hu[2], Weihua Zhang[2], Haibo Chen[3], Jian Li[4], Binyu Zang[2]

[1]The State Key Lab of ASIC & System, Fudan University. [2]Parallel Processing Institute, Fudan University.

{fangzhenman, minqh, zky, yil, huyibin, zhangweihua, byzang}@fudan.edu.cn

[3]Institute of Parallel and Distributed Systems, Shanghai Jiaotong University. haibochen@sjtu.edu.cn

[4]IBM Austin Research Laboratory. jianli@us.ibm.com

ABSTRACT

Full-system simulators are extremely useful in evaluating design alternatives for multicore. However, state-of-the-art multicore simulators either lack good extensibility due to their tightly-coupled design between functional model (FM) and timing model (TM), or cannot guarantee cycle-accuracy. This paper conducts a comprehensive study on factors affecting cycle-accuracy and uncovers several contributing factors ignored before. Based on the study, we propose a loosely-coupled functional-driven full-system simulator for multicore, namely Transformer. To ensure extensibility and cycle-accuracy, Transformer leverages an architecture-independent interface between FM and TM and uses a lightweight scheme to detect and recover from execution divergence between FM and TM. Based on Transformer, a graduate student only needs to write about 180 lines of code and takes about two months to extend an X86 functional model (QEMU) in Transformer. Moreover, the loosely-coupled design also removes the complex interaction between FM and TM and opens the opportunity to parallelize FM and TM to improve performance. Experimental results show that Transformer achieves an average of 8.4% speedup over GEMS while guaranteeing the cycle-accuracy. A further parallelization between FM and TM leads to 35.3% speedup.

Categories and Subject Descriptors

B.2.2 [**Performance Analysis and Design Aids**]: Simulation

General Terms

Design, Measurement, Performance

Keywords

Functional-driven, Multicore simulation, Full-system, Extension

1. INTRODUCTION

Full-system simulation is a key tool to evaluate new ideas in architectural design. Generally, there are two basic models in a full-system simulator: *functional model* (FM), which provides a full-system execution environment to execute operating systems and applications and collects the resulted instruction flow and data access information; and *timing model* (TM), which simulates micro-architectural behavior of the instruction flow generated by FM. Due to the importance of full-system simulator, researchers have designed and implemented a number of FMs such as Simics [8], QEMU [2], and COREMU [15], and TMs such as GEMS [9], MPTLsim [18] and RAMP GOLD [14]. However, FMs and TMs are usually tightly coupled together in a full-system simulator and it is usually hard to extend new FMs or TMs in the simulator. For example, developers have spent years to combine M5 with GEMS (i.e., gem5 [4]) or extend QEMU to PTLsim (MARSS [11]). Further, such a tightly-coupled design also makes it hard to efficiently parallelize FM and TM, resulting in inferior performance.

There is actually a good reason to take the tightly-coupled design in current mainstream full-system multicore simulators. To guarantee cycle-accuracy such as faithful instruction execution behavior and timing, they usually use TM to drive the execution of FM: in each cycle, TM advises FM on which instruction FM should execute; FM will also report to TM with information regarding the executed instruction, to let TM maintain correct architecture states and timing information. Such a tightly-coupled and complex interaction between TM and FM limits both extensibility and performance of full-system simulators. Though there have been some efforts in trying to explore a loosely-coupled design for multicore simulators [7, 6], their solutions cannot guarantee cycle-accuracy and there is no implemented prototype for multicore simulators.

In this paper, we first present a comprehensive study on the limiting factors that lead to execution divergence between FM and TM. We show that besides traditional well-known factors such as branch misprediction and shared data access order, interrupt/exception handling and shared page access order also lead to execution divergence and thus cycle-inaccuracy in a loosely-coupled design. To understand the probability of occurrence of these factors, we profile the proportions of these events in a set of benchmarks and find that these events happen very infrequently (less than 1%). This indicates that for most cases, there is no execution divergence between FM and TM.

Based on the above analysis, we propose Transformer, a loosely-coupled, functional-driven simulation scheme for full-system multicore simulation. In Transformer, FM runs ahead and provides instructions and data access information to TM. TM then uses such information to simulate the detailed timing of micro-architecture. Transformer also provides a lightweight scheme to detect and recover from execution divergence, thus ensures cycle-accuracy. Basically, Transformer rolls back FM to the path indicated by TM. For branch misprediction and interrupt/exception handling, Transformer uses an additional simple FM to generate the instruction flow information in wrong path to feed TM, so as to further reduce the interaction between FM and TM caused by the rollback scheme. Further, to make Transformer extensible, we provide an architecture-independent instruction and data flow interface between FM and TM.

In Transformer, the interaction between FM and TM is much simpler, and thus provides great flexibility to extend with new FMs or TMs. Further, as FM and TM are now loosely-coupled, it also opens the opportunity to parallelize FM and TM to improve the performance.

We have implemented Transformer based on GEMS [9], a widely-used tightly-coupled simulator, and parallelize FM and TM to achieve better performance. And we plan to release the source code of Transformer to the community in future. Based on Transformer, a graduate student only needs to write about 180 LOCs and takes about two months to extend an X86 functional model (QEMU) in Transformer. Furthermore, experiments with SPLASH-2 [17] and PARSEC [3] show that Transformer achieves about 8.4% speedup compared to GEMS while guaranteeing the cycle-accuracy. And the speedup increases to 35.3% after FM and TM are parallelized.

In summary, this paper makes the following contributions:

- The first comprehensive analysis on the factors leading to execution divergence between FM and TM, which uncovers that interrupt/exception handling and shared page access are also limiting factors to cycle-accuracy.
- A loosely-coupled full-system multicore simulation framework that is extensible, fast, and cycle-accurate, as well as a set of techniques to detect and recover from execution divergence.
- An experimental evaluation that confirms the effectiveness and efficiency of Transformer and a case study that extends QEMU in Transformer to demonstrate the extensibility.

The rest of the paper is organized as follows. Section 2 discusses the motivation of the loosely-coupled design and comprehensively analyzes which factors affect cycle-accuracy. Section 3 proposes the Transformer framework, describes the lightweight cycle-accurate solutions and discusses the architecture-independent interface. Section 4 demonstrates an example for extending X86 support and evaluates the performance speedup of Transformer. Section 5 describes related work. Finally, section 6 concludes the paper and discusses possible future work.

2. MOTIVATION

2.1 Limitations with a Tightly-coupled Design

To achieve cycle-accuracy (e.g., guarantee correct interleaving in parallel applications), existing full-system multicore simulators usually exploit a tightly-coupled timing-driven design. As shown in Figure 1, in each cycle, TM directs FM with which instructions should be executed and FM feeds back the executed results to TM to maintain correct architecture states and timing. Moreover, TM has to simulate part of the functional model so as to direct the execution of FM. Such a tightly-coupled and complex interaction between FM and TM makes it very difficult to extend a new FM or TM into those simulator frameworks. For example, the developers spends years to combine M5 with GEMS (gem5 [4]) or extend QEMU into PTLsim (MARSS [11]).

In addition, the complex interaction in current tightly-coupled design limits simulation speed. To illustrate this problem, we profile the execution proportion of FM, TM and their interactions (using the experiment setup in section 4.1). First, to support TM, FM has to execute in instruction-by-instruction model instead of fast

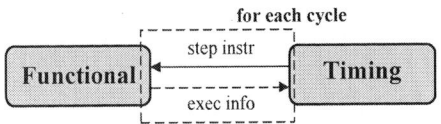

Figure 1: Tightly-coupled Functional and Timing.

binary translation to provide execution information to TM. As a result, FM occupies about 10% of the whole execution time, which cannot be neglected any more. However, it is impossible for a tightly-coupled design to gain performance improvement through parallelizing FM and TM. Moreover, complex interaction produces about 26% overhead due to complex control logic and frequent state transformation with poor locality.

2.2 Factors to Cycle-Accuracy

To gain insight into possible solutions to loosely-coupled cycle-accurate design, we study the factors leading to execution divergence between FM and TM. Besides traditional well-known factors such as branch misprediction and shared data access order, we find that interrupt/exception handling and shared page access order also lead to execution divergence and thus cycle-inaccuracy in a loosely-coupled design.

- *Branch misprediction:* In modern architectures, branch prediction is usually exploited in the pipeline design to avoid stall caused by branch instructions. The branch could be mispredicted to execute a wrong path in TM. However, FM always executes the instructions on the correct path, leading to execution divergence with actual architectural execution (i.e., TM).
- *Shared data access order:* In parallel applications, not all shared data are protected by lock operations to achieve some harmless operations, such as user-level synchronization. Therefore, FM may execute a different write/read order compared with that of TM, which will diverge the execution path.
- *Interrupt/exception handling:* Interrupt or exception is similar to branch misprediction. TM handles the interrupt or exception (i.e., jumps to the interrupt or exception handling path) in the commit stage after squashing the pipeline. Before that, TM will fetch instructions from the wrong path, i.e., next program counter (PC) instead of the interrupt/exception handler code. However, FM directly simulates the interrupt/exception handling path, which leads to execution path divergence.
- *Shared page access order (i.e., MMU miss order):* In full-system multicore simulation, the system behavior has to be simulated. Although such a design guarantees cycle-accuracy, it involves some additional shared data access among different threads, which would further lead to path divergence between FM and TM. The divergence will take place under two conditions. First, two memory operations in different threads may access data within the same page. When this page is not in memory, the first access will result in a MMU miss and its corresponding thread has to include the operations to process the MMU miss. Second, two pages (suppose A and B) accessed by two data accesses might be mapped to the same entry in the page table. Suppose page A is in memory while page B is not present. If the access to page B is executed first, page A will be split out. When page A is accessed again, a MMU miss occurs. However, if page A is executed first, no MMU miss will occur. Since both of these two conditions are related to MMU miss, we will also refer to this factor as MMU miss order.

Although these factors would lead to execution divergence between FM and TM, they occur rarely. To illustrate this problem,

Table 1: Proportion of path diversities.

Path divergence Source	Proportion
Branch Misprediction	5.3E-3
Interrupt/Exception Handling	1.4E-4
Shared Data Access Order Violation	7.9E-6
MMU Miss	1.6E-5

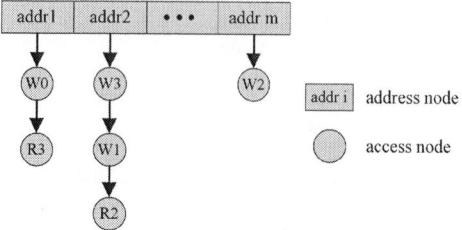

Figure 3: Memory Access Table structure.

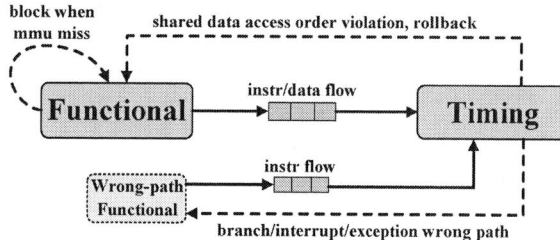

Figure 2: The Transformer framework.

we profile the occurrence proportion of each divergence factor in the total execution (using the configuration in section 4.1). As the data shown in Table 1, branch misprediction occurs most and only occupies about 0.53%. The total proportion occurs less than 1%. Therefore, in most cases (more than 99%), there is no execution divergence between FM and TM. This opens the opportunity to use a loosely-coupled design that may result in better extensibility to support other FMs or TMs and superior performance due to possible parallelization.

3. THE TRANSFORMER FRAMEWORK

This section presents the design of our loosely-coupled framework called Transformer. We first describe a lightweight scheme to detect and recover from execution divergence to guarantee cycle-accuracy. Then, we illustrate an architecture-independent instruction and data flow interface between FM and TM, to make Transformer more extensible. The overall Transformer framework works as follows, as shown in Figure 2.

In Transformer, FM in most cases generates the architecture-independent instruction and data flow information (e.g., pipeline dependence, memory access address) to TM. TM simulates the detailed micro-architecture using instruction and data information provided by FM. When a divergence factor is detected, different strategies (roll back FM and create a wrong-path FM) are applied to revise the divergence execution.

3.1 Divergence Detection

To guarantee cycle accuracy in a loosely-coupled design, the first thing is to detect when and where an execution divergence occurs. Among the four factors, it is easier to detect branch misprediction and interrupt or exception handling. For branch misprediction, we can detect the divergence through checking whether the target address of a branch instruction in TM is the same as that in FM. If they are different, a divergence occurs. For interrupt or exception handling, whenever it occurs, a divergence happens. Therefore, we will mainly focus on how to detect the divergences caused by shared data access order and shared page access order.

3.1.1 Shared Data Access Order

As an important factor affecting cycle-accuracy, prior work [6] detects violation in shared data access order through checking whether the loaded values of shared memory between FM and TM are the same. However, based on such an approach, it is difficult to

know where and when the actual thread interleaving violation occurs. Since the loaded value may be affected by a faraway prior store instruction or two store instructions may have written the same value, the order violation information may have already been lost when the loaded value is detected to be violated. To overcome this problem, we use a more accurate method: when FM executes instructions, it records its access order for each shared datum. When TM commits the memory instruction, it checks whether its access order is the same as that of FM. If it is different, a divergence occurs. To achieve this, we design a data structure called *Memory Access Table (MAT)* to efficiently record and check the shared data access order. As shown in Figure 3, MAT is a two dimensional table. The first level is a hashed list of memory addresses and we will call the node as the memory address node; for each address, it maintains a list of memory accesses from different cores and the node in it will be referred as to the memory access node. Each memory access node records which core it comes from and its operation type (i.e., read or write). The shared data access recording and checking mechanism works as follows:

- *Shared data access order recording:* When FM executes a memory instruction, it first checks whether there is a memory address node in MAT for the accessing address. If not, a new address node is created and inserted into the end of memory address list. Otherwise, an access node for this operation is added to the end of the memory access list for its address.

- *Shared data access order checking:* Order violation is checked by TM. Since the memory operations in a memory access list are inserted based on their execution sequence in FM, it is easier for TM to check the violation. When TM commits a memory instruction, it only needs to check whether there is no store node before it in the memory access list. If so, there is no violation and this node is deleted from MAT. When the memory access list becomes empty, the memory node of this address is also deleted from MAT. Otherwise, the violation is reported.

After the order checking, the node of a memory operation will be deleted from MAT. Therefore, the size of MAT should not be larger than the number of memory instructions that FM executes exceeding TM, which makes MAT relatively small and low-overhead. More detail of MAT design could be found in the appendix section B.

3.1.2 Shared Page Access Order

For shared page order, i.e., MMU miss order, it is instinct to still use MAT to check the divergence. However, the functionality of MMU is only simulated by FM. In order to check whether the order violates, TM has also to be able to check whether MMU miss or hit. As a result, the information of entire page table has to be transferred from FM to TM as well, which will lead to more interactions between FM and TM.

To simplify the design, our solution is to avoid this type of divergence. Whenever a MMU miss is encountered, we block FM execution until TM directs it to advance, i.e., until the MMU miss instruction commits in TM. However, this may bring the danger of draining pipeline in TM, i.e., no instructions are provided by FM. Actually, the pipeline draining will never happen due to the *wrong-path FM* mechanism discussed in section 3.2. In TM, when a MMU miss happens, it raises a MMU miss interrupt. As for interrupt handling, it will fetch instructions from the wrong path until the MMU miss instruction commits. Though we block the execution of FM, we will create a wrong-path FM and provide instruction flow to TM, which can avoid pipeline draining.

3.2 Divergence Revision

When a path divergence is detected, we need to revise the simulation to keep the cycle-accuracy. As discussed in [7, 6], we can always deal with the divergence through rolling back FM when a divergence is detected. However, since FM runs ahead, it's difficult to know when to do a checkpoint. Therefore, it will produce large overhead to frequently save the states for checkpoint. Moreover, the rollback strategy can incur double rollback (from right path to wrong path, and again from wrong path to right path) for branch misprediction and interrupt or exception handling. Therefore, besides the rollback strategy, we will also exploit some other optimized strategy: to create a wrong-path FM to execute the wrong path to provide the instruction information to TM for branch misprediction and interrupt or exception handling.

Basic strategy: roll back FM. To roll back FM, we need the correct architecture states at a rollback point, including registers, memory values, MMU states and I/O states. The direct solution is to checkpoint architecture states. For example, SlackSim [5] uses the *fork* system call to do checkpoint. However, it is difficult to know when a checkpoint is required. Moreover, saving all states will produce large overhead. Therefore, we introduce a lightweight mechanism to roll back FM states.

- For registers, which are lightweight inherently, TM maintains a copy of these states for rollback. At initialization, TM reads these values from FM. Then, when each instruction is executed, FM transfers the changed registers to TM. Finally, TM updates the copy when it commits an instruction.

- For memory values, we record the old value before each store instruction in MAT for rollback. When a divergence is detected, we only need to restore these old values from MAT, which greatly reduces memory checkpoint and rollback overhead.

- As discussed in section 3.1, to avoid shared page access order divergence, we block FM when a MMU miss (note that only MMU miss changes MMU states) occurs until TM directs it to advance its execution. Thus, MMU states are always correct in FM and there is no need for rollback.

- As some I/O operations cannot be rolled back, we simply block the execution of FM until TM commits all instructions before it. This mechanism avoids I/O rollback.

Optimized strategy: create a wrong-path FM. One problem for the rollback strategy is that it would incur double rollback for branch misprediction and interrupt or exception handling. It first rolls back FM to execute the wrong path when a branch predicts a wrong PC or an interrupt or exception instruction is in its fetch stage. Then, it again rolls back FM to execute the right path when branch misprediction is finished or interrupt or exception instruction jumps to the trap handling path in the commit stage.

To further optimize the rollback strategy, we create a *wrong-path FM* to execute the wrong path to provide TM with the instruction information. However, no data information is transferred because the wrong path instructions actually are not committed to change architecture states.

When creating a wrong-path FM, we only initialize the register values for it. During wrong-path execution, it uses its own copy of registers. For memory values, it reads from the main FM and MAT, or the values it stores. While for MMU states and I/O states, it reads directly from the main FM since wrong-path instructions are not committed and cannot change these states. When branch misprediction is finished or interrupt/exception jumps to the trap handling path, Transformer terminates the wrong-path FM and TM gets instruction and data flow information from main FM again.

3.3 Architecture-Independent Interface

For the sake of extensibility, we design an architecture-independent interface in Transformer between FM and TM. FM only needs to map the instructions to the interface and TM only needs to read the interface to do detailed simulation. As TM mainly simulates the pipeline dependence and memory behavior, we abstract each instruction as the following architecture-independent information:

- *Pipeline dependence:* Whether an instruction can issue in the pipeline depends on two conditions: 1) whether the functional unit is ready; 2) whether the source operands are ready. The second type of dependence is maintained by registers for computational instructions and by memory address for memory instructions. Thus, we abstract the pipeline dependence of an instruction as three factors: functional unit for this instruction, source/destination register ID, and memory address.

- *Memory information:* For instruction cache simulation, we need the PC address for each instruction. For data cache simulation, we need memory address for memory instructions. Moreover, to detect shared data access order violation, the interface needs to include shared data access order in MAT.

- *Rollback information:* As discussed in section 3.2, we need changed register values for each instruction. Also, we need to save the old memory value for a store instruction.

Such an architecture-independent interface provides Transformer with more flexibility to extend the state-of-the-art FMs or TMs. Since a loosely-coupled design and clear interface, a new FM only needs to map its instruction information to the interface and support the rollback or the block strategy. It does not need to know other details in TM. Moreover, for a new TM, it only needs to read the interface to do detailed simulation and generate necessary checking information to direct rollback.

4. EVALUATION RESULTS

This section evaluates the extensibility and performance of Transformer. As our performance results show that Transformer guarantees the cycle-accuracy compared with GEMS, we omit the results for cycle-accuracy and only present the performance results of Transformer.

4.1 Experimental Setup

Our baseline processor is a 4-core out-of-order SPARC processor with a MOESI cache coherence protocol. Each core has an out-of-order pipeline with yags branch predictor. Detailed configuration is shown in Table 2. We also evaluate an 8-core configuration. Due to the space constraint, the results are shown in the appendix sections. We use SPLASH-2 [17] and PARSEC [3] benchmark suites for evaluation. The benchmarks run on a Solaris 10 operating system with reference input. The baseline simulator is Simics 3.0.31 + GEMS 2.1.1 [9], a widely-used tightly-coupled multicore simulator. Our Transformer prototype is constructed based on GEMS and it is with about 5.5K LOCs changes in total: about 1.5K modified LOCs in GEMS to decouple FM and TM, and about 4K

Table 2: Baseline 4-core OoO SPARC configuration.

4-core SPARC configuration			
Per-core parameters		**Memory Hierarchy Parameters**	
Pipeline width	4		split I/D cache
Functional units	4 Int add/mul, 2 Int div, 2 Load	L1 Cache	each 64KB
	2 Store, 2 Branch, 4 FP add		2-way set associative
	2 FP mul, 2 FP div/sqrt		64B cache lines
Integer FU latencies	1 add, 4 mul, 20 div		2 cycle latency
FPFU latencies	2 default, 4 mul, 12 div, 24 sqrt	L2 Cache	unified 4MB cache
Reorder buffer size	128		8-way set associative
Instruction window size	64		64B cache lines
Load-store queue	64		20 cycle latency
Branch predictor	yags predictor	Memory	200 cycle latency
Data speculation	no	Cache coherence	MOESI_CMP_directory

Table 3: Extension efforts comparison.

Simulator	Combining Work	Extension Efforts
gem5 [4]	GEMS + M5	Dozens of person-years
MARSS [11]	PTLsim + QEMU	1.5 years by a 4-people group
Transformer	GEMS + QEMU	About two person-months

added LOCs to guarantee cycle-accuracy and provide architecture-independent interface between FM and TM. All the experiments are executed on a 6-core Intel I7 980 CPU (3.33GHz, private L1 and L2 cache, 12M shared L3 cache) with 2GB memory.

4.2 Simulation Extensibility

As Transformer exploits the loosely-coupled design and provides an architecture-independent interface (e.g., pipeline dependence, memory information) between FM and TM, the extension becomes much easier. Extending a FM only needs to map the executed instructions into the interface information, which is generally direct available through instrumentation. While extending a TM only needs to make TM read directly from interface, instead of doing detailed decode itself. As a result, the extension efforts of constructing multicore simulators, measured in man-months, can be significantly reduced.

To demonstrate the extensibility of Transformer, we have extended a functional model QEMU [2] into our framework to construct an X86 simulator. The reason we choose QEMU as FM simulator is that it well supports X86 and plenty of full-system features and it is open-sourced and widely-used. We first decode X86 instructions into RISC-like micro-instructions using the decoder from PTLsim [11] and then directly map the micro-instructions into the architecture-independent interface.

This extension only consists about 180 lines of code. The whole extension work is done by a graduate student, who is familiar with QEMU but new to GEMS, in about two months. Compared to multiple person-year efforts cost in prior extension work such as gem5 and MARSS, shown in Table 3, much less efforts are needed to extend novel models in Transformer to construct a new full-system multicore simulator.

4.3 Performance Speedup

Speedup of sequential Transformer. We first evaluate the performance of the sequential Transformer framework, i.e., loosely-coupled Simics and GEMS, against the tightly-coupled baseline simulator Simics and GEMS. As shown in Figure 4, the sequential Transformer achieves about 8.4% speedup on average, which is mainly from simpler interaction: 1) less interactions only for rare path divergence cases (less than 1%), where TM revises the execution; 2) TM no longer simulates redundant functional execution.

Speedup of parallel Transformer. Due to the loosely-coupled design between FM and TM, we can parallelize FM (i.e.,Simics) and TM (i.e., GEMS) with pipeline parallelism. The parallelized

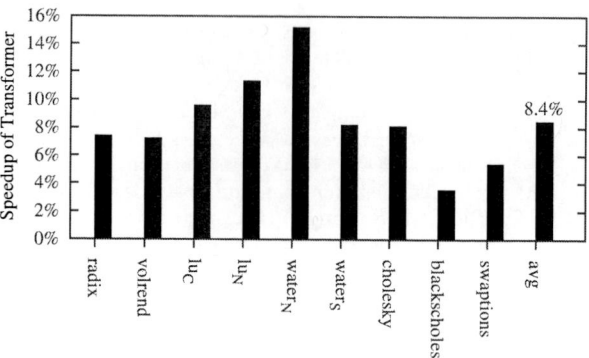

Figure 4: Speedup of sequential Transformer under 4-core configuration.

FM and TM works as two threads: FM thread produces instruction and data flow information to a buffer; TM thread reads the buffer, simulates micro-architecture, and revises FM or creates a wrong-path FM (in the same thread) if necessary.

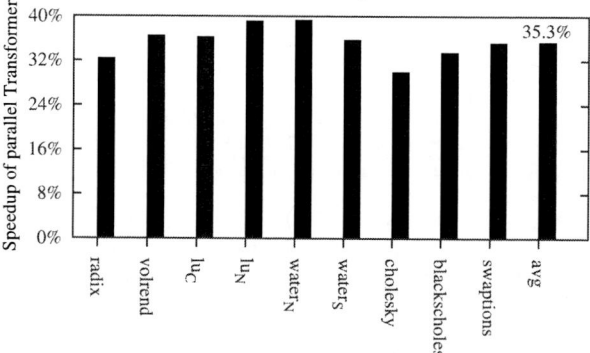

Figure 5: Speedup of parallel Transformer under 4-core configuration.

As the data shown in Figure 5, parallel Transformer achieves about 35.3% speedup against the baseline GEMS simulator. The speedup from parallelizing FM and TM is about 29%. The reason under this speedup is that the parallelization not only distributes the computation into two different cores, but also achieves better instruction and data locality as FM and TM are separated. This speedup is orthogonal to other acceleration techniques such as parallel TM simulation and FPGA-based simulation discussed in section 5. We can combine them to other acceleration techniques to further improve the performance.

5. RELATED WORK

Existing full-system multicore simulators usually exploit a tightly-coupled FM and TM design to achieve cycle-accuracy. A good example is the widely-used Simics + GEMS [9] simulator, which is used in this paper as the baseline simulator. Other mainstream simulators, such as MARSS [11] and gem5 [4], exploit integrated FM and TM design, even more tightly-coupled than Simics + GEMS.

One of the most closest work to Transformer is UT-FAST [7, 6]. UT-FAST exploits a speculative Functional-First simulation design, in which FM (using QEMU) speculatively executes ahead to provide TM (implemented in FPGA) the instruction stream, and TM rolls back FM if branch misprediction. However, UT-FAST only supports single-core simulation and interrupt/exception handling as well as shared page access are not considered. Though it further discusses its extension for multicore simulation in [6], it only considers memory value violation between FM and TM. On one hand, as we discussed in section 2, the factor of shared page access order is not discussed in their paper, which leads to that their solutions cannot achieve cycle-accuracy. On the other hand, they have not implemented their design in a real simulator. While Transformer has done a real implementation with several novel designs such as wrong-path FM and architecture-independent interface, and guarantees cycle-accuracy. To achieve sampling simulation, Cotson [1] also exploits a functional-directed loosely-coupled design, where FM executes ahead for most cases and TM gives feedback to the FM periodically. However, Cotson does not provide cycle-accurate solutions to revise those path divergences and only periodically provides feedbacks to FM.

In contrast, Transformer gives a comprehensive analysis to which factors will affect cycle-accuracy in loosely-coupled design, i.e., branch misprediction, interrupt/exception handling, shared data access order, shared page access order. We further provide several lightweight solutions to detect and revise the simulation instead of simply rolling back FM using heavy-overhead checkpoint mechanism. Moreover, we design an architecture-independent interface between FM and TM to make it more extensible.

Another category of related work is simulation acceleration techniques, including parallel simulation [5, 10], FPGA-based simulation [13, 12], sampling techniques [16, 1], and etc. We use a method orthogonal to the above ones to improve the performance: simple interaction and parallelization between FM and TM.

6. CONCLUSION AND FUTURE WORK

In this paper, we proposed Transformer, an extensible, fast, and cycle-accurate loosely-coupled full-system multicore simulator. We first presented a comprehensive analysis to four factors affecting cycle-accuracy in loosely-coupled design and provided lightweight solutions to detect and revise these divergence factors to ensure cycle-accuracy. Then we further designed an architecture-independent interface between FM and TM, which makes Transformer more flexible to extend state-of-the-art FMs and TMs. As demonstrated, a graduate student only wrote about 180 lines of code and took about two months to extend an X86 functional model (QEMU) based on Transformer. Finally, besides the simple interaction, we further parallelized FM and TM to improve the performance. Experiments showed that it achieved about 8.4% speedup compared to the widely-used tightly-coupled baseline simulator GEMS [9] and 35.3% speedup after parallelizing FM and TM.

There are mainly two directions in our future work. First, we plan to release the source code of Transformer to the public in the near future. Second, we will extend Transformer to support System-on-Chip (SoC) simulation by taking advantage of QEMU and the architecture-independent interface between FM and TM.

7. ACKNOWLEDGMENTS

We thank the anonymous reviewers for their insightful comments. This work was funded by China National Natural Science Foundation under grant numbered 60903015, a joint program between China Ministry of Education and Intel numbered MOE-INTEL-10-04, Key Project of National 863 Program of China under Grant No. 2009AA012201, National 863 Program of China under Grant No. 2012AA010905, Key Project of Major Program of Shanghai Committee of Science and Technology under Grant No. 08dz501600, Opening Project of Architecture Key Laboratory of Institute of Computing Technology in Chinese Academy of Sciences under Grant No. ICT-ARCH2009082009, Fundamental Research Funds for the Central Universities in China and Shanghai

Leading Academic Discipline Project (Project Number: B114).

8. REFERENCES

[1] E. Argollo, A. Falcíon, P. Faraboschi, M. Monchiero, and D. Ortega. Cotson: Infrastructure for full system simulation. *Operating Systems Review*, 43(1):249–261, 2009.

[2] F. Bellard. Qemu, a fast and portable dynamic translator. *USENIX ATC 2005*.

[3] C. Bienia, S. Kumar, J. P. Singh, and K. Li. The parsec benchmark suite: Characterization and architectural implications. *PACT 2008*, pages 72–81.

[4] N. Binkert, B. Beckmann, G. Black, S. K. Reinhardt, A. Saidi, A. Basu, J. Hestness, D. R. Hower, T. Krishna, S. Sardashti, R. Sen, K. Sewell, M. Shoaib, N. Vaish, M. D. Hill, , and D. A. Wood. The gem5 simulator. *Computer Architecture News*, 2011.

[5] J. Chen, L. K. Dabbiru, M. Annavaram, and M. Dubois. Adaptive and speculative slack simulations of cmps on cmps. *Micro 2010*, pages 523–534.

[6] D. Chiou, H. Angepat, N. A. Patil, and D. Sunwoo. Accurate functinal-first multicore simulators. *Computer Architecture Letters*, 8:64–67, 2009.

[7] D. Chiou, D. Sunwoo, J. Kim, N. A. Patil, W. Reinhart, D. E. Johnson, J. Keefe, and H. Angepat. Fpga-accelerated simulation technologies (fast): Fast, full-system, cycle-accurate simulators. *MICRO 2007*, pages 249–261.

[8] P. S. Magnusson, M. Christensson, J. Eskilson, D. Forsgren, G. Hallberg, J. Hogberg, F. Larsson, A. Moestedt, and B. Werner. Simics: A full system simulation platform. *Computer*, 35:50–58, 2002.

[9] C. J. Mauer, M. D. Hill, and D. A. Wood. Full-system timing-first simulation. *SIGMETRICS Perform. Eval. Rev.*, 30:108–116, 2002.

[10] J. E. Miller, H. Kasture, G. Kurian, C. Gruenwald, N. Beckmann, C. Celio, J. Eastep, and A. Agarwal. Graphite: A distributed parallel simulator for multicores. *HPCA 2010*.

[11] A. Patel, F. Afram, S. Chen, and K. Ghose. Marss: a full system simulator for multicore x86 cpus. *DAC 2011*, pages 1050–1055.

[12] M. Pellauer, M. Adlery, M. Kinsy, A. Parashary, and J. Emer. Hasim: Fpga-based high-detail multicore simulation using time-division multiplexing. *HPCA 2011*, pages 406–417.

[13] Z. Tan, A. Waterman, R. Avizienis, Y. Lee, H. Cook, D. Patterson, and K. Asanovic. Ramp gold: An fpga-based architecture simulator for multiprocessors. *DAC 2010*, pages 463–468.

[14] Z. Tan, A. Waterman, H. Cook, S. Bird, K. Asanovic, and D. Patterson. A case for fame: Fpga architecture model execution. *ISCA 2010*, pages 290–301.

[15] Z. Wang, R. Liu, Y. Chen, X. Wu, H. Chen, W. Zhang, and B. Zang. Coremu: a scalable and portable parallel full-system emulator. *PPoPP 2011*, pages 213–222.

[16] T. F. Wenisch, R. E. Wunderlich, M. Ferdman, A. Ailamaki, B. Falsafi, and J. C. Hoe. Simflex: Statistical sampling of computer system simulation. *IEEE Micro*, 26:18–31, 2006.

[17] S. C. Woo, M. Ohara, E. Torrie, J. P. Singh, and A. Gupta. The splash-2 programs: characterization and methodological considerations. *ISCA 1995*, pages 24–36.

[18] H. Zeng, M. Yourst, K. Ghose, and D. Ponomarev. Mptlsim: A simulator for x86 multicore processors. *DAC 2009*, pages 226–231.

APPENDIX

The appendix is organized as follows. Section A gives out the detailed data related to the motivation of the loosely-coupled Transformer framework. The data include the detailed simulation time breakdown of current tightly-coupled simulator design and detailed rate of the factors that would diverge the execution of FM and TM. Section B discusses detailed design of MAT, which is used to efficiently record and check shared data access order violation as discussed in section 3.1. Finally, to further validate the performance of Transformer, section C demonstrates the speedup of sequential and parallel Transformer under the 8-core configuration.

A. MOTIVATION OF TRANSFORMER

This section illustrates detailed data to support the motivation of the loosely-coupled Transformer framework. All these data are evaluated for each benchmark under two different configurations: a 4-core configuration shown in section 4.1 and an 8-core configuration shown in section A.1. We first give out the breakdown of the simulation time in current tightly-coupled simulator design to support that complex interaction produces additional overhead and FM occupies a proportion of the whole execution time that cannot be ignored. Then we demonstrate the detailed rate of four FM and TM path divergence factors, which support that divergences rarely happen and thus loosely-coupled design could achieve simpler interaction.

A.1 8-core Configuration

This section gives the detailed 8-core configuration, which is used to further validate our evaluation results. The detailed configuration is shown in Table 4. Different to the 4-core configuration, the L2 cache is 8M size with 25 cycles latency and the pipeline is simpler: reorder buffer, instruction window and load-store queue are of half size.

Table 4: 8-core OoO SPARC configuration.

8-core SPARC configuration			
Per-core parameters		**Memory Hierarchy Parameters**	
Pipeline width	4		split I/D cache
Functional units	4 Int add/mul, 2 Int div, 2 Load	L1 Cache	each 64KB
	2 Store, 2 Branch, 4 FP add		2-way set associative
	2 FP mul, 2 FP div/sqrt		64B cache lines
Integer FU latencies	1 add, 4 mul, 20 div		2 cycle latency
FPFU latencies	2 default, 4 mul, 12 div, 24 sqrt		unified 8MB cache
Reorder buffer size	64	L2 Cache	8-way set associative
Instruction window siz	32		64B cache lines
Load-store queue	32		25 cycle latency
Branch predictor	yags predictor	Memory	200 cycle latency
Data speculation	no	Cache coherence	MOESI_CMP_directory

A.2 Simulation Time Breakdown

This section evaluates the detailed simulation time breakdown of current tightly-coupled simulator design for each benchmark under 4-core and 8-core configuration. As the data shown in Figure 6, to support TM, the proportion of FM occupies about 10% of the whole time on average for both 4-core and 8-core configuration, which cannot be ignored any more. For interaction time, it occupies about 26% under 4-core configuration and 17% under 8-core configuration on average. Though the interaction percent decreases for 8-core configuration as TM occupies more, the percent is still significant to slowdown the whole performance.

A.3 Rate of Path Divergence Factors

First we illustrate the rate of the four factors for path divergence to validate that they actually occur rarely, thus motivating our loosely-coupled Transformer design with simple interaction.

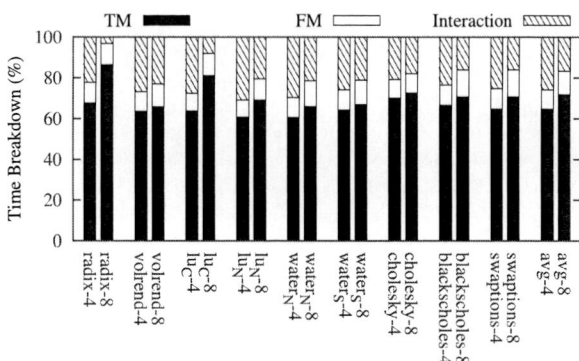

Figure 6: Detailed simulation time breakdown.

Then we further show the I/O rate which involves FM blocking in the "roll back FM" path correcting strategy to validate that it indeed has little influence. All these rate values are shown for each benchmark under both 4-core and 8-core configuration.

Branch Misprediction Rate.

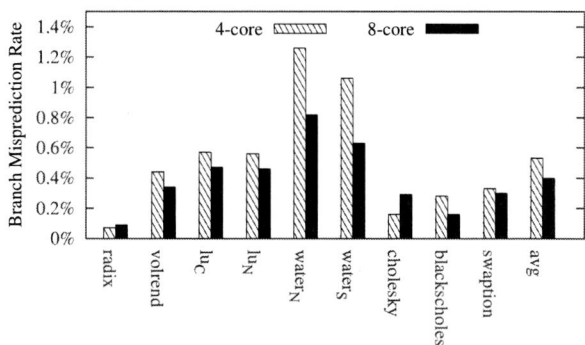

Figure 7: Branch misprediction rate.

Figure 7 shows the branch misprediction rate for each benchmark. As the data shown, the rate of the most occurring branch misprediction factor is only about 0.53% and 0.40% on average under 4-core and 8-core configuration respectively.

Interrupt/Exception Rate.

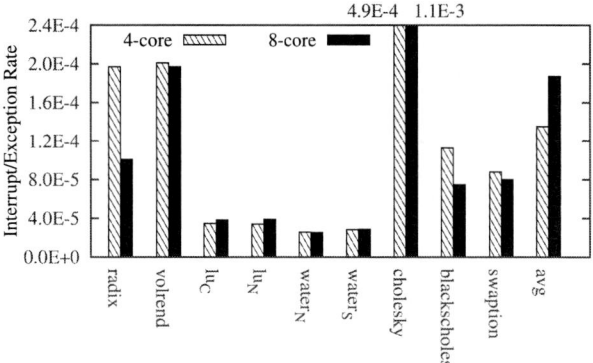

Figure 8: Interrupt/exception rate.

Figure 8 shows the total interrupt and exception rate for each benchmark. As the data shown, the average rate of interrupt and exception is only about 1.3E-4 and 1.9E-4 under 4-core and 8-core configuration respectively. Even for the benchmark (cholesky) with most occurring proportion, it's less than 0.2%.

112

Shared Data Access Order Violation Rate.

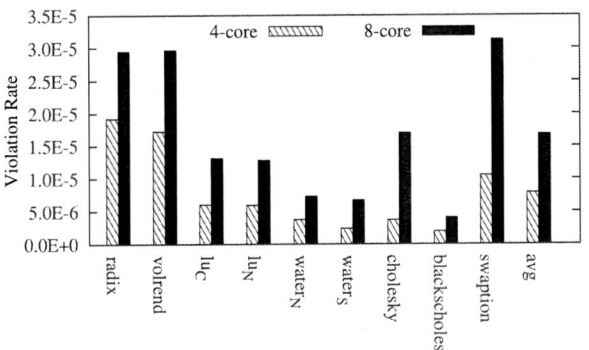

Figure 9: Shared data access order violation rate.

Figure 9 shows the shared data access order rate for each benchmark. As the data shown, the average rate of shared data access order violation is only about 7.8E-5 and 1.7E-5 under 4-core and 8-core configuration respectively.

MMU Miss Rate.

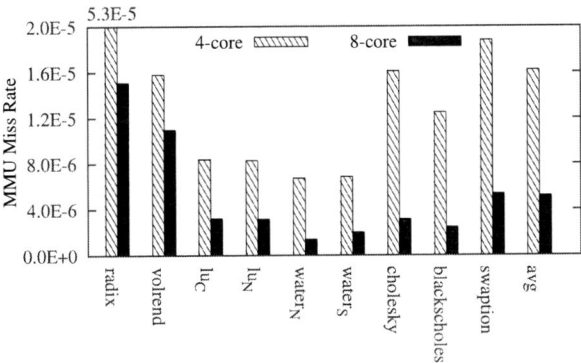

Figure 10: MMU miss rate.

Figure 10 shows MMU miss rate for each benchmark. As the data shown, the average rate of MMU miss is only about 1.6E-5 and 5.2E-6 under 4-core and 8-core configuration respectively. Even for the benchmark (radix) with most occurring proportion, it's only about 5.3E-5.

I/O Rate.

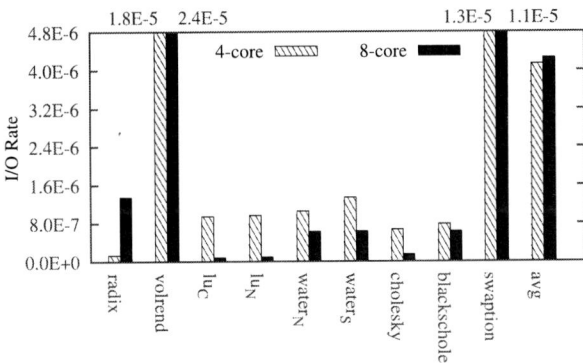

Figure 11: I/O rate.

Figure 11 shows I/O operation rate for each benchmark. As shown, the average rate of I/O operation is only about 4.1E-6 and 4.3E-6 under 4-core and 8-core configuration respectively. Even for the benchmarks (volrend and swaption) with more I/O opera-

tions, the percent is less than 2.5E-5.

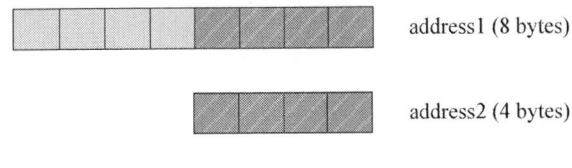

address2 = address1 + 4

Figure 12: Interleaved cases with different memory address.

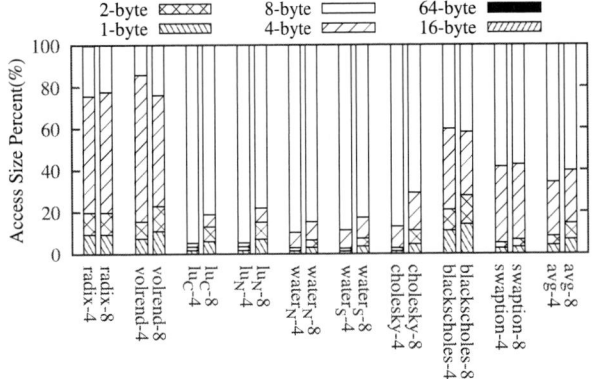

Figure 13: Proportion of each address size.

B. MAT DESIGN

There are two more detailed issues in MAT design, which is used to efficiently record and check shared data access order violation. First, we need to detect access order violation for shared memory access with different addresses. Second, to reduce memory access address list search time in MAT, we hash the memory address.

First, memory accesses with different addresses might access interleaved shared data. As shown in Figure 12, addr1 (access size: 8 bytes) and addr2 (access size: 4 bytes) access 4 bytes shared data. To detect this interleaving order violation, we can divide each memory access into several memory accesses where each accesses only one byte memory. However, this solution is too time-consuming for recording and checking.

Instead, we find most RISC ISAs align memory address (for CISC ISAs, we can divide an unaligned address into two aligned addresses), i.e., n-byte access address is n-byte aligned and it provides us with another opportunity. We have profiled the memory access address and find that more than 99.9% of memory access size is smaller than 8 bytes. The data are shown in Figure 13. Therefore, using 8-byte aligned address can fix in most addresses to avoid dividing the address.

- For memory addresses accessing not larger than 8-byte data, we fixed it into an 8-byte aligned memory address slot and using a 8-bit bitmap to record which bytes it accesses in the corresponding 8 bytes. By doing so, we define shared memory access, i.e., interleaved accesses, as two memory accesses 1) from different cores, 2) fixed into the same 8-byte address slot, and 3) the *and* operation result for two bitmaps does not equal zero, i.e., two addresses are interleaved.
- For memory addresses accessing larger than 8-byte data (e.g., 16-byte, or 64-byte), we divide the address to several 8-byte aligned addresses.

Using this solution, we can efficiently detect all interleaved

113

shared data accesses and detect whether there is order violation.

The second issue is to use hash to reduce memory access address list search time in MAT. The memory address is hashed (e.g., addr mod m) to be an entry of m-entry address array. If two addresses are hashed into the same array entry, they are maintained using a list.

Using these two techniques, to record or check/delete a memory access node, we first map the address into one or more 8-byte aligned addresses. Then for each address, we find the hashed address array entry and search the corresponding address list (much smaller than the design without hash) to find the address in MAT. Finally we record or check the memory access node in the found memory access node list.

C. SPEEDUP UNDER 8-CORE CONFIGU-RATION

This section illustrates the speedup of sequential and parallel Transformer against the baseline GEMS simulator under the 8-core configuration. The speedup data are shown in Figure 14 and Figure 15 respectively. As the data shown, the average speedup of sequential Transformer under 8-core configuration is about 7.0% while the average speedup of parallel Transformer is about 29.7%. They are smaller that of 4-core configuration because TM under the 8-core configuration occupies more proportion of the execution time, as discussed in section A.2.

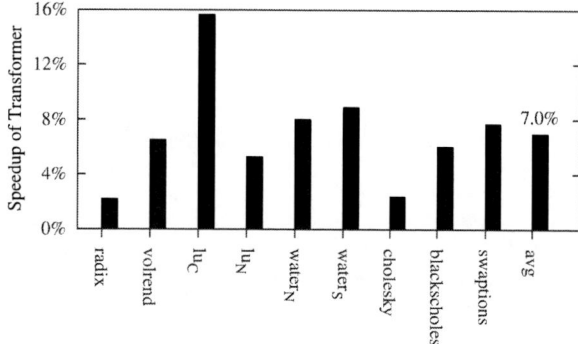

Figure 14: Speedup of sequential Transformer under 8-core configuration.

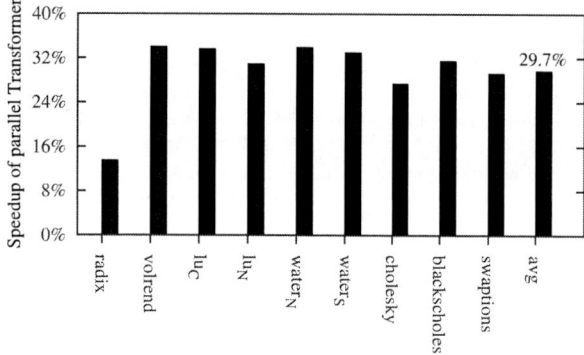

Figure 15: Speedup of parallel Transformer under 8-core configuration.

114

SAGA: SystemC Acceleration on GPU Architectures *

Sara Vinco
Dip. Informatica
Università di Verona, Italy
sara.vinco@univr.it

Debapriya Chatterjee
EECS Department
University of Michigan, USA
dchatt@umich.edu

Valeria Bertacco
EECS Department
University of Michigan, USA
valeria@umich.edu

Franco Fummi
Dip. Informatica
Università di Verona, Italy
franco.fummi@univr.it

ABSTRACT

SystemC is a widespread language for HW/SW system simulation and design exploration, and thus a key development platform in embedded system design. However, the growing complexity of SoC designs is having an impact on simulation performance, leading to limited SoC exploration potential, which in turns affects development and verification schedules and time-to-market for new designs. Previous efforts have attempted to parallelize SystemC simulation, targeting both multiprocessors and GPUs. However, for practical designs, those approaches fall far short of satisfactory performance. This paper proposes *SAGA*, a novel simulation approach that fully exploits the intrinsic parallelism of RTL SystemC descriptions, targeting GPU platforms. By limiting synchronization events with ad-hoc static scheduling and separate independent dataflows, we shows that we can simulate complex SystemC descriptions up to 16 times faster than traditional simulators.

Categories and Subject Descriptors

C.3 [**Special-Purpose and Application-Based Systems**]: Realtime and embedded systems

General Terms

Design, Performance

Keywords

Parallel SystemC, CUDA simulation acceleration

1. INTRODUCTION

Design simulation has traditionally been a key technique to validate digital systems and to conduct early performance and constraints evaluations. However, the increasing complexity of modern designs has been pushing the scalability limits of this technology: as of today its poor performance on complex systems has heavy impacts on the development timeline and ultimately on a system's time-to-market [2].

---

*This work has been partially supported by EU project FP7-ICT-2011-7-288166 (TOUCHMORE) and the Gigascale Systems Research Center

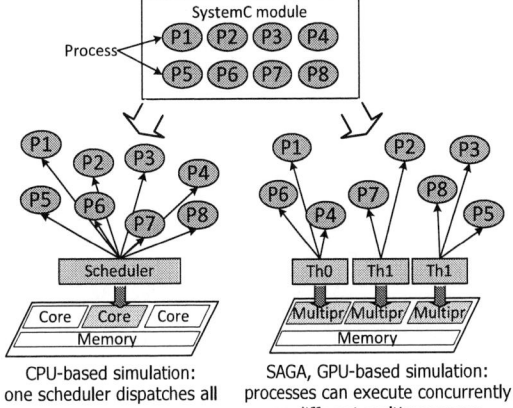

Figure 1: Methodology overview.

One of the most common languages for modeling many digital designs, and particularly embedded systems, is SystemC [8]. SystemC extends C/C++ with libraries to describe HW constructs. It is widely deployed in early-stage analyses and design-space explorations. However, its simulation performance is fairly slow, typically 10x slower than other RTL languages' simulations [2]. To make things worse, the most common SystemC simulation kernel (OSCI) uses application-level threading (co-operative threads), thus it is intrinsically sequential because the operating system cannot dispatch co-operative threads to different processing elements. When simulating transaction-level models (TLMs) these limitations do not have a major impact because the scheduler intervenes rarely and does not introduce heavy overhead. In contrast, RTL simulation requires frequent scheduler operations, leading to a heavy performance impact.

Several works in the literature have attempted to optimize and parallelize SystemC simulation, targeting heterogeneous architectures in order to reduce synchronization overheads and improve performance [1, 4, 10, 9]. Among them, the most promising direction targets GP-GPUs, highly parallel architectures designed for graphical applications and used in a wide range of scientific applications. Early solutions available in this space, such as [6], forego many performance benefits available when simulating SystemC designs on GP-GPU platforms. Looking ahead, a successful GP-GPU based solution for RTL SystemC simulation would bring additional benefits in integrated CPU-GPU architectures [5]. Indeed, the GPU could simulate embedded system's hardware while the CPU would remain available to execute embedded software applications, providing fast simulation of the entire system as a result.

Contributions. This paper proposes *SAGA*, a novel approach for concurrent SystemC simulation that leverages the massive parallelism available on GP-GPUs. Figure 1 overviews our approach

115

and compares it to a traditional SystemC simulation flow. The main contributions of our work are:

- A new concurrent simulation model for SystemC that exploits static scheduling to eliminate the need of frequent synchronization.
- A novel partitioning technique to carve independent data-flows; these are then mapped to distinct threads and multiprocessors to achieve concurrency in the execution.

We show the effectiveness of *SAGA* by applying it to a set of industrial SystemC designs and comparing its performance against that of simulating on chip multiprocessors (CMPs) and against that of previous GPU-based solutions.

The rest of this paper is organized as follows: Section 2 provides background, Section 3 highlights our contributions and the proposed simulation model. Finally, Section 4 presents our experimental evaluation and Section 5 concludes the presentation.

2. RELATED BACKGROUND

This section overviews a typical SystemC simulation flow and the CUDA programming model and architecture. It also discusses state-of-the-art solutions in concurrent SystemC simulation.

2.1 SystemC simulation

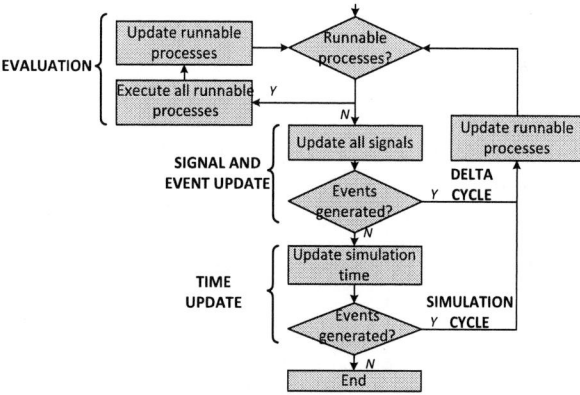

Figure 2: Traditional SystemC simulator scheduler.

SystemC uses an event-based architecture, where a centralized scheduler controls the execution of processes based on events (synchronizations, time notifications or signal value changes). Figure 2 depicts the execution flow of a typical SystemC simulator kernel. The flow is iterated until no event is left to be processed, indicating the end of the simulation. A *simulation cycle* completes at the end of each iteration through the complete flow. Within each cycle, there is first an *evaluation phase*, during which all runnable processes are executed. Signals are updated at the end of the execution of each process. If a signal value change occurs, all processes sensitive to that signal are added to the runnable queue (this is called *signal and event update phase*). Finally, during the *time update phase*, the time of the next simulation cycle is determined by setting it to the earliest of (i) the time at which the simulation ends, (ii) the next time at which an event occurs, or (iii) the next time at which a process is scheduled to resume. If simulation time is not increased, the next simulation cycle will be a delta cycle. When no new event is fired, simulation ends.

The scheduler in a SystemC simulator coordinates the activation of all processes and manages both delta and simulation cycles. Because of this centralized approach, traditional SystemC simulators cannot take advantage of the concurrency of modern CMPs.

2.2 GP-GPU programming through CUDA

NVIDIA's Compute Unified Device Architecture (CUDA) [7] has been proposed to facilitate GP-GPU programming with a general purpose interface. In the CUDA execution model, the GPU is

a co-processor capable of executing many threads in parallel, following the single instruction multiple data (SIMT) model of execution. A data parallel computation process, known as a kernel, can be offloaded to the GPU for execution. The collection of threads represented by a kernel is divided into a grid of thread-blocks.

The CUDA architecture (Figure 3) consists of a number of multiprocessors contained in a single GPU chip. Multiprocessors are responsible for the execution of the thread-blocks that can be mapped to each of them, as dictated by resource limits. Each multiprocessor is comprised of multiple stream processors that have common instruction fetch and support a large number of concurrent threads. Since all resident threads in a multiprocessor execute on a fixed number of stream processors with a common instruction fetch unit, each thread-block executes groups of threads at a time (known as a *warp*) in a time-multiplexed fashion, with frequent context-switches from one warp to another. Because of the shared fetch unit, execution path divergence between threads of a same warp is detrimental to performance as only one branch path can be executed at a time. Thus, if threads in a same multiprocessors must execute different code paths, the least penalizing solution is to map them to different warps.

Each multiprocessor has access to low latency scratchpad memory, divided between local registers and shared memory. All multiprocessors also have access to a region of global memory called device memory, which has higher access latency. Communication with the host CPU's main memory is achieved by means of direct memory access (DMA) transfers. Thus, it is important to keep communication between the host and the GPU to the bare minimum.

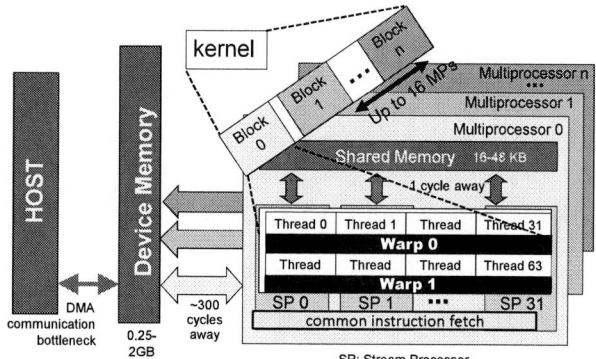

Figure 3: NVIDIA CUDA architecture.

2.3 Concurrent SystemC simulation

Several works in the literature propose to take advantage of the inherent parallelism of SystemC processes to speedup simulation [1, 4, 10, 9]. In SystemC, the order of process execution within a delta cycle does not affect the simulation's output since the simulator presents the same system's status to all those processes. Thus, processes that are activated within the same delta cycle can be executed in parallel, either by using multiple threads or by designing a distributed scheduler. For instance, in [4] SystemC processes are executed as distinct threads on multiple CPUs. Simulation relies on a simulation platform (ArchSim) that introduces heavy overhead, thus making this approach ineffective. In [1], each processing node includes a copy of the scheduler and it simulates a subset of the application modules. All scheduler's copies must synchronize after each delta cycle to update the value of shared signals and of simulation time, thus generating many synchronization events among the separate processors. A different approach is proposed in [9], which transforms the modules' structure. The methodology analyzes SystemC modules and it identifies those blocks within processes that can be executed within one simulator's phase and can be scheduled according to their data and control dependencies. All these solutions rely on code modifications or introduce overhead, as they rely on an existing simulator [1, 10].

A different approach is proposed by the authors of [6], who also target the massive parallelism offered by today's GP-GPUs. In their solution, independent SystemC processes are mapped into parallel threads that synchronize at each iteration of a delta cycle (Figure 2), through a barrier synchronization, to maintain the correct producer-consumer relation among threads. Since typical SystemC processes contain few word-level and arithmetic operations, this can lead to more time spent on synchronization than execution.

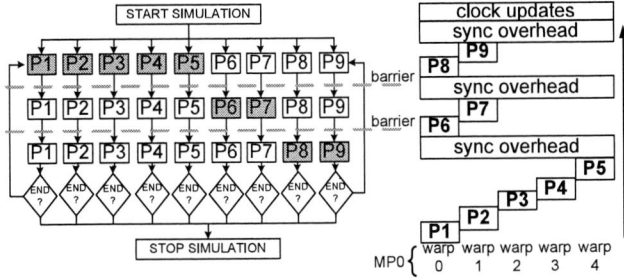

Figure 4: Scheduling example based on [6].

More importantly, the authors of [6] propose to map distinct processes to distinct threads in a same thread-block, so that they can leverage the fast intra-block synchronization mechanisms. However, this is unattainable for most practical SystemC descriptions, since different processes tend not to share the same code. The evaluation in [6] uses unusual designs, such as a 10-stage buffer, that do present lots of inter-process code similarity; however, this is not the common case. Our approach differs from [6] in that we do use distinct multiprocessors to map distinct processes. Then we propose a new scheduler design to minimize the number of synchronizations necessary. As a result we gain concurrent execution even in the common case of processes not sharing any code similarity. Figure 4 illustrate briefly their approach: on the left we show their planned scheduling of processes and on the right the timeline of computation for a same set of non-identical processes on a CUDA platform (*i.e.*, processes are serialized because they do not share the same SystemC code). The evaluation section compares their performance improvement with that of *SAGA*.

3. MAPPING SYSTEMC TO CUDA

Exposing parallelism in a SystemC simulator is not trivial, since the simulation is neither embarrassingly parallel, nor homogeneous. However, some parallelism can be extracted when treating processes active in a same delta cycle as concurrent tasks. *SAGA* exploits this aspect through three steps, as depicted in Figure 5:

1. construction of the *dependency graph* based on the signals read and written by each process, to build a static schedule for the SystemC processes (Section 3.1);
2. partitioning of the static schedule into *parallel dataflows*, that will be executed concurrently in different warps on the CUDA architecture (Section 3.2);
3. levelization of processes within each dataflow based on a *sequential order*. The resulting process blocks will be executed by concurrent thread-blocks in the GP-GPU (Section 3.3).

3.1 Construction of process dependency graph

In our construction, we assume that there is no circular dependency loop between processes and we arrange them in a producer-consumer order based on the I/O direction of their connecting signals. To this end, we build a *process-graph PG = (V;E)* where each process is represented by a vertex V; a directed edge E from vertex v_1 to vertex v_2 represents a process dependency due to a

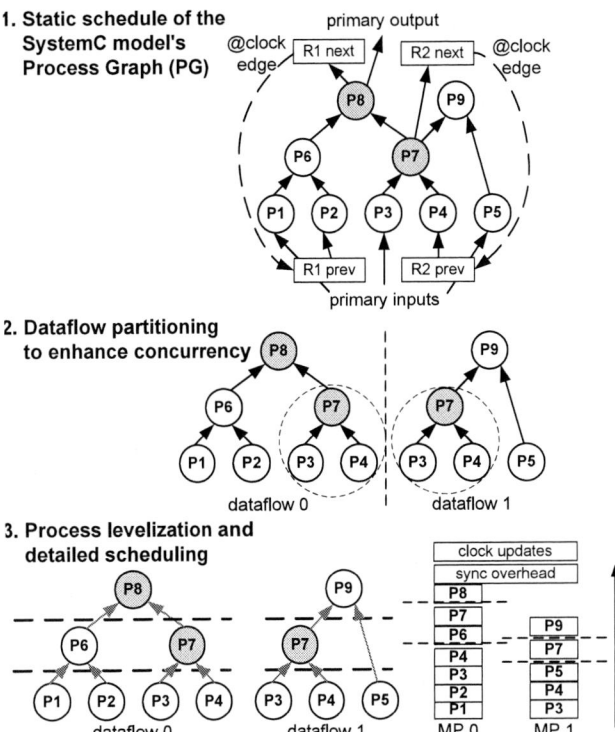

Figure 5: *SAGA* **methodology steps.**

signal generated by v_1 and consumed by v_2. We do not represent synchronous statements in the process-graph, since they create a dependency between present-state values and next-state values through time, while we focus on exploring the concurrency within a same time tick. The PG is a directed acyclic graph (DAG), and thus we can apply a topological sort to it. Processes dependent only on delta events at their primary inputs and at synchronous variables occupy the lowest level; the other levels are established by the edge connections.

Figure 5.1 shows an example of a process graph built for a SystemC module. Nodes in grey represent synchronous processes (*e.g.*, *P8*), while white corresponds to asynchronous ones (*e.g.*, *P6*). Signals *R1* and *R2* are written by synchronous statements, thus they have a current value (*R1 prev* and *R2 prev*) and a future value (*R1 next* and *R2 next*). Their current value will be updated once the dataflow execution has completed (as suggested by the dashed arrows). Steady-state values at the primary output signals and next state values for the synchronous signals can be obtained by executing the processes level-by-level. Because of how delta cycles operate in a traditional simulator, at stable state the PG-based simulator is guaranteed to provide the same results as a traditional one.

Moreover, our construction leveraging static scheduling presents an intrinsic advantage for parallel platforms, since there is no need for a central scheduler to manage events and to activate processes. Note that we can still benefit from the advantages of an event-driven simulation: if we only execute a process conditionally to a change at its inputs, then we are basically using an event-based approach. This optimization brings upon a 10% performance improvement on average over our baseline solution.

3.2 Partitioning into concurrent dataflows

There are several ways of partitioning the process graph obtained in the previous section: we select one based on the constraints of our target GPU platform. A straightforward approach would map different processes to distinct threads, one thread per process. We can then execute all processes in a same schedule level concurrently. However, this could lead to some of the same shortcomings

117

of the previous work discussed in Section 2.3, if the processes do not share the same source code. Thus, to extract as much parallelism as possible, we devise a novel scheme in which the static schedule of the process graph is partitioned into multiple independent dataflows. These are then mapped to distinct multiprocessors for concurrent execution since those have distinct fetch units. The dataflows we create are segments of the scheduled process graph that can be executed independently. When necessary we may replicate some portions of the process graph to attain independence among dataflows.

The partitioning algorithm is outlined Figure 6. First, we select processes in the static schedule that do not activate any other process asynchronously, that is, they are root processes in the PG graph (line 4) (*e.g.*, *P8* and *P9* in Figure 5.1). For each of these nodes, we select their fan-in cone in the PG (line 5–12), as illustrated in the second step in Figure 5. Processes that are common to multiple cones are replicated (*e.g.*, the processes in the dashed circles in the Figure) so to make the cones independent of each other and to enable concurrent execution. Even though we need to replicate some portions of the PG, thus increasing the amount of simulation required, replication ultimately eliminates the need of communicating values among dataflows, thus leading to an important reduction in communication cost through device memory.

```
1:  list queue;
2:  for each node n ∈ V do
3:      list current_dataflow;
4:      if n has no exiting edges then
5:          queue.add(n);
6:          while queue is not empty do
7:              Node current_node = queue.pop();
8:              current_dataflow.add(current_node);
9:              for all incoming edges edge of current_node do
10:                 queue.add(edge.getSource());
11:             end for
12:         end while
13:     end if
14: end for
15: dataflow_list.add(current_dataflow);
```

Figure 6: Dataflow partitioning algorithm for *SAGA*'s step 2.

3.3 Parallel execution in CUDA

The cones built in the previous step are process dependency trees, that must be executed level-by-level to respect the internal dependency constraints. Thus, for each dataflow obtained in the previous step, we now generate a total serial order of processes that satisfies the level-to-level dependencies.

First of all, we levelize the cones by following the algorithm in Figure 7. If the current node has no incoming edges (and thus it is not activated by any other process in the dataflow), then it belongs to the lowest scheduling level (lines 3–4). Otherwise, the node is scheduled at a level higher than that of all its fan-in processes (line 6-11). This step strengthens the dependency relation between processes (*e.g.*, in the example in Figure 5.3, not only *P3* and *P4* execute before *P7*, but also *P5* does). Then, processes in each dataflow are serialized, starting from the lower levels up to the root processes (processes at the same level can be executed in any sequential order). It is advantageous to create such sequential order for each dataflow, since it eliminates the need of frequent synchronization after each level. An example timeline obtained from this example is shown on the right hand side of Figure 5.3.

At this point *SAGA* generates the CUDA code corresponding to the generated process schedule. A *simulation kernel* manages dataflow execution, and it is constructed by listing all the dataflows and predicating each by a thread-block ID condition, so that only a specific thread-block is responsible for executing a certain dataflow. The body of each individual process is replaced by equivalent CUDA code, which might require translation of SystemC datatypes into native datatypes, as reported in Section 4.1. The simulation

```
1:  for each dataflow dataflow in dataflow_list do
2:      for each node n in dataflow do
3:          if n has no incoming edges then
4:              n.setLevel(0);
5:          else
6:              n.setLevel(-1);
7:          end if
8:      end for
9:      while at least one node has not been assigned a non-negative level
        do
10:         for each node n in dataflow do
11:             if for each incoming edge edge, the source node
                edge.getSource() has a non-negative level then
12:                 for each incoming edge edge of n do
13:                     if n.getLevel() ≤ edge.getSource().getLevel() then
14:                         n.setLevel(edge.getSource().getLevel() + 1);
15:                     end if
16:                 end for
17:             end if
18:         end for
19:     end while
20: end for
```

Figure 7: Dataflow levelization algorithm for *SAGA*'s step 3

kernel alternates execution with a *value-update kernel*, responsible for transferring the next-state values into the corresponding present-state values and performing testbench actions. A simulation cycle is completed by one execution of the simulation kernel followed by one execution of the update kernel.

Since device memory accesses are particularly slow, as indicated in Section 2.2, we allocate only variables written by synchronous processes in global memory, since their value must be persistent among different kernel executions. All other variables can be declared as local variables, and will consequently be mapped to registers with much faster access latency.

4. EXPERIMENTAL EVALUATION

In this section we evaluate the performance of *SAGA*, provide insights on its intermediate data structure and compare it against other state-of-the art solutions in this space. Section 4.1 discusses our experimental setup; Section 4.2 compares *SAGA*'s performance against that of a sequential simulator, while Section 4.3 evaluates the performance of code compilation in *SAGA*. A comparison with other available solutions is provided in Section 4.4. Finally, Section 4.5 provides insights on the scalability of our solution.

4.1 Experimental setup

SAGA considers as input a SystemC design, it transforms it as discussed in Section 3, producing all the CUDA code necessary to run the corresponding simulation on a GPU as output. The code can then be off-loaded to a GPU platform and executed. All experiments discussed below were evaluated on a NVIDIA GTX480 GPU and a Intel quad core i7 operating at 2.8Ghz and running Linux RedHat 5.7. In addition, we leveraged the HIFSuite framework [3] to parse the SystemC code and generate an intermediate data structure that is used by *SAGA* for its internal transformations.

The first task in *SAGA* consists of considering a SystemC description and translating it into the HIFSuite's internal format (HIF) by using the HIFSuite sc2hif tool. The code generated at this point is a tree-structured XML-like representation of the original code, where semantic objects are represented with TAGS.

SAGA then applies a number of pre-processing steps to the HIF description. First it extracts all the processes and builds an initial dependency graph, according to signal dependencies among processes. It then applies the 3-step transformation described in Section 3. At this stage SystemC data types are substituted with native C/C++ data types and all corresponding data structures are built. Finally, *SAGA* generates the code for the kernel functions, and outputs the generated HIF description representing the detailed scheduled dataflows obtained with our algorithm. As a last step, the resulting HIF code is converted into C code by means of the HIFSuite *hif2c* tool. This representation is ready to be compiled for the target CUDA architecture.

Table 1 presents our testbench designs. The designs are part of a complex embedded platform that was developed in the context of a European project together with silicon vendor industry partners:

- ECC is an error correction code device.
- ClockGen, ResGen, Sync and RegCtrl are part of a complex DSPI system. ClockGen is a multiple clock generator. ResGen transforms and outputs the computed results in the specified format. Sync is a specialized synchronization function among a number of components. RegCtrl is a register controller for a set of registers.
- 8b10b is a module performing encoding and decoding byte-wide data according to the 8b/10b protocol.

We evaluated *SAGA* on the individual designs and on two more complex SoC design assemblies: Half Platform, comprising ECC, ClockGen, ResGen and Sync; and Platform integrating all the designs previously discussed. For each design, Table 1 reports the number of processes in the original SystemC description (*Processes (#)*), the lines of code (*SystemC (loc)*) and the number of dataflows extracted (*Dataflows (#)*). Column *Replic. proc. (#)* reports the amount of code replication due to our step 2 (see Section 3) as number of replicated processes and the maximum amount of replication for these processes.

| Design | Processes(#) | | SystemC | Dataflows | Replic. |
	Synch.	Asynch.	(loc)	(#)	proc. (#)
ECC	4	7	582	4	4 / 3
ClockGen	6	15	741	12	7 / 3
ResGen	3	6	478	9	0 / 0
Sync	4	22	641	23	0 / 0
RegCtrl	18	32	2677	43	17 / 8
8b10b	7	30	799	7	9 / 3
Half Platform	18	51	2355	48	11 / 3
Platform	42	112	5643	98	37 / 8

Table 1: Characteristics of the designs.

4.2 Performance

Table 2 compares *SAGA*'s performance with that of a SystemC sequential execution as discussed in Section 2. For each design, Table 2 reports simulation time of the SystemC simulation (Column *SystemC simul. (ms)*) and of the *SAGA*-generated CUDA code (Column *SAGA simul. (ms)*). It then reports their comparative performance in terms of *SAGA*'s speedup over sequential execution (Column *Speedup (x)*). The results show that the *SAGA* simulation is always faster than its corresponding SystemC sequential simulation. However, the speedup is moderate when comparing the small, individual component designs, leading to up to a 3.89 times improvement. Note, however, that even in presence of highly heterogeneous and complex processes *SAGA* achieves a respectable performance improvement. In addition the speedup achieved with the two more complex designs is much higher, ranging from 10 to almost 16x. This result suggests that *SAGA* is a promising solution that can extract even more concurrency from the more complex designs, where there are more processes available, leading to a better utilization of the parallel resources available on the GP-GPU.

The speedup achieved by *SAGA* is bounded by the amount of concurrency that can be extracted from each module and by the amount of computation they require. When both these factors are high, the generated code greatly outperforms sequential SystemC simulation. A low level of parallelism (ECC and ClockGen) or non-intensive computation (ResGen and Sync) lead to lower speedups, due to a heavier contribution of synchronization not balanced by computation, or because the limited concurrency is not offset by its setup overhead.

Our performance results also indicate that the benefits of replication far outweigh the costs. Indeed, as indicated in Table 2, even designs with the heaviest replication maintain a good speedup over

a sequential simulator, since replication reduces communication by reducing the need of synchronization.

Design	SystemC simul. (ms)	*SAGA* simul. (ms)	Speedup (x)
ECC	11.99	5.05	2.37
ClockGen	18.00	7.13	2.52
ResGen	8.97	5.22	1.71
Sync	9.98	5.73	1.74
RegCtrl	41.97	13.05	3.21
8b10b	15.99	4.11	3.89
Half Platform	83.98	8.143	10.31
Platform	228.96	14.34	15.97

Table 2: Performance of *SAGA* vs. sequential simulation.

4.3 Compilation

Table 3 compares the costs of compilation for the target platform between a sequential simulator and our proposed *SAGA* flow. Column *SystemC comp. (ms)* reports the time needed to compile the original SystemC code. Column *Code generation* indicates the time needed to generate the target CUDA code in *SAGA*. For this component we report both the time spent in the execution of the *SAGA* algorithm presented in Section 3 (*SAGA (ms)*) and time required for the intermediate language transformations by the HIFSuite tools (*HIFSuite (ms)*). Finally, we show the time required to compile and generate the code to be off-loaded to the GPU.

The time spent in *SAGA* for code generation is a very small fraction of overall compilation time, which is dominated by the CUDA compiler and the HIFSuite transformations. Moreover, the total compilation time is within a factor of 2 of the compilation time of the sequential SystemC simulator. We expect that in a mature version of our solution, *SAGA* could parse and operate directly on the original SystemC source code, thus eliminating the need of resorting to the HIF intermediate format. Furthermore, since many simulations can be run for each model compilation, the value of *SAGA* does not lie in its compilation performance, but rather on the performance of the simulation generated.

| Design | SystemC comp.(ms) | Code generation | | CUDA comp.(ms) |
		SAGA(ms)	HIFSuite (ms)	
ECC	3,893	44	2,356	3,133
ClockGen	3,321	28	878	2,572
ResGen	2,863	16	864	2,457
Sync	3,027	24	180	2,608
RegCtrl	4,154	56	692	3,232
8b10b	3,354	40	3,284	2,936
Half Platform	965	101	3,850	6,431
Platform	10,960	187	7,428	6,824

Table 3: Comparison of compilation times for the sequential SystemC simulator and *SAGA*.

4.4 Architecture comparison

In order to show the effectiveness of the proposed methodology, we compared the performance of *SAGA* against two other concurrent solutions for SystemC simulation. For this study we report results on only two designs for sake of brevity. However, these two designs are representative of typical behavior and we found that the other designs lead to similar outcomes.

We first considered a concurrent SystemC simulator implementing the *SAGA* approach on a CMP architecture, where each dataflow is mapped to a different `pthread`. Furthermore, we compare the *SAGA* approach against the SCGPsim GPU-target simulation solution [6], which we implemented based on the authors' description, as outlined in Section 2. We report our findings in Table 4, where speedups are normalized to the performance of the sequential simulator.

119

The table indicates that *SAGA* is the fastest solution, providing a speedup of 2 to 4x over [6], and even more over the CMP design. Also note that the other solutions do not provide a performance improvement over the sequential simulation for ECC. Upon further inspection we found that the CMP solution does not achieve good concurrency because distinct processes are mapped to co-operative threads, as discussed in Section 1. SCGPsim's performance is limited because design processes do not share the same code, and thus they are executed sequentially when mapped on a same multiprocessor. We believe that the authors of [6] experienced much higher speedups because they evaluated their solution only on SystemC descriptions where processes had identical code. However, this is a very rare situation for any practical design.

Implementation	ECC		ClockGen	
	Time (ms)	Speedup	Time (ms)	Speedup
SystemC	11.99	1x	18.00	1x
Multiprocessor	94.00	0.13x	20.00	0.9x
SCGPsim [6]	20.08	0.59x	14.77	1.22x
SAGA	5.05	2.37x	7.13	2.52x

Table 4: Performance comparison of *SAGA* vs. a CMP concurrent simulator and the SCGPSim simulator.

In order to evaluate the speedup trends of *SAGA* against the solution in [6], we repeated a portion of functionality described in each process of the two designs, ECC and ClockGen, a number of times and then compared the simulation times achieved by all three solutions on these variants of the two designs. Figure 8 plots our findings; on the X axis we report the number of times that the functionality was repeated, and on the Y axis we indicate the corresponding simulation time. It can be noted from the graphs that *SAGA* outperforms SCGPsim and the sequential simulator even in this artificially larger designs with repeated functionality. However, note that the sequential simulator follows an exponential trend, as expected, while SCGPsim appears to follow a fairly constant trend, although with a higher baseline than *SAGA*.

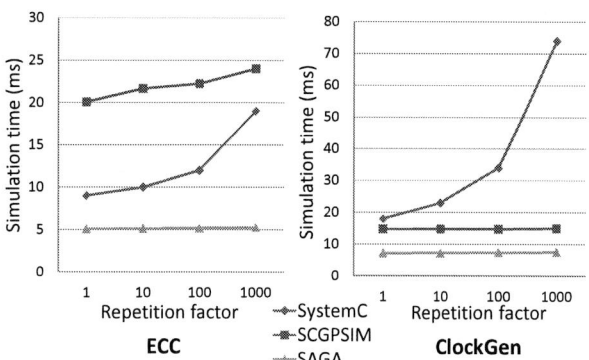

Figure 8: Simulation trends for SCGPsim and *SAGA*.

4.5 Scalability

To estimate the scalability of *SAGA*, we evaluated the GPU resource usage of our solution, so that we could determine when a SystemC design would reach a complexity that would exhaust the resources available on the GPU platform. A required resource in our solution is device memory; however, we only need to store there the values of the synchronous signals in the SystemC design and thus the usage of device memory is negligible, even for the largest designs (less than 1% of the available memory).

All other SystemC signals are stored as local variables and they exclusively require registers. As expected, the demand on registers is thus much more pressing, and these constitute a scarce resource that we need to consider in evaluating the scalability of *SAGA*. To

this end we analyzed the register usage information for each of our testbench designs and used it to determine the maximum amount of dataflow concurrency that can be achieved. Figure 9 plots the fraction of concurrent resources used by our testbeds, based on this analysis. If a design reaches the limit of available concurrent resources, a portion of the computation will be serialized. Note, from Figure 9, that none of our designs reaches this limit, although the Platform testbench, our most complex design, was close, at 80%.

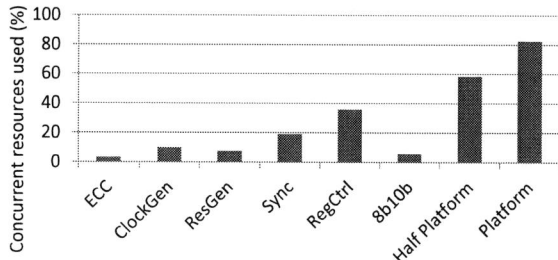

Figure 9: Percentage of available GPU concurrency required by our designs.

5. CONCLUSION

This paper presented *SAGA*, a novel solution for concurrent simulation of SystemC designs on GPU architectures. *SAGA* achieves its goal by extracting independent dataflows from a static schedule of SystemC designs, thus reducing synchronization overheads. As a result, the simulation is more efficient than both a sequential SystemC simulator and other state-of-the-art concurrent approaches. Experimental results show that we achieve the best speedups on complex designs, highlighting the effectiveness of the methodology when targeting complex industrial designs. Future work will focus on developing further optimizations for *SAGA* to further boost its performance and on evaluating the fitness of our solution for other hardware description languages.

6. REFERENCES

[1] P. Combes, E. Caron, F. Desprez, B.Chopard, and J. Zory. Relaxing synchronization in a parallel SystemC kernel. In *Proc. Of ISPA*, 2008.

[2] W. Ecker, V. Esen, L. Schonberg, T. Steininger, M. Velten, and M. Hull. Impact of description language, abstraction layer, and value representation on simulation performance. In *Proc. of DATE*, 2007.

[3] EDALab. *HIFSuite*, 2011. http://www.hifsuite.com/.

[4] P. Ezudheen, P. Chandran, J. Chandra, B. Simon, and D. Ravi. Parallelizing SystemC kernel for fast hardware simulation on SMP machines. In *Proc. of PADS*, 2009.

[5] L. Gwennap. Sandy bridge spans generations. *Microprocessor Report (www.MPRonline.com)*, September 2010.

[6] M. Nanjundappa, H. D. Patel, B. A. Jose, and S. K. Shukla. SCGPSim: A fast SystemC simulator on GPUs. *Proc. of ASP-DAC*, 2010.

[7] NVIDIA. *NVIDIA CUDA Compute Unified Device Architecture - Programming Guide*, 2008. http://developer.download.nvidia.com.

[8] Open SystemC Initiative. *SystemC Language Reference*, 2011. http://www.systemc.org/downloads/standards.

[9] N. Saviou, S. Shukla, and R. Gupta. *Design for Synthesis, Transform for Simulation: Automatic Transformation of Threading Structures in High Level System Models*. University of California at Irvine, 2008. Technical Report TR-01-58.

[10] H. Ziyu, Q. Lei, L. Hongliang, X. Xianghui, and Z. Kun. A parallel SystemC environment: ArchSC. In *Proc. of ICPADS*, 2009.

Synchronization for Hybrid MPSoC Full-System Simulation

Luis Gabriel Murillo, Juan Eusse, Jovana Jovic,
Sergey Yakoushkin, Rainer Leupers and Gerd Ascheid
Institute for Communication Technologies and Embedded Systems
RWTH Aachen University, Germany
{murillo,eusse,jovic,yakoushkin,leupers,ascheid}@ice.rwth-aachen.de

ABSTRACT

Full-system simulators are essential to enable early software development and increase the MPSoC programming productivity, however, their speed is limited by the speed of processor models. Although hybrid processor simulators provide native execution speed and target architecture visibility, their use for modern multi-core OSs and parallel software is restricted due to dynamic temporal and state decoupling side effects. This work analyzes the decoupling effects caused by hybridization and presents a novel synchronization technique which enables full-system hybrid simulation for modern MPSoC software. Experimental results show speed-ups from 2x to 45x over instruction-accurate simulation while still attaining functional correctness.

Categories and Subject Descriptors

I.6.7 [**Simulation and Modeling**]: Simulation Support Systems—*Environments*

General Terms

Design

Keywords

MPSoC, Virtual Platforms, Hybrid Simulation, HySim, Synchronization, Temporal Decoupling

1. INTRODUCTION

The increasing complexity of modern electronic systems and the spread of multi-processor systems-on-chip (MPSoCs) have demanded a drastic design paradigm shift. Platform architectures and software are designed, developed and evaluated from a system perspective, focusing on components interaction, synchronization and communication. This system perspective is of utmost importance to achieve not only better performance-power ratio, but also to support cutting-edge features required by new products.

System simulation plays an important role to support electronic design, as it provides the means to estimate performance, test functionality and guide the development process. Naturally, the complexity of modern systems is reflected in the complexity of their simulators. Current MP-SoC simulators comprise several models of different processing elements (PEs), hardware peripherals, accelerators, and communication infrastructures. Specialized simulation frameworks, such as SystemC-TLM2 [4], OVPSim [1], QEMU [2], Synopsys Virtualizer [3] and Simics [6], are widely used to create models at different levels of abstraction and assemble virtual systems. These full software models of hardware systems, also known as Virtual Platforms (VPs), have gained tremendous popularity among designers because of their early availability.

Accuracy and simulation speed can be traded-off in order to create VPs for different use scenarios (e.g. architectural design, software verification). This is typically done by choosing an appropriate level of abstraction when modeling devices and the communication among them. Models of PEs can be found in the form of Instruction Set Simulators (ISSs), either *cycle-accurate* (ISS-CA) or *instruction-accurate* (ISS-IA), and as *host-compiled* simulators. However, VPs are often many orders of magnitude slower than the real systems they represent and further increasing their speed is a difficult challenge.

Although VPs are composed of several models, PEs remain as a major simulator bottleneck due to their complexity. Several techniques, such as dynamic binary translation [17] and just-in-time compiled simulation [13], have been used to increase ISS speed. However, ISSs remain slow if compared to more abstract processor models. Conversely, *host-compiled* or *native* simulators [9, 14, 15, 16], corresponding to one level of abstraction above ISS-IA, provide faster simulation at the cost of accuracy and a limited view of the underlying hardware. The latter limits the applicability of host-compiled simulators for software development, validation and debugging in a practical context.

1.1 Hybrid Full-System Simulation

Hybrid simulation approaches try to bridge the gap between two different abstraction levels and have been used before with different objectives [11, 12]. *HySim* [10] is a hybrid processor simulator that allows bidirectional dynamic switching between a target ISS-IA and a host-compiled simulator, while keeping the *processor-centric* state synchronized between both simulation modes. The target ISS executes processor specific functions, whereas the host-compiled simulator executes target-independent parts of the application.

The host-compiled part of a hybrid ISS runs outside the context of the simulation kernel, hence leading the PE into a future state with regard to the rest of the system. This decoupling, namely *hybridization-introduced decoupling*, is highly dynamic and might decouple PEs long enough to disturb behavior of systems with interrupt-driven functionality, multi-threading, multi-core synchronization, preemptive scheduling, dynamic task migration and other common features of modern OSs and concurrent run-time environments. In consequence, the use of HySim and other hybrid ISS frameworks is limited to MPSoCs where reactive behavior and inter-processor activity do not define the system correctness and usability.

This paper introduces and describes the aforementioned concept of hybridization-introduced decoupling, and proposes a new synchronization mechanism to attain system coherency when using hybrid ISSs, thus allowing proper interaction among PEs and preserving the functional correctness of the system. The mechanism was used to define an improved hybrid processor simulation architecture and extend the HySim framework in order to enable hybridization, software-centric synchronization, traditional temporal decoupling and reactive behavior to coexist in the simulator.

Paper Outline. The remainder of this paper is organized as follows. In Section 2, the necessary background to better understand the concepts herewith discussed is presented. Section 3 introduces the concept of hybridization-introduced decoupling, and proposes a technique to temporally synchronize hybrid ISSs with the simulation kernel and the rest of the system. This is complemented in Section 4 with an approach to keep a behaviorally correct system-wide state in hybrid systems which run parallel applications and utilize software synchronization functions. Section 5 presents results of our synchronization techniques for hybrid ISSs when applied to MPSoC VPs created in two major commercial frameworks, namely Synopsys Virtualizer and Simics. Results cover complex cases of parallel applications with distributed scheduling mechanisms and complete software stacks. Finally, Section 6 presents conclusions of this work.

2. BACKGROUND

2.1 ISS vs. Host-compiled

ISS and host-compiled simulators cover different use cases of VPs. An ISS emulates execution of a cross-compiled binary, including a one-to-one match of instruction sets and dynamic effects at the instruction level. ISSs of VLIWs, DSPs and ASIPs model also non traditional resources common to these devices, such as irregular register architectures or extended memory interfaces.

On the other hand, host-compiled approaches model computation at the source code level and execute it directly on the host machine. To take into account dynamic effects, parts of the hardware-specific software layers (e.g. HAL, context switching) are usually abstracted away thus losing the visibility of the target architecture, such as in [14, 15].

In a practical context, modifications for host-compiled simulation (e.g. intrusive instrumentation of software and/or simulation models [9]) might be prohibited due to the presence of third-party IPs, legacy code, or even limitations to alter previously verified systems. For this reason, software development often falls back to slow, traditional ISS simulation. Another reason is the necessity to port or de-

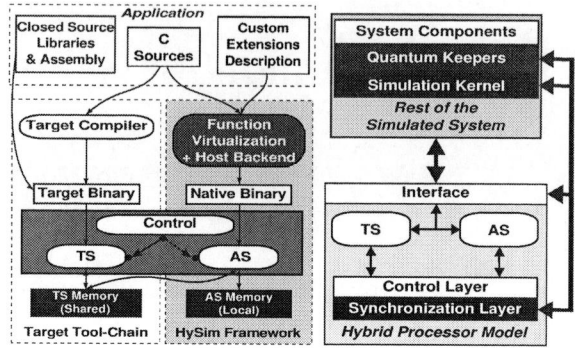

(a) Workflow   (b) New core structure

Figure 1: The HySim Framework.

velop OSs, evaluate target-optimized software and run low-level code and middleware. These are critical tasks in almost all signal processing and multi-media software used to power today's telecommunication and consumer electronic devices.

2.2 HySim - A Hybrid Simulation Framework

HySim combines the advantages of a fast native simulator and a target-specific ISS. It was designed mainly to support embedded software development and debugging. The key idea behind it is partitioned execution at function level, which allows bidirectional run-time switching between a target ISS (TS) and a host-compiled abstract simulator (AS).

The major components of the HySim workflow, shown in Figure 1a, are a compiler-like function virtualizer and a control module which switches simulation modes. Function virtualization analyzes application C source code and selects functions with only target-independent features (i.e. *virtualizable*), which can be executed in AS. The selected functions are extended and instrumented using special synchronization APIs which guarantee processor-centric consistency (i.e. registers and memory view) between TS and AS modes [8, 10]. The remaining functions in an application (i.e. *non-virtualizable*) are executed in TS. This set comprises functions with inline (or written in) assembly, *state-altering* functions (e.g. *fopen*), functions with direct stack manipulation (e.g. *setjmp* and *longjmp*), function pointers, and functions without definition (e.g. closed source libraries). The user, using partitioning algorithms for ultrafast forward breakpoints or manually mapping functions to AS, has the final choice on how to run the simulation. The AS mode relies on dynamic software performance estimation solutions to give a notion of simulated time, which were also introduced in [7]. Although HySim requires the application sources as input, it can be successfully applied to a wide range of practical scenarios because it does not modify the target binary and allows closed source libraries, target-dependent code, and drivers in the application.

2.3 Temporally Decoupled Simulation

Based on the observation that components in a system do not interact with the surrounding environment frequently, some parts might be allowed to run ahead of the rest of the system without consequences. This concept, known as temporal decoupling, has been used in simulation to avoid unnecessary kernel synchronization points and context switches, which cause a significant overhead.

In practice, every PE is assigned a *quantum*, either statically or dynamically, that defines how many simulation steps (i.e. instructions or cycles) it can advance without synchronizing. A small quantum allows to handle external events more accurately but at slow simulation speed, while a big quantum achieves fast speed at the cost of corrupting the timing behavior of the system.

3. HYBRIDIZATION-INTRODUCED DECOUPLING

This section analyzes the decoupling effects in hybrid simulators by (i) introducing the concept of hybridization-introduced decoupling and (ii) proposing a mechanism to ensure proper time-driven behavior in MPSoC system simulators.

In HySim, every function mapped to AS is executed in synchronous mode with the simulated application. Host-compiled execution is incapable of affecting directly the simulated time, neither globally nor locally. Thus, the execution of a virtualized function is performed in *zero time* from the simulator's perspective. Software performance estimation techniques help to obtain timing values for the functions executed natively, which are annotated to the cycle counters of the PEs. However, this causes a hybrid ISS to be temporally decoupled from the rest of the system.

Since switching from TS to AS is performed upon the execution of a function in the application, the introduced temporal decoupling could at best be synchronized at function borders. This is a consequence of having a highly abstracted model for PEs which represents functions as instructions from the ISS perspective. Therefore, interrupts, software-centric synchronization and other events will be delayed to interact with the system state left by the AS execution. Moreover, the loss of accuracy during native mode might disturb timing and change the behavior of OS schedulers. Without any further action, a hybrid ISS might lead to a system crash (e.g. due to unhandled interrupts) or to non-deterministic simulator behavior, thus restricting its use for software development and debugging. This effect, what we call hybridization-introduced decoupling, has some similarities to traditional temporal decoupling scenarios, but poses new constraints that must be handled differently.

3.1 Modified Hybrid Processor Structure

HySim's original hybrid processor model did not consider the necessity to define a synchronization interface for hybridization-introduced decoupling. Therefore, it was necessary to replace HySim's core architecture with a new structure which links the hybrid PE to the simulator kernel's time, as shown by Figure 1b. Our structure features a Synchronization Layer, on top of HySim's Control Layer, that analyzes simulation mode switches and events sent by other system devices. In this way, it is possible to tell the kernel when the hybrid PE should be scheduled again and what simulation mode to be used. Depending on the simulation technology, the link between kernel and PE can be created either directly or through dedicated time manipulation interfaces, like SystemC-TLM2 Quantum Keepers. When used with extensible, API rich simulation frameworks, such as Simics or Synopsys Virtualizer, it is possible to set up the Synchronization Layer in a transparent way and no changes are required in the kernel or other system components.

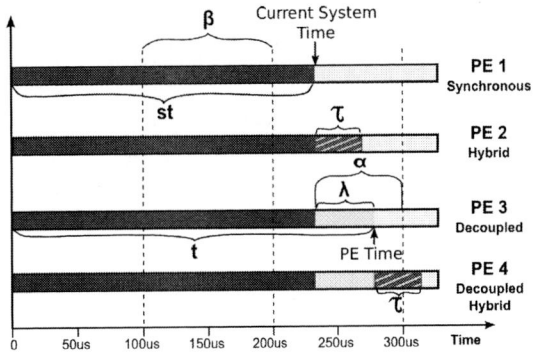

Figure 2: Temporally decoupled simulation.

3.2 Suspension Quantum

The estimated time for a given virtualized function represents at run-time the amount of time a hybrid PE will be ahead of the system. In a PE running synchronously with the system (e.g. triggered by the simulator kernel on every new instruction), a new instruction cannot be executed right after executing a virtualized function without synchronizing with the system. This is because other PEs will stay in the "past" with regard to the PE state that was modified during the virtualized function.

To allow other PEs and system components to reach the new temporal state, it is necessary to introduce a mechanism to suspend a PE for a given amount of time. We define a *suspension quantum* (denoted τ) to be the time a PE will be suspended as system time advances. The suspension quantum is created dynamically upon the execution of a virtualized function, and its length is equal to the estimated time associated to the function. In Figure 2, *PE1* runs always in synchronous mode, whereas *PE2* is synchronized only until it starts executing a virtualized function. This creates a decoupling time equal to τ. To avoid unnecessary kernel synchronizations, a new synchronization point for the suspended PE is set after τ.

In HySim, the larger the code parts executed in native mode, the faster the simulation runs. Functions mapped to AS mode are usually application hotspots that take significant execution time. Therefore, the suspension quantum could possibly take a very large value, causing the PE to lose responsiveness to external events. To avoid this, the suspension quantum must (i) be visible outside the context of a suspended PE and (ii) be breakable by system events.

3.3 Breaking the Suspension Quantum

In an MPSoC, any system component might trigger external events that need interaction with the suspended PE. This is the case for interrupts produced by peripherals, such as timers, accelerators and multi-core mailbox-based communication modules. Losing these interrupts causes systems to change their timing behavior and, in some cases, they even behave incorrectly (e.g. an OS that waits for a timer interrupt to boot). As this situation also happens with dynamic quanta in normal temporal decoupling, some specialized mechanisms allow to break a decoupling quantum and recompute new synchronization intervals. Similarly, the suspension quantum is broken when a PE receives an interrupt or a hardware signal that must be handled "in time".

But, in contrast to normal quanta, the mechanism to break τ needs to take a PE out of its suspension state while

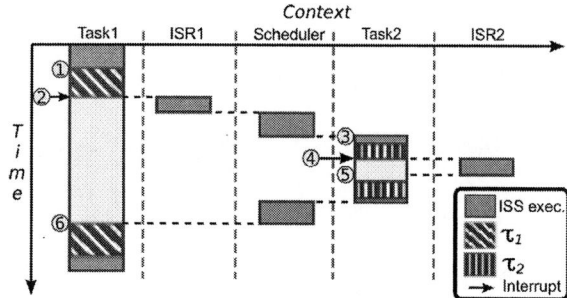

Figure 3: Breaking the suspension quantum.

guaranteeing that the processor jumps into special handling functions (i.e. interrupt service routines (ISR)). The remaining suspension quantum needs to be stored to be consumed later. If the suspension quantum is just canceled and the time is not consumed at all, it will lead to serious inaccuracy errors in the simulation. Moreover, if the remaining suspension quantum is not consumed in the original function execution context, the timing behavior of the application might change considerably.

We introduced a mechanism to ensure that the rest of the suspension quantum is consumed in its proper context, performing the following steps: (i) The hybrid ISS detects an incoming interrupt. (ii) The processor is waken up and the value of the program counter (PC) is taken in the instruction which is aborted by the core interrupt handling mechanisms. (iii) The PC value is associated to a *remaining suspension quantum*. (iv) A breakpoint-like mechanism is activated on the saved PC in order to restore the suspension the next time the processor executes the aborted instruction.

The previous mechanism is specially important in systems with OS and preemptive context switching. Figure 3 shows how a suspension quantum should be broken in an interrupt-triggered scheduling mechanism. In the figure, a task (Task 1) executes one of its functions in native mode (①) and introduces a suspension quantum denoted by τ_1. After some time of suspension, an interrupt (②) arrives, breaking the suspension quantum and triggering the OS scheduler. The scheduler preempts Task 1 and triggers Task 2 (③). The remaining part of τ_1 is only consumed after the context of Task 1 is restored by the scheduler (⑥), because it is tied to the address of the instruction that was interrupted by the incoming signal. In the same way, other nested interrupts that break the quantum in a different context can be supported (as with τ_2 in ④ and ⑤). This approach is limited to applications that do not share functions in their concurrent tasks. However, it can be easily extended by adding OS awareness and detecting a context switch. To avoid errors caused by handling interrupts in a PE with a "future" state (i.e. virtualized function was already executed in zero time and cannot be reverted), this mechanism relies on modifications to the virtualization chain, as discussed later in Section 4.

3.4 Suspension Quantum and Traditional Temporal Decoupling

To achieve maximum simulation speed, it is possible to mix hybridization-introduced decoupling and traditional temporal decoupling, in the same PE. To do so, suspension quanta are used to recompute traditional quanta, and define new synchronization points. Figure 2 shows a processing element (*PE3*) which uses traditional temporal decoupling to

run ahead of other system components. In a given point, the PE is assigned a quantum that defines the amount it is allowed to run decoupled. In the meantime, the global system time remains unchanged, and will be modified only after the next synchronization point (i.e. when the quantum is over). In this situation we use the following definitions:

- **PE Global Quantum (β_i).** Time unit on which PE_i synchronizes.
- **Current System Time (st).** Time elapsed uniformly in all components.
- **Local Time (t_i).** Time elapsed in PE_i. Can be greater than st.
- **Local Quantum (α_i).** Time remaining from t_i to the end of the next β_i.
- **Local Time Offset (λ_i).** Time PE_i is ahead of the system. Difference between t_i and st.

When a hybrid ISS is present, the decoupling parameters need to be dynamically modified depending on the suspension quantum. Therefore, after the execution of a virtualized function the value of t_i is updated, the processor is suspended, and a new β_i is recomputed, according to the following conditions:

1. If the updated local time exceeds the end of the next β_i (i.e. $\tau > \alpha - \lambda$), synchronization is done immediately. The quantum has been overshot, thus a new value for β_i is defined to be used the next time the PE is scheduled. The following operations are performed on the PE timing values:

$$\beta_i' = \beta_i - (\tau - (\alpha - \lambda)) \qquad t_i' = t_i + \tau$$

2. If the updated local time does not exceed the end of the current β_i (i.e. $\tau \leq \alpha - \lambda$), synchronization is performed normally at the next quantum end. In this case, the only operation performed is:

$$t_i' = t_i + \tau$$

These operations need to be performed only when virtualized functions are executed in a given quantum. Otherwise, temporal decoupling is used normally.

4. SYSTEM STATE SYNCHRONIZATION

Allowing PEs to run ahead of others creates momentary inconsistencies in the system. Besides, in shared-memory architectures, the "future" state left by a HySim processor might be wrongly propagated to other processing elements, thus causing a concurrency bug (e.g. atomicity violation, deadlock). This situation is very likely to happen if software synchronization functions (e.g. locks, semaphores, mutexes) or functions unrestrictedly accessing shared memory are virtualized and executed in native mode. Additionally, a memory access in AS mode is not able to trigger behavior in the adjacent peripherals. This is due to the fact that memory accesses do not use the traditional ports or sockets in ISSs, in order to achieve maximum speed. Instead, memory is read by using debug APIs or direct memory interfaces (DMI), thus failing to trigger behavior in other components. Functions that rely on global or static variables which are modified during ISRs cannot be virtualized either.

Because of this, restrictions need to be added to the HySim virtualization chain. In a full system simulation, virtualizable functions are not allowed to:

124

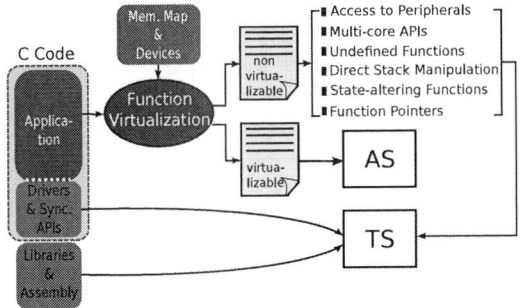

Figure 4: Virtualization chain for state synchronization.

- Perform software synchronization or unrestrictedly access shared memory.
- Interact with peripherals and accelerators.
- Depend on global states modified by ISRs or exception handlers.

All these functions are explicitly excluded from the set of virtualizable functions. Functions without definition are treated in the same way as closed libraries, and are marked as non-virtualizable. In well-formed applications, drivers and the communication and synchronization APIs are clearly separated, and sources can be excluded easily. If the separation is unclear, then the programmer still has the choice to map any virtualizable function to AS or not. Detection of unsafe mappings is done dynamically by address monitoring inside the AS, which uses a memory map description with the location of shared memories and memory-mapped peripherals. Figure 4 shows the modified virtualization flow. Under these constraints, a system with HySim will behave like a temporally decoupled simulator, yet with higher speed and application-defined synchronization.

5. TEST CASES AND RESULTS

To test our synchronization mechanism, we used the HySim framework to simulate different scenarios which are prone to behave wrongly in the presence of decoupling. Platforms for multi-media and signal processing were modeled in Simics and Synopsys Virtualizer, whereas Tensilica[5] Diamond and Xtensa ISSs, wrapped under our hybrid architecture, were selected as PEs. The synchronization layer was implemented using extensibility APIs provided by both simulation frameworks (e.g. *Haps*, execution, cycle and step interfaces in Simics; instrumentation points, quantum observers and simulation context handlers in Synopsys).

For the synchronization, function execution times were estimated by sampling the execution of virtualized functions in the ISS, using the statistical sampling theory from [18]. Although software performance estimation is not the focus of this paper, it is worth to note that it introduces certain error in the number of simulated cycles with and without HySim, which will be presented in the results.

All experiments were executed on a host with a 64-bit AMD Phenom Quad-Core Processor running at 2.4GHz, 8GB of memory and Fedora Core 5.

Scenario 1: 3DES on Single-core System. A Simics platform with one Diamond DC_B_570T core was used to execute a simple 3DES encryption/decryption application. Since it does not depend on reactive behavior, the traditional HySim can be used normally. Thus, this scenario allows to obtain the overhead caused by the new hybrid architecture.

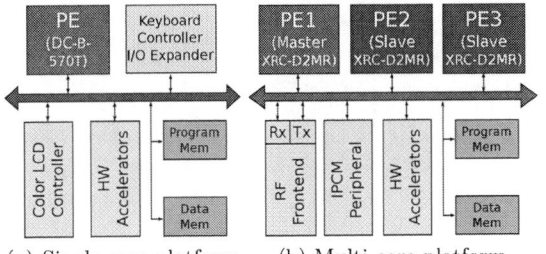

(a) Single-core platform   (b) Multi-core platform

Figure 5: Test systems.

Table 1 compares the elapsed wall-clock time, the amount of ISS executed cycles, and the speed-up values using the normal ISS, HySim and the synchronization-capable HySim (HySim-Sync). Comparing HySim and HySim-Sync, the results show an overhead, traduced to 5x less speed-up, due to the synchronization layer. The speed-up with HySim-Sync is still a significant ~27.2x with respect to the normal ISS.

Scenario 2: MJPEG on Single-core System. An enhanced version of the platform from Scenario 1, shown in Figure 5a, was used to execute a Motion-JPEG (MJPEG) player. This platform includes a set of hardware blocks for multi-media acceleration, a detailed model of a color LCD controller and a peripheral for user interaction. In this system, external interrupts are necessary to enable proper behavior of the device drivers. If the drivers lose interrupts, the system could crash with an unhandled interrupt exception. This is the case when simulating it with a big quantum for traditional temporal decoupling or when using HySim without synchronizations. The last situation is particularly difficult to handle since the hybridization-introduced decoupling might induce a crash only when some specific functions are mapped to HySim's AS mode. Comparative results when running the system are illustrated in Table 1. When HySim-Sync is used, all critical functions are marked as non-virtualizable by the framework and the suspension quantum mechanism enables handling the interrupts as expected by the drivers. If compared to the normal HySim, HySim-Sync guarantees the correct operation at the cost of less speed-up (76.9x vs. 45.2x), however, it ensures the usability of the simulator at a considerable high speed.

Scenario 3: Circular-FFT on Multi-core System. A Synopsys platform consisting of three Xtensa XRC_D2MR cores and an AMBA AXI bus was used to execute a Circular-FFT. This application is a token-passing system in which a core owning the token has to perform an FFT over some data and then pass the token to the next core. The tokens are passed using a simple communication protocol over shared memory upon the reception of a timer interrupt. Cores not holding the token wait while polling the shared memory. This system features a special time-triggered, software-based synchronization and its behavior with HySim depends on the FFT processing time and the timer period:

- If the FFT time is much greater than the timer and the polling (i.e. huge input data set), then HySim yields enormous speed-ups. The results table shows a value of 313x when the FFT size is 1024.
- If the FFT time is less than the timer and the polling, then the time saved by native execution is offset by the ISS execution speed during the polling loops. Thus, HySim might achieve marginal or no speed-up.
- If both times are similar, then the system loses determinism and might lock due to unhandled interrupts.

Application	Normal		HySim				HySim-Sync					
	Simulated ISS Cycles(M)	Wall-clock Time(s)	Simulated ISS Cycles(M)	Wall-clock Time(s)	App. in AS(%)	Speedup (times)	Simulated ISS Cycles(M)	Wall-clock Time(s)	App. in AS(%)	Synchro-nizations	Speedup (times)	Estima-tion Error(%)
3DES	2214.3	1625	26.5	50	98.8	32.5	26.5	59.7	98.8	600000	27.2	-6.3
MJPEG	1705.4	1231	8.8	16	99.4	76.9	10.5	27.2	99.3	7444	45.2	-7.4
Circular-FFT	2360	17238	0.75	55	99.9	313.4	642.6	7494	72.8	200	2.3	-33.3
OFDM-Trans	797.6	5816	71.7	574	91.1	10.1	162.3	3061	79.66	1000	1.9	-12.4

Table 1: HySim speed-up in VPs with and without synchronization.

Figure 6: OFDM transceiver application.

HySim-Sync's suspension quantum and interrupt support are necessary to avoid the last situation, however, at the cost of a significant speed-up reduction of 2 orders of magnitude, as shown by the table (313x vs. 2.3x). In this case, the user has to decide whether such trade-off is acceptable based on his knowledge of the application.

Scenario 4: OFDM Transceiver System. The base platform from Scenario 3 was used to set up a full digital wireless transceiver (OFDM-Trans). The hardware was extended with a mailbox-based Interprocessor Communication Peripheral (IPCM) for multi-core support and peripherals to handle input and output data flows, as shown in Figure 5b. The program features a complete software stack consisting of HAL, a priority-based preemption mechanism, a distributed scheduler, and a task management system. On top, the user application implements an OFDM transceiver algorithm which is divided into sub-tasks corresponding to algorithmic kernels, as illustrated in Figure 6. All sub-tasks are mapped to two cores that act as "Slaves", whereas the job management, launching and scheduling are done in the remaining core ("Master"). The Master launches dynamically a reception or transmission job every time a new packet is reported by the radio frontend. Scheduling a sub-task is done by sending a message through the IPCM which interrupts the destination core. The destination core's scheduler receives the order and manages it locally according to task priorities. Since this application performs priority scheduling and preemption based on incoming interrupts, it is mandatory to synchronize frequently the hybrid ISSs and the system. With the normal HySim, the simulation might still be functionally correct if the application code itself is written to be perfectly synchronized. If not, deadlocks and data races will arise in the system due to unsupported interrupt rate in the drivers or due to the randomization of task interleavings in the scheduler. For this system, HySim-Sync achieves 1.9x speed-up and guarantees correct operation and reproducibility. The level of details of other models (e.g. the bus) prevent to obtain more speed-up.

It is worth to mention that speed-up and execution times are not comparable between different simulation tools because the systems contain models at different levels of abstraction (e.g. systems in Synopsys have a detailed AXI bus model, while Simics uses a point-to-point bus).

6. CONCLUSIONS

Hybrid processor simulators are essential to provide both high speed and target-specific functionality. This paper presented an approach to synchronize hybrid processor simulators within full-system simulators in order to attain correctness. The temporal and the state decoupling problems were addressed by (i) defining a specialized temporal decoupling mechanism and (ii) identifying functions that must be avoided in native execution in order to ensure correctness of parallel applications. The proposed mechanisms were used to refine the internal architecture of a representative hybrid simulator (HySim) and analyze it in four application scenarios. Future work should address the application of hybridization to many-core systems as well as its combination with other advanced simulation techniques (e.g. parallelization).

Acknowledgment

This work has been supported by the FP7 Euretile project, the UMIC Research Center and Huawei Technologies. The authors would like to thank Yao Zhiliang and Guo Can from Huawei for their valuable contributions.

7. REFERENCES
[1] Open Virtual Platforms. http://www.ovpworld.org.
[2] Qemu. http://www.qemu.org.
[3] Synopsys Virtualizer. http://www.synopsys.com.
[4] SystemC. http://www.systemc.org.
[5] Tensilica processors. http://www.tensilica.com.
[6] Windriver Simics. http://www.windriver.com.
[7] L. Gao, K. Karuri, S. Kraemer, R. Leupers, G. Ascheid, and H. Meyr. Multiprocessor performance estimation using hybrid simulation. In *Design Automation Conference (DAC)*, 2008.
[8] L. Gao, S. Kraemer, R. Leupers, G. Ascheid, and H. Meyr. A fast and generic hybrid simulation approach using C virtual machine. In *CASES*, 2007.
[9] P. Gerin, M. M. Hamayun, and F. Pétrot. Native MPSoC co-simulation environment for software performance estimation. In *CODES+ISSS*, 2009.
[10] S. Kraemer, L. Gao, J. Weinstock, R. Leupers, G. Ascheid, and H. Meyr. HySim: a fast simulation framework for embedded software development. In *CODES+ISSS*, 2007.
[11] W. Lee, K. Patel, and M. Pedram. B2sim: a fast micro-architecture simulator based on basic block characterization. In *CODES+ISSS*, 2006.
[12] A. Muttreja, A. Raghunathan, S. Ravi, and N. Jha. Hybrid simulation for energy estimation of embedded software. *IEEE Transactions on Computer-Aided Design of Integrated Circuits and Systems*, 26, 2007.
[13] A. Nohl, G. Braun, O. Schliebusch, R. Leupers, H. Meyr, and A. Hoffmann. A universal technique for fast and flexible instruction-set architecture simulation. In *DAC*, 2002.
[14] P. Razaghi and A. Gerstlauer. Host-compiled multicore RTOS simulator for embedded real-time software development. In *Design, Automation and Test in Europe Conference (DATE)*, 2011.
[15] G. Schirner, A. Gerstlauer, and R. Dömer. Fast and accurate processor models for efficient MPSoC design. *ACM Trans. on Design Automation of Electronics Systems*, 15, 2010.
[16] J. Schnerr, O. Bringmann, A. Viehl, and W. Rosenstiel. High-performance timing simulation of embedded software. In *Design Automation Conference (DAC)*, 2008.
[17] N. Topham, B. Franke, D. Jones, and D. Powell. Adaptive high-speed processor simulation. In *Processor and System-on-Chip Simulation*. Springer-Verlag, 2010.
[18] R. E. Wunderlich, T. F. Wenisch, B. Falsafi, and J. C. Hoe. Statistical sampling of microarchitecture simulation. *ACM Trans. Model. Comput. Simul.*, 2006.

A Non-Intrusive Timing Synchronization Interface for Hardware-Assisted HW/SW Co-Simulation

Yu-Hung Huang, Yi-Shan Lu, Hsin-I Wu, and Ren-Song Tsay
National Tsing Hua University
HsinChu, Taiwan
yhhuang@cs.nthu.edu.tw, {shinsan941501, hiwu.dery}@gmail.com, rstsay@cs.nthu.edu.tw

ABSTRACT

This paper proposes using a non-intrusive timing synchronization interface approach to facilitate shared-data synchronization for fast and accurate hardware-assisted HW/SW co-simulation. Our synchronization interface device is specially designed for non-transparent components. With the device, we can systematically monitor shared-data accesses on a bus and control the progressing time of hardware-assisted components for fast and accurate system co-simulation. Experiments show that our approach is 10 to 140 times faster than the cycle-based hardware-assisted co-simulation approach.

Categories and Subject Descriptors

I.6.7 [**SIMULATION AND MODELING**]: Simulation Support Systems; B.7.2 [**INTEGRATED CIRCUITS**]: Design Aids – *Simulation*

General Terms

Performance, Design, Verification

Keywords

Timing synchronization interface, Hardware-assisted HW/SW co-simulation

1. INTRODUCTION

Due to the increasing manufacturing costs and complexity of System-on-a-Chip (SoC) designs, HW/SW co-simulation has become an essential step prior to production. This indispensible co-simulation process verifies the system integration of concurrent operations of all hardware and software components. Therefore, the performance of HW/SW co-simulation is critical for improving the time-to-market of system designs. Usually, hardware-assisted co-simulation techniques [1][2] are adopted to accelerate the overall HW/SW co-simulation speed.

The challenge of the hardware-assisted co-simulation approach is that it requires two heterogeneous simulation engines: a hardware engine and a software engine. Software-modeled components can be driven by an instruction-set simulator (ISS) or by SystemC modules, while hardware-assisted components rely on an FPGA (Field Programmable Gate Array) or legacy IPs (Intellectual Properties). Although high-performance is the goal, the difficulty arises from the intrinsic speed difference between the hardware and software engines.

Proper timing synchronization is the key to guarantee correct co-simulation results. The synchronization mechanism requires consistent shared-data accesses, with no hazard violations, from all

components. The access time points determine the access order and hence, whether there are violations. Since the hardware and software engines run at different speeds, it is particularly challenging to develop a proper timing synchronization scheme to enforce a consistent shared-data access order.

Although timing synchronization approaches have been well-developed in the past and non-intrusive cycle-based timing synchronization can be achieved in hardware-assisted co-simulation [3], the resultant simulation performance is too slow for practical use.

A recently developed shared-data approach [4] can achieve fast and accurate co-simulation, but it is an intrusive approach that requires inserting timing synchronization points before shared-data accessing codes. Therefore, we cannot easily so implement this approach in hardware-assisted components, whose internal states can only be observed and controlled externally through the input/output (I/O) ports.

Our main contribution in this paper is to propose an innovative *non-intrusive timing synchronization interface* to resolve the *non-transparency* issue, i.e., the limited *observability* and poor *controllability* of hardware-assisted simulation components.

The concept behind our approach is based on two findings. First, we observe that the shared data in an HW/SW co-simulation are mainly from shared memory and hardware I/O registers. For components connected to a bus, we can check for shared-data access operations by monitoring the bus with no intrusion to any component.

Secondly, although the shared-data access order is determined by the chronological order of the access points, the different operating speeds of the different components prevent precise determination of the chronological order. To resolve this issue, we devise a local clock to count the progressing time of each component instead of driving all components under a single global clock. However, a hardware local clock is non-trivial in contrast to a software clock; we will elaborate our proposed solution later.

To summarize, our proposed hardware-assisted co-simulation approach works as like this: First, the local clock of each component model advances independently until it hits a shared-data access point. Then, the least-locally-timed component is executed to achieve synchronization. This process is repeated to create our practical, fast, and accurate hardware-assisted HW/SW co-simulation approach.

The experimental results show that our co-simulation approach performs 10 to 140 times faster than traditional cycle-based approaches. Moreover, the experiment demonstrates that the proposed interface method can easily integrate HW/SW components for hardware-assisted co-simulation.

The rest of this paper is organized as follows: Section 2 discusses related work. We briefly review existing time synchronization techniques, introduce the non-transparency issue, and discuss our solution in Section 3. Section 4 presents our non-intrusive timing synchronization interface. Section 5 presents the experimental results. Finally, we offer a brief conclusion in Section 6.

2. RELATED WORK

In this section, we review existing timing synchronization techniques and compare them to our proposed approach.

Many timing synchronization techniques proposed in the past are intrusive and are only applicable to software-modeled HW/SW co-simulation [4][5][6][7][8]. For hardware-assisted HW/SW co-simulation with components of limited observability and poor controllability, a non-intrusive timing synchronization approach is necessary.

The cycle-based method is an intuitive approach [3][9] since a cycle is the basic time unit. Synchronizing at each cycle certainly guarantees the correctness of the hardware-assisted simulation. However, it requires explicit control of hardware clock signals. Consequently, the simulation performance is severely degraded due to heavy synchronization overhead. On the other hand, our approach synchronizes only at the necessary points and is much more efficient.

To reduce synchronization overhead, others have proposed to synchronize at every few cycles instead of at each cycle [10]. Although this approach improves performance, it cannot guarantee accurate simulation results. On the other hand, our approach dynamically captures synchronization points and ensures simulation correctness.

Wu et al. [4] have proposed an effective shared-data synchronization approach that guarantees correctness. With a much-reduced number of synchronization points at shared-data access points, this approach greatly reduces unnecessary synchronization overhead and achieves fast and accurate co-simulation. However, this work is only for software-modeled HW/SW co-simulation. We have extended this idea of shared-data synchronization to our non-intrusive interface designs and we have dynamically identified interaction points related to hardware-assisted components.

To perform timed emulations, Yang et al. proposed a trace-driven approach to reduce synchronization overhead [11]. However, the major deficiency of the trace-driven approach results from its intrinsic difference with the execution-driven approach. Since there is no relative chronological order among separately generated component data access traces, this approach requires a special simulation kernel to align event traces for correct co-simulation. Conversely, our execution-driven approach does not require any specific simulation kernel and is therefore more portable.

Moreover, in the trace-driven approach [11], a CPU cannot be modeled as a hardware-assisted component. This is because the trace-driven approach only allows local clock counters on slave hardware components, but CPUs are mostly masters. In contrast, our clock counter is implemented on a synchronization interface attached to the target hardware-assisted component. The clock counter senses, counts, and controls the corresponding hardware local clock for each triggered transaction. Our non-intrusive approach allows a master to be hardware-assisted.

In summary, the existing timing synchronization techniques are

either incapable of supporting non-intrusive synchronization for hardware-assisted simulation or are unable to provide efficient simulation performance. Our proposed hardware-assisted co-simulation is proven to be both fast and accurate.

In the following section, we will discuss the challenges of timing synchronization for hardware-assisted co-simulation and then propose a non-intrusive synchronization interface approach to resolve these issues.

3. MANAGING NON-TRANSPARENT COMPONENTS

We adopt and extend Wu's shared-data timing synchronization framework [4] for fast and accurate hardware-assisted co-simulation containing non-transparent components. Essentially, consistent and correct shared-data access order is the only way to guarantee accurate HW/SW co-simulation, whether it is software-modeled or hardware-assisted.

To illustrate the synchronization issue, Figure 1(a) depicts a target system execution with a pair of shared-data access points in the indicated chronological order. Without explicit synchronization, some simulation executions can happen to be correct, as shown in Figure 1(b), but some can also be incorrect, as shown in Figure 1(c). Wu's shared-data synchronization approach considers timing information and synchronizes at every shared-data access point, as shown in Figure 1(d), to guarantee correct simulation results.

However, applying this shared-data synchronization approach to hardware-assisted co-simulation presents three challenges associated with *non-transparent components*. In our discussion, we will refer to software-modeled components as *transparent components*, as in Figure 2(a), and hardware-assisted components as *non-transparent components*, as in Figure 2(b).

The three challenges concerning non-transparent components are: to non-intrusively identify shared-data accesses, to compute the progressing time of the non-transparent components, and to manage the timing synchronization.

In the following, we first review the shared-data synchronization approach and then discuss our proposed solution to the non-transparent components related issues.

3.1 Shared-Data Synchronization

The shared-data synchronization approach simply maintains all shared-data access points in the same chronological order as they appear in the target system. Since the components of a system can interact only during shared-data accesses, it is sufficient to synchronize them at their shared-data access points. Using accurate timing annotations, Wu computes the timing of each shared-data access point and hence, derives the chronological order.

Additionally, an individual local clock is attached to each simulated component to record its local time. Timing calculation techniques

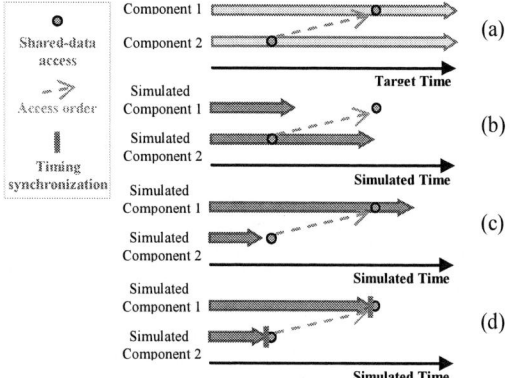

Figure 1: (a) An execution run on a target system. (b) A snapshot of a simulated execution with a correct shared-data access order. (c) A snapshot of a simulated execution with an incorrect shared-data access order. (d) A snapshot of timing synchronization before each shared-data access.

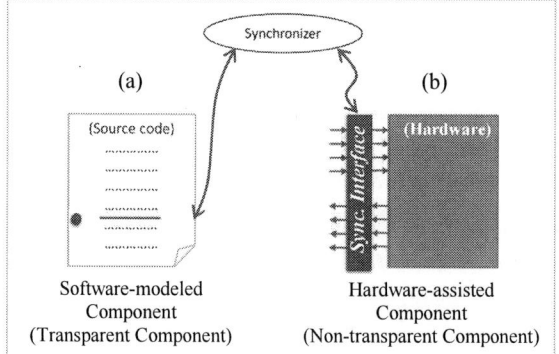

Figure 2: (a) A software-modeled component with a synchronization point. (b) A hardware-assisted component with a synchronization interface to resolve the non-transparency issue.

128

such as those used for CPUs [12][13] and memories [14] are applied. With Wu's algorithm, all simulated components progress together and are synchronized at shared-data access points, as shown in Figure 1(d). The approach is proven to be fast and accurate.

However, Wu's approach cannot be applied directly to hardware-assisted co-simulations mainly due to the issues related to the non-transparent components as discussed below.

3.2 Accurate HW-Assisted Co-Simulation

For software-modeled HW/SW co-simulation, we can easily manipulate the *transparent* software code and achieve timing synchronization. For instance, with *intrusively* annotated timing information in the operation code [12][13][14], an *intrusive* timing synchronization procedure can be invoked before each shared-data access instruction [4], as shown in Figure 2(a).

However, for hardware-assisted co-simulation, the internal states of *non-transparent* components are of limited observability and poor controllability or in a black-box, as illustrated in Figure 2(b). In other words, we can observe or change the internal states of hardware-assisted components only through I/O ports. This key difference between software-modeled and hardware-assisted HW/SW co-simulation leads to difficulties for shared-data synchronization.

The first challenge for non-transparent components is that we cannot pre-analyze or predict when the next shared-data access operation may occur as we can for transparent components. We can only check at runtime on the I/O ports to determine if the active operation is an external shared-data access operation. Therefore, we propose an add-on device, a *Synchronization Interface* (SI), for the I/O ports to monitor and manage the shared-data access operations.

The second challenge is to synchronize the internal accesses with the shared I/O registers of the non-transparent hardware components. Since these are internal operations, we cannot easily detect or control the synchronizations. A practical internal access example is that a hardware timer continuously updates an internal I/O register while an external CPU may access the register value.

If the non-transparent hardware components are hard-sealed legacy components, then we are totally in the dark and can do nothing. However, for hardware-assisted system co-simulation, most non-transparent components are implemented in FPGAs. In this case, we propose a simple probing approach that still performs shared-data synchronization related to internal shared I/O register accesses without changing the target designs.

Once we capture the external and internal shared-data accesses at runtime, we will perform synchronization and confirm the correct data accessing order. Technically, we will have to be able to stop and start the executions of the involved components and know the precise computation time of each transaction in each component. In Section 4, we will discuss in more detail how we have devised the *Synchronization Interface* (SI) for *non-intrusive* timing synchronization, as shown in Figure 2(b).

Before that, we will provide some more insights regarding the interaction behaviors of actual systems.

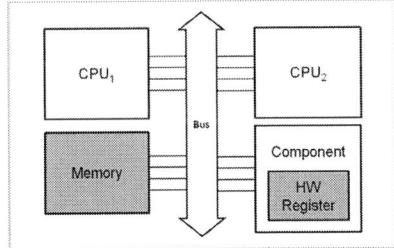

Figure 3: A representative target SoC system with a shared memory and a bus for inter-communication.

3.2.1 Identifying Shared-Data Accesses

We assume that the target system is a shared-memory SoC with a bus for inter-component communication, as shown in Figure 3. A shared-data location is a data location that can be accessed by more than two components. For instance, we assume that the memory and the hardware I/O registers are shared-data locations, as shown in Figure 3. Note that our proposed approach is not limited to the above system and can be applied to more complicated systems.

Since all components connect to the bus, as shown in Figure 3, we can check whether an *external* data access operation is a shared-data access by monitoring the *bus* with no intrusion to any component. Then, we can non-intrusively capture these accesses for synchronization. Here, "non-intrusive" means that no changes are imposed on the original code or design inside a component.

On the other hand, to address the challenging *internal* shared-data accesses issue, we take advantage of the following non-trivial facts.

According to the protocol of software programs communicating with hardware I/O devices [15][16], we categorize the behaviors between them into four phases: (i) Set Data Phase; (ii) Set Control Phase; (iii) Computation Phase; and (iv) Acknowledgment (ACK) Phase, as shown in Figure 4(a). We also denote the "read" (denoted as R) and "write" (denoted as W) behaviors in the four phases, as shown in Figure 4(a), according to the following discussion.

(i) To initiate an I/O request, the software program (denoted as SW) would check the hardware I/O status and feed input data into the

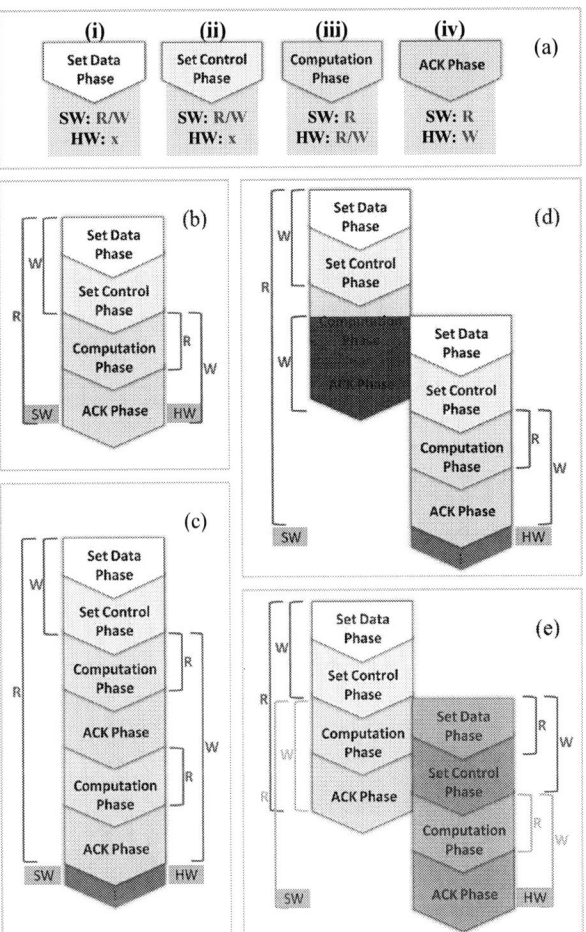

Figure 4: (a) Four generic phases of the communication protocol between software programs (SW) and hardware components (HW). (b) A generic one-run behavioral model for HW according to the HW protocol. (c) An example of one-run HW behavioral model for a timer on execution. (d) An example of one-run HW behavioral model for a timer on reset. (e) A more complex example of one-run HW behavioral model for a device with spooling.

129

hardware I/O data registers (Set Data Phase).

(ii) Immediately before a hardware component (denoted as HW) starts execution, hardware I/O control bits are set by the SW (Set Control Phase).

(iii) The HW starts computation according to the input and control signals while the SW periodically checks the status of the HW without interfering with the hardware execution (Computation Phase).

(iv) After completing an I/O request, the HW sets either a hardware I/O register (polling bits) or an interrupt to notify the SW (ACK Phase).

Using our four-phase model, we will now describe the transaction behaviors of a hardware component. A general hardware transaction behavioral model consists of four execution phases in the order shown in Figure 4(b). In this category, we have many simple hardware I/O devices such as hardware accelerators, IPs (decoder/encoder, etc.), input devices (keyboard, etc.), output devices, and input/output devices (network, etc.).

Some hardware transaction behaviors are more complicated, but these can still be modeled by repeating the phases corresponding to the hardware behaviors. For example, a timer sets an interrupt periodically, as shown in Figure 4(c), or can be reset by a software program in the midst of an execution, as shown in Figure 4(d).

An even more complex hardware transaction behavior model, as shown in Figure 4(e), is for devices capable of feeding input data in a pipeline manner, i.e. performing simultaneous peripheral operations on-line (Spool). In this case, a software application may subsequently feed a few groups of input data to the interacting device while keeping different data groups in distinct registers. Each group operation can be modeled as a separate transaction and these transactions are executed in a pipeline fashion.

With the four-phase modeling technique, we can further analyze the "read" and "write" behaviors of the internal shared-data accesses. An important observation is that the SW "read" operation and the HW "write" operation can concurrently occur at the "Computation" and "ACK" phases and result in a SW-read-after-HW-write or an HW-write-after-SW-read hazard. In contrast, the SW "write" operation to the HW internal shared data never overlaps with the HW "read" and "write" within a transaction.

However, other transactions writing to the same HW internal shared data location always have to wait until the previous transaction finishes according to the polling or interrupt protocol. For pipelined transactions, as in Figure 4(e), different transactions access different data locations and do not conflict. In other words, the SW "write" always occurs before the HW "read" and "write" in a transaction and hence, write-after-write, HW-read-after-SW-write, and SW-write-after-HW-read hazards will never occur. Therefore, we only need to manage the SW-read-after-HW-write and HW-write-after-SW-read hazards to achieve synchronization.

To ensure correct synchronization and avoid hazards or data dependency violations, we also need to know the timing of the

internal shared-data "write" operations of non-transparent components and compare it to the external "read" access time point from other components.

During actual implementation, we will tap *non-intrusive probes* to the I/O registers without changing or affecting users' original designs for the FPGA-assisting hardware components. Normally, the probes are just additional wire connections in the Verilog or VHDL code for FPGA implementation that sense register values. The probed information will be used by the SI to perform synchronization. Details will be discussed in Section 4.

3.2.2 Calculating Timing

In addition to identifying shared-data accesses, the progressing time of components is another critical non-trivial issue for shared-data synchronization. We have discussed the progressing time of transparent components in Section 3. Now we will elaborate the progressing time of non-transparent components in detail.

Unlike the intrusive code insertion, our SI is equipped with a local clock to calculate the progressing time of its associated non-transparent component. A possible hardware local clock solution can be a programmable logic counter. A counter starts and stops at the same time when its associated component starts and stops computing. For instance, the accompanying counter of a slave non-transparent component starts at the request of a master component and ends when the request is fulfilled. This is the implementation used in [11]. In this way, our SI can precisely know the progressing time of each non-transparent component.

Model-based progressing time calculation is an alternative solution. For instance, if we are processing a memory device, the pre-analyzed timing model proposed by Lo et al. [14] can be applied to determine the progressing time for each memory transaction. In that case, we would not need an explicit counter.

Note that the bus timing information is also critical and the bus contention delay has to be included in our considerations since all components communicate with each other through the bus. Although the actual bus design is complicated, an efficient abstract bus timing model [17][18] can be extracted for our purpose.

4. NON-INTRUSIVE SYNCHRONIZATION INTERFACE

In this section, we will elaborate the details of our proposed SI, a critical add-on interface that enables hardware-assisted system co-simulation. We will first introduce the non-intrusive SI for synchronizing shared-data accesses to memory and then for synchronizing hardware I/O registers in Section 4.1, followed by a brief discussion in Section 4.2.

To demonstrate the effectiveness of our proposed approach, we will use an example with mixed transparent and non-transparent components, as shown in Figure 5. This example has a software-modeled *transparent* CPU$_1$, an FPGA-implemented *non-transparent* CPU$_2$, and a *non-transparent* hardware component. Each component has a *local clock (t_x)* as marked. A software-modeled *bus* is placed on the edge, as shown in Figure 5. Additionally, a central system timing synchronizer, named *synchronizer*, determines which component should progress. The simplest scheduling algorithm is to let the component with the earliest local clock time proceed with execution. Details of the synchronization mechanism are discussed below.

4.1 Synchronization

As shown in Figure 6(a), *memory* accesses are always external accesses from either a transparent component or a non-transparent component. For accesses from transparent components, the inserted synchronization code before each shared-data access point is invoked to perform synchronization under the guidance of the

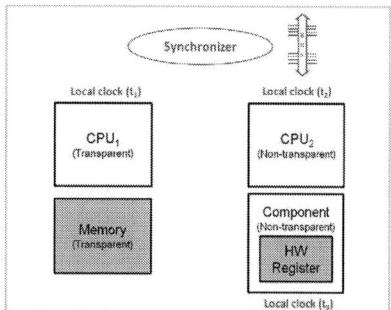

Figure 5: A hardware-assisted HW/SW co-simulation implementation of the target system from Figure 3.

130

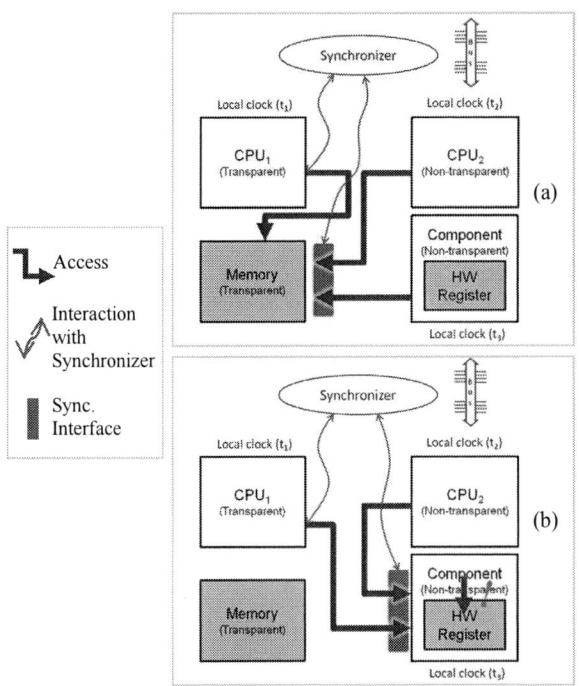

Figure 6: (a) All possible routes for memory accesses. (b) All possible routes for hardware I/O register accesses.

system timing synchronizer. In this case, we essentially follow Wu's approach, as discussed in Section 3.1. Hence, we will focus our discussion on the SI for non-transparent components.

For accesses from non-transparent components, the associated SI will work with the system synchronizer to control timing synchronization as depicted by the detailed synchronization steps in Figure 7. Detailed descriptions of each step are below.

As shown in step ①, the non-intrusive SI, connected between the associated non-transparent component and the bus, will monitor and determine if any shared-data access operation is issued from the component. If so, then in step ②, we buffer the operation for actual access after being cleared by the system synchronizer while simultaneously allowing the component to be frozen in step ③. The SI freezes the component by masking the input clock of the component. Next, in step ④, the SI obtains the progressing time from the local clock of the component. Then SI serves as a synchronization agent to the system synchronizer, as shown in step ⑤. With the timing information of all components and all access requests in the system, the system synchronizer decides which access is safe to proceed and then informs the corresponding SI or transparent component to proceed. Once an SI is granted permission to proceed, it then performs the actual memory access according to

the buffered operation instruction shown in step ⑥. Finally in step ⑦, the SI frees the input clock and lets the component resume its execution. With this careful synchronization procedure, we can guarantee accurate memory access order.

However, it is much more difficult to handle the synchronization of the *hardware I/O register* case shown in Figure 6(b). The challenge is mainly due to the uncontrollable and unobservable internal shared-data accesses discussed in Section 3.2. As with our hardware transaction behavioral model in Section 3.2.1, our proposed solution here is to have the SI record all the time changes of the hardware I/O registers in a table via non-intrusive probing of the hardware I/O registers, as shown in Figure 8. Also, all external "read" accesses to the hardware I/O registers will now go through the corresponding SI, and the SI will check and make sure that the last "write" data is ready before the requested "read" access time point. Then, the SW-read-after-HW-write and HW-write-after-SW-read hazards can be avoided and the hardware I/O register access order can be successfully synchronized.

The detailed steps of the hardware I/O register accesses from non-transparent components are similar to that of Figure 7, except for step ⑤. Instead of serving as an agent, the SI does its own timing synchronization checking. It compares the local clock of the accessing component and that of the accessed component. If the accessing component has an earlier local clock time, i.e., the component accessed has been simulated beyond the access time, then we shall record all the timed register values and the SI will let the accessing component read the corresponding register value on the table; otherwise, the SI will wait until the clock time of the accessed component catches up.

To sum up, when the memory access order and the hardware I/O register access order are synchronized in a chronological order by our synchronization interface, the HW/SW co-simulation is guaranteed to work correctly.

4.2 Discussion

Although the proposed SI is designed for non-transparent components, it can be extended to serve any component, including transparent ones, as long as they can support timing annotation and are connected to the proposed synchronization interfaces.

Another issue that needs to be discussed is the overhead of our SI. Since the SI recording table has a size limitation, faster hardware-assisted components may overflow the SI table. In practice, we freeze the component before it overflows the table. Once other components catch up, then this SI will release past table entry for reuse. The overhead affecting simulation performance is discussed in the next section.

5. EXPERIMENTAL RESULTS

To verify the proposed approach, we use the *WatchDog Timer* benchmark to demonstrate the hardware-assisted HW/SW co-

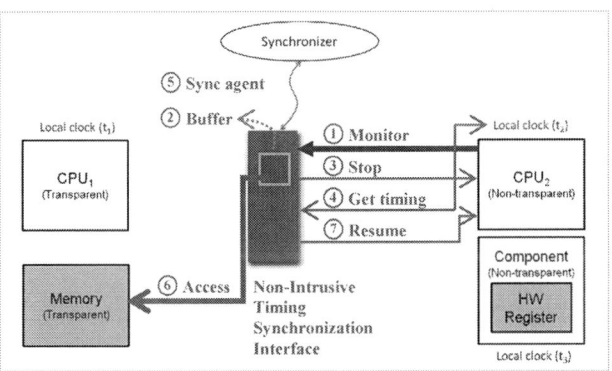

Figure 7: An example of memory access order preservation accessing via a non-transparent component.

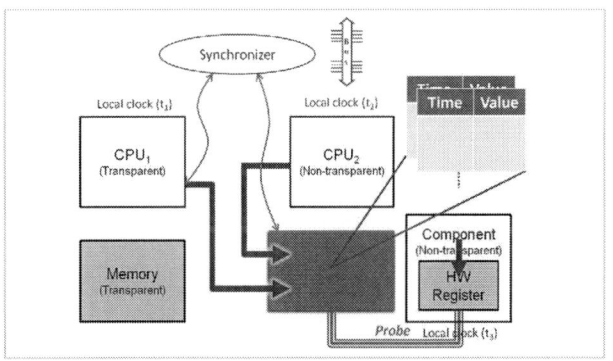

Figure 8: A solution for hardware I/O register preservation.

131

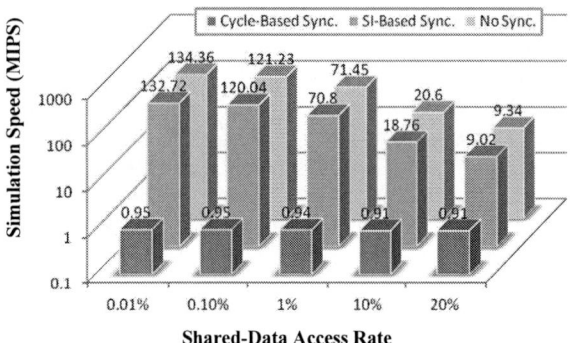

Figure 9: Simulation speeds of different synchronization approaches.

simulation speed and accuracy. The *WatchDog Timer* is a widely-used HW device that avoids SW system stalls by resetting the system when a timeout period is met. The interaction between the HW and SW is through a shared-data WTCNT register in the HW component. The HW will periodically decrease and check the value of the WTCNT register until the value is zero. Then, the HW will notify the SW to reset. By contrast, the SW will periodically set a non-zero value to the WTCNT register to continue system execution.

The SW component is an ISS implemented in a compiled simulation method [4]. The target processor is an Andes 16/32-bit mixed length RISC ISA [19]. Additionally, the HW device is implemented on a Creator PreSoC Development Board [20], which has an Xscale-PXA270 520MHz CPU and an FPGA-EP1C6. We use Creator Bus to bridge the communication between the SW and HW.

We compare the test results of our SI-based approach with that of the cycle-based approach and also of a simulation without synchronization. We keep the same SW program size, but test cases of various shared-data access rates of the *WatchDog Timer* benchmark. The simulation performance comparisons are summarized in Figure 9. Among these cases, the simulation speed of the cycle-based approach is consistently at around 0.93 million instructions per second (MIPS). On the other hand, our SI-based approach synchronizes the HW and SW only at each shared-data access, and hence, the simulation speed is raised up to the range of 9.02 MIPS to 132.72 MIPS.

As shown in Table 1, the speedup are around 10X to 140X. We observe that the speedup is inversely proportional to the shared-data access rate. In a rare case, when we need to synchronize at each cycle (i.e., 100% shared-data access rate), our approach's efficiency is the same as that of the cycle-based approach. In other words, our approach is always faster than the cycle-based approach.

Table 1. The speedup of the SI-based approach.

Shared-Data Access Rate	0.01%	0.1%	1%	10%	20%
Speedup	139.7X	126.5X	75X	20.5X	9.9X

Note that the speed of our approach is very close to the ideal simulation speed, i.e., the one without synchronization. Namely, the synchronization overhead is minimal.

We also compare the trace of shared-data accesses of our proposed SI-based approach with that of the cycle-based approach and verify that our approach is 100% accurate. On the other hand, the results of the approach without synchronization are incorrect. Our approach guarantees correct shared-data access order and co-simulation results.

6. CONCLUSION

In this paper, we have proposed and demonstrated the effectiveness of using a non-intrusive timing synchronization interface for hardware-assisted HW/SW co-simulation. By systematically monitoring shared-data accesses on a bus and controlling the progressing time of hardware-assisted components, we can accurately and non-intrusively co-simulate a system model with transparent and non-transparent components. Our proposal contributes to efficient hardware modeling and synchronization. The experimental results show that our *Synchronization-Interface*-based approach not only guarantees correctness but also performs 10 to 140 times faster than the cycle-based approach for hardware-assisted co-simulation.

7. REFERENCES

[1] William D. Bishop and Wayne M. Loucks, "A Heterogeneous Environment for Hardware/Software Cosimulation", in *ANSS'97*, April 1997.

[2] Stuart Swan, "SystemC transaction level models and RTL verification", in *DAC'06*, July 2006.

[3] Yuichi Nakamura, Kouhei Hosokawa, Ichiro Kuroda, Ko Yoshikawa, and Takeshi Yoshimura, "A fast hardware/software co-verification method for system-on-a-chip by using a C/C++ simulator and FPGA emulator with shared register communication", in *DAC'04*, June 2004.

[4] Meng-Huan Wu, Wen-Chuan Lee, Chen-Yu Chuang, and Ren-Song Tsay, "Automatic generation of software TLM in multiple abstraction layers for efficient HW/SW co-simulation", in *DATE'10*, March 2010.

[5] Luca Formaggio, Franco Fummi, and Graziano Pravadelli, "A timing-accurate HW/SW co-simulation of an ISS with SystemC", in *CODES+ISSS'04*, Sep. 2004.

[6] Meng-Huan Wu, Cheng-Yang Fu, Peng-Chih Wang, and Ren-Song Tsay, "An effective synchronization approach for fast and accurate multi-core instruction-set simulation", in *EMSOFT'09*, October 2009.

[7] Franco Fummi, Stefano Martini, Giovanni Perbellini, and Massimo Poncino, "Native ISS-SystemC Integration for the Co-Simulation of Multi-Processor SoC", in *DATE'04*, February 2004.

[8] Dohyung Kim, Youngmin Yi, and Soonhoi Ha, "Trace-driven HW/SW cosimulation using virtual synchronization technique", in *DAC'05*, June 2005.

[9] Luca Benini, Davide Bertozzi, Davide Bruni, Nicola Drago, Franco Fummi, and Massimo Poncino, "SystemC Cosimulation and Emulation of Multiprocessor SoC Designs", in *Computer*, April 2003.

[10] Franco Fummi, Mirko Loghi, Stefano Martini, Marco Monguzzi, Giovanni Perbellini, and Massimo Poncino, "Virtual Hardware Prototyping through Timed Hardware-Software Co-Simulation", in *DATE'05*, March 2005.

[11] Hoeseok Yang, Youngmin Yi, and Soonhoi Ha, "A timed HW/SW coemulation technique for fast yet accurate system verification", in *SAMOS'09*, July 2009.

[12] Jürgen Schnerr, Oliver Bringmann, Alexander Viehl, and Wolfgang Rosenstiel, "High-performance timing simulation of embedded software", in *DAC'08*, June 2008.

[13] Kai-Li Lin, Chen-Kang Lo, and Ren-Song Tsay, "Source-level timing annotation for fast and accurate TLM computation model generation", in *ASPDAC'10*, January 2010.

[14] Yi-Len Lo, Mao-Lin Li, and Ren-Song Tsay, "Cycle count accurate memory modeling in system level design", in *CODES+ISSS'09*, October 2009.

[15] David A. Patterson and John L. Hennessy, *Computer Organization and Design: The Hardware/Software Interface*, Morgan Kaufmann Publishers Inc., 2007.

[16] A. Silberschatz, P. B. Galvin, and G. Gagne, *Operating System Principles*, John Wiley & Sons (Asia) Re Ltd, 2004.

[17] Sudeep Pasricha, Nikil Dutt, and Mohamed Ben-Romdhane, "Extending the transaction level modeling approach for fast communication architecture exploration", in *DAC'04*, June 2004.

[18] Chen-Kang Lo and Ren-Song Tsay, "Automatic generation of Cycle Accurate and Cycle Count Accurate transaction level bus models from a formal model", in *ASPDAC'09*, January 2009.

[19] AndeStar™ ISA, available at www.andestech.com/p2-2.htm, 2010.

[20] Microtime Computer Inc., www.microtime.com.tw

Can EDA Combat the Rise of Electronic Counterfeiting?

Farinaz Koushanfar
Rice University,
Houston, TX

Saverio Fazzari
Booz Allen Hamilton, Inc.,
Arlington, VA

Carl McCants
Defense Advanced Research
Projects Agency, Arlington, VA

William Bryson
Analytical Solutions, Inc.,
Albuquerque, NM

Matthew Sale
U.S. Naval Surface Warfare
Center, Crane, IN

Peilin Song
IBM Research,
Yorktown Heights, NY

Miodrag Potkonjak
University of California,
Los Angeles, CA

ABSTRACT

The Semiconductor Industry Associates (SIA) estimates that counterfeiting costs the US semiconductor companies $7.5B in lost revenue, and this is indeed a growing global problem. Repackaging the old ICs, selling the failed test parts, as well as gray marketing, are the most dominant counterfeiting practices. Can technology do a better job than lawyers? What are the technical challenges to be addressed? What EDA technologies will work: embedding IP protection measures in the design phase, developing rapid post-silicon certification, or counterfeit detection tools and methods?

Categories and Subject Descriptors

B.7 [**Hardware**]: Integrated Circuits

General Terms

Design, Security

Keywords

Counterfeiting; Reliability; Device and IC aging

1. INTRODUCTION

A counterfeit (fake) product is an illegal forgery or imitation of an original design. The counterfeit parts intend to fraudulently deceive consumers by pretending to be genuine. A 2008 report by the US department of commerce estimated counterfeiting to account for about 8% of the global merchandize trade, equivalent to lost sales of as much as 600B in 2008, and expected to grow to 1.2T in 2009 [30]. Counterfeiting of microelectronics components, embedded systems, and computer peripherals is a common practice in many parts of the world.

Estimates for IC counterfeiting losses vary greatly depending on the source. One of the lowest estimates is provided by

SIA at $7.5 B. Very recently, EE Times estimated that IC counterfeiting losses are as high as $ 169 B annually. Therefore, the fake parts are at least 2.5% of the annual IC sales and are significantly larger than the overall EDA revenues per year. Sources of fake products are diverse, ranging from re-labeling and using defective components to illegal over-building by manufacturers. Conventional chip identification methods such as printing serial numbers and burning fuses can be forged and thus, they have a limited effectiveness in preventing or detecting counterfeit chips.

Counterfeiting is a particularly important problem to address since it has at least four important ramifications: (i) the original IC part providers incur an irrecoverable loss due to the sale of often cheaper counterfeit components, (ii) low performance of counterfeit products (that are often of lower quality and/or cheaper older generations of a chip family) affects the overall efficacy of the integrated systems that unintentionally use them; this could in turn harm the reputation of authentic providers, (iii) unreliability of fake devices could render the integrated systems that unknowingly use the parts unreliable; this potentially affects the performance of weapons, airplanes, cars or other crucial applications that use the fake components [36], and (iv) untrusted fake components may have intentional malware or some backdoors for spying information or remotely controlling critical objects.

The rising trends in chip counterfeiting and its important consequences clearly motivate the urgent need for development of advanced IC anti-counterfeiting techniques. In this paper, we discuss the present state of IC anti-counterfeiting practice and research, technical challenges, and some potential EDA research and development directions for addressing the open problems.

2. PRESENT ANTI-COUNTERFEITING PRACTICES AND EFFORTS

Fake electronic parts have been a known issue since the early days of IC design and fabrication. The IC companies and government organizations have been aware of the occasional counterfeit incidents. Therefore, a preliminary set of guidelines and legal procedures are typically in place for handling the counterfeit subjects. The increasing growth of the number of counterfeit parts and the ascending potential risk of exploits have recently raised the awareness of this prevalent problem. In November 2011, the US Senate Armed

Services Committee held a hearing to address the growing issue of counterfeit parts in the U.S. military supply chain. Senator John McCain of Arizona, the committeeŠs ranking Republican, and Senator Carl Levin of Michigan, committee chairman, were among the officials who investigated defense contractors about the rising number of detected counterfeits in the supply chain. The two senators have used the 2012 Defense Authorization Act to alter acquisition guidelines and make contractors responsible for the expenses of replacing the fake components.

Until now, only a handful of industrial/research labs, and government/military agencies and contractors have a set of more technical procedures for electronic counterfeit prevention and detection, which are often classified. In the remainder of this section, we briefly summarize four major industrial, research lab, government and defense agencies' anti-counterfeiting approaches and initiatives.

The Navy lab defines counterfeit parts in two broad categories: New parts that are misrepresented and old parts that are sold as new. One can envision testability and integrated identification methods and quantify them in terms of practicality, cost, and accessibility beyond factory production tests. The key challenge is to develop methods for at-speed functional test and parametric characterization available and easily implementable for non-factory test screeners as well as methods for better detecting aging and other usage/improper handling induced stresses in parts. One can point out that these challenges exist for digital, mixed-signal, and analog parts. It is important to identify perils from a security point of view of allowing the end user greater insight into the IC through testability.

A different path is taken by Analytical Solutions, Inc., an independent company which provides analyses of complex electronic devices related to commercial, military, medical, security, and space applications [38]. The company offers a "Five Tier Approach" to counterfeit detection to meet the specific needs of the customer, summarized in Table 1. This "Five Tier Approach" gives customers the options that may be required to evaluate and validate that the electronic part or device is not a counterfeit product and provide users the trust they need in the production or manufacturing of high reliability products.

Researchers at IBM have developed a new technique for detecting chip alterations using intrinsic light emission in combination with electrical tests. This method is based on the fact that any active device emits infrared light emission when it is powered on or operational. High sensitivity photon detectors can be employed to capture the weak emission while the chip under evaluation is powered on and electric stimuli are applied to it. In particular, two main families of electrical test modes, static and dynamic, can be applied. It has been demonstrated that combining the optical diagnostics and electrical tests significantly increases the delectability of the malicious circuits. The optical tool of choice that is used for this method is Picosecond Imaging Circuit Analysis (PICA), originally developed for diagnosing time-critical IC failures. This tool can measure time-resolved emission from switching gates, as well as time-integrated and time-resolved from gates in fixed logic state, also known as Light Emission from Off-State Leakage Current (LEOSLC). The approach resulted in many positive results, including high spatial resolution image processing and data interpretation.

IC anti-counterfeiting has been well recognized by DARPA as a strategically important and a necessary technology. For example, the original active hardware metering for IC piracy prevention and anti-counterfeiting was supported by a DARPA/MTO program [1]. As another example, the objectives of the ongoing DARPA/MTO Integrity and Reliability of Integrated Circuits (IRIS) program is to develop the technology to derive the functionality of an IC to determine unambiguously if malicious modifications have been made to that IC, and to accurately determine the IC's useful lifespan and reliability from a physical perspective [39]. While the IRIS program is still in an early phase, it is expected to provide new research results and advance the state-of-the-art of IC trust and anti-counterfeiting techniques.

3. RESEARCH CHALLENGES

Development of newer and more efficient IC anti-counterfeiting techniques is exceedingly interesting from scientific and engineering viewpoints. There are a number of challenging tasks that require not just implementation, design, algorithmic, and modeling innovations, but also conceptual breakthroughs that may greatly benefit future generations of ICs. This difficulty and richness is a consequence of the confluence of several technological, security, and design aspects, including:

- *Ultra large scale of integration.* Modern and future chips have a surprising number of subcomponents, in the order of billions. Failure or performance degradation of each subcomponent may result in an overall system failure. Although the need to simultaneously consider so many components is a nontrivial challenge, it may in some situations facilitate counterfeiting detection. For example, to show that a chip was already significantly used, it suffices to verify that any part of the chip was considerably used.

- *Limited controllability and observability.* It is well known that the ratio of the number of transistors vs. the number of inputs/outputs has been continuously increasing over time. For example, this ratio was around a hundred for the first generations of processors, while it is more than a million for contemporary processors. Thus, it is increasingly more difficult to organize any type of testing in modern ICs.

- *Identification and accessing System-on-a-Chip (SoC) subcomponents.* In case a complex SoC with multiple subcomponents is being investigated, typically optical investigations are needed for identifying and accessing the subparts to be individually tested. Most modern SoC designs are packaged using flip chip technology and a heat spreader which have to be removed for optical evaluations. There is a need to develop more advanced optical methods that not only identify the IP subparts, but also can classify the type of IP (e.g., RF front-end, A2D, memory, etc.) and find the test ports and scan chains to each subcomponent in an automated manner.

- *Functional identification and reverse engineering.* The effective lifetime of an airplane or a tank may span over several decades, while the average lifespan of the underlying electronic components is typically much

Tier 1	Tier 2	Tier 3	Tier 4	Tier 5
External Visual Inspection	External Visual Inspection	External Visual Inspection		
Configuration Marking Permanency	Configuration Marking Permanency	Configuration Marking Permanency		
Lead Finish ID	Lead Finish ID	Lead Finish ID	Construction and Comparative Analysis	Die Analysis and Comparison "Trusted IC"
	X-Ray	X-Ray		
	CSAM	CSAM		
		Electrical Test		
		Internal Visual Inspection		

Table 1: "Five Tier Approach" to counterfeit detection from Analytical Solutions.

shorter. Therefore, to maintain the costly equipment or automotive/avionic systems, failed electronics must be replaced. A standing problem is the lack of sufficient documentation/description of the failed components. This may be because of the long supply chains, or part obsoleteness. In lieu of this information, one must reverse-engineer the target failed component using other working instances. Development of automated methods and tools for functional identification and IC reverse engineering is a major research challenge.

- *Interdisciplinary nature of problems.* Anticounterfeiting requires knowledge and skills in several domains including IC technology, design, EDA, testing, security, statistical analysis, and game theory. Hence, continuous educational efforts are needed.

- *Variety of (unpredictable) attacks.* It is often not easy to develop sound, comprehensive, and practical defenses against known attacks. The situation becomes much more challenging when the attacks are difficult to predict.

- *Process variation and device aging.* Although the security aspects are obviously the most challenging and difficult to address, recent history teaches us that process variation and device aging models are critical for effective research and studies. On one side, they enable development of conceptually new security mechanisms such as physical unclonable functions (PUFs). On the other hand, they also directly invalidate numerous hardware security approaches that assume uniform temporal and spacial characteristics of similar elements across one chip. Luckily, in terms of developing new anti-counterfeiting techniques, such imperfections can be essential and positive.

Furthermore, there is a large spectrum of IC counterfeiting and anti-counterfeiting problems. They are related to issues relevant to contracts between an IC seller and buyer and fulfillment of all contractual agreements, including ones that may not be explicitly stated. Currently popular exploits include reselling old ICs, selling lower performance chips as a higher performance model, and selling untested or defective ICs [30].

4. RELATED WORK

In this Section, we briefly summarize the related concepts which directly influence the research and development of advanced IC anti-counterfeiting methods. A closely related line of research is focused on detection of IC Trojans; such methods are exceptionally relevant when the counterfeit components introduce an exploit in the system. Due to space constraints, we do not discuss Trojan detection and prevention in this paper. Instead, we refer the interested readers to a comprehensive survey on this topic [28].

4.1 IP Watermarking

A watermark embeds a hidden signature in the chip at the design time [3, 29]. The signature is checked against its intended attributes for authenticity verification. Watermarking at different levels of design abstraction is useful and necessary, in particular for designs with multiple IPs where the IP infringement tracking is a challenging task. A watermark can identify a design, and not individual IC instances. It can become useful for tracking stolen design IPs in the supply chain.

4.2 Hardware Metering and Auditing

IC metering or hardware metering refers to tools, methodologies, and protocols that enable post-fabrication tracking of the ICs. Metering can differentiate legitimate hardware from pirated ones. Research efforts have been focused on how to generate a unique ID for a specific device. Hardware metering may be *passive*, or *active*. In passive metering, the ICs are specifically identified, either in terms of their functionality, or by other forms of unique identification [4]. The identified ICs may be matched against their record in a pre-formed database that could reveal unregistered ICs or overbuilt ICs (in case of collisions). In active metering, not only the ICs are uniquely identified, but also parts of the chipŠs functionality can be only accessed, locked (disabled), or unlocked (enabled) by the designer and/or IP rights owners using a high level knowledge of the design not transferred to the foundry [1].

Metering methods may also be classified as *intrinsic* or *extrinsic*: (i) Intrinsic hardware metering leverages process variation to create unique fingerprints by using the existing properties or side channels of the device, such as delay and power. Several approaches have been proposed to characterize the gate-level IC properties for hardware metering pur-

135

pose [5][6][7][8]. Intrinsic metering methods are inherently passive. (ii) Extrinsic hardware metering inserts additional hardware or software components to the device for ID generation [2][4]. The additional components can be configured to produce a unique, difficult to predict or clone fingerprint for each authentic device. Extrinsic metering methods may be passive or active.

There is a natural way to establish a connection between hardware metering and IC anti-counterfeiting techniques. The simple but powerful observation is that once a chip is identified by hardware metering techniques, then the foundry or other reliable sources can be contacted for obtaining more information about the IC such as the manufacturing date and the original buyer.

4.3 Physical Unclonable Functions (PUFs)

Physical unclonable functions (PUFs) are one potential candidate stricture for implementing the unique extrinsic IC identifiers. A PUF is a physical function that provides a mapping between its inputs and outputs based on the unique fluctuations in the unclonable device material properties such as timing or current. The PUF input vector is typically called a *challenge* and the PUF output vector is commonly called a *response*. To ensure security, the mapping should be such that responses can be rapidly evaluated, but they are hard to model, characterize, clone, or reproduce. PUFs have been proposed for both ASIC and FPGAs [22, 23, 26, 37, 27]. Comprehensive summaries and more detailed definition/classification of PUFs can be found in [31, 32]. Very little is known about industrial practices of authentic chip identification. An approach by Sun Microsystems was proposed where they use the unique EM radiation from each chip to establish its authenticity [24].

4.4 Device Aging Models

Natural phenomena such as negative bias temperature instability (NBTI) cause the aging of devices in form of threshold voltage increase. As a result, the IC structural properties such as delay and power would be impacted significantly, which cause degradation in the device lifetime. In order to evaluate and predict the increase of threshold voltages and, consequently the failure time, several quantitative models have been proposed. The time dependence of V_{th} shift follows fractional power law of the stress time [9], as shown in the following equation:

$$\Delta V_{th} = A \cdot exp(\beta V_G) \cdot exp(-E_\alpha/kT) \cdot t^{0.25} \qquad (1)$$

where V_G is the applied gate voltage; A and β are constants; E_α is the measured activation energy of the NBTI process; T is the temperature; and t is the stress time.

Due to the increase of threshold voltages, the aging process significantly impacts IC's structural properties. For example, the leakage energy of a logic gate exponentially decreases with the increase of threshold voltage, as indicated by the leakage current model [10]:

$$I_{leakage} = 2 \cdot n \cdot \mu \cdot C_{ox} \cdot \frac{W}{L} \cdot \left(\frac{kT}{q}\right)^2 \cdot e^{\frac{\sigma \cdot V_{dd} - V_{th}}{n \cdot (kT/q)}} \qquad (2)$$

where L is effective channel length, V_{th} is threshold voltage, W is gate width, V_{dd} is supply voltage, n is subthreshold slope, μ is mobility, C_{ox} is oxide capacitance, ϕ_t is thermal voltage $\phi_t = kT/q$, and σ is drain induced barrier lowering (DIBL) factor.

On the other hand, delay of a logic gate increases in a close-to-linear manner with aging, as shown by the following delay model [10]:

$$Delay = \frac{k_{tp} \cdot k_{fit} \cdot L^2}{2 \cdot n \cdot \mu \cdot \phi_t^2} \cdot \frac{V_{dd}}{(ln(e^{\frac{(1+\sigma)V_{dd} - V_{th}}{2 \cdot n \cdot \phi_t}} + 1))^2}$$
$$\cdot \frac{\gamma_i \cdot W_i + W_{i+1}}{W_i} \qquad (3)$$

where subscripts i and $i+1$ represent the the driver and load gates, respectively; γ is the ratio of gate parasitic to input capacitance; and k_{tp} and k_{fit} are fitting parameters.

4.5 Device Degradation

IC lifetime is influenced by a variety of phenomena that have been studied by the material science and semiconductor communities, including electromigration (EM) [11], stress migration (SM) [12], time dependent dielectric breakdown (TDDB) [13], thermal cycling (TC) [14], oxide breakdown [9], vertical interconnect access (VIA) [15][16], negative bias temperature instability (NBTI) [17], and hot-carrier injection (HCI) [18][19].

Electromigration is the transport of interconnect material (copper in modern designs) due to high density electric currents, i.e., movement of ions that alters the conductivity of an interconnect. The ramifications include physical disconnection of wires or failure of the overall IC to function correctly due to increase in wire delays. Closely related thermomigration has high impact on reliability of vias in particular in new technologies that do not use lead (Pb). Thermomigration moves metal material due to the underlying thermal gradient. Thermomigration is also related to stress migration where the difference in temperature creates thermo-mechanical stress due to different expansion rates of various materials in the IC.

Dielectric breakdown is a process where dielectric around wires develops cracks that drastically change its dielectric properties to the extent that it does not serve as an isolator anymore. Stress migration is a phenomenon where the metal atoms in the interconnects migrate due to mechanical stress, much like electromigration. Stress migration is caused by thermo-mechanical stresses which are caused by differing thermal expansion rates of different materials in the device. Thermal cycling is a phenomenon where the temperature of an IC or its parts is subject to high and rapid changes. It causes permanent damages that accumulate every time there is a cycle in the processor temperature, eventually leading to failures.

4.6 Aging Sensors

Accurate performance-degradation monitoring of CMOS circuits is one of the most critical issues for adaptive design techniques. Therefore, in addition to modeling and studies of aging, additional on-chip aging sensors can be implemented to monitor and report aging and reliability. Because of the increasing importance of this topic, various methods for realizing on-chip aging sensors have been recently proposed, including [33, 34, 35]: [33] proposed a technique to measure the beat frequency of two ring oscillators, one stressed and the other unstressed, to a very high delay sensing resolution for aging differentiation; [34] introduced two compact structures to digitally quantify the change in performance and power of devices undergoing NBTI and defect-induced oxide breakdown. The small size of the sensors makes them amenable to use in a standard-cell design with low area and power overhead; [35] deployed a threshold voltage detector for monitoring the performance degradation of an aged MOSFET. Developing more sophisticated sensing methods with a higher resolution aging differentiation can directly advance the state-of-the-art anti-counterfeiting techniques. The relevant EDA research is to find the best placement of each sensor type on the chip to maximize age sensing coverage while minimizing the overall sensor overhead/cost.

4.7 System Failures

System failure has become a major concern in hardware-based system design, especially with the rapid growth in nanoscale technologies where power and temperature significantly increase due to the transistor scaling. Therefore, prediction and evaluation of system failure has drawn a great deal of attention from both industry and academic communities. Srinivasan et al. [20] introduced a reliability-aware microprocessor (RAMP) design model to predict and evaluate the mean time to failure resulting from different components of the system, such as applications, system architectures, and processor designs. They further extended the RAMP model to evaluate the system failure caused by technology scaling. In particular, estimates the mean times to failure (MTTF) of devices due to various aging phenomena can be found in [20].

5. IC ANTI-COUNTERFEITING TECHNIQUES: CASE STUDIES

Our goal in this section is to provide additional impetus for development of design automation techniques and tools for fighting counterfeiting techniques. While we do not discuss a detail presentation of any of the proposed techniques and do not show any security proofs, we believe that readers will find non-trivial IC counterfeiting ideas and starting points for the development of industrial-strength practical and fundamentally sound methods.

5.1 Techniques for Integrated Circuit Age Characterization

The threshold voltage of each CMOS transistor is a function of the number of dopants injected during manufacturing and the number of bonds that are broken during IC operation. The first component is subject to process variation and essentially follows an exponential distribution. The level of spatial correlation is exceedingly low to the extent that the threshold voltages of different transistors can be considered independent for any pair of transistors. The aging of PMOS transistors is due to chemical structure and is by an order of magnitude faster than the aging of their NMOS counterparts. There are three important observations. The first is that device aging is recoverable to a significant extent but not completely: when a transistor is not under stress or is only under stress for a small percentage of time, its threshold voltage reduces at an exponentially decreasing rate. The second is that a transistor ages when it is under stress, i.e., when its channel acts as an open switch. The third important piece of information is that although there are techniques for extraction of threshold voltage [8], such an extraction is a slow and expensive procedure unless it is restricted to a relatively few transistors.

A particularly important question is whether a chip is essentially new or has been used for a significant amount of time. If used, the distribution of threshold variations would be far from the expected foundry models. Noninvasive IC characterization methods can be used for determining the post-silicon distribution and correlations among the threshold voltages. Statistical outlier detection methods can determine if the measured distributions or correlations significantly deviate from the expected characteristics of new chips.

5.2 Design for Counterfeit Detection (DCD)

In testing and testing-related research and development fields, it is a standard practice that after each successful development of a set of techniques that utilize new technologies or new conceptual insights, the next phase is to develop approaches that incorporate these mechanisms into the design flow. We expect that a major rise of the practical importance of anti-counterfeiting techniques will be provided by the need for management of chips that use ultra-high levels of integration. Cutting edge ICs have already approached about 10 billion transistors per chip; it is unrealistic to expect high yields and long expected lifetimes. For example, one can easily envision that some types of maintenance will be required, at least in form of occasional adjustment of reverse and forward body biasing, to compensate for device aging or high operational temperatures. Thus, we anticipate that DCD approaches will share both the techniques and resources with IC characterization and maintenance methods.

An excellent example of one such maintenance technique has been developed by Mitra's research group at Stanford [25]. They have developed a low cost approach for measuring device aging and therefore slowdown of gates on the critical paths of an IC. This approach is also important because it uses only differential time measurements and induces very low hardware and energy overheads. It is relatively easy to re-target architectural primitives and measurements for detection of old chips using one of the techniques described in the previous subsection.

6. CONCLUSION

Counterfeiting of electronic components is a growing illegal business with important economical, military, governmental, and industrial ramifications. Repackaging the old ICs, selling the failed test parts, as well as gray marketing, are the most dominant counterfeiting practices. Surprisingly, although recently many aspects of hardware related security have been extensively studied, there has been very little IC counterfeiting research. Creation of IC anti-counterfeiting techniques poses numerous and diverse challenges while it also has the potential to address a spectrum of important IC design, management, and maintenance problems. We have analyzed the currently most popular IC counterfeiting attacks and identified the most important IC anti-counterfeiting desiderata. To make the paper self-contained, we presented the most relevant related technologies. The technical highlight of the paper are vignettes of two approaches for fighting IC counterfeiting using EDA techniques and a brief summary of presently available industrial and government methods.

7. REFERENCES

[1] Y. Alkabani and F. Koushanfar. "Active Hardware Metering for Intellectual Property Protection and Security" *USENIX Security*, pp. 291–306, 2007.

[2] A. Caldwell et al., "Effective iterative techniques for fingerprinting design IP," *IEEE T-CAD*, vol. 23, no. 2, pp. 208–215, 2004.

[3] A. Kahng et al., "Copy detection for intellectual property protection of VLSI designs," *ICCAD*, pp. 600–604, 1999.

[4] F. Koushanfar, G. Qu, and M. Potkonjak, "Intellectual property metering," *IH*, pp. 81–95, 2001.

[5] F. Koushanfar and M. Potkonjak, "CAD-based security, cryptography, and digital rights management," *DAC*, pp. 268–269, 2007.

[6] Y. Alkabani et al. "Trusted integrated circuits: a nondestructive hidden characteristics extraction approach," *IH*, pp. 102–117, 2008.

[7] S. Wei, S. Meguerdichian, and M. Potkonjak, "Gate-level characterization: foundations and hardware security applications," *DAC*, pp. 222–227, 2010.

[8] S. Wei, A. Nahapetian, and M. Potkonjak, "Robust passive hardware metering," *ICCAD*, pp. 802–809, 2011.

[9] R. Chau et al., "High-k/metal-gate stack and its MOSFET characteristics," *IEEE EDL*, vol. 25, no.6, pp.408–410, 2004.

[10] D. Markovic et al., "Ultralow-power design in near-threshold region," *Proc. of the IEEE*, vol. 98, no. 2, pp. 237–252, 2010.

[11] J. Black, "Electromigration - a brief survey and some recent results," *IEEE T-ED*, vol. 16, no. 4, pp. 338–347, 1969.

[12] A. Sekiguchi, J. Koike, and K. Maruyama, "Microstructural influences on stress migration in electroplated Cu metallization," *Appl. Phys. Lett.*, vol. 83, no. 10, pp. 1962–1964, 2003.

[13] J. Stathis, "Physical and predictive models of ultrathin oxide reliability in CMOS devices and circuits," *IEEE T-DMR*, vol. 1, no. 1, pp. 43–59, 2001.

[14] J. Pang, D. Chong, and T. Low, "Thermal cycling analysis of flip-chip solder joint reliability," *IEEE T-CPMT*, vol. 24, no. 4, pp. 705–712, 2001.

[15] K. Mistry et al., "A 45nm logic technology with high-k+metal gate transistors, strained silicon, 9 Cu interconnect layers, 193nm dry patterning, and 100% Pb-free packaging," *IEDM*, pp. 247–250, 2007.

[16] R. Havemann and J. Hutchby, "High-performance interconnects: an integration overview," *Proc. of the IEEE*, vol. 89, no. 5, pp. 586–601, 2001.

[17] S. Bhardwaj et al., "Predictive modeling of the NBTI effect for reliable design," *CICC* pp. 189–192, 2006.

[18] E. Takeda and N. Suzuki, "An empirical model for device degradation due to hot-carrier injection," *IEEE EDL*, vol. 4, no. 4, pp. 111–113, 1983.

[19] P. Heremans et al., "Consistent model for the hot-carrier degradation in n-channel and p-channel MOSFETs," *IEEE T-ED*, vol. 35, no. 12, pp. 2194–2209, 1988.

[20] J. Srinivasan et al., "The case for microarchitectural awareness of lifetime reliability," *ISCA*, 2004.

[21] E. Y. Wu et al., "Interplay of voltage and temperature acceleration of oxide breakdown for ultra-thin gate dioxides," *Solid-State Electronics Journal*, 2002.

[22] B. Gassend et al., "Silicon physical random functions," *ACM CCS*, pp. 148–160, 2002.

[23] M. Majzoobi, F. Koushanfar, and M. Potkonjak, "Techniques for design and implementation of secure reconfigurable PUFs," *ACM T-RETS*, vol. 2, no. 1, pp. 1–33, 2009.

[24] K. Gross, R. C. Dhankula, and A. J. Lewis, "Detecting counterfeit electronic components using EMI telemetric fingerprints." US Patent Application US 2009/009830 A1, 2009.

[25] M. Agarwal et al., "Circuit failure prediction and its application to transistor aging," *IEEE VTS*, pp. 277–286, 2007.

[26] N. Beckmann and M. Potkonjak, "Hardware-based public-key cryptography with public physically unclonable functions," *IH*, pp. 206–220, 2009.

[27] M. Potkonjak, S. Meguerdichian, and A. Nahapetian, "Differential public physically unclonable functions: architecture and applications," *DAC*, pp. 242–247, 2011.

[28] M. Tehranipoor and F. Koushanfar, "A survey of hardware Trojan taxonomy and detection," *IEEE D & T of Computers*, vol. 27, no. 1, pp. 10–25, 2010.

[29] G. Qu and M. Potkonjak, "Intellectual Property Protection in VLSI Design," *Kluwer Academic Publisher*, 2003.

[30] Technical report by U.S. Department Of Commerce, Bureau Of Industry And Security, Office Of Technology Evaluation, "Defense industrial base assessment: Counterfeit electronics," 2010.

[31] U. Ruhrmair, S. Devadas, and F. Koushanfar, "Security based on Physical Unclonability and Disorder," Book Chapter in 'Introduction to Hardware Security and Trust', Editors: M. Tehranipoor and C. Wang, *Springer*, 2011.

[32] F. Armknecht, R. Maes, A.-R. Sadeghi, F.-X. Standaert, and C. Wachsmann, "A formalization of the security features of physical functions," *IEEE S & P*, pp. 397–412, 2011.

[33] T. Kim, R. Persaud, and C. H. Kim, "Silicon odometer: An on-chip reliability monitor for measuring frequency degradation of digital circuits," *IEEE JSSC*, vol. 43, no. 4, pp. 874–Ü880, 2008.

[34] E. Karl et al., "Compact in-situ sensors for monitoring negative bias-temperature-instability effect and oxide degradation," *ISSCC*, pp. 410–411, 2008.

[35] K.K. Kim, W. Wang, and K. Choi, "On-chip aging sensor circuits for reliable nanometer MOSFET digital circuits," *T-CAS-II*, vol.57, no. 10, pp.798–802, 2010.

[36] F. Koushanfar, A-R. Sadeghi, and H. Seudie "EDA for Secure and Dependable Cybercars: Challenges and Opportunities," *DAC*, 2012.

[37] M. Majzoobi and F. Koushanfar, "Time-Bounded Authentication of FPGAs," *IEEEE T-IFS*, vol. 6 , no. 3, pp. 1123–1135 , 2011.

[38] Analytical Solutions Inc., *http://www.asinm.com/*.

[39] DARPA Microsystem Technology Office, *http://www.darpa.mil/Our_Work/MTO/*.

Physics Matters: Statistical Aging Prediction under Trapping/Detrapping

Jyothi Bhaskarr Velamala[1], Ketul Sutaria[1], Takashi Sato[2], Yu Cao[1]

[1]School of Electrical, Computer and Energy Engineering, Arizona State University, Tempe, AZ 85287
[2]School of Informatics, Kyoto University, Kyoto, Japan
{jvelamal, kbsutari, ycao}@asu.edu, {takashi}@i.kyoto-u.ac.jp

ABSTRACT

Randomness in Negative Bias Temperature Instability (NBTI) process poses a dramatic challenge on reliability prediction of digital circuits. Accurate statistical aging prediction is essential in order to develop robust guard banding and protection strategies during the design stage. Variations in device level and supply voltage due to Dynamic Voltage Scaling (DVS) need to be considered in aging analysis. The statistical device data collected from 65nm test chip shows that degradation behavior derived from trapping/detrapping mechanism is accurate under statistical variations compared to conventional Reaction Diffusion (RD) theory. The unique features of this work include (1) Aging model development as a function of technology parameters based on trapping/detrapping theory (2) Reliability prediction under **device variations and DVS** with solid validation with using **65nm statistical silicon data** (3) Asymmetric aged timing analysis under NBTI and comprehensive evaluation of our framework in ISCAS89 sequential circuits. Further, we show that RD based NBTI model significantly overestimates the degradation and TD model correctly captures aging variability. These results provide design insights under statistical NBTI aging and enhance the prediction efficiency.

Categories and Subject Descriptors

B.7.2 [**Integrated Circuits**]: Design Aids – *performance analysis and design aids*; B.8.2 [**Performance and Reliability**]: Performance Analysis and Design Aids.

General Terms

Design, Experimentation, Performance, Reliability

Keywords: Negative Bias Temperature Instability, Hole Trapping, Dynamic Voltage Scaling, Timing Violations

1. INTRODUCTION

Negative Bias Temperature Instability (NBTI) is a serious reliability concern in today's digital circuits and becomes predominant with CMOS technology scaling [1]–[4]. NBTI manifests itself as an increase in the threshold voltage (V_{th}) of PMOS devices, resulting in gate and circuit delay shift [3]. Accurate prediction of the aging rate during the design stage is crucial to the decision of guard banding. However, statistical prediction is not trivial since NBTI has strong dependence on device physical parameters and operation conditions. The situation is more complex in today's digital circuits

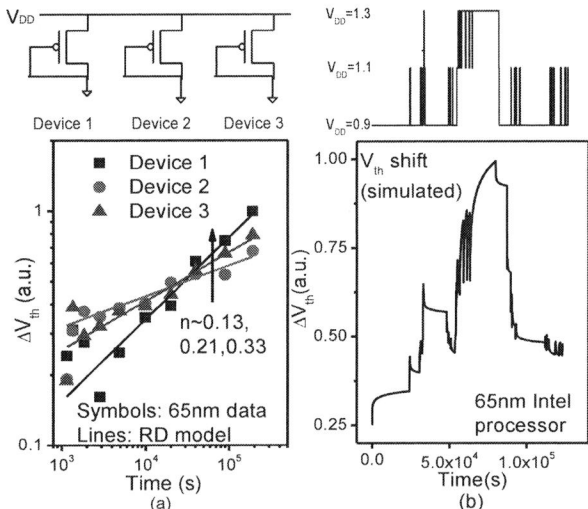

Figure 1. (a) Variation of time exponent extracted from three 65nm devices (b) DVS operation pattern in a Intel 65nm processor and predicted ΔV_{th}.

as they operate under multiple V_{DD} in Dynamic Voltage Scaling (DVS), leading to supply voltage variations. Hence, it is critical to understand, simulate and mitigate the NBTI effect in early design stages to ensure reliable circuit operation for desired lifetime [5]. NBTI mechanism has been explained by the classical Reaction Diffusion (RD) model. According to RD model, ΔV_{th} follows power-er law relation with stress time, where the time exponent (n) should be independent of process and operation parameters. However, Figure 1a illustrates that even from the same process and the same stress condition, the values of n from three different devices do not match with each other, similar to that observed at circuit level [6]. ΔV_{th} prediction from RD model is very sensitive to the time exponent. Hence, even a small change in n leads to dramatic difference in long term prediction of circuit lifetime [7]. On the other hand, recent works show the role of charge trapping/detrapping (TD) in NBTI degradation, where PMOS V_{th} increases if a trap captures the channel carrier. TD based BTI model follows logarithmic relation with time and is less sensitive to variation in model parameters. Further, this work incorporates ΔV_{th} dependence on technology parameters in TD model and correctly predicts statistical aging.

Threshold voltage and gate delay shifts can be determined using TD model based on the circuit operation conditions. Today's circuit operation involves multiple V_{DD} under Dynamic Voltage Scaling (DVS). Figure 1b presents the operation pattern in a 65nm dual core Intel processor and threshold voltage shift under such an operation. When the supply voltage is changed from a higher V_{DD} to lower V_{DD}, the degradation undergoes recovery. TD model correctly predicts the recovery under DVS, whereas RD model predicts a very low degradation when operated under lower V_{DD}, resulting in aging overestimation. Further, the gate delay and circuit delay shifts may turn certain paths into critical paths, resulting in timing violations. Therefore, it is essential to include NBTI in the aged timing analysis to guarantee circuit lifetime. TD model based aging analysis realizes realistic failure rate in digital circuits avoiding overly pessimistic prediction from RD model.

This work leverages trapping/detrapping based compact models for transistor degradation into the VLSI simulation flow at the system level, under device to device and supply voltage variations. The main contributions of this work include:

1. Modeling of NBTI degradation based on trapping/detrapping mechanism and incorporating the technology dependence into the model.

2. Aging prediction under variations at device level and supply voltage using TD model; recovery present in DVS operation is correctly captured avoiding the pessimistic aging prediction.

3. Comprehensive validation with 65nm statistical device data under long term NBTI stress (200ks) and multiple supply voltages in DVS operation.

4. Aging aware timing analysis to capture path delay shifts and timing violations in sequential circuits; RD model significantly overestimates aging rate compared to TD model.

Section 2 presents the trapping/detrapping based V_{th} shift models with dependence on technology and statistical parameters. Section 3 illustrates the statistical aging analysis and timing violations in VLSI circuits. The aging analysis is further demonstrated in ISCAS89 benchmark circuits using 45nm Nangate standard cell library. Section 4 concludes this paper.

2. AGING MODELS

The fundamental step in circuit aging prediction is to estimate V_{th} shift in PMOS device under NBTI. In this section, measurement setup to collect statistical aging data and trapping/detrapping based NBTI model is presented. Further, aging dependence on technology parameters such as V_{DD} and t_{ox} is incorporated into the modeling framework. The TD model accuracy under device level and supply voltage variations are demonstrated in this section.

2.1 Measurement Setup

To analyze the variability and develop an efficient aging prediction, the first step is to collect statistical device data. The measurement time plays a crucial role in NBTI test since even a small measurement time leads to large recovery, resulting in inaccurate aging data. Hence, obtaining degradation data by removing stress from all devices leads to a large measurement error. One solution is to place multiple DUTs on a chip so that stress periods and threshold voltage measurements can be conducted in parallel. This approach is very expensive and needs a larger area. Contrary to parallel measurement method, a parallelize stress period in a pipeline manner is implemented and V_{th} measurements for the DUTs are conducted in this work [8]. Figure 2 shows the microphotograph of test chip implemented in 65nm and 11 metal layer CMOS process. Total area of the test chip is 489x332um² and 128 PMOS devices of four different aspect ratios are implemented as DUTs. Aging measurements are conducted when all the devices are stressed at a voltage of 1.8V and a temperature of 125°C for 200ks. These measurements are required in order to analyze device to device

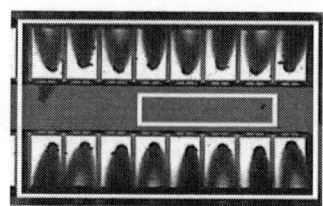

Figure 2. Microphotograph of a 65nm test chip (489x332um²) with a 11 metal layer CMOS; 128 PMOS transistors of 4 different aspect ratios.

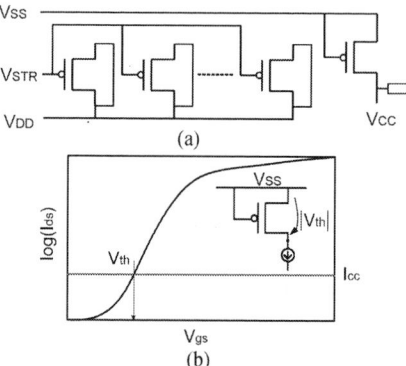

Figure 3. Measurement setup: (a) All devices are under stress except the device under measurement (b) Vth measurement using constant current method

statistical aging behavior in long term. Further, measurements are conducted when all the devices are stressed under multiple V_{DD} to realize the aging in DVS operation. Figure 3 presents the test structure and measurement principle implemented in our work. Except for the device in measurement, all other devices are stressed and V_{th} is measured using constant current method with a resolution of 0.2mV.

2.2 Trapping/Detrapping based BTI Model

Several works show the role of charge trapping/detrapping mechanism in NBTI degradation as opposed to classical RD theory [9]. Clear steps showing single trapping or detrapping events have been reported through discrete V_{th} shifts [10]. Figure 4 presents our measurement in a device under pure recovery. Discrete V_{th} shifts due to trapping/detrapping are observed, confirming the necessity of TD based NBTI models for reliable aging prediction. According to the trapping theory, threshold voltage of a device increases when a trap captures a charge carrier, resulting in reduction of drain current. If the device is not under stress, only localized traps with energy close to Fermi level can change their states, originating into low frequency noise. If NBTI stress leads to a high electric field, the trap energy is modulated and might capture charge carriers. The occupation probability of the trap to be captured is independent of stress time [10]. The probability of trapping depends on capture time constant and that of detrapping depends on emission time constant. The gradual change in number of traps occupied results in the time evolution of V_{th} shift.

The primary assumptions in the modeling framework are (1) the number of traps are Poisson distributed, (2) time constants are uniformly distributed on log scale and (3) trap energy distribution is assumed to be U shaped (key to include V_{DD}, t_{ox} dependence) [11]. The average number of traps that capture charge carriers during the stress phase is given by:

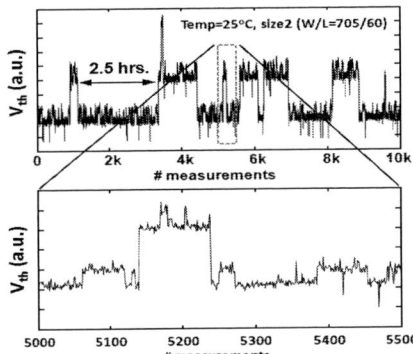

Figure 4. V_{th} shift under pure recovery; discrete V_{th} shifts observed due to trapping/detrapping events [8].

$$n(t) = N.p(t, \tau_c, \tau_e) \qquad (1)$$

where N is the Poisson parameter for the trap distribution and p is the capture probability, which is a function of time constants and stress time. Integrating the number of occupied traps over a period of time gives the V_{th} shift:

$$\Delta V_{th} = \phi \left[A + \log(1 + Ct) \right] \qquad (2)$$

where ϕ is proportional to the number of available traps per device. The variation in V_{th} shift is mainly due to ϕ and parameters, A and C are relatively constant. The V_{th} shift from TD model follows logarithmic relation with time, different from power law time dependence in RD model.

To validate trapping theory, RD and TD model parameters are extracted from stress data<20ks. The parameter ϕ in TD model has σ/μ of 26% and A, C exhibit $\sigma/\mu < 1\%$, indicating that ϕ is the main variation parameter from device to device. Figure 5 shows the distribution of ΔV_{th} prediction using extracted parameters from 20ks and measured ΔV_{th} after stress time of 200ks. The logarithmic model from TD theory correctly estimates mean and variance of V_{th} shift, since the degradation is less sensitive to model parameter variation compared to RD model. Hence, trapping/detrapping based BTI model accurately predicts variability in aging due to randomness in number of traps per device.

2.3 Technology Dependence

As the CMOS technology scales down and gate oxide thickness becomes lesser than 4nm, NBTI is the dominant aging mechanism which limits the device and circuit lifetime [5]. Therefore, it is essential to incorporate technology parameters in aging models

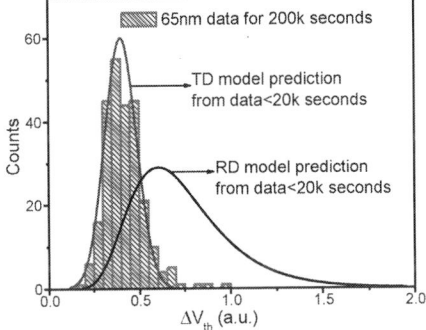

Figure 5. TD model well predicts the long term aging variability compared to RD model.

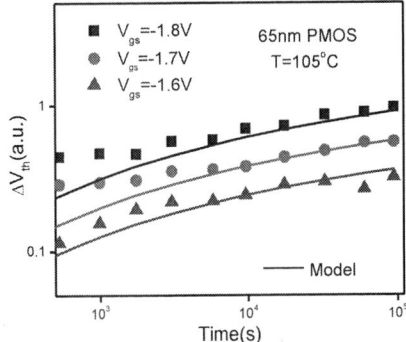

Figure 6: Prediction of V_{th} shift using TD model in 65nm PMOS device under different supply voltages [12].

besides time evolution. In this subsection, we derive a closed form solution for ΔV_{th} dependence on V_{DD} and t_{ox}, facilitating the designers to estimate aging for various technology nodes.

The threshold voltage shift as a function of trap and Fermi energy levels is given by [11]:

$$\Delta V_{th} = N' \left(\int_{Ev}^{Ec} \frac{f(E_T) dE_T}{1 + e^{-(E_T - E_F)/kT}} \right) \left[A + \log(1 + Ct) \right] \qquad (3)$$

where $f(E_T)$ is trap energy distribution, E_T and E_F are trap and Fermi energy levels respectively. The trap energy changes as function of electric field (E_{ox}) and the difference E_T-E_F decreases under high E_{ox}, resulting in a larger V_{th} shift. Trap energy is assumed to follow bucket shape (close to U shape) i.e., energy of traps is 0 from E_F to kT, and linearly increases beyond kT. Since $V_{gs} \sim -V_{DD}$ in digital circuits, performing integration of Eq. (3) leads to:

$$\Delta V_{th} \sim K_1.\exp\left(\frac{-E_0}{kT}\right)\exp\left(\frac{BV_{dd}}{kT.t_{ox}}\right)\left[A + \log(1 + Ct)\right] \qquad (4)$$

The compact model in Eq. (4) predicts the dependence of device degradation as a function of V_{DD}, t_{ox} and temperature (T). Figure 6 presents the validation of our TD model in a 65nm PMOS device stressed under different voltages, showing that our model well matches with experimental data [12]. Further, Eq. (4) is essential to predict aging under multiple V_{DD} in DVS operation.

2.4 Dynamic Voltage Scaling

Today's digital circuit operation is compounded with Dynamic Voltage Scaling (DVS) as shown in Figure 1b. Since the aging rate is a strong function of V_{DD}, it is important to predict the degradation under sequence of V_{DD}s present in DVS operation. The TD based BTI model is capable of correctly predicting the degradation under multiple supply voltages. Our model is evaluated with 65nm device measurement under DVS in this subsection.

Figure 7 illustrates two cases where a PMOS device is stressed under different voltages sequentially. When a digital circuit is operated at V_{DD} of 1.5V and followed by a higher V_{DD} of 1.8V, the number of occupied traps increases and V_{th} shift increases exponentially as shown in Figure 7a. This behavior is captured by both TD and RD models. When V_{DD} is changed back to 1.5V from 1.8V, the degradation undergoes recovery due to traps emitting the charge carriers (Figure 7a). This recovery behavior is captured by TD model and RD model predicts a very low degradation at a lower V_{DD} using the boundary condition [3]. The recovery is significant when operated under a much lower V_{DD} of 1.2V as presented in Figure 7b, which results in a large prediction error in the long term.

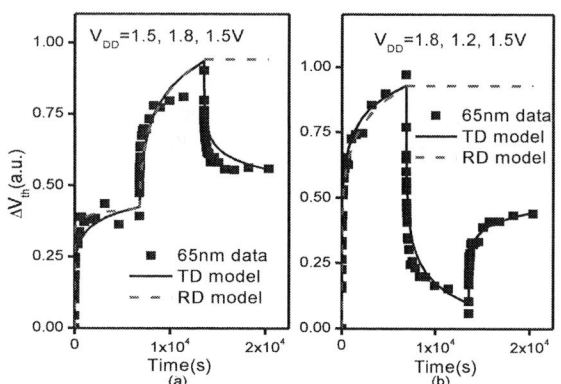

Figure 7. Threshold voltage shift under different V_{DD}; RD model overestimates compared to TD model.

The recovery behavior is also governed by the $log(t)$ function similar to the stress phase. Further, it is important to determine if the device undergoes stress or recovery when V_{DD} is changed. Given the values of ΔV_{th0}, stress time (t') experienced by the device and the supply voltage to be operated (V_{DD}'), we can predict if the degradation increases or recovers. Based on our BTI model, we can predict the V_{th} shift assuming that the device is stressed under V_{DD}' from time $t=0$ to $t=t'$. Using Eq. (4), the degradation increases further if:

$$\Delta V_{th0} < K_1.\exp\left(\frac{-E_0}{kT}\right)\exp\left(\frac{BV_{dd}'}{kT.t_{ox}}\right)\left[A+\log\left(1+Ct'\right)\right] \quad (5)$$

else, the degradation recovers until it reaches equilibrium. This condition at the boundary of supply voltage change allows the accurate aging prediction under DVS operation.

3. STATISTICAL AGING ANALYSIS IN VLSI CIRCUITS

The ΔV_{th} model based on trapping/detrapping proposed in previous section is crucial in statistical aging analysis at the circuit level. In this section, an experimental setup for failure analysis and timing violations in sequential circuits due to NBTI is demonstrated. Further, failure rate in digital circuits is comprehensively estimated under device to device and supply voltage variations.

3.1 Experimental Setup

Figure 8 presents the experimental setup and static timing analysis framework implemented in this work. NBTI aging aware library is used to calculate the delay shift in digital gates. Our framework uses delay information under different V_{DD} in standard library and predicts the delay shift due to change in V_{th} using a simple gate delay model, capable of handling the inherent stack effect present in NOR gates.

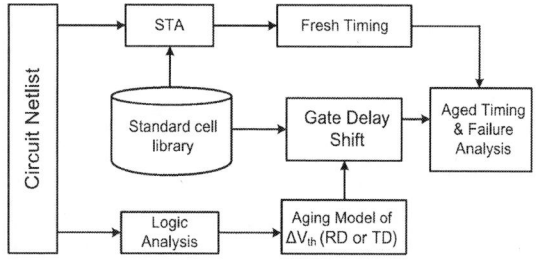

Figure 8. NBTI based statistical timing analysis framework.

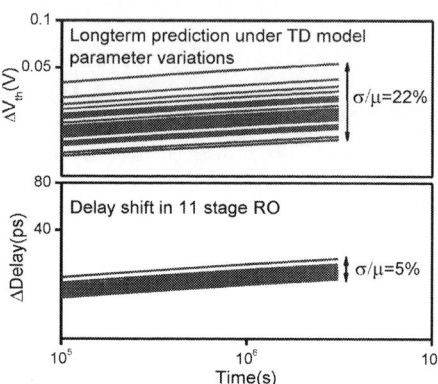

Figure 9. Long term prediction under device level variations of PMOS V_{th} shift and delay change in an 11 stage RO.

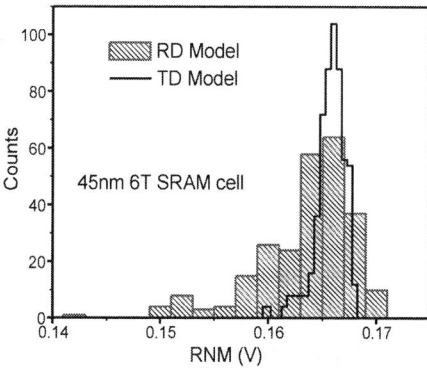

Figure 10. Distribution of read noise margin in a 6T SRAM cell under RD and TD models.

For a given digital circuit, we begin with the standard Static Timing Analysis (STA) which generates fresh timing report with timing information of all the paths in the circuit, without considering the NBTI effect. Logic analysis is performed on the circuit to obtain switching activities (α) in case of AC stress and node voltages in case of static stress. Based on the stress conditions, PMOS V_{th} shift and gate delay shifts are computed using delay information from standard cell library under different slew rates and load capacitances. Aged timing report is then obtained by updating gate and path delay shifts in fresh timing report, thus identifying the paths violating timing requirements. Our framework is general and can be extended to other aging mechanisms such as Positive Bias Temperature Instability (PBTI). The library used in the entire aging analysis is 45nm Nangate standard cell library.

3.2 Long Term Prediction

Device level aging due to trapping/detrapping exhibits large variations due to randomness in number of available traps. The variability at ($t>0$) is also a function of transistor sizing similar to process induced variations ($t=0$) [13]. Aging variability increases with downsizing the devices and hence, it becomes significant with CMOS technology scaling. Further, it is important to understand and estimate the translation of device aging to circuit level degradation. Figure 9 presents the long term prediction of PMOS V_{th} shift and ring oscillator frequency shift under device level variations. Though, the device level prediction has a large variation, circuit aging exhibits moderate variation due to the average effect of multiple transistors in the circuit. Standard deviation (σ) of V_{th} shift is

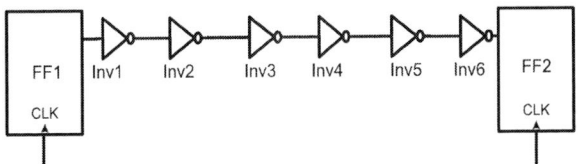

Figure 11. Schematic example of an inverter chain between two sequential elements.

Table 1. Fresh and aged analysis of circuit in Figure 11

Gates in path	Fresh		Tr. type	Aged (2x10⁶ s)			
	Gate delay (ps)	Path delay (ps)		RD Model Gate delay (ps)	TD Model Gate Delay (ps)	RD Model Path delay (ps)	TD Model Path delay (ps)
DFF1	192.6	192.6	Fall	192.6	192.6	192.6	192.6
Inv1	20.7	213.3	Rise	22.8	22	215.4	214.6
Inv2	11.1	224.4	Fall	11.1	11.1	226.5	225.7
Inv3	17.2	241.6	Rise	19.2	18.5	245.7	244.2
Inv4	11	252.6	Fall	11	11	256.7	255.2
Inv5	17.1	269.7	Rise	19.1	18.4	275.8	273.6
Inv6	10.9	280.6	Fall	10.9	10.9	286.7	284.5
Required data arrival time	285			Required data arrival time		285	285
Setup slack	+4.4			Setup slack		-1.7	0.5

22% of mean (μ) from our measurement data. When these V_{th} shifts are induced randomly into an 11 stage Ring Oscillator (RO), σ/μ~5% is observed. Figure 10 presents the distribution of Read Noise Margin (RNM) in a 6T SRAM cell predicted with extrapolation using the model coefficients from the short term data. Lower bound from RD model is much less compared to TD, model indicating the accurate prediction from the trapping theory.

3.3 Timing Violations in Sequential Circuits

NBTI induced gate and circuit delay shifts imply that many circuit paths that are not critical in the design stage may turn critical over time, causing timing violations during the operation [14]. Figure 11 demonstrates a timing violation in an inverter chain between two sequential elements. The output of the inverter chain passes through FF2 at the rising edge of clock signal. The path delay of the inverter chain should be less than required data arrival time. NBTI results in increase of path delay along the inverter chain, causing a setup violation and subsequent logic failure. A setup violation in this circuit is illustrated in Table I. When operated at a clock period of 340ps, the required data arrival time of this circuit is 285ps. The accumulated path delay is 280.6ps, resulting in a setup slack of +4.4ps. Positive setup slack indicates that the data reaches earlier than the required data arrival time and a setup violation do not occur. For a stress time of 2x10⁶s, gate and path delays shift due to NBTI. Since NBTI only shifts low to high delay of a signal, only the rising edges are shifted resulting in asymmetric aging. This is considered in our analysis as only delays of Inv1, Inv3 and Inv5 (rising transitions) are shifted, as highlighted in Table I. Power law time dependence in RD model overestimates aging compared to *log(t)* dependence in TD model. Path delay for the

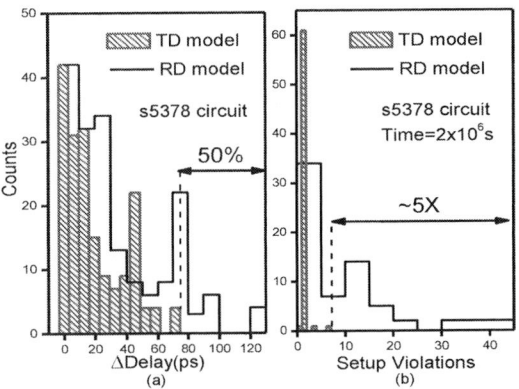

Figure 12. Distribution of (a) path delays and (b) violations under variations in TD and RD model parameters in s5378 circuit.

circuit increases to 286.7ps and 284.5ps under RD and TD models respectively. When RD model is used, the accumulated delay along the logic path is higher than the required data arrival time, causing a setup violation with negative slack of -1.7ps. TD model predicts a positive slack of 0.5ps and the circuit does not encounter any timing violation. Similar to shift in the logic path delay, NBTI induces delay in the clock buffer. If the clock travels slowly between two sequential elements and data is allowed to penetrate through both the registers in the same clock tick, a hold violation occurs. Since modern circuits are well designed to handle clock skew, only NBTI induced setup violations are estimated in this work. Our framework can be further extended to evaluate violations due to aging in sequential benchmark circuits.

The proposed trapping/detrapping based BTI models and asymmetric aging analysis are implemented in ISCAS89 benchmark circuits and PrimeTime, a commercial STA tool is used in our work. A 45nm Nangate open cell library characterized by Predictive Technology Model (PTM) is used in our analysis. ΔV_{th} is predicted using both TD and RD models based on the stress conditions. Gate delay shifts are computed by aging aware library and aged timing report is generated. Figure 12a shows the distribution in shift in path delays when the proposed framework is implemented in s5378 circuit with 179 paths. TD based aging model exhibits a narrow distribution with μ of 19ps and σ of 18.9 ps. RD model predicts a wider distribution of delay shifts and overestimates aging by 50% as illustrated in Figure 12a. The observed behavior is due to fast changing degradation with time in RD model compared to gradual change in TD model due to logarithmic time dependence.

The variations present at the device level are also significant to determine the failure rate due to aging. The predicted ΔV_{th} variations using the extracted model parameters (Figure 5) are used to calculate the variations in gate delay shifts. Aged timing under variations is obtained and number of violations can be predicted under both models. Figure 12b presents the distribution of the number of setup violations in s5378 circuit due to model parameter variations in both TD and RD models. When the circuit is under static operation for 2x10⁶s, RD model has a wider distribution and predicts approximately 5X more number of violations compared to TD model. The main variation parameter in RD model is the time exponent (*n*) and a small change in *n* value leads to huge difference in aging prediction. Hence, the exponential sensitivity of predicted ΔV_{th} to time exponent leads to a wide failure distribution. On the other hand, the main variation parameter in the TD model is ϕ and

143

Table II: Setup Violations under RD and TD models in ISCAS89 sequential circuits.

Design	Clock period (ns)	t=1year		t=5years		t=10years	
		RD	TD	RD	TD	RD	TD
S27	0.48	1	1	1	1	1	1
S382	0.9	1	1	2	2	2	2
S386	0.87	2	2	3	2	3	2
S444	1.05	1	1	1	1	1	1
S510	0.95	1	1	2	1	2	1
S641	2.76	3	3	4	4	4	4
S820	2.3	4	1	4	1	4	1
S832	2.4	4	3	4	3	4	3

ΔV_{th} has a linear dependence on it. Hence, the predicted degradation has a narrow distribution under variations due to less sensitivity of aging model to ϕ. Trapping/detrapping based NBTI model correctly predicts the critical paths under aging, avoiding the need to protect large number of paths during the design stage. Comprehensive demonstration of our aging analysis is demonstrated in different ISCAS89 benchmark circuits and summarized in Table II. Initially the clock period is fixed for each design such that all the paths in the circuit meet the timing requirements. Aging analysis is conducted by setting all the inputs to 0V and operation times of 1, 5 and 10 years. As the number of gates and inputs in the circuit increase, there is an increase in the number of setup violations. Also, RD model overestimates the failure due to aging in majority of the circuits. Therefore, TD based BTI model correctly predicts the degradation and guide designers to correctly predict the guard band under device to device variations.

3.4 Supply Voltage Variations

Along with variations at the device level, supply voltage variations also impact aging due to strong dependence of NBTI to V_{DD}. When a circuit is operated under multiple V_{DD}, the degradation can be computed by using the boundary condition in Eq. (5). Figure 13 presents the absolute shift in frequency of an 11 stage RO when operated under two supply voltages of 1.8V and 1.5V for random operation times. RD model does not predict any recovery when operated under a lower V_{DD} and hence, predicts monotonic RO frequency shift. However, TD model correctly predicts the recovery under a lower V_{DD} and estimates the upper bound of degradation accurately. The situation is more complex when the circuit is operated under more than 2 supply voltages, which can be handled by TD aging model using the appropriate boundary condition. Therefore, our proposed aging prediction under trapping/detrapping

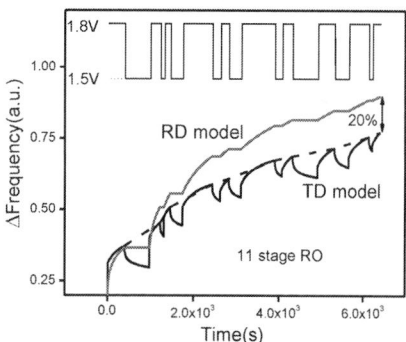

Figure 13. Prediction of delay shift in an 11 stage RO under sequence of V_{DD}=1.8V and 1.5V; RD model overestimates circuit aging.

theory facilitates robust long term device and circuit reliability prediction under V_{DD} variations inherent in DVS operation.

4. CONCLUSION

This paper presents statistical aging prediction under NBTI due to variations at device level and operation conditions. Degradation models based on trapping/detrapping mechanism are presented and the proposed model well predicts the device to device aging variability, which is not captured by conventional RD based BTI model. In this work, technology parameters are incorporated into TD models and thereby, accurately predicting the aging when operated under multiple V_{DD} present in DVS operation. Further, statistical timing analysis is performed under device level variations and demonstrated in ISCAS89 benchmark circuits; RD model overestimates aging and TD model correctly predicts aging under device to device and supply voltage variations. With solid verification with measured 65nm device data, the proposed statistical aging helps designers to accurately monitor and manage circuit lifetime.

REFERENCES

[1] The International Technology Roadmap for Semiconductors (*ITRS*), 2009.

[2] D. K. Schroder and J. A. Babcock, "Negative bias temperature instability: Road to cross in deep submicron silicon semiconductor manufacturing," *J. Applied Physics*, vol. 94, no. 1, pp. 1-18, 2003.

[3] S. Bharadwaj, W. Wang, R. Vattikonda, Y. Cao, and S. Vrudhula, "Predictive modeling of the NBTI effect for reliable design," *CICC*, pp. 189–192, Sep. 2006.

[4] M. A. Alam, H. Kufluoglu, D. Varghese and S. Mahapatra, "A comprehensive model for PMOS NBTI degradation", *Microelectronics Reliability*, vol. 47, no. 6, pp. 853-862, June. 2007

[5] R. Vattikonda, W. Wang, and Y. Cao, "Modeling and minimization of PMOS NBTI effect for robust nanometer design", *DAC*, pp. 1047–1052, Jul. 2006.

[6] W. Zhang and C. H. Kim, "An on-chip monitor for statistically significant circuit aging characterization," *IEDM*, pp. 4.2.1-4.2.4, 2010.

[7] R. Zheng, J. Velamala, V. Reddy, V. Balakrishnan, E. Mintarno, S. Mitra, S. Krishnan, and Y. Cao, "Circuit aging prediction for low-power operation," *CICC*, pp. 427–430, Sep. 2009.

[8] T. Sato, et al., "A device array for efficient bias temperature instability measurements," *ESSDERC*, pp. 143-146, 2011.

[9] T. Grasser et al., "Switching oxide traps as the missing link between negative bias temperature instability and random telegraph noise," *Int. Electron Devices Meeting Tech. Dig.*, pp. 729-732, 2009.

[10] G. I. Wirth, R. da Silva and B. Kaczer, "Statistical model for MOSFET bias temperature instability component due to charge trapping", *IEEE Tran. Electron Devices*, vol. 58, no. 8, pp.2743-2751, Aug, 2011.

[11] R. da Silva and G. Wirth, "Logarithmic behavior of the degradation dynamics of metal oxide semiconductor devices", *Journal of Statistical Mechanics*, vol. P04035, pp. 1-12, 2010.

[12] W. Wang, V. Reddy, B. Yang, V. Balakrishnan, S. Krishnan and Y. Cao, "Statistical prediction of circuit aging under process variations," *CICC*, pp. 13-16, 2008.

[13] S. Pae, J. Maiz and C. Prasad, "Effect of NBTI degradation on transistor variability in advanced technologies," *Integrated Reliability Workshop*, pp.18-21, Oct. 2007.

[14] W. Wang, Z. Wei, S. Yang and Y. Cao, "An efficient method to identify critical gates under circuit aging," *International Conference on Computer Aided Design*, pp. 735-740, 2007.

144

Library-Aware Resonant Clock Synthesis (LARCS)

Xuchu Hu[†‡]
hxcu@soe.ucsc.edu

Walter Condley[†]
jas@soe.ucsc.edu

Matthew R. Guthaus[†]
mrg@soe.ucsc.edu

[†]University of California Santa Cruz
1156 High Street
Santa Cruz, California CA 95064

[‡]Cadence Design Systems, Inc.
2655 Seely Avenue
San Jose, California CA 95134

ABSTRACT

Clock grids are often used in high-performance ASIC designs because of their low skew and robustness to variations. Resonant clock grids have the potential to reduce the power consumption of these high-performance clocks without sacrificing the skew and robustness of a clock grid. We present the first methodology to synthesize high-performance distributed resonant LC tank clock grids that utilize a pre-characterized inductor library. The use of a library reduces designer effort and total inductor area when compared with previous resonant clock grids while still attaining 59% power reduction and competitive skew when compared to traditional buffered clock grids.

Categories and Subject Descriptors

J.6 [**Computer-Aided Engineering**]: Computer-aided design

General Terms

Algorithms, Design

Keywords

Resonant, clock grid, low power

1. INTRODUCTION

Circuits are said to "resonate" when the imaginary components of the inductive and capacitive reactance cancel at a specific frequency. Resonant circuits are widely used in radio transmission and signal processing to filter certain frequencies, but they also have a great potential to lower power consumption of oscillating signals.

Power consumption is one of the major concerns in VLSI design, especially due to the increasing demand of mobile applications. Clock gating, power gating, dynamic frequency scaling, dynamic voltage scaling and multiple threshold voltages are used at different design abstractions to reduce either dynamic or leakage power. However, the power is usually saved by exploiting inactivity or dynamically adjusting performance. Fully active circuits that cannot sacrifice performance, however, can always benefit from additional techniques to lower power consumption. In that regard, resonant circuits can improve the power efficiency in oscillating circuits by recycling energy. Both academia and industry have proposed different types of resonant circuits to save power in clock distribution networks (CDNs) since the on-chip clock is the most active signal and consumes a large percentage of total power.

Previous resonant clock approaches have included standing wave [10], rotary/salphasic [3, 15, 19], and resonant inductor-capacitor (LC) tank clocks [1, 5, 6, 20, 21]. Standing wave resonant clocks provide constant phases but the amplitude of the clock signal varies with position in the CDN. Rotary resonant clocks provide constant amplitude but the phase varies with position in the CDN. The inconsistent phase/amplitude of standing wave and rotary clocks make them difficult to apply in practice since they are dissimilar to existing CDN structures. Resonant LC tank clocks, however, ideally have constant phase and constant magnitude which is similar to previous non-resonant clocks and allows similar structure and design methodologies.

Nearly all high-performance integrated circuits use a buffered clock tree driving a buffered clock grid [12, 18]. Clock grids have better skew and robustness than clock trees because the redundant wires reduce the impact of variation. However, these redundant wires increase the total capacitance and, therefore, the power consumption. Resonant circuits can reduce the power of these high-performance clock grids, but most previous works on distributed LC tank resonant clock require significant manual design. These early resonant CDNs used identical inductors across an entire chip [2, 5, 13], but this does not maximize energy savings or minimize skew.

The first automated resonant clock design methodology [6] focused on local resonant clock tree synthesis, but required impractically large inductors. The most recent methodology, ROCKS [7], is the first to address resonant clock grid synthesis and saves significant power with low skew. However, the ROCKS inductor sizing method produces a continuous range of inductor sizes which makes it impractical to implement. The rounding of these inductors requires either significant overhead for accurate matches or results in high skew due to coarse matches. The design of high-Q inductors requires significant manual design and characterization.

This work presents a Library-Aware Resonant Clock Synthesis (LARCS) algorithm which is the first automated physical design of distributed resonant clock grids using a pre-made library of inductors. We present algorithms to place LC tanks in resonant grids and tune buffers and dummy capacitance to optimize clock performance. Specifically, LARCS is the first methodology to:

- automatically synthesize distributed LC tank resonant clocks with a discrete inductor library;
- consider both amplitude and phase conflict in resonant grids;
- fine-tune phase differences using dummy capacitance to bal-

ance resonant frequencies;

- and consider the impact of inductor quality factor (Q) on resonant efficiency.

In Section 2, we provide a brief review of LC resonant circuits with integrated inductors. In Section 3, we present an overview of LARCS for LC tank placement, buffer sizing and phase tuning. Then, in Sections 4-6, we discuss the details of the tank placement, buffer sizing, and phase tuning, respectively. In Section 7, we present our experimental methodology and results. Finally, we offer conclusions in Section 8.

2. RESONANT THEORY

An LC resonant "tank" circuit is an inductor and capacitor in parallel (or series) as shown in Figure 1(a). At the *resonant frequency* of a tank with sizes L and C, $f_0 = \frac{1}{2\pi\sqrt{LC}}$, the capacitive and inductive reactances cancel which results in a series LC tank with zero impedance or a parallel LC tank with infinite impedance.

When an oscillating signal such as a clock is applied to these LC tanks, they form a *resonant oscillator* that freely exchanges energy between the electric field in the capacitor and a magnetic field in the inductor. A decoupling capacitor (C_d) on the grounded end of each inductor positively biases the oscillating voltage to a compatible CMOS logic range (0 to V_{dd}) [1]. The oscillation of the CDN capacitance with inductance reduces the required size of the grid buffers and saves power. Ideally no grid buffers are needed, however, the parasitic resistance of the CDN causes energy loss in the oscillating system for which small grid buffers must compensate. Distributing multiple LC tanks throughout the CDN rather than a monolithic LC tank helps to minimize the energy lost due to parasitic resistance.

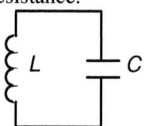

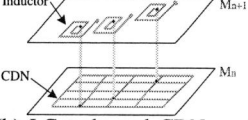

(a) Inductor-capacitor (LC) tank where L is inductor and C is CDN capacitance.

(b) LC tanks and CDN are implemented in adjacent global metal layers.

Figure 1: On-chip LC tank circuits are designed to resonant with the CDN capacitance.

Inductors can be created on-chip using normal metal layers as in many radio frequency (RF) designs. Figure 1(b) shows three spiral inductors on metal M_{n+1} connected to the CDN in metal M_n. For simplicity, we don't show the decoupling capacitance which connects to the open end of inductor. The actual inductance along with their parasitic resistance and capacitance requires specialized EM inductor simulation tools. For this reason, our methods rely on a pre-designed and pre-characterized library of inductors. The on-chip decoupling capacitors which we use are also widely utilized to manage transient power supply noise and are commonly implemented as polysilicon-insulator-polysilicon (PIP), MOS-based or metal-insulator-metal (MIM) capacitors [8].

3. METHODOLOGY

Figure 2(a) shows a resonant grid with a top-level buffered tree and three distributed LC tanks. The tanks are placed and sized so that the distributed clock grid capacitance resonates with the inductors at the operating frequency forming a resonant clock grid. The intersection of grid wires are called *grid nodes*. Clock sinks are connected to grid wires through *stubs*. The intersection nodes of grid wires and stubs are called *stub nodes*. A *node capacitance* is the sum of the capacitance of all wires, sinks and buffers which are

adjacent to a node. The grid wire, grid buffer and sink capacitances form the capacitor (C) in Figure 1(a) and, as such, the inductors (L) and decoupling capacitors (C_d) should be placed/sized according to the capacitive load of the grid. This requires several inductors because the clock capacitance is distributed throughout the die. The grid buffers are large enough only to compensate for the energy loss of the tanks due to parasitic resistance.

As previous authors have described, the main concerns in resonant clock design are the amplitude and phase of the distributed clock signal [7]. In resonant clock design, the amplitude of voltages v at each sink should be large enough for the CMOS sequential elements to fully switch. The different phases among buffers and LC tanks in a resonant clock grid causes *phase conflicts* which introduce extra *phase-conflict induced skew*. The authors of [7] introduced a small signal (AC) analysis methodology to compute the amplitude and phase of clock sinks in resonant clock networks. We utilize and expand on this previous methodology in LARCS.

Figure 2(b) is the flow of our resonant clock network synthesis methodology. Distributed LC tank resonant clocks need a relatively small change in design methodology compared to non-resonant clock grids. We use previous non-resonant clock grid methods to generate the grid [12, 18]. The buffer insertion and sizing, however, is different because smaller buffers are needed in a resonant grid and the buffer driving ability cannot be easily estimated by logical effort due to interaction with the LC tanks.

The resonant clock grid synthesis problem with a pre-made inductor library can be formulated as follows: Given an initial clock grid, an inductor library $\mathcal{L}$, and a maximum total inductor area A, insert LC tanks and insert/size buffers to minimize the power and skew of the resonant clock grid.

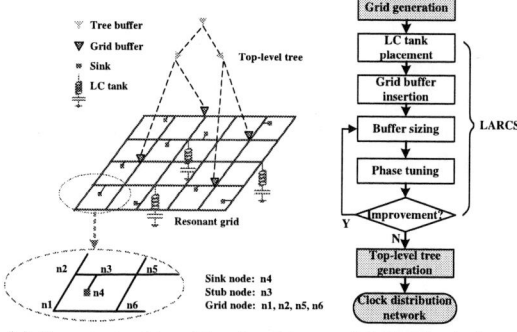

(a) Resonant grid and top-level tree (b) Synthesis flow

Figure 2: Distributed inductor-capacitor (LC) tank circuits are placed and directly attached to a clock grid in order to resonate it at the operating frequency and save power and buffer area

4. LC TANK PLACEMENT

We formulate our LC tank placement problem as a set covering problem of all grid nodes N. Candidate LC tanks are placed at various grid nodes which cover subsets of the nodes W. The objective is to find the optimal subset $T \subset W$ to cover all nodes N with minimum inductor area cost A. Each candidate $w \in W$ represents a set of nearby grid nodes which resonate with an inductor $l_w \in \mathcal{L}$ placed at grid node $p_w \in N$. The cost of the subset T is the total inductor area A required to complete the covering. Some common terms are defined to facilitate the algorithm discussion in Table 1.

4.1 Resistance (R) and Capacitance Mismatch (ΔC) Constraints

The parasitic resistance of the clock grid determines the efficiency of the resonant clock, because power is dissipated as heat when current passes through a resistance. A highly resistive path between a node n and an LC tank (i.e. a large r) means the charging and discharging currents between node n and the LC tank will pass through large resistances and consume more power. We address this by constraining each node capacitance to be less than a maximum resistance R away from the placed inductor in order to properly form an LC tank.

The mismatch between the total clock grid capacitance C_{clk} and the sum of all LC tank resonant capacitances C_T is ΔC. Ideally, ΔC equals zero which means that the clock capacitance is exactly covered by the LC tanks. In other words, each grid node resonates with only one LC tank within resistance R. However, since inductors can only be chosen from the library and they are discrete values, it is hard to guarantee that ΔC equals zero. ΔC is always non-negative because we require a full coverage of all grid nodes N by LC tanks. A positive ΔC, however, shifts the resonant frequency because some nodes resonate with more than one inductor which can increase power and skew. A large ΔC will require a significant amount of dummy capacitance to balance phase in Section 6, so our algorithm minimizes ΔC by setting an upper bound C_{MAX} during the LC placement algorithm and gradually loosening the constraint until a solution is found.

Table 1: Common terms in algorithm

N	the set of all grid nodes (capacitances).
$\mathcal{L}$	inductor library.
r	resistance from an LC tank to a node.
R	resistance limit from an LC tank to a node. When $r > R$, the LC tank is assumed to not resonate with this node capacitance.
R_{MAX}	maximum permissible value of R.
W	set of candidate subsets of grid nodes N. Each element w in W is a subset of grid nodes N. All nodes in w are within resistance R of a sized LC tank at a particular grid node.
l_w	the inductor size of candidate subset w, $l_w \in \mathcal{L}$.
p_w	the inductor position of candidate subset w, $p_w \in N$.
T	subset of W, $(T \subset W)$, that will cover all grid nodes N. Each grid node n will at least resonate with one inductor in T or equivalently $\bigcup_{T_i \in T} T_i = N$.
A_l	area of inductor l.
A	total inductor area used to cover T.
C_{clk}	total grid capacitance, including sink capacitance and wire capacitance.
C_l	ideal capacitance for an inductor l to resonate at the target frequency, $C_l = \frac{1}{(2\pi f)^2 l}$
C_T	total capacitance of subset cover T. $C_T = \sum_{t \in T} C_{l_t}$
ΔC	difference between clock capacitance C_{clk} and total cover capacitance C_T. $\Delta C = C_T - C_{clk}$
C_{MAX}	maximum permissible value of ΔC.

4.2 LC Tank Placement Overview

Algorithm 1 is the pseudo-code of our LC tank placement method. Initially, we set very restrictive bounds on both R and ΔC. After generating all candidate subsets W (Step 2) with the R constraint, the LC tank placement is solved as an integer linear programming (ILP) covering problem (Step 3) with a ΔC constraint. If a feasible solution is found with these tight constraints, the resonant clock grid will have a good performance in both power and skew.

Normally, however, the ILP cannot find a feasible solution with the initial constraints so we iteratively loosen the R and ΔC constraints and re-run step 2 and 3 until the ILP successfully returns a feasible solution. A large ΔC mismatch will induce extra skew, but

we can compensate for this by adding dummy capacitance which is expensive and not preferable as described later in Section 6. Therefore, we instead first increase the R limit which only decreases efficiency but does not add much skew. However, we do not allow the R limit to increase beyond R_{MAX} (Line 5), because a very large resistance means that an LC tank can not resonate efficiently with a node and may not result in a significant power saving as discussed in Section 4.1. We try to keep this resistance limit small, but must increase it to find a feasible solution. If a feasible solution is not found with $R = R_{MAX}$, we must start sacrificing skew by increasing C_{MAX} to find a feasible solution.

Algorithm 1 ILP set covering LC tank placement

1: Initialize R and ΔC with small values
2: Generate candidate subsets for each $\mathcal{L}$ with R constraint
3: Solve ILP set covering problem with ΔC constraint
4: **if** ILP can not successfully find a feasible solution with R and ΔC constraints **then**
5:    **if** R is less than a threshold value R_{MAX} **then**
6:       Loosen R constraint, increase R
7:    **else**
8:       Loosen ΔC constraint, increase C_{MAX}
9:       Reset R constraint, assign a small value to R
10:   **end if**
11: Go to step 2
12: **end if**

4.3 Candidate Subset Generation

To generate the candidate subsets W, we place each inductor in the library $\mathcal{L}$ at every potential grid node in N. When an inductor $l_w \in \mathcal{L}$ is inserted at a grid node $p_w \in N$, it is assumed to resonate with the set of nodes within resistance R. We add the closest (i.e. least resistance away) nodes until the total capacitance is greater than the ideal capacitance C_l for the inductor at the target resonant frequency. The resulting set of grid nodes is the candidate subset $w \subset N$ that is covered by inductor l_w at node p_w. Since the number of inductors $|\mathcal{L}|$ is typically small (3 in our case), the number of candidate subsets $|W|$ is not very large and does not add significant complexity to the ILP.

4.4 Candidate Subset Selection

We solve the set covering problem as $ILP(W, C_{clk}, C_{MAX}, A)$ which is

$$Min : \sum_{t \in T} A_t \times x_t, (t \in T, T \subset W) \qquad (1)$$

subject to:

$$x_t \in \{0, 1\}, t \in W \qquad (2)$$

$$\sum_{t \in T_n} x_t \geq 1, T_n = \{u | n \in u, u \in T\}, \forall n \in N \qquad (3)$$

$$\sum_{t \in T} A_{l_t} \times x_t \leq A \qquad (4)$$

$$0 \leq \Delta C \leq C_{MAX} \qquad (5)$$

$$\Delta C = \sum_{t \in T} (C_t \times x_t) - C_{clk} \qquad (6)$$

The inputs to the ILP are the candidate subsets of nodes (W) covered by sized inductors at grid nodes, the total grid capacitance (C_{clk}), the maximum allowed capacitance difference (C_{MAX}) between C_{clk} and total capacitance resonating with all inductors C_T, and the maximum total inductor area (A). The objective is to find a set T which is the subset of W that satisfies all the constraints (Equation 2-6) and minimizes the total inductor area A (Equation 1).

In Equation 2, x_t is a decision variable (0 or 1) that decides whether inductor t is selected to cover the subset or not. Equation 3 requires that every grid node n resonates with at least one inductor. T_n is a set of all subsets of T which contain node n. Equation 4 is the total inductor area constraint where l_t is the library inductor of subset t and A_{l_t} is its physical area. Equation 5-6 are the capacitance mismatch constraints that ensure the capacitance C_t resonated by each x_t is sufficient to resonate the total clock grid capacitance C_{clk} by limiting ΔC mismatch. If a subset of T can be found to satisfy all these constraints, the LC tanks of specified size are placed in the grid.

5. BUFFER INSERTION AND SIZING

After the placement of LC tanks, buffers must compensate for power loss due to parasitic resistance. In addition, we must reduce the phase difference due to unbalanced capacitance assigned to each inductor. Insufficient total buffer size will not be able to drive the grid and could prevent full voltage swing at the sinks. Unnecessarily large buffers, however, will consume extraneous power and nullify the advantages of a resonant clock. To address this, we insert buffers at each grid node where there is no LC tank and use an AC-based buffer. The AC-based analysis is similar to ROCKS [7].

Algorithm 2 AC-based buffer sizing to make all sinks full voltage swing

---

1: Set all buffers to minimum buffer size
2: AC analysis
3: Update partial-swing sinks set S_u
4: **while** $|S_u| > 0$ **do**
5:     Find all buffers RB that drive S_u
6:     Increase each buffer size in RB
7:     AC analysis
8:     Update partial-swing sinks set S_u
9: **end while**

Algorithm 2 is the pseudo-code of the buffer sizing algorithm. We initially set all buffers to the minimum size on Line 1 and perform an AC analysis to update the voltage amplitude/phase at all sinks. S_u on Line 3 is a subset of sinks that do not achieve full voltage swing. For each sink s in S_u, we find the buffers which are within resistance (R) of s that may influence the voltage swing and define these as the regional buffer set (RB) on Line 5. Note that R is the same parameter used in the tank placement Algorithm 1. Increasing the buffer size of each buffer in RB improves the voltage swing of the violating sinks, S_u. Before starting the next iteration, we update the sink voltage amplitude/phase using AC analysis and repopulate the set S_u on Lines 7 and 8. This buffer sizing procedure is repeated until all sinks have full swing. This buffer insertion and sizing is able to find a solution after several iterations because every grid node is connected to either a buffer or LC tank and the clock grid is mainly driven by LC tanks.

6. PHASE TUNING

At this point, the buffer sizing ensured that all sinks achieve a full voltage swing, but there may be phase conflicts among LC tanks and the newly sized buffers. The phase conflict will appear as skew, but, more importantly, it means that extraneous power is used when buffers and LC tanks conflict while driving the grid. The phase conflicts are caused by both the resistance of the CDN and the remaining ΔC after the LC tank placement.

Though these two effects are minimized, the sink phases do not match perfectly at the resonant frequency and can be improved further. If a capacitance resonates with an inductor at resonant frequency, the phase of this sink will ideally be zero. Some sinks

will have a phase less than zero which means that these sinks see a larger inductor than required for their capacitance. Other sinks will see a greater than zero phase due to long resistive paths to the nearest LC tank which appears as a smaller inductor than required due to resistive shielding. While we cannot fix the phase less than zero, we can reduce the positive phase differences by adding extra dummy capacitance.

Algorithm 3 is used to minimize the phase conflicts in the resonant grid. Lines 1-4 reduce the phase conflict induced by ΔC. For simplicity, we assume a node will only resonate with the closest LC tank (Line 1). After updating the grid capacitance C_g resonating with each LC tank (Line 2), we find the difference between C_g and C_l. C_l is the ideal capacitance value to resonate with inductor l at the target frequency. Adding an extra $C_l - C_g$ of dummy capacitance reduces the phase conflict and improves resonant performance. After this, another buffer resizing (Line 4) reduces the total buffer size required since less compensation by the buffers is required due to the improved efficiency.

Algorithm 3 Resonant clock phase tuning

1: Assign each node in N to the nearest LC tank, l
2: Calculate the capacitance seen by each LC tank, C_g
3: Add extra capacitance to the LC tank if $C_g < C_l$
4: Buffer sizing
5: **repeat**
6:     Add extra cap. to stub nodes of sinks with maximum phase
7:     Buffer sizing
8:     AC analysis
9: **until** No improvement in phase conflict and buffer area

---

In Lines 5-9, the phase of individual sink nodes are adjusted by adding small extra dummy capacitance. After each capacitor is added, the buffers are resized and AC analysis is run (Lines 7-8). This procedure is repeated until no improvement is seen in total buffer area and phase conflict. The buffer area is used as a quick estimate of the power of the grid during this procedure because buffers are used to compensate for the power loss from phase conflicts.

7. EXPERIMENTAL RESULTS

7.1 Setup

We implemented the previous resonant clock synthesis methodology in C++ and used the 45nm technology data from 2010 ISPD Clock Synthesis contest for all experiments [14]. The resonant clock frequency f_0 is set to 1GHz and V_{dd} is set to $1V$ as in previous resonant clock works [7]. HSPICE is used to accurately measure the power and skew of the final clock network.

The inductor library $\mathcal{L}$ contains three inductors: $3nH$, $5nH$ and $8nH$ with areas of $0.16mm^2$, $0.21mm^2$, and $0.30mm^2$, respectively. The on-chip inductors have parasitic resistance that effects the performance of the resonant clock. The quality factor $Q = \frac{\omega L}{R}$ of the inductors is 10 which is reasonable for modern on-chip inductors [4,9] in this frequency range.

The initial clock grids use a uniform spacing of $100\mu m$ for benchmarks 03-08 and $250\mu m$ for 01-02. In order to reduce the resistance in the resonant grids, we use $4\times$ minimum wire width. During the tuning phase, we set a phase limit of $0.12rad$ which is approximately equal to $20ps$ global skew at $1GHz$ in the time domain. A top-level tree driving the clock grid is generated using minimum wire length DME [17]. Equal levels of buffers are inserted to satisfy slew constraints [16] and then buffers are greedily inserted in a top-down manner to minimize skew in the tree. The buffered tree and grid compose the entire CDN.

Table 2: LARCS results on ISPD 2010 benchmarks show power is 41% of non-resonant power while skew is limited to 17ps on average.

	Non-resonant CDN				ROCKS [7]					LARCS				
	Sink	MA	$\frac{S.Cap}{MA}$	Pwr.[1]	Pwr.[2]	$\frac{Pwr.^2}{Pwr.^1}$	$\frac{LA}{MA}$	$\frac{LLA}{MA}$	Skew	Pwr.[3]	$\frac{Pwr.^3}{Pwr.^1}$	$\frac{LA}{MA}$	Skew	Time
	#	mm^2	$\frac{pF}{mm^2}$	mW	mW	%	%	%	ps	mW	%	%	ps	m
01	1107	64.0	0.3	472	81	17	98	196	23	368	78	29	32	130
02	2249	91.0	0.4	708	134	19	97	195	46	364	51	30	33	212
03	1200	1.4	12.9	66	33	50	78	229	34	26	39	92	12	2
04	1845	5.7	2.2	177	90	50	46	130	10	46	26	69	14	58
05	1016	5.8	0.9	85	58	68	28	74	6	48	57	29	16	23
06	981	1.5	8.4	78	26	33	96	258	24	23	29	100	10	2
07	1915	3.5	5.2	156	78	50	31	60	9	39	25	96	14	197
08	1134	2.6	5.0	111	39	35	79	196	16	27	24	81	8	3
Avg.	1431	21.9	4.4	232	67	29	69	167	21	118	41	66	17	78

Pwr.[1]: Switched capacitance CDN power. Pwr.[2]: ROCKS power in HSPICE Pwr.[3]: LARCS power in HSPICE.
$\frac{S.Cap}{MA}$: sink density per area. LA/MA: total inductor area normalized to metal layer area.
MA: single metal layer (chip) area. LLA/MA: total mapped inductor area, normalized to metal layer area.

To compare the power efficiency of our method, we compare the resonant power including dynamic, short-circuit and leakage of all buffers, wires, inductors, and sinks with the switched capacitance power. Since a traditional buffered CDN would require larger grid buffers, this creates conservative estimate of the power saved, but this avoids potential bias of other CDN synthesis methodologies due to the CDN topology, sizing, etc. We also compare our results with an implementation of ROCKS [7] obtained directly from its authors. ROCKS is the only prior work on resonant grid synthesis, but since ROCKS uses continuously sized inductors, we added a discretization step to map to the inductor library. For accurate library mapping, our algorithm considers both parallel and series combinations of inductors to create the best matches. While this consumes extra inductor resources, it is the only feasible method to not horribly degrade both power and skew.

7.2 Experimental Data

We generated resonant clock networks for the ISPD 2010 benchmarks [14] using LARCS. Table 2 lists the simulation results of the resonant grid with top-level tree.

7.2.1 Power

The dynamic power of a non-resonant clock is $P_{dyn} = CV_{dd}^2 f$ and is shown as "Pwr.[1]" in Table 2. The "Pwr.[2]" and "Pwr.[3]" are the resonant clock power of ROCKS and LARCS measured by HSPICE including all buffers, wires, inductors, sinks and short-circuit power. The "Pwr.[2]/Pwr.[1]" and "Pwr.[3]/Pwr.[1]" ratios in Table 2 show the relative power savings of the resonant clock compared to the total switched capacitance (dynamic) power. On average, the power of LARCS is 41% of the switched capacitance dynamic power while ROCKS is slightly lower at 29%, but ROCKS requires drastically more inductor resources.

7.2.2 Skew

The average global skew of our resonant clocks is 17ps. Except for the two largest benchmarks, 01 and 02, all the other skews are significantly less than 20ps. At 1GHz, 20ps global skew is about 2% of clock cycle which is very aggressive.

The skews of benchmark 01 and 02 are just above 30ps. Compared with other benchmarks, these two benchmarks are much larger and have relatively low sink density as shown in the column "$\frac{S.Cap}{MA}$" in Table 2. Because of this, the resistive shielding effect becomes more significant and makes it difficult to tune for phase conflicts. To improve the skew of a big chip like this, we can use wider wires to reduce the wire resistance and put a tighter constraint on the maximum resistance (R_{max}) when generating the subsets for LC tank placement.

7.2.3 Inductor Q factor

The quality factor of the inductors affects the performance of the resonant clock. We simulated benchmark 03 with inductor Q ranging from 1 to 100 and plot the power reduction in Figure 3. The inductor parasitic resistance is significant for low quality inductors ($Q < 10$) and becomes less significant with $Q > 20$. Even with a very poor $Q = 1$, we see more than 25% power savings. The effect of inductor Q on skew is not very significant. Skew changes by only $2 - 5ps$ over the entire range of Q.

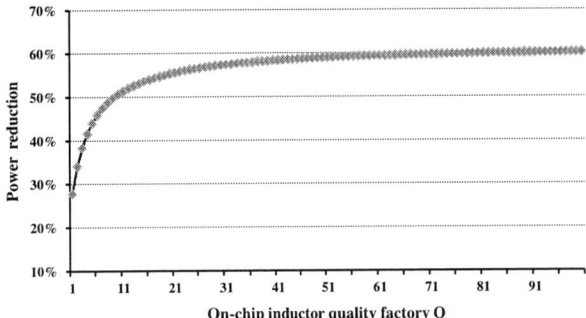

Figure 3: Significant power gains are obtained even with low-Q on-chip inductors

7.2.4 Robustness

We performed Monte Carlo HSPICE simulations with $\pm15\%$ 3-sigma variation of buffer gate lengths, wire widths and sink load variations to assess robustness. We assume ideal inductors as Q was shown to have little effect on skew in Section 7.2.3. Each benchmark was simulated 500 times and the mean (μ) and standard deviation (σ) of global skew were measured. We compare the resonant CDNs with the baseline CDNs which are uniform grids driven by top-level buffered trees. The grid buffers in the baseline grid are placed and sized using a method similar to [18]. The results are listed in Table 3.

The deterministic skew (i.e. without variations) is shown as Det in Table 3. While the average deterministic and mean expected skew (μ) skew of the resonant CDN are slightly worse than baseline CDN, the standard deviation (σ) of the resonant CDN is 2.5ps compared to 4.0ps for the baseline. The resonant clocks have an average $\mu + 3\sigma$ skew of 29ps while the baseline clocks are 32ps. While clock grids are known for their robustness to variation, the resonant grid is able to provide more robustness with less power due to reduced dependence on clock buffers.

7.2.5 Costs of Resonant Grid

The cost of implementing a resonant clock is the time to create and characterize the inductors along with the total metal area occupied by the instances of these on-chip inductors.

Table 3: Resonant clock distributions exhibit decreased worst case skew variation (29ps compared to 32ps) when considering process variations.

Ben.	Baseline CDN				LARCS			
	Det (ps)	μ (ps)	σ (ps)	$\mu+3\sigma$ (ps)	Det (ps)	μ (ps)	σ (ps)	$\mu+3\sigma$ (ps)
01	16	27	5.6	43.8	32	38	5.7	55.1
02	26	39	8.1	63.3	33	41	5.3	57.0
03	15	17	3.1	26.3	12	15	1.6	19.8
04	12	14	3.2	23.6	14	20	1.6	24.5
05	13	15	3.1	24.3	16	18	2.1	23.8
06	13	15	2.5	22.5	10	13	1.1	16.9
07	13	15	3.6	25.8	14	19	1.3	23.1
08	17	19	3.1	28.3	8	10	1.0	13.1
Avg.	15.6	20.1	4.0	32.2	17.2	21.8	2.5	29.1

As on-chip inductors are often implemented with metal, we assume inductors are restricted to one metal layer and set the maximum inductor area A so that inductors can only occupy this metal routing layer. Therefore, we normalize the inductor area to the area of this one metal layer in Table 2. The average inductor area is 66% of one metal layer, but it should be noted that this metal layer can overlap other devices and routing and is not wasted space. When presented with the same constraint, ROCKS uses 69% of the metal layer on average with continuous inductor sizes. The ROCKS inductors, however, range from a few tenths to 10's of nH and must be rounded to the pre-made library inductors. After the replacement, the average inductor area usage is increased to 167% as shown in the column "$\frac{LLA}{MA}$" in Table 2 which requires multiple layers of metal and is not practical.

The decoupling capacitance of the LC tanks are sized to 10 times the resonant capacitance [1, 13] and additional capacitors are used to reduce the phase conflict. From our experiments, ΔC is observed to be between $0.1-0.3\times$ the grid capacitance and the capacitance inserted to compensate for ΔC is typically $10-25pF$ except for benchmarks 01-02 which have odd sink densities as described earlier. Including the both phase tuning and decoupling capacitance, the maximum total capacitance required of all benchmarks is $10.2nF$ for benchmark 02 while the average is just $2.5nF$. To put this in perspective, this maximum capacitance is only 5.7% of the decoupling capacitance used on a Pentium II [11] which is $113mm^2$, a little bit larger than benchmark 02.

7.2.6 CPU Time

The average time to synthesize the resonant clocks is 78.5 minutes. The CPU time in Table 2 is the total run time including grid generation, LC tank placement sizing, phase tuning and top-level tree generation with skew reduction. In the experiments, 5%-30% of the total run time is spent on the actual resonant clock grid generation.

8. CONCLUSIONS

We demonstrated the first Library-Aware Resonant Clock Synthesis (LARCS) algorithm for distributed LC resonant clock grids. This method uses a pre-characterized inductor library which requires less design time than methods with continuous inductor sizes. With a library of three inductors, this method used nearly one-third of the total inductor area while only decreasing power efficiency modestly when compared to a rounded solution of prior continuous-sized inductor synthesis algorithms. Our resonant clock grids, however, still reduced power by 59% over previous non-resonant clock grids and achieved comparable skews with improved robustness.

Acknowledgments

This work was supported in part by the National Science Foundation under grant CCF-1053838. Any opinions, findings, and conclusions or recommendations expressed herein are those of the authors and do not necessarily reflect the views of the NSF.

9. REFERENCES

[1] S. Chan, P. Restle, K. Shepard, N. James, and R. Franch. A 4.6GHz resonant global clock distribution network. *ISSCC*, pages 342 – 343, 2004.

[2] S. C. Chan, P. J. Restle, T. J. Bucelot, J. S. Liberty, S. Weitzel, J. M. Keaty, B. Flachs, R. Volant, P. Kapusta, and J. S. Zimmerman. A resonant global clock distribution for the cell broadband engine processor. *JSSC*, 2009.

[3] V. Chi. Salphasic distribution of clock signals for synchronous systems. *IEEE Transactions on Computers*, 43(5):597 – 602, 1994.

[4] Y.-S. Choi and J.-B. Yoon. Experimental analysis of the effect of metal thickness on the quality factor in integrated spiral inductors for RF ICs. *IEEE Electron Device Letters*, 25:76–79, 2004.

[5] A. Drake, K. Nowka, T. Nguyen, J. Burns, and R. Brown. Resonant clocking using distributed parasitic capacitance. *JSSC*, 39(9):1520 – 1528, Sep 2004.

[6] M. R. Guthaus. Distributed LC resonant clock tree synthesis. In *ISCAS*, pages 1215–1218, 2011.

[7] X. Hu and M. R. Guthaus. Distributed resonant clock grid synthesis (ROCKS). In *DAC*, pages 516–521, 2011.

[8] R. Jakushokas, M. Popovich, A. V. Mezhiba, S. Kose, and E. G. Friedman. *Power Distribution Networks with On-Chip Decoupling Capacitors*. Springer, 2011.

[9] L. Liu, M. Yu, J. Wang, and M. Jiang. A wideband circuit model of on-chip spiral inductor in 0.13um CMOS process. *IEEE International Symposium on Microwave, Antenna, Propagation and EMC Technologies for Wireless Communications*, pages 136–139, 2009.

[10] F. O'Mahony, C. Yue, M. Horowitz, and S. Wong. Design of a 10GHz clock distribution network using coupled standing-wave oscillators. In *DAC*, pages 682–687, 2003.

[11] M. D. Pant, P. Pant, and D. S. Wills. On-chip decoupling capacitor optimization using architectural level prediction. *TVLSI*, 10:319–326, June 2002.

[12] A. Rajaram and D. Z. Pan. Meshworks: an efficient framework for planning, synthesis and optimization of clock mesh networks. In *ASP-DAC*, 2008.

[13] J. Rosenfeld and E. Friedman. Design methodology for global resonant H-tree clock distribution networks. *TVLSI*, 15(2):135–148, February 2007.

[14] C. N. Sze. ISPD 2010 high performance clock network synthesis contest. In *ISPD*, 2010.

[15] B. Taskin, J. Demaio, O. Farell, M. Hazeltine, and R. Ketner. Custom topology rotary clock router with tree subnetworks. *TODAES*, 14(3), May 2009.

[16] G. E. Tellez and M. Sarrafzadeh. Minimal buffer insertion in clock trees with skew and slew rate constraints. *TCAD*, 16(4):333–342, 1997.

[17] R.-S. Tsay. Exact zero skew. In *ICCAD*, pages 336–339, 1991.

[18] G. Venkataraman, Z. Feng, J. Hu, and P. Li. Combinatorial algorithms for fast clock mesh optimization. In *ICCAD*, pages 563–567, 2006.

[19] J. Wood, T. C. Edwards, and S. Lipa. Rotary traveling-wave oscillator arrays: A new clock technology. *JSSC*, 36(11):1654–1664, 2001.

[20] Z. Yu and X. Liu. Implementing multiphase resonant clocking on a finite-impulse response filter. *TVLSI*, 17(11):1593 – 1601, Nov 2009.

[21] C. Ziesler, S. Kim, and M. Papaefthymiou. A resonant clock generator for single-phase adiabatic systems. *ISLPED*, 2001.

Incremental Power Grid Verification*

Abhishek
ECE Department
University of Toronto
Toronto, Ontario, Canada
abhishek@eecg.utoronto.ca

Farid N. Najm
ECE Department
University of Toronto
Toronto, Ontario, Canada
f.najm@utoronto.ca

ABSTRACT

Verification of the on-die power grid is a key step in the design of complex high-performance integrated circuits. For the very large grids in modern designs, *incremental verification* is highly desirable, because it allows one to skip the verification of a certain section of the grid (internal nodes) and instead, verify only the rest of the grid (external nodes). We propose an efficient approach for incremental verification in the context of *vectorless* constraints-based grid verification, under *dynamic* conditions. The traditional difficulty is that the dynamic case requires iterative analysis of both the internal and external sections. This has been previously overcome for *simulation* purposes, but we provide the first solution for *verification*, through two key contributions: 1) a bound on the internal nodes' voltages is developed that eliminates the need for iterative analysis, and 2) a multi-port Norton approach is used to construct a reduced macromodel for the internal section. As a result, we demonstrate significant reductions in runtime, with speed-ups in the range of 3-8x, with negligible impact on accuracy.

Categories and Subject Descriptors

B.7.2 [**Integrated Circuits**]: Design Aids

General Terms

Performance, Algorithms, Verification

Keywords

Power Grid, voltage drop

1. INTRODUCTION

As supply voltages have decreased with technology scaling, the performance and reliability of modern integrated circuits (IC) have become increasingly susceptible to supply voltage fluctuations. Reduced voltage levels degrade the circuit timing performance and can lead to soft errors. Therefore, voltage integrity verification has become crucial for reliable high-speed chip design.

Most grid verification techniques use some form of circuit simulation to simulate the grid. Simulation-based approaches require complete knowledge of current waveforms drawn by the underlying logic circuitry, which are used to simulate the grid and determine the grid node voltage drops. However, verifying the grid in this way is prohibitively expensive, because the number of current traces required to cover all possible circuit behaviors is extremely

---

*This work was supported in part by Advanced Micro Devices (AMD) and the Natural Sciences and Engineering Research Council (NSERC) of Canada.

large. Another disadvantage is that a simulation-based flow does not allow for early grid verification (when changes to the grid can most easily be incorporated) because no current traces may be available at that time.

To overcome these issues, a *vectorless* verification approach based on partial current specification in the form of *current constraints* was proposed in [1], and further developed in subsequent work over the last decade. Grid verification is reduced to a problem of finding the worst-case voltage drop over all possible currents that satisfy certain current constraints. In [2], the authors used an *RC* model of the power grid and gave an upper bound on the worst-case voltage drop using an iterative approach. A closed form expression for the upper bound was later proposed in [3], which involved solving a *linear program* (LP) for every grid node. An efficient way to reduce the size of the LPs based on the sparse approximate inverse technique was proposed in [4].

This previous work is useful for verifying the entire power grid but becomes an overkill when verification of only a part of the grid is required, a scenario that we refer to as *incremental verification*. In large modern grids, incremental verification has become desirable because the grids can be so large that verification on traditional workstations becomes impossible, and a divide-and-conquer approach becomes a necessity. Alternatively, incremental verification is desirable when design changes are made to a local region of a previously-verified grid, and the local impact of these changes needs to be verified. There are also various other cases, such as in case of IP reuse, where a portion of the grid may not need to be verified. In [5], a technique is given for incremental verification but only for the case of a resistive grid, under the influence of DC currents. In this paper, we extend incremental verification to the case of transient currents, i.e., a *dynamic verification* context, for the case of *RC* grids.

In our formulation, the user identifies a part of the grid that does not need to be verified, which we refer to as the *subgrid*. Verification is required only for grid nodes that are outside the subgrid, referred to as *external nodes* (nodes inside the subgrid are referred to as either *internal nodes* or *port nodes*, as we will see). A subgrid can be an arbitrary section of the grid, but must be a connected graph. Strictly speaking, the solution in the dynamic case requires an iterative relaxation-based analysis of both the internal and external grid sections. This difficulty was overcome in [6] for the purpose of circuit simulation (not vectorless verification), through the use of a multi-port Norton theorem. In this work, we provide the first solution in the dynamic case for the purpose of incremental vectorless verification, based on two contributions: 1) upper bounds on the voltage drops at internal nodes are efficiently computed, and used in lieu of the worst-case drops, and 2) a macromodel is constructed for the subgrid based on the movement of internal current sources by adapting the multi-port Norton theorem proposed in [6] to our verification framework, followed by reduction of a passive *RC* circuit by combining the moment-matching based approach as described in [7] and [8] with the nodal elimination based approach of [9].

The remainder of the paper is organized as follows. In section 2, we present the power grid model and the constraints-based approach to vectorless verification. Section 3 describes our pro-

151

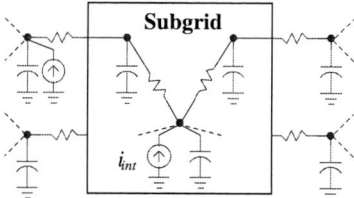

Figure 1: Original Grid

posed incremental verification approach. Implementation details are given in section 4, followed by experimental results in section 5. Finally, the paper is concluded in section 6.

2. BACKGROUND
2.1 The Power Grid Model

Consider an RC model of the grid where each branch is represented by a resistor and where there exists a capacitor from every node to ground. Some nodes have ideal current sources (to ground) to represent the current drawn by underlying circuitry, and some have ideal voltage sources to represent the connections to external power supply. Let the power grid consist of $n + p$ nodes, where nodes $1, 2, \ldots, n$ have no voltage sources attached, and the remaining nodes are nodes where the p voltage sources are attached. Let $i(t)$ be the element-wise non-negative vector of all current sources connected to the grid. We assume that $\forall k = 1, \ldots, n$, $i_k(t)$ is well-defined, so that nodes with no current source attached have $i_k(t) = 0$. Let $i(t)$ be the vector of all current sources $i_k(t)$ and $u(t)$ be the vector of nodal voltages. Applying Modified Nodal Analysis (MNA) to the grid leads to:

$$\mathbf{G}u(t) + \mathbf{C}\dot{u}(t) = -i(t) + \mathbf{G}_0 V_{dd} \qquad (1)$$

where $\mathbf{G}$ and $\mathbf{G}_0$ are $n \times n$ conductance matrices, $\mathbf{C}$ is a $n \times n$ diagonal matrix of node capacitances, and V_{dd} is a constant vector each entry of which is equal to the supply voltage value. The matrix $\mathbf{G}$ is known to be diagonally-dominant, symmetric positive-definite, and an $\mathcal{M}$-matrix (so that $\mathbf{G}^{-1} \geq 0$). Let $v(t) = V_{dd} - u(t)$ be the vector of voltage drops. The RC model for the power grid can then be written as [2]:

$$\mathbf{G}v(t) + \mathbf{C}\dot{v}(t) = i(t) \qquad (2)$$

Note that this equation can be obtained directly by writing the MNA system for a modified network in which all voltage sources are shorted (set to 0) and all current sources are reversed. In the rest of this paper, it will be assumed that we are working with this *modified* topology so that, for example, certain power grid nodes may be connected to ground by a resistor.

In our proposed framework, the nodes inside the subgrid that are connected to external nodes are called port nodes while all the remaining subgrid nodes are referred to as internal nodes. Let n_{ext}, n_{prt}, and n_{int} be the number of external nodes, port nodes, and internal nodes respectively such that $n_{ext} + n_{prt} + n_{int} = n$. Because external nodes connect only to port nodes, the grid equation can now be written as:

$$\begin{bmatrix} \mathbf{G}_{11} & \mathbf{G}_{12} & 0 \\ \mathbf{G}_{12}^T & \mathbf{G}_{22} & \mathbf{G}_{23} \\ 0 & \mathbf{G}_{23}^T & \mathbf{G}_{33} \end{bmatrix} \begin{bmatrix} v_{ext}(t) \\ v_{prt}(t) \\ v_{int}(t) \end{bmatrix} + \begin{bmatrix} \mathbf{C}_{ext} & 0 & 0 \\ 0 & \mathbf{C}_{prt} & 0 \\ 0 & 0 & \mathbf{C}_{int} \end{bmatrix}$$
$$\begin{bmatrix} \dot{v}_{ext}(t) \\ \dot{v}_{prt}(t) \\ \dot{v}_{int}(t) \end{bmatrix} = \begin{bmatrix} i_{ext}(t) \\ i_{prt}(t) \\ i_{int}(t) \end{bmatrix} \qquad (3)$$

where v_{ext} and i_{ext} are sub-vectors corresponding to voltage drops and current sources at external nodes, v_{prt} and i_{prt} correspond to voltage drops and current sources at port nodes, and v_{int} and i_{int} correspond to voltage drops and current sources at internal nodes. The matrices $\mathbf{G}$ and $\mathbf{C}$ are partitioned into sub-matrices of appropriate dimensions.

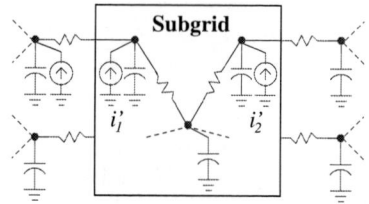

Figure 2: Modified Grid

Using a finite difference approximation as in [2], the system (3) can be written as:

$$\mathbf{A}v(t) = \frac{\mathbf{C}}{\Delta t}v(t - \Delta t) + i(t) \qquad (4)$$

where $\mathbf{A} = \left(\mathbf{G} + \frac{\mathbf{C}}{\Delta t}\right)$ is also a symmetric positive-definite $\mathcal{M}$-matrix, so that $\mathbf{A}^{-1} \geq 0$.

2.2 Current Constraints

We perform verification using a vectorless approach in which the information about the currents drawn by underlying circuitry is not known. Instead, current constraints [1] are used, which capture the uncertainty about circuit behaviors and the fact that one is uncertain about the circuit currents early in the design flow. As in previous work, we use two types of constraints: *local constraints* and *global constraints*.

Local constraints are upper bounds on the individual current sources, where one specifies that the current $i_k(t)$ never exceeds a certain fixed level $i_{L,k}$. We assume that every current source tied to the grid has an upper bound associated with it, so that if a node does not have a current source attached, the upper bound for that current is 0. We can express these constraints as:

$$0 \leq i(t) \leq i_L, \quad \forall t \geq 0 \qquad (5)$$

If only local constraints are provided, the problem is much simplified but the results become overly pessimistic, because it can never be the case that all the components of the chip are simultaneously drawing maximum currents. Global constraints are upper bounds on the sums of currents for groups of current sources. They represent the peak total power dissipation of a group of circuit blocks. Assuming that we have a total of m global constraints, then we can express them in matrix form as:

$$0 \leq \mathbf{S}i(t) \leq i_G, \quad \forall t \geq 0 \qquad (6)$$

where $\mathbf{S}$ is a $m \times n$ matrix that contains only 0s and 1s, which indicate which current sources are present in each global constraint. Together, local and global constraints define a *feasible space* of currents, denoted by $\mathcal{F}$, such that $i(t)$ lies inside the feasible space ($i(t) \in \mathcal{F}$) if and only if it satisfies (5) and (6).

2.3 Power Grid Verification

Given a power grid, we are interested in finding the worst-case voltage drops at all the nodes, under all possible (transient) current waveforms $i(t)$ that satisfy the current constraints. In [3], the authors provide an upper bound v_{ub} on the worst-case voltage drop vector, so that $v(t) \leq v_{ub}, \forall t$, and this bound is given by:

$$v_{ub} = \left(\mathbf{I} + \mathbf{G}^{-1}\frac{\mathbf{C}}{\Delta t}\right) V_a \qquad (7)$$

where V_a is the worst-case voltage drop vector at $t = \Delta t$ in the special case when $i(t) = 0, \forall t \leq 0$. Since $v(0) = 0$ in this special case, it follows from (4) that:

$$v(\Delta t) = \mathbf{A}^{-1}i(\Delta t) \qquad (8)$$

and V_a can be expressed as:

$$V_a = \operatorname*{emax}_{\forall i(\Delta t) \in \mathcal{F}} \mathbf{A}^{-1}i(\Delta t) \qquad (9)$$

152

where emax is an operator that denotes element-wise maximization of its vector argument, under the given constraints. In other words, V_a is the result of the `for`-loop: `for`$(k = 1, \ldots, n)\{$maximize the kth element of the vector $\mathbf{A}^{-1}i(\Delta t)$, over all $i(\Delta t) \in \mathcal{F}\}$. Maximizing each element becomes a linear program (LP). Note that, because the definition of local and global constraints does not depend on time, then $\mathcal{F}$ is the same at every time point. Therefore, V_a is independent of t and we can drop the Δt argument, and write:

$$V_a = \underset{\forall i \in \mathcal{F}}{\mathrm{emax}}\, \mathbf{A}^{-1}i \qquad (10)$$

where, for the purpose of this optimization, i can be viewed as simply a "dummy variable", a $n \times 1$ real vector with units of current. Thus, the problem of finding the worst-case voltage drop is reduced to performing element-wise maximization of $\mathbf{A}^{-1}i$ over all $i \in \mathcal{F}$, to find V_a, followed by a standard linear system solve, to find v_{ub}.

3. PROPOSED APPROACH

In our proposed incremental verification approach, we are interested in the worst-case voltage drops at only the external nodes. As our first contribution in this paper, we propose an efficient way to compute the bounds on the worst-case voltage drops, benefiting from the fact that voltage drops at internal nodes are not required.

3.1 Efficient Bounds Computation

To compute v_{ub}, we need to have an estimate of worst-case voltage drop at internal nodes. From (3), we have:

$$\mathbf{G}_{23}^T v_{prt}(t) + \mathbf{G}_{33} v_{int}(t) + \mathbf{C}_{int}\dot{v}_{int}(t) = i_{int}(t) \qquad (11)$$

which, after time-discretization, leads to:

$$v_{int}(\Delta t) = \mathbf{A}_{int}^{-1} i_{int}(\Delta t) - \mathbf{A}_{int}^{-1}\mathbf{G}_{23}^T v_{prt}(\Delta t) \qquad (12)$$

where $\mathbf{A}_{int} = \left(\mathbf{G}_{33} + \frac{\mathbf{C}_{int}}{\Delta t}\right)$ is a symmetric positive-definite $\mathcal{M}$-matrix, so that $\mathbf{A}_{int}^{-1} \geq 0$. Because $-\mathbf{G}_{23}^T$ and $\mathbf{A}_{int}^{-1}$ are non-negative matrices, then in the special case used above to define V_a, we can write:

$$\underset{\forall i \in \mathcal{F}}{\mathrm{emax}}(v_{int}(\Delta t)) \leq \mathbf{A}_{int}^{-1} \underset{\forall i \in \mathcal{F}}{\mathrm{emax}}(i_{int}(\Delta t)) + \mathbf{T}^T \underset{\forall i \in \mathcal{F}}{\mathrm{emax}}(v_{prt}(\Delta t)) \qquad (13)$$

where $\mathbf{T} = -\mathbf{G}_{23}\mathbf{A}_{int}^{-1}$ is a transformation matrix that will also be useful later. Equation (13) gives an upper-bound on the worst-case voltage drops at $t = \Delta t$ for all internal nodes, so that we can write:

$$V_a = \underset{\forall i \in \mathcal{F}}{\mathrm{emax}}(v(\Delta t)) \leq \begin{bmatrix} \mathbf{I}_{ext} & 0 & 0 \\ 0 & \mathbf{I}_{prt} & 0 \\ 0 & \mathbf{T}^T & \mathbf{A}_{int}^{-1} \end{bmatrix} \underset{\forall i \in \mathcal{F}}{\mathrm{emax}} \begin{bmatrix} v_{ext}(\Delta t) \\ v_{prt}(\Delta t) \\ i_{int}(\Delta t) \end{bmatrix}$$

where $\mathbf{I}_{ext}$ and $\mathbf{I}_{prt}$ are identity matrices of sizes n_{ext} and n_{prt}, respectively. From this, and because $\mathbf{G}^{-1} \geq 0$, we have from (7) that:

$$v_{ub} \leq \left(\mathbf{I} + \mathbf{G}^{-1}\frac{\mathbf{C}}{\Delta t}\right)\begin{bmatrix} \mathbf{I}_{ext} & 0 & 0 \\ 0 & \mathbf{I}_{prt} & 0 \\ 0 & \mathbf{T}^T & \mathbf{A}_{int}^{-1} \end{bmatrix} \underset{\forall i \in \mathcal{F}}{\mathrm{emax}} \begin{bmatrix} v_{ext}(\Delta t) \\ v_{prt}(\Delta t) \\ i_{int}(\Delta t) \end{bmatrix} \qquad (14)$$

Because $\mathrm{emax}_{\forall i \in \mathcal{F}}(i_{int}(\Delta t)) = i_L$ (the vector of local constraint values), this gives a faster way to compute an upper bound on v_{ub} which involves solving LPs for external and port nodes only, followed by standard linear solve. Our second contribution is the macromodeling of the internals of the subgrid, as described in the next sub-section.

3.2 Power Grid Macromodeling

Because the internals of the subgrid do not need to be verified, further performance improvement can be obtained by reducing or eliminating much of the subgrid network. Two steps are involved in this: 1) moving the internal current sources to the port nodes, which benefits from multi-port Norton theorem from previous work [6], and 2) reducing the remaining parasitic RC network inside the subgrid using model order reduction.

3.2.1 Moving Internal Current Sources

Norton's theorem is a fundamental theorem in circuit theory that converts any linear two-terminal network into a simple parallel circuit consisting of an equivalent current source, and an equivalent internal impedance. The equivalent current source value is the current that will flow through a short circuit between the two terminals [10]. In HiPRIME [6], multi-port Norton equivalent circuits were used to move the current sources internal to a block, to the ports. Previous work benefited from the multi-port Norton theorem for simulation purposes. In our work, we adapt this theorem for use in verification, where the current sources are not known, but are instead subject to current constraints.

Norton Equivalent Current Sources.

Applying the multi-port Norton theorem entails removing the current sources internal to the subgrid and replacing them by new current sources attached to the port nodes. The values of the new port current sources are found by a familiar construction in which 1) the subgrid is disconnected from the rest of the grid, 2) each port node is connected to ground via a short circuit, and 3) the current flowing through these short circuit connections (due to the applied internal current sources) is evaluated.

Consider a subgrid with n_{int} internal nodes and which is connected to the external grid through n_{prt} port nodes. The grid equation of the subgrid when the port nodes are shorted to ground is given by:

$$\mathbf{G}_{33}v'_{int}(t) + \mathbf{C}_{int}\dot{v}'_{int}(t) = i_{int}(t) \qquad (15)$$

where v'_{int} is the $n_{int} \times 1$ voltage drop vector at internal nodes. Let us call an internal node that connects to a port node k, a neighbor of k. The current through a port node to ground $i'_k(t)$ is given by:

$$i'_k(t) = \sum_{\text{neighbors } j \text{ of } k} g_{kj}v'_{int_j}(t) \qquad (16)$$

where g_{kj} is the conductance through which port node k is connected to internal node j and $v'_{int_j}(t)$ is the voltage drop for internal node j. In (3), $\mathbf{G}_{23}$ is the $n_{prt} \times n_{int}$ matrix consisting of all the conductance links from port nodes to internal nodes. Therefore, using (16), the Norton current vector $i'(t)$ can be written as:

$$i'(t) = -\mathbf{G}_{23}v'_{int}(t) \qquad (17)$$

Modified Grid.

The grid resulting after the internal current sources of the subgrid have been removed and replaced by the new port current sources will be referred to as the *modified grid*. In this modified grid, the voltage drops at the nodes will be denoted by $\hat{v}(t)$, and the system equation becomes:

$$\mathbf{G}\begin{bmatrix} \hat{v}_{ext}(t) \\ \hat{v}_{prt}(t) \\ \hat{v}_{int}(t) \end{bmatrix} + \mathbf{C}\begin{bmatrix} \dot{\hat{v}}_{ext}(t) \\ \dot{\hat{v}}_{prt}(t) \\ \dot{\hat{v}}_{int}(t) \end{bmatrix} = \begin{bmatrix} \mathbf{I}_{ext} & 0 & 0 \\ 0 & \mathbf{I}_{prt} & 0 \\ 0 & 0 & 0 \end{bmatrix}\begin{bmatrix} i_{ext}(t) \\ i_{prt}(t) \\ i_{int}(t) \end{bmatrix} + \begin{bmatrix} 0 \\ \mathbf{I}_{prt} \\ 0 \end{bmatrix}i'(t)$$

which can also be written as:

$$\mathbf{G}\hat{v}(t) + \mathbf{C}\dot{\hat{v}}(t) = \mathbf{M}i(t) + \mathbf{N}i'(t) \qquad (18)$$

where $\mathbf{M}$ is a $n \times n$ matrix consisting of $\mathbf{I}_{ext}$ and $\mathbf{I}_{prt}$, and $\mathbf{N}$ is a $n \times n_{prt}$ matrix consisting of $\mathbf{I}_{prt}$. Time-discretizing (18) gives:

$$\mathbf{G}\hat{v}(t) + \frac{\mathbf{C}}{\Delta t}\big(\hat{v}(t) - \hat{v}(t - \Delta t)\big) = \mathbf{M}i(t) + \mathbf{N}i'(t) \qquad (19)$$

We now return to the special case situation used to define V_a earlier. In that case, the voltage in the modified grid at time $t = \Delta t$ is given by:

$$\hat{v}(\Delta t) = \mathbf{A}^{-1}\big(\mathbf{M}i(\Delta t) + \mathbf{N}i'(\Delta t)\big) \qquad (20)$$

153

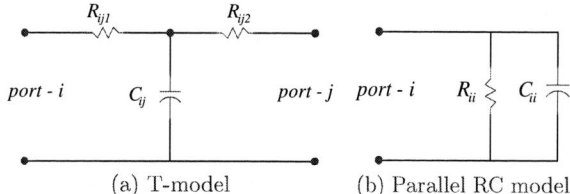

(a) T-model   (b) Parallel RC model

Figure 3: Macromodels for synthesis

Likewise, time discretizing (15) and evaluating at $t = \Delta t$, we get:

$$v'_{int}(\Delta t) = \mathbf{A}_{int}^{-1} i_{int}(\Delta t) \qquad (21)$$

From (17) and (21),

$$
\begin{aligned}
i'(\Delta t) &= -\mathbf{G}_{23} v'_{int}(\Delta t) \\
&= -\mathbf{G}_{23} \mathbf{A}_{int}^{-1} i_{int}(\Delta t) \equiv \mathbf{T} i_{int}(\Delta t)
\end{aligned}
\qquad (22)
$$

which can also be written as:

$$
i'(\Delta t) = \begin{bmatrix} 0 & 0 & \mathbf{T} \end{bmatrix}
\begin{bmatrix} i_{ext}(\Delta t) \\ i_{prt}(\Delta t) \\ i_{int}(\Delta t) \end{bmatrix} = \mathbf{P} i(\Delta t)
\qquad (23)
$$

where $\mathbf{P}$ is a $n_{prt} \times n$ matrix that contains $\mathbf{T}$. Using (23) in (20), we get:

$$\hat{v}(\Delta t) = \mathbf{A}^{-1}(\mathbf{M} + \mathbf{N}\mathbf{P})i(\Delta t) = \mathbf{A}^{-1}\hat{\mathbf{M}} i(\Delta t) \qquad (24)$$

where:

$$
\hat{\mathbf{M}} = \begin{bmatrix} \mathbf{I}_{ext} & 0 & 0 \\ 0 & \mathbf{I}_{prt} & \mathbf{T} \\ 0 & 0 & 0 \end{bmatrix}
\qquad (25)
$$

The above results will be used in the following to efficiently verify the external nodes in the modified grid. From Norton's theorem, the modified grid will exhibit the same voltage response at external and port nodes as the original grid. Therefore, $\forall i \in \mathcal{F}$:

$$v_{ext}(\Delta t) = \hat{v}_{ext}(\Delta t)$$
$$v_{prt}(\Delta t) = \hat{v}_{prt}(\Delta t)$$

so that:

$$\underset{i \in \mathcal{F}}{\mathrm{emax}}(v_{ext}(\Delta t)) = \underset{i \in \mathcal{F}}{\mathrm{emax}}(\hat{v}_{ext}(\Delta t)) \qquad (26)$$

$$\underset{i \in \mathcal{F}}{\mathrm{emax}}(v_{prt}(\Delta t)) = \underset{i \in \mathcal{F}}{\mathrm{emax}}(\hat{v}_{prt}(\Delta t)) \qquad (27)$$

Therefore, any verification that we will do below on the external nodes in the modified grid will also verify the same nodes in the original grid.

3.2.2 Subgrid Reduction

After moving the internal current sources, we are left with a subgrid consisting only of parasitic RC elements. Therefore, we can use a passive Model Order Reduction (MOR) technique to reduce the internals of the subgrid. The reduction approach that we have found applicable and beneficial for this work combines the two standard techniques of moment-matching and node elimination. Because this is a mix-and-match of various content from previous work, it is useful for us to describe some existing techniques, for clarity.

Calculation of Moments.

We use a nodal-formulation based method [8] to compute moments of the system transfer function. The passive subgrid is first isolated from the rest of the grid by removing (i.e., make into an open-circuit) the connections from all port nodes to external nodes. Therefore, the isolated subgrid can be represented in the s-domain by:

$$
\left(\begin{bmatrix} \hat{\mathbf{G}}_{22} & \mathbf{G}_{23} \\ \mathbf{G}_{23}^T & \mathbf{G}_{33} \end{bmatrix} + s \begin{bmatrix} \mathbf{C}_{prt} & 0 \\ 0 & \mathbf{C}_{int} \end{bmatrix} \right) \begin{bmatrix} v_{prt}(s) \\ v_{int}(s) \end{bmatrix}
$$
$$
= \begin{bmatrix} i_{prt}(s) + i'(s) \\ 0 \end{bmatrix} \qquad (28)
$$

where $\hat{\mathbf{G}}_{22}$ is an $n_{prt} \times n_{prt}$ conductance matrix that is derived from $\mathbf{G}_{22}$ by removing the connections from port nodes to external nodes represented by $\mathbf{G}_{12}^T$. The admittance looking into the ports of the subgrid, a matrix $\mathbf{Y}(s)$, can be approximated as [11]:

$$\mathbf{Y}(s) \approx \mathbf{M}_0 + \mathbf{M}_1 s \qquad (29)$$

where $\mathbf{M}_0 = \hat{\mathbf{G}}_{22} - \mathbf{G}_{23}\mathbf{V}$ and $\mathbf{M}_1 = \mathbf{C}_{prt} + \mathbf{V}^T\mathbf{C}_{int}\mathbf{V}$ are the $n_{prt} \times n_{prt}$, zero and first-order moment matrices with $\mathbf{V} = \mathbf{G}_{33}^{-1}\mathbf{G}_{23}^T$. Because of the quadratic form of $\mathbf{M}_1$, it is clear that it is a non-negative matrix and this will be useful below.

Moment-Matching.

For circuits with a non-negative $\mathbf{M}_1$ matrix, a 2π-model between pairs of ports was constructed in [7] by matching the zero and first-order moments. The circuit between a pair of ports is synthesized using a T-model as shown in Fig. 3(a) and port-to-ground elements are modeled with a parallel RC model as shown in Fig. 3(b). The elements of the T-model are given by [7]:

$$R_{ij1} = \frac{-\sqrt{m_1^{jj}}}{m_0^{ij}\left(\sqrt{m_1^{ii}} + \sqrt{m_1^{jj}}\right)}$$

$$R_{ij2} = \frac{-\sqrt{m_1^{ii}}}{m_0^{ij}\left(\sqrt{m_1^{ii}} + \sqrt{m_1^{jj}}\right)} \qquad (30)$$

$$C_{ij} = \frac{\left(\sqrt{m_1^{ii}} + \sqrt{m_1^{jj}}\right)^2 m_1^{ij}}{\sqrt{m_1^{ii} m_1^{jj}}}$$

where m_0^{ij} and m_1^{ij} are the $(i,j)^{th}$ elements of $\mathbf{M}_0$ and $\mathbf{M}_1$, respectively. To macromodel the port-to-ground connections for port i, the authors [7] provide:

$$R_{ii} = \left(m_0^{ii} + \sum_{\substack{j=1 \\ i \neq j}}^{n_{prt}} m_0^{ij} \right)^{-1}, \quad C_{ii} = m_1^{ii} - \sum_{\substack{j=1 \\ i \neq j}}^{n_{prt}} \frac{C_{ij} R_{ij2}^2}{(R_{ij1} + R_{ij2})^2} \qquad (31)$$

Node Elimination.

Note that the above T-model generates an extra node (a new internal node) for each pair of ports. If we have n_{prt} port nodes, generating a T-model for every pair of ports can be expensive because it will result in $n_{prt}(n_{prt}-1)/2$ new nodes. To overcome this issue, we can eliminate many of the new internal nodes by using the nodal elimination based reduction approach proposed in [9]. For every new internal node, the nodal time constant [9] is given by:

$$\tau = \frac{C_{ij} R_{ij1} R_{ij2}}{R_{ij1} + R_{ij2}} \qquad (32)$$

If $\tau < \tau_N$ where τ_N is a user specified nodal time constant value, the internal node can be eliminated by adding capacitors C_i and C_j to port nodes i and j and resistors R_{ij} between i and j. The capacitors and resistors are given by:

$$R_{ij} = R_{ij1} + R_{ij2}$$
$$C_i = \frac{C_{ij} R_{ij2}}{R_{ij1} + R_{ij2}}, \quad C_j = \frac{C_{ij} R_{ij1}}{R_{ij1} + R_{ij2}} \qquad (33)$$

Sparsification.

Once the system matrices of the reduced model are formed, it turns out that they contain large numbers of negligible (near zero) entries. As a final step in the reduction, therefore, we have found it useful to apply a final sparsification step, where entries whose absolute value is below a small value κ are automatically set to 0. We have found the error resulting from this to be insignificant and very much worth the effort.

Algorithm 1 INCR_VERIFY

Input: Partitioned power grid matrices in (3), τ_N, κ, δ_1 and δ_2
Output: Upper bounds on worst-case voltage drops for external nodes
1: Construct subgrid matrices in (28)
2: $(\mathbf{T}, \tilde{\mathbf{G}}_{sub}, \tilde{\mathbf{C}}_{sub}) = $ MACRO(subgrid matrices, τ_N, κ, δ_2)
3: Construct $\tilde{\mathbf{M}}$, $\tilde{\mathbf{G}}$, $\tilde{\mathbf{C}}$ and $\tilde{\mathbf{A}}$
4: **for** $(j = 1, \ldots, n_{ext} + n_{prt})$ **do**
5:     Compute j^{th} row of $\tilde{\mathbf{A}}^{-1}$ using SPAI [4] with $\delta = \delta_1$
6:     Multiply that row by the columns of $\tilde{\mathbf{M}}$, get row vector $\mathbf{d}$
7:     Maximize: $\mathbf{d} \cdot i$, subject to: $i \in \mathcal{F}$
8: **end for**
9: Compute $\tilde{v}_{ub}$ using (37)

Algorithm 2 MACRO(subgrid matrices, τ_N, κ, δ_2)

Output: $\mathbf{T}$ in (13), $\tilde{\mathbf{G}}_{sub}$ and $\tilde{\mathbf{C}}_{sub}$ in (34)
1: Construct $\mathbf{A}_{int}$
2: **for** (every port node k) **do**
3:     Find neighbors of k and g_{kj} in (16)
4:     **for** (every neighbor j of k) **do**
5:         Compute the jth row of $\mathbf{A}_{int}^{-1}$ using SPAI [4] with $\delta = \delta_2$
6:         Multiply the row entries by g_{kj}
7:         Add the row entries to the kth row of $\mathbf{T}$
8:     **end for**
9: **end for**
10: Compute $\mathbf{M}_0$ and $\mathbf{M}_1$
11: **for** (every pair of ports i, j) **do**
12:     Compute R_{ij1}, R_{ij2} and C_{ij} using (30)
13: **end for**
14: **for** (every port node k) **do**
15:     Compute R_{kk} and C_{kk} using (31)
16: **end for**
17: **for** (every new internal node created) **do**
18:     Compute τ using (32)
19:     **if** $\tau < \tau_N$ **then**
20:         Compute R_{ij}, C_i and C_j using (33)
21:         Eliminate the new internal node
22:     **end if**
23: **end for**
24: Drop insignificant connections with conductance less than κ
25: Construct $\tilde{\mathbf{G}}_{sub}$ and $\tilde{\mathbf{C}}_{sub}$

Final Reduced Model.

After applying all the reduction techniques discussed above, we end up with new conductance ($\tilde{\mathbf{G}}_{sub}$) and capacitance ($\tilde{\mathbf{C}}_{sub}$) matrices for the isolated subgrid, given by:

$$\tilde{\mathbf{G}}_{sub} = \begin{bmatrix} \tilde{\mathbf{G}}_{22} & \tilde{\mathbf{G}}_{23} \\ \tilde{\mathbf{G}}_{23}^T & \tilde{\mathbf{G}}_{33} \end{bmatrix}; \quad \tilde{\mathbf{C}}_{sub} = \begin{bmatrix} \tilde{\mathbf{C}}_{prt} & 0 \\ 0 & \tilde{\mathbf{C}}_{int} \end{bmatrix} \quad (34)$$

where $\tilde{\mathbf{G}}_{22}$ and $\tilde{\mathbf{C}}_{prt}$ are the modified port-to-port conductance matrix and capacitance matrix respectively, $\tilde{\mathbf{G}}_{33}$ and $\tilde{\mathbf{C}}_{int}$ are $\tilde{n}_{int} \times \tilde{n}_{int}$ conductance and capacitance matrices for the $\tilde{n}_{int}$ new internal nodes, and $\tilde{\mathbf{G}}_{23}$ and $\tilde{\mathbf{G}}_{23}^T$ are matrices consisting of connections between port nodes and newly formed internal nodes.

3.2.3 Verification after Macromodeling

After macromodeling, the *reduced* grid matrices can be constructed by stitching together the reduced subgrid and external grid matrices using the connections from external nodes to port nodes. The reduced grid matrices are given by:

$$\tilde{\mathbf{G}} = \begin{bmatrix} \mathbf{G}_{11} & \mathbf{G}_{12} & 0 \\ \mathbf{G}_{12}^T & \tilde{\mathbf{G}}_{22}' & \tilde{\mathbf{G}}_{23} \\ 0 & \tilde{\mathbf{G}}_{23}^T & \tilde{\mathbf{G}}_{33} \end{bmatrix}, \quad \tilde{\mathbf{C}} = \begin{bmatrix} \mathbf{C}_{ext} & 0 & 0 \\ 0 & \tilde{\mathbf{C}}_{prt} & 0 \\ 0 & 0 & \tilde{\mathbf{C}}_{int} \end{bmatrix} \quad (35)$$

where $\tilde{\mathbf{G}}_{22}'$ is the updated port-to-port conductance matrix in which the connections from external nodes to port nodes have been added. Since we have used a realizable macromodeling approach, $\tilde{\mathbf{G}}$ is also a $\tilde{n} \times \tilde{n}$ symmetric positive-definite $\mathcal{M}$-matrix, where $\tilde{n} = n_{ext} + n_{prt} + \tilde{n}_{int}$. From (24), the voltage at time $t = \Delta t$ for the modified grid in the special case is given by:

$$\tilde{v}(\Delta t) = \tilde{\mathbf{A}}^{-1}\tilde{\mathbf{M}}i(\Delta t) \quad (36)$$

where $\tilde{v}$ is a $\tilde{n}$-vector of voltage drops at external, port and new internal nodes, $\tilde{\mathbf{A}} = \left(\tilde{\mathbf{G}} + \frac{\tilde{\mathbf{C}}}{\Delta t}\right)$ is a $\mathcal{M}$-matrix and $\tilde{\mathbf{M}}$ is a $\tilde{n} \times n$ matrix with the first $n_{ext} + n_{prt}$ rows equal to $\tilde{\mathbf{M}}$ defined in (25), and the remaining rows have all entries equal to 0. The worst-case voltage drop at external and port nodes can be found by element-wise maximization of $\tilde{v}_{ext}$ and $\tilde{v}_{prt}$. The upper bound on worst-case voltage drops can be computed by using (14) and is given by:

$$\tilde{v}_{ub} = \left(\mathbf{I} + \mathbf{G}^{-1}\frac{\mathbf{C}}{\Delta t}\right) \begin{bmatrix} \mathbf{I}_{ext} & 0 & 0 \\ 0 & \mathbf{I}_{prt} & 0 \\ 0 & \mathbf{T}^T & \mathbf{A}_{int}^{-1} \end{bmatrix} \underset{\forall i \in \mathcal{F}}{\mathrm{emax}} \begin{bmatrix} \tilde{v}_{ext}(\Delta t) \\ \tilde{v}_{prt}(\Delta t) \\ i_{int}(\Delta t) \end{bmatrix} \quad (37)$$

where $\tilde{v}_{ub}$ is the $\tilde{n}$-vector of upper bounds on worst-case voltage drops with the first n_{ext} entries corresponding to upper bounds for external nodes.

4. IMPLEMENTATION

The overall flow of the proposed incremental verification approach is given in Algorithm 1. We start with a user-specified power grid and subgrid, along with parameter values for τ_N, κ, and error tolerance values (δ_1, δ_2) for the sparse approximate inverse (SPAI [4]) engine. The grid matrices are appropriately partitioned and macromodeling of the subgrid is performed. As a result of macromodeling, the size of the original power grid gets reduced and the internal current sources are moved to port nodes. The inverse of the matrix $\tilde{\mathbf{A}}$ is then computed, and every row is multiplied by $\tilde{\mathbf{M}}$ to account for the effect of movement of internal current sources. Next, we maximize the voltage drop at external and port nodes, and compute the upper bounds on worst-case voltage drops by using the efficient bounds computation approach.

The macromodeling algorithm is presented in Algorithm 2. To avoid the cost of constructing the full matrix $\mathbf{A}_{int}^{-1}$, we generate one row of the transformation matrix $\mathbf{T} = -\mathbf{G}_{23}\mathbf{A}_{int}^{-1}$ at a time, and that is possible through the use of SPAI [4] which is inherently parallelizable and can compute a single column/row of the approximate inverse. For a port node, we first identify the neighbors (k), and the connections (g_{kj}) to the neighbors. Then, we compute the corresponding row of the approximate inverse, multiply the row vector by g_{kj}, and add the result to the kth row of $\mathbf{T}$. Calculation of moments is also efficiently done by factorizing $\mathbf{G}_{33}$, followed by standard system solves.

5. EXPERIMENTAL RESULTS

A C++ implementation has been written to test the proposed approach. We use SPAI [4] to compute the approximate inverses, and solve the linear programs using MOSEK [12]. The test grids were generated from user specifications, including grid dimensions, metal layers, pitch and width per layer, and supply voltage sites and current sources distribution. The supply voltages and current sources were randomly placed on the grid. The technology specifications were consistent with 1.1 V 65nm CMOS technology. A global constraint is specified for the subgrid and other global constraints were specified to cover the entire chip. The subgrid nodes are also identified by the user. Computations were done using a 2.6 GHz Linux machine with 24 GB of RAM. A SPAI error tolerance value of $\delta_1 = 0.1$mV is used to compute the approximate inverse for the original and modified power grids. A lower value of tolerance $\delta_2 = 0.01$mV is used to construct $\mathbf{T}$.

Table 1 shows the speed and accuracy of the proposed bounds computation approach (section 3.1). Since we are interested in analyzing only the external nodes, we report maximum error and average percentage error values for the upper bounds on worst-

Table 1: Speed and accuracy after using efficient bounds computation

Power Grid		Subgrid		Max Error	Avg. %	CPU time		Speed
Name	Nodes	n_{int}	n_{prt}	(mV)	Error	Original	Fast v_{ub}	Up
G1	8,413	3,891	118	0.08	0.075	22.92 min.	10.58 min.	2.16x
G2	18,678	10,788	176	0.08	0.102	68.19 min.	24.97 min.	2.73x
G3	32,554	15,714	208	0.07	0.057	2.61 h.	1.12 h.	2.33x
G4	50,444	29,458	290	0.07	0.055	4.5 h.	1.77 h.	2.54x
G5	72,692	42,764	348	0.07	0.047	7.71 h.	3.11 h.	2.47x
G6	98,162	68,972	402	0.08	0.079	11.87 h.	3.10 h.	3.82x
G7	128,241	95,294	413	0.08	0.064	18.71 h.	4.19 h.	4.46x
G8	162,087	124,824	518	0.08	0.078	25.30 h.	5.27 h.	4.8x

Table 2: Speed and accuracy after applying macromodeling

Power Grid	Max Error (mV)	Avg. % Error	CPU time		Speed Up Reduced vs Fast v_{ub}	Total Speed Up Reduced vs Original
			Fast v_{ub}	Reduced		
G1	2.4	1.96	10.58 min.	5.03 min.	2.1x	4.55x
G2	2.06	2.05	24.97 min.	12.26 min.	2.03x	5.56x
G3	1.4	1.20	1.12 h.	0.93 h.	1.2x	2.81x
G4	2.04	1.29	1.77 h.	1.26 h.	1.4x	3.57x
G5	2.01	0.88	3.11 h.	2.36 h.	1.31x	3.26x
G6	2.79	1.14	3.10 h.	2.22 h.	1.39x	5.34x
G7	3.13	1.11	4.19 h.	2.64 h.	1.58x	7.08x
G8	2.6	1.14	5.27 h.	3.14 h.	1.67x	8.05x

case voltage drops at external nodes only. The runtime and accuracy are compared with the original approach based on finding the worst-case voltage drop at every node, and then computing the upper bound using (7). The results show that we are able to achieve significant runtime savings with negligible error values.

The macromodeling approach was tested with user-specified values for nodal time constant ($\tau_N = 5ps$), and conductance threshold ($\kappa = 5 \times 10^{-3}\mho$). Table 2 gives the speed and accuracy obtained after applying macromodeling. The runtime for the reduced grid includes the time taken to perform macromodeling of the subgrid, and then using the efficient bounds computation approach to find the upper bounds. The accuracy is compared with the original approach while speed-up is measured with respect to the efficient bounds computation approach. We also report the total speed-up with respect to the original approach. The results show that we incurred an average error of about 1% for large grids while extracting a total speed-up in the range of 3-8x.

Of course, it is to be expected that, if fewer nodes are to be verified, then corresponding time savings would be the result. However, in our case, the speed-ups are much higher than would be obtained based solely on this argument. For example, if one wants to verify only 15% of the grid nodes, one would expect a speed-up of 100/15=6.67x. However, we can verify 15% of the nodes with a typical total speed-up of 27x. Thus, the benefits of our upper-bounding and macromodeling techniques are quite significant.

6. CONCLUSIONS

We describe an early incremental verification approach for *RC* grids under a constraints-based power grid verification framework, in which only the nodes that are *external* to a subgrid region are to be verified. Our approach gives a fast and accurate way to compute the upper bounds on worst-case voltage drops at external nodes, based on two contributions: 1) an upper-bound method that eliminates the need to perform multiple iterations, and 2) a macromodeling method that drastically reduces the internals of the subgrid. As a result, 3-8x speed-ups are obtained, with negligible 1-2% error. With this proposed approach, it becomes practical to perform early incremental design verification of the on-die power grid under dynamic conditions.

7. REFERENCES

[1] D. Kouroussis and F. N. Najm. A static pattern-independent technique for power grid voltage integrity verification. In *ACM/IEEE DAC*, pages 99–104, Anaheim, CA, Jun. 2-6 2003.

[2] M. Nizam, F. N. Najm, and A. Devgan. Power grid voltage integrity verification. In *ACM/IEEE ISLPED*, pages 239–244, San Diego, CA, Aug. 8-10 2005.

[3] I. A. Ferzli, F. N. Najm, and L. Kruze. A geometric approach for early power grid verification using current constraints. In *ACM/IEEE ICCAD*, pages 40–47, San Jose, CA, Nov. 5-8 2007.

[4] N. H. Abdul Ghani and F. N. Najm. Fast vectorless power grid verification using an approximate inverse technique. In *ACM/IEEE DAC*, San Fransisco, CA, Jul. 26-31 2009.

[5] D. Kouroussis, I. A. Ferzli, and F. N. Najm. Incremental partitioning-based vectorless power grid verification. In *ACM/IEEE International Conference on Computer Aided Design*, pages 358–364, San Jose, CA, November 6-10 2005.

[6] Y.-M. Lee, Y. Cao, T.-H. Chen, J.M. Wang, and C.C.-P. Chen. HiPRIME: hierarchical and passivity preserved interconnect macromodeling engine for RLKC power delivery. *IEEE Transactions on Computer-Aided Design of Integrated Circuits and Systems*, 24(6):797–806, June 2005.

[7] Haifang Liao and W. Wei-Ming Dai. Partitioning and reduction of RC interconnect networks based on scattering parameter macromodels. In *Digest of Technical Papers, ACM/IEEE International Conference on Computer-Aided Design*, pages 704–709, Nov 1995.

[8] P. Miettinen, M. Honkala, J. Roos, C. Neff, and A. Basermann. Study and development of an efficient RC-in-RC-out MOR method. In *IEEE International Conference on Electronics, Circuits and Systems*, pages 1277–1280, Aug 2008.

[9] B. Sheehan. Realizable reduction of RC networks. *IEEE Transactions on Computer-Aided Design of Integrated Circuits and Systems*, 26(8):1393–1407, August 2007.

[10] Samuel L. Oppenheimer, Jean Paul Borchers, and F. Roger Hess. *Direct and Alternating Currents*. McGraw-Hill, New York, 1973.

[11] K.J. Kerns and A.T. Yang. Stable and efficient reduction of large, multiport rc networks by pole analysis via congruence transformations. *IEEE Transactions on Computer-Aided Design of Integrated Circuits and Systems*, 16(7):734–744, 1997.

[12] The MOSEK optimization software (www.mosek.com).

Analysis of DC Current Crowding in Through-Silicon-Vias and Its Impact on Power Integrity in 3D ICs

Xin Zhao[1], Michael Scheuermann[2], and Sung Kyu Lim[1]

[1]School of ECE, Georgia Institute of Technology, Atlanta, GA, USA
[2]IBM T. J. Watson Research Center, Yorktown Heights, NY, USA.
{xinzhao, limsk}@ece.gatech.edu

ABSTRACT

Due to the large geometry of through-silicon-vias (TSVs) and their connections to the power grid, significant current crowding can occur in 3D ICs. Prior works model TSVs and power wire segments as single resistors, which cannot capture the detailed current distribution and may miss trouble spots associated with current crowding. This paper studies DC current crowding and its impact on 3D power integrity. First, we explore the current density distribution within a TSV and its power wire connections. Second, we build and validate effective TSV models for current density distributions. Finally, these models are integrated with global power wires for detailed chip-scale power grid analysis.

Categories and Subject Descriptors

B.7.2 [**Hardware, Integrated Circuit**]: Design Aids

General Terms

Design

Keywords

3D IC, TSV, DC current crowding, power integrity, reliability

1. INTRODUCTION

3D IC power delivery network (PDN) provides power supply to all devices in the entire 3D stack. The inter-die power delivery interconnects, formed by power/ground (P/G) through-silicon-vias (TSVs) or micro-bumps, are unique components in 3D power grids. These vertical connections carry large amounts of current and may suffer from Electro-migration (EM) degradation due to an excessive current density as well as have large IR drops. Therefore, detailed and accurate analysis on the 3D PDN is important to predict the performance and improve the power integrity as necessary.

The purpose of this paper is to investigate DC current crowding in TSVs and its impact on power integrity of generic 3D PDNs. A small cross-section of the global 3D PDN is illustrated in Figure 1. Two dies (top and bottom) are bonded face-to-back and are connected using vias-last TSVs. Voltage is supplied from the package

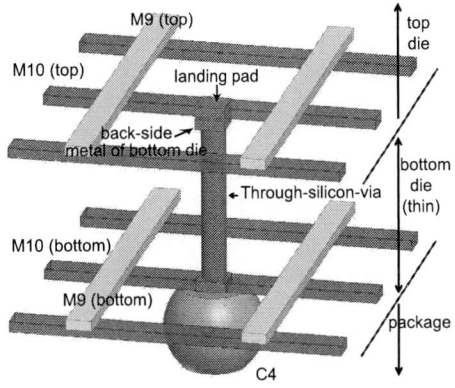

Figure 1: 3D connection in a global power delivery network.

through C4s. For the bottom die, current is delivered directly to Metal 10 and Metal 9; however, for the top die, current is delivered to Metal 10 and Metal 9 through TSVs. Intermediate and local sections of the PDN (using Metal 1 to Metal 8) are connected using local vias to the global PDN. Inter-die connection can be achieved by either directly bonding the landing pads of the bottom-die backside metal with the top-die Metal 10, or through micro-bumps. The TSV has 5um diameter and 30um height, which is similar to the structure described in [3]. The TSV landing pads are 6um×6um [4] and the global power wires are 2um thick. This generic structure is used throughout the paper for both isolated TSV modeling and large-scale 3D PDN modeling.

Some recent papers discussed TSV EM modeling and analysis [8] and TSV-based 3D PDN analysis [6]. However, none of these works investigates detailed current density distribution or current crowding inside P/G TSVs, where some of the edges may suffer from a large current gradient and are subject to a potential EM reliability issue [1]. Moreover, prior works model TSVs and powers wire segments as single resistors, which are insufficient to accurately analyze the detailed current density distribution inside P/G TSVs and 3D PDNs.

In this paper, we present an in-depth investigation of the DC current density distribution in 3D IC PDNs with specific focus on the current crowding inside the TSVs and at the connections between TSVs and power wires. We also study the current crowding impact on the IR drop. In Section 2 we investigate the current density distribution and the voltage drop on a TSV test case. In Section 3 an effective TSV model is implemented and validated. Finally, in Section 4 the proposed TSV models are integrated with power wires

---
[1]EM may become significant with the current density near or higher than 10mA/um^2 [1].

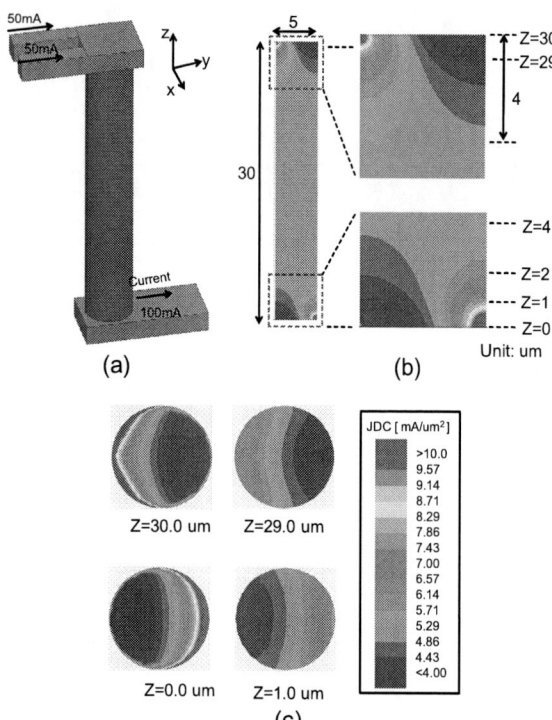

(a)   (b)

(c)

Figure 2: Current crowding in a test case of a TSV and power wires. The current density distribution is shown in a ZY plane (side view) (b), and in top-down XY planes for Z=30.0um, 29.0um, 1.0um, and 0.0um (c).

and simulated using a power simulator (PSIM) for chip-scale TSV-based power grid analysis. The simulation results show that PSIM is able to efficiently analyze detailed current density distribution within TSVs in the context of chip-scale PDNs. PSIM can identify regions of excessive current density or IR drop to help the designer optimize 3D PDNs.

2. CURRENT CROWDING IN 3D IC

2.1 Current Density Distribution Inside a TSV

The test case used to investigate the current density distribution inside a TSV is shown in Figure 2. This corner case was chosen specifically to study a highly asymmetric current distribution and consists of the following components: (1) a TSV with 5um diameter and 30um height, (2) landing pads (6um×6um), (3) two power wires on the top, each 2um wide, (4) one 6um wide large power wire at the bottom[2]. The power wire has 2um thickness with copper resistivity of $18\Omega\cdot$nm. Two current sources are inserted at the top-left corner, each sourcing 50mA current; the current sink is defined at the bottom-right corner. In a 3D PDN, power wires connect to landing pads in both directions. This test case constrains the current flow direction and helps us to investigate the current density distribution in the TSV and its connection points to the PDN. AN-SYS Q3D is used to simulate the DC current density distribution and the voltage drop.

The magnitude of current density is plotted for several cross-sections in Figures 2(b) and 2(c). In Figure 2(b), we observe that a

[2]In PDN designs, depending on the dimension of power wire and landing pads, either multiple power wires or a single power wire can connect to a landing pad.

Table 1: Impact of current crowding on voltage drop through a TSV.

Wire thickness (um):	1.0	2.0	3.0
voltage drop w/ current crowding (mV)	3.33	3.11	3.02
voltage drop w/o current crowding (mV)	2.75	2.75	2.75
Increase by current crowding (%)	21.1	13.1	9.8

large portion of current from the power wires dives into the top-left TSV edge and flows out at the bottom-right edge. Compared with the average current density inside the TSV which is 5.1mA/um^2, the edge current density is approximately 10mA/um^2, nearly twice as large. For the current density distribution along the ZY plane, significant current crowding is observed approximately 4um into the TSV from both the top and bottom interfaces in Z direction. In the center region of the TSV (4um<Z<26um), the current is uniformly distributed inside the TSV. Figure 2(c) is a plot showing the magnitude of current density distribution in the XY planes, when Z=30.0um, 29.0um, 1.0um, and 0.0um. Most of the current is concentrated at the connection between power wires and the TSV.

2.2 Impact of Current Crowding on IR Drop

Current crowding inside the TSV changes the effective resistance of the TSV as well as the voltage drop across the TSV. The spreading resistance [5] is caused by the nonparallel current between two spatially separated contacts. Considering the TSV current crowding, its effective resistance is larger than the value obtained using $R_0 = \rho \cdot l/A$ where ρ is the resistivity, l is the length, and A is the cross sectional area of the TSV.

Q3D is used to find the voltage drop across the TSV taking into account the spreading resistance. We hold the TSV dimensions constant and increase the power wire thickness from 1um to 3um. The resulting voltage drop through the TSV is shown in Table 1. For a 100mA current, the voltage drop through R_0 is $IR_0 = 2.75$mV. However, the current crowding results to is 21.1% to 9.8% greater voltage drop than calculated IR_0 as the power wire thickness increases from 1um to 3um.

3. TSV CURRENT CROWDING MODEL

In traditional PDN modeling, power wire segments and TSVs are modeled as single resistors. This model can only represent uniform current density within a wire segment or a TSV, and is insufficient to accurately capture non-uniform current distributions caused by current crowding. Likewise, it is also insufficient to accurately calculate voltage drop that is related to the spreading resistance and depends on current distributions. Here we describe a TSV model that allows non-uniform current densities within a TSV and its transition regions. The model is simple enough that runtime remains reasonable. Our TSV model has been integrated into a power grid simulator (PSIM) [7]. This simulator is used to solve the power resistance network and analyze current and voltage.

3.1 3D Resistance Network for TSV Modeling

An illustration on our TSV model and the corresponding resistance network is shown in Figure 3. A TSV is composed of rectangular mesh boxes having the basic structure shown in Figure 3(a), where each mesh box consists of six resistors: east, south, west, north, up, and down. The resistance value depends on the geometry of each box and how the TSV overlays the box. The 3D mesh structure of a TSV is generated as follows: (1) Z-mesh: The TSV is divided into multiple short cylinders with the same diameter but various thicknesses; (2) XY-mesh: Each short cylinder is then meshed into a 2D resistance network on a virtual XY plane, which is located at the center of each cylinder.

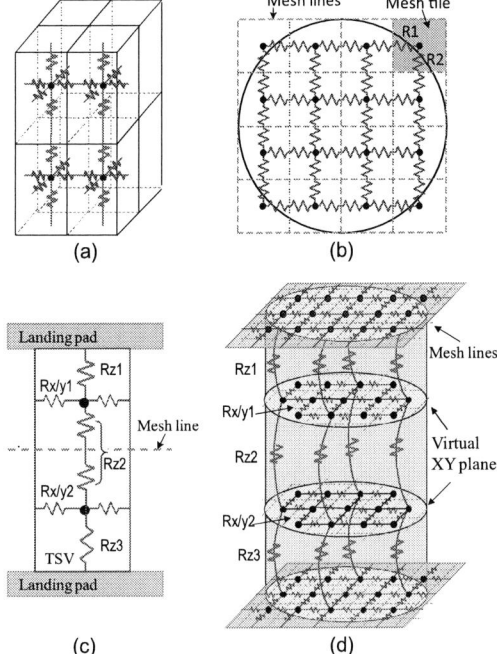

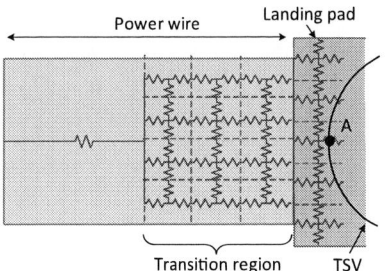

Figure 4: Meshing on the transition region. Using a single resistor in the transition region leads to all the current injected into the Point A.

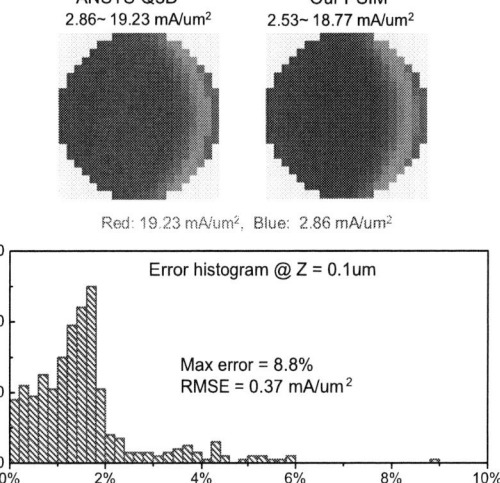

Figure 3: TSV modeling approach of a 3D resistance network. (a) Basic rectangular box after 3D meshing; (b) XY-mesh and partial overlap mesh tiles; (c) side view; (d) 3D view.

Along the Z-dimension, several virtual XY planes are created by partitioning this dimension. The Z locations of these XY-planes, referred to as the Z-mesh, are determined by the current gradient in the ZY-plane shown in Figure 2(b). More cylinders are created in the regions of larger current crowding and fewer in the regions of uniform current density. As a result, more virtual XY planes will be created at the top and bottom of the TSV. A side view of the 3D resistance network is shown in Figure 3(c), where two virtual XY-planes are generated for one Z-mesh.

The resulting model is a non-uniform 3D resistance network consisting of two types of resistors: (1) The resistors along Z-axis connecting the neighboring virtual XY planes (Rz1, Rz2, and Rz3 in Figures 3(c) and 3(d)); (2) the resistors in virtual XY planes (Rx/y1 and Rx/y2 in Figures 3(c) and 3(d)). If a mesh tile is completely covered by the real TSV shape, Rz and Rx/y are directly obtained referring to the XY-mesh and Z-mesh size. However, around the TSV boundary, mesh tiles partially overlap with the real shape as shown in Figure 3(b). For this case, the overlap area is calculated as the cross sectional area for Rz calculation, and the effective length along X-axis and Y-axis is then obtained for R1 and R2 calculation.

Figure 3(d) is a schematic in nature. Most virtual planes and resistors are not shown for readability. The non-uniform Z-mesh actually used in the model is a trade-off between complexity and accuracy. The 30um height TSV is partitioned at Z=0.1, 0.4, 0.9, 2.0, 5.0, 16.0, 27.0, 28.9, 29.4, 29.7, 29.9, 30.0, where the TSV bottom is at Z=0. We implemented three different XY-mesh sizes for comparisons: 0.25um, 0.5um, and 1.0um.

3.2 Modeling of the Transition Region

The transition region is defined as the connection area between the power wires and the TSV landing pad. The meshing on the transition region is shown in Figure 4, where non-uniform current gradient can occur at the connections between power wires and TSV landing pads. Without meshing the transition region, the entire current would entirely flow into the single Point A, which results in a

Figure 5: Current density distributions and the error histogram of ANSYS Q3D and our TSV modeling approach in PSIM at Z=0.1um virtual XY plane. The error in each tile is the absolute difference between Q3D and PSIM.

large but incorrect current at the edge of the TSV. By meshing the transition region, the current spreads evenly along the power wire and then flows into the landing pad and the TSV along the edge, which results in better accuracy. A transition region approximately 6um long is found to be long enough.

3.3 Modeling Accuracy

Detailed comparisons between ANSYS Q3D and PSIM using the new TSV model are shown in Figure 5. The XY-mesh size is 0.25um. For the PSIM results, the current in each mesh tile is extracted and divided by the effective area. For the Q3D results, the current gradient is first simulated using its internal mesh generator and solver. Next, the current values are mapped into our mesh tiles. The current density distribution from Q3D and PSIM are plotted in the top half of Figure 5 for the virtual plane located at Z=0.1um. The error histogram of the current density between Q3D and PSIM for all 357 tiles is shown at the bottom of Figure 5 where the error for each tile is the absolute difference between Q3D and PSIM.

This comparison is for the closest virtual XY plane to the landing pad in this model, where the largest current crowding is observed. PSIM has very good accuracy compared with the Q3D results. The relative error for each tile is less than 10%, and most of them are within 5%. The root-mean-square error (RMSE) is 0.36mA/um^2, defined as the square root of the arithmetic mean of the square of the error for each tile:

Table 2: Impact of the XY-mesh size on the current density (mA/um^2) and the voltage drop (mV).

mesh (um)	#tiles		Max. Current density			Voltage drop		
		RMSE	Q3D	PSIM	err (%)	Q3D	PSIM	err (%)
0.25	4641	0.25	19.2	18.8	-2.1	3.1	3.10	0.3
0.5	1313	0.34	18.0	20.8	15.6	3.1	3.09	0.7
1.0	325	0.55	12.2	15.6	27.9	3.1	2.99	3.9
none	1	–	19.2	5.1	73.4	3.1	2.75	11.3

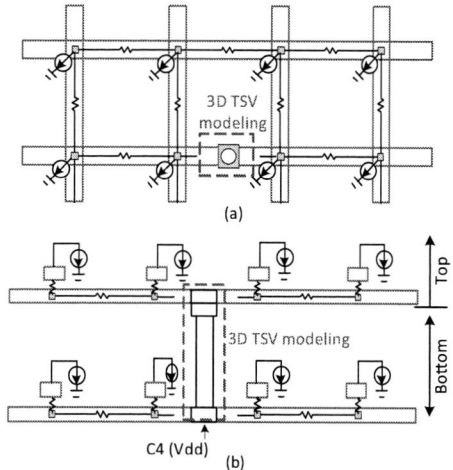

Figure 6: A circuit model for a two-die TSV-based PDN using the proposed 3D TSV modeling approach in top-down view (a) and side view (b).

$$\sqrt{\left(\sum_{i=1 \text{ to } n}(J_i^{Q3D}-J_i^{PSIM})^2\right)/n}$$

where i is the ith tile and n is the total number of tiles. The voltage drop from PSIM is 3.07mV, which is 0.33% different from the Q3D value of 3.08mV.

Differences between Q3D and PSIM are mainly due to constraining the XY-mesh and Z-mesh sizes and mesh structure. While we use low density orthogonal meshing boxes for simplicity, Q3D supports more sophisticated meshing structures such as triangular and tetrahedral shapes. However, PSIM takes less than one second simulation time, whereas Q3D takes up to one hour. This demonstrates that our modeling approach has potential for chip-scale power analysis with reasonable accuracy and acceptable runtime.

3.4 Impact of XY-Mesh Size

The impact of varying the XY-mesh size on the accuracy of current density and voltage drop is shown in Table 2. The Z-mesh size is held constant and the XY-mesh size is increased from 0.25um, 0.5um, to 1.0um. Using larger meshing tiles, the RMSE of the current density increases from 0.25mA/um^2 to 0.55mA/um^2, which is equal to 4.9% to 10.7% of the average current density. Note that, to report the maximum current density in Q3D for a given mesh size, the Q3D simulation result is mapped into each mesh tile. Thus, the maximum current density of Q3D reduces with different mesh sizes in Table 2. The error of the maximum current density increases from 2.1% to 27.9%, and the voltage drop error increases from 0.3% to 3.9%. The cost of using finer size is the increased number of mesh tiles from 325 to 4641. Simulation results using a single resistor are also shown in the table, which results in average current density of 5.1mA/um^2 (73.4% smaller than the maximum current density from Q3D) and lower voltage drop of 2.75mV.

4. CHIP-SCALE 3D PDN ANALYSIS

4.1 Chip-Scale PDN Circuit Model

In this section, we integrate our TSV models with the chip power grid and use PSIM to analyze the entire network. A circuit model of a partial 3D PDN for two dies is illustrated in Figure 6. Both power wire segments and local vias are represented as lumped resistors. The 3D power connection, including TSVs and transition regions, is modeled using the proposed approach. We assume ideal voltage of 1V supplied from C4s. The current sinks are located at the intersections of the power grids in each die. Power wires are 2um thick and 5um wide. The TSV has 30um height, 5um diameter, and 6um×6um landing pads. The goal is to analyze global PDNs that have high current density and contain TSV connections.

4.2 Simulation Results

A global PDN and the power map in the bottom die are shown in Figure 7(c). The footprint area is 1.4mm×1.4mm. Each die has 16×16 power wires (in purple and yellow) and a thick power ring around the boundary with 15um wide. A 4×4 TSV array and C4s are aligned in the bottom die, which are enlarged as the white blocks for readability. Each black box at the intersection of power wires represents the current sink. The XY-mesh size of the TSV model is 0.25um, and the Z-mesh size is kept the same as in Section 3.1.

Power maps in both top die (S2) and bottom die (S1) have a cool spot in the bottom-left corner and a hot spot in the top-right corner. In the center of each die, another two narrow cool spots are placed in the left and right. These power maps result in different current density patterns connecting to the TSVs: 1) symmetric current density, e.g., TSV1 in Figure 7(c), where all the power wires have high current density; 2) asymmetric current density, e.g., TSV2 in Figure 7(c), where left power wires have much lower current density than the right power wires.

The voltage drop maps in S2 and S1 are shown in Figures 7(a) and 7(b). The top-right corner has the maximum IR drop: 23.0mV in top and 19.0mV in bottom. The IR drops in S2 are larger than S1 due to the TSV parasitic resistance. Since TSVs and C4s are aligned, the region closer to TSVs has a smaller IR drop than the region far from TSVs.

Detailed current density distribution in two TSVs (TSV1 and TSV2 in Figure 7(c)) are shown in Figure 8, where TSV1 located in the center of the hot region has a fairly symmetric current density along power wires, and TSV2 located at the boundary between a power hot region and a cool region has asymmetric current densities in the power wires. Figure 8 shows plots of the current density distribution in the XY direction (Jxy) and Z direction (Jz) on different layers. Jxy in metal layers S2-M10, S1-BM, and S1-M10 are plotted in Figures 8(a), 8(c), and 8(f), respectively. Figures 8(b), 8(d), and 8(e) plot Jz through the interface between S1-BM and S2-M10, the top surface of the TSV, and the bottom surface of the TSV, respectively.

First, PSIM is effective to capture the detailed current density distribution inside the 3D power connection. Symmetric current crowding is observed in both edges of TSV1; most of the current crowding occurs at the right edge of TSV2. Second, large current crowding inside TSVs is observed. For TSV1, the maximum current density (J_{max}) in Figure 8(a) is 7.6mA/um^2, where most current concentrates at the connection between the power wire and the landing pad; the J_{max} through the TSV in Z-direction (Figure 8(d)) can reach to 25.6mA/um^2, which is approximately 2.4 times larger than the wire J_{max}. Third, larger current crowding occurs at the TSV top surface than the bottom surface due to the aligned TSVs and C4s. Jz through the TSV bottom (Figure 8(e)) is 14.9mA/um^2 compared with 25.6mA/um^2 through the top surface. Fourth, in the

160

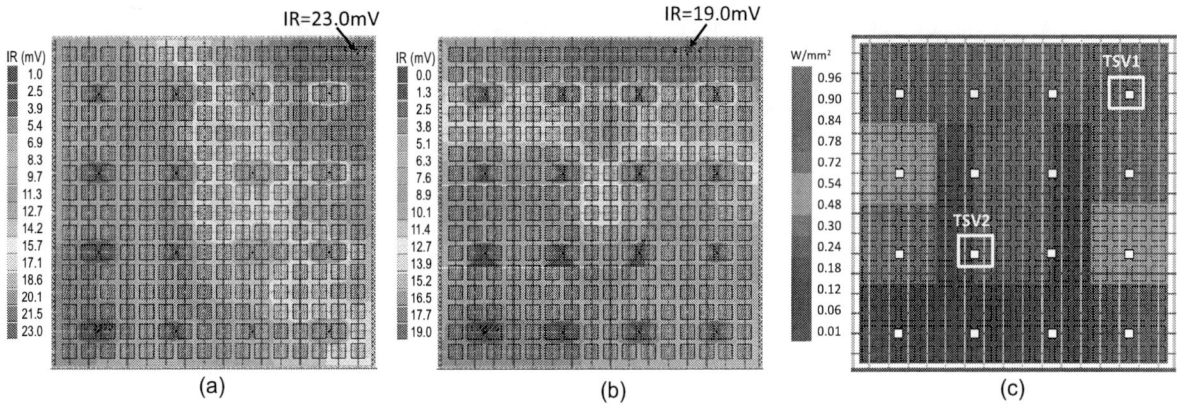

Figure 7: (a) Voltage drop map of the top die (max = 23.0mV), **(b)** voltage drop map of the bottom die (max = 19.0mV), **(c)** power map of the bottom die, where TSVs and C4s are aligned and enlarged for readability.

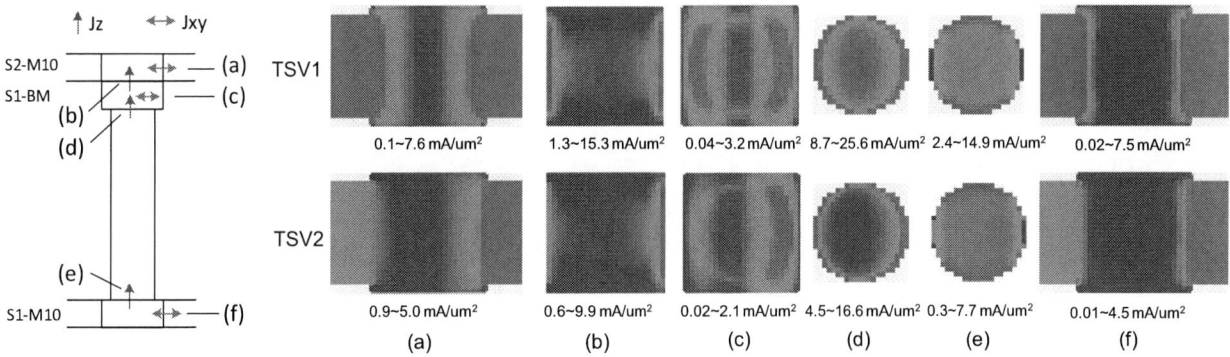

Figure 8: Current density distribution in XY (J_{xy}) and Z (J_z) direction of TSV1 and TSV2 in Figure 7. TSV1 has symmetric current density along power wires; TSV2 has asymmetric current.

bottom TSV surface (Figure 8(e)), J_z current crowds at the top and bottom edges instead of concentrating at the left and right edges. This is because a large amount of J_{xy} currents flow out from the left and right edges to feed the S1 current sinks. As a result, the current, delivered to the S2 power grid, concentrates at the top and bottom edges. Moreover, current crowding leads to 5.7mV IR drop through TSV1, which is 3.7% larger than the IR drop without considering the crowding value of 5.5mV.

The next subsections contain the following results: (1) Maximum current density (J_{max}) along the power wires. (2) Maximum and average current density (J_{avg}) of the TSVs. For multiple TSVs, J_{max} and J_{avg} through each TSV are first measured. The percentage increase (J_{inc}) of J_{max} over J_{avg} is then calculated for each TSV, as $J_{inc}=(J_{max}-J_{avg})/J_{avg}$. The TSV with the maximum J_{max} are reported, as well as the min/max/avg J_{inc} among all the TSVs. (3) Min/max/avg IR drops in top and bottom dies. (4) IR drop through the TSVs. The IR drop with (IR_c) and without (IR_n) considering current crowding are reported. The percentage increase of IR_c over IR_n is also included. The baseline PDN design contains a 16×16 power gird in each die, 4×4 TSVs and C4s, and TSVs with 5um diameter and 0.25um mesh size.

4.3 Impact of TSV Mesh Size

We increase the mesh size of the TSV model from 0.25um, 0.5um, to 1.0um, and fix the other design factors. The results of the current density and the IR drop are shown in Table 3. First, using a larger mesh size in the TSV model results in lower J_{max} in each mesh tile compared with using a finer mesh size. As the mesh size

increases from 0.25um to 1.0um, J_{max} reduces from 25.6mA/um^2 to 14.3mA/um^2 and J_{inc} reduces from 151%-192% to 41%-48%. This is because the coarse mesh averages out the current gradient. Second, the mesh size does not affect the IR drop of the power grid and wire J_{max} very much. In contrast, significant current crowding is observed in the TSVs for smaller mesh sizes. For the mesh size of 0.25um, the J_{max} in the TSV is 25.6mA/um^2, which is 110% larger than the TSV J_{avg} of 10.2mA/um^2.

4.4 Impact of TSV and C4 Offset

Previous simulations assume aligned TSVs and C4s. To study the offset impact on power integrity, we leave 175um distance between the TSV and the C4. The offset design has 12 C4 and 16 TSVs. The simulation results are shown in Table 4. Current crowding has larger impact on TSV IR drop in the offset design than the aligned design. The six columns from the right compare the IR drop through the TSV with (IR_c) and without (IR_n) considering current crowding. Current crowding in the offset design results in 5.9% to 10.6% larger IR drop than the IR_n; whereas, in the aligned design, the TSV IR drop caused by current crowding is 3.4% to 5.2% larger than the IR_n. This is mainly because large current crowding occurs in both the top and bottom surfaces of TSVs in the offset design; whereas, in the aligned design, only the top interface between the TSV and the backside metal has large current crowding, where the voltage at the bottom interface between TSV and S1-M10 is constantly supplied by C4s.

4.5 3D Power Integrity on Large-Scale PDNs

Table 3: Impact of the TSV mesh size on current density (mA/um^2) and IR drop (mV).

Power grid	#TSV &#C4	Mesh (um)	TSV (um)	Wire J_{max}	TSV with max(J_{max})			J_{inc} (%) of TSVs			IR_Bottom			IR_Top		
					J_{max}	J_{avg}	J_{inc}(%)	min	avg	max	min	avg	max	min	avg	max
16×16	4×4	0.25	5	10.5	25.6	10.2	151	151	161	192	2.1	9.5	19.1	3.8	12.7	23.0
16×16	4×4	0.50	5	10.4	20.2	10.1	100	100	105	124	2.4	10.0	19.8	4.1	13.3	23.7
16×16	4×4	1.00	5	10.5	14.3	10.2	41	41	42	48	2.1	9.4	18.9	3.9	12.9	23.1

Table 4: Impact of TSV and C4 offset on current density (mA/um^2) and IR drop (mV) through TSVs.

	# TSVs	# C4	Wire J_{max}	TSV with max(J_{max})			J_{inc}(%) of TSVs			IR_Bottom			IR_Top			TSV w/ max(IR)			Inc (%) of TSV IR		
			J_{max}	J_{max}	J_{avg}	J_{inc}(%)	min	avg	max	min	avg	max	min	avg	max	IR_c	IR_n	Inc.(%)	min	avg	max
Aligned	16	16	10.5	25.6	10.2	151	151	161	192	2.1	9.5	19.1	3.8	12.7	23.0	5.7	5.5	3.7	3.4	4.0	5.2
Offset	16	12	25.2	22.0	8.3	165	149	164	190	10.2	26.5	48.3	29.7	46.7	65.5	4.8	4.5	6.4	5.9	8.0	10.6

Table 5: The footprint (mm^2), power density (W/mm^2), current density (mA/um^2), and IR drop (mV) for large-scale PDNs.

Design	Footprint	Pwr dens.		Power grid	#TSVs	#C4	J_{max}_wire		TSV with max(J_{max})			IR_Bottom			IR_Top			TSV with max(IR)		
		top	bot				top	bot	J_{max}	J_{avg}	J_{inc}(%)	min	avg	max	min	avg	max	IR_c	IR_n	Inc.(%)
PDN1	5×5	0.57	0.57	50×50	144	144	7.0	7.2	9.6	6.8	41	5.1	8.7	15.9	7.9	11.7	19.6	4.1	3.7	11.4
PDN2	6×6	0.40	0.75	60×60	225	225	3.5	6.6	5.0	3.6	40	6.0	9.9	13.3	5.0	7.2	9.2	2.1	1.9	11.4
PDN3	9×9	0.80	0.80	90×90	484	484	13.6	12.1	18.5	13.1	41	4.4	11.3	24.2	6.8	15.6	37.8	7.9	7.1	11.5
PDN4	11×11	0.71	0.91	110×110	729	729	8.7	11.4	11.1	7.5	47	8.1	12.8	25.2	9.7	13.5	24.2	5.1	4.5	11.4
PDN5	15×15	0.47	0.49	150×150	1369	1369	16.2	17.4	23.3	16.3	43	1.8	6.8	34.9	2.7	8.8	49.6	9.9	8.8	12.2

Five large-scale two-die stacked PDNs are designed for 3D power analysis using PSIM. The power wire utilization, defined as the power wire area per die over the footprint area, is set to 5%. Assume the bottom die needs $N_P \times N_P$ power wires, the aligned TSV and C4s count are determined as $(25\%N_P) \times (25\%N_P)$. The power TSVs have 5um diameter with 1.0um mesh size. Power wires have 5um width and 2um thickness. The local and global power density refers to the 3D core-to-memory PDN designs [6][2], Intel microprocessors, and the power density estimation in ITRS 2005.

The results of power analysis on large-scale PDNs are shown in Table 5. First, excessive current density through the TSVs is observed. The J_{max} of the TSV is 40% to 47% larger than the J_{avg} through the TSV. Note that the 1um mesh size is used for this simulation. As previously discussed, using coarser mesh size usually underestimates the J_{max} on 3D power connection compared with using finer mesh or using FEM tools. It is reasonable to expect even larger current density through the TSVs. Second, the wire J_{max} in S1 and S2 are affected by the power density in each die, where larger power density (PDN2 and PDN4) in the bottom die results in comparable J_{max} between wires and TSVs; for other designs with lower power density in the bottom die, TSV J_{max} is 13% larger than wire J_{max}. Third, current crowding also increases the IR drop through the TSVs, which is 11.4% to 12.2% larger than the IR drop without considering current crowding. Moreover, IR drops in S1 and S2 power grid are also affected by the power density in each die. When each die has comparable power density, the maximum IR in the top die is usually larger than the bottom die; allocating the high power density close to C4s (in the bottom die) helps to reduce the IR drops in the top die.

In summary, (1) PSIM with the proposed simple TSV model can analyze the chip-scale 3D PDN for detailed current density distribution and voltage drop; (2) by identifying the current crowding corner inside each TSV, PSIM helps the designer to assign reasonable current limits and voltage drop limits for 3D PDN design and optimization; and (3) PSIM can select a different mesh size depending on the resolution of the power analysis: for a large-scale PDN, 1.0um mesh size can be firstly used to quickly identify the hotspots associated with the maximum current density and IR drop; then in a bounded hotspot region, a finer mesh size is used to identify the detailed current density distribution and optimize the power grid, correspondingly.

5. CONCLUSIONS

In this paper, we studied current crowding inside TSV-based 3D power connections. First, we investigated the current density distribution inside the 3D TSV-based power grids. A large current gradient, current crowding, near the interface between power wires and TSVs has been observed. In addition, the current crowding also increases the effective resistance of the TSV and thus voltage drop in the PDN. Second, a 3D TSV model has been implemented and simulated using PSIM. The model has good accuracy and far less complexity compared with FEM tools. Third, PSIM has been applied on chip-scale 3D power grid analysis, which is able to analyze the current density distribution and the resulting voltage drop.

6. REFERENCES

[1] J. Abella and X. Vera. Electromigration for Microarchitects. *ACM Comput. Surv.*, 42(2):9:1–9:18, March 2010.

[2] M. B. Healy and S. K. Lim. Distributed TSV Topology for 3-D Power-Supply Networks. *IEEE Trans. on VLSI Systems*, PP(99):1–14, October 2011.

[3] C. Huyghebaert, J. Van Olmen, Y. Civale, A. Phommahaxay, A. Jourdain, S. Sood, S. Farrens, and P. Soussan. Cu to Cu interconnect using 3D-TSV and wafer to wafer thermocompression bonding. In *International Interconnect Technology Conference*, pages 1 –3, 2010.

[4] M. Jung, J. Mitra, D. Pan, and S. K. Lim. TSV Stress-aware Full-Chip Mechanical Reliability Analysis and Optimization for 3D IC. In *Proc. ACM Design Automation Conf.*, pages 188–193, 2011.

[5] S. Karmalkar, P. Mohan, H. Nair, and R. Yeluri. Compact Models of Spreading Resistances for ElectricalThermal Design of Devices and ICs. *IEEE Trans. on Electron Devices*, 54(7):1734 –1743, July 2007.

[6] N. Khan, S. Alam, and S. Hassoun. Power Delivery Design for 3-D ICs Using Different Through-Silicon Via (TSV) Technologies. *IEEE Trans. on VLSI Systems*, 19(4):647–658, April 2011.

[7] S. R. Nassif and J. N. Kozhayz. Fast Power Grid Simulation. In *Proc. ACM Design Automation Conf.*, pages 156–161, 2000.

[8] Y. C. Tan, C. M. Tan, X. W. Zhang, T. C. Chai, and D. Q. Yu. Electromigration performance of Through Silicon Via (TSV), A modeling approach. *Microelectronics Reliability*, 50(9-11):1336–1340, Sept.-Nov. 2010.

Tracking Appliance Usage Information in Residential Settings Using Off-the-Shelf Low-Frequency Meters

Deokwoo Jung
Advanced Digital Sciences Center
Illinois at Singapore
1 Fusionopolis Way, Singapore
dj92@illinois.edu

Andreas Savvides
Dept. of Electrical Engineering
Yale University,
New Haven, CT, USA
andreas.savvides@yale.edu

Athanasios Bamis
Dept. of Computer Sci. and Eng.
University of Connecticut,
Storrs, CT, USA
athanasios.bamis@uconn.edu

ABSTRACT

Given the ongoing widespread deployment of low frequency electricity sub-metering devices at residential and commercial buildings, fine-grained usage information of end-loads can bring a new powerful sensing modality in Cyber-Physical Systems (CPS). Motivated by the opportunity, this paper describes an algorithm of estimating the ON/OFF sequences for typical household end-loads in close-to-real-time using an off-the-shelf power meter. Unlike previous algorithms that lacks in scalability to support diverse applications in CPS our algorithm is designed to provide control knobs to support various trade-offs between accuracy and computation load or delay to satisfy the different application requirements. We experimentally verify the proposed algorithm using a collection of home appliances. Our experiment result shows that our algorithm is able to detect ON/OFF sequences of 7 appliances nearly without error and 3 appliances with moderate error rate less than 6% among 12 typical household appliances.

Categories and Subject Descriptors

H.4 [**Information Systems Applications**]: Miscellaneous;
I.2.8 [**Computing Methodologies**]: Artificial Intelligence—
Problem Solving, Control Methods,and Search

General Terms

Algorithms, Design, Performance

Keywords

Cyber Physical System, Load Monitoring, Smart Grid

1. INTRODUCTION

The deregulation of electricity markets and the increasing requirements for greener operation are creating an imminent need for a new breed of Cyber-Physical Systems (CPS) that closely coordinate generation and consumption. Given the ongoing widespread deployment of low frequency electricity sub-metering devices at residential and commercial buildings, fine-grained usage information of end-loads can bring a new type of intelligence in CPS for tighter generation planning and online coordination in smart grid. For example, appliance usage information can be used for understanding electricity consumption [4] or drive decisions for automated load shedding [1]. In this new CPS realm, the data must be obtained and processed as a reusable form of contextual information that can provide scalability to many applications. We argue that ON/OFF traces can provide one such form and we propose a method for obtaining this from central meter information.

Many novel approaches for detecting appliances usage transition states require the analysis of aggregated power consumption measurements in the frequency domain. Although such frequency domain analysis is generally robust, it is not applicable to a large number of typical off-the-shelf electricity meters already deployed by consumers and utilities that do not support high-frequency sampling. To avoid this limitation in this paper we develop a method using time domain electrical event signatures of appliances. This enables the use of off-the-shelf meters. Furthermore, it is also more amenable to emerging web based monitoring systems hosted on the cloud since they don't require the transport and processing of high frequency data.

Solving the problem in the time domain is challenging since appliance signatures inherently suffer from additive noise and interference from other appliances that introduces a lot of ambiguity in their signatures over time. Furthermore, unstable signatures in time domain amplifies this ambiguity. Resolving the ambiguity requires a certain computational loads or delays that need to stay within the tolerances of CPS applications.

To resolve some of theses challenges, in this paper we develop a method of estimating the ON/OFF sequences of appliances, that is the sequence in which they are turned ON and OFF, using the total energy consumption profile measured with a low-frequency central meter. To support various applications in CPS, our method provides control knobs to support various trade-offs between accuracy and computation load or delay to satisfy the different application requirements. To give timely information our method is designed to report the best estimate of ON/OFF sequences to applications at any given time by generating the most likely ON/OFF sequences up to the current observation. This also distinguishes our method from similar existing methods [3, 5, 6] that have little or no freedom of the trade-off and the timing of estimation.

Our approach is validated using two sets of experimental data we collected from a residential deployment using a TED5000 meter with approximately 1 Hz sampling rate. The first data set is used as training data set to characterize the profile of each appliance. We use the second data set to evaluate the performance of our algorithm.

2. RELATED WORK

Previously proposed methods of learning usage information of appliances from a central electricity meter have mainly focused on detecting and classifying ON/OFF event signatures for electricity load monitoring, but less paying attention to estimating sequence of state transition under various near real-time requirements that would apply to emerging CPS applications. Significant work has been done in the area of Non-Intrusive Load Monitoring (NILM) systems, which were first proposed by Hart et al. in [3]. This pioneer work introduced an algorithm for estimating daily usage profiles of each appliance of interest in residential houses by detecting changes in both real and reactive power consumption during their steady state, the *step-change detector* [3].

Since then, many NILM variants have been proposed using additional or new signature features such as the *transient event detector* proposed by Norford et al. in [7] and harmonic signatures in frequency domain proposed by Leeb et al. in [5]. More recently, the *ElectriSense* approach proposed by Gupta et al. in [2] presented a method for detecting appliance events solely from a unique signature of electromagnetic interference (EMI) on frequency domain with a 500 kHz real power measurement equipment. Another approach by, Marchiori et al.[6] revisited Hart et al.'s original method [3] to propose a circuit-level load monitoring method using a low-frequency electricity meter. The method first constructs a two-dimensional probability map of the total active and reactive power consumption for all possible combinations of power consumptions of individual appliances on an electrical sub-circuit. Then the most likely power consumption of each appliance is found using maximum likelihood estimation.

Each of these approaches has its own weaknesses and strengths; however, they overall lack the scalability support needed by large CPS applications. For instance, in the original method by Hart et al.[3], the algorithm has to wait until power consumption is stabilized at steady-state. Algorithms of using signatures in frequency domain [5, 2] is not scalable to typical low-frequency meters. Finally, the methods that use time-domain signatures, such as [6] are not applicable for whole household due to unmanageable computation complexity given the number of appliances.

Our work takes a different approach trying to extract reliable ON/OFF sequences from traces of ambiguous event transitions detected in low-frequency metering data with controllable computational complexity. This is achieved by pruning a set of hypothesis cleverly trading off accuracy and computation load. Instead of detecting a state transition using two consecutive samples, something that can introduce significant error, our method considers a probability of sequences of measurements over a time window. This allows for a more robust detection of state transition that also considers time delay between state transitions.

3. ON/OFF SEQUENCE ESTIMATION

For our algorithm, we model the operation of each appliance as a two state machine that transitions between ON and OFF with time delay. When an appliance changes its state, the transition generates the change in total power consumption measurement. Similarly, an appliance adds its instantaneous power consumption to the total power consumption measurement during the ON or OFF state, which is used as an additional feature of the event signature.

3.1 Our Approach

Our goal is to estimate the ON/OFF sequences of each appliance of interest by detecting their state transitions. We approach the goal based on the key observations that inside a typical household: *a*) there are usually a small number of important appliances that are frequently used; and *b*) the loads of these appliances are very diverse. The diversity of the loads entails that different appliances will have different absolute consumptions and they will also generate different patterns when transitioning between the ON/OFF states. The patterns will differ in the transient spikes they exhibit and also in the time steps required for the power consumption to settle.

The relatively small number of devices ensures that two devices will rarely switch states simultaneously. Thus, by simultaneously combining the information about the absolute value of the current power consumption and the relative change that occurred in consumption we can estimate ON/OFF states of individual appliances for every observed sample from a central meter.

The above detection is a difficult task due to the high variation in the magnitude and the time delay during state transitions of an appliance. Similarly, additive noise caused by the fluctuation in the ON state of different appliances can be frequently confused as an ON/OFF event. To take the uncertainty into account we use a probabilistic model that keeps track of the most likely sequence of ON/OFF events using two variables: *a*) the current power consumption measurement, we call this the *Absolute Magnitude*; and *b*) the magnitude change of the current power consumption during a state transition, we call this the *Differential Magnitude*.

The former is used to compute the likelihood of the current state given the current total consumption value. The latter is used to compute the likelihood that an ON/OFF state transition has taken place.

3.2 Algorithm Description

Our algorithm consists of a one-time training phase and an online detection phase. Fig. 1 provides a high-level overview of its structure. During the training phase, the system learns the typical value of power consumption of each target appliance during its state transition and steady state by having the user perform an experiment. The likelihood functions of absolute and differential magnitudes are computed by kernel density estimator.

During the detection phase, the algorithm performs a two-stage process by first generating a set of *candidate hypothesis* for each appliance in parallel which reduced to *likely hypothesis* by filtering out similar or unlikely hypothesis. Then the algorithm associates hypothesis from each appliance to find *the most likely hypotheses*. To optimally control the trade-off between computation complexity and estimation accuracy for tracking multiple hypothesis, the algorithm provides in-

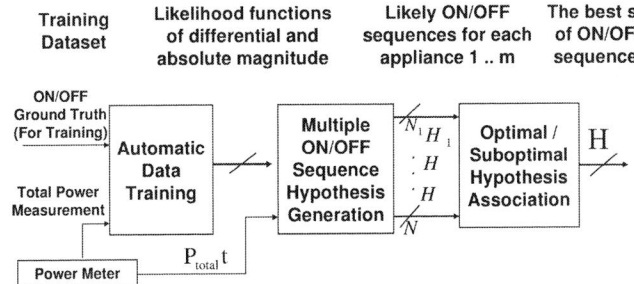

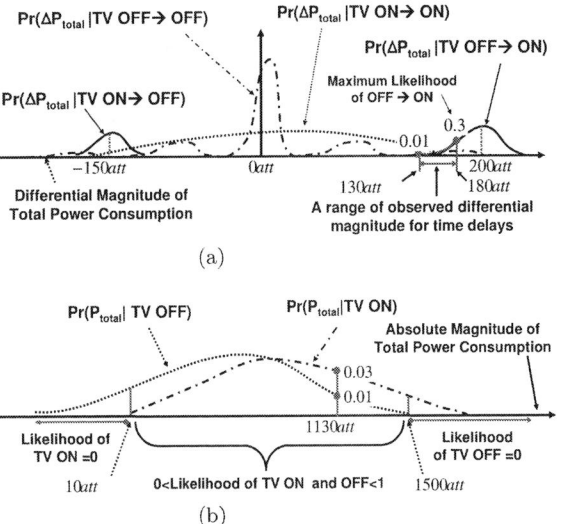

Figure 1: Overall algorithm structure: the algorithm consists of 3 main functions to generate the most likely ON/OFF sequences from the total power consumption measurement at a central power meter

Figure 2: Illustrated examples of likelihood functions of (a) differential magnitude of total power consumption, (b) absolute magnitude of total power consumption

tuitive control knobs: *a)* the maximum number of hypotheses to be retained for each appliance; and *b)* threshold value of the correlation coefficient[1] that quantifies similarity between two hypotheses at different computation stages.

3.2.1 Training Phase

During the training phase, a user performs a simple experiment of turning on and off each target appliance in turn while other appliances at home are off. The *ON and OFF* events in the data set are manually or automatically tagged with the corresponding appliance label. Assuming that each differential and absolute magnitudes for a single appliance follows a truncated gaussian distribution, our algorithm computes the probability density estimate for each appliance using a normal kernel density estimator from the training data set. The possible maximum and minimum time delay of state transition is also learned by searching for the time interval taken to reach a low or a high peak of power consumption.

By superposing the probability density estimates of individual appliances, we can obtain a naive version of a likelihood function for differential magnitude of observed total power consumption, ΔP_{total}. Being said *naive*, we assume that all turning ON and OFF events are mutually exclusive; i.e., only a single state transition occurs at any time instance among all appliances. The assumption simplifies the likelihood function and in practice it holds for most scenarios. Fig. 2(a) shows an illustrative example of the likelihood function of ΔP_{total} given state transitions of Television.

Computing the likelihood function of the absolute magnitude of observed total power consumption, P_{total}, is much more complicated because events of appliances' state visits often overlap each other. For the exact likelihood function, it needs to compute joint probabilities for all possible combinations of appliances' ON/OFF states, which is very computationally intensive and mostly intractable. Instead, we use simulation based technique, *Monte Carlo methods*, in order to obtain their good approximation.

Fig. 2(b) illustrates an example of the likelihood function of the TV being ON and OFF. Note that the likelihood of ON or OFF state for some appliances becomes 0 above or

---

[1]We use the Pearson correlation coefficient that is defined by $\rho(X,Y) = \frac{n \sum x_i y_i - \sum x_i \sum y_i}{\sqrt{n \sum x_i^2 - \sum x_i^2}\sqrt{n \sum y_i^2 - \sum y_i^2}}$ where x_i and y_i are *i*th sample of discrete signal X and Y with n number of samples.

Table 1: Illustrated example of computing likelihood

time	P_{total}	ΔP_{total}	Likelihood of state transition			
			$0 \to 0$	$1 \to 0$	$0 \to 1$	$1 \to 1$
t	1130	-	-	-	-	-
$t-1$	1000	+130	0.1	0	0.01	0.01
$t-2$	990	+140	0.05	0	0.1	0.02
$t-3$	950	+180	0.01	0	0.3	0.05
Max.Trns.Likelihood			0.1	0	0.3	0.05
State Visit Likelihood			0.01		0.03	
Combined Likelihood			0.001	0	0.09	0.015

below a certain threshold value of observed total power consumption, by which we can effectively suppress false alarms. For example, the likelihood of being ON (OFF) becomes 0 when P_{total} is below 10W (above 1500W) in the figure.

3.2.2 ON/OFF Sequence Hypothesis Generator

The *Hypothesis Generator* simultaneously computes the following two likelihood metrics below given the meter measurement, $P_{total}(t)$ at time t.

- Maximum transition likelihood: the maximum value of likelihoods for each possible state transition given differential magnitudes.

- State visit likelihood: a likelihood of ON or OFF state given absolute magnitude, $P_{total}(t)$.

The product of these two likelihoods forms a single state transition likelihood metric, the Combined likelihood that is subsequently multiplied over time for each appliance. A specific sequence of state transitions, e.g. $(0, 0, \cdots, 1, 1)$, constitutes a single hypothesis. Each appliance usually has a large set of hypothesis. A subset of the hypothesis is simultaneously pruned by *zero* likelihood and similarity measure. The maximum likelihood tends to increase the event detection probability by allowing more hypothesis. Meanwhile, the state visit likelihood helps to suppress false alarms with zero likelihood.

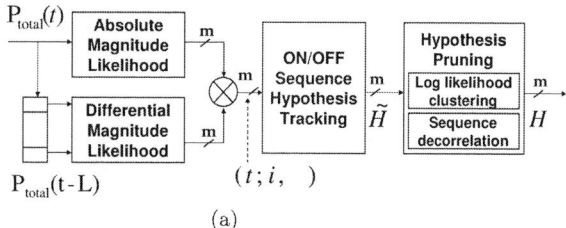

(a)

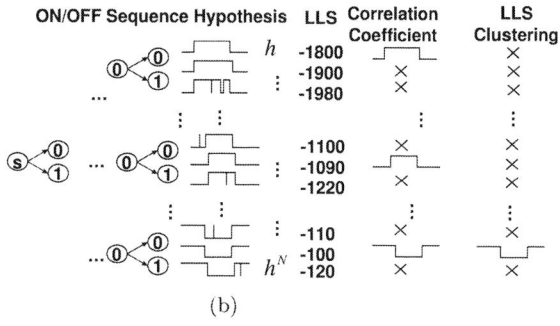

(b)

Figure 3: ON/OFF Sequence Hypothesis Generator (a) Overall algorithm structure , (b) Illustrated example of hypothesis pruning for an appliance

Table 1 illustrates an example of likelihood computation for TV assuming 4 time steps of delay between state transitions given likelihood functions in Fig. 2. In the table, ON and OFF state are represented by 1 and 0, respectively. Given measurement of 1130 Watt at time t, the algorithm evaluates the likelihoods of state visit and the maximum likelihood of each state transition. In Fig.2(a), when the differential magnitude ranges from 130W to 180W in 4 time step delay, the likelihood of state transition from OFF to ON varies from 0.01 to 0.3 where the maximum likelihood is 0.3 at 180W. In the table, the likelihood from ON to OFF state at time t becomes zero that instantly eliminates all hypotheses having that transition. The combined likelihoods are computed by multiplying those two likelihoods for each transition. The largest likelihood is 0.09 for the state transition from OFF to ON.

The overall structure of the hypothesis generator is shown in Fig. 3. Let us assume that there are m appliances at home. For a new measurement $P_{total}(t)$ the hypothesis generator computes the combined likelihoods $L_k(t; i, j)$ of each possible state transition for appliance k from state i to j at time t where $k \in \{1, \cdots, m\}$ and $i, j \in \{0, 1\}$. For each hypothesis Log-Likelihood-Sum (LLS) of the combined likelihoods, $\sum_{\tau \le t} \log L_k(\tau; i, j)$, is updated over time. The hypothesis tracking algorithm generates or updates a candidate set of ON/OFF sequence hypothesis, $\tilde{H}_k = (h_k^1, ..., h_k^{n_k})$ for each appliance k where n_k and h_k^l are the number of hypothesis and lth hypothesis for ON/OFF sequence of appliance k, respectively. Note that n_k could grow exponentially by $O(2^t)$ for each appliance k in the worst case. It, however, rarely occurs because zero likelihood instantly eliminates a large fraction of hypothesis.

To efficiently track the hypothesis we use a pruning algorithm that continuously eliminates nearly duplicated or less likely sequences when the number of hypothesis exceeds a predefined maximum number, N_{max}. To quantify the similarity of two sequences, we use their correlation coefficient. Our algorithm determines two sequences are similar if their correlation coefficient is greater than a predefined threshold value ρ_{min}, which is set to 0.8 by default. Higher ρ_{min} retains more hypotheses.

The pruning process occurs in two steps, which illustrated in Fig. 3(b). In the first step, only one hypothesis with the highest LLS is selected among hypotheses of similar shape of sequences, which are hypotheses with LLS value of $\{-1800, -1090, -100\}$ in the example of the figure. Others are eliminated. In the second step, if the number of the remaining hypotheses exceeds N_{max}, they are clustered[2] into two groups based on their LLS values, and all hypotheses in a group with lower average LLS are eliminated. In Fig. 3(b), hypothesis of $\{-1900, -1090\}$ are removed. The second step is repeated recursively until the number of hypothesis is less than N_{max}. As the result, at most N_{max} number of likely hypotheses H_k are chosen from $\tilde{H}_k$ hypothesis candidates of appliance k. The correlation coefficient threshold value ρ_{min} and maximum number of hypothesis N_{max} can be adaptively adjusted to control computation complexity and estimation performance in real time.

3.2.3 Hypothesis Association

During this stage, the algorithm determines the most likely hypothesis, h_k^* from a set of likely hypothesis, H_k for each appliance k. The simplest way is selecting the one with largest LLS value for each appliance. However, it can result in a large error if a lot of ON and OFF events of appliances occur in a short time. Such error can be significantly reduced by searching all possible associations of hypothesis that minimizes error terms between the reconstructed and the observed total power consumption. It is basically a combinatorial search problem that would cause exponentially heavy computational loads as the number of appliances or the number of retained hypothesis grows. Many generic stochastic search algorithms such as genetic programming or simulated annealing could be used to solve the problem. However, we find that simply associating at most $\tilde{n}_{max}$ number of hypothesis from the highest LLS value works well in our algorithm. When $\tilde{n}_{max} = 1$, our algorithm simply chooses the hypothesis with the largest LLS value.

4. EVALUATION

To validate our approach we performed a short experiment in an 2-story townhouse with 12 typical household end-loads and a TED 5000 meter[3] at the central breaker. For the training phase of our algorithm, shown in Fig. 4(a), we switched OFF most of the appliances in the house, bringing the static power consumption of the house down to approximately 180W. Subsequently, we performed a training sequence consisting of turning each device ON for 30 seconds followed by another 30 seconds of inactivity. The ground truth information was recorded manually. After the training phase a series of ON/OFF experiments was performed, shown in Fig. 4(b). During the experiment, some appliances such as the Refrigerator were constantly ON, sometimes that contributes fluctuation of the background power consumption. We ran our algorithm with those collected data sets for performance evaluation.

In Fig. 5, we estimate the likelihood functions of differen-

---

[2]We use an agglomerative hierarchical clustering algorithm due to its low time complexity.

[3]http://www.theenergydetective.com

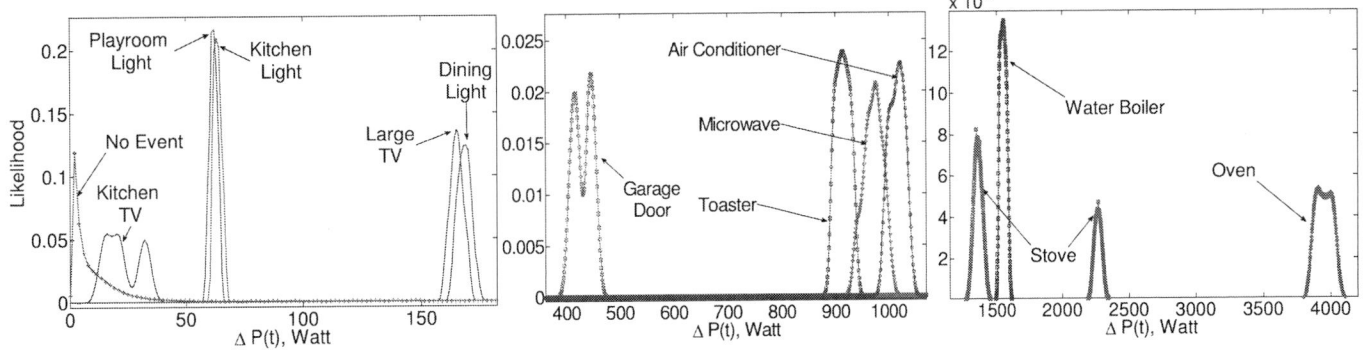

Figure 5: Estimated Likelihood Function of differential magnitude of $OFF \rightarrow ON$ event for appliances

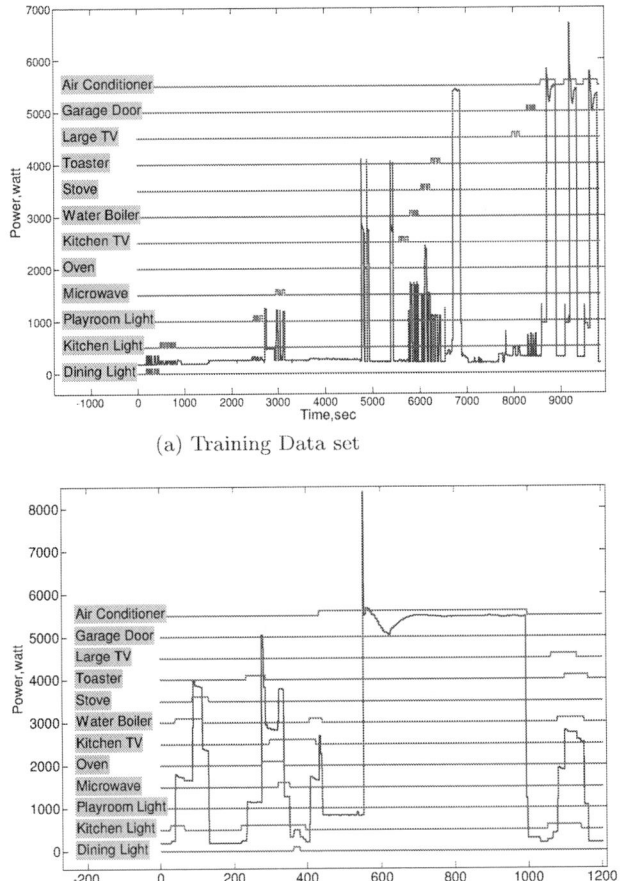

(a) Training Data set

(b) Evaluation Data set

Figure 4: Data sets for experiment: the aggregated power consumption from a power meter (blue line) and ground truth of ON/OFF states (red line)

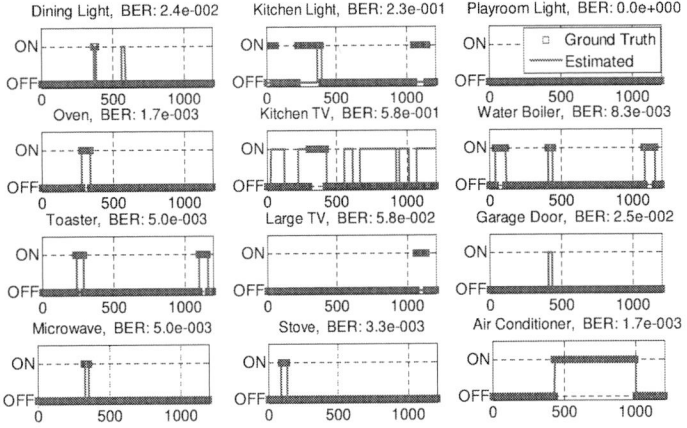

Figure 6: Estimated ON/OFF sequence versus its ground truth of each appliance

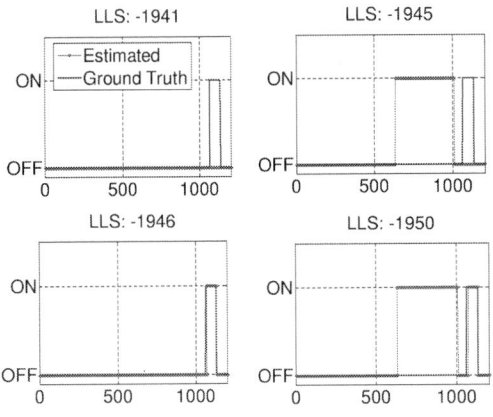

Figure 7: Likely ON/OFF sequences of *Large TV* from Hypothesis Generator

tial magnitude of $OFF \rightarrow ON$ event for 12 appliances using the training data set. We see a large overlap in likelihood functions between *Playroom Light* and *Kitchen Light* and also between *Large TV* and *Dining Light*, including some moderate overlap among *Air Conditioner*, *Microwave*, and *Toaster*. Those ambiguities are reduced when the time delays during state transitions are taken into account. For

example, a typical time delay from OFF to ON state for *Air Conditioner*, *Microwave*, and *Toaster* are 2, 5, and 7 seconds respectively. It gives different maximum likelihood values for similar magnitude changes. We note that a fluctuation of background power consumption, which denoted by *NO Event* often contributes a large false alarm for appliances with low consumption load. In Fig. 5, the likelihood

167

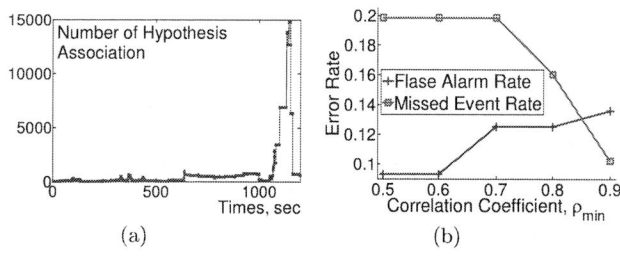

(a)   (b)

Figure 8: (a)Number of the associated hypothesis, (b) False alarm and missed event rate over different correlation coefficient thresholds, ρ_{min} at $N_{max} = 10$

function of *NO Event* significantly overlaps with *Kitchen TV* from 10 to 40 Watt which causes highly unreliable estimates.

In Fig. 6, we evaluate the estimation performance of our algorithm with parameters of $\rho_{min} = 0.8$ and $N_{max} = 10$. For performance metric, we use the bit error rate (BER) which is the number of bit (ON/OFF state) errors divided by the total number of bits generated by an appliance. In Fig. 6, the estimate and ground truth of ON/OFF sequences are plotted as a red line and a blue square mark respectively. It shows that our algorithm can correctly estimate the ON/OFF sequences of most appliance with less than 6% of error, except appliance with relatively small loads, i.e. *Kitchen TV* and *Kitchen Light*. For these smaller loads, the performance of our algorithm is greatly affected by fluctuation of the background consumption that results in more than 50% of error rate for both appliances. Such unreliable performance is caused by low Signal-to-noise ratio(SNR) that is imposed by the hardware limitation of a meter (i.e. a low resolution of signature due to the low sampling rate of a meter) rather algorithm itself.

We note that our algorithm is designed to retain up to N_{max} hypotheses in database. Even if the algorithm cannot determine the correct estimate, it can potentially find one when side information (e.g. reactive power) is available from additional sensors or higher frequency meters. For example, a set of likely ON/OFF sequences for *Large TV* is stored in database that are shown with LLS on top of each plot in Fig. 7. The hypothesis with $LLS = 1941$ is chosen by algorithm, but other closely competing hypothesis including the correct one of $LLS = 1946$ are also retained in database. The algorithm can revisit a set of the retained hypotheses and effectively determine the correct hypothesis.

In Fig. 8(a), we plot a trace of the number of possible hypothesis association that is a product of the number of each appliance's hypotheses. The figure shows that the number stays relatively stable up to 1000 seconds because our algorithm instantaneously eliminates the majority of false alarms and many redundant hypothesis per each sample. The number grows exponentially from 1000 to 1100 seconds when many appliances are turned on in a short intervals, generating many competing hypotheses. Even in such case the individual number of hypotheses of most appliances stays well below $N_{max} (= 10)$. For few appliances whose individual number of hypotheses exceeds N_{max}, LLS clustering prunes less likely hypotheses that instantly drops the number from 15000 to 1000 at $t = 1150$.

To describe a performance trade-off we use a metric of a false alarm rate and a missed event rate. The false alarm rate is defined by a ratio of the number of falsely detected ON states to the total number of OFF states and the missed

event rate is defined by the opposite way of the false alarm rate. Fig. 8(b) shows the trend of average false alarm rate and missed event rate of appliances as we increase the correlation coefficient threshold ρ_{min} from 0.5 to 0.9. For higher ρ_{min}, our algorithm tends to retain more hypothesis because it more strictly applies similarity measures between ON/OFF sequences. Consequently, it reduces the missed event rate but also increases the false alarm rate. Our algorithm is designed to gracefully manage such trade-off. Fig. 8(b), a missed event rate drastically drops from 0.2 to 0.1 as ρ_{min} increases from 0.7 to 0.9. Meanwhile, it modestly raises a false alarm rate from 0.1 to 0.13.

5. CONCLUSIONS AND FUTURE WORK

In this paper, we propose an algorithm for estimating ON/OFF sequences of appliances in residential home from an off-the shelf power meter of 1 Hz sampling rate. Our result shows that our algorithm can successfully track ON/OFF sequences of a majority of loads with higher consumption than background noise. For the future work, we plan to further improve our algorithm to more robustly resolve ambiguity among closely competing hypotheses by utilizing side information from other low cost sensors.

Acknowledgment

This work was supported by generous gifts from The Yale Climate & Energy Institute (YCEI) and a travel grant from Advanced Digital Sciences Center (ADSC), Illinois at Singapore. Finally, authors would like to thank Dr. Jerry T. Chiang at ADSC for his precious feedback.

REFERENCES

[1] David C. Bergman, Dong Jin, Joshua P. Juen, Naoki Tanaka, Carl A. Gunter, and Andrew Wright. Nonintrusive load-shed verification. *IEEE Pervasive Computing*, 10, 2011.

[2] S. Gupta, M.S. Reynolds, and S.N. Patel. Electrisense: single-point sensing using emi for electrical event detection and classification in the home. In *Proceedings of the 12th ACM International Conference on Ubiquitous computing (Ubicomp)*. ACM, 2010.

[3] G.W. Hart. Nonintrusive appliance load monitoring. *Proceedings of the IEEE*, 80(12):1870 –1891, 1992.

[4] D. Jung and A. Savvides. Estimating building consumption breakdowns using on/off state sensing and incremental sub-meter deployment. In *Proceedings of the 8th ACM Conference on Embedded Networked Sensor Systems (SenSys)*. ACM, 2010.

[5] C. Laughman, K.Lee, R. Cox, S. Shaw, S. Leeb, L. Norford, and P. Armstrong. Power signature analysis. *Power and Energy Magazine, IEEE*, 1(2), 2003.

[6] Alan Marchiori, Douglas Hakkarinen, Qi Han, and Lieko Earle. Circuit-level load monitoring for household energy management. *IEEE Pervasive Computing*, 10:40–48, 2011.

[7] L. K. Norford and S. B. Leeb. Non-intrusive electrical load monitoring in commercial buildings based on steady-state and transient load-detection algorithms. *Energy and Buildings*, 24(1), 1996.

Implementing an FPGA System for Real-Time Intent Recognition for Prosthetic Legs

Xiaorong Zhang, He Huang, and Qing Yang
University of Rhode Island
{zxiaorong, huang, qyang}@ele.uri.edu

ABSTRACT

This paper presents the design and implementation of a cyber physical system (CPS) for neural-machine interface (NMI) that continuously senses signals from a human neuromuscular control system and recognizes the user's intended locomotion modes in real-time. The CPS contains two major parts: a microcontroller unit (MCU) for sensing and buffering input signals and an FPGA device as the computing engine for fast decoding and recognition of neural signals. The real-time experiments on a human subject demonstrated its real-time, self-contained, and high accuracy in identifying three major lower limb movement tasks (level-ground walking, stair ascent, and standing), paving the way for truly neural-controlled prosthetic legs.

Categories and Subject Descriptors

C.3 [**Special-purpose and Application-based Systems**]: Real-time and embedded systems.

General Terms

Algorithms, Performance, Design, Experimentation

Keywords

Neural-machine interface, embedded system, prosthetic leg, field-programmable gate array (FPGA)

1. INTRODUCTION

Neural-machine interface (NMI) is a typical example of biomedical cyber physical system (CPS) which utilizes neural activities to control machines. The neural signals collected from nerves, central neurons, and muscles contain a lot of important information that can represent human states such as emotion, intention, and motion. In such a CPS, a computer senses bioelectric signals from a physical system (i.e. human neural control system), interprets these signals, and then controls an external device, such as a power-assisted wheelchair [1], a telepresence robot [2], or a prosthesis [3-5], which is also a physical system.

The neural signals captured from muscles are called electromyographic (EMG) signals. The EMG signals can be picked up with electrodes on the body surface and are effective bioelectric signals for expressing movement intent. In recent years, EMG-based NMI has been widely studied for control of artificial limbs in order to improve the quality of life of people with limb loss.

Researchers have aimed at utilizing neural information to develop multifunctional, computerized prosthetic limbs that perform like natural-controlled limbs. The NMI needs to interface with multiple sensors for collecting neural signals, decipher user intent, and drive the prosthetic joints simultaneously. EMG pattern recognition (PR) is a sophisticated technique for characterizing EMG signals and classifying user's intended movements. It usually contains a training phase for constructing the parameters of a classifier from a large amount of EMG signals, and a testing phase for recognizing user intent using the trained classifier. While the PR algorithm for artificial arm control has been successfully developed and neural-controlled prosthetic arms have already been clinically tested [4, 6-7], there has been no EMG-based NMI commercially available for control of powered prosthetic legs. Challenges in the management of both physical and computational resources have limited the success of a CPS for neural control of artificial legs.

One of the challenges on physical resources is due to the muscle loss of leg amputees. Patients with leg amputations may not have enough EMG recording sites available for neuromuscular information extraction [8]. The non-stationary of EMG signals during dynamic leg movement further increases the difficulty of user intent recognition (UIR). To address this challenge, Huang et al. proposed a phase-dependent PR strategy for classifying user's locomotion modes [8]. This PR algorithm extracted neural information from limited signal sources and showed accurate classification (90% or higher accuracy) of seven locomotion modes when 7-9 channels of EMG signals were collected from able-bodied subjects and leg amputees. The performance of the phase-dependent PR strategy was further improved by incorporating EMG signals with mechanical signals resulting from forces/moments acting on prosthetic legs [9]. The experimental results showed that the classification accuracies of the neuromuscular-mechanical fusion based PR algorithm were 2%-27% higher than the accuracies derived from the strategies using EMG signals alone, or mechanical signals alone [9].

The challenges on computational resources include tight integration of software and hardware on an embedded computer system that is specifically tailored to this environment. It requires high speed classifier training, fast response, real-time decision making, high reliability, and low power consumption. Embedded systems are usually resource constrained and typically have processors with slower system clock, limited memory, small or no hard drives. To make the idea of neural-controlled artificial leg a reality, we need efficiently manage the constrained computational resources to meet all the requirements for smooth control and safe use of prosthetic legs. Our previous study proposed an NMI implemented on a commodity 32-bit microcontroller unit (MCU) for recognizing two non-locomotion tasks of sitting and standing in real-time [10]. It was reported that there was a noticeable delay of 400 ms for producing classifying decisions, implying inadequate computational power of the MCU for real-time control of artificial legs. Furthermore, this NMI implementation realized only the testing phase of the PR algorithm on the MCU. The training algorithm which involved intensive computations was implemented on graphic processing units (GPUs) and showed good speedups over CPU-based implementation [10]. However, currently most of the GPU cards only have PCI Express interfaces and are not portable. Relative high power consumption further makes it more difficult to use GPU as an embedded wearable device.

To tackle these technical challenges, we present a new design of an embedded system that is specifically tailored to the new NMI. A unique integration of hardware and software of the embedded system is proposed that is suitable to this real-time CPS with adequate computational capability, high energy efficiency, flexibility, reliability, and robustness. The NMI on an embedded platform continuously monitors EMG activities from leg muscles as well as mechanical forces/moments acting on prosthetic legs. Information fusion technique is then used to decode and decipher the collected signals to recognize users' intended locomotion modes in real-time. The embedded system contains two major parts: a data collection module for sensing and buffering input signals and an intelligent processing unit for executing the UIR algorithm. The data collection module was implemented on a microcontroller unit (MCU) with multiple on-board analog-to-digital converters (ADCs) for signal sampling. A reconfigurable FPGA device was designed as the main computing engine for this system. There are several reasons for choosing FPGAs for the designed NMI. First, the parallelism of FPGAs allows for high computational throughput even at low clock rates. Secondly, FPGAs are not constrained by a specific instruction set, thus are more flexible and more power efficient than processors. Furthermore, FPGAs can easily generate customized IO interfaces with existing IP cores, and appear to be good choices for real-time embedded solutions. In our design, a high-level synthesis tool was used to help reducing the implementation difficulty of coding with hardware design language (HDL). A special parallel processing algorithm for UIR was designed, realizing the neuromuscular-mechanical fusion based PR algorithm coupled with the real-time controlling algorithm in hardware. A serial peripheral interface (SPI) was built between the MCU and the FPGA to transfer digitized input data from the MCU to the FPGA device. The decision stream of user's intended movements can be output to either control a powered prosthetic leg or drive a virtual reality (VR) system with the purpose of evaluating the NMI. Although our previous research has made the attempt to use FPGA in EMG pattern recognition and has shown high processing speed in the offline analyses [11], the embedded system presented here is the first complete CPS for the NMI that implements both training and testing modules on one single chip. The New CPS integrates all the necessary interfaces and control algorithms for interacting with the physical system in real-time.

The newly designed NMI was completely built and tested as a working prototype. The prototype was then used to carry out real-time testing experiments on an able-bodied subject for classifying three movement tasks (level-ground walking, stair ascent, and standing) in real-time. The system performance was evaluated to demonstrate the feasibility of a self-contained and high performance real-time NMI for artificial legs. Videos of our experiments on the human subject can be found at http://www.youtube.com/watch?v=KNhihjXProU.

This paper is organized as follows. Next section presents the overall system design. Section 3 describes the detailed implementation of the UIR algorithm. The experimental results are demonstrated in Section 4. We conclude our paper in Section 5.

2. SYSTEM DESIGN

The architecture of designed CPS is shown in Figure 1. The embedded NMI samples input signals from two physical systems--a human neuromuscular system and a mechanical prosthetic leg. The sampled signals are then processed to decipher user's intent to control the prosthesis. The NMI consists of two modules: a data

collection module built on an MCU with multiple on-chip ADCs for sensing and buffering input signals, and an FPGA device as the computing engine for fast data decoding and pattern recognition. A serial peripheral interface (SPI) is located between the two devices for transferring digitized input data from the MCU to the FPGA device.

1) Input signals: Multi-channel EMG signals are collected from multiple surface electrodes mounted on patient's residual muscles. Mechanical forces and moments are recorded from a 6 degrees-of-freedom (DOF) load cell mounted on the prosthetic pylon. The EMG signals and the mechanical signals are preprocessed by filters and amplifiers and then simultaneously streamed into the NMI.

2) MCU module: The MCU device does not do any compute-intensive task. It provides multi-channel on-chip ADCs to sample the input signals and convert the analog signals to digital data. The digitized data is then stored in the user-defined result queues allocated in the RAM buffer. In the system RAM, two equal sized result queues are defined. With direct memory access (DMA) support, the MCU core can be insulated from the data acquisition process. Thus these two result queues forms a circular buffer that can continuously receive new data while transmitting old data to the FPGA module for further processing.

3) FPGA module: The FPGA device receives digitized data from the MCU module continuously. In order to fully utilize the computing capacity of the FPGA system and produce dense decisions, the input signals are segmented by overlapped analysis windows with a fixed window length and window increment [4]. The designed FPGA module contains six components: an SPI module that serially receives input signals from the MCU module, a user defined module implementing the UIR algorithm, a high-speed on-chip memory for fast online pattern recognition, an SDRAM controller that interfaces with a large-capacity external SDRAM, parallel IOs for outputting UIR decisions, and a soft processor for managing hardware components and directing data flows. The FPGA module works in two modes: offline training and online pattern recognition. Offline training needs to be performed before using the artificial leg and also whenever a complete re-training is required. During the training procedure, users are instructed to do different movement tasks, and a large amount of data is collected by the NMI to train the classifier. The external SDRAM is only used in the offline training phase to store the training data because FPGAs usually have limited on-chip memory. For online pattern recognition, the input streams are stored in the on-chip memory for fast processing and provided to the classifier for decisions to continuously identify the user's intended movements.

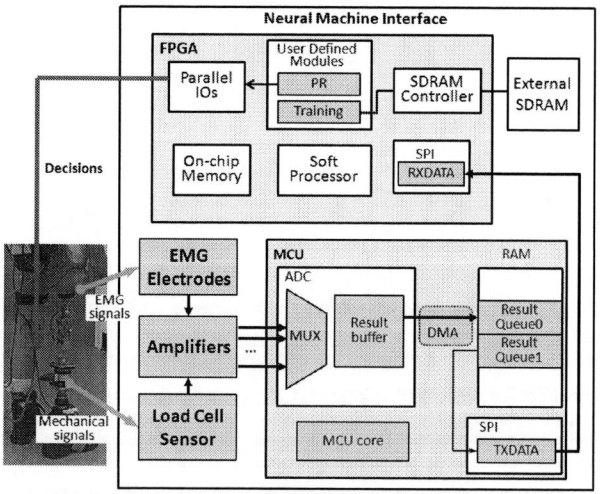

Figure 1. System architecture of the embedded NMI for artificial legs.

3. IMPLEMENTATION OF THE UIR ALGORITHM ON FPGA

3.1 Architecture of the UIR Strategy

The architecture of the UIR strategy based on neuromuscular-mechanical information fusion and phase-dependent pattern recognition (PR) is shown in Figure 2. It is a self-contained architecture that integrates the functions of training and phase-dependent pattern recognition in one embedded system. For every analysis window, features of EMG signals and mechanical signals are extracted from each input channel. A feature vector is formed and normalized by fusing the features from all the input channels. The feature vector is then fed to the classifier for pattern recognition. The phase-dependent classifier consists of a gait phase detector and multiple classifiers. Each classifier is associated with a specific gait phase. During the process of pattern recognition, the gait phase for current analysis window is first determined by the phase detector, and then the corresponding classifier is adopted to do the classification. In this study, four gait phases are defined: initial double limb stance (phase 1), single limb stance (phase 2), terminal double limb stance (phase 3), and swing (phase 4) [9]. The real-time gait phase detection is based on the measurements of the vertical ground reaction force (GRF) sampled from the 6-DOF load cell.

In the real-time embedded system design, to ensure a smooth control of artificial legs, precise timing control is necessary. Figure 3 shows the timing diagram of the control algorithm during the real-time UIR process. In the designed system, the MCU and the FPGA device collaborates to produce a decision at every window increment. While the MCU is sampling data for window $i+1$, the user intent recognition for window i, including the tasks of SPI data transfer, feature extraction, gait phase detection, feature vector formation and normalization, and pattern recognition must be done within the window increment. In other words, the execution time of the UIR algorithm determines the minimum window increment. Larger window increments will introduce longer delay to the NMI decision, which may not be safe to control the prosthesis in real-time. Therefore fast processing speed is very critical to the embedded system design.

3.2 Parallel Implemetations on FPGA

The implementation of the CPS was based on the Altera DE3 education board with a Stratix III 3S150 FPGA device, coupled with the Freescale MPC5566 132 MHz 32 bits MCU evaluation board (EVB) with 40-channel 12-bit on-chip ADCs. The MPC5566 module and the DE3 module are connected with each other via

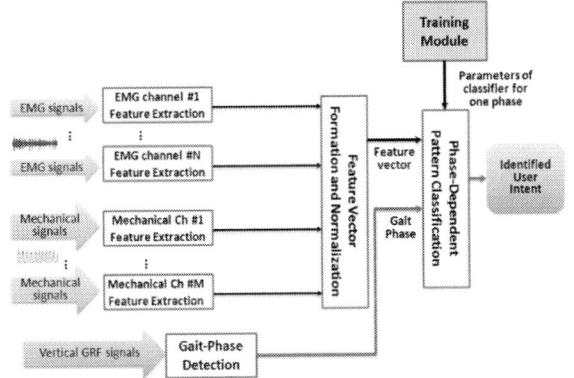

Figure 2. Architecture of UIR strategy based on neuromuscular-mechanical fusion-based phase-dependent pattern recognition.

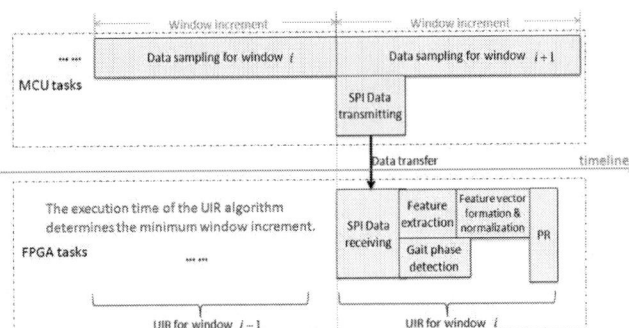

Figure 3. Timing diagram of the control algorithm during online UIR process.

serial peripheral interface (SPI). In this design, DE3 was configured as the SPI master and MPC5566 was the slave. A parallel UIR algorithm tailored to FPGA was designed and implemented on DE3. Fixed-point operations were adopted in this implementation because of their less resource cost and lower latency than floating-point operations. In addition, because the input signals were sampled by ADCs with 12-bit resolution, all the arithmetic operation types in the PR algorithm could be handled by 32-bit fixed-point data formats with careful management.

The UIR algorithm was implemented on the FPGA with the help of a high-level synthesis tool--CoDeveloper from Impulse Accelerated Technologies. The PR algorithm was first developed using C programming language, and then CoDeveloper was used to generate VHDL (VHSIC hardware description language) modules from the C program. The VHDL modules were integrated into the FPGA system as the user defined modules as shown in Figure 1, and worked with other hardware components as a complete NMI. To utilize the parallelism of FPGAs, CoDeveloper provides a multiple process, parallel programming model. In our design, the algorithm was partitioned into a set of processes. These processes can run on the FPGA in parallel if there are no data dependencies. The communications between processes can be done using communication objects, such as streams, signals, and shared memories. Streams are implemented in hardware as dual-port FIFO RAM buffers. A stream connects two concurrent processes (a producer and a consumer), where the producer stores data into and the consumer accesses data from the stream buffer. A single process can be associated with multiple input and output streams. Signals are useful objects to communicate status information among processes. Shared memories are used to store and access large blocks of data from specific external memory locations using block read and block write functions.

1) Feature Extraction: Before offline training or online pattern recognition is performed, features need to be extracted from raw input signals. In every analysis window, four time-domain (TD) features (mean absolute value, number of zero crossings, waveform length, and number of slope sign changes) are extracted from each EMG channel. For the mechanical forces/moments recorded from the 6-DOF load cell, the mean value is calculated as the feature from each individual DOF. The procedure of feature extraction is independent for individual input channel and identical for homogeneous sensors. This property can be utilized to greatly reduce the computation time for feature extraction because all the channels can be processed in parallel. Figure 4 shows the partitioned processes and the data flows of the FPGA implementation of feature extraction. Each white box in the figure represents a small process. The black arrows located between processes are one-way data streams. In this design, $N+6$ parallel threads are generated, where N denotes the number of EMG channels, and the other six threads are assigned for extracting

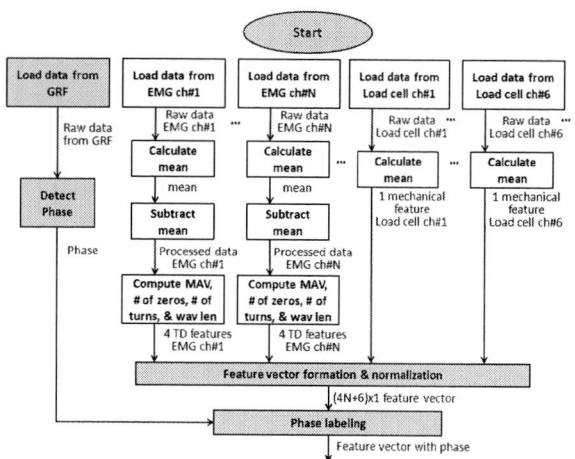

Figure 4. Partitioned processes and data flows of the FPGA implementation of feature extraction.

features from mechanical forces/moments. For each EMG channel, the thread contains four processes: loading raw input data from memory, calculating mean, subtracting mean from the raw data, and extracting four TD features from the processed data. For mechanical forces/moments, each thread fetches raw data from memory and calculates mean as the mechanical feature. After all the features are extracted, the feature streams are sent to the process of feature vector formation and normalization and then fused into a $(4N+6) \times 1$ feature vector. To implement the phase-dependent PR strategy, a thread of gait phase detection loads the vertical GRF measured from the load cell in each analysis window, and then determines current gait phase. This thread is also independent from the threads for feature extraction so that it can run simultaneously with other threads. The detected gait phase is streamed to the phase labeling process, and the feature vector generated in current window is labeled with a specific gait phase. During online pattern recognition, the feature vector with a labeled phase is the input data for pattern classification. In the training procedure, signals are recorded for a period of time under each movement task. Same procedure of feature extraction is performed for every training window. A $(4N+6) \times M_p (p \in [1,4])$ feature matrix is generated as the training data for each gait phase, where M_p is the number of training windows in the p_{th} phase.

2) Pattern Recognition: In this study, linear discriminant analysis (LDA) is adopted for user intent classification because of its computational efficiency for real-time prosthesis control and the comparable accuracy to more complex classifiers [7]. Four gait phases are defined for recognizing user's locomotion mode, giving rise to four LDA-based classifiers. Each classifier is trained for a specific phase. The details of the LDA algorithm can be found in the supplemental material.

Most of the computations involved in the training algorithm are matrix operations. Because a large amount of data need to be processed in the training procedure, the dimensions of the matrices can be very large. Only using on-chip memory is not enough to handle all the computations. External memory with large capacity is required to store the processing data. In our implementation, several external memory buffers are defined to store either large matrices during the training computations or data that might be reused in the online PR phase or the re-training phase. A process is designed to perform a simple task with a small block of data, such as a matrix row/column, and store partial results in the external memory. In this way, the operations of subsequent matrix rows/columns can be efficiently pipelined.

During online pattern recognition, based on the gait phase of current analysis window, the parameters of the corresponding classifier are loaded from memory. The observed feature vector derived from each analysis window is provided to the classifier for intent recognition.

4. PROTOTYPING & EXPERIMENTAL RESULTS

This study was conducted with Institutional Review Board (IRB) approval at our university and informed consent of subjects. To evaluate the performance of the designed NMI, two experiments with different purposes were conducted. First, to evaluate the classification accuracy and the computation speed of the FPGA-based PR algorithm, the performance of the FPGA implementation was compared with our previous software implementation by processing the same dataset offline. Secondly, to evaluate the performance of the entire CPS, a real-time test was carried out on a male able-bodied subject for identifying three movement tasks (level-ground walking, stair ascent, and standing).

4.1 Performance of FPGA vs. CPU

In order to verify the correctness of the FPGA-based PR algorithm and compare the performance of the FPGA design with our previous Matlab implementation, we processed the same dataset on both platforms. The testing dataset was previously collected from a male patient with transfemoral amputation (TF). Seven EMG channels recording signals from the gluteal and thigh muscles and six channels of mechanical forces/moments measured by a 6-DOF load cell were collected in this dataset for identifying three locomotion modes including level-ground walking, stairs ascent, and stairs descent. The dataset was segmented by overlapped analysis windows. The window length and the window increment were set to 160 data points and 20 data points, respectively. The dataset contained 936 analysis windows totally, where 596 of them were used as the training data and the rest 340 windows were testing data. The Matlab implementation was based on a PC with Intel Core i3 3.2 GHz CPU and 6 GB DDR3 SDRAM at 1333 MHz. For the FPGA implementation, a 1GB DDR2-SDRAM SO-DIMM module was plugged into the DDR2 SO-DIMM socket on the DE3 board as the system external memory. The Altera high performance DDR2 SDRAM IP generated one 200 MHz clock as SDRAM's data clock and one half-rate system clock 100 MHz for all other hardware components in the system. The dataset was preloaded into the SDRAM, and the output decisions were printed to the Nios II console [12] for performance evaluation.

It was observed that the classification results of the FPGA system matched very well with the Matlab implementation. Both platforms provided a training accuracy of 98.99% and a testing accuracy of 98.00%. The missed classification points of the two implementations appeared in the same locations. These results clearly demonstrated that the FPGA-based PR algorithm did not lose any computation accuracy as compared to the software implementation.

Table 1 compares the execution time of the LDA-based PR algorithm between the software implementation and the FPGA design. Two configurations with different number of input channels were considered, one with 7 EMG channels and 6 mechanical channels, the other with 12 EMGs and 6 mechanical channels. For the training algorithm that processed 600 analysis windows, the FPGA provided a speedup of around 7X over the software implementation. In the testing phase, the FPGA system took less than 0.3 ms to classify one analysis window. Compared with the Matlab implementation, the FPGA-based PR testing algorithm demonstrated a speedup of 30 times for the configuration of 7

Table 1. Comparison of the execution time of the PR algorithm

	Configuration	FPGA	Matlab	Speedup
Training Algorithm (600 analysis windows)	7 EMGs 6 Mech.	0.46 s	3.2 s	6.96 x
	12 EMGs 6 Mech.	0.64 s	4.7 s	7.34 x
Testing algorithm (classify one analysis window)	7 EMGs 6 Mech.	0.23ms	6.8 ms	29.56 x
	12 EMGs 6 Mech.	0.25 ms	9.5 ms	38.00 x

EMGs and 6 mechanical signals. If more input channels were used (i.e. 12 EMG channels and 6 mechanical channels), a more significant speedup of 38X was observed, which further demonstrated the advantages of FPGA parallelism. From Table 1 we can see that the FPGA implementation of the testing algorithm shows better performance than the training algorithm. This is because the testing algorithm only used fast on-chip memory while the computation complexity of the training algorithm required the FPGA to interact with the external memory. In our experiments it was observed that loading training data from external memory to the FPGA took more than half of the total execution time of the training algorithm. The summary of FPGA resource utilization is listed in Table 2.

4.2 System performance in real-time

The designed NMI prototype was tested on one male able-bodied subject (Figure 5) in real-time. A plastic adaptor was made so that the subject could wear a hydraulic passive knee on the left side. Seven surface EMG electrodes (MA-420-002, Motion Lab System Inc., Baton Rouge, LA) were used to record signals from the gluteal and thigh (or residual thigh) muscles on the subject's left leg. An MA-300 system (Motion Lab System Inc., Baton Rouge, LA) collected seven channels of EMG signals. A ground electrode was placed near the anterior iliac spine of the subject. The mechanical ground reaction forces and moments were measured by a 6-DOF load cell mounted on the prosthetic pylon. The analog EMG signals and mechanical signals were digitally sampled at the rate of 1.1 KHz by the MPC5566 EVB. The intent decisions made by the FPGA device were sent out to 4-bit parallel IO pins on the DE3 board, and displayed by a software GUI. The window length and the window increment were still set to 160 data points and 20 data points, respectively.

Three movement tasks (level-ground walking (W), stair ascent (SA) and standing (ST)) and four mode transitions (ST→W, W→ST, ST→SA and SA→ST) were investigated in this experiment. For the subject's safety, he was allowed to use hand railings and a walking stick. A training session was conducted first to collect the training data for the pattern classification. The

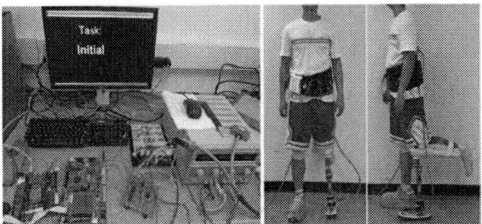

Figure 5. The NMI prototype based on MPC5566 EVB and DE3 education board (left figure) and the experimental setup of the real-time test on a male able-bodied subject (right figure).

subject was instructed to do each movement task for about 10 seconds in one trial. Three trials were collected as the training data. In the real time testing sessions, 10 real-time testing trials were conducted. To evaluate the system performance of real-time intent recognition, we adopted the evaluation criteria as described in our previous study [13]. The testing data were separated into static states and transitional periods. The static state was defined as the state of the subject continuously walking on the same type of terrain (level ground and stair) or performing the same task (standing). A transitional period was the period when subjects switched locomotion modes. The purpose of the UIR system is to predict mode transitions before a critical gait event for safe and smooth switch of prosthesis control mode. In this study the critical timing was defined for each type of transition. For the transitions from standing to locomotion modes (level-ground walking and stair ascent), the critical timing was defined at the beginning of the swing phase (i.e. toe-off). For the transitions from locomotion modes to standing, the critical timing was the beginning of the double stance phase (i.e. heel contact). The real time performance of our embedded system was evaluated by the following parameters.

Classification Accuracy in the Static States: The classification accuracy in the static state is the percentage of correctly classified observations over the total number of observations in the static states.

The Number of Missed Mode Transitions: For the transitions from standing to locomotion modes, the transition period starts one second before the critical timing, and terminates at the end of the single stance phase after the critical timing; for the transitions from locomotion modes to standing, the transition period includes the full stride cycle prior to the critical timing and the period of one second after the critical timing. A transition is missed if no correct transition decision is made within the defined transition period.

Prediction Time of Mode Transitions: The prediction time of a transition in this experiment is defined as the elapsed time from the moment when the decisions of the classifier changes movement mode to the critical timing for the investigated task transitions.

The overall classification accuracy in the static states across 10 testing trials for classifying level-ground walking, stair ascent and standing was 99.31%. For all the 10 trials, no missed mode transitions were observed within the defined transition period. Table 3 lists the average and the standard deviation of the prediction time for four types of transitions. The results show that there was around 104 ms decision delay for the transitions from stair ascent to standing (SA→ST). This is because the subject

Table 2. Stratix III 3S150 Resource Utilization

Resources	Available	Training 12 EMG 6 Mech.	Training 7 EMG 6 Mech.	Online PR 12 EMG 6 Mech.	Online PR 7 EMG 6 Mech.
Combinational ALUTs	113,600	46%	33%	32%	25%
Memory ALUTs	56,800	3%	2%	3%	2%
Registers	113,600	43%	30%	27%	24%
Block memory bits	5,630,976	16%	12%	16%	12%
DSP blocks	384	72%	44%	27%	24%

Table 3. Prediction Time of Mode Transitions Before Critical Timing

Transition	ST→W	W→ST	ST→SA	SA→ST
Prediction Time (ms)	412.8±76.7	124.39±114.2	549.83±139.2	-104.67±54.1

could not perform foot-over-foot alternating stair climbing with a passive knee joint. In our experiments, the subject climbed stairs by lifting the sound leg on one step and then pulled up the prosthetic leg on the same step, which produced the same pattern as the mode transition from stair ascent to standing. Therefore the transition SA→ST was only able to be recognized after the subject was standing still. This problem will be eliminated by replacing the passive device with a powered knee in the near future. Wearing the powered knee, the prosthesis user is able to climb stairs foot-over-foot, which provides a very different pattern from the transition SA→ST. For the other three types of transitions (ST→W, W→ST, and ST→SA), the user intent for mode transitions can be accurately predicted 104-549 ms before the critical timing for switching the control of prosthesis. Figure 6 shows the real-time system performance for one representative testing trial. The white area in Figure 6 denotes the periods of static states (level-ground walking, stair ascent, and standing), the gray area represents the transitional period, and the black vertical dash line indicates the critical timing for each transition. We can see in this trial all the transitions were correctly recognized within the transitional period. No missed classifications occurred in the static states in this trial. The video of our real-time experiments can be found at http://www.youtube.com/watch?v=KNhihjXProU.

5. CONCLUSIONS

This paper presented the design and implementation of the first complete cyber physical system of neural machine interface for artificial legs. The new CPS implemented both training and testing modules on one single chip, and integrated all the necessary interfaces and control algorithms for identifying the user's intended locomotion modes in real-time. The designed NMI incorporated an MCU for sensing and buffering input EMG signals and mechanical signals, and an FPGA device as the computing engine for fast decoding and pattern recognition. A special parallel processing algorithm for UIR was designed and implemented that realized the neuromuscular-mechanical fusion based PR algorithm coupled with the real-time controlling algorithm on the FPGA. The FPGA implementation of the PR algorithm achieved a speedup of 7X over the Matlab implementation for the training phase, and a speedup of more than 30X for the testing phase with no sacrifice of

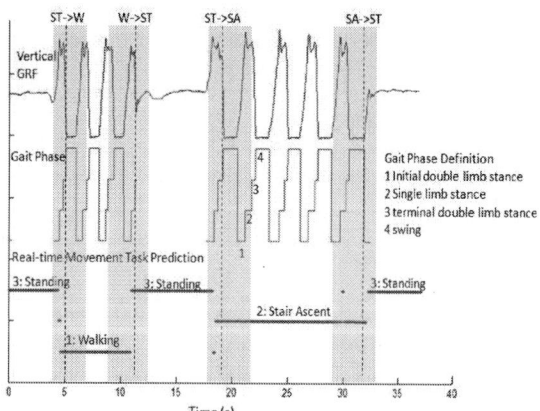

Figure 6. Real-time system performance for one representative testing trial. The white area denotes the periods of static states (level-ground walking, stair ascent, and standing); the gray area represents the transitional period; the black vertical dash line indicates the critical timing for each transition.

computation accuracy. The designed NMI prototype was tested on an able-bodied subject for accurately classifying multiple movement tasks (level-ground walking, stair ascent, and standing) in real-time. The results demonstrated the feasibility of a self-contained and high performance real-time NMI for artificial legs. Our future work includes real-time testing of the designed NMI system on amputee subjects, using the NMI system to control powered prosthetic legs, studying management of power consumption, and increasing the system reliability.

6. ACKNOWLEDGMENTS

This work is partly supported by National Science Foundation NSF/CPS #0931820, NIH #RHD064968A, NSF #1149385, NSF/CCF #1017177, and NSF/CCF #0811333. The authors thank Fan Zhang, Quan Ding, Ding Wang, Lin Du, and Ming Liu at the University of Rhode Island, for their suggestion and assistance in this study.

7. REFERENCES

[1] Oonishi, Y., Oh, S., and Hori, Y., "A New Control Method for Power-Assisted Wheelchair Based on the Surface Myoelectric Signal," Industrial Electronics, IEEE Transactions on, vol. 57, pp. 3191-3196, 2010.

[2] Tonin, L., Carlson, T., Leeb, R., and Millán, J. R., "Brain-Controlled Telepresence Robot by Motor-Disabled People," in 33rd Annual International Conference of the IEEE EMBS, Boston, MA, 2011.

[3] Parker, P. and Scott, R., "Myoelectric control of prostheses," Critical reviews in biomedical engineering, vol. 13, p. 283, 1986.

[4] Englehart, K. and Hudgins, B., "A robust, real-time control scheme for multifunction myoelectric control," IEEE Trans Biomed Eng, vol. 50, pp. 848-54, Jul 2003.

[5] Lin, C. T., Ko, L. W., Chiou, J. C., Duann, J. R., Huang, R. S., Liang, S. F., Chiu, T. W., and Jung, T. P., "Noninvasive neural prostheses using mobile and wireless EEG," Proceedings of the IEEE, vol. 96, pp. 1167-1183, 2008.

[6] Kuiken, T., "Targeted reinnervation for improved prosthetic function," Phys Med Rehabil Clin N Am, vol. 17, pp. 1-13, Feb 2006.

[7] Huang, H., Zhou, P., Li, G., and Kuiken, T. A., "An analysis of EMG electrode configuration for targeted muscle reinnervation based neural machine interface," IEEE Trans Neural Syst Rehabil Eng, vol. 16, pp. 37-45, Feb 2008.

[8] Huang, H., Kuiken, T. A., and Lipschutz, R. D., "A strategy for identifying locomotion modes using surface electromyography," IEEE Trans Biomed Eng, vol. 56, pp. 65-73, Jan 2009.

[9] Zhang, F., DiSanto, W., Ren, J., Dou, Z., Yang, Q., and Huang, H., "A Novel CPS System for Evaluating a Neural-Machine Interface for Artificial Legs," 2011, pp. 67-76.

[10] Zhang, X., Liu, Y., Zhang, F., Ren, J., Sun, Y. L., Yang, Q., and Huang, H., "On Design and Implementation of Neural-Machine Interface for Artificial Legs," IEEE Transactions on Industrial Informatics, 2011.

[11] Zhang, X., Huang, H., and Yang, Q., "Design and Implementation of A Special Purpose Embedded System for Neural Machine Interface," in ICCD'2010, Amsterdam, the Netherlands, October 2010.

[12] Altera. Nios II Processor: The World's Most Versatile Embedded Processor. Available: http://www.altera.com/products/ip/processors/nios2/ni2-index.html

[13] Huang, H., Zhang, F., Hargrove, L., Dou, Z., Rogers, D., and Englehart, K., "Continuous Locomotion Mode Identification for Prosthetic Legs based on Neuromuscular-Mechanical Fusion," Biomedical Engineering, IEEE Transactions on, pp. 1-1, 2011.

Supplemental Material

S1. PATTERN RECOGNITION USING LINEAR DISCRIMINANT ANALYSIS

The principle of the LDA-based PR strategy is to find a linear combination of features which separates multiple locomotion classes $C_g (g \in [1, G])$. G denotes the total number of classes. Suppose μ_g is the mean vector of class C_g and every class shares a common covariance matrix $\sum$, the linear discriminant function is defined as

$$d_{C_g} = \bar{f}^T \sum{}^{-1} \mu_g - \frac{1}{2} \mu_g{}^T \sum{}^{-1} \mu_g . \qquad (1)$$

During the training procedure, $\sum$ and μ_g are estimated based on the feature matrix calculated from the training data. The estimations of $\sum$ and μ_g are expressed as

$$\widetilde{\sum} = \frac{1}{G} \sum_{g=1}^{G} \frac{1}{K_g - 1} (F_g - Mi_g)(F_g - Mi_g)^T$$

and

$$\widetilde{\mu}_g = \frac{1}{K_g} \sum_{k=1}^{K_g} \bar{f}_{C_g, k}$$

where K_g is the number of analysis windows in class C_g; $\bar{f}_{C_g, k}$ is the k_{th} observed feature vector in class C_g; $F_g = [\bar{f}_{C_g,1}, \bar{f}_{C_g,2}, ..., \bar{f}_{C_g,k}, ..., \bar{f}_{C_g,K_g}]$ is the feature matrix of class C_g; $Mi_g = [\widetilde{\mu}_g, \widetilde{\mu}_g, ..., \widetilde{\mu}_g]$ is the mean matrix that has the same number of columns as in F_g. The results of the LDA training procedure can be represented by a weight matrix as $W = [\bar{w}_1, \bar{w}_2, ..., \bar{w}_g, ..., \bar{w}_G]$ and a weight vector as $\bar{c} = [c_1, c_2, ..., c_g, ..., c_G]$. Here

$$\bar{w}_g = \widetilde{\sum}{}^{-1} \widetilde{\mu}_g \qquad (2)$$

and

$$c_g = -\frac{1}{2} \widetilde{\mu}_g{}^T \widetilde{\sum}{}^{-1} \widetilde{\mu}_g . \qquad (3)$$

Therefore (1) can be estimated as

$$\widetilde{d}_{C_g} = \bar{f}^T \bar{w}_g + c_g . \qquad (4)$$

The major task of the training procedure is to calculate the mean vector $\widetilde{\mu}_g$ for each class, the common covariance matrix $\widetilde{\sum}$, and its inverse matrix $\widetilde{\sum}{}^{-1}$. In practice, matrix inversion is a compute-intensive and time consuming task, which should be avoided if possible. From (2) and (3), it can be found that $\widetilde{\sum}{}^{-1}$ does not appear alone. If $\widetilde{\sum}{}^{-1} \widetilde{\mu}_g$ can be calculated in an efficient way, W and $\bar{c}$ can be achieved easily. In our implementation, a more efficient algorithm was adopted to solve this problem. First, a Cholesky decomposition is performed as $\widetilde{\sum} = R^T \times R$, where R is upper triangular. Then $\widetilde{\sum}{}^{-1} \widetilde{\mu}_g$ can be quickly computed with a forward substitution algorithm for a lower triangular matrix R^T, followed by a back substitution algorithm for an upper triangular matrix R. The condition of a successful Cholesky decomposition is that $\widetilde{\sum}$ must be symmetric and has real positive diagonal elements, which can be perfectly satisfied by a covariance matrix. In this way, (2) and (3) can be reformulated as $\bar{w}_g = R \backslash (R^T \backslash \widetilde{\mu}_g)$ and

$$c_g = -\frac{1}{2} \widetilde{\mu}_g{}^T \bar{w}_g .$$

During the testing phase, the observed feature vector $\bar{f}$ derived from each analysis window is applied to calculate $\widetilde{d}_{C_g}$ in (4) for each movement class and is classified into a class $\widetilde{C}_g$ that satisfies

$$\widetilde{C}_g = \arg \max C_g \{\widetilde{d}_{C_g}\}, C_g \in \{C_1, C_2, ..., C_G\} .$$

Statistical Design and Optimization for Adaptive Post-silicon Tuning of MEMS Filters

Fa Wang, Gokce Keskin, Andrew Phelps, Jonathan Rotner, Xin Li, Gary K. Fedder, Tamal Mukherjee and Lawrence T. Pileggi

Carnegie Mellon University, Pittsburgh, PA 15213

{fwang1, gkeskin, rphelps, jrotner, xinli, fedder, tamal, pileggi}@ece.cmu.edu

ABSTRACT

Large-scale process variations can significantly limit the practical utility of microelectro-mechanical systems (MEMS) for RF (radio frequency) applications. In this paper we describe a novel technique of adaptive post-silicon tuning to reliably design MEMS filters that are robust to process variations. Our key idea is to implement a number of redundant MEMS resonators to form an array and then optimally select a subset of these resonators to achieve the desired frequency response. Several new CAD algorithms and methodologies are proposed to optimize and configure the design variables of the proposed MEMS resonator array. A MEMS design example demonstrates that the proposed post-silicon tuning is able to reduce the ripple of the channel filter gain by 7× over other traditional approaches.

Categories and Subject Descriptors

B.7.2 [**Integrated Circuits**]: Design Aids – Verification

General Terms

Algorithms

Keywords

MEMS Filter, Process Variation

1. INTRODUCTION

Advanced radio systems (e.g., software defined radio, cognitive radio, etc.) are of great importance for both commercial and military applications [1]. Designing and manufacturing fully-integrated on-chip radio systems is extremely important to achieve low power consumption and small system size for high-speed wireless communication. This goal, however, is challenging due to the lack of channel filters that can be fully integrated with RF transceivers and their signal processing circuits.

MEMS resonator filters hold great potential for the aforementioned on-chip integration problem [2]-[6], [14]. MEMS resonators made in low-loss materials (e.g., silicon, polysilicon, aluminum nitride, etc.) have been demonstrated to achieve extremely high quality factors (e.g., 3000 ~ 80000). Most importantly, MEMS resonators can be designed with their center frequencies set by layout, thereby enabling on-chip integration of channel filters required by today's radio systems.

While the research and development of MEMS filters has made significant progress in recent years, process variations posed by nanoscale IC technology remain the key bottleneck that limits the practical applications of these MEMS devices. For example, since the quality factor (i.e., Q) of a MEMS resonator is extremely high, a small variation (e.g., less than 1%) of the center frequency

can completely change the frequency response of the resonator and make the MEMS filter ineffective for the intended application.

In prior art, mechanically-coupled MEMS resonators have been proposed to create a tunable resonator array and, hence, achieve the desired frequency response [6]. However, mechanical coupling cannot offer a large tuning range that is sufficient to accommodate the large-scale process variations from run-to-run, lot-to-lot, wafer-to-wafer and die-to-die. In addition, mechanically-coupled MEMS resonators often generate spurious modes out of the intended pass-band, thereby distorting the frequency response of the filter. The technical challenge here is how to develop a new design methodology to make a MEMS filter flexibly tunable and, consequently, match the desired frequency characteristics.

In this paper, we borrow the concept of *statistical element selection* [7]-[8] to develop a novel post-silicon tuning methodology for MEMS filters. Namely, we propose to place a large number (i.e., N) of resonators on a die and select a subset (i.e., M where $M < N$) of these resonators to build a MEMS filter. Other resonators that are not selected are simply turned off. The frequency response of the MEMS filter is equal to the summation of the frequency responses of the M selected resonators. Based on the post-silicon measurement results, the optimal set of M resonators is selected to achieve the desired filter response. As such, the impact of large-scale process variations is effectively minimized. Note that the aforementioned post-silicon tuning is completely different from the traditional N-modular redundancy method where only one device out of N candidates is selected [15].

To make the proposed post-silicon tuning efficient, a number of CAD algorithms and methodologies are developed in this paper to optimize the design variables of the redundant resonators (e.g., the total number of resonators, the center frequencies of resonators, etc.). In particular, a linear programming formulation is derived to determine the optimal values of these design variables. Since the linear programming is convex, it can be solved efficiently (i.e., with low computational cost) [13]. Furthermore, an efficient heuristic algorithm is developed to select the optimal subset of resonators based on the post-silicon measurement results. Compared to the traditional branch and bound algorithm [13], the proposed heuristic method achieves more than 1000× runtime speed-up while converging to a similar solution (less than 8% increase in cost function).

The remainder of this paper is organized as follows. In Section 2, we first describe our proposed post-silicon tuning scheme and then develop a number of statistical CAD algorithms and methodologies in Section 3. The efficacy of the proposed post-silicon tuning is demonstrated by a MEMS mixer example in Section 4. Finally, we conclude in Section 5.

2. ADAPTIVE POST-SILICON TUNING FOR MEMS FILTERS

Figure 1 shows a simplified circuit schematic of a MEMS resonator. The input voltage V_{IN} is placed across electrostatic gap electrodes to create the drive force. As a result, the motional current I_{OUT} is generated at the output electrode. The transfer function from the input V_{IN} to the output I_{OUT} can be approximated as [14]:

$$\frac{I(s)}{V(s)} = \frac{\beta \cdot (\omega_0/Q) \cdot s}{s^2 + (\omega_0/Q) \cdot s + \omega_0^2} \tag{1}$$

where ω_0 represents the center frequency, Q denotes the quality factor and β is a parameter defining the gain of the resonator.

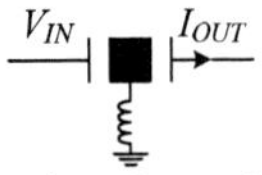

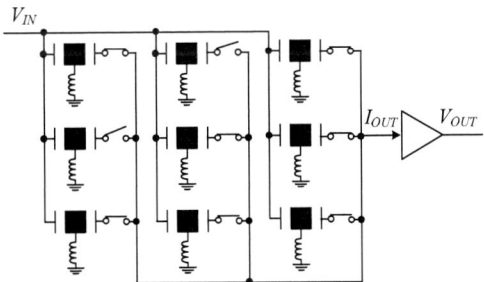

Figure 1. Simplified circuit schematic is shown for a MEMS resonator.

The frequency response of the transfer function in (1) represents a band-pass filter whose center frequency is ω_0 and the bandwidth is ω_0/Q. Since a MEMS resonator is typically made in low-loss materials, its Q value is extremely large (e.g., 3000 ~ 80000). Hence, a MEMS resonator can accurately select the input signals around its center frequency ω_0 and filter out the unwanted signals at other frequencies. More details about MEMS resonator can be found in [14].

In this paper our objective is to design a band-pass MEMS filter instead of a single resonator. In many practical applications (e.g., the channel filter of an RF transceiver) the pass-band of the filter is substantially wider than the pass-band of a single MEMS resonator. Hence, multiple MEMS resonators with different center frequencies should be connected in parallel to form an array. As such, the desired pass-band can be achieved by appropriately designing the resonator array. Due to the high quality factor of MEMS resonators, a sharp transition from the pass-band to the stop-band can be realized and, hence, the resulting MEMS filter can select the signal components within the pass-band and reject the unwanted components outside the pass-band with excellent precision. For illustration purpose, Figure 2 shows an example of the proposed MEMS filter architecture using a resonator array.

Figure 2. The proposed MEMS filter architecture is based upon a resonator array where the switches are used to turn on/off the resonators for post-silicon selection.

To make the proposed MEMS filter of practical utility, all MEMS resonators in the array must be appropriately designed to meet the following three criteria: (i) the ripple of the filter gain should be minimized within the pass-band to avoid in-band distortion; (ii) the filter gain within the stop-band must be minimized to guarantee sufficient signal attenuation; and (iii) the transition band must be minimized to provide excellent frequency selectivity. Without manufacturing variations, these three criteria can be easily satisfied by appropriately designing the center frequencies and gain values for all of the MEMS resonators.

In practice, due to large-scale process variations, the frequency responses of MEMS resonators can significantly differ from their nominal values. In general, process variations can be classified into two broad categories [9]-[11]: (i) inter-die variations, and (ii) intra-die variations. Inter-die variations refer to the common or average variations across the die. Since inter-die variations identically affect all resonators on the same die, they can be easily compensated by using circuit-level design techniques. On the other hand, intra-die variations represent the individual local variations within the die. These intra-die variations introduce random mismatches among MEMS resonators and must be carefully managed in order to achieve the desired frequency response for the proposed MEMS filter. Therefore, we will mainly focus on intra-die variations in this paper.

From (1) we notice that the transfer function of the MEMS resonator is uniquely determined by three parameters: (i) the center frequency ω_0, (ii) the quality factor Q, and (iii) the gain-related parameter β. All three of these parameters can vary due to intra-die variations. However, the variation of the center frequency is most critical to our filter application because the quality factor of a MEMS resonator is extremely high and, therefore, a small perturbation of its center frequency can completely change the frequency response of the filter [12]. Once the intra-die variation (e.g., random mismatch) of center frequency is successfully handled, the frequency response of the proposed MEMS filter can be accurately controlled.

To address the problem of intra-die center frequency variation, we propose to add redundant resonators into the array and then select an optimal subset of the resonators based on the post-silicon measurement results to accurately match the desired filter response in frequency domain. Other resonators that are not selected are simply turned off so that they do not contribute to the frequency response of the filter, as shown in Figure 2. While the proposed idea of post-silicon tuning looks simple, a number of comprehensive CAD algorithms and methodologies must be developed to optimally design the proposed resonator array with redundancy. In particular, the following two critical problems must be solved. First, a statistical optimization method must be developed to determine the optimal number of redundant resonators and their center frequencies. Second, once the post-silicon measurement results (i.e., the post-manufacturing center frequencies of all resonators) are available, an optimal selection algorithm must be used to find the subset of resonators that should be selected to form the MEMS filter. In what follows, we will discuss our proposed approaches to address these two open problems.

3. STATISTICAL DESIGN METHOD
3.1 Statistical Resonator Array Optimization

Without loss of generality, we assume that M MEMS resonators are used to implement the band-pass filter of interest. In the ideal scenario (i.e., without process variations), these M resonators are designed with the same gain and their center frequencies $\{f_m; m = 1,2,\dots,M\}$ can be optimized by the filter designer to achieve the desired frequency response.

In practice, due to process variations, N ($N > M$) resonators are required to form a resonator array with redundancy for post-silicon tuning. Once process variations are considered, the center frequencies of these N resonators are no longer deterministic. Instead, they must be modeled as N random variables:

$$g_n \sim Gaussian\left(\mu_n, \sigma_n^2\right) \quad (n = 1,2,\cdots,N). \tag{2}$$

Eq. (2) assumes that the center frequency of the nth resonator g_n is modeled as a Gaussian distribution. Its mean value μ_n, referred to as the *nominal center frequency* in this paper, is a design variable that can be controlled by the layout of the MEMS resonator. The standard deviation σ_n models the magnitude of the center frequency variation and is determined by the manufacturing process. Since we focus on intra-die variations as described in Section 2, we further assume that the center frequency variations of different MEMS resonators are dominated by random mismatches and they are statistically independent.

The objective of our proposed post-silicon tuning is to select an optimal subset from $\{g_n; n = 1,2,\ldots,N\}$ to accurately match the desired center frequencies $\{f_m; m = 1,2,\ldots,M\}$. Towards this goal, we define the following interval to quantitatively assess whether a given frequency f_m can be accurately matched:

$$[f_m - d, f_m + d] \qquad (3)$$

where $d \geq 0$ is referred to as the *frequency offset* in this paper. The frequency f_m is *successfully matched* if we can find g_n from the set $\{g_n; n = 1,2,\ldots,N\}$ such that:

$$f_m - d \leq g_n \leq f_m + d. \qquad (4)$$

Restating in words, there is at least one resonator whose center frequency closely matches the desired value f_m after fabrication. Otherwise, the frequency f_m is *unsuccessfully matched*, if there exists no g_n from the set $\{g_n; n = 1,2,\ldots,N\}$ to satisfy the inequality in (4).

The aforementioned frequency matching plays an important role in minimizing the ripple of the in-band filter gain. If every frequency f_m in the set $\{f_m; m = 1,2,\ldots,M\}$ is successfully matched with a small offset d, it implies that the desired frequency response can be accurately approximated and, hence, a good MEMS filter can be realized even with large-scale process variations. For this reason, the quality of frequency matching, measured by the offset d, is directly correlated to the quality of the filter we design. When designing the proposed MEMS resonator array, we must optimally determine the number of resonators N and their nominal center frequencies $\{\mu_n; n = 1,2,\ldots,N\}$ so that successful matching can be accomplished for every f_m with a small offset d.

Given the Gaussian distribution in (2), we calculate the probability for the inequality (4) to hold:

$$
\begin{aligned}
P_{mn} &= \mathrm{Prob}\left(f_m - d \leq g_n \leq f_m + d\right) \\
&= \Phi\left(\frac{f_m + d - \mu_n}{\sigma_n}\right) - \Phi\left(\frac{f_m - d - \mu_n}{\sigma_n}\right)
\end{aligned} \qquad (5)
$$

where $\mathrm{Prob}(\bullet)$ denotes the probability of a random event and $\Phi(\bullet)$ represents the cumulative distribution function of a standard Gaussian distribution (i.e., zero mean and unit variance). Eq. (5) estimates the probability for the frequency f_m to be successfully matched by g_n. The matching for f_m is successful, if any g_n from the set $\{g_n; n = 1,2,\ldots,N\}$ satisfies the inequality in (4). Hence, the probability of successful matching for f_m is equal to:

$$P_m = 1 - \prod_{n=1}^{N}\left(1 - P_{mn}\right). \qquad (6)$$

Studying (5)-(6) reveals two important observations. First, the probability P_m increases, as N increases; namely, the chance of successful matching becomes higher as more redundant resonators are added. Second, the probability P_m decreases as d decreases. This simply implies that if the offset specification becomes tight, the parametric yield decreases.

After deriving the probability in (6) we are ready to formulate the following optimization problem:

$$
\begin{aligned}
&\min_{N,\mu_1,\cdots,\mu_N} \quad N \\
&\text{S.T.} \qquad P_m \geq 1 - \varepsilon \quad (m = 1,2,\cdots,M)
\end{aligned} \qquad (7)
$$

where ε is a given specification defining the maximum probability of unsuccessful matching for each frequency f_m. The value of ε should be sufficiently small (e.g., $\varepsilon = 0.01$) in order to guarantee a high probability for successful matching. The objective of (7) is to find: (i) the minimum number of resonators (i.e., N), and (ii) the nominal center frequencies $\{\mu_n; n = 1,2,\ldots,N\}$ for all resonators. Note that directly solving the unknowns $\{\mu_n; n = 1,2,\ldots,N\}$ and N from (7) is not trivial, since the constraint set in (7) is not convex. For this reason, we will propose an alternative optimization formulation in order to solve these problem unknowns efficiently (i.e., with low computational cost).

We discretize the pass-band of the filter and represent it by K discrete frequency values $\{h_k; k = 1,2,\ldots,K\}$. In other words, instead of solving the unknowns $\{\mu_n; n = 1,2,\ldots,N\}$ as continuous variables, we constrain the solution space to a set of discrete values defined by $\{h_k; k = 1,2,\ldots,K\}$. If the step size of the discretization is sufficiently small (i.e., the value of K is sufficiently large), the discrete solution space accurately approximates the original continuous solution space and the approximation error is negligible.

Once the discretization step is complete, we know that the nominal center frequency of the nth resonator (i.e., μ_n) must belong to the finite set $\{h_k; k = 1,2,\ldots,K\}$. Hence, we define a new symbol n_k ($n_k \geq 0$) as the number of resonators whose nominal center frequencies equal h_k. Since we have N resonators in total, the summation of $\{n_k; k = 1,2,\ldots,K\}$ is equal to N:

$$n_1 + n_2 + \cdots + n_K = N \quad \left(n_1 \geq 0 \quad n_2 \geq 0 \quad \cdots \quad n_K \geq 0\right). \qquad (8)$$

Based on this notation, $\{n_k; k = 1,2,\ldots,K\}$ can be considered as a set of new optimization variables that should be solved. Namely, instead of solving $\{\mu_n; n = 1,2,\ldots,N\}$ and N, we propose to derive an alternative-yet-equivalent optimization formulation to find the optimal values of $\{n_k; k = 1,2,\ldots,K\}$.

Given the aforementioned definition of $\{h_k; k = 1,2,\ldots,K\}$ and $\{n_k; k = 1,2,\ldots,K\}$, it is easy to verify that the probability of successfully matching in (6) can be re-written as:

$$P_m = 1 - \prod_{k=1}^{K}\left(1 - Q_{mk}\right)^{n_k} \qquad (9)$$

where

$$Q_{mk} = \Phi\left(\frac{f_m + d - h_k}{\sigma_k}\right) - \Phi\left(\frac{f_m - d - h_k}{\sigma_k}\right). \qquad (10)$$

In (10), σ_k represents the standard deviation of the center frequency variations for the resonators whose nominal center frequencies equal h_k. Note that once $\{f_m; m = 1,2,\ldots,M\}$ (i.e., the desired center frequencies), $\{h_k; k = 1,2,\ldots,K\}$ (i.e., the discrete solution space), and d (i.e., the frequency offset) are known, Q_{mk} in (10) is determined. The probability P_m in (9) is a function of the unknown variables $\{n_k; k = 1,2,\ldots,K\}$ that we need to solve.

Substituting (9) into (7), the constraints in (7) become:

$$1 - \prod_{k=1}^{K}\left(1 - Q_{mk}\right)^{n_k} \geq 1 - \varepsilon \quad (m = 1,2,\cdots,M) \qquad (11)$$

or equivalently:

$$\prod_{k=1}^{K}\left(1 - Q_{mk}\right)^{n_k} \leq \varepsilon \quad (m = 1,2,\cdots,M). \qquad (12)$$

Taking the logarithm for both sides of (12) yields:

$$\sum_{k=1}^{K} n_k \cdot \log(1 - Q_{mk}) \le \log(\varepsilon) \quad (m = 1, 2, \cdots, M). \qquad (13)$$

Combining (7) and (13) results in the following optimization:

$$\min_{n_1, \cdots, n_K} \quad n_1 + n_2 + \cdots + n_K$$

$$\text{S.T.} \quad \sum_{k=1}^{K} n_k \cdot \log(1 - Q_{mk}) \le \log(\varepsilon) \quad (m = 1, 2, \cdots, M). \qquad (14)$$

$$n_k \ge 0 \qquad\qquad (k = 1, 2, \cdots, K)$$

In (14), both the cost function and the constraints are linear. Hence, it is a linear programming problem and the optimal solution can be efficiently found by an interior point algorithm [13].

It is worth mentioning that two practical issues must be carefully considered when applying the optimization formulation in (14). First, the variables $\{n_k; k = 1, 2, \ldots, K\}$ solved by the linear programming in (14) are non-negative and real-valued, while they should be non-negative integers by definition. Therefore, a post-processing step must be applied to "legalize" the solution of (14). In this paper we simply round these variables to the nearest integers during the legalization step.

Second, the frequency offset d must be appropriately determined in order to calculate $\{Q_{mk}; m = 1, 2, \ldots, M; k = 1, 2, \ldots, K\}$ in (10) that are required by the optimization in (14). As previously discussed, the value of d is directly correlated to the ripple of the in-band filter gain. If d is large, the optimization in (14) will result in a small number of resonators to implement the band-pass filter (i.e., N is small). The ripple of the filter gain, however, is expected to be large. Hence, varying the value of d allows us to explore the design trade-offs between the number of required resonators and the ripple of the in-band filter gain. In practice, if the ripple is given as a performance specification, we can repeatedly solve the optimization in (14) with different values of d and then find the appropriate d value that meets the ripple specification.

3.2 Optimal Post-silicon Resonator Selection

In the previous section we formulate the optimization in (14) that enables us to optimally determine the nominal center frequencies $\{\mu_n; n = 1, 2, \ldots, N\}$ for all resonators during the design stage. Once these resonators are manufactured, their actual center frequencies differ substantially from the designed values due to process variations. We therefore must measure the actual center frequencies after fabrication and then optimally select a subset of these resonators for frequency matching. Namely, given the post-manufacturing center frequencies of N resonators $\{\tilde{g}_n; n = 1, 2, \ldots, N\}$, we aim to select M resonators out of these N ($N > M$) candidates such that the M selected center frequencies match the desired values $\{f_m; m = 1, 2, \ldots, M\}$ as closely as possible. Here, we use a new symbol $\tilde{g}_n$ instead of g_n shown in (2), to emphasize that $\{\tilde{g}_n; n = 1, 2, \ldots, N\}$ represent the post-manufacturing center frequencies with deterministic values (i.e., they are not random variables any more). The aforementioned resonator selection, however, is not trivial, because both M and N are typically large (e.g., $100 \sim 1000$), as will be demonstrated by our design example in Section 4. As a result, there are numerous possible choices to select M resonators out of N candidates. Instead of enumerating all possible options, we propose a heuristic algorithm to solve this combinatorial optimization problem efficiently.

To derive our proposed resonator selection algorithm, we first mathematically define the cost function that we aim to minimize. Remember that our objective is to match the desired center frequencies $\{f_m; m = 1, 2, \ldots, M\}$ as closely as possible. Therefore,

we use the following cost function to quantitatively measure the quality of frequency matching:

$$F = \sum_{m=1}^{M} |f_m - \tilde{g}_{sm}| \qquad (15)$$

where $\tilde{g}_{sm}$ denotes the selected center frequency to match f_m. Ideally, if all frequencies $\{f_m; m = 1, 2, \ldots, M\}$ are exactly matched, the cost function in (15) reaches its minimum (i.e., zero).

Next, we define the *distance matrix* $V \in R^{M \times N}$. The (m, n)-th element of the M-by-N matrix V is the distance between the desired center frequency f_m and the post-manufacturing center frequency $\tilde{g}_n$:

$$V_{mn} = |f_m - \tilde{g}_n|. \qquad (16)$$

Our proposed heuristic algorithm iteratively selects the optimal $\tilde{g}_n$ to match f_m based on a greedy search method. At the first iteration step, it finds the minimum element (say, V_{mn}) in the distance matrix V and assign $\tilde{g}_n$ to match f_m. Since the matching between f_m and $\tilde{g}_n$ is complete, the mth row and nth column of the distance matrix V is removed. It simply means that the frequencies f_m and $\tilde{g}_n$ will not be considered any more during the following iteration steps. The dimension of the distance matrix V is reduced to $(M-1)$-by-$(N-1)$. Next, during the second iteration step, we apply the same heuristic to the distance matrix V with reduced dimension and find a new pair of f_m and $\tilde{g}_n$ with minimum distance. The aforementioned procedure is repeated until every desired center frequency f_m is matched to an optimal post-manufacturing center frequency $\tilde{g}_n$.

Algorithm 1: Optimal Post-silicon Resonator Selection

1. Start from a set of desired center frequencies $\{f_m; m = 1, 2, \ldots, M\}$ and a set of post-manufacturing center frequencies $\{\tilde{g}_n; n = 1, 2, \ldots, N\}$.
2. Initialize the distance matrix V based on (16). Set the iteration index $i = 1$.
3. Find the minimum element V_{mn} in the distance matrix V. Assign $\tilde{g}_n$ to match f_m.
4. Remove the mth row and nth column from the distance matrix V.
5. If i is equal to M, stop iteration. Otherwise, set $i = i + 1$ and go to Step 3.

Algorithm 1 summarizes the major steps of the proposed optimal resonator selection method that is based on simple greedy search. The time complexity of this algorithm is M^2N. Compared to the traditional branch and bound algorithm [13], the proposed heuristic method achieves more than $1000\times$ runtime speed-up and converges to a similar solution (less than 8% increase in cost function), as will be demonstrated by our design example in Section 4.

4. DESIGN EXAMPLE

In this section, we demonstrate the efficiency of the proposed adaptive post-silicon tuning by using a MEMS mixer example with channel filters. All numerical experiments are run on a 2.9 GHz Linux server with 4 GB memory.

4.1 MEMS Mixer and Channel Filter Structure

Shown in Figure 3 is the geometrical structure of a CMOS MEMS square frame resonator [3]. An RF signal V_{RF} and a local oscillator (LO) signal V_{LO} are placed across electrostatic gap electrodes to control the motional current I_{OUT}. The polarizing voltage V_P is used to turn on and off the MEMS resonator for

179

post-silicon selection.

In this example, the LO frequency is set to 840 MHz and the RF frequency band is from 850 MHz to 850.1 MHz. The RF signal is down-converted to the intermediate frequency (IF) band from 10 MHz to 10.1 MHz. The gain of the MEMS resonator, measured by the transfer function in (1), is −160 dBS. Its quality factor is 2000 and, hence, the bandwidth of a single resonator is equal to: 10 MHz / 2000 = 5 kHz. As previously described in Section 2, multiple resonators with different center frequencies are connected in parallel to form an array to implement the band-pass channel filter that is required for the given IF band. Here, the in-band gain of the channel filter must be no less than −135 dBS and its bandwidth is equal to: 10.1 MHz − 10 MHz = 100 kHz.

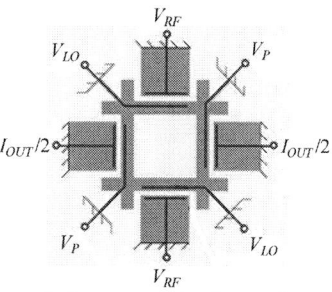

Figure 3. Geometrical structure is shown for a CMOS MEMS square frame resonator [3].

The MEMS resonators were designed based on a commercial 0.35 μm CMOS process. From the device-level variation models in the design kit, it was estimated that the random mismatch of the resonator center frequency has a standard deviation of $\sigma = 21$ kHz. Note that the value of σ (i.e., 21 kHz) is significantly larger than the bandwidth of a single resonator (i.e., 5 kHz). Table 1 summarizes the important parameters for the aforementioned MEMS mixer example. In what follows, we will apply the proposed post-silicon tuning to design and optimize the MEMS mixer to achieve the desired frequency response.

Table 1. Important parameters of the MEMS mixer example

Technology Node	0.35 μm CMOS
RF Frequency Band	[850 MHz, 850.1 MHz]
LO Frequency	840 MHz
IF Frequency Band	[10 MHz, 10.1 MHz]
IF Bandwidth	100 kHz
Channel Filter Gain (I/V)	≥ −135 dBS
Resonator Quality Factor	2000
Resonator Bandwidth	5 kHz
Resonator Center Frequency Variation	$\sigma = 21$ kHz
Resonator Gain (I/V)	−160 dBS

4.2 Resonator Array Optimization

Given the MEMS resonator in Figure 3, we first manually design the desired center frequencies $\{f_m; m = 1,2,...,M\}$ for the ideal scenario (i.e., without process variations). It is determined that 228 resonators in total (i.e., $M = 228$) are needed to implement the specified channel filter. The targeted center frequencies of these 228 resonators are distributed over the pass-band of the channel filter, i.e., [10 MHz, 10.1 MHz].

We next apply the linear programming in (14) to optimally determine the nominal center frequencies of the redundant resonators $\{\mu_n; n = 1,2,...,N\}$. Figure 4 shows the optimization results where the pass-band of the filter is partitioned into 100

discrete frequencies and the frequency offset d is set to different values. Studying Figure 4 we observe that the resonators are allocated at six nominal center frequencies only: 10.018 MHz, 10.019 MHz, 10.049 MHz, 10.050 MHz, 10.080 MHz and 10.081 MHz. At other frequencies the number of resonators is set to zero by the proposed optimization. This implies that we do not need to design many different resonators with different nominal center frequencies during the design stage, thereby reducing the design complexity. Due to random device mismatches, the actual center frequencies of these resonators will naturally spread over a wide frequency range. In the next sub-section we will demonstrate the proposed resonator selection algorithm to optimally find a subset of the redundant resonators and construct the channel filter with the desired frequency response.

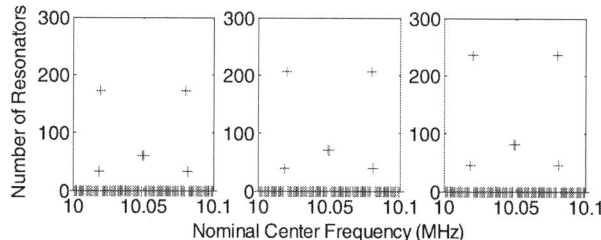

Figure 4. Nominal center frequencies are optimally determined by the linear programming in (14) where the pass-band of the filter is partitioned into 100 discrete frequencies and the frequency offset d is set to different values: (Left) 530 resonators in total, (Middle) 628 resonators in total, and (Right) 724 resonators in total.

4.3 Optimal Resonator Selection

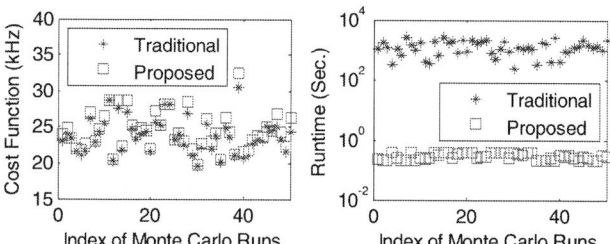

Figure 5. Comparison is shown for the proposed optimal resonator selection algorithm and the traditional branch and bound algorithm [13] based on 50 Monte Carlo runs with 628 resonators in the array: (Left) both algorithms minimize the cost function in (15) and the difference of the minimized cost function values is less than 8%; (Right) the proposed algorithm achieves more than 1000× runtime speed-up over the traditional approach.

Once the MEMS resonator array is manufactured, the center frequencies of all resonators are measured and the proposed heuristic algorithm (i.e., Algorithm 1) is used to select an optimal subset of resonators to minimize the cost function in (15). As such, the response of the selected resonators matches the desired frequency response. For comparison purpose, a traditional branch and bound algorithm [13] is implemented to minimize the same cost function in (15). Figure 5 compares the results arising from these two algorithms when applied to 50 Monte Carlo runs of an array of 628 MEMS resonators.

The data in Figure 5 leads to two important observations. First, the minimum cost function found by Algorithm 1 is close to that found by the branch and bound method (less than 8% difference).

This implies that the proposed heuristic algorithm can reliably find a good sub-optimal solution for the combinatorial optimization problem in (15). Second, but more importantly, the proposed heuristic algorithm is computationally cheaper than the traditional branch and bound method. The branch and bound method often takes more than 15 minutes to converge, while our proposed algorithm can reach convergence within 1 second (more than 1000× runtime speedup). Since Algorithm 1 has substantially lower computational cost, it can be more easily integrated into a practical post-silicon testing and configuration flow to improve the manufacturing yield of MEMS devices.

Table 2. Estimated ripple (dBS) of in-band filter gain for different design methods based on 1000 Monte Carlo runs

	# of Resonators	Mean	Sigma
No Redundancy	228	5.65	1.08
TMR [15]	684	4.56	0.81
Proposed Post-silicon Tuning	530	0.62	0.17
	628	0.51	0.08
	724	0.46	0.06

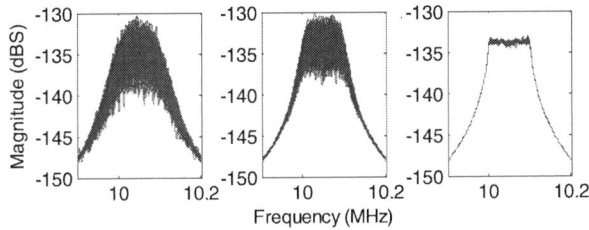

Figure 6. Frequency response (magnitude) is calculated by 1000 Monte Carlo runs: (Left) 228 resonators in total without including redundancy, (Middle) 684 resonators in total with the traditional triple modular redundancy (TMR) [15], and (Right) 628 resonators in total with the proposed post-silicon tuning.

To demonstrate the efficacy of the proposed post-silicon tuning, we further implement two traditional design methodologies for the MEMS channel filter. First, a simple MEMS resonator array is implemented without any redundancy (228 resonators in total). Second, the traditional triple modular redundancy (TMR [15]) approach is implemented where three resonators are designed for each desired center frequency {f_m; m = 1,2,…,228} and one of these three resonators is selected to optimally match the desired value of f_m. Table 2 summarizes the ripple of the in-band filter gain estimated from 1000 Monte Carlo runs for these different methods. Note that the ripple of the two traditional methods is extremely large. Their mean values are greater than 4.5 dBS. On the other hand, the proposed post-silicon tuning reduces the mean value of the ripple from 4.56 dBS to 0.62 dBS (more than 7× reduction). It, therefore, leads to significantly reduced in-band signal distortion when the proposed MEMS resonator array is used as a channel filter.

Figure 6 plots the frequency response (magnitude) of 1000 Monte Carlo runs for three different methods. Note that the two traditional methods result in non-flat in-band filter gain with significant variations. The proposed post-silicon tuning, however, yields a robust filter implementation that is not sensitive to large-scale process variations. These observations are consistent with the results previously shown in Table 2.

5. CONCLUSIONS

In this paper an efficient technique of adaptive post-silicon tuning is proposed to design robust MEMS filters with consideration of large-scale process variations. The proposed post-silicon tuning applies statistical element selection [7]-[8] to optimally select a group of resonators from a large array so that the desired frequency response can be achieved. Several CAD algorithms and methodologies are developed to optimize the design variables of the proposed MEMS resonator array. In particular, a new linear programming formulation is derived for statistical resonator array optimization and a heuristic algorithm is developed for optimal post-silicon resonator selection. Our design example of a MEMS mixer demonstrates that the proposed post-silicon tuning can efficiently reduce the ripple of the channel filter gain by 7× over other traditional approaches. The design methodologies and CAD algorithms proposed in this paper can be further integrated into a practical post-silicon testing and configuration flow to improve the manufacturing yield of MEMS devices.

6. ACKNOWLEDGEMENTS

The authors acknowledge the support of the C2S2 Focus Center, one of six research centers funded under the Focus Center Research Program (FCRP), a Semiconductor Research Corporation entity. This material is based upon work supported by DARPA and SPAWAR under Award No. N66001-09-1-2084. Any opinions, findings, and conclusions or recommendations expressed in this publication are those of the authors and do not necessarily reflect the views of DARPA and SPAWAR.

7. REFERENCES

[1] B. Razavi, "RF CMOS transceivers for cellular telephony," *IEEE Communications Magazine*, pp.144-149, 2003.

[2] F. Chen, J. Brotz, U. Arslan, C. Lo, T. Mukherjee and G. Fedder, "CMOS-MEMS resonant RF mixer-filters," *IEEE MEMS*, pp. 24-27, 2005.

[3] C. Lo, F. Chen and G. Fedder, "Integrated HF CMOS-MEMS square-frame resonators with on-chip electronics and electrothermal narrow gap mechanism," *IEEE Transducers*, pp. 2074-2077, 2005.

[4] E. Quevy, A. Paulo, E. Basol, R. Howe, T. King and J. Bokor, "Back-end-of-line poly-SiGe disk resonators," *IEEE MEMS*, pp. 234-237, 2006.

[5] S. Li, Y. Lin. Z. Ren and C. Nguyen, "Disk-array design for suppression of unwanted modes in micromechanical composite-array filters," *IEEE MEMS*, pp. 866-869, 2006.

[6] H. Chandrahalim and S. Bhave, "Digitally-tunable MEMS filter using mechanically-coupled resonator array," *IEEE MEMS*, pp. 1020-1023, 2008.

[7] G. Keskin, J. Proesel and L. Pileggi, "Statistical modeling and post manufacturing configuration for scaled analog CMOS," *IEEE CICC*, pp. Sep. 2010.

[8] J. Proesel, G. Keskin, J. Plouchart and L. Pileggi, "An 8-bit 1.5GS/s flash ADC using post-manufacturing statistical selection," *IEEE CICC*, Sep. 2010.

[9] Semiconductor Industry Associate, *International Technology Roadmap for Semiconductors*, 2010.

[10] P. Drennan and C. McAndrew, "Understanding MOSFET mismatch for analog design," *IEEE JSSC*, vol. 38, no. 3, pp. 450-456, Mar. 2003.

[11] P. Kinget, "Device mismatch and tradeoffs in the design of analog circuits," *IEEE JSSC*, vol. 40, no. 6, pp. 1212-1224, Jun. 2005.

[12] J. Wang, Y. Xie and C. Nguyen, "Frequency tolerance of RF micromechanical disk resonators in polysilicon and nanocrystalline diamond structural materials," *IEEE IEDM*, pp. 285-289, 2004.

[13] D. Bertsekas, *Nonlinear Programming*, Athena Scientific, 1999.

[14] S. Senturia, *Microsystem Design*, Springer, 2000.

[15] M. Stanisavljevi, A. Schmid and Y. Leblebici, *Reliability of Nanoscale Circuits and Systems: Methodologies and Circuit Architectures*, Springer, 2010.

Generic Low-Cost Characterization of V_{TH} and Mobility Variations in LTPS TFTs for Non-Uniformity Calibration of Active-Matrix OLED Displays

G. Reza Chaji and Javid Jaffari
Ignis Innovation Inc.
Waterloo, ON, Canada N2V 2C5
{gchaji,jjaffari}@ignisinnovation.com

ABSTRACT

Active-matrix organic light emitting diode displays are prone to significant V_{TH} and mobility variations in low-temperature polycrystalline-silicon thin-film transistors. A low-cost characterization of these variations can lead to a practical external calibration and simulation of the display non-uniformity. This paper proposes a generic methodology based on principal component analysis, relying on the display current levels corresponding to applied characterization images. This technique results in simultaneous characterization of the V_{TH} and mobility for the entire active matrix. Measurement results show that taking advantage of spatial correlation leads to 100 times reduction in characterization time with less than 30% relative error.

Categories and Subject Descriptors

B.7.2 [**Integrated Circuits**]: Design Aids—*Verification*;
J.2 [**Physical Sciences and Engineering**]: Electronics

General Terms

Measurement, Algorithms

1. INTRODUCTION

The active-matrix organic light emitting diode (AMOLED) display has been considered as the next generation in display technology because of its advantages on power consumption, viewing angle, color quality, response time, and contrast ratio. Unlike active matrix LCD (AMLCD), OLED is current driven devices and so thin film transistors (TFTs) play a more active role as a programmable current source. A 2-TFT AMOLED pixel circuit is shown in Figure 1. During the programming cycle, a row of pixels is selected by the gate driver, and the proper voltages are stored in the storage capacitor connected to the gate of the drive TFTs of the selected row, through an array of digital-to-analog converters in the source driver. The drive TFT converts the voltage to current and drives the OLED. Subsequently,

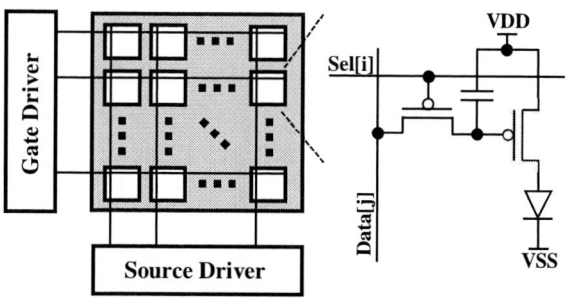

Figure 1: Schematic of a 2T pixel AMOLED display

the luminance of the pixel is proportionally determined by the magnitude of the OLED current. A major candidate for fabrication of AMOLED display is the low-temperature polycrystalline-silicon (LTPS) TFT. Since LTPS backplane offers higher mobility compared to other backplane technologies (e.g. amorphous silicon) [5, 9], it allows manufacturing of higher pixel density displays. However, it is prone to significant process variation as a result of recrystallization techniques [12, 7]. Figure 2(a) shows a contrast-boosted photo of a LTPS AMOELD panel, displaying a flat field image. The non-uniformities are visible as vertical mura, non-uniform background with clouds of dark and bright regions, and shorter range variation (skin mura).

Thus, a compensation technique is needed to eliminate the artifacts of TFT's non-uniformities. These techniques are either based on in-pixel TFT circuits [9] or external calibration of video signals [11, 8, 3]. In-pixel compensation circuits are composed of few TFTs and capacitors which create a correction voltage based on a reference current or discharging path during the operation of the display [9]. The correction voltage then calibrates the programming voltage internally and so controls the non-uniformity artifacts. For high pixel density and/or large-area displays, the use of in-pixel compensation circuits, such as 6TFT+2Cap [14], is limited by small area and yield considerations. On the other hand, external calibration uses the display non-uniformity characterization data stored in an embedded memory to modulate the input video signals. The data tables can be populated using post-fabrication electrical or optical measurements. Despite simple pixel circuits, most of external calibration techniques suffer from long runtime due to few millions of measurement iterations (e.g. electrical measurements) required for

(a) Original   (b) Vertical mura filtered

Figure 2: Photo of an AMOLED panel, displaying a flat field image. The contrast of the photo is increased by 8 times to help observing the vertical and spatial non-uniformities. Note that the horizontal patterns are due to interference between the camera's sensor and the panel's refresh frequencies (not an actual AMOLED panel non-uniformity).

a typical display resolution, expensive electronics and measurement circuits, or significant cross talks (e.g. optical measurements).

Therefore, a characterization technique is needed that provides accurate enough results while it requires significantly shorter characterization time and lower-cost implementation. Besides, the extracted information can be used in a simulation setup for accurate yield analysis of displays. Moreover, the extracted characterization can be used as a flag for process monitoring to find out the divergence from a typical setup [3, 2].

In this paper, a methodology is proposed to efficiently characterize the driving TFT's non-uniformity parameters, the V_{TH} and mobility. This is achieved by employing the spatial correlation information of the process parameters and through statistical learning of their principal components. The initial learning phase is done by a fully electrical characterization of few reference panels. Pattern generated by the principal component analysis (PCA) are applied to the display and the total current of the panel is sensed through the V_{DD} line. The extracted current data is used to simultaneously characterize V_{TH} and mobility variation of each driving TFT.

2. BACKGROUND

In this section, the background required for the development of the non-uniformity extraction method is presented. First, the notion of principal components for a spatially correlated random variable is reviewed. Then, the electrical (I-V) characteristic of an LTPS-TFT 2T pixel is modeled.

2.1 Principal Components of a Spatially Correlated Random Variable

The spatial correlation of a scalar random variable Z on a 2-D plane can be formed by determining the $cov(Z(s_1), Z(s_2))$ at any arbitrary locations of s_1 and s_2. In a second-order stationary process, the spatial covariance is a function of the direction and distance (for an anisotropy process) between the two points rather than their actual position. The correlation generally reduces as the distance increases. The spatial correlation of gate length variation has been measured and

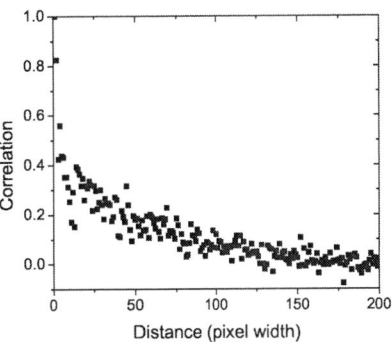

Figure 3: Experimental spatial correlation of the AMOLED panel non-uniformity

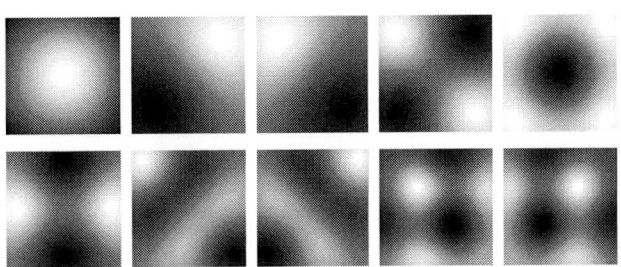

Figure 4: The 10 most important principal components of the experimented spatial correlation matrix

extensively studied in MOS technology [1, 4, 13]. The recent experiments also reveal spatial correlation in threshold voltage and mobility of LTPS TFTs known as long-range variation [12]. The existence of spatial correlation can be verified from the Figure 2(b). In this figure, the vertical mura that is due to vertical thermal annealing of the backplane silicon layer is filtered out. The remaining background image containing dark and light regions shows the similarity (correlation) of light intensity level of pixels that are close together. The experimental spatial correlation of the panel brightness is plotted in Figure 3. As expected, the correlation reduces as the distance between two points increases.

Since the random parameters are spatially correlated, the principal component analysis is very effective in compressing the random parameters. PCA linearly transforms the underlying data to a new coordinate system such that the greatest variance appears on the first coordinate (the first principal component), the second greatest variance on the second coordinate, and so on. If the profile of the random parameter is decomposed to a weighted sum of the principal components, the dimension of the original data (dimension being the number of sub-pixels for each process parameter) can be significantly reduced in the PCA coordinate system by eliminating the less important principal components.

If Σ_Z is the spatial covariance matrix of the process parameter Z, $\Sigma_Z(i,j) = cov(Z(s_i), Z(s_j))$, the m principal components of this process parameter is equivalent to the m eigenvectors of Σ_Z corresponding to its m largest eigen-

183

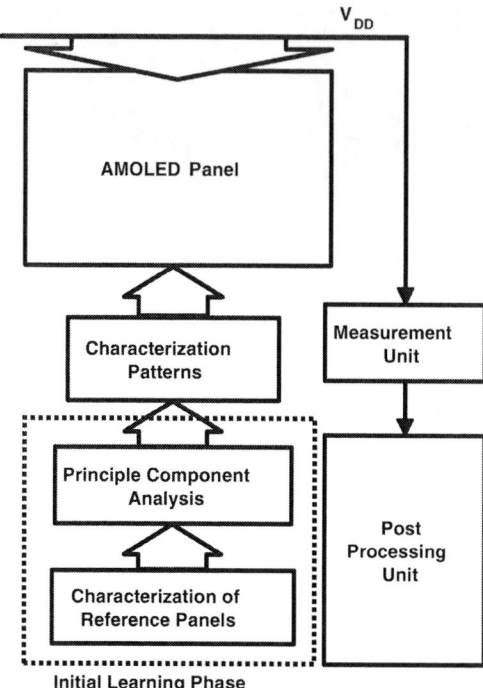

Figure 5: 2T pixel simulation versus I-V model ($V_{DD} = 16v$)

values. Figure 4 shows the first 10 principal components of the spatial correlation matrix according to the experimented spatial statistics earlier in the section. As can be seen, the first important principal components, that capture most of the variance, mostly contain the low spatial frequencies to represent the global non-uniformity trend.

2.2 I-V Characteristic of the 2T Pixel

As discussed earlier in the introduction, the proposed characterization technique is non-invasive and extracts the mobility and threshold variation of each driving transistor through sensing the supply current of the panel. To develop such a methodology the I-V characteristic of a pixel circuit is studied and modeled.

As a voltage programming pixel, the driving transistor must supply a certain amount of current determined by the OLED optical efficiency, for a given gate voltage regardless of the OLED bias. Therefore, the driving transistor of the 2T pixel shown in Figure1 is biased in a way that it remains in strong saturation for the entire range of the gray-scale OLED operation. Consequently, the OLED I-V shift effect, due to electrical aging, on the current of the driving TFT will also be minimized.

Following model is used throughout the paper to capture the process variation effect on the I-V of the pixel.

$$I = \beta(\mu_o + \Delta\mu)(V_{DD} - V_G + V_{TH_o} + \Delta V_{TH})^2 \quad (1)$$

where μ_o is the and $\Delta\mu$ are the nominal and variation of the transistor mobility, V_{TH_o} and ΔV_{TH} are the nominal and variation of the effective threshold voltage.

Figure 5 compares the spice simulations with the quadratic model at the nominal and two extreme process corners. Using this model, the coefficient of determination, R^2, is found to be 0.98 for the gate voltage range of 13-14 v. Therefore, this voltage range is later used as V_{min} and V_{max} in the non-uniformity extraction phase.

3. THE PROPOSED METHOD

In this section, a low cost process variation characterization methodology are proposed to extract the non-uniformity

Figure 6: System-level block diagram of the proposed methodology

profile of an AMOLED panel. Recent advances in the post-fab characterization of silicon wafers have been mainly focused on efficient and minimal probing techniques. This has been partly achieved by novel solutions based on compressive sensing (CS) [15, 16]. Compressive sensing is a technique for finding sparse solutions to an under-determined linear system. For example, the process non-uniformity profile can be considered as a sparse 2D signal under the spatial correlation principal components dictionary or any general energy-compacting transformation such as discrete cosine transformation. However, the major problem in using CS for characterization of the AMOLED process is the scalability of the L1-minimization problem, the computation core of the CS algorithm. For a panel with at least hundreds of thousand pixels, efficient solution of such a large L1-minimization problem is almost unachievable in a reasonable amount of time, even by the use of advanced algorithms such as CoSaMP [10]. In fact, if a long computation time (e.g. hours) is affordable, alternatively one may scan the whole panel, pixel by pixel, to accurately extract the full process profile without a need to applying any acceleration technique. Finally, due the structure and functionality of AMOLED displays, it is impossible to insert probes in predefined locations which eliminate the probing techniques from possible candidates.

However, the unique structure of AMOLED display enables an indirect extraction of non-uniformity profile through V_{DD} line. This can be achieved by applying characterization image patterns to the display, sensing the total current of the panel, and post-processing of the data. These patterns are generated based on major principal components of the background non-uniformity of the mobility and V_{TH}. Figure

6 depicts the system-level block diagram of the proposed methodology.

Following is the total current of a panel of size $R \times C$:

$$I_p = \beta \sum_{i,j=1}^{R,C} (\mu_o + \Delta\mu_{ij}) P_{ij}^2 \left(1 + \frac{\Delta V_{TH_{ij}}}{P_{ij}}\right)^2 \quad (2)$$

where $P_{ij} = V_{DD} - V_{G_{ij}} + V_{TH_o}$ is the drive-in voltage of the pixel at the i-th row and j-th column. For the gate voltage range of 13-14 v, since $\frac{\Delta V_{TH_{ij}}}{P_{ij}} << 1$, the equation is approximated as

$$I_p = \beta \sum_{i,j=1}^{R,C} P_{ij} (\mu_o + \Delta\mu_{ij}) (P_{ij} + 2\Delta V_{TH_{ij}}) \quad (3)$$

The Eq. (3) is the key formula to derive the vertical average and the coefficients of the principal components, all of which being weighted sums of a type of a process parameters.

We start with the extraction of the vertical laser scan impact on the mobility. In this step, the average mobility of each column is computed by displaying two patterns on the column and measuring their current, consequently. While the rest of panel is programmed by full V_{DD} gate voltage (to turn off the drive TFT) the column of interest is driven by two different constant voltages, $V_G^{(1)}$ and $V_G^{(2)}$, sequentially. Note that the choice of the voltages is made in a way that the gate voltage must be set within the range of the I-V model validity. If the measured current of the corresponding patterns are I_1 and I_2, the average mobility variation of the column j can then be obtained from

$$\hat{\Delta\mu_j} = \frac{\sum_{i=1}^{R} \Delta\mu_{ij}}{R} = \frac{I_2 - \frac{p_2}{p_1}I_1 - R\beta\mu_o p_2 (p_2 - p_1)}{R\beta p_2 (p_2 - p_1)} \quad (4)$$

where $p_1 = V_{DD} + V_{TH_o} - V_G^{(1)}$ and $p_2 = V_{DD} + V_{TH_o} - V_G^{(2)}$.

After running the measurement for all columns, the background mobility variation (anything except vertical artifacts) must be efficiently extracted by finding the coefficients of the most important principal components. Suppose $W_{R \times C}$ is a principal component and W_{max} is absolute value of the largest element. For computing each principal component factor, we display 4 patterns sequentially and record the panel current for each. The 4 patterns provide following gate voltage profile:

$$\begin{aligned} V_{G_{ij}}^{(1)} &= V_{DD} + V_{TH_o} - (a - \frac{bW_{ij}}{2})^{\frac{1}{2}} \\ V_{G_{ij}}^{(2)} &= kV_{G_{ij}}^{(1)} \\ V_{G_{ij}}^{(3)} &= V_{DD} + V_{TH_o} - (a + \frac{bW_{ij}}{2})^{\frac{1}{2}} \\ V_{G_{ij}}^{(4)} &= kV_{G_{ij}}^{(3)} \end{aligned} \quad (5)$$

where k is an arbitrary constant close to 1 (e.g. 1.1), and

$$\begin{aligned} a &= \frac{(V_{DD}+V_{TH_o}-V_{min})^2 + (V_{DD}+V_{TH_o}-V_{max})^2}{2} \\ b &= \frac{(V_{DD}+V_{TH_o}-V_{min})^2 - (V_{DD}+V_{TH_o}-V_{max})^2}{W_{max}} \end{aligned} \quad (6)$$

where V_{max} and V_{min} are maximum and minimum applied gate voltages, that are 14 and 13v in this case. Note that,

such values for a and b guarantees the gate voltage, V_G, to stay between the desired maximum and minimum levels.

Consequently, if the panel current for those 4 patterns are measured as $I_1, \cdots, I_4$, then the coefficient of the principal component W of the background mobility non-uniformity can be computed as

$$\sum_{i,j=1}^{R,C} W_{ij} \left(\Delta\mu_{ij} - \hat{\Delta\mu_j}\right) = \frac{\frac{I_4 - I_2 - k(I_3 - I_1)}{k^2 - k} - b\beta\mu_o \sum_{i,j=1}^{R,C} W_{ij}}{b\beta} - \sum_{i,j=1}^{R,C} W_{ij}\hat{\Delta\mu_j} \quad (7)$$

Therefore, the total number of current measurements (number of image frames to be displayed), required for the extraction of the mobility non-uniformity using the average vertical variation and the top m_μ principal components, is $2C + 4m_\mu$.

Once the mobility variation profile is estimated, the threshold voltage variation can be characterized in two steps, similarly by decomposing it into vertical and background variation components. In order to extract the average threshold voltage variation of the column j, only one current measurement is needed by applying the following gate voltage pattern to the column while leave the rest of the panel off:

$$\begin{aligned} if \ (k = j) \quad V_{G_{ik}} &= V_{DD} + V_{TH_o} - \frac{c}{\mu_o + \Delta\mu_{ik}} \\ if \ (k \neq j) \quad V_{G_{ik}} &= V_{DD} \end{aligned} \quad (8)$$

where

$$c = 0.5 \times \left(\begin{array}{c} (V_{DD} + V_{TH_o} - V_{min})(\mu_o + \Delta\mu_{min}) + \\ (V_{DD} + V_{TH_o} - V_{max})(\mu_o + \Delta\mu_{max}) \end{array} \right) \quad (9)$$

to ensure that the gate voltage remains between the V_{min} and V_{max} limits, at the column of interest, so that the condition for the first order approximation model of the pixel I-V holds. Therefore, if the measured current is I, the average threshold variation of the column j is

$$\hat{\Delta V}_{TH_j} = \frac{\sum_{i=1}^{R} \Delta V_{TH_{ij}}}{R} = \frac{I - \beta c^2 \sum_{i=1}^{R} \frac{1}{\mu_o + \Delta\mu_{ij}}}{2\beta cR} \quad (10)$$

Finally, to extract the coefficients of the major principal components of the background threshold voltage variation, two measurements are applied per coefficient, as follows:

$$\begin{aligned} V_{G_{ij}}^{(1)} &= V_{DD} + V_{TH_o} - \left(d - \frac{eW_{ij}}{2(\mu_o + \Delta\mu_{ij})}\right) \\ V_{G_{ij}}^{(2)} &= V_{DD} + V_{TH_o} - \left(d + \frac{eW_{ij}}{2(\mu_o + \Delta\mu_{ij})}\right) \end{aligned} \quad (11)$$

where

$$\begin{aligned} d &= \frac{0.5}{\mu_0} \times \left(\begin{array}{c} (V_{DD} + V_{TH_o} - V_{min})(\mu_o + \Delta\mu_{min}) + \\ (V_{DD} + V_{TH_o} - V_{max})(\mu_o + \Delta\mu_{max}) \end{array} \right) \\ e &= \frac{1}{W_{max}} \times \left(\begin{array}{c} (V_{DD} + V_{TH_o} - V_{min})(\mu_o + \Delta\mu_{min}) - \\ (V_{DD} + V_{TH_o} - V_{max})(\mu_o + \Delta\mu_{max}) \end{array} \right) \end{aligned} \quad (12)$$

Consequently, if the full-panel current for the displayed patterns are measured as I_1 and I_2, the coefficient of the corresponding principal component of the background threshold

voltage variation is

$$\sum_{i,j=1}^{R,C} W_{ij}\left(\Delta V_{TH_{ij}} - \Delta \hat{V}_{TH_j}\right) = -\sum_{i,j=1}^{R,C} W_{ij}\Delta \hat{V}_{TH_j} +$$

$$\frac{\frac{I_2 - I_1}{\beta} - \sum_{i,j=1}^{R,C}\left(\left(d + \frac{eW_{ij}}{2\left(\mu_o + \Delta\mu_{ij}\right)}\right)^2 - \left(d - \frac{eW_{ij}}{2\left(\mu_o + \Delta\mu_{ij}\right)}\right)^2\right)\left(\mu_o + \Delta\mu_{ij}\right)}{2e}$$

$$\tag{13}$$

Consequently, to estimate the threshold voltage and mobility variation profile, the total number of current measurements is $3C + 4m_\mu + 2m_{V_{TH}}$, where C is the number of panel columns, m_μ is the number of principal components used to model mobility variation component other than mura impacts, and $m_{V_{TH}}$ is that of the threshold voltage variation.

Finally, in order to remove the small impact of first degree approximation in the Eq.(3), the computations of Eq.(4,7,10,13) can be repeated by changing the value of current measurements according to the following equation:

$$I_{new} = I - \beta \sum_{i,j=1}^{R,C} \left(\mu_o + \Delta\mu_{ij}\right)\Delta V_{TH_{ij}}^2 \tag{14}$$

where $\Delta\mu$ and ΔV_{TH} are the estimated variation from the last iteration. The subtracted term is equal to the second degree term that has been ignored by applying the first degree approximation.

4. RESULTS

In this section, the efficiency of the proposed methodology is verified by using experimental spatial statistics of an LTPS-AMOLED panel. The following function is used to fit a spatial correlation function with the experimented data [13]:

$$\rho(v) = 2\left(\frac{bv}{2}\right)^{s-1} K_{s-1}(bv)\,\Gamma(s-1)^{-1} \tag{15}$$

where b and s, the fitting parameters, are set to 7.8E-3 and 1.3. K is the modified Bessel function of the second kind, and Γ is the gamma function where v is the distance between two pixels. An isotropy covariance matrix is formed using the Eq.(15).

In an industrial framework, few panels can be fully characterized to directly form a covariance matrix from the sample data. A pre-determined number of the largest eigenvectors of the covariance matrix will then be calculated using ARPACK [6], an iterative Arnoldi algorithm which extracts a few largest eigenvalues/eigenvectors.

Threshold voltage and mobility variation profiles are simulated from the generated covariance matrix, for the verification purposes. Vertical mura defects are also added to both profiles. Figures 7(a) and 7(c) show the original 2D variation profiles. The image is normalized so that the range of $\pm 3\sigma$ of each variation component is demonstrated as 0–255 gray-scale level.

Figures 7(b) and 7(d) depict the extracted profile using the first 1000 principal components out of 320×320 that is less than a percent. The total number of patterns to be displayed is $3 \times 320 + 6 \times 1000 = 6960$ taking at least 116 seconds to run, considering the frame rate of 60 and one current measurement/frame per image pattern. The root mean square

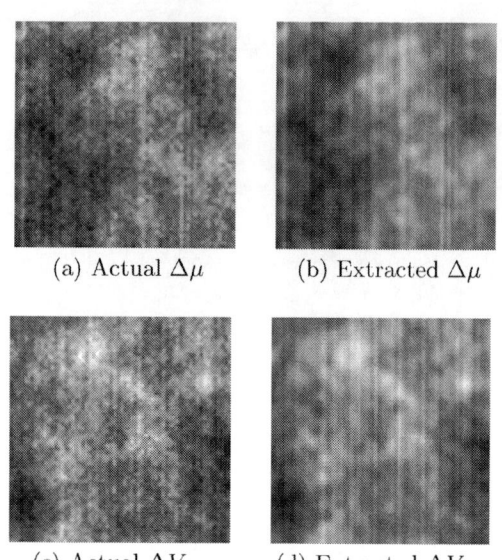

(a) Actual $\Delta\mu$ (b) Extracted $\Delta\mu$

(c) Actual ΔV_{TH} (d) Extracted ΔV_{TH}

Figure 7: The extracted variation profile from a 320×320 panel using the top 1000 principal components and vertical mura characterization versus the actual variation (root mean square error $\approx 0.3\sigma$). Note that, the profiles are scaled so that the range of -3σ to 3σ of each parameter is mapped to 0–255 gray scale.

error of the estimation, compared to the original profile, is close to 0.3σ for both the variation types.

Figure 8 plots the error with respect to the number of principal components. As can be seen, even when no principal components is utilized, the root mean square error is down to 0.5 from full σ. That is due to the part of extracting the average variations of each pixel column to detect the mura-induced effects. However, the extracted data, the average, not only reveals the mura effects but also extracts some part of the background profile which results in a gradient along the x-axis. Figure 8(a) shows the early convergence rate of the characterization using the 1250 most important principal components, while the Fig. 8(a) shows the rest. As expected from the PCA properties, most of the variation data (energy) is captured by the first few components.

5. CONCLUSIONS

While AMOLED displays offer significant performance improvement over other existing technologies, the analog role of the TFTs in pixel along with significant process variations is the major hurdle in the design and implementation of such displays. This becomes more severe in high pixel density and/or large area displays since both in-pixel compensations and external calibration suffers from serious drawbacks. In such applications, the number of TFTs is limited due to the pixel area and yield consideration. Moreover, the characterization of the displays can suffer from long runtime due to few million measurement iterations.

Our studies show a significant spatial correlation of the process parameters which enables the use of principle component analysis for fast non-invasive characterization of TFTs

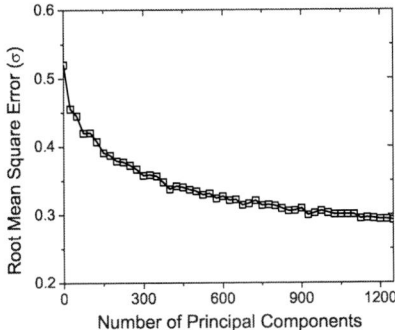

(a) The most important principal components

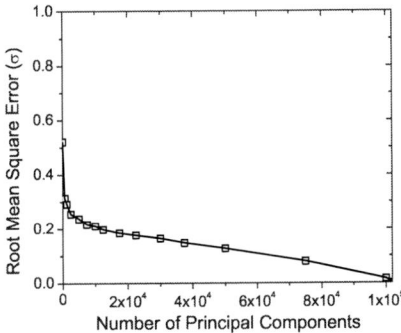

(b) All of the principal components

Figure 8: Root mean square error (normalized by the standard deviation of the process parameter variation) with respect to the number of applied principal components. Note that the vertical mura has been extracted first.

in pixel circuits. The method proposed in this paper benefits from simultaneous characterization of V_{TH} and mobility variation using few measurement iterations and tractable computation. Moreover, it works with any existing display architecture by treating the panel as a black box and applying pre-defined images and measuring display power line current. The results highlighted over 100 times reduction in measurement iteration for a relative error of below 30%.

6. REFERENCES

[1] A. Agarwal, D. Blaauw, V. Zolotov, S. Sundareswaran, M. Zhao, K. Gala, and R. Panda. Statistical delay computation considering spatial correlations. In *Proc. of IEEE/ACM Asia and South Pacific Design Automation Conference*, pages 271–276, 2003.

[2] S. Alexander, G. R. Chaji, J. M. Dionne, C. Church, J. Hamer, and A. Nathan. Unique electrical measurement technology for compensation inspection, and process diagnostics of AMOLED HDTV. In *SID Symposium Digest of Technical Papers*, pages 1356–1359, 2010.

[3] G. R. Chaji, S. Alexander, J. M. Dionne, Y. Azizi, C. Church, J. Hamer, J. Spindler, and A. Nathan. Stable RGBW AMOLED display with OLED degradation compensation using electrical feedback. In

Proc. of IEEE International Solid-State Circuits Conference, pages 118–120, Feb 2010.

[4] P. Friedberg, Y. Cao, J. Cain, R. Wang, J. Rabaey, and C. Spanos. Modeling within-field gate length spatial variation for process-design co-optimization. In *Proc. of SPIE*, pages 178–188, 2005.

[5] M. Kimura, I. Yudasaka, S. Kanbe, H. Kobayashi, H. Kiguchi, S. I. Seki, S. Miyashita, T. Shimoda, T. Ozawa, K. Kitawada, T. Nakazawa, W. Miyazawa, and H. Ohshima. Low-temperature polysilicon thin-film transistor driving with integrated driver for high-resolution light emitting polymer display. *IEEE Trans. Electron Devices*, 46:2282–2288, Dec. 1999.

[6] R. B. Lehoucq, K. Maschhoff, D. Sorensen, and C. Yang. ARPACK software package. *http://www.caam.rice.edu/software/ARPACK/*, 1996.

[7] J. Li, K. Kang, and K. Roy. Variation estimation and compensation technique in scaled LTPS TFT circuits for low-power low-cost applications. *IEEE Trans. Comput.-Aided Design Integr. Circuits Syst.*, 28:46–59, Jan. 2009.

[8] C. L. Lin and Y. C. Chen. A novel LTPS-TFT pixel circuit compensating for TFT threshold-voltage shift and OLED degradation for AMOLED. *IEEE Electron Device Lett.*, 28:129–131, Feb. 2007.

[9] A. Nathan, G. R. Chaji, and S. J. Ashtiani. Driving schemes for a-Si and LTPS AMOLED displays. *IEEE/OSA J. Display Technol.*, 1:267–277, Dec. 2005.

[10] D. Needell and J. Tropp. CoSaMP: Iterative signal recovery from incomplete and inaccurate samples. *Appl. Computat. Harmon. Anal.*, 26:301–321, May 2009.

[11] N. P. Papadopoulos, A. A. Hatzopoulos, and D. K. Papakostas. An improved optical feedback pixel driver circuit. *IEEE Trans. Electron Devices*, 56:229–235, Feb. 2009.

[12] Y. H. Tai, S. C. Huang, W. P. Chen, Y. T. Chao, Y. P. Chou, and G. F. Peng. A statistical model for simulating the effect of LTPS TFT device variation for SOP applications. *IEEE/OSA J. Display Technol.*, 3:426–433, Dec. 2007.

[13] J. Xiong, V. Zolotov, and L. He. Robust extraction of spatial correlation. *IEEE Trans. Comput.-Aided Design Integr. Circuits Syst.*, 26:619–631, Apr. 2007.

[14] J. Yamashita, K. Uchino, and T. Yamamoto. New driving method with current subtraction pixel circuit. In *SID Symposium Digest of Technical Papers*, pages 1452–1455, 2005.

[15] W. Zhang, X. Li, and R. A. Rutenbar. Bayesian virtual probe: minimizing variation characterization cost for nanoscale IC technologies via Bayesian inference. In *Proc. of IEEE/ACM Design Automation Conference*, pages 262–267, 2010.

[16] C. Zhuo, K. Agarwal, D. Blaauw, and D. Sylvester. Active learning framework for post-silicon variation extraction and test cost reduction. In *Proc. of IEEE/ACM International Conference on Computer-Aided Design*, pages 508–515, 2010.

Towards Fault-Tolerant Embedded Systems with Imperfect Fault Detection

Jia Huang
fortiss GmbH, Germany
huang@fortiss.org

Kai Huang
fortiss GmbH, Germany
khuang@fortiss.org

Andreas Raabe
fortiss GmbH, Germany
raabe@fortiss.org

Christian Buckl
fortiss GmbH, Germany
buckl@fortiss.org

Alois Knoll
TU München, Germany
knoll@in.tum.de

ABSTRACT

Many state-of-the-art approaches on fault-tolerant system design make the simplifying assumption that all faults are detected within a certain time interval. However, based on a detailed experimental analysis, we observe that perfect fault detection is not only an impractical assumption but even if implementable also a suboptimal design decision. This paper presents an approach that takes imperfect fault detection into account. Novel analysis and optimization techniques are developed, which distinguish detectable and undetectable faults in the overall workflow. Besides synthesizing the task schedules, our approach also decides which of the available fault detectors is selected for each task instance. Experimental results show that our approach finds solutions with several orders of magnitude higher reliability than current approaches.

Categories and Subject Descriptors

B.8.1 [**Performance and Reliability**]: Reliability, Testing, and Fault-Tolerance; C.3 [**special-purpose and application-based systems**]: Real-time and embedded systems

General Terms

Algorithms, Design, Reliability

Keywords

Embedded Systems, Reliability, Design Optimization

1. INTRODUCTION

To meet the reliability requirements of safety-critical embedded systems, fault-tolerance techniques such as active redundancy are widely adopted. Active redundancy can be implemented in both the space and the time domains. In the space domain, critical components can be replicated into multiple copies to enhance the error resilience. In the time domain, software tasks can be selectively re-executed. Fault-tolerant system design using active redundancy is a very challenging task that involves solving two major problems, namely finding the optimal utilization of temporal and/or spatial redundancy and the scheduling of tasks (including replicas) under timing constraints. Over the past decades, a lot of research efforts have been devoted to this field. A review of related work is presented in Appendix A.

To cope with the high problem complexity, many state-of-the-art studies make simplifying assumption on the fault models and modes. Perfect fail-silent behavior is one assumption that is often used in literature. It is assumed that all faults are detected within a certain time interval and the fault-detection overhead is contained in the tasks' Worst-Case Execution Times (WCETs), e.g., in fault-tolerant task scheduling [8, 15, 4, 17, 6, 5], in reliability-aware energy management [16, 20, 22] and in error-aware system design [9, 10]. With this assumption, each task will produce either a correct output or no output at all. Although fail-silence is a highly desirable property, it is difficult to implement in practice. The prerequisite is the existence of a perfect fault detector that achieves 100% coverage under the given fault hypothesis.

The simplifying assumption of perfect fault detection is problematic. On the one hand, a perfect detector might not exist or is difficult to implement, making the algorithms developed under this assumption less useful in practice. On the other hand, even if implementable, perfect detectors typically come with high resource and timing overheads. In recent work [12, 18] it has been shown that the time needed for high-coverage fault detection may become much longer than the execution time of the task itself (e.g. the timing overhead could be 400% using techniques proposed in [12]). Hence, approaches under this assumption are very pessimistic, as the most expensive fault detector is selected for every task.

This problem can be viewed from a slightly different angle: choosing to implement the perfect fault detector is not only an **assumption** but also an important **design decision**. While making this assumption, all design alternatives with partial fault detectors are ignored without any justification. For example, when active redundancy is concerned, no analysis is performed to find out if it is more efficient to spend the available resources on applying better fault detection or a higher number of replications. Actually, our experimental results show that the answer is highly application and architecture dependent. Detailed discussions are presented in Section 3 and Appendix B.

In this work, we put special emphasis on the effect of imperfect fault detection and present the first approach (to the best of our knowledge) to synthesize fault-tolerant schedules with reliability guarantee using imperfect fault detectors. Besides computing the task schedule and utilizing redundancy, our approach also decides which of the available fault detectors should be selected for each task. The main contributions of this papers are: 1) an experimental analysis on the impact of imperfect fault detection on system-level reliability; 2) a reliability analysis approach that computes the probabilities of both detectable and undetectable faults in the presence of redundancy; 3) an Multi-Objective Evolutionary Algorithm (MOEA) based approach for reliability-aware design optimization.

The remainder of the paper is organized as follows. The system models is first introduced in Section 2. The next Section discusses a motivating example. The main contribution of this work, namely the reliability analysis and optimization approaches are presented in Section 4 and Section 5, respectively. Experimental results are provided in Section 6. Section 7 concludes the paper.

2. PRELIMINARIES

2.1 Fault Model and Fault Tolerance

This paper focuses on tolerating transient faults and adopts the classical fault model that occurrence of transient faults follows a Poisson distribution with a constant failure rate λ. The consideration of permanent faults could be added, e.g. using the technique proposed in [5]. A task may be replicated into multiple copies (or instances) to implement temporal/spatial redundancy. The set of N replicas for a task t_i is denoted as $R(t_i) = \{t_{i,1}, ..., t_{i,N}\}$. For a specific instance, software fault detectors can be implemented. A software fault detector typically transforms the original program into an instrumented version, adding the capability to detect transient faults that occur at runtime of the program. The arithmetic codes [18] and critical variable technique [13] are examples of this kind. We assume that each task instance tries to implement the fail-silent behavior, i.e., as long as the fault detector[1] reports a fault, this specific task instance will not produce any output. This behavior is desirable since the correct outputs from other instances will not be polluted.

As discussed in section 1, fault detectors are typically imperfect in reality. We characterize a fault detector implementation as a pair $d = \{c, o\}$, where c is the fault detection *coverage* in percentage and o is the timing overhead for fault detection. The overhead is defined in percentage with respect to the stand-alone WCET of the task. In this way, a task t_i that implements the fault detector indexed k has the WCET $w_i(1 + o_k)$. We assume that a library of implementable fault detectors are available at design time for each task (denoted as D_i for task t_i).

We assume a voting mechanism with majority voting is

[1] For simplicity, the term fault detector used in the rest of the paper is meant to be software fault detectors unless mentioned otherwise.

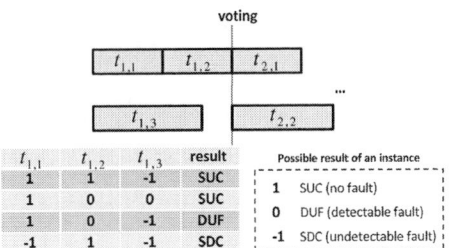

Figure 1: Example Fault Scenario

implemented if redundancy is available. The voter collects results from all instances and produces a single output for the successor tasks. The qualitative execution results of these replicas (i.e. if they deliver a correct output or not) are described by a *fault scenario*. A fault scenario is a vector $x = \{x_1, ..., x_N\}$, which contains a variable $x_l \in \{1, 0, -1\}$ for each task instance, where x_l is 1 if $t_{i,l}$ produces a correct output (no fault occurs); x_l is 0 if $t_{i,l}$ fails silent (a fault occurs and is detected) and x_l is -1 if $t_{i,l}$ produces an incorrect output (i.e., a fault occurs and is not detected). The voter generates an output if and only if a dominating result (or a majority) is found.

The overall execution of a task, considering all its instances, could result in the following 3 scenarios: 1) the task executes successfully (SUC): it experiences no fault or only some faults that are later masked by the voter; 2) Detected Unrecoverable Faults (DUF): the voter fails to find a dominating result and thus produces no output; and 3) Silent Data Corruption (SDC): multiple faults occur and the incorrect outputs mask the correct one. Both DUF and SDC are unwanted behavior that negatively influences the system reliability (see Section 3).

Figure 1 depicts an example of the voting scenario. If the fault scenario is $x = \{1, 1, -1\}$, the incorrect output of $t_{1,3}$ is masked and the overall result is SUC. In the scenario $x = \{1, 0, 0\}$, both $t_{1,2}$ and $t_{1,3}$ produce no result, and the only output from $t_{1,1}$ will be taken. Hence, the overall result is also SUC. In the scenario $x = \{1, 0, -1\}$, a correct and an incorrect output are sent to the voter. However, the voter cannot identify the correct input since no majority is found. In this case, the voter will generate no output and the overall result is DUF. In the last example scenario $x = \{-1, 1, -1\}$, two incorrect outputs are sent to the voter. Note that the fault scenarios model only the qualitative result (0,1, or -1), but the voting is performed based on the real value of the tasks' outputs. Hence, if two outputs are incorrect, two cases might happen: 1) the two incorrect outputs are equal and mask the single correct one, resulting a SDC; 2) the two incorrect outputs are unequal and the voter does not see a dominating value, resulting in a DUF. To stay on the safe side, we have to assume the first case (SDC), because the probabilities of the two cases are very difficult to be quantified, even if possible[2].

2.2 System models

We consider applications modeled as directed acyclic Task Graphs (TGs). The vertices $\mathcal{T} = \{t_0, t_1, ..., t_m\}$ of a TG represent a set of *tasks* to be executed and the edges capture data dependencies. The stand-alone WCET of

[2] The probabilities are highly influenced by the application characteristic, the output data type, common caused errors, etc.

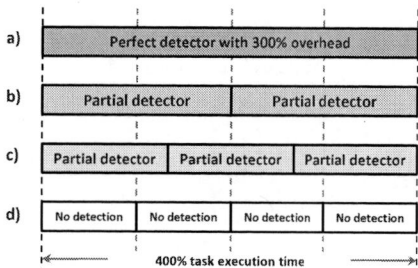

Figure 2: Example Scenario

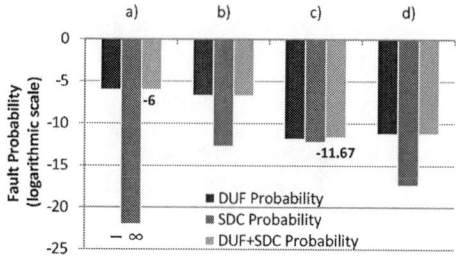

Figure 3: Reliability of the Example Schedules

the task t_i on processor p_j without any fault detection is denoted using $w_{i,j}$. For a instance $t_{i,l} \in R(t_i)$, the processor that will execute $t_{i,l}$ is denoted by $node(t_{i,l})$ and the fault detector ID it implements is denoted by $det(t_{i,l})$. The execution time and fault detection coverage of this instance are therefore $w_i^l = w_{i,node(t_{i,l})}(1 + o_{det(t_{i,l})})$ and $c_{det(t_{i,l})}$, respectively. According to the Poisson fault model, the following formulas could be used to compute the probabilities that an instance is executed successfully without transient faults (denoted by SUC) or experiences detectable/undetectable faults (denoted by DUF/SDC):

$$P_{SUC}(t_{i,l}) = e^{-\lambda_{node(t_{i,l})} w_i^l}$$

$$P_{DUF}(t_{i,l}) = (1 - e^{-\lambda_{node(t_{i,l})} w_i^l}) c_{det(t_{i,l})}$$

$$P_{SDC}(t_{i,l}) = (1 - e^{-\lambda_{node(t_{i,l})} w_i^l})(1 - c_{det(t_{i,l})})$$

Our target architectures are heterogeneous multiprocessor platforms with time-triggered communication, e.g., the GENESYS [3] architecture. The communication between tasks is implemented with messages. The communication can be protected with dedicated techniques (e.g., error correction code) and is therefore assumed as reliable.

3. MOTIVATING EXAMPLE

To understand the impact of imperfect fault detection on the system reliability, we carried out a set of experiments considering two scenarios. In the first one, we fix the amount of redundancy and analyze the influence of detection coverage on the system-level reliability. In the second one, we do it vice-versa, i.e., varying the number of replications with fixed fault detector. The detailed experimental data and discussion is presented in Appendix B. In general, we observe that the selection of fault detector and the utilization of redundancy show a tradeoff. In particular, when the system features only limited amount of resources or the application has tight timing constraints, inappropriate selection of fault detector might disallow certain options for redundancy due to the timing overhead. We explain this issue using the following example.

Consider a simple task running on a single processor system. Similar as the experiments in Appendix B, we reuse the result of [18] and assume that the rate of undetectable faults decreases exponentially with linear fault detection effort. It is further assumed that the perfect fault detection (100% coverage) incurs 300% timing overhead (typical value in [18]). Figure 2a depicts the schedule using the perfect detector. By spending all resources on fault detection, SDCs are completely eliminated. Figure 2b is another possible schedule, in which the task is replicated twice

and the remaining time (200% task execution time in this case) is used to implement two partial fault detectors (each 90% coverage using the 100% detection effort). Figures 2c and 2d show two similar schedules with higher number of replications. When multiple replicas of the same task are available, the results from different instances can be compared to detect or even mask the occurred faults. Figure 3 compares the probability of DUF and SDC for each schedule. For schedule a, although SDCs are avoided completely, the DUF probability is very high, since any transient fault occurring on the single task instance results in a DUF. With imperfect fault detectors (schedule b to d), SDC will not totally disappear but the probability of DUF can be significantly reduced. If both types of faults are considered together, the overall failure probability ($DUF + SDC$) of schedule c is almost six orders of magnitude lower than that of schedule a.

The selection of the best schedule depends on the reliability goal of the application. Many systems have specific requirements concerning DUF and/or SDC. For example, the IBM Power 4 processor-based systems target 10-25 years Mean Time Between Failures (MTBF) for DUF and 1000 years MTBF for SDC [2]. The schedule using perfect fault detectors may not meet the requirements of all applications. Moreover, the criticality of a certain type of faults is application-specific. For systems that require fail-operational behavior, DUFs and SDCs could be equally bad and schedule c is clearly a much better design choice. For other systems, SDCs might be more critical and schedule a or d are more preferable.

From the analysis above, it can be seen that the selection of appropriate fault detectors is critical. The decision has to be made jointly with other design parameters, e.g., task mapping and utilization of redundancy. However, the existing work assuming perfect fault detection prohibits the exploration of design alternatives using partial fault detectors. To tackle this problem, we need 1) a way to evaluate the system quality regarding both DUF and SDC; and 2) an optimization approach for reliability-aware design space exploration. The next two sections present our approach on these issues.

4. RELIABILITY ANALYSIS

Using the voting setup introduced in Section 2, the schedule generated by our algorithm falls into the category of *strict schedules* [4, 1]. Strict schedules obey the rule that if a task t has a data dependency on task t', all replicas of t' should be completed before any replica of t starts. With this restriction, all tasks use exclusively the voter output and the tasks of a TG can be considered independently in the reliability analysis.

For a task t_i, a fault scenario x is *tolerable* if the voter can produce a correct output in the presence of the faults specified in x. This condition can be computed by the following binary function *tolerable()*, which evaluates to *true* if the correct outputs are able to dominate.

$$tolerable(x) = ((\sum_{t_{i,l} \in R(t_i)} x_l) > 0) \qquad (1)$$

Where $R(t_i)$ denotes the set of replicas of task t_i and $x_l \in \{1, 0, -1\}$ is the execution result of task $t_{i,l}$. Similarly, the fault scenario x is *silent* if the voter cannot distinguish a dominating result and x is *faulty* if the incorrect results are majority.

$$silent(x) = ((\sum_{t_{i,l} \in R(t_i)} x_l) = 0) \qquad (2)$$

$$faulty(x) = ((\sum_{t_{i,l} \in R(t_i)} x_l) < 0) \qquad (3)$$

The probability that a task is executed successfully can be computed by summarizing the occurrence probability of all tolerable fault scenarios:

$$P_{SUC}(t_i) = (\sum_{\forall x: tolerable(x) = true} P(t_i, x)) \qquad (4)$$

where $Pr(t_i, x)$ is the probability that the fault scenario x happens. As x specifies the qualitative execution result $(SUC/DUF/SDC)$ of each instance of task t_i, the probability $Pr(t_i, x)$ can be computed as a product of occurrence probability of each task instance:

$$P(t_i, x) =$$
$$\prod_{\substack{t_{i,l} \in R(t_i) \\ \wedge x_l = 1}} P_{SUC}(t_{i,l}) \prod_{\substack{t_{i,l} \in R(t_i) \\ \wedge x_l = 0}} P_{DUF}(t_{i,l}) \prod_{\substack{t_{i,l} \in R(t_i) \\ \wedge x_l = -1}} P_{SDC}(t_{i,l})$$

The instance-level probabilities $P_{SUC}(t_{i,l})$, $P_{DUF}(t_{i,l})$ and $P_{SDC}(t_{i,l})$ are computed from the fault model introduced in Section 2.2. In a similar way as in equation 4, the probability that a task results in a fail-silence ($P_{DUF}(t_i)$) or it produces a faulty output ($P_{SDC}(t_i)$) can be computed.

The complete set of tolerable (or silent or faulty) scenarios can be obtained by systematically enumerating all fault scenarios. Since each task instance has three possible results (1,0, or -1), the overall number of combinations is 3^N, where N is the number of replicas. Although this enumeration has exponential complexity, it is still acceptable in practice since the number of replicas for a task is typically very small, e.g., more than 3 replicas for a task is rarely used in practice. The above step is performed for all tasks in the application so that the task-level probabilities P_{SUC}, P_{DUF} and P_{SDC} are obtained. Then, we can proceed with analyzing the reliability of the entire application. Naturally, an application consisting of tasks $\mathcal{T}$ is successful (i.e. SUC) only if all of its tasks are successful:

$$P_{SUC}(\mathcal{T}) = (\prod_{t_i \in \mathcal{T}} P_{SUC}(t_i)) \qquad (5)$$

The application is silent (i.e. DUF) if at least one of its tasks is silent, because if any task fails to produce an output, the successor tasks cannot proceed due to data dependency and the entire application has to start over. This probability is denoted by $P_{DUF}(\mathcal{T})$. The application is faulty (i.e. SDC,

the corresponding probability is denoted by $P_{SDC}(\mathcal{T})$), if none of its tasks is silent and at least one of its tasks is faulty. Assume t_0 is the first task in $\mathcal{T}$, the application is faulty if t_0 is faulty and the remaining tasks are non-silent (denoted by $P_{\overline{DUF}}(\mathcal{T} \backslash t_0)$), or t_0 is successful and the remaining tasks are faulty.

$$P_{SDC}(\mathcal{T}) = P_{SDC}(t_0)P_{\overline{DUF}}(\mathcal{T} \backslash t_0) + P_{SUC}(t_0)P_{SDC}(\mathcal{T} \backslash t_0)$$

Since $P_{\overline{DUF}}(\mathcal{T} \backslash t_0)$ is the sum of $P_{SUC}(\mathcal{T} \backslash t_0)$ and $P_{SDC}(\mathcal{T} \backslash t_0)$, the above formula can be rewritten as:

$$P_{SDC}(\mathcal{T}) = P_{SDC}(t_0)P_{SUC}(\mathcal{T} \backslash t_0) +$$
$$(P_{SDC}(t_0) + P_{SUC}(t_0))P_{SDC}(\mathcal{T} \backslash t_0) \qquad (6)$$

As can be seen, $P_{SDC}(\mathcal{T} \backslash t_0)$ is the only unknown term. Hence, the SDC probability can be computed in a recursively manner. The complexity is linear with the number of tasks. The DUF probability can then be computed by:

$$P_{DUF}(\mathcal{T}) = 1 - P_{SUC}(\mathcal{T}) - P_{SDC}(\mathcal{T}) \qquad (7)$$

5. OPTIMIZATION PROCEDURE

After having the reliability analysis, the next step is to develop an optimization approach to search for high-quality designs. We identify two major scenarios that the designers may encounter. In the first one, the system is intended to execute a single application, so the design goal is to maximize the reliability while meeting the deadline. We show that this problem can be transformed into a deadline assignment problem that can be solved using Integer Linear Programming. Appendix C details the transformation and ILP formulation. In the second scenario, multiple applications may be executed on the same platform. We add an additional optimization objective that the resource consumption is to be minimized so that more space can be reserved for future applications. A Multi-Objective Evolutionary Algorithm (MOEA) based optimization procedure is presented for this case.

To use MOEA for optimization, the solutions (in our case the task schedules) need to be encoded into chromosome. The proposed encoding scheme maintains a gene (i, M) for each task, where i is the integer index of the task and M is a list of mapping entries. Each mapping entry encodes an instance of the task and is represented as a pair (p, d), where p is the processor that will execute the instance and d is the index of the fault detector to implement. Figure 4 illustrates an example, in which task 1 and 3 are replicated 2 times and task 2 is replicated 3 times. The lower part of the figure depicts the corresponding schedule that the chromosome represents. Since we are targeting on generating strict schedules [4, 1], the reconstruction of the schedule from the chromosome can be done using a simple greedy heuristic. We consider all tasks in the TG in topological order. For each task, the replicas specified in the chromosome are instantiated and scheduled greedily at the earliest possible time. Output messages are scheduled at the end of execution. If the current task has data dependency on previous tasks, a voter is inserted. The failure rate of the voter is added to the failure rate of the current task.

We consider three optimization objectives. The first two are the reliability objectives, one for DUF and one for SDC. The metric is Failure In Time (FIT). One unit FIT specifies one failure in a billion hours. The conversion from failure probabilities computed in section 4 to FIT is

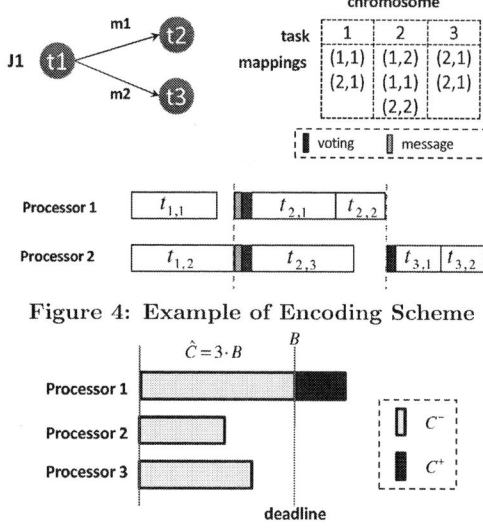

Figure 4: Example of Encoding Scheme

Figure 5: The Resource Consumption Objective

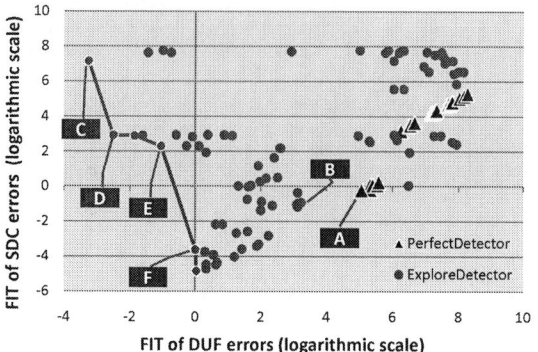

Figure 6: 2D Projection of Optimization Results

straightforward. In the third objective, we intend to encode the design goal of minimizing resource consumption while meeting the deadline. The resource consumption (denoted by C) is defined as the overall processor time that a schedule occupies. Let B be the deadline of the application and N be the number of processors available in the execution platform. The available time budget within the deadline is $\hat{C} = NB$. For a given schedule S, we use C^- to denote the fraction of resource consumption that is within the deadline and C^+ to denote the part above the deadline. Figure 5 depicts an example. The objective function is defined as follows:

$$penalty = \begin{cases} C & \text{iff } C^+ = 0 \\ \hat{C} + C^+ & \text{otherwise} \end{cases} \quad (8)$$

By constructing the objective function as above, each schedule that violates the deadline ($C^+ > 0$) has a higher penalty value than any schedule that meets the deadline. For two schedules that meet the deadline, the one that has less resource consumption will be preferred. Clearly, all three objectives are to be minimized.

6. EXPERIMENTS

We implement the analysis and optimization algorithms in JAVA using the opt4j library [11]. We assume that the target platform consists of two types of Processing Elements (PEs), namely a RISC processor and a DSP. The failure probability of each task on a certain PE is randomly generated between 1×10^{-5} and 1×10^{-7}. We again use the exponential model in [18], i.e., the undetectable faults reduce exponentially with linear fault detection effort. Random fault detectors are generated following this law.

The proposed approach is applied on an mpeg2 decoder example taken from [19]. We compare the performance of two approaches: 1) the proposed approach that explores the optimal utilization of fault detectors (*ExploreDetector*); 2) existing approaches that utilize the perfect fault detector[3]

---

[3]For better visualization in the logarithmic scale, the results we presents are using the detector with 99.9% coverage. If the perfect fault detector is used, the probability of *SDC*

for all tasks (*PerfectDetector*). We use the MOEA optimizer to compute the Pareto optimal results considering the three objectives introduced in section 5. Figure 6 shows the results using an example platform that consists of 2 *RISC*s and 2 *DSP*s. The dots in the figure show the results projected into a 2D plane, with the vertical axis being the FIT of *SDC* and the horizontal axis being the FIT of *DUF*. The Pareto front considering only the two reliability objectives is marked using a solid line. The triangle symbols in Figure 6 show the results of the *PerfectDetector* approach. Clearly, the solutions found by *ExploreDetector* is of much higher quality than those found by *PerfectDetector*. The gap in terms of FIT is several orders of magnitude in this experiment. Moreover, the *ExploreDetector* approach provides a much wider spectrum of solutions, allowing the designers to carefully evaluate the tradeoff between the two classes of faults and select the implementation that fits the application requirements.

We mark some representative implementation alternatives in Figure 6. A is the best solution in terms of reliability found by the *PerfectDetector* approach; B is a solution found by *ExploreDetector* which is close to and dominates A; C to F belong to the Pareto optimal solutions found by *ExploreDetector*. Table 1 compares these implementations in several aspects, e.g., the average number of replications for each task, the average fault detection coverage over all task instances and the resource consumption. It can be seen that implementation A has the lowest number of replications, since a lot of resources are already consumed by fault detection. The solution B has higher quality than A concerning all three objectives. Using fault detectors with average coverage of 63%, it achieves much higher reliability than A and saves 35% resources. The implementation F achieves higher reliability than A as well. By spending 65% more resources, it reduces the FIT of *DUF* by 5 orders of magnitude and the FIT of *SDC* by more than 3 orders of magnitude. It is also worth noticing that, since most of the solutions found by *PerfectDetector* implement 2 replicas, the curve formed by those solutions has similar shape as the curve in Figure 7b.

The optimization results from MOEA can also be viewed from different angles. Extended discussion of this case study is presented in Appendix D. To go one step further, our approach can be used to perform reliability-aware design space exploration (DSE), e.g., to find out the amount and

---

can be reduced to 0, but the probability of *DUF* remains almost the same.

solution	DUF FIT (log.)	SDC FIT (log.)	avg. rep.	avg. cov.(%)	resource (time unit)
A	5.06	-0.23	3.25	99.9	114.0
B	3.24	-0.93	3.67	63.0	74.2
C	-3.25	7.15	3.50	84.4	55.9
D	-2.49	2.93	3.83	74.9	65.5
E	-1.04	2.30	3.92	83.0	150.6
F	0	-3.62	3.92	89.3	189.5

Table 1: Comparing Representative Implementation Alternatives

Application (num. tasks)	200 round	500 round	1000 round	1500 round
mpeg2(13 tasks)	29.0	76.4	120.8	198.3
TG1(50 tasks)	78.3	179.0	395.1	583.0
TG2 (100 tasks)	195.9	442.2	777.0	1692.0

Table 2: Execution Time of Optimization Approach

type of PEs necessary to meet certain reliability goal. The DSE flow is also illustrated in Appendix D.

We measure the execution time (in seconds) of the our approach on a Windows machine with 3GHz CPU. The MOEA is configured to run for 200, 500, 1000 and 1500 rounds. Table 2 presents the results. For a small TG (e.g. mpeg2), the analysis and optimization procedure takes only a few minutes to execute for 1500 iterations. As expected, the execution time grows linearly with the number of iterations. It is also worth mentioning that the execution time also increases roughly linearly with the size of TG. This is because the reliability analysis, as most computational intensive operation, has linear complexity in the number of tasks. For a syntactic TG[4] with 100 tasks, the 1000-iteration EA takes about 13 minutes. In general, the runtime is acceptable for an off-line design space exploration procedure.

7. CONCLUSION

In reliability-aware system design, many existing studies adopt the assumption that fault detection is always perfect to simplify the problem. We observe that this assumption causes several practical issues and may exclude the optimal design alternative. In this paper, we present an approach to synthesizing fault-tolerant design with reliability guarantee applying imperfect fault detectors. The proposed analysis and optimization techniques can be used for reliability-aware DSE. Experimental results verify that our approach finds solutions with several orders of magnitude higher reliability compared to current approaches.

Acknowledgement

This work has been supported in part by the European research project ACROSS under the Grant Agreement ARTEMIS-2009-1-100208.

8. REFERENCES

[1] A. Benoit, L.-C. Canon, E. Jeannot, and Y. Robert. Reliability of task graph schedules with transient and fail-stop failures: complexity and algorithms. *Journal of Scheduling*, 2011.

[4] generated using TGFF http://ziyang.eecs.umich.edu/~dickrp/tgff/

[2] D.C.Bossen. Cmos soft errors and server design. In *Reliability Physics Tutorial Notes, Reliability Fundamentals*, pp. 121.07.1. 2002.

[3] GENESYS. http://www.genesys-platform.eu/.

[4] A. Girault and H. Kalla. A novel bicriteria scheduling heuristics providing a guaranteed global system failure rate. *IEEE Transactions on Dependable and Secure Computing*, 2009.

[5] J. Huang, J. O. Blech, A. Raabe. C. Buckl, and A. Knoll. Analysis and optimization of fault-tolerant task scheduling on multiprocessor embedded systems. In *CODES+ISSS*, Oct 2011.

[6] J. Huang, J. O. Blech, A. Raabe, C. Buckl, and A. Knoll. Reliability-aware design optimization for multiprocessor embedded systems. In *Euromicro Conference on Digital System Design (DSD)*, 2011.

[7] V. Izosimov, I. Polian, P. Pop, P. Eles, and Z. Peng. Analysis and optimization of fault-tolerant embedded systems with hardened processors. In *Design, Automation and Test in Europe (DATE)*, 2009.

[8] V. Izosimov, P. Pop, P. Eles, and Z. Peng. Design optimization of time-and cost-constrained fault-tolerant distributed embedded systems. In *DATE*, 2005.

[9] J. Lee, I. Shin, and A. Easwaran. Online robust optimization framework for qos guarantees in distributed soft real-time systems. In *EMSOFT*, 2010.

[10] A. Lifa, P. Eles, Z. Peng, and V. Izosimov. Hardware/software optimization of error detection implementation for real-time embedded systems. In *CODES+ISSS*, 2010.

[11] M. Lukasiewycz, M. Glaß, F. Reimann, and J. Teich. Opt4J - A Modular Framework for Meta-heuristic Optimization. In *GECCO*, Dublin, Ireland, 2011.

[12] G. Lyle, S. Chen, K. Pattabiraman, Z. Kalbarczyk, and R. Iyer. An end-to-end approach for the automatic derivation of application-aware error detectors. In *DSN*, 2010.

[13] K. Pattabiraman, Z. Kalbarczyk, and R. Iyer. Automated derivation of application-aware error detectors using static analysis: The trusted illiac approach. *Dependable and Secure Computing, IEEE Transactions on*, 8(1):44 –57, 2011.

[14] C. Pinello, L. P. Carloni, and A. L. Sangiovanni-Vincentelli. Fault-tolerant deployment of embedded software for cost-sensitive real-time feedback-control applications. In *DATE*, 2004.

[15] P. Pop, V. Izosimov, P. Eles, and Z. Peng. Design optimization of time- and cost-constrained fault-tolerant embedded systems with checkpointing and replication. *IEEE Trans. VLSI*, 2009.

[16] P. Pop, K. H. Poulsen, V. Izosimov, and P. Eles. Scheduling and voltage scaling for energy/reliability trade-offs in fault-tolerant time-triggered embedded systems. In *CODES+ISSS*, 2007.

[17] P. K. Saraswat, P. Pop, and J. Madsen. Task mapping and bandwidth reservation for mixed hard/soft fault-tolerant embedded systems. In *RTAS*, 2010.

[18] U. Schiffel, A. Schmitt, M. Süßkraut, and C. Fetzer. Software-implemented hardware error detection: Costs and gains. In *Third International Conference on Dependability*, 2010.

[19] L. Thiele, I. Bacivarov, W. Haid, and K. Huang. Mapping applications to tiled multiprocessor embedded systems. In *ACSD*, 2007.

[20] B. Zhao, H. Aydin, and D. Zhu. Enhanced reliability-aware power management through shared recovery technique. In *ICCAD*, 2009.

[21] D. Zhu and H. Aydin. Energy management for real-time embedded systems with reliability requirements. In *ICCAD*, 2006.

[22] D. Zhu and H. Aydin. Reliability-aware energy management for periodic real-time tasks. *IEEE Trans. Computers*, 2009.

APPENDIX

A.   RELATED WORK

In the past decades, much research effort has been devoted to fault-tolerant system design considering transient faults. *Girault et al* [4] combine task scheduling with active spatial redundancy and present a bicriteria heuristic algorithm. Beside scheduling parameters, their algorithm also determines the number of replications that are needed to achieve certain reliability goal. *Izosimov et al* [8] study the design of fault-tolerance systems using both spatial and temporal redundancy. In particular, the technique of sharing re-execution slack among multiple tasks is proposed to improve the efficiency. A tabu-search based optimization procedure is used to find the best schedule with scheduling length being the optimization goal. In [15] *Pop et al* study a similar problem and consider in addition the utilization of check-pointing and roll-back technique. The authors in [17] utilize a hybrid scheduling approach to handle mixed hard and soft real-time tasks. The aforementioned work [8, 15, 17] is based on a simplified fault model. Instead of modeling faults as probabilistic events, they assume that the system may experience at most N faults and those faults may occur in any component of the system. In the follow-up work [7], a more accurate probabilistic analysis is presented. Nevertheless, this analysis considered only temporal redundancy. *Huang et al* further extend the approach and propose a binary tree based approach for probabilistic reliability analysis considering both spatial/temporal redundancy and shared re-execution slack.

Other work also studies the tradeoff between reliability and other design objectives, such as energy [22] and cost [14]. In [16] the authors present a Constraint Logical Programming (CLP) based approach for scheduling and voltage scaling for fault-tolerance systems. *Zhu et al* show that voltage scaling has direct and adverse effects on system reliability [21]. They study static scheduling approaches for energy minimization under reliability constraints [22]. The core idea is, instead of using all available slack time for energy management, a portion of the slack is especially reserved to schedule task re-executions, such that the reliability loss can be recuperated.

In all the work mentioned above, perfect fault detection is assumed. The assumption is that, all faults can be detected when a task is completed and timing overhead of fault detection is contained in the WCETs of tasks. In this paper, we show that certain configuration using imperfect fault detectors combined with replication can outperform those approaches that assumes perfect fault detection.

B.   EXPERIMENTAL ANALYSIS ON THE IMPACT OF IMPERFECT FAULT DETECTION ON SYSTEM RELIABILITY

Figure 7 summarizes the results of the first simulation. We increase the fault detection coverage from 1% to 100% with a step width of 1% while fixing the number of replications. Figure 7a shows the case that a single instance is scheduled. As expected, the probability of SDC decreases linearly with the detection coverage, since all detected faults are converted to $DUFs$. In Figure 7b, two replicas are scheduled. The probabilities of both SDC and DUF decrease with higher coverage. The reason is

as follows: In general, adding redundant components is a recovery technique that migrates faults from the DUF to the DTF class and implementing fault detectors is a detection technique that migrates faults from the SDC to the DUF class. However, if used together, the effects of redundancy and that of fault detection become *correlated*. As an example, assume the first instance generates a correct output whereas the second one encounters a fault. If the fault is undetected, the second one will produces a faulty output. Since the voter cannot distinguish which of the two outputs is correct, the system results in DUF. As the counterpart, if the fault is detected, the faulty instance can fail-silent and the only (and correct) output from the first instance is taken, resulting in a successful scenario. Hence, besides converting $SDCs$ to $DUFs$, fault detection can also convert $DUFs$ to $DTFs$ if voting is available. For this reason, probabilities of both DUF and SDC decrease.

If three replicas are utilized (Figure 7c), the DUF probability first increases and then decreases with higher coverage, whereas the probability of SDC decreases constantly. The reason is that, the effect of SDC-to-DUF dominates when the coverage is still low (upper part of the Figure 7c), and the effect of DUF-to-DTF dominates when the coverage is relatively high. An observation from this set of simulations is that, higher fault detection coverage reduces the amount of $SDCs$ but not necessarily reduces the amount of $DUFs$.

In the second simulation, we increase the number of replications while fixing the detector implementation. We reuse the result of [18] and assume that the rate of undetectable faults decreases exponentially with linear fault detection effort. We further assume the perfect fault detection (100% coverage) incurs 300% timing overhead (typical value in [18]). Several fault detectors with timing overhead ranging from 0% to 300% (corresponds to detection coverage from 0% to 100%) are tested. Figure 8 summarizes the results[5]. As can be seen, when the detection coverage is low, the probability curve shows a zigzag behavior with increasing number of replications (e.g., curve a). This is because the task instances themselves have only poor fault detection and the system relies mainly on the voter to discover the faults. On the one hand, the voter detects a fault when the number of correct and incorrect results breaks even. Hence, when we increment the number of replicas from an odd number and make it even (e.g., from 1 to 2), the fault detection capability of the voter is enhanced, resulting in a reduction of undetected faults (SDC probability drops). On the other hand, the voter recovers a fault when correct results are the majority. Hence, when a new instance is added to an even number of replicas, the amount of recoverable faults increases (DUF probability drops).

As the counterpart, if the task instances have already good fault detectors (e.g., in the case of curve d), the system reliability will be improved more smoothly by inserting more redundancy, i.e., both $DUFs$ and $SDCs$ can be eliminated at the same time. In other words, the effect of active redundancy could be amplified by good fault detection.

C.   ILP BASED OPTIMIZATION FOR SINGLE-OBJECTIVE CASE

---

[5]The figure excludes the case of 0% and 300% by intension, because some of the probabilities are 0 and hard to be visualized in logarithmic scale.

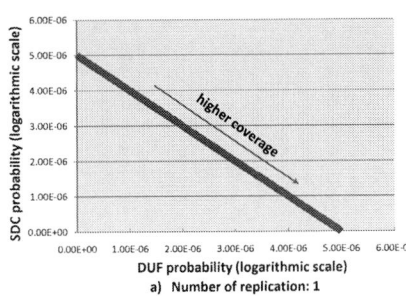
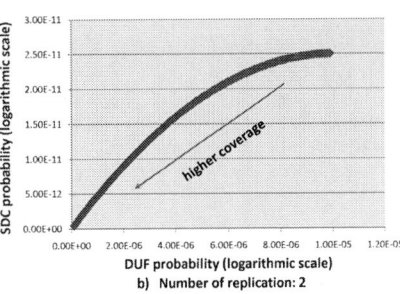
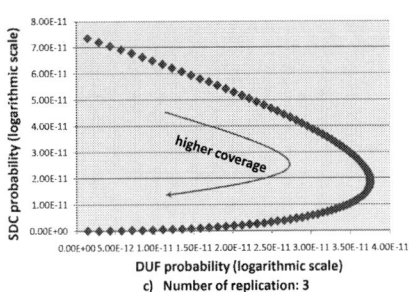

a) Number of replication: 1   b) Number of replication: 2   c) Number of replication: 3

Figure 7: Effect of Fault Detection with Fixed Replication

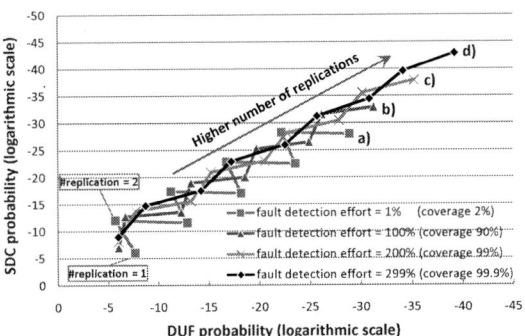

Figure 8: Effect of Replication with Fixed Fault Detection Coverage

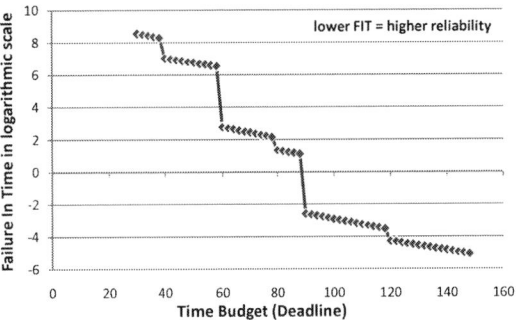

Figure 9: Example of Reliability Function

As mentioned in Section 5, this section presents an ILP based solution to handle the design scenario of maximizing the reliability of a single application. A real-time application typically has an end-to-end deadline that represents the time budget B for the entire application. The total budget can be distributed to individual tasks so that each task t_i has a local deadline b_i. The maximum reliability that can be archived by a task is constrained by the available local time budget. To describe this relationship, we define a Reliability Function (RF) $U_i(b)$, which is a monotonic function that models the achievable reliability of task t_i with given time budget b. Figure 9 depicts an example RF. The metric for reliability is Failure In Time (FIT). To capture both $DUFs$ and $SDCs$, we define the value of $U_i(b)$ to be a weighted sum of the FIT of both fault classes, i.e.:

$$U_i(b_i) = \alpha FIT_{DUF}(b_i) + \beta FIT_{SDC}(b_i) \quad (9)$$

The weighting factors represent the criticality of the type of fault for the application. The RF for a task t_i can be obtained as follows. The possible time budget b_i assigned to t_i is lower-bounded by its execution time and upper-bounded by the available system slack time, i.e., $b_i \in [C_i, C_i + B - \sum_{\forall j} C_j]$. We sample this range with a fixed step width. For each sample value b, we investigate all design alternatives that fit into b, i.e., we try different numbers of replications and all implementable fault detectors[6]. For each design, the DUF and SDC probabilities are analyzed using equation

[6]This procedure is durable since the number of alternatives is very limited. On the one hand, the number of replications for a single task is typically very small. On the other hand,

1-3 and the reliability is evaluated by equation 9. We assign the value of the $U(b)$ to be highest achievable reliability under the budget constraint.

After having the reliability function for all tasks, we can now compute the system reliability. The system-level SDC probability can be computed using equation 6. Since the success probabilities $P_{SUC}(\mathcal{T}\backslash t_0)$ and $P_{SUC}(t_0)$ are typically very close to 1, we approximate equation 6 as follows:

$$P_{SDC}(\mathcal{T}) < P_{SDC}(t_0) + P_{SDC}(\mathcal{T}\backslash t_0) < \dots$$
$$= \sum_i P_{SDC}(t_i)$$

As can be seen, the system-level SDC probability can be overestimated by summarizing the SDC probabilities of all tasks. It can easily be verified that the system FIT can also be computed in an additive manner from the tasks' FITs. Similar approximation exists for the DUF probability. Let $\vec{b}$ be a vector that contains the timing budget for each task. The system reliability can be approximated as $U_{sys}(\vec{b}) = \sum_{t_i \in \mathcal{T}} U_i(b_i)$. The optimization problem becomes a deadline assignment problem stated as follows:

$$Minimize : U_{sys}(\vec{b}) = \sum_{t_i \in \mathcal{T}} U_i(b_i),$$
$$Subject\ to : \sum_{b_i \in \vec{b}} b_i \leq B \quad (10)$$

since fault coverage increases monotonically with detection effort, we can simply choose the best detector that fits into the budget. When the complexity is still too high, methods like Monte Carlo simulation can be used to approximate the RF.

195

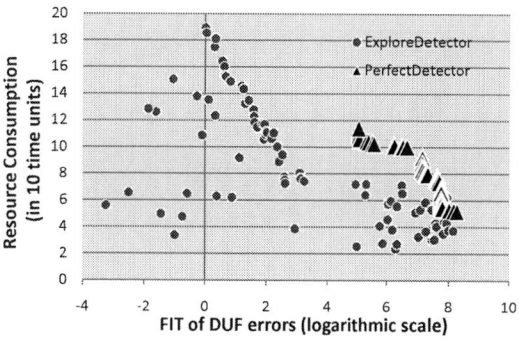

Figure 10: 2D Projection of Optimization Results: FIT of DUF vs Resource Consumption

By restricting the local time budget of each task to be a set of discrete values (as what is done to sample the RF), the above problem can be transformed into an integer linear programming problem and solved using standard solvers. Assume that M samples in the RF are considered for each local deadline value, i.e. $b_i \in \{b_{i,1}, ..., b_{i,M}\}$. We define a set of binary variables to describe the assignment of b_i:

$$x_{i,m} = \begin{cases} 1 & \text{iff } b_i \text{ is assigned to the } m\text{th sample } b_{i,m} \\ 0 & \text{otherwise} \end{cases}$$

Obviously, b_i can only be assigned to exactly one sampling value:

$$\sum_{m \in [1,M]} x_{i,m} = 1, \ \forall t_i \in \mathcal{T}.$$

The actual value of b_i can then be denoted as:

$$b_i = \sum_{m \in [1,M]} x_{i,m} b_{i,m}.$$

The actual reliability of the task i is:

$$u_i = \sum_{m \in [1,M]} x_{i,m} U_i(b_{i,m}).$$

The ILP problem can be stated as:

$$\begin{aligned} Minimize &: \sum_{t_i \in \mathcal{T}} u_i, \\ Subject\ to &: \sum_{t_i \in \mathcal{T}} b_i \leq B \end{aligned} \quad (11)$$

The ILP formulation consists of $M|\mathcal{T}|$ binary variables (the x variables) and $2|\mathcal{T}|$ integer variables (for the b and u variables).

D. EXTENDED EXPERIMENTAL RESULTS

As mentioned in Section 6, this section presents extended experimental results. In Figure 6, the performance of two approaches is compared. For the *PerfectDetector* approach, the FIT of *SDC* can be kept relatively low due to good detection coverage. However, the FIT of *DUF* is always beyond 10^5. Using the *ExploreDetector* approach, we can obtain a wider spectrum of solutions, from the one that achieves very low FIT of *DUF* (C in Figure 6) to the one that achieves very low FIT of *SDC* (F in Figure 6). The designers can select the best implementation according to the application requirements.

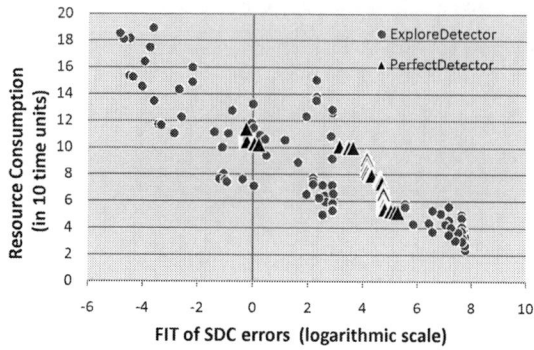

Figure 11: 2D Projection of Optimization Results: FIT of SDC vs Resource Consumption

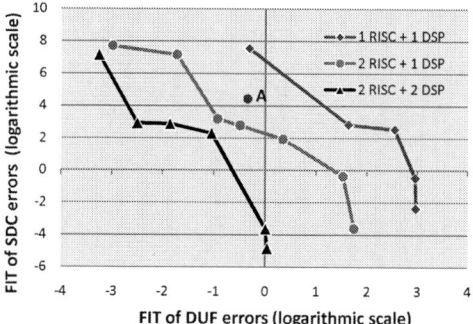

Figure 12: Comparing Results of three Architectures

The optimization results from MOEA can also be viewed from different angles. In Figure 10, the results are projected to in a 2D plane considering the FIT of *DUF* and resource consumption. Similarly, Figure 11 focuses on the FIT of *SDC* and resource consumption. Clearly, for both cases, the solutions found by *ExploreDetector* have better quality than those found by *PerfectDetector*. Concerning *SDC*, the performance of *PerfectDetector* is relatively close to that of *ExploreDetector*. Nevertheless, the performance gap is significant considering *DUF*. In this sense, the main drawback of the *PerfectDetector* approach is that the design objective is biased. It fails to take application-specific reliability requirements into account. Instead, the focus is always on reducing the *SDC*s. For many applications (e.g. those requires fail-operational behavior), this is certainly a suboptimal approach.

The propose approach can be used to perform reliability-aware design space exploration. To show the DSE flow, we apply the approach on several platforms consisting of 2 to 4 processors. In Figure 12, we compare the maximum achievable reliability using different architectures. Clearly, the solutions found using a larger architecture dominate those obtained using a smaller architecture, due to the possibility of implementing more replications and/or better detectors. From these results, the designer may choose the best platform that meets the application requirement. For example, if the reliability goal is point A in Figure 12, the $2RISC + 1DSP$ platform is the cheapest one adhering to the requirement.

Steady-State Dynamic Temperature Analysis and Reliability Optimization for Embedded Multiprocessor Systems

Ivan Ukhov	Min Bao	Petru Eles	Zebo Peng
Linköping University	Linköping University	Linköping University	Linköping University
Sweden	Sweden	Sweden	Sweden
ivan.ukhov@liu.se	min.bao@liu.se	petru.eles@liu.se	zebo.peng@liu.se

ABSTRACT

In this paper we propose an analytical technique for the steady-state dynamic temperature analysis (SSDTA) of multiprocessor systems with periodic applications. The approach is accurate and, moreover, fast, such that it can be included inside an optimization loop for embedded system design. Using the proposed solution, a temperature-aware reliability optimization, based on the thermal cycling failure mechanism, is presented. The experimental results confirm the quality and speed of our SSDTA technique, compared to the state of the art. They also show that the lifetime of an embedded system can significantly be improved, without sacrificing its energy efficiency, by taking into consideration, during the design stage, the steady-state dynamic temperature profile of the system.

Categories and Subject Descriptors

C.3 [**Special-Purpose and Application-Based Systems**]: Microprocessor/microcomputer applications, real-time and embedded systems; G.1.3 [**Numerical Linear Algebra**]: Sparse, structured, and very large systems; G.3 [**Probability and Statistics**]: Reliability and life testing; J.6 [**Computer-Aided Engineering**]: Computer-aided design.

General Terms

Algorithms, Design, Performance, Reliability.

Keywords

Multiprocessor System, Periodic Power Profile, Temperature Analysis, Leakage Power, Thermal Cycling Fatigue.

1. INTRODUCTION AND PRIOR WORK

Due to increasing power densities, temperature has evolved into a major concern for designers of modern embedded systems. Thus, temperature analysis has become an important component of current embedded system design frameworks.

Temperature-aware system-level design methods rely on the availability of temperature modeling and analysis tools. System-level temperature modeling approaches are mostly based on the duality between heat transfer and electrical phenomena [1]. The basic idea is to build an equivalent circuit of thermal resistances and capacitances capturing both the architecture blocks and elements of the thermal package. HotSpot [2], an architecture and system-level model and simulator, is the state of the art choice for system-level temperature analysis, as in [3, 4, 5, 6, 7, 8, 9].

However, temperature analysis time with HotSpot, or other similar approaches, is too long to be used inside a temperature-aware system-level optimization loop. The long thermal simulation time can severely limit the efficiency of the design space exploration. There has been some work on establishing fast system-level temperature analysis techniques. They also build on the duality between heat transfer and electrical phenomena and are based on restrictive assumptions in order to simplify the model. The approaches proposed in [10, 11], for example, are strictly restricted to monocore systems. The method described in [12] is restricted to homogeneous platforms and to applications in which the execution time of individual tasks is long, comparable with the thermal time constant of the package (in the order of 100 s).

Broadly speaking, there are two types of thermal analysis: (1) static temperature analysis, that produces a hypothetical temperature value (the steady-state temperature) at which the circuit is supposed to function if running for a long time under a certain constant (average) input power; (2) dynamic temperature analysis, that produces a transient temperature curve, describing the temperature behavior of the circuit as a function of time, when exposed to an arbitrary power profile.

The steady-state temperature, as produced by static analysis, is an approximation of the thermal behavior with limited applicability. It assumes that, eventually, the circuit will function at one constant temperature. This, however, is very often not the case in reality. In the context of a variable power profile applied periodically, the circuit will not reach a constant steady-state temperature but a steady state in which temperature is varying according to a certain periodic pattern. This pattern is captured by the steady-state dynamic temperature profile (SSDTP).

A typical design task, for which the SSDTP is of central importance, is temperature-aware reliability optimization. The impact of temperature on the lifetime of electronic circuits is well-known [3, 5, 8, 13]. The failure mechanisms commonly considered are electromigration, time-dependent dielectric breakdown, and thermal cycling, which are directly driven by the temperature [3]. What is important in this context is that not only average and maximum temperature, but also the amplitude and frequency of temperature oscillations, have a huge impact on the overall lifetime of the chip. Thus, efficient reliability optimization depends on the availability of the actual SSDTP.

Two approaches have been applied in the literature in order to obtain the SSDTP, as a prerequisite for reliability optimization. An approximate SSDTP can be produced by running a temperature simulator over one or more successive periods of the application until one can assume that a sufficient approximation of the thermal steady state has been reached [3]. Such an approach is both time consuming and potentially inaccurate. A very rough but fast approximation of the SSDTP is proposed in [7]. It constructs a stepwise temperature curve where each step corresponds to the static steady-state temperature that would be reached if a certain constant power was applied for a sufficiently long time. In Sec. 4 we will further elaborate on these two state of the art solutions. As our experiments show, they are too slow and/or too inaccurate in order to efficiently be used inside a temperature-aware system-level optimization loop for, e.g., reliability optimization.

In this paper we consider multiprocessor systems running applications exhibiting a power profile that can be considered periodic. Our contribution is twofold. First, we propose an approach that is both accurate and fast, for SSDTP calculation. Second, we show how our approach makes it possible to efficiently perform reliability optimization, based on the thermal cycling (TC) failure mechanism. More exactly, we propose a temperature-aware task mapping and scheduling technique that addresses the TC ageing effect. Experiments demonstrate the superiority of the proposed techniques, compared to the state of the art.

The rest of the paper is organized as follows. In Sec. 2 we introduce the architecture, power, and thermal models. The problem formulation is given in Sec. 3. The state of the art solutions are discussed in Sec. 4. In Sec. 5 we obtain an analytical formulation for the SSDTP calculation. In Sec. 6 and Sec. 7 we propose a fast and accurate technique to compute the SSDTP. Sec. 8 formulates the temperature-aware reliability optimization problem and proposes a solution based on our fast SSDTP calculation. Experimental results are given in Sec. 9 and Sec. 10 concludes the paper. Supplementary materials are given in the appendix, Sec. S1–S3.

2. ARCHITECTURE, POWER, AND THERMAL MODELS

We consider a heterogeneous multicore architecture with a set of processing elements Π defined as the following:

$$\Pi = \{\pi_i = (V_i, f_i, N_{\text{gate } i}) : i = 0, \ldots, N_p - 1\}$$

where V_i, f_i, and $N_{\text{gate } i}$ are the supply voltage, frequency, and number of gates [4] of the ith core, respectively.

The total power dissipation of a processing element is defined as the sum of the dynamic and leakage power: $P = P_{\text{dyn}} + P_{\text{leak}}$. The dynamic part is modeled as $P_{\text{dyn}} = C_{\text{eff}} \cdot f \cdot V^2$ where C_{eff} is the effective switched capacitance, V and f are the supply voltage and frequency, respectively. The leakage part of the power dissipation is defined as [4]:

$$P_{\text{leak}}(T) = N_{\text{gate}} V I_0 \left[A T^2 e^{\frac{\alpha V + \beta}{T}} + B e^{(\gamma V + \delta)} \right] \quad (1)$$

where T and V are the current temperature and supply voltage, respectively, N_{gate} is the number of gates in the circuit, I_0 is the average leakage current at the reference temperature and supply voltage. A, B, α, β, γ, and δ are the technology-dependent constants found in [4].

Our proposed technique is based on the RC thermal model that employs the analogy between electrical and thermal circuits [1]. Heat transfer is modeled with the following system of differential equations:

$$\mathbf{C} \frac{d\mathbf{T}(t)}{dt} + \mathbf{G} \left(\mathbf{T}(t) - \mathbf{T}_{\text{amb}} \right) = \mathbf{P}(t) \quad (2)$$

where $\mathbf{T}$ is the temperature vector, $\mathbf{T}_{\text{amb}}$ is the ambient temperature vector, $\mathbf{C}$ is the thermal capacitance matrix, $\mathbf{G}$ is the thermal conductance matrix, and $\mathbf{P}$ is the power dissipation vector. The dimensions of the system are $N_n \times N_n$, where N_n is the number of nodes in the equivalent RC thermal circuit, which is further discussed in Sec. S1.

3. PROBLEM FORMULATION

Consider a multicore system that consists of N_p processing elements $\Pi = \{\pi_i : i = 0, \ldots, N_p - 1\}$ and executes a periodic application with a period τ. We construct an equivalent RC thermal circuit of the system that contains N_n thermal nodes. The dynamic power profile of the system is sampled into N_s time intervals of duration Δt, called sampling interval, in such a way that the dynamic power dissipation and temperature of each node are assumed to be constant within an interval. The discrete dynamic power profile is defined as the following:

$$\mathbb{P}_{\text{dyn}} \overset{\text{def}}{=} \{P_{ij} : i = 0, \ldots, N_s - 1; j = 0, \ldots, N_n - 1\}$$

where P_{ij} is the dynamic power dissipation during the ith time interval of the jth thermal node. After the steady state is reached, the corresponding temperature profile becomes periodic and is defined as:

$$\mathbb{T} \overset{\text{def}}{=} \{T_{ij} : i = 0, \ldots, N_s - 1; j = 0, \ldots, N_n - 1\}$$

where T_{ij} is the temperature of the jth node in the ith time interval. The profile is called the steady-state dynamic temperature profile (SSDTP).

Given:

○ A multicore system with a set of processing elements Π executing a periodic application.

○ The discrete dynamic power profile $\mathbb{P}_{\text{dyn}}$ of the system[1] with the sampling interval Δt.

○ The floorplan of the chip corresponding to the level of details at which the thermal modeling is performed.

○ The configuration of the thermal package, i.e., dimensions of the thermal interface material, heat spreader, and heat sink.

○ The thermal parameters of the die and package, e.g., the thermal conductivity and thermal capacitance.

Find:

○ The corresponding periodic temperature profile $\mathbb{T}$ of the system when the steady state is reached.

---

[1]Power dissipation of inactive nodes, i.e., the nodes that belong to the thermal package, is zero.

4. STATE OF THE ART SOLUTIONS

4.1 Iterative Simulation

A rough approximation of the SSDTP can be obtained by running a temperature simulation over successive periods of the application until it can be assumed that the system has reached the thermal steady state. The simulator performs the transient temperature analysis where the common approach is to solve Eq. (2) numerically, for instance, using the fourth-order Runge-Kutta method [14].

The number of iterations required to reach the SSDTP depends on the thermal characteristics of the system. In order to illustrate this aspect, we have considered an application with the period of 0.5 s running on five hypothetical platforms with core areas between 1 and 25 mm^2. The configuration of the die and thermal package can be found in the appendix (Tab. S1). We have run the temperature simulation with HotSpot [2] for 50 successive periods. The temperature profile in each period has been compared with the actual SSDTP, obtained with our analytical approach (Sec. 6), and the normalized root mean square error (NRMSE) has been calculated. The result is shown in Fig. 1a. It can be observed that the number of successive periods over which the temperature simulation has to be performed, in order to achieve a satisfactory level of accuracy, is significant for the majority of configurations. For a 9 mm^2 die, for example, after 15 iterations, the NRMSE is still close to 20%. This leads to large computation times, making it difficult to apply the technique inside an intensive optimization loop.

4.2 Steady-State Approximation (SSA)

An approximation of the SSDTP has been proposed in [7]. Instead of solving the system of equations in Eq. (2), it is assumed that during each time interval Δt_i, in which the power is constant, the system stays in its steady state. The derivative $d\mathbf{T}/dt = 0$ and temperature can be calculated as $\mathbf{T}_i = \mathbf{G}^{-1}\mathbf{P}_i$. The result is a stepwise temperature curve where each step corresponds to the steady-state temperature $\mathbf{T}_i$ that would be reached if the constant power $\mathbf{P}_i$ was applied for a sufficiently long time.

An example of such an approximation (SSA) along with the corresponding SSDTP for an application with 10 tasks and period of 0.1 s is given in Fig. 1b. The die area is 25 mm^2, the configuration of the chip is the same as in Tab. S1. The reduced accuracy of the SSA is due to the mismatch between the actual temperature within each interval Δt_i and the hypothetical steady-state temperature. The inaccuracy depends on the thermal characteristics of the respective platform and on the application itself. To illustrate this, we have generated five applications with periods between 0.01 and 1 s and computed approximated SSDTPs using the SSA for die areas between 1 and 25 mm^2. The NRMSE relative to the correct SSDTP is shown in Fig. 1c. It can be seen that, e.g., for a die area of 10 mm^2 and a period of 100 ms the NRMSE with the SSA is close to 40%.

5. ANALYTICAL SOLUTION

As shown in Sec. 4, the state of the art solutions either produce inaccurate and, in many cases, completely useless results, or they are unacceptably slow. In this section we eliminate the first problem by obtaining an analytical solution for the SSDTP and tackle the second one in Sec. 6 where a fast solution technique is proposed.

In the following explanation, without loss of generality, we assume $\mathbf{T}(t) \equiv \mathbf{T}(t) - \mathbf{T}_{\text{amb}}$. Let the power consumption vector $\mathbf{P}(t)$ be constant and equal to $\mathbf{P}$; then the system given by Eq. (2) is a system of ordinary differential equations (ODE) with the following solution:

$$\mathbf{T}(t) = e^{\mathbf{A}t} \mathbf{T}_0 + \mathbf{A}^{-1}(e^{\mathbf{A}t} - \mathbf{I}) \mathbf{C}^{-1}\mathbf{P} \quad (3)$$

where $\mathbf{A} = -\mathbf{C}^{-1} \mathbf{G}$, $\mathbf{T}_0$ is the initial temperature and $\mathbf{I}$ is the identity matrix. Therefore, given a discrete power profile, the corresponding temperature profile can be found using the following recurrence:

$$\mathbf{T}_{i+1} = \mathbf{K}_i \mathbf{T}_i + \mathbf{B}_i \mathbf{P}_i \quad (4)$$

where $\mathbf{K}_i = e^{\mathbf{A}\Delta t_i}$ and $\mathbf{B}_i = \mathbf{A}^{-1}(e^{\mathbf{A}\Delta t_i} - \mathbf{I})\mathbf{C}^{-1}$. The approach can be used to perform the TTA as it is discussed in the appendix (Sec. S2.1).

198

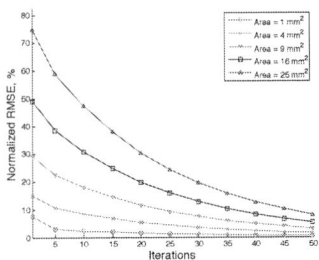

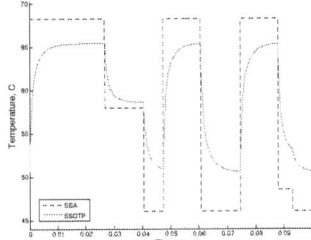

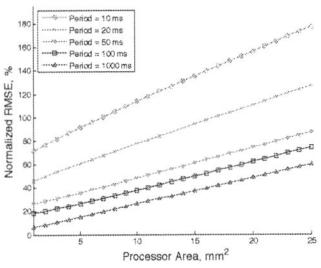

(a) NRMSE with HotSpot. (b) SSA and real SSDTP. (c) NRMSE with SSA.

Figure 1: State of the art solutions.

For the SSDTA calculation the following system of linear equations can be derived from Eq. (4):

$$\begin{cases} \mathbf{K}_0\,\mathbf{T}_0 - \mathbf{T}_1 & = -\mathbf{B}_0\,\mathbf{P}_0 \\ \dots \\ -\mathbf{T}_0 + \mathbf{K}_{N_s-1}\,\mathbf{T}_{N_s-1} & = -\mathbf{B}_{N_s-1}\,\mathbf{P}_{N_s-1} \end{cases}$$

where the last equation enforces the boundary condition, the equality of temperature values on both ends of the period:

$$\mathbf{T}_0 = \mathbf{T}_{N_s} \qquad (5)$$

To get the whole picture, the system can be written as:

$$\underbrace{\begin{bmatrix} \mathbf{K}_0 & -\mathbf{I} & 0 & \cdots & 0 \\ 0 & \mathbf{K}_1 & -\mathbf{I} & & \vdots \\ \vdots & & \ddots & -\mathbf{I} & 0 \\ 0 & & & \mathbf{K}_{N_s-2} & -\mathbf{I} \\ -\mathbf{I} & 0 & \cdots & 0 & \mathbf{K}_{N_s-1} \end{bmatrix}}_{\mathbb{A}} \underbrace{\begin{bmatrix} \mathbf{T}_0 \\ \vdots \\ \vdots \\ \mathbf{T}_{N_s-1} \end{bmatrix}}_{\mathbb{X}} = \underbrace{\begin{bmatrix} -\mathbf{B}_0\,\mathbf{P}_0 \\ \vdots \\ \vdots \\ -\mathbf{B}_{N_s-1}\,\mathbf{P}_{N_s-1} \end{bmatrix}}_{\mathbb{B}} \quad (6)$$

where $\mathbb{A}$ is a $N_n N_s \times N_n N_s$ matrix, $\mathbb{X}$ and $\mathbb{B}$ are vectors with $N_n N_s$ elements. It can be seen that we have obtained a regular system of linear equations. Straight-forward techniques to solve it and their disadvantages are further discussed in the appendix (Sec. S2.2).

6. PROPOSED TECHNIQUE

In this section we propose a fast approach to solve the system in Eq. (6). The approach consists of an auxiliary transformation (Sec. 6.1) and the actual solution (Sec. 6.2).

The major problem with straight-forward techniques (see Sec. S2.2) is that (1) the sparseness of the matrix is not taken into account and/or (2) its specific structure is totally ignored, resulting in inefficiency and inaccuracy of the computations. Using direct dense and sparse solvers, for example, requires a computation time proportional to $N_n^3 N_s^3$ [14]. Our proposed technique considers both features and delivers solutions in time proportional to $N_s N_n^3$ while operating only on a few $N_n \times N_n$ matrices. It is important that the dependency on N_s (the number of steps in the power profile), which is by far dominating ($N_s \gg N_n$), is linear.

Observing the structure of the matrix in Eq. (6), non-zero elements are located only on the block diagonal, on one subdiagonal just above the block diagonal, and on one subdiagonal in the left bottom corner. The block diagonal is composed of $N_n \times N_n$ matrices while all elements of the subdiagonals are equal to -1. Linear systems with the same structure arise in boundary value problems for ODEs where a technique to solve them is to form a so-called condensed equation (CE), or condensed system [15].

6.1 Auxiliary Transformation

The analytical solution in Eq. (3) includes two computationally expensive operations, namely the matrix exponential and inverse involving $\mathbf{A} = -\mathbf{C}^{-1}\,\mathbf{G}$, which is an arbitrary square matrix. It is preferable to have a symmetric matrix to perform these computations, since for a real symmetric matrix $\mathbf{M}$ the following eigenvalue decomposition with independent eigenvectors holds [14]:

$$\mathbf{M} = \mathbf{U}\mathbf{\Lambda}\mathbf{U}^{\mathbf{T}} \qquad (7)$$

where $\mathbf{U}$ is a square matrix of the eigenvectors, $\mathbf{U}^T$ is the transpose of $\mathbf{U}$, and $\mathbf{\Lambda}$ is a diagonal matrix of the eigenvalues

λ_i of $\mathbf{M}$. With such a decomposition, the calculation of the matrix exponential and inverse becomes trivial: $e^{\mathbf{M}} = \mathbf{U}\,e^{\mathbf{\Lambda}}\,\mathbf{U}^{\mathbf{T}}$ and $\mathbf{M}^{-1} = \mathbf{U}\,\mathbf{\Lambda}^{-1}\,\mathbf{U}^{\mathbf{T}}$, where the central matrices are diagonal with elements e^{λ_i} and λ_i^{-1}, respectively.

The conductance matrix $\mathbf{G}$ is a symmetric matrix, since if a node A is connected to B, then B is also connected to A with the same conductance. However, as it is mentioned previously, the product of $\mathbf{G}$ with the inverse of the capacitance matrix $\mathbf{C}$ does not have this property. Since $\mathbf{C}$ is a diagonal matrix, we use the following transformation in order to keep the desired symmetry:

$$\tilde{\mathbf{T}}(t) = \mathbf{C}^{\frac{1}{2}}\mathbf{T}(t) \qquad \tilde{\mathbf{A}} = -\mathbf{C}^{-\frac{1}{2}}\,\mathbf{G}\,\mathbf{C}^{-\frac{1}{2}} \qquad (8)$$

where $\tilde{\mathbf{A}}$ is symmetric[2]. Consequently, the system of ODEs (Eq. (2)) and its solutions (Eq. (3)) can be rewritten as the following:

$$\frac{d\tilde{\mathbf{T}}(t)}{dt} = \tilde{\mathbf{A}}\,\mathbf{Y}(t) + \mathbf{C}^{-\frac{1}{2}}\mathbf{P}$$

$$\tilde{\mathbf{T}}(t) = e^{\tilde{\mathbf{A}}t}\tilde{\mathbf{T}}_0 + \tilde{\mathbf{A}}^{-1}(e^{\tilde{\mathbf{A}}t} - \mathbf{I})\mathbf{C}^{-\frac{1}{2}}\mathbf{P}$$

where $\tilde{\mathbf{A}}$ is a symmetric matrix. Therefore, in the case of, e.g., the matrix exponential we have:

$$e^{\tilde{\mathbf{A}}t} = \mathbf{U}\,e^{\mathbf{\Lambda}t}\,\mathbf{U}^T = \mathbf{U}\,\mathrm{diag}\left(e^{t\lambda_0}, \dots, e^{t\lambda_{N_n-1}}\right)\mathbf{U}^T \quad (9)$$

where $diag$ denotes a diagonal matrix and λ_i are the eigenvalues of $\tilde{\mathbf{A}}$. A similar equation can be obtained for $\tilde{\mathbf{A}}^{-1}$.

The next step is to update the SSDTP system in Eq. (4):

$$\tilde{\mathbf{T}}_{i+1} = \tilde{\mathbf{K}}_i\,\tilde{\mathbf{T}}_i + \tilde{\mathbf{B}}_i\,\mathbf{P}_i \qquad (10)$$

$$\tilde{\mathbf{K}}_i = e^{\tilde{\mathbf{A}}\,\Delta t_i} \qquad \tilde{\mathbf{B}}_i = \tilde{\mathbf{A}}^{-1}\left(e^{\tilde{\mathbf{A}}\Delta t_i} - \mathbf{I}\right)\mathbf{C}^{-\frac{1}{2}}$$

Using the eigenvalue decomposition, the last equation can be computed in the following way:

$$\tilde{\mathbf{B}}_i = \mathbf{U}\,\mathrm{diag}\left(\frac{e^{\Delta t_i\,\lambda_0}-1}{\lambda_0}, \dots, \frac{e^{\Delta t_i\,\lambda_{N_n-1}}-1}{\lambda_{N_n-1}}\right)\mathbf{U}^T\,\mathbf{C}^{-\frac{1}{2}}$$

6.2 Solution with Condensed Equation (CE)

In the recurrence given by Eq. (10) we denote $\mathbf{Q}_i = \tilde{\mathbf{B}}_i\mathbf{P}_i$:

$$\tilde{\mathbf{T}}_{i+1} = \tilde{\mathbf{K}}_i\,\tilde{\mathbf{T}}_i + \mathbf{Q}_i,\ i = 0, \dots, N_s-1 \qquad (11)$$

$$\tilde{\mathbf{T}}_0 = \tilde{\mathbf{T}}_{N_s}$$

Performing the iterative repetition of Eq. (11) leads to:

$$\tilde{\mathbf{T}}_i = \prod_{j=0}^{i-1}\tilde{\mathbf{K}}_j\,\tilde{\mathbf{T}}_0 + \mathbf{W}_{i-1},\ i = 1, \dots, N_s \qquad (12)$$

where $\mathbf{W}_i$ are defined as follows:

$$\mathbf{W}_0 = \mathbf{Q}_0 \qquad \mathbf{W}_i = \tilde{\mathbf{K}}_i\,\mathbf{W}_{i-1} + \mathbf{Q}_i,\ i = 1, \dots, N_s-1 \quad (13)$$

We calculate the final vector $\tilde{\mathbf{T}}_{N_s}$ using Eq. (12) and Eq. (13):

$$\tilde{\mathbf{T}}_{N_s} = \prod_{j=0}^{N_s-1}\tilde{\mathbf{K}}_j\,\tilde{\mathbf{T}}_0 + \mathbf{W}_{N_s-1}$$

$^2\tilde{\mathbf{A}}^T = -(\mathbf{C}^{-\frac{1}{2}}\mathbf{G}\mathbf{C}^{-\frac{1}{2}})^T = -(\mathbf{C}^{-\frac{1}{2}})^T\mathbf{G}^T(\mathbf{C}^{-\frac{1}{2}})^T = \tilde{\mathbf{A}}.$

Taking into account the boundary condition given by Eq. (5), we obtain the following system of linear equations:

$$\left(\mathbf{I} - \prod_{j=0}^{N_s-1} \tilde{\mathbf{K}}_j\right) \tilde{\mathbf{T}}_0 = \mathbf{W}_{N_s-1} \qquad (14)$$

We recall that $\tilde{\mathbf{K}}_i$ is the matrix exponential given by Eq. (9); therefore, the following simplification holds:

$$\prod_{j=0}^{N_s-1} \tilde{\mathbf{K}}_j = \mathbf{U} \operatorname{diag}\left(e^{\tau\lambda_0}, \dots, e^{\tau\lambda_{N_n-1}}\right) \mathbf{U}^T$$

where τ is the application period. Substituting this product into Eq. (14), we obtain the following system:

$$\left(\mathbf{I} - \mathbf{U}\, e^{\tau\mathbf{\Lambda}}\, \mathbf{U}^T\right) \tilde{\mathbf{T}}_0 = \mathbf{W}_{N_s-1}$$

The identity matrix $\mathbf{I}$ can be split into $\mathbf{U}\mathbf{U}^T$, hence:

$$\tilde{\mathbf{T}}_0 = \mathbf{U}\left(\mathbf{I} - e^{\tau\mathbf{\Lambda}}\right)^{-1} \mathbf{U}^T \mathbf{W}_{N_s-1} = \mathbf{Z}\, \mathbf{W}_{N_s-1} \qquad (15)$$

where:

$$\mathbf{Z} = \mathbf{U} \operatorname{diag}\left(\frac{1}{1 - e^{\tau\lambda_0}}, \dots, \frac{1}{1 - e^{\tau\lambda_{N_n-1}}}\right) \mathbf{U}^T$$

The equation gives the initial solution vector $\tilde{\mathbf{T}}_0$; the rest of the vectors $\tilde{\mathbf{T}}_i$ are successively found from Eq. (11).

Since the power profile is evenly sampled with the sampling interval Δt, the recurrence in Eq. (11) turns into:

$$\tilde{\mathbf{T}}_{i+1} = \tilde{\mathbf{K}}\,\tilde{\mathbf{T}}_i + \mathbf{Q}_i = \tilde{\mathbf{K}}\,\tilde{\mathbf{T}}_i + \tilde{\mathbf{B}}\,\mathbf{P}_i$$

where $\tilde{\mathbf{K}} = e^{\tilde{\mathbf{A}}\,\Delta t}$ and $\tilde{\mathbf{B}} = \tilde{\mathbf{A}}^{-1}(e^{\tilde{\mathbf{A}}\,\Delta t} - \mathbf{I})\mathbf{C}^{-\frac{1}{2}}$. Here $\tilde{\mathbf{K}}$ and $\tilde{\mathbf{B}}$ are constants, since they depend on the matrices $\tilde{\mathbf{A}}$, $\mathbf{C}$, and sampling interval Δt, which is fixed. In this case, the block diagonal of the matrix $\tilde{\mathbb{A}}$, similar to Eq. (6), is composed of the same repeating block $\tilde{\mathbf{K}}$ and the recurrent expressions take the following form:

$$\mathbf{W}_i = \tilde{\mathbf{K}}\,\mathbf{W}_{i-1} + \mathbf{Q}_i,\ i = 1, \dots, N_s - 1 \qquad (16)$$

$$\tilde{\mathbf{T}}_{i+1} = \tilde{\mathbf{K}}\,\tilde{\mathbf{T}}_i + \mathbf{Q}_i,\ i = 0, \dots, N_s - 1 \qquad (17)$$

where $\mathbf{Q}_i = \tilde{\mathbf{B}}\,\mathbf{P}_i$, $\mathbf{W}_0 = \mathbf{Q}_0$, and $\tilde{\mathbf{T}}_0$ is given by Eq. (15).

The last step of the solution is to return to temperature by performing the backward substitution opposite to Eq. (8):

$$\mathbf{T}_i = \mathbf{C}^{-\frac{1}{2}}\,\tilde{\mathbf{T}}_i,\ i = 0, \dots, N_s - 1$$

As we see, the auxiliary substitution from Sec. 6.1 allows us to perform the single-time eigenvalue decomposition with orthogonal eigenvectors (Eq. (7)) that later eases the computational process at several stages. In Sec. 6.2 it can be observed that the solution of the system in Eq. (6) has been reduced to two successive recurrences in Eq. (16) and Eq. (17) over N_s steps in the power profile, which implies a linear complexity on N_s mentioned earlier.

It should be noted that the eigenvalue decomposition along with matrices $\tilde{\mathbf{K}}$ and $\tilde{\mathbf{B}}$ are computed only once for a particular RC thermal circuit and can be considered as given together with the RC circuit. It has not to be recalculated when a SSDTP is generated, which significantly decreases the computation time.

7. LEAKAGE POWER

So far, we have assumed that power is independent of temperature. However, due to the leakage component, the power dissipation is a strong function of temperature that cannot be neglected (Sec. 2). Two techniques can be applied to include in our proposed solution temperature-dependent leakage modeling.

7.1 Iterative Computation

In this case, we have an iterative process, depicted in Fig. 2, where the temperature and power profiles are calculated in turns. With each new temperature profile we update the power profile by computing the leakage power and adding it to the dynamic power: $\mathbb{P}_i = \mathbb{P}_{\text{dyn}} + \mathbb{P}_{\text{leak}}(\mathbb{T}_i)$. The process continues until the temperature converges, i.e., the

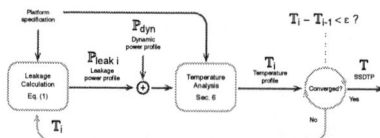

Figure 2: SSDTP with leakage modeling.

difference between two successive temperature profiles is below a predefined bound. In our experiments we used $0.5°C$ as the maximal acceptable difference and observed that the number of required iterations to converge is 4–7.

7.2 Linear Approximation

A linear approximation of the leakage power has the following matrix form: $\mathbf{P}_{\text{leak}}(\mathbf{T}) = \mathbf{A}\,\mathbf{T}(t) + \mathbf{B}$ where $\mathbf{A}$ is a $N_n \times N_n$ diagonal matrix of the proportionality and $\mathbf{B}$ is a vector with N_n elements of the intercept. Both characterize the leakage power for each of the N_n thermal nodes in the system. It can be seen that the approximation keeps Eq. (2) untouched: $\mathbf{C}\,\frac{d\mathbf{T}(t)}{dt} + \bar{\mathbf{G}}\,(\mathbf{T}(t) - \mathbf{T}_{\text{amb}}) = \bar{\mathbf{P}}$ where $\bar{\mathbf{G}} = \mathbf{G} - \mathbf{A}$ and $\bar{\mathbf{P}} = \mathbf{P}_{\text{dyn}} + \mathbf{A}\,\mathbf{T}_{\text{amb}} + \mathbf{B}$. Therefore, all solutions proposed in this paper are perfectly valid with the linearized model. Moreover, in spite of its simplicity, the model provides a good estimation, as shown in [6].

In order to evaluate the linearization, we have constructed a number of hypothetical platforms with 2–32 cores (other parameters are given in Tab. S1) and compared temperature profiles obtained with the linearization and the exponential model (Sec. 2), respectively. For the later, we use the iterative approach described in Sec. 7.1. For the linearization, the power curve fitting with the least squares regression [14] has been employed, targeted at the range between 40 and $80°C$. From the experiments we have observed that the NRMSE is bounded by 1–2%, indicating a good accuracy of the linear approximation.

8. RELIABILITY OPTIMIZATION

The proposed calculation of the SSDTA can be used in a wide range of system optimizations. One of them is reliability optimization that we discuss in this section. We perform a temperature-aware task mapping and scheduling in order to address the thermal cycling fatigue.

8.1 Application Model

The periodic application is modeled as a task graph $G = (V, E, \tau)$ where V is a set of N_t tasks (vertices of the graph), E is a set of data dependencies between tasks (edges), and τ is the period of the application, which we assume to be equal to the deadline. Each pair of a task $v_i \in V$ and processing element $\pi_j \in \Pi$ is characterized by a tuple $(N_{\text{clock}\,ij}, C_{\text{eff}\,ij})$, where $N_{\text{clock}\,ij}$ is the number of clock cycles and $C_{\text{eff}\,ij}$ is the effective switched capacitance.

8.2 Temperature-Aware Reliability Model

We address temperature-driven failure mechanisms with the reliability model presented in [7, 8]. In this paper, our particular focus is on the thermal cycling (TC) fatigue, which is directly connected to the temperature variations. The derivation of the model is given in the appendix (Sec. S3).

Assuming the TC fatigue, the parameters affecting reliability are the amplitude and number of thermal cycles as

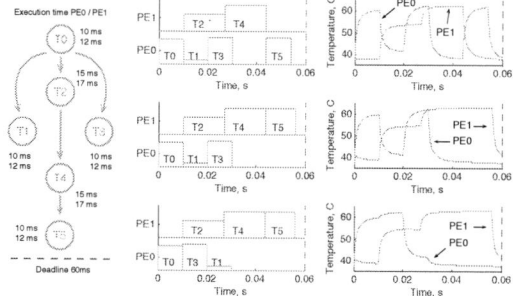

(a) Graph. (b) Mappings, schedules, SSDTPs.

Figure 3: Motivational example.

200

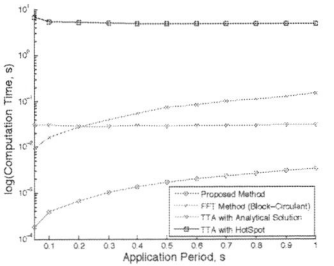

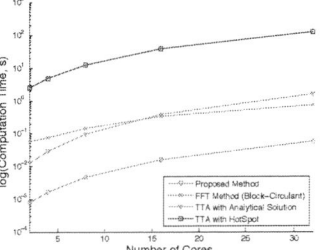

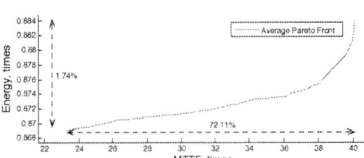

Figure 6: Average Pareto front.

Figure 4: Scaling with τ.   **Figure 5: Scaling with N_p.**

well as the maximal temperature. A thermal cycle is a time interval in which the temperature starts from a certain value and, after reaching an extremum, returns back.

The mean time to failure (MTTF) of one processing element in the system can be estimated as the following:

$$\theta = \frac{\tau}{\sum_{i=0}^{N_m - 1} \frac{1}{N_{c\,i}}} \qquad (18)$$

where N_m is the number of thermal cycles during the application period τ. $N_{c\,i}$ characterizes the ith thermal cycle and is calculated according to the following expression:

$$N_c = A(\Delta T - \Delta T_0)^{-b} e^{\frac{E_a}{kT_{\max}}} \qquad (19)$$

where ΔT is the thermal cycle excursion (the distance between the minimal and maximal temperatures) and $T_{\max}$ is the maximal temperature during the thermal cycle (more details in Sec. S3).

It can be seen that the computation requires the identification of the thermal cycles with their amplitudes and maximal temperatures. All these are captured by the SSDTP, which is needed as an input to the reliability optimization.

8.3 Motivational Example

Consider an application with six tasks, denoted "T0"–"T5", and a heterogeneous architecture with two cores, labeled "PE0" and "PE1". The task graph of the application is given in Fig. 3a along with the execution times for both cores. The period of the application is 0.06 s. A first alternative mapping and schedule, and the resulting SSDTP are shown at the top of Fig. 3b (where the height of a task represents its relative dynamic power consumption). It can be observed that initially PE0 is experiencing three thermal cycles. If we change the mapping of T5 and move it to PE1, we achieve two thermal cycles of PE0 instead of three. Finally, if we vary the schedule as well and change the order of T1 and T3, the number of cycles of PE0 becomes one. Using the reliability model from Sec. 8.2, we observe improvements in the lifetime of 44.69% and 54.53%, respectively, relative to the initial configuration.

8.4 Problem Formulation and Optimization

The problem formulation is the following:
Given:

○ A multiprocessor system Π (Sec. 2).
○ A periodic application G (Sec. 8.1).
○ The floorplan of the chip at the desired level of details, configuration of the thermal package, and thermal parameters.
○ The parameters of the reliability model (Sec. 8.2), i.e., the constants A, ΔT_0, b, E_a (see Eq. (19)).

Maximize:

$$\mathcal{F} = \min_{i=0}^{N_p - 1} \theta_i \quad (20)$$

s.t.
$$t_{\text{end}\,i} \leq \tau, \, \forall i \qquad (21)$$
$$T_{ij} \leq T_{\max}, \, \forall i, j \qquad (22)$$

where θ_i is the MTTF of the ith processing element given by Eq. (18), $t_{\text{end}\,i}$ denotes the end time of the ith task, τ is the period of the application, and T_{ij} are temperature values in the SSDTP. Eq. (21) imposes the application deadline, which we assume to be equal to the period. Eq. (22) enforces the constraint on the maximal temperature in the temperature profile $\mathbb{T} = \{T_{ij}\}$.

The optimization procedure is based on a genetic algorithm (GA) [16] with the fitness function $\mathcal{F}$ given by Eq. (20). The algorithm is outlined in Sec. S3.2.

9. EXPERIMENTAL RESULTS

9.1 SSDTP Calculation

In this subsection we investigate the scaling properties of the proposed solution for the SSDTP calculation and compare it with the approach based on the TTA with HotSpot (Sec. 4.1)[3]. We also include in the comparison two additional techniques described in the appendix, namely the TTA with the analytical solution (Sec. S2.1) and the fast Fourier transform (FFT) (Sec. S2.2). In the cases of the TTA, the simulation over successive iterations is run until the NRMSE relative to the SSDTP obtained with the proposed method is less than 1%.

In the following experiments, the power sampling interval is set to 1 ms and the thermal configuration of the die is the same as in Tab. S1. For the experiments in this subsection, the leakage power has not been considered. If considered according to the linearized model (Sec. 7.2), execution times remain unchanged; if considered according to the iterative model (Sec. 7.1), execution times increase proportionally for all the methods, which does not affect any of the conclusions.

First, we vary the application period τ keeping the architecture fixed, which is a quad-core platform with the core area of 4 mm^2. The comparison is depicted in Fig. 4 on a semilogarithmic scale. It can be seen that the proposed technique is roughly 5000 times faster than calculating the SSDTP by running the TTA with HotSpot and from 9 to 170 times faster than the TTA with the analytical solution.

In the second experiment we evaluate the scaling of the proposed method with regard to the number of processing elements. The application period is fixed to 0.5 s. The results are shown in Fig. 5. It can be observed that the proposed technique provides a significant performance improvement relative to the alternative solutions.

9.2 Reliability Optimization

In this section we evaluate the reliability optimization approach described in Sec. 8, first with a set of synthetic applications and, finally, using a real-life example.

The experimental setup is the following. Heterogeneous platforms and periodic applications are generated randomly [17] in such a way that the execution time of tasks is uniformly distributed between 1 and 10 ms and the leakage power accounts for 30–60% of the total power dissipation[4]. The linear leakage model is used in the experiments, since, as discussed in Sec. 7.2, it provides a good approximation. The area of one core is 4 mm^2, other parameters of the die and thermal package are given in Tab. S1. The temperature constraint $T_{\max}$ (see Eq. (22)) is set to $100°C$. In Eq. (19) the Coffin-Manson exponent b is set to 6, the activation energy E_a to 0.5, and the elastic temperature region ΔT_0 to zero [13]. The coefficient of proportionality A is not significant, since we are concerned about the relative improvement.

In each of the experiments, we compare the optimized solution with an initial temperature-aware solution proposed

---

[3]All the experiments are performed on a Linux machine with Intel® Core™ i7-2600 3.4GHz and 8Gb of RAM.
[4]The parameters of the applications and platforms (task graphs, floorplans, HotSpot configurations, etc.) used in our experiments are available online at [18].

Table 1: Optimization results.

(a) Different numbers of cores.

N_p	N_t	t,s	$\text{MTTF}_\times$	$\text{E}_\times$
2	40	7.84	39.41	0.97
4	80	65.76	37.11	0.99
8	160	759.29	31.36	0.97
16	320	3484.59	13.51	0.98

(b) Different application sizes.

N_p	N_t	t,s	$\text{MTTF}_\times$	$\text{E}_\times$
4	40	9.96	64.53	0.88
4	80	56.57	38.01	0.96
4	160	352.20	18.08	1.07
4	320	408.42	12.92	1.05

(c) Different techniques.

N_p	N_t	$\text{MTTF}_\times^{PM}$	$\text{MTTF}_\times^{HS}$	$\text{MTTF}_\times^{SSA}$
4	40	64.53	1.29	25.10
4	80	38.01	1.67	13.87
4	160	18.08	2.02	5.33
4	320	12.92	1.72	3.82

in [19]. This solution consists of a task mapping and schedule that captures the spatial temperature behavior and tries to minimize the peak temperature while satisfying the real-time constraints. The deadline is set to the duration of the initial schedule extended by 5%.

In the first set of experiments, we change the number of cores N_p while keeping the number of tasks N_t per core constant and equal to 20. For each problem we have generated 20 random task graphs and found the average improvement of the MTTF over the initial solution ($\text{MTTF}_\times$). We also have measured the change in the consumed energy ($\text{E}_\times$). The results are given in Tab. 1a (t indicates the optimization time in seconds). It can be seen that the reliability-aware optimization dramatically increases the MTTF by 13 up to 40 times. Even for large applications with, e.g., 320 tasks deployed onto 16 cores, a feasible mapping and schedule that significantly improve the lifetime of the system can be found in an affordable time. Moreover, our optimization does not impact the energy efficiency of the system.

For the second set of experiments, we keep the quad-core architecture and vary the size (number of tasks N_t) of the application. The number of randomly generated task graphs per application size is 20. The average improvement of the MTTF along with the change in the energy consumption are given in Tab. 1b. The observations are similar to those for the previous set of experiments.

The above experiments have confirmed that our proposed approach is able to effectively increase the MTTF of the system. The efficiency of this approach is due to the fast and accurate SSDTP calculation, which is at the heart of the optimization, and which, due to its speed, allows a huge portion of the design space to be explored. In order to prove this, we have replaced, inside our optimization framework, the proposed SSDTP calculation with the calculation based on HotSpot (Sec. 4.1) and based on the SSA (Sec. 4.2), respectively. The goal is to compare our results with the results produced using HotSpot and the SSA, after the same optimization time as needed with the proposed SSDTP calculation technique. The experimental setup is the same as for the experiments in Tab. 1b. The MTTF obtained with HotSpot and the SSA is evaluated and compared with the MTTF obtained by our proposed method. The results are summarized in Tab. 1c. For example, the lifetime of the platform running 160 tasks can be extended by more than 18 times, compared to the initial solution, using our approach, whereas, the best solutions found with HotSpot and the SSA, using the same optimization time, are only 2.02 and 5.33 times better, respectively. The reason for the poor results with HotSpot is the excessively long execution time of the SSDTP calculation. This allows for a much less thorough investigation of the solution space than with our proposed technique. In the case of the SSA, the reason is different. The SSA is fast but also very inaccurate (Sec. 4.2). The inaccuracy drives the optimization towards solutions that turn out to be of low quality.

We have seen that our reliability-targeted optimizations have significantly increased the MTTF without affecting the energy consumption. This is not surprising, since our optimization will search towards low temperature solutions, which implicitly means low leakage. In order to further explore this aspect, we have performed a multi-objective optimization[5] along the dimensions of energy and reliability. An example of the Pareto front averaged over 20 applications with 80 tasks deployed onto a quad-core platform is given in Fig. 6. It can be observed that the variation of energy is less than 2%. This means that solutions optimized for the MTTF have an energy consumption almost identical to those optimized for energy. At the same time, the difference along the MTTF is huge. This means that ignoring the reliability aspect one may end up with a significantly decreased MTTF, without any significant gain in energy.

Finally, we have applied our optimization technique to a real-life example, namely the MPEG2 video decoder [21] that is deployed onto a dual-core platform. The decoder was analyzed and split into 34 tasks. The parameters of each task were obtained through a system-level simulation using MPARM [22]. The deadline is set to 40 ms assuming 25 video frames per second. The solution found with the proposed method improves the lifetime of the system by 23.59 times with a 5% energy saving, compared to the initial solution. The same optimization was solved using HotSpot and the SSA. The best found solutions are only 5.37 and 11.50 times better than the initial one, respectively.

10. CONCLUSION

In this paper we have proposed an efficient and accurate technique to calculate the SSDTP of an embedded multiprocessor system. Using the proposed approach, we conducted a temperature-aware reliability optimization based on the thermal cycling failure mechanism and have shown that taking into consideration the temperature variations within a multicore platform can significantly prolong its lifetime without affecting its energy efficiency. The improvement, compared using the state of the art, is significant.

11. ACKNOWLEDGMENTS

We would like to thank Prof. Åke Björck from Linköping University for the valuable discussions and suggestions about the analytical solution.

12. REFERENCES

[1] F. Kreith. *CRC Handbook of Thermal Engineering*. CRC Press, 2000.

[2] K. Skadron et al. Temperature-aware microarchitecture. In *ISCA*, pages 2–13, 2003.

[3] J. Srinivasan et al. The impact of technology scaling on lifetime reliability. In *DSN*, pages 177–186, 2004.

[4] W. Liao et al. Temp. and supply voltage aware performance and power modeling at microarchitecture level. *IEEE Trans. CAD*, 24(7):1042–1053, 2005.

[5] A. K. Coskun et al. Analysis and optimization of MPSoC reliability. *J. of Low Power Electronics*, 2(1):56–69, 2006.

[6] Y. Liu et al. Accurate temperature-dependent IC leakage power estimation is easy. In *DATE'07*, 2007.

[7] L. Huang et al. Lifetime reliablity-aware task allocation and scheduling for MPSoCs. In *DATE'09*, 2009.

[8] Y. Xiang et al. System-level reliability modeling for MPSoCs. In *CODES+ISSS'10*, 2010.

[9] L. Thiele et al. Thermal-aware sys. analysis and software synthesis for embedded multi-processors. In *DAC'11*, 2011.

[10] D. Rai et al. Worst-case temperature analysis for real-time systems. In *DATE'11*, 2011.

[11] M. Bao et al. Temp.-aware idle time distribution for energy optim. with dynamic voltage scaling. In *DATE'10*, 2010.

[12] R. Rao et al. Fast and accurate prediction of the steady-state throughput of multicore processors under thermal constraints. *IEEE Trans. CAD of ICs and Systems*, 28(10):1559–1572, 2009.

[13] JEDEC. *Failure Mechanisms and Models for Semiconductor Devices*. JEDEC Publication, 2010.

[14] W. H. Press et al. *Numerical Recipes*. Cambridge University Press, 3rd edition, 2007.

[15] J. Stoer et al. *Introduction to Numerical Analysis*. Springer-Verlag, New York, 3rd edition, 2002.

[16] M. T. Schmitz et al. *Sys.-Level Design Techniques for Energy-Efficient Embed. Sys.* Kluwer Academ. Pub., 2004.

[17] R. P. Dick et al. TGFF: Task graphs for free. In *CODES/CASHE'98*, 1998.

[18] http://www.ida.liu.se/~ivauk83/research/SSDTA.

[19] Y. Xie et al. Temperature-aware task allocation and scheduling for embedded MPSoC design. *Journal of VLSI Signal Processing*, 45(3):177–189, 2006.

[20] K. Deb et al. A fast and elitist multiobjective GA: NSGA-II. *IEEE Trans. Evol. Comp.*, 6(2), 2002.

[21] http://ffmpeg.mplayerhq.hu.

[22] L. Benini et al. MPARM: Expl. MPSoC design space with SystemC. *J. of VLSI Sig. Proc. Sys.*, 41(2):169–182, 2005.

---

[5] The multi-objective optimization is based on NSGA-II [20].

APPENDIX

S1. RC THERMAL CIRCUIT

The equivalent circuit of a multiprocessor system with a thermal package can be built in different ways depending on the intended level of details. Consequently, the number of nodes N_n and structure of the matrices $\mathbf{C}$ and $\mathbf{G}$ in Eq. (2) depend on the particular model. Thermal nodes that belong to the package are called inactive, in the sense that their power dissipation is assumed to be zero.

Without loss of generality, in this paper we use thermal circuits where each of the N_p processing elements is captured by one thermal node. Similar to [2], in the model, three cooling layers are present, namely the thermal interface material, heat spreader, and heat sink captured by N_p, $N_p + 4$, and $N_p + 8$ inactive thermal nodes, respectively. Therefore, the total number of thermal nodes $N_n = 4 \times N_p + 12$. The parameters of the die and thermal package, used throughout this paper, are given in Tab. S1.

A simplified example of such a thermal circuit for a dual-core architecture is depicted in Fig. S1. It can be seen that the inter-core thermal influence is taken into account by modeling the heat flux between the cores (the top two thermal nodes) with the corresponding thermal resistance.

Some or all cores can be also modeled at a finer level of granularity, where caches, ALUs, or registers will be captured as individual thermal nodes.

S2. ANALYTICAL SOLUTION

In this section we further discuss the analytical solution from Sec. 5, its application for the TTA, and possible solution techniques in the case of the SSDTA.

S2.1 Transient Temperature Analysis (TTA)

The recurrence obtained with the analytical solution in Eq. (3) is the following (Eq. (4)):

$$\mathbf{T}_{i+1} = \mathbf{K}_i \, \mathbf{T}_i + \mathbf{B}_i \, \mathbf{P}_i$$

Given the initial temperature $\mathbf{T}_0$, it can be applied to perform the TTA. Our experiments show that, since intervals Δt_i have the same length and matrices $\mathbf{K}_i$ and $\mathbf{B}_i$ become constant, this approach produces a significant performance improvement compared to iterative solutions of ODEs, e.g., the fourth-order Runge-Kutta method used in HotSpot. The same observation is made in [9].

The TTA using the analytical technique given in Eq. (4) can be employed to approximate the SSDTP by applying it over successive application periods, as shown in Sec. 4.1. Since each iteration, with this approach, is much faster than with HotSpot, it will significantly speed up the SSDTP calculation. However, the number of required iterations is similar to the case when HotSpot is used (see Fig. 1a), still keeping the computational process slow (Sec. 9.1).

S2.2 Straight-Forward Solutions (SSDTA)

The first straight-forward way to solve the system in Eq. (6) is to use dense solvers such as the LU decomposition [14]. However, a more advanced approach is to employ sparse solvers since the matrix of the system is a sparse matrix. Therefore, algorithms specially designed for such cases are preferable, e.g., the unsymmetric multifrontal method [S1]. The computational complexity of the solution is proportional to $N_s^3 N_n^3$ [14] where N_n is the number of nodes and N_s is the number of steps in the power profile. The problem here is that the systems to solve can be extremely large, in particular due to N_s. Our experiments have shown that direct

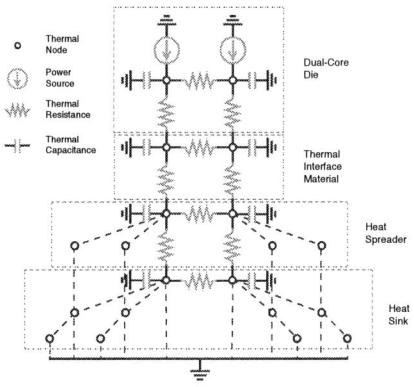

Figure S1: Equivalent RC thermal circuit.

solvers are extremely slow and consume a large amount of memory. Therefore, we do not consider them in the paper.

The overall matrix of the system in Eq. (6) is, in fact, a block Toeplitz matrix. To be more specific, the matrix is a block-circulant matrix where each block row vector is rotated one block element to the right relative to the preceding block row vector. This leads to a wide range of possible techniques to solve the system, e.g., the fast Fourier transform (FFT) [S2] that we include in our experiments in Sec. 9.1.

Another possible technique is iterative methods for solving systems of linear equations (e.g., Jacobi, Gauss–Seidel, Successive Overrelaxation) [14]. These methods are designed to overcome problems of direct solvers and, consequently, they are applicable for very large systems. However, the most important issue with these methods is their convergence. In our experiments we did not observe any advantages of using these methods compared to the others considered in this paper. Therefore, they are excluded from the discussion.

S3. RELIABILITY OPTIMIZATION

This section contains the derivation of the reliability model discussed in Sec. 8 and the description of the actual optimization procedure.

S3.1 Temperature-Aware Reliability Model

In our analysis, we use the reliability model presented in [7, 8]. The model is based on the assumption that the time to failure $\mathcal{T}$ has a Weibull distribution, i.e., $\mathcal{T} \sim Weibull(\eta, \beta)$ where η and β are the scaling and shape parameters, respectively. The expectation of the distribution is the following:

$$\mathbb{E}\left[\mathcal{T}\right] = \eta \, \Gamma(1 + \frac{1}{\beta}) \qquad (23)$$

where Γ is the gamma function. $\mathbb{E}\left[\mathcal{T}\right]$ is the mean time to failure (MTTF) that we denote by θ.

The shape parameter β is independent of the temperature variation [S3], which, however, is not the case with the scaling parameter η. Therefore, the distribution varies with the temperature. We can split the overall period of the application τ into N_m time intervals Δt_i, so that during each time interval Δt_i the corresponding η_i is a constant:

$$\eta_i = \frac{\theta_i}{\Gamma(1 + \frac{1}{\beta})} \qquad (24)$$

where θ_i is the MTTF in the ith time interval as if we had the failure distribution of this interval all the time. For now the values θ_i are unknown and depend on the particular failure mechanism. As it is shown in [8], the reliability function $R(t)$, i.e., the probability of survival until an arbitrary time $t \geq 0$, can be approximated as the following:

$$R(t) = e^{-(\frac{t}{\tau} \sum_{i=0}^{N_m-1} \frac{\Delta t_i}{\eta_i})^{\beta}}$$

The formula keeps the form of the Weibull distribution with the scaling parameter equal to:

$$\eta = \frac{\tau}{\sum_{i=0}^{N_m-1} \frac{\Delta t_i}{\eta_i}} \qquad (25)$$

Table S1: Parameters of the die and package.

Parameter	Value
Ambient temperature	$27\ {}^\circ C$
Convection capacitance	140.4 J/K
Convection resistance	0.1 K/W
Die thickness	0.15 mm
Thermal interface material thickness	0.02 mm
Heat spreader side	20 mm
Heat spreader thickness	1 mm
Heat sink side	30 mm
Heat sink thickness	15 mm

The MTTF with respect to the whole application period can be obtained by combining Eq. (23), Eq. (24), and Eq. (25).

As mentioned previously, in order to compute the MTTF, we need to consider the particular failure mechanism and determine the values θ_i needed in Eq. (24). We focus on the thermal cycling fatigue (Sec. 8.2). Assuming this concrete failure model, the duration Δt_i, during which the corresponding scaling parameter η_i is constant Eq. (24), is exactly a thermal cycle.

When the system is exposed to identical thermal cycles, the number of such cycles to failure can be estimated using a modified version of the well-known Coffin-Manson equation with the Arrhenius term [8, 13]:

$$N_c = A(\Delta T - \Delta T_0)^{-b} e^{\frac{E_a}{kT_{\max}}}$$

where A is an empirically determined constant, ΔT is the thermal cycle excursion, ΔT_0 is the portion of the temperature range in the elastic region which does not cause damage, b is the Coffin-Manson exponent, which is also empirically determined, E_a is the activation energy, k is the Boltzmann constant, and $T_{\max}$ is the maximal temperature during the thermal cycle. Over the application period, the system undergoes a number of different thermal cycles each with its own duration Δt_i and each cycle causes its own damage. Therefore, having N_m thermal cycles characterized by the number of cycles to failure $N_{c\,i}$ and duration Δt_i, we can compute θ_i:

$$\theta_i = N_{c\,i}\,\Delta t_i \tag{26}$$

Taking equations (23), (24), (25), and (26) together, we obtain the following expression to estimate the MTTF of one component in the system:

$$\theta = \frac{\tau}{\sum_{i=0}^{N_m-1} \frac{1}{N_{c\,i}}} \tag{27}$$

In order to identify thermal cycles in the temperature curve, we follow the approach given in [8] where the rainflow counting method is employed.

S3.2   Optimization Procedure

The optimization procedure is based on a genetic algorithm [16] with the fitness function $\mathcal{F}$ given by Eq. (20). Each chromosome is a vector of $2 \times N_t$ elements, where the first half encodes priorities of the tasks and the second represents a mapping. The population contains $4 \times N_t$ individuals that are initialized partially randomly and partially based on the initial temperature-aware solution [19]. In each generation, a number of individuals, called parents, are chosen for breeding by the tournament selection with the number of competitors proportional to the population size. The parents undergo the 2-point crossover with 0.8 probability and uniform mutation with 0.01 probability. The evolution mechanism follows the elitism model where the best individual always survives. The stopping condition is an absence of improvement within 200 successive generations.

The fitness of a chromosome, Eq. (20), is evaluated in a number of steps. First, the decoded priorities and mapping are given to a list scheduler that produces schedules for each of the cores. If the application schedule does not satisfy the deadline, the solution is penalized proportionally to the delay and is not further evaluated; otherwise, based on the parameters of the architecture and tasks, a power profile is obtained and the corresponding SSDTP is computed by our proposed method. If the SSDTP violates the temperature constraint given by Eq. (22), the solution is penalized proportionally to the amount of violation and not further processed; otherwise, the MTTF of each core is estimated according to Eq. (18) and the fitness function $\mathcal{F}$ is computed.

S4.   REFERENCES

[S1] T. A. Davis. Algorithm 832: UMFPACK. *ACM Trans. Mathematical Software*, 30(2):196–199, 2004.

[S2] T. De Mazancourt et al. The inverse of a block-circulant matrix. *IEEE Trans. Anten. Prop.*, 31(5):808–810, 1983.

[S3] S.-C. Chang et al. Electrical characteristics and reliability properties of MOSFET with Dy2O3 gate dielectric. *Applied Physics Letters*, 89(5), 2006.

Considering Diagnosis Functionality during Automatic System-Level Design of Automotive Networks*

Michael Eberl[†], Michael Glaß[†], Jürgen Teich[†], and Ulrich Abelein[‡]

[†]University of Erlangen-Nuremberg, Germany
{michael.eberl, glass, teich}@cs.fau.de

[‡]AUDI AG, Ingolstadt, Germany
ulrich.abelein@audi.de

ABSTRACT

Today, design automation approaches for automotive E/E-architectures focus solely on application functionality, neglecting firmware-related functionalities like diagnostic tests that are of utmost importance for quality features such as dependability or maintenance. However, the latter are typically considered dispensable since they do not provide direct service to the user. This paper proposes a novel approach for integrating optional diagnosis functionality into a holistic design space exploration of automotive E/E-architectures at system-level. Opposed to application functionality, hardware-diagnostics dig deep into the hardware-structures and, hence, require specific tailoring for the employed resources. A case study with Software-Based Self-Tests representing advanced diagnosis functionality gives evidence of the viability and efficiency of the proposed approach, highlighting the importance of a holistic consideration of application as well as firmware-related functionality.

Categories and Subject Descriptors

C.3 [SPECIAL-PURPOSE AND APPLICATION-BASED SYSTEMS]: Real-time and embedded systems

General Terms

Design

Keywords

Diagnosis, Automotive, Design Space Exploration

1. INTRODUCTION

Modern automobiles include a tremendous amount of applications, implemented by means of a network embedded system that consists of up to 100 *Electronic Control Units* (ECUs), several field buses, as well as numerous sensors and actuators. However, the progress of diagnostic capabilities of modern vehicles has not kept up with the increased design complexity and the rise of manufacturing variances and runtime errors caused by shrinking device structures of electronic components. In the automotive domain, *diagnosis* is typically associated with observing operational values on each ECU like monitoring the operating voltage and storing error messages as Diagnostic Trouble Codes (DTCs) [16] into an error memory, see for example [5]. The actual diagnosis is then carried out

---

*Supported in part by the German Ministry for Research and Education (BMBF) within the DIANA project.

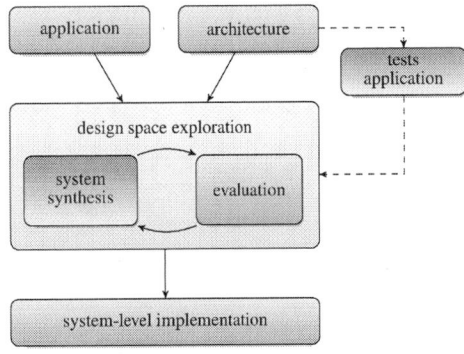

Figure 1: Proposed design flow: A tests application is derived from the architecture and considered during DSE by providing novel steps solely in the system synthesis phase while not affecting the evaluation phase.

offline in the workshop based on the data that is retrieved from the error memory of each ECU, cf. the ISO 15031 standard [6]. Several errors caused by defect electronic components, however, cannot be properly diagnosed with this offline approach because only their consequences, i. e., the operational values exceeding given bounds, is logged, but not their cause, cf. [17].

Opposed to that, the work at hand focuses on advanced diagnosis features that enable structural as well as functional *tests* of the silicon components. These tests in the form of Hardware or Software-Based Self-Tests are already state-of-the-art in other domains but have not yet been successfully applied in automotive E/E-architectures. Moreover, the presence of such tests for the components enables to obtain more system-wide knowledge by implementing online monitoring approaches that collect data or selectively activate more detailed tests. The work at hand aims at closing the existing gap by introducing advanced diagnosis features in the design of automotive systems at system-level as schematically depicted in Fig. 1. First, an approach is shown that allows to model self-tests and online monitoring in a similar scheme as conventional applications. The *tests application* represents all diagnosis features like hardware tests, monitoring functionalities, or data analysis procedures by means of functions that exchange data via messages. However, the tests applications compete with the conventional applications for resources as carrying out tests or collecting data for monitoring increases the load on processors, memories, and buses. Hence, analysis techniques must be applied to ensure that conventional and often safety-critical applications are not influenced to an extend that deadlines or other constraints are violated. Here, it is of utmost importance that the difference between advanced diagnosis features and conventional applications is transparent for the analysis techniques to ensure the applicability of established and typically commercial analysis tools. The proposed design flow addresses this problem by a special treatment of tests applications solely during the process of *system synthesis* where the conventional and tests applications are mapped onto the given architecture platform. Special treatment becomes necessary because conventional functions are *independent* of the executing component, each one having to be implemented in the system either in hardware or executed on an ECU as software to provide correct system service. Opposed to that, each test has to be developed and implemented separately for each individual type of hardware and is thus *architecture-dependent*. Hence, a test function is only available if a component is allocated. Moreover, since this advanced diagnosis is an additional and user-transparent feature, test functions are *optional* and shall, thus, only be implemented if conventional applications are not affected. The outlined requirements and special treatments are all captured within a proposed 0-1 ILP formulation of the system model, enabling a combined system synthesis of conventional and tests applications with a state-of-the-art hybrid optimization approach.

The remainder of this paper is structured as follows: Section 2 presents fundamentals and discusses related work. Section 3 introduces the proposed system model including a novel graph-based model for the consideration of advanced diagnosis features and the 0-1 ILP that enables a combined system synthesis of conventional and diagnosis applications. The efficiency of the proposed design approach is investigated in Section 4 where an example automotive E/E architecture is enriched with an online monitoring system based on Software-Based Self-Tests (SBSTs). The paper is concluded in Section 5.

2. FUNDAMENTALS & RELATED WORK

Considering the boost in the amount of functionality and the up to 100 ECUs that modern automobiles consist of, it is obvious that selecting an optimal platform and selecting an optimal mapping of the functions to the components of the platform becomes more and more complex. *Design Space Exploration* (DSE) supports the system designer in finding optimal implementations. Several methodologies for *system-level design*, typically targeting SoC or MPSoC platforms, are based on the well known Y-Chart model, see [1], where the system is split up in an abstract representation of the functional behavior of the system and a selection of hardware resources on which the functionality can be implemented. Up to now, several tools and approaches for platform-based system-level design have been developed like Sesame [14], SystemCoDesigner [7], MILAN [12], and MESCAL [11]. An overview that discusses various SoC and MPSoC approaches in detail can, e. g., be found in [2]. Though there are individual differences, all presented approaches try to decouple the functional description from the hardware implementation.

To build up the online monitoring system, this work investigates *Software-Based Self-Testing* (SBST) as an example self-test technique, see [15]. SBSTs can be used to carry out periodic online self-tests of the individual silicon units during the operation of the automobile, see [13, 3]. Although [13] and [3] address online SBST scheduling, there is no approach that investigates the concurrent design of conventional applications and SBSTs on an architecture platform including component allocation, function binding, message routing, and scheduling. An important aspect of self-tests is that most of them have to be developed and implemented separately for each individual hardware component. This contrasts with existing approaches for platform-based design mentioned before that decouple functionality from hardware. This work overcomes this drawback by proposing an extension for a symbolic system synthesis approach that has already been successfully applied in the design of automotive E/E architectures, see [9, 4]. To the best of the authors knowledge, this is the first approach that combines a classic Y-chart approach to system design with optional architecture-dependent applications, making the concept applicable also in other domains and for other kinds of applications.

3. AUTOMATIC TESTS INTEGRATION

In this section, the proposed system design flow that integrates advanced diagnosis features into the design of automotive E/E architectures at system-level is presented. First, a short introduction of the employed system design flow is given. On top of the already introduced self-tests of components, this work also takes into account online monitoring approaches that collect operational information derived by the individual tests, may selectively activate detailed tests, or even carry out further analyses, see [19]. Thus, the individual self-tests and the online monitoring itself constitute an advanced diagnosis service. The proposed graph-based tests application model that captures this kind of service is introduced in detail in the middle part of this section. A presentation of the novel system synthesis formulation closes this section.

3.1 System Design

This work extends a system design flow that follows the Y-chart approach and has already been employed in the automotive domain, see [4, 9], as outlined in Fig. 1. As model input, an *application* that represents the algorithmic description of the problem and an *architecture* that represents the available hardware components is given. The aim of the *Design Space Exploration* (DSE) is to find mappings of the application to the architecture that are optimized with respect to multiple and often conflicting design objectives while meeting all given design constraints. During DSE, a step called *system synthesis* derives system *implementations* by (a) *allocating* resources from the architecture, (b) *binding* functions of the application to the selected resources, (c) *routing* messages, and (d) determining scheduling parameters. Each *implementation* is evaluated with respect to given quality measures to determine its design objectives and checked for compliance with given design constraints. The result of the DSE is a set of high-quality system-level implementations from which a designer selects an implementation with the most appropriate trade-off between all design objectives for further refinements on lower levels of abstraction. This whole design flow requires mathematically well-defined models that are introduced in Section 3.1.1.

As already pointed out in Section 2, the conventional approaches cannot handle functionality that is architecture-dependent and optional in the implementation like the activation/execution of self-tests provided by the hardware itself. As a remedy, this work proposes the modeling of test functionality as a special *tests application*. Analogous to conventional functions and messages that form the conventional application, the tests application consists of *test functions* and *test messages*. Test functions represent tests that are provided by selected ECUs. The representation as an application enables to consider more complex test strategies that incorporate data-fusion of several test functions, hierarchical tests, situation-based activation of test functions and so forth. As outlined in Fig. 1, the tests application results from the test capabilities of the given architecture. However, in order not to sacrifice the applicability of standard and often commercial analysis techniques used for the evaluation phase, the work at hand proposes to capture required special treatment of an architecture-dependent application completely within the system synthesis phase. As a result, the difference between conventional functions and messages and test functions and messages becomes transparent during evaluation, ensuring a seamless integration of advanced diagnosis features in existing design flows.

3.1.1 Conventional System Model

The conventional system model, see [10], also termed *specification* is a graph based model that includes an *architecture graph* G_R, an *application graph* G_T, and *mapping edges* E_M:

- The architecture is represented by a directed graph $G_R(R, E_R)$. Vertices R represent different hardware resources like ECUs, sensors, actors, or communication buses. Directed edges $E_R = R \times R$ represent communication connections between resources.

- The application is modeled by a bipartite graph $G_T(T, E_T)$ with $T = P \cup C$. Vertices T represent either functions $p \in P$ or messages $c \in C$. Each directed edge $e \in E_T$ connects a function in P with a message in C, or vice versa. A function may have multiple incoming edges that model its data dependencies to its predecessor vertices. A message can only be preceded by one function though a function is able to be succeeded by several different messages. There can be

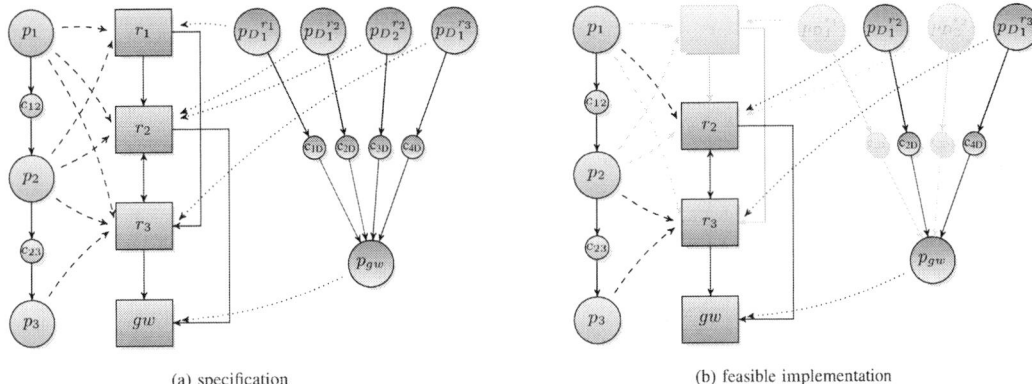

(a) specification   (b) feasible implementation

Figure 2: The example specification in (a) with an application graph (left), an architecture graph (center), and a tests application graph (right). Mapping edges for conventional functions are depicted as dashed edges while mapping edges for test functions are depicted as dotted edges. A corresponding feasible implementation is shown in (b).

multiple functions following each message so that multi-cast communication is enabled.

- The mapping edges E_M provide information on which resource each function may be implemented. At this, each edge $m = (p, r) \in E_M$ represents a possible implementation of function $p \in P$ on resource $r \in R$.

During DSE, different implementations are derived from the specification by a set of steps typically referred to as system synthesis. Such an implementation x consists of an *allocation graph* G_α, a *binding* β that maps functions to allocated resources, a *route* for each message in C, and, optional, several scheduling parameters. A formal definition of *system synthesis* is given as follows:

- During the *allocation*, an induced subgraph $G_\alpha(\alpha, E_\alpha)$ of $G_R(R, E_R)$ is derived. It contains all the resources and the respective communication paths that are available in the current implementation.

- The *binding* β contains one mapping edge of each function in P so that $\beta \subseteq E_M$:

$$\forall p \in P : |\{m | m = (p, r) \in \beta\}| = 1$$

  Each contained mapping $m = (p, r) \in \beta$ indicates that function $p \in P$ is implemented on resource r.
  Additionally, each function is only allowed to be bound onto allocated resources:

$$\forall m = (p, r) \in \beta : r \in \alpha$$

- For each message $c \in C$, a *route*, i.e., a tree-like subgraph $G_{\gamma,c}$ of the allocation α is created. The routing of each message has to meet the following constraints: The root of $G_{\gamma,c}$ has to equal the resource that implements the preceding sending function $(p, c) \in E_T$ of c:

$$\forall (p, c) \in E_T, m = (p, r) \in \beta :$$
$$r \in G_{\gamma,c} \wedge [|\{e | e = (\tilde{r}, r) \in G_{\gamma,c}\}| = 0]$$

  Each message $c \in C$ has to be routed to the same resource as the resource that implements a succeeding function $(c, p) \in E_T$:

$$\forall (c, p) \in E_T, m = (p, r) \in \beta : r \in G_{\gamma,c}$$

An implementation is *feasible* iff all constraints concerning allocation, binding, and routing are fulfilled.

3.1.2 Modeling of Test Functionality

In this work, a monitoring model is created in which test functions are executed directly on the ECUs or their integrated components, respectively. The results of these tests are then propagated to a central monitoring function. There, the incoming data is either stored for later offline failure diagnosis or used for more sophisticated online analysis. To incorporate this monitoring approach, the tests are periodically scheduled on the ECUs and the information is exchanged by sending messages over the communication buses.

The novel *tests application* is modeled as a bipartite graph $G_{T,D}(T_D, E_{T,D})$ with $T_D = P_D \cup C_D$. Vertices T_D represent either a test message $c_D \in C_D$ or a test function $p_D \in P_D$. Each directed edge $e_D \in E_{T,D}$ connects a test function in P_D with a test message in C_D, or vice versa. In the set of test mapping edges $E_{M,D} = R \times P_D$, each test function p_D has exactly one mapping edge $m_D = (p_D, r) \in E_{M,D}$, indicating that the test function is available on resource r. Given that different test features may be available for the same resource like checking the integrity of a processor and checking for faulty memory segments, more than one test function can be mapped to a resource. In particular, activating several test functions on a single ECU is enabled as well. Thus, either none or several test functions may be bound whereas conventional functions are bound exactly once.

To complete the synthesis of the system model, the test allocation is given as follows:

- The *test binding* β_D contains a subset of mapping edges of each test function in P_D so that $\beta_D \subseteq E_{M,D}$. Each contained mapping $m_D = (p_D, r) \in \beta_D$ indicates that function $p_D \in P_D$ is activated on resource r. An activation of a test function is possible iff the corresponding resource is allocated:

$$\forall m_D = (p_D, r) \in \beta_D : r \in \alpha$$

- For each message $c_D \in C_D$ a *route*, i.e., a tree-like subgraph G_{γ,c_D} of the allocation α is created iff:

$$\forall p_D, \widetilde{p_D} \in P_D \text{ with } (p_D, c_D), (c_D, \widetilde{p_D}) \in E_{T,D} :$$
$$m_D = (p_D, r), \widetilde{m_D} = (\widetilde{p_D}, \tilde{r}) \in \beta_D$$

That is, the preceding test function and (at least one) succeeding test function are activated. The feasibility of a route corresponds to the rules presented for the routing of conventional messages.

Fig. 2 (a) depicts an example specification including an application (left), an architecture (middle), and a tests application (right): The test-application models the employed monitoring scenario where all test data is gathered on a central monitoring device. On this central component, in this example the already available gateway gw, a *test gateway function* $p_{gw} \in P_D$ creates error logs and may invoke more sophisticated analysis techniques. Therefore, all test functions send their data to the central test gateway function via test messages.

Given this definition, the test-aware synthesis approach includes the steps of system synthesis (allocation, binding, and routing) and the test allocation (test binding and routing). For the derivation of other system parameters like scheduling priorities, no distinction between test features and conventional applications is required. After these steps, the overall binding is given as $\beta \cup \beta_D$ and the overall routing is given as $\bigcup_{c \in C \cup C_D} G_{\gamma, c_D}$. Thus, the consideration of test features is restricted to the proposed graph-based model and the synthesis step but transparent to all employed evaluation techniques. This, in particular, enables to seamlessly consider the competition of conventional applications and test features for computational and communication resources.

3.2 System Synthesis

Design Space Exploration (DSE) as carried out in the design flow proposed here basically solves the following multi-objective optimization problem:

optimize $f(x)$
on condition that:
x is a feasible implementation

Common optimization methods that are based on integer linear programming or evolutionary algorithms are either limited to only one optimization constraint or do not perform well in scenarios with many constraints and only a few feasible solutions. In this paper, the exploration of the design space is carried out by a hybrid approach that combines a state-of-the-art *SAT-solver* with *Evolutionary Algorithms* (EAs). This method termed *SAT-decoding* [8] combines a *SAT-solver* to generate feasible implementations while an EA enables an efficient exploration of the search space with respect to multiple conflicting and even non-linear design objectives. Such an advanced constraint-handling is particularly important in the context of automotive systems where numerous stringent design constraints are present.

The SAT-decoding approach enables the encoding of linear constraints into the conventional system model. Therefore, by providing a 0-1 Integer Linear Program (ILP) for the test allocation as introduced in Section 3.1.2, the tests application can be fully integrated into this state-of-the-art optimization technique together with the conventional applications. At first, the 0-1 ILP encoding of the conventional system synthesis is briefly introduced. Afterwards, the encoding for the novel test allocation based on the proposed system model is presented in detail.

3.2.1 Conventional Binary Encoding

In [10], a binary search problem for which a solution $x \in \{0, 1\}^n$ has to be found is introduced. Here, x represents a feasible implementation. As part of the symbolic encoding, the following binary variables are introduced:

$\mathbf{r}$ – is created for each resource $r \in R$ and indicates whether the corresponding resource is part of the allocation (1) or not (0).

$\mathbf{p_r}$ – is created for each mapping $m = (p, r) \in E_M$ and indicates whether a function $p \in P$ is bound onto the resource $r \in R$ (1) or not (0).

$\mathbf{c_r}$ – is created for every message $c \in C$ and resource $r \in R$ and indicates whether a message is routed on the resource (1) or not (0).

$\mathbf{c_{r,n}}$ – are additional variables for each communication and resource tuple. They indicate at what communication step $n \in N = \{0, \dots, |N|\}$ a message is routed over that specific resource.

The linear constraints are formulated as follows:
$\forall p \in P$:

$$\sum_{r \in R_p} \mathbf{p_r} = 1 \tag{1a}$$

$\forall c \in C$:

$$\sum_{r \in R_c} \mathbf{c_{r,0}} = 1 \tag{1b}$$

$\forall c \in C, p \in \{\tilde{p} | (\tilde{p}, c) \in E_T\}, r \in R_p \cap R_c$:

$$\mathbf{p_r} - \mathbf{c_{r,0}} = 0 \tag{1c}$$

$\forall p \in P, c \in \{\tilde{c} | (\tilde{c}, p) \in E_T\}, r \in R_p \cap R_c$:

$$\mathbf{c_r} - \mathbf{p_r} \geq 0 \tag{1d}$$

$\forall c \in C, r \in R_c$:

$$\mathbf{c_{r,1}} + \mathbf{c_{r,2}} + \dots + \mathbf{c_{r,n}} \leq 1 \tag{1e}$$

$$\mathbf{c_{r,1}} + \mathbf{c_{r,2}} + \dots + \mathbf{c_{r,n}} - \mathbf{c_r} \geq 0 \tag{1f}$$

$\forall c \in C, r \in R_c, i = \{1, \dots, n\}$:

$$\mathbf{c_r} - \mathbf{c_{r,i}} \geq 0 \tag{1g}$$

$\forall c \in C, r \in R_c, i = \{1, \dots, n-1\}$:

$$-\mathbf{c_{r,i+1}} + \sum_{\tilde{r} \in R_c \wedge e = (\tilde{r}, r) \in E_R} \mathbf{c_{\tilde{r}, i}} \geq 0 \tag{1h}$$

$\forall p \in P, r \in R_p$:

$$\mathbf{r} - \mathbf{p_r} \geq 0 \tag{1i}$$

$\forall c \in C, r \in R_c$:

$$\mathbf{r} - \mathbf{c_r} \geq 0 \tag{1j}$$

$\forall r \in R$:

$$-\mathbf{r} + \sum_{c \in C \wedge r \in R_c} \mathbf{c_r} + \sum_{p \in P \wedge r \in R_p} \mathbf{p_r} \geq 0 \tag{1k}$$

Equation (1a) states that each function will be bound exactly once. Equations (1b) and (1c) limit each message to only have one root that equals the bound resource of the preceding function. Analogously, the preceding message of each function has to be bound on the corresponding resources, see Eq. (1d). Equation (1e) ensures that each resource is only used once for each message so that no cycles are generated. If a message is bound to a resource, the corresponding message has to pass that resource on one communication step. This expression is formulated by Eqs. (1f) and (1g). Equation (1h) states that only adjacent resources may directly exchange information. Equations (1i) and (1j) say that functions or messages can only be bound or routed on resources that are activated and, thus, part of the allocation. In contrast, Eq. (1k) demands a resource only being allocated if at least one function is bound on it or one message is routed on it. This constraint encoding method enables the adding of additional linear or linearized constraints to the model, e.g., limitations regarding the maximum bus load. Section 3.2.2 explains how this mechanism can be used to integrate diagnostic features to the model.

An implementation x is deduced from the result x of the linear problem: The allocation is deduced by the variables r and the binding of the functions by the variables p_r. Variables c_r and $c_{r,n}$ determine the routing of the messages.

3.2.2 Tests Encoding

To encode the monitoring test allocation as introduced in Section 3.1.2, respective constraints have to be provided. For the sake of simplicity, diagnosis-related variables are introduced as:

$\mathbf{p_r^D}$ – is created for each test mapping $m_D = (p_D, r) \in E_{M,D}$ and indicates whether the corresponding test function $p_D \in P_D$ is active (1) or not (0).

$\mathbf{c_r^D}$ – is created for each test message $c_D \in C_D$ and resource $r \in R$ and indicates whether the message is routed on the resource (1) or not (0). The routing steps for test messages, i.e., $\mathbf{c_{r,n}^D}$ are defined accordingly.

The binary encoding for the test allocation is based on the binary encoding for conventional system synthesis as presented previously. As not all of the existing constraints have to be modified, only constraints that have to be omitted or adapted will be explicitly explained here. For all constraints that remain unchanged, variables P are seamlessly interchanged for variables P_D and C for C_D, respectively.

First, Eq. (1a) has to be omitted since it is not required to bind each test function $p_D \in P_D$. In that sense, it sufficient to ensure that a test function may only be activated if the corresponding resource is allocated. This requirement is already encoded in the conventional constraint given in Eq. (1i).

The concept of activating a message distinguishes the test message from the conventional message. Given that either the sending test function or the receiving test gateway function may not be active, it is important to ensure that the respective message is not routed at all. Otherwise, it would consume communication capacities which deteriorates the design objectives and, possibly, the design constraints. This can be ensured by omitting Eq. (1b).

The test functionality as introduced here allows a sending test function to be inactive while the receiving test function (gateway) being active. This would be contradicted by the conventional constraint in Eq. (1d) that requires a routing of data to each receiving function. Thus, Eq. (1d) is exchanged with the following constraints:
$$\forall \widetilde{p_D}, p_D \in P_D, (\widetilde{p_D}, c_D), (c_D, p_D) \in E_T, \widetilde{r}, r \in R_p \cap R_c:$$

$$-\mathbf{c_r^D} + \mathbf{\widetilde{p_{\widetilde{r}}^D}} \geq 0 \tag{2a}$$

$$-\mathbf{c_r^D} + \mathbf{p_r^D} \geq 0 \tag{2b}$$

$$-\mathbf{c_r^D} + \mathbf{\widetilde{p_{\widetilde{r}}^D}} + \mathbf{p_r^D} \leq 1 \tag{2c}$$

They also guarantee that test messages are only activated in an implementation if the corresponding sending test function and the receiving test gateway function are active in the implementation, but enable also the absence of a respective message which is not supported by the conventional model.

In the proposed monitoring system, see again Fig. 2 (a), it should be possible to remove and add several tests from the implementation. However, the test gateway function $p_{gw} \in P_D$ must not be excluded from the implementation as long as there is at least one active test on a resource, because the tests are meant to send their results to a central monitoring device. If there is no active test, the gateway function should not be activatable, though. Therefore, a new binary variable $\mathbf{p_r^{gw}}$ is introduced. $\mathbf{p_r^{gw}}$ is created for the central test gateway function and its corresponding resource $r \in R_p$ and indicates whether the gateway function is active (1) or not (0).

In the required 0-1 ILP formulation, the requirements are formulated as the following constraints:

$$j \cdot \mathbf{p_r^{gw}} - \sum_{p_D \in P_D \setminus \{p_{gw}\}} \mathbf{p_r^D} \geq 0 \tag{3}$$

$$-\mathbf{p_r^{gw}} + \sum_{p_D \in P_D \setminus \{p_{gw}\}} \mathbf{p_r^D} \geq 0 \tag{4}$$

Here, j is given as $j = |P_D \setminus \{p_{gw}\}|$. Fig. 2 (b) shows a feasible implementation based on the specification given in Fig. 2 (a). As only resources r_2 and r_3 are allocated, only $p_{D_1^{r_2}}$, $p_{D_1^{r_3}}$, c_{2D} and c_{4D} are active in the implementation. Though, in this example, $p_{D_2^{r_2}}$ is not part of the implementation, it would be possible to also activate $p_{D_2^{r_2}}$ because the corresponding resource r_2 is allocated. The resulting test binding is $\beta_D = \{(p_{D_1^{r_2}}, r_2), (p_{D_1^{r_3}}, r_3), (p_{gw}, gw)\}$.

4. EXPERIMENTAL RESULTS

In this section the proposed test-aware design approach is applied to an automotive ECU subnetwork. In the case study, a centralized online monitoring approach is implemented. First, details of the case study and the DSE are introduced. Afterwards, the results of the DSE are discussed in detail.

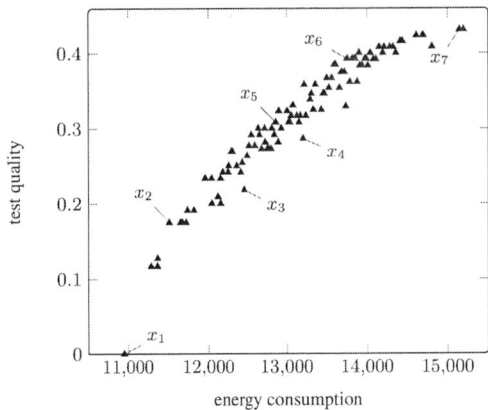

Figure 3: High-quality system implementations: A two-dimensional projection of the three-dimensional objective space depicting test quality and energy consumption for the case study outlined in Section 4.1.

4.1 Case Study

A typical automotive subnet that is made up of 15 ECUs, 2 CAN buses, and 1 FlexRay bus serves as a case study. The buses are interconnected by a special hardware component called *gateway*. 4 applications that consist of 45 functions and 41 periodic messages have to be distributed on the network. Furthermore, a tests application is added to the system specification. In this tests application, individual test functions represent Software-Based Self-Testing (SBST) features. These are provided for 10 out of the 15 ECUs. For each of these 10 ECUs, 4 different tests are provided. Each test requires a certain amount of instructions and offers a stuck-at fault coverage ranging from 70% to 95%. However, the overall fault coverage of multiple tests on an ECU cannot be determined by summing up the fault coverages of the individual tests. For example, a test with a fault coverage of 30% and another test with a fault coverage of 40% typically do not sum up to a fault coverage of 70% because there is a certain amount of overlap between several tests. Thus, the introduced test functions represent a complete test package, i.e., a combination of several individual tests on a single ECU including, e.g., checks of the memory, CPU, and I/O controllers. Hence, in this case study, it is assumed that only one test package, i.e., one test function is activated. All active tests are carried out periodically and the corresponding results are sent periodically to a central test gateway function. The Supplementary Material S1 features a screenshot of the introduced case study represented in a graph based manner as presented in Section 3.

As introduced in the previous sections, the applied DSE is able to vary resource allocation, conventional function as well as test function binding, and message routing. Additionally, a priority-based scheduling is assumed for the ECUs, CAN buses, and the dynamic segment of the FlexRay bus. Thus, the priority of each function and message is varied as well. During DSE, three design objectives, i.e., *test quality*, *energy consumption*, and *area costs*, and one design constraint, i.e., end-to-end latency, is considered. The test quality of an implementation is an *abstract* quality measure defined as the arithmetic mean of all fault coverage values achieved by the activated test functions on all ECUs. If no test function is carried out on an ECU, the fault coverage for that ECU is set to zero. Otherwise, the fault coverage equals the fault coverage of the activated test function. Note that it is important to take the zero coverage of ECUs that do not provide advanced diagnosis functionality into account, as well. That way, the exploration is guided to system implementations that distribute conventional applications predominantly on ECUs that have activated self-tests. Although abstract, the value can be interpreted as an average additional percentage of stuck-at faults that can be covered per ECU compared to a zero coverage in current system implementations. The compliance with the given timing constraints for each application is checked by a performance

Table 1: The test quality, energy consumption increase, and relative hardware costs referring to x_1 of selected implementations from Fig. 3.

	test quality	energy in %	area in %
x_1	0	0	0
x_2	0.17	5.17	1.1
x_3	0.21	13.81	-2.9
x_4	0.28	20.59	-2.9
x_5	0.30	17.47	0.4
x_6	0.39	25.61	4.3
x_7	0.43	38.37	2.9

analysis test using the *Real-Time Calculus* (RTC), see [18]. Given that all functions are activated periodically, a simplified model for the calculation of the energy consumption of the overall system on the average load of each ECU is applied. The load of each ECU is multiplied with its standardized energy consumption for the active state and the energy consumption for the power-down state is multiplied with the idle capacity. The hardware costs are calculated by adding the monetary costs of each ECU and bus in the implementation.

4.2 Results

Fig. 3 shows the test quality and the corresponding energy consumption of the best system implementations found as a two-dimensional projection of the objective space spanned by the three introduced design objectives. Note that implementations with the same test quality but higher energy consumption come at lower hardware costs. It can be seen that the higher the test quality, the higher the energy consumption. This is an obvious trade-off since carrying out test functions increases the load on the ECUs. The highest test quality of 0.43 is achieved by activating test functions on 7 out of 13 allocated ECUs. Interestingly, no implementation is found that achieves a higher test quality without violating timing constraints.

As a reference, the high-quality implementation x_1 is chosen that optimizes energy consumption, in part, because no test functionality is activated. Table 1 depicts the test quality, increase in energy consumption, and increase in area costs of six selected high-quality implementations compared to x_1. These six implementations are also marked in Fig. 3. The Supplementary Material S2 features a screenshot of implementation x_7 and the Supplementary Material S3 a screenshot of implementation x_1. Starting from a test quality of 0, it can be improved to 0.17% at the cost of an increased energy consumption of only 5.17% and a 1.1% area increase. For an additional improvement in test quality of only 0.04, the additional energy consumption rises up to 13.18%. Implementations x_4 and x_5 show that there is a trade-off between test quality, energy consumption, and hardware costs. The test quality of x_4 is lower than the test quality of implementation x_5 despite the higher energy consumption of x_4. However, the hardware costs of implementation x_4 are lower than the hardware costs of x_5. With a total increase of 38.37% in energy consumption and 2.9% in area costs, the test quality can be improved to 0.43.

The presented results show a significant diversity of the found high-quality implementations when considering test quality together with two other common design objectives. Revealing this diversity while at the same time providing high-quality system implementations is only enabled with an approach that considers conventional applications together with test features as proposed in this work. At the same time, meeting given design constraints such as maximum end-to-end latencies is guaranteed by construction.

5. CONCLUSION

In future automobiles, more and more electric/electronic components will carry out various applications. On the other hand, shrinking device structures of electronic components result in a significant increase of manufacturing variances and runtime errors. To cope with this problem, design approaches that integrate advanced diagnosis and test features into such complex networked embedded systems become mandatory. Compared to conventional applications, test features are provided and tailored to hardware components, optional in the final implementation, but compete with conventional application functions for computational and communication resources. This work introduces an automatic design approach at system-level that models test features as a new class of architecture-dependent applications that are considered together with the conventional application and architecture model. The important advantage of the proposed design approach is that test features require special treatment solely during the step of system synthesis, not affecting existing analysis techniques employed for the evaluation. This enables a seamless consideration of conventional applications and (optional) test features regarding design objectives and design constraints. Given the typically huge number of stringent design constraints in automotive E/E architectures, this work presents how the novel model can be encoded as a 0-1 ILP to utilized a state-of-the-art hybrid optimization approach. The result is a combined modeling, analysis, and optimization of the E/E architecture with respect to both conventional applications and test features. A case study taken from the automotive domain with Software-Based Self-Tests (SBSTs) as test features gives evidence of the efficiency of the proposed design approach by delivering several high-quality implementations that trade-off test quality for additional power consumption while respecting all given design constraints. In particular, a maximum gain in diagnosability of 0.43 could be achieved with an increased energy consumption of 38.4% when using energy-intensive SBSTs. Moreover, several high-quality implementations offer trade-offs in between like a 0.17 test quality increase at only 5.2% increased energy consumption.

6. REFERENCES

[1] D. Gajski and R. Kuhn. New VLSI Tools. *Computer*, 16:11–14, 1983.

[2] A. Gerstlauer, C. Haubelt, A. D. Pimentel, T. Stefanov, D. Gajski, and J. Teich. Electronic System-Level Synthesis Methodologies. *Trans. on TCAD '09*, 28(10):1517, 2009.

[3] D. Gizopoulos. Online Periodic Self-Test Scheduling for Real-Time Processor-Based Systems Dependability Enhancement. *IEEE Trans. Dependable Secur. Comput.*, 6:152–158, 2009.

[4] M. Glaß, M. Lukasiewycz, J. Teich, U. Bordoloi, and S. Chakraborty. Designing heterogeneous ecu networks via compact architecture encoding and hybrid timing analysis. In *Proc. of DAC '09*, pages 43–46, 2009.

[5] R. Isermann. Model-based fault-detection and diagnosis - status and applications. *Annual Reviews in Control*, 29(1):71–85, 2005.

[6] ISO 15031. *Road vehicles – Communication between vehicle and external equipment for emissions-related diagnostics.*

[7] J. Keinert, M. Streubühr, T. Schlichter, J. Falk, J. Gladigau, C. Haubelt, J. Teich, and M. Meredith. SYSTEMCODESIGNER - An Automatic ESL Synthesis Approach by Design Space Exploration and Behavioral Synthesis for Streaming Applications. *TODAES '09*, 14(1):1–23, 2009.

[8] M. Lukasiewycz, M. Glaß, C. Haubelt, and J. Teich. Efficient Symbolic Multi-Objective Design Space Exploration. In *Proc. of ASP-DAC '08*, pages 691–696, 2008.

[9] M. Lukasiewycz, M. Glaß, C. Haubelt, J. Teich, R. Regler, and B. Lang. Concurrent topology and routing optimization in automotive network integration. In *Proc. of DAC '08*, pages 626–629, 2008.

[10] M. Lukasiewycz, M. Streubühr, M. Glaß, C. Haubelt, and J. Teich. Combined System Synthesis and Communication Architecture Exploration for MPSoCs. In *Proc. of DATE '09*, pages 472–477, 2009.

[11] A. Mihal, C. Kulkarni, M. Moskewicz, M. Tsai, N. Shah, S. Weber, Y. Jin, K. Keutzer, C. Sauer, K. Vissers, and S. Malik. Developing Architectural Platforms: A Disciplined Approach. *IEEE Design & Test*, 19:6–16, 2002.

[12] S. Mohanty, V. K. Prasanna, S. Neema, and J. Davis. Rapid Design Space Exploration of Heterogeneous Embedded Systems using Symbolic Search and Multi-Granular Simulation. In *Proc. of LCTES/SCOPES '02*, 2002.

[13] A. Paschalis and D. Gizopoulos. Effective Software-Based Self-Test Strategies for On-Line Periodic Testing of Embedded Processors. In *Proc. of DATE '04*, 2004.

[14] A. D. Pimentel, C. Erbas, and S. Polstra. A Systematic Approach to Exploring Embedded System Architectures at Multiple Abstraction Levels. *IEEE Transactions on Computers*, 55:99–112, 2006.

[15] M. Psarakis, D. Gizopoulos, E. Sanchez, and M. S. Reorda. Microprocessor Software-Based Self-Testing. *IEEE Design & Test*, 27:4–19, 2010.

[16] SAE International. Standard J/1979.

[17] A. Unger, K. Lange, D. Peters, and H. Reuss. Methods of a Holistic System View for Function-Oriented Error Detection and Diagnosis in Automotive Networks. In *International Congress on Electronic Systems for Vehicles*, 2005.

[18] E. Wandeler and L. Thiele. Real-Time Calculus (RTC) Toolbox. http://www.mpa.ethz.ch/Rtctoolbox, 2006.

[19] D. Wybranietz and D. Haban. Monitoring and Performance Measuring Distributed Systems during Operation. *SIGMETRICS Perform. Eval. Rev.*, 16:197–206, 1988.

Supplemental Material

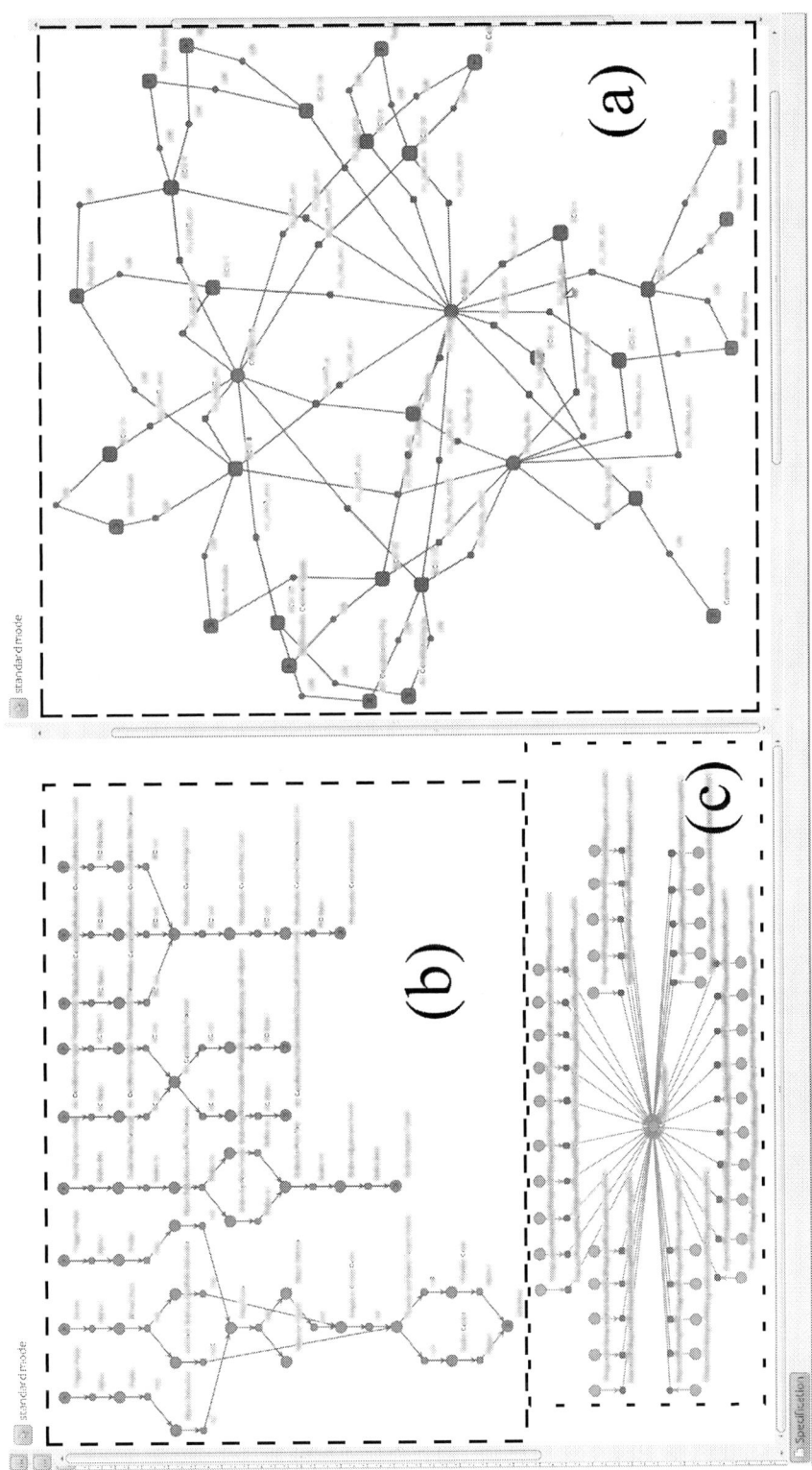

Figure 4: This screenshot shows the specification of the used case study. The architecture in (a) consists of 15 ECUs, 2 CAN buses, 1 FlexRay bus, and 1 Gateway. The conventional application in (b) features 4 separate applications that consist of 45 functions and 41 periodic messages. The tests application in (c) consists of 40 individual tests that send their messages to a central monitoring function.

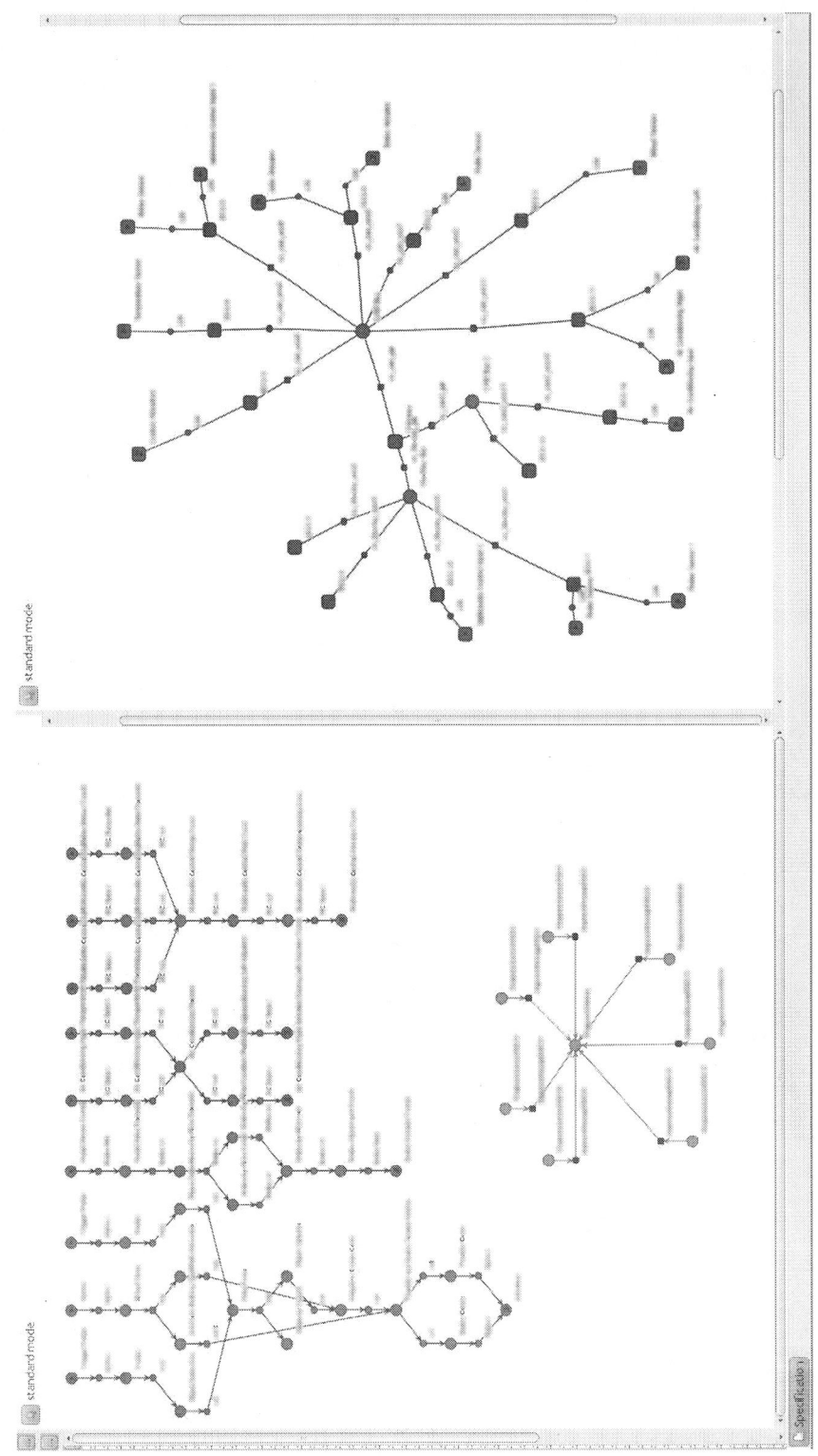

Figure 5: This screenshot shows the implementation x_7 that achieves the highest test quality. There are 13 ECUs and 7 activated test functions. Note that the conventional application that models conventional functionality is completely implemented in the system.

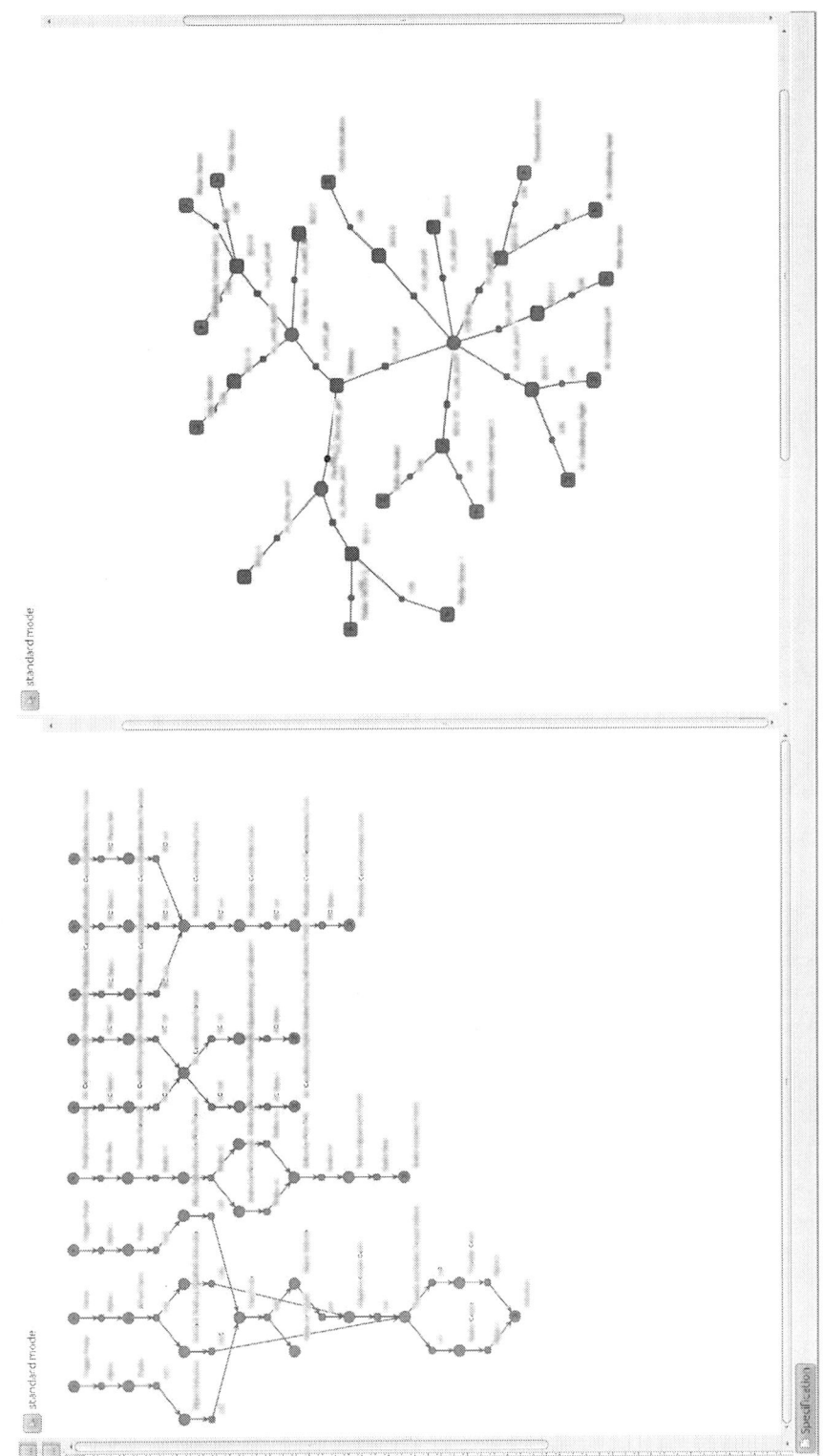

Figure 6: This screenshot shows the implementation x_1 that has no activated test functions. In contrast to implementation x_7 it takes only 11 ECUs to implement the complete conventional functionality. As there is no test function activated on the ECUs, the test gateway function is also removed from the implementation.

Meta-Cure: A Reliability Enhancement Strategy for Metadata in NAND Flash Memory Storage Systems

Yi Wang[†], Luis Angel D. Bathen[§], Nikil D. Dutt[§], Zili Shao[†]

[†]Department of Computing
The Hong Kong Polytechnic University
Hung Hom, Kowloon, Hong Kong
{csywang, cszlshao}@comp.polyu.edu.hk

[§]Center for Embedded Computing Systems
University of California, Irvine
Irvine, CA 92697, USA
{lbathen, dutt}@uci.edu

ABSTRACT

The increasing density of NAND flash memory leads to a dramatic increase in the bit error rate of flash, which greatly reduces the ability of error correcting codes (ECC) to handle multi-bit errors. To ensure the functionality and reliability of flash memory, the pages containing address mapping information and other metadata should be carefully stored in flash memory. This paper presents *Meta-Cure*, a novel hardware and file system interface that transparently protects metadata in the presence of multi-bit faults. Meta-Cure exploits built-in ECC and replication in order to protect pages containing critical data. Redundant pairs are formed at run time and distributed to different physical pages to protect against failures. Meta-Cure requires no changes to the file system, on-chip hierarchy, or hardware implementation of flash memory chip. Experimental results show that the proposed technique can reduce uncorrectable page errors by 92% with less than 1% space overhead in comparison with conventional error correction techniques.

Categories and Subject Descriptors

D.4.2 [**Operating Systems**]: Storage Management—*secondary storage*; B.3.4 [**Memory Structures**]: Reliability, Testing, and Fault-Tolerance—*redundant design*

General Terms

Design, Performance, Reliability

Keywords

NAND flash memory, metadata, reliability, redundancy, ECC

1. INTRODUCTION

During past decades, the capacity of NAND flash memory has been increasing dramatically, leading to the use of non-volatile flash in the system's memory hierarchy. However,

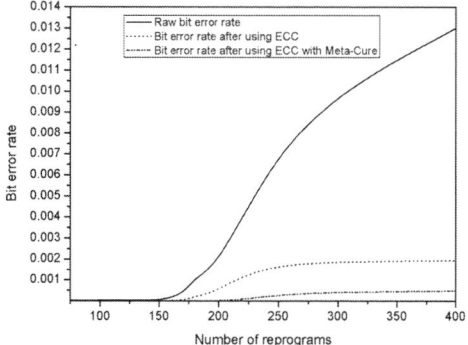

Figure 1: (1) For an 8Gbit MLC NAND flash, a sharp increase in raw bit error rate after about 150 reprograms [4]. (2) Using 8-bit ECC, the bit error rate dramatically decreases. (3) Using ECC and Meta-Cure, the bit error rate drops to a reasonably low rate, such that the reliability is enhanced.

the increasing density of flash causes severe reliability issues. Previous studies have shown that most multi-level-cell (MLC) flash chips experience a sharp increase in bit error rate after a number of reprograms [4]. As shown in Figure 1, raw bit error rate of flash caused by program disturb increases from 10^{-5} to 2×10^{-3} after 200 reprogram operations, and it further climbs to 0.013 after 400 reprograms. With the trend of increasing capacity of flash memory, the bit error rate will become even worse because of closer voltage levels assigned to consecutive logic states.

Error correcting codes (ECC) are widely used in NAND flash memory to ensure data integrity. ECC can effectively protect data when the bit error rate is relatively low (e.g., SECDED ECC can provide single-error correction and double-error detection). Since the strength of ECC is predefined and fixed by chip manufacturer, with the wear of flash, ECC's ability to handle multi-bit errors beyond what it was originally designed for greatly diminishes over the lifetime of the flash. This poses a threat to the integrity of metadata (e.g., file system metadata, page mappings) stored in flash. If a flash memory page contains metadata, the data corruption of the page is very serious, as it may cause an unintended change in functionality of the entire flash. This observation motivates us to design a strategy to enhance the built-in ECC and to provide a more reliable NAND flash memory storage system.

In order to provide a reliable flash memory storage system, the primary objective is to ensure the integrity of metadata

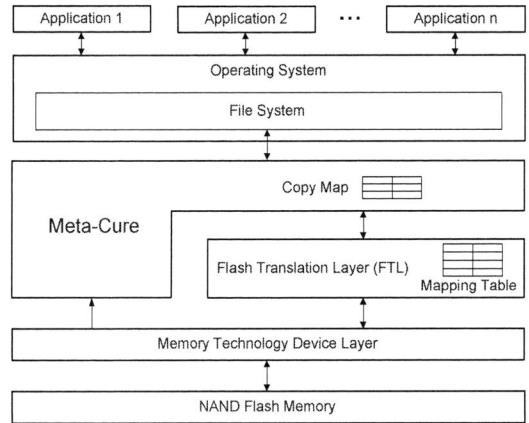

Figure 2: System architecture for Meta-Cure.

stored in flash while considering the distinct characteristics of flash memory (i.e., endurance and "out-of-place update"). Most existing approaches aim to minimize either the amount of critical data or/and the number of block erase counts [5, 9]. Reliability is enhanced through the modification of different components in NAND flash architecture, such as file system [15, 18], hardware implementation of flash memory chip [11, 13], energy consumption [8], or an intermediate software module called flash translation layer (FTL) [12, 16]. These approaches can provide good solutions in terms of endurance, wear-leveling, memory usage, energy consumption, and response time. Nevertheless, they make no specific attempt to cope with faults in either the hardware or the software design.

Only a few approaches have been proposed to achieve fault-tolerant flash memory systems [2, 6, 10, 17]. These approaches apply their fault-tolerant policies at the file system level, which requires significant modifications to existing address mapping schemes and error correction methods. Although recent literature has investigated from the perspective of file system and hardware implementation of flash, they have not yet explored opportunities at the level between file system and flash translation layer, with the resulting advantage that *no changes are required to the file system, cache hierarchy, or hardware implementation of the flash memory chip.*

This paper presents *Meta-Cure*, a new hardware and file system interface that transparently protects *critical data* (e.g., page mappings, and file system metadata) in the presence of faults. Figure 2 shows the system architecture for Meta-Cure. Meta-Cure handles write/read requests from file system and transparently issues the requests to flash translation layer. It cooperates with the built-in ECC to enhance the reliability of flash. Meta-Cure adopts *metadata replication* to dynamically replicate write requests for critical pages and uses *selective update* to prevent the accumulation of errors. Redundant pairs are formed and updated at run time to protect against faults.

We implement Meta-Cure in the Linux kernel and evaluate Meta-Cure using a variety of I/O traces. We use the number of uncorrectable page errors as a performance metric to evaluate the reliability of Meta-Cure. Experimental results show that our approach can reduce the number of uncorrectable page errors by 92.13%, while introducing less than 1% overhead for the total number of block erase counts in comparison with the baseline error correction scheme.

The rest of this paper is organized as follows. Section 2 discusses the motivation and analyzes the problem. Section 3 presents our proposed Meta-Cure in detail. Section 4 presents experimental results on the reliability enhancement and space overheads. Finally, in Section 5, we conclude the paper and describe future work.

2. MOTIVATION AND ANALYSIS

NAND flash memory normally uses ECC to ensure data integrity. ECC typically specifies the number of bit errors they can detect and the number of errors they can correct for one physical page (or one subpage). Given a B-bit physical page, if the bit error rate is p and the ECC scheme can correct N bits and detect M bits errors, the probability with more than N faulty bits on a page is:

$$\sum_{i=N+1}^{B} \binom{B}{i} \times p^i \times (1-p)^{B-i} \qquad (1)$$

In Meta-Cure, we make redundant copies for critical pages to enhance the reliability. By duplicating one page, the probability that both pages have more than N faulty bits is:

$$\sum_{j=N+1}^{B} \sum_{i=N+1}^{B} \binom{B}{i} \times \binom{B}{j} \times p^{i+j} \times (1-p)^{2B-i-j} \qquad (2)$$

We can also get the probability that the original page has more than N faulty bits, while the redundant copy has less than or equal to N faulty bits:

$$\sum_{j=0}^{N} \sum_{i=N+1}^{B} \binom{B}{i} \times \binom{B}{j} \times p^{i+j} \times (1-p)^{2B-i-j} \qquad (3)$$

To illustrate how data integrity is enhanced by duplicating one critical page, Figure 3 shows an example. For the sake of illustration, we assume that the size of each physical page is 2KB (16384 bits) and bit error rate is 10^{-5} [14]. We adopt SECDED ECC to detect and correct errors.

Figure 3(a) illustrates the case that more than one error occur on a page. In this case, page P0 cannot provide correct data. Based on Equation 1, the probability of case (a) is 1.20×10^{-2}. Note that this probability is calculated when bit error rate equals to 10^{-5}, corresponding to the initial bit error rate of a new flash chip. With the wear of flash, bit error rate experiences a sharp increase after several reprogram operations [4]. Then the probability of case (a) will dramatically increase. If an uncorrectable page contains metadata or other critical data, the change in state is very serious, as

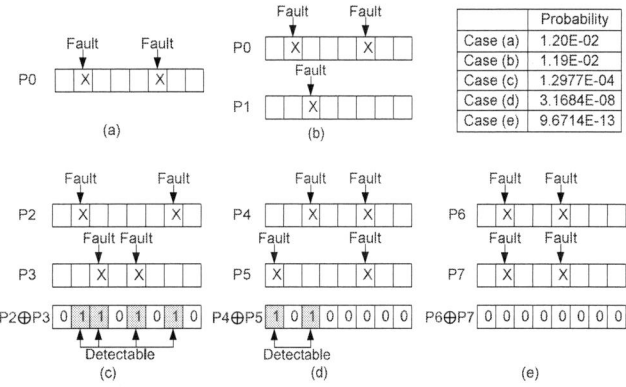

Figure 3: A motivational example of duplicating one critical page.

it may cause an unintended change in functionality of the flash.

Figure 3(b) illustrates the case that page P0 has more than one faulty bit while page P1 has less than two faulty bits. Using a redundant copy, even though the original page P0 is corrupted, page P1 can still provide correct data. Based on Equation 3, the probability of this case is 1.19×10^{-2}, which is similar to the probability of case (a). This result shows that a redundant copy can effectively enhance the data integrity of critical data.

Even when both copies have 2 faulty bits each, there may be opportunities to correct errors, as shown in Figure 3(c) and (d). Figure 3(c) shows the case where the faulty bits in pages P2 and P3 are at different locations; therefore a simple XOR operation still allows us to identify the faulty bits in both pages and recover completely. Similarly, case (d) allows us to recover 2 bits. Furthermore, we note that case (c) has a high probability compared with cases (d) and (e) as shown in Figure 3's probability table, thus indicating that additional opportunities for multi-bit error recovery may be possible.

This example shows that by complementing the built-in ECC and duplicating critical pages, the reliability of the system is enhanced. This observation motives us to propose a fault handling strategy to supplement the built-in ECC to provide a more reliable flash memory.

3. META-CURE: A RELIABILITY ENHANCEMENT STRATEGY FOR METADATA

3.1 Structure and Strategy of Meta-Cure

To facilitate the replication of critical pages, Meta-Cure introduces a new level of indirection between the file system and hardware to protect critical pages from errors and to improve the reliability of flash (see Figure 2). Meta-Cure adopts a series of features to be integrated into the flash memory storage system. First, Meta-Cure handles each write or read request from the file system, and it identifies critical pages. Second, Meta-Cure creates new logical page numbers (LPNs) for critical pages and uses a *Copy Map* to maintain the copy information. Third, Meta-Cure issues write or read operations to the flash translation layer and transparently protects metadata in the presence of faults. Finally, Meta-Cure monitors the errors reported from Memory Technology Device (MTD) layer and selectively updates faulty pages.

Meta-Cure adopts two reliability enhancement strategies, *metadata replication* and *selective update*, to ensure the data integrity of critical pages. *Metadata replication* is a coarse-grained strategy that determines the number of copies of each critical page. *Selective update* is a fine-grained strategy that checks the error rate of a page and decides whether or not to update the faulty page.

Meta-Cure saves redundant copies of each critical page. The number of copies is dependant on three factors: current bit error rate, the error correction strength of the built-in ECC, and the application's tolerable error rate. Since the basic unit for a read or write operation in flash is a page, we define the page error rate after applying ECC and Meta-Cure as *uncorrectable page error rate* (UPER). If an ECC scheme can correct N bits for a B-bit physical page, the raw bit error rate of flash is p, and Meta-Cure preserves k copies

of a critical page, then uncorrectable page error rate is:

$$UPER = \sum_{i_k=N+1}^{B} \cdots \sum_{i_2=N+1}^{B} \sum_{i_1=N+1}^{B} \binom{B}{i_1}\binom{B}{i_2}\cdots\binom{B}{i_k} \cdot p^{\sum_{j=1}^{k} i_j} \cdot (1-p)^{kB - \sum_{j=1}^{k} i_j} \tag{4}$$

The uncorrectable page error rate should be less than the required error rate of the application to ensure the integrity of critical data. Since the raw bit error rate p is constantly increasing through the lifetime of flash memory, the number of copies k for a critical page is configurable by Meta-Cure. Metadata comprises a very small percentage of the total file system capacity, so the redundancy for critical pages incurs very low overhead.

The second reliability enhancement strategy of Meta-Cure is selective update. NAND flash memory suffers from the "out-of-place update" write constraint. When one page is written, it cannot be updated (rewritten) until the block with this page is erased. If a page consists of several faulty bits that can be corrected by ECC, a new page has to be allocated to perform the update operation. Most error correction schemes in flash memory do not provide update operations for faulty pages. Then accumulated faulty bits on a page may eventually go beyond the capability of error correction by ECC.

Meta-Cure selectively updates faulty pages. It compares the number of faulty bits on a page reported from MTD layer, denoted by N_{faulty}, with a predefined threshold $\mathcal{T}$, to decide whether or not to update the faulty page. If the page contains very few faulty bits ($\mathcal{T} \geq N_{faulty}$), then Meta-Cure will not update the faulty page. Otherwise, a new free page is allocated to store the correct data. The threshold value is a tunable parameter that can be set by the designers or the file system depending the degree of fault-tolerance needed.

3.2 Write and Read Operations for Meta-Cure

This section presents the write and read operations for Meta-Cure. Algorithm 3.1 describes the process of a write operation. Algorithm 3.1 has two inputs: LPN and its page

---

Algorithm 3.1 Write operation for Meta-Cure

Input: A logical page number ($LPN\#A$), LPN page content.
Output: Write the content to a physical page.
1: **if** $LPN\#A$ is a critical page **then**
2:     **if** $LPN\#A$ exists in Copy Map **then**
3:         Get the redundant LPN ($LPN\#A'$) from Copy Map.
4:     **else**
5:         Meta-Cure creates a redundant LPN ($LPN\#A'$).
6:         Meta-Cure creates the copy information for the redundant pair $\langle LPN\#A, LPN\#A' \rangle$ in Copy Map.
7:     **end if**
8:     **for** each LPN in $\langle LPN\#A, LPN\#A' \rangle$ **do**
9:         Meta-Cure issues a write request to FTL.
10:     **end for**
11: **else**
12:     Meta-Cure issues a write request to FTL.
13: **end if**
14: **for** each write request **do**
15:     FTL checks mapping table in RAM and flash.
16:     **if** the mapping information for the LPN exists **then**
17:         FTL invalids the physical page and mapping information.
18:     **end if**
19:     FTL writes the page content to flash via MTD and updates mapping information.
20: **end for**

---

content. In the first step, Meta-Cure handles each write request from the file system, and it distinguishes critical pages from non-critical pages (Line 1). Meta-Cure monitors the logical page request and checks the request queue that is populated by filesystem driver. Meta-Cure maintains a small table called Copy Map to track the logical page numbers for critical pages. Note that, Meta-Cure is a general strategy that can be incorporated with various reliability enhancement schemes at different levels of flash memory systems. Users or system developers can also determine the level of reliability and manipulate the differentiation between critical pages and normal pages.

For a critical page containing metadata, if the LPN exists in Copy Map, the redundant LPN can be found in Copy Map (Line 3). Otherwise, Meta-Cure will create a second logical page number and issue two write requests to FTL (Line 8-10). For a non-critical page, Meta-Cure will directly pass the request to FTL. For each write request, FTL will handle the write request and allocate a physical page. It will search and update the mapping table stored in RAM or in flash (Lines 15-18). Based on the mapping information, FTL will issue write operations to the MTD layer, and the MTD layer will write or update data in the flash memory chip (Line 19).

Algorithm 3.2 describes the process of a read operation. The logical page number is used as the input. Similar to the process of a write operation, Meta-Cure will also differentiate the read request for a critical page from that for a non-critical page (Line 1). For a critical page, Meta-Cure checks Copy Map to obtain two LPNs (Line 2). For each LPN of the critical page, Meta-Cure will issue a read operation to the FTL (Line 4). Then it will check the mapping table and read the data from flash (Lines 5-6). Different from write operations, Meta-Cure will monitor the errors reported from the MTD layer (Line 7). Meta-Cure will adopt the selective update strategy, and it will update faulty pages and the corresponding mapping table (Line 9). For a non-critical page, it will directly issue the read operation to the FTL and obtain the page content from flash (Lines 11-13). Meta-Cure

---

Algorithm 3.2 Read operation for Meta-Cure

Input: A logical page number ($LPN\#A$).
Output: Read the content from a physical page.
1: **if** $LPN\#A$ is a critical page **then**
2:     Meta-Cure checks Copy Map and gets the redundant pair $\langle LPN\#A, LPN\#A' \rangle$.
3:     **for** each LPN in $\langle LPN\#A, LPN\#A' \rangle$ **do**
4:         Meta-Cure issues a read request to FTL.
5:         FTL checks mapping table and issues a read request to MTD.
6:         MTD reads the page content from flash and sends it to Meta-Cure through FTL.
7:         Meta-Cure monitors MTD and gets the number of faulty bits on the physical page.
8:     **end for**
9:     Meta-Cure adopts *selective update* strategy to update faulty page(s) and the corresponding mapping information.
10: **else**
11:     Meta-Cure issues a read request to FTL.
12:     FTL checks mapping table and issues a read request to MTD.
13:     MTD reads the page content from flash and sends it to Meta-Cure through FTL.
14: **end if**
15: Meta-Cure returns the page content to file system.

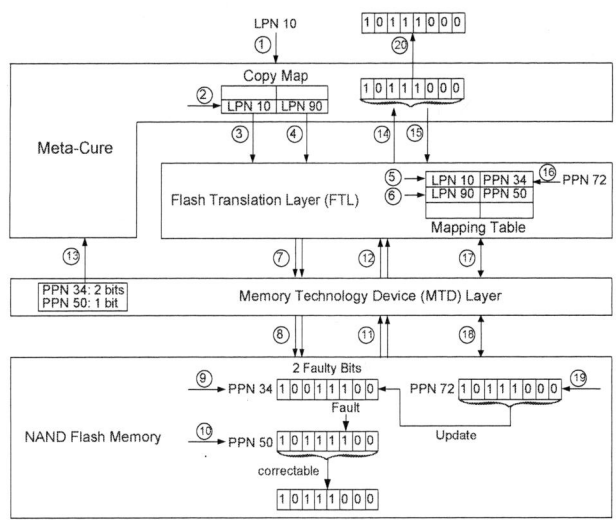

Figure 4: An example of *metadata replication* and *selective update* for a critical page.

will send the results to the file system (Line 15).

Figure 4 shows an example of metadata replication and selective update for a critical page. For the sake of illustration, we assume that each page consists of 8 bits, and an ECC scheme that can provide single-error correction and double-error detection. The threshold $\mathcal{T}$ for selective update is set as one. The following sequence of actions occur in Figure 4. Given a read request (LPN#10) from file system, (1) Meta-Cure finds out that the LPN contributes to a critical page. (2) Meta-Cure checks the Copy Map to obtain the redundant pair $\langle LPN\#10, LPN\#90 \rangle$. (3) and (4) Meta-Cure issues two read requests to FTL. (5) and (6) FTL checks the mapping table and obtain corresponding physical page numbers (PPNs). (7) FTL issues two read requests to MTD. (8) MTD reads data from PPN#34 and PPN#50. (9) and (10) ECC circuit detects two faulty bits in PPN#34 and corrects one faulty bit in PPN#50. (11) and (12) MTD gets the results from flash and reports them to FTL. (13) Meta-Cure monitors the errors reported from MTD. (14) FTL sends the results to Meta-Cure. (15) Since page PPN#34 has two faulty bits and $\mathcal{T}$ is less than 2, this satisfies the condition for selective update. Meta-Cure issues a write operation to FTL to update the faulty page PPN#34. (16) FTL allocates a new physical page PPN#72 to replace PPN#34. FTL updates the mapping table. (17) and (18) FTL issues a write operation to PPN#72 and invalidates the physical pages PPN#34. (19) MTD writes the correct data (10111000) to PPN#72. (20) Meta-Cure sends the correct data to file system.

3.3 Overhead Analysis

Most of the widely used fault detection or correction mechanisms (e.g., triple-modular redundancy) necessarily allocate a portion of the available storage space to implement redundancy, and thus cause the protected part's product specification to be lower than the full capacity of the device [7]. Meta-Cure, on the other hand, only applies data redundancy to metadata in critical pages. Since metadata comprises a very small percentage of the total file system capacity (typically in the low single digits), the redundancy

217

for critical pages incurs very low overhead (from the additional write and erase operations). Our experimental results also demonstrate that Meta-Cure can significantly improve reliability with less than 1% space overhead. Furthermore, Meta-Cure is implemented through a hardware-software interface that supplements the built-in ECC scheme, while requiring zero modifications to the hardware of flash and file system. Therefore Meta-Cure can be implemented with low overhead.

4. EVALUATION

4.1 Experimental Setup

Meta-Cure was implemented on Fedora 7 (Linux kernel 2.6.21). The performance evaluation is based on a trace-driven simulation. The trace of I/O request was collected from a desktop running DiskMon [1]. To eliminate the effects of the operating system's internal operations, a 40GB external hard drive enclosure is used to collect disk access characteristics. Table 1 lists the characteristics of traces.

Table 1: Characteristics of traces.

Trace	# of write operations	# of read operations	% of write	% of read
chatOnline	559,085	472,907	54.18	45.82
copyFiles	3,145,994	2,128	99.93	0.07
p2p	3,695,873	2,101,474	63.75	36.25
unzipFiles	2,425,940	2,441,992	49.84	50.16
zipFiles	2,588,585	2,614,055	49.76	50.24

In our experiments, a 8Gbit NAND flash memory is configured based on the specifications of a flash chip K9F8G08U0M from SAMSUNG. K9F8G08U0M provides 8-bit ECC correction for one physical page. We simulate a set of errors caused by program disturb and read disturb. The distribution and the probability of errors are obtained from Grupp et al. [4]. The page size, the number of pages in a block, the block size, and the size of spare area of a page are set as 4KBytes, 64, 256KBytes, and 128Bytes, respectively. The threshold $\mathcal{T}$ for selective update is set as 6. We adopt a fault-tolerant cache [3], which can ensure its reliability. This configuration aims to guarantee critical information against data loss.

4.2 Results and Discussion

In this section, we present the experimental results with analysis. We adopt the default error correction scheme in the Linux kernel as the baseline scheme, labeled as "Baseline". To make a fair comparison, both the baseline scheme and Meta-Cure handle the traces containing the same set of errors.

4.2.1 Uncorrectable Page Errors

An uncorrectable page error can be caused as the result of data corruption along the path of address mappings from a logical page number to a physical page number, or can be as the result of accumulated multi-bit errors on a page that could not be corrected by ECC. Figure 5 presents the number of uncorrectable page errors. Not surprisingly, Meta-Cure can significantly reduce the number of errors. We observe that Meta-Cure also suffers from several uncorrectable page errors. That is because, the probability of damaging a critical page is almost the same as that for a normal data page. Uncorrectable page errors in Meta-Cure are due to

the accumulated multi-bit errors on normal pages. Different from critical pages, normal pages usually contain non-critical data (e.g., pixel data), which do not influence the functionality of flash.

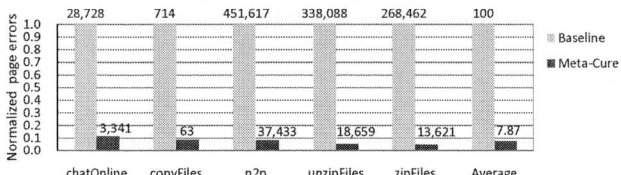

Figure 5: The number of uncorrectable page errors for the baseline scheme and Meta-Cure.

4.2.2 Block Erase Counts

Since Meta-Cure adopts the metadata replication strategy for critical pages, it will issue more write operations and cause more erase operations. This overhead is quantified, and the results are shown in Figure 6. We observe that Meta-Cure causes very low extra erase operations (0.51%) in comparison with the baseline scheme. This is because of two reasons: first, the number of critical pages is relatively small comparing to the total number of pages in a flash; second, Meta-Cure adopts the selective update strategy, such that only the faulty pages with many accumulated bit errors will be updated.

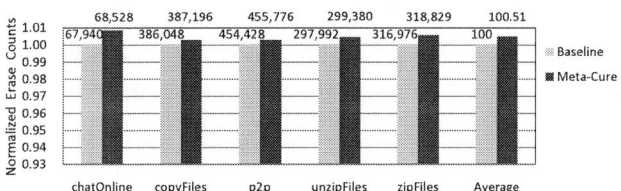

Figure 6: The total number of block erase counts for the baseline scheme and Meta-Cure.

4.2.3 Endurance and Wear-Leveling

The endurance of NAND flash is mainly affected by the worst case erase count of a physical block in the flash. Figure 7 presents the results. We observe that Meta-Cure can achieve similar or even better results in comparison with the baseline scheme. A critical page normally stores metadata that may control several physical pages. Updating one address mapping is normally associated with at least one erase operation. By storing redundant critical pages in flash, Meta-Cure increases the chance to erase the physical block

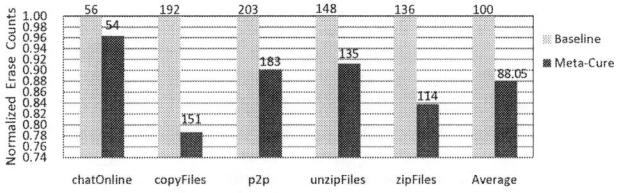

Figure 7: The maximum number of block erase count for the baseline scheme and Meta-Cure.

218

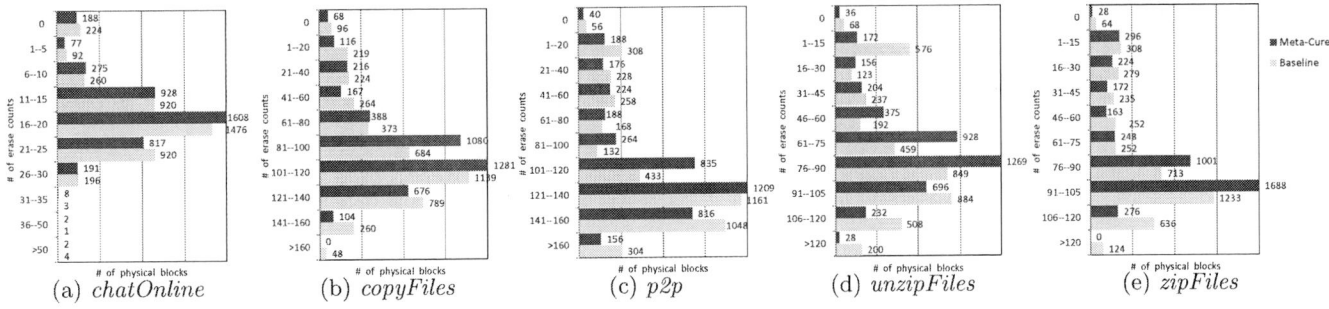

(a) *chatOnline* (b) *copyFiles* (c) *p2p* (d) *unzipFiles* (e) *zipFiles*

Figure 8: The wear-leveling for the baseline scheme and Meta-Cure.

containing critical pages. Although Meta-Cure will introduce low overhead in terms of the total number of block erase counts, it can achieve better worst case erase count by distributing erase operations across different blocks.

Wear-leveling is another factor that influences the endurance of flash memory. Figure 8 shows the distribution of the number of block erase counts for physical blocks in flash. We observe that Meta-Cure can achieve better wear-leveling compared with the baseline scheme. From the results, some physical blocks have very limited numbers of block erase counts. Two types of physical blocks could cause this wear-leveling problem. It could be a data block that is not frequently updated, or an untrackable block due to the data corruption of critical pages. Since untrackable blocks may contain valid pages, a read operation to an existing valid page could point to wrong locations or eventually incur an uncorrectable page error. Meta-Cure can accelerate the process to find these untrackable blocks and improve the wear-leveling of the flash.

5. CONCLUSION

In this paper, we proposed Meta-Cure, the first hardware and file system interface to enhance the reliability of NAND flash memory in the presence of multi-bit errors. Meta-Cure provides protection for critical data using two strategies: metadata replication and selective update. Experimental results show that our Meta-Cure approach reduces the number of uncorrectable errors by 92.13%, improves the endurance of flash by 11.95% in comparison with the baseline scheme, while suffering only a 0.51% degradation in block erase counts. In the future, we plan to investigate the access pattern of flash to build an application specific flash management system, which can improve the effects of our approach.

6. ACKNOWLEDGMENTS

The work described in this paper is partially supported by the grants from NSF Variability Expedition (Grant Number CCF-1029783), the Innovation and Technology Support Programme of Innovation and Technology Fund of the Hong Kong Special Administrative Region, China (ITS/082/10), and the Hong Kong Polytechnic University (G-YK24).

7. REFERENCES

[1] DiskMon for Windows. *http://technet.microsoft.com/en-us/sysinternals/bb896646.aspx*.

[2] M. Balakrishnan, A. Kadav, V. Prabhakaran, and D. Malkhi. Differential RAID: Rethinking RAID for SSD reliability. *Trans. Storage*, 6:4:1–4:22, July 2010.

[3] A. Chakraborty, H. Homayoun, A. Khajeh, N. Dutt, A. Eltawil, and F. Kurdahi. E < MC^2: less energy through multi-copy cache. In *CASES '10*, pages 237–246, 2010.

[4] L. M. Grupp, A. M. Caulfield, J. Coburn, S. Swanson, E. Yaakobi, P. H. Siegel, and J. K. Wolf. Characterizing flash memory: anomalies, observations, and applications. In *MICRO '09*, pages 24–33, 2009.

[5] A. Gupta, Y. Kim, and B. Urgaonkar. DFTL: a flash translation layer employing demand-based selective caching of page-level address mappings. In *ASPLOS '09*, pages 229–240, 2009.

[6] P.-H. Hsu, Y.-H. Chang, P.-C. Huang, T.-W. Kuo, and D. H.-C. Du. A version-based strategy for reliability enhancement of flash file systems. In *DAC '11*, pages 29–34, 2011.

[7] E. Ipek, J. Condit, E. B. Nightingale, D. Burger, and T. Moscibroda. Dynamically replicated memory: building reliable systems from nanoscale resistive memories. In *ASPLOS '10*, pages 3–14, 2010.

[8] Y. Joo, Y. Cho, D. Shin, and N. Chang. Energy-aware data compression for multi-level cell (MLC) flash memory. In *DAC '07*, pages 716–719, 2007.

[9] Y. Joo, Y. Choi, C. Park, S. W. Chung, E. Chung, and N. Chang. Demand paging for OneNAND flash eXecute-in-place. In *CODES+ISSS '06*, pages 229–234, 2006.

[10] Y. Kang and E. L. Miller. Adding aggressive error correction to a high-performance compressing flash file system. In *EMSOFT '09*, pages 305–314, 2009.

[11] S. Kim, C. Park, and S. Ha. Architecture exploration of NAND flash-based multimedia card. In *DATE '08*, pages 218–223, 2008.

[12] T.-W. Kuo, Y.-H. Chang, P.-C. Huang, and C.-W. Chang. Special issues in flash. In *ICCAD '08*, pages 821–826, 2008.

[13] Y. Lee, S. Jung, and Y. H. Song. FRA: a flash-aware redundancy array of flash storage devices. In *CODES+ISSS '09*, pages 163–172, 2009.

[14] N. Mielke, T. Marquart, N. Wu, J. Kessenich, H. Belgal, E. Schares, F. Trivedi, E. Goodness, and L. Nevill. Bit error rate in NAND flash memories. In *IEEE International Reliability Physics Symposium*, pages 9–19, 2008.

[15] S. K. Mylavarapu, S. Choudhuri, A. Shrivastava, J. Lee, and T. Givargis. FSAF: file system aware flash translation layer for NAND flash memories. In *DATE '09*, pages 399–404, 2009.

[16] Z. Qin, Y. Wang, D. Liu, Z. Shao, and Y. Guan. MNFTL: an efficient flash translation layer for MLC NAND flash memory storage systems. In *DAC '11*, pages 17–22, 2011.

[17] Y. Wang, L. Bathen, Z. Shao, and N. Dutt. 3D-FlashMap: A physical-location-aware block mapping strategy for 3D NAND flash memory. In *DATE '12*, pages 1307–1312, 2012.

[18] P.-L. Wu, Y.-H. Chang, and T.-W. Kuo. A file-system-aware FTL design for flash-memory storage systems. In *DATE '09*, pages 393–398, 2009.

EDA for Secure and Dependable Cybercars: Challenges and Opportunities

Farinaz Koushanfar
Electrical& Computer Engineering
Rice University
Houston, TX
farinaz@rice.edu

Ahmad-Reza Sadeghi[†], Hervé Seudie[†,‡]
[†]Fraunhofer SIT, Germany
[†]Intel-TU Darmstadt Security Institute, Germany
[‡]Robert Bosch GmbH, Germany
{ahmad.sadeghi,herve.seudie}@trust.cased.de

ABSTRACT

Modern vehicles integrate a multitude of embedded hard realtime control functionalities, and a host of advanced information and entertainment (infotainment) features. The true paradigm shift for future vehicles (cybercars) is not only a result of this increasing plurality of subsystems and functions, but is also driven by the unprecedented levels of intra- and inter-car connections and communications as well as networking with external entities.

Several new cybercar security and safety challenges simultaneously arise. On one hand, many challenges arise due to increasing system complexity as well as new functionalities that should jointly work on the existing legacy protocols and technologies; such systems are likely unable to warrant a fully secure and dependable system without afterthoughts. On the other hand, challenges arise due to the escalating number of interconnections among the realtime control functions, infotainment components, and the accessible surrounding external devices, vehicles, networks, and cloud services. The arrival of cybercars calls for novel abstractions, models, protocols, design methodologies, testing and evaluation tools to automate the integration and analysis of the safety and security requirements.

Categories and Subject Descriptors

C.3 [**Real-time and embedded systems**]; D.4.6 [**Software**]: OPERATING SYSTEMS—*Security and Protection*

General Terms

Security, Reliability

Keywords

Perspective Article, Automotive Security, CPS Security

1. INTRODUCTION

Modern vehicles include several Electronic Control Units (ECUs) that form an in-vehicle distributed networked embedded system. The ECU networks not only command the hard realtime control of automobile mechanical parts and support infotainment functions [14], but also they provide a gateway between modern cars and their surroundings (e.g., traffic lights), devices (e.g., smartphones), vehicles, and accessible networks. The terms car-to-car and car-to-X (infrastructure or device) are used to refer to cybercars' communication scenarios. The emergence of Intelligent Transport System (ITS) is expected to further reduce the number of road accidents and improve the road traffic conditions. Furthermore, connecting cars to the cloud or smartphones offers a new set of applications and business models. In this context, Infotainment and safety related subsystems may need to interact to provide various information to the driver.

Figure 1 shows the view of ITS depicted by the standardization body ETSI (European Telecommunication Standards Institute). While the introduction of new communication technologies offers an unprecedented number of new opportunities, it increases the complexity of the car system and demands new analysis of security and privacy requirements. A few preliminary attacks that could seriously impact the car's safety and reliability have already been demonstrated in simulations or experiments including [26, 37, 27, 28, 43, 31, 12]. For example, a study by Barisani et al. shows that 2 malicious cars out of 400 vehicles affected 20% of the traffic [9], where this estimation did not even consider system failures due to design errors or poor implementation and testing schemes. A comprehensive summary of related work which includes the description of the demonstrated attacks, is provided in Appendix A.

The cyber-physical attributes of modern automotive systems directly link security vulnerabilities to the cybercar's physical safety and reliability features. Therefore, the scope of the potential vulnerabilities is much more vast than what has been demonstrated, and is far beyond attacking an individual car [42, 29]. Thinking of vehicle safety without considering security, as it was done in the past, is no longer a viable option. In this respect, security vulnerabilities of cars are markedly different from typical security issues in conventional computer and network systems.

A few rather recent projects and initiatives have started investigating the safety and security of modern vehicular electronics, including [51, 52, 20]. The work thus far has mostly centered on identifying protection primitives to be included in the emerging standards, characterizing the attack models and security threats, as well as devising technical and specific protection guidelines to further secure cybercars. Independent of the cybercar security initiatives, the

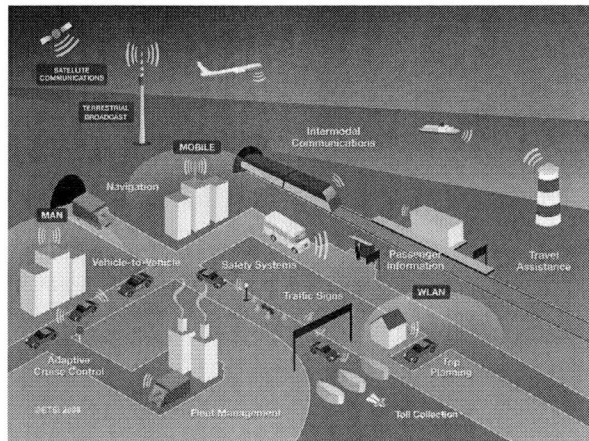

Figure 1: Intelligent Transport Systems. Source: www.etsi.org

EDA and embedded systems communities have been working for years on problems pertaining to modeling, analysis, simulation, and automation of complex vehicular networked embedded systems. Such methods are required not only to ensure design time predictability and composability, but also for architecture selection and design-space exploration [48].

The complexity of the modern cybercar's networked embedded systems, its various interfaces to external entities, together with the scope of emerging attacks, supersedes present knowledge and capabilities pertaining to both lines of efforts in vehicular security and in embedded system communities. The evolving nature of the system complexity and attack possibilities suggests that a continuous flow of research and development is needed before cybercar systems can be efficiently designed, and safely/securely operated.

The situation with cybercars today is similar to the evolution of personal computing and networked communication in the past several decades: One can make an analogy between connecting individual cars to external objects/networks and linking personal computers to the Internet. Since the Internet was not originally designed with an explicit set of security objectives, connected computers still suffer from a range of attacks that could be largely avoided by correct-by-construction methodologies. Due to safety criticality and the vital role of vehicular and transportation systems in personal, business, government, and economic affairs, leaving security as an afterthought is disadvantageous. However, since legacy protocols and hardware take a long time to change, for present and pending cybercar generations, security afterthoughts maybe the only practicable choice.

This perspective article calls for development of novel holistic but systematic EDA methodologies and tools that simultaneously ensure a robust and secure cybercar design flow. The important architecture-evaluation, design, and evaluation process phases need to be automatically augmented with security primitives and flows as defined by rising standards and protection measures. The challenges for realizing such a holistic and systematic automated cybercar security approach are abundant, but so are the research and development opportunities.

2. PRELIMINARIES

Before we delve into more detailed discussions of cybercars security and the pertinent challenges and opportunities, we briefly outline a typical cybercar system architecture and some key evolving standards.

Cybercar system architecture. The term cybercar refers to a generation of vehicles that are fully connected to their surrounding objects, environments, and networks. A cybercar is part of an ecosystem where it could either play the role of a content or a service provider in addition to expressing content and determining applications. Thus, a cybercar can be described as a sophisticated mobile device. As it is connected to external entities, the cybercar has the possibility to rely on outside computing resources, externally available data, and services. For instance, a cybercar without the proper sensors could still run driving assistance applications with the help of a smartphone and a cloud connection. A cybercar may also provide services to connected entities. Cybercars are composed of following domains: (i) the in-car system with the network and the car components, (ii) communication interfaces to external communication partners, (iii) communications partners represent by external devices, cars, infrastructure or cloud. Figure 2 represents an example of a cybercar system architecture.

Emerging automotive standards. With some exceptions [4, 2], automotive standards are only regionally accepted, which might be due to the role of the legislation and the influence of regional automotive industry. However, the AUTOSTAR experience has shown that the global participation of automotive manufacturers and suppliers coming from different countries in the specification process, is a key to the global acceptance of standards [4].

With the emergence of cybercars and the requirement of global acceptance, the automotive industry has created different consortiums to support the specification of standards for ITS. Examples are the GENIVI alliance, the Car Connectivity Consortium and the Car2Car Consortium [7, 13, 55], where the first two are driving the specification of an in-vehicle infotainment platform with different connectivity technologies. The Car2Car Consortium coordinates different investigation results on communication and system architectures for car-to-car and car-to-infrastructure communication. The resulting specifications are provided to standardization bodies. The Sevecom and the EVITA projects are examples of unclassified funded projects [20, 51] that have significantly influenced standardization bodies like the ETSI and the Communication Access for Land Mobiles (CALM) [6, 5].

Typical goals of the automotive industry are guaranteeing interoperability, reliability, dependability and quality. Security has not been a major concern. Hence, standards for the car components and networks have been specified with very limited security requirements. The integration of IT in cars and recent attacks on cars has led to a change of this view [12]. As a result, the security has been integrated both in cybercar related standardization activities and in established specifications such as AUTOSAR, with the introduction of cryptographic interfaces [4]. There is still a need to analyze the overall impact of new technologies emerging on the automotive process, development, and tool chain, and the role of security.

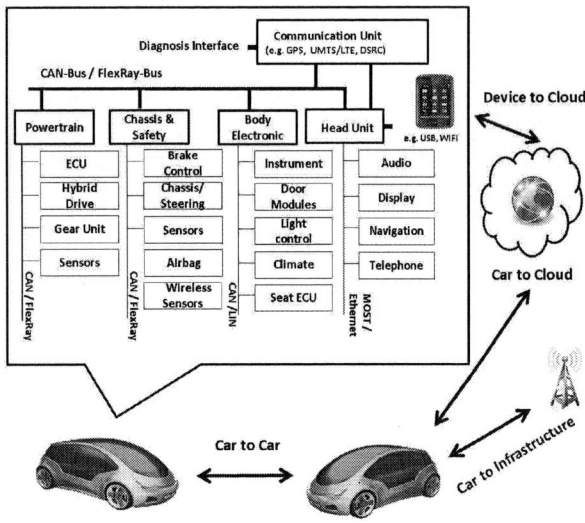

Figure 2: Cybercar System Architecture.

3. CYBERCAR SECURITY CHALLENGES

3.1 Threats

The automobile industry has traditionally focused on providing safety and reliability, under the assumption that a vehicle is an isolated system which is not accessible to adversaries [35], leading to the design of insecure car components and bus systems. Several successful attack scenarios have invalidated traditional models for cybercar security [26, 37, 43, 11, 19, 31, 12, 56]. The vulnerabilities are exasperated by poor implementation of protocols and firmware [30, 39]. The growing connectivity of modern cybercars escalates the range of potential exploits [42, 29, 18]. The existing weaknesses in the automotive domain can be classified as follows:

• Threats caused by physical access, e.g., by connecting to the in-car network through the on-board diagnostic interface, or the in-vehicle network cables.
• Vulnerabilities due to access via infotainment, which introduce the possibility to manipulate cars using multimedia interfaces, e.g., by disabling safety functions.
• Exposures due to the remote access, which describe the possibility to manipulate cars using wireless interfaces.

Exploiting such vulnerabilities may even enable the complete remote control of a car by an attacker, who might possess the expertise to gain access to the in-vehicle network using the on-board diagnostic interfaces or the multimedia interfaces (e.g., USB connection, Bluetooth connection or media players). Such an adversary could remotely disable brakes or manipulate sensor values using typical attacks such as eavesdropping, dropping, modification, spoofing, injection, or message replay on the bus system [43, 12].

3.2 Challenges

The security and dependability challenges of cybercars are closely related to the general engineering challenges of cars and the current state of car networks. These challenges include,

• Automotive components have resource constraints, e.g., limited memory, processing, or number of sensors.
• Automotive networks are insecure and have limited throughput with strict latency requirements.
• Automotive components must be cost effective. Security must be integrated without a high cost overhead.
• The lifecycle of the automotive industry is up to 20 years. Solutions should be capable to hold for a long time.
• Safety critical applications have realtime constraints. Security should not disable the safety function.
• Interference exists between safety and entertainment applications on the infotainment platforms.

In-vehicle security. It seems that there is still no concrete plan to change the specification of the most used automotive bus protocol, the Controller Area Network (CAN) protocol. In fact, CAN is deployed in billions of cars, industry machines and aircrafts. A specification change would impact the complete development and supply chain within the automotive industry. With CAN being the most used protocol for safety critical applications, it has been the most attractive protocol for attackers [27, 26, 37, 31, 12]. The challenge is to embed security in the CAN protocol and ensure that the safety applications are not affected by the changes.

While the CAN specification cannot be changed, using the fields of the CAN frames is a plausible alternative to embed authentication and data integrity. Today's constraints on car applications (e.g., fault tolerance times of about 100ms, message sizes of approximately 2-4 bytes, and asymmetry in the available performance in the different modules) requires the use of efficient cryptography algorithms to fulfill authentication and integrity goals. Thus, researchers have proposed to use a truncated Message Authentication Code (MAC) added in the payload of the CAN frames [50, 37]. This approach can be improved if one considers the error detection attributes of MACs. In the CAN specifications, two bytes are reserved for the Cyclic Redundancy Check (CRC) which the MAC uses. The potential of the combination of MAC and error correcting codes could be investigated as a measure to provide safety and security.

Car-to-X (a.k.a., Car2X) Security and Privacy. The success of Car2X applications depends on their penetration in the car market. Current cars will have to be upgraded for car-2o-X communications. Car2X systems will have to be compatible in terms of messages formats, communications technologies, and security, among other requirements. Moreover, applications will have to react in realtime for safety critical tasks. An example of a Car2X application is an active brake, where a car brakes based on a warning message received from an external communication. This task requires an instant brake manoeuver with a maximum delay of 250ms [21]. In 250ms the car will have to perform plausibility checks on the message content, and verify the authentication and integrity of the message. The challenges can be summarized in the following key words: Upgradeability of car systems, compatibility, realtime performance, trustworthiness, and privacy.

The fact that cybercars may need to deal with huge number of signature verifications per second enforces the use of hardware accelerators to cope with the high computation requirements for authentication and integrity verifications (e.g., of up to 4000 per second [49]). EVITA has made a

222

proposal for such a hardware accelerator enhanced with security features [57]. However, this module still has to be integrated in available architectures [51, 41, 6, 5] to provide a security architecture covering the security requirements of Car2X applications [22]. With the high density of Car2X communications, the common approach is to use public key infrastructure to manage the high number of required keys. An example of a public key infrastructure is proposed in [10]. One important and still open issue is the owner's privacy protection. The project Sevecom proposed the use of short term credentials (e.g., certificates), that are periodically changed by the car users [51]. Nevertheless, other layers in the communication stack may still send sensitive private information such as vehicle position.

Secure integration at the infotainment platform. Recent attacks have demonstrated that the integration surfaces at the infotainment platform (head unit) like the media players or Bluetooth are insecure [12]. The head unit typically runs three types of applications: applications with no access rights to the in-vehicle domains, applications with read access such as diagnostic tasks, and applications with read/write access such as firmware updates. Read access allows collecting privacy-sensitive data, such as location data, fuel consumption, and the vehicle identification number (VIN), which can result in creation of driver profiles. The reuse of such profiles could be interesting for businesses, e.g., insurance companies. Write access allows the control of the car which could potentially cause fatalities. The head unit may run applications with different access rights in parallel, which could lead to confused deputy attacks [25]. Present efforts to isolate the different types of applications are done in the GENIVI consortium but currently seem to rely on desktop standard approaches of virtualization [7] and do not address the unique issues of transportation systems, which is related to the safety of system users.

4. EDA CHALLENGES AND OPPORTUNITIES

Unless security is integrated within the automotive electronics and communication design flows, cybercars will be vulnerable to several nefarious attacks. The scope of the potential exploits is likely much more sophisticated than what has been demonstrated thus far in simulations or in limited practical experiments. The design, realization, and validation of complex networked embedded systems for automotive applications is already a standing challenge, even before the security demands are considered [48]. Security requirements have to be carefully recognized and implemented at each complex step of modeling, simulation, end-to-end design, and implementation of the cybercar systems.

Due to the rising complexity and scale of automotive functionalities, networks, interactions, and the supply chain, stand alone or adhoc protective solutions have a very limited effectiveness. Novel EDA methodologies and tools are required for scalable automated security modeling, integration, verification, checking, and analysis of the cybercar complex systems.

In the remainder of this section, we outline some of the EDA challenges that have to be overcome in order to realize secure cybercars and highlight the great research opportunities in this field.

4.1 Model-based automotive security

Vehicular system integration has been traditionally done based on black-box integrated subsystems along with original equipment manufacturer's high-level specifications and overall performance metrics. The design flow has to be enhanced to better model the secure cybercar system structure and interactions, possibly through refined diagrams consisting of block entities linked by interconnects and flows, which specify a topology and variables for communication among the blocks. A block may represent a logical or physical component, e.g., a hardware or software part, often either representing an abstraction of the full component description or a characterization of the component interface [44, 16, 15]. Like any other proper abstraction, neither description nor interface should contain more information than necessary to realize or connect the components. Let us more carefully investigate the design requirements and security demands for components and interfaces.

Component-models and abstractions. In order to meet the stringent time-to-market constraints, the reuse-based paradigms are a standard practice in design and implementation of complex automotive network embedded systems. Stand-alone component models are typically available and if not, they are attainable at the proper level of abstraction by the original IP owners or manufacturers. However, one has to pay a special attention to component models, especially those for the reusable ready blocks, since their functionality may not be static; their behavior has to be considered in the context of complex dynamic interactions with other system components and environmental/user variables.

Abstraction of the electronic components has traditionally been an integral part of EDA methods and tools. However, the Cyber-Physical nature of cybercar networked embedded systems requires adding a totally new dimension to the component modeling and abstraction. Essentially, there is a need to model and abstract the non-electronic components which include the mechanical parts and user inputs. In the development of such models and abstractions, challenges arise due to the inherent continuous and analog nature of the underlying components, which are far from discrete and (often) binary form of well-known digital or logical abstractions. Attacks based on exploiting the component vulnerabilities that infiltrate the electronic components, the mechanical parts, or the user inputs can be envisioned. Finding the proper level of component abstraction that also captures such potential exposures is a major challenge.

Interface-based design for security. Designers commonly make assumptions about the environment where the component will be employed, or the pairwise interactions between the components. Such descriptions can also be a part of the interface specifications. Abstracting the component security requirements brings upon yet another dimension to the already sophisticated environmental models and interactions among the underlying parts developed by disparate entities. Such secure interface-based models should support compositional refinement as well as different degrees of component abstractions. A model-based approach may profile secure control and dataflow task requirements in a graphical language amenable to graph-theoretic analysis.

According to the recent analyses of automotive security attacks in [12], virtually all exploits outlined in their practical experiments, emerged at the interface between codes written by distinct organizations. A significant advantage of developing sound and proper component and secure interface abstractions is that they can be together used as a precursor to formal verification and correct-by-construction designs. Components and interfaces which expose protocol information about component interactions or secure protocol information can be naturally expressed in an automation-based language. A great introduction to such interface automata and interface languages are outlined in [16, 15]. They have shown that several aspects of interface models, including compatibility and refinement checking for interface interactions can be viewed in a game-theoretic framework.

4.2 Temporal models and constraints

Automotive control systems including the ECUs have traditionally been designed for real-time operations. For example, the CAN protocol implements a deterministic algorithm and assigns priorities to messages. Assuming a fault-free operation and predictable task times, the worst-case timing behavior of such a system can be estimated. However, it has been shown that especially in presence of small changes in the temporal parameters, there are points of discontinuity in the system where the increased number of preemptions adds the execution of one or more tasks to the execution time [48]. This situation is exacerbated for larger and more sophisticated interactions which in turn limit the usefulness of such predictive models.

Priority-based scheduling and security. Practical implementations of security often require a nontrivial overhead on the plaintext processing and tasks such as ciphertext decryption. The complexity would be even worse if keys must be established with external entities, such as on-the-fly Car2X communications which require public key protocols. Priority-based scheduling such as the one implemented by CAN, has very high worst-case latency and discrepancy between the best- and worst- case system timing behavior in the presence of faults or jitters. In particular, a drawback of priority-based resource scheduling is sensitivity to high priority computation or communication flows that can easily take control of the interface or the ECU in order to steal time from lower priority tasks. An intelligent attacker could use this sensitivity to attack the system timing and to violate the system's timing constraints through fault injection in high priority applications.

Additional control layers might be able to avoid the timing faults and could improve the security, but significantly add to the system overhead. Added control layers may very well lead to violations of end-to-end real-time constraints for priority-based schedules, which should be included in the system-level models discussed in Section 4.3.

Isolation of safety-critical and security sensitive tasks by time-based scheduling. Time-based schedulers such as those supported by the FlexRay enforce assignment of the communication bus at predefined time slots and are not always sensitive to system load requests [40]. For example, in FlexRay a cycle is divided into up to four segments: static, dynamic, symbol, and nit. The static part is strictly reserved for transmission of time-critical messages which have a fixed length time slot for each node. The dynamic interval is reserved for non-critical messages with more robust timing requirements. The arbitration among the less critical messages is based on priorities similar to CAN.

Time-triggered communications could allow for better isolation between security sensitive tasks and non-critical applications; they are also better suited for distributed control applications. However, time-triggered protocols are still far from providing a comprehensive security measure. For example, a denial of service attack targeted at fixed time slots could result in significant loss of bandwidth and even in the worst-case, failure in meeting realtime constraints. Thus, along with isolation, attack models as well as proactive and reactive measures to secure against attacks, must be integrated within the system timing analysis framework.

Note that isolation for security may be done by the software control layer interface standards for both priority-based and time-based scheduling policies. For example the AUTOSTAR software standard description allows isolation of timing error for one IP from the other IPs interacting with it. Since such additions may lead to timing violations, a layer in the standard is aimed at addressing performance and delay violations. In a complex control software interactions could generate timing dependencies because of scheduling conflicts, synchronization issues, or buffering. If correct timing models and abstractions are available, it is possible to set up a simulation tool which can model the behavior of tasks and their interactions in a complex environments. An interface-based design methodology could also be used for addressing the security challenges.

4.3 End-to-end secure & reliable integration

An end-to-end design of secure cybercars requires a combination of security requirements along with functional models and abstractions of the architecture and hardware platforms. Novel system-level security abstractions and analysis not only guide partitioning and separation of concerns at the design time, but also accommodate an efficient exploration of the complex design space in early design phases. Such abstractions could lead to a significant increase in efficiency of performance, reliability, delay, and implementation cost of the security solutions. We advocate the use of platform-based designs along with our suggested security flow models and simulation tools for addressing the sophisticated end-to-end system security requirements.

Platform-based design for secure architecture selection. Platform-based design [47, 46] is based upon explicitly characterized layers of abstraction and a design interface between the behavioral specifications and abstractions of possible implementation platforms. Therefore, this methodology decouples application-layer software from variations in the underlying hardware. By decoupling these two components, the same applications are permitted to run across several vehicle platforms without modifications. Once security is abstracted at the proper level, e.g., using the flow model described in Section 4.1, one would be able to use platform-based design principals and tools for system optimization. Essentially, the platform interface and the security requirements should be independent and isolated from lower-level architecture details, while simultaneously allowing design-space exploration with a good predictions of the properties for the system realization.

The design-space-exploration finds the secure system's optimal mapping into a platform candidate instance. There is a need to develop new methods and tools that can provide a measure of the appropriateness of a particular architecture solution for optimizing performance metrics while also satisfying various performance constraints.

Iterative synthesis for security. To find a good optimization solution in the large space of possibilities, often an iterative approach is taken by the EDA community [17]. Such an iterative approach is suitable for the complex mapping between the space of security parameters/flows and the hardware platform. After a set of end-to-end design/security metrics and constraints are specified, a set of initial candidate configurations for the platform architecture are then determined. The candidates are then analyzed and compared for fitting to the design goals and security objectives. If the results are not up to expectations, alternative sets of platform architectures are iteratively evaluated.

Automated robust secure software design. Iterative synthesis and analysis results typically guide the selection of the next set of candidate architectures to evaluate. The analysis include evaluation of delay properties, timing sensitivity, faults, security, and cost. In the cybercar distributed ECU environment, software architecture and mapping can become rather complex. There is an intermediate layer between the specified function and the underlying architecture which is often implemented in software for flexibility reasons. Such tools can be automated to select the best combination of timing, security, and performance constraints.

Counterfeit parts prevention. A possible set of attacks can be launched by the counterfeit automotive parts that are prevalent in the market. The car owners' incentive in buying the fake parts is mostly driven by economics. Counterfeit car electronics not only often have various reliability problems, they could also allow for several nefarious attacks such as Trojan embedding [1, 53]. While legal measures could potentially suppress the rising problem of fake automotive components, they have not been effective in practice. This is largely because of the long and hard-to-track supply chain of fake components, the improved appearance of the cloned components, lack of sufficient reliability and security tests for the fake parts, and the black market nature of the fake parts suppliers [3, 33].

Development of methods for unclonable and secure identification of devices are very relevant for addressing these challenges [24, 36, 8, 45, 32]. Devising possible measures to disable a system when a fake component is identified, are of great interest. For cases where the exact functional description of an electronic part is unknown or a part cannot be found because of its obsoleteness, one may need to reverse engineer the functionality. Thus, research and developments in functional reverse-engineering, and in formal functional verification are important for preventing counterfeits.

4.4 Security validation and testing

System security models and exact characterization of attacks are both necessary steps for proactively or reactively protecting against pertinent vulnerabilities.

Rule checking for secure implementation correctness. There is a need for methods that define and express Security Rule Checks (SRC) based on a set of protection primitives and security protocols at the proper abstraction layer, e.g., at the control and data flow vulnerable interface levels [12]. The SRC can be used to enforce unimplemented constraints of the existing protocols which may not be always in effect, as suggested in [31]. Many widely-used strong protection protocols, including those for access control, encryption, or secure sessions, are available in a standard format. To ensure robustness of interfaces against potential adversaries, implementation of such protocols within the system's realtime constraint is necessary; this may lead to requiring hardware acceleration. Once security community spends the time and effort to develop security protocols, new SRC tools must be developed accordingly to ensure a correct implementation.

Continuous attack analysis and countermeasure development. The attacks targeted at cybercars will likely not be static and will evolve over time, as it is the case with other computer and network attacks. As researchers and practitioners devise new security primitives, rules, and protocols, adversaries could simultaneously find new holes in the system or its implementation. As the protection protocols and security methods become more advanced, the attacks will become more sophisticated. There is a need to continuously and dynamically monitor for potential vulnerabilities and attacks. Once instances of attacks are observed, the corresponding countermeasures should be implemented within the system. Software patches and anti-virus software should be continually updated to limit the spread of any exploits. Online tests for detection of possible exploits should also be implemented and enforced in the cybercar systems.

Counterfeit detection. Since counterfeits provide physical access to the system, malicious fake devices could potentially launch efficient attacks. They may be Trojans that spy information to the outside world or disrupt the system's functionality on a trigger event. Even when the counterfeit parts are not intentionally malicious, they introduce a high risk to the system reliability [33]. This is because the counterfeit components are often lower quality grades, or recycled old ICs. Therefore, the system designers must automatically embed means for testing and detecting the discrepancy and unreliability of the system components, and for monitoring/detecting the Trojan components.

Development of cybercar security benchmarks. Like other areas of EDA and testing, it is necessary to create benchmarks and platforms in order to evaluate and compare the competing methodologies. A flurry of research activities are being directed towards addressing the known and potential vulnerabilities of cybercar systems. There is a need to understand the effectiveness and limitations of each method. This will not be only used as an evaluation tool, but also would help in standardization of the best methodologies and tools. Effective security benchmark development requires outlining the taxonomy and details of the attacks, as well as research and experiments which realistic but challenging hard-to-address attack instances.

5. CONCLUSION

Tight integration of networked computation and physical components in safety-critical modern automotive environments and applications (cybercars) makes them exceptionally vulnerable to security attacks, as confirmed by a number of recent studies. The evolving nature of the complexity of cybercar systems and the severity of possible attacks suggest integration of security within the embedded system design flow. This perspective article highlights the importance of developing novel holistic and systematic EDA methodologies and tools that simultaneously ensure a robust and secure cybercar design flow. We discussed the challenges and opportunities in realizing our proposed vision.

APPENDIX

A. RELATED LITERATURE

Modern vehicular technology is typically comprised of several intercommunicating electronic and software components that enable ever-increasing flexibility, efficiency, safety, and a myriad of new and exciting functionalities [14]. While automotive systems, standards, functions, and components are almost always devised to satisfy strict safety and reliability constraints, requirements for security and protection of the pertinent components and functions have not yet been implemented. Therefore, concerns over the potential risks and vulnerabilities of this complex networked environment were expressed at a high level [59, 39, 58, 34, 35, 54]. Such worries have been exacerbated by the emergence of newer car-to-X technologies [42, 29].

The EVITA project took the first fundamental step towards addressing the rising concerns about the security of modern cars. The project identified a list of automotive use cases with a security impact that could be potentially misused [21]. It also developed a security and trust model for modern vehicles along with attack scenarios [22]. To address identified vulnerabilities, a set of high level requirements along with concrete technical recommendations have been proposed [23]. Architecture solutions that could address the identified vulnerabilities were subsequently suggested and prototyped in hardware.

A majority of the reported attacks have been targeted at the Control Area Network (CAN) protocol. Attacks on simulation models of CAN were suggested in [26, 38]; the first implementation on a real car was demonstrated in [27], where the researchers performed several practical tests on the CAN network and demonstrated their vulnerabilities. The demonstrated attacks included controlling the windows, lights, and airbag systems. Later, they discussed CAN privacy violation issues [28].

The authors in [43] performed an analysis of the Wireless Tire Pressure System (TPMS) in a modern automobile. They outlined methods to manipulate drivers by spoofing the faulty tire pressure measurements. The tire pressure values are typically sent from the tire pressure wireless sensor to the ECU managing the TPMS data. The authors were able to send wrong values to stop the functionality of the ECU managing the TPSM data. Note that our focus in this work is on the intrusions at the network level that are orthogonal to the practical demonstration of attacks on single-device vehicle access control mechanisms such as those targeting the keyless entry system [19] and vehicle immobilizers [11].

A more comprehensive practical security analysis of the CAN vulnerabilities was later demonstrated in [31], where cars' components were tested in isolation in a lab, in controlled settings, and in live road tests. Their attacks confirm that an attacker infiltrating any ECU in the CAN network could circumvent a large array of of automotive functions while ignoring the driver inputs. The attacks included safety critical tasks such as break disabling, engine halting, and light control. Other less critical but vulnerable modules were heating and cooling, infotainment, and instrument panels. Their findings show the extent of potential damages from the existing security holes, ease of attacks, weakness of contemporary vehicular access control, and the ability to delete the traces of infiltration by the attackers. Lastly, they explore considerations and future possible high-level directions for addressing some of the known vulnerabilities.

With the exception of the wireless access experiment in [43], most of the attacks described thus far assume the adversary has physical access to the car's internal network. More recent work in [12] provided a systematic synthesis of possible attack vectors at three modalities: indirect physical access, short-range wireless access, and long-range wireless access. For each modality, the authors reported the practicality of exploitable vulnerabilities, allowing unauthorized control without the need for physical access. They further demonstrated a number of post-compromise control channels that could act like a remotely controllable Trojan. Capabilities of theft and surveillance were also shown. A set of high-level recommendations for raising the security bar were subsequently suggested. We believe that these efforts are a precursor to future research and developments as we are still far from fully secure and protected cybercars.

B. REFERENCES

[1] Defense science board (DSB) study on high performance microchip supply. http://www.acq.osd.mil/dsb/reports/2005-02-hpms_report_final.pdf.

[2] ASAM: Association for standardisation of automation and measuring systems. http://www.asam.net.

[3] Defense industrial base assessment: Counterfeit electronics. Technical report, U.S. Department Of Commerce, Bureau Of Industry And Security, Office Of Technology Evaluation, 2010.

[4] AUTOSAR automotive open software architecture specification. http://www.autosar.org, 2012.

[5] CALM communications access for land mobiles. http://calm.its-standards.eu/, 2012.

[6] ETSI TC ITS standards european telecommunications standards institute. http://www.etsi.org, 2012.

[7] G. Alliance. Genivi. http://www.genivi.org/, 2009.

[8] F. Armknecht, R. Maes, A.-R. Sadeghi, F.-X. Standaert, and C. Wachsmann. A formalization of the security features of physical functions. In *IEEE Symposium on Security and Privacy*, pages 397–412, 2011.

[9] A. Barisani and D. Bianco. Unusual car navigation tricks. In *CanSecWest*, April 2007.

[10] N. Bissmeyer, H. Stuebing, E. Schoch, S. Götz, J. P. Stotz, and B. Lonc. A generic Public Key Infrastructure for securing Car-to-X Communication. In *18th ITS World Congress, Orlando, USA*, Oct. 2011.

[11] S. C. Bono, M. Green, A. Stubblefield, A. Juels, A. D. Rubin, and M. Szydlo. Security analysis of a cryptographically-enabled RFID device. In *USENIX Security Symposium*, pages 1–16, 2005.

[12] S. Checkoway, D. McCoy, B. Kantor, D. Anderson, H. Shacham, S. Savage, K. Koscher, A. Czeskis, F. Roesner, and T. Kohno. Comprehensive experimental analyses of automotive attack surfaces. In *USENIX Security Symposium*, 2011.

[13] C. C. Consortium. Mirror link. http://www.terminalmode.org/, 2010.

[14] J. Cook, I. Kolmanovsky, D. McNamara, E. Nelson, and K. Prasad. Control, computing and

communications: Technologies for the twenty-first century model T. *Proceedings of the IEEE*, 95(2):334 –355, 2007.

[15] L. de Alfaro and T. Henzinger. Interface theories for component-based design. In T. Henzinger and C. Kirsch, editors, *Embedded Software*, volume 2211 of *Lecture Notes in Computer Science*, pages 148–165. Springer Berlin / Heidelberg, 2001.

[16] L. de Alfaro and T. Henzinger. Interface-based design. In M. Broy, J. Grünbauer, D. Harel, and T. Hoare, editors, *Engineering Theories of Software Intensive Systems*, volume 195 of *NATO Science Series*, pages 83–104. Springer Netherlands, 2005.

[17] G. De Micheli. *Synthesis and Optimization of Digital Circuits*. McGraw-Hill Higher Education, 1st edition, 1994.

[18] F. Dressler, F. Kargl, J. Ott, O. Tonguz, and L. Wischhof. Research challenges in intervehicular communication: lessons of the 2010 dagstuhl seminar. *Communications Magazine, IEEE*, 49(5):158 –164, 2011.

[19] T. Eisenbarth, T. Kasper, A. Moradi, C. Paar, M. Salmasizadeh, and M. T. M. Shalmani. On the power of power analysis in the real world: A complete break of the keeloqcode hopping scheme. In *CRYPTO*, pages 203–220, 2008.

[20] EVITA Consortium. The EVITA project: E-safety vehicle intrusion protected applications, 2011. http://www.evita-project.org.

[21] The EVITA project Deliverable 2.1, http://evita-project.org/deliverables.html. *Specification and evaluation of e-security relevant use cases*, 2009.

[22] The EVITA project Deliverable 2.3, http://evita-project.org/deliverables.html. *Security requirements for automotive on-board networks based on dark-side scenarios*, 2009.

[23] The EVITA project Deliverable 3.1, http://evita-project.org/deliverables.html. *Security and trust model*, 2009.

[24] B. Gassend, D. Clarke, M. van Dijk, and S. Devadas. Silicon physical random functions. In *ACM Computer and Communications Security Conference (CCS)*, pages 148–160, 2002.

[25] N. Hardy. The confused deputy (or why capabilities might have been invented). *Operating Systems Review*, 22(4):36–38, 1988.

[26] T. Hoppe and J. Dittman. Sniffing/replay attacks on CAN buses: A simulated attack on the electric window lift classified using an adapted CERT taxonomy. In *Workshop on Embedded Systems Security (WESS)*, 2007.

[27] T. Hoppe, S. Kiltz, and J. Dittmann. Security threats to automotive CAN networks - practical examples and selected short-term countermeasures. In *Computer Safety, Reliability, and Security (SAFECOMP)*, pages 235–248, 2008.

[28] T. Hoppe, S. Kiltz, and J. Dittmann. Automotive IT-security as a challenge: Basic attacks from the black box perspective on the example of privacy threats. In *Computer Safety, Reliability, and Security (SAFECOMP)*, pages 145–158, 2009.

[29] F. Kargl, P. Papadimitratos, L. Buttyan, M. Muter, E. Schoch, B. Wiedersheim, T.-V. Thong, G. Calandriello, A. Held, A. Kung, and J.-P. Hubaux. Secure vehicular communication systems: implementation, performance, and research challenges. *IEEE Communications Magazine*, 46(11):110–118, 2008.

[30] P. Kleberger, T. Olovsson, and E. Jonsson. Security aspects of the in-vehicle network in the connected car. In *IEEE Intelligent Vehicles Symposium (IV)*, pages 528 –533, 2011.

[31] K. Koscher, A. Czeskis, F. Roesner, S. Patel, T. Kohno, S. Checkoway, D. McCoy, B. Kantor, D. Anderson, H. Shacham, and S. Savage. Experimental security analysis of a modern automobile. In *IEEE Symposium on Security and Privacy (S& P)*, pages 447–462, 2010.

[32] F. Koushanfar. Provably secure active ic metering techniques for piracy avoidance and digital rights management. *IEEE Transactions on Information Forensics and Security*, 7(1):51–63, 2012.

[33] F. Koushanfar, S. Fazzari, C. McCants, W. Bryson, M. Sale, P. Song, and M. Potkonjak. Can EDA combat the rise of electronic counterfeiting? In *Design Automation Conference (DAC)*, 2012.

[34] A. Lang, J. Dittmann, S. Kiltz, and T. Hoppe. Future perspectives: The car and its IP-address - a potential safety and security risk assessment. In *International Conference Computer Safety, Reliability, and Security (SAFECOMP)*, pages 40–53, 2007.

[35] U. E. Larson and D. K. Nilsson. Securing vehicles against cyber attacks. In *Cyber Security and Information Intelligence Research Workshop (CSIIRW)*, pages 30:1–30:3, 2008.

[36] M. Majzoobi, F. Koushanfar, and M. Potkonjak. Lightweight secure PUFs. In *International Conference on Computer-Aided Design (ICCAD)*, pages 670–673, 2008.

[37] D. K. Nilsson, U. Larson, and E. Jonsson. Efficient in-vehicle delayed data authentication based on compound message authentication codes. In *VTC Fall'08*, pages 1–5, 2008.

[38] D. K. Nilsson and U. E. Larson. Simulated attacks on CAN buses: vehicle virus. In *IASTED International Conference on Communication Systems and Networks (AsiaCSN)*, pages 66–72, 2008.

[39] C. Paar, A. Weimerskirch, and M. Wolf. Security in automotive bus systems. In *Embedded Security in Cars Workshop*, 2004.

[40] T. Pop, P. Pop, P. Eles, Z. Peng, and A. Andrei. Timing analysis of the FlexRay communication protocol. In *Euromicro Conference on Real-Time Systems*, pages 203–216, 2006.

[41] PRESERVE Consortium. Preparing secure vehicle-to-x communication systems, 2011. http://www.preserve-project.eu.

[42] M. Raya, P. Papadimitratos, and J.-P. Hubaux. Securing vehicular communications. *IEEE Wireless Communications*,, 13(5):8–15, 2006.

[43] I. Rouf, R. Miller, H. Mustafa, T. Taylor, S. Oh, W. Xu, M. Gruteser, W. Trappe, and I. Seskar. Security and privacy vulnerabilities of in-car wireless

networks: a tire pressure monitoring system case study. In *USENIX Security Symposium*, 2010.

[44] J. Rowson and A. Sangiovanni-Vincentelli. Interface-based design. In *Design Automation Conference (DAC)*, pages 178–183, 1997.

[45] U. Ruhrmair, S. Devadas, and F. Koushanfar. *Security based on Physical Unclonability and Disorder, Book Chapter in 'Introduction to Hardware Security and Trust'*. Springer, 2011.

[46] A. L. Sangiovanni-Vincentelli. Defining platform-based design. *EE Times*, Feb 2007.

[47] A. L. Sangiovanni-Vincentelli. Quo Vadis, SLD? reasoning about the trends and challenges of system level design. *Proceedings of the IEEE*, 95(3):467 –506, 2007.

[48] A. L. Sangiovanni-Vincentelli and M. Di Natale. Embedded system design for automotive applications. *IEEE Computer*, 40(10):42–51, 2007.

[49] E. Schoch and F. Kargl. On the efficiency of secure beaconing in vanets. In *Proceedings of the third ACM conference on Wireless network security (WiSec)*, pages 111–116, 2010.

[50] H. Schweppe, M. Idrees, Y. Roudier, B. Weyl, R. El Khayari, O. Henniger, D. Scheuermann, G. Pedroza, L. Apvrille, H. Seudié, H. Platzdasch, and M. Sall. Secure on-board protocols specification. Technical Report Deliverable D3.3, EVITA Project, 2010.

[51] SEVECOM Consortium. Secure vehicular communication, 2008. `http://www.sevecom.org`.

[52] SIMTD Consortium. Sichere intelligente mobilität testfeld deutschland, 2011.

[53] M. Tehranipoor and F. Koushanfar. A survey of hardware trojan taxonomy and detection. *IEEE Design & Test of Computers*, 27(1):10–25, 2010.

[54] P. R. Thorn and C. A. MacCarley. A spy under the hood: Controlling risk and automotive EDR. In *Risk Management*, 2008.

[55] C. to Car Communication Consortium. CAR TO CAR COMMUNICATION CONSORTIUM. `http://www.car-to-car.org/`, 2006.

[56] VDAT. Der deutsche tuningmarkt. `http://www.vdat.org/tuningmarkt_deutschland.php/`, 2010.

[57] B. Weyl, M. Wolf, F. Zweers, T. Gendrullis, M. Idrees, Y. Roudier, H. Schweppe, H. Platzdasch, R. El Khayari, O. Henniger, D. Scheuermann, A. Fuchs, L. Apvrille, G. Pedroza, H. Seudié, J. Shokrollahi, and A. Keil. Secure on-board architecture specification. Technical Report Deliverable D3.2, EVITA Project, 2010.

[58] M. Wolf, A. Weimerskirch, and T. J. Wollinger. State of the art: Embedding security in vehicles. *EURASIP Journal of Embedded Systems*, 2007.

[59] Y. Zhao. Telematics: safe and fun driving. *IEEE Intelligent Systems*, 17(1):10–14, jan/feb 2002.

Software Controlled Cell Bit-Density to Improve NAND Flash Lifetime

Xavier Jimenez, David Novo and Paolo Ienne
Ecole Polytechnique Fédérale de Lausanne (EPFL)
School of Computer and Communication Sciences
CH–1015 Lausanne, Switzerland
xavier.jimenez@epfl.ch, david.novobruna@epfl.ch and paolo.ienne@epfl.ch

ABSTRACT

Hybrid flash architectures combine static partitions in *Single Level Cell (SLC)* mode with partitions in *Multi Level Cell (MLC)* mode. Compared to MLC-only solutions, the former exploits fast and short random writes while the latter brings large capacity. On the whole, one achieves an overall tangible performance improvement for a moderate extra cost. Yet, device lifetime is an important aspect often overlooked. In this paper, we show how a dynamic SLC-MLC scheme provides significant lifetime improvement (up to 10 times) at no cost compared to any classic static SLC-MLC partitioning based on any state of the art Flash Translation Layer policy.

Categories and Subject Descriptors

B.3.1 [**Memory Structures**]: Semiconductor Memories

General Terms

Design, Experimentation, Measurement, Performance, Reliability

Keywords

NAND Flash Memory, Flash Endurance, FTL, SLC, MLC

1. INTRODUCTION

NAND flash memory is the leading data storage technology for mobile devices, such as MP3 players, smartphones, tablets or netbooks. It features low power consumption, high responsiveness and mobility. However, flash technology also comes with several disadvantages. The device has a very specific physical organization, which results in a coarse granularity of data accesses. A flash memory is a type of EEP-ROM and, as such, memory cells need to be erased before being written again. Moreover, a flash memory cell can only be written a limited number of times before wearing out. The severity of those limitations is somehow mitigated by

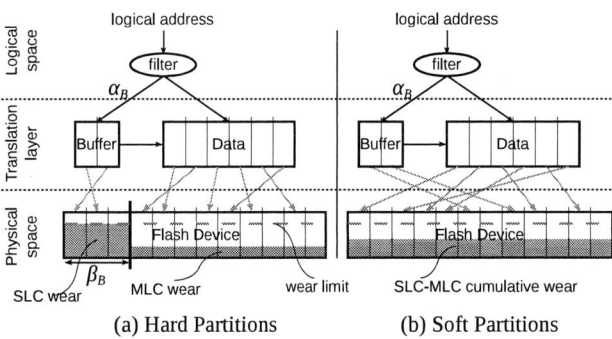

(a) Hard Partitions   (b) Soft Partitions

Figure 1: Hard versus soft partitioning. The FTL redirects random writes to a buffer and sequential writes directly to the data partition. The buffer is mapped to SLC to benefit from speed and low energy, while the data uses MLC for density. When writes are unbalanced across buffer and data, a hard partition will wear faster than the other, while soft partitioning (the contribution of the present paper) allows to balance the wear on the global device.

a software abstraction layer, called *Flash Translation Layer (FTL)*, which interfaces between common file systems and the flash device.

The basic functionality of an FTL is the translation of logical addresses to physical addresses. Such translation can be done following different policies which directly affect read/write performances, device cost and lifetime. The address translation table is typically stored in expensive SRAM cells, which makes the table size a critical design parameter of any FTL. A small translation table is cheap but implies coarse grained data manipulation (block-level) that results in poor performance and lifetime. Instead, a large table enables fine grained data manipulation (page-level), and thus higher performance and lifetime, but at a significant increase in device cost.

Hybrid-FTLs [4, 5, 6, 8, 9, 10] combine the two mappings by dividing the flash memory in two regions: a large data partition, which is addressed at the block-level, and a small log buffer partition, which is addressed at the page-level. The purpose is to direct random writes to the log buffer so that they can be written back to the data partition in-order as big chunks. Such an FLT requires a filter, illustrated in Figure 1a, to decide whether a particular data should be stored in the buffer partition or in the data partition.

Considering that a significant amount of memory accesses go to the small buffer partition, previous work [3, 6, 7, 11]

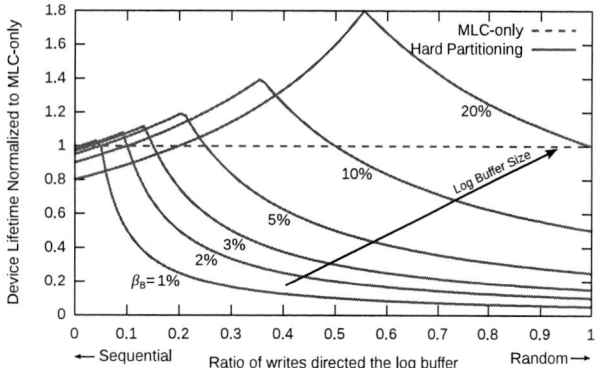

Figure 2: Hard partitions lifetime model for different buffer sizes as a function of the write ratio to the buffer. The model is normalized to a 2-bit MLC-only flash device lifetime. For large sequential accesses, where a FTL will more likely bypass the buffer, the device lifetime is bounded to the data partition. Small and frequently updated random accesses will wear out the buffer first, reducing the device lifetime to the buffer partition one. For reasonable amounts of random writes in practically affordable SLC buffers, lifetime is easily reduced by one order of magnitude.

proposed to build the buffer partition on *Single Level Cell (SLC)* flash, which provides high performance and low energy consumption but poor density, and the larger data partition on *Multiple Level Cells (MLC)* of lower performance but higher density. As a result, the flash device exhibits performances close to SLC (particularly for random data accesses) while keeping the area efficiency of MLC to a great extent. However, this previous work largely disregarded the effect of such SLC-MLC partitioning on the device lifetime.

Managing the SLC and MLC partitions as distinct physical parts of the flash device, as suggested by all the previous pieces of work, can lead to a serious reduction in lifetime. Figure 1a suggests how the extensive use of the buffer partition, due to a particular application write pattern, results in an unbalanced wear causing the device to fail well before most of its cells deteriorate above their maximum wear level (the large data partition is still healthy).

A more quantitative view on the impact of SLC-MLC partitioning to the device lifetime is illustrated in Figure 2, which will be further analyzed later. The figure shows the device lifetime, normalized to an MLC-only device for different buffer sizes as a function of the ratio of writes directed to the log buffer. Localized stress applied to small buffer (on the right of the graph) will easily reduce lifetime by an order of magnitude.

Accordingly, this paper proposes a soft SLC-MLC partitioning which changes the physical allocation of the buffer depending on the wear of the device. Such technique relies on the fact that an MLC can be managed as an SLC while largely keeping the performance benefits of a physical SLC. Figure 1b illustrates the soft partitioning technique, where each cell has a cumulated wear from MLC- and SLC-mode that can be globally balanced. The proposed soft partitioning can be coupled to the existing hybrid FTLs to significantly increase device lifetime while keeping the benefits in

performance, energy and area shown on a hard partitioning.

The rest of the paper is organized as follows. In Section 2, we introduce combined SLC-MLC architectures and analyze their lifetime limitations. The soft partitioning is covered in Section 3. Lifetime measurements for soft and hard partitions are presented in Section 4. The related work is presented in Section 5 and Section 6 concludes our paper.

2. FLASH LOG MANAGEMENT

We principally identify two main NAND flash memory technologies: SLC and MLC. The latter stores multiple bits per memory cell providing a larger bit density and hence a smaller cost per bit. Several bits can be stored in a single cell by using multiple voltage levels; e.g., four voltage levels can store two bits. Manipulating MLCs is trickier than SLCs: a higher precision is now required to differentiate the multiple voltage levels making programming and reading significantly more complex, and therefore resulting in lower performances and higher energy consumption. Furthermore, MLC is more sensitive than SLC to the charge loss and neighboring cell interferences that typically affect flash reliability, which translates into a shorter lifetime. Therefore, MLC offers a higher bit density than SLC at the expense of a lower performance, higher energy consumption and reduced lifetime.

Hybrid FTLs combine an SLC log buffer partition with an MLC data partition to improve the performance of MLC: the more hot data (frequently accessed data) directed to the log buffer, the closer to the SLC performance. However, log buffers need to be carefully dimensioned: The smaller the buffer partition, the higher the bit density of the flash device. Yet, the impact of such partitioning on the device lifetime needs to be carefully considered.

Depending on the application write pattern, an unbalanced wear can occur between the buffer and data partitions. Each partition lifetime is proportional to its technology endurance and size, and inversely proportional to the ratio of writes directed to it. For example, let us take a budget of 100 cells, reserve 5% for an SLC buffer and the rest to the MLC data partition. Consider that the endurance of an SLC is about 10 times larger than of an MLC and, for this particular example, that each partition receives 50% of the writes. The MLC partition will then last $0.95/0.5 = 3.8$ times longer than an MLC-only device taking 100% of the writes. On the other hand, the SLC partition includes only 5% of the cells, which can store half of the bits of an MLC, and is 2.5% of the capacity an MLC-only device; accordingly, the lifetime of the SLC partition corresponds to $0.025 \times 10/0.5 = 0.5$ times the lifetime of an MLC-only device. This indicates that a device with such a hybrid configuration will last half of the time of an MLC-only device, which is already significantly shorter than the lifetime of an SLC-only device.

In order to compute analytically the lifetime of a hybrid flash device, we define α_B and α_D as the write ratios directed to the buffer and data partitions, respectively. Let β_B and β_D respectively be the ratio of the device's cells allocated to the buffer and data, and L_B and L_D be their respective lifetime normalized to MLC-only. Considering an n-bit MLC technology and an SLC endurance γ times larger than MLC, the lifetime of a hybrid device can be expressed as follows:

$$L_B = \frac{\gamma \beta_B}{n \alpha_B} \quad \text{and} \quad L_D = \frac{\beta_D}{\alpha_D}. \quad \text{(1a, b)}$$

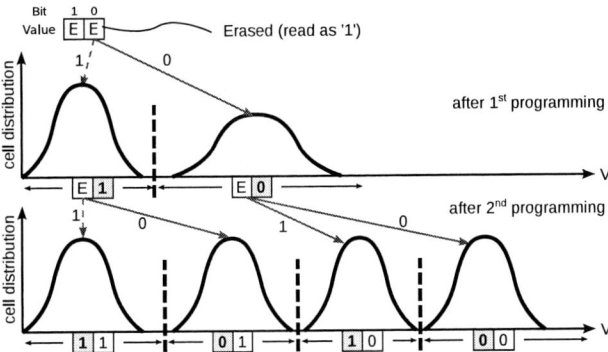

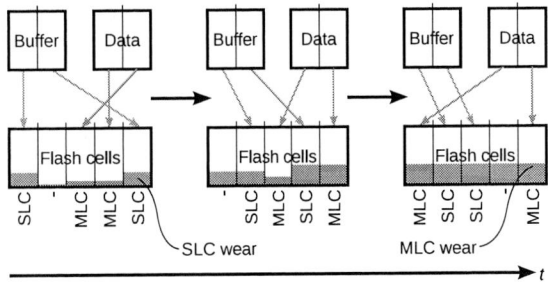

Figure 4: Software-controlled log buffer. A practical scenario of a hybrid-FTL, where cells switch between SLC- and MLC-mode regularly in order to balance the overall wear.

Figure 3: Programming of a 2-bit MLC. Each bit of a cell is programmed separately. Programming the first bit, or LSB, requires to target a single level (staying at the erased level is free) and does not need to be very precise. Programming the MSB, requires to read the current state of the cell and targets potentially 3 different levels, which requires more precision and time. Using cells in SLC-mode consists of programing only the first bit of each cell.

The hard partitions device lifetime corresponds to

$$L_H = min(L_B, L_D). \quad (2)$$

Assuming $n = 2$ and $\gamma = 10$, Figure 2 plots Equation 2 and represents the device lifetime, normalized to an MLC-only device, for different buffer sizes, β_B, and different ratio of writes directed to the log buffer, α_B. We observe that for reasonable buffer sizes (e.g., 5% and less), the lifetime of hybrid devices drops significantly when more than 25% of writes are directed to the buffer. Around one fifth of the cells should be allocated to the buffer to ensure a lifetime close to the MLC-only's.

The inability to balance the wear between its partitions is the main problem of hard partitioning. This, as shown in Figure 2, can seriously compromise the viability of hybrid-FTLs. In the following section we propose a different partitioning that overcomes this issue extending device lifetime.

3. SOFT PARTITIONING

Soft partitioning breaks the rigidity of hard partitioning by changing the physical placement of the log buffer depending on the device wear. This is enabled by the fact that MLC can be manipulated to obtain SLC write performances and that the FTL can keep track of the cumulative wear (SLC and MLC) to decide on the actual physical allocation.

3.1 Faster MLC: Managing MLC as SLC

MLC can also be used to store a single bit and recover the performance, energy consumption and lifetime benefits of SLC [6]. Figure 3 illustrates the programming of a 2-bit MLC. Each cell represents two bits, namely the LSB and the MSB, which are part of different pages and are programmed separately. Before starting the actual programming, the cell needs to be first erased. Then, the LSB is programmed targeting a single voltage level, which does not need to be very precise. In a final step, the MSB is programmed, which requires to first read the current state (i.e., the LSB value) to then push the cell voltage to either of the 3 different levels (see solid arrows in the figure). This second programming

requires higher precision and it is typically three to four times longer than the LSB programming.

Interestingly, programming only the LSB of MLC shows performances very similar to SLC, which motivated previous researchers to propose the use of MLC as SLC for the statically allocated log buffer partition of Figure 1a. Thereby, performance is obtained at the expense of density in an MLC device. Such way of manipulating an MLC will be referred to as *SLC-mode* in the rest of the paper. Here, we propose to go one step further and opportunistically change the physical allocation of the buffer depending on the device wear.

3.2 Software-Controlled Log Buffer

Whereas hard partitions is typically applied on hybrid SLC-MLC hardware, the soft partitions scheme is applied to a completely homogeneous hardware architecture, made only of one or more MLC chips. The FTL is able to configure selectively regions of the flash chip to SLC-mode or MLC-mode at will, with the intention to evenly distribute the wear throughout the whole device. While small buffers are likely to die first for hard partitions, a soft partitioning can spread the localized stress over the complete device.

Figure 4 illustrates the evolution of a device using hybrid-FTL on soft partitions. Focusing on the leftmost physical block, one can see how this is initially allocated to the buffer, thus managed as SLC, then is freed from both partitions to be later allocated to the data partition, managed as MLC. Such transitions are typically triggered by the FTL wear balancing algorithm. Notice that the wear of the block increases through time and results from the times that the block is programmed as SLC and MLC.

Accordingly, the FTL needs to consider a global wear metric that includes the effects of both, the MLC-mode and SLC-mode, to decide the physical allocation of the log buffer. Such metric is detailed in the following subsection.

3.3 Lifetime of Soft Partitions

Let ω_{MLC} and ω_{SLC} be the wear associated to an MLC and SLC, respectively. The lifetime function of the soft partition scheme with respect to MLC-only is a linear function which is proportional to the write ratio directed to the buffer, α_B. In the one extreme, when the MLC-mode is exclusively used ($\alpha_B = 0$), the device lifetime is equal to the MLC-only. In the other extreme, when the SLC-mode is exclusively used ($\alpha_B = 1$), the device lifetime is limited by the wear of the SLC writes and the sensitivity of the MLC reads, which

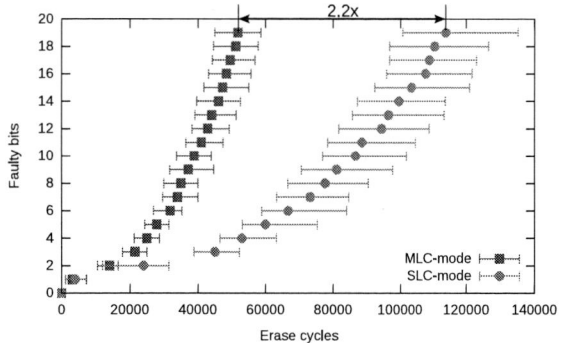

Figure 5: Comparison of SLC-mode and MLC-mode wear speed. Erase cycles are applied on the cells during which, in the case of SLC, only the first bit is programmed. Every hundred cycles, both bits are programmed and are verified for errors. We observed that SLC-mode requires in average 2.2 times more erase cycles than MLC to achieve the same error rate.

corresponds to $\frac{\omega_{MLC}}{n\omega_{SLC}}$, considering that lifetime is inversely proportional to wear. This corresponds to the ratio of the MLC and SLC wear divided by n, as MLC is able to store more bits than SLC. Notice that such lifetime is different with respect to an SLC-only device, where the lifetime is limited by the sensitivity of the SLC read (larger than MLC, as discussed in Section 3.1). When $\frac{\omega_{MLC}}{n\omega_{SLC}} \geq 1$, the soft partitions ensures a lifetime larger or equal to MLC-only. Accordingly, the indicated lifetime function corresponds to

$$L_S = \left(\frac{\omega_{MLC}}{n\omega_{SLC}} - 1 \right) \alpha_B + 1. \qquad (3)$$

However, $\frac{\omega_{MLC}}{n\omega_{SLC}}$ is a parameter that can not be found in the specifications of a standard MLC flash device, as chip manufacturers do not characterize its usage in SLC-mode. We built an FPGA-based platform to interface ONFI [1] compliant NAND flash chips and extracted experimentally this parameter. The latest NAND flash chips generations not being available for average consumers, we unsoldered some from USB sticks. We wore out cells programming them either in SLC- or MLC-mode, while periodically measuring faulty bits. In order to isolate the wear factor from the read reliability, faulty bits are gathered by programming in MLC and reading back the LSB and MSB. On Figure 5, we report the average number of cycles required to observe a certain amount of faulty bits for the first time on a 2-bit MLC. Error bars represent the positive and negative mean differences. We observe for this particular chip that 2.2 times more cycles are needed in SLC-mode to reach the same error rate than MLC. In our experiments, we will assume $\frac{\omega_{MLC}}{2\omega_{SLC}} = 1.1$.

4. RESULTS

In this section, the proposed soft partitioning technique is coupled with three published FTLs and the results of running three different data traces are compared to the hard partitioning reference.

4.1 Experimental Setup

We developed a trace-driven flash simulator in order to measure the execution time and erase counts of several FTL

	financial2	file copy	alix 8 d.
Total Data	1860MB	3606MB	5566MB
Mem Footprint	383MB	2767MB	1030MB
Data/Footprint Ratio	4.86	1.30	5.40
Weighted Req. Avg.	52.0KB	500KB	230KB
Weighted Req. Std. Dev.	65.8KB	21.3KB	177KB

Table 1: Benchmark characteristics.

executing three realistic traces. The 'financial2', obtained from the UMass Trace Repository [2], was gathered from an OLTP application running at a large financial institution. We generated two other disk traces from a tiny home server running a light Linux distribution. The traced storage is a 16 GB Compact Flash, which contains the systems main partition and a swap partition. The first trace, 'alix 8 days' is a 1-week trace running a web server. The second trace, 'file copy' was obtained from writing several GBytes of MP3 files. Some of the characteristics of the selected benchmarks are included in Table 1. The ratio between total data and memory footprint indicates the level of data reuse, while the ratio between the average and the standard deviation of the weighted request size indicates how different are the sizes of the different requests. Accordingly, we can conclude that 'file copy' includes sequential memory requests of similar size while 'financial2' and 'alix 8 days' include significant data reuse with memory requests of varied sizes.

We implemented three different FTLs, namely FAST [9], ROSE [4] and ComboFTL [7]. FAST is one of the first Hybrid-FTL, while ROSE is the most recent amelioration of FAST known by the authors. Although they originally use an MLC buffer, we allocate the buffer to SLC, which, as motivated in Section 3.1, increases performance and extends device lifetime. The only side effect is the increase in area, which we assume to pay off for reasonable buffer sizes. Lastly, ComboFTL includes an SLC buffer that gives multiple chance to victim data upon eviction. If the victim data is considered as being likely to be updated, it can be fed back into the buffer avoiding an expensive migration to the MLC partition.

The simulated flash characteristics were extracted from our measurement on a 4GBytes NAND flash chip. The flash chip shows a program latency of $400\mu s$ and $1600\mu s$ for the first bit and second bit, respectively. It has a read latency of $120\mu s$ and erasing takes $4ms$. We set the number of cells to 80 billion, corresponding to a 20GBytes MLC-only device.

4.2 Soft vs. Hard Partitioned Hybrid FTLs

The traces are executed twice by each FTL for several buffer sizes. The first run serves as a warm up and we collect the result with the second run. We assume a fully allocated logical space. We visit a large spectrum of parameters specific to each FTL and only keep the most effective combination for each trace. For FAST and ROSE, reducing execution time will systematically maximize lifetime, whereas for ComboFTL, originally build on hard partitions, a parameter provides a trade-off between endurance and performance. We show two parameter sets for ComboFTL: $Combo_L$ maximizes lifetime while $Combo_P$ maximizes performance.

The time spent in wear leveling is assumed to be negligible compared to the actual data migration process. Such assumption is supported by Chiao et al. [4], where the performance overhead of aggressive wear balancing applied on a Hybrid-FTL is shown to be between 1-2%. Thus, the execu-

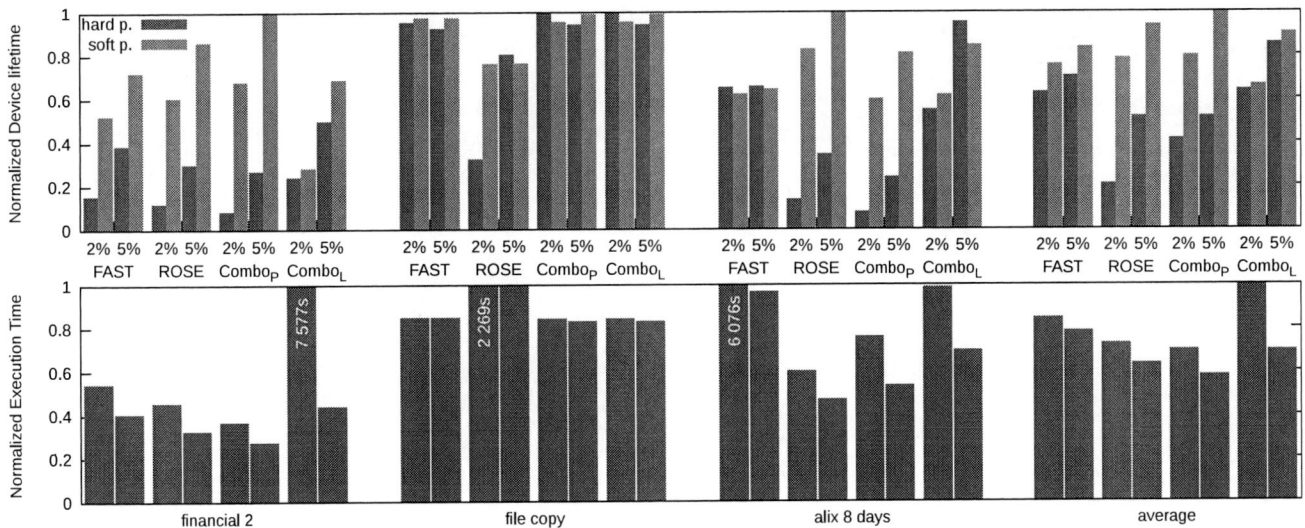

Figure 6: Lifetime and Performance. The results contrast our technique versus hard partitioning for various FTLs implemented with 2% and 5% buffer sizes and normalized for each trace to the largest value. Among a large spectrum of parameters specific to each FTL, only the best results are shown. $Combo_P$ and $Combo_L$ maximize performance and lifetime, respectively. In the case of performance, we assume the difference between hard and soft partitioning to be negligible. Overall, our soft partitioning significantly increases lifetime for practically every considered FTLs and benchmarks.

tion time of both, the soft and the hard partitioning scheme, is assumed to be the same in our experiments.

Figure 6 shows normalized life time (top) and normalized execution time (bottom) of the selected FTLs executing the benchmarks of Table 1 for buffer sizes of 2% and 5% of the total cell budget managed as hard partitions and as the proposed soft partitions. The results are normalized to the largest value for each trace.

The 'financial 2' trace has a large amount of small accesses. Increasing the buffer size will naturally reduce the amount of data evicted from the buffer and results in better performance and lifetime. We observe that the proposed soft partitioning is able to considerably increase the device lifetime with respect to hard partitioning for most of the configurations. We can see that when a lot of stress is put on the buffer, $Combo_L$ is able to extend hard partitions lifetime, sacrificing significantly performance. Interestingly, maximizing lifetime for hard partitions does not improve soft partitions lifetime. Indeed, soft partitions benefits from the fact that SLC-erase cycles wear less the cells while improving performance. Where hard partitioning requires to trade-off lifetime for performance, soft partitioning is able to obtain the best of both.

The 'file copy' trace being mostly made of very large sequential accesses, it bypasses completely the buffer and directs most of the accesses to the data partition, except for ROSE. Having the majority of writes directed to the MLC partition, annihilate most of the benefit of an SLC-MLC combined architecture and it is not surprising to observe similar lifetime between hard and soft partitioning. It also explains why a bigger buffer does not increase performance.

The 'alix 8 days' benchmark is composed of a good balance of both random and large sequential accesses. For FAST, we observed a large amount of sequentiality mispredictions, resulting in poor performances but slightly larger lifetime for hard partitions. Similarly to 'file copy', neither its lifetime

nor execution time will be affected much by the buffer size. For ROSE and ComboFTL, increasing the buffer size has similar effects as 'financial 2': it increases performance and lifetime and we observe a large lifetime improvement with the soft partitioning. Also, we observe a case where $Combo_L$ can maximize lifetime further than soft partitions, however this comes at a performance cost with respect to $Combo_P$.

4.3 Generalization of Experimental Results

Figure 7 plots the lifetime results for the different configurations discussed in the previous subsection in the space introduced in Figure 2. New configurations, corresponding to additional log buffer sizes are also added to the figure.

For applications with random accesses, we observe that increasing the buffer size reduces the pressure on the data partition and results in higher ratios of writes to the buffer. In that spectrum of the plot, hard partitioning is only able to outperform soft partitioning for very large buffer sizes. Such region is shaded in the figure and annotated as *high cost*. However, hard partitioning fails to maximize the device lifetime for sensible buffer sizes. Such region is shaded in the figure and annotated as *typical*. The latter corresponds to scenarios that are likely to target a hybrid-FTL to increase the flash performance and that importantly will largely extend device lifetime when adopting the proposed soft partitioning.

For sequential access patterns, all the buffer sizes present a very low ratio of buffer writes, which results in marginal differences between the different cases, except for large buffer sizes. Such region, shaded in the figure and annotated as *sequential*, does not benefit from the hybrid-FTL schemes targeted in this paper.

5. RELATED WORK

The idea of an SLC-MLC combined architecture has been investigated in previous work. Grupp et al. [6] experimen-

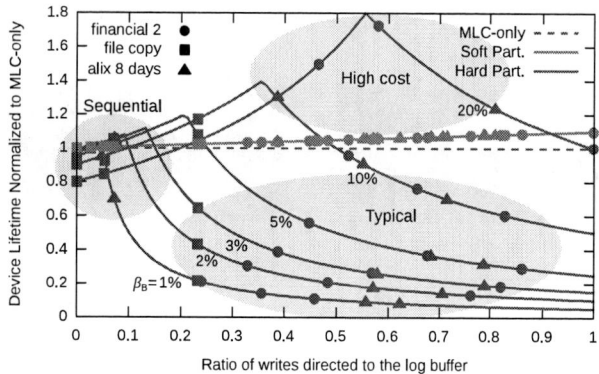

Figure 7: Lifetime models populated by benchmarks results. This graph shows the result of every best combinations of FTLs, traces and buffer sizes. In typical applications including random access patterns, soft partitions do systematically better. Only configurations characterized by a considerably higher cost can reach higher lifetimes by some 50% at most.

tally characterize MLC flash chips reporting performance and energy figures, even to use in SLC-mode. Based on those observations, they propose Mango, an FTL that makes use of SLC-mode to improve response time and reduce the total energy consumed. However, their FTL achieves a lifetime reduction of 35% with respect to MLC. They compensate such reduction by using an error correction code, the benefit of which can also be coupled to our approach.

Chang [3] proposes a Hybrid SSD combining SLC and MLC chips, which is a clear example of hard partitions referred in the paper. In order to extend the lifetime, they adapt the ratio of writes directed to the log buffer to balance the wear on each partition. Thereby, the performance is reduced for most of the cases. Instead, the proposed scheme respects the ratio of writes to the log buffer, which should have been optimized for performance by the FTL, and changes the physical allocation of the log buffer to balance the device wear, obtaining high performance without compromising device lifetime.

Park et al. [11] propose HFTL, an FTL based on an SSD architecture very similar to Chang's. In particular, they propose techniques to exploit multi-banks parallelism and maximize bandwidth. As us, they realize that the device lifetime is limited by the partition with the shortest lifetime, however, they mitigate the problem by sizing the SLC partition to guarantee a lifetime larger than the MLC partition. This oversizing, with 10 to 30% of the cells allocated to the log buffer, significantly increases the cost of the system, not only for the increase in flash cells but mostly for the huge address translation table associated, being only affordable in special niche markets.

Similarly, Im et al. propose ComboFTL [7], which can be tuned to either optimize lifetime at expenses of reducing performance or performance at expenses of reducing lifetime. Figure 6 shows that the combination of ComboFTL optimized for performance with our soft partitions is able to simultaneously achieve the largest lifetime and the higher performance.

Thus, to the best of our knowledge, this is the first work that proposes a soft partitioning of the log buffer present in hybrid FTLs that is continuously reallocated to distribute the device wear extending device lifetime at no cost.

6. CONCLUSIONS

Flash architectures combining SLC and MLC technologies are targeting new cost-sensitive applications displaying frequent random write patterns. The random accesses benefit from SLC performances while devices are primarily in MLC-mode to take advantage of the low cost of flash memory. However, unbalanced pressure on the SLC partition may lead to a premature death of the device. In this paper we have shown an approach that is robust to unbalanced stress and ensures a lifetime at least as long as that of an MLC-only device, showing up to 10 times lifetime improvement compared to the original approach. Interestingly for volume devices, this advantage comes at practically no extra cost.

7. REFERENCES

[1] (2011, Oct.) Open NAND flash interface 2.3. [Online]. Available: http://onfi.org/specifications/

[2] K. Bates and B. McNutt. (2007, Jun.) OLTP application I/O. [Online]. Available: http://traces.cs.umass.edu/index.php/Storage/Storage

[3] L.-P. Chang, "A Hybrid Approach to NAND-Flash-Based Solid-State Disks," *IEEE Trans. Computers*, pp. 1337–1349, Oct. 2010.

[4] M.-L. Chiao and D.-W. Chang, "ROSE: A novel flash translation layer for NAND flash memory based on hybrid address translation," *IEEE Trans. Computers*, pp. 753–766, Jun. 2011.

[5] H. Cho, D. Shin, and Y. I. Eom, "KAST: K-associative sector translation for NAND flash memory in real-time systems," in *Design Automation and Test in Europe*, Nice, France, Apr. 2009, pp. 507–512.

[6] L. M. Grupp, A. M. Caulfield, J. Coburn, S. Swanson, E. Yaakobi, P. H. Siegel, and J. K. Wolf, "Characterizing flash memory: anomalies, observations, and applications," in *ACM/IEEE Int. Symp. Microarchitecture*, New York, NY, USA, Dec. 2009, pp. 24–33.

[7] S. Im and D. Shin, "ComboFTL: Improving performance and lifespan of MLC flash memory using SLC flash buffer," *Journal of Systems Architecture*, pp. 641–653, Dec. 2010.

[8] J. Kim, J. M. Kim, S. Noh, S. L. Min, and Y. Cho, "A space-efficient flash translation layer for CompactFlash systems," *IEEE Trans. Consumer Electronics*, pp. 366–375, May 2002.

[9] S.-W. Lee, D.-J. Park, T.-S. Chung, D.-H. Lee, S. Park, and H.-J. Song, "A log buffer-based flash translation layer using fully-associative sector translation," *ACM Trans. Embedded Computing Systems*, Jul. 2007.

[10] S. Lee, D. Shin, Y.-J. Kim, and J. Kim, "LAST: Locality-aware sector translation for NAND flash memory-based storage systems," *ACM SIGOPS Operating Systems Review*, pp. 36–42, Oct. 2008.

[11] J.-W. Park, S.-H. Park, C. C. Weems, and S.-D. Kim, "A hybrid flash translation layer design for SLC-MLC flash memory based multibank solid state disk," *Microprocessors & Microsystems*, pp. 48–59, Feb. 2011.

Observational Wear Leveling:
An Efficient Algorithm for Flash Memory Management

Chundong Wang and Weng-Fai Wong
School of Computing
National University of Singapore
Email: {wangc, wongwf}@comp.nus.edu.sg

ABSTRACT

In NAND flash memory, wear leveling is employed to evenly distribute program/erase bit flips so as to prevent overall chip failure caused by excessive writes to certain hot spots of the chip. In this paper, we analyze latest wear leveling algorithms, and propose Observational Wear Leveling (OWL). OWL considers the temporal locality of write activities at runtime when blocks are allocated. It also transfers data between blocks of different ages. From our experiments, with minimal additional space and time overhead, OWL can improve wear evenness by as much as 29.9% and 43.2% compared to two state-of-the-art wear leveling algorithms, respectively.

Categories and Subject Descriptors

D.4.2 [**Storage Management**]: Secondary Storage

General Terms

Design, Management, Measurement

Keywords

Flash Management, Wear Leveling

1. INTRODUCTION

The ferromagnetic hard disk drive (HDD) has been the defacto storage device in last several decades. In recent years, flash-based storage is becoming a viable alternative in embedded systems and enterprise servers. Comparatively, flash memory has lower access latency, is more shock-resistant, and consumes less power. However, the issue of *write endurance* continues to be a concern in the large scale deployment of flash memory.

By its very nature, flash cells can only withstand a limited number of *program/erasure flips*, i.e., "writes". There are two types of flash memory, namely NOR and NAND flash. This paper focuses on the latter one that is more prevalent. The unit of programming in NAND flash is a *page*, whose size is 2KB or more [7]. By default, a flash cell stores a logic '1'. To program a page is to write data by selectively setting its cells to '0'. An erase operation can reset a cell to be '1'. The unit of an erasure in NAND flash is a *block* which comprises many pages. Moreover, a page cannot be reprogrammed unless the block it is in has been erased. Thus data can be only updated *out-of-place*. Instead of overwriting the old data, the new data are written to another page with the old one invalidated. The number of flips between the programmed and erased states of a cell is physically limited. Typically it is 100,000 for single-level cell (SLC) flash that has one bit in a cell, and 10,000 for multi-level cell (MLC) flash whose cell can store two or more bits [7]. If a cell is excessively flipped, it is likely to be permanently damaged. The block it is in will be considered to be *worn out*. Worn-out blocks together with another type of *bad* blocks that are caused in manufacturing process will be kept away from regular use [4]. Too many bad blocks would make the entire chip defective.

Wear Leveling is a technique that is employed to distribute erasures as evenly as possible to avoid excessive flips. Data are usually classified to be *hot* or *cold* according to their update frequencies. Also, a physical block is considered to be *old* or *young* depending on its erase counts. Typically wear leveling will transfer cold data to old blocks at the expense of some performance overhead. A wear leveling scheme can be *proactive* or *passive*, or has both manners. Proactive wear leveling aims to put data in suitable blocks, and passive wear leveling will swap data when the distribution of erase counts over all blocks is skewed beyond a certain limit.

Currently wear leveling is performed by an embedded software called the *flash translation layer* (FTL). The FTL also conducts address mapping, garbage collection and other functions. Address mapping translates logical addresses of file system to physical addresses in the flash chips. The state-of-the-art mapping schemes are *hybrid mapping* ones that use *log space*.

In this paper, we shall reconsider the problem of write endurance, and propose a novel wear leveling algorithm called *observational wear leveling* (OWL). OWL exploits hybrid mapping, using proactive methods to avoid the unevenness of erasures through monitoring temporal locality and block utilization. Our experiments show that OWL can outperform the latest passive algorithms by 29.9% and 43.2% respectively on wear evenness with a performance overhead of at most 1.1%. The main ideas of OWL are as follows:

- A *locality-based block allocation* (LBA) scheme is employed within hybrid mapping that leverages on the temporal locality of accesses observed using a *block access table* (BAT).

- A *scan and transfer* (ST) scheme is periodically triggered to transfer cold *or* very hot data to elder blocks. ST can prevent young blocks from being occupied for long periods of time.

The rest of this paper is organized as follows. Section 2 shows hybrid address mapping. Section 3 describes state-of-the-art wear leveling schemes, and motivates the need for an efficient one. Section 4 shows our OWL. Section 5 presents experimental results and analysis. Section 6 will conclude this paper.

2. ADDRESS MAPPING IN FTL

Address mapping is a basic function of flash management. Hybrid mapping [9] is a popular mapping strategy. It combines *page mapping* and *block mapping* whose units of mapping are pages and blocks, respectively. In hybrid mapping, all physical blocks are partitioned into a *data space*, a *log space* and *free blocks*. Blocks in the data space (which we shall call "data blocks") are managed in block mapping. However, block mapping is not flexible because data are written and read in units of pages. If data in a page are to be modified, a free block has to be allocated for out-of-place updates, and data in other pages of the same block have to be moved. Obviously frequent updates will result in continual data movements. Hybrid mapping maintains the log space using page mapping to solve this problem. Upon an update, instead of writing to another block, a page will be allocated from a log block to accept the data. Consecutive updates will be handled by more log page allocations.

FAST [9] is a popular hybrid mapping scheme. In FAST, the log space is "fully associative", which means a log page is not bound to some data block but can accept data from anyone.

Because the log space is managed in page mapping, its capacity cannot be too big since page-level mapping has a significant space overhead. When the log space runs out of pages, a *merge* procedure is called. Merging makes new space by evicting some data in the log space to the data space. In FAST, a victim log block is picked in a round-robin manner, and each page will be checked. If its data are valid, the FTL will merge the page with its corresponding data block to a newly allocated block in the data space. Otherwise, it will be skipped. Thus, during a merge there may be frequent allocation requests for free blocks. After all log pages are merged, the victim log block and original data blocks will be erased for future use. The log space will then be replenished with a free block.

3. PROBLEM FORMULATION

3.1 Passive and Proactive Wear Leveling

Wear leveling attempts to even out erasures across all blocks. It can be either *dynamic* or *static* [2]. Dynamic algorithms consider free blocks and blocks with frequently updated data, i.e., hot data. Static ones make use of all blocks, including blocks of cold data. Table 1 shows four of the latest wear leveling algorithms. They all fall into the static category, although how they perform wear leveling varies significantly. Among them, the dual-pool [1], BET [3] and lazy wear leveling [2] take actions only when the level of unevenness reaches some thresholds. So they are *passive* schemes.

In dual-pool algorithm, hot and cold data stay in the respective hot and cold pool. When the difference on erase count between the head of hot pool and the rear of cold pool exceeds a predefined threshold, the two blocks will swap places.

The BET is a key structure of the algorithm in [3]. We shall use this acronym to refer to this algorithm. Blocks are divided into sets. Each set is associated to a bit in the BET. When a preset interval begins, all bits in the BET are '0'. If a member of a set is erased within the interval, its associated bit will be set to '1'. The total number of erasures in the interval is recorded. If the ratio of the number of erasures over the number of 1's in the BET reaches a predefined threshold, a set whose associated bit is still '0' will be randomly selected. All valid data in this set are moved to a free block set, after which the former set will be erased for future use.

Lazy wear leveling [2] is a recently proposed strategy. It is performed in the merge procedure of hybrid mapping. Before lazy wear leveling, a data block that is involved in merge, say D, will be immediately erased. In lazy wear leveling, if D's erase count is higher than the average by a threshold Δ which is tuned online, besides erasing D, the FTL will find a data block with cold data, say C, transfer C's data to D, erase C, and return C as a free block.

In summary, the dual-pool scheme responds to the widening gap between two blocks' erasure counts, the BET scheme is activated when the erasures are unevenly distributed beyond an extent, and lazy wear leveling works when the block to be reclaimed is much older than the average.

Rejuvenator [6] has both proactive and passive mechanisms. It allocates hot or cold data to young or old blocks respectively in a proactive way. It records recent access frequencies of logical pages, and identifies the temperature of pages accordingly. It also groups blocks that have the same erase count in a list. A list is in the *lower numbered lists* if its erase count is smaller than a dynamic threshold; or it is in *higher numbered lists*. When new write requests arrive, based on the recorded access information, cold data are put into younger blocks of the lower numbered lists using page mapping, and hot data are placed in elder blocks of the higher numbered lists in hybrid mapping. Between the smallest and biggest erase counts is a window. If the number of free blocks in either partition drops below two thresholds (T_L and T_H) respectively, data will be moved out from the lowest list to upper lists, and the window is then adjusted. This is how Rejuvenator performs passive wear leveling.

Table 2: Block Allocation Ratios in FAST

Trace	New Allocation	Merge Allocation
SPC1	3.90%	96.10%
TPC-C	33.76%	67.24%
MSR-hm_0	4.87%	95.13%
MSR-mds_0	13.02%	86.98%
MSR-prn_0	16.07%	83.93%
MSR-prxy_0	7.07%	92.93%
MSR-rsrch_0	18.42%	81.58%
MSR-stg_0	7.30%	92.70%
MSR-ts_0	8.29%	91.71%
MSR-web_0	6.75%	93.25%

3.2 Motivation

In this paper, we shall propose an efficient proactive algorithm to do wear leveling. The key idea of our approach lies in the management of blocks within address mapping. Block allocation is essential in flash management. In general there are two ways to allocate blocks: first-in-first-out (FIFO) and youngest block first [1].

In hybrid mapping, besides supplying log blocks, there are two scenarios in which block allocation needs to be performed, either when a location is being written to for the first time, or in the merge procedure. Table 2 shows the relative frequency of these two scenarios in FAST without wear leveling. Traces from [10], [11] and [5] were used. It is apparent that most of the allocation requests are made in the merge procedure. Note that log space is used to hold the updated copies of data. Some of them may have to be evicted by merging to free up space. However, in terms of temporal locality, some of evicted data may be accessed soon while others may be cold. If at this time we can predict which data are likely to go cold, and allocate elder blocks to them, then future cold data movements can be avoided. Moreover, allocating young blocks to data that are still hot improves wear evenness. Furthermore, blocks with valid data can be organized in an effective way that the utilization of young blocks are more exploited. These are the essence of *Observational Wear Leveling* (OWL) that we are proposing.

Table 1: A summary of the latest wear leveling algorithms

Algorithm	Type	Block Organization	Address Mapping
Dual-pool [1]	Passive	Hot pool and cold pool: a block with valid data is in either pool, where blocks are prioritized upon their erase counts.	Not constrained
BET [3]	Passive	Block sets and BET: A set has one block or several consecutive blocks to correspond a bit in the *block erasing table* (BET).	Not constrained
Rejuvenator [6]	Proactive + Passive	Multiple block lists: blocks that have the same erase count are grouped in a list.	Page mapping + Hybrid mapping
Lazy wear leveling [2]	Passive	Common way: free block pool, valid block pool, etc.	Hybrid mapping
OWL (this paper)	Proactive	Free block pool is ordered on erase count to be a min-heap, and valid block pool is sorted on arrival time.	Hybrid mapping

4. OBSERVATIONAL WEAR LEVELING

4.1 Overview

Observational wear leveling (OWL) attempts to reduce and evenly distribute erasures under hybrid mapping with log blocks in a proactive way. The key idea is to observe the temporal locality of write behaviors, and allocate blocks proactively. To do so, OWL maintains a *block access table* (BAT) that records the footprints of recent logical block accesses. The BAT is then used to perform *locality-based block allocation* (LBA) in the merge process. OWL also detects blocks with valid data and transfers data if necessary to prevent young blocks from being occupied for too long time. These two schemes are made effective by OWL's organization of blocks.

4.2 Block Organization

The organization of blocks is important not just for wear leveling but for all aspects of flash management. For example, DFTL maintains a *free block pool* of clean blocks for address mapping [8]. From Table 1 we see that wear leveling algorithms usually have special ways to organize blocks for better effectiveness.

In this paper, all blocks, excluding log blocks, are grouped in two pools, the *free block pool* and the *valid block pool*. This is a common organization in FTL designs [8]. In OWL we modify it slightly. The free block pool is sorted according to the erase count of each block. Its data structure can be a min-heap in our implementation, or other complicated ones that may consume less space [1]. Using a min-heap, if the number of blocks is n, it will take $O(log(n))$ to enqueue an erased block into the pool. Blocks in the valid block pool are ordered by their arrival time. It is almost like an ordinary FIFO queue, except that a valid block in the middle of the pool may, at the appropriate time, be moved to the head. The valid block pool can be managed in a linear structure, where the cost of insertion and removal is $O(1)$ and $O(n)$, respectively.

4.3 Locality-based Block Allocation

As pointed out earlier, block allocation requests may be issued for new log blocks, arrivals of new data, or in the merge procedure. Traditional wear leveling algorithms usually employ one policy, either FIFO or youngest block first [1, 4]. In OWL, they are handled differently. In particular, allocations for log block and new data are done using the youngest block first policy. This is easy to implement using OWL's free block pool organization as it is just a matter of fetching the top of the min-heap. Requests from merge, however, are serviced by the allocation of a suitable block that is selected in a predictive way according to the data's recent write history.

Data Structure and Overheads

The recent history of writes to logical blocks is recorded in the BAT in the form of write frequencies. Hence, the BAT is a runtime record of the temporal locality of writes. The BAT comprises two components: a hashed table for rapid looking-up, and a linked list to hold blocks' access frequencies. A sketch of BAT is shown in

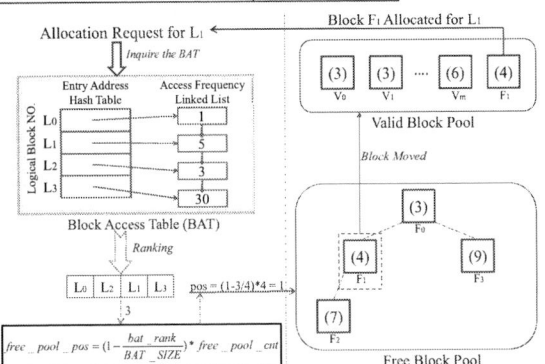

Figure 1: Locality-based Block Allocation with BAT

Figure 1. The hash table maps a logical block number to a linked list entry. In the linked list, an entry can be quickly appended or moved to the end of the list (being the most recently used). On the arrival of a write request with a logical address, the hash table will be checked. If the logical block number does not exist, an entry will be created and appended to the end of the linked list, and the hash table mapping is set up accordingly. Otherwise, the relevant entry will be updated and moved to the end. If space in the linked list is exhausted, the least-recently-used (LRU) entry, i.e., the one in the front, will be deleted to make space for the new arrival. Hence, the BAT keeps the latest information of the temporal locality of recent writes. It will be used by the FTL for the servicing of block allocation requests.

The temporal and spatial overheads of the BAT are fairly small. It is maintained in the RAM with the block and log page mapping tables. The access latency is much smaller than that of flash. It is not necessary to store the BAT in flash because temporal locality is always changing. The spatial overhead is also low. For each entry, an entry address mapped to each logical block number takes 4 bytes, and another 4 bytes are needed for its frequency. Thus, a 2KB table can hold the records of 256 logical blocks. From our experiments, a 2KB BAT is sufficient to support OWL's LBA scheme.

Locality-based Block Allocation

Here we will present the LBA algorithm used in the merge procedure. As mentioned, during a merge, free blocks are needed to accept data from the log page selected as the eviction victim, and its related data blocks. These data were not recently used. However, the situation may change in the near future. LBA aims to put the data to be merged into blocks of suitable ages in a predictive way. In particular, LBA tries to make younger blocks hold hot data, while using elder ones for the cold.

Algorithm 1 presents the skeleton of LBA. It is called in the merge procedure with the logical block number as a parameter and returns the block number of a free physical block. At line 2, the FTL first calculates the "rank" of the logical block in the BAT. In brief, the rank of a logical block is the count of blocks in the BAT

237

that have lower access frequencies than it. At line 3, the FTL computes a position in the free block pool using a heuristic formula. From line 4 to line 11, LBA will find the block at that position in the free block pool, and return it to the merge procedure.

The idea behind Algorithm 1 is as follows. First, the rank of the given logical block is calculated using the recent write history recorded in the BAT. If the logical block is highly accessed, its access frequency will be higher than many others. Then its rank in the BAT will be high too. The LBA puts this rank in the formula at line 3 to get the position in the free block pool, and looks for a free block accordingly. The free block pool is a min-heap sorted with the blocks' erase counts. Hence, LBA can easily locate the one with the suitable age in $O(log(n))$ time.

Computing the rank of a logical block is not straightforward. Since the BAT stores the frequencies of recently referenced blocks, an intuitive way to rank a logical block L is $\frac{BAT[L].freq}{\sum_{l \in BAT} BAT[l].freq}$. However, this is incorrect. In the most recent interval, some blocks may be highly accessed. These hot blocks will have very high frequencies, and they can dominate the total sum. The above fraction will show a bias towards these blocks, and the ranks of other blocks will be inaccurate. Worse, physical blocks cannot be fairly utilized because hot data are unlikely to be merged soon but always occupy younger blocks. In OWL, we first sort the blocks according to their frequencies. The rank is obtained after sorting. Our experiments show that this is a better measure.

Algorithm 1: Locality-based Block Allocation

Input : *logical_blk_no*, logical block number in request
Output: *free_blk_no*, allocated free block number
1 begin
2 $bat_rank := $ CalcBATRank ($logical\_blk\_no$);
3 $free\_pool\_pos := (1 - \frac{bat\_rank}{BAT\_SIZE}) * free\_pool\_cnt$;
4 $blk_pt := $ GetFreePoolHead (**void**);
5 $cnt := 0$;
6   **while** ($cnt < free\_pool\_pos$) **do**
7 $cnt ++$;
8 $blk_pt := $ GetNextFreeBlk (blk_pt);
9   **end**
10 $free\_blk\_no := blk\_pt$;
11   return $free\_blk\_no$;
12 end

Figure 1 gives an example of LBA scheme. There are 4 entries in the BAT, and 4 blocks in the free block pool. The number in the brackets of each block is its erase count. When a request is raised for logical block L_1, the FTL will examine the BAT, and perform sorting. The rank of L_1 is 3, and according to the formula, its position in the free block pool is calculated to be 1. With this number, the FTL finds physical block F_1, and moves it to the valid block pool. Finally, the FTL will return the block number F_1.

OWL differs from Rejuvenator in three important ways. Firstly, Rejuvenator focuses on the block allocation upon the arrival of new write requests; OWL works in the merge procedure that issues much more allocation requests (as shown in Table 2). Secondly, Rejuvenator uses page mapping for hot data and hybrid mapping for cold data. This is quite complicated and interferes too much with the other flash management modules. OWL utilizes hybrid mapping only, and is hence simpler. Thirdly, while they both utilize a structure to record reference counts of logical addresses at runtime, Rejuvenator maintains access information at the granularity of pages. OWL's BAT works at the block-level. With the same amount of RAM space, OWL can store longer historical accesses.

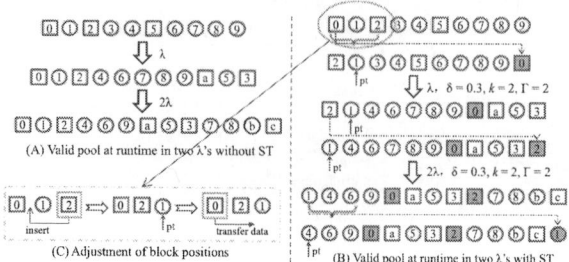

Figure 2: An example of ST scheme

4.4 Scan and Transfer Scheme

LBA works in the merge process, and it may miss two types of data. One is data that are seldom, or possibly never updated after being stored. They have no up-to-date copies in log space. In other words, their data blocks are not related to any log page. Another is ones that are very hot. If data are highly updated, their old copies in log pages will be quickly invalidated. So they can avoid being merged. Evidently blocks occupied by these data are unlikely to be erased. Thus, we use a proactive scheme named *scan and transfer* (ST) to find these data, and efficiently place them in elder blocks.

Many methods have been proposed for hot/cold data identification [6, 4]. Note that here valid blocks are chronologically appended to the valid block pool, and blocks at the head have been there for the longest time. ST exploits the organization of the valid block pool, and periodically scans a small portion through the pool to find a block containing one of the above two types of data. To do so, ST employs two variables, λ and δ. Briefly, ST scans $(\delta \cdot 100\%)$ of the valid block pool after every λ write requests.

In its scanning, ST identifies a young block with cold data using the block's erase count and mapping status. In our implementation, we deem a block to be "young" if its erase count is smaller than half of the average erase count of all blocks, which is more strict than lazy wear leveling that sets such standard to be the average erase count [2]. If a young block is not associated to any log pages, it will be picked. After the scanning, more than one candidate may be found. To minimize the performance overhead, ST will transfer one block's data each time. Let functions $T(b)$ and $Q(b)$ represent block b's residence time in the pool and the quantity of valid pages to be transferred, respectively. The victim should be the one that has stayed for the longest time, and has the least data. Let

$$v(b) = \frac{T(b)}{Q(b)}. \quad (1)$$

The block that has the largest $v(b)$ can be selected as the victim. Given the valid pool's organization, $T(b)$ can be replaced by $\frac{1}{P(b)}$ where $P(b)$ is block b's *position number* in the pool. For example, the head of the pool has a position number 1. Then Eq. (1) will be

$$v(b) = \frac{1}{P(b) \cdot Q(b)}. \quad (2)$$

There are several issues to use Eq. (2), however. Firstly, $P(b)$ can be easily obtained, but to maintain $Q(b)$ for each block requires a large amount of RAM space. Secondly, in Eq. (2), $Q(b)$ has the same weight as $P(b)$. Since ST transfers one block after every λ requests, a larger $Q(b)$ is acceptable, but a block with a big $P(b)$ might be mistakenly identified as cold. Thirdly, computing $v(b)$ for all the candidates may consume too much time.

Based on Eq. (2), ST can be done in a simplified yet efficient way. Besides λ and δ, ST employs a pointer pt and a counter k. Initially, pt points at the first block that is associated to log pages, and k is set to zero. ST will check each block's erase count and

238

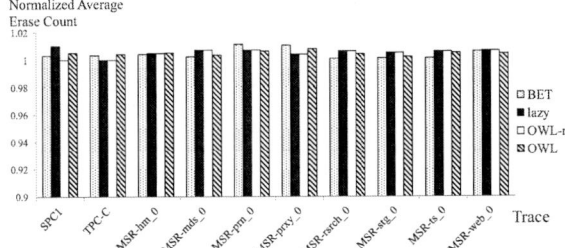

Figure 3: Average Erase Counts of Each Trace

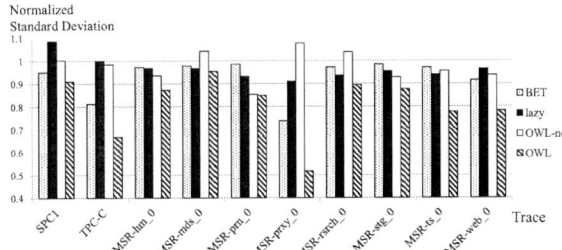

Figure 4: Standard Deviation of Erase Counts

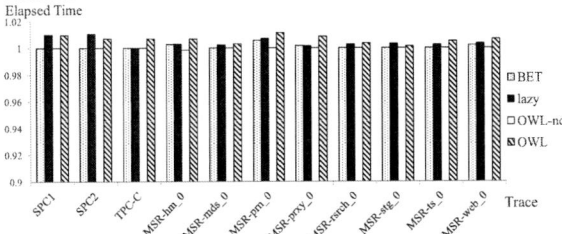

Figure 5: Elapsed Time with Four Algorithms

mapping status through scanning δ blocks of the pool. If a block satisfies the condition mentioned above, i.e. is young and not associated to any log page, it will be selected, and inserted before pt. After scanning, data of the first selected block will be transferred and k will count by one. Before next scan, if the block that pt points at is to be merged, pt will be replaced at the next block that is associated to some log pages, and k will be reset to zero. In the next scan, if blocks found in previous scans exist, ST cancels scanning and just performs data transfer on the first one of these blocks. If after scanning no candidate is found, ST will check k. If k is bigger than a threshold Γ, the block that pt points to has been there for at least $(\Gamma \cdot \lambda)$ requests, and avoided being merged. The data that block holds could be very hot. So ST will select and transfer it; pt and k will be reset accordingly. If $k < \Gamma$, ST just returns.

Obviously ST prefers blocks of cold data to blocks of hot data because the latter still might be merged. It uses pt and k heuristically to identify a block with very hot data. Figure 2 shows an example of ST at runtime. Figure 2(A) is the pool's being in two λ requests without ST. Squares are blocks that are not associated to any log page, and circles are ones that are. The number inside is the logical block number mapped to each block. In Figure 2(B), ST transfers data in logical block 0, 2 and 1 to elder blocks. Figure 2(C) shows a case that a selected block is inserted before pt.

5. EXPERIMENTS

We shall evaluate the effectiveness of the OWL in this section. All the experiments were conducted using the FlashSim [8] simulator in a Linux 64-bit system with GCC 4.6. The address mapping used was FAST [9] that has been modified with our block organizations. We implemented BET, lazy wear leveling and Rejuvenator as comparisons to OWL. In the following texts, baseline refers to a configuration that has no wear leveling, lazy is the one with lazy wear leveling, OWL refers to our proposed OWL algorithm, and OWL-nc has all of OWL except the ST module.

The traces we used came from three sources. They are shown in Table 2. SPC1 and SPC2 were downloaded from [10]. TPC-C is a typical online transaction processing (OLTP) workload from [11]. All others were from Microsoft's data centers [5]. They represent various environments, and the numbers of write requests vary from 1 to 12 million. Note that each write request in the trace may consist of multiple write operations. *Caveat lector*: these traces were recorded at different machines whose configurations were never clearly documented. In our simulations, in order to assess wear evenness, we used a different configuration for each trace so that all physical blocks can be involved. Similarly experiments in [6] were confined to a small area of an SSD disk for the same reason.

We studied three metrics. The average erase count, and its standard deviation are used to measure the effectiveness of the wear leveling algorithms. The overhead is measured by the elapsed time needed to finish processing the trace. All three metrics have to be assessed together in order to obtain a qualitative judgement about the efficacy of the algorithms.

For BET, we configured each block set to be a single block. This is the best case for BET in terms of wear leveling. The threshold Δ of lazy wear leveling was initialized to be 2. It is adaptively tuned online according to [2]. For OWL, the default values of λ, δ and Γ are 1000 requests, 0.4% and 50. All flash parameters, like the latencies of write and erase operations, were obtained from [7].

5.1 Effectiveness of OWL

Figure 3-5 are results on average erase count, standard deviation, and elapsed time for each trace, normalized against baseline. Figure 3 shows OWL can reduce the number of erasures in many cases, while Figure 4 shows that the standard deviation decreases, by as much as 29.9% and 43.2% compared to BET and lazy with MSR-prxy_0 respectively. These lead us to conclude that OWL performs better than BET and lazy in evening out erasures. Figure 5 shows the elapsed time on processing each trace. OWL is at most 1.1% slower than the baseline in the case of MSR-prn_0.

As mentioned earlier, the three metrics should be considered together. Take for example TPC-C. It has 7.7 millions requests in the workload. From Figure 3, we can see OWL has a similar number of erasures as BET and lazy. However, as shown in Figure 4 the difference in standard deviation is significant. This implies OWL achieves better wear evenness with roughly the same erasures.

There are traces in which OWL did not do too well also. Figure 3 shows that OWL has slightly more erasures than lazy for MSR-prxy_0. We analyzed MSR-prxy_0, and found it quite different from other traces. Normally, one would expect a write request to access a number of pages. MSR-prxy_0, however, has a large number of very small write requests, with 77.8% of the requests accessing only one page. Since the BAT works at block-level, such a situation is difficult for the BAT to record access information accurately. This in turn affected LBA's allocations. Even so, OWL was still able to use the ST module to perform wear leveling. This is why OWL has a little more erasures, but the best evenness.

We have also implemented Rejuvenator. However, there were several stumbling blocks. Specifically, two thresholds (T_L and T_H) were not given in their paper. Also, it was said initially all blocks will have a zero erase count, and all will be in the lower numbered lists. However, when and how to migrate from such initial state to the two partitions of the lower and higher numbered lists were not described in [6]. These parameters and process are important for

239

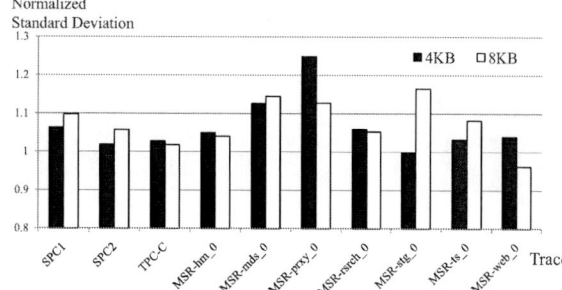

Figure 6: The Effects of Different BAT Size

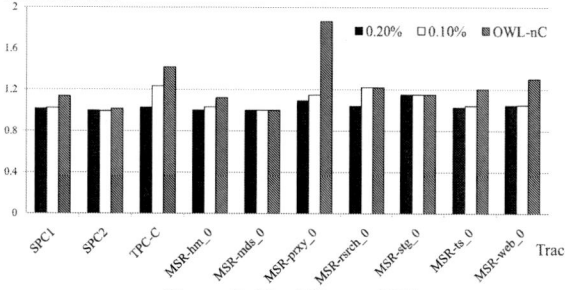

Figure 7: The Effects of ST

Rejuvenator. Nonetheless, we tried to simulated it but the results are not comparable to those for BET, lazy and OWL. Take TPC-C trace for example. It should be easy to identify hot and cold data based on the access information of TPC-C workload. Our simulation of Rejuvenator has a similar erase count as OWL but its standard deviation over all blocks is 44.3% more than OWL. It is worse for other traces. (See the appendix for more comparisons.)

5.2 Effects of BAT Size

The BAT is used to support LBA in the merge procedure. The default size in our experiments is 2KB, allowing for 256 records. We also tried varying the size to 4KB and 8KB. The standard deviations of these normalized against the 2KB configuration are presented in Figure 6. From it we can see in general a larger BAT results in more unevenness. The BAT records the latest write frequencies, and one with a larger capacity is more likely to store outdated information. This will mislead LBA. In terms of overhead, besides saving space, a smaller BAT can also have a lower access time.

5.3 Effectiveness of ST

δ, λ and Γ are three parameters of ST module. We experimented with different values of them to study ST's functions. The results of various δ are shown in Figure 7.

In our default setting, OWL will go through 0.4% of the valid block pool. We also experimented with δ being 0.2% and 0.1%, and normalized their results against those for 0.4%. Figure 7 shows that in general the wear evenness will worsen when a lower proportion of blocks is checked (TPC-C and MSR-rsrch_0). The worst case occurs in OWL-nc that does not have ST. From Figure 7, processing the MSR-prxy_0 will suffer the most from the removal of ST module. Processing less blocks means that ST is less aggressive on moving cold data. This will result in cold or very hot data occupying their blocks longer, preventing these blocks from being utilized. On the other hand, a less aggressive movement would also mean less performance overhead.

We also did experiments to measure the effects of different λ. ST will be activated every λ interval. The default value of λ is 1000 requests. Figure 8 shows the standard deviations in wear evenness

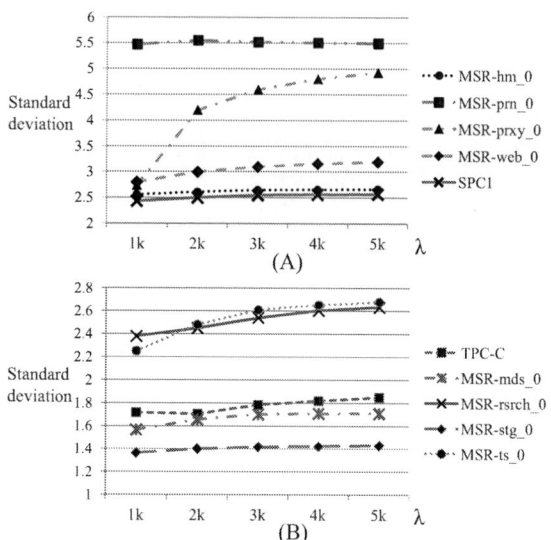

Figure 8: The Effects of λ length

with λ being set at 2000(2k), 3000(3k), 4000(4k) and 5000(5k) requests. From the results we can conclude the effect of λ depends on specific workload. For MSR-prn_0 or MSR-stg_0, the interval length has no significant impact on wear evenness. For others, however, a longer λ will worsen the evenness. This is because ST will be less aggressive on a longer λ. With the same δ, ST will miss blocks that ought to be transferred. Still, for MSR-prxy_0, a more frequently executed ST module can greatly enhance wear evenness.

From our observation, Γ marginally affects OWL's efficacy. The discussion on Γ and more results of δ are attached in the appendix.

6. CONCLUSION

In this paper, we have proposed a novel algorithm for wear leveling of NAND flash called *observational wear leveling* (OWL). OWL records the temporal locality of write activities at runtime, and allocates blocks judiciously in the merge procedure of hybrid mapping. To further even out erasures, OWL also employs a scanning and transfer module to identify and move cold or very hot data. Experimental results show that OWL can improve the evenness of erasures by as much as 43.2% with about 1.1% performance degradation, and a space overhead of 2KB.

7. REFERENCES

[1] L.-P. Chang. On efficient wear leveling for large-scale flash-memory storage systems. In *SAC '07*, pages 1126–1130. ACM, 2007.

[2] L.-P. Chang et al. A low-cost wear-leveling algorithm for block-mapping solid-state disks. In *LCTES '11*, 2011.

[3] Y.-H. Chang et al. Improving flash wear-leveling by proactively moving static data. *IEEE Trans. Comput.*, 59:53–65, January 2010.

[4] C. Wang et al. Extending the lifetime of NAND flash memory by salvaging bad blocks. In *DATE '12*, 2012.

[5] D. Narayanan et al. Write off-loading: Practical power management for enterprise storage. *Trans. Storage*, 4:10:1–10:23, November 2008.

[6] M. Murugan et al. Rejuvenator: A static wear leveling algorithm for NAND flash memory with minimized overhead. *MSST 2011*, 0:1–12, 2011.

[7] Y. Hu et al. MLC vs. SLC NAND flash in embedded systems. Technical report, September 2009.

[8] A. Gupta et al. DFTL: a flash translation layer employing demand-based selective caching of page-level address mappings. In *ASPLOS '09*, 2009.

[9] S.-W. Lee et al. A log buffer-based flash translation layer using fully-associative sector translation. *ACM Trans. Embed. Comput. Syst.*, 6(3):18, 2007.

[10] Storage Performance Council. SPC traces. http://traces.cs.umass.edu/, December 2009.

[11] BYU trace distribution center. TPC-C database benchmark traces. http://tds.cs.byu.edu/tds/, 2001.

APPENDIX

A. EXPERIMENTAL METHODOLOGY

There are three ways to do experiments in order to measure the effectiveness of wear leveling algorithms. They all aim to ensure that all blocks are covered in assessing wear evenness. The first way is what lazy wear leveling did [2]. They configured a 20.5GB SSD (0.5GB was over-provisioned for log space of hybrid mapping) in their simulator, and "replayed the input workload one hundred times". The second was used in the Rejuvenator paper [6]. Their SSD in simulation had 32GB, but they "restrict the active region" for write requests and "the remaining blocks did not participate in the I/O operations". The third way is what we did. For each workload, we assigned a reasonable capacity so that all blocks have the chance to be involved in wear leveling. The capacities for workloads we used are shown in Table 3. Note that the over-provisioning rate for log space is 3% which is the same as other works [8].

Table 3: Capacities for Traces

Trace	Capacity
SPC1	2.06GB
TPC-C	3.09GB
MSR-hm_0	4.12GB
MSR-mds_0	2.06GB
MSR-prn_0	6.18GB
MSR-prxy_0	4.12GB
MSR-rsrch_0	2.06GB
MSR-stg_0	4.12GB
MSR-ts_0	2.06GB
MSR-web_0	2.06GB

All trace in Table 3 are in the public domain, and can be downloaded. The simulator we used, FlashSim [8], is also open-source.

B. RESULTS OF REJUVENATOR

Here we will show the detailed experimental results of Rejuvenator. As previously mentioned, their paper did not show how the values of W (the size of the structure to record access frequencies), T_L and T_H are to be set. In our simulation, W was set to hold 1024 entries, the same as the authors claimed in a presentation[1]. T_L and T_H were set according to our communication with one of the authors. The paper also did not describe how the entire system moves from the initial state where all blocks are in lower numbered lists to the state of two partitions that one block is in either lower numbered lists or higher numbered lists based on its erase count. We approximated this as follows. Initially, blocks are allocated from lower numbered lists as described in their paper because no higher numbered list exists. Then blocks will be erased in the merge procedure. As lists with bigger erase counts start to be populated, we begin to partition lists so that the number of lower and higher numbered lists are adaptively adjusted based on the sliding window.

The results are presented in Table 4. Note that the capacities and Flash parameters of Rejuvenator are the same as that for OWL.

From Table 4 we can see OWL evidently outperforms Rejuvenator. However, since T_L and T_H are workload dependent, it is difficult to draw an absolute conclusion. This is especially so for the case for MSR-hm_0 and MSR-prxy_0 where the results of two algorithms differ significantly.

There are several reasons why OWL can achieve better wear evenness than Rejuvenator. First, with the same RAM space, Rejuvenator has a less complete estimation on access patterns because

[1]http://storageconference.org/2011/Presentations/Research/14.Murugan.pdf

Table 4: Comparison between OWL and Rejuvenator

Trace	Average Erase Count		Standard Deviation	
	OWL	Rejuvenator	OWL	Rejuvenator
SPC1	14.057838	15.264644	2.430129	5.526887
TPC-C	14.09825	14.124719	1.716122	2.476510
MSR-hm_0	9.840271	27.1584	2.554806	10.965194
MSR-mds_0	5.525630	6.778785	1.566230	1.843398
MSR-prn_0	12.085569	13.746652	5.467150	7.568265
MSR-prxy_0	19.043258	25.913277	2.738008	16.781061
MSR-rsrch_0	9.738963	8.495363	2.378286	3.874533
MSR-stg_0	5.729223	13.648533	1.364073	4.930392
MSR-ts_0	10.140859	12.721007	2.250717	3.988268
MSR-web_0	12.527585	13.648533	2.796568	4.930392

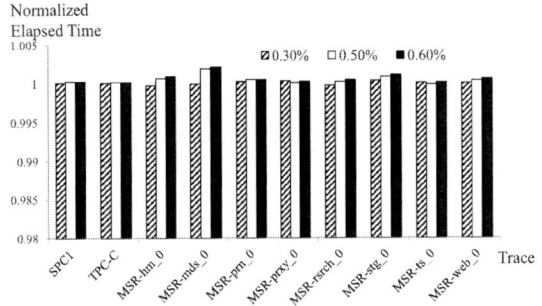

Figure 9: Normalized Elapsed Time with Various δ

it does that at the page-level. Second, Rejuvenator responds to the number of available free blocks by T_L and T_H, and there will be a delay. OWL works in a proactive way, and takes actions more promptly.

Rejuvenator also has a shortcoming in managing mapping tables. Rejuvenator maintains page mapping for both hot data and updates of cold data which are in log blocks. Rejuvenator at any time only has one log block, instead of the usual 3% of all blocks, and it "picks a free block with the least possible erase count in the higher numbered lists". It seems that the entries for this page mapping are very small – about the same as the number of pages in the block. However, Rejuvenator does not immediately merge the log block with data blocks like a standard hybrid mapping strategy [9]. It performs garbage collection in the higher numbered lists only when the number of free blocks in the higher numbered lists drops below T_H. This implies that, as long as there are at least T_H free blocks, a log block can be picked from the higher numbered lists for log space, and more mapping information has to be added to the mapping table. In other words, Rejuvenator can take up a significant amount of RAM space for hybrid mapping. If Rejuvenator merges log space with data blocks like the standard hybrid mapping, the single log block is seriously insufficient and *trashing* will frequently occur. In the paper of Rejuvenator, it was claimed that the proportion of hot logical pages is very small ($< 10\%$). So more than 90% data would be handled with hybrid mapping. This would aggravate the hybrid mapping of Rejuvenator if it has a traditional merge procedure.

C. MORE RESULTS ON OWL

In the experiment section, we reported on the impacts of λ and δ. For the latter, we also experimented with δ being 0.3%, 0.5% and 0.6%. We normalized their results to those of 0.4%, as shown in Figure 9, 10 and 11.

Note that a larger δ value will cause ST to scan more blocks in

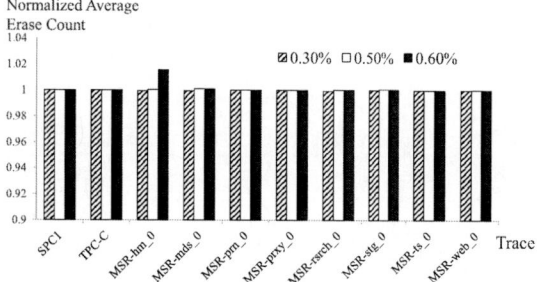

Figure 10: Normalized Average Erase Count with Various δ

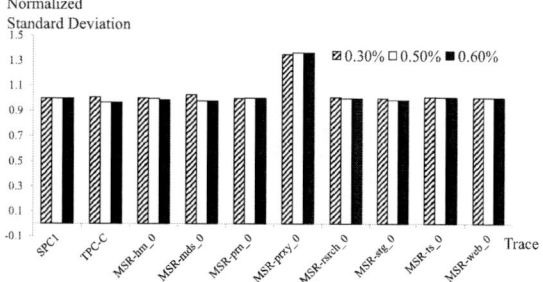

Figure 11: Normalized Standard Deviation with Various δ

the valid block pool. However, the effect is also dependent on the workload. From the three figures, we can see with most traces the impact of various δ is not significant. This is due to the access patterns of these workloads being quite uniform. But for MSR-prxy_0 again, it is obvious in Figure 11 that results of $\delta = 0.4\%$ can be viewed as optimal. That is to say, scanning more blocks will incorrectly classify the data, and transfers based on such erroneous identification will only worsen the wear evenness. On the other hand, scanning less blocks may miss blocks that should be transferred.

Figure 12, 13 and 14 show the results upon various values of Γ. It is obvious that Γ only has a marginal impact on wear evenness in most cases. Note that Γ is the threshold for identifying very hot data. If a block stays in the valid block pool for more than $(\Gamma \cdot \lambda)$ requests, its data are most likely to be very hot. The default value of Γ in previous experiments was 50. We conducted more experiments with Γ being 30, 40, 60 and 70 ($\lambda = 1000$ and $\delta = 0.4\%$). Figure 12 shows that the elapsed time did not change much with various Γ (results of $\Gamma = 30$ are used to normalize other settings). Neither did average erase count in Figure 13 (results of $\Gamma = 30$ are used to normalize other settings). In most cases, the standard deviation of erase counts was not affected as shown in Figure 14. How-

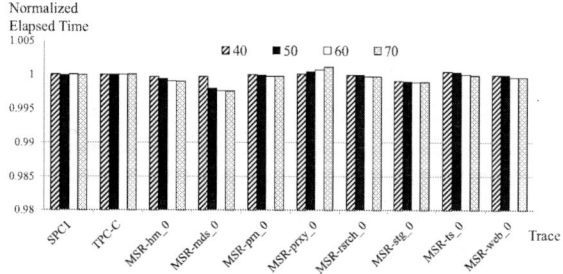

Figure 12: Normalized Elapsed Time with Various Γ

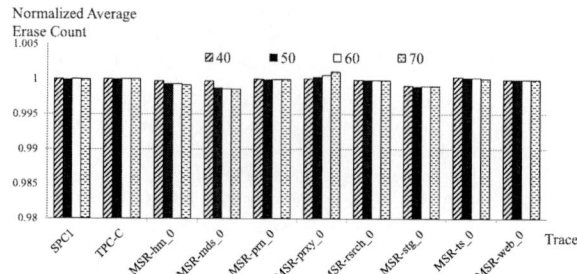

Figure 13: Normalized Average Erase Count with Various Γ

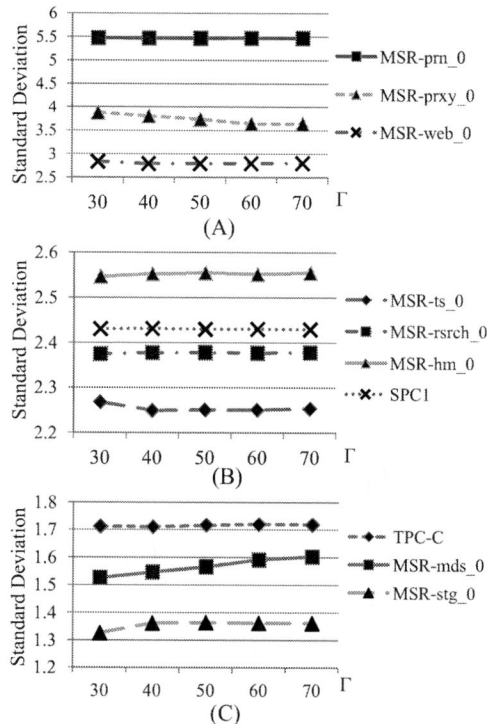

Figure 14: Standard Deviation with Various Γ

ever, with a longer interval, MSR-prxy_0 would result in slightly more erasures for better evenness, while MSR-mds_0 suffered from wear unevenness with slightly less erasures. For the former trace, a bigger Γ could result in more data blocks being identified as very hot. The characteristics of MSR-prxy_0 have been described before. Because small requests were frequently issued, a longer interval (in more requests) would be more suitable and could help to accurately filter out very hot data. Thus ST modules had lower standard deviation with the increase of Γ. This also confirms our argument in Section 5.1 and 5.3 that ST has played an important role in processing MSR prxy_0. For MSR mds_0, which is taken from a media server with a majority of big requests, a longer interval may miss blocks of very hot data with the same δ depth to scan, and less erasures would be performed by the ST module in transferring the data. This explains why as the interval was lengthened, the erase counts decreased and the wear evenness worsened for MSR-mds_0, as shown respectively in Figure 13 and 14(C).

Cache Revive: Architecting Volatile STT-RAM Caches for Enhanced Performance in CMPs

Adwait Jog[†]   Asit K. Mishra[§]   Cong Xu[†]   Yuan Xie[†]
Vijaykrishnan Narayanan[†]   Ravishankar Iyer[§]   Chita R. Das[†]

[†]The Pennsylvania State University
University Park, PA 16802, USA
{adwait,czx102,yuanxie,vijay,das}@cse.psu.edu

Intel Corporation[§]
Hillsboro, OR 97124, USA
{asit.k.mishra,ravishankar.iyer}@intel.com

ABSTRACT

High density, low leakage and non-volatility are the attractive features of Spin-Transfer-Torque-RAM (STT-RAM), which has made it a strong competitor against SRAM as a universal memory replacement in multi-core systems. However, STT-RAM suffers from high write latency and energy which has impeded its widespread adoption. To this end, we look at trading-off STT-RAM's non-volatility property (data-retention-time) to overcome these problems. We formulate the relationship between retention-time and write-latency, and find optimal retention-time for architecting an efficient cache hierarchy using STT-RAM. Our results show that, compared to SRAM-based design, our proposal can improve performance and energy consumption by 18% and 60%, respectively.

Categories and Subject Descriptors

B.3.2 [**Hardware**]: Memory Structures—*Cache memories*; B.7.1 [**Integrated Circuits**]: Types and Design Styles—*Advanced technologies, Memory technologies*

General Terms

Design, Experimentation, Measurement, Performance

Keywords

STT-RAM, Heterogeneous (hybrid) systems

1. INTRODUCTION

The emergence of multicore architectures in both embedded processors as well as general purpose computing domain has started to increasingly stress the demand for the on-chip cache memories. As the number of cores on a chip continues to increase with technology scaling, the demand for the on-chip memory would continue to increase significantly, further worsening the memory wall problem [3]. This memory wall problem is critical both from the performance and power perspectives. Thus, it is imperative to look for novel technology; circuit, and architectural techniques to address the memory wall problem.

Spin-Transfer Torque RAM (STT-RAM) is a promising memory technology for future multi-core general purpose and embedded systems that delivers on many aspects desirable of a universal memory. STT-RAM has the potential to replace the conventional on-chip SRAM caches because of its higher density, competitive read times, and lower leakage power consumption compared to SRAM. Its high-density property (3x-4x denser than conventional SRAM) can provide denser memories at lower area footprint with near-zero leakage energy. However, the latency and energy overhead associated with the write operations are the key drawbacks of this technology for providing competitive or better performance compared to the SRAM-based cache hierarchy. Consequently, recent efforts have focused on masking the effects of high write latencies and write energy both at the architectural [16, 19] level and circuit level [18]. In contrast to these approaches, a recent work explored the feasibility of relaxing STT-RAM data retention times to reduce both write latencies and write energy [15]. This adaptable feature of tuning the data retention time can be exploited in several dimensions. The focus of this paper is to tune this data retention time to closely match the required refresh time of the Last Level Cache (LLC) blocks to achieve significant performance and energy gains. In this context, the paper addresses several design issues such as how to decide an appropriate retention time for the LLC, what the relationship between retention time and write latency is, and how we architect the cache hierarchy with volatile STT-RAM.

The non-volatile nature and non-destructive read ability of STT-RAM are the key differences with regard to traditional on-chip cache design with SRAM technology. However, as our analysis will show, for many emerging applications, it is sufficient to store the valid data in the LLC for a few tens of ms in contrast to μs for the L1 cache [10]. Consequently, the duration of data retention in STT-RAM is an obvious candidate for device optimization in the cache design. We, therefore, conduct an application-driven study to analyze the inter-write times (refresh times) of the L2 cache blocks to determine a suitable data retention time. Although lifetime analysis of cache lines has been conducted earlier to improve performance and reduce power consumption [9, 10], we revisit this topic with a different intention - correlating STT-RAM data-retention time to cache lifetime. An extensive analysis of emerging workloads using the M5 simulator [2] indicates that the average inter-write times for most of the L2 cache blocks is close to $10ms$, and thus, we advocate our STT-RAM design with this retention time.

A key challenge in determining a suitable data-retention time for the STT-RAM is to balance the reduced write latency of STT-RAM cells with lower retention time against the overhead for data refresh or write-back of cache lines with longer lifetimes. In this paper, we compare 3 different STT-RAM based cache designs: (1) STT-RAM cache without retention time relaxation (10+ years of data

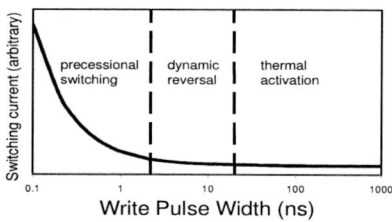

Figure 1: Demonstration of three different switching phases

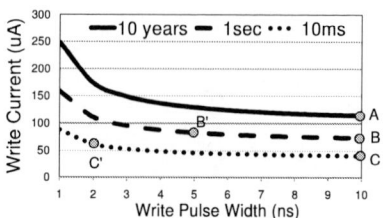

Figure 2: Write current vs. write pulse width for different MTJ retention time

retention time); (2) STT-RAM cache with retention time of $1sec$, which is long enough for the inter-write time of majority of the cache lines, and therefore, no refreshing overhead is incurred; and (3) STT-RAM cache with retention time of $10ms$, which is a more aggressive design with better performance/energy gain, but a data refreshing technique is needed for correct operations, since cache lines that have inter-write times exceeding $10ms$ are likely to lose data. Thus, we propose simple extensions to the L2 cache design for avoiding any data loss. This includes a simple 2-bit counter to keep track of the inter-write times of all the cache blocks and a small buffer to temporarily store the blocks whose inter-write time has exceeded the retention time.

The primary contributions of this paper can be summarized as:
(1) Detailed characterization of STT-RAM volatility property: We present a detailed device characterization of data-retention tunability in STT-RAM cells, providing insight into the underlying principles enabling these tradeoffs.
(2) An application-driven study to determine retention time: With the aid of application level characterization, we propose the design of STT-RAM with the retention time in the range of $10ms$. Also, such a design makes the LLC homogeneous (all same type of STT-RAM cells) leading to lower die cost and ease in fabrication.
(3) Architectural solution to handle STT-RAM volatility: We present a simple buffering mechanism to ensure the integrity of the programs given the volatile nature of our tuned STT-RAM cells. This scheme is simple, yet very energy and performance efficient.

2. STT-RAM DESIGN

In this section, we present a detailed discussion of STT-RAM models which we have developed to guide us in our exploration of suitable STT-RAM device for LLC. Preliminaries for STT-RAM can be reviewed in supplemental Sec. S.1 as well as in [4, 5, 12].

2.1 Write Current vs. Write Pulse Width

In this section, we will establish the relationship between write current and write pulse width. In [4], three distinct switching modes were identified based on the operating range of switching pulse width (τ): thermal activation (TA) ($\tau > 20ns$), precessional switching (PS) ($\tau < 3ns$) and dynamic reversal (DR) ($3ns < \tau < 20ns$). The relationship between switching current density (J_c), and write pulse width (τ) in these three operating ranges are characterized by the following analytical model [14, 18]:
(1) $J_{c,TA}(\tau) = J_{c0}\{1 - (\frac{k_B T}{E_b})ln(\frac{\tau}{\tau_0})\}$
(2) $J_{c,PS}(\tau) = J_{c0} + \frac{C}{\tau^\gamma}$
(3) $J_{c,DR}(\tau) = \frac{J_{c,TA}(\tau) + J_{c,PS}(\tau)e^{-k(\tau-\tau_c)}}{1 + e^{-k(\tau-\tau_c)}}$
where, $J_{c,TA}$, $J_{c,PS}$, $J_{c,DR}$ are the switching current densities for TA, PS and DR respectively. J_{c0} is the critical switching current density, k_B is the Boltzmann constant, T is the temperature, E_b is the thermal barrier, and τ_0 is the inverse of attempt frequency. C, γ, k, and τ_c are fitting constants.

Figure 1 shows write current vs. write pulse width characteristics for PS, DR and TA modes. In TA mode, the switching current increases very slowly with decrease in write pulse width. This sug-

gests that, in TA mode, shorter pulse width is beneficial in reducing write latency and energy without hampering the read latency and energy. On the contrary, in PS mode, write current goes up rapidly as we reduce the write pulse width and minimum write energy of the MTJ can only be obtained at some particular write pulse width in this region, as shorter write pulse width with optimal write current can still provide necessary write energy to switch. DR mode has intermediate characteristics compared to PS and TA. Based on these trade-offs, this paper focuses on PS and DR modes for achieving our overall goal of minimizing write latency and energy.

2.2 Impact of Retention Time on MTJ Characteristics

Retention time of an MTJ is primarily impacted by the thermal stability of its free layer. The relationship between retention time and thermal barrier can be modeled as $t = C \times e^{k\Delta}$ [15], where t is the retention time and Δ is the thermal barrier, while C and k are fitting constants. This relationship suggests that retention time of an MTJ reduces exponentially with reduction in the thermal barrier. As described in [7], the switching current of MTJ decreases as thermal barrier is reduced. Here, we combine this observation with the write current versus write time trade-off described in Sec. 2.1, to conclude that faster write speed or/and small write current/energy can be obtained by lowering the thermal barrier of a MTJ, at the cost of lower retention time.

Thermal barrier of an MTJ can be lowered by reducing the MTJ planar area [15] and thickness of the MTJ [13]. In our study, we use $2F^2$ state-of-the-art in-plane MTJ [7] as our baseline with 10+years of retention time and thermal barrier of $72k_B T$, where k_B is the Boltzmann constant and T is the temperature. Since we use optimized $2F^2$ cell, there is not much leeway to reduce the planar area. Instead, we decrease the thickness of the free layer and lower the saturation magnetization to obtain lower thermal barrier. The minimum thickness of the free layer for the MTJs in our work is $2nm$ and we do not reduce it further to avoid reliability and process variation issues. Our volatile MTJs have thermal barriers of $46k_B T$ and $40k_B T$, which correspond to retention times of $1sec$ and $10ms$ at $125\,°C$, respectively.

We obtained raw experimental data from our device collaborator and further did curve fitting using the in-plane MTJ device equations (1)-(3). The curve-fitted results for three different MTJs (10years, $1sec$, $10ms$) are shown in Figure 2. Operating point A($10ns$, $114\mu A$) serves as our baseline [5]. By fixing the write pulse width at $10ns$, we reduce the write current to obtain volatile MTJs operating at B($10ns$, $73\mu A$) and C($10ns$, $40\mu A$). Then we apply the write current versus write time trade-off on these operating points to reduce the write latency. Specifically, we operate the MTJ with $1sec$ retention time at point B'($5ns$, $82\mu A$) which corresponds to 28% lower write current than that of baseline (point A). Also, we operate $10ms$-retention time MTJ at point C'($2ns$, $61\mu A$) which further cuts down the write current by 25%. Hence, we observe that relaxation of retention time of STT-RAM can reduce both write latency and energy. Based on this analysis, we integrate the cache-level SRAM and STT-RAM models in

244

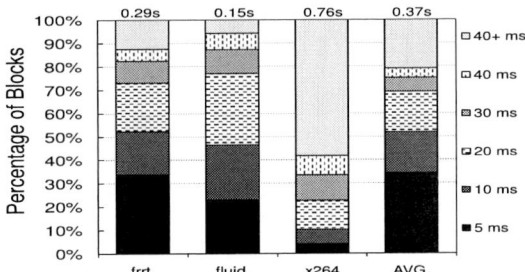

Figure 3: Distribution of blocks showing different revival times (value on the top of a bar show the maximum revival time for that distribution)

NVsim [6] and simulation results are tabulated in Table 3 (supplemental Sec. S.3).

3. AN APPLICATION-DRIVEN APPROACH TO DETERMINE RETENTION TIME

In order to leverage the volatile STT-RAM properties, we need to know what the ideal/feasible retention time should be. Ideally, the STT-RAM write latency should be competitive to SRAM latency and the cache retention time should be high. However, as discussed in the previous section, since the write latency is inversely proportional to the retention time, we need to find a feasible trade-off based on the STT-RAM device characteristics. Thus, we attempt to decide an ideal retention time by analyzing the retention times of LLC blocks in multithreaded PARSEC 2.1 [1] and multiprogrammed SPEC 2006 environment. These suites include emerging workloads from desktop to server domains. `dedup` and `bzip2` are compression schemes and `facesim` simulates face (important in facial recognition and virtual-reality systems). `vips` and `x264` are image processing and video encoding workloads, respectively. All these workloads can be hosted on high end mobile/tablet processors. The main idea is to understand the distribution of the inter-write intervals to an LLC and use the average of these intervals as the STT-RAM retention time.

3.1 Relating Application Characteristics to Retention Time

Application characterization gives the basis for evaluating the impact of retention time on the overall system performance. In order to do this characterization, the first step is to investigate the duration for which the cache block should retain the data. A cache block is only refreshed when the block is written. Thus, we record intervals between two successive writes (refreshes) to the same L2 cache block. We define this interval as the *revival time*. While collecting these results, we ensure that if a block gets invalidated in between two consecutive writes, we do not consider the time in between the invalidation and the next write. Previous work [10] has done similar type of revival time analysis, but it was for the L1 cache. Figure 3 shows the distribution of L2 cache blocks having different revival time intervals. These results are obtained by running emerging workloads on the M5 Simulator [2]. Table 2 contains additional details of the system configuration. Figure 3 shows the results of three applications along with the averages across the entire PARSEC 2.1 suite. We observe similar distribution for SPEC 2006 applications and more details are given in supplemental Sec. S.7. We observe from the figures that, on average, approximately 50% of the cache blocks get refreshed within $10ms$, which is in contrast to the μs reuse for L1 cache studied in [10]. About 20% of blocks remain in the cache for more than $40ms$ and rest of the blocks have intermediate revival times. Blocks that stay longer than the retention time in the cache without being refreshed

Table 1: Retention and Write Latencies for STT-RAM L2 Cache

Retention Time	10years	1sec	10ms
Write Latency @2GHz	22 cycles	12 cycles	6 cycles

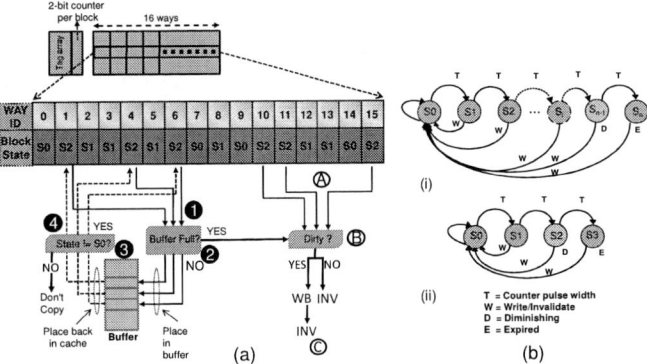

Figure 4: A modified 16-way L2 cache architecture with a 2-bit counter and a small buffer

are assumed to be expired. This distribution also gives us the basis on which we can choose the optimal retention time. Reducing the retention time too much will make the cache highly volatile leading to degraded performance, while increasing the retention time would negatively affect the write latency.

3.2 Low Retention STT-RAM Characteristics

Table 1 summarizes the retention time and write latency tradeoffs based on the analysis of Sec. 2. The results indicate a significant reduction in write latency with reduction in retention time. Note that one can possibly lower the retention time further beyond the ms ranges, but it becomes much harder to control the variations which in turn diminish the benefits of performance/energy, since the number of blocks to be refreshed increases (Figure 3).

From above discussions we conclude that, from the application perspective, it is best to choose a retention time which minimizes the number of unrefreshed blocks and from the technology side, it is ideal to use STT-RAM with minimum write latency and energy.

4. ARCHITECTING VOLATILE STT-RAM

In this section, we propose architectural solutions starting with a naive scheme of writing back all the dirty blocks to a more sophisticated scheme, where we minimize the number of write backs.

4.1 Volatile STT-RAM

In this design, we write back all the unrefreshed dirty blocks which become volatile after the retention time. To identify these blocks, we maintain a counter per cache block. Each cache block uses an n-bit counter (shown in Figure 4(b)(i)). We assume the time between transitions (T) from one state to another equals to the retention time divided by the number of states, where the number of states is 2^n. A block starts in state S_0 when it is first brought to the cache. After every transition time (T), the counter of each block is incremented. When a block reaches state S_{n-1}, it indicates that it is going to expire in time T. We define this time as the *leftover* time and the block in state S_{n-1} as the *diminishing* block. Increasing the value of n will decrease the leftover time at the cost of increased overhead of checking the blocks at a finer granularity. For example, if we use a 2-bit counter, the leftover time is $2.5ms$ and for a 3-bit counter, it is $1.25ms$. A larger bit counter decreases the leftover time and allows more time for a block to stay in the cache before applying any refreshing techniques. This gives the block more opportunity to stay in the cache at the cost of maintaining a counter with high number of bits.

245

Our experimental results show that a 2-bit counter, similar to the one used in [9], suffices to detect the expiration time of the blocks without significantly affecting the performance. With a 2-bit counter, a block can be in one of the four states as shown in Figure 4 (b)(ii). A block moves from state S_0 to state S_3 in steps of $2.5ms$ and at any time the block is written/invalidated, it goes back to the initial state. The counter bits are kept as a part of the SRAM tag array. The overhead of the 2-bit counter is 0.4% for one L2 cache bank. This scheme has a negative impact on the performance for two reasons: (1) There will be a large number of write backs to the main memory for all the dirty blocks at the end of the retention time. (2) The expired block could have been frequently read and losing it will incur additional read misses.

4.2 Revived STT-RAM Scheme

To minimize the write back overhead of the expired blocks at the end of retention time, we propose a different technique, where we use a small buffer to hold a subset of diminishing blocks. We call this design as the *revived* STT-RAM scheme. These dirty blocks are, thus, not written back to the main memory. They are written to the temporary buffer and written back to the cache to start another fresh cycle. Figure 4(a) shows the schematic diagram of the proposed scheme. The main components of this design are:

Buffer: It is a per bank small storage space with a fixed number of entries made up of low-retention time STT-RAM cells. We use these entries to temporarily store the diminishing blocks.

Buffer Controller: The buffer controller consists of a $\log_2 N$ bit buffer overflow detector, where N is the buffer size. The overflow detector is first checked to see the occupancy of the buffer, when a diminishing block is directed to the buffer and the buffer overflow detector is incremented. The block is copied to one of the empty buffer entries along with the set and way ID, if there is buffer space. If the buffer is full, the dirty blocks are written back to the main memory; otherwise they are invalidated.

Implementation Details: Figure 4 (a) shows a 16-way set associative cache bank with the associated tag array. We show the working of our scheme using a 2-bit counter. One of the sets is shown in detail to clarify the details of the scheme. All the blocks in a set are marked with their current state. Each bank is associated with a buffer and the buffer controller. Let us consider that the buffering scheme incorporates eight MRU slots (in supplemental Sec. S.8 we show that eight MRU slots give the best benefits). In Figure 4(a), ❶ shows that three blocks in first eight MRU slots are diminishing and directed to the buffer. Please note that, we apply our scheme only to the diminishing (to-be-expired) blocks, which gives enough time for the scheme to be completed before actual data loss happens. ❷ checks the occupancy of the buffer and if it is not full, each of the diminishing blocks is copied to one of the entries of ❸ along with way and set IDs. Way and set IDs are again used by ❷ to copy back the blocks to the same place in the L2 cache. Ⓐ shows the blocks which are not in MRU slots, but are diminishing. We check these blocks in Ⓑ to see whether they are dirty or not. If they are dirty, we write back (WB) those blocks as shown in Ⓒ and then invalidate (INV). If they are not dirty, they are just invalidated. During this whole refresh process, if a read request for the cache block arrives, it will be successfully completed as the data is still valid during that time. If a write/invalidate request arrives, the cache block goes back to its original state. To make sure we don't copy the stale data from the buffer, we perform a state check before copying back as shown in ❹. Moreover, our implementation assumes the worst-case temperature of $125\,^{\circ}C$, and hence, there is no possibility of early expiration of block because of sudden reduction in STT-RAM retention time.

Table 2: Baseline system configuration

Processor Pipeline	2 GHz processor Fetch/Exec/Commit width 8
L1 Cache (SRAM)	32 KB per-core (private) I/D cache, 4-way 64B block size, write-back, 10 MSHRs
L2 Cache (SRAM or STT-RAM)	1MB (SRAM) or 4MB (STT-RAM) bank, shared, 16-way, 64B block size, 10 MSHRs
Network	Ring network, one router per bank, 3 cycle router and 1 cycle link latency
Main Memory	400 cycle access

5. EXPERIMENTAL EVALUATION

We evaluate our designs using M5 Simulator [2] with PARSEC 2.1 and SPEC 2006 applications. We model a 2GHz processor with four cores. We modified the M5 simulator to model L2 cache banks composed of tunable retention time STT-RAM cells. Table 2 details our experimental system configuration. More details on methodology of collection of results are described in supplemental Sec. S.7. **The design scenarios we evaluate are:**

• **S-1MB:** This is our baseline scheme, where all L2 cache banks are composed of SRAM cells. Capacity of each bank is 1MB.

• **S-4MB:** This is a hypothetical case, where capacity of each bank is 4MB and each bank has the same read and write latency as that of S-1MB. This case is analyzed to see the potential benefit of having a 4x improvement in cache capacity at 4x area density, while still having read/write latencies comparable to SRAM. This hypothetical design has the capacity and area of an STT-RAM but the read/write latencies of an SRAM cache.

• **M-4MB:** This is our baseline scheme for STT-RAM design, where all L2 cache banks are composed of 10 year retention time STT-RAM cells. Capacity of each bank is 4MB and each bank occupies the same area as that of an SRAM bank.

• **Volatile M-4MB(1sec):** This design is used to evaluate our Volatile STT-RAM Scheme described in Sec. 4, where all L2 cache banks are composed of 1sec retention time STT-RAM cells.

• **Volatile M-4MB(10ms):** This design is similar to Volatile M-4MB(1sec) but with 10ms STT-RAM retention time.

• **Revived M-4MB(10ms):** This design is used to evaluate our Revived STT-RAM Scheme, where all L2 cache banks are composed of 10ms retention time STT-RAM cells. All the results are for the design with 8 MRU Slots and 1900 Buffer Slots.

6. ANALYSIS OF RESULTS

In this section, we provide a comparative analysis of the performance and energy results of the proposed six designs.

6.1 Performance Comparison

Figure 5(a) shows speedup results with multithreaded applications along with the average improvements. Only 9 applications are shown to reduce clutter in the plots, however, the average (arithmetic mean) numbers are computed across the entire suite.

(A) Speedup when replacing an SRAM with hypothetical cache: We find that this hypothetical design has an average speedup of 23% over the SRAM cache. This is the maximum performance that any scheme can provide.

(B) Speedup when replacing an SRAM with 4MB, $10 years$ retention time STT-RAM: We find that, for the M-4MB design, all applications to the right of x264 (including x264) exhibit speedup improvements over S-1MB. Most of these applications are read intensive applications (see Table 4) and thus, they benefit from not only the 4x capacity increase of STT-RAM, but also from the presence of a L2 cache write buffer. On average, we find 6% improvement in speedup over S-1MB for these applications. Although ferret and vips are write-intensive applications, they benefit with a 4x improvement in capacity when going from a S-1MB to M-4MB. This is because, the write request to L2 cache banks in

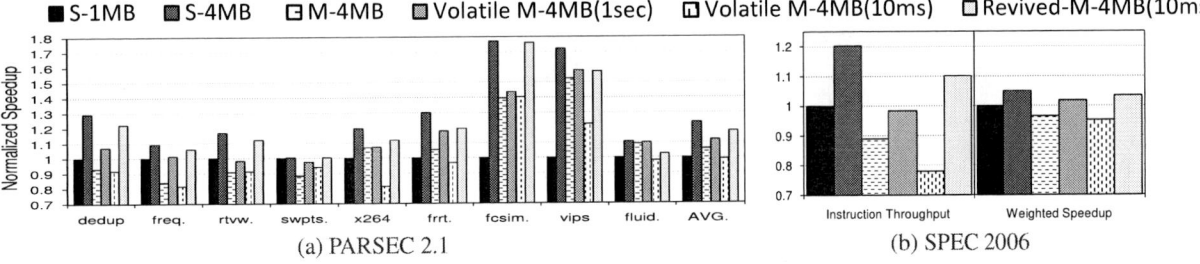

Figure 5: Performance of applications normalized to that of S-1MB

these two applications are staggered in time and an 10-entry write buffer proves to be sufficient to hide the long write latencies.

All applications to the left of x264 are write-intensive and have bursty requests arriving at L2 cache banks. For these applications, we observe significant degradation in speedup (on average 11% degradation) because of the high write latency of STT-RAM. Overall, when averaged across the entire suite, a traditional 10years STT-RAM gives a minimal 5% speedup improvement over S-1MB. However, a 10years STT-RAM cache organization has 14% lower performance when compared to the hypothetical design S-4MB and with write-intensive applications. This is the gap that our proposal seeks to bridge by tuning the retention time.

(C) Speedup when replacing a SRAM with 4MB, 1sec retention time STT-RAM: With such a STT-RAM cache bank, no refreshing schemes are employed. As shown earlier, almost all blocks get refreshed within a 1sec time interval. Reducing the retention time from 10years to 1sec reduced the write latency of a cache bank by 10 cycles (from 22 cycles to 12 cycles). This leads to significant speedup improvements over a 10years retention time STT-RAM cache organization. On average, this reduction in 10 cycles lead to 6% performance improvement (14% for write intensive applications). However, this design is still 9% (11% for write intensive applications) lower in performance than the hypothetical case.

(D) Speedup when replacing a SRAM with 4MB, 10ms retention time STT-RAM without refreshing: This volatile M-4MB (10ms) design also does not have any refreshing scheme, but the retention time of STT-RAM cells used is 10ms. After 10ms, this STT-RAM device will lose its data and hence to keep the integrity of the data a large number of write-backs are forced from the LLC to the main-memory before the actual expiration happens.

(E) Speedup when replacing a SRAM with 4MB, 10ms retention time STT-RAM with refreshing (Revived-M-4MB(10ms)): This is our proposed scheme, which incorporates refreshing of dirty blocks beyond the 10ms retention time. Our scheme significantly improves performance when compared to all realistic design scenarios evaluated. On average, the proposed revived scheme is better than the conventional SRAM design (S-1MB) by 18%, traditional 10years STT-RAM by 15% and over Volatile M-4MB(1sec) by 4.5%. The write latency of this STT-RAM cache bank is 6 cycles and when compared to a 1sec retention time STT-RAM, the difference of 6 cycles reduction in L2 cache bank access time helps in improving performance. This performance improvement is in spite of the increase in the number of write-backs (which increase by over 2x) compared to an SRAM cache. The Revived-M-4MB(10ms) scheme is closest to the hypothetical S-4MB case and is within 5% of it, showing the benefits of our scheme in making the STT-RAM device a choice for universal memory.

(F) Analysis with SPEC 2006: In general, the observations with SPEC applications are consistent with those made with PARSEC 2.1 applications. Figure 5(b) shows Instruction Throughput (IT) and Weighted Speedup (WS) with the multiprogrammed mixes (supplemental Table 5). Simply replacing a SRAM bank with 4MB

STT-RAM would lead to 11% degradation in IT, and 4% degradation in WS. However, employing our refreshing scheme on a 10ms volatile STT-RAM can lead to an average 22%, 11% and 10% improvement over M-4MB, Volatile M-4MB(1sec) and S-1MB, respectively. With a write intensive mix (bzip2, gcc, lbm and hmmer) this improvement can be as high as 35% over the baseline SRAM design. WS also follows a similar trend as WS and we find that our proposed refreshing scheme on a 10ms retention time volatile STT-RAM shows the best results. Overall, our proposed design is within 9% (2%) of the hypothetical device (S-4MB) in terms of IT (WS).

6.2 Energy Usage Comparison

Figure 6 shows normalized leakage, dynamic energy (reads + writes), and total energy for a subset of applications. While computing the energy numbers, we take into account the overheads of our proposed cache block refreshing schemes. We observe that on average, there is 44% improvement in total energy going from S-1MB to M-4MB designs. This improvement is mainly because of the drastic reduction in leakage energy (43%). In general, all volatile STT-RAMs reduce the energy envelope of the LLC. With 1sec volatile STT-RAM, total energy is reduced because of reduction in leakage energy and also nominal performance improvement. However, when comparing Volatile M-4MB(10ms) with Volatile M-4MB(1sec), because of larger number of write-backs with 10ms retention time STT-RAMs, the performance degrades and thus, leakage energy increases. With our proposed cache block refreshment scheme, although there is an increase in the dynamic energy on account of additional back and forth writes to the buffer and cache lines, we consistently observe improvement in total energy. This is mainly attributed to the fact that fraction of dynamic energy to the total energy is not significant because of very high leakage energy and fast switching times of cores. On average, we find 11% energy benefits of using revived M-4MB design over Volatile M-4MB(1sec) and 18% improvement over the baseline STT-RAM design. We observe similar benefits with SPEC 2006 applications, but for the sake of brevity we are not showing them.

7. PRIOR WORK

The work that is most closely related to ours is [15]. In this, the authors relax the retention time of STT-RAM from 10years to $56\mu s$ by reducing the planar area of MTJ. Our application driven analysis shows that the ideal retention time of LLC should be in the range of ms. Recently published work related to STT-RAM [7, 11] uses state-of-the-art MTJ designs in the range of $2\text{-}3F^2$ which do not give the leeway of reducing the retention time by aggressively reducing the MTJ planar area, we reduce the retention time by lowering the saturation magnetization and the thickness of the free layer. Also, our proposed refresh scheme is simple, yet very performance and energy efficient compared to the DRAM-style refreshing proposed in [15] (also see Sec. S.9)

A very recent effort [17] explores the possibility of designing

■ S-1MB ■ M-4MB □ Volatile M-4MB(1sec) ▨ Volatile M-4MB(10ms) ▤ Revived M-4MB(10ms)

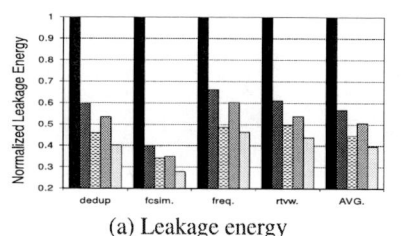
(a) Leakage energy

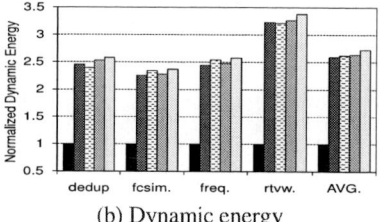
(b) Dynamic energy

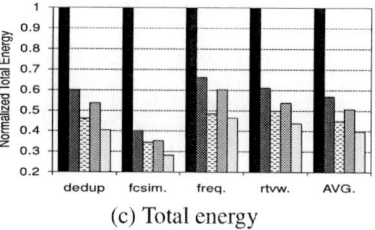
(c) Total energy

Figure 6: Energy of applications normalized to that of S-1MB

LLC with STT-RAM banks of varying retention times and moving dying blocks from lowest retention time (μs) bank to the higher ones [17]. In contrast to this work that populates STT-RAM cells of different retention times, we tune our design for single retention times across the memory hierarchy. Our approach eases the challenge of higher die costs and yield issues associated with heterogeneous retention cells and irregular structures. Further, the μs retention times used in the prior effort are challenging to achieve in newer, scaled MTJ dimensions as indicated earlier.

8. CONCLUSIONS

Spin-Transfer Torque RAM (STT-RAM) is a promising candidate for future multi-core general purpose and embedded systems, due to its high-density, low leakage, and immunity to soft errors. However, its high write latency and dynamic write energy are the disadvantages compared to SRAM based cache design. In this paper, we propose to trade-off the non-volatility (data-retention time) for better write performance/energy in STT-RAM cache design. In this context, we conduct an application-driven study to characterize the lifetime of LLC with the intention of using this time as the optimal retention time for the STT-RAM. We analyze three different scenarios for designing the L2 cache: one with $1sec$ retention time with write back, second with $10ms$ retention time with write back and the third with $10ms$ retention time with buffering, called Revived-M-4MB. The results not only indicate that it is possible to get up to 18% improvement in speedup for PARSEC applications and 60% reduction in total energy consumption over S-1MB design, but also show that our proposed design can be within 5% of the hypothetical case (S-4MB) with an equal capacity SRAM configuration, while being more energy efficient. Furthermore, compared to the prior schemes that are aimed at hiding the high write latency of STT-RAMs, the approach to reduce its write latency seems to be a better solution for designing a performance and power efficient memory hierarchy for multi-core systems.

9. ACKNOWLEDGMENTS

We thank the anonymous reviewers for their reviews and comments towards improving this paper. This work is supported in part by NSF grants CNS-0721479, CNS-1152449, CCF-1147388, CCF-0903432 and DoE grant DE-SC0005026.

10. REFERENCES

[1] C. Bienia. *Benchmarking Modern Multiprocessors*. PhD thesis, Princeton University, 2011.

[2] N. Binkert, R. Dreslinski, L. Hsu, K. Lim, A. Saidi, and S. Reinhardt. The M5 Simulator: Modeling Networked Systems. *Micro, IEEE*, 26(4):52 –60, 2006.

[3] D. Burger, J. R. Goodman, and A. Kägi. Memory bandwidth limitations of future microprocessors. In *ISCA*, 1996.

[4] Z. Diao, Z. Li, S. Wang, Y. Ding, A. Panchula, E. Chen, L.-C. Wang, and Y. Huai. Spin-transfer torque switching in

magnetic tunnel junctions and spin-transfer torque random access memory. *Journal of Physics 2007*.

[5] X. Dong, X. Wu, G. Sun, Y. Xie, H. H. Li, and Y. Chen. Circuit and microarchitecture evaluation of 3D stacking magnetic RAM (MRAM) as a universal memory replacement. In *DAC*, 2008.

[6] X. Dong, C. Xu, Y. Xie, and N. P. Jouppi. NVSim: A Circuit-Level Performance, Energy, and Area Model for Emerging Non-Volatile Memory. *IEEE Transactions on Computer-Aided Design of Integrated Circuits and Systems*, 2012.

[7] A. Driskill-Smith. Latest Advances in STT-RAM. In *2nd Annual NVM Workshop*, 2011.

[8] F. Fishburn, B. Busch, J. Dale, D. Hwang, and et al. A 78nm $6F^2$ DRAM technology for multigigabit densities. In *Proceedings of the Symposium on VLSI Technology*, 2004.

[9] S. Kaxiras, Z. Hu, and M. Martonosi. Cache decay: exploiting generational behavior to reduce cache leakage power. In *ISCA*, 2001.

[10] X. Liang, R. Canal, G.-Y. Wei, and D. Brooks. Process Variation Tolerant 3T1D-Based Cache Architectures. In *MICRO*, 2007.

[11] C. Lin, S. Kang, Y. Wang, K. Lee, X. Zhu, W. Chen, X. Li, W. Hsu, Y. Kao, M. Liu, Y. Lin, M. Nowak, N. Yu, and L. Tran. 45nm low power CMOS logic compatible embedded STT MRAM utilizing a reverse-connection 1T/1MTJ cell. In *IEDM*, 2009.

[12] A. K. Mishra, X. Dong, G. Sun, Y. Xie, N. Vijaykrishnan, and C. R. Das. Architecting on-chip interconnects for stacked 3D STT-RAM caches in CMPs. In *ISCA*, 2011.

[13] A. Nigam, C. Smullen, V. Mohan, E. Chen, S. Gurumurthi, and M. Stan. Delivering on the promise of universal memory for spin-transfer torque RAM (STT-RAM). In *ISLPED*, 2011.

[14] A. Raychowdhury, D. Somasekhar, T. Karnik, and V. De. Design space and scalability exploration of 1T-1STT MTJ memory arrays in the presence of variability and disturbances. In *IEDM*, 2009.

[15] C. Smullen, V. Mohan, A. Nigam, S. Gurumurthi, and M. Stan. Relaxing non-volatility for fast and energy-efficient STT-RAM caches. In *HPCA*, 2011.

[16] G. Sun, X. Dong, Y. Xie, J. Li, and Y. Chen. A novel architecture of the 3D stacked MRAM L2 cache for CMPs. In *HPCA*, 2009.

[17] Z. Sun, X. Bi, H. H. Li, W.-F. Wong, Z.-L. Ong, X. Zhu, and W. Wu. Multi retention level STT-RAM cache designs with a dynamic refresh scheme. In *MICRO*, 2011.

[18] C. Xu, D. Niu, X. Zhu, S. Kang, M. Nowak, and Y. Xie. Device-architecture co-optimization of STT-RAM based memory for low power embedded systems. In *ICCAD*, 2011.

[19] P. Zhou, B. Zhao, J. Yang, and Y. Zhang. Energy reduction for STT-RAM using early write termination. In *ICCAD*, 2009.

SUPPLEMENTAL

In this supplementary section, we present additional details on STT-RAM modeling, application properties and refresh schemes. This section is organized as follows: Sec. S.1 presents the preliminaries on STT-RAM. Sec. S.2 and Sec. S.3 discusses STT-RAM modeling details and simulation results, respectively. Latency and energy numbers of our STT-RAM and SRAM designs are tabulated in Table 3. Sec. S.4 extends the discussion of revival time distribution, started in Sec. 3, for SPEC 2006 applications. Sec. S.5 describes the importance of performing a state check before copying the data back to the buffer. The read and write percentages of our applications taken from PARSEC 2.1 and SPEC 2006 suites are detailed in Sec. S.6 and Table 4. Sec. S.7 describes the evaluation metrics and methodology of collection of results. The sensitivity results to find the optimal number of MRU slots and buffer size are shown in Sec. S.8. In Sec. S.9 we compare our refreshing scheme (Revive) with DRAM-style refresh presented in [15]. Sec. S.10 describes how our refreshing scheme saves number of write backs leading to performance improvements. Finally, we conclude our supplementary section with comparisons to additional prior work (Sec S.11) and concluding remarks (Sec S.12).

S.1 Preliminaries on STT-RAM

STT-RAM uses an Magnetic Tunnel Junction(MTJ) as the memory storage and leverages the difference in magnetic directions to represent a memory bit ("0"/"1" state). As shown in Figure 7, an MTJ contains two ferromagnetic layers. One ferromagnetic layer has a fixed magnetization direction and called the reference layer. The second layer's magnetic direction can be changed by passing write current, and, thus it is called the free layer. The relative magnetization direction of two ferromagnetic layers determines the resistance of the MTJ. If two ferromagnetic layers have different directions, the resistance of MTJ is high, indicating a "0" state; if two layers have the same directions, the resistance of MTJ is low, indicating a "1" state. The current amplitude required to reverse the direction of the free ferromagnetic layer is determined by the size, the aspect ratio of MTJ, and the write pulse duration [4, 11].

S.2 STT-RAM Modeling

In this section, we first introduce models to determine the area of an STT-RAM cell. Typically, the area of each STT-RAM cell would determine the area of a cache bank composed of these cells and in turn influence the read/write latency of the bank. As shown earlier in Figure 7, each 1T1J STT-RAM cell is composed of an NMOS and one MTJ. The NMOS access device is connected in series with the MTJ. The size of NMOS is constrained by both SET and RESET current, which are inversely proportional to the writing pulse width. In order to estimate the current driving ability of MOSFET devices, a small test circuit using HSPICE with PTM 45nm HP model [R3] is simulated. The driving current is obtained by assuming typical TMR (120%) and Low Resistance State (LRS) ($3k\Omega$) value [11] and wordline voltage to be 1.5V (the optimal value is extracted from [19]). Further, we oversize the access transistor width to guarantee enough write current is provided to MTJ using the methodology discussed in [R2]. To achieve high cell density, we model the STT-RAM cell area by referring to the DRAM design rules [8]. As a result, the cell size of a STT-RAM cell is given as:

$$\text{(4)} \qquad \text{Area}_{\text{cell}} = 3\left(W/L + 1\right)(F^2)$$

where, W and L are the channel width and length of the access NMOS transistor, respectively.

Thermal Barrier vs. Retention Time As described in Sec. 2.2, the retention time of an MTJ is largely determined by the *thermal*

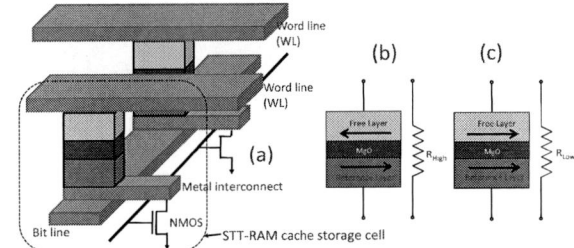

Figure 7: (a) Structural view of an STT-RAM Cache Cell (b) Anti-Parallel High Resistance, Indicating "0" state (c) Parallel Low Resistance, Indicating "1" state

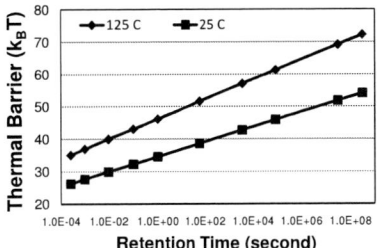

Figure 8: MTJ thermal barrier for different retention times

stability of the MTJ. The relationship between retention time (log scale) and thermal barrier is shown in Figure 8. We observe that MTJ with higher retention time has higher thermal barrier. For higher temperature, thermal barrier decreases at a faster rate, leading to faster reduction in retention time. Since there is a strong dependency of retention time of MTJ on the operating temperature, our implementation assumes the worst-case temperature of $125\,^{\circ}\text{C}$ to avoid any possibility of early expiration of block because of sudden reduction in STT-RAM retention time.

S.3 STT-RAM Simulation Results

Table 3 shows the architectural parameters based on our STT-RAM models. It shows the read, write times and energy numbers of three stable operating points A, B', and C' (shown in Figure 2) for MTJs with different retention times. We find that a 4MB NVM STT-RAM cache occupies similar chip area as 1MB SRAM. This is consistent with previous work [5]. For the leakage simulation, we didn't apply any power gating techniques for the cache banks and hence our leakage numbers are higher than the previously published numbers in recent STT-RAM papers. We observe that STT-RAM consumes almost half of the leakage power than that of SRAM. That is basically because of the fact that, half of STT-RAM die area is occupied by peripheral circuitry, which in turn means that half of the chip is leaky. The switching energy per STT-RAM cell is < 1pJ [7] and only half of the total write energy is consumed on switching the cells. For SRAM numbers, we use balance L2 design for leakage, density and performance (already implemented in CACTI).

S.4 Distribution for SPEC 2006 applications

As described in Sec. 3, the revival time is defined as the time interval between two successive writes. Figure 9 shows the distribution of L2 cache blocks having different revival time intervals. These results are obtained by running multiprogrammed applications on the M5 Simulator [2]. We observe that these results are similar to that of PARSEC 2.1 applications, and we draw similar conclusions as mentioned in Sec. 3.

Table 3: 16-way L2 cache simulation results

		Area (mm^2)	Read Latency (ns)	Write Latency (ns)	Read Energy (nJ)	Write Energy (nJ)	Leakage Power (mW)
1MB SRAM		2.612	1.012	1.012	0.578	0.578	4542
4MB STT-RAM	$t = 10yr$	3.003	0.998	10.61	1.035	1.066	2524
	$t = 1s$	2.904	0.973	5.571	1.015	1.036	2235
	$t = 10ms$	2.901	0.959	2.598	1.002	1.028	2227

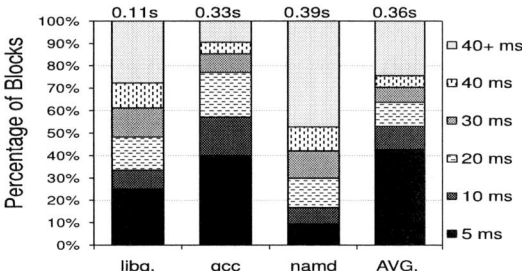

Figure 9: Distribution of blocks showing different revival times (value on the top of a bar show the maximum revival time for that distribution)

S.5 Buffer Architecture

As mentioned in Sec. 4, key idea of our proposal is to copy *important diminishing* blocks to a temporary buffer and immediately copy it back to the cache. Since buffering process for a block starts off in the penultimate state of the counter and not in the final state as usually would have happened in a typical counter [9], the data is still valid for incoming reads in the caches. This scheme cuts the overhead of searching in the buffer. Moreover, it is important to note that incoming writes during buffering process will put back the state of the cache block to S_0 making the corresponding buffer data stale, and hence it is necessary to perform a state check before copying the buffer data back.

S.6 Application Properties

Table 4 shows the properties of various emerging applications. *Read%* denotes the percentage of reads to the L2 cache out of the total L2 accesses and *Write%* denotes the percentage of writes to the L2 cache out of the total L2 accesses. *Intensity* denotes Read/Write intensity based on read%/write%.

S.7 Evaluation Metrics

In this section, we describe our evaluation metrics and methodology for collection of results. For the multithreaded applications, we assume 4 threads are mapped to our modeled processor with four cores. We report normalized speedup for these applications, which is defined as the improvement over the slowest thread. For the multiprogrammed SPEC applications, we report Instruction Throughput and Weighted Speedup. We define instruction throughput (IT) to be the sum of all the number of instructions committed per cycle in the entire chip (Eq. (5)). The weighted speedup (WS) is defined as the slowdown experienced by each application in a multiprogram mix, compared to its run under the same configuration when no other application is running on other cores (Eq.(6)). For analyzing the energy behavior, we measure the leakage energy, dynamic energy and total energy for all designs.

$$(5)\ Instruction\ throughput = \sum_i IPC_i$$

$$(6)\ Weighted\ speedup = \sum_i \frac{IPC_i^{shared}}{IPC_i^{alone}}$$

Collection of Results:
We report results of 12 multithreaded applications(Table 4) and

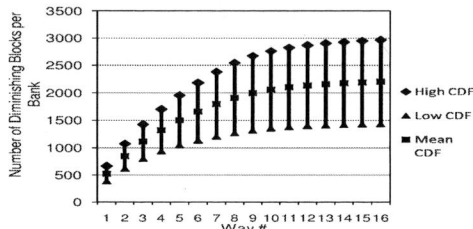

Figure 10: 95% Confidence Intervals of diminishing blocks for each way

14 multiprogrammed mixes (Table 5). We selected these multiprogrammed mixes from read and write intensive categories (Table 4) to get varying percentages of reads and writes. We use *simsmall* input for mulithreaded applications and report the results of only Region of Interest, (ROI) (except for facesim, where we report results for only 2B instructions of ROI) after warming up the caches for 500M instructions and skipping the initialization and termination phases. For the multiprogrammed mixes, we fast forward 1B instructions, warm up caches for 500M instructions and then report results for 1B instructions.

S.8 Sensitivity Analysis

(A) Sensitivity to number of buffer entries:

The number of buffer entries can affect the performance of the Revived-M-4MB scheme in two ways: increasing the buffer size will accommodate more diminishing blocks at a particular instance and decreasing the buffer size will lead to more buffer overflows. Increasing the buffer size, leads to fewer buffer overflows, but this reduction comes at a cost of increase in buffer area and consequent revival overheads. Decreasing the buffer size eventually leads to additional write backs (discussed in Sec. 4).

To find the optimal buffer size, we calculate the 95% Confidence Intervals (CIs) for the cumulative distribution of diminishing blocks per bank. This is shown in Figure 10. We observe that, for the first 8 MRU slots, the mean value of the buffer entries is 1900 blocks, which corresponds to a 3% area overhead per L2 cache bank. Upper limit of the 95% CI corresponds to 2500 blocks, which represents 4% area overhead per L2 cache bank. The lower limit of 95% CI corresponds to 1300 blocks (2% area overhead).

Figure 11(a) shows the normalized speedup to 1300 blocks, with a subset of applications by varying the number of buffer entries. on average, varying the number of buffer entries from 1900 to 2500, results in only 1% speedup improvement. Hence, in all our results, we used 1900 buffer entries (resulting in a 3% area overhead) with the best possible performance per area-overhead.

(B) Sensitivity to number of MRU slots:

In order to calculate the optimal MRU slots for buffering, we collected statistics of MRU positions of diminishing blocks. Figure 10 shows the average cumulative distribution (mean CDF) of diminishing blocks per bank as a function of the number of ways in a set for applications. We observe that the number of diminishing blocks becomes stable after first eight MRU ways. The mean number of blocks corresponding to the first eight ways is 1900 (3% overhead over per L2 cache bank), which is a good initial choice

Table 4: Application characteristics of PARSEC 2.1 and SPEC 2006 applications

#	PARSEC 2.1	Read%	Write%	Intensity	#	SPEC 2006	Read%	Write%	Intensity
1	blackscholes	91.9	8.1	Read	13	bzip2	86.2	13.8	Read
2	bodytrack (btrack.)	92.2	7.8	Read	14	gcc	99.4	0.6	Read
3	dedup	73.8	26.2	Write	15	mcf	94.5	5.5	Read
4	facesim (fcsim.)	78.7	21.3	Read	16	leslie3d	70.7	29.3	Write
5	ferret (frrt.)	46.2	53.8	Write	17	namd	92.7	7.3	Read
6	fluidanimate (fluid.)	82.4	17.6	Read	18	soplex	59.6	40.4	Write
7	freqmine (freq.)	72.1	27.9	Write	19	hmmer	63.6	36.4	Write
8	rtview (rtvw.)	64.1	35.9	Write	20	sjeng	76.6	23.4	Write
9	streamcluster	98.4	1.6	Read	21	libquantum(libq.)	100.0	0.0	Read
10	swaptions (swpts.)	49.9	50.1	Write	22	lbm	15.7	84.3	Write
11	vips	75.0	25.0	Write	23	GemsFDTD	99.2	0.8	Read
12	x264	95.5	4.5	Read	24	omnetpp	97.7	2.3	Read
					25	h264ref	57.8	42.2	Write

Table 5: SPEC 2006 multiprogrammed mixes

Mixes	Applications
mix-1	mcf, leslie3d, bzip2, gcc
mix-2	mcf, gcc, bzip2, namd
mix-3	hmmer, sjeng, gcc, bzip2
mix-4	bzip2, lbm, hmmer, gcc
mix-5	gcc, GemsFDTD, omnetpp, bzip2
mix-6	bzip2, gcc, omnetpp, h264ref
mix-7	h264ref, bzip2, leslie3d, gcc
mix-8	gcc, sjeng, mcf, bzip2
mix-9	leslie3d, gcc, bzip2, hmmer
mix-10	sjeng, omnetpp, gcc, bzip2
mix-11	bzip2, gcc, mcf, hmmer
mix-12	bzip2, leslie3d, gcc, sjeng
mix-13	gcc, hmmer, GemsFDTD, bzip2
mix-14	namd, bzip2, h264ref, gcc

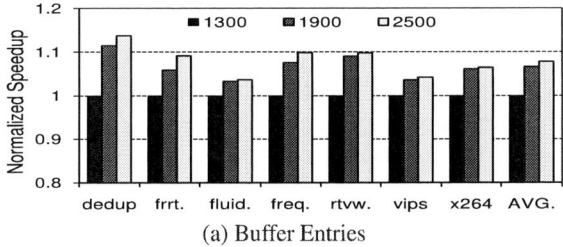
(a) Buffer Entries

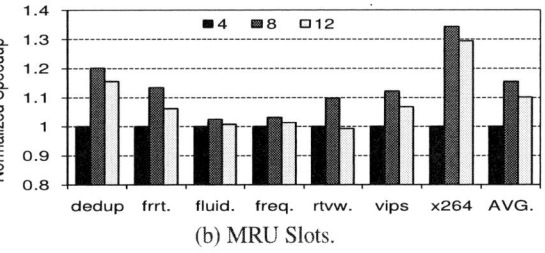
(b) MRU Slots.

Figure 11: Speedup as a function of number of buffer entries and MRU Slots

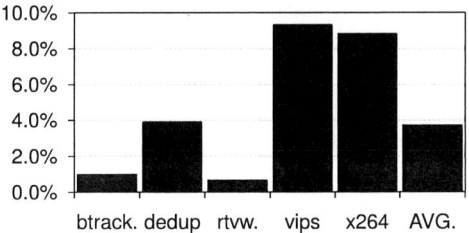

Figure 12: Energy impacts of Revive refresh scheme as a percentage of DRAM-style refresh

for the buffer size.

We find that after 8 MRU slots, the number of diminishing blocks saturates, which suggests that the optimal number of MRU slots is 8. Figure 11(b) shows speedup of a subset of applications along with the average across all mulithreaded applications, with varying number of MRU slots. Buffer size is kept constant at 1900 per bank. We see degradation in performance when we decrease the number of slots from 8 to 4, since buffering 8 MRU slots would have covered more frequently used blocks and hence, reducing write backs of useful blocks. We also see degradation in performance by increasing the number of slots from 8 to 12. This is because, with a constant buffer size, 12 MRU slots increase the probability of buffer overflows, which increases the write backs leading to performance degradation.

S.9   Comparison to DRAM-style refresh

In this section, we compare our scheme, Revive, with the DRAM style in-place refresh scheme proposed in [15]. DRAM style in-place refresh unnecessarily refreshes every block in the LLC after every 10ms, leading to high refresh overheads. In our selective buffering scheme we try to minimize the number of refreshes required. Figure 12 shows the energy comparisons of our revive refresh scheme as a percentage of DRAM-style refresh scheme proposed in [15]. Since, our scheme only refreshes a small number of required blocks as opposed to all the LLC blocks, on average,

it consumes only 4% of energy compared to the DRAM-style refresh. For the same reason, our scheme results in less than 90% of performance penalty taking into our observation that about 10% of the total diminishing blocks need to be written back to the main memory. Although our scheme has 4% buffer area overhead, the significant refreshing benefits obtained justifies our design. Moreover, refreshing benefits will diminish radically if it is done at μs level [15], and hence our ms proposal helps in sustaining the energy benefits.

S.10   Boosting Performance by saving number of write backs

In this section, we describe how reducing the number of write backs can boost the performance. Since our scheme proposes to retain the useful blocks (not writing back) in the cache after their expiration time by copying them to a temporary buffer and copying it back, it is important to see how many write backs our scheme is able to save. Figure 13 shows the number of write backs of all the designs normalized to M-4MB. We observe that the 4MB, $10ms$ retention time STT-RAM design, on average, has 21x more write backs than the traditional STT-RAM design. This leads to significant performance degradation across most applications when compared to simply using a $10years$ retention time STT-RAM cache (8% performance degradation on average). For instance, in *vips*, there is about 20% speedup degradation over M-4MB. It is interesting to see the case of *swaptions*, where there is a slight improvement in speedup over M-4MB, although there is an increase in the number of write backs. The reason for this improvement is due to the fact that the majority of blocks that are not refreshed within $10ms$ interval, are not accessed in future as well leading to a low number of read misses. This helps in reaping benefits from the reduced write latency.

S.11   Additional Prior Work

Sun et al. [16] showed that write buffers can be helpful in hiding the long write latencies of STT-RAM banks. Our analysis shows that, if an application is bursty, write-buffers fail to hide this la-

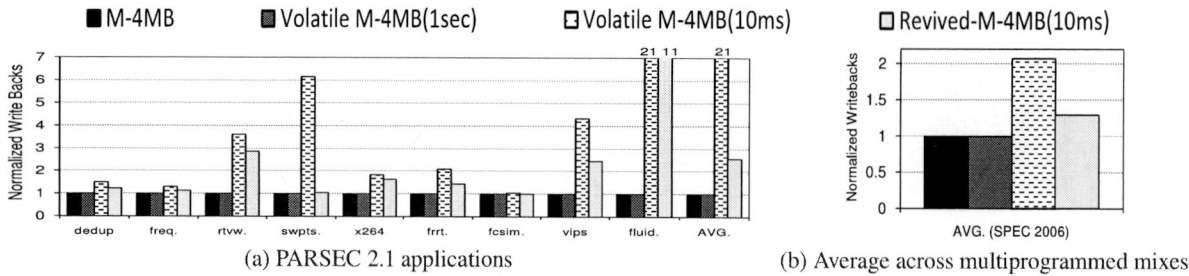

(a) PARSEC 2.1 applications   (b) Average across multiprogrammed mixes

Figure 13: Number of write backs normalized to M-4MB

tency and are rendered in-effective. Out of 25 applications, we found 12 applications to be write intensive and bursty and hence, write-buffering is ineffective for these applications. Moreover, all our results are conservative since we have already assumed a 10-entry (as used in [16]) write-buffer at every STT-RAM bank and our results would be significantly better without the presence of write-buffers.

In a recent work [12], the authors have proposed a network level solution to hide the write latency of STT-RAM banks. This solution requires complex busy/idle bank detection followed by prioritization mechanisms in the network. On a qualitative basis, the network level solution to hide write latency in [12] was shown as the most promising technique compared to any other techniques. The application level performance improvement with this scheme was about 2-4% higher compared to the write buffering technique of Sun et al. [16]. Contrasting this to our work, our scheme provides about 15%/4%(PARSEC IPC/SPEC weighted-speedup) improvement over $10years$ traditional STT-RAM, on top of the write buffering scheme, thereby making it more attractive compared to [12]. Overall, we believe that no prior work makes a case for tuning the retention time of STT-RAM banks based on profiling retention duration of LLC blocks of applications, which our proposal does.

The 3T1D designs proposed in [R1] has typical worst-case retention time in μs region, which makes it incompatible for our LLC design that needs ms retention time. Increasing the retention time of 3T1D cell to ms region will enlarge the size of the gated-diode or will increase the threshold voltage of the access transistor. These choices will incur significant area overhead or performance degradation for our cache design, respectively.

In regard to eDRAM, it also has similar density advantages but has at least 2x higher leakage than STT-RAM. Moreover, eDRAM is considered to have scalability challenges in sub-45nm process nodes due to the difficulty of precise charge placement and data sensing. STT-RAM is believed to at least scale beyond 10nm technology [7]. The retention time of eDRAM is in microseconds making it unsuitable for our design. (We claim the retention time to be in milliseconds). Also, the refresh energy overheads of eDRAM is much higher.

A few other prior works have also proposed architectural and circuit level solutions to handle this long write latency problem in STT-RAMs. Architectural techniques such as early write termination [19], hybrid SRAM/STT-RAM architecture [16] and read-preemptive write-buffer designs have been shown to mitigate write latency/energy. The circuit level techniques such as eliminating redundant bit-writes [19] and data inverting technique [16] have also been shown to be effective in hiding the long write latency. In contrast to all these prior works that attempt to *hide* the write latency, our scheme investigates techniques to *actually* reduce the write latency of STT-RAM banks and make their write latency comparable to SRAM banks. When compared to Zhou et al.'s work [19] that require additional gates for detection and termination of writes inside

each STT-RAM sub-bank, our techniques are simpler to implement since our proposal works at a much coarser granularity.

Unlike recent STT-RAM papers that assume that the non-active STT-RAM banks are power gated, we conservatively assume that these banks leak. Consequently, our absolute leakage numbers are larger.

S.12 Concluding Remarks

Spin-Transfer Torque RAM (STT-RAM) is an emerging non-volatile memory (NVM) technology that has the potential to replace the conventional on-chip SRAM caches for designing a more efficient memory hierarchy for multi-core architectures. Although the high density, low leakage and high endurance are attractive features of STT-RAM, the latency and energy overhead associated with write operations are major obstacles for being competitive with the SRAM. Our study showed that the non-volatility feature with years of data-retention time for STT-RAM technology is not necessary for its usage in on-chip cache, since the refresh times of cache data are usually in μs (for L1 cache) or ms (for L2 cache) range. Thus, we proposed to trade-off the non-volatility (data-retention time) of STT-RAM for better write performance and energy for designing STT-RAM-based L2 cache. The paper addressed several critical design issues such as how we decide on a suitable retention time for last level cache, what the relationship between retention time and write latency is, and how we architect the cache hierarchy with volatile STT-RAM. We studied two data-retention time relaxation cases, one with data-retention time of $1sec$, which satisfies the refresh time requirement of typical cache blocks; and the other one with data-retention time of $10ms$, which is a more aggressive design for better performance and energy gains, but required a data refreshing mechanism. For the aggressive $10ms$ retention time design, we proposed a selective block refreshing scheme for the cache blocks that have a higher refresh time than the STT-RAM retention time to avoid any data loss. Our experiments with a four-core architecture with an SRAM-based L1 cache and volatile STT-RAM-based L2 cache indicated that not only we can eliminate the long write latency overhead of the NVM STT-RAM, but also can provide on an average 18% improvement in performance compared to the traditional SRAM-based design, while reducing the energy consumption by 60%.

S. REFERENCES

[R1] L. Xiaoyao, C. Ramon, W. Gu-Yeon, and B. David. Replacing 6T SRAMs with 3T1D DRAMs in the L1 Data Cache to Combat Process Variability. *IEEE Micro 2008*.

[R2] W. Xu, H. Sun, X. Wang, Y. Chen, and T. Zhang. Design of last-level on-chip cache using spin-torque transfer RAM (STT RAM). *IEEE TVLSI*, 2011.

[R3] W. Zhao and Y. Cao. New generation of predictive technology model for sub-45 nm early design exploration. *IEEE TED*, 2006.

Point and Discard: A Hard-Error-Tolerant Architecture for Non-Volatile Last Level Caches

Jue Wang, Xiangyu Dong, Yuan Xie
Department of Computer Science and Engineering
Pennsylvania State University
{jzw175,xydong,yuanxie}@cse.psu.edu

ABSTRACT

Technology scaling of SRAM and embedded DRAM is increasingly constrained by limitations such as leakage power and silicon area. Emerging non-volatile memory technologies are considered as the potential SRAM/eDRAM alternatives for last-level caches in terms of energy and area savings. Unfortunately, these non-volatile memory technologies usually have limited write endurance. Even worse, process variation causes some cells to wear out much earlier than others. While state-of-the-art error-tolerant techniques such as ECC can handle transient soft errors, we need a new architecture for non-volatile last-level caches whose reliability is mainly challenged by hard errors. This paper presents *Point-and-Discard* (PAD), a hard-failure-tolerant architecture for non-volatile caches. PAD has no initial performance penalty and ensures gradual performance overhead with small storage overhead. By adopting PAD, the lifetime of non-volatile caches can be improved by 4.6X over the conventional architecture under a typical process variation condition.[1]

Categories and Subject Descriptors

B.3.4 [**Memory Structures**]: Reliability, Testing, and Fault-Tolerance

General Terms

Design, Experimentation

Keywords

Cache, non-volatile memories, error tolerance

1. INTRODUCTION

Low energy consumption and high density are two major concerns in designing on-chip last-level caches. However, it becomes more challenging to satisfy these requirements because the scaling of traditional SRAM and embedded DRAM technologies is increasingly constrained by leakage power consumption and cell density.

---

[1]This work is supported in part by SRC grant, NSF 1147388, 0903432 and by DoE under Award Number DE-SC0005026.

Recently, several new non-volatile memory technologies[2], such as *Spin-Torque Transfer RAM (STT-RAM)*, *Phase-change RAM (PCRAM)* and *Resistive RAM (ReRAM)*, are being explored as the potential alternatives of SRAM and DRAM. For example, Toshiba used STT-RAM to replace a 512KB SRAM L2 cache for low power purpose [1], Samsung announced a 58nm 1Gb PCRAM chip equipped with a DRAM-compatible LPDDR2 bus [2], and HP and Hynix plan to launch flash-replacement ReRAM products in 2013 and go into the DRAM market in 2014/2015 and later the SRAM market as well.

It is attractive to substitute XRAM macros for the conventional SRAM or embedded DRAM (eDRAM) ones in the *last level of the cache hierarchy* because of two reasons: First, XRAM has advantages in cell density and process scalability compared to SRAM or eDRAM. While XRAM has disadvantages in long write latency and high write energy, it is still beneficial to use XRAM in last-level caches where memory capacity is more important than write performance. Second, XRAM has non-volatility and zero leakage power consumption from memory cells when idle. Considering leakage energy can be dominant in some cases and previous work [3] shows the leakage energy can be as high as 80% of total energy consumption for an L2 cache in 130nm process, using XRAM last-level caches to mitigate standby leakage power is an attractive choice.

Unfortunately, there is a major obstacle before widely deploying XRAM last-level caches – the limited cache lifetime caused by hard errors. Although many wear-leveling techniques [4–6] can be applied to balance the write traffic to cache lines, their basic assumption that all the memory cells have the same quality and the same write endurance is not true. Due to process imperfection, a more severe problem to XRAM cache lifetime is that there might be a portion of cells with worse quality and much shorter write endurance. This means the actual lifetime might be much shorter than the expected one even with perfect wear leveling if there is no other error-tolerant technique to be combined and the cell with the worst quality might determine the entire cache lifetime.

In addition, state-of-the-art error-tolerant techniques such as Error Correcting Codes (ECC) are designed to handle transient soft errors instead of permanent hard errors. If we use ECC to tolerate permanent wear-out errors, we have to use stronger ECC protection by paying larger hardware overhead. The overhead of *DEC-TED (double-error correction and triple-error detection)* for 64-bit data could be as high as 23%. Moreover, if the bit error count exceeds the ECC protection, we have to discard the affected cache line and waste the healthy bits within it. By doing this, the number of available cache ways in a set might decrease rapidly followed by a quick cache capacity reduction and a significant performance degrada-

---

[2]We use *XRAM* to refer to these emerging non-volatile memory technologies in this paper.

tion. Therefore, ECC cannot effectively tolerate hard errors from limited write endurance and process imperfection, and new cache architectures for hard error tolerance must be explored.

Previous work proposed several hard-error-tolerate architecture for XRAM main memory [7–10] based on error correction, but the problem is different here since main memory and caches are in different levels of hierarchy. In this paper, we propose *Point-and-Discard* (PAD), a hard-error-tolerant architecture for XRAM caches. The principle of PAD is to discard hard errors instead of repairing them. Moreover, it only discards affected bytes instead of the entire cache line when a hard error happens and uses unaffected bytes in the affected cache line to store the locations of affected bytes. The advantages of PAD are:

- *Small storage overhead*: PAD incurs small storage overhead (around 1.4%) to the conventional cache design. This is because it leverages the cache associativity redundancy and uses healthy bytes to record the hard error locations.

- *Gradual performance overhead*: PAD dynamically reduces the number of workable ways in one set along with the failure bit increase in that set. Thus, there is no performance penalty in the system's early life, and it ensures a gradual performance degradation as the memory cells continuously wear out even under a very severe process variation.

Our experimental results show that PAD can improve the lifetime of non-volatile caches by 1.6X-440X compared to the conventional cache architecture under different process variations.

2. ERROR DETECTION AND DISTRIBUTION MODEL

Write endurance is defined as the number of times that a memory cell can be overwritten. Limited write endurance is a common problem for XRAM. For example, PCRAM cells are only expected to sustain 10^8 writes before experiencing permanent errors [11]. The write endurance of ReRAM is recently improved but is still at the level of 10^{11} [12]. While STTRAM is usually predicted to have the write endurance of 10^{15}, the current best test result for STTRAM devices is still less than 4×10^{12} cycles [13].

For last-level caches, some of the XRAM write endurance values (e.g. 10^{11} for ReRAM and 10^{12} for STTRAM) seem to be sufficiently high. However, the real problem is that each cell's lifetime is different due to process imperfection and some weaker cells might wear out much earlier than expected. This problem cannot be solved by wear-leveling technologies [4–6] and has to rely on new cache architecture that can tolerate hard errors.

When a hard error occurs during the runtime of XRAM caches, a cell would stuck at a value and never change after that. In XRAM caches, hard erros can be detected by a "read-write-read" pattern for write operations. This mechanism is common in all the XRAM memory systems to reduce write energy [8]. When there is a write hit, the old data in the cache line is read out at first. Then, only the changed bits are written to the cache line. Finally, the newly-stored data is read out again. If the data is different from the writing data, a hard error is then detected.

In this work, we assume the cell write endurance is under a normal distribution like previous work does [7,8,10] and wear-leveling techniques are adopted into the system to get a uniform write distribution across all cache blocks. In the rest part of this paper, we study the XRAM cache lifetime under a process variation[3] with

---

[3]CoV is coefficient of variation. A sensitivity analysis of CoV from 0.2 to 0.4 is in Section 5.

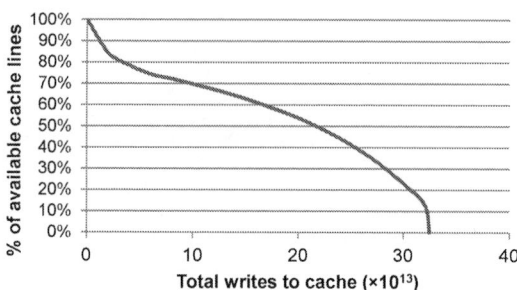

Figure 1: The percentage of available cache lines for a 1MB cache with CoV=0.3 under naive mechanism during runtime.

CoV=0.3. In addition, we assume hard errors are independent and identically distributed.

3. MOTIVATION OF PAD

The simplest method to handle hard erros is to force the affected cache lines tagged as *INVALID*, so that no further data would be written to the cache lines containing hard errors. But, the problem of this mechanism is that even one wear-out cell can cause the waste of the entire cache line, which usually contains 512 bits. In addition, since hard errors are distributed randomly, it is highly possible that almost every cache line has some weak cells. Therefore, the number of unaffected cache lines would rapidly decrease causing significant performance degradation.

We simulate a 1MB ReRAM cache with 8-way associativity and 64-byte cache lines as an example, in which the mean write endurance is 10^{11} and the CoV is 0.3. Figure 1 shows the percentage of unaffected cache lines as the write count increases. The result shows the percentage of unaffected cache lines is quickly reduced to 10% only after 3.25×10^{14} writes to the cache. Assuming there are 400 times/second writes to the cache on average, the lifetime is about 1.5 years, which cannot satisfy the requirement of using ReRAM in last-level caches.

This naive mechanism does not work because it over-discards the entire cache line for every single wear-out bit. In order to solve this problem, the basic idea of PAD is to reduce the discard granularity from cache lines to bytes. In PAD, when there is a hard error, only the affected byte is discarded instead of the entire cache line. The remaining healthy bytes from different cache lines are then reorganized to compose new complete cache lines.

Figure 2 shows a brief example of the PAD architecture, in which we use a simplified cache structure with 4-way associativity and 8-byte cache lines. The shadow bytes are the ones affected by hard errors. The last few healthy bytes in each way are used to store the positions of the hard-error-affected bytes. By knowing which bytes have hard error, we can use the remaining bytes to compose three complete healthy ways.

4. IMPLEMENTATION OF PAD

In this section, we describe the architecture and the operation details of a PAD-structured cache.

4.1 Architecture of PAD

In a cache with 64-byte cache lines (Byte[0] to Byte[63]), we allow PAD to discard up to 32 bytes in one cache line. Thus, 5 bits are added to store the number of pointers for each cache line, and their initial values are zeros. Since these cells are only need to be written 31 times at most, it is not possible for them to wear out.

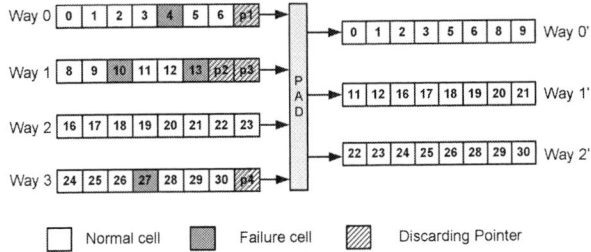

Figure 2: A brief example of the PAD architecture.

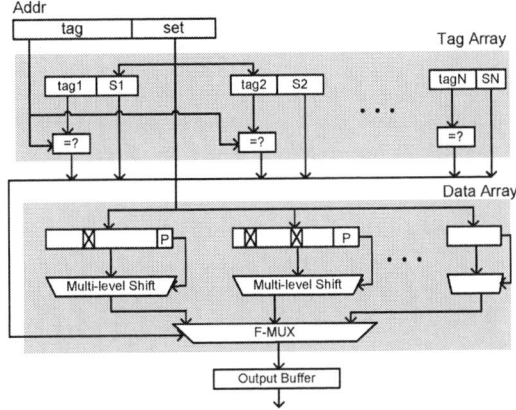

Figure 3: The architecture of PAD mechanism.

Figure 3 shows the diagram of the PAD architecture. In every cache line, if one cell wears out, the address of the affected byte is stored in a pointer and the number of pointers is incremented by 1. Assuming the current number of pointers in a cache line is N, if a new pointer needs to be added, the position of this new pointer is set as Byte[$63 - N$]. In PAD, we use one byte to store one pointer entry although only 6 bits are needed. The remaining 2 bits are used as $VALID\_INFO$. It is set to "00" if this pointer works normally, otherwise it means some bits in this pointer have worn out and this pointer should be ignored. The $VALID\_INFO$ bits themselves are protected by the 2-bit redundancy.

After a byte becomes a pointer, it is impossible for this pointer to wear out because the pointer is read-only once its value is assigned. However, it is possible that the byte to be discarded happens to be the pointer to be designated (i.e. Byte[$63 - N$]), which means the new pointer position is just the same as the affected byte. PAD can also tolerate this type of errors: First, PAD uses $VALID\_INFO$ to indicate this pointer is invalid and should be ignored; Second, if $VALID\_INFO$ itself is stuck at "00", another pointer can be used to discard this pointer. In the worst case, if the designated pointer is healthy at the first place but fails at the same time as the one it points to, PAD can invalid it by setting the $VALID\_INFO$ bits and use another pointer. However, it should be noticed that the possibility of such worst-case scenario is extremely low since it is rare for multiple bits to wear out at the same time.

Unlike conventional cache architectures, it is required for PAD to reconstruct healthy bytes into new cache lines when hard errors occur in a cache set. A multi-level shift component is added for this purpose. To access a cache line, data bytes are first loaded into the shift component, then the information stored in the pointers is used to shift out (discard) the hard-error-affected bytes.

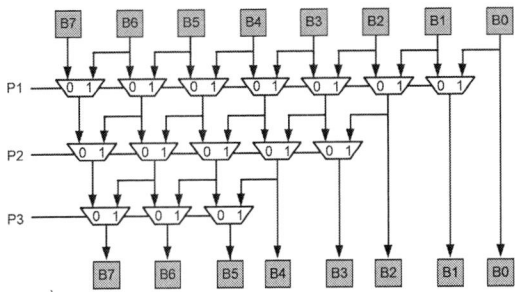

Figure 4: Architecture of the single cycle multi-level shifter using 8-byte cache line as an example.

For demonstration purpose, the structure of the multi-level shift component for 8-byte cache lines is shown in Figure 4. Every pointer implies to discard one hard-error-affected byte and to shift that byte out. Assuming the hard-error-affected bytes in this example are B6 and B3, then B1 and B0 are used to store the pointers for discarding B6 and B3, respectively. When this cache line is accessed, the pointers stored in B1 and B0 are decoded to control the multiplexer signals on each level. In this example: the control bits on Level 1 are "0111111", and the output data is B7 B5 B4 B3 B2 B1 B0 B0; the control bits on Level 2 are "00011", and the data becomes B7 B5 B4 B2 B1 B1 B0 B0; the control bits on Level 3 are "000", and the final output data is B7 B5 B4 B2. Thus, we remove two hard-error-affected bytes (i.e. B6 and B3) by using two pointer (i.e. B1 and B0) and get the remaining healthy bytes (i.e. B7, B5, B4, and B2).

We call this architecture of multi-level shifter as Single-Cycle-MS since all levels of shifters are connected together and the result can be output in one cycle. However, for conventional 64-byte cache lines, using Single-Cycle-MS architecture brings large area and latency overhead. Thus, we propose an improved architecture called as Multi-Cycle-MS which is shown in Figure 5. In each cycle, the data is processed by a n-stage Single-Cycle-MS, in which n is a small number and can be changed by designers according to different target frequencies. In this work, we choose n to be 3. Initially when there is no hard-error-affected bytes, cache lines can be accessed directly and there is no latency overhead. During runtime, when the number of hard-error-affected bytes in one cache line increases, the Multi-Cycle-MS circuit is enabled to reconstruct the correct data by paying latency overhead in an incremental way. Multi-Cycle-MS has two advantages over Single-Cycle-MS:

- *Smaller area*: For Single-Cycle-MS, it needs 1023 2-to-1 multiplexers in every multi-level shift. For Multi-Cycle-MS, if we choose n as 3, it only needs 183 2-to-1 multiplexers. It saves about 82.1% areas.

- *Faster data reconstruction speed*: The latency overhead of Single-Cycle-MS is fixed for every cache access and equals to the sum of 31 2-to-1 multiplexers (more than 10 cycles). On the other hand, the latency overhead of Multi-Cycle-MS only increases gradually as the number of hard-error-affected bytes increases. Although the worst-case latency of Multi-Cycle-MS is the same as that of Single-Cycle-MS, the average latency of Multi-Cycle-MS is much shorter. The analysis of the Multi-Cycle-MS latency overhead is in Section 5.4.

4.2 Operation of PAD

In PAD, two status bits are added to every cache line to represent 1 out of 4 possible states of the corresponding cache line:

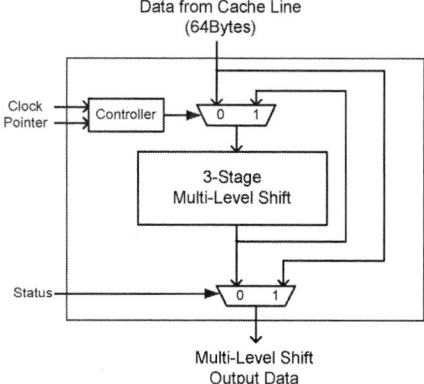

Figure 5: Architecture of the multi-cycle multi-level shifter for 64-byte cache lines.

- *Normal*: This cache line does not have any hard-error-affected bytes and can be treated as a normal line, and shift operation is not needed for accessing this cache line.

- *Fixable*: This cache line has some hard-error-affected bytes but can be fixed by using the valid data bytes in the next cache line of the same set.

- *Patch*: All the healthy bytes in this cache line are used to fix the previous *Fixable* line of the same set. *Patch* cache lines never get a hit because they do not store actual data anymore.

- *Discarded*: After the number of the hard-error-affected bytes in a cache line is more than 31, the cache line is set as *Discarded* and is not used anymore.

In every cache access, the result of comparators in the tag array shows which cache line is hit in the set[4]. Besides the tag, PAD stores a $Start$ information in the tag. For *Normal* ways, $Start$ always equals to 0. For *Fixable* ways, $Start$ denotes which byte is the actual first byte in this cache line, which means that the healthy bytes in Byte[0]–Byte[$Start - 1$] are used to shift into previous ways. When the first hard error occurs in a 8-way set, the last cache line in this set is tagged as *Patch* because this set can only store 7 cache lines instead of 8 lines at most. The affected cache line and its following ones are tagged as *Fixable*. When a new hard-error-affected byte occurs in a *Fixable* line, the number of pointers (N) in this cache line is added by 1. If $Start$ equals to $64 - N$, which means all healthy bytes in this cache line are used to fix the previous line, then the status of this cache line is changed from *Fixable* to *Patch*. During runtime, more ways are tagged to be *Patch* as more weak cells wear out, and the percentage of *Normal* and *Fixable* ways in the set is reduced.

The hit and $Start$ information is loaded into a final MUX (F-MUX) to get the output data. F-MUX outputs the data based on the status of the hit cache line, which must be *Normal* or *Fixable* since hits never occur to *Patch* and *Discarded* lines. If the status of the hit cache line is *Normal*, F-MUX chooses this line and outputs the complete cache line. If its status is *Fixable*, F-MUX chooses this *Fixable* line and all the following lines until the next *Fixable* or *Normal* line or the last line of this set. Assuming $Start$ of a hit *fixable* line is S_n, then F-MUX would choose all the bytes from

Byte[S_n] to Byte[63] in this line, and the remaining bytes from the followed *Patched* or *Fixable* ways. This part is easy to implement since the Multi-Cycle-MS has already discarded the hard-error-affected bytes and F-MUX only needs to assemble them together[5].

5. EXPERIMENTAL RESULTS

In this section, we describe our experiment methodology and evaluate the lifetime improvement and the overhead after applying PAD.

5.1 Experiment Methodology

While it is intractable to simulate the cache behavior over its entire lifetime, we follow the same assumptions used in previous work [7, 8], in which we assume that existing wear-leveling techniques already spread writes evenly over cache lines and every write randomly hits one cache line. In addition, the bit flipping probability is assumed to be 0.5. In this work, we use ReRAM caches as the analysis target. For each simulation run, the simulator assigns a random lifetime to each cell using a normal distribution with a mean of 10^{11} writes [12]. Our experiment tracks the number of available cache lines after each write operation. If a new hard error occurs, the available cache line count is recalculated.

We use the architectural parameters shown in Table 1. The baseline architecture is set to a naive cache architecture in which an entire hard-error-affected cache line (i.e. 64 bytes) has to be disabled for a single hard error. To evaluate the effectiveness of PAD, we also simulate the cache architecture adopted 64-bit $Single\ Error\ Correction$ (SEC_{64}) which is a common type of ECC in today's memory. SEC_{64} corrects up to one error bit for each 64-bit data blocks by adding 7 check bits. The storage overhead of SEC_{64} is 7 bits per 64-bit data (11%).

5.2 Lifetime Improvement

Figure 6 shows the percentage of available cache lines (i.e. *Normal* and *Fixable* cache lines) in PAD compared to baseline and SEC_{64}. Under the assumed write endurance CoV of 0.3, the percentage of available cache lines in baseline quickly decreases to 10% only after 3.25×10^{14} writes to the cache. On the other hand, after adopting PAD architecture, the percentage of available cache lines is reduced to 10% after 1.41×10^{15} writes. Assuming there are 400 times/second writes to the cache on average, the lifetime is about 1.5 years for the baseline system and about 6.8 years after adopting PAD. Thus, PAD can improve the simulated ReRAM cache lifetime by 4.6X.

For SEC_{64}, the percentage of available cache lines is higher than PAD in the system's early life. It is because SEC_{64} pays much larger storage overhead (11% compared to 1.4%) to correct the wear-out bits while PAD uses unaffected bytes to fix the affected lines. However, PAD has larger error-tolerant capability than

Table 1: Experiment settings

Set Number	2048
Way Number	8
Cache Line Size	64 Byte
Cache Size	1MB
Mean Cell Lifetime	10^{11}
Lifetime Variation	0.3

---

[4]For caches, the size of tag array is very small compared to the data array. Thus we can use simple redundant encoding to tolerate the hard errors in tag cells.

[5]Although we focus on reads in the discussion, reads and writes are symmetric. Also, Data-Comparison Write [14] is adopted to avoid the unnecessary writes.

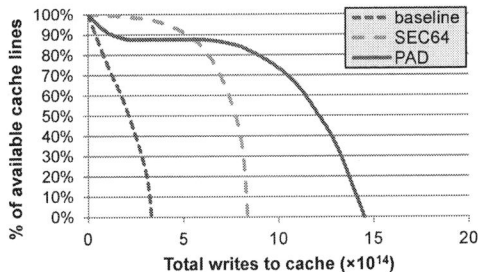

Figure 6: The percentage of available lines of a 1MB cache with CoV=0.3 in PAD compared to baseline and ECC (SEC64).

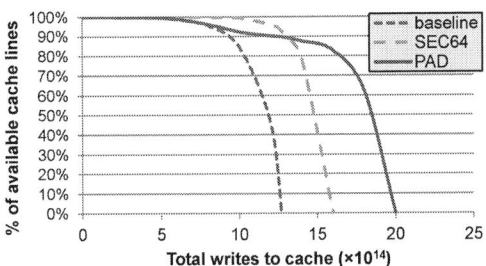

Figure 7: The percentage of available lines of a 1MB cache with CoV=0.2 in PAD compared to baseline and ECC (SEC64).

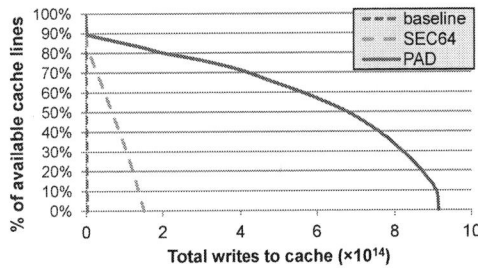

Figure 8: The percentage of available lines of a 1MB cache with CoV=0.4 in PAD compared to baseline and ECC (SEC64).

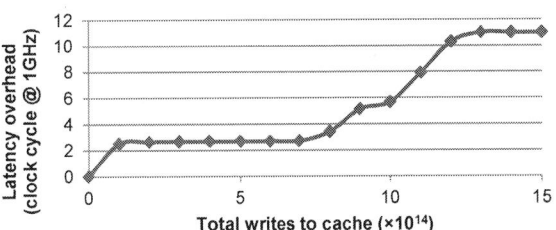

Figure 9: The read latency of PAD for a 1MB ReRAM cache. It is small in the early life and increases during the running time.

SEC_{64}. The result shows that PAD improves the cache lifetime by 1.7X compared to SEC_{64}.

5.3 Sensitivity Analysis of CoV

It is expected that the lifetime of cache architecture is very sensitive to the write endurance distribution. Therefore, we use the same model to simulate a 1MB ReRAM cache with different CoV (higher CoV means more severe process variation, and lower means less). Figure 7 and Figure 8 show the percentage of available cache lines in PAD compared to baseline and SEC_{64} systems when CoV is 0.2 and 0.4, respectively. The results show the 1MB ReRAM cache lifetime is increased by 1.6X and 440X compared to baseline, respectively. Compared to SEC_{64}, PAD can also get 1.3X and 6.1X lifetime improvment. It implies that PAD is more effective under more severe process variation, which means it would be more useful as the process node advances.

5.4 Overhead Analysis

PAD has the capability of tolerating hard errors. However, it is also important to evaluate the area and performance overhead of PAD when we deploy it in low-power high-density XRAM caches.

For each cache line, PAD only needs to add a 5-bit counter (for pointers) and a 2-bit status label. Thus, the storage overhead is about 1.4% for a 64-byte cache line. The major area overhead comes from the multi-level shifters. For a 8-way cache, it needs 8 multi-level shifters, and there are 183 2-to-1 multiplexers in every Multi-Cycle-MS. Thus, the overall area overhead is 1,464 2-to-1 multiplexers. Compared to common error-tolerant schemes, such as SEC_{64}, which has 11% storage overhead and additional encoder/decoder, PAD has much smaller overhead.

The latency overhead of PAD is zero in the system's early life when there is no hard error in the cache. But, the latency increases over time when more cells wear out because Multi-Cycle-MS needs more cycles to reconstruct a valid data. Figure 9 shows the simula-tion result of the latency overhead. It shows that the access latency overhead is smaller than 3 cycles during half of the cache's lifespan, and the maximum latency is 11 cycles which only occurs near the end of the lifetime.

5.5 Performance Analysis

We use Instructions-Per-Cycle (IPC) to compare the performance degradation between PAD and the other cache architectures. In baseline and SEC_{64}, there is no latency overhead[6] but the percentage of available cache lines is reduced rapidly. On the other hand, the latency overhead is increased gradually in PAD, but the percentage of available lines reduces much more slowly as shown in Figure 6.

We used the gem5 simulator for our performance evaluations [15]. Figure 10 shows the normalized IPC of PAD compared to baseline and SEC_{64} of simulating EEMBC 2.0 benchmark [16]. As more write operation conducted to caches, Figure 10 shows the baseline performance is degraded quickly from 100% to 50% after only 3×10^{14} writes. For PAD, the write count is about 1.4×10^{15} when IPC degrades 50%, which is increased by 4.7X. Compared to SEC_{64}, the IPC of PAD is reduced slightly (smaller than 2%) in the cache's early life. However, due to the higher error-toleration, PAD achieves better performance than SEC_{64} with the number of writes increasing. Therefore, compared to baseline and SEC_{64}, PAD architecture can improve the performance of XRAM caches with longer lifetime.

6. RELATED WORK

We compare PAD to some related work in this section. First, there have been many studies on SRAM cache architectures for low Vccmin [17–22]. In low-Vcc SRAMs, process variation induces bit errors increase. This problem is similar with the wear-out hard

---

[6]We use zero latency overhead for SEC_{64} in the simulation. However, depending on the implementation, the encoder and decoder could bring some latency overhead.

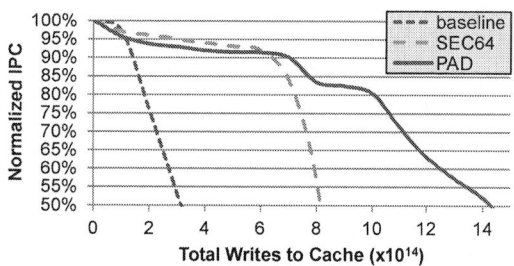

Figure 10: The IPC degradation of PAD architecture compared to baseline and ECC (SEC64).

errors in XRAM caches, but the difference is that the positions of hard errors in SRAM cache are fixed under a specific voltage value. Thus, a defect map for the error correction can be generated at boot time using the memory built-in self-test (BIST) unit. It makes these techniques have significant and fixed performance degradation with large storage overhead. However, for XRAM caches, the number and the position of hard errors are accumulated during runtime. Thus, PAD has no initial performance penalty and ensures gradual performance overhead with small storage overhead.

Second, some previous work focused on tolerating the lifetime variation in the XRAM main memory, such as dynamically replicated memory [7], error correction pointer (ECP) [8], SAFER [9], and FREE-p [10]. However, the major difference when considering the lifetime variation in XRAM caches is that caches have more error-tolerant capacity than main memory since the set-associate cache already has its redundancy. It is easier to sacrifice some blocks to tolerate the hard errors without adding any initial area overhead. But compared to main memory, it cares more about the performance in PAD since it is at the inner level of the memory hierarchy.

Besides, there is another type of techniques to extend the lifetime of XRAM memory which is focused on evenly distributing unbalanced write traffic to every memory blocks [4–6, 23, 24]. These wear-leveling techniques are compatible with our work to enhance the lifetime of XRAM caches from different aspects.

7. CONCLUSION

Adopting non-volatile memory technologies in last-level caches is attractive because they has lower energy consumption and higher density compared to the traditional SRAM and embedded DRAM technologies. However, limited write endurance is one obstacle before adopting non-volatile caches widely. Although the write endurance values of some non-volatile memories seem sufficient for last-level caches, a more severe problem is lifetime variations due to process imperfection. It causes the actual cache lifetime much shorter than expectation because the weakest cell might determine the overall lifetime. However, the current error-tolerant techniques for caches are designed to handle transient faults and cannot effectively tolerate hard errors in non-volatile caches.

This paper presents *Point-and-Discard* (PAD), a hard-error-tolerant architecture for non-volatile caches. PAD has better error-tolerant capacity than some common schemes (such as SEC_{64}) with smaller storage overhead. The experimental results show that PAD improves the lifetime of non-volatile caches by 1.6X-440X under different process variations. Moreover, PAD incurs no performance overhead in the system's early life and ensures a gradual performance degradation as the cells continuously wear out.

8. REFERENCES

[1] K. Nomura *et al.*, "Ultra low power processor using perpendicular-STTMRAM/SRAM based hybrid cache toward next generation normally-off computers," in *MMM*, 2011.

[2] H. Chung *et al.*, "A 58nm 1.8V 1Gb PRAM with 6.4MB/s program BW," in *ISSCC*, 2011.

[3] C. H. Kim *et al.*, "A forward body-biased low-leakage SRAM cache: device and architecture considerations," in *ISLPED*, 2003.

[4] P. Zhou *et al.*, "A durable and energy efficient main memory using phase change memory technology," in *ISCA*, 2009.

[5] M. K. Qureshi *et al.*, "Enhancing lifetime and security of PCM-based main memory with start-gap wear leveling," in *MICRO*, 2009.

[6] N. H. Seong *et al.*, "Security refresh: prevent malicious wear-out and increase durability for phase-change memory with dynamically randomized address mapping," in *ISCA*, 2010.

[7] E. Ipek *et al.*, "Dynamically replicated memory: Building reliable systems from nanoscale resistive memories," in *ASPLOS*, 2010.

[8] S. Schechter *et al.*, "Use ECP, not ECC, for hard failures in resistive memories," in *ISCA*, 2010.

[9] N. H. Seong *et al.*, "SAFER: Stuck-at-fault error recovery for memories," in *MICRO*, 2010.

[10] D. H. Yoon *et al.*, "FREE-p: Protecting non-volatile memory against both hard and soft errors," in *HPCA*, 2011.

[11] S. J. Ahn *et al.*, "Highly manufacturable high density phase change memory of 64Mb and beyond," in *IEDM*, 2004.

[12] Y.-B. Kim *et al.*, "Bi-layered rram with unlimited endurance and extremely uniform switching," in *VLSI*, 2011.

[13] Y. Huai, "Spin-transfer torque MRAM (STT-MRAM): Challenges and prospects," *AAPPS Bulletin*, vol. 18, no. 6, 2008.

[14] W. Zhang *et al.*, "Characterizing and mitigating the impact of process variations on phase change based memory systems," in *MICRO*, 2009.

[15] N. Binkert *et al.*, "gem5:A Multiple-ISA Full System Simulator with Detailed Memory Model," in *Computer Architecture News*, 2011.

[16] The Embedded Microprocessor Benchmark Consortium, "EEMBC 2.0," http://www.eembc.org/.

[17] C. Wilkerson *et al.*, "Trading off cache capacity for reliability to enable low voltage operation," in *ISCA*, 2008.

[18] S. Sasan *et al.*, "A fault tolerant cache architecture for sub-500mV operation: Resizable Data Composer Cache (RDC-Cache)," in *CASES*, 2009.

[19] A. Ansari *et al.*, "ZerehCache: Armoring cache architectures in high defect density technologies," in *MICRO*, 2009.

[20] J. Abella *et al.*, "Low Vccmin fault-tolerant cache with highly predictable performance," in *MICRO*, 2009.

[21] Z. Chishti *et al.*, "Improving cache lifetime reliability at ultra-low voltages," in *MICRO*, 2009.

[22] Y. Choi *et al.*, "Matching cache access behavior and bit error pattern for high performance low Vcc L1 cache," in *DAC*, 2011.

[23] M. K. Qureshi *et al.*, "Scalable high performance main memory system using phase-change memory technology," in *ISCA*, 2009.

[24] Y. Joo *et al.*, "Energy- and endurance-aware design of phase change memory caches," in *DATE*, 2010.

Self-aware Computing in the Angstrom Processor

Henry Hoffmann[1]   Jim Holt[1,2]   George Kurian[1]   Eric Lau[1]   Martina Maggio[3]

Jason E. Miller[1]   Sabrina M. Neuman[1]   Mahmut Sinangil[4]   Yildiz Sinangil[4]

Anant Agarwal[1]   Anantha P. Chandrakasan[4]   Srinivas Devadas[1]

[1]MIT Computer Science and Artificial Intelligence Laboratory [2]Freescale Semiconductor
[3]Lund University [4]MIT Microsystems Technology Laboratories

[1] {hank,jholt,gkurian,elau,jasonm,sneuman,agarwal,devadas}@csail.mit.edu, [2]jim.holt@freescale.com
[3]maggio.martina@gmail.com, [4]{sinangil,koken,anantha}@mit.edu

ABSTRACT

Addressing the challenges of extreme scale computing requires holistic design of new programming models and systems that support those models. This paper discusses the Angstrom processor, which is designed to support a new Self-aware Computing (SEEC) model. In SEEC, applications explicitly state goals, while other systems components provide actions that the SEEC runtime system can use to meet those goals. Angstrom supports this model by exposing sensors and adaptations that traditionally would be managed independently by hardware. This exposure allows SEEC to coordinate hardware actions with actions specified by other parts of the system, and allows the SEEC runtime system to meet application goals while reducing costs (*e.g.,* power consumption).

Categories and Subject Descriptors

C.1.3 [**Other Architectural Styles**]: Adaptable architectures

General Terms

Performance, Design, Experimentation

Keywords

Adaptive Systems, Self-aware Computing

1. INTRODUCTION

The constraints and complexity of extreme-scale computing systems make them extremely difficult to program. This difficulty arises partly from a need to meet multiple – often competing – goals, such as maximizing performance while

---

This work was funded by the U.S. Government under the DARPA UHPC program. The views and conclusions contained herein are those of the authors and should not be interpreted as representing the official policies, either expressed or implied, of the U.S. Government.

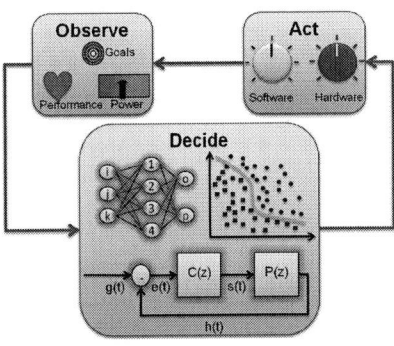

Figure 1: The Observe-Decide-Act loop in SEEC.

minimizing energy consumption. Additional difficulty stems from the fact that these systems are increasingly dynamic and must continue to function in the face of both dramatically changing application workloads and unreliable, failure-prone components.

Programming extreme scale systems to meet multiple constraints and modify behavior in the face of unforeseen events is a challenge beyond most application developers. Doing so requires expertise in the application domain, but also a deep systems knowledge to balance performance against competing goals like power efficiency. In addition, adjusting to dynamic fluctuations, such as workload variance or component failure, requires further knowledge in the design and implementation of adaptive systems.

The Angstrom project [29] addresses the challenge of programming extreme-scale systems by supporting a novel *self-aware computing* model, called SEEC [15]. SEEC makes it easy to create self-adaptive, or autonomic, computing systems which can alter their behavior to meet multiple goals and automatically adapt to environmental changes. Like all self-adaptive systems, SEEC is characterized by the presence of an *observe-decide-act*, or ODA, loop [18, 23]. The Angstrom system continuously monitors its goals (observe) and available resources (actions) using a decision engine to determine how best to use resources to meet goals (decide) given the current state of the system.

One of the unique features of Angstrom compared to prior work in adaptive systems is that every component of the system, from applications to hardware, is designed to support adaptation as a *first-class object*. In practice, this means that these components all contribute to the specification of the observe-decide-act loop as illustrated in Figure 1. In this model, applications use one interface to specify goals (*e.g.,* a video encoder should run at thirty frames per second), while system software and hardware use a separate interface to specify actions (*e.g.*, allocating processing cores, changing cache configuration). A runtime decision system then determines how to use the available actions to meet goals while reducing cost.

This paper provides an overview of the SEEC model and describes the current design of the Angstrom processor, with a focus on those features that explicitly support SEEC. Angstrom is a proposed 1000-core, massively manycore processor that supports self-aware computing by exposing a wide array of actions (in the form of different hardware configurations) and observations (including both traditional performance counters [34] and energy counters [31]) to the SEEC runtime system. Simulations of 256 core Angstrom systems show that exposing hardware adaptation to a software management system has the potential to improve performance per watt by an average of over 100% compared to a non-adaptive system.

The rest of this paper is organized as follows. Section 2 discusses related work in adaptive computing. Section 3 summarizes the SEEC model, while Section 4 describes the Angstrom processor's support for this model. Section 5 presents an evaluation of the SEEC model on both an existing system and on simulations of future Angstrom systems. The paper concludes in Section 6.

2. BACKGROUND

Self-aware, or autonomic, computing has been proposed as one method to deal with the rising complexity of computer systems [18, 23], and adaptive systems have been implemented in both hardware [2, 5, 10, 11] and software [32]. One limitation of these approaches is that they typically do not support adaptation as a first-class object. Instead, they propose *closed* adaptive systems that are not accessible by other components of the system. While this approach completely insulates application programmers from the complexity of adaptive system design, it can lead to other difficulties. Specifically, many hardware-based approaches assume a fixed set of adaptations, which exist exclusively in hardware and are unable to incorporate application specific goals. Similarly, many software-based approaches assume that the hardware is fixed and thus make no attempt to coordinate with lower-level adaptations.

As an example of a closed adaptive hardware system, consider the resource manager described by Bitirgen et al [5]. This system uses a neural network to allocate resources to a multi-programmed multicore system, with a goal of optimizing total system throughput, or total number of programs completed. This system does an excellent job of coordinating resources to meet goals, but it works with a fixed set of hardware resources and optimizes only a fixed goal of total system throughput. This system would not be able to incorporate additional adaptations specified at the software level or work with application specific goals (like achieving a desired frame rate for a video encoder).

To illustrate the problems of composing closed adaptive systems, we present the following experiment. Using the Graphite simulator [28], we run the barnes application from the SPLASH2 benchmark suite on a multicore system with two possible adaptations: the total number of cores assigned to it (from 1-64, by powers of 2), and the size of the L2-cache on each core (from 16-256 KB, by powers of 2). For each combination of core allocation and cache size, we measure

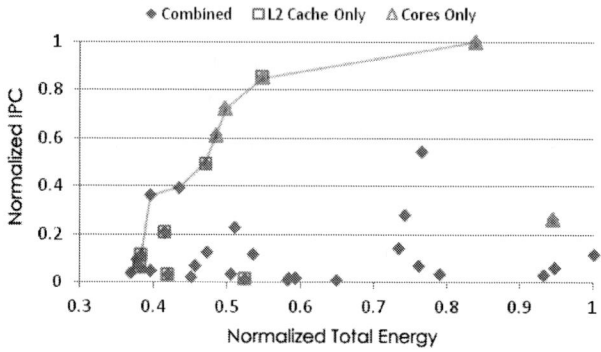

Figure 2: Efficiency of closed adaptive systems.

the performance of the application and the total energy consumed. The results are shown in Figure 2, where the x-axis shows energy and the y-axis shows instructions per second. The solid diamond points represent all tested configurations. The squares show configurations that appear optimal for a closed system which only considers cache adaptations. The triangles show possible configurations for a system that only considers core allocations. The best configurations are the ones with highest performance and lowest total energy; *i.e.*, the Pareto-optimal frontier which is depicted by those diamond points that are connected by a line in the figure. Notice that both triangles and squares appear to the right of the Pareto frontier, and these points represent configurations that closed systems would believe to be optimal, but, in fact, are sub-optimal for the overall system.

To avoid sub-optimal configurations, the SEEC model provides a general interface allowing adaptations supported by different system components to be described by their designer and then manipulated by the SEEC runtime decision system. For example, this interface can be used to describe both operating system-level actions (*e.g.*, allocation of cores to an application [26]) and hardware-level actions (*e.g.*, reconfiguration of the hardware data cache [4]). Given this information, the SEEC runtime system can coordinate adaptation to keep the system on the Pareto optimal curve shown in Figure 2. To support this model, hardware must be explicitly designed to expose adaptations instead of attempting to adapt as a closed system.

3. THE SEEC MODEL

The key feature of the SEEC model is its open ODA loop where different components of the system contribute to the specification of observations, actions, and decisions. This section summarizes the model; more detailed information is available in [15].

3.1 Observe

In order to change something, it is essential to measure the value to be changed. In SEEC, applications explicitly state their goals and other system components measure whether those goals are being met. SEEC uses the Application Heartbeats API [14] to specify application goals and progress. The API's key abstraction is a heartbeat; applications use a function to emit heartbeats at important intervals, while additional API calls specify goals in terms of this heartbeat. SEEC currently supports three application specified goals: performance, accuracy, and power. Performance is specified as a target heart rate or a target latency between specially tagged heartbeats. Accuracy goals are specified as a *distortion*, or linear distance from an application defined nominal value [16], measured over some set of heartbeats. Power and energy goals can be specified as target average power for a given heartrate or as a target energy between tagged heartbeats. All goals are expressed using a C/C++API so that applications can explicitly state these goals. A second API allows other components of the system to observe the current value of any of these metrics and thus evaluate whether or not the goals are being met.

3.2 Act

In the SEEC model, applications provide goals while system components specify *actions* that change the behavior of the system. SEEC supports a range of actions specified from the application-level [3, 16], system software level [12, 25], and the hardware level as exposed by the Angstrom processor (see Section 4).

As discussed in Section 2, SEEC works to avoid suboptimal combinations of actions by providing an interface that all system components use to specify available actions. This interface must be general enough to support actions exposed by different developers working at different levels of the system stack. Actions are specified by describing the *actuators* that implement them. In SEEC, an actuator is a data object with: a name, a list of allowable settings, a function that changes the setting, a set of axes which the actuator affects (*e.g.*, performance and power), and the effects of each setting on each axis. These effects are listed as multipliers over a nominal setting, whose effects are 1 on all axes. Each actuator specifies a *delay*, which is the time between when it is set and when its effects can be observed. Finally, each actuator specifies whether it works on only the application that registered it or if it works on all applications, so that SEEC can distinguish between adaptations specified at application-level (*e.g.*, changing algorithms) and adaptations that affect the whole system (*e.g.*, allocating cores).

3.3 Decide

SEEC's runtime system automatically and dynamically selects actions to meet goals while attempting to minimize cost. The SEEC decision engine is designed to handle general-purpose environments and the SEEC runtime system will often have to make decisions about actions and applications with which it has no prior experience. In addition, the runtime system will need to react quickly to changes in application load and fluctuations in available resources. To meet these requirements for handling general and volatile environments, the SEEC decision engine is designed with multiple layers of adaptation as described in [15]. At the lowest-level, SEEC acts as a classical control system, taking feedback, in the form of heartbeats, and using it to tune actuators to meet goals [26]. The classical control system works well given prior knowledge about the application's behavior. Additional layers of adaptation, including adaptive control and machine learning based techniques [27], allow the SEEC runtime to allocate resources efficiently without prior knowledge of the application, or when the behavior of the actuator diverges from the predicted behavior.

4. ANGSTROM SUPPORT FOR SEEC

Angstrom is a manycore system design targeting the integration of 1000 cores onto a single chip. A full description of the design is beyond the scope of this paper, so we focus on features of the architecture explicitly designed to support the SEEC model.

4.1 Observation

The Angstrom processor design supports SEEC by providing visibility into the hardware in the form of traditional performance counters, event probes, and non-traditional sensors. This information allows SEEC's runtime decision engine to diagnose either why goals are not being met, or whether there might be a lower cost set of actions that would achieve the same goal.

Performance counters provide valuable insight into the behavior of an application on a particular hardware architecture. Unfortunately, many existing systems limit the number of counters that can be read simultaneously by software. This limitation means that application tuning requires multiple profiling runs and prevents dynamic exploitation of performance counter information. The Angstrom design exposes multiple performance counters that are memory-mapped and can be read by any level of the software stack. These count simple events such as: memory operations, cache hits and misses, pipeline stall cycles, network flits sent and received, etc. These are useful for assessing average behavior over a period of time but since they must be polled by software, they cannot be queried too frequently.

To reduce the overhead of monitoring the performance counters and watch for rare events, Angstrom's design includes *event probes*, that can be associated with a performance counter or some other piece of state with the processor. They contain a *trigger* register and a programmable comparator that continually watches for the state to match the trigger. The comparator can be set to different operations including: equal, less than, greater than and their logical inverses. In addition, a mask can be specified to compare only selected bits. When a match occurs, an interrupt can be generated or an event record can be placed in a small hardware queue.

Besides observing processor state, Angstrom includes sensors to monitor things like temperature, voltage, battery charge, and energy consumption [31]. This allows the runtime decision engine to react to changing environmental conditions (such as cooling failures or dying batteries) as well as observe how its actions impact these quantities to handle goals like minimizing power consumption or limiting temperature extremes. We expect some of these sensors to be deployed in a fine-grained manner to measure variations between the 1000 cores.

4.2 Action

As further support for the SEEC model, the Angstrom processor exposes a number of different actions or different hardware configurations. Keeping with the theme of building an open adaptive system, Angstrom provides these "knobs" but relies on the SEEC runtime system to set them in coordination with other adaptations specified at the software level. This section discusses some of the adaptations exposed by the Angstrom processor at both the intra- and inter-core level.

4.2.1 Intra-core Adaptation

In the Angstrom design, each core is capable of running at different voltages and frequencies. Operating the processor designs at lower voltage levels has been shown to increase energy efficiency, as with the voltage-scalable 32-bit microprocessor design demonstrated in [17]. This processor operates at peak performance with nominal voltages while supporting an energy efficient mode at 0.54 V with only 10.2 pJ/cycle energy consumption. Similarly, making each Angstrom core capable of running at different voltage and frequency levels will optimize them for applications with limited energy budgets and time varying processing loads.

Technology scaling is fueling integration of larger on chip caches in processor design (*e.g.,* up to 50 MB [30]). In order to enable ultra-low power consumption, Angstrom cores need to feature voltage-scalable SRAMs. Conventional SRAMs cannot work at low-voltage levels due to stability problems. Thus, recent work has focused on implementing different bit-cell topologies [7, 6] and peripheral assist circuits [21, 33] to enable operation down to sub-VT levels.

Reconfiguration of the local caches is shown to reduce power consumption for the same performance [4]. Disabling unnecessary parts of the Angstrom caches (sets and ways) will help SEEC to optimize power and performance tradeoffs. This adaptation can be beneficial both for applications with small working sets and applications with large working sets that do not achieve much locality on their data.

4.2.2 Inter-core Adaptations

Angstrom supports dynamic adaptation of the on-chip network by enabling software and hardware to interact in achieving goal-driven tradeoffs between performance and efficiency. This is accomplished with three architecture features: express virtual channels (EVC) [8], bandwidth adaptive networks (BAN) [9], and application-aware oblivious routing (AOR) [22].

EVC allows flits to attempt bypassing of buffering and arbitration within a network router by proceeding straight to switch and link traversals. This can significantly reduce latency and can also reduce energy spent in buffering flits. Previous EVC research has demonstrated significant performance/energy tradeoff capability; Angstrom enhances this by introducing a software interface to routing tables maintained in the network hardware. This information is then used by EVC logic to manage virtual channels.

A BAN can rapidly adjust bisection bandwidth to adapt to changing network conditions. This is done by using bidirectional links with arbitration logic and tristate buffers that prevent simultaneous writes to the same wire. A hardware bandwidth allocator governs the direction of a link. Angstrom expands on prior BAN approaches by exposing configuration parameters in the bandwidth allocator to software while still providing for fine-grained bandwidth management in hardware.

AOR algorithms can produce deadlock-free routes while maximizing satisfaction of flow demands on the network, achieving better throughput than traditional oblivious routing approaches because optimization is done using global application knowledge. At the same time the router remains simple because routes are configured via a routing table. Unlike previous offline routing algorithm approaches, Angstrom provides for online routing computation by exposing the routing table to software, allowing the system to adapt to changing application characteristics over time.

Angstrom also supports adaptation of the cache-coherence protocol used between cores. For some applications, directory-based cache-coherence provides the best performance and energy consumption [13]. However, for other applications it is more efficient to use a shared-NUCA (non-uniform cache access) protocols because it provides for a large shared cache capacity and reduces the total number of off-chip accesses [20]. The ARCc architecture has shown that combining these protocols and adaptively selecting the best on a per application basis can improve performance and energy efficiency [19]. Angstrom adopts the ARCc approach of providing multiple coherence protocols and exposes these adaptations to SEEC for management.

4.3 Decision

Although self-aware optimizations are capable of dramatically improving the behavior of applications, they do not come for free. Some resources must be devoted to making runtime decisions to have a dynamic, adaptive system. To help reduce the costs of runtime decision making the Angstrom processor contains specialized, low-power cores called *partner cores*, which we describe below. More detail is available in [24].

Each main core in the Angstrom design has a partner core associated with it. These two cores are tightly integrated so that the partner core can inspect and manipulate state (including performance counters and configuration registers) within the main core. The partner core also has access to the event queues fed by event probes. The partner core targets a lower performance point than the main core and is designed to take much less area and energy. It has a simplified pipeline, smaller caches and fewer functional units. It is designed to run at lower clock frequencies and makes heavy use of low-power circuit techniques, requiring less energy per operation, and making it more efficient to run dynamic optimization code on the partner core than the main core. We estimate that each partner core will consume about 10% of the area and 10% of the power of a main core.

5. EVALUATION

This section presents experiments designed to evaluate the SEEC model and highlight the benefits of Angstrom's support for it. We first use SEEC to control applications on an existing Linux/x86 system. We then collect data on simulations of a 256 core Angstrom processor and use that data to build analytical models that predict how SEEC will behave on such a system.

5.1 Benchmark Applications

Both experiments use the same set of five benchmark applications selected from the SPLASH2 benchmark suite: barnes, ocean (non contiguous), raytrace, water (spatial), and volrend [35]. Each application is instrumented with the Application Heartbeats API to emit goals. In both experiments, applications request a target performance and SEEC is tasked with meeting that performance while minimizing power consumption.

5.2 SEEC on Existing Systems

We begin by presenting results using SEEC to control these benchmarks on an existing architecture with some adaptive capabilities, but which was not explicitly designed to support SEEC. Specifically, we use a Dell PowerEdge R410 server with two quad-core Intel Xeon E5530 processors running Linux 2.6.26. The processors support seven power

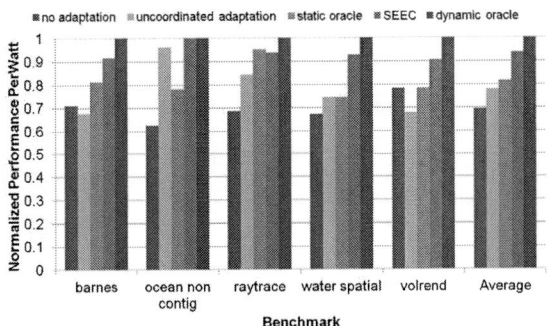

Figure 3: SEEC on a Linux/x86 system.

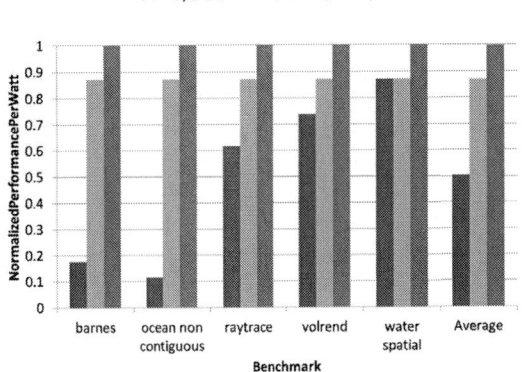

Figure 4: Anticipated SEEC results on Angstrom.

states with clock frequencies from 2.4 GHz to 1.6 GHz. The `cpufrequtils` package enables software control of the clock frequency (and thus the power state). We use a WattsUp device to sample and store the average consumed power over 1 second intervals [1]. All benchmark inputs have been expanded to allow them to run for significantly more than 1 second. The measured power ranges from 220 watts (at full load) to 80 watts (idle), with a typical idle power consumption of approximately 90 watts.

Each benchmark is launched on a single core set to the minimum clock speed and it requests a performance equal to half the maximum achievable. SEEC's runtime attempts to meet this application specified performance goal while minimizing power consumption using the following actions: changing the number of cores assigned to the application, changing the clock speed of the cores assigned to an application, and changing the number of active (or non-idle) cycles assigned to the application. For each application, we measure the average performance and power consumption over the entire execution of the application. We then subtract out the idle power of the processor and compute performance per Watt as the minimum of the achieved and desired performance divided by the power beyond idle.

To show the benefits of SEEC we compare to several other approaches. First, we compute the best that could be achieved in a system with *no adaptation*; *i.e.,* when all applications use the same number of cores and the same clock speed, which provides a baseline that any adaptive system should hope to beat. Next, we compare to a system where adaptation is *uncoordinated*; *i.e.,* separate instances of the SEEC runtime system control cores, clock speed, and idle cycles but do not coordinate with each other. Uncoordinated adaptation represents what happens when multiple closed adaptive systems work together. Third, we compare to a *static oracle*, which adapts resource usage to the individual application, but only assigns resources once per application. The static oracle represents what is possible with an adaptive system which does not make fine-grained adaptation. Finally, we compare to a *dynamic oracle* which adapts resource allocation at every heartbeat to the best configuration for the application's next heartbeat. Obviously, the dynamic oracle cannot be built in practice, but instead its actions are computed after the fact by post processing empirical data for each application. The dynamic oracle represents an upper bound on the possible benefits of adaptation because it has 1) no overhead and 2) perfect knowledge of the future.

The results of this experiment are shown in Figure 3, where benchmarks are labeled on the x-axis and performance per Watt is on the y-axis. For each benchmark, performance

per Watt is normalized to that achieved by the dynamic oracle (so the maximum possible is one). Separate bars show the value achieved for each approach of: no adaptation, uncoordinated adaptation, SEEC, and dynamic oracle.

The results show the benefits of using SEEC to provide coordinated management of adaptive systems. SEEC is able to outperform the uncoordinated adaptive system by over 20% and the static oracle by over 15%. Indeed, for some applications, uncoordinated adaptation is actually worse than not adapting because uncoordinated adaptations oscillate through suboptimal resource allocations. In contrast, the SEEC model exposes these adaptations as first class objects so it can coordinate and avoid suboptimal configurations. SEEC is able to achieve almost 94% of the performance of the dynamic oracle, which demonstrates that SEEC is responsive and low overhead.

5.3 SEEC on Angstrom

This section explores the benefits of running SEEC on a future Angstrom architecture with 256 cores. We use the Graphite simulator [28] to model an Angstrom processor that can adapt its cache size (from 32-128 KB, by powers of 2), the number of cores assigned to an application (from 1-256, by powers of 2), and the voltage (0.4, 0.8 V) and frequency (100, 500 MHz) of those cores. We use simulation because existing processors do not have as many cores or available adaptations.

We run each benchmark in every possible configuration and use this data to compute the performance per watt as described above and construct the performance of a system with no adaptation, and the performance of a system which adapts using the static oracle. We then compute the predicted value of using SEEC by multiplying the value of the static oracle by the multiplier that SEEC achieved on the x86 architecture in the above section.

The results for this experiment are shown in Figure 4. This figure shows the benchmark on the x-axis and the performance per Watt on the y-axis. For each benchmark, performance per Watt is normalized to that achieved by the predicted SEEC result. Separate bars show the value achieved for: no adaptation, the static oracle, and our estimate of how SEEC will behave on the Angstrom architecture.

The results show that the Angstrom architecture indeed benefits more from adaptation than modern architectures, because of the increased number of available adaptations. As a specific example, barnes contains large amounts of parallelism that can be exploited by allocating cores, but that is not the case for all applications. In fact, the non-adaptive system allocates 64 cores out of a possible 256 to the system because it provides the best average performance per watt efficiency across all applications. On the other hand, a static oracle allocates 256 cores for running barnes, outperforming the non-adaptive configuration by over 5x.

Overall, the static oracle outperforms the non-adaptive system by 72% on average. Under the assumption that SEEC outperforms the static oracle by 15%, we predict that SEEC on the Angstrom architecture can achieve an improvement of 100% over a standard non-adaptive system. Indeed, future manycore architectures that are designed explicitly to leverage the self-aware computing model will enjoy SEEC considerably more than current systems.

6. CONCLUSION

This paper has presented the design for the Angstrom architecture. Angstrom is a multicore which addresses the challenge of extreme-scale computing by explicitly supporting a self-aware computational model. This support comes in two forms. First, Angstrom provides additional visibility into the hardware state through performance and energy counters. Second, Angstrom exposes hardware adaptations so that they can be managed by the SEEC runtime system in coordination with other, software based actions. We have demonstrated how self-aware computing provides a benefit on existing architectures and presented evidence that this approach will provide an even greater benefit when implemented on an architecture like Angstrom with explicit support for the model.

7. REFERENCES

[1] Wattsup .net meter. http://www.wattsupmeters.com/.

[2] D. H. Albonesi, R. Balasubramonian, S. G. Dropsho, S. Dwarkadas, E. G. Friedman, M. C. Huang, V. Kursun, G. Magklis, M. L. Scott, G. Semeraro, P. Bose, A. Buyuktosunoglu, P. W. Cook, and S. E. Schuster. Dynamically tuning processor resources with adaptive processing. *Computer*, 36:49–58, December 2003.

[3] J. Ansel, C. Chan, Y. L. Wong, M. Olszewski, Q. Zhao, A. Edelman, and S. Amarasinghe. PetaBricks: A language and compiler for algorithmic choice. In *PLDI*, 2009.

[4] R. Balasubramonian, D. Albonesi, A. Buyuktosunoglu, and S. Dwarkadas. Memory hierarchy reconfiguration for energy and performance in general-purpose processor architectures. In *MICRO*, 2000.

[5] R. Bitirgen, E. Ipek, and J. F. Martinez. Coordinated management of multiple interacting resources in chip multiprocessors: A machine learning approach. In *MICRO*, 2008.

[6] B. Calhoun and A. Chandrakasan. A 256kb sub-threshold SRAM in 65nm CMOS. In *ISSCC*, 2006.

[7] L. Chang, D. Fried, J. Hergenrother, J. Sleight, R. Dennard, R. Montoye, L. Sekaric, S. McNab, A. Topol, C. Adams, K. Guarini, and W. Haensch. Stable SRAM cell design for the 32 nm node and beyond. In *Symposium on VLSI Technology*, 2005.

[8] C.-H. O. Chen, N. Agarwal, T. Krishna, K.-H. Koo, L.-S. Peh, and K. C. Saraswat. Physical vs. Virtual Express Topologies with Low-Swing Links for Future Many-core NoCs. In *NOCS*, 2010.

[9] M. H. Cho, M. Lis, K. S. Shim, M. Kinsy, T. Wen, and S. Devadas. Oblivious Routing in On-Chip Bandwidth-Adaptive Networks. In *PACT*, 2009.

[10] S. Choi and D. Yeung. Learning-Based SMT Processor Resource Distribution via Hill-Climbing. In *ISCA*, 2006.

[11] C. Dubach, T. M. Jones, E. V. Bonilla, and M. F. P. O'Boyle. A predictive model for dynamic microarchitectural adaptivity control. In *MICRO*, 2010.

[12] J. Eastep, D. Wingate, M. D. Santambrogio, and A. Agarwal. Smartlocks: lock acquisition scheduling for self-aware synchronization. In *ICAC*, 2010.

[13] A. Gupta, W. Weber, and T. Mowry. Reducing memory and traffic requirements for scalable directory-based cache coherence schemes. In *ICPP*, 1990.

[14] H. Hoffmann, J. Eastep, M. D. Santambrogio, J. E. Miller, and A. Agarwal. Application heartbeats: a generic interface for specifying program performance and goals in autonomous computing environments. In *ICAC*, 2010.

[15] H. Hoffmann, M. Maggio, M. D. Santambrogio, A. Leva, and A. Agarwal. SEEC: A General and Extensible Framework for Self-Aware Computing. Technical Report MIT-CSAIL-TR-2011-046, MIT, November 2011.

[16] H. Hoffmann, S. Sidiroglou, M. Carbin, S. Misailovic, A. Agarwal, and M. Rinard. Dynamic knobs for responsive power-aware computing. In *ASPLOS*, 2011.

[17] N. Ickes, Y. Sinangil, F. Pappalardo, E. Guidetti, and A. Chandrakasan. A 10 pJ/cycle ultra-low-voltage 32-bit microprocessor system-on-chip. In *ESSCIRC*, sept. 2011.

[18] J. O. Kephart and D. M. Chess. The vision of autonomic computing. *Computer*, 36:41–50, January 2003.

[19] O. Khan, H. Hoffmann, M. Lis, F. Hijaz, A. Agarwal, and S. Devadas. ARCc: A case for an architecturally redundant cache-coherence architecture for large multicores. In *ICCD*, 2011.

[20] C. Kim, D. Burger, and S. W. Keckler. An adaptive, non-uniform cache structure for wire-delay dominated on-chip caches. In *ASPLOS*, 2002.

[21] T.-H. Kim, J. Liu, J. Keane, and C. Kim. A High-Density Subthreshold SRAM with Data-Independent Bitline Leakage and Virtual Ground Replica Scheme. In *ISSCC*, 2007.

[22] M. Kinsy, M. H. Cho, T. Wen, E. Suh, M. van Dijk, and S. Devadas. Application-Aware Deadlock-Free Oblivious Routing. In *ISCA*, 2009.

[23] R. Laddaga. Guest editor's introduction: Creating robust software through self-adaptation. *IEEE Intelligent Systems*, 14:26–29, May 1999.

[24] E. Lau, J. E. Miller, I. Choi, D. Yeung, S. Amarasinghe, and A. Agarwal. Multicore performance optimization using partner cores. In *HotPar*, 2011.

[25] M. Maggio, H. Hoffmann, M. D. Santambrogio, A. Agarwal, and A. Leva. Power optimization in embedded systems via feedback control of resource allocation. *IEEE Transactions on Control Systems Technology*, PP(99):1–8.

[26] M. Maggio, H. Hoffmann, M. D. Santambrogio, A. Agarwal, and A. Leva. Controlling software applications via resource allocation within the heartbeats framework. In *CDC*, 2010.

[27] M. Maggio, H. Hoffmann, M. D. Santambrogio, A. Agarwal, and A. Leva. Decision making in autonomic computing systems: comparison of approaches and techniques. In *ICAC*, 2011.

[28] J. E. Miller, H. Kasture, G. Kurian, C. Gruenwald III, N. Beckmann, C. Celio, J. Eastep, and A. Agarwal. Graphite: A distributed parallel simulator for multicores. In *HPCA*, 2010.

[29] MIT. The MIT angstrom project. http://projects.csail.mit.edu/angstrom, 2012.

[30] R. Riedlinger, R. Bhatia, L. Biro, B. Bowhill, E. Fetzer, P. Gronowski, and T. Grutkowski. A 32nm 3.1 billion transistor 12-wide-issue Itanium processor for mission-critical servers. In *ISSCC*, 2011.

[31] E. Rotem, A. Naveh, D. R. amd Avinash Ananthakrishnan, and E. Weissmann. Power management architecture of the 2nd generation Intel Core microarchitecture, formerly codenamed Sandy Bridge. In *Hot Chips*, Aug. 2011.

[32] M. Salehie and L. Tahvildari. Self-adaptive software: Landscape and research challenges. *ACM Trans. Auton. Adapt. Syst.*, 4(2):1–42, 2009.

[33] M. Sinangil, H. Mair, and A. Chandrakasan. A 28nm high-density 6T SRAM with optimized peripheral-assist circuits for operation down to 0.6V. In *ISSCC*, 2011.

[34] P. Team. Online document, http://icl.cs.utk.edu/papi/.

[35] S. C. Woo, M. Ohara, E. Torrie, J. P. Singh, and A. Gupta. The splash-2 programs: characterization and methodological considerations. *SIGARCH Comput. Archit. News*, 23:24–36, May 1995.

The Case for Elastic Operating System Services in fos

Lamia Youseff[*]
Google, Inc
651 N. 34th Street
Seattle, WA 98103
lyouseff@google.com

Nathan Beckmann
M.I.T CSAIL
32 Vassar Street,
Cambridge, MA 02139
beckmann@csail.mit.edu

Harshad Kasture
M.I.T CSAIL
32 Vassar Street,
Cambridge, MA 02139
harshad@csail.mit.edu

Charles Gruenwald
M.I.T CSAIL
32 Vassar Street,
Cambridge, MA 02139
cg3@csail.mit.edu

David Wentzlaff[†]
Princeton University
Engineering Quadrangle
Princeton, NJ 08544
wentzlaf@princeton.edu

Anant Agarwal
M.I.T CSAIL
32 Vassar Street,
Cambridge, MA 02139
anantagarwal@csail.mit.edu

ABSTRACT

Given exponential scaling, it will not be long before chips with hundreds of cores are standard. For OS designers, this new trend in architectures provides a new opportunity to explore different research directions in scaling operating systems. The primary question facing OS designers over the next ten years will be: *What is the correct design of OS services that will scale up to hundreds or thousands of cores, and adapt to the unprecedented variability in demand of the system resources?* A fundamental research challenge addressed in this paper is to identify some characteristics of such a scalable OS service for next multicore and cloud computing chips.

We argue that the OS services have to deploy elastic techniques to adapt to this variability at runtime. In this paper, we advocate for elastic OS service, illustrate their feasibility and effectiveness in meeting the variable demands through providing elastic technologies for OS services in the fos operating system. We furthermore showcase a prototype elastic file system service fos and illustrate its effectiveness in meeting variable demands.

Categories and Subject Descriptors

D.4.7 [**Operating Systems**]: [Organization and Design]

General Terms

Design, Performance

---

[*]Lamia worked on this project while she was a postdoctoral associate at MIT

[†]David worked on this project while he was a Ph.D. candidate at MIT

Keywords

multicore, scalable operating systems, cloud computing, OS services, self-aware systems

1. INTRODUCTION

Trends in multicore architectures point to an ever-increasing number of cores available on a single chip. Moore's law predicts an exponential increase in integrated circuit density; in the past, this increase in circuit density has translated into higher single-stream performance, but recently single-stream performance has plateaued and industry has turned to adding cores to increase processor performance. In only a few years, multicores have gone from esoteric to commonplace: 12-core single-chip offerings are available from major vendors [3] with research prototypes showing many more cores on the horizon [20, 14], and 64-core chips are available from embedded vendors [24] and 100-core chips available through the TILE-Gx family processors [2].

Given exponential scaling, it will not be long before chips with hundreds of cores are standard, with thousands of cores following close behind. This new architecture trend is providing an exciting opportunity for exploring different research directions in scaling operating systems. In traditional monolithic OSs, OS code executes in the kernel on the same core which makes an OS service request. Corey [7] showed that this led to significant performance degradation for important applications compared to intelligent, application-level management of system services. Our prior work [23] also showed significant cache pollution caused by running OS code on the same core as the application. This becomes an even greater problem if multicore trends lead to smaller per-core cache. We also showed severe scalability problems with OS microbenchmarks, even only up to 16 cores.

A similar, independent trend can be seen in the growth of cloud computing. Rather than consolidating a large number of cores on a single chip, cloud computing consolidates multicore machines within a data center. Current Infrastructure as a Service (IaaS) cloud management solutions require a cloud computer user to manage many virtual machines (VMs). Unfortunately, this presents a fractured and non-uniform view of resources to the programmer. For example, the user needs to manage communication differently

265

depending on whether the communication is within a VM or between VMs. Also, the user of a IaaS system has to worry not only about constructing their application, but also about system concerns such as configuring and managing communicating operating systems. In fos [25], we put a cloud under a single system image OS, thereby improving manageability and resource utilization. There is much commonality between constructing OSs for clouds and multicores, such as large-scale resource management, heterogeneity, and possible lack of widespread shared memory. These similarities allow fos to be designed for both multicore and cloud computers.

The primary question facing OS designers over the next ten years will be: *What is the correct design of OS services that will scale up to hundreds or thousands of cores?* We argue that the structure of monolithic OSs is fundamentally limited in how they can address this problem. In contrast, the structure of fos brings scalability concerns to the forefront by decomposing an OS into services, and then parallelizing within each service. To facilitate the conscious consideration of scalability, fos system services are moved into userspace and connected via messaging. In fos, a set of cooperating system servers which implement a single system service is called a *fleet*.

Monolithic operating systems are designed assuming that computation is the limiting resource. This assumption is proving incorrect as the advent of multicore and clouds is providing abundant computational resources. The abundant computational resources present opportunities to the OS to allocate computational resources to auxiliary purposes, which accelerates application execution, but do not run the application itself. fos's architecture leverages this insight by factoring OS code out of the kernel and running it on cores separate from application code. Once moved out of the kernel, each fos service is implemented as a parallel, cooperating set of servers, called a fleet. A fundamental research challenge in this design is to identify the characteristics of such a fleet. Given the unprecedented variability in demand of the system resources, the OS fleets have to deploy elastic techniques to adapt to this variability at runtime.

In this paper, we advocate for elastic OS services and illustrate their feasibility in meeting the variable demand on system resources. We furthermore showcase a prototype elastic file system service and illustrate its effectiveness in meeting variable demands. The rest of this paper is organized as following. Section 2 overviews the general design of the fos operating system. Section 3 describes the design of elastic operating system services in fos. Section 4 studies the feasibility and the effectiveness of the elastic OS services through a prototype elastic file system service. Section 5 quickly surveys the related work and we conclude our preliminary study in section 6.

2. OVERVIEW OF FOS DESIGN

Current OSs were designed in an era when computation was a limited resource. With the expected exponential increase in number of cores, the landscape has fundamentally changed. The question is no longer how to cope with limited resources, but rather how to make the most of the abundant computation available. fos is designed with this in mind, and takes scalability and adaptability as *the* first-order design constraints. The goal of fos is to design system services that scale from a few to thousands of cores.

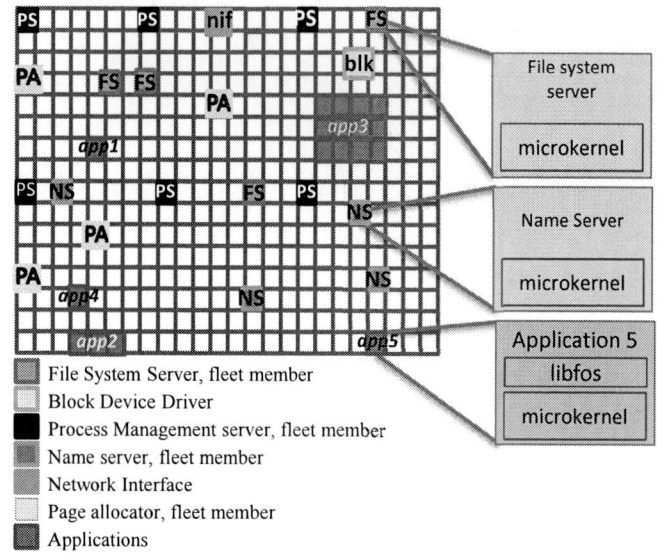

Figure 1: A high-level diagram of the fos servers layout in a multicore machine. The figure illustrates that each OS service fleet consists of several servers which are assigned to different cores.

One key design principle in fos is that *OS is factored into function-specific services, where each is implemented as a parallel, distributed service.* The OS runs independent of the application on separate cores. Applications interact with servers through a library layer that translates requests into messages. This allows fos to *adapt resource utilization to elasticity of system needs.* The utilization of active services is measured, and highly loaded services are provisioned more cores.

Figure 1 shows the high-level architecture of fos. A small microkernel runs on every core. Operating system services and applications run on distinct cores. Applications can use shared memory, but OS services communicate via message passing. A library layer (*libfos*) translates traditional syscalls and a subset of the POSIX API into messages to fos services. A naming service is used to find a message's destination server. The naming service is maintained by a distributed set (fleet) of naming servers. Finally, fos can run on top of a hypervisor and seamlessly span multiple machines, thereby providing a a single system image across a cloud computer.

One unique approach to the organization of multiple communicating processes that fos takes is the use of a naming and lookup scheme. fos processes are able to register a particular name for a mailbox. This namespace is a hierarchical URI much like a web address or filename. This abstraction allows fos to provide great flexibility in load balancing and locality to the operating system. The basic organization for many of fos servers is to divide the service into several independent processes (running on different cores). As a result, when an application messages a particular service, the nameserver will provide a member of the fleet that is best suited for handling the request. To accomplish this, all of the servers within the fleet register under a given name. When a message is sent, the nameserver will provide the server that is optimal based on the load of all of the servers as well as the latency between the requesting process and each server within the fleet.

Several policies can be used to determine the correct server to route the message to. One solution is to have a few fixed policies such as round robin or closest server. Alternatively, custom policies could be set via a callback mechanism or complex load balancer. Meta-data such as message queue lengths can be used to determine the best server to send a message to. As much of the system relies on this naming mechanism, the question of how to optimally build the nameserver and manage caching associated with it is also a challenging research area. This service must be extremely low latency while still maintaining a consistent and global view of the namespace. In addition to servers joining and leaving fleets on the fly, thus requiring continual updates to the name-lookup, servers could also be migrating between machines, requiring the nameserver (and thus routing information) to be updated on the fly as well. The advantage to this design is that much of the complexity dealing with separate forms of inter-process communication in traditional cloud solutions is abstracted behind the naming and messaging API. Each process simply needs to know the name of the other processes it wishes to communicate with, fos assumes the responsibility of efficiently delivering the message to the best suited server within the fleet providing the given service.

3. ELASTIC OS SERVICES

In addition to unprecedented amounts of resources, clouds and multicores introduce unprecedented variability in demand for these resources. Dynamic load balancing and migration of processes go a long way towards solving this problem, but still require over-provisioning of resources to meet demand. This would quickly become unfeasible, as every service in the system claims the maximum amount of resources it will ever need. Instead, the fos architecture allows the OS fleets to be *elastic*, meaning they can grow to meet increases in demand, and then shrink to free resources back to the OS.

Current OSs achieve elasticity by accident, as OS code runs on the same core as the application code. This design relinquishes control of how many cores to provision the service. As discussed earlier, running the OS on the same core as the application can force unnecessary stalls waiting for OS requests to finish and also pollutes application cache. But it also has other downsides in respect to elastic fos services. Running the OS on each core leads to unnecessary sharing of OS data in the cache of each application core. This creates extraneous cache misses every time the service executes on a new core, whereas factoring the service onto distinct cores avoids this. Furthermore, for applications that rely heavily on the OS it may be best to provision *more* cores to the OS service than the application. The servers can then collaborate to provide the service more efficiently.

A fleet is grown by starting a new instance of the service on a new core. This instance joins the fleet by contacting other members (either the coordinator or individual members via a distributed discovery protocol) and synchronizing

its state. Some of the distributed, shared state is migrated to the new member, along with the associated transactions. Transactions could be migrated in any number of ways, for example by sending a redirect message to the client from the "old" server.

Shrinking is in some ways easier, although there are still several possibilities. For services with short-lived or stateless transactions, the fleet member simply de-registers itself from the global namespace and lets all of its transactions complete. Then it flushes the shared data in its distributed state and shuts down. For long-lived transactions, state must be migrated to surviving members of the fleet, and redirect messages must be sent. Some designs may allow for the server to flush its state and then just "disappear". Clients will detect that they are unable to communicate with the server, and restart the transaction with a new member of the fleet. This requires that all necessary state is kept in a distributed data structure.

An interesting problem is determining when to grow and shrink the fleet. This problem is closely related to scheduling, as the new fleet member will consume a core and affect the spatial layout of processes in the system. It is not as simple as detecting when a service is saturated and growing the fleet – doing so may result in performance degradations if it forces migration of important processes. Although we do not fully address this problem in this paper, it is an important area of future work. We are evaluating the efficiency of few possible heuristics and approaches to be used by the system in managing the fleet size, such as a simple low watermark, a more sophisticated fleet-specific scheduler that allows a fleet to decide when to grow, or a more general scheduler that manages the size of all fleets in the system.

4. EXPERIEMENTAL RESULTS

In order to illustrate the feasibility and effectiveness of elastic operating system services, we have implemented a read-only file system service as a prototype of an elastic fleet. The service starts initially with a single server, which acts as the coordinator. This server monitors the utilization of the service, and detects when the service has become saturated. At this point, the fleet grows by adding a new member to meet demand. Similarly, the coordinator detects if utilization has dropped sufficiently to shrink the fleet, freeing up system resources.

Load balancing is performed through the name server. Each file system server is registered under a single alias. At the beginning of each transaction, a name look-up is performed through the name server. This name is cached for the remainder of the transaction. The names are served round-robin, performing dynamic load balancing across the two servers. To simplify our experiment, this load balancing is performed through *libfos*, by counting the number of open files and only refreshing the cache when no files are open. This ensures that there is no local state stored on the file system server, and eliminates the need to distribute the state for the open files.

We conducted a preliminary experiment that focuses on proving the utility of the elastic features of fos fleets. In the experiment, two clients make queries to the read-only file system service. Each client repeats a single transaction (open a file, read one kilobyte, and close the file) 5,000 times. The request rate is throttled initially to one transaction per two million cycles. The throttling gradually decreases until the service is saturated, before gradually increasing again to the initial value.

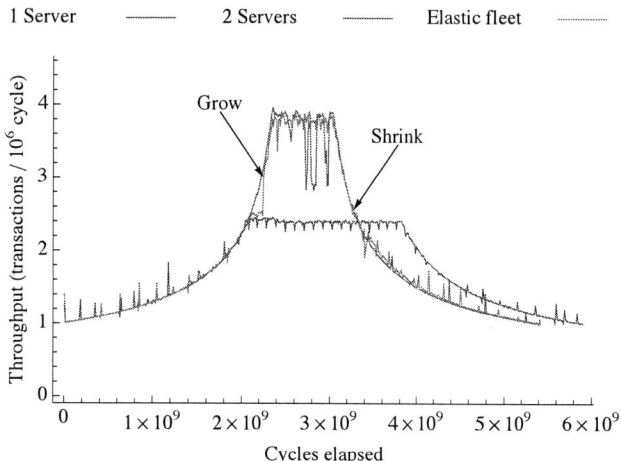

Figure 2: A demonstration of an elastic read-only file system fleet.

Figure 2 shows the results for three different service configurations. Aggregate throughput from both clients is plotted against the number of elapsed cycles in the test. The first configuration is with a single file system server serving both clients. As expected, the service is saturated quickly and peaks at roughly 2 transactions per million cycles. This configuration also takes the longest to complete all transactions. The second configuration has two file system servers, and each client is statically scheduled to communicate with one of the servers. This avoids any contention overhead from dynamic load balancing, so should provide peak performance. This configuration peaks at just below 4 transactions per million cycles. The last configuration is the elastic fleet. This configuration briefly saturates at the same level as a single file system server, before expanding the fleet (marked by the "grow" arrow on the graph). Subsequently, it immediately achieves the throughput of the two server, statically load balanced configuration. Finally, when the request rate again decreases near the end of the experiment, the fleet shrinks to use fewer resources.

Figure 2 shows that our prototype elastic fleet performs equivalently to an over-provisioned service, and that there is negligible overhead for performing dynamic load balancing through the name service. The elastic fleet consumes a single core for the majority of the experiment, only expanding between the "grow" and "shrink" arrows, or for 18.8% of the run-time of the experiment. Note that roughly one-third of the transactions lie within this range, but the fleet processes these transactions at nearly double rate. Because of this, in general a fleet needs to be provisioned extra resources for less than the percentage of requests that saturate the service. Therefore, demand on cores is less than one might otherwise expect. Compared to the two server configuration, this translates into a resource savings of 40.6% at no performance cost, which could be used for other computation or to save power by turning off the core. Although this is a very simple toy-like example of an elastic OS service,

which avoids some of the more difficult issues in expanding and shrinking an OS fleet, it is still indicative of a desirable approach for dealing with variability of demand and the effectiveness of elastic OS services.

5. RELATED WORK

There are several classes of systems which have similarities to fos: traditional microkernels, distributed OSs, and cloud computing infrastructure.

Traditional microkernels include Mach [4] and L4 [16]. fos is designed as a microkernel and extends the microkernel design ideas. However, it is differentiated from previous microkernels in that instead of simply exploiting parallelism between servers which provide different functions, this work seeks to distribute and parallelize within a server for a single high-level function. fos also seeks to exploit the "locality" of massively multicore processors by spatially distributing servers which provide a common OS function.

Like Tornado [11] and K42 [5], fos explores how to parallelize microkernel-based OS data structures. They are differentiated from fos in that they require SMP and NUMA shared memory machines instead of loosely coupled massively multicore machines and clouds of multicores. Also, fos targets a much larger scale of machine than Tornado/K42. Corey [7] OS shares the spatial awareness aspect of fos, but does not address parallelization within a system server and focuses on smaller configuration systems. Also, fos is tackling many of the same problems as Barrelfish [6] but fos is focusing more on how to parallelize the system servers as well as addresses the scalability on chip and in the cloud.

Disco [8] and Cellular Disco [13] run multiple cooperating virtual machines on a single multiprocessor system. fos's spatial distribution of fleet resources is similar to the way that different VM system services communicate within Cellular Disco. Disco and Cellular Disco argue leveraging traditional OSs as an advantage, but this approach likely does not reach the highest level of scalability as a purpose built scalable OS such as fos will. Also, the fixed boundaries imposed by VM boundaries can impair dynamic resource allocation.

fos bears much similarity to distributed OSes such as Amoeba [22], Sprite [17], and Clouds [10]. One major difference is that fos communication costs are much lower when executing on a single massive multicore, and the communication reliability is much higher. Also, when fos is executing on the cloud, the trust model and fault model is different than previous distributed OSes where much of the computation took place on student's desktop machines.

The manner in which fos parallelizes system services into fleets of cooperating servers is inspired by distributed Internet services. For instance, load balancing is one technique taken from clustered webservers. The name server of fos derives inspiration from the hierarchical caching in the Internet's DNS system. In future work, we hope to leverage many distributed shared state techniques such as those in peer-to-peer and distributed hash tables such as Bit Torrent [9] and Chord [21]. fos also takes inspiration from distributed services such as distributed file systems such as AFS [19], OceanStore [15] and the Google File System [12].

fos differs from existing cloud computing solutions in several aspects. Cloud (*IaaS*) systems, such as Amazon's Elastic compute cloud (EC2) [1] and VMWare's VCloud, provide computing resources in the form of virtual machine (VM) instances and Linux kernel images. fos builds on top

268

of these virtual machines to provide a single system image across an IaaS system. With the traditional VM approach, applications have poor control over the co-location of the communicating applications/VMs. Furthermore, IaaS systems do not provide a uniform programming model for communication or allocation of resources. Cloud aggregators such as RightScale [18] provide automatic cloud management and load balancing tools, but they are application-specific, whereas fos provides these features in an application agnostic manner.

6. CONCLUSIONS

Trends in multicore architectures point to an ever-increasing number of cores available on a single chip. A similar, independent trend can be seen in the growth of cloud computing as well. Monolithic operating systems are, however designed assuming that computation is the limiting resource. This assumption is proving incorrect as the advent of multicore and clouds is providing abundant computational resources. However, the abundance in resources is also accompanied by an unprecedented variability in demand of the system resources. In this paper, we advocated for elastic OS service and illustrated their feasibility in meeting the variable demands. We furthermore showcased a prototype elastic file system service in *fos* and illustrated its effectiveness in meeting the variable demands through experimental results.

7. REFERENCES

[1] Amazon Elastic Compute Cloud (Amazon EC2), 2009. http://aws.amazon.com/ec2/.

[2] Tilera Announces the World's First 100-core Processor with the New TILE-Gx Family, Oct. 2009. http://www.tilera.com/.

[3] AMD Opteron 6000 Series Press Release, Mar. 2010. http://www.amd.com/us/press-releases/Pages/amd-sets-the-new-standard-29mar2010.aspx.

[4] M. Accetta, R. Baron, W. Bolosky, D. Golub, R. Rashid, A. Tevanian, and M. Young. Mach: A new kernel foundation for UNIX development. In *Proceedings of the USENIX Summer Conference*, pages 93–113, June 1986.

[5] J. Appavoo, M. Auslander, M. Burtico, D. M. da Silva, O. Krieger, M. F. Mergen, M. Ostrowski, B. Rosenburg, R. W. Wisniewski, and J. Xenidis. K42: an open-source linux-compatible scalable operating system kernel. *IBM Systems Journal*, 44(2):427–440, 2005.

[6] A. Baumann, P. Barham, P.-E. Dagand, T. Harris, R. Isaacs, S. Peter, T. Roscoe, A. Schüpbach, and A. Singhania. The multikernel: a new OS architecture for scalable multicore systems. In *SOSP '09: Proceedings of the ACM SIGOPS 22nd symposium on Operating systems principles*, pages 29–44, 2009.

[7] S. Boyd-Wickizer, H. Chen, R. Chen, Y. Mao, F. Kaashoek, R. Morris, A. Pesterev, L. Stein, M. Wu, Y. D. Y. Zhang, and Z. Zhang. Corey: An operating system for many cores. In *Proceedings of the Symposium on Operating Systems Design and Implementation*, Dec. 2008.

[8] E. Bugnion, S. Devine, and M. Rosenblum. Disco: Running commodity operating systems on scalable multiprocessors. In *Proceedings of the ACM*

Symposium on Operating System Principles, pages 143–156, 1997.

[9] B. Cohen. Incentives build robustness in bittorrent, 2003.

[10] P. Dasgupta, R. Chen, S. Menon, M. Pearson, R. Ananthanarayanan, U. Ramachandran, M. Ahamad, R. J. LeBlanc, W. Applebe, J. M. Bernabeu-Auban, P. Hutto, M. Khalidi, and C. J. Wileknloh. The design and implementation of the Clouds distributed operating system. *USENIX Computing Systems Journal*, 3(1):11–46, 1990.

[11] B. Gamsa, O. Krieger, J. Appavoo, and M. Stumm. Tornado: Maximizing locality and concurrency in a shared memory multiprocessor operating system. In *Proceedings of the Symposium on Operating Systems Design and Implementation*, pages 87–100, Feb. 1999.

[12] S. Ghemawat, H. Gobioff, and S.-T. Leung. The Google file system. In *Proceedings of the ACM Symposium on Operating System Principles*, Oct. 2003.

[13] K. Govil, D. Teodosiu, Y. Huang, and M. Rosenblum. Cellular Disco: Resource management using virtual clusters on shared-memory multiprocessors. In *Proceedings of the ACM Symposium on Operating System Principles*, pages 154–169, 1999.

[14] J. Howard, S. Dighe, Y. Hoskote, S. Vangal, D. Finan, G. Ruhl, D. Jenkins, H. Wilson, N. Borkar, G. Schrom, F. Pailet, S. Jain, T. Jacob, S. Yada, S. Marella, P. Salihundam, V. Erraguntla, M. Konow, M. Riepen, G. Droege, J. Lindemann, M. Gries, T. Apel, K. Henriss, T. Lund-Larsen, S. Steibl, S. Borkar, V. De, R. Van Der Wijngaart, and T. Mattson. A 48-core ia-32 message-passing processor with dvfs in 45nm cmos. In *Solid-State Circuits Conference Digest of Technical Papers (ISSCC), 2010 IEEE International*, pages 108 –109, 7-11 2010.

[15] J. Kubiatowicz, D. Bindel, Y. Chen, S. Czerwinski, P. Eaton, D. Geels, R. Gummadi, S. Rhea, H. Weatherspoon, W. Weimer, C. Wells, and B. Zhao. Oceanstore: An architecture for global-scale persistent storage. In *Proceedings of the Conference on Architectural Support for Programming Languages and Operating Systems*, pages 190–201, Nov. 2000.

[16] J. Liedtke. On microkernel construction. In *Proceedings of the ACM Symposium on Operating System Principles*, pages 237–250, Dec. 1995.

[17] J. K. Ousterhout, A. R. Cherenson, F. Douglis, M. N. Nelson, and B. B. Welch. The Sprite network operating system. *IEEE Computer*, 21(2):23–36, Feb. 1988.

[18] Rightscale home page. http://www.rightscale.com/.

[19] M. Satyanarayanan. Scalable, secure, and highly available distributed file access. *IEEE Computer*, 23(5):9–18,20–21, May 1990.

[20] L. Seiler, D. Carmean, E. Sprangle, T. Forsyth, M. Abrash, P. Dubey, S. Junkins, A. Lake, J. Sugerman, R. Cavin, R. Espasa, E. Grochowski, T. Juan, and P. Hanrahan. Larrabee: A many-core x86 architecture for visual computing. In *SIGGRAPH '08: ACM SIGGRAPH 2008 papers*, pages 1–15, New York, NY, USA, 2008. ACM.

[21] I. Stoica, R. Morris, D. Karger, M. F. Kaashoek, and

H. Balakrishnan. Chord: A scalable peer-to-peer lookup service for internet applications. In *SIGCOMM '01: Proceedings of the 2001 conference on Applications, technologies, architectures, and protocols for computer communications*, pages 149–160, New York, NY, USA, 2001. ACM.

[22] A. S. Tanenbaum, S. J. Mullender, and R. van Renesse. Using sparse capabilities in a distributed operating system. In *Proceedings of the International Conference on Distributed Computing Systems*, pages 558–563, May 1986.

[23] D. Wentzlaff and A. Agarwal. Factored operating systems (fos): the case for a scalable operating system for multicores. *SIGOPS Oper. Syst. Rev.*, 43(2):76–85, 2009.

[24] D. Wentzlaff, P. Griffin, H. Hoffmann, L. Bao, B. Edwards, C. Ramey, M. Mattina, C.-C. Miao, J. F. Brown III, and A. Agarwal. On-chip interconnection architecture of the Tile Processor. *IEEE Micro*, 27(5):15–31, Sept. 2007.

[25] D. Wentzlaff, C. Gruenwald, III, N. Beckmann, K. Modzelewski, A. Belay, L. Youseff, J. Miller, and A. Agarwal. An operating system for multicore and clouds: mechanisms and implementation. In *Proceedings of the 1st ACM symposium on Cloud computing*, SoCC '10, pages 3–14, New York, NY, USA, 2010. ACM.

A Compiler and Runtime for Heterogeneous Computing

Joshua Auerbach   David F. Bacon   Ioana Burcea   Perry Cheng

Stephen J. Fink   Rodric Rabbah   Sunil Shukla

IBM Thomas J. Watson Research Center

{josh, dfb, ioana, perry, sjfink, rabbah, skshukla}@us.ibm.com

ABSTRACT

Heterogeneous systems show a lot of promise for extracting high-performance by combining the benefits of conventional architectures with specialized accelerators in the form of graphics processors (GPUs) and reconfigurable hardware (FPGAs). Extracting this performance often entails programming in disparate languages and models, making it hard for a programmer to work equally well on all aspects of an application. Further, relatively little attention is paid to *co-execution*—the problem of orchestrating program execution using multiple distinct computational elements that work seamlessly together.

We present Liquid Metal, a comprehensive compiler and runtime system for a new programming language called Lime. Our work enables the use of a single language for programming heterogeneous computing platforms, and the seamless co-execution of the resultant programs on CPUs and accelerators that include GPUs and FPGAs.

We have developed a number of Lime applications, and successfully compiled some of these for co-execution on various GPU and FPGA enabled architectures. Our experience so far leads us to believe the Liquid Metal approach is promising and can make the computational power of heterogeneous architectures more easily accessible to mainstream programmers.

Categories and Subject Descriptors

D.3.3 [**Programming Languages**]: Language Constructs and Features; B.6.3 [**Logic design**]: Design aids

General Terms

Design, Languages

Keywords

Heterogeneous, GPU, FPGA, Streaming, Java

1. INTRODUCTION

The mixture of computational elements that make up a processor is increasingly heterogeneous. There are already processors that couple a conventional general purpose CPU with a graphics processor (GPU), and there is increasing use of the GPU for more than graphics processing because of its exceptional computational abilities at particular problems. In this way, a computer with a CPU and a GPU is heterogeneous because it offers a specialized computational element (the GPU) for computational tasks that suit its architecture, and a truly general purpose computational element (the CPU) for all other tasks (*e.g.*, including if needed the computational tasks that are well suited for the GPU). The GPU is an example of a hardware accelerator. Other forms of hardware accelerators are gaining wider consideration, including field programmable gate arrays (FPGAs) and fixed-function accelerators.

Programming technologies exist for CPUs, GPUs, FPGAs, and various accelerators in isolation. For example, programming languages for a GPU include OpenMP, OpenCL [5], and CUDA, all of which can be viewed as extensions of the C programming language. A GPU-specific compiler inputs a program written in one of these languages, and preprocesses the program to separate the GPU-specific code (hereinafter referred to as *device code*) from the remaining program code (hereinafter referred to as the *host code*). The device code is typically recognized by the presence of explicit device-specific language extensions or compiler directives (*e.g.*, pragma), or syntax (*e.g.*, kernel launch with <<<...>>> in CUDA). The device code is further translated and compiled into device-specific machine code (hereinafter referred to as an *artifact*). The host code is modified as part of the compilation process to invoke the device artifact when the program executes. The device artifact may either be embedded into the host machine code, or it may exist in a repository and identified via a unique identifier that is part of the invocation process. Programming solutions for heterogeneous computers that include FPGAs are comparable to GPU programming solutions. However, FPGAs do not enjoy the benefits of a widely accepted C dialect yet.

Regardless of the heterogeneous mix of processing elements in a computer, the programming methodology to date is generally similar and shares the following characteristics. First, the disparate languages or dialects in which different architectures must be programmed make it hard for a single programmer or programming team to work equally well on all aspects of a project. Second, relatively little attention is paid to *co-execution*—the problem of orchestrating a program execution using multiple distinct computational elements that work seamlessly together. Co-execution requires (1) partitioning a program into tasks that can be mapped to the computational elements, (2) scheduling the tasks onto the computational elements, and (3) handling the communication between computational elements, which in itself involves serializing data and preparing it for transmission, routing data between processors, and receiving and deserializing data. Given the complexities associated with orchestrating the execution of a program on a heterogeneous computer, a very early static decision must be made on what will execute where, a decision that is complex and also costly to revisit as a project evolves. This is exacerbated by the fact that some of the accelerators, for example, FPGAs, are difficult to program well and place a heavy engineering burden on programmers.

This paper presents Liquid Metal, a comprehensive compiler infrastructure and runtime system that together enable the use of a single language for programming computing systems with heterogeneous accelerators, and the co-execution of the resultant programs on such architectures. Using Liquid Metal, a program is lowered into an intermediate representation that describes the computation as independent but interconnected computational nodes. Liquid Metal then gives a series of quasi-independent backend compilers a chance to compile one or more groups of computational nodes for different target architectures. Most backend compilers are under no obligation to compile everything. However, the CPU compiler always compiles the entire program, guaranteeing that every node has at least one implementation.

The result of a compilation with Liquid Metal is a collection of artifacts for different architectures, each labeled with the particular computational node that it implements. As a consequence, the runtime can choose from a large number of functionally-equivalent configurations depending on which nodes are activated during execution. In this way, Liquid Metal solves the problem of a premature static partitioning of an application by offering a dynamic partitioning of the program across available processing elements, targeting CPUs, GPUs, and FPGAs. The advantage of runtime-partitioning is that it is not permanent, and it may adapt to changes in the program workloads, program phase changes, availability of resources, and other fine-grained dynamic features.

Liquid Metal source files are programmed using a new language called Lime [1] which is a Java-compatible object-oriented language, with new features that make it feasible to (1) define a unit of migration between CPUs and accelerators, (2) statically compile efficient code for the various accelerators in a given heterogeneous system, and (3) orchestrate the dataflow across the system. We have developed a number of benchmarks in Lime and successfully compiled and run these on CPU+GPU and CPU+FPGA systems. We present an overview of Lime in the following section, with an emphasis on those particular language features that facilitate application development for heterogeneous architectures. Then we describe the Liquid Metal compiler and runtime system, and highlight a development flow using the Lime. We conclude with related work and future directions.

2. LIME LANGUAGE OVERVIEW

The foundation for the Liquid Metal compiler and runtime is a new language called Lime, a general purpose object-oriented language with functional and stream-oriented programming features. Lime is Java-compatible and, as a consequence, it enables an evolutionary migration of existing Java code to Lime, exploiting the new language features where most profitable first. Lime is designed to ease the programming burden for heterogeneous architectures via (1) strong isolation and (2) abstract parallelism with explicit communication and computation operators.

2.1 Strong Isolation

Strong isolation allows the Liquid Metal runtime to safely migrate computation between heterogeneous processors. Isolation is realized using a combination of orthogonal language features called *value* and *local*. The former is a type modifier which declares that any instance of the type is recursively immutable. That is, a value

```
01 public value enum bit {
02     zero, one;
03     public bit ~ this {
04         return this == zero ? one : zero;
05     }
06 }

07 public class Bitflip {
08     local static bit flip(bit b) {
09         return ~b;
10     }
11     local static bit[[]] mapFlip(bit[[]] input) {
12         var flipped = Bitflip @ flip(input);
13         return flipped;
14     }
15     static bit[[]] taskFlip(bit[[]] input) {
16         bit[] result = new bit[input.length];
17         var flipit =  input.source(1)
18                   => ([ task flip ])
19                   => result.<bit>sink();
20         flipit.finish();
21         return new bit[[]](result);
22     }
23 }
```

Figure 1: Lime examples.

cannot mutate once it is created. All primitive types in Lime (*e.g.*, int, float) are also values. The local modifier is attributed to method declarations, and requires that local methods only call other local methods and access fields that are either reachable from the method arguments, fields of the object instance itself (if the method is not static), or compile-time constants.

An example of a value type is shown in Figure 1 lines (1-6). The type bit is an enumeration of two values zero and one. Unlike Java enumerations which are mutable, this type is declared immutable using the value keyword. The methods of a value type are local by default. In contrast, the method flip in Figure 1 line 8 is explicitly declared local because Bitflip is not a value class and its methods are *global* by default. A global method may perform side-effecting operations, including I/O.

The value and local modifiers provide practical invariants for the Liquid Metal compiler and runtime to rely on and exploit. Namely, a local method whose arguments are values is pure (*i.e.*, stateless) if it is either a static method (*e.g.*, flip) or an instance method of a value type (*e.g.*, bit ~). Pure methods are candidates for migration between processors in a heterogeneous architecture because they provide implementation freedom. On a GPU or CPU, the compiler and runtime can invoke threads as needed to apply pure methods and scale-up throughput. Similarly on an FPGA where methods may be embodied as modules, the compiler may instantiate modules as needed to scale-up throughput.

Stateful instance methods are also candidates for co-execution if they are local and the object instance is constructed using an isolating constructor: a local constructor with value arguments. Unlike pure methods which provide *data-parallelism*, stateful methods require the exploitation of *pipeline-parallelism* to achieve higher throughput. Lime provides additional language features to facilitate the expression of data and pipeline parallelism.

2.2 Abstract Parallelism

There are two primary ways of expressing parallelism in Lime. These are illustrated in the methods mapFlip and taskFlip in Figure 1. The first is an example using the Lime *map* operator "@" (line 12) which, in this case, applies the method flip (lines 8-10) to every element of the input array to produce a new bit array of the same size. The argument to the method mapFlip is a value array (denoted using [[]]) and hence its elements are read-only and cannot be assigned. One may call the method using a bit array or bit literal. Lime provides bit literals as syntactic sugar to compactly define bit arrays because of their prevalence in FPGA designs. For example the bit literal `100b` is a 3-bit array where `bit[0]=0` and `bit[2]=1`. The result of `mapFlip(100b)` is a bit array equal to the bit literal `001b`.

When the map operator is a `local static` method applied to `value` arguments, the compiler can infer data-parallelism and optimize the implementation accordingly. In Liquid Metal, the map and *reduce* (not shown) operators are exploited heavily for optimizing code for co-execution on a GPU. We achieved end-to-end speedups of $12\times-431\times$ for a number of benchmarks co-executing between CPU and GPU using an NVidia GTX580 (Fermi) [3].

Pipeline-parallelism in Lime is expressed using explicit operators to create tasks (*i.e.*, pipeline stages) and connect them so data flows between them. A 3-stage pipeline is shown on lines 17-19 in Figure 1 as an example. The first stage is a "source" task that produces a *stream* of bits, one bit at a time. The second stage applies the `flip` method which flips one bit at a time. The third and final "sink" stage accumulates the bits into a new bit array, one bit at a time. The source and sink tasks use utility methods provided by Lime array types. The flip task (line 18) explicitly uses the Lime `task` operator. When applied to a static method as in this case, the result is a dataflow actor that repeatedly applies the named method. The actor consumes data from its input port and produces data to its output stream, applying the named method when the port contains sufficient data to satisfy the argument requirements of the method. The connect operator "`=>`" connects tasks so values flow between them. Hence, every bit produced by the source (line 17) flows to the flip task (line 18) and the result of the flip task flows to the sink task (line 19). Connected tasks are called *task graphs* in Lime.

Only values may flow between tasks. This restriction, which is enforced by the Lime type system, guarantees that data that cross physical boundaries in a heterogeneous system cannot mutate in flight. Hence, the data may be marshaled on one end and unmarshaled on the other without concern for data-races. Furthermore, values are cycle-free and may be marshaled using custom strategies that are tailored to the physical wire-format of the system.

Lime tasks which are either source or sink nodes in the task graph (*i.e.*, have no input connections or output connections, respectively) are allowed to perform I/O and may have side-effects. In contrast, inner tasks, called *filters*, must be strongly isolated in that the task operator can only be applied to a `local` method with `value` arguments. It is the inner tasks that are usually migrated from the CPU and co-executed on accelerators.

Task graph construction is separated from task graph execution. The task graph construction does not cause any of the actors in the graph to execute. Instead, Lime requires an explicit operation to cause the actors to execute. This is accomplished using a `start()` or `finish()` method on tasks (line 20). The latter causes the execution to start and blocks the caller until computation has terminated. In the example, the graph execution terminates when the last bit produced by the source is consumed by the sink.

2.3 Task Relocation and Co-Execution

Lime requires the use of *relocation brackets* (`[ ]`) around task expressions (Figure 1, line 18) in order to inform the compiler and runtime of the programmer's desire to co-execute tasks. When omitted, the compiler and runtime provide no guarantees that the task graph contains any co-executable regions.

The relocation brackets provide a lightweight and convenient mechanism for a programmer to experiment with many different partitions between device code and host code without perturbing the rest of their code. In addition, since the methods that a task

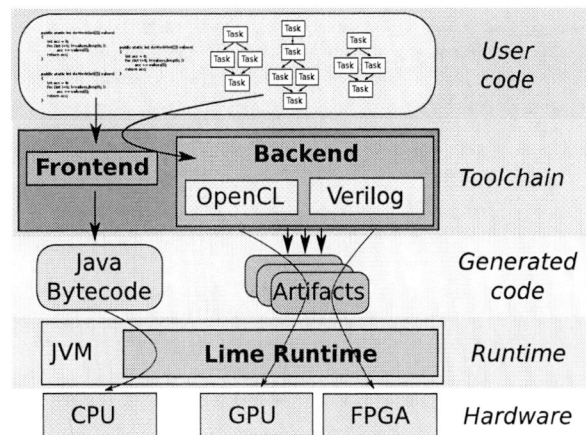

Figure 2: Liquid Metal compiler & runtime.

operator acts on may be used seamlessly in a number of contexts (*e.g.*, map/reduce, task graph, or imperative code), the programmer can develop and debug their code in a single semantic domain and reuse much of their existing code. We believe that programmers will favor using the Lime `value` and `local` modifiers because they are non-intrusive, provide general soundness guarantees that a programmer may wish to assert, and most notably, when combined with task and map/reduce operators, provide a path for exploiting heterogeneous architectures.

3. COMPILING FOR HETEROGENEITY

Figure 2 presents an overview of our compilation and runtime toolchain. Liquid Metal accepts a set of source files and produces artifacts for execution. An artifact is an executable entity that may correspond to either the entire program (as is the case with the bytecode generation) or its subsets (as is the case with the OpenCL/GPU and Verilog/FPGA backends). An artifact is packaged in such a way that it can be replaced at runtime with another artifact that is its semantic equivalent.

The compiler frontend performs shallow optimizations and generates Java bytecode for executing the entire program in a Java virtual machine (JVM). The compiler backend generates code for GPUs and FPGAs. The backend operates on a subset of the input program, focusing solely on compiling Lime task graphs.

The backend consists of architecture-specific *device* compilers; currently, a GPU compiler and an FPGA compiler. The former generates OpenCL for the GPU, while the latter generates Verilog for the FPGA. These are subsequently compiled using device-specific toolflows to complete the artifact generation for each accelerator.

The compiler relies on the presence of relocation brackets around task graphs to learn of the tasks it must compile. In general, task graphs can be constructed using rich control flow (*e.g.*, iterative, recursive). The compiler discovers the shape and other properties of these task graphs statically. As expected, compile-time analysis may not discover *all* possible task graphs that the program might build. If the relocation brackets are present and the compiler fails to determine the shape of the task graph, the programmer is informed at compile time with an appropriate error message. The benchmarks we have developed so far use task construction idioms that our compiler can recognize. We believe these benchmarks are written in a style that is natural to programmers.

Each of the device compilers operates autonomously and independent of the other compilers. It examines the tasks that make up each task graph and decides whether the code that comprises the tasks is suitable for the device. A task containing language constructs that are not suitable for the device is *excluded* from further compilation by that backend. The device compiler produces artifacts for the task (sub)graphs that are not subject to exclusion.

The frontend and backend compilers cooperate to produce a *manifest* describing each generated artifact and labeling it with a unique task identifier. The set of manifests and artifacts are used by the runtime system in determining which task implementations are semantically equivalent to each other. The same task identifiers are incorporated into the final phase of bytecode compilation such that the task graph construction in the Lime program can pass along these identifiers to the runtime when the program is executed.

4. LIQUID METAL RUNTIME

The Lime language runtime is implemented primarily in Java and runs in any modern JVM. It constructs task graphs at run time, elects which implementations of a task to use from the available artifacts, performs task scheduling, and if necessary, orchestrates marshaling and communication with native artifacts. Its structure is best understood by focusing on its phases of execution.

4.1 Task Graph Construction

The runtime contains a class for every distinct kind of task that can arise in the Lime language (*e.g.*, sources, sinks, filters). When the bytecode compiler produces the code for a task, it generates an object allocation for one of these runtime classes. A *connect* operation "=>" creates a FIFO queue between tasks. When the program executes, the task creation and connection operators are reflected in an actual graph of runtime objects. The generated bytecode that instantiates a task also provides the runtime with the unique identifiers that were generated by the backend compilers to advertise the set of available task artifacts.

The start() method, when called on any task graph, activates the scheduling portion of the runtime. Assuming for the moment that the tasks in the graph only have bytecode artifacts, the runtime creates a thread for each task. These threads will block on the incoming connections until enough data is available for the task to produce data on its outgoing connection.

4.2 Task Substitution

The runtime learns about the tasks in the graph that have non-bytecode artifacts while examining the task graph during construction, since the unique identifiers of tasks, which are stored in the task runtime objects, can be looked up efficiently in the *artifact store* populated by the backends. For each task (sub)graph that has an alternative implementation, the runtime is in a position to perform a substitution. At present, the runtime algorithm for doing this substitution is primitive: it prefers a larger substitution to a smaller one. It also favors GPU and FPGA artifacts to bytecode although that choice can be manually directed as well. A more sophisticated algorithm that accounts for communication costs, performs dynamic migration, or runtime adaptation is left to future work. Performing the actual substitution involves the native connection and marshaling issues described in the next section. Substitution precedes the creation and starting of threads as described in the previous section but is otherwise transparent to most of the runtime.

4.3 Native Connections and Marshaling

Although the data and code isolation of Lime tasks makes computation offloading possible, the runtime system must still efficiently transfer data between the main system memory and the device. Because the runtime supports a number of disparate accelerators that

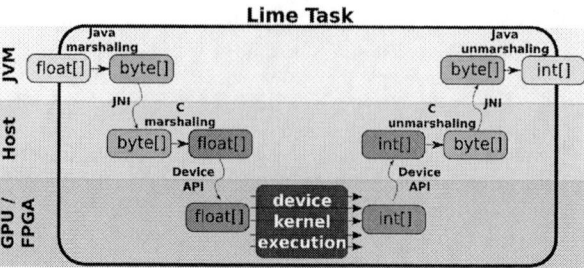

Figure 3: Data transfer between JVM and native device using a float array as input and int array as output.

include GPUs and FPGAs, the runtime implementation adopts a universal "wire" format that relies only on sending a byte stream as shown in Figure 3.

The communication steps between the host JVM and the native device entail (1) serializing a Lime value to a byte array, (2) crossing the JNI boundary, and (3) converting this byte array into a C-style value. The return path is a mirror image in which we convert the native data structure to a byte array, return from the JNI call, and then deserialize from the byte array back into a heap-resident value.

During the task substitution process, the runtime will find a custom serializer based on the task I/O data type. Marshaling on the C side is similar but more specialized because the data is generally densely packed.

Our current design affords a common format as a starting point for a communication layer that supports heterogeneous devices. One might further optimize the protocol by creating specific communication channels so that the sender and receiver are aware of the data format the other party desires. Going even further, one might be able to avoid a low-level memory copy by pinning memory pages and managing memory explicitly. However, these changes come at the cost of OS and JVM portability.

5. DESIGN FLOW USING LIME

In addition to the Liquid Metal compiler and runtime, we developed an interactive development environment (IDE) to aid Lime developers design and implement their applications. The IDE is an Eclipse plugin and provides a number of features that a typical Java developer is accustomed to. Among these is a package explorer, seen in the leftmost panel in Figure 4, and a class outline in the rightmost panel. The Bitflip Lime class is open in the editor (this is the same class used in an earlier part of the paper, albeit here it is more complete). Syntax highlighting is visible in the figure as well. In addition, the reader may notice a green underline at line 18 and a round marker to the left of the line number. These are Lime-specific markers indicating that the compiler generated a device artifact for the corresponding task in the relocation brackets. The console panel toward the bottom of the screen shot shows the result of executing the program, which has emitted two rows of bit values.

The panel labeled "Run Configuration", right of center in the screen shot, shows how a Lime developer interacts with the compiler and runtime to generate artifacts for co-execution. It is currently possible to compile Lime tasks to generate artifacts for native binaries, GPUs and FPGAs. In the case of native binaries, the compiler generates C code and builds shared libraries that are dynamically loaded by the Liquid Metal runtime to co-execute with the remaining Lime bytecodes. For GPUs, the compiler mostly follows the same flow as with native binaries, but the tasks are com-

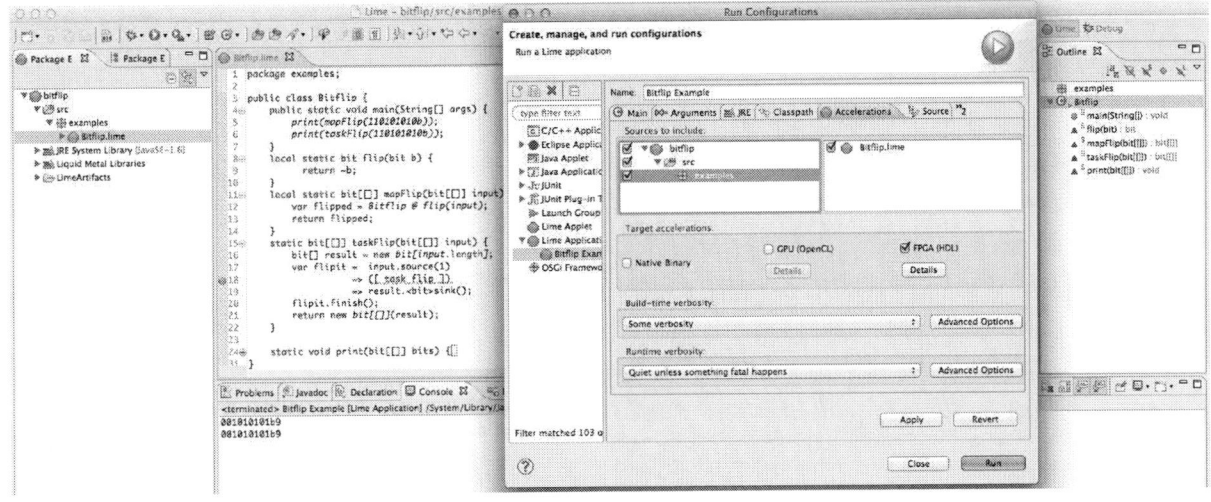

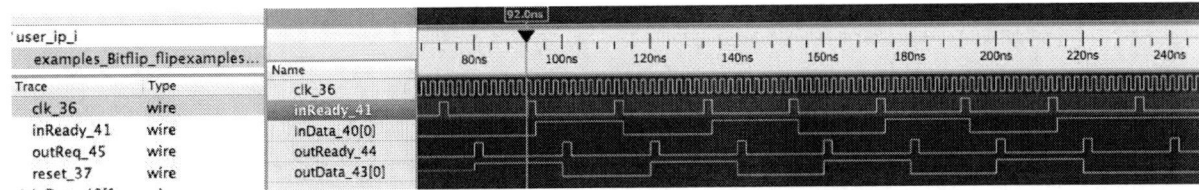

Figure 4: Liquid Metal IDE (top) and co-execution CPU+FPGA simulator (bottom).

piled to OpenCL kernels. Lastly for FPGAs, the compiler generates Verilog and then drives FPGA-specific logic synthesis flows to generate bitfiles that may be loaded into the FPGA at runtime for co-execution. However, it is also possible to simulate the Verilog alongside the Liquid Metal runtime using a Verilog/VHDL simulator.

A typical design flow starts off with the developer writing Lime code that describes the computation, then introducing map/reduce operators and task-graphs, verifying correctness (in bytecodes) using normal debugging methodologies (*e.g.*, using the Eclipse interactive symbolic debugger), and lastly introducing relocation brackets to invoke the backend device compilers. For FPGA development where the Lime source code may affect the behavioral synthesis in a number of ways, it is useful for the developer to rapidly iterate over their designs, and access low-level timing results from a simulator so they may improve their code and achieve better performance. This is especially important at this point in time because our FPGA backend is a work in progress. The bottom half of Figure 4 shows the result of a co-execution between the Liquid Metal runtime and a Verilog simulator, visualized using a waveform viewer.

The waveform captures the entire execution of the `taskFlip` task graph. The example is driven with 9 input bits (line 9 in the editor). In the waveform, these are represented by the 9 transitions on the `inReady` signal. The generated logic uses a FIFO which produces a value on the next rising edge of the clock. This is most clearly visible starting at the clock cursor at 92ns (`input[1]=1`). The `inReady` signal is asserted and `inData[0]` is high one cycle later. Another three cycles later, the output of the module is ready as indicated by the `outReady` signal. In this case, the module I/O is not fully pipelined and the computation essentially consists of one cycle to read, one cycle to compute, and one cycle to publish the result.

6. RELATED WORK

The work described in this paper is an entirely new compilation infrastructure for Lime that includes bytecode, OpenCL and Verilog code generation. This is in contrast to our previous work [4] which compiled a much smaller language lacking explicit task-graph features or map/reduce operations. The current work also includes a runtime that enables co-execution for the generated artifacts on GPUs and FPGAs. The previous paper does not discuss runtime replacement or the kind of compilation methodology described in this paper.

Several research projects address the programming burdens of heterogeneous systems. For example, SoC-C [8] is an extension to the C language that allows the programmer to manage distributed memory, express pipeline parallelism and map different tasks to resources. EXOCHI [12], a research effort from Intel, describes a runtime for integrating accelerators with general purpose processors with shared memory and dynamic allocation of tasks to accelerators. The Quilin system [7] provides a C API that can be used to write parallel programs and an adaptable runtime that dynamically maps computations to processing elements in a CPU+GPU system. Accelerator [10] provides a C# API library that uses a data-parallel model based on parallel arrays to program GPUs. Elastic computing [13] is another approach to support heterogeneous systems where a library of functions is built with several implementations (for different devices or algorithms). During an installation phase, the system profiles each implementation and at runtime, the profile is used to determine the best implementation to use based on the input parameters.

Liquid Metal differs from these approaches in a number of ways: the compiler and runtime are founded upon a new language called Lime. Unlike its C counterparts, Lime is much richer, unifies several forms of parallelism, and makes the computation and communication explicit. All of the programming features in support of heterogeneous computing are first-class language constructs, and as such, they are checkable, provide strong static guarantees, and empower the compiler to perform simple yet highly effective optimizations [3]. Furthermore, the relocation brackets in Lime provide a convenient and lightweight feature for a programmer to experiment with many different program partitions without specializing the bulk of their code for a particular style of computation. While the current Liquid Metal runtime does not employ intelligent heuristics for partitioning and mapping, we view much of the existing work on this particular topic to be complimentary to ours. Our work has focused on identifying a robust unit of migration and co-execution (*e.g.*, task-graphs), and exposing the computation and communication in a way that facilitates compilation and co-execution to multiple targets.

OpenCL [5] provides an interesting comparison point to Liquid Metal. While OpenCL is largely used for programming GPUs today, there is significant interest in extending its reach to FPGAs as well. The main advantage of our work is that Lime was specifically designed with synthesis to FPGAs in mind. Hence, bit-data types are first class, there are no predefined data types that imply a particular vector-width, and complicated synchronization and explicit communication orchestration is not required. The Liquid Metal compiler and runtime automatically generate the synchronization and communication code for GPUs and FPGAs as needed, thus freeing the programmer from tedious and error prone coding.

Compilation of restricted subsets of C or C++ to FPGA code is common (see [2] for a recent survey). There is arguably some degree of co-execution supported by these tools in that they often generate a "testbench" in addition to the FPGA artifact. However, these tools are not oriented toward building complete applications in a co-execution style, and often the extensions for FPGA programming are not well suited for GPU programming.

Virgil [11, 6] is an object-oriented language originally designed for micro-controllers, but unlike Java or Lime, it restricts object creation to an initialization phase. The Kiwi [9] project synthesizes FPGA code for selected library components in C#. These efforts do not emphasize co-execution, rather, they generate entire designs to run wholly in the FPGA.

7. CURRENT AND FUTURE DIRECTIONS

In this paper we highlighted features of the Lime language that we expect to lighten the burden on developers who wish to exploit heterogeneous architectures. The Liquid Metal compiler and runtime leverage the properties of the language to free the programmer from the tedious partitioning of computation between host and device code, and the orchestration of communication for co-execution. Instead, a Lime developer writes their application for GPU and FPGA accelerators in a single semantic domain, uses map/reduce and task graphs to expose parallelism, and then experiments with partitioning their code for co-execution using a lightweight language mechanism, namely relocation brackets.

We have successfully developed a number of applications using Lime and run on CPU+GPU and CPU+FPGA systems. Our current system supports virtually all GPUs, and we have demonstrated significant performance gains on AMD and NVidia GPUs [3]. In contrast, because FPGAs require device-specific synthesis flows,

we currently support the following devices: PCIe attached Nallatech 280 boards with Xilinx FPGA parts, and Xilinx XUP V5 and Spartan LX9 development boards connected to the host via UART. In addition we support Verilog/VHDL simulators including NCSim and Modelsim.

Our current and future emphasis is on improving the quality of our FPGA device compiler and broadening the set of language features that are compiled to Verilog. In addition, we are exploring a number of research topics related to runtime introspection and adaptation of the task-graph partitioning so that tasks run where they are best suited. Lastly, we are exploring applications that can benefit simultaneously from CPU+GPU+FPGA co-execution.

8. REFERENCES

[1] J. Auerbach, D. F. Bacon, P. Cheng, and R. Rabbah. Lime: a Java-compatible and synthesizable language for heterogeneous architectures. In *OOPSLA*, pp. 89–108, Oct. 2010.

[2] J. M. Cardoso and P. C. Diniz. *Compilation Techniques for Reconfigurable Architectures*. Springer, 2008.

[3] C. Dubach, P. Cheng, R. Rabbah, D. F. Bacon, and S. J. Fink. Compiling a high-level language for GPUs. In *PLDI*, June 2012.

[4] S. S. Huang, A. Hormati, D. F. Bacon, and R. Rabbah. Liquid Metal: Object-oriented programming across the hardware/software boundary. In *ECOOP*, pp. 76–103. Springer-Verlag, 2008.

[5] Khronos OpenCL Working Group. *The OpenCL Specification*.

[6] S. Kou and J. Palsberg. From OO to FPGA: fitting round objects into square hardware? In *OOPSLA*, pp. 109–124. ACM, 2010.

[7] C.-K. Luk, S. Hong, and H. Kim. Qilin: exploiting parallelism on heterogeneous multiprocessors with adaptive mapping. In *MICRO*, pp. 45–55. ACM, 2009.

[8] A. D. Reid, K. Flautner, E. Grimley-Evans, and Y. Lin. SoC-C: efficient programming abstractions for heterogeneous multicore systems on chip. In *CASES*, pp. 95–104. ACM, 2008.

[9] S. Singh and D. J. Greaves. Kiwi: Synthesis of FPGA circuits from parallel programs. In *FCCM*, pp. 3–12. IEEE Computer Society, 2008.

[10] D. Tarditi, S. Puri, and J. Oglesby. Accelerator: using data parallelism to program GPUs for general-purpose uses. In *ASPLOS*, pp. 325–335. ACM, 2006.

[11] B. L. Titzer. Virgil: objects on the head of a pin. In *OOPSLA*, pp. 191–208. ACM, 2006.

[12] P. H. Wang, J. D. Collins, G. N. Chinya, H. Jiang, X. Tian, M. Girkar, N. Y. Yang, G.-Y. Lueh, and H. Wang. Exochi: architecture and programming environment for a heterogeneous multi-core multithreaded system. In *PLDI*, pp. 156–166. ACM, 2007.

[13] J. R. Wernsing and G. Stitt. Elastic computing: a framework for transparent, portable, and adaptive multi-core heterogeneous computing. In *LCTES*, pp. 115–124. ACM, 2010.

The HELIX Project: Overview and Directions

Simone Campanoni
Harvard University
Cambridge, USA
xan@eecs.harvard.edu

Timothy Jones
University of Cambridge
Cambridge, UK
timothy.jones@cl.cam.ac.uk

Glenn Holloway
Harvard University
Cambridge, USA
holloway@eecs.harvard.edu

Gu-Yeon Wei
Harvard University
Cambridge, USA
guyeon@eecs.harvard.edu

David Brooks
Harvard University
Cambridge, USA
dbrooks@eecs.harvard.edu

ABSTRACT

Parallelism has become the primary way to maximize processor performance and power efficiency. But because creating parallel programs by hand is difficult and prone to error, there is an urgent need for automatic ways of transforming conventional programs to exploit modern multicore systems. The HELIX compiler transformation is one such technique that has proven effective at parallelizing individual sequential programs automatically for a real six-core processor. We describe that transformation in the context of the broader HELIX research project, which aims to optimize the throughput of a multicore processor by coordinated changes in its architecture, its compiler, and its operating system. The goal is to make automatic parallelization mainstream in multiprogramming settings through *adaptive* algorithms for extracting and tuning thread-level parallelism.

Categories and Subject Descriptors

D.3.4 [**PROGRAMMING LANGUAGES**]: Processors—*Runtime environments*

General Terms

Performance, Languages

Keywords

Coarse grain parallelism extraction, runtime code adaptability, multiple programs

1. INTRODUCTION

By conventional definition, a "parallel program" is either expressed in terms of explicit parallel threads or tasks, or else is heavily annotated to guide compilers in mapping its data and control structures to parallel hardware. Research in recent years, however, has shown that in a very practical sense, every program is a parallel program, even one that has been designed and implemented with sequential semantics. Every long-running program

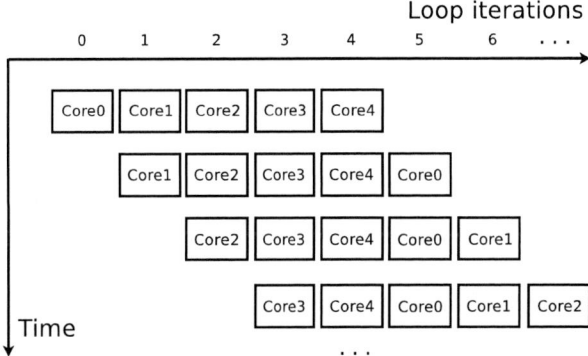

Figure 1: Cores execute loop iterations in round robin order.

depends on loops, and an increasing body of work demonstrates that automatic parallelization of loops, without help from the programmer, can lead to substantial speedup of the overall program [3, 6, 15, 10]. In fact, there have been discussions about whether parallelism should be explicit or not [2]. Since multicore microprocessors are now at the heart of devices from cell phones to supercomputers, it is important that these research demonstrations be translated soon into mainstream compilers and run-time software.

That is the goal of our HELIX project. HELIX starts with a simple idea for loop transformation: to run a loop in parallel, assign its separate iterations to separate processing elements (cores) as shown in Figure 1. In general, the cores that handle separate iterations must communicate, both to synchronize and to exchange data. So successful parallelization of a loop depends on whether the benefit of running it in parallel outweighs the communication costs. When the separate iterations are independent, or nearly so, this simple approach to loop parallelization scales well with the number of available cores.

The reason this approach has not been more widely used is that historically the cost of communication between processing elements has swamped the benefits of running in parallel. Now that a powerful multiprocessor can come on a single chip, intercore communication costs are greatly reduced, and the trend is towards even greater reduction.

To show that the HELIX loop transformation is practical on a current commodity processor, we implemented a prototype compiler that parallelizes ordinary sequential code, including programs with irregular control and data behavior. The prototype

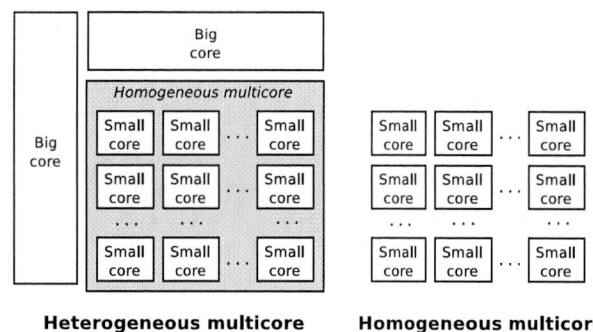

Heterogeneous multicore   **Homogeneous multicore**

Figure 2: Classification of most multicore processor designs into heterogeneous and homogeneous solutions. In both cases, the majority of cores are homogeneous.

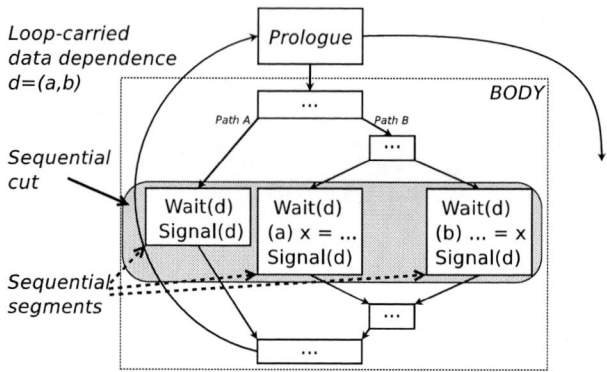

Figure 3: Code produced by the HELIX loop transformation. Depending on the execution path currently hot (path A or path B), the amount of parallelism within the loop iterations can change as the amount of code executed within sequential segments varies.

uses the memory system of the processor for communication between the hardware threads on separate cores that are executing separate loop iterations. To reduce latency, it couples each such iteration thread with a helper thread on the same core to force each inter-core signal to begin its journey (through shared cache) as soon as possible. One reason why the prototype is successful is that helper threads hide much of the cost of using the memory system for signaling.

Another reason that the HELIX prototype succeeds in producing significant overall speedups in workloads like the SPEC CPU2000 suite is that it is good at choosing which loops to parallelize and which to run in their original sequential form. It can select loops efficiently because the basic HELIX loop transformation is so simple. With the aid of a profile obtained as the program runs, HELIX can quickly estimate whether and by how much each loop will speed up if implemented in parallel. Since the speedup model accounts for the overhead of transferring program data between iteration threads, the loop selection heuristic tends to choose loops for parallelization that do not exchange much data between iterations.

Most multicore processor designs can be classified as either heterogeneous or homogeneous (see Figure 2). A homogeneous design is typically an array of relatively simple cores. Their number depends on the intended application (e.g., a sensor, a multimedia processor, an embedded system, or a commodity processor). A heterogeneous multicore processor generally augments such a homogeneous array with a small number of more complex cores, designed to run sequential code as fast as possible. HELIX is well suited to either the symmetric or asymmetric design. Parallelized loops run on the homogeneous array. If more powerful cores are also present, HELIX can use them for the parts of the program that run sequentially.

Our experience with the HELIX prototype is not intended to suggest that using the memory hierarchy and helper threads is the best way for parallel loop iterations to communicate. The experiment shows the benefit of reduced communication latency in a concrete way. Our speedup model is accurate enough to be used for predicting the effects of further overhead reduction. It shows that as ways are found to improve communication, the HELIX approach can achieve even better speedups.

One goal of the HELIX project is to investigate architectural enhancements that enable faster communication between cores. Another is to make HELIX-compiled programs more adaptable at run time. The HELIX prototype assumes that one program at a time uses the cores of its target processor, and the code of

that program does not vary at run time. In reality, programs go through phases and their utilization of parallel resources can vary markedly from phase to phase. Furthermore, contention for such resources from other programs can change with time, as can the overall power target for the enclosing system. While HELIX will still use a static compiler to parallelize programs, a lightweight run-time thread will be added that can detect program phase changes and adjust the program's use of parallel cores accordingly. This lightweight run-time will also interact with the operating system, which will be modified to help schedule HELIX processes according to system load and its user's performance/power guidelines.

Section 2 describes the HELIX prototype in more detail, and Section 3 discusses the project's future directions. Section 4 locates HELIX with respect to related research, and Section 5 draws some conclusions from project so far.

2. HELIX PROTOTYPE

By constraining communication overhead, the HELIX loop transformation makes the simple idea of spawning different loop iterations on different cores efficient. The simplicity of the approach (and the resulting generated code) allows us to define a simple and accurate model for the speedup of a given loop. The transformation chooses the most profitable loops to parallelize automatically by using this speedup model, which relies on a profile obtained using representative input (e.g., the training input of SPEC benchmarks). Parallelized loops run one at a time. The iterations of each parallelized loop run in round-robin order on the cores of a single processor. The generated code can be adapted (even at run time) to use a different number of cores just by changing the mapping between loop iterations and cores.

The following paragraphs describe how HELIX minimizes inefficiencies that arise because certain code segments (known as *sequential segments*) must be executed in loop iteration order, and because of communication overhead, including both data transfer and synchronization. Our paper describing the HELIX prototype [6] implemented in the ILDJIT compilation framework [5] contains more detail.

Parallelism extracted.

Code not related to data dependences across loop boundaries

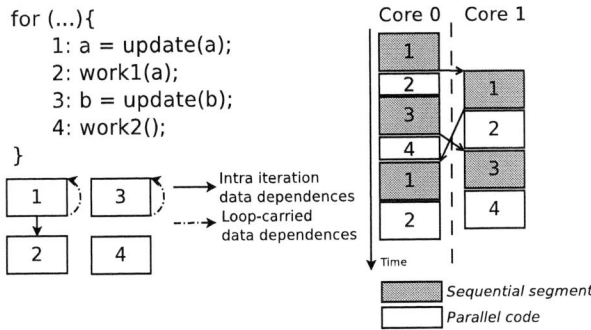

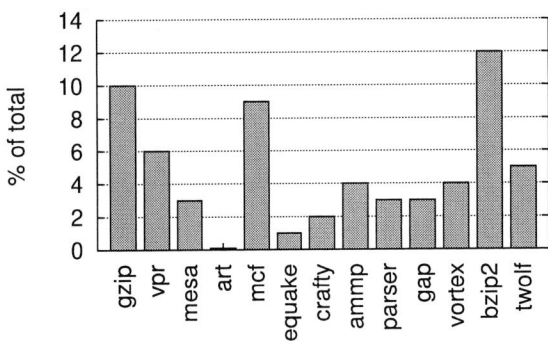

Figure 4: Execution of code produced by HELIX for a dual-core processor. Note that code blocks 1 and 3 must each be executed sequentially, but since they are independent, HELIX overlaps them in time.

Figure 7: In the loops that HELIX chooses for parallelization, the fraction of potential data transfers that must be realized is small.

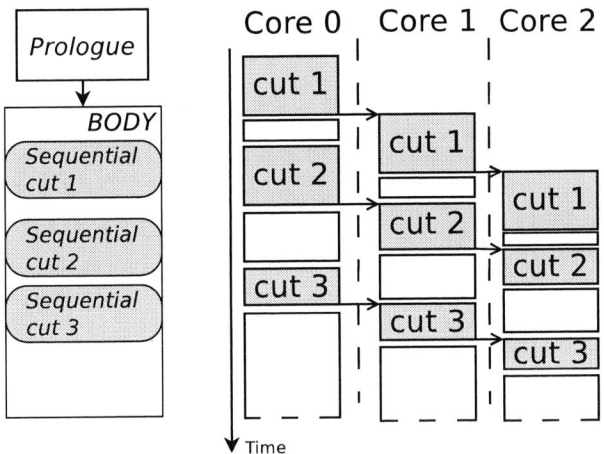

Figure 5: Sequential cuts created in the body of the loop due to loop-carried data dependences. The amount of parallelism among sequential cuts that HELIX is able to extract is shown on the right side.

is executed in parallel by different cores (code outside the sequential segments of Figure 3 and inside the white boxes of Figure 4). On the other hand, HELIX inserts code to ensure that the execution order of the remaining parts, the sequential segments, respects data dependences across loop boundaries, creating sequential cuts in the body (as shown in Figure 5) that must be traversed to end the execution of the body. The boundaries of these cuts are defined by data flow equations specifically designed for this purpose.

While the sequential segments of a given dependence must run in loop-iteration order, those of different dependences may run in parallel. HELIX executes distinct sequential segments concurrently whenever possible, as shown in Figure 4, where sequential segments 1 and 3 overlap.

Figure 6 shows the overall speedups achieved by HELIX. The geometric mean of the resulting speedups on a six core CPU is $2.25\times$, with a maximum of $4.12\times$ (for art). Our experiments use an Intel® Core™ i7-980X with six cores, each operating at 3.33 GHz, with Turbo Boost disabled. The processor has three cache levels. The first two are private to each core and are 32KB

and 256KB each. All cores share the last level 12MB cache, which is used to forward data values across cores of the same processor through the MESIF cache coherence protocol.

Communication overhead.

Execution overhead for the parallelized loops comes from two sources: data transfers and thread synchronizations.

Transferring data between threads to satisfy loop-carried data dependences is potentially a significant component of the overall overhead. However, as shown in [6] and summarized by Figure 7, in the loops that HELIX chooses for parallelization, the fraction of such potential transfers that must actually be realized is surprisingly small. In art, for example, only 0.1% of the data transfers are actually realized, which contributes to its large speedups in Figure 6.

Threads synchronize by sending signals. When a sequential segment ends, for example, it sends a signal to its successor thread to notify it that the corresponding sequential segment is free to start. HELIX minimizes the number of signals sent by exploiting redundancy among them. Moreover, as mentioned earlier, HELIX reduces the effective signal latency by exploiting the simultaneous multi-threading technology of the processor. It couples each thread that is running an iteration with a *helper thread* on the same core, whose role is to force the cache-mediated transmission of every inter-core signal to begin as soon the sending core makes it available.

3. THE FUTURE OF HELIX

The HELIX approach to parallelization aims to answer the following research questions: (i) How can microprocessors can be enhanced to speed inter-core communication? (ii) How can HELIX be extended to take advantage of faster inter-core communication? (iii) How can the parallel code produced by HELIX be used in a multiprogram scenario? (iv) How can HELIX tune the code at run time to accommodate the program's changing needs as it moves through different phases?

Changing the underlying hardware to further reduce inter-core communication overhead is the key to the success of our approach. We will extend HELIX's static compiler to take advantage of this faster inter-core communication. Moreover, we expect to extend the compiler to produce an additional lightweight run-time for each compiled program that makes it adaptable. This run-time collaborates with the OS to either acquire or re-

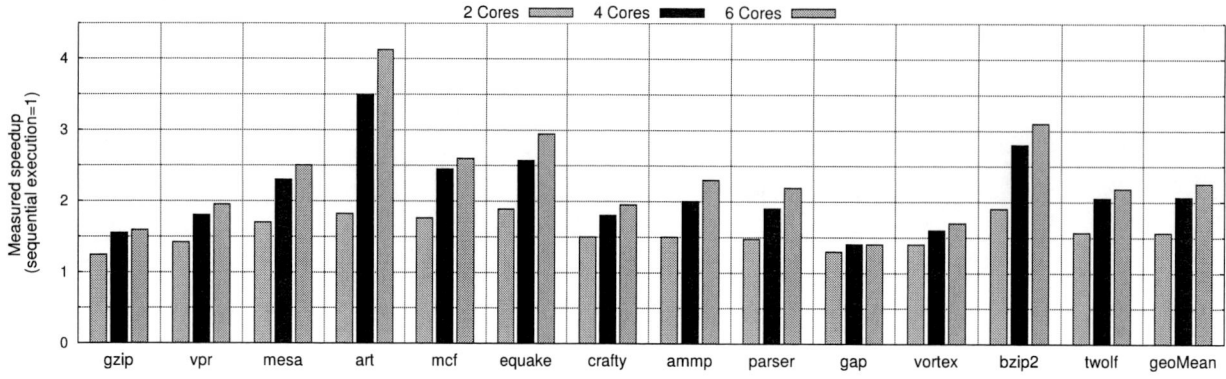

Figure 6: Speedups achieved by HELIX on a real system.

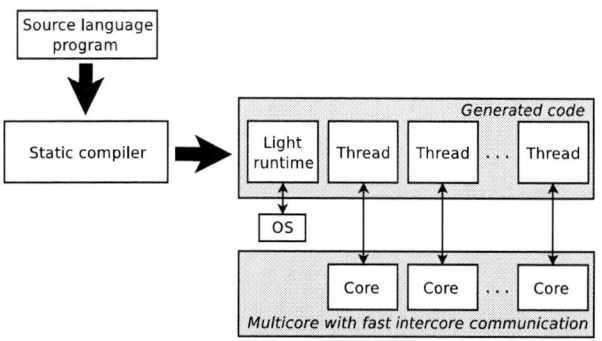

Figure 8: The main elements of the planned HELIX infrastructure and their principal interactions.

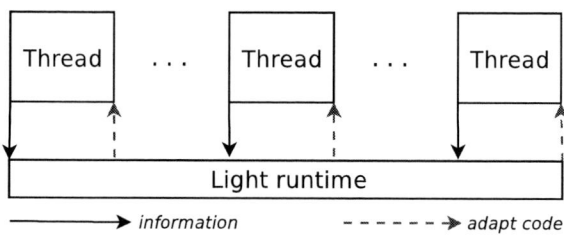

Figure 9: The run-time profiles the execution of the generated code to detect when statically defined rules match (solid lines). When they do, the run-time tunes the code by refining solutions created by the static compiler (dashed lines).

lease system resources. Through these negotiations the OS imposes constraints to balance the load of the overall system. Figure 8 shows the components just outlined.

3.1 Hardware Support

Currently, the HELIX prototype [6] targets commodity processors, which are not designed for frequent thread synchronizations. The key to increasing the number of cores that HELIX can accommodate is to reconsider the design of multicore processors. In particular, it is critical to find a small set of architectural changes that enable fast inter-core communication.

3.2 Static Compilation

The static compiler will extend the HELIX prototype [6] in two ways. First, if inter-core communication becomes faster, the profitability of parallelizing every loop changes, which makes a broader set of solutions available to the compiler. Second, the compiler will analyze the code to produce the most efficient run-time for achieving adaptability, as sketched in the next section.

3.3 Lightweight Run-Time and OS Support

Making parallelized code adaptable at run time depends on lightweight run-time support generated by the static compiler, together with help from the operating system. Code adaptability is important both for performance and for coexistence with other programs on the same system. This is because the resources an application requires and has access to can change at run time. Other researchers have also identified run time adaptation is im-

portant for achieving better overall parallelism [17]. Resource information is held by the operating system, but the compiler has the knowledge about how to best transform the program to make use of the available resources. Therefore, a run-time interaction is required which should be as unobtrusive as possible to avoid introducing overhead into the application.

Lightweight run-time.

It is well known that programs often go through different execution phases [19] that call for different optimizations. For example, the loop in Figure 3 might have a phase in which path A is taken exclusively followed by another phase where path B is taken exclusively. In the former phase, a larger fraction of the loop's execution time is spent running in parallel because path A includes less code that must execute in loop-iteration order. As this example shows, for HELIX, the best way of parallelizing a loop may depend on the phase. Moreover, since the profitability of loops can be different in different phases, loop selection is also affected.

To improve the performance of the running code, the run-time system applies a set of rules defined by the static compiler. The run-time system monitors the patterns of the rules. When it detects a match, it takes the corresponding action. A rule can be as simple as "if execution often leads to a given basic block, execute a certain loop in parallel". In this example, lightweight profiling needs to check the execution frequency of that basic block in case it becomes worth switching the loop to execute in parallel.

Figure 9 shows the use case of the lightweight run-time em-

280

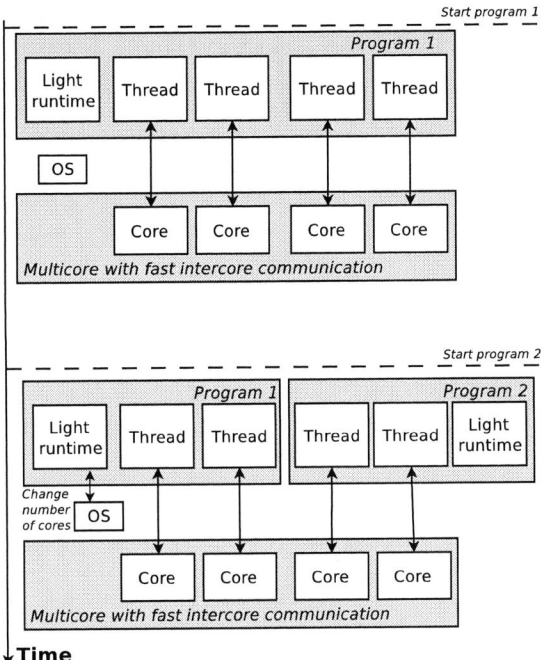

Figure 10: Example of code adaptation where a new program starts (i.e., *Program 2*) and it requires resources in use by a currently running process *Program 1*. In this example, the OS triggers a negotiation with the lightweight run-time to reduce the number of cores assigned to *Program 1* from 4 to 2.

bedded in the generated code when the objective is performance. This run-time can retrieve information such as execution paths taken, data communication patterns used, and thread synchronization patterns. If the run-time detects that is worth increasing the number of cores, it starts a negotiation with the OS.

The objective function of the run-time may not always be performance. For example, the OS may start to turn off parts of the chip if it detects that too much heat is being generated and thermal constraints are likely to be violated. In this situation, the run-time may be instructed to obtain the maximum performance for a given energy budget and will follow a different set of rules to achieve this.

OS support.

Figure 10 shows the use case for HELIX when the objective is coexistence between multiple programs. In this example, the lightweight run-time and the OS interact to adjust the number of cores in use based on the system load.

This interaction is a negotiation triggered by the OS when it decides to either reduce or increase the number of cores assigned to a given process based on the current load. The OS presents the lightweight run-time a hard and a soft threshold representing the maximum and the desired number of cores that the process can use, respectively. The choice of how many cores to use is left to the lightweight run-time, which must choose a number below the hard threshold and as close as possible to the soft threshold.

The lightweight run-time can also trigger negotiation when the needs of its program change. It can either request more cores when it comes to a highly parallel code section, or offer cores back to the OS when it detects a phase with limited parallelism.

In this interaction, each agent controls the resources that it knows best. The OS uses its complete view of the system to define the maximum and desired numbers of cores per process. The lightweight run-time uses the predictability of the HELIX loop transformation and its monitoring of the running code to determine the resource needs of its program.

4. RELATED WORK

There is a rich literature on parallelization of sequential programs by transforming loops into parallel threads of control. There are two main approaches: categorized into two principal paradigms: *pipelined multi-threading* and *cyclic multi-threading*.

Pipelined multi-threading (PMT).

Pipeline multi-threading techniques, the most established of which is called decoupled software pipelining (DSWP) [10, 15, 18], break loops into multiple threads such that cyclic data dependences never cross thread boundaries. The loose coupling of the resulting pipeline of threads allows data transfer between them to be buffered to prevent stalls in one thread from affecting others. The technique can produce significant speedups when this kind of parallelism is available in the program. Speculation has also been used to obtain speedups through the use of software transactional memory [16, 21].

The main drawback of PMT is that it restructures the code in a complex way that makes predicting the impact of this transformation code execution difficult. Therefore, it is unclear how to predict the speedup obtainable by applying these techniques to a given loop. That makes the problem of automatically choosing the most profitable loops for parallelization hard to solve. So for PMT, it is difficult to ensure that applying the transformation will not slow the program down.

Cyclic multi-threading (CMT).

Cyclic multi-threading techniques, such as DOALL and DO-ACROSS, target the parallelism between iterations of a given loop. The main drawback of these techniques comes from their high sensitivity to data communication overhead, which can easily lead to either slowdown or negligible speedup. The HELIX loop transformation belongs to this category, but it is able to constrain communication overhead enough to achieve significant speedups.

The closest approach to HELIX is the DOACROSS parallelization technique [8, 11], which has been studied in depth for regular workloads [1, 13, 20]. DOACROSS executes sequential segments without exploiting TLP between them [8]. Moreover, it does not permit either irregular control flow or irregular memory accesses within the loop [8]. Since HELIX has no such constraints and it considers a broader set of options during loop transformation, it can be seen as a generalization of the DOACROSS scheme that can be applied both to regular and irregular code.

Recent work on DOALL parallelism has used code transformations and thread-level speculation techniques to expose hidden parallelism in general purpose programs [22]. DSWP has also been mixed with DOALL [10, 18] to remove constraints on the number of threads extracted.

Run-time code adaptation.

Adapting code at run time in response to changes in program behavior has been studied deeply for managed code, such as Java or C#, in virtual machines [4]. However, these transformations

281

can change the code quite significantly at run time. In contrast, the HELIX approach will be to fine tune code to adapt its execution, avoiding drastic transformations in order to minimize run-time overhead, which can be substantial for code parallelization techniques. There are also dynamic schemes to execute loop iterations in parallel at run time when they are detected to be independent [9, 23].

Helper threads.

Exploiting SMT to help critical threads was introduced in [7] and adapted to different domains later on [12, 14]. HELIX uses this scheme to solve the specific problem of fetching signals sent from another core.

5. CONCLUSION

The HELIX loop transformation shows that distributing loop iterations among cores of a real multicore processor can be effective even though it is not designed to support the necessary inter-core communication. The broader HELIX research project aims to design hardware more suitable for such transformations. It also adds corresponding enhancements to the static compiler and adds support to make the generated code adaptable to phase changes and availability of system resources. These include a lightweight run-time to adapt the program to the underlying system as its requirements change and allow coexistence with other applications. While HELIX has focused on sequential programs, there is no reason why it cannot work as well for explicitly multi-threaded programs.

Acknowledgements

This work was possible thanks to the sponsorship of Microsoft Research, HiPEAC, the Royal Academy of Engineering, EP-SRC and the National Science Foundation (award number IIS-0926148). Any opinions, findings, and conclusions or recommendations expressed in this material are those of the authors and do not necessarily reflect the views of our sponsors.

6. REFERENCES

[1] J. Allen and K. Kennedy. *Optimizing compilers for modern architectures.* Morgan Kaufmann, 2002.

[2] Arvind, David August, Keshav Pingali, Derek Chiou, Resit Sendag, and Joshua J. Yi. Programming multicores: Do applications programmers need to write explicitly parallel programs? *IEEE Micro*, 30, 2010.

[3] David I. August, Jialu Huang, Thomas B. Jablin, Hanjun Kim, Thomas R. Mason, Prakash Prabhu, Arun Raman, and Yun Zhang. Automatic extraction of parallelism from sequential code. In A. Tabatabai et al., editors, *Fundamentals of Multicore Software Development.* Chapman & Hall / CRC Press, 2011.

[4] Michael G. Burke, Jong-Deok Choi, Stephen J. Fink, David Grove, Michael Hind, Vivek Sarkar, Mauricio J. Serrano, Vugranam C. Sreedhar, Harini Srinivasan, and John Whaley. The jalapeño dynamic optimizing compiler for java. In *Java Grande*, pages 129–141, 1999.

[5] Simone Campanoni, Giovanni Agosta, Stefano Crespi-Reghizzi, and Andrea Di Biagio. A highly flexible, parallel virtual machine: Design and experience of ILDJIT. *Softw. Pract. Exper.*, 2010.

[6] Simone Campanoni, Timothy M. Jones, Glenn Holloway, Vijay Janapa Reddi, Gu-Yeon Wei, and David Brooks.

HELIX: Automatic Parallelization of Irregular Programs for Chip Multiprocessing. *CGO*, 2012.

[7] Robert S. Chappell, Jared Stark, Sangwook P. Kim, Steven K. Reinhardt, and Yale N. Patt. Simultaneous subordinate microthreading (SSMT). *ISCA*, 1999.

[8] R. Cytron. DOACROSS: Beyond vectorization for multiprocessors. *ICPP*, 1986.

[9] Francis Dang and Lawrence Rauchwerger. Speculative parallelization of partially parallel loops. In *5th International Workshop on Languages, Compilers, and Run-Time Systems for Scalable Computers*, pages 285–299, 2000.

[10] Jialu Huang, Arun Raman, Thomas B. Jablin, Yun Zhang, Tzu-Han Hung, and David I. August. Decoupled software pipelining creates parallelization opportunities. *CGO*, 2010.

[11] Ali R. Hurson, Joford T. Lim, Krishna M. Kavi, and Ben Lee. Parallelization of DOALL and DOACROSS loops - a survey. *Advances in Computers*, 1997.

[12] Dongkeun Kim, Steve Shih wei Liao, Perry H. Wang, Juan del Cuvillo, Xinmin Tian, Xiang Zou, Hong Wang, Donald Yeung, Milind Girkar, and John P. Shen. Physical experimentation with prefetching helper threads on Intel's hyper-threaded processors. *CGO*, 2004.

[13] J. T. Lim, A. R. Hurson, K. Kavi, and B. Lee. A loop allocation policy for DOACROSS loops. *SPDP*, 1996.

[14] Chi-Keung Luk. Tolerating memory latency through software-controlled pre-execution in simultaneous multithreading processors. *SIGARCH Comp. Arch. News*, 2001.

[15] Guilherme Ottoni, Ram Rangan, Adam Stoler, and David I. August. Automatic thread extraction with decoupled software pipelining. *MICRO*, 2005.

[16] Arun Raman, Hanjun Kim, Thomas R. Mason, Thomas B. Jablin, and David I. August. Speculative parallelization using software multi-threaded transactions. *ASPLOS*, 2010.

[17] Arun Raman, Hanjun Kim, Taewook Oh, Jae W. Lee, and David I. August. Parallelism orchestration using dope: the degree of parallelism executive. In *PLDI*, 2011.

[18] Easwaran Raman, Guilherme Ottoni, Arun Raman, Matthew J. Bridges, and David I. August. Parallel-Stage decoupled software pipelining. *CGO*, 2008.

[19] Timothy Sherwood, Suleyman Sair, and Brad Calder. Phase tracking and prediction. In *ISCA*, pages 336–347, 2003.

[20] Hong-Men Su and Pen-Chung Yew. Efficient DOACROSS execution on distributed shared-memory multiprocessors. *ACM/IEEE conference on Supercomputing*, 1991.

[21] N. Vachharajani, Ram Rangan, Easwaran Raman, Matthew J. Bridges, Guilherme Ottoni, and David I. August. Speculative decoupled software pipelining. *PACT*, 2007.

[22] Hongtao Zhong, Mojtaba Mehrara, Steve Lieberman, and Scott Mahlke. Uncovering hidden loop level parallelism in sequential applications. *HPCA*, 2008.

[23] Xiaotong Zhuang, Alexandre E. Eichenberger, Yangchun Luo, Kevin O'Brien, and Kathryn M. O'Brien. Exploiting parallelism with dependence-aware scheduling. In *PACT*, pages 193–202, 2009.

Exploring Sub-20nm FinFET Design with Predictive Technology Models

Saurabh Sinha, Greg Yeric, Vikas Chandra, Brian Cline, Yu Cao*

ARM Inc., *Arizona State University, Tempe, AZ

saurabh.sinha@arm.com

ABSTRACT

Predictive MOSFET models are critical for early stage design-technology co-optimization and circuit design research. In this work, Predictive Technology Model files for sub-20nm multi-gate transistors have been developed (PTM-MG). Based on MOSFET scaling theory, the 2011 ITRS roadmap and early stage silicon data from published results, PTM for FinFET devices are generated for 5 technology nodes corresponding to the years 2012-2020 on the ITRS roadmap.

Categories and Subject Descriptors

B.7.1 [**Hardware**]: Integrated Circuits—*Types and Design Styles, Advanced Technologies*; B.8.2 [**Hardware**]: Performance and Reliability—*Performance Analysis and Design Aids*

General Terms

Theory

Keywords

FinFET, multi-gate, scaling theory, predictive models, SPICE

1. INTRODUCTION

CMOS scaling has continued up to the 20nm node through innovative techniques such as incorporating high-k dielectrics in the gate stack, strain engineering, pocket implants and optimization in materials and device structures. However, further scaling of planar devices is proving to be extremely challenging due to degrading short channel effects, process variations and reliability degradation [1].

Multi-gate transistor structures such as FinFETs will be the technology of choice for extending CMOS scaling beyond the 20nm node. Improved short channel control through a fully depleted fin, reduced random dopant fluctuation, improved mobility, lower parasitic junction capacitance and improved area efficiency are some of the primary advantages

of FinFETs [2]. However, FinFETs will be markedly different than planar FETs due to added fringing capacitance, higher access resistance, width-quantization, 3D-factor, and low-field mobility. Hence, it is crucial to develop accurate representative FinFET compact models to be used as tools for design-technology co-optimization, identify key design needs and explore design solutions upfront.

In the area of predictive modeling, the Berkeley Predictive Technology Model (BPTM) [3] and the Arizona State University (ASU) PTM [4] were developed for planar CMOS technology nodes up to 20nm based on the BSIM4 model [5]. BPTM was developed by empirically extracting model parameters from early stage silicon data while ASU PTM further improved the methodology by taking into account significant physical correlations among model parameters. Predictive technology models for bulk planar transistors [4] have enabled a wide variety of exploratory circuit design research by the design community [6–8].

In this work, a new generation of Predictive Technology Model for multi-gate transistors (PTM-MG), specifically FinFETs for sub-20nm technology nodes, is developed. In Section 2, we discuss the development of these model parameters using BSIM-CMG (short for Berkeley Short-channel IGFET Common Multi-Gate) model [9], scaling theory of multi-gate devices, physical models and ITRS projections. In Section 3 we develop separate models aligning to the even years 2012-2020 of the 2011 ITRS roadmap. These years approximately align to technology node names of 20nm, 16nm, 14nm, 10nm and 7nm, respectively. Additionally, ring-oscillator delay metrics and comparison with bulk devices is presented. The new PTMs for sub-20nm multi-gate transistors have been developed in two application-specific versions, high performance (HP) and low-standby power (LSTP).

2. PTM FOR MULTI-GATE TRANSISTORS

2.1 BSIM-CMG: Multi-gate transistor model

Multi-gate transistors have been studied previously using TCAD device simulators, but the speed of TCAD tools limits their use in circuit design exploration. BSIM-CMG is a surface-potential-based compact model that can model different multi-gate structures (double-gate, tri-gate and gate-all-around FETs) [9] [10]. In addition to incorporating the effect of the 3D structure and quantum mechanical effects (QME) on device characteristics and short channel effects, the model retains the standard framework of BSIM4 and BSIM-SOI models for real-device effects such as mobility

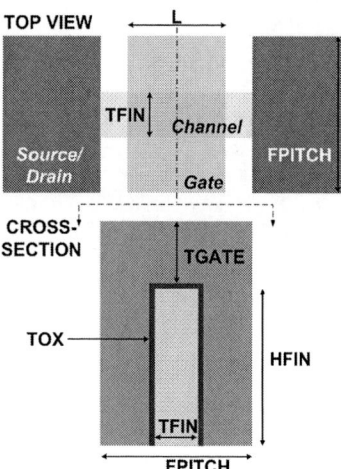

Figure 1: Top view and cross-sectional schematic of a FinFET device.

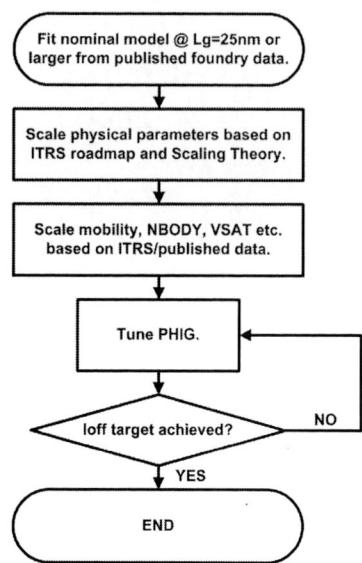

Figure 2: Flowchart describing PTM-MG model generation for FinFETs.

degradation, velocity saturation, series resistance, parasitic capacitance, etc., allowing ease in efficient extraction of model parameters. The model has been verified against TCAD and experimental data [9] and is a standard feature in commercial circuit simulators.

Figure 1 shows the top and cross-sectional view of a FinFET. The dimensions are labeled using the corresponding BSIM-CMG model parameters. The parameters used in PTM-MG development are listed in Table 1. The behavior of a FinFET device is most sensitive to the primary parameters, technology specifications and physical parameters. The secondary parameters are useful to fine-tune a fit to the complete current-voltage characteristics or capture secondary effects.

2.2 Predicting Model Parameters

Since FinFET data for advanced technology nodes (sub-20nm) is not available from foundries, our BSIM-CMG model

Table 1: PTM-MG Parameters

Primary Parameters	
L	Gate length
TFIN	Fin thickness
HFIN	Fin height
FPITCH	Fin pitch
Technology Specifications	
EOT	Equivalent oxide thickness
V_{DD}	Supply voltage
R_{DS}	S/D resistance
Secondary Parameters	
PHIG	Gate work function
NBODY	Channel doping
CDSC	SD-channel coupling
E_{ta0}	DIBL coefficient
Physical Parameters	
μ_0	Low-field mobility
V_{sat}	Saturation velocity

development begins by fitting to previously published transistor data with gate length 25nm or higher. The complete fit of I-V characteristics is discussed in Section 3. This initial step forms the nominal model and parameters for advanced technologies are derived by scaling the PTM-MG parameters based on multi-gate MOSFET physics, guidance from the ITRS road-map and published data. Figure 2 shows a flowchart describing PTM-MG model development. The details of each step will be described in this subsection.

The primary process parameters are determined by multi-gate MOSFET physics. Traditional bulk devices have relied on reducing gate oxide thickness and increased channel/halo doping to reduce Short-Channel Effects (SCE) with scaling. However, for fully depleted devices such as FinFETs, the ratio of channel length to fin thickness affects SCE adversely. The Voltage Doping Transformation (VDT) model [11] is a simple tool that translates the effects of device geometry scaling into electrical parameters. It is important to study how L, TFIN and HFIN influence short channel effects such as Drain-Induced Barrier Lowering (DIBL) and we use the modified VDT model to get its first-order expression as

$$DIBL = 0.80 \frac{\epsilon_{si}}{\epsilon_{ox}} EI \times V_{DS} \tag{1}$$

EI stands for Electrostatic Integrity and it represents the way the electric field from the drain influences the channel region [11]. It is given by

$$EI = \frac{1}{2}\left[1 + \frac{t_{si}^2/4}{L_{el}^2}\right]\frac{t_{ox}}{L_{el}}\frac{t_{si}/2}{L_{el}} \tag{2}$$

Modifying Eq.1 to incorporate the scale length (λ) [9] and plotting for different t_{si} (t_{si} is equivalent to TFIN in MG devices), we see that the fin thickness needs to scale with gate length scaling in order to keep short-channel effects under control (Figure 3). For double-gate transistors it has been demonstrated that the $L_g/TFIN$ of ~ 1.5 is sufficient

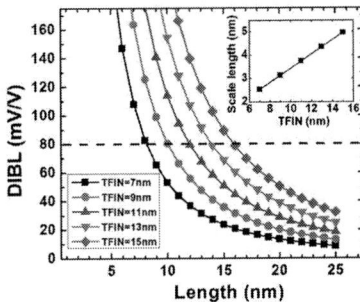

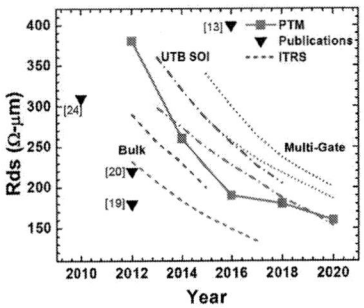

Figure 3: DIBL vs. FinFET gate length for different fin thickness (TFIN). Inset graph shows the scale length (λ) vs. fin thickness (TFIN).

Figure 5: Parasitic Source-Drain Resistance for each technology node.

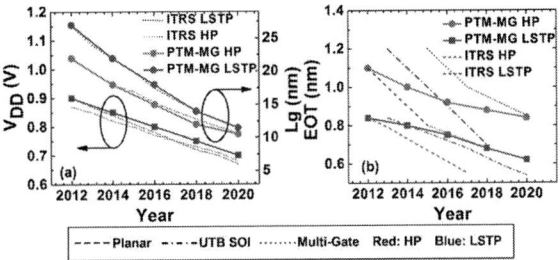

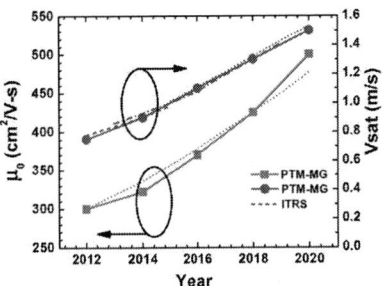

Figure 4: (a) V_{DD} and (b) EOT scaling with technology node. The parameters are scaled according to 2011 ITRS predictions.

Figure 6: Low field mobility and saturation velocity trend. ITRS trends are for HP FETs.

to achieve subthreshold slope less than 90mV/dec [12]. This forms the basis of deriving the transistor dimensions of each technology node as listed in Table 2.

V_{DD} and EOT are scaled according to 2011 ITRS predictions as shown in Figure 4, with the various ITRS device-specific parameters included for reference as the various dashed lines. The ITRS roadmap predicts the introduction of multi-gate transistors in the year 2015 preceded by UTB SOI and planar transistors, but because the leading foundries have announced FinFETs beyond 20nm, we have concentrated our efforts on developing FinFET PTM models. For the sake of simplicity we have used the BSIM-CMG models to fit an average of the planar and UTB SOI values in the years 2012 and 2014. In this way we attempt to best match the likely ITRS device targets as the devices change to improve electrostatics. The outcome for Figure 4 is a relaxed overall trend on EOT as compared to any one ITRS device type. As shown in later figures, this same philosophy has been applied to several of the other device-dependent parameters discussed later.

Parasitic series resistance (R_{DS}) is a major source of performance degradation in nanoscale transistors. It is enhanced in multi-gate structures due to increased spreading resistance in the narrow fins and difficulty in contacting the source-drain regions [13]. R_{DS} is highly dependent on the contact resistivity in addition to fin geometry, length of spacer and doping gradient from S/D region to the channel [14]. These parameters can be instantiated in BSIM-CMG by invoking the geometry based parasitic series resistance model (RGEOMOD=1). Based on ITRS predictions of silicide contact resistivity from the Front End Process

(FEP) tables and published data [15], Figure 5 shows the trend of extracted R_{DS} (normalized to W_{eff}) in these PTM-MG models.

Apart from the gate metal work-function (PHIG), the rest of the secondary process parameters do not have a major influence on the device characteristics. A fixed off-current of 100nA/µm is used for HP devices, which matches the ITRS HP trends. For LSTP devices, we have targeted a 0.1nA/µm off-current that is higher than the 0.01nA/µm of the ITRS LSTP devices. The ITRS V_{TH} set point is indicative of a higher V_{TH} device that would be used in power critical or performance non-critical applications or paths. Both devices are important for circuit prediction, with the higher leakage device more properly tracking performance trends and the lower leakage device more properly tracking the static power of non-critical paths/applications. V_{TH} adjustment in Fin-FETs can be a complicated combination of doping, work function, and fin dimension that can affect other device performance aspects such as mobility, DIBL and subthreshold slope. For the sake of simplicity, we have modeled our different off-current targets using the gate work-function (PHIG) parameter only. The remainder of this paper will use our higher off-current device targets, but a list of 'DVTSHIFT' values targeting the ITRS off-currents for each technology node is available on the PTM website [4].

The transistor channel is assumed to be very lowly doped (NBODY= 10^{17} cm^{-3}) for the 20nm node and is gradually reduced to near intrinsic (NBODY=10^{16} cm^{-3}) at 7nm. Since FinFETs are fully-depleted and the fin thickness determines the short channel effects, channel to S/D coupling (CDSC) and DIBL parameter (E_{ta0}) can be used to model imperfections in real devices such as tapered fins or irreg-

285

Table 2: PTM-MG Summary

Year	2012	2014	2016	2018	2020
Node (nm)	20	16	14	10	7
M1 Pitch (nm)	64	48	38	30	24
Lg (nm)	24	21	18	14	11
V_{DD} (V)	0.9	0.85	0.8	0.75	0.7
TFIN (nm)	15	12	10	9	6.5
HFIN (nm)	28	26	22	21	18
W_{EFF} (nm)	71	64	54	51	42.5
FPITCH (nm)	60	42	32	28	22
3D factor	1.183	1.52	1.68	1.82	1.93

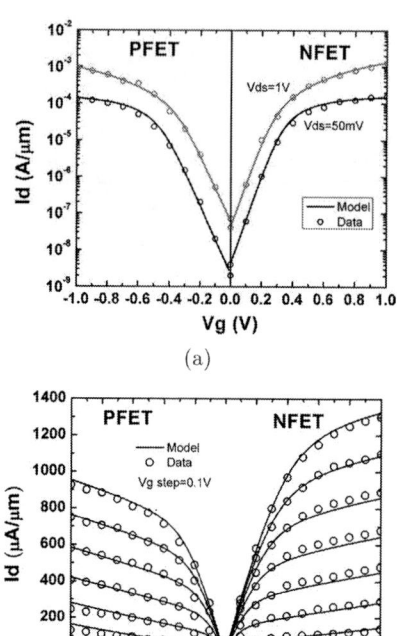

(a)

(b)

Figure 7: PTM-MG fit with measurement data from [20].

ular fin thickness along fin height [16] [17]. With these NBODY assumptions, the gate metal work function is tuned to achieve the off-current target. This results in arbitrary gate work functions that may not be practical to achieve in manufacturing. Thus it is possible that some additional channel doping may be needed in practical FinFET implementations.

Very low channel doping and volume inversion in narrow fins reduce carrier scattering and lead to improved low field carrier mobility (μ_0) and saturation velocity (V_{sat}) [18]. Additionally, strain engineering and ballistic transport in sub-20nm channel lengths can further improve carrier mobility [19]. Figure 6 shows this trend, which is crucial to getting consistent improvement in drive current through subsequent technology nodes.

A summary of the primary process parameters of the PTM-MG models for each technology node are listed in Table 2. Metal 1 pitches are also provided as well as the approximate corresponding technology node values. Since the effective channel width (W_{eff}) of a FinFET is equal to 2×HFIN+TFIN, the total transistor width is quantized. The fin pitch (FPITCH) determines area efficiency compared to a planar device. W_{eff}/ FPITCH is referred to as the "3D factor" in Table 2, demonstrating the additional device width from a FinFET as compared to a planar FET due to fin construction. The ITRS roadmap does not provide guidance on the prediction of these device geometries so our values are tied to the existing published literature for the earlier nodes and then scaled to meet electrostatic requirements and I_{ON}/I_{OFF} values according to ITRS scaling.

Table 3: PTM-MG Verification

Data source	[16]	[20]	[19]	[21]
Foundry	Intel	TSMC	TSMC	IBM
Lg (nm)	40	25	24	25
V_{DD} (V)	1.1	1	1	1
EOT (nm)	1.2	1.1	1.2	1.15
TFIN (nm)	25	15	15	10
HFIN (nm)	29	30	30	30
R_{DS} ($\Omega - \mu m$)	194	220	244	262
I_{on} ($\mu A/\mu m$)	1395	1300	1200	1300
PTM-MG I_{on}	1385	1330	1214	1264
I_{off} (nA/μm)	139	41	100	100
PTM-MG I_{off}	139	43	100	100
Worst Case Error	0.7%	4.87%	1.16%	2.77%

3. PTM-MG EVALUATION

In order to verify accuracy, the models are fitted with measurement data from industrial publications and parameters such as subthreshold slope, DIBL, etc., are plotted with the PTM-MG trends [13, 16, 19–26].

Based on the limited available information and device characteristics, assumptions regarding the geometry are made allowing us to generate I-V curves using the PTM-MG parameters to match with published data. Table 3 summarizes the verification of PTM-MG predictions with published measurement results. Tuning the primary parameters, the I_{ON} and I_{OFF} of FinFET devices with various device geometries have been matched to published results with high accuracy. Figure 7 shows an example fit of the complete I-V characteristics of a FinFET device with L_g=25nm obtained by fine-tuning both primary and secondary parameters.

3.1 PTM-MG Model Trends

The PTM-MG model cards are developed using ITRS as a reference for Lg, V_{DD}, EOT, I_{DSAT} and intrinsic delay trends. Figure 8 shows saturation drive current of PTM-MG models normalized per effective width (W_{eff}) for a constant off-current (I_{off}=0.1nA/μm for LSTP and 100nA/μm for HP). The PTM-MG LSTP devices follow the ITRS LSTP trend but are shifted to be slightly stronger as discussed in Section 2.

Figure 9(a) and 9(b) show subthreshold slope and DIBL of PTM-MG models across technology generations. Selected publication data are plotted for comparison. Short channel effects are determined by the scale length (λ) which

286

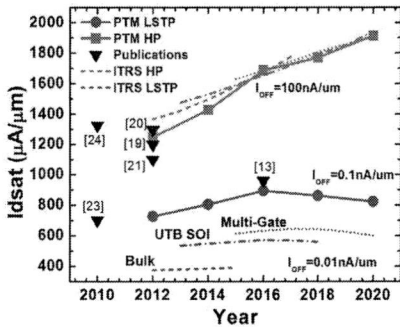

Figure 8: Prediction of I_{DSAT} for PTM-MG HP and LSTP devices. Trends from ITRS and selected publications are annotated for comparison.

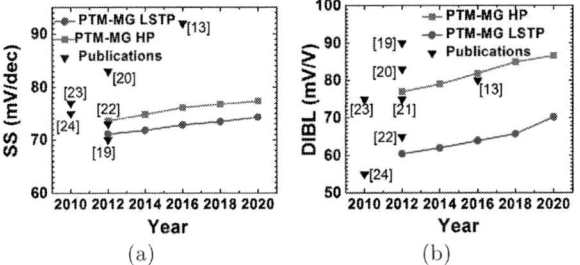

(a)   (b)

Figure 9: Prediction of subthreshold slope and DIBL across technology generations. Selected publication results are included for comparison.

has a strong dependence on fin-width (TFIN). The ratio of L/TFIN is kept greater than 1.5 across technology nodes. However, secondary model parameters CDSC and E_{ta0} are slightly modified to capture real-device effects such has nonuniform fin thickness, trapezoid or notched fins [16].

3.2 Design Benchmarking

FinFETs possess superior electrostatics compared to bulk planar transistors. However, the final power/performance metrics below 20nm will be highly dependent on the total gate capacitance and not just the intrinsic device electrostatics. The prediction of gate capacitance along with ITRS trends are presented in Figure 10. The parameter-based intrinsic capacitance is in agreement with the ITRS predictions, but the total gate capacitance in the PTM FinFETs is much higher due additional fringing capacitance associated with the 3D nature of the device. This fringe capacitance is sensitive to structural dimensions [27] and the BSIM-CMG model has a provision (CGEOMOD=2) to estimate parasitic fringe capacitance based on device geometry. The total capacitance in PTM-MG is estimated using this mode.

Figure 11 shows the scaling of intrinsic delay (τ) calculated from total gate capacitance, supply voltage and saturation current. FO4 inverter delay is plotted in Figure 12. Both metrics match the general ITRS trends showing that FinFETs possess the capability to support the roadmap targets to the 10nm node or below. The ITRS roadmap provides FO4 delay predictions only for HP devices and not for LSTP devices. As was the case for EOT in Figure 4(b), the

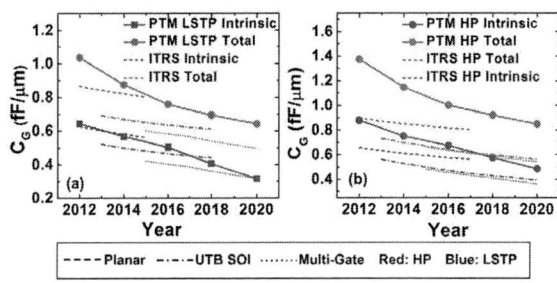

Figure 10: Intrinsic and total gate capacitance of PTM-MG LSTP and HP FinFETs. ITRS prediction trends are plotted for comparison.

2012 and 2014 PTM-MG devices are made to approximate the averaged bulk and UTB-SOI trends.

In Figure 13 we compare the FO4 delay as V_{DD} is scaled for two different technologies: 22nm PTM HP bulk model [4] and 20nm PTM-MG HP model. For fair comparison, the normalized delay with supply voltage scaling is plotted. It is seen that the delay of the 22nm PTM planar transistor increases by about 9 times while scaling V_{DD} from 1V to 0.5V. However, the FinFET transistor shows much better performance with V_{DD} scaling owing to superior electrostatics. This trend matches the results demonstrated in [21].

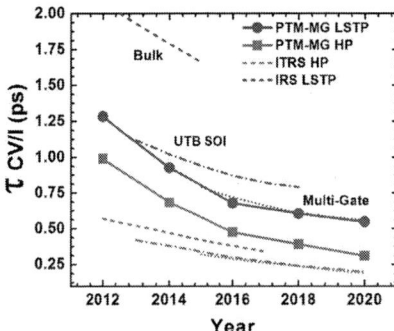

Figure 11: Intrinsic delay (CV/I) of PTM FinFET models and ITRS trend across technology nodes.

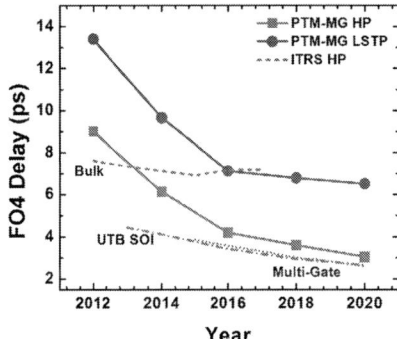

Figure 12: Fan out of 4 (FO4) delay prediction by PTM-MG HP and LSTP models.

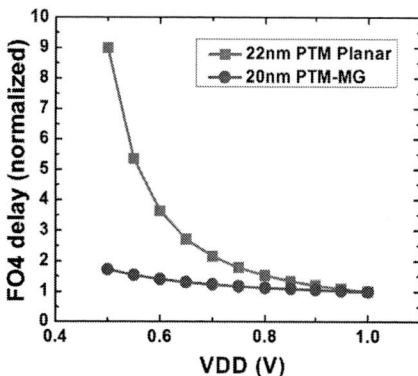

Figure 13: Normalized FO4 delay with V_{DD} scaling comparing 22nm HP planar PTM model to 20nm HP FinFET PTM model.

4. CONCLUSIONS AND FUTURE WORK

A new generation of Predictive Technology Model for Multi-gate (PTM-MG) devices has been developed for early-stage design-technology exploration. Model parameter files for 5 technology nodes corresponding to the years 2012-2020 on the 2011 ITRS roadmap have been developed using BSIM-CMG models. The predictive models have been verified with published results and show excellent scalability across process and design conditions. The PTM-MG models allow the movement beyond time constants to actual circuit simulation in order to explore the impact of technology scaling using multi-gate transistors. For more detailed predictions of circuit design using FinFETs, accurate modeling of process variations, temperature and wire effects are important areas of future work.

5. REFERENCES

[1] [Online]. Available: www.itrs.net/

[2] M. Jurczak et al., "Review of finfet technology," in SOI Conference, oct. 2009, pp. 1 –4.

[3] Y. Cao et al., "New paradigm of predictive mosfet and interconnect modeling for early circuit simulation," in CICC, 2000, pp. 201 –204.

[4] [Online]. Available: ptm.asu.edu

[5] [Online]. Available: www-device.eecs.berkeley.edu/bsim/?page=BSIM4

[6] J. Lee et al., "Analyzing impact of multiple abb and avs domains on throughput of power and thermal-constrained multi-core processors," in ASP-DAC, jan. 2010, pp. 229 –234.

[7] B. Liu, "Spatial correlation extraction via random field simulation and production chip performance regression," in DATE '08, march 2008, pp. 527 –532.

[8] A. Agarwal et al., "A process-tolerant cache architecture for improved yield in nanoscale technologies," TVLSI, vol. 13, no. 1, pp. 27 –38, jan. 2005.

[9] M. Dunga et al., BSIM-CMG: A Compact Model for Multi-Gate Transistors. Springer US, 2008.

[10] S. Yao et al., "Global parameter extraction for a multi-gate mosfets compact model," in ICMTS, march 2010, pp. 194 –197.

[11] T. Skotnicki et al., "The voltage-doping transformation: a new approach to the modeling of mosfet short-channel effects," EDL, vol. 9, no. 3, pp. 109 –112, mar 1988.

[12] J. Kedzierski et al., "High-performance symmetric-gate and cmos-compatible vt asymmetric-gate finfet devices," in IEDM, 2001, pp. 19.5.1 –19.5.4.

[13] H. Kawasaki et al., "Finfet process and integration technology for high performance lsi in 22 nm node and beyond," in IWJT, june 2007, pp. 3 –8.

[14] M. Lundstrom. (2008) Ece 612: Nanoscale transistors. [Online]. Available: nanohub.org/resources/5328.

[15] K.-W. Ang et al., "Advanced contact and junction technologies for improved parasitic resistance and short channel immunity in finfets beyond 22nm node," in ISDRS, dec. 2011, pp. 1 –2.

[16] J. Kavalieros et al., "Tri-gate transistor architecture with high-k gate dielectrics, metal gates and strain engineering," in VLSIT, 2006, pp. 50 –51.

[17] J. Chang et al., "Scaling of soi finfets down to fin width of 4 nm for the 10nm technology node," in VLSIT, june 2011, pp. 12 –13.

[18] C. Young et al., "Critical discussion on (100) and (110) orientation dependent transport: nmos planar and finfet," in VLSIT, june 2011, pp. 18 –19.

[19] C. Wu et al., "High performance 22/20nm finfet cmos devices with advanced high-k/metal gate scheme," in IEDM, dec. 2010, pp. 27.1.1 –27.1.4.

[20] C.-Y. Chang et al., "A 25-nm gate-length finfet transistor module for 32nm node," in IEDM, dec. 2009, pp. 1 –4.

[21] T. Yamashita et al., "Sub-25nm finfet with advanced fin formation and short channel effect engineering," in VLSIT, june 2011, pp. 14 –15.

[22] K. Maitra et al., "Aggressively scaled strained-silicon-on-insulator undoped-body high-kappa/metal-gate nfinfets for high-performance logic applications," EDL, vol. 32, no. 6, pp. 713 –715, june 2011.

[23] R. Rios et al., "Comparison of junctionless and conventional trigate transistors with l_g down to 26 nm," EDL, vol. 32, no. 9, pp. 1170 –1172, sept. 2011.

[24] C.-C. Yeh et al., "A low operating power finfet transistor module featuring scaled gate stack and strain engineering for 32/28nm soc technology," in IEDM, dec. 2010, pp. 34.1.1 –34.1.4.

[25] K. von Arnim et al., "A low-power multi-gate fet cmos technology with 13.9ps inverter delay, large-scale integrated high performance digital circuits and sram," in VLSI Technology, june 2007, pp. 106 –107.

[26] G. Vellianitis et al., "Gatestacks for scalable high-performance finfets," in IEDM, dec. 2007, pp. 681 –684.

[27] M. Guillorn et al., "Finfet performance advantage at 22nm: An ac perspective," in VLSIT, june 2008, pp. 12 –13.

Fast Nonlinear Model Order Reduction via Associated Transforms of High-Order Volterra Transfer Functions

Yang Zhang, Haotian Liu, Qing Wang, Neric Fong and Ngai Wong
Department of Electrical and Electronic Engineering
The University of Hong Kong
Pokfulam Road, Hong Kong
{yzhang,htliu,wangqing,nfong,nwong}@eee.hku.hk

ABSTRACT

We present a new and fast way of computing the projection matrices serving high-order Volterra transfer functions in the context of (weakly and strongly) nonlinear model order reduction. The novelty is to perform, for the first time, the association of multivariate (Laplace) variables in high-order multiple-input multiple-output (MIMO) transfer functions to generate the standard single-s transfer functions. The consequence is obvious: instead of finding projection subspaces about every s_i, only that about a single s is required. This translates into drastic saving in computation and memory, and much more compact reduced-order nonlinear models, without compromising any accuracy.

Categories and Subject Descriptors

B.7.2 [**Hardware**]: Design Aids—*Simulation, Verification*; I.6.5 [**Computing Methodologies**]: Model Development—*Modeling methodologies*; J.6 [**Computer Applications**]: Computer-Aided Engineering—*Computer-aided design (CAD)*

General Terms

Algorithms, Design, Theory, Verification

Keywords

Association of variables, Model order reduction (MOR), Nonlinear system, Analog/RF circuits

1. INTRODUCTION

Simulation techniques for VLSI circuits at the system level are strongly demanded. The analog and radio-frequency (RF) modules, though occupying a small part of a typical mixed-signal chip, are critical while hard to simulate and design due to their nonlinearities. Indeed, model order reduction (MOR) for complex nonlinear systems is often needed whereby the reduced-order model (ROM) inherits the dominant dynamics of the original system while featuring a much smaller dimension for simulation.

So far, MOR techniques have been studied extensively for linear time-invariant (LTI) systems, such as explicit moment matching based asymptotic waveform evaluation (AWE) [12] and projection-based implicit moment matching [9]. The underlying workhorse appears to be the Krylov subspace method (e.g. [11, 9]), which projects the original system onto a ROM via a projection matrix from matching the moments of the original system. Although these methods have made a great success on the LTI systems, the situation becomes complicated in the nonlinear scenario.

As a powerful analysis tool for nonlinear systems, Volterra theory has been studied for decades [15] and successfully applied to nonlinear MOR (NMOR) for years, e.g., [10, 7, 6]. Although it provides an analytical and systematic approach, deployment of the Krylov subspace method (e.g., the NORM algorithm proposed in [7, 6]) suffers from exponentially growing subspace dimensions due to the multiple frequency axes in high-order Volterra transfer functions [15]. Subsequently, the order of the Volterra series used for matching the moments in NMOR is severely limited, rendering the Volterra approach applicable essentially to weakly nonlinear systems. Other NMOR methods, such as the trajectory piecewise-linear (TPWL) approximation [14], which can deal with strongly nonlinear systems, also suffer from training input sequence dependence.

Recently, the MOR of strongly nonlinear systems is transformed into the MOR problem of the quadratic-linear differential algebraic equations (QLDAEs) [4, 5], which are obtained by adding extra states related to strong nonlinearities such as the sine/cosine or exponential (diode-type) curves:

$$C\dot{x} = G_1 x + G_2 x \otimes x + D_1 xu + D_2 x \otimes xu + bu, \quad (1)$$

where $x \in \mathbb{R}^n$ is the state vector and $\otimes$ denotes the Kronecker product. All other matrices are of compatible dimensions and a scalar input u is assumed for notational ease whose multi-input multi-output (MIMO) generalization is possible. The key advantage of QLDAE is that it keeps the strongly nonlinear functions only in quadratic-linear format instead of cubic or higher-order terms.

In the following, we work with a trimmed version of (1) by assuming an invertible C (called a regular system borrowing from the linear system terminology) so that it can be replaced with an identity matrix. Also, we leave out the D_2 term as it seldom appears in electrical circuits of interest. Subsequently, (1) turns into

$$\dot{x} = G_1 x + G_2 x \otimes x + D_1 xu + bu. \quad (2)$$

These assumptions are made mainly for the ease of notation.

For instance, a singular C is analogous to a linear descriptor system whereby the regular (nonsingular) part can be extracted via the canonical projector or a Weierstrass form transformation [16], and the impulsive (singular) part is often immaterial or related algebraically to the regular subsystem. Moreover, all results in this paper are extensible to include the D_2 component and a multi-column B (instead of the vector input matrix b).

The key contribution of this paper is to break the "dimensionality curse" of Krylov subspace method in the NMOR context. Strong nonlinearity is accommodated with the use of QLDAE [4, 5]. Whereas the true computation bottleneck, viz. the transfer matrix moment expansion at multiple frequency axes, is completely avoided by the *first-time proposed* use of association of variables in multivariate MIMO high-order Volterra transfer functions. The consequence is obvious: instead of finding projection subspaces about every s_i which leads to an exponential growth in the overall subspace dimension, only that about a single s is required. This translates into drastic saving in computation and memory, and much more compact nonlinear ROMs, without compromising any accuracy.

This paper is organized as follows. Section 2 briefly reviews the nonlinear system description by Volterra theory, as well as the *association of variables* and the corresponding *associated transform*. Then, the conventional single-input single-output (SISO) association of variables method is extended to its MIMO counterpart through two important theorems. The Krylov subspace generation in associated transfer functions, with practical considerations, is described. In Section 3, the proposed scheme is verified through examples. Section 4 discusses some important remarks. Finally, Section 5 draws the conclusion.

2. ASSOCIATED TRANSFORM IN NMOR

We present the main results of the paper in this section. A succinct account of association of frequency-domain (Laplace) variables is first given, which is mainly applied to SISO systems in the literature, see e.g., [1, 2, 8, 13, 15]. Then, important theorems are devised which allow the natural utilization of associated transform in MIMO scenario to facilitate projection-based NMOR.

2.1 Volterra theory and association of variables

To begin with, the state vector of a Volterra system is progressively approximated with high-order responses, namely,

$$x(t) = x_1(t) + x_2(t) + x_3(t) + \cdots,$$

where

$$x_n(t) = \int_{-\infty}^{\infty} \cdots \int_{-\infty}^{\infty} h_n(\tau_1, \cdots, \tau_n) \cdot$$
$$u(t-\tau_1) \cdots u(t-\tau_n) d\tau_1 \cdots d\tau_n, \quad (3)$$

and $h_n(\tau_1, \cdots, \tau_n)$ is the n-th order Volterra kernel. In particular, x_1 is the usual first-order convolution having its Laplace domain representation $X_1(s) = H_1(s)U(s)$ where $H_1(s) = \int_{-\infty}^{\infty} h_1(\tau)e^{-s\tau}d\tau$ is the impulse response transfer function form. Analogously, the (nonlinear) high-order transfer function counterparts are defined as

$$H_n(s_1, \cdots, s_n) = \int_{-\infty}^{\infty} \cdots \int_{-\infty}^{\infty} h_n(\tau_1, \cdots, \tau_n) \cdot$$
$$e^{-s_1\tau_1} \cdots e^{-s_n\tau_n} d\tau_1 \cdots d\tau_n. \quad (4)$$

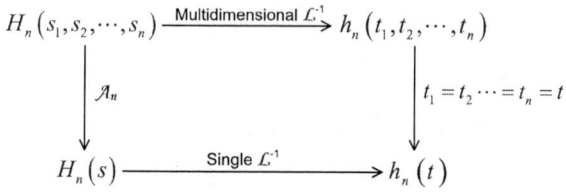

Figure 1: Association of variables for finding $H_n(s)$ from $H_n(s_1, s_2 \cdots, s_n)$.

Nonetheless, unlike the first-order case, there is no direct counterpart in the (multivariate) Laplace domain except if we replace the single time axis in the product of u in (3) by multiple axes as $u(t_1 - \tau_1) \cdots u(t_n - \tau_n)$, yielding [15]

$$X_n(s_1, \cdots, s_n) = H_n(s_1, \cdots, s_n)U(s_1) \cdots U(s_n). \quad (5)$$

The (multidimensional) inverse Laplace transform of (4) is a generalization of the univariate formula, given by

$$h_n(t_1, \cdots, t_n) = \mathscr{L}^{-1}(H_n(s_1, \cdots, s_n))$$
$$= \frac{1}{(2\pi j)^n} \int_{\sigma_n - j\infty}^{\sigma_n + j\infty} \cdots \int_{\sigma_1 - j\infty}^{\sigma_1 + j\infty} H_n(s_1, \cdots, s_n) \cdot$$
$$e^{s_1 t_1} \cdots e^{s_n t_n} ds_1 \cdots ds_n. \quad (6)$$

To restore the required $h_n(t)$, one then evaluates along the *diagonal line* in the multi-time hyperplane, i.e., $h_n(t) = h_n(t_1, \cdots, t_n)|_{t_1 = t_2 = \cdots = t_n = t}$. Of course, the same procedure is used to obtain $x_n(t)$ directly from $X_n(s_1, \cdots, x_n)$ in (5) if the input $u(t)$ or $U(s)$ is known.

The above approach unifies the time variables in the time domain. An alternative is to carry out the unification first in the multi-frequency domain. So its time-domain counterpart is automatically a single-time function. This process is termed the *association of variables* and the corresponding frequency function the *associated transform*, denoted as $H_n(s) = \mathscr{A}_n(H_n(s_1, \cdots, s_n))$, from which $h_n(t)$ can be derived from the conventional inverse Laplace transform of $H_n(s)$. Fig. 1 depicts the relationship between these time- and frequency-domain operations.

A closed-form expression for the association of variables also follows from (6) by setting $t_1 = \cdots = t_n = t$ [13, 15]:

$$H_n(s) = \mathscr{A}_n(H_n(s_1, \cdots, s_n))$$
$$= \frac{1}{(2\pi j)^{n-1}} \int_{\sigma_n - j\infty}^{\sigma_n + j\infty} \cdots \int_{\sigma_2 - j\infty}^{\sigma_2 + j\infty}$$
$$H_n(s - s_2 - \cdots - s_n, s_2, \cdots, s_n) ds_2 \cdots ds_n. \quad (7)$$

Moreover, certain factored forms in $H_n(s_1, \cdots, s_n)$ allow the direct use of (6) and/or (7) to produce useful theorems, e.g., see Chapter 2 of [15]. (Henceforth, the dimensional subscripts in the transfer functions or association operator are sometimes omitted when they are obvious from context.) In particular, an interesting property [1] is that if $H(s_1, \cdots, s_n)$ can be written as $F(s_1 + \cdots + s_n)G(s_1, \cdots, s_n)$, then by (7)

$$H(s) = \mathscr{A}(F(s_1 + \cdots + s_n)G(s_1, \cdots, s_n))$$
$$= F(s)\mathscr{A}(G(s_1, \cdots, s_n)) = F(s)G(s). \quad (8)$$

Nonetheless, to our best knowledge, existing works on association of variables mainly deal with SISO (viz. scalar) systems (see e.g., [1, 2, 8, 13]), even though the formulas in (3)–(7) do not distinguish between SISO or MIMO cases.

290

2.2 MIMO extension

In the following, we propose two important theorems which facilitate the use of the association of variables in the NMOR of general MIMO systems. To start with, two properties of Kronecker product ($\otimes$) and Kronecker sum ($\oplus$), to be used in later proofs, are recalled: i) for compatible dimensions, $(M_1 \otimes M_2)(N_1 \otimes N_2) = (M_1 N_1) \otimes (M_2 N_2)$ and ii) $e^{M \oplus N} = e^M \otimes e^N$, wherein $M \in R^{n_M \times n_M}$, $N \in R^{n_N \times n_N}$ and $M \oplus N = M \otimes I_{n_N} + I_{n_M} \otimes N$. We also use two handy shorthands for multiple Kronecker product or sum of a matrix M, namely, $M \otimes M = M^{②}$ and $M \oplus M = ②M$ etc.

THEOREM 1. *For two square matrices $A_1 \in \mathbb{R}^{n_1 \times n_1}$ and $A_2 \in \mathbb{R}^{n_2 \times n_2}$, associating the Kronecker product of their resolvent matrices in variables s_1 and s_2 is given by*

$$\mathscr{A}_2 \left((s_1 I_{n_1} - A_1)^{-1} \otimes (s_2 I_{n_2} - A_2)^{-1} \right)$$
$$= (s I_{n_1 n_2} - (A_1 \oplus A_2))^{-1}. \quad (9)$$

PROOF. We apply (6) to find the associated time function of the Kronecker product, namely,

$$\frac{1}{(2\pi j)^2} \int_{\sigma_2 - j\infty}^{\sigma_2 + j\infty} \int_{\sigma_1 - j\infty}^{\sigma_1 + j\infty} (s_1 I_{n_1} - A_1)^{-1} \otimes (s_2 I_{n_2} - A_2)^{-1} \cdot$$
$$e^{s_1 t_1} e^{s_2 t_2} ds_1 ds_2$$
$$= \left(\frac{1}{2\pi j} \int_{\sigma_1 - j\infty}^{\sigma_1 + j\infty} (s_1 I_{n_1} - A_1)^{-1} e^{s_1 t_1} ds_1 \right) \otimes$$
$$\left(\frac{1}{2\pi j} \int_{\sigma_2 - j\infty}^{\sigma_2 + j\infty} (s_2 I_{n_2} - A_2)^{-1} e^{s_2 t_2} ds_2 \right)$$
$$= e^{A_1 t_1} \otimes e^{A_2 t_2} = e^{A_1 t_1 \oplus A_2 t_2}. \quad (10)$$

Setting $t_1 = t_2 = t$ in (10) the proof is complete. $\square$

COROLLARY 1. *Repeatedly using Theorem 1 we get the general result*

$$\mathscr{A}_k \left((s_1 I_{n_1} - A_1)^{-1} \otimes \cdots \otimes (s_k I_{n_k} - A_k)^{-1} \right)$$
$$= \left(s I_{n_1 n_2 \cdots n_k} - \oplus_{i=1}^{k} (A_i) \right)^{-1}. \quad (11)$$

THEOREM 2. *The two-variable association of the univariate transfer function $(s_1 I - A)^{-1} b$ is simply b, or*

$$\mathscr{A}_2 \left((s_1 I - A)^{-1} b \right) = b. \quad (12)$$

PROOF. We apply (6) to find the associated time function

$$\frac{1}{(2\pi j)^2} \int_{\sigma_2 - j\infty}^{\sigma_2 + j\infty} \int_{\sigma_1 - j\infty}^{\sigma_1 + j\infty} (s_1 I - A)^{-1} b e^{s_1 t_1} e^{s_2 t_2} ds_1 ds_2$$
$$= \left(\frac{1}{2\pi j} \int_{\sigma_1 - j\infty}^{\sigma_1 + j\infty} (s_1 I - A)^{-1} b e^{s_1 t_1} ds_1 \right) \cdot$$
$$\left(\frac{1}{2\pi j} \int_{\sigma_2 - j\infty}^{\sigma_2 + j\infty} e^{s_2 t_2} ds_2 \right) = e^{A t_1} b \delta(t_2). \quad (13)$$

Setting $t_1 = t_2 = t$ and taking Laplace transform again, the proof follows from the sieving property of the delta function. $\square$

With the above properties in place, we are ready to derive the key results of this paper. Using the growing exponential (also called harmonic probing) method [15], the first three transfer functions of the QLDAE (2) are

$$H_1(s) = (s I - G_1)^{-1} b, \quad (14a)$$

$$H_2(s_1, s_2) = \frac{1}{2} \left((s_1 + s_2) I - G_1 \right)^{-1} \{ G_2 [H_1(s_1) \otimes H_1(s_2)$$
$$+ H_1(s_2) \otimes H_1(s_1)] + D_1 (H_1(s_1) + H_1(s_2)) \}, \quad (14b)$$

$$H_3(s_1, s_2, s_3) = \frac{1}{3} \left((s_1 + s_2 + s_3) I - G_1 \right)^{-1} \cdot$$
$$\{ G_2 [H_1(s_1) \otimes H_2(s_2, s_3) + H_2(s_2, s_3) \otimes H_1(s_1)$$
$$+ H_1(s_2) \otimes H_2(s_1, s_3) + H_2(s_1, s_3) \otimes H_1(s_2)$$
$$+ H_1(s_3) \otimes H_2(s_1, s_2) + H_2(s_1, s_2) \otimes H_1(s_3)]$$
$$+ D_1 [H_2(s_1, s_2) + H_2(s_1, s_3) + H_2(s_2, s_3)] \}. \quad (14c)$$

Taking the G_2 part of (14b) as an example and using the property in (8) and Theorem 1, we immediately get

$$\mathscr{A}_2 \left([(s_1 + s_2) I - G_1]^{-1} G_2 [H_1(s_1) \otimes H_1(s_2) \right.$$
$$\left. + H_1(s_2) \otimes H_1(s_1)] / 2 \right)$$
$$= (s I - G_1)^{-1} G_2 (s I - ②G_1)^{-1} b^{②}. \quad (15)$$

Next, using Theorem 2 the D_1 part is easily checked to be

$$\mathscr{A}_2 \left([(s_1 + s_2) I - G_1]^{-1} D_1 [H_1(s_1) + H_1(s_2)] / 2 \right)$$
$$= (s I - G_1)^{-1} D_1 b. \quad (16)$$

Subsequently, using the often used transfer function notation $\left[\begin{array}{c|c} A & B \\ \hline C & D \end{array} \right] = C(s I - A)^{-1} B + D$ to combine (15) and (16),

$$\mathscr{A}_2(H_2(s_1, s_2)) = (s I - G_1)^{-1} \left(G_2 (s I - ②G_1)^{-1} b^{②} + D_1 b \right)$$
$$= \left[\begin{array}{c|c} G_1 & I_n \\ \hline I_n & 0 \end{array} \right] \cdot \left[\begin{array}{c|c} ②G_1 & b^{②} \\ \hline G_2 & D_1 b \end{array} \right]$$
$$= \left[\begin{array}{cc|c} G_1 & G_2 & D_1 b \\ 0 & ②G_1 & b^{②} \\ \hline I_n & 0 & 0 \end{array} \right]$$
$$= \left[\begin{array}{c|c} \tilde{G}_2 & \tilde{b}_2 \\ \hline \tilde{c}_2 & 0 \end{array} \right], \quad (17)$$

where obviously $\mathscr{A}_2(H_2)$ is recast into a higher order $(n + n^2)$ linear state space. Using similar mechanism, $\mathscr{A}_3(H_3)$ can be carefully derived to be

$$\mathscr{A}_3(H_3) = (s I - G_1)^{-1} \left(G_2 \tilde{H}_3(s) + D_1^2 b \right)$$

where

$$\tilde{H}_3(s) = (I_n \otimes \tilde{c}_2) \left(s I - G_1 \oplus \tilde{G}_2 \right)^{-1} (b \otimes \tilde{b}_2)$$
$$+ (\tilde{c}_2 \otimes I_n) \left(s I - \tilde{G}_2 \oplus G_1 \right)^{-1} (\tilde{b}_2 \otimes b),$$

which can again be put into a linear state space as in (17).

2.3 Krylov subspace for NMOR

We give concise exposition regarding efficient computer implementation. To construct the projection matrix spanning the moment space of the associated $H_1(s)$, $H_2(s) =$

$\mathscr{A}_2(H_2(s_1, s_2))$ etc. (note that these are all $n \times 1$ vectors), one often resorts to finding the Krylov subspace defined as

$$\mathcal{K}_p(G_1, b) = span\left(b, G_1 b, G_1^2 b, \cdots, G_1^{p-1} b\right).$$

The subspace basis construction is popularly done through the Arnoldi iteration, e.g. [3, 9], which then results in a lower-order orthogonal NMOR projection matrix for matching the transfer function moments [7, 6]. Nonetheless, in all practical cases, it is important to respect and exploit the structures pertinent to the state-space matrices to achieve fast speed and high accuracy.

For instance, expanding (14a) differently at $s = \infty$ and $s = 0$ would invoke $K_p(G_1, b)$ and $K_p(G_1^{-1}, G_1^{-1} b)$, respectively. Not surprisingly, the latter is more accurate in matching low-pass responses, though at the expense of computing the matrix factorization (e.g., LU) of G_1 for once. Another implementation issue is in the efficient computation of Krylov subspace projector. Referring to the second last equality in (17), the direct Arnoldi process requires multiplying the large 2×2 upper triangular block matrix (or its inverse as discussed above) with a tall matrix, consuming expensive $O((n + n^2)^2)$ work. Then at the end $[I_n \quad 0]$ is left-multiplied onto the terminated iterate to reduce it to n rows. Apparently, such brute force realization results in poor algorithmic scalability.

To reduce the computational cost, an important insight is to perform an eigenspace decomposition by applying a one-time similarity transform to the state space in (17),

$$\begin{bmatrix} G_1 & G_2 \\ 0 & \textcircled{2}G_1 \end{bmatrix} \begin{bmatrix} I_n & \Pi \\ 0 & I_{n^2} \end{bmatrix} = \begin{bmatrix} I_n & \Pi \\ 0 & I_{n^2} \end{bmatrix} \begin{bmatrix} G_1 & 0 \\ 0 & \textcircled{2}G_1 \end{bmatrix}$$

where Π is solved through the Sylvester equation

$$G_1 \Pi + G_2 = \Pi \textcircled{2}G_1$$

which is always solvable when $\lambda_i(G_1) + \lambda_j(G_1) + \lambda_k(G_1) \neq 0$, $i, j, k = 1, \cdots, n$, where $\lambda_i(\circ)$ denotes the eigenvalue. This is always true, e.g., when G_1 is stable. Subsequently, $H_2(s)$ can be put into

$$\left[\begin{array}{cc|c} G_1 & 0 & D_1 b - \Pi b^{\textcircled{2}} \\ 0 & \textcircled{2}G_1 & b^{\textcircled{2}} \\ \hline I_n & \Pi & 0 \end{array}\right]$$
$$= (sI - G_1)^{-1}(D_1 b - \Pi b^{\textcircled{2}}) + \Pi(sI - \textcircled{2}G_1)^{-1}b^{\textcircled{2}}. \quad (18)$$

Now it becomes obvious that the Krylov subspace for $H_2(s)$ can be found from each of these subsystems. In general, a similar procedure produces k subsystems in $H_k(s)$, implying that *parallelization* is feasible for such Krylov subspace generation from distinct subsystems.

A final note is on accelerating the matrix inversion, say, in the multiplication of $(\textcircled{2}G_1)^{-1}$ onto a vector when computing the Krylov subspace expanded at $s = 0$ in (18). The trick is to first factor G_1 into a convenient form. For example, suppose the Schur form of $G_1 = QRQ^T$ whereby Q is unitary and R is quasi (upper) triangular [3], then $\textcircled{2}G_1 = Q^{\textcircled{2}}(\textcircled{2}R)(Q^{\textcircled{2}})^T$ so that every inversion requires essentially a backward solve as $\textcircled{2}R$ is quasi triangular, too.

Further results on optimized computer implementation, however, are beyond the scope of this work and would be reported elsewhere.

3. NUMERICAL EXPERIMENTS

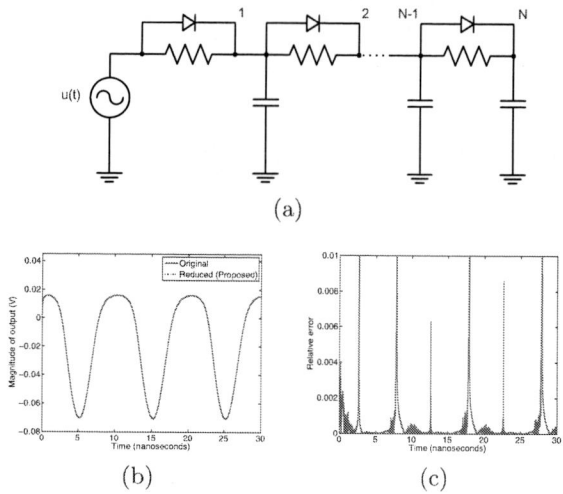

(a)

(b) (c)

Figure 2: A nonlinear transmission line circuit with voltage source. (a) Circuit schematic. (b) Transient responses. (c) Relative errors.

In this section, the proposed associated transform-based NMOR method is applied to the following cases: QLDAEs with and without the D_1 term, multi-input single-output (MISO) QLDAE and ODE with a cubic term. All experiments are performed on a platform of Intel Pentium 4 with 2.8GHz CPU and 2GB RAM.

3.1 QLDAE with D_1 term

We first try the nonlinear transmission line circuit common to many NMOR papers, with a minor modification made to the signal source. As shown in Fig. 2(a), a voltage source is injected into the circuit consisting of 100 stages. All resistors and capacitors are set to 1. The I-V characteristic of the diodes is $i_D = e^{40v_D} - 1$, which has been quadratic-linearized. Using modified nodal analysis (MNA), the circuit can be characterized by a QLDAE of (2). The full model is reduced to a 13th-order ROM by the proposed associated transform approach, with moment matching up to 6 moments of $H_1(s)$, 3 moments of $H_2(s)$ and 2 moments of $H_3(s)$ (by similar order selection as in [7, 6]). The transient simulations of the full model and ROM are shown in Fig. 2(b), and the relative errors in Fig. 2(c). This immediately validates the accuracy of the proposed association of variables NMOR scheme.

3.2 QLDAE without D_1 term

If the above circuit is injected with a current source instead of the voltage source, the resulting QLDAE equation does not have the D_1 term and the final characteristic equation has a form of

$$C\dot{x} = G_1 x + G_2 x^{\textcircled{2}} + u(t)$$

with $x \in \mathbb{R}^{70}$. Compared with NORM [7, 6] which results in a ROM of order 20, the proposed NMOR method requires only 9 to match the same number of moments. Though the NMOR time is longer in the proposed scheme due to the larger-size matrix-vector multiplication in Arnoldi iteration, such MOR process is done *only for once* and the more compact nonlinear ROM (*to be used repeatedly*) brings about a 61% reduction in simulation time compared to the

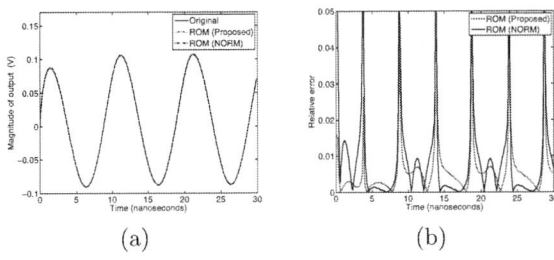

(a)   (b)

Figure 3: A nonlinear transmission line circuit with current source. (a) Transient responses. (b) Relative errors.

Table 1: Runtime comparison between the proposed method and NORM

	Original	Reduced (Proposed)	Reduced (NORM)
Sect. 3.2 Ex.			
Arnoldi	—	268s	88s
ODE solve	2723s	649s	1663s
Sect. 3.3 Ex.			
Arnoldi	—	159s	72s
ODE solve	1876s	182s	381s

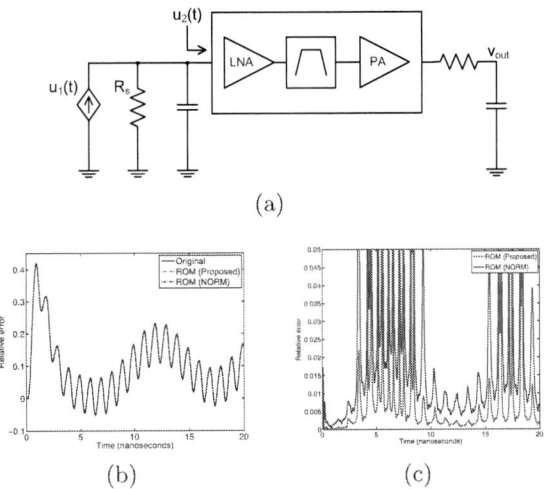

(a)

(b)   (c)

Figure 4: An example of an RF receiver system. (a) Block diagram. (b) Transient responses. (c) Relative errors.

NORM-reduced ROM, as recorded in Table 1. The transient responses are plotted in Fig. 3(a) and the nonlinear ROM from association of variables has almost the same accuracy as that produced by NORM.

3.3 MISO QLDAE

In Fig. 4(a), an RF receiver system with an input signal $u_1(t)$ is interfered by a noise signal $u_2(t)$ coupled from external environment. Considering both the signal and noise sources, the model of this system can be described in an MISO QLDAE form of (2) with $D_1 = 0$. In this experiment,

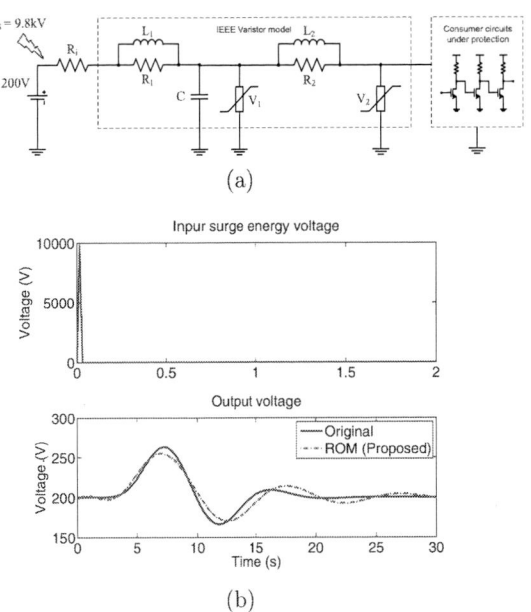

(a)

(b)

Figure 5: An example of a ZnO varistor protection circuit. (a) Equivalent circuit. (b) Transient responses.

with the same moment matching orders, the original full system has 173 voltage/current unknowns and is reduced to 14 and 27 states with the proposed and NORM methods, respectively. Transient responses and relative errors of original model and ROMs are shown in Figs. 4(b) and 4(c). Timing data are also listed in Table 1, with similar observations as in the previous example.

3.4 ODE with cubic terms

The proposed NMOR method not only applies to the QL-DAEs, but is also applicable to other forms of ODEs. In this example, another industrial example verifies the feasibility of the proposed NMOR framework for a nonlinear system with a cubic term. In Fig. 5(a), a surge protection circuit by ZnO varistors is described by an ODE system with a cubic Kronecker product

$$C\dot{x} + G_1 x + G_3 x^{③} = u.$$

The ODE has 102 states which is reduced to only 8 by the proposed method. A sudden, high voltage pulse is fed into the system and generates the dynamic response as shown in Fig. 5(b). Again, a close match in the responses is obtained even with such a low-order ROM.

4. DISCUSSION AND REMARKS

- Compared to the classical Krylov-based NORM approaches [7, 6, 4, 5], the proposed NMOR method via association of variables enjoys a much more compact ROM while matching the same order of moments. For example, for a ROM to preserve up to k_1, k_2 and k_3th-order moments in the first-, second- and third-order transfer functions, the size of the projection matrix of the proposed scheme has a dimension of $\mathcal{O}(k_1 + k_2 + k_3)$, in contrast to the much higher $\mathcal{O}(k_1 + k_2^3 + k_3^4)$

293

in NORM. Moreover, automatic selection of moment numbers in $H_1(s)$, $H_2(s)$, $H_3(s)$ etc. can utilize the Hankel singular values or similar measure inherent to linear MOR [11], again in contrast to the *ad hoc* order choice in NORM.

- As briefly mentioned, a singular C in (1) can proceed with the regular part extraction (viz. the "differential" or ODE part) with respect to the tuple (C, G_1). This can be done by Weierstrass canonical transform or the descriptor-system projector technique [16] taking advantages of circuit structures. In physical circuits, the decoupled "algebraic" part can often be easily handled as they are either immaterial or proportionally related to the regular subsystem.

- Non-DC or multipoint frequency expansion for moment matching is particularly straightforward with this associated transform approach. The resultant Volterra transfer functions all contain one single s and thereby practice from linear system theory follows. Moreover, the decomposition of $H_2(s)$ in (17) or (18) into the cascade of two LTI systems, and $H_3(s)$ into three etc., allows insightful interpretation of stability and passivity of the original nonlinear model. To our knowledge, this kind of simplicity has not appeared in the NMOR literature before.

Results along the 2nd and 3rd bullet points, however, are outside the focus of this paper and would be reported in a separate work.

5. CONCLUSION

This paper has presented an elegant approach for highly efficient Volterra-based NMOR. For the first time, the associated transform is extended to MIMO transfer functions to reduce the high-order multi-frequency-parameter transfer functions into standard single-frequency linear state spaces. Subsequently, linear MOR techniques can be directly utilized for NMOR. Such approach gives rise to remarkable savings in computation and memory, and produces much more compact models than existing NMOR schemes.

6. REFERENCES

[1] C. F. Chen and R. F. Chiu. New theorems of association of variables in multiple dimensional Laplace transform. *Int. J. Systems Sci.*, 4(4):647–664, 1973.

[2] J. Debnath and N. C. Debnath. Associated transforms for solution of nonlinear equations. *Intl. J. Math. & Math. Sci.*, 14(1):177–190, 1991.

[3] G. Golub and C. V. Loan. *Matrix Computations.* JohnsHopkins Univ. Press, Baltimore, 3rd edition, 1989.

[4] C. Gu. QLMOR: a new projection-based approach for nonlinear model order reduction. In *Proc. Int. Conf. Computer Aided Design*, pages 389–396, Nov. 2009.

[5] C. Gu. QLMOR: a projection-based nonlinear model order reduction approach using quadratic-linear representation of nonlinear systems. *IEEE Trans. Comput.-Aided Design Integr. Circuits Syst.*, 30(9):1307–1320, Sept. 2011.

[6] P. Li and L. Pileggi. Compact reduced-order modeling of weakly nonlinear analog and RF circuits. *IEEE Trans. Comput.-Aided Design Integr. Circuits Syst.*, 23(2):184–203, Feb. 2005.

[7] P. Li and L. T. Pileggi. NORM: compact model order reduction of weakly nonlinear systems. In *DAC*, pages 472–477, 2003.

[8] J. K. Lubbock and V. S. Bansal. Multidimensional Laplace transforms for solution of nonlinear equations. *Proc. IEE*, 116(12):2075–2082, Dec. 1969.

[9] A. Odabasioglu, M. Celik, and L. T. Pileggi. PRIMA: Passive reduced-order interconnect macromodeling algorithm. *IEEE Trans. Comput.-Aided Design Integr. Circuits Syst.*, 17(8):645–654, Aug. 1998.

[10] J. R. Phillips. Projection-based approaches for model reduction of weakly nonlinear, time-varying systems. *IEEE Trans. Comput.-Aided Design Integr. Circuits Syst.*, 22(2):171–187, Feb. 2003.

[11] J. R. Phillips, L. Daniel, and L. M. Silveira. Guaranteed passive balancing transformations for model order reduction. *IEEE Trans. Comput.-Aided Design Integr. Circuits Syst.*, 22(8):1027–1041, Aug. 2003.

[12] L. Pillage and R. Rohrer. Asymptotic waveform evaluation for timing analysis. *IEEE Trans. Comput.-Aided Design Integr. Circuits Syst.*, 9:352–366, Apr. 1990.

[13] D. C. Reddy and N. C. Jagan. Multidimensional transforms: new technique for the association of variables. *Electron. Lett.*, 7(10):278–279, May 1971.

[14] M. Rewienski and J. White. A trajectory piecewise-linear approach to model order reduction and fast simulation of nonlinear circuits and micromachined devices. *IEEE Trans. Comput.-Aided Design Integr. Circuits Syst.*, 22(2):155–170, Feb. 2003.

[15] W. Rugh. *Nonlinear System Theory – The Volterra-Wiener Approach.* Baltimore, MD: Johns Hopkins Univ. Press, 1981.

[16] Z. Zhang and N. Wong. An efficient projector-based passivity test for descriptor systems. *IEEE Trans. Comput.-Aided Design Integr. Circuits Syst.*, 29(8):1034–1042, Aug. 2010.

AMOR: An Efficient Aggregating Based Model Order Reduction Method for Many-Terminal Interconnect Circuits

Yangfeng Su[1,2], Fan Yang[1*], and Xuan Zeng[1*]
[1]State Key Lab of ASIC & System, Microelectronics Dept., Fudan University, China
[2]School of Mathematical Sciences, Fudan University, China

ABSTRACT

In this paper, we propose an efficient *Aggregating* based *Model Order Reduction* method (AMOR) for many-terminal interconnect circuits. The proposed AMOR method is based on the observation that those adjacent nodes of interconnect circuits with almost the same voltage can be aggregated together as a "super node". Motivated by such an idea, we propose an efficient spectral partition algorithm in AMOR method to partition the nodes into groups with almost the same voltages. The reduced-order models are then obtained by aggregating the adjacent nodes within the same groups together as "super nodes" in AMOR method. The efficiency of AMOR method is not limited by the numbers of the terminals of the networks. Moreover, noticing that the aggregating procedure can be regarded as mapping the original problem into a coarse-grid problem in multigrid method, we propose a computation-efficient smoothing procedure to further improve the simulation accuracy of the reduced-order models. With such a strategy, the simulation accuracy of the reduced-order models can always be guaranteed. Numerical results have demonstrated that, without the smoothing procedure, the reduced-order models obtained by AMOR can still achieve higher simulation efficiency in terms of accuracy and CPU time than the reduced-order models obtained by the existing elimination based methods. With the smoothing procedure, the simulation accuracy of the reduced-order models can further be improved with several iterations.

Categories and Subject Descriptors:
J.6 [Computer-Aided Engineering]: Computer-Aided Design
General Terms: Algorithm, Design
Keywords: Interconnect, Many-Terminal, Model Order Reduction

1. INTRODUCTION

The interconnect circuits have a large number of terminals intrinsically. Generally, a practical IC can be divided into active components and interconnects. The active components have several pins which connect to the pins of other active components through interconnects. The pins of the active components should be regarded as the terminals of the interconnect circuits. A single interconnect may connect thousands of pins together, and thus may has thousands of terminals. On the other hand, although the number of

terminals of a single interconnect can be small, due to the coupling effects of interconnects, the number of terminals of the network consisting of a large number of coupling interconnects will be very large.

Unfortunately, when dealing with the interconnect circuits with a large number of terminals, the efficiency of the traditional Krylov subspace based Model Order Reduction (MOR) methods degrades as the number of external ports of the circuit increases remarkably [1]. This is because when matching the same number of moments, the size of the reduced-order model is proportional to the number of inputs. The TBR-like methods [2], [3] also encounter difficulties while tackling with the interconnect circuits with a large number of terminals, since the Hankel singular values decay slower as the number of inputs increasing [4].

A variety of methods attempting to improve the efficiency of the traditional Krylov subspace based and TBR-like methods have been proposed in the past decade [1], [4]–[7]. These methods exploit the correlations of the entries of matrix transfer functions [1] [5], the correlations of the input waveforms [4] or the correlations of the terminals [6] [7] to enhance the traditional Krylov subspace based and TBR-like methods, but cannot handle general many-terminal interconnect circuits without correlation information.

An alternative class of methods for MOR of interconnect circuits with large numbers of terminals are the elimination based methods, e.g., PACT [9], TICER [10] and SIP [8]. In these elimination based methods, the reduced-order model is obtained by eliminating the internal nodes, in a manner similar to sparse Gaussian elimination. The reduced-order models generated by these methods aim to match the first two moments of the transfer functions of the original systems. Because the reduction is achieved by eliminating the internal nodes, the efficiency of the elimination based methods is not limited by the number of terminals.

A weakness of the elimination based methods is that the elimination will introduce a large number of fill-ins, which makes the reduced-order models very dense, even if the sparsity control strategy is employed. The efficiency gained by reducing the number of nodes degenerates significantly due to the large number of fill-ins. In our numerical experiments, we find that the simulation time for the reduced-order models by the elimination based methods will be even higher than the original models for many-terminal interconnect circuits, due to the large number of fill-ins.

In this paper, we propose an efficient *Aggregating* based *Model Order Reduction* method (AMOR) for many-terminal interconnect circuits. We focus on the RC network in this paper, and will extend the proposed method to RLC network in the future work. The proposed AMOR method is based on observation that those adjacent nodes of the interconnect circuits with almost the same voltage can be aggregated together as a "super node". Motivated by such an idea, we propose an efficient spectral partition algorithm in AMOR method to partition the nodes into groups with almost the

---

*Corresponding authors. Email: {yangfan, xzeng}@fudan.edu.cn.

same voltages. The reduced-order models are then obtained by aggregating the adjacent nodes within groups together as "super nodes". The values of the resistors and capacitors of the reduced-order models are always positive, and thus the reduced-order models are passive and physically realizable, which makes the reduced-order models more applicable in the downstream simulations.

Moreover, we remark that the aggregating procedure in AMOR method can be regarded as mapping the original problem into a coarse-grid problem in multigrid method. We propose a computation-efficient smoothing procedure to further improve the simulation accuracy of the reduced-order models. With such a strategy, the simulation accuracy of the reduced-order models can always be guaranteed.

Compared with the elimination based methods, the proposed AMOR method employs an "aggregating" strategy, and will never introduce fill-ins. The simulation time will be definitely reduced by AMOR method. Furthermore, the AMOR method exploits the idea of multigrid, and utilizes a computation efficient smoothing procedure to improve the accuracy of the reduced-order models during the simulation procedure. The simulation accuracy of the reduced-order models can always be guaranteed. Numerical results have demonstrated that, without the smoothing procedure, the reduced-order models obtained by AMOR can still achieve higher simulation efficiency in terms of accuracy and CPU time than the reduced-order models obtained by the existing elimination based method SIP [8] and PACT [9]. With the smoothing procedure, the reduction error of the reduced-order models can be further reduced with several iterations.

The rest of the paper is organized as follows. In section 2, the problem formulation of MOR of many-terminal interconnect circuits will be described firstly. The AMOR method and the smoothing procedure will be proposed in section 3 and section 4, respectively. The efficiency of the proposed method is demonstrated by several practical examples in section 5. In section 6, we conclude the paper.

2. PROBLEM FORMULATION

Generally, a linear circuit can be described by the following equation,

$$\begin{bmatrix} C_p & C_c^T \\ C_c & C_i \end{bmatrix} \begin{bmatrix} \dot{x}_p \\ \dot{x}_i \end{bmatrix} + \begin{bmatrix} G_p & G_c^T \\ G_c & G_i \end{bmatrix} \begin{bmatrix} x_p \\ x_i \end{bmatrix} = \begin{bmatrix} BI_p \\ 0 \end{bmatrix}, \quad (1)$$

where $G_p \in R^{M \times M}$ and $C_p \in R^{M \times M}$ represent the contributions of the resistors and capacitors between the ports. $G_i \in R^{N \times N}$ and $C_i \in R^{N \times N}$ represent the contributions of the resistors and capacitors between the internal nodes. $G_c \in R^{N \times M}$ and $C_c \in R^{N \times M}$ describe the branches that connect internal nodes to ports. $B \in R^{M \times p}$ is an incidence matrix, which relates the input sources to the corresponding nodes. I_p represents the input currents.

The MOR methods for the parasitic linear circuit aim to find a projection matrix Q, which transforms the original linear circuit (1) to a $(M+n)th$ order reduced-order model,

$$\begin{bmatrix} C_p & \tilde{C}_c^T \\ \tilde{C}_c & \tilde{C}_i \end{bmatrix} \begin{bmatrix} \dot{x}_p \\ \dot{\tilde{x}}_i \end{bmatrix} + \begin{bmatrix} G_p & \tilde{G}_c^T \\ \tilde{G}_c & \tilde{G}_i \end{bmatrix} \begin{bmatrix} x_p \\ \tilde{x}_i \end{bmatrix} = \begin{bmatrix} BI_p \\ 0 \end{bmatrix}. \quad (2)$$

where $\tilde{C}_i, \tilde{G}_i \in R^{n \times n}$, $\tilde{C}_c, \tilde{C}_c \in R^{n \times M}$ and $\tilde{x}_i \in R^n$. The number of the internal nodes is reduced from N to n in the reduced-order model, while the ports are preserved in the reduced-order model such that the reduced-order model can be simulated together with the nonlinear circuits.

3. THE PROPOSED AMOR METHOD

In this section, we will present the proposed Aggregating Model Order Reduction (AMOR) method, including the spectral partition algorithm and the aggregating procedure.

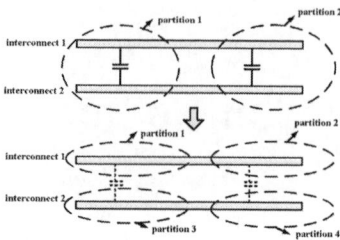

Figure 1: Neglecting the coupling capacitors leads to a more accurate reduced-order model.

3.1 Spectral Partition Algorithm

In the AMOR method, the partition algorithm aims to partition the adjacent nodes of the parasitic RC network into groups with almost the same potentials. Therefore, those partition algorithms developed by Kernighan and Lin [11], and Fiduccia and Mattheyses [12], which emphasize the equal-sized partition, are not suitable for this purpose. Instead, we use the idea of spectral partition [13] [14] in AMOR method to cluster the nodes with almost the same voltages into groups, based on the intuition that the nodes connected together with large admittance tend to be equipotential nodes.

The parasitic RC network can be described by an undirected graph $G = (V, E)$, where V denotes the nodes of the circuit, E represents the resistors and capacitors connecting between the nodes. The weights of edge $e(i, j)$ of the graph can be defined as the admittance between the node i and node j,

$$w(i, j) = \sum_k \frac{1}{r_k},$$

where r_k represent the resistors connecting between node i and node j. The admittance is a good criteria of the similarity of the potential of the nodes, i.e., the larger the admittance, the higher probability the nodes have the same waveforms.

The effects of capacitors are neglected in the weight $w(i, j)$ based on the following considerations. The capacitors can be divided into two categories, i.e., grounded capacitors and coupling capacitors. The grounded capacitors will not affect the relative potentials of the adjacent nodes. The coupling capacitors are connected between different interconnects to model the coupling effects between the interconnects. If the coupling capacitors are not taken into account, we actually divide a group of nodes connected together by the coupling capacitors into several individual smaller partitions, as illustrated in Fig. 1. It will lead to a more accurate reduced-order model.

The spectral partition algorithm can cluster the nodes connected together by edges with high weights together [14]. The spectral partition algorithm is based on the spectral analysis of the Laplacian matrix L of the graph $G = (V, E)$. The entries of the Laplacian matrix $L = (l_{ij})$ is defined as following,

$$\begin{cases} l_{ij} = -w(i, j), & i \neq j, \ (i, j) \in E; \\ l_{ij} = 0, & i \neq j, \ (i, j) \notin E; \\ l_{ij} = \sum_{(k,i) \in E} w(k, i), & i = j. \end{cases}$$

The Laplacian matrix L may be viewed as the discrete analog of the Laplace Δ operator. One can verified that the Laplacian matrix L has the following properties
1). L is a sparse matrix (because the graph $G = (V, E)$ is a sparse graph);
2). L is symmetric;
3). L is non-negative definite, and all the eigenvalues of L is non-negative;

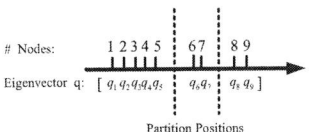

Figure 2: Illustration of eigenvector q placement.

4). The smallest eigenvalue of L is zero, with eigenvecor $(1, 1, \cdots, 1)^T$.

The eigenvector q corresponding to the second smallest eigenvalue of the Laplacian matrix L provides a one-dimensional placement of the nodes of the graph G, as demonstrated in Fig. 2. In this one-dimensional placement, the coordinates of the nodes V are the corresponding values in the eigenvector q. This placement reflects the distribution of the nodes of the graph $G = (V, E)$ with the weights of edges defined earlier in a one-dimensional view. The distance between two nodes in the one-dimensional placement can be viewed as a measure of the impedance between them. Note that the nodes connected together with large admittance or small impedance tend to be equipotential nodes. Therefore, the best partition of the nodes of the graph G is found by examining such a one-dimensional placement [14]. The first k biggest gaps of the one-dimensional placement are select as the best partition positions, as illustrated in Fig. 2. Here, k is a predefined integer.

Algorithm 1 Spectral Partition Algorithm

Input: graph $G = (V, E)$, and the Laplacian matrix L
Output: the partition of the nodes
1: Compute the eigenvector q corresponding to the second smallest eigenvalue of the Laplacian matrix L.
2: Sort the eigenvector q, and obtain the best partition positions of the nodes by finding the biggest k gaps of the sorted eigenvector.

We summarize the spectral partition algorithm in Algorithm 1. It appears that eigenvalue and eigenvector computations of the Laplacian matrix is computation intensive. Actually, the well-developed Lanczos algorithm [15] has greatly speedup the eigenvalue and eigenvector computations of a sparse matrix. The Lanczos algorithm only requires several sparse matrix multiplication with vector to complete the eigenvalue and eigenvector computations.

By recursively call the aforementioned spectral partition algorithm, we can partition the nodes of the graph into groups, and the number of nodes in each group is less than a predefined threshold η. Here, η can be viewed as the "granularity" of the "coarse grid" in the following aggregating procedure. The bigger η is set to, the coarser-grid reduced-order model we obtain.

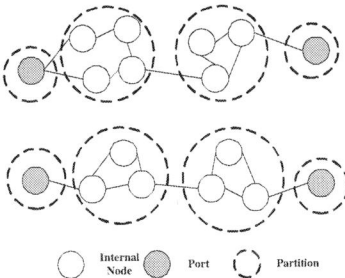

Figure 3: An example of the partition of a graph.

Before we derive the final partition algorithm, there still exist two practical issues to be addressed. Firstly, the graph $G = (V, E)$ which represents the RC network may be di-

vided into several connected components. We can employ the computation efficient breadth-first search or depth-first search algorithm [16] to obtain these connected components, and then use the spectral partition algorithm to partition each connected components. Secondly, the RC network is connected to the nonlinear devices through the ports. We should preserve these ports in the reduced-order models. In order to preserve these ports in the aggregating procedure, we should artificially set those ports as individual partitions. Consequently, we get the final partition algorithm in Algorithm 2.

For illustration, we show an example of the partition obtained by the spectral partition algorithm in Fig. 3. In Fig. 3, the graph is firstly partitioned into two connected components by breadth-first search or depth-first search algorithm. The spectral partition algorithm is then employed to further partition each connected component into several partitions. The ports are artificially set as individual partitions.

Algorithm 2 Partition Algorithm

Input: graph $G = (V, E)$, the weights $w(i, j)$ of the edges and the label of the ports
Output: the partition of the nodes
1: Obtain the connected components of the graph $G = (V, E)$.
2: For each connected component, recursively call the spectral partition algorithm to partition the nodes into groups in which the numbers of nodes are less than a predefined threshold η.
3: Artificially set the ports as individual partitions.

3.2 Aggregating Procedure

Once the partition of nodes is obtained, we employ an aggregating procedure to merge the nodes in the same partitions to "super nodes", and finally derive the reduced-order model.

The aggregating procedure is summarized in Algorithm 3. In the aggregating procedure, the adjacent nodes in the same partition, or equivalently the adjacent nodes tending to be equipotential, are aggregated together as "super nodes". Correspondingly, the capacitors/resistors connected between the nodes of the same partition are neglected. The capacitors/resistors connected between the nodes in a partition and ground are aggregated together and connected between the "super node" of the partition and ground. The capacitors/resistors connected between the nodes of different partitions are aggregated together and connected between the "super nodes" which represent the different partitions.

From the aggregating procedure, we can find that the resistors/capacitors of the reduced-order circuit are all positive. It means that the reduced-order circuit is physically realizable and also passive.

The aggregating procedure in AMOR method can be regarded as mapping the problem into a coarse-grid problem. From our numerical experiments for several industrial parasitic RC network, the coarse-grid reduced-order models achieve much higher accuracy than the reduced-order model obtained by the elimination based MOR method, such as SIP [8] and PACT [9]. The coarse-grid models obtained by the proposed AMOR method are good reduced-order models for circuit simulation.

3.3 Connection to Projection

The aggregating procedure can also be expressed as a projection procedure mathematically. We rewrite the original RC network (1) here.

$$\begin{bmatrix} C_p & C_c^T \\ C_c & C_i \end{bmatrix} \begin{bmatrix} \dot{x}_p \\ \dot{x}_i \end{bmatrix} + \begin{bmatrix} G_p & G_c^T \\ G_c & G_i \end{bmatrix} \begin{bmatrix} x_p \\ x_i \end{bmatrix} = \begin{bmatrix} BI_p \\ 0 \end{bmatrix}. \quad (1)$$

297

Algorithm 3 Aggregating Procedure

Input: the RC network, and the partition result obtained by spectral partition algorithm

Output: the reduced-order circuit

1: **for** Each Partition **do**
2:   Use a "super node" to represent all the nodes in the partition.
3:   Neglect the resistors and capacitors connected between the nodes in the partition.
4:   Denote $\{r_1, r_2, \cdots, r_j\}$, and $\{c_1, c_2, \cdots, c_l\}$ the resistors and capacitors connected between the nodes in the partitions and ground. Add a resistor $r = 1/(\sum_{i=1}^{j} 1/r_i)$, and a capacitor $c = \sum_{i=1}^{l} c_i$ between the "super node" and ground.
5: **end for**
6: The capacitors and resistors connected between the nodes of different partitions are now connected between the "super nodes" which represent the partitions.
7: Use equivalent resistor/capacitor to represent the multiple resistors/capacitors parallel connected between the same pair of nodes.

For simplicity, we assume that equation (1) has been permutated such that

$$x_i = \{x_1^1, \cdots, x_{n_1}^1, x_1^2, \cdots, x_{n_2}^2, \cdots, x_1^\alpha, \cdots, x_{n_\alpha}^\alpha\}^T \quad (3)$$

where $\{x_1^i, \cdots, x_{n_i}^i\}$ represent the nodal voltages of the nodes in the i-th partition of the internal nodes. n_i is the number of nodes in the i-th partition, and α denotes the number of partitions for the internal nodes. It can be verified that the aggregating procedure is equivalent to project the RC network (1) onto the following projection subspace

$$Q = \begin{bmatrix} I & \\ & Q_i \end{bmatrix}, \quad (4)$$

where I is an $M \times M$ dimensional identity matrix, $Q_i \in R^{N \times \alpha}$. The submatrix Q_i is defined as,

$$Q_i = \begin{bmatrix} \frac{1}{\sqrt{n_1}} & & & \\ \vdots & & & \\ \frac{1}{\sqrt{n_1}} & & & \\ & \ddots & & \\ & & \frac{1}{\sqrt{n_\alpha}} & \\ & & \vdots & \\ & & \frac{1}{\sqrt{n_\alpha}} & \end{bmatrix} \begin{array}{l} \left.\rule{0pt}{15pt}\right\} n_1 \\ \vdots \\ \left.\rule{0pt}{15pt}\right\} n_\alpha \end{array}$$

By using the projection basis Q, we can get the reduced-order model in the form of (2).

4. ACCURACY IMPROVEMENT BY SMOOTHING PROCEDURE

Before the discussion of smoothing procedure of the multi-grid method, we briefly review the transient simulation process of a linear circuit. During the transient simulation, the algebraic differential equation (1) is firstly discretized by using backward Euler, forward Euler trapezoidal, or multi-step approximation. After discretization, a linear system has to be solved to obtain the response for this time step.

The reduction procedure presented in the previous section actually builds an explicit mapping between the nodes of the original and the reduced-order circuits. The ports of the linear circuit are preserved in the reduced-order model. Therefore, there exists a one-to-one mapping between the ports of the original and the reduced-order models. For the internal nodes of the original circuit, we aggregate the nodes in the same partition into a "super node". Thus, the internal nodes of the original circuit of a partition are mapped to the corresponding "super node". We can use such an explicit mapping to derive a multigrid iteration during the simulation of the original circuit as described by (1).

Multigrid methods has been widely applied for solving PDE-like problems. Generally, multigrid methods consist of the following complementary components:

1). Coarse grid correction which reduces the low frequency error components on coarse grid.

2). Relaxation (smoothing) which reduces the high frequency error components on fine-grid.

Coarse grid correction first maps the original problem to a coarser grid, solves the mapped problem, and then map the solution back to the fine grid. The mapping between the fine and coarse grids is realized by so called restriction and prolongation operators. The restriction operator R_h^{2h} maps the problem from fine grid to the coarse grid. The prolongation operator P_{2h}^h maps the solution back from coarse grid to fine grid.

As shown in the review of the transient simulation process, the simulation process finally relies on the solution of a linear systems, we assume the linear systems to be solved can be expressed as

$$Ax = b. \quad (5)$$

We use the solution of this linear system to demonstrate the proposed smoothing procedure. For simplicity, we also assume that equation (5) has been permutated such that x satisfies (3).

Firstly, we map this problem to coarse grids by restriction operator R_h^{2h}. The restriction operator R_h^{2h} here is actually the transpose of the projection matrix Q defined in (4), i.e., $R_h^{2h} = Q^T$. The reduced problem on coarse grid can thus be obtained

$$A_r x_r = b_r, \quad (6)$$

where $A_r = Q^T A Q$ and $b_r = Q^T b$. By solving this reduced system (6), we get the solution x_r on the coarse grid.

Afterwards, we map the solution x_r to the fine grid by the prolongation operator P_{2h}^h. The prolongation operator P_{2h}^h here is the projection matrix Q defined in (4), i.e., $P_{2h}^h = Q$. The solution on the fine grid can be obtained by

$$x = Q x_r.$$

The residue can be expressed as $r = b - Ax$. Then, we employ the relaxation (smoothing) procedure to compute the high frequency corrections to the solution on the fine grid. We employ a block Jacobi iterative method in the proposed smoothing procedure. The block Jacobi iterative method is based on splitting matrix A into block diagonal part A_1 and non-block-diagonal part A_2, i.e., $A = A_1 + A_2$ and the iteration can be derived as

$$A_1 x^{n+1} = -A_2 x^n + b,$$

or equivalently

$$A_1 \Delta x = -(A_1 + A_2)x^n + b = b - Ax^n = r^n,$$

where x^{n+1}, x^n represent the solutions at $(n+1)$-th and n-th iterations, respectively. $\Delta x = x^{n+1} - x^n$. The block Jacobi iterative method for the fine grid problem (5) can be summarized in Algorithm 4.

One should note that the size of $A(\mathbf{ind}_j, \mathbf{ind}_j)$ in Algorithm 4 is less than the predefined threshold η as we defined in subsection 3.1. The solution $\Delta x(\mathbf{ind}_j)$ can thus be efficiently obtained. The block Jacobi iterative method presented in Algorithm 4 is therefore very computation efficient.

The previous multigrid iteration can be repeated several times to achieve higher accuracy. An interesting fact of the

Algorithm 4 Block Jacobi Iteration

1: **for** $n \leq number\_of\_iterations$ **do**
2:    **for** $j \leq number\_of\_partitions$ **do**
3:       Find nodes which belong to the j-th partition. Denote $\mathbf{ind}_j$ the indexes of these nodes.
4:       $\Delta x(\mathbf{ind}_j) = A(\mathbf{ind}_j, \mathbf{ind}_j) \backslash r(\mathbf{ind}_j)$
5:    **end for**
6:    $x^{n+1} = x^n + \Delta x$
7:    $r = b - Ax^{n+1}$
8: **end for**

Table 1: Information of the six test cases.

ckt	# ports	# nodes	# MOSFETs	# Rs & Cs	Simulation time (s)
zeni_pe_2	158	1351	90	6942	15.0
C1_2	1326	7314	836	44726	14.0
MUL8x8_2	3685	18112	1956	208497	190.0
testx1	9374	34463	1928	192379	397.8
testx2	4653	15860	848	43569	51.0
testx3	6379	21077	4281	104274	4666.3

multigrid method is that the reduced problem (6) is just the discretized linear systems to be solved for the reduced-order system (2). This means we can start the transient simulation of the original linear system (1) by simulating the reduced-order model (2). The low frequency error components will rapidly vanish on the coarse grid reduced-order model. Afterwards, we map the solution to the original fine grid model, and run the block Jacobi iterative method to reduce the high frequency error components and therefore smooth the solution. The smoothing procedure only requires several block Jacobi iterations, which are very computation efficient as we pointed out earlier. However, the simulation accuracy of the reduced-order models can be significantly improved with such a smoothing procedure.

5. NUMERICAL RESULTS

In this section, we will present several numerical results to demonstrate the efficiency of the proposed AMOR method. All the experiments are conducted on a PC with Intel P8600 2.4GHz CPU and 4GB memory. The test cases are real post-layout simulation cases from industry.

There are six test cases in our experiments. The test case zeni_pe_2 is a decoder for RAM. MUL8x8_2 is an 8x8 multiplier. C1_2 and testx1 are clock driven circuits. testx2 is an amplifier and testx3 is a PLL circuit. We list the information of all the six test cases after parasitic RC extraction in Table 1. The numbers of Rs and Cs in the table denote the number of resistors and capacitors in the linear circuit.

The ports of the linear circuits are preserved in the reduced-order circuits. We check the consistency of the waveforms of these nodes in the original circuits and the reduced-order circuits to verify the accuracy of the MOR methods. The reduction error is defined as

$$reduction\_error = \frac{||v - \tilde{v}||}{||v||},$$

where v represents the voltage of the node in the original circuit, and $\tilde{v}$ represents the voltage of the corresponding node in the reduced-order circuit.

5.1 Comparison between AMOR and SIP

For the first three test case, we employ SIP [8] and AMOR methods to reduce the linear circuits into lower-order models, then synthesize the reduced-order models into equivalent circuits, and finally combine the synthesized equivalent circuits with the nonlinear circuits to formulate the final reduced-order circuits. The final reduced-order circuits are then simulated by HSPICE. The sparsity control techniques [8] are employed in SIP to make the reduced-order models more sparse. The predefined threshold η in AMOR

method is set as 30 for these test cases. We list the information of the final reduced-order circuits by SIP and AMOR in Table 2. From Table 2, we can find that although the sparsity control techniques are employed in SIP, the reduced-order models by AMOR is still more sparse than those by SIP. The numbers of resistors and capacitors of the reduced-order models by AMOR is significantly less then those by SIP. Due to the sparsity of the reduced-order models, the simulation times of the reduced-order models by AMOR are much lower than those by SIP, though the numbers of nodes of the reduced-order models by AMOR is slightly larger than those by SIP.

Especially, for the test case C1_2, the reduced-order model generated by SIP has 84178 resistors and capacitors. Although the sparsity control techniques has been employed in SIP, the reduced-order model is still very dense. These large numbers of resistors and capacitors correspond to the fill-ins introduced during the elimination. Since the reduced-order model is very dense, it takes 128.0 seconds by HSPICE to simulate the reduced-order model, which is even much higher than the simulation time of the original circuit. Contrary to SIP, AMOR method is based on aggregation, and it will never introduce fill-ins. The numbers of resistors and capacitors will be definitely reduced by AMOR method. The efficiency of AMOR method is guaranteed.

We also list the reduction errors in Table 2. From Table 2, we can find that AMOR can achieve remarkably higher accuracy than SIP method.

5.2 Comparison between AMOR and PACT Implementation in HSPICE

HSPICE has an embedded MOR tool, which implements the PACT algorithm [9]. Several sparsity control techniques, such as node reserving and eliminating large resistors and small capacitors, are employed in the PACT implementation in HSPICE (HSPICE-PACT for short). Because HSPICE is a commercial tool, the PACT implementation in HSPICE is well tuned to provide nice reduction efficiency.

We compare the proposed AMOR method with HSPICE-PACT in this subsection. For the last three test cases, we first employ HSPICE-PACT and the proposed AMOR method to generate the reduced-order circuits. The reduced-order circuits are then simulated by HSPICE. The predefined threshold η in AMOR method is set as 30 for all the test cases. We list the information of the final reduced-order circuits by SIP and AMOR in Table 3. HS-PACT represents HSPICE-PACT reduction method in Table 3.

From Table 3, we can find that the reduced-order models by AMOR is more sparse than those by HSPICE-PACT. The numbers of resistors and capacitors of the reduced-order models by AMOR is significantly less then those by HSPICE-PACT. Due to the sparsity of the reduced-order models, the simulation times of the reduced-order models by AMOR are much lower than those by HSPICE-PACT, though the numbers of nodes of the reduced-order models by AMOR is slightly larger than those by HSPICE-PACT.

Especially, for the test case "testx3", the reduced-order model generated by HSPICE-PACT has 1022341 resistors and capacitors, which is about 10 times of those of the original circuits. These large numbers of resistors and capacitors correspond to the fill-ins introduced during the elimination. Since the reduced-order model is very dense, it takes 36729 seconds by HSPICE to simulate the reduced-order model, which is nearly 10x simulation time of the original circuit. Contrary to HSPICE-PACT, AMOR method is based on aggregation, and it never introduces fill-ins. The numbers of resistors and capacitors are definitely reduced by AMOR method. Moreover, AMOR can achieve better accuracy than HSPICE-PACT except the test case "testx3". However, one should note that the simulation time of the reduced-order model of HSPICE-PACT is nearly 10x of the original circuit

Table 2: Information of the reduced-order circuits for the test cases by SIP and AMOR.

ckt	# nodes		# Rs and Cs		Reduction time (s)		Simulation time(s)		Reduction error	
	SIP	AMOR	SIP	AMOR	SIP	AMOR	SIP	AMOR	SIP	AMOR
zeni_pe_2	371	404	694	488	0.557	0.196	4.0	3.3	0.025	0.016
C1_2	3264	3271	84178	4593	4.70	0.886	128.0	4.0	0.22	0.0021
MUL8x8_2	4792	5114	14637	12985	1.70	3.04	36.0	27.0	0.093	0.0087

Table 3: Information of the reduced-order circuits for the test cases by HSPICE-PACT and AMOR.

ckt	# nodes		# Rs and Cs		Reduction time (s)		Simulation time(s)		Reduction error	
	HS-PACT	AMOR	HS-PACT	AMOR	HS-PACT	AMOR	HS-PACT	AMOR	HS-PACT	AMOR
testx1	13675	13035	80823	60691	2.02	9.49	179.3	98.4	0.0073	0.0058
testx2	5379	6299	20492	15173	0.45	1.52	28.5	12.6	9.30e-6	3.39e-6
testx3	14878	14539	1022341	38037	175.5	8.71	36729.0	505.3	1.00e-7	5.90e-6

for this case.

5.3 Accuracy Improvement by Smoothing Procedure

In this subsection, we will use the test case C1_2 to illustrate the efficiency of the smoothing procedure. From section 4, we know that the transient simulation process finally relies on the solution of a set of linear systems. We take the smoothing procedure for solving such a linear system as an example here. In our smoothing procedure, the reduced problem on coarse grid, i.e., (6) is firstly solved. Equation (6) also corresponds to the linear system to be solved for the reduced-order circuits. After the solution x_r of equation (6) is obtained, it is mapped to the fine grid by the prolongation operator P_{2h}^h. Then, the smoothing procedure, i.e., the block Jacobi iterations, are employed to further improve the accuracy. We list the relative errors and the elapsed time during the smoothing procedure in Table 4. The relative error is defined as

$$relative\_error = \frac{||x_{iter} - x_{exact}||}{||x_{exact}||},$$

where x_{iter} denotes the solution during the iterations, x_{exact} represents the exact solution for the problem on the fine grid, i.e., $Ax = b$. From Table 4, we can see that after two smoothing steps, the relative error is reduced from 5.04e-5 to 4.52e-6. The elapsed time for each smoothing procedure is about half of solving the reduced problem $A_r x_r = b_r$ on the coarse grid.

Table 4: Relative errors and the elapsed times during the smoothing procedure.

Step	$A_r x_r = b_r$	$x = P_{2h}^h x_r$	Iter # 1	Iter # 2
Time(s)	0.0207	–	0.0103	0.0106
Relative error	–	5.04e-5	4.35e-5	4.52e-6

6. CONCLUSION

In this paper, we present an Aggregating based Model Order Reduction method (AMOR) for many-terminal interconnect circuits. Compared with the existing elimination based methods, the proposed AMOR method can generate more sparse and accurate reduced-order models. The reduced-order models of the proposed AMOR method can be synthesized as physically realizable linear circuits and the passivity of the reduced-order circuits can also be guaranteed. Furthermore, the aggregating procedure in AMOR method can be regarded as mapping the problem into a coarse-grid problem in multigrid method. We propose a computation-efficient smoothing procedure, which can be embedded in the simulation process of the reduced-order models, to further improve the simulation accuracy of the reduced-order models. With the smoothing procedure, the simulation accuracy of the reduced-order models can further be improved with several iterations. In the future work, we will extend the AMOR method to RLC network and seek a

more efficient smoothing procedure to improve the simulation accuracy of reduced-order circuits.

Acknowledgment

This research is supported partially by National Natural Science Foundation of China (NSFC) research projects 61125401, 61006030, 60976034 and 61076033, E-Institute of Shanghai Municipal Education Commission, N. E03004, National Basic Research Program of China under the grant 2011CB309701, National Major Science and Technology Special Project 2011ZX01035-001-001-003 of China during the 12-th five-year plan period.

7. REFERENCES

[1] P. Feldmann and F. Liu, "Sparse and efficient reduced order modeling of linear subcircuits with large number of terminals," in *Proceedings of IEEE/ACM International Conference on Computer-Aided Design*, Nov. 2004, pp. 88–92.

[2] B. C. Moore, "Principal component analysis in linear systems: Controllability, observability, and model reduction," *IEEE Trans. Automatic Control*, vol. 35, no. 1, pp. 17–32, Feb. 1981.

[3] J. Phillips and L. Silveira, "Poor man's TBR: A simple model reduction scheme," *IEEE Transactions on Computer-Aided Design of Integrated Circuits and Systems*, vol. 24, no. 1, pp. 43–55, Jan. 2005.

[4] L. Silveira and J. Phillips, "Exploiting input information in a model reduction algorithm for massively coupled parasitic networks," in *Proceedings of IEEE/ACM Design Automation Conference*. San Diego, June 2004, pp. 385–388.

[5] P. Feldmann, "Model order reduction techniques for linear systems with large numbers of terminals," in *Proceedings of IEEE/ACM Design, Automation and Test in Europe*, 2004.

[6] P. Li and W. Shi, "Model order reduction of linear networks with massive ports via frequency-dependent port packing," in *IEEE/ACM DAC*, 2006, pp. 267–272.

[7] P. Liu, S. Tan, H. Li, Z. Qi, J. Kong, B. McGaughy, and L. He, "An efficient method for terminal reduction of interconnect circuits considering delay variations," in *Proceedings of IEEE/ACM International Conference on Computer-Aided Design*, 2005.

[8] Z. Ye, D. Vasilyev, Z. Zhu, and J. R. Phillips, "Sparse implicit projection (sip) for reduction of general many-terminal networks," in *Proc. ICCAD' 2008*.

[9] K. J. Kerns and A. T. Yang, "Stable and efficient reduction of large, multiport networks by pole analysis via congruence transformations," *IEEE Trans. CAD*, vol. 16, no. 7, pp. 734–744, July 1997.

[10] B. N. Sheehan, "TICER: Realizable reduction of extracted RC circuits," in *Proc. ICCAD'1999*, pp. 200–203.

[11] B. W. Kernighan and S. Lin, "An efficient heuristic procedure for partitioning graphs," *Bell Syst. Tech. J.*, vol. 49, pp. 291–307, Feb. 1970.

[12] C. M. Fiduccia and R. M. Mattheyses, "A linear-time heuristic for improving network partitions," in *Proc. DAC*, 1982, pp. 175–181.

[13] E. R. Barnes, "An algorithm for partitioning the nodes of a graph," *SIAM J. Alg. Disc. Meth.*, vol. 3, no. 4, pp. 541–550, 1970.

[14] L. Hagen and A. B. Kahng, "Fast spectral methods for ratio cut partitioning and clustering," *IEEE Transactions on Computer-Aided Design of Integrated Circuits and Systems*, vol. 11, no. 9, pp. 1074–1085, 1992.

[15] G. Golub and C. V. Loan, *Matrix Computations*. Baltimore: Johns Hopkins University Press, 1983.

[16] T. H. Cormen, C. E. Leiserson, R. L. Rivest, and C. Stein, *Introduction to Algrorithms*. The MIT Press, 2002, ch. 22.

BLAST: Efficient Computation of Nonlinear Delay Sensitivities in Electronic and Biological Networks using Barycentric Lagrange enabled Transient Adjoint Analysis

Arie Meir*‡, and Jaijeet Roychowdhury*
*Department of Electrical Engineering and Computer Science, The University of California, Berkeley, CA, USA
‡Contact author. Email: ariemeir@berkeley.edu

ABSTRACT

Transient waveform sensitivities are useful in optimization and also provide direct insight into system metrics such as delay. We present a novel method for finding parametric waveform sensitivities that improves upon current transient adjoint methods, which suffer from quadratic complexity, by applying barycentric Lagrange interpolation to reduce computation to near linear in the time-interval of interest. We apply our technique to find sensitivities of a "nonlinear" Elmore-delay like metric in digital logic and biochemical pathway examples. Our technique achieves order-of-magnitude speedups over traditional adjoint and direct sensitivity computation.

Categories and Subject Descriptors

B.8.2 [**Integrated Circuits**]: Performance and Reliability—*Performance Analysis and Design Aids*

General Terms

Algorithms, Design, Reliability, Performance

Keywords

Sensitivity Analysis, Circuit Simulation, Computational Modeling

1. INTRODUCTION

Estimating and optimizing gate and interconnect delays have long been central to IC design. With extreme scaling in transistor feature sizes, variability in every step in the manufacturing process – translating to variability in the delay of individual transistors – has become of growing concern [4, 5]. To enable circuit performance metrics (such as critical path delays) to be optimized over the multidimensional parameter space induced by the manufacturing variability, accurate and effective methods for evaluating delay sensitivities are especially important today. In addition to their use in optimization, sensitivities have an immediate merit of its own, allowing designers to obtain insight about the impact that various system parameters have on delay.

Over more than two decades, many models and algorithms have been devised for accurate prediction of delay and its use for optimal design of ICs (*e.g.*, [6, 10, 13, 19–21]). Most of the work on delay modeling and sensitivity calculation, has focused on estimating delays through the use of linear time invariant (LTI) approximations, such as RC or RLC networks [18, 24]. While LTI techniques are appropriate for estimating delays of individual segments of interconnect [19, 20, 23, 25], they can only *approximate* delays through nonlinear elements such as logic gates or sequential elements. Since logic elements typically involve large signal swings and operation in strongly nonlinear regimes (*e.g.*, involving saturation and hysteresis), the appropriateness of LTI approximations can be very suspect. The same is true for systems that involve many nonlinear logic and interconnect elements. Alternative approaches such as [8] apply proprietary, nonlinear metrics for delay optimization, but rely on piecewise-linear device model approximations [11].

In this work, we introduce a novel method for efficient computation of transient sensitivity waveforms, termed BLAST (**B**arycentric **L**agrange **A**djoint **S**ensitivity **T**ransient). We apply this method to analyze the sensitivity of delay to variations in system parameters. To demonstrate the utility of our method, we propose a nonlinear generalization of the classic Elmore delay metric for LTI systems, *i.e.*, a delay metric defined using waveforms in strongly nonlinear systems, that reduces to Elmore delay for LTI systems. This delay metric is directly motivated by the time-domain definition of Elmore delay (using LTI impulse/step responses); it can be computed as a simple post-processing operation after regular (fully nonlinear) transient simulation. A key feature of this delay metric is that it uses information from waveform values over an entire time interval (*i.e.*, not just a single timepoint, or a few discrete timepoints)[1], thereby taking full account of detailed shapes of waveforms. This makes the metric more representative and more broadly applicable than simple alternatives (such as the 50% rise-time point of a step response).

To find the sensitivities of the proposed delay metric to system parameters, we apply transient sensitivity analysis followed by simple post-processing of the sensitivity waveforms. Existing techniques for "efficient" (*i.e.*, adjoint based) transient sensitivity analysis [9, 14, 17] suffer from *quadratic computational complexity* with respect to the length of the time interval of interest, making their application for computing entire waveforms of transient sensitivity (as opposed to the sensitivity at a single time-point) inefficient.

BLAST solves the problem of quadratic time complexity by applying a quadrature technique known as *barycentric Lagrange interpolation* (BLI) [3] to the transient adjoint computation problem. In its essence, BLI is able to approximate a waveform over an interval, to extremely high accuracy, using only a few carefully chosen samples (see section §2.3). As a result, the quadratic time complexity of existing transient adjoint sensitivity methods is reduced to approximately linear. As noted earlier, the waveform sensitivities generated using BLAST are post-processed to obtain network delay sensitivities. The use of BLI also makes this post-processing step highly accurate and computationally inexpensive.

Being able to compute sensitivities of "nonlinear delays" efficiently makes it possible to obtain design insights regarding, *e.g.*, which parameters have the most impact on delay, thus enabling the designer to focus on the most relevant parameters for delay stability — not only for individual logic or cell library components, but for complex combinational or sequential circuits involving many gates and interconnect segments. The near-linear time complexity of BLAST also facilitates use in delay optimization flows, which make repeated calls to sensitivity routines.

Furthermore, BLAST is equally applicable to the domain of quantitative biology, where recent research has focused on novel synthetic genetic and biochemical pathways such as genetic inverters, toggle switches, oscillators, *etc.* [2]. Since such elements are inherently strongly nonlinear, BLAST is particularly well suited for finding their delay sensitivities.

We demonstrate BLAST on examples from electronics and biology, obtaining average speedups of $17\times$ over adjoint sensitivities without BLI, and $30\times$ over direct transient sensitivity computation. We also demonstrate how non-intuitive design insights into the relative importance of parameters in a delay chain can be observed immediately, from the delay

---

[1]Details are provided in the supplementary material §S5.

sensitivities obtained by BLAST. We emphasize that BLAST can be readily applied to any delay metric that uses multiple time points to estimate delay, i.e., it is not limited to our simple Elmore-like metric.

The remainder of the paper is organized as follows. §2 provides brief background on direct and adjoint sensitivity analysis as well a basic overview of the BLI method. We describe the application of BLI to adjoint sensitivities in §3. §4 presents our "nonlinear" Elmore-like delay metric. Results on examples are presented in §5 .

2. PRELIMINARIES

2.1 Direct sensitivity computation for Differential-Algebraic Equations (DAE) systems

We assume that the system of interest is represented as a set of differential-algebraic nonlinear equations [27]:

$$\frac{d}{dt}[\vec{q}(\vec{x}(t))] + \vec{f}(\vec{x}(t)) + \vec{b}(t) = 0, \qquad (1)$$

where $\vec{x}(t)$ represents the internal state vector of dimension n, $\vec{f}(\cdot)$, $\vec{q}(\cdot)$ capture static and dynamic terms, respectively. $\vec{b}(t)$ represents the time-varying input to the system. As we are interested in the sensitivities of the system to external parameters, we further assume that the internal state, and both static and dynamic terms of the system, depend on a vector of parameters $\vec{p}$ of dimension n_p, allowing us to rewrite the DAE as

$$\frac{d}{dt}[\vec{q}(\vec{x}(t,\vec{p}),\vec{p})] + \vec{f}(\vec{x}(t,\vec{p}),\vec{p}) + \vec{b}(t) = 0. \qquad (2)$$

We assume the initial condition is given at time $t = 0$, and it is $\vec{x}(0) = \vec{x}_0$. Moreover, we assume for simplicity that $\vec{x}_0$ and $\vec{b}(t)$ do not depend on $\vec{p}$; it is simple to extend our analysis if this is not the case.

In transient sensitivity analysis, we are interested in finding the time-varying sensitivities of the solution of (2) with respect to $\vec{p}$. To achieve this goal, we first solve the DAE system for some nominal parameter set $\vec{p}_{nom}$ by running a full transient analysis. Denoting the solution of (2) as $\vec{x}_{nom}(t)$, we let $\Delta\vec{p}$ be a small perturbation to the parameters vector $\vec{p}_{nom}$ and $\Delta\vec{x}$ be a small perturbation to the solution $\vec{x}_{nom}$. We then start the process of linearizing the system (2) around its nominal solution:

$$\frac{d}{dt}[\vec{q}(\vec{x}_{\text{nom}}(t) + \Delta\vec{x}(t), \vec{p}_{\text{nom}} + \Delta\vec{p})]$$
$$+ \vec{f}(\vec{x}_{\text{nom}}(t) + \Delta\vec{x}(t), \vec{p}_{\text{nom}} + \Delta\vec{p}) + \vec{b}(t) = \vec{0}. \qquad (3)$$

Denoting the Jacobian matrices w.r.t state variables by

$$\boldsymbol{C}(t) = \left.\frac{\partial\vec{q}(\vec{x},\vec{p})}{\partial\vec{x}}\right|_{\vec{x}_{\text{nom}}(t),\vec{p}_{\text{nom}}}, \quad \boldsymbol{G}(t) = \left.\frac{\partial\vec{f}(\vec{x},\vec{p})}{\partial\vec{x}}\right|_{\vec{x}_{\text{nom}}(t),\vec{p}_{\text{nom}}}, \qquad (4)$$

and the Jacobian matrices w.r.t parameters by

$$\boldsymbol{S}_q(t) = \left.\frac{\partial\vec{q}(\vec{x},\vec{p})}{\partial\vec{p}}\right|_{\vec{x}_{\text{nom}}(t),\vec{p}_{\text{nom}}}, \quad \boldsymbol{S}_f(t) = \left.\frac{\partial\vec{q}(\vec{x},\vec{p})}{\partial\vec{p}}\right|_{\vec{x}_{\text{nom}}(t),\vec{p}_{\text{nom}}}, \qquad (5)$$

we can now expand $\vec{q}(\cdot)$ and $\vec{f}(\cdot)$ in first-order Taylor series and simplify (3) to

$$\frac{d}{dt}[\boldsymbol{C}(t)\Delta\vec{x}(t) + \boldsymbol{S}_q(t)\Delta\vec{p}] + \boldsymbol{G}(t)\Delta\vec{x}(t) + \boldsymbol{S}_f(t)\Delta\vec{p} \simeq \vec{0}. \quad (6)$$

Denoting the sensitivity matrix $\boldsymbol{M}(t) = \left.\frac{\partial\vec{x}}{\partial\vec{p}}\right|_{\vec{x}_{\text{nom}}(t),\vec{p}_{\text{nom}}}$ and "dividing" by $\Delta\vec{p}$, it can be shown that (6) is equivalent to

$$\frac{d}{dt}[\boldsymbol{C}(t)\boldsymbol{M}(t) + \boldsymbol{S}_q(t)] + \boldsymbol{G}(t)\boldsymbol{M}(t) + \boldsymbol{S}_f(t) = \vec{0}. \qquad (7)$$

For clarity, we note that $\boldsymbol{C}(t), \boldsymbol{G}(t) \in \mathbb{R}^{n\times n}$ and $\boldsymbol{S}_q(t), \boldsymbol{S}_f(t), \boldsymbol{M}(t) \in \mathbb{R}^{n\times n_p}$. $\boldsymbol{M}(t)$ is the matrix of transient sensitivities we are interested in. Rearranging (7) into the form of (2) yields:

$$\frac{d}{dt}[\boldsymbol{C}(t)\boldsymbol{M}(t)] + \boldsymbol{G}(t)\boldsymbol{M}(t) + \underbrace{\left\{\frac{d}{dt}[\boldsymbol{S}_q(t)] + \boldsymbol{S}_f(t)\right\}}_{S(t)} = \vec{0}. \qquad (8)$$

If (8) is solved as a matrix initial value problem with initial condition $\boldsymbol{M}(0) = \boldsymbol{0}$, we obtain transient sensitivities in "direct" fashion. Note that (8) can be solved as n_p separate vector DAE systems, using the columns of $\boldsymbol{S}$ as inputs and solving to obtain the columns of $\boldsymbol{M}$. Once $\boldsymbol{M}(t)$ is available, it is easy to find $\Delta\vec{x}$ in (6) via

$$\Delta\vec{x}(t) = \boldsymbol{M}(t)\Delta\vec{p}. \qquad (9)$$

For a small number of parameters n_p, direct sensitivity computation as above involves the same order of computation as the transient solution itself. However, if the system size n and the number of parameters n_p are both large (furthermore, $n_p \gg n$ for typical real-life systems), one has to keep track of the $n_p \times n$ entries of the sensitivity matrix at every time point of the simulation, increasing computation and memory requirements to the point of infeasibility. Getting around this computational bottleneck is the primary motivation for *adjoint* transient sensitivity, where the sensitivities of a *few* selected "outputs", with respect to *all* parameters, can be obtained more efficiently than by the above direct route.

A practical note concerning the computation of $S(t)$ is appropriate at this point. manual, analytic differentiation of $f(\cdot)$ and $q(\cdot)$ to obtain entries of $S(t)$ tends to be laborious and error prone. Run-time automatic differentiation [12, 22] is an attractive solution for this issue, though it is typically more compute-intensive than the use of hard-coded derivatives. In this work, we have extended [22] and applied the extended automatic differentiation method to compute $S(t)$.

2.2 Adjoint Operator for Linear Differential Equations

A linear differential equation can be written in the form of (2) as:

$$\frac{d}{dt}\boldsymbol{C}(t)\vec{x}(t) + \boldsymbol{G}(t)\vec{x}(t) = \vec{u}(t). \qquad (10)$$

(10) can be viewed as a linear operator $\mathcal{L}: \vec{u}(t) \mapsto \vec{x}(t)$ i.e., a linear mapping between inputs $\vec{u}(t)$ in a domain $\mathcal{D}$ and solutions $\vec{x}(t)$ in a range $\mathcal{R}$. It is a well known fact that any linear mapping has an adjoint operator [15]. The adjoint operator, usually denoted by $\mathcal{L}^\dagger: \vec{y}(t) \mapsto \vec{z}(t)$, takes its input in the range space $\mathcal{R}$ and produces its output in the domain space $\mathcal{D}$. Rewriting equation (6) in the same form as (10), we get

$$\frac{d}{dt}[\boldsymbol{C}(t)\Delta\vec{x}(t)] + \boldsymbol{G}(t)\Delta\vec{x}(t) = \vec{u}(t) = -\boldsymbol{S}(t)\Delta\vec{p}. \qquad (11)$$

Solving (11) defines the mapping $\mathcal{L}: \vec{u}(t) \mapsto \Delta\vec{x}(t)$. It can be shown[2] that the adjoint operator $\mathcal{L}^\dagger: \vec{y}(t) \mapsto z(t)$ of $\mathcal{L}: \vec{u}(t) \mapsto \Delta\vec{x}(t)$ is given by the differential equation

$$-C^*(t)\frac{d}{dt}\vec{z}(t) + G^*(t)\vec{z}(t) = \vec{y}(t). \qquad (12)$$

Assume we are interested in the sensitivities of some scalar output defined as $d(t) = \vec{c}^*\vec{x}(t)$, where $\vec{c}^*$ is a row vector; assume further that we are specifically interested in the sensitivity $d(T_0)$, i.e., limiting our attention to a specific time point $t = T_0$ on the integration interval $[0, T]$. The reason for this will become clearer in the following section, as we demonstrate how finding the sensitivity at a small number of carefully chosen points allows us to approximate the sensitivity for any time point $t \in [0, T]$.

It can be shown[3] that the sensitivity vector of the output $d(t)$ at a point $t = T_0$, denoted $\vec{m}_d(T_0)$, is given by

$$\vec{m}_d(T_0) = \frac{\vec{c}^*\Delta x(T_0)}{\Delta\vec{p}} = -\int_0^T \vec{z}^*_{\vec{c},T_0}(\tau)\,\boldsymbol{S}(\tau)d\tau, \qquad (13)$$

where $\vec{z}_{\vec{c},T_0}(t)$ is obtained by solving the adjoint system

$$-\boldsymbol{C}^*(t)\frac{d}{dt}\vec{z}(t) + \boldsymbol{G}^*(t)\vec{z}(t) = \vec{c}\,\delta(t - T_0), \qquad (14)$$

backwards from $t = T$ to $t = 0$, with initial condition $\vec{z}(T) = \vec{0}$. The reader is invited to follow our analysis in §S3 of the supplementary material; equations (13) and (14) are obtained simply by replacing $\vec{e}_1$ by $\vec{c}$. Note that the sub-indices of $\vec{z}_{\vec{c},T_0}(t)$, namely $\vec{c}$ and T_0, indicate that the solution of the adjoint system (12) depends on the input $\vec{y}(t) = \vec{c}\delta(t-T_0)$.

---

[2]Details are provided in the supplementary material §S2.

[3]See the supplementary material §S3.

Observing equation (14) reveals that it cannot immediately be solved using standard numerical integration methods, because the input on the RHS is a δ-function. Numerical methods are not well suited for δ-function inputs, which involve an infinite value at a single time-point. Therefore, further analytical machinery is needed to re-phrase (14) in a form suitable for numerical integration. To simplify this analysis, we now restrict ourselves to ODEs: *i.e.*, $\vec{q}(\vec{x}) \equiv \vec{x}$ in (2), leading (*w.l.o.g.* for invertible C) to $C(t) \equiv I_{n \times n}$. For the ODE case, (14) thus becomes

$$-\frac{d}{dt}\vec{z}(t) + G^*(t)\,\vec{z}(t) = \vec{c}\,\delta(t - T_0). \qquad (15)$$

It can be shown[4] that finding the solution of (15) over the interval $[0, T]$ is equivalent to solving the homogeneous part of (15) *i.e.*,

$$-\frac{d}{dt}\vec{z}(t) + G^*(t)\,\vec{z}(t) = \vec{0}, \qquad (16)$$

with initial condition $\vec{z}(T_0) = \vec{z}_{\vec{c}, T_0}(T_0) = \vec{c}$, backwards from $t = T_0$ to $t = 0$. Applying this result, we can compute the sensitivity $\vec{m}_d(T_0)$ using standard numerical methods: we first obtain the solution of the homogeneous system (16) using $\vec{z}(T_0) = \vec{c}$ as the initial condition, then compute the integral (13). In section §3, we will define an algorithm for obtaining $\vec{m}_d(T_0)$ based on this analysis.

2.3 Lagrange interpolation and Chebyshev Nodes

Details about Lagrangian interpolation can be found in most textbooks on numerical analysis such as [1], but in the interest of self-sufficiency, we briefly survey the basic concepts of the method here. Given a function $f(x)$ and p sample points $\{c_1, \cdots, c_p\}$, the unique polynomial of degree $p - 1$ that agrees with $f(\cdot)$ at each of these points is

$$L(x) = \sum_{m=1}^{p} u_m(x) f(c_m), \qquad (17)$$

where $u_m(x)$ is the mth Lagrange polynomial (of degree $p - 1$)

$$u_m(x) = \frac{\prod_{\substack{k=1 \\ k \neq m}}^{p}(x - c_k)}{\prod_{\substack{k=1 \\ k \neq m}}^{p}(c_m - c_k)}. \qquad (18)$$

p is referred to as the interpolation order.

In this work, we are interested in approximating the true sensitivity waveform $\vec{m}_d(t)$ by an Lagrangian approximation $L(t)$ which agrees with $\vec{m}_d(t)$ exactly at a small number p of sample points $\{c_1, \cdots, c_p\}$. To estimate the quality of such an approximation, a bound on the error at any point in the interval $t \in [0, T]$ is useful. By choosing the sample points c_i to be Chebyshev nodes over the interval $[0, T]$, it can be shown that the error goes down faster than exponentially with respect to the interpolation order p. Chebyshev nodes on $[a, b]$, are defined as

$$c_m = \frac{a + b}{2} + \frac{b - a}{2}\cos\left(\frac{(2m - 1)\pi}{2p}\right), \quad m = 1, \cdots, p. \qquad (19)$$

With Chebyshev interpolation nodes, the approximation error over $[a, b]$ is bounded as

$$|f(x) - L(x)| \leq \frac{2\left(\frac{b-a}{4}\right)^p}{p!} \max_{\xi \in [a, b]} |f^{(p)}(\xi)|. \qquad (20)$$

Therefore, if the interval $[a, b]$ is fixed, and the derivatives of $f(\cdot)$ over this interval are bounded, the approximation error falls faster-than-exponentially (due to the factorial term) with p [7]. As a rule of thumb, choosing p in the range 15-20 typically leads to double-precision accuracy, *i.e.*, the "approximation" is as good as the original function for all computational purposes. In our case, $[a, b]$ is the integration interval $[0, T]$, which is fixed, and the sensitivity function $\vec{m}_d(t)$ is assumed to have bounded derivatives, the error bound (20) holds. This allows us to choose a relatively small interpolation order p and evaluate $\vec{m}_d(t)$ using the approximation polynomial $L(t)$, which for practical purposes is equivalent to evaluating the original function $\vec{m}_d(t)$. We demonstrate the dependence of the approximation error on p when we discuss the performance of BLAST[5].

---

[4]See the supplementary material §S4.

[5]See the supplementary material §S1.

Using the approximation $L(x)$ instead of the original function $f(x)$ becomes computationally advantageous if a) computing $L(x)$ at each x is expensive and b) values of $f(x)$ are needed at many (*i.e.*, $\gg p$) points x. The key to this efficiency is that only p evaluations of $f(x)$ are needed to set up the approximation $L(x)$ in (17). Once this is done, $L(x)$ can be evaluated very cheaply ($O(p)$ using Barycentric techniques, see §2.4 below) for any value of x.

2.4 Barycentric Lagrange Interpolation

In the form based on (18), each evaluation of $L(x)$ requires $O(p^2)$ additions and multiplications. Another potential impediment is having to recompute all evaluations from scratch once a new sampling point c_{p+1} is been added. An improvement over the classic Lagrange form, described in [3], implements the computation of $u_m(x)$ as

$$u_m(x) = u(x)\frac{w_m}{x - c_m}, \qquad (21)$$

where

$$u(x) = \prod_{k=1}^{p}(x - c_k) \qquad (22)$$

and w_m are the *barycentric weights*, defined as

$$w_m = \frac{1}{\prod_{\substack{k=1 \\ k \neq m}}^{p}(c_m - c_k)}. \qquad (23)$$

Using (22) and (23), the interpolation polynomial can be written as

$$L(x) = u(x)\sum_{m=1}^{p}\frac{w_m}{x - c_m}f(c_m), \qquad (24)$$

which is referred to as the "first form of the barycentric interpolation formula" by Rutishauser in [28]. This formula requires $O(p^2)$ flops for the initial computation of the barycentric weights. The barycentric weights are independent of x, therefore the cost of their computation is amortized over a large number of function evaluations. In addition, each evaluation requires $O(p)$ flops for evaluating $L(x)$, once the weights are known. As shown in [3], incorporating a new node c_{p+1}, also requires $O(p)$ additional operations, an improvement over the direct Lagrange form approach.

To summarize, we have shown that the transient sensitivity waveform $\vec{m}_d(t)$ can be approximated well by the interpolation polynomial $L(t)$. Moreover, using the barycentric form of $L(t)$ outlined in (24) we can now evaluate $L(t)$ at any point, $t \in [0, T]$ at the cost of $O(p)$ flops.

3. BLI BASED TRANSIENT ADJOINT SENSITIVITY COMPUTATION

We now describe how to apply Barycentric Lagrange Interpolation to speed up adjoint-based sensitivity computation when sensitivities are needed not just at one timepoint, but over an entire interval. We first apply the results of §2.2 to present an algorithm for evaluating the sensitivity vector $\vec{m}_d(T_0)$ at a specific point $T_0 \in [0, T]$. Assuming that the sensitivity function is smooth, we then apply the results of section §2.3 and §2.4 to arrive at an algorithm which computes the sensitivity waveform $\vec{m}_d(t)$, over the entire interval $[0, T]$. We compute $\vec{m}_d(t)$ efficiently by evaluating the sensitivity function at a small number of interpolation nodes p, and then apply the BLI method to evaluate the approximation polynomial $L(t)$ at any other point in the interval for a small computational price of $O(p)$ operations.

3.1 Local sensitivity vector using adjoint equation

Assume, as before, that our scalar output of interest is $d(t) = \vec{c}^*\vec{x}(t)$. The sensitivity vector of this output with respect to all parameters was given by equation (13), rewritten here for convenience:

$$\vec{m}_d(T_0) = -\int_0^{T_0} \vec{z}_{\vec{c}, T_0}^{**}(\tau)\,S(\tau)d\tau. \qquad (25)$$

Note that the upper integration limit is effectively T_0, since the adjoint solution $\vec{z}_{\vec{c}, T_0}(t)$ is zero for the interval $t \in [T_0, T]$.

Algorithm 1, presented below utilizes (25) to compute the sensitivity vector $\vec{m}_d(T_0)$:

303

Algorithm 1: *Local Sensitivity*

Input: Circuit DAE D, timepoint of interest T_0, output vector $\vec{c}$, initial circuit state $\vec{x}_0$, final time T
Output: Sensitivity vector $\vec{m}_d(T_0)$

1 $\vec{p} \longleftarrow \vec{p}_{nom}$;
2 $\vec{x}_{nom}(t) \longleftarrow$ Numeric solution of (2) over $t \in [0, T_0]$ with i.c. $\vec{x}_0$;

/* See equation (8) for details */

3 $G(t) \longleftarrow \left. \frac{d\vec{f}(\vec{x},\vec{p})}{d\vec{x}} \right|_{\vec{x}_{nom}(t),\vec{p}_{nom}}$;

4 $S(t) \longleftarrow \frac{d}{dt}[S_q(t)] + S_f(t) =$
$\frac{d}{dt}\left[\left. \frac{d\vec{q}(\vec{x},\vec{p})}{d\vec{p}} \right|_{\vec{x}_{nom}(t),\vec{p}_{nom}} \right] + \left. \frac{d\vec{f}(\vec{x},\vec{p})}{d\vec{p}} \right|_{\vec{x}_{nom}(t),\vec{p}_{nom}}$

/* See §S4 of supplementary material */

5 $\vec{z}_{\vec{c},T_0}(t) \longleftarrow$ solution of equation (16) *backwards from* $t = T_0$ to $t = 0$ *with "initial" condition* $\vec{z}_{\vec{c},T_0}(T_0) = \vec{c}$;

6 $\vec{m}_d(T_0) \longleftarrow -\int_0^{T_0} \vec{z}_{\vec{c},T_0}^*(\tau) S(\tau)\, d\tau$;

7 **return** $\vec{m}_d(T_0)$;

3.2 Sensitivity computation using barycentric Lagrange interpolation (BLI)

In many real-life scenarios, such as for the purpose of computing delay sensitivities, the whole transient sensitivity waveform $\vec{m}_d(t)$ over a period of time $t \in [0, T]$ is of interest.

When using direct sensitivity computation, computing $\vec{m}(t)$ $\forall t \in [0, T]$, is essentially the same effort as computing $\vec{m}(t_0)$: a single, albeit expensive, solution of (8) over $[0, T]$ provides $M(t)$, from which $\vec{m}_d(t)$ can be obtained via $\vec{m}_d(t) = \vec{c}^* M(t)$. Requiring no additional computation for finding all of $\vec{m}_d(t)$ is an attractive feature of direct sensitivity computation.

However, finding all of $\vec{m}_d(t)$ using adjoint sensitivities to compute quantities in in (25) requires the adjoint algorithm to be re-run for each sample point $t \in [0, T]$ - which requires $O((n+n_p)t)$ operations. Hence the total computation needed for finding $\vec{m}_d(t)$ $\forall t \in [0, T]$ is $O((n+n_p)T^2)$. This quantity grows quadratically with T, an undesirable expense that is due to the non-incremental nature of the adjoint transient sensitivity computation algorithm.

In this section, we apply the BLI method outlined in §2.4 to speed up adjoint computation of $\vec{m}_d(t)$, $\forall t \in [0, T]$. The algorithm we present computes $\vec{m}_d(t)$, $\forall t \in [0, T]$ in $O((n + n_p)T)$ floating point operations. Consider (25), which we rewrite as

$$\vec{m}_d(T_0) = -\int_0^{T_0} S^*(\tau)\, \vec{z}_{\vec{c},T_0}(\tau)\, d\tau. \tag{26}$$

We have explicitly used the notation $\vec{z}_{\vec{c},T_0}(t)$, as a reminder that the solution $\vec{z}(t)$ of the adjoint system, depends on the output vector $\vec{c}$ and the time point T_0. In particular, note that the adjoint solution $\vec{z}_{\vec{c},T_0}$ satisfies (see section §S4 of the supplementary material for details):

$$\vec{z}_{\vec{c},T_0}(T_0) = \vec{c}, \text{ and} \tag{27}$$

$$\vec{z}_{\vec{c},T_0}(\tau) \equiv \vec{0} \quad \forall \tau > T_0. \tag{28}$$

If we sample the interval $[0, T_0]$ using N_{T_0} samples $\{\tau_1, \cdots, \tau_{N_{T_0}}\}$, and approximate (26) by a simple discrete summation (for illustration), we obtain

$$\vec{m}_d(T_0) \simeq \vec{m}_{d_{approx}}(T_0) \triangleq -\sum_{i=1}^{N_{T_0}-1} S^*(\tau_i)\, \vec{z}_{\vec{c},T_0}(\tau_i)\, \Delta_i, \tag{29}$$

where $\Delta_i \triangleq \tau_{i+1} - \tau_i$ and $\vec{m}_{d_{approx}}(t)$ denotes a discrete numerical approximation to $\vec{m}_d(t)$.

Suppose we fix a set of time-points

$$\mathcal{T}_T \triangleq \{\tau_0, \tau_1, \cdots, \tau_{N_T}\} \tag{30}$$

for discretizing the interval $[0, T]$, and want to re-use $\mathcal{T}_T$ for calculating (29) for all $t \in [0, T]$. Then it is convenient to re-write (29) as a summation over $\mathcal{T}_T$, using (28) to define $\vec{z}_{\vec{c},T_0}(\tau)$ when $\tau > T_0$, as

$$\vec{m}_{d_{approx}}(t) = -\sum_{i=1}^{N_T} S^*(\tau_i)\, \vec{z}_{\vec{c},T_0}(\tau_i)\, \Delta_i. \tag{31}$$

We would like to find $\vec{m}_{d_{approx}}(t)$ for each $t \in \mathcal{T}_T$. Computing (31) directly for each $t \in \mathcal{T}_T$, would require N_T separate summations, each of which has N_T terms – requiring a total computation of $O(N_T^2 \times (n+n_p))$. To compute (26) efficiently for many values of $t \in [0, T]$, we have applied the BLI method outlined in §2.3 and §2.4. Interpolating $\vec{m}_{d_{approx}}(t)$ we get:

$$\vec{m}_{d_{approx}}(t) \simeq \sum_{j=1}^{p} u_j(t)\vec{m}_{d_{approx}}(c_j), \tag{32}$$

where c_j are the Chebyshev nodes on $[0, T]$, and $u_j(t)$ are the Barycentric Lagrange polynomials (see (21)). Assuming $\vec{m}_d(t)$ is smooth, *i.e.*, its derivatives are bounded over $[0, T]$, we have the approximation error falling faster-than-exponentially with p, as presented in section §2.3. Algorithm 2 presented below, is used to compute the sensitivity waveform $\vec{m}_{d_{approx}}(t)$ in the interval $[0, T]$:

Algorithm 2: Sensitivity Waveform Computation

Input: Circuit DAE D, output vector $\vec{c}$, interpolation order p, time discretization $\mathcal{T}_T$, initial circuit state at time 0 x_0, final time T
Output: Sensitivity vector waveform $\vec{m}_{d_{approx}}(t)$ $\forall t \in [0, T]$

1 $a \longleftarrow 0$;
2 $b \longleftarrow T$;

/* Evaluate sensitivity at Chebyshev nodes */

3 **for** *m=1 to p* **do**

/* See equation (19) */

4 $\quad c_m \longleftarrow \frac{a+b}{2} + \frac{b-a}{2}\cos\left(\frac{(2m-1)\pi}{2p}\right)$;

5 $\quad \vec{m}_{basis}(m) \leftarrow \boldsymbol{LocalSensitivity}(D, c_m, \vec{c}, x_0, T)$;

6 **end**

/* Interpolate sensitivity $\forall t \in [0, T]$ */

7 **for** *i=1 to N_T* **do**

8 $\quad \vec{m}_{d_{approx}}(i) = \sum_{j=1}^{p} u_j(t)\vec{m}_{basis}(j)$

9 **end**

10 **return** $\vec{m}_{d_{approx}}(t)$;

We assume that the Barycentric Lagrange polynomials, $u_j(t)$ (21), are pre-computed in advance and available at step 7.

3.3 Spatial and temporal complexity considerations

Consider the computation and memory requirements of Algorithm 1:

1. Step 2 requires $O(nT)$ in both computation and memory.

2. Step 5 also requires $O(nT)$ in both computation and memory.

3. Computing the $\vec{m}_d(T_0)$ integral numerically in step 6 requires $O((n+n_p)T)$ flops, since $S(t)$ is *sparse* rectangular of size $n \times n_p$.

Hence, the overall time required to compute adjoint sensitivity at a single time point is $O((n + n_p)T)$. For large n_p, this is far superior to the direct sensitivity procedure, where finding $M(t)$, a dense matrix of size $n \times n_p$, requires $O(nn_pT)$ storage and $O(nn_pT)$ computation. For Algorithm 2, the computational cost breakdown is:

1. p iterations of Algorithm 1 in steps $3-6$, which cost $O(pT(n+n_p))$.

2. Evaluating $\vec{m}_{d_{approx}}(t)$ at each point using the BLI polynomials, which requires $O(n_p p)$ flops. Assuming a discretized $[0, T]$ interval with $O(T)$ time points, leads to total computational effort of $O(Tn_p p)$ flops for steps $7 - 9$.

Overall, the total computational effort required for evaluating the entire waveform $\vec{m}_d(t)$ over $[0, T]$ via (32) is $O(pT(n + n_p) + Tn_p p)$. If the interval $[0, T]$ is fixed, p is typically a small constant, hence this computation is much faster that the quadratic complexity of brute-force evaluation of, *e.g.*, (31).

4. A "NONLINEAR" DELAY METRIC

Substantial efforts have been invested by the CAD research community to predict and optimize circuit delay. As mentioned in the introduction, previous works typically assume small-signal conditions when analyzing delay [19, 20, 23, 25], and rely heavily on those assumptions. However, several delay optimization works use proprietary, nonlinear delay metrics, *e.g.*, in the context of underlying piecewise-linear device model functions [8, 11]. Here, we propose an Elmore-delay [10] like metric for nonlinear systems. We then apply the transient sensitivities we have computed using BLAST in order to find delay sensitivities to various system parameters. Given a Linear Time Invariant (LTI) system with impulse response $h_A(t)$ (at some output/node A), the classic Elmore-delay metric is defined as:

$$T_d(A) = \int_0^\infty h_A(t) \cdot t \, dt. \qquad (33)$$

Motivated by this definition, we define a delay metric $T_{d_{NL}}(A)$ that uses the response at A of any system, linear or nonlinear, to a step input. Denoting this response by $g_A(t)$, we define:

$$T_{d_{NL}}(A) = \frac{\int_0^\infty g_A'(t) \cdot t \, dt}{\int_0^\infty g_A'(t) \, dt}. \qquad (34)$$

For LTI systems, this definition reduces to the classic Elmore delay metric, since the derivative of a step response is the impulse response which Elmore's original metric is based upon. However, (34) also applies to nonlinear systems, for which the impulse-response based Elmore delay formula (33) cannot be used.

Computing the sensitivities of this delay using standard adjoint methods runs is quadratic in the time interval of the simulation, as described in §3.2. By using our BLI-based adjoint sensitivity procedure BLAST, however, the complexity is reduced to linear. Note that BLAST can be readily applied to find sensitivities of any other delay metric which uses information from multiple time-points.

5. RESULTS

5.1 Delay sensitivity in a digital inverter chain

We apply BLI-based transient adjoint sensitivity computation to the delay metric in section §4 to analyze delay sensitivities in a chain of three CMOS inverters, shown in Fig. 1. To compute delay sensitivities (in other words, to find the vector $\frac{dT_{d_{NL}}}{d\vec{p}}$), the entire sensitivity waveform is required; we obtain this waveform from the algorithm in section §3.2.

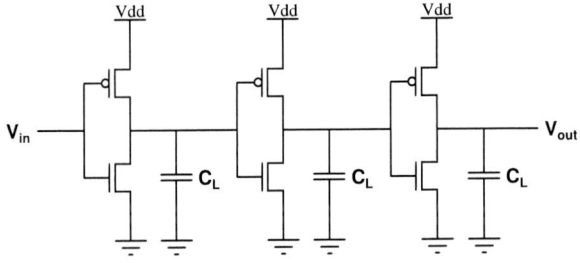

Figure 1: Three stage CMOS 3 inverter chain

An example of the sensitivity waveform for the inverter chain can be seen in Fig. 2, where the direct sensitivity and the adjoint sensitivity are presented side by side.

The sensitivities of the output delay with respect to all model parameters are presented in Table 1, where column I_j lists the sensitivities to parameters in the jth stage of the 3-stage chain.

Parameter	I_1	I_2	I_3
V_{DD}	+10.67E-02	-84.31E-04	-12.09E-04
β_N	+31.19E-07	+12.48E-08	-11.02E-09
β_P	+50.03E-04	-40.31E-06	-12.33E-06
V_{T_N}	-61.90E-03	-15.91E-03	+47.31E-04
V_{T_P}	-53.42E+00	+23.20E-03	+16.02E-04
R_{DS_N}	+34.56E+01	-21.69E+01	+81.04E-01
R_{DS_P}	-99.42E+01	+21.61E+01	-45.86E+00
C_L	-14.56E-09	+12.38E-10	+10.02E-11

Table 1: Digital inverter chain output delay sensitivities to parameters of the different stages

As might be expected given the fact that each nonlinear inverter has an "amplifying" effect on parts of the input waveform, the output delay is much more sensitive to parameters in earlier stages of the chain than to those in later stages, *i.e.*, the impact of the first stage inverter is much larger on the delay, than the impact of the last stage.

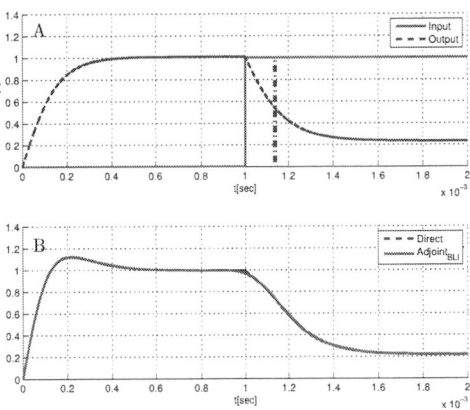

Figure 2: A. Inverter response to a step input, and the delay (vertical line) found by the delay metric. B. Sensitivity of an inverter output to V_{DD} under a step input

5.2 Genetic "inverter-chain" sensitivity analysis

Over more than a decade, research in synthetic biology has focussed on designing and implementing simple, artificial, biological networks. The intention of this research direction is to design standard, modular building blocks in order to build larger artificial biological systems. Another important motivation is to study the behaviour of simple, canonical bio-circuits in order to gain insight into the dynamics of real metabolic pathways [2, 26]. In [16], a synthetic transcriptional cascade is described, and its outputs manually analysed with respect to their sensitivity to design parameters. We demonstrate the applicability of BLAST by computing the delay sensitivities of this transcriptional cascade.

Although functionally similar, the underlying implementation mechanism and the mathematical model of the biological inverter is rather different from its electronic counterpart: a schematic description of a single inverter is presented in Fig. 3. In a very crude approximation, the presence of a transcription factor **tf1** induces the promoter **ip1** and results in the production of a protein **dr1**, which in turn represses the production of a fluorescent report protein **fp1** downstream by blocking its promoter **rp1**. To chain the standalone inverters into a cascade, the second output is engineered to produce a repressor protein for the next stage in all but the last stage, which produces a fluorescent protein for output readout. The input to the system is encoded in the amount of the transcription factor **tf1**.

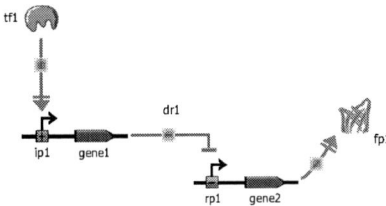

Figure 3: Schematic diagram of a biochemical inverter

Our model for the single biological inverter, based on a simplified version of [16], consists of the following bio-chemical reactions:

$$R + pR_a \rightarrow pR_i, \qquad (35)$$
$$pR_i \rightarrow R + pR_a, \qquad (36)$$
$$pR_a \rightarrow pR_a + mRNA, \qquad (37)$$
$$mRNA \rightarrow mRNA + GFP, \qquad (38)$$
$$mRNA \rightarrow \phi, \qquad (39)$$
$$GFP \rightarrow \phi. \qquad (40)$$

In the above, R stands for repressor, prR_a for active promoter, pR_i for inactivated promoter, *mRNA* stands for the messenger RNA synthesized

305

during transcription and *GFP* represents the **G**reen **F**lourescent **P**rotein, which acts as the output of the inverter. In this reaction system, promoter inactivation is modeled by (35) and (36), transcription is modeled by (37), translation by (38) and the degradation of mRNA and GFP by equations (39) and (40) respectively.

The sensitivities of the output delay to the parameters of the different stages are shown in Table 2; sensitivity waveforms are shown in Fig. 4. Remaining consistent with the digital electronics example, column I_j lists the sensitivities to the parameters of the jth stage.

Parameter	I_1	I_2	I_3
k_{on}	-52.81E-03	+38.21E-04	-41.01E-04
k_{off}	+26.44E-03	-19.33E-04	+20.21E-04
$k_{transcription}$	+41.23E-03	-43.16E-03	-31.29E-04
$k_{translation}$	+38.55E-03	-42.31E-03	-27.91E-04
k_{deg_p}	-18.43E-03	+19.97E-03	-30.96E-05
k_{deg_m}	-13.10E-04	+19.11E-04	+12.90E-06
p_{tot}	+45.04E-01	+45.95E-02	-55.11E-03

Table 2: Biological inverter chain output delay sensitivities to parameters of the different stages

It can be observed that in the biological case too, the sensitivity of the delay at the output is affected more by parameters in the first stage of the cascade. In fact, due to the large variability in the distinct biological implementations of the same basic component, and the inherent, often undesired, inter-component coupling through the reaction medium, sensitivity analysis can assist in estimating compatibility between various components of the same bio-circuit.

This example demonstrates how BLAST is as relevant to the biological domain as it is to electronics. Indeed, we believe that with the rising prominence of synthetic biological design techniques, the rôle of delays in the design of complex biochemical pathways will achieve an importance equalling or surpassing that in electronic/digital design.

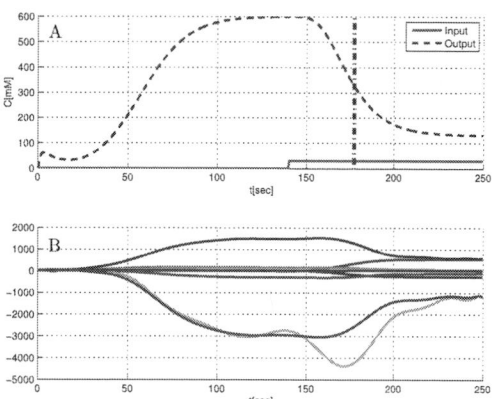

Figure 4: A. Genetic inverter response to a step input and the delay (vertical line) found by the delay metric. B. Sensitivity of genetic inverter output to parameters under a step input

5.3 Performance evaluation

To evaluate computational performance, we measured the run-time of BLAST, and compared it with the direct sensitivity computation as well as the full adjoint computation(without BLI). Sensitivities via BLAST feature average speedups factors of **29.81** and **17.24** when compared to direct and adjoint sensitivities, respectively. Charts and additional performance results are provided in §S1 in the supplementary material.

6. SUMMARY AND CONCLUSIONS

In this paper, we have presented BLAST: an efficient method for computing entire waveforms of transient sensitivity with respect to parameters. We have also proposed an Elmore-delay like metric to estimate delay in nonlinear systems, and as an example, have successfully applied BLAST

to calculate the delay sensitivities of an inverter chain to its system parameters. We have shown that BLAST achieves an average speedup of ~ 30 compared to direct transient sensitivities, and an average speedup of ~ 17 relative to traditional adjoint. We have also shown how BLAST can be readily applied as a design tool in a synthetic biology application. Our hope for the future is that by applying design tools such as BLAST from the CAD community to biological applications, we can contribute to a transformation in the field of quantitative biology, similar to the one that reshaped the field of digital electronics in the 1970s.

7. REFERENCES

[1] F. Acton. *Numerical methods that work*. Spectrum Series. Mathematical Association of America, 1990.

[2] E. Andrianantoandro, S. Basu, D. Karig, and R. Weiss. Synthetic biology: new engineering rules for an emerging discipline. *Mol Syst Biol*, 2:2006, 2006.

[3] J. Berrut and L. Trefethen. Barycentric lagrange interpolation. *SIAM Review*, 46(3):501, 2002.

[4] D. Boning and S. Nassif. Models of process variations in device and interconnect. In *Design of High Performance Microprocessor Circuits, chapter 6*. IEEE Press, 1999.

[5] K. Bowman, S. Duvall, and J. Meindl. Impact of die-to-die and within-die parameter fluctuations on the maximum clock frequency distribution for gigascale integration. *Solid-State Circuits, IEEE Journal of*, 37(2):183–190, Feb 2002.

[6] P. Chan and K. Karplus. Computing signal delay in general rc networks by tree/link partitioning. *Computer-Aided Design of Integrated Circuits and Systems, IEEE Transactions on*, 9(8):898–902, Aug 1990.

[7] W. Cheney and D. Kincaid. *Numerical Mathematics and Computing*. Brooks/Cole, Pacific Grove, CA, 3d edition, 1994.

[8] A. Conn, I. Elfadel, J. Molzen, W.W., P. O'Brien, P. Strenski, C. Visweswariah, and C. Whan. Gradient-based optimization of custom circuits using a static-timing formulation. In *Design Automation Conference, 1999. Proceedings. 36th*, pages 452–459, 1999.

[9] S. Director and R. Rohrer. The Generalized Adjoint Network and Network Sensitivities. *IEEE Trans. Ckt. Theory*, 16:318–323, August 1969.

[10] W. C. Elmore. The transient response of damped linear networks with particular regard to wideband amplifiers. *Journal of Applied Physics*, 19(1):55–63, Jan 1948.

[11] P. Feldmann, T. Nguyen, S. Director, and R. Rohrer. Sensitivity computation in piecewise approximate circuit simulation. *Computer-Aided Design of Integrated Circuits and Systems, IEEE Transactions on*, 10(2):171–183, feb 1991.

[12] A. Griewank. On automatic differentiation. *Mathematical Programming: recent developments and applications*, pages 83–108, 1989.

[13] R. Hitchcock, G. Smith, D. Cheng, et al. Timing analysis of computer hardware. *IBM Journal of Research and Development*, 26(1):100–105, Jan 1982.

[14] D. Hocevar, P. Yang, T. Trick, and B. Epler. Transient sensitivity computation for mosfet circuits. *IEEE Trans. on CAD of Integrated Circuits and Systems*, 4(4):609, 1985.

[15] K. Hoffman and R. Kunze. *Linear algebra*. Prentice-Hall mathematics series. Prentice-Hall, 1971.

[16] S. Hooshangi, S. Thiberge, and R. Weiss. Ultrasensitivity and noise propagation in a synthetic transcriptional cascade. *Proc Natl Acad Sci U S A*, 102(10):3581–3586, Mar. 2005.

[17] Z. Ilievski, H. Xu, A. Verhoeven, E. Maten, W. Schilders, and R. Mattheij. Adjoint transient sensitivity analysis in circuit simulation. *Scientific Computing in Electrical Engineering*, 2006.

[18] T. Lin and C. Mead. Signal delay in general rc networks. *Computer-Aided Design of Integrated Circuits and Systems, IEEE Transactions on*, 3(4):331–349, October 1984.

[19] N. Menezes, R. Baldick, and L. Pileggi. A sequential quadratic programming approach to concurrent gate and wire sizing. *Computer-Aided Design of Integrated Circuits and Systems, IEEE Transactions on*, 16(8):867–881, Aug 1997.

[20] N. Menezes, S. Pullela, F. Dartu, and L. Pillage. Rc interconnect synthesis-a moment fitting approach. In *Computer-Aided Design, 1994., IEEE/ACM International Conference on*, pages 418–425, Nov 1994.

[21] S. Nazarian, M. Pedram, E. Tuncer, and T. Lin. Sensitivity-based gate delay propagation in static timing analysis. In *Quality of Electronic Design, 2005. ISQED 2005. Sixth International Symposium on*, pages 536–541, March 2005.

[22] R. Neidinger. Introduction to Automatic Differentiation and MATLAB Object-Oriented Programming. *SIAM Review*, 52(3):545–563, 2010.

[23] A. Odabasioglu, M. Celik, and L. Pileggi. Prima: passive reduced-order interconnect macromodeling algorithm. *Computer-Aided Design of Integrated Circuits and Systems, IEEE Transactions on*, 17(8):645–654, Aug 1998.

[24] J. Ousterhout. Crystal: A timing analyzer for nmos vlsi circuits. Technical Report UCB/CSD-83-119, EECS Department, University of California, Berkeley, Jan 1983.

[25] L. Pillage and R. Rohrer. Asymptotic waveform evaluation for timing analysis. *Computer-Aided Design of Integrated Circuits and Systems, IEEE Transactions on*, 9(4):352–366, Apr 1990.

[26] P. Purnick and R. Weiss. The second wave of synthetic biology: from modules to systems. *Nature Reviews Molecular Cell Biology*, 10(6):410–422, 2009.

[27] J. Roychowdhury. Numerical simulation and modelling of electronic and biochemical systems. *Foundations and Trends in Electronic Design Automation*, 3(2-3):97–303, 2009.

[28] H. Rutishauser. *Vorlesungen ber numerische Mathematik*. Birkhuser-Verlag, 1976.

Supplementary Material

S1. PERFORMANCE RESULTS

BLAST performance data is presented in Fig. S1.1 where the sensitivity computation runtime is plotted vs. the number of integration points T. For comparison, we present the runtimes of the direct sensitivity and the adjoint sensitivity methods without applying the BLI technique (marked as $Adjoint_{Full}$ in the plots). The speedup factor depends on the choice of the parameters n, n_p and T, and for coherence of presentation, we report performance measurements in Table (S1.1) for a fixed integration interval and varying n and n_p. From the performed experiments, BLAST obtains an average speedup factor of **29.81** when compared to direct sensitivity computation under the same configuration, and **17.24** speedup factor when compared to the adjoint sensitivity method without BLI. A certain pattern can be observed from these measurements, and was confirmed by additional experiments: as we fix the integration interval and make our system larger (by increasing n and n_p), the speedup factor achieved over the direct sensitivity method grows. At the same time, if we compare the speedup factor to the adjoint method that uses no BLI, we see that as the size of the system dominates the length of the integration interval T, the speedup becomes smaller.

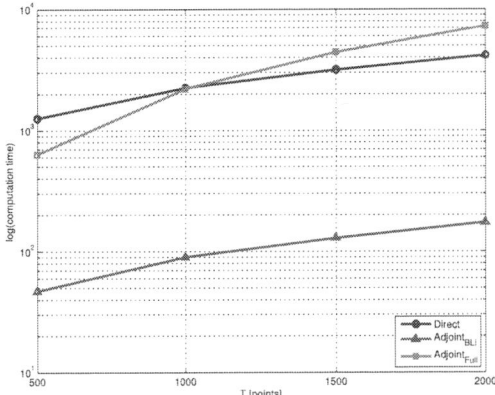

Figure S1.1: Sensitivity map computation time: Direct,Adjoint, Adjoint with BLI

n	n_p	T	$Direct$	$Adjoint_{BLI}$	$Adjoint_{Full}$
1	8	1000	2237.76	89.57	2195.07
2	16	1000	4127.92	145.82	2739.87
4	32	1000	8456.16	268.83	4107.86
8	64	1000	17865.36	517.92	6882.84

Table S1.1: Performance measurements (runtime in seconds)

To use BLAST, one has to choose a value of p, the interpolation order parameter. As usual, this choice represents a trade-off between accuracy and computation time. To test whether the choice of p can become the accuracy bottleneck of the system, we have measured the relative error as a function of the interpolation order p. We present our accuracy measurements for both equispaced and Chebyshev nodes in Fig. S1.2. It can be seen, that for $p = 20$ with Chebyshev nodes, we effectively reach double precision, which means that for a large system (large n and n_p) the factor introduced by p in the time complexity, becomes insignificant, and BLAST effectively performs at linear time. This means that the optimal choice of p will have no effect on the accuracy of the system, as the accuracy bottleneck will be somewhere else.

For completeness, we present the error in the sensitivity computed by BLAST in Fig. S1.3 and compare it to the accuracy of the sensitivity map produced by the direct method.

S2. ADJOINT OPERATOR FOR LINEAR DIFFERENTIAL EQUATIONS

Given an operator $\mathcal{L}$ taking its input from some domain $\mathcal{D}$ into the output range $\mathcal{R}$, the adjoint $\mathcal{L}^\dagger$ of the operator $\mathcal{L}$ can be intuited as an inner-

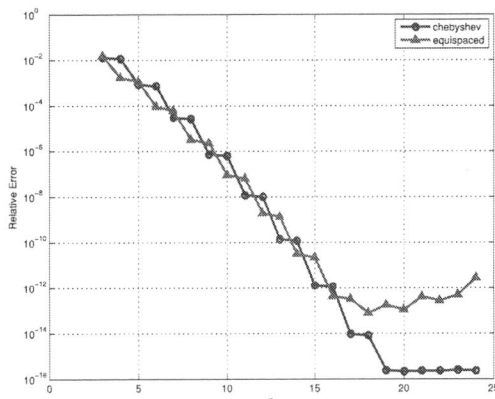

Figure S1.2: Approximation error vs. interpolation order tested on $f(t) = e^{-t} sin(2\pi f t)$

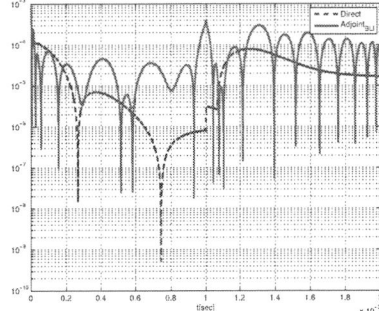

Figure S1.3: Error vs time for direct and adjoint sensitivities

product preserving operator. For a fixed point y in the range, we operate $\mathcal{L}$ on any u in the domain $\mathcal{D}$, and take an inner product between the image of u, i.e. $\mathcal{L}(u)$, and the fixed point y :see Fig. S2.1 for illustration. The adjoint operator, denoted $\mathcal{L}^\dagger$, operating on y in the range, will return z in the domain $\mathcal{D}$, whose inner product with the original u is the same, as the inner product of $\mathcal{L}(u)$ and the fixed point y. If for any choice of the fixed point y, we can find such a z, we say that the mapping $y \mapsto z$ defines the adjoint operator $\mathcal{L}^\dagger$.

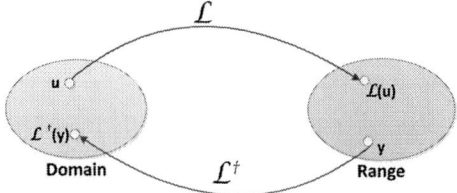

Figure S2.1: Schematic representation of the adjoint operator.

More formally, $\mathcal{L}^\dagger$ would be defined as the adjoint operator of $\mathcal{L}$, if $\forall u \in \mathcal{D}$, $\forall y \in \mathcal{R}$, we have:

$$\langle \mathcal{L}\{u\}, y \rangle_{\mathcal{R}} = \langle u, \mathcal{L}^\dagger\{y\} \rangle_{\mathcal{D}}. \tag{41}$$

Equation (12) which is rewritten here for convenience as

$$-C^*(t)\frac{d}{dt}\vec{z}(t) + G^*(t)\vec{z}(t) = \vec{y}(t), \tag{42}$$

defines a mapping $\vec{y}(t) \mapsto \vec{z}$, i.e. between $\vec{y}(t) \in \mathcal{R}$ and $\vec{z}(t) \in \mathcal{D}$. We will show that this mapping is the adjoint of the mapping $\mathcal{L}$, defined by equation (10), rewritten here for convenience as

$$\frac{d}{dt}C(t)\,\vec{x}(t) + G(t)\,\vec{x}(t) = \vec{u}(t). \tag{43}$$

To show that $\vec{z} = \mathcal{L}^\dagger\{\vec{y}\}$, we need to show, that following the definition in

(41), the inner product is preserved, *i.e.*,

$$\forall \vec{u} \in \mathcal{D}, \vec{y} \in \mathcal{R} \text{ s.t. } \mathcal{L}\{\vec{u}\} = \vec{x} \text{ and } \mathcal{L}^\dagger\{\vec{y}\} = \vec{z},$$
$$\langle \vec{x}(t), \vec{y}(t) \rangle_\mathcal{R} = \langle \vec{u}(t), \vec{z}(t) \rangle_\mathcal{D}. \qquad (44)$$

We use the standard definition of inner product for vector functions:

$$\langle \vec{x}(t), \vec{y}(t) \rangle = \int_A^B w(\tau) \left[\vec{y}^*(\tau) \, \vec{x}(\tau) \right] d\tau, \qquad (45)$$

where $w(t)$ is some appropriate scalar weighting function. We assume $w(t) \equiv 1$ in this work.
We note that

$$\langle \vec{x}(t), \vec{y}(t) \rangle_\mathcal{R} = \left\langle \vec{x}(t), -\boldsymbol{C}^*(t) \frac{d}{dt} \vec{z}(t) + \boldsymbol{G}^*(t) \vec{z}(t) \right\rangle_\mathcal{R}$$
$$= \left\langle \vec{x}(t), -\boldsymbol{C}^*(t) \frac{d}{dt} \vec{z}(t) \right\rangle_\mathcal{R} + \langle \vec{x}(t), \boldsymbol{G}^*(t) \vec{z}(t) \rangle_\mathcal{R}, \qquad (46)$$

and also that

$$\langle \vec{u}(t), \vec{z}(t) \rangle_\mathcal{D} = \left\langle \frac{d}{dt} \left[\boldsymbol{C}(t) \, \vec{x}(t) \right] + \boldsymbol{G}(t) \, \vec{x}(t), \vec{z}(t) \right\rangle_\mathcal{D}$$
$$= \left\langle \frac{d}{dt} \left[\boldsymbol{C}(t) \, \vec{x}(t) \right], \vec{z}(t) \right\rangle_\mathcal{D} + \langle \boldsymbol{G}(t) \, \vec{x}(t), \vec{z}(t) \rangle_\mathcal{D}. \qquad (47)$$

Our purpose is to show that the quantities (46) and (47) are equal. From inner product properties, it is obvious that the second terms are equal. We will show that the first terms are also equal, *i.e.*, that

$$\left\langle \vec{x}(t), -\boldsymbol{C}^*(t) \frac{d}{dt} \vec{z}(t) \right\rangle_\mathcal{R} = \left\langle \frac{d}{dt} \left[\boldsymbol{C}(t) \, \vec{x}(t) \right], \vec{z}(t) \right\rangle_\mathcal{D}.$$

To do this, we use (45) to expand out the inner products and get

$$\left\langle \vec{x}(t), -\boldsymbol{C}^*(t) \frac{d}{dt} \vec{z}(t) \right\rangle_\mathcal{R} = \int_A^B -\left[\frac{d}{dt} \vec{z}^*(t) \right] \boldsymbol{C}(t) \, \vec{x}(t) \, dt.$$

Since integration by parts can be stated as

$$\int_A^B \frac{dr}{dt} s(t) \, dt = \left[r(t) \, s(t) \right]_A^B - \int_A^B r(t) \frac{ds}{dt} \, dt. \qquad (48)$$

we set $s(t) \equiv -\boldsymbol{C}(t) \vec{x}(t)$ and $\vec{r}(t) \equiv \vec{z}^*(t)$, to arrive at:

$$\left\langle \vec{x}(t), -\boldsymbol{C}^*(t) \frac{d}{dt} \vec{z}(t) \right\rangle_\mathcal{R} = \int_A^B -\left[\frac{d}{dt} \vec{z}^*(t) \right] \boldsymbol{C}(t) \, \vec{x}(t) \, dt$$
$$= \int_A^B \frac{dr(t)}{dt} s(t) \, dt$$
$$= \left[r(t) \, s(t) \right]_A^B - \int_A^B r(t) \frac{ds}{dt} \, dt$$
$$= \left[-\vec{z}^*(t) \boldsymbol{C}(t) \vec{x}(t) \right]_A^B + \int_A^B \vec{z}^*(t) \frac{d}{dt} \left[\boldsymbol{C}(t) \vec{x}(t) \right] \, dt$$
$$= \left[-\vec{z}^*(t) \boldsymbol{C}(t) \vec{x}(t) \right]_A^B + \left\langle \frac{d}{dt} \left[C(t) \vec{x}(t) \right], \vec{z}(t) \right\rangle_\mathcal{D}$$

To complete our proof, we are left with the task of making the term $\left[-\vec{z}^*(t) \boldsymbol{C}(t) \vec{x}(t) \right]_A^B$, zero. This term, which we will call the "boundary effect" term, leads to challenges in many situations having to do with adjoints. One way to deal with it, is to make $A \to -\infty$, $B \to \infty$, and restrict the functions $\vec{x}(t)$, $\vec{z}(t)$, to be zero at $\pm\infty$. This is a viable practical solution to this problem for stable systems of differential equations, since we can usually choose inputs that go to zero far away from the regions of time we are interested in. If we cannot (*i.e.*, the system exhibits non-zero asymptotic behaviour), then other technical tricks – like adding weight functions to the inner-product definition, and restricting the domain of definition of the adjoint operator – can be tried to zero out boundary effect terms. In applications, one should be careful to check that such boundary terms are indeed zero, or if non-zero, that they are properly accounted for in the overall analysis.
In our situation, however, getting around this is particularly simple. Since both $\vec{x}(t)$ and $\vec{z}(t)$ are solutions of differential equations, we do need to specify initial conditions. We choose the initial conditions to be $\vec{x}(A) = 0$

and $\vec{z}(B) = 0$. This choice ensures that $\left[-\vec{z}^*(t) \boldsymbol{C}(t) \vec{x}(t) \right]_A^B = 0$. As we will see later, the adjoint differential equation (42) will need to be solved *backwards*, from B to A, so an "initial" condition on $\vec{z}(B)$ is a natural one for the adjoint.

S3.   LOCAL SENSITIVITY USING THE ADJOINT OPERATOR

In section §2.2 we have used the linearized system (11), rewritten here for convenience as

$$\frac{d}{dt} \boldsymbol{C}(t) \, \Delta\vec{x}(t) + \boldsymbol{G}(t) \, \Delta\vec{x}(t) = \vec{u}(t) = -\boldsymbol{S}(t)\Delta\vec{p}, \qquad (49)$$

and its adjoint system, rewritten here for convenience as

$$-C^*(t) \frac{d}{dt} \vec{z}(t) + G^*(t) \vec{z}(t) = \vec{y}(t), \qquad (50)$$

to arrive at the sensitivity vector of a specific output $d(t)$ at a given time point $t = T_0$.
While most technicalities were omitted from the main article, we now provide the detailed derivation, arriving at equation (57), which lays the foundation for Algorithm 1 (*Local Sensitivity Algorithm*).
The main feature of the adjoint operator relevant for our purpose, is the preservation of inner product. We will utilize this property to reduce the computational burden of sensitivity analysis. In this work, we use the standard definition of inner product for vector functions (45) rewritten here for convenience:

$$\langle \vec{x}(t), \vec{y}(t) \rangle = \int_A^B w(\tau) \left[\vec{y}^*(\tau) \, \vec{x}(\tau) \right] d\tau. \qquad (51)$$

In (51), $w(t)$ is some scalar weighting function, which we define to be $w(t) \equiv 1$.
As mentioned in section §S2 of supplementary material, inner product preservation simply means, that for a linear operator $\mathcal{L} : \vec{u}(t) \mapsto \Delta\vec{x}(t)$, and its adjoint operator $\mathcal{L}^\dagger : \vec{y}(t) \mapsto \vec{z}(t)$ the following holds:

$$\forall \vec{u} \in \mathcal{D}, \vec{y} \in \mathcal{R} \text{ s.t. } \mathcal{L}\{\vec{u}\} = \Delta\vec{x} \text{ and } \mathcal{L}^\dagger\{\vec{y}\} = \vec{z},$$
$$\langle \Delta\vec{x}, \vec{y} \rangle_\mathcal{R} = \langle \mathcal{L}\{\vec{u}\}, \vec{y} \rangle_\mathcal{R} = \langle \vec{u}, \mathcal{L}^\dagger\{\vec{y}\} \rangle_\mathcal{D} = \langle \vec{u}, \vec{z} \rangle_\mathcal{D}. \qquad (52)$$

To demonstrate the utility of (52) let us assume that we are interested in a single output $x_1(t)$. Let us further assume that we are interested in sensitivity of $x_1(t)$ at a single time point T_0. Denoting an elementary vector as $\vec{e}_1$, and applying the basic property of the inner product we get:

$$\Delta x_1(T_0) = \langle \Delta\vec{x}(t), \vec{e}_1 \delta(t - T_0) \rangle, \qquad (53)$$

which now allows us to apply (52) and arrive at:

$$\langle \Delta\vec{x}(t), \vec{e}_1 \delta(t - T_0) \rangle =$$
$$\langle \mathcal{L}\{\vec{u}(t)\}, \vec{e}_1 \delta(t - T_0) \rangle_\mathcal{R} = \left\langle \vec{u}(t), \underbrace{\mathcal{L}^\dagger\{\vec{e}_1 \delta(t - T_0)\}}_{\vec{z}_{T_0}(t)} \right\rangle_\mathcal{D}. \qquad (54)$$

Note that the $\mathcal{L}$ represents the linear operator defined by (49), and $\vec{z}_{T_0}(t)$ is the solution of the differential equation defined by the adjoint operator $\mathcal{L}^\dagger : \vec{y}(t) \mapsto \vec{z}(t)$ for an input $\vec{y}(t) = \vec{e}_1 \delta(t - T_0)$. From (49) we observe that the input $\vec{u}(t)$ to our forward equation is $\vec{u}(t) = -\boldsymbol{S}(t)\Delta\vec{p}$, allowing us to rewrite (54), and using (53) arrive at:

$$\Delta x_1(T_0) = \left\langle \vec{u}(t), \underbrace{\mathcal{L}^\dagger\{\vec{e}_1 \delta(t - T_0)\}}_{\vec{z}_{T_0}(t)} \right\rangle_\mathcal{D} = \langle -\boldsymbol{S}(t)\Delta\vec{p}, \vec{z}_{T_0}(t) \rangle_\mathcal{D}. \qquad (55)$$

Finally we can apply the inner product definition from (51) and get:

$$\Delta x_1(T_0) = \langle -\boldsymbol{S}(t)\Delta\vec{p}, \vec{z}_{T_0}(t) \rangle_\mathcal{D} = -\left(\int_A^B \vec{z}_{T_0}^*(\tau) \, \boldsymbol{S}(\tau) d\tau \right) \Delta\vec{p}. \qquad (56)$$

Performing a "division" operation similar to the one in (7), we arrive at $\vec{m}_1(T_0)$, defined as the sensitivity vector of $x_1(t)$ with respect to parameters $\vec{p}$, evaluated at time $t = T_0$:

$$\vec{m}_1(T_0) = \frac{\Delta x_1(T_0)}{\Delta\vec{p}} = -\int_A^B \vec{z}_{T_0}^*(\tau) \, \boldsymbol{S}(\tau) d\tau. \qquad (57)$$

To compute $\vec{m}_1(T_0)$ we first need to obtain $\vec{z}_{T_0}(t)$ by solving the adjoint system for an input $\vec{u}(t) = \vec{e}_1 \delta(t - T_0)$, and then compute the integral in (57). Note that the matrices $\boldsymbol{S}(t)$ can be readily computed from (8).

A subtle, yet important point concerning the integration limits in the definition of inner product(A and B), is in order. As outlined in section §S2 of the supplementary material, we need to zero out the boundary effect term $\left[-\vec{z}^*(t)\boldsymbol{C}(t)\Delta\vec{x}(t)\right]_A^B$. Choosing $A = 0$, will set the term $\left[-\vec{z}^*(t)\boldsymbol{C}(t)\Delta\vec{x}(t)\right]_A$ to zero, since the initial condition of (49), given at $t = 0$, is $\vec{0}$.

As for B, we know from the definition of the inner product it follows that we must have $B \geq T_0$. Moreover, for reasons elucidated in section §S4 of the supplementary material, we choose $B > T_0$. To zero out the term $\left[-\vec{z}^*(t)\boldsymbol{C}(t)\Delta\vec{x}(t)\right]^B$, we must impose the the "initial" condition $\vec{z}(B) = \vec{0}$ when solving the adjoint DAE (50); to obtain $\vec{z}(t)$ over the interval $[A, B]$ (needed for computing (57)), the adjoint DAE must be solved backwards from B to A. In this context, note that if (49) is a stable system (as will typically be the case in applications), then (50) is an unstable system — in other words, its solution will blow up exponentially as t increases. Hence, solving it backwards is *necessary*, from a numerical standpoint, to avoid uncontrolled error blowup; since (12) is unstable going forward in time, it is stable going backward in time, hence standard numerical integration methods can be applied directly.

S4. ANALYSIS OF ADJOINT ODE WITH δ-FUNCTION INPUT

In section §2.2 of our work, we show that a key step in the computation of transient adjoint sensitivities, is the solution of the adjoint equation (14), rewritten here for convenience:

$$-\boldsymbol{C}^*(t)\frac{d}{dt}\vec{z}(t) + \boldsymbol{G}^*(t)\,\vec{z}(t) = \vec{c}\,\delta(t - T_0). \tag{58}$$

As mentioned before, we need to solve (58) backwards from $t = B$ to $t = 0$, with "initial" condition $\vec{z}(B) = \vec{0}$, and $B > T_0$.

However, this solution cannot immediately be obtained using standard numerical integration methods, because the input is a δ-function. Numerical methods are not well suited for δ-function inputs, which involve an infinite value at a single time-point, while being identically zero elsewhere. Therefore, further analysis is needed to re-phrase (58) in a form suitable for numerical integration.

To simplify this analysis, we now restrict ourselves to ODEs: *i.e.*, $\vec{q}(\vec{x}) \equiv \vec{x}$ in (2), leading (*w.l.o.g.* for invertible C) to $\boldsymbol{C}(t) \equiv I_{n \times n}$. For the ODE case, (58) becomes

$$-\frac{d}{dt}\vec{z}(t) + \boldsymbol{G}^*(t)\,\vec{z}(t) = \vec{c}\,\delta(t - T_0), \tag{59}$$

to be solved backwards from $t = B$ to $t = 0$, with "initial" condition given as $\vec{z}(B) = \vec{0}$.

The δ-function form of the input (59), together with its zero initial condition at $t = B$, has the following implications as we integrate backwards from $t = B$ to $t = 0$ (note that $B > T_0$, by assumption above):

1. Over the interval $[T_0^+, B]$, the solution of (59), $\vec{z}(t)$, is identically the zero vector. This is because (59), a linear ODE, has an identically zero input over this interval, as well as an initial condition of zero, at $t = B$.

2. Over the interval $[T_0^-, T_0^+]$, the δ-function in the input is "active".

Integrating (59) over this interval, we obtain

$$\int_{T_0^-}^{T_0^+} \left[-\frac{d}{dt}\vec{z}(t) + \boldsymbol{G}^*(t)\,\vec{z}(t) \right] dt = \int_{T_0^-}^{T_0^+} \vec{c}\,\delta(t - T_0)\,dt$$

$$\Rightarrow \int_{T_0^-}^{T_0^+} \left[-\frac{d}{dt}\vec{z}(t) + \boldsymbol{G}^*(t)\,\vec{z}(t) \right] dt = \vec{c}$$

$$\Rightarrow -\left[\vec{z}(t)\right]_{T_0^-}^{T_0^+} + \int_{T_0^-}^{T_0^+} \boldsymbol{G}^*(t)\,\vec{z}(t)\,dt = \vec{c}$$

$$\Rightarrow \vec{z}(T_0^-) - \vec{z}(T_0^+) + \int_{T_0^-}^{T_0^+} \boldsymbol{G}^*(t)\,\vec{z}(t)\,dt = \vec{c}$$

$$\Rightarrow \vec{z}(T_0^-) = \vec{z}(T_0^+) - \int_{T_0^-}^{T_0^+} \boldsymbol{G}^*(t)\,\vec{z}(t)\,dt + \vec{c}. \tag{60}$$

Note that

- $\vec{z}(T_0^+) \equiv \vec{0}$ (we just established this in 1, above), and
- $\int_{T_0^-}^{T_0^+} \boldsymbol{G}^*(t)\,\vec{z}(t)\,dt \equiv \vec{0}$ (we are integrating finite quantities over an infinitesimally small interval).

Hence (60) becomes

$$\vec{z}(T_0^-) = \vec{c}. \tag{61}$$

3. Over the interval $[0, T_0^-]$, the input to (59) is again identically zero since the δ-function is not active. However, the "initial" condition over this interval is given by (61), *i.e.*, $\vec{z}(T_0^-) = \vec{c}$.

From these observations, we see that finding the solution of (59) over the interval $[0, T_0^-]$, is equivalent to solving the homogeneous part of (59) (*i.e.*, with zero input), *i.e.*,

$$-\frac{d}{dt}\vec{z}(t) + \boldsymbol{G}^*(t)\,\vec{z}(t) = \vec{0}, \tag{62}$$

with initial condition $\vec{z}(T_0^-) = \vec{0}$, backwards from $t = T_0^-$ to $t = 0$. Since T_0^- is only infinitesimally separated from T, we can replace T_0^- with T, since no infinite values or δ-functions arise in (62).

S5. ELMORE DELAY SENSITIVITY

For a system with step response $g_A(t)$ at a node A, we have proposed an Elmore-delay like metric in section §4 of our work. The metric $T_{d_{NL}}(A)$ was defined as

$$T_{d_{NL}}(A) = \frac{\int_0^\infty g_A'(t) \cdot t\,dt}{\int_0^\infty g_A'(t)\,dt}. \tag{63}$$

Assuming that we operate in a finite integration interval $[0, T]$, we can rewrite and simplify (63) as

$$T_{d_{NL}}(A) = \frac{\int_0^T g_A'(t) \cdot t\,dt}{\int_0^T g_A'(t)\,dt} = \frac{\int_0^T g_A'(t) \cdot t\,dt}{g_A(T) - g_A(0)}. \tag{64}$$

We can now write the expression for the delay sensitivity vector:

$$\begin{aligned}
\frac{d}{d\vec{p}}T_{d_{NL}}(A) &= \frac{\frac{d}{d\vec{p}}\left(\int_0^T g_A'(t) \cdot t\,dt\right) \cdot \left[g_A(T) - g_A(0)\right]}{\left(\frac{d}{d\vec{p}}\left[g_A(T) - g_A(0)\right]\right)^2} \\
&\quad - \frac{\left(\int_0^T g_A'(t) \cdot t\,dt\right) \cdot \frac{d}{d\vec{p}}\left[g_A(T) - g_A(0)\right]}{\left(\frac{d}{d\vec{p}}\left[g_A(T) - g_A(0)\right]\right)^2} \\
&= \frac{\left(\int_0^T \frac{dg_A'}{d\vec{p}}(t) \cdot t\,dt\right) \cdot \left[g_A(T) - g_A(0)\right]}{D^2} \\
&\quad - \frac{\left(\int_0^T g_A'(t) \cdot t\,dt\right) \cdot \frac{d}{d\vec{p}}\left[g_A(T) - g_A(0)\right]}{D^2},
\end{aligned} \tag{65}$$

where D^2 denotes the denominator for notational simplicity.

For the purpose of computing (65) numerically, assume we are using the set of time samples

$$\mathcal{T}_T \triangleq \{\tau_0, \tau_1, \cdots, \tau_{N_T}\}, \tag{66}$$

spanning the interval $[0, T]$. Approximating the integral by finite summation, and the temportal derivatives by finite backward differences, we are ready to write the discrete approximation expression for the delay sensitivity vector, but first another notational simplification. Let us assign the aliases α and β, to the first and second terms, in (65), respectively, and rewrite (65) as:

$$\frac{d}{d\vec{p}} T_{d_{NL}}(A) = \frac{\alpha - \beta}{D^2}. \tag{67}$$

Expanding the different terms we arrive at:

$$\alpha = \left[\sum_{i=1}^{N_T} \left(\frac{dg_A}{d\vec{p}}(\tau_i) - \frac{dg_A}{d\vec{p}}(\tau_{i-1}) \right) \cdot \bar{\tau}_i \cdot \Delta_i \right] \cdot \left(g_A(\tau_{N_T}) - g_A(\tau_0) \right),$$

$$\beta = \left(\frac{dg_A}{d\vec{p}}(\tau_{N_T}) - \frac{dg_A}{d\vec{p}}(\tau_0) \right) \cdot \left[\sum_{i=1}^{N_T} \left(g_A(\tau_i) - g_A(\tau_{i-1}) \right) \cdot \bar{\tau}_i \cdot \Delta_i \right],$$

$$D^2 = \left(\frac{dg_A}{d\vec{p}}(\tau_{N_T}) - \frac{dg_A}{d\vec{p}}(\tau_0) \right)^2,$$

$$\tag{68}$$

where $\Delta_i \triangleq \tau_{i+1} - \tau_i$, and $\bar{\tau}_i \triangleq \frac{(\tau_i + \tau_{i-1})}{2}$. Note that $\frac{dg_A}{d\vec{p}}(\tau_i)$ is exactly the transient sensitivity vector, we have computed using BLAST.

To summarize, in this section we have shown a possible application of BLAST, in order to compute the delay sensitivity vector $\frac{dT_{d_{NL}}}{d\vec{p}}(A)$. While we have utilized a particular, Elmore-delay like metric, a similar technique can be used to leverage BLAST in any delay metric, which uses multiple timepoints on the waveform to compute the delay.

DAE2FSM: Automatic Generation of Accurate Discrete-Time Logical Abstractions for Continuous-Time Circuit Dynamics

Karthik .V. Aadithya[*‡] and Jaijeet Roychowdhury[*]

[*]Department of Electrical Engineering and Computer Science, The University of California, Berkeley, CA, USA
[‡]Contact author. Email: aadithya@berkeley.edu

ABSTRACT

We abstract the I/O functionality of continuous-time dynamical systems (e.g., SPICE netlists with combinational and sequential logic) as Finite State Machines (FSMs). This enables efficient simulation of large designs implemented with less-than-perfect devices and components, and also opens the door to formal verification of transistor-level designs against higher-level specifications. In particular, our automatically generated FSMs faithfully capture the behaviour of latches, flip-flops, and circuits constructed from them. Among other technical advances, we generalize an existing (binary-only) FSM-learning approach to arbitrary I/O alphabets, which empowers it to learn high-fidelity abstractions of multi-level-discretized, multi-input/multi-output systems. Our approach, when applied to correctly functioning latches and flip-flops, is able to learn compact, multi-input FSM abstractions whose predictions closely match SPICE simulations. In addition, we have also applied our technique to produce multi-level-discretized FSM representations of digital systems that nevertheless exhibit "analogish" traits, such as an overclocked, error-prone D-flip-flop. For such circuits, the automatically learned FSM abstraction includes additional states that characterise "failure modes" of the circuit for specific input sequences (these failure modes are also confirmed by SPICE simulations). Finally, we demonstrate that our technique is also applicable to larger and more complex multi-input, multi-output systems; for example, we are able to automatically derive an accurate FSM abstraction of a 280-transistor (BSIM4), 0-to-5 increment/decrement counter.

Categories and Subject Descriptors

D.2.2 [Design Tools and Techniques] State Diagrams
B.2.2 [Performance Analysis and Design Aids] Simulation

General Terms

Algorithms, Design, Theory

Keywords

Finite State Machine Learning, Circuit Simulation

1. INTRODUCTION

With technology scaling to 22nm and below, individual devices are increasingly becoming non-ideal, thus compromising the clean Boolean abstractions that underpin the effectiveness and power of the digital design paradigm. Indeed, many components in cutting-edge digital systems today behave more like analog/RF circuits than like digital ones. The design, validation and debugging of digital systems with such components can be challenging because "analog issues" stemming from nonlinear analog dynamics, analog waveshapes, noise/interference, etc., compounded by increased variability, cannot be directly captured within the Boolean modelling and simple delay frameworks that are natural for digital systems.

Purely SPICE-level simulation-based approaches for validation and debugging are impractical, on account of the large sizes of typical digital systems. In existing design methodologies, components in cell libraries are simulated extensively at the SPICE-level, over a range of PVT corners, to verify functionality and to characterize delays. This process does not, however, provide executable abstractions (such as finite state machines) that can reproduce details of analog wave-shapes; nor is it suited for components whose functionality is affected significantly by complex analog effects.

In this work, we develop and demonstrate techniques (collectively dubbed DAE2FSM[1]) to abstract executable Boolean descriptions of transistor-level circuits. The central notion of DAE2FSM is to approximate transistor-level circuits *accurately* as finite state machines (FSMs) by adapting and applying computational machine learning techniques. The resulting FSMs can not only capture the intended logical functionality of the circuit being abstracted, but also take into account analog effects and non-idealities, producing "non-ideal FSMs" that accurately reflect actual (rather than intended) operation (see Fig. 1).

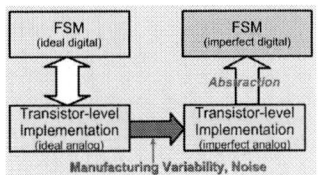

Figure 1: DAE2FSM: Transistor level non-idealities are captured in FSM representations.

We develop and apply an Angluin-based [1] computational learning technique that uses *finite alphabets of $N \geq 2$ symbols*, thus moving beyond the binary symbols used in prior work [2, 3]. This enables us to learn FSMs for transistor-level circuits with *multiple inputs*, by encoding input value or transition combinations using multiple symbols. Multi-symbol learning also enables us to obtain increased fidelity by using *multi-level discretizations* to approximate analog signals better. We also show how *non-deterministic FSMs* can be used to capture situations with unpredictable inputs. We demonstrate the application of these advances on transistor-level latch and flip-flop circuits, as well as on a counter circuit composed of several sequential and combinational components.

DAE2FSM features several points of novelty and promise. FSMs generated by DAE2FSMcan be simulated (in discrete time in the logical domain) much faster than the underlying SPICE-level representations they are derived from, while at the same time capturing the impact of analog/manufacturing imperfections. Indeed, the results of logical FSM simulation can be translated back to analog values that, in many cases, reproduce SPICE-simulated waveforms well. Compromised functionality and failure modes are also captured by the FSMs, which opens possibilities for system-level/post-silicon workarounds.

An important feature of DAE2FSM is that it is suitable for application to practical industrial circuits. Detailed, non-linear SPICE-level circuits can be utilized unchanged, within simulation environments of the user's choice, ensuring that no analog subtlety that SPICE can predict is ignored. At the same time, the I/O-based FSM learning approach behind DAE2FSM ensures that only those details of transistor-level blocks that are "observable from the outside" (i.e., relevant from a system perspective) are captured; in other words, only what is needed is represented in the learned FSM. Indeed, the automated, push-button nature of FSM generation frees the user from understanding in detail how a given transistor-level block functions; all that is needed is a SPICE-level netlist that simulates. From the standpoint of fitting into established design flows, the fact that simulation, validation and debugging can all be performed purely in the logical domain, is very attractive.

Moreover, by representing transistor-level circuits as FSMs, DAE2FSM opens the door to the application of *Boolean formal verification and model checking* techniques (e.g., [4,5]) for determining whether a given transistor-level component with strong analog characteristics satisfies a given property. Unlike alternative formal approaches based on hybrid systems (e.g., [6–9]), DAE2FSM neither suffers from scalability limitations, nor requires *a priori* modelling simplifications of the SPICE-level transistor blocks involved.

The rest of the paper is organized as follows. In §2, we outline the multi-symbol learning technique underlying DAE2FSM. In §3, we present detailed results applying DAE2FSM to latch and flip-flop circuits, generating FSMs for single and multiple inputs, binary and multi-level discretizations, and properly functioning as well as failing circuits. We also demonstrate how DAE2FSM captures the intended functionality of a 280-transistor, 0-to-5 increment/decrement counter from a SPICE-level description.

---

[1]Differential-Algebraic Equation to Finite State Machine.

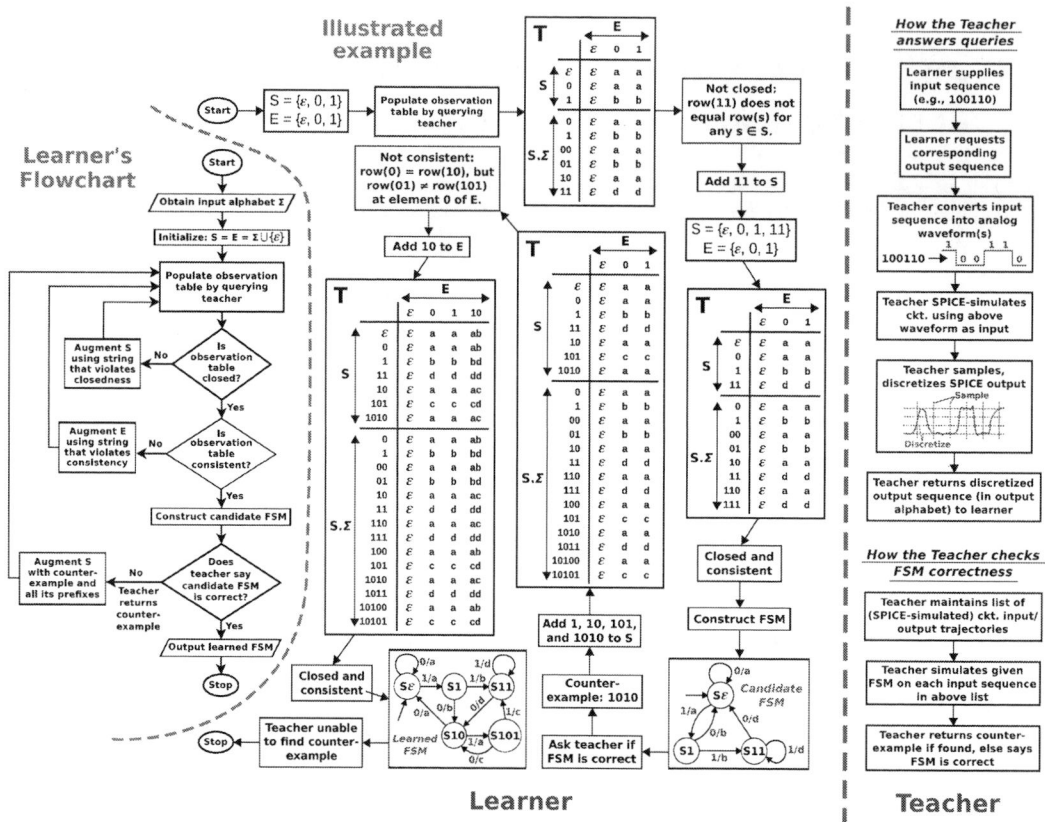

Figure 2: Flowchart and an example that describe our technique for fully automated multi-symbol Mealy machine learning.

2. CORE TECHNIQUE: MULTI-SYMBOL ANGLUIN-STYLE MEALY MACHINE LEARNING

We now describe our core technique for abstracting a given SPICE netlist as a finite state, multi-symbol Mealy machine[2]. Our approach derives from the well-known Anguin's algorithm [1] for Deterministic Finite Automata (DFAs)[3], which we have modified to learn multi-symbol Mealy machine abstractions of continuous dynamical systems such as circuits.

Our FSM abstraction algorithm for circuits has two key entities: a teacher, and a learner. Before the learning begins, both entities decide on an input alphabet Σ, and an output alphabet Γ (these can be any finite sets of symbols).

The teacher has access to the SPICE-level circuit netlist and a SPICE simulator. The learner, on the other hand, operates purely in the discrete domain: it has no knowledge of the circuit, and the only way it learns about the circuit is by asking specific questions of the teacher. These questions can take two forms: (a) I/O queries, and (b) FSM checks.

An I/O query involves the learner presenting the teacher with an input sequence (any word from the input alphabet), to which the teacher responds with an output sequence (a word from the output alphabet of length equal to the input sequence). Given the input sequence, the teacher constructs an input *waveform* from it (as shown in Fig. 2 (right)), and then SPICE-simulates the circuit on this input waveform. The teacher then discretizes the SPICE output into symbols from the output alphabet, and returns the discretized sequence as the "answer" to the learner's query. Thus, with each I/O query, the learner increases its knowledge about the circuit; eventually, it learns enough about the circuit to propose an FSM abstraction (in the form of a multi-symbol Mealy machine) to the teacher. This is the second type of query (an FSM check). Given an FSM proposed by the learner, the teacher runs a number of simulations comparing the outputs produced by the FSM with discretized versions of outputs produced by SPICE-simulating the circuit. If the teacher finds a discrepancy between the FSM and SPICE, it alerts the learner to the counter-example, and the learner uses this information to refine its FSM. If the teacher is unable to find a counter-example, it accepts the FSM as a faithful discrete-domain representation of the continuous-domain circuit behaviour.

We note that the teacher has enormous freedom in the way it converts an input *sequence* into an input *waveform*: for example, the teacher can interpret input symbols as *quantized voltage levels* (e.g., symbol "a" means 0V, symbol "b" means 0.1V, etc.), leading to a *multi-level-discretized* learned FSM. Alternatively, the teacher can interpret input symbols as *bit vectors* specifying boolean values for multiple circuit inputs (e.g., symbol "a" means 000, symbol "b" means 001, etc.); this interpretation results in a *multi-input* Mealy machine abstraction for the circuit. Another possibility is that the teacher can interpret the input symbols as *switching events* (e.g., symbol "a" means the clock switches, symbol "b" means the data switches, etc.), which results in a different kind of multi-input FSM that sometimes offers greater intuition into circuit dynamics (e.g., see our FSMs for latches and flip-flops in §3.2 and §3.3). This *freedom to interpret the input alphabet in multiple ways* is an important aspect of DAE2FSM: it allows us to use the same fundamental framework (automated multi-symbol Mealy machine learning) to generate multi-level, or multi-input, or multi-output, or any combination of these, FSM abstractions, depending on the circuit-driven application at hand. For example, if the application is to characterise a failing flip-flop (§3.4), a multi-level FSM would be the best option. On the other hand, if the application is a combinational/sequential circuit such as a counter, a multi-input, multi-output FSM would be best-suited (see §3.5).

Having presented the teacher's side, we now discuss the learner's algorithm (shown, along with an example, in Fig. 2 (left)). At any point, the learner maintains two sets of words over the input alphabet, S and E. In addition, it maintains an *observation table* T that contains all the information acquired about the circuit thus far. Initially, $S = E = \Sigma \cup \varepsilon$, where ε is the empty string. At any time, for every ordered pair $(s, e) \in (S \cup S.\Sigma) \times E$ (where . denotes concatenation), the observation table T contains an entry[4] $T(s, e)$, which is equal to the last $|e|$ output symbols returned by the teacher for the input sequence $s.e$ (Fig. 2 illustrates how T evolves as the algorithm progresses). For each $s \in S \cup S.\Sigma$, let row(s) denote the row corresponding to s in T.

---

[2]Recall that Mealy machines are FSMs that take in an input sequence, and produce an output symbol at each state transition, thereby returning an output sequence of length equal to the input sequence.

[3]Recall that DFAs are FSMs that take in an input string, and either output a "1" (indicating that the string has been accepted) or output a "0" (indicating that the string has been rejected).

[4]In Anguin's original algorithm for DFAs, T only contained 0/1 entries. However, to extend the algorithm to multi-symbol Mealy machine learning, we store output strings in T instead of binary values.

312

Using the information in T, the learner tries to match an *FSM state* to each word in S; for each $s \in S$, the learner associates with s the *final* state reached by the FSM on input s. For this, T must satisfy two conditions:

Closedness: For each $s_1 \in S.\Sigma$, there must exist $s_2 \in S$ such that row(s_1) = row(s_2). The intuition is that: the set of FSM states associated with S is incomplete if there is no destination state for an input in $S.\Sigma$. If T is not closed, then the strings violating closedness must be added to S, and T recompleted.

Consistency: For every $s_1, s_2 \in S$ such that row(s_1) = row(s_2), it should also be true that row($s_1.\sigma$) = row($s_2.\sigma$) for every $\sigma \in \Sigma$. The intuition is that: one cannot associate identical FSM states with two different strings (s_1 and s_2) in S, unless one can also associate identical states with $s_1.\sigma$ and $s_2.\sigma$, for every $\sigma \in \Sigma$. If T is not consistent, then the string $\sigma.e$ for which $T(s_1.\sigma.e) \neq T(s_2.\sigma.e)$ must be added to E, and T recompleted.

Thus, the learner issues I/O queries to the teacher until T is both closed and consistent. At that point, the learner proposes an FSM (whose states are associated with strings in S). This is repeated until the teacher is unable to find a counter-example to the learner's FSM. In Fig. 2 (left), we present the complete learner's algorithm as a flowchart, and also illustrate it with an example. The example shows how the learner, starting from scratch, learns a multi-output FSM for a failing D-flip-flop (for more details, see §3.4).

3. RESULTS

As mentioned earlier, we have applied the techniques of §2 to generate multi-symbol Mealy machine abstractions of latches, flip-flops, and circuits constructed from them. Here, we discuss these results in detail.

We begin by generating binary FSM abstractions of correctly functioning latches and flip-flops (§3.1), for different timing relationships between the clock (CLK) and data (D) signals. We then use *multi-symbol* FSM learning to construct *multi-input* FSMs for latches and flip-flops (§3.2, §3.3); this encodes all relevant switching patterns of CLK and D using a multi-symbol alphabet (i.e., the algorithm automatically generates a single FSM capturing the circuit's behaviour across all relevant timing scenarios). After this, in §3.4, we apply *multi-level* discretization to realize FSM abstractions of error-prone flip-flops, which fail on certain inputs because of analog effects. Finally, in §3.5, we present a *multi-input, multi-output, combinational/sequential* application: that of automatically learning a state machine abstraction for a 0-to-5 increment/decrement counter implemented with 280 transistors.

3.1 Binary FSMs for correctly functioning latches and flip-flops

We start with the simplest case: generating binary FSMs for latches and flip-flops. We consider six different timing scenarios for the clock and data signals, and we demonstrate that DAE2FSM is able to produce Mealy machines whose predictions match well with SPICE-level simulations.

At this stage, the target FSMs are limited to a single input (i.e., the data input D). The other input (the clock CLK) is fixed, and assumed to be a periodic pulse, whose one period is shown in Fig. 3 (right). It is assumed that D transitions exactly once per clock cycle. Also, the input/output sequences required for DAE2FSM are sampled exactly once per clock cycle (either before, or after D transitions).

Input changes	Input, output are sampled	Scenario #	Latch FSM	Flip-Flop FSM
Only when CLK is low	When CLK is low (before input changes)	1	Buffer	Buffer
	When CLK is low (after input changes)	2	Delay	Delay
	When CLK is high	3	Buffer	Delay
Only when CLK is high	When CLK is low	4	Buffer	Buffer
	When CLK is high (before input changes)	5	Buffer	Buffer
	When CLK is high (after input changes)	6	Buffer	Delay

Figure 3: Binary FSMs learned for latches and flip-flops in various operating modes.

This gives rise to six possible timing scenarios, tabulated in Fig. 3. For each scenario, we used DAE2FSM to learn FSMs for both a D-latch and a D-flip-flop[5]. Fig. 3 shows that, depending on the scenario and the circuit, the learned FSM can be either a *buffer* or a *delay*. These FSMs are shown in Figs. 4 (a) and 4 (b) respectively; it is seen that the buffer FSM simply relays

---

[5]We note that a D-latch FSM has also been learned by the authors of [2]. However, that work considers only one of the scenarios outlined in Fig. 3, whereas we have meticulously analysed all possible scenarios. Also, the authors of [2] consider only latches, whereas we have developed FSMs for latches, flip-flops and beyond.

its input directly to the output (i.e., its input and output sequences are always identical), whereas the delay FSM shifts the input to the right by one element. Together, the two FSMs capture the behaviour of ideal latches and ideal flip-flops in their various operating modes (Fig. 3).

For example, we know that an ideal D-latch can operate in two modes: it is transparent when CLK is high, but retains its output (even if the input changes) when CLK is low. Thus, if the input transitions (once a clock cycle) when CLK is low, and input/output sequences are sampled just before CLK turns high (scenario 2 in the above table), the output would reflect the previous input (applied 1 clock cycle earlier), which corresponds exactly to a delay FSM (as indeed, the table above shows).

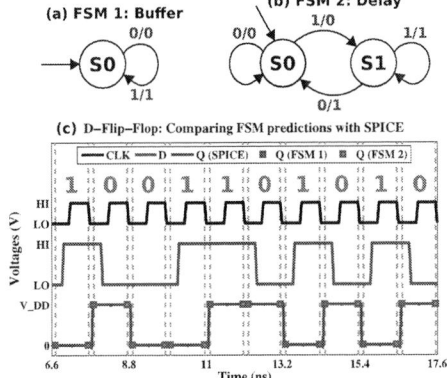

Figure 4: The *buffer* and *delay* FSMs returned by DAE2FSM accurately reflect the behaviour of an ideal D-flip-flop.

Similarly, we know that an ideal D-flip-flop captures the value of D precisely at the negative edge of each clock cycle (i.e., during the interval when CLK transitions from high to low), and remains opaque to changes in D at all other times. Fig. 4 (c) shows a SPICE simulation[6] of such a flip-flop's output (the blue waveform) where the input D (the green waveform) transitions, once per clock cycle, when CLK (the black waveform) is low.

From Fig. 3, we see that a buffer FSM predicts the flip-flop's output at uniformly spaced time points just *after* the negative edge of the clock (i.e., scenario 1). For example, given the input sequence of Fig. 4 (c), the buffer FSM predicts the output sequence 0100110101 at these time points. This digital output sequence can now be mapped back to an *analog* output sequence, by tagging each output symbol with a specific analog voltage (determined at the time of sampling I/O sequences for the FSM learning algorithm). For instance, in this example, we tag the output symbol "0" with the analog voltage 0V, and the output symbol "1" with the voltage V_{DD} (which, in this case, is 0.8V). Hence we obtain a sequence of analog voltages (in this case, [0V, 0.8V, 0V, 0V, 0.8V, 0.8V, 0V, 0.8V, 0V, 0.8V]) associated with uniformly spaced time points. We now overlay this analog time series on top of the SPICE waveform (see the magenta markers on top of the blue SPICE waveform in Fig. 4 (c)), to judge how well the discrete-domain FSM is able to predict the circuit's continuous-domain behaviour. In this case, just from a visual examination, it is clear that the FSM's predictions do in fact, tally closely with the SPICE-simulated waveform. Similarly, we have also plotted (with green markers in Fig. 4) the analog time series version of the delay FSM's predictions (which we know, from Fig. 3, to be valid just *before* the negative edge of the clock). From the figure, we see that this set of predictions also closely match the SPICE waveform at these time points (the grey vertical lines).

3.2 Multi-input FSMs for correctly functioning latches

The FSMs of the previous subsection handle only one circuit input (i.e., D); the other input (CLK) is taken to be a fixed waveform that is known *a priori*. This necessitates many separate learnings, one for every possible switching pattern of D relative to CLK (as tabulated in Fig. 3). Moreover, if the input can transition twice or more per clock cycle, none of the FSMs learned above would be valid, and a fresh set of FSMs would need to be learned. The learned FSMs must then be "pieced together" to understand the functionality of the given circuit. This process can be tedious and inconvenient.

Instead, DAE2FSM offers the capability of learning *multi-input FSMs* (as outlined in §2), thereby dispensing with the need to generate/piece together many one-input FSMs. By considering D and CLK as two separate circuit inputs, the algorithm automatically learns a multi-input FSM that fully takes

---

[6]All SPICE simulations in this paper have been carried out with 22nm devices, using BSIM4 device models. We obtained device parameters from [10], and the SPICE engine from [11].

into account all relevant switching combinations of both inputs. We now demonstrate the multi-input capability of DAE2FSM on a D-latch.

To produce the multi-input latch FSMs (explained in §2 and in §1), we first encode all relevant switching patterns of both inputs, using a 4-symbol input alphabet $\Sigma = \{w,x,y,z\}$ for the multi-symbol Angluin procedure outlined in §2. Input symbol w indicates that both CLK and D are held constant (until the next sampling instant). Similarly, input symbol x (y) indicates that only CLK (D) switches (becomes high if it was low earlier, and vice-vera), while D (CLK) is held constant. Finally, symbol z indicates that both CLK and D switch their values: z therefore has two different meanings, depending on whether CLK switches first or D switches first. The two meanings lead to two different multi-input FSMs, as we show below.[7] Clearly, this 4-symbol input alphabet can represent all possible sequences of (legal) switching events in both inputs. For example, if one wants to determine the latch output following three switches of D before a single switch of CLK, and then a clock switch, and then two more switches of D, one simply passes the input sequence $[y,y,y,x,y,y]$ to the Mealy machine learned by DAE2FSM.

Fig. 5 (b) shows the multi-symbol Mealy machine automatically learned by DAE2FSM for the D-latch, for the case that CLK switches ahead of D on input symbol z. This FSM has two *transparent* states (TR1 and TR2), and four *opaque* states (OP1 to OP4). These states offer intuition into how the latch functions: every switch in CLK (i.e., input symbol x or z) causes the FSM state to toggle between transparent and opaque. By contrast, a switch in D does not affect the transparency or opacity of the current FSM state. Closer examination reveals that when CLK goes low (i.e., the latch transitions from transparent to opaque), the FSM always reaches the opaque state with the correct polarity (i.e., OP1 if D is low at the instant the clock switches, OP4 otherwise). Transitions from opaque to transparent states also reflect precisely how one would expect an ideal latch to behave. Indeed, as Fig. 5 (a) shows, the latch output (Q) predictions made by this FSM closely tally with SPICE simulations, even for a complicated sequence of input switches that cannot be handled by any binary-only FSM derived earlier.

Fig. 5 (d) depicts the FSM derived when D switches ahead of CLK at input symbol z (with the meaning of the other input symbols unchanged). As expected, the FSMs in Figs. 5 (b) and 5 (d) are identical except for transitions on input z: their states are in one to one correspondence for all z-less input sequences. This FSM's predictions are also in excellent agreement with SPICE simulations (as seen from Fig. 5 (d)).

Thus, our multi-input DAE2FSM technique has automatically produced Mealy machines that accurately mimic the latch's behaviour under all relevant switching conditions. The only caveat is that the algorithm does not (yet) automatically handle illegal race conditions in the input[8]; for example, if CLK switches to low, and D switches at exactly the same time (a well-known "illegal" situation that can produce unpredictability, and even metastability), the output of a D-latch can become unpredictable, which the learned FSM, being a deterministic automaton, fails to capture. This unpredictability is illustrated in Fig. 5 (e): when both CLK and D switch simultaneously, the latch sometimes behaves as though CLK switched first, and sometimes as though D switched first. To account for such conditions, we combine the FSMs in Figs. 5 (b) and 5 (d) to arrive at a non-deterministic FSM. The rationale is that CLK in Fig. 5 (b) switches ahead of D at input symbol z, while Fig. 5 (d) applies when D switches ahead of CLK at input symbol z. Therefore, if CLK and D switch at the same time, the latch could (in theory) choose to behave according to either of these FSMs; in practice, the latch's "choice" of FSM would depend on many factors, including the exact shapes of the switching input waveforms, clock jitter, voltages at internal nodes, device parameters, noise processes (e.g., thermal noise, shot noise), etc. Since most of these factors are inherently unpredictable, it is convenient to abstract them by introducing non-deterministic transitions in the learned FSM. This results in the Mealy machine of Fig. 5 (f), whose non-z transitions are identical to the original FSMs, but whose z-transitions include non-determinism.

3.3 Multi-input FSMs for correctly functioning flip-flops

We now repeat the analysis of the previous subsection, but for a D-flip-flop[9] instead of a D-latch. The results (Fig. 6) roughly mirror those of the previous subsection (Fig. 5); however, there are interesting differences, as noted below.

As with the D-latch, we have generated FSM abstractions of the D-flip-flop using the multi-symbol input alphabet $\{w,x,y,z\}$ (where the symbols have the same meaning as before). The auto-generated FSMs are shown in Figs. 6 (b) and 6 (d).

Unlike the D-latch, however, there is no concept of a "transparent" or "opaque" state in the D-flip-flop's FSMs. Rather, the intuition is that each state can be viewed as an ordered 3-tuple, whose dimensions are the stored flip-flop value Q, the clock CLK, and the input D. For example, state S101 indicates that: (a) the flip-flop currently stores a value $Q = 1$ (captured at the most recent negative clock edge), (b) the clock is low, and (c) D is high. With this intuition, it is readily seen that the FSMs in Figs. 6 (b) and 6 (d) capture the precise functionality expected of a D-flip-flop: the input is captured and relayed to the output exactly once per "cycle" of the clock, and this happens only when CLK transitions from high to low. Moreover, as shown in Figs. 6 (a) and 6 (c), the predictions made by these FSMs match very well with values sampled from SPICE simulations.

Finally, Fig. 6 (e) shows that the flip-flop's output can be unpredictable when CLK switches from high to low, and D also switches at exactly the same time. In such a situation, the flip-flop behaves at times as if CLK switched first, and at other times as if D switched first. Moreover, as before, we observe that the two FSMs learned for the different meanings of z, in this case also, behave identically for all z-less sequences, with their states being in one-to-one correspondence. Therefore, as before, it is possible to combine these FSMs by introducing non-deterministic transitions. The resulting combined FSM is shown in Fig. 6 (f).

3.4 Multi-level FSMs for failing flip-flops

Having shown how DAE2FSM produces correct abstractions of properly functioning latches and flip-flops, we now demonstrate another crucial feature of DAE2FSM: that it can abstract useful FSMs even for circuits that suffer from such significant analog imperfections that digital functionality is compromised. The motivation is that it is often important to characterise the behaviour of latches and flip-flops functioning under non-ideal operating conditions (e.g., under lowered supply voltages, extreme overclocking, a particularly unfavourable process variability corner, etc.). Such characterisation, for example, plays a central role in the design of error-resilient communication systems. We note that the generation of (binary) FSMs for non-ideal latches has already been demonstrated in [2]. In this work, we focus on *multi-level discretizations* applied to flip-flops and demonstrate how DAE2FSM captures novel failure modes.

Fig. 7 shows two possible failure modes (discovered by DAE2FSM) that a D-flip flop can exhibit: both result from overclocking the flip-flop (i.e., as the clock frequency increases, the flip-flop has less time to capture the input value at the negative clock edge, eventually being unable to do so for some input sequences). One failure mode (Fig. 7 (a)) can be adequately captured by a binary FSM, whereas the other (Fig. 7 (c)) needs a multi-level output alphabet (supported only by the multi-symbol approach).

In the first mode of failure[10] (Fig. 7 (a)), a single "1" applied at the input of the flip-flop fails to register at the output; the flip-flop's response is too slow to capture the fleeting bit. However, if the applied "1" remains in place for two or more consecutive clock cycles, the flip-flop is able to register it, because its internal (analog) state has already been nudged in the right direction by the first "1". This is illustrated by the SPICE plots in Fig. 7 (a). As shown in Fig. 7 (b), DAE2FSM is able to capture this failure mode. The FSM in Fig. 7 (b) starts at the "zero" state marked Sε. In this state, if the FSM encounters a "1", it moves to an *intermediate state* S1, where it waits for the next input. If the next input is "0", the FSM goes back to its initial state without registering the previously applied "1" (reflecting the failing flip-flop's behaviour). If the next input is also "1", the FSM enters the "one" state (and stays there for as long as the input remains "1") because it has witnessed at least two consecutive "1s" at the input. Thus, the binary FSM of Fig. 7 (b) is adequate to capture this failure mode.

The second failure mode, illustrated in Fig. 7 (c), occurs when the flip-flop is clocked at 9.26 GHz. The output of the flip-flop clearly shows 4 distinct levels: at 0, V_{DD}, and two intermediate levels indicated by the horizontal, dashed red lines (marked L1 and L2). These intermediate levels appear in the output only when specific sequences are detected at the input. For example, level L1 appears when the input sequence contains a "1" that is preceded by neither a "1" nor the sequence "10". On the other hand, level L2 is reached whenever the input sequence "101" is applied at the input. This can be explained by recognizing that the failing flip-flop has a memory longer than one clock cycle: each time a "1" is applied, the failing flip-flop remembers it for the next two

---

[7] Alternatively, one could also learn a 5-symbol FSM, where the two meanings of z are encoded by two different input symbols z_1 and z_2; however, the 4-symbol approach has the added advantage that the resultant FSMs can be combined via non-deterministic transitions to model race conditions in the input (Fig. 5 (f)).

[8] We are currently working on improvements to DAE2FSM that can handle race conditions, metastability, etc.

[9] The flip-flop we have used is a master-slave, negative-edge triggered D-flip-flop built from two D-latches.

[10] This failure is observed at a clock frequency of 10.42 GHz, a reasonable frequency at which to expect the flip-flop to fail by its own, i.e., without the added effect of combinational delays from external sources.

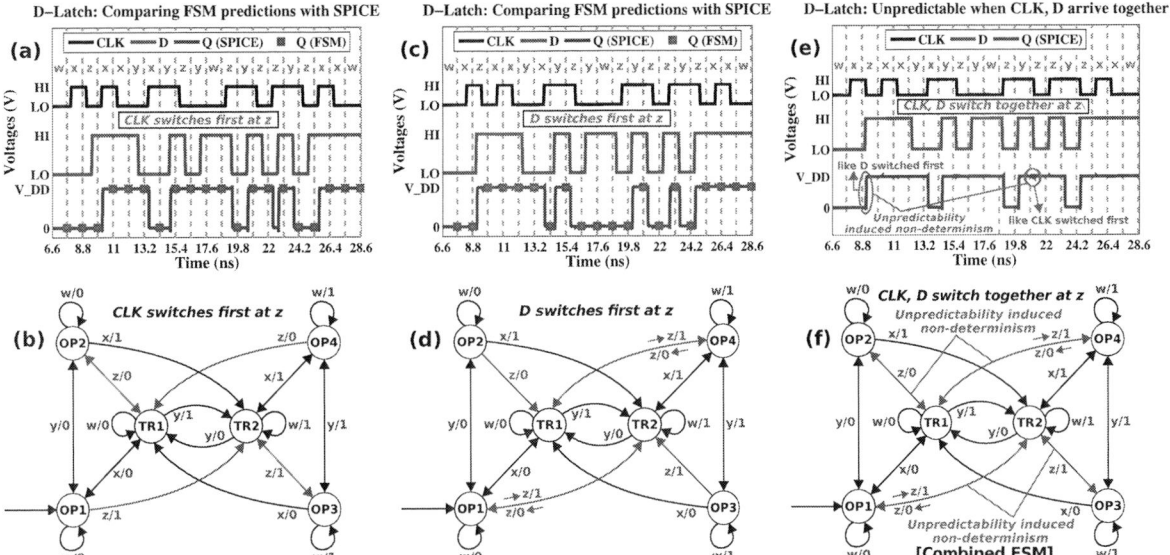

Figure 5: Multi-input DAE2FSM applied to construct multi-input FSM abstractions of a D-Latch.

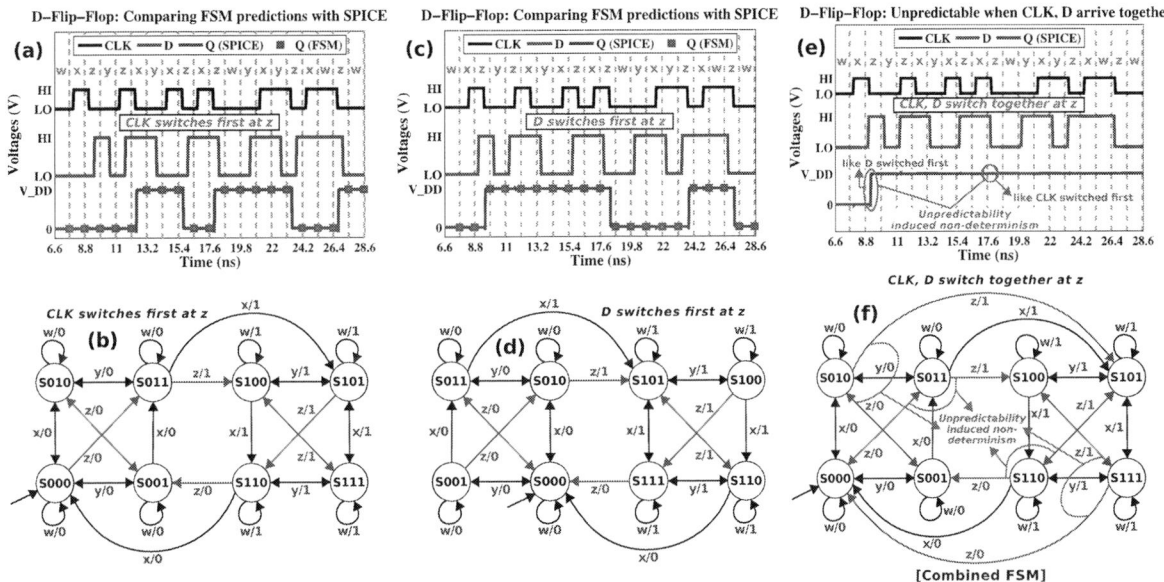

Figure 6: Multi-symbol DAE2FSM applied to construct unified FSM abstractions of a D-flip-flop.

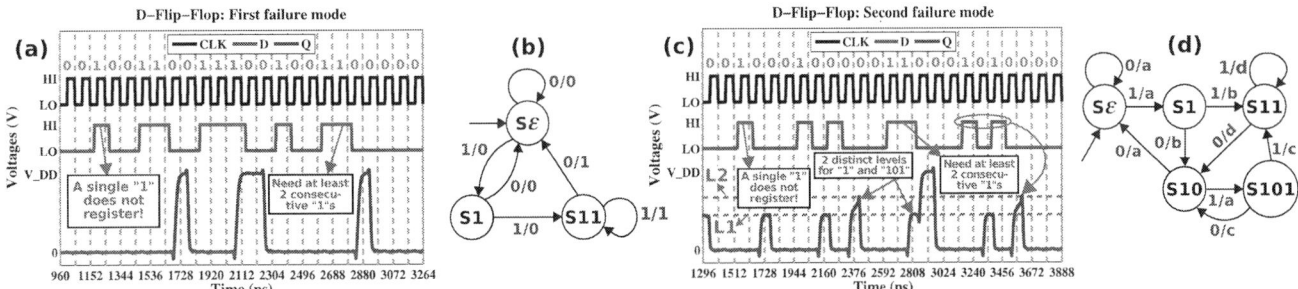

Figure 7: Failure modes observed for flip-flops, and FSMs learned for failing flip-flops.

315

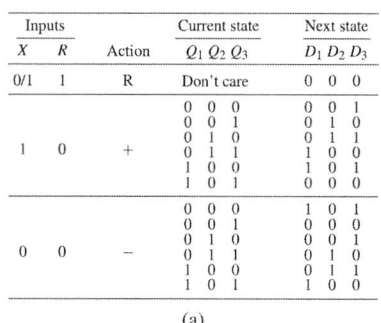

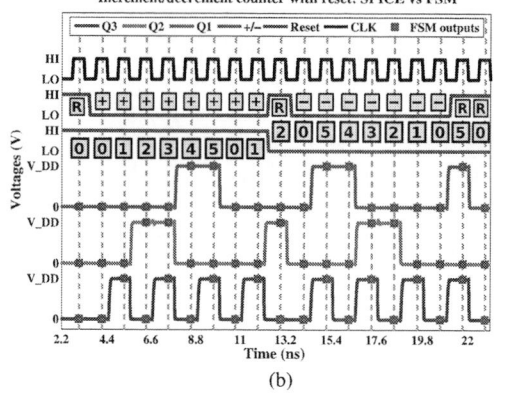

Increment/decrement counter with reset: SPICE vs FSM

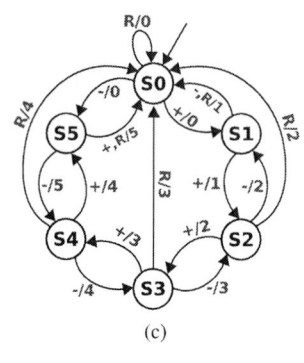

Inputs			Current state	Next state
X	R	Action	$Q_1\ Q_2\ Q_3$	$D_1\ D_2\ D_3$
0/1	1	R	Don't care	0 0 0
1	0	+	0 0 0 0 0 1 0 1 0 0 1 1 1 0 0 1 0 1	0 0 1 0 1 0 0 1 1 1 0 0 1 0 1 0 0 0
0	0	−	0 0 0 0 0 1 0 1 0 0 1 1 1 0 0 1 0 1	1 0 1 0 0 0 0 0 1 0 1 0 0 1 1 1 0 0

(a)

(b)

(c)

Figure 9: (Left) Table showing the state transitions of a 0-to-5 increment/decrement counter with reset (please see Fig. 8 for a circuit schematic). The counter takes the increment (decrement) action + (−) when $X = 1$ ($X = 0$), unless the reset bit R is set, in which case the counter is reset to 0. (Right) Multi-symbol Mealy machine automatically learned by DAE2FSM for the counter. (Middle) SPICE simulation showing a complete increment cycle and a complete decrement cycle of the counter. The top row of yellow boxes indicate the next "action" that will be taken by the counter, while the bottom row indicates the current count.

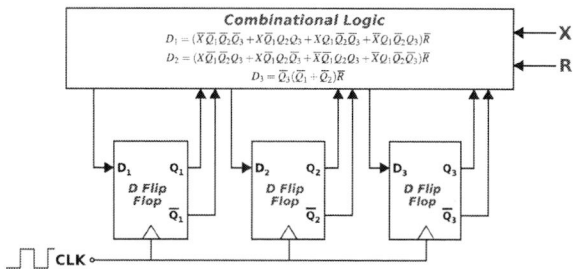

Figure 8: Schematic of a 0-to-5 increment/decrement counter with reset (please see Fig. 9 (a) for the corresponding state transition table).

clock cycles, which has an effect on its output during that time-span. However, because the output is now a 4-level signal, it cannot be reproduced by a binary FSM. In this case, therefore, one needs an FSM with an output alphabet of at least 4 symbols, one for each output level (the input alphabet can still be binary). Applying multi-level DAE2FSM to this circuit results in the multi-symbol Mealy machine of Fig. 7 (d). The example of Fig. 2 shows in detail how DAE2FSM was able to learn this Mealy machine from scratch. It can be verified that this FSM produces output "c" (corresponding to level L2) on input sequence "101", and output "b" (level L1) on a "1" that is preceded by neither a "1" nor the sequence "10". The output is "a" (the symbol for 0) for a "0" at the input, and "d" (the symbol for V_{DD}) only if the input has two or more consecutive "1s". Thus, multi-level DAE2FSM can be applied to find state machines that model failing flip-flops.

3.5 A combinational/sequential, multi-input/multi-output case

Having demonstrated FSM abstraction of basic units such as latches and flip-flops, we now apply DAE2FSM to a much larger design: a 280-transistor, multi-input, multi-output circuit that includes both combinational and sequential logic elements. Although the circuit is considerably more complex than the examples above, the learned FSM reproduces its behaviours perfectly.

The circuit is a 0-to-5 increment/decrement counter with reset (schematic shown in Fig. 8), which takes two (digital) inputs (not counting CLK) X and R, and returns three (digital) outputs Q_1, Q_2 and Q_3. The output bits Q_1 to Q_3 encode a whole number (the count) in the range 0-to-5 (0 and 5 included), with Q_1 (Q_3) being the most (least) significant bit. At the negative edge of each clock cycle, the count is either incremented (where 5 "increments" to 0), decremented (where 0 "decrements" to 5) or reset to 0, depending on the inputs supplied. The "reset to 0" action (denoted R) is taken whenever the reset input (also R) is set; otherwise, the count is incremented (denoted $+$) if $X = 1$, and decremented (denoted $-$) if $X = 0$. Fig. 9 (a) shows the state transition table associated with the counter, while Fig. 9 (b) shows SPICE simulations of a complete increment cycle and a complete decrement cycle of the counter (with a reset in between). These simulations confirm that the counter functions exactly as intended.

We applied DAE2FSM to learn a Mealy machine representation of this counter automatically. Each of the two inputs and three outputs was discretized using

two levels; they were then encoded using 4 and 6 symbols[11], respectively, for multi-symbol Angluin-based learning (see §2).

Fig. 9 (c) depicts the Mealy machine learned by DAE2FSM; it consists of six states S0 to S5 (arranged in a circle in the figure), corresponding to the count values 0-to-5. Increment actions, taken when $(X, R) = (1, 0)$, result in clockwise traversal of the circle, while decrement actions, taken when $(X, R) = (0, 0)$, result in anti-clockwise traversal. At each state, the reset action (at $R = 1$) results in a state transition back to S0. This state machine captures the intended logical functionality of the counter exactly. Also, the output sequence predicted by the learned Mealy FSM, when translated to analog values (as described earlier), matches SPICE-level simulations well (see the magenta markers in Fig. 9 (b)).

4. SUMMARY, CONCLUSIONS AND FUTURE WORK

In this paper, we have demonstrated DAE2FSM, a technique for automatically learning multi-symbol discrete-domain FSM abstractions of continuous-domain dynamical systems such as circuits. We have extended Angluin's algorithm to multi-symbol Mealy machine learning, which enables the generation of different classes of FSMs (including multi-symbol, multi-input, multi-output, and any combination of these) supporting different circuit-level applications. We have applied our technique to automatically produce multi-input FSMs for properly functioning latches and flip-flops, by encoding all relevant input switching patterns with a single 4-symbol input alphabet. The auto-generated FSMs are able to produce output sequences that match well with SPICE simulations. We have also used the multi-symbol framework to learn multi-level FSM abstractions of error-prone flip-flops, where DAE2FSM was able to identify two failure modes in overclocked D-flip-flops. We have also generated a multi-input, multi-output Mealy machine abstraction of a 0-to-5 increment/decrement counter, which illustrates the applicability of our technique to a larger and more complex system than an individual latch/flip-flop.

In future, we would like to extend our framework to automatically learn non-deterministic FSMs that characterise circuit behaviour for inputs that create race conditions. Closely related to this, we would also like to auto-generate FSMs that capture the effects of metastability on latches and flip-flops.

5. REFERENCES

[1] D. Angluin. Learning regular sets from queries and counter-examples. *Information and Computation*, 75:87–106, November 1987.

[2] C. Gu and J. Roychowdhury. FSM model abstraction for analog/mixed-signal circuits by learning from I/O trajectories. In *ASP-DAC '11: Proceedings of the 16th Asia and South Pacific Design Automation Conference*, pages 7–12, 2011.

[3] C. Gu. *Model order reduction of non-linear dynamical systems*. PhD thesis, The University of California, Berkeley, 2011.

[4] E. M. Clarke, T. A. Henzinger, and H. Veith, editors. *Handbook of model checking*. Springer-Verlag, 2011.

[5] R. K. Brayton and A. Mishchenko. ABC: An academic industrial-strength verification tool. In *CAV '10: Proceedings of the 22nd International Conference on Computer Aided Verification*, pages 24–40, 2010.

[6] R. Alur, T. A. Henzinger, G. Laferriere, and G. Pappas. Discrete abstractions of hybrid systems. *Proceedings of the IEEE*, 88(7):971–984, July 2000.

[7] C. L. Guernic and A. Girard. Reachability analysis of hybrid systems using support functions. In *CAV '09: Proceedings of the 21st International Conference on Computer Aided Verification*, pages 540–554, 2009.

[8] A. Girard, C. L. Guernic, and O. Maler. Efficient computation of reachable sets of linear time-invariant systems with inputs. In *HSCC '06: Proceedings of the 9th International Conference on Hybrid Systems: Computation and Control*, pages 257–271, 2006.

[9] C. Tomlin, I. Mitchell, A. M. Bayen, and M. Oishi. Computational techniques for the verification of hybrid systems. *Proceedings of the IEEE*, 91(7):986–1001, 2003.

[10] http://ptm.asu.edu/modelcard/HP/22nm_HP.pm.

[11] http://www.spiceopus.si/.

---

[11]Ordinarily, 8 symbols are needed to encode three digital outputs; however, two of these (bit vectors 110 and 111) never appear in this counter's output.

Chip/Package Co-Analysis of Thermo-Mechanical Stress and Reliability in TSV-based 3D ICs

Moongon Jung[1], David Z. Pan[2], and Sung Kyu Lim[1]
[1] School of ECE, Georgia Institute of Technology, Atlanta, GA, USA
[2] Department of ECE, University of Texas at Austin, Austin, TX, USA
moongon@gatech.edu, dpan@ece.utexas.edu, limsk@gatech.edu

ABSTRACT

In this work, we propose a fast and accurate chip/package thermo-mechanical stress and reliability co-analysis tool for TSV-based 3D ICs. We also present a design optimization methodology to alleviate mechanical reliability issues in 3D IC. First, we analyze the stress induced by chip/package interconnect elements, i.e., TSV, μ-bump, and package bump. Second, we explore and validate the principle of lateral and vertical linear superposition of stress tensors (LVLS), considering all chip/package elements. This linear superposition principle is utilized to perform full-chip/package-scale stress simulations and reliability analysis. Finally, we study the mechanical reliability issues in practical 3D chip/package designs including wide-I/O and block-level 3D ICs.

Categories and Subject Descriptors

B.7.2 [**Hardware, Integrated Circuit**]: Design Aids

General Terms

Design

Keywords

3D IC, TSV, stress, mechanical reliability, chip/package co-analysis

1. INTRODUCTION

Most previous works on the thermo-mechanical stress and reliability of TSV-based 3D ICs have been done separately in chip or package domain. The impact of TSV-induced stress due to coefficient of thermal expansion (CTE) mismatch between TSV and substrate materials on device performance [1] and crack growth in TSV [7] were studied in the chip domain. As for the package domain, many works focused on the reliability of package bump (= C4 bump) [9]. Recently, authors in [8] showed the significant impact of package components on the chip domain stress. They proposed a stress exchange file to transfer the boundary conditions from package-level to silicon-level analysis. However, all of these approaches require FEA methods which are computationally expensive or infeasible for full-chip or -package analysis.

To overcome the limitation of FEA method, linear superposition of stress tensors [5] and response surface method [4] were utilized. However, all of these are limited to the chip domain analysis. In this paper, we propose a full-chip/package-scale thermo-mechanical stress and reliability co-analysis flow as well as a de-

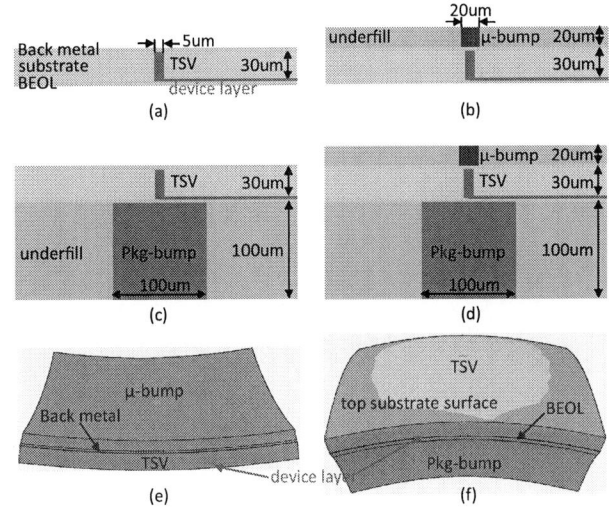

Figure 1: Impact of bumps and underfill on the stress of device layer (= red line). (a) TSV only [5] (b) TSV + μ-bump (c) TSV + package-bump (d) TSV + μ-bump + package-bump. (e) Deformed structure of (b). (f) Deformed structure of (c). Both (e) and (f) are drawn with 10X deformation scale factor.

sign optimization methodology to reduce the mechanical reliability problems in TSV-based 3D ICs. We show the impact of design parameters such as the size and pitch of chip/package interconnect elements and the number of dies in the stack on thermo-mechanical stress and reliability.

The main contributions of this work include the following: (1) Modeling: Compared with existing works, we simulate more detailed 3D IC structures including both chip and package components and study their interaction and impact on thermo-mechanical stress and reliability. (2) Full-chip/package co-analysis: We, for the first time, validate the principle of lateral and vertical linear superposition of stress tensors induced by each chip/package interconnect element such as TSV, μ-bump, and package-bump against FEA simulations. We apply this methodology to generate a stress map and a reliability metric map in full-chip scale. (3) Case study: we study the mechanical stress and reliability issues in practical 3D chip/package designs including wide-I/O and block-level 3D ICs.[1]

2. MOTIVATION

We first examine how various chip/package interconnect components interact and alter the thermo-mechanical stress distribution on the *device layer around TSV* caused by the CTE mismatch between

---

[1] We also explore the impact of TSV/bump placement, size, and pitch on the overall system, and the materials are moved to the Supplemental Section due to the space limit.

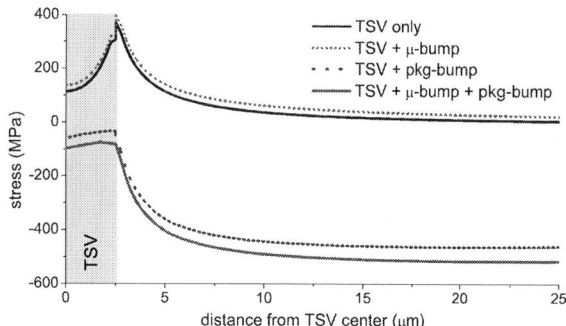

Figure 2: Impact of package components on the stress (σ_{rr}) around TSV on device layer (FEA results).

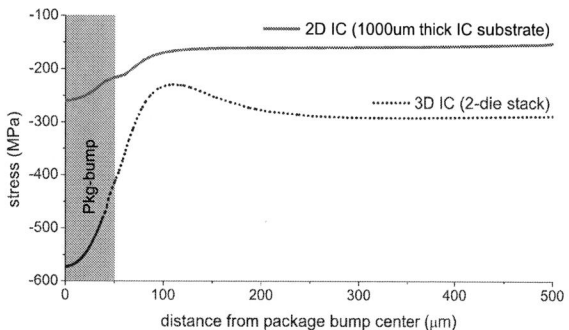

Figure 3: Comparison of impact of package-bump on the device layer stress (σ_{rr}) between 2D IC and 3D IC (2-die stack) (FEA results).

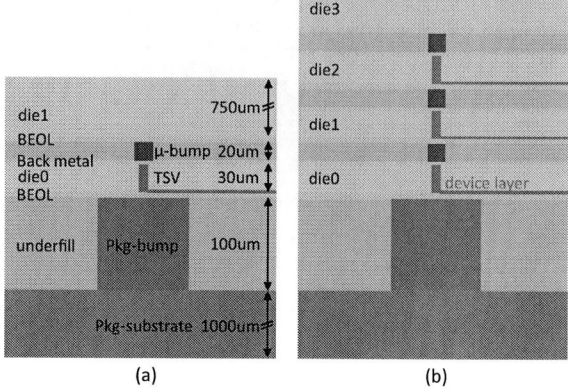

Figure 4: Side view of baseline chip/package simulation structures. (a) 2-die stack (b) 4-die stack.

TSV and substrate materials. First, we only consider TSV and substrate which most previous works studied. We employ the same simulation structure used in [5] as shown in Figure 1(a). Then, we add a μ-bump and underfill layer *above* the substrate as shown in Figure 1(b). All structures undergo $\Delta T = -250°C$ of thermal load (annealing/reflow 275°C → room temperature 25°C). As Figure 2 shows, by adding the μ-bump layer (= dotted red line), we see slightly more tensile (= positive) stress than the TSV-only case (= solid black line). This is because ΔCTE of μ-bump and underfill is 24 ppm/K, while that of TSV and substrate is 14.7 ppm/K, hence the deformation of the entire structure is largely determined by the μ-bump and underfill layer. Since the top side of μ-bump layer is free surface, the entire structure easily bends upward as all the elements shrink from the negative thermal load as shown in Figure 1(e). Thus, the materials on device layer stretch outward, which results in more tensile stress.

On the other hand, if we add a package-bump (= C4 bump) layer *below* the substrate as shown in Figure 1(c), now the entire structure bends downward as shown in Figure 1(f) because package elements are shrinking more than chip elements. The ΔCTE of package bump and underfill is 22 ppm/K. This generates highly compressive (= negative) stress on the device layer. Comparing Figure 1(b) and Figure 1(c), we see that the bending direction depends on which layer shrinks more: in both cases, the bump layers shrink more than the silicon substrate.

Lastly, we include both bump layers as shown in Figure 1(d). In this case, the ΔCTE is almost the same (24 ppm/K on the top, 22 ppm/K on the bottom). However, the overall structure bends down in a similar fashion as shown in Figure 1(f) because of the sheer volume of package bump layer (= shrinking more than the μ-bump layer). This in turn causes compressive stress in the device layer. However, the magnitude is slightly *more* (= solid green line in Figure 2) than the package-bump layer only case (= dotted blue line). One might expect the overall compressive stress would

be less because the μ-bump layer tries to bend upward while the package-bump layer tries to bend downward (= canceling effect). However, this additive effect is because the μ-bump layer eventually bends down and adds more compressive stress to the device layer. Remember that the bending direction of the μ-bump layer is affected by adjacent layers. Since now the deformation of the entire structure is dominated by the package-bump layer, the flexible underfill material in the μ-bump layer easily bends downward. These basic simulations clearly show the importance of considering package element impact on the chip-domain stress distribution.

Figure 3 shows the stress contributions of package bump and underfill layer to the chips (2D vs. 3D) mounted on it. For the 3D IC/package structure, we build a two-die stack chip/package structure similar to Figure 4(a) excluding TSV and μ-bump. This was to examine the impact of package-bump solely. The bottom die (= die0) is thinned, and we examine the device layer of this thin die. One 2D IC/package structure is also created, where we use a single un-thinned die of 1000 μm thickness. We examine the device layer of this un-thinned die. We apply the same thermal load (ΔT = -250°C) for both cases. Figure 3 shows that the 3D IC experiences more severe compressive stress than the 2D IC case. The main reason is the thickness and the flexibility of the die that we are monitoring. Even though the thickness of the entire structure is thicker in 3D IC, the thin die (30 μm thick) and the underfill material above the thin die is much more flexible than the un-thinned substrate in 2D IC. Thus, this thin die is highly affected by the package-bump underneath it. This indicates that the impact of package-bump is more significant in 3D IC.

3. 3D IC/PACKAGE STRESS MODELING

We use the von Mises yield criterion [10] as a mechanical reliability metric for TSVs, which is explained in Section S.1. However, we do not use a specific threshold value for the von Mises criterion in this work, since it is greatly affected by fabrication process.

3.1 3D IC/Package Simulation Structure

Figure 4 shows our simulation structure, where the dimensions of our baseline simulation structures are based on the fabricated and/or published data [2, 8]. In this work, we specifically examine the stress distribution on device layer for each die shown in red lines in Figure 4. Our baseline TSV diameter, height, landing pad size, Cu diffusion barrier thickness, and dielectric liner thickness are 5 μm, 30 μm, 6 μm, 50 nm, and 125 nm, respectively. We use Ti and SiO$_2$ as Cu diffusion barrier and liner materials. Also, diameter/height of μ-bump and package-bump are 20 μm and 100 μm, respectively, unless otherwise specified. Material properties used for our experiments are as follows: CTE (ppm/K) / Young's modulus (GPa) for Cu = (17/110), Si = (2.3/188), SiO$_2$ =

318

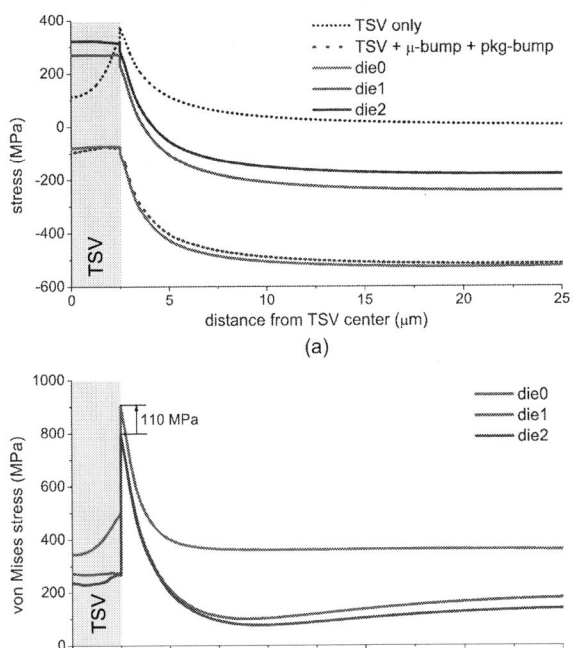

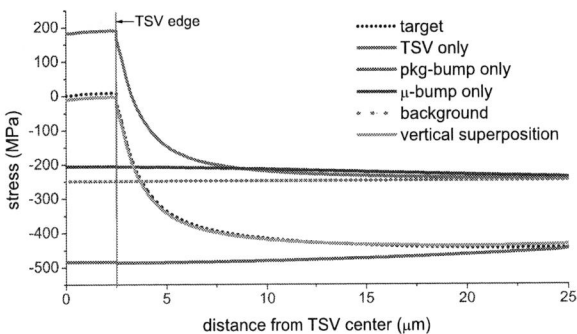

Figure 7: Vertical linear superposition of σ_{rr} stress in a 2-die stack shown in Figure 6

die without TSVs) is almost flat (-110 ± 5 MPa). Since die3 does not contain any TSVs, there is no local von Mises stress peak (= dangerous region) caused by TSVs. Thus, we only consider the dies containing TSVs in this work.

Moreover, we observe that the mechanical reliability problem is most severe in die0 shown in Figure 5(b). The maximum von Mises stress at TSV edge in die0 is about 110 MPa higher than the upper two dies. This is again mostly due to the package-bump that induces large deformation at the nearest die.

4. HANDLING FULL-CHIP/PACKAGE

FEA simulation for multiple TSVs, μ-bumps, and package-bumps require huge computing resources and time, thus it is not feasible for a full-system-scale analysis. In this section, we present a chip/package thermo-mechanical stress co-analysis flow in full-chip/package scale. We use the principle of lateral and vertical linear superposition of stress tensors from individual TSVs, μ-bumps, and package-bumps to enable a full-system-level analysis. We validate our approach against FEA simulation results. Based on the linear superposition method, we build full-chip stress maps and then compute the von Mises yield metric to assess the mechanical reliability problems in TSV-based 3D ICs.

4.1 Lateral and Vertical Linear Superposition

In [5], authors used the principle of linear superposition of stress tensors to perform a full-chip stress and reliability analysis considering many TSVs. In that case, all stress contributors (= TSVs) are on the same layer, hence we call this lateral linear superposition. However, as we consider the impact of μ-bump and package-bump, which are not in the same layer where TSVs are located, this lateral linear superposition cannot be used alone. Fortunately, the principle of linear superposition is not limited to 2D plane, but applicable to any linearly elastic structures including 3D structures.

Figure 6 illustrates our vertical linear superposition method, which enables us to consider the stress induced by elements which are not in the same layer. We first decompose the target structure into four separate structures: TSV only, package-bump only, μ-bump only, and background which does not contain TSV and bumps. Next, we obtain stress tensors along the red line on device layer affected by each interconnect element separately from FEA simulations. Then, we add up the stress tensors from TSV only, package-bump only, and μ-bump only structures, and subtract twice the magnitude of the background stress tensors since this background stress is already included in previous three structures. If the point under consideration is affected by n components, then we need to subtract $n-1$ times the background stress.

Figure 7 shows the stress distributions from each structure as well as the stress obtained by the vertical linear superposition. We see that μ-bump induces more tensile stress than background and package-bump generates much more compressive stress than background, which is discussed in Section 2. We also observe that even without interconnect elements (= background) device layer is in

---

Figure 5: Impact of die stacking on device layer stress (FEA results). (a) σ_{rr} **stress on device layer in each die in 4-die stack. (b) von Mises stress in each die in a 4-die stack.**

(0.5/71), Ti = (8.6/116), package-bump (SnCu)= (22/44.4), μ-bump (Sn$_{97}$Ag$_3$) = (20/26.2), underfill = (44/5.6), package substrate (FR-4) = (17.6/19.7).

We use a FEA simulation tool ABAQUS to perform experiments, and all materials are assumed to be linear elastic and isotropic. We also discuss the impact of the anisotropic Si material property on the thermo-mechanical stress and reliability in Section S.4. The entire structure undergoes ΔT = -250°C of thermal load (annealing/reflow 275°C → room temperature 25°C) to represent a fabrication process. In addition, all materials are assumed to be stress free at the annealing/reflow temperature.

3.2 Impact of Die Stacking

Previous works on the full-chip thermo-mechanical stress analysis used the same stress pattern for different dies in a multiple-die stack [1, 5]. In this section, for the first time, we examine how the thermo-mechanical stress distribution on the device layer around TSV differs across strata. We employ the four-die stack structure for this purpose. Also, we use only one TSV, μ-bump, and package-bump for each die or layer, respectively, and their center locations are aligned as shown in Figure 4.

First of all, the stress level, the extent of compression or tension, differs significantly across dies as shown in Figure 5(a). The overall stress trend remains similar: the stress is highest at TSV edge and decays then saturates as distance increases from the TSV center. However, the bottom-most die (= die0, solid red line), which is closest to the package-bump layer, shows most compressive stress among three dies containing TSV. This is because the impact of package-bump is most significant in die0 due to their proximity.

Also, as we go to the upper dies, the stress level becomes closer to the case considering only TSV and substrate. We also see that the stress curve of die0 is very close to the case of TSV + μ-bump + package-bump (= dotted purple line), which does not contain the package substrate and un-thinned top die shown in Figure 1(d). This also indicates that the stress level in die0 is mostly determined by package-bump. The stress distribution in die3 (un-thinned top

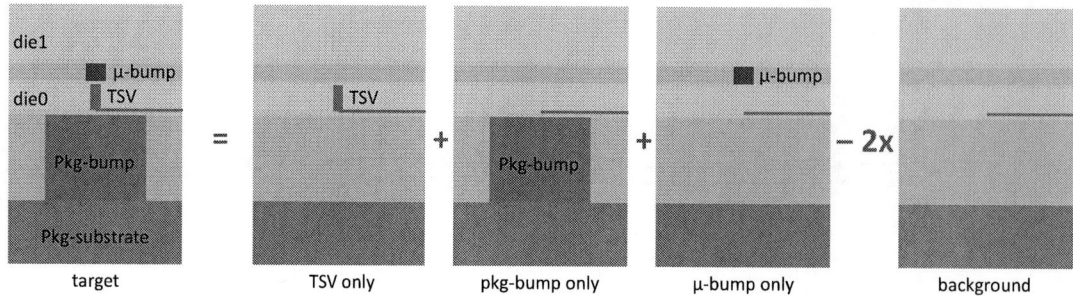

Figure 6: Illustration of vertical linear superposition with a 2-die stack structure. Stress is extracted along the red line on device layer from each structure using FEA tool.

compression due to the shrinking of the underfill material which has the highest CTE (= 44 ppm/K) among all materials in the simulation structure. Most importantly, our vertical linear superposition method matches well with the target stress distribution. Although we see the maximum error (11 MPa) occurs inside TSV, this is inevitable since we ignore the direct vertical interaction between TSV, μ-bump, and package-bump by decomposing the structure. Nonetheless, this error is acceptable for a fast full-system-scale analysis.

To obtain the stress tensor at a point affected by multiple TSVs, μ-bumps, and package-bumps, we apply both lateral and vertical linear superposition (LVLS) as follows:

$$
S = \sum_{i=1}^{n_{SV}} S_{TSVi} + \sum_{j=1}^{n_{\mu}} S_{\mu B_j} + \sum_{k=1}^{n_{pkg}} S_{pkgB_k}
$$
$$
-(n_{TSV} + n_{\mu B} + n_{pkgB} - 1) \times S_{bg} \quad (1)
$$

where, S is the total stress at the point under consideration and S_{TSVi}, $S_{\mu B_j}$, and S_{pkgB_k} are individual stress tensor at this point due to i^{th} TSV, j^{th} μ-bump, and k^{th} package-bump, respectively. S_{bg} indicates the background stress at that point.

4.2 Full-Chip/Package Stress Analysis Flow

We briefly explain how we perform a full-chip/package stress analysis based on the LVLS method. We first build a stress library from FEA simulations. This library contains stress tensors along an arbitrary radial line on the device layer induced by each interconnect element, i.e., TSV, μ-bump, and package-bump, separately. Given locations of TSVs, μ-bumps, and package-bumps, we find a stress influence zone for each element. Beyound this stress influence zone of each interconnect element, the stress induced by the element under consideration is negligible [5]. In our work, we use five times the diameter of each component as a stress influence zone, where stress level is saturated to the background stress level from FEA simulations.

Then, we associate each grid point with all the interconnect elements whose stress influence zone overlaps with the point. Next, we apply the LVLS method at the point under consideration to obtain the stress tensor induced by every component found in the association step. Finally, we compute the von Mises stress to assess the mechanical reliability problem in TSVs. More details of our algorithm is discussed in Section S.6.

4.3 Validation of LVLS

In this section, we validate our LVLS method against FEA simulations by varying the number of TSVs, μ-bumps, and package-bumps as well as their arrangement. We set the minimum pitch of TSV, μ-bump, and package-bump as 10 μm, 20 μm, and 200 μm for all test cases. Stress tensors along the radial line on device layer induced by each interconnect element (stress tensor library) are obtained through FEA simulation with 0.25 μm interval. In our linear superposition method, simulation area is divided into uniform array style grid with 0.1 μm pitch. If the stress tensor at the grid

Table 1: Von Mises stress comparison between FEA and LVLS for a 4-die stack structure (die0). Error = LVLS - FEA. (At TSV edge, typical von Mises stress level is around 900 MPa.)

# TSV /μ-B /pkg-B	FEA		LVLS		max error (MPa)		
	# node	run time	# grid	run time	inside TSV	TSV edge	outside TSV
1/1/1	754K	1d2h	1M	23s	-11.4	-12.6	7.9
2/2/1	812K	1d2h	1M	26s	-12.7	-13.2	7.3
5/5/2	902K	1d6h	6M	2m43s	-14.1	-15.3	8.2
10/10/4	1.3M	1d20h	9M	6m44s	-23.1	-19.8	9.4
10/10/9	1.4M	2d0h	16.8M	11m11s	-22.5	-20.5	11.9

point under consideration is not obtainable directly from the stress library, we compute the stress tensor using linear interpolation with adjacent stress tensors in the library.

Table 1 shows some of our comparisons in die0 in a four-die stack, which shows the largest errors among three dies containing TSVs due to its proximity to package-bumps. Also, we only list the cases with the minimum pitches for each component, which again shows maximum errors. First, we observe a huge run time reduction in our LVLS method. Note that we perform FEA simulations using 8 CPUs while only one CPU is used for our linear superposition method. Even though the LVLS method performs stress analysis on a 2D plane (= device layer), whereas FEA simulation is performed on the entire 3D structure, we can perform stress analysis for other planes in a similar way if needed.

The error between FEA simulations and LVLS is very small. Results show that our LVLS method underestimates stress magnitude inside TSV and TSV edge, and overestimates outside TSV, as shown in Figure 7. In general, the most critical region for the mechanical reliability is the interface between different materials, hence TSV edge is most important in our case. Even though the maximum error at TSV edge is as high as -20.5 MPa, its % error is only -2.24 %. Figure 8 shows one test case comparison of von Mises stress between FEA and LVLS. The structure has 10 TSVs (5 μm diameter and 10 μm pitch), 10 μ-bumps (20 μm diameter and 40 μm pitch), and 9 package-bumps (100 μm diameter and 200 μm pitch). It clearly shows our LVLS method matches well with the FEA simulation result.

5. SIMULATION RESULTS

We implement a chip/package thermo-mechanical stress and reliability co-analysis flow based on LVLS in C++/STL. More details can be found in Section S.6. We explore the impact of package-bump and μ-bump on the reliability in full-system scale. Also, we examine the reliability concerns in wide-I/O DRAM and block-level 3D IC designs.

In our experiments, we adopt a regular TSV placement style in which TSVs are placed uniformly across each die or inside TSV blocks with pre-defined pitch. In all cases, the pair of TSV and μ-bump is vertically aligned. Default diameter/height (μm) of TSV, μ-bump, and package-bump are 5/30, 10/10, and 100/100, respec-

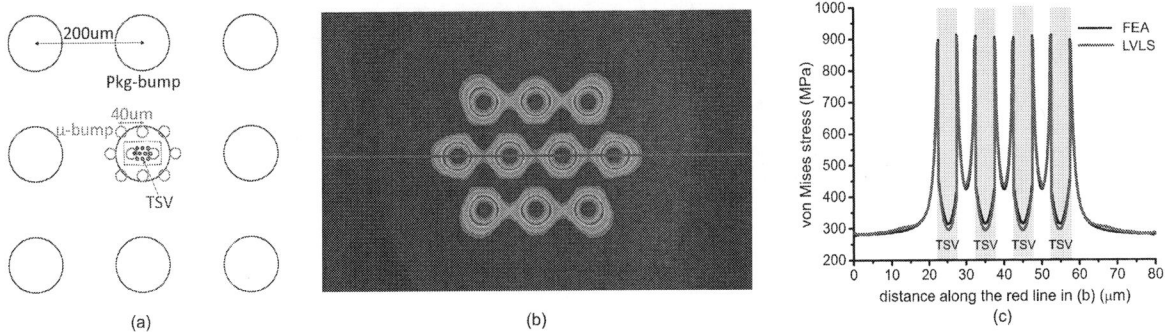

Figure 8: Sample stress comparison between FEA and LVLS. (a) Test structure. (b) Close-up shot of von Mises stress map (using LVLS) taken from the red box in (a) on the device layer in die0 in a 4-die stack. (c) FEA vs. LVLS along the red line in (b).

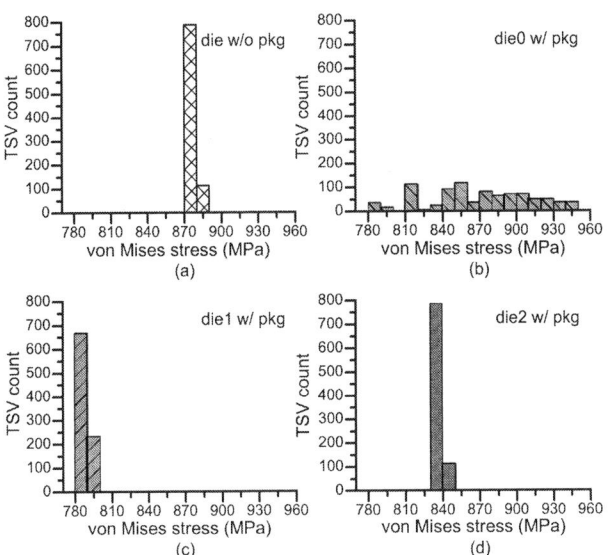

Figure 9: Impact of package components and die stacking on the mechanical reliability of TSVs (900 TSVs in each die).

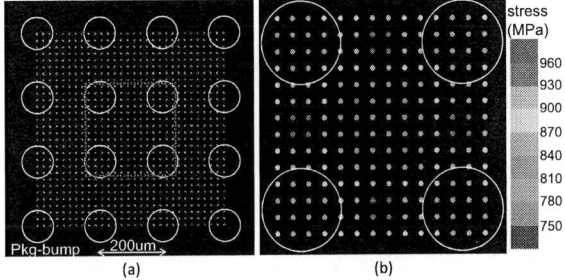

Figure 10: Von Mises stress map for TSVs (die0 in a 4-die stack). Colored dots are TSVs and white circles are package-bumps. (a) Test structure. (b) Close-up shot of red box in (a)

Table 2: Reliability in wide-I/O DRAM.

case	von Mises stress distribution (MPa)					median (MPa)
	780-810	810-840	840-870	870-900	900-930	
(a)	30	114	52	220	608	944.8
(b)	182	842	0	0	0	856.2

center of Figure 10(b), if the distance between TSV and package-bumps is long enough, the von Mises stress of TSV becomes low.

Interestingly, die1 shows lowest von Mises stress level among all cases even though die2 is farthest from package-bumps. This is due to the fact that die2 is affected by the rigid un-thinned top silicon substrate above it. Since die0 is most problematic in terms of the mechanical reliability, we only consider die0 in a four-die stack in the subsequent experiments.

5.2 Case Study I: Wide-I/O DRAM

Wide-I/O based 3D DRAM is fast becoming the first mainstream product that utilizes TSV in 3D ICs, mainly targeting mobile computing applications such as smart phones which need lower power consumption and high data bandwidth. In this section, we evaluate the reliability concerns of TSVs in wide-I/O DRAM. We follow the TSV placement style similar to the work in [6], where TSV arrays are placed in the middle of a chip. We assume that 2x128 TSV array (per memory bank) is placed in the middle of a chip shown in Figure 11. We employ four memory banks and 1024 TSVs in total. We set the pitch of TSV/μ-bump and package-bump as 15 μm and 200 μm, respectively. We compare two cases; (a) Package-bumps are placed right underneath TSV arrays; (b) Package-bumps are placed with 200 μm spacing from TSV arrays. This 200 μm distance is chosen since we see that the effect of package-bump on the TSV reliability is negligible beyond 200 μm in case of the 100 μm diameter package-bump shown in Figure 16.

Table 2 clearly shows that the chip/package co-design can greatly reduce the mechanical reliability concerns in TSV-based 3D ICs.

tively, unless otherwise specified.

5.1 Impact of Package-Bump and μ-Bump

We first study the impact of package-bump and μ-bump on the mechanical reliability of different dies in a four-die stack. We also compare this to the case without these components as in the previous work [5] as shown in Figure 1(a). In this experiment, the pitch of TSV/μ-bump and package-bump are 20 μm and 100 μm, respectively; the total number of TSV/μ-bump and package-bump are 900 and 16, respectively, as shown in Figure 10(a).

We first observe that unlike the die without package-bumps and μ-bumps (Figure 9(a)) and the upper dies with package components (Figure 9(c) and (d)), TSVs in die0 (Figure 9(b)) experience large variations of von Mises stress across the die. This is because die0 is highly affected by package-bumps underneath it, and hence depending on the relative position between TSVs in die0 and package-bumps the von Mises stresses of TSVs change noticeably.[2]

We also identify that higher von Mises stress occurs around package-bump edge and in between package-bumps due to constructive stress interference shown in Figure 10(b). However, as we see in the

---

[2]Note that we see higher von Mises stress level in (Figure 9(a)) than the previous work [5] even with the same simulation structure. This is because we use the Young's modulus of 188 GPa for Si instead of 130 GPa in [5] as a worst case scenario. More details are discussed in Section S.4.

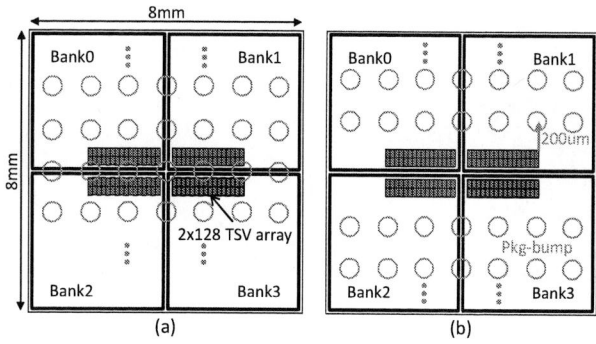

Figure 11: Mechanical reliability in wide I/O DRAM. 1024 TSVs are placed in the middle of a chip. (a) Package-bumps are placed underneath TSV arrays. (b) Package-bumps are placed 200 μm apart from TSV arrays. (not drawn in scale.)

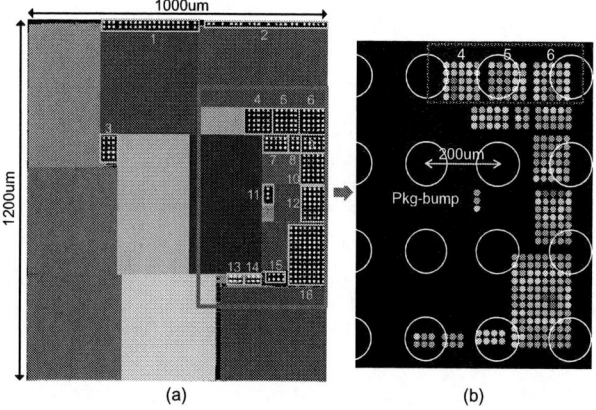

Figure 12: Mechanical reliability in block-level 3D IC. (a) Sample layout of block-level design. (b) Von Mises stress map for TSVs in red box in (a).

With a safe margin of 200 μm (= case(b)), von Mises stress magnitude reduces significantly. Thus, given the TSV placement, we can find safe locations for package-bumps without affecting the package design much, or vice versa.

5.3 Case Study II: Block-Level 3D IC

In this section, we study the reliability issues in block-level 3D designs. 3D block-level designs are generated using an in-house 3D floorplanner which treats a group of TSVs as a block shown in Figure 12. Total 16 TSV blocks (368 TSVs) are used and the TSV pitch is 15 μm. Package-bumps are regularly placed with 200 μm pitch.

Table 3 shows von Mises stress level in selected TSV blocks. We first observe that larger TSV blocks experience more variation of von Mises stress within the TSV block. This is because the distance between each TSV in the block and package-bumps can vary more than small TSV blocks, which is a key factor that affects the reliability of TSVs. We also see that TSV blocks with the same size can show quite different characteristics depending on the distance to the nearest package-bump. For example, although TSV block 4, 5, and 6 are all 5x5 TSV blocks and are located side-by-side, TSV block 5 shows the lowest von Mises stress level. However, its standard deviation of von Mises stress is highest among three blocks. We observe lower von Mises stress if TSV is placed near the package-bump center or far away from it; however, we see higher stress in TSV located around package-bump edge shown in Figure 16 in Section S.5. In case of TSV block 5, most TSVs are near the package-bump center, which lowers von Mises stress level.

Table 3: Mechanical reliability in block-level 3D IC. TSV blocks are shown in Figure 12.

TSV block #	# TSV	von Mises stress (MPa)				blk-bump dist (μm)
		max	min	avg	std dev	
3	5x3	901.0	811.1	859.5	26.0	96.4
4	5x5	939.6	853.5	902.6	24.0	67.6
5	5x5	908.6	816.0	858.7	33.3	24.1
6	5x5	942.3	874.4	910.4	22.0	91.4
11	3x1	896.6	855.9	871.0	18.2	39.3
16	12x8	943.7	806.0	877.2	33.6	90.7

However, at the same time a few TSVs are around the package-bump edge, which increases the standard deviation of von Mises stress inside the TSV block.

From this experiment, we observe two possible ways to reduce the mechanical reliability problems in block-level 3D designs: (1) Assign TSV blocks right above package-bump center locations if possible. (2) Place package-bumps outside the TSV block locations with a safe margin such as outside the red box in Figure 12(a). However, other design constraints such as package area and the required number of pins sholud be carefully considered as well.

6. CONCLUSIONS

In this work, we showed how package elements affect the stress field and the mechanical reliability on top of the TSV-induced stress in 3D ICs. We observed that the mechanical reliability of TSVs in the bottom-most die in the stack are highly affected by packaging elements, and that effect decreases as we go to the upper dies. We also presented an accurate and fast full-chip/package stress and mechanical reliability co-analysis flow based on the principle of lateral and vertical linear superposition of stress tensors (LVLS), considering all chip/package elements.

7. ACKNOWLEDGMENTS

This work is supported in part by the National Science Foundation under Grants No. CCF-1018216, CCF-1018750, the SRC Interconnect Focus Center (IFC), SRC Task 2239.001 and 2238.001 (CADTS), and SRC task 2244.001 and 2243.001 (SEMATECH 3D EC).

8. REFERENCES

[1] K. Athikulwongse, A. Chakraborty, J.-S. Yang, D. Z. Pan, and S. K. Lim. Stress-Driven 3D-IC Placement with TSV Keep-Out Zone and Regularity Study. In *Proc. IEEE Int. Conf. on Computer-Aided Design*, 2010.

[2] G. V. der Plas et al. Design Issues and Considerations for Low-Cost 3D TSV IC Technology. In *IEEE Int. Solid-State Circuits Conf. Dig. Tech. Papers*, 2010.

[3] M. A. Hopcroft, W. D. Nix, and T. W. Kenny. What is the Young's Modulus of Silicon. In *J. Microelectromechanical Systems*, 2010.

[4] M. Jung, X. Liu, S. K. Sitaraman, D. Z. Pan, and S. K. Lim. Full-Chip Through-Silicon-Via Interfacial Crack Analysis and Optimization for 3D IC. In *Proc. IEEE Int. Conf. on Computer-Aided Design*, 2011.

[5] M. Jung, J. Mitra, D. Z. Pan, and S. K. Lim. TSV Stress-aware Full-Chip Mechanical Reliability Analysis and Optimization for 3D IC. In *Proc. ACM Design Automation Conf.*, 2011.

[6] J.-S. Kim et al. A 1.2V 12.8GB/s 2Gb Mobile Wide-I/O DRAM with 4x128 I/O Using TSV-Based Stacking. In *IEEE Int. Solid-State Circuits Conf. Dig. Tech. Papers*, 2011.

[7] K. H. Lu, S.-K. Ryu, J. Im, R. Huang, and P. S. Ho. Thermomechanical Reliability of Through-Silicon Vias in 3D Interconnects. In *IEEE Int. Reliability Physics Symposium*, 2011.

[8] M. Nakamoto et al. Simulation Methodology and Flow Integration for 3D IC Stress Management. In *Proc. IEEE Custom Integrated Circuits Conf.*, 2010.

[9] S. R. Vempati et al. Development of 3-D Silicon Die Stacked Package Using Flip Chip Technology with Micro Bump Interconnects. In *IEEE Electronic Components and Technology Conf.*, 2009.

[10] J. Zhang et al. Modeling Thermal Stresses in 3-D IC Interwafer Interconnects. In *IEEE Trans. on Semiconductor Manufacturing*, 2006.

S. SUPPLEMENTAL MATERIAL

In this section, we provide basic concepts and thorough modeling results of thermo-mechanical stress and reliability analysis. We first introduce the concept of stress tensor and von Mises yield criterion. Then, we discuss the impact of the thickness of both package substrate and un-thinned top substrate on the stress. We also model how the alignment of TSV, package-bump, and μ-bump affect the mechanical reliability, and observe that relative distance between TSV and package-bump is the key factor that determines the reliability of TSV. In addition, we examine how the anisotropic Si material property affects the stress and reliability compared with the isotropic Si, and why we use the isotropic Si material property as a worst case scenario in our work.

We also present details of our full-chip/package stress and reliability analysis flow. Then, we provide extensive full-chip/package analysis results which show the impact of TSV/bump size and pitch on the reliability. In general, smaller size and larger pitch of each interconnect element help reduce the mechanical reliability problem of TSV-based 3D ICs. However, other design constraints such as the area of chip and package should be carefully considered.

S.1 Stress Tensor & Von Mises Criterion

To help understand stress modeling results, we introduce the concept of a stress tensor. Stress at a point in a body can be described by the nine-component stress tensor:

$$\sigma = \sigma_{ij} = \begin{bmatrix} \sigma_{11} & \sigma_{12} & \sigma_{13} \\ \sigma_{21} & \sigma_{22} & \sigma_{23} \\ \sigma_{31} & \sigma_{32} & \sigma_{33} \end{bmatrix}$$

where, the first index i indicates that the stress acts on a plane normal to the i axis, and the second index j denotes the direction in which the stress acts. If index i and j are same we call this a normal stress, otherwise a shear stress. Since we adopt a cylindrical coordinate system in this modeling for the cylindrical TSV, μ-bump, and package-bump, index 1, 2, and 3 represent r, θ, and z, respectively.

In order to evaluate if computed stresses indicate possible reliability concerns, a critical value for a potential mechanical failure must be chosen. The von Mises yield criterion is known to be one of the most widely used mechanical reliability metric [10]. If the von Mises stress exceeds a yielding strength, material yielding starts. Prior to the yielding strength, the material will deform elastically and will return to its original shape when the applied stress is removed. However, if the von Mises stress exceeds the yield point, some fraction of the deformation will be permanent and non-reversible even if applied stress is removed [5].

There is a large variation of yield strength of Cu in the literature, from 225 MPa to 3.09 GPa, and it has been reported to depend upon thickness, grain size, and temperature [10]. In this work, rather than selecting a specific value of yield stress for Cu TSV, we show how von Mises stress level changes under various circumstances. The yield strength of silicon is 7000 MPa, which will not be reliability concerns for the von Mises yield criterion.

The von Mises stress is a scalar value at a point that can be computed using components of a stress tensor shown in Equation (2).

S.2 Impact of Thickness of Substrate

In this section, we study the impact of the thickness of package substrate and un-thinned top silicon substrate on the thermo-mechanical stress. We use a 1 mm thick package substrate and a 750 μm thick un-thinned top die as a baseline structure.

We first vary the package substrate thickness from 0.75 mm to 3 mm, and monitor the stress around TSV on device layer in die0 in a four-die stack structure. We observe that stress becomes more

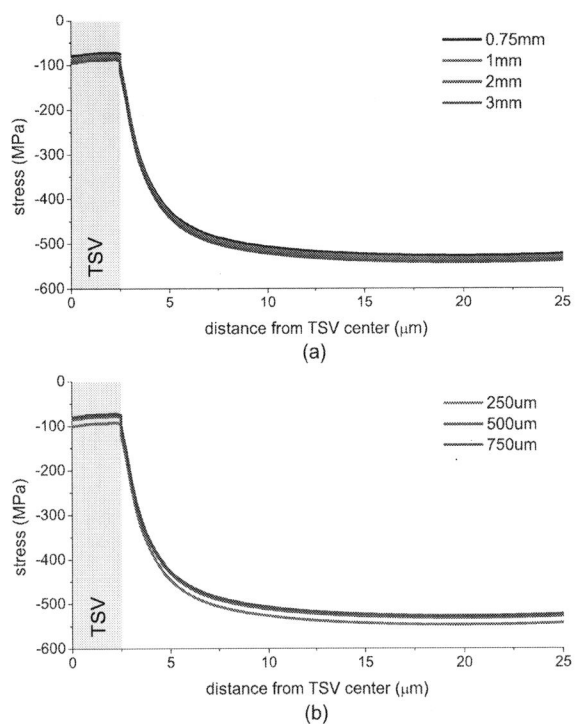

Figure 13: Impact of package substrate and un-thinned top die thickness on stress (FEA results). (a) Impact of package substrate thickness. (b) Impact of un-thinned top die thickness.

compressive as thickness increases, but the difference is almost indistinctive shown in Figure 13(a). This is mainly because this package substrate is already much thicker than other layers, hence its increased thickness impact on device layer is negligible.

We also change the thickness of the un-thinned top die from 250 μm to 750 μm, and observe that thinner die induces more compressive stress. This is because thinner die is more flexible as we see in the thin die case, and hence helps the entire structure bend more easily. However, still the differene is not significant. Thus, we use the baseline 1 mm thick package substrate and 750 μm thick un-thinned top die in our experiments.

S.3 Impact of Multiple Die Stacking

We now examine the stress magnitude in each die with a different number of die stacking. Figure 14 shows stress distributions in die0 with a two-die, a three-die, and a four-die stack. As more dies are stacked, more compressive stress occurs in die0 due to the additional stress from dies above. However, we see that this difference becomes smaller as we go to the upper dies, e.g., die1 stress in a three-die and a four-die stack.

S.4 Isotropic vs. Anisotropic Si Property

Up to this point, all materials are assumed to be isotropic for simplicity. However, Si is an anisotropic material with elastic behavior that depends on which crystal direction the structure is being stretched. The possible values of Young's modulus (E), which is a measure of stiffness of a material, for Si range from 130 to 188 GPa, and those for Poisson's ratio (ν) range from 0.048 to 0.4. Thus, the choice of this value can affect analysis results significantly [3]. In this section, we examine the impact of anisotropic material property of Si on the stress distribution compared with the isotropic Si material property.

$$\sigma_v = \sqrt{\frac{(\sigma_{xx} - \sigma_{yy})^2 + (\sigma_{yy} - \sigma_{zz})^2 + (\sigma_{zz} - \sigma_{xx})^2 + 6(\sigma_{xy}^2 + \sigma_{yz}^2 + \sigma_{zx}^2)}{2}} \qquad (2)$$

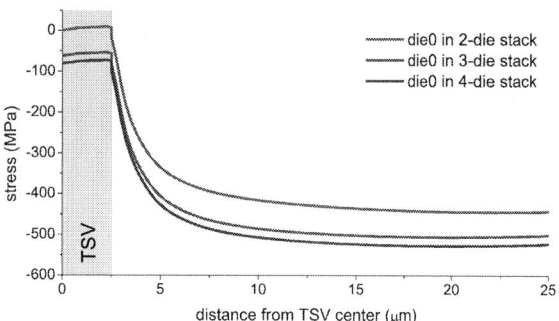

Figure 14: σ_{rr} **stress on die0 with a different number of die stacking.**

Elasticity is the relationship between stress (σ) and strain (ϵ). Hooke's law describes this relationship in terms of stiffness C, i.e., $\sigma = C\epsilon$. For isotropic uniaxial cases, stiffness C can be represented by a single value of Young's modulus E, and the equation takes the form of $\sigma = C\epsilon$. In an anisotropic material, a fourth rank stiffness tensor with $3^4 = 81$ terms is required to describe the elasticity. Fortunately, due to the cubic symmetry of Si, the elastic properties can be expressed in terms of orthotropic material constants. An orthotropic material is one which contains at least two orthogonal planes of symmetry, and Si, with cubic symmetry, can be described this way. The orthotropic elasticity of Si can be expressed with reference axes of a standard (100) Si wafer, which are [110], [$\bar{1}$10], and [001],

$$\begin{bmatrix} \sigma_{xx} \\ \sigma_{yy} \\ \sigma_{zz} \\ \sigma_{yz} \\ \sigma_{zx} \\ \sigma_{xy} \end{bmatrix} = \begin{bmatrix} c1 & c5 & c6 & 0 & 0 & 0 \\ c5 & c1 & c6 & 0 & 0 & 0 \\ c6 & c6 & c2 & 0 & 0 & 0 \\ 0 & 0 & 0 & c3 & 0 & 0 \\ 0 & 0 & 0 & 0 & c3 & 0 \\ 0 & 0 & 0 & 0 & 0 & c4 \end{bmatrix} \begin{bmatrix} \epsilon_{xx} \\ \epsilon_{yy} \\ \epsilon_{zz} \\ \epsilon_{yz} \\ \epsilon_{zx} \\ \epsilon_{xy} \end{bmatrix}.$$

where, orientation specific constants c1, c2, c3, c4, c5, c6 are 194.5, 165.7, 79.6, 50.9, 35.7, and 64.1, all in GPa, respectively. This stiffness tensor translates to $E_x = E_y = 169$ GPa, $E_z = 130$ GPa, $\nu_{yz} = 0.36$, $\nu_{zx} = 0.28$, and $\nu_{xy} = 0.064$ [3].

Figure 15 shows the stress comparison between anisotropic and isotropic Si (Young's modulus = 188 GPa for all directions) material properties. We see that the normal stress component becomes less compressive and the von Mises stress is lower with the anisotropic Si compared with the isotropic Si case. This is largely due to the fact that we use the maximum Young's modulus for the isotropic Si case. With higher Young's modulus Si substrate becomes stiffer, hence higher stress builds up at the TSV/substrate interface. In this work, even though anisotropic Si property is more realistic, we use the isotropic Si property as a worst case scenario.

S.5 Impact of TSV and Bump Alignment

In this section, we explore the impact of alignment between TSV, μ-bump, and package-bump on the mechanical reliability of TSVs. We first examine the impact of relative position between TSV/μ-bump and package-bump. We use a two-die stack structure in which center locations of TSV, μ-bump, and package-bump are aligned as shown in Figure 16(a). Then we shift both TSV and

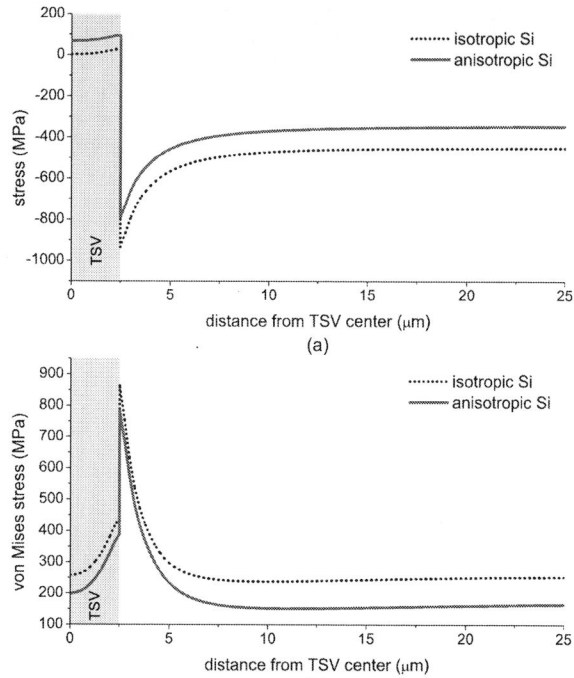

Figure 15: Impact of silicon material property on stress (FEA results). (a) $\sigma_{\theta\theta}$ **stress. (b) Von Mises stress.**

μ-bump together from the package-bump center with a 25 μm step and monitor the von Mises stress at the right edge of TSV.

Figure 16(c) shows that the von Mises stress is maximum around package-bump edge region and then decreases and saturates as distance increases. The difference between minimum and maximum is as high as 11.1 %. As Figure 3 shows, the highest stress gradient occurs around package-bump edge which results in the highest deformation of the structure near this region. Hence, this higher deformation causes more severe mechanical reliability problem in TSV.

We also see the decrease in von Mises stress near the package-bump center. This is because the material around this area is the same (= package-bump material), hence its deformation is relatively smaller than the edge which is the interface between two different materials.

We also examine whether relative position between μ-bump and TSV/package-bump affects the mechanical reliability of TSV. We fix the location of TSV and package-bump whose centers are aligned, then move μ-bump only with a 5 μm step up to 30 μm and monitor the von Mises stress at TSV edges. We observe the similar trend as before. However, the difference between minimum and maximum is only 6.5 MPa (0.8 %), which is negligible. Thus, we identify that the relative position between TSV and package-bump is a critical factor that affects the mechanical reliability of TSV.

S.6 Full-Chip/Package Analysis Algorithm

In this section, we discuss details of our full-system-scale thermo-mechanical stress and reliability analysis flow. First, based on the

324

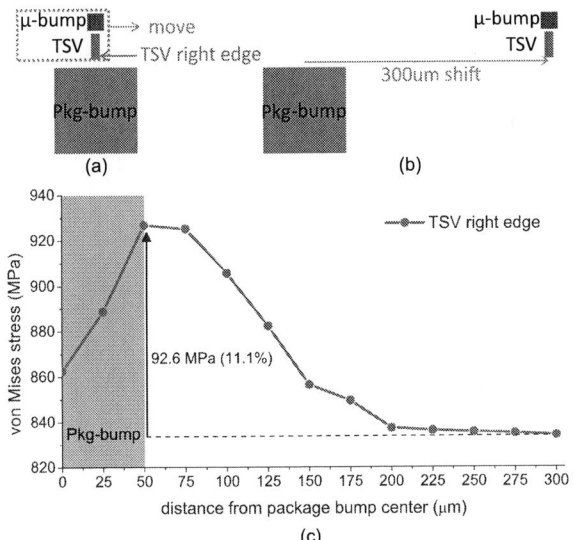

Figure 16: Impact of relative position between TSV/μ-bump and package-bump on von Mises stress. (a) Initial position. (b) Final position where TSV/μ-bump are shifted by 300 μm from package bump center. (c) Von Mises stress at TSV edge along the distance between TSV/μ-bump and package-bump (FEA results).

observation that the stress field induced by a single TSV, a μ-bump, and a package-bump in isolation is radially symmetrical due to their cylindrical shape, we obtain stress tensors for each interconnect component along an arbitrary radial line on device layer from their center location in a cylindrical coordinate system. To evaluate the stress tensor at a point affected by multiple interconnect elements, a conversion of a stress tensor to a Cartesian coordinate system is required. This is due to the fact that we extract stress tensors from these interconnect components whose center position is the origin in the cylindrical coordinate system; hence we cannot perform a vector sum of stress tensors from each component which has a different center location.

Then, we compute the stress tensor at the point of interest by adding up the stress tensors from TSVs, μ-bumps, and package-bumps that affect this point. We set a stress influence zone of TSV, μ-bump, and package-bump 25 μm, 100 μm, 500 μm from the center of each component, which is five times the diameter of each component, respectively. This is because the magnitude of each stress tensor component saturates well before this distance, hence there is a negligible impact from the interconnect element beyond this stress influence zone.

Let the stress tensor in Cartesian and cylindrical coordinate system be S_{xyz} and $S_{r\theta z}$, respectively.

$$S_{xyz} = \begin{bmatrix} \sigma_{xx} & \sigma_{xy} & \sigma_{xz} \\ \sigma_{yx} & \sigma_{yy} & \sigma_{yz} \\ \sigma_{zx} & \sigma_{zy} & \sigma_{zz} \end{bmatrix}, S_{r\theta z} = \begin{bmatrix} \sigma_{rr} & \sigma_{r\theta} & \sigma_{rz} \\ \sigma_{\theta r} & \sigma_{\theta\theta} & \sigma_{\theta z} \\ \sigma_{zr} & \sigma_{z\theta} & \sigma_{zz} \end{bmatrix}$$

The transform matrix Q is the form:

$$Q = \begin{bmatrix} \cos\theta & -\sin\theta & 0 \\ \sin\theta & \cos\theta & 0 \\ 0 & 0 & 1 \end{bmatrix}$$

where, θ is the angle between the x-axis and a line from the center of each interconnect element to the simulation grid point. A stress tensor in a cylindrical coordinate system can be converted to

---

input : TSV list T, pkg-bump list P, μ-bump list M, stress library
output: stress map, von Mises stress map
for *each TSV t, pkg-bump p, and μ-bump m in T, P, and M* **do**
    $(it, ip, im) \longleftarrow$ FindStressInfluenceZone(t, p, m)
    for *each point it', ip', and im' in it, ip, and im* **do**
        $it'.TSV \longleftarrow it$
        $ip'.pkg\text{-}bump \longleftarrow ip$
        $im'.\mu\text{-}bump \longleftarrow im$
    end
end
for *each simulation point r* **do**
    if $r.TSV \neq \emptyset \,||\, r.pkg\text{-}bump \neq \emptyset \,||\, r.\mu\text{-}bump \neq \emptyset$ **then**
        for *each $(t, p, m) \in (r.TSV, r.pkg\text{-}bump, r.\mu\text{-}bump)$* **do**
            $(dt, dp, dm) \longleftarrow distance(t, p, m, r)$
            $S_{cyl}(t, p, m) \longleftarrow$ GetStressTensor(dt, dp, dm)
            $S_{cyl}(t, p, m) \longleftarrow S_{cyl}(t, p, m) - BGstress$
            $\theta(t, p, m) \longleftarrow$ GetAngle$(line\ tr, pr, mr, x\text{-}axis)$
            $Q_{(t, p, m)} \longleftarrow$ SetConversionMatrix$(\theta_t, \theta_p, \theta_m)$
            $S_{Cart}(t, p, m) \longleftarrow$
            $Q_{(t, p, m)}S_{cyl}(t, p, m)Q_{(t, p, m)}^T$
            $r.S_{Cart} \longleftarrow r.S_{Cart} + S_{Cart}(t, p, m)$
        end
    end
    $r.S_{Cart} \longleftarrow r.S_{Cart} + BGstress$
    $vonMises(r) \longleftarrow$ ComputeVonMises$(r.S_{cart})$
end

Algorithm 1: Full-Chip/Package Stress and Reliability Analysis Flow (LVLS)

---

a Cartesian coordinate system using conversion matrices: $S_{xyz} = QS_{r\theta z}Q^T$ [5].

Our full-system-scale thermo-mechanical stress and reliability analysis flow is shown in Algorithm 1. We first start to find a stress influence zone from each TSV, μ-bump, and package-bump. Then, we associate the points in the influence zone with the affecting interconnect elements. Next, for each grid point under consideration, we look up the stress tensors from each interconnect component found in the association step, and subtract background stress from the stress tensor. Then, we use the coordinate conversion matrices to obtain stress tensors in the Cartesian coordinate system. We visit an individual TSV, μ-bump, and package-bump affecting this simulation point and add up their stress contributions. After visiting all the components effecting this point, we add one background stress back. Once we finish the stress computation at the point, we obtain the von Mises stress value using Equation (2).

S.7 Impact of Bump Size

In this section, we study the impact of package-bump and μ-bump size on the reliability. First, we vary the package-bump diameter/height from 100 μm to 300 μm, while fixing the package-bump pitch and the TSV/μ-bump count and pitch as 400 μm, 1600, and 20 μm, respectively. Table 4 shows that the number of TSVs experiencing higher von Mises stress increases with larger package-bumps due to the larger deformation of a stack and the increased package-bump circumference where highest von Mises stress occurs. However, in the 300 μm package-bump case, there are more TSVs with lower von Mises stress (780 - 870 MPa) than the 200 μm package-bump case. As discussed in Section S.5, TSVs lo-

325

Table 4: Maximum von Mises stress distribution of TSVs with different size of package-bump and μ-bump. (die0 in four-die stack with 1600 TSVs)

von Mises stress (MPa)	pkg-bump size (μm)			μ-bump size (μm)		
	100	200	300	10	20	30
780-810	31.2%	17%	17.8%	4.2%	5.6%	0%
810-840	33.8%	18%	27.8%	6.9%	6.3%	6.9%
840-870	19%	14%	17.2%	22.9%	22.2%	15.3%
870-900	14%	28.5%	12.2%	21.5%	22.9%	18.1%
900-930	2%	20.5%	13.5%	27.1%	22.9%	22.9%
930-960	0%	2%	10%	17.4%	20.1%	36.8%
960-	0%	0%	1.5%	0%	0%	0%
median (MPa)	824.6	871.7	848.2	893.3	890.1	908.0

Table 5: Impact of package-bump and TSV/μ-bump pitch on von Mises stress. (die0 in four-die stack with 900 TSVs)

von Mises stress (MPa)	pkg-bump pitch (μm)			TSV/μ-bump pitch (μm)		
	200	250	300	15	25	35
780-810	4.7%	6.3%	19.5%	0.6%	4.6%	7.1%
810-840	4.7%	21.5%	27.0%	3.1%	4.6%	6.9%
840-870	21.9%	27%	31.6%	19.1%	21.6%	22.9%
870-900	19.5%	33.2%	20.3%	23.5%	20.9%	21.5%
900-930	24.2%	12.1%	1.6%	26.4%	23.7%	17.5%
930-960	25.0%	0%	0%	23.4%	24.6%	24.1%
960-	0%	0%	0%	3.9%	0%	0%
median (MPa)	897.9	863.9	844.1	901.8	897.9	893.2

Table 6: Details of the mechanical reliability in TSV blocks in Figure 12.

TSV block #	# TSV	von Mises stress (MPa)				blk-bump dist (μm)
		max	min	avg	std dev	
1	2x19	909.0	798.5	839.6	34.0	96.4
2	1x20	921.9	805.2	846.5	35.2	97.9
3	5x3	901.0	811.1	859.5	26.0	96.4
4	5x5	939.6	853.5	902.6	24.0	67.6
5	5x5	908.6	816.0	858.7	33.3	24.1
6	5x5	942.3	874.4	910.4	22.0	91.4
7	3x5	915.2	855.3	891.3	16.6	61.0
8	3x2	887.2	854.3	865.4	11.2	78.4
9	3x5	889.3	802.5	856.3	24.6	106.0
10	6x5	933.6	812.7	857.8	36.0	111.2
11	3x1	896.6	855.9	871.0	18.2	39.3
12	7x5	952.7	797.1	871.3	43.9	98.8
13	2x3	879.4	807.4	836.9	24.4	100.7
14	2x3	834.7	800.7	820.4	10.9	114.8
15	2x4	909.6	888.5	895.3	7.1	73.9
16	12x8	943.7	806.0	877.2	33.6	90.7

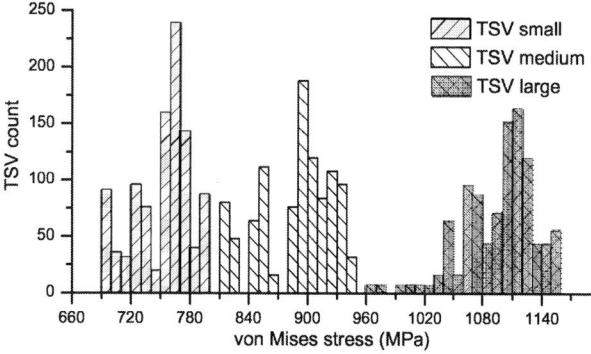

Figure 17: Impact of TSV size on von Mises stress distribution of TSVs. (die0 in four-die stack with 1024 TSVs)

cated near the package-bump center region show lower von Mises stress than those around package-bump edge. Hence, with larger package-bumps more TSVs reside near the package-bump center, which results in lower von Mises stress level for these TSVs.

We now vary the μ-bump size, and use a 100 μm package-bump with a 200 μm pitch. Note that since we align center locations of TSV and μ-bump, we set the TSV pitch as 35 μm to accommodate the largest μ-bump diameter of 30 μm. We observe that larger μ-bump causes more TSVs to experience higher von Mises stress. However, this μ-bump size impact is less significant than the package-bump size.

S.8 Impact of TSV Size

In general, package-bumps and μ-bumps generate global stress distribution, while TSVs create local stress distribution. Therefore, TSV size and pitch are still critical factors that affect the mechanical reliability problem in TSVs even with the presence of other interconnect elements. In this section, we investigate the effect of TSV size. We use three different sizes of TSV with the same aspect ratio of 6; TSV small ($H/D = 15/2.5\ \mu m$), TSV medium ($H/D = 30/5\ \mu m$), and TSV large ($H/D = 60/10\ \mu m$), where H/D is TSV height/diameter. We set the pitch of TSV and package-bump as 25 μm and 200 μm, respectively for all cases.

Figure 17 shows that smaller TSVs reduce the von Mises stress level significantly. This is mainly because larger TSV induces higher stress level at TSV edge due to the sheer volume of TSV. Also, the magnitude of normal stress components decay proportional to $(D/2r)^2$, where r is the distance from the TSV center. In other words, larger TSV affects larger area, hence increases stress level around it more than smaller TSV.

S.9 Impact of Pitch

In this section, we explore the effect of package-bump and TSV/μ-bump pitch on the reliability. We employ a 100 μm package-bump and change its pitch from 200 μm to 300 μm. The pitch of TSV/μ-bump is set to 25 μm. Table 5 shows that larger package-bump pitch reduces the von Mises stress level noticeably by reducing constructive stress interference between package-bumps. However, we cannot arbitrarily increase the package-bump pitch considering the package size increase given the required number of pins.

We also examine the impact of TSV pitch on the von Mises stress. In this case, we set the package-bump pitch as 200 μm. In Table 5, we see that larger TSV pitch reduces von Mises stress level. However, there is not much difference between 25 μm and 35 μm pitch cases. This is because the stress influence zone of a 5 μm diameter TSV is 25 μm, hence there is a negligible difference between these two cases in terms of the stress induced by TSVs solely. Thus, in this case, the von Mises stresses of TSVs are largely determined by relative position between TSVs and package-bumps. Therefore, the proper TSV placement considering the locations of package-bumps is a key design knob to mitigate the reliability concerns in TSV-based 3D ICs.

S.10 Full Details of Table 3

Table 6 shows the details of the missing TSV blocks in Table 3.

Symbolic Model Checking on SystemC Designs[*]

Chun-Nan Chou[†], Yen-Sheng Ho[‡], Chiao Hsieh[‡], and Chung-Yang (Ric) Huang[†‡]

[†]Graduate Institute of Electronics Engineering, National Taiwan University, Taipei 10617, Taiwan

[‡]Department of Electrical Engineering, National Taiwan University, Taipei 10617, Taiwan

ABSTRACT

SystemC is a de-facto standard for modeling system-level designs in the early design stage. Verifying SystemC designs is critical in the design process since it can avoid error propagation down to the final implementation. Recent works exploit the software model checking techniques to tackle this important issue. But they abstract away relevant semantic aspects or show limited scalability. In this paper, we devise a symbolic model checking technique using bounded model checking and induction to formally verify SystemC designs. We introduce the notions of behavioral states and transitions to guarantee the soundness of our approach. The experiments show the scalability and the efficiency of our method.

Categories and Subject Descriptors

D.2.4 [**Software Engineering**]: Software/Program Verification—*Model checking*

General Terms

Verification

Keywords

Formal Verification, Symbolic Model Checking, SystemC.

1. INTRODUCTION

Since the cost of fixing a design error increases significantly in the later design stages, it becomes a must to develop modern System-on-Chips (SoCs) in the early design cycle with a higher level of abstraction. Among all the high-level hardware description languages, SystemC [1] has been promoted as an industry standard for its high simulation speed.

The fast simulation of SystemC enables the simulation-based verification to work with the system-level complexity. However, it has a potential problem of false-positive since the SystemC scheduler only simulates one possible scheduling of the concurrent threads. That is, unless we can enumerate all the possible permutations of the interleaving threads, which have the factorial complexity, we may miss a design error from an unsimulated thread scheduling. To resolve this problem, the authors in [2,3] utilize Partial Order Reduction (POR) techniques to explore the alternative schedulings during simulation and thus to identify the bugs that the traditional SystemC simulation cannot catch. Nevertheless, to compromise the factorial complexity, they limit the exploration with a fixed input value and thus the input-dependent corner-case errors may be unwittingly ignored.

On the other hand, formal verification of SystemC seems a plausible alternative. If the formal engine can completely capture all possible input sequences and all possible schedulings, the mathematical certainty of the properties under verification can be assured. There are a number of works applying model checking techniques for this purpose [4]-[7]. These approaches differ on the models they use to interpret the SystemC semantics and yet they all resort to certain symbolic techniques like UPPAAL [8] for formal verification. Nevertheless, they either fail to precisely interpret the SystemC scheduler, or are limited in scalability due to the excessive complexity in checking the underlying models with the symbolic techniques. For example, the work in [4] does not handle channel updates and time delays.

Recently, there have been some works adopting the software model checking approaches to verify SystemC designs. For example, the authors of [9] translate SystemC designs into threaded sequential C programs and apply Explicit-Scheduler/Symbolic Threads (ESST) to the translated C programs for unbounded model checking. In [10], they further improve ESST by employing the POR method. Although they do present some promising results on SystemC formal verification, there are two potential problems. Firstly, the slow refinement by the formal engine makes it inefficient to verify the unsafe designs, and secondly, the underlying state-based paradigm may limit its scalability for large designs.

In contrast, stateless methods have been proposed to cope with the abovementioned limitations [11,12]. Their methods are stateless [13] in the sense that they store no state representations in memory but only information about which transitions and traces have been executed so far. The work in [11] utilizes the bug hunting capability of Bounded Model Checking (BMC) [14] to verify the deadlock properties and thus can be scaled well for industrial designs. However, it has the inherited limitation that it can only prove the properties up to a certain bound. On the other hand, the work in [12] proposes a complete method based on induction to formally verify the properties. Nevertheless, they cannot support SystemC designs with timed language constructs such as *wait(t)*, which means waiting for *t* time units with $t \geq 0$.

In this paper, we introduce a formulation interpreting SystemC designs with timed language constructs as Kripke structures. With this formulation, we propose a reachability analysis for this Kripke structure. Based on the concepts of BMC and induction, we devise an effective symbolic model checking approach utilizing the proposed reachability analysis for SystemC designs.

The paper is organized as follows. In Section 2, the preliminaries are provided. We translate SystemC designs with timed language constructs to Kripke structures in Section 3. Our symbolic model checking approach is described in Section 4 and the experiments are given in Section 5. Finally, we conclude this paper and discuss some directions for future work in Section 6.

2. PRELIMINARIES

In this section, we first review some background and notational conventions in the conventional model checking. We then briefly discuss the difficulties of applying the model checking techniques to SystemC designs. A simple SystemC example is presented in Subsection 2.2 to highlight these issues and provide the intuitions for the formal settings as defined in Section 3.

---

[*]This work is partially supported by Springsoft Inc. We are grateful for their support on SystemC front-end and valuable advices on the implementation. This work is also sponsored in part by the National Science Council of Taiwan ROC under grants NSC 99-2221-E-002-211-MY3.

2.1 Model Checking

Model checking [15] is a technique for automatically verifying finite-state concurrent systems. In this verification technique, the specification is expressed in temporal logic and the system is modeled as a Kripke structure. A Kripke structure M over a set of atomic propositions $\mathcal{A}$ is a 4-tuple $M = (S, I, T, \ell)$ with S a finite set of states, $I \subseteq S$ the set of initial states, $T \subseteq S \times S$ a transition relation between states, and $\ell: S \to \mathcal{P}(\mathcal{A})$ the labeling function of the states. Please note that there are two important properties of a Kripke structure: (1) S must be a finite set and (2) T must be total, that is, for every state $s \in S$ there is a state s' such that $T(s, s')$.

Model checking has been successfully adopted for hardware and software verification. Without loss of generality, the core techniques of model checking rely on the reachability analysis of its set of states. Therefore, it is required that the states and the corresponding transitions of the design under verification should be clearly defined. For hardware, the states are the valuations of the flip-flops and the transitions are the combinational logic in the circuit; for software, they are the valuations of variables and the statements in the program, respectively.

However, a design described in SystemC is a more complex entity, and thus its states and transitions are much more difficult to define. In essence, SystemC itself is based on the C++ object-oriented software language, and it also contains a rich set of constructs to represent the concurrent, synchronous and asynchronous hardware behaviors. Albeit the designers can use SystemC to easily model their designs on a higher abstraction level, this will pay the price in the difficulties of formal modeling and verification on the SystemC designs.

For example, it is not sufficient to describe the state of a SystemC design by the valuations of the variables. Please note that different schedulings of the concurrent threads may lead to different behaviors of the design. Therefore, to characterize the state space of a SystemC design, we also need to take the status and scheduling of the threads into account. However, this will substantially increase the number of states.

Besides, it is hard to derive the transitions between the states. Unlike the traditional hardware and software model checking where the transition functions are the logic between the state variables, there is no clear structural construct to define the timeframe boundary in SystemC. This is mainly because, first, SystemC contains both synchronous and asynchronous semantics, and second, the behavior of a SystemC design also hinges on the scheduling of its concurrent threads. For instance, the *wait* function can be used to wait for some clock/time to elapse (synchronous), or wait for an event to occur (asynchronous). Therefore, if the *wait* functions in a design are involved in certain control statements, the design may be halted at different time stamps by different execution paths. Thus no repeated structure can be defined as the transition function. Moreover, the scheduling of the concurrent threads can influence execution paths of the design and thus alter the transition functions between the states. Therefore, we need to permute the thread ordering in order to construct the complete transition functions. In the next subsection, we will give a simple example to further illustrate the difficulties.

2.2 A SystemC Example

Figure 1 depicts two concurrent threads, t1 (lines 3-11) and t2 (lines 12-22), of a simple SystemC example. For clarity, the syntactic details of SystemC such as SC_MODULE and the sc_main function are not shown. In this example, each thread has an infinite while loop to ensure the thread can keep on running. b_in and a are input and global variables, respectively. In

t1, the value of a is determined by b_in while in t2, the time to resume the thread is dependent on the different values of a.

If we assume that both t1 and t2 are runnable at the start of the simulation, to fully explore the behavior of the design, we should consider the value of the input variable b_in and the scheduling orders on t1 and t2. This will constitute the 4 combinations as shown in Figure 2, and we denote the corresponding transition relations as TR_{ti}^0 to TR_{ti}^3 for thread ti. Note that the first three cases will halt the design at different time stamps. For example, if the scheduling order is t1→t2 and the input b_in is true, t2 will wait for 20 time units. Otherwise if b_in is false, t2 waits for 30 time units. For the scheduling order t2→t1, no matter what the value of b_in is, t2 waits for 10 time units. Clearly, these different waiting time units will result in different transition functions.

The example above sufficiently demonstrates how the scheduling order can affect the executions of the design, and the stop locations of the threads also have an effect on the sequel behaviors. Intuitively, if we include the scheduling orders to define the states and transitions of a SystemC design, we should be able to construct its corresponding Kripke structure. This opens up the possibility to perform model checking on SystemC designs and we will present the basis for the transformation and the symbolic model checking approach in the following sections.

```
1.  sc_in<bool> b_in;      12.  void t2(){
2.  sc_int<2> a=1;         13.    while(true){
3.  void t1(){             14.      if(a==2)
4.    while(true){         15.        wait(20);
5.      if(b_in)           16.      else if(a==3)
6.        a=2;             17.        wait(30);
7.      else               18.      else
8.        a=3;             19.        wait(10);
9.      wait(e);           20.      e.notify();
10.   }                    21.    }
11. }                      22.  }
```

Figure 1. Example 1: The excerpt of a simple SystemC design

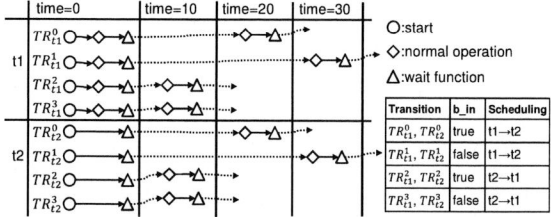

Figure 2. Different transitions of Example 1 with different inputs and schedulings

3. TRANSLATING SYSTEMC TO KRIPKE STRUCTURE

In this section, we propose several definitions, lemmas, and theorems to translate a SystemC design to a state transition system, i.e., the Kripke structure. With this transformation, we will be able to adopt any of the conventional symbolic model checking techniques on SystemC designs.

Note that there are three types of processes in SystemC: SC_METHOD (method), SC_CTHREAD (clocked thread), and SC_THREAD (thread). Because both methods and clocked threads in SystemC are special cases of threads [4], for the clarity of presentation, we will focus the discussion on threads only.

3.1 Assumptions

In developing this formal model, we impose several assumptions on the SystemC design under verification (DUV). Basically, our assumptions are based on the principle of treating the DUV as a reasonable hardware. Otherwise, verifying SystemC designs is intractable in theory. Please note that these assumptions do not limit the applicability of this formal model since most SystemC designs satisfy our assumptions.

First, we treat the variables in the design as bit-vectors. This implies that each variable takes a finite set of values. Then, we assume no dynamic creations of objects or threads in a design and thus the number of components is known statically (i.e. during compilation). Besides, we assume the time arguments passed to *wait* and *notify* functions are all constants instead of variables. We also assume that each function can be inlined statically and there are no infinite loops or infinite delta cycles between two consecutive timed notification phases. Finally, we only sample the new values of the input variables at the timed notification phase.

3.2 Behavioral States

While the conventional model checking defines the states structurally as the valuations of the variables, we call our states the *behavioral* states because they involve the behavior from both the functionalities in the design and the scheduling in the simulation process. In the following, we first define the behavioral boundary from which the behavioral states can later be derived.

Definition 1: A **behavioral boundary** is the time point when the scheduler enters the evaluation phase right after the initialization phase or the timed notification phase.

In short, we can treat the behavioral boundaries as the time points when the design encounters time advancements during the simulation. They are marked with *stars* as shown in Figure 6 of the appendix, which illustrates the OSCI-standard simulation flow [1]. Please note that there might exist multiple delta cycles between two consecutive behavioral boundaries according to the above definition. Besides, at the behavioral boundaries, no thread is running. They are suspended by certain wait functions and some of them can be runnable in the coming evaluation phase. These runnable threads will be selected to run iteratively by the SystemC scheduler. On the other hand, threads that are not runnable would have to wait until the time is advanced or is notified by the event. To describe this situation precisely, we define the *suspended status* of a thread as:

Definition 2: The **suspended status** of a thread is an ordered pair $ss = <loc, ttr>$, where loc is the program location at which the thread is suspended, and ttr is the time to resume (ttr) describing how long it takes for the thread to become runnable from the current time point.

For a thread i suspended by a timed *wait*, ttr_i is a nonnegative value. Specially, if this thread is *runnable* immediately, ttr_i is zero. Otherwise, it should be suspended by an event and we denote ttr_i as U (unknown).

To capture the behavior of the SystemC design when the scheduler enters the evaluation phase, we define the behavioral state at behavioral boundary, quantifying out the intermediate statuses resulted from the delta-cycle events, for example.

Definition 3: If there are N threads in a SystemC design, a **behavioral state** is an ordered pair $bs = <<ss_1, ss_2,..., ss_N>, \varphi>$ at the behavioral boundary, where the first element describes the suspended statuses of the threads, and the second element φ is the valuation of all variables in the design.

For example, Figure 3 presents partial behavioral states of the design in Figure 1. At the behavioral boundary *time*=0, the initial behavioral state is $bs_0=<<ss_{t1}^0, ss_{t2}^0>, \varphi_0>$ where $ss_{t1}^0=<3, 0>$, $ss_{t2}^0=<12, 0>$, and $\varphi_0=\{a=1\}$. If the scheduling is $t1 \rightarrow t2$ and b_in is true, then t1 and t2 will be suspended by wait(e)(line 9) and wait(20) (line 15) after running the simulation, respectively. Hence, the scheduler performs a timed

notification and the simulation time is advanced to 20 based on the simulation semantics of SystemC. According to our definition of behavioral states, the next behavioral state is at behavioral boundary *time*=20 and is denoted as $bs_1=<<ss_{t1}^1, ss_{t2}^1>, \varphi_1>$ where $ss_{t1}^1=<9, U>$, $ss_{t2}^1=<15, 0>$, and $\varphi_1=\{a=2\}$.

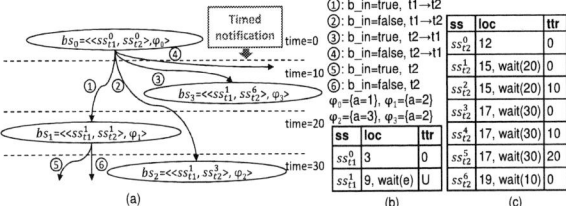

Figure 3. Some partial state transitions of Example 1: (a) One step transition from the initial state; (b)(c) Possible suspended statuses of t1 and t2

3.3 Finite Behavioral State Space

As mentioned in Subsection 2.1, the set of states in a Kripke structure must be finite. If we want to apply model checking techniques to verify a SystemC design, we should guarantee that the set of all possible behavioral states in a design is finite. In this subsection, we will prove this important attribute.

Lemma 1: For the wait function with the specific *loc*, the set of the possible *ttr*s at behavioral boundary is finite.

Proof. There are two possible cases for wait functions:

(a) wait for some clock/time: Since we assume that the time argument is a constant, the number of the possible *ttr*s is limited by this constant at behavioral boundary.

(b) wait for an event: If there are only immediate and delta notifications related to this wait function, the only possibility of *ttr*s is U (unknown). Otherwise, the number of the possible *ttr*s is limited by the constant which is the biggest time argument of the related timed event notifications.

According to the two cases above, this lemma is proven. □

Lemma 2: For each thread, the set of the possible suspended statuses at behavioral boundary is finite.

Proof. We know the fact that, for a finite number of finite sets, the Cartesian product of these sets is finite. From Lemma 1, we can complete this proof because there are a finite number of wait functions in each thread. □

Theorem 1: The set of the possible behavioral states in a SystemC design is finite.

Proof. Since dynamic creation of threads is not allowed, the set of threads in a design is finite. Then by Lemma 2, the set of the possible suspended statuses of the threads is also finite. Moreover, since the values of variables are bit-vectors, the set of their valuations is finite. Hence, we can conclude the proof. □

3.4 Behavioral Transition Relations

Another key property of a Kripke structure is that the state transitions must be total. That is, given any state, there must be a next state. In this subsection, we will define the behavioral transition relation to complete the transformation from SystemC to Kripke structure.

As shown in Subsection 2.2 and in Figure 3, the scheduling of the runnable processes has an effect on the next behavioral state. Therefore, we should define the behavioral transition function with the scheduling as well as the current behavioral state and the valuation for input variables. Given a SystemC design, let $\mathcal{BS}$ denote the set of all the possible behavioral states, Γ denotes the set of all the possible valuations for the input variables, and Ω represents the set of all the possible schedulings at behavioral boundaries. Please note that Γ and Ω are both finite at behavioral boundaries based on our assumptions. We have the definition of the behavioral transition function as:

Definition 4: A **behavioral transition function** in SystemC is δ: $\mathcal{BS} \times \Gamma \times \Omega \rightarrow \mathcal{BS}$.

Similarly, we can define transition relations from one behavioral state to another as:

Definition 5: A **behavioral transition relation** in SystemC is $\mathcal{BTR} = \{<bs, bs'> \in \mathcal{BS} \times \mathcal{BS} \mid \exists \gamma \in \Gamma. \exists \omega \in \Omega.\ bs' = \delta(bs, \gamma, \omega)\}$

In Figure 3, given $bs_0 = <<ss_{t1}^0, ss_{t2}^0>, \varphi_0>$, $\gamma_0 = \{\text{b\_in=true}\}$, and $\omega_0 = <\text{t1}, \text{t2}>$, the next behavioral state is $bs_1 = <<ss_{t1}^1, ss_{t2}^1>, \varphi_1>$. Thus, we can know that $bs_1 = \delta(bs_0, \gamma_0, \omega_0)$ and $<bs_0, bs_1> \in \mathcal{BTR}$. It is worth noting a special case of $bs = <<ss_1, ss_2, ..., ss_N>, \varphi>$ where $\forall i\ ss_i = <loc_i, U>$. According to SystemC specification, this special case results in the termination of the simulation. However, in our formal model, we interpret this situation as follows. $\forall \gamma \in \Gamma. \forall \omega \in \Omega.\ bs = \delta(bs, \gamma, \omega)$ if bs belongs to this special case. That is, the next behavioral state of this special state is itself. This interpretation makes $\mathcal{BTR}$ total.

Let $\mathcal{A}$ be a set of atomic propositions. Now we can view a SystemC design as a Kripke structure $BM = (\mathcal{BS}, I, \mathcal{BTR}, \ell)$ where $I \subseteq \mathcal{BS}$ is the set of initial behavioral states, and ℓ: $\mathcal{BS} \rightarrow \mathcal{P}(\mathcal{A})$ is the labeling function of the behavioral states. Although our formulation opens up the possibility to perform model checking, the number of behavioral states can be very large, and the explicit traversal of the behavioral state space may be infeasible. To overcome this obstacle, we will propose our symbolic approach in the next section.

4. SYMBOLIC MODEL CHECKING

In this section, we present a novel approach to apply symbolic model checking to the Kripke structure BM defined in the previous section. Our approach explores the reachability of BM as shown in Figure 4(a). In our reachability analysis, we use a symbolic behavioral state (Section 4.3) to represent a set of behavioral states and conduct symbolic simulation to compute the behavioral transition function. In a symbolic behavioral state, we use bit-vector formulas to encode the possible values of variables and keep track of possible suspended statuses of each thread symbolically. In the sequel, we refer to this technique as Symbolic Data and Suspended Status (SDSS) reachability analysis.

Besides, we do not interpret the logic of SystemC scheduler as formulas directly. Instead, we embed the algorithm of SystemC scheduler in SDSS to further enumerate all possible schedulings of runnable threads. Please note that the idea of embedding the scheduler in the model checking algorithm is similar to ESST in [9]. However, our symbolic approach is totally different from ESST utilizing lazy predicate abstraction in its back-end checker.

For ease of exposition, we assume that there is only one statement in a line and a statement cannot be split into two lines. Based on this assumption, we will use the statement and the line (program location) interchangeably in the sequel. In the following, we first use Example 1 to illustrate how our symbolic simulation works in Subsection 4.1 and 4.2. With the concept of symbolic simulation, we introduce the symbolic behavioral state in Subsection 4.3. Finally, Subsection 4.4 outlines our proof process.

4.1 Symbolic Valuation of Variables

To conduct the reachability analysis of BM, we employ the symbolic simulation technique to compute behavioral transition function. Our tool invokes the public API of Boolector [16] to generate bit-vector formulas and applies Boolector for the constraint satisfiability checks. In the following, we describe how to update the possible values of the variables symbolically.

In our symbolic simulation, we use a bit-vector formula to represent the possible values of each variable in a SystemC design. To execute the codes symbolically, our symbolic simulation engine conducts a level-ordered traversal in the extended Petri Net [11] of the SystemC design. When encountering an assignment operator, our symbolic simulation engine uses the If-Then-Else (ITE) operator to update the bit-vector formula of the left-hand-side variable. This ITE mechanism can be achieved by properly maintaining the statement formula defined as:

Definition 6: The **statement conditions** (SC) of the statement is a set of constraints. We call the conjunction of *SC* the **statement formula** (SF) which is satisfiable iff the statement can be reached.

The *SC* is maintained by using a stack. Our symbolic simulation engine pushes and pops the corresponding constraint when it enters and leaves the statement block of the branch, respectively. For example, we assume that f_{b_in} is the bit-vector formula of b_in. If the symbolic simulation engine starts from the line 3 in Figure 1, the *SC* and *SF* of the line 8 are $\{\text{true}, \neg f_{b\_in}\}$ and $\neg f_{b\_in}$, respectively. When our engine leaves the line 8, the *SC* becomes $\{\text{true}\}$. That is, the line 9 waiting for event can be reached no matter what the value of b_in is.

Based on the concept of the *SF*, we can update the bit-vector formulas of the variables by using the ITE operator $f_{new\_lhs} = f_{ass\_SF}$: f_{rhs}: f_{old_lhs} where f_{new_lhs} is the new bit-vector formula of left-hand-side variable, f_{ass_SF} is the *SF* of the assignment statement, f_{rhs} is the bit-vector formula of right-hand-side variable, and f_{old_lhs} is the old bit-vector formula of left-hand-side variable. We also take Figure 1 as the example and assume that the symbolic simulation engine starts from the line 3. After our engine simulates the line 6, the bit-vector formula of a becomes $f_a^0 = f_{b\_in}? 2: 1$ where f_{b_in} is the bit-vector formula of b_in. f_a^0 means the possible values of a are 2 and 1 so far. When our engine stops the line 9, the bit-vector formula of a is $f_a^1 = \neg f_{b\_in}? 3: f_a^0$ and the resulted ITE structure is shown in Figure 4(b). As the ITE structure indicates, the possible values of a are 2 and 3 since either f_{b_in} or $\neg f_{b\_in}$ must be true.

Our symbolic simulation engine supports all the operators stated in [16]. Please note that, if pointer dereferencing operators are used, this requires a standard points-to analysis. Therefore, our symbolic simulation engine does not yet support pointer. However, support for the pointer is future extensions of our symbolic simulation engine. Besides, it is noteworthy that there is an important property in our symbolic simulation:

Property 1: The conjunction of SF_T and SF_F is unsatisfiable, where SF_T and SF_F are the *SFs* for the true and false statement blocks of the same branch, respectively. We call this property the **mutual exclusion** of the **branch** (MEB).

Intuitively, if the *MEB* property did not hold in the symbolic simulation, there would be an input combination which can reach both sides of the same branch. But this situation is not allowed in the semantic of SystemC language.

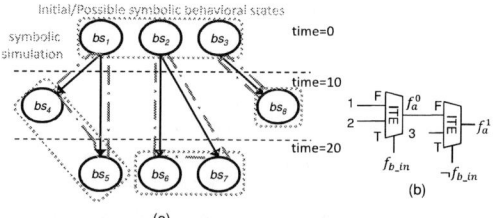

(a)

(b)

Figure 4. (a)Overall idea of our reachability analysis; (b) The ITE structure of variable a

4.2 Symbolic Mechanism for Wait Statements

After discussing the ITE structures of the variables, we now illustrate how our symbolic simulation handles the synchronous and asynchronous semantics in SystemC. The motivation of our mechanism here is to keep the *MEB* property when we deal with these two semantics. We describe them separately as follows.

(a) wait for some clock/time: We assume that the program location and the *ttr* of this wait statement are loc_i and ttr_i, respectively. If our symbolic simulation engine reaches this kind of statement, loc_i and ttr_i are recorded. Besides, the related *SF*—called the *timed SF* (TSF)—is also recorded. When ttr_i elapses, our engine places the *TSF* into the *SC* of loc_i and resumes the symbolic simulation from loc_i. The *TSF* would remain in the *SC* of the following statements and never be popped until the symbolic simulation from loc_i is finished. This means the discontinuing symbolic simulation only occurs under the constraint, *TSF*.

(b) wait for an event: We assume that the program location of this wait statement is loc_j. If our symbolic simulation engine reaches this kind of statement, loc_j and the *SF*—called the *event SF* (ESF)—are recorded. In addition, we assume that the program location of the statement notifying the corresponding event is the loc_k and the *SF* of loc_k is SF_k. When our symbolic simulation engine reaches the event notification at loc_k, we check whether the conjunction of SF_k and the corresponding *ESF* is satisfiable or not. This check can be achieved by applying the incremental satisfiability check in Boolector [16]. If the conjunction is unsatisfiable, we do nothing. Otherwise, our engine updates the *ESF* of loc_j to *ESF'* where $ESF': (ESF \land \neg SF_k)$. This update is a key step to make the *MEB* property hold. After all the other locs scheduled to simulate symbolically have been finished, the *SC* of loc_j is updated to the conjunction of the original *ESF* and SF_k. Then our engine resumes the symbolic simulation from loc_j. This means the event notification only happens under the constraint, $ESF \land SF_k$.

Again, we take Figure 1 as the example. We assume that the scheduling is t1→t2 and the start points for symbolic simulation are the line 3 and the line 12. As illustrated before, Figure 4(b) is the ITE structure of a after our engine symbolically simulates t1. Besides, the *ESF* of the line 9 is ESF_9: true. This means t1 stops at the line 9 without a constraint. Based on the ITE structure of a, the engine continues to symbolically simulate t2. When the engine stops, the *TSF*s of the line 15 and the line 17 are TSF_{15}: $f_a^1 == 2$ and TSF_{17}: $f_a^1 \neq 2 \land f_a^1 == 3$, respectively. This means t2 can stop at line 15 with TSF_{15} or at line 17 with TSF_{17}. Please note that the line 19 cannot be reached because TSF_{19}: $f_a^1 \neq 2 \land f_a^1 \neq 3$ is unsatisfiable. When the simulation time is advanced to 20, our engine starts to symbolically simulate t2 from the line 15 and TSF_{15} is placed into the *SC*. Then the event notification statement is met and our engine checks the satisfiability of $ESF_9 \land TSF_{15}$. Because $ESF_9 \land TSF_{15}$ is satisfiable, ESF_9 is updated to ESF_9': $\{f_a^1 \neq 2\}$. After the symbolic simulation of t2 is finished, our engine places $ESF_9 \land TSF_{15}$ into the *SC* and begins to symbolically simulate t1 from the line 9.

4.3 Symbolic Behavioral States

In this subsection, we first observe the situation after applying the symbolic simulation to the initial behavioral state, bs_0, in Figure 3. Based on this observation, we devise the definition of the symbolic behavioral state.

If we apply the symbolic simulation to bs_0 based on the scheduling t1→t2, the result of the symbolic simulation is the same as the example described in the previous subsection. Thus, we can know that ESF_9: true, TSF_{15}: $f_a^1 == 2$, and TSF_{17}: $f_a^1 \neq 2 \land f_a^1 == 3$. If we keep track of suspended status of each

thread explicitly, this still requires two states in the reachability analysis as shown in Figure 7 of the appendix. The first state represents that t2 stops at the line 15. The other state stands for that t2 stops at the line 17. Obviously, the state explosion problem remains even if we use bit-vector formulas to encode the possible values of variables. However, by benefiting from the *MEB* property, we can use the *TSF* or *ESF* as the unique identification of the suspended status. Thus, we can represent the suspended status symbolically and define the symbolic suspended status as:

Definition 7: The **symbolic suspended status** of a thread is an ordered pair $sss = <ss, sf>$, where ss is the suspended status of the thread, and sf is the *SF* of the suspended status. For a thread suspended by wait(e), sf is its *ESF*. Otherwise, sf is its *TSF*.

Due to the nature of the symbolic approach, there might be more than one suspended status for a thread at the specific behavior boundary. Therefore, for a thread i, we use S_i to denote the set of the symbolic suspended statuses and define the symbolic behavioral state as:

Definition 8: If there are N threads in a SystemC design, a **symbolic behavioral state** is an ordered pair $sbs = << S_1, S_2, ..., S_N>, \sigma>$ at behavioral boundary, where the first element describes the set of the symbolic suspended statuses of each thread and the second element σ is the symbolic valuation of all variables.

For example, Figure 5 shows the effect after we apply SDSS to Example 1. $f_{b\_in\_0}$, $f_{b\_in\_10}$, $f_{b\_in\_20}$, and $f_{b\_in\_30}$ are the free bit-vector variables which represent the input variable, b_in, at different behavior boundaries. The symbols of all suspended statuses are the same as those in Figure 3. The initial symbolic behavioral state is $sbs_0 = <<S_{t1}^0, S_{t2}^0>, \sigma_0>$ where $S_{t1}^0 = \{<ss_{t1}^0, true>\}$, $S_{t2}^0 = \{<ss_{t2}^0, true>\}$, and $\sigma_0 = \{a=1\}$. The *SF*s for ss_{t1}^0 and ss_{t2}^0 are trues because the line 3 and line 12 can be reached without a constraint. If the scheduling is t1 → t2, t1 is suspended by wait(e) and t2 is suspended by wait(20) or wait(30) after applying symbolic simulation. We can quickly derive sbs_1 from the result of the example described in Subsection 4.2.

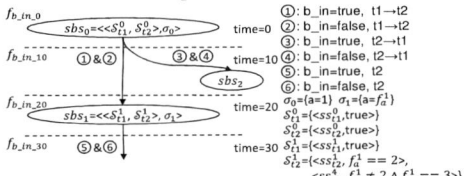

Figure 5. Partial symbolic behavioral states of Example 1

4.4 BMC and Induction Based Proof Process

Our symbolic model checking approach which is outlined in Algorithm 1 of the appendix performs BMC [14]. That is, we conduct SDSS and prove the property up to a given bound. Based on the definition of the symbolic behavioral state, the bound is the global time in SystemC. Currently, our tool incrementally increases the bound until Boolector exceeds its time or memory bound.

On the other hand, the fundamental concept of the inductive step in the induction is that if an arbitrary state satisfies the property, then every next state also satisfies it [17]. In terms of SDSS, the major difference between BMC and induction is the initial symbolic behavioral state, the beginning set of behavioral states. To adopt induction, we need to construct an over-approximation, all the possible behavioral states, as the start point of SDSS.

All the possible behavioral states can be derived as follows. By Lemma 2, we know that, for each thread, the set of the possible suspended statuses is finite. For example, Figure 3 shows the sets of the possible suspended statuses for t1 and t2. Besides, for each variable in the design, we can use a free bit-vector variable to represent its possible values. The remaining issue is how we encode the possible combinations of suspended statuses symbolically instead of enumerating the combinations explicitly. To this end, we introduce a free bit-vector variable to represent the choice of suspended statuses for each thread. We denote this free bit-vector variable as CSS_i for the thread i. For j-th suspended status in the thread i, we use the equality $CSS_i==j$ as its sf and put this symbolic suspended status into S_i in the initial symbolic behavioral state. Since the introduced equalities adhere to the *MEB* property, SDSS can explore the reachability from all possible behavioral states properly.

5. EXPERIMENTAL EVALUATION

We implement the proposed symbolic model checking method in the framework [11]. We use benchmarks taken and adapted from [9,11] to experiment with our approach. Due to limited space, we only select some representative test cases. All experiments are carried out on an Intel-Xeon 2GHz system running Linux, equipped with 32GB RAM. We fix the time limit to 1000 seconds and the memory limit to 2GB. Before showing the results, we firstly describe the checked properties in each design.

5.1 Checked Properties

In general, the safety properties supported in our prototype tool can be expressed as: at every behavioral boundary, the processes concerned should (not) stop at the specific program locations, and/or the variables concerned should (not) have certain values. This kind of property is enough to express some important system-level properties in SystemC designs like the deadlock condition defined in [11]. In contrast, the traditional program assertion is hard to express the property at behavioral boundary.

Except `kundu`, `toy`, and `pipeline`, the safety properties we check in our test cases are the deadlock conditions. We consider the following properties of these three designs:

    (a) `kundu`: the value of the array index should be 0 or 1.

    (b) `toy`: the computation of the variables should be correct.

    (c) `pipeline`: the value of the variable in the final stage stays at 0 after 5 clocks.

5.2 Results and Comparison

To the best of our knowledge, KRATOS [9] is the only available tool for comparison. We used ESST in KRATOS with enabling both the persistent and the sleep set reductions. The results of experiments are shown on Table 1. In the second column, S and U indicate the verification status of the benchmark is safe or unsafe, respectively. We use the code size to roughly demonstrate the complexity of each design. The verification time and total memory usage are given. T.O. represents time out, and M.O. stands for memory out.

When verifying unsafe designs, our approach outperforms ESST because ESST conducts over-approximations involving symbolic computations and requiring many refinements. The comparison has shown that our approach is efficient for bug hunting. For safe designs, ESST efficiently derives the results since ESST is an abstraction-based method. However, ESST faces the state explosion issue when handling larger designs. By comparison, our induction taking advantage of stateless paradigm can prove the safeness in those big designs.

6. CONCLUSION

In this paper, we translate SystemC designs with timed language constructs to Kripke structures and propose a symbolic model checking approach for verifying SystemC designs. The experiments successfully show the scalability and efficiency of our approach. As future work, we will enhance our implementation to support different kinds of properties. Besides, we will sur-

vey optimization techniques to reduce the complexity of all the possible schedulings in SDSS.

Table 1. Results for experimental evaluation

Case Name	V	Code Size	KRATOS		Ours	
			Time(s)	Mem(M)	Time(s)	Mem(M)
kundu	S	141	0.49	21.4	5.16	17.6
toy	S	162	0.49	22.7	0.09	15.6
pipeline	S	315	T.O.	579.3	0.06	16.9
token_ring5	U	124	2.48	30.5	0.00	14.9
token_ring10	U	204	T.O.	771.0	7.17	624.0
pc_sfifo1	S	128	0.09	18.7	0.01	15.3
mem_slave5	S	371	86.27	107.2	14.43	18.4
alarm_bug	U	1876	N/A	N/A	77.53	27.0
partial_bus	S	678	540.82	M.O.	10.76	20.1
partial_bus_bug	U	673	27.88	122.3	4.62	17.9
usb_fifo	U	442	912.5	1592.5	0.16	17.0
usb_fifo_bug	U	442	T.O.	1745.4	0.19	17.0
usb_fifo_re	S	461	T.O.	1934.8	121.52	22.3

7. REFERENCES

[1] IEEE Standard 1666 SystemC Language Reference Manual, 2005. www.systemc.org.

[2] C. Helmstetter, F. Maraninchi, L. Maillet-Contoz, and M. Moy, "Automatic Generation of Schedulings for Improving the Test Coverage of Systems-on-a-Chip," In *Proc. of FMCAD*, 2006.

[3] S. Kundu, M. Ganai, and R. Gupta, "Partial Order Reduction for Scalable Testing of SystemC TLM Designs," In *Proc. of DAC*, 2008.

[4] D. Kroening and N. Sharygina, "Formal Verification of SystemC by Automatic Hardware/Software Partitioning," In *Proc. of MEMOCODE*, 2005.

[5] D. Karlsson, P. Eles, and Z. Peng, "Formal Verification of SystemC Designs Using a Petri-Net Based Representation," In *Proc. of DATE*, 2006.

[6] P. Herber, J. Fellmuth, and S. Glesner, "Model Checking SystemC Designs using Timed Automata," In *Proc. of CODES+ISSS*, 2008.

[7] N. Blanc and D. Kroening, "Race Analysis for SystemC using Model Checking," In *Proc. of ICCAD*, 2008.

[8] G. Behrmann, A. David, K. G. Larsen, J. Håkansson, P. Pettersson, W. Yi, and M. Hendriks, "UPPAAL 4.0," In *Proc. of QEST*, 2006.

[9] A. Cimatti, A. Micheli, I. Narasamdya, M. Roveri, "Verifying SystemC: a software model checking approach," In *Proc. of FMCAD*, 2010.

[10] A. Cimatti, I. Narasamdya, and M. Roveri, "Boosting Lazy Abstraction for SystemC with Partial Order Reduction," In *Proc. of TACAS*, 2011.

[11] C.-N. Chou, C.-H. Hsu, Y.-T. Chao, and C.-Y. Huang, "Formal deadlock checking on high-level SystemC designs," In *Proc. of ICCAD*, 2010.

[12] D. Große, H.M. Le, R. Drechsler, "Proving transaction and system-level properties of untimed SystemC TLM designs," In *Proc. of MEMOCODE*, 2010.

[13] P. Godefroid, "Model checking for programming languages using VeriSoft," In *Proc. of POPL*, 1997.

[14] A. Biere, A. Cimatti, E. M. Clarke, and Y. Zhu, "Symbolic model checking without BDDs," In *Proc. of TACAS*, 1999.

[15] E. Clarke, I. Grumberg, and D. Peled, "Model Checking," The MIT Press, 1999.

[16] R. Brummayer and A. Biere, "Boolector: An Efficient SMT Solver for Bit-Vectors and Arrays," In *Proc. of TACAS*, 2009.

[17] M. Sheeran, S. Singh, and G. Stålmarck, "Checking safety properties using induction and a SAT-Solver," In *Proc. of FMCAD*, 2000.

S1. SYSTEMC SCHEDULER

There are several phases in the SystemC scheduler as shown in Figure 6. In the initialization phase, all processes are added to the set of runnable processes except the processes declaring dont_initialize. It selects a process from the set of runnable processes non-deterministically and executes this process in its evaluation phase until there is no runnable process. Then, the scheduler updates channels in the update phase and some processes could become runnable due to the updates or delta notifications. After this step, if there are still runnable processes, the scheduler goes back to the evaluation phase. Otherwise, in the timed notification phase, the scheduler would advance the simulation time to the earliest time where there exists an event to be notified.

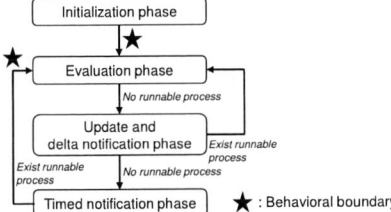

Figure 6. Phases of the SystemC scheduler

S2. SYMBOLIC VALUATION WITH EXPLICIT SUSPENDED STATUS

If we use the symbolic valuation to encode variables while keeping track of possible suspended statuses of each thread explicitly, this reachability analysis is called the symbolic valuation with explicit suspended status. Figure 7 shows the result applying this kind of reachability analysis to Example 1. In Figure 7, the symbols of all suspended statuses are the same as those in Figure 3. Besides, σ is the symbolic valuation of all variables.

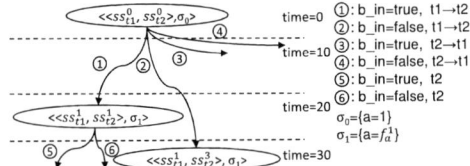

Figure 7. Partial states of Example 1 in symbolic valuation with explicit suspended status

S3. OUR SYMBOLIC MODEL CHECKING ALGORITHM

Algorithm 1 Our Symbolic Model Checking
1: **procedure** SMC()
2:      $T \leftarrow 0$;
3:      **while** TURE **do**
4:          **if** BMC(T) == CEX_FOUND **then**
5:              **return** UNSAFE;
6:          **end if**
7:          $T' \leftarrow T + t_M$;
8:          **if** InductiveStep(T, T') == NO_CEX **then**
9:              **return** SAFE;
10:          **end if**
11:          $T \leftarrow T'$;
12:      **end while**
13: **end procedure**

In our proposed algorithm, there are two procedures, BMC and InductiveStep. BMC(T) performs SDSS which starts from the symbolic behavioral state containing the initial behavioral states of the design. When the time exceeds the specified bound T, BMC terminates and evokes Boolector to check the satisfiability of the safety property. On the other hand, InductiveStep takes two arguments, T as well as T', and performs SDSS which starts from the symbolic behavioral state containing all possible behavioral states. Our inductive step turns into checking whether, during the interval t_M, a counterexample can be reachable from an arbitrary behavioral state satisfying the property. Our algorithm increases T until it obtains any result or runs out of resources. Our algorithm can derive a result in two circumstances. First, a counterexample is found in BMC. In this situation, our algorithm can conclude that the safety property does not hold and thus return UNSAFE. The other one is that no counterexample can be found in both procedures. In this case, the property is satisfied by all reachable behavioral states according to induction. Therefore, our algorithm returns SAFE.

Please note that our inductive step becomes trivially unsatisfiable if there is no behavioral state between T and T'. This trivially unsatisfiable situation makes it unable to perform induction. To avoid this situation, we let T' always larger than T by t_M, *the maximum value of time arguments in wait/notify functions*. For example, $t_M = 30$ in Figure 1.

System Verification of Concurrent RTL Modules by Compositional Path Predicate Abstraction

Joakim Urdahl, Dominik Stoffel, Markus Wedler, Wolfgang Kunz
Dept. of Electrical and Computer Eng., Univ. of Kaiserslautern, Germany

ABSTRACT

A new methodology for formal system verification of System-on-Chip (SoC) designs is proposed. It does not only ensure correctness of the system-level models but also of the concrete implementation at the Register-Transfer-Level (RTL). For each SoC module at the RTL an abstract description is obtained by *path predicate abstraction*. Since this leads to time-abstract system models the main challenge is to deal with the concurrency between the individual RTL components. We propose a compositional scheme describing the communication between SoC modules independently of their individual processing speed. The composed abstract system is modeled as an *asynchronous composition* and can be verified using the SPIN model checker. We demonstrate the practical feasibility of our approach by a comprehensive case study based on Infineon's FPI Bus.

Categories and Subject Descriptors:

B.5.2 [**Hardware**] RTL Implementation — *Design Aids*

General terms: Verification

Keywords: Formal System Verification, Abstraction

1. INTRODUCTION

New design methodologies based on Electronic System Level (ESL) descriptions impose new challenges and offer new opportunities for formal verification technology. System-level models for SoC components are created for different reasons including design exploration, evaluation and verification. Since the productivity of the subsequent Register Transfer Level (RTL) design phase benefits only little from the ESL descriptions the creation of system-level models in most practical cases constitutes extra design and verification efforts that must be added to the efforts required in a conventional design flow.

One of the main bottlenecks in SoC design is the verification of the RTL implementation. Unfortunately, this also remains true at the presence of ESL descriptions. A comparison with verification practices at lower design levels may illustrate the problem: as a matter of course, when verifying the logic input/output behavior of a large combinational circuit we consider its gate-level description. Simulating its transistor-level description instead, in principle, is possible but would be considered highly inefficient and inappropriate. Similarly, when verifying the micro-architecture of an SoC design we base this verification on its RTL description rather than its gate netlist. This is standard procedure because we can take it for granted that the gate netlist and the RTL description are *sound*

abstractions for the next lower levels.

On the other hand, when evaluating the global system behavior of an SoC this verification is still done at the RTL. Chip-level simulations of the RTL design are running for weeks and months on large workstation clusters even when ESL descriptions are available. The reason is, obviously, that the system-level models are usually not considered to be sound abstractions of the RTL implementation.

This paper intends to contribute a theoretical basis and the sketch of a methodology to create *sound abstractions* for system-level models. It is shown in [1] that a sound abstraction called *path predicate abstraction* can be created from an RTL description based on *Complete Interval Property Checking (C-IPC)* [2, 3]. In this paper, we show how path predicate abstraction can be made *compositional*. This makes it applicable to large designs and can provide the basis for a verification flow that verifies system behavior at the system level and restricts RTL verification to examining local *operation properties*. The main challenge to be addressed in this paper is how to deal with the concurrency of the RTL modules that are each represented by time-abstract models at the system level.

It has been pointed out in [1] that path predicate abstraction is related to the notion of stuttering bisimulation [4, 5]. For the latter, a so-called refinement map which needs to be determined by the verification engineer establishes this relation. Each state visited in a run of the abstract model corresponds to a finite state sequence in the corresponding run of the concrete model. This correspondence is proven using a theorem prover. In this paper, we extend the notion of path predicate abstraction which relates abstract states to sets of concrete states and abstract transitions to sets of concrete state sequences. Unlike in [4, 5] not all concrete states are mapped to abstract states. Instead they are only considered as elements of state sequences which are mapped to abstract transitions. We show in this paper that this relationship between the abstract and the concrete model can be described in terms of a complete set of operation properties that can be proven automatically by a property checker. Leveraging established property checking technology, we are able to handle industrial RTL design implementations as concrete models in our compositional system verification approach.

In the context of ESL design, most research in property checking is concerned with developing new methods for proving properties on abstract descriptions, e.g. based on SystemC [6, 7]. There is little research, however, on compositional approaches that create sound, time-abstract system-level descriptions based on property checking. A notable exception is [8]. Their work describes an approach from mathematical logic based on interpretation of theories to relate time-abstract system descriptions to cycle-accurate implementations. Moreover, a criterion is stated that identifies special cases in which abstractions obtained from C-IPC are compositional. Another promising approach to reduce the verification costs in ESL-based flows is to employ high-level synthesis and/or high-level equivalence checking [9, 10]. High-level equivalence checking is very appropriate in applications where a notion of equivalence can be established between the system level and the RTL. On the other hand, in general, we believe that the expressiveness of modern property languages and property checking techniques like C-IPC can provide a more general framework to close the semantic gap between the different levels of abstraction. However, high-level equivalence checking, when applicable, can nicely complement the envisioned design and verification flow.

The paper is organized as follows: In Sec. 2 the formal definitions and terminology of [1] describing C-IPC-based abstraction are extended to make path predicate abstraction compositional. Sec. 3 is dedicated to modeling communication between SoC modules and how to obtain a sound model for a system with concurrency. Our experimental case study is summarized in Sec. 4.

2. PATH PREDICATE ABSTRACTION

We first introduce some notations. The basic building block of a concrete or abstract system model is the deterministic finite state machine $(S, I, X, Y, \delta, \lambda)$ with a set of states S, a set of initial states $I \subseteq S$, an input alphabet X, an output alphabet Y, a transition function $\delta : S \times X \mapsto S$ and an output function $\lambda : S \mapsto Y$.

We will also refer to the *transition relation* $T \subseteq S \times X \times S$ of a finite state machine (FSM) which is the set of all valid transitions $(s, x, s') \Leftrightarrow s' = \delta(s, x)$.

A state sequence π_l of length l is a sequence of $l + 1$ states $(s_0, s_1, \ldots, s_l)$. (Note that s_0 only denotes the zero-th element of the sequence and is not necessarily an initial state of the system.) A state sequence *predicate* $\sigma(\pi_l) = \sigma((s_0, s_1, \ldots, s_l))$ of length l is a Boolean function characterizing a set of state sequences. A state sequence predicate σ of length l can be applied to a state sequence π_m of arbitrary (non-negative) length m. If $m >= l$ then σ is evaluated on the l-prefix of π_m (beginning with s_0). If $m < l$ then σ is evaluated on the set of sequences obtained by extending π_m with all possible $(l - m)$-suffixes. The result is *true* if there exists an extended sequence where σ evaluates to *true*, otherwise it is *false*.

We also define input and output sequences and predicates for such sequences. An input sequence $\rho_l = (x_0, x_1, \ldots, x_{l-1})$ describes l inputs which may be associated with the l transitions between the $l + 1$ states in a state sequence. Accordingly, an output sequence $\gamma_l = (y_0, y_1, \ldots, y_{l-1})$ describes l outputs. Input and output sequences may be characterized using I/O sequence predicates which are defined analogously to state sequence predicates. The sequence predicates $\mu(\rho_l)$ and $\mu(\gamma_l)$ are Boolean functions characterizing a set of input sequences and output sequences, respectively. (As a convention, we use Greek letters for sequences, predicates and abstraction functions and Roman letters for most of the other objects.)

Since sequence predicates return Boolean values, we can apply the usual Boolean operators $\vee, \wedge, \neg, \Rightarrow$ to sequence predicates. The length of the resulting predicate is the maximum length of the operand predicates.

We define the general path predicate $ispath(\pi_l, \rho_l) =$

$$ispath((s_0, \ldots, s_l), (x_0, \ldots, x_{l-1})) = \bigwedge_{i=1}^{l} (\delta(s_{i-1}, x_{i-1}) = s_i)$$

We also define the output sequence predicate $isoutput(\pi_l, \gamma_l) =$

$$isoutput((s_0, \ldots, s_l), (y_1, \ldots, y_l)) = \bigwedge_{i=1}^{l} (\lambda(s_i) = y_i)$$

It characterizes output sequences as computed by the output function for a given state sequence. Together with *ispath* it can be used to characterize valid output sequences of the FSM.

We can shift predicates in time using the *next* operator:

$$next(\sigma_l, n)((s_0, s_1, \ldots, s_{n-1}, s_n, s_{n+1}, \ldots, s_{n+l}))$$
$$:= \sigma_l((s_n, s_{n+1}, \ldots, s_{n+l}))$$

The $next(\sigma_l, n)$ operator syntactically extends the length of the sequence predicate to $(n + l)$ but semantically it merely shifts the starting point of the evaluation of a predicate σ_l by n positions.

We also define concatenation $\odot$ for l-sequence predicates:

$$\sigma_l \odot \sigma_k = \sigma_l \wedge next(\sigma_k, l)$$

This predicate evaluates to true for all sequences that begin with a sequence of length l characterized by σ_l and continue with a sequence of length k characterized by σ_k, where the last state in the l-sequence is the first state in the k-sequence.

In our methodology, Complete Interval Property Checking (C-IPC) is used to create *path predicate abstractions* [1] for RTL modules. The basic idea is that the behavior of a module is completely described by a set of *interval properties*, also called *operation properties*. Each operation describes a piece of design behavior over a number of clock cycles starting and ending in so called *important states*. Every operation is "triggered" by specific input conditions and produces certain outputs.

Based on C-IPC an abstract FSM model called *path predicate abstraction* can be created whose states correspond to the important states of the RTL implementation and whose edges correspond to the operations. The trigger conditions and outputs of the operations are abstracted into abstract FSM inputs and outputs. The completeness of the property set establishes the soundness of the abstract model.

We consider a concrete model $(S, I, X, Y, \delta, \lambda)$ and a path predicate abstraction $(\hat{S}, \hat{I}, \hat{X}, \hat{Y}, \hat{\delta}, \hat{\lambda})$. Both are represented as deterministic finite state machines. The two are related to each other as described in the following.

The abstract input and output alphabets, $\hat{X}$ and $\hat{Y}$ are related to the concrete behavior through a set of *message predicates*, $\{\mu_i\}$. We use the notion of messages, message sequences and predicates to define such sequences for modeling the communication between modules (finite state machines) in a system. In such a system, the input alphabet X is global, meaning that every participating machine potentially reads the output of every other machine. For an FSM $M_i = (S_i, I_i, X, Y_i, \delta_i, \lambda_i)$ in a (concrete) system of n communicating finite state machines the input space is equal to $X = Y_1 \times Y_2 \times \ldots \times Y_n$ (see Sec. 3).

Message predicates are used to define output sequences that serve as input sequences in other modules. These sequences trigger operations, i.e., behaviors of the design between *important states*. They are defined by referring to the outputs y_j of the machines M_j that produce the output sequences.

DEFINITION 1: A *message predicate* $\mu_j((x_0, \ldots, x_{l-1}))$ is an I/O sequence predicate characterizing sequences of symbols from the (global) input alphabet, X. □

Note that, for simpler notation, we write $\mu_j((y_0, \ldots, y_{l-1}))$ when characterizing a sequence of *outputs* sent by a specific machine M_j in a system. The same predicate written as $\mu_j((x_0, \ldots, x_{l-1}))$ characterizes all global system *input* sequences where M_j produces the specified output sequences. In the former notation the message predicate describes outgoing messages, in the latter the same messages are considered as ingoing. In practice, such message predicates may be defined as *macros* in a property language, relating to logic values on the interconnect lines between communicating machines. The same macro can be used without modification for ingoing and outgoing message specification.

Important states are identified by state predicates $\eta_i(s)$. The vector of important state predicate values $\alpha(s) := (\eta_1(s), \eta_2(s), \ldots)$ is the state abstraction function. It defines an abstract state value for every concrete state s.

DEFINITION 2: An *important-state predicate* $\eta_i(s)$ is a predicate evaluating to *true* for a set of concrete important states s and to *false* for all other states. A concrete state cannot belong to more than one important state. The disjunction of all $\eta_i(s)$ is a state predicate $\Psi(s) = \eta_1(s) \vee \eta_2(s) \vee \ldots$ characterizing the set of all important states. Finally, we require that the η_i satisfy the *abstraction requirements* stated in Def. 5, below. □

Operations are finite sequences of behavior of a module moving from an important state to another important state, visiting an arbitrary number of unimportant states along the path. These operations are triggered by input sequences that are composed of messages received from other modules.

335

DEFINITION 3: An *operational path* of length l between two important states, $s_0 \in S$ and $s_l \in S$, is a pair, (π_l, ρ_l), of a state sequence $\pi_l = (s_0, s_1, \ldots, s_{l-1}, s_l)$ and an input sequence $\rho_l = (x_0, x_1, \ldots, x_{l-1})$ with $l > 0$ such that

- $\Psi(s_0) = true$,
- $\Psi(s_l) = true$,
- if $l > 1$: all intermediate states $s_1, \ldots, s_{l-1}$ are unimportant states, i.e., $\Psi(s_1) = \ldots = \Psi(s_{l-1}) = false$,
- the input sequence ρ_l drives the FSM along the path π_l, i.e., it is $ispath(\pi_l, \rho_l) = true$. $\square$

A key element of our abstraction methodology is the relationship between operations and messages. The important state predicates and the message predicates are chosen such that every operation between two sets of important states can be described based on the messages. Every machine in the system produces output sequences that are described by a set of message predicates. In the following, the set $Q_j = \{\mu_{j,1}, \mu_{j,2}, \ldots\}$ is the set of message predicates chosen for describing the output of the j-th machine, M_j, in the system.

DEFINITION 4: An *operational input predicate* $\iota(\rho_l)$ is a conjunction of message predicates, one from each machine M_j in a system of n machines:

$$\iota(\rho_l) = \bigwedge_{j=1}^{n} \mu_j(\rho_l), \text{ where } \mu_j \in Q_j \qquad \square$$

An operation in our methodology is characterized by the set of important beginning states, η_B, the set of important ending states, η_E, and a set of operational input predicates, ι_j, that each describe input sequences leading from any state in η_B to some state in η_E. We call the disjunction of these input predicates, $\bigvee \iota_j$, the *trigger* of the operation. Note that a trigger of an operation, in practice, usually has a compact representation since only a small subset of machines in a system has an influence on the trigger of a specific operation. Similarly like an implicant of a Boolean function can compactly describe a large disjunction of minterms, in practice, a trigger condition for an operation can compactly describe a large disjunction of operational inputs.

As can be seen from these definitions, the important state predicates η_i and the message predicates μ_j cannot be chosen arbitrarily. Instead, the choice must satisfy the following requirements so that the proposed abstraction is created.

DEFINITION 5: The important-state predicates η_i and the message predicates μ_j are chosen to fulfill the following *abstraction requirements*:

1. For any pair of message predicates, $\mu_{j,1}$ and $\mu_{j,2}$, from the same sending machine M_j there does not exist an (arbitrary) sequence ρ_l such that both predicates become *true*.

2. For every operational path (π_l, ρ_l) between two sets of important states given by $\eta_B(s)$ and $\eta_E(s)$ there exists an operational input predicate ι_j that is satisfied by ρ_l: $\iota_j(\rho_l) = true$.

3. For all pairs of (concrete) important states $s_B, s_E \in S$ between which there exists an operational path there is an l_{max} such that every operational l-path between s_B and s_E is of length l_{max} or shorter: $l \leq l_{max}$.

4. For every pair of important-state predicates, η_B, η_E, such that there exists an operational path $(\tilde{s}_B, \ldots, \tilde{s}_E)$ with $\eta_B(\tilde{s}_B) = true$ and $\eta_E(\tilde{s}_E) = true$ it holds that there also exists an operational path $(s_B, \ldots, s_E)$ for *every* state s_B satisfying $\eta_B(s_B) = true$ and *some* state s_E satisfying $\eta_E(s_E) = true$. $\square$

The first condition ensures that no operational path "is forgotten" in the abstract model. We fulfill this condition using the completeness check in C-IPC. Together with the second condition we make sure that the message predicates chosen for describing the inputs from a given sending machine M_j disjointly partition the set of input sequences received from M_j.

The third condition requires that *all cyclic paths in the concrete model intersect an important state*, i.e., there are only finite operational paths between important states. This is fulfilled "automatically" by the C-IPC methodology.

The fourth requirement makes sure that states with the same "operational behavior" are grouped into important states: when the FSM receives a set of messages it reacts in the same way for all concrete important states grouped into the same abstract state. Since every operational path is mapped to an abstract transition this also ensures that abstract paths assembled from abstract transitions can always be mapped to some concrete path, i.e., there are no false abstract paths. Also this condition is "automatically" fulfilled by the C-IPC methodology.

The abstraction is created by the abstraction function $\alpha(s)$ for the states and by an abstraction function $\beta(\mu)$ for message predicates.

DEFINITION 6: [Message Abstraction]
For every message predicate $\mu \in Q_j$ there is a distinct, unique, *abstract message* symbol $\hat{q} \in \hat{Q}_j$. There is a one-to-one mapping between the message predicates in Q_j and the abstract message symbols in $\hat{Q}_j$ by the mapping function: $\beta_j : Q_j \mapsto \hat{Q}_j$. $\square$

Using β_j we can map every operational input predicate ι (which is a conjunction of n message predicates according to Def. 4) to a corresponding n-tuple of abstract messages, $\hat{x}$ (and vice versa).

DEFINITION 7: The abstract input alphabet is a set of n-tuples, $\hat{X} = \hat{Q}_1 \times \ldots \times \hat{Q}_n$ where $\hat{Q}_j$ is the set of abstract message symbols of the j-th sending machine according to Def. 6. $\square$

For readability, we introduce the following notation:

$$input(\hat{x}, \rho_l) = input((\hat{q}_1, \ldots, \hat{q}_n), \rho_l) = \bigwedge_{j=1}^{n} \mu_i(\rho_l)\beta_j(\mu_j)$$

where $\hat{q}_j = \beta_j(\mu_j)$. The predicate $input(\hat{x}, \rho_l)$ is true iff the input sequence ρ_l satisfies the operational input predicate corresponding to the abstract input $\hat{x}$. It is used in the definitions of the abstract transition relation and the abstract output function.

DEFINITION 8: We consider an abstraction function α such that the important-state predicates and the message predicates fulfill the requirements of Def. 5. The abstract transition relation $\hat{T}$ is given by:

$$(\hat{s}, \hat{x}, \hat{s}') \in \hat{T} \Leftrightarrow$$
$$\forall \pi_l = (s_0, s_1, \ldots, s_l), \forall \rho_l = (x_0, x_1, \ldots, x_{l-1}):$$
$$\Psi(s_0) \wedge (\alpha(s_0) = \hat{s})$$
$$\wedge\, input(\hat{x}, \rho_l)$$
$$\wedge\, ispath(\pi_l, \rho_l)$$
$$\rightarrow \Psi(s_l) \wedge \alpha(s_l) = \hat{s}' \qquad \square$$

This definition states that an abstract transition from $\hat{s}$ to $\hat{s}'$ under $\hat{x}$ exists iff, in the concrete system, *all* operational paths beginning at states corresponding to $\hat{s}$ and triggered by messages corresponding to $\hat{x}$ make the FSM move to a state corresponding to $\hat{s}'$.

LEMMA 1. *The abstract transition relation of Def. 8 implicitly defines a function, i.e., it holds that*

$$(\hat{s}, \hat{x}, \hat{s}'_1) \in \hat{T} \wedge (\hat{s}, \hat{x}, \hat{s}'_2) \in \hat{T} \Rightarrow \hat{s}'_1 = \hat{s}'_2 =: \hat{\delta}(\hat{s}, \hat{x})$$

Proof: Assume $(\hat{s}, \hat{x}, \hat{s}'_1) \in \hat{T} \wedge (\hat{s}, \hat{x}, \hat{s}'_2) \in \hat{T}$ and $\hat{s}'_1 \neq \hat{s}'_2$. Assume two concrete states, s_0 and $\tilde{s}_0$ such that $\alpha(s_0) = \alpha(\tilde{s}_0) = \hat{s}$, and two input sequences ρ_l and $\tilde{\rho}_l$ with $input(\hat{x}, \rho_l) = true$ and $input(\hat{x}, \tilde{\rho}_l) = true$ such that that the concrete FSM moves from s_0 to a state s_l with $\alpha(s_l) = \hat{s}'_1$ and from $\tilde{s}_0$ to a state $\tilde{s}_l$ with $\alpha(\tilde{s}_l) = \hat{s}'_2$. The input sequence $\tilde{\rho}_l$ violates the implication in Def. 8 (because under $\tilde{\rho}_l$ it is not $\alpha(s_l) = \hat{s}'_1$), hence $(\hat{s}, \hat{x}, \hat{s}'_1) \notin \hat{T}$. This violates our initial assumption, proving the lemma. $\square$

Based on the operational view on the concrete FSM we also define abstract output symbols and the abstract output function.

DEFINITION 9: The *abstract output alphabet* of an abstract FSM $\hat{M}$ is $\hat{Y} = \hat{Q}$ where $\hat{Q}$ is a set of abstract messages according to Def. 6. □

An abstract output symbol denotes a message sent by an abstract module during an abstract transition. The abstract output alphabet of an abstract FSM $\hat{M}$ is $\hat{Y} = \hat{Q}$. This set of messages, $\hat{Q}$ is in one-to-one correspondence with a set of message predicates Q, by the mapping function β of Def. 6. An operation in a FSM M produces a sequence of output symbols along an operational l-path that satisfies exactly one message predicate in Q. On the abstract level, this corresponds to one abstract message symbol. The abstract output function $\hat{\lambda}(\hat{s}, \hat{x})$ computes for a given abstract state $\hat{s}$ and a given input symbol $\hat{x} \in \hat{X}$ an abstract output symbol $\hat{y} \in \hat{Y}$, corresponding to the message predicate $\{\mu_j\}$ that is true on every operational l-path triggered in the corresponding operation.

For readability, again, we introduce some more notation:

$$output(\hat{y}, \gamma_l) = \mu_j(\gamma_l), \text{ where } \beta(\mu_j) = \hat{y}$$

The predicate $output(\hat{y}, \gamma_l)$ identifies the message predicate that corresponds to the abstract output $\hat{y}$.

DEFINITION 10: The abstract output function $\hat{\lambda}$ is given by:

$$\hat{y} = \hat{\lambda}(\hat{s}, \hat{x}) \Leftrightarrow$$
$$\forall \pi_l = (s_0, s_1, \ldots, s_l), \forall \rho_l = (x_0, x_1, \ldots, x_{l-1}),$$
$$\forall \gamma_l = (y_1, y_2, \ldots, y_l):$$
$$\Psi(s_0) \wedge (\alpha(s_0) = \hat{s})$$
$$\wedge \, input(\hat{x}, \rho_l)$$
$$\wedge \, ispath(\pi_l, \rho_l)$$
$$\wedge \, isoutput(\pi_l, \gamma_l)$$
$$\rightarrow output(\hat{y}, \gamma_l) \qquad \square$$

According to this definition, the abstract output function produces an abstract output symbol $\hat{y}$ representing exactly that message that is produced by the concrete FSM on every operational path beginning from a state in $\hat{s}$ and triggered by $\hat{x}$.

Our definition of an FSM allows for a single abstract output per transition. In practice, our state machines may have several communication partners and produce several outputs simultaneously. We could model this with an extended FSM model allowing for more than one λ function (and several output alphabets). Instead, for keeping our notations simple when modeling such behavior we instantiate several FSMs sharing the same δ function but having different λ functions.

Supplement S1 illustrates how the developed theoretical notions can be used in practice. In order to implement the compositional system verification methodology proposed in this paper we use the C-IPC verification methodology. Path predicate abstractions are created by following coding guidelines for macro definitions in operation properties that are expressed using standard property languages (SVA). These macros correspond to the objects of our formalism.

3. COMPOSITION OF ABSTRACTIONS

In this section we discuss how path predicate abstractions of RTL modules can be composed into an abstract system model that is sound with respect to certain LTL properties. Path predicate abstraction leads to *time-abstract* models — the abstract model contains no information on how long operations in the components of the system actually take. The abstract system model therefore needs to be an *asynchronous composition* (to be defined later) of path predicate abstractions that models all possible interleavings of operations in the individual modules.

The system behavior is determined by the communication between the modules. We therefore need to make sure that this is soundly modeled. Every communication between concrete ma-

chines uniquely corresponds to a communication between the corresponding abstract machines, and vice versa.

Communication in digital hardware relies on a few basic principles. The communicating parties need to synchronize with each other before messages can be transmitted. Also, there needs to be an agreement on how the message is actually transferred from the sender to the receiver(s).

For reasons of space, we cannot elaborate on all communication schemes used in digital hardware design. With our methodology we can model asynchronous and synchronous communication, both with unilateral or bilateral synchronization. Supplement S2 discusses these schemes in more detail.

3.1 Synchronization and stuttering

A basic element of communication in hardware is synchronization using special *synchronization signals*. SoC modules remain in a waiting state until a synchronization signal to which they are sensitive becomes asserted. In our methodology, we capture the de-asserted signal value that keeps an FSM waiting in a message predicate of length 1. We view this message as a "trigger" for the wait operation. However, the FSM driving such a synchronization signal may keep the signal deasserted also during operations that are of length l greater than 1. We therefore need to make a small extension to our formalism of Sec. 2 in order to accomodate for repetitions of messages when we are not interested in the actual number of times a message is repeated. This is needed in two cases: either the repeated message triggers wait operations or the repeated message is not used as a trigger at all. We use the following notation for a finite number of repetitions of a sequence predicate:

$$stutter(\mu_j, l) = \underbrace{\mu_j \odot \mu_j \odot \ldots \odot \mu_j}_{l \text{ times}}$$

We change the definition of the abstract output function by redefining the *output* predicate used in Def. 10 in the following way:

$$output_S(\hat{y}, \gamma_l) = stutter(\mu_j, l)(\gamma_l), \text{ where } \beta(\mu_j) = \hat{y}$$

The predicate $output(\hat{y}, \gamma_l)$ is true iff the output sequence ρ_l satisfies l concatenations of the message predicate corresponding to the abstract abstract output $\hat{y}$.

Obviously, this leads to a sound abstraction of the composed system only if, in the receiving machines, the effect of l repetitions of the sequences characterized by μ_j is indistinguishable from a single occurrence of μ_j. This is only the case if the receiving machines do not change state upon receiving μ_j. In other words, μ_j must not trigger any operation except for a waiting operation.

DEFINITION 11: A *synchronization signal* is a (concrete) signal with the following property: If the signal's assertion triggers an operation in an important state $\eta_i()$ of an FSM then its de-assertion triggers a *waiting* operation, i.e., an operation of length 1 that leads back to $\eta_i()$. □

Using this additional concept we are now able to cover all elements of standard communication schemes between finite state machines as described in Supplement S2:

- Synchronization signals are abstracted using stuttering of message predicates μ_j of length 1. Def. 10 of the abstract output function is based on the $output_S()$ predicate introduced above.

- Data exchange operations of lengths $l > 1$ are abstracted using message predicates of length l. As elaborated in Supplement S2, such operations must be implicitly synchronized using a common clock. Abstraction of such message predicates is based on the unmodified $output()$ predicate introduced in Sec. 2. Communication works only if such implicit synchronization is preceded by explicit synchronization (based on signaling).

- Data exchange operations of length $l = 1$ may or may not be directly combined with explicit synchronization. In our formalism, they are abstracted as introduced in Sec. 2.

3.2 Modelling digital H/W communication

In the following, we discuss how the standard communication schemes as described in Supplement S2 are modeled in our methodology. A path predicate abstraction of a module represents its operations including those for communication. A communication operation is represented by a transition from an abstract synchronization state to another abstract state. The abstract finite state machine contains no represention of the concrete timing of the operations. A system model composed from individual abstract machines is therefore an asynchronous one, as will be formally defined later. Communication between the abstract finite state machines relies solely on event signaling.

Path predicate abstraction of communicating modules identifies the beginning and ending important states of each communication operation. It also identifies and represents unilateral or bilateral synchronization signals if such explicit synchronization is used in the implementation. The additional self-transitions and guard conditions needed for a complete four-cycle handshake are introduced in the course of composing the path-predicate abstracted modules into a system, as discussed in detail in Supplement S3.

3.3 Abstract System Model

In the concrete system, every finite state machine, M_i, corresponds to a path predicate abstraction $\hat{M}_i$. An operation in a machine M_i may comprise a sequence of state transitions but it corresponds to a single transition in the abstract model $\hat{M}_i$. The temporal relationship between the operations in different machines based on a common clock is lost in the abstraction. Hence, in our abstract system model the abstract FSMs communicate *asynchronously* with each other by exchanging messages. The unknown temporal relationship between the modules is modeled using non-determinism: While each abstract FSM $\hat{M}_i = (\hat{S}_i, \hat{I}_i, \hat{X}, \hat{Y}_i, \hat{\delta}_i, \hat{\lambda}_i)$ is still a deterministic FSM the composed model $\hat{M}$ has a non-deterministic transition behavior modeled by a relation $\hat{T}$ rather than a function $\hat{\delta}()$.

DEFINITION 12: [Asynchronous composition]

The *asynchronous composition* $\hat{M}$ of n path-predicate-abstracted FSMs $\hat{M}_i$ is given by $\hat{M} = (\hat{S}, \hat{I}, \hat{X}, \hat{Y}, \hat{T}, \hat{\lambda})$, where

- $\hat{S} = \hat{S}_1 \times \hat{S}_\times \ldots \times \hat{S}_n$, is the set of states,

- $\hat{I} = \hat{I}_1 \times \hat{I}_2 \times \ldots \times \hat{I}_n$, is the set of initial states,

- $\hat{X} = \hat{Y}_0 \times \hat{Y}_1 \times \ldots \times \hat{Y}_n$ is the input alphabet where $\hat{Y}_0$ is the set of abstract primary input messages and, for all i, $1 \leq i \leq n$, $\hat{Y}_i$ is the set of messages produced by the abstract FSM $\hat{M}_i$,

- $\hat{Y} = \hat{Y}_1 \times \ldots \times \hat{Y}_n$ is the output alphabet where $\hat{Y}_i$ is the set of messages produced by the abstract FSM $\hat{M}_i$,

- $\hat{T}$ is the transition relation:
$\big((\hat{s}_1, \ldots, \hat{s}_n), \hat{x}, (\hat{s}'_1, \ldots, \hat{s}'_n)\big) \in \hat{T}$ iff $\exists i, 1 \leq i \leq n$ such that $\hat{s}'_i = \hat{\delta}_i(\hat{s}_i, \hat{x})$ and $\forall j, 1 \leq j \leq n, j \neq i : \hat{s}'_j = \hat{s}_j$.

- $\hat{\lambda} : \hat{S} \mapsto \hat{Y}$ is the output function, labeling every state with the output messages produced by all sub-modules:
$\hat{\lambda}((\hat{s}_1, \ldots, \hat{s}_n)) = (\hat{\lambda}_1(\hat{s}_1), \hat{\lambda}_2(\hat{s}_2), \ldots \hat{\lambda}_n(\hat{s}_n))$. □

This notion of an asynchronous composition is illustrated in Supplement S4. All possible interleavings of operations between machines are represented. This model is closely related to the asynchronous product of ω-automata in SPIN [11]. Note that the transitions in the asynchronous composition represent single transitions of sub-modules, i.e., modules never transition simultaneously but always "one after the other".

When checking liveness properties based on this asynchronous composition fairness constraints need to be added to make sure that every sub-FSM $\hat{M}_i$ infinitely often makes a transition (cf. *finite progress assumption* in [11]). This is implemented as *weak fairness* in SPIN.

3.4 Model checking on abstract system

Let us now discuss what properties can be formulated so that their validity on the abstract model implies the validity of the corresponding property on the concrete model. Obviously, since the concrete timing is lost in the abstraction, the X operator must not appear in a property. We consider here LTL and discuss properties based on the operators F and G. The properties we formulate for the abstract system correspond to concrete properties exclusively relating to important states. Path predicate abstraction of a single module has been shown to be sound with respect to certain CTL properties [1]; this proof also holds for the LTL formulas considered here.

When we integrate path predicate abstractions into a system then the set of abstract product states is a superset of the product states that have corresponding concrete product states. Not every abstract product state has a concrete counterpart. We therefore do not formulate LTL properties that relate to more than one module in a given product state. The allowed abstract LTL formulas are defined recursively:

DEFINITION 13: The following are LTL formulas allowed on our abstract model:

- $F \hat{p}_i$, where $\hat{p}_i$ relates to the atomic formulas of only one sub-FSM, $\hat{M}_i$, of the abstract model,

- $G \hat{p}_i$, where $\hat{p}_i$ relates to the atomic formulas of only one sub-FSM, $\hat{M}_i$, of the abstract model,

- $F \hat{f}$ if $\hat{f}$ is an allowed LTL formula,

- $G \hat{f}$ if $\hat{f}$ is an allowed LTL formula,

- $\hat{f} \wedge \hat{g}$ and $\hat{f} \vee \hat{g}$ and $\neg \hat{f}$ if $\hat{f}$ and $\hat{g}$ are allowed LTL formulas.

THEOREM 1. *Consider an LTL formula $\hat{\varphi}$ for the abstract model according to Def. 13. The corresponding LTL formula φ for the concrete model is obtained by replacing every sub-formula $F \hat{p}_i$ with $F(\Psi_i \wedge p_i)$ and every sub-formula $G \hat{p}_i$ with $G(\Psi_i \Rightarrow p_i)$. The Boolean formula p_i is obtained by replacing all occurrences of abstract atomic state variables $\hat{a}_i$ by their corresponding important-state predicates: $p_i = \hat{p}_i(\hat{a}_1 := \eta_1(s), \hat{a}_2 := \eta_2(s), \ldots)$.*
If the formula $\hat{\varphi}$ holds on the abstract model then the formula φ holds also for the concrete model. □

Supplement S5 gives a proof of the theorem.

Note that a property in the abstract model may fail if the corresponding property on the concrete level only holds for a specific processing speed in between synchronization events. This is a consequence of describing the system model as an asynchronous composition. Only system behavior that is *speed independent*, i.e., it does not depend on the processing speed of the individual modules, can be proved on the abstract level.

We use LTL rather than CTL formulas since CTL existentially quantifies over the paths. Since not *every* abstract path must have a concrete counterpart, the asynchronous composition, in general, is not a sound model with respect to CTL.

4. EXPERIMENTAL RESULTS

The proposed methodology has been evaluated by an experimental case study based on an industrial implementation of the on-chip bus protocol called *Flexible Peripheral Interconnect (FPI) bus*. The protocol and the implementation are owned by Infineon Technologies and are used in numerous industrial designs. The main objective of our experiments was to check the applicability of our abstraction method in a realistic setting. For this purpose we considered the FPI bus which is a complex and highly optimized industrial design.

Our abstraction method results in time-abstract modules communicating asynchronously using a four-cycle handshake. This means that both the receiver and the sender are blocked during such a communication until certain conditions are fulfilled. This matches well with the communication mechanisms implemented in the SPIN model checker. In SPIN, no order of execution is enforced among the modules. Instead, synchronization is done using explicit blocking mechanisms [11].

Module	#inputs	#outputs	#state bits	#LOC
MasterAgent	199 / 17	202 / 13	292 / 17	3568 / 91
SlaveAgent	199 / 10	202 / 5	292 / 7	3568 / 49
BCU	258 / 9	215 / 4	941 / 14	8966 / 41

Table 1: Sizes of concrete/abstract system components

The modules of the FPI bus are listed in Table 1. For each module we created path predicate abstractions, as described in this paper, based on C-IPC using the OneSpin 360 MV [3] property checker. The size of each module in its concrete and path-predicate-abstracted version is shown in Table 1. Column LOC refers to the number of lines of code in VHDL for the concrete model and in PROMELA for the abstract SPIN model.

The effort to create a complete set of properties for SoC module verification is estimated to be around 2000 LOC per person month [3]. This matches also with our experiences in this project. Since the abstract model results from these properties based on macro coding conventions as illustrated in Supplement S1 the additional effort for creating the abstract model is small. Coding conventions similar to the ones described, anyway, are considered "good practice" among verification engineers.

The FPI bus is a modular system — every master and slave interfaces the bus using a dedicated master agent or slave agent, respectively. For our experiments, we constructed a system where two peripherals act as masters and two simple memories act as slaves. As an implementation of such a system we instantiated the FPI bus with two master agents and two slave agents.

In our industrial implementation, MasterAgent and SlaveAgent are designed as a single module which can be configured at system startup to act as an agent either for the master or the slave. Our abstractions of the MasterAgent and SlaveAgent, however, cover only the relevant behavior of either role. We estimate that the module configured as SlaveAgent covers about 2/3 of the inputs, outputs and states listed in Tab. 1, while the module configured as MasterAgent covers nearly all of them. Similarly, for the Bus Control Unit (BCU) we ignored certain features like debugging mode and starvation protection, and abstracted only the behavior relevant for address decoding and bus arbitration. We estimate that about 1/3 of the BCU inputs, outputs and states contribute to these functions. But even taking this into account, our composed abstract system soundly models concrete system functions involving approximately 750 inputs, 750 outputs and 1300 states. Proving global safety and liveness properties on a system of this size is clearly beyond the capacities of today's model checking technology.

Note that separating out the considered system functionality is practically impossible in the highly optimized concrete system. It is an advantage of our methodology that it can selectively extract functionality which is in the focus of the intended verification plan.

In our experiments we have attempted to verify that the FPI bus is conforming to its protocol in any environment, independently of the specific system into which it is integrated. Therefore, we implemented a peripheral which behaves in as general a way as possible. In order to achieve this we exploited the non-determinism native to SPIN and implemented a peripheral that randomly makes requests from a set containing all types of FPI bus transactions.

SPIN features a number of pre-defined checks that can be run on a PROMELA model. In particular, since we model the communication between modules using the PROMELA notion of a channel, SPIN globally proves freedom of deadlock in the composed system. In our experiments this global proof took 3 minutes of CPU time on a state-of-the-art PC using 3 GB of main memory .

Moreover, we formulated a representative LTL property of type $G\,(a \Rightarrow F\,c)$ to verify that a requested transaction is executed correctly in the system. This and similar properties to verify the correctness of transactions can be proven in about 1 minute using

0.5 GB of main memory.

As a consequence of our compositional methodology, proving these properties on the abstract level does not only establish the correctness of our abstract model but implies valid proofs also for the actual implementation of the FPI bus. We are not aware of any other technology that could prove system-level properties of a similarly large system with concurrency using a largely automated approach based on property checking.

5. CONCLUSION

This paper proposes a compositional approach to system-level verification of systems with concurrent modules. It is based on creating sound abstractions for concrete RTL implementations. The large semantic gap between the system level and the RTL description is closed leveraging the expressiveness of modern property languages such as SVA, and the power of state-of-the-art property checking technology such as C-IPC. The paper demonstrates that the model checking paradigm of SPIN is not only useful for verifying concurrent processes in software but, in the context of the proposed methodology, can also verify complex hardware implementations. Ideally, a front-end should be created that links transaction-level modelling in SystemC to the computational models of SPIN. This would increase the scope of the proposed approach in an industrial environment creating new and promising opportunities to integrate formal verification in ESL-based design flows.

6. REFERENCES

[1] J. Urdahl, D. Stoffel, J. Bormann, M. Wedler, and W. Kunz, "Path predicate abstraction by complete interval property checking," in *Proc. Intl. Conf. on Formal Methods in Computer-Aided Design (FMCAD)*, 2010, pp. 207–215.

[2] M. D. Nguyen, M. Thalmaier, M. Wedler, J. Bormann, D. Stoffel, and W. Kunz, "Unbounded protocol compliance verification using interval property checking with invariants," *IEEE Transactions on Computer-Aided Design*, vol. 27, no. 11, pp. 2068–2082, November 2008.

[3] Onespin Solutions GmbH, "Germany. OneSpin 360MV," http://www.onespin-solutions.com.

[4] P. Manolios and S. K. Srinivasan, "A refinement-based compositional reasoning framework for pipelined machine verification," *IEEE Transactions on VLSI Systems*, vol. 16, pp. 353–364, 2008.

[5] M. Abadi and L. Lamport, "The existence of refinement mappings," *Theoretical Computer Science*, vol. 82, no. 2, pp. 253–284, 1991.

[6] D. Kroening and N. Sharygina, "Formal verification of SystemC by automatic hardware/software partitioning," *Formal Methods and Models for Co-Design*, 2005.

[7] D. Große, H. M. Le, and R. Drechsler, "Proving transaction and system-level properties of untimed SystemC TLM designs," in *Proc. ACM/IEEE International Conference on Formal Methods and Models for Codesign (MEMOCODE)*, 2010, pp. 113–122.

[8] H. Eveking, T. Dornes, and M. Schweikert, "Using SystemVerilog assertions to relate non-cycle-accurate to cycle-accurate designs," in *Proc. IEEE Intl. High Level Design Validation and Test Workshop (HLDVT)*, 2011.

[9] A. Koelbl, R. Jacoby, H. Jain, and C. Pixley, "Solver technology for system-level to RTL equivalence checking," in *Design, Automation Test in Europe Conference (DATE)*, april 2009, pp. 196 –201.

[10] P. Chauhan, D. Goyal, G. Hasteer, A. Mathur, and N. Sharma, "Non-cycle-accurate sequential equivalence checking," in *Proc. Design Automation Conference (DAC)*, 2009, pp. 460–465.

[11] G. J. Holzmann, *The SPIN Model Checker*. Addison-Wesley, 2004.

7. SUPPLEMENTS

S1. Practical Methodology

The example system depicted in Fig. 1 mimics the behavior of a bus. Transactions are requested by a peripheral and are executed

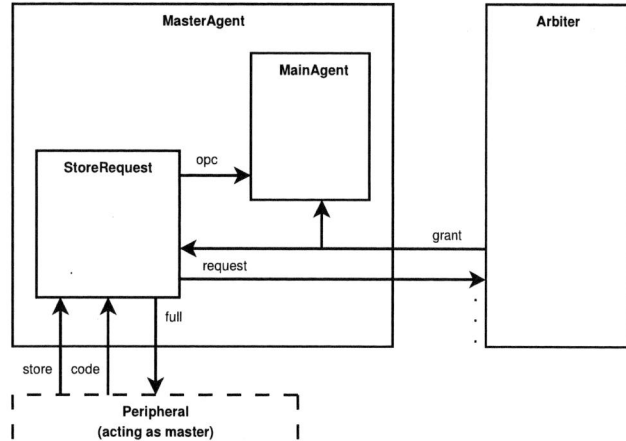

Figure 1: Block diagram of example system

by the bus. A dedicated MasterAgent implements the interface towards the bus peripheral. The signals *store*, *code* and *full* connecting our system with the peripheral are the *primary inputs* and *primary output*, respectively. The Arbiter grants access to the shared bus lines (omitted in this example).

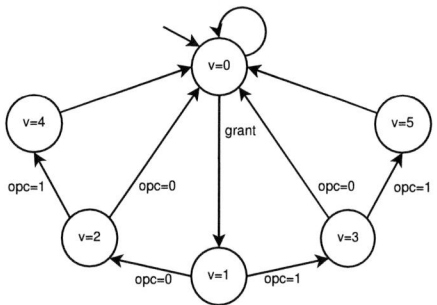

Figure 2: MainAgent

The MasterAgent consists of the sub-modules MainAgent and StoreRequest. MainAgent executes the transaction stored by StoreRequest when granted by the Arbiter. Its FSM is shown in Fig. 2. When granted it takes some action according to the sequence received for *opc*. StoreRequest is described in Fig. 3. When *store* goes active the current and next value of *code* is stored in two subsequent steps, each represented by a unique state. The request to the Arbiter is set and the FSM waits until a grant from Arbiter is received. After a grant has occurred *opc* outputs a sequence identifying the stored transaction while the FSM moves back to its initial state. StoreRequest also sets the *full* flag to indicate that no new transaction can be stored. (The peripheral needs to comply with this.) For reasons of space we do not present the full behavior of the Arbiter. However, for correct system behavior it is required that the arbiter only grants to a requesting MasterAgent, i.e., *grant* → *request*.

A complete set of properties can be formulated and proved for MainAgent and StoreRequest. The resulting abstract FSMs are shown in Fig. 4 and Fig. 5.

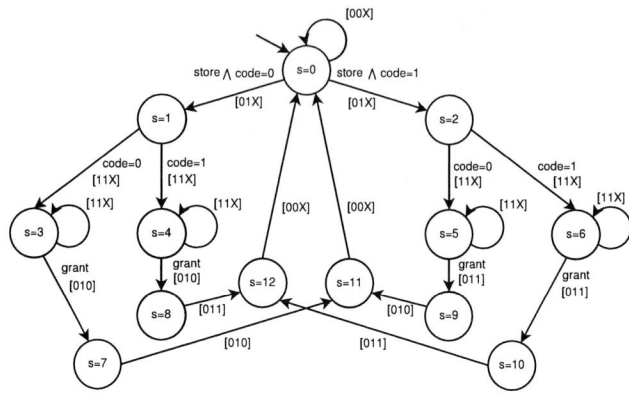

Figure 3: StoreRequest (output encoding [*request, full, opc*])

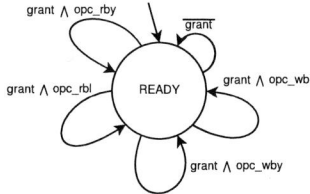

Figure 4: Abstract MainAgent

As an example, one of the operation properties written for MainAgent is presented in the following.

```
property granted_with_blockwrite is
 assume :
   at t : READY;
   at t : grant;
 prove :
   at t+1: opc_wbl;
   during [t+1,t+2]: not_free ;
   during [t+1,t+2]: not_request ;
   at t+1: not IMPORTANT_STATE ;
   at t+2: STORE_WBL;
end property ;
```

READY and STORE_WBL are the names of macros describing the important states, i.e., they correspond to the important state predicates η_i as described in Sec. 2.

Specifically, in this example it is

```
macro READY      { s=0}
macro STORE_WBL  { s=6}
```

In more realistic examples, the concrete design often consists of several FSMs and combinational blocks, i.e., several processes and combinatorial statements in an HDL. Ideally, the abstract state machine is created from the specification and purely reflects the control path. Therefore, in contrast to our simple example, the structure of the concrete and abstract machines differ greatly. In an industrial design, important state predicates therefore rarely describe only single concrete states as in this example but rather sets of concrete states as given by complex expressions over the internal variables of a module. This allows us to prove operations spanning the entire module across several FSMs.

IMPORTANT_STATE in the above operation property is a macro containing the disjunction of all important state predicates Ψ of the module.

The inputs and outputs of the abstract FSM are defined by message predicates. Just like for the important state predicates, in practice, macros are used to define them. In our example, the primary inputs are abstracted by the following macros. Note that if an abstract peripheral was to be included in our system the same macros would have to be used for the abstraction of its outputs.

340

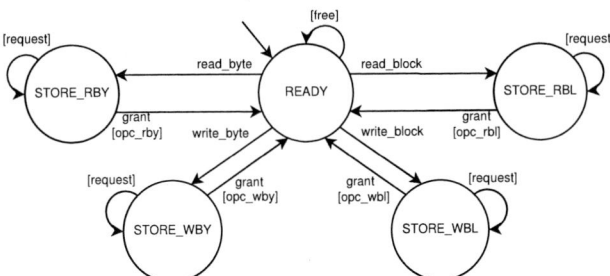

Figure 5: Abstract StoreRequest. (If no output is written the values *not_free*, *not_request* and *opcX* are assumed for their respective output functions.)

```
macro read_byte    {store and code=0 and next(code)=0}
macro read_block   {store and code=0 and next(code)=1}
macro write_byte   {store and code=1 and next(code)=0}
macro write_block  {store and code=1 and next(code)=1}
macro no_store     {not store}
```

The abstraction functions α and β are defined by the mapping of objects in the concrete FSM to these macro names. The abstract alphabet for the primary input $\hat{Y}_0$ becomes {*read_byte*, *read_block*, *write_byte*, *write_block*, *no_store*}.

The abstract transition function and the abstract output function are given by the abstract FSM. Every property describes an abstract transition between the abstract states for a certain trigger condition and abstract outputs. Following our coding guidelines, such a property is described by the important state macros, the input macros, and the output macros.

The abstraction fulfills the requirements defined Sec. 2. Every operation is described by an IPC property proved for the module. The set of operations is proven to be complete. Additionally, we explicitly check for each operation that no important state is traversed (using macro IMPORTANT_STATE) and make sure that our message predicates never can be fulfilled simultaneously. Finally, the modules are verified only in terms of our predicates. Practically, this means that all properties are completely described in terms of the developed macros. Direct references to the RTL variables are prohibited. Note that in the manual process of formulating properties and creating the abstraction it is always possible that a mistake is made by the verification engineer. This, however, will always be discovered either syntactically, by a violation of the described composition rules, or semantically, by a failing property or a failing completeness check.

Before concluding the example the output of the abstract FSM still needs to be discussed. In our example, we have chosen to abstract *request*, *full* and *opc* as separate output functions of StoreRequest. This is modeled in our formalism by introducing new FSMs differing only in the output function.

```
macro opcX         { not grant}
macro opc_rby      { grant and opc=0 and next(opc)=0 }
macro opc_rbl      { grant and opc=0 and next(opc)=1 }
macro opc_wby      { grant and opc=1 and next(opc)=0 }
macro opc_wbl      { grant and opc=1 and next(opc)=1 }

macro request      { request }
macro not_request  { not request }

macro free         { not full }
macro not_free     { full }
```

We realize that *opc* is only relevant after a grant and we exploit this to simplify the abstract system. A general *don't care* message would, however, contradict our requirement 1 of Def. 5, namely, that a sequence must not satisfy more than one message predicate. Since all inputs and outputs in our system are defined as elements of the same global alphabet we can specify the value of the *grant* signal also in output message predicates. In order to make our mes-

sages unambiguous we explicitly state the don't care situation using the input *grant* in the output message predicates. In this way, we ensure also that we only observe the value of *opc* when it is relevant.

Def. 10 describes message predicates that have the length of the operation considered. Observe that this is not the case for the predicates *opcX*, *not_request* and *not_free* in this example. Note that these predicates never trigger a transition to another state. Predicates *free* and *request* are used for synchronization. The peripheral waits for the *free* flag to be set before starting a communication. Similarly, the Arbiter only issues a grant when *request* is set. In our timing-inaccurate abstraction we do not want to consider how long the machines are waiting to be synchronized. Therefore, in Sec. 3.1, we have extended our terminology by the concept of *stuttering* for such predicates. In the case of the *free* predicate, for example, using this notion we can prove that *free* is kept unset for the entire operation and abstract it by the message *not_free* for every message length. Similarly, we handle *opcX* which is not part of any communication, i.e., it never triggers any operation. Therefore, we do not care how many times the message is sent. The resulting abstraction is shown in Fig. 5.

S2. Communication schemes in digital hardware

Communication in digital hardware relies on a few basic principles. The communicating parties need to synchronize with each other before messages can be transmitted. Also, there needs to be an agreement on how the message is actually transferred from the sender to the receiver(s).

Let us discuss the different communication schemes that we address in this paper. A first fundamental distinction is between *asynchronous* communication, relying on dedicated event signaling, and *synchronous* communication, relying on a common clock. Another distinction is to be made between implementations of communication systems that rely on timing constraints/guarantees (*implicit timing*) and those that do not.

In asynchronous communication the synchronization of sender and receiver is carried out through event signaling: one or more communication partners signal their being ready for communication by asserting a synchronization signal. If only one partner sends a synchronization signal then local timing constraints must guarantee that the other is ready to communicate when the synchronization signal comes. The message is then transferred either through implicit timing, meaning that that the communication partners comply to timing constraints such as latency periods or setup/hold times. Or, if no implicit timing information is used, proper transmission is signaled through a handshake.

In synchronous communication the situation is similar, however, all communication partners rely on a common clock. Before a message is sent the communication partners synchronize by asserting synchronization signals (*explicit synchronization*). If only one communication partner sends such a signal (*unilateral synchronization*) then it must be guaranteed (through local timing constraints) that the other partner(s) are ready to receive the signal and the message. Because of the common clock, in synchronous communication the actual data exchange may comprise several steps (such as the beats in a burst operation on a bus). Proper reception of the data may be signaled explicitly (e.g., to accomodate for access latencies) or it may not need to be signaled because of the implicit synchronization through the clock.

S3. Modeling the Communication Schemes

Fig. 6 shows an example of a four-cycle handshake between two Moore machines M_1 and M_2. The handshake is carried out using signals *s* and *r*. The signal *s*, produced by M_1, is asserted only

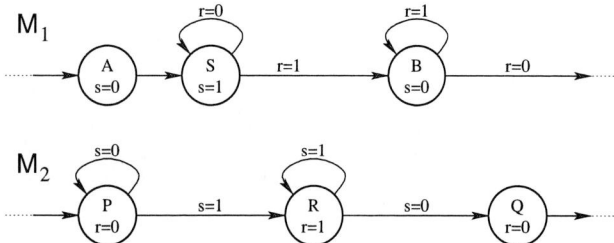

Figure 6: Asynchronous communication based on four-cycle handshake

in state S and de-asserted in all other states. Likewise, signal r, produced by M_2, is asserted in no state other than R.

Before data can actually be transferred both machines need to synchronize. Assume that M_2 is waiting in P. When M_1 moves from A to S it sends a synchronization signal $s = 1$, possibly together with some data. M_2 is triggered by this signal and moves into R. Because there are no timing guarantees machine M_1 needs to wait in its sending state S until M_2 has actually received the message, moved into state R and acknowledged back to M_1 by sending the signal $r = 1$, again possibly together with some data. Machine M_1 then de-asserts s and waits for M_2 to de-assert r as well. Note that M_1 needs to wait for M_2 in state B, otherwise a new message sent during some state sequence $(B, \ldots, A, S, B)$ could go unrecognized if machine M_2 remains in state R during that time. The four-cycle handshake built with the signals s and r ensures certain reachability constraints according to the four cycles: state P is not left unless state S is taken, S is not left while in P, R is not left while in S and B is not left while in R.

In order to obtain a composed model of path-predicate-abstracted state machines, each of the communication schemes discussed in Supplement S2 needs to be mapped to this four-cycle handshake. This is trivial in the case of asynchronous communication without local timing guarantees. In this case a four-cycle handshake communication is present in the concrete implementation as well as in its path predicate abstraction. In the other schemes discussed above only parts of the four-cycle handshake are present in the abstraction. We then need to extend the abstract models of sender and receiver(s) with the missing elements.

As an example, consider the unilateral synchronous scheme: two machines M_1 and M_2 communicating in a synchronous system with M_1 sending the synchronization signal and M_2 receiving it. Fig. 7

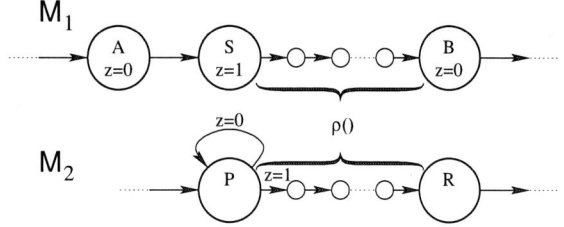

Figure 7: Synchronous communication with unilateral synchronization

shows parts of their state transition graphs. The system relies on an implicit timing guarantee stating that machine M_2 is always in state P whenever machine M_1 enters state S. (Such implicit timing guarantees result from the communication protocol and can usually be identified easily.) Machine M_1 sends the synchronization signal, $z = 1$, in state S to indicate that a communication operation begins. Machine M_2 waits in P until the synchronization signal triggers a communication operation. The operation lasts for several clock

cycles in which data can be exchanged between the machines. During this operation, both machines remain in synchrony due to the common clock while they each traverse the non-important states of the operation (indicated by the smaller circles). A message predicate $\rho()$ is used to characterize the I/O sequences exchanged in the operation.

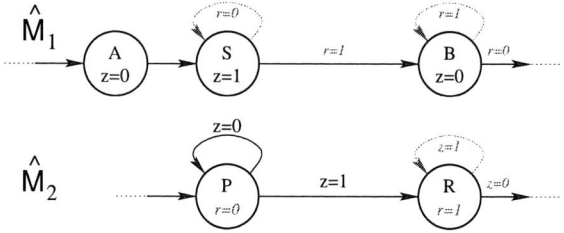

Figure 8: Path predicate abstractions of M_1 and M_2 with extensions

The path predicate abstractions of M_1 and M_2 as described in Sec. 2 are given by the state transition graphs in Fig. 8, however without the signal r and the dotted arcs. The abstract states S and P correspond to the important starting states of the communication operation between M_1 and M_2, the abstract states B and R mark its end. Comparing this with the four-cycle handshake of Fig. 6 we see that states with the same names correspond to each other. The synchronization signal z serves as one of the handshake signals, namely s in Fig. 6. However, for a full four-cycle handshake, the dotted/italic elements of Fig. 8 need to be added so that we soundly model the communication in an abstract system that does not rely on timing guarantees. In this example, an abstract handshake signal r needs to be introduced that is asserted only in state R. Self loops and guard conditions are added to the state transition graphs of the abstract machines $\hat{M}_1$ and $\hat{M}_2$ as shown in Fig. 8.

When are we allowed to add these elements to the state transition graphs of the path-predicate-abstracted models? As stated above, the elements (states, transitions, guard conditions) of a four-cycle handshake produce a behavior with certain reachability constraints on the product states of the composed abstract system. We may extend path predicate abstractions to full four-cycle handshake communication if and only if the corresponding concrete system has the same reachability constraints on the involved important states as the extended abstract system. This must be shown for all communication schemes considered in our methodology.

Let us begin with the case of unilateral synchronization as shown in the above example. Referring to Fig. 6, the first reachability constraint requires that state S must not be entered before P is entered, (i.e., M_2 is always ready for M_1 in P). This reachability constraint must be guaranteed by the implicit timing constraints used in the implementation. (Note that the soundness of our model relies on the validity of this constraint; see discussion below.) The other three reachability constraints (S not left while in P, R not left while in S and B not left while in R) are fulfilled through the synchronous communication operation following state S. They are verified by the fact that the machines transition synchronously throughout the communication operation and that this operation is unambiguously described by the message predicate $\rho()$. This same predicate is used in the formal property proofs of the communication operations in both, the sending and the receiving machine.

The discussion of the remaining communication schemes is analogous. For the case of synchronous communication with bilateral synchronization both signals, s and r, exist in the concrete implementation and therefore also in the path predicate abstractions, as do the self-loops in state S and P. The extension towards a full four-cycle handshake requires only the self-loops in the communication ending states, B and R. This is justfied in the same way as for the unilateral synchronous case.

For the case of asynchronous communication we identify two cases: bilateral synchronization and unilateral synchronization with an implicit timing guarantee. The first case yields a four-cycle handshake on the concrete as well as on the abstract level, as mentioned before, and needs no extension. The second case is similar to the synchronous unilateral scheme in the following respect: Instead of having a feedback signal r from machine M_2 to M_1 we have implicit timing constraints (enforced, e.g., through timer circuits or counters) that enable state transitions in machine M_1 only if M_2 is guaranteed to have moved into the corresponding communication states. In other words, the timing constraints enforce the reachability constraints of the four-cycle handshake abstraction.

The soundness of these models for the considered communication schemes can also be established by the following argument. If the reachability constraints are fulfilled by the implementation then the following construction yields a concrete system which is functionally equivalent with the original design and has a path predicate abstraction with a four-cycle handshake: Assume we extend the original concrete implementation of M_2 with an additional output signal r that is evaluated by M_1 as in Fig. 8. Obviously, the operations corresponding to the dotted arcs are never triggered if the implementation fulfills the set of reachability constraints. Therefore, the extended implementation which has a full four-cycle handshake communication abstraction is functionally equivalent to the original implementation.

In all communication schemes, the extended path predicate abstractions composed in an asynchronous system with communication through four-cycle handshakes soundly models the concrete system. In those cases where implicit timing constraints are used in the implementation the soundness of our model relies on the validity of the timing constraints. In practice, however, these timing constraints are often obvious by inspection because the respective state machines have only very few states and may be always in a state ready for communication.

S4. Example of Asynchronous Composition

Fig. 9 shows the asynchronous composition of the machines in Fig. 6. As can bee seen from the state transition graph the four-

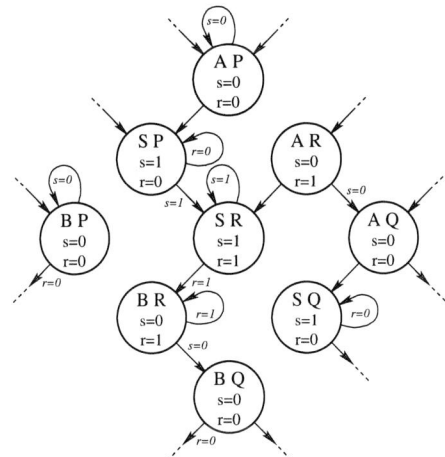

Figure 9: Asynchronous composition of machines in Fig. 6

cycle handshake between the two machines is capable of modeling a synchronous communication as, for example, in Fig. 8: The starting states of the communication operation are S and P, the ending states are B and Q. In a synchronous communication the product state BQ is reached always some time after product state SP. This is reflected in the asynchronous composition of Fig. 9: All fair paths

leaving SP always reach BR. (Fairness forbids infinite cycling in SP, SR or BR.)

S5. Proof of Theorem 1

In order to ensure soundness of the proposed abstraction it must be possible to translate the abstract LTL formulas to concrete formulas such that these characterize concrete paths that are represented by the abstract paths characterized by the abstract formulas. LTL formulas in the concrete model can therefore only refer to important states.

As a result of time abstraction, product states in the abstract model cannot always be represented as products of important states in the concrete model. (In the specific concrete timing of the implementation the considered important states may not occur simultaneously. A certain important state of one module may only be combined with a certain unimportant state of another module.) Therefore, we must exclude Boolean sub-formulas relating to the atomic propositions of more than one module. For similar reasons, due to time abstraction, the X operator is not allowed. This is why the LTL formulas allowed for the abstract model are restricted as given by Def. 13.

The soundness of the asynchronous composition with respect to the LTL formulas as by Def. 13 relies on the fact that every path in the concrete system is represented by one or several paths in the abstract system. This means that if there is a counterexample for a property in the concrete model then there also exists a counterexample on the abstract model.

We state the proof for the communication scheme of synchronous communication on the concrete level. For the other communication schemes the proof is similar.

Consider an arbitrary (possibly infinite) path π from the initial state in the concrete system. In every product state on the path an arbitrary number of machines may be in a respective important state. In a synchronous communication operation, the sender and the receiver begin and end the communication in synchronized start and end states, i.e., at product states that are important in both, sender and receiver. We now split up the path π into fragments between such synchronized states. A communication fragment is one that begins at a communication start state and that ends at a communication end state. Obviously, this fragment has a corresponding abstract path fragment, because the start and end states, by construction (cf. Sec. 2 and Supplement S3), have corresponding product states in the asynchronous composition.

A non-communication fragment is a path fragment that begins at the end of some communication and that ends at the beginning of the next communication. In between these states the machines do not communicate and the specific product states occurring on the path fragment are a result of the specific speed at which each module actually runs in the implementation. The asynchronous model abstracts from concrete timing and represents all possible interleavings of operations in the two modules. In every individual module, an operation on the concrete level always corresponds to an abstract transition, as proved in [1]. Hence, for such a concrete path fragment at least one abstract path fragment exists.

Each path in the concrete and in the abstract machine consists of numerous path fragments and represents a sequence of communications, in the concrete and in the abstract model, respectively. As a result of the asynchronous composition, the sequence of communications contained in the abstract model is a superset of the sequence of communications in the concrete model. Therefore, every path in the concrete system beginning at the initial state has a representation in the abstract system. If a concrete LTL formula does not hold on a specific path leaving the initial state in the concrete model then there is an abstract path leaving the initial state where the abstract LTL formula is violated, also.

343

Equivalence Checking for Behaviorally Synthesized Pipelines

Kecheng Hao
Dept. of Computer Science
Portland State University
kecheng@cs.pdx.edu

Sandip Ray
Dept. of Computer Sciences
University of Texas at Austin
sandip@cs.utexas.edu

Fei Xie
Dept. of Computer Science
Portland State University
xie@cs.pdx.edu

ABSTRACT

Loop pipelining is a critical transformation in behavioral synthesis. It is crucial to producing hardware designs with acceptable latency and throughput. However, it is a complex transformation involving aggressive scheduling strategies for high throughput and careful control generation to eliminate hazards. We present an equivalence checking approach for certifying synthesized hardware designs in the presence of pipelining transformations. Our approach works by (1) constructing a provably correct pipeline reference model from sequential specification, and (2) applying sequential equivalence checking between this reference model and synthesized RTL. We demonstrate the scalability of our approach on several synthesized designs from a commercial synthesis tool.

Categories and Subject Descriptors

B.6.3 [**Logic Design**]: Design Aids—*automatic synthesis, optimization, verification*

General Terms

Algorithms, Performance, Reliability, Verification

Keywords

Equivalence checking, behavioral synthesis, pipeline

1. INTRODUCTION

As hardware complexity increases, it is getting increasingly difficult to develop a high-quality hardware system via hand-crafted RTL. Electronic System Level (ESL) designs provide a promising solution to this problem, by facilitating more abstract design description (*e.g.*, with SystemC). The adoption of this approach, however, is crucially dependent on the correctness of *behavioral synthesis* [13, 16, 11, 5, 3], *viz.*, the compilation of an ESL description to RTL.

A critical transformation in behavioral synthesis is loop pipelining, producing temporal overlap of successive loop iterations. It is available in most state-of-the-art tools (*e.g.*,

AutoESL [17], CatapultC [12], and Cynthesizer [4]) and is crucial to the synthesis of high-throughput hardware. However, it induces retiming and out-of-order executions; furthermore, the mapping of internal operations is lost between the sequential description and the pipelined RTL. This rules out standard sequential equivalence checking (SEC) techniques for their comparison. In particular, some key optimizations (*e.g.*, cutpoints) become inapplicable.

We present an SEC framework for certifying synthesized designs with pipelined loops. We have applied our tool on industrial-size designs with thousands of lines of RTL, synthesized by AutoESL. This scalability is derived from tight integration with the synthesis flow. Instead of directly comparing the synthesized RTL with the sequential description, we develop an intermediate *pipeline reference model*. This model provably preserves the semantics of the sequential description. However, our model generation algorithm is parameterized by pipeline parameters, whose values are obtained from the synthesis tool; this ensures that the structure of the generated model is similar to that of the synthesized RTL, and enables internal operation mapping between the reference model and the RTL.

The rest of this paper is organized as follows. Section 2 provides relevant background. Sections 3 and 4 present our approach. We present experimental results in Section 5. We discuss related work in Section 6 and conclude in Section 7.

2. BACKGROUND

2.1 Behavioral Synthesis

A behavioral synthesis tool applies a sequence of transformations to an ESL specification to transform it into RTL. Figure 1(a) illustrates an ESL description with a simple loop structure and Figure 1(b) shows the schematics of the pipelined RTL design synthesized by AutoESL. The following transformation phases are involved in this synthesis.

- First, *compiler transformations* are applied to the ESL description. For instance, constant propagation is used in the example to remove unnecessary variables.

- The second phase is *scheduling*, which optimizes the clock cycle for each operation. In this phase, operations are chained across conditional blocks and decomposed into smaller multi-cycle sub-operations. Loop pipelining is employed as part of this phase.

- The third phase is *resource binding and control synthesis*. This phase binds operations to hardware units,

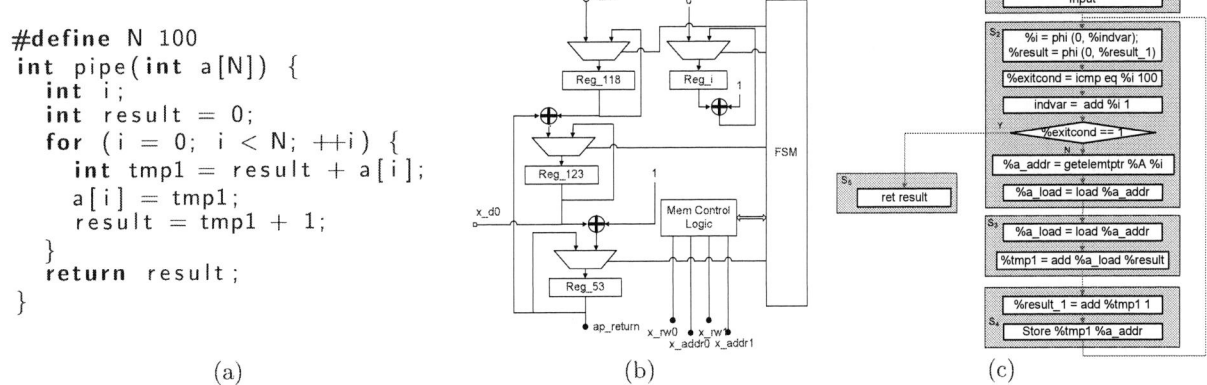

```
#define N 100
int pipe(int a[N]) {
    int i;
    int result = 0;
    for (i = 0; i < N; ++i) {
        int tmp1 = result + a[i];
        a[i] = tmp1;
        result = tmp1 + 1;
    }
    return result;
}
```

(a)   (b)   (c)

Figure 1: (a) C Code with Loop Design. (b) Schema of RTL Synthesized by AutoESL. (c) CCDFG

and allocates registers. For instance, the "+" operation is bound to a hardware adder. Furthermore, a control circuit is generated (typically as a finite-state machine module) to implement the schedule.

After the transformations, the design is expressed as RTL.

2.2 A Certification Framework

In [14], we proposed a verified/verifying framework for certifying behaviorally synthesized RTL. The key idea was to compare the RTL with the design representation after the high-level (compiler and scheduling) transformations have been applied to the ESL description. To achieve this, the framework introduced CCDFG, a formalization of design specification that augments the traditional Control/Data Flow Graph (CDFG) with a schedule. High-level transformations can be certified offline (by theorem proving) to preserve CCDFG semantics. The transformed CCDFG is compared with RTL through SEC via dual-rail symbolic simulation. In [6], we also developed three optimizations to optimize SEC performance. The resulting framework was scalable to industrial-size designs.

However, this previous work ignored pipelining. In particular, the optimizations were critically dependent on ready availability of mapping information for internal operations between the CCDFG and the RTL, which was determined from the resource bindings performed by the synthesis tool; pipelining destroys this ready correspondence, making direct mapping inapplicable.

2.3 CCDFG Formalization

Figure 1(c) illustrates the CCDFG corresponding to the C code shown in Figure 1(a). Details of the formal semantics of CCDFG are presented in previous paper [14]. Formally, a CCDFG $G \triangleq \langle G_{CD}, M, T \rangle$, where G_{CD} is the control/data flow graph, M is a *microstep partition*, and T is a schedule. The formalization assumes that the underlying language provides the semantics for *primitive operations* (*e.g.*, arithmetic operations, comparison, etc.). The operations are partitioned into *microsteps* that stipulate the operations that can be executed concurrently. Finally, microsteps are grouped into a schedule which specifies the microsteps that are completed within a single clock cycle. Following standard conventions, the control flow is broken up into basic blocks; data dependencies follow the "read after write"

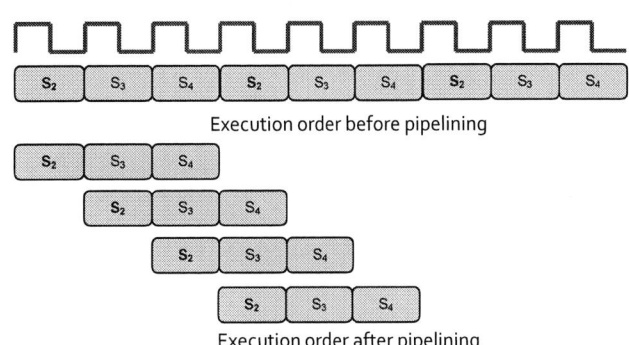

Execution order before pipelining

Execution order after pipelining

Figure 2: Execution Orders Before and After Pipelining. The rounded boxes S_2, S_3, and S_4 are scheduling steps in the sequential design.

paradigm: op_j is dependent on op_i if op_j occurs after op_i in a control path and computes an expression over some variable v that is assigned most recently by op_i in the path. A CCDFG execution is formalized by a state-based semantics. A *CCDFG state* (resp., *CCDFG input*) is a valuation of the state (resp., input) variables. Given a sequence of inputs, an *execution* of a CCDFG G with microstep partition M and schedule T is a sequence of CCDFG states that corresponds to an evaluation of the microsteps of M respecting T.

3. CHALLENGES WITH LOOP PIPELINES

Loop pipelining allows multiple successive iterations of a loop to operate in parallel by executing a new iteration before the previous iteration completes. Consider pipelining the loop in Figure 1(a). Figure 2 shows the execution orders of the scheduling steps in the loop body before and after pipelining. In the sequential design, execution of iteration i involves reading the value of $a[i]$ from the memory in S_2, adding i and $a[i]$ in S_3, and storing new value to the memory and computation of $result$ in S_4. However, with pipelining, iteration $i + 1$ is initiated before iteration i completes.

The result of overlapping executions is a significant difference in the schedule of operations between the CCDFG of the sequential design and the RTL generated from the pipeline. Each scheduling step of the pipeline is composed of

345

a number of scheduling steps of the sequential design; there is no longer a direct operation mapping between the CCDFG and RTL. Furthermore, due to the difference in the execution order of the scheduling steps, the controlling finite-state machines are also different. A direct SEC between the two reduces to comparison of their input-output relations, which is prohibitively expensive for loops with many iterations.

4. SEC WITH REFERENCE MODEL

Our solution to the above problem is to develop a *reference pipelining transformation* on CCDFGs. Given a CCDFG G and certain pipeline parameters (see below), we generate a new CCDFG G' by pipelining the loops. Note that our transformation is different from that used by the synthesis tool to generate the pipelined RTL. The synthesis tool transformation includes algorithms and heuristics to *determine* how many iterations to pipeline, when to introduce stalls and bubbles, etc.; on the other hand, our algorithm merely *takes* such information as parameters to create G'. In fact, we obtain this information from the synthesis tool itself. Thus the output CCDFG G', if successfully generated by our algorithm,[1] is guaranteed to have close structural correspondence with the synthesized RTL. On the other hand, irrespective of the actual value of these parameters, G' is guaranteed to be semantically equivalent to G and can therefore be soundly used instead of G for SEC.

The following definition characterizes the loops handled by the algorithm.

REMARK 1 (CONVENTIONS). *For a given CCDFG $G \triangleq \langle G_{CD}, M, T \rangle$ and a set $t \in T$, we use the term "projection of G on t" to mean the CCDFG $G_t \triangleq \langle G'_{CD}, M', \{t\} \rangle$ where G'_{CD} and M' contain only the operations in G_{CD} and M respectively, that are members of t. For a set $T_0 \subseteq T$, we use "projection of G on T_0" to denote the following graph G'. The nodes of G' are given by the set $\mathcal{N} \triangleq \{G_t : t \in T_0\}$; given $g_0, g_1 \in \mathcal{N}$, there is an edge from g_0 to g_1 if there are operations o_1 and o_2 such that $o_1 \in g_0$, $o_2 \in g_1$ and there is an edge from o_1 to o_2 in G_{CD}.*

DEFINITION 1 (PIPELINABLE LOOP). *For a CCDFG $G \triangleq \langle G_{CD}, M, T \rangle$ and for $T_0 \subseteq T$, we say that T_0 induces a "pipelinable loop" if (1) the projection of G on T_0 is a cycle C, and (2) in the projection of G on T there is a unique node (called the "entry node") in C with a predecessor outside C and a unique node (called the "exit node") in C with a successor outside C.*

REMARK 2. *The notion of pipelinable loops is more restrictive than the common loop definition in programming languages. In particular, a pipelinable loop has a single exit and loop nesting is disallowed. Our definition is based on the kind of loops that can be pipelined during behavioral synthesis. For instance, if a design contains nested loops, then the inner loop can be unrolled completely (possibly by compiler transformations) before the outer loop can be pipelined.*

---

[1] Our algorithm does not use semantic invariants of the program being transformed. Thus we may fail to pipeline a loop for a given number of iterations (and report spurious hazard) when in fact such a pipeline is hazard-free. However, in practice we have not seen a case where the synthesis tool generates a pipeline with specific parameters and our algorithm reports a spurious hazard on those parameters.

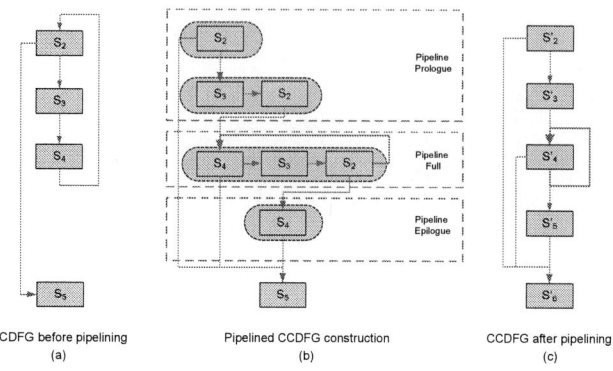

Figure 3: Input and Output CCDFGs of Loop Pipelining Transformation

REMARK 3. *Since a schedule is a partition of microsteps, T_0 induces a partition of G_{CD} such that if $t_0 \neq t_1$ the partition induced by t_0 is disjoint from that induced by t_1. Given a set T of scheduling steps, one can describe the CCDFG $G \triangleq \langle G_{CD}, M, T \rangle$ uniquely as the triple $\langle S, E, M \rangle$ where S and E denote the nodes and edges of the projection of G on T, and M is the set of microstep partitions refined by T. We use this view in the rest of the paper for pipelinable loops.*

Given CCDFG G, our reference transformation replaces each loop L in G with the pipelined refinement of L as described in Algorithm 1. Here I is iteration interval, which indicates how many clock cycles later a new iteration is to be "fed" into the pipeline, and N is the number of scheduling steps in L. Values of these parameters are readily available from AutoESL.

Algorithm 1 PIPELINELOOP($L = \langle S, E, M \rangle, I, N$)

1: $S'_1 \leftarrow GenerateSchedulingSteps(S, I, N)$
2: $\langle S'_2, M'_1 \rangle \leftarrow GeneratePipelineRegs(S'_1, M, E, I)$
3: $E'_1 \leftarrow GenerateEdges(S'_2, E, I, N)$
4: $\langle S'_3, M'_2 \rangle \leftarrow GenerateForwarding(S'_2, M'_1, E'_1, I)$
5: **return** $\langle S'_3, E'_1, M'_2 \rangle$

Figure 3 illustrates the use of the algorithm on our simple example. We now discuss the different steps of the algorithm in greater detail.

Algorithm 2 GenerateSchedulingSteps (S, I, N)

1: $S_G \leftarrow \emptyset$;
2: $iter \leftarrow 0$ /*loop iteration*/
3: **while** $iter * I < N$ **do**
4: $\quad S_G \leftarrow mergeIteration(S_G, S, I, iter)$
5: $\quad iter \leftarrow iter + 1$
6: **end while**
7:
8: /*build new edges within one single scheduling step */
9: **for** each step s' in S_G **do**
10: $\quad$ **for** each step pair $(s'[pos], s'[pos + 1])$ in s' **do**
11: $\quad\quad e' \leftarrow buildEdge(s'[pos], s'[pos + 1])$
12: $\quad\quad s' \leftarrow append(s', e')$
13: $\quad$ **end for**
14: **end for**
15: **return** S_G

346

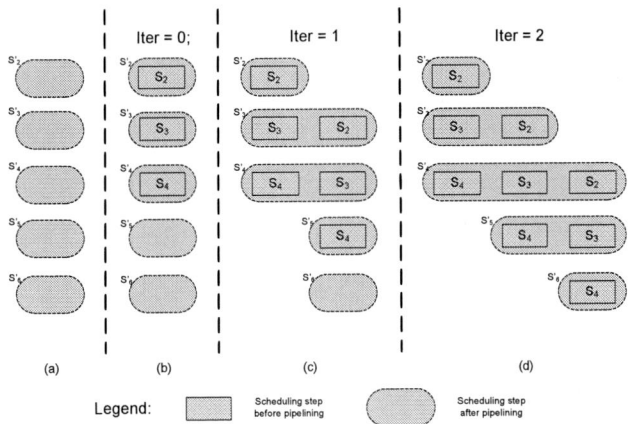

Figure 4: Construction of Scheduling Steps

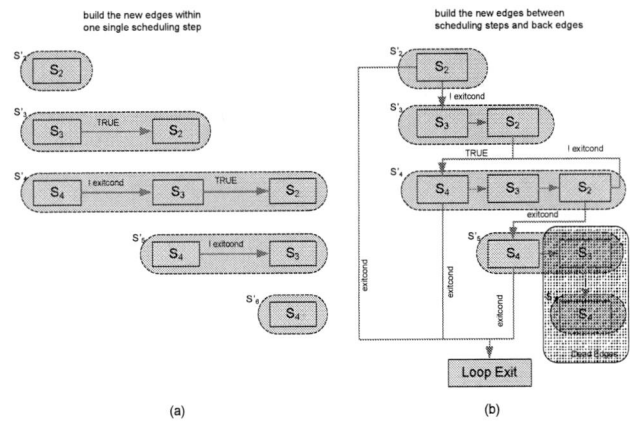

Figure 5: Construction of Edges

Algorithm 2 describes the construction of scheduling steps of the pipelined CCDFG. The algorithm simulates the process of "feeding" a new loop iteration into the pipeline until the pipeline is full. Consider the sequence of iterations shown in Figure 4. The output is an array (initially empty) of graphs. Each graph represents the projection of the reference pipeline CCDFG at a single scheduling step. We first build the nodes of each graph in the array (Lines 3-6); we then compute the edges within each graph (Lines 8-14). The set of nodes of each graph in S_G is determined by I and N. The algorithm updates S_G for every iteration. If the pipeline is not yet full, *i.e.*, can accept a new iteration but no iteration is completed yet (Line 3), then a new iteration is introduced and merged with the existing iterations in the pipeline by subroutine *mergeIteration*. Subroutine *mergeIteration* merges each scheduling step in the new iteration with the corresponding steps already in pipeline, returns new scheduling steps as shown in Figure 4(b), (c), (d). To model the exit, the pipeline enters the "flushing" stage in which iterations are completed without new iteration being introduced. The pipeline full stage corresponds to the new loop body for the pipelined CCDFG while the prologue and epilogue correspond to the entry and exit.

We now build the edges for each graph in S_G. The goal is to ensure that the new control flow respects that of the input loop. The process is demonstrated in Figure 5 (a). Recall that a scheduling step of the pipeline involves a number of scheduling steps of the original CCDFG (across several iterations). To ensure that the original control flow is respected, a scheduling step s' of the pipeline is executed following the iteration order. This is achieved by adding edges enforcing the evaluation of microsteps from left to right. For instance, in S_4' shown in Figure 5 (a), an edge is created to connect S_4 and S_3. Since S_4 is from an earlier iteration, the direction is from S_4 to S_3. The edge condition $!exitcond$ states that loop does not exit. If the loop exits at iteration i, all iterations from $(i + 1)$ must be skipped: this is ensured by inserting the exit condition on all such edges. Subroutine *buildEdge* creates the correct edge condition according to the control flow.

Algorithm 3 inserts "pipeline registers" between iterations to facilitate correct data flow and prevent variables from being overwritten before being consumed. In a CCDFG, the

effect of pipeline registers is mimicked using temporary variables as follows. We first compute all program variables that may be overwritten before being consumed(Line 2); this constitutes the variables that potentially require pipeline registers. To find such variables, we compare the distance between the producer ms_p and the last consumer ms_c; if the distance is greater than I, v is assigned the new data value of the next iteration before current iteration's value has been fully consumed; this warrants insertion of pipeline variables in every scheduling step between ms_p and ms_c. The value is propagated every clock cycle following the CCDFG data flow. In Figure 6, variable $\%a\_addr$ is computed in S_2 and the last use scheduling step is S_4. The distance is greater than $I = 1$, therefore, temporary variables a_addr_pipe1 and a_addr_pipe2 are inserted. Subroutine $addPipelineReg$ generates new microsteps for assignments of the pipeline variables, create new edges to integrate these microsteps into the data path, and updates the schedule.

Algorithm 3 GeneratePipelineRegs (S, M, E, I)

1: $S' \leftarrow S; M' \leftarrow M$
2: $V_{pr} \leftarrow getPipelineRegisterVars(S, M, E, I)$
3: **for** each variable v in V_{pr} **do**
4:    $ms_p \leftarrow getProducer(v)$
5:    $ms_c \leftarrow getLastComsumer(v)$
6:    $\langle S', M' \rangle \leftarrow addPipelineReg(S', M', E, ms_p, ms_c)$
7: **end for**
8: **return** $\langle S', M' \rangle$

Algorithm 4 shows the construction of edges governing the control flow of the pipelined CCDFG. Figure 5 (b) shows how to build edges between new scheduling steps (Lines 3-6). One example is the edge from S_2 in S_2' to S_4 in S_3'. Because the pipeline is still in prologue stage, the edge condition is that loop does not exit.

The back edge of the new loop connects the last scheduling step of the pipeline full stage to the first one. $S'[N - 1]$ is the last one and $S'[N - I]$ is the first step in the pipeline full stage. Finally, for an unbounded loop, exit can occur in any iteration. Thus, we must allow the pipeline to start flushing in any iteration, even when the pipeline is not full (Lines 12-17). In the example shown in Figure 5(b), the exit point of the loop is in S_2, therefore in pipeline epilogue, the edge from S_4 to S_3 will never be valid. This is because the loop would have already exited and the S_3 and S_4 of the new iteration will not execute. The dead edges will be removed to simplify the final CCDFG.

Algorithm 4 GenerateEdges (S, E, I, N)

1: $E' \leftarrow \emptyset$
2: /*build the edges between new scheduling steps*/
3: **for** each step pair $(S[i], S[i+1])$ in S **do**
4:    $e' \leftarrow buildEdge(S[i], S[i+1])$
5:    $E' \leftarrow append(E', e')$
6: **end for**
7: /*build the back edge*/
8: $s_{src} \leftarrow S[N-1]$; $s_{dst} \leftarrow S[N-I]$
9: $e_{backedge} \leftarrow buildEdge(s_{src}, s_{dst})$
10: $E' \leftarrow append(E', e_{backedge})$
11: /*build the early exit edge*/
12: $i \leftarrow N-1$
13: **while** $i < sizeof(S) - 1$ **do**
14:    $e' \leftarrow buildEdge(S[i], s_{loopexit})$
15:    $E' \leftarrow append(E', e')$
16:    $i \leftarrow i + I$
17: **end while**
18: **return** $\langle E' \rangle$

Algorithm 5 GenerateForwarding (S, M, E, I)

1: /*find all loop carried dependencies*/
2: $D_{lc} \leftarrow getLoopCarriedDependencies(S, M, E)$
3: $S' \leftarrow S$; $M' \leftarrow M$
4: **for** each pair (o_w, o_r) in D_{lc} **do**
5:    **if** $checkForwarding(o_r, I, S')$ **then**
6:       $\langle S', M' \rangle \leftarrow moveOp(o_w, o_r, S', E, M')$
7:    **else**
8:       **return** $ERROR$
9:    **end if**
10: **end for**
11: **return** $\langle S', M' \rangle$

A critical puzzle is computation of data forwarding paths along pipeline iterations (Algorithm 5). Data forwarding is critical to achieving aggressive pipelining and eliminate the data hazards. The first key observation is that forwarding is only necessary for loop carried dependencies, which extend back to the previous iteration. D_{lc} denotes a list of dependencies and Subroutine $getLoopCarriedDependencies$ finds all loop carried dependencies. Each dependency is pair of operations (o_w, o_r), o_w is the last write operation in the loop body and o_r is the first read operation. Subroutine $checkForwarding$ checks if the data forwarding is possible (*i.e.*, whether the value is computed before use) for these variables in the scheduling steps of the pipeline. We then implement forwarding using so-called "Φ nodes". Φ nodes are special operators in compiler transformations and are widely used in resolving conditional branches in a number of compilers, and used to postpone computation of control flow to run time. In particular, a Φ node is introduced in a basic block which has multiple predecessors; the values of variables in a Φ node for a specific execution are given by the specific block which actually precedes the node in that execution. To understand its utility for data forwarding,

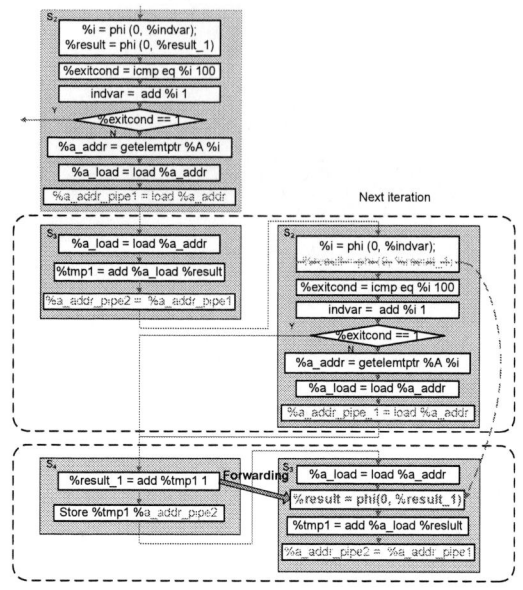

Figure 6: Pipeline Registers and Forwarding

consider Figure 6. In the non-pipelined design Φ operators can occur only in scheduling step S_2. The valid value of variable *%result* is computed by the Φ node in scheduling step S_2. Since we desire to execute scheduling steps S_2 and S_3 within a single scheduling step, we move the Φ from S_2 to S_3 and forward the value directly from the producer to the consumer. In general, to implement pipeline forwarding, we need to relocate the position of the Φ operator for a variable to immediately before its first consumer, also update the assignment of Φ node according to the new control flow. The "move" is implemented in $moveOp$, which will generate new scheduling S' and new microstep partition M'.

5. EXPERIMENTAL RESULTS

We implemented the loop pipelining algorithm on top of our verified/verifying certification framework for behavioral synthesis [14]. SEC is implemented by cycle-by-cycle dual-rail, word-level symbolic simulation between CCDFG and RTL that utilizes CVC3 SMT engine, as well as several optimizations including cutpoints, cut-loop, and modular analysis [6]. We ran our tool on a collection of pipelined designs synthesized by AutoESL. All experiments were conducted on a workstation with 3GHz Intel Xeon processor with 2GB memory.

Table 1 illustrates the results. Our framework could successfully handle SEC for synthesized designs with pipelined loops involving several thousand lines of RTL within reasonable time and memory bounds. Note that this success on pipelines depends on the applicability of other optimizations during SEC. The reason is that because of the presence of non-trivial loops, SEC without cut-loop optimization requires expensive fixed-point computation which runs out of memory and time. For all designs, brute-force SEC between the unpipelined CCDFG and the RTL times out. SEC between the pipelined CCDFG and the RTL can mostly finish. With the optimizations applied, SEC finishes with reduced memory and time usages. The results thus support

Table 1: Loop Pipelining Experimental Results

Design	RTL #line	App. Domain	Loop Info.			Pipeline Info.		Without Opt.		With Opt.	
			Inter-val	Depth	Oper-ations	Forw-arding	Pipeline Register	Mem. (MB)	Time (Sec)	Mem. (MB)	Time (Sec)
MemoryOp	291	Memory operation	1	4	18	2	2	24	38	4	0.3
TEA	383	Cryptography	1	4	28	4	2	-	-	40	6.2
XTEA	483	Cryptography	1	3	37	4	1	-	-	52	7.8
CORDIC	485	Data processing	1	3	31	4	0	38	7.9	5	0.9
SmithWater	517	Data processing	2	3	73	3	0	-	-	134	50.2
FIR	610	Signal processing	3	5	27	3	1	763	127.4	63	10.8
YUVToRGB	756	Image processing	2	6	77	1	5	-	-	335	128.9
MotionComp	1248	Image processing	1	3	53	3	0	434	132.2	50	11.4
DES	3292	Cryptography	1	3	17	2	2	468	364.7	257	163.3

our preference to compare the RTL with a closely resembling pipelined CCDFG that facilitates the optimizations, rather than develop a specialized SEC algorithm for pipelines.

6. RELATED WORK

Koelbl *et al.* [8] provide a tutorial introduction on methods of comparing high-level designs with RTL. Chauhan *et al.* [2] propose a technique for SEC between non-cycle-accurate designs by constructing a pair of normalized cycle-accurate designs from the original designs. Kundu *et al.* [9] propose the use of bisimulation correspondence to validate designs generated by behavioral synthesis. However, neither approach provides pipelining-specific equivalence checking strategies that effectively integrate with behavioral synthesis flows.

There is a significant literature on verifying pipelined microprocessors [1, 7, 10, 15], which has parallels with our work. However, there has been very little published work on formal verification of pipelines generated by behavioral synthesis. Nevertheless, any viable SEC framework for behavioral synthesis (*e.g.*, Synopsys Hector tool) must handle loop pipelining. To our knowledge current implementations either involve cost-prohibitive input-output comparison or require the user to provide the requisite mappings.

7. CONCLUSIONS AND FUTURE WORK

We have presented an approach to equivalence checking of pipelined designs generated by behavioral synthesis. Its efficiency and scalability has been attested by application to industrial-size case studies. The key insight is that a parameterized, synthesis-guided reference transformation on CCDFG permits comparison with RTL even after mappings with the original sequential specification has been destroyed by an aggressive transformation such as pipelining. Furthermore, the approach permits smooth integration with pipeline-oblivious transformations such as cut-loop.

In future work, we plan to handle more industrial examples. We also plan to handle more diverse pipelines, including function pipelines and pipelines for nested loops.

Acknowledgment

This research was partially supported by National Science Foundation Grants #CCF-0916772 and #CCF-0917188 and by a research grant from Intel Corporation. We sincerely thank Disha Gandhi, Naren Narasimhan, Jin Yang, and Zhenkun Yang for their help.

8. REFERENCES

[1] J. R. Burch and D. L. Dill. Automatic verification of pipelined microprocessor control. In *Proc. of CAV*, 1994.

[2] P. Chauhan, D. Goyal, G. Hasteer, A. Mathur, and N. Sharma. Non-cycle-accurate sequential equivalence checking. In *Proc. of DAC*, 2009.

[3] J. Cong, Y. Fan, G. Han, W. Jiang, and Z. Zhang. Behavioral and Communication Co-Optimizations for Systems with Sequential Communication Media. In *Proc. of DAC*, 2006.

[4] Forte Design Systems. *Cynthesizer Manual*, 2011.

[5] D. Gajski, N. D. Dutt, A. Wu, and S. Lin. *High Level Synthesis: Introduction to Chip and System Design.* Kluwer Academic Publishers, 1993.

[6] K. Hao, F. Xie, S. Ray, and J. Yang. Optimizing equivalence checking for behavioral synthesis. In *Proc. of DATE*, 2010.

[7] R. B. Jones, D. L. Dill, and J. R. Burch. Efficient validity checking for processor verification. In *Proc. of ICCAD*, 1995.

[8] A. Koelbl, Y. Lu, and A. Mathur. Formal Equivalence Checking between System-level Models and RTL. In *Proc. of ICCAD*, 2005.

[9] S. Kundu, S. Lerner, and R. Gupta. Validating High-Level Synthesis. In *Proc. of CAV*, 2008.

[10] J. Levitt and K. Olukotun. A scalable formal verification methodology for pipelined microprocessors. In *Proc. of DAC*, 1996.

[11] Y.-L. Lin. Recent developments in high-level synthesis. *ACM Trans. Des. Autom. Electron. Syst.*, 2(1), 1997.

[12] Mentor Graphics. *Catapult C Reference Manual*, 2011.

[13] A. Pnueli, M. Siegel, and E. Singerman. *A Survey of High-Level Synthesis Systems.* Kluwer Academic Publishers, 1991.

[14] S. Ray, K. Hao, F. Xie, and J. Yang. Formal verification for high-assurance behavioral synthesis. In *Proc. of ATVA*, 2009.

[15] M. N. Velev and R. E. Bryant. Verification of pipelined microprocessors by correspondence checking in symbolic ternary simulation. In *Proc. of ACSD*, 1998.

[16] R. Walker and R. Camposano. *A Survey of High-Level Synthesis Systems.* Kluwer Academic PublishersBoston, MA, USA, 1991.

[17] Xilinx. *AutoESL Reference Manual*, 2011.

Proving Correctness of Regular Expression Accelerators

Mitra Purandare
IBM Research Zurich

Kubilay Atasu
IBM Research Zurich

Christoph Hagleitner
IBM Research Zurich

ABSTRACT

A popular technique in regular expression matching accelerators is to decompose a regular expression and communicate through instructions executed by a post-processor. We present a complete verification method that leverages the success of sequential equivalence checking (SEC) to proving correctness of the technique. The original regular expression and the system of decomposed regular expressions are modeled as net-lists and their equivalence is proved using SEC. SEC proves correct handling of 840 complex patterns from the Emerging Threats open rule set in 50 hours, eliminating altogether informal simulation and testing.

Categories and Subject Descriptors

B.6.3 [**Logic Design**]: Design aids—*Verification*

General Terms

Verification, Algorithms

Keywords

Regular Expression Accelerators, Formal Verification

1. INTRODUCTION

Network security and monitoring applications have become exceedingly important due to the ever increasing security threats. Packet content scanning, an essential part of such applications, is commonly used for various purposes, such as network intrusion detection, unwanted traffic control. Commercially popular intrusion detection systems such as Snort [3] are signature based and rely heavily on regular expressions to express complex attack patterns. Incoming packets (flows) are parsed and when a flow matches a certain signature, preventive/blocking action is taken against the incoming attack.

Regular expressions are typically implemented as deterministic finite automata (DFAs) or non-deterministic finite automata (NFAs). An incoming packet of characters is simulated on the DFA or NFA of the regular expression to detect an intrusion. Implementing a regular expression as a DFA is the preferred method as it yields higher scanning rates. Regular expressions can be compiled into DFAs using well-known techniques [7]. In addition, DFA architectures that can be programmed to realize one or more regular expressions by loading a set of state transition rules into standard off-chip or specialized on-chip memories are available [15, 8, 12, 13].

Implementing a regular expression as a DFA suffers from potential state space explosion, especially when a number of patterns are combined [12]. Hence, performance of the DFA-based approaches critically depends on the storage efficiency of the compiled patterns. As a result, a number of research efforts target compact encoding of DFAs using various transformations. A technique that is known to be very effective in compressing the state space involves decomposition of complex regular expressions into simpler ones communicating through bitwise instructions and counters [8, 12]. Such a scheme can be implemented using a post-processor that contains local storage [15, 12]. The resulting architecture can be classified as a Finite State Machine with Datapath (FSMD), which is a common architectural template in hardware design [6]. Figure 1 illustrates an FSMD model that combines multiple DFAs with a datapath that involves combinational logic and registers.

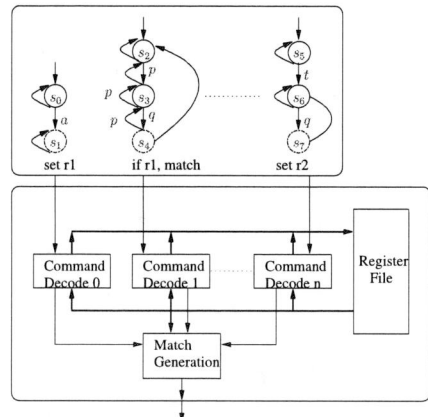

Figure 1: DFAs generate instructions to be executed by a post-processor involving combinational logic and registers.

For a complex set of regular expressions and a post-processor with complex instructions, correct use of instructions is quite involved. Correct functioning of the network security application necessitates that the original regular expression and the system of decomposed regular expressions detect exactly the same set of strings in incoming packets as intrusions. This entails that the transformed system generates matches for only the language of the original regular expression. Verifying that the set of transformations applied on the regular expressions preserve the language of the original set of regular expressions is a challenging task. Note that even if the transformation is mathematically proved to be correct, its implementation needs to be verified. Manual solutions are often incomplete and not scalable, e.g., applicable to a small and simple set of regular expressions only. Thus, automated verification techniques are needed. Automated simulation and testing is useful, however it is incomplete.

We present a complete verification method that uses sequential equivalence checking (SEC) for proving correctness of the transformations and use of post-processor instructions. Sequential equivalence checking (SEC) tools are used for determining functional equivalence of register transfer level (RTL) models at various levels of abstraction (e.g., post-synthesis, post-retiming) during hardware development. Current SEC tools benefit from modern fast SAT solver technology, logic rewriting techniques etc. reducing the verification overhead considerably. When two designs are proved not to be equivalent, most SEC tools generates error trace. The error trace demonstrates functional mismatch between the two designs. To the best of our knowledge, this work presents the first application of SEC technology to verification of pattern matching accelerators. When the original regular expression and the system of decomposed regular expressions are proved to be inequivalent, the error trace can be effectively utilized to understand and fix the cause of the mismatch, reducing the time spent on debugging.

The rest of the paper is organized as follows. In Section 2, we briefly explain the preliminaries of regular expressions and SEC. In Section 3, we provide examples of regular expression transformations that can be applied given a post-processor that supports some basic bitwise operations. In Section 4, we present some simulation and testing approaches and their shortcomings. We then show how the problem of verifying correctness of regular expression transformations can be solved by SEC. Details of implementation and its experimental evaluation are presented in Section 5. Section 6 presents our conclusion.

2. BACKGROUND

2.1 Moore Machine

A sequential circuit is a deterministic finite state machine (DFSM) for which our basic model is a Moore finite state machine with fixed initial states. A Moore machine M is a 6-tuple $(S, S_0, I, O, \delta, \lambda)$ where S is a finite set of states, $S_0 \subseteq S$ is the set of initial states, I is a finite set of input alphabet, O is a finite set of output alphabet, $\delta : S \times I \to S$ is a state transition function that maps a pair of state and an input alphabet to a corresponding next state, and $\lambda : S \to O$ is an output function mapping a state to the corresponding output alphabet.

2.2 Regular Expressions

A regular expression can be converted into a NFA which can in turn be converted into a DFA. Regular expression matching using DFAs is faster than using NFAs. An NFA needs needs $O(n)$ time to process a single character, whereas a DFA takes $O(1)$ time to process a single character. However, the exponential blow up in time and memory required during the NFA to DFA conversion is a bottleneck. A DFA is usually defined as a 5-tuple (S, S_0, I, δ, F) where S, S_0, I, and δ are as defined above and $F \subseteq S$ is a set of accepting states. Figure 2 shows a DFA for the regular expression $abc.^*def$. The accepting state is shown in dotted circle. A DFA can also be represented as a DFSM $(S, S_0, I, O, \delta, \lambda)$ with one output alphabet $accept$ and λ maps F to $accept$.

2.3 Sequential Equivalence Checking

Given two finite state machines with same input and output alphabet, equivalence checking addresses if the machines have identical output behavior in all executions and for all possible input sequences. Combinational equivalence checking (CEC) is applicable to designs in which a 1:1 state element pairing exists. Sequential equivalence checking (SEC) overcomes this limitation of CEC, i.e., a 1:1 latch mapping need not exist between the two designs, making SEC computationally expensive. Techniques like redundancy removal [10], binary decision diagram (BDD)-based and propositional satisfiability (SAT)-based reachability analysis [5, 9] are used to reduce the computational complexity of the equivalence check. Most sequential equivalence checking tools generate an error trace demonstrating a behavioral mismatch between two inequivalent designs. The error trace is useful in debugging and fixing the cause of the mismatch.

3. REGULAR EXPRESSION TRANSFORMATIONS

Various regular expression transformation are possible; we briefly explain one most commonly used one.

Let E_1 and E_2 be two regular expressions. Any regular expression of the form $E_1.^*E_2$ is split into E_1 and E_2. If E_1 matches in the input string, a bit r1 is set. If E_2 matches in the input string and r1 is set, a match is reported.

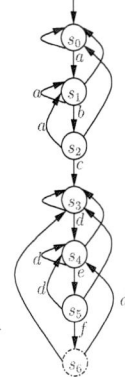

Figure 2: DFA of the regular expression abc.*def. State S_6 is the accepting state.

Consider a regular expression $abc.^*def$. The corresponding DFA is shown in Figure 2. The initial state of the DFA is labeled as S_0. The state transitions are shown using directed edges, and the set of input characters resulting in the transitions are given alongside each transition.

Transforming *abc.*def* as explained above, the two decomposed subexpressions are *abc* and *def*. A transformed DFA to match each of the subexpressions independently is built as shown in Figure 3. In the transformed DFA, whenever *abc* is matched in the input stream, a 1-bit register (r1) is set. When the input stream matches *def*, r1 is checked. If r1 is set, a match is reported. It is straightforward to see that only for the set of input strings accepted by the DFA in Figure 2, does the transformed system report a match. However, as the regular expressions and the instructions get more complex, manual verification is not possible.

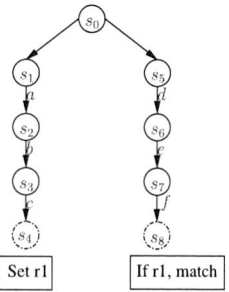

Figure 3: Transformed DFA making use of bitwise instructions. The back edges to states s_1, s_2, s_5, and s_6, called as default transitions are omitted as these are generally optimized for storage efficiency.

The above transformation does not guarantee correctness for any E_1 and E_2. Consider a regular expression *abc.*cde*. Following the above approach, *abc.*cde* is split into *abc* and *cde*. A 1-bit register r2 is set whenever *abc* matches. A match is reported whenever *cde* matches and r2 is set. This transformation results in generating matches for input strings that are not in the language of the original regular expression. The transformation results in a match for the string *abcde* whereas *abcde* does not belong to the language of the regular expression *abc.*cde*. In this case, it can still be possible to reduce the state space by applying further transformations on the DFA, and making use of more complex instructions.

The above example highlights the need to verify that the system of the decomposed subexpressions and post-processor generates matches only for the input strings in the language of the original regular expression. Any missed match or additional match is an error. Verifying this manually is impractical as the regular expressions can be very large. Automated verification methods are required to establish correctness of the transformations.

4. PROVING CORRECTNESS OF TRANSFORMATIONS

Proving correctness of the transformations is necessary. A transformation is correct if and only if the transformed system reports a match only for the set of input strings accepted by the DFA of the original regular expression. One possible but insufficient way to check correctness is via simulation and testing.

4.1 Simulation and Testing

In our preliminary verification effort, various ways of generating input strings for testing were tried.

- Given a DFA of the original regular expression, input strings can be generated that reach/cover accepting states only. This generates only positive test input strings, implying that negative test inputs like *abcde* for pattern *abc.*cde* will not be generated.

- Test input strings covering each state in the DFA can be generated. However, a state can be reached along multiple paths in the DFA, necessitating traversal of each path.

- Test input strings that simulate each path in the DFA can be generated. It is easy to see that the set of test strings grows very quickly, resulting in long simulation times.

- Focussing only on correct execution of bit-wise instructions, input strings triggering such instructions can be generated.

- The information about how a regular expression is transformed can be put to use instead of traversing any DFA. Consider a regular expression *abcp{2}def*. Assume that it is split into *abc* and *def*. The postprocessor takes care of counting number of *p*'s between *abc* and *def*. Some useful input strings that can be generated without traversing the DFA are *abcpdef* (negative test), *abcppdef* (positive test), and *abcpppdef* (negative test), covering positive and negative test input strings.

- Lastly there is also an option of generating random test input strings.

It is easy to see that the above approaches may result in large number of input strings, requiring extremely large simulation times. The next section explains how sequential equivalence checking can be used to prove correctness of the transformed DFAs.

4.2 Proving Correctness using SEC

SEC tools to verify equivalence of two designs are widely used in electronic design automation (EDA) industry. The novelty of our work lies in using SEC to prove equivalence of two regular expressions by synthesizing the regular expressions into net-lists that an SEC tool can handle.

Given a transformed DFA M^T, and a post-processor M^P, let M^C represent the FSMD that combines M^T and M^P as shown in Figure 1. Let M denote the reference DFA. We model M^C and M as Moore machines. We then show that M and M^C are output equivalent, i.e., for any possible sequences of inputs, the same sequences of outputs are produced by the machines.

4.2.1 Representation as DFSM

A character in the incoming packets, i.e., input to the DFA, typically consists of 8-bits (256 characters) [15]. If the reference DFA implements a single regular expression, a single output, say *accept*, is sufficient: the output function maps the accepting states F to *accept*. If a DFA implements multiple regular expressions, multiple output al-

352

phabets can be introduced, one per each DFA. [1] The reference DFA represented as a Moore machine is a 6-tuple $M = (S, I, O, s_0, \delta, \lambda)$, where

- S is a finite set of states;
- $I \in \{0,1\}^N$ is the set of inputs (N is the length of the input in bits);
- $O \in \{0,1\}^K$ is the set of outputs (K is the length of the output in bits);
- $s_0 \in S$ is the initial state;
- $\delta : S \times I \to S$ is the transition function;
- $\lambda : S \to O$ is the output function.

The output alphabet of the transformed DFA consists of the bit-wise instructions to the post-processor. The transformed DFA represented as a Moore machine is a 6-tuple $M^T = (S^T, I, P, s_0^T, \delta^T, \lambda^T)$, where

- S^T is a finite set of states;
- $I \in \{0,1\}^N$ is the set of inputs (N is the length of the input in bits);
- P is a finite set of instructions supported by the post-processor;
- $s_0^T \in S^T$ is the initial state;
- $\delta^T : S^T \times I \to S^T$ is the transition function;
- $\lambda^T : S^T \to P$ is the output function.

The input alphabet of the post-processor is the set of bit-wise instructions it supports. The post-processor is a 6-tuple $M^P = (R, P, O, r_0, \delta^R, \lambda^R)$, where

- R is a finite set of states;
- P is a finite set of instructions supported by the post-processor;
- $O \in \{0,1\}^K$ is the set of outputs (K is the length of the output in bits);
- $r_0 \in R$ is the initial state;
- $\delta^R : R \times P \to R$ is the transition function;
- $\lambda^R : R \to O$ is the output function.

The cross product of M^T and M^P is another Moore 6-tuple $M^C = (S^T \times R, I, O, (s_0^T, r_0), \delta^C, \lambda^C)$ representing the combined architecture M^C, where

- $S^C = S^T \times R$ is a finite set of states;
- $I \in \{0,1\}^N$ is the set of inputs (N is the length of the input in bits);
- $O \in \{0,1\}^K$ is the set of outputs (K is the length of the output in bits);
- (s_0^T, r_0) is the initial state;
- δ^C is the transition function such that $((s,r), i, (s', r')) \in \delta^C$ iff $(s, i, s') \in \delta^T$, $\lambda^T(s) = p$, and $(r, p, r') \in \delta^R$;
- $\lambda^C : S^T \times R \to O$ is the output function such that $((s,r), o) \in \lambda^C$ iff $(r, o) \in \lambda^R$.

---

[1]The length of the input can also be increased to support DFAs processing multiple characters per clock cycle.

Multiple DFAs.

For many regular expressions, combination of their DFAs results in state space explosion. A frequently used optimization is to identify such regular expressions and create multiple separate DFAs for them. This gives rise to multiple DFAs operating in parallel and sending instructions to the post-processor [15, 11]. Figure 1 shows one such architecture. The formalization presented above can be easily extended to support multiple DFAs ($M_1^T, M_2^T, \cdots$) sending instructions to a single M^P.

Next, we explain how SEC tools can be incorporated for proving correctness of the regular expression transformations.

4.2.2 Sequential Equivalence Checking

As shown in the preceding section, the original DFA and the system of transformed DFAs can be represented as Moore machines, which can be easily expressed in any hardware description language. These can further be converted into net-lists using state-of-the-art synthesis tools. Algorithm 1 depicts the usage of SEC in our verification tool chain. Steps 5 and 8 are performed by SEC tools. A useful feature of SEC tools is their ability to generate an error trace that exhibits the mismatch between two inequivalent designs. In our case, an error trace consists of sequence of input characters, i.e., an input string that is accepted by one design but not by the other. This input can be used to analyze and fix the cause of the mismatch.

Algorithm 1 Check soundness of RegX transformations

Input: A regular Expression E
Output: A Yes/No answer to whether E is transformed soundly and an error trace if transformation is unsound
1: Synthesize E generating M.
2: Transform E and synthesize it generating M^C.
3: Express M and M^C in a hardware description language, e.g., VHDL, called H and H^C respectively.
4: Synthesize H and H^C into gate-level net-lists, called N and N^C.
5: **if** Equivalent(N,N^C) **then**
6:     **return** Yes
7: **else**
8:     Generate an error trace corresponding to the mismatch between N and N^C
9:     **return** NO along with the error trace

4.2.3 Soundness of DFA optimizations

The transformations usually adopted in regular expression accelerators are not limited to decomposition of regular expressions. A variety of DFA optimizations have been proposed in the literature [14, 15] to keep the memory requirement low.

One commonly used optimization exploits default transitions. The transitions in a DFA are prioritized. Consider the DFA of *abc* in Figure 4. The prioritized transitions are shown in Table 1.

The lower priority transition $(s_0, *, s_0)$ is taken iff (s_0, a, s_1) is not taken. Observe that all the states in this DFA have a default lowest priority transition $(*, *, s_0)$. The DFA memory usage can be optimized by storing this transition only once. We emphasize that SEC is also applicable to identification of unsound optimizations. The DFSM model needs

353

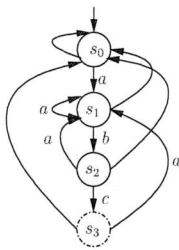

Figure 4: The DFA of the regular expression abc.

Current State	Input	Next State	Priority
s_0	a	s_1	1
s_1	b	s_2	1
s_2	c	s_3	1
*	a	s_1	2
*	*	s_0	3

Table 1: Prioritized transitions on DFA in Figure 4. Highest priority=1, Lowest priority=3

to be adapted accordingly.

5. EXPERIMENTS

We have implemented the proposed verification method (Algorithm 1) using IBM's formal verification tool called SixthSense [4]. The patterns represented as regular expressions in PCRE format [2] are a complex subset of the open rule set from Emerging Threats [1] and our proprietary pattern set. Note that transformation and verification is done for a single pattern at a time.

Figure 5 illustrates a simplified model of the architecture of the accelerator. When a regular expression is split into a number of sub-expressions, the corresponding DFAs are combined producing four DFAs, keeping the storage requirement of the resulting DFAs as low as possible. These DFAs are synthesized into net-lists and are mapped on four engines of the accelerator. Each engine sends instructions to the post-processor. The post-processor comprises instruction decoding units, execution units and a 128-bit register file. The engines have built-in mechanisms to support various DFA optimizations, e.g., they have a special memory bank to store the default rules.

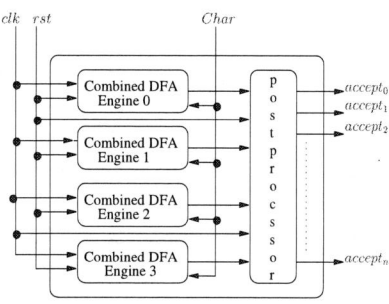

Figure 5: Simplified model of the accelerator

Since there is no 1:1 state element pairing between M and M^C, equivalence checking of the two designs is computationally expensive unlike CEC. We make effective use of a variety of techniques available in SixthSense, keeping the verification overhead low. Combinational redundancy removal (via BDD-sweeping and SAT sweeping), constant propagation, and redundant register removal are some of the techniques applied to reduce the size of the designs. If all verification targets, i.e., inequality of all pairs of outputs are proved unreachable, the two designs are equivalent. One of the engines employed in our methodology fragments targets into simpler sub-targets that may be easier to solve. Structurally similar targets are identified restricting the verification to only one of them. A cut-based abstraction engine to reduce the number of input variables is also employed. The state space is explored exhaustively using a BDD-based engine.

5.1 Emerging Threats Open Rule Set

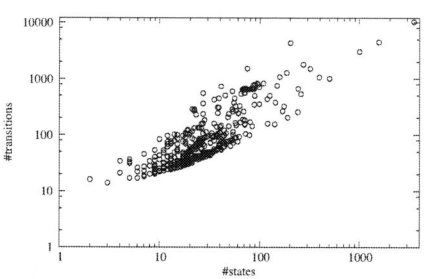

Figure 6: Emerging Threats Pattern Set Characteristics: #transitions vs. #states in original DFA

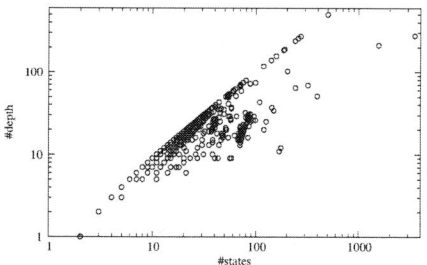

Figure 7: Emerging Threats Pattern Set Characteristics: Depth vs. #states in original DFA

Figure 6 is a log-scale scatter plot comparing the number of transitions and the number of states in the original DFAs of the regular expressions from Emerging Threats on which the proposed method is run. Figure 7 is a log-scale scatter plot comparing the depth and number of states in the original DFA of each regular expression. The patterns result in DFAs that have very large number of transitions, drastically increasing the compilation time of their VHDL representations. Verification of 840 patterns took 50 hours. Out of the 840 patterns, 335 are decomposed to use the post-processor. Out of these, 10 patterns are reported to be incorrectly transformed. The DFA optimizations, e.g., exploitation of default rules (see Section 4.2.3) are performed

354

on all the patterns. We observe that the VHDL compilation time dominates the total run time. Only 10% of the total time is spent in SEC.

5.2 Proprietary Pattern Set

We have also run our verification tool on our proprietary pattern set consisting of 6445 patterns. Out of these, 564 patterns were decomposed to use the post-processor. Figures 8 and 9 depict the characteristics of this pattern set in terms of the number of states, number of transitions, and the depth of the original DFAs. Out of the 564 patterns, the transformations on 17 of them are proved to be incorrect, which simulation and testing failed to uncover even when ran for about a week. The overall process of VHDL compilation and running SEC on these transformations took just 17 hours. We note that years of simulation and testing, devising various test generation options described in Section 4.1 failed to detect these errors.

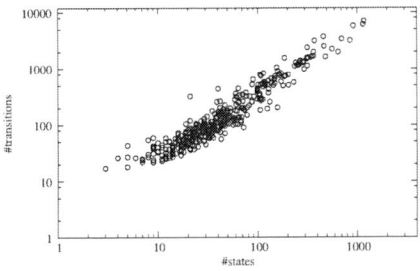

Figure 8: Proprietary Pattern Set Characteristics: #transitions vs. #states in original DFA

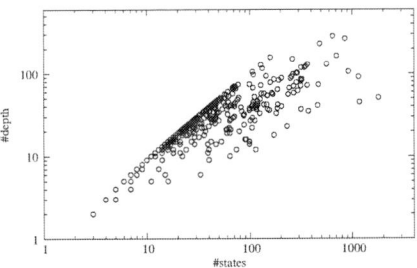

Figure 9: Proprietary Pattern Set Characteristics: Depth vs. #states in original DFA

6. CONCLUSION

Unverified regular expression transformations employed in fast packet content scanning pose a security threat in network security and monitoring applications. However, automated verification of regular expression transformations has received very little attention. We have presented a formal verification methodology for verifying correctness of the regular expression transformations commonly employed in accelerating packet context scanning. As complexity of patterns grows, techniques like propositional satisfiability solving, interpolation, induction etc. can be applied to increase

the scalability of SEC. As future work, we intend to transform and verify multiple patterns all together instead of transforming and verifying patterns one by one.

7. ACKNOWLEDGEMENT

The authors would like to thank Jason Baumgartner and Hari Mony for their discussions on the work and help with SixthSense.

8. REFERENCES

[1] Emerging Threats. http://www.emergingthreats.net/.

[2] PCRE - Perl Compatible Regular Expressions. http://www.pcre.org/.

[3] SNORT network intrusion detection system. http://www.snort.org/.

[4] J. Baumgartner, H. Mony, V. Paruthi, R. Kanzelman, and G. Janssen. Scalable sequential equivalence checking across arbitrary design transformations. In *ICCD'06*, pages 259–266, 2006.

[5] J. R. Burch, E. M. Clarke, K. L. McMillan, D. L. Dill, and L. J. Hwang. Symbolic model checking: 1020 states and beyond. *Inf. Comput.*, 98:142–170, June 1992.

[6] D. D. Gajski, N. D. Dutt, A. C.-H. Wu, and S. Y.-L. Lin. *High-level synthesis: Introduction to Chip and System Design*. Kluwer Academic Publishers, Norwell, MA, USA, 1992.

[7] J. E. Hopcroft, R. Motwani, and J. D. Ullman. *Introduction to Automata Theory, Languages, and Computation*. Addison Wesley, 2000.

[8] S. Kumar, B. Chandrasekaran, J. Turner, and G. Varghese. Curing regular expressions matching algorithms from insomnia, amnesia, and acalculia. In *ANCS '07*, pages 155–164. ACM, 2007.

[9] K. McMillan. Interpolation and SAT-based model checking. In *Computer Aided Verification*, volume 2725 of *Lecture Notes in Computer Science*, pages 1–13. Springer Berlin / Heidelberg, 2003.

[10] H. Mony, J. Baumgartner, A. Mishchenko, and R. Brayton. Speculative reduction-based scalable redundancy identification. In *DATE '09*, pages 1674–1679, 2009.

[11] J. Rohrer, K. Atasu, J. van Lunteren, and C. Hagleitner. Memory-efficient distribution of regular expressions for fast deep packet inspection. In *CODES+ISSS*, pages 147–154, 2009.

[12] R. Smith, C. Estan, S. Jha, and S. Kong. Deflating the big bang: fast and scalable deep packet inspection with extended finite automata. In *SIGCOMM '08*, pages 207–218. ACM, 2008.

[13] L. Tan and T. Sherwood. A high throughput string matching architecture for intrusion detection and prevention. In *ISCA '05*, pages 112–122, 2005.

[14] N. Tuck, T. Sherwood, B. Calder, and G. Varghese. Deterministic memory-efficient string matching algorithms for intrusion detection. In *IEEE Infocom*, pages 2628–2639, 2004.

[15] J. van Lunteren. High-performance pattern-matching for intrusion detection. In *INFOCOM 2006*, pages 1–13, 2006.

Sciduction: Combining Induction, Deduction, and Structure for Verification and Synthesis

Sanjit A. Seshia
UC Berkeley
sseshia@eecs.berkeley.edu

ABSTRACT

Even with impressive advances in formal verification, certain major challenges remain. Chief amongst these are environment modeling, incompleteness in specifications, and the complexity of underlying decision problems.

In this position paper, we contend that these challenges can be tackled by integrating traditional, deductive methods with inductive inference (learning from examples) using hypotheses about system structure. We present *sciduction*, a formalization of such an integration, show how it can tackle hard problems in verification and synthesis, and outline directions for future work.

Categories and Subject Descriptors

B.5.2 [**Design Aids**]: Verification; I.2.6 [**Learning**]: Concept Learning

General Terms

Algorithms, Design, Theory, Verification

Keywords

Formal verification, synthesis, learning, deduction, induction

1. INTRODUCTION

> "Formal methods research is about mechanizing creativity" — J. Strother Moore, FMCAD 2011 keynote.

> "To be creative requires divergent thinking (generating many unique ideas) and then convergent thinking (combining those ideas into the best result)." – Bronson & Merryman, Newsweek, July 2010.

Formal verification has made enormous strides over the last few decades. Verification techniques such as model checking [12, 38, 15] and theorem proving (see, e.g. [25]) are used routinely in computer-aided design of integrated circuits and have been widely applied to find bugs in software and embedded systems. However, certain problems in system verification remain very challenging, stymied by computational hardness or requiring significant human input into the verification process. This position paper seeks to outline these challenges, discuss recent promising trends, and generalize these trends into an approach for tackling these challenges.

Let us begin by examining the traditional view of verification, as a decision problem with three inputs (see Figure 1):

1. A model of the system to be verified, S;
2. A model of the environment, E, and
3. The property to be verified, Φ.

The verifier generates as output a YES/NO answer, indicating whether or not S satisfies the property Φ in environment E. Typically, a NO output is accompanied by a counterexample, also called an error trace, which is an execution of the system that indicates how Φ is violated. Some formal verification tools also include a proof or certificate of correctness with a YES answer. The first point to note

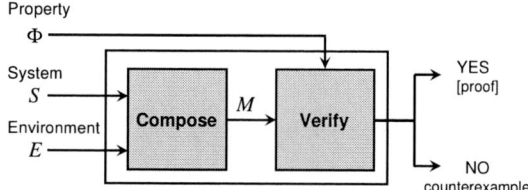

Figure 1: Formal verification procedure.

is that this view of verification is high-level and a bit idealized; in particular, it somewhat de-emphasizes the challenge in generating the inputs to the verification procedure. In practice, one does not always start with models S and E — these might have to be generated from implementations. To create the system model S, one might need to perform automatic abstraction from code that has many low-level details. Similarly, the generation of an environment model E is usually a manual process, involving writing constraints over inputs, or a state machine description of the parts of the system S communicates with. Bugs can be missed due to incorrect environment modeling. In systems involving third-party intellectual property (IP), not all details of the environment might even be available. Finally, the specification Φ is rarely complete and sometimes inconsistent, as has been noted in industrial practice (see, e.g., [6]). Indeed, the question "when are we done verifying?" often boils down to "have we written enough properties (and the right ones)?"

The second point we note is that Figure 1 omits some inputs that are key in successfully completing verification. For example, one might need to supply hints to the verifier in the form of inductive invariants or pick an abstract domain for generating suitable abstractions. One might need to break up the overall design into components and construct a compositional proof of correctness (or show that there is a bug). These tasks requiring human input have one aspect in common, which is that they involve a *synthesis sub-task* of the overall verification task.[1] This sub-task involves the synthesis of *verification artifacts* such as inductive invariants, auxiliary lemmas, abstractions, environment assumptions or input constraints for compositional reasoning, etc. One often needs human insight into at least the form of these artifacts, if not the artifacts themselves, to succeed in verifying the design.

---
[1] Here we use the term "*synthesis*" in a different way from what is standard in the EDA community. We do not mean "logic synthesis". Rather, we mean the synthesis of formal artifacts from very high-level specifications or as part of the verification procedure.

Finally, it has been a long-standing goal of the fields of electrical engineering and computer science to automatically synthesize systems from high-level specifications. The EDA community has been in the forefront of work towards this goal. In fact, the genesis of model checking lies in part in the automatic synthesis problem; the seminal paper on model checking by Clarke and Emerson [12] begins with this sentence:

> "We propose a method of constructing concurrent programs in which the synchronization skeleton of the program is automatically synthesized from a high-level (branching time) Temporal Logic specification."

In automatic synthesis, one starts with a specification Φ of the system to be synthesized, along with a model of its environment E. The goal of synthesis is to generate a system S that satisfies Φ when composed with E. Figure 2 depicts the synthesis process. Modeled thus, the essence of synthesis can be viewed as a *game*

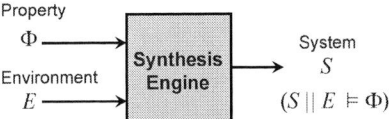

Figure 2: Formal synthesis procedure.

solving problem, where S and E represent the two players in a game; S is computed as a winning strategy ensuring that the composed system $S\|E$ satisfies Φ for all input sequences generated by the environment E. If such an S exists, we say that the specification (Φ and E) is *realizable*. Starting with the seminal work on automata-theoretic and deductive synthesis from specifications (e.g. [31, 37]), there has been steady progress on automatic synthesis. In particular, many recent techniques (e.g. [49, 50]) build upon the progress in formal verification in order to perform synthesis. However, there is a long way to go before automated synthesis is practical and widely applicable. One major challenge, shared with verification, is the difficulty of obtaining complete, formal specifications from the user. Even expert users find it difficult to write complete, formal specifications that are realizable. Often, when they do write complete specifications, the effort to write these is arguably more than that required to manually create the design in the first place. Additionally, the challenge of modeling the environment, as discussed above for verification, also remains for synthesis. Finally, synthesis problems typically have greater computational complexity than verification problems for the same class of specifications and models. For instance, equivalence checking of combinational circuits is NP-complete and routinely solved in industrial practice, whereas synthesizing a combinational circuit from a finite set of components is Σ_2-complete and only possible in very limited settings in practice. In some cases, both verification and synthesis are undecidable, but there are still compelling reasons to have efficient procedures in practice; a good example is hybrid systems — systems with both discrete and continuous state — whose continuous dynamics is non-linear, which arise commonly in embedded systems and analog/mixed-signal circuits.

To summarize, there are two main points. First, the main challenges facing formal verification and synthesis include system and environment modeling, creating good specifications, and the complexity of underlying decision problems. Second, the key to efficient verification is often in the synthesis of artifacts such as inductive invariants or abstractions, and thus, at the core, the challenges facing automatic verification are very similar to those facing automatic synthesis. Some of these challenges — such as dealing with computational complexity — can be partially addressed by

advances in *computational engines* such as Binary Decision Diagrams (BDDs) [9], Boolean satisfiability (SAT) [29], and satisfiability modulo theories (SMT) solvers [5]. However, these alone are not sufficient to extend the reach of formal methods for verification and synthesis. New *methodologies* are also required.

As J. Strother Moore observed in his FMCAD 2011 keynote talk, formal methods involve the mechanization of human creativity in design. Arguably, the design of circuits and programs ranks amongst the most creative activities performed by humankind, alongside finding proofs in mathematics, conducting basic scientific research, composing music, creating artistic paintings, and more. The question naturally arises: do current automated formal methods follow the same approach as these other creative endeavors? Upon reflection, we find that this is not entirely the case. We contend that the "creativity gap" arises from an imbalance between the use of *deduction* and *induction* in formal methods, that recent trends are closing the gap, and that one can build upon those trends to develop a new paradigm for formal verification and synthesis.

Induction is the process of inferring a general law or principle from observation of particular instances.[2] Machine learning algorithms are typically inductive, generalizing from (labeled) examples to obtain a learned *concept* or *classifier* [34, 3]. *Deduction*, on the other hand, involves the use of general rules and axioms to infer conclusions about particular problem instances. Traditional formal verification and synthesis techniques, such as model checking or theorem proving, are deductive. This is no surprise, as formal verification and synthesis problems (see Figures 1 and 2) are, by their very nature, deductive processes: given a particular specification Φ, environment E, and system S, a verifier typically uses a rule-based decision procedure for that class of Φ, E, and S to deduce if $S\|E \models \Phi$. On the other hand, inductive reasoning may seem out of place here, since typically an inductive argument only ensures that the truth of its premises make it *likely or probable* that its conclusion is also true. However, one observes that humans often employ a combination of inductive and deductive reasoning while performing verification or synthesis. For example, while proving a theorem, one often starts by working out examples and trying to find a pattern in the properties satisfied by those examples. The latter step is a process of inductive generalization. These patterns might take the form of lemmas or background facts that then guide a deductive process of proving the statement of the theorem from known facts and previously established theorems (rules). Similarly, while creating a new design, one often starts by enumerating sample behaviors that the design must satisfy and hypothesizing components that might be useful in the design process; one then systematically combines these components, using design rules, to obtain a candidate artifact. The process usually iterates between inductive and deductive reasoning until the final artifact is obtained.

In this paper, we present a methodology, *sciduction*, that formalizes such a combination of inductive and deductive reasoning.[3] This methodology is inspired by recent successes in automatic abstraction and invariant generation such as counterexample-guided abstraction refinement (CEGAR), as well as the enormous advances in machine learning over the past two decades. The key in integrating induction and deduction is the use of *structure hypotheses*. These are mathematical hypotheses used to define the class of artifacts to be synthesized within the overall verification or synthesis problem. Sciduction constrains inductive and deductive reasoning

---

[2] The term "induction" is often used in the verification community to refer to *mathematical induction*, which is actually a deductive proof rule. Here we are employing "induction" in its more classic usage arising from the field of Philosophy.

[3] sciduction stands for structure-constrained induction and deduction. An early version of this position paper appeared as a UC Berkeley technical report [43].

using structure hypotheses, and actively combines inductive and deductive reasoning: for instance, deductive techniques generate examples for learning, and inductive reasoning is used to guide the deductive engines. We explain how one can combine inductive and deductive reasoning to obtain the kinds of soundness/completeness guarantees one desires in formal verification and synthesis.

The rest of this paper is organized as follows. We describe the methodology in detail, with comparison to related work, in Section 2. Some new instances of this methodology are presented in Section 3. Future applications and further directions are explored in Section 4.

2. SCIDUCTION: FORMALIZATION AND RELATED WORK

We begin with a formalization of the sciduction methodology, and then compare it to related work. This section assumes some familiarity with basic terminology in formal verification and machine learning — see the relevant books by Clarke et al. [15], Manna and Pnueli [30], and Mitchell [34] for an introduction.

2.1 Verification and Synthesis Problems

As discussed in Section 1, an instance of a verification problem is defined by a triple $\langle S, E, \Phi \rangle$, where S denotes the system, E is the environment, and Φ is the property to be verified. Here we assume that S, E, and Φ are described formally, in mathematical notation. Similarly, an instance of a synthesis problem is defined by the pair $\langle E, \Phi \rangle$, where the symbols have the same meaning. In both cases, as noted earlier, it is possible *in practice* for the descriptions of S, E, or Φ to be missing or incomplete; in such cases, the missing components must be synthesized as part of the overall verification or synthesis process.

A *family of verification or synthesis problems* is a triple $\langle \mathcal{C}_S, \mathcal{C}_E, \mathcal{C}_\Phi \rangle$ where $\mathcal{C}_S$ is a formal description of a class of systems, $\mathcal{C}_E$ is a formal description of a class of environment models, and $\mathcal{C}_\Phi$ is a formal description of a class of specifications. In the case of synthesis, $\mathcal{C}_S$ defines the class of systems to be synthesized.

2.2 Elements of sciduction

An instance of sciduction can be described using a triple $\langle \mathcal{H}, \mathcal{I}, \mathcal{D} \rangle$, where the three elements are as follows:

1. *A structure hypothesis*, $\mathcal{H}$, encodes our hypothesis about the form of the *artifact to be synthesized*, whether it be an abstract system model, an environment model, an inductive invariant, a program, or a control algorithm (or any portion thereof);

2. *An inductive inference engine*, $\mathcal{I}$, is an algorithm for *learning from examples* an artifact h defined by $\mathcal{H}$, and

3. *A deductive engine*, $\mathcal{D}$, is a *lightweight decision procedure* that applies deductive reasoning to answer queries generated in the synthesis or verification process.

We elaborate on these elements below. For concreteness, the context of synthesis is used to explain the central ideas in the approach. Note however that all of the discussion applies also to verification, in the context of synthesis of *verification artifacts*. We will note points specific to verification or synthesis as they arise.

2.2.1 Structure Hypothesis

The structure hypothesis, $\mathcal{H}$, encodes our hypothesis about the form of the artifact to be synthesized.

Formally $\mathcal{H}$ is a (possibly infinite) set of *artifacts*. $\mathcal{H}$ encodes a hypothesis that the system to be synthesized falls in a subclass $\mathcal{C}_\mathcal{H}$ of $\mathcal{C}_S$ (i.e., $\mathcal{C}_\mathcal{H} \subseteq \mathcal{C}_S$). Note that $\mathcal{H}$ needs not be the same as $\mathcal{C}_\mathcal{H}$, since the artifact being synthesized might just be a portion of the full system description, such as the guard on transitions of a state machine, or the assignments to certain variables in a program. Each

artifact $h \in \mathcal{H}$, in turn, corresponds to a unique set of *primitive elements* that defines its semantics. The form of the primitive element depends on the artifact to be synthesized.

More concretely, here are two examples of a structure hypothesis $\mathcal{H}$:

1. Suppose that $\mathcal{C}_S$ is the set of all finite automata over a set of input variables V and output variables U satisfying a specification Φ. Consider the structure hypothesis $\mathcal{H}$ that restricts the finite automata to be synchronous compositions of automata from a finite library L. The artifact to be synthesized is the entire finite automaton, and so, in this case, $\mathcal{H} = \mathcal{C}_\mathcal{H}$. Each element $h \in \mathcal{H}$ is one such composition of automata from L. A primitive element is an input-output trace in the language of the finite automaton h.

2. Suppose that $\mathcal{C}_S$ is the set of all hybrid automata [1], where the guards on transitions between modes can be any region in $\mathbb{R}^n$ but where the modes of the automaton are fixed. A structure hypothesis $\mathcal{H}$ can restrict the guards to be hyperboxes in $\mathbb{R}^n$ — i.e., conjunctions of upper and lower bounds on continuous state variables. Each $h \in \mathcal{H}$ is one such hyperbox, and a primitive element is a point in h. Notice that $\mathcal{H}$ defines a subclass of hybrid automata $\mathcal{C}_\mathcal{H} \subset \mathcal{C}_S$ where the guards are n-dimensional hyperboxes. Note also that $\mathcal{H} \neq \mathcal{C}_\mathcal{H}$ in this case.

A structure hypothesis $\mathcal{H}$ can be syntactically described in several ways. For instance, in the second example above, $\mathcal{H}$ can define a guard either set-theoretically as a hyperbox in $\mathbb{R}^n$ or using mathematical logic as a conjunction of atomic formulas, each of which is an interval constraint over a real-valued variable.

2.2.2 Inductive Inference

The inductive inference procedure, $\mathcal{I}$, is an algorithm for learning from examples an artifact $h \in \mathcal{H}$.

While any inductive inference engine can be used, in the context of verification and synthesis we expect that the learning algorithms $\mathcal{I}$ have one or more of the following characteristics:

- $\mathcal{I}$ performs *active learning*, selecting the examples that it learns from.

- Examples and/or labels for examples are generated by one or more *oracles*. The oracles could be implemented using deductive procedures or by evaluation/execution of a model on a concrete input. In some cases, an oracle could be a human user.

- A deductive procedure is invoked in order to synthesize a concept (artifact) that is consistent with a set of labeled examples. The idea is to formulate this synthesis problem as a decision problem where the concept to be output is generated from the satisfying assignment.

2.2.3 Deductive Reasoning

The deductive engine, $\mathcal{D}$, is a lightweight decision procedure that applies deductive reasoning to answer queries generated in the synthesis or verification process.

The word "lightweight" refers to the fact that the decision problem being solved by $\mathcal{D}$ must be easier, in theoretical or practical terms, than that corresponding to the overall synthesis or verification problem.

In theoretical terms, "lightweight" means that at least one of the following conditions must hold:

1. $\mathcal{D}$ must solve a problem that is a strict special case of the original.

2. One of the following two cases must hold:

(i) If the original (synthesis or verification) problem is decidable, and the worst-case running time of the best known procedure for the original problem is $O(T(n))$, then $\mathcal{D}$ must run in time $o(T(n))$.

(ii) If the original (synthesis or verification) problem is undecidable, $\mathcal{D}$ must solve a decidable problem.

In practical terms, the notion of "lightweight" is fuzzier: intuitively, the class of problems addressed by $\mathcal{D}$ must be "more easily solved in practice" than the original problem class. For example, $\mathcal{D}$ could be a finite-state model checker that is invoked only on abstractions of the original system produced by, say, localization abstraction [26] — it is lightweight if the abstractions are solved faster in practice than the original concrete instance. Due to this fuzziness, it is preferable to define "lightweight" in theoretical terms whenever possible.

$\mathcal{D}$ can be used to answer queries generated by $\mathcal{I}$, where the query is typically formulated as a decision problem to be solved by $\mathcal{D}$. Here are some examples of tasks $\mathcal{D}$ can perform and the corresponding decision problems:

- Generating examples for the learning algorithm.
  Decision problem: "does there exist an example satisfying the criterion of the learning algorithm?"
- Generating labels for examples selected by the learning algorithm.
  Decision problem: "is L the label of this example?"
- Synthesizing candidate artifacts.
  Decision problem: "does there exists an artifact consistent with the observed behaviors/examples?"

2.2.4 Discussion

We now make a few remarks on the above formalism.

In the above description of the structure hypothesis, $\mathcal{H}$ only "loosely" restricts the class of systems to be synthesized, allowing the possibility that $\mathcal{C}_{\mathcal{H}} = \mathcal{C}_S$. We argue that a tighter restriction is often desirable. One important role of the structure hypothesis is to reduce the search space for synthesis, by restricting the class of artifacts $\mathcal{C}_S$. For example, a structure hypothesis could be a way of codifying the form of human insight to be provided to the synthesis process. Additionally, restricting $\mathcal{C}_{\mathcal{H}}$ also aids in inductive inference. Fundamentally, the effectiveness of inductive inference (i.e., of $\mathcal{I}$) is limited by the examples presented to it as input; therefore, it is important not only to select examples carefully, but also for the inference to generalize well beyond the presented examples. For this purpose, the structure hypothesis should place a strict restriction on the search space, by which we mean that $\mathcal{C}_{\mathcal{H}} \subsetneq \mathcal{C}_S$. The justification for this stricter restriction comes from the importance of *inductive bias* in machine learning. Inductive bias is the set of assumptions required to *deductively* infer a concept from the inputs to the learning algorithm [34]. If one places no restriction on the type of systems to be synthesized, the inductive inference engine $\mathcal{I}$ is unbiased; however, an unbiased learner will learn an artifact that is consistent only with the provided examples, with no generalization to unseen examples. As Mitchell [34] writes: "a learner that makes no a priori assumptions regarding the identity of the target concept has no rational basis for classifying any unseen instances." Given all these reasons, it is highly desirable for the structure hypothesis $\mathcal{H}$ to be such that $\mathcal{C}_{\mathcal{H}} \subsetneq \mathcal{C}_S$. We present in Sec. 3 two applications of sciduction that have this feature.

Another point to note is that it is possible to use randomization in implementing $\mathcal{I}$ and $\mathcal{D}$. For example, a deductive decision procedure that uses randomization can generate a YES/NO answer with high probability.

Next, although we have defined sciduction as combining a single inductive engine with a single deductive engine, this is only for simplicity of the definition and poses no fundamental restriction. One can always view multiple inductive (deductive) engines as a being contained in a single inductive (deductive) procedure where this outer procedure passes its input to the appropriate "sub-engine"

based on the type of input query.

Finally, in our definition of sciduction, we do not advocate any particular technique of combining inductive and deductive reasoning. Indeed, we envisage that there are many ways to "configure" the combination of $\mathcal{H}$, $\mathcal{D}$, and $\mathcal{I}$, perhaps using inductive procedures within deductive engines and vice-versa. Any mode of integrating $\mathcal{H}$, $\mathcal{I}$, and $\mathcal{D}$ that satisfies the requirements stated above on each of those three elements is admissible. We expect that the particular requirements of each application will define the mode of integration that works best for that application. We present illustrative examples in Section 3.

2.3 Soundness and Completeness Guarantees

It is highly desirable for verification or synthesis procedures to provide *soundness* and *completeness* guarantees. In this section, we discuss the form these guarantees take for a procedure based on sciduction.

A verifier is said to be *sound* if, given an arbitrary problem instance $\langle S, E, \Phi \rangle$, the verifier outputs "YES" only if $S \| E \models \Phi$. The verifier is said to be *complete* if it outputs "NO" when $S \| E \not\models \Phi$.

The definitions for synthesis are similar. A synthesis technique is *sound* if, given an arbitrary problem instance $\langle E, \Phi \rangle$, if it outputs S, then $S \| E \models \Phi$. A synthesis technique is *complete* if, when there exists S such that $S \| E \models \Phi$, it outputs at least one such S.

Formally, for a verification/synthesis procedure $\mathcal{P}$, we denote the statement "$\mathcal{P}$ is sound" by $\text{sound}(\mathcal{P})$.

Note that we can have probabilistic analogs of soundness and completeness. Informally, a verifier is *probabilistically sound* if it is sound with "high probability." We leave a more precise discussion of this point to a later stage in this paper when it becomes relevant.

2.3.1 Validity of the Structure Hypothesis

In sciduction, the existence of soundness and completeness guarantees depends on the validity of the structure hypothesis. Informally, we say that the structure hypothesis $\mathcal{H}$ is *valid* if the set of correct artifacts to be synthesized, if any exist, is guaranteed to include one in $\mathcal{C}_{\mathcal{H}}$.

Let us elaborate on what we mean by the phrase "correct artifacts to be synthesized":

- In the context of a synthesis problem, this is relatively easy: one seeks an element c of $\mathcal{C}_S$ that satisfies a specification Φ. If Φ is available as a formal specification, this phrase is precisely defined. However, as noted earlier, one of the challenges with synthesis can be the absence of good formal specifications. In such cases, we use Φ to denote a "golden" specification that one would have in the ideal scenario.
- For verification, there can be many artifacts to be synthesized, such as inductive invariants, abstractions, or environment assumptions. Each such artifact is an element of a set $\mathcal{C}_S$. The specification for each such "synthesis sub-task", generating a different kind of artifact, is different. For invariant generation, $\mathcal{C}_S$ is the set of candidate invariants, and the specification is that the artifact $c \in \mathcal{C}_S$ be an inductive invariant of the system S. For abstractions, $\mathcal{C}_S$ defines the set of abstractions, and the specification is that $c \in \mathcal{C}_S$ must be a sound *and* precise abstraction with respect to the property to be verified, Φ; here "precise" means that no spurious counterexamples will be generated. We will use Ψ to denote the cumulative specification for all synthesis sub-tasks in the verification problem.

Thus, for both verification and synthesis, the existence of an artifact to be synthesized can be expressed as the following logical formula:

$$\exists c \in \mathcal{C}_S . c \models \Psi$$

where, for synthesis, $\Psi = \Phi$, and, for verification, Ψ denotes the cumulative specification for the synthesis sub-tasks, as discussed above.

Similarly, the existence of an artifact to be synthesized that additionally satisfies the structure hypothesis $\mathcal{H}$ is written as:

$$\exists c \in \mathcal{C}_{\mathcal{H}} \,.\, c \models \Psi$$

Given the above logical formulas, we define the statement "the structure hypothesis is *valid*" as the validity of the logical formula `valid(`$\mathcal{H}$`)` given below:

$$\texttt{valid}(\mathcal{H}) \triangleq (\exists c \in \mathcal{C}_S \,.\, c \models \Psi) \implies (\exists c \in \mathcal{C}_{\mathcal{H}} \,.\, c \models \Psi) \tag{1}$$

In other words, if there exists an artifact to be synthesized (that satisfies the corresponding specification Ψ), then there exists one satisfying the structure hypothesis.

Note that `valid(`$\mathcal{H}$`)` is trivially valid if $\mathcal{C}_{\mathcal{H}} = \mathcal{C}_S$, or if the consequent $\exists c \in \mathcal{C}_{\mathcal{H}} \,.\, c \models \Psi$ is valid. Indeed, one extremely effective technique in verification, counterexample-guided abstraction refinement (CEGAR), can be seen as a form of sciduction where the latter case holds (Sec. 2.4.1 has a more detailed discussion of the link between CEGAR and sciduction.) However, in some cases, `valid(`$\mathcal{H}$`)` can be proved valid even without these cases; see Sec. 3.1 for an example.

2.3.2 Conditional Soundness

A verification/synthesis procedure following the sciduction paradigm must satisfy a conditional soundness guarantee: procedure $\mathcal{P}$ must be *sound if the structure hypothesis is valid.*

Without such a requirement, $\mathcal{P}$ is a heuristic, best-effort verification or synthesis procedure. (It could be extremely useful, nonetheless.) With this requirement, we have a mechanism to formalize the assumptions under which we obtain soundness — namely, the structure hypothesis.

More formally, the soundess requirement for $\mathcal{P}$ can be expressed as the following logical expression:

$$\texttt{valid}(\mathcal{H}) \implies \texttt{sound}(\mathcal{P}) \tag{2}$$

Note that one must prove `sound(`$\mathcal{P}$`)` under the assumption `valid(`$\mathcal{H}$`)`, just like one proves unconditional soundness. The point is that making a structure hypothesis can allow one to devise procedures and prove soundness where previously this was difficult or impossible.

Where completeness is also desirable, one can formulate a similar notion of conditional completeness. We will mainly focus on soundness in this paper.

2.4 Context and Previous Work

In both ancient and modern philosophy, there is a long history of arguments about the distinction between induction and deduction and their relationship and relative importance. This literature, although very interesting, is not directly relevant to the discussion in this paper.

Within computer science and engineering, the field of artificial intelligence (AI) has studied inductive and deductive reasoning and their connections (see, e.g., [41]). As mentioned earlier, Mitchell [34] describes how inductive inference can be formulated as a deduction problem where inductive bias is provided as an additional input to the deductive engine. *Inductive logic programming* [35], an approach to machine learning, blends induction and deduction by performing inference in first-order theories using examples and background knowledge. Combinations of inductive and deductive reasoning have also been explored for synthesizing programs (plans) in AI; for example, the SSGP approach [19] generates plans by sampling examples, generalizing from those samples, and then proving correctness of the generalization.

Our focus is on the use of combined inductive and deductive reasoning in *formal verification and synthesis.* While several techniques for verification and synthesis combine subsets of induction,

deduction, and structure hypotheses, there are important distinctions between many of these and the sciduction approach. Below, we highlight a representative sample of related work; this sample is intended to be illustrative, not exhaustive.

2.4.1 Instances of Sciduction

We first survey prior work in verification and synthesis that has provided inspiration for formulating the sciductive approach. Sciduction can be seen as a "lens" through which one can view the common ideas amongst these techniques so as to extend and apply them to new problem domains.

Counterexample-Guided Abstraction-Refinement (CEGAR). In CEGAR [14], depicted in Fig. 3, the key problem is to synthesize an abstract model so as to eliminate spurious counterexamples. CEGAR solves a synthesis sub-task of generating abstract models

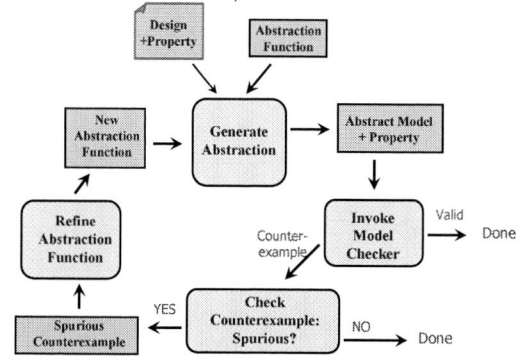

Figure 3: Counterexample-guided abstraction refinement (CEGAR).

that are *sound* (they contain all behaviors of the original system) and *precise* (any counterexample for the abstract model is also a counterexample for the original system). The synthesized artifact is thus the abstract model. CEGAR has been successfully applied to hardware [14], software [4], and hybrid systems [13]. One can view CEGAR as an instance of sciduction as follows:

- The *abstract domain*, which defines the form of the abstraction function, is the structure hypothesis. For example, in verifying digital circuits, one might use localization abstraction [26], in which abstract states are cubes over the state variables.
- The inductive engine $\mathcal{I}$ is an algorithm to learn a new abstraction function from a spurious counterexample. Consider the case of localization abstraction. One approach in CEGAR is to walk the lattice of abstraction functions, from most abstract (hide all variables) to least abstract (the original system). This problem can be viewed as a form of learning based on version spaces [34], although the traditional CEGAR refinement algorithms are somewhat different from the learning algorithms proposed in the version spaces framework. Gupta, Clarke, et al. [21] have previously observed the link to inductive learning and have proposed version of CEGAR based on alternative learning algorithms (such as induction on decision trees).
- The deductive engine $\mathcal{D}$, for finite-state model checking, comprises the model checker and a SAT solver. The model checker is invoked on the abstract model to check the property of interest, while the SAT solver is used to check if a counterexample is spurious.

In CEGAR, usually the original system is in $\mathcal{C}_{\mathcal{H}}$, and since it is a sound and precise abstract model, the consequent $\exists c \in \mathcal{C}_{\mathcal{H}} \,.\, c \models \Psi$ is valid. Thus, the structure hypothesis is valid, and the notion of soundness reduces to the traditional (unconditional) notion.

There are other counterexample-guided techniques that are also instances of sciduction. Programming by sketching is a novel approach to synthesizing software by encoding programmer insight in the form of a *partial program*, or "sketch" [49]. An algorithmic approach central to this work is counterexample-guided inductive synthesis (CEGIS) [49, 48], which is analogous to CEGAR. The structure hypothesis is the sketch, and the inductive and deductive procedures are similar to those in CEGAR.

Invariant Generation. One of the important steps in verification based on model checking or theorem proving is the construction of *inductive invariants*. (Here "inductive" refers to the use of mathematical induction.) One often needs to strengthen the main safety property with auxiliary inductive invariants so as to succeed at proving/disproving the property.

In recent years, an effective approach to generating inductive invariants is to assume that they have a particular structural form, use simulation to prune out candidates, and then use a SAT solver or model checker to prove those candidates that remain. This is an instance of the sciduction approach, and is very effective. For example, these strategies are implemented in the ABC verification and synthesis system [8] and described in part in Michael Case's PhD thesis [11]. The structure hypothesis $\mathcal{H}$ defines the space of candidate invariants as being either constants (literals), equivalences, implications, or in some cases, random clauses or based on k-cuts in the and-inverter graph. The inductive inference engine is very rudimentary: it just keeps all instances of invariants that match $\mathcal{H}$ and are consistent with simulation traces. The deductive engine is a SAT solver. Clearly, in this case, the structure hypothesis is restrictive in that the procedure does not seek to find arbitrary forms of invariants. However, the verification procedure is still sound, because if a suitable inductive invariant is not found, one may fail to prove the property, but a buggy system will not be deemed correct.

This idea has also been explored in software verification, by combining the Daikon system [17] for generating likely program invariants from traces with deductive verification systems such as ESC/Java [18].

Learning for Compositional Verification. The use of learning algorithms has been investigated extensively in the context of synthesizing environment assumptions for compositional verification. Most of these techniques are based on Angluin's L^* algorithm [2] and its variants; see [16] for a recent collection of papers on this topic. These techniques are an instance of sciduction in the following sense. The artifact being synthesized is an environment model in the form of a state machine. For finite-state model checking, one typically assumes that the environment is also a finite-state transducer whose inputs and outputs are defined by those of the system. The structure hypothesis is thus not restrictive, i.e., $\mathcal{C}_{\mathcal{H}} = \mathcal{C}_E$. The L^* algorithm is a learning algorithm based on queries and counterexamples. The counterexamples are generated by the model checker, which forms the deductive procedure in this case.

2.4.2 Other Related Work

Program Analysis Using Relevance Heuristics. McMillan [33] describes the idea of verification based on "relevance heuristics", which is the notion that facts useful in proving special cases of the verification problem are likely to be useful in general. This idea is motivated by the similar approach taken in (CDCL) SAT solvers. A concrete instance of this approach is interpolation-based model checking [32], where a proof of a special case (e.g., the lack of an assertion failure down a certain program path) is used to generate facts relevant to solving the general verification problem (e.g., correctness along all program paths). Although this work generalizes from special cases, the generalization traditionally is not inductive, and no structure hypothesis is involved.

However, one might note the possibility of using inductive infer-

ence in the generalization step as follows: since an interpolant is a specific type of logical formula that is consistent with states reachable at a particular program point (positive examples), but inconsistent with the particular path extension from that point generated by the model checker (negative examples). Indeed, this is the approach taken very recently by Sharma et al [47]. This variant, based on inductive inference, can be viewed as an instance of sciduction, where the structure hypothesis is a particular assumption on the syntactic form of the interpolant (which is the artifact being synthesized).

Automata-Theoretic Synthesis from Linear Temporal Logic (LTL). One of the classic approaches to synthesis is the automata-theoretic approach for synthesizing a finite-state transducer (FST) from an LTL specification, pioneered by Pnueli and Rosner [37]. The approach is a purely deductive one, with a final step that involves solving an emptiness problem for tree automata. No structure hypothesis is made on the FST being synthesized. Although advances have been made in the area of synthesis from LTL, for example in special cases [36], some major challenges remain: (i) writing complete specifications is tedious and error-prone, and (ii) the computational complexity for general LTL is doubly-exponential in the size of the specification. It would be interesting to explore if inductive techniques can be combined with existing deductive automata-theoretic procedures to form an effective sciduction-based approach to some class of systems or specifications.

Verification-Driven Software Synthesis. Srivastava et al. [50] have proposed a verification-driven approach to synthesis (called VS3), where programs with loops can be synthesized from a scaffold comprising of a logical specification of program functionality, and domain constraints and templates restricting the space of synthesizable programs. The latter is a structure hypothesis. However, the approach is not sciduction since the synthesis techniques employed are purely deductive in nature.

3. NEW INSTANCES OF SCIDUCTION

In this section, we discuss, in somewhat more depth, two newer applications of sciduction. The first of these, controller synthesis for hybrid systems (Sec. 3.1), tackles the problem of high complexity of the underlying decision problem. The second, timing analysis of embedded software (Sec. 3.2), tackles the difficulty of environment modeling.

3.1 Controller Synthesis

We present here a new approach to the synthesis of *switching logic* in hybrid systems [23]. It differs from CEGAR in that $\mathcal{C}_{\mathcal{H}} \subset \mathcal{C}_S$; yet, there are reasonable conditions under which the structure hypothesis is valid. Since the problem area might be a unfamiliar to many readers in the EDA community, we provide more background for this problem.

Many embedded and control systems are conveniently modeled as multi-modal dynamical systems (MDSs). An MDS is a physical system (also known as a "plant") that can operate in different modes. The dynamics of the plant in each mode is known, and is usually specified using a continuous-time model such as a system of ordinary differential equations (ODEs). However, to achieve safe and efficient operation, it is typically necessary to switch between the different operating modes using carefully construed *switching logic*: guards on transitions between modes. The MDS along with its switching logic consitutes a *hybrid system*. Manually designing switching logic so as to ensure that the hybrid system satisfies its specification can be tricky and tedious.

While several techniques for switching logic synthesis have been proposed (see [23] for a survey), it remains quite challenging to handle systems with a combination of rich discrete structure (in the form of multiple modes) and complex non-linear dynamics within modes. One representative switching logic synthesis problem is for ensuring that specified *safety* properties are satisfied. More precisely, the problem is stated as follows:

⟨SLS⟩ Given a safety property, a multimodal dynamical system (MDS), and a set of initial states, synthesize switching logic for the MDS so that the resulting hybrid system is safe.

There are no constraints on the intra-mode continuous dynamics in the MDS, other than it be deterministic and locally Lipschitz at all points [23].

An example of such a switching logic synthesis problem is the 3-gear automatic transmission system [28] depicted in Figure 4 as a hybrid automaton [1]. This example has seven modes. The tran-

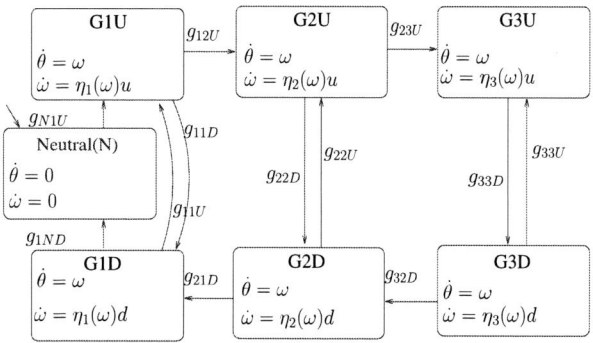

Figure 4: Automatic Transmission System

sitions between modes are labeled with guard variables: g_{ij} labels the transition from Mode i to Mode j. Such a guard is termed an *entry guard* for Mode j and an *exit guard* for Mode i.

Note that for this example, the dynamics in each mode are *nonlinear differential equations*. u and d denote the throttle in accelerating and deaccelerating mode. The transmission efficiency η is η_i when the system is in the ith gear, given by:

$$\eta_i = 0.99e^{-(\omega-a_i)^2/64} + 0.01$$

where $a_1 = 10, a_2 = 20, a_3 = 30$ and ω is the speed. The distance covered is denoted by θ. The acceleration in mode i is given by the product of the throttle and transmission efficiency.

The synthesis problem is to find the guards between the modes such that the efficiency η is high for speeds greater than some threshold, that is, $\omega \geq 5 \Rightarrow \eta \geq 0.5$. Also, ω must be less than an upper limit of 60. So, the safety property ϕ_S to be enforced would be

$$(\omega \geq 5 \Rightarrow \eta \geq 0.5) \wedge (0 \leq \omega \leq 60)$$

Note that for the class of hybrid automata with nonlinear dynamics within modes, even reachability analysis is undecidable. Synthesizing safe switching logic is therefore undecidable too, unless additional assumptions are imposed. While one cannot expect to have a synthesis procedure that works in all cases, finding safe switching logic is a non-intuitive task that can be tricky and involve a lot of trial-and-error for human designers to get right.

Instance of sciduction. A new approach to the ⟨SLS⟩ problem has been presented by the author and colleagues (Jha et al. [23]). This approach is as an instance of sciduction, as follows:

- *Structure Hypothesis:* A particular syntactic form is imposed on the guards of the hybrid system: they are *hyperboxes*. More precisely, the structure hypothesis includes the following two properties:

1. The safe switching logic, if one exists, has all guards as n-dimensional hyperboxes with vertices lying on a known discrete grid.[4]
2. For each mode, if all exit guards and all but one entry guard are fixed as hyperboxes, then for the remaining entry transition to that mode, the safe switching states constitute a hyperbox on the above-mentioned discrete grid.

While the above structure hypothesis may not be valid for general multi-modal dynamical systems, it can be proved valid under two additional properties: (i) the continuous dynamics within a mode is such that state variables vary *monotonically* within a mode [23], and (ii) the discrete grid reflects the finite-precision with which values of continuous system variables can be recorded. These are reasonable assumptions that hold in many embedded control systems.

To summarize, $\mathcal{C}_{\mathcal{H}}$ is the set of all hybrid automata in which the guards satisfy the above structure hypothesis.

- *Inductive Inference:* This routine is an algorithm to learn hyperboxes in $\mathbb{R}^n$ from labeled examples. An example is a point in $\mathbb{R}^n$. Its label is positive if the point is inside the box, and negative otherwise.

The main idea is to view safe switching states as positive examples and unsafe switching states as negative examples. Jha et al. [23] show how, given such labels, one can learn hyperboxes using the results of Goldman and Kearns [20]. The positive/negative labels on states, required by the inductive routine, are generated by a deductive engine, as described below.

- *Deductive Reasoning:* In order to label a switching state s for a mode m as safe or unsafe, we need a procedure to answer the following question: if we enter m in state s and follow its dynamics, will the trajectory visit only safe states until some exit guard becomes true?

This is a reachability analysis problem for purely continuous systems modeled as a system of ordinary differential equations (ODEs) with a single initial state. This problem is known to be undecidable in general [40]. However, in practice, this reachability problem can be solved for many kinds of continuous dynamical systems (including the intra-mode dynamics for the example shown in Fig. 4) using state-of-the-art techniques for *numerical simulation*. Thus, the deductive engine in this approach is a numerical simulator that can handle the dynamics in each mode of the multi-modal dynamical system. The numerical simulator must be *ideal*, in that it must always return the correct YES/NO answer to the above reachability question. Since this reachability problem is a strict special case of reachability for the entire hybrid systems model, the deductive engine is "lightweight" as per the requirement in Sec. 2.

The reader might wonder why a numerical simulator is termed as a deductive engine. Indeed, on the surface a numerical simulator seems quite different from a deductive theorem prover. However, on closer inspection one finds that both procedures employ similar deductive reasoning: they both solve systems of constraints using axioms about underlying theories and rules of inference, and they both involve the use of rewrite and simplification rules.

The overall approach of Jha et al. [23] operates within a fixpoint computation loop that initializes each guard with an overapproximate hyperbox, and then iteratively shrinks entry guards using the hyperbox learning algorithm that selects states, queries the simula-

---

[4]Recall that a hyperbox corresponds to a conjunction of interval constraints over the continuous variables. The requirement for the vertices of the hyperbox to lie on a discrete grid is equivalent to requiring the constant terms in the hyperbox to be rational numbers with known finite precision.

tor for labels, and then infers a smaller hyperbox from the resulting labeled states. They show that their approach can efficiently synthesize safe switching logic for many kinds of systems with monotonic continuous dynamics [23], including the automotive transmission system shown in Fig. 4.

3.2 Timing Analysis of Sofware

The analysis of quantitative properties, such as bounds on timing and power, is central to the design of reliable embedded software and systems. Fundamentally, such properties depend not only on program logic, but also on details of the program's environment. The environment includes several elements — the processor, characteristics of the memory hierarchy, the operating system, the network, etc. Moreover, in contrast with many other verification problems, the environment must be modeled with a relatively high degree of precision. Most state-of-the-art approaches to worst-case execution time (WCET) analysis employ significant manual modeling, which can be tedious, error-prone and time consuming, even taking several months to create a model of a relatively simple microcontroller. See [42] for a more detailed description of the challenges in quantitative analysis of software.

There are several variants of the timing analysis problem. We will consider the following representative timing analysis problem in this paper:

⟨TA⟩ Given a terminating program P, its execution platform (environment) E, and a fixed starting state of E, is the execution time of P on E always at most τ?

If the execution time can exceed τ, it is desirable to obtain a test case (a state of P) that shows how the bound of τ is exceeded.

The main challenge in solving this problem, as noted earlier, is the generation of a model of the environment E. While the problem statement ⟨TA⟩ includes E, complete details of the environment may not be available due to intellectual property issues. Even if a complete description of E is available, it would require substantial abstraction to facilitate timing analysis.

Additionally, the complexity of the timing analysis arises from two dimensions of the problem: the *path dimension*, where one must find the right computation path in the task, and the *state dimension*, where one must find the right (starting) environment state to run the task from. Moreover, these two dimensions interact closely; for example, the choice of path can affect the impact of the starting environment state.

We show in the next section that sciduction offers a promising approach to address this challenge of environment modeling.

Instance of sciduction: GAMETIME. Automatic inductive inference of models offers a way to mitigate the challenge of environment modeling. We have created a new approach, termed GAME-TIME [46, 45, 44], in which a *program-specific timing model* of the platform is inferred from observations of the program's timing that are automatically and systematically generated. The program-specificity is an important difference from traditional approaches, which seek to manually construct a timing model that works for *all* programs one might run on the platform. GAMETIME only requires one to run end-to-end measurements on the target platform, making it easier to port to new platforms. The GAMETIME approach, along with an exposition of theoretical and experimental results, including comparisons with other methods, is described in existing papers [46, 45, 44]. We only give a brief summary here to describe how it is an instance of sciduction.

The central idea in GAMETIME is to view the platform as an adversary that controls the choice and evolution of the environment state, while the tool has control of the program path space. The problem is then a formulated as a game between the tool and the platform. GAMETIME uses a sciductive approach to solve this game based on the following elements:

- *Structure hypothesis:* The platform E is modeled as an adversarial process that selects weights on the edges of the control-flow graph of the program P in two steps: first, it selects the path-independent weights w, and then the path-dependent component π. The edge weights represent execution times of the basic blocks corresponding to those edges. Formally, w cannot depend on the program path being executed, whereas π is drawn from a distribution which is a function of that path. Both w and π are elements of $\mathbb{R}^m$, where m is the number of edges in the CFG after unrolling loops and inlining function calls. We term w as the *weight* and π as the *perturbation*, and the structure hypothesis as the *weight-perturbation model*

  More specifically, the structure hypothesis $\mathcal{H}$ used by GAME-TIME for problem ⟨TA⟩ is to define the space of environment models $\mathcal{C}_{\mathcal{H}}$ to be all processes that select a pair (w, π) every time the program P runs, where additionally the pair satisfies the following constraints (see [45] for details):

  C1: The mean perturbation along any path is bounded by a quantity $\mu_{\max}$.

  C2: The worst-case (longest) path is the unique longest path by a specified margin ρ.

  In general, this structure hypothesis is restrictive: i.e., $\mathcal{C}_{\mathcal{H}} \subset \mathcal{C}_E$, where $\mathcal{C}_E$ is the space of all environments. The restrictions come from conditions C1 and C2, since any environment can be viewed as picking times for basic blocks in terms of a path-independent component w and a path-dependent component π. The implications of C1 and C2 will be discussed later.

- *Inductive inference:* The inductive inference routine is a learning algorithm that operates in a *game-theoretic online setting* The task of this algorithm is to learn the (w, π) model from measurements. The idea in GAMETIME is to measure execution times of P along so-called *basis paths*; these paths are those that form a basis for the set of all paths, in the standard linear algebra sense of a basis. GAMETIME chooses amongst these basis paths uniformly at random over a number of trials, recording the length (time) for each one. A (w, π) model is inferred from the end-to-end measurements of program timing along each of the basis paths.

- *Deductive reasoning:* Timing measurements for basis paths constitute the examples for the inductive learning algorithm. These examples are generated using SMT-based test generation. More precisely, an SMT solver combined with an integer linear programming (ILP) engine generates a set of feasible basis paths. The ILP engine generates candidate basis paths. From each candidate basis path, an SMT formula is generated such that the formula is satisfiable if and only if the path is feasible. Thus, the deductive procedure for GAMETIME is the SMT solver combined with the ILP solver. The procedure is lightweight, as it solves an NP-hard problem, whereas the overall timing analysis is PSPACE-hard, requiring reachability analysis for the composition of the program with an abstract timing model of the platform, typically a finite-state system.

The operation of GAMETIME is sketched in Figure 5. As shown in the top-left corner, the process begins with the generation of the control-flow graph (CFG) corresponding to the program (possibly at the binary level), where all loops have been unrolled to a maximum iteration bound, and all function calls have been inlined into the top-level function. The CFG is assumed to have a single source node (entry point) and a single sink node (exit point); if not, dummy source and sink nodes are added. A feasible set of basis paths (along with corresponding test cases) are then extracted using a combination of ILP and SMT solving. This is the main deductive step of GAMETIME. The program is then compiled for the target platform, and executed on these test cases. In the basic GAMETIME algo-

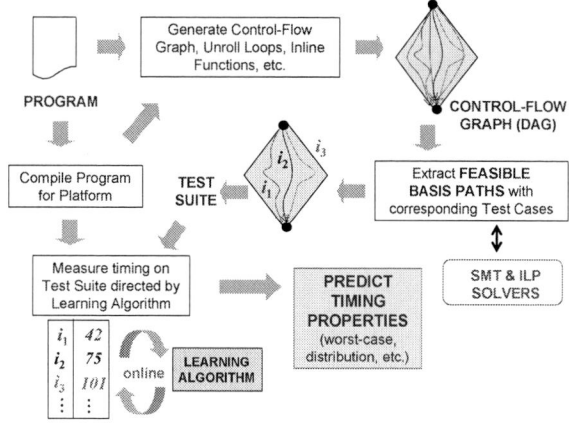

Figure 5: GAMETIME **overview**

rithm (described in [46, 45]), the sequence of tests is randomized, with basis paths being chosen uniformly at random to be executed. The overall execution time of the program is recorded for each test case. From these end-to-end execution time measurements, GA-METIME's learning algorithm — the inductive inference engine — generates the (w, π) model that can then be used for timing analysis. In contrast with most existing tools for timing analysis (see, e.g., [39]), GAMETIME can not only be used for WCET estimation, it can also be used to predict execution time of arbitrary program paths, and certain execution time statistics (e.g., the distribution of times). For example, to answer the problem $\langle TA \rangle$ presented in the preceding section, GAMETIME would predict the longest path, execute it to compute the corresponding timing τ^*, and compare that time with τ: if $\tau^* \leq \tau$, then GAMETIME returns "YES", otherwise it returns "NO" along with the corresponding test case.

Theoretical Guarantees and Empirical Results. Assuming the structure hypothesis holds, GAMETIME answers the timing analysis question $\langle TA \rangle$ with high probability. In other words, if the structure hypothesis is valid, GAMETIME is *probabilistically sound and complete* in the following sense:

> Given any $\delta > 0$, if one runs a number of tests that is polynomial in $\ln \frac{1}{\delta}$, $\mu_{\max}$, and the program parameters, GAMETIME will report the correct YES/NO answer to Problem $\langle TA \rangle$ with probability at least $1 - \delta$.

See the theorems in [46, 45] for details. Every part of the structure hypothesis, with the possible exception of condition C2, is valid in practice. Condition C1 holds since the mean perturbation cannot be unbounded for physical platforms. However, C2 must be validated; this has been done experimentally.

Experimental results indicate that in practice GAMETIME can accurately predict not only the worst-case path (and thus the WCET) but also the distribution of execution times of a task from various starting environment states. These results have been obtained on pipelined processors with instruction and data caches as well as branch prediction [45].

4. DISCUSSION AND FUTURE DIRECTIONS

This paper posits that sciduction, a tight integration of induction and deduction with structure hypothesis, is a promising approach to addressing challenging problems in formal verification and synthesis. Our proposal seeks to mirror the approach a human might take to a verification or design problem, by combining inductive reasoning with systematic deductive processes. Some of the recent

successes in the formal methods area can be seen as instances of sciduction. The structure hypothesis provides a way for a human to provide creative input into the verification process without getting mired in tedious details.

We conclude with a discussion of how the ideas herein can be applied to other problems, and outline directions for future work.

4.1 Insights for Verification and Synthesis

How can one apply sciduction to a new problem in verification or synthesis?

At present, we feel that more experience is needed before particular combinations of induction and deduction are deemed more useful than others. However, one can use a general prescription for applying sciduction in cases where a purely deductive approach falls short, as follows:

1. Identify the hard synthesis sub-task(s) within the overall synthesis or verification problem.
2. Formalize suitable structure hypotheses, if needed, for each sub-task.
3. For each sub-task, devise a top-level inductive or deductive procedure to solve it under the structure hypothesis. This procedure may in turn invoke other inductive or deductive procedures: to generate examples for learning, synthesize candidate artifacts, verify properties of synthesized artifacts, etc.
4. Prove the validity of the structure hypotheses, ideally theoretically, otherwise empirically.

In addition to the work described in Sec. 3, we have applied sciduction in other settings. For example, a major challenge for automatic synthesis from linear temporal logic (LTL) is in writing complete and consistent specifications, of which the environment assumptions are a large part. In recent work [27], we have demonstrated that environment assumptions can be mined from traces and counter-strategies. We have also recently used a combination of induction on decision trees (see [34]) and SMT-based ("term-level") model checking using UCLID [10] to perform conditional term-level abstraction of register-transfer level (RTL) designs [7]. We have used sciduction in the automatic synthesis of loop-free programs, with applications to synthesizing high-performance code and deobfuscating malware [22]. Finally, in the area of controller synthesis for hybrid systems, initial results have been obtained on synthesizing switching logic for optimality, rather than just safety [24].

4.2 Future Directions

There are several directions for future work.

First, recall that the soundness guarantees of sciduction only hold when the structure hypothesis is valid. When this is not trivially true (e.g., $\mathcal{C_H} = \mathcal{C_S}$), one has to use other properties of the problem, such as monotonicity of dynamics in the hybrid systems problem of Sec. 3.1. It would be useful to develop a general, systematic approach for checking the validity of the hypothesis $\mathcal{H}$.

Second, we note that sciduction offers ways to integrate inductive reasoning into deductive engines, and vice-versa. It is intriguing to consider if SAT and SMT solvers can benefit from this approach — for example, using inductive reasoning to guide the solver for specific families of SAT/SMT formulas. Similarly, how can one effectively use deductive engines as oracles in learning algorithms? Are there new concept learning problems that can be effectively solved using this approach?

Finally, the landscape of applications is yet to be fully explored. An interesting direction is to take problems that have classically been addressed by purely deductive methods and apply the sciductive approach to them. As an example, consider again the problem of synthesis from LTL specifications. Another challenge for this problem is to deal with the doubly-exponential computational complexity. It would be interesting to see if the synthesis algorithms themselves can be made more scalable using sciduction. Another direction is to generalize the ideas used for timing analysis to other quantitative properties of cyber-physical systems, and also for verification problems at the hardware-software interface ("hardware-software verification"). In both settings, generating environment models can be quite challenging, and, from our experience with timing analysis, it appears that sciduction can be effectively brought to bear on these problems.

Acknowledgments

This article is a result of ideas synthesized and verified (!) over the last few years in collaboration with several students and colleagues. The contributions of Susmit Jha, in particular, are gratefully acknowledged. This paper has benefited from feedback on talks on this work given by the author at several venues in 2009-11. This work has been supported in part by several sponsors including the National Science Foundation (CNS-0644436, CNS-0627734, and CNS-1035672), Semiconductor Research Corporation (SRC) contracts 1355.001 and 2045.001, an Alfred P. Sloan Research Fellowship, the Hellman Family Faculty Fund, the Toyota Motor Corporation under the CHESS center, and the Gigascale Systems Research Center (GSRC) and MultiScale Systems Center (MuSyC), two of six research centers funded under the Focus Center Research Program (FCRP), a Semiconductor Research Corporation entity.

5. REFERENCES

[1] R. Alur, C. Courcoubetis, N. Halbwachs, T. A. Henzinger, P.-H. Ho, X. Nicollin, A. Olivero, J. Sifakis, and S. Yovine. The algorithmic analysis of hybrid systems. *Theoretical Computer Science*, 138(1):3–34, February 1995.

[2] D. Angluin. Queries and concept learning. *Machine Learning*, 2:319–342, 1988.

[3] D. Angluin and C. H. Smith. Inductive inference: Theory and methods. *ACM Computing Surveys*, 15:237–269, Sept. 1983.

[4] T. Ball, R. Majumdar, T. D. Millstein, and S. K. Rajamani. Automatic predicate abstraction of C programs. In *Proc. ACM SIGPLAN 2001 Conference on Programming Language Design and Implementation (PLDI)*, pages 203–213, June 2001.

[5] C. Barrett, R. Sebastiani, S. A. Seshia, and C. Tinelli. Satisfiability modulo theories. In A. Biere, H. van Maaren, and T. Walsh, editors, *Handbook of Satisfiability*, volume 4, chapter 8. IOS Press, 2009.

[6] I. Beer, S. Ben-David, C. Eisner, and Y. Rodeh. Efficient detection of vacuity in ACTL formulas. *Formal Methods in System Design*, 18(2):141–162, 2001.

[7] B. Brady, R. E. Bryant, and S. A. Seshia. Learning conditional abstractions. In *Proceedings of the IEEE International Conference on Formal Methods in Computer-Aided Design (FMCAD)*, pages 116–124, October 2011.

[8] R. Brayton and A. Mishchenko. ABC: An Academic Industrial-Strength Verification Tool. In *Computer Aided Verification (CAV)*, 2010.

[9] R. E. Bryant. Graph-based algorithms for Boolean function manipulation. *IEEE Transactions on Computers*, C-35(8):677–691, August 1986.

[10] R. E. Bryant, S. K. Lahiri, and S. A. Seshia. Modeling and verifying systems using a logic of counter arithmetic with lambda expressions and uninterpreted functions. In E. Brinksma and K. G. Larsen, editors, *Proc. Computer-Aided Verification (CAV'02)*, LNCS 2404, pages 78–92, July 2002.

[11] M. Case. *On Invariants to Characterize the State Space for Sequential Logic Synthesis and Formal Verification*. PhD thesis, EECS Department, UC Berkeley, Apr 2009.

[12] E. M. Clarke and E. A. Emerson. Design and synthesis of synchronization skeletons using branching-time temporal logic. In *Logic of Programs*, pages 52–71, 1981.

[13] E. M. Clarke, A. Fehnker, Z. Han, B. H. Krogh, O. Stursberg, and M. Theobald. Verification of hybrid systems based on counterexample-guided abstraction refinement. In *TACAS*, pages 192–207, 2003.

[14] E. M. Clarke, O. Grumberg, S. Jha, Y. Lu, and H. Veith. Counterexample-guided abstraction refinement. In *12th International Conference on Computer Aided Verification (CAV)*, volume 1855 of *Lecture Notes in Computer Science*, pages 154–169. Springer, 2000.

[15] E. M. Clarke, O. Grumberg, and D. A. Peled. *Model Checking*. MIT Press, 2000.

[16] Dimitra Giannakopoulou and Corina S. Pasareanu, eds. Special issue on learning techniques for compositional reasoning. *Formal Methods in System Design*, 32(3):173–174, 2008.

[17] M. Ernst. *Dynamically Discovering Likely Program Invariants*. PhD thesis, University of Washington, Seattle, 2000.

[18] C. Flanagan, K. R. M. Leino, M. Lillibridge, G. Nelson, J. B. Saxe, and R. Stata. Extended static checking for Java. *SIGPLAN Notices*, 37:234–245, May 2002.

[19] H. Fox. *Agent problem solving by inductive and deductive program synthesis*. PhD thesis, Massachusetts Institute of Technology, Dept. of Electrical Engineering and Computer Science, 2008.

[20] S. A. Goldman and M. J. Kearns. On the complexity of teaching. *Journal of Computer and System Sciences*, 50:20–31, 1995.

[21] A. Gupta. *Learning Abstractions for Model Checking*. PhD thesis, Computer Science Department, Carnegie Mellon University, 2006.

[22] S. Jha, S. Gulwani, S. A. Seshia, and A. Tiwari. Oracle-guided component-based program synthesis. In *Proceedings of the 32nd International Conference on Software Engineering (ICSE)*, pages 215–224, 2010.

[23] S. Jha, S. Gulwani, S. A. Seshia, and A. Tiwari. Synthesizing switching logic for safety and dwell-time requirements. In *Proceedings of the International Conference on Cyber-Physical Systems (ICCPS)*, pages 22–31, April 2010.

[24] S. Jha, S. A. Seshia, and A. Tiwari. Synthesis of optimal switching logic for hybrid systems. In *Proceedings of the International Conference on Embedded Software (EMSOFT)*, pages 107–116, October 2011.

[25] M. Kaufmann, P. Manolios, and J. S. Moore. *Computer-Aided Reasoning: An Approach*. Kluwer Academic Publishers, 2000.

[26] R. Kurshan. Automata-theoretic verification of coordinating processes. In *11th International Conference on Analysis and Optimization of Systems – Discrete Event Systems*, volume 199 of *LNCS*, pages 16–28. Springer, 1994.

[27] W. Li, L. Dworkin, and S. A. Seshia. Mining assumptions for synthesis. In *Proceedings of the Ninth ACM/IEEE International Conference on Formal Methods and Models for Codesign (MEMOCODE)*, July 2011.

[28] J. Lygeros. Lecture notes on hybrid systems. 2004.

[29] S. Malik and L. Zhang. Boolean satisfiability: From theoretical hardness to practical success. *Communications of the ACM (CACM)*, 52(8):76–82, 2009.

[30] Z. Manna and A. Pnueli. *The Temporal Logic of Reactive and Concurrent Systems:Specification*. Springer-Verlag, 1992.

[31] Z. Manna and R. Waldinger. A deductive approach to program synthesis. *ACM TOPLAS*, 2(1):90–121, 1980.

[32] K. L. McMillan. Interpolation and SAT-based model checking. In *Proc. 15th International Conference on Computer-Aided Verification (CAV)*, pages 1–13, July 2003.

[33] K. L. McMillan. Relevance heuristics for program analysis. In *Proceedings of the 35th ACM SIGPLAN-SIGACT Symposium on Principles of Programming Languages (POPL)*, pages 145–146. ACM Press, 2008.

[34] T. M. Mitchell. *Machine Learning*. McGraw-Hill, 1997.

[35] S. Muggleton and L. de Raedt. Inductive logic programming: Theory and methods. *The Journal of Logic Programming*, 19-20(1):629–679, 1994.

[36] N. Piterman, A. Pnueli, and Y. Sa'ar. Synthesis of reactive(1) designs. In *7th International Conference on Verification, Model Checking, and Abstract Interpretation (VMCAI)*, volume 3855 of *Lecture Notes in Computer Science*, pages 364–380. Springer, 2006.

[37] A. Pnueli and R. Rosner. On the synthesis of a reactive module. In *ACM Symposium on Principles of Programming Languages (POPL)*, pages 179–190, 1989.

[38] J.-P. Queille and J. Sifakis. Specification and verification of concurrent systems in CESAR. In *Symposium on Programming*, number 137 in LNCS, pages 337–351, 1982.

[39] Reinhard Wilhelm et al. The Determination of Worst-Case Execution Times—Overview of the Methods and Survey of Tools. *ACM Transactions on Embedded Computing Systems (TECS)*, 2007.

[40] K. Ruohonen. Undecidable event detection problems for ODEs of dimension one and two. *Informatique Théorique et Applications*, 31(1):67–79, 1997.

[41] S. Russell and P. Norvig. *Artificial Intelligence: A Modern Approach*. Prentice Hall, 2010.

[42] S. A. Seshia. Quantitative analysis of software: Challenges and recent advances. In *Proc. Formal Aspects of Component Software (FACS)*, 2010.

[43] S. A. Seshia. Sciduction: Combining induction, deduction, and structure for verification and synthesis. Technical Report UCB/EECS-2011-68, EECS Department, University of California, Berkeley, May 2011.

[44] S. A. Seshia and J. Kotker. GameTime: A toolkit for timing analysis of software. In *Proc. Tools and Algorithms for the Analysis and Construction of Systems (TACAS)*, 2011.

[45] S. A. Seshia and A. Rakhlin. Quantitative analysis of systems using game-theoretic learning. *ACM Transactions on Embedded Computing Systems (TECS)*. To appear.

[46] S. A. Seshia and A. Rakhlin. Game-theoretic timing analysis. In *Proceedings of the IEEE/ACM International Conference on Computer-Aided Design (ICCAD)*, pages 575–582. IEEE Press, 2008.

[47] R. Sharma, A. V. Nori, and A. Aiken. Interpolants as classifiers. Technical Report MSR-TR-2012-13, Microsoft Research, January 2012.

[48] A. Solar-Lezama, G. Arnold, L. Tancau, R. Bodík, V. A. Saraswat, and S. A. Seshia. Sketching stencils. In *PLDI*, pages 167–178, 2007.

[49] A. Solar-Lezama, L. Tancau, R. Bodík, S. A. Seshia, and V. Saraswat. Combinatorial sketching for finite programs. In *ASPLOS*, 2006.

[50] S. Srivastava, S. Gulwani, and J. S. Foster. From program verification to program synthesis. In *Proceedings of the 37th ACM SIGPLAN-SIGACT Symposium on Principles of Programming Languages (POPL)*, pages 313–326, 2010.

Cost-Efficient Buffer Sizing in Shared-Memory 3D-MPSoCs Using Wide I/O Interfaces

Sahar Foroutan
TIMA Laboratory
46 Avenue Félix Viallet
38031, Grenoble, France
+334-76-57-43-02

Sahar.Foroutan@imag.fr

Abbas Sheibanyrad
TIMA Laboratory
46 Avenue Félix Viallet
38031, Grenoble, France
+334-76-57-48-64

Abbas.Sheibanyrad@imag.fr

Frédéric Pétrot
TIMA Laboratory
46 Avenue Félix Viallet
38031, Grenoble, France
+334-76-57-48-70

Frederic.Petrot@imag.fr

ABSTRACT

This paper addresses link-buffer capacity allocation in the design process of best-effort 3DNoCs holding hotspot memory ports. We show that in 3DSoCs with integrated wide I/O DRAMs, the congestion spreading is different from SoCs with external DRAMs: the bottlenecks are not anymore the external memory ports but the network links that become saturated and retro-propagate the congestion. The distribution of bottleneck links is directly affected by the traffic directed to the hot memory ports. Using an analytical performance evaluation method, we determine network link buffer capacities according to the given workload composed of regular and hotspot traffics.

Categories and Subject Descriptors

B.8.2 [Performance and Reliability]: Performance Analysis and Design Aids; B.3.3 [Memory Structures] Performance Analysis and Design Aids--Formal models; B.4.3 [Input/Output and Data Communications]: Interconnections (subsystems)--Topology (eg. bus, point-to-point)

General Terms

Design, Performance

Keywords

Multiprocessor Systems-on-Chip (MPSoCs), Networks-on-Chip (NoCs), Performance Analysis.

1. INTRODUCTION

In typical multi-core SoC (System-on-Chip)-based devices, many applications (e.g. HD video, image processing and 3D gaming) request data from shared memories leading to a drastically increasing demand on memory bandwidth. In current SoC architectures a significant part of the required shared memory is located outside the chip (i.e. off-chip DRAMs). Next-generation handheld devices with 3DHD video and 3D graphics will require a total memory bandwidth of 10's of GB/s. It is even possible that these requirements reach 1 TB/s in the coming years [1]. Fast memory interfaces (e.g. DDR and LPDDR) have emerged to cope with the increasing demanding memory bandwidth. However, these standards have limitations when the bandwidth must be further increased, mainly due to high frequency in a noisy environment. Reducing the frequency is possible if the bus width

can be enlarged. But, current MPSoCs cannot afford this solution due to limitations on the number of external pads [2].

Emerging 3D-Integration [3], by enabling the stacking of dies fabricated in different technologies, allows the migration of dense dynamic memories inside chips. Using Through-Silicon-Vias (TSV), 3D-Integration provides a unique opportunity to propose new wide data bus interface, called *Wide I/O*, between the MPSoC and the stacked memory die. With wider and slower memory interfaces, Wide I/O improves the bandwidth and reduces power compared to existing memory interfaces [2, 4].

In spite of all advantages of on-chip stacked-memories, hotspot dealing, as a common concern of MPSoCs with a centralized shared memory, remains but as shown in this paper in a different shape. Traditionally a hotspot is defined as a module receiving traffic from the majority of network sources with an aggregate rate exceeding the rate at which it can absorb data [5, 6]. When the arriving traffic persists indefinitely the hotspot causes a serious congestion situation in the network, called *tree saturation* [6]: when a memory buffer becomes saturated it forces that router's link-level flow control to throttle back all the inputs feeding the memory module (since the network is lossless), that in turn cause the previous routers to fill their buffer and so on back to the traffic sources. An external shared memory module is a hotspots and causes a serious bandwidth bottleneck in the system [7], while this concern is less important with Wide I/O 3D-DRAMS as their bandwidth is not less than the NoC bandwidth. This paper shows that with Wide I/O memories, as opposed to off-chip memories, the root of a tree saturation is not necessarily the memory port. Instead, because of NoC limited buffering capacity, network links saturate and become system bottlenecks.

Several dynamic software or hardware methods (mostly for off-chip networks) have been proposed to detect the hotspot contention at runtime and thus to avoid, prevent or remove the saturation tree as soon as detected (e.g. in [6, 8, 9]). Hardware techniques are more efficient since according to [8] the congestion tree is filled in less than 10 traversal times of the network – far too quickly for software to react in time to the problem. Most of the hardware techniques are too costly when applied to NoCs that are extremely cost-constrained architectures. *Hardware combining* [7] in which each router combines hotspot requests into a single one, *Multipath networks* [7, 9, 10] in which extra paths are added to divert the regular traffic around the hotspot (requires large number of buffers, virtual channels, multiplexors, demultiplexors, etc.), *Discard strategies* [11] that forwards only one request to the hot memory and discards others, and the technique of *Deflecting* packets away of loaded locations [12] (requires adaptive routing), are some of these techniques.

This work is supported by Catrene Project CT105-3DIM3

Among dynamic techniques, *rate-regulation* strategies have been widely implemented in both off-chip networks (e.g. [7]) and NoCs. (e.g. [2, 5]). Generally speaking, the hot memory is monitored dynamically and processors are notified (usually by using an end-to-end flow control mechanism) to stop or regulate the data injection if memory is at a congestion state. Since these strategies need to detect the congestion sate at runtime, they are usually expensive to implement and thus more appropriate for real-time applications that need guaranteed services. Also, in end-to-end flow control strategies, the problem is not solved at the network layer; instead, it's transferred to upper layers such as network-interface or application layers. While for the best-effort networks a simple and cost-efficient solution must be provided at the network layer.

In this paper we do not intend to remove the hotspot problem dynamically. Instead, by properly sizing the NoC buffers we try to alleviate the impact of hotspot traffic on the network performance in terms of network average latency and saturation point. The proposed buffer sizing method is static and applied at design time. It addresses best-effort general purpose NoCs or those applications specific NoCs in which the traffic distribution (in terms of the average data rate transferred between different communicating IP pairs) is known at design time. Using the analytical method proposed in [13], we have developed a tool to automate the buffer sizing process. Hu et al. in [14] and Guz et al. in [15] take similar approaches in terms of using an analytical delay model as the heart of the buffer allocation method. However their buffer allocation algorithms are completely different from ours. They both use greedy algorithms while we distribute the buffering budget simply according to the distribution of the workload on network links (resulting in a less complex method).

The rest of this paper is organized as follows. Section 2 discusses how the presence of hotspot nodes deteriorates the whole NoC performance. It also presents the assumptions of this work. The impact of 3D-specific wide-interfaces and NoC architectural parameters on the congestion is described in section 3. Section 4 shows how the system bottlenecks are distributed over the network. In order to alleviate the system performance, in Section 5 we propose a buffer sizing method to assigns non-uniform buffer capacities to network links. We also compare the NoC cost-performance, before and after buffer sizing. Finally we summarize our contributions in section **Error! Reference source not found.**.

2. PERFROMANCE DEGRADATION DUE TO HOTSPOTS

The performance metric taken as reference of comparisons in this paper is the *network saturation point* as defined in the following.

2.1 Saturation Point as a Performance Metric

The network saturation point is the offered-load at which the network gets saturated and reaches its *maximum sustainable throughput* and thus cannot accept more offered-load from traffic sources. The term *offered-load* is used as the rate of flit injection by each core. Therefore the same unit of measurement for both offered-load and saturation point is used which is flits/cycle (shown in %). We use average-latency/offered-load curves to determine the network saturation point. When the network is saturated the average latency tends to infinity [16, 17]. Thus the offered-load at which the average latency grows infinitely equals to the NoC saturation point. The average latency is defined as the average latency of all possible paths of the network.

To obtain the latency/load curves we use an ameliorated version of the analytical method proposed in [13] (see the appendix added to the end of this paper). The method is based on a router delay model that determines different delays inside a best-effort wormhole router (e.g. link acquisition and link transfer delays) within an acceptable accuracy. The model deals with the direct contentions between different flows coming to a router. It also handles the back pressure impact that happens in sequences of routers. For a given NoC (in terms of topology, routing, and link capacities) and its workload consisting of a regular traffic (determined by the traffic spatial distribution and data generation rate) and a hotspot traffic (determined by the number of hotspots, their location, and the fraction of each source load sent to each hotspot), the method provides an average delay analysis of the network layer. More precisely it provides average per-path latency (i.e. the latency of any desired path in the network), average router latency (the latency between any I/O ports of a router), average buffer utilizations and the probability of contention for accessing any shared link of the network. Using the method we obtain the latency/load curves and thus determine the network saturation point. The method is very fast and converges in less than one minute for experiments that require up to two days of Cycle-Bit-Accurate simulation. Therefore regarding the number of analyses and the dimension of the 3D-NoC used in our experiments, the use of the analytical method is more time-efficient than using simulations.

2.2 Saturation-Point as a Function of Hotspot-Fraction

When the tree saturation occurs even the normal packets (i.e. packets whose destinations are not hotspot modules) are affected by congestion. This is more serious in best-effort networks where hotspot and non hotspot packets compete for the same resources. Hence the network as a whole suffers a catastrophic loss of performance. This conclusion applies to any network but hotspot effects in a wormhole network are more severe than in a store-and-forward one, as packets are blocked across multiple routers and buffering space is limited. To better understand the importance of the hotspot impact on the NoC performance, Figure 1 shows the saturation points of a 3D-NoC obtained for different amounts of hotspot fraction (h). In this experiment (and other experiments of the paper) it is assumed that the 3D-SoC contains a NoC with a fully connected 3x5x5 3D-Mesh topology. Since the emergence of 3D-SoCs, other topologies have been proposed for 3D interconnection that may exhibit better performance for 3D shared-memory MPSoCs (e.g. the network-in-memory architecture proposed in [18]). However, in this work we do not aim to investigate the impact of topology since the occurrence of tree saturation is independent from the network topology [8]. In general in any network holding at least one hotspot a tree saturation appears [8]. Only the shape of the tree may change form one topology to another. In the experiment of Figure 1 the NoC contains only one hotspot located on the center of its upper die.

We assume that two kinds of traffic are simultaneously running namely *hotspot* and *regular* traffic. To generate hotspot traffic we use the model proposed in [6]. A new generated packet has a probability h, called the *hotspot fraction*, of being directed to the hotspot node and probability ($1-h$) of being directed, according to the regular traffic, to the other network nodes. Nodes generate traffic independently of each other following a Poisson process with a mean rate of λ flits/cycle (λ = offered-load), consisting of hotspot and regular portions of $h\lambda$ and ($1-h$)λ, respectively. When there are more than one hotspot modules, h signifies the total hotspot fractions sent to all hotspots. For example when four hotspots exist and each node sends h_i% of its offered load (λ) to the *i-th* hotspot, h is equal to $\sum_{i=1}^{4} h_i$. The regular traffic is distributed uniformly in the following experiments, but it can follow any

367

spatial distribution determining the average data rate (λ) transferred between each communicating IP pairs. The hotspot memory controllers are assumed to absorb data flits at a rate equal or greater than the data arrival rate (the network bandwidth <= memory bandwidth). Regarding the high bandwidth of the new generation of stacked 3D-DRAM protocols (e.g. Wide I/O) this assumption is realistic. Similarly other destinations (non hotspot ones) are also assumed to behave as sinks.

As it is shown in Figure 1, when the hotspot fraction increases the saturation point decreases considerably. For a hotspot fraction greater than 11% (i.e. each node sends only 11% of its total offered-load to the hot memory) the saturation point falls under 10%. This is whereas the same network is capable to sustain up to 45% of offered-load when the traffic is uniformly distributed (h=0). This example shows how drastically a hotspot module can deteriorate the overall NoC performance.

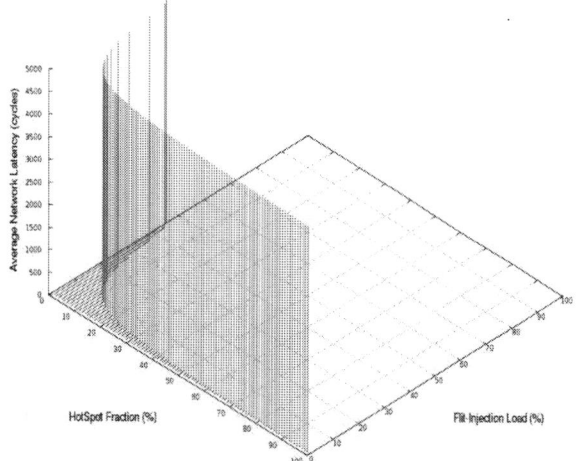

Figure 1. The NoC saturation point determined by latency/load curves for various amounts of hotspot fraction

3. THE EFFECTS OF SoC WORKLOAD AND ARCHITECTURAL PARAMETERS

This section investigates the impact of the parameters related to the SoC workload (e.g. number and location of memory controllers shaping the hotspot traffic) and also its architectural parameters (e.g. routing algorithm) on the SoC performance.

3.1 4-Port Wide I/O Memory Interface

Memory vendors and JEDEC (JC-42.6) are currently developing standards for Wide I/O memory interface with TSV interconnects stacked on SoC [2, 4]. Wide I/O calls for a 512-bit wide data interface providing the total peak bandwidth of 12.8GB/second (with 200 MHz I/O bus clock). The chip architecture of a Wide I/O DRAM is made up of four independent memory partitions (channels or ranks) which are symmetric with respect to the chip center. Each channel has 4 Banks, with a density per channel between 256Mb and 8Gb.

Although there are some works in the literature focusing on the location of memory controllers in on-chip many-core systems [19], few of them concern directly 3D-NoCs and Wide I/O memories. According to [2, 4] the Wide I/O memory controllers are located on the logic die and not on the DRAM die. To access the four channels either a single centralized memory controller is considered as a unique resource shared among the sub-systems (similar to Figure 2.A) or, four independent controllers manage parallel traffic streams to the four independent memory channels (like Figure 2.B). In this case each channel has its own 128-bit

wide PHY interface working at 200MHz, thus providing 3.2 GByte/s bandwidth which is approximately in the range of the link bandwidth of current NoCs [20]. In our experiments presented in the remaining of the paper we consider both cases.

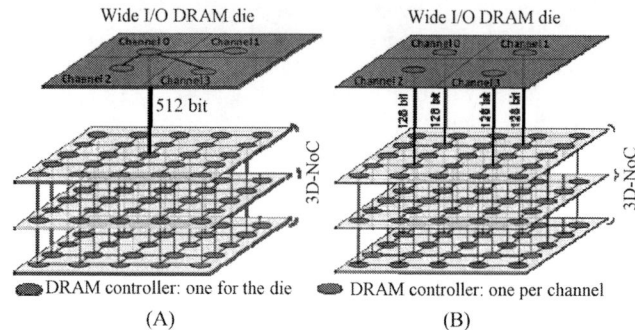

Figure 2. Wide I/O DRAM controller schemes: (A) one 512-bit wide controller for the DRAM die and (B) four separate 128-bit wide controllers for the four channels

3.2 Hotspot and Architectural Parameters

Figure 3 shows the saturation points (y-axis) obtained for different amounts of hotspot fraction in the 3D-NoC with one and four DRAM controllers located like Figure 2 (note that Figure 3 gives a 2D-view of the x-y plane of the 3D curves similar to Figure 1 when looking down from above). It is shown that having four memory controllers instead of one, results in a significant increase in saturation point. This is because the hotspot traffic is distributed more homogenously. As a result collision and congestion of the network decrease. Therefore the maximum sustainable throughput increases. However, even with four controllers the saturation point degrades considerably when the hotspot fraction rises. Low saturation points signify that such workloads (that mimic the behavior of applications requiring a huge amount of the shared memory bandwidth), are practically unfeasible on the NoCs in which the total buffering budget is uniformly distributed between network links (link have identical buffer length).

An interesting point in Figure 3 is that the peak saturation point with four memory ports is achieved when the hotspot fraction is equal to 6% and not 0%. This signifies that in order to have a better performance, what is important is the lower contentions over the whole network, due to the combination of both hotspot and regular traffics. In this experiment the best case achieves when 6% of the load of each node is sent to the four hotspots and the rest sent uniformly to other destinations.

Apart from the hotspot traffic parameters, the NoC architectural parameters are also of crucial importance on the propagation of congestion, and thus the network saturation-point. Figure 4 compares latency/load curves related to different configurations of DRAM controllers and different routing algorithms. Three configurations are shown: 1) a single memory controller (1-port) located in the center of the upper die, 2) four memory controllers (4-port) located in the center of each corresponding partition of the upper die (Figure 2.B), and 3) four memory controllers on the four corners of the upper die. The dimension-ordered routing algorithms of XYZ (first X, then Y, and last Z) and ZXY are applied to each configuration. The hotspot fraction is set to 25%. As we can observe these parameters have influence on the network performance. When the controllers (either one or four) are nearer to the center of the NoC, the saturation point is higher with ZXY routing (the effect is more important for the case of one single memory controller), whereas, with four controllers located on the corner nodes, the highest saturation point is achieved when the XYZ routing is applied.

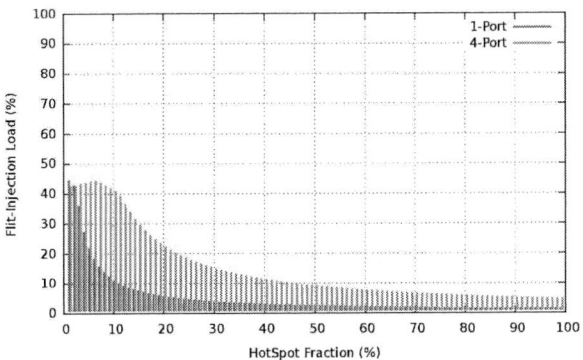

Figure 3. Saturation point as a function of hotspot fraction

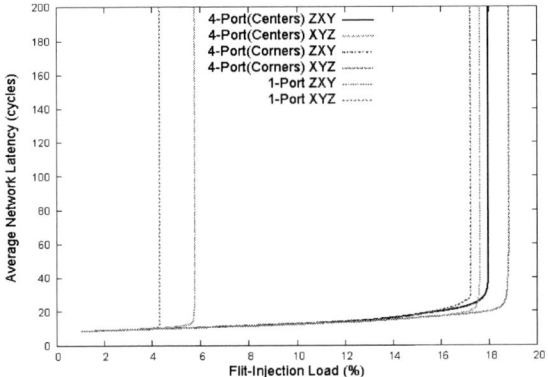

Figure 4. The impact of routing algorithm and hotspot locations on the performance of a NoC containing one or four hotspots (h=25%)

4. BOTTLENECK DISTRIBUTION

In NoCs with external memory the hot memory port is the main bottleneck since it cannot absorb data with the request arrival rate (i.e. the network bandwidth > the memory bandwidth). However, the new generation of 3D wide-interface DRAMs makes the memory interfaces to work with approximately the same rate as the NoC (even much faster if one 512-bit port is used instead of four 128-bit ports). In such systems even with a high rate of memory bandwidth demand, the memory port is not the main root of the tree saturation in the system. Instead, network contention due to high traffic toward the hotspot node causes some of the network links become congested and perform tree saturations behind them. It means that the system hotspots are not necessarily the network bottlenecks. Instead, because of the network limitations (e.g. high contentions and limited buffering capacity), any of the network links can become a system bottleneck. Thus, let's distinguish henceforth a hotspot from a bottleneck such that a *hotspot* signifies a hot destination (here a DRAM controller), and a *network bottleneck* is a congested buffer that starts to and thus throttles all upstream buffers such that a tree saturation appears.

The distribution of bottlenecks depends on the NoC topology, traffic shape, and routing algorithm. For example Figure 5 (A) and (B), presenting the spectrum of buffer-utilizations, show that the distribution of congested links can be completely different in the same NoC when only one parameter changes. The same NoC configuration is considered in both figures: uniform regular traffic, 4 memory controllers on the centers of the upper die, ZXY routing, 8-flit buffer per link, and 16-flit packets. The only parameter that differs is the hotspot fraction that is 5% in (A) and 25% in (B).

- With *h*=25% the hotspot traffic is too high. So the most congested buffers are the input ports of the routers connected to the 4 hotspots. Note that all of the hotspots' input links are

not congested and they are not equally loaded (because of the effect of the routing algorithm). This testifies that the roots of saturation trees are these links, and not the memory ports.

- With *h*=5% the workload is distributed more uniformly and thus the buffers saturate in a more uniform way. Just before the NoC saturation point, the most loaded buffers are those that are located on the middle die (shown by arrows) far from hotspots. More interesting, these links do not even belong to the saturation trees due to any of the four hotspots (regarding the ZXY routing).

4.1 How to find Bottlenecks

Since bottlenecks severely affect the network throughput and average latency, we would like to recognize the regions of the network that are more sensitive to the accumulation of data flits. In order to do this we have implemented a tool that determines the distribution of traffic in the buffers of the network by obtaining the *average buffer utilizations* (also called *average buffer occupancy*) for all buffers (like Figure 5). The tool uses the aforementioned analytical method [13] (described in the appendix) to determine average buffer utilizations for a given workload. It selects an operational offered load very close to the saturation point and obtains buffer utilizations and determines the links almost saturated at this offered load. These buffers are the bottlenecks and cause retro-propagation of congestion to upstream buffers and thus a saturation tree is build behind any of them.

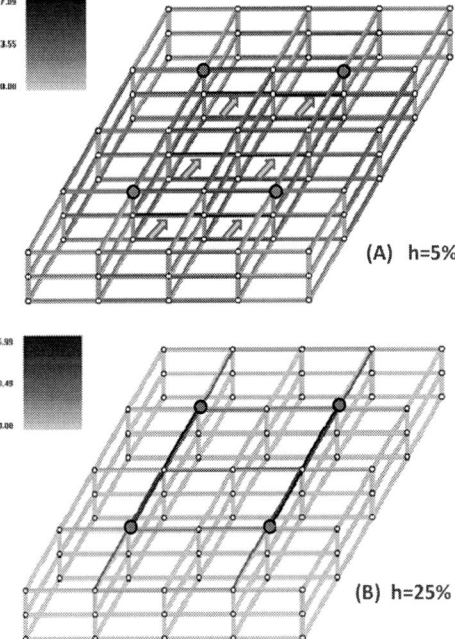

Figure 5. The distribution of congested buffers: In the same NoC (the ZXY routing, buffer length=8 flits, Uniform background traffic), the congested buffer distribution is completely different in (A) with h=5%, and (B) h=25%.

5. BUFFER SIZING

As known, if the capacity of link buffers in a wormhole network increases the saturation point of the network increases. In a wormhole packet-switching network when a packet is traversing the network, all resources in the path between the header (first flit) and the trailer (last flit) are allocated to the packet and no other packet can use those resources. This is the reason that contentions rises in the network as there are less free resources to be allocated to new packets. Now by a higher capacity for link buffers we let

369

the packets occupy a shorter part of a path and thus the resources of the rest of the path are released. Reminding that more than 80% of the area of a router belongs to its buffers [20], due to cost constraints assigning large buffers to all network links is not practically feasible. On the other hand, as seen in Figure 5, the distribution of workload is not homogeneous in all links. For example with higher value of hotspot fraction most of the links have little buffer utilization and there are only few highly loaded buffers in the networks. Using large buffers (more than the need of links) is a waste of area and power. In the following we propose a cost-efficient buffer capacity allocation method that resizes buffers proportionally to their average occupancy:

- Performing an initial analysis to determine the network saturation point. For this initial analysis the buffer length is assumed to be infinite (in practice a very large buffer, e.g. 200 times the average packet length which is 3200-flit buffer in our experiments is assigned to each link).

- Once the saturation point is known, determining the average buffer utilization for all network links at a load just before the NoC saturation point using infinite FIFOs (e.g. 3200 flits).

- Determining the normalizing scale factor as follow:

$$Scale = \frac{Max_{Len} - Min_{Len}}{Bottleneck}$$

where *Bottleneck* is the buffer utilization of the most congested link, and Max_{Len} is the largest buffer the designer intends to allocate to the bottleneck link and is given to the tool as an input parameter. Min_{Len} is the buffer length (determined also by the system designer) allocated to buffers that are too slightly loaded (i.e. the buffers that do not receive important traffic but because of the architectural reasons and in order to absorb occasional bursts their existence is needed).

- Optimizing the buffer length of each network link as follow:

$$Buffer = \lfloor Utilization \times Scale \rfloor + Min_{Len}$$

Where *Utilization* = the average buffer utilization of each link

We believe that distributing the buffering budget according to the average buffer utilization is one of the most straightforward and efficient ways for link capacity allocation. Although more complex methods like the methods proposed in [14, 15] can be used for this purpose, our experimental results show how proper buffer allocating at design time can help to mitigate the hotspot impacts, even when a simple and intuitive method is used.

Besides, one can argue that buffer sizing methods need to iterate as removing bottlenecks can just move them somewhere else, or even results to spontaneously remove some others according to the link dependencies. It is true but please remember that the use of huge FIFOs, as described in the first step of our algorithm, is exactly for this reason. In fact using theoretically infinite FIFOs (e.g. 3200 flits in our experiments) does not allow the congestion to be retro-propagated to upstream buffers and consequently there is no any dependency between congested links. Our experiments also proved that doing iterations does not change the value of normalized FIFOs significantly and we will have almost the same results.

Figure 6 shows the saturation-point gain achieved by our buffer sizing tool. Two sets of curves are demonstrated: the first four curves from left correspond to 1-port memory controller and the next four correspond to the case of four separate memory

controllers. For both sets we obtain the saturation point by setting the size of all buffers to 8 flits. See the curves of "F8 1-port" and "F8 4-port". Then we replace all buffers with very large ones (3200-flit) to imitate the infinite link capacity and observe how the network saturation point improves (the most right hand curves in each sets). As can be seen the performance improvement is significant. For normalizing the buffer lengths we have chosen Min_{Len}=4 flits and Max_{Len}=320 flits. Maybe the buffer length of 320 flits seems too costly at the first glance but after the assignment of buffer lengths (according to their average utilization) most of the links are assigned with short buffers. For each buffer length between 4 and 320 flits Table 1 shows how many links are assigned with that buffer length. The average buffer length after optimizing between 4 and 320 is 5.74 for 1-port memory and 14.67 for 4-port memory and the corresponding curves in Figure 6 are "FA5.74(4~320) 1-port" and "FA14.67(4~320) 4-port". As can be seen the buffer sizing results in a significant gain in the network saturation that is quite near to the saturation point of the network with 3200-flit buffers.

In order to show the importance of buffer sizing we perform the fourth analysis for each set, in which the buffer length is identical for all buffers and equal to the average of buffer lengths after optimization between 4 and 320 (i.e. equal to 6 for 1-port memory in the curve of "F6 1-port", and 15 for four-port memory in the curve of "F15 4-port"). The poor saturation points of the last experiment emphasize the importance of distributing the link capacities with regard to their buffer utilizations at equilibrium, rather than distributing the same total amount of capacity uniformly between all links (i.e. identical length for all buffers).

As a last point please note that the choice of the NoC architecture and workload was just an example for the experiments of this paper. Our method is completely generic for arbitrary topology, deterministic routing algorithms, and target applications that their traffic characteristics are determined at design time.

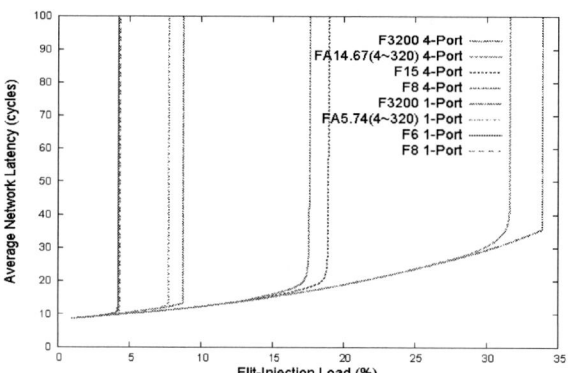

Figure 6. Performance enhancement due to buffer sizing with h=25%

5.1 Cost-Performance Analysis

Table 2 and Table 3 present rough cost-performance comparisons between different configurations with one and four memory controllers respectively. The cost of a configuration signifies the normalized average buffer length and the performance of a configuration signifies the normalized network saturation point in the configuration. The cost and performance of the configuration with identical 8-flit buffers everywhere are assumed to be 1, thus

Table 1. Buffer length distribution

Buffer Length	4	5	6	7	8	9	12	13	14	17	18	19	20	21	22	23	26	36	39	42	43	48	58	59	97	98	128	131	318	319	320
Number of links (4-port memory)	-	124	46	46	12	46	10	-	4	3	4	13	1	4	2	1	-	1	3	2	2	4	2	2	-	-	1	3	1	2	1
Number of links (1-port memory)	328	-	4	-	-	-	-	2	-	-	-	-	-	-	-	1	2	-	-	-	-	-	-	-	1	1	-	-	-	-	1

370

the reference of the comparisons. The most considerable point of these comparisons is the last line of Table 2 that shows with a reduction in the cost (about 29%), if the buffer capacities are distributed according to the buffer utilizations, we can obtain a gain of 78% in performance.

In order to show the cost-performance analysis of all configurations and the impact of our buffer sizing method for XYZ and ZXY routings or location of hotspots, we present a normalized comparison summary of all configurations in Table 4. The reference configuration is that of one hotspot, XYZ routing, and identical 8-flit buffer length. Considering the best case of the table, it shows that if we split the memory controller in four separate parts located at four corners of the upper die, and if we distribute a normalized buffer capacities between 4 and 320 (averagely 13.80) flits according to our method, we can improve the performance by a factor of 775 %, while the cost is increased only 72%.

Table 2. Cost-Performance comparison between different buffer capacity allocations when the SoC contains one DRAM controller

Configuration	Cost	Performance
F8 1-Port	1	1
F6 1-Port	0.75	0.97
F3200 1-Port	400	2.01
FA5.74 (4~320) 1-Port	0.71	1.78

Table 3. Cost-Performance comparison between different buffer capacity allocations when the SoC contains four DRAM controller

Configuration	Cost	Performance
F8 4-Port	1	1
F15 4-Port	1.87	1.07
F3200 4-Port	400	1.92
FA14.67 (4~320) 4-Port	1.83	1.79

Table 4. Cost-Performance comparison between all experiments

Configuration	Cost	Performance
F8 1-Port XYZ	1	1
FA5.74 (4~320) 1-Port XYZ	0.71	1.78
F8 1-Port ZXY	1	1.33
FA6.47 (4~320) 1-Port ZXY	0.80	1.69
F8 4-Port (centers) XYZ	1	4.06
FA14.67 (4~320) 4-Port (centers) XYZ	1.83	7.30
F8 4-Port (centers) ZXY	1	4.14
FA20.08 (4~320) 4-Port (centers) ZXY	2.51	6.42
F8 4-Port (corners) XYZ	1	4.34
FA13.80 (4~320) 4-Port (corners) XYZ	**1.72**	**7.75**
F8 4-Port (corners) ZXY	1	3.97
FA17.43 (4~320) 4-Port (corners) ZXY	2.17	7.33

6. CONCLUSION

In this paper we studied the impact of hotspot shared memory (DRAM) on the system-level performance (in terms of average packet latency and network saturation point) of a 3D-SoC, when the DRAM is stacked on top of the 3D-NoC. We have shown that as opposed to the current external shared memories, with the high bandwidth provided by 3D specific wide-interfaces (such as wide I/O), the memory port is not itself the bottleneck of the system. Instead the bottlenecks are the network links that, affected by the heavy traffic addressing the hot memory, become saturated and retro-propagate very rapidly the congestion to upstream links. This results in a drastic degradation of system performance. To deal with this problem in wormhole best-effort NoCs, we have proposed a simple method to distribute the total buffering budget according to the distribution of the traffic on the network links, at design time. Using a mix of uniform random and hotspot traffics, we have shown that the buffer capacity assignment method significantly improves the NoC performance compared to a uniform buffer capacity assignment.

As the future direction of the work we plan to use more complex buffer sizing algorithms such as those proposed in [14] and [15] and analyze their efficiency and accuracy against our method.

REFERENCES

[1] P. D. Franzon, et al., "Creating 3D specific systems: architecture, design and CAD," in *Proceedings of the Conference on Design, Automation and Test in Europe (DATE10)* 2010, pp. 1684-1688.

[2] B. Akesson, et al., "Memory Controllers for High-Performance and Real-Time MPSoCs," presented at the CODES+ISSS, Taipei, Taiwan, 2011.

[3] A. Sheibanyrad, et al., *3D Integration for NoC-based SoC Architectures*: Springer Publishing Company, Incorporated.

[4] J. S. Kim, et al., "A 1.2 V 12.8 GB/s 2Gb mobile Wide-I/O DRAM with 4× 128 I/Os using TSV-based stacking," in *International Solid-State Circuits Conference*, 2011, pp. 496-498.

[5] A. C. Isask'har Walter, et al., "«Curing Hotspots in Wormhole NoCs»," presented at the Design, Automation and Test in Europe (DATE06), 2006.

[6] G. F. Pfister and V. Norton, "Hot spot" contention and combining in multistage interconnection networks," *IEEE TRANS. COMP.*, vol. 34, pp. 943-948, 1985.

[7] S. P. Dandamudi, "Reducing hot-spot contention in shared-memory multiprocessor systems," *Concurrency, IEEE*, vol. 7, pp. 48-59, 1999.

[8] G. Pfister, et al., "Solving hot spot contention using infiniband architecture congestion control," *Proceedings HP-IPC 2005*, 2005.

[9] J. Duato, et al., "A new scalable and cost-effective congestion management strategy for lossless multistage interconnection networks," in *11th International Symposium on High-Performance Computer Architecture (HPCA'05)*, 2005.

[10] M. C. Wang, et al., "Using a multipath network for reducing the effects of hot spots," *IEEE Transactions on Parallel and Distributed Systems*, vol. 6, pp. 252-268, 1995.

[11] W. S. Ho and D. L. Eager, "A novel strategy for controlling hot spot congestion," in *International Conference on Parallel Processing*, 1989, pp. 14-18.

[12] P. T. Gaughan and S. Yalamanchili, "Adaptive routing protocols for hypercube interconnection networks," *IEEE Transactions on Computers*, pp. 12-23, 1993.

[13] S. Foroutan, et al., "An analytical method for evaluating network-on-chip performance," in *DATE '10 Proceedings of the Conference on Design, Automation and Test in Europe* 2010, pp. 1629-1632.

[14] J. Hu and R. Marculescu, "Application-specific buffer space allocation for networks-on-chip router design," 2004, pp. 354-361.

[15] Z. Guz, et al., "Efficient link capacity and QoS design for network-on-chip," in *Design Automation & Test in Europe Conference* 2006, p. 11.

[16] E. Salminen, et al., "On network-on-chip comparison," in *DSD 2007. 10th Euromicro Conference on Digital System Design Architectures, Methods and Tools*, 2007.

[17] P. P. Pande, et al., "Performance Evaluation and Design Trade-Offs for Network-on-Chip Interconnect Architectures," *IEEE TRANSACTIONS ON COMPUTERS*, pp. 1025-1040, 2005.

[18] F. Li, et al., "Design and management of 3D chip multiprocessors using network-in-memory," presented at the 33rd International Symposium on Computer Architecture (ISCA'06), Boston, Massachusetts 2006.

[19] D. Abts, et al., "Achieving predictable performance through better memory controller placement in many-core cmps," *ACM SIGARCH Computer Architecture News*, vol. 37, pp. 451-461, 2009.

[20] A. Sheibanyrad, et al., "Multisynchronous and fully asynchronous NoCs for GALS architectures," *IEEE Design and Test of Computers*, pp. 572-580, 2008.

Appendix: An Iterative Computational Technique for Performance Evaluation of NoCs

The design of a NoC-based system usually starts with a design space exploration phase whose goal is to find the optimum NoC instance by evaluating the performance of different candidates. To date the NoC performance evaluation relies largely on simulation based approaches. It is true that such approaches can offer the highest accuracy but each simulation run may take considerable time to evaluate only a single NoC configuration. Therefore, it is practically impossible to use simulation for automatic design space exploration. A more efficient way would be to use analytical methods during the initial stages of the design to accelerate the design space exploration and then, simulation-based methods when the design space is reduced to a relatively small set. In this 4-page appendix we densely describe a generic and parametric NoC performance analysis method which addresses best-effort NoCs with wormhole packet-switching strategy. Given the NoC topology, routing algorithm, router characteristics, target application and its mapping on the network, our method provides average per-path latency (or per-flow latency, i.e. the latency of any desired path in the network), average router latency (the latency between any I/O ports of a router), average buffer utilizations and the probability of contention for accessing any shared link of the network.

1. THE ANALYTICAL METHOD

The method aims to analyze the latency of any desired path. Other performance metrics are obtained either during the latency analysis procedure (such as buffer utilization and router delay components) or are results of it (such as saturation threshold).
Path latency is defined as the average latency of a *tagged packet* crossing a user-specific path, obtained from the sum of the average latencies of the path routers:

$$\textbf{\textit{Average path latency}} = \boldsymbol{l_{c2e}^{r_1} + l_{w2e}^{r_2} + \cdots + l_{w2e}^{r_{n+1}} + l_{w2c}^{r_n}}$$

where l_{i2o}^r depicts the average latency of crossing router r from its input port i to output o (for 2D-meshes, the I/O ports are identified by their directions e.g. c for core, e for east etc.).
Due to resource sharing in BE NoCs, the tagged packet may have contention with packets coming from other flows. These packets are called *disrupting packet*. We distinguish two kinds of contention called *direct* and *indirect* contentions:

- Direct contention happens in *one router* and causes a *cyclic dependency* in the computation of latencies of different flows incoming to the router. This is physically explained by the reciprocal impact that different flows produce on the average latency of each other when they are in direct contention. The cyclic dependency is resolved by an *iterative technique*.

- Indirect contention happens in a *sequence of routers*. In a BE wormhole network a chain of packets with different destinations may stay blocked one after the other over a sequence of routers. It means that the latency of each router is a function of the latency of its following routers (downstream routers in the sequence). To deal with this *acyclic dependency*, we build the *dependency tree* and then recursively compute router latencies from the leaves of the tree backward to its root.

The dependency tree is to compute the latency of the source router represented by the root of the tree. A node of the tree, illustrated by the couple (r, i2o), represents the latency l_{i2o}^r. The edges going out of a node determine its dependencies. Thus the children of each node represent the latencies of the following router to which the node is dependent (i.e. the latencies between the corresponding input port and all possible outputs of the

downstream router toward which the disrupting packets can be forwarded). The children of node (r, i2o) are determined from the topology and routing algorithm. We determine the set of all output ports of the router r' that a packet coming from input port i' can be forwarded to, called *Forwarding set* ($\mathcal{F}_{i'}^{r'}$):

$$\mathcal{F}_{i'}^{r'} = \{o' \in O_{r'} | \Gamma_{r'}(i', o') \neq 0\}$$
$$\Gamma_r(i, o) = \sum_s \sum_d \mathcal{R}_{s \to d}(i, o) \qquad (1\text{-}1)$$

where for a given router r, O_r depicts the set of its output ports, $\Gamma_r(i,o)$ the number of flows crossing trough input port i and output port o of r, and $\mathcal{R}_{s \to d}(i, o)$ the routing algorithm which is equal to 1 if the data flow from source s to destination d is routed through input port i and output port o of r, and equals to 0 otherwise (s and d belong to the set of all routers of the NoC):

$$\mathcal{R}_{s \to d}(i, o) = \begin{cases} 1, & \textit{if flow "s to d" passes through ports i \& o} \\ 0, & \textit{otherwise} \end{cases} \qquad (1\text{-}2)$$

Since we address deadlock (and livelock) free routing algorithms the dependency tree doesn't contain any cycle and they eventually terminate to cores, simply because cores are not followed by any router. So using a Depth First Search and by retaining the reverse order of the dependency, we traverse each node of the tree and compute its corresponding latency and use it for computing the mean latency of its (backward) upstream node.

2. ROUTER DELAY MODEL

Through this section we assume that P_i represents the tagged packet coming from input port i of router r addressing output o of that router and P_j represents a disrupting packet coming from any input port j of r (other than i) addressing the same output port in competition with P_i. Router latency l_{i2o}^r is the average latency of the header of P_i to cross r and the buffer located on the output link o (i.e. the buffer between r and its downstream router r'). In our abstract router model, the physical implementation of buffers in output or input of routers is not visible. What is important is the total buffer space, located between two neighbor routers r and r'. Therefore depending on situation we may use the term output or input buffer.

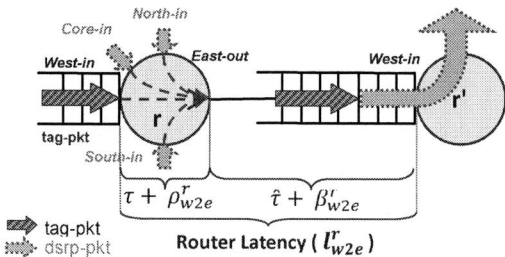

Fig. 1. Router delay model related to a 2D-mesh

As it is shown in Fig. 1, the router latency consists of two major delays. The first is related to crossing the router i.e. from the moment the header of P_i is present at the input port until the moment its output is allocated and thus the header can be written in the output buffer. This delay is composed of *router service time* (τ) and *port acquisition delay* (ρ_{i2o}^r) as defined below. The second is related to transmission of the header of P_i through the output buffer and thus arriving to the input port of r' and is the sum of two delay components namely *buffer constant delay* ($\hat{\tau}$) and *buffer transfer delay* (β_{i2o}^r) as defined in the following. As a result, the router latency can be expressed as the sum of the four aforementioned delay components:

$$l_{i2o}^r = \tau + \rho_{i2o}^r + \hat{\tau} + \beta_{i2o}^r \qquad (2\text{-}1)$$

- τ and $\hat{\tau}$ are constant delays representing respectively *router service time* as the pure delay a router takes to route a packet even when there is no contention (depends on the router architecture) and *Buffer Constant delay* as the delay required for a flit to be transferred through a buffer when it is empty (depends on the buffer architecture).

- ρ_{i2o}^r: *Port acquisition delay* is the average delay P_i experiences at the input port of r while waiting for other disrupting packets (demanding simultaneously the same output port) to pass through the router. In other words it is the router waiting time.

- β_{i2o}^r: *Buffer transfer delay* is the delay caused by the flits previously accumulated in the output buffer at the moment the header of P_i is written in the buffer. In other words it is the time duration the header passes in the buffer until it arrives to the input port of r.

Assuming that τ and $\hat{\tau}$ are given to the method as input, the challenging part of the router model is the computation of ρ_{i2o}^r and β_{i2o}^r described in the two following subsections.

2.1 ρ_{i2o}^r: Port Acquisition Delay

ρ_{i2o}^r is a function of *contention probability* (π_{j2o}^r) between the P_i and any P_j and also *contention delay* (δ_{j2o}^r) which is the delay that P_j takes until it releases shared output port o (i.e. until the tail of P_j leaves the router):

$$\rho_{i2o}^r = \sum_{j \in I_r} \pi_{j2o}^r \times \delta_{j2o}^r \qquad (2\text{-}2)$$

where I_r is set of input ports of r and π_{j2o}^r and δ_{j2o}^r are defined as:

2.1.1 δ_{j2o}^p: Contention Delay

δ_{j2o}^p is defined as the average delay P_j produces in front of P_i, before releasing shared output port o. P_i may arrive at any moment of the *presence time* of P_j (γ_{j2o}^r) that is the total time during which output port o is blocked by P_j. γ_{j2o}^r is composed of the following delay components:

$$\gamma_{j2o}^r = \mathcal{L}_{j2o}^r + \rho_{j2o}^r + \varphi_o^r \qquad (2\text{-}3)$$

- $\mathcal{L}_{j2o}^r$: is the average length of Pj in flit. Since the unit of time is equivalent to the transmission time of one flit, every flit of Pj (scheduled before Pi) takes at least one unit of time to be transferred. Thus Pi is at least blocked by the flits of Pj.

- ρ_{j2o}^r: is the port acquisition delay met by Pj due to contentions with other packets (including packets coming from i) which are reciprocally considered as disrupting packets for Pj. This delay must be considered because Pi may arrive to r just when the header of Pj is waiting at the router input port to be served.

- φ_o^r: is called back pressure delay related to output buffer o. It represents the back pressure impact which arises when contention in the network is very high. In this case data flits accumulated in the output buffer o may exceed the buffer size. Since we assume lossless link level flow control these extra flits retro propagate in the preceding input buffers. Therefore the waiting packet (here Pj) has to wait until the retro-propagated flits are transferred. In subsection 2.2.4 we will see how φ_o^r is computed.

However, P_i doesn't always meet all the presence time of a pre-scheduled P_j. Instead, according to its arrival time, it meets the *average* delay taken by P_j, called contention delay (δ_{j2o}^r). To compute δ_{j2o}^r we separate two cases according to the arrival time of P_i:

$$\delta_{j2o}^r = \left(\frac{\mathcal{L}_{j2o}^r + \varphi_o^r}{\gamma_{j2o}^r} \cdot \frac{\mathcal{L}_{j2o}^r + \varphi_o^r + 1}{2} \right) + \left(\frac{\rho_{j2o}^r}{\gamma_{j2o}^r} \cdot \left(\mathcal{L}_{j2o}^r + \varphi_o^r \right) \right) \qquad (2\text{-}4)$$

Case A: With the probability of $\left(\mathcal{L}_{j2o}^r + \varphi_o^r \right) / \gamma_{j2o}^r$, P_i falls on the body of P_j ($\mathcal{L}_{j2o}^r$) or on the remaining flits of previous

transmissions because of back pressure impact (φ_o^r). In this case blocking flits ($\mathcal{L}_{j2o}^r + \varphi_o^r$) are in moving mode so when P_i arrives it meets a stream of flits crossing the router in a pipeline fashion. Thus the delay experienced by P_i is equivalent to the average number of flits left to be transferred which is equal to the mean value of $\mathcal{L}_{j2o}^r + \varphi_o^r$. The first parentheses in (2-4) represent this case.

Case B: With the probability of $\rho_{j2o}^r / \gamma_{j2o}^r$, P_i arrives while the header of P_j is waiting for output allocation. In this case P_i is exactly stalled by the total flits of $\mathcal{L}_{j2o}^r + \varphi_o^r$. The second parentheses of (2-4) represent this case.

As it is deduced from equations (2-2) and (2-4), the cyclic dependency appears here between the delay components related to flows coming from different inputs of r. (e.g. ρ_{i2o}^r in (2-2) is a function of δ_{j2o}^r which according to (2-4) is a function of ρ_{j2o}^r, and the latter is similarly a function of δ_{i2o}^r and thus function of ρ_{i2o}^r. This is because like P_j that is considered as a disrupting packet for P_i, a packet arriving from i is considered as disrupting packet for P_j). This dependency exists between all inputs of r which compete to obtain output port o and is solved by an iterative fixed point technique.

2.1.2 π_{j2o}^p: Contention Probability

In general the probability of contention between P_i and P_j for obtaining output o is represented by $\pi_{\{i,j\}2o}^p$:

$$\pi_{\{i,j\}2o}^r = \alpha_{i2o}^r \cdot \alpha_{j2o}^r \qquad (2\text{-}5)$$

where α_{i2o}^r and α_{j2o}^r depict the probabilities of the presence of P_i and P_j on input ports i and j both toward o. Since in our model we assume that the tagged packet (i.e. P_i) is already present, α_{i2o}^r is always equal to 1 and thus:

$\pi_{\{i,j\}2o}^r = \alpha_{j2o}^r$ (At the presence of the P_i)

For the reason of coherency with other variables of the model, we denote $\pi_{\{i,j\}2o}^r$ simply with π_{j2o}^r as used in (2-2). Note that $\pi_{i2o}^r = 0$ (i.e. $\pi_{\{i,i\}2o}^r = 0$) since i cannot be in contention with itself (packets arrive to the router in a fifo mode). In quasi zero loads when the contention in the network is very low, a good approximation for π_{j2o}^r would be the *forwarding rate j to o* (λ_{j2o}^r) that is the data rate passing through input j to output o. But in practice when the load increases, all delays imposed to the header of a disrupting packet P_j such as ρ_{j2o}^r and φ_o result in the increase of the presence time of P_j and consequently the increase of the contention probability:

$$\pi_{j2o}^r = \lambda_{j2o}^r \frac{\gamma_{j2o}^r}{\mathcal{L}_{j2o}^r} \qquad (2\text{-}6)$$

where $\gamma_{j2o}^r / \mathcal{L}_{j2o}^r$ gives the average presence time per flit for P_j which is multiplied by λ_{j2o}^r to give the probability of the presence of data on port j addressing port o.

Forwarding rate λ_{j2o}^r is obtained by the following equation:

$$\lambda_{i2o}^r = \sum_s \sum_d \mathcal{R}_{s \to d}(i, o) \cdot t_{sd} \qquad (2\text{-}7)$$

where t_{sd} is the $(s,d)^{th}$ element of traffic distribution matrix T and $\mathcal{R}_{s \to d}(i, o)$ is the routing indicator.

2.2 β_{i2o}^r: Buffer Transfer Delay

β_{i2o}^r is the time duration counted from the moment the header of P_i is written into the output buffer until the moment it arrives to the input port of the next router (r'). In a wormhole flow control strategy, as soon as the last flit of any P_j is transferred through router r, the header of P_i could be written in the output buffer (in the condition the output buffer has enough space to accept at least one flit of P_i). Then the header of P_i shifts after the accumulated flits of previous transmissions. This pipeline shifting takes an amount of time (i.e. β_{i2o}^r) which is equivalent to the average number of flits accumulated in the buffer just when header of P_i is written in the buffer:

$$\beta_{i2o}^r = \sum_{j \in I_r} \pi_{j2o}^r \cdot \hat{\delta}_{j2o}^r + \sum_{j \in I_r} \hat{\pi}_{j2o}^r \cdot \frac{\hat{\delta}_{j2o}^r}{2} \qquad (2\text{-}8)$$

where $\hat{\delta}_{j2o}^r$ is the *buffer occupancy* and $\hat{\pi}_{j2o}^r$ the *buffer occupancy rate* related to any previous P_j, both explained later, and π_{j2o}^r is the contention probability (equation (2-6)). In the computation of β_{i2o}^r two cases are considered (two sigmas of the above equation):

- The first sigma covers the cases in which Pi is in direct contention with any Pj. So when the output is allocated to Pi , Pj has already crossed the router and some of its flits are still in the buffer. Pi is shifted behind the tail of Pj and thus until arriving to the input port of r' it experiences a delay caused by the buffer space occupied by Pj (i.e. $\hat{\delta}_{j2o}^r$). This case is demonstrated in Fig. 2. This case happens with the probability of contention between Pi and Pj (i.e. π_{j2o}^r).

- The second sigma covers the cases in which Pi arrives after a transmission already accomplished and so crosses the router without any contention. When the header of Pi is written into the output buffer, the buffer may be empty or there may be some flits from previous transmissions. This happens with the probability equals to the rate of buffer occupancy by packets coming from any j ($\hat{\pi}_{j2o}^r$). In this case Pi meets the mean buffer space occupied by Pj (i.e. $\hat{\delta}_{j2o}^r / 2$). This case is demonstrated in Fig. 3.

2.2.1 $\hat{\delta}_{j2o}^r$: Buffer Occupancy

Buffer occupancy is the buffer space occupied after the transmission of one packet from any input port j and is obtained by the following equation:

$$\hat{\delta}_{j2o}^r = \beta_{j2o}^r + \nu_o^{r'} \qquad (2\text{-}9)$$

where β_{j2o}^r is the buffer transfer delay, experienced by P_j and $\nu_o^{r'}$ is the delay due to the blocking of the header of P_j at r', called *next node blocking delay* (explained in the following). This is because, similarly to P_i, when P_j is written into the output buffer, its header meets the average number of flits already accumulated in the buffer (i.e. β_{j2o}^r). Thus its header shifts toward r' after the accumulated flits and experience a delay equal to β_{j2o}^r. Simultaneously, its body flits are written and shifted behind it. When the header arrives to the input port of r', there are exactly a number of body flits equal to β_{j2o}^r, accumulated after it. Then at the input port of r', the header of P_j is stalled for a time duration equal to ν_o^r to be routed. Meanwhile more body flits are accumulated behind it. Therefore, as shown in Fig. 2.(A) at this instant the buffer space occupied by P_j is equal β_{j2o}^r plus ν_o^r. Then the header of P_j is routed toward the proper output and its body flits follow it. Finally when the last flit of P_j crosses r (Fig. 2.(B)), P_i is written into the buffer and meets $\hat{\delta}_{j2o}^r$ number of flits.

Note that the cyclic dependency appears between the delay components of buffer transfer delay (β, obtained from equation (2-8)) and buffer occupancy ($\hat{\delta}$, obtained from equation (2-9)) related to different competing inputs.

2.2.2 $\hat{\pi}_{j2o}^r$:Buffer Occupancy Rate

Buffer occupancy rate represents the probability of having some flits remaining from a previous transmission when P_i crosses the router without any contention. Contrary to π_{j2o}^r, $\hat{\pi}_{j2o}^r$ is not zero for input port i because P_i may confront with the residual flits left from a previous packet from its own input port. $\hat{\pi}_{j2o}^r$ is proportional to λ_{jo} and obtained as below:

$$\hat{\pi}_{j2o}^r = \lambda_{j2o}^r \frac{\hat{\delta}_{j2o}^r}{\mathcal{L}_{j2o}^r} \qquad (2\text{-}10)$$

where, the fraction $\hat{\delta}_{j2o}^r / \mathcal{L}_{j2o}^r$ determines the occupied buffer space per flit.

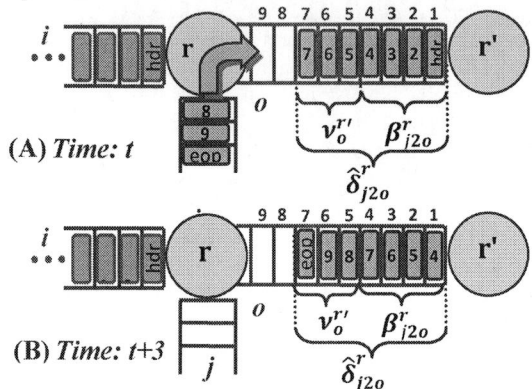

Fig. 2 . Buffer occupancy caused by P_j at time instant t (A), and $t+3$ (B) when P_j is transferred and P_i can be written in the buffer

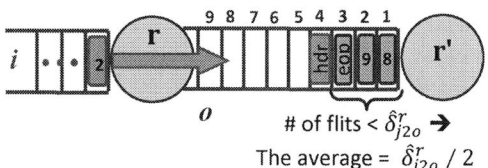

Fig. 3. The average number of accumulated flits in the output buffer at the arrival of P_i when there is no header contention

2.2.3 $\nu_o^{r'}$: Next Node Blocking Delay

$\nu_o^{r'}$ is equal to the average service time needed for P_j to be routed through r' and thus depends on the delay components of r'. As P_j can be routed to different output ports of r', we have:

$$\nu_o^{r'} = \sum_{o' \in \mathcal{F}_{i'}^{r'}} f_{i'2o'}^{r'} \cdot (\tau + \rho_{i'2o'}^{r'} + \varphi_{o'}^{r'}) \qquad (2\text{-}11)$$

where, assuming that P_j arrives to input port i' of r', $\mathcal{F}_{i'}^{r'}$ is the forwarding set of i' (equation I-1) and $f_{i'2o'}^{r'}$ is the *forwarding probability* indicating with which probability P_j is forwarded to output port o' of r':

$$f_{i'2o'}^{r'} = \frac{\lambda_{i'2o'}^{r'}}{\sum_{(o \in \mathcal{F}_{i'}^{r'})} \lambda_{i'2o}^{r'}} \qquad (2\text{-}12)$$

The sum of the three delay components in (2-11) is equivalent to the waiting time of P_j to acquire its output port at r'. τ and ρ have been explained earlier, in the following we describe how the back pressure delay is computed.

2.2.4 φ_o^r : Back pressure Delay

φ_o^r is equivalent to the number of flits retro-propagated in the preceding input buffers and thus the waiting packet is blocked until these flits be transferred. This arises under high traffic load when the network is beginning to saturate. In our computation, the *average buffer utilization* (θ_o^r) can be theoretically larger than the buffer size ($\mathcal{B}_o^r$). This imaginary extra buffer space is interpreted as the back pressure delay:

$$\varphi_o^r = \begin{cases} 0, & \theta_o^r \le \mathcal{B}_o^r \\ \theta_o^r - \mathcal{B}_o^r, & otherwise \end{cases} \qquad (2\text{-}13)$$

where: $\theta_o^r = \sum_{i \in I_r} \lambda_{i2o}^r \times (1 + \hat{\delta}_{i2o}^r)$

In practice φ_o^r indicates the aggregate time durations during which the output port is allocated to the demanding packet but the output buffer is full and thus the packet flits cannot be written into the buffer.

2.3 Cyclic Dependency: Iterative Technique

The cyclic dependencies that exist between the delay components of a router r is demonstrated in Fig. 4 in which the order of computation of delay components is shown by numbers beside the related equation. Assuming that all the downstream nodes in the dependency tree have been traversed and thus all of the required delay components of the following router (r') are available, we can use a fixed-point iteration technique to compute delay components of r.

In Fig. 4 $\Pi_o^r = \left[\pi_{j2o}^r\right]_{1 \times k}$, $\Delta_o^r = \left[\delta_{j2o}^r\right]_{1 \times k}$, $\widehat{\Pi}_o^r = \left[\widehat{\pi}_{j2o}^r\right]_{1 \times k}$, $\widehat{\Delta}_o^r = \left[\widehat{\delta}_{j2o}^r\right]_{1 \times k}$ are row vectors representing respectively contention probabilities, contention delays, buffer occupancy rates, and buffer occupancies for all input ports of r addressing output o, $j \in I_r, |I_r| = k$. Therefore equations (2-2) and (2-8) become:

$$\rho_{i2o}^r = \Pi_o^r \cdot \Delta_o^{r^T},$$

$$\beta_{i2o}^r = \Pi_o^r \cdot \widehat{\Delta}_o^{r^T} + \widehat{\Pi}_o^r \cdot \frac{\widehat{\Delta}_o^{r^T}}{2} \qquad (2\text{-}14)$$

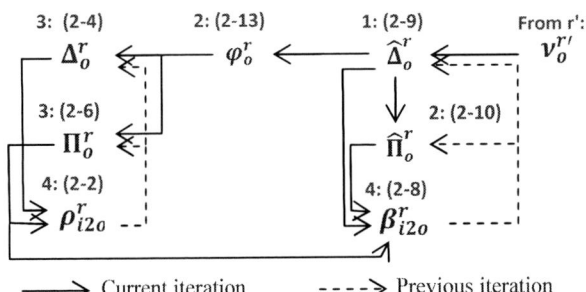

Fig. 4. The order of delay component computation in one iteration.

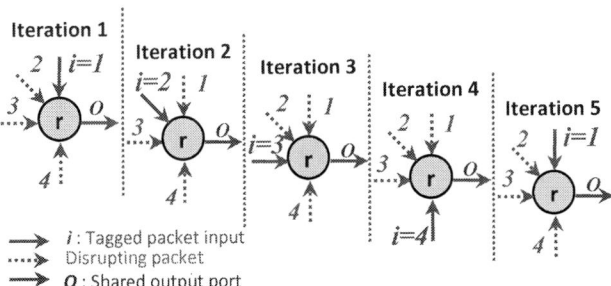

Fig. 5. Iterative computation for inputs {1,2,3,4} of router r

As shown in Fig. 5 in each iteration one of the inputs is marked 'i' as the tagged packet (P_i) input and the others are considered as the disrupting packet inputs. Thus port acquisition and buffer transfer delays are computed for i (i.e. ρ_{i2o}^r and β_{i2o}^r) which are used in the next iteration to update the aforementioned row vectors. Then the position of i changes in the next iteration. For example in Fig. 5 input 1 is the tagged input in iteration 1 (i.e. $i=1$) while it is considered as a disrupting packet input in iteration 2, 3, and 4, until iteration 5 that it is again marked i. Before beginning the iterations, $\nu_o^{r'}$ is computed from the delay components of the downstream router. In iteration 1 we build the row vectors (according to the order shown in Fig. 4) where β_{i2o}^r and ρ_{j2o}^r are initialized to 0 for all $j \in I_r$. Then ρ_{12o}^r and β_{12o}^r are obtained. In the second iteration $i=2$ and input 1 is a disrupting input so we first obtain $\widehat{\delta}_{12o}^r$ and update $\widehat{\Delta}_o^r$ with new $\widehat{\delta}_{12o}^r$ while other elements of $\widehat{\Delta}_o^r$ do not change. In the same way the first element of the other row vectors are updated and finally ρ_{22o}^r and β_{22o}^r are computed. In this way we fill the row vectors with proper values

obtained from iterations. The more we progress in iterations, the more the result becomes accurate. As soon as the difference between the results of two successive iterations for the same input i is less than a given constant ε (determined by the desired accuracy) the computation stops for that input. Since the equations of the router in the dependency cycle of Fig. 4 make a monotonically increasing sequence when the network gets saturated, in this situation the iterative computation converges to infinity.

The iterative approach enables us to determine, for a given router, the latencies between all inputs and the specified output port o. For example in Fig. 5 the iterative computation gives ρ_{j2o}^r and β_{j2o}^r for all $j \in I_r$.

3. VALIDATION OF THE METHOD

We have compared and validated the results of our analytical method against the results obtained by a SystemC CABA (Cycle Accurate Bit Accurate) simulator. However because of the lack of space here we show only results for the average buffer utilization which is the main target of the paper itself.

Uniform distribution is the simplest traffic pattern that can mimic the homogenous communication in general purpose NoCs. However, from a design perspective, it is more efficient to place frequently communicating resources close to each other. This leads to also the use of the localized traffic pattern which is closer to real application behavior.

Fig. 6 shows the average utilization of buffer $r_{3,4} \rightarrow r_{4,4}$ for the uniform (A) and localized (B) traffics. It demonstrates that the results obtained by the analytical method almost match the simulation results. As it is observed the localized traffic loads the buffer as half of the uniform traffic does. The average buffer utilization gives helpful insights into optimal link capacity allocation and the distribution of the traffic over network buffers.

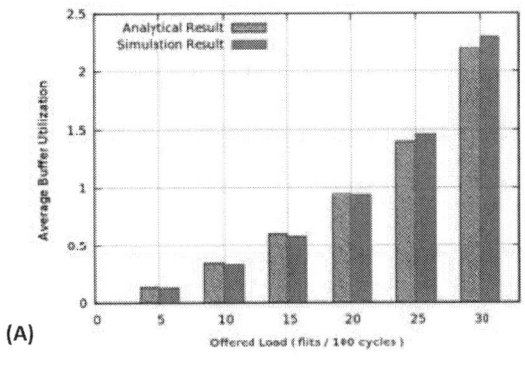

(A)

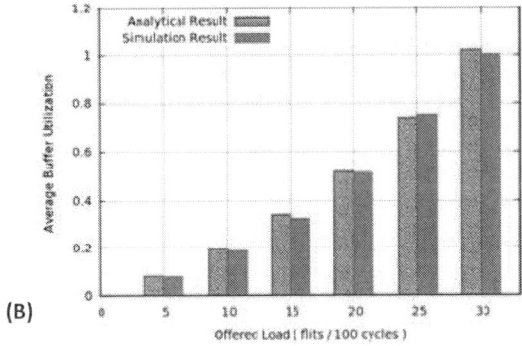

(B)

Fig. 6. The average utilization of buffer $r_{3,4} \rightarrow r_{4,4}$ under the uniform (A) and localized (B) traffic distributions

Attackboard: A Novel Dependency-Aware Traffic Generator for Exploring NoC Design Space

Yoshi Shih-Chieh Huang, Yu-Chi Chang, Tsung-Chan Tsai,
Yuan-Ying Chang, and Chung-Ta King
Department of Computer Science, National Tsing Hua University, Hsinchu, Taiwan
{yoshi, yuchi, tctsai, elmo, king}@cs.nthu.edu.tw

ABSTRACT

Network-on-chip (NoC) is very important for many applications, such as many-core architectures and application-specific usages. For exploring the design space, several approaches have been proposed with different considerations. In this paper, inspired by bloom filters, we propose *Attackboard*, a novel design for exploring the design space of NoC, which satisfies accuracy, space efficiency, and simplicity. To justify the usage of Attackboard, a parallel object detection program is used as the benchmark program to evaluate the performance of a specific NoC. By comparing the results with an execution-based simulator, it shows that Attackboard simultaneously achieves the requirements of fast speed, simplicity, and accuracy.

Categories and Subject Descriptors

B.4 [**Input/output and data communications**]: Processors

General Terms

Performance, Design

Keywords

Network-on-chip, Many-core, Dependency, Traffic generator, Table-driven

1. INTRODUCTION

Network-on-chip (NoC) has become the de facto of the substrate of many-core architecture due to its simplicity and scalability. Exploring the design space of NoC is therefore increasingly important. To study a novel NoC architecture, architects usually exercise it with realistic workloads and measure quantitatively how close the design objective is approached. Consequently, a model of the interested architecture needs to be established and evaluated first, even when the detailed implementations are not available yet.

Early-stage models are essential for architects working on novel architectures, in order to evaluate the proposed architecture even before the detailed RTL or circuit design is available. Existing early stage models for NoC architectures come in different flavors, and cover a wide range of accuracy, simulation speed, and flexibility. One of the most confident solution to early stage models are full-system simulators, which have been success in flexibility as it takes moderate efforts to develop and configure at early design stage. For the mature full-system simulators including NoC, a wide range of topologies, routing algorithms, and router architectures can be modeled by changing the simulation parameters, such as Simics with Garnet, GEMS, M5, SESC, and etc [3, 10, 5, 8]. Full-system simulators have been successfully deployed to study various architectures, due to their high flexibility and accessibility. However, these simulators have relatively lower simulation performance, and does not scale with the progress of modern many-core architectures.

In contrast to simulate the full system, one another way to address the simulation complexity is through trace-driven simulator, which takes the trace log generated in execution as input. A second simulation run, usually with cycle-accuracy, is driven by the previously generated trace. Based on similar methodology, a number of trace-driven simulators for NoCs have been proposed, such as BookSim [1] and [9]. Comparing to full-system simulation, trace-driven simulation is simple and fast. However, the trace-driven simulation is open-loop, which does not consider the backpressure from NoC to the processing elements. Therefore its accuracy is low and may not be tolerable for all studies.

As a result, improving the accuracy of trace-driven simulation has been addressed in state-of-the-art research. In [11], dependencies among injected packets are inferred and embedded into the raw trace logs. Embedding dependencies into traces absolutely brings more confident results of evaluation. However, once dependencies are embedded into original trace logs, two following problems arise. First, it makes the trace logs much complicated and require more storage space, ranging from megabytes to gigabytes according to the granularity. Second, each entry in trace-log becomes correlated and therefore need to be processed while evaluating systems with dependency-aware traces.

In this paper, we propose *Attackboard*, a new methodology to help explore the design space of network-on-chip. Attackboard strikes a balance between speed and accuracy, while keeping the simplicity as trace-driven simulation. The essence of Attackboard is multiple bloom filters. By taking advantages of the property of denial for sure, permitting for high confidence, Attackboard[1] can rebuild the application behavior with a simple pattern-oriented table.

Since each attackboard is a dependency-driven table rather than a time-driven table, redundant dependency patterns only take one line in an attackboard. This property helps reduce the size of table since a program is usually intrinsically repeating program because of loops. By keeping the causality-domain instead of time-domain, the execution can be logged with the pattern-oriented method.

Our evaluations show the accuracy and space overhead of Attackboard by comparing with an execution-based simulator. To identify the contribution of this paper, we listed them as follows:

- A novel design with multiple bloom filters for NoC evaluation is proposed.

- Experiments for comparing Attackboard with an execution-based simulator are evaluated.

- Analysis for the practical implementation is discussed.

The rest of paper is organized as follows. Related works are given in Section 5. Next, Section 2 provides a formal definition of Attackboard. In Section 3, we firstly start with the overview of Attackboard, and followed by the detailed operations and design options. In Section 4, we evaluate Attackboard by comparing with execution-based simulation. Finally, conclusion is drawn in Section 6.

2. PROBLEM FORMULATION

Assume that there are N routers in an NoC. To each router k, it contains its own *attackboard*, denoted as ab_k, which is a two-dimension table, and in which each row stands for a unique dependency pattern, and each column represents the necessity of *must having data from this router*, i.e., a necessary predecessor. For example, if there is a **1** in the cell of row i and column j, it means that for satisfying pattern i, router j must have already injected data to router k. In contrast, if the number in the cell is **0**, it means that this pattern is not necessary to be after the injection of router j to k.

Each router k contains a *Current Status* (CS_k) with length N in terms of bit. By using CS_k as the index to match an entry in ab_k, once a pattern is satisfied, the corresponding injection is generated. This problem can be modeled as a *Multiple Bloom Filters* problem. Each row represents a bloom filter, and CS_k is an entry which is trying to pass one or more bloom filters.

3. SYSTEM DESIGN

3.1 Overview

There are mainly two stages in Attackboard. In the first stage, as Figure 1(a) shows, the dependencies of the injections from each source node are firstly discovered, and then represented with a table structure. The basic idea of this stage is to *periodically capture the relationships among injections*. Once the stage one is done, the characteristics of the benchmarks are now represented by the attackboards.

---

[1]In the following context, *Attackboard* with initial capital stands for the whole design, and *attackboard* with the lower case stands for the table structure.

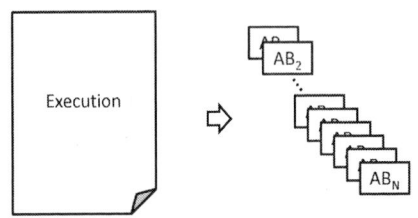

(a) Capturing dependencies and represented by attackboards

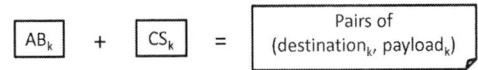

(b) Each router generates its traces by feeding current status to its AB

Figure 1: The system flow of Attackboard. (a) Execution logs are broken to attackboards and distributed to each router. (b) A router generates traffic by feeding its current status to its attackboard.

In the second stage, as shown in Figure 1(b), while evaluating an NoC configuration, each router only needs to look up its own attackboard to generate traffic. Similarly, the basic idea of this stage is to *periodically generate the traffic according to the indications of attackboards*.

3.2 Stage 1 - Creation of Attackboard

For each send event, we observe the received traffic for an interval. Suppose that a send event n is observed at cycle x, and an interval size I is selected as our observing window size, i.e., any traffic occurs between cycle $I - x$ to x would be granted as a dependent receive event to the send event n. As the representation between these dependent events and send event n, a corresponding row is inserted into the attackboard. For each dependency event, the relationship would be marked as **1** in the corresponding cell in the table, indicating if the source core of a dependent receive event had communication with the core of the observed send event. For example, in the scenario shown in Figure 2, the receive event $n - 1$ and $n - 2$ would be granted as the dependent events to send event n. And then a dependency pattern could be built as $< 1, 0, 1 >$, which implies that core 0 and core 2 have communicated with core 3 before the send event n occurs.

The attackboard is able to semantically reveal the program behavior, thus serve as a good evaluation tool to exploit the performance of NoCs. As previous works [11, 7], it is important to collect *correct* dependency information for attackboard so as to discover the semantic meaning of programs. The selection of dependency-extraction interval size I is thus critical. The size of I decides how many dependent (or independent) events of a send event could be seen, and therefore strongly correlates to the accuracy of the recorded dependency information.

Take the function `MPI_Allreduce` as an example. `MPI_Allreduce` performs a reduce operation after core 0 collects data from all other cores and then broadcast the results back to all cores. Since the broadcast can only happen after core 0 receives all data from other cores, the corresponding attackboard of this operation should be like $< 0, 1, 1, 1 >$ with send events $P_1(4), P_2(4), P_3(4)$. This attackboard indicates that only after receiving from core 1, 2, 3, can core 0 sends

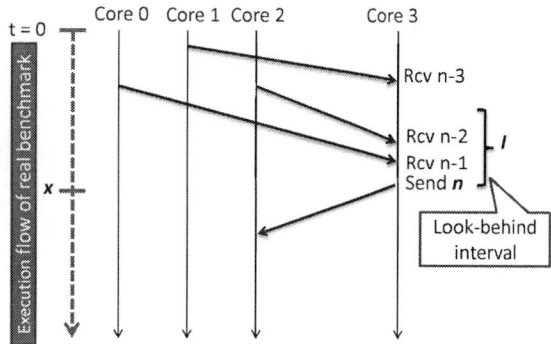

Figure 2: Recognizing the dependent receives of send n.

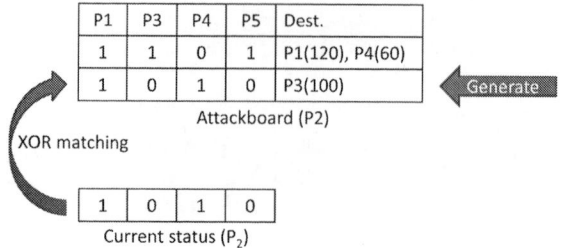

Figure 3: A sample of an attackboard and the procedure of entry matching and traffic generating

data to core 1, 2, 3, each with data quantity 4, which is exactly how `MPI_Allreduce` works. By using this small and scalable attackboard, the semantic meaning of program behaviors can be well presented. In Section 4.1, we have an experiment to discuss the choices of dependency extracting interval.

3.3 Stage 2 - Usage of Attackboard

3.3.1 Initialization

The dependency status of each router k is initially set as all **0**s. Note that there is always an attackboard with dependency pattern as all **0**s, since the very first send event of the program depends on no other send events but the startup of the execution. Therefore, the attackboard need nothing to be propelled but the startup of the simulation. After the first send event generated by the attackboard, the simulation would keep generating dependency-aware traffic to cause the following injections as a chain reaction.

3.3.2 Entry Matching

As Figure 3 shows, each router k keeps its *Current Status* (CS_k) which is caused by others. CS_k is a bitwise vector with length $|N| - 1$, where N is the number of processors. For a bit in CS_k at column y, **0** means that currently processor k has not received any packet from processor y. In contrast, **1** represents that processor k has received data from processor y.

In the end of each interval I', CS_k is used as an index to match its own $attackboard_k$ (ab_k) to find the matches. As the matches are found, the recorded send events of this dependency pattern entry will be generated. Note that the generated traffic of an entry will be evenly distributed to the incoming interval for avoiding all the send events injecting at the starting point of the next interval. This is for simplifying the simulation without recording the computation time before an injection.

3.4 Table Minimization

Since the design of Attackboard are based on bitwise bloom filters, the number of entries are proportional to the number of processors, i.e., $2^{(N-1)}$, where N is the number of processors. For future many-core architecture, the number of processors is expected to be thousands of cores in a chip. Considering such a situation, we propose two methods to reduce the size of attackboards.

3.4.1 Merging duplicated entries

During our gathering the dependency information, those send events with the same dependency pattern would be compressed into the same dependency pattern category. Upcoming send events would be appended to the end of the dependency pattern category, except for those send events with same quantity and destination. The send events of same quantity and destination would be granted as the repetition of one send event, and thus only one send event would be recorded. However, the payloads of two or more send events may be different. Denote a set S as set of the send events which have the same predecessors and injecting destination, and denote the corresponding payloads as $payload_i, i \in S$. We proposed three solutions: First, by averaging the payloads and only record the averaged value as the constant payload. That is, $\frac{\sum_{i \in S} payload_i}{|S|}$. Second, record the payloads as a sequence, i.e., in the cell of destination, keep the sequence as $P_i(payload_1, payload_2, ..., payload_{|S|})$. While generating traffic, the sequence acts as a circular queue. Third, based on the second solution, instead of using the round-robin strategy to select, different payloads are tagged with probability values according to the number of occurrences, and the selection is based on the probability.

3.4.2 Merging similar entries

For further minimizing each attackboard, similar entries can be merged into one. For identifying those entries which have similar patterns, *eXclusive OR* (XOR) operation can calculate the Hamming distance of two entries. For those entries which have closest Hamming distance are considered to be merged first. Once two entries are considered as high similarity, denoted as E_i and E_j, the two entries are removed and replaced by a new entry $E_k = E_i \cdot E_j$. For example, if $E_i =< 1, 0, 1, 1 >$ and $E_j =< 1, 1, 0, 1 >$, then it can be replaced with $E_k =< 1, 0, 0, 1 >$. This operation relaxes the condition of the necessity of predecessors, but the new entry still keeps the coexisting predecessors.

4. EVALUATION

The evaluation of NoC architectures usually involves performance of different NoCs during the executions of real programs. If the average network delay of NoC A is more than that of NoC B, we assert that the NoC B is more suitable than NoC A for target programs. In the context of this evaluation standard, we would evaluate how close the average network delay is by comparing the results of attackboard with real benchmark executions on an execution-based simulator, since it is a representative of the semantic meaning

Table 1: Default simulation setup

Simulation Platform	
Native processor element	Tilera TILE64 [4]
Native processor frequency	700 MHZ
Simulated topology	4×4 mesh network
Routing algorithm	Dimension-order
Bandwidth	1 flit/cycle per port

Core ID	Attackboard dependency pattern entry	send event
0	0 0 0 0 0 0 0 0 0 0 0 0 0 0 0 0	$P_1(4)$
	0 1 0 0 0 0 0 0 0 0 0 0 0 0 0 0	$P_1(4)$
1	0 0 0 0 0 0 0 0 0 0 0 0 0 0 0 0	$P_2(4)$
	0 0 1 0 0 0 0 0 0 0 0 0 0 0 0 0	$P_0(4)$
15	0 0 0 0 0 0 0 0 0 0 0 0 0 0 1 0	$P_{14}(4)$

(a) Attackboards of odd-even sort (16 cores)

Core ID	Attackboard dependency pattern entry	send event
0	0 0 0 0 0 0 0 0 0 0 0 0 0 0 0 0	$P_1(2, 4, 30)$
1	0 0 0 0 0 0 0 0 0 0 0 0 0 0 0 0	$P_2(2)$
	1 0 0 0 0 0 0 0 0 0 0 0 0 0 0 0	$P_2(4, 30), P_0(30), P_3(30)$
15	0 0 0 0 0 0 0 0 0 0 0 0 0 0 1 0	$P_7(30)$

(b) Attackboards of object detection (16 cores)

Figure 4: Attackboards of odd-even sort and object detection.

of real programs. The accuracy is considered higher if the simulation results of attackboard is closer to the results of real benchmark executions. Besides accuracy, the size of attackboard is also evaluated compared with communication traces to meet the needs of easy distribution.

In the following paragraph, we evaluate Attackboard with execution-based simulation in terms of accuracy, size, and how representative the attackboard is for semantic meaning of real programs. The parameters of the simulated environment are shown in Table 1. The utilization of Attackboard involves two stages. At the first stage, we run real benchmarks to generate Attackboard by retrieving dependency information, and examine if attackboard could successfully portray the semantic meaning of the behaviors of real programs. Afterwards, at the second stage, we would run the simulation to generate dependency-aware traffic with Attackboard to examine the accuracy. Finally, we would compare the size of attackboard with trace.

4.1 Dependencies Extracting

We use an execution-based simulator to execute the instrumented programs and capture the dependencies for a micro-benchmark from Intel MPI Benchmark (IMB) suites [2] and a parallel object detection program. The instrumentations are done by hand and therefore guaranteeing the true-dependencies. Other strategies to automatically capture the dependencies are discussed in recent works [7, 11, 13].

Note that the dependent events could be repetitive among send events. That is, if an upcoming send event $n+1$ occurs right after send event n, the observing window of these two send events might overlap, which results in the same dependency pattern, the reason is that these two send events are just dependent on the same receive events simultaneously (such as the MPI group communication as `MPI_allgather`, `MPI_alltoall`, etc.) in real benchmarks. Furthermore, as a dependency pattern driven traffic generation, Attackboard would generate two send events once the receiving dependency pattern is satisfied, which is compatible with the semantic meaning of programs.

Figure 4 shows the derived attackboard of odd-even sort and object detection. In the odd-even sort part, the attackboards of core number 0, 1, and 15 are shown. The first entry of attackboards shows that core 0 would send data (with quantity 4) to core 1 without depending on any send events, and the core 1 would send data to core 2 in the same situation. Meanwhile, the core 15 would only generate data to core 14 after receiving data from core 14. This shows the odd phase of the odd-even sort, while the even phase is shown in the second dependency entry of core 1, and the first entry of core 15. In the even phase, the core 1 would pass data back to core 0, which describes the "swap" operation in the odd-even sort. And the second entry of core 0 conveys that if the even phase ends, there should be another upcom-

ing odd phase. The attackboards of the unmentioned cores also obey this communication behavior as discussed.

In the other hand, the object detection part shows that, initially core 0 would pass data to core 1, while core 1 passes data to core 2. And after the receiving data from core 0, core 1 would send (handled) data back to core 0 and also pass data to core 2 and core 3. Meanwhile, the core 15 would send data to core 7 after receiving data from core 14. This behavior indicates group communication. The object detection algorithm here uses data parallelism to detect if object exists in a frame. To achieve data parallelism, some data must be exchanged between cores to learn acknowledge of other part of data. The attackboard also presents the behavior of data parallelism, which is compatible of the semantic meaning of the programs of object detection.

To conclude, it is feasible to use the derived attackboards to reveal the semantic meaning of the corresponding programs.

4.2 Traffic Generating

As discussed in section 4.1, the retrieval of Attackboard involves the dependency extraction. The dependency extraction methodology has been proposed, with an interval size I introduced. However, it may be questionable that if a fixed dependency-extraction interval size could really embrace all essential dependency information for different send events occurred in execution. Thus, during the generation of dependency-aware traffic, instead of matching the recorded dependency patterns of attackboard, we grant these dependency patterns as bloom filters. By using the characteristic of bloom filter – *denial for sure, permitting for high confidence* – we could generate the traffic only with essential dependent send events, and still provide chances for those unrecorded send events (which is also likely to be a predecessor, but somehow lost during creation of attackboard).

With the Attackboard composed of bloom filters, we could then generate confident dependency-aware traffic. The traffic is generated periodically. Every time the setup simulation interval I' expires, an attackboard would be thoroughly examined. During the examination, we compare the current status and the dependency entries of attackboard. Only the 1s in the dependency pattern entries matter. That is, if we encounter a **0** in the dependency patterns, we simply treat it as a "don't care". For example, with current status $< 0, 1, 1 >$ comparing a dependency pattern $< 0, 1, 0 >$, the generation of traffic is permitted. Contrarily, if a dependency pattern $< 1, 1, 1 >$ is encountered, the traffic genera-

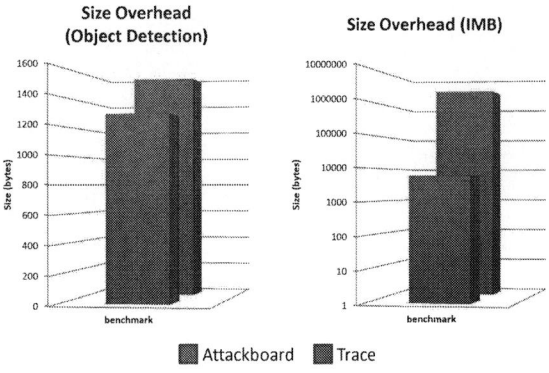

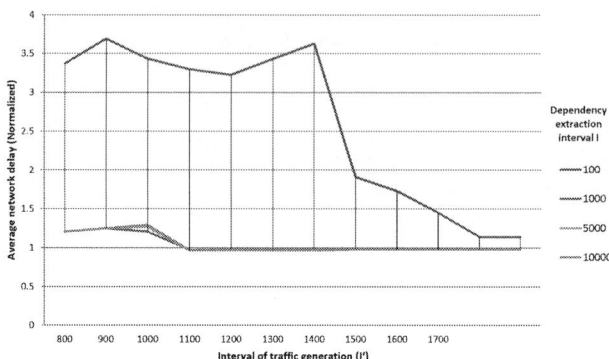

Figure 5: The space overhead of attackboards compared to trace files.

Figure 6: The accuracy of object detection under different I and I'.

tion is not permitted.

Once a suitable dependency pattern is found, corresponding traffic would be generated. If no suitable dependency pattern is found, then no traffic would be generated. Since the generated traffic would obey the form of dependency information revealed from real benchmarks, the generated traffic behaves like the real benchmarks.

As described, the attackboard would periodically generate dependency-aware traffic with simulation interval I'. Since the traffic injection rate is a critical factor to evaluate NoCs, the selection of I' (i.e., how often Attackboard generates traffic) plays an important role of attackbard simulation. An appropriate I' could attackboard behave as real programs. The selection of I' heavily depends on the NoC architectures, so a tuned I' for a simulating NoC architecture is always needed in the attackboard simulation. In the following experiments, we would show that with the easily tuned I', the attackboard could be a representative of real benchmark. After the tuned I' is found, the average network delay could then be generated with attackboard simulations.

4.3 Space Overhead

As shown in Figure 5, the space requirement of Attackboard is quite small. In the size comparison for object detection, the size of attackboard is comparatively small than the size of communication traces. Note that the improvement is not big because of the communication events are relatively lesser and execution time is shorter compared to other real programs. A great improvement can be observed in the size comparison for IMB, it is because the long execution time and intensive communications, thus the size of communication traces is very large, and it would grow even larger if the execution time is set longer. Compared with the large size of IMB traces, attackboard is much smaller. Attackboard is scalable, since Attackboard is a timeless table, so the size of Attackboard would not grow linearly with time, but depends on *how many distinct dependency patterns exist in the programs*. Further, the dependency pattern that Attackboard records are in essence the semantic meaning of program, the repetition of behaviors would be folded naturally with Attackboard, which makes Attackboard more scalable to use.

4.4 Case Study

The accuracy of Attackboard is evaluated by the comparison of average network delay. We compared Attackboard with the average network delay of the execution of parallel object detection and IMB Broadcast. The accuracy is presented as the normalized average network delay with that of execution of real benchmarks. Also, multiple I and I' are tested to derive the best result.

Figure 6 and Figure 7 show the statistics of normalized average network delay Attackboard achieved with different I and I' (in terms of cycle). The x-axis represents simulation interval length I' and the y-axis represents the normalized results. Different curves in the figure represent an attackboard with different simulation interval length I. In the following paragraphs, we discuss different cases, respectively.

4.4.1 Parallel Object Detection

As we can see in Figure 6, there are two groups of the curves. The first group is the curve with I set as 100. The curve is initially very high, and falls down slowly. After I' is set larger than 1600, it would arrive a steady value while still remain a comparatively high value. The reason could be found obviously in the content of the attackboards. There is only one dependency entry – all zeros – of the attackboard with I set as 100. This shows that the length of dependency-extraction interval is too small, so that the attackboard could not capture any useful dependency information for send events. Therefore, during the simulation of Attackboard, the send events would inject every interval, which causes severe congestion.

The second group is the curve with I set as 100, 1000, 5000, and 10000. The curves of second group would startup high and then flatten out afterwards. It is because when the length of I' is too small, congestion would occur since the injection rate is too high. In contrast, after I' is fairly set, the injection would then balance with routers' receiving, thus generate a steady average network delay value, which matches the result of the real program. In this group, the results almost fit the ideal case.

4.4.2 Intel IMB Broadcast

Different from the simulation behaviors of object detection, the simulation results of attackboard of IMB broadcast is very centralized. As Figure 7 shows, the accuracy (normalized average network delay) of the simulation results ranges from 0.8 to 1.1, showing the high feasibility of the

380

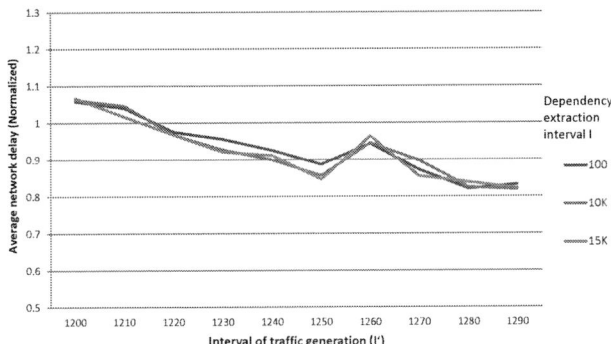

Figure 7: The accuracy of IMB under different I and I'.

dependency-aware traffic generation by Attackboard. However, the selection of I does not really matter in this case. As discussed in section 4.4.1, the interval size has a strong correlation with the dependency information. If the smaller the dependency-extraction interval is chosen, the less dependency information. But in this case study, attackboards of various interval sizes seem to behave in the same way. It is thanks to the power of bloom filters that suffice the distortion of dependency information. Since Attackboard only checks the confident recorded predecessor (the 1s in the dependency entry of Attackboard), and considers others as "don't cares", the distorted result is automatically corrected by the natural operations of simulator. Because Attackboard will not reject those "high-confident" traffic generation, the behavior of broadcast is appropriately replayed, thus achieve high accuracy.

In summary, since Attackboard could achieve accurate results with easily tuned parameter, it becomes a good substitute for evaluating NoCs. Using Attackboard as the simulation tool thus much alleviates the evaluation of NoC for real programs. The easy implementation, small size, and the capability of exploiting the semantic meaning of real programs even make Attackboard a better tool to evaluate NoCs.

5. RELATED WORKS

Research on the on-chip interconnection network greatly uses a trace-driven simulator [1, 6, 13]. Unfortunately, the accuracy of trace-driven simulation is not convincing, because in most cases, the dependencies of the transmission are broken, i.e., each entry in the trace log are not correlated. As a result, improving the accuracy of trace-driven simulation has been important in state-of-the-art research. In [11], dependencies among injected packets are inferred and embedded into the original trace logs. Similar idea can also be found in [7]. Once dependencies are embedded, it absolutely improves the accuracy of trace-driven simulation; however, it makes the trace logs much complicated and require more storage space, ranging from megabytes to gigabytes.

Many other methodologies to help explore design space of NoC are proposed, including full-system simulation [3, 5], mathematical models, FPGA-based acceleration [12], etc. Each of them has their own pros and cons. Among these existing works, Attackboard strikes the balance between full-system simulation and trace-driven simulation.

6. CONCLUSION AND FUTURE WORKS

In this paper, we propose Attackboard, a new methodology to help explore the design space of network-on-chip. Attackboard takes advantages of repetitive behaviors and dependencies among injections. We use IMB and a parallel object detection program to evaluate the performance of Attackboard. The results show that Attackboard has high accuracy as programs directly run on the execution-based simulator. On the other hand, we also compare Attackboard with trace-driven simulation in terms of space requirements. The results show that the space overhead is much smaller while keeping the accuracy.

The future works are twofold. First, the intervals for extracting and generating rely on empirical rule to find. We are developing an automatic process to find the suitable intervals. Second, Attackboard currently is evaluated with message passing programs. In our ongoing work, the traffic of shared-memory programs will be included.

7. REFERENCES

[1] BookSim 2.0.

[2] Intel MPI Benchmarks.

[3] N. Agarwal, T. Krishna, L.-S. Peh, and N. Jha. Garnet: A detailed on-chip network model inside a full-system simulator. In *Proceedings of IEEE International Symposium on Performance Analysis of Systems and Software, 2009. ISPASS 2009.*, pages 33 –42, April 2009.

[4] S. Bell, B. Edwards, J. Amann, R. Conlin, K. Joyce, V. Leung, J. MacKay, M. Reif, L. Bao, J. Brown, M. Mattina, C.-C. Miao, C. Ramey, D. Wentzlaff, W. Anderson, E. Berger, N. Fairbanks, D. Khan, F. Montenegro, J. Stickney, and J. Zook. TILE64 - processor: A 64-core soc with mesh interconnect. In *Proceedings of Internation Conference on Solid-State Circuits, 2008. ISSCC 2008.*, pages 88 –598, feb. 2008.

[5] N. Binkert, R. Dreslinski, L. Hsu, K. Lim, A. Saidi, and S. Reinhardt. The m5 simulator: Modeling networked systems. In *Proc. of the 39th Int'l Symposium on Microarchitecture*, volume 26. pages 52–60, 2006.

[6] F. Fazzino, M. Palesi, and D. Patti. Noxim: Network-on-chip simulator, 2008.

[7] J. Hestness, B. Grot, and S. W. Keckler. Netrace: Dependency-driven trace-based network-on-chip simulation. In *Proceedings of the Third International Workshop on Network on Chip Architectures*, pages 31–36, New York, NY, USA, 2010. ACM.

[8] C. Hughes, V. Pai, P. Ranganathan, and S. Adve. Rsim: Simulating shared-memory multiprocessors with ilp processors. *Computer*, 35(2):40–49, 2002.

[9] A. B. Kahng, B. Lin, K. Samadi, and R. S. Ramanujam. Trace-driven optimization of networks-on-chip configurations. In *Proceedings of the 47th Design Automation Conference*, pages 437–442, New York, NY, USA, 2010. ACM.

[10] M. M. K. Martin, D. J. Sorin, B. M. Beckmann, M. R. Marty, M. Xu, A. R. Alameldeen, K. E. Moore, M. D. Hill, and D. A. Wood. Multifacet's general execution-driven multiprocessor simulator (gems) toolset. *SIGARCH Comput. Archit. News*, 33:92–99, November 2005.

[11] C. Nitta, M. Farrens, K. Macdonald, and V. Akella. Inferring packet dependencies to improve trace based simulation of on-chip networks. In *Proceedings of the Fifth ACM/IEEE International Symposium on Networks-on-Chip*, NOCS '11, pages 153–160, New York, NY, USA, 2011. ACM.

[12] Z. Tan, A. Waterman, H. Cook, S. Bird, K. Asanović, and D. Patterson. A case for FAME: FPGA architecture model execution. In *Proc. of the 37th Annual Int'l Symposium on Computer Architecture*, pages 290–301, 2010.

[13] F. Trivino, F. J. Andujar, F. J. Alfaro, J. L. Sanchez, and A. Ros. Self-related traces: An alternative to full-system simulation for nocs. In *High Performance Computing and Simulation (HPCS), 2011 International Conference on*, pages 819 –824, july 2011.

Towards Graceful Aging Degradation in NoCs Through an Adaptive Routing Algorithm

Kshitij Bhardwaj   Koushik Chakraborty   Sanghamitra Roy

USU BRIDGE LAB
Electrical and Computer Engineering, Utah State University
kshitij.bhardwaj@aggiemail.usu.edu,{sanghamitra.roy, koushik.chakraborty}@usu.edu

ABSTRACT

Continuous technology scaling has made aging mechanisms such as Negative Bias Temperature Instability (NBTI) and electromigration primary concerns in Network-on-Chip (NoC) designs. In this paper, we model the effects of these aging mechanisms on NoC components such as routers and links using a novel reliability metric called *Traffic Threshold per Epoch (TTpE)*. We observe a critical need of a robust aging-aware routing algorithm that not only reduces power-performance overheads caused due to aging degradation but also minimizes the stress experienced by heavily utilized routers and links. To solve this problem, we propose an aging-aware adaptive routing algorithm and a router microarchitecture that routes the packets along the paths which are both least congested and experience minimum aging stress. After an extensive experimental analysis using real workloads, we observe a 13%, 12.7% average overhead reduction in network latency and Energy-Delay-Product-Per-Flit (EDPPF) and a 10.4% improvement in performance using our aging-aware routing algorithm.

Categories and Subject Descriptors

C.4 [**Performance of Systems**]: Design Studies, Fault Tolerance

General Terms

Reliability, Algorithms, Design

Keywords

NoC, Aging, NBTI, Electromigration, Routing algorithms

1. INTRODUCTION

With the proliferation of on-chip cores allowed through rapid technology scaling, Network-on-Chips (NoCs) are becoming a critical determinant of overall system power-performance characteristics. Consequently, the growing reliability challenges, which are continuously reshaping the system design considerations, must now be thoroughly analyzed in the context of NoC designs [14]. In this work, we study two primary mechanisms responsible for circuit wear-outs in an NoC design: *Negative Bias Temperature Instability (NBTI)* and *Electromigration*.

An NoC architecture comprises two major components: NoC router and link. A pipelined NoC router consists of both combinational logic structures (e.g. virtual channel allocation logic) and storage-cell structures (e.g. virtual channels). Due to the presence of these structures, NBTI is the major aging mechanism associated with NoC routers [6]. NoC links, on the other hand, are implemented using repeated copper interconnects [8]. Therefore, NBTI (repeaters) and electromigration (copper interconnects) are the two primary aging problems associated with NoC links. Unfortunately, previous works on NoCs have completely ignored the role of links in their reliability analysis, focusing solely on the routers [10, 6]. We demonstrate that such limitations can grossly underestimate the NoC lifetime by nearly a factor of 2.

In the context of reliability in NoCs, another critical design challenge stems from the asymmetric usage of NoC components. Mishra et al. have shown this non-uniform pattern of router buffer and link utilization [13]. They observed that the routers in the center of the mesh are highly (75%) utilized, while the peripheral routers have low (35%) utilization. Similarly, our experiments with multi-threaded workloads on a 4×4 mesh indicates a wide disparity in buffer utilization of different routers. Due to such asymmetric utilization, each router and link will also suffer from different amounts of aging degradation. Therefore, there is a need for an aging-aware routing algorithm for NoCs that considers the aspect of asymmetric aging while routing packets, so as to improve the system reliability.

At the system level, solely improving reliability may hurt the power-performance. For example, to alleviate the aging degradation on a heavily utilized path, it may be necessary to use an alternate route. However, employing such an alternate route can increase the network latency, thereby degrading the system level power-performance. Thus, efficient ways to improve the system robustness requires design space exploration techniques that simultaneously optimize multiple objectives. Such an optimization problem must effectively model several NoC design aspects in the context of the overall system: (a) routing topology, (b) network traffic during the execution of real programs, (c) device level models capturing the effect of NBTI and electromigration, and (d) latency and energy consumption of the routing policies. Ad-hoc analysis and optimization of these complex objectives can lead to sub-optimal solutions, with limited insight for future improvements.

To effectively model multiple device level aging characteristics, system level asymmetric usage patterns and runtime traffic variations, we introduce a specific reliability metric for NoC components: *Traffic Threshold per Epoch (TTpE)* (Section 3). *TTpE* is defined as the amount of traffic that a stressed link or router should accept in a particular epoch[1] during the runtime. The purpose of this metric is two folds: (a) allow formal analysis of reliability impact on an NoC, and (b) a means to dynamically analyze the traffic patterns and adapt the routing algorithms. Subsequently, we present an adaptive aging-aware routing algorithm that a) reduces the aging-induced power and performance overheads by routing through paths that experience least aging effects and congestion, and (b) minimizes the stress experienced by heavily utilized routers and links by constraining them to meet their respective *TTpEs* for different epochs throughout the total running time.

We make the following contributions in this paper:

- We show the effects of NBTI and electromigration on NoC links and its significance in reliability analysis and fault tolerance of NoCs (Section 2).
- We formulate a comprehensive system-level aging model for NoC routers and links. This model considers the effects of asymmetric aging in routers and links during the program execution. Subsequently we show the impact of NBTI and electromigration on the performance of an NoC-based multicore system (Section S2). Our work integrates SPICE level process variation and aging analysis, circuit-level statistical timing analysis, and full system architectural simulation (Section 5).
- We propose an aging-aware adaptive routing algorithm and router micro-architecture that not only mitigates the impact of aging on the NoC's power-performance characteristics but also minimizes the stress experienced by heavily utilized routers and links. This way, the algorithm is able to provide robustness at the system-level design of NoCs (Section 4).
- An extensive experimental analysis using the GARNET NoC simulator [3] and real workloads (PARSEC benchmarks [2]) indicates an average of 13% and 12.17% reduction in the network latency and Energy-Delay-Product-Per-Flit (EDPPF) [11] in a typical NoC undergoing aging stress. We also obtain an average improvement of 10.4% in performance using our proposed algorithm (Section 6).

2. MOTIVATION

In this section, we present a brief robustness analysis of an NoC that not only considers impact of aging on routers but also considers the effects of aging mechanisms such as NBTI and electromigration on links.

To show the importance of considering the aging effects on links, we perform an experiment using an NoC architecture that comprises two routers connected by two unidirectional links. We use the flexible numerical model of NBTI degradation based on reaction-diffusion [5] and wire resistance based electromigration stress model in our calculations [15]. These models are discussed in more details in Section S2.2. We analyze the network latency under the following schemes:

Table 1: Different Degradation Schemes

Scheme	Degradation in Routers	Degradation in Links
A	NBTI	NONE
B	NBTI	NBTI
C	NBTI	Electromigration
D	NBTI	NBTI and Electromigration

The above schemes are evaluated under tornado synthetic traffic pattern for three different injection rates: a) low (0.1 flits/cycle), b) medium (0.3 flits/cycle), c) high (0.5 flits/cycle).

[1] To capture runtime traffic variations, total time taken to route the flits is divided into equal intervals of time called epochs.

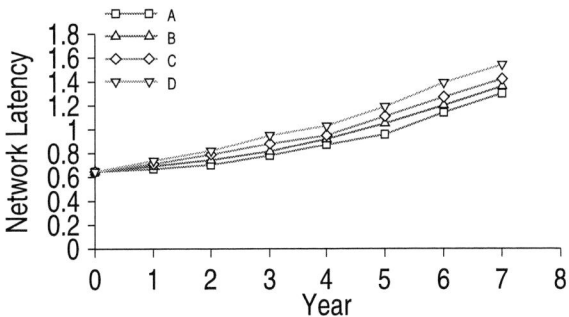

Figure 1: Latency variation with time (low injection rate).

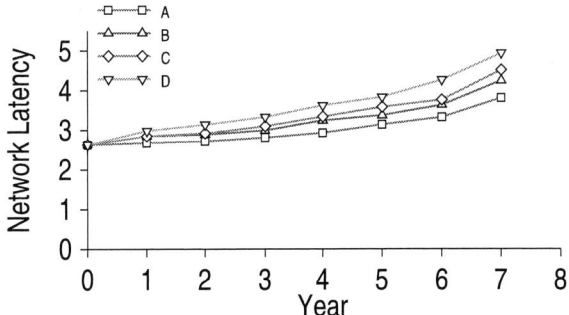

Figure 2: Latency variation with time (medium injection rate).

2.1 Analyzing Network Latency

Figures 1, 2 and 3 show the variation of network latency with time for different injection rates. As is evident, *D* estimates the highest network latency. The consideration of both NBTI and electromigration in *scheme D* degrades the link more, thereby causing this substantial increase. Network latency is least affected in *scheme A* as there is no link degradation and the small increase is only due to NBTI degradation in routers.

2.2 Effect on Fault-Tolerance of NoC

We assume that a network becomes faulty when the increase in network latency exceeds a pre-defined threshold (20% in our study). Figure 4 shows the time taken for the network to become faulty in the case of high injection rate traffic. For example, under *scheme D*, a network can be rendered faulty in almost three years. However, using *scheme A* grossly over-estimates the time to failure (almost six years). In reality, due to the combined effects of NBTI and electromigration, the copper wires are likely to degrade beyond the threshold by that time.

3. TTPE: NEW RELIABILITY METRIC F-OR RUNTIME ADAPTATIONS IN NOCS

In an NoC, every router and link is utilized in different amounts, some more than others. These variations stem from two key factors: (a) asymmetric communication patterns between NoC nodes; (b) runtime traffic variation due to policies of the routing algorithm. Section S1 illustrates this property through our rigorous analysis with the GARNET NoC simulator.

Based on the above observations, we derive the system-level aging model to find the relationship between router/link utilization and the amount of stress experienced from NBTI/electromigration degradation. For this purpose, we introduce a novel metric called

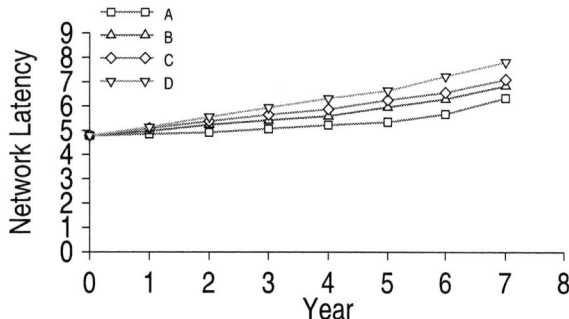

Figure 3: Latency variation with time (high injection rate).

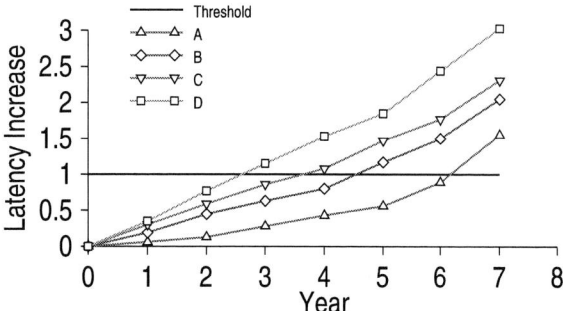

Figure 4: Time taken for the network to become faulty under various aging models (high injection rate).

Traffic Threshold per Epoch (TTpE), defined as the fraction of the nominal traffic[2] that a stressed router/link should accept during a particular epoch. The significance of using *TTpE* as a reliability metric for an aging-stressed NoC design lies in the following facts:

- It determines an upper limit on the amount of traffic that a router or link should accept so as to keep the variation in network latency below a pre-defined threshold for a particular aging period (7 years in this work). If the limit imposed by *TTpE* is exceeded in a router undergoing maximum degradation, it will be rendered faulty.
- *TTpEs* are derived from continuous monitoring of the traffic, and are used to adapt the routing policies for every epoch to mitigate the long-term degradation in the NoC.

The *TTpEs* vary over the runtime with different values during different epochs for each stressed router and link. In order to calculate these thresholds, we first profile a congestion-aware routing algorithm to estimate the router/link utilization for every epoch during runtime. Depending on the utilization, *TTpEs* are calculated for the stressed routers and links for every epoch using our system-level aging model. *TTpEs* are then stored in each router in the form of lookup tables so that the router can select the appropriate threshold depending on the epoch during runtime. These steps are discussed in more details next.

3.1 TTpE calculation

The calculation of *TTpE* involves the following stages:

- **Threshold calculation:** This stage mainly involves profiling of a state-of-the-art congestion-aware routing algorithm [7] to cal-

---

9[2]Nominal traffic is the traffic across routers/links when they are unstressed.

culate the *TTpE* of different stressed links and routers.
- **Using *TTpE* Estimation in Routing:** During this stage, we build the traffic threshold tables, which are then stored inside each router.

We now discuss these stages in more detail.

3.1.1 Threshold Calculation

This stage can be further divided into two steps:

- **Profiling:** We first profile the congestion-aware routing algorithm that routes the flits based on both local and global congestion information. The total time taken to route these flits is then divided into several epochs. The significance of adding epochs lies in the fact that an application's communication characteristics may change during the runtime and therefore the traffic must be monitored continuously. This way we can keep track of the link and the router utilization during runtime and take additional measures if the utilization reaches *TTpE* for the epoch under consideration.
- **TTpE Calculation:** For each epoch, we find the n most stressed links and routers based on their utilization. The *TTpEs* for these routers and links are calculated as follows:

  1. **Router TTpE:** We use the aging model developed in Section S2.1 to calculate the *TTpE* for routers. This aging model considers only NBTI degradation in routers.
  2. **Link TTpE:** To calculate the *TTpEs* for links, we use the NBTI and electromigration degradation model as described in Section S2.2. Also we know that the traffic across a router can be controlled only by controlling the traffic across the links input to the router. Therefore, we transfer the stress experienced by the router to the input links and calculate their *TTpEs* for the given epoch. For example, if a stressed router R has four input links ($l0...l3$) then the utilization of the router is given by:

$$util(R) = util(l0) + util(l1) + util(l2) + util(l3) \quad (1)$$

where $util()$ estimates the utilization of a router or a link. Using the above equation, we can calculate the amount of traffic each link should accept (or *TTpE* of each link) for the router to meet its threshold. Therefore, there are two kinds of stressed links: i) links that are directly under aging stress, and ii) input links that experience stress because of the stressed router.

3.1.2 Using TTpE Estimation in Routing

Here, we store the computed *TTpEs* for different epochs in the form of lookup tables (SL_{set}) in each router. The router at runtime can then select the appropriate *TTpE* depending on the epoch. During this stage, we also compute the routing tables for each router. In order to minimize network latency and communication energy, we select only the deadlock-free shortest paths for each flow.

4. AGING-AWARE ADAPTIVE ROUTING

In this section, we present a robust aging-aware routing algorithm that not only reduces the stress experienced by heavily utilized NoC components but also minimizes the overall aging induced power-performance overheads. The algorithm involves the following two stages (Table 2):

  1. **Congestion and aging-aware routing**: For each flow at runtime, the routing algorithm selects the best shortest path from the routing table that i) suffers from least aging degradation

Table 2: Algorithm for Aging-aware adaptive routing

ALGORITHM Aging_Adaptive:

For each flow,

    1. Select the best shortest path from the routing
table which:

        a) suffers from least delay variation due to
aging (sc_{age} is minimum).

        b) is least congested based on global and local
congestion information (sc_{cong} is minimum).

    2. For each stressed link in SL_{set} of each epoch:

        a) Check if the link meets its *TTpE*:

          - If the link has already reached
its *TTpE*, keep the link idle for the
rest of the epoch (insert *recovery cycles*).

          - If link utilization is safely below its
TTpE then there is no need for
inserting *recovery cycles*.

---

i.e. the path that suffers from least delay variation due to aging (**1-a**); and ii) is least congested (**1-b**). We give a higher priority to a path that is least degraded as compared to a path with the least congestion. For example, in case of an arbitrary flow F, if the available number of deadlock-free shortest paths is four ($path0....path3$) then the algorithm maintains congestion and aging scores (sc_{cong} and sc_{age}) for each of these paths. These scores are calculated based on both local and global aging and congestion information obtained from different routers and links present in the paths. Now if the sc_{age} is least for $path0$ but sc_{cong} is least for $path3$ then the algorithm selects $path0$ for routing. In a different situation, if sc_{age} is same for all paths but sc_{cong} is least for $path3$ then $path3$ is only selected for routing.

2. **Honoring TTpE by employing recovery cycles**: During the execution of the routing algorithm, each stressed link in SL_{set} is checked to see if it meets its respective *TTpE* for every epoch (**2-a**). There can be two possible cases: i) In the epoch, if the link has already reached its *TTpE* then the link must be kept idle for the rest of the epoch so that its utilization does not exceed its *TTpE*; and ii) If the link operates safely inside its *TTpE* for that epoch, then there is no need for inserting idle cycles. The physical significance of inserting these idle cycles is that they provide additional time to the links and routers to recover from the aging stress. Therefore, we call these additional idle cycles as *recovery cycles*. This procedure also avoids unnecessary insertion of *recovery cycles* in the epoch and thus keeps the network latency in check.

4.1 Aging-aware adaptive router microarchitecture

In order to implement the proposed routing algorithm, we extend a congestion-aware router such that it computes the best route that is both aging and congestion-aware. Moreover, this aging-aware router must also ensure that the best allocated output link is operating under its respective *TTpE*. Figure 5 shows a detailed microarchitecture of the aging-aware adaptive router. Based on the functionality of each module of the router, we can divide the routing unit of the aging-aware adaptive router into two different stages:

- **Route Computation:** This stage involves selection of the output link towards the least stressed and least congested shortest path.

During this stage both the global and local congestion information from different routers is aggregated to calculate the congestion score (sc_{cong}) of the paths. Similarly, based on the number of stressed links in SL_{set} for each epoch, an aging score (sc_{age}) is calculated for each path. The *Route Computation* unit then selects the output link corresponding to the shortest path which has the least sc_{age} and sc_{cong}. Note that the additional logic-based circuitry is introduced in parallel paths rather than in sequence, for example sc_{age} and sc_{cong} are calculated in parallel.

- **Recovery Cycles:** During this stage of the routing unit, utilization of the selected output link corresponding to the current epoch is evaluated. If the link is stressed, its utilization is compared with the threshold for the epoch (*TTpE*), stored inside the lookup tables (SL_{set}). In case the link has reached its *TTpE* then this stage inserts recovery cycles for this link during the given epoch. If the link has not reached its *TTpE*, then there is no need for the recovery cycles. Also, an unstressed link can skip this stage of the routing unit.

5. EXPERIMENTAL METHODOLOGY

Our experimental setup combines SPICE level analysis for process variation and NBTI aging, statistical timing analysis using synthesized Verilog for NoC routers, and full system architectural simulation. The effect of process variation and NBTI aging in basic logic gates are performed through Synopsys *HSPICE*, using Predictive Technology Models (PTM), and long term degradation due to NBTI [4]. On each of these gates, we run 10K Monte Carlo simulations to obtain respective statistical distributions of their performance characteristics. Using these gates at the 45nm technology, we synthesize the NoC router RTL obtained from Stanford University's open-source NoC router resources [1]. Subsequently, we perform a statistical timing analysis to find various critical paths in the router, and their delay distributions under the combined effect of process variation and NBTI aging.

For architectural simulation, we use the GARNET NoC simulator, embedded inside GEMS [12]. GARNET uses the ORION power model [9] to calculate power consumptions of the routers and the links. In our experiment, we consider a system with 16 processors in a 4×4 mesh topology. Each processor has a dual issue 32 entry out-of-order issue window and a private L1 cache (2-way, 32 KB, response latency: 3 cycles) and a shared L2 cache (4-way, 2 MB, response latency: 15 cycles). For traffic generation and system-level analysis, we use PARSEC benchmarks with 16 threads pinned to cores: Canneal (can), Dedup (ded), Facesim (fac), Ferret (fer), Fluidanimate (flu), Freqmine (fre) and Raystone (ray).

6. EXPERIMENTAL RESULTS

To study the power-performance impact of aging on NoC designs, we conduct a set of experiments on a 4×4 NoC mesh shown in Figure 10. Different schemes implemented for comparison are discussed next.

6.1 Comparative Schemes

We use three different schemes to show the importance of a robust aging-aware adaptive routing algorithm:

- **RCA-1D**: In this scheme, we use a state-of-the-art congestion-aware routing algorithm (Section S4) to route the flits in an NoC system comprising aging-stressed routers and links. Without aging awareness, this scheme continues to use heavily degraded links/routers, thereby incurring power-performance overhead. Delay degradations in the stressed routers and links are modeled according to Section S2.

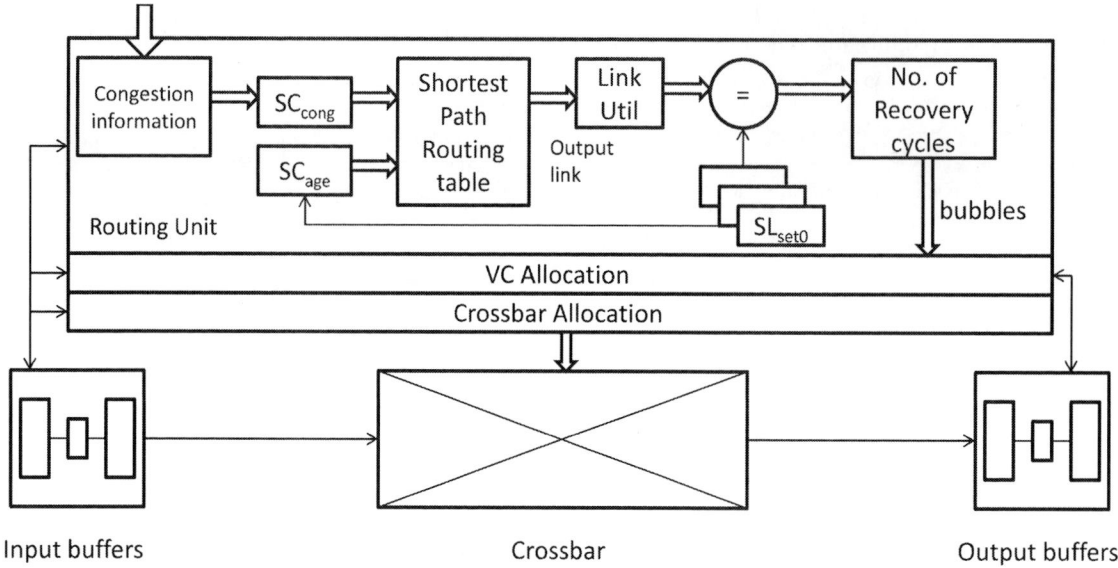

Figure 5: Aging-aware adaptive router micro-architecture: In addition to the global congestion information in the form of sc_{cong}, the router also uses the aging information given by sc_{age} to select the best output route. The additional logic circuitry for this route computation is implemented in parallel, e.g. sc_{age} and sc_{cong} are computed simultaneously.

- **AGE-ADAP**: Here we implement our proposed congestion and aging-aware adaptive routing algorithm to route flits in a stressed NoC design. This scheme employs the step 1 of our proposed routing algorithm (Section 4) but does not honor the *TTpE* limits and therefore does not insert any *recovery cycles* in the epochs during runtime. Delay variation in the routers and links are based on the model in Section S2.
- **AGE-ADAP-REC**: This scheme extends AGE-ADAP such that it inserts *recovery cycles* for stressed links/routers during an epoch if the utilization has reached *TTpE*. Therefore, this scheme ensures that none of the stressed routers/links operates beyond their calculated *TTpE*.

6.2 Robustness Evaluation

In order to show the robustness degradation in a 4×4 NoC mesh that does not consider aging aware-routing (RCA-1D), we compare the reliability of this scheme with the AGE-ADAP and AGE-ADAP-REC schemes that model aging-awareness.

Figure 6 shows the reliabilities of all three schemes for an aging period of 7 years, calculated using the reliability's dependence on failure rate and *TTpE* (Section S3). As expected, RCA-1D shows substantially higher failure rate compared to AGE-ADAP and AGE-ADAP-REC, as its design does not adapt to the wear-out degradation of NoC components. Also the traffic utilization per epoch of stressed routers and links is always above the *TTpE* for RCA-1D which further reduces its reliability. As stressed routers and links are well below their *TTpE* limits in AGE-ADAP-REC, its reliability is even better than AGE-ADAP.

6.3 Overhead Analysis

6.3.1 Network Latency

Figure 7 shows the network latency for various schemes normalized to the RCA-1D scheme. As RCA-1D does not employ any aging-awareness in its congestion-aware routing, it only selects those paths which are least congested. However, frequent heavy utilization in these paths results in large aging stress, which man-

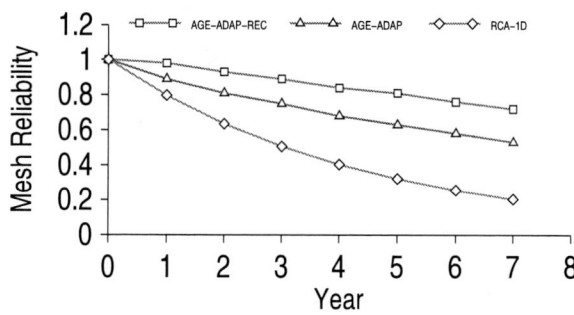

Figure 6: NoC Robustness Degradation Over Time.

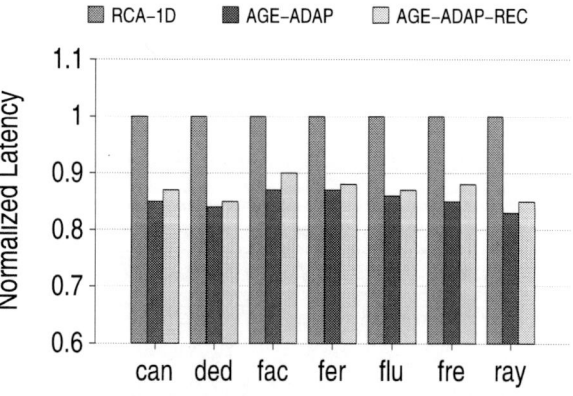

Figure 7: Normalized network latency (lower is better).

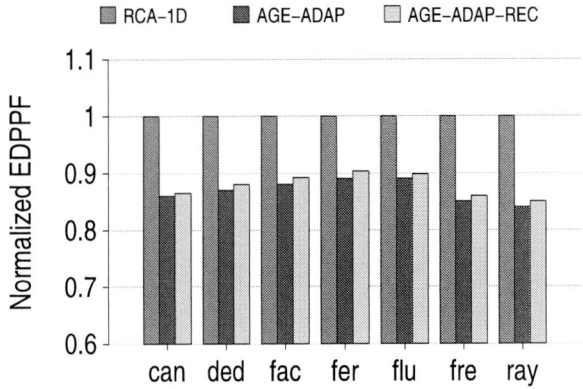

Figure 8: Normalized EDPPF (lower is better).

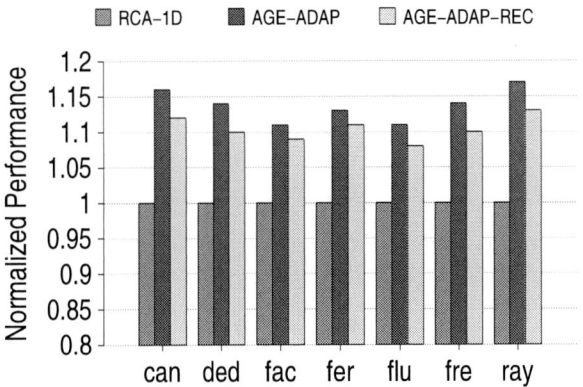

Figure 9: Normalized performance (higher is better).

ifests as performance degradation in routers and links along this path. Consequently, packets transmitted along these paths suffer from increased latency. On the other hand, AGE-ADAP and AGE-ADAP-REC route packets intelligently using paths that are least stressed and least congested (Section 4) and therefore incur lower overheads. Among AGE-ADAP and AGE-ADAP-REC, additional *recovery cycles* in case of AGE-ADAP-REC to meet *TTpEs* for each epoch leads to a higher latency as compared to AGE-ADAP. On an average, AGE-ADAP-REC reduces the latency by 13% relative to RCA-1D.

6.3.2 Energy-Delay-Product-Per-Flit

To show the impact of aging stress on power-performance characteristics of NoC based multicore system, EDPPF is also evaluated for each of the schemes. Figure 8 shows the EDPPF for different benchmarks across various schemes normalized to the RCA-1D scheme. AGE-ADAP and AGE-ADAP-REC are able to achieve reduced overhead as compared to RCA-1D. Due to additional *recovery cycles*, AGE-ADAP-REC incurs a higher EDPPF as compared to AGE-ADAP. On an average, AGE-ADAP-REC reduces EDPPF by 12.17% relative to RCA-1D.

6.3.3 Performance

Figure 9 shows the system performance for the schemes normalized to RCA-1D. Here also we observe that RCA-1D shows lower performance as compared to AGE-ADAP and AGE-ADAP-REC schemes. As RCA-1D is only congestion aware, it selects the least congested paths over the least stressed ones and therefore incurs

higher performance overheads at the system-level. For these multithreaded benchmarks, we use fair speedup as a metric for performance as it provides a more accurate estimate (see Section S6). Across different benchmarks, AGE-ADAP-REC shows 10.4% performance improvement over RCA-1D, demonstrating its effectiveness.

7. CONCLUSION

In this paper, we introduce a new reliability metric called *Traffic Threshold per Epoch (TTpE)* to model the effects of aging on NoC routers and links. Using this metric, we propose an aging-aware adaptive routing algorithm and router micro-architecture. Extensive experimental analysis incorporating power-performance impact of aging demonstrate 13%, 12.17% improvements in network latencies and EDPPF for our algorithm, respectively. At the system level, our algorithm shows 10.4% performance improvement for real workloads.

Acknowledgments

This work was supported in part by National Science Foundation grant CNS-1117425 and Micron Research Foundation.

8. REFERENCES

[1] *Open Source NoC Router RTL.* https://nocs.stanford.edu/cgi-bin/trac.cgi/wiki/Resources/Router.
[2] *PARSEC.* http://parsec.cs.princeton.edu/.
[3] N. Agarwal, T. Krishna, L.-S. Peh, and N. K. Jha. Garnet: A detailed on-chip network model inside a full-system simulator. pages 33–42, 2009.
[4] S. Bhardwaj, W. Wang, R. Vattikonda, Y. Cao, and S. Vrudhula. Predictive modeling of the nbti effect for reliable design. In *IEEE Custom Integrated Circuits Conference*, pages 189 –192, sept. 2006.
[5] T.-B. Chan, J. Sartori, P. Gupta, and R. Kumar. On the efficacy of nbti mitigation techniques. In *Proc. of DATE*, pages 1–6, 2011.
[6] X. Fu, T. Li, and J. A. B. Fortes. Architecting reliable multi-core network-on-chip for small scale processing technology. In *Proc. of DSN*, pages 111–120, 2010.
[7] P. Gratz, B. Grot, and S. W. Keckler. Regional congestion awareness for load balance in networks-on-chip. In *HPCA*, pages 203–214, 2008.
[8] C. Hernandez, F. Silla, and J. Duato. A methodology for the characterization of process variation in noc links. In *Proc. of DATE*, pages 685–690, 2010.
[9] A. B. Kahng, B. Li, L.-S. Peh, and K. Samadi. Orion 2.0: A fast and accurate noc power and area model for early-stage design space exploration. In *DATE*, pages 423–428, 2009.
[10] A. K. Kodi, A. Sarathy, A. Louri, and J. M. Wang. Adaptive inter-router links for low-power, area-efficient and reliable network-on-chip (noc) architectures. In *Proc. of ASP-DAC*, pages 1–6, 2009.
[11] B. Li, L.-S. Peh, and P. Patra. Impact of process and temperature variations on network-on-chip design exploration. In *NOCS*, pages 117–126, 2008.
[12] M. M. K. Martin, D. J. Sorin, B. M. Beckmann, M. R. Marty, M. Xu, A. R. Alameldeen, K. E. Moore, M. D. Hill, and D. A. Wood. MultifacetâĂŹs general execution-driven multiprocessor simulator (gems) toolset. *SIGARCH Comput. Archit. News*, 33, 2005.
[13] A. K. Mishra, N. Vijaykrishnan, and C. R. Das. A case for heterogeneous on-chip interconnects for cmps. In *ISCA*, pages 389–400, 2011.
[14] J. D. Owens, W. J. Dally, R. Ho, D. N. Jayasimha, S. W. Keckler, and L.-S. Peh. Research challenges for on-chip interconnection networks. *IEEE Micro*, 27(5):96–108, 2007.
[15] M. Sun, M. G. Pecht, and D. Barbe. Lifetime rc time delay of on-chip copper interconnect. In *IEEE Tran. on Semiconductor Manufacturing*, volume 15, pages 253–259, 2002.

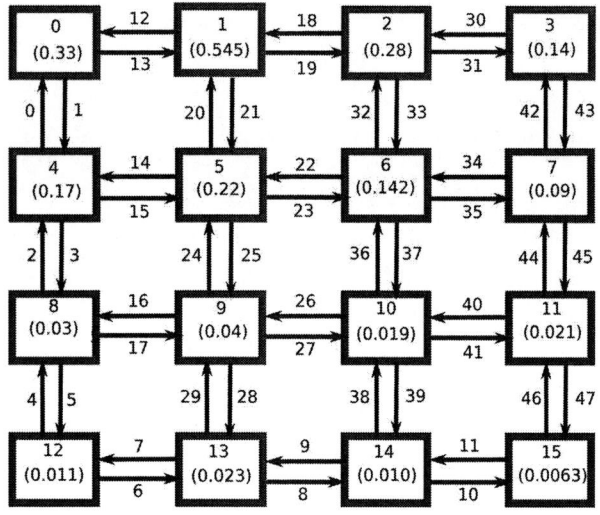

Figure 10: Different router utilization in a 4×4 NoC Mesh.

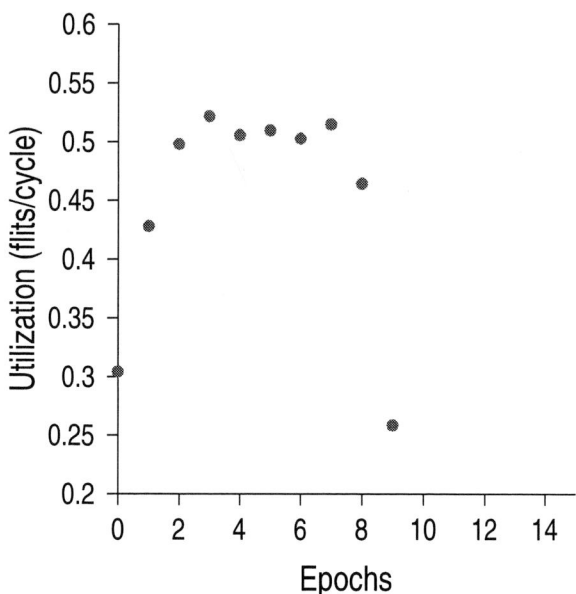

Figure 11: Buffer Utilization of router R1 for Different Epochs.

Supplemental Materials

S1. DISPARITY IN ROUTER UTILIZATION

Figures 10 and 11 show the disparity in NoC router utilization. Figure 10 shows the variation in buffer utilization (flits/cycle) in a 4×4 NoC mesh. These utilizations were obtained using the GARNET NoC simulator when traffic is generated by the *canneal* benchmark run. Similarly, the utilization of routers and links also varies at the runtime. We conducted another experiment using the same benchmark and NoC mesh to find the runtime variation in buffer utilization of router R1 (marked as 1 in Figure 10). We divided the total runtime into 10 different epochs of equal width. Figure 11 shows the different utilizations obtained for different epochs. Therefore, we observe two kinds of variations in utilization: a) every router and link has different utilization (asymmetric); and b) every router and link also experiences variation in utilization across different time windows (runtime).

S2. MODELING AGING IMPACT ON NOC ROUTERS AND LINKS

In this section, we present the system-level aging model that models the impact of aging mechanisms such as NBTI and electromigration on NoC routers and links. Here, we also derive the dependency of *TTpE* of stressed links and routers on the aging-induced delay variation.

S2.1 Modeling Delay Variations due to NBTI Degradation in NoC Routers

Due to the presence of both combinational and storage circuitry in NoC routers, the effects of NBTI on the performance of these routers cannot be ignored [SR5]. In this section, we model these effects using the flexible numerical model of NBTI degradation based on reaction-diffusion [SR6]. According to this model, V_{th} shift is given by:

$$\Delta V_{th} = \frac{q N_{it}(t)}{C_{ox}} \qquad (2)$$

where q is the elementary charge, $N_{it}(t)$ is the number of interface traps per unit area at time t and C_{ox} is the PMOS gate capacitance.

This model calculates $N_{it}(t)$ using the different parameters mentioned in Table 3.

As the architectural level techniques such as dynamic voltage scaling and activity and power management mechanisms are not applicable for network-on-chip design, we can also use the device level model [SR7] for NBTI modeling. We analyzed the reaction-diffusion model using the 65 nm technology parameters and obtained a similar degradation as the device level model.

To find the *TTpE* of a stressed router, we use a similar analysis as in [SR10] that estimates the workloads across the stressed cores by considering delay variations, discussed next.

S2.1.1 Analyzing Delay Variation in a Stressed Router

In an NoC system, different routers can experience a wide variation in performance degradation due to the combined effect of process variation and NBTI aging. Fundamentally, the *TTpE* of a stressed router during an epoch is estimated by comparing its performance degradation, measured as the delay variation, with that of the router experiencing the maximum performance degradation for the same epoch.

To estimate the delay variation in a stressed router, we extend the gate delay model in [SR8] to the critical path delay model. After perturbing V_{th} as $V_{th} = V_{th0} + \Delta V_{th}$, the i-th critical path delay can be written as:

$$d_i = d_i(V_{th0}, L_{eff}) + \left(\frac{\delta d_i}{\delta V_{th}}\right) \Delta V_{th} \qquad (3)$$

where delay $d_i(V_{th0}, L_{eff})$ is modeled as a Gaussian distribution with V_{th0} and L_{eff} as the nominal threshold voltage and channel length. As there can be many critical paths in a single router, the critical path with the biggest variation is used in the calculation of *TTpE*. We analyze all the routers in the system and estimate their biggest critical path variation. For a particular epoch, the *TTpE* of a given router is estimated by comparing its delay variation with that of the router with the worst variation. However, in case the worst router experiences more than $3\sigma_{delay_r}$ variation, which statistically covers 99.7% of all delay variations in the system [SR9], we simply use the $3\sigma_{delay_r}$ variation for *TTpE* estimation. Section 5 describes our experimental setup in more detail.

$$\Delta d_r = min(max_i((\delta d_i/\delta V_{th})\Delta V_{th}), 3\sigma_{delay_r})[SR10] \quad (4)$$

To relate the delay variation with *TTpE*, we use the percentage model proposed in [SR10]. According to this model, when the delay has zero variation, the value of *TTpE* is 100% and when the delay variation is maximum (at $3\sigma_{delay_r}$), the router must not accept any traffic (*TTpE* = 0). Hence,

$$TTpE_r = 1 - \left(\frac{\Delta d_r}{3\sigma_{delay_r}}\right) \quad (5)$$

where $TTpE_r$ is the Traffic Threshold per Epoch of the stressed router due to delay variation. We use a similar approach to model *TTpE* of NoC links next.

S2.2 Modeling Delay Variations due to NBTI and Electromigration Degradations in NoC Links

NoC links are modeled as repeated copper interconnects and therefore suffer from two different types of stresses: a) NBTI stress that increases the repeater resistance [SR11] and b) electromigration stress due to the use of barrier layers in copper interconnects that increases the wire resistance [SR12].

S2.2.1 Analyzing Delay Variation in Stressed Links

To model the propagation delay of a repeated interconnect in the presence of NBTI and electromigration stress, we include the increase in wire resistance due to electromigration in the NBTI-aware delay model given in [SR11]. Therefore, the propagation delay of the link under both NBTI and electromigration is:

$$d_{l_s} = kT_d + p(0.69)\left(C_d + \frac{C_w}{k} + C_g\right)\Delta R_o$$
$$+p(0.69)\left(\frac{C_g}{k}\right)\Delta R_w + p(0.38)\left(\frac{C_w}{k^2}\right)\Delta R_w \quad (6)$$

where k is the number of repeaters, R_o is the repeater resistance, C_d is the output drain diffusion capacitance of the repeater, C_g is the input gate capacitance of the repeater, R_w is the wire resistance, C_w is the wire capacitance, p is the number of stressed repeaters, T_d is the original unstressed delay, ΔR_o is the increase in repeater resistance due to NBTI and ΔR_w is the increase in wire resistance due to electromigration.

The variability of repeater resistance with the threshold voltage (ΔV_{th}) due to NBTI is given as [SR11]:

$$\frac{\delta R_o}{\delta V_{th}} = g\left[\frac{2 + (V_{GS} - |V_{th}| - \Delta V_{th})(\frac{\mu}{2.v_{sat}.L} + \theta)}{\frac{1}{2}\mu C_{ox}\frac{W}{L}(V_{GS} - |V_{th}| - \Delta V_{th})^3}\right] \quad (7)$$

where

$$g = \frac{3}{4}V_{dd}\left(1 - \frac{7}{9}\lambda V_{dd}\right) \quad (8)$$

ΔV_{th} is given by Equation 2 and rest of the symbols are similar to those used in [SR11].

The variability of the wire resistance due to electromigration stress is modeled as [SR12]:

$$\Delta R_w = \frac{2R_w \frac{\gamma}{A_0}D_0^{\frac{1}{2}}t^{\frac{1}{2}}e^{\frac{-Q_a}{2RT_a}}}{1 - 2\frac{\gamma}{A_0}D_0^{\frac{1}{2}}t^{\frac{1}{2}}e^{\frac{-Q_a}{2RT_a}}} \quad (9)$$

Table 3: Parameters used in NBTI stress modeling

Parameter	Value
$D0$	1e8
Ea	0.49 eV
k	8.617e-5
q	1.6e-19
T	Temperature
D	$D0 * exp(-Ea/(k*T))$
KH	1
Kf	1
Kr	1
tox	2.2 nm
$e0$	8.85e-21 F/nm
k	8.617e-5 eV/K (Boltzman constant)
eox	$3.9 * e0$
Cox	eox/tox
$powerFactor$	2
t	stress duration

Table 4: Parameters used in electromigration stress modeling

Parameter	Value
D_0	6.5E-7 m^2/s
R	8.31 J/mole K
A_0	400E-9 m
Q_a	1.64E5 J/mole

where different parameters are shown in Table 4.

We find the effective delay variation by comparing the delay variation with its $3\sigma_{delay_l}$ value.

$$\Delta d_l = min(d_{l_s} - T_d, 3\sigma_{delay_l}) \quad (10)$$

After calculation of the effective delay variation, we use the percentage model to evaluate the $TTpE_l$ for the stressed link:

$$TTpE_l = 1 - \left(\frac{\Delta d_l}{3\sigma_{delay_l}}\right) \quad (11)$$

S3. EFFECT OF TTPE: RELIABILITY ANALYSIS

In this section, we show the effect of *TTpE* on the reliability of the NoC mesh.

Reliability of a component in a system is defined as the probability that the component will perform its normal operation for a set period of time. Similarly, reliability of an NoC router ($\alpha(t)$) can be defined as the probability that the router is able to operate correctly from time 0 to time t [SR3]. We use the exponential failure law to define the reliability of a router [SR4]:

$$\alpha(t) = e^{-tk(t)} \quad (12)$$

where $k(t)$ is called the failure rate, measured by the number of failures per unit time.

Since reliability of a system is a probability, it can also be fun-

389

damentally expressed as a ratio:

$$\alpha(t) = \frac{N(t)}{N(0)} \tag{13}$$

where $N(0)$ is the number of components operating at time 0 and $N(t)$ is the number of components operating at time t.

As is evident from our system-level aging model, *TTpE* signifies a limit on the amount of traffic that can be accepted by a stressed router in any epoch. If the router continues to accept more traffic than this limit, it must be considered faulty. Therefore, the number of routers operating correctly at some particular time are the routers that are operating under their respective *TTpE* limits. Or,

$$N(t) = \sum_{i=1}^{R_{stress}} S_i(t) + \sum_{m=1}^{R_T - R_{stress}} N_m(t) \tag{14}$$

where R_T is the total number of routers, R_{stress} is the number of stressed routers and $S_i(t)$ for i^{th} stressed router at time t is defined as:

$$S_i(t) = \prod_{j=1}^{E} x_{ij}(t) \tag{15}$$

where j is an epoch, E is the total number of epochs and x_{ij} is given by:

$$x_{ij}(t) = \begin{cases} 1 & \text{if } U_{ij}(t) < lim_{ij}(t) \\ 0 & \text{if } U_{ij}(t) \geq lim_{ij}(t) \end{cases}$$

where $U_{ij}(t)$ is the utilization of ith stressed router at time t for jth epoch, in flits/cycle and $lim_{ij}(t)$ is the limit on the amount of traffic that the ith stressed router can accept at time t for jth epoch, also in flits/cycle. Here, $lim_{ij}(t)$ for i^{th} router is calculated as *TTpE* percentage of $U_{ij}(t)$.

In this analysis, we assume that faults are only caused by aging stress and therefore $N_m(t)$ is 1 for all the unstressed routers. For stressed routers, we can show the dependence of the number of routers surviving at time t on traffic limits due to *TTpE* by expressing $x_{ij}(t)$ in terms of router utilization and traffic limit:

$$x_{ij}(t) = 1 - \left\lceil \frac{\left\lfloor \frac{U_{ij}(t)}{lim_{ij}(t)} \right\rfloor}{\frac{U_{ij}(t)}{lim_{ij}(t)}} \right\rceil \tag{16}$$

S3.1 Calculation of Failure Rate

Failure rate of an NoC router can be derived by differentiating both sides of Equation 12. Therefore, failure rate $k(t)$ can be calculated by using the following equation:

$$k(t) = -\frac{\frac{d}{dt} N(t)}{N(t)} \tag{17}$$

In order to obtain $\frac{d}{dt} N(t)$, we first replace $S_i(t)$ in Equation 14 by its expression given in Equation 15. Equation 14 is now transformed to:

$$N(t) = \sum_{i=1}^{R_{stress}} \prod_{j=1}^{E} x_{ij}(t) + \sum_{m=1}^{R_T - R_{stress}} N_m(t) \tag{18}$$

Equation 18 is then differentiated w.r.t. time to obtain the value of $\frac{d}{dt} N(t)$. The complexities involved in the differentiation can be stated as:

1. Equation 18 has floor and ceiling functions which are discontinuous functions. These functions are differentiated using

their Fourier series continuous expansions:

$$\lfloor f(t) \rfloor = f(t) - \frac{1}{2} + \frac{1}{\pi} \sum_{k=1}^{\infty} \frac{Sin(2\pi k f(t))}{k} \tag{19}$$

$$\lceil f(t) \rceil = f(t) + \frac{1}{2} - \frac{1}{\pi} \sum_{k=1}^{\infty} \frac{Sin(2\pi k f(t))}{k} \tag{20}$$

2. Router utilization ($U_{ij}(t)$) and traffic limit ($lim_{ij}(t)$) are both time-dependent. Differentiation of $N(t)$ also involves differentiating the above mentioned time-dependent metrics. These derivatives are calculated as follows:

   - We first plot both the time-dependent functions w.r.t. time t.
   - We select a very small time interval for which the functions are linear (Δt) and find the slope for this interval after plotting both the metrics w.r.t. time for each stressed router i.

Once $\frac{d}{dt} N(t)$ is obtained, Equation 17 is used to calculate the failure rate. This failure rate is used in the calculation of reliability as described next.

S3.2 Reliability

After calculating the failure rate, we modify Equation 12 to obtain the variation of reliability of an NoC mesh ($\alpha_{mesh}(t)$) with time. Since a $P \times P$ mesh comprises $P \times P$ routers, therefore reliability of the entire mesh can be calculated as:

$$\alpha_{mesh}(t) = \left(e^{-tk(t)P^2} \right) \tag{21}$$

S4. CONGESTION-AWARE ROUTING ALGORITHM

In our proposed aging-aware adaptive routing algorithm, we have employed the Regional Congestion Awareness technique in one-dimension (RCA-1D), presented in [SR2]. Our algorithm uses the count of free input buffers present in the router as a measure of congestion. The RCA-1D technique aggregates and propagates this congestion metric along each dimension independently. The source router uses this congestion information to select the least congested path from the routing table to route the flits. In order to generate this congestion information, RCA-1D uses the *Aggregation* and *Propagation* modules, described next:

- **Aggregation**: This module is basically used to aggregate the local and global congestion information. Inputs to the aggregation module come from the downstream routers and the local Congestion Value Registers (CVRs). The Aggregation module then combines the two congestion values with equal weights of 50-50.

- **Propagation**: This module is responsible for the transmission of congestion information to adjacent routers. The propagation module combines the congestion values generated by the source router's aggregation unit. For RCA-1D, the congestion information is forwarded upstream, reflecting the conditions along a particular dimension.

S5. ROUTING TABLE: AN EXAMPLE

In this section, we present an example of a generic routing table. Table 5 shows a routing table stored inside a source router where the first column consists of the destination nodes and the second column gives a list of deadlock-free shortest paths available to this destination.

Table 5: An example of a generic routing table

Destination Nodes	Deadlock-free Shortest Paths
Node 0	(path A0, Output link La0), (path A1, Output link La1), ..
Node 1	(path B0, Output link Lb0), (path B1, Output link Lb1), ..
...	...
...	...
Node n	(path Z0, Output link Lz0), (path Z1, Output link Lz1), ..

S6. FAIR SPEEDUP

To measure the performance in our multi-programmed workload, we use a metric called *Fair Speedup* (FS) [SR1]. FS is defined as the harmonic mean of per-thread Speedups. We use FS as it follows the Pareto efficiency principle where improvements on the system should increase the performance of all running threads.

$$FS(scheme) = \frac{n}{\sum \frac{IPC_i(base)}{IPC_i(scheme)}}$$

Additionally, FS is a fair metric because the harmonic mean of speedup rewards uniform improvements across all threads while at the same time penalizing slowdowns. Such fairness is not exhibited by other metrics such as speedup of aggregate IPCs (Agg_{ipc}).

We give an example below to illustrate this point clearly. In this example, there are 4 threads running with IPC values indicated in Table 6. Suppose we introduce a scheme that degrades the performance of the first three threads by 10% while increasing the last one by 100%.

Setup	IPC1	IPC2	IPC3	IPC4	Agg_{ipc}	FS
Baseline	1	2	3	4	-	-
NewScheme	0.9	1.8	2.7	8	**1.34**	**1.17**

Table 6: Sample IPC values

Both performance improvements from Agg_{ipc} and FS are shown in the table (last two columns). Agg_{ipc} reports this as an impressive 34% performance increase while FS will report this at a modest 17% improvement. We believe FS gives a more accurate picture of the actual improvement because outlier values cannot heavily influence the final speedup as compared to other metrics based on arithmetic averages.

S7. REFERENCES

[SR1] CHANG, J., AND SOHI, G. S. Cooperative cache partitioning for chip multiprocessors. In *International Conference on Supercomputing* (2007).

[SR2] GRATZ, P., GROT, B., AND KECKLER, S. W. Regional congestion awareness for load balance in networks-on-chip. In *HPCA* (2008).

[SR3] CHANG, Y. C., CHIU, C. T, AND LIN, S. Y. On the design and analysis of fault tolerant noc architecture using spare routers. In *ASP-DAC* (2011).

[SR4] WANG, L. T., STROUD, C. E, AND TOUBA, N. A. System-on-Chip test architectures: Nanometer design for testability. *Morgan Kaufmann* (2008).

[SR5] FU, X., TAO, L., AND FORTES, J. A. Architecting reliable multi-core network-on-chip for small scale processing technology. In *DSN* (2010).

[SR6] CHAN, T., SARTORI, J., GUPTA, P., AND KUMAR, R. On the efficacy of NBTI mitigation techniques. In *DATE* (2011).

[SR7] BHARDWAJ, S., WANG, W., VATTIKONDA, R., CAO, Y., AND VRUDHULA, S. Predictive modeling of the NBTI effect for reliable design. In *CICC* (2006).

[SR8] CHANG, H., AND SAPATNEKAR, S. Statistical timing analysis considering spatial correlations using a single pert-like traversal. In *ICCAD* (2003).

[SR9] NAVIDI, W. Statistics for engineers and scientists. *Mc Graw Hill* (2010).

[SR10] SUN, J., KODI, A. K., LOURI, A., AND WANG, J. M. NBTI aware workload balancing in multi-core systems. In *ISQED* (2009).

[SR11] DATTA, B., AND BURLESON, W. Analysis and mitigation of NBTI-impact on PVT variability in repeated global interconnect performance. In *GLSVLSI* (2010).

[SR12] SUN, M., PECHT, M. G., AND BARBE, D. Lifetime RC time delay of on-chip copper interconnect. In *IEEE Tran. on Semiconductor Manufacturing* (2002).

Explicit Modeling of Control and Data for Improved NoC Router Estimation

Andrew B. Kahng[†‡], Bill Lin[†] and Siddhartha Nath[‡]

UC San Diego ECE[†] and CSE[‡] Departments, La Jolla, CA 92093

abk@ucsd.edu, billlin@ece.ucsd.edu, sinath@cs.ucsd.edu

ABSTRACT

Networks-on-Chip (NoCs) are scalable fabrics for interconnection networks used in many-core architectures. ORION2.0 is a widely adopted NoC power and area estimation tool; however, its models for area, power and gate count can have large errors (up to 110% on average) versus actual implementation. In this work, we propose a new methodology that analyzes netlists of NoC routers that have been placed and routed by commercial tools, and then performs explicit modeling of control and data paths followed by regression analysis to create highly accurate gate count, area and power models for NoCs. When compared with actual implementations, our new models have average estimation errors of no more than 9.8% across microarchitecture and implementation parameters. We further describe modeling extensions that enable more detailed flit-level power estimation when integrated with simulation tools such as GARNET.

Categories and Subject Descriptors

C.2 [**Computer-Communications Networks**]: Network Architecture and Design

General Terms

Algorithms, Design, Performance

Keywords

network-on-chip, flit-level power modeling, parametric regression

1. INTRODUCTION

Networks-on-Chip (NoCs) have proven to be a highly scalable and low-latency interconnection fabric in the era of many-core architectures, as evidenced by in commercial chips such as the Intel 80-core [31], IBM Blue Gene [32] and Tilera TILE-Gx [33] processors. Because of their growing importance, NoC implementations must be optimized for latency and power [7, 9, 11, 15, 20]. To aid architects and designers in early design-space exploration, accurate NoC power and area estimators are required. Previous approaches to modeling are of two kinds, (1) based on templates at the architecture level, such as ORION2.0 [3], and (2) based on regression analysis on post-P&R data, such as [2]. ORION2.0 is widely used as a stand-alone tool as well as with full-system NoC simulators such as GARNET [13].

Both template- and regression-based modeling approaches, however, are in need of improvement. ORION2.0 has large estimation errors [2] for two fundamental reasons: (1) models are incomplete because control path resources are not modeled, even though they contribute significantly to power and area, and (2) models are not refined using post-P&R power and area data. Kahng et al. [2] and Jeong et al. [10] proposed non-parametric regression models to overcome the limitations in ORION2.0; [2] further concluded that parametric regression can be very inaccurate. In Figure 1(a), we show power estimation errors at 65nm in ORION2.0 and the previous regression approach [2], as a function of the number of virtual channels in the router. The maximum errors are 185% and 75%. Sim-

ilarly, in Figure 1(b), we show power estimation errors at 90nm in ORION2.0 and the previous regression approach [2] when the flit-width is changed.

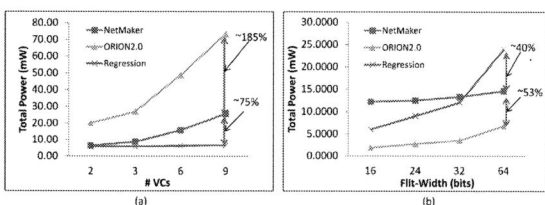

Figure 1: Poor estimations by ORION2.0 [3] and previous regression approach [2]. NetMaker vs. ORION2.0 vs. regression at (a) 65nm. (b) 90nm.

In this work, we propose a new model, ORION_NEW, which improves the ORION2.0 models by explicitly modeling control and data path resources. We perform parametric regression analysis with post-P&R area and power data to refine ORION_NEW models such that the estimates for area and power are highly accurate across multiple router RTLs, microarchitectures and implementation parameters. We demonstrate that accurate parametric models lead to better minimization of error in least-squares regression, and the worst-case errors are significantly better than the worst-case errors of non-parametric regression approaches [2]. We further describe modeling extensions that enable more detailed flit-level power estimation when integrated with simulation tools such as GARNET [13].

Our main contributions are as follows.

1. We explicitly model control and data paths to create ORION_NEW models that are highly accurate and robust across multiple router RTLs, and across microarchitecture and implementation parameters.

2. We demonstrate that parametric regression with accurate models can significantly reduce the worst-case error compared to non-parametric regression approaches for NoC routers.

3. We are the first to propose a detailed, efficient and fine-grained flit-level power estimation model that seamlessly integrates with full-system NoC simulators.

The remainder of this paper is organized as follows. Section 2 presents related work. Section 3 describes the ORION_NEW model. Section 4 describes our modeling methodology. Section 4.3 presents our new flit-level power estimation model. Section 5 presents experimental results to validate and compare ORION_NEW models with ORION2.0 and the non-parametric regression approach in [2]. Section 6 concludes and outlines future work.

2. RELATED WORK

Previous works have focused primarily on two broad modeling paradigms: (1) architecture-level models using templates for each router component block (input and output buffer, crossbar, switch and VC arbiter) and (2) RTL and gate-level simulation-driven models. For the first approach, Patel et al. [4] propose a transistor count-based analytical model for NoC power. However, their models have large errors because they do not consider any router microarchitecture parameter. ORION [7] and ORION2.0 [3] are architectural models that use microarchitecture and technology parameters for the router component blocks. However, from our experimental studies as well as from [2], ORION2.0 estimates have very large errors.

The other approach is based on pre-layout (RTL or post-synthesis gate-level) [6, 8, 5, 16] or post-layout [12, 1, 17, 14] simulations.

Banerjee et al. [1] report accurate power for a range of routers, but do not present any analytical models for router power. Chan et al. [8] develop cycle-accurate power models with reported average errors up to 20%. Meloni et al. [14] and Lee et al. [16] perform parametric regression analysis on post-layout and RTL simulation results, respectively. Their models, however, are fairly coarse-grained as they cannot explain how power dissipates in each router block with change in load, microarchitecture or implementation parameters. Kahng et al. [2] use non-parametric regression to model NoCs.

Our methodology uses accurate parametric models along with non-negative least-squares regression analysis to provide accurate area and power estimates, with average error of no more than 9.8% across microarchitecture and implementation parameters. Our models calculate area and power on a per-instance basis but avoid the overhead of slow gate-level simulations.

Furthermore, we significantly extend our models to achieve flit-level power estimations. Ye et al. [18] and Penolazzi et al. [17] estimate power dissipation using bit-level model, and Penolazzi et al. [17] propose a static bit-based model to estimate Nostrum NoC power. However, each of these models is tied to a specific router implementation and cannot explain how different bit encodings affect the power consumption in each block within the router. Our flit-level power estimation methodology estimates the power impact for each component block and reports accurate power numbers across different bit encodings in flits.

3. MODEL DESCRIPTION

We now describe the ORION_NEW modeling of each component in a modern on-chip network router. We have developed these models by analyzing post-synthesis and post-P&R netlists of two RTL generators, NetMaker [28] from Cambridge and the Open Source NoC router from Stanford [29]. (Our methodology is described in detail in Section 4.) Figure 2 shows the component blocks in a router, i.e., input buffer, switch and VC (virtual channel) arbiter, crossbar and output buffer [11]. We model instances (or gates) in each component block because our studies show that accurate estimations of area and power are possible only if the instance modeling is accurate. The microarchitecture parameters used are #Ports (P), #VCs (V), #Buffers (B) and Flit-width (F).

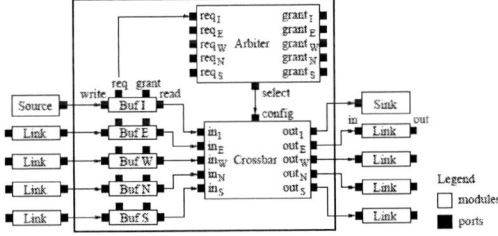

Figure 2: Router architecture [7].

3.1 New Model Elements

The new model explicitly accounts for control and data resources in the router. The new modeling elements are:

1. Control resources such as FIFO select and decode logic signals in the input and output buffers.

2. Tri-state crossbar model.

3. Additional input buffer resources for delay-optimized arbiters [11].

4. Output buffer model to store only head flits.

5. Clock frequency dependent scaling.

3.2 Crossbar (XBAR) Model

This component block is responsible for connecting input ports to output ports so that all flit bits are transferred to output ports [19]. The ORION2.0 models for router crossbar consider two implementations, matrix [11] and multiplexer tree [3]. The multiplexer tree is the smaller of these in terms of instance count and area and is modeled as $P \times P \times F$ multiplexers at each level of the tree. .

Modern router RTLs such as NetMaker and Stanford NoC use a simpler and smaller crossbar implementation where each flit bit is controlled using a tri-state buffer, which can be modeled as a $2:1$ MUX. Hence, the total number of such MUXes required are: $P \times P \times F$. This new model reduces the instance count by a factor of $[2^{\log_2 P} - 1]$ when compared to the multiplexer tree implementation.

3.3 Switch and VC Arbiter (SWVC) Model

This block is responsible to generate control signals for the crossbar such that a connection is established between input buffers to output ports [19]. ORION2.0 adds an overhead of 30% to the arbiter by default. Our analysis indicates that this overhead is not needed with frequency ranges $400MHz$-$900MHz$ for process nodes $45nm$ to $130nm$. Beyond this range of frequency a derating factor must be applied, which is discussed in Section 3.7. The ORION_NEW model for switch and virtual channel arbiter is: $9 \times (P \times (P \times (V^2 + 1) + (V^2 - 1))$. The constant factor 9 arises because six 2-input NOR gates, two INVerters and one D-FlipFlop are used to generate one grant signal on each path.

3.4 Input Buffer (InBUF) Model

This block holds the entire incoming payload of flits at the input stage of the router for decode [19]. ORION2.0 models only the buffer instances and does not take into account control signals which are needed at this stage for decode such as FIFO select, buffer enable control signals and logic for housekeeping, such as the number of free buffers available per VC, VC identification tag per buffer, etc. As a result, ORION2.0 underestimates the instances at the input stage of the router.

In our new model, we model control signals and housekeeping logic in addition to the actual FIFO buffers. Modern routers implement the same stage VC and SW allocation to optimize delay [11], leading to doubling of input buffer resources. Hence, the number of FIFO buffers are $2 \times P \times V \times B \times F$. The control signals for decoding the housekeeping logic are modeled as: $180 \times P \times V + 2 \times P^2 \times V \times B + 3 \times P \times V \times B + 5 \times P^2 \times B + P^2 + F \times P + 15 \times P$ (as analyzed from the post-synthesis and post-P&R netlists). Each constant factor in the model denotes the number of instances per path. For example, the 180 factor accounts for instances to generate FIFO select signals and flags for each buffer in the $P \times V$ path. The smaller constant factors $2, 3, 5$ account for instances for local flags in the decode logic. The factor 15 denotes the number of buffers in each FIFO select path of an input port.

3.5 Output Buffer (OutBUF) Model

This block holds the head flits between the switch and the channel for a switch with output speedup [19]. ORION2.0 models the output buffers in exactly the same way as input buffers; this is inaccurate for modern routers that use hybrid output buffers, and leads to an overestimate of the instance count. The output buffers need to only store enough flits to match the speed between the switch and the channel. At the output, these buffers are used to stage the flits between the switch and channel when channel and switch speeds mismatch. Instead of $P \times V \times B \times F$ used in ORION2.0, output buffers are proportional to $P \times V$. There are several control signals per port and VC associated with each buffer, which makes the overall instance count grow as $P \times (80 \times V + 25)$. The constant factor 80 accounts for the instances used to generate flow control credit signals for each VC, while the constant factor 25 accounts for buffers and flags.

3.6 Clock and Control Logic (CLKCTRL) Model

ORION2.0 does not accurately model clock buffers and control logic routing resources as clock frequency scales. ORION_NEW models these resources as 2% of the sum of instances in the SWVC, InBUF and OutBUF component blocks.

3.7 Frequency Derating Model

As frequency changes, timing constraints change. To meet setup time at higher frequencies, buffers are inserted leading to an overall increase in instance count in the design. ORION2.0 scaling is agnostic to implementation parameters such as clock frequency. This causes large errors in area and instance counts at higher frequencies for component blocks such as SWVC, InBUF and OutBUF where there are several logic signals which consume routing resources. The number of instances in the crossbar does not vary much with frequency because there are no critical paths. So, we can ignore the effects of frequency on the crossbar.

To derate for frequency, we find the frequency below which the instance counts change by less than 1%. In $65nm$ technology, this is $400MHz$ for both NetMaker and Stanford NoC routers. We derate instance counts based on this frequency as: $\Delta Instance = \Delta Frequency \times Constant Factor$. The constant factor is dependent on the amount of control logic versus FIFO for each component block. In SWVC and InBUF, the $control/FIFO \approx 1$, so the constant factor value is 1. In OutBUF, $control/FIFO \approx 0.16$, and a fitted constant factor of 0.03 is used to account for setup buffers.

4. ORION_NEW METHODOLOGY

In this section, we describe how we estimate power and area using the two approaches described in Sections 4.1 and 4.2. We extend our methodology to flit-level power estimation in Section 4.3. We use:

- Multiple parametrized NoC RTL generators, NetMaker [28] from Cambridge University and the Open Source NoC from Stanford [29] to make the ORION_NEW models robust.
- Range of values of microarchitecture parameters, #Ports (*P*), #VCs (*V*), #Buffers (*B*) and Flit-width (*F*) and implementation parameters such as clock frequency and technology node.
- Operational parameters for power calculation: switching activity (*TR*) and static probability of 1's in the input (*SP*).
- Multiple commercial tools, Synopsys DesignCompiler (DC) [22] and Cadence RTL Compiler (RC) [21], with options to preserve module hierarchy after synthesis because we analyze each router component block. We compare instance counts, area and power reported by each tool to ensure that for a given RTL these results do not vary by more than 10%.
- Cadence SOC Encounter (SOCE) [21] with die utilization of 0.75 and die aspect ratio of 1.0 to place and route the synthesized router netlist.
- Synopsys PrimeTime-PX (PT-PX) [23] to run power analysis of the post-P&R netlist, SPEF [26] and SDC [27].
- MATLAB [30] function *lsqnonneg* for regression analysis.

Table 1 summarizes these details.

Table 1: ORION_NEW Methodology: Tools and Parameters

Stage	Tool	Options
RTL	NetMaker	ISLAY config
	Stanford NoC	default
μarch	Ports; VCs;	$P = \{5, 6, 8, 10\}; V = \{2, 3, 6, 9\}$
	BUFs; Flit-Width	$B = \{8, 10, 15, 22\}; F = \{16, 24, 32, 64\}$
Impl	Clock Freq	$Freq = \{400, 700, 1200, 2000\}$ MHz
		Switching Activity $(TR) = \{0.2, 0.4, 0.6, 0.8\}$
		Static Prob of 1's $(SP) = \{0, 0.25, 0.5, 0.75, 1.0\}$
	Tech Nodes	$45nm$ = OpenPDK45 from NCSU/OSU
		$65nm, 90nm, 130nm$ = TSMC GP, G, GHP resp.
Syn	Synopsys DC	*compile_ultra -exact_map*
	(v2009.06-SP2)	*-no_autoungroup -no_boundary_optimization*
		report_area -hierarchy; report_power -hierarchy
	Cadence RC	default synthesis flow
	(vEDI09.12)	
Power	Synopsys PT-PX	*set power_enable_analysis true*
	(v2009.06-SP2)	*set power_analysis_mode averaged*
		set_switching_activity -toggle_count TR
		-static_probability SP -type inputs
		read_sdc router.sdc; read_parasitics router.spef
Regression	MATLAB	*lsqnonneg*

Figure 3 shows the flow we use to develop ORION_NEW models for each component block of the router. In Table 2, we summarize the ORION_NEW instance count model of each component block.

Table 2: ORION_NEW model for Instances

Component	Equation
XBAR	P^2F
SWVC	$9(P^2V^2 + P^2 + PV - P)$
InBUF	$180PV + 2PVBF + 2P^2VB + 3PVB + 5P^2B + P^2 + PF + 15P$
OutBUF	$25P + 80PV$
CLKCTRL	$0.02 \times (SWVC + InBUF + OutBUF)$

There are two ways to estimate NoC area and power using the ORION_NEW models as shown in Figure 4. The manual approach is described in Section 4.1, and the regression analysis approach is described in Section 4.2. The benefits of each are described below.

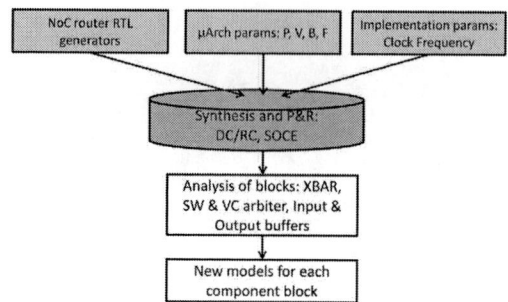

Figure 3: High-level flow used to arrive at ORION_NEW models.

- Both the approaches have minimum estimation error when the router RTLs are modular so that instance count and area numbers per component block can be calculated.
- The manual approach requires knowledge of process node and finer implementation details such as (HP, LSTP, LOP) × (HVT, NVT, LVT) × (bc, wc) to correctly select a technology library file. The regression analysis approach, on the other hand, is agnostic of implementation details. It only depends on a training set of data. More data points help the tool to minimize the sum of square error.
- The manual approach leads to faster estimation since it only involves technology library look-ups and plugging-in of library values into the ORION_NEW model. In contrast, the regression analysis approach requires synthesis and P&R to be performed on the router RTL for at least six data points. On an Intel Core *i3* $2.4GHz$ processor, the runtime of the manual approach when used with ORION2.0 code is less than $10ms$, whereas the regression analysis approach takes about $140ms$, when 64 test data points are used.
- It is extremely difficult to capture fine-grained implementation details in ORION_NEW models, e.g., area and power contribution of wires after routing, and change in coupling capacitance and power after metal fill. These missing details cause estimation errors versus actual implementation when the manual approach is used. In order to reduce errors with respect to implementation, the regression analysis approach with post-P&R area and power is preferred.

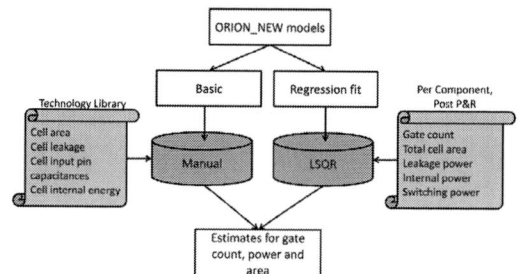

Figure 4: High-level view of power and area estimation methodology using Manual and Regression Analysis (LSQR) approaches.

4.1 Manual Approach to Estimate NoC Power and Area

This approach uses ORION_NEW models along with the technology library file of the process node in which the router is going to be fabricated. The key ingredients of this approach are:

- Microarchitecture parameters {*P, V, B* and *F*} and implementation parameter (clock frequency).
- Cell areas, leakage, internal energy and load capacitance.
- Switching activity.

ORION_NEW simplifies design of a NoC, using only a few standard cells. Instance count for each component block for a given set of router microarchitecture parameters is calculated from Table 2. Cell area is obtained from technology files. The area calculation, along with TSMC standard-cell names in parentheses, is shown in Table 3.

394

Table 3: Area Models using Instance count

Component	Logic (TSMC Cell Name)	Area
XBAR	MUX2 (MUX2D0)	$Area_{MUX} \times XBAR_{insts}$
SWVC	6 NOR2, 2 INV, 1 DFF (NR2D1, INVD1, DFQD1)	$\left(\frac{6Area_{NOR} + 2Area_{INV} + Area_{DFF}}{9}\right)$ $\times SWVC_{insts}$
InBUF + OutBUF	1 AOI, 1 DFF (AOI22D1, DFQD1)	$\left(\frac{Area_{AOI} + Area_{DFF}}{2}\right)$ $\times (In + Out)BUF_{insts}$
CLKCTRL	1 INV, 1 AOI (INVD1, AOI22D1)	$\left(\frac{Area_{AOI} + Area_{INV}}{2}\right)$ $\times (CLKCTRL)_{insts}$

Power has three components, that is, leakage, internal and switching. Leakage power is static power when the cell is not transitioning between logic states. It is dependent on current state of the input pins of the cell as well as process corner, voltage and temperature. Switching and internal power together constitute dynamic power, which varies with operating voltage, capacitive load and frequency of operation. Switching power is the power consumed when a load capacitance on a net is charged and discharged; internal power is the power dissipated inside a cell and consists of short-circuit power and switching power of internal nodes.

In ORION_NEW, toggle rate (*TR*) is equal to the input switching activity for all nets in the crossbar, arbiters and buffer control logic. We assume that buffer cells toggle at 25% of the input switching activity, since multiple VCs do not require buffer contents to change in every cycle.

Leakage power calculation: For leakage power, the model uses the weighted average of the state-dependent leakage of the cells. Equations (1)-(4) are used to calculate the leakage power of each component block.

$$P_{leak\_XBAR} = MUX_{leak} \times XBAR_{insts} \quad (1)$$

$$P_{leak\_SWVC} = \left(\frac{6NOR_{leak} + 2INV_{leak} + DFF_{leak}}{9}\right) \times SWVC_{insts} \quad (2)$$

$$P_{leak\_BUF} = \left(\frac{AOI_{leak} + DFF_{leak}}{2}\right) \times (In + Out)BUF_{insts} \quad (3)$$

$$P_{leak\_CLKCTRL} = \left(\frac{AOI_{leak} + INV_{leak}}{2}\right) \times (CLKCTRL)_{insts} \quad (4)$$

Internal power calculation: For internal power, table look-ups in technology library files return the internal energy of given standard cells with load capacitance of fanout pins and slew value of $\approx 5 \times FO4 \; delay$.[1] Internal energy for a pin is the minimum of the rise and fall energies. Equations (5)-(8) are used to calculate internal power of each component block.

$$P_{int\_XBAR} = MUX_{int} \times TR \times XBAR_{insts} \quad (5)$$

$$P_{int\_SWVC} = (6NOR_{int} + 2INV_{int} + DFF_{int}) \times TR \times SWVC_{insts} \quad (6)$$

$$P_{int\_BUF} = (AOI_{int} + 0.25DFF_{int}) \times TR \times (In + Out)BUF_{insts} \quad (7)$$

$$P_{int\_CLKCTRL} = (AOI_{int} + INV_{int}) \times TR \times (CLKCTRL)_{insts} \quad (8)$$

Switching power calculation: For switching power, the load capacitance is calculated as the sum of the input capacitances of pins that are driven by a net and the wire capacitance on the net. The wire capacitance is approximately calculated as a constant factor times the total pin capacitances. This constant factor is 1.4 at 65*nm* and is assumed to decrease by 14% with for each successive process node shrink. Equations (9)-(12) are used to calculate switching power of each component block.

$$P_{sw\_XBAR} = XBAR_{load} \times TR \times XBAR_{insts} \quad (9)$$

$$P_{sw\_SWVC} = SWVC_{load} \times TR \times SWVC_{insts} \quad (10)$$

$$P_{sw\_BUF} = (In + Out)BUF_{load} \times TR \times (In + Out)BUF_{insts} \quad (11)$$

$$P_{sw\_CLKCTRL} = (CLKCTRL)_{load} \times TR \times (CLKCTRL)_{insts} \quad (12)$$

---

[1] The *FO4* delay is the delay of a minimum-sized INV and is a standard proxy for switching speed in a given process technology. The resulting slew time values are 80 – 100ps for 45nm and 65nm technologies.

Flow details: The steps below describe how total area and power are estimated using the ORION_NEW models and equations above.

1. Choose microarchitecture parameters (P,V,B,F), clock frequency and average switching activity at inputs.

2. Use models in Table 2 to calculate the instance count of each component block of the router.

3. Use models in Table 3 to calculate the area of each router component block. Total area is calculated as the sum of areas of all blocks.

4. Obtain state-dependent leakage of cells from technology library files. Use Equations (1)-(4) to calculate leakage power of each component block. Total router leakage power is calculated as the sum of leakage power of all component blocks.

5. Obtain internal energy of cells from technology library files. Use Equations (5)-(8) to calculate internal power of each component block. Total internal power is calculated as the sum of internal power of all component blocks.

6. Obtain input pin capacitances of cells from technology library files. Use Equations (9)-(12) to calculate switching power of each component block. Total switching power is calculated as the sum of switching power of all component blocks.

7. The total power dissipated by the router is calculated as the sum of total leakage, total internal and total switching power.

4.2 Regression Analysis Approach to Estimate NoC Power and Area

As another approach to estimation of router area and power, we use parametric regression to fit parameters for cell area, leakage, internal energy and load capacitance into ORION_NEW models. This approach requires instance counts, area, and total leakage, internal and switching power of each component block of the router from post-P&R tools. Options are set in synthesis to preserve module hierarchy and names. Constrained least-squares regression (LSQR) is used to enforce non-negativity of coefficients (cell area, leakage, internal energy, load capacitance). We use the MATLAB [30] function *lsqnonneg* for this purpose, and tool options as given in Table 1.

Flow Details: LSQR is applied to fit a model of post-P&R instance count for each router component block. At least six data points are needed in the training set because there are four microarchitecture parameters and two implementation parameters (clock frequency and toggle rate). Our parametric LSQR setup is as follows.

$$a_1 \cdot Insts_{model\; <component>} + a_0 = Insts_{tool\; <component>} \quad (13)$$

$Insts^R_{model\; <component>}$ is the refined instance count of each component block after LSQR. The refined instance count is used to fit models of post-P&R area and power as follows:

$$b_1 \cdot Insts^R_{model\; <component>} + b_0 = Area_{tool\; <component>} \quad (14)$$

In Equation (14), b_1 is the fitting coefficient for cell area and the coefficient b_0 accounts for the routing overhead.

We model leakage, internal and switching power as:

$$\{c_5, d_5, e_5\} \cdot Insts^R_{model\; XBAR} + \{c_4, d_4, e_4\} \cdot Insts^R_{model\; SWVC} +$$
$$\{c_3, d_3, e_3\} \cdot Insts^R_{model\; InBUF} + \{c_2, d_2, e_2\} \cdot Insts^R_{model\; OutBUF} +$$
$$\{c_1, d_1, e_1\} \cdot Insts_{model\; CLKCTRL} = \{P_{leak\; tool}, P_{int\; tool}, P_{sw\; tool}\} \quad (15)$$

where coefficients $\{c_5, \cdots, c_0\}$ are used to fit cell leakage power, and similarly $\{d_5, \cdots, d_0\}$ and $\{e_5, \cdots, e_0\}$ are respectively used to fit internal energy and load capacitance.

It is possible to skip the instance count refinement step (Equation (14)) and directly perform LSQR for area and leakage, internal and switching power using the above equations. We observe that average error can change by 3% in either direction by omitting the instance count refinement step. Note that it is necessary to perform per-component LSQR; if LSQR is performed for the entire router's area or power, large errors result because multiple components have the same parametric combination of (P,V,B,F). Failing to separate these contributors to area or power results in large errors: at 65nm, we have experimentally observed worst-case errors of 296%

395

for power and 557% for area. Thus, it is important to preserve module hierarchy during synthesis in the flow.[2]

4.3 Extension to Flit-Level Power Modeling

The dynamic power models used in ORION2.0 and ORION_NEW do not consider bit encodings in a flit, which can lead to significant errors in dynamic power estimation. As an example, consider an 8-bit flit with four bits as 1. This flit can either be $8b'11110000$ or $8b'10101010$. In the first encoding, there is only one toggle per flit, whereas in the second encoding there are seven toggles per flit. Clearly, the second flit will lead to higher dynamic power than the first one. To model this effect, we devise a flow as shown in Figure 5. Before using a testbench, the netlists must pass an equivalence check using tools such as Synopsys Formality [24]. We inject different bit encodings in the input during simulation over 10000 cycles and the resultant VCD (Value Change Dump) is validated using a waveform analyzer such as Synopsys DVE [25]. A satisfactory VCD is used as input to Synopsys PrimeTime-PX [23] to obtain power values. Regression analysis is performed using the tool-reported power values with the ORION_NEW estimates to obtain an enhanced ORION_NEW model for flit-level power estimation. These models may be invoked by NoC full-system simulators such as GARNET [13] to obtain very accurate estimates.

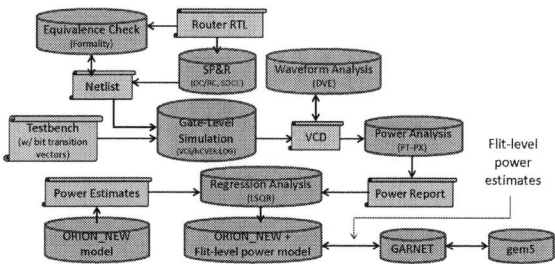

Figure 5: Methodology to enhance ORION_NEW dynamic power models with flit-level power estimation.

5. VALIDATION AND RESULTS

We set up experiments as described in Table 1 of Section 4. We use parameters and tools for our experiments as listed in Table 1. We discuss the results in two parts - (1) ORION2.0 versus ORION_NEW comparisons for area and power, and (2) impact of results with our regression analysis approach versus the approach used in prior work of [2]. We compare the results of our methodology with post-P&R instance count, power and area outcomes for two router RTL generators, Netmaker [28] and Stanford NoC [29].

5.1 ORION2.0 versus ORION_NEW Comparisons

Since the instance count per component is at the core of the ORION_NEW model, we compare ORION2.0 estimates of instance (or gate) counts, as well as the ORION_NEW model estimates with implementation (post-P&R) for each component block. Figures 6(a), 6(c) and 6(e) show the large errors in ORION2.0 in the crossbar, output buffer and input buffer respectively, and Figures 6(b), 6(d) and 6(f) show the significant reduction in estimation error for these components with ORION_NEW models. ORION2.0 and ORION_NEW are plotted in different graphs because of the large errors in instance counts in ORION2.0.

ORION2.0 modeling of instance count for a component does not consider implementation parameters such as clock frequency. As a result, the instance count does not scale when frequency is changed, even though at higher frequencies several buffers are inserted to meet tight setup time constraints. ORION_NEW models apply a frequency derating factor on the instance models for component blocks as described in Section 3.7. Figures 7(a) and 7(b) show the incorrect estimates by ORION2.0; by contrast, the estimates from

---

[2]Use of hierarchical synthesis in general leads to lower instance count, standard-cell area, and total power as compared with flat synthesis results. This comes at the cost of frequency (timing slack), since flat optimization across module boundaries can sometimes achieve better timing results. For our selection of microarchitecture and implementation parameters, hierarchical synthesis on average has 35% fewer instances, 48.8% less standard-cell area and 49.4% less total power – along with 8% less timing slack – compared with flat synthesis. The runtimes for hierarchical and flat synthesis are within 5% of each other.

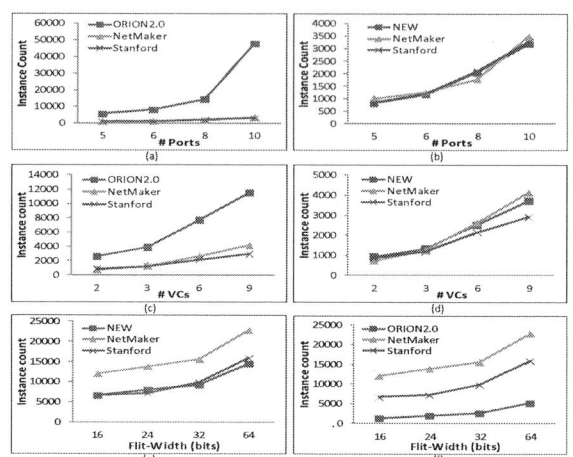

Figure 6: (a) XBAR with #Ports: ORION2.0 vs. Implementation. (b) XBAR with #Ports: ORION_NEW vs. Implementation. (c) Output Buffer with #VCs: ORION2.0 vs. Implementation. (d) Output Buffer with #VCs: ORION_NEW vs. Implementation. (e) Input Buffer with Flit-Width: ORION2.0 vs. Implementation. (f) Input Buffer with Flit-Width: ORION_NEW vs. Implementation.

ORION_NEW are very close to actual implementation for output and input buffer component blocks respectively.

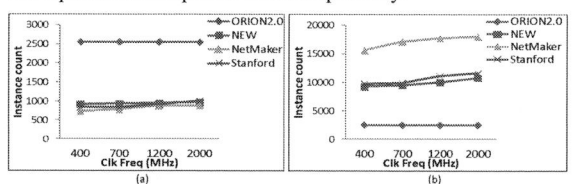

Figure 7: (a) Output buffer with Clock Frequency: ORION2.0 vs. ORION_NEW. (b) Input buffer with Clock Frequency: ORION2.0 vs. ORION_NEW.

Component	Avg Error: #Instances		Max Error: #Instances		Avg Error: Total Area		Max Error: Total Area	
	2.0	NEW	2.0	NEW	2.0	NEW	2.0	NEW
XBAR	86.10%	2.10%	93.10%	3.00%	86.20%	0.90%	93.20%	1.80%
SWVC	12.30%	12.30%	35.40%	35.40%	15.90%	20.80%	39.10%	66.80%
InBUF	270.70%	8.00%	417.30%	19.30%	134.40%	6.50%	199.40%	20.20%
OutBUF	69.00%	13.60%	80.60%	27.80%	74.70%	24.80%	86.40%	60.10%
Overall	109.50%	8.80%	156.60%	21.40%	77.80%	13.30%	104.50%	37.20%

Figure 8: Instance and Area error comparison of ORION2.0 vs. ORION_NEW. Error% = ABS((TOOL - MODEL) / MODEL * 100).

Table 8 summarizes the error in estimates of ORION2.0 and ORION_NEW when compared with NetMaker and Stanford NoC router post-P&R area. Higher values of error among the two models are highlighted in red. Figures 9(a) and 9(b) plot the estimation errors in power and area respectively at 45nm and 65nm technology nodes after applying the regression fitting approach described in Section 4.2. We see that ORION_NEW estimates are very close to implementation (average error of 9.8% in estimating NetMaker power at 45nm) and are robust across multiple microarchitecture and implementation parameters as well as router RTLs.

Next, we analyze the impact of flit-level power modeling as described in Section 4.3. To capture the effect of running simulations with input vectors having different bit encodings (shown in Figure 5), we use options in Synopsys PrimeTime-PX [23] to vary toggle rate and bit encodings in the input. We run simulations using four different toggle rates (0.2, 0.4, 0.6, 0.8) and four different encodings of 1's in 32-bit input flits, and observe that leakage power is not dependent on bit encodings (changes by less than 2%). However, dynamic power varies by up to 30% (on average) depending on bit encodings in each flit. ORION2.0 models are incomplete because they consider only the flit arrival rates in the dynamic power estimation models. In Figure 10 we compare error in dynamic power estimations in ORION2.0, only ORION_NEW, and ORION_NEW with flit-level power models. We observe that by using flit-level power models, dynamic power estimations can be within 12% on average.

396

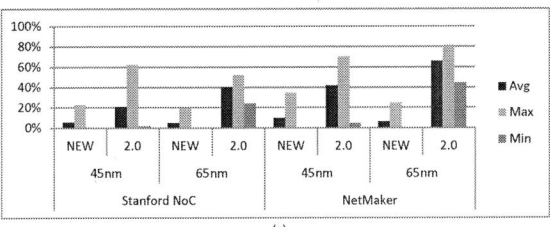

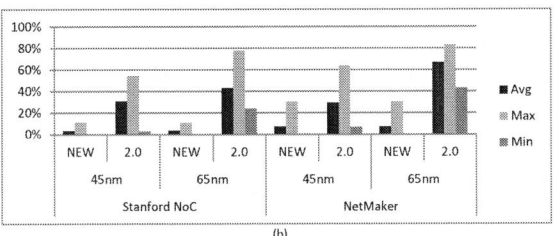

Figure 9: ORION_NEW with regression fit vs. ORION2.0: (a) Power estimation error. (b) Area estimation error.

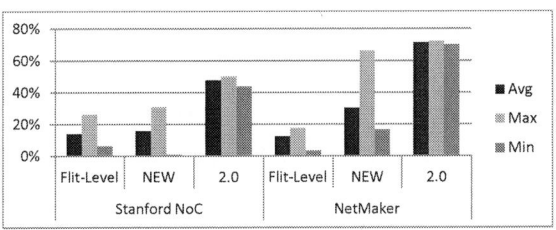

Figure 10: Comparison of dynamic power estimation error using (1) flit-level power model and ORION_NEW, (2) only ORION_NEW, and (3) ORION2.0.

We use these enhanced models (i.e., ORION_NEW with flit-level power models) in the full-system NoC simulator, GARNET [13]. We run simulations with synthetic uniform-random traffic for 10000 cycles and observe the difference in power estimates with the enhanced models and the default ORION2.0 models.

5.2 Impact of our regression analysis approach

In Section 4.2, we describe our parametric regression analysis approach using the ORION_NEW models. As seen from the results in Section 5.1, the ORION_NEW models are accurate across microarchitecture and implementation parameters because they explicitly model control and data path elements. With these accurate models, regression analysis can minimize errors and generate accurate fitting coefficients. The previous parametric regression approach [2] reports large errors because underlying ORION2.0 models do not model control path elements. The non-parametric regression approach of [2] using MARS (Multi-variate Adaptive Regression Splines) achieves reduced average power modeling errors of 5.82% at 65nm and 5.65% at 90nm, and reduced average area errors of 5.41% at 65nm and 5.01% at 90nm. In our work, we use parametric regression analysis but with accurate ORION_NEW models. Our average errors are similar to [2]; however, our maximum error for power (resp. area) is reduced by 58.89% (resp. 51%) at 65nm. At 90nm the reduction of maximum power (resp. area) error is 67.77% (resp. 53.38%). The reduction of maximum estimation error is significant because designers and architects of NoC care about worst-case accuracy.

6. CONCLUSIONS AND FUTURE WORK

Accurate modeling for NoC area and power estimation is critical to successful early design-space exploration in the era of many-core computing. ORION2.0, while very popular, has large errors versus actual implementation because it does not model control path resources. In this work, we propose ORION_NEW models that explicitly account for control and data path resources; we further refine the resulting area and power models by performing parametric regression analysis on post-P&R data. We are also the first to propose a detailed flit-level power estimation model that can seamlessly

integrate with full-system NoC simulators such as GARNET. We validate robustness of our models across multiple router RTLs, and across microarchitecture and implementation parameters, and show that the ORION_NEW models are highly accurate with average error $\leq$ 9.8%. We also demonstrate that accurate models and parametric regression can reduce the worst-case estimation errors by more than 50% as compared to previous non-parametric regression models for NoC routers. We plan to extend our work to more accurately model link power by incorporating link signaling elements such as differential signaling, scrambling, serdes, equalization and 3D routing.

7. REFERENCES

[1] A. Banerjee, R. Mullins and S. Moore, "A power and energy exploration of network-on-chip architecture", *Proc. NOCS*, 2007, pp. 163-172.

[2] A. B. Kahng, B. Lin and K. Samadi, "Improved on-chip router analytical power and area modeling" *Proc. ASPDAC*, 2010, pp. 241-246.

[3] A. B. Kahng, B. Li, L.-S. Peh and K. Samadi, "ORION 2.0: A fast and accurate NoC power and area model for early-stage design space exploration", *Proc. DATE*, 2009, pp. 423-428.

[4] C. S. Patel, S. M. Chai, S. Yalamanchili and D. E. Schimmel, "Power constrained design of multiprocessor interconnection networks", *Proc. IEEE ICCD*, 1997, pp. 408-416.

[5] G. Guindani, C. Reinbrecht, T. Raupp, N. Calazans and F. G. Moraes, "NoC power estimation at the RTL abstraction level", *Proc. IEEE ASVLSI*, 2008, pp. 475-478.

[6] G. Palermo and C. Silvano, "PIRATE: A framework for power/performance exploration of network-on-chip architectures", *Proc. PATMOS*, 2004, pp. 521-531.

[7] H.-S. Wang, L.-S. Peh and S. Malik, "Orion: A power-performance simulator for interconnection networks", *Proc. MICRO*, 2002, pp. 294-305.

[8] J. Chan and S. Parameswaran, "NoCEE: Energy macro-model extraction methodology for network-on-chip routers", *Proc. IEEE ICCAD*, 2005, pp. 254-259.

[9] K. Chang, J. Shen and T. Chen, "A low-power crossroad switch architecture and its core placement for network-on-chip", *Proc. DATE*, 2005, pp. 375-380.

[10] K. Jeong, A. B. Kahng, B. Lin and K. Samadi, "Accurate machine learning-based on-chip router modeling", *IEEE ESL 2(3)*, 2010, pp. 62-66.

[11] L.-S. Peh, "Flow control and micro-architectural mechanisms for extending the performance of interconnection networks" *PhD Thesis*, Stanford University, 2001.

[12] N. Banerjee, P. Vellanki and K. S. Chatha, "A power and performance model for network-on-chip architectures", *Proc. DATE*, 2004, pp. 1250-1255.

[13] N. Agarwal, T. Krishna, L.-S. Peh and N. K. Jha, "GARNET: A detailed on-chip network model inside a full-system simulator", *Proc. IEEE ISPASS*, 2009, pp. 33-42.

[14] P. Meloni, I. Loi, F. Angiolini, S. Carta, M. Barbaro, L. Raffo and L. Benini, "Area and power modeling for network-on-chip with layout awareness", *Proc. IEEE VLSI Design*, 2007, pp. 1-12.

[15] R. Mullins, A. West and S. Moore, "The design and implementation of a low-latency on-chip network", *Proc. ASPDAC*, 2006, pp. 164-169.

[16] S. E. Lee and N. Bagherzadeh, "A high level power model for network-on-chip (NoC) router", *Integration, the VLSI journal* 35(6), 2009, pp. 1-7.

[17] S. Penolazzi and A. Jantsch, "A high level power model for the Nostrum NoC", *Proc. Digital System Design*, 2006, pp. 673-676.

[18] T. T. Ye, G. de Micheli and L. Benini, "Analysis of power consumption on switch fabrics in network routers", *Proc. DAC*, 2002, pp. 524-529.

[19] W. J. Dally and B. Towles, *Principles and practices of interconnection networks*, Morgan Kaufmann, 2004.

[20] X. Chen and L.-S. Peh, "Leakage power modeling and optimization in interconnection networks", *Proc. IEEE ISLPED*, 2003, pp. 90-95.

[21] Cadence Encounter RTL Compiler User Guide. *http://www.cadence.com/products/ld/rtl_compiler/pages/default.aspx*

[22] Synopsys Design Compiler User Guide. *http://www.synopsys.com/Tools/Implementation/RTLSynthesis/DCUltra/pages/default.aspx*

[23] Synopsys PrimeTime User Guide. *http://www.synopsys.com/Tools/Implementation/SignOff/PrimeTime/pages/default.aspx*

[24] Synopsys Formality User Guide. *http://www.synopsys.com/tools/verification/formalequivalence/pages/formality.aspx*

[25] Synopsys VCS and DVE User Guide. *http://www.synopsys.com/tools/verification/functionalverification/pages/vcs.aspx*

[26] Standard Parasitic Exchange Format. *http://www.edaboard.com/thread37705.html*

[27] SDC User's Guide. *http://www.actel.com/documents/SDC_AN.pdf*

[28] Netmaker. *http://www-dyn.cl.cam.ac.uk/~rdm34/wiki*

[29] Stanford NoC. *https://nocs.stanford.edu/cgi-bin/trac.cgi*

[30] MATLAB. *http://www.mathworks.com/*

[31] Intel 80-core Report. *http://techresearch.intel.com/ProjectDetails.aspx?Id=151*

[32] IBM Blue Gene processor. *http://www.research.ibm.com/journal/rd49-23.html*

[33] Tilera TILE-Gx processor. *http://www.tilera.com/products*

Approaching the Theoretical Limits of a Mesh NoC with a 16-Node Chip Prototype in 45nm SOI[*]

Sunghyun Park, Tushar Krishna, Chia-Hsin Chen, Bhavya Daya,
Anantha Chandrakasan, Li-Shiuan Peh
Massachusetts Institute of Technology, Cambridge, MA

ABSTRACT

In this paper, we present a case study of our chip prototype of a 16-node 4x4 mesh NoC fabricated in 45nm SOI CMOS that aims to simultaneously optimize energy-latency-throughput for unicasts, multicasts and broadcasts. We first define and analyze the theoretical limits of a mesh NoC in latency, throughput and energy, then describe how we approach these limits through a combination of microarchitecture and circuit techniques. Our 1.1V 1GHz NoC chip achieves 1-cycle router-and-link latency at each hop and energy-efficient router-level multicast support, delivering 892Gb/s (87.1% of the theoretical bandwidth limit) at 531.4mW for a mixed traffic of unicasts and broadcasts. Through this fabrication, we derive insights that help guide our research, and we believe, will also be useful to the NoC and multicore research community.

Categories and Subject Descriptors

B.4 [Hardware]: Input/Output and Data Communications

General Terms

Design, Performance, Measurement

Keywords

Network-on-Chip, Theoretical Mesh Limits, Virtual Bypassing, Multicast Optimization, Low-Swing Signaling, Chip Prototype

1. INTRODUCTION

Moore's law scaling and diminishing performance returns of complex uniprocessor chips have led to the advent of multicore processors with increasing core counts. Their scalability relies highly on the on-chip communication fabric connecting the cores. An ideal communication fabric would incur only metal-wire delay and energy between the source and destination core. However, there is insufficient wiring for dedicated global point-to-point wires between all cores [8], and hence, packet-switched Networks-on-Chip (NoCs) with routers that multiplex wires across traffic flows are becoming the de-facto communication fabric in multicore chips [5].

These routers, however, can impose considerable overhead. Latency wise, each router can take several pipeline stages to perform

---

[*]The authors acknowledge the support of the Gigascale Systems Research Center and Interconnect Focus Center, research centers funded under the Focus Center Research Program (FCRP), a Semiconductor Research Corporation entity, and DARPA under Ubiquitous High-Performance Computing.

the control decisions necessary to regulate the sharing of wires across multiple flows. Inefficiency in the control also frequently leads to poor link utilization on NoCs. Buffers queues have been used to improve flow control and link utilization, but come with overhead in energy consumption. Conventional wisdom is that NoC design involves trading off latency, bandwidth and energy.

In this paper, we describe our design of a NoC mesh chip that aims to simultaneously approach the theoretical latency, bandwidth and energy limits of a mesh, for all kinds of traffic (unicasts, multicasts and broadcasts). We first derive such theoretical limits of a mesh NoC for unicasts and broadcasts. This analysis closely guided us in our design which leverages virtual bypassing to approach the theoretical latency limit of a single cycle per hop for unicasts, multicasts and broadcasts. This, coupled with the speed benefits of low-swing signaling, enabled us to swiftly reuse buffers and approach theoretical throughput without trading off energy or latency. Finally, low-swing signaling applied to the datapath helps us towards the theoretical energy limit.

Contributions. In this paper, we make the following contributions:

- We present a mesh NoC chip prototype that shows 48-55% latency benefits, 2.1-2.2x throughput improvements and 31-38% energy savings as compared with an equivalent textbook baseline NoC described in Section 3.1. To the best of our knowledge, this is the first mesh NoC chip with multicast support.
- We define the theoretical mesh limits for unicasts and broadcasts, in terms of latency, throughput and energy. We also characterize several prior chip prototypes' performance relative to these limits.
- We present lessons learnt from our prototyping experience:
  - Virtual bypassing can enable 1GHz single-cycle router pipelines and 32% buffering energy savings with negligible area overhead (5% only). It comes at the expense of a 21% increased critical path, though this timing overhead can be masked in multicore processors where cores limit the clock frequency rather than routers. More critically, virtual bypassing does not address non-data-dependent power.
  - Low-swing signaling can substantially reduce datapath energy (3.2x less energy in 1mm links compared to a full-swing datapath) as well as realize high frequency single-cycle traversal per hop (5.4GHz with a 64bits 5×5 crossbar and 1mm links), but comes with increased process variation vulnerability and area overhead.
  - System-level NoC power modeling tools like ORION 2.0 [12] can be way off in absolute accuracy (~5x of measured chip power) but maintain relative accuracy. RTL-based post-layout power simulations (post-layout) are much closer to measured power numbers, but post-layout timing simulations are still off.

The rest of the paper is organized as follows: Section 2 defines our baseline router, derives the theoretical limits of a mesh NoC, and characterizes prior chips performance relative to these limits. Section 3 describes our fabricated NoC prototype, while Section 4 details measurement results. Finally, we conclude in Section 5.

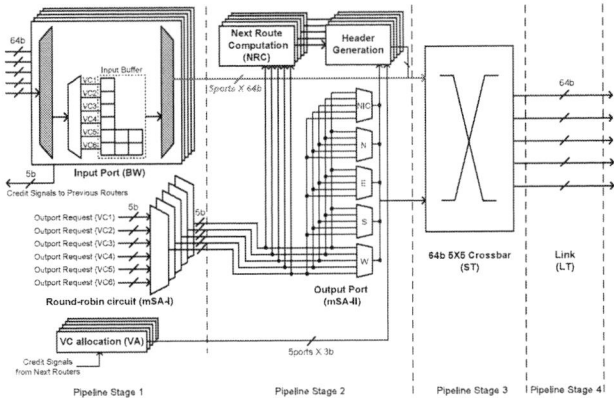

Figure 1: Baseline router microarchitecture.

2. BACKGROUND AND RELATED WORK

2.1 Baseline Mesh NoC

The mesh [6] is the most popular NoC topology for a general-purpose multicore processor, as it is scalable, is easy to layout, and offers path diversity [7, 10, 11, 21, 23]. Each core in a multicore processor communicates with other cores by sending and receiving messages through a network interface controller (NIC) that connects the core to a router (hence the network). Before a message is injected into the network, it is first segmented into packets that are then divided into fixed-length flits, short for flow-control units. A packet consists of a head flit that contains the destination address, body flits, and a tail flit that indicates the end of a packet. If the amount of information the packet carries is little, single-flit packets are also possible, *i.e.* where a flit is both the head and tail flit. Because only the head flit carries the destination information, all flits of a packet must follow the same route through the network.

Figure 1 shows the microarchitecture of an input-buffered virtual channel router. Before an incoming flit is forwarded to the next router, it needs to go through several actions in order: buffer write (BW), route computation (NRC) (only for head flits), switch allocation (SA), virtual channel allocation (VA) (only for head flits), buffer read (BR), switch traversal (ST), and link traversal (LT). Out of all these actions, only ST and LT actually move the flits toward the destination. Thus, we consider all other actions as overhead. We will refer to this as the baseline router throughout the paper.

2.2 Latency, Throughput and Energy Limits

A mesh topology by itself imposes theoretical limits on latency, throughput and energy (*i.e.* minimum latency and energy, and maximum throughput). We derive these theoretical bounds of a $k \times k$ mesh NoC for two traffic types, unicast and broadcast traffic, as shown in Table 1. Specifically, each NIC injects flits into the network according to a Bernoulli process of rate R, to a random, uniformly distributed destination for unicasts, and from a random, uniformly distributed source to all nodes for broadcasts. All derived bounds are for a complete action: from initiation at the source NIC, till the flit is received at all destination NIC(s). More details on the derivation of the bounds is shown in Appendix A.

2.3 Related Work

There have been few chip prototypes with mesh NoCs as the communication fabric between processor cores or nodes, as listed in Table 2. Other NoCs, *e.g.* KAIST [2], Spidergon [4], Pleiades [24], are targeted for heterogeneous topologies and architectures, making it difficult to characterize them against the theoretical mesh limits. The prototypes range from full multicore processors to stand-alone NoCs. Of these, three chips were selected for comparison, that differ significantly with respect to targeted design goals and optimizations: Intel Teraflops which is the precursor of the Intel IA-32 NoC, Tilera TILE64 which is the successor of the MIT RAW, and SWIFT, a NoC with low-swing signaling. Each processor is described further in Appendix B.

Flit Size	64 bits
Request Packet Size	1 flit *(coherence requests and acknowledges)*
Response Packet Size	5 flits *(cache data)*
Router Microarchitecture	10 X 64b latches per port (6VCs over 2MCs)
Bypass Router-and-link Latency	1 cycle
Operating Frequency	1GHz
Power Supply Voltage	1.1V and 0.8V
Technology	45nm SOI CMOS

Figure 2: Die photo and overview of our fabricated 4×4 mesh NoC.

We calculated zero-load latency and channel load of these networks for both unicast-only and broadcast-only traffic. Zero-load latency is calculated by multiplying the average hop-count by the number of pipeline stages to traverse a hop, with serialization latency added on to model pipelining of all flits. We computed channel load based on an flit injection rate per core of R, following the methodology of [6]. The results are shown in the Table 2. We can see that our proposed router optimizes for broadcast (multicast) traffic and has much lower zero-load latency and channel load compared to all other networks.

TILE64 attempts to optimize for all three metrics, by utilizing independent simple networks for different message types. The simple router design, with no virtual channels, improves unicast zero-load latency but broadcast traffic latency is poor as its lack of multicast support forces the source NIC to duplicate $k^2 - 1$ copies of a broadcast flit and send a copy to every destination NIC. This increases channel load by $k^2 - 1$ times, causing contention at all routers along the shared route, making it impossible to meet the single-cycle per hop. TILE64's static partitioning of traffic across 5 networks may also lead to poor throughput when exercised with realistic uniform traffic. Similar effect on broadcast latency and channel load is observed for the Teraflops and SWIFT NoCs as none of these chip prototypes have multicast support. The SWIFT NoC with a single-cycle pipeline for unicasts performs better on zero-load latency, albeit at a lower operating frequency. The TeraFLOPS NoC has poor zero-load latency in terms of cycles due to a 5-stage pipeline, which is aggravated with broadcasts.

In the rest of this paper, we will describe how we designed a NoC chip specifically to approach the theoretical limits.

3. PROPOSED NOC CHIP DESIGN

This section describes the design of our chip prototype. Figure 2 shows our fabricated 16-node 4x4 NoC. The network is packet-switched, and all routers are connected to network interface circuits (NICs) to generate and receive packets. Each router has 5 I/O ports: North, East, South, West and NIC. Each input port has two message classes (MCs), request and response, to avoid message-level deadlocks in cache-coherent multicores.

3.1 Overview of Proposed Router Pipeline

Our design essentially evolves the original textbook router pipeline (Fig. 1) into a strawman 4-stage router pipeline tailored for multicasts so multicasts/broadcasts do not require multiple unicast packets to be injected. Next, we add features pushing latency towards the theoretical limit of a single cycle per hop, throughput towards the theoretical limit of maximum channel load, and energy towards the theoretical limit of just datapath traversal.

In the first pipeline stage, (1) flits entering the router are first buffered (BW). (2) Each input port chooses one output port request (mSA-I) out of the requests from all VCs at that input port with a round-robin logic that guarantees fair and starvation-free arbitration. Since multicast flits can request multiple output ports, the request is a 5b vector. (3) The next router VC is selected (VA) for each neighbor from a free VC queue at each output port. These

Table 1: Theoretical Limits of a $k \times k$ mesh NoC for unicast and broadcast traffic.

Metric	Unicasts (one-to-one multicasts)	Broadcasts (one-to-all multicasts)
Average Hop Count ($H_{average}$)	$2(k+1)/3$	$(3k-1)/2$, for k even $(k-1)(3k+1)/2k$, for k odd
Channel Load on each bisection link ($L_{bisection}$)	$k \times R/4$	$k^2 \times R/4$
Channel Load on each ejection link ($L_{ejection}$)	R	$k^2 \times R$
Theoretical Latency Limit given by $H_{average}$	$2(k+1)/3$	$(3k-1)/2$, for k even $(k-1)(3k+1)/2k$, for k odd
Theoretical Throughput Limit given by $\max\{L_{bisection}, L_{ejection}\}$	R, for $k <= 4$ $k \times R/4$, for $k > 4$	$k^2 \times R$
Theoretical Energy Limit E_{xbar}: energy of crossbar traversal E_{link}: energy of link traversal	$2(k+1)/3 \times E_{xbar}$ $+ E_{xbar}$ $+ 2(k+1)/3 \times E_{link}$	$k^2 \times E_{xbar}$ $+ (k^2-1) \times E_{link}$

Table 2: Comparison of mesh NoC chip prototypes

	Intel Teraflops [10] 8×10, 65nm	Tilera TILE64 [23] 5 8×8, 90nm	SWIFT [14] 2×2, 90nm	This work 4×4, 45nm SOI	
Clock frequency	5GHz	750MHz	225MHz	1GHz	
Power supply	1.1-1.2V	1.0V	1.2V	1.1V	
Power consumption	97W	15-22W	116.5mW	427.3mW	
Latency Metrics	Modeled as 8×8 networks			4×4 network	
Delay per hop	1ns	1.3ns	8.9-17.8ns	1-3ns	
Zero-load latency (cycles)	30 (unicast) 120.5 (broadcast)	9 (unicast) 77.5 (broadcast)	12 (unicast) 86 (broadcast)	6 (unicast) 11.5 (broadcast)	3.3 (unicast) 5.5 (broadcast)
Throughput Metrics	Modeled as 8×8 networks			4×4 network	
Channel width	39b	5×32b	64b	64b	
Bisection bandwidth	1560Gb/s	937.5Gb/s	112.5Gb/s	512Gb/s	256Gb/s
Channel load (R:injection rate/core)	64R (unicast) 4096R (broadcast)	64R (unicast) 4096R (broadcast)	64R (unicast) 4096R (broadcast)	64R (unicast) 64R (broadcast)	16R (unicast) 16R (broadcast)

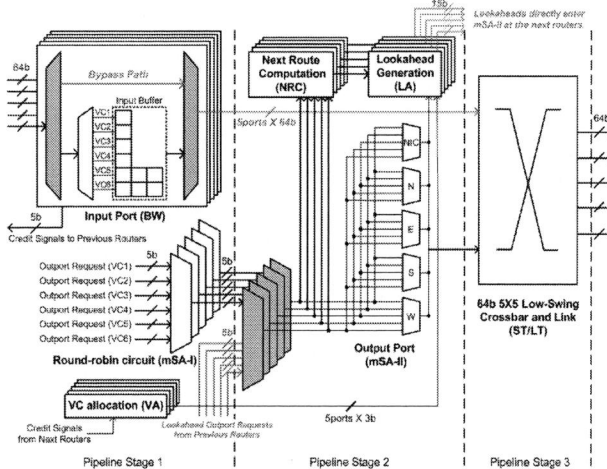

Figure 3: Proposed router microarchitecture and pipeline.

3 operations are executed in parallel without decreasing operating frequency as they are not dependent on each other. In the second stage, output port requests for the next routers are computed (NRC) for the winners of mSA-I, and concurrently, a matrix arbiter at each output port grants the crossbar ports to the input port requests (mSA-II). Multicast requests get granted multiple output ports. In the third stage flits physically traverse the crossbar (ST) and reach the next router through the link (LT) in the fourth stage.

At this point, our strawman router can simultaneously send a broadcast packet to all 16 nodes of a NoC. The baseline textbook router (Fig. 1), on the other hand, needs to generate multiple unicasts at each cycle to implement the broadcast packet and such unicasts takes 4 cycles per hop. The proposed design (Fig. 3) will completely be described through the following subsections.

3.2 Towards Theoretical Latency Limits

We push our strawman towards the limit by adding two key features: (1) virtual bypassing [15–17] to remove/hide delays due to buffering and arbitration and (2) low-swing circuits on the datapath to achieve single cycle ST+LT without lowering clock frequency.

Single-stage pipeline with lookaheads. In stage 2 of the strawman, we add and generate 15b lookahead signals from the results of NRC and mSA-II, and send them to the next router. The lookaheads try to pre-allocate the crossbar ahead of the actual flit, thus hiding mSA-II from the router delay. The lookahead takes priority over requests from buffered flits at the next router, and directly enters mSA-II. If the lookahead wins an output port, this pre-allocation allows the following flit to bypass the first two pipeline stages and go into the third stage directly, reducing the router pipeline depth from 4 to 2. Active pre-allocation by lookaheads enables incoming flits to bypass routers at all loads, in contrast to a naive approach of bypassing only at low-loads when the input queues are empty.

Single-cycle ST+LT with low-swing circuits. We apply a low-swing signaling technique, which can reduce the charging / discharging delay and dynamic energy when driving capacitive parasitics [20], to the highly-capacitive datapath. As will be described later in Section 3.4, the proposed low-swing circuits obtain higher current driving ability (or lower linear drive resistance) even at small V_{ds} than the reduced-swing signaling generated by simply lowering supply voltage, and hence, our low-swing datapath enables single-cycle ST+LT at higher clock frequency. Such single-cycle ST+LT can operate at up to 5.4GHz with 1mm 0.15um-width 0.30um-space links as demonstrated with measurement results in Section 4.3.

These two optimizations achieve a single-cycle-per-hop delay for unicasts and multicasts, exactly matching the theoretical latency limits. The caveat is that in case of contention for the same output port from multiple lookaheads, one of them will have to be buffered and then forced to go through the 3-stage pipeline. In addition, critical path delay is stretched, which will be analyzed in Section 4.

3.3 Towards Theoretical Throughput Limits

We take two steps towards the throughput limit for both unicasts and broadcasts (1) multicast support inside routers, and (2) single-cycle hop latency for fast buffer reuse.

Multicast support inside routers. We design a router that can replicate flits, allowing one multicast/broadcast flit to be sent from the source NIC, and get routed to all other routers in the network via a tree. This allows a broadcast flit to share bandwidth till it does not require an explicit forking into different directions. This dramatically reduces contention compared to the baseline design where multiple flits would have be sent as unicasts which are guaranteed to create contention at along the shared routes. We use a dimension ordered XY-tree in our design as it is deadlock free, and simplifies the routing algorithm. The ability to replicate flits in the router is implemented in the form of our broadcast-optimized

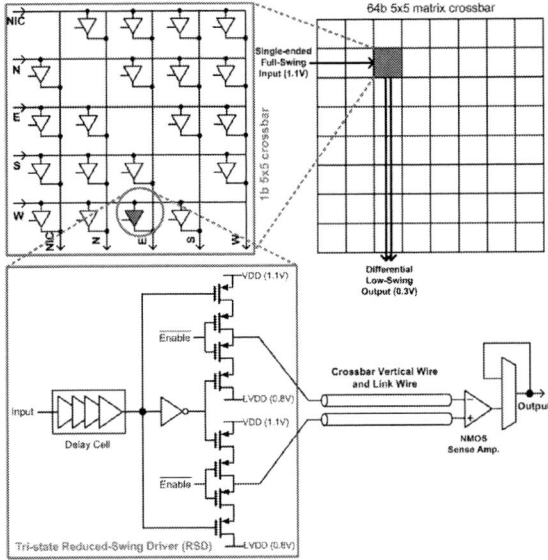

Figure 4: Proposed low-swing crossbar and link circuits.

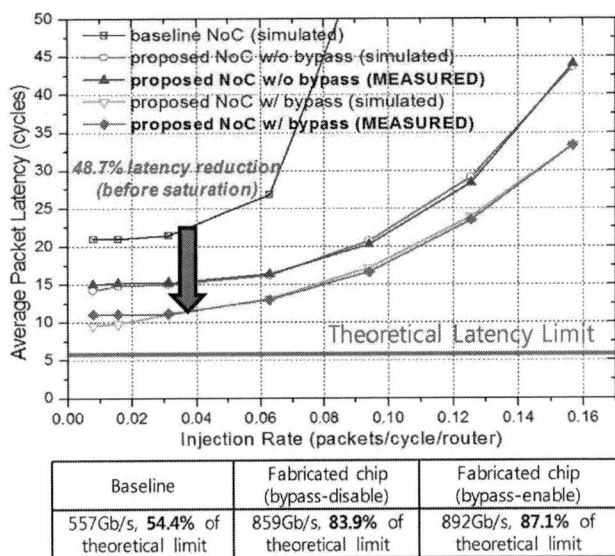

Baseline	Fabricated chip (bypass-disable)	Fabricated chip (bypass-enable)
557Gb/s, **54.4%** of theoretical limit	859Gb/s, **83.9%** of theoretical limit	892Gb/s, **87.1%** of theoretical limit

Maximum throughput comparison (received)

Figure 5: Throughput-latency performance evaluation with mixed traffic at 1GHz.

crossbar and mSA-II (switch allocation for multiple output ports).

Single-cycle-per-hop latency. The number of buffers/VCs required at every input port to sustain a particular throughput depends upon the buffer/VC turnaround time, *i.e.* the number of cycles for which the buffer/VC is occupied. This is where our optimizations for latency in Section 3.2 come in handy here since they reduce the pipeline depth, thus reducing buffer turnaround time, thereby increasing throughput given the same number of buffers. For our single-cycle pipeline, the turnaround time for buffers/VCs is 3: one cycle for ST+LT to the downstream router, one cycle for the free VC/buffer signal to return from the downtsream router (if the flit successfully bypassed), and one cycle for it to be processed and ready to be used for a new flit. We thus choose 4 VCs in the request message class, each 1-flit deep (since requests packets in our design are 1-flit wide) to satisfy VC turnaround time and sustain high throughput for broadcasts. We chose 2 VCs in our response message class, each 3-flit deep, for the 5-flit response packets. This number was chosen to be less than the turnaround time to shorten the critical path, and reduce the total buffers (which increase power consumption). We thus chose a total of 6 VCs per port, with a total of 10 buffers.

3.4 Towards Theoretical Energy Limits

Section 2 reveals a significant energy gap between the baseline router energy and the theoretical energy limit (which is just clocking and datapath energy, E_{xbar} and E_{link}). Such a gap is due to buffering energy (E_{buff}), arbitration logic energy (E_{arb}) and silicon leakage energy (E_{lkg}). Conventionally, these energy overheads are traded off against latency and throughput as follows: (1) Fewer buffers reduce E_{buff} and E_{lkg}, but stretch latency due to contention and lower throughput. (2) Simple routers like wormhole routers reduce E_{arb} and E_{lkg}, and increase operating frequency f, but these come at the expense of poorer latency and throughput.

Our proposed NoC first includes multicast support so even broadcasts and multicasts can approach the theoretical energy limit. Then, it incorporates two new features that permits different tradeoffs of latency, throughput and energy. First, our multicast virtual bypassing reduces E_{buff}, while improving both latency and throughput. The hidden cost lies in increased E_{arb} and decreased f. As shown in Section 4.1, the savings in E_{buff} outweigh the E_{arb} overheads, and operating frequency can still be in GHz. Second, our chip employs low-swing signaling to reduce dynamic energy in the datapath (E_{xbar} and E_{link}) which is unavoidable and part of the theoretical energy limit. Low-swing signaling provides an opportunity to break the conventional trade-offs that achieve dynamic energy savings at the cost of latency and throughput penalties; In fact, low-

swing optimizes both energy and latency. Its downsides lie in its area overheads and reduced process variation immunity.

Figure 4 shows the circuit implementation of the low-swing crossbar directly connected to links with tri-state reduced-swing drivers (RSDs). This crossbar enables low-swing signaling in the datapath (crossbar vertical wires and links). The tri-state RSD disconnects horizontal and vertical wires and only drives the corresponding vertical wire and link, thereby providing energy-efficient multicasting capability. With an additional supply voltage (LVDD), the 4-PMOS stacked RSD design generates more reliable low-swing signaling in the presence of wire capacitance and resistance variation than equalized interconnects [9, 13, 18] where low-swing signaling is obtained by wire channel attenuation. A delay cell aligns an input signal (which drives only a 1b crossbar) to an enable signal (which drives all of 64 1bit crossbars). It reduces mismatch between charging and discharging time, thus decreasing inter-symbol interference (ISI). The 64bits links are designed with 0.15um-width 0.30um-space fully shielded differential wires, to eliminate noise coupling of crosstalk effects and supply voltage variation.

4. EVALUATION

In this section, we first evaluate the measured energy-latency-throughput of our fabricated NoC against that of the baseline mesh and theoretical limits defined in Section 2. Armed with our chip measurements, we then delve into three specific case studies on virtual bypassing, low-swing signaling and power modeling and estimation to dissect our design choices.

4.1 Energy-Latency-Throughput Performance

We measured average packet latency of our NoC as a function of packet injection rate, with two different traffic patterns: mixed traffic (50% broadcast request, 25% unicast request and 25% unicast response messages) and broadcast-only traffic (100% broadcast request messages), at 1GHz operating frequency. For brevity, Figure 5 only shows the results for mixed traffic along with the baseline performance and theoretical mesh limits. Here, we chose a more aggressive baseline that has single-cycle ST and LT stages shown in Fig. 1. Since even the full-swing baseline can support single-cycle ST+LT at 1GHz, this baseline is a fairer model of an equivalent unicast full-swing NoC. Except for the the single-cycle ST+LT, the baseline used in this section is identical as that described in Section 2.1. The theoretical latency limits (cycles/packet) include two extra cycles for NIC-to-router and router-to-NIC traversals which are indispensable since traffic injects and ejects through the NICs. Theoretical throughput

401

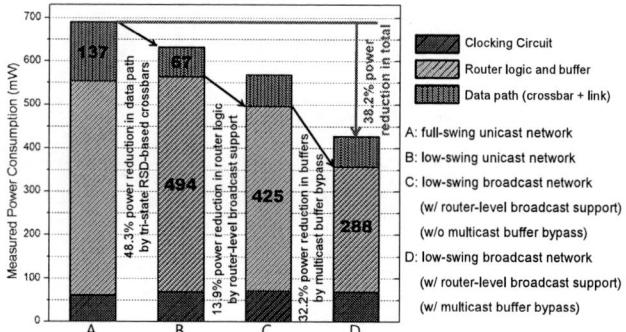

Figure 6: Measured power reduction at 653Gb/s at 1GHz.

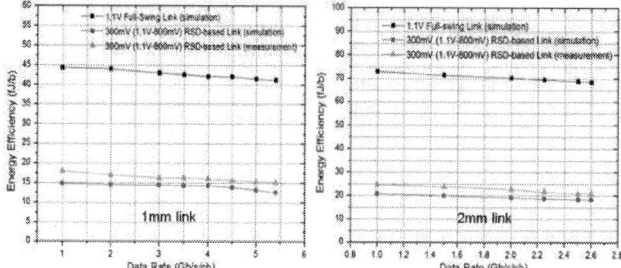

Table 3: Critical path analysis results.

Pre-layout simulations	
Baseline router design	549ns
Our virtual bypassed router design	593ns (1.08x overhead)
Post-layout simulations	
Baseline router design	658ns
Our virtual bypassed router design	793ns (1.21x overhead)
Measured critical path	
Our virtual bypassed router design	961ns (1/1.04GHz)

limits are calculated based on received flits, then converted into Gb/s to factor in the 1GHz clock frequency and 64-bit flit size ($16\times64b\times1/1GHz=1024Gb/s$). Simulation results were obtained from pre-layout synthesis with sufficient simulation cycles (10^4 cycles) to make scan-chain warmup (128 cycles) negligible.

For latency, our design enables 48.7% (mixed traffic) and 55.1% (broadcast-only) reductions before the network saturates[1] as compared to the baseline. The low-load latency gap from the theoretical latency limit is 5.7 (6.3) cycles for mixed (broadcast) traffic, *i.e.* only 1.03 (1.14) cycles of contention latency per hop for mixed (broadcast) traffic. This can be further improved to 0.04 (0.05) cycles of contention latency per hop (obtained through RTL simulations) by removing the artifact in our chip whereby all NICS had identical pseudo-random generators that caused contention which lowers the amount of bypassing even at low injection rates.

Throughput wise, the fabricated NoC approaches the theoretical limits: 87% (mixed traffic) and 91% (broadcast-only) of the theoretical throughput limits. In addition, our NoC design has 2.1x (mixed traffic) and 2.2x (broadcast-only) higher saturation throughput than the baseline. In other words, the proposed NoC can obtain the same throughput as the baseline with fewer buffers or VCs. The throughput gap between the theoretical mesh and the fabricated chip is due to imperfect arbitration (like all prior chips, we use separable allocators, mSA-I and mSA-II, to lower complexity) and routing (XY routing can lead to imbalance in load).

Figure 6 shows the measured power reduction at 653Gb/s broadcast delivery at 1GHz at room temperature. The low-swing signaling enables 48.3% power reduction in the datapath. In addition, the single-cycle multicast capability and virtual bypassing result in 13.9% and 32.2% power reduction in router logics and buffers, respectively. Overall, our chip prototype achieves 38.2% power reduction compared to the baseline. To compare against the theoretical power limit, we performed a post-layout power simulation of a router in the middle of the mesh to further breakdown data-dependent power from non-data-dependent components like clocking. We then calculate the theoretical power limit to comprise just clocking and a full-swing datapath: 5.6mW/router, at close to zero-load injection rate (3/255). Compared to our NoC power consumption at the same low injection rate (13.2mW/router), our overhead comes largely from VC bookkeeping state (1.9mW/router) and buffers (2.0mW/router), whereas the allocators (0.7mW/router) and additional lookahead signals (0.2mW/router) contribute little additional power. The data-dependent power (*e.g.* buffers, allocators) is due to our identical PRBS generators at NICs that limited bypassing at low loads and can be removed by virtual bypassing, but the non-data-dependent power (*e.g.* VC state) will remain. Also, since our chip consumes nontrivial leakage power (76.7mW measured, 18% of overall chip power consumption at 653Gb/s), power gating will help to further close the gap, at the expense of a decrease in operating frequency.

---

[1]To enable precise comparisons, we define the saturation point as the injection rate at which NoC latency reaches 3 times the average no-load latency; most multi-threaded applications run within this range.

Figure 7: Measured energy efficiency of the proposed low-swing circuit on pseudo-random binary sequence data.

4.2 Virtual bypassing

Virtual bypassing of buffering to achieve single-cycle routers has been proposed in various forms [3, 15–17] in research papers. The aggressive folding of multiple pipeline stages into a single cycle naturally raises the question of whether that comes at the expense of router frequency f. While our chip is the first prototype to demonstrate a single-cycle virtual bypassed router at GHz frequency, it begs the question of how much f is affected. To quantify the timing overhead, we performed critical path analysis on pre- and post-layout netlists of the baseline and our design. Table 3 shows such estimates along with the actual measured timing.

The critical paths of both the baseline and the proposed router occur in the second pipeline stage where mSA-II is performed. The overhead of lookaheads lengthens the critical path by 8% in pre-layout simulations and 20% in post-layout simulations. It should be pointed out though that if the operating frequency is limited by the core rather than the NoC router, which is typically the case, this 20% critical path overhead can be hidden. In the Intel 48 core chip, nominal operation is 1GHz core and 2GHz router frequencies, allowing any network overhead to be masked [11].

Also notable is the fact that while the critical path of the post-layout simulation is 793ns, the maximum frequency of our chip prototype is 1.04GHz (*i.e.* the actual critical path is 961ns). This is mainly due to nonideal factors (*e.g.* a contaminated clock, supply voltage fluctuation, unexpected temperature variations, and *etc.*) whose effects cannot be exactly predicted in design phase.

4.3 Low-Swing Signaling

Low-swing signaling has demonstrated substantial energy gains in domains such as off-chip interconnects and SRAMs. However, in NoCs, there are few chip prototypes employing low-swing signaling [2, 14]. So a deep understanding of its trade-offs and its applicability to NoCs can be useful. To investigate such effects with longer links (necessary in a multicore processor as cores are much larger than routers), and at higher data rates than the network clock frequency (which is limited by synthesized router logic), an identical low-swing crossbar with longer link wires (1mm and 2mm) is separately implemented as shown in Figure 2.

Energy savings and 1-cycle ST+LT. The measured energy efficiency (Fig. 7) shows that the 300mV-swing tri-state RSD consumes up to 3.2x less energy as compared to a equivalent full-swing repeater. Experimental results also demonstrates that the tri-state RSD-based crossbar supports single-cycle ST+LT at up to 5.4GHz and 2.6GHz clock frequency with 1mm and 2mm links, respectively. The tri-state RSDs enables a reduction in the total amount of charge and delay required for data transitions, thereby resulting in these energy and latency benefits.

402

Synthesized full-swing crossbar	$26,840 um^2$
Proposed low-swing crossbar	$83,200 um^2$ (3.1x overhead)
Router with the full-swing crossbar	$227,230 um^2$
Router with the low-swing crossbar	$318,600 um^2$ (1.4x overhead)

Table 4: Area comparison with full-swing signaling.

Area overheads. Table 4 shows the area overhead of our 5×5 64bits low-swing crossbar against an equivalent full-swing crossbar. The low-swing crossbar has a high area overhead (3.1x) compared to a synthesized full-swing crossbar, as the proposed RSDs employ differential signaling while the full-swing crossbar uses single-ended signaling. In addition, since our low-swing crossbar was carefully laid out due to noise coupling issues, such restricted placement and wiring of tri-state RSDs exacerbate the area overhead. However, at the router level, the relative area overhead goes down to 1.4x, and naturally, it will again diminish when compared against an entire tile with a core, cache and router.

Process variation effects. The critical drawback of low-swing signaling is reduced noise margin. In our circuit, the primary noise source is a sense amplifier offset caused by process variation. While low-swing signaling enables more dynamic energy savings as voltage swing decreases, the process variation effect worsens. Based on 1000-run Monte-Carlo Spice simulations, we chose 300mV-swing for above 3-σ reliability, but the voltage swing can be further decreased by offset compensation circuit techniques [1, 19, 22] at the cost of design complexity. Appendix C delves into further evaluation of our low-swing datapath.

4.4 Power Modeling and Estimation

Architectural power models such as ORION have been extensively adopted by researchers for early-stage evaluation of research ideas, while RTL-based energy estimates have also been widely used. With our chip, we can now study the gap between silicon-proven energy and different levels of energy modeling.

We compare our chip power measurements with two power estimates obtained from ORION 2.0 and post-layout netlists. The experiments (or simulations) are conducted with 1.1V supply voltage, 1GHz clock frequency, 653Gb/s throughput at room temperature. Figure 8 summarizes the results.

ORION 2.0 substantially over-estimates power (4.8-5.3x of measured chip power), but its estimate of relative power reduction between the baseline and our design (32% reduction) is not far from the measurements (38% reduction). This is because the transistor sizes assumed in ORION are much larger than the actual sizes in the chip. Thus, while ORION can be used for comparison of various system-level optimizations or early-stage design space exploration, its estimates should not be the basis of absolute power budgets.

On the other hand, the post-layout simulation gives us fairly accurate power estimates (6-13% deviation from measurements). Specifically, it slightly under-estimates the power of buffers and arbitration logic but over-estimates clocking and datapath power. Relative power reduction (34%) also matches well with measurements (38%). However, such accurate estimates come at the cost of tremendous simulation time overheads (several days for an entire NoC simulation) because the post-layout simulation calculates its estimates at the transistor-level along with parasitic effects. Moreover, since the post-layout estimation requires complete extracted netlists, it is difficult to apply to early-stage NoC evaluation.

5. CONCLUSION

This chip prototype offered us insights that may help guide future research. While virtual bypassing can effectively skip buffering and arbitration energy, there is a need for architectural techniques that tackle the non-data-dependent energy components as well without trading off latency and throughput. Similarly, though sophisticated off-chip signaling techniques have been shown to deliver substantial energy savings when applied to NoC interconnects, circuit or system-level solutions to their increased vulnerability to process variations need to be developed to ensure viability in future

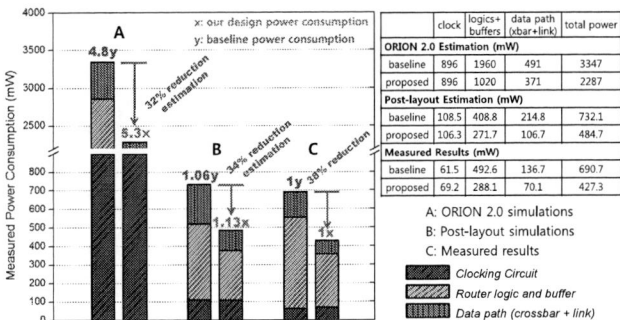

Figure 8: Comparison of power estimates with measurements.

technology nodes. Finally, accurate timing and power models for early-stage NoC design are still sorely needed.

6. REFERENCES

[1] I. Arsovski and R. Wistort. Self-referenced sense amplifier for across-chipvariation immune sensing in high-performance content-addressable memories. In *IEEE Custom Integrated Circuits Conf.*, pages 453–456, 2006.

[2] S. Bell et al. A 118.4 gb/s multi-casting network-on-chip with hierarchical star-ring combined topology for real-time object recognition. *IEEE Journal of Solid-State Circuits*, 45:1399–1409, 2010.

[3] C.-H. O. Chen et al. Physical vs. virtual express topologies with low-swing links for future many-core nocs. In *Int'l Symp. on Networks-on-Chip*, May 2010.

[4] M. Coppola et al. Spidergon: a novel on-chip communication network. In *Int'l Symp. on System-on-Chip*, page 15, 2004.

[5] W. J. Dally and B. Towles. Route packets not wires: On-chip interconnection networks. In *DAC*, June 2001.

[6] W. J. Dally and B. Towles. *Principles and Practices of Interconnection Networks*. Morgan Kaufmann Publishers, 2004.

[7] P. Gratz et al. On-chip interconnection networks of the trips chip. *IEEE Micro*, 27(5):41–50, 2007.

[8] S. Heo and K. Asanovic. Replacing global wires with an on-chip network: A power analysis. In *Int'l Symp. on Low Power Elect. and Design*, pages 369–374, 2005.

[9] R. Ho et al. High-speed and low-energy capacitive-driven on-chip wires. In *Int'l Solid-State Circuits Conf.*, pages 412–413, 2007.

[10] Y. Hoskote et al. A 5-ghz mesh interconect for a teraflops processor. *IEEE Micro*, 27(5):51–61, 2007.

[11] J. Howard et al. A 48-core ia-32 message-passing processor with dvfs in 45nm cmos. In *Int'l Solid-State Circuits Conf.*, pages 108–109, 2010.

[12] A. Kahng et al. Orion 2.0: A fast and accurate noc power and area model for early-stage design space exploration. In *Proc. Design, Automation and Test in Europe*, pages 423–428, 2009.

[13] B. Kim and V. Stojanovic. A 4gb/s/ch 356fj/b 10mm equalized on-chip interconnect with nonlinear charge-injecting transmit filter and transimpedance receiver in 90nm cmos. In *Int'l Solid-State Circuits Conf.*, pages 66–67, 2009.

[14] T. Krishna et al. Swift: A swing-reduced interconnect for a token-based network-on-chip in 90nm cmos. In *Int'l Conf. on Computer Design*, pages 439–446, 2010.

[15] T. Krishna et al. Towards the ideal on-chip fabric for 1-to-many and many-to-1 communication. In *MICRO*, 2011.

[16] A. Kumar et al. Express virtual channels: Towards the ideal interconnection fabric. In *Int'l Symp. on Computer Architecture*, June 2007.

[17] A. Kumar et al. Token flow control. In *MICRO*, Nov 2008.

[18] E. Mensink et al. A 0.28pj/b 2gb/s/ch transceiver in 90nm cmos for 10mm on-chip interconnects. In *Int'l Solid-State Circuits Conf.*, pages 314–315, 2000.

[19] M. Qazi et al. A 512kb 8t sram macro operating down to 0.57v with an ac-coupled sense amplifier and embedded data-retention-voltage sensor in 45nm soi cmos. In *Int'l Solid-State Circuits Conf.*, pages 350–351, 2010.

[20] J. M. Rabaey et al. *Digital Integrated Circuits: A design perspective*. Prentice Hall, 2nd Edition, 1998.

[21] M. B. Taylor et al. The raw microprocessor: A computational fabric for software circuits and general-purpose programs. *IEEE Micro*, 22(2):25–35, 2002.

[22] N. Verma and A. P. Chandrakasan. A high-density 45nm sram using small-signal non-strobed regenerative sensing. In *Int'l Solid-State Circuits Conf.*, pages 380–381, 2008.

[23] D. Wentzlaff et al. On-chip interconnection architecture of the tile processor. *IEEE Micro*, 27(5):15–31, 2007.

[24] H. Zhang et al. A 1 v heterogeneous reconfigurable processor ic for baseband wireless applications. In *Int'l Solid-State Circuits Conf.*, pages 68–69, 2000.

APPENDIX

A. DERIVATION OF THEORETICAL MESH LIMITS

First, we detail assumptions made in our analysis of the theoretical latency, energy and throughput limits, and explain our derivation process.

Assumptions:

1. Perfect routing: A router would route all packets with minimal hop-counts, balancing injected packets (termed channel load in our analysis) across multiple routes perfectly, thus keeping the load on all links optimally balanced.

2. Perfect flow control: A router maintains maximum utilization of the links; in other words, a link is never left idle when there is traffic routed across it.

3. Perfect router microarchitecture: All flits only incur the delay and energy of the datapath (ST and LT), that is, the router arbitrates between competing flits, performs crossbar and link traversal all in a single cycle and do not expend extraneous energy for buffering and control.

Assumption (1) and (2) are conventionally assumed in theoretical analysis of NoCs [6] while we further add assumption (3) as that is the minimum energy-delay per hop with synchronous NoCs.

Based on these assumptions, we derived the theoretical limits for unicast and broadcast traffic.

For unicasts, we analyze the theoretical limits for latency and throughput using the same technique as in [6]. We then derive the energy limit by multiplying hop count with crossbar and link energy costs.

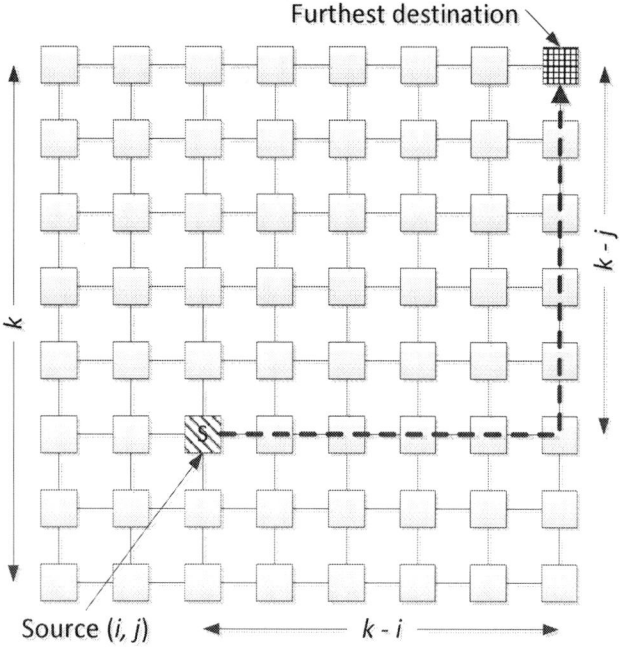

Figure 9: Latency calculation for broadcast traffic.

For broadcast traffic, to the best of our knowledge, no prior theoretical analysis exists. Here, we define the time till a flit is received by all destination NICs as equivalent to when this flit is received by the furthest NIC relative to the source NIC (Fig. 9). Hence, we derived the theoretical latency limit for received packets by averaging the hop delay from each source NIC to its furthest destination

NIC. We obtained the theoretical throughput limit by analyzing the channel load across the ejection links and bisection links [6], and observed that the maximum throughput for broadcast traffic is limited by the ejection links. This differs from unicast traffic where throughput is always limited by the bisection links. As for the theoretical energy limit, intuitively, due to the nature of broadcasting, a broadcast flit needs to visit all k^2 routers in the network and traverse k^2 crossbars/links connecting them. Therefore, the energy limit grows quadratically with the number of routers in the network.

B. BACKGROUND ON PRIOR MESH CHIP PROTOTYPES

Here, we describe in detail these three other chips and corresponding NoC architecture.

Tilera TILE64 is a multiprocessor consisting of 64 tiles interconnected by five 2D mesh networks, where each tile contains a CPU, cache and a router, fabricated on the TSMC 90nm process and running at a speed of 700 to 866 MHz [23]. Four of the five networks are dynamically routed, each servicing a different type of traffic: user dynamic network (UDN) for user-level messages, I/O dynamic network (IDN) for I/O traffic, memory dynamic network (MDN) for traffic to/from the memory controllers, and tile dynamic network (TDN) for cache-to-cache transfers. The dynamic networks are packetized, wormhole routed, with a one cycle pipeline for straight-through traffic and two cycles for turning traffic. The static network is software scheduled, and has a single-cycle pipeline.

Intel Teraflops had a more complex NoC architecture, but the cores are much simpler than a standard RISC processor. Since simpler cores are more area- and energy-efficient than larger ones, more functional units can be supported within a single chip's area and power budget. Teraflops is a demonstration of the possibility of including an on chip interconnect, operating at 5 GHz, and achieving performance in excess of teraflops while maintaining a power usage of less than 100W [10]. Teraflops NoC has a five-port, two-lane, five-pipeline-stage router with a double pumped crossbar used to interconnect the tiles in a 2D mesh network. Each input port is connected to two 16 entry deep FIFO buffers, one for each lane. A single crossbar for both lanes is double pumped in the fourth pipeline stage using dual-edge triggered flip-flops, allowing the switch to transfer data at both edges of the clock signal.

SWIFT is a 2x2 standalone NoC research chip demonstrating the practicality of implementing token flow control [17] and low voltage swing crossbars and links. The bufferless traversal of flits through a reduce-swing datapath is demonstrated to perform at 400 MHz and obtain latency and power reductions of approximately 40 percent each [14]. The token flow control microarchitecture pre-allocates buffers and links in the network by using tokens. Many flits are then able to bypass buffering, improving link utilization and reducing the buffer turnaround time. Dual voltage supply differential reduced-swing drivers and sense-amplifier receivers sustain the low-swing signaling necessary to reduce the dynamic power consumption.

C. LOW-SWING CIRCUIT EVALUATION

Here, we present additional measured and simulated results of our low-swing signaling circuits.

Figure 10 shows energy efficiency and link failure probability of the 1mm 5Gb/s tri-state RSD as a function of voltage swing level. The normalized probability was calculated from 1000 Monte-Carlo Spice simulations. These results explicitly reveal the low-swing signaling energy gain trade-off against process variation vulnerability, as discussed earlier in Section 4.3.

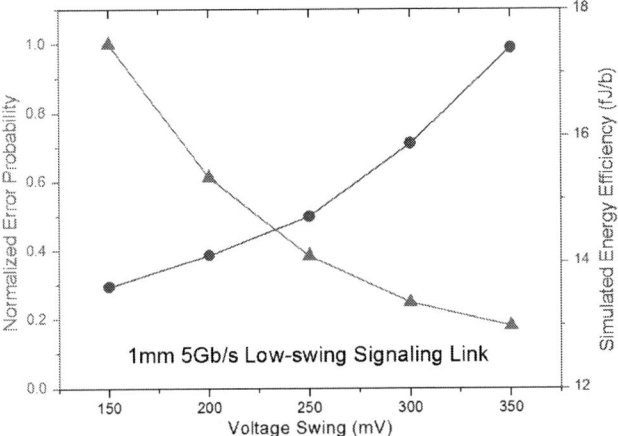

Figure 10: Low-swing signaling trade-off between reliability and energy efficiency.

We also measured power consumption of the 1b 5×5 tri-state RSD-based crossbar connected with 1mm link wires, with various multicast counts: a unicast, 2-multicast, 3-multicast and broadcast. Figure 11 show such results. As described in Section 3.4, the proposed low-swing crossbar drives only the corresponding vertical wires and link wires according to the multicast counts, and hence, it can provide energy-efficient multicasting capability (*i.e.* linearly increasing power consumption as a function of multicast counts).

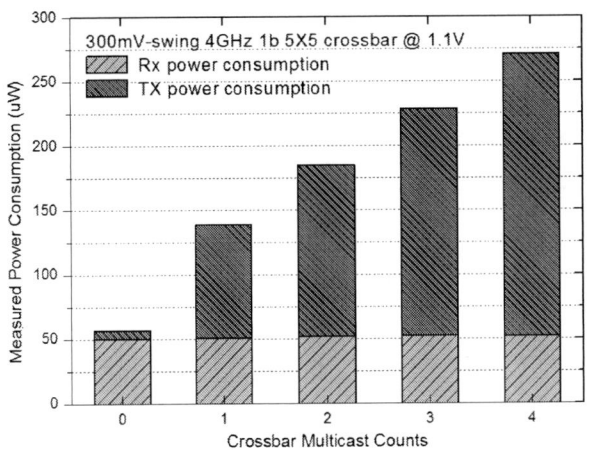

Figure 11: Measured dynamic power of the tri-state RSD-based crossbar.

We present another interesting trade-off between repeated and directly-transmitted (*i.e.* repeaterless) low-swing signaling. Figure 12 shows the 2.5Gb/s simulated vertical eye values with wire resistance variation at two 2mm-LT configurations: 1mm-repeated tri-state RSD and 2mm-repeaterless tri-state RSD. The results show that the 1mm-repeated low-swing link has a larger noise margin

but takes an additional cycle and 28% more energy than the 2mm-repeaterless low-swing link.

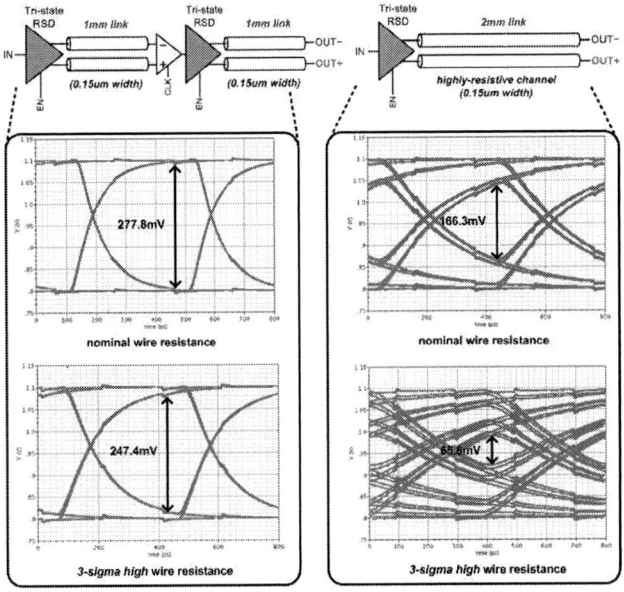

Figure 12: Reduced noise margin comparison of repeated and directly-transmitted low-swing signaling for 2mm-LT.

D. NETWORK PERFORMANCE WITH BROADCAST-ONLY TRAFFIC

As a case study, we evaluated the broadcast-only traffic performance of the fabricated NoC. Figure 13 shows the measured network performance and comparison with the simulated baseline performance. Compared to the mixed traffic performance (Fig. 5), the proposed NoC achieves more latency reduction and throughput improvement. In general, performance benefits of the proposed NoC get larger as network traffic becomes more broadcast-intensive (*i.e.* cache coherence protocols incorporate more broadcast messages with increasing core counts).

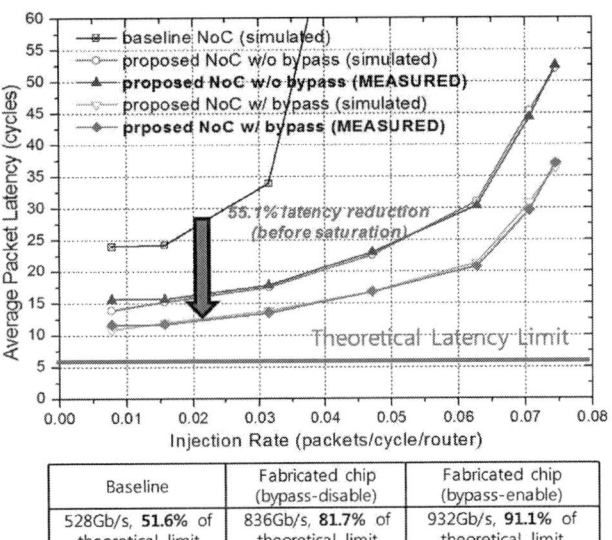

Baseline	Fabricated chip (bypass-disable)	Fabricated chip (bypass-enable)
528Gb/s, **51.6%** of theoretical limit	836Gb/s, **81.7%** of theoretical limit	932Gb/s, **91.1%** of theoretical limit

Maximum throughput comparison *(received)*

Figure 13: Throughput-latency performance evaluation with broadcast-only traffic at 1GHz.

High Radix Self-Arbitrating Switch Fabric with Multiple Arbitration Schemes and Quality of Service

Sudhir Satpathy, Reetuparna Das, Ronald Dreslinski, Trevor Mudge, Dennis Sylvester, David Blaauw
University of Michigan, Ann Arbor
sudhirks@umich.edu

ABSTRACT

A scalable architecture to design high radix switch fabric is presented. It uses circuit techniques to re-use existing input and output data buses and switching logic for fabric configuration and supporting multiple arbitration policies. In addition, it integrates a 4-level message-based priority arbitration for quality of service. Fine grain clock gating, tiled fabric topology and self-regenerating bit-line repeaters enable scaling the router to 8k wires. A 64×64(128b data) switch fabric fabricated in 45nm SOI CMOS spans 4.06mm^2 and achieves a throughput of 4.5Tb/s at 3.4Tb/s/W at 1.1V with a peak measured efficiency of 7.4Tb/s/W at 0.6V.

Categories and Subject Descriptors

B.4.0 [**INPUT/OUTPUT and Data Communications**]: General

General Terms

Algorithms, Design

Keywords

Switch Fabric, Radix, Arbitration, Quality of Service

1. INTRODUCTION

Technology scaling has made billion transistors design feasible on a single die. With transistors getting cheaper and faster, the core count in multi-processor systems has been steadily increasing [1,2]. High end servers [3], gigabit Ethernet routers [4,5] and multimedia processors [6,7] now serve workloads dealing with terabytes of data flow every second. Even medium throughput applications now prefer multi-core architectures over a single core implementation for better energy efficiency and fault tolerance [8]. These system need a network to communicate data among processing and storage elements in the chip. Although processing units are getting smaller and simpler, the dramatic rise of their number in a single die has resulted in the growing complexity of interconnect. As a result, the interconnect fabric has become a bottleneck in improving overall system efficiency. Thus, the design paradigm for multi-core chips is gradually shifting from a core-centric architecture towards an interconnect-centric architecture, where overall system performance is limited by the bandwidth of the interconnect fabric rather than the processing ability of any individual core.

A generic switch fabric comprises of three key modules: 1) A data routing module to transfer information among different IPs connected to the fabric. This could be a single or a collection of shared buses for systems where processing units rarely communicate [9] or could be a fully connected crossbar [10]. 2) An arbiter that receives requests from processing units and configures the fabric to ensure data sent from a source reaches the appropriate destination. 3) A priority management module that monitors traffic flow pattern within the fabric and assists the arbiter to ensure fairness in resource allocation. The overall efficiency of the switch fabric relies on how efficiently each of these independent modules function and how seamlessly they communicate among one another. With a growing number of input and output ports, each module gets physically bigger and hence farther from the others. Beyond a size, the latency and energy overhead due to communication between these modules starts limiting overall fabric efficiency. Existing circuit techniques to build high radix switch fabrics rely on assembling together smaller switches that are usually 5×5 in dimension [11]. This approach has certain limitations: 1) The elementary switches are built using multiplexers that select bits rather than buses. Hence, as buses get wider, routing at the fabric input ports involves a lot of interleaving among the wires incurring additional area penalty. 2) A switch with ports far exceeding 5 would require multiple of these switches to be connected in stages. Data has to traverse through multiple stages to reach its destination thereby increasing latency and energy dissipation. They would also require additional data storage elements in the data routing path for higher throughput resulting in further latency and power overhead. 3) Each source sends requests to the arbiter before it could access a destination, and eventually the arbiter sends an acknowledgement back to the source after setting up the routing path. Configuring all the switches along the routing path incurs latency. In systems that have well defined communication patterns and usually operate on massive sets of data, the latency and energy cost for configuring the fabric can be amortized by setting up the routing path ahead of time or by sending multiple chunks of data once the path is established. However, in generic multiprocessor systems most traffic patterns are not pre-defined. Hence, the fabric configuration cost becomes a bottleneck. 4) A non-blocking switch supports all possible permutations and hence can guarantee starvation free communication for all applications. However many applications (like FFT, LDPC, Color space conversion etc.) can be sped up significantly by incorporating multicast and broadcast features in the switch fabric [12].

Arbitration policies have a noticeable impact on throughput and fairness of interconnection networks. Some arbitration policies like the greedy allocation policy tend to maximize network throughput at the cost of quality of service. At the other extreme, some complex schemes like probabilistic distance [13] based arbitration can guarantee quality of service at the price of more complex logic and hence additional arbitration latency and power overhead. Hence, adaptive and hybrid resource allocation schemes are preferred in general over static schemes because of their ability to mitigate congestion.

In this paper, we propose a fabric architecture called swizzle switch network (SSN) to accomplish a variety of arbitration policies with very minimal overhead as shown in Fig. 1. For proof of concept, we fabricated a 64x64 SSN prototype with 128b data bus in 45nm SOI CMOS with the following key features: 1) A novel, single cycle least recently granted (LRG) priority arbitration technique that re-uses the already present input and output data buses and their drivers and sense amps. 2) An additional 4-level message-based priority arbitration for quality of service (QoS) with 2% logic and 3% wiring overhead. 3) A new bidirectional bit-line repeater that allows the router to scale to >8000 wires. These features result in a compact fabric (4.06mm²) with throughput gain of 2.1× over [5] at 3.4Tb/s/W efficiency which improves to 7.4Tb/s/W at 600mV.

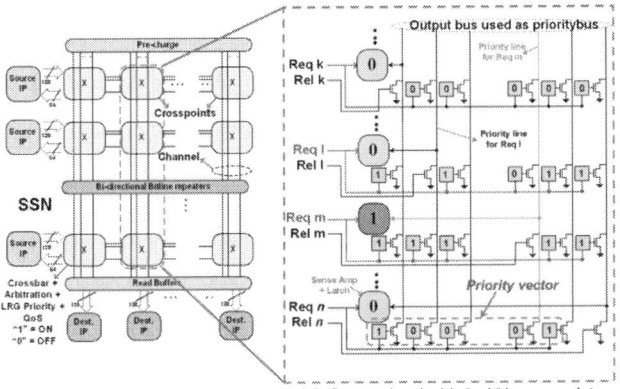

Data routing, arbitration, priority update and QoS control embedded within crosspoints

Figure 1. Swizzle Switch Network

The rest of the paper is organized as follows: In section 2, we present the SSN fabric architecture. Section 3 explains the various arbitration policies that can be implemented in SSN. In section 4, we present SSN's message based QoS arbitration scheme. Detailed circuit level implementation and design choices are then presented in section 5. Measurement results from SSN test prototype fabricated in 45nm have been described in section 6 before conclusion in section 7.

2. SSN ARCHITECTURE

SSN is a matrix-type fabric as shown in Fig. 1 with input buses running horizontally and output buses vertically. When data is routed, the input and output buses transfer data traffic. During arbitration the input bus routes a multi-hot code indicating which output channel(s) are requested by that input, and the output bus is used for conflict detection and arbitration [14]. Each crosspoint stores a connectivity status bit indicating whether the input bus was granted access to the output channel. An *n-1* bit priority vector is also stored to represent the priority of the input bus with respect to all other inputs for that output bus. Fig. 1 shows the priority vector at each crosspoint in a blow-up of a *single* output channel. Each input bus is assigned a unique bit line from the channel as its *priority line* which, if high, indicates it as the winner in a particular arbitration cycle. Similarly, each bit in the priority vector at a crosspoint indicates whether the input bus at that crosspoint has higher or lower priority than the input bus associated with the priority line. For instance, in Fig. 2 priority line *m* corresponds to input bus *m* while the *m*-th priority bit of bus *n* is a 1, indicating that *n* has higher priority than *m*. When input *n* requests the output channel this high bit results in the

discharge of priority line *m*, suppressing access by input *m*. In contrast, input *l* stores a 0 at its *m*-th priority bit and hence does not suppress an access request from input *m*, meaning that *l* has lower priority that input *m*.

Priority vectors need to be set consistently and indicate the same priority order. In Fig. 2, the 0 at bit *m* of input *l* must be mirrored with a 1 at bit *l* of input *m*. Furthermore, the priority bits need to be correctly updated after each arbitration cycle to implement LRG policy. We propose a new, simple mechanism to accomplish this. In Fig. 1, inputs *l* and *m* request the output channel in an arbitration cycle. Input *m* wins owing to its higher priority and its

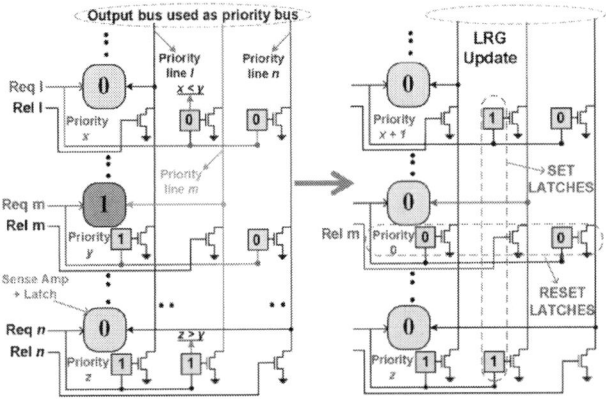

Figure 2. Arbitration and priority update.

connectivity status bit is set to 1. After data transfer, input *m* releases its channel during a channel release cycle. In this cycle, input *m* first *resets* all its priority bits. This guarantees that *m* now has lowest priority, as required by the LRG algorithm. At the same time, input *m* also lowers its *priority line m*, which is a signal to other crosspoints in the output channel to *set* their *m*-th priority bit. This ensures that all other input buses now have higher priority than *m*. Input buses with higher priority than *m* remain unchanged and *only* inputs with lower priority than input *m* are increased in their priority by exactly one level. This simple and fast update mechanism provably guarantees both consistency of all priority vectors and correct implementation of the LRG arbitration scheme, which enables efficient and deadlock-free routing.

3. ARBITRATION SCHEMES

The SSN is capable of supporting multiple arbitration schemes. The priority vectors at various crosspoints along a single output bus form a priority matrix. A priority matrix with 6 inputs (arranged top to down numerically from *input1* to *input6*) is shown in Fig. 3. Here, the priority line connections are denoted as Xs. The priority matrix satisfies the following criteria: 1) The total number of 0s equals the total number of 1s. 2) Each row has a unique number of 1s which represents the corresponding input's priority. 3) Each column has a unique number of 1s. 4) At any priority line connection (denoted as X in Fig. 3.), the sum of the number of 1s (or 0s) in its corresponding row and column must add to one less than the total number of inputs.

3.1 Least Recently Granted (LRG) scheme

Though priority lines can be randomly assigned to inputs without limiting the generality of the priority update schemes, a diagonal assignment as shown in Fig. 3 makes the priority matrix skew (or anti) symmetric and easy to understand. As shown in Fig. 3, the

407

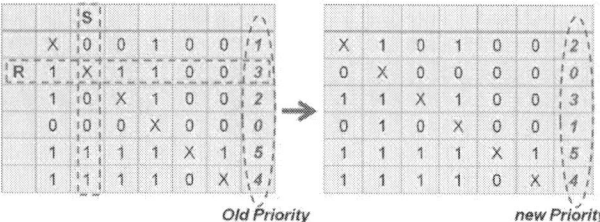

Figure 3. LRG based priority update

input corresponding to second row used the channel most recently. It is assigned a priority level 3. An LRG priority update is accomplished by resetting all priority bits along the second row (denoted by R) and by setting all priority bits (denoted by S) along the second column (which is the priority line for *input2*). *Input2* is thus downgraded to have the least priority while all inputs with lower priorities get upgraded by exactly one level. In the rest of this section, the other priority update schemes will be explained using the priority matrix notation.

3.2 Most Recently Granted (MRG) scheme

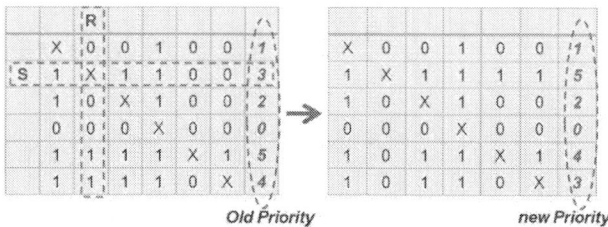

Figure 4. MRG based priority update

For accomplishing an MRG based update, we set all priority bits along the second row (denoted by S) and reset all priority bits (denoted by R) along the second column (which is the priority line for the *input2*) as shown in Fig. 4. By setting bits in the second row, *input2* now gets the highest priority. Inputs that previously had higher priorities than *input2* get downgraded by exactly one level. Inputs that previously had lower priority than *input2* retain their old priorities.

3.3 Round Robin schemes

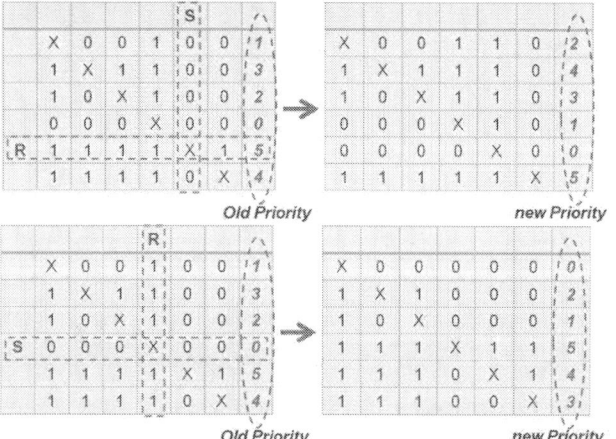

Figure 5. Incremental (top) and decremental (bottom) round robin based priority updates

For accomplishing an incremental Round Robin based update, we pick the row with the highest priority. This can be identified by a logical AND operation of all the priority bits. In this case *input5* has the highest priority. We reset all priority bits along the fifth row (denoted by R) and set all priority bits (denoted by S) along the fifth column (which is the priority line for *input5*) as shown in Fig. 5 top. By resetting bits in the fifth row, *input5* now gets the least priority. All other inputs get upgraded by exactly one level.

For accomplishing a decremental Round Robin based update, we pick the row with the lowest priority. This can be identified by a logical OR operation of all the priority bits. In this case *input4* has the highest priority. We set all priority bits along the fourth row (denoted by S) and reset all priority bits (denoted by R) along the fourth column (which is the priority line for *input4*) as shown in Fig. 5 bot. By setting bits in the fourth row, *input4* now gets the highest priority. All other inputs get downgraded by exactly one level.

3.4 Priority swap and reversal

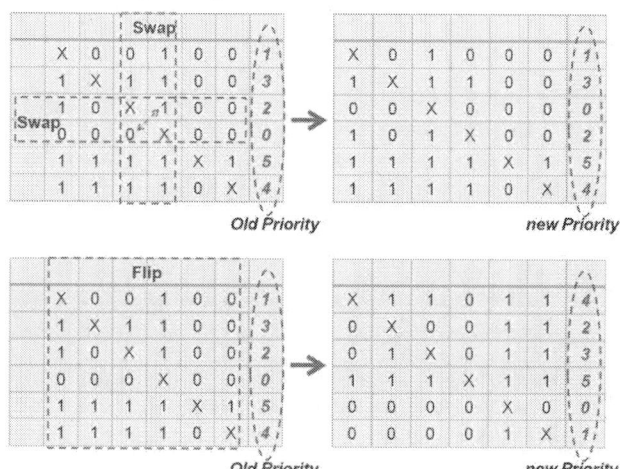

Figure 6. Priority swap (top) and priority reversal (bottom)

The priorities of 2 inputs can be swapped (without affecting priorities of other inputs) by swapping the priority bits in their corresponding rows and those in the columns corresponding to their priority lines as shown in Fig. 6 top. Here, we intend to swap the priorities of *input3* and *input4*. In the physical realization of this technique, already existing word-lines will be used to swap priority bits between columns and bit-lines to swap priority bits between rows. In a single cycle, any two priorities can be swapped.

The unique priority encoding scheme also allows reversing the priority of all inputs instantaneously by flipping all the priority bits as shown in Fig. 6 bottom. In physical circuit level implementation, rather than flipping all the bits, a multiplexer can be used to select the inverted priority. Hence, this functionality can be achieved without the expense of a clock cycle. The consistency of the new priority vectors is guaranteed because this transformation ensures that the priority matrix still satisfies all the criteria mentioned before.

3.4 Selective LRG and MRG

In this scheme LRG update is applied to a selective section of inputs. In the standard LRG scheme, the input that used the output bus most recently is downgraded to have the least priority while all inputs with lower priorities get upgraded by exactly one level. In this case *input6* with a priority level 4 used the channel most recently. However, in the selective LRG scheme, instead of downgrading *input6* all the way down to 0, we intend to downgrade it to some intermediate priority (say priority level of *input0* which is 1). To accomplish this, before setting/resetting priority bits we identify certain rows and columns that need to be frozen. In this case, all columns corresponding to priority bits that are high in the first row are frozen as shown in Fig. 7 top. Simultaneously, all rows corresponding to priority bits that are low in the first column (which is the priority line for *input0*) are also frozen. Following this the priority bits in the sixth row are reset (except the bits in frozen columns) and those in the sixth column are set (except the bits in frozen rows). This ensures that the new priority matrix is consistent and the intended priority update is achieved.

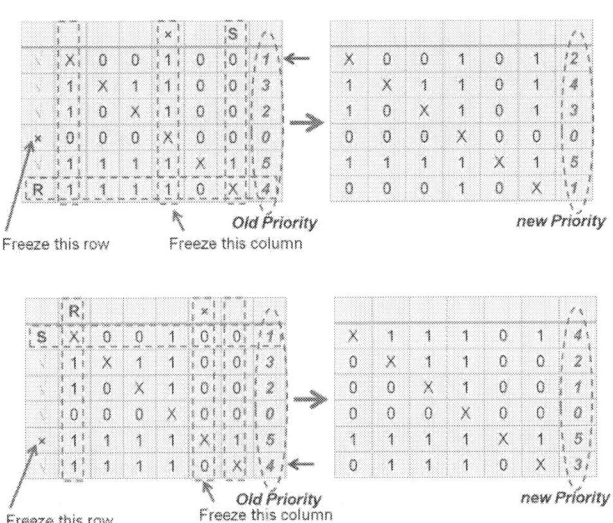

Figure 7. Selective LRG (top) and selective MRG (bot)

In this scheme MRG update is applied to a selective section of inputs. In the standard MRG scheme, the input that used the output bus most recently is upgraded to have the highest priority while all inputs with higher priorities get downgraded by exactly one level. In this case *input1* with a priority level 1 used the channel most recently. However, in the selective MRG scheme, instead of upgrading *input1* all the way up to 5, we intend to upgrade it to some intermediate priority (say priority level of *input6* which is 4). To accomplish this, before setting/resetting priority bits we identify certain rows and columns that need to be frozen. In this case, all columns corresponding to priority bits that are low in the sixth row are frozen as shown in Fig. 7 bot. Simultaneously, all rows corresponding to priority bits that are high in the sixth column (which is the priority line for *input6*) are also frozen. Following this the priority bits in the first row are set (except the bits in frozen columns) and those in the first column are reset (except the bits in frozen rows). This ensures that the new priority matrix is consistent and the intended priority update is achieved.

4. QoS ARBITRATION

In a 64×64 SSN it might take a message 64 cycles to win arbitration in the worst case when all inputs collide. To assist critical messages to reach destination early, SSN also features a 4-level message-based QoS arbitration technique that allows only input buses with the highest message priority to arbitrate for the channel as shown in Fig. 8. A 2-bit *message priority* is decoded into a 4-bit thermometer code at the crosspoint, which is used to selectively discharge priority bit-lines comprising the *QoS priority bus*. A multiplexer samples one of those priority bit-lines using its own *message priority* and the input bus progresses to the LRG arbitration cycle if the monitored priority bit is not discharged. Using separate wires for QoS arbitration incurs 3% area overhead. However, the additional QoS arbitration cycle can be overlapped with the prior routing operation for the output bus, avoiding a latency penalty.

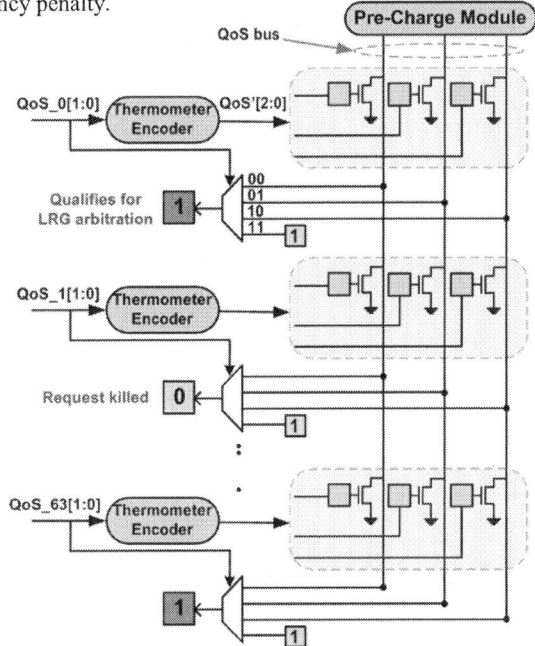

Figure 8. QoS arbitration technique

5. PROTOTYPE IMPLEMENTATION

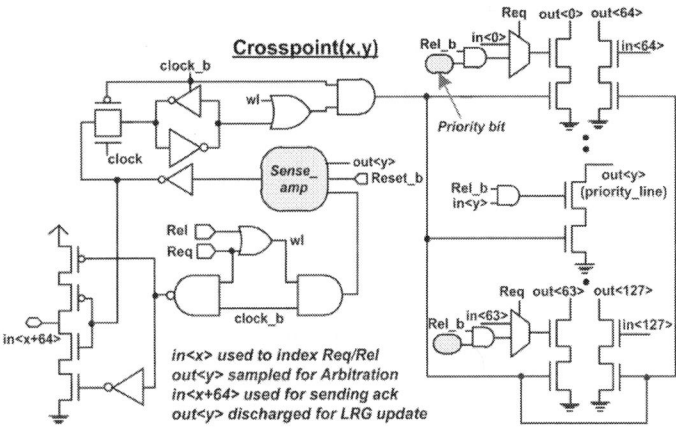

Figure 9a. Crosspoint circuit

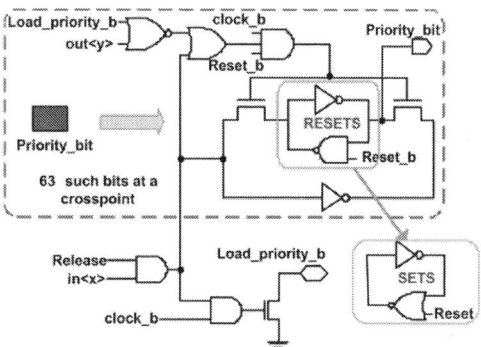

Figure 9b. Priority storage latch

Fig. 9 shows the SSN crosspoint circuit and the priority storage latch. *Load_priority_b* is an additional bit-line provided per channel that is discharged during the release cycle. This triggers the priority update mechanism. During a request/release cycle the channels are indexed using the lower 64 bits from the input bus. Crosspoints send acknowledgements over the upper 64 bits.

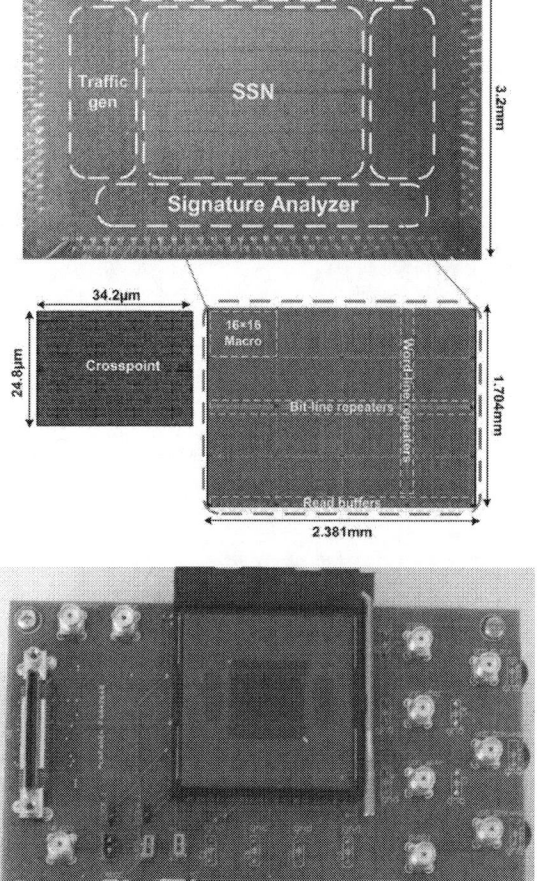

Figure 10. SSN die photo (top) and printed circuit board hosting SSN test prototype (bottom)

Fig. 10 shows the test prototype fabricated in 45nm SOI CMOS with a 64×64 SSN as the communication network between the traffic generators and traffic analyzers. SSN is laid out using a semi custom design flow. A generic crosspoint cell is laid out as a parameterized cell. A skill script parses the generic crosspoint by taking in the x coordinate, y coordinate and the priority vector (at reset) as the arguments and generates crosspoint specific to each location. These crosspoints are then tiled using a compiler that appropriately sizes the word line drivers and precharge transistors to generate the switch fabric.

The SSN features 8448 word-lines and 8576 bit-lines spread across 4096 crosspoints. The integration of the LRG and QoS control within this fabric with very low overhead greatly improve SSN's scalability and makes it possible to realize a fabric of this large size. In addition, new bi-directional repeaters (Fig. 11) are used for bit-lines that use a regenerative sensing element to improve delay despite high slew rates on long bit-lines. The proposed repeater uses a thyristor element to detect and amplify a transition on the bit-line. Once a transition is detected the repeater enters a self regeneration mode where it decouples itself from the slow transitioning bit-line. This allows the internal nodes in the thyristor to switch faster and reduces delay. The regeneration and self-decoupling mechanism improves bit-line delay by 32% and allows for a 50% smaller bit-line driver compared to a conventional repeater. Simulated fabric latency with increasing SSN size shows 1.6× performance improvements over an SSN with un-repeated bit-lines as shown in Fig. 11 due to the near-linear latency increase with radix size rather than quadratic dependency without repeaters. Bit-lines are pre-charged within every 16×16 SSN macro. This improves pre-charge time by 59% over a similar sized lumped driver and results in more uniform current drawn from the power grid. Bit-line delay degrades more rapidly than word-line delay under voltage scaling. Hence, the bit-cell aspect ratio (1:0.73) is chosen to shorten bit-lines, improving fabric latency at low V_{dd}. Fine grain clock gating reduces clock power by 94% at each crosspoint. A crosspoint is clocked only if its connectivity status is ON, a request is asserted, or an LRG priority update occurs. These events are registered in the positive clock phase, allowing gating of the negative (active) phase with 2.3% delay penalty. Adjacent SSN input ports are driven from opposite directions, reducing routing congestion and local Ldi/dt drop when repeaters on the 2.5mm long word-lines switch.

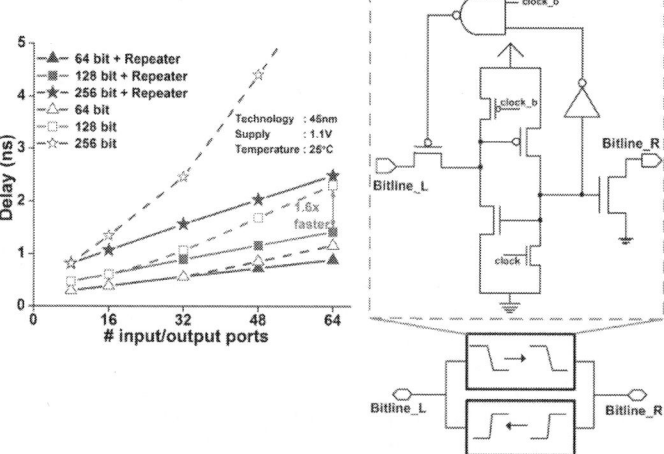

Figure 11. Self-regenerating bit-line repeater improves SSN delay by 1.6×

6. MEASUREMENT RESULTS

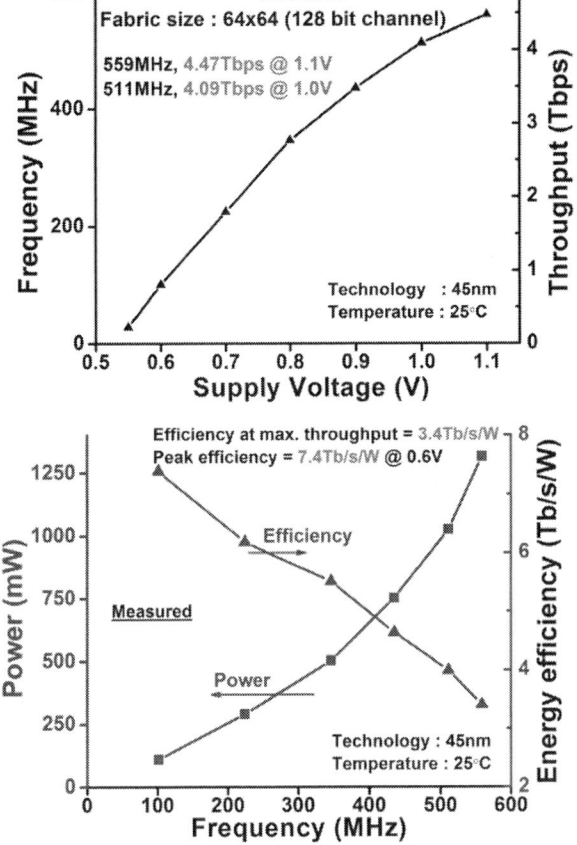

Figure 12. Measured performance and power for 64×64 SSN

The traffic generators in the test prototype can be tuned to produce traffic patterns with varying switching activity and collision patterns. The SSN is tested for functionality by streaming various data streams through it and verifying the signatures. Fig. 12 (top) shows the measured SSN's operating frequency and the aggregate throughput at varying supply voltage. SSN's power consumption at different operating frequencies is shown in Fig. 12 (bottom). At 1.1V, the SSN operates at 559MHZ with a throughput of 4.47Tb/s while consuming 1.32W. This translates into an efficiency of 3.4Tb/s/W which is which is 3.7× higher than [4] at similar bandwidth. The work in [4] uses an 8×8 mesh topology based on 5×5 routers at each node to connect 64 units, whereas the SSN uses a 64×64 single-stage fabric. The SSN is fully functional down to 550mV with a measured peak efficiency of 7.4Tb/s/W at 0.6V.

7. CONCLUSION

In this paper, we present a self-arbitrating fabric called SSN that leverages a novel priority encoding scheme that re-uses existing logic and interconnect resources in switch fabric to locally store priorities at router crosspoints resulting in a compact implementation. A 64×64 SSN with 128b data bus achieves a peak throughput 4.5Tb/s at an energy efficiency of 3.4Tb/s/W while spanning only 4.06mm^2 in 45nm SOI CMOS. It features a single cycle least recently granted arbitration technique that re-

uses data buses and switching logic, a 4-level message based priority arbitration for quality of service and unique bidirectional bit-line repeaters to aid scalability. The unique priority encoding scheme also allows seamless implementation of many other arbitration policies in addition to LRG with very minimal overhead.

8. ACKNOWLEDGEMENTS

The authors gratefully acknowledge funding and support from ARM Ltd.

9. REFERENCES

[1] D.Truong et al., "A 167-processor 65nm Computational Platform with per-Processor Dynamic Supply Voltage and Dynamic Clock Frequency Scaling," International symposium of VLSI Circuits, pp. 22-23, 2008.

[2] S. Vangal et al., "An 80-Tile 1.28TFLOPS Network-on-Chip in 65nm CMOS," International Solid State Circuits Conference, pp. 98-99, 2007.

[3] S.Tremblay et al., "A Third-Generation 65nm 16-Core 32-thread Plus32-Scout-Thread CMY SPARC Processor," International Solid State Circuits Conference, pp. 82-83, 2008.

[4] S. Vangal et al., "A 5.1 GHz 0.34mm2 Router for Network-on-Chip Applications," International Symposium on VLSI Circuits, pp. 42-43, 2007.

[5] M.Anders et al., "A 4.1Tb/s Bisection-Bandwidth 560Fb/s/W Streaming Circuit-Switched 8×8 Mesh Network-on-Chip in 45nm CMOS," International Solid State Circuits Conference, pp. 110-111, 2010.

[6] K.Kim et al., "A 211 GOPS/W Dual-Mode Real-time Object Recognition Processor with Network-on-Chip," European Solid State Circuits Conference, pp. 462-465, 2008.

[7] J.Kim et al., "A 118.4 GB/s multi-casting Network-on-Chip with hierarchical star-ring combines topology for real-time object recognition," Journal of Solid State Circuits, vol.45 pp. 1309-1409, 2010.

[8] E. Karl et al., "ElastIC: An Adaptive Self-Healing Architecture for Unpredictable Silicon," IEEE Design and Test of Computers, Vol. 23, No. 6, pp. 484-490, 2006.

[9] D. Flynni et al., "AMBA: enabling reusable on-chip designs," IEEE Micro, pp. 20-27, 1997.

[10] P. Kongetira et al., "Niagara: A 32-way multithreaded Sparc processor," IEEE Micro, pp. 21-29, 2005.

[11] P. Salihundam et al. ,"A 2Tb/s 6×4 Mesh Network with DVFS and 2.3Tb/s/W router in 45nm CMOS," International Symposium on VLSI Circuits, pp. 79-80, 2010.

[12] S. Satpathy et al., "A 1.07 Tbit/s 128×128 Swizzle Network for SIMD Processors," International Symposium on VLSI Circuits, pp. 81-82, 2010.

[13] M. Lee et al., "Probabilistic distance-based arbitration: Providing equality of service for many-core CMPs," IEEE MICRO, 2010.

[14] S. Satpathy et al., "SWIFT: A 2.1Tb/s 32x32 self-arbitrating many-core interconnect fabric," International Symposium on VLSI Circuits, pp. 81-82, 2010.

411

WCET-Centric Partial Instruction Cache Locking

Huping Ding[1], Yun Liang[2] and Tulika Mitra[1]
[1]School of Computing, National University of Singapore
[2]Advanced Digital Sciences Center, Illinois at Singapore
{d-huping,tulika}@comp.nus.edu.sg, eric.liang@adsc.com.sg

ABSTRACT

Caches play an important role in embedded systems by bridging the performance gap between high speed processors and slow memory. At the same time, caches introduce imprecision in Worst-case Execution Time (WCET) estimation due to unpredictable access latencies. Modern embedded processors often include cache locking mechanism for better timing predictability. As the cache contents are statically known, memory access latencies are predictable, leading to precise WCET estimate. Moreover, by carefully selecting the memory blocks to be locked, WCET estimate can be reduced compared to cache modeling without locking. Existing static instruction cache locking techniques strive to lock the entire cache to minimize the WCET. We observe that such aggressive locking mechanisms may have negative impact on the overall WCET as some memory blocks with predictable access behavior get excluded from the cache. We introduce a partial cache locking mechanism that has the flexibility to lock only a fraction of the cache. We judiciously select the memory blocks for locking through accurate cache modeling that determines the impact of the decision on the program WCET. Our synergistic cache modeling and locking mechanism achieves substantial reduction in WCET for a large number of embedded benchmark applications.

Categories and Subject Descriptors

C.3 [**Special-purpose and Application-based Systems**]: [Real-time and embedded systems]

General Terms

Algorithm, Design, Performance

Keywords

WCET, Partial Cache Locking

1. INTRODUCTION

Cache memories are often employed in embedded systems to hide the main memory access latency. Caches are quite effective in improving the average-case performance due to the temporal and spatial locality of memory accesses in a program. For hard real-time systems, however, caches are problematic due to timing unpredictability — specially in the context of Worst-case Execution Time (WCET) estimation. WCET is an important metric for hard real-time systems. It is defined as the upper bound on the maximum execution time of the program on a particular hardware platform across all the possible inputs. In the presence of caches, it is challenging to determine the cache behavior for a memory access (hit or miss) for WCET estimation through static program analysis. If a memory access cannot be guaranteed as a cache hit, it is conservatively estimated to be a miss in WCET analysis. This leads to significant imprecision in WCET estimation.

In this paper, we focus on instruction caches, which are present in almost all embedded systems today. There exist many static program analysis techniques that model the instruction cache for tight WCET estimation [19, 10]. For example, Theiling et al. [19] model the cache states at each program point and classify the cache behavior (hit or miss) of a memory access based on the cache state. However, in the presence of complex control flow, cache modeling may fail to accurately determine the cache behavior for some memory accesses. Such unclassified accesses are conservatively assumed to be cache misses in WCET analysis due to the safety critical nature of hard real-time systems. This over-estimation can lead to serious over-dimensioning of the processor resources.

To improve timing predictability, modern embedded processors often feature cache locking mechanisms. Memory blocks can be locked in the cache using special lock instructions. Once a memory block is locked, it cannot be evicted from the cache under replacement policy. Thus, locking the entire cache resolves the problem of timing unpredictability. More importantly, by carefully choosing the memory blocks to be locked, WCET estimate can be reduced compared to cache modeling techniques without locking [6, 14].

Embedded processors also provide the option of partial cache locking through two different mechanisms: way locking and line locking. In way locking, particular ways are entirely locked for all the cache sets. Way-locking is employed in several ARM processor series. Line locking allows different number of lines to be locked in different cache sets and is employed in Intel Xscale, ARM9 family and Blackfin 5xx family. In partial cache locking, the unlocked portion of the cache behaves as a normal cache. For example, if in a 4-way set associative cache, two cache ways are locked, then the other two cache ways serve as a 2-way set associative cache. Clearly, line locking is more flexible than way locking, which in turn is more flexible than full cache locking.

Recently, a heuristic [6] and an optimal solution [14] have been proposed to minimize the WCET via static instruction cache locking. These existing techniques make an *implicit but important de-*

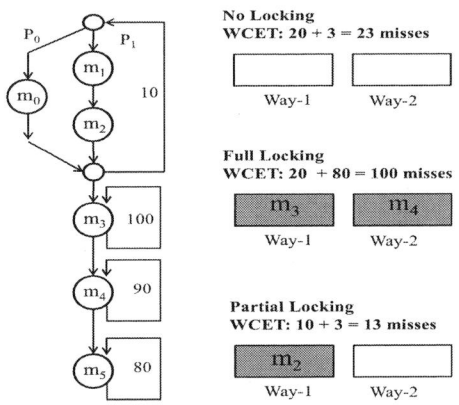

Figure 1: Advantage of partial cache locking over full cache locking and cache modeling with no locking. The program consists of four loops. The first loop contains two paths (P_0 and P_1) and the other three loops contain only one path. The loop iteration counts appear on the back edges.

cision of locking the entire cache. This crucial decision arises from the assumption that instruction cache modeling for WCET analysis is quite imprecise. By employing full cache locking, [6, 14] can completely bypass cache modeling in WCET analysis phase and thereby achieve tight WCET estimation. Indeed, as these techniques are oblivious to cache modeling, they assume the worst-case behavior with empty cache (where all the accesses are serviced from main memory) as the reference point and improve upon it through locking of memory blocks along the WCET path. In this context, it is guaranteed that locking the entire cache will provide maximum WCET reduction compared to the baseline empty cache. In other words, the cache locking problem becomes equivalent to the scratchpad memory allocation problem [5, 18].

In this paper, we argue (and experimentally validate) that aggressive full cache locking as proposed in [6, 14] may have substantial negative impact on WCET reduction. State-of-the-art instruction cache modeling techniques for WCET analysis are quite mature. Most memory accesses can thus be successfully classified as hit/miss through WCET analysis techniques. Consider a memory block m originally classified as cache hit in a normal cache through static WCET analysis. But m is not selected for locking under full cache locking scenario. Thus m does not have any opportunity to reside in the cache and all its accesses incur cache misses. Now consider an alternative scenario where partial cache locking is employed. Again m is not selected for locking. However, as the cache has some unlocked lines, m may still be brought into the cache at runtime and the cache misses can be avoided.

In summary, full cache locking does not exploit the entire spectrum of opportunities presented by cache locking. Thus, in this paper we propose a partial cache locking technique that explores in conjunction with accurate cache modeling the entire spectrum of choices. Partial cache locking problem is more challenging compared to full cache locking as it requires careful cost-benefit analysis to decide between locking a cache line with a single memory block versus keeping it unlocked so that more than one memory blocks can benefit from it. This synergistic interaction between cache modeling and memory block selection for locking sets apart our technique from the state-of-the-art.

Motivating Example. We illustrate the benefit of partial cache locking over full cache locking with a concrete example shown in Figure 1. The program consists of four loops and we assume that all the memory blocks are mapped to the same cache set in a 2-way set associative cache.

Cache modeling with no locking: Let us first estimate the WCET via cache modeling with no locking. Theiling et al. [19] models the cache states at all program points. All the memory blocks in the first loop ($m0$, $m1$, $m2$) are cache misses in the worst case because alternate execution of the two program paths ($P0$ and $P1$) can lead to mutual eviction of the blocks. Thus, program path $P1$ with 2 cache miss is the worst case path in the first loop. For the other three loops, cache modeling techniques can easily determine that the first access is a cold miss and the subsequent accesses are cache hits via persistence analysis or virtual unrolling [15, 19]. Therefore, cache modeling estimates 23 cache misses in the worst case — 20 misses for the first loop and 3 misses for the other loops.

Full cache locking: Existing cache locking techniques [6, 14] first build the worst case path (e.g., $(m1m2)^{10}m3^{100}m4^{90}m5^{80}$) assuming that all accesses are serviced from the main memory (i.e., there is no cache). Now memory blocks are selected for locking along the worst-case path so as to improve the WCET until the cache is fully locked. Both cache locking techniques [6, 14] model the fact that the WCET path may change after locking some memory locks. For this example, the heuristic [6] and the optimal [14] approach return the same solution. $m3$ and $m4$ are chosen to be locked as they contribute most towards WCET reduction. After locking, we get 100 cache misses in total in the worst case — 20 misses in the first loop and 80 misses in the last loop. Thus, cache locking performs worse than cache modeling in this example.

Partial cache locking: Our partial locking technique can determine that it is beneficial to keep one cache line free so that accesses to $m3$, $m4$, and $m5$ can be cache hits after the first cold miss. It only chooses to lock $m1$ or $m2$ in the cache. Thus we get 13 cache misses in the worst case — 10 misses in the first loop and 3 cold misses for the other loops. Thus partial cache locking improves upon cache modeling and full locking.

From the example above, we first observe that full locking techniques [6, 14] are not guaranteed to perform better than cache modeling (with no locking) specially when some memory accesses can be easily classified as cache hits ($m3$, $m4$, $m5$ in our example). Locking these memory blocks with deterministic access pattern does not yield any benefit. On the other hand, if the cache is fully locked and these memory blocks with deterministic access pattern are not chosen for locking, it can have serious impact on the WCET.

Our partial locking mechanism integrates cache locking with cache modeling. We model the cache content at all program points and select the memory blocks for locking based on the cache state and their impact on the WCET. In particular, we use the concrete cache states or the abstract cache states to model the cache content. Concrete cache state captures the exact path behavior while abstract cache state is a compact representation that merges multiple concrete cache states together. For concrete cache state, we use integer linear programming (ILP) approach to optimally select the memory blocks for locking. As no cache locking and full cache locking are just two extreme instances of partial cache locking, partial locking is guaranteed to be equivalent to or better than them. To improve the efficiency, we also propose a heuristic partial locking strategy based on abstract cache states. Experimental results show that our partial cache locking techniques improve WCET substantially for a large number of embedded benchmark applications.

2. RELATED WORK

Cache locking is used to improve timing predictability in real-time systems. Puaut and Decotigny [16] explore static cache locking in multitasking real-time systems. Two content selection algorithms have been proposed in their work to minimize the utilization and inter-task interferences. Campoy et al. [4] employ genetic algorithm to perform instruction cache locking. However, both [16] and [4] do not model the change in worst-case path after locking.

Falk et al. [6] perform cache locking by taking into account the change of worst-case path and achieve better WCET reduction. Their greedy algorithm computes the worst-case path and selects the procedure with maximum WCET reduction for locking. This process continues until the cache is fully locked. Liu et al. [14] present an optimal solution to minimize WCET via cache locking. However, their approach is optimal on the premise that the cache is fully locked. It may not be optimal towards minimizing WCET as shown in our motivating example. More importantly, they do not consider the cache mapping function at all in the locking algorithm. They simply assume that any memory block can be locked in any cache set (as if the cache is a scratchpad memory). After locking decisions are taken, they have to use code placement/layout technique [7, 12] that force the locked memory blocks to be mapped to the appropriate cache sets. This can lead to serious code size blowup, which has not been addressed.

Vera et al. [20] combine compile-time cache analysis and data cache locking in order to estimate a safe and tight worst-case memory performance. This work also assume full cache locking. Arnaud and Puaut [1] propose dynamic instruction cache locking for hard real-time systems. In their approach, the program is partitioned into regions, and static cache locking is performed for each region. In [17], cache locking is explored for predictable shared caches on multi-core systems. Cache locking is also shown to be quite effective for improving average-case execution time [13]. Finally, optimal on-chip scratchpad memory allocation to improve the WCET has been explored in [5, 18].

3. CACHE MODELING

Cache design depends on a few parameters: line (block) size L, which defines the unit of transfer of instructions or data between the cache and the main memory; number of sets K that the cache is divided into; associativity A, which determines the number of lines (blocks) in a set. Then the capacity of a cache is $L \times A \times K$. We assume LRU (Least Recently Used) cache replacement policy.

Given a memory block m, it is mapped to only one cache set. Thus, the different cache sets are independent and can be modeled independently. In the following, we describe our modeling technique for one cache set. The same modeling techniques can be repeated for other cache sets. We use M to denote the set of memory blocks mapped to a cache set and use $\perp$ to indicate the absence of any memory block in a cache line.

3.1 Cache States

DEFINITION 1 (**Concrete Cache State**). *A concrete cache state c is a vector* $\langle c[0], ..., c[A-1] \rangle$ *of length A where $c[i] \in M \cup \{\perp\}$. If $c[i] = m$, then m is the ith most recently used memory block in the cache set. We also define a special concrete cache state $c_\perp = \langle \perp, ..., \perp \rangle$ called the empty concrete cache state.*

DEFINITION 2 (**Concrete Cache State Hit**). *Given a concrete cache state c and a memory access $m \in M$*

$$c\_hit(c, m) = \begin{cases} 1 & if \; \exists i \; (0 \leq i \leq A-1) \; s.t. \; c[i] = m \\ 0 & otherwise \end{cases}$$

We use Ω to denote the set of all possible concrete cache states of a program. Note that a program point can be reached via multiple paths and these paths may lead to different concrete cache states. We use $\mathcal{P}$ to denote the set of all possible concrete cache states at

a program point, i.e., $\mathcal{P} \in 2^\Omega$. We can easily compute $\mathcal{P}$ at each program point through static program analysis as shown in [11].

Given the set of all possible cache states $\mathcal{P}$ at a program point and a memory access $m \in M$,

$$p\_hit(\mathcal{P}, m) = \begin{cases} 1 & if \; \forall c \in \mathcal{P} \; c\_hit(c, m) = 1 \\ 0 & otherwise \end{cases}$$

That is, an access m is a hit at a program point with the set of all possible concrete cache states $\mathcal{P}$ if and only if m is hit in all the concrete cache states of $\mathcal{P}$.

Maintaining the set of all possible cache states may not be feasible (and scalable) for large programs with complex control flows where a program point can potentially have hundreds or even thousands of cache states. Thus we also employ abstract interpretation to compute the abstract cache state at every program point [19]. An abstract cache state is derived by joining all possible concrete cache states at a program point.

DEFINITION 3 (**Abstract Cache State**). *An abstract cache state a is a vector* $\langle a[0], ...a[A-1] \rangle$ *of length A where $a[i] \in 2^M$.*

An abstract cache state maps a cache line to a set of memory blocks. Must analysis and may analysis [19] are usually employed to compute abstract cache states for WCET analysis. Given a program point, must analysis determines the set of memory blocks that are guaranteed to be present in the cache, while may analysis determines the set of memory blocks that are never in the cache. Must analysis uses abstract cache states where the position of a memory block is an upper bound of its age. In may analysis, the lower bound of the age of a memory block is used as its position in the abstract cache state, in order to capture the set of all memory blocks that may be in the cache. Figure 2 shows the relationship between a set of concrete cache states and the corresponding abstract cache states.

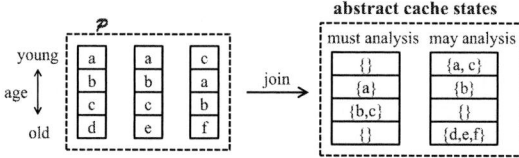

Figure 2: Concrete cache states and abstract cache states.

4. CACHE LOCKING

In this paper, we consider static cache locking, where the selected memory blocks are locked into the cache before the program starts execution and remain unchanged throughout the execution. Furthermore, we consider line locking mechanism, where different number of lines can be locked in different cache sets. As discussed before, for our purposes, we can treat each cache set independently because the memory blocks mapped to different cache sets do not interfere. Each cache set can be considered as a fully associative cache containing A lines, where A is the associativity. Once a memory block is locked in a cache line, it can not be evicted from the cache. The remaining unlocked lines in the cache set serve as a fully associative cache with reduced capacity.

Note that the mapping of instructions to the cache sets depends on the code memory layout. Inserting additional code for cache locking may tamper this layout. To avoid this problem, we use the trampolines [2] approach. The extra code to fetch and lock the memory blocks in the cache are inserted at the end of the program as a trampoline. We leave a dummy NOP instruction at the entry point of the program that gets replaced by a call to this trampoline.

The main challenge is in selecting the memory blocks for locking so as to minimize the WCET. In the following, we propose two solutions. The first one is an optimal solution employing Integer Linear Programming (ILP) formulation based on concrete cache states and the second one is a heuristic approach based on abstract cache states.

4.1 Optimal solution with concrete cache states

The set of concrete cache states at any program point captures the exact set of cache states resulting from all possible program paths. Based on this accurate set of cache states, we formulate an ILP problem to optimally select the memory blocks for partial locking. In the following, we first show the ILP formulation for a loop and then extend it to the whole program.

4.1.1 ILP Formulation for Loop

We represent the loop body as a Directed Acyclic Graph (DAG). Each DAG is associated with a unique source and sink node. We compute the set of possible concrete cache states $\mathcal{P}$ at any point of the program through static program analysis [11]. Given the set of all possible cache states $\mathcal{P}$ and a memory block access m, $p\_hit(\mathcal{P}, m)$ determines whether the access is a cache hit or miss before locking. Next, we proceed to determine the cache access behavior of m after locking.

For each memory block m, we define a 0-1 decision variable L_m, which indicates whether m is locked in the cache. Thus,

$$0 \le L_m \le 1$$

There are only A (associativity) cache lines available for locking in each cache set. Thus for each cache set i

$$\sum_{m \in M_i} L_m \le A$$

where M_i is the set of memory blocks mapped to cache set i.

The accesses to the locked memory blocks are cache hits. Let $Lock_i$ denote the set of memory blocks locked in cache set i. For an unlocked memory block m mapped to cache set i ($m \in M_i, m \notin Lock_i$), its access can be classified as hit or miss depending on the concrete cache states $\mathcal{P}$ at that program point and $Lock_i$.

For a concrete cache state $c \in \mathcal{P}$, we define age_m^c as the age of the memory block m in c, where $age_m^c = 0$ ($age_m^c = A - 1$) if m is the most (least) recently accessed memory block in c. If $m \notin c$ ($c\_hit(c, m) = 0$), then $age_m^c = A$. Thus, $0 \le age_m^c \le A$. If $m \in M_i$ and $m \notin Lock_i$, then given a concrete cache state c, the access to m is cache hit if

$$\left(age_m^c + \sum_{m' \in Lock_i \wedge age_{m'}^c > age_m^c} L_{m'} \right) < A \qquad (1)$$

In other words, if a locked memory block $m' \in Lock_i$ is younger than m in the cache state c, then locking m' does not change the hit classification of m. However, if $m' \in Lock_i$ is older than m in cache state c (i.e., $age_{m'}^c > age_m^c$), then locking m' essentially increases age of m by 1. If the number of such older memory blocks added to age_m^c exceeds the associativity, then m becomes a cache miss due to locking.

We define a 0-1 variable h_m^c, which specifies whether m is a cache hit in c after locking. Based on Equation 1,

$$h_m^c = \begin{cases} 1 & if \left(A - age_m^c - \sum_{m' \in Lock_i \wedge age_{m'}^c > age_m^c} L_{m'} \right) > 0 \\ 0 & otherwise \end{cases}$$

However, the above equation is not linear. We substitute it with the equivalent linear equations as follows.

$$A - \sum_{m' \in Lock_i \wedge age_{m'}^c > age_m^c} L_{m'} - age_m^c - U \times h_m^c \le 0$$

$$A - \sum_{m' \in Lock_i \wedge age_{m'}^c > age_m^c} L_{m'} - age_m^c + U - U \times h_m^c > 0$$

where U is a large constant ($U \ge A$).

The set of concrete cache states $\mathcal{P}$ at a program point usually contains more than one concrete cache states ($|\mathcal{P}| > 1$). Memory block access m is guaranteed as cache hit if and only if it is cache hit for every concrete cache state $c \in \mathcal{P}$. We define a 0-1 variable $h_m^\mathcal{P}$, which specifies whether m is a cache hit in $\mathcal{P}$ after locking.

$$h_m^\mathcal{P} = \begin{cases} 1 & if \sum_{c \in \mathcal{P}} h_m^c = |\mathcal{P}| \\ 0 & otherwise \end{cases}$$

We linearize the above equation as follows.

$$\sum_{c \in \mathcal{P}} h_m^c - h_m^\mathcal{P} \le |\mathcal{P}| - 1$$

$$\sum_{c \in \mathcal{P}} h_m^c - |\mathcal{P}| \times h_m^\mathcal{P} \ge 0$$

Finally, for each memory block access m, we define a 0-1 decision variable hit_m, which specifies whether m is cache hit or miss after locking. Locked memory blocks are guaranteed to be cache hits. On the other hand, for an unlocked memory block m, we rely on its corresponding cache state $\mathcal{P}$ to determine the cache behavior.

$$hit_m = \begin{cases} 1 & if\ L_m = 1 \\ h_m^\mathcal{P} & otherwise \end{cases}$$

We linearize the above equation as follows.

$$hit_m \ge L_m,\ hit_m \ge h_m^\mathcal{P}\ and\ hit_m \le L_m + h_m^\mathcal{P}$$

Thus, the access latency of basic block B after cache locking is calculated as follows

$$T_B = \sum_{m \in B} \left(miss\_lat - (miss\_lat - hit\_lat) \times hit_m \right)$$

where $miss\_lat$ and hit_lat are the cache miss penalty and cache hit latency, respectively.

We also define a variable W_B for each basic block B in the loop, which represents the latency of the worst-case path rooted at basic block B in the DAG after cache locking. Then

$$W_B = \max_{B' \in imsucc(B)} \{ W_{B'} + T_B \}$$

where $imsucc(B)$ is the set of immediate successors of B in DAG. Therefore, for any outgoing edge from node B to node B' ($B \to B'$) in the DAG, we have the following constraint

$$W_B \ge W_{B'} + T_B$$

Since there is no outgoing edge for the sink node of the loop, it is defined specially

$$W_{sink} = T_{sink}$$

Obviously, W_{src} will capture the latency of the worst-case acyclic path in the DAG (src is the source node of DAG). Let lb be the loop bound of this loop (maximum number of iterations of this loop). Then, $W_{src} \times lb$ is the WCET of this loop after cache locking. Thus, the optimal cache locking result for this loop can be obtained by minimizing $W_{src} \times lb$ (the objective function of ILP formulation).

415

4.1.2 Extension to the Whole Program

In the previous section, we present an ILP formulation to obtain the optimal cache locking for a loop. In order to obtain the ILP formulation for the whole program, we are required to start from the innermost loops of the program. We first generate the ILP formulation for the innermost loops, and then each innermost loop is treated as a dummy basic block of the outer loop. Therefore, we can construct the ILP formulation for the next level of loop. We continue this way until we reach the outmost loop in the program. Clearly, W_{entry} represents the WCET of the whole program under cache locking, where $entry$ denotes the entry node of program. Finally, the locking overhead (e.g., the execution of the locking instructions) are included in the WCET of the whole program.

4.2 Heuristic with abstract cache states

In the previous section, we develop an optimal ILP formulation using concrete cache states. However, programs with complex control flow may have hundreds or even thousands of cache states at a program point. For such programs, maintaining all possible concrete cache states may not be feasible. Also ILP formulation may take very long to reach a solution specially for larger programs and larger associativity. Thus, we propose a heuristic approach based on abstract cache states. Abstract cache state is a more compact representation compared to the set of concrete cache states.

We first perform WCET analysis with cache modeling based on abstract interpretation [19]. Then we can easily determine cache hit/miss classification for each memory access based on the abstract cache states. As a by-product of the WCET analysis, we obtain the abstract cache states under *must analysis* at all program points. Meanwhile, we also collect the execution frequency of each basic block along the worse-case path. Then we iteratively lock some memory blocks on the worse-case path to improve the WCET.

Suppose memory block m is on the worst-case path. Let lat_m be the access latency of m according to the hit/miss classification in WCET analysis, and f_m be its execution frequency along the worst-case path. By locking memory block m, all accesses to m will be cache hits. Therefore, we define the benefit of locking m as

$$benefit_m = (lat_m - hit\_lat) \times f_m$$

where hit_lat is the cache hit latency. Thus, locking a memory block guaranteed to be hit before locking does not give any benefit.

On the other hand, locking memory block m in cache may have negative impact for the memory blocks mapped to the same set as the associativity for this set is reduced by 1. Similar to concrete cache state, we define the age of a memory block m in abstract must cache state $\mathcal{C}$ as $age_m^{\mathcal{C}}$. When $m \in \mathcal{C}$, $0 \le age_m^{\mathcal{C}} \le A - 1$, where A is the associativity. Otherwise, we set its age to A.

Suppose we choose to lock memory block m in the cache and its $benefit_m > 0$. In other words, m is not in the abstract must cache state before locking. Then, locking m will downgrade the memory block m' from cache hit to cache miss if $age_{m'}^{\mathcal{C}} = A - 1$. Note that the associativity A here refers to the current associativity of the set. That is A refers to the original associativity of the cache minus the number of memory blocks locked in the set so far. Therefore, we define the cost of locking m as follows.

$$cost_m = \sum_{m' \in M_i \wedge age_{m'}^{\mathcal{C}} = A-1} (miss\_lat - hit\_lat) \times f_{m'}$$

where as before $m \in M_i$. Then, the overall gain of locking m is

$$gain_m = benefit_m - cost_m$$

We compare different memory blocks in terms of their gain and select the most beneficial memory block m to be locked. However,

Table 1: Characteristic of benchmarks & analysis time

Benchmarks	Code Size (bytes)	Optimal (sec)	Heuristic (sec)	Speedup
adpcm	12,480	313.37	1.28	245
cnt	1,648	0.43	0.05	9
compress	4,864	145.61	0.33	441
crc	2,048	1.44	0.10	14
edn	7,296	1.07	0.16	7
fir	1,152	0.10	0.02	5
jfdctint	5,520	0.35	0.06	6
matmult	1,632	0.37	0.07	5
minver	6,256	114.20	0.35	326
qurt	2,048	1.20	0.13	9

$gain_m$ may not be the actual WCET reduction because the worst-case path may change after locking m. Thus, we update cache state for instructions mapped to the affected cache set and perform WCET analysis again to obtain the exact WCET after locking m. If the WCET is actually reduced, we lock m in the cache. We continually select memory blocks for locking until either there is no actual WCET improvement after locking any memory block or there is no gain in the cost-benefit analysis for any memory block m (i.e., $gain_m \le 0$). Finally, the locking overhead is included. The detail algorithm is shown in Appendix A.

5. EXPERIMENTAL EVALUATION

In this section, we present the experimental evaluation of partial cache locking. We compare both the optimal and the heuristic solutions with static cache analysis [19] and the full cache locking approach proposed by Falk et al. [6].

5.1 Experimental Setup

We use the benchmarks from MRTC benchmark suite [8] as shown in Table 1. We compile our benchmarks for SimpleScalar PISA (Portable ISA) instruction set [3] — a MIPS like instruction set architecture — with gcc cross-compiler. The control flow graphs of these benchmarks are extracted and provided as input to our cache locking analysis. Our framework is built on top of the open-source WCET analysis tool Chronos [9]. The binary code size of each program is shown in the second column of Table 1. We perform all the experiments on 2.53GHz Intel Xeon CPU with 24GB memory. IBM CPLEX is used as the ILP solver.

We assume only one level of instruction cache in the architecture. In other words, an instruction access is either cache hit or it has to be fetched from memory. The cache hit latency is 1 cycle, while a cache miss takes 30 cycles. As we are modeling the instruction cache, we assume a simple in-order processor with unit-latency for all data memory references.

5.2 Partial Cache Locking vs. Static Analysis

Figure 3 shows the WCET improvement of partial cache locking over static analysis with no locking based on abstract interpretation [19]. The instruction cache is 4-way set associative with block size of 32 bytes, and its capacity is varied from 512B to 1KB (see Appendix for other settings).

Our partial cache locking technique significantly improves the WCET over static analysis with no locking for many benchmarks (e.g., cnt, crc and $qurt$) for different cache sizes. However, some benchmarks show limited improvement of WCET via partial cache locking, especially when the cache size is small. This is mainly due to the fact that locking memory blocks destroys the deterministic access pattern for some unlocked blocks. Therefore, our partial locking technique decides not to lock these memory blocks and the result of partial locking is close to that of static analysis.

416

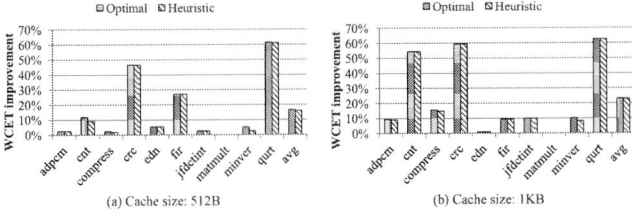

(a) Cache size: 512B (b) Cache size: 1KB

Figure 3: WCET improvement of partial cache locking (optimal and heuristic solution) over static cache analysis with no locking (cache: 4-way set associative, 32-byte block).

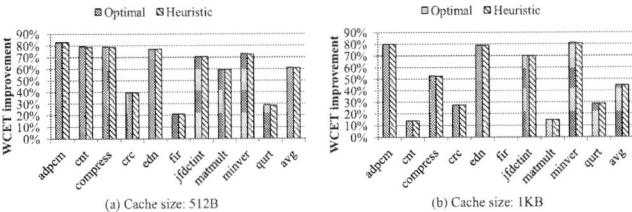

(a) Cache size: 512B (b) Cache size: 1KB

Figure 4: WCET improvement of partial cache locking (optimal and heuristic solution) over Falk et al.'s method (cache: 4-way set associative, 32-byte block).

For most of the benchmarks, the improvement increases as the cache size increases, because there is more space for locking and more memory blocks can be locked into the cache. However, for some benchmarks, the improvement decreases as cache size increases, for example fir. For fir, when the cache size increases, more memory accesses become deterministic, which can be successfully identified by static cache analysis. Thus, cache locking may not help to improve the WCET much compared to static cache analysis. Overall, more WCET improvement is observed as the cache size increases. On an average, 16% and 23% improvement are achieved for 512B and 1KB size cache, respectively.

5.3 Partial versus Full Cache Locking

There exist two full cache locking techniques as mentioned in Section 2 [14, 6]. Even though Liu et al. [14] show that their approach can achieve better WCET reduction compared to [6], it has several limitations. Liu et al. do not consider the cache mapping function in the locking algorithm. They simply assume that any memory block can be locked in any cache set (as if the cache is a scratchpad memory). After locking decisions are made, they employ code placement techniques that force the locked memory blocks to be mapped to the appropriate cache sets. This can lead to code size blowup, which has not been addressed in their work.

Thus we decide to compare our partial locking results with Falk et al.'s method [6] as both approaches do not require any subsequent code placement technique. We choose memory blocks as locking granularity instead of procedures originally used in Falk et al.'s method for a fair comparison. This choice of granularity does not change the greedy heuristic algorithm proposed in [6]. The instruction cache is 4-way associative with 32-byte blocks and capacity varying from 512B to 1KB (see Appendix for other settings).

The WCET improvement of partial cache locking over Falk et al.'s method is shown in Figure 4. Both optimal and heuristic partial locking approaches outperform Falk et al.'s method for different cache sizes. Our partial cache locking techniques usually lock part of the cache (see Appendix for detailed results). Thus, after locking, there are still some cache lines left for the unlocked memory blocks to exploit their locality of accesses. However, in Falk et al.'s method, the cache is fully locked and all the accesses to the

unlocked memory blocks are cache misses. On an average, partial cache locking improves the WCET by 61% and 45% over full cache locking for 512B and 1KB caches, respectively.

5.4 Optimal vs. Heuristic Approach

As shown in Figure 3 and 4, our heuristic approach obtains nearly the same results as the optimal solution. Table 1 presents the average analysis time of different algorithms for all the benchmarks. Clearly, our heuristic approach produces comparable results to the optimal solution while it is more efficient in analysis time.

6. CONCLUSION

In this paper, we propose partial cache locking for WCET reduction. We have proposed an optimal partial locking solution based on concrete cache states as well as a heuristic approach based on abstract cache states. Our partial cache locking significantly reduces the WCET compared to the static cache analysis and the state-of-the-art cache locking techniques that fully lock the cache. Our heuristic achieves comparable WCET reduction to the optimal solution but it is more efficient in terms of runtime. In the future, we will consider data cache locking and explore dynamic cache locking for further improvement.

7. ACKNOWLEDGMENTS

This work was partially supported by Singapore Ministry of Education Academic Research Fund Tier 2, MOE2009-T2-1-033.

8. REFERENCES

[1] A. Arnaud and I. Puaut. Dynamic instruction cache locking in hard real-time systems. In *RTNS*, 2006.

[2] B. Buck and J. K. Hollingsworth. An API for runtime code patching. *Int. J. High Perform. Comput. Appl.*, 14(4), 2000.

[3] D. Burger and T. M. Austin. The simplescalar tool set, version 2.0. *SIGARCH Comput. Archit. News*, 25(3), 1997.

[4] A. M. Campoy, I. Puaut, A. P. Ivars, and J. V. B. Mataix. Cache contents selection for statically-locked instruction caches: An algorithm comparison. In *ECRTS*, 2005.

[5] H. Falk and J. C. Kleinsorge. Optimal static WCET-aware scratchpad allocation of program code. In *DAC*, 2009.

[6] H. Falk, S. Plazar, and H. Theiling. Compile-time decided instruction cache locking using worst-case execution paths. In *CODES+ISSS*, 2007.

[7] C. Guillon, F. Rastello, T. Bidault, and F. Bouchez. Procedure placement using temporal-ordering information: Dealing with code size expansion. *J. Embedded Comput.*, 1(4), 2005.

[8] J. Gustafsson, A. Betts, A. Ermedahl, and B. Lisper. The Mälardalen WCET benchmarks – past, present and future. In *WCET*, 2010.

[9] X. Li, Y. Liang, T. Mitra, and A. Roychoudury. Chronos: A timing analyzer for embedded software. *Science of Computer Programming*, 69(1-3), 2007.

[10] Y.-T. S. Li, S. Malik, and A. Wolfe. Cache modeling for real-time software: beyond direct mapped instruction caches. In *RTSS*, 1996.

[11] Y. Liang and T. Mitra. Cache modeling in probabilistic execution time analysis. In *DAC*, 2008.

[12] Y. Liang and T. Mitra. Improved procedure placement for set associative caches. In *CASES*, 2010.

[13] Y. Liang and T. Mitra. Instruction cache locking using temporal reuse profile. In *DAC*, 2010.

[14] T. Liu, M. Li, and C. J. Xue. Minimizing WCET for real-time embedded systems via static instruction cache locking. In *RTAS*, 2009.

[15] F. Martin, M. Alt, R. Wilhelm, and C. Ferdinand. Analysis of loops. In *CC*, 1998.

[16] I. Puaut and D. Decotigny. Low-complexity algorithms for static cache locking in multitasking hard real-time systems. In *RTSS*, 2002.

[17] V. Suhendra and T. Mitra. Exploring locking & partitioning for predictable shared caches on multi-cores. In *DAC*, 2008.

[18] V. Suhendra, T. Mitra, A. Roychoudhury, and T. Chen. WCET centric data allocation to scratchpad memory. In *RTSS*, 2005.

[19] H. Theiling, C. Ferdinand, and R. Wilhelm. Fast and precise WCET prediction by separated cache and path analyses. *Real-Time Syst.*, 18(2/3), 2000.

[20] X. Vera, B. Lisper, and J. Xue. Data cache locking for higher program predictability. *SIGMETRICS Perform. Eval. Rev.*, 31(1), 2003.

APPENDIX

A. HEURISTIC ALGORITHM

Algorithm 1 presents the details of our heuristic approach described in Section 4.2. This approach is based on abstract cache states. The algorithm iteratively selects the most beneficial memory block for locking based on the cost and benefit metrics defined in Section 4.2. This process continues until there is no WCET improvement with further locking.

Algorithm 1: Heuristic with abstract cache states

Input: Cache configuration cfg and binary executable $prog$
Output: Set of locked memory blocks $lock\_set$ and WCET after locking $wcet$

```
 1  begin
 2  |   stop_locking := false; lock_set := null;
 3  |   analyze_abstract_cache_states(prog, cfg);
 4  |   wcet := analyze_wcet();
 5  |   while (!stop_locking) do
 6  |   |   /* select candidate memory block to lock */
    |   |   cnd := null; gain_cnd := 0;
 7  |   |   foreach m ∈ M do
 8  |   |   |   Suppose m is mapped to cache set s;
 9  |   |   |   Let assoc be the current associativity of s;
10  |   |   |   if (m ∉ lock_set ∧ assoc > 0) then
11  |   |   |   |   benefit_m := calculate_benefit();
12  |   |   |   |   cost_m := calculate_cost(assoc);
13  |   |   |   |   gain_m := benefit_m - cost_m;
14  |   |   |   |   if gain_m > gain_cnd then
15  |   |   |   |   |   cnd := m; gain_cnd := gain_m;
16  |   |   |
17  |   |
18  |   |   if cnd ≠ null then
19  |   |   |   lock_to_cache(cnd);
20  |   |   |   /* update cache states for affected cache set */
    |   |   |   update_cache_state(prog, cfg, cnd);
21  |   |   |   new_wcet := analyze_wcet();
22  |   |   |   if new_wcet < wcet then
23  |   |   |   |   wcet := new_wcet;
24  |   |   |   |   lock_set := lock_set ∪ cnd;
25  |   |   |   |   update associativity for the affected cache set;
26  |   |   |   else
27  |   |   |   |   stop_locking := true;
28  |   |
29  |   |   else
30  |   |   |   stop_locking := true;
31  |   |
32  |
33  end
```

The input to the algorithm is the cache configuration cfg and the binary executable $prog$. First, we perform cache modeling based on abstract interpretation [19] for this binary executable (line 3). The output of this analysis are the abstract cache states at each program point. Next we perform WCET analysis of the binary executable (line 4) where memory accesses are categorized into always hit, always miss, and unclassified based on abstract cache states. The $wcet$ obtained in this step is the original WCET obtained through static cache analysis and no cache locking.

Now, we iteratively select the most beneficial memory block for locking into the cache. Let M be the set of all memory blocks. We perform cost-benefit analysis for each memory block $m \in M$ where m is not yet locked ($m \notin lock\_set$) and the cache

set s where m is mapped to still has some unlocked cache lines ($assoc > 0$). We gain benefit from locking m if m was not guaranteed to be a hit after static cache analysis (see Section 4.2). However, there is a cost associated with locking m. The other memory blocks mapped to cache set s but not yet locked will have one less cache block available in the cache set s. As discussed in Section 4.2, some of these blocks now may incur cache miss (even though their accesses were hits under static cache analysis) depending on their relative age with respect to the age of m in cache set s. The additional latency incurred due to these cache misses will contribute to the cost of locking m. The difference between benefit and cost of locking m is the gain. We identify the memory block cnd with maximum gain.

If we cannot identify any memory block with positive gain, then the locking algorithm terminates. Note that the cost-benefit analysis is approximate in nature because it depends on the frequency of memory accesses along the worst-case path before locking memory block cnd. After locking memory block cnd, the worst-case path may change. So we update the abstract cache states for the cache set where cnd is mapped to and repeat WCET analysis with this new abstract cache states. If the new WCET is indeed lower than the previous WCET, then we add the memory block cnd to $lock\_set$. We also need to decrease the associativity of the corresponding cache set. If the actual WCET after locking m is not lower than the previous WCET, then we terminate the algorithm.

B. EVALUATION WITH DIFFERENT CACHE CONFIGURATIONS

In this section, we perform an extensive comparison of our partial cache locking solutions (optimal and heuristic) with static cache analysis (no locking) [19] and Falk et al.'s method (full cache locking) [6] for various cache configurations.

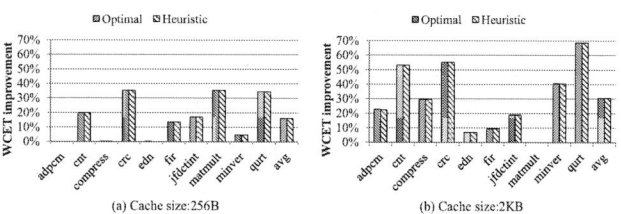

Figure 5: Comparison between partial cache locking and static cache analysis with no locking for cache size of 256B and 2KB (cache:4-way set associative, 32-byte block).

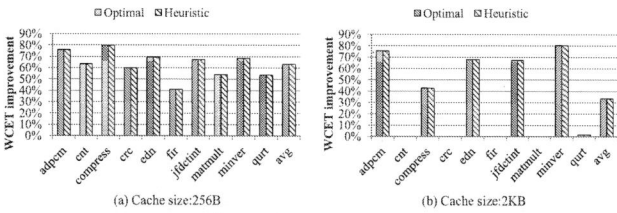

Figure 6: Comparison between partial cache locking and Falk et al.'s method for cache size of 256B and 2KB (cache:4-way set associative, 32-byte block).

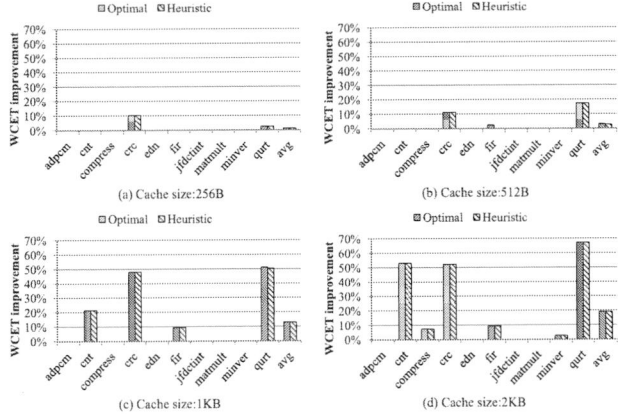

Figure 7: WCET improvement of partial cache locking over static cache analysis (no locking) for direct mapped cache, 32-byte block.

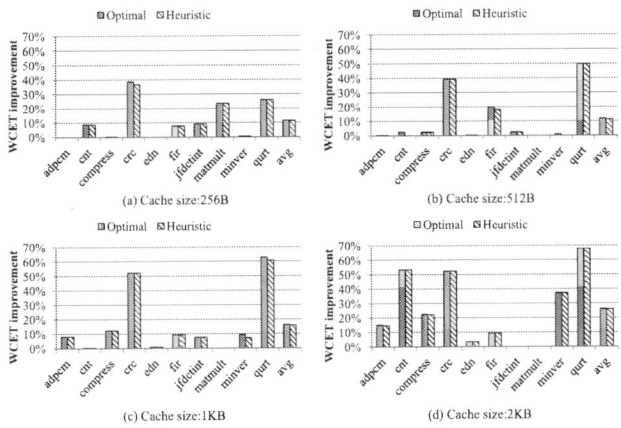

Figure 8: WCET improvement of partial cache locking over static cache analysis (no locking) for 2-way set-associative cache, 32-byte block.

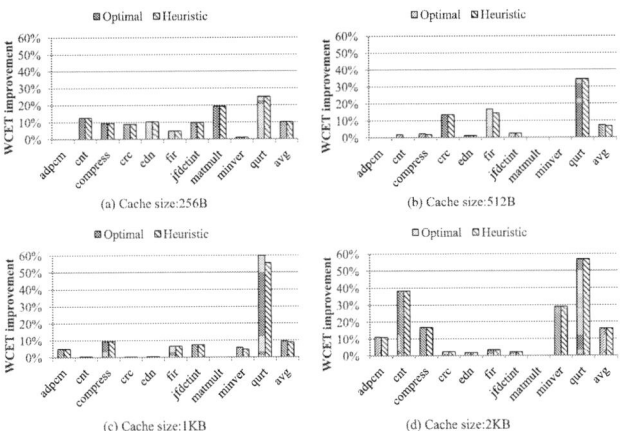

Figure 9: WCET improvement of partial cache locking over static cache analysis (no locking) for 2-way set-associative cache, 64-byte block.

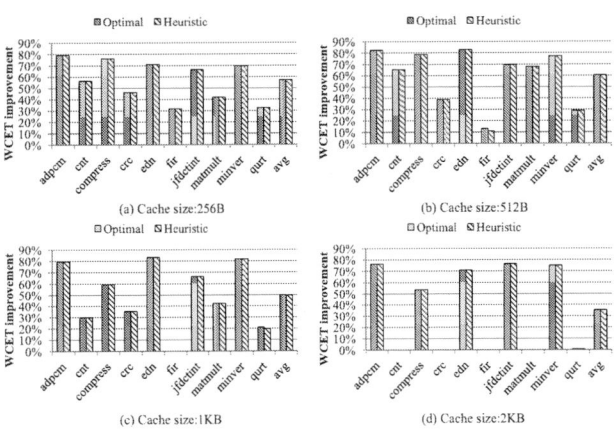

Figure 10: WCET improvement of partial cache locking over Falk et al.'s method (full locking) for direct mapped cache, 32-byte block.

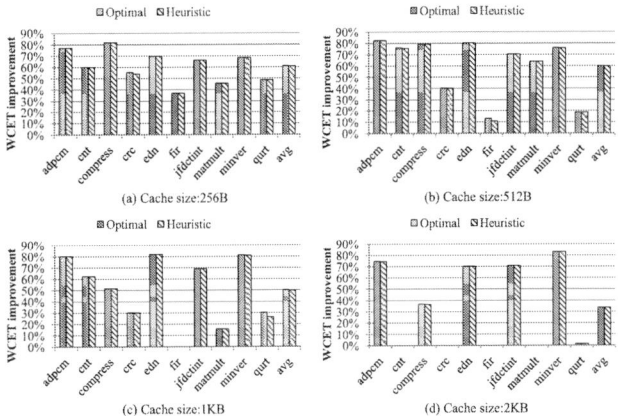

Figure 11: WCET improvement of partial cache locking over Falk et al.'s method (full locking) for 2-way set-associative cache, 32-byte block.

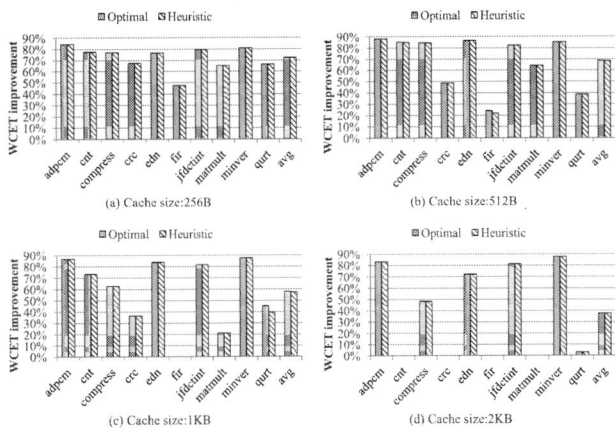

Figure 12: WCET improvement of partial cache locking over Falk et al.'s method (full locking) for 2-way set-associative cache, 64-byte block.

419

B.1 Different Cache Sizes

In this subsection, we present the evaluation results for 256B and 2KB size caches, respectively. The block size (32-byte) and associativity (4-way) are the same as the cache configuration presented in Section 5. Figure 5 shows the WCET improvement of partial cache locking over static cache analysis with no locking and Figure 6 shows the WCET improvement of partial cache locking over full locking (Falk et al.'s method). As expected, WCET improvement of partial cache locking over static analysis is much higher with bigger cache size as more space is available for locking memory blocks. As compared to Falk et al.'s method, when the cache is large enough to hold the entire program, all the memory blocks can be locked to achieve the minimum WCET. In that scenario, partial and full cache locking obtain identical solutions for some benchmarks (e.g., *cnt*, *crc*, *fir*, and *matmult*).

B.2 Different Associativity

In this subsection, we evaluate our partial cache locking for different cache associativity values. In Section 5, 4-way cache associativity results are presented. Here we show the results of direct mapped and 2-way set associative caches, while the block size remains constant at 32 bytes. Figure 7 and 8 present the improvement of partial cache locking over static cache analysis with no locking for direct mapped cache and 2-way set associative cache, respectively. Figure 10 and 11 present the improvement over full locking (Falk et al.'s method) for different cache associativity.

It is observed that the WCET improvement of direct mapped cache is not as good as that of 2-way and 4-way set associative cache, especially when the cache size is small. For direct mapped cache, there is only one cache line available in each set. Locking a memory block in a cache set implies that all the accesses to the other memory blocks in the cache set will be cache miss. Thus our partial cache locking method decides not to lock any memory block for most of the benchmarks. Therefore, the partial cache locking results are similar to that of static cache analysis with no locking, especially when the cache size is small. 2-way set associative caches provide more opportunities for partial cache locking. Thus, more WCET improvement is achieved compared to direct mapped cache. Finally, partial cache locking outperforms full locking (Falk et al.'s method) for different associativity.

B.3 Different Block Sizes

In this section, we evaluate our partial cache locking for different block size. In Section 5, we present results for 32 byte block size. Here we evaluate the benefits of partial cache locking for 64 bytes block size. Figure 9 and 12 present the WCET improvement with partial cache locking over static cache analysis with no locking and full locking (Falk et al.'s method), respectively. As shown, our partial locking still achieves significant improvement.

C. PERCENTAGE OF LINES LOCKED

As mentioned before, the main strength of partial cache locking lies in the fact that cache lines are locked judiciously after performing careful cost-benefit analysis. If it is beneficial to keep a cache line unlocked so that multiple memory blocks can benefit from it, partial cache locking can identify such situations. In this subsection, we present the cache locking solutions derived by our partial cache locking mechanisms (optimal and heuristic) for the cache configurations in Section 5. That is, the cache is 4-way set associative with 32-byte block size, and its capacity is varied from 512B to 1KB. Table 2 presents the percentage of lines locked in the cache for different cache configurations. As we can observe, for all the benchmarks, our partial cache locking algorithms (optimal and

Table 2: Percentage of lines locked in cache (cache: 4-way set associative, 32-byte block).

Benchmarks	512B cache (%)		1KB cache (%)	
	optimal	heuristic	optimal	heuristic
adpcm	25.00	25.00	56.25	56.25
cnt	25.00	25.00	65.63	75.00
compress	43.75	43.75	59.38	68.75
crc	68.75	68.75	71.88	71.88
edn	18.75	18.75	12.50	37.50
fir	75.00	75.00	40.63	87.50
jfdctint	50.00	75.00	75.00	75.00
matmult	18.75	18.75	40.63	46.88
minver	12.50	18.75	25.00	28.13
qurt	81.25	75.00	75.00	75.00

heuristic) lock only a fraction of the cache lines. The percentage of lines locked is generally lower when the cache size is small as the unlocked memory blocks need the remaining cache lines. As the cache size increases, partial cache locking chooses to lock higher percentage of cache lines. These results clearly confirm that partial cache locking is indeed important to minimize WCET compared to the two extreme ends of the spectrum of choices, namely, full cache locking and no cache locking.

Worst-Case Execution Time Analysis for Parallel Run-Time Monitoring

Daniel Lo and G. Edward Suh
Cornell University
Ithaca, New York, USA
{dl575,gs272}@cornell.edu

ABSTRACT

The increasing safety-critical role of real-time systems requires increased attention to their security and reliability. Several recent studies have shown that parallel run-time monitoring of programs can significantly improve the security and reliability of computing systems. However, these techniques cannot be applied to real-time systems without first estimating their impact on worst-case execution time (WCET). In this paper, we present a method for determining the impact of parallel monitoring on WCET using a mixed integer linear programming (MILP) formulation. We use our method to estimate the WCET for seven benchmark programs and two possible monitoring techniques. This estimate is compared against observed execution times from simulation and an upper bound based on sequential monitoring. The results show that our method estimates a WCET within 71% of worst-case observed execution times and up to 74% lower than the sequential bound.

Categories and Subject Descriptors

C.3 [**Special-purpose and Application-based Systems**]: Real-time and embedded systems; C.4 [**Performance of Systems**]: Modeling techniques

General Terms

Measurement, Performance, Reliability, Security

Keywords

WCET analysis, run-time monitoring, real-time systems

1. INTRODUCTION

Embedded real-time systems are becoming increasingly prevalent as we deeply integrate computing devices into the physical world. For example, many mechanical systems including automobiles and planes are now electronically controlled by computers. Because electronic systems can provide more intelligent control and coordination through networks, such cyber-physical integration is expanding into even more systems including buildings, medical systems, and power grids. Secure and reliable computation is critical in these systems because a malfunction may cause physical damage or loss of life.

In this context, recent studies have shown that parallel monitoring of run-time program behavior can significantly improve the security and reliability of a computing system with minimal overheads. As an example, Dynamic Information Flow Tracking (DIFT) is a recently proposed security technique that tracks and restricts the use of untrusted I/O inputs, and has been shown to be able to effectively detect a large class of common software attacks [17]. While software DIFT on a single core can incur a significant slowdown even with optimizations (3.6x on average) [14], parallel DIFT on multiple processing modules can reduce overheads to tens of percents on average [4]. Similarly, run-time monitoring can enable many new capabilities such as fine-grained memory protection [19], array bound checks [5], hardware error detection [11], etc.

However, today's parallel monitoring techniques cannot be easily applied to critical real-time systems due to their lack of timing guarantees. The development of a safety-critical real-time system requires an estimate of the worst-case execution time (WCET) of each task in order to ensure that tasks meet the system's real-time deadlines. Yet, previous studies on parallel monitoring have only focused on average slowdowns through simulations with no worst-case guarantee. Unfortunately, estimation of the worst-case performance overhead of parallel monitoring is not straightforward because of its loosely coupled nature. In the best case, the monitoring happens in parallel to the main task and does not cause any slowdown. However, parts of the main task with heavy monitoring may be required to slow down in order to allow the parallel monitor to keep up.

In this paper, we present a method for estimating the increase in WCET of programs running on a system with parallel monitoring. We first investigate how to mathematically model the loosely-coupled relationship between the main processing core and parallel monitoring hardware, which are often connected through a FIFO buffer with a fixed number of entries. The resulting model is non-linear but can be transformed into a mixed integer linear programming (MILP) formulation. The MILP formulation produces the maximum number of cycles for each basic block that the main core may be stalled due to monitoring. These monitoring stalls can be incorporated into popular WCET analysis methods based on implicit path enumeration techniques (IPET) [9] which use integer linear programming (ILP).

We evaluate the effectiveness of the proposed method by

comparing its WCET estimates with a conservative estimate for sequential monitoring, simulation results, and the WCET without monitoring. The experiments use the Mälarden WCET benchmark suite [7] and two monitoring techniques: uninitialized memory check (UMC) and control flow protection (CFP). The results indicate that our WCET formulation can provide a bound that is up to 74% lower than a straightforward estimate from a sequential formulation. These bounds are within 71% of observed worst-case run times from simulations for the selected benchmarks. This is similar to the results when no monitoring is present which show up to a 52% difference between simulations and WCET estimates. As a result, the proposed WCET estimation method enables parallel monitoring techniques to be applied to hard real-time systems with worst-case timing guarantees without being excessively conservative.

This paper is organized as follows. Section 2 discusses related work. Section 3 describes the parallel monitoring architecture that is modeled in this paper. Section 4 develops the MILP formulation for WCET analysis and Section 5 presents experimental evaluation results. Finally, Section 6 concludes the paper.

2. RELATED WORK

This paper aims to enable parallel monitoring techniques on real-time systems by providing a general WCET analysis framework that can be applied to a broad range of monitoring techniques. While there exist many parallel monitoring schemes where our WCET analysis can be applied to, we briefly discuss some recent parallel platforms here as examples. For example, INDRA [15] uses a checker core to monitor coarse-grained events on a computation core such as function call/return, code origin inspection, and control flow inspection. Nagarajan et al. studied implementing DIFT on multi-cores [12]. Chen et al. proposed hardware acceleration techniques for multi-core systems and showed that a set of parallel monitoring techniques for security and software debugging can be realized with low performance overheads (tens of percents) [4]. FlexCore [6] shows that parallel monitoring can be made even more efficient by utilizing heterogeneous accelerators implemented on FPGA fabric. These previous studies demonstrate that parallel monitoring can significantly improve system security and reliability with minimal overheads.

Estimating the worst-case execution time of a sequential program on a single-core system is a well studied problem. A survey paper by Wilhelm et al. [18] provides an overview of existing methods and tools in this context. However, to the best of our knowledge, this paper represents the first study on the WCET of parallel monitoring. Researchers have recently started studying the WCET problem for multi-core systems. For example, Paolieri et al. proposed a multi-core hardware architecture for hard real-time systems and analyzed its WCET behavior [13]. McAiT is a tool that has been developed for WCET analysis of multi-core real-time software [10]. These studies focused on the contention between parallel programs for shared resources such as memory. However, the loosely coupled link between the main core and parallel monitoring hardware represents a producer-consumer relationship rather than shared resources. Thus, we found that previously developed techniques were not directly applicable or easily adaptable to provide a tight WCET bound for a system with parallel monitoring.

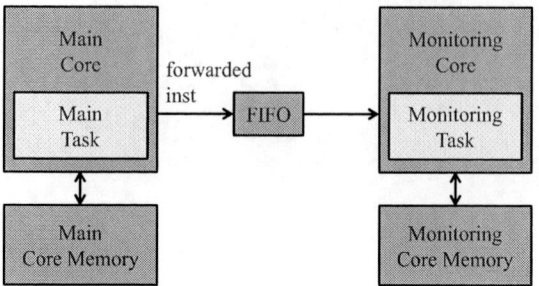

Figure 1: Parallel monitoring architecture model.

3. ARCHITECTURE MODEL

Figure 1 shows the model of run-time parallel monitoring architecture that is assumed in this paper. The architecture consists of two parallel processing elements, main and monitoring cores, which are loosely coupled with a FIFO buffer. The main core runs a computation task, called the *main task*, which performs the original function of the real-time system. The monitoring core receives a trace of certain main task instructions through a FIFO, and performs a *monitoring task* in parallel. We refer to the main task instructions that need to be sent to the monitoring core as *forwarded instructions* or *monitored instructions*. The forwarded instructions are determined based on a particular monitoring technique, and are often sent by the main core transparently without explicit instructions added to the main task. The FIFO allows the monitoring core to operate in a decoupled manner by buffering forwarded instructions. However, if the FIFO is full, the main core needs to wait on a forwarded instruction until a FIFO entry becomes available. We refer to these stalls of the main core due to monitoring as *monitoring stalls* and the number of cycles stalled as *monitoring stall cycles*. The forwarded instruction triggers the monitoring core to execute a series of monitoring instructions.

There are many possible monitoring techniques that can be implemented on this monitoring architecture. One example is to use the monitoring core to detect a software bug that reads a memory location before writing a valid value. We call this monitor an uninitialized memory check (UMC). In UMC, the main core forwards load and store instructions to the monitoring core once they happen in the main task. On a store, the monitoring task sets a tag bit corresponding to the store's memory location, indicating that the location has been initialized. On a load, the monitoring task checks the tag bit, and raises an exception if the bit is not set.

The analysis in this paper focuses on the interaction between the main core and the monitoring core through a FIFO with an assumption that each core has its own memory, as shown in Figure 1. Therefore, there is no interference between the two cores on memory accesses. This configuration applies for typical multi-core embedded microprocessors or a small monitor with a dedicated memory that is attached to a large core. The main core is assumed to not exhibit timing anomalies. This is required so that the worst-case monitoring stall cycles can be assumed to produce the WCET on the main core. This paper does not make any other assumptions on the microarchitecture of each processing core. However, we assume that the WCET of a main task and a monitoring task on the given processing cores can be estimated individually using traditional WCET techniques.

4. WCET ANALYSIS

This section presents our method for estimating the impact of parallel run-time monitoring on WCET. We first review the traditional ILP-based analysis for a sequential program execution, and show that this analysis can be extended to incorporate the overhead of monitoring if the worst-case increase in execution time can be estimated for each basic block. Then, we discuss how to estimate the worst-case monitoring stalls, which happen when the FIFO is full and cannot take a forwarded instruction. We start this discussion with a simple yet rather conservative bound based on the case when a monitoring task always stalls the main task (Section 4.2). Then, we show how the FIFO decoupling can be modeled analytically (Section 4.3), and formulated using MILP (Section 4.4) to create a tighter WCET bound.

4.1 Implicit Path Enumeration

Most of the WCET analysis techniques today rely on an ILP formulation that is obtained from implicit path enumeration techniques [9]. In this method, a program is converted to a control flow graph (CFG). From the control flow graph, an ILP problem is formulated that seeks to maximize

$$t = \sum_{B \in \mathcal{B}_{CFG}} N_B \cdot c_{B,max}$$

where $\mathcal{B}_{CFG}$ is the set of basic blocks in the control flow graph. N_B is the number of times block B is executed and $c_{B,max}$ is the maximum number of cycles to execute block B. The maximum value of t is the WCET of the task. To account for the fact that only certain paths in the graph will be executed, a set of constraints are placed on N_B. For example, on a branch, only one of the branches will be taken on each execution of the block. A variable can be assigned to each edge corresponding to the number of times that edge is taken. The number of times edges out of the block are taken must equal the number of times the block is executed. Similarly, the number of times edges into the block are taken must equal the number of times the block is executed. Various methods have been developed to create additional constraints to convey other program behavior [9, 18].

Integer linear programming is an attractive optimization technique for this problem because the solution found is a global optimum. In addition, many aspects of program and architecture behavior can be described by adding constraints to the ILP problem. Several open source and commercial ILP solvers exist which can solve the formulated ILP problem. Thus, in developing a method for estimating the WCET of parallel run-time monitoring, we look to build upon this ILP framework.

The IPET-based ILP formulation can be extended in a straightforward fashion to incorporate run-time monitoring overheads if we have the maximum (worst-case) monitoring stall cycles for each basic block by maximizing

$$t = \sum_{B \in \mathcal{B}_{CFG}} N_B \cdot (c_{B,max} + s_{B,max})$$

Here, $s_{B,max}$ represents the maximum number of cycles that block B is stalled due to monitoring. In this sense, the challenge in WCET analysis with monitoring lies in determining $s_{B,max}$. The rest of this section addresses this problem.

4.2 Sequential Monitoring Bound

One way to determine a conservative bound on the worst-case monitoring stall cycles is to consider sequential moni-

toring. In sequential monitoring, the monitoring task is run in-line with the main task on the same core rather than in parallel. That is, after each instruction that would be forwarded, the monitoring task is run on the main core before the main task resumes execution. In this case, the WCET estimate can be obtained from a traditional method by analyzing one program that contains both main and monitoring tasks. The resulting WCET can be considered as a simple bound for parallel monitoring because it models the case where every forwarded instruction causes the main core to stall. However, this bound is extremely conservative as it does not account for the FIFO buffering or the parallel execution of the monitoring core. These features are critical to utilizing run-time monitoring techniques while maintaining low performance overheads.

4.3 FIFO Model

To obtain tighter WCET bounds, we need to model the FIFO. The main task can continue its execution as long as a FIFO entry is available, but needs to stall on a forwarded instruction if the FIFO is full. The WCET model needs to capture the worst-case (maximum) number of entries in the FIFO at each forwarded instruction and determine how many cycles the main task may be stalled due to the FIFO being full. Here, we propose a mathematical model to express the load in the FIFO and estimate the worst-case stalls.

In this approach, the original control flow graph must be transformed so that each node contains at most one forwarded instruction which is located at the end of the code sequence represented by the node. This transformed graph is called a *monitoring flow graph (MFG)*. Intuitively, the analysis needs to consider one forwarded instruction at a time in order to model the FIFO state on each forwarded instruction and capture all potential stalls from monitoring.

To model how full the FIFO is, we define the concept of *monitoring load*. The monitoring load is the number of cycles required for the monitoring core to process all outstanding entries in the FIFO at a given point in time. The monitoring load increases when a new instruction is forwarded by the main task, and decreases as the monitoring core processes forwarded instructions. For simplicity, the increase in monitoring load for any forwarded instruction is conservatively assumed to be the worst-case (maximum) monitoring task execution time among all possible forwarded instructions. This maximum, $t_{M,max}$, can be obtained from the WCET analysis of the monitoring tasks. We make this simplification because it is difficult to model the FIFO mathematically at an entry-by-entry level. With this simplification, each FIFO entry is identical and so the monitoring load fully represents the state of the FIFO. The monitoring load cannot be negative and is upper-bounded by the maximum monitoring load the FIFO can handle, l_{max}. The maximum monitoring load is the number of FIFO entries, n_F, multiplied by the increase in monitoring load for one forwarded instruction, $t_{M,max}$.

In our context, we need to determine the worst-case (maximum) monitoring load at the node boundaries in the MFG. For a given node, M, in the MFG, we define li_M as the monitoring load coming into the node and lo_M as the monitoring load exiting the node. The change in monitoring load for the node is denoted by Δl_M. The maximum Δl_M can be calculated as the difference between the WCET of a monitoring task that corresponds to M and the minimum execution cycles of the node, $c_{M,min}$:

$$\Delta l_M = \begin{cases} t_{M,max} - c_{M,min}, & \text{forwarded inst.} \in M \\ -c_{M,min}, & \text{no forwarded inst.} \in M \end{cases}$$

In order to ensure that the analysis is conservative in estimating the worst-case (maximum) stalls, we use the best-case (minimum) execution time for the main task here.

Because the monitoring load is bounded by zero and the maximum load that the FIFO can handle, l_{max}, the monitoring load coming out of a node is

$$lo_M = \begin{cases} 0, & li_M + \Delta l_M < 0 \\ li_M + \Delta l_M, & 0 \le li_M + \Delta l_M \le l_{max} \\ l_{max}, & li_M + \Delta l_M > l_{max} \end{cases}$$

$$l_{max} = n_F \cdot t_{M,max}$$

The worst-case monitoring load entering node M, li_M, is the largest of the output monitoring loads among nodes with edges pointing to node M. Let $\mathcal{M}_{prev}$ represent the set of nodes with edges pointing to node M. Then,

$$li_M = \max_{M_{prev} \in \mathcal{M}_{prev}} lo_{M_{prev}}$$

The above equations describe the worst-case monitoring load at each node boundary. A monitoring stall occurs when a forwarded instruction is executed but there is no empty entry in the FIFO buffer. In terms of monitoring load, if a node would add monitoring load that would cause the resulting total load to exceed l_{max}, then a monitoring stall occurs. The number of cycles stalled, s_M, is the number of cycles that this total exceeds l_{max}. That is,

$$s_M = \begin{cases} 0, & li_M + \Delta l_M < l_{max} \\ (li_M + \Delta l_M) - l_{max}, & li_M + \Delta l_M \ge l_{max} \end{cases}$$

Then, the worst-case monitoring stall cycles for each MFG node can be obtained by maximizing the sum of the s_M across all possible execution paths:

$$\max \sum_{M \in \mathcal{M}_{MFG}} s_M$$

where $\mathcal{M}_{MFG}$ is the set of nodes in the MFG. Once the worst-case stalls for each MFG node is found, the worst-case stalls for a CFG node, $s_{B,max}$, can be computed by simply summing the stalls from the corresponding MFG nodes. We note that since the monitoring load is always conservative in representing the FIFO state, no timing anomalies are exhibited by this analysis. That is, determining the individual worst-case stalls results in the global worst-case stalls.

4.4 MILP Formulation

The proposed FIFO model requires solving an optimization problem to obtain the worst-case stalls, where the input and output monitoring loads, li_M and lo_M, and the monitoring stalls, s_M, need to be determined for each node. Here, we show how the problem can be formulated using MILP. Although the equations for lo_M and s_M are non-linear, they are piecewise linear. Previous work has shown that linear constraints for piecewise linear functions can be formulated using MILP [16]. In the following constraints, all variables are assumed to be lower bounded by zero unless otherwise specified, as is typically assumed for MILP.

First, a set of variables, lo' and s', are created to represent the unbounded versions of lo and s. For readability, the per block subscript M has been omitted.

$$s' = li + \Delta l - l_{max}, \; s' \in (-\infty, \infty)$$
$$lo' = li + \Delta l, \; lo' \in (-\infty, \infty)$$

The following piecewise linear function calculates s from s'.

$$s = f(s') = \begin{cases} 0, & s' < 0 \\ s', & s' \ge 0 \end{cases}$$

This function can be described in MILP using the following set of constraints.

$$a_s \lambda_0 + b_s \lambda_2 = s'$$
$$\lambda_0 + \lambda_1 + \lambda_2 = 1$$
$$\delta_1 + \lambda_2 \le 1$$
$$\delta_2 + \lambda_0 \le 1$$
$$\delta_1 + \delta_2 = 1$$
$$b_s \lambda_2 = s$$

where a_s is chosen to be less than the minimum possible value of s' and b_s is chosen to be greater than the maximum possible value of s'. The choice of a_s and b_s is arbitrary as long as it meets these requirements. λ_i are continuous variables and δ_i are binary variables. In this set of constraints, s' is expressed as a sum of the endpoints of a segment of the piecewise function. The δ_i variables ensure that only the segment corresponding to s' is considered. $\delta_1 = 1$ corresponds to the $s' < 0$ segment of $f(s')$ and $\delta_2 = 1$ corresponds to the $s' \ge 0$ segment of $f(s')$. The λ_i variables represent exactly where s' falls on the domain of that segment. s can be calculated using this information and the values of the function at the segment endpoints.

Similarly, lo can be bound between 0 and l_{max} by using the following set of constraints.

$$a_l \lambda_3 + l_{max} \lambda_5 + b_l \lambda_6 = lo'$$
$$\lambda_3 + \lambda_4 + \lambda_5 + \lambda_6 = 1$$
$$2\delta_3 + \lambda_5 + \lambda_6 \le 2$$
$$2\delta_4 + \lambda_3 + \lambda_6 \le 2$$
$$2\delta_5 + \lambda_3 + \lambda_4 \le 2$$
$$\delta_3 + \delta_4 + \delta_5 = 1$$
$$l_{max} \lambda_5 + l_{max} \lambda_6 = lo$$

As before, a_l and b_l are chosen such that $lo' \in (a_l, b_l)$. Again, λ_i are continuous variables and δ_i are binary variables.

Finally, for each node, the input monitoring load li_M must be determined. li_M depends on the previous nodes, $\mathcal{M}_{prev}$. If there is only one edge into the node, then li_M is simply

$$li_M = lo_{M_{prev}}$$

When there is more than one edge into node M, one set of constraints is used to lower bound li_M by all $lo_{M_{prev}}$.

$$li_M \ge lo_{M_{prev}}, \; \forall M_{prev} \in \mathcal{M}_{prev}$$

Then, another set of constraints upper bounds li_M by the maximum $lo_{M_{prev}}$,

$$li_M - b \cdot \delta_{M_{prev}} \le lo_{M_{prev}}, \; \forall M_{prev} \in \mathcal{M}_{prev}$$
$$\sum_{M_{prev} \in \mathcal{M}_{prev}} \delta_{M_{prev}} = |\mathcal{M}_{prev}| - 1$$

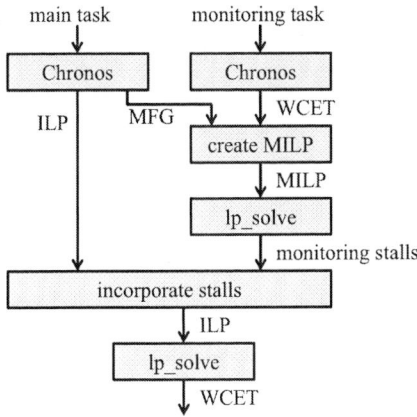

Figure 2: Toolflow for WCET estimation of parallel monitoring.

where b is chosen to be greater than $[\max(lo_{M_{prev}}) - \min(lo_{M_{prev}})]$ and $|\mathcal{M}_{\text{prev}}|$ is the number of nodes with edges pointing to M. δ_i are binary variables. The use of the binary variables δ_i and the second constraint ensure that li_M is only upper bound by one of the $lo_{M_{prev}}$. In order for all constraints to hold, this must be the maximum $lo_{M_{prev}}$. Together with the lower bound constraints, these constraints result in $li_M = \max(lo_{M_{prev}})$. For an example of this complete analysis with numbers, see Appendix A.

5. EVALUATION

5.1 Experimental Setup

Our toolflow for the proposed WCET method is shown in Figure 2. We first use Chronos [8], an open source WCET tool, to estimate the WCET for the main task and the monitoring tasks. We also modified Chronos to produce a MFG of the main task. This MFG and the monitoring task WCET are used to produce an MILP formulation as in Section 4. This MILP problem is solved using lp_solve [3], which produces the worst-case monitoring stall cycles for each forwarded instruction. These monitoring stalls are combined into the ILP formulation that is originally generated for the main task to estimate the overall WCET with parallel run-time monitoring. Although we use Chronos and lp_solve for our implementation, these components can be replaced with any WCET estimation tool and LP solver respectively.

To evaluate the effectiveness of our WCET scheme, we compared its estimate with a simple WCET bound from sequential monitoring (Section 4.2) as well as simulation results using the SimpleScalar tool [2]. In addition to the WCET estimates with monitoring, we also compared the results with the WCET of the main task without monitoring, using both Chronos and simulations.

For the experiments, we configured Chronos and SimpleScalar to model simple processing cores that execute one instruction per cycle for both main and monitoring cores and used an 8-entry FIFO. This configuration represents typical embedded microcontrollers, and is designed to focus on the impact of parallel run-time monitoring by removing complex features such as branch prediction and caches. In the evaluation, we used seven benchmarks from the Mälarden WCET benchmark suite [7] and two monitoring techniques: uninitialized memory checks (UMC) and control flow protection (CFP). UMC detects a software bug that reads memory without a write as briefly explained in Section 3. CFP pro-

tects a program's control flow by checking a target address on each control transfer [1]. In this technique, a compiler determines a set of valid targets for each branch and jump in the main task. This information is stored on the monitoring core. On a branch or jump, the monitoring core ensures that the target is contained in the list of valid targets.

5.2 Results

Table 1 shows the experimental results for each benchmark under different configurations. The first set of rows show the WCET estimate from Chronos (`wcet-none`) and actual run-times from simulations (`sim-none`) without monitoring. The remaining rows show the WCET for the UMC and CFP monitoring extensions. The results are shown for three different approaches: a bound from sequential monitoring (`sequential`), our approach (`wcet`), and simulations (`sim`). The numbers indicate the number of clock cycles. Appendix B includes running times for these experiments.

Table 2 shows relative comparisons between different configurations or WCET methods. The first set of rows compare the WCET estimates from ILP or MILP formulations with the worst-case simulation cycles for each monitoring setup. The results show that the analytical WCET estimates from our proposed scheme are larger than the observed WCET by 0% to 52% for UMC and 0% to 71% for CFP, depending on the main task. This difference is comparable to the case without parallel run-time monitoring, where the analytical WCET from Chronos is larger than simulation results by 0% to 52%. In fact, for `expint`, the majority of the difference is from the WCET estimate of the main task rather than the effects of monitoring. This result suggests that our WCET approach is not significantly more conservative than the baseline WCET tool for the main task.

The second set of rows compare the bound from sequential monitoring and the WCET from our proposed method. For UMC, our approach shows up to a 74% reduction in WCET estimates over the simple bound. Similarly, for CFP, our method shows up to a 73% improvement. These results demonstrate that modeling the FIFO decoupling between the main and monitoring tasks is important for obtaining tight WCET estimates of parallel monitoring.

Finally, the last two rows in Table 2 compare the WCET estimates with and without run-time monitoring. The results show that the increase in WCET varies significantly depending on benchmark and monitoring technique. Benchmarks with infrequent monitoring events (forwarded instructions) show minimal overheads while ones with frequent monitoring can see significant impacts. Also, the benchmarks with large WCET increases differ between UMC and CFP. Therefore, when applying parallel run-time monitoring techniques to real-time systems, a careful WCET analysis for the given tasks and monitoring techniques needs to be performed.

The impact of run-time monitoring on the execution time in our experiments (up to 3.48x in UMC and 2.58x in CFP) is roughly in line with previous studies on multi-cores without any hardware support [4, 12]. The performance overheads will be much lower for multi-cores with optimizations [4] or heterogeneous monitors [6]. Our analysis technique does not depend on any specific monitoring core microarchitecture and is applicable to more optimized architectures.

6. CONCLUSION

Parallel run-time monitoring techniques are an attractive

Monitoring	Experiment	Benchmark						
		cnt	expint	fdct	fibcall	insertsort	matmult	ns
None	wcet-none	64531	3483	1805	245	598	133668	5951
	sim-none	62931	2293	1805	245	598	133668	5951
UMC	sequential-umc	103052	3591	4382	257	2489	357453	10338
	wcet-umc	64550	3498	3035	245	2083	256120	5953
	sim-umc	62931	2297	2564	245	1864	235120	5951
CFP	sequential-cfp	151732	11669	1976	794	1174	231507	18623
	wcet-cfp	93544	8984	1805	547	677	133668	13614
	sim-cfp	72540	5247	1805	382	598	133668	9824

Table 1: Estimated and observed WCET (clock cycles) with and without monitoring.

Ratio			Benchmark							min	max	geomean
			cnt	expint	fdct	fibcall	insertsort	matmult	ns			
wcet-none	:	sim-none	1.03	1.52	1.00	1.00	1.00	1.00	1.00	1.00	1.52	1.07
wcet-umc	:	sim-umc	1.03	1.52	1.18	1.00	1.12	1.09	1.00	1.00	1.52	1.12
wcet-cfp	:	sim-cfp	1.29	1.71	1.00	1.43	1.13	1.00	1.39	1.00	1.71	1.26
sequential-umc	:	wcet-umc	1.60	1.03	1.44	1.05	1.19	1.40	1.74	1.03	1.74	1.33
sequential-cfp	:	wcet-cfp	1.62	1.30	1.09	1.45	1.73	1.73	1.37	1.09	1.73	1.45
wcet-umc	:	wcet-none	1.00	1.00	1.68	1.00	3.48	1.92	1.00	1.00	3.48	1.41
wcet-cfp	:	wcet-none	1.45	2.58	1.00	2.23	1.13	1.00	2.29	1.00	2.58	1.55

Table 2: Ratios comparing results from different experiments.

solution for improving the safety and reliability of future real-time systems. Before these solutions can be applied, the WCET impact of these techniques must be analyzed. In this paper we have presented a method for estimating the WCET for tasks running on a parallel monitoring system. We have shown how the non-linear FIFO behavior can be modeled as an MILP problem to produce the worst-case monitoring stall cycles. These can then be incorporated into traditional IPET methods for WCET estimation. Our evaluation of the method shows significant improvements over an estimate assuming sequential execution of the monitoring. In addition, the amount of overestimation is comparable to the overestimation for a system without parallel monitoring. Appendix C discusses some future directions for this work.

Acknowledgments

This work was partially supported by the National Science Foundation grants CNS-0746913 and CNS-0708788, the Air Force grant FA8750-11-2-0025, the Office of Naval Research grant N00014-11-1-0110, the Army Research Office grant W911NF-11-1-0082, and an equipment donation from Intel.

7. REFERENCES

[1] D. Arora, S. Ravi, A. Raghunathan, and N. K. Jha. Secure Embedded Processing Through Hardware-Assisted Run-Time Monitoring. In *Proceedings of the Conference on Design, Automation and Test in Europe*, 2005.

[2] T. Austin, E. Larson, and D. Ernst. SimpleScalar: An Infrastructure for Computer System Modeling. *IEEE Computer*, 2002.

[3] M. Berkelaar, K. Eikland, and P. Notebaert. lp_solve Version 5.5. http://lpsolve.sourceforge.net/5.5/.

[4] S. Chen, M. Kozuch, T. Strigkos, B. Falsafi, P. Gibbons, T. Mowry, V. Ramachandran, O. Ruwase, M. Ryan, and E. Vlachos. Flexible Hardware Acceleration for Instruction-Grain Program Monitoring. In *Proceedings of the 35th International Symposium on Computer Architecture*, 2008.

[5] J. Clause, I. Doudalis, A. Orso, and M. Prvulovic. Effective Memory Protection Using Dynamic Tainting. In *Proceedings of the 22nd International Conference on Automated Software Engineering*, 2007.

[6] D. Deng, D. Lo, G. Malysa, S. Schneider, and G. Suh. Flexible and Efficient Instruction-Grained Run-Time Monitoring Using On-Chip Reconfigurable Fabric. In *Proceedings of the 43rd International Symposium on Microarchitecture*, 2010.

[7] J. Gustafsson, A. Betts, A. Ermedahl, and B. Lisper. The Mälardalen WCET Benchmarks – Past, Present and Future. In *Proceedings of the 10th International Workshop on Worst-Case Execution Time Analysis*, 2010.

[8] X. Li, Y. Liang, T. Mitra, and A. Roychoudury. Chronos: A Timing Analyzer for Embedded Software. *Science of Computer Programming*, 2007.

[9] Y.-T. S. Li and S. Malik. Performance Analysis of Embedded Software Using Implicit Path Enumeration. In *Proceedings of the 32nd Conference on Design Automation*, 1995.

[10] M. Lv, W. Yi, N. Guan, and G. Yu. Combining Abstract Interpretation with Model Checking for Timing Analysis of Multicore Software. In *Proceedings of the 31st Real-Time Systems Symposium*, 2010.

[11] A. Meixner, M. E. Bauer, and D. Sorin. Argus: Low-Cost, Comprehensive Error Detection in Simple Cores. In *Proceedings of the 40th International Symposium on Microarchitecture*, 2007.

[12] V. Nagarajan, H.-S. Kim, Y. Wu, and R. Gupta. Dynamic Information Flow Tracking on Multicores. In *Proceedings of the Workshop on Interaction Between Compilers and Computer Architectures*, 2008.

[13] M. Paolieri, E. Quiñones, F. J. Cazorla, G. Bernat, and M. Valero. Hardware Support for WCET Analysis of Hard Real-Time Multicore Systems. In *Proceedings of the 36th International Symposium on Computer Architecture*, 2009.

[14] F. Qin, C. Wang, Z. Li, H. seop Kim, Y. Zhou, and Y. Wu. LIFT: A Low-Overhead Practical Information Flow Tracking System for Detecting Security Attacks. In *Proceedings of the 39th International Symposium on Microarchitecture*, 2006.

[15] W. Shi, H.-H. S. Lee, L. Falk, and M. Ghosh. INDRA: An Integrated Framework for Dependable and Revivable Architectures Using Multicore Processors. In *Proceedings of the 33rd International Symposium on Computer Architecture*, 2006.

[16] G. Sierksma. *Linear and Integer Programming*, pages 237–239. Marcel Dekker, Inc., 2002.

[17] G. E. Suh, J. Lee, D. X. Zhang, and S. Devadas. Secure Program Execution via Dynamic Information Flow Tracking. In *Proceedings of the 11th International Conference on Architectural Support for Programming Languages and Operating Systems*, 2004.

[18] R. Wilhelm, J. Engblom, A. Ermedahl, N. Holsti, S. Thesing, D. Whalley, G. Bernat, C. Ferdinand, R. Heckmann, T. Mitra, F. Mueller, I. Puaut, P. Puschner, J. Staschulat, and P. Stenström. The Worst-Case Execution-Time Problem – Overview of Methods and Survey of Tools. *ACM Transactions on Embedded Computing Systems*, 2008.

[19] E. Witchel, J. Cates, and K. Asanovic. Mondrian Memory Protection. In *Proceedings of the 10th International Conference on Architectural Support for Programming Languages and Operating Systems*, 2002.

APPENDIX

A. EXAMPLE OF MILP-BASED METHOD

In this section we show a detailed example of applying our MILP-based method for estimating the WCET of a task running on a system with parallel run-time monitoring.

A.1 Example Setup

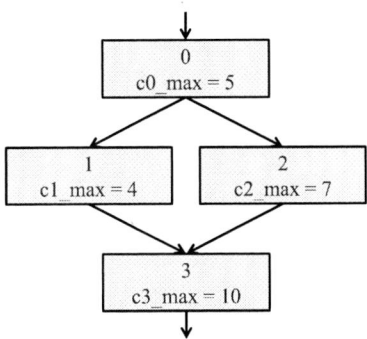

Figure 3: Control flow graph of a main task.

The control flow graph for an example main task is shown in Figure 3. We assume that the execution time for each node has already been calculated using previous methods. These execution times are labeled as `cB_max` in the figure.

In this example, let us assume that the monitoring technique requires loads and stores to be forwarded, as in the case of UMC. The monitoring task requires 5 cycles to handle a load and 7 cycles to handle a store. Thus, the maximum execution time of the monitoring task, $t_{M,max}$, is 7 cycles.

Because of the simplicity of the example, we assume that the FIFO only holds one entry ($n_F = 1$). Thus, $l_{max} = n_F \cdot t_{M,max} = 7$.

A.2 Creating the MFG

The first step is to create the monitoring flow graph. For each node in the CFG, the code represented by that node is analyzed. After any forwarded instruction, in this case any load or store instructions, an edge is created, dividing a node into 2 new ones. For example, the assembly-level code for node 1 in the CFG is shown below.

	node 1
1	add $t0, $t1, $t2
2	add $t3, $t4, $t5
3	lw $t4, 0($t3)
4	add $t0, $t0, $t4

Since the third instruction is a load instruction, node 1 must be split into two nodes in the MFG. The first node represents the first three instructions and the second node represents the last instruction.

The full MFG is shown in Figure 4. Nodes that are blue (dark) include a forwarded instruction, which is located at the end of the node. Nodes that are yellow (light) do not include a forwarded instruction. The nodes in the graph are labeled with minimum (`cB_min`) rather than maximum execution times. It can be seen that node 1 from the CFG corresponds to nodes 1.0 and 1.1 in the MFG. In this example, nodes in the CFG were only transformed into at most 2 nodes in the MFG. However, in general, a CFG node will

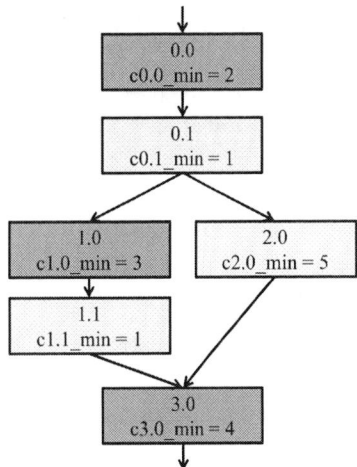

Figure 4: Monitoring flow graph of the main task. Blue (dark) nodes indicate ones with a forwarded instruction at the end. Yellow (light) nodes indicate ones without a forwarded instruction.

be transformed into a number of nodes in the MFG equal to the number of forwarded instructions plus one.

A.3 Calculating the Monitoring Load

Once the MFG is constructed, a set of MILP constraints is generated for each node. This process can be automated, but for this example we will construct the constraints for one node by hand. Specifically we will consider node 3.0 in the MFG. We will also calculate, by hand, the MILP solution for the node using some assumed values for variables associated with other nodes. Note that all variables are assumed to be non-negative unless otherwise specified.

Calculating input monitoring load: First, we will determine the worst-case input monitoring load for node 3.0, $li_{3.0}$. One set of constraints lower bounds the monitoring load by all possible incoming monitoring loads.

$$li_{3.0} \geq lo_{1.1}$$
$$li_{3.0} \geq lo_{2.0}$$

Then, a set of constraints upper bounds this input monitoring load.

$$li_{3.0} - 1000\delta_{1.1} \leq lo_{1.1}$$
$$li_{3.0} - 1000\delta_{2.0} \leq lo_{2.0}$$
$$\delta_{1.1} + \delta_{2.0} = 1$$

Here, the value 1000 is chosen arbitrarily but is known to be greater than $|lo_{2.0} - lo_{1.1}|$. A different value could have been chosen as long as this condition was true. $\delta_{1.1}$ and $\delta_{2.0}$ are binary variables which can only assume values of 0 or 1. To see how these constraints work, suppose that $li_{2.0} = 7$ and $li_{1.1} = 4$. The constraints are then evaluated as

$$li_{3.0} \geq 4$$
$$li_{3.0} \geq 7$$
$$li_{3.0} - 1000\delta_{1.1} \leq 4$$
$$li_{3.0} - 1000\delta_{2.0} \leq 7$$
$$\delta_{1.1} + \delta_{2.0} = 1$$

The first pair of constraints ensures that $li_{3.0} \geq 7$. This

427

means that for the third constraint to hold, $\delta_{1.1} = 1$. If $\delta_{1.1} = 1$, then by the last constraint, $\delta_{2.0} = 0$. Plugging this value into the fourth constraint gives $li_{3.0} \leq 7$. Thus the only possible solution is $li_{3.0} = 7$.

Calculating output monitoring load: In order to determine the output monitoring load for node 3.0, we must first calculate the change in monitoring node, $\Delta l_{3.0}$. Since there is a forwarded instruction in node 3.0,

$$\Delta l_{3.0} = t_{M,max} - c_{3.0,min}$$
$$= 7 - 4 = 3$$

We first create a variable, $lo'_{3.0}$ to represent the unbounded output monitoring load.

$$lo'_{3.0} = li_{3.0} + \Delta l_{3.0}, \ lo'_{3.0} \in (-\infty, \infty)$$

Using the example input monitoring load previously calculated of $li_{3.0} = 7$, this unbounded output monitoring load is $lo'_{3.0} = 7 + 3 = 10$. Then, the following set of constraints determines the bounded output monitoring load, $lo_{3.0}$.

$$-1000\lambda_3 + 7\lambda_5 + 1000\lambda_6 = 10 \tag{1a}$$
$$\lambda_3 + \lambda_4 + \lambda_5 + \lambda_6 = 1 \tag{1b}$$
$$2\delta_3 + \lambda_5 + \lambda_6 \leq 2 \tag{1c}$$
$$2\delta_4 + \lambda_3 + \lambda_6 \leq 2 \tag{1d}$$
$$2\delta_5 + \lambda_3 + \lambda_4 \leq 2 \tag{1e}$$
$$\delta_3 + \delta_4 + \delta_5 = 1 \tag{1f}$$
$$7\lambda_5 + 7\lambda_6 = lo_{3.0} \tag{1g}$$

The -1000 and 1000 values were chosen arbitrarily and only require that $lo'_{3.0}$ to fall between them. δ_3, δ_4, and δ_5 are binary variables. By Constraint 1b, it can be seen that all λ_i are less than or equal to 1. Thus, in order for Constraint 1a to hold, $\lambda_6 > 0$. Since $\lambda_6 > 0$, Constraints 1c and 1d force δ_3 and δ_4 to both be zero. From this, by Constraint 1f, $\delta_5 = 1$. Then, by Constraint 1e, λ_3 and λ_4 are both forced to be zero. If we now go back to the first two constraints, they are reduced to

$$7\lambda_5 + 1000\lambda_6 = 10$$
$$\lambda_5 + \lambda_6 = 1$$

Solving this system of equations gives the solution $(\lambda_5, \lambda_6) = (0.997, 0.003)$. Plugging these values into Constraint 1g,

$$lo_{3.0} = 7\lambda_5 + 7\lambda_6$$
$$= 7 \cdot 0.997 + 7 \cdot 0.003$$
$$= 7$$

Thus, the output monitoring load is indeed bound by the maximum monitoring load of 7. Although this may seem to be a complicated series of calculations to determine this obvious result, this set of constraints is required in order for the piecewise linear, and thus non-linear, bounding function to be expressed in an MILP problem.

Calculating the monitoring stall cycles: The one remaining value that needs to be determined for node 3.0 is the monitoring stall cycles. Based on our previous calculations, the worst-case input monitoring load ($li_{3.0}$) is 7, the change in monitoring load ($\Delta l_{3.0}$) is 3, and the maximum monitoring load (l_{max}) is 7. Thus, we expect the worst-case monitoring stall cycles to be $(7 + 3) - 7 = 3$. To handle this

as an MILP problem, first the unbounded monitoring stall cycles, s', is calculated.

$$s'_{3.0} = li_{3.0} + \Delta l_{3.0} - l_{max}, \ s'_{3.0} \in (-\infty, \infty)$$
$$= 7 + 3 - 7 = 3$$

In this case, since $s'_{3.0}$ is positive, we expect $s_{3.0} = s'_{3.0}$. The MILP problem determines $s_{3.0}$ using the following set of constraints.

$$-1000\lambda_0 + 1000\lambda_2 = 3 \tag{2a}$$
$$\lambda_0 + \lambda_1 + \lambda_2 = 1 \tag{2b}$$
$$\delta_1 + \lambda_2 \leq 1 \tag{2c}$$
$$\delta_2 + \lambda_0 \leq 1 \tag{2d}$$
$$\delta_1 + \delta_2 = 1 \tag{2e}$$
$$1000\lambda_2 = s_{3.0} \tag{2f}$$

The -1000 and 1000 values are chosen arbitrarily, only requiring that $s'_{3.0}$ is between them. From Constraint 2a, λ_2 must be positive. Since δ_i are binary variables, Constraint 2c then implies that $\delta_1 = 0$. Constraints 2d and 2e then force $\delta_2 = 1$ and $\lambda_0 = 0$. The first two constraints then reduce to

$$1000\lambda_2 = 3$$
$$\lambda_1 + \lambda_2 = 1$$

Solving this system of equations leads to $(\lambda_1, \lambda_2) = (0.997, 0.003)$ and thus calculating $s_{3.0}$ using Constraint 2f:

$$s_{3.0} = 1000\lambda_2$$
$$= 1000 \cdot 0.003 = 3$$

This is the value for s that we expected. If s' had instead been negative, then δ_1 would be forced to 1 and λ_2 would be forced to 0. From the last constraint, it can be seen that if λ_2 is 0, then s is also 0.

A.4 MILP Optimization

In the previous subsection, the monitoring loads for one node were calculated in detail. However, note that the output monitoring load for each node with an edge pointing to node 3.0 was assumed to be a certain value. In an actual MILP problem, these would be variables that are also being solved for. Solving for these inter-related variables and determining the global maximum number of cycles stalled due to monitoring is impractical to do by hand. While the amount of calculations may seem excessive for these simple examples, the ability to formulate the problem in MILP is essential in order to solve large problems.

B. TIME TO SOLVE LINEAR PROGRAMMING PROBLEM

The most time intensive portion of the WCET analysis is the actual solving of the linear programming (LP) problem. For our experiments, we used lp_solve 5.5.2.0 [3] as our LP solver. These experiments were run on a 2.67 GHz Xeon E5430 quad-core processor with 4 GB of RAM. The running times for lp_solve are shown in Table 3. The first set of rows show the running time for determining the worst-case stalls from the monitoring flow graph (stall). The second set of rows show the lp_solve running time for finding the sequential bounds. The final set of rows show the running

428

Solver Target	Benchmark							min	max	geomean
	cnt	expint	fdct	fibcall	insertsort	matmult	ns			
stall-umc	17.789	6.256	21.733	0.043	0.39	161.796	3.655	0.043	161.796	4.224
stall-cfp	3.691	97.93	0.038	0.024	0.025	14.209	1.474	0.024	97.930	0.778
sequential-umc	0.006	0.004	0.004	0.005	0.002	0.004	0.006	0.002	0.006	0.004
sequential-cfp	0.007	0.001	0.003	0.002	0.003	0.006	0.003	0.001	0.007	0.003
wcet-none	0.003	0.003	0.004	0.002	0.002	0.002	0.001	0.001	0.004	0.002
wcet-umc	0.004	0.004	0.003	0.001	0.004	0.005	0.002	0.001	0.005	0.003
wcet-cfp	0.002	0.007	0.005	0.004	0.003	0.005	0.004	0.002	0.007	0.004

Table 3: Running time of lp_solve in seconds to determine worst-case stalls (stall), sequential bound (sequential), and worst-case execution times (wcet).

time for determining the overall WCET (wcet). For wcet-umc and wcet-cfp, this is for the ILP problem given the worst-case stalls .

The running times for the sequential cases and the wcet cases are very similar. This is because these cases are all solving essentially the same problem with different numbers. That is, for a given benchmark, these different cases are all solving a linear programming problem for the same control flow graph (CFG). As a result, the number of variables and the set of constraints is the same, though the WCET for each basic block changes depending on the extension and the estimation method. The stall cases have a longer running time. This is due to the fact that a MFG has more nodes than its corresponding CFG. The increased number of nodes also implies more variables and more constraints.

C. FUTURE WORK

There are two main directions for future work. One direction for future work is to tighten the WCET bound and the other is to improve the time needed to solve the linear programming (LP) problem. The WCET bound could be improved by incorporating more detailed information about the main task. Program behavior such as infeasible paths and loop bounds have previously been studied in the IPET context [18]. Incorporating this information into the WCET analysis for parallel monitoring can decrease the worst-case stall cycles found. For example, the current formulation does not include any notion of loop bounds. As a result, for a loop that increases the monitoring load, the worst-case conclusion is that there are enough loop iterations for the FIFO to become full. With information about loop bounds, it may be the case that certain loops do not cause the FIFO to fill completely. We believe that since our formulation uses a linear programming approach similar to IPET, additional program behavior can be added in a similar manner using additional constraints. Along these lines, another possible direction for future work is to extend this work for addi-

tional architectural features. For example, one assumption in this work was that the main and monitoring cores had separate memory spaces. It would be interesting to extend this WCET analysis to a system where the main and monitoring cores shared memory.

Appendix B shows the running time for solving the LP problems created. Determining the worst-case stalls requires a longer run time than determining the WCET. This is due primarily to the larger graph size of the MFG compared to the CFG. The larger graph size means that there are more variables to optimize over. It may be possible to model the monitoring load for each basic block in such a way that the optimization problem can remain at the CFG graph size. Since the monitoring load calculations for a series of MFG nodes, without branch entries or exits, is relatively straightforward, it may be possible to "collapse" them into a set of equations for one node. However, care must be taken that these simplifications do not remove any worst-case possibilities.

Finally, we mention the possibility of combining the worst-case stall cycles MILP problem and WCET ILP problem into a single LP problem. Combining the two problems into a single optimization may provide improvements in tightening the WCET bound. It may also improve the LP running time by requiring only one LP problem to be solved. At first glance, this may seem possible by combining the constraints from both problems and maximizing the objective

$$t = \sum_{B \in \mathcal{B}_{CFG}} N_B \cdot (c_{B,max} + s_{B,max})$$

from Section 4.1. However, combining the problems means that this is optimizing over both N_B and $s_{B,max}$. These variables form a product term in the equation for t and so the optimization objective is no longer linear. There may exist a method to formulate a combined problem that has a linear objective. Alternatively, non-linear programming techniques may serve as a solution.

Conforming the Runtime Inputs for Hard Real-Time Embedded Systems

Kai Huang
fortiss GmbH, Germany
khuang@fortiss.org

Gang Chen
Technical University Munich, Germany
gangchen1170@tum.edu

Christian Buckl
fortiss GmbH, Germany
buckl@fortiss.org

Alois Knoll
Technical University Munich, Germany
knoll@in.tum.de

ABSTRACT

Timing is an important concern when designing an embedded system. While lots of researches on hard real-time systems focus on design-time analysis, monitoring the corresponding runtime behaviors are seldom investigated. In this paper, we investigate the conformity problem for runtime inputs of a hard real-time system. We adopt the widely used arrival curve model which captures the worst/best-cases event arrivals in the time interval domain and propose an algorithm to on-the-fly evaluate the conformity of the system input w.r.t. given arrival curves. The developed algorithm is lightweight in terms of both computation and memory overheads, which is particularly suitable for resource-constrained embedded systems. We also provide proofs and an FPGA implementation to demonstrate the effectiveness of our approach.

Categories and Subject Descriptors

C.3 [**SPECIAL-PURPOSE AND APPLICATION-BASED SYSTEMS**]: Real-time and embedded systems; B.8.0 [**Hardware**]: Performance and Reliability—*General*

General Terms

Algorithms, Design, Performance

Keywords

Real-Time Calculus, Leaky Bucket, Greedy Shaper

1. INTRODUCTION

Guaranteeing timing properties is an important aspect for building embedded systems. In particular for the class of real-time embedded systems, meeting timing constraints, e.g., worst-case response time and end-to-end latencies, is a major design concern. Researchers on hard real-time

timing analysis in the literature [1, 5, 6, 12] often focus on design-time analysis, trying to compute worst-case bounds on timing properties at an early phase of the system design. The validity of the design-time analysis and the safeness of the derived bounds rely on the assumption that the system input follows certain specifications. In order to not harm the safeness of the analysis results, the runtime inputs of the system (or components) need to be conformed to the specifications used by the design-time analysis.

The conformity verification, however, is non-trivial. On the one hand, the verification has to cover the worst cases in order to be in consistence with the offline analysis. On the other hand, the verification and a possibly preceeding regulation mechanism have to be lightweight due to the stringent timing and resource budgets of the system. Therefore, directly applying the commonly used techniques which usually rely on expensive numerical computation may not be suitable for the runtime monitoring.

In this paper, we investigate the runtime conformity problem. Targeting hard real-time embedded systems, we try to provide on-the-fly verification for the worst-case conformity of system inputs. We adopt the widely used arrival-curve model which captures the worst/best-cased system inputs in the time interval domain and propose an algorithm to evaluate the conformity of input traffic with respect to given arrival curves. In case too many events are detected, our algorithm regulates the traffic such that the traffic complies again with the curve specifications assumed at design time. In case too few events are detected, no generic solution can be offered, but the applications can be notified.

Based on the results in [9] that an arrival curve can be conservatively approximated by a set of staircase functions each of which can be modeled by a leaky bucket, we use a dual-bucket mechanism to monitor each staircase function during runtime, one for conformity verification and one for traffic regulation. By tracking the fill level of buckets, the computationally expensive (de-)convolutions used by the design-time analysis are eliminated. Our approach is thus lightweight in terms of both computational overhead and memory footprint, particularly suitable for embedded systems with limited resources. We also provide formal proofs and an FPGA prototype for our algorithm to demonstrate the effectiveness of our approach.

The rest of this paper is organized as follows: Section 2 reviews related work in the literature. Section 3 presents the system models and analyzes the problem. Section 4 presents our algorithm and the proofs. Experimental results are presented in Section 5 and Section 6 concludes the paper.

2. RELATED WORK

The analysis of traffic regulators is not new. In the domain of classical networking flow control, the studies of lossless greedy regulators by means of network calculus can be found in [2, 11]. Such traffic regulators are usually modeled as leaky-bucket shapers. To model lossy systems, traffic clippers [4, 3] are introduced to regulate network packets. Unlike shapers that delay network packets, a traffic clipper actively discards non-conformed packets. The modeling of leaky-bucket greedy shapers in the context of real-time calculus (RTC) [12] is presented in [13]. The latest work on this direction [7] uses a greedy shaper to optimally reduce the peak temperature of a real-time system. All aforementioned work is offline analysis. Whether such analysis can be applied for online monitoring is not clear.

In [9], a methodology for coupling timed automata-based [1] and RTC-based models are proposed, which enables hybrid analysis of a system containing both state-based (timed automata) and state-less (RTC) components. The basic idea underlying the proposed methodology is to use a set of leaky-bucket event generators as an interface between the state-based and stateless abstractions, such that models can be interchanged between these two abstractions. This technique is applied in [8] to predict tighter worst-case bounds for the arrivals of future workload. Taking the idea in [9, 8] to the level of runtime, this paper investigates traffic conformance, trying to develop lightweight routines for on-the-fly traffic verification and regulation. We also prototyped an FPGA implementation of our approach on an ALTERA Cyclone III development board.

3. MODELS AND PROBLEM

Event Arrival Curves.

Event streams in a system can be described using a cumulative function $R(s, t)$, defined as the number of events seen in the time interval $[s, t]$. While any R always describes *one* concrete trace, a 2-tuple $\alpha(\Delta) = [\alpha^u(\Delta), \alpha^l(\Delta)]$ of upper and lower *arrival curves* [10] provides an abstract event stream model that characterizes a whole class of (non-deterministic) event streams. $\alpha^u(\Delta)$ and $\alpha^l(\Delta)$ provide an upper and a lower bound on the number of events seen on an event stream in *any* time interval of length Δ:

$$\alpha^l(t - s) \leq R(t) - R(s) \leq \alpha^u(t - s), \ \forall\, 0 \leq s \leq t, \quad (1)$$

with $\alpha^l(\Delta) = \alpha^u(\Delta) = 0$ for $\Delta \leq 0$.

The concept of arrival curves unifies many other common timing models of event streams. For example, a periodic event stream can be modeled by a set of step functions where $\alpha^u(\Delta) = \lfloor \frac{\Delta}{p} \rfloor + 1$ and $\alpha^l(\Delta) = \lfloor \frac{\Delta}{p} \rfloor$. For a sporadic event stream with minimal inter arrival distance p and maximal inter arrival distance p', the upper and lower arrival curve is $\alpha^u(\Delta) = \lfloor \frac{\Delta}{p} \rfloor + 1$, $\alpha^l(\Delta) = \lfloor \frac{\Delta}{p'} \rfloor$, respectively. A widely used model to specify an arrival curve is the PJD model by which an arrival curve is characterized with a period p, jitter j, and minimal inter arrival distance d. The upper arrival curve is thus $\alpha^u(\Delta) = \min\{\lceil \frac{\Delta+j}{p} \rceil, \lceil \frac{\Delta}{d} \rceil\}$. For details, please refer to [12].

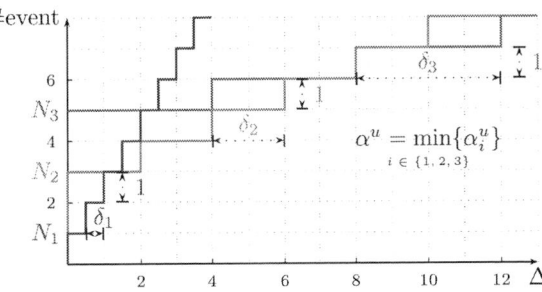

Figure 1: An example for an arrival curve as the combination of staircase functions.

Complex Arrival Patterns.

In this work, we deal with discrete numbers of event arrivals and their arrival patterns. In principle any (discrete) complex arrival pattern can be bounded by a set of upper and lower staircase functions, as long as the system under consideration is monotone and time-invariant [9]. The *monotone* property means that a higher number of input events seen in an interval yields a higher number of output events in intervals of equal or larger sizes. The *time-invariant* property means that the system behavior depends on the system states only. No matter when this state is reached, the possible set of the reactions of the systems is always the same, independently upon the concrete time when the actual state is reached.

An upper arrival curve thereby can be conservatively approximated as the minimum on the set of staircase functions of the form $\alpha_i(\Delta) = N_i^u + \lfloor \frac{\Delta}{\delta_i^u} \rfloor$:

$$\forall \Delta \in \mathbb{R}_{\geq 0} : \alpha^u(\Delta) \leq \min_{i \in n}(\alpha_i^u(\Delta)). \quad (2)$$

An example for such approximation is depicted in Fig. 1, where α^u is given as the minimum of three staircase functions $\alpha_1^u = 1 + \lfloor \frac{\Delta}{0.5} \rfloor$, $\alpha_2^u = 3 + \lfloor \frac{\Delta}{2} \rfloor$, and $\alpha_3^u = 5 + \lfloor \frac{\Delta}{4} \rfloor$.

Analogously, a set of staircase functions and maximum of which, i.e., $\alpha^l(\Delta) \geq \max_{j \in m}(0, \alpha_j^l(\Delta))$, can be employed for approximating a lower curve, where $\alpha_j^l(\Delta) = -N_j^l + \lfloor \frac{\Delta}{\delta_j^l} \rfloor$.

Problem Statement.

Given a trace R, checking its conformity w.r.t to an arrival curve is theoretically not a problem. One can simply inspect the trace by the definition in Eqn. (1). In the case that $\exists s, t, 0 \leq s < t, R(t) - R(s) > \alpha^u(t - s)$ or $R(t) - R(s) < \alpha^l(t - s)$, a violation occurs. Alternatively, one can apply the min-plus de-convolution:

$$\sup_{u \geq 0}\left\{ R(t + u) - R(u) \right\} \stackrel{\text{def}}{=} R(t) \oslash R(t) > \alpha^u(t) \quad (3)$$

according to [10],

Once a violation is detected, the traffic can be regulated to re-conform again to the specified arrival curves, e.g., by imposing a certain delay for the over-bursty input events. A usual way is to use a greedy shaper σ such that[1]

$$(R \oslash R) \otimes \sigma \leq \alpha^u \quad (4)$$

The shaper σ can be simply the convex hull of α^u.

One might notice that above approaches require numerical computation for the min-plus (de-)convolution, which demands intensive computing power as well as large

---

[1]min-plus convolution: $f \otimes g \stackrel{\text{def}}{=} \inf_{0 \leq s < t}\left\{ f(t - s) + g(s) \right\}$

431

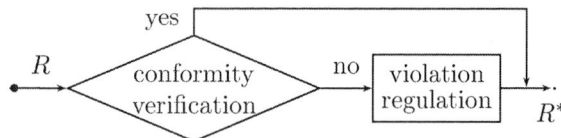

Figure 2: The flow of the approach.

memory footprint. Directly applying these approaches for online monitoring is thus prohibited, in particular for the class of embedded systems with stringent resource constraints. Therefore, lightweight alternatives are needed to conduct efficient conformance verification as well as violation recovery. In the next section, we will present an approach that solves this problem in a particular way.

4. OUR APPROACH

The idea of our approach is based on the knowledge that an arrival curve can be conservatively approximated by a set of staircase functions [9], each of which can be modeled by a leaky-bucket kind event generator. Rather than generating events, we use the leaky bucket mechanism for online monitoring. In this context, the bucket capacity corresponds to the maximally tolerable number of bursty events and the leak rate models the period of the staircase function. The fill level of the bucket is used as an indicator for the remaining capacity of the tolerable burst.

For each staircase function, we employ two leaky buckets, namely V-bucket for input conformity verification and R-bucket for input regulation. The fill level of V-bucket indicates how many bursty incoming events can still be tolerated while the fill level of R-bucket shows how many bursty events are allowed to release. By simply tracking the fill levels of the buckets, the computationally expensive min-plus (de-)convolution normally used by the offline analysis can be eliminated for the online monitoring, resulting in a lightweight software or hardware implementation.

The flow of our approach is shown in Fig. 2. Upon each event arrival, the conformity of the event is tested. If the arrival of this event conforms to the specification, this event will be immediately released. If not, certain regulation is applied to enforce the conformity. For the current version of our algorithm, we delay the release time of non-conformed events. Discarding events due to deadline violation or buffer overflow can be easily adapted based on the proposed scheme. We will discuss their solutions in Section 4.3. Note that we only present the algorithm and proofs for the upper bound α^u. The conformity verification of the lower bound works in a similar manner.

4.1 Algorithm

Assume an arrival curve is approximated by n staircase functions S_i, $i \in n$. Each S_i is defined by a leaky bucket with two parameters, namely bucket capacity N_i^u and period δ_i^u. During runtime, the status of a bucket is tracked by two variables, namely the fill level BFL and a timer CLK. The timer CLK records the time passed within a period δ_i^u. Thus our algorithm maintains a 4-tuple $<BFL_i^v, CLK_i^v, BFL_i^r, CLK_i^r>$ during runtime, BFL_i^v, CLK_i^v for V-bucket and BFL_i^r, CLK_i^r for R-bucket. The algorithm is invoked when a signal comes. A signal can be triggered by the arrival of an event or the timeout of a timer. Initially,

---

Algorithm 1 On-the-fly traffic verification and regulation for an n-staircase arrival curve.

Input: signal s ▷ tuple $<BFL_i^v, CLK_i^v, BFL_i^r, CLK_i^r>$ and event queue q are global variables

```
 1: for i ← 1 to n do                                    ▷ timeout
 2:     if s = CLKᵢᵛ_timeout then
 3:         BFLᵢᵛ ← min(BFLᵢᵛ + 1, Nᵢᵘ)
 4:         reset_timer(CLKᵢᵛ)
 5:     end if
 6:     if s = CLKᵢʳ_timeout then
 7:         BFLᵢʳ ← min(BFLᵢʳ + 1, Nᵢᵘ)
 8:         reset_timer(CLKᵢʳ)
 9:     end if
10: end for
11: if s = event_arrival then                      ▷ event arrival
12:     for i ← 1 to n do
13:         if BFLᵢᵛ = Nᵢᵘ then
14:             reset_timer(CLKᵢᵛ)
15:         end if
16:         BFLᵢᵛ ← BFLᵢᵛ − 1
17:     end for
18:     q.enqueue()
19: end if
20: if minᵢ∈ₙ(BFLᵢᵛ) < 0 then                    ▷ nonconformity
21:     report_violation()
22: end if
23: while q.length()>0 ∧ minᵢ∈ₙ(BFLᵢʳ)>0 do         ▷ regulation
24:     q.dequeue()
25:     for i ← 1 to n do
26:         if BFLᵢʳ = Nᵢᵘ then
27:             reset_timer(CLKᵢʳ)
28:         end if
29:         BFLᵢʳ ← BFLᵢʳ − 1
30:     end for
31: end while
```

$BFL_i^v = BFL_i^r = N_i^u$ and $CLK_i^v = CLK_i^r = \delta_i^u$.

The pseudo code of the algorithm is shown in Algo. 1. BFL_i^v of V-bucket indicates the number of bursty events that can still be tolerated at current time. Its value is decreased by 1 when an event arrives (Line 16) and increased by 1 when CLK_i^v is timeout (Lines 2–5). BFL_i^v has a limit of N_i^u (Line 3). Based on the algorithm, BFL_i^v can be computed as a function of the trace R:

$$BFL_i^v(t) = \min(N_i^u + \lfloor \frac{t}{\delta_i^u} \rfloor - R(t), N_i^u) \quad (5)$$

For any time interval $(s, t]$, BFL_i^v can be computed as:

$$BFL_i^v(t) = \min\left(BFL_i^v(s) + \lfloor \frac{t-s}{\delta_i^u} \rfloor - (R(t) - R(s)), N_i^u\right) \quad (6)$$

Since BFL_i^v records the remaining capacity of the bucket, conformity violation occurs when $BFL_i^v < 0, \exists i \in n$ (Lines 20–22). In other words, a nonconformity occurs when the number of events arrived in the interval Δ is larger than $N_i^u + \lfloor \frac{\Delta}{\delta_i^u} \rfloor$.

The CLK_i^v of V-bucket notifies when the budget can be recharged. It is reset when BFL_i^v reaches its limit, i.e., $BFL_i^v = N_i^u$ (Lines 13–15). When BFL_i^v is equal to N_i^u, a burst of maximal N_i^u events can be tolerated from this time on. This case can also be considered as a renew point of the bucket. We will use this property in the later-on proofs.

The R-bucket works similarly. BFL_i^v controls when and how many events can be released (Lines 23–31). Only when all $BFL_i^r > 0$, an event can be released. It is decreased by 1 when an event is released (Line 29). Otherwise events will be postponed until every BFL_i^r turns nonzero. To release buffered events, we use a first-come first-out scheme.

4.2 Correctness

This section proves the correctness of our algorithm. For simplicity, we provide the formal proof for the case of $n = 1$, i.e., $\alpha^u(\Delta) = N^u + \lfloor \frac{\Delta}{\delta^u} \rfloor$. Proofs for $n > 1$ cases follow a similar scheme.

To prove the algorithm, we divide the time axis for the system execution into a set of consecutive time segments. A time segment is defined as follows.

DEF. 1. *A time segment F is the time interval for the value of BFL^v changing from N^u back to N^u, i.e., between two renew points in the trace (Lines 2–5, Algo. 1).*

The starting and ending time instants for F_i are denoted by t_{S_i} and t_{E_i}, respectively. For the arrival of the nth event e_n in the trace, we also designate t_n and x_n the time instant and the value of BFL^v, respectively. Based on above definitions, we have following lemmas.

LEMMA 1. *Within any F_i, $i \in \mathbb{N}$, at time t_1, i.e., the first arrived event of this segment, $BFL^v = N^u - 1$. At each time instant t_i of the arrivals of subsequent events within F_i, $BFL^v \leq N^u - 2$.*

PROOF. According to Def. 1, BFL^v always starts from N^u for any F_i. When the first event arrives, BFL^v turns to $N^u - 1$. The arrivals of subsequent events will further decrease BFL^v. Assume at some point of time, BFL^v is recharged back to $N^u - 1$ due to the timeouts of CLK^v (Lines 2–5). There are only two cases: a) a next timeout of CLK^v comes, BFL^v reaches N^u, and the segment ends, or b) an event arrives, BFL^v decreases back to $N^u - 2$. Therefore, the lemma holds. $\square$

LEMMA 2. *At time t_n for the arrival of event e_n, if $t_{m+n} - t_n \geq (m - (N^u - 1))\delta^u$, $\forall m$, the trace between $[t_n, t_{m+n}]$ conforms α^u.*

PROOF. Consider an arbitrary interval $[s, t]$. There are $m + 1$ events arrived within this interval and these events are numbered as $n, n+1, \ldots, m+n$, so that $t_{n-1} \leq s < t_n \leq t_{n+1} \leq \ldots \leq t_{m+n} \leq t$. We get $\alpha^u(t-s) \geq \alpha^u(t_{m+n} - t_n) = N^u + \lfloor \frac{t_{m+n} - t_n}{\delta^u} \rfloor \geq m + 1$. Because $R(t) = m + n$ and $R(s) = n-1$, we get $\alpha^u(t-s) \geq R(t) - R(s)$. From Eqn. (1), the lemma holds. $\square$

LEMMA 3. *Let T_{F_i} denote the length of segment F_i and $|F_i|$ the number of events arrived within F_i, $T_{F_i} \geq |F_i| \cdot \delta^u$.*

PROOF. Fig. 3(a) illustrates such an example. According to the algorithm, the timer CLK^v is cleared at time instants t_n, t_{S_i}, and t_{E_i}. With Eqn. (6), we have $N^u - 1 + \frac{t_{E_i} - t_n}{\delta^u} - (|F_i| - 1) = N^u$. Therefore, $t_{E_i} - t_{S_i} \geq t_{E_i} - t_n = |F_i| \cdot \delta^u$. $\square$

THEOREM 1. *Given a trace R and a staircase α^u, in the case that $BFL^v \geq 0$, Algo. 1 guarantees R conform to α^u.*

PROOF. What we need to prove is for $\forall 0 \leq s \leq t$, $R(t) - R(s) \leq \alpha^u(t - s)$. We consider two cases, i.e., s and t are located within one segment and in two different segments.

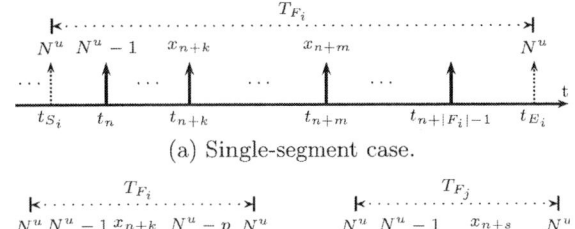

(a) Single-segment case.

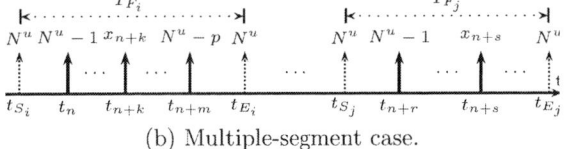

(b) Multiple-segment case.

Figure 3: Graphical illustration of the proof.

First we consider the special case of $N^u = 1$. In this case, each segment allows only one event to guarantee $BFL^v \geq 0$, i.e., there is only one event in each segment. Without loss of generality, let s and t the arrival time of event e_n and e_m which arrive at segment F_n and F_m, respectively. Form Lem. 3, we have $t_{E_n} - t_n = \delta^u$ and $t_{S_m} - t_{E_n} = (m-1-n)\delta^u$. Thus $t_m - t_n = (t_m - t_{S_m}) + (t_{S_m} - t_{E_n}) + (t_{E_n} - t_n) \geq (m-n)\delta^u = (m - n - (N^u - 1))\delta^u$. Based on Lem. 2, the theorem holds for $N^u = 1$.

In the following, we provide the proof for $N^u \geq 2$.

• Single-segment case:

For any given segment F_i, assume e_n is the first event arrived in F_i, as shown in Fig. 3(a). Obviously, BFL^v is $N^u - 1$ at time instant t_n. We further assume that events e_{n+k} and e_{n+m} $(m > k \geq 0)$ arrive within F_i at time instants t_{n+k} and t_{n+m}, the corresponding values of BFL^v being x_{n+k} and x_{n+m}, respectively. According to Eqn. (6), we have

$$N^u - 1 + \lfloor \frac{t_{m+n} - t_n}{\delta^u} \rfloor - m = x_{n+m} \tag{7}$$

$$N^u - 1 + \lfloor \frac{t_{n+k} - t_n}{\delta^u} \rfloor - k = x_{n+k} \tag{8}$$

For $k > 0$, we know $0 \leq x_{n+m}, x_{n+k} \leq N^u - 2$ (Lem. 1). Therefore, we have

$$
\begin{aligned}
t_{m+n} - t_{n+k} &= (t_{m+n} - t_n) - (t_{n+k} - t_n) \\
&= \lfloor \frac{t_{m+n} - t_n}{\delta^u} \rfloor \delta^u + \sigma_{m+n}\delta^u \\
&\quad - \lfloor \frac{t_{k+n} - t_n}{\delta^u} \rfloor \delta^u - \sigma_{k+n}\delta^u \\
&= (m + x_{n+m} - k - x_{n+k})\delta^u + (\sigma_{m+n} - \sigma_{k+n})\delta^u \\
&\geq (m - k - (N^u - 2))\delta^u + (\sigma_{m+n} - \sigma_{k+n})\delta^u \\
&\geq (m - k - (N^u - 1))\delta^u
\end{aligned}
$$

where[2] $\sigma_{m+n} = \frac{t_{m+n} - t_n}{\delta^u} - \lfloor \frac{t_{m+n} - t_n}{\delta^u} \rfloor$.

For $k = 0$, we have

$$
\begin{aligned}
t_{m+n} - t_n &= \lfloor \frac{t_{m+n} - t_n}{\delta^u} \rfloor \delta^u + \sigma_{m+n}\delta^u \\
&= (m + x_{n+m} - (N^u - 1))\delta^u + \sigma_{m+n}\delta^u \\
&\geq (m - (N^u - 1))\delta^u + \sigma_{m+n}\delta^u \\
&\geq (m - (N^u - 1))\delta^u
\end{aligned}
$$

Based on Lem. 2, the theorem holds for this case.

• Multiple-segment case:

We consider segments F_i and F_j with $j > i$. As shown in Fig. 3(b), event e_n is the first event in F_i with $BFL^v = N^u - 1$ and event e_{n+m} is the last event in F_j with $BFL^v = N^u - p$ $(p \geq 2)$. In F_j, event e_{n+r} is the first event with

---

[2]σ_{k+n}, σ_{n+s}, and σ_{n+k} in the subsequent text follow similar definition.

$BFL^v = N^u - 1$. Therefore, there are $r - m - 1$ events arrived in the interval $[t_{E_i}, t_{S_j}]$. From Lem. 3, we have

$$t_{S_j} - t_{E_i} \geq (r - m - 1)\delta^u \qquad (9)$$

$$t_{E_i} - t_n = (m + 1)\delta^u \qquad (10)$$

Considering events e_{n+s} of F_j and e_{n+k} of F_i, we have following equations based on Eqn. (6):

$$N^u - 1 + \lfloor \frac{t_{n+s} - t_{n+r}}{\delta^u} \rfloor - (s - r) = x_{n+s} \qquad (11)$$

$$N^u - 1 + \lfloor \frac{t_{n+k} - t_n}{\delta^u} \rfloor - k = x_{n+k} \qquad (12)$$

For $k \neq 0, s \neq r$, we have $0 \leq x_{n+s}, x_{n+k} \leq N^u - 2$ (Lem. 1). Together with Eqns. (9)–(12), we get

$$\begin{aligned}
t_{n+s} - t_{n+k} &= (t_{n+s} - t_{n+r}) + (t_{n+r} - t_{S_j}) + (t_{S_j} - t_{E_i}) \\
&\quad + (t_{E_i} - t_n) - (t_{n+k} - t_n) \\
&\geq (t_{n+s} - t_{n+r}) + (t_{S_j} - t_{E_i}) + (t_{E_i} - t_n) \\
&\quad - (t_{n+k} - t_n) \\
&\geq \lfloor \frac{t_{n+s} - t_{n+r}}{\delta^u} \rfloor \delta^u + \sigma_{n+s}\delta^u + (r - m - 1)\delta^u \\
&\quad + (m + 1)\delta^u - (\lfloor \frac{t_{n+k} - t_n}{\delta^u} \rfloor \delta^u + \sigma_{n+k}\delta^u) \\
&= (x_{n+s} - x_{n+k} + s - k)\delta^u + (\sigma_{n+s} - \sigma_{n+k})\delta^u \\
&\geq (s - k - N^u + 1)\delta^u
\end{aligned}$$

For $k \neq 0, s = r$, from Lem. 3, Eqns. (10), (9), and (12) as well as $N^u \geq 2$, we have

$$\begin{aligned}
t_{n+r} - t_{n+k} &\geq (t_{S_j} - t_{E_i}) + (t_{E_i} - t_n) - (t_{n+k} - t_n) \\
&\geq (r - k + 1)\delta^u + (\sigma_{n+m} - \sigma_{n+k})\delta^u \\
&\geq (r - k - (N^u - 1))\delta^u
\end{aligned}$$

For $k = 0$, we get $t_{n+s} - t_{n+r} \geq (s - r - N^u + 1)\delta^u$ and $t_{E_i} - t_n = (m+1)\delta^u$ from single-segment case and Eqn. (10). Then we have

$$\begin{aligned}
t_{n+s} - t_n &\geq (t_{n+s} - t_{n+r}) + (t_{S_j} - t_{E_i}) + (t_{E_i} - t_n) \\
&\geq (s - N^u + 1)\delta^u
\end{aligned}$$

From above cases, the theorem holds. $\square$

COROLLARY 1. *At the time instant when BFL^v turns small than 0, a violation to α^u occurs.*

PROOF. As shown in Fig. 3(a), assume $BFL^v < 0$ occurs when event e_{m+n} arrives in F_i, i.e., $x_{n+m} \leq -1$. According to Eqn. (7), $t_{m+n} - t_n = (m + x_{n+m} - (N^u - 1))\delta^u + \sigma_{m+n}\delta^u$. Consider the interval [s,t] with $s = t_n - \lambda$ and $t = t_{m+n}$, where $0 < \lambda < (1 - \sigma_{m+n})\delta^u$. Thus, $\alpha^u(t - s) = N^u + \lfloor \frac{t_{m+n} - t_n + \lambda}{\delta^u} \rfloor = m + x_{n+m} + 1 \leq m$. Because $R(t) = m + n$ and $R(s) = n - 1$, $\alpha^u(t - s) < R(t) - R(s)$. A violation to α^u occurs. $\square$

THEOREM 2. *The resulting trace regulated by Algo. 1 conforms to α^u.*

PROOF. The output traffic is modulated by the R-bucket. The R-bucket works in the same mechanism as V-bucket. In addition, variable *backlog* (Lines 23) guarantees that BFL^r will never go below zero. Therefore, the theorem holds according to Thm. 1. $\square$

4.3 Discussion

The algorithm and proof in the previous section are for the conformance verification and regulation of the upper bound of an arrival curve. Similar technique can be used for the violation detection of the lower bound. The regulation of input traffic to re-conform to a lower curve is however not possible in this context. As violation of the lower bound basically means no sufficient number of events occurs for

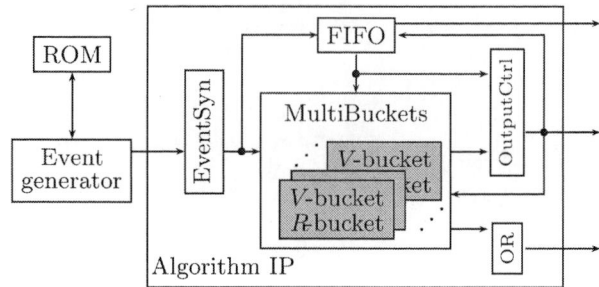

Figure 4: The block diagram of the FPGA testbed.

a certain time interval, a regulation by injecting artificial events into the system would violate our basic assumption that we only consider time-invariant systems. Nevertheless, a warning can be issued to the application, so that the application itself might be able to react to the violation.

Note that another assumption of our approach is that the violated events will be stored in a buffer and released at a later point of time. Too many buffered events may lead to buffer overflow of our algorithm. Although it is unavailable, we nevertheless can detect such occurrence by modulating the event queue q in the algorithm. Another fact is that delaying the input events may result in deadline violation of input events. Detecting the deadline violations can also be included based on the current scheme.

5. EXPERIMENTS

This section demonstrates the effectiveness of our approach by empirical case studies. We implement our algorithm both in MATLAB and Verilog HDL. The Verilog HDL code is synthesized in ALTERA Cyclone III FPGA.

We adopt the PJD model (Section 3) for the specification of event streams. The upper bound α^u for such a model can be represented as the minimum of two staircase functions. The parameters of the two staircase functions can be computed as follows [9]:

- Case $d = 0 \lor d \leq p - j$:
  $N^u = \lceil \frac{i}{p} \rceil + 1$; $\qquad N^l = -\lceil \frac{i}{p} \rceil$; $\qquad \delta^u = \delta^l = p$

- Case $d > 0 \land d > p - j$:
  $N_1^u = 1$; $\qquad \delta_1^u = d$; $\qquad N_2^u = \lceil \frac{i}{p} \rceil + 1$;
  $N^l = -\lceil \frac{i}{p} \rceil$; $\qquad \delta_2^u = \delta^l = p$

To generate traces with different patterns, the RTC/RTS-toolbox [14] is used. We first generate a worst-case trace that conforms to the upper bound. Then we inject random events to artificially create violations. In our experiment, we employ an arrival curve with period of $100us$, jitter of $300us$, and delay of $20us$.

In order to validate our algorithm, we implement a discrete-time simulation in MATLAB and an FPGA testbed. The MATLAB simulation is implemented using the RTC/RTS-toolbox. The testbed is comprised of an event generator IP and the algorithm IP, as the block diagram shown in Fig. 4. The event generator IP is used to generate events that comply with those used in the MATLAB simulation. The algorithm IP itself consists of four modules. As shown in the figure, the MultiBuckets module contains a reconfigurable number of bucket pairs, each of which contains a V-bucket and R-bucket. The EventSyn module synchronizes the FIFO

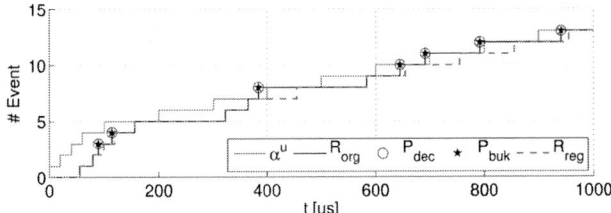

(a) Detection and traffic regulation.

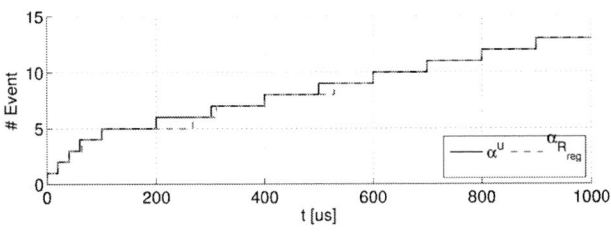

(b) Upper bound of the regulated trace.

Figure 5: Results for verification and regulation of a trace with seven violation points.

# Bucket Pair	Logic Elements	Register	LUTs	Memory (Bit)	Delay (cycle)
2	134	107	103	64	6
4	245	195	191	64	6
6	353	283	279	64	6

Table 1: Resource and timing overhead for the FPGA implementation.

performance.

6. CONCLUSION

This paper presents an online algorithm for the traffic conformity and regulation of hard real-time systems. Our algorithm can detect input violation and regulate the violated traffic to comply again with the specifications. We also present formal proofs, simulation results, and an FPGA implementation to demonstrate the effectiveness of our algorithm. The experiment results show that the resource and timing overheads of our algorithm are lightweight, particularly suitable for embedded systems with stringent resource constraints.

7. REFERENCES

[1] R. Alur and D. Dill. Automata for modeling real-time systems. In M. Paterson, editor, *Automata, Languages and Programming*, volume 443 of *Lecture Notes in Computer Science*, pages 322–335. Springer Berlin / Heidelberg, 1990. 10.1007/BFb0032042.

[2] C.-S. Chang. On deterministic traffic regulation and service guarantees: a systematic approach by filtering. *IEEE Transactions on Information Theory*, 44(3):1097–1110, may 1998.

[3] C.-S. Chang, R. Cruz, J.-Y. Le Boudec, and P. Thiran. A min, + system theory for constrained traffic regulation and dynamic service guarantees. *IEEE/ACM Transactions on Networking*, 10(6):805–817, dec 2002.

[4] R. L. Cruz and M. Taneja. An analysis of traffic clipping. In *Princeton University*, 1998.

[5] M. González Harbour, J. J. Gutiérrez García, J. C. Palencia Gutiérrez, and J. M. Drake Moyano. MAST: Modeling and Analysis Suite for Real Time Applications. In *Proc. Euromicro Conference on Real-Time Systems*, pages 125–134, Delft, The Netherlands, June 2001.

[6] R. Henia, A. Hamann, M. Jersak, R. Racu, K. Richter, and R. Ernst. System Level Performance Analysis — The SymTA/S Approach. *IEE Proceedings Computers and Digital Techniques*, 152(2):148–166, Mar. 2005.

[7] P. Kumar and L. Thiele. Cool shapers: Shaping real-time tasks for improved thermal guarantees. In *Proc. of Design Automation Conference (DAC)*, San Diego, 2011. ACM.

[8] K. Lampka, K. Huang, and J. J. Chen. Dynamic counters and the efficient and effective online power management of embedded real-time systems. In *the International Conference on Hardware-Software Codesign and System Synthesis (CODES+ISSS)*, 2011.

[9] K. Lampka, S. Perathoner, and L. Thiele. Analytic real-time analysis and timed automata: A hybrid methodology for the performance analysis of embedded real-time systems. *Design Automation for Embedded Systems*, 14(3):193–227, 2010.

[10] J. Le Boudec and P. Thiran. *Network Calculus: A Theory of Deterministic Queuing Systems for the Internet*. Springer, 2001.

[11] J.-Y. Le Boudec. Application of network calculus to guaranteed service networks. *IEEE Transactions on Information Theory*, 44(3):1087–1096, may 1998.

[12] L. Thiele, S. Chakraborty, and M. Naedele. Real-time calculus for scheduling hard real-time systems. In *Proc. IEEE International Symposium on Circuits and Systems (ISCAS)*, volume 4, pages 101–104, 2000.

[13] E. Wandeler, A. Maxiaguine, and L. Thiele. Performance analysis of greedy shapers in real-time systems. In *Design, Automation and Test in Europe (DATE)*, pages 444–449, Munich, Germany, 2006.

[14] E. Wandeler and L. Thiele. Real-Time Calculus (RTC) Toolbox. http://www.mpa.ethz.ch/Rtctoolbox, 2006.

and MultiBuckets modules when events arrive. The output module controls the release of events. The FIFO module is used to buffer regulated events. The testbed is simulated using ModelSim. Details of the implementation are given in Fig. 6 in the appendix.

We compare the theoretical and experimental results in Fig. 5. The solid line R_{org} and dashed line R_{reg} represent the original and regulated traces, respectively. The star dots P_{dec} and round dots P_{buk} indicate the violation events computed by Eqn. (3) and detection by our algorithm, respectively. As expected, the two sets of dots match (Fig. 5(a)). We also compute the upper bound $\alpha_{R_{reg}}$ for the regulated trace R_{reg} (by Eqn. (3)) and compare with the input specification α^u. As the results shown in Fig. 5(b), $\alpha_{R_{reg}}$ is bounded by α^u. From Fig. 5, we experimentally confirm that our algorithm performs correctly.

We also report the resource consumption and latency for the Verilog HDL implementation. We synthesis the implementations for 2, 4, and 6 pairs of buckets using ALTERA Cyclone III EP3C120F780 development kit. The resource consumption is shown in Tab. 5. As shown in the table, the used logic elements even for the case of 6-pair buckets are still under 0.3% of the total resources (in total 119,088 logic elements in the FPGA board). Furthermore, the resource usage is linear w.r.t the number of bucket pairs, which indicates our algorithm can be used to regulate the runtime traces for complex arrive curves. Note that, in this implementation, the same FIFO buffer is used for all buckets as events of the trace belong to the same arrival curve. Therefore, the size of the FIFO module is independent on the number of buckets and is decided by the over-burst events that is intended to tolerate.

Regarding timing overhead, 6 cycles are needed to transfer an event from the input to the output of our IP in the case that no regulation is employed. As the working frequency of the FPGA is set to 50 Mhz, this latency corresponds 120 ns. This result indicates the timing overhead of our algorithm is considerably small, which can be integrated into the WCET of the events without significant side-effects for the timing

APPENDIX

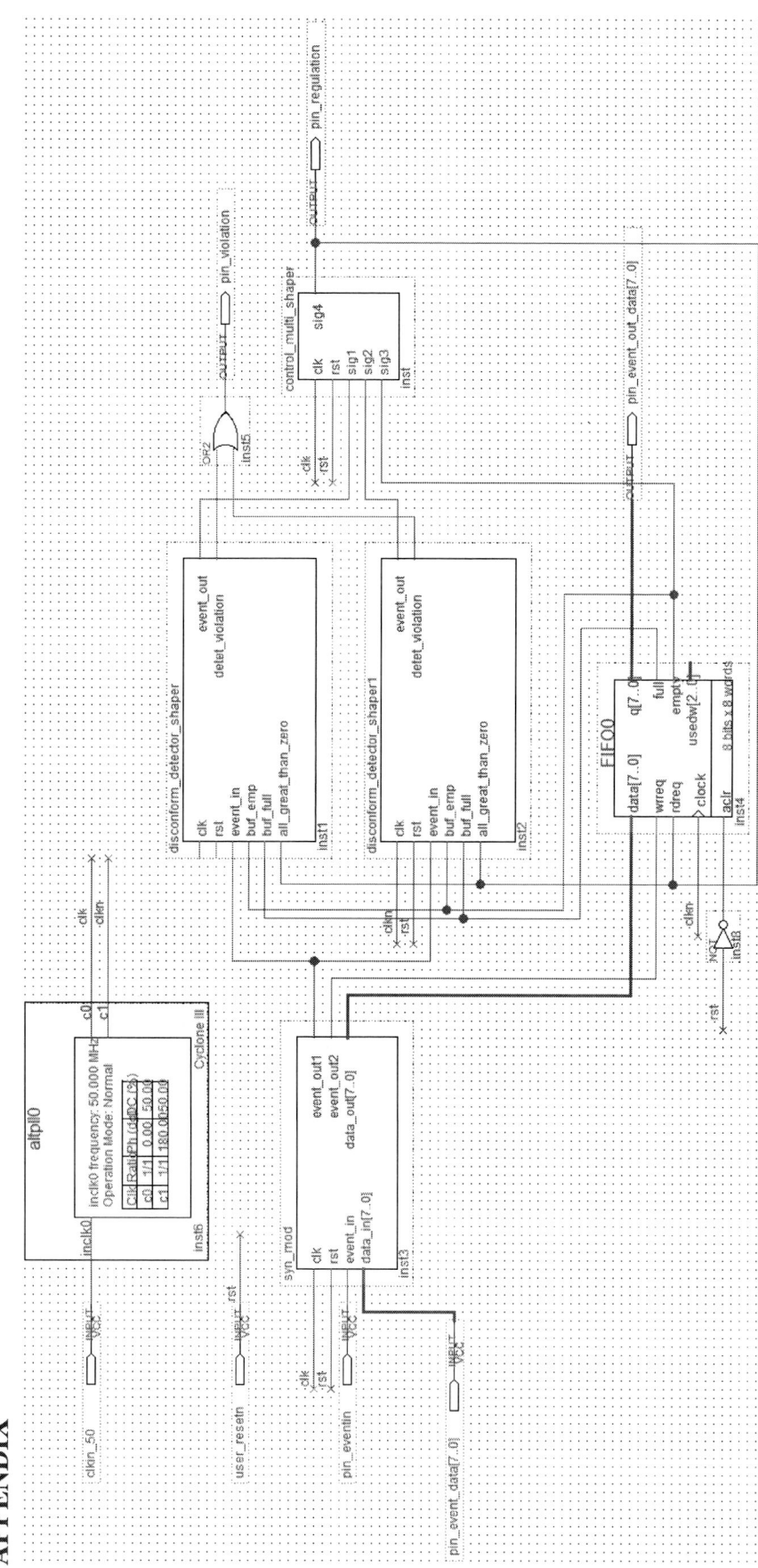

Figure 6: The detailed block diagram of algorithm IP (Fig. 4), implementing two bucket pairs. Pins pin_eventin and pin_event_data, respectively, are event synchronization signal and event data bus. Pin pin_violation represents a pulse signal when violation occurs. Regulation signals are generated at pin pin_regulaion and pin_event_out_data. When BFL_r value is larger than zero, pin_regulation asserts a high level and event data is put on pin_event_out_data at the same time.

436

STM Concurrency Control for Embedded Real-Time Software with Tighter Time Bounds

Mohammed El-Shambakey
ECE Dept., Virginia Tech
Blacksburg, VA 24060, USA
shambake@vt.edu

Binoy Ravindran
ECE Dept., Virginia Tech
Blacksburg, VA 24060, USA
binoy@vt.edu

ABSTRACT

We consider software transactional memory (STM) concurrency control for multicore real-time software, and present a novel contention manager (CM) for resolving transactional conflicts, called length-based CM (or LCM). We upper bound transactional retries and response times under LCM, when used with G-EDF and G-RMA schedulers. We identify the conditions under which LCM outperforms previous real-time STM CMs and lock-free synchronization. Our implementation and experimental studies reveal that G-EDF/LCM and G-RMA/LCM have shorter or comparable retry costs and response times than other synchronization techniques.

Categories and Subject Descriptors

C.3 [**Special-Purpose and Application-based Systems**]: Real-time and embedded systems

General Terms

Design, Experimentation, Measurement

Keywords

Software transactional memory (STM), real-time contention manager

1. INTRODUCTION

Lock-based concurrency control suffers from programmability, scalability, and composability challenges [12]. These challenges are exacerbated in emerging multicore architectures, on which improved software performance must be achieved by exposing greater concurrency. Transactional memory (TM) is an alternative synchronization model for shared memory objects that promises to alleviate these difficulties. With TM, programmers organize code that read/write shared objects as transactions, which appear to execute atomically. Two transactions conflict if they access the same object and one access is a write. When that happens, a contention manager (or CM) resolves the conflict by aborting one and allowing the other to commit, yielding (the illusion of) atomicity. In addition

to a simple programming model, TM provides performance comparable to highly concurrent fine-grained locking and lock-free approaches, and is composable. TM has been proposed in hardware, called HTM, and in software, called STM, with the usual tradeoffs: HTM has lesser overhead, but needs transactional support in hardware; STM is available on any hardware. See [11] for an excellent overview on TM.

Given STM's programmability, scalability, and composability advantages, we consider it for concurrency control in multicore real-time software. Doing so requires bounding transactional retries, as real-time threads, which subsume transactions, must satisfy time constraints. Retry bounds in STM are dependent on the CM policy at hand. Thus, real-time CM is logical.

Past research on real-time CM have proposed resolving transactional contention using dynamic and fixed priorities of parent threads, resulting in Earliest-Deadline-First-based CM (ECM) and Rate Monotonic Assignment-based CM (RCM), respectively [7–9]. These works show that, ECM and RCM, when used with the Global EDF (G-EDF) and Global RMA (G-RMA) multicore schedulers, respectively, achieve higher schedulability than lock-free synchronization techniques only under some ranges for the maximum atomic section length. This raises a fundamental question: is it possible to increase the atomic section length by an alternative CM design, so that STM's schedulability advantage has a larger coverage?

We answer this question by designing a novel CM that can be used with both dynamic and fixed priority (global) multicore real-time schedulers: length-based CM or LCM (Section 4.1). LCM resolves conflicts based on the priority of conflicting jobs, besides the length of the interfering atomic section, and the length of the interfered atomic section. We establish LCM's retry and response time upper bounds, when used with G-EDF (Section 4.2) and with G-RMA (Section 4.5) schedulers. We identify the conditions under which G-EDF/LCM outperforms ECM (Section 4.3) and lock-free synchronization (Section 4.4), and G-RMA/LCM outperforms RCM (Section 4.6). We implement LCM and competitor CM techniques in the Rochester STM framework [14] and conduct experimental studies (Section 5). Our study reveals that G-EDF/LCM and G-RMA/LCM have shorter or comparable retry costs and response times than competitors.

Thus, the paper's contribution is LCM with superior timeliness properties. This result thus allows programmers to reap STM's significant programmability and composability benefits for a broader range of multicore embedded real-time software than what was previously possible.

2. RELATED WORK

Transactional-like concurrency control without using locks, for real-time systems, has been previously studied in the context of non-blocking data structures (e.g., [1]). Despite their numerous advantages over locks (e.g., deadlock-freedom), their programmability has remained a challenge. Past studies show that they are best suited for simple data structures where their retry cost is competitive to the cost of lock-based synchronization [3]. In contrast, STM is semantically simpler [12], and is often the only viable lock-free solution for complex data structures (e.g., red/black tree) [10] and nested critical sections [15].

STM concurrency control for real-time systems has been previously studied in [2, 7, 9, 10, 13, 16, 17].

[13] proposes a restricted version of STM for uniprocessors. Uniprocessors do not need contention management.

[9] bounds response times in distributed systems with STM synchronization. They consider Pfair scheduling, limit to small atomic regions with fixed size, and limit transaction execution to span at most two quanta. In contrast, we allow transaction lengths with arbitrary duration.

[16] presents real-time scheduling of transactions and serializes transactions based on deadlines. However, the work does not bound retries and response times. In contrast, we establish such bounds.

[17] proposes real-time HTM. The work does not describe how transactional conflicts are resolved. Besides, the retry bound assumes that the worst case conflict between atomic sections of different tasks occurs when the sections are released at the same time. However, we show that this is not the worst case. We develop retry and response time upper bounds based on much worse conditions.

[10] upper bounds retries and response times for ECM with G-EDF, and identify the tradeoffs with locking and lock-free protocols. Similar to [17], [10] also assumes that the worst case conflict between atomic sections occurs when the sections are released simultaneously. The ideas in [10] are extended in [2], which presents three real-time CM designs. But no retry bounds or schedulability analysis techniques are presented for those CMs.

[7] presents the ECM and RCM contention managers, and upper bounds transactional retries and response times under them. The work also identifies the conditions under which ECM and RCM are superior to lock-free synchronization. In particular, they show that, STM's superiority holds only under some ranges for the maximum atomic section length. Our work builds upon this result.

3. PRELIMINARIES

We consider a multiprocessor system with m identical processors and n sporadic tasks $\tau_1, \tau_2, \ldots, \tau_n$. The kth instance (or job) of a task τ_i is denoted τ_i^k. Each task τ_i is specified by its worst case execution time (WCET) c_i, its minimum period T_i between any two consecutive instances, and its relative deadline D_i, where $D_i = T_i$. Job τ_i^j is released at time r_i^j and must finish no later than its absolute deadline $d_i^j = r_i^j + D_i$. Under a fixed priority scheduler such as G-RMA, p_i determines τ_i's (fixed) priority and it is constant for all instances of τ_i. Under a dynamic priority scheduler such as G-EDF, a job τ_i^j's priority, p_i^j, differs from one instance to another. A task τ_j may interfere with task τ_i for a number of times during an interval L, and this number is denoted as $G_{ij}(L)$.

Shared objects. A task may need to read/write shared, in-memory data objects while it is executing any of its atomic sections, which are synchronized using STM. The set of atomic sections of task τ_i is denoted s_i. s_i^k is the kth atomic section of τ_i. Each object, θ, can be accessed by multiple tasks. The set of distinct objects accessed by τ_i is θ_i without repeating objects. The set of atomic sections used by τ_i to access θ is $s_i(\theta)$, and the sum of the lengths of those atomic sections is $len(s_i(\theta))$. $s_i^k(\theta)$ is the kth atomic section of τ_i that accesses θ. $s_i^k(\theta)$ executes for a duration $len(s_i^k(\theta))$. The set of

tasks sharing θ with τ_i is denoted $\gamma_i(\theta)$.

Atomic sections are non-nested (supporting nested STM is future work). Each section is assumed to access only one object; this allows us to be consistent with the assumptions in [7], enabling a comparison with retry-loop lock-free synchronization [5], which is an important goal of this paper. The maximum-length atomic section in τ_i that accesses θ is denoted $s_{i_{max}}(\theta)$, while the maximum one among all tasks is $s_{max}(\theta)$, and the maximum one among tasks with priorities lower than that of τ_i is $s_{max}^i(\theta)$.

STM retry cost. If two or more atomic sections conflict, the CM will commit one section and abort and retry the others, increasing the time to execute the aborted sections. The increased time that an atomic section $s_i^p(\theta)$ will take to execute due to a conflict with another section $s_j^k(\theta)$, is denoted $W_i^p(s_j^k(\theta))$. If an atomic section, s_i^p, is already executing, and another atomic section s_j^k tries to access a shared object with s_i^p, then s_j^k is said to "interfere" or "conflict" with s_i^p. The job s_j^k is the "interfering job", and the job s_i^p is the "interfered job." The total time that a task τ_i's atomic sections have to retry over T_i is denoted $RC(T_i)$. The additional amount of time that a task τ_j causes to response time of τ_i when interfering with τ_i during L, without considering retries due to atomic sections, is denoted $W_{ij}(L)$.

4. LENGTH-BASED CM

LCM resolves conflicts based on the priority of conflicting jobs, besides the length of the interfering atomic section, and the length of the interfered atomic section. This is in contrast to ECM and RCM [7], where conflicts are resolved using the priority of the conflicting jobs. This strategy allows lower priority jobs, under LCM, to retry for lesser time than that under ECM and RCM, but higher priority jobs, sometimes, wait for lower priority ones with bounded priority-inversion.

4.1 Design and Rationale

Algorithm 1: LCM

Data: $s_i^k(\theta) \rightarrow$ interfered atomic section.
$s_j^l(\theta) \rightarrow$ interfering atomic section.
$\psi \rightarrow$ predefined threshold $\in [0, 1]$.
$\delta_i^k(\theta) \rightarrow$ remaining execution length of $s_i^k(\theta)$
Result: which atomic section of $s_i^k(\theta)$ or $s_j^l(\theta)$ aborts

1 **if** $p_i^k > p_j^l$ **then**
2 | $s_j^l(\theta)$ aborts;
3 **else**
4 | $c_{ij}^{kl} = len(s_j^l(\theta))/len(s_i^k(\theta));$
5 | $\alpha_{ij}^{kl} = ln(\psi)/(ln(\psi) - c_{ij}^{kl});$
6 | $\alpha = (len(s_i^k(\theta)) - \delta_i^k(\theta))/len(s_i^k(\theta));$
7 | **if** $\alpha \leq \alpha_{ij}^{kl}$ **then**
8 | | $s_i^k(\theta)$ aborts;
9 | **else**
10 | | $s_j^l(\theta)$ aborts;
11 | **end**
12 **end**

For both ECM and RCM, $s_i^k(\theta)$ can be totally repeated if $s_j^l(\theta)$ — which belongs to a higher priority job τ_j^b than τ_i^a — conflicts with $s_i^k(\theta)$ at the end of its execution, while $s_i^k(\theta)$ is just about to commit. Thus, LCM, shown in Algorithm 1, uses the remaining length of $s_i^k(\theta)$ when it is interfered, as well as $len(s_j^l(\theta))$, to decide which transaction must be aborted. If p_i^k was greater than p_j^l, then $s_i^k(\theta)$ would be the one that commits, because it belongs to a higher priority job, and it started before $s_j^l(\theta)$ (step 2). Otherwise, c_{ij}^{kl} is calculated (step 4) to determine whether it is worth aborting $s_i^k(\theta)$ in favor of $s_j^l(\theta)$, because $len(s_j^l(\theta))$ is relatively small compared to the remaining execution length of $s_i^k(\theta)$ (explained further).

We assume that:

$$c_{ij}^{kl} = len(s_j^l(\theta))/len(s_i^k(\theta)) \tag{1}$$

where $c_{ij}^{kl} \in\,]0,\infty[$, to cover all possible lengths of $s_j^l(\theta)$. Our idea is to reduce the opportunity for the abort of $s_i^k(\theta)$ if it is close to committing when interfered and $len(s_j^l(\theta))$ is large. This abort opportunity is increasingly reduced as $s_i^k(\theta)$ gets closer to the end of its execution, or $len(s_j^l(\theta))$ gets larger.

On the other hand, as $s_i^k(\theta)$ is interfered early, or $len(s_j^l(\theta))$ is small compared to $s_i^k(\theta)$'s remaining length, the abort opportunity is increased even if $s_i^k(\theta)$ is close to the end of its execution. To decide whether $s_i^k(\theta)$ must be aborted or not, we use a threshold value $\psi \in [0,1]$ that determines α_{ij}^{kl} (step 5), where α_{ij}^{kl} is the maximum percentage of $len(s_i^k(\theta))$ below which $s_j^l(\theta)$ is allowed to abort $s_i^k(\theta)$. Thus, if the already executed part of $s_i^k(\theta)$ — when $s_j^l(\theta)$ interferes with $s_i^k(\theta)$ — does not exceed $\alpha_{ij}^{kl}len(s_i^k(\theta))$, then $s_i^k(\theta)$ is aborted (step 8). Otherwise, $s_j^l(\theta)$ is aborted (step 10).

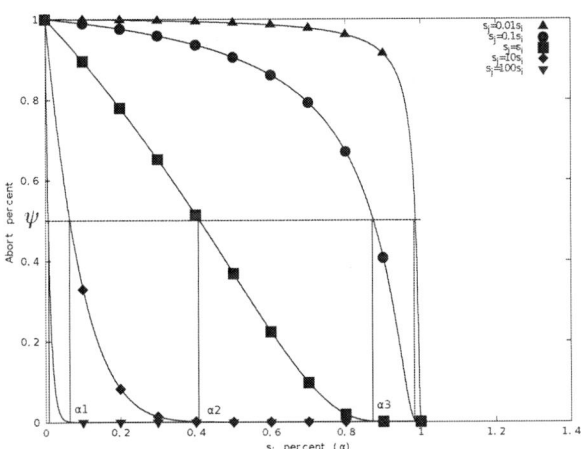

Figure 1: Interference of $s_i^k(\theta)$ by various lengths of $s_j^l(\theta)$

The behavior of LCM is illustrated in Figure 1. In this figure, the horizontal axis corresponds to different values of α ranging from 0 to 1, and the vertical axis corresponds to different values of abort opportunities, $f(c_{ij}^{kl},\alpha)$, ranging from 0 to 1 and calculated by (2):

$$f(c_{ij}^{kl},\alpha) = e^{\frac{-c_{ij}^{kl}\alpha}{1-\alpha}} \tag{2}$$

where c_{ij}^{kl} is calculated by (1).

Figure 1 shows one atomic section $s_i^k(\theta)$ (whose α changes along the horizontal axis) interfered by five different lengths of $s_j^l(\theta)$. For a predefined value of $f(c_{ij}^{kl},\alpha)$ (denoted as ψ in Algorithm 1), there corresponds a specific value of α (which is α_{ij}^{kl} in Algorithm 1) for each curve. For example, when $len(s_j^l(\theta)) = 0.1 \times len(s_i^k(\theta))$,

$s_j^l(\theta)$ aborts $s_i^k(\theta)$ if the latter has not executed more than $\alpha3$ percentage (shown in Figure 1) of its execution length. As $len(s_j^l(\theta))$ decreases, the corresponding α_{ij}^{kl} increases (as shown in Figure 1, $\alpha3 > \alpha2 > \alpha1$).

Equation (2) achieves the desired requirement that the abort opportunity is reduced as $s_i^k(\theta)$ gets closer to the end of its execution (as $\alpha \to 1$, $f(c_{ij}^{kl},1) \to 0$), or as the length of the conflicting transaction increases (as $c_{ij}^{kl} \to \infty$, $f(\infty,\alpha) \to 0$). Meanwhile, this abort opportunity is increased as $s_i^k(\theta)$ is interfered closer to its release (as $\alpha \to 0$, $f(c_{ij}^{kl},0) \to 1$), or as the length of the conflicting transaction decreases (as $c_{ij}^{kl} \to 0$, $f(0,\alpha) \to 1$).

LCM is not a centralized CM, which means that, upon a conflict, each transactions has to decide whether it must commit or abort.

CLAIM 1. *Let $s_j^l(\theta)$ interfere once with $s_i^k(\theta)$ at α_{ij}^{kl}. Then, the maximum contribution of $s_j^l(\theta)$ to $s_i^k(\theta)$'s retry cost is:*

$$W_i^k(s_j^l(\theta)) \le \alpha_{ij}^{kl}len\left(s_i^k(\theta)\right) + len\left(s_j^l(\theta)\right) \tag{3}$$

(Proofs of all claims are provided in the Supplementary Material section at the end of the paper.)

CLAIM 2. *An atomic section of a higher priority job, τ_j^b, may have to abort and retry due to a lower priority job, τ_i^a, if $s_j^l(\theta)$ interferes with $s_i^k(\theta)$ after the α_{ij}^{kl} percentage. τ_j's retry time, due to $s_i^k(\theta)$ and $s_j^l(\theta)$, is upper bounded by:*

$$W_j^l(s_i^k(\theta)) \le \left(1 - \alpha_{ij}^{kl}\right)len\left(s_i^k(\theta)\right) \tag{4}$$

CLAIM 3. *A higher priority job, τ_i^z, suffers from priority inversion for at most number of atomic sections in τ_i^z.*

CLAIM 4. *The maximum delay suffered by $s_j^l(\theta)$ due to priority inversion is caused by the maximum length atomic section accessing object θ, which belongs to a lower priority job than τ_j^b that owns $s_j^l(\theta)$.*

4.2 Response Time of G-EDF/LCM

CLAIM 5. *$RC(T_i)$ for a task τ_i under G-EDF/LCM is upper bounded by:*

$$
\begin{aligned}
RC(T_i) = &\left(\sum_{\forall \tau_h \subset \gamma_i} \sum_{\forall \theta \subset \theta_i \wedge \theta_h} \left(\left\lceil \frac{T_i}{T_h} \right\rceil \sum_{\forall s_h^l(\theta)} len\left(s_h^l(\theta)\right) \right.\right. \\
& \left.\left. + \; \alpha_{max}^{hl}len\left(s_{max}^h(\theta)\right) \right) \right) \\
& + \sum_{\forall s_i^i(\theta)} \left(1 - \alpha_{max}^{iy}\right)len\left(s_{max}^i(\theta)\right) \tag{5}
\end{aligned}
$$

where α_{max}^{hl} is the α value that corresponds to ψ due to the interference of $s_{max}^h(\theta)$ by $s_h^l(\theta)$. α_{max}^{iy} is the α value that corresponds to ψ due to the interference of $s_{max}^i(\theta)$ by $s_i^y(\theta)$.

Response time of τ_i is calculated by (11) in [7].

439

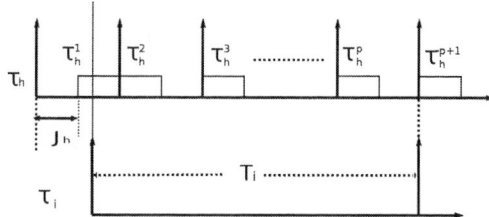

Figure 2: τ_h^p **has a higher priority than** τ_i^x

4.3 Schedulability of G-EDF/LCM and ECM

We now compare the schedulability of G-EDF/LCM with ECM [7] to understand when G-EDF/LCM will perform better. Toward this, we compare the total utilization of ECM with that of G-EDF/LCM. For each method, we inflate the c_i of each task τ_i by adding the retry cost suffered by τ_i. Thus, if method A adds retry cost $RC_A(T_i)$ to c_i, and method B adds retry cost $RC_B(T_i)$ to c_i, then the schedulability of A and B are compared as:

$$\sum_{\forall \tau_i} \frac{c_i + RC_A(T_i)}{T_i} \quad \leq \quad \sum_{\forall \tau_i} \frac{c_i + RC_B(T_i)}{T_i}$$

$$\sum_{\forall \tau_i} \frac{RC_A(T_i)}{T_i} \quad \leq \quad \sum_{\forall \tau_i} \frac{RC_B(T_i)}{T_i} \qquad (6)$$

Thus, schedulability is compared by substituting the retry cost added by the synchronization methods in (6).

CLAIM 6. *Let s_{max} be the maximum length atomic section accessing any object θ. Let α_{max} and α_{min} be the maximum and minimum values of α for any two atomic sections $s_i^k(\theta)$ and $s_j^l(\theta)$. Given a threshold ψ, schedulability of G-EDF/LCM is equal or better than ECM if for any task τ_i:*

$$\frac{1 - \alpha_{min}}{1 - \alpha_{max}} \leq \sum_{\forall \tau_h \in \gamma_i} \left\lceil \frac{T_i}{T_h} \right\rceil \qquad (7)$$

4.4 G-EDF/LCM versus Lock-free

We consider the retry-loop lock-free synchronization for G-EDF given in [5]. This lock-free approach is the most relevant to our work.

CLAIM 7. *Let s_{max} denote $len(s_{max})$ and r_{max} denote the maximum execution cost of a single iteration of any retry loop of any task in the retry-loop lock-free algorithm in [5]. Now, G-EDF/LCM achieves higher schedulability than the retry-loop lock-free approach if the upper bound on s_{max}/r_{max} under G-EDF/LCM ranges between 0.5 and 2 (which is higher than that under ECM).*

4.5 Response Time of G-RMA/LCM

CLAIM 8. *Let*

$$\lambda_2(j, \theta) = \sum_{\forall s_j^l(\theta)} len(s_j^l(\theta)) + \alpha_{max}^{jl} len(s_{max}^j(\theta))$$

where α_{max}^{jl} is the α value corresponding to ψ due to the interference of $s_{max}^j(\theta)$ by $s_j^l(\theta)$. The retry cost of any task τ_i under G-RMA/LCM during T_i is given by:

$$RC(T_i) = \sum_{\forall \tau_j^*} \left(\sum_{\theta \in (\theta_i \wedge \theta_j)} \left(\left(\left\lceil \frac{T_i}{T_j} \right\rceil + 1 \right) \lambda_2(j, \theta) \right) \right)$$

$$+ \sum_{\forall s_i^y(\theta)} \left(1 - \alpha_{max}^{iy} \right) len\left(s_{max}^i(\theta) \right) \qquad (8)$$

where $\tau_j^* = \{\tau_j | (\tau_j \in \gamma_i) \wedge (p_j > p_i)\}$.

The response time is calculated by (17) in [7] with replacing $RC(R_i^{up})$ with $RC(T_i)$.

4.6 Schedulability of G-RMA/LCM and RCM

CLAIM 9. *Under the same assumptions of Claims 6 and 8, G-RMA/LCM's schedulability is equal or better than RCM if:*

$$\frac{1 - \alpha_{min}}{1 - \alpha_{max}} \leq \sum_{\forall \tau_j} \left(\left\lceil \frac{T_i}{T_j} \right\rceil + 1 \right) \qquad (9)$$

5. EXPERIMENTAL EVALUATION

Having established LCM's retry and response time upper bounds, and the conditions under which it outperforms ECM, RCM, and lock-free synchronization, we now would like to understand how LCM's retry and response times in practice (i.e., on average) compare with that of competitor methods. Since this can only be understood experimentally, we implement LCM and the competitor methods and conduct experimental studies.

5.1 Experimental Setup

We used the ChronOS real-time Linux kernel [4] and the RSTM library [14]. We modified RSTM to include implementations of ECM, RCM, G-EDF/LCM, and G-RMA/LCM contention managers, and modified ChronOS to include implementations of G-EDF and G-RMA schedulers.

For the retry-loop lock-free implementation, we used a loop that reads an object and attempts to write to the object using a compare-and-swap (CAS) instruction. The task retries until the CAS succeeds.

We use an 8 core, 2GHz AMD Opteron platform. The average time taken for one write operation by RSTM on any core is $0.0129653375 \mu s$, and the average time taken by one CAS-loop operation on any core is $0.0292546250 \, \mu s$.

We used the periodic task set shown in Table 1. Each task runs in its own thread and has an atomic section. Atomic section properties are probabilistically controlled (for experimental evaluation) using three parameters: the maximum and minimum lengths of any atomic section within the task, and the total length of atomic sections within any task. All task atomic sections access the same object, and do write operations on the object (thus, contention is the highest).

5.2 Results

Figure 3 shows the retry cost (RC) for each task in the three task sets in Table 1, where each task's atomic section length is equal to half of the task length. Each data point in the figure has a confidence level of 0.95. We observe that G-EDF/LCM and G-RMA/LCM achieve shorter or comparable retry cost than ECM and RCM. Since all tasks are initially released at the same time, and due to the specific nature of task properties, tasks with lower IDs somehow have higher priorities under the G-EDF scheduler. Note that tasks with lower IDs have higher priorities under G-RMA, since tasks are ordered in non-decreasing order of their periods.

Thus, we observe that G-EDF/LCM and G-RMA/LCM achieve comparable retry costs to ECM and RCM for some tasks with lower IDs. But when task ID increases, LCM — for both schedulers — achieves much shorter retry costs than ECM and RCM. This is because, higher priority tasks in LCM suffers blocking by lower priority tasks, which is not the case for ECM and RCM. However, as task priority decreases, LCM, by definition, prevents higher priority tasks from aborting lower priority ones if a higher priority task

440

Table 1: Task sets. (a) Task set 1: 5-task set; (b) Task set 2: 10-task set; (c) Task set 3: 12-task set

(a)

	$T_i(\mu s)$	$c_i(\mu s)$
τ_1	500000	150000
τ_2	1000000	227000
τ_3	1500000	410000
τ_4	3000000	299000
τ_5	5000000	500000

(b)

	$T_i(\mu s)$	$c_i(\mu s)$
τ_1	400000	75241
τ_2	750000	69762
τ_3	1200000	267122
τ_4	1500000	69863
τ_5	2400000	152014
τ_6	4000000	286301
τ_7	7500000	493150
τ_8	10000000	794520
τ_9	15000000	1212328
τ_{10}	20000000	1775342

(c)

	$T_i(\mu s)$	$c_i(\mu s)$
τ_1	400000	58195
τ_2	750000	53963
τ_3	1000000	206330
τ_4	1200000	53968
τ_5	1500000	117449
τ_6	2400000	221143
τ_7	3000000	290428
τ_8	4000000	83420
τ_9	7500000	380917
τ_{10}	10000000	613700
τ_{11}	15000000	936422
τ_{12}	20000000	1371302

interferes with a lower priority one after a specified threshold. In contrast, under ECM and RCM, lower priority tasks abort in favor of higher priority ones. G-EDF/LCM and G-RMA/LCM also achieve shorter retry costs than the retry-loop lock-free algorithm.

Figure 4 shows the response time of each task in the Table 1 task sets with a confidence level of 0.95. (Again, each task's atomic section length is equal to half of the task length.) We observe that G-EDF/LCM and G-RMA/LCM achieve shorter response time than the retry-loop lock-free algorithm, and shorter or comparable response time than ECM and RCM.

We repeated the experiments by varying three parameters: the relative total length of all atomic sections to the length of the task, the maximum relative length of any atomic section to the length of the task, and the minimum relative length of any atomic section to the length of the task. Full set of results are omitted here due to space constraints; however additional results are included in the Supplementary Material section. Full set of results are given in Appendix B in [6].

6. CONCLUSIONS

In ECM and RCM, a task incurs at most $2s_{max}$ retry cost for each of its atomic section due to conflict with another task's atomic section. With LCM, this retry cost is reduced to $(1+\alpha_{max})s_{max}$ for each aborted atomic section. In ECM and RCM, tasks do not retry due to lower priority tasks, whereas in LCM, they do so. In G-EDF/LCM, retry due to a lower priority job is encountered only from a task τ_j's last job instance during τ_i's period. This is not the case with G-RMA/LCM, because, each higher priority task can be aborted and retried by any job instance of lower priority tasks. Schedulability of G-EDF/LCM and G-RMA/LCM is better or equal to ECM and RCM, respectively, by proper choices for

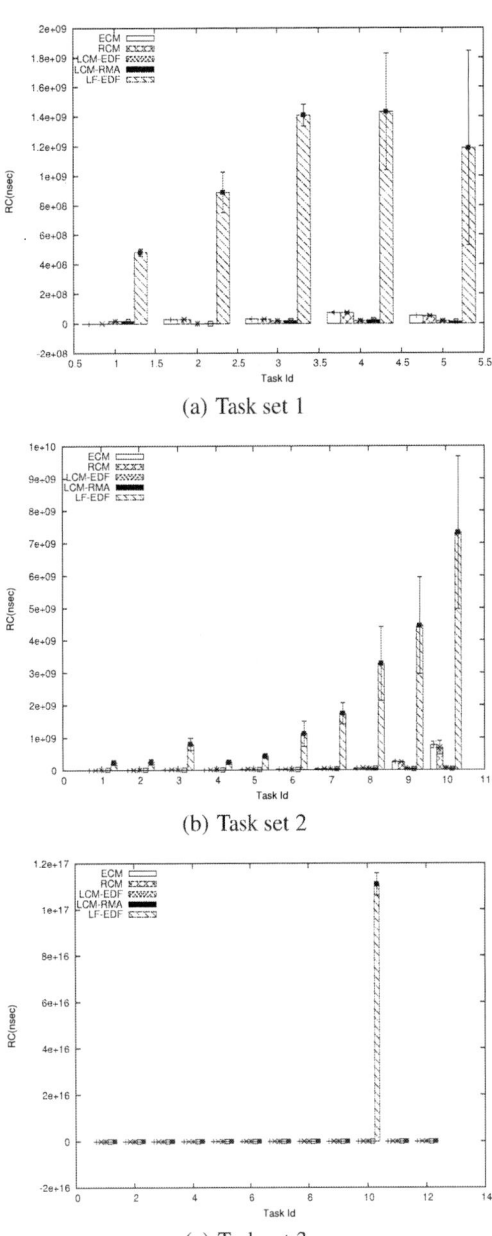

(a) Task set 1

(b) Task set 2

(c) Task set 3

Figure 3: Task retry costs under LCM and competitor synchronization methods

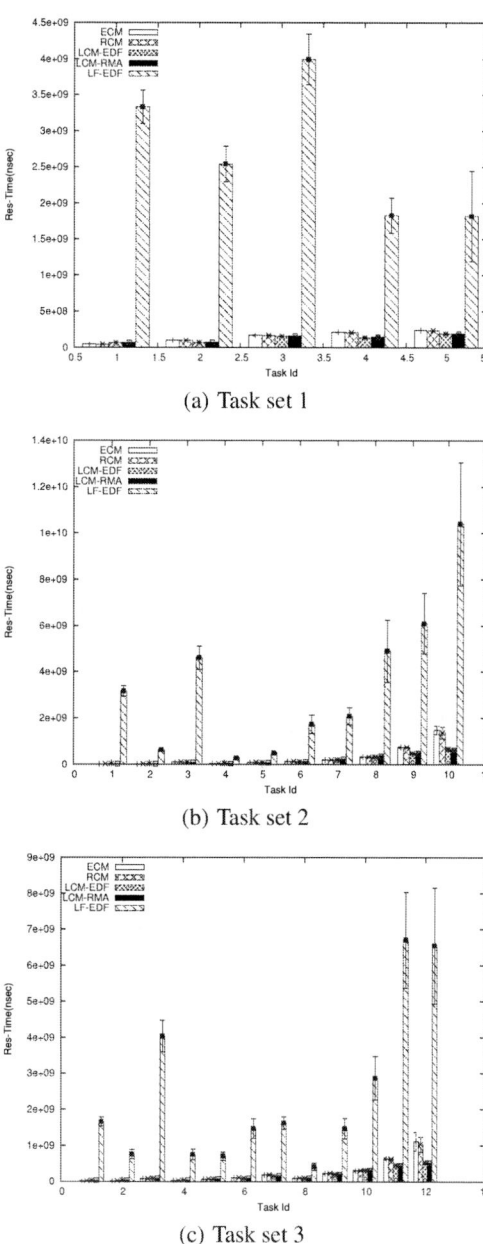

(a) Task set 1

(b) Task set 2

(c) Task set 3

Figure 4: Task response times under LCM and competitor synchronization methods

α_{min} and α_{max}. Schedulability of G-EDF/LCM is better than retry-loop lock-free synchronization for G-EDF if the upper bound on s_{max}/r_{max} is between 0.5 and 2, which is higher than that achieved by ECM.

Acknowledgments

This work is supported in part by NSF CNS 0915895, NSF CNS 1116190, and NSF CNS 1130180.

7. REFERENCES

[1] J. Anderson, S. Ramamurthy, and K. Jeffay. Real-time computing with lock-free shared objects. In *RTSS*, pages 28–37, 1995.

[2] A. Barros and L. Pinho. Managing contention of software transactional memory in real-time systems. In *IEEE RTSS, Work-In-Progress*, 2011.

[3] B. B. Brandenburg et al. Real-time synchronization on multiprocessors: To block or not to block, to suspend or spin? In *RTAS*, pages 342–353, 2008.

[4] M. Dellinger, P. Garyali, and B. Ravindran. Chronos linux: a best-effort real-time multiprocessor linux kernel. In *Design Automation Conference (DAC), 2011 48th ACM/EDAC/IEEE*, pages 474–479. IEEE, 2011.

[5] U. C. Devi, H. Leontyev, and J. H. Anderson. Efficient synchronization under global EDF scheduling on multiprocessors. In *ECRTS*, pages 75–84, 2006.

[6] M. El-Shambakey and B. Ravindran. On the design of real-time stm contention managers. Technical report, ECE Department, Virginia Tech, 2011. Available as `http://www.real-time.ece.vt.edu/tech-report-rt-stm-cm11.pdf`.

[7] M. Elshambakey and B. Ravindran. Stm concurrency control for multicore embedded real-time software: Time bounds and tradeoffs. In *SAC*, 2012.

[8] S. Fahmy, B. Ravindran, and E. Jensen. Response time analysis of software transactional memory-based distributed real-time systems. In *ACM SAC*, pages 334–338, 2009.

[9] S. Fahmy, B. Ravindran, and E. D. Jensen. On bounding response times under software transactional memory in distributed multiprocessor real-time systems. In *DATE*, pages 688–693, 2009.

[10] S. F. Fahmy. *Collaborative Scheduling and Synchronization of Distributable Real-Time Threads*. PhD thesis, Virginia Tech, 2010.

[11] T. Harris, J. Larus, and R. Rajwar. *Transactional Memory*. Morgan & Claypool Publishers, 2nd. edition, December 2010.

[12] M. Herlihy. The art of multiprocessor programming. In *PODC*, pages 1–2, 2006.

[13] J. Manson, J. Baker, et al. Preemptible atomic regions for real-time Java. In *RTSS*, pages 10–71, 2006.

[14] V. Marathe, M. Spear, C. Heriot, A. Acharya, D. Eisenstat, W. Scherer III, and M. Scott. Lowering the overhead of nonblocking software transactional memory. In *Workshop on Languages, Compilers, and Hardware Support for Transactional Computing (TRANSACT)*, 2006.

[15] B. Saha, A.-R. Adl-Tabatabai, et al. McRT-STM: a high performance software transactional memory system for a multi-core runtime. In *PPoPP*, pages 187–197, 2006.

[16] T. Sarni, A. Queudet, and P. Valduriez. Real-time support for software transactional memory. In *RTCSA*, pages 477–485, 2009.

[17] M. Schoeberl, F. Brandner, and J. Vitek. RTTM: Real-time transactional memory. In *ACM SAC*, pages 326–333, 2010.

Supplementary Material

This section includes proofs of all Claims.

S.1 Proof of Claim 1

PROOF. If $s_j^l(\theta)$ interferes with $s_i^k(\theta)$ at a Υ percentage, where $\Upsilon < \alpha_{ij}^{kl}$, then the retry cost of $s_i^k(\theta)$ is $\Upsilon len(s_i^k(\theta)) + len(s_j^l(\theta))$, which is lower than that calculated in (3). Besides, if $s_j^l(\theta)$ interferes with $s_i^k(\theta)$ after α_{ij}^{kl} percentage, then $s_i^k(\theta)$ will not abort. □

442

S.2 Proof of Claim 2

PROOF. It is derived directly from Claim 1, as $s_j^l(\theta)$ will have to retry for the remaining length of $s_i^k(\theta)$. □

S.3 Proof of Claim 3

PROOF. Assuming three atomic sections, $s_i^k(\theta)$, $s_j^l(\theta)$ and $s_a^b(\theta)$, where $p_j > p_i$ and $s_j^l(\theta)$ interferes with $s_i^k(\theta)$ after α_{ij}^{kl}. Then $s_j^l(\theta)$ will have to abort and retry. At this time, if $s_a^b(\theta)$ interferes with the other two atomic sections, and the LCM decides which transaction to commit based on comparison between each two transactions. So, we have the following cases:-

- $p_a < p_i < p_j$, then $s_a^b(\theta)$ will not abort any one because it is still in its beginning and it is of the lowest priority. So. τ_j is not indirectly blocked by τ_a.

- $p_i < p_a < p_j$ and even if $s_a^b(\theta)$ interferes with $s_i^k(\theta)$ before α_{ia}^{kb}, so, $s_a^b(\theta)$ is allowed abort $s_i^k(\theta)$. Comparison between $s_j^l(\theta)$ and $s_a^b(\theta)$ will result in LCM choosing $s_j^l(\theta)$ to commit and abort $s_a^b(\theta)$ because the latter is still beginning, and τ_j is of higher priority. If $s_a^b(\theta)$ is not allowed to abort $s_i^k(\theta)$, the situation is still the same, because $s_j^l(\theta)$ was already retrying until $s_i^k(\theta)$ finishes.

- $p_a > p_j > p_i$, then if $s_a^b(\theta)$ is chosen to commit, this is not priority inversion for τ_j because τ_a is of higher priority.

- if τ_a preempts τ_i, then LCM will compare only between $s_j^l(\theta)$ and $s_a^b(\theta)$. If $p_a < p_j$, then $s_j^l(\theta)$ will commit because of its task's higher priority and $s_a^b(\theta)$ is still at its beginning, otherwise, $s_j^l(\theta)$ will retry, but this will not be priority inversion because τ_a is already of higher priority than τ_j. If τ_a does not access any object but it preempts τ_i, then CM will choose $s_j^l(\theta)$ to commit as only already running transactions are competing together.

So, by generalizing these cases to any number of conflicting jobs, it is seen that when an atomic section, $s_j^l(\theta)$, of a higher priority job is in conflict with a number of atomic sections belonging to lower priority jobs, $s_j^l(\theta)$ can suffer from priority inversion by only one of them. So, each higher priority job can suffer priority inversion at most its number of atomic section. Claim follows. □

S.4 Proof of Claim 4

PROOF. Assume three atomic sections, $s_i^k(\theta)$, $s_j^l(\theta)$, and $s_h^z(\theta)$, where $p_j > p_i$, $p_j > p_h$, and $len(s_i^k(\theta)) > len(s_h^z(\theta))$. Now, $\alpha_{ij}^{kl} > \alpha_{hj}^{zl}$ and $c_{ij}^{kl} < c_{hj}^{zl}$. By applying (4) to obtain the contribution of $s_i^k(\theta)$ and $s_h^z(\theta)$ to the priority inversion of $s_j^l(\theta)$ and dividing them, we get:

$$\frac{W_j^l(s_i^k(\theta))}{W_j^l(s_h^z(\theta))} = \frac{\left(1 - \alpha_{ij}^{kl}\right)len(s_i^k(\theta))}{\left(1 - \alpha_{hj}^{zl}\right)len(s_h^z(\theta))}$$

By substitution for αs from (2):

$$= \frac{(1 - \frac{ln\psi}{ln\psi - c_{ij}^{kl}})len(s_i^k(\theta))}{(1 - \frac{ln\psi}{ln\psi - c_{hj}^{zl}})len(s_h^z(\theta))} = \frac{(\frac{-c_{ij}^{kl}}{ln\psi - c_{ij}^{kl}})len(s_i^k(\theta))}{(\frac{-c_{hj}^{zl}}{ln\psi - c_{hj}^{zl}})len(s_h^z(\theta))}$$

$\because ln\psi \le 0$ and $c_{ij}^{kl}, c_{hj}^{kl} > 0, \therefore$ by substitution from (1)

$$= \frac{len(s_j^l(\theta))/(ln\psi - c_{ij}^{kl})}{len(s_j^l(\theta))/(ln\psi - c_{hj}^{zl})} = \frac{ln\psi - c_{hj}^{zl}}{ln\psi - c_{ij}^{kl}} > 1$$

Thus, as the length of the interfered atomic section increases, the effect of priority inversion on the interfering atomic section increases. Claim follows. □

S.5 Proof of Claim 5

PROOF. The maximum number of higher priority instances of τ_h that can interfere with τ_i^x is $\left\lceil \frac{T_i}{T_h} \right\rceil$, as shown in Figure 2, where one instance of τ_h and τ_h^p coincides with the absolute deadline of τ_i^x.

By using Claims 1, 2, 3, and 4, and Claim 1 in [7] to determine the effect of atomic sections belonging to higher and lower priority instances of interfering tasks to τ_i^x, claim follows. □

S.6 Proof of Claim 6

PROOF. Under ECM, $RC(T_i)$ is upper bounded by:

$$RC(T_i) \le \sum_{\forall \tau_h \in \gamma_i} \sum_{\forall \theta \in (\theta_i \wedge \theta_h)} \left(\left\lceil \frac{T_i}{T_h} \right\rceil \sum_{\forall s_h^z(\theta)} 2len(s_{max}) \right) \quad (10)$$

with the assumption that all lengths of atomic sections of (4) and (8) in [7] and (5) are replaced by s_{max}. Let α_{max}^{hl} in (5) be replaced with α_{max}, and α_{max}^{iy} in (5) be replaced with α_{min}. As α_{max}, α_{min}, and $len(s_{max})$ are all constants, (5) is upper bounded by:

$$RC(T_i) \le \left(\sum_{\forall \tau_h \in \gamma_i} \sum_{\forall \theta \in \theta_i \wedge \theta_h} \left(\left\lceil \frac{T_i}{T_h} \right\rceil \sum_{\forall s_h^z(\theta)} (1 + \alpha_{max}) \right. \right.$$
$$\left. \left. len\left(s_{max}\right) \right) \right) + \sum_{\forall s_i^z(\theta)} \left(1 - \alpha_{min}\right) len\left(s_{max}\right)$$
$$(11)$$

If β_1^{ih} is the total number of times any instance of τ_h accesses shared objects with τ_i, then $\beta_1^{ih} = \sum_{\forall \theta \in (\theta_i \wedge \theta_h)} \sum_{\forall s_h^z(\theta)}$. Furthermore, if β_2^i is the total number of times any instance of τ_i accesses shared objects with any other instance, $\beta_2^i = \sum_{\forall s_i^y(\theta)}$, *where θ is shared with another task*. Then, $\beta_i = max\{max_{\forall \tau_h \in \gamma_i}\{\beta_1^{ih}\}, \beta_2^i\}$ is the maximum number of accesses to all shared objects by any instance of τ_i or τ_h. Thus, (10) becomes:

$$RC(T_i) \le \sum_{\tau_h \in \gamma_i} 2 \left\lceil \frac{T_i}{T_h} \right\rceil \beta_i len(s_{max}) \quad (12)$$

and (11) becomes:

$$RC(T_i) \le \beta_i len(s_{max}) \left((1 - \alpha_{min}) + \sum_{\forall \tau_h \in \gamma_i} \left\lceil \frac{T_i}{T_h} \right\rceil (1 + \alpha_{max}) \right) \quad (13)$$

We can now compare the total utilization of G-EDF/LCM with

that of ECM by comparing (11) and (13) for all τ_i:

$$\sum_{\forall \tau_i} \frac{(1-\alpha_{min}) + \sum_{\forall \tau_h \in \gamma_i}\left(\left\lceil\frac{T_i}{T_h}\right\rceil(1+\alpha_{max})\right)}{T_i}$$

$$\leq \sum_{\forall \tau_i} \frac{\sum_{\forall \tau_h \in \gamma_i} 2\left\lceil\frac{T_i}{T_h}\right\rceil}{T_i} \qquad (14)$$

(14) is satisfied if for each τ_i, the following condition is satisfied:

$$(1-\alpha_{min}) + \sum_{\forall \tau_h \in \gamma_i}\left(\left\lceil\frac{T_i}{T_h}\right\rceil(1+\alpha_{max})\right) \leq 2\sum_{\forall \tau_h \in \gamma_i}\left\lceil\frac{T_i}{T_h}\right\rceil$$

$$\therefore \frac{1-\alpha_{min}}{1-\alpha_{max}} \leq \sum_{\forall \tau_h \in \gamma_i}\left\lceil\frac{T_i}{T_h}\right\rceil$$

Claim follows. $\square$

S.7 Proof of Claim 7

PROOF. From [5], the retry-loop lock-free algorithm is upper bounded by:

$$RL(T_i) = \sum_{\tau_h \in \gamma_i}\left(\left\lceil\frac{T_i}{T_h}\right\rceil + 1\right)\beta_i r_{max} \qquad (15)$$

where β_i is as defined in Claim 6. The retry cost of τ_i in G-EDF/LCM is upper bounded by (13). By comparing G-EDF/LCM's total utilization with that of the retry-loop lock-free algorithm, we get:

$$\sum_{\forall \tau_i} \frac{\left((1-\alpha_{min}) + \sum_{\forall \tau_h \in \gamma_i}\left(\left\lceil\frac{T_i}{T_h}\right\rceil(1+\alpha_{max})\right)\right)\beta_i s_{max}}{T_i}$$

$$\leq \sum_{\forall \tau_i} \frac{\sum_{\forall \tau_h \in \gamma_i}\left(\left\lceil\frac{T_i}{T_h}\right\rceil + 1\right)\beta_i r_{max}}{T_i}$$

$$\therefore \frac{s_{max}}{r_{max}} \leq \frac{\sum_{\forall \tau_i}\frac{\sum_{\forall \tau_h \in \gamma_i}\left(\left\lceil\frac{T_i}{T_h}\right\rceil + 1\right)\beta_i}{T_i}}{\sum_{\forall \tau_i}\frac{\left((1-\alpha_{min}) + \sum_{\forall \tau_h \in \gamma_i}\left(\left\lceil\frac{T_i}{T_h}\right\rceil(1+\alpha_{max})\right)\right)\beta_i}{T_i}} \qquad (16)$$

Let the number of tasks that have shared objects with τ_i be ω (i.e., $\sum_{\tau_h \in \gamma_i} = \omega \geq 1$ since at least one task has a shared object with τ_i; otherwise, there is no conflict between tasks). Let the total number of tasks be n, so $1 \leq \omega \leq n-1$, and $\left\lceil\frac{T_i}{T_h}\right\rceil \in [1,\infty[$. To find the minimum and maximum values for the upper bound on s_{max}/r_{max}, we consider the following cases:

- $\alpha_{min} \to 0, \alpha_{max} \to 0$

$\therefore$ (16) will be:

$$\frac{s_{max}}{r_{max}} \leq 1 + \frac{\sum_{\forall \tau_i}\frac{\omega - 1}{T_i}}{\sum_{\forall \tau_i}\frac{1 + \sum_{\forall \tau_h \in \gamma_i}\left\lceil\frac{T_i}{T_h}\right\rceil}{T_i}} \qquad (17)$$

By substituting the edge values for ω and $\left\lceil\frac{T_i}{T_h}\right\rceil$ in (17), we derive that the upper bound on s_{max}/r_{max} lies between 1 and 2.

- $\alpha_{min} \to 0, \alpha_{max} \to 1$

(16) becomes

$$\frac{s_{max}}{r_{max}} \leq 0.5 + \frac{\sum_{\forall \tau_i}\frac{\omega - 0.5}{T_i}}{\sum_{\forall \tau_i}\frac{1 + 2\sum_{\forall \tau_h \in \gamma_i}\left\lceil\frac{T_i}{T_h}\right\rceil}{T_i}} \qquad (18)$$

By applying the edge values for ω and $\left\lceil\frac{T_i}{T_h}\right\rceil$ in (18), we derive that the upper bound on s_{max}/r_{max} lies between 0.5 and 1.

- $\alpha_{min} \to 1, \alpha_{max} \to 0$

This case is rejected since $\alpha_{min} \leq \alpha_{max}$.

- $\alpha_{min} \to 1, \alpha_{max} \to 1$

$\therefore$ (16) becomes:

$$\frac{s_{max}}{r_{max}} \leq 0.5 + \frac{\sum_{\tau_i}\frac{\omega}{T_i}}{2\sum_{\tau_i}\frac{\sum_{\forall \tau_h \in \gamma_i}\left\lceil\frac{T_i}{T_h}\right\rceil}{T_i}} \qquad (19)$$

By applying the edge values for ω and $\left\lceil\frac{T_i}{T_h}\right\rceil$ in (19), we derive that the upper bound on s_{max}/r_{max} lies between 0.5 and 1, which is similar to that achieved by ECM.

Summarizing from the previous cases, the upper bound on s_{max}/r_{max} lies between 0.5 and 2, whereas for ECM [7], it lies between 0.5 and 1. Claim follows.

$\square$

S.8 Proof of Claim 8

PROOF. Under G-RMA, all instances of a higher priority task, τ_j, can conflict with a lower priority task, τ_i, during T_i. (3) can be used to determine the contribution of each conflicting atomic section in τ_j to τ_i. Meanwhile, all instances of any task with lower priority than τ_i can conflict with τ_i during T_i. Claims 2 and 3 can be used to determine the contribution of conflicting atomic sections in lower priority tasks to τ_i. Using the previous notations and Claim 3 in [7], the claim follows. $\square$

S.9 Proof of Claim 9

PROOF. Under the same assumptions as that of Claims 6 and 8, (8) can be upper bounded as:

$$RC(T_i) \leq \sum_{\forall \tau_j^*}\left(\left(\left\lceil\frac{T_i}{T_j}\right\rceil + 1\right)(1+\alpha_{max})len(s_{max})\beta_i\right)$$

$$+ \quad (1-\alpha_{min})len(s_{max})\beta_i \qquad (20)$$

For RCM, (16) in [7] for $RC(T_i)$ is upper bounded by:

$$RC(T_i) \leq \sum_{\forall \tau_j^*}\left(\left\lceil\frac{T_i}{T_j}\right\rceil + 1\right)2\beta_i len(s_{max})$$

By comparing the total utilization of G-RMA/LCM with that of RCM, we get:

$$\sum_{\forall \tau_i}\frac{len(s_{max})\beta_i\left((1-\alpha_{min}) + \sum_{\forall \tau_j^*}\left(\left(\left\lceil\frac{T_i}{T_j}\right\rceil + 1\right)(1+\alpha_{max})\right)\right)}{T_i}$$

$$\leq \sum_{\forall \tau_i}\frac{2len(s_{max})\beta_i\sum_{\forall \tau_j^*}\left(\left\lceil\frac{T_i}{T_j}\right\rceil + 1\right)}{T_i} \qquad (21)$$

(21) is satisfied if $\forall \tau_i$ (9) is satisfied. Claim follows. $\square$

S.10 EXTENDED RESULTS

The three parameters x, y, z for each figure specify respectively the relative total length of all atomic sections to the length of the task, the maximum relative length of any atomic section to the length of the task, and the minimum relative length of any atomic section to the length of the task.

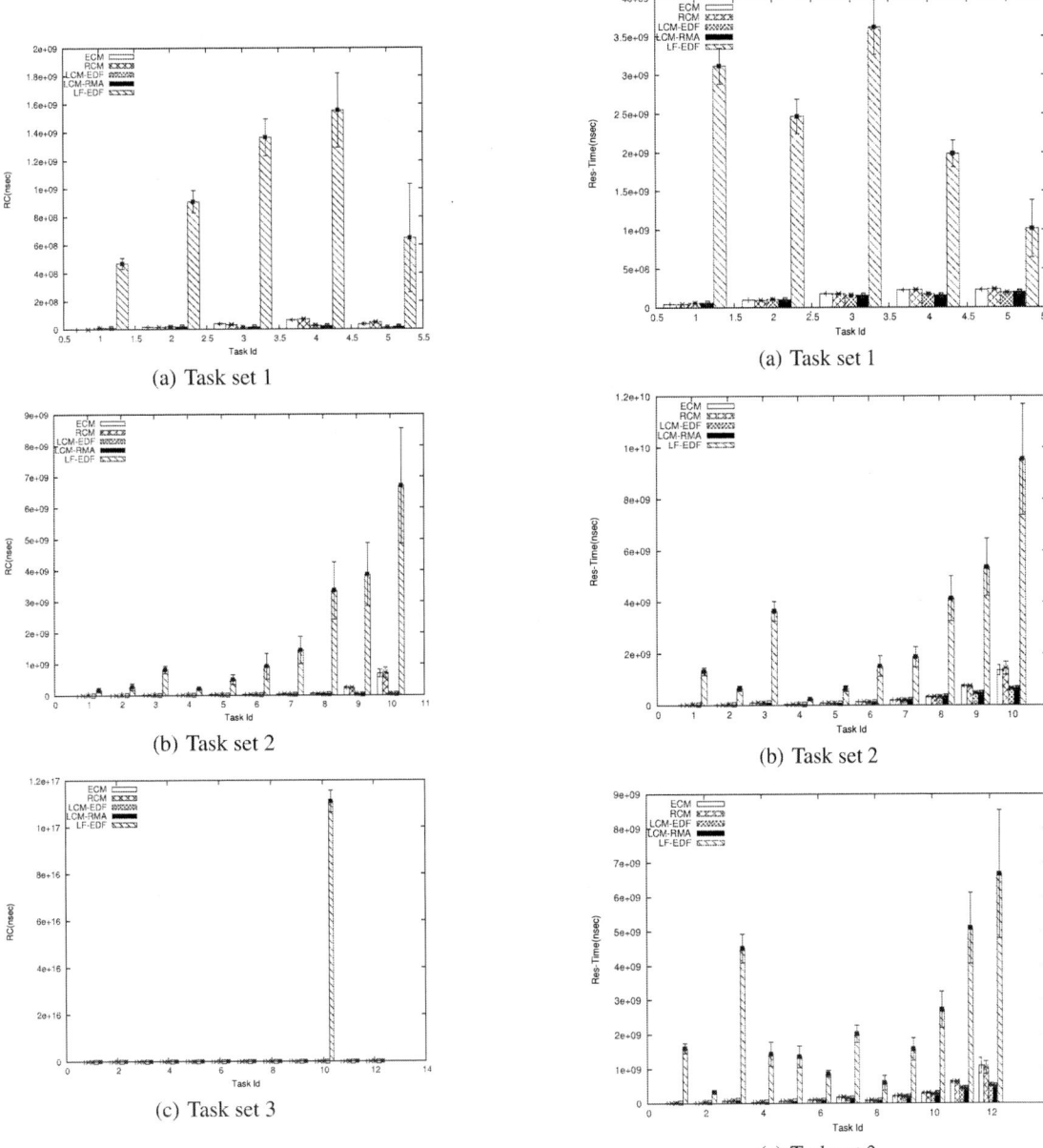

(a) Task set 1

(b) Task set 2

(c) Task set 3

Figure 5: Task retry costs under LCM and competitor synchronization methods (0.5,0.2,0.2)

(a) Task set 1

(b) Task set 2

(c) Task set 3

Figure 6: Task response times under LCM and competitor synchronization methods (0.5,0.2,0.2)

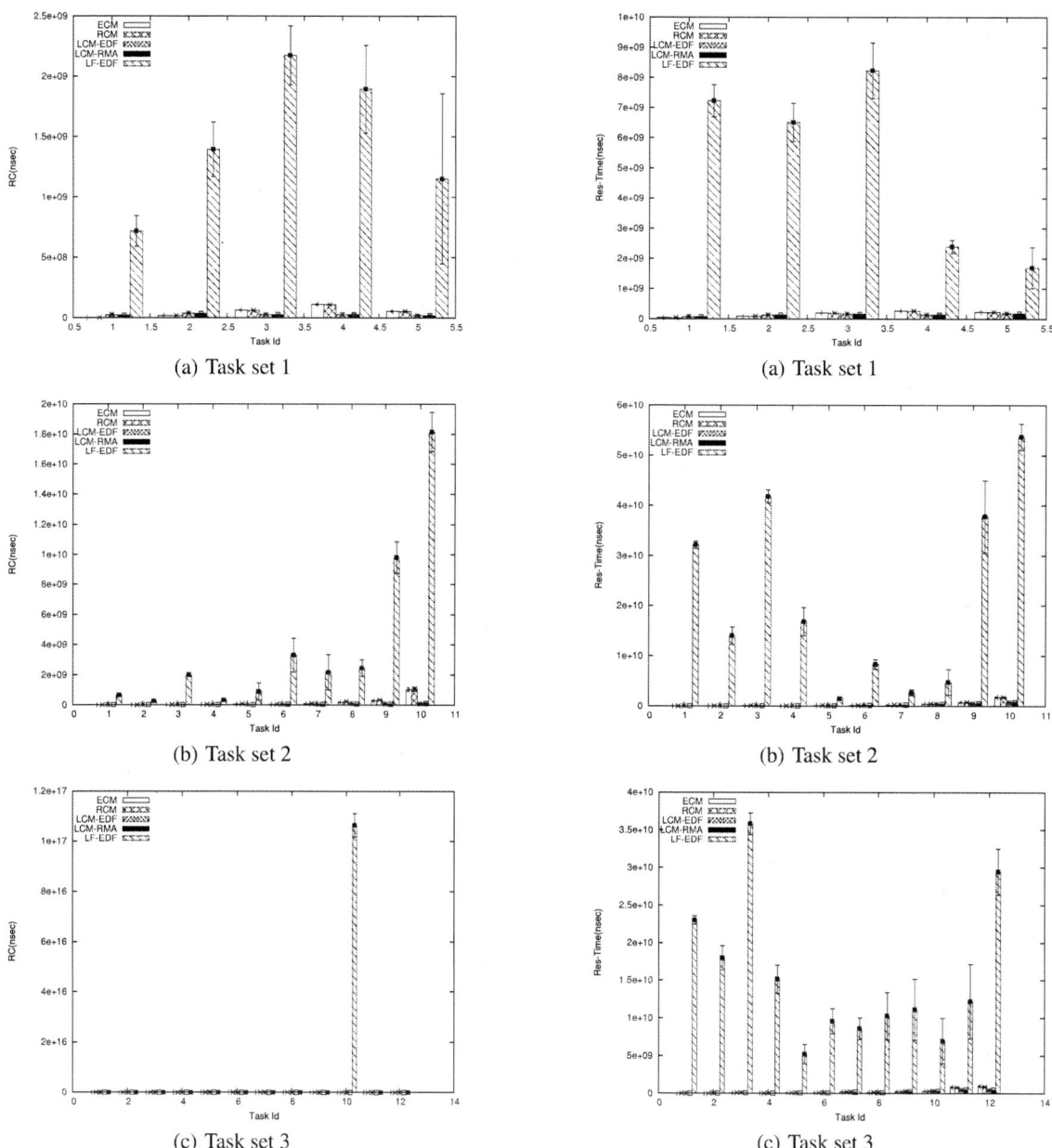

Figure 7: Task retry costs under LCM and competitor synchronization methods (0.8,0.5,0.2)

Figure 8: Task response times under LCM and competitor synchronization methods (0.8,0.5,0.2)

HaVOC: A Hybrid Memory-aware Virtualization Layer for On-Chip Distributed ScratchPad and Non-Volatile Memories

Luis Angel Bathen; Nikil Dutt

Center for Embedded Computer Systems
University of California, Irvine
Irvine, CA 92697, USA
{lbathen, dutt}@uci.edu

ABSTRACT

Hybrid on-chip memories that combine Non-Volatile Memories (NVMs) with SRAMs promise to mitigate the increasing leakage power of traditional on-chip SRAMs. We present HaVOC: a run-time memory manager that virtualizes the hybrid on-chip memory space and supports efficient sharing of distributed ScratchPad Memories (SPMs) and NVMs. HaVOC allows programmers and the compiler to partition the application's address space and generate data/instruction block layouts considering virtualized hybrid address spaces. We define a data volatility metric used by our hybrid memory-aware compilation flow to generate memory allocation policies that are enforced at run-time by a filter-inspired dynamic memory algorithm. Our experimental results with a set of embedded benchmarks executing simultaneously on a Chip-Multiprocessor with hybrid NVM/SPMs show that HaVOC is able to reduce execution time and energy by 60.8% and 74.7% respectively with respect to traditional multitasking based SPM allocation policies.

Categories and Subject Descriptors

C.3 [**Special-purpose and Application-based systems**]: Real-time and embedded systems; B.3 [**Design Styles**]: Virtual Memory; D.4 [**Storage Management**]: Distributed memories

General Terms

Algorithms, Design, Management, Performance

1. INTRODUCTION

The ever increasing complexity of embedded software and adoption of open-environments (e.g., Android) is exacerbating the deployment of multi-core platforms with distributed on-chip memories [12, 18]. Traditional memory hierarchies consist of caches, however, it is known that caches may consume up to 50% of the processor's area and power [3]. As a result, ScratchPad Memories (SPMs) are rapidly being adopted and incorporated into multi-core platforms for their high predictability, low area and power consumption. Efficient SPM management can significantly reduce power consumption [17, 29, 13], and may be a good alternative to caches for applications with high levels of regularity (e.g., Multimedia).

As sub-micron technology continues to scale, leakage power will overshadow dynamic power consumption [19, 14]. Since SRAM-based memories consume a large portion of the die, they are a major source of leakage in the system [2], which is a major issue for multi-core platforms. In order to reduce leakage power in SRAM-based memories, designers have proposed emerging Non-Volatile Memories (NVMs) as alternatives to SRAM for on-chip memories [27, 15, 24]. Typically, NVMs (e.g., PCRAM [20]) offer high densities, low leakage power, comparable read latencies and dynamic read power with respect to traditional embedded memories (SRAM/e-DRAM). One major drawback across NVMs is the expensive write operation (high latencies and dynamic energy). To mitigate the drawbacks of the write operation in NVMs, designers have made the case for deploying hybrid on-chip memory hierarchies (e.g., SRAM, NVM) [27], which have shown up to 37% reduction in leakage power [11], and increased IPC as a byproduct of the higher density provided by NVMs [24]. Orthogonal to traditional hybrid on-chip memory subsystems which have been predominately focused on caches, Hu et al. [11] showed the benefits of exploiting hybrid memory subsystems consisting of SPMs and NVMs.

In this paper, we present *HaVOC*, a system-level HW/SW solution to efficiently manage on-chip hybrid memories consisting distributed ScratchPad Memories (SRAM) and Non-Volatile Memories (e.g., MRAMs) to support multitasking Chip-Multiprocessors. *HaVOC* allows programmers to partition their application's address space into virtualized SRAM address space and virtualized NVM address space through a minimalistic API. Programmers (through annotations) and compilers (through static analysis) can then specify hybrid memory-aware allocation policies for their data/instruction blocks at compile-time, while *HaVOC* dynamically enforces them and adapts to the underlying memory subsystem. The **novel contributions** of our work are that we:

- Explore distributed shared on-chip hybrid memories consisting of SPMs and NVMs and virtualize their address spaces to facilitate the management of their physical address spaces

- Introduce the notion of data volatility analysis to drive efficient compilation and policy generation for hybrid on-chip memories

- Present a filter-driven dynamic allocation algorithm that exploits filtering and volatility to find the best memory placement

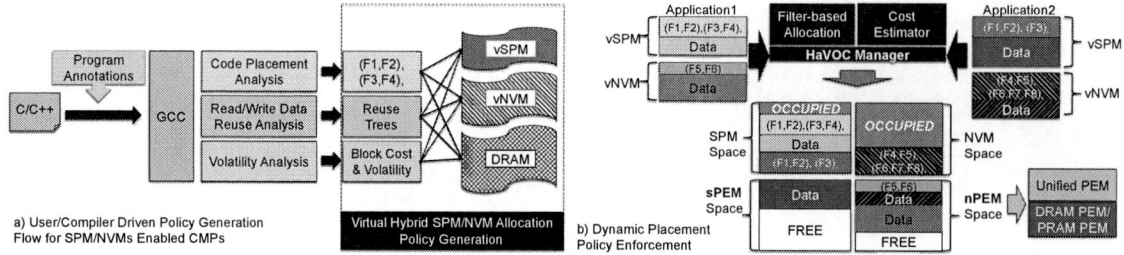

Figure 1: HaVOC-aware Policy Generation (a) and Enforcement (b).

2. MOTIVATION

Unlike caches, SPMs are software controlled memories as their management is completely left to the programmer and compiler. At first glance, we would need to take traditional SPM management schemes and adapt them to manage hybrid memories. However, SPM based allocation schemes (e.g., [29, 13]) assume physical access to the memory hierarchy; consequently, the traditional SPM based programming model would require extensive changes to account for the different characteristics of the NVMs. This motivates the need for a simplified address space to minimize changes to the SPM programming model.

The challenge of programming and managing SPM/NVM-based hybrid memories is aggravated by the adoption of open environments (e.g., Android OS), where users can download applications, install them, and run them on their devices. In these environments, it is possible that many of the running processes will require access to the physical SPMs, therefore, programmers and compilers can no longer assume that their applications are the only ones running on the system. Traditional SPM-sharing approaches [9, 26, 28] would either allocate part of the physical address space to each process (spatial allocation) or time-share the SPM space (temporal allocation). Once the entire SPM space has been allocated, all remaining data is then mapped to off-chip memory. In order to reduce the overheads of sharing the SPM/N-VMs, our scheme exploits programmer/compiler-driven policies obtained through static analysis/annotations (Sec. 3.3) and uses the information to efficiently manage the memory resources at run-time (Sec. 3.4).

3. HAVOC OVERVIEW

Figure 1 (a) shows our proposed compilation flow, which takes annotated source code, and performs various types of SPM/NVM-aware static analysis techniques (e.g., instruction placement, data reuse analysis, data volatility analysis); the compiler then uses this information to generate allocation policies assuming the use of virtual SPMs (vSPMs) and virtual NVMs (vNVMs). Figure 1 (b) shows our proposed dynamic policy enforcement mechanism for multitasking CMPs. The *HaVOC* manager (black box) takes in the vSPM/vNVM allocation policies provided by each application (Application 1 & 2), and decides how to best utilize the underlying memory resources. The rest of this section will go over each of the different components at a high level, for more details please refer to our technical report [4]. In the following discussion we use data/instruction blocks interchangeably as our approach supports placement of both data and instructions onto the hybrid memories.

3.1 Target Platform and Assumptions

Figure 2 shows a high level diagram of our SPM/NVM

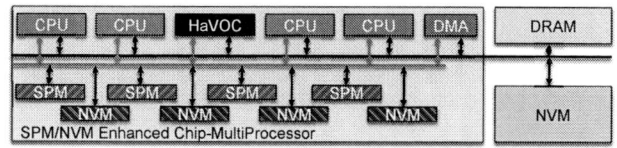

Figure 2: HaVOC-Enhanced CMP.

enhanced CMP, which consists of a number of OpenRISC-like cores, the *HaVOC* manager, a set of distributed SPMs and NVMs, a DRAM/NVM main memory hierarchy, and an AMBA AHB bus-based communication fabric.

We make the following assumptions: 1) The application can be statically analyzed/profiled so that data/instruction blocks can be mapped to SPMs [12, 13, 28]. 2) We operate over blocks of data (e.g., 1KB mini-pages). 3) We can map all data/instructions to on-chip/off-chip memory and do not use caches (e.g., [12]). 4) Part of off-chip memory can be locked in order to support the virtualization of the on-chip SPM/NVM memories.

3.2 Virtual Hybrid Memory Space

In order to present the compiler/programmer with an abstracted view of the hybrid memory hierarchy and minimize the complexity of our run-time system we propose the use of virtual SPMs and virtual NVMs. We leverage the concept of vSPMs [5], which enables a program to view and manage a set of vSPMs as if they were physical SPMs. In order virtualize SPMs, a small part of main memory (DRAM) called protected evict memory (*PEM*) space is locked and used as extra storage. The run-time system would then prioritize the data mapping to SPM and PEM space based on a utilization metric. In this work we introduce the concept of virtual NVMs (vNVMs), which behave similarly to vSPMs, meaning that the run-time environment transparently allows each application to manage their own set of vNVMs. Management of virtual memories is done through a small set of APIs [4], which send management commands to the *HaVOC* manager. The *HaVOC* manager then presents each application with intermediate physical addresses (*IPAs*), which point to their virtual SPMs/NVMs. Traditional SPM-based memory management requires the data layout to use physical addresses by pointing to the base register of the SPMs, as a result, the same is expected of SPM/NVM-based memory hierarchies [11]. In our scheme, all policies use virtual SPM and NVM base addresses, so any run-time re-mapping of data will remain transparent to the initial allocation policies as the IPAs will not change.

3.3 Hybrid Memory-aware Policy Generation

The run-time system needs compile-time support in order to make efficient allocation decisions at run-time. In this paper we present various ways by which designers may gen-

448

erate policies (manual through annotations or through static analysis). These policies are then enforced (currently in best effort fashion) by the run-time system in order to prioritize the access to SPM/NVM space for the various applications running on the system. Each policy attempts to map data to virtual SPMs/NVMs, while the *HaVOC* manager dynamically maps the data to physical memories.

3.3.1 Volatility Analysis

$$Data_{lifetime} \leftarrow \bigcup_{i=0}^{n} ST_i \qquad (1)$$

$$Write_{freq.}^{i} \leftarrow Writes_i \div ST_i, i = 0 \cdots n \qquad (2)$$

$$Data_{volatility} \leftarrow STDEV(Write_{freq}^{i}), i = 0 \cdots n \qquad (3)$$

$$C(D_i) = C_{load}(D_i^f) + C_{evict}(D_i^t) + \\ C_{util}(D_i^r, D_i^w) + \Delta_{leak}(D_i^r, D_i^w, D_i^{lifetime}) \qquad (4)$$

We introduce a new metric, *data volatility*, to facilitate efficient loading of data on the hybrid on-chip memory configurations. Data volatility is defined as the write frequency of a piece of data over its accumulated lifetime. In order to estimate the volatility of a data block we first define a sampling time (ST_i), which can be in cycles, so that the union of all sample times equals the block's lifetime (Eq. 1). Next, we calculate the write frequency for each sample time (Eq. 2). Finally, we estimate the volatility of the data as the variation in its write frequency (Eq. 3). This metric is useful when deciding whether data is worth (cost effective) being mapped onto NVM. Highly volatile data implies that at some point the cost of keeping data in NVM during its entire lifetime might be greater than leaving it in main memory. As a result, when two competing applications request NVM space, the estimated cost function (e.g., energy savings) will be used to prioritize allocation of on-chip space, while volatility can be used as a tie breaker and prediction metric of cost fluctuation. Volatility may also be used to decide the granularity at which designers might do their data partitioning. We define the expected cost metric ($C(D_i)$) for a given data block (D_i) as shown in Eq. 4, which takes into account the cost of transferring data between memory type D_i^f and memory type D_i^t (C_{load}, C_{evict}), the utilization cost (C_{util}), and the extra leakage power consumed by mapping the given data to the preferred memory type. The cost represents *energy or latency*.

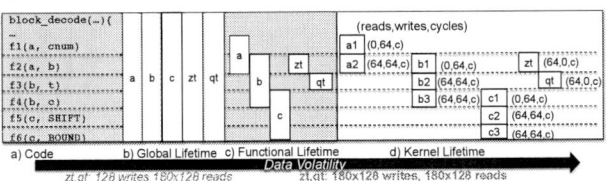

Figure 3: Data volatility across various lifetimes.

Figure 3 (a) shows sample JPEG [21] code and how partitioning its data's lifetime may affect its data's volatility. Figure 3 (b) shows the global life time of the data arrays (*a, b, c, zt, qt*), where the number of accesses to NVM would be (128 rd, 23K wr) for *qt/zt* if we map and keep them in NVM during the entire execution of the program. To accommodate other data structures onto SPM space, arrays *qt/zt*'s lifetime may be split, resulting in finer life-time granularities (Figure 3 (c-d)). Though the rd/wr ratio of data remains the same (*qt/zt* have 23K reads to 0 writes), finer granularity

lifetimes might yield higher volatility (*qt/zt* now have 23K writes to NVM since they are loaded every time *block_decode* executes), making *qt/zt* poor candidates for NVM.

3.3.2 Memory Allocation Policy Generation

Programmers can embed application-specific insights into source code through annotations [10] in order to guide the compilation process. Since we are working with virtualized address spaces, programmers can create hybrid memory-aware policies that define the placement for a given data structure by simply defining the following parameters: < *preferred memory type, reads, writes, lifespan, volatility* >. These annotations are used at run-time by the *HaVOC* manager to allocate the data onto the *preferred* memory type.

Instruction blocks are very good candidates for mapping onto on-chip NVMs since their volatility is quite low (e.g., write once and use many times). In this work, we borrow traditional SPM-based instruction-placement schemes [16] and enhance them to account for the possibility of mapping the instructions to NVM memories by introducing volatility analysis into the flow. Like we discussed in Sec. 3.3.1, the granularity of the code partitioning (e.g, function, basic block, etc.) will affect how volatile the placement will become. As a result, when mapping a block of instructions onto vNVM/vSPM, we need to partition our code such that Eq. 5 is met, where $C(D_i)$ represents the cost in Eq. 4. Our goal is to partition the application such that we can minimize the number of instruction replacements ([16]) in order to minimize energy and execution time.

$$C(D_i)_{Off-Chip} \ll C(D_i)_{On-Chip} \qquad (5)$$

Data placement candidates are obtained from static analysis (e.g., data reuse analysis [13]) or profiling. The idea is to map highly read-reused data with a long access distance onto vNVM to minimize number of fetches from off-chip memory, while highly reused read-data with short lifetimes will be mapped to vSPM preferably. Highly reused-read-modify data with low write-volatility should be mapped to vNVM, while highly reused write-data and read-modify data with high write-volatility should go to vSPM.

The last step in our flow involves the generation of near-optimal hybrid memory layouts we define as *allocation policies*. This process takes as input data/instruction blocks with various pre-computed costs (e.g., Eq. 4) obtained from static analysis, profiling, and annotations, which are combined and fed as inputs to an enhanced hybrid memory-aware allocator based on [11].

3.4 HaVoC Manager

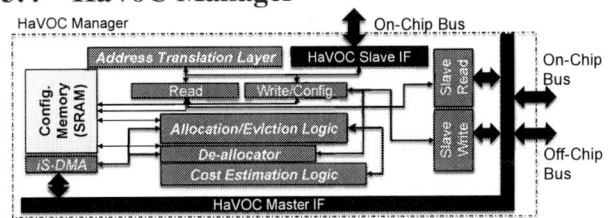

Figure 4: *HaVOC* Manager

The *HaVOC* manager may be implemented in software as an extended memory manager embedded within a hypervisor/OS or as a hardware module (e.g., [9, 5]). The software implementation is quite flexible and does not require modifying existing platforms. The hardware version requires additional hardware and the necessary run-time support, but

449

the run-time overheads will be much lower than the software version. In this work, we present a proof-of-concept embedded hardware implementation (Figure 2). Figure 4 shows a block diagram of the *HaVOC* manager. It consists of a memory-mapped slave interface, which is used by the system's masters (e.g., CPUs) and handles the read/write/configuration requests. The address translation layer module converts IPAs to physical addresses (PAs) in one cycle [7, 5]). The manager consists of 1KB to 256KB of configuration memory used to keep block metadata information (e.g., volatility, # accesses, etc.). The allocation/eviction logic uses the cost estimation (e.g., efficiency) logic to prioritize access to on-chip storage. Finally, the internal DMA (iS-DMA) allows the manager to asynchronously transfer data between on-chip and off-chip memory. In order to use the *HaVOC* manager, the compiler generates two things: 1) the creation of the required virtual SPM/NVMs through the use of our APIs and 2) The use of IPAs instead of PAs during the data/instruction layout stage (e.g., memory maps using purely virtual addresses for SPMs/NVMs). Any read/write to a given IPA is translated and routed to the right memory. Any write to *HaVOC* configuration memory space is then used to manage the virtualized address space. The goal is to allow each application to manage its virtual on-chip memories as if it had full access to the on-chip real-estate.

3.5 HaVOC's Dynamic Policy Enforcement

Table 1: Filter Inequalities and Preferred Memory Type

Filter	Pref.	Inequalities
F1	sram	$E(D_i^{spm}) > E(D_i^{dram}) \bigwedge E(D_i^{nvm}) < E(D_i^{dram}) \bigwedge V > T_{vol}$
F2	nvm	$E(D_i^{nvm}) > E(D_i^{dram}) \bigwedge V < T_{vol}$
F3	either	$E(D_i^{spm}) > E(D_i^{dram}) \bigwedge E(D_i^{nvm}) > E(D_i^{dram})$
F4	dram	$E(D_i^{spm}) < E(D_i^{dram}) \bigwedge E(D_i^{nvm}) < E(D_i^{dram})$

```
Algorithm 1 FilterDynamic Allocation Algorithm
───────────────────────────────────────────────
   Require: req{size, cost, volatility}
      pref_mem ← filter(req, volatility)
2: if allocatable(req, pref_mem) then
         return  ipa ← update_alloc_table(req)
4: end if
   min_set ← sort_Hi2LowEff(alloc_table, size)
6: if E(min_set) < C_evict(min_set) + E(req) then
      evict(min_set)
8:       return  ipa ← update_alloc_table(req)
   else
10:      return  ipa ← mm_malloc(req)
      end if
```

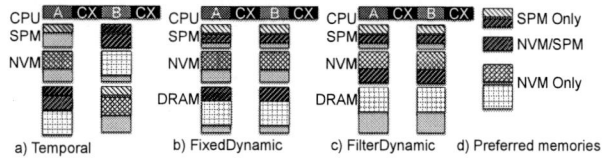

a) Temporal   b) FixedDynamic   c) FilterDynamic   d) Preferred memories

Figure 5: Dynamic Hybrid Memory Allocation Policies

We define *three* block-based allocation policies: 1) *Temporal* allocation, which combines temporal SPM allocation ([28]) and hybrid memory allocation ([11]), and adheres to the initial layout obtained through static analysis (Sec. 3.3); however, the application's SPM and NVM contents must be swapped on a context-switch to avoid conflicts with other tasks (Fig. 5 (a)). 2) *FixedDynamic* allocation, which combines dynamic-spatial SPM allocation ([9]) and hybrid memory allocation [11], and maps the data block to the preferred memory type (adhering to the initial layout) as long as there is space, otherwise, data is mapped to DRAM (Fig. 5 (b)). 3) *FilterDynamic* allocation (Alg. 1), which exploits the

concept of filtering and volatility to find the best placement. Each request is filtered according to a set of inequalities (shown in Table 1) which determine the preferred memory type (Fig. 5 (d)). The volatility of the data block (V) and its mapping efficiency ($E(D_i) = C(D_i)/|D_i|$) are used to determine what memory type would minimize the block's energy (or access latency). For instance, data with low volatility and high energy efficiency could potentially benefit more from being mapped to NVM than SRAM (e.g., filter *F2* in Table 1). If there is enough preferred memory space (e.g., SPM or NVM), the dynamic allocator adheres to Eq. 4 prior to loading the data. If there is not enough space, then the allocator follows Alg. 1 and sorts the allocated blocks from highest to lowest efficiency (e.g., energy per bit). It then compares the cost of evicting the least important blocks (MIN_{Set}) with the cost of dedicating the space to the new block. If the efficiency of bringing the new block offsets the eviction cost and efficiency of the data already mapped to the preferred memory type ($E(MIN_{Set})$), then *HaVOC* evicts the MIN_{Set} blocks and updates the allocation table with the new block ($|new\ block| \leq |MIN_{Set}|$). In the event the preferred memory type is either NVM or SPM (filter *F3* in Table 1), the allocator evicts the $min(MIN_{Set}^{spm}, MIN_{Set}^{nvm})$. At the end, *HaVOC* allocates either on-chip space or off-chip space (unified PEM space ($sPEM$)), resulting in the allocation shown in Fig. 5 (c), where a data block originally intended for SPM is mapped to NVM.

4. RELATED WORK

Most efforts have focused on replacing and/or complementing main memory (DRAM) or caches (SRAM) with a combination of various NVMs to reduce leakage power and increase throughput. Joo et al. [15] proposed PCM as an alternative to SRAM for on-chip caches. Sun et al. [27] introduced MRAM into the cache hierarchy of a NUCA-based 3D stacked multi-core platform. Mishra et al. [24] followed up by introducing STT-RAM as an alternative MRAM memory and hid the overheads in access latencies by customizing how accesses are prioritized by the the interconnect network. Wu et al. [31] presented a hybrid cache architecture consisting of SRAM (fast L1/L2 accesses), eDRAM/MRAM (slow L2 accesses), and PCRAM (L3). Hybrid main memory has also been studied [32]. Mogul et al. [25] and Wongchaowart et al. [30] attempted to reduce write overheads in main memory by exploiting page migration and block content signatures. Ferreira et al. [8] introduced a memory management module for hybrid DRAM/PCM main memories. Static analysis has been explored to efficiently map application data on off-chip hybrid main memories [23, 22].

HaVOC is different from approaches that address hybrid cache/main memories in that we primarily focus on hybrid SPM/NVM-based hierarchies, however, our scheme can be complemented by both existing schemes that leverage the benefits of both on-chip SRAMs and NVMs as well as hardware/system-level solutions that address hybrid off-chip memories (e.g., DRAM/eDRAM and PCM). Our work is different from [11] in that we consider: 1) shared distributed on-chip SPMs and 2) dynamic support for multitasking systems, however, our scheme can benefit from compile-time analysis schemes that provide our runtime system with allocation hints (e.g., [22, 23, 11]). Like [5], we use part of off-chip memory to virtualize on-chip memories, however, our approach differs in that our primary focus is the ef-

450

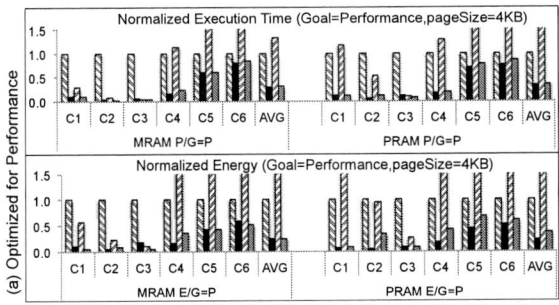
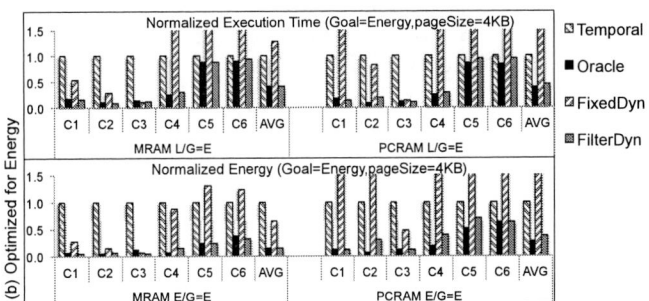

Figure 6: Normalized Execution Time and Energy for Performance Optimized (a) and Energy Optimized (b) Policies

ficient management of hybrid on-chip memories, and as a result, *HaVOC*'s programming model and run-time environment account for the different physical characteristics of SRAMs, DRAMs and NVMs (MRAM and PCRAM). Moreover, we believe that we can complement our static-analysis/allocation policy generation with other SPM management techniques [17, 29, 9, 26, 28].

5. EXPERIMENTAL RESULTS

Table 2: Configurations

Config.	Applications	CPUs	vSPM/vNVM Space	SPM/NVM Space
C1	adpcm,aes	1	32/128 KB	16/64 KB
C2	adpcm,aes,blowfish,gsm	1	64/256 KB	16/64 KB
C3	C2 & h263,jpeg,motion,sha	1	128/512 KB	16/64 KB
C4	same as C2	2	64/256 KB	32/128 KB
C5	same as C3	2	128/512 KB	32/128 KB
C6	same as C3	4	128/512 KB	64/256 KB

5.1 Experimental Setup and Goals

Our goal is to show that *HaVOC* is able maximize energy savings and increase application throughput in a multitasking environment under various scenarios. First, we generate two sets of hybrid memory aware allocation policies (Sec. 3.3.2), one set of policies minimizes execution time (Sec. 5.2) and the other minimizes energy (Sec. 5.3). These policies are generated at compile-time and enforced at run-time by a set of dynamic allocation policies (Sec. 3.5) under various system configurations (Table 2). Next, we show the effects of the allocation policy's block-size on execution time (Sec. 5.4). We built a trace-driven simulator that models a light-weight OS, with a round-robin scheduler and context-switching enabled (window = 50K cycles). We model CMPs consisting of an AMBA AHB bus, OpenRISC-like in-order cores, distributed SPMs and NVMs, and the *HaVOC* manager (Fig. 2). We bypassed the cache and mapped all data to either SPM, NVM, or main memory (see Sec. 3.3.2). We obtained traces from Mediabench II [21] by using SimpleScalar [1]. We model on-chip SPMs (SRAMs), MRAMs and PCRAM by interfacing our simulator with NVSim [6] and set leakage power as the optimization goal. To virtualize SPMs/NVMs we use the unified PEM space model discussed in Sec. 3.2 (sPEM). The *HaVOC* manager consists of 4KB low power SRAM memory.

5.2 Enforcing Performance Optimized Policies

For this experiment we generated allocation policies that minimized execution time for each application. We then executed each application on top of our simulated RTOS/CMP. Table 2 shows each configuration (C1-6), which has a set of applications running concurrently over a numberof CPUs, and a predefined hybrid memory physical space. To show the benefit of our approach we implemented four policies:

the three described in Sec. 3.5 (*Temporal, FixedDynamic, FilterDynamic*), and a policy we call *Oracle* (black bar in Fig. 6), which is a near-optimal policy because on every block-allocation request, it feeds the entire memory map to the same policy generator the compiler uses to generate policies statically (see Sec. 3.3.2). The idea is to show that our *FilterDynamic* policy (backward-slashed bars in Fig. 6) achieves competitive quality allocation solutions as the more complex *Oracle* policy. Fig. 6 (a) shows the normalized execution time and energy for each of the different configurations (*C1-6*, Goal=Min Execution Time denoted as *G=P*) using 4KB blocks and different memory types with respect to the *Temporal* policy. The *FixedDynamic* policy (forward-slashed bars in Fig. 6 (a)) suffers the greatest impact on energy and execution time as memory space increases (*C4-6*) since it adheres to the decisions made at compile-time and does not efficiently allocate memory blocks at run-time. In general, we see that the *FilterDynamic* policy performs almost as good as the *Oracle* policy (within 8.45% execution time). Compared with the *Temporal* policy, *HaVOC*'s *FilterDynamic* policy is able to reduce execution time and energy by an average 75.42% and 62.88% respectively when the initial application policies have been optimized for execution time minimization.

5.3 Enforcing Energy Optimized Policies

Fig. 6 (b) shows the the normalized execution time and energy for each of the different configurations and memory types (Goal=Min Energy denoted as *G=E*) with respect to the *Temporal* policy. Similar to the case of *G=P*, both the *Temporal and FixedDynamic* policies are unable to efficiently manage the on-chip real-estate. The *FilterDynamic and Oracle* policies are able to greatly reduce execution time and energy, with the *FilterDynamic* within 3.54% of the execution time achieved by the *Oracle* policy. Compared with the *Temporal* policy, *HaVOC*'s *FilterDynamic* policy is able to reduce execution time and energy by an average 85.58% and 61.94% respectively when the initial application policies have been optimized for energy minimization. The goal of this experiment was to show that regardless of the initial optimization goal, *HaVOC*'s *FilterDynamic* policy is able to achieve as good results as the *Oracle* policy.

5.4 Block Size Effect on Allocation Policies

So far we have seen that the *Oracle* policy appears to be a feasible dynamic allocation solution, which would potentially enhance *HaVOC*'s virtualization engine. However, the *Oracle* policy is very complex as it runs in $O(Blks * spm_{size} * nvm_{size})$. The *FixedDynamic* policy on the other extreme runs in $O(Blks)$, however, its efficiency may be even worse than the *Temporal* policy. *HaVOC*'s *FilterDy-*

451

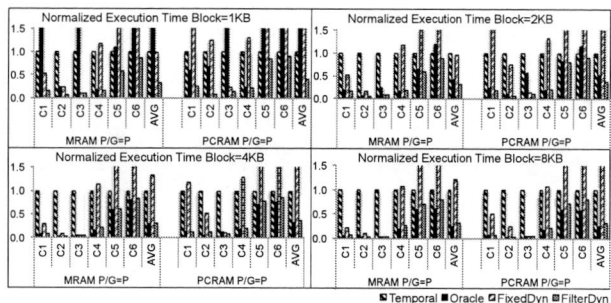

Figure 7: Effects of Varying Block Size.

namic policy on the other hand, keeps a semi-sorted list of data blocks, as a result it can be $O(Blks)$ best case or $O(Blks\ log\ Blks)$ worst case for sorting, and the final filtering decision runs in $O(Blks_{spm} + Blks_{nvm} + Blks_{dram})$, which results in $O(Blks\ log\ Blks) + O(Blks_{spm} + Blks_{nvm} + Blks_{dram})$. Thus, the complexity and execution time of the *Oracle* policy will increase orders of magnitude as the number of data/instruction blocks to allocate increases (*Blks*) or as the available resources increases (spm_{size}, nvm_{size}). This is validated in Fig. 7, where we have varying block size (as a result, number of blocks to allocate increases) and increase in available resources (*C3-C6*). As we can see, for block size = 1KB, where the number of blocks to allocate is in the hundreds, the allocation time of the *Oracle* is orders of magnitude greater than the *FilterDynamic* policy. For block size = 2KB, we see that as resources increase, the *Oracle* allocation time once again prevents it from being a feasible solution. On average, we observe that *HaVOC*'s *FilterDynamic* is capable of achieving as good solutions as the *Oracle* policy (within 10% margin) with much lower complexity. On average, across all test scenarios (varying page sizes, different optimization goals, and different configurations) we see that the *FilterDynamic* is able to reduce execution time and energy by 60.8% and 74.7% respectively.

6. CONCLUSION

We presented *HaVOC*, a hybrid memory aware virtualization layer for dynamic memory management of applications executing on CMP platforms with hybrid on-chip NVM/SPMs. We introduced the notion of data volatility analysis and proposed a dynamic filter-based memory allocation scheme to efficiently manage the hybrid on-chip memory space. Our experimental results for embedded benchmarks executing on a hybrid memory-enhanced CMP show that our dynamic filter-based allocator greatly minimizes execution time and energy by 60.8% and 74.7% respectively with respect to traditional multitasking-aware SPM allocation policies. Future work includes: 1) Integrating *HaVOC*'s concepts in a hypervisor to support full on-chip hybrid memory virtualization, 2) Adding off-chip NVM memories to support the virtualization of on-chip NVMs (*nPEM*), and 3) Designing a scalable hybrid-memory aware virtualization layer.

7. ACKNOWLEDGMENTS

This work was partially supported by NSF Variability Expedition Grant Number CCF-1029783.

8. REFERENCES

[1] T. Austin et al. Simplescalar: an infrastructure for computer system modeling. *Computer*, 35(2), Feb. 2002.

[2] N. Azizi et al. Low-leakage asymmetric-cell sram. *TVLSI*, Vol. 11, aug. 2003.

[3] R. Banakar et al. Scratchpad memory: design alternative for cache on-chip memory in embedded systems. In *CODES '02*, 2002.

[4] L. Bathen and N. Dutt. A hybrid-memory-aware virtualization layer for cmps. In *UCI TR #12-03*, 2011.

[5] L. Bathen et al. SPMVisor: dynamic scratchpad memory virtualization for secure, low power, and high performance distributed on-chip memories. In *CODES+ISSS '11*, 2011.

[6] X. Dong et al. Pcramsim: system-level performance, energy, and area modeling for phase-change ram. In *ICCAD '09*, 2009.

[7] B. Egger et al. Dynamic scratchpad memory management for code in portable systems with an mmu. *ACM TECS*, 7, January 2008.

[8] A. Ferreira et al. Using pcm in next-generation embedded space applications. In *RTAS '10*, april 2010.

[9] P. Francesco et al. An integrated hardware/software approach for run-time scratchpad management. In *DAC '04*, 2004.

[10] S. Guyer et al. An annotation language for optimizing software libraries. In *DSL '99*, 1999.

[11] J. Hu et al. Towards energy efficient hybrid on-chip scratch pad memory with non-volatile memory. In *DATE '11*, 2011.

[12] IBM. Cell. *http://www.research.ibm.com/cell/*, 2005.

[13] I. Issenin et al. Multiprocessor system-on-chip data reuse analysis for exploring customized memory hierarchies. In *DAC '06*, 2006.

[14] ITRS. Process integration, device and structures. *http://www.itrs.net/*, 2007.

[15] Y. Joo et al. Energy- and endurance-aware design of phase change memory caches. In *DATE '10*, march 2010.

[16] S. Jung et al. Dynamic code mapping for limited local memory systems. In *ASAP '10*, july 2010.

[17] M. Kandemir et al. Dynamic management of scratch-pad memory space. In *DAC '01.*, 2001.

[18] S. Kaneko et al. A 600mhz single-chip multiprocessor with 4.8gb/s internal shared pipelined bus and 512kb internal memory. In *SSCC '03*, 2003.

[19] N. Kim et al. Leakage current: Moore's law meets static power. *Computer*, 36(12), dec. 2003.

[20] B. Lee et al. Architecting phase change memory as a scalable dram alternative. *SIGARCH Comput. Archit. News*, 37, June 2009.

[21] C. Lee et al. Mediabench: a tool for evaluating and synthesizing multimedia and communicatons systems. In *MICRO '97*.

[22] K. Lee et al. Application specific non-volatile primary memory for embedded systems. In *CODES+ISSS '08*, 2008.

[23] T. Liu et al. Power-aware variable partitioning for dsps with hybrid pram and dram main memory. In *DAC '11*, 2011.

[24] A. Mishra et al. Architecting on-chip interconnects for stacked 3d stt-ram caches in cmps. In *ISCA '11*, 2011.

[25] J. Mogul et al. Operating system support for nvm+dram hybrid main memory. In *HotOS'09*, 2009.

[26] V. Suhendra et al. Scratchpad allocation for concurrent embedded software. In *CODES+ISSS '08*, 2008.

[27] G. Sun et al. A novel architecture of the 3D stacked MRAM L2 cache for CMPs. In *HPCA '09*, feb. 2009.

[28] H. Takase et al. Partitioning and allocation of scratch-pad memory for priority-based preemptive multi-task systems. In *DATE '10*, 2010.

[29] M. Verma et al. Data partitioning for maximal scratchpad usage. In *ASP-DAC '03*, 2003.

[30] B. Wongchaowart et al. A content-aware block placement alg. for reducing pram storage bit writes. In *MSST '10*.

[31] X. Wu et al. Hybrid cache architecture with disparate memory technologies. In *ISCA '09*, 2009.

[32] P. Zhou et al. A durable and energy efficient main memory using phase change memory technology. In *ISCA '09*, 2009.

Age-Based PCM Wear Leveling with Nearly Zero Search Cost

Chi-Hao Chen[1], Pi-Cheng Hsiu[2,3], Tei-Wei Kuo[1,2,3,4], Chia-Lin Yang[1,4], and Cheng-Yuan Michael Wang[5]

[1] Department of Computer Science and Information Engineering, National Taiwan University, Taipei, Taiwan
[2] Research Center for Information Technology Innovation, Academia Sinica, Taipei, Taiwan
[3] Institute of Information Science, Academia Sinica, Taipei, Taiwan
[4] Graduate Institute of Networking and Multimedia, National Taiwan University, Taipei, Taiwan
[5] Macronix Int'l Co., Ltd., Hsinchu, Taiwan

r98922075@csie.ntu.edu.tw, pchsiu@citi.sinica.edu.tw, {ktw, yangc}@csie.ntu.edu.tw, michaelwang@mxic.com.tw

ABSTRACT

Improving the endurance of PCM is a fundamental issue when the technology is considered as an alternative to main memory usage. In the design of memory-based wear leveling approaches, a major challenge is how to efficiently determine the appropriate memory pages for allocation or swapping. In this paper, we present an efficient wear-leveling design that is compatible with existing virtual memory management. Two implementations, namely, bucket-based and array-based wear leveling, with nearly zero search cost are proposed to tradeoff time and space complexity. The results of experiments conducted based on popular benchmarks to evaluate the efficacy of the proposed design are very encouraging.

Categories and Subject Descriptors

D.4.2 [**Operating Systems**]: Storage Management—*Virtual memory*

General Terms

Design, Experimentation, Management, Performance

Keywords

Phase change memory, memory management, wear-leveling, endurance

1. INTRODUCTION

Energy efficiency has become a critical design issue for embedded systems as well as for servers. A number of recent studies have revealed the potential energy savings in replacing DRAM with non-volatile memory. For example, a recent study predicted that energy savings of 65% could be achieved if 4GB DRAM is replaced by 4GB non-volatile memory under intensive memory write access on desktops and web servers [17]. In addition to the potential energy savings, the advantages in adopting non-volatile memory include improvements in cost, density, and non-volatility. With proper system designs, even greater performance gains may be possible because a larger non-volatile memory could help reduce page faults during program executions, compared to DRAM, which is approaching its scalability limit [12]. Based on the above observations, a number of researchers have explored various proposals for replacing DRAM with non-volatile memory, especially *Phase Change Memory* (PCM) [4, 9, 12, 17]. However, such proposals must address a major challenge, namely, the endurance of PCM. This motivates us to explore solutions that could enhance the endurance of PCM-based memory systems.

Permission to make digital or hard copies of all or part of this work for personal or classroom use is granted without fee provided that copies are not made or distributed for profit or commercial advantage and that copies bear this notice and the full citation on the first page. To copy otherwise, to republish, to post on servers or to redistribute to lists, requires prior specific permission and/or a fee.

In recent years, a number of promising mechanisms have been proposed to improve the endurance of PCM. Generally, the mechanisms can be categorized into three types of techniques, which are orthogonal to, and can cooperate with, one another. The first type is based on *write reduction*, and is designed to reduce the amount of data that needs to be written over in PCM, e.g. [3, 7, 9, 12, 17]. The second type is based on *recycling*. It tries to increase the space utilization in a page-based memory management system when only a small number of bits on a physical page are worn out, e.g., [13, 15]. The last type is based on *wear leveling*, e.g. [8, 11, 14, 16, 17]. It improves PCM endurance by spreading writes evenly over the managed PCM space to prevent the PCM pages from being worn out quickly by extensive memory access; thus, it prolongs the lifetime of PCM.

Wear leveling is an effective way to improve the lifetime of systems with non-volatile memory technologies. Many excellent approaches based on *dynamic* and/or *static wear leveling* have been proposed for flash memory, e.g., [1, 2, 6]. However, they cannot be applied to PCM-based systems directly due to the different characteristics of PCM and flash memory. Moreover, most of the work on wear leveling for flash memory is based on studies of storage systems. Research on PCM wear leveling for main memory usage can be roughly classified into two categories: *random-based* and *age-based* wear leveling (WL). Random-based WL approaches usually try to use the PCM space evenly by swapping data randomly, e.g., [5, 10, 11, 14]. The approaches relies on the assumption that applications often have a highly skewed distribution of writes to main memory [5]. In addition, a huge number of extra writes might be triggered because of a high swapping rate. On the other hand, age-based WL approaches usually keep some housekeeping information, such as the ages of PCM pages, in order to move hot (resp. cold) data to young (resp. old) PCM pages so that the wear in the PCM space is even, e.g., [16, 17]. A typical approach is *segment swapping*, which searches for the most frequently written segment and the least frequently written segment periodically and swaps them [17]. The main concern with age-based WL is the time and space overheads incurred by maintaining housekeeping information.

In this paper, we propose an age-based wear-leveling design, which is compatible with existing virtual memory management systems, to improve the endurance of PCM. To avoid the overheads involved in searching for the most frequently and least frequently written pages, the design is based on a simple yet effective concept of "placing old pages far away so that they are less likely to be used". Consequently, a page acquired immediately is likely to be young. Two implementations with time and space complexity considerations are proposed. *Bucket-based WL* uses a number of buckets linked in a circular format to implement different distances so that a young page (when needed) can be acquired immediately without searching or sorting pages. *Array-based WL* reduces the overheads involved in maintaining sophisticated data structures at the cost of a limited search overhead. For the performance evaluation, we collect real traces with *Simics* by running several *SPEC2000* benchmarks with

453

large memory footprints and intensive memory access. To provide useful insights, we compare the proposed approaches with segment swapping [17] and random swapping [5]; and also estimate two baseline schemes, no wear leveling and an ideal case. The results of experiments demonstrate that the proposed approaches are very effective in improving the endurance of PCM, yet only incur low overheads.

The remainder of this paper is organized as follows. In Section 2, we present the system architecture and explain the motivation for this work. In Section 3, we propose an efficient wear-leveling design for PCM-based memory management. Section 4 reports the results of experiments on improving the endurance and incurring extra overheads. Section 5 contains some concluding remarks.

2. SYSTEM ARCHITECTURE AND MOTIVATION

2.1 System Architecture

In recent years, researchers have explored the possibility of utilizing PCM for main memory and/or storage because of its non-volatile and byte-addressable nature. PCM exploits *chalcogenide glass*, a phase-change material, to record bit-information in a cell. The material can stay in two stable phases[1], i.e., *crystalline* and *amorphous*, which represent 1 and 0, respectively. Unlike DRAM, in which the stored information eventually fades unless the capacitor charge is refreshed periodically, PCM does not require standby power and it only consumes a small amount of energy for read operations. However, the phase transition of PCM is driven by a heating process. A cell transforms into the crystalline phase if it is heated above the crystalline temperature for a certain period of time; and it goes into the amorphous phase if it is heated above the (higher) melting temperature and quenched quickly. The adoption of a heating process in PCM results in much higher power consumption and longer latency for a write operation, compared with such operations in DRAM. Most importantly, because of the heating process, a PCM cell can only sustain $10^6 \sim 10^9$ writes before a failure occurs; hence, replacing DRAM with PCM raises some challenging issues. Note that a DRAM cell can sustain more than 10^{15} writes in practice. The attributes of PCM and DRAM are summarized in Table 1.

Attributes	PCM	DRAM
Scalability	20nm	32nm
Byte Addressability	Yes	Yes
Read Latency	$50ns$	$10ns$
Write Latency	$150ns$	$10ns$
Endurance (Write Cycles)	$10^6 \sim 10^9$	$> 10^{15}$

Table 1: Comparison of PCM and DRAM

Virtual memory management in modern operating systems (OS) plays an important role in designing a PCM-based main memory system. Any memory usage requested by user processes should be granted by the OS. On receipt of a request, the OS will maintain the mapping between the virtual and physical addresses in the *page table* according to its free space management policy. During a program's execution, the translation of virtual addresses issued by user processes into physical addresses in DRAM (or PCM in this paper) is performed by the memory management unit (MMU), as shown in Figure 1. To speed up the translation time, a small and fast cache, called the translation look-aside buffer (TLB), is deployed to cache the page and frame numbers of the most recently used pages. If PCM is adopted as the main memory, we may have PCM only or PCM and DRAM together in the paging architecture. Although PCM is byte-addressable, the management unit of PCM would be one page in the proposed design to comply with the virtual memory management format. The size of a page is defined by the OS.

---

[1]Multiple levels per cell are also possible technically.

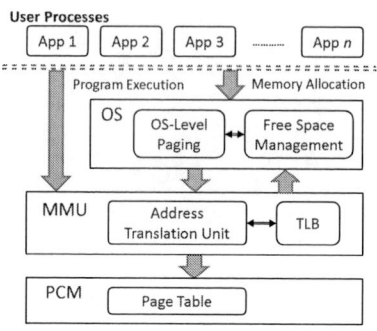

Figure 1: An illustration of the address translation mechanism with demand paging

2.2 Motivation

One of the technical challenges in adopting PCM for main memory usage is the wear-leveling issue because the number of possible writes per PCM cell before a failure is far less than in a DRAM cell. Existing wear-leveling approaches designed for storage devices, such as those based on flash memory, cannot be applied to PCM management directly because memory-related approaches must be extremely efficient in terms of latency, and the extra cost should be justifiable. In general, wear leveling can be classified as dynamic or static. When *dynamic wear leveling* updates data, it tends to write frequently updated data, referred to as *hot data*, to management units (i.e., pages in this paper) with less writes in the past (referred to as *young pages*) to avoid the endurance problem of PCM. In contrast, *static wear leveling* might move less frequently updated data (referred to as *cold data*) from time to time to pages with more writes in the past (referred to as *old pages*). Note that dynamic and static wear leveling can be applied at the same time.

Wear-leveling approaches designed for PCM must rely on some *swapping or allocation* policies, which involve determining the appropriate pages for swapping or allocation, preferably youngest pages. The cost of maintaining the age information of physical pages and the search of young pages (or their ordering) might introduce significant performance/space overheads if an age-based WL approach is adopted for PCM-based memory management. Furthermore, the required swapping mechanisms might need an additional translation layer with potential overheads in cell/page location management and extra latency in access. There is a strong demand for PCM management methodology with low overheads and an efficient and effective wear-leveling capability. In addition, the methodology should be compatible with existing memory management designs or only need minor modifications.

3. AGE-BASED PCM WEAR LEVELING WITH NEARLY ZERO SEARCH COST

3.1 Overview

In this section, we present a wear-leveling design that can be incorporated into the OS's virtual memory management systems to improve the endurance of PCM-based main memory. A major technical challenge in the design is how to select an appropriate page efficiently when a free page is needed during a program's execution or when another page should be used instead to prevent an old page from being worn out.

The proposed wear-leveling design for PCM-based memory management is illustrated in Figure 2. Three components, namely, a *lifetime-aware management unit*, a *counter cache*, and *counters*, are incorporated into the memory management architecture shown in Figure 1. The counters record the ages (representing the numbers of write counts) of the PCM physical pages and provide the OS with the information to perform wear level-

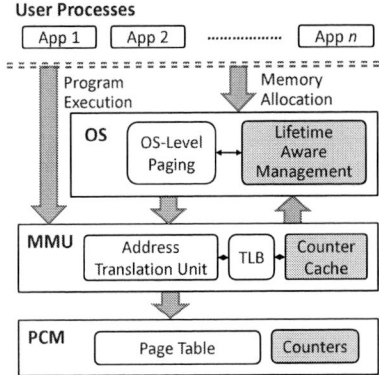

Figure 2: The proposed design for PCM-based memory management

ing. Note that the PCM pages that store the counters are also protected by wear leveling. The counter cache is employed to avoid intensive writes over the PCM due to counter updates, and it can be in the MMU, as shown in Figure 2, or in the DRAM if available. Any cache design can be adopted as long as it can determine whether a page's counter is in the cache efficiently. This work simply assumes a *direct-mapped cache*. We evaluate the impact of the cache size on the proposed approach's performance in Section 4. The focus of this work is the lifetime-aware management unit, which decides how to allocate and swap pages to enhance the PCM endurance. With the existing OS-level paging mechanism, address remapping incurred by page swapping/allocation can be performed by simply modifying the corresponding entries in the page table. In this way, address translation is still performed by the page table, rather than by an additional translation layer, so as to avoid extra overheads and latency.

For the lifetime-aware management, a young page should be acquired immediately when page allocation or page swapping needs to be performed. The idea is to place older pages far away so that they are less likely to be used. Two implementations, called *bucket-based WL* and *array-based WL*, are developed to realize the idea. Bucket-based WL relies on bucket-based data structures to find a young page quickly without searching or sorting pages (Section 3.2). Array-based WL reduces the overheads incurred by maintaining the bucket-based data structures, and achieves a comparable performance at the cost of a limited search overhead (Section 3.3).

3.2 Bucket-Based WL

3.2.1 Data Structures

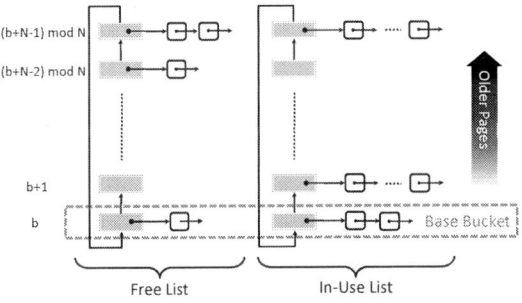

Figure 3: Bucket-based data structures

Bucket-based WL uses two bucket lists, as shown in Figure 3, to help classify pages according to their ages and to identify the youngest pages quickly. One bucket list, called the *free list*, manages all the free pages; while the other list, called the *in-use list*, manages the pages that are in use. Each bucket list employs N buckets linked in a circular format to implement "different distances", and the ages of pages in the same bucket are similar. Once a page has been written R times already, it will be moved to the next bucket with older pages. Note that each bucket on one list corresponds to a bucket on the other list. The buckets that contain the current youngest free pages and in-use pages are referred to as the *free base bucket* and the *in-use base bucket* respectively. To find a free page for allocation or swapping efficiently without searching, it can always be acquired from one of the two base buckets. When both of the base buckets become empty eventually (because their pages will gradually be moved to other buckets), the bucket lists rotate so that the current base buckets become the buckets maintaining the oldest pages, and the next buckets on the respective lists become the free and in-use base buckets.

3.2.2 Wear-Leveling Algorithms

Now, we present two procedures, FREE-PAGE-ALLOCATION() and WORN-OUT-AVOIDANCE(), which manage PCM physical pages based on the bucket-based data structures. Procedure FREE-PAGE-ALLOCATION() always returns a young page when a free page is needed during the execution of a user process or when invoked by the OS. Procedure WORN-OUT-AVOIDANCE(), which prevents old pages from being worn out, is invoked whenever an in-use page is to be updated. If the page is still young, the update is performed on the page directly. However, if the page is comparatively old and likely to be updated frequently, it is swapped with a young free page to facilitate wear leveling.

Algorithm 1 Free Page Allocation

Require: bucket indexes b and h
Ensure: a free page p
1: **if** $F[b] \neq \emptyset$ **then**
2:    $p \leftarrow$ the last page in $F[b]$
3:    move p to the head of $I[b]$
4: **else**
5:    $p \leftarrow$ the last page in $I[b]$
6:    $q \leftarrow$ the last page in $F[h]$
7:    copy the data of p to q
8:    $c[q] \leftarrow c[q] + 1$
9:    move q to the head of $I[h]$
10: **return** p

Procedure FREE-PAGE-ALLOCATION(), as shown in Algorithm 1, employs two bucket indexes b and h, where b indicates the current base buckets and h always indicates the nonempty bucket with the oldest free pages. When a free page is needed, it is always acquired from the base buckets. The procedure first checks if the free base bucket $F[b]$ is empty. If the bucket is not empty, the last page in $F[b]$ is moved to the head of the in-use base bucket and returned (Lines 1-3). In contrast, the last page in the in-use base bucket $I[b]$ is released and returned (Lines 4-5). However, before releasing the young page, we should pick a free page to store its data. Because the page is relatively young, it might not have been updated frequently. In other words, the data should be cold data. Thus, an oldest free page in $F[h]$ is moved to the head of $I[h]$ and used to store the cold data instead (Lines 6-9). Note that this procedure allocates free pages, simultaneously along with static wear leveling by moving cold data to old pages.

Procedure WORN-OUT-AVOIDANCE() is shown in Algorithm 2. Given a page q to be updated and its corresponding bucket index s, if q is not in the bucket with the oldest in-use pages, the update is performed on q directly, and the write count $c[q]$ is increased by 1 (Lines 1-2). Furthermore, if page q has been written R times already, it is moved to the next bucket $I[s+1]$

455

Algorithm 2 Worn Out Avoidance

Require: bucket indexes b and s, and a page q to be updated
Ensure: none
1: **if** $s \neq (b + N - 1) \bmod N$ **then**
2:     perform update on q
3:     $c[q] \leftarrow c[q] + 1$
4:     **if** $c[q] \geq R$ **then**
5:         move q to the head of $I[s + 1]$
6:         $c[q] \leftarrow 0$
7: **else**
8:     **if** $0 < c[q] < \frac{R}{2}$ **then**
9:         perform update on q
10:         $c[q] \leftarrow c[q] + 1$
11:     **else**
12:         $p \leftarrow$ invoke free page allocation
13:         copy the data of q to p
14:         perform update on p
15:         move q to the head of $F[s]$
16:         $c[p] \leftarrow c[p] + 1$
17:         **if** $c[p] \geq R$ **then**
18:             move p to the head of $I[b + 1]$
19:             $c[p] \leftarrow 0$

with older in-use pages, and $c[q]$ is reset as 0 (Lines 4-6). On the other hand, if q is already in the bucket with the oldest pages, we delineate two cases, depending on whether $0 < c[q] < \frac{R}{2}$. If $c[q] = 0$, it implies that page q was moved from the previous bucket $I[s-1]$ to $I[s]$ due to intensive updates, and thus the data it stores are likely to be hot data. In contrast, if $c[q] \neq 0$, then q was moved from a bucket of free pages to $I[s]$ when FREE-PAGE-ALLOCATION() was invoked. Recall that, during free page allocation, the page q is used to store the cold data of the young page to be released. Thus, $c[q]$ is not supposed to increase quickly, unless the cold data is changed into hot data. Based on the above observations, we update q directly when $0 < c[q] < \frac{R}{2}$, because it contains cold data (Lines 8-10). In contrast, if $c[q] = 0$ (i.e., q contains hot data) or $c[q] \geq \frac{R}{2}$ (i.e., the cold data may have turned into hot data), we should find a young free page to store the hot data so as to prevent q from being worn out (Lines 11-19). Note that this procedure performs dynamic wear leveling (by moving hot data to young pages) and static wear leveling (when free page allocation is invoked) simultaneously.

3.3 Array-Based WL

3.3.1 Data Structures

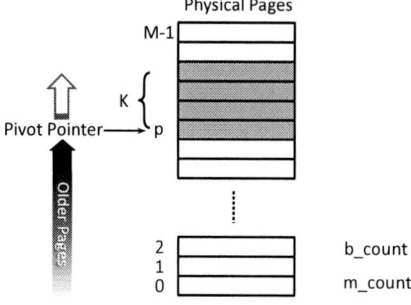

Figure 4: Array-based data structures

In this section, we present an alternative, array-based WL, which implements the idea of "placing older pages far away so that they are less likely to be used". Unlike bucket-based WL, which relies on sophisticated data structures, array-based WL simply uses a pivot pointer and two global counters to manage PCM physical pages, as shown in Figure 4. The pivot points to a page p around which to partition the array, such that the pages closer to the pivot (from the bottom of the array) are farther away and are less likely to be used soon. Once any page q has been written R' times already, we try to find a "young enough" page y from the

K pages after the pivot, and then swap the data of y and q to prevent q from being worn out. Then, the pivot pointer moves ahead so that page y locates behind the pivot and will not be used soon. To determine whether a page is young enough, we employ two global counters, m_count and b_count. During the moving of the pivot pointer, m_count is used to record the minimum write count (i.e., the age of the youngest pages). After the pivot pointer rotates back to the beginning of the array, a new round starts, and b_count is set as m_count to approximate the age of the youngest pages when the new round starts.

3.3.2 Wear-Leveling Algorithms

Based on the array-based data structures, array-based WL manages the PCM physical pages. Unlike bucket-based WL, array-based WL does not involve free page allocation. When a free page is needed, the OS simply returns a page according to its original free space management policy, irrespective of whether the page is young or old. If the page has been written R' times already, Procedure WORN-OUT-AVOIDANCE() is invoked to swap the page with a young page so that intensive writes can be spread evenly over the PCM.

Algorithm 3 Worn Out Avoidance

Require: a page q to be updated
Ensure: none
1: perform update on q
2: $c[q] \leftarrow c[q] + 1$
3: **if** $c[q] \bmod R' = 0$ **then**
4:     $y \leftarrow p$
5:     **for** $i \leftarrow 0$ to $K - 1$ **do**
6:         **if** $c[(p + i) \bmod M] \leq$ b_count $+ R'$ **then**
7:             $y \leftarrow (p + i) \bmod M$
8:             **break**
9:         **else if** $c[(p + i) \bmod M] \leq c[y]$ **then**
10:             $y \leftarrow (p + i) \bmod M$
11:         **if** $c[(p + i) \bmod M] \leq$ m_count **then**
12:             m_count $\leftarrow c[(p + i) \bmod M]$
13:     swap y and q
14:     $p \leftarrow p + i + 1$
15:     **if** $p \geq M$ **then**
16:         $p \leftarrow p \bmod M$
17:         b_count $\leftarrow$ m_count
18:         m_count $\leftarrow c[p]$

Procedure WORN-OUT-AVOIDANCE() is shown in Algorithm 3. Given a page q to be updated, the update is performed on q directly (Lines 1-2). Then, the procedure checks if q has been written R' times already (Line 3). If so, the procedure tries to find a "young enough" page y from the K pages after the pivot, namely, pages $(p + i)^2 \bmod M$, $\forall 0 \leq i \leq K - 1$ (Line 5). A page is deemed young enough if its write count is not more than b_count$+R'$, i.e., the difference in age between the page and the youngest page (when the current round starts) is not more than R'. If any pages are young enough, the first one is selected to avoid searching all the K pages (Lines 6-8); otherwise, the youngest page among the K pages is used for swapping (Lines 9-10). After being selected, page y is swapped with page q (Line 13). Then, the pivot moves ahead to the first page that has not been examined (Line 14). The design is motivated by the following observation. The probability that this procedure will be invoked by hot data is much higher than the probability that it will be invoked by cold data. Thus, the selected page y tends to become old and should be placed far away (which means right behind the pivot). By shifting the pivot pointer sequentially, the pages used recently will not be used soon, and all the pages could be used evenly. Moreover, during the shifting of the pivot pointer, m_count is used to record the age of the youngest pages (Lines 11-12). After the pivot pointer rotates back, b_count is set as m_count and m_count is set as the age of the page currently pointed by the pivot when the new round starts (Lines 15-18).

---

[2] A page is indexed by its physical address, so page $p + i$ denotes the ith page next to page p.

4. PERFORMANCE EVALUATION

4.1 Experimental Setup and Performance Metrics

In this section, we evaluate the performance of the proposed bucket-based WL and array-based WL, in terms of the lifetime and overhead. The lifetime metric is based on the failure time of the first PCM cell; and the overhead is based on the percentage of extra writes due to wear leveling. We compare two approaches: *segment swapping* [17] and *random swapping* [5]. Segment swapping is a coarse-grained mechanism that swaps the oldest and youngest segments periodically. To reduce the global search cost, the segment size is set at 1MB with a swap interval of 2×10^6 writes, as suggested in [17]. Random swapping swaps the page currently being written with a randomly selected page for every 512 writes to PCM pages, as adopted in [5]. To better identify the motivation for the adoption of wear leveling, we also estimate two baseline schemes: *no wear leveling* and an *ideal case* when all the writes are spread evenly over the PCM space in terms of cachelines.

One 4GB PCM memory (4KB per page and 10^6 write cycles per cell) is investigated. We collected the traces with *Simics* on a 3.0GHz in-order core with a 2MB 16-way private L2 cache. Eight benchmarks with large memory footprints and intensive memory accesses, namely *ammp, art, bzip2, equake, gcc, gzip, mcf*, and *twolf*, were chosen from *SPEC2000*. In addition, the benchmarks were grouped into three benchmark sets based on their memory writes per second, namely *high* (equake, gcc, gzip, and mcf), *low* (ammp, art, bzip2, and twolf), and *mixed* (gcc, equak, ammp, and twolf), and executed together to simulate multi-programming workloads. The collected traces (of each benchmark) were applied repeatedly in the simulations until a page cell was damaged. The lifetime (years) is computed based on how many runs were required to wear out the first page and the simulated time required for each run [5].

We investigate two scenarios. The first scenario does not involve process reallocation; however, in the second scenario, the processes terminate and are reallocated to different memory addresses according to the OS's free memory management policy. A memory allocation algorithm similar to the *buddy system* is implemented to cooperate with all the investigated wear-leveling approaches (except bucket-based WL, which adopts its free page allocation procedure, as shown in Algorithm 1). To ensure a fair comparison, all the (inter-page) wear-leveling approaches cooperate with the same intra-page wear-leveling mechanism, called *row shifting* [17]. First, the counter cache is assigned a size of 4MB with 4B per entry (which implies that each PCM page corresponds to a cache entry) to eliminate its impact on the performance. Then, we explore the impact of different cache sizes. In bucket-based WL, the number of used buckets N is set at 500, and the threshold R is set at 10. In array-based WL, the threshold R' is set at 5000 ($= N \times R$), and the maximum number of look-ahead pages K is set at 32. Note that searching K entries only needs $\frac{K \times \text{entry size}}{\text{cacheline size}}$ memory accesses instead of K times. The cacheline size of the last level on-chip cache is usually 64B, which implies $\frac{32 \times 4B}{64B} = 2$ memory accesses. We also explore the impact of K on the performance of array-based WL.

4.2 Experimental Results

Figures 5(a) shows the lifetime achieved by the investigated approaches under various benchmarks when process reallocation is not simulated. The lifetime varies significantly and depends to a large extent on how memory-intensive the benchmarks are. It is observed that the first cell is worn out in only few days if no wear-leveling approach is applied, and all the wear-leveling approaches delay the first failure time significantly. The results show that bucket-based WL, array-based WL, segment swapping, and random swapping achieve, respectively, 86%, 76%,

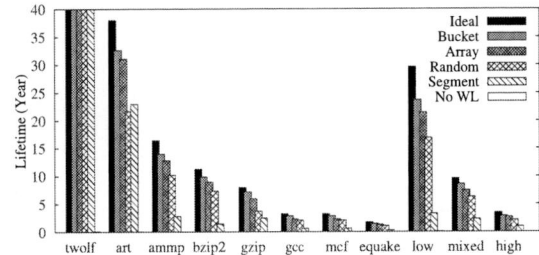

(a) The lifetime without reallocation

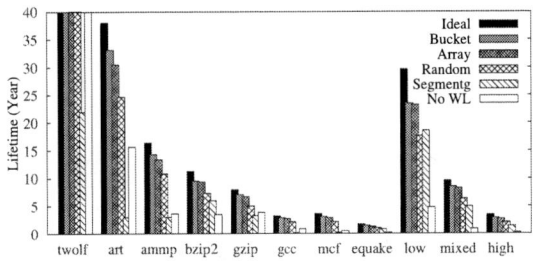

(b) The lifetime with reallocation

Figure 5: The lifetime under various benchmarks

60%, and 29% of the lifetime of the ideal case. Moreover, different benchmarks do not have a significant impact on the normalized lifetime achieved by bucket-based WL and array-based WL, but that is not the case for the segment swapping and the random swapping. The reason for the phenomenon is that, in bucket-based WL and array-based WL, whether wear leveling is invoked depends on the number of writes to a page, instead of the total number of writes to the PCM. As a result, the performance is comparatively insensitive to the locality of the benchmarks.

Figure 5(b) shows the lifetime achieved by the investigated approaches when process reallocation is simulated. We observe that the behavior of process reallocation in the OS could spread the memory writes over the PCM space naturally, especially for the benchmarks that are not memory-intensive; and even if no wear-leveling approach is applied, 59% of the lifetime of the ideal case is achieved for twolf. This is because twolf only invokes a few writes for each run, and it may be allocated to execute in different memory addresses for different runs. In general, with process reallocation, a wear-leveling approach could further improve the endurance of PCM. The results show that bucket-based WL, array-based WL, segment swapping, and random swapping achieve, respectively, 86%, 81%, 64%, and 32% of the lifetime of the ideal case. Interestingly, for some benchmarks, e.g. twolf, art, and gcc, segment swapping with process reallocation results in poorer endurance than without process reallocation. This is because segment swapping assumes that hot data are stored in old segments and cold data are stored in young segments, so that wear leveling can be achieved by swapping the data in the youngest and oldest segments. The assumption becomes invalid when processes are reallocated to somewhere else after termination.

Figure 6 shows the overhead incurred by the investigated approaches under various benchmarks. We observe that the overhead incurred by wear leveling is similar irrespective of whether process reallocation is simulated; thus, for brevity, we only report the results with process reallocation. As shown in the figure, random swapping incurs approximately 25% more writes than no wear leveling. The overhead is obviously higher than that incurred by the other wear-leveling approaches. The reason for the high overhead is that random swapping has to swap

457

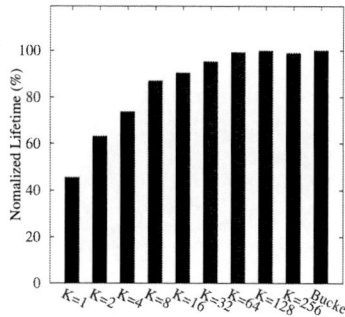

Figure 7: The impact of K on the lifetime

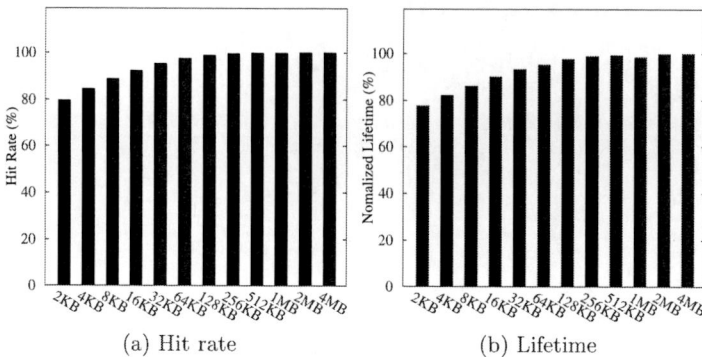

(a) Hit rate   (b) Lifetime

Figure 8: The impact of the cache size on the hit rate and lifetime

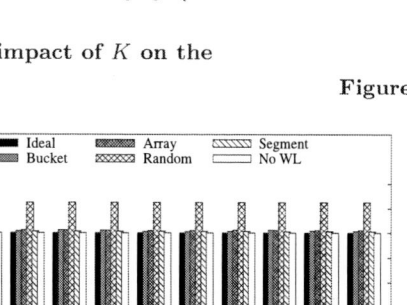

Figure 6: The overhead under various benchmarks

pages frequently (i.e., every 512 writes to PCM pages) so as to spread the writes over the PCM evenly. The overhead incurred by the other approaches is about 2%, which means one extra write for every 50 normal writes on average.

Figure 7 shows the impact of the maximum number K of look-ahead pages on the lifetime achieved by array-based WL, compared with that achieved by bucket-based WL. Array-based WL reduces the overhead of maintaining the bucket-based data structures at the cost of a limited search overhead. As expected, the lifetime increases as K increases because a "young enough" page is more likely to be found. The results show that the lifetime achieved by array-based WL is saturated when $K = 64$, which implies that a young enough page can be found within 4 memory accesses. The reported results are the average values under low, high, and mixed benchmarks with process reallocation. It is observed that array-based WL could achieve a comparable performance when $K = 32$, which implies only one additional memory access.

Figures 8(a) and 8(b) show, respectively, the impact of the counter cache size on the hit rate and the lifetime achieved by bucket-based WL. As expected, the hit rate increases as the cache size increases, as shown in Figure 8(a). When a cache miss occurs, the data in the corresponding cache entry must be written back to the PCM. The extra writes degrade the lifetime of the PCM, as shown in Figure 8(b). The reported results are the average values under low, high, and mixed benchmarks with process reallocation. It is observed that a cache size of 128KB is sufficient to achieve a reasonable performance, and the counter cache captures the locality of the most frequently written pages.

5. CONCLUSION

This paper considers the endurance issue when PCM is adopted as the main memory. We propose two wear-leveling approaches, namely, bucket-based WL and array-based WL, which can be incorporated into existing designs of virtual memory management. Both approaches implement the concept of placing older pages far away so that they are less likely to be used. We conduct ex-

tensive experiments based on popular benchmarks to evaluate the endurance improvement derived by the proposed approaches as well as the extra overhead incurred. The experimental results show that the approaches could achieve 80% of the lifetime of an ideal case, while incurring 2% extra writes compared to no wear leveling. In addition, the results of experiments conducted to compare the proposed approaches with segment swapping and random swapping provide useful insights into efficient wear-leveling designs for PCM-based memory management.

Acknowledgement

This work was supported in part by Macronix, by the Excellent Research Projects of National Taiwan University under grant No. 10R80919-2, and by the National Science Council under grant No. 100-2628-E-001-003-MY2 and 101-2219-E-002-002.

6. REFERENCES

[1] L.-P. Chang and T.-W. Kuo. An Adaptive Striping Architecture for Flash Memory Storage Systems of Embedded Systems. In *Proc. of the IEEE RTAS*, pages 187–196, 2002.

[2] Y.-H. Chang, J.-W. Hsieh, and T.-W. Kuo. Endurance Enhancement of Flash-Memory Storage Systems: An Efficient Static Wear Leveling Design. In *Proc. of the IEEE/ACM DAC*, pages 212–217, 2007.

[3] S. Cho and H. Lee. Flip-N-Write: A Simple Deterministic Technique to Improve PRAM Write Performance, Energy and Endurance. In *Proc. of the IEEE/ACM MICRO*, pages 347–357, 2009.

[4] A. P. Ferreira, B. Childers, R. Melhem, D. Mossé, and M. Yousif. Using PCM in Next-generation Embedded Space Applications. In *Proc. of the IEEE RTAS*, pages 153–162, 2010.

[5] A. P. Ferreira, M. Zhou, S. Bock, B. Childers, R. Melhem, and D. Mossé. Increasing PCM Main Memory Lifetime. In *Proc. of the IEEE/ACM DATE*, pages 914–919, 2010.

[6] E. Gal and S. Toledo. Algorithms and Data Structures for Flash Memories. *ACM Computing Surveys*, 37(2):138–163, 2005.

[7] J. Hu, C. J. Xue, W.-C. Tseng, Y. He, M. Qiu, and E. H.-M. Sha. Reducing Write Activities on Non-volatile Memories in Embedded CMPs via Data Migration and Recomputation. In *Proc. of the IEEE/ACM DAC*, pages 350–355, 2010.

[8] L. Jiang, Y. Du, Y. Zhang, B. R. Childers, and J. Yang. LLS: Cooperative Integration of Wear-Leveling and Salvaging for PCM Main Memory. In *Proc. of the IEEE/IFIP DSN*, pages 221–232, 2011.

[9] B. C. Lee, E. Ipek, O. Mutlu, and D. Burger. Architecting Phase change Memory as a Scalable DRAM Alternative. In *Proc. of the IEEE/ACM ISCA*, pages 2–13, 2009.

[10] M. Qureshi, A. Seznec, L. Lastras, and M. Franceschini. Practical and Secure PCM Systems by Online Detection of Malicious Write Streams. In *Proc. of the IEEE HPCA*, pages 478–489, 2011.

[11] M. K. Qureshi, J. Karidis, M. Franceschini, V. Srinivasan, L. Lastras, and B. Abali. Enhancing Lifetime and Security of PCM-Based Main Memory with Start-Gap Wear Leveling. In *Proc. of the IEEE/ACM MICRO*, pages 14–23, 2009.

[12] M. K. Qureshi, V. Srinivasan, and J. A. Rivers. Scalable High Performance Main Memory System Using Phase-Change Memory Technology. In *Proc. of the IEEE/ACM ISCA*, pages 24–33, 2009.

[13] S. Schechter, G. H. Loh, K. Straus, and D. Burger. Use ECP, not ECC, for Hard Failures in Resistive Memories. In *Proc. of the IEEE/ACM ISCA*, pages 141–152, 2010.

[14] N. H. Seong, D. H. Woo, and H.-H. S. Lee. Security Refresh: Prevent Malicious Wear-Out and Increase Durability for Phase-Change Memory with Dynamically Randomized Address Mapping. In *Proc. of the IEEE/ACM ISCA*, pages 383–394, 2010.

[15] D. H. Yoon, N. Muralimanohar, J. Chang, P. Ranganathan, N. Jouppi, and M. Erez. FREE-p: Protecting Non-volatile Memory against both Hard and Soft Errors. In *Proc. of the IEEE HPCA*, pages 466–477, 2011.

[16] W. Zhang and T. Li. Exploring Phase Change Memory and 3D Die-Stacking for Power/Thermal Friendly, Fast and Durable Memory Architectures. In *Proc. of the IEEE PACT*, pages 101–112, 2009.

[17] P. Zhou, B. Zhao, J. Yang, and Y. Zhang. A Durable and Energy Efficient Main Memory Using Phase Change Memory Technology. In *Proc. of the IEEE/ACM ISCA*, pages 14–23, 2009.

Algorithms and Data Structures for Fast and Good VLSI Routing

Michael Gester, Dirk Müller, Tim Nieberg, Christian Panten, Christian Schulte, Jens Vygen*

ABSTRACT

We present advanced data structures and algorithms for fast and high-quality global and detailed routing in modern technologies. Global routing is based on a combinatorial approximation scheme for min-max resource sharing. Detailed routing uses exact shortest path algorithms, based on a shape-based data structure for pin access and a two-level track-based data structure for long-distance connections. All algorithms are very fast. We demonstrate their superiority over traditional approaches by a comparison to an industrial router (on 32 nm and 22 nm chips). Our router is over two times faster, has 5 % less netlength, 20 % less vias, and reduces detours by more than 90 %.

Categories and Subject Descriptors

B.7.1 [**Integrated Circuits**]: Types and Design Styles—*VLSI (very large scale integration)*; B.7.2 [**Integrated Circuits**]: Design Aids—*Placement and Routing*

General Terms

Algorithms

Keywords

VLSI Design, Global Routing, Detailed Routing, Routing Optimization

1. INTRODUCTION

We restrict ourselves to *Manhattan routing*, meaning that all wires run parallel to the x- or y-axis. On a single routing layer, either almost all wires are horizontal or almost all are vertical (this is called the *preferred direction* of a layer; wires running orthogonally are called *jogs*). Horizontal and vertical layers alternate. This still is common design practice today.

Routing the most complex chips in the current technologies poses several challenges (cf. also [1]):

- Finding any feasible solution is often very difficult.
- The design rules tend to become more complicated with each new technology generation. While *diff-net rules*, requiring a certain minimum distance of wires that belong to different nets, are most important, *same-net rules* also require more and more attention. Ideally, the computed routing should pass all design rule checks (DRC).
- Pins often have irregular geometries and many blockages around them, which makes pin access a challenging packing problem in itself.
- Mostly for timing reasons, many nets require nonstandard wire widths, increased wire spacing, or restriction to a subset of routing layers.
- The quality of the overall routing solution is extremely important. Large detours often lead to violated timing constraints. The manufacturing yield and power consumption is significantly influenced by the routing.
- Huge instance sizes contrast with expected turn-around times of just a few hours. This requires extremely efficient algorithms and data structures.

1.1 Outline

Our global router, which we describe in Section 2, is based on an efficient fully polynomial approximation scheme for the min-max resource sharing problem. It produces a provably near-optimal fractional solution, rounds it, and removes the local congestion caused by rounding.

For detailed routing, we pre-compute routing tracks that we will use for the majority of the wires in order to pack them efficiently. The available routing space is modeled by an efficient two-level data structure (called fast grid and shape grid). A specialized version of Dijkstra's algorithm exploits this data structure to find shortest paths extremely fast even for long distances. Off-track paths are computed by a shape-based shortest path algorithm to access pins, also taking same-net rules into account. We present these routing algorithms and data structures in Section 3, and explain how we combine them with an industrial router for DRC cleanup.

The experimental results in Section 4 show that competitive numbers of DRC violations are obtained by this approach, while improving considerably wire length, via count, detours, and overall runtime.

---

*The authors are with the Research Institute for Discrete Mathematics, University of Bonn, Lennéstr. 2, D-53113 Bonn (e-mail: {gester,mueller,nieberg,panten,schulte,vygen}@or.uni-bonn.de).

2. GLOBAL ROUTING

2.1 Modeling the Global Routing Problem

Global routing is an abstraction of the actual routing problem to a coarser model of the routing space. The chip area is divided into an array of *tiles*. The size of our tiles is chosen such that approximately 50 to 100 parallel wires (of minimum width) would fit into one tile on each layer. For each tile and each routing layer we have a vertex. The vertices thus constitute a partition of the routing space. Two vertices (t, l) and (t', l') are connected by an edge if $t = t'$ and $|l - l'| = 1$, or if $l = l'$ and t and t' are two tiles that are adjacent in the preferred direction of routing layer l. This defines an undirected graph G. The edges of G are assigned capacities $u : E(G) \rightarrow \mathbb{R}_{\geq 0}$, estimating the number of standard wires that can go from any vertex to any neighbor while obeying the minimum distance requirements.

Each pin p has shapes in one or more tiles, and is represented by the respective set V_p of vertices. Each net is a set of pins, and $\mathcal{N}$ denotes the set of nets. Let $\mathcal{T}_n$ denote the set of feasible *Steiner forests* for net n, i.e., minimal edge sets $F \subseteq E(G)$ such that $F \cup \bigcup_{p \in n} K(V_p)$ connects all pins of n, where $K(V_p)$ denotes the clique on V_p. Any net n for which $\emptyset \in \mathcal{T}_n$ (e.g. because all pins are in the same tile on the same layer) can be removed.

For a net n and an edge $e \in E(G)$ let $w(n, e)$ denote the minimum required width of a wire for n along e plus the minimum required distance to any neighbor.

A major difference to most other global routers is that we allow to allocate extra space $s(n, e) \geq 0$ as this may reduce coupling capacitance and hence delay and power consumption, and may also lead to better yield.

Then the global routing task is to find for each $n \in \mathcal{N}$ a Steiner forest $T_n \in \mathcal{T}_n$ and an extra space assignment $s(n, e) \geq 0$ for $e \in T_n$ such that $\sum_{n \in \mathcal{N}: e \in T_n} (w(n, e) + s(n, e)) \leq u(e)$ for each $e \in E(G)$.

However, we are not satisfied with an arbitrary feasible solution. While a traditional objective function considers only wire length and number of vias, we can also optimize other objective functions like power consumption or expected yield, and we can also deal with constraints bounding e.g. detours of certain nets or weighted sums of capacitances on critical paths. The objective function and each of such constraints will be modeled as a *resource*. We denote by $\mathcal{R}$ the set of resources.

Each wire consumes a certain amount of some of the resources modeled by functions $\gamma_{n,e}^r : \mathbb{R}_{\geq 0} \rightarrow \mathbb{R}_{\geq 0}$ for $r \in \mathcal{R}$, $n \in \mathcal{N}$ and $e \in E(G)$. Here, $\gamma_{n,e}^r(s)$ is the estimated use of resource r if e is used by net n with allocated space $w(n, e) + s$. For all relevant resources that we consider, including electrical capacitance (with coupling), power consumption, and estimated wiring yield loss, these functions are convex (cf. [7] and [14]).

For each constraint, we have an upper bound $u^r > 0$ of how much we may use of the corresponding resource r. For the objective function, we do not have an upper bound a priori, but we can guess a value that we expect to be achievable and adapt it if needed (we could apply binary search to find approximately the optimum value, but this is usually not needed in practice).

We also model the edge capacities as resources. If $r(e)$ denotes the resource corresponding to edge e, we have $u^{r(e)} =$ $u(e)$, $\gamma_{n,e}^{r(e)}(s) = w(n, e) + s$, and $\gamma_{n,e'}^{r(e)}(s) = 0$ for any other edge $e' \neq e$ and $s \geq 0$.

Now the task is to find for each $n \in \mathcal{N}$ a Steiner forest $T_n \in \mathcal{T}_n$ and an extra space assignment $s(n, e) \geq 0$ for $e \in T_n$ such that $\sum_{n \in \mathcal{N}} \sum_{e \in T_n} \gamma_{n,e}^r(s(n, e)) \leq u^r$ for all $r \in \mathcal{R}$.

2.2 Reduction to Min-Max Resource Sharing

Let $\chi(T) \in \{0, 1\}^{E(G)}$ denote the incidence vector of a Steiner forest T (i.e. $(\chi(T))_e = 1$ for $e \in T$ and $(\chi(T))_e = 0$ for $e \notin T$). Then

$$\mathcal{B}_n^{\text{int}} := \{ (\chi(T), s) \mid T \in \mathcal{T}_n, \, s \in \mathbb{R}_{\geq 0}^{E(G)}, \, s_e = 0 \text{ for } e \notin T \}.$$

represents the set of feasible solutions for net n. In our relaxation we consider the convex hull $\mathcal{B}_n := \text{conv}\left(\mathcal{B}_n^{\text{int}}\right)$ for each $n \in \mathcal{N}$. For $(x, s) \in \mathcal{B}_n$ we define

$$g_n^r(x, s) := \frac{1}{u^r} \sum_{e \in E(G): x_e > 0} x_e \, \gamma_{n,e}^r(s_e / x_e).$$

Note that we have $g_n^r(\chi(T), s) = \frac{1}{u^r} \sum_{e \in T} \gamma_{n,e}^r(s_e)$ for $(\chi(T), s) \in \mathcal{B}_n^{\text{int}}$. If each $\gamma_{n,e}^r$ is convex (as in our applications), then each g_n^r is also convex. With this notation, the global routing problem asks for an element $b_n \in \mathcal{B}_n^{\text{int}}$ for each $n \in \mathcal{N}$ such that $\sum_{n \in \mathcal{N}} g_n^r(b_n) \leq 1$ for all $r \in \mathcal{R}$. As in previous works, we now consider a fractional relaxation first and look for $b_n \in \mathcal{B}_n$ $(n \in \mathcal{N})$ approximately attaining

$$\lambda^* := \inf \left\{ \max_{r \in \mathcal{R}} \sum_{n \in \mathcal{N}} g_n^r(b_n) \,\middle|\, b_n \in \mathcal{B}_n \, (n \in \mathcal{N}) \right\}.$$

If the instance is feasible and the guess of an achievable value of the objective function is realistic (see above), then λ^* will be close to 1.

This relaxation is known as the (block-angular) MIN-MAX RESOURCE SHARING PROBLEM. In its standard formulation, the sets $\mathcal{B}_n$ are not given explicitly. Rather we assume to have an oracle, denoted by $f_n : \mathbb{R}_{\geq 0}^{\mathcal{R}} \rightarrow \mathcal{B}_n$, for $n \in \mathcal{N}$, which for $y \in \mathbb{R}_{\geq 0}^{\mathcal{R}}$ returns an element $b \in \mathcal{B}_n$ with $\sum_{r \in \mathcal{R}} y_r g_n^r(b) \leq \sigma \, \text{opt}_n(y)$, where $\text{opt}_n(y) := \inf_{b \in \mathcal{B}_n} \sum_{r \in \mathcal{R}} y_r g_n^r(b)$ and $\sigma \geq 1$ is a constant. In other words, we assume that we can optimize linear functions over each $\mathcal{B}_n$ efficiently with an approximation factor σ. This is indeed true in our case:

THEOREM 1. [9] *The oracle functions f_n can be implemented by an approximation algorithm for the Steiner tree problem in weighted graphs.*

We use the fastest known algorithm for the MIN-MAX RESOURCE SHARING PROBLEM, due to Müller, Radke and Vygen [9]. The core algorithm works in t phases, where t has to be chosen depending on the desired approximation guarantee. In each phase, a solution for each net is computed by applying Theorem 1, and resources become more expensive as they are used. The output of the algorithm is a convex combination $\sum_{b \in \mathcal{B}_n^{\text{int}}} x_{n,b} b$ of the elements in $\mathcal{B}_n^{\text{int}}$ for each net n, i.e. $\sum_{b \in \mathcal{B}_n^{\text{int}}} x_{n,b} = 1$. See [9] for a detailed description and a proof of the following Theorem:

THEOREM 2. [9] *If our oracle computes solutions within a factor σ of optimal in time θ, then a $\sigma(1 + \omega)$-approximate solution can be computed in $O(\theta \log |\mathcal{R}|(|\mathcal{N}| + |\mathcal{R}|)(\log \log |\mathcal{R}| + \omega^{-2}))$ time for any $\omega > 0$.*

1	2	3	4	5	6	7	0	0	0	0	0
0	0	0	0	0	0	0	0	0	0	0	0
0	0	0	0	8	8	8	8	8	8	8	8
0	0	9	10	11	12	12	12	12	12	12	12
0	0	0	0	0	0	0	0	0	1	2	3

Figure 1: Shape grid cell configurations and their configuration numbers for a given wiring. Sequences of identical numbers in preferred direction are merged to intervals. In this example, we store 15 intervals with a configuration other than 0.

The algorithm can be parallelized very well using *volatility-tolerant* block solvers [9] to guarantee the desired approximation ratio even with concurrent access to the same resources by multiple threads, and achieves a very good scaling with the number of processors [9].

As we need an element of $\mathcal{B}_n^{\text{int}}$ for each net n, we apply randomized rounding to the output of our resource sharing algorithm, choosing each $b \in \mathcal{B}_n^{\text{int}}$ with probability $x_{n,b}$, for each net n independently. In practice, this results only in few violations, i.e. resources $r \in \mathcal{R}$ with $\sum_{n \in \mathcal{N}} (g_n(\hat{b}_n))_r > 1$, if $\hat{b}_n$ is the solution picked by randomized rounding for $n \in \mathcal{N}$. These violations can be eliminated easily by postprocessing ("ripup and reroute") using standard heuristic techniques.

3. DETAILED ROUTING

Most of the wiring on a routing layer has the minimum possible width. Such *standard wires* can be packed best with a net-by-net routing approach if they are aligned on parallel *tracks* in preferred direction. Normally, the optimal distance of two adjacent tracks is the minimum wire width plus the minimum required distance of two wires, but in the presence of blockages, total usable track length can be maximized only with non-uniform track spacings in general. Optimum routing tracks can be computed in $O(n \log n)$ time for n axis-parallel rectangular blockages [8]. The intersection points of routing tracks with tracks projected from neighboring wiring layers define the vertices of a *track graph* in a natural way.

As most pins are not aligned with tracks, we combine a very fast on-track path search for long distance connections (cf. Section 3.3), obyeing only diff-net rules, with an off-track path search that also takes same-net rules into account and is used for pin access (cf. Sections 3.4, 3.5, and 3.6). The on-track path search operates on a two-level data structure for representing routing space, which we describe in Sections 3.1 and 3.2. Section 3.7 explains our parallelization approach for detailed routing. Finally, we describe how we combine our router with an industrial router for DRC cleanup in Section 3.8.

3.1 Shape Grid

The shape grid efficiently stores all relevant data about blockage, wire, and via shapes. The information in the shape grid allows to decide whether a wire can be placed somewhere without violating minimum distance rules. If not, it allows to find out if there is a set of shapes that can be removed such that the answer becomes positive.

The shape grid partitions the chip area on each wiring

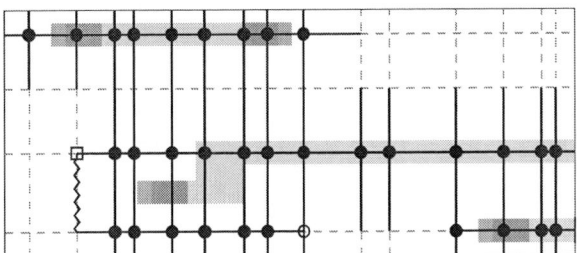

Figure 2: The fast grid: dashed lines represent x- and y-edges of the track graph, with vertices at their intersections. For some given wiretype, circles mark vertices at which no jog can start (unless ripup is allowed); if a circle is filled, then no wire in preferred direction can start at this vertex either. Thick black edges represent edges that are unusable with the given wiretype. Usability of an edge can be deduced from the vertex information in all cases except the zigzag edge: For this edge, we set a bit at one of its incident vertices (marked by a square) to indicate that usability of its incident edges must be determined by querying the shape grid.

layer and on the via layers into axis-parallel rectangular *cells*, such that each of them has at most one neighbor cell to the left, right, bottom and top, respectively. The size of the cells is small enough such that shapes of different nets cannot be present in the same cell while obeying all distance rules.

In each cell, all intersections of shapes with its area are stored with coordinates relative to an anchor point (e.g. its center), plus their respective type and minimum distance requirements. Since this *cell configuration* is typically identical in a large number of cells, we store this data indirectly by a *cell configuration number* which is used as an index into a lookup table that stores the actual data.

The data volume is further reduced by grouping neighboring cells in preferred routing direction to *intervals* which are stored in an AVL-tree in each row or column of cells, depending on the preferred routing direction (see Figure 1 for an example).

Moreover, for each nonempty interval we store the net that the shapes of this interval belong to if they are removable. The type and net identifier of a shape are used to assign a *ripup level* to it. The ripup-and-reroute algorithm can then be restricted to rip out shapes of at most some specified ripup level in order to avoid re-routing of critical connections.

3.2 Fast Grid

As most of the routing is done on predefined routing tracks, we obtain faster query times by storing pre-computed data for a restricted set of locations and *wire types* (i.e. wire and via geometries with spacing requirements) based on these tracks. This data is maintained in the *fast grid* data structure. Queries for other, less frequently used, wire types are transferred to the shape grid.

If there is only on-track wiring, legality of a wire whose stick figure (usually the center line) connects two neighboring vertices of the track graph is implied by legality of the zero-dimensional stick figures at each of the vertices. Because of this, the fast grid stores information for vertices, but not for edges. This allows to group longer sequences of vertices with identical data to intervals. If data was stored for vias or jogs, more changes would occur along a track in preferred direction, resulting in a higher interval count.

If legality of a wire on an edge $\{v, w\}$ cannot be deduced from legality at v and w because off-track wires or other shapes are present in the vicinity, an extra bit at one of the vertices v and w encodes this (see Fig. 2 for an illustration). If this bit is set, a query to the fast grid is transferred to the shape grid.

We store information on the usability of up to five different wire types in the fast grid. In our experiments, this allowed to answer 97.9% of the legality queries using the information stored in the fast grid, resulting in an average overall speedup of 5.29 of on-track path search.

3.3 On-Track Path Search

The on-track path search finds a shortest path between two sets of vertices (corresponding to connected components of a net) in a graph G which results from the track graph by removing all edges whose usage would introduce a diff-net violation. We do not store G explicitly, but rather query the fast grid for usable edges as needed. If the fast grid does not store legality information for the wire type we want to use, a query to the shape grid is performed.

When routing a modern chip, millions of path searches have to be performed. Thus even a runtime linear in $|V(G)|$ would be too slow. Our on-track path search uses a Dijkstra-based algorithm with two major speed-up features:

1) Merging sequences of usable vertices on a track, and labeling whole intervals instead of single vertices. This was proposed by Hetzel [3] for the special case of equidistant routing tracks which match in all layers with the same preferred direction, generalized by Humpola [4] to arbitrary routing tracks, and further generalized by Peyer et al. [11] to more general vertex sets.

2) Using a *future cost* (similar to the A* heuristic [12]) to reduce the number of labeling steps by providing a lower bound of the distance to the target for each vertex. Peyer et al. [11] proposed a blockage-aware future cost which computes shortest paths to the target from all nodes in a supergraph of G with a simpler structure. As this also consumes non-negligible runtime, we do it only if the global routing for this connection already contains a large detour, and use a weaker future cost based on l_1-distance otherwise, which is cheaper to compute.

With edge costs

$$c(\{v, v'\}) = \begin{cases} \gamma_{\{z, z'\}} & \text{if } \{v, v'\} \text{ is a via} \\ \beta_z \cdot ||v - v'||_1 & \text{if } \{v, v'\} \text{ is a jog} \\ ||v - v'||_1 & \text{otherwise} \end{cases}$$

for two neighboring vertices $v := (x, y, z)$ and $v' := (x', y', z')$ of G, where $\beta_z, \gamma_{\{z, z'\}} \in \mathbb{Z}_{>0}$ are parameters that encode penalty costs for wires running in non-preferred direction and for vias, respectively, the following performance guarantee can be given:

THEOREM 3. [3, 4, 11] *A shortest path P in a track graph G with edge costs $c : E(G) \to \mathbb{N}$ as above can be found in time $O(\min\{(\Lambda + 1)|\mathcal{I}| \log |\mathcal{I}|, |V(G)| \log |V(G)|\})$, where $\mathcal{I}$ is the set of intervals representing G, and Λ is the difference of the cost of P w.r.t. c and the minimum future cost of a source vertex (i.e., Λ measures how good this lower bound is).*

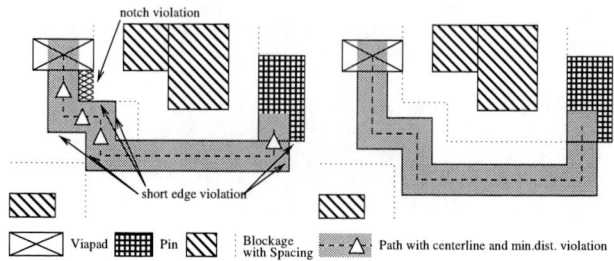

Figure 3: Example showing geometric design rule violations for geometric shortest paths without minimum segment length (left). The path is sought to connect the viapad to the pin. Taking minimum segment lengths into account, a feasible path (right) obeys these rules.

According to recent experiments on 22 nm chips, labeling and storing intervals instead of single nodes speeds up the path search by at least a factor of 6.

To support *ripup-and-reroute*, the on-track path search can be driven in a mode where vertices of the track graph that are unusable only because of minimum distance violations to modifiable shapes are flagged as usable, but high extra costs are imposed on intervals containing such vertices.

If there is unused space in a region, *wire spreading* can improve timing and manufacturing yield (both by reducing the probability of extra material defects [7], and by increasing the number of vias that can be replaced by larger, i.e. more robust, vias in postprocessing). To do this, the on-track path search imposes extra costs on intervals that should be kept free, based on congestion observed by global routing.

3.4 Same-Net Rules

Most same-net rules are in place to avoid geometric configurations with features below the lithographic capabilities, to ensure space for optical proximity correction (OPC), and to improve manufacturing yield. The list of these rules is often very long, we briefly discuss the most important ones:

- Notch violations: Even within the same path, non-adjacent segments have to obey distance requirements.

- Short edge violations: Two short adjacent edges of any polygon bounding metal area are forbidden.

- Minimum area violations: Every connected metalized polygon on a layer must have a certain minimum area, independent of its actual shape.

These rules are often violated in geometric shortest paths and Steiner trees, see also Figure 3. It is particularly important to obey them in pin access because of the often irregular geometries of pins and limited space available for fixing errors by postprocessing.

3.5 Blockage Grid for Off-Track Path Search

The main goal of the blockage grid is to support fast search for shortest (off-track) paths that avoid the most important classes of same-net rule violations. Most same-net rules can be mapped to requirements on minimum segment lengths in the constructed path [10]. Our approach to DRC-clean off-track wiring hence is as follows.

Suppose that we only have a single requirement stating that each segment of a wire must have length at least $\tau > 0$.

Call a rectilinear path τ-*feasible* if each of its segments has length at least τ and does not intersect the interior of any obstacle. We construct a data structure called *blockage grid*, which allows for finding shortest τ-feasible paths.

Starting with a Hanan grid, we add additional vertical and horizontal lines at distances $k\tau$, $k_1 \leq k \leq k_2$ and $k \in \mathbb{Z}$, from each original line, where k_1 is minimal and k_2 maximal such that each added line is at most 2τ from any of the original vertical or horizontal lines, respectively. It is not difficult to see that the total number of vertices, i.e. crossing points of lines, is then bounded by $O(n^4)$, where n denotes the number of rectangles representing the blockages. The following result shows that these vertices suffice.

THEOREM 4. [6] *Consider a set of n rectilinear obstacles in the plane, a distance $\tau > 0$, and $s, t \in \mathbb{R}^2$. If there exists a τ-feasible path from s to t, then there is also a shortest τ-feasible path from s to t all whose segments have vertices of the blockage grid as endpoints.*

In order to respect the minimum segment length τ during the path search, we construct a path-preserving digraph G on which we can run a regular shortest path algorithm. We construct G from the blockage grid as follows. G contains up to four vertices for each vertex of the blockage grid; one collecting incoming arcs of each direction. Arcs of G connect neighboring vertices of the blockage grid where this does not induce a bend together with the incoming direction. Additional arcs connect each vertex to the (at most two) nearest vertices at distance $\geq \tau$ perpendicular to the incoming direction. This way, each bend is followed by a long arc as G does not contain short arcs after bends. More sophisticated rules like short edge avoidance can be added as well by introducing corresponding edges in G or removing edges that induce a respective violation.

Storing the blockage grid coordinates only takes $O(n^2)$ space. For a multi-layer path search, we have a blockage grid for each wiring and via layer. Coordinates are also induced by blockages of neighboring layers. Vertices of two adjacent wiring layers are connected if a via is allowed here.

3.6 Off-Track Pin Access

Pins that are not aligned with routing tracks require off-track pin access [10], which is based on the off-track path search and blockage grid presented in Section 3.5. For each such pin we construct a *catalogue* of several DRC clean paths connecting it to on-track points within a small radius. Among these, we compute one primary access path for each pin such that the set of these paths for the pins of each circuit forms a *conflict-free solution* (see also Figure 4), i.e. is DRC-clean also w.r.t. diff-net rules. We add the primary access paths to the routing space as reservations before actually starting to route: this way, newly added wires do not invalidate the precomputed conflict-free solution. We do all this in a preprocessing step, but also have the functionality to construct additional paths on-the-fly when needed.

Although there may be millions of circuits placed on the chip, containing the vast majority of pins, there are only a few thousand prototype circuits in a library. Furthermore, non-circuit blockages and wires, e.g. stemming from power supply or pre-designed clock nets, often have a regular structure. The preprocessing step of constructing the catalogues for the pins exploits this.

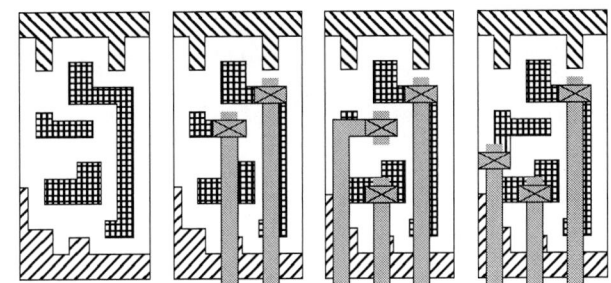

Figure 4: Examples of different pin access solutions. On the left, a circuit is given with its pins (on layer 1) and blockage structure. The following three figures show three different situations for pin access wiring (grey, on layer 2). First, a situation that may occur by a greedy approach is shown: after connecting the first two pins, the third pin is blocked. The two figures on the right show two conflict-free solutions (using on-track via positions on metal 2 as endpoints). The one on the right is superior and will be chosen by our algorithm.

We partition the set of circuits into *circuit classes*. Expanding the bounding area of a circuit and taking all shapes influencing this area and the track coordinates into account, we can identify geometrically equal situations on the chip area (up to translation, mirroring, and rotation), and we thus collect these configurations into equivalence classes. The preprocessing is then done based on these.

To anticipate the effects of local congestion, we are not interested in feasible conflict-free solutions alone, but also take (local) spreading of the paths into account. Knowing that the ongrid path-search adds wiring around the endpoints of the access paths, we evaluate a conflict-free solution based on spreading of endpoints, number of blocked tracks, directions of feasible on-track continuation, and length.

We use a branch-and-bound based enumeration technique called destructive bounding to compute a good conflict-free solution for each circuit class.

3.7 Parallelization

The parallelization approaches we use in global and detailed routing are fundamentally different, although both are shared-memory approaches. While our global router allows different threads to work in the same region, in detailed routing we partition the chip area into regions assigned to threads in order to strictly obey minimum distance rules without needing a collision detection mechanism.

Hence, in detailed routing, each thread is allowed to make only changes that do not affect regions assigned to other threads. E.g., if a wire of a certain type has large minimum distance requirements to other objects, it may not be put close to the border of the region assigned to a thread. Because this allows only a subset of connections to be closed, a sequence of partitions is defined such that in each of them the estimated workload in each region is balanced. The number of regions, and hence the number of threads used, is reduced in later partitions to allow longer connections to be closed. In addition, there is an initial critical net routing step. Nets can be considered critical here from a timing perspective, or because they use wide wires that would be hard to route when the majority of nets has already been routed.

463

3.8 Combination with External DRC Cleanup

The algorithms described above constitute the core parts of BonnRoute, which is the routing solution of the University of Bonn, developed within the scope of our cooperation with IBM. It is part of the BonnTools [5]. BonnRoute has been used by IBM and its customers for the design of more than thousand of the most complex chips. It focuses on near-optimum packing of wires and avoids only the most important classes of DRC violations: first, with very few exceptions, it leaves no violations of diff-net spacing rules; second, DRC violations that need additional space for fixing, e.g. minimum area errors, are avoided as much as possible. In addition, many other types of rules are handled by a post-processing step. Cleanup of the remaining DRC violations is left to an external tool. The experimental results in Section 4 demonstrate that this approach works very well in practice.

4. EXPERIMENTAL RESULTS

We compare routing results of an industry standard router (ISR) with our combined flow ("BR+ISR"), in which we use the same ISR to clean up design rule violations left by BonnRoute. ISR is a current router which is used in practice on many industry designs of former and recent technologies. In contrast to BonnRoute, it uses traditional rip-up and reroute [13] and a track assignment step to cover long distances.

All experiments were done on a 3.47GHz Intel Xeon machine. Both routers used 12 threads. Table 1 reports runtime, wire length, via counts, scenic nets and error counts, which are the sums of the numbers of DRC violations and opens (i.e. number of connected components minus number of nets). Chips 5 and 8 are 32nm designs, all others are 22nm. Out of 23:08 hours total runtime of our combined "BR+ISR" flow on these instances, BonnRoute takes only 7:12 hours. DRC cleanup takes considerably longer although only local changes are made. Nevertheless, the total runtime of our "BR+ISR" flow is less than half compared to "ISR". Global routing takes only 24 minutes of the total BonnRoute runtime, 17 minutes of which are spent in the resource sharing algorithm.

We call a net *scenic* if it has routed wiring length of at least $100\,\mu m$ and a detour of at least 25% or 50%, respectively, over the length of a Steiner tree with minimum length (for nets with at most 9 terminals [2]) or approximately minimum for nets with 10 or more terminals (obtained by a Steiner tree heuristic). The Steiner trees used as baseline for defining scenic nets are of course identical in the "ISR" and "BR+ISR" rows. The table shows that the quality of the routing obtained by BonnRoute is far superior.

Acknowledgment

The authors would like to thank our cooperation partners at IBM, in particular Karsten Muuss, Sven Peyer, and Gustavo Tellez. Moreover, we would like to acknowledge the contributions of former members and students of the BonnRoute team, in particular Asmus Hetzel, Christoph Albrecht, André Rohe, Sven Peyer, Jesco Humpola, Lars Bellinghausen, Corinna Gottschalk, Daniel Joachimi, Niko Klewinghaus, Felix Nohn, Thomas Petig, and Rudolf Scheifele.

Chip (Nets)	Runtime	Wire len. (m)	Vias	Scenics (≥25%)	Scenics (≥50%)	Err.
1	2:20:10	3.08	1,266 k	587	103	18
(121k)	0:35:53	2.97	932 k	7		28
2	2:09:32	3.56	1,113 k	1,545	1,071	124
(126k)	1:28:38	3.40	904 k	16		194
3	2:04:40	3.07	1,240 k	695	286	14
(130k)	0:35:23	2.97	925 k	8	1	25
4	2:28:17	3.37	1,248 k	502	232	15
(135k)	0:42:19	3.24	945 k	5		3
5	2:49:58	9.91	2,936 k	2,401	1,157	50
(384k)	1:46:37	9.57	2,399 k	105	6	61
6	12:42:42	14.01	4,253 k	14,829	10,668	459
(438k)	5:28:50	12.47	3,234 k	1,568	520	506
7	8:51:05	12.31	4,039 k	3,853	2,310	154
(466k)	2:31:47	11.87	3,185 k	350	193	183
8	14:44:44	38.87	7,769 k	11,516	6,539	111
(962k)	9:59:18	37.31	6,239 k	2,619	1,285	117
Sum	48:11:08	88.18	23,864 k	35,928	22,366	945
	23:08:45	83.80	18,763 k	4,678	1,285	1,117

Table 1: Runtimes, wire length, vias, scenic nets and error counts for ISR (light gray) and our combined BR+ISR flow (dark gray)

5. REFERENCES

[1] C. Alpert, Z. Li, M. Moffitt, G. Nam, J. Roy, and G. Tellez. What makes a design difficult to route. In *Proc. ISPD*, pages 7–12, 2010.

[2] C. Chu and Y.-C. Wong. FLUTE: Fast lookup table based rectilinear steiner minimal tree algorithm for VLSI design. *IEEE Trans. Computer-Aided Design Integr. Circuits Syst.*, 27:70–83, 2008.

[3] A. Hetzel. A sequential detailed router for huge grid graphs. In *Proc. DATE*, pages 332–339, 1998.

[4] J. Humpola. Schneller Algorithmus für kürzeste Wege in irregulären Gittergraphen. *Diploma Thesis, University of Bonn*, 2009.

[5] B. Korte, D. Rautenbach, and J. Vygen. BonnTools: Mathematical innovation for layout and timing closure of systems on a chip. *Proc. IEEE*, 95:555–572, 2007.

[6] J. Maßberg and T. Nieberg. Rectilinear paths with minimum segment lengths. *Discrete Appl. Math., to appear*.

[7] D. Müller. Optimizing yield in global routing. In *Proc. ICCAD*, pages 480–486, 2006.

[8] D. Müller. *Fast Resource Sharing in VLSI Routing*. PhD thesis, University of Bonn, 2009.

[9] D. Müller, K. Radke, and J. Vygen. Faster min-max resource sharing in theory and practice. *Math. Prog. Comp.*, 3:1–35, 2011.

[10] T. Nieberg. Gridless pin access in detailed routing. In *Proc. DAC*, pages 170 – 175, 2011.

[11] S. Peyer, D. Rautenbach, and J. Vygen. A generalization of dijkstra's shortest path algorithm with applications to VLSI routing. *J. Discrete Algorithms*, 7:377–390, 2009.

[12] P. E. Hart and N. J. Nilsson and B. Raphael. A formal basis for the heuristic determination of minimum cost paths. *IEEE Trans. Syst. Sci. Cybern.*, 2:100–107, 1968.

[13] J. Salowe. Rip-up and reroute. In C. Alpert, D. Mehta, and S. Sapatnekar, editors, *Handbook of Algorithms for Physical Design Automation*, pages 615–626. CRC Press, 2009.

[14] J. Vygen. Near-optimum global routing with coupling, delay bounds, and power consumption. In *Proc. IPCO*, pages 308–324, 2004.

Guiding a Physical Design Closure System to Produce Easier-to-Route Designs with More Predictable Timing

Zhuo Li[1], Charles J. Alpert[1], Gi-Joon Nam[1], Cliff Sze[1], Natarajan Viswanathan[2], Nancy Y. Zhou[2]

[1] IBM Austin Research Laboratory, 11501 Burnet Road, Austin, TX 78758.
[2] IBM Systems & Technology Group, 11400 Burnet Road, Austin, TX 78758.
{ lizhuo, alpert, gnam, csze, nviswan, nancyz }@us.ibm.com

ABSTRACT

Physical synthesis has emerged as one of the most important tools in design closure, which starts with the logic synthesis step and generates a new optimized netlist and its layout for the final sign-off process. As stated in [1], "it is a wrapper around traditional place and route, whereby synthesis-based optimization are interwoven with placement and routing." A traditional physical synthesis tool generally focuses on design closure with Steiner wire model. It optimizes timing/area/power with the assumption that each net can be routed with optimal Steiner tree. However, advanced design rules, more IP and hierarchical design styles for super-large billion-gate designs, serious buffering problems from interconnect scaling and metal layer stacks make routing a much more challenging problem [2]. This paper discusses a series of techniques that may relieve this problem, and guide the physical design closure system to produce not only easier to route designs, but also better timing quality. Open challenges are also overviewed at the end.

Categories and Subject Descriptors

B.7.2 [**Integrated Circuits**]: Design Aids—*Placement and Routing*

General Terms

Design, Experimentation

Keywords

Physical Synthesis, Timing Driven Routing

1. INTRODUCTION

Routability has become an increasingly challenging issue in VLSI design flow in advanced technology nodes (65 nm and beyond). High number of metal layers, wide range of metal thickness, the explosion of design rules, complicated logic structures like crossbars and the ever-increasing design complexity, of course, are just a few of the many reasons that make routing closure much more difficult. The situation is that without conscious attention to routability through the entire physical synthesis flow, it's more likely that the design cannot be fully routed without any routing violations. Even

if the design can be routed clean, how to prevent the timing degradation from the routing is another challenge task.

A typical physical synthesis flow may consist of the following steps [3], though the order may be different for various systems: 1) global placement, 2) optimization, 3) timing-driven placement, 4) optimization, 5) clock insertion and optimization, 6) routing and post-routing optimization. Traditionally, each step focuses on timing and power optimization with Steiner wire model, but as illustrated later in the paper, without paying attention to the congestion, it is easy to create unroutable designs even if the timing with Steiner wire model closes. To produce a routable instance and reduce the gap between the timing before and after the routing, various congestion mitigation techniques and routing techniques are discussed in this paper. With these techniques, a new physical synthesis system is proposed with the following steps, and one may combine them in different orders,

1. logic cleanup,
2. global placement,
3. high fanout buffering and wirelength reduction,
4. global and detail congestion aware spreading,
5. clock insertion,
6. global buffering,
7. optimization,
8. constraints driven global routing and optimization,
9. constraints driven detail routing and optimization.

In the rest of the paper, we first classify the different types of congestion that occur during a physical synthesis flow, and the different factors that cause the timing degradation in the routing stage. Then, each component in the proposed flow is discussed, followed by the experimental results. Finally, open challenges are discussed followed by the final conclusion.

2. CONGESTION MODELS AND METRICS

In practice, different flavors of congestions can occur during physical synthesis optimization. To guide the tools to mitigate the congestion, congestion models and metrics are necessary. In general, congestion modeling is performed in a regular grid structure created on top of a given design. For each grid tile, a relevant congestion metric is calculated.

Placement Cell Congestion and Metrics. This is also known as cell brick wall in design community. Typical placement objective function, the total wire length minimization, can sometimes create locally congested areas where cells are packed tightly as shown in Fig. 1(a). Subsequent timing optimization such as buffering and gate sizing in these congested areas might be problematic due to lack of free space. Particularly when the brick-walled cell area occurs in narrow alleys between fixed macros, the situation gets worse. Additional buffers or gate-sized cells have to be spiraled off over the big macros to find legally placed locations.

To model the placement cell congestion, cell utilization per tile typically used. It is defined as the sum of cells whose outlines overlap with the grid tile divided by the area to a grid tile. Or the number of brick-walled cells (i.e., two adjacent cells that are abutting each other) can be counted as well.

Global Routing Congestion and Metrics. Once all the cells are legally placed without any overlapping, wiring congestion can be measured. Previously, probabilistic congestion estimation approaches or rent rule based models are used to estimate the congestion at the placement stage, but they are highly inaccurate. With recent advancement of global routing algorithms [4–9], one can directly invoke the global router to measure the routing congestion and get much more accurate information. Once a global routing is performed, a set of congested tiles is identified where more routing resources are on demand than the available ones. Figure 1(b) shows the example of routing congestion map on ISPD placement benchmark. Global routing accurately captures the wires that pass between routing edges and it has to be completed successfully without any overflow violation. If global routing congestion exists, signal nets passing the congested area need to be detoured to avoid the increase of congestion, which can lead to degraded signal delay. Thus, global routing congestion not only affects the routability of a design, it can also significantly affect the timing of a design.

While the congestion maps capture the details of global congestion, it is hard to be used as a metric to compare different techniques or guide the optimization tools. Several metrics to measure the global congestion have been proposed before [2], such as total over flow (TOF), average worse X% net-based congestion, total routed wirelength (TWL), the number of scenic nets and the number of nets over X% congested regions. Lately, the Average Congestion of g-cell Edges (ACE) is proposed in [10], which is based on the histogram of g-edge congestion.

Detailed Routing Congestion and Metrics. Global routing does not focus on the congestion internal to global routing tiles. Even though global routing constraints are reasonably well satisfied, local peaks of pin density often create detailed routing congestion. A design may appear to be easily globally routable, but may fail detailed routing leaving several *opens* (wires are not electrically connected) and *shorts* (metals from two different electrical nets are connected). Each short and open requires additional iterations of rip-up and re-routings leading to significant increase of detailed routing runtime. These local pin access congestions happen when relative high number of pins are populated within a small region. For example, when high number of signal pins are fully contained within a global routing tile, it is a good candidate area for potential local pin access problem.[1] Fig. 1(c) shows the example of pin density map of a design. One interesting observation is that the pin density map is quite discrete compared to the global routing congestion map shown in Fig. 1(b). In global routing congestion map, the level of congestion changes gradually while in pin density map, one hot congested tile sits right next to a cold one with no congestion at all.

Detailed routing congestion mitigation is a relatively new research area and still several different techniques are being explored. A few of recent research works demonstrated pin density-based metrics can be effective to address local pin access issue. In this work, then pin density metric proposed in [11] is used for detail routing congestion aware spreading. In addition, the k-factor idea proposed in [10] is used to model the intra-gcell congestion and pin density factor inside the global router, and global congestion aware spreading will mitigate the detail routing congestion to some ex-

---

[1]This situation is popular when complex logic gates such as AOI or OAI types are heavily populated within a tile.

tent as well. After detailed routing, the number of opens and shorts are the typical quality metrics to measure the success of detailed routing.

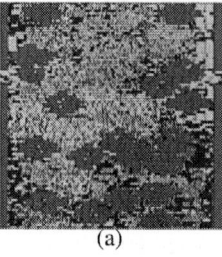
(a)

(c)

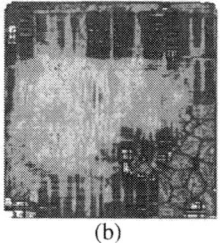
(b)

Figure 1: Examples of different congestion styles: (a) cell-brick-wall, (b) global routing congestion (from [13]), and (c) pin density local congestion.

3. TIMING AND WIRE MODELS

To get efficient turn-around-time for design closure system, physical synthesis may start with large design changes early in the flow, and require quick incremental timing analysis. Since the Steiner topology is fast to be built [12], Steiner wire model is commonly used during the initial timing closure.

However, even if a design has zero slack using Steiner wire model, and can be routed without any opens and shorts, the timing after the routing may still have big degradation.

At the global routing level, the timing degradation is from the scenic nets due to the global routing congestion. One may perform further optimization based on global wires with the incremental global routing.

Global wires modellack the accuracy of modeling the intra-tile congestion, track assignment and exact pin locations and shapes. Therefore, wires after detail routing may have more detours and vias, or different layer assignment than global wires. Coupling is another big factor that would degrade the timing after the track assignment and detail routing. Therefore, a design with good timing based on global wire model could get further degradation after detail routing. Some examples of such timing degradation are

1. Sub-marine effect. Wires after detail routing do not share the same layer assignment as wires from global routing. For example, as shown in Fig. 2(a), after global routing, 70% of the net is routed on higher metal layers from fifth to seventh layers (higher layer has better RC characteristics [14]), while 30% of net is routed on lower metal layers from first to fourth layers. After detail routing, however, 30% of the net is routed on higher metal layers, and 70% of net is routed on lower metal layers as shown in Fig. 2(c). While sub-marine effect may not adds more capacitance, it significantly increases interconnect resistance and causes more timing and electrical violations.

2. Via impact. Even if detail wires and global wires could have same percentage on higher metal layers, detail routing may

add more vias and break a long global wire on higher metal layer into many small segments of up-down jogging patterns shown in Fig. 2(d). More vias add interconnect resistance can cause more timing and electrical violations.

3. Scenics from detail routing: While global wires will natually have more detours than Steiner wires, detail wires could be even worse as shown in Fig. 2(b) and Fig. 2(e). Such wirelength increase adds loads for the driver and causes more timing and electrical violations, especially when the driver may be tuned to have smaller gate size based on Steiner wire model and be sensitive to the load variation.

4. routing order. The routing order also impacts the final timing result when we have multiple nets with limited routing tracks. In Fig. 3, if the non-timing critical net is routed first, the final worst slack is -8 ps. If the timing critical net is routed first, as shown in Fig. 4, the final worst slack is 2 ps. The net ordering clearly has the big impact on the timing.

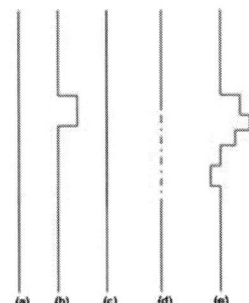

Figure 2: Same net with the different routing. Red line represents higher layers 5 to 7. Blue line represents lower layers 1 to 4.

While the timing after detail routing has the best accuracy, optimization with detail wires may not be the best approach since incremental detail routing is much more complicated and expensive than incremental global routing. In practical, optimization after detail routing is limited to small changes only. Some routing problems, such as long scenic nets, could be seen at global routing stages and one may want to fix it right away by buffering or other approaches.

4. LOGIC CLEANUP

As stated in [2], logic synthesis could easily create structures that are good for timing closure but difficult to route. For the tangled logic structures, such as datapaths, ROMs and crossbar switches, one need techniques to identify such structures and also optimize them in a more global way than traditional incremental logic synthesis used in the physical synthsis system. The structures need to be cleaned before they are fed into global placement. Some recent work try to address such issues, such as tangled logic idenficiation [15], datapath straightening [16], and fan-in tree wirelength minimization [17]. This is still an open research area, where one could also perform global routing directly inside the logic synthesis transforms, though the runtime and QOR trade-off is not clear.

5. GLOBAL PLACEMENT

Global placement is one of the most important steps in a physical synthesis flow. The last few years have witnessed a renaissance in the research on wirelength-driven global placement, largely due to the availability of challenging benchmarks derived from real industrial designs [18, 19]. Traditionally, optimizing wirelength has been

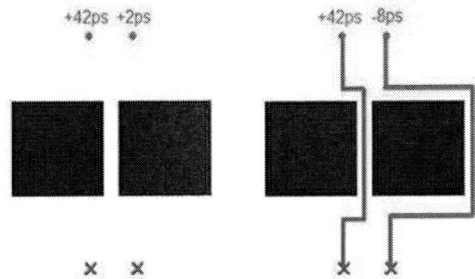

Figure 3: Routing the non-timing critical net first

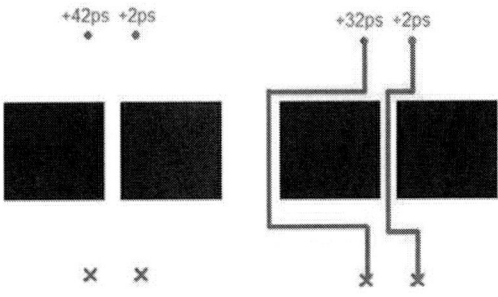

Figure 4: Routing the timing critical net first

the key objective during global placement, with routability being addressed via post-placement refinement techniques [13, 20, 21]. There have been attempts to alleviate congestion via routability-driven global placement [22, 23], but they rely on rudimentary and often inaccurate techniques like pin-density or probabilistic congestion estimation. With the latest fast global routers [4–9] and benchmarks containing information to perform both placement and global routing [24, 25], it is now possible to perform routability-driven global placement using accurate congestion estimation techniques. Some work in this area [26, 27] have shown promising results, and this approach is anticipated to become more prevalent.

6. CONGESTION AWARE SPREADING

Spreading techniques are common in the literature for analytical placement algorithms. One such technique is iterative local refinement (ILR) used by FastPlace [28]. ILR also creates a regular grid for a given placement and performs many rounds of movement for every cell in a design. During each round, each movable cell is examined once. A cell may move from its current grid tile to one of its eight neighboring grid tiles. The choice of destination for each cell is based on a cost function which is a linear combination of the change in wire length and the congestion model function between the source and destination tiles caused by the move. Fig. 5 shows the general congestion mitigation technique via spreading-based placement optimization. For a given design, congestion hot spots are detected with a given congestion model. All the standard movable cells in those regions are artificially inflated to bigger sizes. These inflations create a temporarily overlapped illegal placement. They also violate the target placement density constraint as well and force ILR algorithm to spread movable cells. Once the placement is legalized with these inflated cells, they are restored to the original sizes which introduces additional white spaces around those cells. These additional white spaces open up spaces for additional buffering or gate sizing for placement cell congestion mitigation, or add to more resources for global and detailed routing congestion mitigation. Some detail placement techniques, such as dynamic programming could be used as well. In this work, the spreading techniques similar to [11, 13] are used to mitigate the global and detail routing congestion.

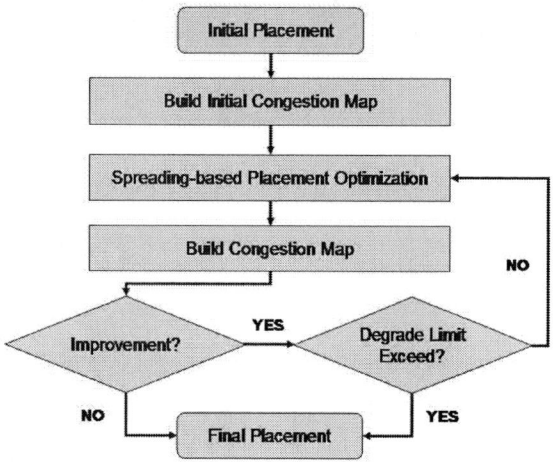

Figure 5: The flow of spreading-based congestion mitigation placement technique.

7. REPEATER INSERTION

Repeater insertion is a critical step in any physical synthesis flow to reduce interconnect delay. After repeater insertion, the long interconnect wire is breaking into small wire segments and the quadratic delay increase along the wire is reduced to linear increase. Repeaters also decouple the capacitive loads of non-critical sinks from the critical paths. For top level designs with tens or even hundreds of macros, memories and IPs, hundreds of thousands of repeaters are needed.

To compensate the technology impact that causes significant increase of wire resistance, more complex metal stacks are used with wiring pitches ranging from $1x$ to $20x$ across 12 metal layers for the 32 nm node. Significant speed and area advantage can be achieved if one combine the repeater insertion and layer assignment [14].

While there exists extensive literatures on repeater insertion and layer assignment lately, most of them focus on improving the timing performance [30–37], and not much work have been focused on the congestion. However, as shown in [2], congestion could happen at the higher metal layers when the nets are blindly promoted to achieve the best timing. If the router obey the layer directives given by the layer assignment, it may not able to finish the wire connection. On the other hand, if the router can not maintain the layer directive, significant timing degradation happens after the routing. Also, repeaters themself could cause the corona effect [2] if they are not placed in a congestion aware way.

To solve this problem, one could invoke global router first, and put down the repeaters along the global wires. However, the layer assignment and routing topology chosen by the router may not be timing aware, and the slack of every single net may be negative before repeater insertion.

It is still an open challenge to address repeater insertion and layer assignment at the same time while meeting both routing and timing constraints. It is a complicated resource sharing problem. Currently, we combine all different techniques and put an iterative flow to address this problem, but we clearly see big room for the improvement.

Another problem, which is often ignored by the literature, is the repeater insertion for the nets with high fanouts ranging from few hundreds sinks to hundreds of thousands of sinks. These nets could be scan, reset or control nets, and are generally not timing critical. However, due to the amount of wiring resources they consume, it is critical to buffer these nets in a wire-length efficient way while still be careful to not create a daisy chain topology.

8. CLOCK NETWORK SYNTHESIS

Traditionally, clock network synthesis (CNS) is performed after placement and timing optimization because latches and flip-flops have to be well-placed as an input to CNS. However, the process of CNS is usually very disruptive to the routing congestion because it can insert thousands of buffers and clock gates and routing for CNS requires a set of wires with shields and wide widths. All these would degrade the routability of a well-optimized layout produced by physical synthesis tool. On the other hand, performing CNS after placement and timing optimization restricts the tool such that optimization would hardly employ useful skew for better timing and power. As a result, it is crucial to consider CNS during placement and optimization stage of physical synthesis suite. In fact, once the latches and flip-flops are placed, an initial clock network should be constructed to estimate its effect to timing and congestion. Timing and congestion optimization can then take advantage of the information from the clock network, and incremental CNS can be done during the whole physical synthesis flow [29].

9. TIMING OPTIMIZATION TECHNIQUES

With Static Timing Analysis (STA) tool, one can query the timing at any point of the physical synthesis flow and get a list of timing critical nets or paths. Timing optimization techniques, such as buffering, gate sizing, cell movement, vt swapping, cloning, logic decomposition, inverter merging, connection reordering, local logic remapping, can then be applied to fix the timing problems. Most of the timing optimization can be done incrementally without disturbing the majority of placed cells and have little impact on routability. This in general requires one to give up the timing-driven placement at the global level, which models the net weight with timing information and throw all cells in the air. More incremental approaches are needed, such as the work shown in [38].

One can perform these timing optimization techniques with different wire models, such as global and detail wires. To perform more surgical changes, incremental global and detail routing are necessary to make the right design changes.

10. CONSTRAINTS DRIVEN ROUTING

The main goal of the router is to finish all metal wire connections without any opens and shorts. It is impossible for the router to generate each single topology with the best timing aware Steiner tree, and sometimes just adding too many constraints, such as layer directives or strict ordering, may choke the router. On the other hand, if a design is easier to route, but the router is not smart enough to make the right choices, the timing after global routing can still be very poor comparing to the Steiner wire model, though they are possible to be fixed. One example is shown in Table 1, where WSLK refers to the worst slack of the design, FOM is the figure of the merit, Avg 20% is the Avg 20% net congestion, NO90 refers to the number of nets over 90% congested regions, and NO100 refers to the number of nets over 100% congested regions. Even though the global congestion metric looks reasonable, the timing variation after global routing is high.

One can perform the optimization after detail routing to fix post-routing timing "jump" [39]. However, it could be prohibitive due to complicated incremental detail routing, or adding too much power and area to the design without controling router itself, and many times abort without space to insert additional buffers or size the gates.

Table 1: The timing information before/after global routing.

Metrics	WSLK	FOM	Avg 20%	NO90	NO100
Steiner Timing	0.06	0.0	82.29	7742	287
Timing After Global Routig	-0.84	-301	-	-	-

We propose a new routing flow to solve this problem. It consists the following steps: 1) constraints generation and constraints driven global routing; 2) optimization based on global wires; 3) constraints generation and constraints driven detail routing; 4) optimization based on detail wires. This flow basically separate the global and detail routing, add the reasonable "soft" constraints to the router, and perform optimization after global routing, as well as the detail routing stage.

The key part of this new flow is the step to generate the constraints. For the industrial router used in our system, the following two type of constraints are used and generated. Similar concepts are also seen for other academic and industrial routers, or could be added without much difficulty.

1. Scenic constraint. This constraint associates a scenic ratio or wire length bound for every net, such that no route is permitted to go more than the given wirelength bound. It could be modeled as either hard or soft constraints, where soft constraints imply the router may violate the constraints with a higher cost to finish the wire connection. It is easier to support this constraint inside global router rather than the detail router since detail routing has more constraints to meet.

2. Net priority or net ordering. This constraint is commonly seen in the literature where all nets are routed in a given order one by one. The granuality of the ordering can change with the tradeoff of the runtime and QOR. A set of five to ten priorities could be enough to separate the timing critical or timing sensitive nets from those with more margin or tolerance to interconnect RC changes. Both global and detail router can easily support this constraint.

In our flow, these constraints are generated based on the timing and wirelength from Steiner wire model, or more accurate wire models if one perform iteration of the proposed flow. For example, tighter constraints are set for top X% percentage of timing critical nets with length greater than a given threshold, or timing sensitive nets that are under given slack threshold. The constraints are relaxed if the net has more timing margin or tolerance of interconnect variation. For the nets with very short wirelength, one can assign a bigger ratio. For the nets with layer directive or wide wire width, one may assign tigher constraints since they are more susceptible to sub-marine and via effects.

Timing optimization after global routing is necessary. By buffering along a big detour nets, or sizing the gates that are sensitive to the load change could easily fix some messy timing problems and help generate reasonable constraints for the detail routing.

11. EXPERIMENTAL RESULTS

A series of experiments are done to show the impact of several techniques proposed in the paper.

Congestion aware spreading. The first experiment is done to show the benefit of global and detailed routing congestion aware spreading techniques on a commercial ASIC design which has approximately 110k objects with high connectivity cells such as AOI or OAI gates. Two placement instances are compared with and without the spreading techniques. Table 2 reports the comparison of congestion statistics. Column 2 and 3 report Avg 20% Nets and

Table 2: Global and detailed routing congestion mitigation results on commercial ASIC design. (For all metrics, smaller is better)

Technique used?	Global Routing Stats		Deailed Routing Stats			
	Avg20%	90% nets	Shorts	Opens	Route Time	Routed WL
No	91.93	10296	367	3	10.9 h	4.11e6
Yes	86.08	5119	1	40	7.4 h	4.00e6

the number of nets over 90 congested regions. Column 4 to 7 report the detailed routing congestion statistics such as the number of shorts, opens, the runtime in hours and the routed wire length. The spreading-based placement optimization is extremely effective in reducing both global and detailed routing congestion metrics. Moreover, the detailed routed wire length improved after the congestion mitigation techniques.

New flow prior to the routing. The second experiment is done to show the benefit of new flow prior the routing stage versus the traditional timing-driven physical synthesis flow on a commercial ASIC design with approximately 600k objects. The design instances after both flows are fed into the same global router and measure the global routing congestion statistics. Fig. 6 shows the comparison of the congestion maps. The Avg 20% is reduced from 93.3% to 84.3%. The number of nets over 100% is reduced from 29157 to 654, and the number of nets over 90% is reduced from 96172 to 17182. The detail routing runtime has also been improved over 2X. Cearly, the new flow has significant benefit than the old flow.

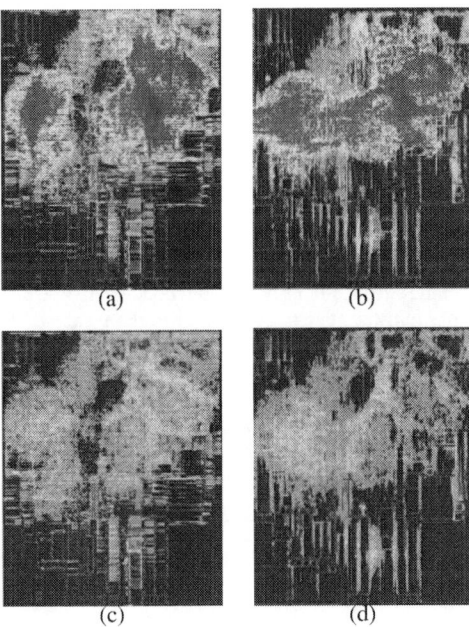

Figure 6: The comparison of congestion maps with old and new flow. (a) Horizontal congestion map with old flow; (b) vertical congestion map with old flow; (c) horizontal congestion map with new flow; (d) vertical congestion map with new flow.

New routing flow. We apply the new routing flow on several 65, 45 and 32 nm ASIC designs, with 598K to 1.3M gates. For all designs, we use the new flow prior to the routing stage to get the placed and optimized instances which has good timing with Steiner wire model and also good congestion statistics. Then the instance is fed to the new routing flow, and a flow where no constraints are generated and no optimization is done after the global routing. The results are shown in Table 3. In all testcases, the new routing flow

Table 3: The comparison of RBO vs. traditional Wire Flow

	# gates	Router	WLSK (ns)	FOM (ns)	Avg 20%	WCI90	WCI100	#opens	#shorts	#Slew Violations	#Cap Violations
Circuit1	966900	default routing flow	-0.979	-1418	84.23	30028	760	2	0	4330	6737
		new routing flow	-0.309	-94	84.57	35758	529	5	0	475	570
Circuit2	598004	default routing flow	-0.889	-416	82.29	7742	287	0	164	5540	2294
		new routing flow	-0.096	-16	82.41	9675	299	2	89	647	161
Circuit3	1382607	default routing flow	-1.196	-841	82.78	27818	489	5	0	24999	5498
		new routing flow	-0.258	-240	82.93	31032	530	13	0	6653	361

achieves much better timing and electrical violation (slew and cap violations) results than the the other one, while maintaining similar routing metrics (Avg 20%, NO90, NO100,#opens, #shorts).

12. OPEN CHALLENGES

As discussed previously, modeling the detail wiring impact inside global routing, global buffering and layer assignment, incremental timing-driven placement without hurting congestion, clock network synthesis without hurting congestion, all remain unsolved problems. Also, it is not clear if one needs congestion driven logic synthesis techniques, or to merge the congestion aware placement into global placement directly. While the techniques discussed in this paper could help produce a design instance with much better timing at the end, the extra time spent in congestion analysis and additional techniques need to be balanced for the overall flow. How to come up with the "best" physical synthesis flow is still an ad hoc approach in the industry, while academic researchers lack the infrastructure and all the components to systematically study the flow. Even if they do, flow optimization is a much harder problem than any single objective oriented optimization.

13. CONCLUSION

Physical synthesis needs to produce designs that are easy to route and meet timing constraints. Traditional placement, logic synthesis, repeater insertion, clock network synthesis may use congestion estimation techniques, such as fast global routing, to mitigate the congestion as a post-fix, or correct-by-construction. Different congestion models and metrics are ncessary for different optimizations, and detail routing effects need to be modeled inside the global router. Proper constraints can be applied to the router to achieve good timing QOR without failing the wire connections. A congestion driven physical synthesis flow is described in this paper and proven to be effective. There are still many open problems in the field, leaving exciting research oppotunities along the way.

14. REFERENCES

[1] Z. Li and C. J. Alpert, "What is Physical Synthesis," *ACM/SIGDA E-Newsletter*, vol. 41, no. 1, Jan 2011.

[2] C. J. Alpert et al., "What makes a design difficult to route," in *Proc. ISPDPD*, 2010, pp. 7 – 12.

[3] C. J. Alpert et al., "Techniques for fast physical synthesis," *Proceedings of the IEEE*, vol. 95, no. 3, pp. 573–599, 2007.

[4] Y. Xu et al., "FastRoute 4.0: Global router with efficient via minimization," in *Proc. ASPDAC*, 2009, pp. 576–581.

[5] Y.-J. Chang et al., "NTHU-Route 2.0: A fast and stable global router," in *Proc. ICCAD*, 2008, pp. 338–343.

[6] C. Minsik et al., "BoxRouter 2.0: Architecture and implementation of a hybrid and robust global router," in *Proc. ICCAD*, 2007, pp. 503–508.

[7] H.-Y. Chen et al., "High-performance global routing with fast overflow reduction," in *Proc. ASPDAC*, 2009, pp. 582–587.

[8] H. Shojaei et al., "Congestion analysis for global routing via integer programming," in *Proc. ICCAD*, 2011, pp. 256–262.

[9] J. Hu, J. A. Roy, and I. L. Markov, "Completing high-quality routes," in *Proc. ISPD*, 2010, pp. 35–41.

[10] Y. Wei et al., "GLARE: Global and Local wiring Aware Routability Evaluation," in *Proc. DAC*, 2012, to appear.

[11] T. Taghavi et al.,"New Placement Prediction and Mitigation Techniques For Local Routing Congestion," in *Proc. ICCAD*, 2010, pp. 621-624.

[12] C. C. N. Chu, Y. C. Wong, "FLUTE: Fast lookup table based rectilinear Steiner minimal tree Algorithm for VLSI Design," *IEEE Trans. on CAD*, vol. 27, no. 1, pp. 70–83, 2008.

[13] J. Roy et al., "CRISP: Congestion reduction by iterated spreading during placement," in *Proc. ICCAD*, 2009, pp. 357–362.

[14] Z. Li et al., "Fast interconnect synthesis with layer assignment," in *Proc. ISPD*, 2008, pp. 71 – 77.

[15] S. I. Ward et al., "Keep it straight: teaching placement how to better handle designs with datapaths," in *Proc. ISPD*, 2012, pp. 79–86.

[16] T. Jindal et al., "Detecting tangled logic structures in VLSI netlists," in *Proc. DAC*, 2010, pp. 603–608.

[17] H. Xiang et al., "Logical and physical restructuring of fan-in trees," in *Proc. ISPD*, 2010, pp. 67–74.

[18] G.-J. Nam et al., "The ISPD2005 placement contest and benchmark suite," in *Proc. ISPD*, 2005, pp. 216–220.

[19] G.-J. Nam,C. J. Alpert, and P. Villarrubia, "ISPD 2006 placement contest: Benchmark suite and results," in *Proc. ISPD*, 2006, pp. 167–167.

[20] Y. Zhang and C. Chu, "CROP: Fast and effective congestion refinement of placement," in *Proc. ICCAD*, 2009, pp. 344–350.

[21] C. Li et al., "Routability-driven placement and white space allocation," *IEEE Trans. on CAD*, vol. 26, no. 5, pp. 858–871, 2007.

[22] U. Brenner and A. Rohe, "An effective congestion-driven placement framework," *IEEE Trans. on CAD*, vol. 22, no. 4, pp. 387–394, 2003.

[23] W. Hou et al., "A new congestion-driven placement algorithm based on cell inflation," in *Proc. ASPDAC*, 2001, pp. 723–728.

[24] N. Viswanathan et al., "The ISPD-2011 Routability-driven placement contest and benchmark suite, " in *Proc. ISPD*, 2011, pp. 141–146.

[25] N. Viswanathan et al., "The DAC 2012 Routability-driven placement contest and benchmark suite," in *Proc. DAC*, 2012, to appear.

[26] M. Pan and C. Chu, "IPR: An integrated placement and routing algorithm," in *Proc. DAC*, 2007, pp. 59–62.

[27] M.-C. Kim et al., "A SimPLR method for routability-driven placement," in *Proc. ICCAD*, 2011, pp. 67–74.

[28] N. Viswanathan, M. Pan and C. C. N. Chu, "FastPlace 3.0: A fast multilevel quadratic placement algorithm with placement congestion control," in *Proc. ASP-DAC*, 2007, pp. 135-140.

[29] D. A. Papa et al., "Physical Synthesis with Clock-Network Optimization for Large Systems on Chips," *IEEE Micro*, vol. 31, no. 4, pp. 51–62, 2011.

[30] W. Shi and Z. Li, "A fast algorithm for optimal buffer insertion," *IEEE Trans. on CAD*, vol. 24, no. 6, pp. 879–891, June 2005.

[31] Z. Li, Y. Zhou and W. Shi, "$O(mn)$ Time Algorithm for Optimal Buffer Insertion of Nets With m Sinks", *IEEE Trans. on CAD*, vol. 31, no. 3, pp. 437-441, March 2012.

[32] S. Hu, Z. Li and C. J. Alpert, "A fully polynomial time approximation scheme for timing driven minimum cost buffer insertion," in *Proc. DAC*, 2009, pp. 424–429.

[33] S. Hu, Z. Li and C. J. Alpert, "A faster approximation scheme for timing driven minimum cost layer assignment," in *Proc. ISPD*, 2009, pp. 167–174.

[34] Z. Li and W. Shi, "An $O(bn^2)$ time algorithm for optimal buffer insertion with b buffer types", *IEEE Trans. on CAD*, vol. 25, no. 3, pp. 484-489, March 2006.

[35] M. Waghmode, Z. Li and W. Shi, "Buffer insertion in large circuits with constructive solution search techniques," in *Proc. DAC*, 2006, pp. 296–301.

[36] C. C. N. Sze et al., "Path based buffer insertion," in *Proc. DAC*, 2005, pp. 509–514.

[37] Y. Zhou et al., "Shedding Physical Synthesis Area Bloat," *VLSI Design*, Article ID 503025, 2011.

[38] N. Viswanathan et al., "ITOP: integrating timing optimization within placement," in *Proc. ISPD*, 2010, pp. 83–90.

[39] B. Dougherty and M. Kazda, "Rapids: Post-Routing Timing Closure," in *DAC 2010 User track Poster 1U.5*.

Rule Agnostic Routing by Using Design Fabrics

Gyuszi Suto
Core CAD Technologies

Intel Corporation
2501 NW 229th Avenue
Hillsboro, Oregon
+01.971.214.2392

gyuszi.suto@intel.com

ABSTRACT

Moore's law requires the shrinking of physical dimensions of the transistors to roughly half their area every two years. This poses a tremendous challenge on how to print and manufacture these ever-shrinking physical components that make up the transistors and the interconnect - generation after process generation. One aspect of this challenge is that the process rules are exploding in complexity – directly translating into physical design EDA (Electronic Design Automation) tool complexity. Traditional design rules governed the spacing, overlap or alignment of any two layout objects from this set: diffusion, poly, via cut, wire, etc. In this work we propose a solution that relies on grids (aka. Fabrics), models the design rules on those grids and presents them to the EDA tools in such a way that it minimizes the complexity cost on the tools' side. In an ideal situation, the proposed solution can completely decouple the tools from the process rules, i.e. even if the tools don't change at all, they'll still be able to support new process nodes.

Categories and Subject Descriptors

J.6 **[Computer-Aided Engineering]**: Computer-aided design (CAD); B.7.2 **[Design Aids]**: Layout, Placement and Routing.

General Terms

Algorithms, Design.

Keywords

Grids, Fabrics, Design rules, Abstraction, API

1. INTRODUCTION

The Current EDA tools and models traditionally store, model and do layout synthesis based on physical design rules between two objects positioned in a free, geometrical space with respect to each other. These models usually build on edge to edge, corner to corner, intersection and containment pairwise relationships,

Permission to make digital or hard copies of all or part of this work for personal or classroom use is granted without fee provided that copies are not made or distributed for profit or commercial advantage and that copies bear this notice and the full citation on the first page. To copy otherwise, to republish, to post on servers or to redistribute to lists, requires prior specific permission and/or a fee.

between the layout objects in the physical domain. These systems and models are stretched far beyond their planned scope and capability, resulting in either a) dumbing down of the process capabilities so that the tools can somehow comprehend and deal with the rules, or b) requiring significant effort to customize the EDA tools and models with each generation – with negative impact on turn-around-time, productivity and design space exploration.

2. GRID-BASED MODEL

The new generation of more complex design rules include relationships built from one or more of the following (or similar) constraints C_i between layout objects: 1) objects must be at closer, equal or greater distance from each other; 2) objects must be aligned in certain ways; 3) configurations (groups) of objects in a certain relative placement to each other are designated to be design rule correct or incorrect 4) groups of objects in a certain configuration constrain the relative position of other groups of objects in certain (possibly different) configurations; 5) the presence of some objects require the presence of (or absence) of some other (specific) objects, 6) others. In this work, we propose a simple language that will be able to express all the constraints C_i from above, and also future (speculative) constraints with little or no change. This language is built upon the premise that there is always a design grid G present. The grid G has payloads (layout objects associated to the various grid points) assigned and the process rules are expressed as logic relationship of required presence or required absence of these payloads. The examples presented in this paper use made-up layout object sizes, shapes and rules. Their purpose is to convey the idea behind the work.

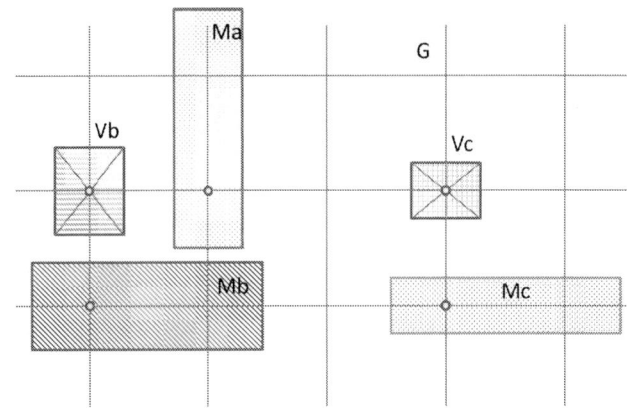

Figure 1

In Figure 1 we define a uniform grid (G). On each of the grid points (grid-line intersections) of G, we can potentially have any of the five layout objects shown (wires Ma, Mb, Mc and via cuts Vb, Vc). Those objects are drawn relative to the gridpoint highlighted with a small circle. The grid defined here sets the potential (discretized) locations and the associated layout object types that can be laid down at those locations.

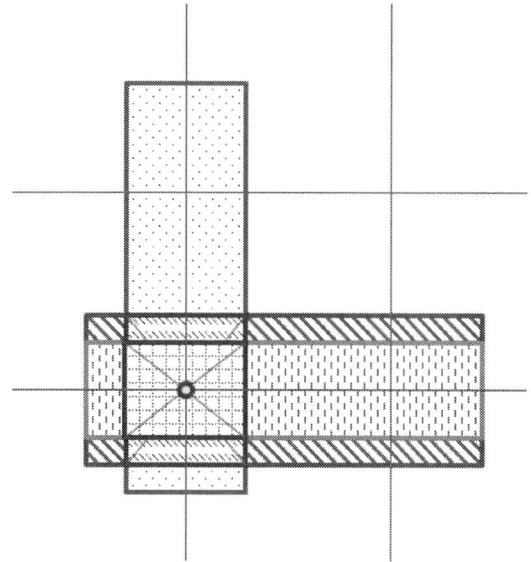

Figure 2

In Figure 2 , we show all five layout objects superimposed on one gridpoint. We now define the manufacturing rules (aka. Design rules) as logic relationships between these objects at given gridpoint locations. Each of the objects can be required present (P) or required absent (A) at some location. A manufacturing rule is defined as a logic relationship between two or more of these entities.

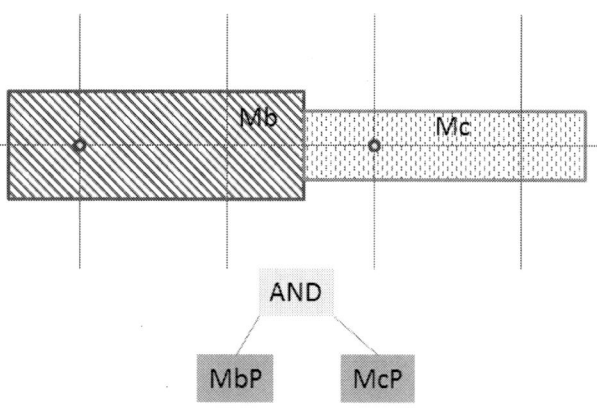

Figure 3

In Figure 3 we define a manufacturing rule that prohibits tapering from wide wire Mb to narrow wire Mc. The logic AND ties the two entities together and essentially states the following: "If you have an Mb wire present AND have an Mc wire present two grid points to its right, then this is not-manufacturable". The leaves in the logic tree represent specific layout object instances at relative grid point coordinates. In the example above both leaves are qualified as must-be-present for this rule to kick in, i.e. if there is layout that matches this pattern then it's a design rule violation.

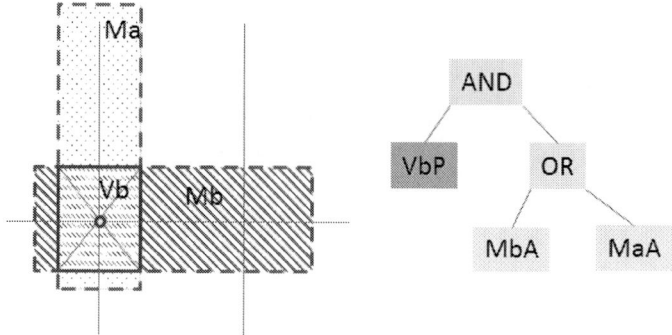

Figure 4

In Figure 4. we state the following manufacturing rule: "If a via Vb is present AND either is not covered by horizontal metal Mb OR not covered by vertical metal Ma, then it's not manufacturable". The P and A suffixes denote must-be-present (green leaf) or must-be-absent (pink leaf in the logic tree – the layout object drawn with dashed outline) properties on the layout objects. The AND and OR are mathematical logic relationships.

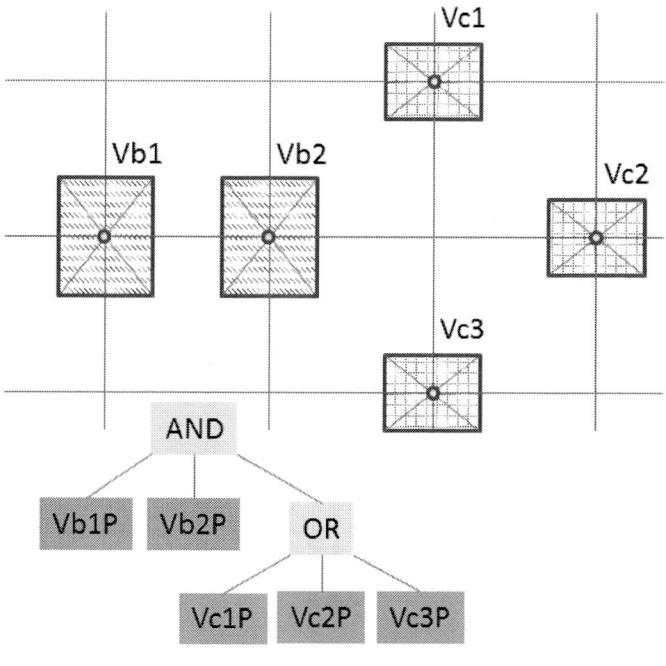

Figure 5

In Figure 5 the rule states that if you have two, horizontally-adjacent vias Vb1 and Vb2 both present, and any of the three vias Vc1 or Vc2 or Vc3 present, it cannot be manufactured. Other neighboring objects are irrelevant for this rule. If they would be relevant, they would've been included in the logic expression tree.

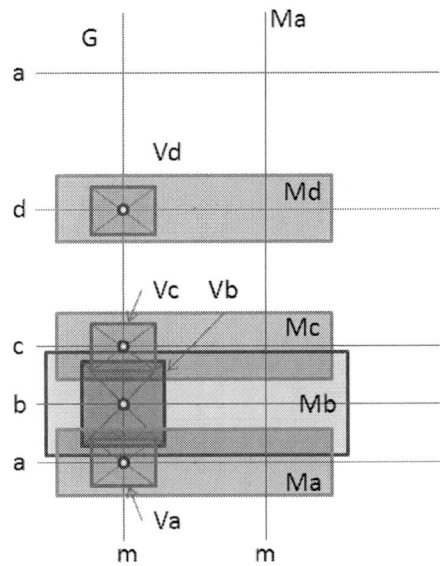

Figure 6

In Figure 6 we define a different, non-uniform grid G which has four repeating horizontal lines a, b, c, d, and one vertical line m. The payloads associated to the grid points are the four metal wires Ma, Mb, Mc, Md, and respectively the four vias: Va, Vb, Vc and Vd. The rule in Figure 7 expresses the fact that we don't want to have Mb or Vb short with any of the adjacent neighbors Ma, Va, Mc or Vc.

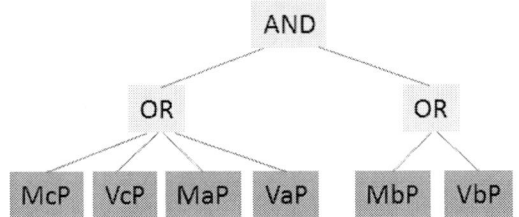

Figure 7

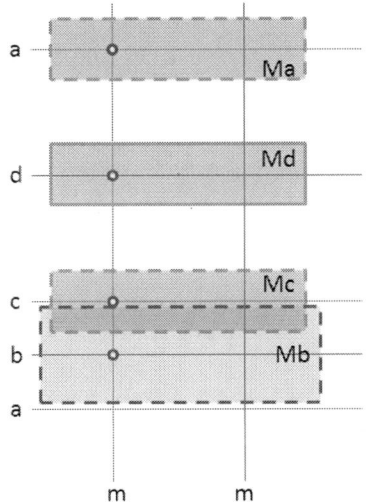

Figure 8

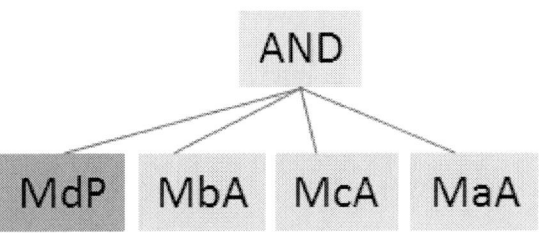

Figure 9

The rule in Figure 9 expresses the rule that a wire Md needs to have at least one of the three adjacent neighbors (Ma, Mc or Mb) wires present, or else it's not manufacturable. The relative positions of these wires are shown in Figure 8 – with the must-be-absent layout objects shown with dashed outline.

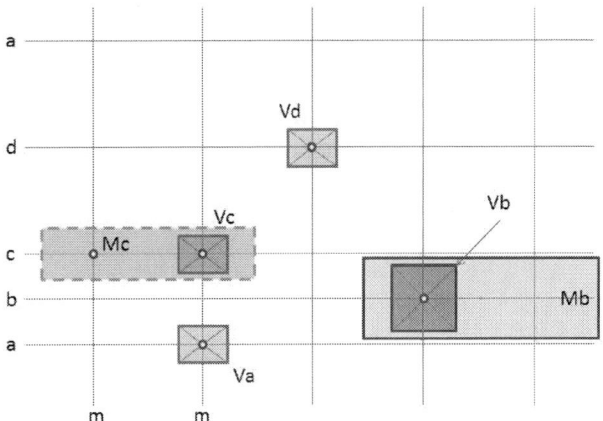

Figure 10

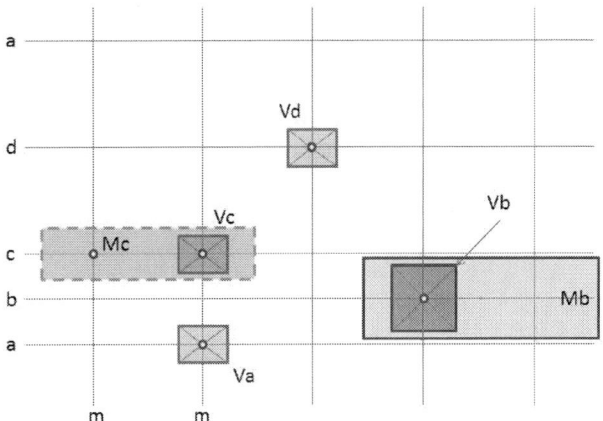

Figure 11

In Figure 10 and Figure 11 we show a rule stating the following: if you have a wire Mb and via Vb, and via Vc is present or the wire Mc is absent (drawn with dashed outline), and in addition any one of the vias Va or Vd are present, then we have a layout that cannot be manufactured.

2.1 Relative geometric positions of leaves

In each one of the examples from above, the leaves of the trees have additional information about the relative positions of the layout objects. In Figure 12 below we are showing the previous example from Figure 10, with the added gridline numbers on both dimensions.

473

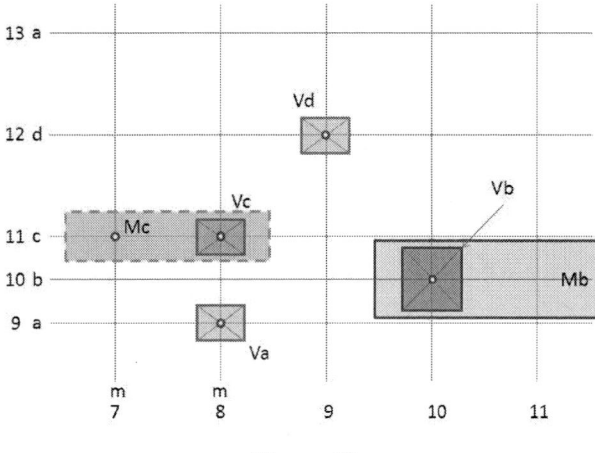

Figure 12

The Via Va is at position (x:8, y:9). The type of that gridpoint is (x:m, y:a). There is another, equivalent point at (x:8, y:13), which is not part of this rule. The leaves of the tree contain the grid point coordinates to make the spatial relationship precisely specified.

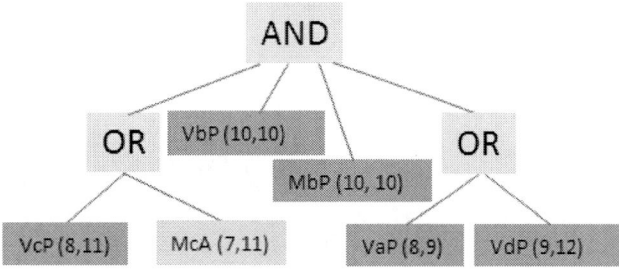

Figure 13

Since the periodicity of this grid is four lines on the y dimension, one could define a rule equivalent to the one above by shifting up each coordinate by four units on the y dimension.

2.2 Logically minimized versus fully enumerated patterns

The patterns presented in the previous examples are captured as a logic relationship between two or more layout objects on the grid. The logic tree can be arbitrarily deep, with nodes representing AND, OR, NOT, XOR relationships. Each leaf is must-be-present (P), or must-be-absent (A) constraint on one particular layout object (payload) at some relative grid point location. Other neighboring layout objects that have no effect on the rule are simply ignored and not part of the logic tree. These patterns could be fully enumerated, in which case each tree would have only one root node (AND) and two or more leaves. If we would want to convey the information in Fig.11 in a fully enumerated fashion, then we would need to combine the left OR node at true state with any of these three combinations: VcP=0, McA=1; VcP=1, McA=0; VcP=1, McA=1 with the three combinations of the right OR node at true: VaP=0, VdP=1; VaP=1, VdP=0; VaP=1, VdP=1 totaling nine individual patterns. These two modes are logically equivalent, but we chose to present them in the compressed tree form for obvious space saving and expressivity. If the rules are automatically generated based on process/lithography simulations/studies, they may be generated in fully enumerated fashion and then logically minimized. This paper does not attempt to solve the logic minimization - or any potential scaling - issue.

2.3 Good versus Bad patterns

The patterns we presented in this paper were assumed to be "bad", meaning that they cause design rule violations, and if they would occur, they would make the layout non-manufacturable. We can attach other properties to these patterns, like low yield, desired, less desired pattern, etc. We can even have "good" patterns that provide waivers to bad patterns. Say we have a bad pattern Pb that states that two vias cannot be placed in straight line on adjacent grid points (11 – two adjacent vias in a line represented as 1s). However 6 vias in a straight line are manufacturable, and that could be expressed with a pattern Pg (111111). The system can be set up such that layout is incorrect if it is covered by a bad rule, unless it is covered by a good rule. Otherwise, the constraint has to be enumerated as several bad patterns: 0110, 01110, 011110, 0111110 expressing the fact that two adjacent vias with no left or right neighboring vias is bad, and same for groups of three, four and five vias. One can see how the number of enumerated rules can blow up, so we propose curbing that with the concept that good rules waive contained bad rule matches.

3. RULE AGNOSTIC ROUTING

In the model we're proposing in this paper, the grid is pre-characterized with all the payloads (layout objects) that are associated to its grid points, and all patterns are expressed as shown above as logic relationships between the payloads on the grid. This enables us to efficiently cross-reference and cache all relationships. Each grid point is associated with a set of discrete payloads. Each payload at each grid point is associated with a finite set of patterns. Each pattern is associated with a finite number of payloads on specific grid points at well-defined relative locations with respect to each other.

3.1 Proposed API

We're then able to write an API (application programming interface) similar to the following list of functions. n –is the number of grid points in a region, k – is the average number of patterns (rules) relevant to a grid point. Next to each function we noted the estimated order of algorithmic complexity:

- Is this layout object correct O(k)

- What rules are governing this layout object O(k)

- What rules are violated by this layout object O(k)

- What rules refer to this grid point O(k)

- Is the layout in this region correct O(n*k)

- Given this layout object, what are all the possible neighboring objects that may be interacting with it O(k)

- Where can I place a layout object in a given region without causing violation O(n*k)

- Where can I remove a layout object from - in a given region - without causing violation O(n*k)

- How can I correct this incorrect layout by addition O(k)

- How can I correct this incorrect layout by removal O(k)

474

- Given a region, enumerate the legal layout options based on an objective like density, sparseness, yield, presence or absence of certain objects, etc. $O(k^n)$

Given this API, it is possible to write physical design EDA tools like: router, layout verification, fixer, placer, optimizer, etc. These tools would operate directly on the grid, the patterns or the API. As long as the rules of a process generations can be expressed as patterns on grids, the API stands as is and the tools can support the new process generation with any new type of rules without the need of code re-write. It is also possible to feed the patterns directly into SAT solvers to find optimal layout configurations – this latter solution obviously would not scale well, but for small sizes it would be feasible (see [1]).

4. RESULTS

We implemented the data model for grids, payloads and patterns as well as a subset of the proposed API from section 3.1. We implemented proof-of-concept tools for: online design rule checking, online layout editing with instant correctness (layout verification) feedback and guidance (place object here – and the layout will be correct, place object there – and layout will be less correct). We also implemented a simple router that exercises the (partial) API. The router is rule agnostic and very simplistic – we cannot present any data about it at this point.

5. CONCLUSIONS

In this paper we presented a method to model the physical design rules as logic relationships between layout objects that are associated as payloads on a grid (aka. Fabric). We also presented an API that enables the implementation of basic physical design tools and one that in-effect decouples the tool implementation from the rules themselves. The advantage of this method is as follows:

- Provides a formal (mathematical) means to express most manufacturing rules of current and (projected) future process generations

- Makes the definition, communication and validation of the design rules much easier.

- Enables automatic definition of the majority of manufacturing rules by exhaustive characterization of layout clusters using lithography simulation.

- Enables physical design (PD) EDA tools like layout verification, placers, routers to support most manufacturing rules without the need to modify the software code base – hence dramatically reducing the need for code re-write for new process nodes.

- Enables better design space exploration by providing the rules directly to SAT solvers and providing novel, discrete algorithm API functions to EDA tools

6. REFERENCES

[1] Nikolai Ryzhenko and Steven Burns. 2012. Standard Cell Routing via Boolean Satisfiability. *DAC 2012*

Making Non-Volatile Nanomagnet Logic Non-Volatile

Aaron Dingler*[1], Steve Kurtz*[1], Michael Niemier[1], Xiaobo Sharon Hu[1], Gyorgy Csaba[2], Joseph Nahas[1], Wolfgang Porod[2], Gary Bernstein[2], Peng Li[2], Vjiay Karthik Sankar[2]

University of Notre Dame Departments of Computer Science and Engineering[1], and Electrical Engineering[2]

{adingler, skurtz, mniemier, shu, gcsaba, jnahas, porod, bernstein.1, pli, vsankar}@nd.edu

ABSTRACT

Field-coupled nanomagnets can offer significant energy savings at iso-performance versus CMOS equivalents. Magnetic logic could be integrated with CMOS, operate in environments that CMOS cannot, and retain state without power. Clocking requirements lead to inherently pipelined circuits, and high throughput further improves application-level performance. However, bit conflicts – *that will occur in defect free, pipelined ensembles* – can make non-volatile logic volatile. Assuming a field-based clock, we present hardware designs to improve steady state non-volatility, and explain how design enhancements could increase clock energy. We then suggest materials-related design levers that could simultaneously deliver non-volatility and low clock energy.

Categories and Subject Descriptors

B.6.1 [**Design Styles**]: *Cellular arrays and automata*, B.8.2 [**Performance and Reliability**]: *Performance Analysis and Design Aids*

General Terms

Performance, Design, Reliability

Keywords

Nanomagnet logic, nanotechnology, NML, MQCA

1. INTRODUCTION

Many device technologies are being developed in an effort to replace or augment CMOS, as both the size and performance scaling trends associated with Moore's Law are becoming more difficult to achieve. Per the ITRS, nanomagnet logic (NML) uses device-to-device coupling between lithographically defined magnets, with nanometer feature sizes, to perform Boolean logic operations. Potential advantages of NML include extremely low power information processing, the ability to function in harsh environments, and non-volatility (even at the gate level).

NML is commonly compared to subthreshold CMOS where the focus is on optimizing power over speed ([1]). This is more poignant because of the upper limit (~1Ghz) on NML speeds as dictated by magnetic reversal times. **Active energy** associated with an NML-based computation could be 96 times less than a CMOS equivalent – even after accounting for clock overhead [2, 3]. With the exception of any drive circuitry, standby power should be 0.

NML could process information robustly in **harsh environments**. Notably, magnetic devices are intrinsically radiation hard. Experimental results by our group (not yet published) have considered NML ensembles that were subjected to radiation doses

*Both A Dingler and S Kurtz contributed equally to this work, are listed alphabetically, and should be considered co-first authors.

of up to 10 Mrad. No state change was observed post-dose, and follow-on switching experiments are planned. Additionally, devices could retain their magnetic properties below the Curie temperature of a material (e.g., 1100° C for Co), and operate equally well at cryogenic temperatures.

In principle, NML ensembles should be **non-volatile** (NV). At the application-level, this is beneficial for multiple reasons. **(i)** With supply voltage scaling, leakage power rivals dynamic power in state-of-the-art CMOS, and magnetic logic devices could mitigate standby power. As examples, [4] proposed using STT-RAM based lookup tables for both memory and combinational logic in multi-core processors, which could reduce chip-level power by 70%. [5, 6] proposed using magnetic tunnel junctions (MTJs) as tunable resistors to implement gate-level storage in dynamic current mode logic (DyCML). [6] considered a DyCML adder where one input took the form of an MTJ. For applications such as a sum of absolute differences (e.g., common in MPEG encoding for motion detection), pattern matching [7], etc., a given input can change infrequently and be stored at the gate. The benefit of NML over other approaches of this sort is that both logic and storage can be accomplished in NML. Thus, storage is intrinsic to the logic and no read/write penalties are incurred. Furthermore, this can simplify the manufacturing process. **(ii)** Systems with unreliable power supplies could also benefit. While computational state could be backed up to NV storage in the event of a power failure – and restored when power is restored – gate level state should be preserved in NML ensembles, as the logic devices themselves are magnetic. Thus, hardware for wireless applications (e.g., RFID), hardware that relies on energy scavenging, etc., could benefit. **(iii)** While closely related to (ii), any logic should be instant off/instant on – and computation could resume whenever power is restored.

For an NML circuit to be truly NV, *every* magnet should maintain its state when computation ceases. Therefore, the energy difference between magnetization states should be on the order of N kT, where N (typically 40-60) is determined by the number of devices and how long one wishes information to be preserved. While steady state stability is a function of a given device's size, shape, and material, it can also be affected by the state of neighboring devices – and may be threatened by inherent pipelining even if circuits are defect free. Here, we consider tradeoffs between design enhancements to preserve information post computation, and clock energy.

After discussing relevant background regarding the experimental state-of-the-art related to both NML ensembles and clocking (**Sec. 2**), we will discuss our technical contributions in more detail.

In **Sec. 3**, we show that even circuits that are free of hard errors (e.g., from fabrication variation [8]) and soft errors (e.g., from thermal noise [9]) are at risk of losing information in the steady state. Bit conflicts (e.g., an anti-ferromagnetic (AF) line with state ↑↓↓↑↓ instead of ↑↓↑↓↑↓) that might occur due to soft errors can initiate "random walks" that can destroy the state of an NML ensemble post re-evaluation [10]. Regardless of clock mechanism, bit conflicts are unavoidable in error free, inherently pipelined NML. Also, structures proposed in previous work (e.g., [2]) that could operate with acceptably low energy clocks may not retain state as computation ceases. We discuss stability qualitatively and quantitatively, and identify two approaches to improve stability.

Using a field-based clock as context, in **Sec. 4**, we discuss design changes (i.e., to magnet aspect ratio, thickness, etc.) to enhance steady state stability in ensembles of polycrystalline devices. Case studies (i) quantify clock energy overhead, (ii) demonstrate correct operation, and (iii) illustrate that stability can be improved despite thermal noise. However, increased stability will (not surprisingly) come at the cost of higher clock energy. To reduce clock energy to acceptable levels, and enable pipelined ensembles with improved steady state stability, in **Sec. 5**, we discuss the benefits of materials-centric design levers – enhanced permeability dielectrics (EPDs) [11] (to more efficiently generate clock fields) and biaxial anisotropy [9] (originally suggested for preserving hard axis stability during ensemble re-evaluation). New experiments with EPDs and consideration of clock field requirements for biaxial devices are presented. Notably, biaxial devices could lead to NV NML with just a 2-5X increase in clock energy (with no other design enhancements). **Sec. 6** concludes.

2. BACKGROUND
2.1 Devices, gates, and interconnect

Single domain magnets can represent and store binary data. Ensembles of magnets can also move and process information. A "wire" (Fig. 1) can be formed from a line of magnets that are anti-ferromagnetically coupled with each other. Ferromagnetic (F) interconnect is also possible [12]. A functionally complete logic set can be realized with combinations of majority voting gates. By setting one input to a logic '0' or '1', the gate can execute a NAND/NOR (or AND/OR) function [13]. This "parts library" has been expanded to include programmable majority gates, fanout, and non-majority AND/OR logic [8].

2.2 Field-driven clocking

Externally supplied switching energy is needed to re-evaluate a magnet ensemble (e.g., Fig. 1a) with new inputs (Fig. 1b,c) to overcome (i) both the intrinsic demagnetizing energy barrier (EB) (Fig. 2a), and (ii) any Zeeman energies contributed by fringing fields from neighboring magnets that reinforce the original state.

To modulate the EBs of magnet ensembles "on-chip," [14] proposed using hard axis directed magnetic fields from current driven wires. Clock energy could be amortized over 100,000s of devices as a single clock line could control many parallel ensembles. Clock lines can be placed in series and in multiple planes to minimize driver overhead [8]. Fringing fields from devices themselves also help with the transition to a 0° state (Fig. 2b). As such, the magnitude of the required clock field (and current) need not be excessively high [2, 14]. If a driving neighbor provides a y-directed field, a target magnet's energy landscape shifts (Fig. 2c), and it rotates toward a preferred easy axis (while the clock is applied). When the clock is removed, the magnets will completely relax to a new, low energy ground state (Fig. 1c).

[14] proposed CMOS compatible clocks that have been (i) fabricated [15], (ii) used to switch the state of individual magnetic islands [15], and (iii) used to re-evaluate line and gate structures with new inputs [16]. Looking to larger systems, parallel ensembles of magnets over a given wire (Fig. 1d) could (a) rest in a ground state to drive a second, adjacent group (i.e., N-1 drives N in Fig. 1b), (b) be placed into the metastable, 0° state required for re-evaluation, (Fig. 1b, Fig. 2b), or (c) relax into a new ground state (Fig. 1c) thus transmitting information. Devices controlled

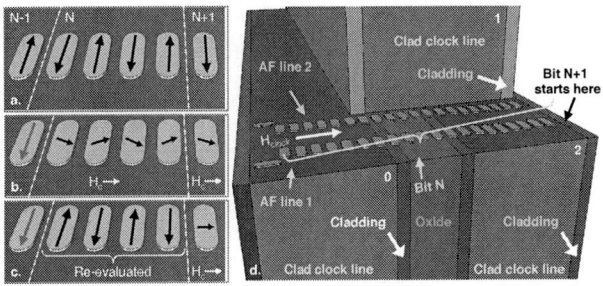

Figure 1: (a) AF-lines move information; (b) an AF-line has a new input, and an external clocking field is used to facilitate re-evaluation of the line; (c) as the field is removed, devices relax along their easy axes; (d) clock lines control many devices, and create an inherent pipeline.

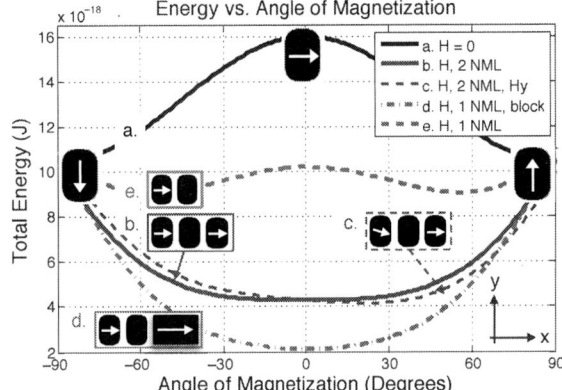

Figure 2: Energy landscape for a 60x90x30 nm³ supermalloy device (a) with clock fields (H) equal to 0; (b) when device is subjected to a clock and fringing fields from 2 neighbors (NML) at 0°; (c) if a neighbor provides a y-bias; (d) a block mimics a neighbor in a 0° state; (e) a device coupled to just one other is in a high energy state when hard-axis biased at 0°.

by a third wire could be hard axis biased to prevent the second group from being driven from two directions (group N+1 in Fig. 1b). This ensures unidirectional dataflow, and creates an inherent pipeline. (In simulations, we employ a block of magnetic material whose easy axis is parallel to the direction of the clock field to mimic this effect – note the similarities between Fig. 2d and Fig. 2b – and avoid the high-energy state at 0° in Fig. 2e.)

2.3 Alternative clocking mechanisms

Multiferroics [17], magnetostriction [18], and STT have also been proposed as NML clocks. While a potential disadvantage of the aforementioned clocks is that every device may need to be contacted individually, potential advantages include additional energy reductions [18], and more fine-grained control of an NML ensemble (useful architecturally per [7]). However, any circuit will still be inherently pipelined [18], bit conflicts can still occur, and our insights are relevant to other clocking approaches too.

3. ORIGINS OF ENSEMBLE VOLATILITY

Non-volatility in NML has largely been quantified by considering the EB of a single device in isolation. However, device-to-device coupling will also affect a circuit ensemble's overall stability.

To begin, consider work published in [10] – which suggests that NML devices could operate near thermal equilibrium *without* the assistance of externally applied magnetic fields. [10] initially considered micromagnetic simulations of 60x120x5 nm³ devices with a device-to-device spacing of 20 nm. This line was initialized

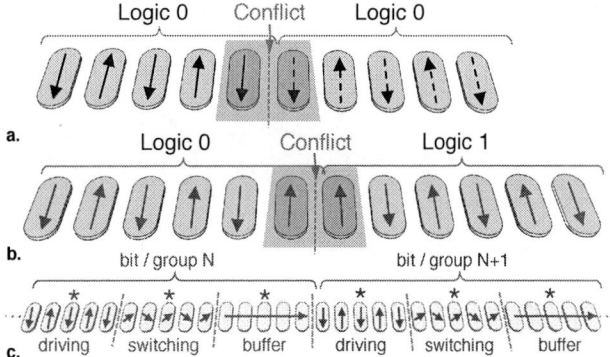

Figure 3: Conflicts occur if (a) an odd number of magnets per clock group or (b) an even number of magnets per clock group is used; (c) bit definition in an AF-line (3-phase clock).

such that all devices were initially in a 0°, metastable state, and *was* subjected to a hard axis magnetic field (37 mT) such that the EB between ↑ and ↓ states was less than the thermal energy kT. While all devices in the line relaxed to a magnetization state defined by an easy axis (i.e., ↑ or ↓), thermal noise caused some devices in the line to switch prematurely. The net result was a line with defects (where "defect" suggests improper signal ordering – e.g., ↑↓↓↑↓ – rather than manufacturing variation). With continued application of the critical switching field, said defects began to randomly walk through the line (i.e., ↑↓↓↑↓ might change to ↑↓↑↑↓, etc.) until all mis-orderings were eventually annihilated. This phenomenon was repeated experimentally where temperature elevations alone induced bit flips (detected by photoemission electron microscopy). In essence, conflicts migrated, and in turn changed the state of surrounding devices.

3.1 Impact of bit conflicts in the steady state

Clocked NML is usually inherently pipelined (e.g., see Fig. 1d, [18], etc.). While this improves throughput, pipelining will lead to bit conflicts during error free operation (i.e., even with no premature switching). Consider a line clock where an odd number of magnets span the width of each line. From Fig. 3a, if a group stores a binary 0, and an adjacent group also stores a binary 0, there cannot be complete AF ordering in the ensemble. Similarly, if an even number of magnets spans each group, a binary 1 next to a binary 0 also results in a steady state conflict (Figure 3b).

To see the potential impact of conflicts in pipelined circuits, we initially discuss micromagnetic simulations of a line of 40, $40\times60\times20$ nm^3 supermalloy magnets spaced 8 nm apart. We use the OOMMF LLG solver [19] to model device interactions. Thermal noise is modeled with random fields, whose strength is a function of temperature, that are applied to each mesh point for each simulation time step. Note that lines of $40\times60\times20$ nm^3 magnets with 8 nm between devices (i) allow for a reasonable comparison to 15 nm CMOS when considering minimum feature size, (ii) should not be superparamagnetic, (iii) could be controlled by a line clock with non-uniform field distributions [2], and (iv) the currents required to generate sufficiently high clocking fields should allow for clock energy budgets to make NML competitive with low power CMOS equivalents [2, 7].

In our simulations, fields were applied to groups of 5 magnets to create a pipelined, AF-line. (With a 3-phase clock, the line could store 3 bits of information as suggested by a subset in Fig. 3c.) Simulations at both 0 and 300 K suggest that multiple bits can simultaneously move through the line in a pipelined fashion, and (as in [2]) a 5 mT clock field was sufficient for re-evaluation.

Now, consider the state of this line of magnets if adjacent magnet groups contain logic 0s. With this correct "architectural state", bit conflicts occur. Lines with this conflict state were again considered via OOMMF simulation. No hard axis fields were applied, and the ensemble was only subjected to the effects of thermal noise representative of a 300K environment. In all simulations, random walks occurred. As an example, after approximately 15 ns, the final state of the 40 magnet line that originally contained conflicts is a completely AF-ordered line, and information originally contained in the line is lost. Simulations of other devices/spacing (e.g., $60\times90\times30$ nm^3 magnets with 10 nm spacing) also exhibit state-destroying random walks.

3.2 Quantifying ensemble stability

For a given device size/spacing, field coupling and thermal noise *alone* could be sufficient to initiate a state destroying random walk. To design pipelined ensembles that preserve state, we must quantify ensemble stability. While it is relatively simple to consider the stability of a device in isolation, the task becomes much more difficult when multiple coupled devices are introduced. At any moment, the stability of a single magnet depends not only on the device's intrinsic EB, but also on the fringing fields from neighboring devices – which continually change due to the effects of thermal noise.

We present an approach for predicting the stability of subcomponents within an ensemble. This will be used to evaluate design space enhancements to preserve steady state stability. Enhancements can then be studied via simulation to estimate the potential impact on clock energy (Sec. 4). (Promising design candidates can then be fed to experimentalists to significantly narrow the design space for temperature-accelerated experiments.)

To estimate ensemble stability, we introduce a simulation-based approach that considers 3-magnet ensembles where a "target" device is subjected to fringing fields from two neighboring devices. Consideration of just three devices is sufficient as conflicts are local to groups of this size (e.g., conflicts can be ↑↓↓, ↓↓↑, ↓↓↓) and the stability of each target depends solely on the state of its two neighbors. (While 2 neighbors are considered here as AF-lines are used as a case study, this approach could be extended to N neighbors to consider majority gates, fanout, etc.)

We use OOMMF to measure the fringing fields that a target magnet could experience to determine how the EB – and hence stability – of a target magnet might be impacted by neighboring device state. Assuming an initial/desired ↑ or ↓ state for a target device, we consider the impact on the EB between the initial/desired state, and the opposite/mistake state when the target is subjected to fringing fields from two neighbors in every possible magnetization state. Device angles between ±90°, in 1° increments are considered (resulting in 32,761 possible magnetization states), and average fields are measured *x* nm away from the simulated neighbor device to account for device-to-device spacing in a circuit. (This data could be reused if one were interested in the effects of different spacing between devices.)

We then calculate the EB of the target magnet (in isolation), which can include energy from shape anisotropy (demagnetizing energy) and crystalline anisotropy (to be discussed in Sec. 5). This curve includes angles between ±90°, in 0.1° increments (1801 total data points), and requires ~30 minutes of simulation time. This data is used to determine the "net" EB of the target (the base EB of the target plus the energy from neighboring devices). This combined EB curve is then used to quantify target stability (the EB between a desired state and a mistake state) for each of the 32,761 possible neighbor states described above.

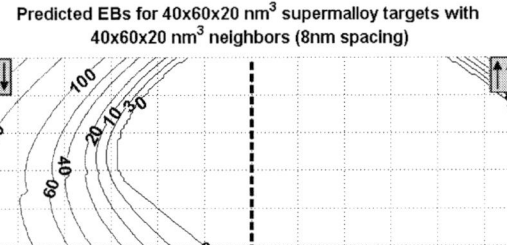

Figure 4: Predicted EB (40x60x20 nm³ device, 8 nm spacing).

As an example, assume the desired state of the target is such that M_y is positive (i.e., +90° or ↑). Beginning at +90°, we sweep the EB curve from +90° to -90° until we either (i) encounter a local energy minimum, or (ii) reach -90 degrees and encounter no minimum. This local minimum represents the state the target would change to if it begins with $+M_y$ (↑) and is subjected to the fields from the neighbors.

There are now three possible cases to consider. **First**, the local minimum is such that M_y is negative, which suggests the target would undergo a state transition to a state such that the y-component of magnetization would be negative (↓). In this case, the EB is predicted to be 0. **Second**, if the local minimum is associated with a positive angle (i.e., such that M_y is still positive), and there exists another minimum (found by sweeping the curve starting from *this* minimum to -90°), the EB is the height of the curve between these two minima. **Finally**, if the local minimum is associated with a positive angle, but there is no other minimum, we consider the EB to be the difference between this minimum and -1° (i.e., the energy to transition from ↑ to ↓). The net result is a stability prediction for all possible neighbor states.

The contour plot in Fig. 4 is a visual representation of the data generated by the process described above for the core of the AF-line in Sec. 3.1 (i.e., three, 40x60x20 nm³ devices with 8 nm spacing). (Lines are representative of 'N' in N kT, for T=300K.) The magnetization state described by quadrant I represents a true worst case when considering the stability of a line of magnets. We would like a target to retain a positive, y-component of magnetization (↑), but fringing fields from both neighbors are directed in the opposite direction. While statistically unlikely, this case could occur if random walks from two adjacent clock groups migrate to a device in a middle group. A bit flip is all but certain, as our model predicts essentially no stability in this quadrant.

Quadrants II and IV represent "architectural conflicts," and suggest that if the angle of magnetization of neighboring devices is approximately +45°/-45° (left/right in quadrant II), the probability of a state transition is ~13% per ns (predicted EB is 2). (By Arrhenius-Neel theory of thermally activated magnetization reversal, the probability per unit time of reversal over the EB is $1/\tau$, where $\tau = t_o \times \exp(EB/k_b T)$, and t_o is the time for a thermally activated reversal – assumed to be 1 ns per [20]). Notably, 0K simulations of the same line suggest that, due to additional

fringing field coupling associated with 8 nm device spacing, the angle of magnetization of each device in an *AF-line with no conflicts* is approximately ±45-50°. (Experiments in [21] also suggest increased ferromagnetic ordering at closer spacing.) This explains the random walk described in Sec. 3.1. Additionally, per supplement Sec. S1, thicker devices in line ensembles can require lower clock fields because of increased device-to-device coupling.

Finally, quadrant III represents the case where there is no conflict, and a line is completely AF-ordered. Given the ~200 kT stability, information should be retained (fringing fields from neighboring devices reinforce a target's state even when at ±45-50° angles).

In the context of these contour plots, we note two approaches to provide N kT stability in a circuit. **First**, one can mitigate conflict migration if the desired N kT stability is achieved for all points in quadrants II and IV for all magnets, as the worst case (quadrant I) should not occur. **Second**, if the stability of selective "latch" magnets (see Sec. 4) is at least N kT in quadrant I; random walks would be allowed in the remaining, unstable magnets, and the latches would preserve state. (See Sec. S2 for additional discussion of this case.)

Finally, we note that our method for quantifying stability could trend toward the pessimistic (we assume static neighbor state, whereas thermal noise may only result in the worst case for a relatively short time). However, it may be possible to relax our model. For example, Fig. 4 suggests that if the angle of magnetization for the left and right neighbors is sufficiently restricted in the steady state, the ensemble will remain stable. (See supporting data for this assumption in Sec. 4.) For example, per Fig. 4 (quadrant II) if the final magnetization states of the left and right neighbors did not deviate beyond +60° and -60° respectively, the probability of the target flipping in a given year is ~2.8x10⁻¹⁰, and in our first approach we would not need to design for N kT for *all* angles in quadrants II, IV.

4. IMPROVING NON-VOLATILITY

Here, we explore ways to tune EBs in polycrystalline devices for improved stability. Lowering inter-device coupling effectively raises the energy barrier of each device, e.g., compare EBs in Fig. 2e where a device couples with 1 neighbor vs. 2b where it couples with 2 neighbors. Coupling can be decreased by increasing inter-device spacing, decreased device thickness (see quantitative discussion in Sec. S1), or changing magnet shapes. Alternatively, raising the intrinsic EB of each device can increase ensemble stability. The EB is proportional to the demagnetizing factor, volume, and saturation magnetization of a magnet. Thus the EB can be raised by, for example, increasing device aspect ratio, volume, or tuning device EB with different materials.

We now present two test cases where we leverage aspect ratio alone to demonstrate the two approaches for improving steady state stability (over the original design in Sec. 3.1). The discussion is not intended to represent an exhaustive study, but rather to identify design options that should be explored to achieve the best balance between low energy clocking and non-volatility. Any potential increases to clock energy are also captured quantitatively. (Additional levers are considered in Sec. 5.)

4.1 Approach 1: Eliminate Random Walk

An obvious first attempt to improve stability is to raise the EB of all devices in an ensemble by increasing their aspect ratio. Improved stability can be explained by considering the demagnetizing energy of a given device – illustrated as a function of angle of magnetization for 40x60x20 nm³ and 40x80x20 nm³

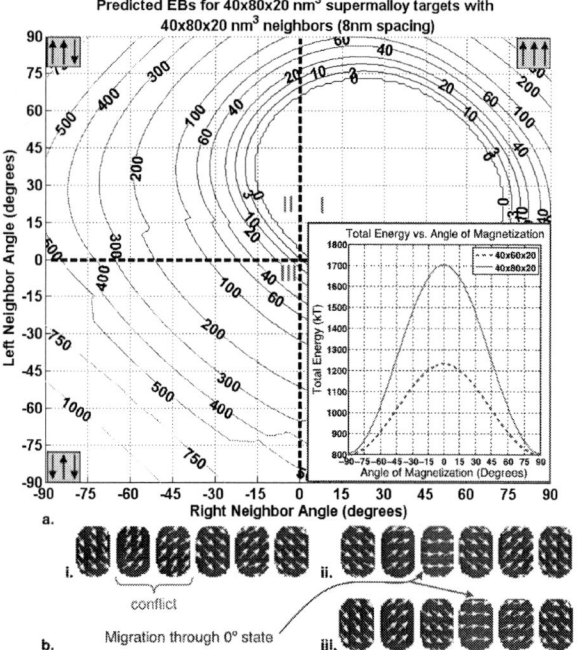

Figure 5: (a) predicted EB of a 40x80x20 nm³ device at 8 nm spacing; (b) snapshots of random walk in 40x60x20 nm³ line.

devices at 300K in the Fig. 5a inset. As the aspect ratio of a device increases, the energy required for a magnet to transition to a 0° state also increases. This is especially important as the state-changing random walks that occurred in the line of 40x60x20 nm³ devices were facilitated by a transition through a lower energy, F-ordered state (Fig. 5b). Given the higher/steeper energy landscape for the 40x80x20 nm³ device (Fig. 5a inset), hard axis directed fringing fields that can impact stability should be less significant.

As expected, lines of 40x80x20 nm³ devices could be quite stable even if conflict states occur. With one conflict (quadrants II, IV in Fig. 5), if devices remain magnetized along their easy axes, an EB of >500 kT is predicted. Micromagnetic simulations (and the energy landscape in Fig. 5a) reinforce this assertion, as the average angle of magnetization observed for the configuration captured by quadrant II is not less than ±86° (averages of 0K and 300K simulations sampled over 40ns.).

However, not surprisingly, the clock energy required to re-evaluate an ensemble will increase. Simulations suggest that required fields will increase from 5 mT (for a line of 40x60x20 nm³ devices) to 50 mT (for a line of 40x80x20 nm³ devices). As the magnitude of a current generated field is directly proportional to the current (i.e., B ~ µI), and line energy is a function of I², Ohmic losses would increase by a factor of 100. Studies of reduced aspect ratio devices (40x70x20 nm³) suggest that AF-lines can be re-evaluated with a 25 mT field (Ohmic losses would increase by 25), and could be sufficiently stable (e.g., 500 kT with angles of ±83° as observed in simulation). However, additional tuning of aspect ratio could be impeded by lithographic resolution, as only 10 nm of "height" remain (to 40x60 nm² footprints).

4.2 Approach 2: Selective Latching

As suggested in Sec. 3.2, not all devices within a clock group must be non-volatile in order to preserve pipe stage state. Instead, we can preserve information within a pipe stage by improving the steady state stability of select "latch" devices. One approach is to selectively introduce higher AR devices within a clocked segment

(i.e., the *'ed magnets in Fig. 3c). For example, one might consider a line of 40x60x20 nm³ devices where select devices are replaced by higher aspect ratio, 40x80x20 nm³ devices. Regarding fabrication, this would not be challenging, as all devices could still be made with a single mask.

Regarding stability, our analysis suggests this latch should increase stability in quadrant I by an average of 83X (compare contour in Sec. S3, Fig. S1 to Fig. 5). When considering clock energy, higher fields were still needed to move data through the higher aspect ratio device during re-evaluation. However, we can modify the waveforms associated with a three-phase clock such that the first phase now operates at 50 mT (to place latches in a 0° state), and the second phase operates at 5 mT (to clock the low-AR devices). Simulations show that this facilitates data movement through the line. Total energy is thus increased by a factor of 50X. Again, lower aspect ratio, 40x70x20 nm³ latches could operate with 5mT/25mT fields (leading to an energy increase of 13X and an average stability increase of 19X in quadrant I). While better than Sec. 4.1, the increase in clock energy is still significant.

5. LOWER ENERGY ALTERNATIVES

Work in Sec. 4 suggests that pipelined, polycrystalline magnet ensembles can be engineered to be NV, but that clock field requirements for re-evaluation could become unacceptably high. Here, we discuss additional (materials centric) approaches for low energy clocking of field driven, NV NML circuits.

5.1 Enhanced permeability dielectrics

As suggested in [2] the ratio of flux density to magnetic field strength (µ=B/H) could be increased by surrounding devices with superparamagnetic CoFe magnetic nanoparticles (~2-5 nm in diameter) [11]. Switching currents could be reduced by a factor of μ_r, and I^2R losses by μ_r^2. Recently, we have reproduced EPD films in [11], and are now working to integrate NML devices. Preliminary films demonstrate relative permeabilities of ~4.5. This could improve clock energy by ~20X – although EPDs may increase steady-state device-to-device coupling as well. (EPDs / new experiments are discussed further in Sec. S4.)

5.2 Biaxial anisotropy

Devices with biaxial anisotropy ([9]) were proposed to increase hard axis metastability during ensemble re-evaluation, and make *soliton*-based switching more reliable. (For a detailed discussion of soliton-based operation and a comparison to our approach, see Sec. S5.2.) Recently, devices with biaxial anisotropy were added to our design space exploration efforts, which emphasize the identification of device features to minimize field-based clock energy. Initial results – presented here for the first time – are encouraging, and could impact both clock energy and stability.

Consider first micromagnetic simulations of AF-lines comprised of 5, 60x90x5 nm³ devices spaced 10 nm apart. The material parameters chosen were essentially identical to those used in [10], (see Sec. S5.4). Also, a detailed discussion justifying the use of devices with a larger footprint appears in Sec. S5.6 – but recall that random walks in lines with devices with the same footprint and spacing were also observed. With K_1 equal to 15 and 30 kJ/m³, a 10 mT field is sufficient to re-evaluate the AF-ordered line with a new input (see Fig. S6a for 30 kJ/m³ data). Per Sec. S5.5, additional energy reductions appear possible (via additional field reductions to 7.5 mT or alternative clock waveforms). In identical lines with $K_1 = 0$ kJ/m³, signal propagation failed in 8 of 10 different simulations (to capture effects of thermal noise) – *even when a 15 mT clock field was employed.*

These trends can be explained by considering the energy landscapes for the biaxial device (K_1=15 kJ/m^3) and polycrystalline device (K_1=0 kJ/m^3) when each device is subjected to a 33 mT field in the 0° direction. (The devices are otherwise similar.) This field is representative of (i) the fringing fields produced by two neighboring devices on a target (~28 mT with 10 nm spacing between devices) and a 5 mT clock field. If we assume that each device originally had a positive, y-component of magnetization, the energy landscape (see Fig. S5a) suggests the biaxial device will transition to the metastable state (for re-evaluation), while the polycrystalline device would retain some of its state. (Dynamic simulations in Sec. S5.7 also show this.)

While biaxial anisotropy has the potential to benefit *all* NML ensembles – especially if devices are thinned for integration with I/O structures (see [8, 21] and Sec. S5.6) – additional analysis suggests that we can also use it to improve steady state stability. As device thickness decreases, the magnetization state of an individual device is more aligned along its easy axis (toward ±90°) as a result of less x-directed coupling. Per Sec. 4, this should help to improve stability, while the biaxial component could help to keep clock energy low (see Fig. S5a).

Consider the case of 3, 60x90x5 devices (K_1=15 kJ/m^3) spaced 10 nm apart (see Fig. S5b for contour plot of data). Our analysis suggests that if the angles of magnetization of the left/right neighbor do not move beyond +75°/-75° respectively, the probability of architectural conflict migration is low (the predicted EB is > 60). (Simulation data sampled over ~5 ns suggests the angle of magnetization does not deviate by more than -86.6° for the left neighbor, and 85° for the right neighbor.) Finally, simulations to quantify clock field requirements of 60x90x30 nm^3 devices with 10 nm spacing, and where random walks *were* observed, suggested that 5 mT fields were needed for re-evaluation. Simulations of biaxial ensembles (including those that were pipelined) suggest that 7.5-12 mT fields are needed to re-evaluate – which would lead to a net increase in clock energy of ~2.3-5.8X with no enhancements. This merits further study.

6. SUMMARY AND CONCLUSIONS

In this paper, we have explained how architectural-level bit conflicts can occur in defect free NML ensembles, and said conflicts can make supposedly NV circuits volatile. When studying magnet ensembles proposed in previous work [2] – that could be re-evaluated with clock fields that would offer energy savings over state-of-the-art low power CMOS – we found that pipelined circuits would *not* retain their state post re-evaluation. To combat this problem, we first developed a methodology for quantifying ensemble stability, and showed that properly designed, 3-magnet ensembles can mitigate volatility introduced by architectural-level conflicts. Moreover, while our work primarily focused on experimentally demonstrated, field-based clocking, the conflicts described in this paper will occur in any pipelined NML circuit, regardless of what clock is employed. As such, our tools should be applicable to other clocking approaches as well – e.g., as tighter device spacing might be required to help preserve hard axis metastability in soliton-based switching with a multiferroic clock. When discussing design alternatives to improve the stability of polycrystalline ensembles, we identified what the NML design space must "deliver" to ensure non-volatility. However, increased clock energy or fine-grained lithography may be needed. We concluded by discussing materials-related design levers. Most notably, biaxial anisotropy could minimize clock energy in *any* NML circuit (especially when considering integration with I/O). An added benefit appears to be improved steady state stability. Future work will further investigate this space (and other NML circuit constructs) to significantly narrow the design space for temperature-accelerated experiments to further study non-volatility.

7. REFERENCES

1. Chandrakasan, A.P., et al. Technologies for Ultradynamic Voltage Scaling. *Proceedings of the Ieee*, 98 (2). 191-214, 2010.
2. Dingler, A., et al. Performance and Energy Impact on Locally Controlled NML Circuits. *ACM Journal on Emerging Technologies in Computing*, 7 (1). 1-24, 2011.
3. Abelson, L.A., et al. Superconductor integrated circuit fabrication technology. *Proceedings of the Ieee*, 92 (10). 1517-1533, 2004.
4. Guo, X., et al. Resistive computation: avoiding the power wall with low-leakage, STT-MRAM based computing. *Proc. of 37th Int. Symp. on Comp. Arch. (ISCA)*. 371-382, 2010.
5. Matsunaga, S., et al. Fabrication of a Nonvolatile Full Adder Based on Logic-in-Memory Architecture Using Magnetic Tunnel Junctions. *Applied Physics Express*, 1. 091301, 2008.
6. Mochizuki, A., et al. TMR-Based Logic-in-Memory Circuit for Low-Power VLSI. *IEICE Trans. Fundam. Electron. Commun. Comput. Sci.*, E88-A (6). 1408-1415, 2005.
7. Crocker, M., et al. Design and Comparison of NML Systolic Architectures. *IEEE NANOARCH*. 29-34, 2010.
8. Niemier, M.T., et al. Nanomagnet Logic: Progress Toward System-Level Integration. *J. Phys. Con. Mat.*, 23. 493202, 2011.
9. Carlton, D.B., et al. Simulation Studies of Nanomagnet-Based Logic Architecture. *Nano Letters*, 8 (12). 4173-4178, 2008.
10. Carlton, D.B., et al. Computing in Thermal Equilibrium With Dipole-Coupled Nanomagnets. *IEEE TNANO*, 10 (6). 1401-1404, 2011.
11. Pietambaram, S.V., et al. Low-power switching in magnetoresistive random access memory bits using enhanced permeability dielectric films. *App. Phys. Let.*, 90. 143510, 2007.
12. Pulecio, J.F., et al. Magnetic cellular automata coplanar cross wire systems. *J. of App. Phys.*, 107 (3). 034308, 2010.
13. Imre, A., et al. Majority logic gate for Magnetic Quantum-dot Cellular Automata. *Science*, 311 (5758). 205-208, 2006.
14. Niemier, M.T., et al. Clocking Structures and Power Analysis for Nanomagnet-Based Logic Devices. *Int. Symp. on Low Power Elec. and Design (ISLPED)*. 26-31, 2007.
15. Alam, M.T., et al. On-Chip Clocking for Nanomagnet Logic Devices. *IEEE Transactions on Nanotechnology*, 9 (3). 348-351, 2010.
16. Alam, M.T., et al. On-chip Clocking of Nanomagnet Logic Lines and Gates. *to appear in IEEE TNANO*, 2011.
17. Chu, Y.H., et al. Electric-field control of local ferromagnetism using a magnetoelectric multiferroic. *Nat. Mat.*, 7 (6). 478-482, 2008.
18. Salehi, F.M., et al. Magnetization dynamics, Bennett clocking and associated energy dissipation in multiferroic logic. *Nanotechnology*, 22 (15). 155201, 2011.
19. Donahue, M. OOMMF User's Guide, Version 1.0, Interagency Report NISTIR 6367.
20. Rizzo, N.D., et al. Thermally activated magnetization reversal in submicron magnetic tunnel junctions for magnetoresistive random access memory. *App. Phys. Let.*, 80 (13). 2335-2337, 2002.
21. Lyle, A., et al. Probing dipole coupled nanomagnets using magnetoresistance read. *J. App. Phys.*, 98 (9). 092502, 2011.

Paper Supplement

1. DEVICE THICKNESS

Intuitively, assuming a constant aspect ratio, as both the volume and demagnetizing factor N_d are roughly proportional to the thickness of a device, the energy barrier (EB) (and thus stability) of an individual magnet decreases with the square of the thickness. Thus, in isolation, thinner devices should (i) require lower fields to flip, (ii) require lower fields to hard axis bias, and (iii) be less stable.

In previous design space exploration efforts [1], the trends described by items (i) and (ii) were also observed when considering the magnitude of the clock field required to re-evaluate ensembles of devices of varying size, aspect ratio, thickness, etc. For example, fields needed to saturate (such that $M_x = M_s$) a line of 40x60x20 nm^3 devices spaced 16 nm apart were ~10% higher than those needed to saturate a line of 40x60x10 nm^3 devices with the same spacing. However, as spacing was decreased, these trends were *reversed*. As a result of increased device-to-device coupling, clock fields required to saturate the line of 10 nm thick devices spaced 8 nm part were ~17% *higher* than those needed for a line of 20 nm thick devices.

2. DESIGN POINT DISCUSSION

Using Fig. 3c in the main manuscript as context, assume there is a power failure when devices in the switching and buffer groups are in the metastable, 0° state. If these devices relax randomly, the situation described by quadrant I could occur. The latch could preserve state in the un-clocked driving group. If we simply save the current clock period in non-volatile memory – just 2 bits would be required – computation could be restarted.

3. LATCH STABILITY

Fig. S1 depicts the stability of the 40x80x20 nm^3 latch magnets with 40x60x20 nm^3 neighbors (discussed in Sec. 4.2 of the text).

4. EPDs

Enhanced permeability dielectrics (EPDs) with embedded magnetic nano-particles (e.g., CoFe particles 2-5 nm in diameter) were proposed in [2] to increase the field from a word or bit line in field MRAM without increasing current. The authors suggest that absolute permeability ($\mu = \mu_o \times \mu_r$) could improve by as much as 30X, although μ_r's of 2-6 were most common experimentally. As illustrated in Fig. S2, we have reproduced the films described in [2], and are now working to integrate NML devices.

As a representative example, a film comprised of 6-to-9 nm diameter CoFe particles appears in Fig. S2a. Vibrating sample magnetometer (VSM) characterization (Fig. S2b) suggests that (i) this film has a relative permeability of ~4.5 (determined by measuring the slope of the low field magnetization data – as was done in [2]), and (ii) that the particles are superparamagnetic (no coercivity or remanence was observed). Notably, this "boost" to μ_r could improve clock energy by ~20X – although EPDs could also increase device-to-device coupling as well. How this affects stability remains to be investigated.

5. BIAXIAL ANISOTROPY

5.1 Concept

Depending on its value, the biaxial anisotropy term flattens and/or introduces a local minimum in the energy landscape of an individual magnetic island at the 0° point (see Fig. S3).

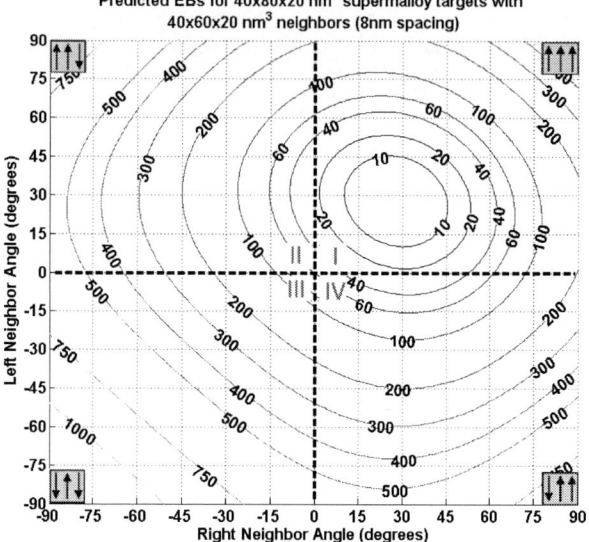

Figure S1: Stability of a 40x80x20 nm^3 latch neighbored by 40x60x20 nm^3 devices with 8 nm spacing.

Figure S2: (a) SEM image of CoFe particles that comprise EPD film; (b) VSM characterization suggests μ_r is ~4.5.

5.2 Soliton-based switching

With soliton-based switching, all devices in a clocked ensemble are simultaneously placed into a 0° state. The clock field is then removed, and dipole fields from neighboring magnets alone are expected to preserve metastable, 0°-ordering until a device is set by an appropriate neighbor. To prevent premature switching due to thermal noise, [5] proposed devices with a magnetocrystalline biaxial anisotropy such that $U(\theta)$ becomes $K_u\cos^2(\theta) + \frac{1}{4}K_1\sin^2(2\theta)$ (K_1 is the biaxial anisotropy constant). The biaxial anisotropy term can impact the energy landscape of an NML device at the 0° point (e.g., per Sec. S5.1, Fig. S3, a local minimum can appear).

Although there are challenges to soliton-based switching, additional advantages could include further reductions in clock energy, as the clock could be turned off after ensemble metastability is achieved. In simulations described in Sec. 3 and Sec. 4 of the main manuscript, devices are biased into a new, logically correct state before an external field is removed.

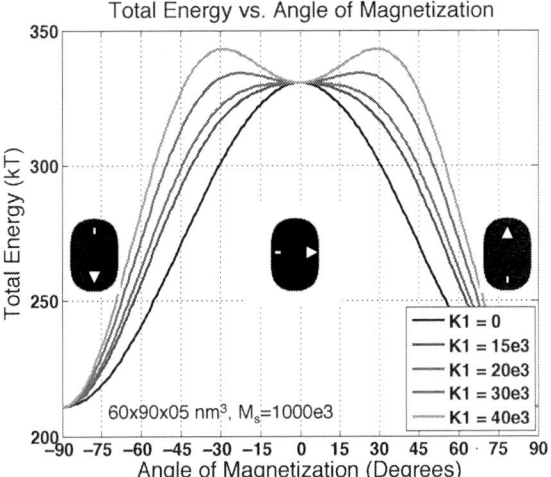

Figure S3: As K_1 increases, the depth of the local energy minimum at 0° increases as well. Per [1], the energy minimum must be tuned such that it is high enough to prevent premature relaxation, and low enough so that a neighbor can still bias a device to a new, and logically correct state.

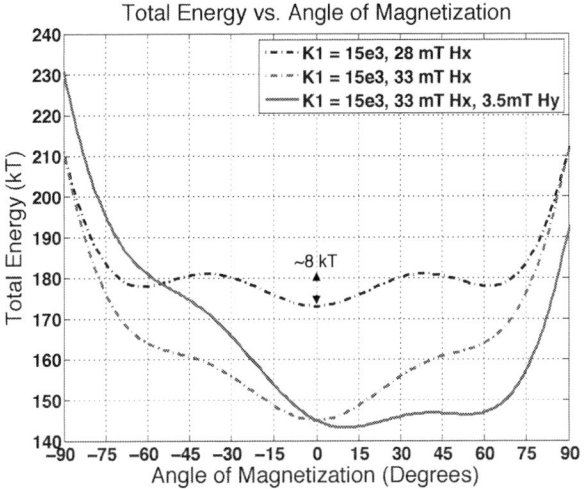

Figure S4: Representative energy landscape for a device in the middle of an anti-ferromagnetic line.

While this will (i) result in an increase in clock energy, and (ii) turn successive devices in a group into weaker drivers (eventually limiting the width of a clock line), potential advantages include a decreased likelihood of premature switching (as switching dynamics are similar to those described by Figs. 2b,c of the main manuscript where there is an absolute energy minimum). Performance wins projected by [3] account for this increased clock energy.

While a detailed discussion is beyond the scope of even this supplement, we believe that switching in the presence of an applied field could help to reduce error rates during the switching process (i.e., after a device ensemble has been placed into a metastable state). Briefly, assuming the soliton-based switching approach discussed at the beginning of this section, only device-to-device fringing fields preserve 0°, metastability during the re - evaluation process. For the devices considered in the simulations captured by Fig. S6, the average field that an internal device

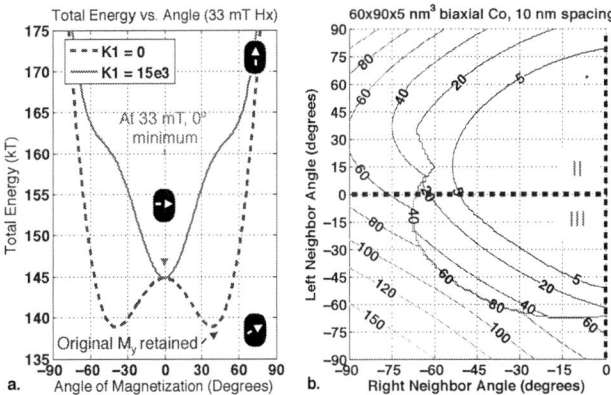

Figure S5: (a) energy landscape for polycrystalline, biaxial devices with 33 mT 0° field; **(b)** predicted EBs (quadrant II, III) for $60x90x5$ nm^3 devices spaced 10 nm apart.

would experience from two nulled neighbors is approximately 28 mT. The energy landscape for a given device (subject to a 28 mT, hard-axis directed field) is illustrated in Fig. S4. While there is a local energy minimum at the 0° / metastable state, only an 8 kT barrier separates this metastable state from a potential mistake state. (As such, thermal noise, fabrication variation, etc. could induce a transition to a mistake state.) However, given a nominal clock field (e.g. 5 mT), the 0° / metastable state represents an *absolute energy minimum*. Moreover, given a y-bias (presumably from a neighboring device), any energy minima are on the logically correct side of the 0° point – suggesting that information can propagate through the ensemble in the presence of an applied field.

Additionally, in the simulation efforts in [4], the authors note that dipole-to-dipole coupling alone could be insufficient to keep an anti-ferromagnetically ordered line (for example) hard-axis-biased even if devices with biaxial anisotropy are employed. Thermal noise could induce premature, random, and unwanted switching. Thus, while appealing, the viability of true, soliton-based switching does need additional study.

5.3 Energy landscape and stability

Fig. S5a depicts the energy landscapes for polycrystalline and biaxial devices in isolation (with a field along the x-axis). Fig. S5b depicts the predicted EBs for 60x90x5 nm3 devices spaced 10 nm apart. Both figures are discussed in Sec. 5.2 of the main manuscript.

5.4 Material parameters

The material parameters chosen were essentially identical to those first proposed in [5] – i.e., as the saturation magnetization (M_s) is also 1,000,000 A/m. Additionally, as in [5], the exchange stiffness A was set at $13x10^{-11}$ J/m. One minor difference between our simulations and those in [5] is the damping parameter α – where a lower, more realistic value of 0.05 was used in our simulations instead of 0.1 used in [5]. Per [6], using even lower values requires additional reductions in simulation time step which can in turn make simulations excessively long. Additionally, differences in projected field requirements for device switching were negligible with lower damping coefficients (e.g., 0.01).

5.5 Simulations of biaxial, AF-lines

Here, we provide a more detailed discussion of simulations of biaxial, AF-ordered lines. Simulation setup is described, and representative results are illustrated graphically in Fig. S6.

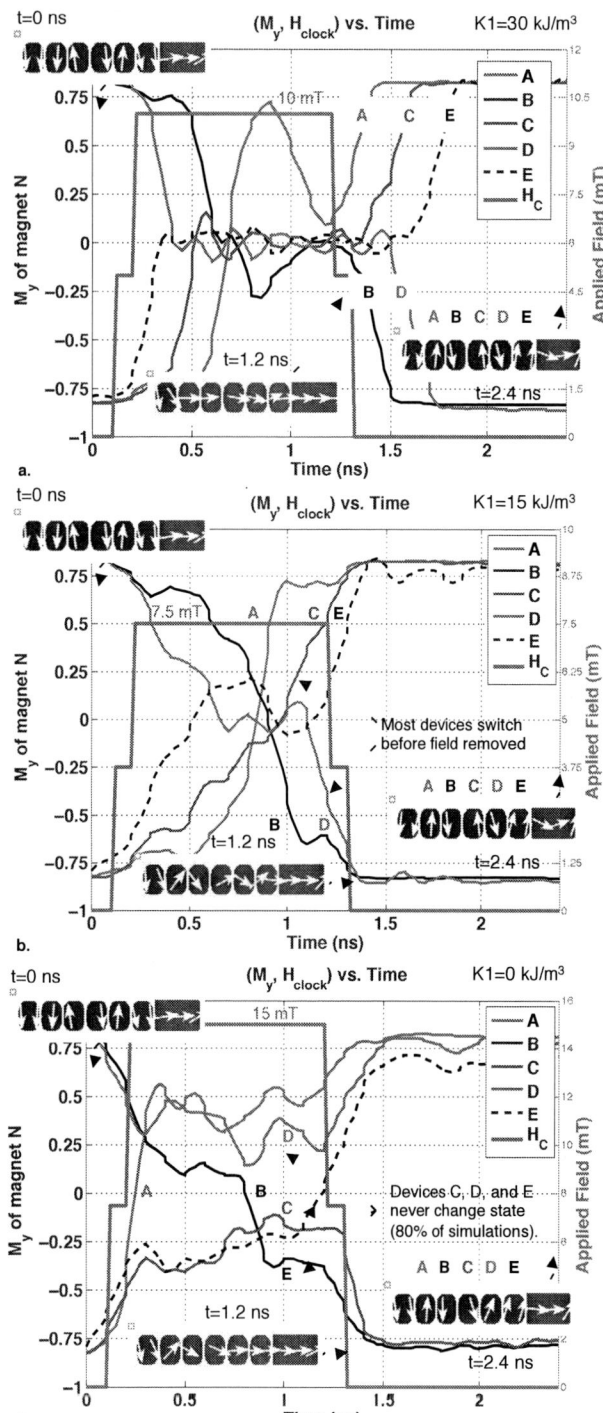

Figure S6: (a) Magnetization state (M_y) for every device ($K_1 = 30$ kJ/m³) in an AF-ordered line as a function of time. When the applied field is removed, devices switch (in the correct order) to an AF-ordered state associated with a new input; (b) with $K_1 = 15$ kJ/m³, fields of 7.5 mT can facilitate re-evaluation (40% success); (c) with no biaxial anisotropy, *15 mT* fields only facilitate re-evaluation in 2 of 10 cases.

Note that at the beginning of a simulation, the state of the input device has been flipped, while the remaining devices are in their previous, logically correct, AF-ordered state (see devices A-through-E in Fig. S6 insets). A 0° directed field was applied to this

ensemble for 1.0 ns (increasing from 0 mT to a peak value in 0.2 ns). The hard axis clock field was then removed (transitioning from a peak value to 0 mT in 0.2 ns). The ensemble was then allowed to completely relax into a final/ground state for 1.0 ns. (Devices were only subjected to fringing fields from neighbors and thermal noise during this last window.) For all cases considered, 10 simulations were performed, each with a different random seed. (Our intention is not to make claims about the statistical reliabilities of line switching, but to ensure that a single result is not an artifact of a chosen seed.)

Figs. S6 a-c illustrate how the magnetization state of each device in the line changes as a function of time. The desired simulation outcome (post clocking) is for the remaining devices in the line to change state in accordance with the new/fixed input, in the correct order. With K_1 equal to 30 kJ/m³, a 10 mT field is sufficient to re-evaluate the AF-ordered line with a new input (see Fig. S6a). When K_1 was reduced to 15 kJ/m³, again, a 10 mT field was sufficient to facilitate the re-evaluation of an identical line (not shown). In both instances, assuming a 10 mT clock field, (i) all 10 simulations exhibited logically correct functionality and (ii) signal propagation occurred after the field was removed – i.e. in a soliton-like fashion.

How biaxial lines responded to *lower* clock fields was also studied. When K_1 was reduced to 15 kJ/m³, a 7.5 mT field was sufficient to facilitate the re-evaluation of the line with a new input, and switching occurred in the presence of an applied field.

While the results illustrated in Fig. S6b suggest that (i) field reductions of 33% are possible by tuning K_1 and (ii) that switching can occur even in the presence of an applied field, we must note that when the results of 10 simulations were considered, 4 of 10 were successful. In all instances, the first device in the line did not flip. Again, while a detailed discussion is beyond the scope of this supplement, alternative clock waveforms – e.g. a shorter, higher magnitude 10 mT pulse, followed by a 5 mT pulse also produced desired switching dynamics with a 100% success rate (similar to the approach suggested in Sec. 4.2 of the main manuscript). This will be studied in future work.

That said, the obvious comparison to be made is to contrast the results discussed above with an identical line of devices with *no* biaxial anisotropy. Representative simulations results are illustrated in Fig. S6c ($K_1 = 0$ kJ/m³). As one can see, even after being subjected to a 15 mT field, devices C and D never reach the metastable state required for re-evaluation, and retain their original state. (Device E – which couples to the terminating block – is biased into a 0° state. Its state change is a result of random thermal noise post field removal.) Note that 8 of the 10 simulations performed resulted in a final, incorrect state of the line. Thus, the addition of biaxial anisotropy at the device-level seems to result in lower energy clocks – at least when considering 5 nm thick devices.

5.6 Biaxial implementations and device sizes

Before discussing *why* the addition of biaxial anisotropy can help to lower clock energy, for completeness, we briefly address (i) manufacturability and (ii) other aspects of this design space.

Regarding manufacturability, additional conversations with the authors of [5] suggest that devices with K_1 values ranging from 0-50 kJ/m³ could be engineered.

Regarding the design space itself, it is probably obvious to the reader that when considering the NML devices discussed in Sec. 4 (40x60x20 nm³ polycrystalline supermalloy devices spaced 8 nm apart) and the devices discussed here (60x90x5 nm³ biaxial cobalt devices spaced 10 nm apart) that several aspects of an AF-line design have changed. We address these design differences in a more controlled fashion below.

First, while no graphical results are included, simulations of biaxial supermalloy devices (K_1 = 15 kJ/m³, M_s = 800,000 A/m) exhibit similar trends to those that were illustrated in Fig. S6. For example, all 10 simulations of the 5 magnet AF-line studied exhibited correct, soliton-like signal propagation when subjected to a 7.5 mT clock field.

Second, while some simulations of thicker devices exhibited correct signal propagation, simulations also suggested that several, undesirable switching characteristics were possible. For example, consider a line of 60x90x20 nm³ devices (10 nm spacing between devices, K_1 = 40 kJ/m³, M_s = 1,000,000 A/m).

Simulations show (graphical results not included) that with the larger K_1 value, devices are almost too strongly coupled, and signal propagation proceeds slowly. (In essence, computational latency will increase.) Consider also a line of 60x90x30 nm³ devices (10 nm spacing between devices, K_1 = 15 kJ/m³, Ms = 1,000,000 A/m). Given this device configuration (again, graphical results not included), the first device in the line enters a relatively stable vortex state [7], and information never propagates down the line. The aforementioned issues were not observed with lines of thinner, biaxial devices.

Third, magnet thickness can impact chip-level energy dissipation. Each magnetic island will dissipate an amount of energy given by the EB between the hard and easy axes of the magnet. This can be tuned precisely by altering the aspect ratio, thickness, and saturation magnetization of the islands, as well as by adiabatic switching [8]. An upper bound on the energy dissipated by the magnets is then simply the number of magnets in a circuit ensemble (N) times the barrier height. The energy difference between magnetization states for the 60x90x5 nm³ devices (~120kT) is more than an order of magnitude *less* that the energy difference between states for the thicker, 60x90x30 nm³ devices (~2100kT). If we assume that 10^9 devices switch every nanosecond, the power dissipation from just the magnets could be ~17X higher (8.7W vs. 0.5W at 300K). As such, a shift to thinner devices can be favorable energetically.

Fourth, moving to thinner devices is also desirable from the standpoint of manufacturability – particularly when considering the integration of NML devices with structures to be used for the purposes of input and output. More specifically, [9] has proposed (and simulated) the use of magnetic tunnel junctions (MTJs) to provide magnetic-to-electrical and electrical-to-magnetic interfaces. For output, fringing fields from an NML device could be used to bias the free layer of an MTJ. The resistance of a given MTJ could then be sensed to determine whether an output of an NML ensemble is a 1 or a 0. For input, one could leverage spin transfer torque (STT) to set the state of an MTJ's free layer. Fringing fields from the free layer could then be used to set the state of successive devices in an NML ensemble.

That said, challenges to this approach include (i) whether fringing fields from a thinner free layer are of a sufficient magnitude to set the state of a thicker device (at the inputs), (ii) the use of STT to set the state of a (relatively) thick free layer, and (iii) the feasibility of integrating thinner devices (free layers) with thicker NML devices (as mask alignments would be required). As such, uniform device / free layer thickness is desirable from the standpoint of manufacturing NML ensembles that are integrated with needed I/O circuitry.

Finally, as noted in the main manuscript, random walks were also observed in thicker, polycrystalline devices with the same footprint and spacing

5.7 Dynamic data

Simulation results in Fig. S7 add further credence to the trends identified in Sec. 5 of the main manuscript. Here, identical, 60x90x5 nm³ biaxial and polycrystalline devices are subjected to (i) a hard axis field that is increased in 1mT/100ps and 1mT/200ps increments and (ii) a constant, easy axis directed, 2 mT biasing field opposite to the original direction of magnetization (representative of a neighbor's fringing field). The biaxial device transitions to a new state (-M_y), through the 0° state in a non-linear fashion. There is no (binary) state change in the polycrystalline device.

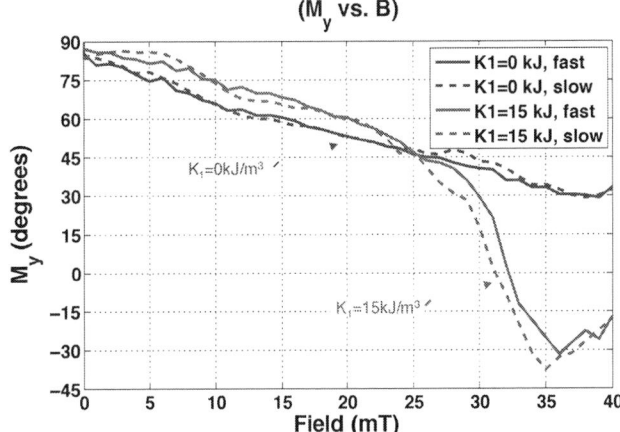

Figure S7: Switching dynamics of polycrystalline and biaxial devices when subjected to a 2 mT downward biasing field and a 0° field that increases in magnitude. The biaxial device experiences a state transition (with lower, 0° fields) while the polycrystalline device does not even change state.

6. REFERENCES

1. Niemier, M.T., et al. Nanomagnet Logic: Progress Toward System-Level Integration. *J.Phys.Con.Mat.*, *23*. 493202, 2011.
2. Pietambaram, S.V., et al. Low-power switching in magnetoresistive random access memory bits using enhanced permeability dielectric films. *App.Phys.Let.*, *90*. 143510, 2007.
3. Dingler, A., et al. Performance and Energy Impact on Locally Controlled NML Circuits. *ACM Journal on Emerging Technologies in Computing*, 7 (1). 1-24, 2011.
4. Spedalieri, F.M., et al. Performance of Magnetic Quantum Cellular Automata and Limitations Due to Thermal Noise. *IEEE TNANO, 10* (3). 537-546, 2011.
5. Carlton, D.B., et al. Simulation Studies of Nanomagnet-Based Logic Architecture. *Nano Letters*, 8 (12). 4173-4178, 2008.
6. Alam, M.T., et al. On-chip Clocking of Nanomagnet Logic Lines and Gates. *to appear in IEEE TNANO*, 2011.
7. Cowburn, R.P., et al. Single-Domain Circular Nanomagnets. *Physical Review Letters*, *83* (5). 1042-1045, 1999.
8. Csaba, G., et al., Power dissipation in nanomagnetic logic devices. in *IEEE Conf. on Nanotechnology*, 2004, 346-8.
9. Liu, S., et al. Magnetic-Electrical Interface for Nanomagnet Logic. *IEEE T. on Nanotechnology, 10* (4). 757-763, 2011.

mLogic: Ultra-Low Voltage Non-Volatile Logic Circuits Using STT-MTJ Devices

Daniel Morris, David Bromberg, Jian-Gang (Jimmy) Zhu, Larry Pileggi
Carnegie Mellon University, Dept. of ECE, 5000 Forbes Avenue, Pittsburgh, PA 15213

ABSTRACT
This paper introduces the design of logic circuits based exclusively on novel magnetoelectronic devices. Current signals are steered by 2x resistance change switching while operating with sub-100 mV voltage pulses for power and synchronization. The inherent memory of the devices results in fully pipelined nonvolatile logic. We demonstrate that co-optimization of the devices, circuits and logic can achieve ultra-low energy-per-operation for design examples.

Categories and Subject Descriptors
B.7.1 [Integrated Circuits] Types and Design Styles
– Advanced Technologies

General Terms
Design

Keywords
Magnetic Logic, MRAM, Spin-Transfer Torque, Emerging Circuits and Devices, Spintronics

1. INTRODUCTION
A number of new magnetoelectronic devices and circuits have been proposed that hold the promise of low power and nonvolatile computation and memory. Magnetic quantum cellular automata (MQCA), for example, have demonstrated promising results regarding logic computation [1] and driving fanout [2]. Although this initial work has met with some success, challenges remain in making commercially-viable ICs with this technology. Many of these MQCA and magnetic nanowire schemes require the use of an external magnetic field, which is either static for biasing or dynamic for clocking. On-chip generation of this field by means of passing current through a clad copper wire has been shown to require currents as high as 680mA [3].

Other approaches have paired CMOS and magnetic tunnel junctions (MTJs) in schemes where fanout is driven electrically without external clocking fields. For example, logic gates have been built with switchable MTJs. But these gates have required CMOS circuits to detect and amplify small resistance changes [4]. Other hybrid approaches have used CMOS for computation paired with MRAM-based retention latches [5]. The schemes that use variants of MRAM cells require tight integration with CMOS and raise the cost of fabrication compared to a non-heterogeneous technology. Furthermore, powering CMOS in these hybrid circuits prevents the low operating voltages possible with magnetoelectronics alone. It is recognized that many existing approaches to magnetoelectronics may have worse energy-delay products than CMOS once the circuit overhead is considered [6].

Other styles of magnetic logic based exclusively on spintronic devices have been proposed [7]. Here, we present *mLogic*, one such nonvolatile circuit scheme. Tight integration with CMOS is not necessary because CMOS would only be used off-chip to generate global clocking signals. The proposed circuits do not use external magnetic fields for powering or clocking. In fact, the device is insensitive to ambient fields. Our device design ensures digital operation and thermal stability at room temperature. We also demonstrate low energy mLogic circuits even with low tunnel magnetoresistance (TMR) switching ratios. Additionally, we show that low energy and high throughput systems are possible by exploiting the inherent non-volatility of the devices in the circuits and system architecture.

In section 2 we describe a novel magnetoelectronic device that is co-optimized with circuits and logic gates. This device enables the low-energy transistor-independent mLogic circuits. The operation of mLogic circuits is explained in section 3. The device and circuit properties enable new logic and datapath architecture optimizations. We describe these in section 4, and discuss our design methodology and some key opportunities for CAD research in this area.

2. MCELL DEVICE
mLogic uses a magnetoelectronic device called the mCell. The mCell is a slightly modified MRAM device [8] and acts as a current-controlled switchable resistor. Like MRAM, this device is nonvolatile so the resistance state is memorized even when the controlling current is shut off. It is novel due to electrical insulation between separate read and write paths.

Figure 1 shows the four-terminal mCell device that consists of a write path (w^+-w^-) and electrically-isolated read path (R-R'). Its write path consists of a low-impedance ferromagnetic metal connecting the bottom electrodes. These electrodes are magnetic, with the moments at the two opposite ends oriented in opposite directions by a pinning mechanism. This opposite magnetization in the write path causes a region of rotating magnetization called a domain wall to form.

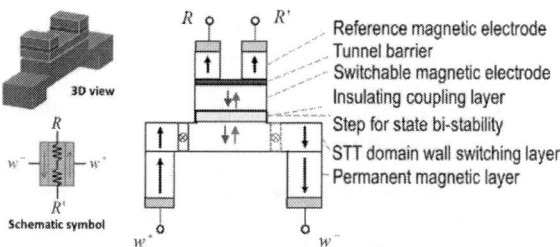

Figure 1. Schematic symbol, 2D and 3D drawings of mCell

The domain wall can be moved through the write path by a mechanism called spin-transfer torque. When a current is sent through the write path, the electrons become spin-polarized by the source terminal. In Figure 1, for example, conventional current flow from w^- to w^+ is equivalent to electron flow from w^+ to w^-. The electrons' spins are polarized "up," by the fixed magnetization of the w^+ terminal. When the electrons reach a region in the write path of opposing magnetization, a torque is exerted on the local moments that causes them to flip 180 degrees, which moves the domain wall in the direction of electron flow. As a result, given the direction of the current, this layer can be "programmed" to have magnetization oriented "up" or "down".

The programming of the write path also programs the magnetization of the switchable layer in the read path via magnetic coupling. The coupling is mediated by exchange through the magnetic oxide, such that the mechanism is not predominantly dipolar. The programming orients the free layer's magnetization parallel or antiparallel to the fixed R and R' terminals. This results in two stable resistance states. These resistance states are nonvolatile, as removing power has no effect on the orientation of any layer's magnetization.

The domain wall position in the write path determines the resistance of an MTJ in the read path. In the read path, a fixed magnetic electrode and switchable magnetic electrode sandwich a tunnel barrier. Flipping the magnetization of the switchable magnetic electrode switches the resistance from its lowest stable value (R_{low}) to its highest stable value (R_{high}), or vice versa. This switching resistance phenomenon is known as tunnel magnetoresistance. The orientation of the magnetization on either side of a tunnel barrier affects electron scattering to produce a minimum resistance when the moments are parallel and a maximum resistance when antiparallel. Given today's technology, the resistance can change by a factor of 2-3x (i.e. 100%-200% TMR) [9]. Even with MTJ device advances, the switching ratio is not expected to surpass 10x with today's materials [10]. As a result, designing circuits that operate with low switching ratios are essential. Note that for the device configuration shown (Figure 1), the resistance is actually two tunnel junction resistances in series. However, the tunnel barrier under one of the contacts may be etched through and a purely ohmic contact deposited to lower the resistance (Figure 2).

Figure 1 also shows that the write path has a step to a thicker region, due to the deposition of the magnetic coupling layer and free layer. This change in cross-section introduces an energy barrier that ensures bi-stability of the domain wall position. This bi-stability results in the digital switching of device resistance and stable operation at room temperature. For a 10 nm wide device, the energy barrier is greater than $50k_BT$. Increasing step height or anisotropy strength could maintain a sufficient energy barrier for even narrower devices.

In summary, the resistance of the output read path is switched by the directionality of current through the input write path (Figure 2). Micromagnetic simulation of state switching is shown in Figure 3. The simulator solves the Landau-Lifshitz-Gilbert equation and uses the finite difference method.

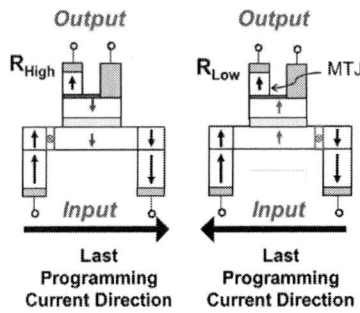

Figure 2. Programming nonvolatile resistance state is a function on electron flow direction

Figure 3. Micromagnetic simulation of domain wall motion and state switching. The color indicates magnetic polarization.

This device is very similar to existing MRAM devices. The key improvement is that there is electrical isolation between the write path and read path [8]. Like previous devices, the mCell can be manufactured with planar processes on sputtered metals.

Micromagnetic simulations of the mCell demonstrate a minimum switching current density of approximately 4 MA/cm² with a 2 ns pulse width. For a 3 nm thick, 35 nm wide nanowire, this corresponds to a switching current of 4 μA. Higher write currents lead to faster switching times (Figure 4). Key parameters of exemplary devices are listed in Table 1.

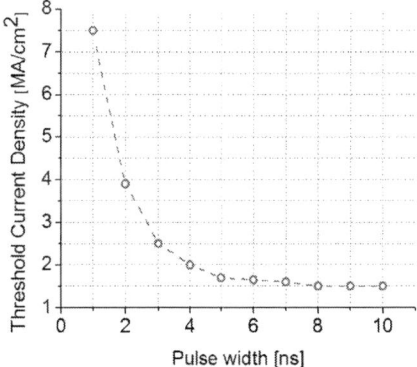

Figure 4. Micromagnetic simulation results for minimum required current pulse amplitude for switching as a function of pulse width

Table 1. Exemplary Device Parameters

Threshold Current Density [MA/cm²]	4
MTJ Resistance*Area [ohm*μm²]	2
Tunnel Magnetoresistance Ratio	100%
Read Path Low Resistance [Ω]	1.25K
Read Path High Resistance [Ω]	2.5K
Write Path Resistance [Ω]	120

A Verilog-A compact model of the mCell was written to allow SPICE simulation of circuits. The base of the model was derived from a first-order approximation of the underlying physics and Verilog-A model parameters were adjusted so that the SPICE simulation would closely match the more accurate micromagnetic model (Figure 5). Simulated current and resistance behavior closely matches with measured data from other groups [8], ensuring the simulation results are reasonable.

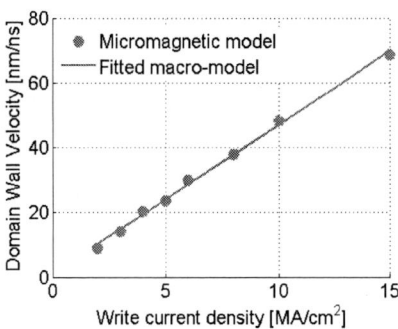

Figure 5. Matching micromagnetic and SPICE simulated domain wall speed as a function of write current density

3. mLogic Circuits

3.1 Current Steering

Switching resistance ratios for MTJs are six to seven orders of magnitude lower than in CMOS devices. With worse R_{high}/R_{low} ratios, the leakage power of CMOS style circuits increases dramatically. A new approach to circuit design using current steering is needed. mLogic current steered circuits are still energy efficient even with switching ratios of 2x-3x. Figure 6 illustrates a mLogic buffer with a logic '1' and logic '0' output. The pull-up and pull-down networks are connected to symmetric positive or negative voltages. The output node of this circuit is connected through a low impedance path to ground, represented in Figure 6 as an mCell with write path resistance of 120 ohms. The output current is

$$I = \frac{V\left(\frac{1}{R_{PU}} - \frac{1}{R_{PD}}\right)}{2k}, \quad k = 1 + R_{outpath}\left(\frac{1}{R_{PD}} + \frac{1}{R_{PU}}\right) \quad (1)$$

where V/2 is the magnitude of the matching positive and negative power supplies. R_{PU} and R_{PD} are the resistance of the pull-up and pull-down mCells in the driving gate. These resistances are controlled by the input logic value and differ by the R_{high}/R_{low} ratio to steer the output current into or out of the fanout load. The direction of this current is the output logic value.

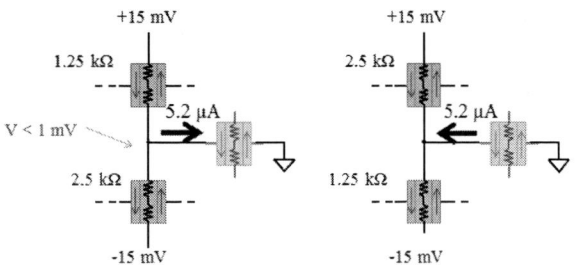

Figure 6. mLogic steering fanout current direction. Fanout is represented by greyed mCell.

By design, the magnitude of the output currents representing logic '1' and logic '0' are large enough to ensure writing of the fanout devices in the presence of variation. It is not necessary to have fine control over write current magnitudes as logical value is based on current direction, as long as the magnitude is above the threshold. Margining against variation can be achieved by raising power supply voltages which raises the output currents. As long as the power supply voltage is below the breakdown

voltage of the tunnel barrier, margin can be added to overcome variation in the MTJ, STT write path, and supply noise.

In addition to the buffer shown in Figure 6, a variety of other pull-up and pull-down networks can be designed with MTJs to steer the directionality of currents. A NAND gate is shown in Figure 7. MTJ areas are sized so that the gate can steer the necessary currents.

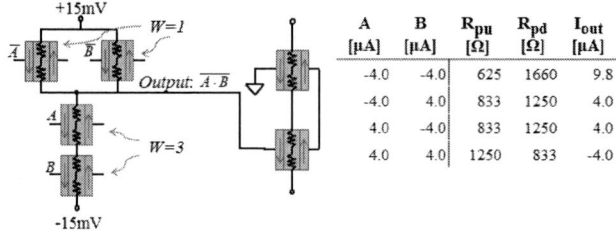

A [μA]	B [μA]	R_{pu} [Ω]	R_{pd} [Ω]	I_{out} [μA]
-4.0	-4.0	625	1660	9.8
-4.0	4.0	833	1250	4.0
4.0	-4.0	833	1250	4.0
4.0	4.0	1250	833	-4.0

Figure 7. NAND circuit driving fanout inverter and associated truth table. The pulldown mCells are sized three times larger than the pullup mCells.

The output current is used to cause magnetic state and resistance switching in the fanout mCells. The fanouts are connected in series through their write paths so that each receives the full programming current. The serial connection of fanout also prevents the shunting of current through unbalanced loads in parallel paths.

3.2 Gain

The magnitude of the output current is designed to be greater than that of the input current. This characteristic of mLogic gates is the equivalent of "gain" in CMOS circuits. This gain ensures a gate is capable of driving fanout and that the circuit operates properly under noise and variation. Figure 8 demonstrates the gain of an inverter in the form of a hysteresis loop. For input currents below the threshold, the mCell states are maintained and the output current is unchanged. When the switching threshold is reached, after a brief transient period the output current settles to a value determined by the supply voltage and the MTJ resistances. For a designed range of input currents, I_{OUT}/I_{IN} is greater than one. As noted earlier, this "gain" may be raised by raising the supply voltage, which increases the output current.

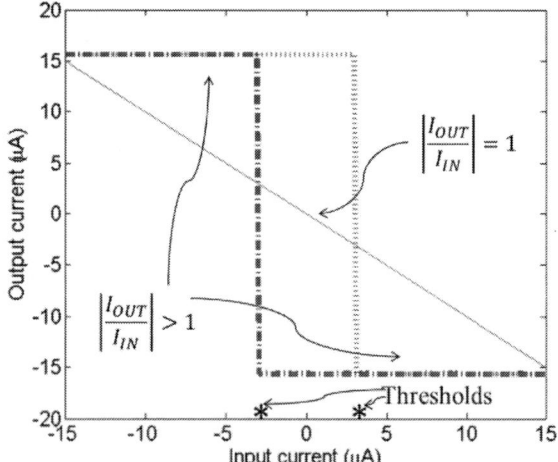

Figure 8: Hysteresis loop of an inverter with VDD=±45mV. Switching thresholds are denoted by "*".

3.3 Powering and Synchronization

A CMOS transistor in the off-state is nearly an open switch and has little static power consumption. In contrast, current steering with mCell MTJs results in the power supply sourcing 10's of microamps of static current per gate, albeit from sub-100mV supplies. For maximum power efficiency, mLogic gates should not be powered until after each input has received enough current to switch. mLogic gates are non-volatile and do not need to be powered during a write. Similarly, the gates can cease to be powered once their outputs source enough current to switch fanout. Selectively powering the gates is accomplished by strobing the matched positive and negative power supplies. These clocked power supplies are called *pClocks*. When a pClock is enabled, current is steered through the gate to produce an output current to program the fanout.

Figure 9 illustrates the switching of a NAND gate evaluated on the 'A' phase of pClock. A separate non-overlapping pClock powers the fanout inverter. The sign of the output current represents the logical sense of the signal (positive current is a "1" and negative current a "0"). The resistance state switching of a fanout mCell due to this output current is also shown. Note that the resistance state remains stable even between pClock pulses. Like conventional MRAM, the mCells have practically infinite retention time.

The NAND gate and the inverter on its fanout are clocked by separate pClocks. It is preferable for alternating stages of logic to be clocked and powered on non-overlapping phases of pClocks. This is shown in Figure 10 where the three NAND gates in the second stage of logic all are connected to pClk2. In this way, many gates share a common set of pClock signals and as few as two pClocks could power an entire chip.

When none of the pClocks are enabled, the nonvolatile circuit retains state and enters a zero power sleep mode. When the pClock is reenergized, the circuit instantaneously exits the sleep mode and logic computation continues as normal. No additional overhead is required to support this sleep mode as it would be in a CMOS chip.

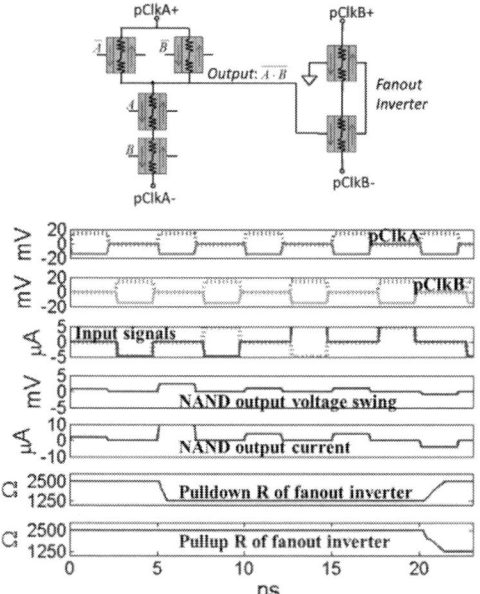

Figure 9. SPICE simulation of NAND gate driving fanout load (top); evaluation of gate with power clocks, input currents, output current, and resistance state of fanout mCell (bottom).

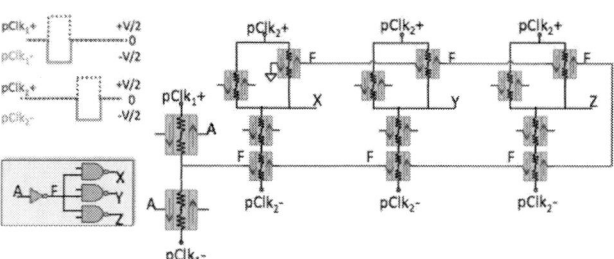

Figure 10. Inverter steers current through NAND fanout gates; driving gate and fanout powered by non-overlapping pClocks.

pClock signals are distributed globally and shared by many gates, much like VDD and ground are common to all gates in a power domain on a CMOS chip. The global nature of the pClocks allows one to use off-chip power regulation circuitry to generate strobed supplies, thus avoiding the need to tightly integrate CMOS with the magnetoelectronic devices. Two separate die, one for CMOS and one for mLogic, may be used because only limited interconnections are necessary. The interconnections can be done with wire bonding, flip-chip packaging, or perhaps other 3D integration techniques. Each die is simpler and less expensive to manufacture than one heterogeneous die due to the reduced number of process steps. If a heterogeneous die is preferred, integration is possible as well because the materials and processing are similar to that of conventional MRAM, which is generally considered to be CMOS compatible.

3.4 mLogic Gate Level Behavior

mLogic gates have a number of characteristics that enable efficient implementation of logic functions. Inversion can be done without any additional gates by altering the routing to swap the direction of current on the w^+ and w^- terminals. This is possible because the logical sense of a signal is based on the direction of write current. Free inversion can simplify implementation of logic functions.

For some logic gates, CMOS consumes less energy. For example, a two-input NAND gate with a fanout of four consumes 1.03 fJ/switch in 32 nm CMOS and 1.71 fJ/switch in mLogic (with VDD = 0.9V in CMOS and ±35 mV in mLogic). The 10-20 ps switching speed of a transistor is also significantly higher than that of an mCell. However, these numbers understate the energy advantages of nonvolatile logic. mLogic dissipates significantly less energy in interconnect because with current-based signaling the voltage swing on an output node is generally no greater than a few millivolts. For example, a CMOS inverter consumes 23 fJ to drive 100 μm of metal. This is dominated by a 23 fF load and a tapered buffer chain to drive the load. But an mLogic inverter consumes just 1.4 fJ. The output swings just 3.2 mV with a ±25 mV pClock driving the 700 ohm wire load.

With memory inherent to every mCell, state storage is free. This enables pipelined digital blocks. Simple deep pipelining of mLogic offsets the moderate switching speed of individual mCells to provide high system-level throughput (Figure 11). In CMOS, pipelining registers are generally inserted about every 12 logic stages [11]. In mLogic, it is possible to pipeline every other logic stage, achieving 6x the speedup of coarser pipelining. This aggressive pipelining capability comes with zero additional area, power, and timing overhead. In contrast, a CMOS flip flop may have a timing overhead of three gate

489

delays, use nearly 20 transistors, and consume almost 3 fJ per switch in a 32 nm process.

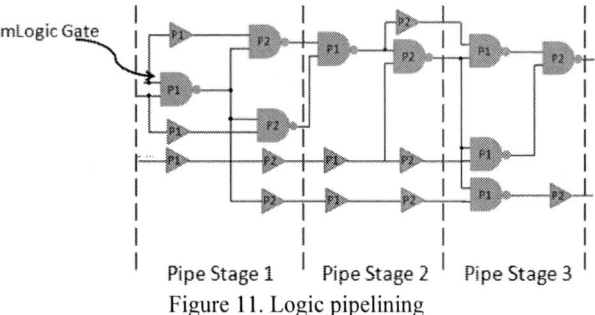

Figure 11. Logic pipelining

4. mLogic Architectures

Optimizing logic and architecture for the capabilities of mLogic allow us to obtain system level energy and speed benefits above what is achievable by the gates alone. The ability for each gate to store state is even more significant if this granular memory can be interspersed with the logic, enabling new types of datapaths and architectures. Ultimately, this would necessitate a new approach to logic synthesis. In this section we discuss the specific design of a hardware architecture that takes advantage of the memory state in each mLogic gate to achieve low energy-per-operation at high throughput.

Bit-serial logic styles are particularly suited to exploit the state present in each mLogic gate. Traditionally, adders are implemented with all bits of the summands added concurrently. The critical path of this circuit is generally from the lowest order bit to the highest order bit. In contrast, bit-serial arithmetic [12] operates on a single bit position of the summands per clock and all the bits pass through the same addition gate in sequential order (Figure 12). The carry-out from the addition of one bit position is held in a sequential delay element and fed back into the gate to compute the next bit position.

The advantage of bit-serial logic is that an adder for an arbitrarily wide data word can be fully implemented by one single bit wide adder, yielding significant area savings. Further, multiple adders can be efficiently composed into larger datapaths because only a single wire per adder needs to be routed, instead of a bus as wide as the data word. This arithmetic style is possible in CMOS, but flip-flops and the high timing overhead for clocking relative to the short switching time of a CMOS gate makes this approach inefficient with traditional logic technologies.

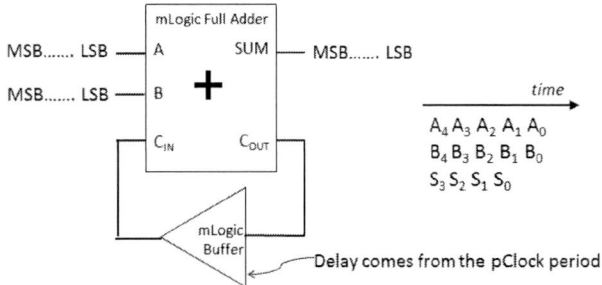

Figure 12. Bit-serial adder with mLogic

Multiplication can also be done compactly using constant coefficient bit-serial logic (Figure 13). Standard constant coefficient multipliers bit-shift the multiplicand to multiply by a power of two and add these products together. mLogic bit-serial bit-shifting is equivalent to a delay element such as a buffer. Multiplication is equivalent to delay because in the bit-serial representation the bits of the data word pass through the same gates sequentially.

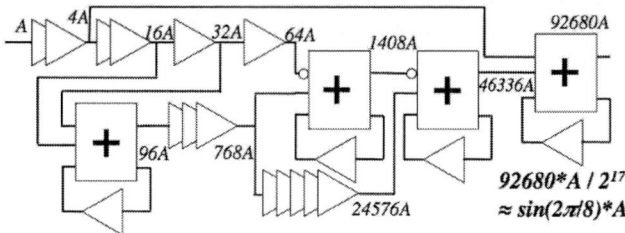

Figure 13. 16b precision bit-serial twiddle multiplier with 22 gates.

Using the building blocks of a bit-serial adder and multiplier, highly efficient FFT hardware can be assembled. The core computation unit of an FFT architecture is the butterfly. This butterfly consists of complex-valued twiddle factor multiplication and an addition and subtraction. In general, FFT hardware architectures use a number of butterfly units to match the application requirements for throughput, power and cost. Higher throughput is enabled by streaming reuse of many duplicated butterfly units per chip while area efficiency is enabled by iterative reuse of fewer butterfly units over several cycles [13].

With mLogic circuits, the FFT hardware tradeoffs between area and latency are recast, and architectures, datapaths, and arithmetic circuits with improved area, throughput and power are possible. Extremely compact pipelined butterfly adders and multipliers can be built with mLogic allowing hardware to efficiently exploit the data parallelism available. The bit-serial datapaths enable an entire 512 point FFT to be implemented without any iterative reuse. This approach would be area inefficient without the sequential bit-serial circuits. With mLogic, the datapath building blocks are simpler and just as importantly the interconnections between the blocks are reduced to single bit wide buses that have swings under 5 mV due to current-based signaling.

To evaluate the efficacy of the mLogic bit-serial FFT hardware, a virtual prototype was built using commercial EDA tools and custom scripts. This illuminated several shortcomings of standard design flows. As discussed prior, there are no tools to synthesize logic with state, including bit-serial datapaths. As a result, the automated design flow started with a gate level Verilog description. In order to fully pipeline the design, the length of all logic paths must match because each pClocked gate delay is a pipestage. A Perl script was used to post process the Verilog netlist to balance the paths. Second, automatic placement, routing and physical synthesis tools are not suited for mLogic. mLogic fanout is connected serially, not in parallel as with voltage based signaling. Hollow LIB abstracts were created for the tools because the LIB format could not express the true electrical and logical behavior of the gates, nor could the existing placement and routing tools do timing or gate sizing optimizations. All verification was done manually using SPICE on layout extracted netlists. SPICE simulations of the mLogic hardware revealed reduced write currents due to the extracted routing resistances. pClock voltages were raised to compensate.

Two voltage domains were used for energy savings, with ±50 mV strobed to the adders and ±20 mV to the buffers.

The simulations showed that bit-serial streaming reuse in mLogic with mCell-enabled pipelining allows throughput of 6.5 million FFT/sec at 23.1 nJ/FFT with extraction (Figure 14). To put these numbers in context, this is 6.7 times less energy and 1.7 million times the throughput of a seminal sub-threshold CMOS implementation optimized for minimum energy [14]. This increase in throughput comes at just over twice the device count (1.5 million mCells to 627,000 transistors). Sub-threshold CMOS is selected as a point for comparison because like mLogic it is optimized for extremely energy constrained systems. The comparison has not been normalized for the lithographic node because of the numerous differences between silicon and magnetic fabrication.

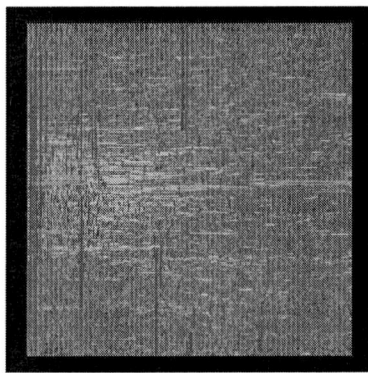

Figure 14. mLogic FFT layout

These design exercises have illustrated the significant role that CAD research plays in the success of any new nano-device or implementation technique. The advantages of the nanotechnologies, such as mLogic, must be viewed in a system context rather than in a gate/circuit level comparison. New front-end design tools are needed to best exploit granular logic and memory. With state in each gate, new synthesis approaches are possible. To verify these sequential gates, new tools would be needed as well. For back-end design, the new electrical properties of the mLogic gates require advances in physical synthesis algorithms and electrical verification software.

5. CONCLUSION

We have demonstrated mLogic, a new logic family composed of modified MRAM devices. This logic family is unique because it performs computation and memory entirely without tightly-integrated transistors. Energy efficiency is achieved despite 2X switching ratios by using current-steering and operating at sub-100 mV supplies. Energy*delay benefits at a system level are achieved by co-optimizing logic and architectures to fully exploit device properties. Nonvolatile memory state in each gate enables full pipelining with minimum overhead as well as new efficient datapath structures. Though not a replacement for all CMOS, these results indicate that mLogic is a prospective technology for low energy and low voltage niches that are unmet by current implementation technology.

Acknowledgements

This material is based upon work supported by the National Science Foundation Graduate Research Fellowship under Grant No. 0750271 and a Qualcomm Innovation Fellowship.

References

[1] Imre, A.; Csaba G.; Ji, L.; Orlov, A.; Bernstein, G.H.; Porod, W. "Majority Logic Gate for Magnetic Quantum-Dot Cellular Automata," *Science*, 311 (5758), pp.205-208, Jan. 2006

[2] Varga, E.; Orlov, A.; Niemier, M.T.; Hu, X.S.; Bernstein, G.H.; Porod, W.; , "Experimental Demonstration of Fanout for Nanomagnetic Logic," *Nanotechnology, IEEE Transactions on* , vol.9, no.6, pp.668-670, Nov. 2010

[3] Alam, M.T.; Siddiq, M.J.; Bernstein, G.H.; Niemier, M.; Porod, W.; Hu, X.S.; , "On-Chip Clocking for Nanomagnet Logic Devices," *Nanotechnology, IEEE Transactions on* , vol.9, no.3, pp.348-351, May 2010

[4] Matsunaga, S.; Hayakawa, J.; Ikeda, S.; Miura, K.; Endoh, T.; Ohno, H.; Hanyu, T.; , "MTJ-based nonvolatile logic-in-memory circuit, future prospects and issues," *Design, Automation & Test in Europe Conference & Exhibition, 2009. DATE '09.* , vol., no., pp.433-435, 20-24 April 2009

[5] Ohno, H.; Endoh, T.; Hanyu, T.; Kasai, N.; Ikeda, S.; , "Magnetic tunnel junction for nonvolatile CMOS logic," *Electron Devices Meeting (IEDM), 2010 IEEE International* , vol., no., pp.9.4.1-9.4.4, 6-8 Dec. 2010

[6] Fengbo Ren; Markovic, D.; , "True Energy-Performance Analysis of the MTJ-Based Logic-in-Memory Architecture (1-Bit Full Adder)," *Electron Devices, IEEE Transactions on* , vol.57, no.5, pp.1023-1028, May 2010

[7] Behin-Aein, B.; Datta, D.; Salahuddin, S.; Datta, S; , "Proposal for an all-spin logic device with built-in memory," *Nature Nanotechnology*, 5, 266 – 270, Feb. 2010

[8] Fukami, S.; Suzuki, T.; Nagahara, K.; Ohshima, N.; Ozaki, Y.; Saito, S.; Nebashi, R.; Sakimura, N.; Honjo, H.; Mori, K.; Igarashi, C.; Miura, S.; Ishiwata, N.; Sugibayashi, T.; , "Low-current perpendicular domain wall motion cell for scalable high-speed MRAM," *VLSI Technology, 2009 Symposium on* , vol., no., pp.230-231, 16-18 June 2009

[9] Parkin, S.S.P.; Kaiser, C.; Panchula, A.; Rice, P.M.; Hughes, B.; Samant, M.; Yang, S.-H., "Giant tunnelling magnetoresistance at room temperature with MgO (100) tunnel barriers," *Nature Materials*, 3, pp.862-867, Dec. 2004

[10] Butler, W.H.; Zhang, X.-G.; Schulthess, T.C.; MacLaren, J.M., "Spin-dependent tunneling conductance of Fe|MgO|Fe sandwiches", *Phys. Rev. B.*, 63, 054416, Jan. 2001

[11] Chandrakasan, A.; Bowhill, W.J.; Fox, F., *Design of High-Performance Microprocessor Circuits*, Wiley-IEEE Press, Oct. 2001

[12] H J Sips. 1986. "Bit sequential arithmetic for parallel processors," In *Advanced computer architecture*, Dharma P Agrawal (Ed.). IEEE Computer Society Press, Los Alamitos, CA, USA 93-106.

[13] Milder, P.A.; Franchetti, F.; Hoe, J.C.; Puschel, M.; , "Formal datapath representation and manipulation for implementing DSP transforms," *Design Automation Conference, 2008. DAC 2008. 45th ACM/IEEE* , vol., no., pp.385-390, 8-13 June 2008.

[14] Wang, A.; Chandrakasan, A. , "A 180-mV subthreshold FFT processor using a minimum energy design methodology," *Solid-State Circuits, IEEE Journal of* , vol.40, no.1, pp. 310- 319, Jan. 2005.

Future Cache Design using STT MRAMs for Improved Energy Efficiency: Devices, Circuits and Architecture

Sang Phill Park, Sumeet Gupta, Niladri Mojumder, Anand Raghunathan, Kaushik Roy

Purdue University, West Lafayette, IN 47907

{sppark,guptask,niladri,raghunathan,kaushik}@purdue.edu

ABSTRACT

Spin-transfer torque magnetic RAM (STT MRAM) has emerged as a promising candidate for on-chip memory in future computing platforms. We present a cross-layer (device-circuit-architecture) approach to energy-efficient cache design using STT MRAM. At the device and circuit levels, we consider different genres of MTJs and bitcells, and evaluate their impact on the area, energy and performance of caches. In addition, we propose micro-architectural techniques *viz.* sequential cache read and partial cache line update, which exploit the non-volatility of STT MRAM to further improve energy efficiency of STT MRAM caches. A detailed comparison of STT MRAM caches with SRAM-based caches is also presented. Our results indicate that the proposed optimizations significantly enhance the efficiency of STT MRAM for designing lower level caches.

Categories and Subject Descriptors

B.3.2 [**Hardware**]: Memory Structures—*Cache Memories*

General Terms

Design, Performance, Experiments

Keywords

Cache, Memory, Emerging devices, STT MRAM, Spin

1. INTRODUCTION

The ever-increasing gap between processor speed and main memory latency has driven the demand for larger on-chip caches in processors. Traditionally, on-chip caches in modern processors are implemented using static random access memories (SRAM). However, limited scalability, susceptibility to soft errors and high leakage power of SRAM pose challenges to high-density on-chip cache implementation. In order to address the limited scalability of SRAMs, several recent processors have adopted embedded dynamic RAM (EDRAM) in lower level caches. However, vulnerability to soft errors and significant standby power of EDRAM caches due to high cell leakage are still major bottlenecks in on-chip cache design [4]. To cope with the above problems, there has been significant research directed towards several alternative embedded memory technologies [3].

Among various candidates, spin-transfer torque magnetic RAM (STT MRAM) is considered as a promising technology that can offer desirable memory attributes such as high endurance, non-volatility, soft error immunity, zero standby power and high integration capability. More importantly, its compatibility with CMOS processes makes it an attractive

vehicle to realize high-density low-power embedded memories in scaled technologies [6]. However, higher write latency and write energy requirements, compared to the traditional embedded memories such as SRAM, are major issues with STT MRAM [17]. These drawbacks can preclude direct deployment of STT MRAM in level-1 (L1) caches that require fast read and write operations. However, in lower level caches such as the level-2 (L2) or last-level (LL) caches, the low leakage and high density of STT MRAMs can be more effectively utilized to replace SRAMs [7, 17].

Some previous works have explored the use of STT MRAM in the cache hierarchy, primarily through architectural techniques such as hybrid caches, write buffers, *etc.* [7, 18]. While these efforts have proven the potential of STT MRAMs, we believe that deriving highest benefits from STT MRAM requires device/circuit/architecture co-design. In this work, we explore different genres of magnetic tunnel junction (MTJ) stacks and bitcell configurations, and analyze their implications on the energy consumption and performance of STT MRAM caches under different cache utilizations. Furthermore, we present circuit/architecture co-design techniques that exploit the non-volatility of STT MRAM, not only in the standby mode, but also during dynamic cache operations. One of the consequences of non-volatility of STT MRAMs is that during column selection, the unselected columns do not consume any energy. This is unlike SRAM caches, in which the unselected bitcells need to be biased with voltages identical to the for the SRAM read operation to prevent disturb failures (known as half-select problem) [12]. Hence, in STT MRAMs, column selection is half-select-free, as a result of which only the selected columns consume energy. Based on this observation, we present two techniques — sequential tag-data access for reads and partial line update for writes — that significantly improve the energy-efficiency of STT MRAM-based caches.

In summary, we utilize device/circuit/architecture co-design to make STT MRAMs an attractive option for high density on-chip memory and to enhance the energy efficiency of STT MRAM caches. The key contributions of this work are as follows:

- We investigate the impact of various STT MRAM bitcells with different genres of MTJ stacks and bitcell configurations on total cache area, energy consumption and performance. We perform a detailed comparison of STT MRAM caches with respect to SRAM caches, based on *physical layouts* of the STT MRAM and SRAM bitcells.

- Exploiting the non-volatility of STT MRAMs, we propose a cache architecture that performs *partial cache line update* for cache writeback energy reduction. The technique does not incur any extra cache misses since it does not change the data flow between different cache levels.

- We propose a read energy reduction technique exploiting the non-volatility of STT MRAM. The technique is based on *sequential tag-data access*, and does not require signif-

- icant architectural modification.
- We also analyzed the total energy consumption of STT MRAM caches considering cache utilization during processor operations. We show that for lower level caches, STT MRAM caches are significantly more energy efficient than SRAM-based caches due to low utilization and low standby power.

2. RELATED WORK

Researchers have studied the power and performance characteristics of STT MRAM caches for general purpose processors. Wu *et al.* [18] have proposed SRAM-STT MRAM hybrid-cache architectures based on partitioning of caches into fast SRAM and slow STT MRAM regions. Xu *et al.* [17] have proposed an STT MRAM-based last level cache. In [17], the analysis is limited to the performance benefits arising from the higher integration density of STT MRAM. There have been efforts that address the large write energy of STT MRAM caches. Zhou *et al.* [10] proposed an early termination of write operations to reduce write energy of STT MRAMs. The value stored in the bitcell is sensed at the beginning of the write cycle, and if found to be identical to the new data, the write operation is terminated. In [7], a write biasing technique has been proposed in order to address the high write energy of STT MRAM in cache operation. This technique reduces the number of writebacks from L1 to lower level caches by biasing dirty cache lines to reside in L1 for a longer time. However, write-biasing in a typical size L1 cache can result in noticeable performance penalty due to an increase in L1 read miss rates. In previous work, the reported power and performance evaluations of STT MRAM caches are based on approximate estimation of STT MRAM characteristics such as area, energy and latency. In our work, the total cache area is accurately calculated from bitcell layouts based on λ-based design rules. The performance and energy consumption are evaluated using bitcell and circuit parameters obtained using physics based simulator that can comprehend different types of MTJ stacks [16] along with a circuit simulator. In addition, we consider different genres of MTJs and analyze their impact on cache performance and energy consumption. For addressing excessive write energy of STT MRAMs, we propose *partial cache line update* to avoid unnecessary overwrite of the same data, thus achieving write energy savings. Our technique does not incur cache miss increase or pre-evaluation of the stored data by exploiting the non-volatility of STT MRAMs and data redundancy in a multi-level cache hierarchy. Moreover, we show that read energy reduction can also be achieved by utilizing the non-volatility of STT MRAMs.

3. STT MRAM CACHE DESIGN

In this section, we first describe the basic operation of STT MRAM bitcells. Next, we explore alternative MTJ stacks and bitcell configurations, and evaluate their impact on the area, performance and energy of caches. Finally, we explore the dependence of the energy benefit of STT MRAM based caches on cache utilization, making a case for the use of STT MRAMs in the lower levels of the cache hierarchy.

3.1 STT MRAM Preliminaries

A conventional STT MRAM cell comprises of a magnetic tunnel junction (MTJ) and an access transistor in series (Figure 1 (a-b)). The MTJ contains a pinned layer and a free layer separated by a dielectric layer (e.g. MgO). The

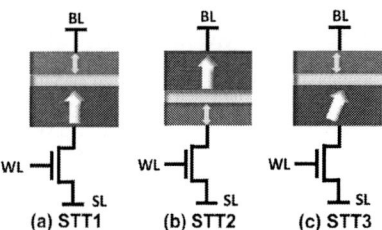

Figure 1: Schematics of an STT MRAM bitcell (a) in the standard-connected configuration (b) in the reverse-connected configuration and (c) with tilted magnetic anisotropy

pinned layer has a fixed magnetization, and the free layer is programmable by changing its magnetic orientation. The resistance of the MTJ depends on the relative magnetization of the free layer with respect to the pinned layer. Parallel magnetization of the free layer with respect to the pinned layer leads to a lower resistance (R_P) compared to the resistance in the anti-parallel state (R_{AP}). The two resistances of the MTJ define the binary states of the memory cell. A read operation is performed by sensing resistance difference of the two binary states. A write operation is performed by passing a current (I_W) through the bitcell that exceeds a critical current (I_C). The direction of (I_W) determines the final magnetization of the free layer (*i.e.*, parallel or anti-parallel states of the MTJ) [16].

3.2 STT MRAM Bitcell Design: Devices and Circuits

Different types of MTJ stacks [1, 9] and bitcell configurations [2] provide several design choices, and can result in substantially different bitcell characteristics. Before exploring these choices, we first discuss design considerations of STT MRAM bitcells. A conventional MTJ [1] has a large switching current density requirement, and the requirement increases dramatically with lower switching delay [2]. The large switching current requirement for fast write operation is one of the major challenges for energy-efficient STT MRAM design. In order to address the excessive switching current requirement, an MTJ with tilted magnetic anisotropy (TMA) has been proposed in [9]. Tilting the direction of the pinned layer, by a larger angle than what stochastic thermal noise can provide, leads to a thermal-noise-independent non-zero initial angle for precessional switching. As a result, the switching current overdrive and switching delay can be reduced significantly [9]. In our work, we consider three different bitcell designs shown in Figure 1: (i) a standard-connected configuration where the access transistor is connected to the pinned layer, (ii) a reverse-connected configuration where the access transistor

Table 1: Bitcell parameters of three STT MRAMs in Figure 1

Bitcell Type	STT1	STT2	STT3
Area*($F^2, F = 32nm$)	56.0625	44.85	34.5
P('0') read current(μA)	143	144	62
AP('1') read current(μA)	71	82	20
Read voltage(V)	0.19	0.24	0.32
P->AP write current(μA)	367	159	126
P->AP critical current(μA)	140	140	40
AP->P write current(μA)	316	316	90
AP->P critical current(μA)	287	287	82
Access transistor width(nm)	263	202	144**
Bitcell layout aspect ratio*	1.70	1.36	1.04**

* 6T-SRAM bitcell area and aspect ratio are $176F^2$ and 2.75, respectively.

** Cell area is limited by metal to metal pitch

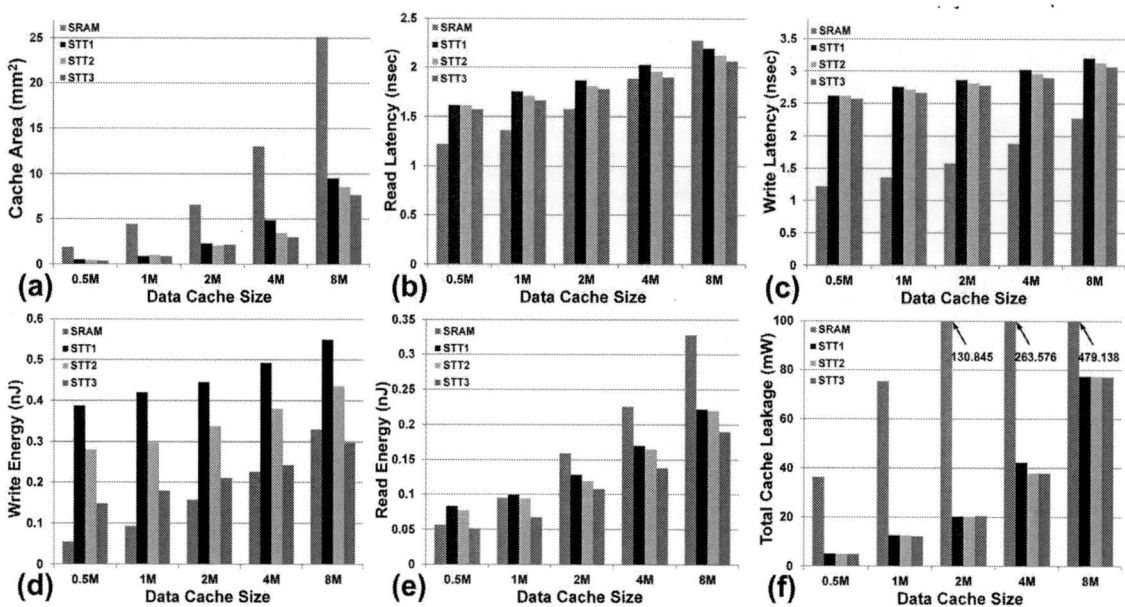

Figure 2: (a) Area requirement of SRAM and STT MRAM based caches (4-way, 64B cache line, B=Byte, M=Mega Byte) in mm^2, (b) read latency and (c) write latency in ns, (d) read energy and (e) write energy in nJ per operation, and (f) leakage power in mW

is connected to the free layer, and (iii) a configuration with tilted magnetic anisotropy. The bitcells are designed to meet the same specification with write error rate of 10^{-9}, switching time of $2ns$, 10% write margin (defined as $(I_W - I_C)/I_C$) and 50% read disturb margin (defined as $(I_C - I_{READ})/(I_C)$). We compare the characteristics of these three design options using manual layout and device simulations. In the simulations, MTJs with a free layer size of 64x64x3 nm^3 are modeled using the Non-equilibrium Green's function (NEGF) formalism [16] to obtain the electronic transport and spin transfer torque characteristics. Modeling of switching dynamics of the MTJ is carried out using the Landau-Liftshitz-Gilbert (LLG) equation with an STT term, and the MTJ models are calibrated against measurements [16, 14]. For the access transistor, $32nm$ predictive technology models (PTM) are used. In addition, a conventional thin-cell layout [11] of an SRAM bitcell is designed for comparison. The SRAM cell is designed to achieve a read access time less than $200ps$ with 128 bitcells per bitline.

Table 1 presents the bitcell parameters obtained from our evaluations. In comparison to the standard-connected bitcell, substantial reduction in write/read current along with reduction in the size of access transistors is observed in the reverse-connected bitcell and the cell with TMA. In the case of the reverse-connected bitcell, a smaller sized access transistor can provide sufficient current drive-ability due to non-source-degenerated transistor operation during P to AP switching [2]. Furthermore, the source degeneration of the transistor operation during AP to P switching reduces excessive overdrive current. In the case of the cell with TMA, the access transistor size can be further reduced due to the significantly lower critical switching current requirement of the MTJ with TMA. The size of the access transistor is a critical parameter in determining the unit cell area of STT MRAM. In the case of the two STT MRAM cells with conventional MTJs, the bitcell width is determined by the width of the access transistor. However, the bitcell width of the STT MRAM with TMA is limited by the metal to metal pitch rather than the size of the access transistor.

3.3 STT MRAM Cache vs. SRAM Cache

In this section, we evaluate STT MRAM caches based on the various bitcells presented in Figure 1 and Table 1. Performance and energy consumption of the arrays can vary, not only with different bitcell characteristics, but also with array parameters, such as capacity, the number of rows and columns, *etc.* [8]. A cache comprises of multiple arrays for storing tags and data bits. In conventional on-chip caches, both the tag and data arrays are implemented using SRAM. On the other hand, in the proposed STT MRAM caches, the tag arrays are implemented using SRAM and data arrays are implemented using STT MRAM. This is due to the fact that the write latency of STT MRAM may not be suitable for tag array operation, which requires frequent and fast updates of status bits and history bits [7]. In order to estimate the overall cache latency, area and energy consumption of the STT MRAM cache, we modified the CACTI 6.5 simulator [8] to consider (i) analog read circuits in STT MRAM data arrays (ii) SRAM-based tag arrays along with STT MRAM data arrays, and (iii) the bitcell layout geometries to optimize the array aspect ratio.

Figure 2 (a) compares the area requirements of caches designed with SRAM and STT MRAM. It is clearly shown that STT MRAM caches have a much higher integration density than SRAM cache. However, the total cache area does not fully reflect the area advantage of STT MRAM bitcells shown in Table 1, due to the area required for SRAM-based tag arrays and peripheral circuits in the STT MRAM caches. For instance, the STT MRAM bitcell with TMA has an approximately 5X smaller footprint in comparison to the SRAM bitcell. However, the 2MB caches based on the three types of STT MRAMs require 2 to $2.3mm^2$, which is slightly larger than the area requirement of 0.5MB SRAM cache ($1.9mm^2$).

The higher integration density of an STT MRAM-based data array can enable improved cache access latency and energy. As the cache area increases with capacity, the impact of wire delay on cache latency becomes larger. The SRAM cache latency increases rapidly with the capacity of the cache due to longer delays in the metal-lines such as the wordlines, the bitlines, and the data bus. However, the latency increase of STT MRAM-based caches is more graceful due to smaller cache area (Figure 2 (b,c)). It can be seen that the STT MRAM cache can be faster in read access when the cache capacity is larger than 4MB. Similarly, as a result of the graceful increase in the write latency of STT MRAM cache, the write latency gap between STT MRAM cache and SRAM cache becomes smaller with increasing capacity.

A similar trend can be observed in the case of dynamic energy. The energy dissipated in read operations in STT1 and STT2 is higher than that of SRAM due to power dissipation in the analog read circuits, despite 4X smaller total cache area. However, for larger capacity (above 1MB), the energy dissipation due to interconnects becomes dominant. Therefore, read dynamic energy is significantly lower in STT MRAM caches. During write operations, STT MRAM caches with conventional bitcells (STT1,STT2) dissipate significantly larger energy than SRAM based caches. However, the STT MRAM cache using TMA bitcells (STT3) shows significantly lower energy dissipation due to the lower write current requirement of the bitcell. As a result, write energy dissipation of STT3 is comparable to that of SRAM cache at a capacity of 4MB.

Figure 2(f) shows that the leakage power increases with cache capacity for both SRAM and STT MRAM caches. In case of SRAM cache, the leakage power increases drastically due to bitcell leakage. On the other hand, for STT MRAM cache, the increase in leakage is much lower due to zero standby power of the bitcells. The leakage power contribution in STT MRAM cache is primarily due to SRAM-based tag arrays and peripherals.

3.4 Cache Utilization and Energy Consumption

The contribution of active and leakage energy to total energy consumption is different for SRAM- and STT MRAM-based caches. The leakage energy in an STT MRAM cache is smaller than an SRAM cache even with 4 times larger capacity (at iso-area). On the other hand, the dynamic energy for a write operation is higher in an STT MRAM cache compared to an SRAM cache. It is important to note that the total energy dissipation in a cache depends on factors such as cache access patterns (number of read and write operations) and cache utilization (number of times a processor accesses the cache per unit cycle). The cache utilization is lower than 30% in today's processors [13]. Moreover, for lower levels of the cache hierarchy, the cache utilization is significantly lower than 30%. We have measured L2 cache utilizations for various SPEC2000 benchmarks based on the Simplescalar framework [15] with a 32KB L1 cache config-

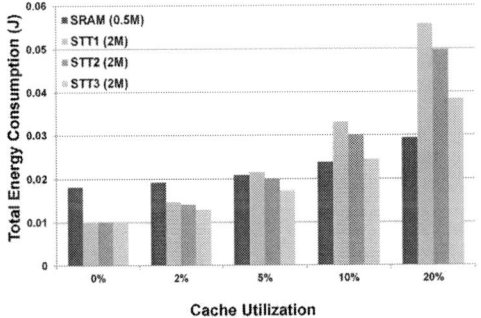

Figure 3: Total energy consumption of L2 caches at iso-area (0.5MB SRAM vs. 2MB STT MRAM)

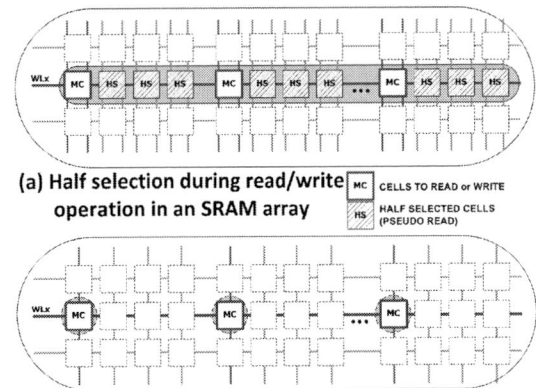

(a) Half selection during read/write operation in an SRAM array

MC CELLS TO READ or WRITE
HS HALF SELECTED CELLS (PSEUDO READ)

(b) No half selection problem in an STT-MRAM array

Figure 4: Column-selective read/write operations in SRAM and STT MRAM arrays

uration. Our simulation results also confirm low L2 cache utilization. For a majority of the benchmarks, L2 cache utilization is lower than 3%. The highest utilization, observed for the AMMP benchmark, is about 13%, and the average utilization across 16 benchmarks is only 2.2%.

As shown in Figure 3, a 2MB STT MRAM cache shows similar or lower energy consumption than a 0.5MB SRAM cache when the utilization is lower than 10%. Although the STT MRAM cache has significantly lower energy consumption at 0% utilization (leakage only), the energy dissipation increases drastically due to excessive write energy as the utilization increases. The results are obtained using the following conditions: read and write operation ratio of 2:1, 2GHz processor speed, and total simulation time of 1 billion processor cycles. Therefore, an STT MRAM cache can achieve high energy-efficiency along with high capacity in comparison to an SRAM cache, especially in lower levels of the cache hierarchy due to the low cache utilization.

4. ENERGY-EFFICIENT STT MRAM CACHE DESIGN

One of the distinct advantages of STT MRAM, compared to SRAM, is *non-volatility* of bitcells. Interestingly, non-volatility can further improve energy efficiency in dynamic operation of STT MRAM caches. In this section, we first investigate the difference in array operations (in particular column selections) for STT MRAM and SRAM arrays. We then propose dynamic energy reduction techniques exploiting the non-volatility for read and write operations of STT MRAM cache, and evaluate their impact on the overall energy consumption.

4.1 Column Selection: SRAM vs. STT MRAM

In a conventional SRAM array, column selection is required for storing multiple words in a single row [12]. Since set associativity is common in modern caches, column selection in SRAM arrays is imperative. Furthermore, bit-interleaving can only be achieved by employing column selection. Bit-interleaving is a commonly adopted technique in SRAM arrays (1) to mitigate soft errors [12], and (2) to increase array density by bitline multiplexing [8]. In the column selection operation of an SRAM array, all unselected bitcells in the accessed row have to be under read mode to prevent unexpected bit flips, when a wordline is asserted. This phenomenon is commonly known as pseudo read or half selection [12]. Note that, in an STT MRAM array, the non-volatility of bitcells can eliminate the half selection problem.

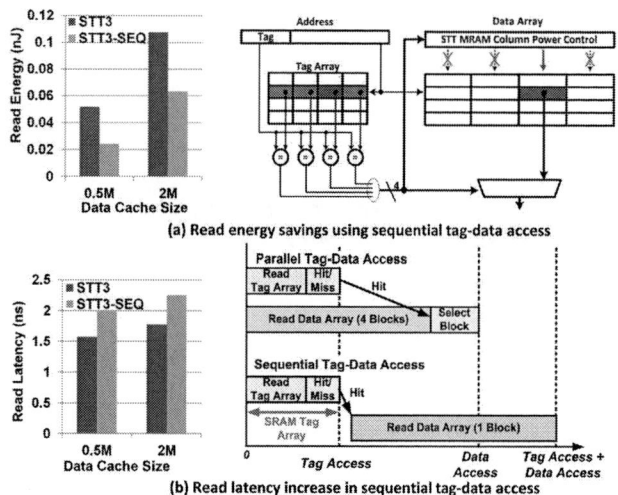

(a) Read energy savings using sequential tag-data access

(b) Read latency increase in sequential tag-data access

Figure 5: (a) Read energy savings and (b) Read latency increase in sequential tag-data access

As presented in Figure 4, the unselected bitcells can remain in standby mode, and hence, consume no energy during both read and write column selection operations. In the next two sub-sections, we will describe read and write energy saving techniques that are based upon this insight.

4.2 Read Energy Reduction in STT MRAM Cache

One challenge to enable energy-efficient column selection can be to identify the selected column address during cache read operation with minimal performance penalty. We observed that the proposed technique can be easily adopted in a cache implementing sequential tag-data access. Sequential tag-data access is often employed in large, lower-level caches to improve energy-efficiency during operation [8, 19]. In sequential tag-data access, a cache probes the tag array first, and identifies a hit or miss. Access to the data array occurs only when there is a cache hit, and only the sub-array storing the corresponding cache line in the data array is accessed. As a result, significant energy savings can be achieved.

This technique can be more energy-efficient in an STT MRAM cache due to half-selection-free column selection. In general, each sub-array in SRAM-based cache stores multiple cache lines in a row, in order to improve area efficiency or to employ bit-interleaving [8, 12]. Due to the half-selection issue, all cache lines in the row of the SRAM sub-array dissipate dynamic energy during read operations (due to precharging/discharging of bitlines). On the other hand, in a sub-array of an STT MRAM-based cache, only the bit-columns storing a single cache line consume energy as discussed previously. Figure 5 illustrates the proposed sequential tag-data cache access for an STT MRAM-based cache. The column address from the tag array is used to enable the selected bit-columns. The single cache line read access in STT MRAM can substantially lower read dynamic energy as shown in Figure 5 (a) (STT3-SEQ). The sequential tag-data access in cache increases the overall cache access latency [19]. However, the tag array has much smaller latency than the data array due to the smaller size of the tag array. Note that, in our proposed STT MRAM cache, the tag arrays are implemented using SRAMs and are much faster than STT MRAM data arrays. Hence, the overall latency increase due to the sequential access is not significant as shown in Figure 5 (b). Moreover, the latency increase in L2 cache does not have significant impact on the overall

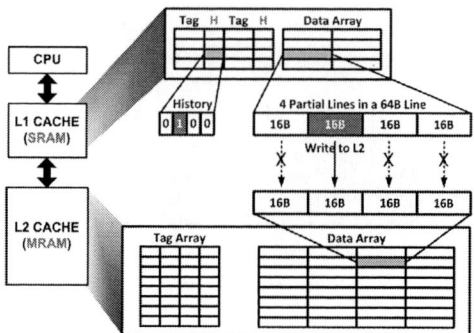

Figure 6: Partial cache line update

processor performance. Our simulation results show that, for a 2MB L2 cache, the increased latency due to sequential tag-data access results in less than 1% IPC (Instructions per cycle) reduction on average for 16 SPEC2000 benchmarks.

4.3 Write Energy Reduction in STT MRAM Cache

Similar to the read energy reduction technique described above, improvement in write energy efficiency of STT MRAM cache can also be achieved by exploiting half-selection-free column selection. We propose partial cache line update (PLU) to reduce cache writeback [5] energy consumption. This technique exploits data redundancy in a multi-level cache hierarchy as well as non-volatility of STT MRAM bitcells. In a writeback cache, writeback is performed when a dirty cache line in the L1 cache needs to be replaced by a new cache line. Hence, the dirty line has to be written into the L2 cache. In general, a cache line consists of multiple processor words in order to take advantage of spatial locality, and the size of a cache line is the unit data size in a cache. As a result, the entire cache line is written during writeback operation, even if there is only a single word that might have changed in the cache line.

Figure 6 presents the proposed PLU STT MRAM cache architecture. Each cache line is partitioned into n partial lines in order to utilize the energy-efficient column selection of STT MRAM arrays ($n = 4$ in the given example). During writeback from the SRAM L1 cache, only the partitions in the cache line that have been updated by the processor (1 out of 4 partitions in the example) are written to the STT MRAM L2 data array. The data in the remaining partitions are identical to the data already stored in the L2 cache. Therefore, writing the unchanged partitions into the L2 cache is unnecessary. The change of partitions can be tracked by using a history bit per partition. In the given example, 4 history bits to support 4 partitions are added into each tag in the L1 SRAM cache. The history data is used and reset whenever the corresponding cache line is written back into the L2 STT MRAM cache. During the PLU in the STT MRAM L2 cache, only the bitcells belonging to the updated partitions are written, while the other bitcells in the unchanged partition remain in standby mode. Note that

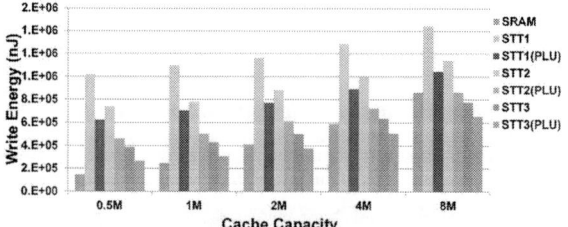

Figure 7: Average cache writeback energy reduction using PLU for 8 SPEC2000 integer benchmarks.

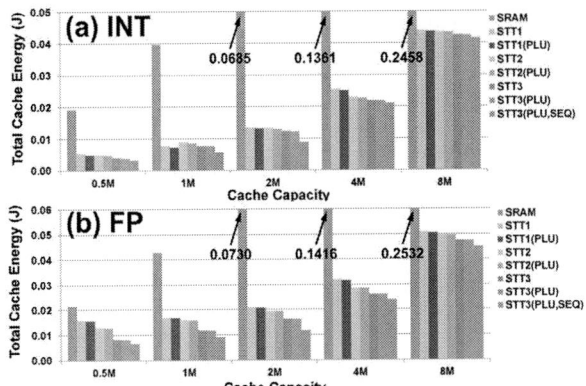

Figure 8: Total energy consumption of SRAM and STT MRAM L2 caches

an SRAM cache may not be able to take advantage of the proposed PLU technique due to the half selection problem.

We performed architectural simulation using 16 SPEC 2000 benchmarks with a processor configuration having 32KB L1 and 512KB L2 cache for 1 billion processor cycles. The results show that, during writeback operations, only 70% of cache line partitions are utilized on average when 4 partitions are used. In the case of 8 partitions per cache line, approximately 60% of the partitions are utilized. Figure 7 presents the total cache writeback energy consumption without and with PLU (for 8 partitions). STT MRAM caches with PLU show significant improvements in writeback energy compared to conventional STT MRAM caches (STT{1,2,3}). The results show that, for large STT MRAM caches (e.g., 4MB and 8MB) exploiting the PLU, the total writeback energy is comparable to an SRAM-based cache.

4.4 Total L2 Energy Consumption

In order to analyze the energy efficiency of STT MRAM caches in comparison to SRAM caches, we measured the total energy consumption of L2 cache including leakage, read and write energy over 1 billion cycles of processor execution. The results presented in Figure 8 are obtained by averaging L2 cache energy consumption across 8 integer and 8 floating point benchmarks. SRAM-based L2 cache shows the largest energy consumption compared to STT MRAM caches with the same capacity, due to the significant leakage energy of SRAM bitcells. The energy difference is further improved for larger cache capacities. Moreover, under iso-area comparison (e.g., 0.5MB SRAM and 2MB STT MRAM caches), STT MRAM caches show significant energy benefit along with larger cache capacity (note that larger capacity improves processor performance by lowering cache misses). Our results show that a processor with 2MB STT MRAM L2 outperforms one with 0.5MB SRAM L2 by 10% in IPC. The energy efficiency is further improved by employing the proposed sequential tag-data access in read and partial-line-update in write operations. In particular, the cache employing the proposed sequential tag-data access in addition to PLU (STT3(PLU,SEQ)) in Figure 8) shows substantial total energy reduction ranging from 2% to 28% across various benchmarks.

5. CONCLUSION

In this work, we performed a comprehensive analysis of the performance, energy consumption and integration density of STT MRAM caches in comparison to conventional SRAM cache. We considered different genres of MTJ stacks

and STT MRAM bitcell configurations in this study. Based on the detailed analysis of various bitcell characteristics including accurate area estimation from physical layout, we showed that, for large cache capacity, STT MRAM caches can have lower dynamic energy consumption and read latency compared to SRAM caches with the same capacity. Moreover, the low leakage energy consumption and high integration density of STT MRAM are highly beneficial for lower level caches (due to low utilization), and improve energy efficiency and processor performance. We also proposed read and write energy reduction techniques, namely sequential tag-data access in reads and partial cache line update in writes, which exploit the non-volatility of STT MRAM bitcells. The results show that the proposed techniques further improve the energy efficiency of STT MRAM caches.

6. ACKNOWLEDGMENTS

This research was supported in part by NRI, INDEX, Intel Corporation, and Qualcomm.

7. REFERENCES

[1] C. Augustine et al. Numerical analysis of typical STT-MTJ stacks for 1T-1R memory arrays. In Proc. IEDM, 2010.

[2] C. J. Lin et al. 45nm low power CMOS logic compatible embedded STT MRAM utilizing a reverse-connection 1T/1MTJ cell. In Proc. IEDM, pages 1 –4, Dec. 2009.

[3] D. Sandre et al. A 90nm 4Mb embedded phase-change memory with 1.2V 12ns read access time and 1MB/s write throughput. In Proc. ISSCC, Feb. 2010.

[4] K. Itoh. Embedded memories: Progress and a look into the future. IEEE Design & Test, 28(1):10 –13, Jan.-Feb. 2011.

[5] J. L. Hennessy et al. Computer Architecture: A Quantitative Approach. Morgan Kaufmann, May 2002.

[6] K. Lee et al. Development of Embedded STT-MRAM for Mobile System-on-Chips. IEEE Trans. Magnetics, 2011.

[7] M. Rasquinha et al. An energy efficient cache design using Spin Torque Transfer (STT) RAM. In Proc. ISLPED, 2010.

[8] N. Muralimanohar et al. Optimizing NUCA Organizations and Wiring Alternatives for Large Caches with CACTI 6.0. In Proc. MICRO, pages 3–14, 2007.

[9] N. N. Mojumder. Design of Hybrid Spintronic Devices at Scaled Technologies for non-Volatile Memory Applications. PhD thesis, Purdue University, Dec. 2011.

[10] P. Zhou et al. Energy reduction for STT-RAM using early write termination. In Proc. ICCAD, Nov. 2009.

[11] S. Ohbayashi et al. A 65-nm SoC Embedded 6T-SRAM Designed for Manufacturability With Read and Write Operation Stabilizing Circuits. IEEE JSSC, Apr. 2007.

[12] S. Park et al. Column-selection-enabled 8T SRAM array with ∼1R/1W multi-port operation for DVFS-enabled processors. In Proc. ISLPED, pages 303 –308, Aug. 2011.

[13] S. Ramaswamy et al. An utilization driven framework for energy efficient caches. In Proc. HiPC, pages 583–594, 2008.

[14] S. Yuasa et al. Giant room-temperature magnetoresistance in single-crystal Fe/MgO/Fe magnetic tunnel junctions. Nat. Mater., 3, Dec. 2004.

[15] Simplescalar LLC. http://www.simplescalar.com.

[16] X. Fong et al. Bit-cell Level Optimization for Non-volatile Memories Using Magnetic Tunnel Junctions and Spin-Transfer Torque Switching. IEEE Trans. Nanotechnology, 2011.

[17] X. Wei et al. Design of Last-Level On-Chip Cache Using Spin-Torque Transfer RAM (STT RAM). IEEE Trans. VLSI, 2011.

[18] X. Wu et al. Hybrid cache architecture with disparate memory technologies. In Proc. ISCA. ACM, 2009.

[19] Z. Chishti et al. Distance associativity for high-performance energy-efficient non-uniform cache architectures. In Proc. MICRO, pages 55–66, 2003.

Hardware Realization of BSB Recall Function Using Memristor Crossbar Arrays

Miao Hu and Hai Li
Polytechnic Institute of New York University
6 Metrotech Center, Brooklyn, NY, USA
mhu01@students.poly.edu,
hli@poly.edu

Qing Wu and Garrett S. Rose
Air Force Research Laboratory
Rome Site, 525 Brooks Road
Rome, NY, USA
qing.wu@rl.af.mil, garrett.rose@rl.af.mil

ABSTRACT

The Brain-State-in-a-Box (BSB) model is an auto-associative neural network that has been widely used in optical character recognition and image processing. Traditionally, the BSB model was realized at software level and carried out on high-performance computing clusters. To improve computation efficiency and reduce resources requirement, we propose a hardware realization by utilizing memristor crossbar arrays. In this work, we explore the potential of a memristor crossbar array as an auto-associative memory. More specificly, the recall function of a multi-answer character recognition based on BSB model was realized. The robustness of the proposed BSB circuit was analyzed and evaluated based on massive Monte-Carlo simulations, considering input defects, process variations, and electrical fluctuations. The physical constrains when implementing a neural network with memristor crossbar array have also been discussed. Our results show that the BSB circuit has a high tolerance to random noise. Comparably, the correlations between memristor arrays introduces directional noise and hence dominates the quality of circuits.

Categories and Subject Descriptors

C.1.3 [**PROCESSOR ARCHITECTURES**]: Other Architecture Styles — *Neural nets*

General Terms

Design, Performance, Reliability

Keywords

neural network, BSB model, memristor, crossbar array, process variation.

1. INTRODUCTION

As demand on high performance computation continuously increases, the traditional Von Neumann computer architecture becomes less efficient. In recent years, neuromorphic hardware systems have gained great attentions. Such systems can potentially provide the capabilities of biological perception and information processing within a compact and energy-efficient platform [1, 2]. Many research activities have been carried out on neural network algorithm enhancement [3] and/or system implementations built upon the conventional CPU, GPU, or FPGA [4].

As a highly generalized and simplified abstract of a biological system, an artificial neural network usually uses a *connection matrix* to represent a set of synapse networks. Accordingly, the net inputs of a group or groups of neurons can be transformed into matrix-vector multiplication(s). Similar to the biological systems, the neural network algorithms inherently are adaptive to the environment and resilience to random noise. As a consequence, hardware realizations of neural networks require a large volume of memory and are associated with high design complexity and hardware cost [2]. Algorithm enhancement can alleviate the situation but cannot fundamentally resolve it. More efficient hardware-level solutions become necessary.

The *Brain-state-in-a-box* (BSB) model is a simple, auto-associative, nonlinear, energy-minimizing neural network [5, 6]. A common application of the BSB model is *optical character recognition* (OCR) for printed text [7]. Recently, a multi-answer character recognition method based on the BSB model has been developed to improve reliability and robustness for noisy or occluded text images [8]. An input character image is processed through the BSB models in parallel for the **recall** (pattern recognition) operation. When all recalls are completed, a set of candidates are selected based on the convergence speed.

The existence of the memristor was predicted in circuit theory nearly forty year ago [9]. However, it wasn't until 2008 that the first physical realization was demonstrated by HP Lab through a TiO_2 thin-film structure [10]. Afterward, many memristor materials and devices have been reported or rediscovered. The memristor has many promising features, such as non-volatility, low-power consumption, high integration density, and excellent scalability [11, 12]. More importantly, the unique property to record the historical profile of the excitations on the device makes it an ideal candidate to realize synapse behavior in electronic neural networks [13, 14].

In this paper, we demonstrate a BSB recall circuit built on the memristor crossbar array. The crossbar architecture can naturally transfer the weighted combination of input signals to output voltages. a fast approximation mapping method is proposed so that the connection matrix can be mapped to pure circuit element relations. Key design parameters and physical constraints have been extracted and studied. Furthermore, we carried out a detailed analysis to study the weight of each and all noise contributors on the accuracy and robustness of the BSB circuit. Interestingly, even if a large process variation exists in memristor devices [15], it will not affect the performance much due to the inherent random noise tolerance of the BSB model. However, the correlation between two

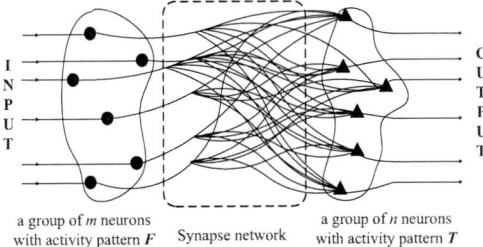

Figure 1: A simple example of neuron network.

a group of m neurons with activity pattern F — Synapse network — a group of n neurons with activity pattern T

memristor arrays within one BSB circuit dominates the robustness of the circuit since it introduces directional noise.

The remainder of the paper is organized as follows. In Section 2 we provide background information. Section 3 describes the hardware requirement to realize the BSB model and explains the details of the circuit implementation. Section 4 classifies the types of noise that can affect the quality of the BSB recall circuit. Section 5 analyzes the robustness of the BSB circuit design based on the simulation results. At the end, we conclude the paper in Section 6.

2. PRELIMINARY

2.1 Neural Network and BSB Model

Figure 1 illustrates a simple example of a neural network, in which two groups of neurons are connected by a set of synapses. We define $a_{i,j}$ as the synaptic strength of the synapse connecting the ith neuron in the input group and the jth neuron in the output one. The relationship of the activity patterns $\mathbf{F}$ of input neurons and $\mathbf{T}$ of output neurons can be described in matrix form:

$$\mathbf{T}_n = \mathbf{A}_{n \times m} \times \mathbf{F}_m, \qquad (1)$$

where matrix $\mathbf{A}$, denoted as the *connection matrix*, consists of the synaptic strengthes between the two neuron groups. The matrix-vector multiplication of Eq. (1) is a frequent operation in neural network theory to model the functionally associated with neurons in brains.

The BSB model is a simple auto-associative neural network with two main operations – *training* and *recall* [5, 6]. In this paper, we will focus on the hardware realization of the BSB recall operation. Its mathematical model can be represented as [8]

$$\mathbf{x}(t+1) = S(\alpha \cdot \mathbf{A} \times \mathbf{x}(t) + \lambda \times \mathbf{x}(t)), \qquad (2)$$

where, $\mathbf{x}$ is an N dimensional real vector, and $\mathbf{A}$ is an N-by-N connection matrix. $\mathbf{A} \times \mathbf{x}(t)$ is a matrix-vector multiplication, which is the main function of the recall operation. α is a scalar constant feedback factor. λ is an inhibition decay constant. $S(y)$ is the "squash" function defined as follows:

$$S(y) = \begin{cases} 1, & \text{if } y \geq 1 \\ y, & \text{if } -1 < y < 1 \\ -1, & \text{if } y \leq -1 \end{cases}. \qquad (3)$$

For a given input pattern $\mathbf{x}(0)$, the recall function computes Eq. (2) iteratively until *convergence*, that is, when all the entries of $\mathbf{x}(t+1)$ are either "1" or "−1".

Recently, Wu *et al.* [8] developed a multi-answer character recognition method based on the BSB model for occluded text images. It processes an input character image through all the BSB models and selects multiple candidates with the fastest convergence speed for word-level or sentence-level language model. This method will be used to evaluate the reliability and robustness of the memristor-based BSB recall circuit proposed in this paper.

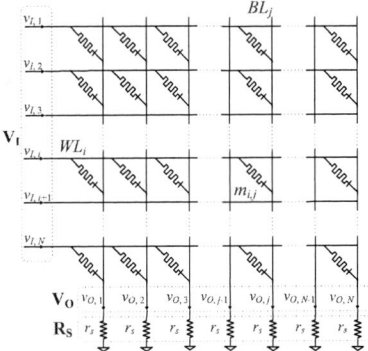

Figure 2: A memristor crossbar array.

2.2 Memristor

Many materials have demonstrated memristive behavior in theory and/or by experimentation via different mechanisms. In general, a certain energy (or threshold voltage) is required to enable the state change in a memristor [16]. When the electrical excitation through a memristor is greater than the threshold voltage, *e.g.*, $v_{in} > v_{th}$, the memristance changes (training). Otherwise, the memristor behaves like a resistor.

By nature, the memristor crossbar array is attractive for the implementation of neural networks. First, it supports a large number of signal connections within a small footprint, which is a basic requirement of synapse networks. Secondly, a frequent operation within neuromorphic circuits is the weighted combination of input signals, which mimic the so-called dendritic potential [17]. As we shall show in Section 3.1, the crossbar array inherently provides capabilities for this operation. Moreover, the adaptability of the connection matrix is particularly important in neural networks. Memristors are good at "learning" for the historical behavior [18].

In this work, we will focus on a hardware realization of the BSB recall function by assuming all ofthe memristors have been pre-programmed or already trained. During recall operations, the voltage across the memristors are constrained below the threshold voltage so that all the memristance values remain unchanged.

3. METHODOLOGY

3.1 Crossbar Array vs. Connection Matrix

Let's use the N-by-N memristor crossbar array illustrated in Figure 2 to demonstrate its the matrix computation functionality.

Here, we apply a set of input voltages $\mathbf{V_I^T} = \{v_{I,1}, v_{I,2}, \ldots, v_{I,N}\}$ on the *word-lines* (WL) of the array, and collect the current through each *bit-line* (BL) by measuring the voltage across a sensing resistor. The same sensing resistors are used on all the BLs with resistance r_s, or conductance $g_s = 1/r_s$. The output voltage vector is $\mathbf{V_O^T} = \{v_{O,1}, v_{O,2}, \ldots, v_{O,N}\}$. Assume the memristor sitting on the connection between WL_i and BL_j has a memristance of $m_{i,j}$. The corresponding conductance is $g_{i,j} = 1/m_{i,j}$. Then the relation between the input and output voltages can be represented by

$$\mathbf{V_O} = \mathbf{C} \times \mathbf{V_I}. \qquad (4)$$

Here, $\mathbf{C}$ can be represented by the memristances and the load resistors as

$$\mathbf{C} = \mathbf{D} \times \mathbf{G} = diag(d_1, \ldots, d_N) \times \begin{bmatrix} g_{1,1} & \cdots & g_{1,N} \\ g_{2,1} & \cdots & g_{2,N} \\ \vdots & \ddots & \vdots \\ g_{N,1} & \cdots & g_{N,N} \end{bmatrix}, \qquad (5)$$

where, $d_i = 1/(g_s + \sum_{k=1}^{N} g_{i,k})$. The conductance matrix $\mathbf{G}$ and the memristor matrix $\mathbf{M}$ have a relation of $\mathbf{G} = 1 \cdot / \mathbf{M}^{1}$.

Eq. (4) indicates that a trained memristor crossbar array can be used to construct the connection matrix $\mathbf{C}$, and transfer the input vector $\mathbf{V_I}$ to the output vector $\mathbf{V_O}$. However, Eq. (5) shows that $\mathbf{C}$ is not a direct one-to-one mapping of conductance matrix $\mathbf{G}$ of the memristor crossbar array, since the diagonal matrix $\mathbf{D}$ is related to $\mathbf{G}$. Though we can use a numerical iteration method to obtain the exact mathematical solution of $\mathbf{G}$, it is too complex and impractical. We assume any $g_{i,j} \in \mathbf{G}$ satisfies $g_{min} \leq g_{i,j} \leq g_{max}$, where g_{min} and g_{max} respectively represent the minimum and the maximum conductances of all memristors in the crossbar array. Instead, we propose a simple and fast approximation to the mapping problem by directly allowing

$$g_{i,j} = c_{i,j} \cdot (g_{max} - g_{min}) + g_{min}. \tag{6}$$

In the following, we will prove that by using this mapping method, a decayed version of the connection matrix $\hat{\mathbf{C}}$ can be approximately mapped to the conductance matrix $\mathbf{G}$ of the memristor array.

PROOF. By plugging Eq. (6) in Eq. (5), we have

$$\hat{c_{i,j}} = \frac{c_{i,j} \cdot (g_{max} - g_{min}) + g_{min}}{g_s + (g_{max} - g_{min}) \cdot \sum_{k=1}^{N} c_{i,j} + N \cdot g_{min}}. \tag{7}$$

Note that many memristor materials, such as TiO_2 memristor, demonstrate a large g_{max}/g_{min} ratio [10]. Thus, a memristor at the high resistance state under a low voltage excitation can be regarded as an insulator, that is, $g_{min} \simeq 0$. Moreover, the BSB recall matrix $\mathbf{A}$ is a special matrix with a small $\sum_{k=1}^{N} c_{i,j}$. For example, all the BSB models used for character recognition in our experiments have $\sum_{k=1}^{N} c_{i,j} < 5$ when $N = 256$. And $\sum_{k=1}^{N} c_{i,j}$ can be further reduced by increasing the ratio of g_s/g_{max}. As a result, the impact of $\sum_{k=1}^{N} c_{i,j}$ can be ignored. These two facts indicate that Eq. (7) can be further simplified as

$$\hat{c}_{i,j} \approx c_{i,j} \cdot \frac{g_{max}}{g_s}. \tag{8}$$

In a summary, with the proposed mapping method, the memristor crossbar array performs as a decayed connection matrix $\hat{c_{i,j}}$ between the input and output voltage signals. $\square$

3.2 Transformation of BSB Recall Matrix

To construct a memristor-based BSB recall circuit, our first task is to transfer the matrix $\mathbf{A}$ in the mathematical BSB recall model to a memristor array with physical meaning. A memristor is a physical device with conductance $g > 0$. Therefore, all elements in matrix $\mathbf{C}$ must be positive as shown in Eq. (5). However, in the original BSB recall model, $a_{i,j} \in \mathbf{A}$ could be positive or negative. We propose to split the positive and negative terms of $\mathbf{A}$ into two matrixes $\mathbf{A}^+$ and $\mathbf{A}^-$ as

$$a_{i,j}^{+} = \begin{cases} a_{i,j}, & \text{if } a_{i,j} > 0 \\ 0, & \text{if } a_{i,j} \leq 0 \end{cases}, \quad \text{and} \tag{9a}$$

$$a_{i,j}^{-} = \begin{cases} 0, & \text{if } a_{i,j} > 0 \\ -a_{i,j}, & \text{if } a_{i,j} \leq 0 \end{cases}. \tag{9b}$$

As such, Eq. (2) becomes

$$\mathbf{x}(t+1) = S\left(\mathbf{A}^+ \times \mathbf{x}(t) - \mathbf{A}^- \times \mathbf{x}(t) + \mathbf{x}(t)\right), \tag{10}$$

Here for the default case we set $\alpha = \lambda = 1$. The two connection matrices $\mathbf{A}^+$ and $\mathbf{A}^-$ can be mapped to two memristor crossbar ar-

---

[9] [1] In the the paper, we use $(\cdot *)$ and $(\cdot /)$ to represent the element wise multiplication and division operations between two matrices, respectively.

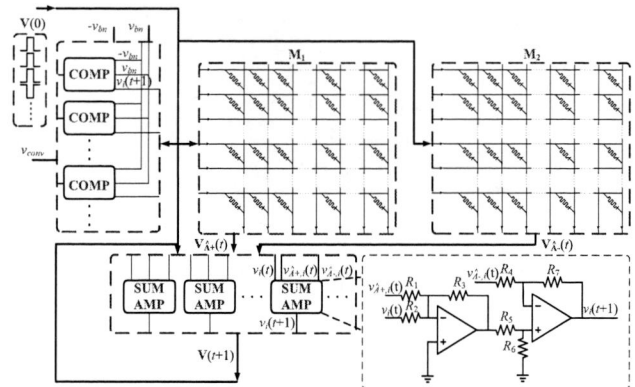

Figure 3: The conceptual diagram of the BSB recall circuit.

rays $\mathbf{M_1}$ and $\mathbf{M_2}$ in a decayed version $\hat{\mathbf{A}}^+$ and $\hat{\mathbf{A}}^+$, respectively, by following the mapping method in Eq. (6).

3.3 Circuit Realization

To realize the BSB recall function at the circuit level, we first convert the normalized input vector $\mathbf{x}(t)$ to a set of input voltage signals $\mathbf{V}(t)$. The corresponding function description of the voltage feedback system can be expressed as

$$\begin{aligned} \mathbf{V}(t+1) &= S'\left(\mathbf{A}^+ \times \mathbf{V}(t) - \mathbf{A}^- \times \mathbf{V}(t) + \mathbf{V}(t)\right) \\ &= S'\left(\mathbf{V_{A+}}(t) - \mathbf{V_{A-}}(t) + \mathbf{V}(t)\right) \end{aligned}. \tag{11}$$

We use v_{bn} to represent the input voltage boundary, that is, $-v_{bn} \leq v_i(t) \leq v_{bn}$ for any $v_i(t) \in \mathbf{V}(t)$. The new saturation boundary function $S'()$ need to be modified accordingly. In implementation, v_{bn} can be adjusted based on requirements of convergence speed and accuracy. Meanwhile, v_{bn} must be smaller than v_{th} so that the memristance values will not change during the recall process.

Figure 3 illustrates the BSB recall circuit built based on Eq. (11). The design is an analog system consisting of three major components. Below is the detailed description.

Memristor crossbar arrays: As the key component of the overall design, the memristor crossbar arrays are used to realize the matrix-vector multiplication function in the BSB recall operation. To obtain both positive and negative weights in the original BSB algorithm in Eq. (2), two memristor crossbar arrays $\mathbf{M_1}$ and $\mathbf{M_2}$ are required in the design to represent the connection matrices $\hat{\mathbf{A}}^+$ and $\hat{\mathbf{A}}^-$, respectively. The memristor crossbar array has the same dimension as the BSB weight matrix $\mathbf{A}$.

Summing amplifier: In our design, the input signal $v_i(t)$ along with $v_{\hat{A}+,i}(t)$ and $v_{\hat{A}-,i}(t)$, the corresponding voltage outputs of two memristor crossbar arrays, are fed into a summing amplifier. The conceptual structure of the summing amplifier can be found in the inner set of Figure 3.

Resulted by the decayed mapping method proposed in Section **3**, the required $v_{A+,i}(t)$ should be g_s/g_{max} times of the generated $v_{\hat{A}+,i}(t)$. $v_{A-,i}(t)$ has the same requirement too. In our design, we set $R_1 = R_4 = R_6 = 1/g_s$ and $R_2 = R_3 = R_5 = R_7 = 1/g_{max}$. The resulting output of the summing amplifier

$$\begin{aligned} v_i(t+1) &= \frac{g_s}{g_{max}} \cdot v_{\hat{A}+,i}(t) - \frac{g_s}{g_{max}} \cdot v_{\hat{A}-,i}(t) + v_i(t) \\ &= v_{A+,i}(t) - v_{A-,i}(t) + v_i(t) \end{aligned}, \tag{12}$$

which indicates that the decayed effect has been canceled out. The N dimensional BSB model requires N summing amplifiers to realize the addition/subtraction operation in Eq. (11). Also, for sum-

500

ming amplifiers, we should adjust their power signals to make their maximum/minimum output voltages equal $\pm v_{bn}$, respectively. In implementation, we can adjust the resistances $R_1 \sim R_7$ to match the required α and λ in Eq. (2), if they are not the default value 1.

Comparator: Once a new set of voltage signals $\mathbf{V}(t+1)$ is generated from the summing amplifiers, we send them back as the input of the next iteration. Meanwhile, every $v_i \in \mathbf{V}$ is compared to v_{bn} and $-v_{bn}$ to determine if path i has "converged". The recall operation stops when all the N paths reach convergence. Totally N comparators are needed to cover all the paths.

4. ROBUSTNESS OF BSB RECALL CIRCUIT

Running the BSB recall circuit constructed in Section 3 under ideal conditions should lead to the exact same results as the BSB mathematical algorithm. Unfortunately, the noise induced by process variations and signal fluctuations in implementation can significantly affect circuit performance. In this section, we will address the modeling of this noise at the component level. The impact of physical design constrains will also be discussed.

4.1 Process Variations

4.1.1 Memristor Crossbar Arrays

Due to process variations, the real memristance matrix $\mathbf{M}'$ of a memristor crossbar array could be quite different from the theoretical $\mathbf{M}$. Their difference can be represented by a *noise matrix* $\mathbf{N_M}$, which includes two contributors – the systematic noise $\mathbf{N_{M,sys}}$ and the random noise $\mathbf{N_{M,rdm}}$. Consequently, $\mathbf{M}'$ can be expressed by

$$\mathbf{M}' = \mathbf{M} \cdot *\mathbf{N_M} = \mathbf{M} \cdot *(1 + \mathbf{N_{M,sys}} + \mathbf{N_{M,rdm}}). \quad (13)$$

The impact of $\mathbf{N_M}$ on the connection matrix $\mathbf{C}$ is too complex to be expressed by a mathematical closed-form solution. But numerical analysis shows that it can be approximated by

$$\mathbf{C'_M} = \mathbf{C} \cdot *\mathbf{N_{CM}} = \mathbf{C} \cdot *\frac{1}{\mathbf{N_M}} \cdot *\frac{1}{\mathbf{N_M}}. \quad (14)$$

Here, $\mathbf{C'_M}$ is the connection matrix after including memristance process variations. $\mathbf{N_{CM}}$ is the corresponding noise matrix.

In the following analysis, we assume $\mathbf{N_{M,sys}}$ follows a normal distribution. To fully demonstrate the impact of the random process variations, the lognormal distribution is used to generate $\mathbf{N_{M,rdm}}$. Coefficient $\mathrm{Corr_M}$ is used to represent the correlation degree between the two memristor crossbar arrays in the same BSB circuit. When $\mathrm{Corr_M} = 1$, the two arrays have the same systematic noise.

4.1.2 Sensing Resistance

Similar to the analysis of memristance variation, we classify the noise induced by $\mathbf{R_S}$ variations into the systematic noise $\mathrm{N_{R,sys}}$ and the random noise $\mathrm{N_{R,rdm}}$. The actual sensing resistance vector becomes

$$\mathbf{R'_S} = \mathbf{R_S} \cdot *\mathbf{N_{Rs}} = \mathbf{N_S} \cdot *(1 + \mathrm{N_{R,sys}} + \mathrm{N_{R,rdm}}). \quad (15)$$

$\mathbf{C'_R}$, the connection matrix after including $\mathbf{N_{Rs}}$, is

$$\mathbf{C'_R} = \mathbf{C} \cdot *\mathbf{N_{CR}} = \mathbf{C} \cdot * [\mathbf{N_{Rs}} \, \mathbf{N_{Rs}} \, \ldots \, \mathbf{N_{Rs}}]. \quad (16)$$

Here, $\mathbf{N_{CR}}$ is the noise matrix of $\mathbf{C}$ after including the process variation of the sensing resistors. The mean value of r_s distribution, which reflects the impact of systematic noise, can be obtained during the post-fabrication testing. When training the memristances in BSB circuit, $\mathrm{N_{R,sys}}$ should have been included. Hence, in the following analysis, we only consider the random noise $\mathrm{N_{R,rdm}}$, which follows a normal distribution.

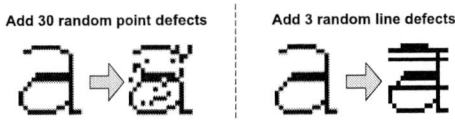

Figure 4: (a) Random line defects; (b) Random point defects.

4.2 Signal Fluctuations

The electrical noise from the power supplies and the neighboring wires can significantly degrade the quality of analog signals. Different from the process variations that remain unchanged after the circuit is fabricated, these signal fluctuations vary during circuit operation. Without loss of generality, we assume the run-time noise of the summing amplifier's output signals follows a normal distribution, same as that of the output of the comparators.

4.3 Physical Constrains

There are three major physical constrains in the circuit implementation: (1) For any $v_i(0) \in \mathbf{V}(0)$, The voltage amplitude of initial input signal $v_i(0)$ is limited by the input circuit, ; (2) boundary voltage v_{bn} must be smaller than v_{th} of memristors; and (3) the summing amplifier has finite resolution.

In the BSB recall function, the ratio between boundaries of $S(y)$ and the initial amplitude of $x_i(0), x_i(0) \in \mathbf{x}(0)$, determines the learning space of the recall function. If the ratio is greater than the normalized value, the recall operation takes more iterations to converge with a higher accuracy. Otherwise, the procedure converges faster by lowering stability. Thus, minimizing the ratio of $|v_i(0)|$ and v_{bn} can help obtain the best performance. However, the real amplifier has a finite resolution and v_{bn} is limited within v_{th} of the memristor. Continuously reducing $|v_i(0)|$ eventually will lose enough information in the recall circuit. So the resolution of the summing amplifier is a key parameter to determine the optimal ratio of $|v_i(0)|$ and v_{bn} in circuit implementation. Certainly it also affects the design cost of amplifier and the overall design.

5. SIMULATION RESULTS

The robustness of the BSB recall circuit was analyzed based on Monte-Carlo simulations conducted with MATLAB. We tested 26 BSB circuits corresponding to the 26 lower case letters from "a" to "z". Each character image consists of 16×16 points and can be converted to a 256-entry vector. Accordingly, the BSB recall matrix has a dimension of 256×256. In each test, we created 500 design samples for each BSB circuit and ran 13,000 Monte-Carlo simulations. Two types of input pattern defects, random point defects and random line defects (see Figure 4), have been evaluated.

5.1 BSB Circuit Under Ideal Condition

For an input pattern, the different BSB circuits have different convergence speeds. Figure 5 shows an example when processing a perfect "a" image through the BSB circuits trained for all 26 lower case letters. The BSB circuits for "a", "l", and "s" reach convergence with the least iteration numbers. The multi-answer character

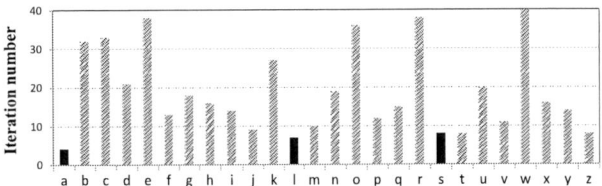

Figure 5: Iterations of 26 BSB circuits for a perfect "a" image.

501

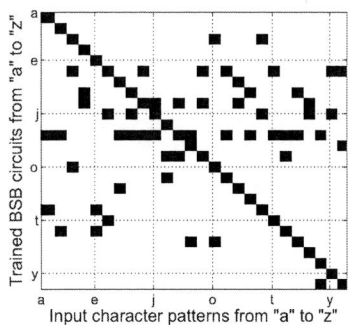

Figure 6: The ideal performance of 26 BSB circuits

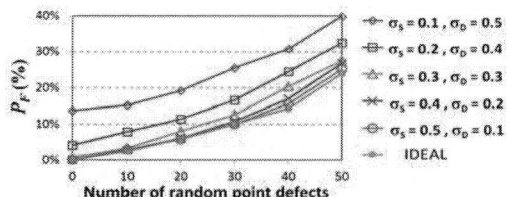

Figure 7: Static noise vs. dynamic noise.

recognition method consider these three letters as winners and send them to word-level language model [8].

Figure 6 summarizes the performance of the BSB circuit design under ideal condition without input defects, process variations, or signal fluctuations. The x-axis and y-axis represent input images and the BSB circuits, respectively. All the winners are highlighted by the black blocks. Figure 6 shows that a BSB circuit corresponding to its trained input pattern always wins under the ideal condition. However, after injecting noise to input pattern or circuit design, some BSB circuits might *fail* to recognize its trained input pattern. In this work, we use the probability of failed recognitions P_F to measure the performance of BSB circuits.

5.2 Process Variations and Signal Fluctuations

Impact of random noise:

The random noise in the BSB circuit could come from process variations as well as electrical signal fluctuations. We summarize the impact of every single random noise component in Table 1, based on Monte-Carlo simulation results. Here, we assume two memristor crossbar arrays are fully correlated, *i.e.*, $\text{Corr}_M = 1$. The simulation results show that BSB circuit design has a high tolerance on the random noise: compared to the ideal condition without any fluctuation ("IDEAL"), these random noise of circuits cause slight performance degradation. This is because resilience to random noise is one of the most important inherent features for the BSB model as well as other neural networks.

Static Noise vs. Dynamic Noise:

The noise matrices $\mathbf{N_M}$ and $\mathbf{N_{Rs}}$ mainly affect the mapping between connection matrix and memristor crossbar array. Physically noise components come from process variations and remain unchanged during the recall operation. Thus, they can be regarded as *static noise* N_S. On the contrary, the noise from the summing amplifiers and comparators are induced by electric fluctuations, which

demonstrates a dynamic behaviour during the iteration process. We classify them as *dynamic noise* N_D.

We can adjust N_S and N_D and observe the combined impact on BSB circuit performance. For simplicity, we set $\sigma_{rdm}(M) = \sigma(R_S) = \sigma_S$ and $\sigma(AMP) = \sigma(COMP) = \sigma_D$. And $\text{Corr}_M = 1$ to exclude the impact of correlations between the two memristor arrays. The result in Figure 7 shows that the dynamic noise dominates P_F. For example, when $\sigma_D = 0.5$ and $\sigma_S = 0.1$, P_F is high even with a clean input image. Decreasing σ_D but increasing σ_S results in P_F reduction in all regions.

Impact of Corr_M:

The BSB circuit implementation uses two memristor crossbar arrays to split the positive and negative elements of $\mathbf{A}$. Reducing Corr_M and hence increasing the difference in the systematic noise of two memristor arrays can be regarded as $\mathbf{A}^+$ and $\mathbf{A}^-$ having different overall shifts. This is *directional noise* in the recall function. As a consequence, Corr_M demonstrates a higher impact. As shown in Figure 8, when decreasing Corr_M from 1 to 0, the average P_F dramatically increases.

5.3 Impact of Summing Amplifier Resolution

To achieve the same learning space as the normalized BSB model, we set $v_{bn} = 1.6V$ and all elements of $\mathbf{V(0)}$ to be $\pm0.1V$. Then we vary the summing amplifiers' resolution under different static and dynamic noise configurations. Corr_M was fixed at 0.6. The simulation results are shown in Figure 9.

Again the simulation results demonstrate the BSB circuit's high tolerance for random noise: when $\sigma_S = \sigma_D \leq 0.4$, P_F is close to the ideal condition of $\sigma_S = \sigma_D = 0$. A 200mV resolution for the summing amplifier is too coarse to be acceptable: the BSB circuit cannot have zero P_F even under the ideal condition when neither input defects nor random noise are included. The resolution of 100mV is acceptable when the noise is not significant (*e.g.*, $\sigma_S = \sigma_D \leq 0.2$) and the input pattern defect number is small (*e.g.*, less than 20 random point defects). For the given physical constraint configuration, the 50mV and 25mV resolutions show similar

Table 1: $P_F(\%)$ of 26 BSB circuits for 26 input patterns.

random point numbers	0	10	20	30	40	50
IDEAL	0	2.1	4.2	5.3	10.0	20.8
$M(\sigma_{sys}=0.1 \& \sigma_{rdm}=0.1)$	0	1.9	4.6	6.5	14.2	24.7
$R_S(\sigma=0.1)$	0	1.8	4.3	6.2	13.7	24.1
SUM-AMP$(\sigma=0.1)$	0	1.9	4.4	7.7	13.5	23.1
COMPARATOR$(\sigma=0.1)$	0	2.3	5.5	5.4	11.1	22.0
$Corr_M=0.6$	5.6	10.2	17.2	22.7	30.8	38.6
OVERALL $(Corr_M=0.6)$	4.6	8.2	15.2	20.7	32.8	36.6
random line numbers	0	1	2	3	4	5
IDEAL	0	7.3	13.8	21.5	35.8	50.2
$M(\sigma_{sys}=0.1 \& \sigma_{rdm}=0.1)$	0	7.4	14.8	25.5	38.8	53.6
$R_S(\sigma=0.1)$	0	7.4	14.8	23.3	35.1	51.8
SUM-AMP$(\sigma=0.1)$	0	7.7	15.3	23.4	34.7	52.6
COMPARATOR$(\sigma=0.1)$	0	6.9	14.5	23.3	33.7	53.2
$Corr_M=0.6$	5.1	14.4	24.7	34.6	44.2	55.1
OVERALL $(Corr_M=0.6)$	6.3	15.4	24.2	34.1	44.0	58.2

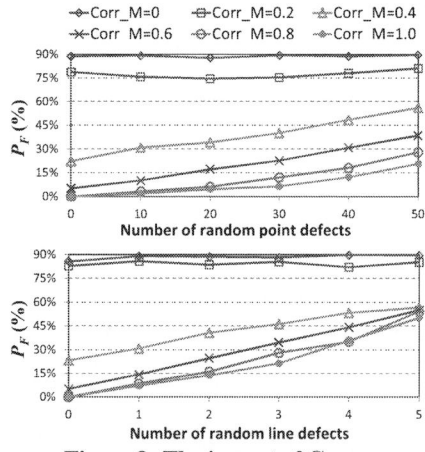

Figure 8: The impact of Corr_M.

502

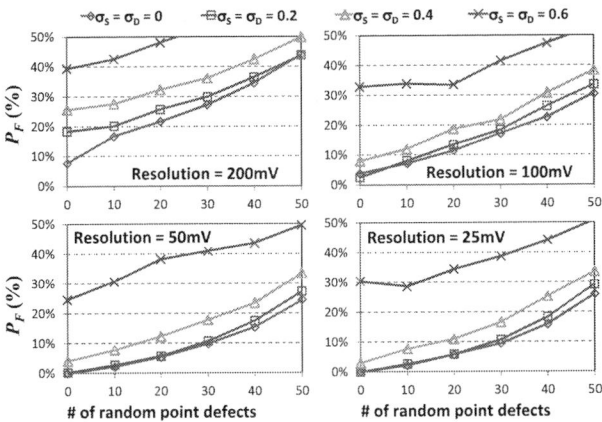

Figure 9: Impact of resolutions of summing amplifiers.

results when $\sigma_S = \sigma_D \leq 0.2$.

5.4 Overall Performance

In the previous analysis, we use the averaged P_F of all 26 BSB circuits for performance evaluation. One thing of particular interest is whether all BSB circuits degrade in the same way as we inject defects and noise into the system, or perhaps certain BSB circuits perform much better or worse than the others. In this test, we set $\text{Corr}_M = 0.8$ and inject 0 or 30 random points defects for each input image. Figure 10 shows the comparison of P_F of each input character pattern under ideal condition (noise free) and under the scenario including all process variations and signal fluctuations ($\sigma_S = \sigma_D = 0.1$).

The simulation shows that the performance degradation induced by process variations and signal fluctuations have a constant impact on all BSB circuits. When processing a perfect image under the ideal condition, no BSB circuits fails and hence $P_F = 0$. After including static and dynamic noise, P_F ranges from 1% to 7% for different input characters. When increasing the random point defects to 30 for input images, the range of P_F increases from 0~10% under ideal condition to 4~16% after including all the noise sources.

6. CONCLUSION

In this work, we firstly introduce a framework for a hardware realization of neural network algorithms using memristor crossbar arrays. More specific, we transfer the mathematical expression of the BSB recall model to a pure physical device relation and design the corresponding circuit architecture. The multi-answer character recognition method [8] was used to perform robustness analysis for the proposed circuit. The impact of various noise components

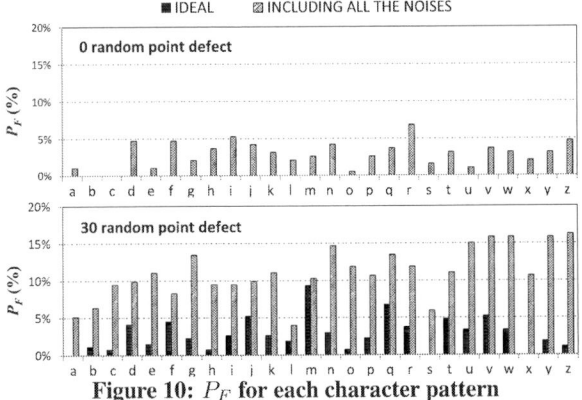

Figure 10: P_F for each character pattern

induced by process variations and electrical fluctuations have been studied. The physical constrains in circuit implementation have also been discussed. We found that the correlation between the two memristor crossbar arrays within a BSB recall circuit dominates the quality of the circuit. The resolution of the summing amplifier is another important factor, which is related to the the physical constrains in circuit implementation. Interestingly, the random noise do not have obvious correlation with the character pattern which is "trained" and stored in the BSB connection matrix.

7. ACKNOWLEDGMENT OF SUPPORT AND DISCLAIMER

Received and approved for public release by AFRL on 01/10/2012, case number 88ABW-2012-0192. Any Opinions, findings, and conclusions or recommendations expressed in this material are those of the authors and do not necessarily reflect the views of AFRL or its contractors.

8. REFERENCES

[1] P. Camilleri, M. Giulioni, V. Dante, D. Badoni, G. Indiveri, B. Michaelis, J. Braun, and P. del Giudice, "A neuromorphic avlsi network chip with configurable plastic synapses," in *International Conference on Hybrid Intelligent Systems*, 2007, pp. 296–301.

[2] J. Partzsch and R. Schuffny, "Analyzing the scaling of connectivity in neuromorphic hardware and in models of neural networks," *Neural Networks, IEEE Transactions on*, vol. 22, no. 6, pp. 919–935, 2011.

[3] M. Wang, B. Yan, J. Hu, and P. Li, "Simulation of large neuronal networks with biophysically accurate models on graphics processors," in *Neural Networks (IJCNN), The 2011 International Joint Conference on*. IEEE, 2011, pp. 3184–3193.

[4] H. Shayani, P. Bentley, and A. Tyrrell, "Hardware implementation of a bio-plausible neuron model for evolution and growth of spiking neural networks on fpga," in *NASA/ESA Conference on Adaptive Hardware and Systems*. IEEE, 2008, pp. 236–243.

[5] J. Anderson, J. Silverstein, S. Ritz, and R. Jones, "Distinctive features, categorical perception, and probability learning: Some applications of a neural model." *Psychological Review*, vol. 84, no. 5, p. 413, 1977.

[6] E. M. H. Hassoun, "Associative neural memories: Theory and implementation," in *Oxford University Press*, 1993.

[7] A. Schultz, "Collective recall via the brain-state-in-a-box network," *Neural Networks, IEEE Transactions on*, vol. 4, no. 4, pp. 580–587, 1993.

[8] Q. Wu, M. Bishop, R. Pino, R. Linderman, and Q. Qiu, "A multi-answer character recognition method and its implementation on a high-performance computing cluster," in *FUTURE COMPUTING 2011, The Third International Conference on Future Computational Technologies and Applications*, 2011, pp. 7–13.

[9] L. Chua, "Memristor-the missing circuit element," *IEEE Transaction on Circuit Theory*, vol. 18, pp. 507–519, 1971.

[10] D. B. Strukov, G. S. Snider, D. R. Stewart, and R. S. Williams, "The missing memristor found," *Nature*, vol. 453, pp. 80–83, 2008.

[11] Y. Ho, G. Huang, and P. Li, "Nonvolatile memristor memory: device characteristics and design implications," in *Proceedings of the 2009 International Conference on Computer-Aided Design*. ACM, 2009, pp. 485–490.

[12] D. Niu, Y. Chen, C. Xu, and Y. Xie, "Impact of process variations on emerging memristor," in *Design Automation Conference (DAC)*, 2010, pp. 877–882.

[13] D. Strukov, J. Borghetti, and S. Williams, "Coupled ionic and electronic transport model of thin-film semiconductor memristive behavior," *SMALL*, vol. 5, pp. 1058–1063, 2009.

[14] Y. Pershin and M. Di Ventra, "Experimental demonstration of associative memory with memristive neural networks," *Neural Networks*, vol. 23, no. 7, pp. 881–886, 2010.

[15] M. Hu, H. Li, Y. Chen, X. Wang, and R. Pino, "Geometry variations analysis of tio 2 thin-film and spintronic memristors," in *Asia and South Pacific Design Automation Conference*. IEEE Press, 2011, pp. 25–30.

[16] Y. Pershin and M. Di Ventra, "Practical approach to programmable analog circuits with memristors," *Circuits and Systems I: Regular Papers, IEEE Transactions on*, vol. 57, no. 8, pp. 1857–1864, 2010.

[17] U. Ramacher and C. V. D. Malsburg, *On the Construction of Artificial Brains*. Springer, 2010.

[18] T. Hasegawa, T. Ohno, K. Terabe, T. Tsuruoka, T. Nakayama, J. K. Gimzewski, and M. Aono, "Learning abilites achieved by a single solid-state atomic switch," *Advanced Materials*, vol. 22, no. 16, pp. 1831–1834, 2010.

A Methodology for Energy-Quality Tradeoff Using Imprecise Hardware

Jiawei Huang	John Lach	Gabriel Robins
Computer Engineering	Electrical and Computer Engineering	Computer Science
University of Virginia	University of Virginia	University of Virginia
jh3wn@virginia.edu	jlach@virginia.edu	robins@cs.virginia.edu

ABSTRACT

Recent studies have demonstrated the potential for reducing energy consumption in integrated circuits by allowing errors during computation. While most proposed techniques for achieving this rely on voltage overscaling (VOS), this paper shows that *Imprecise Hardware (IHW)* with design-time structural parameters can achieve orthogonal energy-quality tradeoffs. Two IHW adders are improved and two IHW multipliers are introduced in this paper. In addition, a simulation-free error estimation technique is proposed to rapidly and accurately estimate the impact of IHW on output quality. Finally, a quality-aware energy minimization methodology is presented. To validate this methodology, experiments are conducted on two computational kernels: DOT-PRODUCT and L2-NORM – used in three applications – Leukocyte Tracker, SVM classification and K-means clustering. Results show that the Hellinger distance between estimated and simulated error distribution is within 0.05 and that the methodology enables designers to explore energy-quality tradeoffs with significant reduction in simulation complexity.

Categories and Subject Descriptors

G.1.6 [**Numerical Analysis**]: Constrained optimization

General Terms

Algorithms

Keywords

Imprecise hardware, energy-quality tradeoff, static error estimation

1. INTRODUCTION

High power consumption is one of the greatest challenges currently facing IC designers. Although circuit-level techniques such as dynamic voltage and frequency scaling (DVFS), as well as sub- and near-threshold operation have proved effective in power reduction, they are fundamentally limited by the critical path of the circuit. Recently, a new design philosophy has emerged that relaxes the absolute correctness requirement to achieve further power reductions. For example, application noise tolerance [1] combines a voltage-overscaled computation core with a low-precision error-compensation core. Significance driven computation [2] identifies functionally non-critical parts of an algorithm and employs VOS to save power. Both techniques exploit the error-tolerant nature of the algorithms being implemented and use V_{dd} as the leverage to tradeoff quality for power. However, a good understanding of the algorithm is usually required to identify functionally non-critical components that could be "imprecisely" implemented without excessively degrading the output quality. In addition, the system must be *simulated* under a range of V_{dd} in order to find the optimal power-quality tradeoff, which is typically a time-consuming process.

This paper presents a generalized methodology for energy[1]-quality tradeoffs with two unique features. First, it incorporates "variables" for imprecise computation other than V_{dd}; namely RTL structural parameters for deterministic design-time energy-quality tradeoffs. The specific IHW components introduced here are parameterized ALUs. IHW is orthogonal to existing V_{dd}-lowering techniques, as VOS can be applied on top of IHW to achieve even higher energy reduction. Second, the methodology utilizes a novel static error estimation method that models the output error distribution based on the input distribution and design parameters. This method enables the automated exploration of the energy-quality space without computational-intensive simulations at each design point. It is also general enough to be used by VOS designs for rapid quality evaluation to speed up V_{dd} selection.

Table 1 lists two kernel functions common in multimedia, recognition and mining applications [3] and three examples of such applications. These kernels and applications will be used to demonstrate and validate the proposed methodology. In principle, this methodology can be used to explore energy-quality tradeoffs in any error-resilient applications with computational kernels that can be implemented with IHW.

Table 1. % application runtime spent in computation kernels

Kernel	Application	Runtime%
$DOT-PRODUCT(X,Y) = \sum_{i=1}^{n} x_i * y_i$	Leukocyte Tracker	22%
$L2-NORM(X,Y) = \sum_{i=1}^{n} (x_i - y_i)^2$	SVM	98%
	K-means	49%

Major contributions of this work include:

- a generalized quality-aware energy minimization methodology,
- a fast and accurate static error estimation method, and
- design of imprecise multipliers based on imprecise adders.

The rest of the paper is organized as follows. Section 2 reviews related work in this area and highlights the motivation of this work. Section 3 introduces two existing imprecise adders as well as some improvements and adaptations to use them to build imprecise multipliers. Section 4 introduces the static error estimation method. The quality-aware energy-minimization methodology is described in Section 5, followed by application-

---

[1] This paper will focus on *energy per operation (E/op)* instead of power, but the methodology is applicable to any hardware metric such as power, area, energy-delay product, etc.

level energy-quality tradeoff results in Section 6. Section 7 concludes the paper.

2. BACKGROUND AND RELATED WORK

Most of the prior work on imprecise computation focuses on power reduction through VOS [1, 2]. Since traditional circuits are designed such that most paths have delays close to those of critical paths, naive VOS will likely induce massive timing violations and circuit failure. These techniques attempt to either correct errors with redundant circuits [1] or delay the onset of massive errors through timing path rebalancing [4]. Mohapatra et. al. [3] suggest a way to memorize the timing errors in a counter and make corrections over longer time intervals. However, these techniques do not fundamentally change the circuit structure to enable energy-quality tradeoffs at design time. Another problem is finding the optimal V_{dd}. There is no easy way to predict the output quality at a certain V_{dd} level except through time-consuming detailed circuit simulation.

The correctness requirement in error-tolerant applications can be further relaxed, leaving errors *uncorrected*. For example, users are unlikely to notice small/rare degradations in multimedia quality, and computation errors often do not affect the results of recognition or data mining analyses. Therefore, many high-energy circuit structures could be simplified (such as breaking long adder carry propagation chains with constant 1s or 0s) with a tolerable impact on application-level quality. Such design-time techniques could be used in conjunction with runtime techniques (e.g., VOS) to achieve more desirable energy-quality tradeoffs.

Since IHW inevitably leads to some loss of accuracy, it is particularly important to be able to evaluate its effect on output quality. The static error estimation technique proposed in Section 4 achieves this goal by leveraging statistical analysis to propagate the error distribution through a system of arithmetic operators. Although the application-level quality impact still needs to be evaluated through simulation, arithmetic kernel-level quality estimation can significantly reduce the number of design points that need to be simulated. Most suboptimal design points are eliminated at the kernel level by the static error estimator. With the exception of the initial simulation to characterize IHW components, no simulation is required at the kernel-level, and the same characterization data can be reused for arbitrary input distributions.

3. IMPRECISE ADDERS AND MULTIPLIERS

Adders and multipliers are used extensively in multimedia and data mining applications. Imprecise implementations of adders and multipliers have the most direct impact on system energy and output quality. This section presents two imprecise adder designs in the literature, and introduces new imprecise multiplier designs.

3.1 ACA Adder

The Almost Correct Adder (ACA) [5] is a modified version of the traditional Kogge-Stone adder (KSA). ACA leverages the fact that under random inputs, the vast majority of the actual timing paths are much shorter than the worst-case critical path. Table 2 gives the probability of two random 64-bit inputs triggering a critical path longer than K. Even with K much smaller than 64, the probability of critical path violation is quite small and that probability decreases rapidly with larger K. ACA then uses a tree structure to compute the *propagate* and *generate* signals similar to KSA but assumes the longest run of *propagate* never exceeds K, i.e., Sum_i is computed using only $A_i \cdots A_{i-K+1}$ and $B_i \cdots B_{i-K+1}$. Its

worst case delay is $\log_2(K)$. ACA's structure is essentially a trimmed KSA tree. A smaller tree translates to lower delay, smaller area and less energy per addition.

Table 2. Prob. of a random propagate chain exceeding K bits

K	12	16	24	30
Probability	0.0024	1.22×10^{-4}	2.4×10^{-7}	9.1×10^{-10}

Errors occur in ACA when the inputs trigger a *propagate* chain longer than K. For example, when A and B are exactly complementary, the *propagate* chain will extend the full adder's length. To produce the correct Sum_i, all the bits from both inputs will be needed, but ACA speculates and approximates it with the *propagate* chain from bit i down to $i-K+1$ with the carry-in set to constant 0. In case of incorrect speculation, a large error will appear in Sum_i. The largest error occurs when bit i is the MSB. Errors with such characteristics are called **infrequent large-magnitude (ILM)** errors [6]: they occur rarely, but whenever they do, their magnitude tends to be large. Energy-quality tradeoffs can be achieved by tuning the design parameter K.

3.2 ETAIIM Adder

The Modified Error-Tolerant Adder Type II (ETAIIM) [7] is another type of imprecise adder based on the Ripple-carry adder (RCA). RCA has a simple linear *propagate* chain. ETAIIM works by partitioning the *propagate* chain into segments of variable widths. The carry bits across two segments are truncated to zero. In order to provide higher precision for higher-order bits, segments are wider (i.e., contain more bits) on the MSB side than on the LSB side. ETAIIM has two parameters: BPB (bits per block) and L (the number of blocks used for generating the MSB). A block refers to the smallest segment, which is usually located at the LSB. The maximum error magnitude of ETAIIM is limited by $BPB \times L$. However, carry generation across blocks is common; therefore, errors occur quite frequently in ETAIIM. These errors are called **frequent small-magnitude (FSM)** errors [6] because their magnitudes are bounded by the design parameters and are usually small compared to ILM errors.

3.3 Improving Imprecise Adders

The original ACA and ETAIIM designs do exhibit a weakness. For simplicity, both designs use a constant 0 as the carry-in at the cut-off point of the critical path, but this leads to negatively-biased errors because 0 is an underestimation of the carry-in bit. Similarly, constant 1 will produce positively-biased errors. One possible improvement is to take the carry-in from the bit immediately before the *propagate* chain. For ACA, this means the propagate chain formed by $A_i \cdots A_{i-K+1}$ and $B_i \cdots B_{i-K+1}$ will take A_{i-K} (or B_{i-K}) as the carry-in. For ETAIIM, it means the carry bit across blocks will be taken from the highest bit in the previous block. If the inputs are random during the computation, every bit has a 50% probability of being 0 or 1. This will eventually produce an unbiased error distribution in the sum. Table 3 is obtained from simulating the summation of 20 numbers randomly drawn from [-0.5, 0.5] using ETAIIM adder ($BPB=8$, $L=4$). The anti-biasing technique notably improves statistical error metrics.

Table 3. Error metrics improvement with anti-biasing

Metrics	Original	w. Anti-biasing
Error Rate	12.3%	6.9%
Mean Error Magnitude	6.3×10^{-8}	3.3×10^{-8}

3.4 Imprecise Multipliers

Despite the lack of imprecise multipliers in the literature, it is possible to build imprecise multipliers based on imprecise adders. A typical multiplier consists of three stages: partial product generation, partial product accumulation and a final stage adder [8]. The idea of building an imprecise multiplier is simple: replace the final stage adder with an imprecise adder. The ACA and ETAIIM adders will thus yield corresponding ACA and ETAIIM multipliers. For the other two stages, we adopt the popular simple partial product generation (shifted versions of the multiplicand without recoding) [8] and Wallace-tree partial product accumulator (3:2 compressor tree) [9]. These choices will influence the actual energy numbers but they do not affect the ability to perform energy-quality tradeoffs.

Table 4 compares the energy-delay product (EDP) of various precise and imprecise adders and multipliers. They are all synthesized to their respective critical path delays in 130nm technology, and imprecise ALUs are operated at lower voltages to match the speed of their precise counterparts. As seen from the table, imprecise ALUs consume significantly less E/op than their precise counterparts at the same delay due to their simplified logic structures.

Table 4. E/op and area of precise and imprecise ALUs

ALU	E/op (pJ)	Delay (ns)	EDP (pJ·ns)
KSA64	8.47	0.8	6.776
ACA64 (K=16)	4.96		3.968
RCA64	5.48	5.3	29.044
METAII (BPB=4, L=4)	0.527		2.793
MULT64_KSA	413.18	2.6	1074.268
MULT64_ACA (K=32)	365.98		951.548
MULT64_RCA	174.8	11.1	1940.28
MULT64_METAII (BPB=4, L=4)	82.56		916.416

4. STATIC ERROR ESTIMATION

While many CAD tools exist to evaluate the energy consumption of an integrated circuit, quality evaluation capability is far less common – i.e., determining how much the imprecise output differs from the precise output. In all VOS techniques, quality is evaluated by running Monte Carlo simulations, since the relationship between the circuit variables and the output cannot be easily derived. There is a fundamental drawback to this approach: the simulation time grows exponentially with data width and computation length. For example, a length-10 DOT-PRODUCT with 32-bit numbers would require $32^{20} \approx 1.3 \times 10^{30}$ different input vectors to cover the entire input space.

This section presents a static error estimation technique that eliminates the need for simulation during quality evaluation at the kernel level. We make two assumptions here: 1) the only operations involved are additions and multiplications, and 2) input data (X and Y) are independent. Assumption 1 is satisfied in both kernel functions in Table 1 and in many error-tolerant application domains. Assumption 2 is necessary to prevent, for example, the product $X * Y$ reducing to certain forms of X^2. If a squaring operation is treated as a normal two-operand multiplication, the estimation accuracy will be significantly lower. In the DOT-PRODUCT kernel, the probability of any $X_i = Y_i$ is quite low so

this assumption is usually satisfied. Estimation of the squaring operation in L2-NORM will be discussed in Section 4.3. All the adders and multipliers in this discussion will be 64-bit wide. The number representation is 2's complement 4_60, with 4 bits (including sign bit) before the decimal point and 60 bits after. In multiplication, the product format is 8_120. All input data are scaled to prevent overflow during computation.

4.1 Probability Mass Function (PMF)

Probability Mass Function (PMF) is a way of representing the statistical distribution of any discrete data/error. It can be visualized as a bar chart on the magnitude vs. frequency plane as shown in Figure 1.

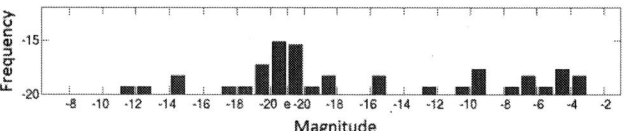

Figure 1. PMF examples

Each bar indicates a non-zero data probability. The location of a bar on the x-axis indicates the magnitude range of the data and the height indicates its frequency of occurrence. The taller a bar is, the more frequent the data occur. Both the x-axis and y-axis are logarithmic-scaled in order to cover a wider frequency-magnitude range. For example, a bar bounded by marker -8 and -7 with a height -10 means that the probability of observing the data between magnitude 2^{-8} and 2^{-7} is 2^{-10}. The e symbol in the middle of the x-axis represents zero; thus, bars to the left have negative magnitude and those to the right have positive magnitude. The sum of the heights of all the bars in a PMF is equal to the probability of the data being non-zero (P_{NZ});. The probability of zero is therefore implicitly obtained by $1-P_{NZ}$. When PMF is used to represent an error distribution, it is possible that $P_{NZ} < 1$. In this case P_{NZ} represents the total error probability P_e and $1-P_e$ gives the error-free probability. Within each bar, the data is assumed to be uniformly distributed.

4.2 Modified Interval Arithmetic (MIA)

Interval Arithmetic (IA) [10] is a classical method to estimate variable ranges during numerical computations. It uses a single interval $[x_l, x_r]$ to represent each variable. When the variable takes part in computation, its interval goes through corresponding IA to produce the output interval. Provided that data are not correlated, the bounds given by IA are tight. However, in many cases data and error distributions such as in Figure 1 cannot be represented by a single uniform distribution.

Modified Interval Arithmetic (MIA) [11] extends IA by using *multiple* intervals to represent a distribution to enhance accuracy. MIA can be easily mapped to PMF: each PMF bar corresponds to one interval in MIA. The entire MIA can be formalized as

$$MIA(x) = P(a \leq x \leq b), \text{if } a \leq x \leq b$$

When an error distribution is represented in MIA, the total error probability is given by $\int MIA(x)$.

When two intervals operate with each other, their resulting interval observes the following rules:

$$[x_{l1}, x_{r1}] + [x_{l2}, x_{r2}] = [x_{l1} + x_{l2}, x_{r1} + x_{r2}]$$
$$[x_{l1}, x_{r1}] * [x_{l2}, x_{r2}] = [\min(x_{l1}x_{l2}, x_{l1}x_{r2}, x_{r1}x_{l2}, x_{r1}x_{r2}),$$
$$\max(x_{l1}x_{l2}, x_{l1}x_{r2}, x_{r1}x_{l2}, x_{r1}x_{r2})]$$

For operations between two MIAs, each IA from the first MIA must perform that operation with each IA from the second MIA and the resulting IAs are merged into a single MIA. While merging, IAs of the same intervals are combined into one IA with its probability being equal to the sum of the constituent IA probabilities.

4.3 Propagating MIA across IHW

Rules in the previous subsection assume precise operation. They must be modified to account for imprecise operators.

The first step is to use a common data structure (MIA_d, MIA_e) to represent any data during imprecise operation. MIA_d is the error-free MIA obtained assuming all operators are precise, while MIA_e is the pure error MIA introduced by imprecise operators. The sum of MIA_d and MIA_e gives the actual data MIA.

It then becomes necessary to build a model to obtain the output (MIA_{d_out}, MIA_{e_out}) from input (MIA_{d_in}, MIA_{e_in}). The imprecise operator (marked with *) will also introduce MIA_{e_op}, which can be regarded as additive noise to the system. We have derived the relationships between these quantities (Figure 2).

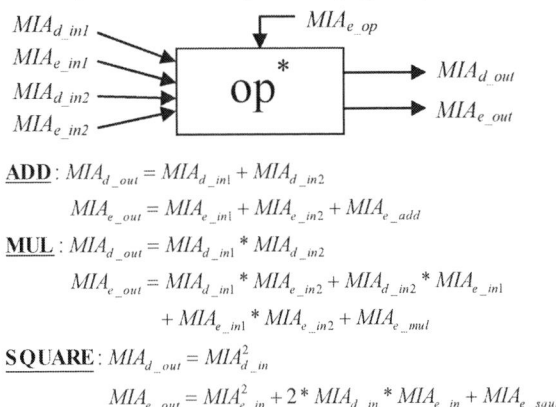

ADD : $MIA_{d\_out} = MIA_{d\_in1} + MIA_{d\_in2}$

$MIA_{e\_out} = MIA_{e\_in1} + MIA_{e\_in2} + MIA_{e\_add}$

MUL : $MIA_{d\_out} = MIA_{d\_in1} * MIA_{d\_in2}$

$MIA_{e\_out} = MIA_{d\_in1} * MIA_{e\_in2} + MIA_{d\_in2} * MIA_{e\_in1}$
$+ MIA_{e\_in1} * MIA_{e\_in2} + MIA_{e\_mul}$

SQUARE : $MIA_{d\_out} = MIA_{d\_in}^2$

$MIA_{e\_out} = MIA_{e\_in}^2 + 2 * MIA_{d\_in} * MIA_{e\_in} + MIA_{e\_square}$

Figure 2. MIA propagation rules for ADD/MUL/SQUARE

Operations between MIAs follow the rules given in Section 4.2. Notice that SQUARE is separated from MUL because it cannot be obtained from simple MIA operations such as * and +. Even if X and Y have the same distribution, the distribution of $X * Y$ and X^2 will be different. The modeling of SQUARE will rely on characterization.

MIA_{e_add}, MIA_{e_mul} and MIA_{e_square} are attributes of the operator determined by the circuit design parameters. They can be obtained by simulation. The process of obtaining MIA_{e_op} through simulation is called *characterization of IHW*. To characterize ADD and MUL, we randomly draw data from single bars from both inputs' MIAs (i.e., draw first operand from $[2^i, 2^{i+1}]$ and draw second operand from $[2^j, 2^{j+1}]$) and perform the imprecise operation. Simulation is made possible by creating a functional model of the imprecise adders and multipliers written in C. The resulting MIA_d and MIA_e are then stored into a matrix at index (i, j). When the entire matrix is populated, we can later use it to quickly retrieve MIA_{e_op} during MIA propagation. For the unary operator SQUARE, the result is stored in a vector instead of a matrix and we need two vectors for SQUARE: one for looking up errors (MIA_{e_square}), and the other for looking up squared data ($MIA_{d\_in}^2$ and $MIA_{e\_in}^2$). IHW can be characterized *a priori* and each IHW configuration (i.e., a unique setting of *BPB, L* and *K*) needs to be characterized only once. The characterization data can then be reused many times for different kernel input workloads.

In summary, kernel-level MIA propagation follows three steps:

1) Construct the characterization vector/matrix by simulating the IHW with inputs being drawn from various $[\pm 2^i, \pm 2^{i+1}]$ intervals.

2) During propagation, use the input MIAs to look up the characterization vector/matrix to obtain MIA_{e_op}.

3) Apply rules in Figure 2 to obtain output MIA.

Step 2 and 3 may need to be repeated because the output MIA normally becomes the input MIA of the next round of computation. The final MIA_d and MIA_e accurately describe the data and error distribution of the kernel output and they can be used to evaluate output quality. Common quality metrics such as error rate and mean error magnitude are computed as follows:

$$error\ rate = \int MIA_e(x)$$
$$mean\ error\ magnitude = \int |x \cdot MIA_e(x)|$$

Static MIA propagation is much faster than Monte Carlo simulation because no actual computation is performed. It is the distributions (in the form of MIA) rather than actual data that are being propagated.

4.4 Experimental Results

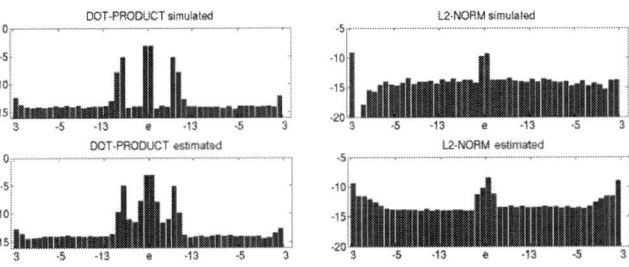

Figure 3. Error MIAs of DOT-PRODUCT and L2-NORM

Figure 3 shows the final error MIAs after performing a size-25 DOT-PRODUCT and a size-49 L2-NORM using both Monte Carlo simulation and static estimation. DOT-PRODUCT contains an ACA adder with K=16 and an ETAIIM multiplier with BPB=8 and L=4; L2-NORM contains an ACA adder with K=16 and an ACA multiplier with K=24. Table 5 compares the speed and accuracy of simulated and estimated error MIAs. All experiments are run on a dual-core Xeon 2.4GHz with 32GB memory. The simulation size is 500,000 and is regarded as the ground truth. As seen in the table, the speed improvement is dramatic and the simulated and estimated error distributions are very close. For example, a Hellinger distance[2] of 0.05 is comparable to 1 million random samples from two uniform distributions between [-1, 1] generated by Matlab's default Mersenne Twister algorithm [13].

Table 5. Speed and accuracy comparison between simulation and static estimation

Kernel	Sim. time	Est. time	Hellinger distance
DOT-PRODUCT	565 hr	13 s	0.05
L2-NORM	620 hr	6 s	0.02

5. QUALITY-AWARE ENERGY MINIMIZATION FLOW

The energy-quality optimization problem can be formulated in many different ways, such as *energy^a/quality^b* cost minimization or quality maximization subject to an energy constraint. This paper focuses on solving the quality-constrained energy minimization problem:

```
minimize:     E(x₀, x₁, ..., xₙ)
subject to:   Q(x₀, x₁, ..., xₙ) >= Q₀
```

---

[2] Statistical measure of similarity between two distributions – smaller values indicate higher similarity [12].

where E denotes the energy consumed while performing a kernel computation; Q denotes the resultant quality. $x_0, x_1, \ldots, x_n$ are circuit structural parameters such as *BPB*, *L* and *K*. Assuming the adders and multipliers are restricted to 64-bit, then the x vector for the DOT-PRODUCT kernel is as follows:

$$[\text{add}_{mode}\ BPB_{add}\ L_{add}\ K_{add}\ \text{mul}_{mode}\ BPB_{mul}\ L_{mul}\ K_{mul}]$$

$\text{add}_{mode}/\text{mul}_{mode}$ is an integer representing IHW type: 0=KSA, 1=ACA, 2=ETAIIM, 3=RCA. L2-NORM needs four additional parameters for its subtractor. Circuit operating conditions such as V_{dd} and frequency can also be included in the x vector, and it is part of the ongoing work of combining IHW with VOS. There are certain restrictions on each parameter, such as the requirement that the adder width (64) must be divisible by *BPB* and *BPB* × *L* cannot exceed 64. Parameters will be swept in their valid ranges only.

Including precise (KSA/ACA) designs, there are a total of 39 adder designs and 101 multiplier designs. DOT-PRODUCT needs 1 adder and 1 multiplier, forming a space of 8 variables and 3939 design points. L2-NORM needs 2 adders and 1 multiplier, forming a space of 12 variables and 154,000 points.

Since all the parameters must be integers, this is an integer programming problem. Matlab offers a genetic algorithm function (GA) to solve these types of problems. It requires two routines to calculate E and Q respectively. For energy calculation, parameterized RTL models were developed for ACA/ETAIIM adders and multipliers and the RTL for KSA/RCA was obtained online [14]. We then synthesized the models into netlists using Cadence RC Compiler in ST 130nm CMOS technology and simulated 1000 random additions and 100 random multiplications using Cadence Ultrasim. Energy per operation can be extracted from the simulation waveforms. An energy model is subsequently built using curve-fitting to extrapolate to the entire parameter space. For simplicity, the energy consumed in the control logic is ignored and the sum of ALU energies is used to represent the energy of the kernel.

For calculation of quality, MIA propagation was implemented in C++ as an extension to the *libaffa* project [15]. The workload is written into a text file with each line in the following format:

```
MUL ETAIIM 8 4 0 4 60 -1 1 -1 1
```

This specifies the operator's parameters (ETAIIM multiplier with *BPB*=8, *L*=4), input format (4_60), and input data ranges ([-1, 1]). A program parses this file and the characterization vector/matrix files, performs the MIA propagation, and writes the output data and error MIA into a result file. A final Matlab script extracts error rate and mean error magnitude metrics from the result file.

5.1 Experimental Results

The methodology was tested on two kernels: size-8 DOT-PRODUCT with inputs in [-1, 1] and size-10 L2-NORM with inputs in [-0.25, 0.25]. Their sizes and dynamic ranges are based on the actual computation and data range profiled while running their corresponding applications. Two quality metrics are evaluated: error rate and mean error magnitude. By setting the quality constraint at different values between $[2^{-10}, 2^{-1}]$, the optimizer is able to produce the energy-quality tradeoff curves in Figure 4. As a comparison, we also show curves obtained by running an exhaustive search on all possible design points. In all four figures, the optimizer curves follow the exhaustive-search curves with a maximum deviation of 2%. Both kernels enjoy a region of about 10% energy reduction with graceful quality degradation. All the curves are significantly lower than the lowest

energy achievable by precise designs (136.44pJ for DOT-PRODUCT and 140.4pJ for L2-NORM).

6. APPLICATION-LEVEL ANALYSIS

Since the application-level quality can only be obtained through simulation, it is difficult to extend the previous methodology to the application level. Simulating the application with IHW is usually 2-3 orders of magnitude slower than with precise hardware, because the host machine cannot use a single ALU instruction to perform an imprecise operation. However, kernel-level solutions can facilitate the application-level exploration process. The first step is to solve the kernel-level problem multiple times using static analysis, each time with a different quality constraint value. Then, assuming application-level quality is a monotonic function of kernel-level quality, the application can be simulated using only the points identified during the kernel-level exploration. The same genetic algorithm (GA) can then be applied to obtain the minimum-energy point given an application-level quality requirement. This section presents experimental results at the application level assisted by kernel-level exploration. The goal of these experiments is to demonstrate the energy-quality behavior of different applications under IHW implementation and the benefits of the proposed methodology.

The three applications chosen to evaluate the proposed methodology are shown in Table 1. Leukocyte Tracker implements an object-tracking algorithm [16] in which an important step is to compute the sum of gradients on the 8 neighboring pixels. SVM is a classification algorithm that consists of a training stage and a prediction stage. The training stage involves computing the Euclidean distance of two data points (called radial basis function) in order to map them into a higher dimensional space. K-means is a data clustering algorithm; the basic operation is calculating the distance between two data points. The Euclidian distance is commonly used. Both K-means and SVM use the L2-NORM kernel, whereas Leukocyte Tracker uses the DOT-PRODUCT kernel. In each application, the corresponding kernel represents a significant percentage of the runtime (Table 1). The source code for Leukocyte and K-means is obtained from the Rodinia benchmark suite [17] and SVM from libSVM [18]. All benchmarks provide sample input data. In Leukocyte tracker we tracked 36 cells in 5 frames; in SVM we attempted to classify 683 breast cancer data points with 10 features into 2 classes; in K-means, we tried to cluster 100 data points with 34 features into 5 clusters.

Quality metrics for the three applications are defined as follows. For Leukocyte, the center locations of the tracked cells are compared with the locations returned by the precise implementation. The *average cell-center deviation* serves as a good negative quality metric. *Classification accuracy* is a well established quality metric for SVM. Finally, for K-means, *mean centroid distance* [3] is used.

Before simulation, the programs are first profiled to determine the dynamic range of data during kernel computation. If the dynamic range is greater than the characterized data range, it is necessary to perform scaling on the input and output data. Certain applications, such as SVM and Leukocyte, already incorporate data normalization into their algorithm so no scaling is necessary.

The design points returned during kernel-level optimization are then used to rewrite the kernel portions of the three applications using those imprecise designs.

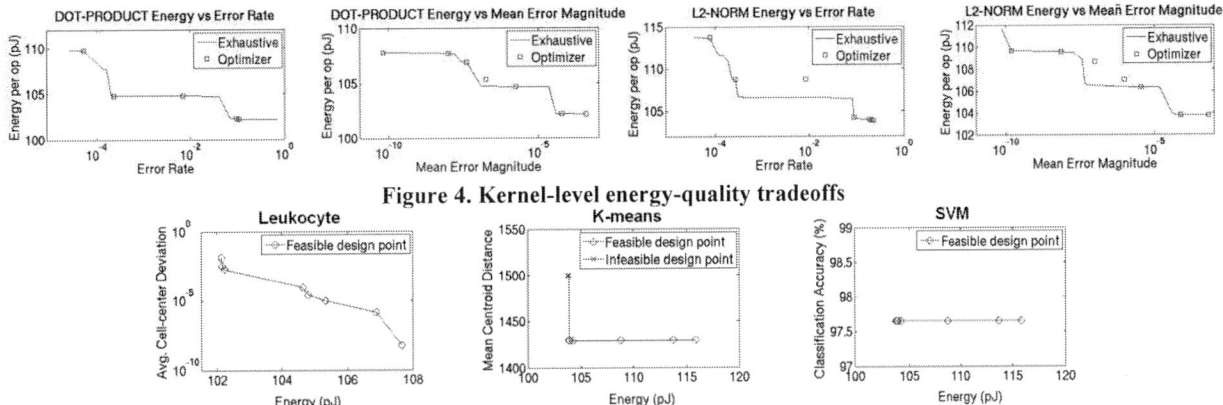

Figure 4. Kernel-level energy-quality tradeoffs

Figure 5. Application-level energy-quality tradeoffs

The final application-level energy-quality tradeoff curves are shown in Figure 5. Since running a SPICE simulation of the entire application to obtain its energy is prohibitively slow, the kernel's energy was used to represent the entire application's energy. Among the three applications, Leukocyte has a smooth quality-energy transition region. At its lowest-energy point (102.24pJ), the mean deviation from precise outputs is merely 0.1 pixels. Its energy is 25% lower than the 136.44pJ precise design. For K-means, the mean centroid distance remains unchanged (1429.22) above the 103.8pJ energy point (i.e., a 26% reduction over precise design). Any design below that energy point failed to converge during simulation. A similar situation is observed in SVM where the critical energy point is 103.76pJ.

Table 6. Number of designs points simulated

Search method	Leukocyte Tracker	SVM	K-means
Exhaustive search	3,939	153,621	153,621
GA (app-level)	887	1,343	1,343
Proposed methodology	**15**	**17**	**17**

Table 6 compares the number of design points that needed to be simulated in order to generate the application-level energy-quality tradeoff curves in Figure 5. Exhaustive search simulates all the design points once, while applying GA at the application-level simulates only a subset. The proposed methodology simulates the least number of design points because it only chooses those points on the optimal kernel-level energy-quality curves.

7. CONCLUSIONS AND FUTURE WORK

This paper presents a methodology to find the lowest-energy design for certain computation kernels given a quality-constraint. This methodology leverages IHW with design-time structural parameters to achieve energy-quality- tradeoffs. It requires no simulation at the kernel level, and the simulation effort at the application level is significantly reduced. Experiments show that the methodology can produce results close to exhaustive search and the runtime is orders-of-magnitude shorter than Monte Carlo simulation. Extending this methodology to support VOS and peak error bounding estimation are valuable future research projects.

8. ACKNOWLEDGMENTS

This work was supported in part by the National Science Foundation, under grants IIS-0612049 and CNS-0831426.

9. REFERENCES

[1] Shim, B., Sridhara, S., Shanbhag, N. 2004, Reliable low-power digital signal processing via reduced precision redundancy, *IEEE Transactions on VLSI Systems*, 12(5):497-510.

[2] Mohapatra, D., Karakonstantis, G., Roy, K. 2009, Significance driven computation: a voltage-scalable, variation-aware, quality-tuning motion estimator, *ISLPED*, pp. 195-200.

[3] Mohapatra, D., Chippa, V.K., Raghunathan, A., Roy, K. 2011, Design of voltage-scalable meta-functions for approximate computing, *DATE*, pp. 1-6.

[4] Kahng, A., Kang, S., Kumar, R., Sartori, J. 2010, Slack redistribution for graceful degradation under voltage overscaling, *ASP-DAC*, pp. 825-831.

[5] Verma, A.K., Brisk, P., Ienne, P. 2008, Variable latency speculative addition: A new paradigm for arithmetic circuit design, *DATE*, pp. 1250-1255.

[6] Huang, J., Lach, J. 2011, Exploring the fidelity-efficiency design space using imprecise arithmetic, *ASP-DAC*, pp.579-584.

[7] Zhu, N., Goh, W.L., Yeo, K.S. 2009, An enhanced low-power high-speed adder for error tolerant application, *ISIC*, pp. 69-72.

[8] Ercegovac, M.D., Lang, T. 2004, *Digital Arithmetic*. Morgan Kaufmann Publishers.

[9] Wallace, C.S. 1964, A suggestion for fast multipliers. *IEEE Trans. Electron. Comput.* EC-13(1):14-17.

[10] Moore, R.E. 1966, *Interval Analysis*, Prentice-Hall.

[11] Huang, J., Lach, J., Robins G. 2011, Analytic error modeling for imprecise arithmetic circuits, *SELSE*.

[12] Nikulin, M.S. 2001, Hellinger distance, *Encyclopaedia of Mathematics*, Springer, ISBN 978-1556080104.

[13] Matsumoto, M., Nishimura, T. 1998, Mersenne twister: a 623-dimensionally equidistributed uniform pseudo-random number generator, *ACM Transactions on Modeling and Computer Simulation*, 8(1):3-30.

[14] http://www.aoki.ecei.tohoku.ac.jp/arith/mg/index.html

[15] http://savannah.nongnu.org/projects/libaffa

[16] Ray, N., Acton, S.T. 2004, Motion gradient vector flow: an external force for tracking rolling leukocytes with shape and size constrained active contours, *IEEE Transactions on Medical Imaging*, 23(12):1466-1478.

[17] Che, S., Boyer, M., Meng, J., Tarjan, D., Sheaffer, J.W., Lee, S.-H., Skadron, K. 2009, Rodinia: A benchmark suite for heterogeneous computing, *IISWC*, pp. 44-54.

[18] Chang, C., Lin, C. 2011, LIBSVM: a library for support vector machines, *ACM Transactions on Intelligent Systems and Technology*, 2(1)27:1-27:27.

On the Exploitation of the Inherent Error Resilience of Wireless Systems under Unreliable Silicon

Georgios Karakonstantis[†], Christoph Roth[‡], Christian Benkeser[‡], Andreas Burg[†]

[†]Telecommunications Circuits Lab (TCL), EPFL, Lausanne, VD 1015, Switzerland
[‡]Integrated Systems Lab (IIS), ETH, Zurich, ZH 8092, Switzerland
{georgios.karakonstantis, andreas.burg}@epfl.ch, {rothc, benkeser}@iis.ee.ethz.ch

ABSTRACT

In this paper, we investigate the impact of circuit misbehavior due to parametric variations and voltage scaling on the performance of wireless communication systems. Our study reveals the inherent error resilience of such systems and argues that sufficiently reliable operation can be maintained even in the presence of unreliable circuits and manufacturing defects. We further show how selective application of more robust circuit design techniques is sufficient to deal with high defect rates at low overhead and improve energy efficiency with negligible system performance degradation.

Categories and Subject Descriptors

B.8.2 [**PERFORMANCE AND RELIABILITY**]: Performance Analysis and Design Aids.

General Terms

Algorithm, Design, Reliability.

Keywords

Error-Resiliency, Memory Failures, Wireless Communication Systems, Energy-Efficiency, Reliability, Yield.

1. INTRODUCTION

With the enormous success of wireless communication systems in the last decade, users are asking for ever higher data rates and better quality of service (QoS). However, sophisticated algorithms/systems with increasing percentage of memory components are required in the transceiver IC to meet the throughput requirements of latest wireless communication standards [1]. Unfortunately, this algorithm-complexity increase and especially the exploding memory requirements of the latest communication standards lead to a paramount increase of power consumption, making energy efficiency one of the main challenges in the design of emerging wireless systems.

Though several schemes exist that try to address the increased power consumption, one of the most effective techniques for low power implementation is still considered to be voltage scaling (VS) due to the quadratic dependency of power consumption on voltage [1,2]. However, VS reduces the circuit performance and increases circuit sensitivity to parametric variations that originate from nanometer device sizes and inaccuracies in the delicate fabrication processes [2-5]. Such variations not only lead to delay and memory failures, that could even worsen over time (i.e., due to aging), but also increase the spread in leakage current, making it more difficult to meet today's strict throughput and energy

requirements with decent yield. In addition, the shrinking of dimensions to 65nm and below increases layout density, which is of particular importance for area-efficient memory design, but at the same time, reduces the amount of charge required to upset a circuit node and raises the likelihood of having a large number of soft errors on chip [5, 6].

Several approaches exist today that try to address both power consumption and parametric variations simultaneously. However, they often lead to significant area overhead and limit the gains in power consumption since they are based, for instance, on the addition of redundant hardware [2,7]. Interestingly, as the percentage of memory components in wireless systems increases (crucial for supporting the large data load), the overhead of such techniques makes them prohibitive, reducing their viability. For instance, error correction coding (ECC) and novel bit-cell architectures (i.e., 8T) might tackle the high failure probability of traditional 6 transistor (6T) bit-cells (under variations and VS) but can lead to more than 50% power overhead [5-9]. However, although in general purpose processors/systems the overhead of such techniques might still be acceptable due to the equal significance of all data, it might be possible in application-specific systems to depart from the 100% error-free computing paradigm. By accepting dies even with a number of defects or restricting application of robust techniques to only the most critical parts of the system we could improve yield and energy efficiency at no/limited overhead even in the presence of hardware errors (due to VS and/or variations). While several approaches tried to take advantage of such an observation in order to address the issues of power and variations in multimedia systems and individual DSP blocks [2, 8, 9], to the best or our knowledge, no such effort has targeted wireless communication systems so far. Therefore, there is a need to study the impact of hardware errors induced by parametric variations and VS on the performance of such systems, which are ubiquitous components of all today's portable devices. Interestingly, the main characteristic of such systems is that corresponding receivers with sophisticated communication algorithms are able to recover the transmitted data even when the received signal has been heavily distorted by noise and interference due to bad wireless channel conditions. This robustness of such systems motivates the investigation of their inherent resilience against unreliable silicon implementations and raises questions regarding the limits of this error resilience and how it could be improved at low cost.

To this end, in this paper we investigate the impact of hardware defects/errors induced for example by VS and parametric variations on the performance and yield of wireless communication systems using the latest high-speed packet access evolution of the 3G mobile cellular standard HSPA+ [10]. Our study focuses on a large and power hungry memory required for the hybrid automatic repeat request (HARQ) block that is critical for the correct and high throughput operation of the overall system. Our contributions can be summarized as follows:

- Develop a system-level fault simulation approach for capturing the effects of errors on the system performance and relating the

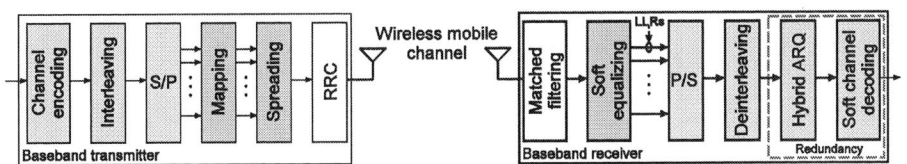 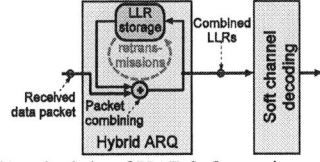

Figure 1: (a) Wireless Communication System (HSPA),   (b) Principle of HARQ Operation

results to the yield in a meaningful way.

- Exploit the resilience limits of communication systems moving away from the 100% reliable computing paradigm. Interestingly, we find that the system is able to operate correctly even in the presence of hardware errors, but as such errors increase beyond a critical rate (making them comparable to channel-induced errors) the system throughput deteriorates significantly. This finding allows us to actually accept dies with up-to a specific number of defects leading either to a better yield or enabling energy reduction (since such dies may operate at low voltages).

- Explore low-overhead techniques that can improve robustness and facilitate aggressive VS, thus improving energy efficiency under very high defect rates. Specifically, we show how selective application of robust circuit techniques, such as 8T cells, only on some critical parts of the system (that are identified by our study) can reduce the overhead of conventional conservative robustness techniques that aim at restoring 100% reliable operation while ensuring minimum system throughput at high yield loss under high defect rates.

The rest of this paper is organized as follows. In Section 2 we briefly present the basic characteristics of a modern HSPA+ communication system which serves as an excellent and commercially relevant test vehicle for our study. Section 3 discusses the various failure mechanisms and their impact on memory cells/arrays and yield. Section 4 presents our approach for studying the impact of errors on throughput and yield. Section 5 then reveals the resilience limits of communication systems to hardware defects and discusses the achieved yield improvement. Section 6 proposes low overhead techniques for improving the robustness of the system. Conclusions are drawn in Section 7.

2. ERROR RESILIENCE OF WIRELESS SYSTEMS

In the following we briefly summarize an HSPA+ system as specified by the 3GPP standard suite [10], which serves as a challenging vehicle to verify our findings and to demonstrate the effectiveness of our proposed low-overhead techniques for modern wireless communication systems. Arguing that a communication system is designed to cope with noisy data, we will then highlight the inherent error resilience of such a system and exploit its characteristics that may help in tolerating hardware induced errors.

2.1. HSPA+ System Model

HSPA+ is based on code division multiple access (CDMA), a channel access method where a single wireless transmission channel is simultaneously shared by several users. A simplified block diagram of an HSPA+ baseband transmitter and receiver separated by a noisy mobile channel is shown in Fig. 1(a).

Baseband transmitter: On the transmit side, a sequence of data bits, referred to as data packet, is encoded using a high-performance error-correction code and then passed through an interleaver which generates a pseudo-random permutation of the input bit stream. This serial bit stream is then converted into parallel streams and each of these streams is individually modulated (with either 16QAM or 64QAM) before they are spreaded and multiplexed to a single stream in the spreading unit.

Finally, this stream of multiplexed data symbols modulates a root-raised cosine (RRC) pulse-train which is then transmitted over the mobile channel.

Baseband Receiver: The main task of the receiver is to extract the originally transmitted bit stream from the distorted received signal using sophisticated equalization and channel decoding algorithms, which are the most challenging blocks in terms of implementation complexity. While the equalizer attempts to undo the destructive effects of the mobile channel, the decoder corrects errors in the equalized data packet, exploiting the redundancy and structure in the transmitted bit stream imposed by the channel encoder. Rather than deciding on hard bits, a soft-decision equalizer produces reliability-indicators, referred to as log-likelihood ratios (LLRs), representing the probability for each bit being logic-0 or logic-1. The magnitude of an LLR reflects the confidence, and the sign shows whether a decision would be in favor of logic-0 or logic-1, respectively. A soft-decision channel decoder works on LLRs instead of simple bits. Clearly, a soft receiver (based on LLRs) implies higher implementation complexity in terms of silicon area and power consumption compared to a hard receiver but the considerable gain in performance, required to fulfill the demanding 3GPP specifications, justifies the overhead. An important performance metric in such a system is the block-error rate (BLER) which is the probability that the channel decoder fails to decode a data package. This metric is usually measured as a function of the signal-to-noise ratio (SNR) at the input of the receiver, representing the ratio of the user signal power over the noise and interference power.

Hybrid automatic repeat request (HARQ): A key feature on the terminal side of an HSPA+ downlink is the HARQ operation, which allows for rapid retransmission of erroneously received data packets. HARQ is a crucial mechanism to enable *high average throughput*, the ultimate performance metric of such systems, over a wide range of rapidly varying mobile channel conditions. The main principle of HARQ is depicted in Fig. 1(b). The received data packets are buffered in the LLR storage prior to decoding. In case the channel decoder fails to decode a data packet, a retransmission is requested by the receiver. In contrast to traditional ARQ-based communication systems where simply the retransmitted data packet is decoded, the HARQ operation combines the retransmitted data packet with the (stored) information (i.e., LLRs) of previous transmissions, increasing the probability of correct decoding. The higher the quality of the combined LLRs used by the soft-decision channel decoder, the lower the average number of retransmissions required to successfully deliver a data packet even under channel errors.

2.2. Error Resilience to Channel Noise

The above functionality reveals the main characteristic of such systems; their ability to operate reliably under channel noise. This is clearly indicated in Fig. 2 that depicts the decoding-failure probability of a data packet (i.e., BLER) evolving over the incremental HARQ retransmissions for three different SNR regimes. In the high SNR (29dB) regime, the channel decoder is able to decode roughly 95% of all data packets already after the initial transmission. For the medium SNR (11dB) regime, the channel decoder is still able to deliver a considerable fraction of

511

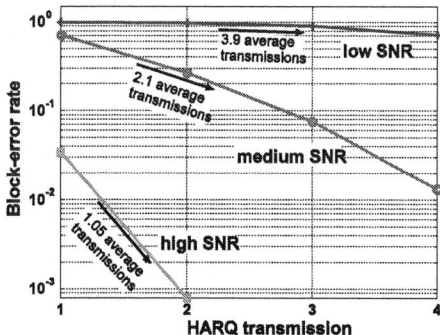

Figure 2: Decoding Error Probability.

the data packets in the initial transmission, revealing the inherent resilience of the system to noisy input data. However, in the low SNR (3dB) regime, the channel-induced noise corrupts the data too severely, and virtually all data packets are scheduled for retransmission. While in a traditional ARQ-based system this would drive the throughput performance to zero at this low SNR, the LLR combination in the HARQ unit increases the decoding probability after each retransmission due to more reliable LLRs as shown in Fig. 2. It is apparent that more retransmissions reduce the throughput.

3. HARDWARE ERRORS AND IMPACT ON YIELD AND MEMORY OPERATION

As explained in the previous section, the receiver's ability to decode the received stream correctly heavily depends on the operation of the HARQ unit. The main component of this block as shown on Fig. 1(b) is the LLR memory that stores the received data packets and combines them with the corresponding retransmissions. Striving for a fully integrated baseband solution, this storage is typically implemented with SRAM memory cells, which account for a considerable fraction of both silicon area and power consumption of the overall system besides the equalizer and channel decoder. The latency in such a complex wireless system combined with the high data rates involved thereby inflate the required HARQ storage size, which can range up to 253 Kb times the number of bits used to quantize an LLR. Unfortunately, while the continuous scaling of devices allows for the realization of such high density memories in a single chip, the small sizes beyond the sub-65nm node make devices more prone to variation-induced defects. To better understand the nature of such defects we briefly explain the basic sources of hardware errors and their impact on yield and SRAM operation. In general, memory failures can be persistent (i.e., failures due to difference in transistor characteristics causing yield loss) or non-persistent (e.g., soft errors due to radiation) and the probability of both of them increases as supply voltage decreases [2, 6, 7].

Parametric Variations: The primary source of device mismatch, which is the dominant failure mechanism in memory cells, is the intrinsic fluctuation of the threshold voltage (V_{th}) of different transistors due to random dopant fluctuations (RDF) [2-8]. Any mismatch in V_{th} of neighboring transistors in an SRAM cell can result in a failure of the corresponding bit cell. For instance, a cell failure can occur due to, i) unstable read (flipping of the cell while reading) and/or write (inability to successfully write to the cell), ii) increase in the cell access time (access time failure), and iii) failure in the data hold capability of the cell in the standby mode. Since these failures are caused by the variations in the device parameters, they are known as parametric failures [11]. The degree of such failures depends on the size/type of the memory bit-cell, but also on the array organization and strongly on the supply voltage (V_{dd}). Specifically, as the on-chip memory

density increases, lowering the supply voltage is the most effective approach for low power operation in order to meet the tight power budgets in wireless communication systems. However, a supply voltage below its nominal value increases the sensitivity of circuits to RDF and thus leads to higher number of failures. Fig. 3 depicts the failure probability of a memory array implemented by medium-sized 6T bit-cells, 15% upsized 6T cells and 8T bit-cells under various voltages in case of slow-fast corner, which was found to be the worst corner for RDF induced memory failures in the 65nm technology node [5, 9]. Such failure rates are directly related to the yield of a memory block. It is apparent that as the effect of intra-die variations increases with technology scaling and lower voltages the memory failures increase and thus yield decreases accordingly. Conventionally, the addition of redundant rows/columns could help to recover from such defects, but as the size of memory and the number of defects increases they are insufficient to avoid yield loss. Moreover, the number and the location of failures due to process variations changes depending on operating condition (e.g., applied V_{dd} and frequency) which cannot be handled efficiently by redundant rows/columns.

Soft Errors: The small size of transistors have made it also easier to upset the stored charge in a node giving rise to soft errors with a rate that is almost constant across technology generations [5]. Such errors do not damage the cell permanently, and studies have shown that they do not depend so much on voltage since they only increase by a factor of 3x for every 500mV decrease in supply voltage as opposed to RDF induced errors that increase by billion times for such a voltage decrease (Fig. 3).

In general purpose systems, techniques such as transistor up-sizing, novel bit-cell configurations (8T) and error correcting codes (ECC) can decrease the failure probability of a memory array to improve yield or enable operation at lower supply voltage. Unfortunately, all these techniques come at an increased cost in terms of silicon area and power consumption overhead. Such additional costs may be prohibitive for wireless communication systems that need to deliver large amounts of data rates at very low energy as part of battery-operated consumer-electronics devices.

The proven inherent resilience of the considered system to channel noise, as discussed in Section 2.2, suggests that such systems may also be able to cope with additional distortions introduced by unreliable hardware. We therefore propose to depart from the 100% error-free computation paradigm and accept hardware-induced errors up to a certain defect rate. This paradigm-change would not only enable more aggressive voltage scaling in wireless communication systems, but it would also facilitate achieving the demanding manufacturing yield targets that are critical for today's cost sensitive applications. In the next sections we investigate the limits of the inherent error resilience of wireless systems by considering the effect of hardware failure

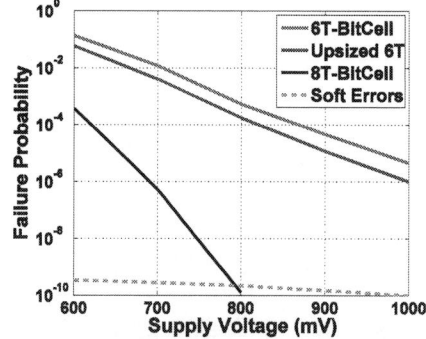

Figure 3: Memory Failure Probability (65nm).

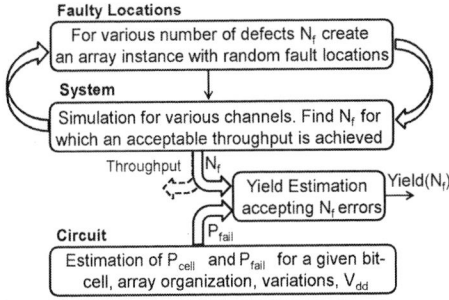

Figure 4: System Level Fault-Simulation Approach.

mechanisms in system-level simulations. Based on these results, we further identify techniques to maintain acceptable system operation beyond this point with minimum additional cost.

4. SYSTEM LEVEL FAULT-SIMULATOR

In the following, we describe our approach for jointly considering circuit misbehavior and the consequences on yield together with the impact on the overall system performance. This analysis will later also be instrumental in identifying the few most critical parts for the operation of the system that may need protection under high hardware failure rates. The primary challenge in estimating the system-performance impact of errors in not-100% operational dies is that meaningful throughput evaluation requires a vast amount of Monte-Carlo simulations averaging over various wireless channel conditions. Unfortunately, as we have departed from the 100% error-free processing paradigm, individual devices may be different since they may be affected by a different number of errors distributed across the storage array according to one out of billions of possible error patterns. To nevertheless capture the effects of the number and the location of defects on the system performance in a meaningful way and to relate the results to yield, we developed the system level fault-simulation methodology depicted in Fig. 4.

Estimation of Cell-Failure probability (P_{cell}): Initially, the failure probability (P_{cell}) of the desired bit-cell type under various degrees of variations and voltage scaling are obtained through Monte-Carlo circuit simulations.

Yield Estimation - (Y): In a conventional, 100% defect-free design the cell-failure probability immediately leads to the failure probability of the overall memory array (P_{fail}) using simple methods that can also consider some advanced robustness techniques such as redundant columns, error correction codes (ECC) and array organizations [6, 7, 11-12]. For example, by assuming that all cell failures are independent, one can easily estimate the array failure probability for an array size of M cells and thus the yield (Y) for this part of the circuit according to:

$$Y(100\% \; reliable) = (1 - P_{cell})^M \qquad (1)$$

However, as discussed above, the inherent error resilience of wireless systems may allow tolerating a limited number of failing cells. This relaxation makes the acceptance of faulty dies possible, which otherwise would be discarded, thus improving the yield. Alternatively, the relaxed selection criterion enables meeting the yield target at a reduced nominal supply-voltage which leads to the desired power reduction. Keeping in mind the impact on the throughput we can investigate the yield that can be achieved by tolerating a number or percentage of faulty cells for a given memory array with size M and cell failure propability P_{cell}. To this end, we redefine yield (Y) for the case where chips with at most N_f faulty cells pass the inspection:

$$Y(N_f) = \sum_{i=0}^{N_f} \binom{M}{i} \times (P_{cell})^i (1 - P_{cell})^{M-i} \qquad (2)$$

The above equation reveals the yield improvement that a manufacturer can get by not discarding chips with a specific

number of faulty cells. Fig. 5 plots $Y(N_f)$ for various P_{cell} and various numbers of N_f. Note that each P_{cell} corresponds to probability of defects due to voltage scaling and parametric variations as we discussed in Section 3. From such a figure we can determine the number of defects that we need to accept for achieving the yield target. For instance in case of $P_{cell} = 10^{-4}$ and $M = 200Kb$, chips with 0.1% defects need to be accepted for meeting the target yield (95%). For determining how many faulty cells N_f can be tolerated we need to evaluate the impact on the throughput of the overall system as we describe next.

Wireless System Simulation - (Throughput): Since we do not require zero defects, we need to assess the worst-case system performance for the dies that pass the selection process (each of which can be affected by different defects within the specified selection criterion). To this end, we consider only the case of N_f defects distributed across the array using random fault-location maps. For a given wireless channel realization, each bit of the received LLRs is mapped to a specific memory cell in the LLR memory array. If the mapped location of the 'bit' indicates a fault in the fault location map, the 'bit' is inverted to indicate a bit-error. These bit flips are considered in the MATLAB Monte-Carlo system simulations and the impact of circuit misbehavior is evaluated using the appropriate system metrics (i.e., average throughput), as also prescribed by the corresponding communication standards.

5. EXPLOITING THE RESILIENCE LIMITS

In this section we evaluate the impact of defects in the LLR storage on the throughput performance of a fully standard-compliant HSPA+ system. We present worst case simulation results for the most noise-sensitive, high throughput 64QAM modulation mode and for a maximum of three retransmissions per data packet over a standard-compliant multipath channel. A minimum mean-square error (MMSE) equalizer is used for the generation of LLRs which are quantized with 10 bits to avoid any throughput-loss due to quantization noise over a wide range of SNR points (according to our simulations of a defect-free system).

Having set the above parameters in our simulation framework we use the approach discussed in Section 4 for injecting errors at random locations of the LLR storage (assuming a medium sized 6T based memory) . In our simulations we cover various choices for N_f. Note that the system-performance results reflect the worst-case behavior of dies with exactly N_f failing cells (i.e., for a given selection criterion) and are thus independent of the failure probability P_{cell}. However, N_f and P_{cell} together define the impact on yield and, due to the dependence of P_{cell} on the supply voltage, also the potential for power savings.

Results: Fig. 6(a) depicts the throughput performance of the considered system for various choices of N_f, specified in % of the size of the LLR storage array. We observe that the throughput is roughly the same as that of the defect-free system (up-to a 0.1% defects), highlighting the inherent resilience of wireless communication systems to unreliable storage. Furthermore, the simulations reveal that the described system is able to meet the required (normalized) throughput (0.53 at 18dB) specified by the standard for this mode of operation (64QAM) withstanding even 10% of defects in LLR storage (corresponding to 2000 defective cells). This indicates that there is no need for protective mechanisms in the LLR storage up to that amount of defects. This resilience allows not only to avoid the cost for protective mechanisms, but also to lower the supply voltage since for example a memory based on conventional 6T cells can function at 0.8V (lower by 200mV compared to 1V in 65nm (Fig. 3, 5)).

513

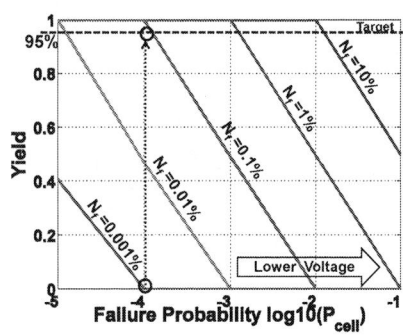
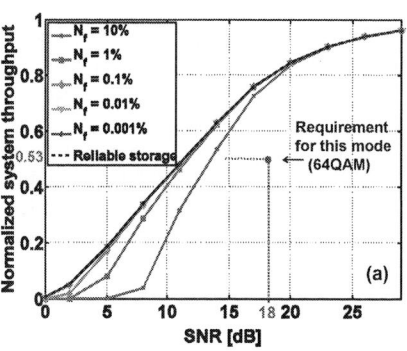
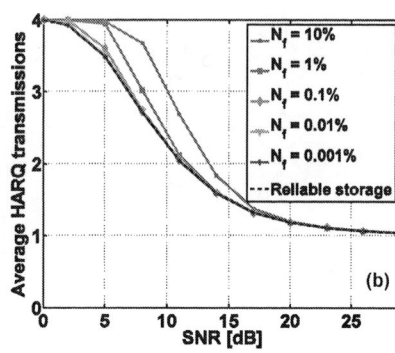

Figure 5: Yield estimation (200Kb array).   Figure 6: (a) Throughput, (b) Average number of transmissions under various defect rates.

As the number of tolerated defects increases beyond 0.1%, the quality of the LLRs deteriorates to a point that becomes dominant over the effects of the signal-distortions due to the wireless channel, increasing the average number of transmissions required to successfully decode a data packet as shown in Fig. 6(b). As outlined in Section 2, this in turn reduces the throughput and increases the overall energy required to deliver a data packet since the entire transmitter/receiver chain is forced to handle the incurred overhead. Hence, a further yield increase under severe process variations or further voltage scaling (increasing P_{cell}) without degrading yield, requires more sophisticated measures to maintain good system performance (throughput) while tolerating more faults.

6. IMPROVING RESILIENCE AND YIELD

In this section we discuss how selective application of more robust circuit design techniques to the LLR storage is sufficient for allowing the wireless system to operate reliably at a low cost in case of high defect rates.

6.1. Proposed Storage Approach

As discussed above, conventionally, designers would apply expensive methods in terms of area and power to the complete memory array such as ECC or larger transistor/different type of cells (i.e., 8T) in order to enhance the robustness of the memory. However, not all bits are of equal weight (e.g., the sign information is of higher importance than the rest bits for the channel decoder). Hence, such expensive techniques for the protection against failures may not be required for all bit-cells. In order to determine the number of LLR bits that need to be protected in order to obtain an acceptable throughput even with a large number of faulty cells in the remaining bits (corresponding to a better yield), we performed a sensitivity analysis by utilizing the approach discussed in Section 4. Specifically, we consider zero or a very low number of tolerated defects ($N_f^{MSB} \leq 0.01\%$)) in the well protected bit locations starting from the most

significant bit (MSB), while in turn considering a high number of tolerated defects in the less well physically protected rest of the bits ($N_f \geq 0.1\%$). In other words, for the sensitive parts of the data we propose to meet the yield target by reducing the cell-failure probability using for example 8T SRAM cells. In exchange, we continue to use area- and energy- efficient, but unreliable 6T cells for the less sensitive parts (bits). We speculate, that we can now tolerate an even higher number of faulty cells for these less significant bits to achieve the yield target even under more severe process variations or at lower voltages. The corresponding analysis reveals that protection of few MSBs is sufficient and allows for a high number of accepted defective cells in the remaining bits without jeopardizing throughput. This is evident by comparing Fig. 7 to Fig. 6, where it is shown that by protecting only the 3-4 MSB bits (rather than all bits) is sufficient for limiting the throughput loss even under 10% defect rates.

6.2. Efficiency of Protection

Of course by protecting more bits, a higher number of defects can be accepted improving the throughput and yield. However, a higher number of protected bits increases the associated area penalty proportionally. In Fig. 8 we plot the throughput gain (Throughput(N_f)/ defect-free Throughput) achieved by protecting various number of bits divided by the area overhead needed by using more robust cells in the case of $N_f = 10\%$. We assume the use of 8T cells for the protection of bits and we plot the overhead of a hybrid array (8T and 6T cells) over the area of a 6T-based array. By focusing on the point where the system with unprotected storage cells experiences the worst-case throughput penalty compared to the error-free case (here at an SNR=8dB), we observe that protecting 4 bits is optimum. The protection of more bits causes further increase in silicon area without any significant throughput improvement. This observation also proves that the conventional approach of using equal protection for all bits cells is not as efficient as protecting few MSB bits only.

Similarly we can argue that the use of ECC protection of all

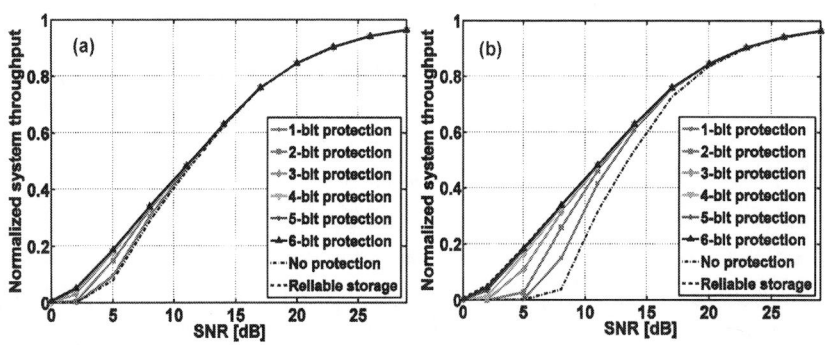
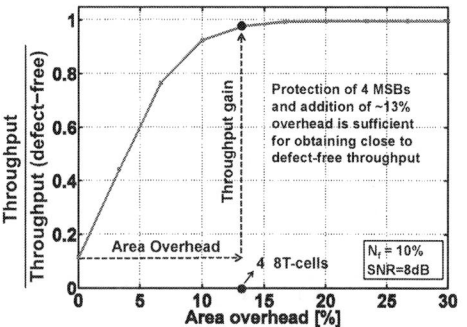

Figure 7: (a) Throughput after protecting various numbers of bits under various   Figure 8: Protection efficiency.
defect rates (a) N_f=1% in 6T cells, (b) N_f=10% in 6T cells.

514

bits is not efficient either. Specifically, a single detection and correction ECC method could be used for the protection of all 10 bits. However, this would result in 35% area overhead compared to the 6T-based array, since 4 redundant bits are required for a single error correction according to Hamming based ECC [5, 6, 8]. Furthermore, the use of higher order ECC for the protection of more bits increases the area and power by more than 50% [6].

Note that for the implementation of such a hybrid memory we could also utilize upsized 6T cells for the protection of the required bits. However, it was shown recently that 8T cells lead to lower area and power overhead for the same stability improvement [9]. In any case, the same techniques applied in the design of such a hybrid memory for multimedia applications [9] can be utilized also here. Nonetheless, a selective protection of LLR bits is very efficient since it protects only what is necessary for obtaining good throughput limiting any unnecessary overhead of conventional techniques (i.e. use of 8T cells in whole memory).

6.3. Potential for Power Reduction

The improvement in throughput achieved by selective protection of significant data bits translates into improvement of yield and offers the potential for power savings through aggressive voltage over-scaling beyond i) the limit of reliable operation of 6T memory cells and ii) the limits imposed by the inherent error resilience of the fully unprotected system described in Section 5. Specifically, as discussed in Section 6.1 the proposed storage scheme can limit the throughput loss, providing acceptable performance even under high number of defects (1%-10%) which could be induced by aggressive voltage scaling down to 0.6V (Fig. 3, 5, 7(b)). In other words, our proposed storage scheme can allow the operation of wireless system with acceptable throughput even at a low supply voltage for the HARQ memory block that can translate to 30% power savings for that block under an iso-area comparison with a conventional 6T SRAM array [9].

Furthermore, the proposed prefential storage does not only reduce the power locally in the HARQ block but also in the overall system. This is evident if we consider Fig. 6(b) and the number of retransmissions. For instance in case of SNR=9dB we can observe that with the utilized partial protection we need 2.4 retransmissions as opposed to 3.5 retransmissions in case of no memory protection under N_f=10%. Therefore, the preferential storage scheme increases the ability of decoder for correct decoding thus reducing the retransmission rate which has an immediate impact on the whole system energy efficiency.

6.4 Joint Consideration of Bit-Width and Defects

One of the main system level decisions that need to be taken into account, while designing a wireless system is the degree of quantization. Traditionally designers tend to use more bits for the quantization for ensuring minimum quantization noise and thus minimum impact on throughput. However, a high number of bits increase the size of the required storage, making memories not only larger, but also more prone to hardware errors. This reveals that when deviating from the paradigm of 100% correct operation, circuit level limitations should also be considered when making decisions on quantization. The necessity for such considerations is suggested also by Fig. 9. Although the 10-bit quantization introduce more noise than using more bits at the high SNR points, it actually results in a better throughput compared to 11/12 bits (which would be the selection of designers in case that only channel noise is considered) with cell failures. This can be attributed to the fact that the system becomes more sensitive to failures in the memory which due to its larger size (in case of 11 and 12 bits), becomes more prone to hardware errors.

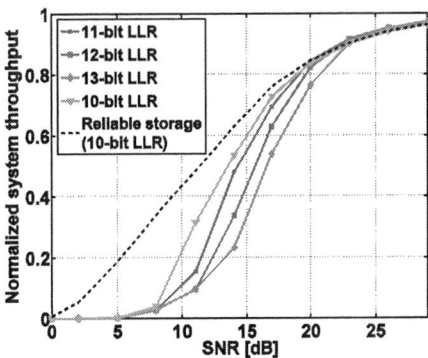

Figure 9: Throughput under various bit-widths (no protection with 10% defects).

7. CONCLUSION

The paper proposes the departure from the paradigm of 100% reliable circuit operation for the design of wireless communication systems. Our study reveals that wireless systems are able to tolerate a considerable number of defects allowing for the acceptance of defective dies. Focusing on the large storage array in the Hybrid ARQ subsystem of the 3GPP HSPA+ standard we show that this not only translates directly to a yield improvement, but also to power savings since circuits can be operated at lower supply voltage. We further show that only partially protecting the memory content ensures reliable operation even under high number of defects at low cost. This preferential storage scheme enables further power savings (through voltage scaling) and reduces the circuit-level overhead required to provide robust operation in sub-65nm process nodes. Overall, our study suggests that taking hardware errors into account already in the system-level design of future wireless systems can be beneficial for achieving robust low power solutions.

8. ACKNOWLEDGMENTS

This research was supported by Swiss National Science Foundation under the project number PP002-119052.

9. REFERENCES

[1] J. M. Rabaey, "Low Power Design Essentials", Springer, 2009.
[2] S. Bhunia, et al, "Low-Power Variation-Tolerant Design in Nanometer Silicon," Springer 2011.
[3] S. Borkar, et al., "Design and reliability challenges in nanometer technologies," *IEEE DAC*, pp.75, 2004.
[4] A. Shrivastava, et al., "*Statistical Analysis and Optimization for VLSI: Timing and Power*", Springer, 2005.
[5] Z. Chishti, et al., "Improving Cache Lifetime Reliability at Ultra-low Voltages," *IEEE MICRO*, 2009.
[6] C. Wilkerson, et al., "Trading off Cache Capacity for Reliability to Enable Low Voltage Operation," *IEEE ISCA*, 2008.
[7] Shi-Ting Zhou, et al. "Minimizing Total Area of Low-Voltage SRAM Arrays through Joint Optimization of Cell Size, Redundancy, and ECC," *IEEE ICCD*, 2010.
[8] Y. Emre, et al., "Memory Error Compensation Techniques for JPEG2000," *IEEE SiPS*, 2010.
[9] I. J. Chang, et al., "A Priority-Based 6T/8T Hybrid SRAM Architecture for Aggressive Voltage Scaling in Video Applications," *IEEE Trans. on CSVT*, 2011.
[10] High speed downlink packet access (HSDPA), Third Generation Partnership Project TS 25.308, Rev. 10.5.0, Jun. 2011.
[11] S. Mukhopadhyay, "Statistical design and optimization of SRAM cell for yield enhancement", *IEEE ICCAD*, 2004.
[12] P. Zuber, et al., "Statistical SRAM analysis for yield enhancement," *IEEE DATE*, 2011.

Near-Optimal, Dynamic Module Reconfiguration in a Photovoltaic System to Combat Partial Shading Effects

Xue Lin[1], Yanzhi Wang[1], Siyu Yue[1], Donghwa Shin[2], Naehyuck Chang[2], and Massoud Pedram[1]

[1]Department of Electrical Engineering, University of Southern California, Los Angeles, CA, USA
[2]Seoul National University, Seoul, Korea

[1]{xuelin, yanzhiwa, siyuyue, pedram}@usc.edu, [2]{dhshin, naehyuck}@elpl.snu.ac.kr

ABSTRACT

Partial shading is a serious obstacle to effective utilization of photovoltaic (PV) systems since it can result in significant output power degradation for the system. A PV system is organized as a series connection of PV modules, each module comprising of a number of series-parallel connected cells. This paper presents modified PV cell structures with integrated switches, imbalanced cell connection topologies for PV modules, and a dynamic programming algorithm to produce near-optimal reconfigurations of each PV module with the goal of maximizing the system output power level under any partial shading patterns. Through simulations, we have demonstrated up to a factor of 2.3X improvement in the output power level of a PV system comprised of 3 PV modules with 60 PV cells per module.

Categories and Subject Descriptors

B.8.2 [**Performance and Reliability**]: Performance Analysis and Design Aids.

General Terms

Algorithms, Management, Performance, Design.

Keywords

Photovoltaic System, Partial Shading, Photovoltaic Module Reconfiguration, Dynamic Programming.

1. INTRODUCTION

Due to increasing appetite for energy sources and environmental concerns about fossil fuels, there has been a growing demand for renewable energy resources, which are clean and eco-friendly (pollution-free.) Among renewable resources, photovoltaic (PV) energy generation techniques have received significant attention, since solar energy is abundant, and can be easily scaled up. Thanks to extensive research efforts on PV energy generation technologies, various scales of PV-powered energy generation systems (PV systems) have been deployed for practical applications, such as PV power stations, solar-powered vehicles, and solar power heating and lighting appliances.

Like most other renewable energy systems, PV systems have a major weakness in that their power output level is greatly affected by environmental conditions. More precisely, solar irradiation received by a PV module is changing frequently, according to the time of day and weather conditions (e.g., a passing cloud.) To provide a rather steady power output level, standalone PV systems are equipped with electrical energy storage (EES) elements. Furthermore, PV modules exhibit highly non-linear current-voltage (I-V) characteristics that change with solar irradiance. Therefore, a maximum power point tracking (MPPT) technique is mandated to extract maximum power from PV modules [1][2]. Recently, the maximum power transfer tracking (MPTT) method, which accounts for changes in the efficiency of the charger, can be more effective than the MPPT methods [3][4].

In a PV system, a string of PV modules (i.e., a PV string) are connected in series to produce a desired voltage level, which is then fed to a charger. This architecture reduces the cost of the PV system due to sharing of the charger among PV modules. We call such a structure the *string charger interface*. Solar irradiations received by PV cells in the PV system, may be different, and such a phenomenon is known as *partial shading*. Unfortunately, the string charger interface is vulnerable to partial shading, which not only reduces the maximum output power of the shaded PV cell itself, but also makes non-shaded PV cells that are in series with the shaded one deviate from their MPPs[1]. This makes the maximum output power of a PV module string much lower than the sum of the MPP power values of all PV cells in the string. Partial shading may also result in multiple power peaks in the power-voltage (P-V) characteristics of a PV string. Therefore the MPPT techniques must be modified to track a global optimum operating point instead of a local optimum [5][6] because the unimodality assumption of the PV string P-V characteristics is the basis for conventional MPPT techniques. But modified MPPT techniques increase the complexity of the PV system controller, while non-shaded PV cells still suffer from power loss due to the deviation from their MPPs caused by shaded cells.

PV module reconfiguration techniques, which have the potential of exploiting MPPs of both non-shaded and shaded PV cells in a partially shaded PV string, may maintain the power output level of PV system under partial shading. Various PV reconfiguration techniques have been proposed [7]~[9]. However, they suffer from one or more of the following limitations.

(1) To compensate the power loss from shaded PV cells, (many) extra PV cells are needed for performing reconfiguration.
(2) There is a lack of systematic and scalable structural support or effective control mechanism.
(3) Variations in the efficiency of the charger or inverter are overlooked, which may result in a sizeable degradation in the system energy conversion efficiency.
(4) The PV system employs an individual charger interface, in which each PV module has an individual charger for

---

[1] MPP stands for maximum power point. On I-V characteristics of a PV cell/module/string, there is a point (V, I) where the power is maximized. This point is the MPP of a PV cell/ module/string.

operating point setting, thereby, increasing the hardware cost of a PV system.

In this paper, we present a dynamic imbalanced PV module reconfiguration method with both a scalable structural support, as well as a systematic and near-optimal control algorithm, to overcome the power output degradation in a PV system under partial shading. We realize imbalanced reconfiguration for PV modules. Different from [10], which always maintains an $n \times m$ configuration for the PV module (where m is the number of parallel-connected elements in each *PV group* [2] and n is the number of groups that are connected in series), our imbalanced reconfiguration method allows more flexible PV module configurations in which there can be an arbitrary number of PV groups connected in series in the PV module and PV groups may have different numbers of parallel-connected PV cells. Our reconfiguration method can utilize both non-shaded and shaded PV cells in a PV module to the largest extent under partial shading.

We present the imbalanced reconfiguration method on the PV system with the more widely-used and cost-effective string charger interface. We introduce an effective reconfiguration control algorithm to realize adaptive and near-optimal PV module reconfiguration for each PV module according to partial shading pattern and charger efficiency variation. The proposed control mechanism is based on dynamic programming with polynomial time complexity. Therefore, it can be incorporated into PV systems with negligible extra computational overhead. Experimental results demonstrate that our proposed reconfiguration method can result in up to a 2.3X output power level improvement, compared with the baseline PV system without PV module reconfiguration method.

2. COMPONENT MODELS
2.1 PV Cell Model and Characterization

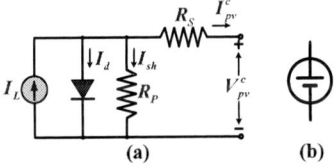

Figure 1. (a) Equivalent circuit and (b) symbol of a PV cell.

We use V_{pv}^c and I_{pv}^c to denote the output voltage and current of a PV cell, respectively. Figure 1(a) shows PV cell equivalent circuit model, with I-V characteristics given by

$$I_{pv}^c = I_L - I_d - I_{sh}$$
$$= I_L(G) - I_0(T)\left(e^{(V_{pv}^c + I_{pv}^c \cdot R_s)\cdot \frac{q}{AkT}} - 1\right) - \frac{V_{pv}^c + I_{pv}^c \cdot R_s}{R_p}, \quad (1)$$

where

$$I_L(G) = \frac{G}{G_{STC}} \cdot I_L(G_{STC}), \quad (2)$$

and

$$I_0(T) = I_0(T_{STC}) \cdot \left(\frac{T}{T_{STC}}\right)^3 \cdot e^{\frac{qE_g}{Ak}\cdot\left(\frac{1}{T_{STC}} - \frac{1}{T}\right)}. \quad (3)$$

For parameters in (1)~(3), G is irradiance level; T is cell temperature; q is charge of the electron; E_g is bandgap and k is Boltzmann's constant. STC stands for standard test condition where $G_{STC} = 1000$ W/m^2 and $T_{STC} = 25$ °C. For rest parameters, i.e., photo-generated current at STC $I_L(G_{STC})$, dark saturation current at STC $I_0(T_{STC})$, PV cell series resistance R_s, PV cell

---

[2] A PV module consists of series-connected PV groups. A PV group composes of a number of parallel-connected PV cells.

parallel (shunt) resistance R_p, and diode ideality factor A, we adopt the method in [11] to extract their values. Therefore, based on this PV cell model, we can obtain I-V characteristics of a PV cell under any given environmental condition (G, T).

2.2 Charger Model

Figure 2 shows the model of a pulse width modulation buck-boost switching converter, which is used as the charger in PV system [12]. The input ports of the charger are connected to the PV string. The output ports are connected to the load. Then the operating point of the PV module string can be regulated by the charger through controlling its output current. The input voltage, input current, output voltage and output current of the charger are denoted by V_{in}, I_{in}, V_{out}, and I_{out}, respectively. The power loss of the charger P_{conv} satisfies

$$V_{in} \cdot I_{in} = P_{conv} + V_{out} \cdot I_{out}. \quad (4)$$

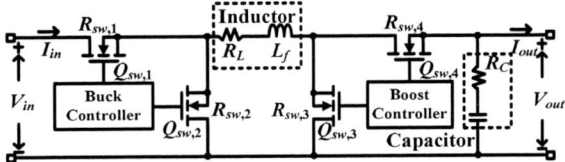

Figure 2. Buck-boost converter architecture.

When the charger is at the buck mode $(V_{in} > V_{out})$, its power loss P_{conv} is given by

$$P_{conv} = I_{out}^2 \cdot (R_L + D \cdot R_{sw,1} + (1-D) \cdot R_{sw,2} + R_{sw,4})$$
$$+ \frac{(\Delta I)^2}{12} \cdot (R_L + D \cdot R_{sw,1} + (1-D) \cdot R_{sw,2} + R_{sw,4} + R_C) \quad (5)$$
$$+ V_{in} \cdot f_s \cdot (Q_{sw,1} + Q_{sw,2}) + V_{in} \cdot I_{controller},$$

where $D = V_{out}/V_{in}$ is the PWM duty ratio and $\Delta I = V_{out} \cdot (1 - D)/(L_f \cdot f_s)$ is the maximum current ripple; f_s is the switching frequency; $I_{controller}$ is the current of the micro-controller of the charger; R_L and R_C are the internal series resistances of the inductor L and the capacitor C, respectively; $R_{sw,i}$ and $Q_{sw,i}$ are the turn-on resistance and gate charge of the i-th MOSFET switch shown in Figure 2, respectively. The charger power loss P_{conv} at the boost mode $(V_{in} \le V_{out})$ is given by

$$P_{conv} = \left(\frac{I_{out}}{1-D}\right)^2 \cdot (R_L + D \cdot R_{sw,3} + (1-D) \cdot R_{sw,4} +$$
$$R_{sw,1} + D \cdot (1-D) \cdot R_C) + \frac{(\Delta I)^2}{12}(R_L + D \cdot R_{sw,3} + (1- \quad (6)$$
$$D)R_{sw,4} + R_{sw,1} + (1-D)R_C) + V_{out} \cdot f_s \cdot (Q_{sw,3} +$$
$$Q_{sw,4}) + V_{in} \cdot I_{controller},$$

where $D = 1 - V_{in}/V_{out}$ and $\Delta I = V_{in} \cdot D/(L_f \cdot f_s)$. Moreover, we use $I_{out} = Chg\_Out\_I(V_{in}, I_{in}, V_{out})$ to denote the function which calculates the charger output current I_{out}, given its input voltage and current, as well as its output voltage.

3. IMBALANCED PV CELL CONNECTION TOPOLOGY

This paper proposes an *imbalanced PV cell connection topology* (imbalanced topology) for each PV module in the system. An N-cell PV module has n series-connected PV groups. PV cells in a PV group are parallel-connected. The imbalanced topology is very flexible because n could be any nonzero integer less than or equal to N. And the number of parallel-connected PV cells $m_j (> 0)$ in the j-th $(1 \le j \le n)$ PV group satisfies

$$\sum_{j=1}^{n} m_j = N. \quad (7)$$

We denote an imbalanced topology or configuration of a PV module by $C(n; m_1, m_2, \cdots, m_n)$. Figure 3 shows an example imbalanced topology of $C(3; 4, 3, 5)$.

Figure 4 shows the reconfigurable PV module architecture for implementing imbalanced topologies of a PV module. The architecture is already provided in [10]. However we attempt the imbalanced reconfiguration for the first time. Each PV cell is integrated with three switches, i.e., a series switch (S-switch) and two parallel switches (P-switches), except for the last one. The P-switches connect PV cells in parallel, and the S-switches connect the PV groups in series. We denote S-switch of the i-th PV cell by $S_{S,i}$, and the top and bottom P-switches of the i-th PV cell by $S_{PT,i}$ and $S_{PB,i}$, respectively. Both $S_{PT,i}$ and $S_{PB,i}$ must be closed or open together. $S_{PT,i}$ and $S_{PB,i}$ are closed exactly if $S_{S,i}$ is open, and vice versa. Refer to Figure 3 on how an imbalanced topology is implemented, in which the first PV group contains PV cells 1 to 4, and these four PV cells are connected in parallel by the P-switches of PV cells 1 to 3. The first PV group is connected in series with the second PV group by the S-switch of PV cell 4.

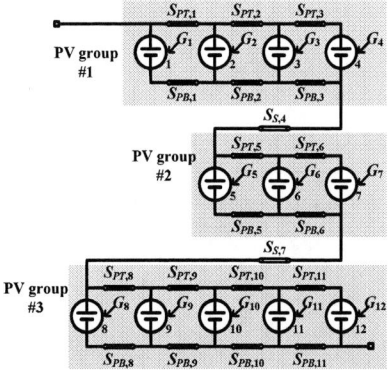

Figure 3. An imbalanced topology of a 12-cell PV module.

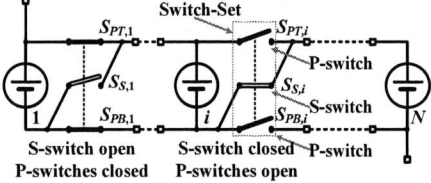

Figure 4. The N-cell reconfigurable PV module architecture.

An imbalanced configuration $\mathcal{C}(n; m_1, m_2, \cdots, m_n)$ of an N-cell PV module can be viewed as a partitioning of the PV cell index set $\mathbf{A} = \{1, 2, \cdots, N\}$. A partitioning is denoted by n subsets $\mathbf{B}_1, \mathbf{B}_2, \cdots, \mathbf{B}_n$, which correspond to n PV groups consisting of $m_1, m_2, \cdots, m_n$ parallel-connected PV cells, respectively. Therefore, the subsets $\mathbf{B}_1, \mathbf{B}_2, \cdots, \mathbf{B}_n$ should satisfy

$$\bigcup_{j=1}^n \mathbf{B}_j = \mathbf{A}, \tag{8}$$

and

$$\mathbf{B}_j \cap \mathbf{B}_k = \emptyset, \text{ for } \forall j, k \in \{1, 2, \cdots, n\} \text{ and } j \neq k. \tag{9}$$

Due to the structural characteristics of the reconfigurable PV module architecture shown in Figure 4, we also have

$$i_1 < i_2, \text{ for } \forall i_1 \in \mathbf{B}_j, \forall i_2 \in \mathbf{B}_k \text{ and } 1 \leq j < k \leq n. \tag{10}$$

A partitioning satisfying (8)~(10) is an *alphabetical partitioning*.

We can develop I-V characteristics of a PV module with the configuration $\mathcal{C}(n; m_1, m_2, \cdots, m_n)$. We denote the current and voltage of the i-th PV cell by $I_{pv,i}^c$ and $V_{pv,i}^c$, respectively, the voltage of the j-th PV group by $V_{pv,j}^g$, the current and voltage of the PV module by I_{pv}^m and V_{pv}^m, respectively. We have

$$I_{pv}^m = \sum_{i \in \mathbf{B}_j} I_{pv,i}^c, \; \forall j \in \{1, 2, \cdots, n\}, \tag{11}$$

$$V_{pv,j}^g = V_{pv,i}^c, \text{ for } \forall i \in \mathbf{B}_j \text{ and } j \in \{1, 2, \cdots, n\}, \tag{12}$$

and

$$V_{pv}^m = \sum_{j=1}^n V_{pv,j}^g. \tag{13}$$

$I_{pv,i}^c$ and $V_{pv,i}^c$ satisfy the PV cell I-V characteristics (1)~(3), given environmental condition (G_i, T_i), where G_i and T_i are the irradiance level and temperature on the i-th PV cell, respectively.

4. PROBLEM STATEMENT

Figure 5 shows the architecture of a PV system with the string charger interface, which consists of M series-connected reconfigurable PV modules, a charger and a supercapacitor as the EES element. The PV modules share one charger. The string charger interface is cost-effective compared with the individual charger interface, and is widely used. Each PV module in the string has the reconfigurable architecture shown in Figure 4. We restrict the freedom of reconfiguration within each PV module, but it is a reasonable compromise of feasibility and performance.

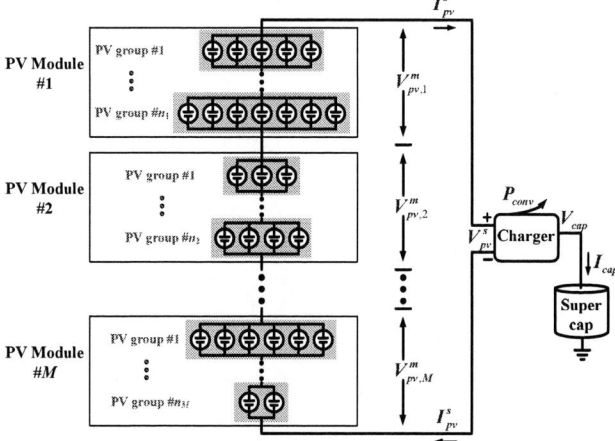

Figure 5. Architecture of a PV system with the string charger interface.

The PV system operation starts at time T_0 and ends at time T_d. We use $V_{pv}^s(t)$ and $I_{pv}^s(t)$ to denote the output voltage and current of the PV module string at $t \in [T_0, T_d]$, respectively. We denote the output voltage of the k-th $(1 \leq k \leq M)$ PV module by $V_{pv,k}^m(t)$. The output currents of the PV modules are the same, namely $I_{pv}^s(t)$, and we have

$$V_{pv}^s(t) = \sum_{k=1}^M V_{pv,k}^m(t). \tag{14}$$

We denote the irradiance level and the temperature on the i-th $(1 \leq i \leq N)$ PV cell of the k-th $(1 \leq k \leq M)$ PV module at time t by $G_{k,i}(t)$ and $T_{k,i}(t)$, respectively, and denote the imbalanced configuration of the k-th PV module by $\mathcal{C}_k(t)$. Obviously $\mathcal{C}_k(t)$ $(1 \leq k \leq M)$ are the control variables of the PV reconfiguration algorithm. We assume all PV cell temperatures are the same and constant for the period $[T_0, T_d]$, denoted by T_{pv}, to focus on the partial shading problem. The PV reconfiguration algorithm proposed in this paper runs in an *online manner*, i.e., the system controller is not aware of $G_{k,i}(t)$ $(1 \leq k \leq M, 1 \leq i \leq N)$ until time t. The relationship of $V_{pv,k}^m(t)$ and $I_{pv}^s(t)$ depends on $\mathcal{C}_k(t)$, $G_{k,i}(t)$, and $T_{k,i}(t) = T_{pv}$ $(i \in \{1, 2, ..., N\})$, as given in (11)~(13).

We use $V_{cap}(t)$ and $I_{cap}(t)$ to denote the terminal voltage and charging current of the supercapacitor, respectively. The power loss of the charger $P_{conv}(t)$ is determined by its input voltage, input current, output voltage, and output current, i.e., $V_{pv}^s(t)$,

$I_{pv}^s(t)$, $V_{cap}(t)$, and $I_{cap}(t)$, respectively, according to (5)~(6). According to (4), we also have

$$V_{pv}^s(t) \cdot I_{pv}^s(t) = P_{conv}(t) + V_{cap}(t) \cdot I_{cap}(t). \quad (15)$$

The objective of the PV system controller is to find the optimal configuration $\mathcal{C}_k^{opt}(t)$ $(1 \le k \le M)$ and the optimal operating point $(V_{pv}^{s,opt}(t), I_{pv}^{s,opt}(t))$ of the PV module string to maximize $I_{cap}(t)$ at any time $t \in [T_0, T_d]$.

5. RECONFIGURATION ALGORITHM

5.1 Motivation

We denote the power at MPP of a PV cell/group/module/string by *MPP power*. Similarly, there are *MPP voltage* and *MPP current* of a PV cell/group/module/string. The utmost solar power harvested by a PV system is the sum of MPP power values of all PV cells. The reconfiguration algorithm aims at making all PV cells operate at or close to their MPPs simultaneously. We update the optimal configurations for PV modules according to current shading patterns to maximize the supercapacitor charging current.

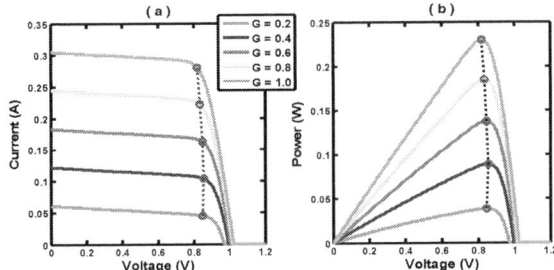

Figure 6. (a) Current-voltage and (b) power-voltage characteristics of a PV cell under different irradiance levels with MPPs labeled by red circles.

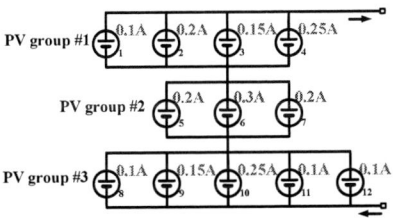

Figure 7. An example of imbalanced reconfiguration according to the PV cell MPP currents at their own MPPs.

One observation made from I-V and P-V characteristics of PV cells in Figure 6 is that MPP voltage values of a PV cell under different irradiance levels are very close to each other, while MPP current values vary significantly. It enables us to use a constant voltage V_{avg}^{MPP} to approximate MPP voltage. In Figure 7, PV cell MPP current values are calculated from irradiance levels on PV cells and labeled beside PV cells, where the sum of PV cell MPP current values in every PV group is 0.7 A. Then with the configuration in Figure 7, all PV cells can operate close to their own MPPs simultaneously if we set the output voltage of this module at $3 \cdot V_{avg}^{MPP}$. Similarly, MPP currents of all PV modules in the PV string shall be close to each other, so that the output power of the PV string can be optimized.

5.2 Algorithm

We omit time index t for convenience, since the reconfiguration algorithm will be performed at every decision epoch $t \in [T_0, T_d]$. We denote the MPP voltage and current values of the i-th $(1 \le i \le N)$ PV cell in the k-th $(1 \le k \le M)$ PV module by

$V_{pv,k,i}^{c,MPP}$ and $I_{pv,k,i}^{c,MPP}$, respectively, which are calculated from the condition $(G_{k,i}, T_{pv})$. These $V_{pv,k,i}^{c,MPP}$ and $I_{pv,k,i}^{c,MPP}$ values are inputs of our reconfiguration control algorithm. All $V_{pv,k,i}^{c,MPP}$ can be approximated by V_{avg}^{MPP} due to the observation in Section 5.1. The objective of our algorithm is finding the optimal configurations $\mathcal{C}_k^{opt}(n_k; m_{k,1}, m_{k,2}, \cdots, m_{k,n_k})$ $(1 \le k \le M)$. It consists of an *initial allocation procedure* and a *kernel algorithm*. The initial allocation procedure determines the number of PV groups n_k for each k-th PV module. The kernel algorithm finds the optimal $m_{k,1}$, $m_{k,2}$, ..., m_{k,n_k} values corresponding to the optimal alphabetical partitioning $B_{k,1}$, $B_{k,2}$, $\cdots$, B_{k,n_k} of the set $A = \{1, 2, \cdots, N\}$ for each k-th PV module, given $I_{pv,k,i}^{c,MPP}$ $(1 \le i \le N)$ and n_k.

The initial allocation procedure determines the number of PV groups n_k for each k-th PV module in a one shot manner. It can improve PV system output power due to the following reasons:

(1) The MPP currents of PV modules in the PV string should be close to each other.

(2) The MPP voltage of the PV module string should match supercapacitor voltage to reduce power loss in the charger.

We do not look into detailed configuration of each PV module in the initial allocation procedure. Detailed procedures of the initial allocation are described as follows. First, we use $\hat{P}_{pv}^{s,MPP}$ to denote the sum of MPP power values of all PV cells in the PV module string, given by

$$\hat{P}_{pv}^{s,MPP} = \sum_{1 \le k \le M} \sum_{1 \le i \le N} V_{pv,k,i}^{c,MPP} \cdot I_{pv,k,i}^{c,MPP}. \quad (16)$$

$\hat{P}_{pv}^{s,MPP}$ is an optimistic estimation of the MPP power of the whole PV module string. We use $\hat{P}_{pv,k}^{m,MPP}$ to denote the sum of MPP power values of all PV cells in the k-th PV module, given by

$$\hat{P}_{pv,k}^{m,MPP} = \sum_{1 \le i \le N} V_{pv,k,i}^{c,MPP} \cdot I_{pv,k,i}^{c,MPP}. \quad (17)$$

It is an optimistic estimation of MPP power of the k-th PV module.

Next we find the estimated optimal PV string voltage and current, denoted by $\hat{V}_{pv}^{s,opt}$ and $\hat{I}_{pv}^{s,opt}$, respectively, satisfying $\hat{V}_{pv}^{s,opt} \cdot \hat{I}_{pv}^{s,opt} = \hat{P}_{pv}^{s,MPP}$, so that the estimated supercapacitor charging current, denoted by $\hat{I}_{cap}^{opt}$, is maximized. We have:

$$\left(\hat{V}_{pv}^{s,opt}, \hat{I}_{pv}^{s,opt}\right) = \underset{V_{pv}^s \cdot I_{pv}^s = \hat{P}_{pv}^{s,MPP}}{\arg\max} Chg\_Out\_I(V_{pv}^s, I_{pv}^s, V_{cap}), \quad (18)$$

$$\hat{I}_{cap}^{opt} = \max_{V_{pv}^s \cdot I_{pv}^s = \hat{P}_{pv}^{s,MPP}} Chg\_Out\_I(V_{pv}^s, I_{pv}^s, V_{cap}), \quad (19)$$

where the function $Chg\_Out\_I()$ is defined in Section 2.2.

Next we find a proper range of the estimated PV string output voltage $[\hat{V}_{pv}^{s,min}, \hat{V}_{pv}^{s,max}]$, so that for any (V_{pv}^s, I_{pv}^s) pair satisfying $V_{pv}^s \in [\hat{V}_{pv}^{s,min}, \hat{V}_{pv}^{s,max}]$ and $V_{pv}^s \cdot I_{pv}^s = \hat{P}_{pv}^{s,MPP}$, the supercapacitor charging current calculated through $Chg\_Out\_I(V_{pv}^s, I_{pv}^s, V_{cap})$ is within a small range of $\hat{I}_{cap}^{opt}$, i.e.,

$$\max_{V_{pv}^s \in [\hat{V}_{pv}^{s,min}, \hat{V}_{pv}^{s,max}], V_{pv}^s \cdot I_{pv}^s = \hat{P}_{pv}^{s,MPP}} Chg\_Out\_I(V_{pv}^s, I_{pv}^s, V_{cap})$$
$$\ge (1 - \varepsilon)\hat{I}_{cap}^{opt}, \quad (20)$$

where ε is a predefined small value. Furthermore we calculate the range of the estimated number of PV groups in the PV module string $[\hat{n}_{total}^{min}, \hat{n}_{total}^{max}]$, where $\hat{n}_{total}^{min}$ and $\hat{n}_{total}^{max}$ satisfy

$$\hat{n}_{total}^{min} = \left\lceil \frac{\hat{V}_{pv}^{s,min}}{V_{avg}^{MPP}} \right\rceil, \quad \hat{n}_{total}^{max} = \left\lfloor \frac{\hat{V}_{pv}^{s,max}}{V_{avg}^{MPP}} \right\rfloor. \quad (21)$$

For each possible estimated number of total PV groups in the PV string $\hat{n}_{total}^{(l)}$ $(\hat{n}_{total}^{(l)} \in [\hat{n}_{total}^{min}, \hat{n}_{total}^{max}]$ and $1 \le l \le \hat{n}_{total}^{max} - \hat{n}_{total}^{min} + 1)$, we calculate the *possible numbers of PV groups for each PV module*. To reduce the exploration space, we assume that there are only two possible numbers of PV groups for each k-th PV module $n_k^{(l),-}$ and $n_k^{(l),+}$, respectively, given by

$$n_k^{(l),-} = \left\lceil \frac{\hat{n}_{total}^{(l)} \cdot \hat{P}_{pv,k}^{m,MPP}}{\hat{P}_{pv}^{s,MPP}} \right\rceil, \quad n_k^{(l),+} = n_k^{(l),-} + 1. \quad (22)$$

The underlying principle of (22) is that MPP currents of different PV modules in the PV string should be close to each other. We define a vector variable $\boldsymbol{n}^{(l)} = \left(n_1^{(l)}, n_2^{(l)}, ..., n_M^{(l)}\right)$, in which $n_k^{(l)}$ ($1 \le k \le M$) denotes the possible number of PV groups in the k-th PV module. There are 2^M (vector) values $\boldsymbol{n}^{(l)}$ could assume, since $n_k^{(l)}$ is either $n_k^{(l),-}$ or $n_k^{(l),+}$. Hence, we can find the optimal $\boldsymbol{n}^{(l)}$ among all possible l and $n_k^{(l)}$ ($1 \le k \le M$), such that the estimated supercapacitor charging current

$$Chg\_Out\_I \left(\sum_{1 \le k \le M} n_k^{(l)} \cdot V_{avg}^{MPP}, \min_{1 \le k \le M}\left(\frac{\hat{P}_{pv,k}^{m,MPP}}{n_k^{(l)} \cdot V_{avg}^{MPP}}\right), V_{cap}\right),$$

can be maximized, i.e., the optimization objective would be

$$\max_{l, \boldsymbol{n}^{(l)}} Chg\_Out\_I \left(\sum_{1 \le k \le M} n_k^{(l)} \cdot V_{avg}^{MPP}, \min_{1 \le k \le M}\left(\frac{\hat{P}_{pv,k}^{m,MPP}}{n_k^{(l)} \cdot V_{avg}^{MPP}}\right), V_{cap}\right). \quad (23)$$

Then we assign the number of PV groups in each k-th PV module n_k, to be the value $n_k^{(l)}$ in the optimal vector $\boldsymbol{n}^{(l)}$. An outline of the initial allocation procedure is shown in Algorithm 1.

Algorithm 1: The Initial Allocation Procedure.

1. Calculate the $\hat{P}_{pv}^{s,MPP}$ value using (16).
2. Calculate the $\hat{P}_{pv,k}^{m,MPP}$ values for $1 \le k \le M$ using (17).
3. Find the $\hat{V}_{pv}^{s,opt}$, $\hat{I}_{pv}^{s,opt}$, and $\hat{I}_{pv}^{opt}$ values using (18), (19).
4. Find the proper range of estimated PV string output voltage $\left[\hat{V}_{pv}^{s,min}, \hat{V}_{pv}^{s,max}\right]$ using (20).
5. Find the proper range of estimated number of total PV groups $\left[\hat{n}_{total}^{min}, \hat{n}_{total}^{max}\right]$ using (21).
6. Calculate $n_k^{(l),-}$ and $n_k^{(l),+}$ for $1 \le l \le \hat{n}_{total}^{max} - \hat{n}_{total}^{min} + 1$ and $1 \le k \le M$ using (22).
7. Find optimal l and $\boldsymbol{n}^{(l)}$ using (23).

Then the number of PV groups in each PV module, i.e., n_k ($1 \le k \le M$), is determined, where n_k is the k-th component of the optimal $\boldsymbol{n}^{(l)}$.

After the initial allocation procedure, the kernel algorithm is applied once to each PV module with given n_k. The kernel algorithm finds the optimal $m_{k,1}$, $m_{k,2}$, ..., m_{k,n_k} values for the k-th PV module. We use $\hat{I}_{pv,k,j}^{g,MPP}$ to denote the sum of PV cell MPP currents in the j-th PV group of the k-th PV module, given by

$$\hat{I}_{pv,k,j}^{g,MPP} = \sum_{i \in B_{k,j}} I_{pv,k,i}^{c,MPP}. \quad (24)$$

$\hat{I}_{pv,k,j}^{g,MPP}$ is an estimation of MPP current of the j-th PV group in the k-th PV module. Inspired by motivations stated in Section 5.1, we should make all $\hat{I}_{pv,k,j}^{g,MPP}$ values ($1 \le j \le n_k$) as close to each other as possible. Equivalently, we maximize the minimal $\hat{I}_{pv,k,j}^{g,MPP}$ value of the k-th PV module, denoted by $Min\_Sum\_I_k$, since MPP current of the k-th PV module is restricted by that minimal value. The objective of kernel algorithm is finding optimal alphabetical partitioning $B_{k,1}, B_{k,2}, \cdots, B_{k,n_k}$ of the k-th PV module, so that $Min\_Sum\_I_k$ is maximized, i.e., the objective would be

$$\max_{B_{k,1}, B_{k,2}, \cdots, B_{k,n_k}} Min\_Sum\_I_k = \max_{B_{k,1}, B_{k,2}, \cdots, B_{k,n_k}} \min_{1 \le j \le n_k} \hat{I}_{pv,k,j}^{g,MPP}. \quad (25)$$

Then all PV cells in the k-th PV module under the configuration corresponding to the optimal alphabetical partitioning $B_{k,1}, B_{k,2}, \cdots, B_{k,n_k}$ could work very close to their MPPs, if the k-th PV module output voltage is set at $n_k \cdot V_{avg}^{MPP}$.

Consider a general problem of finding optimal configuration for an l_1-cell ($l_1 \le N$, corresponding to the first l_1 cells of the original N cells in the k-th PV module) PV module composed of l_2 ($l_2 \le n_k$) PV groups, given $I_{pv,k,i}^{c,MPP}$ ($1 \le i \le l_1$) values and l_2. This is equivalent to finding the optimal alphabetical partitioning $B_{k,1}^{l_1,l_2}, B_{k,2}^{l_1,l_2}, \cdots, B_{k,l_2}^{l_1,l_2}$ of the set $A^{l_1} = \{1, 2, \cdots, l_1\}$, which is optimal in that the value $Min\_Sum\_I_k^{l_1,l_2} = \min_{1 \le j \le l_2} \sum_{i \in B_{k,j}^{l_1,l_2}} I_{pv,k,i}^{c,MPP}$ is maximized. We call this problem (l_1, l_2) *reconfiguration problem*. When $l_1 = N$ and $l_2 = n_k$, the (l_1, l_2) reconfiguration problem becomes the original reconfiguration problem of the k-th PV module as stated in (25). We find optimal substructure property of (l_1, l_2) reconfiguration problem below, implying the applicability of dynamic programming.

The optimal substructure observation: Suppose that (l_1, l_2) reconfiguration problem has been optimally solved, and that the last (l_2-th) PV group consists of $m_{k,l_2}^{l_1,l_2}$ PV cells. Then the sub-problem of finding the optimal configuration for the first $l_1 - m_{k,l_2}^{l_1,l_2}$ PV cells within $l_2 - 1$ PV groups, which corresponds to the $(l_1 - m_{k,l_2}^{l_1,l_2}, l_2 - 1)$ reconfiguration problem, has to be solved optimally. From this observation, we have Algorithm 2 based on dynamic programming as the kernel algorithm for finding the optimal configuration for the k-th PV module with given n_k.

Algorithm 2: The Kernel Algorithm.

Maintain $N \times n_k$ matrixes **Min_Sum_Opt** and **Last_Par**.
Initialize **Min_Sum_Opt**$(l_1, 1) \leftarrow \sum_{1 \le i \le l_1} I_{pv,k,i}^{c,MPP}$, **Last_Par**$(l_1, 1) \leftarrow 0$.
For l_2 from 2 to n_k:
    For l_1 from l_2 to N:
        Min_Sum_Opt$(l_1, l_2) \leftarrow$
        $\max_{l_2-1 \le l < l_1} \min\{$**Min_Sum_Opt**$(l, l_2 - 1), \sum_{l < i \le l_1} I_{pv,k,i}^{c,MPP}\}$.

        Last_Par$(l_1, l_2) \leftarrow$
        $\arg\max_{l_2-1 \le l < l_1} \min\{$**Min_Sum_Opt**$(l, l_2 - 1), \sum_{l < i \le l_1} I_{pv,k,i}^{c,MPP}\}$.
    End
End
Trace back using the matrix **Last_Par** to find the optimal imbalanced configuration C_k^{opt} of the k-th PV module.

---

6. EXPERIMENTAL RESULTS

We compare performances of the PV system with reconfiguration technique and the baseline system without reconfiguration. Both systems consist of 3 PV modules ($M = 3$) with 60 PV cells each ($N = 60$), a charger and a supercapacitor. PV modules in the proposed system have the reconfigurable architecture in Figure 4, while PV modules in the baseline system have a fixed 10x6 configuration, where 10 PV groups are series-connected with 6 PV cells per PV group. The charger model is Linear Technology LTM4607 converter, and the supercapacitor is 100 F. In the baseline system, we incorporate improved MPPT technique [5] and bypass diodes with PV cells [7] to enhance power output under partial shading. In the proposed system, we incorporate the MPTT technique to feed maximum power into the supercapacitor.

First we test instantaneous power output level of the two systems under the shading pattern shown in Figure 8. For the proposed system, Figure 8 shows physical locations of PV cells in the 3 PV modules, instead of the real imbalanced configuration of PV modules. Table 1 summarizes the power output improvement of the proposed system compared to the baseline system, given the shading pattern in Figure 8 and the V_{cap} values. As shown in Table 1, the proposed system achieves up to a 2.31X output power improvement compared with the baseline system with a V_{cap} of 45 V demonstrating effectiveness of the reconfiguration technique.

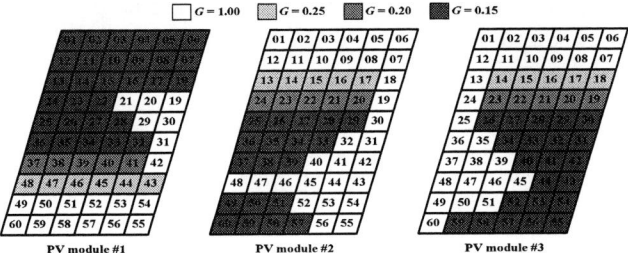

Figure 8. The shading pattern on the 3-module PV string.

Table 1. The improvement of instantaneous power level of the PV system with the imbalanced reconfiguration technique.

V_{cap} (V)	5	25	45
Power output improvement	1.91X	2.04X	2.31X
PV module #1 configuration	$\mathcal{C}(3; 30,22,8)$	$\mathcal{C}(6; 20,10,16,6,4,4)$	$\mathcal{C}(4; 28,18,8,6)$
PV module #2 configuration	$\mathcal{C}(4; 9,21,15,15)$	$\mathcal{C}(9; 4,4,4,11,9,10,4,7,7)$	$\mathcal{C}(6; 6,6,18,12,6,12)$
PV module #3 configuration	$\mathcal{C}(4; 8,16,21,15)$	$\mathcal{C}(8; 4,4,4,12,12,9,4,11)$	$\mathcal{C}(5; 6,6,23,11,14)$

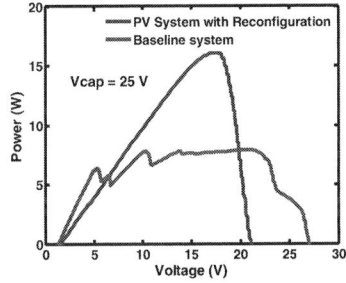

Figure 9. P-V characteristics of the PV strings in the PV system with imbalanced reconfiguration and baseline system.

Table 1 also provides near-optimal configuration obtained by the reconfiguration control algorithm for each PV module in the PV string. Figure 9 plots the P-V curves of the PV module strings in the two systems with the shading pattern in Figure 8 and a V_{cap} of 25 V. The proposed system achieves a peak power much higher than that of the baseline system. Moreover our PV reconfiguration technique makes P-V curve of the proposed system unimodal under partial shading, while P-V curve of the baseline system has multiple peaks. Therefore standard MPTT technique can be incorporated into the proposed system without extra modifications. But in baseline system, the MPPT technique must be modified to track a global optimal operating point instead of a local one.

We also test the overall efficiency of the two systems in a time period of 300 minutes, with timing parameters $T_0 = 0$ min and $T_d = 300$ min. About 2/3 of all PV cells in the 3 PV modules are shaded and irradiance levels on these shaded cells are randomly generated within the range [0, 0.5] which also change with time. The proposed system updates its module configurations once per minute according to current shading pattern and charger efficiency variation. We compare the electrical energy stored into the supercapacitors in the two systems during time $[T_0, T_d]$, and the proposed system achieves a 1.59X improvement compared to the baseline system, demonstrating that the reconfiguration technique can effectively combat the partial shading effects.

7. CONCLUSION

This paper addresses the power output loss problem of a PV system with the string charger interface under partial shading. We introduce the imbalanced reconfiguration technique to combat the partial shading effects. We propose the imbalanced PV cell connection topology for the first time and also provide an effective reconfiguration control algorithm, which realizes adaptive and near-optimal PV module reconfiguration for each PV module in the PV string according to the partial shading pattern and the charger efficiency variation. The proposed reconfiguration control algorithm is based on dynamic programming with polynomial time complexity.

8. ACKNOWLEDGEMENT

This work is sponsored in part by a grant from the National Science Foundation, the Brain Korea 21 Project, and the National Research Foundation of Korea (NRF) grant funded by the Korea government (MEST) (No. 2011-0016480). The ICT at Seoul National University provides research facilities for this study.

9. REFERENCES

[1] N. Femia, G. Petrone, G. Spagnuolo, and M. Vitelli, "Optimization of perturb and observe maximum power point tracking method," *IEEE T. on Power Electronics*, 2005.

[2] F. Liu, S. Duan, F. Liu, B. Liu, and Y. Kang, "A variable step size INC MPPT method for PV systems," *IEEE T. on Industrial Electronics*, 2008.

[3] Y. Kim, N. Chang, Y. Wang, and M. Pedram, "Maximum power transfer tracking for a photovoltaic-supercapacitor energy system," in *ISLPED*, 2010.

[4] C. Lu, V. Raghunathan, and K. Roy, "Maximum power point considerations in micro-scale solar energy harvesting systems," in *IEEE ISCAS*, 2010.

[5] H. Patel, and V. Agarwal, "Maximum power point tracking scheme for PV systems operating under partially shaded conditions," *IEEE T. on Industrial Electronics*, 2008.

[6] R. Bruendlinger, B. Bletterie, M. Milde, and H. Oldenkamp, "Maximum power point tracking performance under partially shaded PV array conditions," in *Proc. 21st EUPVSEC*, 2006.

[7] D. Nguyen, and B. Lehman, "An adaptive solar photovoltaic array using model-based reconfiguration algorithm," *IEEE T. on Industrial Electronics*, 2008.

[8] G. Velasco-Quesada, F. Guinjoan-Gispert, R. Pique-Lopez, M. Roman-Lumbreras, and A. Conesa-Rosa, "Electrical PV array reconfiguration strategy for energy extraction improvement in grid-connected PV systems," *IEEE T. on Industrial Electronics*, 2009.

[9] M. A. Chaaban, M. Alahmad, J. Neal, J. Shi, C. Berryman, Y. Cho, S. Lau, H. Li, A. Schwer, Z. Shen, J. Stansbury, and T. Zhang, "Adaptive photovoltaic system," in *IECON*, 2010.

[10] Y. Kim, S. Park, Y. Wang, Q. Xie, N. Chang, M. Poncino, and M. Pedram, "Balanced reconfiguration of storage banks in a hybrid electrical energy storage system," in *ICCAD*, 2011.

[11] W. Lee, Y. Kim, Y. Wang, N. Chang, M. Pedram, and S. Han, "Versatile high-fidelity photovoltaic module emulation system," in *ISLPED*, 2011.

[12] Y. Wang, Y. Kim, Q. Xie, N. Chang, and M. Pedram, "Charge migration efficiency optimization in hybrid electrical energy storage (HEES) systems," in *ISLPED*, 2011.

Networked Architecture for Hybrid Electrical Energy Storage Systems *

Younghyun Kim, Sangyoung Park, and Naehyuck Chang
Seoul National University, Seoul, Korea
{yhkim, sypark, naehyuck}
@elpl.snu.ac.kr

Qing Xie, Yanzhi Wang, and Massoud Pedram
University of Southern California, Los Angeles, CA, USA
{xqing, yanzhiwa, pedram}@usc.edu

ABSTRACT

A hybrid electrical energy storage (HEES) system that consists of multiple, heterogeneous electrical energy storage (EES) elements is a promising solution to achieve a cost-effective EES system because no storage element has ideal characteristics. The state-of-the-art HEES systems are based on a shared-bus charge transfer interconnect (CTI) architecture. Consequently, they are quite limited in scalability which is a function of the number of EES banks. This paper is the first introduction of a HEES system based on a networked CTI architecture, which is highly scalable and is capable of accommodating multiple, concurrent charge transfers. The paper starts by presenting a router architecture for the networked CTI and an effective on-line routing algorithm for multiple charge transfers. In the proposed algorithm, negotiated congestion (NC) routing for multiple charge transfers is performed and any lack of routing resources is addressed by merging two or more charge transfers while maximizing the overall energy efficiency by setting the optimal voltage level for the shared CTI. Examples of the proposed networked CTI are presented and the efficacy of the routing algorithm is demonstrated on a mesh-grid networked CTI.

Categories and Subject Descriptors

C.3 [**Special-Purpose and Application-Based Systems**]: Real-time and embedded systems

General Terms

Algorithms, Design

Keywords

Hybrid electrical energy storage, Charge transfer interconnect, Routing algorithm

---

*This work was supported by the Brain Korea 21 Project and the National Research Foundation of Korea (NRF) grant funded by the Korea government (MEST) (No. 2011-0016480). The ICT at Seoul National University provides research facilities for this study.

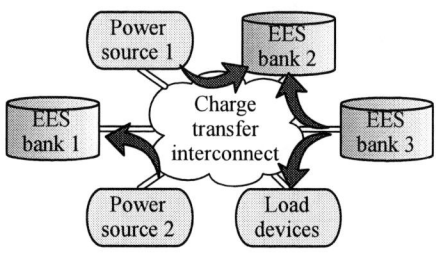

Figure 1: A hybrid electrical energy storage system with three EES banks.

1. INTRODUCTION

Hybrid electrical energy storage (HEES) systems consist of multiple, heterogeneous electrical energy storage (EES) elements [1, 2]. They utilize the strengths of each EES element type while hiding its disadvantages by performing appropriate charge management processes, including charge replacement, charge allocation, and charge migration [3, 4, 5]. HEES systems are one of the key resources in a smart grid enabling energy management, peak shaving, load balancing, and the like [6, 7]. They also improve the generation efficiency of renewable power sources such as solar cells and windmills by enabling maximum power extraction from these sources because of the ability to store the excess generated energy for future use, and by decoupling the load demand variation from the power generation [8].

Figure 1 shows a conceptual diagram of a HEES system composed of three EES banks. The EES banks can freely exchange energy among each other by transferring charge over the charge transfer interconnect (CTI). Power sources and load devices are also connected to the CTI. Previously reported HEES systems rely on a simple CTI architecture e.g., a single shared-bus CTI. However, a shared-bus CTI has a major limitation in terms of the HEES system scalability, that is, the number of EES banks is limited to a rather small count because of the potential contention on the CTI bus. Note that the CTI bus contention is different from contention in computer network in that the objects to be routed (charges) can be merged together into a single object. The contention of a CTI bus is thus defined in terms of the overall charge transfer efficiency. Each charge transfer task has its own optimal CTI voltage level to achieve the maximum transfer efficiency [3, 4, 5]. Merging the charge transfers is possible only at the expense of charge transfer efficiency degradation since transfer tasks have to share a CTI bus which can only have single (likely sub-optimal) CTI voltage. Severe efficiency degradation may thus offset the benefits of merging charge transfer tasks in the HEES system if the optimal CTI voltage levels of candidate charge transfer tasks are manifestly different.

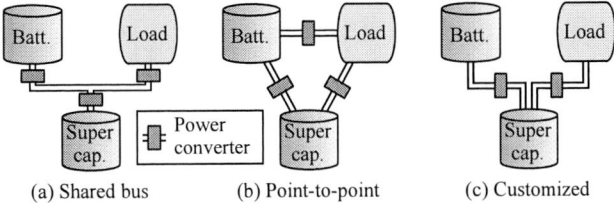

Figure 2: Various CTI architectures in view of the power path.

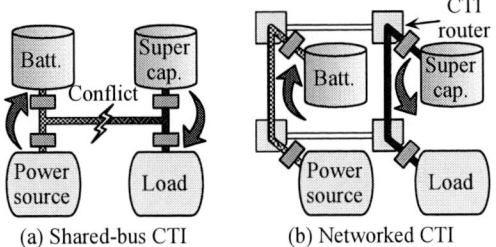

Figure 3: Shared-bus CTI and networked CTI of four nodes.

The solution to the above-mentioned problem is to use multiple CTIs in a HEES system. Multiple CTIs would then enable concurrent charge transfers with high energy efficiency by providing more routing resources and by relieving the single CTI voltage constraint. This paper is the first to introduce a networked CTI architecture for HEES systems to accommodate multiple EES banks and multiple, concurrent charge transfers while ensuring scalability. Since this is the first paper of a networked CTI for HEES systems, we consider the most basic mesh network topology.

This paper addresses the following important issues in charge transfers in networked CTI HEES systems. The first issue is routing of charge transfer paths in mesh networks as in traditional field-programmable gate array (FPGA) signal interconnects. We devise a CTI routing algorithm exploiting the similarity between our routing problem and the FPGA routing problem. We also discuss merging of charge transfer tasks and its impact on the charge transfer energy efficiency. Although the CTI routing in a networked architecture has similarities with the on-chip signal interconnects routing for an FPGA, there is a fundamental difference such that the charge transfers can share a routing path while on-chip signals cannot. We perform merging of charge transfers considering energy efficiency to eliminate routing failures due to lack of routing resources. The efficiency of a charge transfer task depends on the efficiency of converters, which are in turn affected by the CTI voltage and amount of charge to be transferred. We devise an algorithm to select charge transfer tasks to be merged based on their optimal CTI voltage values and congestions among tasks. We formulate the CTI routing problem and devise a systematic method to effectively perform energy-efficient charge transfers in networked CTI HEES systems.

The contributions of this paper are summarized as follows: i) we introduce a networked CTI architecture for HEES systems; ii) we define and precisely formulate the CTI routing problem on a networked CTI; and iii) we propose a computationally efficient online algorithm for the problem with the objective of maximizing the charge transfer efficiency. We demonstrate examples of the proposed networked CTI architecture and show the efficacy of the devised algorithm.

2. RELATED WORK

2.1 HEES CTI Architectures

Figure 2 shows three representative examples of the CTI architectures proposed in the previous HEES systems in view of the power paths. Shared-bus CTI architectures (typically called DC bus) are commonly used when the number of EES banks is limited. Recent works on the HEES system management methodologies [3, 4, 5, 9] assume a general shared-bus CTI architecture (Figure 2(a)). The shared-bus CTI is analogous to an on-chip shared bus on a system-on-chip (SoC) and their advantages and disadvantages are similar. Another architecture is a complete connection among the nodes [7] (Figure 2(b)). This architecture is comparable

to a point-to-point connection in an SoC. Both the shared-bus and point-to-point connection architectures are feasible as long as the number of EES banks is small, but they certainly lack scalability to accommodate a large-scale HEES system. The other architecture is a customized network architecture for a particular application and operation policy. For example, a supercapacitor buffer efficiently mitigates the rate-capacity effect of a Li-ion battery especially for pulsed load demand [6] (Figure 2(c)). As the control policy is to use the supercapacitor as a buffer of the battery, the path from the battery bank to the load device is not necessary. This architecture is similar to a network-on-chip (NoC) architecture with irregular connectivity which is fully dependent on the application. In short, none of the previously introduced CTI architectures can be used to accommodate a large number of EES banks for general applications.

2.2 Conventional Routing Problems

The CTI routing problem in a networked CTI has similarity to the conventional FPGA signal routing problem. In the problem of CTI routing, each task competes for routing resources such as converters and CTI links, whereas each signal competes for wires and connection points in FPGA routing. The FPGA routing is a highly complex combinatorial optimization problem, and thus it is usually done by iterative rip-up and reroute of signals. The success of routing is dependent not just on the choice of which nets to reroute, but also on the order in which rerouting is done as shown in traditional rip-up and reroute methods [10, 11]. The negotiation-based FPGA router successfully relieves the signal ordering problem and provides a systematic rip-up and reroute capability [12]. This routing algorithm allows initial sharing of the routing resources among signals, but subsequently makes them negotiate for the shared resource with other signals until no resource is shared. The negotiation-based routing algorithm is further enhanced in terms of compilation time by incorporating delay-driven routing [13]. More recent works such as [14] focus on the new architecture or technology scaling, but the core of the routing algorithm is still based on [12].

3. CHARGE TRANSFER INTERCONNECT

3.1 Charge Transfer Conflicts

The charge management of a HEES system is achieved by charge allocation, replacement, and migration operations [1]. The operations are basically charge transfers among EES banks using the CTI as charge transfer medium. The previous works on the charge management of HEES [3, 4, 5] assumed that the charge transfer path is always available for a given charge transfer task. They focused on maximizing the energy efficiency by setting a proper value for CTI voltage of the charge transfers. However, it is not always true that a charge transfer path is available whenever it is required. Two or more charge transfer tasks can have a conflict by competing for the shared-bus. Figure 3(a) demonstrates an example where the

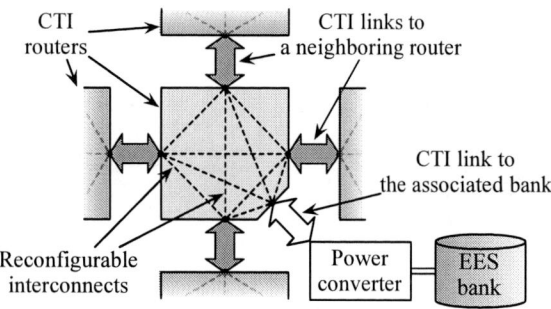

Figure 4: Architecture of a CTI router. An associated EES bank is connected via a power converter. The arrows denote the CTI links.

power supply charges the battery bank and the supercapacitor bank supplies power to the load at the same time. Two charge transfer tasks have different optimal CTI voltage values, which maximize the charge transfer energy efficiency of each task, and there is only one CTI link.

We define that two or more charge transfers *conflict* when they try to occupy the same CTI link and have different optimal CTI voltage values. Such a conflict enforces the charge transfer tasks to use the same CTI voltage, and thus at least one of them has to suffer possibly severe degradation in energy efficiency. Separating the two charge transfers in time domain by scheduling may mitigate the conflict, but we leave this as a future work.

3.2 Networked CTI Architecture

We introduce a *networked CTI architecture* as shown in Figure 3(b) to fundamentally solve the charge transfer conflict problem, which ensures scalability to a large number of EES banks. Specifically, we use a mesh interconnect architecture to ensure flexibility and scalability of networked CTI architecture. One important component to realize the networked CTI architecture is the CTI router. We propose a CTI router that connects CTI links, an associated component (i.e., an EES bank, a power source, or a load device), and a power converter. Figure 4 shows the detailed architecture of the CTI router. Each CTI router is connected with the adjacent CTI routers through the CTI links. The CTI router consists of reconfigurable interconnects which are denoted as dashed lines in Figure 4. We dynamically connect or disconnect the reconfigurable interconnects inside the router to setup a path from one CTI link to another. The reconfigurable interconnects form a complete graph so that the signal can be routed in any direction. The CTI router in Figure 4 has five CTI links, and thus it has ten interconnects each of which is implemented as a pair of back-to-back MOSFET switches. We adopt the switching power converter efficiency model from [15] in this paper.

The networked CTI architecture is comparable to a general NoC architecture. As the number of processing elements in an SoC increases, the single-level on-chip bus architecture is no longer able to handle increased data exchanges between the processing elements. Similar to the NoC which requires packet routing, a HEES system with a networked CTI architecture requires routing of the charge transfers. However, CTI routing on a networked CTI is not the same as the conventional NoC packet routing, conventional signal routing for FPGA, nor application-specific integrated circuit (ASIC). To efficiently describe the networked CTI architecture routing problem, we compare it with the conventional signal routing problem as shown in Table 1.

Table 1: CTI routing and signal routing problem mapping.

	Networked CTI routing	Signal routing
Nodes	EES banks, power sources, load devices	Processing elements
Links	CTI links	On-chip interconnects
Flows	Charge flows	Signal flows
Objective	High efficiency	Low latency
Output	Charge routing trees w/ voltage and current	Signal routing trees w/ buffer size
Resource sharing	Allowed	Not allowed
Routability	Guaranteed w/ resource sharing	Not guaranteed

4. PROBLEM STATEMENT

4.1 Formal Definitions

We present a formal definition of the CTI routing problem in this section. We first define a *node* as a combination of a CTI router and either of an EES bank, a power source, or a load device associated with it. A CTI network is a graph $G = (V, E)$ where V is a set of vertices that corresponds to nodes, and E is a set of edges that corresponds to CTI links between two elements in V. It is an undirected graph as the CTI links are bidirectional electrical conductors. The link between the CTI router and the associated EES element (an EES bank, a power source or a load device) is a dedicated resource, and thus we do not consider this in the routing algorithm.

We define a set of charge transfer tasks such that $\tau = \{T_1, T_2, \ldots, T_k\}$, where k is the number of tasks. Each charge transfer task is a two-tuple such as $T_i = (\sigma_i, \delta_i)$, where σ_i is the set of source nodes and δ_i is the set of destination nodes. The task describes the nodes that should be connected by the routing algorithm. The deadline of the transfer and the amount of charge provided by the source nodes or received by destination nodes are defined separately because they are not related to the routing process. They are used for charge transfer optimization discussed in Section 4.3. We add a newly arriving task to τ or remove a finished task from τ, update remaining time until the deadline of the existing or remaining tasks, and perform routing again.

The CTI routing problem is to find routing paths for a given transfer task set τ, that connects all the nodes in σ_i and δ_i for each $T_i \in \tau$. A node of T_i participates in only one charge transfer, and it is either a source or a destination, not both. That is,

$$\bigcup_{T_i \in \tau} (\sigma_i \cap \delta_i) = \varnothing \quad \text{and} \quad \bigcap_{T_i \in \tau} (\sigma_i \cup \delta_i) = \varnothing. \quad (1)$$

As a result of the CTI routing, a disjoint subset of edges in E that forms an acyclic routing tree is assigned to each T_i. We set each CTI router configuration (make connections of the internal interconnects) according to the edges in the routing trees. An individual routed charge transfer is equivalent to a charge transfer on an independent shared-bus CTI. Therefore, it enables us to apply any previous HEES charge management methods that are based on a shared-bus CTI to each routed charge transfer task.

4.2 Charge Transfer Interconnect Routing

The CTI routing problem resembles to the FPGA routing problem discussed in Section 2.2 in the aspect that the routing resources are discrete and scarce. The routing process allocates limited resources to the nets (the set of charge transfer tasks or signals), and each net is allowed to use the resource for a designated period.

Routing charge transfer tasks requires iterative execution of two steps; i) the CTI routing and ii) charge transfer optimization. The CTI routing operation is to determine a routing path of the charge

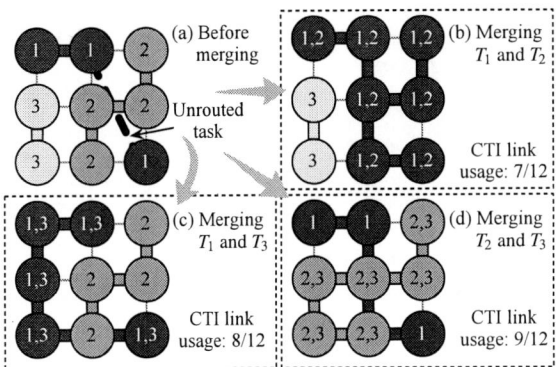

Figure 5: Example routing of three tasks after merging. T_1 **is unrouted in (a), and routing after three possible merging combinations are presented in (b), (c), and (d).**

transfer, and the charge transfer optimization operation is to determine the voltage level of the routing path and the amount of current through the routing path. We discuss the routing and optimization one after another in this section and in Section 4.3, respectively.

The CTI routing problem should tackle limitation in the routing resources (the CTI links) like the conventional FPGA routing problems. Signal routing of FPGA fails if there are unrouted nets which are not routable with remaining routing resources. The workaround is either increasing the resource, i.e., using a larger device or optimizing placement so that the congestion is reduced.

On the other hand, redoing placement is not an option for the CTI routing problem because the nodes are at a fixed location in the HEES system and cannot be moved. Instead, we perform *merging* in order to mitigate the routing congestion. This is a unique feature of the CTI routing for HEES systems. Merging is combining two charge transfer tasks into one to produce a new task set. Two or more migration tasks can be merged and share resources unlike signal routing.

If one task has a longer deadline than the other, the combined task uses the CTI links for whichever the shorter deadline. After the deadline expires, the task with a shorter deadline releases the CTI links and the task with a longer deadline solely occupies the CTI links after rerouting. Merging $T_i = (\sigma_i, \delta_i)$ and $T_j = (\sigma_j, \delta_j)$ results in a new task $T_{i,j} = (\sigma_i \cup \sigma_j, \delta_i \cup \delta_j)$. Then T_i and T_j are removed from τ and $T_{i,j}$ is added to τ. After T_i or T_j that has a shorter deadline is finished, the remaining task is added back to τ with the remaining deadline.

Figure 5 is an example of the merging to improve routability of three tasks. In Figure 5(a), T_2 and T_3 are routed, but T_1 is not routed. There are three possible combinations to merge two tasks out of three as shown in Figures 5(b), (c), and (d). The CTI link usage out of 12 CTI links is different depending on the combinations. The number of unused CTI links directly affects the routability of the other charge transfer tasks.

Most importantly, merging is not free. A merged task suffers efficiency degradation due to single CTI voltage constraint. Therefore, we have to consider not only the routabiltiy but also the efficiency at same time. We discuss this matter in the Section 4.3.

4.3 Charge Transfer Optimization

The energy efficiency of a charge transfer is defined as ratio between the sum of the energy in all the EES banks after and before the transfer. Recent works on EES systems show that the energy efficiency of charge transfers is significantly affected by the CTI voltage and amount of transfer current from/to participating nodes [3, 4, 5]. The goal of charge transfer optimization is to mitigate the

energy loss of charge transfer tasks by finding the best-suited CTI voltage and current. The optimization process considers the efficiency variation of the power converters according to the CTI voltage as well as electrical characteristics of all the EES elements. The input of the optimization process is a merged task, and the results are the CTI voltage, current from/to the nodes in σ and δ, and the overall energy efficiency.

When two or more tasks are merged into one, they have to share the CTI links that has a single voltage level. In case a task had substantially different CTI voltage from the CTI voltage after merging, the charge transfer efficiency would be greatly deteriorated. We choose the best task pairs to merge based on the optimal CTI voltage level similarity to avoid severe transfer efficiency degradation.

The power converter efficiency model and electrical characteristics of the EES elements are complex nonlinear functions that make the optimization process difficult. We adopt the derivation of the optimal CTI voltage and transfer current similar to the methods in the previous work [3, 4, 5] such as fractional optimization for efficiency and ternary search for the optimal CTI voltage. We do not discuss the detailed implementation of the optimization process because it is out of focus of this paper.

5. SOLUTION METHOD

5.1 CTI Link Cost Evaluation

We present the proposed networked CTI routing algorithm in Algorithm 1. The input of Algorithm 1 is the CTI network G and a set of charge transfer tasks τ. Algorithm 1 iteratively performs rip-up and rerouting the charge transfer tasks until all the tasks are routed. The kernel of the routing algorithm is based on the negotiated congestion (NC) routing algorithm in [12]. The cost of resources (CTI links) gradually increases over iterations, and each charge transfer task competes with others to occupy the resource. Only one charge transfer task that is willing to pay the cost occupies the resource, and the other tasks detour via other less-costly resources.

We first define the cost of resources taking into account the distinctive characteristics of the CTI routing problem. An edge $e = (u, v)$ is associated with a congestion cost $c[e]$ that is defined as

$$c[e] = (b[e] + h[e]) \cdot p[e], \qquad (2)$$

where $b[e]$ is the base cost of the edge e, $h[e]$ is the congestion history cost and $p[e]$ is the penalty due to the congestion at the current iteration. The base cost $b[e]$ is related with the unit cost of charge transfer from u to v, and we set the base cost to 1. The penalty $p[e]$ is defined as

$$p[e] = 1 + p_{gradient} \cdot u[e], \qquad (3)$$

where $p_{gradient}$ is a constant, and $u[e]$ is the number of charge transfer tasks that share the edge e. The congestion history cost $h[e]$ increases gradually after each iteration to increase cost of congested edge and make the conflicting nets to avoid it. That is,

$$h[e] = \begin{cases} h[e]' & \text{if } u[e] = 0 \\ h[e]' + h_{gradient} \cdot (u[e] - 1) & \text{if } u[e] \geq 1 \end{cases}, \qquad (4)$$

where $h[e]'$ is $h[e]$ of the previous iteration, $h_{gradient}$ is a constant, and $h[e]$ is initially 0.

Only the congestion history cost is dependent on the number of iterations by (4), and it is a non-decreasing function of the number of iterations. This is because the nets to be routed do not change over iterations in the signal routing, and so the congested resources are likely to be congested again in subsequent iterations. This is not the case for the CTI routing problem because we merge conflicting tasks into one, and then the shared resources are not congested any more. The cost of the previously shared edges are overestimated if we do not decrease $h[e]$ after they are merged. This leads to other charge transfers to avoid using the released edges and results in non-optimal routing results. Therefore, we reduce $h[e]$ of edges that have been congested by the merged tasks.

5.2 Conflict Graph

We define a *conflict graph* as $G^c = (V^c, E^c)$. There are $k = |\tau|$ nodes in $V^c = \{v_1^c, v_2^c, \ldots, v_k^c\}$, and each v_i^c is mapped to T_i. A conflict graph G^c is a complete graph, and each edge $e^c = (v_i^c, v_j^c) \in E^c$ is assigned with $d[e^c]$ which is *conflict count* between tasks T_i and T_j. Initially, $d[e^c]$ is set to zero, and we increase $d[e^c]$ by n if the tasks T_i and T_j share n CTI links. We define the sum of the conflict counts of all the edges in a conflict graph G^c to be a *conflict degree* $D[G^c]$ such that

$$D[G^c] = \sum_{e^c \in E^c} d[e^c], \qquad (5)$$

which is the metric of routability of a given task set.

We also use this conflict graph to prune away the task pairs that do not increase the routability after merging. This is important to efficiently find task pairs to merge by avoiding a situation of trying all the pairs in every iteration. We try merging a pair of tasks, and accept it if it increases the routability. We define that the routability is improved if the conflict degree is reduced after merging by the conflict count between merged transfer tasks or more. That is, we accept the merging of T_i and T_j if

$$D[G^{c\prime}] \leq D[G^c] - d[(v_i^c, v_j^c)], \qquad (6)$$

or reject it otherwise. We mark an edge of rejected task pair with $r[e^c] = 1$ to indicate the task pair is previously rejected, and $r[e^c] = 0$ otherwise.

We merge a pair of tasks T_i and T_j that have the least difference in the optimal CTI voltage if they conflict $(d[(v_i^c, v_j^c)] > 0)$ and have not been rejected previously $(r[(v_i^c, v_j^c)] = 0)$. Merging the two tasks results in a new conflict graph because two tasks T_i and T_j is removed and a new task $T_{i,j}$ is added. The new task is marked not-to-be-merged $(r[e^c] = 1)$ with existing tasks if both the merged tasks were marked not-to-be-merged with the tasks. The conflict count $d[e^c]$ is reset to zero after merging. (Refer to Appendix A.1 for an example of the conflict graph management.)

5.3 Algorithm Description

The algorithm starts from initialization of the cost of $e \in E$ based on (2), (3), and (4) in Line 1. Initially, $u[e] = 0$ for all e. It also initializes the conflict graph G^c in Line 2. We try routing and merging until all the CTI links are not shared by multiple charge transfer tasks in the loop through Lines 3–15. The loop in Lines 4–6 attempts to route the given task set with the NC-router. The NC-router repeats rip-up and rerouting for all the charge transfer tasks while updating the edge cost $c[e]$ in Line 5. We update the conflict graph after one trial for the rip-up and rerouting for all the charge transfer tasks in Line 6. These procedures are repeated until the current task set τ is fully routed. The algorithm is terminated and returns the routing results after the charge transfer optimization for each task in Line 7 if the routing is successful.

We perform merging through Lines 8–15 if the routing fails. We judge that the task set is not routable if routing attempt fails for a certain number of iterations or a certain amount of runtime. The previous merging is rejected in Line 10 if it fails to improve the routability. If the previous merging is rejected, we restore the previous states of τ, G^c, and edge costs of E. We mark rejected pairs of tasks at the edges $(r[e^c] = 1)$ in Line 11 so that they are not explored in the future attempts for merging.

Algorithm 1: Networked CTI routing algorithm

Input: CTI graph G, Charge transfer task set τ
Output: Routing tree for each task with the optimal voltage
1   Initialize cost c
2   Initialize conflict graph G^c
3   **while** shared resource exists **do**
4   **while** routing retry conditions hold **do**
5   NC-route τ on G_{cti} with cost c
6   Update conflict graph G^c
7   Solve the charge transfer optimization problem for each task
8   **if** routing failed **then**
9   **if** previous merging is not successful **then**
10   Reject the previous merging and restore τ, G^c, and costs of E
11   Mark rejected pair of tasks in G^c
12   Save the current τ, G^c, and costs of E
13   Merge the two tasks and update τ
14   Update conflict graph G^c
15   Update costs of E

Merging tasks begins with saving the current states of τ, G^c, and edge costs of E so that we can restore them when the merging is rejected in Line 12. We utilize the conflict graph G^c to find candidate tasks to be merged as described in Section 5.2. We update G^c and reset the conflict count $d[e^c]$ to zero after merging in Line 14. We also update the cost c of CTI links based on the new CTI link utilization after merging in Line 15.

6. EXPERIMENTS

6.1 Experimental Setup

We demonstrate examples of the proposed networked CTI architecture and evaluate the proposed CTI routing algorithm compared with the state-of-the-art shared-bus CTI architecture in this section. The proposed CTI routing algorithm is not restricted to a specific topology, but we assume a CTI network of a regular-shape mesh-grid for the demonstration purpose. All the EES banks are supercapacitor banks, and thus the terminal voltage of each bank is linearly proportional to the state of charge and is initially different to each other. The initial terminal voltage of the EES banks is randomly determined between 15 V and 200 V.

The performance metric to be evaluated is the energy efficiency of charge transfer tasks. The baseline method is the shared-bus CTI architecture. We first begin with the charge transfers tasks that are single-source-single-destination (SSSD) ($|\sigma_i| = 1$ and $|\delta_i| = 1$ for all T_i). The SSSD transfers become multiple-source-multiple-destination (MSMD) transfers after merging. We do not lose any generality by assuming SSSD transfer tasks because the proposed algorithm can handle arbitrary number of nodes in σ and δ.

We assume the followings in charge transfers in the experiments. i) An SSSD transfer task defines the amount of energy into the destination node. The amount of charge transfer is defined from the destination side in an SSSD transfer task. The amount of energy from the source node is determined accordingly by the power converter efficiency. ii) The amount of energy into the destination nodes is kept the same in an MSMD transfer task after merging. We keep the ratio of the amount of energy to be discharged from each source the same.

Table 2: Routing results and efficiency of charge transfers in networked CTIs. All the tasks are initially SSSD transfers.

No.	Network grid size	Number of participating nodes	Number of tasks change	Energy efficiency	
				Networked CTI archi.	Shared-bus CTI archi.
1	3-by-3	4 out of 9	$2 \rightarrow 2$	88.5%	76.1%
2	3-by-3	8 out of 9	$4 \rightarrow 2$	79.2%	73.4%
3	5-by-5	12 out of 25	$6 \rightarrow 6$	81.2%	74.2%
4	5-by-5	18 out of 25	$9 \rightarrow 6$	57.7%	57.3%
5	7-by-7	18 out of 49	$9 \rightarrow 8$	81.6%	74.8%
6	7-by-7	38 out of 49	$19 \rightarrow 15$	75.9%	68.7%

6.2 Experimental Results

We use 3-by-3 to 7-by-7 mesh-grid CTI networks with different number of initial charge transfer tasks as benchmarks. Table 2 shows the number of nodes that participate in the charge transfer, total number of nodes, number of tasks in the initial and final output task sets, and the energy efficiency improvement compared with the shared-bus CTI architecture.

Figure 6 shows the input and output of the proposed algorithm with the benchmark No. 4 having a 5-by-5 CTI network and an initial task set of nine SSSD tasks. The CTI algorithm performs five times of routing and four times of merging (the last routing is not followed by merging). Three merges are accepted and one is rejected, and so the initial nine tasks are merged into six tasks as a result of the routing. Figure 6(b) shows that tasks T_1, T_6, and T_9 in Figure 6(a) are merged into T_1, and tasks T_4 and T_5 are merged into T_4. (Refer to Appendix A.2 for the detailed steps of routing.)

The experimental results show that the proposed routing algorithm successfully routes the charge transfer tasks even the number of tasks is large and the routing resources are limited. For example, there are initially 19 SSSD tasks in the benchmark No. 6, and so 38 nodes out of total 49 nodes participate in the charge transfers, which results in a very congested CTI network routing. The proposed algorithm merges only 4 of 19 tasks into other tasks and achieves a 7.2% higher energy efficiency compared with the charge transfers on a shared-bus CTI architecture. The energy efficiency improvement is up to 12.4% for the example benchmark set.

It is shown that the efficiency improvement diminishes as more number of tasks are merged. Benchmarks No. 1 and 2 end up with two tasks in the end, but efficiency improvement is less significant in No. 2 because we merge more number of tasks. It is the same for the benchmarks No. 3 and 4 that both end up with six tasks after merging. This is because more participating nodes in the same number of tasks imply that there are more nodes that do not have the optimal CTI voltage.

The energy efficiency improvement is significant when we consider the initial voltages of the EES banks are totally randomly generated. In fact, the energy efficiency improvement may be minor as in the benchmark No. 4 if the initial SSSD transfer tasks have a large voltage difference between the source and destination nodes. However, the benefit of the networked CTI architecture is larger when the voltage difference between the source and destination nodes is small in the initial charge transfer tasks before merging.

7. CONCLUSIONS

This paper introduced a networked charge transfer interconnect (CTI) architecture for hybrid electrical energy storage (HEES) systems. The networked CTI architecture is capable of accommodating an increased number of EES banks with an excellent scalability. We also described a CTI router, which is the basic building block of the proposed networked CTI architecture. Since HEES systems with networked CTI architecture require sophisticated control

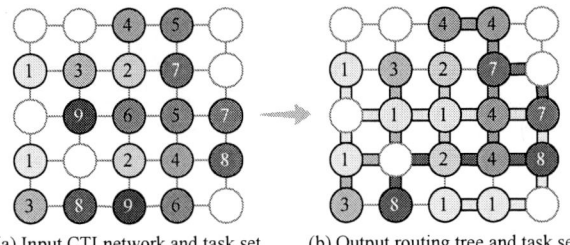

(a) Input CTI network and task set   (b) Output routing tree and task set

Figure 6: An example of routing result of nine tasks on a 5-by-5 regular-shaped mesh-grid CTI network.

mechanisms, we presented a novel CTI routing algorithm that guarantees routability (despite routing congestion due to limited routing resources) by developing the charge transfer merging technique. The proposed CTI routing algorithm finds the routing paths for charge transfers, performing energy-efficient merging of tasks not to significantly degrade the overall energy efficiency. We showed that the proposed networked CTI architecture significantly improves the energy efficiency of the charge transfers by enabling multiple, concurrent transfers with optimal CTI voltages in comparison with the state-of-the-art shared-bus CTI architecture that all the charge transfers should be merged on the bus. The experimental results show that the proposed networked CTI architecture achieves up to 12.4% of energy efficiency improvement.

8. REFERENCES

[1] M. Pedram, N. Chang, Y. Kim, and Y. Wang, "Hybrid electrical energy storage systems," in *ISLPED*, 2010, pp. 363–368.

[2] F. Koushanfar, "Hierarchical hybrid power supply networks," in *DAC*, 2010, pp. 629–630.

[3] Q. Xie, Y. Wang, Y. Kim, N. Chang, and M. Pedram, "Charge allocation for hybrid electrical energy storage systems," in *CODES+ISSS*, 2011, pp. 277–284.

[4] Q. Xie, Y. Wang, M. Pedram, Y. Kim, D. Shin, and N. Chang, "Charge replacement in hybrid electrical energy storage systems," in *ASP-DAC*, 2012, pp. 627–632.

[5] Y. Wang, Y. Kim, Q. Xie, N. Chang, and M. Pedram, "Charge migration efficiency optimization in hybrid electrical energy storage (HEES) systems," in *ISLPED*, 2011, pp. 103–108.

[6] D. Shin, Y. Kim, J. Seo, N. Chang, Y. Wang, and M. Pedram, "Battery-supercapacitor hybrid system for high-rate pulsed load applications," in *DATE*, 2011, pp. 875–878.

[7] A. Mirhoseini and F. Koushanfar, "HypoEnergy. hybrid supercapacitor-battery power-supply optimization for energy efficiency," in *DATE*, 2011, pp. 887–890.

[8] Y. Kim, Y. Wang, N. Chang, and M. Pedram, "Maximum power transfer tracking for a photovoltaic-supercapacitor energy system," in *ISLPED*, 2010, pp. 307–312.

[9] Y. Kim, S. Park, Y. Wang, Q. Xie, N. Chang, M. Poncino, and M. Pedram, "Balanced reconfiguration of storage banks in a hybrid electrical energy storage system," in *ICCAD*, 2011, pp. 624–631.

[10] D. Hill, "A CAD system for the design of field programmable gate arrays," in *DAC*, 1991, pp. 187–192.

[11] Y.-W. Chang, S. Thakur, K. Zhu, and D. F. Wong, "A new global routing algorithm for FPGAs," in *ICCAD*, 1994, pp. 356–361.

[12] L. McMurchie and C. Ebeling, "PathFinder: a negotiation-based performance-driven router for FPGAs," in *FPGA*, 1995, pp. 111–117.

[13] J. S. Swartz, V. Betz, and J. Rose, "A fast routability-driven router for FPGAs," in *FPGA*, 1998, pp. 140–149.

[14] J. Luu, I. Kuon, P. Jamieson, T. Campbell, A. Ye, W. M. Fang, and J. Rose, "VPR 5.0: FPGA cad and architecture exploration tools with single-driver routing, heterogeneity and process scaling," in *FPGA*, 2009, pp. 133–142.

[15] Y. Choi, N. Chang, and T. Kim, "DC–DC converter-aware power management for low-power embedded systems," *IEEE T. on CAD*, pp. 1367–1381, 2007.

APPENDIX

A.1 Conflict Graph Management Example

We give an example of task selection of tasks to be merged using a conflict graph and conflict graph update discussed in Section 5.3. Figure A1 shows an example of conflict graph update with four tasks. Figure A1(a) is the initial conflict graph G^c, and Figure A1(b) is the updated conflict graph $G^{c\prime}$ after merging. Each node corresponds to a charge transfer task. The three-tuple on each edge is $(\Delta V_{opt}^{cti}, d[e^c], r[e^c])$, where ΔV_{opt}^{cti} is the difference of the optimal CTI voltages of two tasks, $d[e^c]$ is the conflict count between two tasks, and $r[e^c]$ is the marking whether if the corresponding merging is rejected previously.

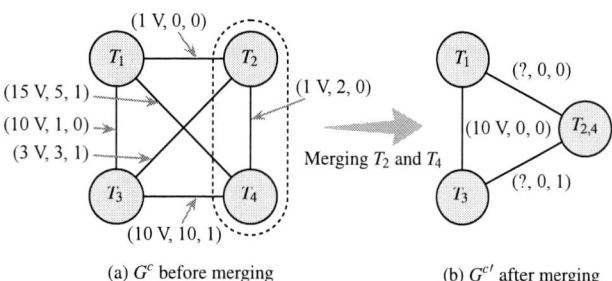

(a) G^c before merging   (b) $G^{c\prime}$ after merging

Figure A1: Example of a conflict graph (a) before and (b) after merging tasks T_2 and T_4. Each edge is annotated with a three-tuple $(\Delta V_{opt}^{cti}, d[e^c], r[e^c])$.

We find a pair of tasks to merge from G^c as follows:

- We do not consider the task pairs for merging if they had been rejected before ($r[e^c] = 1$). This condition excludes the task pairs (T_1, T_4), (T_2, T_3) and (T_3, T_4).

- We do not consider the task pairs for merging if they are not conflicting ($d[e^c] = 0$). This condition excludes the task pair (T_1, T_2).

- Remaining task pairs are (T_1, T_3) and (T_2, T_4).

- We pick (T_2, T_4) for merging because the CTI voltage difference is 1 V, which is smaller than 10 V of (T_1, T_3).

We update the conflict graph to $G^{c\prime}$ after the merging as follows:

- T_2 and T_4 are removed and $T_{2,4}$ is added.

- All the conflict count $d[e^c]$ is reset to zero.

- Task pair (T_1, T_3) is not affected by the merging, and so $\Delta V_{opt}^{cti} = 10$ V and $r[(v_1^c, v_3^c)] = 0$ for this pair remain the same.

- We increase $d[e^c]$ during the routing in Line 5 in Algorithm 1.

- ΔV^{cti} for $(T_1, T_{2,4})$ and $(T_3, T_{2,4})$ is set to unknown because the optimal CTI voltage for of $T_{2,4}$ is not calculated yet. It is calculated in Line 7 in Algorithm 1 after routing.

- We mark $T_{2,4}$ merge-able with T_1, i.e., $r[(v_1^c, v_{2,4}^c)] = 0$ because T_2 was merge-able before merging. On the other hand, we mark $T_{2,4}$ not-to-be-merged with T_3, i.e., $r[(v_3^c, v_{2,4}^c)] = 1$ because neither T_2 nor T_4 was merge-able with T_3 before merging.

Since $D[G^c] = 21$ and $d[(v_2^c, v_4^c)] = 2$, merging T_2 and T_4 is accepted if $D[G^{c\prime}] \leq 21 - 2 = 19$ by (6), and the new conflict graph in Figure A1(b) is used in the next iteration. Otherwise (if rejected), we restore G^c, mark $r[(v_2^c, v_4^c)]$ to 1, and try merging the last remaining pair (T_1, T_3).

A.2 Routing Example

We show the detailed steps of CTI routing of Figure 6. The initial charge transfer task set has nine SSSD tasks. The steps illustrated in Figure A2 explain the iterations of the loop through Lines 3–15 of Algorithm 1 until the routing finishes. Figure A2(a) is the initial state, and it goes through five iterations as shown in Figures A2(b)–(f). The first merge is rejected and the following three merges are accepted, resulting in six tasks.

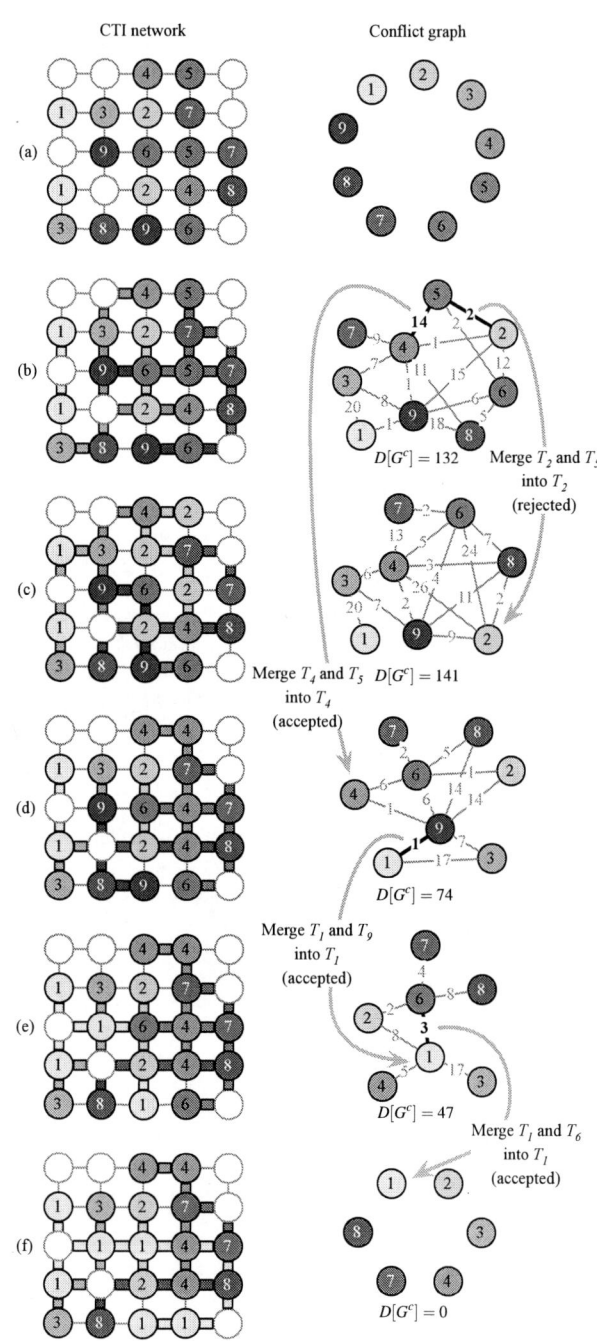

Figure A2: Detailed steps of the CTI routing of Figure 6.

A New Uncertainty Budgeting Based Method for Robust Analog/Mixed-Signal Design

Jin Sun
Orora Design Technologies, Inc.
Issaquah, WA
jinsun@orora.com

Priyank Gupta
Cirrus Logic, Inc.
Tucson, AZ
priyank.gupta@cirrus.com

Janet Roveda
The University of Arizona
Tucson, AZ
wml@ece.arizona.edu

ABSTRACT

This paper proposes a novel methodology for robust analog/mixed-signal IC design by introducing a notion of budget of uncertainty. This method employs a new conic uncertainty model to capture process variability and describes variability-affected circuit design as a set-based robust optimization problem. For a pre-specified yield requirement, the proposed method conducts uncertainty budgeting by associating performance yield with the size of uncertainty set for process variations. Hence the uncertainty budgeting problem can be further translated into a tractable robust optimization problem. Compared with the existing robust design flow based on ellipsoid model, this method is able to produce more reliable design solutions by allowing varying size of conic uncertainty set at different design points. In addition, the proposed method addresses the limitation that the size of ellipsoid model is calculated solely relying on the distribution of process parameters, while neglecting the dependence of circuit performance upon these design parameters. The proposed robust design framework has been verified on various analog/mixed-signal circuits to demonstrate its efficiency against ellipsoid model. An up to 24% reduction of design cost has been achieved by using the uncertainty budgeting based method.

Categories and Subject Descriptors

B.7.2 [**INTEGRATED CIRCUITS**]: Design Aids

General Terms

Algorithms, Design, Performance

Keywords

Robust Design, Performance Yield, Uncertainty Set, Budget of Uncertainty, Process Variations

1. INTRODUCTION AND MOTIVATION

Due to the decreasing feature sizes, the advanced sub-wavelength semiconductor fabrication techniques fail to control precisely dopant diffusion [1][2] and are unsuccessful in printing geometric features accurately [3]. Consequently, a significant amount of process variations are introduced into the integrated circuits. Process parameters of the fabricated devices, such as effective channel length and oxide thickness, can be very different from their nominally designed values. These variations have caused substantial uncertainties in circuit

behaviors and significantly increased the difficulties in analog/mixed-signal circuit design [4][5][6]. The parametric yield of manufacturing process and circuit performance are thus in jeopardy. The motivation of this paper is to develop an efficient method for robust circuit design that optimizes circuit performance while meeting the yield requirement.

In general, process variations can be separated into two components: inter-die (global) variations and intra-die (local) variations. Global variations are generally modeled as Gaussian distributed random variables, while local variations are not easy to capture. One reason is that local variations, also random in nature, exhibit strong spatial correlations. In addition, some variation sources are known to be non-Gaussian with asymmetric distributions [7]. The formulation of variability-affected circuit design depends on the models of parameter variations. With process variations modeled as random variables, a nature way is to formulate design with uncertainty problem as stochastic optimization. However, stochastic optimization assumes that parameter variability has an accurate probabilistic description for computing the statistical features of circuit performance [8], which is very difficult in early design stage. Another difficulty in stochastic optimization is the high computational complexity.

A lot of efforts have been made to formulate the stochastic design problem into its deterministic counterpart. Robust optimization [9][10] is one of the most popular categories. Robust optimization employs a uncertainty set to capture the variability in design parameters. The resulting set-based design problem can be further reduced to a deterministic one by constructing a solution that is robust to the variations within the uncertainty set:

$$\min \; f\left(P_0\right) \;\; \text{s.t.} \;\; \max_{u \in \mathcal{U}} g_i\left(P_0, u\right) \le g_{\text{limit}}^{(i)} \qquad (1)$$

where P_0 denotes the nominal design variables, $f\left(P_0\right)$ is a certain circuit performance to be optimized, and performance limits $g_{\text{limit}}^{(i)}$'s impose a set of performance constraints. The process variations associated with design parameters are described by an uncertainty set $\mathcal{U}$. Traditional robust optimization may lead to conservative designs because it assumes a complete guarantee of performance constraint instead of seeking to immunize the design solution in probabilistic sense [8]. This shortcoming restricts its applicability in variability-affected circuit design. In this work, we propose to use a "budget of uncertainty" in robust optimization to explore the trade-off between parametric yield protection and design parameter variability. The solutions obtained by uncertainty budgeting based robust design can be expected to be close to those produced by stochastic methods.

In [11][12] the authors proposed to use an ellipsoidal uncertainty model for robust circuit design with a a guaranteed probability. The ellipsoidal uncertainty set is defined as:

$$\mathcal{E}_\gamma = \left\{ P \,|\, \left(P - P_0\right)^T \Sigma^{-1} \left(P - P_0\right) \le \gamma \right\} \qquad (2)$$

where Σ stands for the covariance matrix of parameter variations. Since random variable $\left(P - P_0\right)^T \Sigma^{-1} \left(P - P_0\right)$ follows a χ_n^2 distribution, a pre-specified probability η can be captured by an ellipsoid set $\mathcal{E}_\gamma$ with $\eta = F_{\chi_n^2}(\gamma)$ where $F_{\chi_n^2}$ is the χ_n^2 CDF. Figure 1 illustrates a 2-D example of ellipsoid method. The performance limit g_{limit} forms a feasible region in the design space. Each point in the feasible region denotes a nominal design candidate. The process variations of design candidates are captured by an ellipsoid set.

Note that in ellipsoid model the uncertainty set is solely determined by the distribution information in design space, without looking into performance space. For a pre-specified yield η, the structure of the ellipsoid set is determined by the covariance matrix. However, neglecting the dependence of performance metric on design parameters may lead to conservative estimation of process variations. As shown in Figure 1, at design point P_A the ellipsoid set captures parameter variations δP_A with a certain probability. However, due to the high nonlinearity of analog circuit behaviors, the actual performance yield may be very different after a nonlinear mapping from design space to performance space. The conservative estimation will incur considerable over-design especially at low yield specifications.

In addition, at a given probability, ellipsoid model uses an uncer-

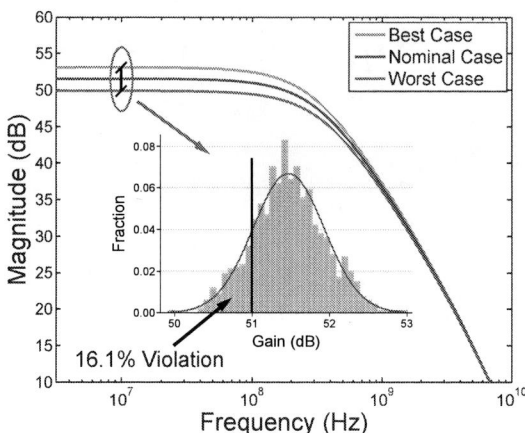

Figure 2: The experimental results on a differential pair show that ellipsoid model may lead to unreliable designs.

attempts to associate the performance yield with the size of conic uncertainty set, and therefore allows the designer a level of flexibility in choosing the tradeoff between robustness and performance. Finally, we thoroughly verify the accuracy of the proposed model against ellipsoid model on several circuits in 90nm technology. Experimental results show promising improvements in terms of design cost reduction by using the uncertainty budgeting based method.

The rest of the paper is organized as follows. Section 2 introduces the concept of uncertainty budgeting and the conic representation of uncertainty model for process variations. Section 3 details the proposed uncertainty budgeting based framework for robust circuit design. Section 4 demonstrates experimental results, and Section 5 concludes this paper.

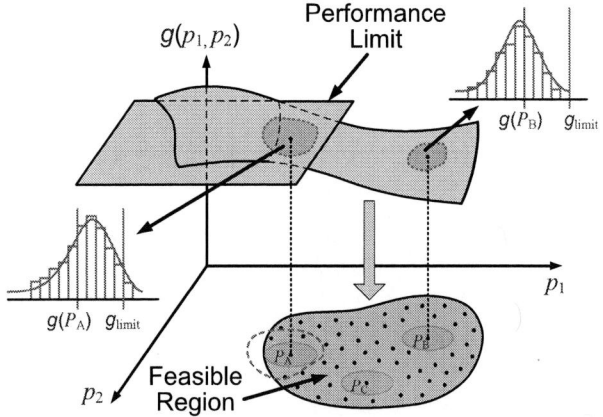

Figure 1: An illustrative example of a 2-dimensional ellipsoidal uncertainty set

tainty set with fixed size and structure for all different nominal designs. In practice the process variations associated with a specific design are highly dependent on the nominal design values [13]. For example, the Pelgrom's model [14] shows that the standard deviation of threshold voltage has a $1/\sqrt{(WL)}$ dependence on transistor area. If the optimal solution falls onto a boundary point where its variation range is much greater than the ellipsoid set, the actual yield may be significantly different from its expected value. We use a CMOS differential pair circuit as an example to describe this problem. The schematic diagram and experiment setup are detailed in Appendix S1. The robust design is formulated as minimizing the power consumption with a minimum gain limit at a particular frequency. For a 100% yield specification, the results show that the design solution by ellipsoid method has caused a 16.1% performance violation, and therefore does not meet the yield requirement.

The contributions of this paper can be summarized as follows. First, we propose to use a conic representation of uncertainty set for characterizing process variations. This conic uncertainty model extracts the spatial correlations among parameter variations and captures the dependence of variation range on design parameters. In addition, we discuss the relationship between budget of uncertainty and uncertainty model in circuit design with variability. Given a yield specification for circuit performance, we show how to incorporate uncertainty budgeting into a robust design flow to achieve optimal design with the required performance yield. Uncertainty budgeting

2. UNCERTAINTY BUDGETING WITH CONIC UNCERTAINTY MODEL

Traditional robust design is set-based and has no probabilistic description of variations in circuit performance. Budget of uncertainty was first introduced in [10][15] to establish a probabilistic guarantee for the robust design that can be computed *a priori*, i.e. as a function of the structure and size of uncertainty set. To be specific, in robust design considering budget of uncertainty, the performance constraint can be rewritten as:

$$\max \left\{ g_i(P_0, u) \,\middle|\, u \in \mathcal{U}(\eta) \right\} \leq g_{\text{limit}}^{(i)} \qquad (3)$$

where $\mathcal{U}(\eta)$ is the uncertainty set, of which the size is now linked with performance yield η. For example, a most simplistic and straightforward way is to define the budget of uncertainty as the number of design parameters that are allowed to have variations. If this number is small, the variations in circuit performance may lead to a low yield, and vice versa. If we can establish a valid relationship between performance yield and uncertainty set, we may achieve robust designs close to the results by stochastic modeling and avoid complicated stochastic computation. It is worth emphasizing that this concept is different from the ellipsoid model (2) in which the probability is captured from the distribution information of design parameters.

We start with the probabilistic constraint in stochastic design fashion, which is given by:

$$\text{Prob}\left\{ g_i(P_0 + \delta P_0) \leq g_{\text{limit}}^{(i)} \right\} \geq \eta \qquad (4)$$

where δP_0 represents the process variations associated with the nominal design P_0. The constraint in (4) identifies a search space consisting of design candidates that are feasible for stochastic optimization.

Apparently if the amount of probability (i.e. the yield requirement for circuit performance) changes, such search space will change accordingly. However, as explained previously, the computation of probabilistic constraint is in general difficult as it requires explicit distribution information and an intensive computation cost. We observe that in a similar manner, the search space confined by the robust constraint (3) is also sensitive to the change of yield requirement as the uncertainty set is now associate with the probability η. To be specific, if a relatively lower yield level is required, it is reasonable that we accordingly choose a relatively smaller size of uncertainty set to model process variations. This will lead to a relatively larger search space containing possibly better design solution(s), which is in accordance with the situation in stochastic optimization. In this sense, we avoid intensive computation of probabilistic constraint, and need not to break the robust optimization framework. If we are able to determine the size of uncertainty set which is closely associated with performance yield, the resulting robust solution could be a good approximation to that of stochastic design.

Two main problems in budgeting uncertainty for robust design are: 1) to set up an appropriate uncertainty model for parameter variations, and 2) to establish the budget of uncertainty that allows the trade-off between yield requirement and uncertainty characterization. In this work we propose to use a conic representation of uncertainty set based on second order cone (SOC) to capture the relationship between process variations and design parameters. Mathematically, a unit SOC of dimension k is defined as [16]:

$$\left\{ \left[\begin{array}{c} X \\ s \end{array} \right] \middle| X \in R^{k-1}, \ s \in R, \ \|X\|_2 \le s \right\} \quad (5)$$

where X is a vector of dimension $k-1$. For a nominal design vector P_0, by introducing an auxiliary variable s, we have the following conic uncertainty set to characterize process variations associated with P_0:

$$\mathcal{S} = \left\{ (\delta P, s) \middle| \|\delta P\|_2 = \|P - P_0\|_2 \le s, s \le s_{\max} \right\} \quad (6)$$

where $\|\delta P\|_2 = \delta P^T \delta P$ is the 2-norm of variations δP. A general-form uncertainty model can be extended from the unit case in (6):

$$\mathcal{S} = \left\{ (\delta P, s) \middle| \|A\,\delta P + b\|_2 \le c^T \delta P + d \right\} \quad (7)$$

where $A \in R^{l \times n}$, $b \in R^n$, $c \in R^n$ and $d \in R^n$ are constant coefficients. For interpretive purpose, in what follows we use the simplified case of unit cone to explain the conic uncertainty model. Referring to (6), the value of s variable in fact restricts how far the parameter variations can deviate from their nominal values. The s value can be varied with a bounding value $s_{\max}$, where $s_{\max}$ captures the furthest variation point away from the nominal location. By continuously varying s all parameter variations will be enclosed by the uncertainty model, and $s_{\max}$ defines the size of the uncertainty set. In ellipsoid model the uncertainty set has a fixed size for all nominal points, whereas in the proposed model the size of uncertainty set varies at different points as we allow a changeable variable s.

If we extend the unit-case conic set to the general case (7), by changing the entries of A matrix we can further adjust the shape of conic uncertainty set (considering that parameter variations are distributed around the nominally designed point, it is reasonable to assume that $b = 0$). On the contrary, the shape of ellipsoid set is fixed as its radius lengths and directions are determined by the eigenvalues of covariance matrix. We thus draw the conclusion that ellipsoid model is in fact a special case of conic model. Such a conic set can be regarded as an ellipsoid set with changing size as well as changing shape. The conic uncertainty model provides better flexibility and accuracy in estimating parameter variations, especially in case of

non-Gaussian local variations with asymmetric distributions. As illustrated in Figure 3, at design point P_A, s varies from 0 to $s_{\max,A}$ to estimate variations around P_A, with $s_{\max,A}$ determining the size of uncertainty set for variations δP_A. At a different location P_B, there exists another conic set whose size is determined by $s_{\max,B}$. The

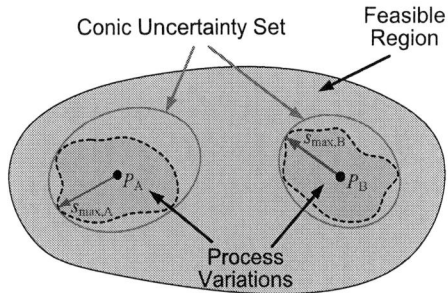

Figure 3: The conic uncertainty model allows changing size and shape of uncertainty set at different locations

size ($s_{\max}$) and shape (matrix A) of conic uncertainty set depends on the location of nominal design point. This property addresses the limitation of ellipsoid model discussed in Section 1.

Having explained the formulation of conic uncertainty set, a nature question is how to determine an explicit expression of $s_{\max}$ in terms of nominal design parameters. We use a fitting technique to find out the relationship between $s_{\max}$ and nominal design values by sampling process variations based on their distribution information. As introduced before, local variations are difficult to capture, especially in the presence of strong spatial correlations. Following [17][18], the entire die can be partitioned into a number of grids. Each grid field on the chip is denoted by its coordinate position: $l = (x, y)$. The process variations at any two grids l_i and l_j on the same chip will be correlated. The variations in close grids are strongly correlated, while those in far-away grids are weakly correlated. The correlation function between two grids l_i and l_j only depends on the Euclidean distance between them:

$$\rho(l_i, l_j) = \rho(\|l_i - l_j\|) = \rho\left(\sqrt{(x_i - x_j)^2 + (y_i - y_j)^2}\right). \quad (8)$$

The covariance function of correlated process variations can be then determined as:

$$\text{cov}(l_i, l_j) = \sigma_i \sigma_j \rho(\|l - l'\|). \quad (9)$$

where σ_i and σ_j denote the standard deviation of process variations at grid l_i and grid l_j, respectively. Note that standard deviations σ_l and $\sigma_{l'}$, as well as the correlation function ρ may be distinct for different chip designs [17]. Therefore, it will be problematic to use a unique uncertainty set for all design candidates.

In the fitting procedure, we employ the Matern model [19] to model the correlation function $\rho(\cdot)$. We perform random sampling around a nominal design P_0 and capture the furthest variation deviated from the nominal values. This distance will be identified as the $s_{\max}$ value for this particular design. Having a set of simulation data pairs (P_0, $s_{\max}$) collected throughout the range of possible parameter values, we assume $s_{\max}$ has: 1) a linear relationship ($s_{\max} = \sum_{i=1}^{n} \alpha_i p_i + \sum_{i,j}^{n} \beta_{ij} p_i p_j + k$), and 2) a quadratic relationship ($s_{\max} = \sum_{i=1}^{n} \lambda_i p_i^2 + \sum_{i=1}^{n} \alpha_i p_i + \sum_{i,j}^{n} \beta_{ij} p_i p_j + k$) with nominal design values, and performed linear regression and nonlinear regression respectively. Note that the cross terms in the fitting function are necessary for capturing the correlations among parameter variations. The

531

results show that linear fitting causes significant approximation errors (up to 16%) compared with quadratic fitting (less than 2%). The results also indicate that the linear assumption of s_{max} expression is potential to generate overly optimistic estimation of parameter variations, i.e. to produce a smaller size of uncertainty set than required. The results validate the necessity of quadratic fitting in the modeling of conic uncertainty set. The details about the correlation function and fitting results are described in Appendix S2.

3. UNCERTAINTY BUDGETING BASED ROBUST DESIGN FRAMEWORK

This section describes how to incorporate uncertainty budgeting into the robust design flow based on the conic uncertainty model. Figure 4 demonstrates the flow of the uncertainty budgeting based methodology that is applicable for general analog circuits. In the

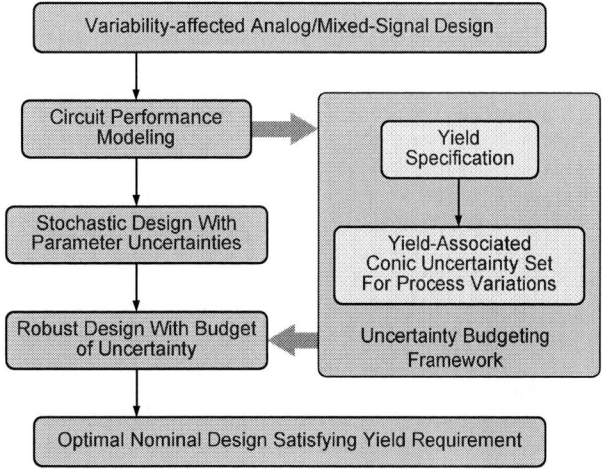

Figure 4: The general flow of uncertainty budgeting based robust circuit design

presence of process variations, the first step of variability-affected circuit design is to model the dependence of circuit performance upon design parameters, which will be required in the uncertainty budgeting flow. Having established the response surface model, we turn to deal with an intractable stochastic optimization with random parameter variations. The key contribution of the proposed design flow is the uncertainty budgeting framework based on conic uncertainty model. The budgeting method employs a conic uncertainty set to capture process variations associated with design parameters. For a pre-specified yield specification, the budgeting method maps the yield information back onto design space and explores the yield-associated conic uncertainty set for process variations. By associating performance yield with the size of the uncertainty set the stochastic design with variability can be formulated as general robust optimization. The end result is the optimal nominal design that satisfies performance constraint with a required yield.

In circuit performance modeling, since process parameters are spatially correlated, a principal component analysis (PCA) [20] is required to transform the correlated parameters into a set of independent variables with standard Gaussian distributions, as denoted by $\tilde{P}$. Following [21], we use a response surface model to approximate a specific circuit performance as a linear combination of a set of basis functions [22][23]:

$$g(\tilde{P}) \approx \sum_{l=1}^{L} k_l \cdot h_l(\tilde{P}) \tag{10}$$

where k_l's are the model coefficients, and $h_l(\cdot)$'s are the basis functions usually selected as linear and/or quadratic polynomials. As indicated in [21], only a few of basis functions are necessary to approximate a specific performance function. In other words, the vector of model coefficients only contains a small number of non-zero items. In this situation, L_0-norm regularization scheme [24] can be utilized to determine the non-zero coefficients. The details of circuit performance modeling are described in [21].

We now discuss how to associate performance yield with conic uncertainty set and incorporate it into the robust design flow. With consideration of budget of uncertainty, we rewrite the conic uncertainty set for process variations as follows:

$$\|A \, \delta P(\eta) + b\|_2 \leq s_{max}(P)\Omega(\eta) \tag{11}$$

where $\Omega(\cdot)$ is a scaling function of the conic set size for a specified yield requirement. We define the representation in (11) as the yield-associated conic uncertainty set. The scaling function $\Omega(\eta)$ is restricted to lie in the interval $[0, 1]$, and is defined as the budget of uncertainty. If $\Omega(\cdot) = 0$, there is no protection against parameter variability. If $\Omega(\cdot) = 1$, the performance constraint is completely protected against variability. If $\Omega(\cdot)$ is between 0 and 1, there exists a trade-off between yield protection and parameter variability.

We assume process variations of design parameters are captured by this yield-associated conic set. Note that the parameter variations defined by (11) is dependent on probability value η, therefore we use $\delta P(\eta)$ to distinguish it from the physical parameter variations δP. For simplicity we let $b = 0$ in this model. Assume that the physical parameter variations follow a multivariate Gaussian distribution $\delta P \sim N(\mu, \Sigma)$ where Σ denotes the covariance matrix. We can derive that the variations defined by the yield-associated uncertainty set, i.e. $\delta P(\eta)$, are also Gaussian distributed:

$$\delta P(\eta) \sim N\left(\Omega\mu, \Omega^2\Sigma\right) \tag{12}$$

We are now able to approximate the budget of uncertainty Ω in the representation of yield-associated conic model. Applying a first-order Taylor series expansion to the performance function yields:

$$\text{Prob}\left\{g\left(P_0 + \delta P(\eta)\right) \leq g_{limit}\right\} \tag{13}$$

$$= \text{Prob}\left\{g(P_0) + \sum_{i=1}^{n} \underbrace{\left(\frac{\partial g}{\partial p_i}\right)\bigg|_{p_{i_0}}}_{q_i(P_0)} \delta p_i(\eta) \leq g_{limit}\right\} \geq \eta$$

where $q_i(P_0)$ is the derivative of performance function (10) with regard to an individual parameter p_i. By substituting (12) into (13), it is possible to approximate the probabilistic constraint by a Gaussian CDF (Cumulative Distribution Function) and establish the mapping relationship between η and Ω. We start with analyzing the uncorrelated case. If the design parameters are all uncorrelated, since $\delta p_i(\eta) \sim N(\Omega\mu_i, \Omega\sigma_i)$, based on probability theory we can derive that the variations in performance metric obeys the following Gaussian distribution:

$$g(P_0, \eta) = g(P_0) + \sum_{i=1}^{n} q_i(P_0)\delta p_i(\eta) \tag{14}$$

$$\sim N\left(\underbrace{g(P_0) + \left(\sum_i q_i(P_0)\right)\Omega\mu_i}_{\mu_g}, \underbrace{\left(\sum_i q_i(P_0)\right)\Omega\sigma_i}_{\sigma_{d_j}}\right).$$

Therefore the probabilistic constraint function (13) can be approximated as follows:

$$\text{Prob}\left\{g(P_0, \eta) \leq g_{\text{limit}}\right\}$$
$$= \text{Prob}\left\{\frac{g(P_0, \eta) - \mu_g}{\sigma_g} \leq \frac{g_{\text{limit}} - \mu_g}{\sigma_g}\right\}$$
$$= \Phi\left(\frac{g_{\text{limit}} - \mu_g}{\sigma_g}\right) \geq \eta \qquad (15)$$

where $\Phi(\cdot)$ is the CDF for a standard Gaussian variable. Therefore, for a required yield specification, we can refer to Gaussian distribution table to find $\Phi^{-1}\left(\frac{g_{\text{limit}} - \mu_g}{\sigma_g}\right)$ and further calculate the budget of uncertainty Ω. Note that μ_g and σ_g both are functions of the nominal design parameters and yield specification η, so as the budget of uncertainty Ω. The resulting function Ω will be utilized in the robust optimization framework for analog circuit design.

The derivation above is based on the assumption of uncorrelated parameters. In the presence of spatial correlations, a Principle Component Analysis (PCA) [20] procedure will be first performed to transform the correlated parameters In addition, we made another simplification that a first-order Taylor expansion is applied for Gaussian distribution approximation. If the constraint function is highly nonlinear, a second-order expansion and chi-square distribution approximation (or even higher-order estimates) would be necessary.

4. EXPERIMENTAL RESULTS

This section presents the experimental results of the robust design framework on several circuits with different complexities. The proposed algorithm was implemented by MATLAB programming integrated with HSPICE simulations. All simulations and experiments were performed on a quad-core 2.8-GHz machine with 4-GB memory. We assume 30% process variations for all design parameters: channel length, channel width, threshold voltage etc.

We first use the same example of differential pair circuit to verify the correctness of the proposed method. The design objective is to minimize the power consumption with a pre-specified gain limit. The schematic diagram and robust design formulation are presented in Appendix S1. We have shown that ellipsoid model leads to a non-robust design solution (refer to Section 1 and Figure 2). Following the same procedure, with a yield requirement of 100% and gain limit of 51dB, we run 5,000 Monte-Carlo simulations to verify the robustness of the design solution generated by the uncertainty budgeting method. Figure 5 shows the distribution information of dB gain at a

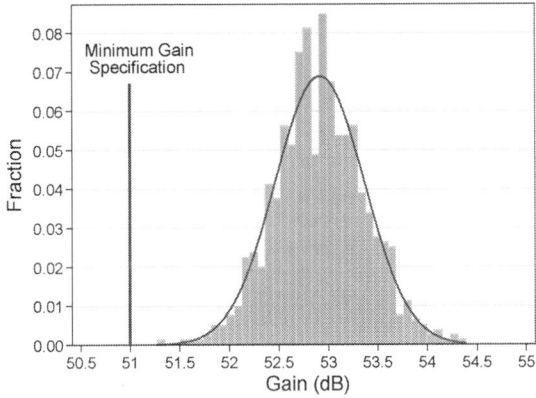

Figure 5: The histogram of gain distribution for robust differential pair design

selected frequency. All fluctuations propagated from process variations are well bounded by the minimum gain limit. Compared with

ellipsoid method, the yield requirement is completely satisfied in this example. This is because the proposed method uses an adaptively varying size of conic uncertainty set to capture process variations and generates more reliable design solutions.

As mentioned before, the uncertainty budgeting method starts with exploring the performance space, and maps the performance yield back onto the design space to form a yield-associated uncertainty set for robust optimization. On the contrary, the uncertainty model solely relies on the distribution of process variations in design space. Due to the high nonlinearity of analog circuit behaviors, neglecting the mapping relationship between parameter and performance may lead to a conservative estimation even if the yield specification can be guaranteed. We compare the results obtained by both methods on a ring oscillator circuit (refer to Appendix S3 for the schematic diagram). Robust design explores the optimal design parameters such that the power consumption is minimized and the phase noise is bounded by a maximum phase noise limit. In this example, four types of process variations from design variables, channel width, channel length, threshold voltage and oxide thickness, are taken into consideration. At different values of yield specifications, we solve the robust design problem and obtain the minimized design cost (i.e. power consumption in this example). Figure 6 shows the power consumption achieved by ellipsoid method and the proposed method, respectively. At a 100% yield specification, a power reduction of

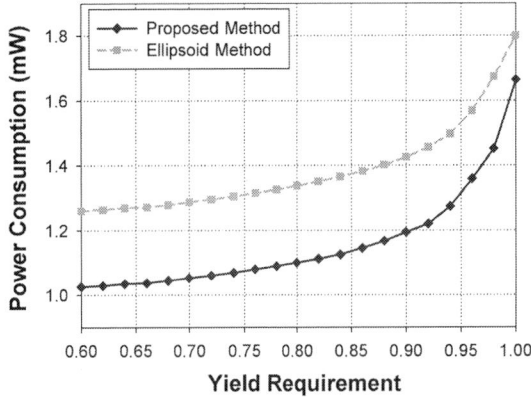

Figure 6: Power comparison between ellipsoid method and the proposed method for robust ring oscillator design

about 8% by using the proposed method can be observed. The improvement increases significantly when lower yield is required, and becomes stable with yield specification below 80%. On average the proposed method has achieved a 18% reduction of design cost.

The uncertainty budgeting framework has been further verified on a CMOS high-gain opamp circuit. The schematic diagram is provided in Appendix S4. In this example, robust design tries to minimize the total area cost while satisfying a bandwidth constraint. We consider three sets of design variables and process variations: channel width, channel length and oxide thickness. Channel width and length have been normalized to their minimum required values. The comparison of area cost by both methods is presented in Figure 7. Similar conclusion can be drawn as in the ring oscillator example. The area reduction by the proposed method becomes obvious as the yield requirement is relaxed. By conducting uncertainty budgeting the proposed method has improved the design cost by up to 23% compared with ellipsoid method. It is worth emphasizing that a drastic increase in the design cost will be incurred when the yield specification is close to 100%. In addition, we perform another set of experiments to show more comparison results. At a fixed yield specification 80%, we set different performance limits and observe the

533

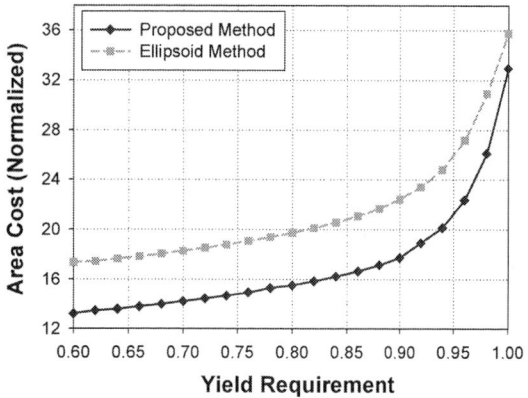

Figure 7: Area cost comparison between ellipsoid method and the proposed method for robust opamp design

improvement of area reduction. Circuits with tighter performance specifications wound be more sensitive to process variations. Figure 8 shows the area costs for the opamp circuit with different values of bandwidth limits. The results show that for tighter performance

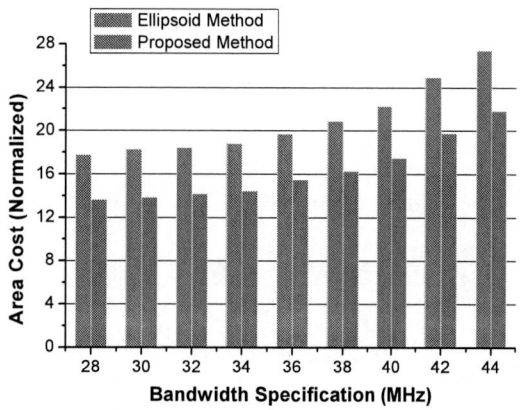

Figure 8: Area costs for the opamp circuit with different bandwidth limits

constraints, the design cost increases considerably. However, a slight difference in area reduction (ranging from 21% to 24%) can be observed as the bandwidth limit varies.

5. CONCLUSIONS

This paper presents a novel robust design framework for general analog/mixed-signal circuits in the presence of process variations. By employing a new conic uncertainty model, the new methodology conducts a uncertainty budgeting procedure by associating yield requirement with the size of conic uncertainty set. With process variations characterized in the conic representation, robust circuit design can be formulated as a regular robust optimization problem.

ACKNOWLEDGMENT

The authors would like to thank Mr. Kiran Potluri for helping design the differential pair circuit and the operational amplifier circuit in 90nm technology.

6. REFERENCES

[1] D. Boning and S. Nassif, "Models of process variations in device and interconnect," in *Design of High-Performance Microprocessor Circuits, chapter 6*, A. Chandrakasan, W. J. Bowhill, and F. Cox, Eds. IEEE Press, 2001, pp. 98–115.

[2] S. Roy and A. Asenov, "Where do the dopants go?" *Science*, vol. 309, no. 5733, pp. 388–390, 2005.

[3] M. Orshansky, L. Milor, and C. Hu, "Characterization of spatial intrafield gate cd variability, its impact on circuit performance, and spatial mask-level correction," *IEEE Transactions on Semiconductor Manufacturing*, vol. 17, no. 1, pp. 2–11, 2004.

[4] H. Yu, X. Liu, H. Wang, and S. Tan, "A fast analog mismatch analysis by an incremental and stochastic trajectory piecewise linear macromodel," in *Proc. ASPDAC*, 2010, pp. 211–216.

[5] M. Orshansky, S. Nassif, and D. Boning, *Design for Manufacturability and Statistical Design: A Constructive Approach.* New York, NY: Springer, 2008.

[6] F. Gong, H. Yu, Y. Shi, D. Kim, J. Ren, and L. He, "Quickyield: an efficient global-search based parametric yield estimation with performance constraints," in *Proc. DAC*, 2010, pp. 392–397.

[7] L. Cheng, J. Xiong, and L. He, "Non-linear statistical static timing analysis for non-gaussian variation sources," in *Proc. DAC*, 2007, pp. 250–255.

[8] D. Bertsimas, D. B. Brown, and C. Caramanis, "Theory and applications of robust optimization," *SIAM Review*, vol. 53, no. 3, pp. 464–501, 2011.

[9] A. Ben-Tal and A. Nemirovski, "Robust convex optimization," *Mathematics of Operations Research*, vol. 23, no. 4, pp. 769–805, 1998.

[10] D. Bertsimas and A. Thiele, "Robust and data-driven optimization: Modern decision-making under uncertainty," in *Tutorials in Operations Research: Models, Methods, And Applications for Innovative Decision making.* INFORMS, 2006.

[11] J. Singh, V. Nookala, Z. Luo, and S. Sapatnekar, "A geometric programming-based worst case gate sizing method incorporating spatial correlation," *IEEE Transactions on Computer-Aided Design of Integrated Circuits and Systems*, vol. 53, no. 3, pp. 464–501, 2011.

[12] Y. Xu, K.-L. Hsiung, X. Li, L. T. Pileggi, and S. P. Boyd, "Regular analog/RF integrated circuits design using optimization with recourse including ellipsoidal uncertainty," *IEEE Transactions on Computer-Aided Design of Integrated Circuits and Systems*, vol. 28, no. 5, pp. 623–637, 2009.

[13] M. Anis and M. H. Aburahma, "Leakage current variability in nanometer technologies," in *Proc. IDEAS*, 2005, pp. 60–63.

[14] M. Pelgrom, A. C. Duinmaijer, and A. P. Welbers, "Matching properties of mos transistors," *IEEE Journal of Solid-State Circuits*, vol. 24, no. 5, pp. 1433–1440, 1989.

[15] D. Bertsimas and M. Sim, "Robust convex optimization," *The Price of Robustness*, vol. 52, no. 1, pp. 35–53, 2004.

[16] M. Lobo, L. Vandenberghe, S. Boyd, and H. Lebret, "Applications of second-order cone programming," *Linear Algebra and its Applications*, vol. 284, no. 1-3, pp. 193–228, 1998.

[17] B. Hargreaves, H. Hult, and S. Reda, "Within-die process variations: How accurately can they be statistically modeled?" in *Proc. ASPDAC*, 2008, pp. 524–530.

[18] H. Chang and S. Sapatnekar, "Statistical timing analysis considering spatial correlation using a single pert-like traversal," in *Proc. ICCAD*, 2003, pp. 621–625.

[19] M. L. Stein, *Interpolation of Spatial Data.* New York, NY: Springer, 1999.

[20] G. Seber, *Multivariate Observations.* Hoboken, NJ: Wiley Series, 1984.

[21] X. Li, "Finding deterministic solution from underdetermined equation: large-scale performance variability modeling of analog/RF circuits," *IEEE Transactions on Computer-Aided Design of Integrated Circuits and Systems*, vol. 29, no. 11, pp. 1661–1668, 2010.

[22] X. Li, J. Le, L. Pileggi, and A. Strojwas, "Projection-based performance modeling for inter/intra-die variations," in *Proc. ICCAD*, 2005, pp. 721–727.

[23] T. McConaghy and G. Gielen, "Template-free symbolic performance modeling of analog circuits via canonical-form functions and genetic programming," *IEEE Transactions on Computer-Aided Design of Integrated Circuits and Systems*, vol. 28, no. 8, pp. 1162–1175, 2009.

[24] R. Tibshirani, "Regression shrinkage and selection via the lasso," *Journal of the Royal Statistical Society*, vol. 58, no. 1, pp. 267–288, 1996.

SUPPLEMENTAL MATERIAL

S1. DIFFERENTIAL PAIR CIRCUIT

Figure 1 shows the schematic diagram of a CMOS differential pair circuit in 90nm technology.. The robust design for this circuit is formulated as

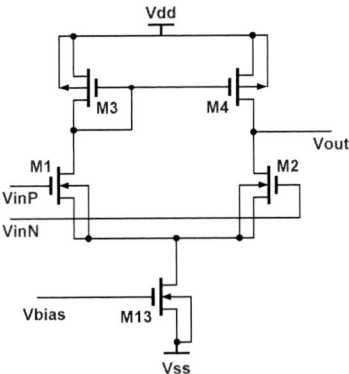

Figure 1: Schematic of a CMOS differential pair circuit

minimizing the power consumption while satisfying a gain constraint:

minimize: $\text{Power}(W, L, V_{\text{th}})$

subject to: $A_v(W + \delta W, L + \delta L, V_{\text{th}} + \delta V_{\text{th}}) \leq A_{\text{limit}}$ (1)

The objective function is the total power consumption for all transistors. A_v stands for the dB gain at frequency 10MHz, with a lower bound limit $A_{\text{limit}} = 51\text{dB}$. The design variables include the following process parameters: channel width (W), channel length (L) and threshold voltage (V_{th}). In this example, we set a 100% yield requirement, and perform robust design with ellipsoid model and robust design by uncertainty budgeting respectively. The results show that the optimal design achieved by ellipsoid method has caused about 16% performance violation (as shown in Figure 2) by Monte-Carlo simulations, while the solution generated by the proposed method perfectly meets the required gain specification (as shown in Figure 5).

S2. CONIC UNCERTAINTY SET FITTING

In the fitting procedure, we use a flexible class of correlation functions called Matern class [19] to model the correlations among process variations. In Matern model the correlation function is parameterized by two parameters, θ_1 and θ_2, and is dependent on the Euclidean distance h:

$$\rho(h) = \frac{1}{2^{\theta_2 - 1}\Gamma(\theta_2)}\left(\frac{2h\sqrt{\theta_2}}{\theta_1}\right)^{\theta_2}\mathcal{K}_{\theta_2}\left(\frac{2h\sqrt{\theta_2}}{\theta_1}\right), \quad (2)$$

where $h = \|l_i - l_j\|$ denotes the distance between two grids. $\mathcal{K}_\alpha(\cdot)$ and $\Gamma(\cdot)$ denote the modified Bessel function and Gamma function, respectively. The parameter θ_1 can be regarded as the decay rate of correlation coefficient as a function of distance, and θ_2 determines the shape of the correlation function (refer to [17] for a detailed discussion). When fitting the conic uncertainty model, we choose a set of three typical values for θ_1 and θ_2 respectively, and perform linear fitting and quadratic fitting for all possible combinations of (θ_1, θ_2) values. Fitting results presented in Figure 2 demonstrate that in all cases linear fitting has caused considerable approximation errors, ranging from 14.3% to 15.9%, while the errors by quadratic fitting are less than 1.5% for different (θ_1, θ_2) values.

S3. RING OSCILLATOR CIRCUIT

Figure 3 shows the schematic diagram of a simple ring oscillator circuit. The robust design for this circuit is formulated as minimizing the power consumption while satisfying a phase noise constraint:

minimize: $\text{Power}(W, L, V_{\text{th}}, T_{\text{ox}})$ (3)

subject to: $\text{PN}(W + \delta W, L + \delta L, V_{\text{th}} + \delta V_{\text{th}}, T_{\text{ox}} + \delta T_{\text{ox}}) \leq \text{PN}_{\text{limit}}$

The maximum phase noise limit is specified as $\text{PN}_{\text{limit}} = -106\text{dBc/Hz}$. In this example, four types of process variations are taken into consideration: channel width, channel length, threshold voltage and oxide thickness. For different yield specifications, the design results by ellipsoid method and the proposed method are presented for comparison.

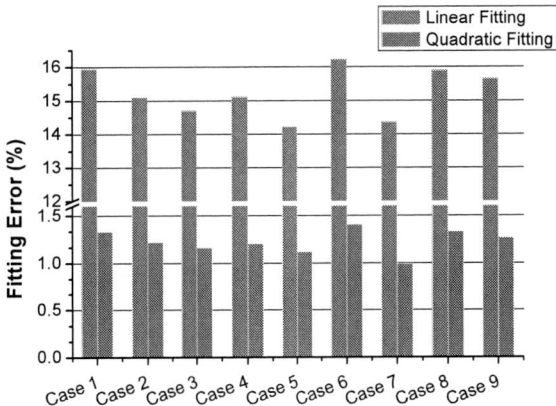

Figure 2: Schematic of a ring oscillator circuit

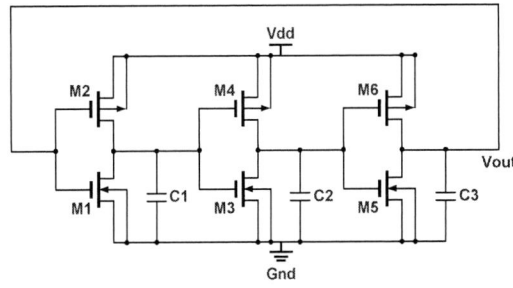

Figure 3: Schematic of a ring oscillator circuit

S4. OPERATIONAL AMPLIFIER CIRCUIT

Figure 4 shows the schematic diagram of a CMOS operational amplifier circuit in 90nm technology. The robust design for this circuit is formulated as

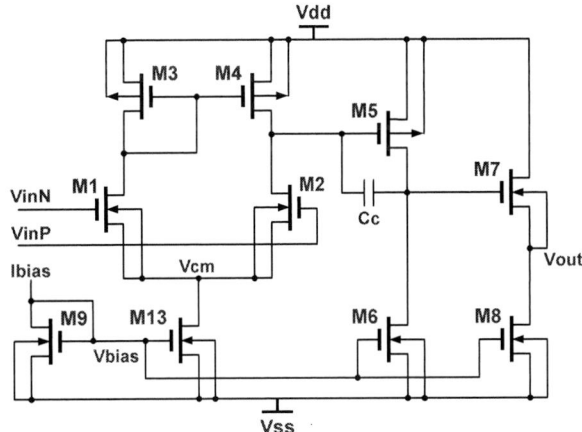

Figure 4: Schematic of an operational amplifier circuit

minimizing the area cost while satisfying a bandwidth constraint:

minimize: $\text{Area}(W, L)$

subject to: $\text{BW}(W + \delta W, L + \delta L, T_{\text{ox}} + \delta T_{\text{ox}}) \leq \text{BW}_{\text{limit}}$ (4)

In this example, we consider three types of process variations associated with the design parameters: channel width, channel length and oxide thickness. There are also minimum required values for channel width and length. In the experimental results these two parameters have been normalized to their minimum values. A series of bandwidth limits, ranging from 28MHz to 44MHz, are selected to perform robust design and compare optimization results between ellipsoid method and the proposed method.

535

Variability-Aware, Discrete Optimization for Analog Circuits

Seobin Jung, Yunju Choi, and Jaeha Kim

Inter-university Semiconductor Research Center, Seoul National University, Seoul, Korea

seobin@mics.snu.ac.kr, yunju@mics.snu.ac.kr, jaeha@snu.ac.kr

ABSTRACT

This paper proposes the use of discrete optimization techniques for variability-aware analog circuit synthesis. Starting from an observation that the continuous design space of analog circuits can be effectively covered by a finite number of discrete points in presence of variations, new algorithms are explored to address three aspects of discrete optimization: discretizing a continuous design space, selecting candidate points on the discretized grid, and comparing the statistical measures between multiple candidate points. A digitally-controlled oscillator (DCO) example demonstrates that the proposed optimization technique can efficiently find the design that balances between the resolution, linearity, phase noise, and power dissipation.

Categories and Subject Descriptors

B.7.2 **[Integrated Circuits]**: Design Aids – *simulation*

General Terms

Algorithms, Design

Keywords

Discrete Optimization, Analog Circuit Synthesis, Variability

1. INTRODUCTION

The 2009 ITRS report identifies the lack of systematic design flows and automation tools for analog circuits as one of the key obstacles to sustaining the complexity growth of mixed-signal system-on-chips (SoC's) [1]. Analog circuit optimizers, that can automatically size the transistors in a circuit according to a prescribed cost function, can be an effective productivity tool for analog, yet the reality is that their adoption into the mainstream flows has been slow. This paper aims to address the shortcomings of the existing analog optimizers. In particular, the focus is to investigate the benefits of using discrete optimizers instead of the conventional continuous optimizers.

Unlike digital logic synthesis which involves searching a solution within a finite, discrete set, analog circuit design involves finding an optimal solution in a continuous, high-dimensional design space. Due to this reason, most analog circuit synthesis tools employ numerical optimization techniques such as simulated annealing and convex optimization [2-4]. Nonetheless, they tend to consume long execution times and require a large number of simulation runs. In fact, many numerical optimizers tend to

repeatedly evaluate the similar design points when refining a solution to a certain precision or when finding a global optimum in presence of surface roughness and/or local optima. Furthermore, the ever-increasing complexities and variabilities in the deeply-scaled devices such as well proximity effects, stress effects, and aging effects pose the greater challenges to these optimizers.

Despite these adverse trends, this paper demonstrates that such aggravating process variation and uncertainty can be in fact *leveraged* to enable a much simpler tool for analog synthesis. A key observation is that in presence of process variation and uncertainty, designers have limited control over the circuit's characteristics with the design parameters (e.g., transistor sizes). In other words, unless the transistor sizes of one design are sufficiently different from those of another, process variation can render them essentially equivalent with the similar performance ranges. This has an important ramification – the design space of an analog circuit need not be continuously explored. Instead, the space can be covered effectively with a finite number of samples on a discrete grid.

The minimum change required in the design parameters to arrive at a sufficiently different performance can be derived based on the Shannon's channel capacity theory [5]. The problem is transformed into a communication model where the linearized sensitivity of the performance with respect to the design parameter is regarded as the transfer gain of the channel and the variation in performance due to process uncertainties is considered as the additive, white Gaussian noise (AWGN) with the equivalent variance. It has been shown that the minimum grid spacing required to explore the design space with discrete samples is quite coarse for a few common circuits [6].

Inspired by this promising result that motivates a grid-based analog circuit optimizer, this paper explores the use of discrete optimization techniques for variability-aware analog circuit synthesis. Specifically, the paper addresses three main challenges: an efficient grid scheme to discretize a continuous design space, a search algorithm to find the optimal design point on the discretized grid, and a way to reduce the computational burden of comparing the statistical measures among the multiple candidate points.

To address the first challenge, the paper proposes a grid scheme we call "Polka-dots" that distributes the candidate points in a multi-dimensional space in a way that keeps the distances to the nearest neighbors uniform, hence achieving the best packing efficiency.

Despite the fact that the number of candidate points has become finite, it is still computationally burdensome to try all the points on the grid to find the global optimum. To address this challenge, this paper explores the local search algorithms. That is, the current design point is compared with its neighbor points and is subsequently updated when a better point is found among them. When the number of neighbors is too large for high-dimensional spaces, the comparison can be made only with a few randomly selected neighbors. It can save the number of design points to be evaluated while still guaranteeing the convergence to a local optimum.

The third challenge applies to all yield-aware circuit optimizers. The candidate points must be compared while considering the circuit's global and local variations without incurring high simulation costs. The key is to realize that for an optimization purpose, one does not need to know the full statistical characteristics of a design point, as long as she can determine which point has the highest yield among the selected candidates. This observation opens the possibility of adopting the incremental Monte-Carlo simulation approaches [8].

The rest of this paper is organized as follows. Section 2 describes the minimum change required in order for a design to be sufficiently different from the others. Section 3 discusses our proposed solutions to the above-mentioned challenges, namely the Polka-dot grid, local search algorithm with randomly selected neighbors, and incremental Monte-Carlo methods. Section 4 demonstrates the proposed optimization algorithm on a digitally-controlled oscillator (DCO) design example. Finally, Section 5 concludes the paper.

2. The Minimum Change Required for Distinguishable Designs

The minimum change required in the design parameter (ΔD_{min}) to arrive at a sufficiently different performance can be derived based on the Shannon's channel capacity theory [5]. The problem is transformed into a communication model where the linearized sensitivity of the performance (P) to the design parameter (D) is regarded as the transfer gain of the channel ($\partial P/\partial D$) and the variation in performance due to process uncertainties is considered as the additive, white Gaussian noise (AWGN) with the equivalent variance (σ_P^2). In this case, Shannon derived the required signal-to-noise ratio (SNR_{min}) to transmit N-bit digital information error-freely:

$$N = \frac{1}{2}\log_2(1 + SNR) \qquad (1)$$

$$SNR_{min} = \left(\frac{(\Delta D_{min}/2)\cdot(\partial P/\partial D)}{\sigma_P}\right)^2 = 2^{2N} - 1 \qquad (2)$$

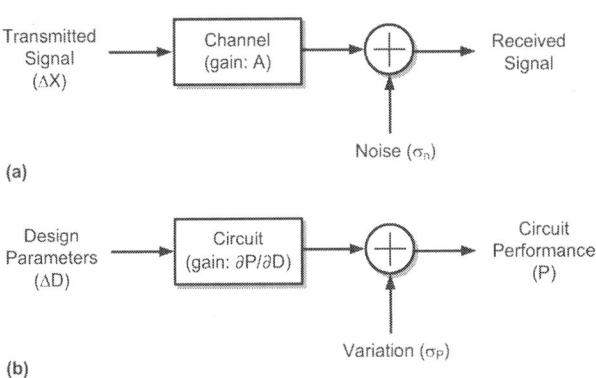

Fig 1. (a) Communication channel in presence of noise and (b) considering a circuit as an effective channel medium in presence of PVT variations and device mismatches

Since we would like to distinguish the two design points by their difference in the performance metrics, it corresponds to an information of one bit (N=1). Then it follows that the minimum difference required is:

$$\Delta D_{min} = \sqrt{12}\cdot\frac{\sigma_P}{\partial P/\partial D} \qquad (3)$$

It suggests that the larger design change is required in presence of larger statistical variations.

3. Variability-Aware, Discrete Optimization Algorithms

3.1 Grid Discretization

Before discussing the proposed Polka dot grid, let's think about one of the most basic discrete grids, the Cartesian grid. As shown in Fig.3, on a Cartesian grid, a design point requires the diagonal, facial, and corner neighbors not to miss the design points which may differ substantially. It can be shown that the number of neighbors to a given point grows exponentially as $O(3^N)$ with the number of parameters N (i.e., the space dimension). As the dimension increases, the computational load to evaluate all the neighbor candidates would become excessive.

In addition, the neighbor points on the Cartesian grid have uneven distances. For example, in a 3-dimensional unit cell shown in Fig. 3-(b), one can find different distances; 1, $\sqrt{2}$, and $\sqrt{3}$. For a general N-dimensional design space, the number of distinct distance values is N. This not only indicates inefficient packing density but also causes difficulty in determining the grid size.

To address these problems, this paper suggests a new grid scheme called "Polka-dot", which is illustrated in Fig. 3 for the three-dimensional case. As shown in the figure, the three-dimensional polka grid has the same structure with the face-centered cubic lattice in chemistry that achieves the most efficient packing of spheres.

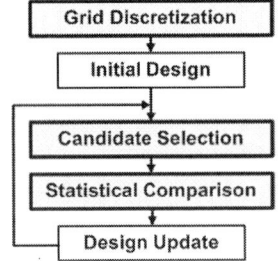

Fig 2. A discrete optimizer

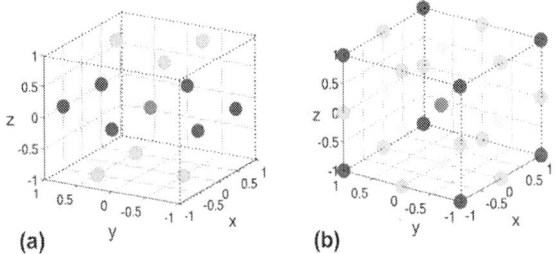

Fig 3. (a) Polka dot grid (b) Cartesian grid

The polka dot grid can be generated in a recursive way; polka dots in an N-dimensional space are newly generated based on the dots in the N-1 dimensional space. To be specific, all dots in the N-1 dimensional space are first shifted upwards and downwards by one unit along the new N-th axis, and then they are shifted side ways to keep the neighbor distances uniform. For example, in the three-dimensional space, a hexagon in the x-y plane is both raised and lowered along the z-axis, and then shifted until it is located at the middle of the hexagon.

This grid scheme has two properties that make it well-suited for discrete optimization even when the dimension gets high. The first property is that the number of the nearest neighbors grows with $O(N^2)$, the far milder increase compared to the exponential growth in the Cartesian case. It can be shown that the number of neighbors is exactly:

$$NB_N = NB_{N-1} + 2N = N(N+1) \ where \ NB_1 = 2 \qquad (4)$$

The second property is that the distance between neighbors is almost uniform, which makes it easy to determine the size of grid.

It should be noted that the proposed Polka-dot grid has some similarities with the recently proposed permutohedral lattice in computer graphics [7] and shares the described favorable properties. The permutohedral lattice is made by tessellating uniform simplices, so the distance between two nearest points is uniform, and the number of the enclosing simplex of a point grows as the order of $O(N^2)$.

Algorithm 1. Local search algorithm for discrete optimizer with randomly selected candidates among the neighbors

---

1: Start from a center

2: While the termination criteria is unsatisfied:

3: Get the nearest neighbors of the center

4: Randomly shuffle the nearest neighbors

5: For each unvisited neighbor:

6:   Compare the selected neighbor and the center

6:   if the neighbor is superior than the center:

7:    the neighbor becomes a new center

8:    break the for loop

9:   else:

10:    if this is the last neighbor:

11:     Termination criteria is satisfied

12:    else: Continue the for loop

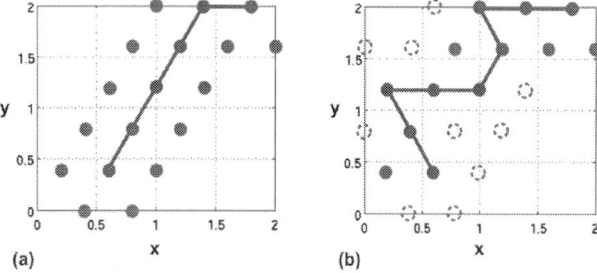

Fig 4. Local searches with (a) all neighbors evaluated and (b) only randomly selected neighbors evaluated.

3.2 Candidate Selection

Although an advantage of using a discrete grid is that one can explore the entire design space with a finite number of evaluations, it can become practically infeasible as the design space dimension increases.

One possible alternative is to evaluate only the local neighbors to the current design point. This is analogous to the local gradient search in continuous optimizers. Once a better point is found among the neighbors, the current best solution is updated and the process is repeated. The proposed Polka-dot grid can be advantageous here as the number of neighbors scales mildly with the dimension.

Interestingly, it is possible not to evaluate all the neighbors to reach the optimum. It has been formally proven [9] that for continuous optimizers, the optimum can be found by evaluating the sensitivity along a single, randomly-chosen perturbation vector. Analogous to the continuous optimization algorithm, a random selection of neighbors to be compared in the discrete design space is suggested as in Algorithm 1. The idea is quite simple; (1) choose one of the neighbors randomly, (2) if the chosen one is better than the center, move to it without testing the other neighbors, (3) iterate this process until the optimizer arrives at a point where there exists no better neighbor than the center. Intuitively, unless the cost surface is discontinuous, following in a direction of increasingly better trace would end up at a local optimal point, so that the total number of comparisons will be lower than the full conservative search.

3.3 Statistical Comparison of the Candidates

As noted before, the quality of the candidate points must be compared in a statistical sense. For example, one may want to optimize the yield of a circuit that satisfies a set of specifications. While it is possible to first estimate the yields of the points via full-blown Monte-Carlo simulations and then compare them, the incurred computational cost can be very high.

It is noteworthy that for the purpose of finding an optimum, the optimizer need not know the absolute yield value for each candidate point, as long as it can determine which point has the highest yield. This observation opens the possibility of adopting the incremental Monte-Carlo test approaches that try to minimize the number of Monte-Carlo samples required to carry out the comparison [8]. The core idea of the incremental Monte-Carlo test, or sometimes called the fully sequential procedure for selecting the best point above some candidates, is that the inferior candidate points can be screened out as soon as possible by updating the statistical information of the candidates increasingly.

In case when the difference between the two designs is large, the comparison can be concluded with far fewer samples than the ones required to estimate the absolute yield value. On the other hand, when the difference between the designs is small, a large number of Monte-Carlo samples may be still required to determine the superiority. However, this should happen only when the grid spacing is finer than the prescribed ΔD_{min}.

A simple problem, designing a 5-stage ring oscillator in Fig.5-(a) is used to explain the details of this algorithm. For simplicity, let's assume the only design parameter for the ring oscillator is the width of PMOS and the width of NMOS is fixed at 30λ (1λ equals to 30nm). The oscillation period (T_P) is the performance metric to optimize. Fig. 5-(b) shows the relationship between the performance metric and the design parameter in presence of PVT variations and device mismatches. After a heavy-duty Monte-Carlo simulation (with 100,000 samples in total), the full statistical characteristics are plotted as in Fig.5-(b). The solid line traces the average and the dashed lines indicate the standard deviation of the oscillation period (i.e., they correspond to one σ above and below the mean values.) From the graph, it can be found that the design achieves the minimum oscillation period around $W_p=40\lambda$.

The same problem is solved by using a discretized grid and the incremental Monte-Carlo tests. Both the Table 1 and Table 2 show the progression of the number of Monte-Carlo runs as the comparison proceeds. After the initial sampling of 10 Monte-Carlo runs for each candidate, the statistical information such as

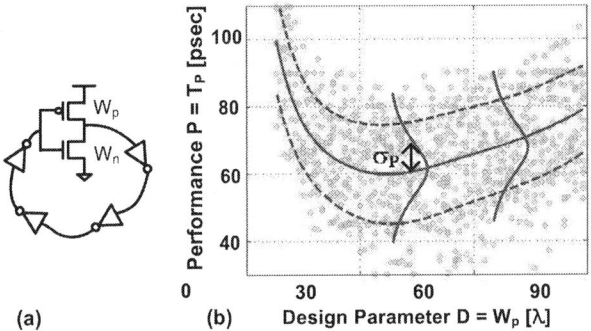

(a) (b)

Fig 5. (a) A 5-stage ring oscillator with W_p being the design parameter and the oscillation period (T_P) being the performance (b) a plot of T_P vs. W_p of (a) in presence of PVT variation and device mismatches. A 65 nm CMOS process is assumed.

Table 1. Experimental results of the incremental Monte-Carlo approach of the ring oscillator in Fig.5. As the number of Monte-Carlo runs increases, the variance of Tp at each Wp decreases so that the required number of Monte-Carlo runs decreases. The optimal point is found in less than 30 Monte-Carlo runs for each.

# of Monte-Carlo runs	Average of Tp at each Wp			Required # of Monte-Carlo runs
	10 λ	50 λ	90 λ	
10	91ps	70ps	78ps	760
27	89ps	66ps	74ps	552
28		65ps	74ps	190

Table 2. Experiment results of the incremental Monte-Carlo approach where the grid size is adjusted to a smaller value than the one in Table 1. Almost 100 Monte-Carlo runs were required to find the optimal point.

# of Monte-Carlo runs	Average of Tp at each Wp			Required # of Monte-Carlo runs
	70 λ	75 λ	80 λ	
10	64ps	70ps	75ps	323
67	65ps	70ps	71ps	352
91	66ps	70ps		342

the averages, variances, and covariances among different results are calculated. Then an iteration loop begins that runs one Monte-Carlo simulation at a time and updates the results to find any inferior candidates to screen. The loop terminates when there is only one candidate left or the maximum number of simulation runs is reached.

For the case with a coarse grid size that gives candidate points of $W_p=10\lambda$, 50 λ, and 90 λ listed in Table 1, the iteration loop terminates by selecting $W_p=50\lambda$ as the optimal point after running less than 30 Monte-Carlo runs for each candidate. This is a significant reduction in the number of Monte-Carlo simulations compared to that needed to characterize the precise statistical distribution in Fig. 5-(b).

For the case with a finer grid size that gives the denser distribution of candidate points, e.g., $W_p=70\lambda$, 75 λ, and 80 λ as in Table 2, the iteration loop terminates by selecting $W_p=70\lambda$ as the optimal point after running 67 Monte-Carlo runs for one candidate and 91 Monte-Carlo runs for the other two. Although the total number of Monte-Carlo simulations is still low, it is 2.7 times larger than the previous case in Table 1. However, it was resulted by the grid size finer than the prescribed $\Delta D_{min}=20\lambda$ (3). Therefore, it is necessary to check the grid size against ΔD_{min}, once enough Monte-Carlo samples are collected.

4. Circuit Description

The core part of the DCO is the inverter-based ring oscillator shown in Fig. 6-(a). The frequency of the oscillator is adjusted by changing its regulated supply V_{reg}; as V_{reg} gets high, the oscillation frequency rises. The oscillation frequency is determined by its terminal I-V characteristics as in (6) where I_{ring} is the load current of the oscillator and V_{reg} is the regulated supply.

$$f_{osc} \propto \frac{I_{ring}}{V_{reg}} = R_{ring} \tag{6}$$

Hence the key idea of this DCO is to control the effective output resistance of the ring oscillator R_{ring} with a digitally-adjusted, passive resistor reference. The constant relative DCO gain can be achieved since the relative change in the resistance relies only on the ratio between the resistances but not on their absolute values. The ratio between the identical elements is perhaps the only thing in IC that can remain constant against the global PVT variations.

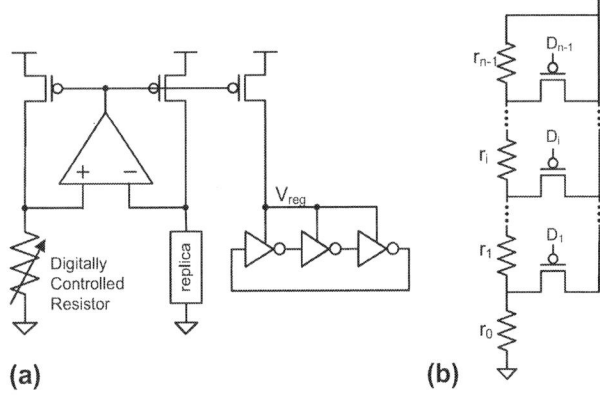

(a) (b)

Fig 6. (a) A digitally-controlled oscillator (b) a digitally-controlled resistor

539

Another subcircuit of the DCO design is the replica cell, of purpose is to have the same I-V characteristics as the ring oscillator. It is used to stabilize the feedback loop while allowing the V_{reg} node to have a large decoupling capacitance for noise rejection.

4.1 Optimization Formulation

4.1.1 Design Parameters

A DCO in general has many design parameters. One design parameter is the DCO resolution, i.e., the number of digital control bits, as it influences the minimum DCO relative gain as well as the area and stability of the feedback loop. The other design parameters may include the transistor sizes in the ring oscillator, replica cell, and feedback amplifier. Especially, the size of the ring oscillator dictates the phase noise and power dissipation.

We have made special efforts to reduce the number of design parameters. For example, for the DCR, instead of using all the switch sizes and resistances as the design parameters, we used a single independent parameter N, the number of control bits, and made the rest of the DCR parameters dependent on it. For example, once the number of bits is given, the switch sizes are determined to make their on-resistances sufficiently small compared to the corresponding series resistances. The resistance values are determined to minimize the area while considering the mismatch effects.

The other parameter is the compensation capacitor connected at the output port of the regulator. Its main purpose is to stabilize the loop, but too large capacitance consumes area and lowers the regulation bandwidth.

4.1.2 Performance Metrics

The performance metrics considered during the optimization of this DCO are listed in Table 3 along with their specifications. In this example, the relative DCO gain, or the differential nonlinearity is selected as the metric to be minimized, while the other metrics are considered as constraints. Note that the constraints demand a very wide frequency range as well as very low power consumption.

4.2 Optimization Results

Table 4 lists the found optimal point and Table 5 lists the achieved performance metrics at the optimal point. A 45nm CMOS process is used here. All the constraints are satisfied and the minimum differential nonlinearity achieved is 0.1LSB.

Fig. 7 graphically illustrates the progression of the center points according to our proposed algorithm. We omitted the compensation capacitor size as it was found nearly constant during the process of optimization. All the visited points by the local search algorithm are marked with filled circles while the candidate points that do not satisfy the constraints are marked as empty circles. In this example, the power budget and stability constraint act as the dominant factors that determine the feasible region; the power budget limits the size of the ring oscillator and the stability constraint limits the DCO resolution.

In Fig.7 (a), the optimizer starts at a point with a low DCO resolution, 2bits, and moves in the direction to increase the resolution while decreasing the replica size until converging. The

incremental Monte-Carlo tests approach is used and the total number of Monte-Carlo evaluation number is 350. On the other hand, when the optimizer starts from a point with a higher DCO resolution, 6 bits, it first searches for a feasible point and moves towards it. Because of the limited feasible region posed by the strict constraints, the number of hops made by the optimizer is relatively small. Note that the path (b) converges to the same optimal point as the path (a).

Table 6 summarizes the results that demonstrate the soundness of our local search algorithm based on randomly selected neighbors. Starting from the initial points as in Fig. 7(a) and (b), the optimizer converges to the same optimal point despite the random optimization traces. Note that the average number of Monte-Carlo simulation runs was reduced by about 30~40% using this approach.

Table 3. Specifications and Constraints for Design Example

Constraint	Specification
Frequency range	min=0.6GHz, max=2GHz
Power	≤ 1mW (at 1.5GHz)
Area	≤ 22,500 μm^2
Closed-loop bandwidth	≥ 1MHz
Phase margin	≥ 45°
Gain margin	≥ 5dB
Supply sensitivity	≤-10dB (at 5MHz)
Phase noise	≤ -100dB/c (at 100MHz offset)
Effective number of bits	≥ 0.8·(DCO resolution)
Differential nonlinearity	Minimize

Table 4. Optimal Design for Design Example (1λ =24nm)

Variable	Value
Resolution	4bits
Ring oscillator	40λ
Replica cell	8λ
Compensation capacitor	m=30,000

Table 5. Performance of Optimal Design for Design Example

Specification/Constraint	Performance
Frequency range	min=0.62GHz, max=2.0GHz
Power	0.89mW (at 1.5GHz)
Area	9,000 μm^2
Closed-loop bandwidth	3MHz
Phase margin	91°
Gain margin	24dB
Supply sensitivity	-12.4dB (at 5MHz)
Phase noise	-137dB/c (at 100MHz offset)
Effective number of bits	3.9bits
Differential nonlinearity (worst case)	0.1LSB

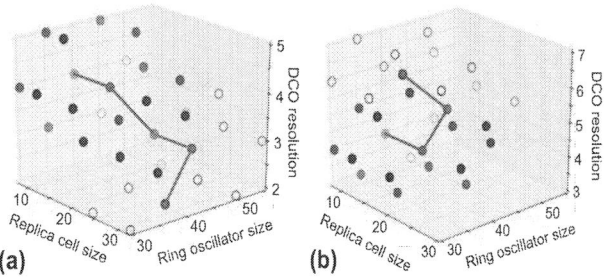

(a)   (b)

Fig 7. (a) Trace of the optimizer when the starting point is (DCO resolution: 2bits, ring oscillator size: 40λ , replica cell size: 28λ). It converges at (4bits, 40λ, 8λ) (b) Trace of the optimizer when the starting point is (6bits, 40 λ, 12 λ). It converges at (4bits, 40λ, 8λ), the same optimal point in (a).

Table 6. Comparison of the Evaluation Number of Two Optimization Algorithms

Path (a)	Deterministic comparison	Random comparison (100 trials)
Convergence rate	100%	100%
Monte-Carlo Evaluation number	350	average = 205 standard deviation = 39
Path (b)	**Deterministic comparison**	**Random comparison (100 trials)**
Convergence rate	100%	100%
Evaluation number	330	average = 236 standard deviation = 57

5. Conclusions

This paper proposed the use of discrete optimization techniques for variability-aware analog circuit synthesis. After justifying the use of discrete grids to cover the continuous design space, the paper described the novel algorithms to address three aspects of discrete optimization: discretizing a continuous design space, selecting the candidate points on the discretized grid, and efficiently comparing the statistical measures between multiple candidate points. A digitally-controlled oscillator (DCO) example demonstrates that the proposed optimization technique can efficiently find the design that balances between the resolution, linearity, phase noise, and power dissipation.

6. ACKNOWLEDGEMENTS

This work has been funded by Semiconductor Research Corporation (SRC) under GRC Task 1836.093 and the authors would like to thank the industry liaisons at Texas Instruments, Inc. including Tom Vrotsos. CAD tool licenses have been supported by IC Design Education Center (IDEC) in Korea.

7. REFERENCES

[1] International Technology Roadmap For Semiconductors (ITRS): 2009 Edition, Available: http://www.itrs.net/Links/2009ITRS/Home2009.hht.

[2] G. Gielen and R. Rutenbar, "Computer-Aided Design of Analog and Mixed-Signal Integrated Circuits," *Proceedings of IEEE*, pp. 1825-1852, Dec. 2000.

[3] R. Phelps, et al., "ANACONDA: Simulation-based Synthesis of Analog Circuits via Stochastic Pattern Search," *IEEE Trans. Computer-Aided Design of Integrated Circuits and Systems*, pp. 703-717, June 2000.

[4] W. Daems, et al., "Simulation-based Generation of Posynomial Performance Models for the Sizing of Analog Integrated Circuits," *IEEE Trans. Computer-Aided Design of Integrated Circuits and Systems*, pp. 517-534, May 2003.

[5] T. Cover and J. Thomas, *Elements of Information Theory*, 2nd Ed., Wiley Interscience, New Jersey, 2006.

[6] S. Jung, S. Youn, and J. Kim, "Analysis on Performance Controllability under Process Variability: A Step Towards Grid-Based Analog Circuit Optimizers," *Frontiers in Analog Circuit (FAC) Synthesis and Verification*, July 2011.

[7] A. Adams, J. Baek and M. Davis, "Fast High-Dimensional Filtering Using the Permutohedral Lattice," *Computer Graphics Forum*, vol. 29, pp. 753–762, May 2010.

[8] S. Kim and L. Nelson, "A fully sequential procedure for indifference-zone selection in simulation," *ACM Trans. Modeling and Computer Simulation*, vol. 11, Issue 3, July 2001.

[9] J. Spall, "Multivariate stochastic approximation using a simultaneous perturbation gradient approximation," *IEEE Trans. Automatic Control*, vol. 37, pp. 332-341, March 1992

[10] R. Staszewski, et al., "All-digital PLL and Transmitter for Mobile Phones," *IEEE J. Solid-State Circuits*, vol.40, no.12, pp. 2469-2482, Dec. 2005

[11] J. Kim, et al. "Design of CMOS adaptive-bandwidth PLL/DLLs: a general approach," *IEEE Trans. Circuits and System II: Analog and Digital Signal Processing*, vol.50, no.11, pp. 860-869, Nov. 2003

Efficient Multi-Objective Synthesis for Microwave Components Based on Computational Intelligence Techniques

Bo Liu

ESAT-MICAS, KU Leuven, Belgium

Bo.Liu@esat.kuleuven.be

Hadi Aliakbarian

ESAT-TELEMIC, KU Leuven, Belgium

Hadi.Aliakbarian@esat.kuleuven.be

Soheil Radiom

ESAT-MICAS, KU Leuven, Belgium

Soheil.Radiom@esat.kuleuven.be

Guy A. E. Vandenbosch

ESAT-TELEMIC, KU Leuven, Belgium

Guy.Vandenbosch@esat.kuleuven.be

Georges Gielen

ESAT-MICAS, KU Leuven, Belgium

Georges.Gielen@esat.kuleuven.be

ABSTRACT

Multi-objective synthesis for microwave components (e.g. integrated transformer, antenna) is in high demand. Since the embedded electromagnetic (EM) simulations make these tasks very computationally expensive when using traditional multi-objective synthesis methods, efficiency improvement is very important. However, this research is almost blank. In this paper, a new method, called Gaussian Process assisted multi-objective optimization with generation control (GPMOOG), is proposed. GPMOOG uses MOEA/D-DE as the multi-objective optimizer, and a Gaussian Process surrogate model is constructed ON-LINE to predict the results of expensive EM simulations. To avoid false optima for the on-line surrogate model assisted evolutionary computation, a generation control method is used. GPMOOG is demonstrated by a 60GHz integrated transformer, a 1.6GHz antenna and mathematical benchmark problems. Experiments show that compared to directly using a multi-objective evolutionary algorithm in combination with an EM simulator, which is the best known method in terms of solution quality, comparable results can be obtained by GPMOOG, but at about 1/3-1/4 of the computational effort.

Categories and Subject Descriptors

D. 2.2. Design Tools and Techniques

General Terms

Algorithms, Performance, Design, Theory

Keywords

Multi-objective microwave components synthesis, Transformer synthesis, Antenna synthesis, Efficient global optimization, MOEA/D, Gaussian Process, Differential evolution

1. INTRODUCTION

In microwave component design, the designers are typically interested in more than one performance and try to make a trade-off between them. For instance, for integrated transformer design, RF designers often want to find a tradeoff between efficiency and area. For antenna design, designers are often interested in both the realized gain and the circular polarization axis ratio, as well as other characteristics such as return loss, efficiency, antenna size. Multi-objective optimization is a useful method for these problems. Hence, a growing number of works were reported in recent years focusing on introducing multi-objective optimization algorithms to microwave engineering [1-4], covering microwave filters, antennas, microwave passive components and circuits.

In most of the works, multi-objective evolutionary algorithms (MOEAs) are selected as the search engine, due to their high ability to approximate the Pareto front. Several existing MOEAs based on different evolutionary algorithms (e.g. genetic algorithm, differential evolution, particle swarm optimization) are compared in [1] using the example of a microwave filter; methods based on differential evolution (DE) [5] show the best performance.

Besides the research on investigating better optimizers, a more critical issue is the efficiency improvement needed for many multi-objective microwave component synthesis problems. The computationally cheap equivalent circuit models [6] can only be used when handling low-frequency (a few GHz) integrated passive components or very simple antennas, and expensive electromagnetic (EM) simulation is often a must for most microwave component synthesis, especially those working at mm-wave frequencies (e.g. 60GHz), where parasitic-aware equivalent circuit models are no longer accurate [7]. For single-objective EM simulation-included synthesis, an on-line surrogate model assisted evolutionary algorithm is proposed in [7], which can obtain a highly optimized result in a practical computational time. Compared with existing off-line surrogate model-based global optimization methods [8] and fine-coarse model mapping-based methods [9] solving the same problem, clear advantages on efficiency and optimization ability have been shown.

Although there is a solution for single-objective synthesis problems, multi-objective microwave component synthesis is much more difficult. The goal of multi-objective optimization is to generate the Pareto front, which includes a bunch of non-dominated points distributed in different areas of the design space, while only one of them is needed for single objective optimization. Even in the computational intelligence field, the available efficient surrogate model assisted multi-objective optimization algorithms can obtain promising results for some benchmark test problems, but not for all of them [10,11]. In the multi-objective microwave component synthesis area, the research on efficiency improvement is still an open research topic. Most of the available works directly use MOEAs as the search engine and EM simulators as the objective function evaluator [1,2,12]. Hence, the synthesis or optimization process is typically very CPU time expensive. Efficiency improvement based on hardware resources has been investigated in [2]. Because the evaluation of different candidate designs in a population is independent from each other in most MOEAs, parallel computation is used. However, to the best of our knowledge, there are very few efficient software algorithm for multi-objective EM-simulation-included synthesis.

To address this problem, a new method, called Gaussian Process assisted multi-objective optimization with generation control (GPMOOG), is proposed. The method aims to:

- achieve comparable results as the traditional method (directly using MOEA with an EM simulator) for multi-objective EM-simulation-included synthesis;
- while highly improving the efficiency of the traditional method and making the computational time practical (a couple of hours to about one day).

The remainder of the paper is organized as follows. Section 2 introduces the basic concepts and techniques used in GPMOOG. Section 3 describes the key ideas and algorithm of GPMOOG. Section 4 demonstrates GPMOOG by practical examples and mathematical benchmark problems. The concluding remarks are presented in Section 5.

2. BASIC CONCEPTS AND TECHNIQUES

2.1 Multi-objective optimization and MOEA/D

Most multi-objective optimization evolutionary algorithms (MOEAs) aim to find a reasonable number of solutions to approximate the Pareto front. A multi-objective optimization problem can be stated as follows [13]:

$$\min\{f_1(x),\dots f_m(x)\} \, , \, x \in \Omega \qquad (1)$$

where $x = (x_1,\dots x_n)$ is the decision variable vector (design variables of the microwave component in this application) and $f_i(x)$ are the objective functions, which are the considered desired performances of the microwave component, such as gain, efficiency or area. Ω is the decision space. A solution x_j is said to dominate solution x_k if and only if $f_i(x_j) \le f_i(x_k)$ for every $i \in \{1,\cdots,m\}$ and $f_i(x_j) < f_i(x_k)$ for at least one index $i \in \{1,\cdots,m\}$. A point $x^* \in \Omega$ is Pareto optimal to (1) if there is no point $x \in \Omega$ such that $f(x)$ dominates $f(x^*)$. $f(x^*)$ is

then a Pareto-optimal objective vector. The set of all the Pareto-optimal points is called the Pareto Set (PS). The set of all the Pareto-optimal objective vectors is called the Pareto Front (PF). The PF shows the trade-off curve of different performances.

A state-of-the-art MOEA is the multi-objective evolutionary algorithm based on decomposition (MOEA/D) [14]. It decomposes a multi-objective optimization problem into a set of scalar optimization sub-problems with neighborhood relations. The neighborhood relations are defined by the distances between their aggregation coefficient vectors. In this way, the fitness assignment is the same as single objective optimization, and the diversity is maintained by the diverse search directions determined by the uniformly distributed weight vectors. The first version of MOEA/D used simulated binary crossover and polynomial mutation as the search engines. Later, a new version using the mutation operator (DE/best/1/bin [5]) in DE as the main search engine was proposed [15] and was shown to outperform MOEA/D and NSGA-II, especially for complex problems. New population updating mechanisms were reported in [15], [16] to improve the diversity of the generated PF. MOEA/D-DE is used as the search engine in the GPMOOG algorithm in this paper. More details of MOEA/D and DE are in [14, 15].

2.2 Surrogate model assisted evolutionary algorithms

To boost the efficiency for computationally expensive optimization problems, a promising way is developing surrogate model assisted evolutionary algorithms (SAEAs). The key of SAEA is to employ efficient surrogate models to replace the computationally expensive exact function evaluations (such as EM simulations). As the construction of the surrogate model and its use to predict the function values cost much less effort than directly embedding the expensive exact function evaluator with the optimizer, the computational cost can be reduced significantly. In recent years, many surrogate model construction methods and the corresponding SAEAs have been investigated. Among them, Gaussian Process (GP) or Kriging, artificial neural networks (ANN), support vector machines (SVM) and radial basis function (RBF) show good performances and are widely used [17].

2.3 Basics of Gaussian Process

This work uses the Gaussian Process surrogate model, which is introduced briefly in this sub-section. More details are in [18]. GP machine learning not only has very good prediction ability, but also can provide an estimation error with a solid mathematical background. Compared to some other machine learning techniques, the advantages of GP are discussed in section 3.1.

GP predicts a function value $y(x)$ at some design point x by modeling $y(x)$ as a stochastic variable with mean μ and variance σ. For two points x_i and x_j, their correlation is defined as:

$$Corr(x_i, x_j) = \exp(-\sum_{l=1}^{d} \theta_l \, | \, x_{il} - x_{jl} \, |^{p_l}), \theta_l > 0, p_l \in [1, 2] \, (2)$$

where d is the dimension of x and θ_l is the correlation parameter which determines how fast the correlation decreases when x_{il} moves in the l direction. p_l is related to the smoothness of the function with x_l. The optimal values of μ, σ and θ are determined by maximizing the likelihood function of the observed data. The function value $y(x^*)$ at a new point x^* can be predicted as (3):

$$\hat{y}(x^*) = \hat{\mu} + r^T R^{-1}(y - I\hat{\mu})$$
$$\hat{\mu} = (I^T R^{-1} I)^{-1} I^T R^{-1} y$$
$$R_{i,j} = Corr(x_i, x_j), \ i, j = 1, 2, \cdots n \qquad (3)$$
$$r = [Corr(x^*, x_1), Corr(x^*, x_2), \cdots, Corr(x^*, x_n)]^T$$

where $x = (x_1, x_2 \cdots, x_n)$ and $y = (y_1, y_2 \cdots, y_n)$ are already evaluated data points and their objective function values. I is a $n \times 1$ vector of ones.

The prediction uncertainty is shown to be:

$$\hat{s}^2(x^*) = \hat{\sigma}^2[I - r^T R^{-1} r + (I - r^T R^{-1} r)^2 (I^T R^{-1} I)^{-1}]$$
$$\hat{\sigma}^2 = (y - I\hat{\mu})^T R^{-1}(y - I\hat{\mu})n^{-1} \qquad (4)$$

In this work, we use the DACE toolbox [19] to implement the Gaussian process-based surrogate model.

3. THE GPMOOG ALGORITHM
3.1 Key Ideas of GPMOOG

The surrogate model used in GPMOOG is the GP model. Unlike many other surrogate models, GP does not fit the function to a predefined kernel or structure. The prediction is based on the correlations of the available data. Therefore, GP does not have the problem of over-fitting. In contrast, the over-fitting of the ANN-based surrogate model has been shown in [7].

GPMOOG is an on-line SAEA. The reason why we do not use the off-line SAEA framework is as follows. Off-line SAEA first constructs a good surrogate model which covers the whole design space and then uses it. To obtain a good surrogate model, the training data need to cover the whole design space with a reasonably high density. Hence, a lot of expensive EM simulations are necessary to generate the training data. On the other hand, only a small part of the design space is used in the optimization. The reason is that EA is not based on enumeration, but based on iteration, so many of these expensive EM simulations are wasted. In contrast, in GPMOOG, the surrogate model construction and improvement are performed on-line, so the expensive EM simulations are used only in the necessary area of the design space, which is controlled by MOEA/D-DE. Consequently, GPMOOG is more efficient in terms of the number of EM simulations than methods using an off-line surrogate model.

However, the risk of on-line SAEA is incorrect convergence. Because the surrogate model is constructed based on the available data, which are often not sufficient in the beginning, the corresponding surrogate model may not be reliable. In other words, when using the current surrogate model to predict the

newly generated candidate solutions, the prediction values of some points may be far from their true objective function values. In this way, the search may converge to some false optimal points after several iterations [17], and the "good" candidates selected to perform EM simulations to update the surrogate model are also not in the promising area. An intuitive solution to address this problem is to make use of the prediction uncertainty in (4). For newly generated candidates with large prediction uncertainty, EM simulation is used; for those with small prediction uncertainty, the prediction value is used. However, the solution is not trivial. The threshold value for the prediction uncertainty to judge if EM simulation should be used is difficult to decide and is different from problem to problem. Therefore, GPMOOG still makes use of the prediction uncertainty $\hat{s}$ (from (4)) but in a different way.

The GPMOOG method holds two basic ideas. (1) The total number of iterations is equally divided to several groups (e.g. for 100 iterations, we can divide them into 1-10 iterations, 11-20 iterations, …, 90-100 iterations). In each group, the population in some iterations use expensive EM simulations for candidate evaluation, while for others, prediction values are used. (2) The number of iterations using EM simulation in each group is adaptively adjusted based on the prediction uncertainty of the current surrogate model. Hence, when there are more available samples and the quality of the surrogate model is improved (as reflected by the prediction uncertainty), more iterations use prediction values, which enhances the efficiency considerably. On the other hand, when the quality of the surrogate model is not good enough, more EM simulations are used to maintain correct convergence and to improve the surrogate model. This is the main idea of generation control for the ANN-based single objective SAEA [20]. The rule how to adjust the number of iterations using EM simulation in each group is as follows:

$$N_{EM}(k+1) = N_{EM(\min)} + \left\lfloor \frac{S(k)}{S_{\max}} \right\rfloor (N_{EM(\max)} - N_{EM(\min)}) \qquad (5)$$

where $N_{EM(\max)}$ is the maximum number of iterations using EM simulation in a group of iterations. $N_{EM(\min)}$ is the minimum number of iterations using EM simulation in a group. Note that in each group, at least 1 iteration needs to use EM simulation in order to calibrate possible false optima. The S value is the maximum of all prediction uncertainties $\hat{s}$ for the current population using the current surrogate model. Using the maximum $\hat{s}$ value emphasizes the surrogate model quality, which is a safe setting. The search mechanism of GPMOOG is shown in Fig. 1.

3.2 Handling multiple objectives in GP model

In MOEA/D-DE, Tchebycheff aggregation is used. The scalar function is as follows:

$$\text{minimize } g^{te}(x \mid \lambda, z^*) = \max_{1 \le i \le m} \{\lambda_i \mid f_i(x) - z_i^* \mid\}, x \in \Omega \qquad (6)$$

where $\lambda = (\lambda_1, \cdots, \lambda_m)$ is a weight vector and $\Sigma_{i=1}^m \lambda_i = 1$, $z^* = (z_1^*, \cdots, z_m^*)$ is the reference point, m is the number of objectives. In MOEA/D-DE evolution, different weight vectors will be used for each $f_i(x)$ to calculate the corresponding aggregation value. If GP model is directly used to predict

544

$g^{te}(x\,|\,\lambda,z^*)$, a large number of surrogate models will be generated, because for different λ and z^* (z^* is updated in search) the objective function is different. [10] provides an approximation method to predict $g^{te}(x\,|\,\lambda,z^*)$ for different weight vectors and reference points by only using the surrogate model for the prediction of $f_i(x)$. The derived formula is as follows, more details are in [10]. Consider two objectives $f_i(x) \sim N(\hat{y}_i(x), \hat{s}_i^2(x))$, and $g^{te}_i(x) \sim N(\hat{y}_i^{te}(x), (\hat{s}_i^{te}(x))^2)$.

$$\hat{y}_i^{te}(x) = \mu_1 \Phi(\alpha) + \mu_2 \Phi(-\alpha) + \tau\phi(\alpha) \tag{7}$$

$$\hat{s}_i^{te}(x))^2 = (\mu_1^2 + \sigma_1^2)\Phi(\alpha) + (\mu_2^2 + \sigma_2^2)\Phi(-\alpha) \\ +(\mu_1 + \mu_2)\phi(\alpha) - (\hat{y}_i^{te}(x))^2 \tag{8}$$

where $\sigma_i^2 = [\lambda_i \hat{s}_i(x)]^2, \mu_i = \lambda_i(\hat{y}_i(x) - z_i^*), i = 1,2$.

$\tau = \sqrt{\sigma_1^2 + \sigma_2^2}, \alpha = (\mu_1 - \mu_2)/\tau$. $\phi(\cdot)$ is the standard normal density function, and $\Phi(\cdot)$ is the standard normal distribution function.

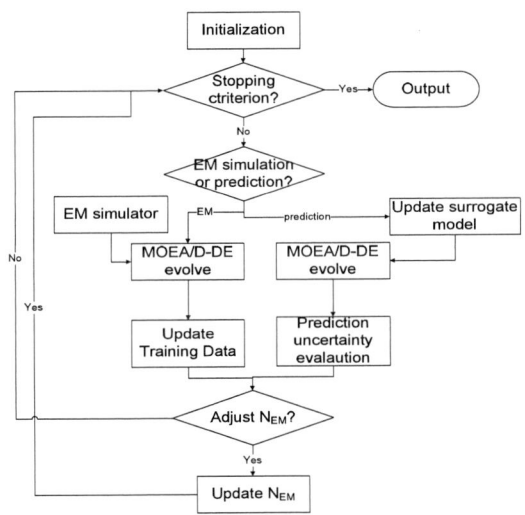

Figure 1. Search mechanism of GPMOOG

3.3 The GPMOOG algorithm

The GPMOOG algorithm works as follows.

Input:

(1) a multi-objective EM-simulation-included synthesis problem with m objectives

(2) a stopping criterion (e.g. maximum number of iterations)

(3) MOEA/D parameters: N: the number of sub-problems; T: the neighborhood size; δ : the probability that parent solutions are selected from the neighborhood; n_r : the maximum number of solutions replaced by a child solution; λ : weight vector (the

generation method is in [14])

(4) evolutionary search parameters: CR: crossover rate in DE, F: the scaling factor in DE (see [5]);

(5) Gaussian Process parameters: a correlation function

(6) Generation control parameters: C: the number of iterations in each group; $N_{EM(init)}$: the number of iterations using EM simulation in the first group. $N_{EM(min)}, N_{EM(max)}$ (see section 3.1)

Output: (1) approximated PF, (2) approximated PS

Procedure:

Step 1: Initialization

Step 1.1: Compute the Euclidean distances between the weight vectors and work out the T closest weight vectors to each weight vector (the set is B). For $i = 1, \cdots N$, set $B(i) = \{i_1, \cdots, i_T\}$. $\lambda^{i_1}, \cdots, \lambda^{i_T}$ are the T closest vectors to λ^i (each sub-problem corresponds to a weight vector).

Step 1.2: Randomly generate an initial population $x_1, \cdots, x_n$. Calculate the fitness values of the population by EM simulations.

Step 1.3: Initialize $z = \{z_1, \cdots, z_m\}$, where $z_j = \min_{1 \le i \le N} f_j(x^i)$.

Step 2: Update

For $i = 1, \cdots N$,

Step 2.1: Selection of the mating pool:

Generate a random number which is uniformly distributed in $[0,1]$. Set

$$P = \begin{cases} B(i) & \text{if rand} < \delta \\ \{1, \cdots, N\} & \text{otherwise} \end{cases} \tag{9}$$

Step 2.2: Reproduction:

Set $r_1 = i$ and randomly select two indexes r_2 and r_3 from P, and generate a new solution $\overline{y}$ by DE mutation (see [15]). Then, perform a polynomial mutation [14] on $\overline{y}$ to produce a new solution y.

Step 2.3: Repair:

If an element of y is out of the bound of Ω , its value is reset to be a randomly selected value inside the boundary.

Step 2.4: Evaluation of the new candidate (y)

The current iteration is noted as *iter*.

(2.4.1) If $remainder(iter, C) < N_{EM}$ and $remainder(iter,$ $C) \ne 0$, use EM simulation; otherwise, use GP surrogate model prediction (see equations (3-4)).

(2.4.2) Evaluate the newly generated candidate y by the selected method. When using EM simulation, update the training data set by adding y and the corresponding performances from EM simulation. When using GP surrogate

545

model, update the S_{max} value.

Step 2.5: Update of the reference point:

For $j = 1, \cdots, m$, if $z_j > f_j(y)$, set $z_j = f_j(y)$.

Step 2.6: Replacement of solutions:

(2.6.1) For each j in P, calculate $g(y \mid \lambda^j, z)$ and $g(x^j \mid \lambda^j, z)$ by (7) and the $\hat{s}_i^{te}(y)$ value by (8).

(2.6.2) Set $c=0$. If $g(y \mid \lambda^j, z) \leq g(x^j \mid \lambda^j, z)$, c=c+1

(2.6.3) If $c \leq n_r$, for each j with $g(y \mid \lambda^j, z) \leq g(x^j \mid \lambda^j, z)$, set $x^j = y$. If $c > n_r$, for each j with $g(y \mid \lambda^j, z) \leq g(x^j \mid \lambda^j, z)$, calculate the Euclidean distances between $f(y)$ and $f(x^j)$ and then rank them. Choose n_r solutions with the smallest distances. Set $x^j = y$.

Step 2.7: Adjustment of N_{EM}:

If $remainder(iter, C) = 0$,

(2.7.1) Using all the $\hat{s}_i^{te}(y)$ from step (2.6.1), calculate the S value of the current surrogate model for the current population by the method described in section 3.1.

(2.7.2) Adjust N_{EM} using (5).

Step 3: Stopping Criterion:

If the stopping criterion is satisfied, then stop the algorithm and output $\{x^1, \cdots, x^N\}$ to Step 4. Otherwise, go to **Step 2**.

Step 4: Output:

Perform EM simulation to the last population $\{x^1, \cdots, x^N\}$ to obtain their performances $\{f(x^1), \cdots, f(x^N)\}$. Delete dominated points and output the final PF.

4. EXPERIMENTAL RESULTS AND COMPARISONS

In this section, the GPMOOG algorithm is demonstrated for a 60GHz transformer in a 90nm CMOS technology, a 1.6GHz antenna in RT5880 technology and mathematical benchmark problems. The parameters are as follows: (1) MOEA/D-DE parameters: N: 80, T: 8, δ: 0.9, n_r: 1; λ: generated by the method in [14], CR: 1, F: 0.5. These are common settings of MOEA/D-DE [15]. (2) Generation control parameters: C: 5, $N_{EM(init)}$: 4, $N_{EM(min)}$: 1, $N_{EM(max)}$: 3. (3) Gaussian Process parameters: the exponential correlation function is used [19]. GPMOOG runs 50 iterations in example 1, and 90 iterations in example 2. These iteration numbers are determined by the improvement of the PF in each iteration. If no clear improvement is shown in 10 consecutive iterations, GPMOOG is stopped. As the PF is visible for two or three objective multi-objective synthesis, human observation can

also be a stopping criterion. The examples are run on a PC with 4 cores, 12GB RAM and Linux operating system. For the two real-world problems, ADS-Momentum is used as the EM simulator. Because EM simulation is included in example 1 and 2, directly using the traditional method (MOEA/D-DE with the EM simulator without surrogate models) costs a very long time for comparison. Hence, we run the traditional method for example 1 and compare both the generated PF and the efficiency. For example 2, we only analyze the efficiency enhancement since using the traditional method to example 2 will cost a week. Then, the optimization ability comparison is reinforced by mathematical benchmark problems with known PF.

4.1 Example 1

The first example is a 60GHz overlay transformer with octagonal shape in a 90nm CMOS process. The design variables are the inner diameter of the primary inductor ($dinp$), the inner diameter of the secondary inductor ($dins$), the width of the primary inductor (wp) and the width of the secondary inductor (ws). The ranges of the design variables are $dinp, dins \in [20, 150]$, $wp, ws \in [5, 10]$ (in μm). The output impedance is $25\,\Omega$. The two objectives are the maximization of the power transfer efficiency (PTE) and the minimization of the square root of the area. The generated PFs of both GPMOOG and the traditional method are shown in Fig.2. Each point in the PF corresponds to a design. An example of the design is shown in Fig. 3.

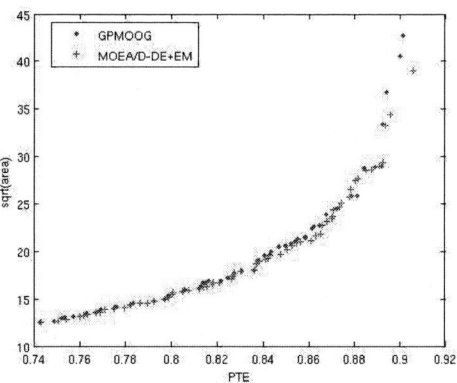

Figure 2. PF generated by GPMOOG and the traditional method (example 1)

From Fig. 2, it can be seen that the PF generated by directly using MOEA/D-DE with the EM simulator is only a slightly better than that of GPMOOG. To quantify the difference, we use:

$$err(PF_{GPMOOG}, PF_{MOEA/D+EM}) - err(PF_{MOEA/D+EM}, PF_{GPMOOG}) \quad (10)$$

where $err(A, B)$ is defined as: for each point a in A which is dominated by some points in B, we calculate $e = (\sum_{i=1}^{m} |a_i - \tilde{b}_i| / |\tilde{b}_i|) / m$, where m is the number of objectives, and $\tilde{b}$ is the nearest nondominated point to a in B. err is the average of the e values to all the points. The result is the PF generated by the traditional method is 2.88% better than GPMOOG. This shows the surrogate model in GPMOOG has good performance. In the first two groups (1-5 iterations, 6-10 iterations), 4 and 3 iterations use EM simulations, respectively.

546

After that, the number of iterations that use EM simulation falls to 1 in all the other groups. This shows the adaptive adjustment of N_{EM} and the final results verify the adjustment mechanism. The time consumption of GPMOOG (clock time) is 28.2 hours, which is reasonable for practical use. In the 50 iterations, 16 of them use EM simulation. When directly using MOEA/D-DE with EM simulation, the time consumption is 3.6 days. Hence, a more than 3 times speed enhancement is achieved.

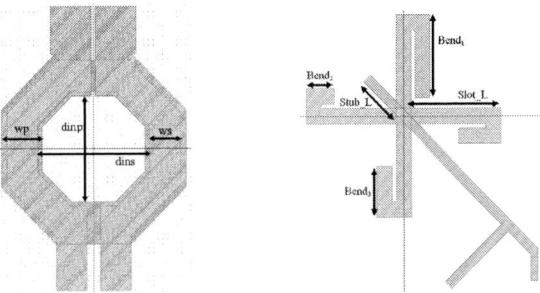

Figure 3. Transformer example Figure 4. Antenna example

4.2 Example 2

One of the characteristics often required in antennas is circular polarization (CP). The second example is a miniaturized circular polarized crossed slot antenna fed by using a single feed line based on a work by Vlastis [21]. The antenna has miniaturized crooked arms instead of the Vlastis design, as illustrated in Fig.4. At the same time, because of the unsymmetrical structure, the radiation pattern can rotate and reduce the broadside gain. Therefore, the realized gain and the CP axial ratio (ARCP) are two important characteristics of the antenna which are taken as the goals of the PF. The antenna is designed at 1.6 GHz frequency. The five main parameters affecting the antenna performances are as follows: lengths of the main arm: Slot_L; length of the third part of each arm (in total four arms, one is set to be fixed and the three others are selected as design variables): bend_1, bend_2 and bend_3; the length of the feed stub, Stub_L. The ranges of the design variables are: Slot_L $\in [25, 35]$, bend_1, bend_2 and bend_3 $\in [5, 25]$ and Stub_L $\in [5, 25]$ (all in mm).

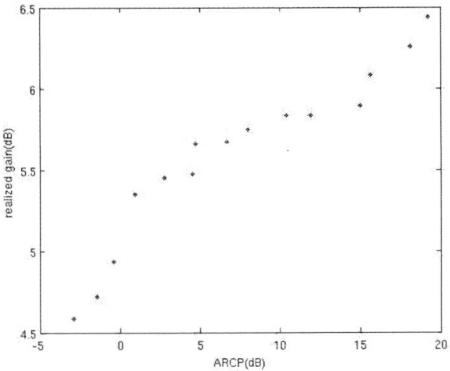

Figure 5. PF showing the trade-off of gain and ARCP (example 2)

The generated PF by GPMOOG is shown in Fig. 5. The result shows the gain versus ARCP tradeoff of the antenna simulated in ADS-Momentum. Because the ARCP is very sensitive to the five critical parameters, the generated PF is not so smooth. As we would like to have the AR as close as possible to 0 dB, we find out from the PF that the achievable gain range is approximately 5 to 5.5 dB with small ARCP, which is a very good gain value for a slot antenna having optimized ARCP (slot antenna often has small gain in such size, e.g. 1-2dB).

The synthesis time is 41.5 hours (clock time). In this example, 24 of the total 90 iterations use EM simulation. Considering that the computational cost is dominated by the EM simulation, the speed enhancement is nearly 4 times.

4.3 Benchmark Tests

To further verify the optimization ability of GPMOOG, two benchmark problems (ZDT1, ZDT2) [10] for testing MOEAs are used. The number of decision variables is 5. N is set to 100, and 60 iterations are used. All the other settings are the same as the previous examples. The median results (by IGD value [10]) over 10 runs for both methods are shown in Fig. 6.

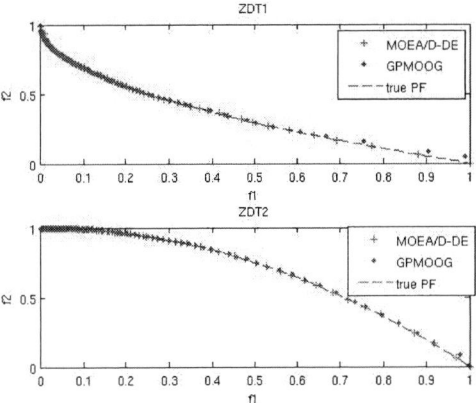

Figure 6. Results of benchmark functions

From Fig. 6, it can be seen that GPMOOG can obtain comparable results as directly using MOEA/D-DE, and the results are also very close to the true PF. In the two examples, 18 of 60 iterations use exact function evaluations. For expensive optimization problems, the efficiency is enhanced for more than 3 times.

5.CONCLUSIONS

In this paper, the GPMOOG algorithm has been proposed for the multi-objective synthesis of microwave components. GPMOOG is the first attempt to enhance the efficiency of EM simulation-included multi-objective synthesis by software algorithms. GPMOOG gets comparable optimization quality and costs 1/3-1/4 of the computational effort as the traditional method, yet. The gain comes from the new surrogate model assisted multi-objective evolutionary algorithm. Future works will focus on expanding the dimensionality of the synthesis problems.

6. ACKNOWLEDGMENTS

This research was supported by a special bilateral agreement scholarship of Katholieke Universiteit Leuven, Belgium and Tsinghua University, P. R. China.

7. REFERENCES

[1] Goudos S. K. et al., 2010. Pareto optimal microwave filter design using multiobjective differential evolution. *IEEE Transactions on Antennas and Propagation*, 132-144.

[2] Poian M. et al., 2008. Multi-objective optimization for antenna design. *Proc. of IEEE conference on microwaves, communications, antennas and electronic systems*, 1-9.

[3] Oliveria D. et al., 2011. Design of a Microwave Applicator for Water Sterilization Using Multiobjective Optimization and Phase Control Scheme. *IEEE Transactions on Magnetic*, 1242-1245.

[4] Brito L. C. et al., 2003. A general and robust method for multi-criteria design of microwave oscillators using an evolutionary strategy. *Proc. of IEEE international microwave and optoelectronics conference*, 135-139.

[5] Price K., et al., 2005. *Differential Evolution. A Practical Approach to Global Optimization.* Springer, Berlin, Heidelberg, New York.

[6] Nieuwoudt A. et al., 2006. Variability-Aware Multilevel Integrated Spiral Inductor Synthesis. *IEEE TCAD*, 2613-2625.

[7] Liu B. et al., 2011. Synthesis of Integrated Passive Components for High-Frequency RF ICs Based on Evolutionary Computation and Machine Learning Techniques. *IEEE TCAD*, 1458-1468.

[8] Mandal S. K. et al., 2008. ANN- and PSO-Based Synthesis of On-Chip Spiral Inductors for RF ICs. *IEEE TCAD*, 188-192.

[9] Bandler J. et al., 2004. Space Mapping: The State-of-the-Art. *IEEE MTT*, 337-361.

[10] Zhang Q. et al., 2010. Expensive multiobjective optimization by MOEA/D with Gaussian Process Model. *IEEE Transactions on Evolutionary Computation*, 456-474.

[11] Knowles J., 2005. ParEGO: A hybrid algorithm with on-line landscape approximation for expensive multiobjective optimization problems. *IEEE Transactions on Evolutionary Computation*, 50-66.

[12] Chung K. L. et al., 2008. Particle swarm optimization of wideband patch antennas. *Proc. of Asia-Pacific microwave conference*, 1-4.

[13] Deb K., 2001. *Multiobjective optimization using evolutionary algorithms*, New York: Wiley.

[14] Zhang Q. et al., 2007. MOEA/D: A multiobjective evolutionary algorithm based on decomposition. *IEEE Transactions on Evolutionary Computation*, 1-20.

[15] H. Li et al., 2008. Multiobjective optimization problems with complicated Pareto sets, MOEA/D and NSGA-II. *IEEE Transactions on Evolutionary Computation*, 1-19.

[16] Liu B. et al., 2010. An enhanced moea/d-de and its application to multiobjective analog cell sizing. *Proc. of IEEE Congress on Evolutionary Computation*, 1-7.

[17] Jin Y., 2005. A comprehensive survey of fitness approximation in evolutionary computation. *Soft computing*, 3-12.

[18] Jones D. et al., 1998. Efficient Global Optimization of Expensive Black-Box Functions. *J. of Global Optimization*, 455-492.

[19] Lophaven S. N. et al., 2002. *DACE: A MATLAB Kriging Toolbox*, Technical Report. Technical University of Denmark.

[20] Jin Y. et al., 2002. *A framework for evolutionary optimization with approximate fitness functions. IEEE Transactions on Evolutionary Computation*, 481-494.

[21] Vlasits T. et al., 1996. Performance of a cross-aperture coupled single feed circularly polarised patch antenna. *Electronics Letters*, 612–613.

Non-Uniform Multilevel Analog Routing with Matching Constraints *

Hung-Chih Ou[1], Hsing-Chih Chang Chien[1], and Yao-Wen Chang[1,2]

[1]Graduate Institute of Electronics Engineering, National Taiwan University, Taipei 106, Taiwan
[2]Department of Electrical Engineering, National Taiwan University, Taipei 106, Taiwan
{howard, lichee626}@eda.ee.ntu.edu.tw; ywchang@cc.ee.ntu.edu.tw

ABSTRACT

Symmetry, topology-matching, and length-matching constraints are three major routing considerations to improve the performance of an analog circuit. Symmetry constraints are specified to route matched nets symmetrically with respect to some common axes. Topology-matching constraints are commonly imposed on critical yet asymmetry nets with the same number of bends, vias, and wirelength. Length-matching constraints are specified to route the nets which have limited resources with the same wirelength. These three constraints can reduce current mismatches and unwanted electrical effects between two critical nets. In this paper, we propose the first work to simultaneously consider the three constraints for analog routing while minimizing total wirelength, bend numbers, via counts, and coupling noise at the same time. We first present an integer linear programming (ILP) formulation to simultaneously consider the three constraints for analog routing, and employ effective reduction techniques to further reduce the numbers of ILP variables and constraints. Then, a non-uniform multilevel routing framework is presented to enhance the performance of our routing algorithm. Experimental results show that our approach can obtain better routing results and satisfy all specified routing constraints while optimizing circuit performance.

Categories and Subject Descriptors

B.7.2 [**Integrated Circuits**]: Design Aids [Placement and Routing]

General Terms

Algorithms, Performance

Keywords

Physical Design, Routing, Analog ICs

1. INTRODUCTION

Routing is one of the most important processes in analog layout synthesis for modern analog and mix-signal circuit designs. It determines all the wire connections of an analog circuit while minimizing the total wirelength and satisfying all the user-specified routing constraints for better circuit performance.

To improve the performance of an analog circuit, *symmetry, topology-matching,* and *length-matching* constraints are three major considerations for analog routing. Symmetry constraints are specified to route

---

*This work was partially supported by IBM, SpringSoft, TSMC, and NSC of Taiwan under Grant No's. NSC 100-2221-E-002-088-MY3, NSC 99-2221-E-002-207-MY3, NSC 99-2221-E-002-210-MY3, and NSC 98-2221-E-002-119-MY3.

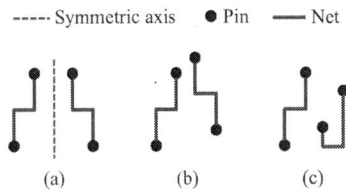

Figure 1: (a) Symmetric constraint. (b) Topology-matching constraint. (c) Length-matching constraint.

matched nets symmetrically with respect to some common axes (see Figure 1(a)). The symmetry constraint can reduce unwanted electrical effects and balance the parasitic resistors and capacitors. Topology-matching constraints are commonly imposed on the critical, yet asymmetry nets with the same number of bends, vias, and wirelength (see Figure 1(b)) to reduce current mismatches. The topology-matching constraint is similar to the exact-matching constraint addressed in [10, 5], but focuses on *non-preferred direction* routing. Symmetry and topology-matching constraints are two most stringent constraints to achieve the best circuit performance; however, they might demand excessive routing resource to satisfy the constraints. For less critical nets, length-matching constraints are imposed to route nets with the same wirelength (see Figure 1(c)) to achieve reasonably good performance.

Existing analog routers focus on one of the two matching constraints alone: symmetry constraints, or exact-matching constraints (similar to the topology-matching constraint addressed in this paper, but mainly for preferred direction routing). For symmetry constraints, Piguet et al. proposed a stretch method for electrical symmetry [11], and Malavasi et al. simply mirrored the routing result from the one side to the other side for matched nets [9]. Later works in [4, 8, 12, 13, 14] proposed a maze-based router to further consider the non-symmetric obstacles. These approaches consider both the real obstacles and the virtual obstacles obtained by mirroring each obstacle with respect to the symmetry axis to determine the locations of all the obstacles. For analog channel routing, Choudhury et al. and Malavasi et al. worked on the constraint-graph-based algorithms to handle the mirror symmetry [2, 8]. Recently, Lin et al. proposed a pattern-based router to consider matching devices [7]. Aside from the symmetry constraint, Ozdal and Hentschke proposed a mathematical model for the exact-matching constraint [10], and Gao et al. proposed a two-stage framework to further consider the obstacles for preferred direction routing in a circuit [5].

The performance of an analog circuit is sensitive to the layout parasitic, so it is desirable to consider these three matching constraints during routing. However, the maze-based analog routers have to define the routing priority for different routing constraints, and this ordered routing style might result in congested regions and performance degradation to consider symmetry, topology-matching, and length-matching constraints at the same time. Figure 2(a) shows a routing result of the ordered routing style. The unordered routing style can simultaneously consider the three matching constraints and obtain a less congested result with better circuit performance as shown in Figure 2(b). In addition, experienced analog designers usually route one net on the same layer to reduce the total wire resistance. However, the previous works focus on preferred direction routing and pre-assign routing layers for each routing direction. As a result, unnecessary vias and bends might be introduced, and thus circuit performance might be degraded.

To simultaneously consider symmetry, topology-matching, and length-matching constraints during routing for better circuit performance,

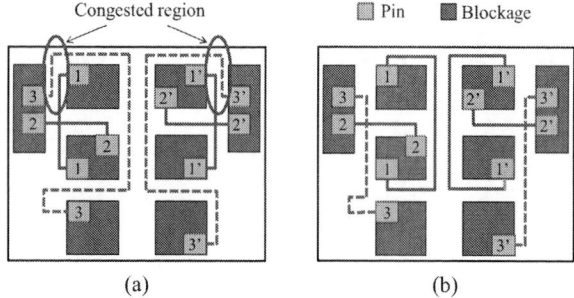

(a)   (b)

Figure 2: Nets $1, 1'$, $2, 2'$, **and** $3, 3'$ **are three net pairs to be routed with the symmetry, topology-matching, and length-matching constraints, respectively. (a) A more congested routing resulting from considering the three constraints in sequence. (b) A less congested routing resulting from considering the three constraints simultaneously.**

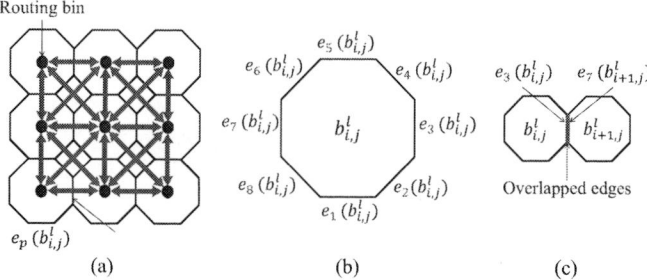

(a)   (b)   (c)

Figure 3: (a) An octagonal grid model, where $e_p(b_{i,j}^l)$ **represents the edge of the bin** $b_{i,j}^l \in B$ **. (b) Definition of the edge** $e_p(b_{i,j}^l)$ **. (c) Overlapped edges** $e_3(b_{i,j}^l)$ **and** $e_7(b_{i+1,j}^l)$ **between the bins** $b_{i,j}^l$ **and** $b_{i+1,j}^l$ **.**

we propose an integer linear programming (ILP) based routing algorithm with an octagonal routing grid model to handle the three matching constraints while minimizing total wirelength, bend numbers, via counts, and coupling noise. Effective reduction techniques are employed to further reduce the numbers of ILP variables and constraints. To enhance the routing performance, we also propose a non-uniform multilevel framework and integrate it into our routing algorithm.

The main contributions of this paper are summarized as follows:

- A new ILP-based analog routing algorithm which simultaneously considers symmetry, topology-matching, and length-matching constraints, is proposed. To the best of our knowledge, this is the first work that handles these three constraints simultaneously.

- Effective reduction techniques are employed to further reduce the numbers of ILP variables and constraints.

- A non-uniform multilevel framework for analog routing is proposed. The non-uniform multilevel framework enhances routing performance and make the routing algorithm conform to the conventional design manner for analog circuits.

- In addition to wirelength minimization, bend numbers, via counts, and coupling noise are simultaneously optimized in our analog routing.

- A non-preferred direction with the 45-degree routing style for analog circuits is presented. Our approach requires no layer assignment process before routing.

- Experimental results show that our method can satisfy all the three matching constraints and achieve shorter wirelength and fewer vias than previous works. Simulation results even show that our approach can achieve higher circuit performance than tape-out manual designs.

The remainder of this paper is organized as follows. Section 2 formulates the analog routing problem. Section 3 presents our analog routing algorithm to simultaneously consider the three matching constraints. Section 4 introduces our non-uniform multilevel routing framework. Section 5 reports the experimental results. Finally, Section 6 concludes this paper.

2. PROBLEM FORMULATION

The analog routing problem can be formally defined as follows:

- **The Analog Routing Problem:** Given a set of placed modules $M = \{m_k | 1 \leq k \leq |M|\}$, a set of nets $N = \{n_k | 1 \leq k \leq |N|\}$, a set of symmetry pairs $N^S = \{n_k^S | 1 \leq k \leq |N^S|\}$, a set of topology-matching pairs $N^T = \{n_k^T | 1 \leq k \leq |N^T|\}$, a set of length-matching pairs $N^L = \{n_k^L | 1 \leq k \leq |N^L|\}$, and the design rules, route all the nets in N to minimize the total wirelength, bend numbers, via counts, and coupling noise such that no design rule is violated and all the pairs in N^S, N^T, and N^L satisfy the symmetry, topology-matching, and length-matching constraints, respectively.

3. THE ANALOG ROUTING ALGORITHM

In this section, we propose a routing grid model for the analog routing problem, and give an ILP formulation for the routing. First, we show an octagonal routing grid model to consider both 45- and 90-degree routing styles in Section 3.1, and then a basic ILP formulation for the analog routing problem is proposed in Section 3.2. Finally, we analyze the computational complexity of the basic ILP formulation in Section 3.3 and provide ILP reduction techniques in Section 3.4.

3.1 Octagonal Routing Grid Model

In addition to the 90-degree routing style, the 45-degree one is also an important style for practical analog routing [6]. This non-Manhattan routing style can reduce routing wirelength, electrical effects, and power consumption. For these reasons, we propose an *octagonal routing grid model* to simultaneously consider both 45- and 90-degree routing styles in a single grid. Figure 3(a) gives an example. For the ILP formulation for analog routing, we introduce a node $b_{i,j}^l$ in the center to represent the bin, and define $B = \{b_{i,j}^l | 0 \leq i \leq w, 0 \leq j \leq h, l \in L\}$ to be the set of all the nodes, where L is the set of routing layers, w and h are the respective numbers of grids in the width and height of the chip. We also define $E = \{e_p(b_{i,j}^l) | 1 \leq p \leq 8, b_{i,j}^l \in B\}$ to be the set of the eight edges for the routing bin $b_{i,j}^l \in B$. The index p for the corresponding edge is illustrated in Figure 3(b). The overlapped edges, such as $e_3(b_{i,j}^l)$ and $e_7(b_{i+1,j}^l)$ are just for the explanation of our algorithm (see Figure 3(c)). In practical implementations, we will not generate any overlapped edge for adjacent bins. To satisfy the design rules, we assume that each edge allows only one net to pass. The advantage of the octagonal routing grid model is that both 45- and 90-degree routing styles can easily and directly be formulated at the same time.

3.2 Basic ILP Formulation

With the octagonal routing grid, we can use an ILP formulation to handle the analog routing problem. For a multi-terminal net, we split it into two-terminal ones by FLUTE [3]. In order not to cross the active regions of devices as mentioned in [13], we simply move the Steiner points out of the active regions for unmatched nets. For matched nets, we move two Steiner points simultaneously out of the active regions. To achieve better circuit performance, our objective focuses on simultaneously minimizing total routing wirelength, bend numbers, via counts, and coupling noise while satisfying the symmetry, topology-matching, and length-matching constraints at the same time.

The notations used in the ILP formulation are as follows:

- $A_{sym}(n_k^S)$: symmetry axis of the symmetry pair $n_k^S \in N^S$.

- $x_p^k(b_{i,j}^l)$: 0-1 integer variable that denotes a wire segment of net n_k passing the corresponding edge $e_p(b_{i,j}^l)$ of the routing bin $b_{i,j}^l \in B$. $x_p^k(b_{i,j}^l) = 1$ if net n_k passes the corresponding edge $e_p(b_{i,j}^l)$; $x_p^k(b_{i,j}^l) = 0$, otherwise.

- $T_{b_{i,j}^l}(n_k)$: 0-1 integer variable that denotes the existence of the terminals of net n_k in bin $b_{i,j}^l \in B$. If the terminal is in $b_{i,j}^l$, $T_{b_{i,j}^l}(n_k) = 1$; otherwise, $T_{b_{i,j}^l}(n_k) = 0$.

- $V_{b_{i,j}^l}(n_k)$: 0-1 integer variable that denotes the existence of a via

550

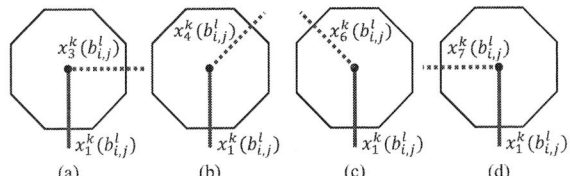

Figure 4: Wire-bend examples. The wire segment $x_1^k(b_{i,j}^l)$ will result in a bend with any of the wire segments $x_3^k(b_{i,j}^l)$, $x_4^k(b_{i,j}^l)$, $x_6^k(b_{i,j}^l)$, and $x_7^k(b_{i,j}^l)$.

of net n_k in bin $b_{i,j}^l \in B$. If the bin $b_{i,j}^l$ has a via, $V_{b_{i,j}^l}(n_k) = 1$; otherwise, $V_{b_{i,j}^l}(n_k) = 0$.

- L: set of routing layers.
- $E(b_{i,j}^l)$: set of edges related to the routing bin $b_{i,j}^l$.
- $O_w(n_k)$: wirelength function of net n_k.
- $O_b(n_k)$: bend-number function of net n_k.
- $O_v(n_k)$: via-count function of net n_k.
- $O_c(n_k)$: estimated coupling-noise function of net n_k.
- $C(e_p(b_{i,j}^l), e_p(b_{i',j'}^{l'}))$: coupling-noise function between two parallel edges $e_p(b_{i,j}^l)$ and $e_p(b_{i',j'}^{l'})$.

Based on the notations, the analog routing problem can be formulated in the following subsections.

3.2.1 Objective Function

The goal of the analog routing problem is to simultaneously minimize the total routing wirelength, bend numbers, via counts, and coupling noise. Therefore, the objective is defined as follows:

$$\min \sum_{n_k \in N} \Big(\alpha O_w(n_k) + \beta O_b(n_k) + \gamma O_v(n_k) + \delta O_c(n_k) \Big),$$

where the functions $O_w(n_k)$, $O_b(n_k)$, $O_v(n_k)$, and $O_c(n_k)$ denote the respective total wirelength, bend numbers, via counts, and coupling noise of net n_k, and α, β, γ, and δ are user-specified constants.

The total wirelength $O_w(n_k)$ of the net n_k is defined as follows:

$$\sum_{b_{i,j}^l \in B} \sum_{1 \leq p \leq 8} \Big(W\big(e_p(b_{i,j}^l)\big) x_p^k(b_{i,j}^l) \Big), n_k \in N, \quad (1)$$

where $W\big(e_p(b_{i,j}^l)\big)$ is the length of edge $e_p(b_{i,j}^l)$.

The bend number $O_b(n_k)$ of the net n_k is defined as follows:

$$\sum_{b_{i,j}^l \in B} \sum_{\substack{l' \in L \\ l' \geq l}} \sum_{1 \leq p \leq 8} \sum_{\substack{1 \leq p' \leq 8 \\ p' > p \\ p' \neq \{p+1, p+4, p+7\}}} \Big(x_p^k(b_{i,j}^l) \bigwedge x_{p'}^k(b_{i,j}^{l'}) \Big), n_k \in N. \quad (2)$$

Figure 4 shows a wire-bend example for the wire segment $x_1^k(b_{i,j}^l)$. The wire segment $x_1^k(b_{i,j}^l)$ will result in a bend with any of the wire segments $x_3^k(b_{i,j}^l)$, $x_4^k(b_{i,j}^l)$, $x_6^k(b_{i,j}^l)$, and $x_7^k(b_{i,j}^l)$.

The via number $O_v(n_k)$ of the net n_k is defined as follows:

$$\sum_{b_{i,j}^l \in B} \sum_{\substack{l' \in L \\ l' > l}} \sum_{1 \leq p \leq 8} \sum_{\substack{1 \leq p' \leq 8 \\ p+2 \leq p' \leq p+6}} \Big(x_p^k(b_{i,j}^l) \bigwedge x_{p'}^k(b_{i,j}^{l'}) \Big), n_k \in N. \quad (3)$$

In our ILP formulation, we treat both the stack via and the single via as one unit via similar to the previous work [1].

The coupling noise $O_c(n_k)$ of the net n_k in one direction is defined as follows:

$$\sum_{\substack{n_k' \in N \\ 1 \leq p \leq 8}} \sum_{|d| \leq D} \left(C\big(e_p(b_{i,j}^l), e_p(b_{i+d,j}^l)\big) P\big(x_p^k(b_{i,j}^l), x_p^{k'}(b_{i+d,j}^l)\big) \right),$$

$$n_k, n_{k'} \in N, b_{i,j}^l \in B, \quad (4)$$

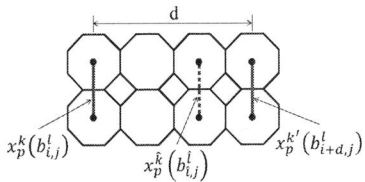

Figure 5: The shielding condition for the wire segments $x_p^k(b_{i,j}^l)$ and $x_p^{k'}(b_{i+d,j}^l)$. If there exists a wire segment $x_p^{\hat{k}}(b_{i,j}^l)$ in between them, the coupling noise between $x_p^k(b_{i,j}^l)$ and $x_p^{k'}(b_{i+d,j}^l)$ will be shielded, and Expression (5) will be zero.

where D is the maximum distance between two parallel wire segments with coupling. The function $P\big(x_p^k(b_{i,j}^l), x_p^{k'}(b_{i+d,j}^l)\big)$ denotes the shielding condition of two wire segments. If one of the two segments equals zero, no coupling exists between the two wire segments, and the function equals zero. If another parallel wire segment exists in between them, which means that the wire segment is shielded by the other wire segment, the value of the function will also be zero. Otherwise, the function will be 1. The function $P\big(x_p^k(b_{i,j}^l), x_p^{k'}(b_{i+d,j}^l)\big)$ can be defined as follows:

$$x_p^k(b_{i,j}^l) \bigwedge \left(x_p^{k'}(b_{i+d,j}^l) \bigwedge \neg \Big(\bigvee_{\substack{n_{\hat{k}} \in N \\ \hat{i} \in (i, i+d)}} x_p^{\hat{k}}(b_{i,j}^l) \Big) \right). \quad (5)$$

Figure 5 gives an example of the shielding condition for two parallel wire segments defined in Expression (5). The coupling noise between $x_p^k(b_{i,j}^l)$ and $x_p^{k'}(b_{i+d,j}^l)$ will be shielded if there exists a wire segment $x_p^{\hat{k}}(b_{i,j}^l)$ in between them, and Expression (5) will be 0. The coupling noise for other directions are defined similarly.

3.2.2 Constraints

To guarantee 100% routability and satisfy all the user-specified routing constraints, we formulate these constraints into four categories: symmetry, topology-matching, length-matching, and routing conservation constraints.

Symmetry constraint: To satisfy the symmetry constraint, two nets of the same symmetry pair should be routed symmetrically with respect to a common axis. The Y-axis symmetry constraint can be modelled as follows:

$$x_1^k(b_{i,j}^l) - x_1^{k'}(b_{i',j}^l) = 0, i' = \Big(2A_{sym}(n_k^S) - i\Big),$$
$$\forall n_k, n_{k'} \in n_k^S, \forall b_{i,j}^l, b_{i',j}^l \in B, \quad (6)$$

$$x_2^k(b_{i,j}^l) - x_8^{k'}(b_{i',j}^l) = 0, i' = \Big(2A_{sym}(n_k^S) - i\Big),$$
$$\forall n_k, n_{k'} \in n_k^S, \forall b_{i,j}^l, b_{i',j}^l \in B, \quad (7)$$

$$x_3^k(b_{i,j}^l) - x_7^{k'}(b_{i',j}^l) = 0, i' = \Big(2A_{sym}(n_k^S) - i\Big),$$
$$\forall n_k, n_{k'} \in n_k^S, \forall b_{i,j}^l, b_{i',j}^l \in B, \quad (8)$$

$$x_4^k(b_{i,j}^l) - x_6^{k'}(b_{i',j}^l) = 0, i' = \Big(2A_{sym}(n_k^S) - i\Big),$$
$$\forall n_k, n_{k'} \in n_k^S, \forall b_{i,j}^l, b_{i',j}^l \in B. \quad (9)$$

Figure 6 gives an example of the symmetry constraint. The wire segment $x_p^k(b_{i,j}^l)$ of the net n_k is symmetrical to the wire segment $x_p^k(b_{i',j}^l)$ of the net $n_{k'}$ with respect to the common axis $A_{sym}(n_k^S)$, where $i' = \big(2A_{sym}(n_k^S) - i\big)$. The symmetry constraints for other symmetry axes are defined similarly.

Topology-matching constraint: The topology-matching constraint is to route both nets of the same topology-matching pair with the same wirelength, bend numbers, and via counts. The topology-matching constraint can be further decomposed into three constraints: the length-matching, bend-matching, and via-matching constraints. The length-matching constraint is to guarantee both nets of the same topology-matching pair to be routed with the same wirelength. The length-matching constraint can be formulated as follows:

$$O_w(n_k) - O_w(n_{k'}) = 0, \forall n_k, n_{k'} \in n_k^T, \quad (10)$$

551

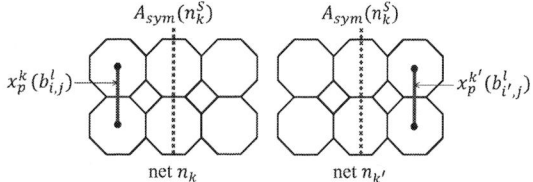

Figure 6: An example of the symmetry constraint. The wire segment $x_p^k(b_{i,j}^l)$ of the net n_k is symmetrical to the wire segment $x_p^k(b_{i',j}^l)$ of the net $n_{k'}$, where $i' = \left(2 \times A_{sym}(n_k^S) - i\right)$.

the total wirelength $O_w(n_k)$ of the net n_k is defined in Expression (1).

The bend-matching constraint is to guarantee both nets of the same topology-matching pair to be routed with the same number of wire bends. The bend-matching constraint can be formulated as follows:

$$O_b(n_k) - O_b(n_{k'}) = 0, \forall n_k, n_{k'} \in n_k^T, \qquad (11)$$

the bend number $O_b(n_k)$ of the nets n_k is defined in Expression (2).

The via-matching constraint is to ensure both nets of the same topology-matching pair to be routed with the same number of vias. The via-matching constraint can be formulated as follows:

$$O_v(n_k) - O_v(n_{k'}) = 0, \forall n_k, n_{k'} \in n_k^T, \qquad (12)$$

the via number $O_v(n_k)$ of net n_k is defined in Expression (3).

Length-matching constraint: The length-matching constraint is used by the nets which have limited routing resource to ensure both nets of the same length-matching pair to be routed with the same wirelength. The length-matching constraint can be formulated as follows:

$$O_w(n_k) - O_w(n_{k'}) = 0, \forall n_k, n_{k'} \in n_k^L, \qquad (13)$$

the total wirelength $O_w(n_k)$ of net n_k is defined in Expression (1).

Routing conservation constraint: The routing conservation constraint guarantees that the route of each net is consecutive and non-branched. The routing conservation constraint can be formulated as follows:

$$\sum_{l \in L} \sum_{1 \leq p \leq 8} x_p^k(b_{i,j}^l) = 1,$$
$$\text{if } T_{b_{i,j}^l}(n_k) = 1, \forall n_k \in N, \forall b_{i,j}^l \in B, \qquad (14)$$

$$\bigoplus_{\substack{l \in L \\ 1 \leq p \leq 8}} x_p^k(b_{i,j}^l) = 0,$$
$$\text{if } T_{b_{i,j}^l}(n_k) = 0, \forall n_k \in N, \forall b_{i,j}^l \in B, \qquad (15)$$

$$\sum_{l \in L} \sum_{1 \leq p \leq 8} x_p^k(b_{i,j}^l) \leq 2,$$
$$\text{if } T_{b_{i,j}^l}(n_k) = 0, \forall n_k \in N, \forall b_{i,j}^l \in B, \qquad (16)$$

$$\sum_{l \in L} x_p^k(b_{i,j}^l) \leq 1, \forall n_k \in N, \forall b_{i,j}^l \in B, 1 \leq p \leq 8 \qquad (17)$$

$$\left(x_1^k(b_{i,j}^l) \bigwedge x_2^k(b_{i,j}^{l'})\right) + \left(x_2^k(b_{i,j}^l) \bigwedge x_3^k(b_{i,j}^{l'})\right) +$$
$$\left(x_3^k(b_{i,j}^l) \bigwedge x_4^k(b_{i,j}^{l'})\right) + \left(x_4^k(b_{i,j}^l) \bigwedge x_5^k(b_{i,j}^{l'})\right) +$$
$$\left(x_5^k(b_{i,j}^l) \bigwedge x_6^k(b_{i,j}^{l'})\right) + \left(x_6^k(b_{i,j}^l) \bigwedge x_7^k(b_{i,j}^{l'})\right) +$$
$$\left(x_7^k(b_{i,j}^l) \bigwedge x_8^k(b_{i,j}^{l'})\right) + \left(x_8^k(b_{i,j}^l) \bigwedge x_1^k(b_{i,j}^{l'})\right) = 0,$$
$$\forall l, l' \in L, \forall n_k \in N, \forall b_{i,j}^l, b_{i,j}^{l'} \in B, \qquad (18)$$

$$\sum_{n_k \in N} \left(\sum_{1 \leq p \leq 8} x_p^k(b_{i,j}^l) + V_{b_{i,j}^l}(n_k)\right) \leq 2,$$
$$\forall l \in L, \forall b_{i,j}^l \in B. \qquad (19)$$

Constraint (14) ensures that only one of the edges in the set $\bigcup_{l \in L} E(b_{i,j}^l)$ can be passed if one of the terminals of net n_k exists in bin $b_{i,j}^l$. Constraints (15) and (16) ensure that either no edge or two edges can

be passed if none of the terminals of net n_k exists in bin $b_{i,j}^l$. Constraint (17) ensures no wire segment overlap between any two layers. Constraint (18) guarantees no routing path with an acute corner. A path with any acute corner will result in design rule violations. Constraint (19) guarantees that no routing path and via overlap in the same routing bin $b_{i,j}^l$.

3.3 Complexity Analysis

Since one routing bin allows only one net to pass, the number of variables is dominated by Constraint (18). The number of variables induced by Constraint (18) is $O(|G||N||L|^2)$, where $|G|$ is the number of routing bins in one layer.

As for the constraint complexity, it can be verified that Constraint (18) introduces the maximum number of constraints, since the constraint checks the existence of an acute corner for any two layers. The number of constraints introduced by Constraint (18) is thus $O(|G||N||L|^2)$.

THEOREM 1. *Given an analog design with $|N|$ nets, $|G|$ routing bins on each layer, and $|L|$ routing layers, the numbers of ILP variables and constraints in our basic ILP formulation are $O(|G||N||L|^2)$ and $O(|G||N||L|^2)$, respectively.*

3.4 ILP Reduction

We present three ILP reduction techniques to reduce the numbers of variables and constraints for our basic ILP formulation.

Since the wire connections usually cannot cross the active regions of devices, the placed modules can be viewed as blockages in the circuit. As a result, the number of variables can be reduced to $O((|G| - |\hat{M}|)|N||L|^2)$, where $|\hat{M}|$ is the number of routing bins occupied by the placed modules in one layer.

In addition, we can apply the following formulations to reduce the constraints induced by Expressions (2) and (3) and Constraint (18). Since Constraint (17) guarantees no wire segment overlap between any two layers, we define

$$F_p^k(i,j) = \sum_{l \in L} x_p^k(b_{i,j}^l), n_k \in N, b_{i,j}^l \in B. \qquad (20)$$

Then, Expression (2) and Constraint (18) can be transformed as follows:

$$\sum_{\substack{0 \leq i \leq w \\ 0 \leq j \leq h}} \sum_{1 \leq p \leq 8} \sum_{\substack{1 \leq p' \leq 8 \\ p' > p \\ p' \neq \{p+1, p+4, p+7\}}} \left(F_p^k(i,j) \bigwedge F_{p'}^k(i,j)\right), n_k \in N, \qquad (21)$$

$$\left(F_1^k(i,j) \bigwedge F_2^k(i,j)\right) + \left(F_2^k(i,j) \bigwedge F_3^k(i,j)\right) +$$
$$\left(F_3^k(i,j) \bigwedge F_4^k(i,j)\right) + \left(F_4^k(i,j) \bigwedge F_5^k(i,j)\right) +$$
$$\left(F_5^k(i,j) \bigwedge F_6^k(i,j)\right) + \left(F_6^k(i,j) \bigwedge F_7^k(i,j)\right) +$$
$$\left(F_7^k(i,j) \bigwedge F_8^k(i,j)\right) + \left(F_8^k(i,j) \bigwedge F_1^k(i,j)\right) = 0,$$
$$\forall n_k \in N, \forall i \in [0, w], \forall j \in [0, h]. \qquad (22)$$

Therefore, any acute corner between different layers can be checked at the same time, and the number of constraints induced by Expression (2) and Constraint (18) can be reduced to $O(|G||N^T|)$ and $O(|G||N|)$, respectively. To reduce the constraints induced by Expression (3), we further define

$$\hat{F}^k(b_{i,j}^l) = 1 - \left|\sum_{1 \leq p \leq 8} x_p^k(b_{i,j}^l) - 1\right|, n_k \in N, b_{i,j}^l \in B. \qquad (23)$$

Then, Expression (3) can be transformed as follows:

$$\sum_{\substack{0 \leq i \leq w \\ 0 \leq j \leq h}} \left(\bigvee_{l \in L} \hat{F}^k(b_{i,j}^l)\right), n_k \in N, b_{i,j}^l \in B. \qquad (24)$$

Therefore, the number of constraints induced by Expression (3) can be reduced to $O(|G||N||L|)$.

After applying the above techniques, the number of constraints is dominated by Expressions (24) and (4), and the number of constraints induced by Expression (4) is $O(D^2|G||N|)$, where D is the maximum distance that two wire segments would have coupling noise. In general, D is smaller than $|L|$.

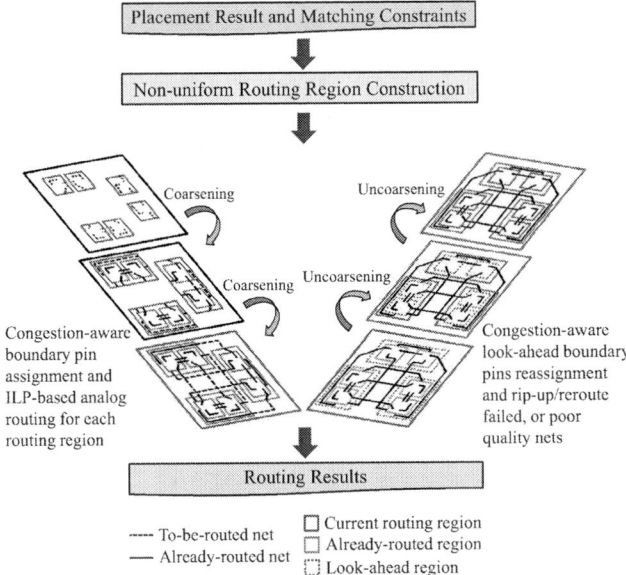

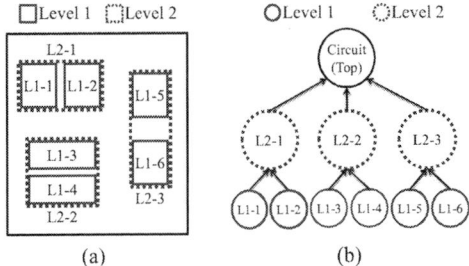

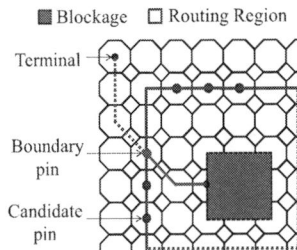

Figure 7: The non-uniform multilevel routing algorithm flow.

THEOREM 2. With the above ILP reduction, the number of ILP variables can be reduced from $O(|G||N||L|^2)$ to $O((|G|-|\hat{M}|)|N||L|^2)$, and that of ILP constraints from $O(|G||N||L|^2)$ to $O(|G||N|(D^2 + |L|))$.

4. NON-UNIFORM MULTILEVEL ROUTING

In this section, we propose a non-uniform multilevel routing algorithm to enhance the performance of our routing algorithm proposed in Section 3. Figure 7 shows our non-uniform multilevel routing flow. Given a placement and design specifications, we first construct a routing-hierarchy tree and non-uniform routing regions according to the design hierarchy. This step is followed by the V-shaped non-uniform multilevel routing framework of coarsening followed by uncoarsening. In the coarsening stage, congestion-aware boundary pin assignment for each routing region is performed before ILP-based analog routing. Each routing region in the previous level is regarded as a blockage in the current level. In the uncoarsening stage, rip-up and reroute with congestion-aware look-ahead boundary pin reassignment are performed level by level for failed or poor-quality nets to further improve the routing quality.

4.1 Non-uniform Routing Region Construction

Figure 8 shows an example of the design hierarchy and its routing-hierarchy tree. According to the routing-hierarchy tree and the design hierarchy, non-uniform local routing regions can be constructed level by level. Since analog designers usually consider part of the layout at one time, our non-uniform multilevel routing framework conforms to the conventional design manner of analog circuits. As mentioned in [13], the active regions of devices should be considered as blockages in analog routing for better performance. In addition, each routing region in the previous level is also regarded as a blockage in the current level. To acquire suitable routing resource and compensate the routing resource which is blocked, we expand the original routing region according to the blockage ratio of the region and the whitespace ratio of the design. Assume that w is the whitespace ratio of the design, and b_i is the blockage ratio of the routing region i. The expansion ratio for each routing region can be defined as follows:

$$\delta_i = 1 + w \times b_i. \tag{25}$$

4.2 Boundary Pin Assignment and Reassignment

In our multilevel routing framework, each level consists of several routing regions according to the routing-hierarchy tree. As a result, local (two terminals in the same routing region) and global (two terminals in different routing regions) connections will co-exist in one routing region. To avoid routing congestion and reserve routing resource, we split the global connection into two parts, the global and local parts. We assign a boundary pin between these two parts, and pre-route the local part in the coarsening stage and reserve routing resource. For these reasons, we propose a congestion-aware boundary pin assignment method to assign a pin to the boundary of the routing region with minimized routing congestion and wirelength.

Figure 9 shows an example of the boundary pin assignment method. To select the most appropriate location of a boundary pin, we first estimate the routing congestion for the routing level by a probability-based breadth first search (BFS) method. Then, the candidates for boundary pins can be selected according to the estimated routing congestion. After selecting the candidates of boundary pins, the cost of each candidate can be calculated by the distance from the candidate pin to the target terminal.

For the uncoarsening stage, rip-up and reroute with congestion-aware look-ahead boundary pin reassignment will be performed. The boundary pins for failed, non-matched, or poor-quality nets will be reassigned according to the real congestion caused by the routed nets in the current level and the next level.

Figure 8: (a) A circuit with design hierarchy statistics. (b) Routing-hierarchy tree of the design.

Figure 9: An example of boundary pin assignment. The boundary pin will be selected from the candidate pins with the smallest cost.

Table 1: Device statistics of OP1 and OP2.

Circuits	Device Number				Specification		
	MOS	Capacitor	Resistor	Total	DC Gain (dB)	Unity Gain Bandwidth (MHz)	Phase Margin (°)
OP1	31	0	7	38	50	100	45
OP2	52	4	0	56	40	5	45

5. EXPERIMENTAL RESULTS

We implemented our routing algorithm and the algorithm in [13] in C++ programming language. All the experiments were performed on an Intel Xeon 2.93GHz Linux workstation with 48GB memory. We used the CPLEX12.3 [15] library to solve the ILP problems. To evaluate our approach, we used two industrial analog circuits OP1 and OP2 which have been taped-out by TSMC 0.18 μm process manually by expert analog designers. The statistics of the two circuits are shown in Table 1. We also applied the TSMC 0.18 μm library for the analog design environment, and used two layers to route the circuits. After generating the layout automatically, we used Calibre nmDRC and nmLVS to check the design rules and verify the layout versus schematic.

553

Table 2: Comparison of the routing results between the work [13] and our approach with ILP reduction.

Circuits	[13]				Ours with ILP reduction			
	Area (um^2)	Wirelength (um)	Via Number	Runtime (s)	Area (um^2)	Wirelength (um)	Via Number	Runtime (s)
OP1	1455.87	327.16	45	16.97	1455.87	273.66	36	42.35
OP2	N/A	N/A	N/A	N/A	7165.22	418.62	80	254.97
Comp.	1	1.20	1.25	0.40	1	1	1	1

Table 3: Comparison of DC Gain, Unity Gain Bandwidth, and Phase Margin between [13] and our approach.

Circuits	DC Gain (dB)				Unity Gain Bandwidth (MHz)				Phase Margin ($^\circ$)			
	Schematic	Manual	[13]	Ours	Schematic	Manual	[13]	Ours	Schematic	Manual	[13]	Ours
OP1	62.81	60.74	60.73	60.92	654.90	182.70	209.60	272.3	59.23	54.30	51.90	50.68
OP2	43.57	43.56	N/A	43.56	9.16	8.97	N/A	8.97	38.44	38.90	N/A	38.96

Table 4: Comparison of running time.

Circuit	Without Multilevel Framework		With Multilevel Framework	
	w/o Reductions	w/ Reductions	w/o Reductions	w/ Reductions
OP1	>1 hr	>1 hr	114.11 s	42.35 s
OP2	>1 hr	>1 hr	490.10 s	254.97 s
Comp.			2.31	1

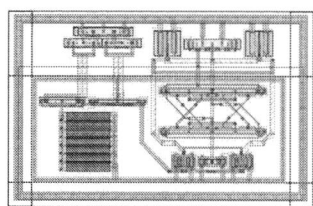

Figure 10: Layout of OP1.

Since there is no existing work simultaneously handling the symmetry, topology-matching, and length-matching constraints, we compared our method with [13] for area, wirelength, via number, and running time. As shown in Table 2, our method with the non-uniform multilevel framework and ILP reduction techniques obtained an average wirelength reduction of 20% and an average via reduction of 25% with a little overhead in running time. These results show that our ILP and the non-preferred direction with 45-degree routing style can reduce the wirelength and the number of vias effectively. Since there are some congested regions in OP2, the algorithm in [13] failed to complete the routing because of the preferred direction routing style.

We also ran the post-layout simulation to verify the quality of our routing results. To obtain the netlist with extracted parasitic RC, we performed post-layout extraction by using Calibre xRC. After RC extraction, we used HSPICE to simulate the netlist with extracted parasitic RC, measured DC gain, unity gain bandwidth, and phase margin. Table 3 shows the simulation results. Our approach can satisfy all the specified routing constraints and better fit the specification. Since the work [13] handles only symmetry constraints, it is predictable that our approach with simultaneously considering symmetry, topology-matching, and length-matching constraints results in much fewer current mismatches and balanced parasitic resistors and capacitors. Moreover, our work can obtain better DC gain and unity gain bandwidth than both the manual design and the work [13]. Besides, our phase margin is the closest to the specification (45°). Figure 10 shows the resulting layout of OP1.

With the ILP reduction, we can reduce the average number of variables by 2.6X and the average number of constraints by 6.5X. As shown in the Table 4, with the reduction techniques, our ILP routing algorithm can reduce the running time by 2.3X. In addition, Table 4 also shows the efficiency of our non-uniform multilevel routing framework. Since the number of variables and constraints in a local routing region are smaller than those in the whole chip, our multilevel approach can reduce the running time significantly. In the two cases, the running time of our ILP routing algorithm without the multilevel framework is more than one hour. In contrast, the ILP routing algorithm with the multilevel routing framework solved the problem efficiently.

6. CONCLUSIONS

In this paper, we have proposed an analog routing algorithm with a non-uniform multilevel routing framework to simultaneously consider symmetry, topology-matching, and length-matching constraints for analog circuit designs. Our algorithm also minimizes the total wirelength, bend numbers, via counts, and coupling noise at the same time. Experimental results have shown that our proposed analog routing algorithm and non-uniform multilevel framework are effective and efficient for analog routing.

7. ACKNOWLEDGMENTS

The authors would like to thank the Nanoelectronics and Gigascale Systems Laboratory of National Chiao Tung University and Mr. Jhao-Yan Liu of National Tsing Hua University for their valuable help.

8. REFERENCES

[1] H.-Y. Chen, M.-F. Chiang, Y.-W. Chang, L. Chen, and B. Han, "Full-chip Routing Considering Double-via Insertion," *IEEE TCAD*, vol 27, no. 5, pp. 844–857, May. 2008.

[2] U. Choudhury, and A. Sangiovanni-Vincentelli, "Constraint-based Channel Routing for Analog and Mixed Analog/Digital Circuits," *IEEE TCAD*, vol 12, no. 4, pp. 497–510, Apr. 1993.

[3] C. Chu, and Y.-C. Wong; "FLUTE: Fast Lookup Table Based Rectilinear Steiner Minimal Tree Algorithm for VLSI Design," *IEEE TCAD*, vol 27, no. 1, pp. 70–83, Jan. 2008.

[4] J. M. Cohn, D. J. Garrod, R. A. Rutenbar, and L. R. Carley, "KOAN/ANAGRAM II: New Tools for Device-level Analog Placement and Routing," *IEEE JSSC*, vol. 26, pp. 330–342, Mar. 1991.

[5] Q. Gao, H. Yao, Q. Zhou, and Y. Cai, "A Novel Detailed Routing Algorithm with Exact Matching Constraint for Analog and Mixed Signal Circuits," *Proc. of ISQED*, Mar. 2011.

[6] S. Kumar, J. D. Carothers, R. D. Newbould, and B. V. Krishnan, "Candidate Generation for 45 Degree Routing for Mixed-Signal Layout," *Proc. of SSMSD*, Feb. 2003.

[7] P.-H. Lin, H.-C. Yu, T.-H, Tsai, and S.-C. Lin, "A Matching-based Placement and Routing System for Analog Design," *Proc. of VLSI-DAT*, Apr. 2007.

[8] E. Malavasi, E. Charbon, E. Felt, and A. Sangiovanni-Vincentelli, "Automation of IC Layout with Analog Constraints," *IEEE TCAD*, vol 15, no. 8, pp. 923–942, Aug. 1996.

[9] E. Malavasi, M. Chilanti, and R. Guerrieri, "A General Router for Analog Layout," *Proc. of CompEuro*, May. 1989.

[10] M. M. Ozdal, and R. F. Hentschke, "Exact Route Matching Algorithms for Analog and Mixed Signal Integrated Circuits," *Proc. of ICCAD*, Nov. 2009.

[11] S. Piguet, F. Rahali, M. Kayal, E. Zysman, and M. Declercq, "A New Routing Method for Full Custom Analog ICs," *Proc. of CICC*, May. 1990.

[12] K. Sajid, J. D. Carothers, J. J. Rodriguez, and W. T. Holman, "Global Routing Methodology for Analog and Mixed-signal Layout," *Proc. of ASIC/SOC Conference*, Sep. 2001.

[13] L.F. Xiao, Evangeline F.Y. Young, X.-Y. He, and K.P. Pun, "Practical Placement and Routing Techniques for Analog Circuit Designs," *Proc. of ICCAD*, Nov. 2010.

[14] L. Zhang, U. Kleine, and Y. Jiang, "An Automated Design Tool for Analog Layouts," *IEEE TVLSI*, vol 14, no. 8, pp. 881–894, Aug. 2006.

[15] IBM ILOG CPLEX Optimizer, http://www-01.ibm.com/software/integration/optimization/cplex-optimizer/

X-Tracer: A Reconfigurable X-Tolerant Trace Compressor for Silicon Debug

Feng Yuan[†‡], Xiao Liu[†] and Qiang Xu[†‡]

[†]CUhk REliable Computing Laboratory (CURE)
Department of Computer Science & Engineering
The Chinese University of Hong Kong, Shatin, N.T., Hong Kong
[‡]Shenzhen Institutes of Advanced Technology, Chinese Academy of Sciences
Email: {fyuan,xliu,qxu}@cse.cuhk.edu.hk

ABSTRACT

The effectiveness of at-speed silicon debug is constrained by the limited trace buffer size and/or trace port bandwidth, requiring highly-efficient trace data compression solutions. As it is usually inevitable to have unknown 'X' values during silicon debug, trace compressor should be equipped with X-tolerance feature in order not to significantly degrade error detection capability. To tackle this problem, this paper presents a novel reconfigurable X-tolerant trace compressor, namely X-Tracer, which is able to tolerate as many X-bits as possible in the trace streams while guaranteeing high compression ratio, at the cost of little extra design-for-debug hardware. Experimental results on benchmark circuits demonstrate the effectiveness of the proposed technique.

Categories and Subject Descriptors

B.7.3 [Integrated Circuits]: Reliability and Testing

General Terms

Design, Verification, Reliability, Algorithm.

Keywords

Silicon Debug, Trace Data Compression, X-Tolerance

1. INTRODUCTION

The ever-increasing design complexity of integrated circuits (ICs) and the inherent inaccuracy of circuit models at high abstraction levels significantly challenge the effectiveness of pre-silicon verification techniques, and it is not uncommon that IC products need to go through multiple re-spins to be error-free [1], despite the fact that more than half of the resources are devoted to verification tasks [2]. Consequently, to reduce expensive re-spins and time-to-market, silicon debug (also known as post-silicon validation) cannot be an afterthought and has become an essential step in today's IC design flow.

1.1 Related Work

Since the circuit under debug (CUD) is a piece of silicon that has already been fabricated, the main challenge in silicon debug is the limited visibility of internal signals. To tackle this problem, usually dedicated design-for-debug (DfD) circuitries are added to the design to improve its observability.

Trace-based debug [3] that allows designers to real-time observe a set of signals in consecutive cycles, being non-intrusive to the circuit's normal operation, is one of the most effective silicon debug techniques and has been widely adopted by the industry [4, 5]. To be specific, in this technique, a set of "key" signals in the CUD are tapped and they can be traced after being triggered. The sampled data are then sent to internal trace buffers and/or external trace ports via trace interconnection fabric [6], for later analysis by debug software and physical probing tools to further root cause and fix the bug (e.g., [7, 8, 9]).

Once a bug is activated, it leaves its erroneous effects in one or more state elements of the circuit at some cycles. The objective of trace-based debug is to observe and localize such errors with as few debug runs as possible. Since it is not possible for us to trace all internal signals in the circuit, on one hand, the effectiveness of trace-based debug depends on the quality of the selected trace signals, which may include both manually-picked signals by experienced designers and signals selected via automated solutions guided by some visibility-enhancement metrics (e.g., [10, 11, 12, 13]). On the other hand, even with pre-determined trace signals that can capture a bug, it will only manifest itself at some specific time and it is crucial to ensure the signals at the "right" time are indeed traced.

Clearly, the more trace data that we can acquire, the higher possibility for us to catch a bug's erroneous effects in them and the less time and effort to identify the bug. Unfortunately, what we can trace in each debug run is usually quite limited. This is because, trace-based debug involves non-trivial overhead and we are only given limited trace buffer size and/or few external pins as trace ports.

Because of the above, it is not quite economical to store the"raw" trace data. In [14], Park and Mitra compressed the execution states of microprocessor into a small amount of *footprints*, taking advantage of the fact that the locality feature of instruction sequence and redundant information in monitored data that can be easily identified with the executed instructions. [15, 16] utilized the data locality feature when accessing cache and adopted dictionary-based compression to improve the compression ratio.

The above trace compression solutions focused on debugging microprocessors. Several compression techniques have also been presented for signal tracing in general logic circuits to improve their error detection capability, and they can be broadly classified into the following three types:

- *lossless trace compressors*, which take advantage of the locality of trace data for lossless compression. In [17], Anis and Nicolici presented several dictionary-based compressors to trace repeatable data. Based on the observation that toggling rate of state values is usually low, Prabhakar *et al.* [18] proposed to compress the differential data to achieve higher compression quality.

- *spatial lossy trace compressors*, which compacts a set of N signals into M parity signals ($N > M$) with XOR network before signal tracing starts [19]. To reduce routing overhead, such spatial compressors are usually organized as a tree-like structure as part of the trace interconnection fabric.

- *temporal lossy trace compressors*, which compacts a number of cycles (e.g., 1k) of the raw data into a signature during signal tracing [20, 21], with the help of multiple-input signature register (MISR), originally used for test response compaction in VLSI testing domain. As shown in Fig. 1, with the assumption that the CUD behaves repeatable in different debug iterations, [20] consecutively zooms-in the failure signatures by reconfiguring the compaction ratio in their MISR-based compressor for each debug run to localize the error.

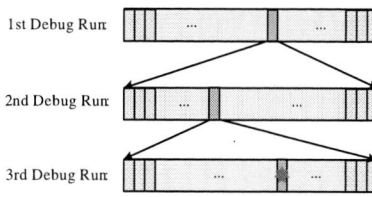

Figure 1: Iterative Debug Flow in [16].

In this work, we focus on the temporal trace compressor design, which provides significantly higher compression ratio when compared to the other types of compressors. Please note, at the same time, the above three trace compression methodologies are not contradictory, and in fact, they can be combined together to further enhance the real-time observability of the CUD.

1.2 Motivation and Summary of Contributions

From the above, it is clear that temporal lossy trace compressors are quite appealing due to their impressive compression ratio. However, the effectiveness of such MISR-based compressors relies on the existence of *clean* "golden vector" to generate reference signatures for comparison. This is usually not the case during silicon debug, rendering the lossy compression technique less effective on error detection. This is because: (i). it is often too time-consuming to run gate-level simulation for failed silicon test, and hence designers often resort to high-level simulator to generate "golden vectors" and many unknown bits (X-bits) are obtained when they are mapped onto gate-level vectors; (ii). asynchronous clock domains and uninitialized state elements also result in many X-bits in functional patterns.

In test response compaction techniques, X-bits will also corrupt the signature and hence result in fault coverage loss. Various techniques were presented to tackle this problem in the VLSI testing domain. X-masking hardware can be introduced to mask X-bits before they are fed into response compactor [22]. X-tolerant techniques are able to inherently tolerate X-bits by connecting each compactor input with redundant channels [23]. X-canceling solution identifies linear dependency of the signature with Gauss-Jordan elimination and cancels X-bits by XORing relevant signatures together [24].

The above techniques, however, cannot be effectively reused to tolerate X-bits in trace-based silicon debug. This is because: (i). signal tracing is conducted in functional mode and hence those solutions that rely on blocking X-bits in test mode cannot be used; (ii). unlike test patterns that are generated before the circuit is fabricated, the vectors used in silicon debug cannot be pre-calculated and hence we do not know the exact distribution of X-bits in advance; (iii). different from manufacturing test that only needs to guarantee fault detection capability, in silicon debug we are required to obtain as much information as possible to root-cause the bug.

Motivated by the above, in this paper, we propose a novel reconfigurable X-tolerant trace compressor, namely *X-Tracer*, which is able to tolerate as many X-bits as possible in the trace streams while guaranteeing high compression ratio, at the cost of little extra design-for-debug hardware. The major contributions of this work include:

- We propose a novel reconfigurable MISR-based trace compressor with redundancy that is able to effectively tolerate X-bits in trace-based debug;

- We develop a trace data extraction algorithm that is able to extract as much useful trace information as possible;

The remainder of this paper is organized as follows. Section 2 presents the proposed X-tracer design. The trace data extraction algorithm is then detailed in Section 3. Next, experimental results are shown in Section 4. Finally, Section 5 concludes this work.

2. PROPOSED X-TRACER DESIGN

The proposed X-Tracer design is constructed in two steps. First, we add redundancy into conventional MISR structure to equip it with dedicated X-tolerance feature. Then, we introduce reconfigurability into the compressor to further improve its X-tolerance capability. The details are given in the following.

2.1 MISR-Based X-Tolerant Trace Compressor with Explicit Redundancy

Conventional MISR design is typically constructed with a primitive polynomial feedback loop. Fig. 2(a) and Fig. 2(b) show two example MISR circuits with different polynomials. In this example, input data are compressed for five cycles and the corresponding MISR outputs are represented as linear combinations of inputs.

Clearly, with the above architecture, any unknown input bit will render the MISR output that utilize it to be unknown as well. However, as an input bit may present itself on multiple outputs, it is possible to cancel some X-bits with the technique shown in [25]. For example, given the MISR circuit in Fig. 2 (a), if only I_{03} appears to be an X-bit in the compression trace window, its effect can be canceled by XORing O_7 with O_5 (with external software support, detailed in Section 3), thus enabling us to preserve some of the traced information, i.e., I_{01}, I_{10}, I_{14}, I_{32} and I_{34} in this example.

The above X-canceling possibility is due to the implicit redundancy for certain trace bits, which, however, is not guaranteed. Let us consider the MISR circuit shown in Fig. 2 (a) again.

- It is inevitable that some information bits are only delivered to a single output bit (e.g., I_{00}). Consequently, if such bits are unknown, they will contaminate all the other information bits located in the same output (e.g., O_6) and results in significant trace information loss;

- Even for those information bits that are delivered to multiple outputs, the by-product of X-canceling operation may result in other trace information loss. For example, consider again the case of $I_{03} = X$ in O_7 and O_5, while I_{01}, I_{10}, I_{14}, I_{32} and I_{34} are reserved, the trace bits I_{12}, I_{21} and I_{30} are lost since they are also canceled being appeared in both O_7 and O_5 and at the same time cannot be recovered from other MISR outputs;

- Large number of X-bits in trace data will make it fail to find any X-canceling combination. As for this example, when I_{00}, I_{02}, I_{03} and I_{23} are all X-bits, there will be no uncontaminated information bits left by conducting X-canceling.

To overcome the above limitations, we propose to construct a MISR-based trace compressor with explicit redundancy, by combining multiple MISRs with different primitive polynomials and input mapping orders. An example is shown in Fig. 2, wherein we simply combine the two MISR circuits together to construct our X-tolerant trace compressor. With this design, all the traced bits will be included in at least two outputs. Then, when I_{03} is unknown, previously-lost information bits I_{12}, I_{21} and I_{30} can be recovered by $O_5 \oplus O_3$ and O_0. Even when we have many X-bits as the earlier example, i.e., when I_{00}, I_{02}, I_{03} and I_{23} are all X-bits, there are choices to cancel their effects with combinations such as $O_7 \oplus O_3 \oplus O_2$, $O_6 \oplus O_3 \oplus O_1$.

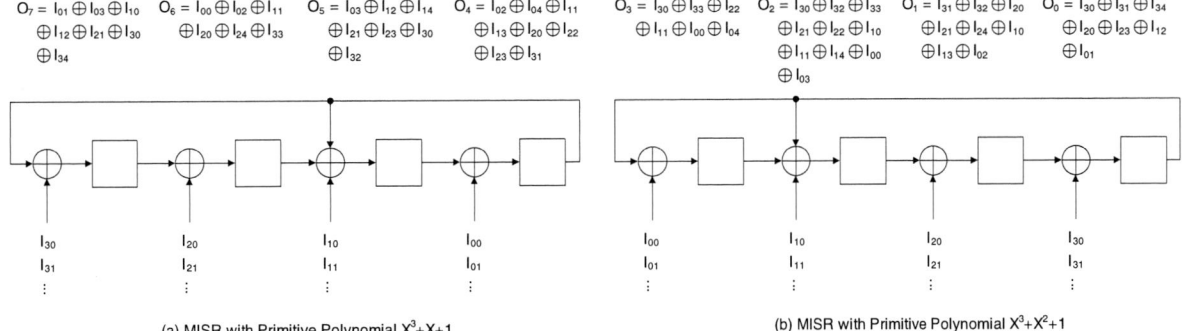

$O_7 = I_{01} \oplus I_{03} \oplus I_{10}$ $\quad$ $O_6 = I_{00} \oplus I_{02} \oplus I_{11}$ $\quad$ $O_5 = I_{03} \oplus I_{12} \oplus I_{14}$ $\quad$ $O_4 = I_{02} \oplus I_{04} \oplus I_{11}$
$\oplus I_{12} \oplus I_{21} \oplus I_{30}$ $\qquad$ $\oplus I_{20} \oplus I_{24} \oplus I_{33}$ $\qquad$ $\oplus I_{21} \oplus I_{23} \oplus I_{30}$ $\qquad$ $\oplus I_{13} \oplus I_{20} \oplus I_{22}$
$\oplus I_{34}$ $\qquad\qquad\qquad\qquad\qquad\qquad$ $\oplus I_{32}$ $\qquad\qquad\qquad$ $\oplus I_{23} \oplus I_{31}$

$O_3 = I_{30} \oplus I_{33} \oplus I_{22}$ $\quad$ $O_2 = I_{30} \oplus I_{32} \oplus I_{33}$ $\quad$ $O_1 = I_{31} \oplus I_{32} \oplus I_{20}$ $\quad$ $O_0 = I_{30} \oplus I_{31} \oplus I_{34}$
$\oplus I_{11} \oplus I_{00} \oplus I_{04}$ $\qquad$ $\oplus I_{21} \oplus I_{22} \oplus I_{10}$ $\qquad$ $\oplus I_{21} \oplus I_{24} \oplus I_{10}$ $\qquad$ $\oplus I_{20} \oplus I_{23} \oplus I_{12}$
$\qquad\qquad\qquad\qquad$ $\oplus I_{11} \oplus I_{14} \oplus I_{00}$ $\qquad$ $\oplus I_{13} \oplus I_{02}$ $\qquad\qquad$ $\oplus I_{01}$
$\qquad\qquad\qquad\qquad$ $\oplus I_{03}$

(a) MISR with Primitive Polynomial X^3+X+1 $\qquad\qquad\qquad$ (b) MISR with Primitive Polynomial X^3+X^2+1

Figure 2: Example MISR-Based Trace Compressors.

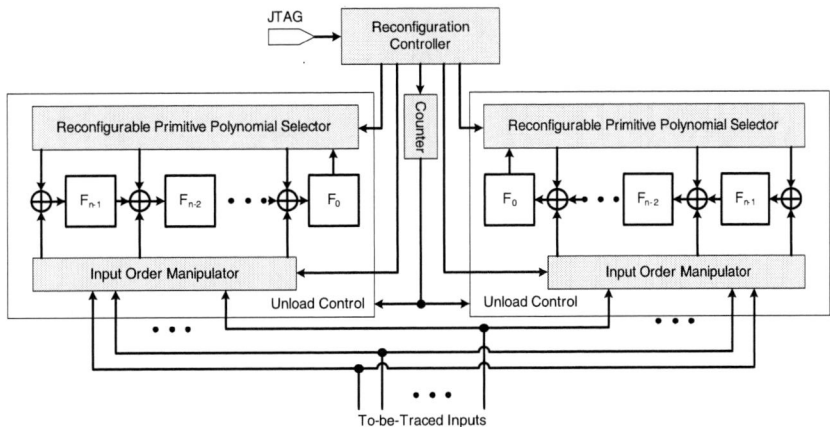

Figure 3: Reconfigurable X Tolerant Trace Compressor Architecture.

We attribute the benefits provided by our X-tolerant trace compressor structure to the following reasons:

1. The increased number of observing points for each and every traced bits provide more information redundancy and hence higher possibility to cancel X-bits.

2. As demonstrated by the symbolic expression of each MISR output (see Fig. 2), MISR with distinct primitive polynomial results in unique information bit distribution at the corresponding observing outputs. By combining MISRs with diversified information bit distributions, the proposed trace compressor thus provides additional opportunities to tolerate X-bits and reduce the possibility of useful information loss. For example, if two identical MISRs shown in Fig. 2(a) are used to compose the trace compressor, there will be only two observing outputs for I_{00} and it may cause X-canceling difficulty and information loss if it is an X-bit. Replacing one of them with the MISR shown in Fig. 2(b), however, gives us three observing outputs for I_{00} all together.

3. Distinct input orders in redundant MISRs facilitate to prevent some fixed combination of information bits. In the example design, $I_{01} \oplus I_{10}$ would be appearing in both MISRs if their input orders are kept the same, thus leading to information loss if one of them is unknown.

It is important to note that, while the proposed X-tolerant trace compressor design involves little area overhead, it does lead to significant compression ratio loss. With two MISRs to compose the proposed X-tolerant trace compressor, the compression ratio is only half of that of the single MISR design with the same trace buffer size. This effect has been taken into consideration for fair comparison in our experimental results shown in Section 4.

2.2 Reconfigurable Trace Compressor Design

As discussed earlier, while more redundancy is helpful to recover more trace information, the compression ratio is reduced and it may also involve high control complexity. Consequently, in our trace compressor design, we keep the redundancy ratio to be two. In order to further enhance the capability of the trace compressor, we introduce reconfigurability into our trace compressor design. By doing so, we are able to flexibly change the compressor's structure for each debug run, which enables us to extract more trace information.

Fig. 3 presents the overall structure of the proposed X-tracer design, which can be configured externally via JTAG interface. Three configuration options are provided to debug engineers. Firstly, the primitive polynomial can be reconfigured for each MISR, and it is implemented by selectively turning on/off the feedback loop from each output. Secondly, the module *Input Order Manipulator* is utilized to change MISR's input order, also individually-controllable. Finally, we introduce an internal *counter* to determine the cycle number to unload the MISR signatures. Adjustment of the counter value enables us to tradeoff compression ratio and X-tolerant capability, similar to [20]. Finally, our proposed trace compressor can be easily reconfigured to simply use a single MISR for trace compression as well, when there is nearly no X-bits and hence redundant tracing is not required.

3. TRACE INFORMATION EXTRACTION

After acquiring the X-contaminated signatures from the trace buffer in each debug run, we rely on an off-line processing step to extract as many useful trace bits as possible.

Given the X-contaminated signatures, we can construct the corresponding X-matrix and employ the X-canceling solution in [25] to extract useful trace data. An example X-matrix is shown in the following example, wherein each row corresponds to a MISR observing bit and each column represents a specific X-bit (entry '1' denotes that the corresponding X-bit contaminates the specific observing bit). Next, Gauss-Jordan elimination is utilized to identify all-zero rows, which represent X-canceling combinations. Assuming I_{00}, I_{02}, I_{03} and I_{23} in Fig. 2 are X-bits, one possible solution with [25] is shown as follows:

$$
\begin{pmatrix}
 & I_{00} & I_{02} & I_{03} & I_{23} \\
O_7 & 0 & 0 & 1 & 0 \\
O_6 & 1 & 1 & 0 & 0 \\
O_5 & 0 & 0 & 1 & 1 \\
O_4 & 0 & 1 & 0 & 1 \\
O_3 & 1 & 0 & 0 & 0 \\
O_2 & 1 & 0 & 1 & 0 \\
O_1 & 0 & 1 & 0 & 0 \\
O_0 & 0 & 0 & 0 & 1
\end{pmatrix}
\rightarrow
\begin{pmatrix}
 & I_{00} & I_{02} & I_{03} & I_{23} \\
O_3 & 1 & 0 & 0 & 0 \\
O_6+O_3 & 0 & 1 & 0 & 0 \\
O_7 & 0 & 0 & 1 & 0 \\
O_6+O_4+O_3 & 0 & 0 & 0 & 1 \\
O_7+O_6+O_5+O_4+O_3 & 0 & 0 & 0 & 0 \\
O_7+O_3+O_2 & 0 & 0 & 0 & 0 \\
O_6+O_3+O_1 & 0 & 0 & 0 & 0 \\
O_6+O_4+O_3+O_0 & 0 & 0 & 0 & 0
\end{pmatrix}
$$
(1)

As a test response compaction technique, the objective of [25] is to find a solution that X-bits do not contaminate the bits used to fault detection. In silicon debug, however, our objective is to extract as many useful trace bits as possible. Consequently, we cannot directly use [25] to generate our solution.

Since different solutions lead to very different X-canceling combinations, the corresponding extracted trace data may vary significantly. We can in fact employ multiple solutions to extract more trace information for silicon debug. This, however, is a challenging problem because the number of solutions for an X-matrix can be huge (even with the rather constrained Gauss-Jordan elimination procedure). In this section, we propose novel techniques that are able to effectively explore X-canceling combination space and we present an algorithm to maximize the number of extracted trace bits.

3.1 X-Canceling Solution Space Exploration

Again, let us take the example shown in Eq. 1 to illustrate our proposed technique for X-canceling solution space exploration.

We first select a targeted MISR observing bit (e.g., O_6), and move the corresponding row down to the last position. By doing so, we aim at finding the combination of remaining observing bits to cancel the unknown targeted bit. Then, instead of conducting row operations, we perform column operations to reduce the modified X-matrix to *column echelon form* as follows:

$$
\left(
\begin{array}{c|cccc}
O_7 & 0 & 0 & 1 & 0 \\
O_5 & 0 & 0 & 1 & 1 \\
O_4 & 0 & 1 & 0 & 1 \\
O_3 & 1 & 0 & 0 & 0 \\
O_2 & 1 & 0 & 1 & 0 \\
O_1 & 0 & 1 & 0 & 0 \\
O_0 & 0 & 0 & 0 & 1 \\
\hline
O_6 & 1 & 1 & 0 & 0
\end{array}
\right)
\rightarrow
\left(
\begin{array}{cccc}
\mathbf{1} & 0 & 0 & 0 \\
0 & \mathbf{1} & 0 & 0 \\
0 & 0 & \mathbf{1} & 0 \\
0 & 0 & 0 & \mathbf{1} \\
1 & 0 & 0 & 1 \\
1 & 1 & 1 & 0 \\
1 & 1 & 0 & 0 \\
\hline
0 & 1 & 1 & 1
\end{array}
\right)
$$
(2)

With the *column echelon form* of the X-matrix, the first non-zero entry is called a pivot (in bold), and its corresponding row is *pivot row*, which is guaranteed to contain only one non-zero entry. In addition, we define all-zero row as the *free rows*, and the other rows as *stack rows*. According to linear algebra, if the last row (representing the targeted observing bit) is not a pivot row, there must exist at least one combination of the remaining bits to cancel the unknown targeted bit. For the sake of easy explanation, we use a vector S to denote a X-canceling combination, where 1 in S means that the corresponding observing bit is involved in the combination. Based on the above definition for the reduced X-matrix, we define the corresponding bits in the S as *pivot bits*, *stack bits* and *free bits*. Therefore, an initial solution S_{init} can be found in the following manner: we first identify non-zero entries on the last row of the reduced X-matrix; then find the pivots on the same column; finally, we fill the related pivot bits in S_{init} with 1s, and left the other as 0s. For the example in Eq. 2, the initial solution could be $S_{init} = \{0, 1, 1, 1, 0, 0, 0, 1\}$.

Next, starting from the initial solution, we try to traverse the solution space and generate new solutions by transforming existing solutions. To guarantee the obtained solution is legal, the transformation obeys three well-defined bit flipping rules: (i). free bits can

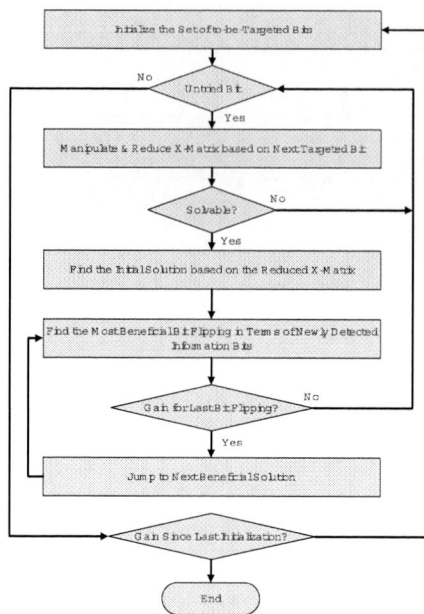

Figure 4: X-Free Information Extraction Algorithm.

be freely flipped to generate a new valid solution; (ii). to flip stack bit, we need to further flip all the pivot bits whose corresponding pivots are on the same columns of non-zero entries of stack row correlated with to-be-flipped stack bit. For example, if we flip the fifth bit in S_{init}, the related stack row is $\{1, 0, 0, 1\}^T$ and thus the first and forth pivots bit need be flipped at the same time. In this case, the new solution could be $S_{sec} = \{1, 1, 1, 0, 1, 0, 0, 1\}$. (iii). the pivot bits and the last bit in each row cannot be flipped. With the above solution space exploration procedure, the X-canceling combinations can be acquired by simply changing different targeted observing bits.

3.2 X-Free Information Extraction

With the flexibly to explore the X-canceling combination solution space, our objective is to extract the maximum number of useful trace data (i.e., those X-free bits that are known).

Our proposed algorithm is depicted in Fig. 4. It starts from putting all the observing outputs into a to-be-targeted bit set, from which we select one output as the targeted bit a time. Based on the original X-matrix, we first move the targeted bit column to the last position, and then perform row operations to reduce the matrix to *row echelon form*. Next, we try to find the next targeted bit if the reduced matrix is solvable, otherwise we try to restore the initial solution. Then, we search the solution space by greedily flipping the most beneficial bit, in which the *gain* is defined as the increased number of extracted information bits, and the search procedure is stopped when there is no more *gain*. After all the targeted bits have been tried, we check whether these is still *gain* since the last initialization. The whole procedure is re-initialized if it is not zero, otherwise it is terminated.

4. EXPERIMENTAL RESULTS

4.1 Experimental Setup

To evaluate the performance of our proposed X-tracer solution, in our experiments, we perform simulation studies under various X-bit distribution in the "raw" trace data, with X-bit ratio ranging from 0.05% to 5.5% injected randomly[1]. The trace buffer size is fixed as $32 \times 1k$ in our experiments, i.e., 32-bits are fed to the temporal compressor in every cycle and the number of trace entries is $1k$. Experiments are conducted under various compression ratio, ranging from 50 to 500. Since the proposed X-tracer architecture results in

---

[1]The simulation study is independent of the specific circuit under debug with random X-bit injection.

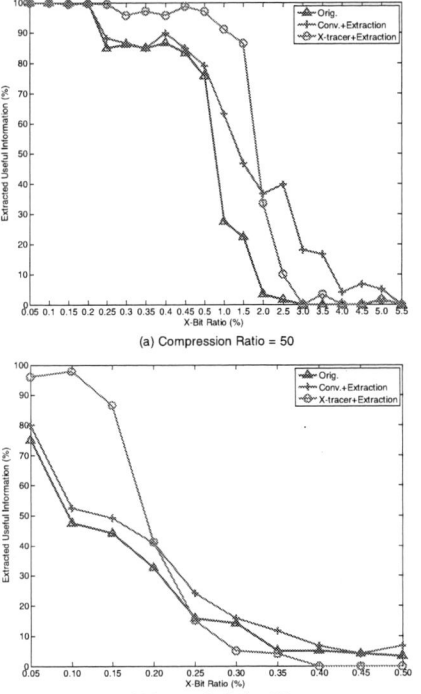

(a) Compression Ratio = 50

(b) Compression Ratio = 500

Figure 5: X-Tracer vs. Conventional MISR-Based Compressor.

compression ratio to be half of the conventional MISR-based compressor without redundancy when the compression window is the same, for fair comparison, as an example, this means when the compression ratio is 50, we compress 50 cycles of "raw" trace data into a signature with conventional compressor while we compress 100 cycles of data with X-tracer.

4.2 Results and Discussion

We first compare the extracted useful trace data with the proposed X-tracer against conventional MISR-based compressor without redundancy, as depicted in Fig. 5. Here, Y-axis represents the percentage of extracted information bits out of the useful information bits in the "raw" trace data (denoted as "Extracted Useful Information"), and we report the ratio from data obtained with X-tracer and the proposed trace information extraction algorithm (denoted as "X-tracer+Extraction"), the ratio from data obtained with conventional MISR-based compressor and the proposed extraction algorithm (denoted as "Conv.+Extraction") and the ratio from data obtained conventional MISR-based compressor and extracted by simply counting those signatures that are not contaminated (denoted as "Orig.").

The gap between "Conv.+Extraction" curve and "Orig." curve in Fig. 5 clearly shows the effectiveness of the proposed extraction algorithm, which does not involve any design overhead. In addition, as can be observed from the figure, in most cases, when the X-bit ratio is higher, the percentage of the extracted useful trace information gets smaller. We attribute the few exceptional cases to the fact that X-bits are injected randomly and hence it is possible to have some outliers.

As shown in Fig. 5(a), when the compression ratio is 50 and the X-bit ratio in the "raw" trace data is low (smaller than 0.25%), almost 100% useful information bits can be obtained by all the three solutions. With the increase of X-bit ratio and when it ranges between 0.25% and 2.0%, the proposed X-tracer design significantly outperforms conventional compressor, which proves the effectiveness of the proposed solution with the help of redundancy. When the X-bit ratio is even higher, none of the solutions can extract more than 50% of the useful trace information. From the results, under these circumstances, it can be seen that conventional compressor with the

proposed extraction algorithm lead to more extracted trace information than X-tracer. This is due to the fact that the proposed X-tracer needs to compress twice cycles of the data when compared to the conventional compressor in order to keep the same compression ratio, the number of X-bits to be dealt with X-tracer in each compression window is thus also roughly twice. Consequently, when the X-bit ratio is very high, it is a lot harder to extract useful trace information with X-tracer.

When the compression ratio is 500, as shown in Fig. 5(b), most of the above conclusions hold true (for less X-bit ratio). We can observe the significant benefits provided with X-tracer when X-bit ratio is low. It is important to note that, what designers are more concerned about is the lost information during silicon debug in order not to miss the bug's erroneous effects. Hence, the proposed design, being able to recover more trace information, is highly desirable.

It is important to emphasize that, we usually would not try to debug a circuit under the situation when the extracted useful information is very limited (e.g., for the case when X-bit ratio is larger than 2.0% with compression ratio of 50). Instead, we would rather reduce the compression ratio and rely on more debug runs to have better visibility for the circuit under debug. Our next experiment is thus conducted to evaluate the impact of compression ratio on extracted trace information, as shown in Table 1. Generally speaking, when the compression ratio is low, X-bits will contaminate less useful information, and hence more information bits can be extracted by our algorithm for both structures. When X-bit ratio is relatively high, X-tracer can easily achieve high information bit ratio by slightly decreasing compression ratio. For example, when X-bit ratio is 0.2%, the extracted useful information grows from 41.18 to 86.89 with the drop of 20% compression ratio (i.e., from C.R.=500 to C.R.=400). Even for the case with X-bit ratio to be 0.5%, our solution is able to collect about 97.54% useful information data with $100\times$ compression ratio, which can be very helpful for silicon debug. Meanwhile, reducing compression ratio with conventional compressor will also increase the extracted useful information, but the improvement is not as significant as that of X-tracer.

With repeatable errors, we can rely on more debug runs to incrementally obtain more useful information. By collecting all the data from multiple debug runs, we can again rely on our trace extraction algorithm to acquire more useful trace information. Fig. 6 describes the result when the compression ratio is fixed as 500 and we conduct 1, 2 and 3 debug runs to collect trace data. To be specific, we configure our X-tracer with different primitive polynomials and different input orders in each debug run. From the results, it is easy to observe that when X-bit ratio is not too high (i.e., less than 0.4%), our solution can achieve very high information ratio (more than 90%) by analyzing the combined trace data from just 3 debug runs. Even for the case with 0.45% and 0.5% X-bit ratios in "raw" trace data, we can extract useful information from barely 0% with one debug run to 73.8% and 57.3%, respectively, when two more debug runs are conducted. The above results prove the effectiveness of the reconfigurable trace compressor design.

The computational time of the proposed extraction algorithm is acceptable. It only takes up to hundreds of seconds to analyze the traced signatures with X-tracer. On the other hand, the extra hardware overhead of the proposed X-tracer design is negligible, because the extra MISR and some control circuitries are much smaller when compared to the trace buffer, which occupies most of the hardware cost in trace-based debug solution.

5. CONCLUSION

In this paper, we propose a reconfigurable trace data compressor design with explicit redundancy for silicon debug, in the presence of many X-bits during signal tracing. The proposed *X-tracer* design, together with the novel algorithms to extract useful trace data out of contaminated trace signatures, facilitates to obtain as much trace information as possible while guaranteeing high compression ratio, as demonstrated in our experimental results.

X-Bit Ratio (%)	Extracted Useful Information (%)									
	C.R.=500		C.R.=400		C.R.=300		C.R.=200		C.R.=100	
	Conv.	X-tracer	Conv.	X-tracer	Conv.	X-tracer	Conv.	X-tracer	Conv.	X-tracer
0.05	80.00	96.06	88.33	98.45	88.33	98.66	100.00	99.39	100.00	100.00
0.1	52.51	92.11	66.66	92.71	88.33	97.92	100.00	98.31	100.00	98.79
0.15	49.15	86.70	54.14	95.04	70.83	97.48	75.03	97.51	84.94	98.32
0.20	40.82	41.18	44.98	86.89	65.00	96.49	69.13	95.75	91.59	98.98
0.25	24.15	15.15	36.63	45.44	50.00	89.03	61.65	96.47	80.80	97.81
0.30	15.83	5.00	36.66	7.20	45.83	65.36	59.98	95.97	78.26	97.48
0.35	11.68	4.17	18.37	8.32	32.50	26.59	52.50	93.08	79.99	93.93
0.40	6.68	0.00	20.00	0.00	27.50	14.98	50.82	88.65	74.14	96.48
0.45	4.17	0.00	17.48	0.00	26.67	8.33	44.16	73.35	72.40	98.34
0.50	6.68	0.00	14.99	0.00	13.33	1.67	34.94	32.01	61.58	97.54

C.R.: Compression ratio;

Conv.: Useful information obtained with conventional MISR and the proposed extraction algorithm;

X-tracer: Useful information obtained with X-tracer and the proposed extraction algorithm;

Table 1: Experimental Results with Different Compression Ratio.

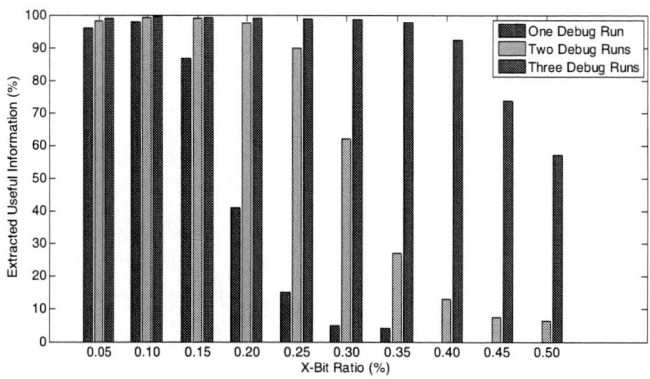

Figure 6: Experimental Results with Reconfigurable Compressors under Different Number of Debug Runs.

6. ACKNOWLEDGEMENTS

This work was supported in part by the General Research Fund CUHK418111 from Hong Kong SAR Research Grants Council (RGC), by National Science Foundation of China (NSFC) under grant No. 60901052, and in part by RGC Grant Direct Allocation under grant No. 2050488.

7. REFERENCES

[1] M. Abramovici. In-System Silicon Validation and Debug. *IEEE Design & Test of Computers*, 25(3):216–223, May-June 2008.

[2] Semiconductor Industry Association (SIA). *The International Technology Roadmap for Semiconductors (ITRS): 2003 Edition.* http://public.itrs.net/Files/2003ITRS/Home2003.htm, 2003.

[3] X. Liu and Q. Xu, "On signal tracing in post-silicon validation," in *Proceedings IEEE/ACM Asia and South Pacific Design Automation Conference (ASP-DAC)*, pp. 262–267, 2010.

[4] ARM Ltd. How CoreSight Technology Gets Higher Performance, More Reliable Product to Market Quicker. http://www.arm.com.

[5] R. Leatherman and N. Stollon. An Embedded Debugging Architecture for SOCs. *IEEE Potentials*, Feb.-Mar. 2005.

[6] X. Liu and Q. Xu. Interconnection fabric design for tracing signals in post-silicon validation. In *Proc. ACM/IEEE Design Automation Conference (DAC)*, pp. 352–357, 2009.

[7] R. H. Livengood and D. Medeiros. Design for (Physical) Debug for Silicon Microsurgery and Probing of Flip-Chip Packaged Integrated Circuits. In *Proc. IEEE International Test Conference (ITC)*, pp. 877–882, 2007.

[8] K. H. Chang, I. L. Markov, and V. Bertacco. Fixing Design Errors with Counterexamples and Resynthesis. In *Proc. IEEE Asia South Pacific Design Automation Conference (ASP-DAC)*, pp. 944–949, 2007.

[9] S. Tang and Q. Xu, "A multi-core debug platform for NoC-based systems," in *Proceedings IEEE/ACM Design, Automation, and Test in Europe (DATE)*, 2008, pp. 870–875.

[10] H. F. Ko and N. Nicolici. Automated Trace Signals Identification and State Restoration for Improving Observability in Post-Silicon Validation. In *Proc. Design, Automation, and Test in Europe (DATE)*, pp. 1298–1303, 2008.

[11] X. Liu and Q. Xu. Trace signal selection for visibility enhancement in post-silicon validation. In *Proceedings Design, Automation, and Test in Europe (DATE)*, pp. 1338–1343, 2009.

[12] J.-S. Yang and N. A. Touba. Automated Selection of Signals to Observe for Efficient Silicon Debug. In *Proc. IEEE VLSI Test Symposium (VTS)*, pp. 79–84, 2009.

[13] X. Liu and Q. Xu, "On Multiplexed Signal Tracing for Post-Silicon Debug," in *Proceedings IEEE/ACM Design, Automation, and Test in Europe (DATE)*, pp. 1–6, 2011.

[14] S. B. Park and S. Mitra. IFRA: Instruction footprint recording and analysis for post-silicon bug localization in processors. In *Proc. ACM/IEEE Design Automation Conference (DAC)*, pp. 373–378, 2008.

[15] C.H. Lai, et al. A trace-capable instruction cache for cost efficient real-time program trace compression in SoC. In *Proc. ACM/IEEE Design Automation Conference (DAC)*, pp. 136 –141, 2009.

[16] A. Vishnoi, P.R. Panda, and M. Balakrishnan. Cache Aware Compression for Processor Debug Suppport. In *Proc. Design, Automation, and Test in Europe (DATE)*, 2009.

[17] E. Anis and N. Nicolici. On Using Lossless Compresstion of Debug Data in Embedded Logic Analysis. In *Proc. IEEE International Test Conference (ITC)*, pp. 1–10, October 2007.

[18] S. Prabhakar, R. Sethuram, and M.S. Hsiao. Trace Buffer-Based Silicon Debug with Lossless Compression. In *Proc. International Conference on VLSI Design*, pp. 358–363, 2011.

[19] J.-S. Yang and N. A. Touba. Enhancing Silicon Debug via Periodic Monitoring. In *Proc. IEEE International Symposium on Defect and Fault Tolerance in VLSI Systems (DFT)*, pp. 125–133, 2008.

[20] E. Anis and N. Nicolici. Low cost debug architecture using lossy compression for silicon debug. In *Proc. Design, Automation, and Test in Europe (DATE)*, pp. 225–230, 2007.

[21] J.S. Yang and N.A. Touba. Expanding Trace Buffer Observation Window for In-System Silicon Debug through Selective Capture. In *Proc. IEEE VLSI Test Symposium (VTS)*, pp. 345–351, 2008.

[22] M. C.-T. Chao, et al. Response Shaper: A Novel Technique to Enhance Unknown Tolerance for Output Response Compaction. In *Proc. International Conference on Computer-Aided Design (ICCAD)*, pp. 80–87, 2005.

[23] S. Mitra, M. Mitzenmacher, S. S. Lumetta, and N. Patil. X-Tolerant Test Response Compaction. *IEEE Design & Test of Computers*, 22(6):566–574, 2005.

[24] J.S. Yang, N.A. Touba, S.Y. Yang, and T.M. Mak. Industrial Case Study for X-Canceling MISR. In *Proc. IEEE International Test Conference (ITC)*, 2009.

[25] N. A. Touba. X-Canceling MISR-An X-Tolerant Methodology for Compacting Output Responses with Unknowns Using a MISR. In *Proc. IEEE International Test Conference (ITC)*, pp. 1–10, 2007.

Quick Detection of Difficult Bugs for Effective Post-Silicon Validation

David Lin[1], Ted Hong[1], Farzan Fallah[1], Nagib Hakim[3], Subhasish Mitra[1,2]

[1]Department of EE and [2]Department of CS
Stanford University, Stanford, CA, USA

[3]Intel Corporation
Santa Clara, CA, USA

Abstract

We present a new technique for systematically creating post-silicon validation tests that quickly detect bugs in processor cores and uncore components (cache controllers, memory controllers, on-chip networks) of multi-core System on Chips (SoCs). Such quick detection is essential because long error detection latency, the time elapsed between the occurrence of an error due to a bug and its manifestation as an observable failure, severely limits the effectiveness of existing post-silicon validation approaches. In addition, we provide a list of realistic bug scenarios abstracted from "difficult" bugs that occurred in commercial multi-core SoCs. Our results for an OpenSPARC T2-like multi-core SoC demonstrate: 1. Error detection latencies of "typical" post-silicon validation tests can be very long, up to billions of clock cycles, especially for bugs in uncore components. 2. Our new technique shortens error detection latencies by several orders of magnitude to only a few hundred cycles for most bug scenarios. 3. Our new technique enables 2-fold increase in bug coverage. An important feature of our technique is its software-only implementation without any hardware modification. Hence, it is readily applicable to existing designs.

Categories and Subject Descriptors

B.6.2 Reliability and Testing – Error-checking, Test generation, B.6.3 Design Aids – Verification, B.7.2 Design Aids – Verification, B.7.3 Reliability and Testing – Test generation, B.8.1 Reliability, Testing, and Fault-Tolerance.

General Terms

Reliability, Verification.

Keywords

Debug, Post-Silicon Validation, Quick Error Detection, Verification

1. Introduction

The purpose of post-silicon validation is to test manufactured ICs in actual systems to detect and fix design flaws (bugs). Traditional pre-silicon verification alone is no longer adequate because it is too slow and is often incapable of detecting electrical bugs. *Electrical bugs* manifest themselves only under specific operating conditions, such as voltage, frequency, and/or temperature corners [Patra 07]. Existing post-silicon validation approaches are *ad hoc*, and their costs are rising [Keshava 10]. Massive integration of a wide variety of components in complex System-on-Chips (*SoCs*) consisting of processor cores, accelerators, adaptive power / thermal / reliability control, and *uncore components* such as cache / memory / network controllers significantly exacerbate post-silicon validation challenges [Adir 11, Mitra 10, Singerman 11].

Post-silicon validation involves three activities: detecting a problem by applying proper stimuli, localizing the problem to a small region inside the chip, and fixing the problem through software patches, circuit editing, or silicon re-spin. The effort to localize the problem from an observed failure often dominates the cost of post-silicon validation [Josephson 06]. Long *error detection latency*, the time elapsed between the occurrence of an error due to a bug and its manifestation as an observable failure, limits the effectiveness of existing bug localization techniques. Bugs with error detection latencies longer than a few thousand clock cycles are highly challenging because it is extremely difficult to trace too far back in history for bug localization.

This paper makes the following contributions:

1. We present a list of bug scenarios obtained by analyzing "difficult" bugs (from proprietary bug databases) that occurred in latest commercial multi-core SoCs. Researchers can use these bug scenarios to quantify the benefits and drawbacks of existing and new validation techniques. Such quantification is essential in advancing the field of design validation.

2. We demonstrate that bugs in uncore components of SoCs can result in very long error detection latencies of several millions to billions of clock cycles unless special attention is paid to shorten these long error detection latencies. We also demonstrate that "typical" post-silicon validation tests with "end result checks" (that check results upon test completion) or "typical" self-checking tests are inadequate in shortening these long error detection latencies.

3. We present a new Proactive Load and Check (*PLC*) technique for **systematically** creating post-silicon validation tests with very short error detection latencies to quickly detect bugs in both uncore components and processor cores of multi-core SoCs. Simulation results of our PLC tests for a complex OpenSPARC T2-like multi-core SoC [OpenSPARC] demonstrate:

(a) Several orders of magnitude improvement in error detection latencies for bugs inside processor cores and uncore components. The error detection latencies of PLC tests are within a few hundred cycles for most bug scenarios we simulated.

(b) 2-fold improvement in the coverage of bug scenarios.

Our PLC tests can be implemented entirely in software with no hardware modifications. Hence, they can be readily applied to existing designs.

Section 2 presents a comprehensive list of bug scenarios. Section 3 describes our new technique for systematically creating post-silicon validation tests. In Sec. 4, we present simulation results to demonstrate the effectiveness of our technique. Related work is discussed in Sec. 5, followed by conclusion.

2. Bug Scenarios

A comprehensive list of bug scenarios is critical for understanding the limitations of existing validation techniques and for evaluating new approaches. Toward this end, we compiled a list of realistic bug scenarios by analyzing reports of "difficult" bugs (primarily "logic" bugs) detected during validation of actual multi-core SoCs. These bug scenarios are considered "difficult" because of very long debug times as indicated in bug reports. (A more precise definition of "difficulty" is desirable).

We performed extensive analysis of various bug reports, and worked closely with validation teams to abstract these bugs into bug scenarios using higher-level descriptions while removing product-specific details. As a result, several actual bugs are abstracted into a single bug scenario. Each bug scenario is decomposed into a bug activation criterion (Table 1a) and a bug effect (Table 1b).

Bug activation criterion is the condition that must be satisfied to activate a bug. In Table 1a, criteria 1-4 correspond to cache controller bugs, criterion 5 corresponds to bugs inside cache / memory controllers and on-chip networks, and criteria 6-8 correspond to processor core bugs. Since we abstracted the activation criteria from "informal" bug reports, it is possible that all activation conditions might not have been completely captured.

Bug effect is defined as the incorrect behavior resulting from bug activation. In Table 1b, effects A-E correspond to cache controller bugs, effect F corresponds to memory controller bugs, effect G corresponds to interconnection network bugs, and effects H-J correspond to bugs inside processor cores.

Tables 1a and 1b allow us to implement each bug scenario using various micro-architectural and RTL simulators (details in Sec. 4). One can create families of bug scenarios by adjusting integer parameters X and Y in Tables 1a and 1b. For example, pairing bug activation criterion 2, for $X=10$, with bug effect A produces the following bug scenario:

> Two stores in 10 cycles to the same cache line cause cache coherence message for that line to be dropped.

Table 1a. Bug activation criteria.

Uncore components	1. Two stores in X clock cycles to different cache lines.
	2. Two stores in X clock cycles to the same cache line.
	3. Two stores in X clock cycles to adjacent cache lines.
	4. Two cache misses in X clock cycles.
	5. A sequence of load and/or store operations in X clock cycles.
Processor core	6. Data forwarding between pipeline stages.
	7. Two branch instructions in X clock cycles.
Other	8. Any clock cycle chosen at random.

Table 1b. Bug effects.

Uncore components	A. Next received cache* coherence message dropped.
	B. Next received cache* coherence message delayed.
	C. Next store operation not allocated a cache* line.
	D. Next store update to cache* delayed by Y clock cycles.
	E. Next data accessed from cache* corrupted.
	F. Next data coming from main memory to cache* / core* corrupted.
	G. Processor core's* load value corrupted.
Processor core	H. Core* jumps to incorrect (random) address in the next cycle.
	I. Error in decoding next instruction's operand inside core*.
	J. Processor core* incorrectly decodes next instruction to a NOP instruction.

* where activation criterion is satisfied.

While there is on-going work on understanding electrical bug behaviors [Gao 11, Hong 10, McLaughlin 09], there exists little consensus on what constitutes accurate logic bug models [ITRS 09]. Earlier researchers analyzed logic bugs using research chips, class projects, and published errata pages [Constantinides 08, DeOrio 08, 09, Ho 95, Van Campenhout 00, Velev 03]. Bug scenarios resulting from Tables 1a and 1b subsume bugs in [DeOrio 08, 09, Ho 95, Velev 03]. It is difficult to compare bugs in [Constantinides 08, Van Campenhout 00] vs. our bug scenarios. This is because

[Constantinides 08] provides RTL-specific bug examples, and [Van Campenhout 00] focuses on implementation-dependent root-cause analysis, e.g., missing inputs, incorrect signal sources.

3. Post-Silicon Validation Tests Targeting Uncore Components

Traditional post-silicon validation tests are inadequate for shortening error detection latencies, especially for bugs in uncore components of complex SoCs. Consider the example in Fig. 1a. When Core 1 stores 1 to memory location A (*mem*[A]), a bug in an uncore component, e.g., cache or memory controller, can result in data corruption. The value stored in *mem*[A] itself may be corrupted, or a local (cached) copy of *mem*[A] for Core 2 may not be properly updated. After a very long time, Core 2 uses the corrupt value from *mem*[A] and produces incorrect outputs. Tests with "end result checks" that check for expected output values upon test completion result in very long error detection latencies (Fig. 1a).

Following the above example, the corrupt value read by Core 2 can result in a livelock / deadlock (Fig. 1b). Techniques for detecting livelocks / deadlocks [Bayazit 05, Chandy 83] are inadequate because it is already too late when the livelock / deadlock happened (long after *mem*[A] got corrupted).

Self-checking tests [Aharon 95, Raina 98, Wagner 08], including QED tests that shorten error detection latencies for bugs inside processor cores, are not sufficient either. As shown in Fig. 1c, it is already too late when checking occurs in Core 2. Assertions for post-silicon validation also have similar challenges (more details in Sec. 5). A self-checking test variant, referred to as *store readback* test (Fig. 1d), performs a load operation on the same core that performed the store operation (Core 1 in this case) to check if the loaded value matches the stored value. However, a bug which does not correctly update a local (cached) copy of *mem*[A] in Core 2 may not be detected by this check.

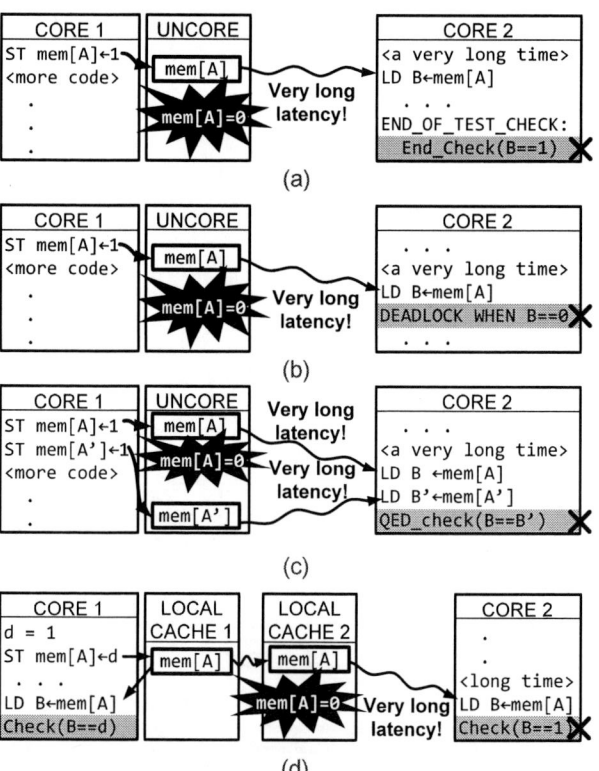

Figure 1. (a) End result check. (b) Deadlock detection. (c) QED test for processor cores. (d) Store readback test.

Checkpoint-based tests, that periodically checkpoint a system and compare checkpointed values with expected values (obtained from simulations), are also insufficient. System-level checkpointing [Silas 03] of processor registers and memory states is complicated for complex many-core SoCs. Hence, frequent system-level checkpointing may not be feasible. To reduce complexity, only processor registers may be locally checkpointed. In that case, our bug example in Fig. 1 will have a long error detection latency (detected only after Core 2 uses *mem*[A]).

3.1. Proactive Load and Check (PLC) Transformation

We overcome the above error detection latency challenges by statically transforming existing validation tests into new tests (during compile time) using a special **Proactive Load and Check (*PLC*) transformation**. Unlike QED tests targeting processor cores [Hong 10], the PLC transformation does not solely rely on (identical or diverse) re-execution of instructions in the original test. Instead, it inserts special PLC operations at very fine granularities across memory (and I/O) address spaces using targeted instructions. The idea is to "strategically" perform loads on all threads executing on all processor cores from selected variables and to insert self-consistency checks on those variables. The PLC transformation is compatible with QED. Hence, bugs inside processor cores can also be detected with short error detection latencies.

In this paper, PLC is implemented as a software-only technique. Hence, it does not require any hardware modification and can readily fit into existing post-silicon validation flows. However, PLC can be further enhanced with hardware support, a complete description of which is beyond the scope of this paper.

Next, we provide a detailed overview of the steps required to implement PLC for an OpenSPARC T2-like SoC.

Step 1: Initialization

We first transform an existing validation test into a QED test, e.g., using the EDDI-V transformation [Hong 10] (Fig. 2). EDDI-V bounds error detection latencies for bugs inside processor cores by strategically inserting duplicated instructions and checks in the original test. EDDI-V uses different (but identically initialized) registers, memory, and variables for the duplicated instructions. The results of the duplicated instructions are compared with the original ones, and an error is indicated upon mismatch.

A global array variable called *PLC_List* is created. Each *PLC_List* entry consists of the tuple <*original variable pointer, EDDI-V variable pointer*>. The *original variable pointer* points to a variable selected for PLC. Variable selection strategies for PLC are discussed later. The *EDDI-V variable pointer* points to the corresponding duplicated variable created by EDDI-V (e.g., <&a, &a'> and <&b, &b'> in Fig. 2). For dynamically allocated variables, the pointer values are determined during runtime by the memory allocator. For statically allocated variables, the pointers are known *a priori* via source code labels.

All variables listed in the *PLC_List* must be protected against race conditions between stores to these variables by one thread and PLC operations (details later) by another thread. Such race conditions can occur due to unexpected interleaving of the four operations: store to an original variable in the *PLC_List*, store to the corresponding EDDI-V variable, load from an original variable in the *PLC_List*, and load from the corresponding EDDI-V variable. We achieved this by locking the original variable and the corresponding EDDI-V variable pair during store and load operations to the variable pair (Fig. 2). More efficient techniques to protect against race conditions exist using architectural support (e.g., store double word) or software-only techniques [Larsson 04]. The details of such techniques are beyond the scope of this paper.

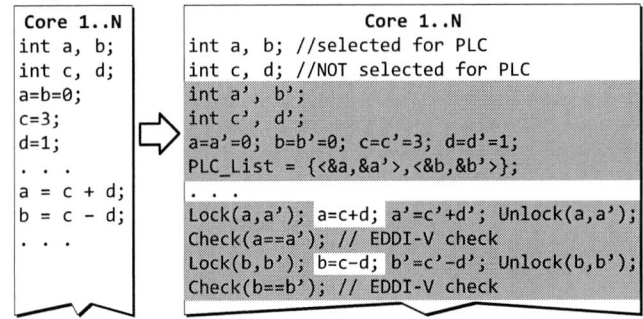

Figure 2. PLC transformation Step 1: initialization.

Step 2: PLC Operation Insertion

After initialization, the PLC transformation inserts proactive load and check operations in each thread in each processor core. Each *Proactive Load and Check (PLC) operation* (Fig. 3) performs the following function. It iterates once over all tuples in *PLC_List*. For each tuple in the list, it locks the tuple and then loads the values from both the *original variable pointer* and the *EDDI-V variable pointer*. After the values are loaded, it unlocks the tuple. Next, the two loaded values are compared with each other. Under bug-free situations, the two values should **exactly** match. An error is indicated upon mismatch. Upon error detection, the system can be halted for bug localization and debug using a variety of existing techniques. Bugs affecting the *PLC_List* itself can be quickly detected because it is highly unlikely that the corrupted addresses corresponding to the *original variable pointer* and the *EDDI-V variable pointer* fields will contain the same data values.

PLC operations are inserted at periodic intervals in **each thread in each processor core**. This is necessary because bugs can affect various pathways between processor cores and uncore components. Furthermore, PLC operations check all variables in the *PLC_List* to cover situations in which some arbitrary variable (not necessarily a recently modified or used variable) is affected by a bug. For example, a store / load operation to one variable can trigger an electrical bug that creates an error in another variable.

Intuitively, highly frequent PLC operations are preferred for quick error detection. However, excessive PLC operations may result in excessive *intrusiveness*, i.e., the deviation in the execution behavior of the transformed test compared to the original test, which can adversely impact coverage of the transformed test. Our results in Sec. 4 do not show any such coverage degradation.

To minimize possible intrusiveness due to PLC operations, we introduce a *PLC_inst_min* parameter, defined as the minimum number of instructions in the same thread that must execute before a PLC operation is inserted. Similar to [Hong 10], code transformations, such as loop unrolling and loop splitting, may be required to satisfy the *PLC_inst_min* constraint.

Variable Selection Strategies for PLC

There can be several strategies to select variables to be included in the *PLC_List*:

1. One can include all variables in a given test. However, the resulting error detection latencies can be long, especially when PLC is implemented in software. This is because PLC operations must iterate through all variables, which can take a long time.

2. One can create a **family** of tests from a given test, where each test in the family selects a subset of variables (in the given test) for its *PLC_List*. The resulting tests are referred to as *PLC family tests*. There is a possibility that some bugs may not be detected quickly with short error detection latencies when PLC family tests are used. This is because a bug may or may not be activated by each

563

individual test in the family. Our results in Sec. 4 demonstrate that PLC family tests are highly effective in significantly shortening error detection latencies for the bug scenarios in Sec. 2.

3. If the inputs to a test are known *a priori* (which is often the case for validation tests [Bentley 01]), one can perform profiling to determine variables with store-to-load or load-to-load latencies longer than the desired error detection latency bounds, and include only those variables in the *PLC_List*.

PLC operation insertion

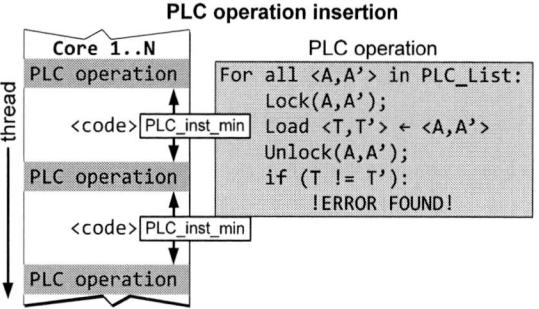

Figure 3. PLC transformation Step 2: PLC operation insertion.

4. Results

We evaluated the effectiveness of our PLC transformation technique by inserting the bug scenarios in Sec. 2 into a micro-architectural simulator. We used Multifacet's General Execution-driven Multiprocessor Simulator (GEMS) [Martin 05] to simulate an OpenSPARC T2-like SoC [OpenSPARC], a 500 million-transistor design with 8 processor cores, 64 threads, private split L1 data and instruction caches, crossbar-based interconnects, 8-way banked L2 cache using directory-based cache coherence protocol, and 4 memory controllers (Fig. 4).

Memory Controller		Memory Controller		Memory Controller		Memory Controller	
L2$	L2$	L2$	L2$	L2$	L2$	L2$	L2$
Crossbar							
L1$D	L1$D	L1$D	L1$D	L1$D	L1$D	L1$D	L1$D
L1$I	L1$I	L1$I	L1$I	L1$I	L1$I	L1$I	L1$I
core 0	core 0	core 0	core 0	core 0	core 0	core 0	core 0

Figure 4. OpenSPARC T2-like SoC [OpenSPARC].

We implemented the bug scenarios in the simulator in the following way: for a selected bug scenario, we modified the source code of the simulated system to include two additional routines: one that constantly monitors the simulated system for the activation criterion, and another that inserts the bug effect in the system (initially disabled). When the activation criterion is satisfied, the routine that inserts the bug effect in the system is enabled to insert a single instance of the bug effect. Since the activation criterion can be satisfied multiple times during a simulation run, the bug effect can be inserted multiple times as well (i.e., once for each time the activation criterion is satisfied), which is consistent with the behavior of (logic) bugs in actual systems.

We ran a separate simulation experiment for each of the 80 bug scenarios (cross product of 8 bug activation criteria and 10 bug effects in Tables 1a and 1b). For each experiment, only one bug scenario is inserted. For bug activation criteria and bug effect parameters, we set $X=10$ clock cycles and $Y=100$ clock cycles because these values gave us "close" representations (on our simulation platform) of the "difficult" bugs found in the bug databases. Our PLC transformation is independent of X and Y, and is not engineered based on the bug scenarios. For each experiment, the selected bug scenario affects any one of the cores, L2 caches, and memory controllers (i.e., no specific component was pre-selected).

The results are summarized in Fig. 5 for the following tests: FFT and LU programs from the SPLASH-2 benchmark suite [Woo 95] in Figs. 5a and 5b, respectively, and a proprietary industrial post-silicon validation test targeting memory bugs (details omitted for confidentiality) in Fig. 5c. The results correspond to 8-threaded (single thread per core) versions of the tests. 64-threaded versions of the tests were not used because of slow simulation speed.

In Figs. 5a-c, the *original* tests are instrumented with "end result checks" only. Since the inputs to these tests are known *a priori*, the expected end results can be calculated.

For *PLC family tests* in Figs. 5a and 5b, we split all the heap variables in the original test into 1,024 groups. This is done by aggregating a list of all heap variables (known *a priori* since the inputs to the tests are known) and dividing the list into 1,024 groups. This results in 1,024 individual tests in the PLC family, each performing PLC operations for only the variables in its corresponding group. Each test in the family is run one after another. For each bug scenario, the error detection latency reported in Fig. 5a (5b) corresponds to the error detection latency when the bug scenario was detected for the first time by the PLC family tests.

For the industrial test, there is a single PLC family test consisting of all variables allocated on the memory. Register variables are not included for PLC because errors in these variables can be quickly detected by EDDI-V alone. Since this test targets memory bugs only, Fig. 5c considers memory bug scenarios (i.e., activation criteria 1-5 in Table 1a and effects A-G in Table 1b).

We performed PLC transformations at the C source code level with *PLC_inst_min* to be approximately 10 lines of code. Figures 5a-c also report results for *QED* tests obtained by transforming the original tests using the QED EDDI-V transformation only.

Table 2 shows the relative runtimes of various tests (normalized to the corresponding original test). The runtimes of PLC family tests are significantly longer than the original tests. That raises the question: Are the significant benefits of PLC tests due to their PLC operations or due to their longer runtimes? To answer this question, we performed a controlled set of experiments. For an original test, we created a version of the test whose runtime is approximately the same as that of the PLC family tests. This is done by keeping most of the PLC family tests intact but removing the error reporting code when an EDDI-V or PLC check detects an error. These tests are referred to as *OERT* tests (Original Equivalent RunTime tests). An OERT was not created for the industrial test because the runtime of the original test is already very long.

The following observations can be made from the results.

Observation 1: Post-silicon validation tests created using our PLC transformation technique shorten error detection latencies by several orders of magnitude for all bug scenarios in Sec 2. Error detection latencies of original tests can be extremely long: several millions or billions of cycles. QED tests targeting bugs inside processor cores have long error detection latencies for uncore bugs. In contrast, PLC tests have error detection latencies of only a few hundred cycles for most bug scenarios. These benefits come from the PLC operations and not from the longer runtimes of PLC tests.

Observation 2: Post-silicon validation tests created using our PLC transformation technique continue to detect bug scenarios detected by the original tests. There is not a single bug scenario that the original tests (or the QED tests or the OERT tests) detected but the PLC tests didn't. Note that, the activation criteria in Table 1a are fairly rare with only tens of activations per ten million cycles.

Observation 3: Post-silicon validation tests created using our PLC transformation technique detect up to 2-fold more bug scenarios that would otherwise remain undetected by the original tests, the QED tests, or the OERT tests. Our simulation experiments confirmed that each bug scenario was activated at least once by the original tests, the QED tests, and the OERT tests.

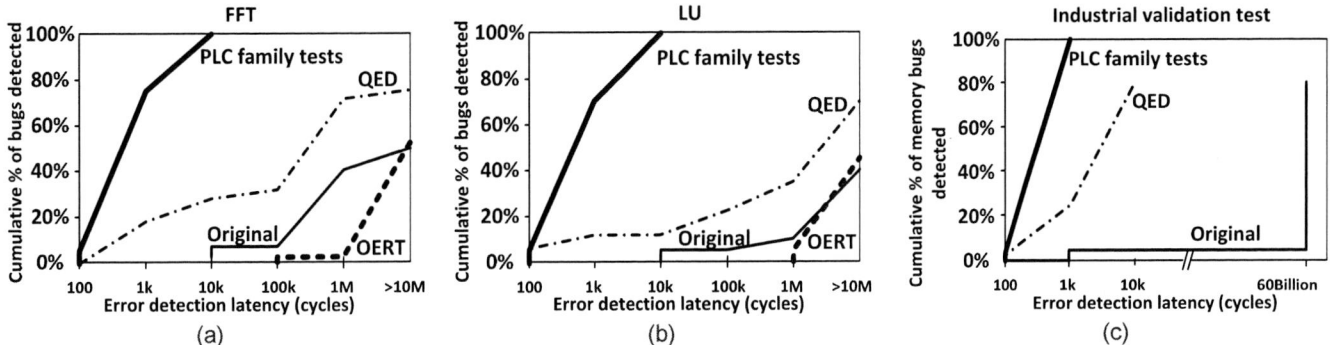

Figure 5. Error detection latencies and coverage of post-silicon validation tests. (a) FFT. (b) LU. (c) Industrial validation test.

Table 2. Runtimes normalized to corresponding original tests.

	FFT	LU	Industrial validation test
Original	1	1	1
QED	≈ 2	≈ 3	≈ 3
PLC family tests	≈ 1,024 x 28	≈ 1,024 x 32	≈ 35
OERT	≈ 1,024 x 28	≈ 1,024 x 32	NA

Since the PLC family tests have long runtimes, we experimented with an alternative variable selection technique for PLC. This variable selection technique first performs test code profiling to determine variables with long store-to-load latencies (since test inputs are known *a priori*). Each PLC operation only checks the most recently stored variables with long store-to-load latencies. Note that, unlike the store readback test in Fig. 1d, all threads executing on all cores (not just the thread which performed the store) perform these PLC operations. Figure 6 compares error detection latency and coverage of this PLC variable selection technique vs. store readback for the FFT test. As expected, the PLC test achieves better error detection latency and higher coverage compared to the store readback test. The runtime of this PLC test is 25 times longer than the original test (which is significantly shorter than the PLC family tests in Fig. 5).

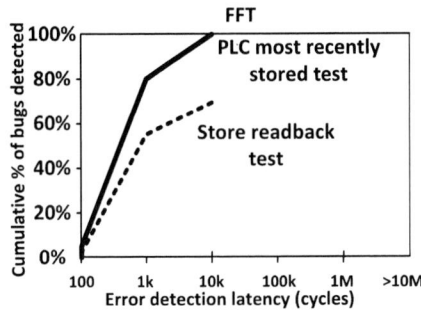

Figure 6. PLC test using most recently stored variables with long stored-to-load latencies vs. store readback.

5. Related Work

Existing work related to this paper can be classified into: post-silicon validation stimuli generation, memory scrubbing for fault-tolerant computing, and assertions for post-silicon validation.

Automatic Post-Silicon Validation Stimuli Generation

We demonstrated that test techniques that target bugs inside processor cores (e.g., [Aharon 95, Hong 10, Wagner 08]) alone are not sufficient for bugs in uncore components. Such tests can have

very long error detection latencies for uncore bugs. [Raina 98] presented a technique for automatically generating random self-tests for microprocessor caches only, without bounds on error detection latencies. Our new technique ensures bounded error detection latencies for bugs in a variety of uncore components as well as processor cores. Furthermore, our technique is compatible with existing test generation techniques such as those using constraint satisfaction [Katz 12].

Memory Scrubbing for Fault-Tolerant Computing

Memory scrubbing [Abraham 83, Shirvani 00] is used in fault-tolerant computing to detect and correct errors inside memory arrays. The PLC transformation technique is inspired by the concept of memory scrubbing. However, there are significant differences:

1. Scrubbing generally uses error-correcting codes to target errors inside memory arrays, but may not detect errors due to bugs outside memory arrays such as cache or memory controllers.

2. Memory scrubbing generally occurs infrequently compared to PLC operations in post-silicon validation.

3. In post-silicon validation, reducing error detection latency is very important because debug time, rather than test execution time, often dominates overall post-silicon validation costs [Josephson 06]. Therefore, some test execution time penalties can be tolerated if error detection latencies can be significantly improved.

4. For post-silicon validation tests, test program inputs may be known *a priori* [Bentley 01]. This allows special transformations, e.g., through profiling (e.g., Fig. 6), to improve both error detection latencies and coverage while reducing test execution time penalties.

5. Failure containment and recovery are not primary concerns during validation.

Assertions for Post-Silicon Validation

The generation and use of assertions for post-silicon validation [Boule 07] are non-trivial. As noted in [Bentley 01], assertions have to be carefully crafted, and it is difficult to keep them up-to-date and to validate their correctness. While there are methods for automated assertion generation [Ernst 07, Hangal 05], these techniques are not widely applicable to all SoC components. In contrast, our approach allows a **structured** way of performing extensive checks, and can be automatically generated and validated.

6. Conclusion

Our new PLC technique for **systematically** creating post-silicon validation tests is highly effective in quickly detecting difficult bug scenarios inside uncore components as well as processor cores in multi-core SoCs. Our results demonstrate several orders of magnitude improvement in error detection latencies and 2-fold improvement in coverage simultaneously. Such short error detection latencies can significantly improve the productivity of post-silicon validation. Moreover, our technique is readily applicable to existing post-silicon validation flows because it doesn't require any hardware changes. The list of bug scenarios, derived from difficult bugs that occurred in commercial multi-core SoCs, can act as excellent benchmarks to advance validation research.

Several opportunities exist to further enhance validation methodologies using our approach. Examples include: 1. Bug localization by analyzing the checks in our PLC tests that detected (or did not detect) errors. 2. Hardware support for PLC tests to reduce their runtimes. 3. Continued collection and analysis of difficult bugs in actual SoCs for better bug benchmarks. 4. Extending our approach to system verification using emulators, hardware accelerators, and prototyping and virtual platforms.

7. Acknowledgment

This research was supported in part by FCRP, GSRC, NSF, SRC, and Stanford Graduate Fellowship. The authors thank Eswaran S. and Sharad Kumar of Freescale, Jagannath Keshava and Sandip Ray of Intel, and Christine Cheng and Subhasis Das of Stanford University for their valuable inputs and support.

8. References

[Abraham 83] Abraham, J. A., E. S. Davidson, and J. H. Patel, "Memory System Design for Tolerating Single Event Upsets," *IEEE Trans. Nuclear Science*, Vol. 30, Issue 6, pp. 4339-4344, December, 1983.

[Adir 11] Adir, A., *et al.*, "A Unified Methodology for Pre-Silicon Verification and Post-silicon Validation," *Proc. IEEE/ACM Design, Automation and Test in Europe Conf.*, pp. 1-6, 2011.

[Aharon 95] Aharon, A., *et al.*, "Test Program Generation for Functional Verification of PowerPC Processors in IBM," *Proc. IEEE/ACM Design Automation Conf.*, pp 279-285,1995.

[Bayazit 05] Bayazit, A. A., and S. Malik, "Complementary Use of Runtime Validation ad Model Checking," *Proc. IEEE/ACM Intl. Conf. Computer-aided Design*, pp. 1049-1056, 2005.

[Bentley 01] Bentley, B., and R. Gray, "Validating the Intel Pentium 4 Processor." *Intel Technology Journal*, Vol. 5 Issue 1, pp. 1-8, February, 2001.

[Boule 07] Boule, M., J.-S. Chenard, and Z. Zilic, "Assertion Checkers in Verification, Silicon Debug and In-Field Diagnosis," *Proc. IEEE Intl. Symp. Quality Electronic Design*, pp. 613-620, 2007.

[Chandy 83] Chandy, K. M., J. Misra, and L. M. Haas, "Distributed Deadlock Detection," *ACM Trans. Computer Systems*, Vol. 1, Issue 2, pp 144-156, May 1983.

[Constantinides 08] Constantinides, K., O. Mutlu, and T. Austin, "Online Design Bug Detection: RTL Analysis, Flexible Mechanisms, and Evaluation," *Proc. IEEE/ACM Intl. Symp. Microarchitecture*, pp. 282-293, 2008.

[DeOrio 08] DeOrio, A., A. Bauserman, and V. Bertacco, "Post-Silicon Verification for Cache Coherence," *Proc. IEEE Intl. Conf. Computer Design*, pp.348-355, 2008.

[DeOrio 09] DeOrio, A., I. Wagner, and V. Bertacco, "DACOTA: Post-silicon Validation of the Memory Subsystem in Multi-Core Designs," *Proc. IEEE Intl. Symp. High-Performance Computer Architecture*, pp. 405-416. 2009.

[Ernst 07] Ernst, M. D., *et al.*, "The Daikon System for Dynamic Detection of Likely Invariants," *Science of Computer Programming*, Vol. 69, Issues 1-3, pp. 35-45, December, 2007.

[Gao 11] Gao, M., P. Lisherness, and K.-T. Cheng, "Post-silicon Bug Detection for Variation Induced Electrical Bugs" *Proc. IEEE/ACM Asia and South Pacific Design Automation Conf.*, pp 273-273, 2011.

[Hangal 05] Hangal S., *et al.*, "IODINE: A Tool to Automatically Infer Dynamic Invariants," *Proc. IEEE/ACM Design Automation Conf.*, pp. 775-778, 2005.

[Ho 95] Ho, R. C., *et al.* "Architecture Validation for Processors," *Proc. ACM/IEEE Intl. Symp. Computer Architecture*, pp. 404-413, 1995.

[Hong 10] Hong, T. *et al.*, "QED: Quick Error Detection Tests for Effective Post-Silicon Validation," *Proc. IEEE Intl. Test Conf.*, pp. 1-10, 2010.

[ITRS 09] http://www.itrs.net/Links/2009ITRS/Home2009.htm.

[Josephson 06] Josephson, D., "The Good, the Bad, and the Ugly of Silicon Debug," *Proc. IEEE/ACM Design Automation Conf.*, pp. 3-6, 2006.

[Katz 12] Katz, Y., M. Rimon, and A. Ziv, "Generating Instruction Streams Using Abstract CSP," *Proc. IEEE/ACM Design, Automation and Test in Europe Conf.*, pp. 15-20, 2012.

[Keshava 10] Keshava, J., N. Hakim, and C. Prudvi, "Post-silicon Validation Challenges: How EDA and Academia Can Help," *Proc. IEEE/ACM Design Automation Conf.*, pp. 3-7, 2010.

[Larsson 04] Larsson, A., *et al.*, "Multi-Word Atomic Read/Write Registers on Multiprocessor System," *Lectures Notes in Computer Science*, Vol. 3221, pp. 736-748, 2004.

[Martin 05] Martin, M., *et al.*, "Multifacet's General Execution-Drive Multiprocessor Simulator (GEMS) Toolset," *ACM SIGARCH Computer Architecture News*, Vol. 33, Issue 4, pp. 92-99, November, 2005.

[McLaughlin 09] McLaughlin, R., S. Venkataraman, and C. Lim, "Automated Debug of Speed Path Failures Using Functional Tests," *Proc. IEEE VLSI Test Symp.*, pp. 91-96, 2009.

[Mitra 10] Mitra, S., *et al.*, "Post-Silicon Validation Opportunities, Challenges and Recent Advances," *Proc. IEEE/ACM Design Automation Conf.*, pp. 12-17, 2010.

[OpenSPARC] "OpenSPARC: World's First Free 64-bit Microprocessor," http://www.opensparc.net.

[Patra 07] Patra, P., "On the Cusp of a Validation Wall," *IEEE Design & Test of Computers*, Vol. 24, Issue 2, pp. 193-196, March, 2007.

[Raina 98] Raina, R., and R. Molyneaux, "Random Self-Test Method Applications on PowerPC^{TM} microprocessor cache," *Proc. ACM/IEEE Great Lakes Symp. VLSI*, pp 222-229, 1998.

[Shirvani 00] Shirvani, P. P., N. R. Saxena, and E. J. McCluskey, "Software-Implemented EDAC Protection Against SEUs," *IEEE Trans. on Reliability*, Vol. 49, Issue 3, pp 273-284. September, 2000.

[Silas 03] Silas, I., *et al.*, "System-Level Validation of the Intel® Pentium® M Processor," *Intel Technology Journal*, Vol. 7 Issue 2, pp. 37-43, May 2003.

[Singerman 11] Singerman, E., Y. Abarbanel, and S. Baartmans, "Transaction Based Pre-To-Post Silicon Validation," *Proc. IEEE/ACM Design Automation Conf.*, pp. 564-568, 2011.

[Van Campenhout 00] Van Campenhout, D., *et al.*, "Collection and Analysis of Microprocessor Design Errors," *IEEE Design & Test of Computers*. Vol. 17, Issue 4, pp. 51-60, Oct-Dec, 2000.

[Velev 03] Velev, M. N., "Collection of High-Level Microprocessor Bugs from Formal Verification of Pipelined and Superscalar Designs", *Proc. IEEE Intl. Test Conf.*, pp. 138-147, September, 2003.

[Wagner 08] Wagner, I., and V. Bertaco, "Reversi: Post-Silicon Validation System for Modern Microprocessors," *Proc. IEEE Intl. Conf. Computer Design*, pp. 307-314, October, 2008.

[Woo 95] Woo, S. C., *et al.*, "The SPLASH-2 Programs: Characterization and Methodological Considerations," *Proc. ACM/IEEE Intl. Symp. Computer Architecture*, pp. 24-36, 1995.

Test-Data Volume Optimization for Diagnosis

Hongfei Wang, Osei Poku, Xiaochun Yu, Sizhe Liu,
Ibrahima Komara and R. D. (Shawn) Blanton
ECE Dept., Carnegie Mellon University
Email: {hongfeiw, blanton}@ece.cmu.edu

ABSTRACT

Test data collection for a failing integrated circuit (IC) can be very expensive and time consuming. Many companies now collect a fix amount of test data regardless of the failure characteristics. As a result, limited data collection could lead to inaccurate diagnosis, while an excessive amount increases the cost not only in terms of unnecessary test data collection but also increased cost for test execution and data-storage. In this work, the objective is to develop a method for predicting the precise amount of test data necessary to produce an accurate diagnosis. By analyzing the failing outputs of an IC during its actual test, the developed method dynamically determines which failing test pattern to terminate testing, producing an amount of test data that is sufficient for an accurate diagnosis analysis. The method leverages several statistical learning techniques, and is evaluated using actual data from a population of failing chips and five standard benchmarks. Experiments demonstrate that test-data collection can be reduced by > 30% (as compared to collecting the full-failure response) while at the same time ensuring >90% diagnosis accuracy. Prematurely terminating test-data collection at fixed levels (e.g., 100 failing bits) is also shown to negatively impact diagnosis accuracy.

Categories and Subject Descriptors

B.8.1 [**Performance and Reliability**]: Testing

General Terms

Algorithms, Design, Reliability

Keywords

Test Cost; Diagnosis; Test Data Collection; Statistical Learning

1. INTRODUCTION

Given the presence of process variations and defects during production, integrated circuits (ICs) are typically tested to verify they satisfy performance specifications. Test patterns are generated and applied to the IC for this purpose. The output responses of failing ICs are recorded into failure-log files in case diagnosis is later performed. Generally, failure diagnosis achieves its best possible outcome when a large amount of test measurement data from a failing IC is available for analysis.

Different test-data collection strategies are adopted in industry. Some companies collect no test data, whereas others collect a finite amount of test data, irrespective of the characteristics of the failing IC under testing. Other companies even attempt to collect data for the entire production test set, despite the number of erroneous bits exceeding 1M. However, the cost of collecting test-measurement data can be significant due to the extra test time

Permission to make digital or hard copies of all or part of this work for personal or classroom use is granted without fee provided that copies are not made or distributed for profit or commercial advantage and that copies bear this notice and the full citation on the first page. To copy otherwise, or republish, to post on servers or to redistribute to lists, requires prior specific permission and/or a fee.

incurred for going beyond the first failing pattern. The data-storage cost can also be significant especially for high-volume products. Also, even when it is desirable to collect ample amounts of test data, it may not be possible due to the limited memory that modern ATE (automatic test equipment) has for storing test-response data.

An adjacent area of research focuses on reducing test-execution cost. One related approach adopted in practice is multi-site testing, where several ICs are tested simultaneously. Multi-site testing improves test throughput and thus saves expensive ATE time, although it does not necessarily reduce test-data volume. The other commonly used and effective approach for reducing test-execution cost is test compaction [1-4]. This area is related since if fewer tests are performed, then less test data is produced. Instead of applying the entire set of generated tests, test compaction identifies a subset of tests and then uses them to determine the overall pass/fail status of an IC. Since a smaller number of tests are applied when compaction is employed, the total volume of the test data that can be collected is also reduced.

Reducing test data via test compaction may not be beneficial for diagnosis however since the objective is to determine chip pass-fail status at minimum cost. To understand the incongruence between test-execution cost reduction and diagnosis further, consider a "stop-on-first-fail" test strategy. For this case, test data is minimized but the negative impact on diagnosis is likely quite significant since there is minimal information for distinguishing the various sources that could be responsible for failure. Moreover, once a subset of tests is selected from the original set, it is fixed and little consideration is given to chip to chip variation. This means that different failing chips may require varying amounts of test-data for an accurate diagnostic analysis. A dynamic method can therefore be quite useful and could counter any negative effects on diagnosis due to test-cost reduction.

The focus of this work is on reducing the cost of test-data collection from a different perspective. In our analyses, we have observed for a substantial number of failing ICs that their diagnosis results do not change with an increasing amount of test data. Although it may seem intuitive that more test data would improve diagnostic resolution and accuracy, it turns out that it may not, or even worse, it may degrade. It is even possible that test data from the first few failing test patterns is sufficient for obtaining an accurate diagnosis result with optimal resolution. If the testing procedure can be terminated after a sufficient amount of data is collected for diagnosis, test-data collection can be reduced.

The histogram in Fig. 1 further illustrates the relation between diagnostic accuracy (defined in Section 2.4) and the amount of test data collected from an actual fabricated IC. Accuracy improves with more test data collected as evidenced by the overall increasing trend of peak heights in the histogram, though the slope of this trend decreases. About 78% of the total sample of 456 failing ICs examined required only 50 bits to perform an accurate diagnosis. On the other hand, if the test-data collection is terminated at a fixed level, e.g., 100 bits, insufficient information is then gathered for an accurate diagnosis for some failing ICs, 51 for the data of Fig. 1.

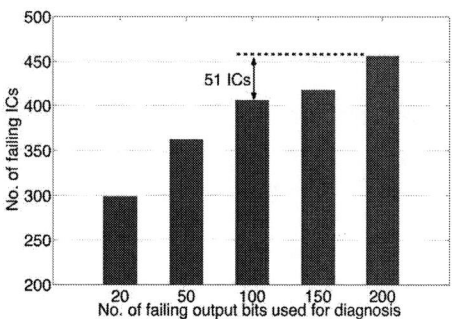

Fig. 1: Number of failing ICs that result in an accurate diagnosis when using the amount of test data shown on the x-axis.

In this work, the objective is to predict the minimal amount of test data necessary to produce both a precise and accurate diagnosis. We propose a method that dynamically predicts test termination for producing an amount of test data that is sufficient for obtaining a quality diagnosis result. The prediction model is learned from the test data produced by a sample of fully-tested ICs that have failed. The learned model is then deployed in production to predict the termination point when testing future ICs of the same type. Specifically, when an IC is tested, for each new failing pattern observed, the prediction model is invoked to determine if sufficient data has been collected for producing a quality diagnosis. The method is dynamic since test termination can vary from one failing IC to next rather than being fixed. In this way, the variation in individual ICs is taken into account in order to produce an optimized test-data volume for each, ensuring that the overall cost from collecting test data is minimized without sacrificing diagnosis quality. The test-termination prediction model is based on analysis of the failing tests and outputs of the ICs, which exhibit certain patterns that can be interpreted as a signal for judging the termination point of testing. Several statistical learning techniques are employed to discover these patterns from full-fail test data.

The work in [5] has the same objective as the work presented here. Specifically, their work dynamically determines test termination by evaluating the benefits of executing a subsequent test, based on the additional fault model coverage it would achieve. This work therefore assumes there is a positive correlation between model coverage and diagnostic resolution/accuracy. In our work, test termination is based on a model that is formulated using characteristics from real defective circuits, thus removing the barrier of abstraction that exists between fault models and actual defects. Moreover, our work complements the test-set compaction [1-4] and test-response compaction [6] based approaches. In other words, our work can be applied after these procedures to further minimize total cost.

The rest of this paper is organized as follows: Section II describes how the raw test-data is pre-processed for further analysis. Section III introduces several statistical learning techniques explored in this work for deriving a test-termination prediction model. Experiment validation is presented in Section IV and Section V summarizes our contributions.

2. DATA EXTRACTION
We begin with a description of the various data sources used in this work, and then address the issue of extracting useful information from the raw test data collected. Different metrics for data preparation and processing are also presented.

2.1 Sources
The first data set stems from a fabricated test chip from the LSI Corporation which consists of 64 identical ALUs (arithmetic logic units). A total of 233 test patterns were used to test this chip, including its scan-chains. Test data and diagnosis results for 493 chips were collected. Some chips were tested more than once, resulting in multiple test-data files. The test outcomes and diagnosis results for a particular chip, however, do not greatly differ among the various test executions. Therefore we arbitrarily decide to use the test data generated by the first test execution.

Besides the above real-world data set, experiments that use five benchmark circuits are also performed to demonstrate the viability of our method for a variety of circuit types. The circuits include c499 and c7552 from the ISCAS'85 benchmarks, s5378 and s7552 from the ISCAS'89 sequential circuits, and b12 from the ITC'99 benchmark suite.

The defect simulation framework from [7] is used to create a population of realistic failures from various benchmark designs, which are the largest circuits available that have simulation responses from layout-injected defects. In [7], randomly-selected defects are generated for each possible defect type that includes open, bridge, polysilicon, transistor stuck-open, and transistor stuck-closed. Defects are injected, one at a time, into the layout of each benchmark. An extracted netlist of each defective circuit is then simulated at the circuit-level using 100% stuck-at test sets generated by a commercial ATPG tool. The resulting simulation responses form the virtual test data for the failing population of circuits created.

2.2 Data Set Overview
In a typical statistical learning scenario, the data set used to learn a prediction model can be represented as an $n \times p$ matrix (denoted by X), where each row is an input instance described by p features. A $n \times 1$ vector (denoted by Y) is also provided, which contains the labels (e.g., quality diagnosis versus non-quality diagnosis) for each corresponding row $X(i,\cdot)$.

The remaining part of this section describes the data scrubbing process used for obtaining X and Y from the test data and diagnosis results, respectively.

2.3 Feature Extraction
A defect in an IC, once activated, produces one or more errors that may be propagated to one or more outputs. Since each output has a fan-in logic cone, back-cone tracing is employed during diagnosis to locate possible defect sites. Consequently, determining the test-termination point for a given IC under test is based on simple observations made from the failing outputs. It is difficult however to use this raw data directly to discover underlying patterns and trends that suggest a termination point for testing. Therefore, the raw test data is processed and then organized into a set of features that are more readily usable for forming a test-termination prediction model. Although "passing patterns" (test patterns that do not cause any outputs to fail) can be useful in some forms of diagnosis [8], they are less focused upon since they can only provide indirect information concerning the source of failure. Table 1 summarizes the features derived from test data.

2.4 Evaluation Metrics
To perform a diagnosis for a failing IC, the circuit design, the corresponding test patterns, and the measured test response are

provided to a software-based diagnostic tool [9]. In order to create training data for forming a test-termination prediction model, an IC that has failed N test patterns is used to produce N diagnostic results. Specifically, test patterns that include the first one (either passing or failing) through the first failing pattern are used to produce the first diagnostic result, test patterns that include the first through the second failing pattern are used to produce the second diagnostic result, and so on until N diagnostic results are created using all N failing test patterns. Failing patterns are focused upon since the impact of passing patterns (patterns applied to failing IC that produce expected results) on the diagnostic result is typically less significant than failing patterns. Each of the N diagnoses produces a list possible solutions or candidates, each of which typically includes a fault type (stuck-at, bridge, etc.), the possible location, along with a percentage score, where 100% indicates a perfect match between the fault simulation response and the tester response. A *golden* diagnosis result for a failing IC refers to the one generated when using all the applied tests, that is, from the first test pattern through the last failing test. The golden result does not have to have exact one fault candidate however. An *intermediate* result refers to any outcome generated using a subset of the test patterns corresponding to diagnoses 1 through N-1. Terminating the test of an IC early may still produce a diagnosis result that is the "same" as the golden one.

Table 1. Seven features extracted from raw test data. Each feature becomes a column vector in the matrix X, denoted by $X(\cdot, j)$.

Feature	Feature description
$X(\cdot,1)$	Number of test patterns (both passing and failing) that have been applied thus far.
$X(\cdot,2)$	Number of failing test patterns that have been applied thus far.
$X(\cdot,3)$	Total number of erroneous output bits that have been accumulated thus far.
$X(\cdot,4)$	Number of erroneous output bits for the current test pattern.
$X(\cdot,5)$	Total number of different erroneous output bits that have been accumulated thus far.
$X(\cdot,6)$	Number of unique erroneous output bits produced by the current test pattern.
$X(\cdot,7)$	Indicates the first failing test pattern.

Our objective is to predict the termination point during the test of an IC that will produce the golden result. In other words, for each failing pattern, we want to predict whether or not enough test data has been obtained to achieve an acceptable diagnosis result. To accomplish this objective, it is desirable to quantitively measure how close an intermediate result matches the golden result. To accomplish this, we examine two metrics:

$$m_A = \frac{\text{No.(intermediate candidates} \cap \text{golden candidates)}^2}{\text{No.(intermediate candidates)} \times \text{No.(golden candidates)}}$$

and

$$m_B = \begin{cases} 0, & \text{intermediate candidates} \cap \text{golden candidates} = \varnothing \\ 1, & \text{otherwise} \end{cases}.$$

The numerator in the equation for m_A is the square of the number of common candidates from the golden result and an intermediate result. Each candidate resulting from the diagnosis is associated with a rank (i.e., a percentage between 0~100%), representing the confidence in the result produced by the diagnosis tool. The highest ranked candidates are selected and used for set intersection. It should also be noted that for a specific common

candidate, its rank may differ in the golden and intermediate results. The value range for metric m_A is [0,1], with 1 denoting a perfect match among the candidates from the intermediate diagnosis result and the golden one. For such a case, the ideal test-termination point is the failing test pattern that produces this intermediate diagnosis result.

Metric m_B takes binary values only: 0 for an intermediate diagnosis result that is distinctly different from the golden, and 1 if any candidates are in common. m_B is essentially a relaxed formed of m_A since

$$m_B = \begin{cases} 0, & m_A = 0 \\ 1, & m_A \neq 0 \end{cases}.$$

m_B may also be useful since any highly-ranked candidate may be sufficient for a follow-on application. For example, physical failure analysis of a failing IC typically only needs one highly-ranked candidate. In summary, m_A looks for an exact match of the golden result, whereas m_B captures a more relaxed notion of correctness. The metric values for intermediate diagnoses, obtained by employing either m_A or m_B are stored in Y, sorted in the same order of the IC indices in X.

3. LEARNING TECHNIQUES

Predicting test termination can be formulated as a classification task in statistical learning: for each failing test pattern, a decision is made about whether to terminate testing, or to continue applying further test patterns, by observing and analyzing the test data thus far collected. To this end, an adopted learning technique is expected to produce a model that is capable of classifying a new data instance by X_i^*, representing a failure response for the most-recent test applied to an IC, with an estimated binary value $\hat{Y}_i^*$, denoting whether to terminate or continue testing. The asterisk notation (*) means the data instances are from an IC that is currently being tested. The learning process is considered a success if $\hat{Y}_i^*$ precisely matches the true Y_i^*, which could be obtained from using the entire test response for diagnosis once the testing of this IC is finished. A binary classification problem of this nature can be handled by various statistical learning techniques, one of which is chosen based on effectiveness.

3.1 Linear Regression

The most widely and frequently used statistical learning technique is linear regression [10]. In particular, linear regression with least-squares estimation builds a model that represents how the response vector Y depends on the feature matrix X. This dependence is studied, and then generalized to predict future instances of X^* whose Y^* is unknown. The learned model can be expressed by

$$Y = X \cdot \beta + \varepsilon, \ \beta = \left[\beta_1, \beta_2, \cdots, \beta_p\right]^\mathrm{T}, \ \varepsilon = \left[\varepsilon_1, \varepsilon_2, \cdots, \varepsilon_p\right]^\mathrm{T},$$

where β is a $p \times 1$ vector of regression coefficients and ε is the error to be minimized by the least-squares estimation. β_i weighs the impact of the corresponding feature on the value of Y. It allows hypothesis testing of the correlation between selected features and the response vector, which will be investigated in Section IV. For classification problems, a threshold is usually pre-specified to discretize the output values.

The specific framework for linear regression employed here is LASSO (Least Absolute Shrinkage and Selection) [11]; it minimizes the objective function

$$\left\| X \cdot \beta - Y \right\|_2^2 + \lambda \|\beta\|_1.$$

The first term in the above expression minimizes the sum-of-square errors. λ in the second term is a tuning parameter that controls the weight of penalty, formulated in the L1 norm for the sum of the absolute values of the coefficients. A large λ decreases the absolute value of β's, and may eventually forces some to 0. This beneficial property helps to yield a parsimonious model, in addition to achieving high accuracy.

It may seem that the use of a linear regression technique such as LASSO is misguided since we have not shown that the test-termination problem is inherently linear. But many linear regression approaches weakly depend on linearity, and employ techniques such as regularization to produce models that effectively deal with the relationships between X and Y. Evidence of this being the case here is demonstrated by the high accuracy produced by the LASSO model presented in Section 4.3.

3.2 *k*NN

Unlike other statistical learning techniques, kNN (k-nearest neighbor) [12] does not require a separate training process. It searches X and finds k instances that have the nearest distance from X_i^*. These k instances are the k-nearest neighbors of X_i^* in this distance-based approach. Using their corresponding values from the response vector Y, kNN applies majority voting to compute the $\hat{Y}_i^*$. kNN is simple but can be very accurate in many cases [13].

There are two factors concerning the implementation of kNN. One is the choice of "k", the number of nearest neighbors to be included when making classification. For the application considered here, $k = 1$ is found to produce the best results for an analysis that considered all integers $k < 16$. The other factor is the distance measurement. One particular metric is based on cosine similarity between two instances (e.g., X_p and X_q), which can be expressed by

$$\theta = \cos^{-1} \frac{X_p \cdot X_q}{\|X_p\| \|X_q\|}.$$

According to [14], the cosine-based distance metric is more appropriate than Euclidean or Mahalanobis distances in many applications. It is thereby employed here given the task of test data collection.

3.3 Support Vector Machines

SVM is becoming more and more popular in recent years, due to its robust and accurate performance [13], and is favored by many researchers from disciplines such as electrical engineering and computational biology. We therefore consider SVM for this application as well.

For binary classification, SVM (support vector machines) aim to find a boundary in the hyperspace defined by the features of the data (column vectors in X), according to their response vector Y. The boundary creates a margin between the two classes of instances so that the minimal distance is maximized. In some cases, instances of the opposite classes are allowed to be mistakenly separated by the boundary. Tolerance of such errors produces a less accurate boundary describing X and its associated Y, but may provide a better classification strategy for a future instance X_i^*. The search for the aforementioned boundary can be formulated as an optimization problem of

$$\min_{w, \xi} \quad \frac{1}{2} w^T w + \lambda \sum \xi_i,$$

where w defines the direction of the boundary, and minimizing $\frac{1}{2} w^T w$ is equivalent to maximizing the margin. The parameter λ (>0) is a tuning parameter that penalizes the error ξ_i, i.e., the distance associated with misclassification. Solving for the vector w can be achieved by employing kernel functions, as explained in [13]. Typical kernels include linear, polynomial, radial, and sigmoid. In our experiments, the linear kernel produced the best tradeoff between accuracy and efficiency.

3.4 Decision Tree

Decision-tree models are straightforward to create and can be easily interpreted by users with even limit statistical background. In addition, they perform well as compared to other statistical learning techniques for classification tasks.

In this work, we apply CART (Classification And Regression Trees) [15] to build a decision-tree prediction model. Unlike other tree-constructed algorithms, the splitting rules in CART are guided by the Gini measure of impurity [13,15]. CART recursively partitions the data, described by X and its associated response Y, until the stopping criteria is satisfied which includes, for example, a minimal number of instances in the resulting partitions. The decision-tree paths to leaf nodes, representing the binary classification decision (terminate or continue testing), are used to predict the failing pattern for test termination.

A large number of experiments are conducted for each aforementioned statistical learning technique in order to improve their classification accuracy, reduce run-time and memory space required. In addition, we scrutinize the parameter settings, such as the choice of kernel for SVM, the minimum node size to terminate partitioning for a decision tree, and the distance-measuring metric for kNN to maximize accuracy. Details of these analyses are omitted however due to limited space.

4. EXPERIMENT RESULTS

In this section, correlations among the output responses of failing ICs, the diagnosis results that identify possible failure locations, and the amount of test data needed to produce an ideal/acceptable diagnosis results are explored and studied.

4.1 Data Sets

Each failing test pattern is transformed into a data instance, based on the feature extraction procedure described in Section 2. In other words, multiple data instances are derived for each failing chip. The total number of chips for the LSI test chip and the benchmarks are 493 and 1,745, respectively, which produces 35,829 and 57,981 data instances, respectively. The raw test data are further processed before being given as input to the statistical learning techniques.

4.2 Feature Effectiveness

In Section II, we described the seven features extracted from the raw data. However, naively and intuitively, one could perform a coin toss to decide the test-termination point after observing a failing test. Thus it would be tempting to know if these features are chosen reasonably and are effectively correlated with the label vector storing test-termination points (denoted by Y). Therefore, a hypothesis testing is employed to examine correlation. Specifically, two models are compared: one with all seven features, and the other one with a random variable drawn from a binomial distribution (with $p = 0.5$) to simulate a coin-tossing process. Both models are constructed using linear regression. An F-test is conducted under the null hypothesis that the coin-tossing model fits the data well. The test statistic is

$$F = \frac{(RSS_{ct} - RSS_{full})/(df_{ct} - df_{full})}{RSS_{full}/df_{full}}$$

where RSS is the residual sum of square errors, and df is the degrees of freedom. Subscripts ct and $full$ denote the "coin-tossing" model, and the model using all the features, respectively. The resultant p-value is quite small (< 0.05) for this test, suggesting strong evidence against the null hypothesis. In other words, the seven features extracted from the raw test data are proved to be closely correlated with test-termination points that produce accurate diagnosis results, thereby providing more useful information for predicting the Y labels than a random coin toss.

4.3 Accuracy and Data Volume Reduction

For each data set, 90% of the IC population is randomly chosen for learning a statistical model. The failure responses, together with the diagnosis results, are processed into $\langle X, Y \rangle$. The remaining 10% of ICs, whose failure output behaviors are denoted by X^*, are reserved and not included in the learning process. Their response vector Y is assumed unknown and are used to gauge the effectiveness of the learned model. In other words, $\langle X_i^*, \hat{Y}_i^* \rangle$ denotes a failure response for any of the ICs in the reserved 10% population, and its corresponding action of whether to continue or terminate testing suggested by the learned model base on $\langle X, Y \rangle$. We repeat this procedure 10 times as to perform a 10-fold cross-validation for each experiment to evaluate the average performance.

Test-termination prediction is inaccurate if the resulting diagnosis result does not match the full-data one as measured by one of the metrics m_A or m_B. Inaccuracy is typically due to early termination, which simply means insufficient test data has been collected to produce a quality diagnosis. Some inaccuracy is expected from statistical learning techniques since they produce data-driven, nondeterministic solutions. To prevent early test termination, a guard-band is imposed. Different from the guard-banding approach employed in [1,16], where multiple models are required to make identical decisions, the guard band invoked here uses a single model. Specifically, testing is only terminated after the model predicts test termination for M successive failing patterns, where M is an integer that can be 0 (no guard banding), 1, 2, etc. A sequence of consistently identical test-termination decisions increases the confidence that a learned model will produce a quality diagnosis result. Guard banding is expected to reduce the chances of early termination but the trade off is that test data-volume reduction (DVR) will be reduced.

Of course, besides accuracy, the other key factor affecting the performance of the model is the level of DVR, which can be measured by

$$DVR = \frac{1}{n}\sum(1 - \frac{|t_i^T|}{|T_i|})$$

where $|T_i|$ is the test-data volume for the ith IC analyzed. This is obtained by accumulating the number of failing output bits for all failing test patterns, from the first to the last. $|t_i^T|$ is the optimized volume calculated in a similar way, from the first pattern to the test-termination point predicted for this IC. The DVR for one IC is not statistically significant; thus it is averaged over the entire population of n failing chips per design.

All four statistical learning techniques, kNN, linear regression (LASSO), SVM, and decision tree, are applied to the industrial data set of a test chip from the LSI Corporation to measure both the accuracy and DVR. Metric m_A is used to measure the

diagnosis results. The results are presented in Fig. 2 which shows that a model produced by any of the statistical-learning techniques investigated is able to effectively predict test termination with high accuracy, which can be further improved with guard banding. For example, in Fig. 2(a), with a guard band >16, the accuracy of kNN is increased by over 30% comparing to a guard band of two. On the other hand, although guard banding proves to increase the accuracy in most of the experiments, it is not always guaranteed that using a large guard band will necessarily improve accuracy of a learned statistical model. For example, in Fig. 2(c), when using SVM with guard bands from 16 to 20, the accuracy is slightly degraded. This non-monotonic behavior of accuracy is caused by using "too much" collected data, as opposed to early test-termination. Choosing a proper guard band is therefore critical to further improve the accuracy of a statistical model.

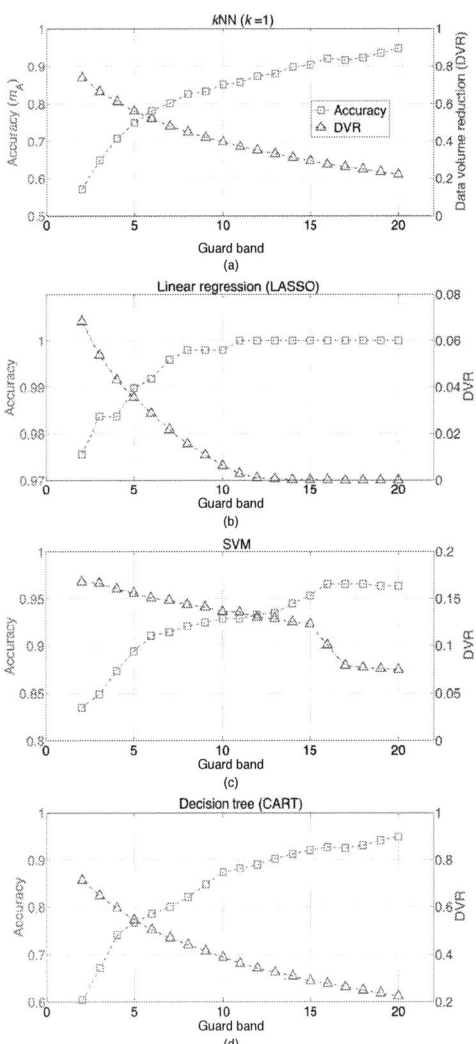

Fig. 2: Experiment results that compare test-termination prediction for an industrial test chip using: (a) kNN, (b) linear regression (LASSO), (c) SVM and (d) decision tree. Dashed lines with squares illustrate the accuracy of the models based on the metric m_A; dash-dot lines with triangles are the levels of DVR achieved. Different guard bands are applied, ranging from 2 to 20.

Unlike accuracy, DVR drops monotonically with the increase of guard band. There is a trade-off between accuracy and DVR. LASSO achieves the best accuracy. With a guard band greater than 11, it reaches 100%. The cost for this perfect accuracy however is that there is little to no DVR benefit. SVM has lower accuracy than LASSO, but is better than kNN and a decision tree. However, the improved accuracy of SVM over kNN and a decision tree becomes negligible when the guard band is large. Moreover, although significantly greater than LASSO, the DVR achieved by SVM cannot compete with either kNN or a decision tree. The performance of a decision tree and kNN are very similar, with a decision-tree model being slightly better in terms of accuracy. For kNN, it is reported to lack robustness when a small k is adopted [13]. Consequently, considering the balance between accuracy and DVR, the best learning technique among those examined is the decision tree. Specifically, a decision tree exhibits, on average, an accuracy of over 90% when achieving a DVR of 32.4%. Tree models are also more easily constructed than SVM and LASSO, both of which require data scaling. The remaining experiments and discussion therefore focus on using decision-tree models.

It should be noted that, once a statistical model is constructed, it takes little time (less than a millisecond) to predict the test-termination point, especially for a decision-tree model. They therefore can easily be incorporated into a production test flow where prompt termination decisions are desired in the course of testing ICs to produce an optimized test-data volume.

4.4 Viability

To demonstrate the general applicability of this methodology, standard benchmarks are analyzed in addition to the test chip from the LSI Corporation. The results are summarized in Table 2.

Table 2. Decision-tree results using 10-fold cross-validation to obtain average accuracy and DVR.

Design	Performances (%) and guard bands (GB) used					
	Accuracy	DVR	GB	Accuracy	DVR	GB
c499	82.9	26.0	1	90.6	18.3	3
c7552	84.7	33.0	3	91.7	21.3	6
s5378	83.1	64.5	2	90.2	50.9	5
s9234	87.0	54.0	2	91.4	46.4	4
b12	83.7	21.5	1	90.1	13.1	3

Guard banding (GB) is expected to prevent early termination but the tradeoff is that DVR will degrade as evidenced by Fig. 2. Determining a GB level for test termination is a function of the design, the tests applied, the defects occurring, the acceptable accuracy, and the desired level of DVR. These areas of uncertainty are naturally handled however by employing a well-known statistical estimation method called cross-validation (Section 4.3). It is therefore straightforward to use cross-validation to identify the optimal level of GB value for a given design, test, and fabrication process. For the designs considered here, GB ≤ 6 ensures accuracy is greater than 80%. A GB > 20 is not required since a lower value has equivalent or superior results.

Including the experiment results from the LSI chip, 38.9% of the test-data volume can be reduced on average with an accuracy that exceeds 85.3%. For an average accuracy greater than 90%, the DVR decreases to 30.4%, which is still significant. The experiment results for standard benchmarks also indicate our method can be effectively applied to a variety of designs.

Generally, a higher accuracy level results in a smaller DVR, which means test cost is not minimized. In practice, the acceptable accuracy level should be determined considering both the test economics and the desired diagnosis accuracy.

4.5 Evaluation Using Different Metrics

As a final experiment, we repeat the analysis of the LSI test chip but instead employ metric m_B. The results are illustrated in Fig. 3.

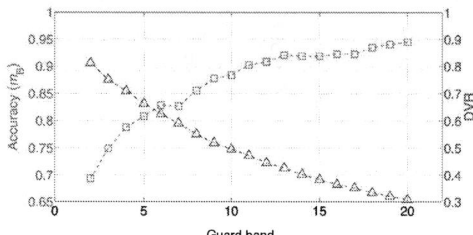

Fig. 3: Decision tree outcome using metric m_B for the LSI test chip.

Comparing with Fig. 2(d), both the accuracy and DVR are increased, as expected, due to using a less stringent metric m_B. The experiment is provided to illustrate that: 1) the performance of a learned model is subject to change when a different metric is used to evaluate diagnosis results and 2) an increase in DVR and accuracy can be expected if partially correct diagnosis results are deemed acceptable.

5. SUMMARY

This paper explored various aspects of using statistical-learning techniques to optimize the cost of collecting test data for producing a high-quality diagnostic result. Test data collected from full-tested ICs that fail are used to learn a model that allows the test-termination points to be accurately predicted. The learned model uses easily obtainable data (number of erroneous outputs, identification of unique erroneous outputs, number of failing test patterns, etc.) and thus can be used in real-time to decide when testing should be terminated. Experiment results from both industrial and simulated data sets demonstrate that the amount of test data needed to obtain an accurate diagnosis can vary from failing IC to failing IC.

6. REFERENCES

[1] S. Biswas and R. D. Blanton, "Statistical Test Compaction Using Binary Decision Trees,"*IEEE Des. Test. Comput.*, vol. 23, no. 6, pp. 452-462, 2006.

[2] L. Amati, C. Bolchini, and F. Salice, "Optimal Test Set Selection for Fault Diagnosis Improvement," *IEEE DFT*, pp.93-99, 2011.

[3] M. Shukoor and V. Agrawal, "A Two Phase Approach for Minimal Diagnostic Test Set Generation," *IEEE ETS*, pp. 115-120, 2009.

[4] Y. Higami et al., "Compaction of pass/fail-based diagnostic test vectors for combinational and sequential circuits," *IEEE ASPDAC*, 2006.

[5] L. Amati et al., "A Formal Condition to Stop an Incremental Automatic Functional Diagnosis," *IEEE DSD*, pp. 637-643, 2010.

[6] W. Cheng, K. Tsai, Y. Huang, N. Tamarapalli, J. Rajski, "Compactor independent direct diagnosis," *IEEE ATS*, pp. 204-209, 2004.

[7] W. C. Tam and R. D. Blanton, "SLIDER: A Fast and Accurate Defect Simulation Framework," *IEEE VLSI*, pp. 172-177, 2011.

[8] R. Desineni, O. Poku, and R. D. Blanton, "A Logic Diagnosis Methodology for Improved Localization and Extraction of Accurate Defect Behavior," *IEEE ITC*, 2006.

[9] TetraMAX ATPG [Online]. Available: http://www.synopsys.com/Tools/ Implementation/RTLSynthesis/Pages/TetraMAXATPG.aspx.

[10] S. Weisberg, *Applied Linear Regression,* 3rd edition. New York: Wiley, 2005.

[11] R. Tibshirani, "Regression Shrinkage and Selection Via the Lasso," Technical report, University of Toronto, 1994.

[12] T. Cover and P. Hart, "Nearest Neighbor Pattern Classification," *IEEE Trans. Inf. Theory* , IT-11, 21-27, 1967.

[13] X. Wu and V. Kumar, *The Top Ten Algorithms in Data Mining.* Chapman & Hall/CRC Press, 2009.

[14] A.M. Qamar et al., "Similarity Learning for Nearest Neighbor Classification," *IEEE ICDM*, pp.983-988, 2008.

[15] L. Breiman et al., *Classification and Regression Trees.* Wadsworth, Belmont, 1984.

[16] S. Biswas and R. D. Blanton, "Specification Test Compaction for Analog Circuits and MEMS," *IEEE DATE*, pp. 164-169, 2005.

Invariance-Based Concurrent Error Detection for Advanced Encryption Standard

Xiaofei Guo and Ramesh Karri
Polytechnic Institute of New York University
6 Metrotech Center, Brooklyn, USA
xguo02@students.poly.edu, rkarri@poly.edu

ABSTRACT

Naturally occurring and maliciously injected faults reduce the reliability of Advanced Encryption Standard (AES) and may leak confidential information. We developed an invariance-based concurrent error detection (CED) scheme which is independent of the implementation of AES encryption/decryption. Additionally, we improve the security of our scheme with Randomized CED Round Insertion and adaptive checking. Experimental results show that the invariance-based CED scheme detects all single-bit, all single-byte fault, and 99.99999997% of burst faults. The area and delay overheads of this scheme are compared with those of previously reported CED schemes on two Xilinx Virtex FPGAs. The hardware overhead is in the 13.2-27.3% range and the throughput is between 1.8-42.2Gbps depending on the AES architecture, FPGA family, and the detection latency. One can implement our scheme in many ways; designers can trade off performance, reliability, and security according to the available resources.

Categories and Subject Descriptors

B.5.3 [**Logic Design**]: Reliability and Testing—*error-checking*; K.6.5 [**Management of Computing and Information Systems**]: Security and Protection—*physical security*

General Terms

Design, Security

Keywords

Concurrent error detection, Reliability, Fault injection attack

1. INTRODUCTION

The Advanced Encryption Standard (AES) is used in a variety of applications, including smart cards, mobile phones, WWW servers, automated teller machines, and digital recorders. The decreasing cost of VLSI chips and increasing user throughput requirements make hardware implementation of AES necessary.

Faults that occur in VLSI chips are classified into two categories: transient faults that eventually die away and permanent faults. The origin of these faults could be internal phenomena in the system, such as threshold changes, shorts, opens, etc., or external influences, such as electromagnetic radiation. These faults affect the memory as well as the combinational parts of a circuit and are detected using concurrent error detection (CED) [1]. Cryptographic chips are sensitive to natural faults in the hardware [2]. A small number of excited faults can cause a large number of output bits of AES to be faulty [3]. Recently, attackers have injected faults into cryptographic circuits to steal secret information as well [4,5].

1.1 Related Work

Previous work on CED can be classified into four types of redundancy: hardware, time, information, and combined redundancy.

Hardware redundancy duplicates the function and detect faults by comparing the outputs of two copies.

Time redundancy: The function is computed twice on the same input and the results are compared. A variation of the time redundancy is in [6]. The function is computed on both clock edges. This speeds up the computation. Under some conditions, this scheme allows the encryption to be computed twice without affecting the global throughput. This scheme is complex and delicate to implement as technology scales.

Information redundancy: Many fault detection schemes are based on error detecting codes. A few check bits are generated from the input message; then they propagate along with the input message and are finally validated when the output message is generated. The basic parity scheme is proposed in [3]. In this scheme, each predicted parity bit is generated from an input byte. Then, the predicted parity bits, and actual parity bits of output are compared to detect the faults. This scheme incurs large hardware overhead. [7] proposes a scheme in which only single-bit parity is used for the entire 128-bit output, and the parity bit is checked once for the entire round. However, the above two schemes only apply to lookup table-based (LUTs) substitution-box (S-box) implementation. In [8], parity is obtained for S-box implementation that uses logic gates. In [9], a general parity-based scheme is proposed to protect the S-box regardless of its implementation. All these parity schemes share the same limitation. If an even number of faults occur in the same byte, none of these schemes can detect them.

Combined redundancy: In [10], the authors consider CED at the operation, round, and algorithm levels for AES. In these schemes, an operation, a round, or the entire encryption and decryption are followed by their inverses, and the results are compared with the original input to detect faults. Although these schemes detect most of the faults, they require both encryption/decryption to be on chip and can suffer from more than 100% throughput overhead.

1.2 Contributions

We propose a low overhead, implementation independent, and invariance-based CED scheme, which uses combined redundancy for obtaining a reliable AES implementation. The scheme achieves a fault detection capability that is close to [10] and hardware redundancy, the state-of-the-art countermeasures, but with much lower cost. Our contributions are as follows:

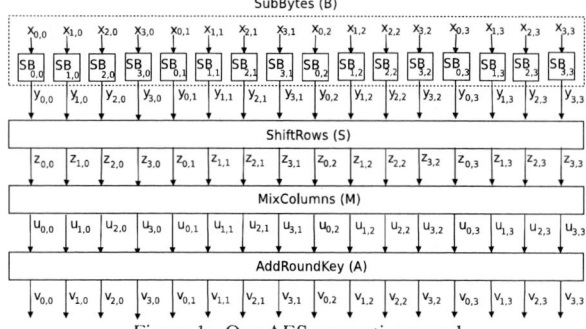

Figure 1: One AES encryption round

- We present the invariance-based[1] CED with Randomized CED Round Insertion and adaptive checking for AES to improve security.
- We prove that the invariance-based CED scheme detects all single-bit and single-byte faults.
- The invariance-based CED achieves an order of 10^5 lower fault miss rate than the best parity schemes for multiple burst and multiple random faults.

This paper is organized as follows: In Section 2, we introduce the AES algorithm. In Section 3, we propose the key idea of our invariance-based CED scheme, and we show the fault coverage, hardware overhead, and detailed analyses of the scheme.

2. ADVANCED ENCRYPTION STANDARD

In this paper, we consider 128-bit AES as specified by the National Institute of Standards and Technology (NIST). AES encrypts a 128-bit input plaintext into a 128-bit output ciphertext with a user key using 10 nearly identical rounds plus an initial special round (round 0). One AES encryption round consists of SubBytes, ShiftRows, MixColumns, and AddRoundKey denoted by B, S, M, and A, respectively, as shown in Figure 1. In round 0, only AddRoundKey is performed and in round 10, MixColumns is not performed. Each operation in every round acts on a 128-bit input state, where each state element is a byte in $GF(2^8)$. In this paper, each byte is denoted by $s_{r,c}$ ($0 \leq r, c \leq 3$), and it indicates that this byte is in row r and column c in state matrix.

$$S = \begin{bmatrix} s_{0,0} & s_{0,1} & s_{0,2} & s_{0,3} \\ s_{1,0} & s_{1,1} & s_{1,2} & s_{1,3} \\ s_{2,0} & s_{2,1} & s_{2,2} & s_{2,3} \\ s_{3,0} & s_{3,1} & s_{3,2} & s_{3,3} \end{bmatrix} = [s_{r,c}]_{r,c=0}^3 \quad (1)$$

In SubBytes, all the bytes are processed separately by 16 S-boxes (SBs). Each S-box performs a nonlinear transformation of the input byte. Let X be the input to the SBs. The resulting output is:

$$Y = B(X) = [y_{r,c}]_{r,c=0}^3 \quad (2)$$

In ShiftRows, the rows of the state are shifted cyclically bytewise using a different offset for each row. Row 0 is not shifted, while rows 1, 2, and 3 are cyclically shifted to the left by 1 byte, 2 bytes, and 3 bytes, respectively. The resulting output is:

$$Z = S(Y) = \begin{bmatrix} y_{0,0} & y_{0,1} & y_{0,2} & y_{0,3} \\ y_{1,1} & y_{1,2} & y_{1,3} & y_{1,0} \\ y_{2,2} & y_{2,3} & y_{2,0} & y_{2,1} \\ y_{3,3} & y_{3,0} & y_{3,1} & y_{3,2} \end{bmatrix}$$

[1]Let $f : \{0,1\}^{128} \rightarrow \{0,1\}^{128}$ denotes an operation on the state space of Rijndael which operates completely on the Galois field $GF(2^8)$. A property $P \subseteq \{0,1\}^{128}$, $P \neq 0$, is called an invariance of f, if P is preserved by f, i.e., for every $x \in P$ it follows that $f(x) \in P$. [11]

$$= [y_{r,(r+c) \, mod \, 4}]_{r,c=0}^3 = [z_{r,c}]_{r,c=0}^3 \quad (3)$$

In MixColumns, the output state is obtained by multiplying a constant matrix with the output of ShiftRows. The resulting output is:

$$U = M(Z) =$$

$$\begin{bmatrix} 02 & 03 & 01 & 01 \\ 01 & 02 & 03 & 01 \\ 01 & 01 & 02 & 03 \\ 03 & 01 & 01 & 02 \end{bmatrix} \begin{bmatrix} z_{0,0} & z_{0,1} & z_{0,2} & z_{0,3} \\ z_{1,0} & z_{1,1} & z_{1,2} & z_{1,3} \\ z_{2,0} & z_{2,1} & z_{2,2} & z_{2,3} \\ z_{3,0} & z_{3,1} & z_{3,2} & z_{3,3} \end{bmatrix} = [u_{r,c}]_{r,c=0}^3$$

$$(4)$$

In AddRoundKey, the input state is added (modulo-2) to the round key, i.e., the 128-bit state U with matrix $K = [k_{r,c}]_{r,c=0}^3$. The resulting output is:

$$V = A(K)U = [u_{r,c}]_{r,c=0}^3 + [k_{r,c}]_{r,c=0}^3 = [v_{r,c}]_{r,c=0}^3 \quad (5)$$

3. INVARIANCES OF AES

In [11], the authors have shown that AES exhibits various mapping invariance properties. They were investigating these as possible source of weakness in AES. In contrast, we use these invariances for CED to protect the AES against random faults and malicious attacks. We analyze three round level invariances of AES. We give a formal proof of the invariance property P_1, which is the most effective invariance according to our experimental results. We analyze the effectiveness of the remaining invariances in Section 4.

THEOREM 1. *An AES round can be represented as*

$$A(K)(M(S(B(X))))$$

where X is the 128-bit input to the round. Exist byte permutation P, so that the following hold true:

$$A(K)(M(S(B(X)))) = P^{-1}(A(P(K))(M(S(B(P(X)))))) \quad (6)$$

where P^{-1} denotes the inverse function of P.

One of the byte permutation is:

$$P_1(X) = P_1([x_{r,c}]_{r,c=0}^3) = \begin{bmatrix} x_{0,3} & x_{0,0} & x_{0,1} & x_{0,2} \\ x_{1,3} & x_{1,0} & x_{1,1} & x_{1,2} \\ x_{2,3} & x_{2,0} & x_{2,1} & x_{2,2} \\ x_{3,3} & x_{3,0} & x_{3,1} & x_{3,2} \end{bmatrix}$$

$$= [x_{r,(c+3) \, mod \, 4}]_{r,c=0}^3 \quad (7)$$

$$P_1^{-1}([x_{r,(c+3) \, mod \, 4}]_{r,c=0}^3) = [x_{r,c}]_{r,c=0}^3 \quad (8)$$

PROOF. Let us start from the right-hand side of the equation. First, we apply permuted input $X' = P_1(X)$ to SubBytes, and from (2) and (7), we get:

$$Y' = [y'_{r,c}]_{r,c=0}^3 = [y_{r,(c+3) \, mod \, 4}]_{r,c=0}^3 = P_1(Y) \quad (9)$$

Given Y' as the input to ShiftRows, we get:

$$Z' = S(Y') = \begin{bmatrix} y'_{0,0} & y'_{0,1} & y'_{0,2} & y'_{0,3} \\ y'_{1,1} & y'_{1,2} & y'_{1,3} & y'_{1,0} \\ y'_{2,2} & y'_{2,3} & y'_{2,0} & y'_{2,1} \\ y'_{3,3} & y'_{3,0} & y'_{3,1} & y'_{3,2} \end{bmatrix}$$

$$= \begin{bmatrix} y_{0,3} & y_{0,0} & y_{0,1} & y_{0,2} \\ y_{1,0} & y_{1,1} & y_{1,2} & y_{1,3} \\ y_{2,1} & y_{2,2} & y_{2,3} & y_{2,0} \\ y_{3,2} & y_{3,3} & y_{3,0} & y_{3,1} \end{bmatrix} = [z'_{r,c}]_{r,c=0}^3 \quad (10)$$

From (3) and (10), we find that:

$$Z' = [z_{r,(c+3) \, mod \, 4}]_{r,c=0}^3 = P_1(Z) \quad (11)$$

Applying Z' to MixColumns, we get:

$$U' = \begin{bmatrix} 02 & 03 & 01 & 01 \\ 01 & 02 & 03 & 01 \\ 01 & 01 & 02 & 03 \\ 03 & 01 & 01 & 02 \end{bmatrix} \begin{bmatrix} z'_{0,3} & z'_{0,0} & z'_{0,1} & z'_{0,2} \\ z'_{1,3} & z'_{1,0} & z'_{1,1} & z'_{1,2} \\ z'_{2,3} & z'_{2,0} & z'_{2,1} & z'_{2,2} \\ z'_{3,3} & z'_{3,0} & z'_{3,1} & z'_{3,2} \end{bmatrix}$$

$$= [u_{r,(c+3) \bmod 4}]_{r,c=0}^3 = P_1(U) \tag{12}$$

Then we apply the permuted round key matrix $K' = P_1(K)$ resulting in:

$$V' = [u'_{r,c}]_{r,c=0}^3 + [k'_{r,c}]_{r,c=0}^3$$

$$= [u_{r,(c+3) \bmod 4}]_{r,c=0}^3 + [k_{r,(c+3) \bmod 4}]_{r,c=0}^3 = P_1(V) \tag{13}$$

Finally we apply inverse permutation P^{-1} to the output. We get:

$$P_1^{-1}(V') = P_1^{-1}(P_1(V)) = V \tag{14}$$

$\square$

3.1 Invariance-based CED Architecture

We design two invariance-based CED architectures for AES: Fully pipelined and iterative.

Fully pipelined: There are 11 stages in our pipelined architecture. For each pipeline stage in Figure 2(a), we add two muxes (mux_x and mux_k) and a comparator (cmp). P is a permutation of wires based on the invariance property. P^{-1} is the inverse permutation. We need two encryption cycles to detect the faults, and let us call them C1 and C2. In C1, let the input and the key of round 1 be $X1$ and $K1$. We run the encryption with mux_s and mux_k, selecting $X1$ and $K1$ as the inputs. The round result $V1$ is stored in the data register. Then we run C2; we run the encryption with mux_s and mux_k, selecting permuted inputs $X1'$ and $K1'$, respectively. At the end of C2, we inverse permute output $V1'$ and compare it with the value $V1$ stored in the data register. If the results are equal, no fault is detected. Otherwise, the comparator will assert the fault indication flag. The comparator does not add delay to the critical path because the comparison can be performed when the next round input is executed. We see that C1 can be any normal encryption cycle, and C2 is the corresponding extra cycle, which selects the permuted inputs to be performed after every C1. One can add a C2 after R rounds; we call R the checking ratio. R can be changed based on the tradeoff between performance, reliability, and security specified by the designer. For a detailed analysis, please see Section 3.3.

Iterative: As shown in Figure 2(b), we add mux_x and mux_k and a comparator. There is one security benefit of iterative implementation. Each ciphertext takes 10 cycles to generate in an iterative architecture. If the designer chooses to add a C2 for every ten cycles, then the faulty ciphertext will not be sent to the output because the fault is detected before it is generated. This will further prevent an attacker from stealing the secret key. In the pipeline architecture, a ciphertext is generated every cycle. Therefore, if the faults are generated before the comparison, faulty outputs will be obtained by the attacker.

3.2 Fault Analysis

Invarianced-based CED detects all single-bit and single-byte faults. Our simulations show that this CED scheme detects 99.99999997% of multiple burst faults and close to 100% of multiple random faults. Fault coverage (FC) is calculated as:

$$FC = 1 - FMR$$

where FMR is the fault miss rate calculated as:

$$FMR = \frac{T_{undetected}}{T_{total} - T_{correct}}$$

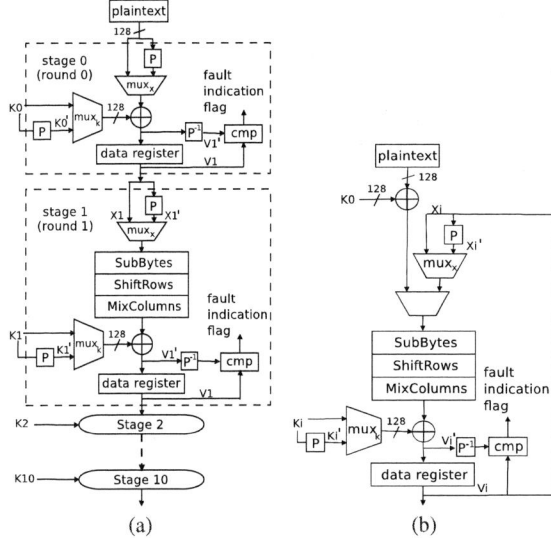

Figure 2: AES hardware architectures (a) Pipeline. (b) Iterative. cmp stands for comparator. We assume that the comparator is fault tolerant.

where $T_{undetected}$ is the number of tests in which faults are excited but not detected. T_{total} is the total number of tests we applied. $T_{correct}$ represents the tests in which the faults are not excited.

3.2.1 100% Fault Coverage of Single-Bit Faults

THEOREM 2. *If a single-bit fault in any of the steps in a round affects the outputs of the final result of that round, our scheme will detect it.*

PROOF. **Case 1: A single-bit fault in S-box (SB).** In Figure 1, let the $SB_{i,j}$ ($0 \le i,j \le 3$) have a single-bit stuck-at fault. If the SBs are implemented using ROMs, the considered fault corresponds to an address fault of the ROM, a fault in memory location, or a fault in the output data lines. If the SBs are implemented using combinational logic, the considered fault can appear in any gate of the implementation.

In C1, the $SB_{i,j}$ generates faulty output $y_{i,j}$. After ShiftRows, the outputs are $[z_{r,c}]_{r,c=0}^3 = [y_{r,(r+c) \bmod 4}]_{r,c=0}^3$, and the faulty state element is $z_{i,(j-i) \bmod 4} = y_{i,j}$. In MixColumns, a single faulty input causes four bytes within the same column to be faulty. The faulty state elements are represented as $[u_{r,(j-i) \bmod 4}]_{r=0}^3$. After AddRoundKey, $[v_{r,(j-i) \bmod 4}]_{r=0}^3$ are the faulty state elements. In C2, we apply X' and K' as the permuted inputs. Using the same steps shown above, faulty state elements are represented as $[v'_{r,(j-i) \bmod 4}]_{r=0}^3$. From (8), we know that

$$[v'_{r,(j-i) \bmod 4}]_{r=0}^3 = [v_{r,((j-i)+3 \bmod 4)}]_{r=0}^3$$

Therefore, the faulty column in C1 is $(j-i) \bmod 4$, but the faulty column in C2 corresponds to column $((j-i) \bmod 4)+3) \bmod 4$ in C1; note that $0 \le i,j \le 3$ and $(j-i) \bmod 4 \ne (((j-i) \bmod 4)+3) \bmod 4$.

Since faulty $SB_{i,j}$ affects different columns in C1 and C2, we always compare a faulty column with a nonfaulty column, and our scheme detects the fault as long as it affects the output. For a concrete example, let $SB_{1,2}$ be faulty, thus, $y_{1,2}$ is the faulty output byte. After ShiftRows, $z_{1,1} = y_{1,2}$. After MixColumns, the faulty state elements are shown as $[u_{r,1}]_{r=0}^3$. Then we apply this as the input of AddRoundKey, so faulty state elements are shown as $[v_{r,1}]_{r=0}^3$. Then we run C2, and faulty state elements are represented as $[v'_{r,1}]_{r=0}^3$. Because $[v'_{r,1}]_{r=0}^3 = [v_{r,0}]_{r=0}^3$, the faulty

575

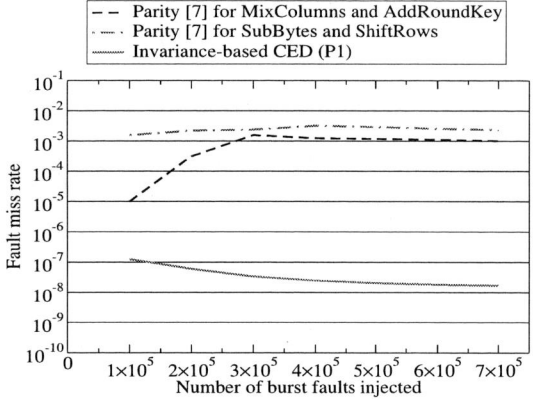

Figure 3: Simulation results show that the fault miss rate of the invariance-based CED is superior to that of the parity scheme in [9].

columns in C1 and C2 are different, and we detect the fault by comparing the outputs.

Case 2: A single-bit fault in ShiftRows. A fault in ShiftRows is equivalent to a fault at the input of MixColumns, and thus, we prove this in case 3.

Case 3: A single-bit fault in MixColumns. Since MixColumns is mainly implemented with XOR and a few other basic gates, we consider a fault in MixColumns in three scenarios: the input, the internal logic gates, and the output. If there is a stuck-at fault in the input, the fault will propagate to all four bytes in the same column. Assuming that column $[u_{r,j}]_{r=0}^{3}$ is faulty in C1, we know that the column $[v_{r,j}]_{r=0}^{3}$ is faulty in the final output of the current round. In C2, we know that the column $[u'_{r,j}]_{r=0}^{3}$ is faulty, and the column $[v'_{r,j}]_{r=0}^{3}$ of the output of the round is also faulty. From (12) and (14), we know that $j' = (j+3) \mod 4$. Because the faulty columns of the two outputs are different, we detect the fault.

Case 4: A single-bit fault in AddRoundKey. AddRoundKey is mainly implemented as bit-wise XOR gates. We consider a fault in AddRoundKey as stuck-at fault in the input or output. The fault at the input is equivalent to the fault in the output of MixColumns. Let us prove the theorem true for a single-bit fault at the output of AddRoundKey. Let the faulty bit affects byte $v_{i,j}$ in the C1 and $v'_{i,j}$ in C2. From (14), we know that $v'_{i,j} = v_{i,(j+3) \mod 4}$. Again, the faulty columns in C1 and C2 are different, and thus we detect the fault. □

3.2.2 100% Fault Coverage for Single-Byte Faults

Recent experiments have shown that high-energy ions can energize two or more adjacent memory cells in a circuit [12]. Because an attacker can choose light [13] or electromagnetic radiation [14] to inject faults, this is a realistic attack model. We define single-byte fault as faults that affect at most a single byte, i.e., they can affect from one bit upto eight bit of the byte.

THEOREM 3. *If multiple faults in any of the processing steps in a round affect a byte quantity, the invariance scheme will detect it.*

PROOF. We prove this theorem for single-byte fault in either S-box, ShiftRows, MixColumns, or AddRoundKey.

Case 1: A single-byte fault in S-box (SB). Let multiple faults in S-box affect one byte output $y_{i,j}$ in C1. From the proof of our theorem 1, after AddRoundKey, faulty state elements are shown as $[v_{r,(j-i) \mod 4}]_{r=0}^{3}$. After C2, the faulty state elements are represented as $[v'_{r,(j-i) \mod 4}]_{r=0}^{3}$. Theorem 1 proves that these faults are detected.

A single-byte fault in ShiftRows, MixColumns, and AddRound-Key can also be similarly detected. □

Table 1: Comparison of fault coverage. a. burst faults b. random faults

FDS	Fault coverage		
	Single bit	Single byte	Multiple bit
HW red.	100%	100%	100%
Parity 1 [3]	100%	50%	99.997%
Parity 2 [7]	98.7%	50%	48-53%
Parity 3 [9]	100%	50%	99.996%
Alg. level [10]	100%	100%	100%
Invariance	**100%**	**100%**	**99.99999997% [a]** **100% [b]**

In order to have a 100% single-byte fault detection rate, one need hardware redundancy or combined redundancy [10], both of which have more than 100% hardware overhead. Our fault simulation confirms that the invariance-based CED detects all single-bit and single-byte faults.

3.2.3 Fault Coverage for Multiple Faults

We simulated multiple faults for the invariance scheme and compared it with the one proposed in [9]. These models cover both natural faults and fault attacks [15].

Due to technology constraints, an attacker may not be able to inject a single-bit fault [15]. Multiple faults are actually injected in the process. We use burst and random fault models [15]. We use Fibonacci Linear Feedback Shift Register (LFSR) with 128-bit output taps to inject faults. The maximum sequence length polynomial for the LFSR is selected as $L(X) = X^{128} + X^{29} + X^{27} + X^2 + 1$.

Burst faults occur at the output of only one operation at a time, i.e., the faults are injected at the 128-bit output of one operation in the AES encryption. This includes both stuck-at zero and one faults. The size, location, and type of the burst are randomly generated.

The results of our simulations for the burst faults in the AES encryption are shown in Figure 3. We compare our miss rate with that of [9]. The dot-dash line respresents the fault miss rate of Sub-Bytes and ShiftRows in the AES encryption in [9]. The dash line represents the fault miss rate of MixColumns and AddRoundKey in the AES encryption in [9]. The solid line represents the fault miss rate of the invariance-based CED. We have injected up to 700,000 burst faults at the operation outputs and monitored the faults that are detected by the fault indication flag. The fault miss rate for [9] is between 10^{-2} and 10^{-3}. The miss rate of our scheme is between 10^{-7} and 10^{-8}; a reduction of 10^5.

Random faults are injected at random locations, i.e., four 128-bit outputs of the operations. In another simulation, we saw 0% fault miss rate after injecting up to 700,000 random faults.

3.2.4 Comparison of Fault Coverage

As shown in Table 1, for single-bit and single-byte faults, the invariance-based CED provides 100% fault coverage, the same as [10] and hardware redundancy. It is note that [10] requires both encryption/decryption to be on chip to achieve such fault coverage. While most of the parity schemes achieves 100% fault coverage for single-bit fault, they can only provide 50% fault coverage for single-byte fault [8]. For multiple faults, [10] and hardware redundancy provide 100% fault coverage. The invariance-based CED provides 99.99999997% fault coverage, and much higher than the parity-based schemes. The tradeoff between performance and detection latency can be explored by varying the checking ratio R, which is the ratio of the number of results computed without invariance to the number of results computed with invariance-based CED.

3.3 Security Analysis

In this section, we propose two policies that the designers can

Table 2: Comparisons of implementation of CED schemes on two Xilinx FPGAs. We use the metrics, FPGA platform, and results from [9]. Our pipeline implementation are shown in bold, and we implement the iterative architectures.   a. the latency is 2x the original AES encryption/decryption   b. using two (256×9) distributed memories for CED of each S-box or inverse S-box   c. using (256×9) distributed memories for CED of each S-box or inverse S-box   d. checking ratio is 11   e. checking ratio is 10   f. checking ratio is 1

FPGA (Device)	Arch.	Scheme	Encryption				Decryption			
			Slice (overhead)	Freq. (MHz)	Thro. (Gbps)	Eff(Mbps /slice)	Slice (overhead)	Freq. (MHz)	Thro. (Gbps)	Eff.(Mbps /slice)
$Virtex^{TM}$-4 (xc4vlx160-12)	Pipe.	Original	18335(-)	240.5	30.8	1.7	19322(-)	203.5	26.0	1.3
		HW Red.	36684(100.1%)	240.5	30.8	1.1	38658(100.1%)	203.5	26.0	0.7
		Parity 1 [3][b]	39104(113.3%)	163.5	20.9	0.5	40244(108.3%)	145.4	18.6	0.5
		Parity 2 [7][c]	21211(15.7%)	240.5	30.8	1.4	22280(15.3%)	203.5	26.0	1.1
		Parity 3 [9]	20127(9.8%)	240.5	30.8	1.5	20909(8.2%)	203.5	26.0	1.2
		Alg. level [10]	38273(108.7% [9]) (1.6%)	194.9	24.9[a]	0.6	38273(98.1% [9]) (1.6%)	194.9	24.9[a]	0.6
		Invariance	**21253(15.9%)**	**232.3**	**27.2[d]– 14.9[f]**	**1.28[d]– 0.7[f]**	**22240(15.1%)**	**197.6**	**23.1[d]– 12.6[f]**	**1.0[d]– 0.6[f]**
	Iter.	Original	1905(-)	224.6	2.9	1.5	2002(-)	192.0	2.5	1.2
		Invariance	**2170(13.9%)**	**217.4**	**2.55[e]– 1.4[f]**	**1.2[e]– 0.7[f]**	**2267(13.2%)**	**186.7**	**2.2[e]– 1.2[f]**	**1.0[e]– 0.6[f]**
$Virtex^{TM}$-5 (xc5vlx110-3)	Pipe.	Original	2960(-)	371.7	47.6	16.1	3906(-)	296.3	37.9	9.7
		HW Red.	5934(100.5%)	371.7	47.6	10.2	7826(100.4%)	296.3	37.9	5.5
		Parity 1 [3][b]	5590(88.9%)	282.8	36.2	6.5	6680(71.2%)	260.2	33.3	4.9
		Parity 2 [7][c]	3619(22.3%)	304.0	38.9	10.7	4426(13.3%)	277.0	35.5	8.0
		Parity 3 [9]	3757(26.9%)	371.7	47.6	12.7	4286(9.7%)	296.3	37.9	8.8
		Alg. level [10]	5849(97.6% [9]) (-14.8%)	284.4	36.4[a]	6.2	5849(49.7% [9]) (-14.8%)	284.4	36.4	6.2
		Invariance	**3664(23.9%)**	**358.9**	**42.2[d]– 23.0[f]**	**12.0[d]– 6.6[f]**	**4434(13.5%)**	**288.9**	**33.9[d]– 18.5[f]**	**7.6[d]– 4.2[f]**
	Iter.	Original	344(-)	347.0	4.4	12.8	462(-)	286.4	3.7	7.9
		Invariance	**438(27.3%)**	**335.8**	**3.9[e]– 2.2[f]**	**7.9[e]– 4.4[f]**	**539(16.7%)**	**273.1**	**3.2[e]– 1.8[f]**	**5.9[e]– 3.7[f]**

employ to further secure their design.

Randomized control: The invariance-based CED technique cannot detect the faults that are not excited in C1 and C2 rounds. If an attacker determines the architecture of the AES with the proposed CED implementation, this feature can be used as a weakness to insert faults in such a way that they do not exist during the CED rounds but only during the normal rounds. To prevent this, we proposed using Randomized CED Round Insertion (RCRI). In this method, the positions of the CED rounds C1 and C2 are randomized during the 11-round AES encryption process for pipelined architecture. This can be implemented as shown in the state diagram of Figure 4. A random number can be obtained using the randomness property of the AES algorithm.

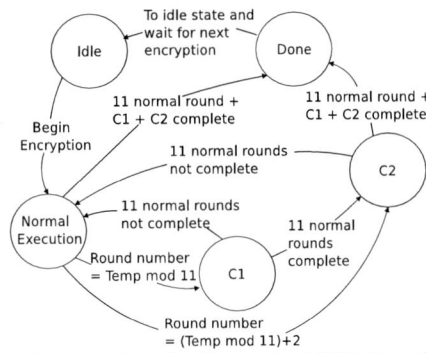

Figure 4:   State machine for Randomized CED Round Insertion (RCRI)

For example, a Rand register can be incorporated into the circuit with some random number stored in it at manufacture time and for every subsequent encryption performed, the resulting ciphertext is xored with Rand to get a Temp number. When an encryption is performed, the algorithm enters the normal execution state. Normal encryption rounds are performed until the value of the Temp modulo 11 equals the round number. Once this condition is satisfied, the CED round C1 is performed. Depending on whether 11 normal rounds have been performed, either C2 or the remaining

normal rounds are performed. The encryption process is complete when 11 normal rounds and the randomly inserted C1 and C2 CED round are complete. For the mod operation, we can take the last four bits of Temp and apply it to a lookup table which contains modulo 11 results from input 0 to 15. For the iterative architecture, since the encryption takes 10 rounds, we use Temp modulo 10.

Adaptive checking: If the designer chooses $R \neq 1$, some of the transient faults injected by the attacker may go undetected. However, an attacker needs to obtain a large number of faulty outputs to steal the secret key. We suggest the designer deploy an adaptive approach. When a fault is detected, the chip changes its checking ratio from the R specified by the designer to 1, and thus almost all the faults will be detected. If faults are detected in a number of consecutive checks, the chip can stop its function or self-destruct.

If $R = 1$, all results will be checked. The fault miss rate remains the same for permanent and transient faults. If $R > 1$, every R^{th} result will be checked. Let us assume the transient faults appear for N cycles. When $R \leq N$, the fault coverage remains the same, because the results of C1 and C2 will be checked before the faults disappear. When $R > N$, the probability of detecting a single-bit and single-byte fault is $\frac{N}{R} \times 100\%$ and that of multiple burst faults is $\frac{N}{R} \times 99.99999997\%$.

3.4   Implementation and Comparison

The implementation results shown in Table 2 are all post place-and-route. We implement fully pipelined and iterative architectures. We use pipelined distributed memories for S-boxes and inverse S-boxes similar to [9]. Hardware redundancy, information redundancy [3, 7, 9], combined redundancy [10], and invariance-based CED are compared. The metrics include (1) slice utilization (the number of occupied slices), (2) slice overhead (ratio of number of slices for CED schemes over the number of slices for AES), (3) maximum clock frequency, (4) throughput, and (5) efficiency (raitio of (4) over (1)).

For pipeline architecture, the hardware overhead of the invariance-based CED is much lower than that of hardware redundancy for both encryption and decryption. The invariance-based CED provides flexible throughput. If the designer checks all rounds, checking ratio R is 1 and the throughput overhead is 100%. If the designer performs one check per ciphertext, the checking ratio R is 11 for pipeline architecture and 10 for iterative. The throughput overhead is $\frac{1}{11}$ for the pipeline architecture and $\frac{1}{10}$ for the iterative architecture. If R is large, then the throughput is unaffected. The invariance-based CED has higher efficiency when the checking ratio is 11 compared to hardware redundancy. Because the scheme in [7] uses 1-bit signatures for the 128-bit block of data, it has lower hardware overhead and higher efficiency compared to invariance-base scheme. However, from Table 1, the fault coverage of this scheme is the lowest. [3] and [9] use 16 bits for each 128-bit block, and this leads to much higher fault coverage. The invariance-based CED has much smaller hardware overhead and higher efficiency than [3], but provides higher fault coverage. Another limitation of [3] and [7] is that they are only applicable to S-box implementation using LUT. On Virtex-5, the invariance-based CED has higher efficiency than most of the CED schemes except [9]. Although the invariance-based CED has approximately the same hardware overhead compared to [9], it detects all single-byte faults and lowers the fault miss rate of multiple burst faults by an order of 10^5. The schemes in [10] are only applicable when both encryption and decryption are both on the same chip. Therefore, if only encryption or decryption is on chip, the hardware overhead of [10] is in the 49.7–108.7% range [9], e.g., 108.7% for AES encryption on Virtex-4 FPGA. If both encryption and decryption are on the same chip, the hardware overhead of [10], which is from the comparator, is very low. For Virtex-5, the overhead of this scheme is -14.8%, because the slice utilization of this scheme is smaller than the total slice utilization of encryption and decryption. However, the efficiency of the invariance-based CED is higher than that of [10]. Most importantly, the invariance-based CED and all other CED schemes do not require both encryption and decryption to be on chip.

For iterative architecture, since round 0 is performed in the same clock cycle as round 1, an extra delay is added in the critical path. The hardware overhead of the invariance-based CED as a 16.7-27.3% is slightly higher than that of the pipeline architecture on Virtex-5.

4. CONCLUDING REMARKS

There are several other mapping invariances that can be used for CED [11]. Most of them restrict the pattern of the inputs and thus are not effective when realistic random inputs are provided. However, there are two other invariances that allow us to perform CED on any inputs:

$$P_2(X) = \begin{bmatrix} x_{0,2} & x_{0,3} & x_{0,0} & x_{0,1} \\ x_{1,2} & x_{1,3} & x_{1,0} & x_{1,1} \\ x_{2,2} & x_{2,3} & x_{2,0} & x_{2,1} \\ x_{3,2} & x_{3,3} & x_{3,0} & x_{3,1} \end{bmatrix} \quad (15)$$

$$P_3 = P_1(P_{pre}(X)) = P_1(\begin{bmatrix} x_{0,0} & x_{0,3} & x_{0,2} & x_{0,1} \\ x_{1,0} & x_{1,3} & x_{1,2} & x_{1,1} \\ x_{2,0} & x_{2,3} & x_{2,2} & x_{2,1} \\ x_{3,0} & x_{3,3} & x_{3,2} & x_{3,1} \end{bmatrix}) \quad (16)$$

P_2 swaps the first and third columns and also the second and fourth columns in the state matrix. P_3 is the same as P_1 except that the initial input X is permuted by P_{pre} before being applied to the input. Fault miss rate for CED using invariances P_1, P_2, and P_3 are compared in Figure 5. The fault miss rate dropped sharply to below 10^{-7} after we injected 11 faults using P_1 and P_3, and 15 faults for P_2. After we injected 15 faults, the fault miss rate was very low and thus not shown in Figure 5. The fault miss rates of invariance P_1 and P_3 are lower than the fault miss rate of invariance P_2 with any number of faults injected. Compare to P_2, the area and

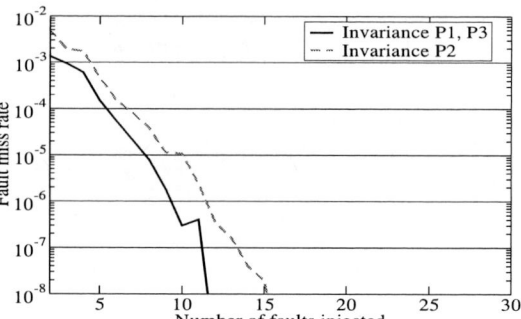

Figure 5: Fault miss rate for invariance P_1, P_2, and P_3

performance overheads of P_1 is the same. For P_3, one first need to apply $P_{pre}(X)$ as input to run C1, and store the result V_{pre}. Then use $P_1(P_{pre}(X))$ as input to run C2, and apply P_1^{-1} to the result V'_{pre} to compare with V_{pre}. Both C1 and C2 are extra overhead for performance. Compare with P_3, C2 is the only performance overhead for P_1. Therefore, P_1 is the most effective invariance.

5. ACKNOWLEDGMENTS

This material is based upon work supported by the NSF CNS program under grant 0831349.

6. REFERENCES

[1] D. P. Siewiorek and R. S. Swarz, "Reliable Computer Systems: Design and Evaluation," *A K Peters/CRC Press; 3 edition*, 1998.

[2] D. Boneh, R. DeMillo, and R. Lipton, "On the Importance of Checking Cryptographic Protocols for Faults," in *Eurocrypt*. Lecture Notes in Computer Science, 1997, pp. 37 – 51.

[3] G. Bertoni, L. Breveglieri, I. Koren, P. Maistri, and V. Piuri, "Error Analysis and Detection Procedures for a Hardware Implementation of the Advanced Encryption Standard," *IEEE Trans. Computers*, vol. 52, no. 4, pp. 492 – 505, 2003.

[4] G. L. P. Dusart and O. Vivolo, "Differential Fault Analysis on AES," in *ACNS*, Oct 2003, pp. 293–306.

[5] G. Piret and J. Quisquater, "A Differential Fault Attack Technique against SPN Structures, with Application to the AES and Khazad," in *CHES*, Sept 2003, pp. 77–88.

[6] P. Maistri and R. Leveugle, "Double-Data-Rate Computation as a Countermeasure against Fault Analysis," *IEEE Transactions on Computers*, vol. 57, no. 11, pp. 1528 – 1539, Nov 2008.

[7] K. Wu, R. Karri, G. Kuznetsov, and M. Goessel, "Low Cost Concurrent Error Detection for the Advanced Encryption Standard," in *ITC*, Oct 2004, pp. 1242–1248.

[8] M. Mozaffari-Kermani and A. Reyhani-Masoleh, "A Lightweight High-Performance Fault Detection Scheme for the Advanced Encryption Standard Using Composite Field," *IEEE Transactions on VLSI*, vol. 19, no. 1, pp. 85–91, 2011.

[9] ——, "Concurrent Structure-Independent Fault Detection Schemes for the Advanced Encryption Standard," *IEEE Transactions on Computers*, vol. 59, no. 5, pp. 608–622, 2010.

[10] R. Karri, W. Kaijie, P. Mishra, and K. Yongkook, "Fault-Based Side-Channel Cryptanalysis Tolerant Rijndael Symmetric Block Cipher Architecture," in *DFT*, Oct 2001, pp. 418–426.

[11] T. V. Le, R. Sparr, R. Wernsdorf, and Y. Desmedt, "Complementation-Like and Cyclic Properties of AES Round Functions," *In: Dobbertin, H., Rijmen, V., Sowa, A. (eds.) AES Springer, Heidelberg (2005)*, vol. 3373, no. 6, pp. 128 – 141, 2005.

[12] R. Reed, M. Carts, P. Marshall, C. Marshall, O. Musseau, P. McNulty, D. Roth, S. Buchner, J. Melinger, and T. Corbiere, "Heavy Ion and Proton Induced Single Event Multiple Upsets," *IEEE Transactions on Nuclear Science*.

[13] S. P. Skorobogatov and R. J. Anderson, "Optical Fault Induction Attacks," in *proceedings of CHES*, Aug 2002, pp. 2–12.

[14] D. Samyde, S. Skorobogatov, R. Anderson, and J.-J. Quisquater, "On a New Way to Read Data from Memory," in *proceedings of IEEE Security in Storage Workshop*, Dec 2002, pp. 65–69.

[15] L. Breveglieri, I. Koren, and P. Maistri, "An Operation-Centered Approach to Fault Detection in Symmetric Cryptography Ciphers," *IEEE Transactions on Computers*, vol. 56, pp. 635 – 649, May 2007.

Accelerating Neuromorphic Vision Algorithms for Recognition

Ahmed Al Maashri[*]   Michael DeBole[†]   Matthew Cotter[*]   Nandhini Chandramoorthy[*]

Yang Xiao[*]   Vijaykrishnan Narayanan[*]   Chaitali Chakrabarti[‡]

*Microsystems Design Lab, The Pennsylvania State University
{maashri, mjcotter, nic5090, yux106, vijay}@cse.psu.edu

†IBM System and Technology Group
mvdebole@us.ibm.com

‡School of Electrical, Computer and Energy Engineering, Arizona State University
chaitali@asu.edu

ABSTRACT

Video analytics introduce new levels of intelligence to automated scene understanding. Neuromorphic algorithms, such as HMAX, are proposed as robust and accurate algorithms that mimic the processing in the visual cortex of the brain. HMAX, for instance, is a versatile algorithm that can be repurposed to target several visual recognition applications. This paper presents the design and evaluation of hardware accelerators for extracting visual features for universal recognition. The recognition applications include object recognition, face identification, facial expression recognition, and action recognition. These accelerators were validated on a multi-FPGA platform and significant performance enhancement and power efficiencies were demonstrated when compared to CMP and GPU platforms. Results demonstrate as much as 7.6X speedup and 12.8X more power-efficient performance when compared to those platforms.

Categories and Subject Descriptors

C.3 [SPECIAL-PURPOSE AND APPLICATION-BASED SYSTEMS]: Signal processing systems

General Terms

Design, Experimentation, Performance

Keywords

Recognition, Domain-Specific Acceleration, Heterogeneous System, Power Efficiency

1. INTRODUCTION

The visual cortex of the mammalian brain is remarkable in its processing and general recognition capabilities providing inspiration for complex, power-efficient systems and architectures. Researchers [1,2,3] have made advances in understanding the processing that occurs in the visual cortex. These advances have a profound impact on a range of application domains used for image recognition tasks such as surveillance, business analytics, and the study of cell migration.

As a step towards exploring how the brain efficiently processes visual information, a brain-inspired feed-forward hierarchical model (HMAX) [2] has become a widely accepted abstract representation of the visual cortex. HMAX models are mainly implemented on general-purpose processors (CPUs) and graphics processing units (GPUs), which do not attain the power and computational efficiency that can be achieved by custom hardware implementations [3,4,5,6]. To achieve the performance, power, and flexibility to support computations used in neuromorphic applications, the ideal platform is the heterogeneous integration of domain-specific accelerators with general-purpose processor architectures.

This paper proposes a neuromorphic system, based on HMAX, for universal recognition. The system is composed of customized hardware accelerators that target four applications; namely, object recognition, face identification, facial expression recognition, and action recognition. This neuromorphic system is evaluated on a prototype heterogeneous CMP system composed of multi-FPGA system interfaced to quad-core Intel Xeon processor. The results indicate that the proposed architecture achieves recognition accuracies ranging from 70-90% across the four recognition algorithms. A detailed comparison of the power and performance with respect to HMAX variants executed on GPUs, multi-core CPUs, and FPGAs is performed. The results reveal that the proposed FPGA prototype provides frames-per-second(fps)-per-watt improvement as much as 12.8X over CPUs, and 9.7X over GPUs.

The rest of this paper is organized as follows; Section 2 provides an overview of HMAX model. Section 3 describes the micro-architecture of the accelerators developed for HMAX, while Section 4 discusses the evaluation platforms and presents results from the FPGA-based emulation system. Section 5 presents Related Work. Finally, Section 6 concludes the paper.

2. HMAX COMPUTATIONAL MODEL

Figure 1 shows a computational template of HMAX [2,3,5]. The model primarily consists of two distinct types of computations: convolution and pooling (non-linear subsampling), corresponding to the Simple, S, and Complex, C, cell types, respectively. The first S-layer (S_1) is comprised of fixed, simple-tuning cells, represented by oriented Gabor filters. Following the S_1 layer, the remaining layers alternate between max-pooling layers and template-matching layers tuned by a dictionary encompassing patterns representative of the categorization task.

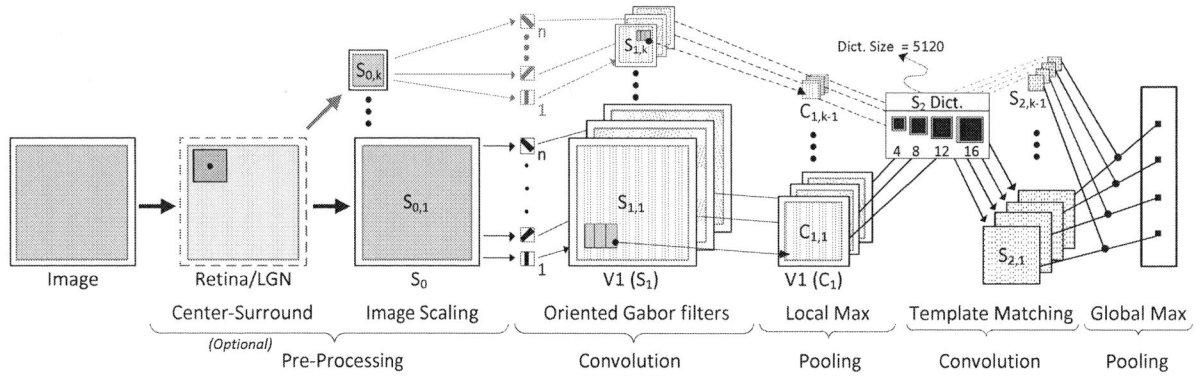

Figure 1. Typical Structure of HMAX Model: Center-Surround can be optionally applied to gray-scale input image. Image Scaling (S_0) produces k scales, forming a pyramid out of downsampled input image. A bank of Gabor filters ($S_{1,1}$ $S_{1,2}$ ... $S_{1,k}$) are used to detect edges at n orientations, while local maximum pooling ($C_{1,1}$ $C_{1,2}$ ... $C_{1,k-1}$) finds the maximum response over the S_1 output. Template matching is performed in S_2. Finally, C_2 performs global max operation producing a feature vector.

2.1 HMAX for Object Recognition

Our work uses a specific implementation for the object recognition task developed by [7], as it represents the current understanding of the ventral stream and produces good results when used for classification. This model is represented by a total of five layers: an image layer and four layers corresponding to the alternating S and C units.

Image layer: In this layer, the image is converted to grayscale and then downsampled to create an image pyramid of 12 scales, with the largest scale being 256x256.

S_1 (Gabor filter) layer: The S_1 layer corresponds to the V1 [2] simple cells in the ventral stream and is computed by performing a convolution with the full range of orientations at each position and scale. The number of orientations used in this model is 12, producing 12 outputs per scale for a total of 144 outputs.

C_1 (Local invariance) layer: This layer provides a model for the V1 complex cells and pools over nearby S_1 units with the same orientation. Within a scale, each orientation is convolved with a 3D max filter of size 10x10x2 (10x10 units in position and 2 in scale). This layer provides scale invariance over large local regions.

S_2 (Tuned features) layer: This layer models the V4 [2] by matching a set of prototypes against C_1 output. These prototypes have been randomly sampled from a set of representative images.

C_2 (Global invariance) layer: This layer provides global invariance by taking the maximum response from each of the templates across the scales. The layer removes all position and scale information, leaving only a complex feature set which can then be used later for classification.

2.2 Extensions of the HMAX Model

Neuroscientists have observed that the primates' visual system often shares a general, early-level processing structure, which eventually branches off into more specific higher-level representations. This serves as a motivation to configure the HMAX model to implement a variety of recognition problems beyond object classification.

2.2.1 HMAX for Face Processing

In order to support face identification and facial expression recognition, Meyers et al. [4] add a center-surround stage of processing to model the 'center-on surround-off' processing that

is present in the retinal and LGN of the thalamus. In addition, the model does not perform S_2 and C_2 stages in order to maintain visual features localized to a particular region in space.

LGN/Retinal (Center-Surround) Layer: The center-surround is computed prior to pyramid generation and helps to eliminate intensity gradients due to shadows. The processing is done by placing a 2D window at each position in the input image that is identical in size to the filter used for S_1. The output is then computed by dividing the current pixel's intensity by the mean of the pixel intensities within the window.

2.2.2 HMAX for Action Recognition

While HMAX was originally limited to model the ventral stream, a model of the dorsal stream is useful for analyzing motion information. Jhuang et al. [6] have proposed augmenting the original HMAX model with the dorsal path as it can then be applicable to motion-recognition tasks, such as identifying actions in a video sequence. Computationally, this is done by integrating spatio-temporal detectors into S_1, while adding two additional layers, S_3 and C_3, which track features over time, providing time-invariance to the structure.

Space-Time Oriented S_1 (Gabor filter) layer: S_1 units for motion are extended by adding a third temporal dimension to the receptive fields. Computationally, this layer becomes an $n \times 2D$ convolution across a sliding window of past, present, and future frames, where n is total number of frames.

Space-Time Oriented S_3/C_3 (Tuning/Pooling) layers: S_3 unit responses are obtained by temporally matching the output of C_2 features to a dictionary, similar to S_2, where each patch represents a sampled sequence of frames. C_3 unit responses are the maximum response over the duration of a video sequence.

Based on these observations, the design of an accelerator based on the HMAX model should also contain additional hardware accelerators designed for spatio-temporal detection, retinal (center-surround) processing, and time-invariance operations, with an option for preserving localized spatial features.

3. HMAX SYSTEM ARCHITECTURE

The proposed HMAX accelerators are interconnected using a communication infrastructure that provides a number of features, including: high-bandwidth communication, run-time re-configurability and inter-accelerator message-passing for synchronization. The infrastructure accepts two types of accelerators; namely, stream-based and compute-based. Stream-based accelerators are more suited for on-the-fly processing of streaming data, while the compute-based accelerators are more suitable for iterative processing on non-contiguous blocks of data. Due to space limitation, this paper does not discuss the details of this infrastructure. A more in-depth treatment of the infrastructure can be found in [8].

3.1 HMAX Accelerators

Table 1 lists the HMAX accelerators and their functions. In this paper, we focus on the S_2/C_2 accelerator since this is the most time consuming stage in the HMAX model.

The S_2/C_2 accelerator combines the S_2 and C_2 stages into a single accelerator, which allows pooling to occur immediately following the computation of a current S_2 feature. Also, this can effectively decrease the amount of data required to be sent across the network by $\sum_{s=0}^{S-1}(N_{prototypes}[X_s, Y_s]) \cdot (X_s \cdot Y_s)/N_{prototypes} * 2$. Here, S is the number of scales at the S_2 stage, X_S (Y_S) is the dimension of scale S in the x-direction (y-direction). Practically, with 5120 prototypes using 12 scales and 12 orientations, this reduces the data transferred by 4,135X.

The accelerator consists of one or more instances of systolic 2D filters, which are the most computationally demanding components of the S_2 unit. These filters are designed to enable data reuse and take advantage of available parallel computational resources. Additionally, the accelerator makes efficient use of available memory hierarchies to improve performance. For instance, the per-scale outputs of the 2D filters in the convolver must be accumulated across all orientations. In order to avoid the network traffic associated with buffering these results in an external network-attached memory, the accelerator uses a scratchpad memory to store and accumulate these outputs immediately as they are produced. This results in increased performance of up to nX, where n is the number of orientations. However, scratchpad memories are not suitable for large volumes of data. As an example, in this implementation of HMAX for object recognition, the S_2 uses 5120 prototypes which require approximately 24 MB of storage. A more suitable storage solution is the use of off-chip memory. While the communication infrastructure supports network-attached memories, the large increase in network traffic will degrade performance. Instead, the S_2/C_2 accelerator integrates an optimized memory controller that interfaces directly to an off-chip memory.

Table 1. HMAX accelerators and their functions

Stream-based	Function(s)
Downsampling (DS)	Generates multiple scales by subsampling input
Normalization (Norm)	Computes windowed average for normalizing S_2 output
	Computes center-surround
S_1	Streaming 2D convolution
S_3	Streaming 1D convolution

Compute-based	Function (s)
C_1	Windowed pooling
S_2/C_2	Prototype correlation and global max
C_3	Global max operation

3.2 HMAX Data Processing Flow

The HMAX accelerators, listed in Table 1, are interconnected using a communication infrastructure [8]. This acceleration system can be loosely coupled to a CMP within a heterogeneous system, as illustrated in Figure 2. In this heterogeneous system, the processor, executing the main application, offloads the entire HMAX computation to the accelerators. However, prior to offloading the computation, the CMPs will configure these accelerators to reflect the current structure of the model (e.g. number of orientations, scales, pooling window sizes, etc...). Figure 2 also demonstrates an example application, action recognition, executed on the accelerators. First, the processor copies the data to the image buffer memory, *Img_Mem*, and a notification message to C_1 through the interface. C_1 performs several reads from the image buffer for every pooled output through *flow0*, calculating the proper scale and S_1 tuned cells each time, on the fly. As each pooled output is computed, it is written to the S_2/C_2 accelerator unit through the Normalization, Norm, accelerator using *flow1*. C_1 then messages S_2/C_2 notifying the latter that all scales have been written and initiating the computation of the S_2/C_2 tuned and pooled cells. Once that is completed, the S_2/C_2 writes the outputs to the C_3 unit, which calculates the S_3 tuned output during data-movement using *flow2*. Finally, when C_3 has completed the computations; the output is returned via the interface to the invoking processor. Similarly, other recognition applications are executed according to their computational structure.

4. EXPERIMENTAL EVALUATION

To evaluate the neuromorphic accelerators, we use a Multi-FPGA system as an emulation platform. Also, the four vision application domains (i.e. object recognition, face identification, facial expression recognition and action recognition) were implemented on multi-core CPU and GPU platforms for performance and accuracy comparisons. The following subsections discuss the datasets used for testing the evaluated platforms and provide a comparative analysis of the performance of each platform.

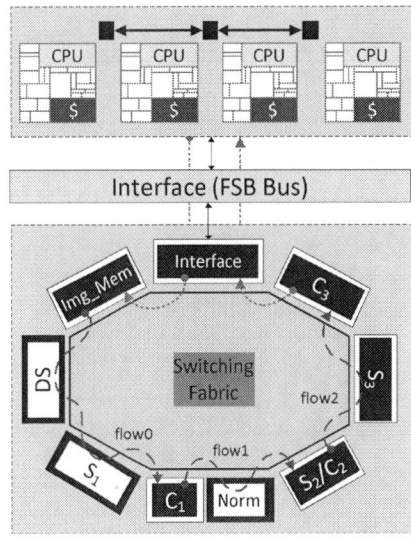

Figure 2: A heterogeneous system showing the interactions between the CMP and neuromorphic accelerators. The neuromorphic accelerators require access to three memory banks used to store the feature dictionaries used for S_2 and S_3 (not shown) and intermediate frames required for action recognition (*Img_Mem*).

4.1 Datasets for Evaluation

Table 2 shows the datasets used for evaluating the recognition accuracy of the neuromorphic accelerators. The Caltech 101 dataset [9] is used to test the accuracy of object classification using all 102 different categories. The *vehicles* dataset, prepared in-house, consists of 16 categories with a variety of objects including vehicles, aircrafts, military equipment and background scenery. The ORL dataset [10], used for testing face identification accuracy, contains a collection of close-up images of the faces of 40 different individuals from varying viewing angles. The FERET dataset [11] includes 1208 individuals from which a random subset of 10 individuals was chosen for evaluation. The JAFFE dataset [12] is used for testing the accuracy of facial expression recognition. Six different expressions were tested—anger, disgust, fear, happiness, sadness, and surprise. Finally, the Weizmann dataset [13] is used for testing human action recognition. This dataset includes 10 different categories of actions: bending, jumping jacks, vertical jumping, horizontal jumping, skipping, running, side-stepping, walking, one-hand-waving, and two-hand-waving.

4.2 Experimental Setup

The accelerators were prototyped on a Multi-FPGA platform that mimics a heterogeneous multi-core system. The platform contains a quad-core Xeon processor running at 1.6 GHz interfaced to an FPGA acceleration system through a Front-Side Bus, FSB. The FPGA system contains four Virtex-5 SX-240T FPGAs [14], all operating at 100 MHz. The quad-core processor is mainly used to transfer data to the accelerators through the FSB and to retrieve the output.

The reference CPU platforms contained 4- and 12-Core Xeon CPU systems with the total number of threads executed on each shown in Figure 3. The 4-Core CPU is clocked at 1.6 GHz, while the 12-Core CPU is clocked at 2.4 GHz. All CPU platforms utilized SSE instruction set extension. The 12-core Xeon processor configuration with 12 threads serves as the CPU reference when comparing implementations as it provides the best performance across CPU platforms and configurations. The GPU platform consists of an Nvidia Tesla M2090 board [15], which houses three 1.3 GHz Tesla T20A GPUs, and using CUDA as the programming language [5].

4.3 Performance

Accuracy: The fifth column in Table 2 shows the classification accuracy across all datasets using the feature vector produced by the accelerated HMAX. The recognition accuracy across all the platforms was similar; however, since accelerators use 32-bit fixed-point representation (1 bit for sign, 7 for integer and 24 for fraction), a slight degradation in accuracy was observed (i.e. $\leq$ 2%). This degradation is due to the truncation of the fixed-point representation during the multiply-accumulate operation.

Speed: We use frames (segments) processed per second (fps) as a metric to compare the speedup gained by each platform. In this paper, we use the term "segment" to refer to a group of 20 frames extracted from a video sequence for action recognition application. **Figure 4** shows a speedup comparison between the three platforms in terms of fps for the four recognition applications. The FPGA prototyping platform demonstrates a speedup ranging from 3.5X to 7.6X (1.5X to 4.3X) when compared to the CPU (GPU) platform. The FPGA platform exhibits increased performance improvement in the action recognition application. This is due to the cumulative effect of per-frame performance of the S_1 stage. Since each video segment

consists of 20 frames, the FPGA accelerator sees a linear increase in performance with the number of frames.

Power Efficiency: **Figure 5** compares the power efficiency (fps/Watt) across the three platforms. For the GPU, the command tool 'nvidia-smi -q' is used to probe the power consumption from a power sensor found on the GPU board. For the CPU and FPGA platforms, power consumption was measured using a power meter. The meter provides continuous and instantaneous reading of power drawn by the platform with 99.8% accuracy.

The power consumption for all platforms is measured only after the platform reaches steady-state to obtain the baseline power measurement. Then, the workload is executed and peak power is measured throughout the duration of the workload execution. For example, the power measurements show that when running HMAX for object recognition, the GPU, CPU and FPGA platforms consume 144, 116 and 69 Watts, respectively. Using these measurements, the power efficiency of each platform is computed as shown in **Figure 5**. The results show that the HMAX accelerators demonstrate a significant performance-per-watt benefit, ranging from 5.3X to 12.8X (3.1X to 9.7X) when compared to CPU (GPU) platform.

Configurability/Tradeoffs: It is often desirable to trade off accuracy for higher performance. We performed further experimentation with the accelerated HMAX to analyze impact of reduced overall accuracy on the execution time. For example, we experimented with changing the number of orientations processed by the HMAX model from 12 to 4. Reducing the number of orientations improved speed by 2.2X, while producing only a 1.1% difference in accuracy for the *vehicles* dataset. In another experiment, the numbers of input scales was varied, while observing its influence on accuracy and speedup using the *vehicles* dataset. **Figure 6** (left) shows that as the number of scales decreases the classification accuracy decreases until it reaches ~70% when using 5 input scales. On the other hand, **Figure 6** (center, right), shows a consistent improvement in speedup and power efficiency as number of scales is decreased, effectively reaching 15.4X better speedup and power efficiency when using only 5 scales compared to 12-scale configuration. Permitting such trade-off analysis makes the proposed accelerator very suitable for studies in modeling refinements and vision algorithm tuning.

4.4 Discussion of Results

There are a number of factors that contribute to the increased performance of the accelerator-based system. First, the underlying framework provides up to 1.6 GB/s (3.2 GB/s) bandwidth when

Table 2: Datasets used for evaluation. Note that there is no overlap in training and testing samples.

Application Domain	Dataset	# Classes	# Test samples	Accuracy (%)
Object recognition	Caltech 101	102	4543	70
	vehicles	16	1382	83
Face ID	ORL	40	200	85
	FERET	10	60	70
Facial expr. recognition	JAFFE	6	60	86.7
Action recog.	Weizmann	10	40	77.7

clocked at 100 MHz (200 MHz), supporting high transfer rates across the network. Second, the parallelism exploited by the architecture is enabled by the large number of resources (multipliers, registers, etc) available on the FPGA. For example, this allows up to 256 multiply-and-add operations to be performed simultaneously providing a 256X increase in performance over sequential operation. Third, the accelerators implement customized processing pipelines, taking advantage of data reuse. Finally, the ability to instantiate multiple processing units of the same type (e.g. S_2/C_2 units), leverages task-level parallelism to the user.

Similarly, the power efficiency benefits are the result of customized, application-specific architectures that are able to process incoming data in fewer cycles (compared to CPU/GPU) at a lower frequency. The use of custom numerical representations also contributes to the performance gain. It should be noted that our FPGA was fabricated with an older 65nm technology, compared to 45nm and 40nm technologies used with CPU and GPU platforms, respectively. It is expected that implementing the neuromorphic accelerators in silicon rather than on an FPGA platform will accentuate such benefits. For instance, Kuon et al. [16] show that at 90nm fabrication process, moving from SRAM-based FPGA to CMOS ASIC architectures improves critical path delay by $3X - 4.8X$, and dynamic power by $7.1X - 14X$.

5. RELATED WORK

The effort demonstrated in this work is synergistic with recent efforts aimed at domain-specific computing with configurable accelerators [17,18,19,20,21,22]. In [17] the authors detail the implementation of a multi-object recognition processor on SoC. They present a biologically inspired neural perception engine that exploits analog-based mixed-mode circuits to reduce area and power. However, except for the visual attention engine and the vector matching processors, all other algorithm acceleration is performed on multiple SIMD processors executing software kernels. Tsai et al. [18] propose a neocortical computing processor interconnected with high-bandwidth NoC. The processor consists of 36 cores; each contains multiple processing elements for performing the actual computations. Unlike the accelerators proposed in this paper, these processing elements are generic and not customized for any specific stage in the HMAX model. Other works [19,22] have proposed an architecture for image processing using Convolutional Neural Networks, CNN. These architectures can configure and train the neural network to support a variety of recognition algorithms. The convolution engine forms the critical component of these accelerators. While the authors in [19,22] indicate that mapping HMAX models to CNN structures is straightforward, the authors do not describe modifications necessary to implement large convolution windows or the n-dimensional convolutions that are required in the S_2 layer.

6. CONCLUSIONS

This work proposed reconfigurable neuromorphic accelerators for universal recognition that can be fabricated within a heterogeneous system. An FPGA prototyping platform is implemented as an emulation of the heterogeneous system. The prototyping platform exhibits a remarkable performance gain of up to 7.6X (4.3X) compared to the CPU (GPU). Moreover, this prototyping platform shows a superior power efficiency of 12.8X (9.7X) compared to the CPU (GPU) platform.

7. ACKNOWLEDGMENTS

This work was funded in part by an award from the Intel Science and Technology Center on Embedded Computing, NSF Awards 1147388, 0916887, 0903432, 0829607.

8. REFERENCES

[1] L. Itti, C. Koch, and E. Niebur, "A Model of Saliency-Based Visual Attention for Rapid Scene Analysis," *IEEE Transactions on Pattern Analysis and Machine Intelligence*, vol. 20, no. 11, pp. 1254-1259, Nov 1998.

[2] M. Riesenhuber and T. Poggio, "Hierarchical Models of Object Recognition in Cortex," *Nature Neuroscience*, vol. 2, no. 11, pp. 1019-1025, November 1999.

[3] T. Serre et al., "Robust Object Recognition With Cortex-Like Mechanisms," *IEEE PAMI*, vol. 29, no. 3, March 2007.

[4] E. Meyers and L. Wolf, "Using Biologically Inspired Features for Face Processing," *International Journal of Computer Vision*, vol. 76, no. 1, pp. 93-104, January 2008.

[5] J Mutch, U Knoblich, and T Poggio, "CNS: A GPU-Based Framework for Simulating Cortically-Organized Networks," Massachusetts Institute of Technology, Cambridge, MA, MIT-CSAIL-TR-2010-013 / CBCL-286 2010.

[6] H. Jhuang, T. Serre, L. Wolf, and T. Poggio, "A Biologically Inspired System for Action Recognition," in *International Conference on Computer Vision (ICCV)*, 2007, pp. 1-8.

[7] J. Mutch and D. G. Lowe, "Object Class Recognition and Localization Using Sparse Features with Limited Receptive Fields," *International Journal of Computer Vision (IJCV)*, vol. 80, no. 1, October 2008.

[8] S. Park et al., "System-On-Chip for Biologically Inspired Vision Applications," *Information Processing Society of Japan*, 2012, [In Press].

[9] L. Fei-Fei et al., "Learning Generative Visual Models from Few Training Examples: An Incremental Bayesian Tested on 101 Object Categories," in *IEEE CVPR 2004, Workshop on Generative-Model Based Vision*, 2004.

[10] F. Samaria and A. Harter, "Parameterisation of a Stochastic Model for Human Face Identification," in *2nd IEEE Workshop on Applications of Computer Vision*, 1994.

[11] P. J. Phillips et al., "The FERET Evaluation Methodology for Face Recognition Algorithms," *Trans. of Pattern Analysis and Machine Intelligence*, vol. 22, no. 10, October 2000.

[12] M. Lyons et al., "Coding Facial Expressions with Gabor Wavelets," in *Third IEEE International Conference on Autmatomatic Face and Gesture Recognition*, 1998.

[13] M. Blank et al., "Actions as Space-Time Shapes," in *International Conference on Computer Vision*, 2005.

[14] Xilinx, "Virtex-5 Family Overview," DS100(v5.0) 2009.

[15] Nvidia. (2011) Tesla M2090 Board Specification. [Online]. http://www.nvidia.com/docs/IO/43395/Tesla-M2090-Board-Specification.pdf

[16] I. Kuon and J. Rose, "Measuring the Gap Between FPGAs and ASICs," *IEEE Transactions on Computer-Aided Design of Integrated Circuits and Systems*, vol. 26, no. 2, pp. 203-

215, February 2007.

[17] J.-Y. Kim et al., "A 201.4 GOPS 496 mW Real-Time Multi-Object Recognition Processor With Bio-Inspired Neural Perception Engine," *IEEE Journal of Solid-State Circuits*, vol. 45, no. 1, pp. 32-45, Jan 2010.

[18] C.-Y. Tsai et al., "A 1.0TOPS/W 36-Core Neocortical Computing Processor with 2.3Tb/s Kautz NoC for Universal Visual Recognition," in *IEEE Int. Conference Digest of Technical Papers*, San Francisco, 2012, pp. 480-482.

[19] C. Farabet et al., "Hardware Accelerated Convolutional Neural Networks for Synthetic Vision Systems," in *ISCAS*, Paris, 2010, pp. 257-260.

[20] R. Iyer et al., "CogniServe: Heterogeneous Server Architecture for Large-Scale Recognition," *IEEE Micro*, vol. 31, no. 3, pp. 20-31, May-June 2011.

[21] J. Clemons et al., "EFFEX: An Embedded Processor for Computer Vision Based Feature Extraction," in *The 48th ACM/EDAC/IEEE DAC*, San Diego, 2011, pp. 1020-1024.

[22] S. Chakradhar et al., "A Dynamically Configurable Coprocessor for Convolutional Neural Networks," in *ISCA*, 2010, pp. 247-257.

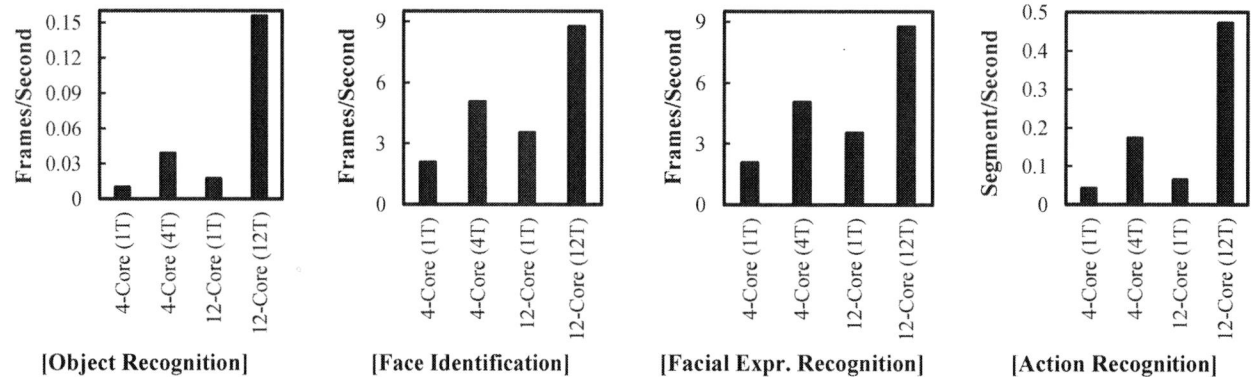

Figure 3: Performance of reference CPU platforms across four application domains. Number of threads is indicated within brackets on the x-axis. The metric used to measure performance is frames/second (segment/second) for the first three applications (action recognition application). Segment in this context refers to a group of twenty frames.

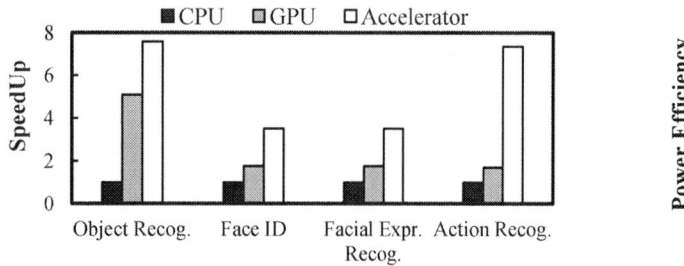

Figure 4: Speedup (FPS): A comparison across the three platforms for each application domain. Figures are normalized to the CPU platform

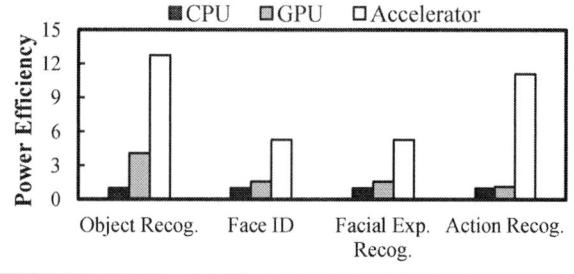

Figure 5: Power Efficiency (fps/Watt): A comparison across the three platforms for each application domain. Figures are normalized to the CPU platform

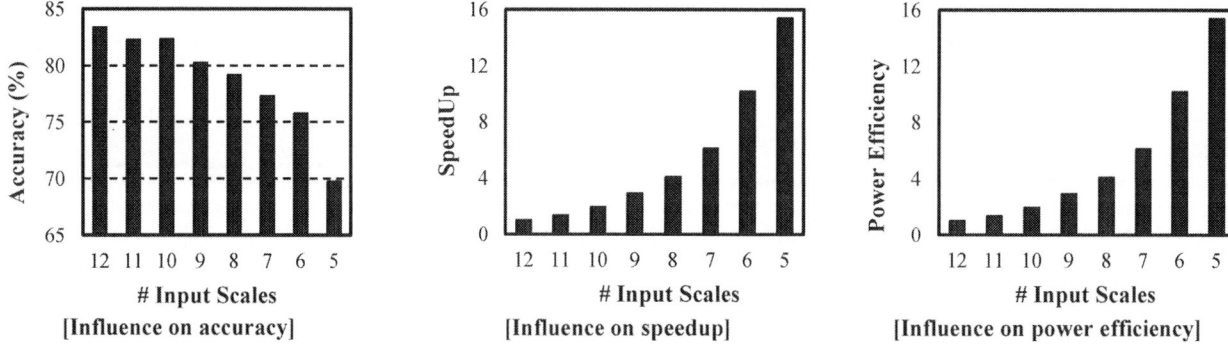

Figure 6: The influence of number of scales on classification accuracy and performance. As the number of input scales decreases, the classification accuracy decreases and power efficiency increases. Figures in "Speedup" & "Power Efficiency" are normalized to the 12-input-scale configuration

Statistical Memristor Modeling and Case Study in Neuromorphic Computing

[1]Robinson E. Pino, [2]Hai (Helen)Li, [3]Yiran Chen, [2]Miao Hu, and [3]Beiye Liu
[1]Air Force Research Laboratory, Rome, NY, USA
[2]ECE Dept., Polytechnic Institute of New York University, Brooklyn, NY, USA
[3]ECE Dept., University of Pittsburgh, Pittsburgh, PA, USA
robinson.pino@rl.af.mil, hli@poly.edu, yic52@pitt.edu, mhu01@students.poly.edu, bel34@pitt.edu

ABSTRACT

Memristor, the fourth passive circuit element, has attracted increased attention since it was rediscovered by HP Lab in 2008. Its distinctive characteristic to record the historic profile of the voltage/current creates a great potential for future neuromorphic computing system design. However, at the nano-scale, process variation control in the manufacturing of memristor devices is very difficult. The impact of process variations on a memristive system that relies on the continuous (analog) states of the memristors could be significant. We use TiO_2-based memristor as an example to analyze the impact of geometry variations on the electrical properties. A simple algorithm was proposed to generate a large volume of geometry variation-aware three-dimensional device structures for Monte-Carlo simulations. A neuromorphic computing system based on memristor-based bidirectional synapse design is proposed as case study. We analyze and evaluate the robustness of the proposed system in pattern recognition based on massive Monte-Carlo simulations, after considering input defects and process variations.

Categories and Subject Descriptors

C.2.6 [**Computing Methodologies**]: Artificial Intelligence–Learning

General Terms

Design, Performance, Reliability

Keywords

Memristor, process variation, neural network, pattern recognition.

1. INTRODUCTION

In 1971, Professor Leon Chua predicted the existence of the memristor [1]. However, the first physical realization of memristors was first demonstrated in 2008 by HP Lab, in which the memristive effect was achieved by moving the doping front along a TiO_2 thin-film device [2]. Soon, memristive systems on spintronic devices were proposed [3].

The unique properties of memristors create great opportunities in future system design. For instance, the non-volatility and excellent scalability make it a promising candidate as the next-generation high-performance high-density storage technology [4]. More importantly, memristors have an intrinsic and remarkable feature called a "pinched hysteresis loop" in the $i - v$ plot, that is, memristors can "remember" the total electric charge flowing through them by changing their resistances (memristances) [5]. For example, the applications of this memristive behavior in electronic neural networks have been extensively studied [6][7].

As process technology shrinks down to decananometer (sub-50nm) scale, device parameter fluctuations incurred by process variations have become a critical issue affecting the electrical characteristics of devices [8]. The situation in a memristive system could be even worse when utilizing the analog states of the memristors in design: variations of device parameters, *e.g.*, the instantaneous memristance, can result in the shift of electrical responses, *e.g.*, current. The deviation of the electrical excitations will affect memristance because the total charge through a memristor indeed is the historic behavior of its current profile. In this work, we explore the implications of the device parameters of memristors to the circuit design by taking into account the impact of process variations.

The device geometry variations significantly influence the electrical properties of nano-devices [10]. For example, the random uncertainties in lithography and patterning processes lead to the random deviation of line edge print-images from their ideal pattern, which is called line edge roughness (LER) [11]. Thickness fluctuation (TF) is caused by deposition processes in which mounds of atoms form and coarsen over time. We propose an algorithm to generate a large volume of three-dimensional memristor structures to mimic the geometry variations for Monte-Carlo simulations. Here, we mainly focus on the impacts of geometry variations because previous experimental results showed that the geometry variations are the dominate fluctuation source as process technology further scales down [8].

Memristive function can be achieved by various materials and device structures. For its popularity, TiO_2-based memristor [3] is analyzed and evaluated in our work. However, our proposed modeling methodologies and design philosophies are not limited by the specific types of devices and can be easily extended to the other structures/materials with necessary modifications.

To demonstrate the impact of process variations on neuromorphic systems, we proposed a bidirectional synapse design and build a computing system using for pattern recognition. After embedding input defects and process variations, Monte-Carlo simulations were conducted to analyze and evaluate the system robustness. Interestingly, our experiments show that even if a large process variation exists in memristor devices, the performance of the memristor-based neuromorphic system is not affected much.

The organization of this paper is as follows: Section 2 introduces

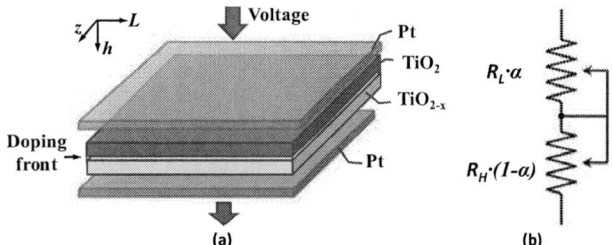

Figure 1: TiO$_2$ thin-film memristor. (a) structure, and (b) equivalent circuit.

the physical mechanisms of TiO$_2$ thin-film memristors; Section 3 analyzes the memristor model under process variations; Section 4 explains the three-dimensional memristor structure algorithm; Section 5 presents and analyzes the impact of geometry variation on the memristor electrical properties; Section 6 describe the neuromorphic system composed of bidirectional synapses and analyze its performance for pattern recognition; at last, Section 7 concludes our work.

2. TIO$_2$ THIN-FILM MEMRISTOR

In 2008, HP Lab demonstrated the first intentional memristive device by using a Pt/TiO$_2$/Pt thin-film structure [2]. The conceptual view is illustrated in Figure 1(a): two metal wires on Pt are used as the top and bottom electrodes, and a thick titanium dioxide film is sandwiched in between. The stoichiometric TiO$_2$ with an exact 2:1 ratio of oxygen to titanium has a natural state as an insulator. However, if the titanium dioxide is lacking a small amount of oxygen, its conductivity becomes relatively high like a semiconductor. We call it oxygen-deficient titanium dioxide (TiO$_{2-x}$) [9]. The memristive function can be achieved by moving the doping front: A positive voltage applied on the top electrode can drive the oxygen vacancies into the pure TiO$_2$ part and therefore lower the resistance continuously. On the other hand, a negative voltage applied on the top electrode can push the dopants back to the TiO$_{2-x}$ part and hence increase the overall resistance. For a TiO$_2$-based memristor, R_L (R_H) is used to denote the lowest (highest) resistance of the structure.

Figure 1(b) illustrates a coupled variable resistor model for a TiO$_2$-based memristor, which is equivalent to two series-connected resistors. The overall resistance can be expressed as

$$M(\alpha) = R_L \cdot \alpha + R_H \cdot (1 - \alpha). \qquad (1)$$

Here α ($0 \leq \alpha \leq 1$) is the relative doping front position, which is the ratio of doping front position over the total thickness of TiO$_2$ thin-film.

The velocity of doping front movement $v(t)$, which is driven by the voltage applied across the memristor $V(t)$ can be expressed as

$$\frac{v(t)}{h} = \frac{d\alpha}{dt} = \mu_v \cdot \frac{R_L}{h^2} \cdot \frac{V(t)}{M(\alpha)} \qquad (2)$$

where, μ_v is the equivalent mobility of dopants, h is the total thickness of the TiO$_2$ thin-film; and $M(\alpha)$ is the total memristance when the relative doping front position is α.

Filamentary conduction has been observed in nano-scale semiconductors, such as TiO$_2$. It shows that the current travels through some high conducting filaments rather than passes the device evenly [16][17]. However, there is no device model based on filamentary conduction mechanism yet. Considering that the main focus of this work is the process variation analysis method of the memristor, which can be separated from the explicit physical model of

Table 1: The device dimensions of TiO$_2$ memristor.

	Length(L)	Width(z)	Thickness(h)
Thin-film	50 nm	50 nm	10 nm

memristor, the popular bulk model of TiO$_2$ is applied. Recent experiments showed that μ_v is not a constant but grows exponentially when the bias voltage goes beyond certain threshold voltage [18]. Nevertheless, the structure of TiO$_2$ memristor model, i.e., Eq. (1), still remains valid.

3. MATHEMATICAL ANALYSIS

The actual length (L) and width (z) of a memristor is affected by LER. The variation of thickness (h) of a thin film structure is usually described by TF. As a matter of convenience, we define that, the impact of process variations on any given variable can be expressed as a factor $\theta = \dfrac{\omega'}{\omega}$, where ω is its ideal value, and ω' is the actual value under process variations. The ideal geometry dimensions of the TiO$_2$ thin-film memristor used in this work are summarized in Table 1.

In TiO$_2$ thin-film memristors, the current passes through the device along the thickness (h) direction. Ideally the doping front has an area $S = L \cdot z$. To simulate the impact of LER on the electrical properties, the memristor device is divided into many small filaments between the two electrodes. Each filament i has a cross-section area ds and a thickness h. Figure 2 demonstrates a non-ideal 3D structure of a TiO$_2$ memristor (i.e., with geometry variations in consideration), which is partitioned into many filaments in statistical analysis.

As shown in Figure 2, ideally, the cross-section area of a filament is ds/S of the entire device area and its thickness is h. Thus, for filament i, the ideal upper bound and lower bound of the memristance can be expressed as

$$R_{i,H} = R_H \cdot \frac{S}{ds}, \text{and } R_{i,L} = R_L \cdot \frac{S}{ds}. \qquad (3)$$

Here, $\theta_{i,s}$ represents the variation ratio on the cross-section area ds, which is caused by 2-D LER. Similarly, $\theta_{i,h}$ is the variation ratio on thickness h due to TF. The resistance of a filament is determined by its section area and thickness, i.e., $R = \rho \cdot \frac{h}{s}$, where ρ is the resistance density. Therefore, the actual upper and the lower bound under the process variations can be expressed as

$$R'_{i,H} = R_{i,H} \cdot \frac{\theta_{i,h}}{\theta_{i,s}}, \text{and } R'_{i,L} = R_{i,L} \cdot \frac{\theta_{i,h}}{\theta_{i,s}}. \qquad (4)$$

If a filament is small enough, we can assume it has a flat doping

3−D model for TiO2 memristor

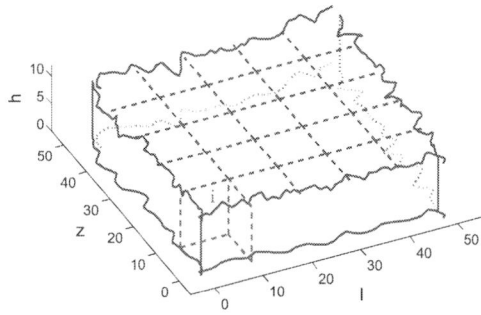

Figure 2: An example of 3D TiO$_2$ memristor structure, which can be partitioned into many filaments in statistical analysis.

front. Then, the actual doping front velocity in filament i considering process variations can be calculated by replacing the ideal values with actual values in Eq.(2). We have

$$v'_i(t) = \mu_v \cdot \frac{R'_{i,L}}{h'^2} \cdot \frac{V(t)}{M'_i(\alpha'_i)}. \tag{5}$$

Here h' and M'_i are the actual thickness and memristance of filament i. Then, we can get a set of related equations for filament i, including the doping front position

$$\alpha'_i(t) = \int_0^t v'(\tau) \cdot d\tau, \tag{6}$$

the corresponding memristance

$$M'_i(\alpha'_i) = \alpha'_i \cdot R'_{i,L} + (1 - \alpha'_i) \cdot R'_{i,H}, \tag{7}$$

and the current through the filament i

$$I'_i(t) = \frac{V(t)}{M'_i(\alpha'_i)}. \tag{8}$$

By combining Eq. (5) – (8), the doping front position in every filament i under process variations $\alpha'_i(t)$ can be obtained by solving the differential equation

$$\frac{d\alpha'_i(t)}{dt} = \mu_v \cdot \frac{R'_{i,L}}{h'^2} \cdot \frac{V(t)}{\alpha'_i(t) \cdot R'_{i,L} + (1 - \alpha'_i(t)) \cdot R'_{i,H}}. \tag{9}$$

Eq. (9) indicates that the behavior of the doping front movement is dependent on the specific electrical excitations, e.g., $V(t)$.

For instance, applying a sinusoidal voltage source to the TiO_2 thin-film memristor such as

$$V(t) = V_m \cdot \sin(2\pi f \cdot t), \tag{10}$$

the corresponding doping front position of filament i can be expressed as:

$$\alpha'_i(t) = \frac{R_{i,H} - \sqrt{R_{i,H}^2 - A \cdot B(t) \cdot \frac{2}{\theta_{i,h}^2} + 2C \cdot A \cdot \frac{\theta_{i,s}}{\theta_{i,h}}}}{A}. \tag{11}$$

Where, $A = R_{i,H} - R_{i,L}$, $B(t) = \mu_v \cdot R_{i,L} \cdot V_m \cdot \cos(2\pi f \cdot t)$, and C is an initial state constant.

The term $B(t)$ accounts for the effect of electrical excitation on doping front position. The terms $\theta_{i,s}$ and $\theta_{i,h}$ represent the effect of both LER and TF on memristive behavior. Moreover, the impact of the geometry variations on the electrical properties of memristors could be affected by the electrical excitations. For example, we can set $\alpha(0) = 0$ to represent the case that the TiO_2 memristor starts from $M(0) = R_H$. In such a condition, C becomes 0, and hence, the doping front position $\alpha'_i(t)$ can be calculated as:

$$\alpha'_i(t) = \frac{R_{i,H} - \sqrt{R_{i,H}^2 - A \cdot B(t) \cdot \frac{2}{\theta_{i,h}^2}}}{A}, \tag{12}$$

which is affected only by TF and electrical excitations. LER will not disturb $\alpha'_i(t)$ if the TiO_2 thin-film memristor has an initial state $\alpha(0) = 0$.

The overall memristance of the memristor can be calculated as the total resistance of all n filaments connected in parallel. Again, i denotes the ith filament. When n goes to ∞, we can have

$$R'_H = \frac{1}{\int_0^\infty 1/R'_{i,H} \cdot di} = R_H \cdot \frac{1}{\int_0^\infty \theta_{i,h}/\theta_{i,s} \cdot di}; \tag{13}$$

and

$$R'_L = \frac{1}{\int_0^\infty 1/R'_{i,L} \cdot di} = R_L \cdot \frac{1}{\int_0^\infty \theta_{i,h}/\theta_{i,s} \cdot di}. \tag{14}$$

The overall current through the memristor is the sum of the current through each filament:

$$I'(t) = \int_0^\infty I'_i(t) \cdot di. \tag{15}$$

The instantaneous memristance of the overall memristor can be defined as

$$M'(t) = \frac{V(t)}{I'(t)} = \frac{1}{\int_0^\infty 1/M'_i \cdot di}. \tag{16}$$

Since the doping front position movement in each filament will not be the same because h'_i varies due to TF (and/or the roughness of the electrode contact), we define the average doping front position of the whole memristor as:

$$\alpha'(t) = \frac{R'_H - M'(t)}{R'_H - R'_L}. \tag{17}$$

4. 3D MEMRISTOR MODELING

Analytic modeling is a fast way to estimation the impact of process variations on memristors. However, we noticed that in modeling some variations analytically, e.g. simulating the LER, may be beyond the capability of analytic model [12]. The data on silicon variations, however, is usually very hard to obtain simply due to intellectual property protection. To improve the accuracy of our evaluations, we create a simulation flow to generate 3-D memristor samples with the geometry variations including LER and thickness fluctuation. The correlation between the generated samples and the real silicon data are guaranteed by the sanity check of the LER characterization parameters. The flow is shown in Figure 3.

Many factors affecting the quality of the line edges show different random effects. Usually statistical parameters such as the auto-correlation function (ACF) and power spectral density (PSD) are used to describe the property of the line edges.

ACF is a basic statistical function of the wavelength of the line profile, representing the correlation of point fluctuations on the line edge at different position. PSD describes the waveform in the frequency domain, reflecting the ratio of signals with different frequencies to the whole signal.

Considering that LER issues are related to fabrication processes, we mainly target the nano-scale pattern fabricated by electron beam lithography (EBL). The measurements show that under such a condition, the line edge profile has two important properties: (1) the line edge profile in ACF figure demonstrates regular oscillations,

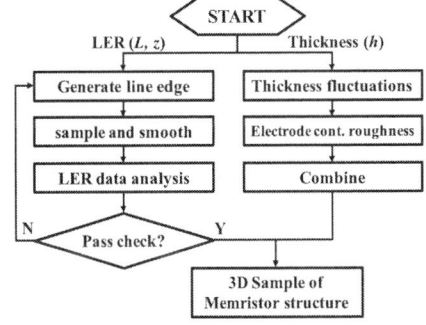

Figure 3: The flow of 3D memristor structure generation including geometry variations.

587

Table 2: The parameters/constraints in LER characterization.

Parameters		Constraints	
L_{LF}	0.8 nm	σ_{LER}	2.5nm ~ 3.5nm
f_{max}	1.8 MHz	σ_{LWR}	4.0nm ~ 5.0nm
L_{HF}	0.4 nm	Sk	0.1nm ~ 0.2nm
/	/	Ku	2.5nm ~ 3.5nm

which are caused by periodic composition in the EBL fabrication system; and (2) the line edge roughness mainly concentrates in a low frequency zone, which is reflected by PSD figure [12].

To generate line edge samples close to the real cases, we can equally divide the entire line edge into many segments, say, n segments. Without losing the LER properties in EBL process, we modified the random LER modeling proposed in [19] to a simpler form with less parameters. The LER of the ith segment can be modeled by

$$\text{LER}_i = L_{LF} \cdot \sin(f_{max} \cdot x_i) + L_{HF} \cdot p_i. \quad (18)$$

The first term on the right side of Eq. (18) represents the regular disturbance at the low frequency range, which is modeled as a sinusoid function with amplitude L_{LF}. f_{max} the mean of the low frequency range derived from PSD analysis. Without loss of generality, a uniform distribution $x_i \in U(-1, 1)$ is used to represent an equal distribution of all frequency components in the low frequency range. The high frequency disturbances are also taken into account by the second term on the right side of Eq. (18) as a Gaussian white noise with amplitude L_{HF}. Here p_i follows the normal distribution $N(0, 1)$ [12]. The actual values of L_{LF}, L_{HF} and f_{max} are determined by ACF and PSD.

To ensure the correlation between the generated line edge samples with the measurement results, we introduce four constraints to conduct a sanity check of the generated samples:

- σ_{LER}: the root mean square (RMS) of LER;
- σ_{LWR}: the RMS of line width roughness (LWR);
- Sk: skewness, used to specify the symmetry of the amplitude of the line edge; and
- Ku: kurtosis, used to describe the steepness of the amplitude distribution curve.

The above four parameters are widely used in LER characterization and can be obtained from measurement results directly [12]. Only the line edge samples that satisfy the constraints will be taken as valid device samples. Table 2 summarizes the parameters used in our algorithm, which are correlated with the characterization method and experimental results in [12]. And Figure 4 shows the LER characteristic parameters distribution among 1000 Monte-Carlo simulations.

Even the main function has captured the major features of LER, it is not enough to mimic all the LER characteristics. The difference

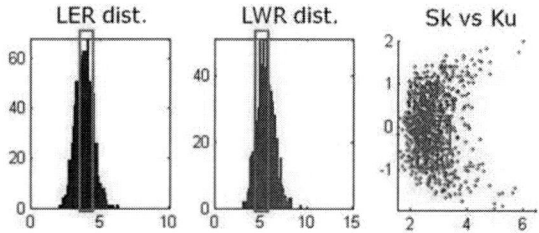

Figure 4: LER characteristic parameters distribution among 1000 Monte-Carlo simulations. Constraints are shown in red rectangles.

between real LER distribution and our modeling function results in the fact that some generated samples are not qualified compared to the characteristic parameters, or the constraints of the real LER profile. Thus, sanity check which screens out the unsuccessful results is necessary. Only those samples in red rectangles shown in Figure 4 satisfy the constraints and will be used for the device electrical property analysis. The criteria of the sanity check are defined based on the measurement results of real LER data.

The thickness fluctuation is caused by the random uncertainties in sputter deposition or atomic layer deposition. It has a relatively smaller impact than the LER and can be modeled as a Gaussian distribution. Since the memristors in this work have relatively bigger dimensions in the horizontal plane than the thickness direction (shown in Table 1), we also considered roughness of electrode contact in our simulation: The means of the thickness of each memristor is generated by assuming it follows the Gaussian distribution. Each memristor is then divided into many filaments between the two electrodes. The roughness of electrode contacts is modeled based on the variations of the thickness of each filament. Here, we assume that both thickness fluctuations and electrode contact roughness follow Gaussian distributions with a deviation $\sigma = 2\%$ of thin film thickness.

Figure 2 is an example of 3D structure of a TiO_2 thin-film memristor generated by the proposed flow. It illustrates the effects of all the geometry variations on a TiO_2 memristor device structure. According to Section 3, a 2-D partition is required for the statistical analysis. In the given example, we partition the device into 25 small filaments with the ideal dimensions of $L = 10nm$, $z = 10nm$, and $h = 10nm$. Each filament can be regarded as a small memristor, which is affected by either only TF or both LER and TF. The overall performance of device can be approximated by paralleled connecting all the filaments.

5. IMPACT ON MEMRISTOR PROPERTIES

To evaluate the impact of process variations on the electrical properties of memristors, we conducted Monte-Carlo simulations with 10,000 qualified 3-D device samples generated by our proposed flow. A sinusoidal voltage source shown in Eq. (10) is applied as the external excitation. The initial state of the memristor is set as $M(\alpha = 0) = R_H$. The device and electrical parameters used in our simulations are summarized in Table 3. Both separate and combined effects of geometry variations on various memristor properties are analyzed, including:

- the distribution of R_H and R_L;
- the change of memristance $M(t)$ and $M(\alpha)$;
- the velocity of wall movement $v(\alpha)$;
- the current through memristor $i(t)$; and
- the I-V characteristics.

The $\pm 3\sigma$ (minimal/maximal) values of the device/electrical parameters as the percentage of the corresponding ideal values are summarized in Table 4. For those parameters that vary over time, we consider the variation at each time step of all the devices. The simulation results considering only either LER or TF are also listed.

Table 4 shows that the static behavior parameters of memristors, i.e., R_H and R_L, are affected in a similar way by both LER and thickness fluctuations. This is consistent to the analytical results

Table 3: TiO_2 memristor electrical parameters.

$R_L(\Omega)$	$R_H(\Omega)$	$\mu_v(m^2 \cdot s^{-1} \cdot V^{-1})$	V_m (V)	f (Hz)
100	16000	10^{-14}	1	0.5

Table 4: 3σ min./max. of TiO_2 memristor parameters

Sinusoidal Voltage	LER only		TF only		overall	
	$-3\sigma(\%)$	$+3\sigma(\%)$	$-3\sigma(\%)$	$+3\sigma(\%)$	$-3\sigma(\%)$	$+3\sigma(\%)$
$R_H \& R_L$	-5.4	4.1	-5.5	4.8	-6.4	7.3
$M(\alpha)$	-5.4	4.1	-37.1	20.8	-36.5	24.1
$\alpha(t)$	0.0	0.0	-13.3	27.5	-14.7	27.4
$v(\alpha)$	0.0	0.0	-9.3	15.6	-10.4	16.9
$i(\alpha)$	-4.7	5.7	-9.3	15.7	-10.7	17.2
Power	-4.7	5.7	-8.8	14.1	-10.1	15.6

Square wave Voltage	LER only		TF only		overall	
	$-3\sigma(\%)$	$+3\sigma(\%)$	$-3\sigma(\%)$	$+3\sigma(\%)$	$-3\sigma(\%)$	$+3\sigma(\%)$
$R_H \& R_L$	-5.3	3.7	-6.2	5.2	-6.6	6.9
$M(\alpha)$	-5.3	3.7	-17.8	13.2	-15.4	14.4
$\alpha(t)$	0.0	0.0	-12.1	16.6	-13.0	15.6
$v(\alpha)$	0.0	0.0	-11.6	17.7	-12.5	16.7
$i(\alpha)$	-4.0	5.2	-11.7	17.7	-12.6	17.6
Power	-4.0	5.2	-7.7	9.8	-8.5	10.1

in Eq. (13) and (14), which show that θ_s and θ_h have the similar effects on the variation of R'_H and R'_L.

However, thickness fluctuation shows a much more significant impact on the memristive behaviors such as $v(t)$, $\alpha(t)$ and $M(\alpha)$, than LER does. It is because the doping front movement is along the thickness direction: $v(t)$ is inversely proportional to the square of the thickness, and $\alpha(t)$ is the integral of $v(t)$ over time as shown in Eq. (5) and (6). For the same reason, thickness fluctuations significantly affect the instantaneous memristance $M(\alpha)$ as well.

Because the thickness of the TiO_2 memristor is relative small compared to other dimensions, we assume the doping front cross-section area is a constant along the thickness direction in our simulation. The impact of LER on $\alpha(t)$ or $v(t)$, which is relatively small compared to that of the thickness fluctuations, is ignored in Table 4.

An interesting observation is that as the doping front α moves toward 1, the velocity v regularly grows larger and reaches its peak at the half period of the sinusoidal excitation. This can be explained by Eq. (7): the memristance is getting smaller as α moves toward 1. With the same input amplitude, a smaller resistance will result in a bigger current and therefore a bigger variation on $v(t)$. Similarly, memristance $M(\alpha)$ reaches its peak variance when α is close to 1.

We also conduct $10{,}000\times$ Monte Carlo simulations on the same samples by applying a square wave voltage excitation. The amplitude of the voltage excitation is ± 0.5V. The simulation results are also shown in Table 4. The results of the static behavior parameters, i.e., R_H and R_L, are exactly the same as those with sinusoidal voltage inputs because they are independent of the external excitations, The results of the memristive behavior parameters such as $v(t)$, $\alpha(t)$ and $M(\alpha)$ show similar trends as those with the sinusoidal voltage inputs. Based on Eq. (11), $\alpha(t)$'s variance is sensitive to the type and amplitude of electrical excitation, because $B(t)$ greatly affects the weight of the thickness fluctuation parameter. That is why the thickness fluctuation has a significantly impact on the electrical properties of memristors under sinusoidal and square voltage excitations.

6. A CASE STUDY

A memristor behaves similarly to a synapse in biological systems and hence can be easily used as the weighted connections in neural networks. Based on the memristor-based bidirectional synapse design, we implement a network serving as neuromorphic computing system with *units* (artificial neurons) and *weighted connections* (synapses). The neuron in this network is a binary threshold unit that produces only two different values to represent its state. A synapse works as a weighted connection to transmit a signal from

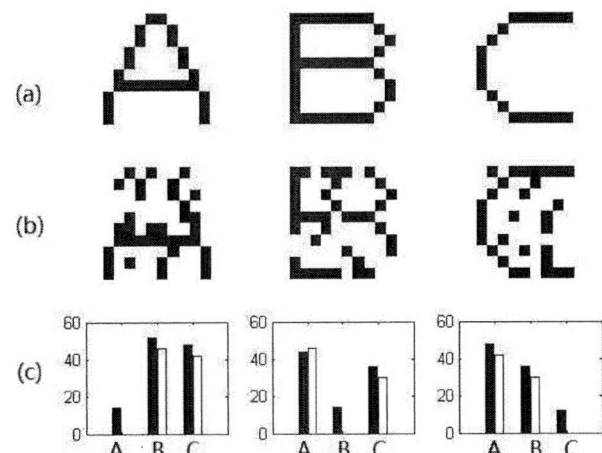

Figure 5: The neural network in pattern recognition: (a) the standard patterns; (b) the noised input patterns; (c) the comparison of the input noised images (black bars) or the output converged images (white bars) from their corresponding standard patterns.

one neuron to another. The activation function can be described as:

$$N_0 = \begin{cases} 1, & \text{if } \sum_{i=1}^{n}(N_i \times W_i) \geq threshold \\ 0, & \text{otherwise} \end{cases} \quad (19)$$

Here, the neuron N_0 collects signals from all the other neurons N_i through the weighted connections W_i. The state of N_0 could be *excitation* ($N_0 = 1$) or *inhibition* ($N_0 = 0$) that is determined by the relation between the summed weighted signals and the threshold. Here, we use bidirectional synapses in the design to build a fully connected neural network, in which any two connected neurons interact each other.

The proposed neural network can be used for pattern recognition: first, multiple standard input images are used to train the connection weights of the system till they reach convergence; after that, any input pattern will produce to a local minimum, which is a stable state corresponding to one the stored standard patterns. Such a network system can even be used to recognize the input image with defects.

In our experiment, we build a network with $100\,(10 \times 10)$ neurons and store the character images 'A', 'B' and 'C' shown in Figure 5(a) as the standard patterns. Each neuron in the network represents a pixel of the image. Then the defected images in Figure 5(b) are applied as inputs to initialize the network's state. Figure 5(c) show that each input has 13 defects compared to its corresponding stan-

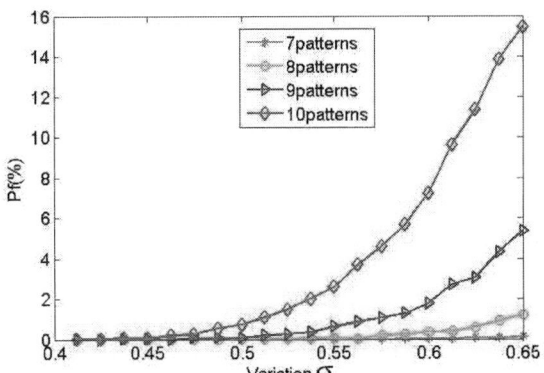

Figure 6: The impact of memristance variations on the probability of failure (P_f).

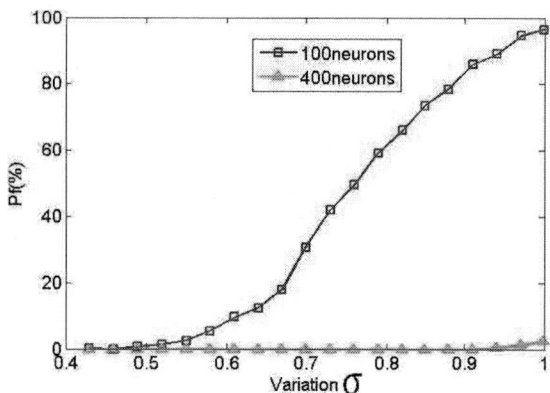

Figure 7: Increasing the network size can reduce P_f.

dard images (see black bars). The proposed system can completely eliminate the difference to zero and converge to one of the standard patterns, as demonstrated by the write bars in Figure 5(c).

The maximal allowed stored standard patterns (*capacity*) of this neural network design is determined by the amounts of neurons and connections. Moreover, the more patterns stored in the system, the higher precision of the connection weights is needed. Therefore, a large number of stored patterns and the high process variation on memristances will result in a higher failure probability (P_f).

To quantitatively evaluate the impact of memristance variations and robustness of the proposed neural network design, we conducted Monte-Carlo simulations for the network with 100 (10×10) neurons. Random variations following Gaussian distribution have been injected to the memristors. And σ is the standard deviation of the memristance. The system could fail to recognize the noised patterns or mismatch an input with other standard patterns due to the inaccurate connection weights. To test the failure probability under different conditions, we ran 10,000 Monte-Carlo simulations by varying the memristance variation σ when 7, 8, 9, or 10 patterns are stored in the system. In this experiment, each input image contain 21 defects among 100 pixels.

The simulation results in Figure 6 demonstrate that the proposed memristor-based neuromorphic system has a high tolerance on memristance variations. When $\sigma < 0.4$, which already exceeds the upper bound of memristance variation in Table 4, P_f of all the four configuration are close to the ideal condition at $\sigma = 0$. This indicates that even a large process variation exists in memristor devices, the performance of the proposed neuromorphic system is not affected much. Further increasing $\sigma > 0.5$, P_f grows significantly. As expected, under the same process variation condition, the system suffers a higher P_f when more patterns are stored.

For the same amount of stored patterns, a larger network with more neurons is more robust to process variations. Figure 7 compares the performance of the systems with 100 neurons (the blue line) and with 400 neurons (the green line). Both systems have 10 standard patterns. And the input defect rate remains at 21% for the two designs. The simulations show that the impact of process variations is smaller and therefore the required precision of connection weights is lower in a bigger network. Hence, in a neural network system design, the tradeoff between network capacity and robustness need to considered.

7. CONCLUSION

In this work, we evaluate the impact of geometry variations on the electrical properties of TiO_2-based memristors by conducting analytic modeling analysis and Monte-Carlo simulations. The responses of the static and memristive parameters under various pro-

cess variations are evaluated and their implication for the electrical properties are analyzed. At the end, we propose a memristor-based neuromorphic computing system and use it as the case study of robustness analysis. Our experiment results show that the proposed design demonstrates high tolerance on process variation and input defects, which is consistent to the intrinsic property of neuromorphic systems.

Acknowledgments and Disclaimer

Contractor acknowledges Government support in the publication of this paper. This material is based upon work funded by AFRL under contract No. FA8750-11-2-0046 and FA8750-11-1-0271, and by NSF under contract No. CNS-1116684 and CNS-1116171.

Any opinions, findings and conclusions or recommendations expressed in this material are those of the author(s) and do not necessarily reflect the views of AFRL.

8. REFERENCES
[1] L. Chua, "Memristor-the missing circuit element," *IEEE Transaction on Circuit Theory*, vol. 18, pp. 507–519, 1971.

[2] D. B. Strukov, G. S. Snider, D. R. Stewart, and R. S. Williams, "The missing memristor found," *Nature*, vol. 453, pp. 80–83, 2008.

[3] X. Wang, Y. Chen, H. Xi, H. Li, and D. Dimitrov, "Spintronic memristor through spin-torque-induced magnetization motion," *IEEE Electron Device Letters*, vol. 30, pp. 294–297, 2009.

[4] Y. Ho, G. M. Huang, and P. Li, "Nonvolatile memristor memory: device characteristics and design implications," in *International Conference on Computer-Aided Design*, 2009, pp. 485–490.

[5] D. Strukov, J. Borghetti, and S. Williams, "Coupled ionic and electronic transport model of thin-film semiconductor memristive behavior," *SMALL*, vol. 5, pp. 1058–1063, 2009.

[6] Y. V. Pershin and M. D. Ventra, "Experimental demonstration of associative memory with memristive neural networks," in *Nanotechnology Nature Proceedings*, 2009, p. 345201.

[7] H. Choi, H. Jung, J. Lee, J. Yoon, J. Park, D.-J. Seong, W. Lee, M. Hasan, G.-Y. Jung, and H. Hwang, "An electrically modifiable synapse array of resistive switching memory," *Nanotechnology*, vol. 20, no. 34, p. 345201, 2009.

[8] A. Asenov, S. Kaya, and A. R. Brown, "Intrinsic parameter fluctuations in decananometer MOSFETs introduced by gate line edge roughness," *IEEE Transaction on Electron Devices*, vol. 50, pp. 1254–1260, 2003.

[9] D. Niu, Y. Chen, C. Xu, and Y. Xie, "Impact of process variations on emerging memristor," in *Design Automation Conference (DAC)*, 2010, pp. 877–882.

[10] G. Roy, A. Brown, F. Adamu-Lema, S. Roy, and A. Asenov, "Simulation study of individual and combined sources of intrinsic parameter fluctuations in conventional nano-MOSFETs," *IEEE Transactions on Electron Devices*, vol. 53, no. 12, pp. 3063–3070, 2006.

[11] P. Oldiges, Q. Lin, K. Petrillo, M. Sanchez, M. Ieong, and M. Hargrove, "Modeling line edge roughness effects in sub 100 nanometer gate length devices," in *SISPAD*, 2000, pp. 131–134.

[12] Z. Jiang and et.al, "Characterization of line edge roughness and line width roughness of nano-scale typical structures," in *International Conference on Nano/Micro Engineered and Molecular Systems*, 2009, pp. 299–303.

[13] A. Asenov, A. Cathignol, B. Cheng, K. McKenna, A. Brown, A. Shluger, D. Chanemougame, K. Rochereau, and G. Ghibaudo, "Origin of the asymmetry in the magnitude of the statistical variability of n-and p-channel poly-si gate bulk mosfets," *IEEE Device Letters*, vol. 29, no. 8, pp. 913–915, 2008.

[14] A. Asenov, S. Kaya, and J. Davies, "Intrinsic threshold voltage fluctuations in decanano mosfets due to local oxide thickness variations," *Electron Devices, IEEE Transactions on*, vol. 49, no. 1, pp. 112–119, 2002.

[15] X. Wang and Y. Chen, "Spintronic memristor devices and applications," in *Design, Automation & Test in Europe Conference and Exhibition (DATE)*, 2010.

[16] D. Kim, S. Seo, S. Ahn, D. Suh, M. Lee, B. Park, I. Yoo, I. Baek, H. Kim, E. Yim *et al.*, "Electrical observations of filamentary conductions for the resistive memory switching in NiO films," *Applied physics letters*, vol. 88, no. 20, pp. 202 102–202 102, 2006.

[17] K. Kim, B. Choi, Y. Shin, S. Choi, and C. Hwang, "Anode-interface localized filamentary mechanism in resistive switching of TiO thin films," *Applied physics letters*, vol. 91, p. 012907, 2007.

[18] D. Strukov and R. Williams, "Exponential ionic drift: fast switching and low volatility of thin-film memristors," *Applied Physics A: Materials Science & Processing*, vol. 94, no. 3, pp. 515–519, 2009.

[19] Y. Ban, S. Sundareswaran, R. Panda, and D. Pan, "Electrical impact of line-edge roughness on sub-45nm node standard cell," in *Proc. SPIE*, vol. 7275, 2009, pp. 727 518–727 518–10.

Triple Patterning Aware Routing and Its Comparison with Double Patterning Aware Routing in 14nm Technology [*]

Qiang Ma   Hongbo Zhang   Martin D. F. Wong

Department of Electrical and Computer Engineering, University of Illinois at Urbana-Champaign
qiangma1@illinois.edu, hzhang27@illinois.edu, mdfwong@illinois.edu

ABSTRACT

As technology continues to scale to 14nm node, Double Patterning Lithography (DPL) is pushed to near its limit. Triple Patterning Lithography (TPL) is a considerable and natural extension along the paradigm of DPL. With an extra mask to accommodate the features, TPL can be used to eliminate the unresolvable conflicts and minimize the number of stitches, which are pervasive in DPL process, and thus smoothen the layout decomposition step. Considering TPL during routing stage explores a larger solution space and can further improve the layout decomposability. In this paper, we propose the first triple patterning aware detailed routing scheme, and compare its performance with the double patterning version in 14nm node. Experimental results show that, using TPL, the conflicts can be resolved much more easily and the stitches can be significantly reduced in contrast to DPL.

Categories and Subject Descriptors

B.7.2 [**Integrated Circuits**]: Design Aids

General Terms

Algorithm, Design

Keywords

Triple patterning aware routing, Double patterning aware routing, Maze routing, 14nm technology

1. INTRODUCTION

As the minimum feature size keeps shrinking, and the availability of the next generation lithography methods (EUV, e-beam direct write, etc.) is further delayed, Double Patterning Lithography (DPL) is commonly recognized as a feasible lithography process for 20nm technology nodes [3, 5, 6, 12]. DPL increases pitch size and enhances resolution by decomposing the features on a critical layer into two masks. The conflicting features with spacing between them less than a predefined threshold d_{min} have to be assigned onto different masks. Whenever necessary, a feature can be

---

[*]This work was partially supported by the National Science Foundation under grant CCF-1017516 and a grant from the semiconductor Research Corporation (SRC).

further sliced to resolve conflicts, which introduce *stitches*. However, in the circumstance of high density layout, it is still possible that some conflicts cannot be resolved with stitches [7]. Moreover, the introduced stitches can lead to yield loss due to overlay error [3]. The problem of layout decomposition with conflicts and stitches minimization for DPL has been extensively studied [7, 13]. Recently, Tang *et al.* have shown in [11] that this problem is polynomial time solvable and provided an optimal algorithm for it.

As technology continues to scale to 14nm node, the scalability of DPL is further challenged. The complexity of layout decomposition grows higher, and stitches become more costly as they are more sensitive to overlay error. As DPL is being pushed to its limit, Triple Patterning Lithography (TPL) is a considerable and natural extension along the paradigm of DPL to alleviate the situation. Industry has already explored the test-chip patterns with triple patterning or even quadruple patterning [1]. Yu *et al.* studied the layout decomposition problem for TPL in [14] and formulated the problem as Integer Linear Programming. With an extra mask to accommodate the features, TPL can be used to (1) eliminate the unresolvable conflicts and reduce the number of stitches while maintaining the same pitch size as DPL (for 14nm node); (2) further increase the pitch size and improve the depth of focus (DOF) (for 10nm node and beyond).

A layout configuration without considering TPL/DPL at design stage can make the layout hard to decompose, and redesigning an indecomposable layout requires high ECO efforts. It is observed that most hard-to-decompose patterns originate from routing wires [4], so considering TPL/DPL during design time, especially detailed routing stage, can significantly benefit the layout decomposition step. Figure 1 gives an example for illustration in the context of TPL. Figure 1(a) shows an indecomposable layout consisting of four nets. All four nets conflict with each other so they need to be assigned to different masks (colors), however, only three masks (colors) are available. If net 2 is implemented with an alternative route, the layout can be decomposed at the expense of introducing a stitch on net 3, as shown in Figure 1(b). In Figure 1(c), net 3 is rerouted, and the layout can be decomposed without stitching. Several previous works studied the problem of optimizing routing for double patterning. Cho *et al.* proposed the first double patterning aware detailed router in [4]. They developed a heuristic which is a modified Dijkstra's shortest path algorithm to take into account the conflicts and stitches when routing one net. Lin and Li developed a double patterning aware gridless router in [8]. They maintain a conflict graph during routing and use it to guide the routes of the incoming nets. Sun *et al.* explored in [10] post-routing layer assignment for DPL optimization. As far as we know, no previous work has addressed triple patterning aware routing. Note that throughout this paper, we interchangeably use the terminologies

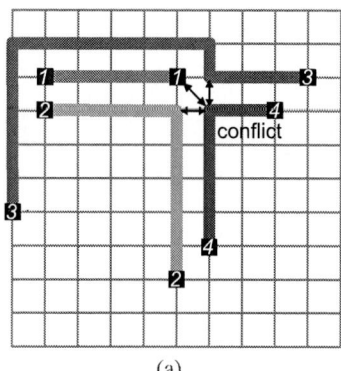
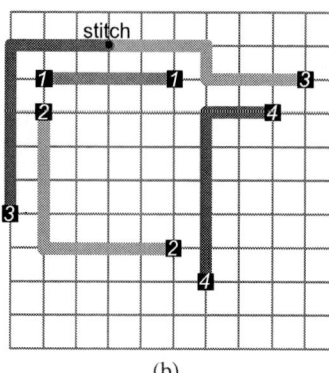
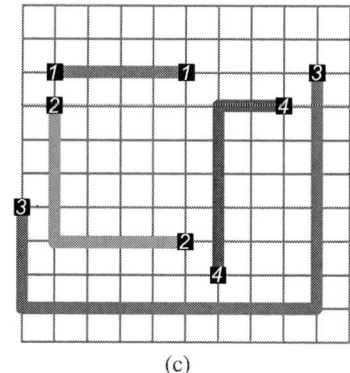

(a)   (b)   (c)

Figure 1: (a) The routing solution cannot be successfully decomposed due to the conflict; (b) The routing solution can be decomposed at the expense of introducing a stitch; (c) The routing solution can be decomposed without stitching.

triple patterning and triple patterning lithography (TPL), as well as double patterning and double patterning lithography (DPL).

In this paper, we study triple patterning aware routing and compare it with double patterning aware routing in 14nm technology node. We first propose a graph model that correctly models the cost of conflicts and stitches in TPL. By replacing each vertex in the routing grid with the graph model and performing shortest path algorithm on the expanded graph, the optimal path with mask/color assignment for one net can be computed, in presence of previously routed and colored nets. Our proposed graph model is a unified model that can be extended to handle multiple patterning and can also be tailored for double patterning. We then develop a negotiated congestion based scheme to resolve conflicts. The regions with conflicts are penalized and the nets are iteratively rerouted and re-colored. Experimental results show that this scheme is very effective and all conflicts can be resolved within a small number of iterations in our test cases. An effective heuristic is also developed to ensure approximately balanced utilization of the three masks. To the best of our knowledge, this is the first triple patterning aware detailed router developed in literature. We also implement a double patterning aware detailed router by adapting our graph model for double patterning, and compare it with the triple patterning version in 14nm node. Experimental results show that using TPL the conflicts can be resolved much more easily and the stitches can be significantly reduced in contrast to DPL.

The rest of this paper is organized as follows: Section 2 states the triple patterning aware routing problem; Section 3 introduces our proposed graph model and describes how to route a single net on the expanded routing graph; Section 4 presents our negotiated congestion based scheme to resolve conflicts, as well as the heuristic to balance the routing on the three masks; Section 5 reports the experimental results, and Section VI concludes this paper.

2. PROBLEM FORMULATION

In TPL, we have three masks available to accommodate all the features within one layer. For the ease of discussion, we use three colors, namely, RED, GREEN and BLUE, to represent the three masks. In TPL aware detailed routing, we perform simultaneous routing and color assignment. If the spacing between two pieces of wires assigned with the same color is smaller than a predefined minimum spacing requirement d_{min}, a conflict is caused and the two pieces of wires will contribute to the total conflicting wire length. If a piece of wire changes its color assignment at some point, as net 3 in Figure 1(b), a stitch is introduced. We want to minimize the total conflicting wire length as well as the number of stitches. The triple patterning aware detailed routing problem is stated as follows.

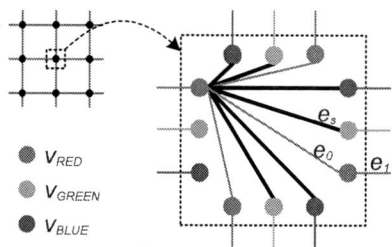

Figure 2: Graph model for a non-pin vertex in the routing grid.

DEFINITION 1. **Triple Patterning Aware Detailed Routing -** *Given a netlist, a routing grid, a minimum spacing requirement d_{min} and three colors (RED, GREEN and BLUE), detailed routing with simultaneous color assignment is performed such that the total conflicting wire length and number of stitches are minimized.*

An immediate subproblem is how to perform maze routing for a single net on the routing grid in the presence of a set of previously routed nets. When we route a net, it is desired to compute a path p with color assignment that produces minimum conflicting wire length and minimum number of stitches. Of course, the wire length of the net, as a conventional metric, also needs to be minimized. Therefore, the weighted sum $l_w^p + \alpha l_{con}^p + \beta n_s^p$ is a good cost metric to minimize when we route a single net, where l_w^p, l_{con}^p and n_s^p denote the wire length, the conflicting wire length and the number of stitches produced by the path p computed, respectively, and α and β are user defined parameters that specify the relative importance between them. We define the triple patterning aware maze routing problem below.

DEFINITION 2. **Triple Patterning Aware Maze Routing -** *Given a set of previously routed nets together with their color assignment on a routing grid, as well as two pins of a net, the objective is to compute a path p with color assignment between the two pins such that the weighted sum $l_w^p + \alpha l_{con}^p + \beta n_s^p$ is minimized.*

3. ROUTING A SINGLE NET

In this section, we propose a graph model that correctly captures the cost of conflicts and stitches, and show that the triple patterning aware maze routing problem can be optimally solved by performing shortest path algorithm on an expanded routing graph constructed using the graph model. We then discuss a practical extension of the proposed graph model to forbid stitch at corner.

3.1 Triple Patterning Aware Maze Routing

Suppose we are given a routing grid G, which can be viewed as a routing graph if we regard every intersection of four line segments

592

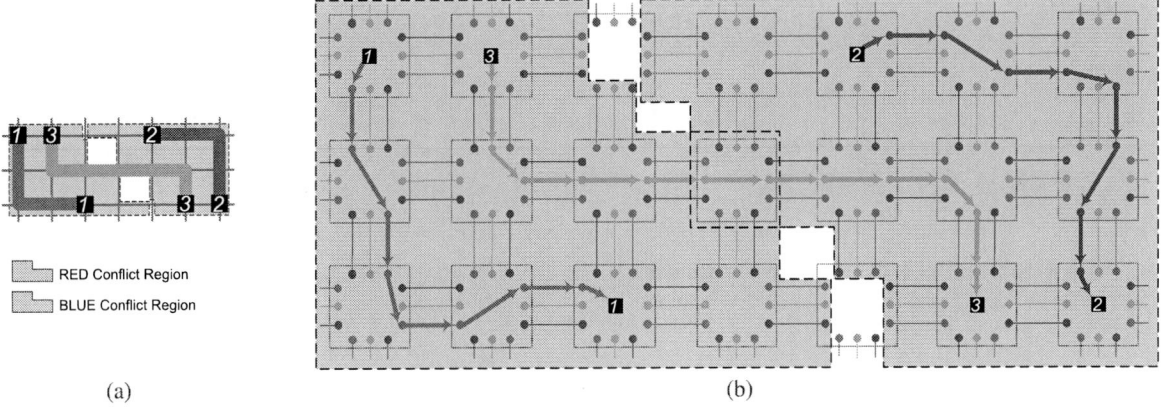

(a)   (b)

Figure 4: (a) Routes on the original routing grid; (b) Routes on the expanded routing graph.

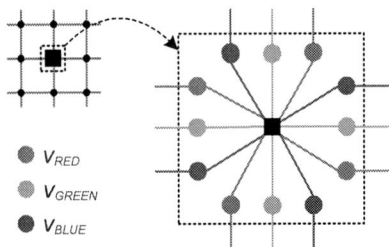

Figure 3: Graph model for a pin vertex in the routing grid.

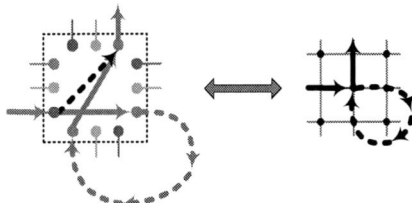

Figure 5: A path in the expanded routing G' may correspond to a walk with loop in the routing grid G.

as a vertex and the short segments between vertices as edges. In the triple patterning aware maze routing problem, the cost of wire length and conflicting wire length can be easily captured by assigning cost to the edges of G. However, the cost of stitches cannot be directly captured. In order to capture the cost of stitches as well, we split each vertex v of G into 12 vertices and construct a graph model on them, as shown in Figure 2. The detailed construction is described as follows:

- The 12 vertices fall into three categories, namely, v_{RED}, v_{GREEN} and v_{BLUE}, which correspond to the three colors RED, GREEN and BLUE, respectively. On each of the four boundaries of the graph model, there are one v_{RED}, one v_{GREEN} and one v_{BLUE}, as shown in Figure 2. On the two adjacent boundaries of two neighboring graph models, the two vertices of the same color are connected.

- Within the graph model, two vertices are connected by an edge if and only if they are not lying on the same boundary. Note that in Figure 2, only the edges adjacent to the vertex v_{RED} on the left boundary are displayed. This graph model works like a switch box. A route can come into the graph model at any boundary and go out at any other boundary. The color of the route can be changed within the model.

- All the edges can be categorized into three types, namely, e_0, e_s and e_1. A type e_0 edge with cost 0 connects two vertices of the same color within the graph model. A type e_s edge with cost β (the cost of a stitch), as indicated by a thick edge in Figure 2, connects two vertices of different colors within the graph model. It corresponds to a stitch. A type e_1 edge connects two same color vertices of two neighboring graph models. It corresponds to an edge in the original routing grid. The base cost of a type e_1 edge is 1, which indicates the cost of the unit wire length. However, when routing on a type e_1 edge causes a coloring conflict, the cost of this edge will be updated to $(1+\alpha)$, where α is the cost of the unit conflicting

wire length. Note that the type e_1 edges are shared by the neighboring graph models.

Note that if a vertex v in G is a pin of a net, we will construct the graph model for v as shown in Figure 3 instead, where v is still split into the 12 vertices, and another vertex representing the pin will be added and connected to the 12 vertices.

We now obtain an expanded routing graph G' from the original routing grid G through the construction described above. When we route one net, we simply apply Dijistra's shortest path algorithm on G'. Figure 4 gives an example for illustration. Figure 4(a) displays the routes of three nets on the original routing grid, while Figure 4(b) demonstrates the corresponding routes on the expanded routing graph. Suppose, without loss of generality, the net ordering during routing is net 1, net 2, then net 3. Net 1 and net 2 are routed and assigned to RED and BLUE, respectively. The two shaded regions are respectively RED conflict region and BLUE conflict region. The type e_1 edges connecting two v_{RED} (v_{BLUE}) vertices within the RED (BLUE) conflict region will have their cost be updated as $1+\alpha$, indicating that using these edges for routing will cause conflicts. Therefore, when net 3 is routed, the shortest path algorithm will find the path colored GREEN as shown in Figure 4.

To show the optimality of our graph expansion based approach, we first show that the shortest path p' between two pins on G' must correspond to a path p between the two pins on G. Suppose the shortest path p' on G' does not correspond to a path p on G, and it corresponds to a walk with loop on G instead. It follows that p' enters a graph model at least twice on G'. Let us assume that, w.l.o.g., p' enters a graph model twice as shown in Figure 5. However, the part of path p' that produces a loop can always be shortcut by an edge inside the graph model. This indicates that p' is not a shortest path, which is contradictory to the assumption. Note that the shortest path p' on G' also determines the color assignment of its corresponding path p on G. Then, from our graph model construction and cost setting, it is easy to see that the $l_w^p + \alpha l_{con}^p + \beta n_s^p$ value of a path p with color assignment on G is the same as the cost of its corresponding path p' on G'. Thus, the shortest path p'

593

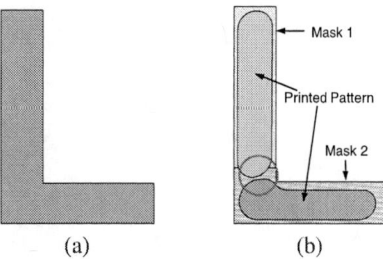

(a)   (b)

Figure 6: (a) Pattern to be printed; (b) Stitching at corner results in significant printability degradation due to overlay error and line-end effect.

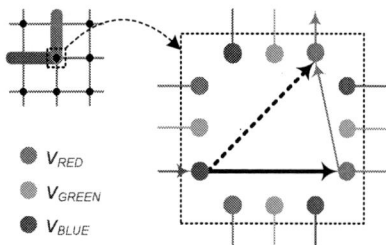

- v_{RED}
- v_{GREEN}
- v_{BLUE}

Figure 7: Corner stitches cannot be prevented by simply removing the edges that directly generate corner stitches.

between two pins on G' corresponds to a colored path p of smallest $l_w^p + \alpha l_{con}^p + \beta n_s^p$ value between the two pins on G. Based on the above analysis, we can conclude that the triple patterning aware maze routing problem on a routing grid G can be optimally solved by computing the shortest path between the two pins on the expanded routing graph G'.

3.2 Forbidding Stitch at Corner

A routing path can change its color assignment at some point to avoid conflicts, which introduces a stitch. Stitches are sensitive to overlay errors and thus lead to printability degradation. In particular, when a stitch occurs at a turning point, or corner, of a route, the situation is even worse and the degradation of printability can be much more significant [2], as illustrated in Figure 6. Therefore, stitch at corner is highly undesirable. In this subsection, we show that our graph model can be extended to disallow stitch at corner.

Intuitively, stitch at corner can be prohibited by removing the type e_s edges that introduce corner stitches in the graph model. For example, in Figure 7, the edge connecting v_{BLUE} on the left boundary and v_{RED} on the top boundary (shown in dashed line) will be removed. However, this does not work as v_{BLUE} on the left boundary can still reach v_{RED} on the top boundary by taking a detour as shown by the solid edges in Figure 7. To address this issue, we further split each vertex v in the graph model into two vertices v^{in} and v^{out}. v_{RED} (v_{GREEN}, v_{BLUE}) is split into v_{RED}^{in} (v_{GREEN}^{in}, v_{BLUE}^{in}) and v_{RED}^{out} (v_{GREEN}^{out}, v_{BLUE}^{out}). We also make the edges directed. The edges are coming into the graph model through v_{in}, and going out of the graph model through v^{out}. Within the graph model, an edge will be directed from v^{in} to v^{out}. The new graph model that disallows stitch at corner is shown in Figure 8. Note that only the edges adjacent to the vertices on the left boundary are displayed. A minor flaw of using this new graph model is that a loop similar to the one shown in Figure 5 may be produced by applying the shortest path algorithm. A path may take a detour outside the graph model to reach v_{RED}^{out} on the top boundary from v_{BLUE}^{in} on the left boundary, but this time the detour can no longer be shortcut as there is no connection between the two vertices within the new graph model. However, according to our experiment, this situation rarely happens. When it happens, we simply block the place where

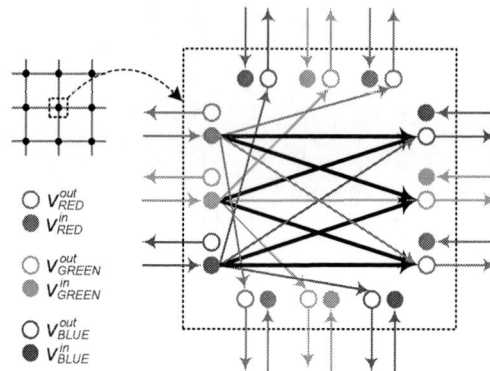

- ○ v_{RED}^{out}
- ● v_{RED}^{in}
- ○ v_{GREEN}^{out}
- ● v_{GREEN}^{in}
- ○ v_{BLUE}^{out}
- ● v_{BLUE}^{in}

Figure 8: Graph model for TPL that disallows stitch at corner.

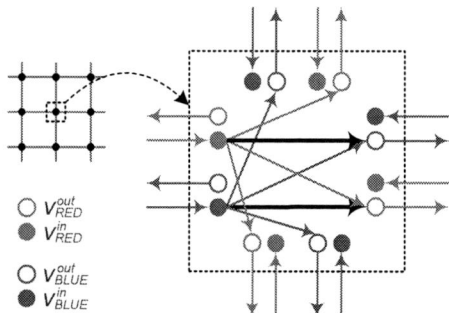

- ○ v_{RED}^{out}
- ● v_{RED}^{in}
- ○ v_{BLUE}^{out}
- ● v_{BLUE}^{in}

Figure 9: Graph model for DPL that disallows stitch at corner.

the detour occurs and run the shortest path algorithm again. This technique turns out to be very effective.

Our proposed graph model is a unified graph model, which can be extended for multiple patterning lithography, and can also be tailored for double patterning lithography. The graph model for DPL is shown in Figure 9, which uses two colors *RED* and *BLUE*.

4. OVERALL ROUTING SCHEME

In this section we present our overall routing scheme for the triple patterning aware detailed routing problem. We adopt a negotiated congestion based scheme to resolve the coloring conflicts over iterations of rip-up and reroute/re-color. Since the nets are routed and colored one by one sequentially, the solution obtained within one single pass is not good enough, and coloring conflicts may exist. The negotiated congestion based scheme is able to dynamically refine the routing and coloring solution over iterations. This scheme can significantly reduce the adverse effect of improper net ordering. We also propose a way to balance the features over the masks, which is beneficial for manufacturing. Our overall routing scheme also works for double patterning aware detailed routing.

4.1 Negotiated Congestion based Scheme to resolve Coloring Conflicts

The negotiated congestion based routing scheme [9] has been widely used in FPGA routing and global routing to resolve routing congestions. In this routing scheme, routability is achieved by forcing all the nets to negotiate for a resource and thereby determine which net needs the resource most. Some nets may use shared resources that are in high demand if all alternative routes utilize resources in even higher demand; other nets will tend to spread out and use resources in lower demand. All the nets are iteratively rerouted until no more resources are shared.

594

We adapt this negotiated congestion based scheme to resolve coloring conflicts in our triple patterning aware detailed routing problem. We let the nets negotiate for the color assignment by adding a history cost to the type e_1 edges in the expanded routing graph G'. The cost of a type e_1 edge is computed by the following formula:

$$cost(e_1) = 1 + \alpha \times (isConflict?\ h_c : 0),$$

where $isConflict$ is a boolean variable indicating if this edge lies in the conflict region produced by the previously routed nets, and h_c denotes the history cost. This means that if routing on this type e_1 edge causes a conflict, the cost of this edge will be set to $1 + \alpha \times h_c$; otherwise, the cost will be set to 1. h_c is initialized as 1.

This scheme works as follows. We start with a global routing solution without color assignment. In the initial iteration, each net is rerouted on the expanded routing graph G'. When the initial iteration terminates, the routes on G' provide a color assignment of all the nets. Note that during routing one net, the access to the graph models occupied by other nets is denied, so that no crossing will be generated and routability is guaranteed. If coloring conflicts exist, iterations of rip-up and reroute will be performed. When a net i is rerouted, we first remove its current route, as well as the conflict region(s) it produces. If a type e_1 edge was lying in net i's conflict region of the same color, its flag $isConflict$ will be updated as $false$ since the conflict region is now gone. The shortest path algorithm is then performed to compute a new path for net i, and the conflict region(s) it produces will be updated. If a type e_1 edge falls into a conflict region representing the same color, its flag $isConflict$ will be set as $true$. In addition, if the path of this net causes coloring conflicts with previously routed nets, the type e_1 edges that are responsible for the conflicts will have their history cost h_c incremented by 1. In this way, the regions with coloring conflicts grow more expensive over iterations, and those nets with more options will tend to choose alternative routes or colors in subsequent iterations, so that the conflicts can potentially be resolved. In our implementation, this procedure will terminate when either no more conflicts exist or enough iterations of rerouting have been performed. Our experimental results show that this scheme is very effective in resolving conflicts.

The example in Figure 10 demonstrates how this negotiated congestion based scheme works. There are 6 nets, where net 1-5 are previously routed with color assignment, and net 6 is the current net to be routed, as shown in Figure 10(a). It is easy to see that net 6 will cause conflict with other nets no matter what color it is assigned. Net 6 is colored in RED by the shortest path algorithm, which causes conflicts with net 4, as shown in Figure 10(b). As a result, the type e_1 edges that are responsible for the conflicts have their history cost h_c incremented. Therefore, in the next iteration, net 4 is rerouted and colored in BLUE, so that the conflicts are effectively resolved, as shown in Figure 10(c).

4.2 Balancing Features on Three Masks

TPL provides three masks to accommodate the features. Balancing the features on the three masks ensures that each mask is fully utilized, so that none of the masks is unnecessarily dense. This helps the printability enhancement during manufacturing. We develop a heuristic that can effectively control the balancing on the fly. During routing, we maintain three variables l_{RED}, l_{GREEN} and l_{BLUE}, which respectively keep track of the wire length colored in RED, GREEN and BLUE. The cost of the type e_1 edges will be adjusted according to the current distribution of the total wire length on the three masks. For example, when relatively more wires are assigned to GREEN, the cost of the type e_1 edges of $GREEN$ color should be increased, so that the router tends to favor the routes

using other colors. In our implementation, we let l_{RED} be the reference. The cost of the type e_1 edges of GREEN color is scaled by $\frac{l_{GREEN}}{l_{RED}}$, and the cost of those of BLUE color is scaled by $\frac{l_{BLUE}}{l_{RED}}$. This technique is shown to be effective through the experiments.

The pseudocode of our overall routing scheme is listed below.

ALGORITHM TPL AWARE DETAILED ROUTING(G, N):
    Construct the expanded routing graph G' from G;
    Set h_c as 1 for all the type e_1 edges;
    for each net i in N **do**   //initial iteration
        Reroute net i using shortest path algorithm;
        Update flag $isConflict$ for the affected e_1 edges;
        Update l_{RED}, l_{GREEN} and l_{BLUE};
    while $\exists$ conflicts **do**
        for each net i in N **do**
            Remove the route of net i;
            Update flag $isConflict$ for the affected e_1 edges;
            Update l_{RED}, l_{GREEN} and l_{BLUE};
            Reroute net i using shortest path algorithm;
            Update flag $isConflict$ for the affected e_1 edges;
            Increment h_c for the conflicting e_1 edges;
            Update l_{RED}, l_{GREEN} and l_{BLUE};

---

5. EXPERIMENTAL RESULTS

We implement a TPL aware detailed router as well as a DPL version, and compare their performance in 14nm node on two sets of benchmarks. The graph model that disallows stitch at corner is adopted for both triple patterning and double patterning. Our program is implemented in C++ and all the experiments are performed on a Linux machine with 2.0GHz CPU and 2GB RAM.

We first compare the two routers on a set of benchmarks derived from industrial data. The result of comparison is shown in Table 1. The column "Init Con. WL" shows the conflicting wire length after the initial iteration. The column "#iter" shows the total number of rerouting iterations performed. The column "Con. WL" reports the final conflicting wire length. We can see that our negotiated congestion based routing scheme is very effective in resolving conflicts. All the conflicts can be resolved within a few iterations for both DPL and TPL. The column "#Stitch" and the column "#Via" report the number of stitches and the number of vias in the final solution, respectively. We can see that TPL, with an extra mask to accommodate the features, can remove almost all of the stitches (99.9% on average) that are necessary for DPL. TPL can also reduce the number of vias by 37% on average compared with DPL. The column "Wirelength" reports the total wire length together with the ratio of the wire length on each mask. The results show that our approach can balance the utilization of the masks very well. We then compare the two routers in terms of ability to resolve conflicts on a set of denser benchmarks. The result is displayed in Table 2. The maximum number of rerouting iterations is set to be 30. We can see from the result that the double patterning router generates a lot more conflicts from the beginning and cannot resolve all of them after 30 iterations, while the triple patterning router can easily bring the conflicting wire length down to zero within a small number of iterations.

By comparing DPL and TPL under the same printing condition, we would like to answer the multiple choice question about which patterning technique to use in 14nm technology node. During the experiment and with help of our proposed algorithm, we have set up the comparison between DPL and TPL fairly enough: (1) the same test benches; (2) the same technology node setup; (3) the routing-coloring co-optimized designs for both DPL and TPL. Note that since the stitches can be almost completely avoided (in Table 1),

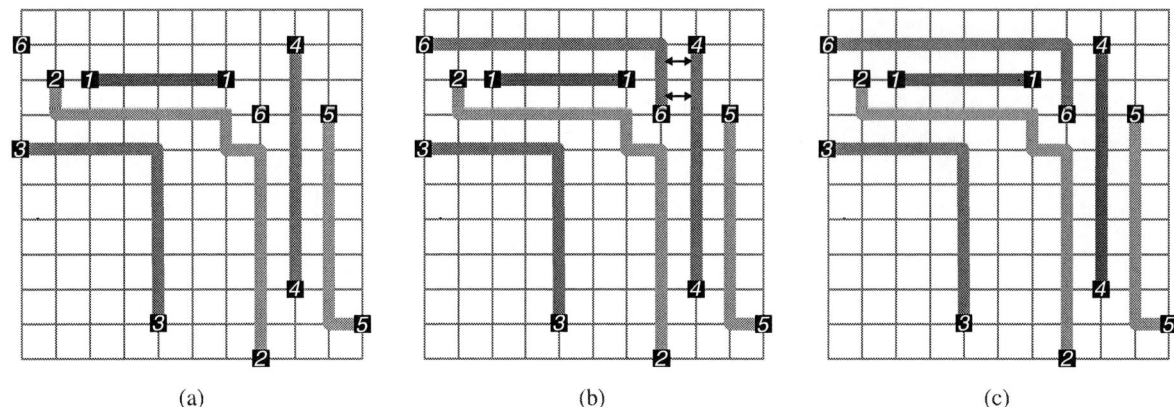

Figure 10: An example of using the negotiated congestion based scheme to handle conflicts. (a) Net 1-5 are routed and colored, and net 6 is not routed yet; (b) Net 6 is routed in color red, causing conflicts with net 4; (c) In the next iteration, net 4 is rerouted in color blue due to the penalty exerted on the red edges.

Table 1: Comparison of TPL and DPL in 14nm technology

Test Cases	#Net	Size (um^2)	Init Con. WL(um)		#Iter		Con. WL		#Stitch		#Via		Wirelength (um)		Runtime (s)	
			DP	TP	DP	TP	DP	TP	DP	TP	DP	TP	DP (R:B)	TP (R:G:B)	DP	TP
test1	1K	31.36	7.39	0.28	16	3	0	0	49	0	648	476	354(1:1.03)	354(1:1.005:0.99)	57	17
test2	2K	70.56	9.07	0.39	9	3	0	0	96	0	1076	750	720(1:1.04)	723(1:1.016:1.0007)	143	68
test3	4K	165.9	8.23	0.28	4	3	0	0	185	0	1622	878	1444(1:1.02)	1461(1:1.02:1.04)	273	301
test4	6K	245.9	12.15	0.56	16	2	0	0	247	0	2252	1496	2130(1:1.012)	2162(1:0.97:1.003)	2431	438
test5	8K	321.1	19.08	0.67	7	3	0	0	365	0	3426	2036	2873(1:0.999)	2911(1:1.01:1.02)	1790	1194
test6	10K	384.2	30.41	2.13	15	4	0	0	419	1	4452	2782	3593(1:1.016)	3625(1:0.998:1.01)	5673	2210
	Avg		14.56	0.716	11.2	3	0	0	227	0.17	2246	1403	1852	1873	1728	705
	Diff		1	-95%	1	-73%	0	0	1	-99.9%	1	-37%	1	+1.13%	1	-59%

Table 2: Comparison of TPL and DPL in resolving conflicts

Test Cases	#Net	Size (um^2)	Init Con. WL(um)		#Iter		Con. WL(um)	
			DP	TP	DP	TP	DP	TP
test1d	1.2K	31.36	18.47	0.82	30	3	7.68	0
test2d	2.5K	70.56	24.50	1.15	30	5	8.35	0
test3d	5K	165.9	28.40	0.96	30	4	6.24	0
test4d	7.5K	245.9	48.92	1.32	30	4	12.46	0
test5d	10K	321.1	71.26	1.89	30	5	20.79	0
test6d	12K	384.2	87.41	4.26	30	8	37.61	0
Avg			46.49	1.73	30	4.83	15.52	0
Diff			1	-96%	1	-84%	1	-100%

the overlay in TPL will not be as harmful as in DPL where stitches are commonly seen. From the comparison (Table 1 and Table 2), we can see that choosing between DPL and TPL in 14nm node is intrinsically a trade-off between mask/process cost and printability. In the circumstances where stitches are highly likely to cause yield loss and the extra mask cost from TPL is still affordable, with the help of our proposed algorithm, the solution of TPL with simultaneous routing and coloring would be a wise choice in 14nm technology node.

6. CONCLUDING REMARKS

In this paper, the problem of TPL aware routing is studied. A unified graph model is proposed to solve the maze routing problem in the context of multiple patterning lithography. An overall routing scheme is developed to resolve the coloring conflicts and balance the utilization of the masks. Triple patterning aware routing is compared with double patterning aware routing in 14nm node. The results show that TPL can resolve the conflicts more easily and significantly reduce the number of stitches compared with DPL.

7. REFERENCES

[1] International Technology Roadmap for Semiconductors 2011, http://www.itrs.net.

[2] D. Abercrombie, P. Lacour, O. El-Sewefy, A. Volkov, E. Levine, K.

Arb, C. Reid, Q. Li, and P. Ghosh. Double patterning from design enablement to verification. In *Proc. SPIE vol. 8166*, 2011.

[3] G. E. Bailey, A. Tritchkov, J.-W. Park, L. Hong, V. Wiaux, E. Hendrickx, S. Verhaegen, P. Xie, and J. Versluijs. Double pattern EDA solutions for 32nm HP and beyond. In *Proc. SPIE vol. 6521*, 2007.

[4] M. Cho, Y. Ban, and D. Z. Pan. Double patterning technology friendly detailed routing. In *Proc. ICCAD*, pages 506–511, 2008.

[5] J. Huckabay, W. Staud, R. Naber, A. van Oosten, P. Nikolski, S. Hsu, R. J. Socha, M. V. Dusa, and D. Flagello. Process results using automatic pitch decomposition and double patterning technology (DPT) at $k_{1eff} < 0.20$. In *Proc. SPIE vol. 6349*, 2006.

[6] Y. Inazuki, N. Toyama, T. Nagai, T. Sutou, Y. Morikawa, H. Mohri, N. Hayashi, M. Drapeau, K. Lucas, and C. Cork. Decomposition difficulty analysis for double patterning and the impact on photomask manufacturability. In *Proc. of SPIE, vol. 6925*, 2008.

[7] A. B. Kahng, C. H. Park, X. Xu, and H. Yao. Layout decomposition for double patterning lithography. In *Proc. ICCAD*, pages 465–472, 2008.

[8] Y.-H. Lin, and Y.-L. Li. Double patterning lithography aware gridless detailed routing with innovative conflict graph. In *Proc. DAC*, pages 398–403, 2010.

[9] L. McMurchie, and C. Ebeling. Pathfinder: a negotiation-based performance-driven router for fpgas. In *Proc. FPGA*, pages 111–117, 1995.

[10] J. Sun, Y. Lu, H. Zhou, and X. Zeng. Post-routing layer assignment for double patterning. In *Proc. ASPDAC*, pages 793–798, 2011.

[11] X. Tang and M. Cho. Optimal layout decomposition for double patterning technology. In *Proc. ICCAD*, 2011.

[12] V. Wiaux, S. Verhaegen, S. Cheng, F. Iwamoto, P. Jaenen, M. Maenhoudt, T. Matsuda, S. Postnikov, and G. Vandenberghe. Split and design guidelines for double patterning. In *Proc. of SPIE, vol. 6924*, 2008.

[13] Y. Xu, and C. Chu. GREMA: Graph reduction based efficient mask assignment for double patterning technology. In *Proc. ICCAD*, pages 601–606, 2009.

[14] B. Yu, K. Yuan, B. Zhang, D. Ding, and D. Z. Pan. Layout decomposition for triple patterning lithography. In *Proc. ICCAD*, 2011.

GDRouter: Interleaved Global Routing and Detailed Routing for Ultimate Routability

Yanheng Zhang
Placement Technology Group
Cadence Design Systems
San Jose, CA 95134 USA
yhzhang@cadence.com

Chris Chu
Department of ECE
Iowa State University
Ames, IA 50010 USA
cnchu@iastate.edu

ABSTRACT

Improving detailed routing routability is an important objective of a global router. In this paper, we propose GDRouter, an interleaved global routing and detailed routing algorithm for the ultimate routability i.e., detailed routing routability. The newly proposed router makes the global routing aware of detailed routing routability by correctly setting global capacity to reduce the inconsistency between the two stages. The final result contains both the detailed routing guided global routing and deailed routing solutions.

Fast and efficient academic global routing and detailed routing tools *FastRoute* [1] and *RegularRoute* [2] are interleaved in GDRouter. In the *Initial Capacity and Routing Weight Esitmation* (ICRWE) phase, the weight for each global and detailed routing grid is calculated to make GDRouter aware of pin distribution based on a *Gridded Voronoi Diagram* method. Then the algorithm generates initial global capacity based on both local usage and global segment usage. In particular, Spine routing is utilized to estimate local usage. And a virtual routing i.e. fast implementations of FastRoute and RegularRoute, is performed to estimate global segment usage. The initial global capacity is applied in *Full Routing* phase to obtain detailed routing routability i.e., number of unassigned global segment. To further improve routability, in the following *Iterative Test Routing* (ITR) phase, GDRouter incrementally applies the interleaved global routing and detailed routing to adjust the global capcity based on detailed routing solution. To save runtime, GDRouter quits the loop if detailed routing routability stops improving or it reaches maximum iteration.

Experimental results reveal that the newly proposed algorithm is capable of enhancing detailed routing routability. In particular, GDRouter reduces number of unassigned global segments by 90% for ISPD98 [3] derived testcases and around 60% for ISPD05/06 [4,5] derived testcases with $2.9\times$ runtime overhead.

Categories and Subject Descriptors

B.7.2 [**Hardware**]: Integrated Circuits—*Design Aids*

General Terms

Algorithms, Design, Performance

Keywords

Routing, Physical Design, VLSI CAD

1. INTRODUCTION

VLSI routing is an important design stage where module and cell pins are connected by over-the-cell metal wires. As the fabrication technology enters the nanometer scale, the routability issue is becoming increasingly challenging. First, there are more and more transistors integrated on chip and the size of routing problem is growing

much bigger. Second, the integration of reusable unit of logic i.e., Intellectual Property (IP) cores due to the increasing design complexity poses more constraints on routing resources.

It has been proved that the routing problem, even the small case containing only a couple of two-pin nets, is NP-hard [6]. Due to the enormous computational complexity, routing is typically carried out through consecutive global routing and detailed routing stages. In global routing, the entire routing region is divided into regular global cells(i.e., G-Cells) and routing is performed based on these G-Cells. The obtained global routing results are used to generate detailed routing solution considering exact metal shapes and positions.

Name	FastRoute [1]		RegularRoute [2]	
	O.F.	CPU(sec.)	unassiged	CPU(sec.)
adaptec1	0	195	3233	566
adaptec2	0	48	1038	442
adaptec3	0	324	3352	1285
adaptec4	0	66	4027	1330
adaptec5	0	559	8826	3782

Table 1: Detailed routing results generated by Regular-Route on five reportedly routable global routing benchmarks in ISPD07/08 [7,8] global routing contest

The primary objective for global routing is to generate a congestion free global routing solution on G-Cells where the wiring demands across the G-Cell boundary is below its capacity. The global capacity is an estimation of how many wires can be accommodated during detailed routing. There have been many research conducted on improving global routing routability since the two consecutive ISPD 2007 and 2008 global routing contests [7,8]. For instance, contest-winning routers like BoxRouter [9], NTHU-R [10], FGR [11] and FastRotue [1,12] are proposed to drive the global routing congestion lower with consistent effort. However, the ultimate routability i.e., detailed routing routability is not consistently pursued. In Table 1, we present the detailed routing results by taking the global routing solution of five routable testcases generated by the *FasRoute*. According to Table 1, all five testcases are easily routable by FastRoute in global routing stage. We use recently proposed detailed router *RegularRoute* [2] to generate the detailed routing solution. Nevertheless, none of the testcases can be easily routed by RegularRoute based on the pre-defined pitch size. The detailed routing routability i.e., number of unassigned global segment reaches several thousands. The results reveal the inconsistency between global routing and detailed routing stages. Or in other words, global capacity is not set in agreement with the detailed routing routability.

Traditionally, global capacity is estimated based on empirical methods. For instance, to reserve routing resources for local usage and pin escape routing, global capacity is reduced for global edges on first horizontal and vertical metal layers. To simulate the effect of macro porosity for big macros, global capacity of the covered global edges is scaled by a fixed percentage. However, these methods are not accurate enough to estimate the detailed routing routability since they simply overlook the exact pin distribution and actual detailed routing usage.

In this paper we will present a novel algorithm called GDRouter for the overall routability in routing problem. We intend to address the inconsistency between global and detailed routing by setting the global capacity more accurately. It is a systematic approach interleaving both global routing and detailed routing beyond the scope defined in conventional routing flow. As far as we know, it is the first attempt in academia to interleave the global routing and detailed routing for the detailed routing routability.

Efficient global routing and detailed routing tools *FastRoute* [1] and *RegularRoute* [2] are interleaved. In the the *Initial Capacity and Routing Weight Estimation* (ICRWE) phase, we calculate the weight for each global and detailed routing grid to make GDRouter aware of pin distribution. We propose a *Gridded Voronoi Diagram* method to evaluate pin distribution. We also generate initial global capacity based on local usage and global segment usage. In particular, we use a Spine routing technique to estimate the local usage. And we apply a virtual routing based on fast implementations of FastRoute and RegularRoute to estimate global segment usage. The initial global capacity is applied in the *Full Routing* phase to obtain detailed routing routability. In the following *Iterative Test Routing* phase, we run interleaved global routing and detailed routing to further improve routability. Detailed routing solutions are adaptively used to update global capacity. To speed up the whole algorithm, we propose incremental global routing and a history based detailed routing for FastRoute and RegularRoute respectively. To save runtime, GDRouter terminates when it reaches maximum iteration or detailed routing routability stops improving.

Therefore, the novel techniques that will be presented in this paper are as follows:

- Novel routing algorithm interleaving global routing and detailed routing for detailed routing routability

- Effective initial capacity and routing weight estimation capturing pin distribution, local usage and global segment usage

- Efficient mechanism to adaptively update the global capacity based on detailed routing solutions

- Useful incremental global routing technique and history based detailed routing technique to reduce runtime overhead

We implemented GDRouter and tested its performance on routing testcases drived from ISPD98 [13] and ISPD05/06 [4, 5] benchmarks respectively. Experimental results show that GDRouter is capable of improving detailed routing routability in terms of number of unassigned global segments. In particular, the number is improved by 90% for ISPD98 and around 60% for ISPD05/06 with 2.9 × runtime overhead.

The rest of the paper is organized as follows: In Section 2, we first review the global routing and detailed routing tools *FastRoute* and *RegularRoute* and present the overview of GDRouter. Section 3 discusses the initial capacity and routing weight estimation (ICRWE) phase. In Section 4, we present techniques in iterative test routing (ITR) phase for how to update global capacity based on detailed routing solution. Experimental results are presented in Section 5 and we will make conclusion of this paper in Section 6.

2. GDROUTER OVERVIEW

In this section, we will first present necessary definitions and problem formulations for global routing and detailed routing respectively. We then present review for global and detailed routing tools *FastRoute* and *RegularRoute*. Finally, we will present the flow of GDRouter.

2.1 Problem Formulation

We will introduce the formulation for global routing and detailed routing respectively.

2.1.1 Global Routing Formulation

In global routing, layout region on each metal layer is divided into *3-D global routing cells* (3-D G-Cells). The *3-D global routing grid graph* is drived where each grid denotes one 3-D G-Cell and each *3-D global edge* represents the common boundary between two 3-D G-Cells. Each edge is assigned with *3-D global capacity* representing the maximum allowable global routing usage. The overflow is the exceeding amount of global routing usage over global capacity.

The major objective in global routing is to minimize total overflow. Since FastRoute generates 2-D global routing solution, we lump the 3-D G-Cells on different layers into a 2-D global routing cells (2-D G-Cell or simply G-Cell). The capacity, usage and overflow will be accumulated as the capacity, usage and overflow of the G-Cell respectively. The 3-D grid graph is therefore projected to become a 2-D grid graph. The global routing solution is generated using global path based on the 2-D grid graph. [1]

2.1.2 Detailed Routing Formulation

The detailed routing resource is modeled as a 3-D regular grid graph. Each grid edge can accommodate exact one wire detailed routing usage. Please note that the grid defined here are not of the same concept with the global routing counterpart. We call the grid detailed routing grid or finest grid. It is defined by the pitch size so the detailed routing path on the grid would not cause spacing rule violations. Each routing layer has a *preferred routing direction* and the preferred direction iterates between adjacent layers. The *routing track* is defined as

---
[1]In the following contexts, G-Cell, global edge and global capacity mean 2-D G-Cell, 2-D global edge and 2-D capacity.

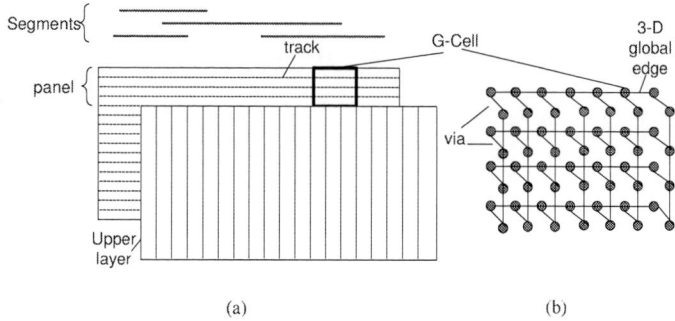

Figure 1: Problem formulation for global routing and detailed routing (a) Detailed routing with panel, track, segments with two layers (b) Corresponding global routing grid graph

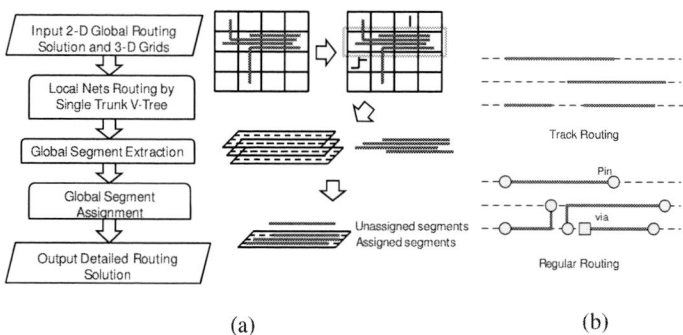

Figure 2: (a) Overview of RegularRoute (b) Comparison between track routing and regular routing

a sequence of grid edges along the preferred routing direction of each layer.

The input is 2-D global routing solution We extract global segments from the global routing solution. Each global segment spans multiple G-Cells that the global route goes through. *Panel* is defined to be the collection of parallel tracks in a row of column of G-Cells for each metal layer. It is introduced to favor the global segment assignment for paralleism. Spacing rule related design rule violations account for the majority of design rule violations in a typical VLSI design. We mainly consider the spacing rules during detailed routing since they have been captured by the application of detailed routing grid. The congestion of detailed routing is evaluated by the number of *unassigned global segments*, meaning such global segments cannot be properly assigned due to conflicts with other usage.

In Figure 1, we show the definitions we have mentioned for global routing and detailed routing respectively. Simply speaking, global routing generates global routing solution based on 2-D global routing grid. The obtained results are used to generate the detailed routing solution based on detailed routing grid. Hence the ultimate routability should be pertinent to the congestion (number of unassigned segments) of detailed routing solution.

2.2 FastRoute and RegularRoute

Next we introduce the global routing and detailed routing algorithms that are interleaved in GDRouter.

FastRoute is very fast and effective global router. It incorporates a couple of ideas for solving the challenging global routing testcases. It contains a number of works: FastRoute [14] is the first work which presented congestion-driven Steiner tree construction and edge shifting. FastRoute 2.0 [15] introduced a novel monotonic routing to enhance the pattern routing and a multi-source multi-sink maze routing to improve maze routing. In FastRoute 3.0 [12], a virtual capacity technique is proposed to achieve fast convergence of maze routing. FastRoute [1] introduces routing techniques for reducing via count.

RegularRoute [2] is a recently proposed grid-based detailed routing technique. The router strives to generate regular routing patterns (less turning) to benefit design rule satisfaction. In general, it adopts a bottom-up layer-by-layer and panel-by-panel scheme. It employs a spine routing technique to route the local nets. The global segments are assigned panel-by-panel and a Maximum Weighted Independent

598

Set (MWIS) problem is formulated for each panel. RegularRoute employes a fast and effective heuristic to solve the problem for each panel. For better detailed routing routability, the router introduces the terminal promotion and partial assignment. RegularRoute looks similar to the track routing problem [16] but there are great difference between the two. One important factor is that RegularRoute generates a valid detailed routing solution where terminal connections are considered. General track routing simply overlooks terminal connection. In Figure 2(a), we show the overview of RegularRoute and its algorithmic flow. In Figure 2(b), we present the example of solutions for both RegularRoute and general track routing. We could easily discover the difference between the two problems.

2.3 Algorithm Flow

The general flow of GDRouter is shown in Figure 3. Basically the flow contains three main phases: (1) Initial Capacity and Rouing Weight Estimation (2) Full Routing (3) Iterative Test Routing. The flow of GDRouter interleaves the flow of FastRoute and RegularRoute but it is not simply an addition of the two. There are many new techniques introduced for better detailed routing routability and less runtime overhead.

Phase 1: Initial Capacity and Routing Weight Estimation
    1. Pin distribution analysis based on Gridded Voronoi Diagram of detailed routing grids
    2. Local usage estimated by Spine routing
    3. Virtual routing by fast implementations of FastRoute and RegularRoute
Phase 2: Full Routing
    4. FastRoute
    5. Global segment extraction
    6. RegularRoute
Phase 3: Iterative Test Routing
    7. Global capacity update based on RegularRoute's solution
    8. Incremental FastRoute to reroute nets with unassigned global segments
    9. History based RegularRoute to tune number of choice of global segments
    10. Repeat Phase 3 until reaching maximum iteration or detailed routing routability stops improving

Figure 3: Flow of GDRouter

In particular, in ICRWE phase, pin distribution is analyzed based on Gridded Voronoi Diagram. The weight for each global and detailed routing grid is calculated and will be applied in the following global and detailed routing. Spine routing and a virtual routing are performed to generate the initial global capacity based on the estimated local usage and global segment usage. Based on the initial global capacity computed in Phase 1. GDRouer applies FastRoute and RegularRoute in full-routing mode to obtain detailed routing soluiton. Next, GDRouter enters routing (ITR) phase to iteratively update the global capacity to make sure the global routing is guided to avoid detailed routing hotspots. To speed up the whole algorithm, an incremental global routing technique and a history based global segment assignment technique are applied for FastRoute and RegularRoute. They are proposed to replace the time-consuming full routing. It should be noted that in GDRouter, detailed routing has been fully integrated in our full-chip routing framework. The final routing solution contains not only the detailed routing guided global routing solution, but also the very detailed routing solution accompanied with it.

3. INITIAL CAPACITY AND ROUTING WEIGHT ESTIMATION

In this section, we will introduce techniques in ICRWE phase in detail. In the first part, we will discuss a Gridded Voronoi Diagram method to analyze pin distribution and how to compute the weight for each global routing and detailed routing grid. Next, we introduce the Spine routing to estimate local usage. Finally, we discuss virtual routing to capture the global segment usage by fast implementations of FastRoute and RegularRoute and how we generate initial global capacity based the two techniques.

3.1 Pin Distribution Analysis based on Gridded Voronoi Diagram

Pin distribution is an important indicator of potential routing congestion. In the past, people usually use pin density i.e., number of pins in one G-Cell to estimate pin related local hotspots. However, pin density cannot fully capture detailed routing routability. Consider one G-Cell containing 5 pins with even distribution while the other G-Cell with 4 pins but with all pins concentrated in the G-Cell. The latter one has lower pin density but more likely to be unroutable.

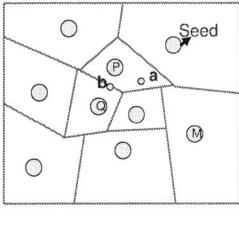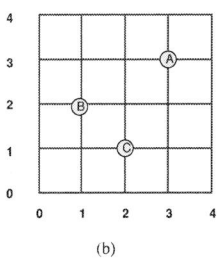

Figure 4: (a) General Voronoi diagram with seeds and proximity cells (b) Gridded Voronoi Diagram of detailed routing grid

Therefore we propose to adopt pin distribution analysis along with pin density in GDRouter.

In the work of [17], the paper utilizes voronoi diagram [18] to estimate the *wire distribution* and the corresponding critical area. In this paper, we utilize *Gridded Voronoi Diagram* to estimate pin distribution. In a general Voronoi Diagram, the *entire region* indicates the whole area under analysis, a *seed* indicates the subject of which we are investigating the distribution (e.g., pin in our case). The *proximity cell* of a seed is the sub-region in which each point has closest proximity to the seed. The *proximity cell boundary* consists of the points along the cell boundary, which have equivalent cloest proximity to multiple seeds. Figure 4 illustrates a general voronoi diagram for a region with nine seeds. For instance, point a belongs to seed P, and point b is along the cell boundary of seed Q and has equal proximity to P and Q.

The size of proximity cell indicates the distribution in the area around the particular seed. For the case with evenly distributed seeds, the voronoi diagram proximity cell is equivalent in size. For a case with non-evenly distributed seeds, the relative size of the proximity cell indicates the seeds' distribution i.e., the smaller the size of a cell, the denser seed's distribution will be. For instance, in Figure 4(a), seed P has smaller proximity cell than seed M, so points around P have denser seeds' distribution than M.

Based on this intuition, we could extend the general voronoi diagram to detailed routing grid, and we call it *Gridded Voronoi Diagram*. The main difference is the region is now grid based and the seed has become the pin. In our formulation, since each pin is located on one detailed routing grid of M1 (metal 1), we will only look at the 2-D detailed routing grid in M1 instead of other metal layers.

We define the *proximity grids* for a pin as the detailed routing grids with closest proximity to a pin. And we define the *peripheral grids* as the grids with equivalent cloest proximity with multiple pins. To analyze the pin distribution for each grid, we first calculate the average count of proximity grids (θ) for all pins, which is equal to total grids divided by total pins. For any pin, we suppppose the pin has g proximity grids, the distribution for the pin's proximity grids is calculated as follows,

$$D_g^{px} = \beta \times (\frac{\theta}{g})^\alpha \qquad (1)$$

$$\beta = \frac{\theta \times (p+1)}{W^2} \qquad (2)$$

where α is a parameters that need be tuned for optimization. In our experiemnt, alpha equals 1.5. β is another factor tuning the distribution value. W is the length of square window which contains number of grids cloest to g i.e., $W^2 \approx g$. p is the number of pins inside the window.[2] For each peripheral grid, its distribution is equal to the average of all pins it has closest proximity.

$$D_g^{pr} = \frac{\sum D_g^{px}}{N_{pr}} \qquad (3)$$

In the equation N_{pr} is the total number of pins the peripheral grid has closest proximity.

A gridded voronoi diagram for detailed routing grids is illustrated in Figure 4, which contains a 5×5 grid with three pins. We list out the closest proximity pin for each grid. For instance, grid $(1, 1)$ in coordinate is a peripheral grid which has equivalent closest proximity to pin B and pin C. While grid $(2, 0)$ and $(2, 1)$ are two proximity grids of pin C. And the pin distribution for each grid can ben computed accordingly.

---

[2] The window is introduced to partly capture the impact of pin density and to defferentiate grids in the same proximity cell.

The pin distribution of one G-Cell is calculated by averaging all grids inside the G-Cell including both proximity grids and peripheral grids, which is

$$D_{cell} = \frac{\sum D_g}{G} \qquad (4)$$

In the equation, D_{cell} is the average pin distribution value of G-Cell, and G is the number of grids inside the G-Cell.

Based on these computations, we assign the weight of each G-Cell in global routing and each detailed routing grid in detailed routing. More specifically, in global routing, the global path generation is weighed by D_{cell}. And during detailed routing, the cost of assigning each global segment is also weighed by D_g.

3.2 Local Usage Estimation

One factor that may affect global capacity is the local usage inside one G-Cell. A net or part of a net that resides inside a G-Cell is a local net. The local usage is treated as blockages when assigning global segments. Hence global capacity needs to be adjusted. In RegularRoute, *spine routing* is applied to route local nets. Spine routing construct routing tree using one vertical trunk and several horizontal branches. The major advantage in terms of routability is the preservation of routing resources. We apply spine routing for estimating local usage.

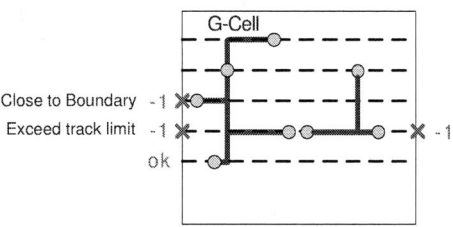

Figure 5: (a) Spine routing for estimating local usage

We reduce the global capacity for a global edge if local usage is close to the boundary of the G-Cell (smaller than 3 grids) or the total usage of one track inside the G-Cell is over 50%. Local usage that close to G-Cell boundary is likely to block global segment usage (routing across the G-Cell). And the track is likely to be blocked when the total local usage is high, in which case we need to reduce capacity for both global edges (e.g., left and right edges). In Figure 5, we show global capacity adjustment when local usage is close to G-Cell boundary. It is also illustrated that one track is assigned with usage more than 50%, the capacity is reduced for both left and right global edges.

3.3 Virtual Routing for Estimating Global Segment Usage

The mere local usage estimation is not capable of accurately finding detailed routing hotspots. In many cases, congestion occurs simply when global segment cannot be assigned. In Figure 6, we show a congested global edge with the global capacity of four. And the global edge is used by four nets, which does not cause overflow in global routing. But in reality only three of the segments can be assigned in detailed routing. In this case congestion is caused by global segment usage and the mismatch between global capacity and detailed routing routability.

Since it is computationally hard to predict global segment usage and its impact on global capacity, we propose to apply a virtual routing for global capacity adjustment. The virtual routing consists of fast implementations of global and detailed routing. In particular, in FastRoute, we only apply very fast pattern routing in "L" and "Z"

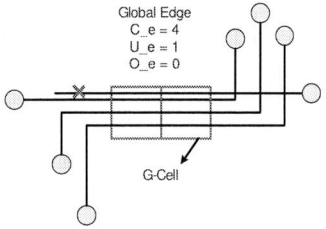

Figure 6: Global segment derived congestion in detailed routing without pin and local usage

shapes [19]. Although we may not obtain overflow free global routing solution, it provides with the information of global segment usage with actual global routing paths. Next, the global routing solutions are imported into RegularRoute. Likewise, we employ a fast mode of RegularRoute with restricted solution space instead of trying to generate the optimized solution. In particular, each global segment is allowed to try fewer regular routing shapes. Terminal promotion and partial assignment are disabled for saving runtime. As a matter of fact, in such configuration, we only perform track routing.

After virtual routing, we obtain the detailed routing routability in terms of the unassigned global segments. For each global edge e along the unassigned global segment, suppose we have the global capacity C_e, the global routing overflow O_e and the total number of unassigned segments is U_e. Although global edge e has global routing overflow O_e, in detailed routing, the actual unassigned global segment is U_e. The mismatch of global capacity and detailed routing is actually $U_e - O_e$. We adjust the global capacity by the following equation, C'_e is the new capacity, γ is a coefficient to suppress the over-adjustment of the global capacity. γ equals 0.5 in our experiment.

$$C'_e = C_e - \gamma \times (U_e - O_e) \qquad (5)$$

Based on the spine routing and the virtual routing, we obtain the inital global capacity. With the initial global capacity, GDRouter enters *full routing* phase in which FastRoute and RegularRoute are applied in full-routing modes.

4. ITERATIVE TEST ROUTING

In this section we present techniques in the *iterative test routing* (ITR) phase. The main idea is to update global capacity iteratively based on interleaved global and detailed routing. Detailed routing routability in terms of unassigned global segments is utilized to adjust the global capacity. To improve runtime, an incremental global routing technique is proposed for the interleaved global routing. And a novel history based detailed routing technique is proposed for the interleaved detailed routing. ITR terminates when it reaches maximum iteration or the number of unassigned global segments stops improving. In the end, the routing solution contains not only a detailed routing guided global routing solution, but also the very detailed routing solution accompanied with it.

4.1 Adaptive Global Capacity Update

As the flow shown in Figure 3, after ICRWE, we apply global routing and detailed routing using *FastRoute* and *RegularRoute* with full-routing features. During the global routing stage, the 2-D global routing solution in terms of the global route on the global routing grids is obtained. We extract global segments based on the solution. In the subsequent detailed routing stage, RegularRoute is applied to route local nets and assign global segments. After detailed routing, the detailed routing routability is obtained in terms of the number of unassigned global segments, with the indication of where the actual detailed routing congestion is. We need to update the global capacity to make sure global routing is properly guided by actual detailed routing bottlenecks.

As defined in Section 2.1.2, a global segment represents an interval spanning multiple G-Cells. It should be assigned to the tracks to the panels it belongs to. If a segment is unassigned, it suggests the panels are unable to accommodate the segment. We update global capacity based on the global edges along the unassigned global segment. For each global edge e after detailed routing, A_e is the number of assigned global segment, C_e is the original global capacity, U_e is the number of unassigned segments, and R_e is the accumulated amount of capacity adjustment (positive for reduction, negative for increase). We will adaptively update the global capacity based on the following cases (we use C'_e to represent the new capacity): i) If $A_e + U_e > C_e + R_e$ and $U_e > 0.2 \times (C_e + R_e)$, when the global capacity is excessively overestimated, $C'_e = C_e \times 0.9$; ii) If $A_e + U_e > C_e + R_e$ and $0.2 \times (C_e + R_e) \geq U_e > 0$, when global capacity is moderately undertimated, $C'_e = C_e - 1$; iii) If $A_e + U_e \leq C_e + R_e - 2$ and $U_e = 0$, when global capacity is underestimated, $C'_e = C_e + 1$; iv) In all other cases, when global capacity is roughly accurate, $C'_e = C_e$. In general, capacity reduction and increase is carried out without too large amount since (1) The detailed routing hotspot may change over-time, it is not proper to drastically reduce the global capacity. (2) We do not intend to disturb the original solution, otherwise it takes more computational effort in the subsequent global routing and detailed routing in ITR.

4.2 Techniques to Reduce Runtime Overhead

Runtime is the major overhead in ITR if we apply FastRoute and RegularRoute in full-routing features. As a matter of fact, in very congested designs global capacity might become very restricted, and the global routing could become too congested to achieve routing convergence. In such case, RegularRoute would spend longer runtime due to the extra global segment usage introduced by the detoured global path in FastRoute. To reduce runtime overhead, we propose techniques for FastRoute and RegularRoute to improve runtime in ITR. More specifically, we propose incremental global routing technique for FastRoute and history based global segment assignment for RegularRoute.

4.2.1 Incremental Global Routing

Instead of applying FastRoute with full features in ITR, we propose an incremental global routing technique. As introduced in Section 2.2, FastRoute is a sequential global router which is rip-up and reroute based. Nets are routed sequentially from step to step. Therefore, we rip-up and reroute nets using over-capacity edges. This approach would reduce the number of routed nets significantly. Moreover, most global routers like FastRoute strive to minimize global routing overflow and spend long runtime on maze routing. We propose to terminate maze routing when the congestion improvement is less than a threshold value, which is 5% of the initial total global routing overflow. After obtaining incremental global routing solution, we only re-extract the global segments of the nets that are rerouted.

4.2.2 History Based Detailed Routing

In *RegularRoute*, solution space of assigning global segment is explored based on a number of regular routing shapes i.e., the specific track, layer and terminal routing, of each global segment. Each candidate shape is defined as one *Choice* of the segment. Basically one segment is more flexible given more choices. As noted earlier, to perform RegularRoute in full features inside ITR would be very computationally expensive. We propose the history based detailed routing to reduce runtime overhead. In particular, for the nets that are not rerouted in incremental global routing, we will recommend to use the choices of global segments of last iteration, or in other words, we rank the recommended those choices with higher priority. If one segment is consistently assignable, we gradually decrease the number of choice or even just provide the choice of last iteration. This technique is especially effecitve for the panels that are not congested. Conversely, for the tough-to-handle global segments, we increase their number of choices to improve routability. But with the algorithm continues, fewer global segments are unassignable and the runtime for each iteration of ITR is further reduced.

5. EXPERIMENTAL RESULTS

We implement GDRouter in C and all our experiments are performed on a machine with 2.67GHz Intel Xeon CPU and 32G memory. We derive the testcases from ISPD98 [3] and ISPD05/06 [4, 5] placement benchmarks. In original ISPD98 benchmarks, pins are set to be at the center of each standard cell, we develop a program to randomly set the pin coordinates and make sure they satisfy the spacing requirements at the bottom layer.[3] We use Dragon [20] to generate the placed testcases for ISPD98 benchmarks and use FastPlace [21] to generate those for ISPD05/06 benchmarks. Other placement methods [22] [23] [24] are worthy alternatives but it is out of the content of this paper. We will then derive the global routing benchmark based on placed testcase with the same format as ISPD07/08 global routing contest benchmarks [7, 8].

5.1 Initial Capacity Estimation Results

Name	EM1		EM2		EM3		ICRWE	
	unass-igned	CPU (sec.)	unass-igned	CPU (sec.)	unass-igned	CPU (sec.)	unass-igned	CPU (sec.)
ibm01	0	9.7	0	7.3	4	18.6	0	11.2
ibm02	25	19.2	36	15.6	78	44.3	21	18.4
ibm07	55	37.6	66	34.3	108	98.5	43	42.5
ibm08	19	62.4	23	41.6	26	122.1	20	57.1
ibm09	24	78.2	32	52.5	42	175.2	19	55.4
ibm10	55	98.4	65	73.2	99	188.4	34	103.6
ibm11	25	93.4	19	68.9	46	155.9	16	87.2
ibm12	145	206.5	168	114.5	255	355.2	102	137.2
Sum	348	605.4	409	407.9	658	1158.2	255	512.6
Norm	1.36	1.48	1.60	1	2.58	2.84	1	1.26

Table 2: Results comparison with three empirical methods for estimating global capacity up to full routing phase (Phase 2) in GDRouter on ISPD98 derived testcases.

We present experimental results of techniques in the initial capacity and routing weight estimation phase on ISPD98 derived testcases. The results are summarized in Table 2. We compare ICRWE against three empirical methods, namely EM1, EM2 and EM3 respectively. In particular, EM1 assumes the first two metal layers (M1 and M2) have

---

[3]We assume all pins are on M1.

0% of full capacity and the rest of layers have 50% of the full capacity. EM2 assumes the first two metal layers have 20% of full capacity and the rest of layers have 80%. And EM3 assumes the first two metal layers have 40% of full capacity and the rest of layers have 100%. In our experiments, all parameters are identical for the four cases.

In Table 2, we present the results for each case respectively. We could notice that ICRWE is capable of achieving best routability among the four methods. EM1 tends to underestimate the global capacity. It makes the global routing much harder to solve and creates more extra detours. The extra detoured usage is likely to incur more global segment usage and thus affect detailed routing routability. On the other hand, EM3 tends to overestimate the global capacity. The global routing problem in EM3 is easier to solve, but it leads to a harder-to-solve detailed routing problem. The runtime becomes much worse since the runtime spent on detailed routing increases dramatically. EM2 generates reasonably good solution. But the same success cannot be guaranteed for all design cases. Overall, ICRWE provides more accurate global capacity estimation than the empiricle methods and it generates best routability.

5.2 GDRouter Results on ISPD98 Derived Test-cases

We present experimental results for GDRouter on the eight ISPD98 derived testcases. We compare our results with three variations of GDRouter. All the results are summarized in Table 3. In the table, the first three columns provides us the basic statistics of the testcases. #Nets is the total number of nets in the testcase, *Grid* is the scale of global routing grid, *Avg.Deg.* is the average net degree for the entire netlist. These statistics provide an overview of the complexity for each testcase.

The next columns show the results on three different variations of GDRouter. In particular, the first one is GDRouter without ICRWE phase. It applies EM2 (20% M1-M2, 80% All other layers). Second variation is GDRouer without ITR phase. It is the same as reported in Table 2 for the "ICRWE" case. The third variation is GDRouter without incremental global routing and history based detailed routing techniques. And the last three columns show results for GDRouter with all proposed techniques.

From the table, it is noticeable that all proposed techniques are indispensable in GDRouter. In particular, ICRWE provides a good starting point with more accurate initial global capacity. Likewise, ITR is an important component since the number of unassigned segments is consistently reduced in this phase. And the incremental global routing technique and history based detailed routing technique significantly reduce runtime overhead (roughly 2 times), though FastRoute and RegularRoute in full features generate slightly better QoR. In terms of runtime, GDRouer without ITR is close to conventional routing flow (except some runtime spent in ICRWE phase). It shows GDRouter is capable of reducing number of unassigned segments by 90% with 2.9× runtime overhead.

5.3 GDRouter Results on ISPD05/06 Derived Test-cases

We present experimental results for GDRouter on the eight ISPD05 /06 derived testcases. Similarlly, we compare our results with three variations of GDRouter. All the results are summarized in Table 4. The circuit statistics report the similar metrics of each testcase. They are typically much larger in design size than the testcases derived from ISPD98 benchmarks. In ISPD07/08 [7, 8] global routing contests, the pitch size is set to 2 to make the global routing benchmarks more challenging. They are not easy to handle in detailed routing. Hence we perform our experiments on testcases with two cases (pitch size equals one and two respectively).

Like the experiments on ISPD98 derived testcases, we compare GDRouter with three variations, GDRouter without ICRWE, GDRouter without ITR and GDRouter without incremental global routing and history based detailed routing. We could notice for large designs, GDRouter is also effective in improving detailed routing routability. In particular, it shows that GDRouter reduces over 60% unassigned global segments when pitch size is two and over 90% when pitch size is one. And it is also noticeable that all techniques are indispensable in GDRouter. In terms of runtime, GDRouter spends 2.8× and 2.1× of conventional routing flow (GDRouter without ITR) for the two cases with pitch size equals 2 and 1 respectively. Overall, GDRouter is capable of generating cost efficient detailed routing results.

6. CONCLUSION

In this paper, we propose GDRouter: an interleaved global routing and detailed routing algorithm for the ultimate routability. It contains three phases: initial capacity and routing weight estimation (ICRWE), full routing and iterative test routing (ITR). Pin distribution is analyzed based on the novel Gridded Voronoi Diagram and each global grid and detailed routing grid is assigned weight based on

	statistics			w/o ICRWE			w/o ITR			w/o Runtime			GDRouter		
Name	#Nets	Grid	Avg. Deg.	unass-igned	CPU (sec.)	iter	unass-igned	CPU (sec.)	iter.	unass-igned	CPU (sec.)	iter.	unass-igned	CPU (sec.)	iter.
ibm01	11507	133×132	3.85	0	7.3	0	0	11.2	0	0	11.2	0	0	11.2	0
ibm02	18427	152×151	4.23	2	54.6	5	21	18.4	0	0	95.5	5	0	60.5	5
ibm07	44394	229×228	3.70	0	104.2	4	43	42.5	0	0	202.3	4	0	111.4	4
ibm08	47944	239×238	4.13	0	130.9	4	20	57.1	0	0	288.6	4	0	142.0	4
ibm09	50393	243×242	3.73	8	162.3	4	19	55.4	0	0	302.6	5	0	177.6	5
ibm10	64227	316×315	4.19	6	221.6	5	34	103.6	0	0	523.2	5	2	220.4	5
ibm11	66994	276×275	3.54	8	213.6	4	16	87.2	0	0	402.6	4	0	235.1	4
ibm12	67739	341×340	4.34	30	466.7	7	102	137.2	0	12	988.1	7	21	498.5	7
Sum	-	-	-	54	1361.2	-	255	512.6	-	12	2814.1	-	23	1456.7	-
Norm	-	-	-	2.3	0.93	-	11.1	0.35	-	0.5	1.9	-	1	1	-

Table 3: Results comparison for GDRouter for variations (GDRouter w/o ICRWE, GDRouter w/o ITR, GDRouter w/o incremental global routing and history based detailed routing in ITR) on ISPD98 derived testcases

	statistics			Pitch	w/o ICRWE			w/o ITR			w/o Runtime			GDRouter		
Name	#Nets	Grid	Avg. Deg.		unass-igned	CPU (min.)	iter	unass-igned	CPU (min.)	iter.	unass-igned	CPU (min.)	iter.	unass-igned	CPU (min.)	iter.
adaptec1	219243	893×892	4.28	2	1244	42	4	3019	16	0	1044	81	4	1259	44	4
adaptec2	257659	1174×1172	4.09		221	24	3	972	11	0	189	55	3	198	26	3
adaptec3	466293	1935×1946	4.01		1462	85	5	3177	29	0	993	186	5	1088	89	5
adaptec4	515300	1933×1945	3.70		1755	67	4	3822	27	0	1450	110	4	1562	71	4
adaptec5	867344	1935×1946	3.99		3345	205	5	8467	74	0	2844	420	5	3177	211	5
newblue1	331106	934×932	3.68		45	18	4	376	7	0	32	31	4	34	19	4
newblue5	1257334	2122×2132	3.87		2821	197	5	7718	71	0	2444	409	5	2593	206	5
newblue6	1286448	2310×2318	4.09		3672	220	5	8892	84	0	3012	449	5	3411	226	5
Sum	-	-	-		14565	858	-	36443	319	-	12008	1741	-	13322	892	-
Norm	-	-	-		1.09	0.96	-	2.7	0.36	-	0.90	1.95	-	1	1	-
adaptec1	/	/	/	1	0	12	0	0	14	0	0	14	0	0	14	0
adaptec2	/	/	/		0	8	0	0	10	0	0	10	0	0	10	0
adaptec3	/	/	/		2	42	2	16	26	0	0	94	2	0	45	2
adaptec4	/	/	/		0	48	3	44	23	0	0	90	3	0	52	3
adaptec5	/	/	/		30	155	4	156	62	0	10	298	4	18	160	4
newblue1	/	/	/		0	5	0	0	6	0	0	6	0	0	6	0
newblue5	/	/	/		24	139	4	81	55	0	0	266	4	4	144	4
newblue6	/	/	/		16	98	3	42	63	0	0	212	3	0	102	3
Sum	-	-	-		72	507	-	339	259	-	10	990	-	22	533	-
Norm	-	-	-		3.27	0.95	-	15.4	0.49	-	0.45	1.86	-	1	1	-

Table 4: Results comparison for GDRouter for variations on ISPD05/06 derived testcases

pin distribution. Spine routing and virtual routing are applied to generate initial global capacity. In ITR, GDRouter further polishes the initial global capacity based on detailed routing results. Experimental results on both ISPD98 and ISPD05/06 derived testcases demonstrate the effectiveness and efficiency of our algorithm.

We will continue to improve GDRouter's performance and scalability. We will propose more systematic frameworks to perform capacity update. Meanwhile, we will also enhance the solution quality of the interleaved global routing and detailed routing algorithms.

7. REFERENCES

[1] Y.Xu, Y.Zhang, and C.Chu. FastRoute 4.0: Global router with efficient via minimization. In *Proc. Asia and South Pacific Design Automation Conf.*, pages 576–581, 2009.

[2] Y. Zhang and C. Chu. RegularRoute: An efficient detailed router with regular routing patterns. In *Proc. ACM/SIGDA Intl. Symp. on Physical Design*, pages 146–151, 2011.

[3] ISPD98 global routing benchmarks. http://www.ece.ucsb.edu/~kastner/labyrinth.

[4] ISPD05 placement contest benchmarks. http://www.sigda.org/ispd2005/contest.htm.

[5] ISPD06 placement contest benchmarks. http://www.sigda.org/ispd2006/contest.htm.

[6] M. R. Garey and D. S. Johnson. *Computers and Intractability: A Guide to the Theory of NP-Completeness.* Freeman, NY, 1979.

[7] ISPD07 global routing contest benchmarks. http://www.sigda.org/ispd2007/contest.htm.

[8] ISPD08 global routing contest benchmarks. http://www.sigda.org/ispd2008/contest.htm.

[9] K. Yuan M. Cho, K. Lu and D. Z. Pan. Boxrouter 2.0: Architecture and implementation of a hybrid and robust global router. In *Proc. Intl. Conf. on Computer-Aided Design*, pages 503–508, 2007.

[10] P.-C. Wu J.-R. Gao and T.-C. Wang. A new global router for modern designs. In *Proc. Asia and South Pacific Design Automation Conf.*, pages 232–237, 2008.

[11] M. M. Ozdal and M. D.F. Wong. High-performance routing at the nanometer scale. In *Proc. Intl. Conf. on Computer-Aided Design*, pages 496–502, 2007.

[12] Y. Zhang, Y. Xu, and C. Chu. FastRoute 3.0: A fast and high quality global router based on virtual capacity. In *Proc. Intl. Conf. on Computer-Aided Design*, pages 344–349, 2008.

[13] IBM-Place 1.0 benchmark suites. http://er.cs.ucla.edu/benchmarks/ibm-place/.

[14] M. Pan and C. Chu. Fastroute: A step to integrate global routing into placement. In *Proc. Intl. Conf. on Computer-Aided Design*, pages 464–471, 2006.

[15] M.Pan and C.Chu. FastRoute 2.0: A high-quality and efficient global router. In *Proc. Asia and South Pacific Design Automation Conf.*, pages 250–255, 2007.

[16] S. Batterywala, N. Shenoy, W. Nicholls, and H. Zhou. Track assignment: A desirable intermediate step between global routing and detailed routing. In *Proc. Intl. Conf. on Computer-Aided Design*, pages 59–66, 2002.

[17] H. Chen, S. Chou, S. Wang, and Y. Chang. Novel wire density driven full-chip routing for cmp variation control. In *Proc. Intl. Conf. on Computer-Aided Design*, pages 831–838, 2007.

[18] M. de Berg, M. van Kreveld, M. Overmars, and O. Schwarzkoph. *Computational Geometry: Algorithms and Applications.* Springer, 1997.

[19] E. Bozorgzadeh R. Kastner and M. Sarrafzadeh. Pattern routing: Use and theory for increasing predictability and avoiding coupling. *IEEE Trans. on Computer-Aided Design and Integrated Circuits and Systems*, 21(7):777–790, July 2002.

[20] X.Yang, B.Choi, and M.Sarrafzadeh. Routability-driven white space allocation for fixed-die standard-cell placement. *IEEE Trans. on Computer-Aided Design and Integrated Circuits and Systems*, 22(4):410–419, April 2003.

[21] N.Viswanathan, M.Pan, and C.Chu. FastPlace 3.0: A fast multilevel quadratic placement algorithm with placement congestion control. In *Proc. Asia and South Pacific Design Automation Conf.*, pages 135–140, 2007.

[22] Y. Zhang and C. Chu. CROP: Fast and effective congestion refinement of placement. In *Proc. Intl. Conf. on Computer-Aided Design*, pages 344–350, 2009.

[23] J.Z. Yan and C. Chu. Handling complexities in modern large-scale mixed-size placement. In *Proc. ACM/IEEE Design Automation Conf.*, pages 436–441, 2009.

[24] J.Z. Yan and C. Chu. DeFer: defered decision making enabled fixed-outline floorplanner. In *Proc. ACM/IEEE Design Automation Conf.*, pages 161–166, 2008.

Standard Cell Routing via Boolean Satisfiability

[1]Nikolai Ryzhenko, [2]Steven Burns

Strategic CAD Labs, Intel Corporation, [1]Moscow, Russia, [2]Hillsboro, OR, US

nikolai.v.ryzhenko@intel.com, steven.m.burns@intel.com

ABSTRACT

We propose a flow for routing nets within a standard cell that 1) generates candidate routes for point-to-point segments; 2) finds conflicts (electrical shorts and geometric design rule violations) between candidate routes; and 3) solves a SAT instance producing a legal and complete routing for all nets in the standard cell. This approach enables routing automation for cutting-edge process technology nodes. We present how to make this technique more effective by introducing pruning techniques to reduce the work required in all three steps. We also show how we can further optimize routing quality within the SAT formulation through the use of successively more stringent constraints. Recent improvements in the speed of SAT solvers make such a formulation practical for even complex standard cells. A routing tool based on our SAT formulation is currently being used to route real industrial standard cell layouts. It demonstrates acceptable runtime and 89% coverage of our industrial standard cell library, including scan flip-flops, adders, and multiplexers. We also observe a significant reduction in amount of metal2 routing in comparison with industrial hand-crafted standard cells.

Categories and Subject Descriptors

J.6 [**Computer-Aided Engineering**]: Computer-aided design (CAD); B.7.2 [**Design Aids**]: Layout, Placement and Routing.

General Terms

Algorithms, Design.

Keywords

Boolean Satisfiability, Standard Cells, Layout Automation

1. INTRODUCTION

Automation of standard cell layout has been practiced in industry for decades. The notion "automated layout" includes: transistor placement, dirty and clean routing, routing completion, compaction, auto-fixing of design rule violations (DRV), etc. However, current reality is that the creation of cell libraries still demands a significant amount of human efforts. At Intel, manual layout work constitutes up to 50% of the library production cycle. Routing constitutes a significant part of this work.

Practical cell-level routing algorithms are limited to traditional sequential and interleaving routing approaches. Channel routing is used in [1][2]. Maze algorithm is implemented in [3]. A greedy sequential approach is applied in [4] to route an experimental layout fabric. One of the major problems of these heuristics is scenic routing. Ordering nets by criticality as well as using pattern routing and/or pre-routes mitigates this negative effect. These routing heuristics use a rip-up-and-reroute technique both to complete the routing and to increase its quality. In [3], this is done using an approach based on simulated annealing. Regular layout routing techniques have introduced new approaches. In one, via-programmable cells are constructed on litho-friendly regular layout fabrics (most recent works are [5][6]). Programmable vias between adjacent layers define the logic function on the pre-fabricated routing. This approach results in overhead in the cell area. Also, enumerative approaches are used to route limited class of cells with a specific layout style [7].

In this work, we propose a standard cell routing flow that uses the solution to a Boolean satisfiability problem (SAT) as a key step in constructing correct and optimized routing.

Given a Boolean formula, SAT determines if the variables can be assigned in such a way to make the formula true. SAT is ideally suited for tasks where there are multiple conflicts among a discrete set of objects [4][8][9][10][11]. In [4], SAT is used to complete a greedy routing approach. In [8], SAT is formulated for transistor placement. It also determines, under some simplistic assumptions, whether the placement is routable or not. In [9], SAT is applied for FPGA routing. In [10], SAT is used for routing regular logic bricks. In [11], nets in a manufacturing hotspot region are simultaneously ripped up and rerouted. The layout patterns, which are stored in a pre-built library, are forbidden to appear in the rerouted design through SAT constraints.

By construction, our SAT formulation does not compromise design rule cleanliness. Another key advantage of our SAT formulation in application to detailed routing and in comparison with the conventional algorithms is its independence on net ordering. By finding a satisfying solution to a Boolean formula, the solver finds a complete detailed routing *simultaneously* for the whole netlist, including power/ground nets, ports, and signal nets. The solution, by construction, guarantees routing completion and satisfies other imposed electrical and quality constraints. Another key advantage of a SAT is that it either: 1) finds a legal solution or, 2) reports precisely that no solution exists. Rip-up-and-reroute is not needed to complete routing.

SAT runtime cost is a major limiting factor. The SAT problem in [8] is limited to minimum-width cells, special layout style, and a coarse routing grid. The SAT formulations in [4][9][10] are applied to very regular layout structures. In [4], SAT is applied to as yet unfinished nets with no more than 4 subnets. In [11], SAT rerouting is performed in small local regions.

The industry trend is to more regular and discrete layouts [12] and recent dramatic improvements in the speed of SAT solvers make such a formulation practical for industrial standard cell routing. In this paper, we present several pruning techniques that make the SAT problem more efficient. The basic idea is that the number of possible layout realizations for a net may be large but very few of them actually may end up in the final layout. We do efficient pruning of these unfeasible layouts. Another key procedure of our routing flow determines conflicts between otherwise feasible layout patterns. If we have N layout patterns, in the worst case, we would have to call a design rule checking (DRC) tool $O(N^2)$ times to find all the conflicts between pairs of routes. In our flow we do this in near-linear time.

After pruning, we formulate the final routing problem for feasible layout patterns in a large Boolean formula. The formula also includes conflicts between layout objects, quality constraints, and cell architecture limitations. This formula, represented in conjunctive normal form (CNF), is then presented to a SAT solver [13]. The SAT formulation is design-rule, layout style, and cell architecture independent. These process-dependent details are coded in the feasible route generation and conflicts between the routes determination stages of the flow.

Key contributions of this work are following. 1) To the best of our knowledge, this work is the first practical implementation of SAT in application to industrial standard cell routing. 2) As a separate part of the flow, several pruning techniques are presented that allow a more efficient SAT formulation. 3) Additional SAT-based techniques are presented that provide the ability to optimize routing quality. The technique imposes quality constraints via a unary counter of true Boolean literals and iterates through the list of constraints via external assumptions.

A routing tool based on our SAT formulation is being currently used to route real layouts. It demonstrates acceptable runtime and high coverage of the industrial standard cell library, including scan flip-flops, full adders, and multiplexers. We also observe a significant reduction in amount of metal2 routing in comparison with industrial hand-crafted standard cells.

The outline of the paper is as follows. We formulate the routing problem in Section 2. We present the routing flow in Section 3. In Section 4, we describe in details the stages preceding our SAT formulation. In Section 5, we formulate the SAT problem for detailed routing. Techniques to improve routing quality are detailed in Section 6. In Section 7, we present experimental results and conclude in Section 8.

2. PRELIMINARIES

Each net in a cell has two types of terminals: poly gate and source/drain diffusion contact (Fig. 1). We distinguish three types of nets: power/ground (vcc/vss) nets, internal signal nets, and port nets. A port net can be either an input or an output. Ports are used for inter-cell connections. The purpose of routing is to create design rule (DR) clean and electrically-correct layout connections among all terminals of signal and port nets, and to connect vcc/vss terminals with their appropriate power grids. Each port net must have a sufficient number of legal intersections (via locations) with the upper layer used for inter-cell routing.

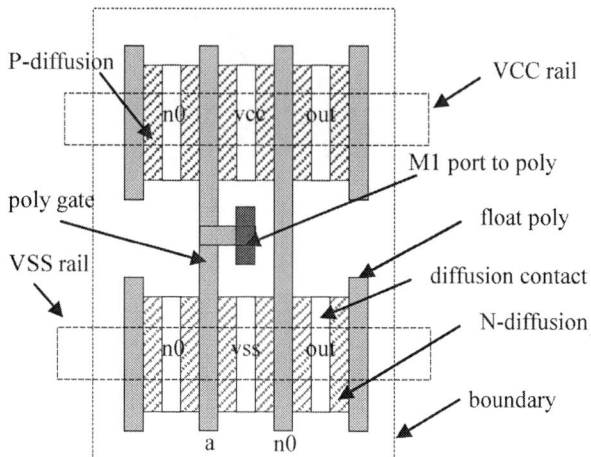

Figure 1. Layout basics.

In our flow, the SAT formulation does not depend on the process technology, the layout style, and cell architecture. The only information at the SAT level is the logical netlist, the list of legal layout routes, and the list of conflicts among routes.

Thus, without loss of generality we use in our exhibition of this work the following assumptions. 1) Metal1 is the upper layer for cell routing. Horizontal metal2 is used for inter-cell routing. Metal2 usage for intra-cell routing is considered as the exception and should be avoided when possible. 2) Port nets must have at least one metal1 wire. Each port must have a sufficient number of intersections with metal2. Usually, this number is 3-4 unique metal2 tracks. 3) The poly layer is vertical and unidirectional [14]-[15]. Vertical diffusion contact (diffcon) is used for routing. 4) Vcc/vss metal2 rails go across the cell in the corresponding P and N regions. Direct hookups to metal2 rails from vcc/vss terminals are preferable but short connections between adjacent terminals within the cell are also allowed. 5) Poly wires adjacent in vertical direction are pre-routed (nets 'a' and 'n0' in Fig. 1).

3. THE ROUTING FLOW

The routing flow is presented in Figure 2. Process-independent, process-dependent, and SAT-related stages have different colorings in the drawing.

In Stage 1, the router creates connections (ties) for each pair of terminals. A tie has a logical meaning. It can be either connected or disconnected. All terminals must be connected by tree of ties (no cycles) and a path must exist between each pair of terminals.

Legal layout realizations (*routes*) for each tie are created in Stage 2. We use a maze routing algorithm with some process-dependent modifications to find routes for all 2-point connections. The maze algorithm is applied to vcc/vss nets as well. Another procedure creates routes for single-terminal ports. We find several possible routes for each tie and store them for the next stages.

In stage 3, we prune *redundant routes*.

A special procedure in the Stage 4 determines *conflicts* between created routes. It considers design rules, electrical rules, and cell architecture specifications. We do an effective search of conflicts via a library of unique layout objects. We also build rectangular bin structures to minimize the number of useless computations.

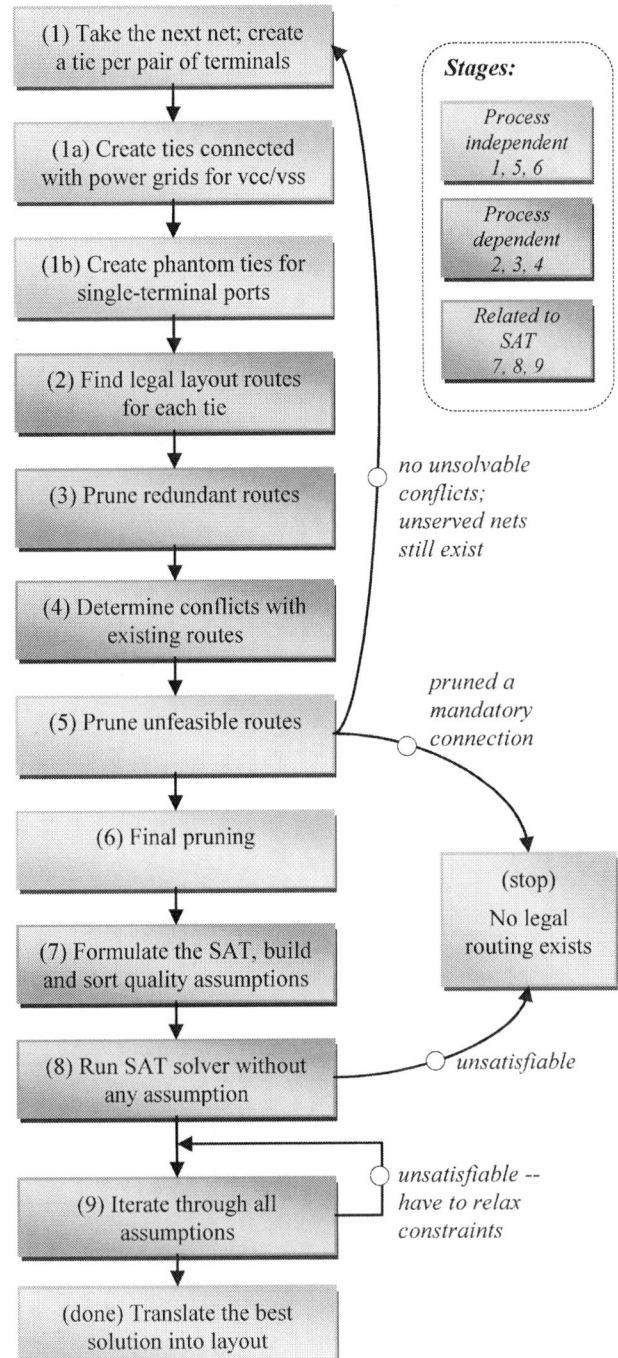

Figure 2. The routing flow.

Unfeasible routes are pruned in Stage 5. A DR-correct and electrically-correct route is termed "unfeasible" if it cannot be applied in the final cell layout because of conflicts with other routes. Multiple conflicts between layout patterns may result in a complete pruning of a mandatory tie. In such a case, we report that no legal routing exists for this transistor placement.

Final pruning is done in Stage 6. Additional pruning technique can be applied when all ties and routes are created.

We formulate the *SAT problem* for the detailed routing in Stage 7. The formulation includes routing completion, conflicts between routes, and additional assumptions on routing quality.

In Stage 8, we run the SAT solver for the first time. All quality assumptions are disabled in this run. If there is a satisfiable solution, we proceed to the next stage. After this stage, we either have a particular instance of a legal routing or we report that no legal routing exists.

In Stage 9, we iterate through the quality assumptions. The SAT solver is called several times at this stage. If we cannot find a satisfiable solution, we relax the constraints and run the solver again. At the end of this stage we either have a default initial solution obtained on the previous Stage 8 or we have another solution with better quality metrics.

Finally, the best satisfiable solution is translated to layout completing the detailed routing.

4. STAGES BEFORE SAT FORMULATION

4.1 Creation of ties

Figure 3 illustrates the routing task for the internal net 'n0'. The net has 4 terminals: 2 diffcon wires and 2 poly gate wires. There are 6 possible connections (ties) among these terminals. To get a complete valid routing for the net 'n0', at least 3 of these 6 ties must be connected. These 3 ties must not form a cycle.

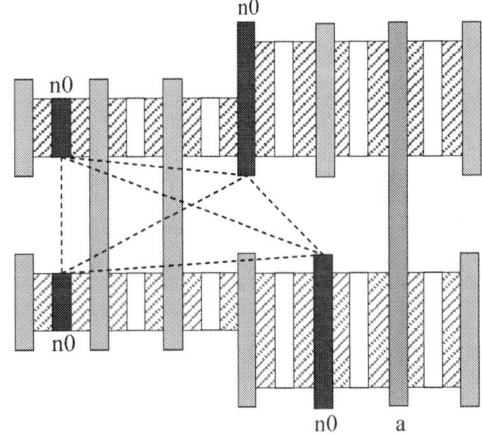

Figure 3. Ties of the internal signal net n0.

Additional ties are created for vcc/vss nets. A tie is created between each terminal and the appropriate power rail. In such a case the power rail is considered as another net terminal. Indeed, all power terminals can be connected within the cell but the routing must have at least one hookup to the power rail.

If a port net has only one terminal (e.g. net 'a' in Fig. 3) another phantom tie is created. Logically, the tie connects the real terminal and some phantom terminal. Physically, each layout realization consists of a metal1 wire connected to the poly gate.

These modifications for vcc/vss nets and for single-terminal ports are made to have a consistent connectivity model for all nets in the cell. All net terminals including "phantom terminals" must be connected so that a path exists between each pair of terminals.

605

4.2 Enumeration of possible routes

In Stage 2, the router finds for each tie a number of DR-clean layout routes. A *route* is a unique layout implementation of a logical tie. A connected tie may have only one applied route.

We use a maze routing algorithm to find all possible routes within a bounding box around appropriate terminals. Maze routes are computed independent of each other. Layouts for vcc/vss hookups to metal2 rails are enumerated in a small bounding box around the corresponding terminal. The number of wires on different layers is limited to 6-8 segments per connection. The maze algorithm considers obstacles (i.e. terminals of other nets and other pre-routes if any), so the created layouts by default have no electrical shorts. The implementation of maze algorithm is process-dependent. We check on-the-fly some via rules and spacing rules to stop a wave front propagation early and not produce obviously illegal layouts. When enumeration is completed, the design rule checker (DRC) leaves only DR-clean routes.

Layouts for single-terminal ports are created separately. When a metal1 wire is connected with a poly gate, the poly-contact positions are enumerated. In the general case, layouts *a* and *b* in Figure 4 are ideal from an electrical perspective. Cell architectures may constrain the feed-through length in the poly layer (Fig. 4c).

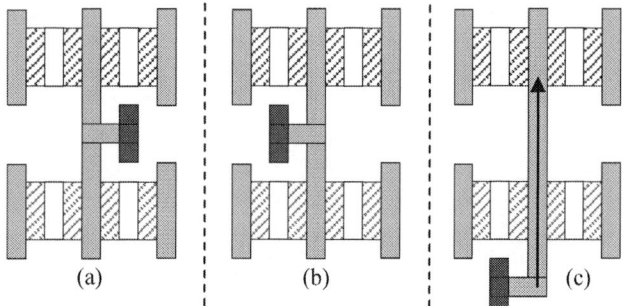

(a)   (b)   (c)

Figure 4. Possible layouts for single-terminal input port.

4.3 Pruning of redundant routes

In stage 3, we prune *redundant routes*. In our flow, they appear because the diffcon layer is used for routing. If a route between two terminals has a diffcon wire, this wire may overlap a third diffcon terminal. This situation is both DR and electrically correct but redundant. We prune the route between terminals A and B if this route can be replaced by two other routes connecting A and B through the third terminal C. These two routes after merging must have the same layout footprint as the redundant route.

4.4 Detection of conflicts between routes

In Stage 4, we determine conflicts between each pair of routes. We say that there is a *conflict* between two routes if these routes cannot be applied simultaneously in the layout. We use conflicts to prune unfeasible routes in the next Stage 5. Electrical shorts constitute the majority of conflicts between routes. Other conflicts appear because of design and electrical rules, and cell architecture constraints. Routes of one tie cannot have conflicts by default, because only one route per tie may be applied in the layout. Routes of different ties on the same net have no electrical shorting issues but can have conflicts of the other types. Complex rules may involve layout from more than 2 routes. Multi-route conflicts are coded into SAT separately. We describe this case in Section 5.

Given N routes, in the worst case we need $R * W^2 * (N^2 - N)/2$ checks, where R is the number of active rules, and W is the average number of wires and vias per route. To reduce the task, we define a collection of unique wires and vias that constitute all routes. The more regular and discrete the technology process is [12], the less variety of unique wires and vias. If a wire W_1 belongs to P routes and a wire W_2 belongs to Q routes, then by one call to a checking procedure, we check $P * Q$ conflicts. We use a geometrical bin grid structure to further minimize the number of useless computations. A bin grid is constructed for each layer separately and wires/vias are put into appropriate bins. Wires from different bins are checked for design rule violations only if these bins are close enough to violate any of the active rules. We only need to check a few rules; the majority of rules are correct by construction.

Cell constraints specify forbidden layout situations at the cell borders. An otherwise DR-correct layout may potentially violate design rules with the layout in an adjacent cell. Pairs of routes leading to such forbidden layouts are marked as conflicting.

The high-resistance poly layer imposes *electrical constraints*. In our flow, a poly jumper between two diffusions is not allowed (routes p and q in Fig. 5). Using bin grids, we find all unmatched contacts W_1 and W_2 for each poly and mark corresponding P and Q routes as conflicting. Thus, we find $P * Q$ conflicts by 1 check.

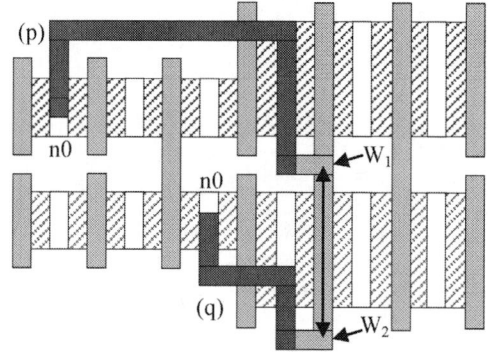

Figure 5. The poly jumper.

4.5 Pruning of unfeasible routes

In Stage 5, we prune *unfeasible routes* that can never be applied in the final layout because of conflicts with other routes. A *mandatory tie* is a tie that is required to be connected. For example, a net with two terminals has only one tie, and this tie is mandatory by default. We prune a route if it fully blocks a mandatory tie. Also, for each terminal we combine all its routes from different ties to a common collection. If a route has conflicts with all routes from such collection – we prune this route because the terminal must be connected with at least one another terminal.

If we find that two non-conflicting routes block together either a mandatory tie or a terminal we mark them as conflicting.

A tie is pruned if it loses all its routes. An *optional tie* of a terminal may become mandatory if all other ties connected to this terminal are pruned. We repeat the pruning until all mandatory routes are determined and all infeasible routes are deleted.

Pruning procedures are process-independent. They operate only with ties and pre-computed conflict information between each pair of routes. New conflicts may appear. However, these conflicts arise from existing conflicts rather than from DRC.

4.6 Pruning routes by signature

Additional pruning is applied in Stage 6 when all ties and routes are created. For each route, we define a *signature*, a set of conflicting routes. If two routes r_i and r_j of one tie have signatures such that $s_i\{r_1, r_2, ..., r_N\} \subset s_j\{r_1, r_2, ..., r_N, ..., r_M\}$, we prune the route r_j. The procedure does not affect the routability of the tie because route r_j has the same conflicts as r_i plus several other conflicts. Pruning by signature is repeated multiple times to eliminate a significant fraction of redundant routes.

5. SAT FORMULATION

The SAT problem is to find such an assignment of variables so that a Boolean formula asserts true. Once a problem is formulated, it can be solved by a SAT solver, in our case MiniSAT [13]. As with many other SAT solvers, MiniSAT demands the formula in conjunctive normal form (CNF). CNF is a conjunction of clauses, where a clause is a disjunction of literals. A literal is either a Boolean variable or a negated Boolean variable. We use a convention translation of Boolean formulas and *always false* and *always true* statements into CNF, as first presented in [16].

In stage 7, we have the netlist (nets, terminals, and ties), a list of legal layout routes per tie, and a list of conflicts between each pair of routes.

Ties and routes. We create a Boolean literal for each route. When r_i is set to 1, it means that the associated route is applied in the layout. Another literal is created per tie (1). A tie clause t_k^n can be either true (connected tie) or false (disconnected tie). A connected tie may have only one applied route, so the formula (2) is stated as *always false*. In this formula, $\{r_i\}$ are all R routes of the tie t_k^n.

$$
\begin{cases}
t_k^n = \bigvee_{i=1}^{R} r_i & (1) \\
\bigvee_{\substack{i \leq R;\, j \leq R \\ i=1;\, j=i+1}} (r_i \wedge r_j) = 0 & (2)
\end{cases}
$$

Connectivity. Given N terminals, we construct a two-dimensional $N \times N$ matrix M. Initially, all entries of M are set to constant zero. If two terminals i and j can be connected by a tie t_{ij} with non-zero number of routes, we set $M(i,j) = M(j,i) = t_{ij}$. Diagonal entries are defined as $M(i,i) = \bigvee_{k=1}^{T_i} t_k$, where $\{t_k\}$ are all T_i ties of the terminal i. Then, each entry of M is transformed by the $O(N^3)$ Floyd–Warshall algorithm [17] into a Boolean formula. Each entry formula becomes a disjunction of all possible paths between the corresponding pair of terminals. We state each entry as *always true*. It means that at least one path between each pair of terminals must be presented in the solution. This formulation does not limit the number of connected ties. However, it must be $N - 1$ to get a cycle-free routing tree for N terminals [18]. Using a unary counter [19], we impose additional constraint on the number of connected ties. We describe the counter in Section 6.

Conflicts and complex design rules. Pair conflicts between routes are presented in the same formula as routes of one tie (2). However, rules may involve layout objects from more than two routes. If v is such an object, we create a literal: $v = \bigvee_{i=1}^{R} r_i$, where $\{r_i\}$ are all R routes containing v. Then, we enumerate the corresponding *always false* statements: $\bigwedge v_i = 0$ to forbid all such multi-object violations.

Ports. Ports must have at least one piece of metal1. So we add another *always true* constraint $\bigvee_{k=1}^{R_n} r_k^{n/m1} = 1$, where $\{r_k^{n/m1}\}$ are the R_n routes of port net n with at least one *metal1* wire.

Routing quality. We formulate quality constraints and add them to CNF. We describe them in details in Section 6. In Stage 8, we run the solver for the first time. Quality clauses do not affect the SAT problem at this run because they are stated neither as *always true* nor as *always false*. After this run either we have a variant of routing or the solver reports that the problem is unsatisfiable.

6. QUALITY ASSUMPTIONS

Some layout situations may degrade electrical characteristics. These situations are not forbidden (as poly jumpers are in our flow, Fig. 5) but their appearance should be as minimal as possible. Below we list the some of these cases.

Firstly, undesired layouts arise from the high-resistance of poly and contact layers (e.g. Fig. 4c) and high coupling capacitance between diffcon and poly wires. Secondly, the more direct hookups between vcc/vss diffcons and the power grid the better. Also, if metal2 is used for intra-cell routing, the number of metal2 routes should be minimal. The list may include other aspects of routing quality. In our flow, we explicitly minimize the number of metal2 routes and the amount of diffcon routing, and explicitly maximize the number of direct hookups to the power rails. Below, we describe the minimization of metal2 usage.

We find all the unique vias $\{v_k\}$ between metal1 and metal2 and create a literal per via: $v_k = \bigvee_{i=1}^{R_k} r_i$, where $\{r_i\}$ are all R_k routes containing via v_k. The number of allowed vias between metal1 and metal2 is small. Metal2 is unidirectional, so 6 unique vias result in at most 3 metal2 wires. We use this limitation to construct for $\{v_k\}$ a partial unary counter.

A full unary counter [19] for 4 Boolean literals is illustrated as a circuit in Figure 6. Outputs c_1, c_2, c_3, and c_4 are true when the number of true literals among $\{x_1, x_2, x_3, x_4\}$ are ≥ 1, ≥ 2, ≥ 3, and $= 4$ respectively. Using negated values of literals as inputs, we get an inverted unary counter. In this case, outputs c_1, c_2, c_3, and c_4 are true when the number of true literals among $\{x_1, x_2, x_3, x_4\}$ are ≤ 3, ≤ 2, ≤ 1, and $= 0$ respectively. The logic as presented in Figure 6 can be extended to any arbitrary number of literals. A partial counter can be built if we need only a limited range of values (e.g. c_2 only). Useless nodes are trimmed from the full counter (dotted objects in Fig. 6); new nodes are not added.

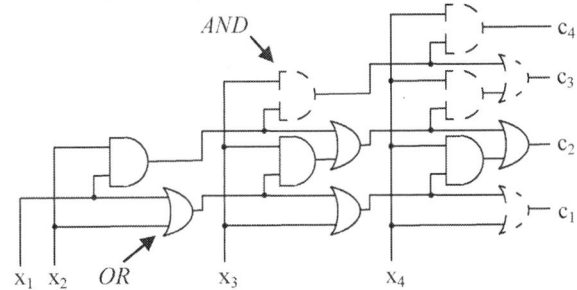

Figure 6. An unary counter for 4 literals.

We restrict the maximal number of vias, so that the counter is inverted. Gates and outputs of the counter are translated to CNF and new clauses are appended into the common SAT formula.

MiniSAT has an ability to solve a problem with assumptions. In Stage 9, we iterate through 4 assumptions: the number of vias = 0, ≤ 2, ≤ 4, and ≤ 6. Corresponding clauses are externally stated as *always true* one by one. Before each new SAT run, we check whether the current SAT solution satisfies the new assumption (we have at least one solution after the initial SAT run in the Stage 8). If we find a satisfiable solution, we store the current assumption as a mandatory statement and proceed to another quality constraint. If there is no satisfiable solution even with the number of vias up to 6 we do not abort the flow, and proceed to the next quality constraints without any restrictions on metal2 routes. However, it is a signal to place transistors differently.

Other routing quality metrics are optimized in a similar way. We construct ordered groups of undesired routes, build corresponding unary counters, and iterate through several assumptions starting with the strongest one.

7. EXPERIMENTAL RESULTS

The formulation explained in Sections 3-6 enables us to route standard cells from a given transistor placement. To test the efficiency of the SAT-based standard cell detailed router, we re-routed hand-crafted cells from an industrial library. We did not compare our flow with any tool because it is hard to extend existing works to handle complex design rules for real cells.

The library has 188 different cell families. We selected one single-height (SH) cell and one double-height (DH) cell from each family. Not all families have both types of cells, so there are 112 SH and 177 DH cells in this selection, totaling 269 cells. 78 cells (29%) have metal2 routes. We extracted transistor placement and re-routed each cell from scratch. 89% cells (239 from total 269) were successfully routed. We observed a significant elimination of metal2 wires. Metal2 was completely eliminated in 59% of cases: for 46 cells from total 78. All results are in Table 1.

Table 1. Routing results.

Cell type	#cells	#routed	routed %	m2 elim.
Combinational	120	115	96%	6 / 11
Delay cell	11	11	100%	0 / 0
Latch	14	14	100%	1 / 1
Non-scan DFF	14	13	93%	9 / 9
Scan DFF	29	21	72%	16 / 24
Adder	16	9	56%	5 / 13
Multiplexer	25	21	84%	7 / 13
Clock driver	40	35	88%	2 / 7
Total	**269**	**239**	**89%**	**46 / 78**

Runtime results are presented in Table 2. We limit the maximal number of wires W per route as described in Section 4.2. First of all, we tried $W = 7$, then $W = 8$, and finally $W = 9$. For example, both routes p and q in Figure 5 consist of 6 wires: 1 diffcon, 1 polycon, 1 poly, and 3 metal wires. The majority of combinational cells and several latches, multiplexers, adders, DFFs, and scan DFFs were routed with $W = 7$. A non-scan DFF demanded the largest runtime ~17 hours with $W = 9$.

Table 2. Runtime statistic.

	< 1 min	< 5 min	< 1 hours	> 1 hour
#cells	169	20	37	14
% from all routed	~70%	~8%	~15%	~6%

The capacity of the routing tool depends on the number of devices as presented in Table 3. For cells with up to 30 transistors, ~5% of

the runs are unsuccessful. This number grows and achieves 50% for big cells with 51-70 devices. The largest routed cell in our experiment was a 63-transistor combinational cell. Runs for 16 unrouted cells were aborted because they exceeded the machine memory limit. Seven of these cells already had an initial SAT solution but the solver exceeded its memory limit iterating through the quality assumptions.

Table 3. Statistic of unrouted cells depending on #devices.

#devices	2-10	11-20	21-30	31-40	41-50	51-70	Total
#cells	99	81	27	31	23	8	269
unrouted	5	3	1	9	8	4	30
%	5%	4%	4%	29%	35%	50%	11%

8. CONCLUSION

In this work, we proposed a standard cell routing flow formulated as a Boolean satisfiability problem (SAT). As part of the flow, we presented several pruning techniques that make the SAT formulation more effective. We also demonstrated how to improve routing quality while remaining in the SAT framework. The routing tool based on our SAT formulation is currently being used to route real layouts. It demonstrated acceptable runtime and 89% coverage of the industrial standard cell library, including scan flip-flops, adders, and multiplexers. We also observed a significant reduction in the amount of metal2 routing in comparison with industrial hand-crafted standard cells.

We consider tool capacity increase and improving the efficiency of the SAT formulation as next steps.

9. REFERENCES

[1] Sanjay Rekhi, J. Donald Trotter, Daniel H. Linder, Automatic layout synthesis of leaf cells. DAC 1995.

[2] Gupta A., J. P. Hayes, CLIP: Integer-Programming-Based Optimal Layout Synthesis of 2D CMOS Cells. ACM Tr. on DA El. Sys., V. 5, № 3, July 2000.

[3] Mohan G. et al, CELLERITY: a fully automatic layout synthesis system for standard cell libraries. DAC 1997.

[4] Lin Y.-W., Marek-Sadowska M., Maly W. Transistor-level layout of high-density regular circuits. ISPD 2009.

[5] Li M.-C. et al, Standard cell like via-configurable logic block for structured ASICs. Proc. of IEEE Computer Society Annual Symp. on VLSI, 2008.

[6] M. Pons, F. Moll, A. Rubio, J. Abella, X. Vera, and A. González, VCTA: a via-configurable transistor array regular fabric. Proc. of VLSISoC, 2010.

[7] N. Ryzhenko, S. Burns, Physical Synthesis onto a Layout Fabric with Regular Diffusion and Polysilicon Geometries. DAC 2011.

[8] T. Iizuka, M. Ikeda, K. Asada, High Speed Layout Synthesis for Minimum-Width CMOS Logic Cells via Boolean Satisfiability. ASP-DAC 2004.

[9] Nam G.-J. et al, Satisfiability-Based Layout Revisited: Detailed Routing of Complex FPGAs via Search-Based Boolean SAT. FPGA 1999.

[10] B. Taylor, L. Pileggi, Exact Combinatorial Optimization Methods for Physical Design of Regular Logic Bricks. DAC 2007.

[11] F. Yang, Y. Cai, Q. Zhou, J. Hu, SAT Based Multi-Net Rip-up-and-Reroute for Manufacturing Hotspot Removal. DATE'10

[12] T. Jhaveri et al, Co-Optimization of Circuits, Layout and Lithography for Predictive Technology Scaling Beyond Gratings. IEEE Trans. of ICs and S., vol. 29, №4, pp. 509-527, Apr. 2010.

[13] MiniSAT, http://minisat.se

[14] http://download.intel.com/newsroom/kits/22nm/pdfs/22nm-Announcement_Presentation.pdf

[15] http://download.intel.com/newsroom/kits/22nm/pdfs/22nm-Details_Presentation.pdf

[16] Tseitin G.S., On the Complexity of Derivation in Propositional Calculus. In Studies in Constructive Mathematics and Mathematical Logic, Part 2, 1968.

[17] Warshall S., A theorem on Boolean matrices. J. of the ACM, Jan. 1962. Floyd W. R., Algorithm 97: Shortest Path. Comm. of the ACM, June 1962.

[18] Graham R. L. and Hell P., On the history of the minimum spanning tree problem. Annals of the History of Computing, 7(1): 43-57, 1985.

[19] Claude E. Shannon, A symbolic analysis of relay and switching circuits. Thesis (M.S.), MIT, Dept. of Electrical Engineering, 1940.

10. SUPPLEMENTAL MATERIALS

We would like to take this opportunity to illustrate key stages of the routing flow. An outline of the supplemental materials is the following. In section 11, we illustrate creation of routes, detection of conflicts, pruning techniques, and SAT formulation. In Section 12, we illustrate the full variety of pruning techniques. In Section 13, we discuss some aspects of global routing in application to standard cell routing.

11. EXAMPLE OF THE ROUTING

An artificial routing task is presented in Figure 7. Seven terminals of three nets a, b, and c are placed on a uniform routing grid. Nets a and b have two terminals. Net c has three terminals.

In stage 1, we create ties. Tie t_{12}^a connects terminals a_1 and a_2 of a 2-terminal net a and this tie is mandatory by default. Similarly, a mandatory tie t_{12}^b connects terminals b_1 and b_2 of net b. Net c has three optional ties t_{12}^c, t_{13}^c, and t_{23}^c. To get a complete routing, two and only two of these ties must be connected.

In Stage 2, we create legal layout routes for each tie within a bounding box around the appropriate terminals. By using bounding boxes, we restrict the number of layout routes and also prevent scenic routing. In the flow, we use similar bounding boxes with small extensions to improve routability. Given this bounding box, all possible routes are shown in Figure 7. They are constructed by a maze algorithm independently from one another.

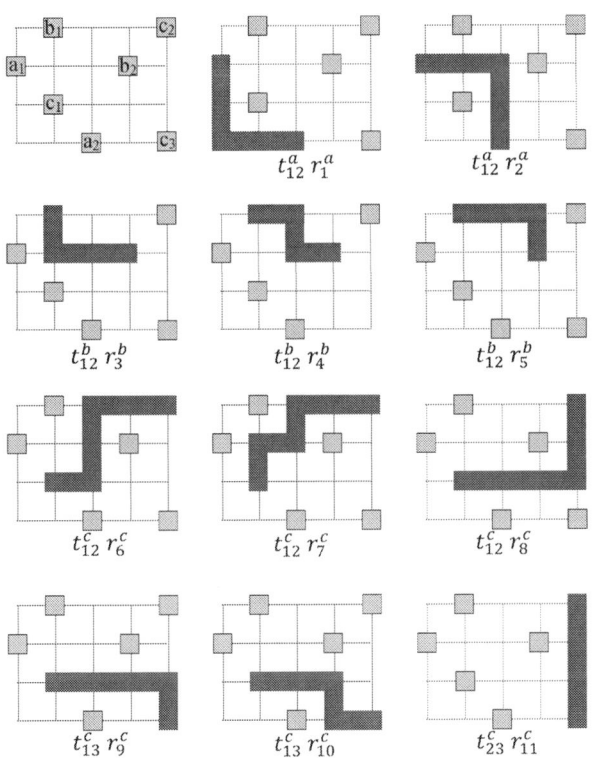

Figure 7. The routing task and possible routes.

In Stage 3, we prune redundant routes. Such routes appear when a route between two terminals touches a third terminal. Our example in Figure 7 does not have such routes. A redundant route is shown in Figure 8. A route r_1 between terminals e_1 and e_3 can be pruned because it can be replaced by two routes r_2 and r_3

between terminals e_1 and e_2 and terminals e_2 and e_3 respectively. It is important, that the layout after merging of r_2 and r_3 must be identical to r_1. These two routes have equivalent influence on all other routes as the route r_1 and thus we can safely prune r_1.

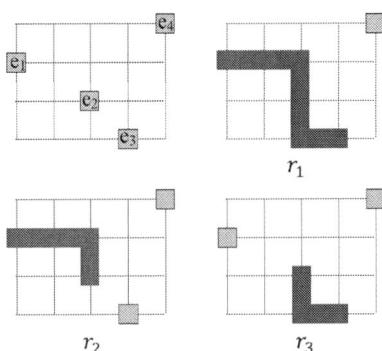

Figure 8. Redundant route.

In Stage 4, we determine conflicts between newly created routes for the current net and routes created previously. This is done to prune unfeasible routes early. To illustrate this, we demonstrate detection of conflicts simultaneously for all routes in Figure 7.

To simplify explanation, for this artificial example, we state that electrical shorts are the only conflicts between pairs of routes. Electrical shorts are the simplest conflicts and constitute the majority of conflicts among routes in industrial layouts. Other conflicts include geometric design rules such as minimum spacing rules and end-to-end attacker rules and will not be discussed in detail here.

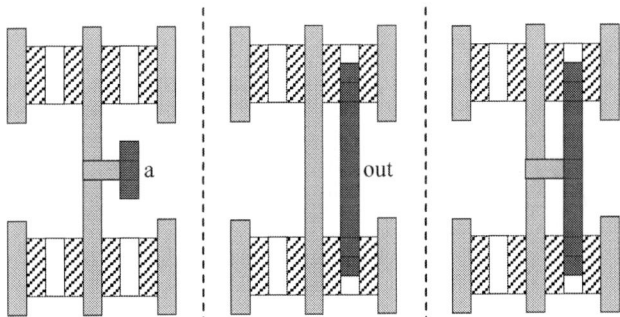

Figure 9. The electrical short.

We determine all the unique wires constituting all the routes and put them into rectangular bins. The bins are located at intersections of the routing grid. Each bin contains all wires passing through the appropriate intersection. In such a formulation, all wires in a bin intersect by construction. An interesting bin is illustrated in Figure 10.

This bin contains 7 unique wires belonging to 5 routes $\{r_2, r_3, r_4, r_6, r_7\}$, 3 ties $\{t_{12}^a, t_{12}^b, t_{12}^c\}$, and 3 nets.

There are 8 conflicts here. They are $\{r_2, r_3\}$, $\{r_2, r_4\}$, $\{r_2, r_6\}$, $\{r_2, r_7\}$, $\{r_3, r_6\}$, $\{r_3, r_7\}$, $\{r_4, r_6\}$, and $\{r_4, r_7\}$. $\{r_3, r_4\}$ is not a conflict because these routes belong to the same tie t_{12}^b. Similarly, $\{r_6, r_7\}$ is also not a conflict.

In a real layout, all design rules are checked in the same way. Wires from adjacent bins are checked for design rule violations only if these bins are close enough to violate any of the active

609

rules. Bins are constructed to minimize the number of DR checks needed between pairs of layout objects.

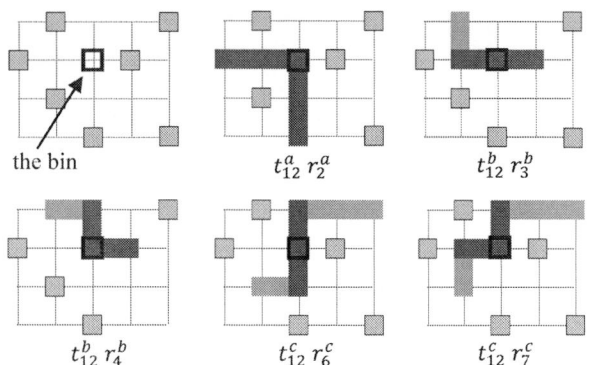

Figure 10. The routing task and possible routes.

In this bin, only routes r_4 and r_7 have a common identical vertical wire. In real standard cells the number of unique wires is much less than the number of possible maze routes. This fact and the imposed bin grids allow us to avoid most of the useless DRC computations.

All conflicts between routes are presented in Table 4. Conflicting pairs of routes are marked by 'x'.

Table 4. Initial conflicts.

	t_{12}^a (m)		t_{12}^b (m)			t_{12}^c			t_{13}^c		t_{23}^c
	r_1	r_2	r_3	r_4	r_5	r_6	r_7	r_8	r_9	r_{10}	r_{11}
r_1											
r_2			x	x		x	x	x	x	x	
r_3		x				x	x				
r_4		x				x	x				
r_5						x	x				
r_6		x	x	x	x						
r_7		x	x	x	x						
r_8		x									
r_9		x									
r_{10}		x									
r_{11}											

In Stage 5, we first prune unfeasible routes r_6 and r_7 because they block the mandatory tie t_{12}^b. Table 4 is transformed into Table 5.

Table 5. Conflicts and routes after 1st pruning.

	t_{12}^a (m)		t_{12}^b (m)			t_{12}^c	t_{13}^c		t_{23}^c
	r_1	r_2	r_3	r_4	r_5	r_8	r_9	r_{10}	r_{11}
r_1									
r_2			x	x		x	x	x	
r_3		x							
r_4		x							
r_5									
r_8		x							
r_9		x							
r_{10}		x							
r_{11}									

Then we prune route r_2 because it completely blocks terminal c_1 (i.e. ties t_{12}^c and t_{13}^c). Table 5 is transformed into Table 6.

Table 6. Conflicts and routes after 2nd pruning.

	t_{12}^a	t_{12}^b			t_{12}^c	t_{13}^c		t_{23}^c
	r_1	r_3	r_4	r_5	r_8	r_9	r_{10}	r_{11}
r_1								
r_3								
r_4								
r_5								
r_8								
r_9								
r_{10}								
r_{11}								

In our small example, after pruning we have no conflicts between the remaining routes. Mandatory tie t_{12}^b has 3 possible routes and any of these routes can be applied in the layout. Optional tie t_{13}^c has 2 possible routes, and either of these routes can be applied in the layout. To minimize the size of the SAT formulation, we prune these redundant routes by signature. For each route we define a *signature* as the set of conflicting routes. If two routes r_i and r_j of one tie have signatures such that $s_i\{r_1, r_2, ..., r_N\} \subset s_j\{r_1, r_2, ..., r_N, ..., r_M\}$, then we can prune the route r_j. In our case, all routes have the identical primitive signature $\{\emptyset\}$. So we can prune any of the available routes considering additional parameters, for example, the number of segments per route. Here we use route r_3 for tie t_{12}^b and route r_9 for tie t_{13}^c (see Table 7).

Table 7. Conflicts and routes after pruning by signature.

	t_{12}^a	t_{12}^b	t_{12}^c	t_{13}^c	t_{23}^c
	r_1	r_3	r_8	r_9	r_{11}
r_1					
r_3					
r_8					
r_9					
r_{11}					

In Stage 7, we formulate the SAT problem. A direct SAT formulation as it was described in Section 5 for this task is following:

$$t_{12}^a = r_1 \tag{3}$$

$$t_{12}^b = r_3 \tag{4}$$

$$t_{12}^c = r_8 \tag{5}$$

$$t_{13}^c = r_9 \tag{6}$$

$$t_{23}^c = r_{11} \tag{7}$$

$$t_{12}^a = 1 \tag{8}$$

$$t_{12}^b = 1 \tag{9}$$

$$t_{12}^c \vee (t_{13}^c \wedge t_{23}^c) = 1 \tag{10}$$

$$t_{13}^c \vee (t_{12}^c \wedge t_{23}^c) = 1 \tag{11}$$

$$t_{23}^c \vee (t_{12}^c \wedge t_{13}^c) = 1 \tag{12}$$

$$\overline{t_{23}^c} \vee (\overline{t_{12}^c} \vee \overline{t_{13}^c}) = 1 \tag{13}$$

Formulas (3)-(7) define Boolean literals for available ties.

Formulas (8)-(9) are *always true* statements for mandatory ties. These formulas also determine connectivity for corresponding 2-terminal nets a and b.

610

Formulas (10)-(12) are constructed by the Floyd–Warshall algorithm [17] and they determine all possible paths between each pair of terminals. These formulas are also stated as *always true* because each terminal must be connected with each other terminal of the net.

Formula (13) is an inverted partial counter that is stated as *always true* to allow no more than two connected ties for net c. Formulas (10)-(13) together limit the number of connected ties for net c to exactly 2.

We translate formulas to CNF and run the MiniSAT solver [13] for this CNF. A final layout is presented in Figure 11.

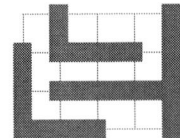

Figure 11. The final layout of routes r_1, r_3, r_8, and r_{11}.

12. PRUNING TECHNIQUES

In this section, we illustrate the full variety of pruning techniques on a more sophisticated example. Let's imagine that the routing task in Figure 7 has not only electrical shorts but also additional design rules. And these rules resulted in several new conflicts as presented in Table 8. New conflicts are marked by the green color.

Table 8. Sophisticated conflicts.

	t_{12}^a (m)		t_{12}^b (m)			t_{12}^c			t_{13}^c		t_{23}^c
	r_1	r_2	r_3	r_4	r_5	r_6	r_7	r_8	r_9	r_{10}	r_{11}
r_1					X						
r_2			X	X		X	X	X	X	X	
r_3		X				X	X				X
r_4		X				X	X		X		
r_5	X					X	X				
r_6		X	X	X	X						
r_7		X	X	X	X						
r_8		X							X		
r_9		X		X				X			
r_{10}		X									
r_{11}			X								

As in the previous example, we prune unfeasible routes r_6 and r_7 because they block the mandatory tie t_{12}^b, and prune route r_2 because it completely blocks terminal c_1 (i.e. ties t_{12}^c and t_{13}^c).

Table 9. Pruned r_2, r_6, and r_7.

	t_{12}^a				t_{12}^b			t_{13}^c		t_{23}^c
	r_1	r_3	r_4	r_5	r_8	r_9	r_{10}	r_{11}
r_1				X				
r_3								X
r_4						X		
r_5	X							
r_8						X		
r_9			X		X			
r_{10}								
r_{11}		X						

After pruning r_2, we have to prune r_5 because it has conflict with r_1 and therefore blocks mandatory tie t_{12}^a (see Table 10). We could not prune this route earlier because it had no conflict with r_2, another route of mandatory tie t_{12}^a.

Table 10. Pruned r_5.

	t_{12}^a	t_{12}^b			t_{12}^c	t_{13}^c	t_{23}^c
	r_1	r_3	r_4	r_8	r_9	r_{10}	r_{11}
r_1							
r_3							X
r_4				X			
r_8				X			
r_9		X	X				
r_{10}							
r_{11}		X					

After pruning r_5, routes r_9 and r_{11} together block mandatory tie t_{12}^b. Route r_9 has a conflict with r_4 and route r_{11} has a conflict with r_3. Routes r_9 and r_{11} cannot be applied in the layout simultaneously so we mark these routes as conflicting (Table 11).

Table 11. Found a conflict between r_9 and r_{11}.

	t_{12}^a	t_{12}^b			t_{12}^c	t_{13}^c	t_{23}^c
	r_1	r_3	r_4	r_8	r_9	r_{10}	r_{11}
r_1							
r_3							X
r_4				X			
r_8				X			
r_9		X	X				X
r_{10}							
r_{11}		X			X		

After adding a conflict between r_9 and r_{11}, we see that r_9 now blocks terminal c_2. It blocks both optional ties t_{12}^c and t_{23}^c connected to c_2. We prune route r_9 (see Table 12).

Table 12. Pruned r_9.

	t_{12}^a	t_{12}^b		t_{12}^c	t_{13}^c	t_{23}^c
	r_1	r_3	r_4	r_8	r_{10}	r_{11}
r_1						
r_3						X
r_4						
r_8						
r_{10}						
r_{11}		X				

At this state, pruning is finished. We cannot prune more routes and cannot discover new conflicts. We prune the rest of routes by signature.

$$signature(r_3) = \{r_{11}\} \qquad (16)$$

$$signature(r_4) = \{\emptyset\} \qquad (17)$$

We see that route r_3 has the same conflicts as route r_4 plus additional conflict with r_{11}, so we can safely prune r_3. It is important, that 1) both routes belong to one tie t_{12}^b and 2) conflicts of r_4 are fully covered by conflicts of r_4. Only in this case we guarantee that we do not compromise the routability of the tie t_{12}^b in particular and net b in general.

In the flow, pruning by signature is repeated several times. When we remove a route from consideration, signatures of other routes can be changed. It can take several iterations to prune all routes by signature.

The final conflicts and routes are presented in Table 13.

Table 13. Pruned r_3 by signature.

	t_{12}^a	t_{12}^b	t_{12}^c	t_{13}^c	t_{23}^c
	r_1	r_4	r_8	r_{10}	r_{11}
r_1					
r_4					
r_8					
r_{10}					
r_{11}					

The resulted list of routes and conflicts is similar to the previous example (Table 7) so we omit the SAT formulation for this task. A new layout is presented in Figure 12.

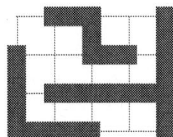

Figure 12. The final layout of routes r_1, r_4, r_8, and r_{11}.

In this example, we did not illustrate another case. An optional tie may lose all its routes. When it happens we prune this tie. An optional tie of a terminal may become mandatory if all other optional ties connected to this terminal are pruned. For example, by pruning any of three ties t_{12}^c, t_{13}^c, or t_{23}^c, both remaining ties become mandatory. As we demonstrated, mandatory ties are used efficiently both to prune routes and to find new conflicts.

13. GLOBAL ROUTING

The concept of ties (2-terminal connections) does not disallow global routing. However, the value of upfront global routing is lessened by the downstream SAT formulation. The SAT solver finds a complete detailed routing simultaneously for the whole netlist. The complete independence on net ordering is a key difference between SAT-based routing and traditional sequential routing approaches.

Figure 13 illustrates a routing task and its optimal routing.

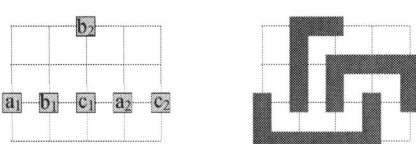

Figure 13. The routing task and the optimal routing.

A few worst-case scenarios for sequential routing are shown in Figure 14. Routes a) and b) lead to the incomplete routing. Routes c) and d) result in significant scenic routing for net a (Fig. 14e). Routes f) and g) result in scenic routing for net b (Fig. 14h). Sequential routing algorithms demand global routing, ordering nets by criticality, rip-up-and-reroute, and other techniques to avoid or to mitigate negative effects presented in Figure 14.

In our SAT-based formulation, a route principally cannot block other routes. This route will be either pruned before the SAT

formulation or this route will never be applied according to the imposed Boolean constraints. Scenic routing is also impossible. SAT operates with pre-created routes. Undesired routes cannot appear in the layout unless we explicitly create such routes. We control the quality of created routes varying bounding boxes and the number of allowed segments per route. If the SAT solver reports that particular problem is unsatisfiable, we can increase the number of segments per route. In some cases we also extend slightly the bounding box for a route.

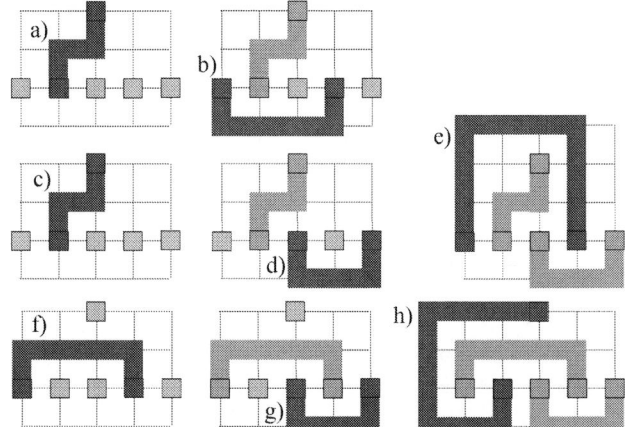

Figure 14. Several worst-case scenarios for sequential routing.

These considerations do not mean that global routing of global assignment of ties is useless for the SAT formulation. Reasonable restrictions both on the variety and the length of routes make the routing flow more effective. These restrictions limit the work required in applying the various pruning techniques: pruning redundant routes, pruning unfeasible routes, and pruning routes by signature.

In our flow, we skip ties that will be redundant with very high probability. For example, if three or more diffusion contacts are placed in a line we connect only adjacent terminals and skip such ties as tie t_{13} in Figure 15. This situation usually takes place for vcc/vss terminals and for multi-legged transistors. We also do not create redundant ties for three or more polys placed in a line and for the analogous situation in the vertical direction for double-height standard cells.

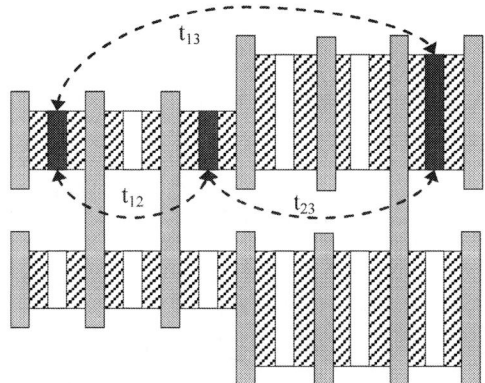

Figure 15. Redundant tie t_{13}.

We do not use Steiner trees because the length of the terminals is comparable with the length of the routes. Terminals cannot be replaced by points in the construction of Steiner trees.

An Efficient Algorithm for Multi-Layer Obstacle-Avoiding Rectilinear Steiner Tree Construction *

Chih-Hung Liu[1], I-Che Chen[1], and D. T. Lee[1,2,3]

[1]Research Center for Information Technology Innovation and [2]Institute of Information Science, Academia Sinica
[3]Department of Computer Science and Engineering, National Chung Hsing University
chliu_10@citi.sinica.edu.tw, ycchen@gmail.com, dtlee@ieee.org

ABSTRACT

We consider the multi-layer *obstacle-avoiding rectilinear Steiner minimal tree* (OARSMT) problem and propose a reduction to transform a multi-layer instance into a 3D instance. Based on the reduction we apply computational geometry techniques to develop an efficient algorithm, utilizing existing OARSMT heuristics. Experimental results show that our algorithm provides a solution with excellent quality and has a significant speed-up compared to previously known results.

Categories and Subject Descriptors

B.7.2 [**Integrated Circuits**]: Design Aids

General Terms

Algorithms, Theory, Design

Keywords

Physical Design, Rectilinear Steiner Minimal Tree (RSMT), Routing, Obstacle-Avoidance

1. INTRODUCTION

The *rectilinear Steiner minimal tree* (RSMT) problem is a fundamental research topic in VLSI layout design and routing since the routes for signal nets are usually represented by rectilinear Steiner trees. However, as the IC technology significantly increases the density of an IC chip, a modern IC design contains more and more hard IP cores, macro blocks, and pre-routed nets, which are referred to as obstacles in the routing process. Therefore, the *obstacle-avoiding RSMT* (OARSMT) problem has become very important and received lots of attention recently [1]–[8].

In addition, the advance of IC technology also offers an abundance of routing layers to provide more routing resources. Therefore, Lin et al. [9] first formulated the *multi-layer OARSMT* (ML-OARSMT) problem and discussed the

---

*This works was supported by National Science Council, Taiwan under grants No's. NSC-98-2221-E-001-007-MY3, NSC- 98-2221-E-001-008-MY3, and NSC-99-2911-I-001-506.

difficulty of extending the methods for the OARSMT problem to the ML-OARSMT problem. They extended the spanning graph in [1] to construct a multi-layer spanning graph and developed an effective algorithm.

Moreover, considering signal integrity and IC manufacturing, the orientation of routing in a single layer tends to be the same (either horizontal or vertical), and thus *preferred directions* are assigned to each routing layer. Liu et al. [10] further considered preferred directions and formulated the *obstacle-avoiding preferred direction Steiner tree* (OAPD-ST) problem. They proposed an $O(n^2)$-space routing graph which contains at least one optimal solution, and developed a factor-2 approximation algorithm with $O(n^2 \log n)$ time complexity. Recently, Chuang and Lin [11] proposed an effective algorithm which does not construct a routing graph.

According to the experimental results in [9], [10], and [11], those algorithms still behave like a quadratic-time algorithm in practice. In our opinion, the inefficiency of the algorithms in [9] and [10] results from the quadratic size of a routing graph, and the inefficiency of the algorithm in [11] results from their backtracking procedure of finding Steiner points. Since the Steiner tree construction will be invoked many times during the physical design automation, it is necessary to develop a more efficient algorithm for the ML-OARSMT problem and the OAPD-ST problem which behaves like a subquadratic-time algorithm for practical applications.

However, the obstacle-avoidance shortest path problems were of major interest in computational geometry in the 1980s and the 1990s, and have been well-studied in both the 2D plane and the 3D space. Under these circumstances, it is probably beneficial to apply those computational geometry techniques to the VLSI routing. For example, Liu et al. [5] made good use of shortest path maps to obtain an excellent algorithm for the OARSMT problem.

We therefore employ computational geometry techniques to develop an efficient algorithm for the ML-OARSMT problem and the OAPD-ST problem, which not only gives excellent solutions but also behaves like a subquadratic-time algorithm for practical applications. We first proved that the size of a *multi-layer obstacle-avoiding rectilinear minimum spanning tree* (ML-OARMST) is $\Omega(n^2)$, indicating that the multi-layer model is very different from its 2D counterpart. Hence, we propose a reduction which transforms a multi-layer instance into a 3D instance and thus enable us to employ computational geometry techniques.

Based on the reduction, we compute the visibility graph as defined in [12] and utilize the Steiner-point based framework in [5] to develop a 4-step algorithm for the ML-OARSMT problem. Specifically, we adopt the algorithm in [12] to construct *multi-layer visibility graph* (ML-VG). We then use the concept of Kruskal algorithm [13] to select good Steiner points from the ML-VG. After that, we take those selected Steiner points to generate a Steiner tree, and finally, we refine the generated tree to reduce the total cost.

Furthermore, we make use of the geometric transformation in [14] to extend our ML-OARSMT algorithm and make it amenable for the OAPD-ST problem.

Experimental results show that our algorithm provides a solution with excellent quality and has a significant speed-up compared to previously known results. Compared with [9], our algorithm for the ML-OARSMT problem improves the solution quality in terms of total cost by 2.95% on average, and also achieves 46.12 times speed-up on average at the same time. Compared with [10], our algorithm for the OAPD-ST problem improves the solution quality in terms of total cost by 5.55% on average, and also achieves 16.61 times speed-up on average at the same time. Moreover, the speed-up over [9] and [10] increases with the input size n and is like a linear function of n, so we claim that our algorithm behaves like a subquadratic-time algorithm for practical applications, and is very close to a loglinear-time algorithm.

The rest of the paper is organized as follows. Section 2 formulates the two problems. Section 3 presents the theoretical results, and Section 4 describes the algorithm. Section 5 shows the experimental results, and Section 6 gives the conclusion. We move all the proofs to the appendices.

2. PROBLEM FORMULATION

An *obstacle* is a rectangle in a layer. Any two obstacles cannot overlap with each other except the boundary. A *pin-vertex* is a vertex in any layer which will be connected in a signal net. No pin-vertex can be inside any obstacle except the boundary. A *via* is an edge connecting two points (x, y, z) and $(x, y, z + 1)$. Each endpoint of a via must not be inside an obstacle except the boundary.

We assume that the costs of vias are the same, and the unit cost of a wire is identical in all the layers [1]. Let $P = \{p_1, p_2, \ldots, p_m\}$ be the set of pin-vertices in an m-pin net, $O = \{o_1, o_2, \ldots, o_k\}$ be a set of k obstacles, and n be the size of $P \cup \{corners\ in\ O\}$ ($n \leq m + 4 * k$). Let N_l be the number of routing layers and C_v be the cost of a via. N_l is a small constant in practice, and n is the input size.

- **The ML-OARSMT Problem:**
  Given a via cost C_v, a set P of pin-vertices, a set O of obstacles, and N_l routing layers, construct a rectilinear routing tree T connecting all the pin-vertices in P possibly through some additional points (called Steiner points) such that no tree edges intersect any obstacles in O and the total cost is minimized, where the cost of T, cost(T), is defined to be the total wirelength $+ C_v$ * the number of vias.

The OAPD-ST problem is a more restricted version of the ML-OARSMT problem in which the orientation of a tree edge must follow the *preferred direction* (PD) constraints. Without loss of generality, we assume the PD constraints as follows: only *vertical* edges are allowed in *odd* number layers, and only *horizontal* edges are allowed in *even* number layers. Hereafter, all the mentioned trees/distances/edges/paths are obstacle-avoiding and rectilinear.

---
[1]Our setting is a little different from that in [10], which assume wires in different layers may have different unit costs.

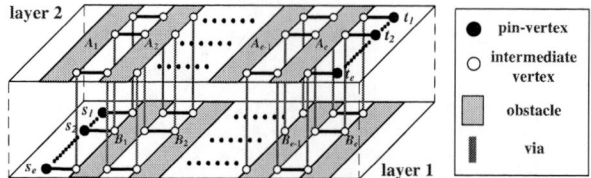

Figure 1: A worst-case ML-OARMST instance.

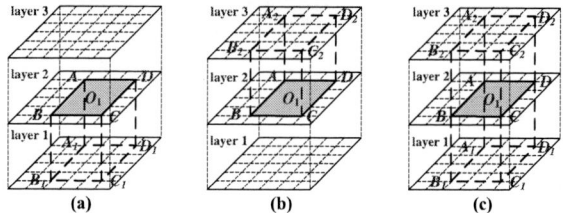

Figure 2: Examples for the properties of an obstacle. (a) projection in layer 1. (b) projection in layer 3. (c) union of (a) and (b).

3. THEORETICAL RESULTS

3.1 Lower Bound of an ML-OARMST

DEFINITION 1. *Given an ML-OARSMT problem instance, a* **multi-layer obstacle-avoiding rectilinear minimum spanning tree (ML-OARMST)** *is a tree connecting all the pin-vertices in P using a set of shortest paths between pairs of those pin-vertices such that the sum of costs of those paths is the minimum.*

In Fig. 1, there are $m = 2e$ pin-vertices ($s_i = p_{2i-1}$ and $t_i = p_{2i}$ for $1 \leq i \leq e$) and $k = 2e$ obstacles ($A_i = O_{2i-1}$ and $B_i = O_{2i}$ for $1 \leq i \leq e$). Thus, $n = m + 4k = 10e$. For $1 \leq i \leq e$, there exists only one shortest path between s_i and t_i, and this path consists of at least $2e$ segments. For two points p and q, let $d(p, q)$ be the distance between them. We assume $d(s_i, t_i)$ for $1 \leq i \leq e$ is smaller than $d(s_j, s_{j+1})$ and $d(t_j, t_{j+1})$ for $1 \leq j \leq e - 1$. In other words, if a shortest path between two pin-vertices is viewed as an edge, the e shortest paths between s_i and t_i for $1 \leq i \leq e$ are the e edges of least cost.

According to the cut property of minimum spanning trees [13], an edge crossing a cut with minimum cost must be included in a minimum spanning tree. Therefore, the e shortest paths between s_i and t_i for $1 \leq i \leq e$ must be included in the ML-OARMST. Since each of those paths has at least $2e$ segments and $e = \frac{n}{10}$, we conclude the following theorem.

THEOREM 1. *The size of an ML-OARMST is $\Omega(n^2)$.*

Theorem 1 indicates that the multi-layer model is very different from its 2D counterpart for the reason that the size of a 2D OARMST is linear. Moreover, Theorem 1 also shows that it is impossible to develop a subquadratic-time algorithm for the ML-OARSMT problem using the ML-OARMST construction in the worst case.

3.2 3D Reduction

For an obstacle o in the multi-layer model, no path can go through the region between o and the projections of o

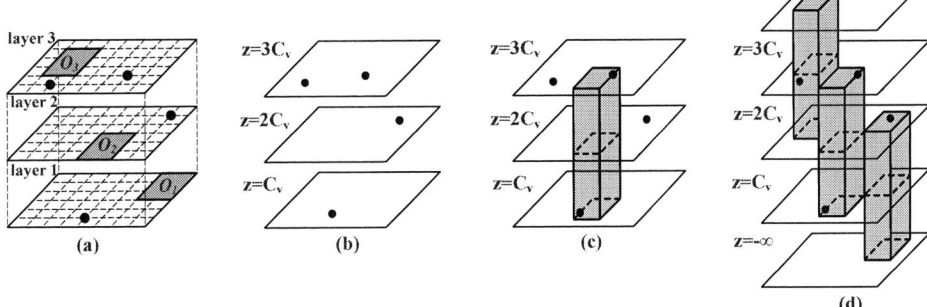

Figure 3: Reduction from a 3-layer instance I to a 3D instance $R(I)$. (a) a 3-layer instance I. (b) transformed pin-vertices. (c) a transformed obstacle from layer 2. (d) transformed obstacles from layer 1 and layer N_l.

in adjacent layers. For instance, in Fig 2(a), there is an obstacle O_1, rectangle $ABCD$, in layer 2, and the projection of O_1 in layer 1 is rectangle $A_1B_1C_1D_1$. No path can go through the interior of box $ABCDD_1A_1B_1C_1$, but a path could go through each face of this box except rectangle $ABCD$. Similarly, we have box $ABCDD_2A_2B_2C_2$ as shown in Fig. 2(b). By merging the two boxes, we have box $A_2B_2C_2D_2D_1A_1B_1C_1$ in Fig. 2(c) such that no path can go through the interior of this box, but a path can go through each face of this box. Therefore, we conclude that a rectangle in the multi-layer model can be viewed as a specific rectilinear box in the 3D space.

Based on the observation, we propose a reduction $R(\cdot)$ to transform a multi-layer instance I into a 3D instance $R(I)$:

1. For each pin-vertex $p = (x, y, l)$ in P, $R(p)$ is $(x, y, l * C_v)$ as shown in Fig. 3(b).

2. For each obstacle o in O, let l be the layer number of o. If $2 \leq l \leq N_l - 1$, $R(o)$ represents a rectilinear box constructed by projecting o to $z = (l + 1) * C_v$ and $z = (l - 1) * C_v$, and then connecting the vertices of those two projections using line segments parallel to the z-axis. For instance, O_2 in layer 2 in Fig. 3(a) is transformed into a box between $z = C_v$ and $z = 3 * C_v$ in Fig. 3(c). If $l=1$, $R(o)$ represents a rectilinear box constructed by applying the above method on $z = -\infty$ and $z = 2 * C_v$; if $l=N_l$, $R(o)$ represents a rectilinear box by applying the above method on $z = (N_l - 1) * C_v$ and $z = \infty$. For instance, O_1 and O_3 in Fig. 3(a) are transformed into two boxes in Fig. 3(d). The purpose of projections on $z = -\infty$ and $z = \infty$ is to prevent shortest paths from passing above layer N_l or below layer 1.

Let $R(P)$ be $\{R(p) | p \in P\}$ and $R(O)$ be $\{R(o) | o \in O\}$. If several boxes in $R(O)$ intersect, we combine them into a rectilinear obstacle. For $1 \leq i \leq N_l$, let Z_i denote plane $z = i * C_v$, and let $Z_{-\infty}$ and Z_∞ denote planes $z = -\infty$ and $z = \infty$ respectively. The following theorem can be simply proved from the reduction $R(\cdot)$. Please refer to Appendix A for a formal proof.

THEOREM 2. *There exists an ML-OARSMT T in a multi-layer instance I if and only if there exists a 3D-OARSMT T' in $R(I)$ such that $cost(T) = cost(T')$.*

4. ALGORITHM

4.1 The ML-OARSMT Problem

Liu et al. [5] adopted a fact that an optimal OARSMT is an OARMST connecting all pin-vertices and appropriately selected Steiner points to propose a Steiner-point based framework for the OARSMT problem: (1) graph construction, (2) Steiner point selection (3) minimum terminal spanning tree (MTST) construction (Definition 3), and (4) refinement. We follow their Steiner point framework to develop a 4-step algorithm for the ML-OARSMT problem.

4.1.1 Multi-layer Visibility Graph Construction

To deal with shortest paths among rectilinear obstacles in the 3D space, Clarkson et al. [12] proposed a *3D visibility graph* (3D-VG) which embeds a shortest path for each pair of vertices in the set of pin-vertices and obstacle corners. Based on our reduction, we use the algorithm in [12] to construct the *multi-layer visibility graph* (ML-VG) $G(V, E)$ for a multi-layer instance I. In detail, we first compute the 3D-VG $G'(V', E')$ for $R(I)$, and then delete vertices in Z_∞ and $Z_{-\infty}$ from G' to construct G. Note that the 3D-VG algorithm can handle rectilinear obstacles each of which is a union of intersecting boxes, and thus is feasible for our reduction.

The 3D-VG construction is summarized as follows:

1. Let V' be the union of $R(P)$ and the obstacle corners in $R(O)$. Let $(x_1, x_2, \cdots, x_{|V'|})$, $(y_1, y_2, \cdots, y_{|V'|})$, and $(z_1, z_2, \cdots, z_{|V'|})$ be the x, y, and z-coordinates of the vertices in V'. Let V_I be the set of vertices obtained by intersecting the edges of obstacles in $R(O)$ with planes $x = x_1, \cdots, x = x_{|V'|}$, $y = y_1, \cdots, y = y_{|V'|}$, and $z = z_1, \cdots, z = z_{|V'|}$. Let V_S be $V' \cup V_I$.

2. Let z_m be the median of the z-coordinates of vertices in V_S, and P_{zm} be the plane $z = z_m$. For each vertex $v = (x, y, z)$ in V_S, create a vertex $u = (x, y, z_m)$ if $\overline{vu}$ does not intersect any obstacle. Apply the 2D-VG algorithm in [12] on P_{zm} to create essential vertices.

3. Let V_H and V_L be the set of vertices in V_S higher than P_{zm} and the set of vertices in V_S lower than P_{zm} respectively.

4. Repeat Step 2 and Step 3 on V_H and V_L separately.

5. Add V_S and the vertices created during step 2–step 4 to V'. For each two vertices $v, u \in V'$, if (1) $\overline{vu}$ is

rectilinear and (2) $\overline{vu}$ does not intersect any obstacle in $R(O)$ or any vertex in V', add $\overline{vu}$ to E'.

THEOREM 3. *The ML-VG has $O(|V_S|\log n)$ vertices and edges, and can be constructed in $O(|V_S|\log^2 n)$ time, where $|V_S|$ is $O(n^2)$ in the worst case and $O(n)$ for practical applications.*

Theorem 3 is mainly from the proofs in [12], and the major difference is that most obstacles are regular in practice and only generate $O(1)$ vertices to V_S, implying that $|V_S|$ is $O(n)$ in practice. Please see Appendix B for a formal proof.

4.1.2 Steiner Point Selection

We will select Steiner points from the ML-VG. In [5], the authors used the concept of Prim algorithm [13] to select Steiner points, i.e., they started from one pin-vertex to find Steiner points until all pin-vertices are reached. However, the choice of the starting pin-vertex will affect the quality of selected Steiner points. Therefore, in order to obtain a stable result, we use the concept of Kruskal algorithm [14] to simultaneously start from all pin-vertices and to select Steiner points.

DEFINITION 2. *For two vertices v and u in a graph $G(V, E)$, their **shortest path component** $SPC(v, u)$ is the minimal subgraph of G such that $SPC(v, u)$ contains all the shortest paths between v and u in G.*

For the ML-VG $G(V, E)$ with a set $P \subset V$ of pin-vertices, our Kruskal-based Steiner point selection is stated as follows.

1. Initialize each $p_i \in P$ as a connected component C_i.

2. Find the nearest pair of connected components C_i and C_j. Let the shortest path between C_i and C_j be from a vertex $s_i \in C_i$ and a vertex $s_j \in C_j$.

3. Construct $SPC(s_i, s_j)$ and combine C_i, $SPC(s_i, s_j)$, and C_j into one connected component. If $s_i \notin P$ ($s_j \notin P$), select s_i (s_j) as a Steiner point.

4. Repeat Step 2 and Step 3 until there exists only one connected component.

We use the Dijkstra shortest path algorithm [13] to implement the selection. We view the vertices of all the connected components as sources, i.e. their weight are zero, and then manipulate the priority queue to find the nearest source for each vertex in V. For each edge $(u, v) \in E$, let s_1 and s_2 be the nearest sources of u and v, respectively. If s_1 and s_2 belong to different components, (u, v) corresponds to a path between s_1 and s_2 whose length is $d(s_1, u) + |(u, v)| + d(v, s_2)$. Finding the shortest one of such "paths" will determine the nearest pair of two connected components. Note that when Step 2 is repeated, we directly set the vertices of $SPC(s_i, s_j)$ as new sources without re-initializing the whole priority queue, which increases the efficiency.

Moreover, in order to construct $SPC(s_i, s_j)$, we find all the edges each of which corresponds to a shortest path between s_i and s_j, and then backtrack from the endpoints of those edges to compute $SPC(s_i, s_j)$, which is similar to the backtracking procedure for a shortest path region in [5]. Besides, we use the operations of disjoint sets in [13] to maintain the relationships between vertices and connected components. Please see Appendix C for more details.

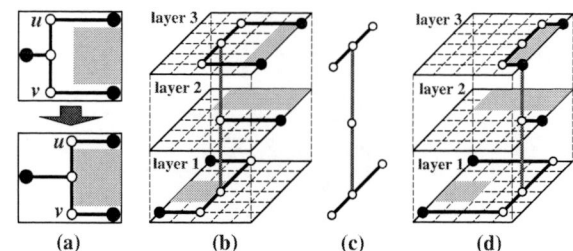

Figure 4: (a) U-shaped pattern. (b)–(d) a multi-layer pattern.

4.1.3 MTST Construction

DEFINITION 3. *For a graph $G(V, E)$ and a terminal set $S \subseteq V$, a **minimum terminal spanning tree (MTST)** connects all the vertices in S using a set of terminal paths among vertices in S such that the sum of lengths of those terminal paths is minimized, where a **terminal path** is a path in G between two vertices in S passing through vertices in $V \setminus S$ [2].*

We view all the pin-vertices and selected Steiner points as terminals and thus construct the MTST of the ML-VG as an initial solution. Mehlhorn [15] proposed an $O(|E| + |V| \log |V|)$-time MTST algorithm. By Theorem 3, this step takes $O(n^2 \log^2 n)$ time in the worst case and $O(n \log^2 n)$ time for practical applications.

4.1.4 Refinement

We refine the initial solution in Section 4.1.3 to reduce more wirelength. For the OARSMT problem, the traditional way is to move the U-shaped patterns [1] as shown in Fig. 4(a). In general, a U-shaped pattern results from a line segment which has different numbers of edges incident on its two sides, e.g., $\overline{uv}$ in Fig. 4(a).

However, in the multi-layer model, line segments in different layers may affect each other. Therefore, we should replace the line segment of a 2D U-shaped pattern with a collection of line segments.

DEFINITION 4. *For a routing tree T in a multi-layer instance, a **preferred direction component** (PDC) is a **maximal connected component** of T whose elements either share the same x-coordinate or share the same y-coordinate. For example, Fig. 4(c) shows a PDC of Fig. 4(b). A PDC is also known as a consecutive line segment in [10].*

As shown in Fig. 4(b)–(d), if a PDC has different numbers of edges incident on its two sides, moving this PDC possibly reduces the wirelength. As a result, we greedily move a PDC to reduce the wirelength until no PDC can be moved to reduce the wirelength. Please see Appendix D for more details.

4.1.5 Time Complexity

THEOREM 4. *Our ML-OARSMT construction takes $O(mn^2 \log^2 n)$ time in the worst-case and $O(n \log^2 n)$ time for practical applications.*

PROOF. Please see Appendix E. □

616

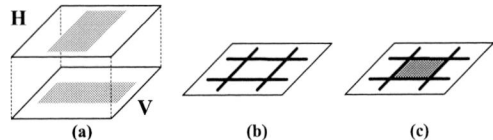

Figure 5: (a) a 2-layer OAPD-ST instance. (b) the transformed 1-layer instance from (a) without PD constraint. (c) the augmented instance from (b).

4.2 The OAPD-ST Problem

We extend our 4-step ML-OARSMT algorithm to deal with the OAPD-ST problem. In fact, the last three steps are directly applicable, and thus we only discuss the graph construction. Without loss of generality, we assume that N_l is an even number.

Lee and Yang [14] proposed an $O(n \log n)$-time algorithm to transform a 2-layer instance with PD constraint into a 1-layer instance without PD constraint. As shown in Fig. 5(a)–(b), a rectangle in the horizontal layer is transformed into its two vertical boundary segments to prevent invalid horizontal routing, and a rectangle in vertical layer is transformed into its two horizontal boundary segments. They proved that a valid path in the transformed 1-layer instance is directly a valid path in the original 2-layer instance where a bend of the path in the transformed 1-layer instance corresponds to a via in the original 2-layer instance.

We combine the transformation in [14] and the ML-VG construction in Section 4.1.1 to construct *preferred direction visibility graph* (PD-VG) for an OAPD-ST problem instance I as follows:

1. Transform I into a multi-layer instance I' by transforming layer (2i-1) and layer 2i of I into layer i of I' for $1 \le i \le \frac{N}{2}$. In order to apply the ML-VG construction, as shown in Fig. 5(c), we further augment each region bounded by 4 transformed segments with a solid rectangle to prevent invalid projections.

2. Apply the ML-VG construction on I' to obtain $G'(V', E')$.

3. Transform the ML-VG $G'(V', E')$ back to the PD-VG $G(V, E)$. For a vertex $v = (x, y, z)$, let v_1 be $(x, y, 2z - 1)$ and v_2 be $(x, y, 2z)$. For each $v \in V'$, insert v_1 and v_2 into V and (v_1, v_2) into E. For each edge $e = (u, v) \in E'$, if e is vertical, insert (u_1, v_1) into E; if e is horizontal, insert (u_2, v_2) into E; if e is perpendicular to routing layers, insert either (u_1, v_2) or (u_2, v_1) into E depending on whether u is higher than v or not. Remove all invalid vertices and edges from G.

Note that the ML-VG embeds all the shortest paths among pin-vertices and obstacle corners, but the PD-VG does not, since the visibility graph does not consider the bends, which represent vias in the transformed layers. Since the transformation takes $O(n \log n)$-time [14], the PD-VG construction takes the same time with the ML-VG construction. Moreover, since the last three steps of the algorithm are the same, we conclude the following by Theorem 4.

THEOREM 5. *Our OAPD-ST construction takes* $O(mn^2 \log^2 n)$ *time in the worst-case and* $O(n \log^2 n)$ *time for practical applications.*

5. EXPERIMENTAL RESULTS

We have implemented our algorithm in C language, and conducted all the experiments on an IBM x3550 server with 12 3.3-GHz processors and 64-GB memory. All the results are generated using one processor instead of twelve. We requested binaries from Lin et al. [9], Liu et al. [10], and Chuang and Lin [11]. However, the authors in [11] could not provide us with their binary executable, so we could not make a direct comparison with them.

5.1 The ML-OARSMT problem

There are totally ten benchmark circuits, five test cases (ind1–ind5) from Synopsys and five random test cases (rt1–rt5) generated by Lin et al. [9].

Table 1 lists the total cost, the number of vias, and the CPU time of [9] and our algorithm under the condition $C_v = 3$. Compared with [9], our algorithm improves the solution quality in terms of total cost by 2.95% on average, and also achieves 46.12 times speed-up on average at the same time as shown in Table 1. For a large test case (ind5), the speed-up is up to about 228 times.

Moreover, the speed-up seems to increase with the input size n. We generate ten vertices whose x-coordinates represent the input sizes n of those test cases and whose y-coordinates represent the CPU times of our algorithm. By least squares fitting on the log-log-axes of those vertices, the respective slope of the fitting line is 0.91, implying that the complexity of the speed-up is similar to $\Theta(n^{0.91})$. Since the approach in [9] is a quadratic-time algorithm, our algorithm behaves like a subquadratic-time algorithm for practical applications, and is very close to a loglinear-time algorithm.

5.2 The OAPD-ST problem

In order to address our motivations, we need to conduct experiments on test cases similar to practical conditions. However, the random test cases generated by Liu et al. [10] are not feasible since the number of pin-vertices is too large and the obstacles are too large and too overcrowded. As a result, we take the shapes and sizes of obstacles in the industrial test cases (ind1–ind5) for the ML-OARSMT problem to randomly generate a new set of test cases for the OAPD-ST problem. The detailed setting of those test cases is shown in Table 2. For each kind of test cases, we generate 100 samples, and the reported results (total cost, number of vias and CPU time) are average results.

Table 2 lists the total cost, the number of vias, and the CPU time of [10] and our algorithm under the conditions $C_v = 3$ and $N_l = 6$. Compared with [10], our algorithm improves the solution quality in terms of total cost by 5.55% on average, and also achieves 16.61 times speed-up on average at the same time as shown in Table 2. For the largest test case ($m = 200$ and $k = 2000$), the speed-up is 54 times.

More importantly, the speed-up obviously increases with the input size n. Similar to Section 5.1, we use the least squares fitting method to analyze the speed-up, and show that the complexity of the speed-up seems to be $\Theta(n^{0.90})$, implying that our algorithm seems to be almost faster than [10] by a factor of n. Since the approach in [10] is a quadratic-time algorithm, our algorithm behaves like a subquadratic-time algorithm for practical applications, and is very close to a loglinear-time algorithm.

We will give more experimental results and analyses in Appendix F to support the claims made in Theorem 3.

Table 1: Comparison on the Total Cost and Cpu Time between [9] and Ours, where C_v is 3.

Test Cases	m / k / N_l	n	Total Cost(# via)		Imp. (%) $(\frac{A-B}{A})$	Time		speed-up $(\frac{C}{D})$
			[9] (A)	ours (B)		[9] (C)	ours (D)	
ind1	50 / 6 / 5	74	55537(49)	54207(49)	2.39	0.067	0.009	7.44x
ind2	200 / 85 / 6	540	12512(224)	12008(206)	4.03	1.863	0.076	24.51x
ind3	250 / 13 / 10	302	10973(359)	10555(348)	3.81	1.973	0.081	24.36x
ind4	500 / 100 / 5	900	77033(0)	77292(0)	-0.34	4.613	0.043	107.28x
ind5	1000 / 20 / 5	1080	14515511(0)	14599961(0)	-0.58	24.024	0.105	228.8x
rt1	25 / 10 / 10	65	4334(76)	4169(70)	3.81	0.067	0.018	3.72x
rt2	100 / 20 / 10	180	9434(215)	9132(209)	3.2	0.596	0.07	8.51x
rt3	250 / 50 / 10	450	15569(490)	14750(478)	5.26	3.737	0.413	9.05x
rt4	500 / 50 / 10	700	22034(918)	21013(936)	4.63	9.286	0.619	15x
rt5	1000 / 100 / 5	1400	27890(869)	26970(814)	3.3	27.865	0.857	32.51x
Avg.	–	–	–	–	2.95	–	–	46.12x

Table 2: Comparison on the Total Cost and Cpu Time between [10] and Ours, where $C_v = 3$ and $N_l = 6$.

Test Cases			Total Cost(# via)		Imp. (%) $(\frac{A-B}{A})$	Time		speed-up $(\frac{C}{D})$
m	k	n	[10] (A)	ours (B)		[10] (C)	ours (D)	
10	50	210	5,272(19)	4,882(22)	7.40	0.053	0.029	1.83x
10	100	410	10,309(19)	9,965(23)	3.34	0.238	0.065	3.66x
20	100	420	14,617(44)	14,141(49)	3.26	0.272	0.069	3.94x
20	200	820	30,215(44)	28,038(48)	7.21	0.748	0.124	6.03x
50	250	1,050	58,149(111)	55,281(124)	4.93	1.573	0.206	7.64x
50	500	2,050	116,266(114)	109,717(132)	5.63	9.574	0.589	16.25x
100	500	2,100	164,850(226)	155,801(261)	5.49	11.525	0.662	17.41x
100	1,000	4,100	333,198(221)	313,109(273)	6.03	29.725	1.302	22.83x
200	1,000	4,200	459,861(457)	429,896(539)	6.52	47.385	1.463	32.39x
200	2,000	8,200	912,896(448)	861,407(546)	5.64	156.605	2.894	54.11x
Average			–	–	5.55	–	–	16.61x

6. CONCLUSION

For the ML-OARSMT problem and the OAPD-ST problem, we have employed computational geometry techniques to develop the first subquadratic-time algorithm for practical applications. In order to employ computational geometry techniques, we have proposed a reduction to transform a multi-layer instance into a 3D instance, and we believe that this reduction will provide a way to employ more computational geometry techniques. Moreover, the computational geometry techniques employed may also be useful for other routing problems. Experimental results show that our algorithm behaves like a subquadratic-time algorithm for practical applications, and when compared with existing approaches, our algorithm provides a solution with excellent quality and achieves a significant speed-up.

7. REFERENCES

[1] C.-W. Lin, S.-Y. Chen, C.-F. Li, Y.-W. Chang and C.-L. Yang, "Obstacle-avoiding rectilinear Steiner tree construction based on spanning graphs," *IEEE Trans. Computer-Aided Design*, vol. 27, no.4, pp. 643–653, 2008.

[2] J. Long, H. Zhou, and S. Memik, "EBOARST: an efficient edge-based obstacle-avoiding rectilinear Steiner tree construction algorithm ," *IEEE Trans. Computer-Aided Design*, vol. 27, no.12, pp. 2169–2182, 2008.

[3] L. Li and Evangeline F.-Y. Young, "Obstacle-avoiding rectilinear Steiner tree construction," *in Proc. ICCAD*, pp. 133–138, 2008.

[4] C.-H. Liu, S.-Y. Yuan, S-Y. Kuo, and Y.-H. Chou, "An $O(n \log n)$ path-based obstacle-avoiding algorithm for rectilinear Steiner tree construction," *in Proc. DAC*, pp. 314–319, 2009.

[5] C.-H. Liu, S.-Y. Yuan, S.-Y. Kuo, and J.-H. Weng, "Obstacle-avoiding rectilinear Steiner tree construction based on Steiner point selection," *in Proc. ICCAD*, pp. 26–32, 2009.

[6] T. Huang and Evangeline F.-Y. Young, "Obstacle-avoiding rectilinear Steiner minimum tree construction: An optimal approach," *in Proc. ICCAD*, pp. 610–613, 2010.

[7] G. Ajwani, C. Chu, and W.-K. Mak, "FOARS: FLUTE based obstacle-avoiding rectilinear Steiner tree construction," *IEEE Trans. Computer-Aided Design*, vol. 30, no. 2, pp. 194–203, 2011.

[8] T. Huang and Evangeline F.-Y. Young, "An exact algorithm for the construction of rectilinear Steiner minimum trees among complex obstacles," *in Proc. DAC*, pp. 164–169, 2011.

[9] C.-W. Lin, S.-L. Hunag, K.-C. Hsu, M.-X. Lee, Y.-W. Chang, "Multilayer obstacle-avoiding rectilinear Steiner tree construction based on spanning graph," *IEEE Trans. Computer-Aided Design*, vol. 27, no. 11, pp.2007–2016, 2008.

[10] C.-H. Liu, Y.-H. Chou, S.-Y. Yuan, and S.-Y. Kuo, "Efficient multilayer routing based on obstacle-avoiding preferred direction Steiner tree," *in Proc. ISPD*, pp. 118–125, 2008.

[11] J.-R. Chuang and J.-M. Lin, "Efficient multi-layer obstacle-avoiding preferred direction rectilinear Steienr tree construction," *in Proc. ASP-DAC*, pp. 527–532, 2011.

[12] K. L. Clarkson, S. Kapoor, And P. M. Vaidya, "Rectlinear shortest paths through polygonal obstacles in $O(n \log^2 n)$ time," *in Proc. SCG*, pp. 251–257, 1987.

[13] T. Cormen, C. Leiserson, R. Rivest, and C. Stein, *Introduction to Algorithms*, 2nd edition, The MIT Press, 2001.

[14] D. T. Lee and C. D. Yang, "Finding rectilinear paths among obstacles in a two-layer interconnection model," *Internat. J. Comput. Geom. Appl.*, vol. 7, no.6, pp. 581–598, 1997.

[15] K. Mehlhorn, "A faster approximation algorithm for the Steiner problem in Graphs," *Inf. Process Let.*, vol. 27, pp. 125–128, 1988.

[16] B. Chazelle, "An algorithm for segment-dragging and its implementation," *Algorithmica*, vol. 3, pp. 205–221, 1988.

[17] M. de Berg, O. Cheong, M. van Kreveld, and M. Overmars, *Computational Geometry: Algorithms and Applications*, 3rd edition, Springer-Verlag, 2008.

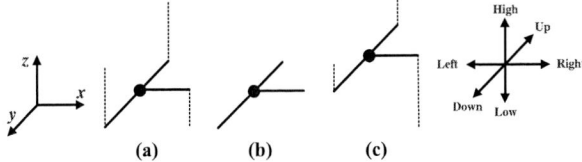

Figure 6: Solid segments represent the z-component and dash segments represent the incident edges. (a) a part of a tree. (b) the z-component of (a). (c) the topology refined from (a) by moving (b) along the high direction.

APPENDIX

A. THE REDUCTION

Theorem 2. There exists an ML-OARSMT T in a multilayer instance I if and only if there exists a 3D-OARSMT T' in $R(I)$ such that $\text{cost}(T) = \text{cost}(T')$.

Proof.

Necessity:

For each edge $e = (u_1, u_2)$ in I, let $R(e)$ be $(R(u_1), R(u_2))$. It is clear that for each edge e of T, $R(e)$ is valid in $R(I)$ and $\text{cost}(e)=\text{cost}(R(e))$. Let $R(T)$ be $\{R(e)|$ an edge $e \in T\}$. Therefore, there exists a 3D-OARSMT $T' = R(T)$ in $R(I)$ such that $\text{cost}(T') = \text{cost}(T)$.

Sufficiency:

Let $R^{-1}(\cdot)$ be the inverse function of $R(\cdot)$ such that for each point p in I, $R^{-1}(R(p)) = p$. If $R^{-1}(T')$ is valid in I, $R^{-1}(T')$ is directly T. Otherwise, we move parts of T' to construct another 3D-OARSMT T'' without increasing the wirelength such that $\text{cost}(T') = \text{cost}(T'')$ and $T = R^{-1}(T'')$. Since each pin-vertex of $R(I)$ is located on Z_i where $1 \le i \le N_l$, T' has no edge higher than Z_N or lower than Z_1; otherwise, T' can be refined to reduce the wirelength, contradicting that T' is minimized. Therefore, if T' has some "invalid" edges which make $R^{-1}(T')$ invalid, each of them must be between Z_i and Z_{i+1} where $1 \le i \le N_l - 1$.

Let a z-component be a maximal connected component whose vertices share the same z-coordinate. For example, Fig. 6(b) shows a z-component of Fig. 6(a). Those "invalid" edges of T' can be grouped into several z-components. For each invalid z-component C of T', the number of incident edges higher than C must be equal to the number of incident edges lower than C; otherwise, moving C along the direction in which the number of incident edges is larger will reduce the wirelength, contradicting that T' is minimized. For example, the z-component of Fig. 6(a) has two high incident edges and one low incident edge, so moving it along the high direction will reduce the wirelength and result in a topology with less wirelength as shown in Fig. 6(c). Therefore, we can move each z-component of T' located between Z_i and Z_{i+1} to Z_{i+1} without increasing the wirelength and finally transform T' into T'' such that $R^{-1}(T'')$ is valid and $\text{cost}(T'')$ is equal to $\text{cost}(T)$. Those movements will not be denied by any obstacle since if an obstacle in $R(I)$ will deny the movement of a z-component located between Z_i and Z_{i+1}, the bottom face of this obstacle must be located between Z_i and Z_{i+1}, which violates the reduction $R(\cdot)$. As a result, there exists an ML-OARSMT $T = R^{-1}(T'')$ in I such that $\text{cost}(T) = \text{cost}(T')$. $\square$

```
Algorithm: Steiner_Point_Selection(P, V, E)
Input:  P /* the set of pin-vertices */
        V /* the set of vertices */
        E /* the set of edges*/
Output: S /* the set of Steiner points */
1    for each vertex v ∈ V
2        (v.wt, v.sr) ← (∞, ∅)
/* v.wt: the weight of v */
/* v.sr: the nearest source of v */
3    let H_V be a priority queue of vertices v whose key is v.wt
4    let H_E be a priority queue of edges e = (u, v)
         whose key is (u.sr).wt + |(u, v)| + (v.sr).wt
5    for each pin-vertex p in P
6        (p.wt, p.sr) ← (0, p)
7        Make-Set(p)
8        insert p into H_V
9    while not all pin-vertices are in the same set
10       (u, v) ← Nearest_Components(H_V, H_E, V, E)
11       (s_1, s_2) ← (u.sr, v.sr)
12       SPC(s_1, s_2) ← Compute_SPC(u, v, V, E)
13       for each vertex v' in SPC(s_1, s_2)
14           Union(Find-Set(s_1), Make-Set(v')))
15           (v'.wt, v'.sr) ← (0, v')
16           insert v' into H_V
17       if s_1 ∉ P
18           S ← S ∪ {s_1}
19       if s_2 ∉ P
20           S ← S ∪ {s_2}
21   return S
```

Figure 7: The Steiner point selection Algorithm.

B. THE ML-VG CONSTRUCTION

Theorem 3. The ML-VG has $O(|V_S| \log n)$ vertices and edges, and can be constructed in $O(|V_S| \log^2 n)$ time, where $|V_S|$ is $O(n^2)$ in the worst case and $O(n)$ for practical applications.

Proof. We first show that the ML-VG has $O(|V_S| \log |V_S|)$ vertices and edges, and can be constructed in $O(|V_S| \log^2 |V_S|)$ time, where V_S is the set of vertices generated in the Step 2 of the 3D-VG construction in Section 4.1.1. Then we explain that $|V_S|$ is $O(n^2)$ in the worst case and $O(n)$ for practical applications.

In [12], the authors had shown that the 3D-VG has $O(|V_S| \log^2 |V_S|)$ vertices and edges, and can be constructed in $O(|V_S| \log^3 |V_S|)$ time. One $O(\log |V_S|)$ factor of the number of vertices, the number of edges, and the time complexity results from the depth of recursion for Step 2–Step 4 of the 3D-VG construction. Since the number of distinct z-coordinates in $R(P) \cup \{$obstacle corners in $R(O)\}$ is at most $N_l + 2$ according to the reduction, the depth of recursion is also at most $N_l + 2$. As a result, we can remove one $O(\log |V_S|)$ factor from the number of vertices, the number of edges, and the time complexity since N_l is a small constant under the problem formulation in Section 2.

Since there are $O(n)$ vertices in $R(I)$, the $O(n)$ coordinates of those vertices will make $O(n)$ axis-parallel planes. Therefore, since there are $O(n)$ boxes in $R(I)$, there are $O(n^2)$ intersections between those axis-parallel planes and the edges of those boxes, implying that that $|V_S|$ is $O(n^2)$ in the worst case. However, I represents the routing environ-

```
Algorithm: Nearest_Components(H_V, H_E, V, E)
Input: H_V /* the priority queue of vertices*/
       H_E /* the priority queue of edges*/
       V /* the set of vertices */
       E /* the set of edges*/
Output: (u_m, v_m) /* the edge corresponds to
       the nearest pair of components */
1   while H_V is nonempty
2     u ← Extract_Min(H_V)
3     for each (u, v) ∈ E
4       if v.wt > u.wt + |(u, v)|
5         (v.wt, v.sr) ← (u.wt + |(u, v)|, u.sr)
6         if v ∉ H_V
7           insert v into H_V
8       else if v ∉ H_V
9         if Find-Set(u.sr) ≠ Find-Set(v.sr)
10          if (u, v) ∉ H_E
11            insert (u, v) into H_E
12          else
13            update the key of (u, v) in H_E
                with (u.sr).wt + |(u, v)| + (v.sr).wt
14  while H_E is nonempty
15    (u_m, v_m) ← Extract_Min(H_E)
16    if Find-Set(u_m.sr) ≠ Find-Set(v_m.sr)
17      return (u_m, v_m)
```

Figure 8: The computation of the nearest pair of connected components. (This algorithm must return an edge except that all the pin-vertices are in the same connected component.)

ment for a practical IC chip. Since there are only N_l routing layers, the number of extremely large obstacles which generate $O(n)$ vertices to V_S is originally limited. More importantly, in practice, some obstacles may be extremely large to generate $O(n)$ vertices to V_S, but most obstacles are regular and only generate $O(1)$ vertices to V_S. Besides, a small but extremely long obstacle would generate $O(n)$ vertices to V_S, while the number of such obstacles is a constant in practice. As a result, $|V_S|$ is $O(n)$ for practical applications. □

C. THE STEINER POINT SELECTION

We write three pseudocodes in Fig. 7, Fig. 8 and Fig. 9 to show the detailed implementation of the Kruskal-based Steiner point selection in Section 4.1.2. Lines 1–8 of Fig. 7 indicate Step 1, lines 10–11 indicate Step 2, and lines 12–20 indicate Step 3. (u_m, v_m) in Fig. 8 and Fig. 9 represents an edge which correspond to a shortest path between the nearest two connected components. Lines 1–13 of Fig. 8 are modified from the Dijkstra shortest path algorithm to compute the current nearest source for each vertex and to select several edges which corresponds to "paths" between possible nearest pairs of connected components; Lines 14–17 of Fig. 8 compute the nearest pair of connected components. Fig. 9 mainly uses a backtracking procedure to compute the corresponding SPC. Make-Set, Union, and Find-Set functions are the operations of disjoint sets in [13].

D. TECHNIQUES OF THE REFINEMENT

In Section 4.1.4, we refine the initial solution by moving some of its PDCs to reduce the wirelength. In order to move

```
Algorithm: Compute_SPC(u_m, v_m, V, E)
Input: (u_m, v_m) /* the edge corresponds to
       the nearest pair of components */
       V /* the set of vertices */
       E /* the set of edges*/
Output: C /* SPR(u_m.sr, v_m.sr) */
1   for each edge (u, v) ∈ E
      where u.sr = u_m.sr and v.sr = v_m.sr
2     if u.wt + v.wt + |(u, v)| =
          u_m.wt + v_m.wt + |(u_m, v_m)|
3       Q ← {u, v}
4       while Q is nonempty
5         u_1 ← Pop(Q)
6         C ← C ∪ {u_1}
7         for each edge (u_1, u_2) ∈ E
8           if u_1.sr = u_2.sr and
                u_2.wt + |(u_1, u_2)| = u_1.wt
9             Q ← Q ∪ {u_2}
10  return C
```

Figure 9: Computation of shortest path component.

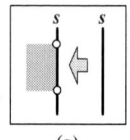 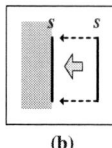 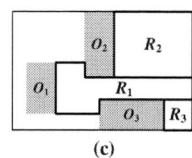

(a)   (b)   (c)

Figure 10: Segment dragging. (a) s touches the obstacle corners. (b) s only touches the obstacle boundary. (c) A planar subdivision for the segment dragging query of (b).

a PDC, it is very important to decide the possible moving offset, and the possible moving offset probably depends on the closest obstacle in the moving direction. Determining the the closest obstacle for a PDC is equivalent to determining the closest obstacles for its segments and its vertices and selecting the closest one. Here, we employ computational geometry techniques to determine the closest obstacle for a segment, and similar for a vertex. For simplicity, we only describe how to compute the *left* closest obstacle for a *vertical* segment, and similar for the other cases.

The left closest obstacle for a segment is exactly the first-touched obstacle by dragging this segment left. As shown in Fig. 10(a)–(b), there are two cases in dragging a segment s to touch an obstacle: (1) s touches at least one obstacle corner, and (2) s only touches the obstacle boundary. As a result, the left closest obstacle for a segment can be determined by the left closest obstacle corner and the left closest obstacle boundary.

For the first case, Chazelle [16] proposed a data structure which can answer the first-touched point for each orthogonal segment dragging query in $O(\log n)$ time after $O(n \log n)$-time preprocessing. Therefore, after preprocessing all the obstacle corners, we can solve the first case in $O(\log n)$ time. For the second case, it is clear that the closest obstacle boundary belongs to the closer one of the first-touched obstacles by shooting a ray from the two endpoints of the segment. As shown in Fig. 10(c), we can partition the plane into

620

a subdivision such that each obstacle O_i is associated with a region R_i and O_i is the first-touched obstacle by shooting a left ray from an arbitrary point in R_i. Such a planar subdivision can be done by a simple plane-sweep algorithm within $O(n \log n)$ time. Moreover, a point location query in a planar subdivision can be answered in $O(\log n)$ time after $O(n \log n)$-time preprocessing [17], implying that the second case can also be solve in $O(\log n)$ time.

To conclude, we can find the closest obstacle for a segment in $O(\log n)$ time after $O(n \log n)$-time preprocessing. For a PDC consisting of i elements, we can determine the possible moving offset in $O(i \log n)$ time.

E. THE TOTAL TIME COMPLEXITY

Theorem 4. Our ML-OARSMT construction takes $O(mn^2 \log^2 n)$ time in the worst-case and $O(n \log^2 n)$ time for practical applications.

PROOF. By Theorem 3 and the discussion in Section 4.1.3, both the ML-VG construction and the MTST construction take $O(n^2 \log^2 n)$ time in the worst case and $O(n \log^2 n)$ time for practical applications. We only analyze the Steiner point selection and the refinement.

We use the Dijkstra shortest path algorithm in [13] to implement our Kruskal-based Steiner point selection. Since the Dijkstra algorithm extracts each vertex from the priority queue at most once, it takes $O(|V| \log |V| + |E|)$ time for a graph $G(V, E)$. However, unlike the Dijkstra algorithm, since we let several vertices be new sources during the selection, an extracted vertex may also be re-inserted into the priority queue and extracted again later. In detail, each time an SPC has been constructed, we will set all the vertices of the SPC as sources, i.e., let their weights be zero, and re-insert them into the priority queue. There are m pin-vertices, so $m - 1$ SPCs will be constructed, implying that a vertex can be extracted at most m times in the worst case. As a result, since the ML-VG has $O(n^2 \log n)$ vertices and edges in the worst case (Theorem 3), the Steiner point Selection takes $O(mn^2 \log^2 n)$ time in the worst case.

Fortunately, for a practical instance, the distribution of pin-vertices and obstacles is more regular, and thus those constructed SPCs will be distributed uniformly. Under these circumstances, a vertex will be extracted only $O(1)$ times on average. Moreover, we don't re-initiate the priority queue after each SPC construction but only re-insert the vertices of the SPC into the queue. As a result, since the ML-VG has $O(n \log n)$ vertices and edges for practical applications, the Steiner point selection takes $O(n \log^2 n)$ time for practical applications.

The refinement moves several PDCs of the initial solution to reduce the wirelength. Since the ML-VG has $O(n^2 \log n)$ vertices and edges in the worst case and $O(n \log n)$ vertices and edges for practical applications, by Definition 4, the total complexity of all PDCs is $O(n^2 \log n)$ in the worst case and $O(n \log n)$ for practical applications. As discussed in Appendix D, if a PDC has i elements, its possible moving offset can be determined in $O(i \log n)$ time after $O(n \log n)$-time preprocessing. Therefore, the refinement takes $O(n^2 \log^2 n)$ time in the worst case and $O(n \log^2 n)$ time for practical applications. Note that during the refinement, two PDCs may be merged into one PDC, and the possible moving offset of the new PDC can be determined from the original two PDCs.

To conclude, our ML-OARSMT algorithm takes $O(mn^2 \log^2 n)$ time in the worst-case and $O(n \log^2 n)$ time for practical applications. $\square$

F. EXTENSIVE EXPERIMENTAL RESULTS

Theorem 3 demonstrates that for practical applications, both the ML-VG and the PD-VG have $O(n \log n)$ vertices and edges, and can be computed in $O(n \log^2 n)$ time. In order to support the claims made in Theorem 3, we use more experimental results to analyze the space complexity of the ML-VG and PD-VG and the time complexity of the ML-VG and PD-VG construction.

We take the following method to make those analyses. For a multi-layer instance I with input size n, let $T(I)$ be the size of a graph or the run time of an algorithm, and let $f(n)$ be a function of n. The relationship between $T(I)$ and $f(n)$ is summarized as follows:

- If $\frac{T(I)}{f(n)}$ seems to increase with n, the complexity of $T(I)$ may be higher than $f(n)$, implying that $T(I)$ seems to be $\omega(f(n))$.

- If $\frac{T(I)}{f(n)}$ seems to decrease with n, the complexity of $T(I)$ may be lower than $f(n)$, implying that $T(I)$ seems to be $o(f(n))$.

- If $\frac{T(I)}{f(n)}$ neither increases nor decreases with n, $\frac{T(I)}{f(n)}$ may depend on the layout of I more than n, implying that $T(I)$ seems to be $\Theta(n)$.

Table 3 lists $|V_S|$, $|V|$, and $|E|$ of the PD-VG and the CPU time of the PD-VG construction. As shown in Table 3, since all $\frac{|V_S|}{n^2}$, $\frac{|V|}{n^2}$, $\frac{|E|}{n^2}$, and $\frac{\text{Time}}{n^2}$ decrease with the input size n, both the space complexity of the PD-VG and the time complexity of the PD-VG construction are subquadratic for practical applications. Moreover, since $\frac{|V_S|}{n}$ neither increases nor decreases with the input size n, $|V_S|$ seems to be $\Theta(n)$ for practical applications. For the same reason, $|V|$, $|E|$, and the time complexity of the PD-VG construction seem to be $\Theta(n \log n)$, $\Theta(n \log n)$, and $\Theta(n \log^2 n)$ for practical applications, respectively, which significantly supports Theorem 3.

Table 4 lists $|V_S|$, $|V|$, and $|E|$ of the ML-VG and the CPU time of the ML-VG construction. The results in Table 4 are similar to those in Table 3, but are not so regular in industrial test cases (ind1–ind5). This is because ind1, ind2, and ind3 have different values of N_l, and both ind4 and ind5 has only one layer for routing, i.e., both ind4 and ind5 are equivalent to a 2D instance. As shown in Table 4, the analyses on the random test cases (rt1–rt5) still show that both the space complexity of the ML-VG and the time complexity of the ML-VG construction are subquadratic and close to loglinear for practical applications, which supports Theorem 3.

In addition, unlike the test cases for the OAPD-ST problem, the benchmark circuits used in [9] for the ML-OARSMT problem have only one sample for each kind of test case. Therefore, in order to obtain better analyses, we also generate more random test cases in which each kind of test case has 100 samples, and those additional experimental results are very similar to those in Table 3.

G. COMMENTARY

Here, we discuss why the algorithms in [9], [10], and [11] behave like a quadratic-time algorithm.

Table 3: Analyses on $|V_S|$, $|V|$, and $|E|$ of the PD-VG and the CPU Time of the PD-VG Construction, where C_v and N_l 6.

Test Cases			PD-VG				$\frac{|V_S|}{n}$	$\frac{10^3|V_S|}{n^2}$	$\frac{|V|}{n\log n}$	$\frac{10^2|V|}{n^2}$	$\frac{|E|}{n\log n}$	$\frac{10|E|}{n^2}$	$\frac{10^6*\text{Time}}{n\log^2 n}$	$\frac{10^7*\text{Time}}{n^2}$
m	k	n	$|V_S|$	$|V|$	$|E|$	Time								
10	50	210	2,853	14,600	24,096	0.018	13.59	64.69	9.01	33.11	14.87	5.46	1.44	4.08
10	100	410	5,835	37,294	61,103	0.045	14.23	34.71	10.48	22.19	17.17	3.63	1.46	2.68
20	100	420	6,159	39,771	64,951	0.046	14.66	34.91	10.87	22.55	17.75	3.68	1.44	2.61
20	200	820	9,320	70,609	115,550	0.077	11.37	13.86	8.90	10.50	14.56	1.72	1.00	1.15
50	250	1,050	12,665	103,386	167,778	0.118	12.06	11.49	9.81	9.38	15.92	1.52	1.12	1.07
50	500	2,050	28,029	268,547	430,862	0.323	13.67	6.67	11.91	6.39	19.10	1.03	1.30	0.77
100	500	2,100	29,566	286,468	459,123	0.349	14.08	6.70	12.36	6.50	19.81	1.04	1.36	0.79
100	1,000	4,100	46,871	515,714	821,879	0.668	11.43	2.79	10.48	3.07	16.70	0.49	1.13	0.40
200	1,000	4,200	50,199	554,546	883,648	0.740	11.95	2.85	10.97	3.14	17.48	0.50	1.22	0.42
200	2,000	8,200	78,562	993,183	1,575,383	1.421	9.58	1.17	9.32	1.48	14.78	0.23	1.03	0.21

Table 4: Analyses on $|V_s|$, $|V|$, and $|E|$ of the ML-VG and the CPU Time of the ML-VG Construction, where C_v .

Test Cases	$m/k/N_l$	n	ML-VG				$\frac{|V_S|}{n}$	$\frac{10^2|V_S|}{n^2}$	$\frac{|V|}{n\log n}$	$\frac{10|V|}{n^2}$	$\frac{|E|}{n\log n}$	$\frac{10|E|}{n^2}$	$\frac{10^6*\text{Time}}{n\log^2 n}$	$\frac{10^7*\text{Time}}{n^2}$
			$|V_S|$	$|V|$	$|E|$	Time								
ind1	50 / 6 / 5	74	397	1384	2886	0.007	5.36	7.25	3.01	2.53	6.28	5.27	2.45	12.78
ind2	200 / 85 / 6	540	8485	23749	52854	0.054	15.71	2.91	4.85	0.81	10.78	1.81	1.21	1.85
ind3	250 / 13 / 10	302	5484	31908	68789	0.055	18.16	6.01	12.82	3.5	27.65	7.54	2.68	6.03
ind4	500 / 100 / 5	900	3298	10418	17384	0.03	3.66	0.41	1.18	0.13	1.97	0.21	0.35	0.37
ind5	1000 / 20 / 5	1080	13218	28877	49995	0.073	12.24	1.13	2.65	0.25	4.59	0.43	0.67	0.63
rt1	25 / 10 / 10	65	871	4844	10515	0.013	13.4	20.62	12.37	11.47	26.86	24.89	5.51	30.77
rt2	100 / 20 / 10	180	3122	30842	66760	0.057	17.34	9.64	22.87	9.52	49.51	20.6	5.64	17.59
rt3	250 / 50 / 10	450	18328	144940	319873	0.265	40.73	9.05	36.54	7.16	80.65	15.8	7.58	13.09
rt4	500 / 50 / 10	700	22875	206379	450542	0.412	32.68	4.67	31.19	4.21	68.1	9.19	6.59	8.41
rt5	1000 / 100 / 5	1400	32843	259594	554775	0.51	23.46	1.68	17.74	1.32	37.92	2.83	3.34	2.60

There are two factors leading the algorithm in [9] to the quadratic-time performance. First, the algorithm uses the *obstacle-avoiding spanning tree* (OAST) construction in [1] to compute a *multi-layer obstacle-avoiding spanning tree* (ML-OAST). However, since the OAST construction in [1] computes a shortest path for each pair of pin-vertices, the ML-OAST construction in [9] takes more than quadratic time. Second, the algorithm adopts a parameter T_n to make the number of vertices be $O(n)$. According to the *Vertex-Projection-between-Layers* algorithm and the *ML-OASG-Construction* algorithm in [9], the number of vertices of their routing graph is actually $O(4^N n)$. Since N_l is a small constant, for an extremely large n, the $O(4^N)$ factor of $O(4^N n)$ can be viewed as a constant, and $O(4^N n)$ is $O(n)$. Nevertheless, for practical applications ($n < 10000$), the $O(4^N)$ factor still affects the practical size of the routing graph, and thus lowers the efficiency of the algorithm in [9].

The algorithm in [10] uses an $O(n^2)$-space routing graph. Since their routing graph is actually an obstacle-avoiding Hanan grid following the PD constraint, the size of their routing graph is still quadratic for practical applications, and thus their algorithm takes more than quadratic time.

The authors in [11] claimed that since the number of backtracking is $O(m)$ in the experiment, their *multi-layer obstacle-avoiding rectilinear full Steiner tree* (ML-OAPDRFST) construction takes $O(m)$ time, and their whole algorithm takes $O(n \log n)$ time in practice. However, their speed-up over [10] does not increase with the input size n, implying their time complexity is similar to that of [10] in practice. Since the algorithm in [10] is a quadratic-time algorithm, their algorithm would still behave like a quadratic-time algorithm. This is because in the presence of obstacles, their backtracking algorithm will construct an obstacle-avoiding shortest path between a new pin-vertex and a point of the current full Steiner tree. Nevertheless, without a routing graph, it takes $\Theta(n \log n)$ time to compute an obstacle-avoiding shortest path between two points. As a result, even if the number of backtracking is $O(m)$, the ML-OAPDRFST construction may still take $O(mn \log n)$ time in the worst case rather than $O(m)$ time, and thus their whole algorithm would take at least quadratic time.

On the other hand, the time complexity of our algorithm depends on the size of the ML-VG and the PD-VG. There do exist some instances in which the size of the ML-VG and the PD-VG is quadratic. However, in practice, the size of the ML-VG and the PD-VG is probably close to loglinear. As a result, our algorithm takes subquadratic time for practical applications.

Avoiding Game Over: Bringing Design to the Next Level

Ofer Shacham

Sameh Galal

Sabarish Sankaranarayanan

Megan Wachs

John Brunhaver

Artem Vassiliev

Mark Horowitz

Andrew Danowitz

Wajahat Qadeer

Stephen Richardson

VLSI Research Group, Stanford University

ABSTRACT

Technology scaling has created a catch-22: technology now can do almost anything we want, but the NRE design costs are so high, that almost no one can afford to use it. Our current situation is reminiscent of the 1980's, when only a few companies could afford to produce custom silicon. Synthesis and placement and routing tools changed this, by providing modular tools with well defined interfaces that codified designer knowledge about the physical design of chips. Now we need a new set of tools that can codify designer knowledge about how to construct software, hardware, and validation to again enable application designers to produce chips. Researchers are developing methodologies that allow users to create hardware constructors, or *generators*. These include Genesis2, which extends SystemVerilog and enables the designer to encode hierarchical system construction procedurally. To demonstrate some of the capabilities that these languages and tools provide, we describe FPGen, a complete floating point generator written in Genesis2, that also generates the needed validation collateral and hints for the backend processes.

Categories and Subject Descriptors

B.6 [**Hardware**]: Logic Design

General Terms

Design, Economics, Languages, Performance

Keywords

HDL, SystemVerilog, Generator, Genesis2, Optimization, Power, Floating Point

1. INTRODUCTION

Digital design is at a crossroads. Technology scaling over the past fifty years has created technologies of almost unimaginable capability. Today we can produce chips with transistor features that are measured in tens of nanometers on silicon dies that are a couple of hundred square millimeters, fabricated on wafers almost half a meter in diameter. This means our chips can contain billions of transistors, yet can be produced at a cost per area that is only marginally higher than chips a decade ago. Thus the cost per function continues to decline, enabling systems that used to sell for hundreds or thousands of dollars to be available today for around a dollar.

Unfortunately, the rising non-recurring engineering (NRE) costs associated with a modern chip design, estimated to be $60-$80 million for a 32nm design [4, 8], only makes sense if you have a billion dollar market. Unless something radical is done, we will again be in a situation where only "chip companies" have the resources to create new silicon, and the high design costs will even limit the diversity of solutions that they can produce. While the performance of these solutions will be impressive, most of the capability of the underlying technology will be lost.

We have been in this situation before. In the 1980s, the extreme cost of custom design limited access to the technology. Back then, the solution was to automate the flow from logic description to a chip physical layout. A set of tools encoded the expertise and knowledge of how to map and optimize the gates primitives, and then place and route these gates to form the final layout. This automation allowed logic designers to create chips. Many people believe the solution to today's ASIC problem is to once again encode even more designer knowledge into our tools and raise the abstraction level.

While many have attempted to do so, no solution has yet been as powerful as the automated synthesize, place and route flows from the 1980s. One reason is that synthesis and place-and-route solve a mapping problem which is very difficult, but context-free: it can be used in all designs. In contrast, higher level synthesis seems to be context sensitive: effective approaches require application specific design knowledge [10]. High level flows that create efficient signal processing applications are unlikely to also create elegant packet processors. Creating one high-level synthesis tool for all seems unlikely.

Recently, a promising research movement has taken a different tack. Rather than creating compilers that incorporate all of the application knowledge directly, it advocates creating tools that allow each designer to incorporate their knowledge about their system in the block they create[1]. This

---

Permission to make digital or hard copies of all or part of this work for personal or classroom use is granted without fee provided that copies are not made or distributed for profit or commercial advantage and that copies bear this notice and the full citation on the first page. To copy otherwise, to republish, to post on servers or to redistribute to lists, requires prior specific permission and/or a fee.

---

[1]Somewhat equivalent to the software "constructor" and

produces "smarter" modules that hide internal implementation details, and provide the user with application level parameters to adjust. Designers can later configure these *generators*, simulate, optimize, and output custom hardware blocks or systems at a much lower cost. Genesis2 [15] is one such tool for creating generators. It extends SystemVerilog, and enables one to hierarchically generate human-readable, synthesizable RTL and its verification collateral, from module template files and a configuration file. Genesis2 has enabled us (and collaborators at other institutions) to create designs with improved power and performance. We demonstrate this through a Floating Point Unit generator which produces designs up to 40% better than the leading libraries.

2. THE EARLY YEARS

Encoding designer knowledge into tools to reduce design complexity and designer error is an old idea that has been used since the 1970's [9]. Early in that decade the first layout and circuit design tools were created. Chip layout, originally the laborious and error prone process of cutting rubylith by hand to produce masks, was automated by digitizing the layout drawings, enabling computer controlled mask generation. This innovation required a primitive abstraction; a machine needed a set of x, y coordinates telling it exactly where each rectangle would begin and end [16]. The early abstraction of x, y coordinates eventually became a sophisticated means of encoding layers, labels and shapes, culminating in an industry standard known as "GDSII."

This interface, also enabled a new set of tools, including design rule checkers (DRC). Checking rules by hand was tedious and error prone; Intel designers, reminiscing about the early days, have stated a belief that *most* of Intel's early chips "shipped with at least some design rule violations [16]." By the mid 1970s, some researchers had embedded knowledge about the rules into a DRC tool that fixed this problem, and by the early 1980's was widely used in IC design houses.

A successful design rule check (DRC) ensures that the layout is electrically sound and ready for fabrication. Simulation proved that the design "works". Now there was a need to validate the layout and the schematic actually describe the same system. It was not long before knowledge about how to extract netlists from layout were encoded into a tool. This enabled the creation of layout-versus-schematic (LVS), which check that the extracted netlist matches the circuit netlist you simulated.

While these tools made it more likely that the chip you fabricated worked when it came back, the design of the circuit and the layout were all done manually. By the end of the 1970s it was clear that design complexity was growing too rapidly for this method to continue. More automation was needed, and many different approaches were applied.

As the level of integration increased, designers increased the level of abstraction once again by creating a class of domain specific computer languages that could formally describe both the *netlist* and the *binary logic* of the electronic circuit—the birth of Hardware Description Languages (HDLs). Early HDLs such as the Instruction Set Processor (ISP) language [2] and TEGAS, were used for *simulating* designs. Verilog, introduced in the mid 1980's, also began as a logic simulation language.

Many groups tried to simplify the layout task by generat-

"factory" approaches.

ing layouts from stick diagrams. A compactor would then use this description to generate the layout. In contrast, others advocated removing the transistor layout task completely and constraining the design to only use predefined logic cells with predefined layout. This group also argued that routing algorithms, already popular for PCB layout tools, could help with the tedious and time consuming task of wiring up cells by hand. Adding placement would take this automation a step further, by automatically laying out the cells in a way that was routable, and, ideally, somewhat efficient. An even more aggressive approach advocated creating silicon compilers: software that uses a high-level design description to generate a customized design from that description. This led to the formation in 1981 of Silicon Compilers, a company started to fulfill this vision [3].

During the early 1980's, another approach to design automation was brewing. It started as automatic netlist sizing and optimization. Then, a few people thought it would be possible to create gate level implementations directly from the logic simulation description. In ways analogous to a normal software compiler, the tools would first map the simulation primitives to the existing standard cells, and then optimize the resulting logic to reduce cost and improve performance. This was called logic synthesis. Synopsys founder, David Gregory reminisces that engineers accepted netlist-to-netlist synthesis first, because it provided a transformation that improved results, and because, once shown the solution, engineers could repeat and/or verify that transformation and thus learn to trust it. That trust opened the door to acceptance of HDL synthesis.

In the end, synthesis, place and route tools were a winning combination, and formed the basis of the ASIC design flow [6, 9]. In hindsight it seems this combination won for a couple of reasons. First, it was restrictive enough that the designer could ignore a large class of design issues: the cells were pre-characterized, pre-validated, and came with layout. Approaches that only tried to simplify the layout step did not go far enough and left designers working at the transistor level, requiring additional expertise. Second, the successful flow consisted of a number of independent steps, done by different tools, each with clean interfaces and understandable transformation. This meant that one company did not need all the design knowledge to make an ASIC. Companies with expertise in one area, such as cell layout, could still push the flow forward because end users could use these cell libraries with their own placers and routers. In addition, since the interfaces are exposed, designers could use the ASIC flow for specific portions of the chip if they preferred to map, place, and route some parts of the circuit by hand.

3. GROWING COMPLEXITY

The ASIC flow allowed small companies to build chips. They no longer needed silicon expertise, or a fab; thus chip design was no longer the sole purview of chip manufacturers. This enabled a new paradigm of *fabless semiconductor companies*, and led to a booming number of logic design starts in the 90s [12]. However, as the amount of logic that could be placed on a silicon chip continued to increase with Moore's law, the complexity of designs grew constantly, concentrating the real cost of building a modern chip not in the backend processes (mask, etc.) but rather in accurately describing and validating the logic.

The design and verification cost for a state-of-the-art ASIC today is reported to be well over \$30M, and the total non-recurring engineering (NRE) costs are more than twice that, due to the custom software required for these custom chips [4, 8]. As a result, the number of ASICs being produced is actually decreasing [7, 12].

As circuit complexity continued to grow, one of the first stress points was validation. Companies started building verification environments, principally using C functions linked into simulators, that required increasing levels of expertise to use. Once again it made sense to incorporate domain specific best practices into tools for use by the industry at large, to ease the job of building complex environments. In particular two new languages were invented: the 'e' language from Verisity (now Cadence) and the Vera language from System Science (now Synopsys). These languages simplified key verification tasks like automatic constrained random test generation and coverage to track how much of the design gets exercised. Unlike synthesis that in essence restricted the HDL language to its synthesizable subset, because verification environments became extensive software engineering projects, Vera and 'e' actually extended Verilog and VHDL with high-level programming features to allow modern software engineering practices, such as object-oriented programming. A third practice of inline assertions was also handled by a newly developed language, Property Specification Language (PSL), which mathematically describes assertions and removes the need for hundreds of lines of error prone assertion code in the RTL. In the early-mid 2000s, all these verification techniques were integrated with Verilog as SystemVerilog. These tools made it successively easier for any designer to construct an expert-grade validation environment, at higher and higher levels of the design.

Today, the coupling of increased complexity and diminishing power and performance returns from scaling makes creating power efficient systems ever more difficult. Like with the design bottlenecks of the early 1980's, many approaches are vying to solve this problem.

One approach has been to build circuits out of pre-made IP blocks, which can be assembled using a System-on-Chip (SoC) methodology. The designer now only needs to verify the interactions of the whole system. In order to make it easier to combine these blocks, they had to have well-defined and understood interfaces. The development of these standardized interfaces led to platform-based design [13].

An orthogonal effort aims to embed designer knowledge at a higher level by raising the level of language abstraction that describes these IP blocks. While RTL descriptions represent a near-structural description of the hardware that is ultimately built, it may not be an intuitive way for DSP application developers to describe their complex WiFi algorithms. For these cases, SystemC was developed as a set of C++ classes and macros that provide an event-driven simulation kernel. Then, high level synthesis (HLS) tools were developed to automatically tranlsate algorithms from SystemC into Verilog [14], or even directly to FPGA-specific mappings [17]. One example of a language highly tuned to its particular domain, the CMU Spiral project, automates hardware development and optimization for digital signal processing (DSP) algorithms [5].

These two attempts to raise the level of abstraction take two different approaches to embed designer knowledge. The pre-verified IP block contains a tremendous amount of application knowledge used to create it, but that knowledge is not encoded in a particularly malleable or extensible form. The knowledge in the HLS tools is much more accessible, since it is used to translate code into hardware, but it cannot possibly cover all the specific domain knowledge that one might find in different IP blocks. The next section provides an intriguing method designed to address this issue.

4. A NEW AGE

Comparing our current situation to the early 1980's, using IP blocks and current high-level synthesis tools make the design task easier, but to build a chip using them still requires the users to have expert designer knowledge. They do not enable a higher-level design framework. These tools would be analogous to stick diagrams and compactors in the 1980's. One could argue that it should be possible to create high-level synthesis that does raise the abstraction level by incorporating a large amount of domain expertise, but creating such a system would be analogous to creating a silicon compiler back in the 1980's: the effort needed to create a useful tool would be enormous, and would need to be done before getting any real users. For example, the necessary design knowledge for creating an image processor is very different from understanding the trade-offs for a network switch. We need a way to capture domain knowledge incrementally, by creating useful domain specific tools, with the hope of eventually being able to aggregate these tools to allow design at a higher level.

One promising approach borrows from SoC methodology the idea of allowing the designer to embed their knowledge into the design, but allows them to incorporate multiple architectural choices, essentially creating templates for the design. At the same time, it retains the transformation capability of the high-level compiler, converting the user's high-level description of what they want into a hardware instance. Using software jargon, the "tool" designer would create a constructor that can then produce the different designs for the user. We call this hardware-oriented constructor a *generator*. Furthermore, like an HLS compiler, the generator must encapsulate as much of the designer knowledge as possible, even including knowledge about placement, validation and software.

The first problem encountered when trying to implement this approach is the limitations of current hardware description languages. Since the introduction of *generate* blocks in Verilog 2001, IP blocks have become somewhat parameterized, meaning that one IP block can cover a wider range of circuit implementations. Unfortunately, parameters are calcified once IP blocks are instantiated with a certain choice for values. Similarly, more often than not, the dependencies between IP blocks, and between their parameter choices are calcified at instantiation. This does not allow the flexibility required for an optimization round as once the system is assembled there are no lower level optimization knobs. History has shown that the ability to optimize through known but semi-hidden knobs[2], was crucial to every successful raise of the abstraction level (e.g., cell layout, gate sizes, Boolean equivalence of circuits).

Even more worrisome, HDLs, historically and by definition, were created to describe hardware, not to construct

---

[2]"Semi-" because when crucial, a user could always cut through the abstraction and instantiate a particular library cell at the HDL level, somewhat like adding assembly code directives in C/C++.

it. With the increasing design costs, there is a tendency to use the same base design to make chips for different market segments or to fabricate on a newer and smaller technology by adding more capabilities as allowed by the technology. Variants of the base design can differ from each other in the number of cores, number and/or type of memory interfaces, cache sizes, etc. Even established "big-iron" companies producing CPUs or GPUs use this technique to amortize the cost of design. However, many of these variants are created by manual code branching and human translations (and/or ad-hoc scripts), a tedious approach. The reason is that we still *describe IP and system instances,* rather than *encoding how instances should be created* given a set of constraints.

Among the first to tackle the constructor problem, Professor Arvind's group from MIT created *Bluespec* in the early 2000's [11] as a new HDL with higher level software constructs. Leveraging the power of a software (functional) programming language, Bluespec allows its users to manipulate fundamental building blocks using conditionals, loops and recursion. It also allows for flexible parameterization and use of polymorphism to enable the generation of multiple different flavors of the hardware without having to rewrite code each time. In Bluespec, a module is in fact a constructing function that returns (or "provides" in Bluespec jargon) an interface.

While Bluespec's atomic rules behavioral synthesis method has its pros and cons, through the use of Bluespec's constructs, we began to understand the need for enabling a designer to write procedural code that describes how the circuit should be built and integrated into the system, and how to resolve inter-block dependences. This enables the designer to create IP blocks that cover a much wider range of implementations and embed his/her understanding of how the system works into the system that they build.

Another clean-slate HDL, UC Berkeley's *Chisel* emphasizes the constructing aspect even further [1]. Chisel uses a Scala embedded language to design hardware but does not claim to do behavioral synthesis, just enable easier construction of the system. Both Bluespec and Chisel introduce a new design language, completely discarding legacy HDLs and associated design/validation methodologies.

Sometime after the introduction of Bluespec and before the introduction of Chisel, our group at Stanford developed the idea of using hardware constructors to encapsulate designer knowledge about a particular design problem. We call these constructors *generators* because, given a set of parameters and constraints, these executable design templates generate the desired module. Unlike Bluespec and Chisel, we did not want to give up SystemVerilog, since we wanted to leverage its tools, in particular its validation support described in Section 3. Yet SystemVerilog lacked the necessary capabilities for the elaboration flexibility that one needed to really build module- or system-generators. To overcome these limitations, we gradually developed scripts that extended, and eventually replaced, the limitation of SystemVerilog parameters and *generate* blocks, as well as enable system introspection during the construction phase.

This suite of scripts became Genesis2, a prototype tool for what SystemVerilog could become. It built on top of SystemVerilog without modifying its formal syntax. The introduction of Genesis2 achieved two goals: First, it showed that it is possible, and even quite simple, to incorporate modern constructs, concepts and methodologies prevalent in the software world for two decades into hardware design. Second, it demonstrates the usefulness of this extra functionality, including how it can facilitate the construction of highly flexible designs, and how that flexibility can be preserved as the system is hierarchically built, to enable later optimization rounds.

Genesis2 provides hardware designers with a rich software language for writing instructions that specify how to generate modules from a set of input parameters. These instructions can be seen as an explicit "elaboration program" or as an object oriented constructor for generating elaborated instances. The behavioral description, however, remains in SystemVerilog. Granted, in software coding, the semantics of coding a constructor for a class is the same as the semantics of coding the rest of that class's functionality. In contrast, the description of the functionality of a hardware module must obey strict rules of synthesizability. This historically resulted in HDLs that enforced strict rules on the construction of the system, also known as its *elaboration*. With Genesis2, we remove this artificial limitation, by allowing the designer to code in two languages simultaneously and interleaved: one that describes the hardware proper (SystemVerilog), and one that decides what hardware to use for a given instance (Perl). However, Genesis2 maintains the notion of modules, hierarchy and system, by forcing the two language layers to share the same scopes.

Genesis2 builds upon the simple realization that during elaboration time, any software construct should be allowed and every computation that can lead to forming the hardware should be possible: Hardware can be generated by software evaluation of procedural code. This simple yet powerful insight is the reason that almost all modern designs already use some kind of code generator as part of the flow, be it the simple Verilog preprocessor, or complex scripts in a higher level software language such as Perl, Python, or C++. A common design construct that often relies on code generation is the register file, whose details these days might be generated from an Excel spreadsheet. Following these same generator concepts, Genesis2 simply takes it to the next level by incorporating the generators and scripts into a single hierarchical system, such that a register file generator can accept the list of registers as a parameter, and a bus controller generator can accept a pointer to its connected slaves to query their parameters. Thus all parameters can be late-bound by optimization and the system is still generated coherently.

To overcome some of the limitations in SystemVerilog's parameter use, Genesis2 refines the way parameters are defined and bound. First, and much like SystemVerilog, Genesis2 enables designers to define and give default values to parameters. Then, it provides a simple mechanism for overwriting these values from external configuration files. This simple mechanism on its own already enables an optimization process to pick the best binding for internal parameters of an IP without ever exposing them to the next level IP integrator. An example might be the aspect ratio of a block RAM, or the number of virtual queues in a network switch.

However, some parameters should be exposed, such as bus widths, or parameters that alter the functionality of the IP block. Therefore Genesis2 allows parameters to be assigned during instantiation (again, much like SystemVerilog does), and it puts this assignment at a higher priority than both the local definition and the external input. We further add one more level of parameters, a declaration or export layer,

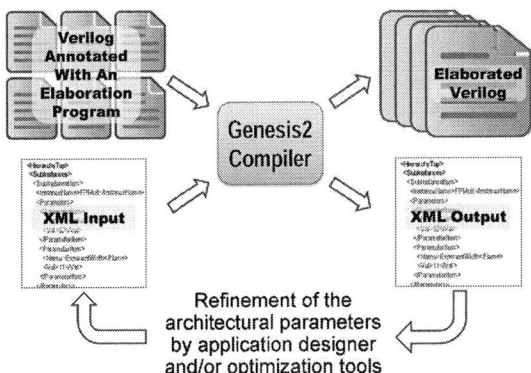

Figure 1: Iterative Generation And Configuration. Genesis2 processes elaboration programs along with an input configuration file. Genesis2 then produces all the Verilog files, along with an output configuration file that represents the newly generated design. The output configuration file can then be further refined and fed back as input.

to allow a designer of an IP block to declare a value to the world, in case some other component will need that information later. This declaration receives the highest priority and cannot be overwritten. Put together, we redefined the priorities of parameter assignments as follows:

1. Parameters can be declared and defined in the module to which they belong.

2. Parameter values which were defined using method 1 can be overruled by external input (using XML format as explained later).

3. Parameter values which were defined using methods 1 or 2 can be overruled at the instantiation.

4. Parameters can alternatively be declared and defined immutably in the module to which they belong.

Because we consider parameters key to enabling the efficient coding of flexible modules, Genesis2 allows parameters to be of complex types, from strings and scalars to structures of arrays and hashes. It also adds a documentation and range field to the declaration of parameters to encourage designers to use them intelligently. Finally, every time the Genesis2 compiler runs, it not only generates code, it also extracts the entire parameterization space, hierarchically, into an XML-formatted description file. Using this XML file, other tools (e.g. the Genesis2 GUI, optimizers, software code generators, etc.) can easily read out the machine configuration. This formal representation of design choices facilitates, for example, the creation of header files for register banks and drivers for the various IP blocks. Genesis2 also uses the same XML schema as hierarchical input for all parameters. Thus by iteratively making changes to the XML configuration file, one can influence the next round of code generation, as shown in Figure 1.

5. EXAMPLE

To continue to refine Genesis2, and to better understand the strengths and weaknesses of this approach, we have used

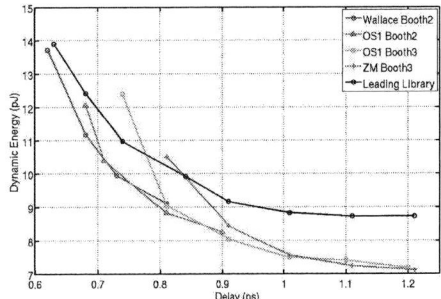

Figure 2: Microarchitecture sweep of double precision multiplier. We were genuinely surprised, as we expected Wallace Booth2 to dominate. It turns out however, that for the low power points Booth3 arrayed type designs perform better.

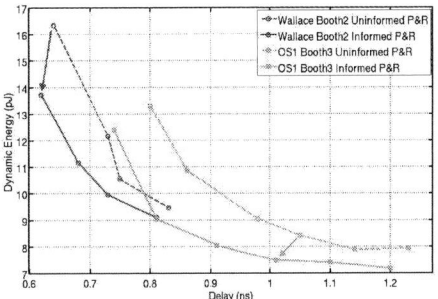

Figure 3: Advantages of encoding layout hints into a double precision multiplier generator. "Informed" layout acheives 10%-30% better in power and performance.

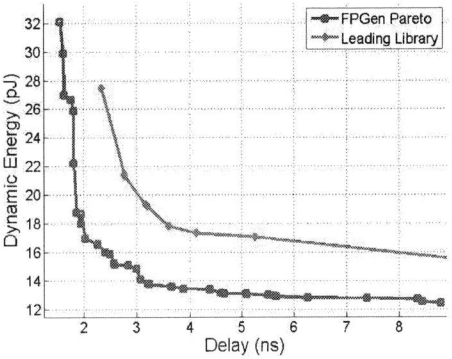

Figure 4: FPGen generated designs for complete double precision FMA vs. a leading library. FPGen achieves 20% to 40% less energy for the same latency, or 30% to 60% less delay for the same energy budget.

it for most of the design work in our group for the past two years. Some of the design projects from our group and other early adopters include small mixed-signal chips; IP blocks for DSP Radar applications [18]; a complete heterogeneous chip multiprocessor system; and an implementation back-end to a domain-specific language for describing memory coherency and consistency protocols. While not perfect, we have been pleased with its performance. For example, we created a chip interface generator for mixed signal test chips that generates all the needed JTAG pads, controller, and register files, to support, control and debug analog components.

To further demonstrate the ability of a generator to encapsulate designer knowledge, this section will describe some of the features and implementation issues of a floating point unit (FPU) generator we have constructed. The goal was for the FPU generator, made up of FPAdd, FPMult, and/or FPMAdd, to handle any type of design, be it throughput oriented for GPU design or latency oriented for CPU design. The FPU should be able to support full IEEE compliant rounding modes if desired and be able to handle any resolution: half, single, double, quad or custom precision.

Creating an FPU is usually a difficult task, due to the extensive work on adders, multipliers, and FP algorithms. Just looking at the multiplier research, there are many different methods of encoding the partial products (Booth 2, 3 and redundant), reducing the partial-products (Wallace, Dadda, overturned staircase, ZM trees) and even the specific interconnection between the underlying carry save adders. Adders have a similarly rich design history. There are many carry tree formulations (Sklansky, Kogge Stone, ripple carry adders) as well as specific tricks for reduced FP adder latency (divided mantissa datapath based on exponent difference). For even an experienced designer, it is hard to know which of these techniques "really work" unless they have built FPUs before.

To address this problem, we created a constructor that is able to use many of these design variants (and their cross products) to create working FPAdd, FPMult, or FPMAdd units using Genesis2. We then combine this constructor with general optimization scripts that explore the parameter space so we can choose the lowest energy implementation for a given performance specification. Furthermore, since we were able to encode the design procedurally, it was both easier and cleaner to code. For example, in a Wallace tree reduction algorithm, each step of the reduction depends on how many outputs the previous level of the tree generated.

The results for a double precision multiplier are shown in Figure 2, and indicate the advantage of embedding the different options into the generator. For example, while Wallace/Booth-2 is the canonic multiplier design, Booth-3 encoding tends to save energy within the lower performance range, as does using a hybrid tree/array partial product reduction network like overturned stairs – the best design changes with performance. With the knowledge we have encoded into the tool, it can find the best structure, and the result is much better than an industry leading library. This type of advantage is replicated for FPUnits of different bit widths, although the "best" configuration changes with size.

An advantage of using a generator is that one can embed many different types of information. For example, while current placement tools are good for modest sized blocks, as the blocks get larger, adding placement "hints" can improve performance. Our generator takes advantage of this knob as

well, as Figure 3 shows. Here instead of showing curves from all multiplier organizations, it shows the performance curve for just two, with and without the placement hints for how to layout the multiplier. In this example adding placement hints improves energy by 10%-30%.

The true advantage of creating generators is that it allows one to use them to create generators for even more complex blocks. We created the higher level FPU generator by embedding floating point unit design knowledge into a tool, hierarchically calling the adder and multiplier generators to create the needed adds and multiplies. At this level we did not need to worry about all the design options for these units, since we knew we could count on the generators to create the best implementation for our needs. Figure 4 shows the power of this approach: one can give the designers the high-level flexibility they want, and still generate implementations that are state-of-the-art.

6. CONCLUSIONS

The continued scaling of technology has enabled us to manufacture systems of enormous complexity for very low cost. The dark side of this scaling is the cost of designing these complex systems. To address this cost, we need to raise the level of abstraction; the dilemma is how to accomplish this goal. As in the early 1980's, there are many possible approaches. We believe having designers create generators and not instances, therefore incorporating their knowledge of a design problem into the parts they build, is a promising approach. Each generator development has immediate incremental value, and can be done in parallel. Like synthesis, place and route in the 1980's, it enables a whole community to work on the problem, and has an incremental insertion strategy. Once enough of these generators are in place, the interface to chip design will be at the level of setting the parameters of these blocks, raising the chip design interface enough to once again allow application experts to create ASICs. Tools like Genesis2 and its cousins, which enable designers to create generators, are a critical first step in enabling this vision.

7. REFERENCES

[1] J. Bachrach, H. Vo, B. Richards, K. Asanovic, and J. Wawrzynek. Chisel: Constructing hardware in a Scala embedded language. In Proceedings of the 49th Design Automation Conference (DAC), 2012.

[2] M. R. Barbacci. Instruction set processor specifications (ISPS): the notation and its applications. IEEE Trans. Comput., 30(1):24–40, January 1981.

[3] E. Cheng. Silicon compilation at 21 years young. Berkeley Wireless Research Center Seminar, 2002.

[4] R. E. Collett. Executive session: How to address today's growing system complexity. DATE '10: Conference on Design, Automation and Test in Europe, March 2010.

[5] P. D'Alberto, P. A. Milder, A. Sandryhaila, F. Franchetti, J. C. Hoe, J. M. F. Moura, M. Püschel, and J. Johnson. Generating FPGA accelerated DFT libraries. In IEEE Symposium on Field-Programmable Custom Computing Machines (FCCM), pages 173–184, 2007.

[6] D. Gregory, K. Bartlett, A. deGeus, and G. Hachtel. SOCRATES: A system for automatically synthesizing and optimizing combinational logic. In Papers on Twenty-five years of electronic design automation, pages 580–586, New York, NY, USA, 1988. ACM.

[7] D. Grose. Keynote: From Contract to Collaboration Delivering a New Approach to Foundry. DAC '10: Design Automation Conference, June 2010.

[8] M. Keating. Third revolution: The search for scalable code-based design. http://www.synopsys.com/ apps/community/ university/video/third$_r$evolution.html.

[9] D. MacMillen, R. Camposano, D. Hill, and T. Williams. An industrial view of electronic design automation. Computer-Aided Design of Integrated Circuits and Systems, IEEE Transactions on, 19(12):1428–1448, December 2000.

[10] G. Martin and G. Smith. High-level synthesis: Past, present, and future. IEEE Des. Test, 2009.

[11] R. Nikhil. Bluespec system verilog: efficient, correct rtl from high level specifications. In Formal Methods and Models for Co-Design, 2004. MEMOCODE '04. Proceedings. Second ACM and IEEE International Conference on, pages 69 – 70, june 2004.

[12] W. C. Rhines. Keynote: World Semiconductor Dynamics: Myth vs. Reality. Semicon West '09, July 2009.

[13] A. Sangiovanni-Vincentelli. Defining platform-based design. EEDesign of EETimes, February 2002.

[14] J. Sanguinetti. A different view: Hardware synthesis from systemc is a maturing technology. IEEE Design and Test of Computers, 23:387, 2006.

[15] O. Shacham. Chip Multiprocessor Generator: Automatic Generation of Custom and Heterogeneous Compute Platforms. PhD thesis.

[16] A. M. Volk, P. A. Stoll, and P. Metrovich. Recollections of Early Chip Development at Intel. Intel Technology Journal, Q1 2001.

[17] Z. Zhang, Y. Fan, W. Jiang, G. Han, C. Yang, and J. Cong. AutoPilot: A Platform-Based ESL Synthesis System. In High-Level Synthesis: From Algorithm to Digital Circuit, chapter 6, pages 99–112. Springer, 2008.

[18] Q. Zhu, E. L. Turner, C. R. Berger, L. Pileggi, and F. Franchetti. Polar format synthetic aperture radar in energy efficient application-specific logic-in-memory. In Proceedings of International Conference on Acoustics, Speech, and Signal Processing (ICASSP), 2012.

PowerField: A Transient Temperature-to-Power Technique based on Markov Random Field Theory

Seungwook Paek[1]
swpaek@mvlsi.kaist.ac.kr

Seok-Hwan Moon[2]
shmoon@etri.re.kr

Wongyu Shin[1]
wgshin@mvlsi.kaist.ac.kr

Jaehyeong Sim[1]
jhsim@mvlsi.kaist.ac.kr

Lee-Sup Kim[1]
lskim@ee.kaist.ac.kr

[1]Department of Electrical Engineering
KAIST
Daejeon, 305-701, Korea

[2]Convergence Components & Materials Research Lab.
ETRI
Daejeon, 305-700, Korea

ABSTRACT

Transient temperature-to-power conversion is as important as steady-state analysis since power distributions tend to change dynamically. In this work, we propose PowerField framework to find the most probable power distribution from consecutive thermal images. Since the transient analysis is vulnerable to spatio-temporal thermal noise, we adopted a maximum-a-posteriori Markov random field framework to enhance the noise immunity. The most probable power map is obtained by minimizing the energy function which is calculated using an approximated transient thermal equation. Experimental results with a thermal simulator shows that PowerField outperforms the previous method in transient analysis reducing the error by half on average. We also applied our method to a real silicon achieving 90.7% accuracy.

Categories and Subject Descriptors: C.4 [**Performance of Systems**]Measurement Techniques

General Terms: Algorithms, Verification, Experimentation, Measurement

Keywords: Power, thermal imaging, post-silicon verification, Markov random field

1. INTRODUCTION

Low power consumption and thermal safety are becoming major design considerations [1, 2]. For these goals, designers should estimate the power distribution and thermal behavior at design time in an acceptable accuracy. Those physical phenomina are getting hard to be predicted as fabrication technology evolves. To resolve the problem of these uncertainties, the power model has to be validated with post-silicon power verification. However, the direct post-silicon power measurement is quite expensive since it requires a number of on-chip measurement circuitries inside a chip[3]. In practice, it is infeasible to measure the dense power map

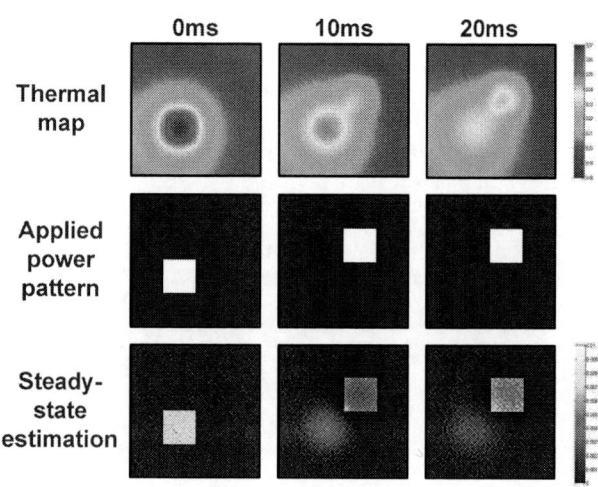

Figure 1: Motivating example: results of a steady-state temperature-to-power technique on transient thermal images.

directly.

This is the motivation of an indirect power measurement using thermal imaging devices. Since power dissipation directly affects thermal map, it is reasonable to find an inverse transformation from temperature to power. Several approaches were proposed to extract the power map from thermal images. The first work is refered to as *Spatially-resolved Imaging of Microprocessor Power* (SIMP)[4]. In this technique, the relationship between the steady-state temperature map and the power map is described as the following equation:

$$A \cdot P = T \qquad (1)$$

where T, P and A are temperature, power and thermal resistance matrices respectively. Each entry of the matrix T and P represents a unit silicon segment which is called as a *thermal node*. SIMP method also proposed a well-organized measurement setup using a scanning laser beam to obtain the matrix A. Once we get A and T, P is obtained by solving (1) using a linear least-square technique[4]. Based on SIMP method, several works had brought novel improvements by adopting regularization theory[5], constrained Levenberg-

Marquardt(LM) algorithm [6] and AC thermography technique[7].

However, the previous works have a limitation in common that all the thermal images have to be steady-state. Indeed, these techniques cannot capture the correct power map if the power pattern changes in runtime. For example, many microprocessors support chip-level dynamic power and thermal management for low-power consumption and thermally safe operation[3, 8]. As the trend of runtime power optimization in chip's operation grows, a need for correct verification of temporal power changes also increases. Thus, to be more useful, a temperature-to-power converter should support not only steady-state but also transient analysis. While Martinez et al. proposed a transient power analysis of a microprocessor by using genetic algorithm[9], the resolution was limited to a functional block level.

Fig. 1 shows a stark difference between a steady-state estimation and the actual power map when transient thermal images are provided. In this example, a significant residual image of power hinders us from recognizing the actual power pattern. For many designs with more complex time-varying power patterns, the steady-state technique is not likely to produce the correct power distribution. Obviously, the fundamental reason of this misestimation is that the intermediate thermal images are assumed to be steady-state.

Motivated by this observation, we aim to design a new framework based on a transient heat transfer equation. In order to cope with high noise sensitivity of transient analysis[6], we borrowed maximum-a-posteriori Markov random field (MAP-MRF) framework[10] including physical modeling and optimization method from computer vision which is already proven to be robust to spatio-temporal noise in many inverse estimation applications such as stereo matching[11] and video denoising[12]. Using this framework, we find the most *probable* solution by regarding the power map as a two-dimensional field of random variables. When we construct a random field of power, two intrinsic properties of planar thermal systems are considered - the temperature distribution calculated by the power estimation has to be smooth in spatial domain and the temperature change of a thermal node is affected only by the node itself and the neighboring nodes.

Taking these into account, we propose a new framework for transient temperature-to-power conversion referred to as *PowerField*. Given two consecutive thermal images with time interval Δt, we calculate an energy function which measures how the current power estimation is *not* probable. The energy function reflects the abovementioned thermal properties. Then the PowerField algorithm finds the most probable power distribution by minimizing the energy function using a global optimization algorithm. In this work, we used graph-cuts optimizer[13] which efficiently minimizes the MRF-based energy function.

We demonstrate the performance of PowerField in two different ways. The first experiment is done with HotSpot thermal simulator[14]. In this experiment, we show that PowerField outperforms previous methods in transient analysis. Then we move on to an analysis of an FPGA chip with controlled power generation patterns in a similar way to the previous works[5, 7].

The major contributions of this paper are summarized as follows.

- A new formulation of transient temperature-to-power

conversion problem based on MAP-MRF framework (Section 2,3).

- An approximate transient heat transfer formula which greatly simplifies the problem by exploiting the thermal measurement data(Section 3.3).

- Demonstration of the effectiveness of PowerField, an implementation of proposed framework, in transient analysis using thermal simulator and real world measurement data (Section 4).

The remainder of this paper is organized as follows. In Section 2, we introduce MAP-MRF framework and show how it is used to solve the temperature-to-power inversion problem. Section 3 describes the problem formulation of transient analysis by defining the energy function and the objective. In addition, we derive an approximate transient heat equation to efficiently compute the energy function. Then we show the evaluation results in Section 4. Finally, we conclude and discuss the future work in Section 5.

2. BACKGROUND

MRF is a probability theory mostly used for analyzing the spatio-temporal dependencies of physical phenomena. Many image analysis and interpretation problems can be posed as *labeling problems* in which the solution is a set of labels assigned to image pixels. In temperature-to-power problem, the power value of each thermal node corresponds to a label. A labeling problem is specified in terms of a set of sites and a set of labels. A site can be a pixel or a set of pixels depending on the formulation of the thermal node network. For spatial representation, a set of site S and a neighborhood system N are defined:

$$S = \{1, ..., m\}, N = \{N_i | \forall i \in S\} \qquad (2)$$

in which $1, ..., m$ are indices. Let L be a set of labels which can have one of M discrete values:

$$L = \{0, ..., M - 1\} \qquad (3)$$

Then we can define a family of random variables:

$$F = \{F_1, ..., F_m\} \qquad (4)$$

on the set S. In F, each random variable F_i takes a value f_i in L. $f = \{f_1, ..., f_m\}$ is called a labeling. The family F becomes an MRF on S with respect to N if and only if the following two conditions are satisfied:

$$P(f) > 0, \forall f \in F \qquad (positivity) \qquad (5)$$

$$P(f_i | f_{S-\{i\}}) = P(f_i | f_{N_i}) \qquad (Markovianity) \qquad (6)$$

The Markovianity depicts the local characteristics of F, i.e. for each site, only neighboring sites have direct interactions with each other. With an aproximation that the heat flows through the neighboring thermal nodes for a small Δt, the power distribution over the thermal node network can be modeled as an MRF. Once the thermal system is modeled into MRF, what we want is to find the labeling f^* which maximizes the posterior probability $P(f|d)$ where d is the observation. This estimation technique is called a MAP-MRF framework. According to the MAP-MRF theory [10], $P(f|d)$ is represented as follows:

$$P(f|d) \propto e^{-U(f|d)} \qquad (7)$$

where

$$U(f|d) = U(d|f) + U(f) \qquad (8)$$

$$= \sum_{p \in S} D_p(d_p, f_p) + \lambda \sum_{\{p,q\} \in N} V_{p,q}(f_p, f_q) \qquad (9)$$

is the *posterior energy* in which λ controls the ratio of two energy components. In (9), $D_p(d_p, f_p)$ measures how the label f_p is *unlikely* to produce the observation d_p and $V_{p,q}(f_p, f_q)$ denotes how f_p disagrees with the intrinsic characteristics of the random field such as smoothness of the temperature distribution. $D_p(d_p, f_p)$ and $V_{p,q}(f_p, f_q)$ are also known as a data energy and a smoothness energy respectively. (9) implies that maximizing the posterior probability $P(f|d)$ is equivalent to minimizing the posterior energy $U(f|d)$. Thus, we can apply energy minimization algorithms to $U(f|d)$ to find the optimal labeling f^*.

3. PROBLEM FORMULATION

In this section, we formulate the problem by defining each energy term $D_p(d_p, f_p)$ and $V_{p,q}(f_p, f_q)$ in (9). The data energy and the smoothness energy jointly contribute to the final solution by considering error reduction and thermal smoothness at the same time. The objective of the problem is to find the power configuration (labeling) f^* that minimizes the posterior energy $U(f|d)$.

3.1 Data Energy

As introduced in Section 2, the data energy depicts the unlikeliness of the current labeling when a set of observation is given. For each thermal node p, our data energy is defined as an absolute difference between the measured temperature and the calculated temperature at $t = \Delta t$ as follows:

$$D_p(d_p, f_p) = |T_{p\Delta t}(d_p) - T_p(\Delta t, f_p)| \qquad (10)$$

where $T_{p\Delta t}(d_p)$, $T_p(\Delta t, f_p)$ are the observed and the calculated temperatures respectively.

3.2 Smoothness Energy

Minimizing only the data energy may not lead to the optimal solution since the measurement data contains considerable noise. In order to cope with this problem, a smoothness energy is introduced which drives the solution to satisfy our expectation. Even if the algorithm gets non-smooth thermal images, it is expected that the calculated temperature distribution with the estimated power has to be smooth which is the prior knowledge of the thermal system. Thus, the smoothness energy is defined to be proportional to a difference between the calculated temperature of a thermal node and its neighbors at $t = \Delta t$:

$$V_{p,q}(f_p, f_q) = |T_p(\Delta t, f_p) - T_q(\Delta t, f_q)|^\alpha \qquad (11)$$

where α controls the shape of the smoothness function. Since (11) is defined in a pairwise manner, it is computed for every pair of thermal nodes in the neighborhood system N.

3.3 Approximate Heat Transfer Formula

In this section, we describe a thermal modeling to compute the temperature of a thermal node p at $t = \Delta t$ which is an essential part of energy computation. Specifically, the objective of this section is to derive a formula of T_p as a function of f_p and t given the observed temperatures of every thermal nodes at time 0 and Δt. As shown in Fig. 2,

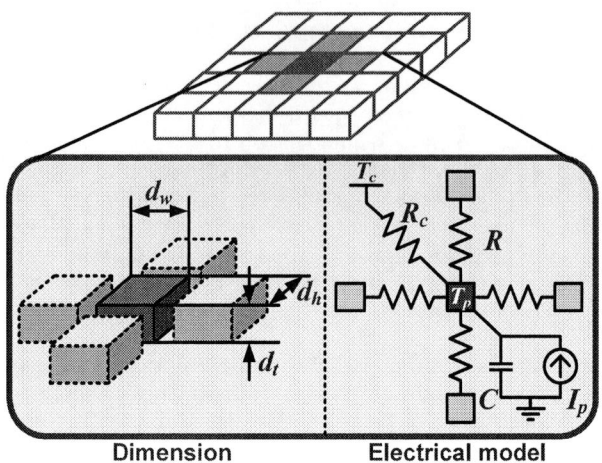

Figure 2: A thermal model used in this work.

our thermal model is based on a well-known lumped RC network model. Each pair of the die segments is connected by thermal resistors R. Each thermal node has its own power source and thermal capacitor which are denoted by I_p and C respectively. The secondary heat dissipation paths such as air and C4 pads are merged into a single thermal node with R_c having constant temperature T_c during the time interval Δt. This approximation is reasonable since they have relatively high heat capacity and the Δt is very short (typically milisecond order). We assume that the thermal characteristics are uniform over the entire die, so every thermal resistance and capacitance have the same values of R_c, R and C as follows:

$$R_c = \frac{1}{k}\frac{d_t}{d_w d_h}, R = 0.5\frac{1}{k}\frac{d_w}{d_h d_t}, C = 0.333 c d_w d_h d_t \qquad (12)$$

where k and c denote the thermal conductance and the specific heat capacity of silicon respectively. The scaling factors of 0.5 and 0.333 in R and C are borrowed from the HotSpot thermal model[14]. Once the thermal behavior is modeled in a lumped RC network, we can apply various circuit analysis techniques to it. Based on Kirchhoff's current law, a heat equation is described as:

$$\sum_{n \in N_p} \left(\frac{T_n - T_p}{R} \right) + \frac{T_c - T_p}{R_c} + I_p - C\frac{dT_p}{dt} = 0 \qquad (13)$$

where N_p and T_n denote the set of thermal node p's neighbors and the neighbor node n's temperature. Then we can rewrite this equation to construct a first-order linear equation of T and t as:

$$\frac{dT_p}{dt} + \frac{1}{C}\left(\frac{4}{R} + \frac{1}{R_c}\right)T_p = \frac{1}{C}\left\{ \sum_{n \in N_p}\left(\frac{T_n}{R}\right) + \frac{T_c}{R_c} + \hat{I}f_p \right\} \qquad (14)$$

Note that I_p is replaced by $\hat{I}f_p$ where $\hat{I}$ denotes the power per unit label. Directly solving (14) induces heavy computational cost because T_n is also a function of T_p. Since we know the observed temperature of the neighbor nodes, we can approximate the temperature of neighbor nodes as a linear function of t:

$$T_n(t) = T_{n0} + \frac{T_{n\Delta t} - T_{n0}}{\Delta t}t \qquad (15)$$

632

where T_{n0} and $T_{n\Delta t}$ are the measured temperature of neighbor node n at $t = 0$ and Δt respectively. This approximation allows T_p to be independent to the temperature of the neighboring nodes and simplify the problem since T_p becomes a function of f_p and t only. Applying (15) to (14) yields the following:

$$\frac{dT_p}{dt} + \frac{1}{C}\left(\frac{4}{R} + \frac{1}{R_c}\right)T_p = \frac{1}{RC}\sum_{n \in N_p}(T_{n0} + T_{n\Delta t} - T_{n0})\frac{t}{\Delta t}$$
$$+ \frac{T_c}{R_c} + \hat{I}f_p \quad (16)$$

Solving (16), we get the solution of T_p as:

$$T_p(t, f_p) = T_{p0}e^{-t/\tau} + T^*(1 - e^{-t/\tau})$$
$$+ \frac{\tau}{RC}\sum_{n \in N_p}(T_{n\Delta t} - T_{n0}) \quad (17)$$

where

$$\tau = \left(\frac{4}{CR} + \frac{1}{CR_c}\right)^{-1}$$

$$T^* = \frac{\tau}{C}\left\{\sum_{n \in N_p}\frac{T_{n0}}{R} - \frac{\tau}{\Delta t}\sum_{n \in N_p}\frac{T_{n\Delta t} - T_{n0}}{R} + \frac{T_c}{R_c} + \hat{I}f_p\right\}$$

3.4 Energy Minimization

Once we have constructed the energy function, finally we can find the most plausible power map by energy minimization techniques. In other words, the objective is to find the best labeling f^* minimizing the energy function among all the possible configurations F:

$$f^* = \arg\min_{f \in F} U(f|d) \quad (18)$$

Since the MAP estimation for a discrete MRF is NP-hard in general, we used an approximate technique called graph-cuts since it greatly reduces the computation time by using fast iterative expansion and swap algorithms[13]. The algorithm is guaranteed to find a strong local minimum of the energy within $|L|$ iterations. In practice, the optimization process is generally finished in 5 - 10 iterations.

4. EVALUATION

In this section, we performed two experiments with simulation results and real world measurement data. The errors are calculated by using rounded power maps which is proposed in previous works[5, 7].

4.1 PowerField Framework

All the features described in previous sections are integrated into our PowerField implementation. As shown in Fig. 3, the system receives two consecutive thermal images from a thermal imaging device or a thermal simulator. Initially, all the power estimation values are set to zero and the system computes the corresponding energy function. Then the graph-cuts algorithm generates a power estimation by minimizing the energy function. This process is iterated until the energy reaches the lowest value.

4.2 HotSpot Simulation Results

Before we apply PowerField to real world thermal images, a fully controlled experiment with an accurate thermal simulator is performed. We used Hotspot thermal simulator

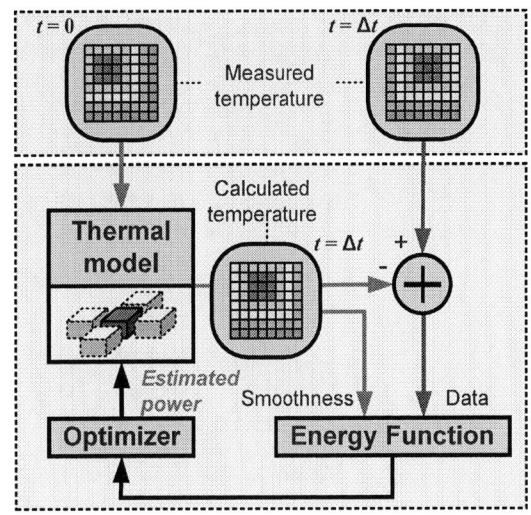

Figure 3: Overview of PowerField framework.

Table 1: Important Parameters for Evaluation using Hotspot Simulator

Hotspot Parameters	
d_t	$500\mu m$
d_w, d_h	$250\mu m$
k	100.0 (W/m-K)
c	1.75×10^6 J/m^3-K
Simulated Runtime	20 msec
Δt	10 msec
Floorplan	checker (64×64)
Power Trace	spot, dacpower
PowerField Parameters	
M	100
$\hat{I}$	1.0×10^{-3} W
α	2.0
λ	1.0×10^{-6}

which provides reliable results compared to accurate finite elements analysis[14]. The simulated runtime is set to 20ms with time interval of 10ms that reflects the microprocessors' thermal time constant (5ms to 300ms)[15]. The detailed configuration of Hotspot simulator is shown in Table 1. Two power traces are applied in this experiment. The first one has a simple one-point heat source (Fig. 4-(a)) which helps to understand the difference between the steady-state method and PowerField intuitively and another one, named dacpower (Fig. 4-(b)), is used to demonstrate the superiority of our method for complex power patterns. Initially, we let the silicon to be steady-state with initial power distribution. Then we abruptly (at 0ms) switch the distribution to a very different one. As shown in Fig. 4, the estimation results with previous method contain an amount of afterimage which disrupts identifying the correct power patterns. In contrast, PowerField outperforms the previous method in transient analysis of one-point and dacpower patterns with average errors of 0%, 6.6% respectively.

4.3 FPGA Measurement Results

We applied PowerField technique to a Xilinx Virtex-4

633

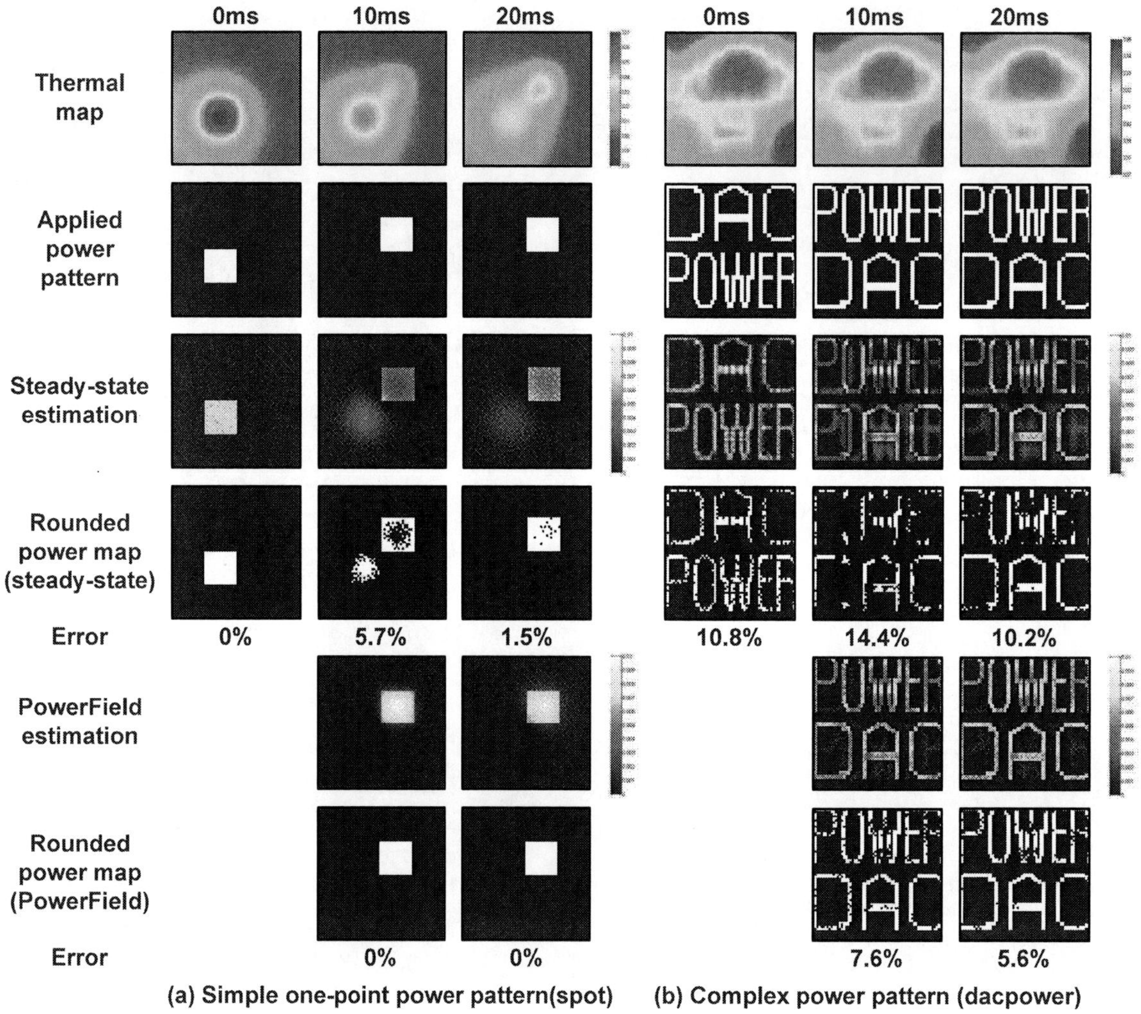

Figure 4: Transient analysis using HotSpot[14] thermal simulator.

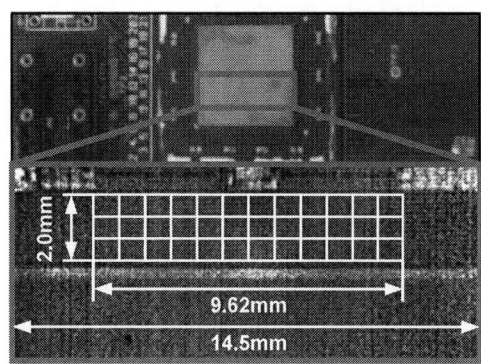

Figure 5: Floorplan of the micro heaters.

FPGA chip with pre-defined power generation patterns. The thermal imaging device (FLIR A325) captures the infrared signals from the backside of the chip. The signal is stored as a raw data of the sensor response and converted to a temperature value using the calibration method described in [7]. Then the images are filtered using spatial low-pass filter to remove the significant high frequency noise.

Fig. 5 shows our floorplan which consists of 3 × 12 micro heater blocks containing 36 free-running ring oscillators. Each micro heater consumes 46.5mW on the average. Two 36-bit registers are used to store two different enable-signals for micro heaters to construct the desired power patterns. Using an external switch, we toggle between two power patterns for transient analysis. For the experiment, a transition of power pattern from 'COLD' to 'HOT' is used. The thermal imaging device captures the transient thermal maps with 16ms time interval which is less than half of the entire thermal map transition. As shown in Fig. 6, our algorithm effectively estimates the correct power pattern even when the thermal map is changing and 90.7% accuracy on average was achieved with rounded power maps.

5. CONCLUSIONS

In this work, we propose the use of computer vision technique, MAP-MRF framework, for transient temperature-to-power inversion problem that is more general situation in real world post-silicon verification. By exploiting its robustness against spatio-temporal noise, PowerField succesfully

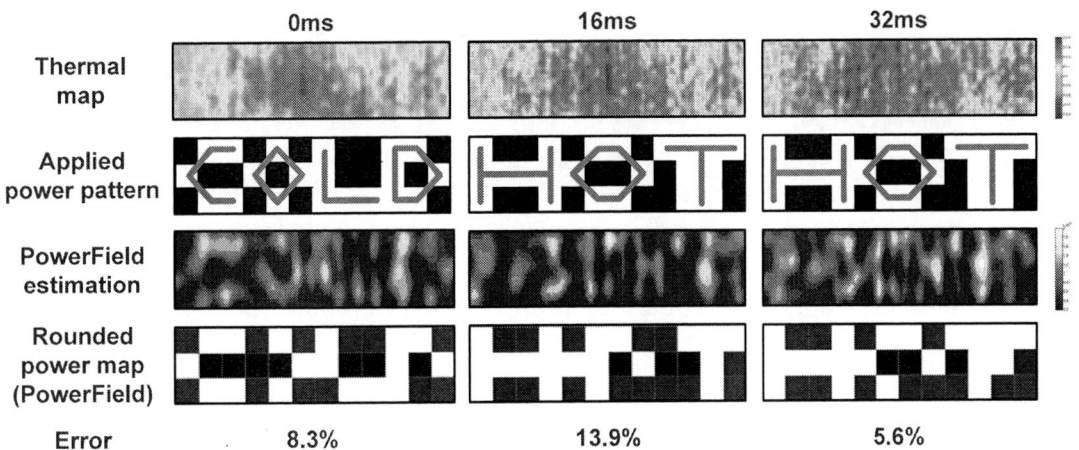

Figure 6: Transient analysis using an FPGA with a transition of power pattern from 'COLD' to 'HOT'.

estimates the correct power map from time-varying thermal images. Experimental results with a thermal simulator shows that PowerField outperforms the previous method in transient analysis reducing the error by half on average. We also applied our method to a real silicon achieving 90.7% accuracy. Our future research direction is to devise a way to obtain the model parameters such as R, R_c and C directly from the measurement similar to the scanning laser beam in [4] for more reliable estimates.

6. ACKNOWLEDGEMENTS

This work was supported by Basic Science Research Program through the NRF of Korea funded by the MEST (No.2011-0000320, No.2011-0018357), IDEC, and the IT R&D program of MKE/KEIT. [KI002134, Wafer Level 3D IC Design and Integration]

7. REFERENCES

[1] D. Brooks, R. P. Dick, R. Joseph, and L. Shang, "Power, thermal, and reliability modeling in nanometer-scale microprocessors," *IEEE Micro*, vol. 27, pp. 49–62, June 2007.

[2] W. Huang, M. R. Stant, K. Sankaranarayanan, R. J. Ribando, and K. Skadron, "Many-core design from a thermal perspective," in *Proc. of Design Automation Conf.*, pp. 746–749, June 2008.

[3] C. Poirier, R. McGowen, C. Bostak, and S. Naffziger, "Power and temperature control on a 90nm itanium family processor," in *Int. Solid-State Circuits Conf.*, pp. 304–305, Feb. 2005.

[4] H. Hamann, J. Lacey, A. Weger, and J. Wakil, "Spatially-resolved imaging of microprocessor power (simp): hotspots in microprocessors," in *The Tenth Intersociety Conf. on Thermal and Thermomechanical Phenomena in Electronics Systems*, pp. 121–125, June 2006.

[5] R. Cochran, A. N. Nowroz, and S. Reda, "Post-silicon power characterization using thermal infrared emissions," in *Proc. Int. Symp. on Low Power Electronics and Design*, pp. 331–336, 2010.

[6] Z. Qi, B. H. Meyer, W. Huang, R. J. Ribando, K. Skadron, and M. R. Stan, "Temperature-to-power mapping," in *Proc. Int. Conf. on Computer Design*, pp. 384–389, Oct. 2010.

[7] A. Nowroz, G. Woods, and S. Reda, "Improved post-silicon power modeling using ac lock-in techniques," in *Proc. Design Automation Conf.*, pp. 101–107, June 2011.

[8] P. Salihundam, S. Jain, T. Jacob, S. Kumar, V. Erraguntla, Y. Hoskote, S. Vangal, G. Ruhl, and N. Borkar, "A 2 tb/s 6 4 mesh network for a single-chip cloud computer with dvfs in 45 nm cmos," *IEEE Journal of Solid-State Circuits*, vol. 46, pp. 757–766, Apr. 2011.

[9] F. J. Mesa-Martinez, J. Nayfah-Battilana, and J. Renau, "Power model validation through thermal measurements," in *Proc. Int. Symp. on Computer Architecture*, pp. 302–311, June 2007.

[10] R. Kindermann and J. L. Snell, *Markov Random Fields and Their Applications*. American Mathematical Society, 1980.

[11] V. Kolmogorov and R. Zabih, "Computing visual correspondence with occlusions using graph cuts," in *Proc. of Int. Conf. on Computer Vision*, vol. 2, pp. 508–515, Jul. 2001.

[12] J. Chen and C.-K. Tang, "Spatio-temporal markov random field for video denoising," in *Conf. on Computer Vision and Pattern Recognition*, pp. 1–8, June 2007.

[13] Y. Boykov, O. Veksler, and R. Zabih, "Fast approximate energy minimization via graph cuts," *IEEE Trans. on Pattern Analysis and Machine Intelligence*, vol. 23, pp. 1222–1239, Nov. 2001.

[14] W. Huang, S. Ghosh, S. Velusamy, K. Sankaranarayanan, K. Skadron, and M. Stan, "Hotspot: A compact thermal modeling methodology for early-stage vlsi design," *IEEE Trans. on Very Large Scale Integration Systems*, vol. 14, pp. 501–513, May 2006.

[15] F. J. Mesa-Martinez, E. K. Ardestani, and J. Renau, "Characterizing processor thermal behavior," in *Proc. of Int. Conf. on Architectural Support for Programming Languages and Operating Systems*, pp. 193–204, Mar. 2010.

EigenMaps: Algorithms for Optimal Thermal Maps Extraction and Sensor Placement on Multicore Processors

Juri Ranieri[*], Alessandro Vincenzi[†], Amina Chebira[*], David Atienza[†], Martin Vetterli[*]
[*]LCAV, [†]ESL
École Polytechnique Fédérale de Lausanne, Lausanne (Switzerland)
name.surname@epfl.ch

ABSTRACT

Chip designers place on-chip sensors to measure local temperatures, thus preventing thermal runaway situations in multicore processing architectures. However, thermal characterization is directly dependent on the number of placed sensors, which should be minimized, while guaranteeing full detection of all hot-spots and worst case temperature gradient. In this paper, we present EigenMaps: a new set of algorithms to recover precisely the overall thermal map from a minimal number of sensors and a near-optimal sensor allocation algorithm. The proposed methods are stable with respect to possible temperature sensor calibration inaccuracies, and achieve significant improvements compared to the state-of-the-art. In particular, we estimate an entire thermal map for an industrial 8-core industrial design within $1°C$ of accuracy with just four sensors. Moreover, when the measurements are corrupted by noise (SNR of 15 dB), we can achieve the same precision only with 16 sensors.

ACM Categories & Subject Descriptors
B.7.1 [Integrated Circuits]: Types and Design Styles.
General Terms: Design, Performance, Algorithms
Keywords: Thermal characterization, principal component analysis, sensor allocations, least-square estimation

1. INTRODUCTION

The continuous evolution of process technology enables the inclusion of multiple cores, memories and complex interconnection fabrics on a single die [7]. Although many–core architectures potentially provide increased performance, they also suffer from increased IC power densities and thermal issues are a serious concern in latest designs with deep submicron process technologies [5, 8]. In particular, it is key to design many–core designs that prevent hot spots and large on-chip temperature gradients, as both conditions severely affect system's characteristics. In fact, thermal stress increases the overall failure rate of the system [13], reduces performance due to an increasing operating temperature [1], significantly increases leakage power consumption (due to its exponential dependence on temperature) and cooling costs [13, 3]. Designers organize the floorplan to limit these thermal phenomena, for example, by placing the highest power density components closer to the heat sink [5]. However, the workload execution patterns are fundamental to determine the transient on-chip temperature distribu-

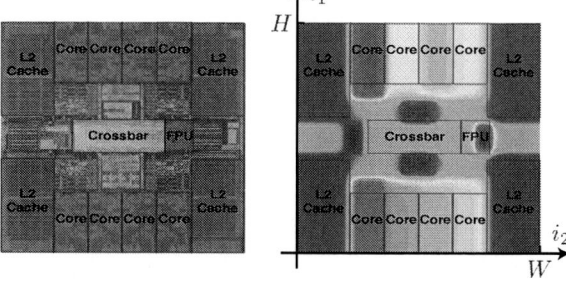

Figure 1: Left: a simplified floorplan of the Ultrasparc T1, the considered 8-core processorn. Right: an example of thermal map [3].

tion in multicore designs and, unfortunately, these patterns are not fully known at design time. Furthermore, these issues are amplified in many–core designs, where thermal hot-spots are generated without a clear spatio-temporal pattern due to the dynamic task set execution nature, based on external service requests, as well as the dynamic assignment to cores by the many–core OS [3, 4].

Therefore, latest many–core designs include dynamic thermal management approaches that incorporate thermal information into the workload allocation strategy to obtain the best performance while avoiding peaks or large gradients of temperature. Nowadays, a few sensors are already deployed on the chip to obtain this thermal information. However, their number is limited by area/power constraints and their optimal placement to detect all the worst case temperature scenarios is a very complex problem that has received significant attention in the last years [2, 10, 12, 14, 15].

Unfortunately, the reconstruction of the entire thermal map from a limited number of thermal sensors poses many — and still unresolved — questions. In particular, for a specific many–core architecture, the two fundamental questions are the possible trade-offs regarding the number of sensors to place and the reachable degree of temporal and spatial thermal precision, as well as the sensor placement to maximize the reconstruction performance.

In this paper, we propose to recover the entire thermal map using a new method, which we call *EigenMaps*. Our method estimates an entire temperature map using a limited set of measurements collected by sensors, as inspired by [12]. First, we reduce the complexity of the thermal map by considering an optimal low–dimension approximation. Then, we use the sensors measurements to estimate the parameters of the approximated thermal map. Specifically, our contributions are:

- A reliable low–dimensional approximation of thermal maps based on a thermal analysis done at design time.
- A reconstruction algorithm that recovers the thermal map approximation from the sensors measurements. The quality of reconstruction can be adjusted according to the number of

sensors and the quality of the measurements by adapting the precision of the aforementioned approximation.

- A theoretical derivation of a sensor allocation method minimizing the reconstruction error.
- An algorithm that finds a near-optimal allocation in polynomial time.

Our methods are stable in the presence of noise: the error corrupting the measurements is not amplified by the reconstruction algorithm. Moreover, the allocation algorithm can easily integrate sensor location constraints, such as the unfeasibility of placing a sensor into the cache [11]. The reconstruction algorithm achieves the highest precision when compared to the available literature. The price of this precision is the necessity of storing a matrix in the memory/cache. The size of this matrix, and therefore the necessary space in memory, grows linearly with the desired precision.

We substantiate our theoretical findings with extensive numerical evidence, based on thermal simulation of an eight–core Niagara T1 Ultrasparc processor, shown in Fig. 1. These experiments show that EigenMaps achieve a significant improvement when compared to the state-of-the-art. In particular, we can estimate an entire thermal map with high precision ($<1°$) with only four sensors. Moreover, we can achieve the same precision with measurements that have a SNR of 15 dB using only 16 sensors.

2. BACKGROUND AND RELATED WORK

The thermal map of a processor can be estimated using two dual strategies:

- Solution of the direct problem, given the heat sources and the physical model of the temperature diffusion (e.g. a nonlinear diffusion equation),
- Solution of the inverse problem, given the value of the temperature in some locations and some a-priori information about the thermal map.

The first approach is limited by its requirements: the knowledge of the heat sources can be ascribed to the knowledge of the detailed power consumption of the different components. This information cannot be known exactly at run time and moreover the computation of a solution would require an excessive computational power.

On the other hand, it is impossible to solve the inverse problem from few, spatially localized, imprecise measurements without some a-priori constraints on the thermal map, such as limited bandwidth [2]. The performance is significantly impacted by the small number of available sensors and the *structure* we consider for the thermal map, i.e. the a-priori information. Nowroz et al. [12] proposed a low-pass approximation strategy to reduce the number of sensors that are placed using an energy-oriented algorithm. This sensor allocation algorithm has been improved by Reda et al. [14] using a heuristic iterative approach to approximate an NP-hard problem. The authors in [9] proposed a grid-based uniform sensor placement followed by interpolation to approximate the temperature. These works estimated entire thermal maps, but the precision of the estimates is limited by the sub-optimality of the proposed a-priori information.

Other works have notable performance but are not focused on the estimation of the entire thermal map. Namely, the approach in [19] employs the correlation in power distribution to estimate the expected value of temperature at different locations of the chip using a dynamically tuned Kalman filter. The problem of noisy measurements has also been already considered. In particular, the correlation between the different sensor measurements has been exploited to denoise the measurements [18].

The remainder of the paper is organized as follows. In Section 3, we describe our three main findings: the optimal approximation, the reconstruction algorithm and the sensor allocation algorithm.

Then, the experimental setup is described in Section 4, followed by the experimental results in Section 5.

3. RECOVERY OF THERMAL MAPS

In this work, we use the two dual estimation strategies to optimize the reconstruction. First, we use the direct problem to define an optimal low–dimensional approximation. Then, we use this approximation to recover thermal maps solving a simpler inverse problem. The inverse problem is simplified because the approximation reduces the number of parameters that must be estimated by the reconstruction algorithm.

The performance of the reconstruction algorithm depends on the approximation quality and the conditioning of the inverse problem— a complicated function of the sensor locations. Therefore, we conclude with a sensor allocation algorithm that minimizes the conditioning of the inverse problem.

We consider a processor with an N–dimensional discrete temperature map t. The temperature at coordinates i_1 and i_2 is defined as $t[i_1, i_2]$, for $0 \leq i_1 \leq H-1$ and $0 \leq i_2 \leq W-1$, where W and H are the width and the height of the discretized thermal map, respectively. We vectorize the thermal map as $x[i]$, for $0 \leq i \leq N-1$ and $N = WH$, that is

$$x[i] = t\left[i \mod H, \left\lfloor \frac{i}{W} \right\rfloor\right].$$

In other words, we stack the columns of the discrete thermal map to transform the matrix into a vector. For the remainder of the paper, a bold symbol, $\boldsymbol{x}$, indicates vectors or matrices while the normal symbols, x are reserved for scalars or elements of vectors.

3.1 Approximation of thermal maps

In this section, we derive the approximation method as a projection onto the low–dimensional linear subspace that minimizes the mean squared error (MSE). It allows us to describe the N–dimensional thermal map with only K coefficients, where $K \ll N$.

We define the subspace at design time, exploiting the set of T thermal maps generated by a numerical solution of the direct problem. These are considered as realizations of the N–dimensional random vector $\boldsymbol{x}$. We assume that the elements of $\{\boldsymbol{x}_j\}_{j=0}^{T-1}$ have zero mean to keep a simple notation[1]. Any vector $\boldsymbol{x}$ can be represented using a basis $\boldsymbol{\Phi}$ as,

$$x[i] = \sum_{n=0}^{N-1} \Phi[i,n]\alpha[n],$$

where $\alpha[n]$ are the coefficients of the expansion over the basis $\boldsymbol{\Phi}$. Note that once we define a basis for the data, knowing the coefficients $\boldsymbol{\alpha}$ is equivalent to knowing the thermal map $\boldsymbol{x}$. We can describe the approximated thermal maps with a linear combination of K columns of $\boldsymbol{\Phi}$ with K elements of $\boldsymbol{\alpha}$ out of N as coefficient. More precisely, the approximated thermal map $\widehat{\boldsymbol{x}}$ is given by the following overdetermined system of equations

$$
\widehat{\boldsymbol{x}} = \begin{bmatrix} \Phi[0,0] & \cdots & \Phi[0, K-1] \\ \vdots & \ddots & \vdots \\ \Phi[N-1, 0] & \cdots & \Phi[N-1, K-1] \end{bmatrix} \begin{bmatrix} \alpha[0] \\ \vdots \\ \alpha[K-1] \end{bmatrix}
$$

$$= \boldsymbol{\Phi}_K \boldsymbol{\alpha}_K, \tag{1}$$

where the subscript K indicates the selection of the first K columns for a matrix or the first K elements for a vector. This approximation is equivalent to a projection onto the K–dimensional subspace

---

[1]Note that we can always subtract the mean to get zero-mean vectors.

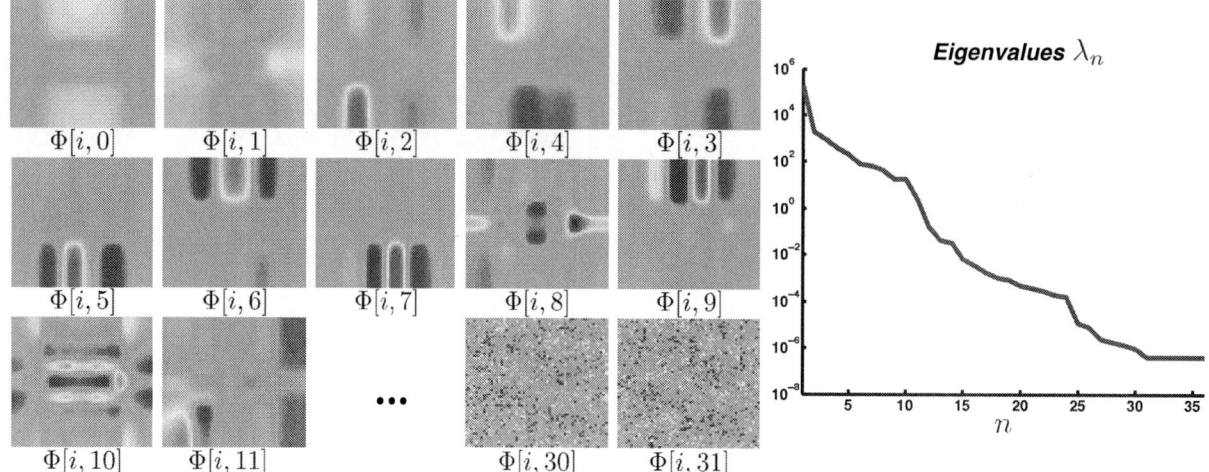

Figure 2: Left: a selection of the first 32 EigenMaps for the Niagara T1 Ultrasparc. Note that the informative content decays rapidly to just noise (the last two EigenMaps). This analysis is confirmed by the decay of the eigenvalues given in the right plot.

spanned by the columns of Φ_K. We suggest that the optimal subspace is the K–dimensional one introducing the smallest error in the MSE sense. We define the following optimization problem to find this basis and the relative optimal subspace we are looking for.

PROBLEM 1. *Find the set of basis vectors in Φ such that the approximation $\widehat{x}$ with the first $K < N$ components, $\widehat{x} = \Phi_K \alpha_K$, minimizes the following error*

$$\xi = \mathbb{E}\left[|\boldsymbol{x} - \widehat{\boldsymbol{x}}|^2 \right] = \mathbb{E}\left[\left| \sum_{n=K}^{N-1} \Phi[i,n]\alpha[n] \right|^2 \right].$$

This dimensionality reduction technique is well known in other fields under different names, such as *Principal Component Analysis* (PCA) and *Karhunen-Loeve Transform*. It has an analytic solution and it requires the covariance matrix C_x, that is defined for real zero-mean random variables as

$$C_x[i,j] = \mathbb{E}\left[x[i]x[j] \right].$$

We estimate this matrix using the set of T thermal maps simulated at design time. The quality of the available dataset impacts the quality of the estimate C_x. This estimation is a well studied topic [6] and will not be discussed here.

We give the solution to Problem 1 in the following proposition. The proof is a well-known result.

PROPOSITION 1. *Optimal Approximation Let us consider a set of T thermal maps $\{\boldsymbol{x}_j\}_{j=0}^{T-1}$ with zero mean and covariance matrix C_x. The orthonormal basis Φ_K that defines the approximation $\widehat{x}$ with the minimum error ξ, is formed by the K eigenvectors of C_x with the largest eigenvalues $\{\lambda_n\}_{n=0}^{K-1}$. Moreover, the approximation error is decreasing when K grows as*

$$\xi = \sum_{n=K}^{N-1} \lambda_n. \tag{2}$$

The connection between C_x and the optimal basis has an intuitive explanation. In fact, if the temperatures at different spatial points are statistically correlated, then C_x has nonzero elements outside its diagonal. These elements can be used to infer the temperature at points without sensors. Moreover, if the correlation is strong, then the eigenvalues λ_n of C_x decay fast and we can precisely approximate the temperature $\boldsymbol{x}$ with a smaller K, see (2). Recall that K is the number of parameters we need to estimate from the sensor measurements; having the optimal approximation

with the minimum K is fundamental to have a truthful reconstruction with just few sensors. Note that this optimal approximation *pays* the increased precision with more space occupied in memory; namely, we need to store the matrix Φ_K in the memory, with an occupation that is proportional to the desired resolution. Other methods, such as the one proposed in [12], use standard basis, avoiding the occupation of memory but obtaining a lower precision.

Inspired by a classical work in computer vision [17], we call *EigenMaps* the eigenvectors of C_x. An example for EigenMaps of the Ultrasparc T1 multicore architecture is given in Fig. 2. Note that each EigenMap represents a particular structure of the processor, such as cores, FPU and cache.

3.2 Reconstruction of thermal maps

Thermal maps are now defined only by their K coefficients α_K in the basis Φ_K. Here, we explain how to estimate them from the sensors measurements.

In principle, we can find the coefficients by inverting the overdetermined linear system of equations given in (1). However, this would require the knowledge of the temperature $x[i]$ at every spatial location i. Let us assume that we can place only M sensors at locations $\mathcal{S} = \{j_i\}_{i=1}^{M}$. Considering (1), it is equivalent to

$$\boldsymbol{x}_{\mathcal{S}} = \begin{bmatrix} \Phi[j_1,0] & \cdots & \Phi[j_1,K-1] \\ \vdots & \ddots & \vdots \\ \Phi[j_M,0] & \cdots & \Phi[j_M,K-1] \end{bmatrix} \begin{bmatrix} \alpha[0] \\ \vdots \\ \alpha[K-1] \end{bmatrix}$$
$$= \widetilde{\Phi}_K \alpha_K, \tag{3}$$

where $\widetilde{\Phi}_K$ is a matrix formed by the rows of Φ_K corresponding to the sensor locations $\mathcal{S}$, $\boldsymbol{x}_{\mathcal{S}}$ is a vector containing our sensor measurements and α_K is the unknown vector.

Before we characterize the solution of (3), we need to introduce the concept of noise into the model. More precisely, we have two different noise sources affecting our measurements. First, we have the approximation error $\epsilon = \widehat{x} - x$ that is systematic and is due to the approximation on the K–dimensional subspace. Second, the measurements are corrupted by a significant amount of noise due to many factors, such as thermal noise, quantization and calibration inaccuracies [15]. Therefore, we consider the following modification of (3),

$$\boldsymbol{x}_{\mathcal{S}} + \boldsymbol{w} = \widetilde{\Phi}_K \alpha_K, \tag{4}$$

where $\boldsymbol{w}$ is the noise term. There is no exact solution to (4). However, we can find the coefficients $\widehat{\boldsymbol{\alpha}}_K$ such that the error w.r.t. the measured temperature is minimized. Namely, we solve the following least square problem,

$$\min_{\widehat{\boldsymbol{\alpha}}_K} \|\boldsymbol{x}_{\mathcal{S}} - \widetilde{\boldsymbol{\Phi}}_K \widehat{\boldsymbol{\alpha}}_K\|_2^2.$$

We reconstruct thermal maps using the K–dimensional approximation given in (1). This leads to the theorem for the reconstruction of a thermal map from noisy measurements.

THEOREM 1. *Noisy Reconstruction Consider an N–dimensional thermal map x, with zero mean and covariance matrix C_x. Choose a basis Φ, such as the one in Proposition 1. Define a new matrix $\widetilde{\boldsymbol{\Phi}}_K$ according to (3) to represent the approximation on the K– dimensional subspace and the sensing with M noisy sensors located at $\mathcal{S}$ as in (4). If $M \geq K$ and $\mathrm{rank}(\widetilde{\boldsymbol{\Phi}}_K) = K$, then the reconstruction $\widetilde{x}$ of the thermal map x is **unique** and equal to*

$$\widetilde{x} = \Phi_K \left(\widetilde{\boldsymbol{\Phi}}_K^* \widetilde{\boldsymbol{\Phi}}_K \right)^{-1} \widetilde{\boldsymbol{\Phi}}_K^* x_{\mathcal{S}}.$$

Moreover, the reconstruction MSE is bounded by the condition number κ of $\widetilde{\boldsymbol{\Phi}}_K$ and the noise energy $\|w\|_2$ as

$$\frac{\|\widetilde{x} - x\|_2}{\|x\|_2} = \mathcal{O}\left(\kappa^2(\widetilde{\boldsymbol{\Phi}}_K) \right) \|w\|_2. \tag{5}$$

This theorem highlights a focal point of our work: *given M sensors and an optimal K–dimensional subspace Φ_K, the optimal sensor location is the one that minimizes the condition number of $\widetilde{\boldsymbol{\Phi}}_K$.* If this condition number is minimal, the reconstruction error $\epsilon_r = \widetilde{x} - x_{\mathcal{S}}$ is minimal for the given amount of noise w. In other words, the condition number is an excellent metric to evaluate different sensing patterns and find the optimal one. Note that, once we have fixed M, increasing K will in general increase the reconstruction error ϵ_r (worse conditioning) and decrease the approximation error ϵ (better approximation). Therefore, we should pick an optimal K such that the sum of ϵ and ϵ_r is minimal.

3.3 A greedy algorithm for sensor allocation

In what follows, we present a greedy polynomial algorithm to find the solution to the sensor allocation problem, i.e. choosing M rows from Φ_K such that the resulting $\widetilde{\boldsymbol{\Phi}}_K$ is full rank and has minimal condition number.

Intuitively, we are looking for M rows that form the best orthonormal basis for the M–dimensional subspace. The optimal solution can be obtained by computing the condition number of all possible sets of M rows out of the N original rows. This is equivalent to computing $\binom{N}{M}$ singular value decompositions, which is computationally impossible, as this number is proportional to $N!$.

We propose a sensor allocation algorithm that has polynomial complexity and achieves the best performance when compared to the state-of-the-art. First, we compute the correlation between all rows of $\widetilde{\boldsymbol{\Phi}}_K$, then we remove one by one the rows that show the highest correlation with the other ones. Intuitively, we do not consider the sensor locations that would add the least informative content. Eventually, the remaining M rows indicate where we should place the sensors. The structure of the algorithm is given in Algorithm 1 and an example of the sensor allocation algorithm output is given in Fig. 6 (a).

4. EXPERIMENTAL SETUP

We tested the proposed methods on an Ultrasparc T1 and we simulate its thermal behavior using **3D-ICE** [16]. This simulator is based on a compact transient thermal model; it can be used for thermal simulations of 2D or 3D chips cooled with conventional or liquid cooling. The simulator has been validated against

---

Algorithm 1 Sensor allocation

Require: Subspace Φ_K, number of sensors M
Ensure: Sensing matrix $\widetilde{\boldsymbol{\Phi}}_K$, sensor locations $\mathcal{S}$

1. Normalize the rows of Φ_K such that they are unit-norm. Call U the normalized matrix.
2. Compute $G = UU^* - I$, where I is the identity.
3. **Repeat until M rows are left:**
   (a) Find the maximum element, i.e. $G[i, j] = \max G$.
   (b) Remove the i-th row and column from G.
   (c) Build $\mathcal{S}$ from the remaining rows, build the sensing matrix $\widetilde{\boldsymbol{\Phi}}_K$.
   (d) **If** $\mathrm{rank}(\widetilde{\boldsymbol{\Phi}}_K) < K$, **then**
      i. Restore $\widetilde{\boldsymbol{\Phi}}_K$ from the previous iteration.
      ii. Break.

---

computational fluid dynamics simulations, it is easily configurable and publicly available. The input of the simulation are the power traces given in [7]. These traces describe the power consumed by the elements of the processor while running different scenarios/workload. The output is a set of thermal snapshots at each time interval: namely, we have $T = 2652$ thermal maps with $N = 3360$, since the thermal maps are discretized with $W = 60$ and $H = 56$. Note that our thermal maps are rather coarse–grained since we consider large blocks having the same average power consumption, but we expect to obtain similar performance on more detailed thermal maps.

As a reference, we choose in the literature the reconstruction algorithm that shows the best performance for the entire thermal reconstruction. Specifically, we consider the k-LSE algorithm for reconstruction and the energy-center technique for sensor allocation, both from [12]. k-LSE is also one of the first algorithms estimating the entire thermal map from few measurements. Moreover, our reconstruction methods are conceptually close to their approach, the main difference being the choice of the approximation subspace.

We consider two main figures of merit when comparing the reconstruction techniques. The MSE of the reconstruction, defined as the average error over all thermal maps, that is

$$\mathrm{MSE} = \frac{1}{TN} \sum_{i=0}^{N-1} \sum_{j=0}^{T-1} |x_j[i] - \widehat{x}_j[i]|^2 ,$$

where the index j points to all T thermal maps available in the dataset. We also consider the maximum error among the maps, that is defined as

$$\mathrm{MAX} = \max_{i,j} |x_j[i] - \widehat{x}_j[i]|^2 ,$$

because localized peaks of error can lead to thermal runaway.

5. RESULTS

In what follows, we present and discuss the results of the numerical experiments. All the experiments have been run on all the 2652 thermal maps generated during the simulation phase.

5.1 Reconstruction Performance

First, we show the impact of the choice of the subspace. We compute the difference between a thermal map x and its approximation $\widehat{x}$ as a function of K for the two different methods, k-LSE and EigenMaps. The results are given in Fig. 3 (a). The theoretical optimality of the EigenMaps basis is confirmed by this experiment, where we note how the error is exponentially lower than for the DCT basis used in k-LSE. Note that this error has a strong impact

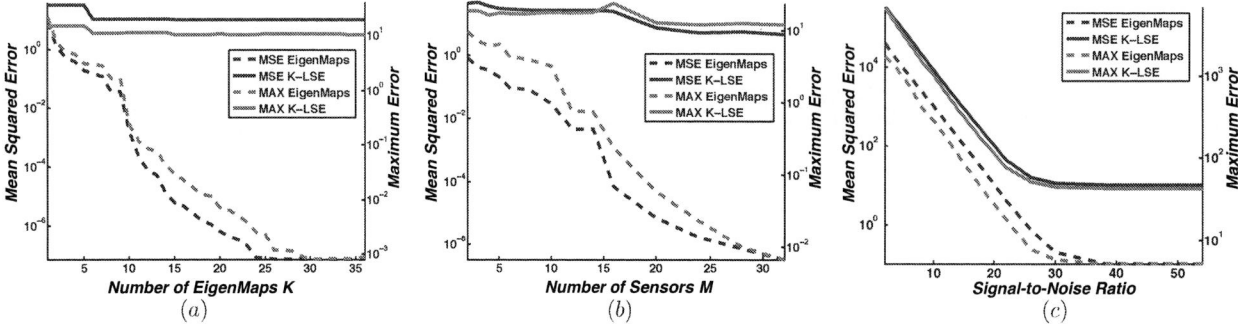

Figure 3: Comparison between EigenMaps (dotted) and k-LSE (continuous). In all the plots, the MSE (dark green) is on the right and the MAX (orange) is on the left. Note the difference of scale between the two y-axes. (a) The approximation error as a function of the number of EigenMaps K. (b) The reconstruction error as a function of the number of sensors used. (c) The reconstruction error in presence of measurement noise as a function of the SNR using 16 sensors.

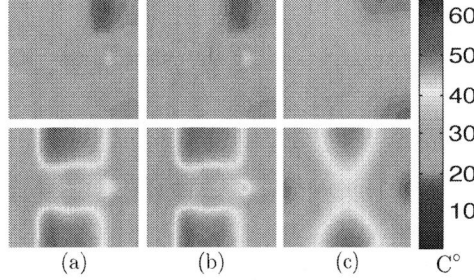

Figure 4: Visual comparison between EigenMaps and k-LSE two different thermal maps (top and bottom row) using 16 sensors. (a) The original thermal maps. (b) Reconstruction using EigenMaps. (c) Reconstruction using the k-LSE method.

in the reconstruction phase. In fact, this error is noise for the reconstruction algorithm and is amplified by the conditioning of $\widetilde{\Phi}_K$ according to (5).

Therefore, we expect to have the same exponential decay for the reconstruction error. To numerically confirm this expectation, we present our first core result: a direct comparison of the reconstruction performance in terms of MSE and MAX between EigenMaps and k-LSE. The results are given in Fig. 3 (b). We observe that the reconstruction error is approximately decaying as fast as the approximation error. Here is our intuitive explanation: we can estimate the parameters of more EigenMaps by increasing the number of sensors M, while keeping a low conditioning. Therefore, the total error follows closely the approximation error.

As further proof of the quality of our reconstruction, in Fig. 4 we give a visual comparison between the original thermal map and the reconstruction with the two methods using 16 sensors for each.

Lastly, we consider the realistic scenario of presence of noise and/or measurements errors. We compare the reconstruction performance using 16 sensors with EigenMaps and with k-LSE as a function of the SNR of the measurements, defined as

$$\text{SNR} = \frac{\|\boldsymbol{x}\|_2}{\|\boldsymbol{w}\|_2},$$

where $\boldsymbol{w}$ is the noise vector. The results are depicted in Fig. 3 (c) and we note that EigenMaps performs better than k-LSE even when there is noise introduced by sensor calibration inaccuracies. We believe that the performance of k-LSE is negatively impacted by two main factors: we consider a very small number of sensors and the processor taken in consideration here (Ultrasparc T1 instead of Athlon dual–core) generates more high frequency content, which is not well-approximated by their choice of basis.

Among all the proposed results, we would like to underline the

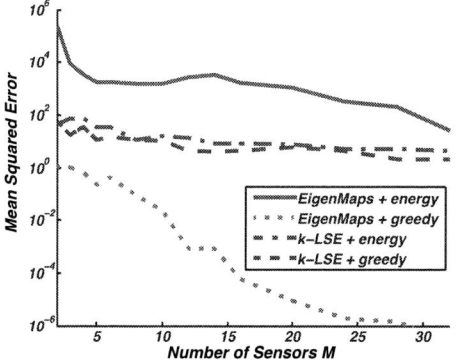

Figure 5: Comparison of the two sensor allocation techniques with both reconstruction algorithms. For both reconstruction algorithms, the greedy approach obtains better reconstruction performance in terms of MSE.

most important ones: we can recover with few sensors (4-5) entire thermal maps while keeping the MSE and the MAX below $1^\circ C$, see Fig 3 (b). Moreover, if we consider a very noisy environment, 15dB of SNR, we can keep the same excellent reconstruction performance with just 16 sensors, see Fig. 3 (c).

5.2 Sensor allocation performance

To underline the effectiveness of the sensor allocation algorithm, we propose the following experiment. We compute the MSE for four different combinations of reconstruction algorithms and sensor allocation algorithms. In particular, we consider EigenMaps, k-LSE, our greedy algorithm for sensor allocation and the energy-center algorithm [12] (referred as "energy" in the figures). The results are depicted in Fig. 5. Note that whichever reconstruction method is chosen, the greedy algorithm improves the performance w.r.t. to the energy-center algorithm. Hence, the greedy algorithm leads to a better condition number of the inverse problem.

Finally, we look at the stability of our sensor allocation algorithm when we have design constraints. This experiment is motivated by the fact that we cannot place sensors in a very regular and/or critical structure, such as a cache [11]. To analyze this scenario we compare the reconstruction performance between the unconstrained and the constrained cases. The constraints are defined using a mask of allowed zones (black) that is given in Fig. 6 (b). The reconstruction error is given in Fig. 6 (d), while examples of sensor allocations are given in Fig. 6 (a) and (c).

6. CONCLUSION

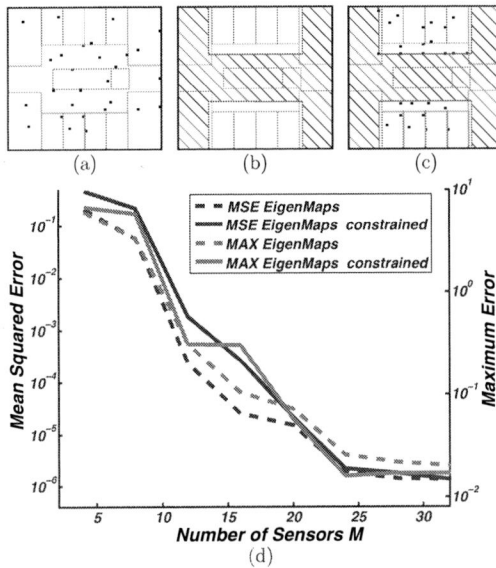

Figure 6: Comparison of the reconstruction error of the proposed method when the sensor locations are free (dotted line) and constrained by design (continuous line). (a) Location of 32 sensors without constraint. (b) Mask of the contraint: sensors can not be placed on the striped red zone. (c) Location of 32 sensors with constraint. (d) MSE (dark green) and MAX (orange) as a function of the number of sensors.

In this work, we proposed a framework to optimally reconstruct thermal maps of multicore processors using a small number of sensors. We defined an optimal approximation of thermal maps to reduce the number of parameters to estimate, without loosing precision. We reconstructed the thermal maps using a least square approach and we exposed the critical role of the sensor location for the conditioning of the inverse problem. We concluded proposing a sensor allocation algorithm that minimizes the reconstruction error by minimizing the conditioning of the inverse problem.

We compared EigenMaps with k-LSE [12] using extensive numerical experiments. We demonstrated the higher fidelity of our reconstruction using a smaller number of sensors. We showed how the reconstruction performance is stable w.r.t. the noise introduced by the electronics or by sensor calibration inaccuracies. Moreover, even if we constrain the locations of the sensors, the reconstruction degrades only slightly. To the best of our knowledge, this is one of the first works that recovers the entire thermal map while considering noise measurements and constrained allocation.

To summarize, our work improves significantly the precision and the stability achievable for the thermal monitoring of a multicore processor. The price for the increased precision is the memory needed to store the matrix Φ_K, that is the main limitation of the proposed method. We also introduce the concept of minimizing the condition number of the inverse problem for the sensor allocation problem, that is the critical figure of merit of the estimation problem.

7. ACKNOWLEDGEMENTS

The work of Juri Ranieri and Martin Vetterli has been supported by *ERC Advanced Grant - Support for Frontier Research - SPARSAM Nr : 247006*, while the work of Alessandro Vincenzi has been partly supported by the *Swiss National Science Foundation (SNF), grant number 200021-130048*.

8. REFERENCES

[1] BROOKS, D., DICK, R., JOSEPH, R., AND SHANG, L. Power, Thermal, and Reliability Modeling in Nanometer-Scale Microprocessors. *Micro, IEEE 27*, 3 (Jan. 2007), 49–62.

[2] COCHRAN, R., AND REDA, S. Spectral techniques for high-resolution thermal characterization with limited sensor data. *Des. Aut. Con.* (June 2009), 478–483.

[3] COSKUN, A. K., ROSING, T., WHISNANT, K., AND GROSS, K. C. Static and Dynamic Temperature-Aware Scheduling for Multiprocessor SoCs. *IEEE Trans. VLSI Syst. 16*, 9 (2008), 1127–1140.

[4] COSKUN, A. K., ROSING, T. S., AND GROSS, K. C. Utilizing Predictors for Efficient Thermal Management in Multiprocessor SoCs. *IEEE Trans. Comput. Aided Des. Integr. Circuits Syst. 28*, 10 (Dec. 2009), 1503–1516.

[5] HUANG, W., STAN, M. R., SANKARANARAYANAN, K., RIBANDO, R. J., AND SKADRON, K. Many-Core Design from a Thermal Perspective. In *DAC* (2008).

[6] KAY, S. M. *Fundamentals of statistical signal processing: estimation theory.* Prentice-Hall, Inc., Mar. 1993.

[7] LEON, A. S., TAM, K. W., SHIN, J. L., WEISNER, D., AND SCHUMACHER, F. A Power-Efficient High-Throughput 32-Thread SPARC Processor. *IEEE J. Solid-State Circuits 42*, 1 (Jan. 2007), 7–16.

[8] LIN, S.-C., AND BANERJEE, K. Cool Chips: Opportunities and Implications for Power and Thermal Management. *IEEE Trans. Electron Devices 55*, 1 (Jan. 2008), 245–255.

[9] LONG, J., MEMIK, S. O., MEMIK, G., AND MUKHERJEE, R. Thermal monitoring mechanisms for chip multiprocessors. *ACM Trans. Archit. Code Optim. 5*, 2 (Aug. 2008), 1–33.

[10] MUKHERJEE, R., AND MEMIK, S. Systematic temperature sensor allocation and placement for microprocessors. In *DAC* (2006).

[11] MUKHERJEE, R., AND MEMIK, S. O. Systematic temperature sensor allocation and placement for microprocessors. In *DAC* (2006).

[12] NOWROZ, A. N., COCHRAN, R., AND REDA, S. Thermal monitoring of real processors: techniques for sensor allocation and full characterization. In *DAC* (2010).

[13] PEDRAM, M., AND NAZARIAN, S. Thermal Modeling, Analysis, and Management in VLSI Circuits: Principles and Methods. *Proc. IEEE 94*, 8 (Aug. 2006), 1487–1501.

[14] REDA, S., COCHRAN, R., AND NOWROZ, A. N. Improved Thermal Tracking for Processors Using Hard and Soft Sensor Allocation Techniques. *IEEE Trans. Comput. 60*, 6 (Nov. 2011), 841–851.

[15] SHARIFI, S., AND ROSING, T.Š. Accurate Direct and Indirect On-Chip Temperature Sensing for Efficient Dynamic Thermal Management. *IEEE Trans. Comput.-Aided Des. Integr. Circuits Syst. 29*, 10 (Oct. 2010), 1586–1599.

[16] SRIDHAR, A., VINCENZI, A., RUGGIERO, M., BRUNSCHWILER, T., AND ATIENZA, D. 3D-ICE: Fast Compact Transient Thermal Modeling for 3D ICs with Inter-tier Liquid Cooling. *IEEE ICCAD 1* 2010.

[17] TURK, M. A., AND PENTLAND, A. P. Face recognition using eigenfaces. In *1991 IEEE Computer Society Conference on Computer Vision and Pattern Recognition* (Nov. 1991), IEEE Comput. Soc. Press, pp. 586–591.

[18] ZHANG, Y., AND SRIVASTAVA, A. Accurate temperature estimation using noisy thermal sensors . In *DAC* (2009).

[19] ZHANG, Y., AND SRIVASTAVA, A. Adaptive and autonomous thermal tracking for high performance computing systems. In *DAC* (2010).

An Information-theoretic Framework for Optimal Temperature Sensor Allocation and Full-chip Thermal Monitoring

Huapeng Zhou and Xin Li
Carnegie Mellon University
Pittsburgh, PA 15213
{huapeng, xinli}@cmu.edu

Chen-Yong Cher, Eren Kursun and Haifeng Qian
IBM T. J. Watson Research Center
Yorktown Heights, NY 10598
{chenyong, ekursun, qianhaifeng}@us.ibm.com

Shi-Chune Yao
Carnegie Mellon University
Pittsburgh, PA 15213
sy0d@andrew.cmu.edu

ABSTRACT

Full-chip thermal monitoring is an important and challenging issue in today's microprocessor design. In this paper, we propose a new information-theoretic framework to quantitatively model the uncertainty of on-chip temperature variation by differential entropy. Based on this framework, an efficient optimization scheme is developed to find the optimal spatial locations for temperature sensors such that the full-chip thermal map can be accurately captured with a minimum number of on-chip sensors. In addition, several efficient numerical algorithms are proposed to minimize the computational cost of the proposed entropy calculation and optimization. As will be demonstrated by our experimental examples, the proposed entropy-based method achieves superior accuracy (1.4× error reduction) for full-chip thermal monitoring over prior art.

Categories and Subject Descriptors

B.7.2 [**Integrated Circuits**]: Design Aids – Verification

General Terms

Algorithms

Keywords

Thermal Monitoring, Integrated Circuit

1. INTRODUCTION

As integrated circuit (IC) technology continues to scale to the nanoscale era, power and thermal issues become increasingly important for microprocessor design. In general, modern microprocessors can be classified into two broad categories [1]: (a) power-limited microprocessors, and (b) hotspot-limited microprocessors. The operation of power-limited microprocessors is constrained by the availability of power supply such as the limited battery life and/or the capacity of the power delivery network. Most embedded microprocessors fall into this category. On the other hand, the performance of hotspot-limited microprocessors, such as the high-performance microprocessors used for enterprise servers, is constrained by their cooling capability. The high temperature at hotspots directly impacts the device lifetime and the circuit reliability [2].

During the past decade, a large number of thermal management approaches have been proposed [2]-[8] to apply circuit-level and/or architecture-level techniques (e.g., power gating, dynamic voltage and frequency scaling, clock throttling, etc.) to optimally explore the trade-offs between microprocessor performance and hotspot temperature. These techniques have been

successfully integrated into many commercial microprocessors and thermal management has become a critical component for advanced microprocessor design.

The efficacy of thermal management, however, heavily relies on the accuracy of on-chip temperature estimation and prediction. Due to localized heating, the temperature variation within a single chip can reach up to 10s of degrees. For instance, within-die temperature variation up to 50 °C was reported in [9]. In order to accurately capture the within-die temperature variation for the emerging processors that become increasingly complex, a large number of temperature sensors must be deployed throughout the chip to collect thermal data in real time. These temperature sensors along with their peripheral circuits introduce substantial overhead in silicon area, power consumption and design complexity. Hence, it is extremely important to minimize the design overhead associated with temperature sensors without surrendering the accuracy of thermal monitoring. In other words, the open question here is how to accurately capture the full-chip thermal map with a *minimum* number of temperature sensors.

To answer this question, two different issues must be carefully addressed. First, an efficient methodology must be developed to identify the optimal spatial locations to deploy a minimum number of on-chip temperature sensors. This goal can be achieved by modeling and analyzing the thermal structure of the microprocessor design, the workload statistics and/or the historical thermal measurement data. Second, once the temperature data are measured by on-chip sensors, an efficient algorithm should be developed to estimate the full-chip thermal map. In the literature, various works have been proposed to address these two issues [10]-[17]. These techniques have been successfully applied to a broad range of practical applications.

In this paper, we propose an information-theoretic framework to address the thermal monitoring problem. Our goal is to fundamentally re-think the on-chip temperature variation from the information point of view and, consequently, develop an optimal solution to solve the thermal monitoring problem. To this end, we propose to accurately model and then quantitatively measure the uncertainty of on-chip temperature variation by differential entropy. Based on this information-theoretic framework, an optimization algorithm is developed to select a set of optimal sensor locations by minimizing the uncertainty (i.e., the differential entropy) of on-chip temperature variation. In other words, once the temperature is measured by on-chip sensors at these selected locations, the full-chip thermal map can be accurately estimated by the proposed information-theoretic framework. As will be demonstrated by the experimental examples in Section 4, the proposed entropy-based approach achieves superior accuracy (around 1.4× error reduction) for full-chip thermal monitoring over other traditional methods.

Another contribution of this paper is to develop an efficient numerical algorithm to reduce the computational complexity for the proposed entropy calculation and minimization. In particular, the proposed numerical algorithm is facilitated by two core techniques: (a) Schur complement, and (b) incremental Cholesky decomposition. The combination of these two techniques results in 69× runtime speedup, compared to a simple implementation without using these fast algorithms.

The remainder of this paper is organized as follows. In Section 2, we first develop the proposed information-theoretic framework for optimal sensor allocation and thermal map estimation. Next, several numerical algorithms are proposed in Section 3 to facilitate an efficient implementation with low computational cost. The efficacy of the proposed method is demonstrated by both simulation results and industrial measurement data in Section 4. Finally, we conclude in Section 5.

2. INFORMATION FRAMEWORK

2.1 Optimal Sensor Allocation

The proposed thermal monitoring is motivated by the assumption that on-chip temperature variation is spatially correlated. Such spatial correlation exists, because the active and leakage power generated at one spatial location can affect the temperature at multiple locations [2]. Due to the spatial correlation, it is possible to measure the temperature at a few locations and then predict the temperature at other locations.

Mathematically, the spatial temperature variation of a given chip can be modeled by a set of random variables:

$$X = \begin{bmatrix} x_1 & x_2 & \cdots & x_N \end{bmatrix}^T \tag{1}$$

where x_i denotes the temperature at the ith spatial location and N represents the total number of spatial locations of interest. In this paper, we further assume that the random variable X follows a multivariate Normal distribution:

$$pdf(X) = \frac{1}{\sqrt{(2 \cdot \pi)^N det(\Sigma_X)}} \cdot exp\left[-\frac{1}{2}(X - \mu_X)^T \Sigma_X^{-1} (X - \mu_X) \right] \tag{2}$$

where $pdf(\bullet)$ denotes the probability density function (PDF) of a random variable, $det(\bullet)$ represents the determinant of a matrix, and $\mu_X \in R^N$ and $\Sigma_X \in R^{N \times N}$ are the mean value and the covariance matrix of the multivariate random variable X respectively. A similar assumption of Normal distribution has been used by several previous works such as [12] and [17]. In practice, the mean μ_X and the covariance Σ_X of the Normal distribution can be estimated by a set of simulated and/or measured thermal maps, as will be illustrated by our experimental examples in Section 4.

Based on the information theory, the uncertainty posed by the multivariate random variable X can be quantitatively measured by its differential entropy [20]:

$$H(X) = -\int pdf(X) \cdot log[pdf(X)] \cdot dX . \tag{3}$$

Since X follows a multivariate Normal distribution, substituting (2) into (3) yields:

$$H(X) = \frac{N}{2} \cdot (1 + log 2\pi) + \frac{1}{2} \cdot log[det(\Sigma_X)] . \tag{4}$$

Note that Eq. (4) equals the summation of two different terms. The first term is proportional to N (i.e., the dimensionality of X). If a lot of random variables are required to model the variation (i.e., N is large), the differential entropy is large. The second term in (4) is proportional to the determinant $det(\Sigma_X)$ that directly measures the variance of the distribution. If the variance is large, the differential entropy (i.e., the uncertainty of the variation) is large. Finally, it is worth mentioning that the differential entropy in (4) is independent of the mean value of the Normal distribution. This observation is consistent with our intuition. Namely, the mean value only determines the "center" of the distribution, but it does not measure the "uncertainty" of the variation.

In our application of thermal monitoring, the objective is to measure a subset of (say, M) random variables and then estimate the values of other $N - M$ random variables. Such estimation is possible, since these random variables are correlated. Without loss of generality, we re-order the random variables in the vector X:

$$Y = W \cdot X = \begin{bmatrix} X_S \\ X_{\tilde{S}} \end{bmatrix} \tag{5}$$

where $X_S \in R^M$ contains the random variables that are measured, $X_{\tilde{S}} \in R^{N-M}$ contains the other random variables that should be estimated, and $W \in R^{N \times N}$ is a permutation matrix. After the permutation, the mean value and the covariance matrix of the new multivariate random variable Y are equal to:

$$\mu_Y = W \cdot \mu_X \tag{6}$$

$$\Sigma_Y = W \cdot \Sigma_X \cdot W^T . \tag{7}$$

Once the multivariate random variable X_S is measured:

$$X_S = T_S \tag{8}$$

the conditional probability $pdf(X_{\tilde{S}} | X_S)$ models the uncertainty of $X_{\tilde{S}}$ after X_S is known. As $X_{\tilde{S}}$ and X_S are jointly Normal, it can be proven that $pdf(X_{\tilde{S}} | X_S)$ is a multivariate Normal distribution [20]. Its mean value and covariance matrix are [20]:

$$\mu_{\tilde{S}|S} = \mu_{\tilde{S}} + \Sigma_{\tilde{S}S} \cdot \Sigma_{SS}^{-1} \cdot (T_S - \mu_S) \tag{9}$$

$$\Sigma_{\tilde{S}|S} = \Sigma_{\tilde{S}\tilde{S}} - \Sigma_{\tilde{S}S} \cdot \Sigma_{SS}^{-1} \cdot \Sigma_{S\tilde{S}} \tag{10}$$

where $\mu_S \in R^M$ and $\mu_{\tilde{S}} \in R^{(N-M)}$ are the elements of μ_Y:

$$\mu_Y = \begin{bmatrix} \mu_S \\ \mu_{\tilde{S}} \end{bmatrix} \tag{11}$$

and $\Sigma_{SS} \in R^{M \times M}$, $\Sigma_{\tilde{S}S} \in R^{(N-M) \times M}$, $\Sigma_{S\tilde{S}} \in R^{M \times (N-M)}$ and $\Sigma_{\tilde{S}\tilde{S}} \in R^{(N-M) \times (N-M)}$ are the four sub-matrices of Σ_Y:

$$\Sigma_Y = \begin{bmatrix} \Sigma_{SS} & \Sigma_{S\tilde{S}} \\ \Sigma_{\tilde{S}S} & \Sigma_{\tilde{S}\tilde{S}} \end{bmatrix} . \tag{12}$$

In (9)-(10), only the mean value $\mu_{\tilde{S}|S}$ depends on the measurement T_S. The covariance matrix $\Sigma_{\tilde{S}|S}$ is independent of T_S. This is an important property. It implies that the uncertainty of $pdf(X_{\tilde{S}} | X_S)$ is *not* dependent on the measured temperature value. Similar to (4), we can use differential entropy to quantitatively measure the uncertainty of $X_{\tilde{S}}$ once X_S is known:

$$H\left(X_{\tilde{S}}\middle|X_S\right) = \frac{N-M}{2} \cdot (1 + log 2\pi) + \frac{1}{2} \cdot log\left[det\left(\Sigma_{\tilde{S}|S}\right)\right] . \tag{13}$$

Based on the covariance matrix in (10), the block partition in (12) and the theory of Schur complement [21], we have:

$$det\left(\Sigma_{\tilde{S}|S}\right) = \frac{det(\Sigma_Y)}{det(\Sigma_{SS})} . \tag{14}$$

Substituting (7) into (14) yields:

$$det\left(\Sigma_{\tilde{S}|S}\right) = \frac{det\left(W \cdot \Sigma_X \cdot W^T\right)}{det(\Sigma_{SS})} = \frac{det(W) \cdot det(\Sigma_X) \cdot det(W^T)}{det(\Sigma_{SS})} . \tag{15}$$

Since W is a permutation matrix and $det(W) = det(W^T) = 1$, Eq. (15) can be further simplified as:

$$det\left(\Sigma_{\tilde{S}|S}\right) = \frac{det(\Sigma_X)}{det(\Sigma_{SS})} . \tag{16}$$

Combining (13) and (16), we have:

$$H\left(X_{\tilde{S}}\middle|X_S\right) = \frac{N-M}{2} \cdot (1 + log 2\pi) + \frac{1}{2} \cdot log\left[\frac{det(\Sigma_X)}{det(\Sigma_{SS})}\right] . \tag{17}$$

Based on the differential entropy in (17), our optimal sensor allocation problem can be mathematically stated as follows: *Given a set of correlated random variables $X = [x_1, x_2, ..., x_N]^T$, we want to identify and then measure an optimal subset of (i.e., M) random variables in X_S such that the differential entropy $H(X_{\bar{S}} \mid X_S)$ (i.e., the uncertainty posed by the remaining random variables in $X_{\bar{S}}$) is minimized.*

Studying (17), we would have two important observations. First, the differential entropy $H(X_{\bar{S}} \mid X_S)$ is independent of the measured value of X_S. This observation is consistent with the fact that the covariance matrix of the conditional probability $pdf(X_{\bar{S}} \mid X_S)$ is independent of the measurement of X_S, as shown in (10). It, in turn, facilitates us to minimize $H(X_{\bar{S}} \mid X_S)$ without knowing the measured value of X_S. Second, only the denominator of the second term in (17) depends on the random variables that are selected for measurement. Hence, in order to minimize the differential entropy $H(X_{\bar{S}} \mid X_S)$, we should maximize the determinant $det(\Sigma_{SS})$.

Algorithm 1: Optimal Sensor Allocation

1. Start from a set of correlated random variables $X = [x_1, x_2, ..., x_N]^T$ representing the temperature at different spatial locations, and a given number M representing the total number of temperature sensors.
2. Set $X_S = [\ ]$ and $X_{\bar{S}} = [x_1, x_2, ..., x_N]^T$, implying that no random variable is initially selected. Set the iteration index $m = 0$.
3. Find the optimal random variable x_k from the vector $X_{\bar{S}}$ such that the determinant of the covariance matrix Σ_{SS} is maximized after adding x_k to the vector X_S.
4. Remove x_k from the vector $X_{\bar{S}}$ and add it to the vector X_S.
5. Update the iteration index $m = m + 1$.
6. If $m = M$, stop iteration. Otherwise, go to Step 3.

Optimally finding the M random variables out of the N candidates, however, is a combinatorial optimization problem that is not trivial to solve. In this paper, we propose to apply a greedy algorithm, as shown in Algorithm 1. It selects one random variable to maximize the determinant $det(\Sigma_{SS})$ and, equivalently, minimize the differential entropy $H(X_{\bar{S}} \mid X_S)$ at each iteration step. Such an iterative process continues until M random variables are selected (i.e., the spatial locations of M temperature sensors are determined). The proposed greedy heuristic does not necessarily converge to the global optimum. However, as will be demonstrated by the experimental examples in Section 4, the proposed greedy algorithm reliably finds a set of good sensor locations and it substantially outperforms other traditional algorithms for sensor allocation.

Note that Algorithm 1 requires the total number of temperature sensors (i.e., M) as its input. In practice, the appropriate value of M should be determined by exploring the trade-offs between the accuracy of thermal monitoring and the overhead of circuit design. In other words, if a lot of temperature sensors are used (i.e., M is large), we can achieve high accuracy of thermal map estimation; however, the design overhead (e.g., the silicon area and the power consumption associated with the temperature sensors and their peripheral circuits) is also large. These design trade-offs will be further illustrated by our experimental examples in Section 4.

2.2 Thermal Map Estimation

Once the sensor locations are determined, we can measure the temperature $X_S = T_S$ at these locations. Our next goal is to estimate the value of $X_{\bar{S}}$, i.e., the temperature at other spatial locations that are not measured.

To this end, we need to re-visit the condition probability

$pdf(X_{\bar{S}} \mid X_S)$ that has been discussed in Section 2.1. Remember that $pdf(X_{\bar{S}} \mid X_S)$ models the distribution of $X_{\bar{S}}$ after X_S is known. As previously mentioned, $pdf(X_{\bar{S}} \mid X_S)$ is a multivariate Normal distribution. Its mean value and covariance matrix are shown in (9) and (10) respectively. The distribution $pdf(X_{\bar{S}} \mid X_S)$ implies that the value of $X_{\bar{S}}$ is not deterministic. In other words, since X_S and $X_{\bar{S}}$ are not fully correlated, it is impossible to exactly know $X_{\bar{S}}$ after X_S is measured.

While $X_{\bar{S}}$ cannot be uniquely determined, it is possible to statistically estimate $X_{\bar{S}}$ based on its probability distribution. In this paper, we aim to find the maximum-likelihood estimation (MLE) of $X_{\bar{S}}$, i.e., the value of $X_{\bar{S}}$ that is most likely to occur. Since $pdf(X_{\bar{S}} \mid X_S)$ follows a multivariate Normal distribution and it reaches the maximum value at its mean, the MLE of $X_{\bar{S}}$ is simply equal to the mean value of $pdf(X_{\bar{S}} \mid X_S)$, as shown in (9). Namely, substituting the measurement $X_S = T_S$ into (9) yields the estimated temperature $X_{\bar{S}}$ at other spatial locations.

Finally, it is worth mentioning that a similar statistical method has been proposed in [18] and applied to low-cost characterization of wafer-level process variation. In this paper, we extend the information-theoretic framework to full-chip thermal monitoring. Note that our thermal monitoring problem is substantially different from the variation characterization problem addressed in [18]. For example, due to the high dimensionality of the thermal monitoring problem, it is necessary to further develop efficient numerical algorithms to make the proposed entropy calculation and minimization practically tractable. These implementation details will be discussed in Section 3.

3. IMPLEMENTATION DETAILS

The proposed entropy calculation and minimization are made computationally efficient by implementing a number of fast numerical algorithms. In this section, we describe these implementation issues in detail.

3.1 Schur Complement

Studying Algorithm 1, we notice that Step 3 is the most computationally expensive part of Algorithm 1. During this step, the optimal random variable x_k should be selected from the vector $X_{\bar{S}}$. Towards this goal, the determinant of the following matrix Φ_k must be repeatedly calculated for each random variable x_k that belongs to the vector $X_{\bar{S}}$:

$$\Phi_k = \begin{bmatrix} \Sigma_{SS} & \Sigma_{Sk} \\ \Sigma_{kS} & \Sigma_{kk} \end{bmatrix} \quad (18)$$

where Σ_{SS} is the covariance matrix for all random variables belonging to the vector X_S, Σ_{kS} and Σ_{Sk} are the row and column vectors representing the covariance between the random variable x_k and those random variables in X_S, and Σ_{kk} is a scalar denoting the variance of the random variable x_k. Once the determinant $det(\Phi_k)$ is calculated for different x_k's, the optimal x_k with the maximum determinant will be selected.

In general, the determinant of the symmetric, positive semi-definite covariance matrix Φ_k can be computed by using Cholesky decomposition [21]. Such a simple implementation can be expensive, because the determinant $det(\Phi_k)$ must be repeatedly calculated for a large number of different x_k's. To minimize the computational cost, we expand $det(\Phi_k)$ by using the theory of Schur complement [21]:

$$det(\Phi_k) = det(\Sigma_{SS}) \cdot det(\Sigma_{kk} - \Sigma_{kS} \cdot \Sigma_{SS}^{-1} \cdot \Sigma_{Sk}). \quad (19)$$

Studying (19), we would have two important observations.

First, Σ_{SS}, a symmetric and positive semi-definite matrix, is independent of the random variable x_k that is selected. Hence, finding the optimal x_k to maximize the determinant $det(\Phi_k)$ is equivalent to maximizing the determinant $det(\Sigma_{kk} - \Sigma_{kS}\cdot\Sigma_{SS}^{-1}\cdot\Sigma_{Sk})$. In other words, only $det(\Sigma_{kk} - \Sigma_{kS}\cdot\Sigma_{SS}^{-1}\cdot\Sigma_{Sk})$ should be repeatedly calculated for different x_k during the search process.

Second, the matrix inverse Σ_{SS}^{-1} can be pre-computed. Once Σ_{SS}^{-1} is known, the determinant $det(\Sigma_{kk} - \Sigma_{kS}\cdot\Sigma_{SS}^{-1}\cdot\Sigma_{Sk})$ can be efficiently calculated by the following three numerical operations:

- *Matrix-vector multiplication*: The matrix Σ_{SS}^{-1} is multiplied by the column vector Σ_{Sk}, resulting in a column vector $\Sigma_{SS}^{-1}\cdot\Sigma_{Sk}$.
- *Vector-vector multiplication*: The row vector Σ_{kS} is multiplied by the column vector $\Sigma_{SS}^{-1}\cdot\Sigma_{Sk}$, resulting in a scalar $\Sigma_{kS}\cdot\Sigma_{SS}^{-1}\cdot\Sigma_{Sk}$.
- *Scalar subtraction*: The scalar $\Sigma_{kS}\cdot\Sigma_{SS}^{-1}\cdot\Sigma_{Sk}$ is subtracted from another scalar Σ_{kk}, resulting in a new scalar $\Sigma_{kk} - \Sigma_{kS}\cdot\Sigma_{SS}^{-1}\cdot\Sigma_{Sk}$. The determinant of a scalar is simply equal to the scalar itself, i.e., $det(\Sigma_{kk} - \Sigma_{kS}\cdot\Sigma_{SS}^{-1}\cdot\Sigma_{Sk}) = \Sigma_{kk} - \Sigma_{kS}\cdot\Sigma_{SS}^{-1}\cdot\Sigma_{Sk}$.

These numerical operations are substantially cheaper than the traditional Cholesky decomposition, as will be demonstrated by the experimental examples in Section 4.

The key idea of our proposed determinant calculation is to efficiently calculate a large number of determinant values by Schur complement. When implementing the proposed fast algorithm, the matrix inverse Σ_{SS}^{-1} must be efficiently calculated. In what follows, we will further develop an incremental algorithm for Cholesky decomposition that enables us to compute Σ_{SS}^{-1} with low computational cost.

3.2 Incremental Cholesky Decomposition

In this sub-section, we describe the detailed algorithm to calculate the matrix inverse Σ_{SS}^{-1} by using incremental Cholesky decomposition. As shown in Algorithm 1, an optimal random variable x_k is selected and added to the vector X_S at each iteration step. Initially, when X_S only contains one element and Σ_{SS} is a scalar, its inverse Σ_{SS}^{-1} can be easily calculated. Hence, our focus of this sub-section is to develop an efficient numerical algorithm to incrementally update Σ_{SS}^{-1} as more and more random variables are added to the vector X_S over the iterations of Algorithm 1.

Without loss of generality, we consider the case where a random variable x_k is added to the vector X_S at the mth iteration step. After adding x_k to the vector X_S at the end of this iteration step, the matrix Φ_k in (18) is exactly the matrix Σ_{SS} that is required for the next iteration step. Hence, we should calculate the matrix inverse Φ_k^{-1} and assign its value to Σ_{SS}^{-1}.

Our proposed numerical algorithm to calculate Φ_k^{-1} consists of two major steps. First, the matrix Φ_k is incrementally factorized by Cholesky decomposition:

$$\Phi_k = L_\Phi \cdot L_\Phi^T \qquad (20)$$

where L_Φ is a lower triangular matrix. During this step, we represent the lower triangular matrix L_Φ by a sub-matrix L_Σ, a row vector $L_{k\Sigma}$ and a scalar L_k:

$$L_\Phi = \begin{bmatrix} L_\Sigma & 0 \\ L_{k\Sigma} & L_k \end{bmatrix}. \qquad (21)$$

Substituting (21) into (20) yields:

$$\Phi_k = \begin{bmatrix} L_\Sigma L_\Sigma^T & L_\Sigma L_{k\Sigma}^T \\ L_{k\Sigma} L_\Sigma^T & L_{k\Sigma} L_{k\Sigma}^T + L_k^2 \end{bmatrix}. \qquad (22)$$

Comparing (18) and (22), it is easy to verify that the matrix

multiplication $L_\Sigma \cdot L_\Sigma^T$ is simply the Cholesky decomposition of the matrix Σ_{SS}. When an incremental algorithm is applied, the matrix L_Σ should already be calculated during the previous iteration step. The row vector $L_{k\Sigma}$ and the scalar L_k should be solved in order to incrementally calculate the new Cholesky decomposition $L_\Phi \cdot L_\Phi^T$ for the matrix Φ_k in (20).

Combining (18) and (22) yields the following linear equations:

$$L_\Sigma L_{k\Sigma}^T = \Sigma_{Sk} \qquad (23)$$

$$L_{k\Sigma} L_{k\Sigma}^T + L_k^2 = \Sigma_{kk}. \qquad (24)$$

In (23), the matrix L_Σ is a lower triangular matrix, since $L_\Sigma \cdot L_\Sigma^T$ is the Cholesky decomposition of Σ_{SS}. Therefore, the vector $L_{k\Sigma}$ can be easily solved from (23) by forward substitutions. Once $L_{k\Sigma}$ is found, the scalar L_k can be directly calculated from (24). After L_Σ, $L_{k\Sigma}$ and L_k are known, the lower triangular matrix L_Φ in (21) is determined.

Next, given the Cholesky decomposition $L_\Phi \cdot L_\Phi^T$ in (20), the matrix inverse Φ_k^{-1} is calculated by solving the following linear equations:

$$L_\Phi \cdot P = I \qquad (25)$$

$$L_\Phi^T \cdot \Phi_k^{-1} = P \qquad (26)$$

where I denotes an identity matrix. Since L_Φ and L_Φ^T are lower and upper triangular matrices respectively, the linear equations in (25)-(26) can be easily solved by forward and backward substitutions, resulting in the inverse matrix Φ_k^{-1} that we need. As Algorithm 1 proceeds to the next iteration step, the aforementioned procedure will be repeatedly applied and the matrices Φ_k, L_Φ and Φ_k^{-1} will be further updated incrementally.

3.3 Summary

Algorithm 2: Incremental Sensor Selection

1. Start from two given vectors X_S (containing the random variables that have already been selected) and $X_{\bar{S}}$ (containing the random variables that have not been selected), the covariance matrix Σ_{SS} for all random variables belonging to the vector X_S, the Cholesky decomposition $\Sigma_{SS} = L_\Sigma \cdot L_\Sigma^T$, and the inverse matrix Σ_{SS}^{-1}.

2. For each random variable x_k in the vector $X_{\bar{S}}$, calculate the determinant $det(\Sigma_{kk} - \Sigma_{kS}\cdot\Sigma_{SS}^{-1}\cdot\Sigma_{Sk})$ where Σ_{kS}, Σ_{Sk} and Σ_{kk} are defined in (18).

3. Select the optimal random variable x_k with the maximum determinant value.

4. Given the optimal random variable x_k, solve the row vector $L_{k\Sigma}$ and the scalar L_k from the equations in (23)-(24).

5. Form the matrix L_Φ in (21) and the Cholesky decomposition $L_\Phi \cdot L_\Phi^T$ for the matrix Φ_k in (20).

6. Calculate the matrix inverse Φ_k^{-1} by solving the linear equations in (25)-(26).

7. Remove x_k from the vector $X_{\bar{S}}$ and add it to the vector X_S. Let $\Sigma_{SS} = \Phi_k$, $L_\Sigma = L_\Phi$ and $\Sigma_{SS}^{-1} = \Phi_k^{-1}$.

Algorithm 2 summarizes the numerical method that is developed for incremental sensor selection. It should be used to select the optimal random variable x_k from the vector $X_{\bar{S}}$ at Step 3 and Step 4 of Algorithm 1. As shown in Algorithm 2, the proposed numerical computation starts from two given vectors X_S and $X_{\bar{S}}$. For each random variable x_k in the vector $X_{\bar{S}}$, it calculates the determinant $det(\Sigma_{kk} - \Sigma_{kS}\cdot\Sigma_{SS}^{-1}\cdot\Sigma_{Sk})$. The optimal random variable x_k with the maximum determinant is selected. Next, the Cholesky decomposition $\Phi_k = L_\Phi \cdot L_\Phi^T$ and the matrix inverse Φ_k^{-1}

are calculated. Finally, the optimal random variable x_k is moved from the vector $X_{\bar{S}}$ to X_S and the matrix Σ_{SS} is replaced by Φ_k, before the next iteration step of Algorithm 1 starts. Relying on these incremental matrix updates, the numerical operations required by Algorithm 2 can be performed with low computational cost. As will be demonstrated by our experimental examples in Section 4, the proposed fast algorithm (i.e., Algorithm 2) is able to achieve 69× runtime speedup over a simple implementation without using incremental updates.

4. EXPERIMENTAL RESULTS

In this section, we demonstrate the efficiency of the proposed sensor allocation and thermal monitoring method by using several experimental examples with both simulation and measurement data. All numerical experiments are run on a 2.53 GHz Linux server with 16 GB memory.

4.1 Simulation Experiment

We first set up a simulation study by using an 8-core microprocessor designed in a 45 nm process. The die area of this microprocessor is 2×2 cm^2. Its workload is simulated by using 11 SPEC2000 benchmarks (7 integer benchmarks and 4 floating point benchmarks). Wattch [19] is used to estimate the power consumption of each microarchitecture unit (e.g., branch predictor, integer queue, etc.). Next, given the microprocessor floorplan, HotSpot [4] is used to run thermal simulation to generate a set of full-chip thermal maps with the resolution of 128×128. For testing and comparison purpose, these thermal maps are partitioned into two non-overlapped groups: the training set and the testing set. The thermal maps in the training set are used to calculate the mean value μ_X and the covariance matrix Σ_X in (2), while those in the testing set are used to calculate the error for the proposed thermal map estimation.

We implement four different methods to determine the sensor allocation: (a) the clustering method [10], (b) the partition method [11], (c) the Bayesian method [12], and (d) the proposed entropy method. Once the sensor locations are determined, the maximum-likelihood estimation method in Section 2.2 is used to estimate the full-chip thermal map for any given temperature data collected at the sensor locations. By studying the accuracy of the estimated thermal maps, the efficacy of different sensor allocation algorithms is compared. In addition, to consider the measurement error of each temperature sensor, a small random noise is added to each sensor reading. In this example, the sensor error is modeled as a Gaussian distribution where the mean value is 0 °C and the standard deviation is 1 °C.

Figure 1 shows the temperature estimation error for four different sensor allocation algorithms. Here, the estimation error is calculated by using 5000 thermal maps in the testing set:

$$Error = \frac{1}{\sqrt{5000}} \cdot \sqrt{\sum_{k=1}^{5000} \max_i \left(\tilde{x}_{i,k} - x_{i,k}\right)^2} \qquad (27)$$

where $\tilde{x}_{i,k}$ and $x_{i,k}$ represent the exact and the estimated temperature values at the ith spatial location of the kth thermal map, respectively. Studying Figure 1, we would have two important observations. First, the proposed entropy method out-performs other traditional approaches. Our entropy method achieves 1.4× error reduction, when 9 sensors are selected. Second, the estimation error of the entropy method does not quickly decrease, if we use more than 9 sensors. In other words, deploying 9 sensors for thermal monitoring is a good choice to

explore the trade-offs between estimation accuracy and design complexity in this example.

Figure 2 further plots the spatial locations of 9 temperature sensors selected by different algorithms. In this example, the proposed entropy method arranges all sensors in a regular pattern. Namely, 8 sensors are optimally located at the hotspots of every core, and the other sensor is located at the center of the die to measure the environmental temperature variation. The other three traditional methods, however, cannot find the optimal sensor locations, as shown in Figure 2 (a)-(c). This is the reason why the traditional methods result in large estimation error in Figure 1. In this example, while the optimal sensor allocation is trivial, it provides an excellent test case for us to extensively compare different sensor allocation algorithms.

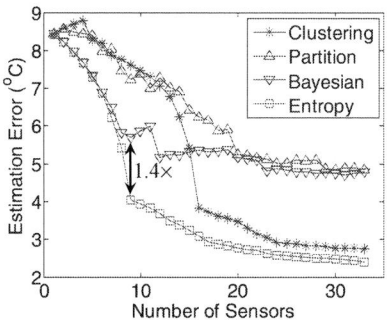

Figure 1. Full-chip thermal estimation error is compared for four different sensor allocation algorithms.

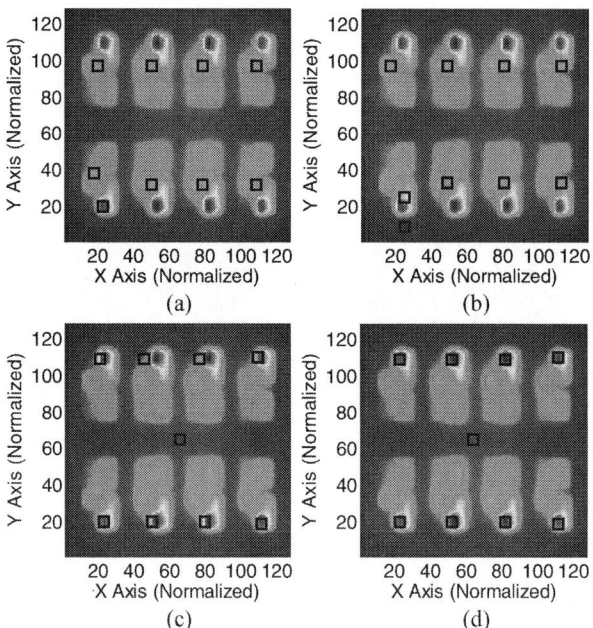

Figure 2. Temperature sensor locations are selected by different allocation algorithms: (a) the clustering method, (b) the partition method, (c) the Bayesian method, and (d) the entropy method.

Table 1. Runtime (Sec.) for different sensor allocation algorithms

Clustering [10]	Partition [11]	Bayesian [12]	Entropy (Simple)	Entropy (Fast)
18.78	0.74	7.42	160.24	2.31

Finally, Table 1 shows the runtime for different methods to determine the sensor locations for our experiment in Figure 1.

Note that our proposed fast algorithms achieve 69× runtime speedup over the simple implementation without using these fast algorithms. While the problem size in this example is small (i.e., only 33 sensors are selected in Figure 1), the runtime speedup of the proposed fast algorithms would be more pronounced when applied to larger-size problems.

4.2 Measurement Experiment

In this sub-section, we further demonstrate the efficacy of the proposed entropy method by collecting the temperature sensor data from an industrial dual-core microprocessor that contains 24 temperature sensors distributed in both cores and caches. In this experiment, all 24 temperature sensors are first calibrated using infrared imaging. Next, temperature readings are recorded from these 24 sensors when running a subset of SPEC2006 benchmarks. Similar to the previous example, the measurement data are partitioned into two non-overlapped groups: the training set and the testing set. Our objective in this example is to identify the most important temperature sensors out of these 24 candidates. Once the important sensors are found, the temperature values of other sensors can be estimated and, hence, these unimportant sensors may be eliminated in future design to reduce silicon area and/or power consumption.

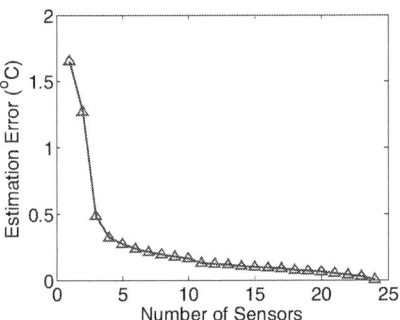

Figure 3. Thermal estimation error decreases, as the number of selected sensors increases.

Figure 3 shows the thermal estimation error as a function of the number of selected temperature sensors. Here, the estimation error is calculated by (27) using the measurement data in the testing set. Note that the estimation error is about 0.3 °C, when 5 sensors are selected. Selecting more than 5 sensors has diminishing returns in information collected. It, in turn, demonstrates that the optimal choice is to deploy 5 sensors for thermal monitoring in this example. The aforementioned analysis has potential benefits in both system and design areas. First, it reduces the number of sensors and information processing for thermal-aware OS (i.e., operating system) task scheduling [5]. Second, it demonstrates that our proposed sensor allocation technique is effective in real world and can be used for future microprocessor design.

5. CONCLUSIONS

In this paper, we propose a new information-theoretic framework to efficiently find the optimal spatial locations of on-chip temperature sensors for full-chip thermal monitoring. The key idea is to quantitatively model the uncertainty of on-chip temperature variation by differential entropy. Our experimental results demonstrate that the proposed entropy-based method achieves superior accuracy (around 1.4× error reduction) for full-

chip thermal monitoring over other traditional methods. The techniques developed in this paper can be further incorporated into various thermal management schemes for hotspot-limited microprocessors.

6. ACKNOWLEDGEMENTS

The authors acknowledge the support of the C2S2 Focus Center, one of six research centers funded under the Focus Center Research Program (FCRP), a Semiconductor Research Corporation entity. This work is also supported in part by the National Science Foundation under contract CCF–0915912.

7. REFERENCES

[1] H. Hamann, A. Weger, J. Lacey, Z. Hu and P. Bose, "Hotspot-limited microprocessors: direct temperature and power distribution measurements," *IEEE JSSC*, vol. 42, no. 1, pp. 56-65, 2007.

[2] M. Pedram and S. Nazarin, "Thermal modeling, analysis, and management in VLSI circuits: principles and methods," *Proceedings of the IEEE*, vol. 94, no. 8, pp. 1487-1501, 2006.

[3] D. Brooks and M. Martonosi, "Dynamic thermal management for high-performance microprocessors," *IEEE HPCA*, pp. 171-182, 2001.

[4] K. Skadron, M. Stan, H. Wei, S. Velusamy, K. Sankaranarayanan and D. Tarjan, "Temperature-aware microarchitecure," *IEEE ISCA*, pp. 2-13, 2003.

[5] J. Choi, C. Cher, H. Franke, A. Weger, and P. Bose, "Thermal-aware task scheduling at the system software level" *IEEE ISLPED*, pp. 213-218, 2007.

[6] Y. Wang, K. Ma and X. Wang, "Temperature-constrained power control for chip multiprocessors with online model estimation," *IEEE ISCA*, pp. 314-324, 2009.

[7] T. Ebi, M. Faruque and J. Henkel, "TAPE: Thermal-aware agent-based power economy multi/many-core architectures," *IEEE ICCAD*, pp. 302-309, 2009.

[8] C. Cher and E. Kursun, "Exploring the effects of on-chip thermal variation on high-performance multicore architectures," *ACM TACO*, vol. 8, no. 1, pp. 2:1-2:22, 2011.

[9] S. Bokar, T. Karnik, S. Narendra, J. Tschanz, A. Keshavarzi and V. De, "Parameter variations and impact on circuits and microarchitecture," *IEEE DAC*, pp. 338-342, 2003.

[10] S. Memik, R. Mukherjee, M. Ni and J. Long, "Optimizing thermal sensor allocation for microprocessors," *IEEE TCAD*, vol. 27, no. 3, pp. 516-527, 2008.

[11] A. Nowroz, R. Cochran and S. Reda, "Thermal monitoring of real processors: techniques for sensor allocation and full characterization," *IEEE DAC*, pp. 56-61, 2010.

[12] Y. Zhang, B. Shi and A. Srivastara, "A statistical framework for designing on-chip thermal sensing infrastructure in nano-scale systems," *IEEE ISPD*, pp. 169-176, 2010.

[13] F. Zanini, D. Atienza, C. Jones and G. Micheli, "Temperature sensor placement in thermal management systems for MPSoCs," *IEEE ISCAS*, pp. 1065-1068, 2010.

[14] S. Sharifi and T. Rosing, "Accurate direct and indirect on-chip temperature sensing for efficient dynamic thermal management," *IEEE TCAD*, vol. 29, no. 1, pp. 1586-1599, 2010.

[15] Y. Zhang, A. Srivastava and M. Zahran, "Chip level thermal profile estimation using on-chip temperature sensors," *IEEE ICCD*, pp. 432-437, 2008.

[16] R. Cochran and S. Reda, "Spectral techniques for high-resolution thermal characterization with limited sensor data," *IEEE DAC*, pp. 478-483, 2009.

[17] Y. Zhang and A. Srivastava, "Accurate temperature estimation using noisy thermal sensors," *IEEE DAC*, pp. 472-477, 2009.

[18] W. Zhang, X. Li and R. Rutenbar, "Bayesian virtual probe: minimizing variation characterization cost for nanoscale IC technologies via Bayesian inference," *IEEE DAC*, pp. 262-267, 2010.

[19] D. Brooks, V. Tiwari and M. Martonosi, "Wattch: a framework for architecture-level power analysis and optimizations," *IEEE ISCA*, pp. 83-94, 2000.

[20] C. Bishop, *Pattern Recognition and Machine Learning*, Prentice Hall, 2007.

[21] R. Horn and C. Johnson, *Matrix Analysis*, Cambridge Press, 2007.

Optimizing Energy Efficiency of 3-D Multicore Systems* with Stacked DRAM under Power and Thermal Constraints

Jie Meng   Katsutoshi Kawakami   Ayse K. Coskun

Electrical and Computer Engineering Department, Boston University, Boston, MA, USA

{jiemeng, kkawakam, acoskun}@bu.edu

ABSTRACT

3D multicore systems with stacked DRAM have the potential to boost system performance significantly; however, this performance increase may cause 3D systems to exceed the power budget or create thermal hot spots. This paper introduces a framework to model on-chip DRAM accesses and analyzes performance, power, and temperature tradeoffs of 3D systems. We propose a runtime optimization policy to maximize performance while maintaining power and thermal constraints. Our policy dynamically monitors workload behavior and selects among *low-power* and *turbo* operating modes accordingly. Experiments with multithreaded workloads demonstrate up to 49% energy efficiency improvements compared to existing thermal management policies.

Categories and Subject Descriptors

C.4 [**Performance of System**]: Modeling techniques

General Terms

Design, Experimentation, Management, Performance

Keywords

energy efficiency, thermal management, 3D multicore system

1. INTRODUCTION

3D stacking is a promising technique to increase transistor density per footprint without scaling the technology node, and it also enables stacking different technologies into a single chip. Using 3D stacking, it is possible to place a sizable DRAM layer within the chip, reducing the delays associated with accessing off-chip memory [1, 2]. On the other hand, 3D systems exacerbate the already existing thermal challenges because of higher thermal resistivities and power densities per chip footprint. Thermal hot spots and large temporal or spatial temperature variations adversely affect reliability and performance while increasing the cooling costs [3]. In addition to the temperature rise on the logic layers, temperature of the DRAM layers substantially increases because of the high memory access rate and the heat transfer from the logic layer. High DRAM temperatures severely affect memory reliability and performance [4, 5].

---

*This work has been funded by DAC A. Richard Newton Scholarship and NSF CAREER grant #1149703.

The thermal challenges in 3D systems require a joint assessment of performance, energy, and temperature tradeoffs. Also, as workload dynamics change during the lifetime of a system, it is imperative to have runtime optimization techniques that monitor and actively manage the interplay among performance, power, and temperature of 3D systems. Prior energy and temperature management methods for 3D systems include workload scheduling, dynamic voltage-frequency scaling (DVFS), thermally-aware floorplanning, and job allocation (e.g., [6, 7, 8, 3]). However, these techniques do not jointly evaluate and optimize performance, power, and temperature profiles at run-time for logic and DRAM layers in 3D systems simultaneously.

This paper focuses on optimizing the energy efficiency and temperature of 3D multicore systems with on-chip DRAM. We first model the performance, power, and thermal impacts of the on-chip DRAM and analyze how the reduced memory access latency changes runtime dynamics. We then propose a novel optimization technique that dynamically monitors application behavior through performance counters and adjusts the operating points for adapting to varying application phases. Our policy selects among *low-power* and *high-performance*, or *"turbo"*, execution modes from available voltage-frequency (V-F) settings by utilizing predictions from a regression-based model. In this way, we maximize throughput while maintaining the power and temperature constraints.

The optimization policy is motivated by two observations derived from our analysis of 3D systems with on-chip DRAM. First, we observe that memory-bound benchmarks have significant performance improvements when running on 3D systems with on-chip DRAM compared to the 2D baseline with off-chip memory. However, power and temperatures on both logic and DRAM layers rise significantly. In this case, our policy selects a *low-power* V-F setting to maximize throughput under power and thermal constraints. Second, for CPU-bound benchmarks, we observe limited performance improvement compared to the 2D baseline. However, for CPU-bound applications, stacking the DRAM layer with the logic layer provides a *temperature slack* as the DRAM layer is much cooler than the logic layer and helps maintain low temperature. In this case, we boost system performance using high-frequency *turbo* modes without creating thermal problems. Our specific contributions are as follows:

- We design a simulation framework to model the on-chip DRAM accesses and jointly analyze performance, power, and temperature for both logic and memory layers on 3D systems with stacked DRAM. Using the framework, we analyze on-chip DRAM accesses at various bandwidths. Enabling parallel access to the DRAM improves performance by up to 86.9% compared to single-bus access.

- We propose a novel runtime optimization policy for selecting V-F settings to maximize system performance subject to power and thermal constraints. Our experiments demonstrate that our policy achieves an average performance improvement of 36.1% for a 16-core 3D system with stacked DRAM compared to a statically optimized 3D system with fixed V-F settings. We reduce the energy-delay product (EDP) by up to 49.4% compared to a 3D system managed by a temperature-triggered DVFS policy.

The rest of the paper starts with an overview of the related work. Section 3 introduces the experimental methodology. Sections 4 and 5 propose the runtime optimization policy and present the experimental results, respectively. Section 6 concludes the paper.

2. RELATED WORK

Recent literature has studied the performance and energy benefits of 3D systems with on-chip DRAM. However, most prior work considers the evaluation of performance, power, and temperature separately. Loi et al. analyze 3D system performance with thermal considerations using a standard heat flow model [9], and Loh explores 3D-stacked memory architectures [2] with temperature analysis using HotSpot [10]. However, their thermal simulations are based on coarse-grained power estimates instead of using power traces obtained from detailed performance statistics.

Prior research on 3D system energy and thermal management includes design-time optimization methods and runtime management polices based on task scheduling and DVFS techniques. Cong et al. introduce a thermally-aware 3D placement approach based on transformation techniques [8]. Healy et al. propose a microarchitectural floorplanning algorithm for 3D ICs using linear programming and simulated annealing [11]. These static optimization methods are implemented at design stage, and do not address dynamic changes in workload profiles.

Dynamic power management on multicore 2D systems has been well studied. Isci et al. present a runtime phase prediction methodology to control DVFS based on frequency of memory operations [12]. Cochran et al. propose a scalable method for determining the optimal V-F settings under power constraints [13]. For dynamic management on 3D systems, Zhu et al. propose a runtime thermal management approach using task migration and DVFS [6]. Zhou et al. introduce an OS-level scheduling algorithm for optimizing 3D system temperature using dynamic workload scheduling [14]. These methods targeting 3D systems, however, do not consider detailed performance analysis of the workloads.

Our research differentiates from prior work as we provide a modeling and management methodology to jointly analyze and optimize performance, power, and temperature for 3D systems with on-chip DRAM. We analyze the performance impact of 3D stacked DRAM for single-bus or parallel access scenarios, and design a detailed on-chip DRAM performance and power model. We then propose a runtime optimization method that selects low-power or turbo operating modes based on processor and DRAM utilization in the 3D stack.

3. METHODOLOGY

Our research targets 3D multicore processors with on-chip DRAM. Figure 1 provides an illustration of the logic layer of a 16-core 3D system with stacked DRAM. In the 3D system, all the processing cores and caches are on one layer and a 3D DRAM layer is stacked below it. Through-silicon vias (TSVs) are used for vertically connecting the core and DRAM layers. We assume face-to-back, wafer-to-wafer bonding for building the 3D systems. Wafer-to-wafer bonding allows for reliably manufacturing larger 3D systems. Both the target 3D system with on-chip DRAM and the 2D baseline with off-chip memory have the same core architecture and the same floorplan for the logic layer. The core architecture of the target system is modeled based on the AMD Family 10h microarchitecture used in AMD Magny-Cours processors. Each core has multiple-issue and out-of-order execution. The architectural parameters for cores and caches are listed in Appendix $S2$. We assume the target processor is manufactured with 45nm technology, has a total die area of $376mm^2$, and can be operated under five different V-F settings, as listed in Table 2.

3.1 Modeling Memory Accesses

In order to accurately quantify the performance improvements of our target 3D systems, we model the memory access latency by examining the different components that contribute to the latency. For multicore systems, there are three main components of the memory access latency from the last-level caches to main memory: the propagation delay between last-level caches to memory controller (LLC-to-controller delay), the data request time spent at the memory controller (memory controller processing latency), and the data retrieval time spent at the DRAM.

To model the LLC-to-memory controller delay, we assume that all the private L2 caches are connected to the memory controllers through a shared bus. Figure 1 illustrates the physical layout of the logic layer. We assume that the global bus interconnect is routed around the chip in a serpentine fashion. For modeling the bus interconnect, we use energy-optimized repeater-inserted pipelined channels to reduce the global wire delay. The wire propagation delay is linear with respect to the wire length, owing to the repeaters that are inserted to partition the wire into smaller segments. Each pipeline stage is designed using predictive technology model for 45nm and has a propagation delay of 183ps per mm. We estimate the average distance from an L2 cache to a memory controller block as 9.4mm based on the layout. Thus, the round trip LLC-to-memory controller latency is 4ns (rounded up).

Memory access latency is strongly governed by the memory controller processing time. Modern memory controllers typically consist of a memory request queue that buffers the pending requests waiting to get scheduled, and a scheduler that selects the next request to be serviced. The memory controller processing latency, thus, refers to the time spent

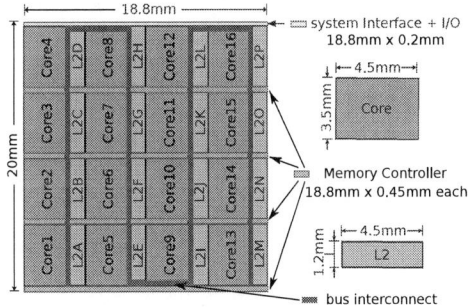

Figure 1: The layout for the logic layer of target 3D system.

Table 1: DRAM access latency

	2D-baseline design	3D system with regular memory access
memory controller	4ns LLC-to-controller delay, 48ns memory controller processing time	4ns LLC-to-controller delay, 24ns memory controller processing time
main memory	off-chip DRAM, $t_{RAS} = 36ns$, $t_{RP} = 15ns$	on-chip DRAM, $t_{RAS} = 36ns$, $t_{RP} = 15ns$
memory bus	off-chip bus, 200MHz, 8-Byte bus width	on-chip bus, 2GHz, 64-Byte bus width
total delay	103ns	79ns

by a memory request waiting to get scheduled. We set the overall memory controller processing latency as 100 cycles from the simulation results reported in prior work [15], where the memory controller latency of a 16-core processor with 4 memory controllers running PARSEC benchmark suite is studied. For the 3D system, we assume that the memory controller latency is reduced by 50% [16].

We use the same DRAM structure for the off-chip DRAM in 2D baseline and for the DRAM layer in 3D system, where we consider a 1GB DRAM consisting of 4 ranks, each of which has 4 banks (16 banks total). We use the row active time $t_{RAS} = 36ns$ and row precharge time $t_{RP} = 15ns$ as reported by MICRON's DDR3 SDRAM. We use the same timing parameters for the DRAM layer of the target 3D system, which is consistent with the assumptions used in earlier studies [2, 9]. Table 1 summarizes the memory access times for 2D and 3D systems.

To simulate the data transfer between logic layer and DRAM layer, we consider `regular memory access` and `parallel memory access`, both with a fast memory bus at 2G-Hz. As illustrated in Figure 2, the `parallel memory access` allows the four on-chip memory controllers to access the four DRAM ranks at the same time. We deploy 512 TSVs on each memory controller, which provide a 64-Byte bus width for each memory controller with only 0.2% chip area overhead. From our simulation results for the NAS and PARSEC benchmarks, we observe the accesses to the main memory are evenly distributed among the four ranks, as shown in Appendix $S3$. Therefore, we assume the memory access latency with `parallel access` is one-fourth of the latency of `regular access`. Note that this is a conservative assumption as the simultaneous accesses also enable faster processing at the memory controller because of fewer pending requests in the request queues.

3.2 Performance Simulation

We use M5 full-system simulator [17] to conduct the performance simulation for our target systems. We use the Alpha instruction set architecture (ISA) as it is the most stable ISA currently supported in M5. The full-system mode in M5 models a DEC Tsunami system to boot Linux OS. We model the 3D system with on-chip DRAM in M5 by configuring the main memory access latency and the bus width/speed between L2 caches and main memory to mimic the high data

transfer bandwidth provided by the TSVs. The architectural configuration parameters and memory access latencies are shown in Appendix $S2$ and in Table 1, respectively.

We select parallel applications from the PARSEC [18] and NAS Parallel Benchmarks (NPB) suite [19] as our target workloads. We run PARSEC benchmarks in M5 with sim-large input sets and NAS with class B problem sets. For each benchmark, we fast-forward to get past the serial initialization phase. Then, we execute each benchmark with the out-of-order CPUs with detailed memory access simulations, and collect statistics at every 100 million instructions for 100 sampling steps. In order to collect the access statistics for the 3D stacked DRAM, we track the memory accesses to each DRAM bank by observing the least significant bits for the physical memory addresses at every interval. The performance statistics collected from M5 simulations are used as inputs for the processor and DRAM power models.

3.3 Power Model

We use McPAT 0.7 [20] for 45nm process to obtain the run-time dynamic power of the cores. In our McPAT simulations for 2D baseline, we set V_{dd} to 1.1V and operating frequency to 2.1GHz. For our target 3D system, we use five V-F settings as shown in Table 2 (see Appendix $S4$ for average core power results). The L2 cache power is calculated using Cacti 5.3 [21], where the dynamic L2 power is scaled using L2 access rates. The average L2 cache power is 0.62W.

We calibrate the McPAT run-time dynamic core power using measurements that we collect on an AMD Magny-Cours processor. We derive the average dynamic core power values from power simulation across the benchmark suite, and compute the calibration factor, R, to translate the McPAT raw data to the target power scale. Then, we use R to scale dynamic core power consumption. A similar calibration approach has been introduced in prior work [22]. At nominal temperature of 343K, we assume the leakage power for the cores is 35% of the total core power, which matches the measurements on the AMD Mangy-Cours system. We also take the temperature and voltage impact on leakage power into account. The impact of temperature on leakage power is exponentially dependent on the temperature [23]. Prior work shows close-to linear relation between V_{dd} and leakage when variation of V_{dd} is small [24]. As voltage change is limited to 10% of the default setting in our system, we model leakage dependence on V_{dd} as linear.

The DRAM power in the 3D system is calculated using MICRON's DRAM power calculator [25], which takes the memory read and write access rates as inputs. We obtain detailed DRAM power traces for each of the DRAM banks at every sampling interval. The average on-chip DRAM bank power across all the benchmarks in 3D system with parallel access is 1.44W. The on-chip memory controller power for both 2D and 3D systems is estimated based on Intel's 48-core single-chip cloud computer as 5.9W [26]. We assume the system interface and I/O power as well as the on-chip bus

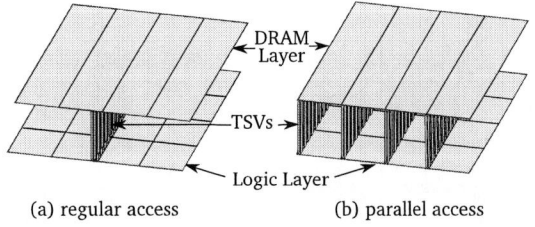

(a) regular access   (b) parallel access

Figure 2: An illustration of the 3D system with DRAM stacking that has (a) `regular memory access` and (b) `parallel memory access`.

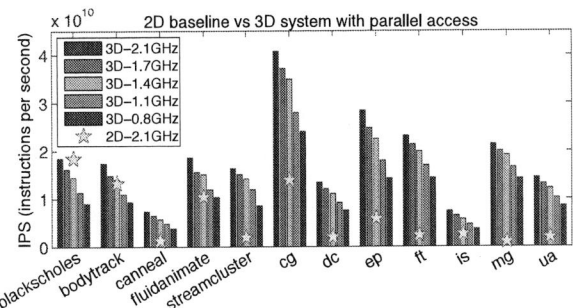

Figure 3: IPS for PARSEC and NAS benchmarks running on 2D baseline and the 3D system with parallel access.

power are negligible with respect to the total chip power. It has been shown that the total on-chip bus power for running PARSEC and NAS workloads is less than 2.0W even for a 64-core system [27].

3.4 Thermal Model

We use HotSpot 5.0 [10] for thermal simulations. We extend HotSpot (see Appendix $S1$) to account for the TSVs in 3D systems by utilizing the methodology for modeling the interlayer material heterogeneity introduced in prior work [3]. Appendix $S5$ provides the thermal parameters for the HotSpot simulations.

Our simulation framework is able to periodically sample runtime events at every fixed time interval or at a certain number of instructions. In this way, we observe the dynamically changing performance patterns that cannot be captured by average or coarse-grained performance estimates. For each benchmark and for each V-F setting, we record the power and performance data at every 100 million instructions into a database. Thermal simulator polls this database to gather power traces based on a fixed or dynamically set V-F setting, as determined by the policy running. Instruction-based intervals are converted to time-based samples as required by thermal simulations. This approach enables decoupling thermal simulation from lengthy performance-power simulations and achieves significant speedups. Even when applying DVFS policies, we are able to maintain accuracy due to two main reasons: (1) we change V-F settings of all the cores together; (2) for all the benchmarks in our evaluation set, the distribution of executed instructions among all the cores is very similar when running at different V-F settings, which allows the runtime V-F setting changes in our modeling methodology.

4. RUN-TIME OPTIMIZATION POLICY

The goal of our runtime optimization policy is to select operating points maximizing performance while maintaining the power and temperature constrains for both logic and DRAM layers. Our optimization policy is motivated by the observations of running PARSEC and NAS benchmarks on our simulation framework under different V-F settings. Figures 3, 4, and 5 present the performance, temperature, and power results of the 2D baseline and the target 3D system with stacked DRAM, respectively.

We notice that, for most of the benchmarks, the average IPS of 3D systems running at 0.8GHz are sufficiently high to match the performance of the 2D baseline. We also observe that applications dramatically differ in their performance behavior. For the memory-intensive benchmarks, such as

streamcluster and *mg*, the high memory access rates result in significant performance improvements when running on 3D systems with stacked DRAM in comparison to 2D baseline; however, the peak temperature also considerably increases. Thus, we run such benchmarks at the *low-power* mode by exploiting the *performance slack*. Figure 3 shows that, even at *low-power* mode, the memory-intensive benchmarks running on the 3D system still have significant performance improvements in comparison to running on 2D baseline. For CPU-intensive workloads, on the other hand, the low memory access rate result in a cooler DRAM layer that shares the temperature of the hotter core layer. For benchmarks such as *blackscholes*, we switch to the *turbo* mode with higher V-F settings for boosting the performance by taking advantage of the *temperature slack*.

The basic concept of our optimization method is presented in Equation (1), where (F, V) is the set of available V-F settings. Our goal is to maximize throughput (instructions per second, IPS) under power and thermal constraints. P_{cap} is the power budget of the target system, and T_{thld} is the peak temperature threshold to ensure reliable operation. As shown in Figure 4, we set T_{thld} at $85^\circ C$. Figure 5 shows three P_{cap} settings. Our policy satisfies T_{thld} and P_{cap} at the same time. For example, at a loose P_{cap} of 200W, T_{thld} at $85^\circ C$ dominates the optimization decisions. A more strict P_{cap} at 175W or 155W requires taking peak power into account. Peak power management is an increasingly important feature owing to power supply limitations and potential energy cost reduction opportunities at large computer clusters.

$$\begin{aligned} \underset{(f,v)\in(F,V)}{\text{maximize}} \quad & IPS(f,v) \quad\quad\quad\quad\quad\quad (1)\\ \text{subject to} \quad & power(f,v) \le P_{cap},\ temperature(f,v) \le T_{thld}. \end{aligned}$$

Our runtime optimization policy is illustrated in Figure 6. We start running the application with the lowest V-F setting to ensure reliable operation, and collect the performance statistics at regular intervals of 100 million instructions. Based on a model we construct offline, we predict the highest V-F setting satisfying the constraints using the performance statistics as inputs. We continue running the application with the predicted V-F setting. This process is repeated at every interval.

We choose instructions per cycle (IPC) and memory access per instruction (MA) to construct a regression-based model for selecting the V-F settings. This is because IPC is a good indicator of the power of the logic layer and MA is a good indicator of the power of the DRAM layer. Power densities on both layers affect chip peak temperature on the 3D system. Our V-F prediction model is in the form of $VF = c_0 + c_1 \cdot MA + c_2 \cdot IPC + c_3 \cdot MA*IPC$. We train

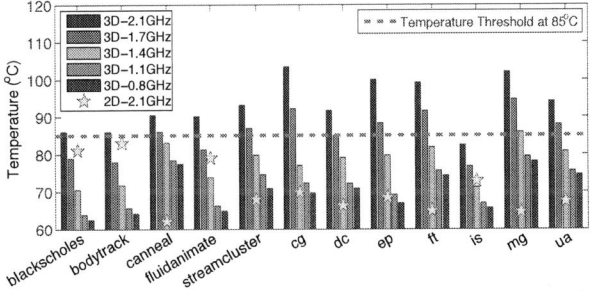

Figure 4: Peak chip temperature on the 3D system with parallel access running at different V-F settings.

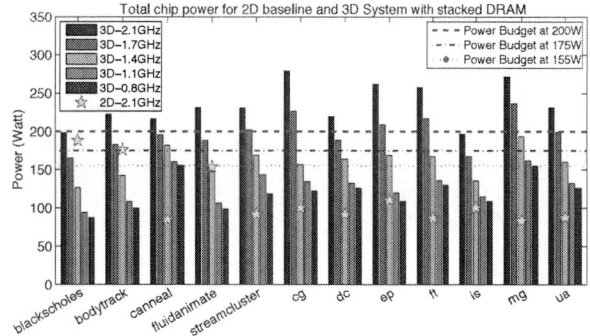

Figure 5: Total chip power on the 3D system with parallel access running at different V-F settings.

the regression model with power and performance statistics from simulations across all benchmarks. Note that we need to use different coefficients in the model depending on the current V-F setting, as MA and IPC vary with the V-F setting. As an example of the V-F prediction for a 3D system with $85^{o}C/175W$ constraints, we list the coefficients of the regression-based model for all the V-F settings in Table 2. The regression model provides accurate prediction as shown in Figure 10, and can be refined at runtime if needed. The overhead of the run-time prediction is negligible, since computing a simple equation at every interval has very low computational cost.

Table 2: Regression coefficients for a target 3D system with $85^{o}C/175W$ constraints for all the V-F settings.

V-F setting	c_0	c_1	c_2	c_3
2.1GHz/1.1V	3.68	-147.95	-0.059	0.19
1.7GHz/1.06V	3.74	-141.77	-0.071	0.23
1.4GHz/1.02V	3.76	-145.71	-0.075	0.36
1.1GHz/1.0V	3.80	-147.08	-0.087	0.41
0.8GHz/0.98V	3.87	-152.01	-0.072	0.58

5. EXPERIMENTAL RESULTS

This section evaluates our technique on 3D systems with parallel access, and compares our optimization policy against using static V-F settings, a temperature-triggered DVFS policy, and a DVFS policy guided by memory accesses.

Figure 7 demonstrates the performance improvement of the 3D system with parallel on-chip DRAM access running at 2.1GHz and 0.8GHz. We show that enabling parallel access to the 3D DRAM layer improves IPS by up to 86.9% compared to using regular access. `streamcluster` and `mg` show higher IPS improvements than the other benchmarks, since they have higher memory access rates and thus benefit more from reduced average memory access time.

Table 3 presents the performance and energy-efficiency improvements for 3D systems running our runtime optimization policy compared to using static V-F settings. We notice that the peak temperatures go over the thermal constraint

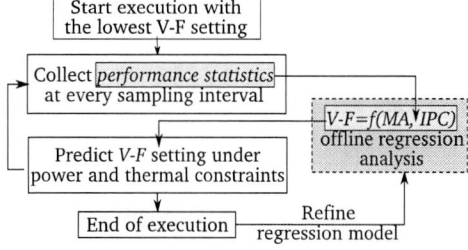

Figure 6: The flowchart of our runtime optimization policy.

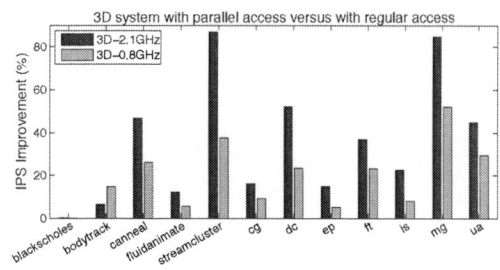

Figure 7: Performance improvement on 3D system with parallel access compared to 3D system with regular access.

of $85^{o}C$ for applications running on the 3D systems with frequency settings higher than 1.1GHz. With a loose power constraint of 200W, we compare our policy with the static V-F setting at 1.1GHz/1.0V which maintains temperature below $85^{o}C$ for all the benchmarks. Our policy achieves an average IPS improvement of 24.6% and EDP reduction of 22.4% across all the benchmarks.

We present the runtime V-F selection process of our optimization policy in Figure 8. For *ua*, 1.4GHz/1.02V is the reliable static operating point, maintaining the temperature below $85^{o}C$. However, the phase change of *ua* creates a temperature slack periodically. Our policy takes advantage of the temperature slack and switches to 1.7GHz during periods of low temperature.

We demonstrate the advantage of our runtime optimization policy over applying temperature-triggered DVFS in Figure 9. Temperature-triggered DVFS is a well-known policy for thermal management on 2D systems [10, 28]. It tracks chip peak temperature and selects the operating point based on temperature sensor readings. For safe operation while maintaining system performance, we choose two temperature thresholds as $80^{o}C$ and $70^{o}C$. Temperature-triggered DVFS reduces/increases the V-F setting when temperature goes above/below $80^{o}C/70^{o}C$. Our policy improves EDP by up to 61.9% and IPS by 32.2% on average across all the benchmarks in comparison to the temperature-triggered DVFS policy. The performance of *blackscholes* and *is* does not differ between our policy and the temperature-triggered DVFS policy. This is because they have low temperature while running at 2.1GHz/1.1V. The benchmarks that have high temperatures when running on 3D systems with stacked DRAM, such as *streamcluster*, show larger performance improvement using our runtime policy. Our policy selects the highest V-F settings to operate at safe temperatures, while temperature-triggered DVFS may oscillate around the high temperature threshold.

We also compare our optimization policy against memory access driven DVFS, in which V-F selections are mainly guided by the memory access rate (e.g., [29]). For implementing memory access driven DVFS, we construct a regression-based model for selecting V-F setting with only

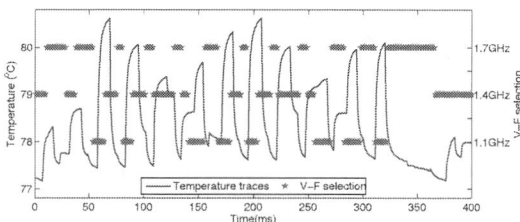

Figure 8: Temperature trace of *ua* on the 3D system running at 1.4GHz/1.02V and the V-F setting selected by our runtime management policy.

652

Table 3: Comparison of our runtime optimization policy against 3D systems with static settings.

Policy	Static V/F settings (GHz/V)					Runtime optimization		
	0.8/0.98	1.1/1.0	1.4/1.02	1.7/1.06	2.1/1.1	85°C/155W	85°C/175W	85°C/200W
Peak P (W)	154.72	161.53	193.37	236.79	279.25	154.85	168.63	189.62
Peak T ($°C$)	78.10	79.46	85.85	94.65	103.39	77.97	80.81	83.32
EDP*★ ($J{\cdot}s$)	246.42	167.63	135.18	132.19	119.82	185.67	145.11	130.03
IPnS**★	10.63	12.86	15.73	16.93	18.93	14.47	15.68	16.02

* EDP per 10billion instructions, ** IPnS stands for instructions per nanosecond, ★ Average across all benchmarks.

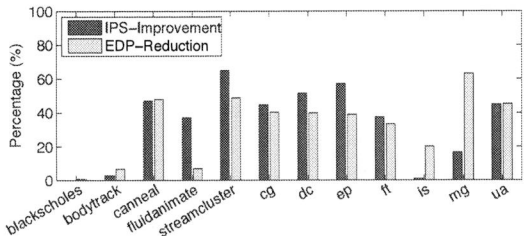

Figure 9: 3D system with runtime management policy in comparison to temperature-triggered DVFS policy.

MA. We show the V-F prediction for 3D system with 85°/ 175W constraints in Figure 10. By only using MA, three out of twelve benchmarks end up with different V-F settings than the optimal ones; while the predictions are all accurate using both IPC and MA as in our policy. The benchmarks that are predicted incorrectly using only MA are *blackscholes*, *is*, and *mg*. *blackscholes* has low MA but high IPC, *is* has both low MA and low IPC, and *mg* has high MA and relatively higher IPC than the other memory-bound benchmarks. Our policy provides accurate prediction as we take the power and temperature constraints on both logic and DRAM layers into account on 3D systems with stacked DRAM, where both high IPC and memory access rate could result in high chip power and peak temperature.

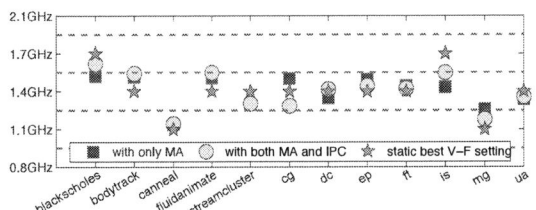

Figure 10: Prediction accuracy of our runtime management policy versus memory access (MA) driven DVFS.

6. CONCLUSION

In this paper, we have provided a methodology to evaluate 3D systems with stacked DRAM and proposed a runtime management policy for dynamically selecting operating points. We have evaluated various access bandwidths to DRAM and demonstrated up to 86.9% performance improvement using parallel access to the DRAM layer compared to regular memory access. Our experiments show that our optimization policy achieves performance improvement of 36.1% for 3D systems with stacked DRAM in comparison to using static V-F settings and EDP reduction of up to 49.4% compared to a temperature-triggered DVFS policy.

REFERENCES

[1] B. Black *et al.*, "Die stacking (3D) microarchitecture," in *MICRO*, pp. 469–479, 2006.

[2] G. H. Loh, "3D-stacked memory architectures for multi-core processors," in *ISCA*, pp. 453–464, 2008.

[3] A. K. Coskun, D. Atienza, T. S. Rosing, T. Brunschwiler, and B. Michel, "Energy-efficient variable-flow liquid cooling in 3D stacked architectures," in *DATE*, pp. 111–116, 2010.

[4] M. Ghosh and H.-H. S. Lee, "Smart refresh: An enhanced memory controller design for reducing energy in conventional and 3D die-stacked DRAMs," in *MICRO*, pp. 134–145, 2007.

[5] S. Liu *et al.*, "Hardware/software techniques for DRAM thermal management," in *HPCA*, pp. 515–525, 2011.

[6] C. Zhu *et al.*, "Three-dimensional chip-multiprocessor run-time thermal management," *IEEE Transactions on CAD (TCAD)*, vol. 27, no. 8, pp. 1479–1492, 2008.

[7] W. Hung *et al.*, "Interconnect and thermal-aware floorplanning for 3D microprocessors," in *ISQED*, pp. 98–104, 2006.

[8] J. Cong *et al.*, "Thermal-aware 3D IC placement via transformation," in *ASP-DAC*, pp. 780–785, 2007.

[9] G. L. Loi *et al.*, "A thermally-aware performance analysis of vertically integrated (3-D) processor-memory hierarchy," in *DAC*, pp. 991–996, 2006.

[10] K. Skadron *et al.*, "Temperature-aware microarchitecture," in *ISCA*, pp. 2–13, 2003.

[11] M. Healy *et al.*, "Multiobjective microarchitectural floor-planning for 2-D and 3-D ICs," *TCAD*, vol. 26, no. 1, pp. 38–52, 2007.

[12] C. Isci, G. Contreras, and M. Martonosi, "Live, runtime phase monitoring and prediction on real systems with application to dynamic power management," in *MICRO*, pp. 359–370, 2006.

[13] R. Cochran, C. Hankendi, A. K. Coskun, and S. Reda, "Identifying the optimal energy-efñAcient operating points of parallel workloads," in *ICCAD*, pp. 608–615, 2011.

[14] X. Zhou *et al.*, "Thermal management for 3D processors via task scheduling," in *ICPP*, pp. 115–122, 2008.

[15] M. Awasthi *et al.*, "Handling the problems and opportunities posed by multiple on-chip memory controllers," in *PACT*, pp. 319–330, 2010.

[16] C. C. Liu *et al.*, "Bridging the processor-memory performance gap with 3D IC technology," *IEEE Design Test of Computers*, vol. 22, no. 6, pp. 556–564, 2005.

[17] N. L. Binkert *et al.*, "The M5 simulator: Modeling networked systems," *IEEE Micro*, vol. 26, no. 4, pp. 52–60, 2006.

[18] C. Bienia, *Benchmarking Modern Multiprocessors*. PhD thesis, Princeton University, January 2011.

[19] D. Bailey *et al.*, "The NAS parallel benchmarks," tech. rep., 1994.

[20] S. Li *et al.*, "McPAT: An integrated power, area, and timing modeling framework for multicore and manycore architectures," in *MICRO*, pp. 469–480, 2009.

[21] S. Thoziyoor *et al.*, "CACTI 5.1," tech. rep., April 2008.

[22] R. Kumar *et al.*, "Single-ISA heterogeneous multi-core architectures: the potential for processor power reduction," in *MICRO*, pp. 81–92, 2003.

[23] J. Srinivasan *et al.*, "The case for lifetime reliability-aware microprocessors," in *ISCA*, pp. 276–287, 2004.

[24] H. Su *et al.*, "Full chip leakage estimation considering power supply and temperature variations," in *ISLPED*, pp. 78–83, 2003.

[25] "DRAM power calculations." http://www.micron.com/. Micron Technology Inc.

[26] J. Howard *et al.*, "A 48-core IA-32 message-passing processor with DVFS in 45nm CMOS," in *ISSCC*, pp. 108–109, 2010.

[27] J. Meng, C. Chen, A. K. Coskun, and A. Joshi, "Run-time energy management of manycore systems through reconfigurable interconnects," in *GLSVLSI*, pp. 43–48, 2011.

[28] A. K. Coskun *et al.*, "Evaluating the impact of job scheduling and power management on processor lifetime for chip multiprocessors," in *SIGMETRICS*, pp. 169–180, 2009.

[29] C. Isci *et al.*, "An analysis of efficient multi-core global power management policies: Maximizing performance for a given power budget," in *MICRO*, pp. 347–358, 2006.

SUPPLEMENTAL MATERIAL

S1. Modeling the Thermal Impact of TSVs on 3D Systems with Stacked DRAM

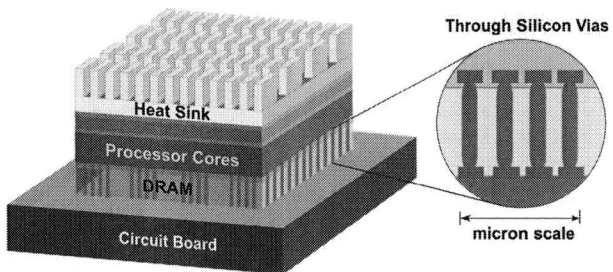

Figure 11: An illustration of the 3D multicore processor with stacked DRAM. TSVs are used to connect the on-chip DRAM layer with the logic layer.

We extend HotSpot to enable the modeling of TSVs in 3D multicore systems with stacked DRAM. The HotSpot extension utilizes the methodology for modeling the interlayer material heterogeneity introduced in prior work [3].

HotSpot 5.0 [10] has the functionality of modeling stacked 3D chips through a layer configuration file which provides the set of layers and the physical properties for each layer. In the default HotSpot tool, the properties across a single layer of the chip are homogenous with same resistivity and capacitance values for all the units. Each layer of the 3D chip is resolved to a grid and the temperature responses are calculated for each grid cell using the parameters from the layer configuration file.

In order to model the thermal effect of the TSVs in 3D stacked systems, our HotSpot extension allows the user to model the heterogeneity in the layer by modifying the resistivity and capacitance for any unit on the chip. For modeling this heterogeneity, we add an additional data structure to each grid to hold grid-specific resistivity and capacitance values. When the temperature responses are being calculated, the tool then uses the grid-specific parameters rather than the layer-specific ones.

Figure 11 provides an illustration of the 3D multicore processor with stacked DRAM. The TSVs connect the on-chip DRAM layer with the logic layer in the 3D multicore systems. Note that TSVs go through both the DRAM and thermal interface layers to connect the active regions of the DRAM and the logic layers. In our target 3D system with parallel access, there are 512 TSVs on each memory controller block. The TSVs have a diameter of $10\mu m$ and a center-to-center pitch of $20\mu m$. To calculate the thermal resistivity of the blocks with TSVs, we assume that the TSVs are evenly spread throughout the memory controller. As we know the dimensions of a single Copper TSV, we can calculate the area the TSVs cover in the memory controller block ($Area_{TSV}$) as well as the area of the memory controller block without TSVs. The joint parallel resistivity (of Copper and thermal interface material, TIM) can be calculated as follows:

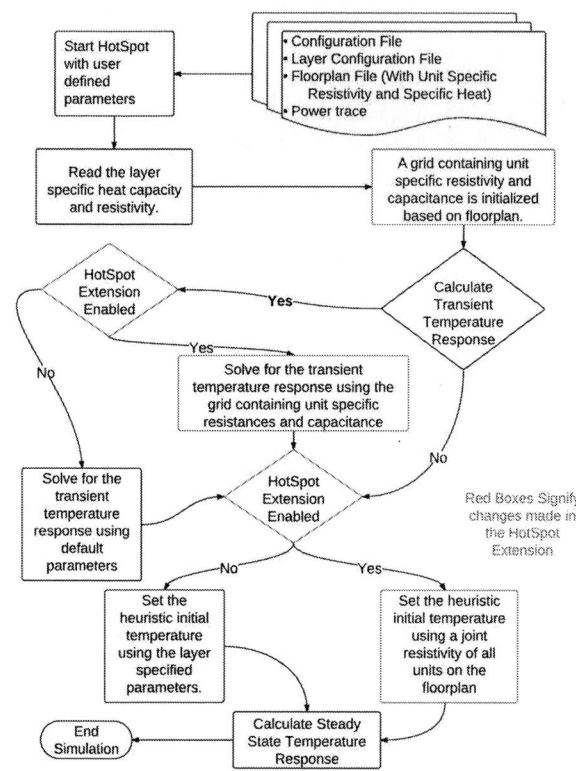

Figure 12: A flow chart for illustrating the HotSpot extension for modeling 3D systems with TSVs by enabling the interlayer material heterogeneity.

$$R_{Joint} = \frac{Area}{\dfrac{Area - Area_{TSV}}{R_{TIM}} + \dfrac{Area_{TSV}}{R_{Copper}}} \quad (2)$$

where $Area$ is the area of a memory controller block where TSVs are located at, $Area_{TSV}$ is the area of the memory controller block with TSVs, R_{TIM} is the thermal resistivity of TIM, and R_{Copper} is the thermal resistivity of Copper. Thus, we get the thermal resistivity for the memory controller block with TSVs as $0.156mK/W$, which is lower than the original TIM resistivity of $0.25mK/W$. We also model the TSVs going through the DRAM layer, and compute the joint thermal resistivity of silicon and Copper as $0.0098mK/W$.

We present a flow chart for the implementation of our HotSpot extension in Figure 12. The black boxes are based on the default HotSpot implementation, while the red boxes indicate changes made in the HotSpot tool. In addition to reading the parameters from the layer configuration file, the tool also reads unit-specific parameters from each of the floorplan files. We use the resistivity and capacitance values specified in the floorplan file and assign them to specific grids on each layer. The addition of grid-specific values changes the heuristic initial temperature for steady-state temperature computations. HotSpot will set the initial temperature using only the layer-specific vertical resistance for each layer while ignoring any lateral resistances. Our extension will find the weighted mean of all the vertical resistances

654

in each grid in the layer. When calculating the initial temperature for each layer, the extension will use this weighted mean instead of using the layer-specific vertical resistance. The Hotspot extension described above has been recently released by our group at: http://lava.cs.virginia.edu/HotSpot-/links.htm.

S2. Target System Modeling Parameters

We model the core architecture of our target system based on the AMD Family 10h microarchitecture used in AMD Magny-Cours processors. Each core has multiple-issue, out-of-order execution, and a 512 KB private L2 cache. All the L2 caches are located on the same layer as the cores and connected by a shared bus. MESI cache coherence protocol is used for communication. The architectural parameters for cores and caches are listed in Table 4. These parameters are used for the target system configuration in our performance simulations, as described in Section 3.2.

Table 4: Core Architecture Parameters

Architectural Configuration	
CPU Clock	2.1 GHz
Branch Predictor	tournament predictor
Issue	out-of-order
Reorder Buffer	84 entries
Issue Width	3-way
Functional Units	3 IntALU, 1 IntMult, 3 FPALU, 1 FPMultDiv
Physical Regs	128 Int, 128 FP
BTB size	2048 entries
RAS size	24 entries
Load Queue	32 entries
Store Queue	32 entries
L1 I/DCache	64 KB @2 ns, 2-way, 64B block
L2 Cache(s)	16 private L2 Caches, each L2: 16-way set-associative, 64B block 512 KB @6 ns

S3. Additional Details for Parallel Memory Access Modeling

We consider both `regular memory access` and `parallel memory access` to simulate the data transfer between logic

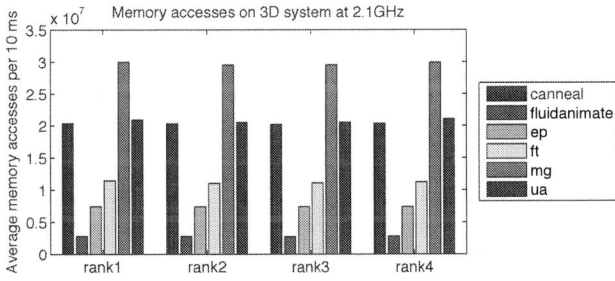

Figure 13: Average memory accesses per 10ms on different DRAM ranks on the 3D system with stacked DRAM.

layer and DRAM layer in our target 3D system. The `parallel memory access` allows the four on-chip memory controllers accessing the four DRAM ranks at the same time. We deploy 512 TSVs on each memory controller, and these TSVs provide a 64-Byte bus width for each memory controller. In our experiments, we consider TSVs with a diameter of $10\mu m$ and a center-to-center pitch of $20\mu m$. Thus the total TSV area only takes up less than 0.2% of the chip area overhead. From our simulation results for the NAS and PARSEC benchmarks as shown in Figure 13, we observe the main memory accesses are evenly distributed between the four ranks. This provides the justification for assuming the memory access latency with `parallel access` is one fourth of the latency with `regular access`.

S4. Power Modeling Parameters

We use five V-F settings in our power model for the target 3D system, matching the five V-F settings in AMD 10h processors. The V-F settings and the corresponding average core power across all the benchmarks for the 3D system with parallel access are shown in Table 5.

Table 5: V-F settings and average per core power values for the 3D system with parallel access.

Frequency(GHz)	2.1	1.7	1.4	1.1	0.8
Voltage(V)	1.10	1.06	1.02	1.0	0.98
Core Power(W)	10.57	8.98	6.98	5.30	4.86

S5. Temperature Modeling Parameters

We provide the thermal parameters used in the HotSpot simulations for 2D and 3D architectures in Table 6. In HotSpot, we set chip thickness at 0.1mm, DRAM thickness at 0.05mm, thermal conductivity of DRAM at 100W/mK (thermal conductivity of silicon), and sampling interval at 1ms. All the other parameters are the same as the default HotSpot configuration to represent efficient packages in high-end systems. The power traces are the inputs for the thermal model. All simulations use the HotSpot grid model for higher accuracy and are initialized with the steady-state temperatures.

Table 6: Thermal simulation configurations in HotSpot.

Parameters	Values
Chip thickness	0.1mm
Silicon thermal conductivity	100 W/mK
Silicon specific heat	1750 kJ/m^{3}K
Sampling interval	0.001s
Spreader thickness	1mm
Spreader thermal conductivity	400 W/mK
DRAM thickness	0.05mm
DRAM thermal conductivity	100 W/mK
Interface material thickness	0.02mm
Interface material conductivity	4 W/mK
Heatsink thickness	6.9mm
Heatsink resistance	0.1K/W

Static Dataflow with Access Patterns:
Semantics and Analysis

Arkadeb Ghosal*, Rhishikesh Limaye*, Kaushik Ravindran*, Stavros Tripakis**
Ankita Prasad*, Guoqiang Wang*, Trung N Tran*, Hugo Andrade*
* National Instruments Corp., Berkeley, CA, USA, {firstname.lastname}@ni.com
** University of California, Berkeley, CA, USA, stavros@eecs.berkeley.edu

ABSTRACT

Signal processing and multimedia applications are commonly modeled using Static/Cyclo-Static Dataflow (SDF/CSDF) models. SDF/CSDF explicitly specifies how much data is produced and consumed per firing during computation. This results in strong compile-time analyzability of many useful execution properties such as deadlock absence, channel boundedness, and throughput. However, SDF/CSDF is limited in its ability to capture how data is accessed in time. Hence, using these models often leads to implementations that are sub-optimal (i.e., use more resources than necessary) or even incorrect (i.e., use insufficient resources). In this work, we advance a new model called Static Dataflow with Access Patterns (SDF-AP) that captures the timing of data accesses (for both production and consumption). This paper formalizes the semantics of SDF-AP, defines key properties governing model execution, and discusses algorithms to check these properties under correctness and resource constraints. Results are presented to evaluate these analysis algorithms on practical applications modeled by SDF-AP.

Categories and Subject Descriptors: C.3 [**Special-purpose and Application-based Systems**]: Signal processing systems
General Terms: Theory, Algorithms, Experimentation
Keywords: Dataflow, semantics, access patterns

1. INTRODUCTION

Static Dataflow (SDF) is a model of computation to specify, analyze, and implement multi-rate computations that operate on infinite streams of data [13]. An SDF model is represented as a directed graph of computational actors interconnected by FIFO channels. The SDF semantics requires that the number of tokens consumed and produced by an actor per firing is fixed and pre-specified. This guarantees decidability of key model properties: existence of deadlock-free and memory-bounded infinite computation, throughput, latency, and execution schedule [1, 13]. The expressiveness of the SDF model in naturally capturing streaming applications, coupled with its strong compile-time pre-

dictability properties, has made it popular in the domains of multimedia, digital signal processing, and communications.

While the standard SDF model is untimed, it is a common practice to associate worst-case execution time (WCET) models to analyze the timing behavior of applications [7, 12, 14, 15, 20]. These timing annotations enable static analysis of SDF models and mapping solutions to specific platforms under resource and performance constraints. Worst-case timing models have been applied to capture execution behavior of SDF actors for software and hardware implementations.

However, these timing models suffer a key deficiency: they lose information about the precise timing of consumption and production of tokens by an actor during a firing cycle. The problem is particularly evident when SDF models are used to capture hardware implementations. Many hardware IP blocks require that data tokens be delivered to them at precisely specified clock cycles from the start of execution. This loss of timing information in SDF models results in sub-optimal analysis and implementations that conservatively estimate the resources needed.

For example, consider a design connecting a producer P to a consumer C. P produces 1 token per firing and executes in 1 time unit, and C consumes 8 tokens per firing and executes in 8 clock cycles. Suppose that the IP block implementing C requires 8 tokens to be delivered in 8 consecutive cycles. Unfortunately, the SDF timing model is not sufficiently expressive to capture this behavior. The semantics of SDF assumes that an actor cannot start firing until sufficient tokens are present at the inputs. As a result, if the above example is modeled with SDF, C cannot start firing until after P completes eight firings. Therefore, a buffer of size at least 8 must be added between P and C; C may start its execution only after the buffer has collected 8 tokens from P. While this is a valid implementation, it is sub-optimal in terms of allocation of buffer resources. In contrast, a better implementation can exploit knowledge about the behavior of C and determine that a buffer of size one is sufficient.

Cyclo-Static Dataflow (CSDF) [2] is a generalization of SDF that appears to resolve the problem. CSDF "breaks" a firing into finer-grained phases, and specifies consumptions and productions of tokens for each phase. But CSDF still relies on the same basic hypothesis as SDF, i.e., that an actor will wait until sufficient tokens have accumulated at the input channels before beginning a phase. Unfortunately, this hypothesis violates requirements related to the precise timing of token accesses. In the example above, C requires that it receive 8 tokens in 8 consecutive clock cycles once it commences firing. CSDF cannot capture this constraint and as a result can lead to incorrect implementations [19].

For example, consider an alternate producer P' with an execution time of 2. Then a CSDF model would conclude that a buffer of size 1 between P' and C is sufficient, but this would violate the timing requirement of C.

It may be argued that the requirement of precise timing of tokens is artificial, since actors can be *stalled* or turned off. Actors implemented in software can easily be disabled or context switched. For hardware IP blocks, there is typically a "clock enable" signal that regulates their execution. Setting this signal to "false" freezes the actor when inputs are unavailable. However, this solution is not satisfactory in practical designs. The area overhead due to the enable logic is undesirable. Also, any logic that regulates the clock contributes additional delay to timing-critical paths. The increased distribution of "clock enable" signals further adversely impacts the achievable frequency. Hence, it is important to capture precise timing of token accesses to generate resource optimal implementations. Both SDF and CSDF models are not equipped for this.

To remedy the expressiveness problems of SDF/CSDF, a new model, called *SDF with Access Patterns* (SDF-AP), is introduced informally in [19]. SDF-AP strikes a balance between the analyzability of SDF/CSDF while accurately capturing the interface timing behavior. The latter is achieved by specifying *access patterns* that capture the precise timing behavior of token productions and consumptions. The original motivation for SDF-AP comes from modeling hardware IP blocks, where access patterns are precisely characterized and presented as timing diagrams. Nevertheless, the timing extensions that access patterns provide are general and applicable to actors implemented in software as well.

The goal of [19] is to justify that choosing the right model is important for generating correct and non-defensive implementations from high level component abstractions. It informally introduces the SDF-AP model and advocates a general methodology based on Finite State Machines to reason about performance and resource trade-offs. However, [19] does not define the semantics of the SDF-AP model. It also does not develop analysis methods to reason about model properties. This paper closes this gap. Our main contributions are: (a) a formal definition of the SDF-AP model with its operational semantics, (b) formal definitions of key model properties, such as executability and throughput, (c) algorithms for efficient static analysis of these properties, and (d) case studies to evaluate these algorithms.

2. RELATED WORK

Real-time streaming applications are widely deployed on embedded platforms. Model-based design is a well-tested approach for the implementation of these systems. A comprehensive survey on concurrent models of computation can be found in [11]. Prior research has shown that dataflow and its variants are sufficiently expressive enough to capture the task and data parallelism in streaming applications. SDF and CSDF models enable compile time analysis of key execution properties, e.g., absence of deadlocks and consistency of execution rates, via efficient algorithms [1,12,13]. Recent variants like Heterochronous Dataflow (HDF) [6], Scenario Aware Dataflow (SADF) [18], and Core Functional Dataflow (CFDF) [7] extend SDF/CSDF with specifications for control. Design frameworks like Ptolemy-II [5], SDF^3 [17], and OpenDF [8] deliver hardware and software implementations.

Though SDF/CSDF models have many advantages, they are limited in their ability to capture precise timing information of data production and consumption. This is particularly evident when dataflow models are targeted for hardware implementations. Prior efforts are conservative in their implementation of the glue logic to stitch SDF actors in hardware [4,8–10]. SDF-AP is introduced in [19] to remedy that deficiency. Model properties like consistency, absence of deadlock, bounded execution, and throughput need to be checked before the model can be implemented. There are existing techniques to check the properties for SDF/CSDF models. However, they cannot be directly used for SDF-AP models due to differences in semantics. In this paper, we present the formal semantics of SDF-AP models and algorithms to efficiently check key model properties.

3. SDF-AP: SYNTAX AND SEMANTICS

An SDF-AP model consists of actors connected over channels. Actors read tokens from incoming channels and write to outgoing channels. Once an actor has fired, it consumes (resp. produces) a fixed number of tokens from (resp. to) input (resp. output) channels over the execution time. An actor associates each channel with a pattern represented as a binary word of length equal to the execution time of the actor. The pattern denotes whether the actor reads (resp. writes) a token or not from (resp. to) the incoming (resp. outgoing) channel at a particular cycle in the execution. The access pattern can be provided by the user or derived from the timing diagrams accompanying the documentation of the IP block [19]. Given the application domain and hardware implementation, we will restrict reading and writing of at most one token per channel at any clock transition. Nevertheless, the model semantics can be easily generalized to allow multiple tokens to be read or written.

Fig. 1 shows an SDF-AP model with a build stream actor, bs, which takes two input streams and merges them in one. Actor bs is fed by two source actors $i1$ and $i2$: $i1$ generates 1 token every 2 cycles, and $i2$ generates 3 tokens every 4 cycles. At each firing, bs consumes 2 tokens produced by $i1$, and 4 tokens produced by $i2$, and places them in a merged stream of 6 tokens at the output (tokens from $i1$ preceding those of $i2$). Actor bs is connected to a sink actor o. The net token count and respective pattern are shown separated by ":", e.g., "3:1101" on $c2$ denotes that $i2$ produces 2 tokens with the pattern 1101 on channel $c2$. Channels $c1$, $c2$, $c3$ connect $i1$ with bs, $i2$ with bs, and bs with o, respectively.

3.1 Syntax

An SDF-AP model is a pair $\mathcal{M} = (aset, cset)$, where $aset$ is a set of actors, and $cset$ is a set of channels. For the example in Fig. 1, $aset = \{bs, i1, i2, o\}$, and $cset = \{c1, c2, c3\}$.

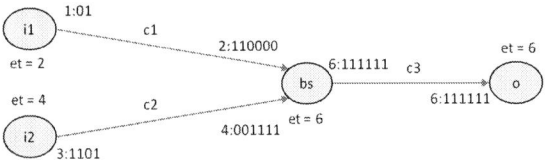

Figure 1: SDF-AP model for build stream actor interacting with two sources and one sink

An actor $a \in aset$ is a tuple $(ic, oc, it, ot, et, cp, pp)$ where $ic \subseteq cset$ (resp. $oc \subseteq cset$) is the set of input (resp. output) channels of a, it (resp. ot) is a map $it : ic \to \mathbb{N}$ (resp.

$ot : oc \to \mathbb{N}$) mapping each input (resp. output) channel to the total number of tokens read from (resp. written to) that channel per firing of a^1, $et \in \mathbb{N}$ is the time in clock cycles it takes to complete one firing of a, cp (resp. pp) is a map of input (resp. output) channels to consumption (resp. production) patterns. The pattern cp (resp. pp) maps each input (resp. output) channel to a binary word of length et, i.e., $cp : ic \to \mathbb{B}^{et}$ and $pp : oc \to \mathbb{B}^{et}$. The i-th letter of the word is denoted as $cp(c,i)$ (resp. $pp(c,i)$). The sum of the 1's in $cp(c)$ (resp. $pp(c)$) equals the input (resp. output) token count for the channel $it(c)$ (resp. $ot(c)$). For source actors, $ic = \emptyset$, $it = cp = \varnothing^2$; for sink actors, $oc = \emptyset$, $ot = pp = \varnothing$.

A channel $c \in cset$ is a unique id, and must appear exactly once in the input channel set of an actor, and exactly once in the output channel set of an actor. This ensures no dangling channels and no non-determinism in channel access.

3.2 Semantics

The operational semantics of an SDF-AP model $\mathcal{M} = (aset, cset)$ is defined as a state transition system. A state of the system tracks the number of tokens on each channel, the set of running instances of each actor, and for each instance, the number of clock cycles it has been executing. Formally, a state s is a pair (γ, υ) where $\gamma : cset \to \mathbb{Z}$ is a *channel quantity* [16] (we allow negative values for token counts, see below for the interpretation), and $\upsilon : aset \to \mathcal{MS}(\mathbb{N}_{+0} \times \{w, r, \bot\})$ maps each actor to a multiset of pairs of the form $(\eta, \kappa) \in \mathbb{N}_{+0} \times \{w, r, \bot\}$. If $\upsilon(a) = \emptyset$ then actor a has no *active* (i.e., running) instances currently. Otherwise, each pair $(\eta, \kappa) \in \upsilon(a)$ represents an active instance of a: η denotes the number of clock cycles the instance has been executing, and κ is a flag denoting the *stage* the instance within the current clock cycle. There are three possible stages: beginning of clock cycle $\bot$ (idle stage), reading r, and writing w. The meaning will become clear in what follows.[3] A state s is called *stable* if $\forall a \in aset, \forall(\cdot, \kappa) \in \upsilon(a), \kappa = \bot$. The *initial state* $s_0 = (\gamma_0, \upsilon_0)$ where $\forall a \in aset$, $\upsilon_0(a) = \emptyset$, and γ_0 maps each channel to a given number of initial tokens. The initial state (which gets modified with different set of initial tokens) determines the behavior of the model.

Following [16], we define operations on channel quantities. If γ_1, γ_2 are channel maps from sets of channels $cset_1, cset_2$, with $cset_2 \subseteq cset_1$, then $\gamma_2 \preceq \gamma_1$ if $\forall c \in cset_2$, $cset_2(c) \leq cset_1(c)$. The operation $\gamma_1 + \gamma_2$ is defined as pointwise addition. If $\gamma_2 \preceq \gamma_1$, then the operation $\gamma_1 - \gamma_2$ is defined as pointwise subtraction. We will use $\gamma = 0$ to denote that token counts on all channels are 0, $\gamma \geq 0$ to denote that all channels map to $\mathbb{N}_{+0}$, and $\gamma \leq \beta$ where $\beta \in \mathbb{N}_{+0}$ to denote that channel counts are bounded by β. For actor a and $i \in \{1, ..., et(a)\}$, we define the following channel quantities: $\gamma_{a,i}^R$ (resp. $\gamma_{a,i}^W$) maps every input (resp. output) channel c of a to $cp(c,i)$ (resp. $pp(c,i)$). For source actors, $\gamma_{a,i}^R = 0$ and for sink actors, $\gamma_{a,i}^W = 0$, for all i. A transition $\delta = (s, l, s')$ of $\mathcal{M}$ from state $s = (\gamma, \upsilon)$ to

---

[1] We denote integers by $\mathbb{Z}$, natural numbers (without 0) by $\mathbb{N}$, $\mathbb{N} \cup \{0\}$ by $\mathbb{N}_{+0}$, and binary numbers by $\mathbb{B} = \{0, 1\}$.

[2] $\emptyset$ and $\varnothing$ denote empty set and empty mapping, respectively.

[3] The definition of state is inspired by the definition used in [16], but differs in several respects. In particular, the flag κ is necessary to track read and write activities at each clock cycle. This is not an issue in [16] since in CSDF reads and writes occur at the beginning and at the end of firings and not at arbitrary times during a firing, as in SDF-AP.

---

state $s' = (\gamma', \upsilon')$ labeled with label l, also denoted $s \xrightarrow{l} s'$, can be any one of those shown in Table 1. s' is called a *successor* of s. A transition labeled $begin(a)$ adds a new instance of a to the set of active actor instances. The clock counter of the new instance is initialized to 0 and the instance is idle (i.e., not ready to read or write). A transition labeled $end(a)$ removes an instance of a from the set of active instances, provided the instance has finished its firing, i.e., its clock counter has reached $et(a)$. A transition labeled *clock* marks the beginning of a clock cycle: all active actor instances increase their clock counter by 1 and move from the idle stage $\bot$ to the reading stage r. A transition labeled $read(a)$ (resp. $write(a)$) corresponds to a reading from (resp. writing to) its input (resp. output) channels. Once it has read, an actor instance moves from reading stage r to writing stage w. Once it has written, it moves back to stage $\bot$, until the beginning of the next clock cycle.

Note that read transitions may result in channel capacities becoming negative. This is because no precondition on having enough tokens in the channel is imposed for taking a read transition. Similarly, nothing prevents writing, which means that no a-priori bounds on channel size are imposed. This approach makes the semantics easier to formalize. We identify below situations where a negative token count models non-executable vs. transient behaviors as well as distinguish between bounded and unbounded executability.

Also note that reads and writes occur *asynchronously* between actor instances (i.e., different instances interleave) while for a given instance, the read always occurs before the write. The latter is done to model causality where a consuming actor needs to wait till a producing actor places a token in the channel. A *synchronous* semantics is also possible, where all actors read simultaneously, then write simultaneously, to complete a clock cycle. The synchronous semantics results in far fewer transitions than the asynchronous semantics. However, the synchronous semantics does not allow to distinguish between non-executable and certain executable models (see discussion on Figure 2 in Section 4).

An *execution trace* τ is an infinite sequence of transitions $\tau = s_0 \xrightarrow{l_1} s_1 \xrightarrow{l_2} s_2 \cdots$, where s_0 is the initial state. Any subsequence $\tau' = s_i \xrightarrow{l_{i+1}} \cdots \xrightarrow{l_n} s_n$ for some $i, n \in \mathbb{N}_{+0}$ and $i \leq n$ is a *sub-trace*. The set of traces of $\mathcal{M}$ is denoted $traces(\mathcal{M})$. The set of states visited along a trace τ is denoted $states(\tau)$. Refer to Supplemental Section S1 for traces from the running example. A state s is called *reachable* from initial state s_0 if $s \in states(\tau)$ for some trace τ. Note that our semantics guarantees that any reachable state s has a successor state s'. State s is a *post clock transition (PCT)* state if $s' \xrightarrow{clock} s$ for some state s'. Given a PCT state s, a stable state s' is a *next stable state* of s, denoted $NSS(s)$, if there exists a sub-trace $\tau' = s \xrightarrow{l_1} \cdots \xrightarrow{l_n} s'$ (for some $n \in \mathbb{N}$) such that none of the labels $l_1, l_2 \cdots l_n$ are of the types $begin(a)$, $end(a)$ or *clock* for all actors $a \in aset$. Our semantics guarantees that *for any PCT state s there is a unique next stable state $NSS(s)$* (refer to Supplemental Section S2 for formal reasoning). A PCT state s corresponds to the beginning of a clock cycle, and $NSS(s)$ corresponds to the end of that cycle. If $s = (\cdot, \upsilon)$ is a PCT state where $\forall a \in aset, \upsilon(a) = \emptyset$, then s is a stable state, and is a next stable state of itself. This corresponds to a situation when no actor has fired, and hence no read transition is enabled.

Given a trace τ, $all(\tau)$ is the set of traces generated by

Table 1: State Transitions (transition $\delta = s \xrightarrow{l} s'$, $s = (\gamma, v)$, $s' = (\gamma', v')$)

Type	Label l	Precondition	Action
begin fire	$begin(a)$	s is stable	$v'(a) = v(a) \uplus \{(0, \bot)\}, v'(a' \neq a) = v(a')$
end fire	$end(a)$	s is stable, and $(et(a), \bot) \in v(a)$	$v'(a) = v(a) \setminus \{(et(a), \bot)\}, v'(a' \neq a) = v(a')$
clock	$clock$	s is stable $\not\exists (\gamma, v) \xrightarrow{end(a)}$	$\forall a \in aset$, if $v(a) = \emptyset$, then $v'(a) = \emptyset$ else each $(i, \bot) \in v(a)$ is updated to $(i+1, r) \in v'(a)$
read	$read(a)$	$(i, r) \in v(a)$	$\gamma' = \gamma - \gamma_{a,i}^R, v'(a) = v(a) \setminus \{(i, r)\} \uplus \{(i, w)\}, v'(a' \neq a) = v(a')$
write	$write(a)$	$(i, r) \in v(a)$	$\gamma' = \gamma + \gamma_{a,i}^W, v'(a) = v(a) \setminus \{(i, w)\} \uplus \{(i, \bot)\}, v'(a' \neq a) = v(a')$

model $\mathcal{M} = (aset, cset)$, actors $a, a' \in aset$, and $\uplus$ and $\setminus$ denote multiset union and difference

combining τ with all possible sub-traces between all the PCT states of τ and their corresponding NSS states. $all(\tau)$ can be seen as a set of traces, but also as a transition system, which is a part of the transition system of the model. The set of states in $all(\tau)$ is denoted as $states(all(\tau))$.

SDF-AP actors are auto-concurrent, i.e., multiple instances of an actor can execute simultaneously. However this may not be feasible in practice due to restrictions like finite resources, IP block properties etc. Such constraints are captured through initiation interval $ii \in \mathbb{N}_{+0}$ which specifies the minimum time between two firings of an actor. If $ii \geq et$, then actor execution cannot be concurrent; otherwise actors can execute in parallel. If ii is specified for an actor, then enabling condition of a begin fire transition should check that the state is stable, and ensure that a minimum of ii clock cycles has passed after the latest firing of the actor.

4. MODEL PROPERTIES

Interesting properties for standard SDF/CSDF models are *deadlock/livelock-freedom* (can the model execute with some/all actors firing infinitely often?), *boundedness* (can the model execute forever with finite buffers?), etc. In this section we define properties similar in spirit for SDF-AP.

DEFINITION 4.1. *A trace τ is* live *if both $\xrightarrow{begin(a)} \forall a \in aset$ and $\xrightarrow{clock}$ appear infinitely often in τ.*

The semantics of SDF-AP allows token counts to be negative. Hence, every model has live traces. In reality buffers cannot have negative token count. However, there is an interesting situation where a trace models an implementable behavior, even though the trace visits states with negative token counts. Consider a channel c whose token count becomes -1 between a PCT state s and $NSS(s)$. This implies a situation c is empty and an actor writes to c while another actor reads from c at the same clock cycle. If the read happens before the write (our asynchronous semantics allows that) c will have a (transient) negative token count of -1. This scenario *can* however be implemented with a *fast buffer* that allows writing and reading a token in the same cycle. A model that can be executed without a fast buffer is *strongly executable*, otherwise, it is *weakly executable*.

DEFINITION 4.2. *An SDF-AP model $\mathcal{M}$ is* weakly executable *if there exists a live trace $\tau \in traces(\mathcal{M})$ such that $\forall s = (\gamma, \cdot) \in states(\tau), \gamma \geq 0$. $\mathcal{M}$ is* strongly executable *if there exists a live trace $\tau \in traces(\mathcal{M})$ such that $\forall s = (\gamma, \cdot) \in states(all(\tau)), \gamma \geq 0$.*

We distinguish between executability and *bounded* executability. The former only captures problems of negative token counts (i.e., deadlocks or livelocks in standard SDF/CSDF parlance). Bounded executability is stronger and requires in addition ability to execute with bounded buffers.

DEFINITION 4.3. *An SDF-AP model $\mathcal{M}$ is* bounded weakly *(resp.* strongly*)* executable *if $\exists \beta \in \mathbb{N}_{+0}$ and live trace τ such that (1) $\mathcal{M}$ is weakly (resp. strongly) executable with respect to τ, and (2) $\forall s \in states(\tau)$ (resp. $states(all(\tau))$), $\gamma(s) \leq \beta$.*

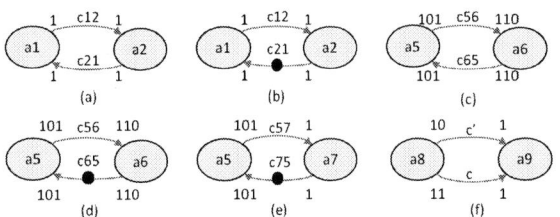

Figure 2: Liveness, Executability and Boundedness

Fig. 2(a) is not executable as any trace generated will always have negative token counts on channels. Note that the synchronous semantics would be unable to detect the deadlock, as the channel counts are 0 at all stable states. With an initial token in one of the channels (Fig. 2(b)) the model is strongly executable. Similarly Fig. 2(c) is not executable. However with a token on channel $c65$, the model (Fig. 2(d)) is weakly executable for the following reason. Consider a cycle when $a5$ fires (with the token from $c65$) and $a6$ is idle; $a5$ consumes 1 token and produces 1 token, i.e., token count on $c65$ and $c56$ are 0 and 1, respectively. In the next cycle $a5$ can continue executing as it does not need any token, and $a6$ starts firing by consuming the token from $c56$; at the end of the cycle, token count on $c65$ and $c56$ are 1 and 0, respectively. In the following cycle, $a5$ consumes 1 token from $c65$, and produces 1 token on $c56$. Actor $a6$ (in the second cycle of its execution) consumes the token and continues execution. Note that this scenario is possible as the token produced by $a5$ can be consumed by $a6$ in the same cycle thus making the model weakly executable. Fig. 2(e) is strongly executable as $a7$ has execution time of 1, and hence can wait indefinitely between firings until required amount of token has been generated in the incoming channel. Fig. 2(b), (d) and (e) are bounded. Fig. 2(f) is strongly executable but not bounded.

We define throughput for bounded executable traces and models. Throughput $\Gamma(\mathcal{M}, a, \tau)$ of an actor a, for a bounded executable trace τ of a model $\mathcal{M}$, is the average rate of firing of the actor a in τ. In a model, one is typically interested in the throughput of certain actors, e.g., certain sources or sinks. We assume that a is such a fixed actor for given model $\mathcal{M}$, and we denote $\Gamma(\mathcal{M}, \tau) = \Gamma(\mathcal{M}, a, \tau)$ to be the throughput of the model for the trace τ. The optimal throughput of the model is then $\Gamma(\mathcal{M}) = \max_{\tau \in T} \{\Gamma(\mathcal{M}, \tau)\}$, where T is the set of all bounded executable traces of the model.

659

5. ALGORITHMS FOR STATIC ANALYSIS

We now show that bounded strong executability is *decidable* by providing algorithms to check the property. As we are concerned with strong executability, synchronous semantics suffice, with the benefit of making analysis efficient.

5.1 Boundedness

A model is bounded if it can be executed without termination using buffers with finite capacity. In SDF models, bounded execution is verified by proving that the model is sample rate consistent [13]. An SDF model is sample rate consistent if there exists a fixed non-zero number of firings for each actor, called the *repetitions vector*, such that executing these firings reverts the model to its original state.

The concept of sample rate consistency can be applied to check boundedness of SDF-AP models. If the underlying SDF model is sample rate consistent, then there exists a non-zero repetitions vector $r : aset \rightarrow \mathbb{N}$ such that the number of tokens produced and consumed on each channel is balanced, i.e. $\forall c \in cset, r(a)ot(c) = r(a')it(c)$, where a and a' are the producing and consuming actors of c. The repetitions vector provides a recipe for a non-terminating periodic execution of the SDF-AP model in bounded memory.

5.2 Bounded Executability

The concept of bounded executability can be translated to the model being bounded and deadlock free. An SDF model is deadlock free if it can be executed without interruption for one full *iteration* (in which each actor fires as many times as specified in the repetitions vector). The algorithmic solution is to compute a *self timed schedule* for one iteration (in which an actor fires as soon as all its input tokens are available) [1].

However, for SDF-AP models, the underlying SDF being deadlock free is a sufficient but not necessary condition. For SDF-AP models, it may be necessary to fire an actor before all tokens are available. Consider the SDF-AP model in Fig. 2(d). The underlying SDF model is deadlocked. But in the SDF-AP model, $a5$, which has a consumption access pattern of [101], can begin firing and consume the initial token. Hence, the SDF-AP model is bounded executable though the underlying SDF model is deadlocked.

We formalize the problem of checking bounded executability of SDF-AP models. The objective is to determine start times for actor firings that respect data dependence and access patterns. Let r be the repetitions vector of a bounded SDF-AP model $\mathcal{M}$. An actor a must produce $r(a)ot(c)$ tokens on output channel $c \in oc$ in one iteration. For each token, we associate a firing index fp and time offset op to characterize when it is produced: $\forall a \in aset, c \in oc(a), n \in \mathbb{N},\ fp(a, c, n) = \lceil n/ot(c) \rceil$, and $op(a, c, n) = \theta(pp(c), (n \bmod ot(c)) + 1)$, where $\theta : \{pp(c)\} \times \{1..ot(c)\} \rightarrow \mathbb{N}_{+0}$ is the offset from the start of a firing when a token is produced. E.g., given access pattern $pp = [11001]$ for an actor that produces 3 tokens in 5 cycles, $\theta(pp, 1) = 0, \theta(pp, 2) = 1, \theta(pp, 3) = 4$. We similarly characterize the firing index $fc(a, c, n)$ and offset $oc(a, c, n)$ at which a token on an input channel is consumed by substituting $ot(c)$ by $it(c)$ and $pp(c)$ by $cp(c)$ in the prior equations.

We present a constraint system to determine if an SDF-AP model is bounded executable. The variables are the start times of actor firings: $x(a, i) \in \mathbb{Z}$, $\forall a \in aset, \forall i \in \{1 \ldots r(a)\}$. The dependencies in start times are encoded as

($\gamma_0(c)$ is the number of initial tokens on channel c):

$$\forall c \in cset, \forall n \in \{\gamma_0(c) + 1, \ldots, r(a)ot(c)\}$$
$$x(a, fp(a, c, n - \gamma_0(c))) + op(a, c, n) + 1 \leq$$
$$x(a', fc(a', c, n)) + oc(a', c, n)$$

These constraints are all of the form $x_1 - x_2 \leq k$, where x_1 and x_2 are variables and k is a constant. Such a system of difference constraints can be solved by encoding it as a problem of finding shortest paths in a weighted directed graph [3]. The Bellman-Ford algorithm is applied to solve the shortest path problem. Two outcomes are possible: (a) Bellman-Ford returns the delay of the shortest path to each vertex, or (b) Bellman-Ford detects a negative cycle proving that the constraint system is infeasible. Outcome (a) corresponds to the SDF-AP model being bounded executable. The shortest path delays correspond to valid start times for all actor firings in one iteration. Outcome (b) proves that the SDF-AP model is not bounded executable. Thus, this translation to a well-known graph theoretic problem provides an effective mechanism to check bounded executability.

The number of constraints is equal to the total number of firings in one iteration, which is exponential in the worst case in the number of actors in the model. However, the problem of checking whether an SDF model deadlocks also incurs the same complexity [12]. As our experiments show, this is a feasible solution method for practical SDF-AP models.

5.3 Throughput and Buffer Sizing

We address the problem of checking if a bounded executable SDF-AP model meets a specified throughput Γ. The constraint formulation can be extended to solve this problem. Intuitively, the throughput constraint is an upper bound on the time between successive firings of an actor. This can be expressed as a difference equation of the form $x(a, i+1) - x(a, i) \leq 1/\Gamma$. The constraint system can still be solved as a shortest path problem. The optimal throughput of the model can be computed by a binary search over the range of feasible values for Γ.

Further, the constraint formulation can be repeatedly applied explore buffer sizes for channels to meet a specified throughput. The buffer size of a channel can be encoded as a back edge with initial tokens corresponding to the size [16]. The solution approach is a search algorithm in which an outer loop fixes buffer sizes and the constraint formulation is analyzed to check throughput of each configuration. One direction of future work is to find efficient heuristics to guide exploration of buffer sizes for SDF-AP models.

6. EXPERIMENTAL RESULTS

We evaluate the benefits of the SDF-AP model on several streaming applications (see Table 2). The first seven applications are SDF models of realistic FPGA implementations consisting of streaming hardware IP blocks. We compute the access patterns for the IP blocks from the cycle-level timing information in their datasheets. The other applications are benchmarks from the SDF³ [17] analysis tool. For our experiments, we choose different amounts of buffer space for each application, and determine throughput (an average rate of firing of the actors) and latency (total duration of a single iteration of the model) using: (1) traditional SDF analysis using eager, self-timed symbolic simulation [16], and (2) our SDF-AP analysis using difference constraints.

Table 2: Throughput and latency analysis for different applications for given buffer spaces

Name	#Actors, #Channels	Firings/ Iteration	Optimal Throughput(Hz)	Buffer Space	Throughput (Hz) SDF	SDF-AP	Latency (μsec) SDF	SDF-AP	Run-time (seconds) SDF	SDF-AP
OFDM Tx 2msps	11, 14	585	833	14210	496	833	2051	1251	0.86	0.16
				15760	833	833	1314	1249	0.70	0.14
OFDM Tx 5msps	11, 14	585	2083	14210	783	2083	1309	537	0.70	0.11
				17928	2083	2083	594	535	0.67	0.11
OFDM Tx 25msps	9, 12	6723	6250	13858	1590	6250	682	205	0.99	0.23
				27294	6250	6250	282	203	0.97	0.22
OFDM Rx Full	46, 66	107054	1667	8350	390	422	2684	2362	40.18	378.9
				52716	1437	1573	760	699	39.65	308.9
				53584	1437	1667	734	697	39.85	305.65
OFDM Tx Full	17, 22	40400	1667	3572	510	1667	1994	658	9.69	25.29
ZeroPad 600	4, 4	2	48804	7	-	48804	-	20	0.01	0.003
Van de Beek	19, 28	28164	39032	12820	662	685	1510	1480	7.78	54.30
				21059	36141	39032	77	52	6.39	0.32
				12059	-	39032	-	52	0.18	0.34
				11112	-	-	-	-	0.15	11.25
MP3Decoder	14, 18	911	27	9264	17	27	66103	37017	0.16	0.28
				10500	27	27	54344	37497	0.15	0.27
H263Decoder	4, 3	1783	15	596	14	15	68243	64831	0.15	0.01
				600	15	15	65361	65361	0.12	0.01
H263Encoder	5, 5	201	120	299	54	120	18548	8347	0.02	0.03
				301	81	120	12409	8314	0.02	0.05

Table 2 summarizes the results. For each model, we specify the number of actors, channels, firings per iteration, and optimal throughput (assuming unbounded buffers). The throughput is relative to a sink actor which has a repetition count of one. Then for each model, throughput and latency analysis is done by bounding the buffer space. Rest of the columns compare throughput, latency and CPU run-time for the two models. We observe that in all cases SDF-AP models have higher throughput and lower latency. In two cases, the SDF model is deadlocked (denoted by "-"), though the SDF-AP model is not.

While run-time for a majority of examples is better for the SDF-AP model, there are instances for which the run-time is worse than the underlying SDF model. The SDF-AP analysis uses a novel algorithmic method based on difference constraints, which can be further optimized for better performance. Independent of how efficiently SDF-AP can be analyzed, the main value lies in its expressiveness which allows to obtain better throughput and latency than SDF.

7. CONCLUSIONS

The SDF-AP model aims to strike a balance between the analyzability of SDF-like models while accurately capturing the interface timing behavior by including access patterns. In this paper, we formalize the SDF-AP model, discuss its operational semantics, and define executability and boundedness properties. We also present algorithms to check these properties. The experimental results validate their performance. As future work, we will develop an analysis framework to automatically reason about these properties, and further investigate the effectiveness of SDF-AP as an abstraction for hardware synthesis.

8. REFERENCES

[1] S. S. Bhattacharyya, P. K. Murthy, and E. A. Lee. *Dataflow Graphs*. Kluwer Academic Press, Norwell, MA, 1996.

[2] G. Bilsen, M. Engels, R. Lauwereins, and J. Peperstraete. Cyclo-static data flow. In *IEEE Intl. Conf. Acoustics, Speech, and Signal Processing*, 1995.

[3] T. H. Cormen, C. Stein, R. L. Rivest, and C. E. Leiserson. *Introduction to Algorithms*. McGraw-Hill Higher Education, 2nd edition, 2001.

[4] M. Edwards and P. Green. The Implementation of Synchronous Dataflow Graphs Using Reconfigurable Hardware. In *FPL*, 00.

[5] J. Eker, J. W. Janneck, E. A. Lee, J. Liu, X. Liu, J. Ludvig, S. Neuendorffer, S. Sachs, and Y. Xiong. Taming Heterogeneity - The Ptolemy Approach. *Proc. of IEEE*, 91(1):127–144, 2003.

[6] A. Girault, B. Lee, and E. Lee. Hierarchical finite state machines with multiple concurrency models. *IEEE Trans. Computer-Aided Design*, 18(6):742–760, 1999.

[7] C.-J. Hsu, M.-Y. Ko, and S. S. Bhattacharyya. Software Synthesis from the Dataflow Interchange Format. In *Intl. Worksop on Software and Compilers for Embedded Processors*, 2005.

[8] J. W. Janneck, I. D. Miller, D. B. Parlour, G. Roquier, M. Wipliez, and M. Raulet. Synthesizing hardware from dataflow programs. *J. Signal Process. Syst.*, 2009.

[9] H. Jung, H. Yang, and S. Ha. Optimized RTL Code Generation from Coarse-Grain Dataflow Specification for Fast HW/SW Cosynthesis. *J. Signal Process. Syst.*, 52(1):13–34, 2008.

[10] R. Lauwereins, M. Engels, M. Adé, and J. A. Peperstraete. Grape-II: A System-Level Prototyping Environment for DSP Applications. *Computer*, 28(2):35–43, 1995.

[11] E. A. Lee. Concurrent models of computation for embedded software. Technical Report UCB/ERL M05/2, EECS Department, University of California, Berkeley, Jan 2005.

[12] E. A. Lee and D. G. Messerschmitt. Static Scheduling of Synchronous Data Flow Programs for Digital Signal Processing. *IEEE Trans. on Computers*, 36(1):24–35, 1987.

[13] E. A. Lee and D. G. Messerschmitt. Synchronous Data Flow. *Proc. of the IEEE*, 75(9):1235–1245, 1987.

[14] O. M. Moreira and M. J. G. Bekooij. Self-Timed Scheduling Analysis for Real-Time Applications. *EURASIP Journal on Advances in Signal Processing*, 2007(83710):1–15, April 2007.

[15] P. Poplavko, T. Basten, M. Bekooij, J. van Meerbergen, and B. Mesman. Task-level Timing Models for Guaranteed Performance in Multiprocessor Networks-on-Chip. In *CASES*, 2003.

[16] S. Stuijk, M. Geilen, and T. Basten. Throughput-buffering trade-off exploration for cyclo-static and synchronous dataflow graphs. *IEEE Trans. Computers*, 57(10):1331–1345, 2008.

[17] S. Stuijk, M. C. Geilen, and T. Basten. SDF^3: SDF For Free. In *Proc. of ACSD*, 2006.

[18] B. D. Theelen, M. C. W. Geilen, T. Basten, J. P. M. Voeten, S. V. Gheorghita, and S. Stuijk. A scenario-aware data flow model for combined long-run average and worst-case performance analysis. In *Proc. of MEMOCODE*, 2006.

[19] S. Tripakis, H. Andrade, A. Ghosal, R. Limaye, K. Ravindran, G. Wang, G. Yang, J. Kornerup, and I. Wong. Correct and non-defensive glue design using abstract models. In *Proc. of the CODES+ISSS*, 2011.

[20] M. H. Wiggers, M. J. G. Bekooij, and G. J. M. Smit. Efficient Computation of Buffer Capacities for Cyclo-Static Dataflow Graphs. In *Proc. of DAC*, 2007.

S1: Supplemental Section 1

The SDF-AP model $\mathcal{M}$ of the example introduced in Section 3 is $\mathcal{M} = (aset, cset)$ where $aset = \{i1, i2, bs, o\}$ and $cset = \{c1, c2, c3\}$. The actors are defined as follows:

- $i1 = (\emptyset, \{c1\}, \varnothing, \{c1 \mapsto 1\}, 2, \varnothing, \{c1 \mapsto 01\})$
- $i2 = (\emptyset, \{c2\}, \varnothing, \{c2 \mapsto 3\}, 4, \varnothing, \{c2 \mapsto 1101\})$
- $bs = (\{c1, c2\}, \{c3\}, \{c1 \mapsto 2, c2 \mapsto 4\}, \{c3 \mapsto 6\}, 6, \{c1 \mapsto 110000, c2 \mapsto 001111\}, \{c3 \mapsto 111111\})$
- $o = (\{c3\}, \emptyset, \{c3 \mapsto 6\}, \varnothing, 6, \{c3 \mapsto 111111\}, \varnothing)$

In the running example, there are no initial tokens. So the initial state $s_0 = (\gamma_0, v_0)$ where $\gamma_0 = (\{c1 \mapsto 0, c2 \mapsto 0, c3 \mapsto 0\}$, and $v_0 = \{i1 \mapsto \emptyset, i2 \mapsto \emptyset, bs \mapsto \emptyset, o \mapsto \emptyset\})$.

Figure 3 shows a possible sub-trace with one complete firing of actor $i1$. The start state remains the same as above. For the rest of the states, only the token count for $c0$ and the instance information for $i1$ change; hence only those values have been shown. The trace starts with a $begin(i1)$ transition which adds one instance of actor $i1$ to the instance information in state $s1$. A clock transition updates the instance of $i1$ to read stage. $i1$ being a source actor, no token is consumed, and the read transition between states $s2$ and $s3$ does not change token counts. A write transition for $i1$ from state $s3$ does not change token count as the output pattern for $i1$ is 01, and one clock cycle has passed after the begin fire transition. A clock transition from $s4$ increases the execution time info for running instance of $i1$ to 2. A read transition from $s5$ does not change any token count, but a write transition from $s6$ produces one token on channel $c1$, and the token count for the channel is updated to 1. An end transition at state $s7$ removes the running instance of $i1$. Figure 4 shows a possible sub-trace where both $i1$ and $i2$ fires. The start state $s0$ remains the same. In the remaining states, only the token counts for $c0$ and $c1$, and the instances for $i1$ and $i2$ change.

S2: Supplemental Section 2

LEMMA 8.1. *Any reachable state has a successor state.*

PROOF. Consider a reachable state s which does not have a successor state. Given that s is reachable, there must be a transition $s'' \xrightarrow{l} s$ for some transition label l. If s is stable (i.e., either $l = begin(a)|end(a)$ for some $a \in aset$, or $l = clock$ but none of the actors are executing, or $l = write(a)$ with no further write transitions possible), then a begin fire transition, or an end fire transition (if precondition is met), or a clock transition (if no end transitions are enabled) are possible from state s. If $l = read(a)|write(a)$ and s is unstable, then at least one read or write transition is enabled at s; if s is stable, then see above. Hence at least one transition is enabled at s, and thus s must have a successor state. $\square$

THEOREM 8.2. *For any PCT state s, $NSS(s)$ is unique.*

PROOF. Let PCT state $s = (\gamma, v)$ has multiple next stable states. The possible scenarios are discussed below.

Scenario 1: If $\forall a \in aset, v(a) = \emptyset$, then s is a $NSS(s)$ by definition. Hence $NSS(s)$ is unique for this scenario.

Scenario 2: If $v(a) = \{(e, r)\}$ $(e \in \mathbb{N})$, and $\forall a' \in aset \setminus \{a\}, v(a') = \emptyset$, then a possible sub trace from s is $\tau' = s \xrightarrow{read(a)} s' \xrightarrow{write(a)} s''$. The first stable state in the trace is

s'' and is $NSS(s)$. There are no other transitions possible for s and s'. Hence $NSS(s)$ is unique for this scenario. Without loss of generality, the same argument can be made for any other actor $a' \in aset \setminus \{a\}$.

Scenario 3: If $v(a) = \{(e, r), (e', r)\}$ $(e, e' \in \mathbb{N})$, and $\forall a' \in aset \setminus \{a\}, v(a') = \emptyset$, there are several possible sub traces starting from s. We will first consider the traces where all read transitions precede all the write transitions. Consider the trace $\tau : s \xrightarrow{read(a)} s_1 = (\gamma_1, v_1) \xrightarrow{read(a)} s_2 = (\gamma_2, v_2) \xrightarrow{write(a)} s_3 = (\gamma_3, v_3) \xrightarrow{write(a)} s_4 = (\gamma_4, v_4)$ where $v_1(a) = \{(e, w), (e', r)\}$, $v_2(a) = \{(e, w), (e', w)\}$, $v_3(a) = \{(e, \perp), (e', w)\}$, and $v_1(a) = \{(e, \perp), (e', \perp)\}$. The channel quantity is computed as $\gamma_4 = \gamma + \gamma_{a,e}^R + \gamma_{a,e'}^R - \gamma_{a,e}^W - \gamma_{a,e'}^W$. The states s_1, s_2, s_3 are unstable, and s_4 a $NSS(s)$. Note that different order of the two reads and the two writes (but all reads preceding all writes) will lead to three other traces starting from s; all of which will have three unstable states (excluding s) and one stable state at the end. The value of v in the stable states will be identical to v_4. And given that ordering of addition and subtraction does not matter, the channel quantities for the stable states will be identical to that of γ_4. Next we will consider the traces where the reads and writes are interspersed. Without loss of generality, consider the trace $\tau' : s \xrightarrow{read(a)} s_1' = (\gamma_1', v_1') \xrightarrow{write(a)} s_2' = (\gamma_2', v_2') \xrightarrow{read(a)} s_3' = (\gamma_3', v_3') \xrightarrow{write(a)} s_4' = (\gamma_4', v_4')$ where $v_1'(a) = \{(e, w), (e', r)\}$, $v_2'(a) = \{(e, \perp), (e', r)\}$, $v_3'(a) = \{(e, \perp), (e', w)\}$, and $v_4'(a) = \{(e, \perp), (e', \perp)\}$. The channel quantity is computed as $\gamma_4' = \gamma + \gamma_{a,e}^R - \gamma_{a,e}^W + \gamma_{a,e'}^R - \gamma_{a,e'}^W$. The states s_1', s_2', s_3' are unstable, and s_4' is a $NSS(s)$. Using the same logic as above, $s_4' = s_4$. Note that if the order of the read-write combination is changed, there will be a new sub trace with three unstable state followed by a stable state. However the value of channel quantity and execution map at the unstable state will be identical to s_4' and thus to s_4. Hence $NSS(s)$ is unique under this scenario, but there are many possible sub traces between the states s and $NSS(s)$. The argument remains the same, if more than two instances of the actor is executing, or actor a is replaced by any other actor $a' \in aset \setminus \{a\}$.

Scenario 4: If $a, a' \in aset, v(a) = \{(e, r)\}, v(a') = \{(e', r)\}$ $(e, e' \in \mathbb{N})$, and $\forall a'' \in aset \setminus \{a, a'\}, v(a'') = \emptyset$, then the argument above can be extended to show that there are multiple possible sub traces starting from s. In each such sub trace, there are three unstable states (excluding s) followed by a stable state. The four states are generated by two read transitions and the two corresponding write transitions. The transitions can be interspersed in any order, but the channel quantity and the execution map would be identical for the stable state along all of the sub traces implying that $NSS(s)$ is unique for this scenario.

The arguments discussed for the last three scenarios can be extended for arbitrary number of active actors with arbitrary number of instances. $\square$

COROLLARY 8.3. *For a trace τ, the set of stable states of τ is identical to the set of stable states in $all(\tau)$.*

PROOF. Every PCT state s has an unique $NSS(s)$ (Lemma 3.3). This implies that generating sub traces between s and $NSS(s)$ do not affect any state before s or any state after $NSS(s)$. States on sub traces between s and $NSS(s)$ are all unstable. Hence generating the sub traces neither adds nor removes any stable state. $\square$

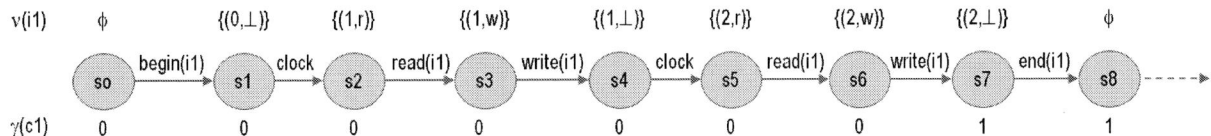

Figure 3: A sub trace with one complete firing of actor $i1$

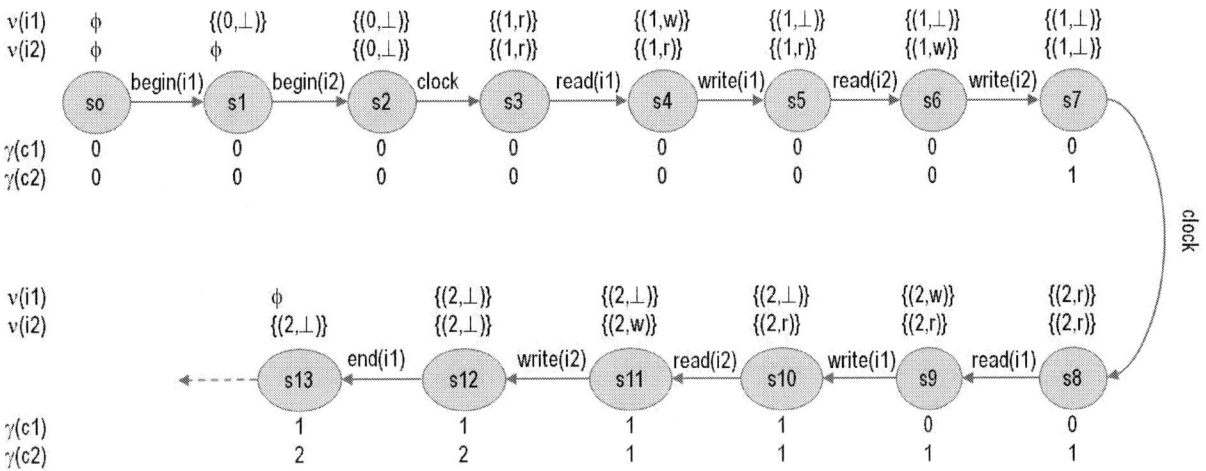

Figure 4: A sub trace where actors $i1$ and $i2$ fire

Executing Synchronous Dataflow Graphs on a SPM-Based Multicore Architecture

Junchul Choi
School of Electrical Engineering
and Computer Science
Seoul National University
Seoul, Korea
hinomk2@iris.snu.ac.kr

Hyunok Oh
Department of information
Systems
Hanyang University
Seoul, Korea
hoh@hanyang.ac.kr

Sungchan Kim
Department of Computer
Science and Engineering
Chonbuk National University
Jeonbuk, Korea
sungchan.kim@chonbuk.ac.kr

Soonhoi Ha
School of Electrical Engineering
and Computer Science
Seoul National University
Seoul, Korea
sha@snu.ac.kr

ABSTRACT

In this paper we are concerned about executing synchronous dataflow (SDF) applications on a multicore architecture where a core has a limited size of scratchpad memory (SPM). Unlike traditional multi-processor scheduling of SDF graphs, we consider the SPM size limitation that incurs code and data overlay overhead. Since the scheduling problem is intractable, we propose an EA(evolutionary algorithm)-based technique. To hide memory latency, prefetching is aggressively performed in the proposed technique. The experimental results show that our approach reduces the overlay overhead significantly compared to a non-optimized approach and the previous approach.

Categories and Subject Descriptors

D.1.3 [**Concurrent Programming**]: Synchronous Dataflow graphs, Multiprocessor Scheduling

General Terms

Algorithm, Performance, Design, Experimentation

Keywords

Multiprocessor scheduling, synchronous dataflow, scratch pad memory, multicore architecture, memory overlay, prefetching

1. INTRODUCTION

With the continuous evolution of semiconductor process technology, hundreds of, even a thousand processors will be integrated in a single chip. There are two types of many-core architectures envisioned in the future. One is a GPU-type architecture where multiple cores that share memories cooperate in processing an application exploiting data parallelism maximally. The other is a distributed memory architecture where each processor core, simply core, will have a limited size of local memory and shared memories will be accessed through an interconnection network. The latter is the target concerned in this paper.

In a distributed memory architecture, if a local memory plays a role of cache, it is called a cache-based multicore system. It is unlikely to use coherent caches in such a system since the overhead of maintaining cache coherence increases dramatically as the number of cores increases. If we use a local memory as a scratchpad memory (SPM), it is called an SPM-based multicore system. SPM has several advantages over cache. SPM consumes less area to store the same number of bits and less power than cache since no tag memory is needed. Since it is controlled purely

by software, we can predict the performance at compile-time. Hence SPM-based multicore architectures are preferred for embedded real-time applications [1][2][3].

We assume that an application is specified by a synchronous dataflow (SDF) graph that is regarded as a viable programming model for stream-based application programming [4]. In an SDF graph, a node represents a computation task and an arc represents a FIFO communication channel between two adjacent nodes. The number of data to be produced or consumed per node execution, which is called sample rate, can be any integer. A node becomes executable only after it has no fewer number of samples accumulated on all input arcs than the specified sample rates. An SDF graph reveals task-level parallelism of an application explicitly by specifying the true dependency only between nodes. Thus parallelizing an application can be readily performed by mapping nodes to cores.

Since the sample rates are fixed in the SDF model, we can construct a static schedule of node executions on a given target architecture [5]. If no schedule can be found, the input specification has an error of buffer overflow or deadlock. From the static schedule, we can estimate the execution latency of an application at compile time. Such static analyzability is a very desirable feature for embedded parallel software design.

In this paper, we focus on minimizing the execution latency of applications specified by an SDF graph in an SPM-based multicore architecture. Unlike traditional multi-processor scheduling of SDF graphs, we have to consider the SPM size limitation that incurs code and data overlay overhead when building a static schedule. When we schedule a task on a processor, we check the available space on the SPM and compare it with the required space of the task. If there is not enough space, we have to move some data in the SPM to a shared memory to secure the sufficient space for task execution; moving of data from an SPM to a shared memory is called *flushing*. Before executing a task, we also have to *fetch* the task code, the input data and internal states from a shared memory. When we make a scheduling decision, we have to consider such code and data overlay overhead that is caused by flushing and fetching.

In order to reduce the overlay delay by hiding memory access latency, we can use prefetching that can be performed concurrently with computation. In the proposed technique, we assume that prefetching is used as much as possible.

Recently a series of researches have been performed to solve the same problem, scheduling of an SDF graph onto an SPM-based many-core architecture [6][7][8]. Since they considered only a subset of factors that we consider in this paper, the proposed technique shows significant overlay overhead reduction over the state-of-the-art technique, which will be demonstrated by experiments in Section 6.

Since multiprocessor scheduling of an SDF graph is an intractable problem, we use an EA(evolutionary algorithm)-based technique when finding a sub-optimal schedule. In an EA-based technique, the key elements are encoding of candidate solutions, evaluation, and evolution strategy, which will be explained in Section 5.

2. RELATED WORK

There are numerous approaches proposed to schedule a task graph onto a multiprocessor system. Since an SDF graph can be converted into a homogeneous synchronous dataflow graph, these approaches can be used for scheduling of the SDF graph. They are distinguished from each other, in terms of scheduling objectives, architecture characteristics, and design constraints. As for scheduling objectives, some aim to minimize the execution latency of the graph while the other aim to maximize the throughput. In this paper, we aim to minimize the execution latency of the graph. Other objectives such as area minimization and energy minimization can also be considered as well to make a multi-objective scheduling problem, which is left as a future work.

As for architecture characteristics, some assume unlimited number of processors while others assume a given target architecture. Some assume homogeneous processors while the others allow heterogeneous processors. We assume that the target architecture is a general heterogeneous multicore architecture. Only recently, an SPM-based multicore architecture gained research attention as the target architecture of multiprocessor scheduling.

Che *et al.* proposed the region-segment overlay scheme and used ILP formulations to minimize the latency on SPM-based processors [6]. Since they configured the memory layout to store all internal data and channel buffers locally in SPM, they simply focused on the code overlay overhead minimization within the remaining capacity. The same authors solved the same problem using heuristic algorithms to overcome the exponential complexity of ILP solution in [8]. Even though we aim to solve the same problem in the paper, the proposed technique is different from theirs as follows. 1) Since they do not consider the data overlay, significant portion of an SPM should be reserved for the data area of all assigned tasks. It may decrease the memory utilization significantly. To overcome this limitation, we propose to perform data overlay as well as code overlay. 2) Prefetching technique is used in their approaches to hide the code overlay latency. But they assume that prefetching is performed during execution of the previous task only. But we use more *aggressive* prefetching scheme if there is a possibility of hiding the overlay latency more effectively by perform pre-fetching earlier. 3) Most importantly, they did not consider communication overhead between cores because they focused on the behavior of a single core only. Since they consider only the code region that is not shared or communicated with the other cores, their solution did not have to consider inter-core communication. On the other hand, we consider the communication overhead in details.

In another work, Che *et al.* proposed a retiming based technique to maximize the throughput on an SPM-based multicore architecture [7]. Here they assumed a single appearance schedule (SAS) that usually requires larger channel buffers than a more memory-efficient schedule. And they used a smart double buffering scheme to minimize the inter-stage communication overhead. On the other hand, we have no restriction on the scheduling strategy, so exploring wider search space of mapping and scheduling. Our scheduling objective is not to maximize the throughput, but to minimize the latency, which is another difference.

Damavandpeyma *et al.* proposed a scheduling technique to minimize the execution latency of a task graph on an SPM-based MPSoC [9]. Since they considered both code and data overlay, their work is closest to ours. But they have a strict restriction that memory objects of two different tasks may not coexist in an SPM at the same time. This assumption might be valid on when the SPM size is too small to accommodate more than one task. With

this assumption, they could not use aggressive prefetching nor hide output data flushing delay.

In summary, the problem complexity of this paper is significantly higher than that of the previous work, considering more factors that affect the execution time when executing an SDF graph on a multicore architecture; we consider both code and data overlay and use aggressive prefetching with no restriction on the scheduling type and the SPM size. As a result, the proposed technique can be applied to more general SPM-based multicore architectures than the previous approaches, to our best knowledge.

3. BACKGROUND

3.1 SPM-based Multicore Architecture

Figure 1 shows the generic structure of an SPM-based multicore architecture we are targeting in this paper. A core and the associated local memory compose a *core* tile and a shared memory module lies in a memory tile. Core tiles and memory tiles are connected to an on-chip network. A core tile has a DMA module inside so that computation and communication can be overlapped in time.

A core has a small-footprint of run-time system that manages the list of assigned tasks according to a given static schedule, controls overlay before task execution, and initiates prefetching at the scheduled times. We assume that this supervisory region is reserved separately from the SPM region that we are concerned about in task scheduling. Another possible option is to designate a core as a central controller that plays the role of supervisor of computation cores, based on a given static schedule. As will be explained later, a static schedule has all timing information when to perform prefetching, flushing and fetching as well as task execution.

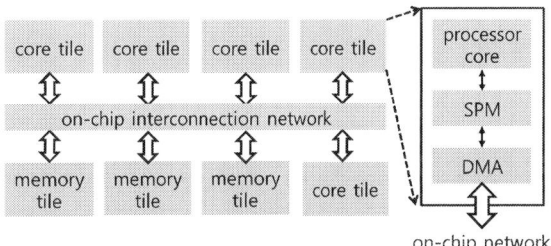

Figure 1. The generic structure of an SPM-based multicore architecture.

3.2 Execution Scenario

We assume that the memory space required for each task execution is known. It includes the code size and the data size. In order to execute an SDF node, the input data should be available. Moreover, temporary data space for the stack and the local variables is also required. And before starting the task execution, the task code and the input data as well as internal states should be fetched into the SPM.

If there is not enough empty space in the SPM, some data in the SPM should be expelled to secure the required size of space. If the data corresponds to a code region or a temporary data region, no flushing is required. We can simply overwrite the region without paying any flushing overhead. If it corresponds to channel data or internal states, we have to flush the data to the shared memory. Since the flushing overlay depends on the types of the secured region, where to locate the required size is also an important factor that affects the performance.

In case there is an empty space large enough to accommodate future tasks, we can hide the overlay overhead by prefetching the code, input data, and the internal states. The DMA (direct memory access) module in a core tile performs prefetching without interrupting the current task execution of the core.

4. PROBLEM FORMULATION

In this section we present the scheduling problem in a formal way. Formal presentation of the scheduling problem is one of the main contributions of this paper since we have to consider many factors in a single framework. An SDF graph is denoted as $G = (V, E)$ where V is a set of nodes and E is a set of edges. For a node $v \in V$, the size of the following three memory objects is assumed given: code, internal state, and temporary data. These memory objects are denoted as $Code_v$, $Internal_v$, and $Temp_v$ respectively. The execution time of node v is denoted as $Exec_v$. For an edge $e \in E$, a channel buffer is associated as a memory object, denoted by $Token_e$.

We define a set I of memory objects of an SDF graph:

$I = \{$ for all $v \in V$ and $e \in E$, $| Code_v, Internal_v, Temp_v, Token_e \}$

For each memory object $i \in I$, we define a size function $Sz(i)$ such that $Sz: I \rightarrow Z$ where Z is a set of positive integers.

And we define a life-time function $Life_S(i)$ such that $Life_S: I \rightarrow \{ t \mid t \in Time \}$ for a given schedule S. If a schedule is given, we can obtain the life-time of each memory object. Suppose that we are given a schedule, AABAB for a simple SDF graph with two nodes A and B. Since three instances of node A appears in the schedule, the code memory of node A has a life-time that consists of three time intervals. The life time interval indicates when to fetch or flush a memory object. At the end of the life time, for example, a channel buffer is flushed to the shared memory. It is fetched into SPM in the beginning of the life time.

Scheduling is to determine on which processor to map a node and on which order to schedule the mapped nodes on each core. We represent a mapping function $M(v)$ such that $M: V \rightarrow N$ where N is a set of processor indices. To represent a schedule, we assign a unique priority to each node instance. If we have three instances of the same node in a schedule, then they are assigned different priorities. Then a static schedule can be determined among the mapped nodes on each processor in the order of their priorities. Hence we define a priority assignment function $Pr(v^*)$ such that $Pr: V^* \rightarrow Z$ where V^* is a set of all node instances. In this function, no two node instances may have the same priority.

When we map a node to a core, we also have to allocate the associated memory objects to SPM. So we define another function for memory allocation $P(i)$ such that $P: I \rightarrow Z$ where Z is a set of offset values on the SPM.

In the proposed technique, we can prefetch memory objects to hide the communication latency. Similarly we can delay flushing memory objects than their life times. It implies that the life time of a memory object can be lengthened by prefetching or *post-flushing*, which means that we delay the flushing of a memory object. To express prefetching and post-flushing in a formal way, we define a life-time extension function $E(i)$ such that $E: I \rightarrow Z$ where Z is an integer indicating how far its life time can be prolonged. If we denote the prolonged life time as $Life_S^*(i)$ for a memory object i, the following is satisfied: $Life_S^*(i) \supseteq Life_S(i)$.

For memory allocation, we have the following constraints.

(C1) No overlapping constraint: If $[P(i), P(i)+Sz(i)) \cap [P(j), P(j)+Sz(j)) \neq \varnothing$ for $i \neq j$ then $Life_S^*(i) \cap Life_S^*(j) = \varnothing$.

(C2) Size constraint: $P(i)+Sz(i) \leq$ the SPM size for all i.

Now we summarize the scheduling problem as follows.

Input: An SDF graph, $G = (V, E)$, with the following profile information: execution time $\{Exec_v\}$, size of memory objects $\{ Sz(i) \mid i \in I \}$

Constraints: (C1) and (C2)

Objective: Minimize the execution latency of an SDF graph

Problem: Find an optimal schedule, which is to determine the following four functions: mapping function $M(v)$, priority assignment function $Pr(v^*)$, life time extension function $E(i)$, and memory allocation $P(i)$.

5. PROPOSED GA-BASED TECHNIQUE

The scheduling problem tackled in this paper is a combinatorial optimization problem that may be solved using various evolutionary algorithms. Among them, the proposed technique is based on a Genetic Algorithm (GA).

5.1 Solution Representation

As explained in the previous section, our scheduling problem can be represented as a set of four sub-problems. It is noteworthy that the four sub-problems are closely inter-related so that they should be solved at once. Hence, in the proposed approach we encode a solution of each sub-problem separately and combine four encoded solutions into a single chromosome, as shown in Figure 2; a chromosome is divided into four sub-chromosomes associated with each sub-problem.

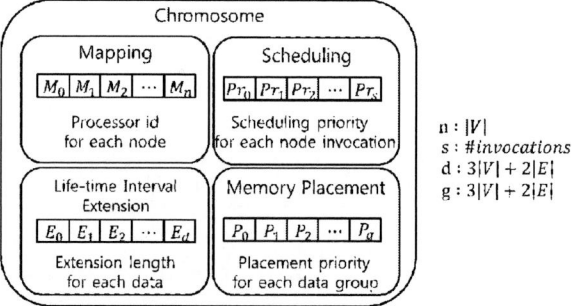

Figure 2. Chromosome organization of the proposed GA.

The first sub-chromosome represents the mapping of SDF nodes to cores and its size is the number of nodes in the SDF graph. Note that we assign all instances of a node to the same core. If there is a loop-level parallelism, or data parallelism, in the SDF graph, multiple instances of a node can be assigned to different cores. In this paper, however, we do not consider this case and leave it as a future work to exploit loop-level parallelism. If a node has internal states, there is dependency between node invocations so that they had better be mapped to the same core as we assume.

The second sub-chromosome represents a static schedule of the nodes in all cores at once, combined with the first sub-chromosome as explained in the previous section. It is a very efficient way of representing a static schedule in an evolutionary algorithm. This representation is inspired by the *invocation-based priority assignment* [10] of dynamic scheduling. The sub-chromosome size is the total number of node invocations in the SDF schedule. In case the number of invocations is huge in a complex multi-rate application, we can reduce the sub-chromosome size by shrinking the solution space of static scheduling by adopting the *node-based assignment* where the priority of a node invocation is expressed by an affine function. For detailed discussion on the scheduling representation, refer to [10].

The third sub-chromosome represents life-time extension of memory objects. This determines the schedule of prefetching and flushing of memory objects, or the life time of them. Figure 3 shows a simple example how an integer solution is related with prefetching. Suppose we are interested in the code memory object of the second invocation of node A. If the value is 0, it means that

666

the life time is not prolonged. Thus we have to fetch the code of node A paying the code overlay overhead. If the value is between 1 and 5, the code is prefetched at the start position of the prolonged interval. Since the core is executing another task, prefetching by DMA is overlapped with computation to hide the code overlay latency. If the value becomes 6, the life time meets the previous invocation of the same node. Then, we do not need to prefetch the code since the code object can stay in the SPM. If we consider an internal state object instead of code object, the last case implies that we do not need to flush the object after the first invocation of the node. As illustrated in this example, life-time extension is a compact way to represent prefetching, post-flushing, and object pinning. It is another novel idea in the proposed technique.

The size of the third sub-chromosome is equal to $3|V| + 2|E|$, where $|V|$ and $|E|$ are the numbers of nodes and edges in an SDF graph respectively. We assign separate entries for prefetching and flushing for the same memory object. For internal states of a node, we should consider both prefetching and flushing. On the other hand, there is no need of life-time extension for temporary data. For a code object, we do not need flushing. Thus we need 3 entries per node. For a channel buffer, we need to consider both prefetching and flushing, requesting 2 entries per edge. In total we need $3|V| + 2|E|$ entries in this sub-chromosome.

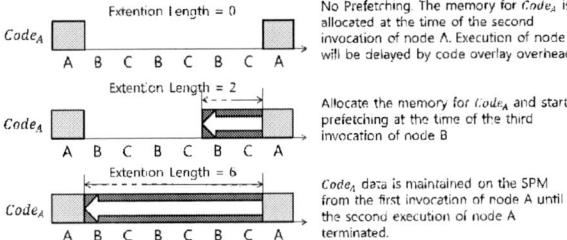

Figure 3. Life-time extension of a code object by prefetching

The last sub-chromosome represents the allocation of memory objects in the SPM in an indirect way. It does not solve the memory allocation function $P(i)$ directly. Similar to the second sub-chromosome that represents a static schedule by node priorities, we determine the priorities of the memory objects. Based on the priorities, we use Move-Down Algorithm (MDA) [11] that is proposed as a heuristic to produce overlay-aware data layout of an SPM. The MDA algorithm requires the order of data segments to be placed in memory. Thus this sub-chromosome represents the placement orders of data. For a node, we need to allocate three memory objects on the assigned core. For an edge, we need two entries for a channel buffer since two end nodes may be allocated to different cores. Then, we have to allocate the channel buffer in a pair of SPMs. As a result, the size of the sub-chromosome becomes $3|V| + 2|E|$.

5.2 Overall Flow of the Proposed GA

The overall flow of the proposed GA is illustrated in Figure 4. The procedure begins with creating initial populations randomly. At each generation, we select two chromosomes (solutions) as parents and apply conventional one-point crossover or mutation separately to each of four sub-chromosomes with the selected parents. A mutation operation simply changes the value of any randomly chosen entry at each sub-chromosome with a given mutation probability. For the third sub-chromosome corresponding to the life-time extension, however, we give higher probability to increasing life-time over decreasing in order to produce better solutions with less DMA transfers.

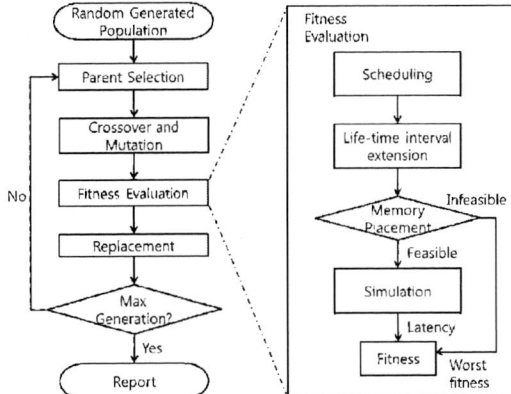

Figure 4. Overall flow of the proposed GA-based technique.

In the next step, the fitness of a newly generated chromosome is evaluated in terms of application latency and the total size of SPM required. This step first constructs the static schedule based on the first and the second sub-chromosomes. Second, by combining the life-time of data objects and the schedule information, we compute the start time of prefetching/flushing of memory objects, if applicable. Next, these life-time intervals and the memory placement priorities of the fourth sub-chromosome are given to the memory placement algorithm [11] to obtain the required SPM size.

If the required size satisfies the SPM size constraint, the solution is feasible and the application latency is calculated, which is defined as the fitness value of the current solution. When an infeasible solution is found, we enforce subsequent siblings to have shorter life-time intervals, by introducing a *life-time reduction factor* R in order to increase the probability of making feasible solutions. In the current implementation, we reduce all life-times in the current chromosome by multiplying R to the current value. The reduction factor itself also varies according to the following formula:

$$R = 1 - \frac{(c_{max} - c)^2}{c_{max}^2}$$

where c and c_{max} is the current and maximum count value. The value of c_{max} is empirically determined. As it gets small, the decreasing rate increases. In our experiments, we set it to 20. Initially c is set to c_{max} and decreases by 1 whenever an infeasible solution is found. It increases by 1 if a feasible solution is found, but limited by c_{max}. Since we will have many infeasible solutions in the beginning stage of evolution, R would decrease rapidly, limiting the life-time extension of memory objects severely. But as the solutions converge, the memory allocation will be successful with high probability. Also the reduction factor itself will become stable as c approaches to c_{max}.

The execution latency of an application is obtained from the schedule and execution time information of nodes. We also consider the data communication delay for inter-core communication as well as code/data overlay overhead what could not be hidden by prefetching and post-flushing. Remind that all communication between SPM and the outside is performed by DMA in our model. We can simulate the interconnection network to compute the network delay more accurately

6. EXPERIMENTAL RESULTS

In order to validate the efficiency of the proposed approach, we have conducted using some applications in StreamIt Benchmarks [13]. As the target SPM-based multicore architecture, we adopt the IBM Cell BE architecture [3]. While the channel buffer sizes of benchmark programs are small except MP3 and MPEG applications, we increase the buffer size by running multiple iterations to reduce the communication overhead, which is known as scalable synchronous dataflow [14]. The iteration factors are 302.85 on average, by which the buffer sizes become KB unit. Table 1 shows the detailed information of the benchmark applications. Every node in all applications except MPEG is assigned to SPE in Cell processor. Since two nodes of MPEG, MotionPrediction and PictureReorder, require larger data memory than the SPM size, 256KB, they cannot be assigned to SPE; they are assigned to PPE.

Table 1. The SDF applications used in the experiments.

Benchmarks	#Nodes	#Edges	#Invocations	Average execution time(μs)
BeamFormer	52	84	80	7.1038
DCT	19	25	30	3.0454
DES	101	132	117	3.5912
FFT	12	11	95	1.5393
FilterBank	26	28	80	3.7361
FMRadio	44	62	48	3.7903
Serpent	101	100	119	5.4124
MP3	29	39	977	1.5524
MPEG	20	27	3809	1.6384

Since the target architecture is the IBM Cell BE, the DMA communication is modeled based on the IBM Cell BE. Typically the latency of DMA communication is modeled as follows:

$$\text{DMACost(D)} = T_{base} + \frac{Size(D)}{V_{DMA}}$$

where T_{base} is the start-up time for DMA transaction, $Size(D)$ is the size of data object D to transfer, and V_{DMA} is the bandwidth of the communication network. We should consider a contention between DMA transfers from different PEs. The bus in the IBM Cell BE consists of four data rings which allow concurrent data transfers. However, in our execution scenario, almost all data transfers occur between SPM in each SPE and a remote shared memory, in which the data port of shared memory becomes a bottleneck. In contrast, direct transfers between SPEs, which rarely occur, are likely to be progressed concurrently. Thus, data transfers using same data ports should be arbitrated. The transfer bandwidth for such transfers is divided by the number of transfers using same data ports. We revised the equation to consider the bus contention for the experiment. We use parameters available in [8]. The speed of SPE is 3.2 GHz, T_{base} is 0.21us, and V_{DMA} is 13.3 GB/s.

We compare the proposed approach with two approaches: "NoOpt" and "code-overlay". The NoOpt approach is the baseline approach of the proposed technique but without prefetching and post-flushing. It does not extend lifetime intervals without hiding code/data overlay delays. The code-overlay approach is the approach adopted from [8]; it aims to minimize the code overlay delays only while data objects are statically allocated in SPM. Unlike [8] that aims to maximize the throughput, however, we implemented the approach to minimize the latency of the graph.

The experimental results are obtained by simulation. In the first set of experiments, we measure the end-to-end latency of benchmark applications assuming that the SPM size of each core is 256 KB and the number of cores is 8. Figure 5 shows the comparison results among three approaches. The y-axis indicates the latency of the applications in micro seconds. The graph shows how much the code/data overlay overhead contributes to the execution latency of application. In order to compute the upper bound of performance improvement by the prefetching and post-flushing, we compute ideal latency assuming that the overlay delays and the communication time between cores is zero, which

is labeled as "Computation" in Figure 5. The results show that the proposed technique could hide the most overlay delay efficiently. The impact of the overlay delay onto the end-to-end latency of applications is negligible throughout all applications, which is about 2.52 % on average. On the other hand, the other approaches result in additional latency due to the disclosed overlay overhead. The "Serpent" application shows the worst case, where the overlay overhead occupies 30% of the end-to-end latency if no optimization is applied. Our approach reduces the overlay overhead by 98.25 % and 95.27% compared with the NoOpt approach and the code-overlay approach on average respectively. Note that the detailed graphs are available in a supplementary material.

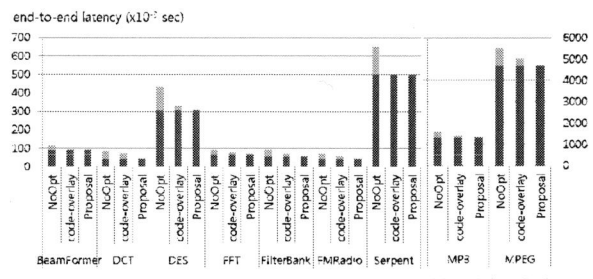

Figure 5. Latency comparison between three approaches, NoOpt, code-overlay only, and the proposed, for StreamIt benchmark applications

In the next set of experiments, we change the number of cores from 1 to 8 as while the SPM size is fixed to 256 KB. Figure 6 shows the latency results. For DCT, FFT and MP3 applications on a single processor, the code-overlay approach provides equivalent results to our approach since the total size of code and data is less than the SPM size and, thus, no overlay is occurred. However, as the number of processors increases, our technique outperforms the code overlay approach since communication delay between cores can be hidden by prefetching or post-flushing in the proposed technique.

Furthermore, the code overlay approach often fails to find a feasible solution when the number of processors is less than 2 or 4 for several applications since the SPM size is not large enough to hold all code and data objects for the nodes mapped on a core. On the other hand, our approach is found to be robust by providing feasible solutions even with the small number of cores, paying negligible overlay overhead by aggressive prefetch and post-flushing. Note that no latency graph is plotted at the configuration points where the algorithm does not find a feasible solution.

In the next set of experiments, we vary the SPM size from 32KB to 1MB while the number of SPEs is fixed to 8. Figure 7 shows the overlay delays. The similar tendency is observed as the previous experiments. The proposed approach finds a feasible solution even though the solution requires larger transferring delays as the SPM size becomes smaller while the code-overlay approach cannot find any feasible solution. For instance, the DES and Serpent applications cannot be executed with the code-overlay approach when the SPM size is 128 KB. Note that after the SPM size is large enough to hold all code and data, the increment of the SPM size does not improve performance any more.

In summary, the proposed approach is able to efficiently utilize SPM compared to the previous approaches. It consistently provides well-optimized solutions with negligible overlay overhead even in case frequent prefetch and flushing are required due to small SPM sizes where the previous approaches could not be successful.

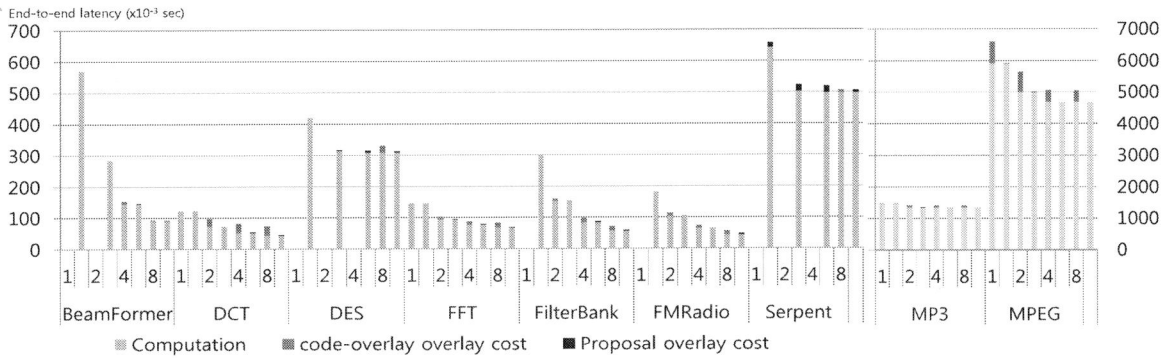

Figure 6. Latency improvement as the number of cores changes.

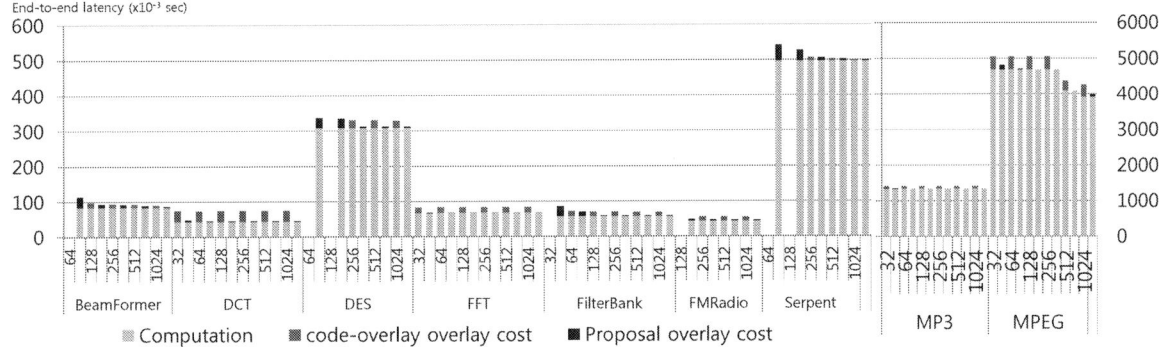

Figure 7. Overlay delays as the SPM size varies.

7. CONCLUSION

In this paper, we proposed a GA-based scheduling technique of an SDF graph, aiming to minimize the execution latency on a SMP-based multicore architecture. Unlike traditional multi-processor scheduling of SDF graphs, we consider the SPM size limitation that incurs code and data overlay overhead. We organized the scheduling problem as a combination of four sub-problems: mapping of nodes to cores, scheduling of nodes on each core, stretching of life-time intervals for memory objects (code, channel buffer, internal states and temporary variables), and placement of the memory objects onto SPMs. The problem is formally formulated and solved by the proposed EA (evolutionary algorithm)-based technique. To hide memory latency, prefetching and post-flushing are aggressively performed in the proposed technique. The experimental results show that our approach reduces the overlay overhead by 98.25 % and 95.07% compared to a non-optimized approach and the previous code-overlay approach on average, respectively.

8. ACKNOWLEDGEMENT

This work was supported by the MKE(The Ministry of Knowledge Economy), Korea, under the ITRC(Information Technology Research Center) support program supervised by the NIPA(National IT Industry Promotion Agency) (NIPA-2012-H0301-12-1011), research Grant of Education and Research Foundation College of Engineering, Seoul National University, and the Basic Science Research Program through the National Research Foundation of Korea (NRF) funded by the Ministry of Education, Science and Technology (2010-0005982 and 2011-0026105). The ICT at Seoul National University provides research facilities for this study.

9. REFERENCES

[1] S. Vangal, *et al.*, "An 80-tile 1.28 TFLOPS network-on-chip in 65 nm CMOS," in IEEE ISSCC, 2007.

[2] J. Owens, "Gpu architecture overview," ACM SIGGRAPH 2007 courses, 2007.

[3] D. Pham, *et al.*, "Overview of the architecture, circuit design, and physical implementation of a first-generation cell processor," IEEE Journal of Solid-State Circuits, Jan. 2006.

[4] E. A. Lee and D. G. Messerschmitt, "Synchronous data flow," Proceedings of IEEE, Sep. 1987.

[5] E. Lee and D. Messerschmitt, "Static scheduling of synchronous data flow programs for digital signal processing," IEEE Trans. Computers, Jan. 1987

[6] W. Che and K. Chatha, "Scheduling of synchronous data flow models on scratch pad memory based embedded processors", in ICCAD, 2010

[7] W. Che and K. Chatha, "Compilation of stream programs onto scratchpad memory based embedded multicore processors through retiming," in DAC, 2011

[8] W. Che and K. Chatha, "Scheduling of Stream Programs onto SPM Enhanced Processors with Code Overlay," in ESTIMedia, 2011.

[9] M. Damavandpeyma, S. Stuijk, T. Basten, M. Geilen, and H. Corporaal, "Hybrid Code-Data Prefetch-Aware Multiprocessor Task Graph Scheduling," in DSD, 2011

[10] J. Kim, T.-H. Shin, S. Ha and H. Oh, "Resource Minimized Static Mapping and Dynamic Scheduling of SDF Graphs," in ESTIMedia, 2011.

[11] M. H. Foroozannejad, M. Hashemi, T. L. Hodges, and S. Ghiasi, "Look into details: the benefits of fine-grain streaming buffer analysis," in LCTES, 2010.

[12] W. Thies, M. Karczmarek, and S. Amarasinghe, "StreamIt: A language for streaming applications," in CC, 2002.

[13] StreamIt benchmarks, http://groups.csail.mit.edu/cag/streamit/shtml/benchmarks.shtml.

[14] S. Ritz, M. Pankert, and H. Meyr, "Optimum vectorization of scalable synchronous dataflow graphs," in ASAP, 1993

S1. A SIMPLE EXAMPLE

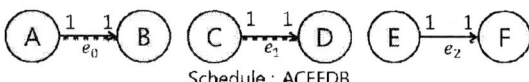

Schedule : ACEFDB

SDF Graph					
$Code_A$	300B	$Code_B$	800B	$Code_C$	200B
$Code_D$	100B	$Code_E$	100B	$Code_F$	500B
$Token_{e_0}$	200B	$Token_{e_1}$	200B	$Token_{e_2}$	200B
$Exec_A$	3 μs	$Exec_B$	2 μs	$Exec_C$	7 μs
$Exec_D$	2 μs	$Exec_E$	1 μs	$Exec_F$	2 μs
Architecture					
SPM size		1300B			
DMA speed		100B/μs			

Figure 8. Example SDF graph and input table

In this section, we show how the proposed technique is applied to an example scenario. Consider Figure 8 which is a simple SDF graph where all sample rates are unity with "ACEFDB" schedule. We only consider life time intervals of code and channel buffer in this example in order to simplify the problem representation. Information of SDF graph and architecture is also shown in the figure. We have 1300B scratchpad memory and DMA operates with the speed of 100B/us and have no start-up time.

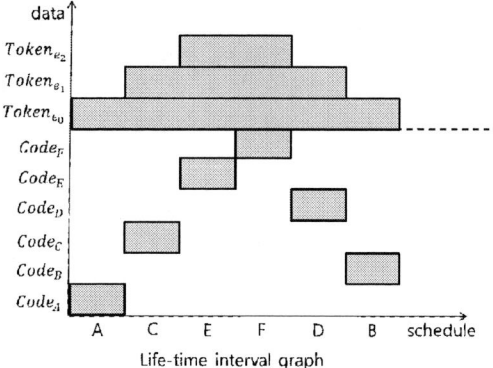

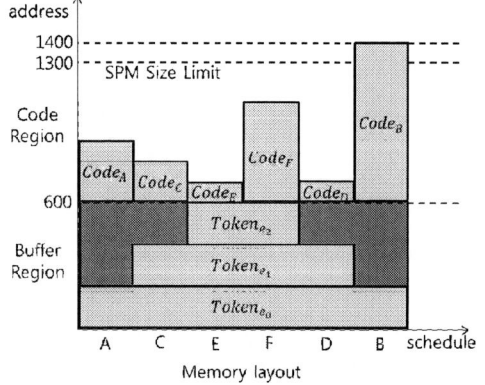

Figure 9. Life-time interval graph and memory layout of code-overlay approach.

Figure 9 illustrates the life-time interval graph and the memory layout that the code-overlay approach will produce. 600B must be allocated to buffer region to store all channel buffers in SPM locally. Since code data for node B cannot be allocated in the SPM, schedule "ACEFDB" is infeasible in the code-overlay approach.

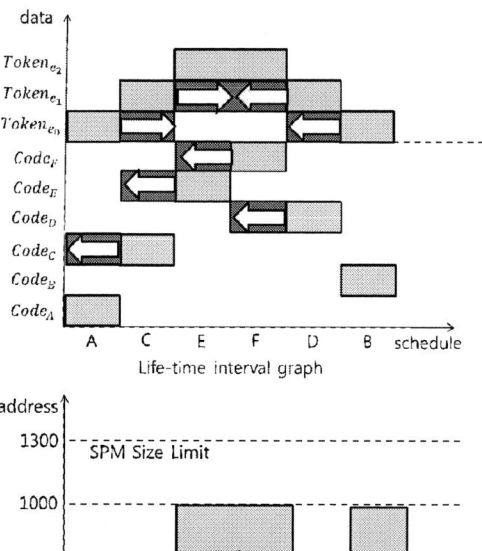

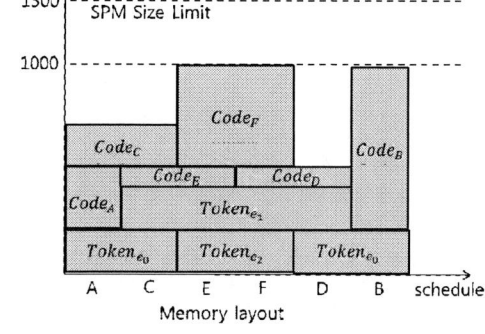

Figure 10. Life-time interval graph and memory layout of code/data overlay approach.

However, we can find a feasible solution with the same schedule when data overlay is also considered as the proposed technique. Memory layout in Figure 10 illustrates that memory allocation can be successful with data overlay. Note that channel buffers don't have to be stored during entire schedule any more. $Token_{e_0}$ is flushed to the shared memory and fetched again before execution of node B. another data block can be allocated to $Token_{e_0}$'s region in the meantime.

Now let us consider the life-time interval extension. Life time intervals of code object of node C, D, E and F are extended by one to left, causing pre-fetching during the previous node execution. The life time interval of $Token_{e_0}$ is extended by one to right, causing post-flushing after node execution. The effect of concurrent fetching/flushing with executions is shown with a time line graph in Figure 12. Note that extension for $Token_{e_1}$ causes two life time intervals to be merged into one interval. In this case no fetching/flushing is needed.

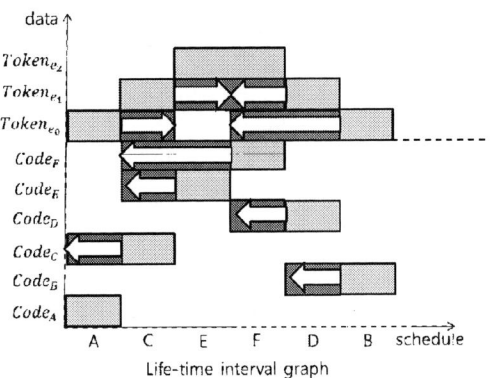

Life-time interval graph

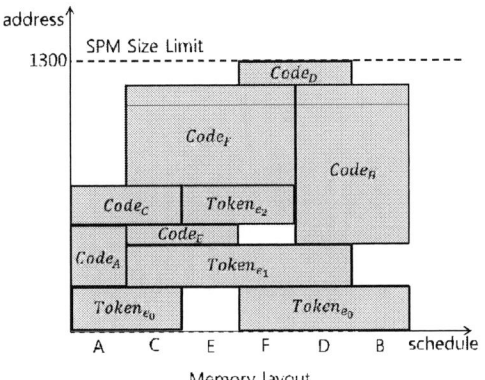

Memory layout

Figure 11. Life-time interval graph and memory layout of aggressive pre-fetching example.

We can optimize latency in Figure 12 by prefetching the code object earlier. Figure 11 shows how it can be achieved. Since we have limited SPM size(1300B), we can extend life time intervals as far as the memory space permits. In Figure 11, we use full SPM space and extend $Code_F$ and $Token_{e_0}$ more. As a result overlay cost of $Code_F$ is fully hidden with node C execution and part of overlay cost of $Code_B$ is removed. The overall latency is reduced by 5 μs, as can be seen in Figure 13.

S2. Experimental Results

In this section, we show the comparison result in terms of the communication overhead only. Figure 15 shows the normalized performance over the NoOpt solution.

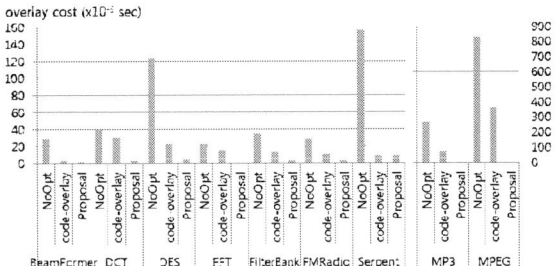

Figure 14. Overlay cost comparison between three approaches, NoOpt, code-overlay, and the proposed, for StreamIt benchmark applications

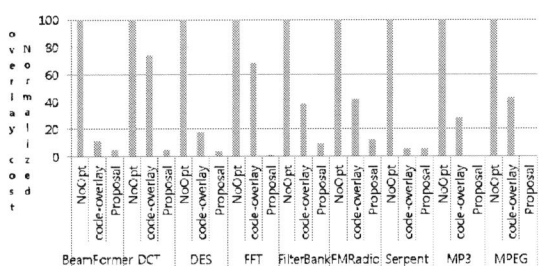

Figure 15. Overlay cost comparison between three approaches, NoOpt, code-overlay, and the proposed, for StreamIt benchmark applications. Normalized by NoOpt

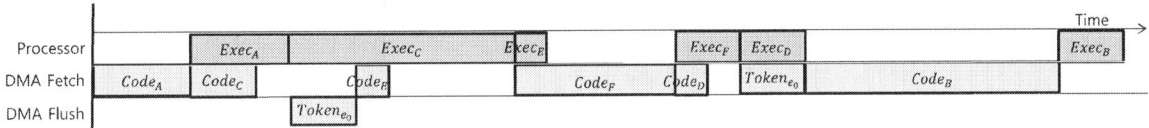

Figure 12. Time line graph of the example in Figure 10.

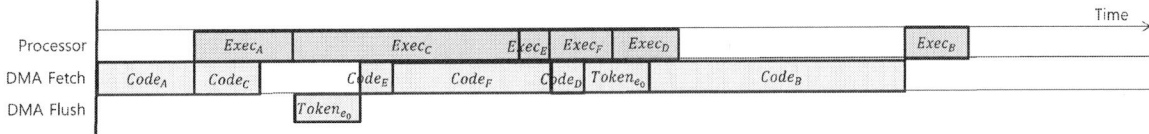

Figure 13. Time line graph of the example in Figure 11.

671

System-level Synthesis of Memory Architecture for Stream Processing Sub-Systems of a MPSoC

Glenn Leary, Weijia Che, and Karam S. Chatha
Arizona State University
Department of CSE, P.O. Box 875406, Tempe, AZ, 85287-5406
{gleary, wche2, kchatha}@asu.edu

ABSTRACT

Many embedded processor chips aimed at high performance and low power application domains are implemented as multi-processor System-on-Chip (MPSoC) devices. The multi-media and communication sub-systems of an MPSoC perform some of the most computation intensive and performance critical tasks, and are key determinants of the system-level performance and power consumption. This paper presents an automated technique for synthesizing the system-level memory architecture (both code and data) for the streaming sub-systems of an embedded processor. The experimental results evaluate effectiveness of the proposed technique by synthesizing the system-level memory architecture for benchmark stream processing applications and comparisons against an existing approach.

Categories and Subject Descriptors

B.3.0 [**Memory Structures**]: General

General Terms

Performance, Algorithms, Design, Experimentation

Keywords

Memory Synthesis, Code Overlay, Data Minimization, SDF

1. INTRODUCTION

The past decade has seen the emergence of smart mobile devices (smart phones, tablets) as the new technology drivers. Present day versions of these devices support a multitude of applications (voice/data communication, HD camera, media player, GPS, and 3D displays) on the same device. The processors aimed at such devices must support the desired performance while literally "sipping" energy from the battery pack. Further, as smart devices fall in the realm of embedded computing the processors must be designed with a short turn around time. Consequently, chip designers have adopted a heterogeneous System-on-Chip architecture for these processors where each sub-system (application, graphics/media, communication, peripheral) is designed with an

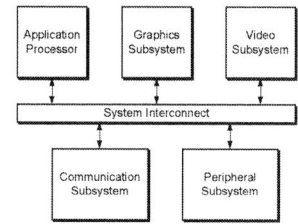

Figure 1: Generic MPSoC architecture

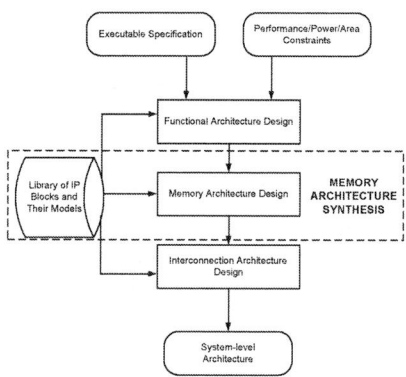

Figure 2: System-level design flow

optimal constellation of processors, hardware accelerators, memory hierarchy and interconnection network.

Figure 1 shows a top-level view of a generic MPSoC aimed at smart mobile devices along with its major sub-systems. The application processor is a general purpose processor such as a dual core ARM Cortex-A9 with cache coherent memory hierarchy. The other sub-systems of the MPSoC are composed of highly optimized instruction processors (such as ARM M3), graphics/DSP processors, and custom hardware accelerators. Further, the non-application sub-systems do not typically support a cache hierarchy and have scratchpad memories for both code and data. The overall architecture is an integration of the various sub-systems via a high bandwidth system-level interconnect.

The focus of this paper is on the system-level architecture design of a sub-system for such a MPSoC. Figure 2 depicts the three primary design stages in developing the system-level architecture. The inputs to the system-level design flow are the executable specification, performance/area/power constraints and a library of characterized IP blocks (performance/power/area models). The functional architecture

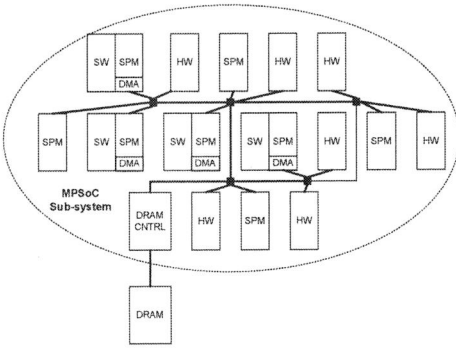

Figure 3: Architecture of MPSoC sub-system

design stage selects the processor core(s) and hardware accelerator(s), and maps the functionality on the processing elements (PE). The memory architecture design stage selects the number and configuration (sizes, ports) of the various memory elements. Finally, the interconnection architecture design stage specifies the topology of the interconnect for the architecture. *The paper focuses on the design automation of the memory architecture design stage for a domain specific sub-system of a MPSoC.*

The graphics, multimedia and communication sub-systems of the MPSoC depict classical streaming behavior. Consequently, the functionalities of these sub-systems can be most naturally described by stream programming formats. For the purposes of this paper we assume that the functionality is described by a synchronous dataflow (SDF) specification [1]. As the focus of the paper is on memory-interconnection architecture synthesis, we assume that the designer performs the functional architecture design stage (selection of PEs and mapping of the SDF actors onto the PEs). *Thus, the objective of our synthesis flow is to select the number and configuration (sizes, ports) of the memory elements such that the performance and area constraints are satisfied, and the power consumption is minimized.*

Figure 3 shows the generic architecture of a sub-system belonging to the MPSoC. In the figure, the instruction processors are denoted by SW, hardware accelerators as HW and the scratch-pad memories as SPM. Each SW PE has a local SPM, and a DMA controller. There may be other SPMs distributed in the architecture that act as shared resources. The various compute nodes and memory elements are connected together by a Network-on-Chip (NoC). The overall performance (and consequently power consumption) of the architecture is a consequence of several design decisions and trade-offs. As the same SPM is shared by code and data for the SW PE, its performance is dictated by the SDF schedule and code overlay (if required). Code overlay schemes are utilized to minimize the memory required for actor code by mapping the code of multiple actors to the same region of memory. Thus, if the code for the actor to be executed next is not in the memory, it is fetched from DRAM and the currently resident actor code is overlayed. At the system-level the interconnect delays are dictated by the topology and the DMA schedules. As the objective is to minimize the power consumption subject to both performance and area constraints the number (and sizes) of memory elements, and router nodes that can be utilized are limited. In our approach we utilize an existing NoC synthesis technique [2]. This paper presents a novel automated

memory synthesis approach that is able to effectively perform the various trade-offs, and generate a highly optimized memory and NoC architecture for the sub-system.

The paper is organized as follows: Section 2 formally defines the problem, Section 3 describes the related work, Section 4 discusses our synthesis approach, Section 5 presents the experimental results and finally Section 6 concludes the paper.

2. PROBLEM DEFINITION

The formal definition of the problem is as follows. *Given*:

1. *a synchronous dataflow specification of a streaming application*: A Directed Graph $G(V, E)$, where $v \in V$ is a set of filters or actors, and the set of directed edges $e(u, v) \in E$ denotes that the data produced by u is consumed by v. Each directed edge $e(u, v) \in E$ is annotated with the size of the data block, $\delta(u)$, produced by filter u and the size of the data block, $\phi(v)$, consumed by filter v.

2. *a set PEs and a mapping of filters to the PEs*: A bipartite Graph $G(V, P, M)$, where $v \in V$ is the set of filters pertaining to the streaming application, $p \in P$ is the set of PEs (HW or SW), and the set of undirected edges $e(v, p) \in M$ denotes the mapping of filter v onto PE p. In the case of SW PEs more than one filter may be mapped to it. Each filter $v \in V$ is annotated with the code size of the filter, $\omega(v)$, and the execution time of the filter, $\tau(v)$.

3. *performance and area constraints*: Designer specified throughput constraint on the SDF, and area constraint on the sub-system.

4. *library of characterized memory elements*: A library consisting of memory elements parameterized in terms of size and number of ports, and characterized in terms of power consumption, area requirement, and access latencies.

5. *library of characterized NoC router architectures*: A library of NoC IP components (routers and network interfaces) characterized in terms of power consumption, area requirement, and no load latency.

synthesize

1. *a memory architecture for the sub-system*: The memory architecture specifies the number and configuration of distinct SPM elements in the sub-system.

2. *a NoC topology*: The NoC topology specifies the number and configuration of the routers used in the architecture, and their interconnection to the PEs and memory elements.

3. *a memory usage description for the sub-system*: The memory usage description describes the utilization of the various SPMs for actor code and data blocks. The description specifies if a code overlay scheme has been utilized for the SPM, and if indeed it has been utilized, the description includes a mapping of actors to region and segments in the SPM. The usage description also defines a mapping of the actor data blocks to memory regions of various SPMs. As the memory usage is minimized by utilizing shared SPM for ephemeral data, more than one data block may be assigned to same region of a SPM.

4. *a execution schedule for SDF and DMA*: The execution schedule gives the global schedule for firing of various actors, and launching of DMA operations for code overlays, and data transfers.

such that the performance and area constraints are satisfied, and the power consumption of the design is minimized.

3. PREVIOUS WORK

System-level MPSoC memory synthesis has been attracting growing attention over the past few years. A representative selection of existing work is discussed in this section. Meftali et al. [3] presented an integer linear programming approach for memory synthesis that focused only on data blocks. Pasricha et al. [4] proposed an integrated heuristic approach for memory and bus matrix synthesis that was also primarily aimed at data blocks. Pandey et al. [8] presented a bus and data memory architecture co-synthesis approach based on slack allocation. Issenin et al. [5] proposed a MILP and heuristic memory synthesis approaches that utilized a fixed topology bus architecture and aimed at minimizing data memory usage. An extension of the same work for mesh based NoC was also proposed [6]. Monchiero et al. [7] presented the results for design space exploration of a non-uniform memory access architecture interconnected with a parameterized (ring, spidergon or mesh) NoC fabric. Recently, Lee et al. [9] presented an approach for integrated MPSoC synthesis for SDF specification that considered pre-selected bus templates. In contrast to these approaches we consider NoC aware memory architecture design for streaming applications. Further, we not only optimize and account for data block memory usage but also consider the impact of code memory optimization. Specifically, we consider the design trade-offs for partitioning the same SPM between code and data. We also consider the performance and power overheads of code overlay schemes that can reduce the memory requirements (and consequently the MPSoC area). To the best of our knowledge the system-level memory synthesis approach presented in the paper is the only technique that considers the impact of both data and code memory requirements during design space exploration.

4. SYSTEM-LEVEL MEMORY SYNTHESIS

The top-level view of our memory synthesis technique is shown in Figure 4. The overall strategy of our technique is to begin with the highest performance and lowest power consuming solution. We then iteratively arrive at a solution that satisfies an area constraint with minimal decrease in performance and increase in power consumption. The inputs to the memory synthesis design stage are the i) performance and area constraints, ii) functional architecture description, and iii) the library of memory and interconnect IP blocks along with their power, performance, and area models. The memory synthesis technique broadly consists of two stages, an initial solution generation step followed by an iterative improvement stage.

Initial solution generation stage: As a first step we generate a minimum buffer usage multi-core SDF schedule. We utilize a well known heuristic approach to generate the schedule [10]. We then consider a maximal memory architecture for the sub-system. In the maximal memory architecture the local SPM of each SW PE is large enough to host the entire code base of all the actors assigned to the PE. Further, there is sufficient memory for inter-PE transfers. Finally, we do not perform any memory optimization for ephemeral data blocks. Thus, the maximal memory architecture represents the maximum SPM memory that is required for the design. Consequently, the design also depicts the best possible performance and minimal power consumption[1]. We

---

[1]Power consumption is minimal because the number of accesses to DRAM is minimal.

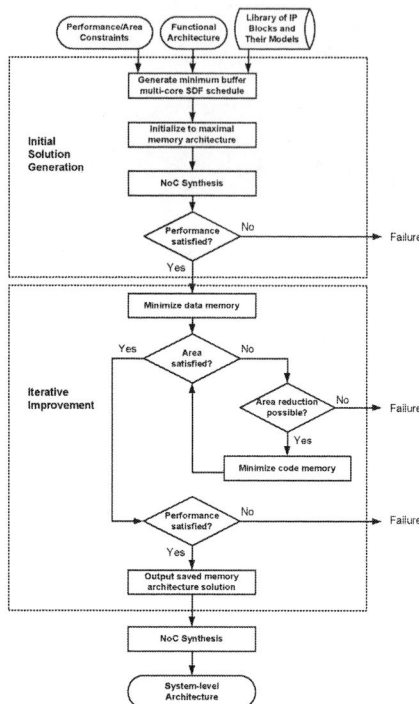

Figure 4: Top-level view of memory synthesis

then synthesize the NoC architecture for the sub-system. As mentioned earlier we utilize an existing approach to synthesize the NoC [2]. The NoC synthesis technique supports guaranteed throughput traffic which is ideal for streaming applications. The synthesis technique includes a system-level floorplanning stage, and is thus able to generate very good estimates for communication latencies and power consumption. The NoC synthesis technique minimizes both the power consumption (primary goal) and resource requirement (secondary goal) of the interconnection architecture subject to the communication bandwidth requirements. Finally, we evaluate the performance of the design and verify if the performance constraint is satisfied. As the initial design represents the best performance design, we declare failure if the performance constraint is not satisfied. Alternatively, if the performance constraint is satisfied we enter the iterative improvement stage in which we aim to satisfy the area constraints, and minimize power consumption.

Iterative improvement stage: The objective of the iterative improvement stage is to satisfy the area constraints and minimize power consumption. As a first step we minimize the memory required for ephemeral data by analyzing their lifetimes and mapping them to the same memory region wherever possible. We introduce shared SPM into the memory architecture if the data blocks that share the memory region are from different PEs. Notice, that data memory reduction does not have an appreciable impact on the performance[2]. However, the power consumption is expected to increase due to an increase in NoC communication. We next check if the area constraint is satisfied. If it is we have successfully

---

[2]Mapping the data blocks to remote SPM may introduce additional communication delays. However, the NoC synthesis technique is able to generate designs with minimum latency, and consequently the performance impact is minimal.

```
1: minimize_data_memory()
2: {
3:   G = generate_interference_graph()
4:   clique_partitioning(G)
5:   for each clique C in G do
6:       combine_data_blocks(C)
7:   end for
8: }
```

Figure 5: Data memory minimization pseudo-code

```
1:  clique_partitioning(G)
2:  {
3:  while vertex exists with degree greater than zero do
4:       V = smallest_non_zero_degree_vertex()
5:       U = smallest_degree_attached(V)
6:       N = combine_vertices(U, V)
7:       for each vertex P attached to V do
8:           for each vertex L attached to U do
9:               if P equals L then
10:                  add edge from N to P
11:              end if
12:          end for
13:      end for
14:      update_degrees()
15: end while
16: return partitioning
17: }
```

Figure 6: Clique partitioning pseudo-code

synthesized the memory architecture. Alternatively, we try to further reduce the memory requirement by introducing code overlays. Introduction of code overlay involves periodic fetching of code from the off-chip DRAM memory, and it increases the power consumption and reduces the performance. The impact of performance reduction can be amortized to some extent by scheduling code pre-fetch DMAs whenever possible. Code overlay is only introduced if the area constraints are not satisfied. We iteratively reduce the code memory usage (increase code overlay overheads) until either the area constraints are satisfied or no further reduction in memory can be achieved. In the case of the later we declare failure as the area constraints are not satisfied. If they are satisfied we again evaluate the performance constraint. If the performance constraint is still satisfied we declare success and output the memory architecture. Alternatively, we declare failure due to non-satisfaction of the performance constraint. In the following two sub-sections we discuss the data and code memory minimization stages in further detail.

4.1 Data memory minimization

The objective of the data memory minimization stage is reduce the memory requirement for ephemeral data blocks by analyzing their lifetimes, and assigning them to the same memory region. We utilize a classical clique partitioning algorithm to achieve our goal. The pseudo-code for data memory minimization stage is shown in Figure 5. We first generate an interference graph (Line 3, Figure 5). The interference graph is specified as $G(V, E)$ where V is the set of data_blocks and E is the set of edges from (u, v) where u and v are vertices in V. An edge (u, v) exists when there is no interference between data blocks u and v. Interference is defined as both data blocks being alive during a portion of the same time frame. A data block is alive from the time when it first begins to be written to, and up to (and including) the last time instance that it is read from. As the data blocks may be present in distinct SPMs, we annotate each edge $(u, v) \in E$ with the physical distance between the two distinct SPMs, d. Notice that we do synthesize a

```
1: minimize_code_memory()
2: {
3:   Initialize each filter to occupy its own region
4:   calculate_IF()
5:   while area constraint not met and | R | greater than 1 per SPM
     do
6:       collapse_smallest_IF()
7:       update_IF()
8:   end while
9:   perform_segmentation()
10: }
```

Figure 7: Code memory minimization pseudo-code

NoC as part of the initial solution, and our NoC synthesis technique generates a floorplan as part of its design flow. Consequently, we can deduce the distance between two distinct SPMs. The distance is used as a tie breaker during the clique partitioning stage.

As a next step (Line 4, Figure 5) we invoke the clique partitioning algorithm (Figure 6). The algorithm begins by finding the vertex with the smallest non-zero degree (Line 4, Figure 6). The degree of a vertex is equal to the number of edges incident on the vertex. The algorithm then finds the smallest degree vertex that is attached to the previously found vertex (Line 5, Figure 6). If there is a tie between vertices the algorithm will choose the vertex with the highest common neighbors as the first vertex. If there is still a tie the algorithm will choose the vertex that has the smallest physical distance d (remember this is annotated on the edge). The algorithm will then combine these two vertices into a single vertex. Next the algorithm updates the edges of the graph. An edge will exist from the new compound vertex to another vertex if and only if the vertex was connected to both the vertices that have been collapsed into the compound node (Line 10, Figure 6). The degrees of the vertices are updated and the algorithm repeats until all vertices have a degree of zero.

The $minimize\_data\_memory()$ algorithm then collapses the data blocks which are part of a clique into a single SPM (Line 4, Figure 5). Notice that at this stage we might introduce new shared SPMs if the data blocks were originally resident on local SPMs of distinct PEs.

4.2 Code memory minimization

We invoke the code memory minimization stage only if the area constraints are not satisfied. The objective of the code memory minimization stage is reduce the code memory requirements for SW PEs by off loading code to DRAM. We would like to emphasize that the code is always resident in the DRAM. In the initial solution generated by our approach the entire code base is fetched in to the on-chip SPM before the start of the first iteration. Consequently, in the initial solution for we do not need to fetch code from DRAM for any subsequent iteration of the SDF. In the code memory minimization stage we assign code bases of two or more filters to the same region of the memory. Thus, during an iteration of SDF execution, we would have to fetch code for one or more filters from the DRAM. Therefore, there is both a performance and power (as accessing DRAM consumes a lot more power) penalty associated with code memory reduction.

The pseudo-code for the code minimization algorithm is given in Figure 7. The algorithm begins by initializing each filter to its own unique region (Line 3). We next calculate the interaction factor (IF) for each region pair (Line 4). The IF is first initialized to zero for all region pairs. Next we step

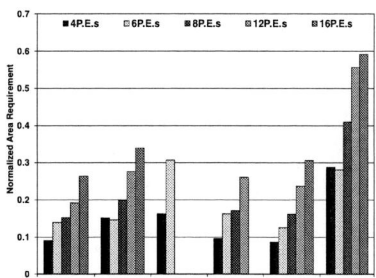 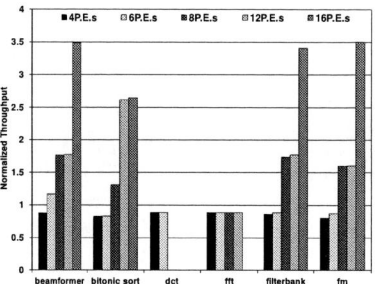 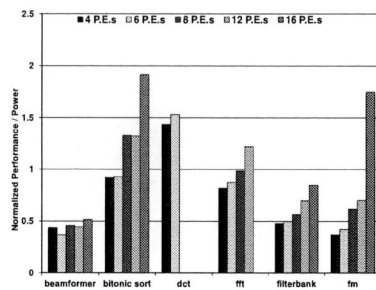

Figure 8: Normalized area Figure 9: Normalized throughput Figure 10: Normalized Perf./Watt

Benchmarks	#Actors	#Edges	#Executions
Beamformer	40	72	64
Bitonicsort	26	31	68
DCT	15	22	28
FFT	17	16	58
Filterbank	51	65	94
Fmradio	29	39	58
Average	30	41	62

Table 1: Benchmark Specifications

through the SDF execution schedule, and for each switch from region r_i to region r_j or vice versa the IF(r_i, r_j) is increased by one. The IF for regions on distinct SPMs is initialized to infinity. Next, the algorithm enters a loop if the area constraint has not been met, and there is at least one SPM with 2 or more regions. Within the loop the algorithm collapses the region pair with the smallest IF. The IF of the regions is then updated and the loop repeats. Upon exiting the loop, the algorithm performs segmentation on the regions where two or more filter belonging to a single region are assigned to the same segment. Segmentation amortizes the DMA cost for fetching the code bases of the filters from the DRAM.

4.3 Time Complexity Analysis

The time complexity of finding the minimum buffer schedule is $O(n)$, where n is the number of actors. The time complexity of the minimizing the data is $O(b^3)$, where b is the number of data blocks. And lastly, the time complexity of minimizing the code is $O(n^4)$, where n is the number of actors. Therefore, the total time complexity for the memory architecture synthesis is $O(n + b^3 + n^4)$. Typically, the number of data blocks is substantially larger than the number of actors and therefore in practice the time is dominated by $O(b^3)$.

5. RESULTS

We evaluated the efficacy of our proposed memory synthesis approach by considering the design of sub-systems that implemented six benchmarks from the StreamIt [11] suite. The benchmarks are described in Table 1. In the table the second and third columns denote the number of actors and edges in each benchmark, and the last column denotes the total number of actor firings in one iteration of the SDF. We generated MPSoC designs for each benchmark by considering 4, 6, 8, 12, and 16 PEs. For each number of PEs we set the throughput constraint to be 0.75 times the throughput of the initial best performance baseline solution. We then iteratively reduced the area constraint until we had the tightest area constraint for each benchmark in which our technique was able to generate a valid design. We compared the solutions generated by our technique with the initial baseline

solutions as well as with the designs generated by the existing 2-stage technique proposed in [6]. Our technique took on average 15 minutes to generate the designs which is reasonable considering we perform NoC synthesis which contains a floorplanning stage.

5.1 Comparison against Baseline Solution

The first set of experiments we compared the designs generated by our technique after the final NoC synthesis stage with the baseline initial solution for each benchmark. Figures 8, 9, and 10 plot the normalized area, throughput, and performance per watt of the various designs. For each benchmark in the plot, the results are normalized to the initial baseline (or maximal area) solutions of the 4 PE design. For example, the area plots for the beamformer benchmark designs are normalized to the area of the initial baseline solution for the beamformer implemented with 4 PEs. For some benchmarks (dct and fft) we do not plot results for all PEs as the benchmarks were too small to be mapped onto the larger number of PEs.

In Figure 8, we see that our technique is able to generate designs with very tight area constraints. With the smallest area constraint at 4 PEs being 10% for the 'beamformer' benchmark and the largest constraint being 30% for 'fm-radio' benchmark. On average, across all benchmarks our technique is able to generate designs that require 75.3% less area than the initial baseline solutions for a 25% loss in performance. We also see that the area requirement compared to the intial 4 PE design increases as we increase the number of cores. This is due to the increase in the required amount of SPM memory for the cores (each core requires a minimal amount). In Figure 9, we notice that for the initial 4 PE design the throughput of the designs generated by our technique is slightly lower than the initial solution. This is to be expected due to the code overlay overhead from DRAM accesses to retrieve code. However, we see as the number of PEs increases we gain a substantial increase in throughput over the initial baseline solution. Figure 10 illustrates that the designs generated by our technique have higher performance per watt than the initial baseline solutions. At 16 PEs the performance per watt of our design is almost 2 times the intitial 4 PE baselin solution for both 'bitonic sort' and 'fm.' And in the other three benchmarks at 16 PEs, our designs had higher performance per watt than the initial baseline 4 PE solutions.

5.2 Impact of Code Overlay Optimization

Figure 11 demonstrates the impact of the code overlay optimization for two benchmarks with the maximum number of actors (namely beamformer and filterbank). The plot

676

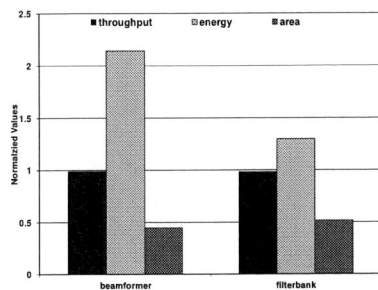

Figure 11: Impact of code overlay

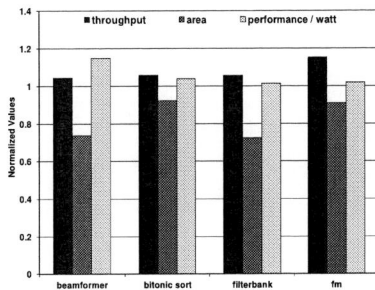

Figure 12: Existing approach

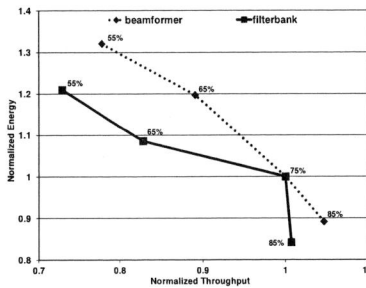

Figure 13: Area impact

depicts normalized throughput, energy and area for 16 PE designs. The plots are normalized to the solutions that only apply data optimizations and do not apply code overlay. As is depicted in the plot, the code overlay optimization is able to considerably reduce the area requirements (by over 50%) for comparable performance. The trade-off is the increase in energy due to code overlay accesses to DRAM. Particularly for the filterbank application the increase in energy is only about 30%. Area minimization is critical as the silicon real estate determines the cost of manufacturing. Code overlay optimization is able to generate design alternatives for tight area constraints that would not be otherwise possible.

5.3 Comparison with Existing Approach

Figure 12 compares the designs generated by our technique against a 2-stage synthesis technique presented in [6]. The technique proposed in [6] only accounts for data memory optimization (at the fine grain). Also, the technique considers a mesh (template) topology for the NoC network. The technique generates a data reuse graph consisting of data buffers in a hierarchical manner with each higher level buffer containing all of the data in the buffers below it in the hierarchy. The technique then greedily selects buffers to add to the design based on the energy savings of using the buffer. We modified the technique to use the larger data blocks present in SDF specifications. We also modified the technique to use the same NoC synthesis tool that we use. This will ensure a fair comparison between the techniques.

Figure 12 plots the normalized throughput, area, and performance per watt for 4 StreamIt benchmarks. The plots are normalized to the respective values for the designs generated by the existing approach [6]. As can be observed in the figure, our technique consistently gives better performaning designs that utilize lower area and have higher performance per watt. On an average our designs show 7.8% increase in performance, 17.7% reduction in area and 5.6% increase in the performance per watt. Our technique is able to give better results because of more comprehensive data minimization methods and incorporation of code overlay optimizations.

5.4 Impact of Area Constraint

In our last experiment, we evaluated our approach by varying the area constraints for the 12 PE designs. In this experiment we only considered 2 benchmarks, and Figure 13 plots the results. In the plot each point (energy and throughput) depicts the design obtained for the respective area constraints (55%, 65%, 75%, 85%). The area constraint is achieved by percentage scaling the area for the initial baseline (maximal area) solution. The plots are normalized to the 75% area constraint design. From the plot, we can see

that as the area constraint is made tighter the throughput of the designs decreases and the energy consumption increases. This is expected due to the increase in code overlay overheads as more code is forced into main memory.

6. CONCLUSION

We presented an approach for synthesizing the system-level memory of a MPSoC sub-system that demonstrates streaming characteristics. The approach accounts and optimizes for the memory requirements for both code and data. We evaluated our approach by extensive experimentation with streaming application benchmarks through comparisons with an existing approach and the initial baseline solution. Our technique performed superiorly to the existing approach and clearly demonstrated the ability to generate high quality designs meeting the area and performance constraints while maintaining a low energy consumption. Future work will address automation of the functional architecture design stage.

7. ACKNOWLEDGMENT

The research presented in this paper was supported in part by grants from the National Science Foundation (Career CCF-0546462).

8. REFERENCES

[1] E. A. Lee and D. G. Messerschmitt, "Synchronous Data Flow", *Proceedings of the IEEE*, Vol. 75, No. 9, September, 1987.

[2] G. Leary and K. S. Chatha, "A Holistic Approach to Network-on-Chip Synthesis", *Proceedings of CODES+ISSS*, 2010.

[3] S. Meftali, F. Gharsalli, F. Rousseau and A. A Jerraya, "An Optimal Memory Allocation for Application-Specific Multiprocessor System-on-Chip", *Proceedings of International Symposium in System Synthesis (ISSS)*, 2001.

[4] S. Pasricha and N. Dutt, "COSMECA: Application Specific Co-Synthesis of Memory and Communication Architectures for MPSoC", *Proceedings of DATE*, 2006.

[5] I. Issenin, E. Brockmeyer, B. Durinck and N. Dutt, "Data-Reuse-Driven Energy-Aware Cosynthesis of Scratch Pad Memory and Hierarchical Bus-based Communication Architecture for Multiprocessor Streaming Applications", *IEEE Transactions on Computer-Aided Design of Integrated Circuits and Systems*, Vol. 27, No. 8, August 2008.

[6] I. Issenin and N. Dutt, "Data Reuse Driven Memory and Network-on-Chip Co-Synthesis", *IFIP Embedded System Design: Topics, Techniques and Trends*, Eds. A. Rettberg, M. Zanella, R. Domer, A. Gerstlauer and F. Rammig, 2007.

[7] M. Manchiero, G. Palermo, C. Silvano and O. Villa, "Exploration of Distributed Shared Memory Architectures for NoC-based Multiprocessors", *Journal of Systems Architecture: the EUROMICRO Journal* Volume 53 Issue 10, October, 2007.

[8] S. Pandey and R. Drechsler, "Slack Allocation Based Co-synthesis and Optimization of Bus and Memory Architectures for MPSoCs", *Proceedings of DATE*, 2008.

[9] C. Lee, S. Kim and S. Ha, "A Systematic Design Space Exploration of MPSoC Based on Synchronous Data Flow Specification", *Journal of Signal Processing Systems*, Vol 58, 2010.

[10] A. Jantsch, *Modeling Embedded Systems and SoCs: Concurrency and Time in Models of Computation*, Morgan Kaufmann Publishers, 2004.

[11] W. Thies, M. Karczmarek and S. Amarasinghe, "Streamit: A language for streaming applications", *Proceedings of International Conference on Compiler Construction*, 2002.

AUTHOR INDEX

Aadithya, Karthik . V. 311
Abelein, Ulrich 205
Agarwal, Amit 1149
Agarwal, Anant 265
Agosta, Giovanni 77
Ahmadyan, Seyed Nematollah 1018
Akesson, Benny 988
Al Maashri, Ahmed....................579
Aliakbarian, Hadi 542
Alpert, Charles J. 465, 768, 774
Anders, Mark 1149
Anderson, Charles 1274
Andrade, Hugo 656
Ascheid, Gerd 121, 1262
Ashouei, Maryam 962
Atasu, Kubilay 350
Athikulwongse, Krit 741
Atienza, David 636
Auerbach, Joshua 271
Augustine, Charles 1258
Bachrach, Jonathan 1212
Bacon, David F. 271
Baj-Rossi, Camilla 6
Bamis, Athanasios 163
Bampi, Sergio 866
Bao, Min 197
Barenghi, Alessandro 77
Bartolini, Davide B.....................856
Bathen, Luis Angel D. 214, 447
Beckmann, Nathan 265
Beigne, E. 1049
Belhadj, Bilel 1260
Benini, Luca 1137
Benkeser, Christian....................510
Beretta, Ivan 1043
Bernstein, Gary 476
Bertacco, Valeria....................115, 729, 955
Bhardwaj, Kshitij 382
Bhuyan, Laxmi 1006
Blaauw, David 406, 980, 1037
Bobba, Shashikanth 42
Bobda, Christophe 1203
Boero, Cristina 6
Borkar, Shekhar....................1149
Brisk, Philip 26
Bromberg, David....................486
Brooks, David 277
Browy, Chris 936
Brunhaver, John 623
Bryson, William 133
Buckl, Christian 188, 430
Burcea, Ioana 271
Burg, Andreas 510
Burleson, Wayne 12
Burns, Steven....................603

Butler, Kenneth M....................808
Campanoni, Simone 277, 856
Cancare, Fabio....................856
Cao, Yu 139, 283
Carrara, Sandro 6
Castrillon, Jeronimo....................1262
Chaji, G. Reza 182
Chakrabarti, Chaitali....................579
Chakrabarty, Krishnendu 18, 1024
Chakraborty, Koushik 382, 1074
Chakraborty, Samarjit 688
Chan, Carven 1222
Chan, Ya-Chung 1163
Chandra, Vikas....................283
Chandramoorthy, Nandhini 579
Chandrasekar, Karthik....................988
Chang, Chen-Feng 1088
Chang, Hua-Yu 802
Chang, Kai-Hui 936
Chang, Leland 1155
Chang, Naehyuck 516, 522
Chang, Yao-Wen 549, 762, 802, 1082,
1088, 1175, 1181
Chang, Yuan-Hao....................882
Chang, Yu-Chi 376
Chao, Mango C.-T.....................1012
Charkrabari, Chaitali 980
Chatha, Karam S.....................672, 1268
Chatterjee, Debapriya 115, 955
Che, Weijia 672, 1268
Chebira, Amina 636
Chen, Chia-Hsin 398
Chen, Chi-Hao 453
Chen, Fu-Wei 1094
Chen, Gang 430
Chen, Haibo 106
Chen, I-Che 613
Chen, Wei-Yu 1181
Chen, Wu-Hsin 1254
Chen, Xiang 1000
Chen, Xiaoming 1125
Chen, Xinke 876
Chen, Yi-Hung 1113
Chen, Yiran....................585, 1000, 1187
Cheng, Perry 271
Cheng, Yi-Kan 1113
Cher, Chen-Yong....................642
Chiang, Charles 1163, 1169
Chiang, Patrick....................974
Chien, Hsing-Chih Chang....................549
Chiou, Derek 790
Choi, Junchul....................664
Choi, Yunju 536
Chou, Chun-Nan 327
Chou, Sheng....................762

AUTHOR INDEX

Christmann, J. F. 1049
Chu, Chris 597
Chung, Jaewoong..............................888
Chung, Sung Woo 1193
Chung, Yi-Ting.................................1055
Clark, Shane S. 12
Cline, Brian 283
Condemine, C. 1049
Condley, Walter 145
Cong, Jason 843, 1229, 1235
Coskun, Ayse K. 648
Cotter, Matthew 579
Crop, Joseph 974
Csaba, Gyorgy 476
Danowitz, Andrew.............................623
Das, Reetuparna 406
Davoodi, Azadeh 709
Daya, Bhavya....................................398
De Gyvez, Jose Pineda 962
De Marchi, Michele............................42
De Micheli, Giovanni 6, 42
Debole, Michael 579
Derrien, Steven 48
Dey, Sujit 826
Ding, Duo 756
Ding, Huping 412
Ding, Wei 834
Dingler, Aaron...................................476
Dong, Xiangyu 253
Donkoh, Eric 62
Donohoo, Brad 1274
Dreslinski, Ronald G...........406, 980, 1143
Du, Yang 1160
Duranton, Marc 1260
Dutt, Nikil D. 214, 447
Eberl, Michael 205
Edith, Beigne 994
Eles, Petru 197
El-Shambakey, Mohammed 437
Erez, Mattan 850
Eusse, Juan 121
Fallah, Farzan 561
Fang, Shao-Yun........................1175, 1181
Fang, Zhenman 106
Fazzari, Saverio................................133
Fedder, Gary K. 176
Feng, Zhuo 1119
Fey, Gorschwin..................................941
Fick, David......................................1143
Finder, Alexander 941
Fink, Stephen J. 271
Flamand, Eric 1137
Fong, Neric 289
Foreman, Eric 1061
Foroutan, Sahar 366

Forte, Domenic 96
Fu, Kevin 12
Garg, Siddharth................................697
Gester, Michael 459
Ghasemi, Hamid Reza 56
Ghodrat, Mohammad Ali 843
Ghosal, Arkadeb 656
Gill, Michael....................................843
Glaß, Michael 205
Goossens, Kees 988
Goswami1, Dip 688
Grassi, Paolo Roberto 1043
Grigorian, Beayna 843
Grissom, Daniel 26
Gruenwald, Charles 265
Guo, Jing 1169
Guo, Xiaofei 573
Gupta, Priyank 529
Gupta, Sumeet 492
Guthaus, Matthew R. 145
Ha, Soonhoi 664
Haensch, Wilfried 1155
Hagleitner, Christoph 350
Hakim, Nagib 561
Hamzeh, Mahdi 1280
Hao, Kecheng 344
Harris, Ian G....................................1252
Haubelt, Christian 1203
Haugou, Germain 1137
Hayes, Jerry 1107
Heliot, Rodolphe 1260
Hemmett, Jeffrey 1061
Henkel, Jorg............................ 866, 1288
Ho, Yen-Sheng 327
Ho, Yuan-Kai 1088
Hodges, Ben R. 723
Hoffmann, Henry 259, 856
Holloway, Glenn 277
Holt, Jim 259
Hong, Ted.......................................561
Horowitz, Mark 623, 783
Hsieh, Chiao 327
Hsiu, Pi-Cheng 453
Hsu, Meng-Kai.................................762
Hsu, Steven 1149
Hu, Miao 498, 585
Hu, Wei-Yi 1113
Hu, Xiaobo Sharon 476
Hu, Xuchu 145
Hu, Yibin 106
Huang, Chung-Yang 327
Huang, He 169
Huang, Jia 188
Huang, Jiawei 504
Huang, Kai 188, 430

AUTHOR INDEX

Huang, Po-Chun.....................................882
Huang, Rei-Fu.....................................1012
Huang, Shi-Yu.....................................1031
Huang, Shuai.....................................876
Huang, Yoshi Shih-Chieh.....................................376
Huang, Yu-Hung.....................................127
Huisken, Jos.....................................962
Hwang, Tingting.....................................1094
Ienne, Paolo.....................................229
Irwin, Mary Jane.....................................678
Jaffari, Javid.....................................182
Jane, Mary.....................................834
Jang, Ohyoung.....................................834
Jego, Bruno.....................................1137
Jeong, Min Kyu.....................................850
Jha, Somesh.....................................1250
Jiang, Iris Hui-Ru.....................................802, 1163
Jiang, Jie-Hong Roland.....................................1055
Jiang, Lei.....................................907
Jimenez, Xavier.....................................229
Jog, Adwait.....................................243
Jones, Timothy.....................................277
Joshi, Rajiv.....................................1107
Joubert, Antoine.....................................1260
Jovic, Jovana.....................................121
Jung, Byunghoo.....................................1254
Jung, Deokwoo.....................................163
Jung, Moongon.....................................317
Jung, Seobin.....................................536
Kahng, Andrew B..... ....392, 820
Kandemir, Mahmut.....................................678, 834
Kang, Seokhyeong.....................................820
Kanj, Rouwaida.....................................1107
Karakonstantis, Georgios.....................................510
Karri, Ramesh.....................................83, 573
Kasture, Harshad.....................................265
Kaul, Himanshu.....................................1149
Kawakami, Katsutoshi.....................................648
Kelley, Kyle.....................................783
Keng, Brian.....................................947
Keskin, Gokce.....................................176
Khaled, Nadia.....................................1043
Khatri, Sunil P..... ....734, 1256
Kim, Dae Hyun.....................................888
Kim, Dongki.....................................888
Kim, Heesoo.....................................808
Kim, Jaeha.....................................536
Kim, Lee-Sup.....................................630
Kim, Myung-Chul.....................................747
Kim, Nam Sung.....................................56
Kim, Sungchan.....................................664
Kim, Yejoong.....................................1037
Kim, Younghyun.....................................522
Kim, Youngsik.....................................897
Kirsch, Christoph M..... 913

Klefstad, Raymond.....................................1006
Knoll, Alois.....................................188, 430
Kong, Joonho.....................................1193
Koushanfar, Farinaz.....................................68, 90, 133, 220
Koyfman, Anatoly.....................................955
Kozhikkottu, Vivek.....................................796, 826
Krishna, Tushar.....................................398
Krishnaswamy, Smita.....................................814
Kuan, Jui-Feng.....................................1113
Kuang, Jilong.....................................1006
Kuehlmann, Andreas.....................................814
Kumar, Jayanand Asok.....................................808, 1018
Kumar, Pratyush.....................................688
Kumar, Rakesh.....................................918, 1297
Kunz, Wolfgang.....................................334
Kuo, Chin-Cheng.....................................1113
Kuo, Tei-Wei.....................................453, 882
Kurian, George.....................................259
Kursun, Eren.....................................642
Kurtz, Steve.....................................476
Lach, John.....................................504
Lau, Eric.....................................259
Lauwereins, Rudy.....................................1
Leary, Glenn.....................................672
Leblebici, Yusuf.....................................42
Lee, D. T..... 613
Lee, Hsu-Chieh.....................................1082, 1088
Lee, Po-Wei.....................................1088
Lee, Sunggu.....................................897
Lee, Sungkwang.....................................888
Lee, Yoonmyung.....................................1037
Lee, Yunsup.....................................1212
Lepley, Thierry.....................................1137
Leupers, Rainer.....................................121, 1262
Li, Hai.....................................498, 585
Li, Jian.....................................106
Li, Kai.....................................90
Li, Peng.....................................476, 876
Li, Wenchao.....................................1250
Li, Xin.....................................176, 642
Li, Zhuo.....................................465, 768, 774, 1107
Liang, Yun.....................................412
Lim, Sung Kyu.....................................157, 317, 741
Limaye, Rhishikesh.....................................656
Lin, Bill.....................................392
Lin, David.....................................561
Lin, I-Jye.....................................1088
Lin, Pey-Chang Kent.....................................734, 1256
Lin, Shih-Chin.....................................1012
Lin, Xue.....................................516
Lin, Yu-Hsiang.....................................1031
Lingamneni, Avinash.....................................924
Lionel, Vincent.....................................994
Liu, Beiye.....................................585
Liu, Bin.....................................1235

AUTHOR INDEX

Liu, Bo	542, 962
Liu, Chih-Hung	613
Liu, Frank	723
Liu, Haotian	289
Liu, Sizhe	567
Liu, Xiao	555
Lo, Daniel	421
Lowery, Alicia	62
Lu, Yi	106
Lu, Yi-Shan	127
Luo, Yan	18
Ma, Qiang	591
Maggio, Martina	259
Mahlke, Scott	980
Malburg, Jan	941
Malik, Sharad	1222
Mandal, Ayan	1256
Marek-Sadowska, Malgorzata	1100
Markov, Igor L.	747
McCants, Carl	133
Mei, Arie	301
Melpignano, Diego	1137
Meng, Jie	648
Middendorf, Lars	1203
Milder, Peter	1241
Miller, D. Michael	36
Min, Qinghao	106
Mirhoseini, Azalia	68
Mishra, Asit K.	243
Mitra, Subhasish	561
Mitra, Tulika	412
Mojumder, Niladri	492
Moon, Seok-Hwan	630
Morad, Ronny	955
Morris, Daniel	486
Mudge, Trevor	406, 1143
Mukherjee, Tamal	176
Muller, Dirk	459
Murillo, Luis Gabriel	121
Nahas, Joseph	476
Najm, Farid N.	151
Nam, Gi-Joon	465
Narayanan, Vijaykrishnan	579
Nassif, Sani	1107
Nath, Siddhartha	392
Neuman, Sabrina M.	259
Nieberg, Tim	459
Niemier, Michael	476
Novo, David	229
Oh, Hyunok	664
Ohlsen, Chris	1274
Ou, Hung-Chih	549
Paek, Seungwook	630
Palem, Krishna	924
Pan, David Z.	317, 756

Panagopoulos, Georgios	1258
Panten, Christian	459
Park, Sang Phill	492
Park, Sangyoung	398, 522
Park, Yongjun	980
Pasricha, Sudeep	1274
Patel, Hiren D.	697
Pathak, Mohit	741
Paver, Nigel	850
Pawlowski, Robert	974
Payer, Hannes	913
Pedram, Massoud	516
Pelosi, Gerardo	77
Peng, Zebo	197
Petrot, Frederic	366
Phelps, Andrew	176
Philippe, Maurine	994
Piguet, C.	1049
Pileggi, Larry	486
Pileggi, Lawrence T.	176
Pinckney, Nathaniel	1143
Pino, Robinson E.	585
Pino, Yougok	83
Poku, Osei	567
Porod, Wolfgang	476
Potkonjak, Miodrag	68, 90, 133, 133
Prasad, Ankita	656
Purandare, Mitra	350
Püschel, Markus	1241
Qian, Haifeng	642
Qiu, Xiang	1100
Raabe, Andreas	188
Rabbah, Rodric	271
Radiom, Soheil	542
Raghunathan, Anand	492, 796, 826
Rajendiran, Aravindkumar	697
Rajendran, Jeyavijayan	83
Ramanathan, Parmeswaran	709
Rana, Vincenzo	1043
Ranieri, Juri	636
Ransford, Benjamin	12
Ravindran, Binoy	437
Ravindran, Kaushik	656
Ray, Sandip	344
Reddy, Lakshmi	768
Rehman, Semeen	1288
Reinman, Glenn	843
Ren, Ling	1125
Richards, Brian	1212
Rinard, Martin	930
Rincon, Francisco	1043
Roa, Elkim	1254
Robins, Gabriel	504
Rose, Garrett S.	498
Roth, Christoph	510

AUTHOR INDEX

Rotner, Jonathan 176
Roveda, Janet 529
Roy, Kaushik 492, 796, 1258
Roy, Sanghamitra 382, 1074
Roychowdhury, Jaijeet 301, 311
Ryzhenko, Nikolai 603
Sabne, Amit 796
Sadeghi, Ahmad-Reza 220
Sankar, Vjiay Karthik 476
Sartori, John 918, 1297
Sasanian, Zahra 36
Sato, Takashi 139
Satpathy, Sudhir 406
Savvides, Andreas 163
Scheffer, Louis K. 717
Scheuermann, Michael 157
Schulte, Christian 459
Schulte, Michael J. 56
Schwartz-Narbonne, Daniel 1222
Sciuto, Donatella 856
Sentieys, Olivier 48
Seo, Sangwon 980
Seok, Mingoo 968
Seshia, Sanjit A. 356, 1250
Sethi, Divjyot 1222
Seudie, Herve 220
Severson, Matt 1160
Sewell, Korey 1143
Shacham, Ofer 623
Shafique, Muhammad 866, 1288
Shao, Zili 214
Sharad, Mrigank 1258
Sharifi, Akbar 678
Sheibanyrad, Abbas 366
Shen, Chin-Fang 1088
Shin, Donghwa 516
Shin, Wongyu 630
Shojaei, Hamid 709
Shrivastava, Aviral 1280
Shriver, Emily 62
Shukla, Sunil 271
Sim, Jaehyeong 630
Sinangil, Mahmut 259
Sinangil, Yildiz 259
Sinanoglu, Ozgur 83
Sinha, Debjit 1061, 1067
Sinha, Saurabh 283
Sinha, Subarna 1163, 1169
Sinkar, Abhishek A. 56
Sironi, Filippo 856
Sloan, Joseph 918
Srikantaiah, Shekhar 678
Srivastava, Ankur 96
Stevenson, John 783
Stoffel, Dominik 334

Su, Yangfeng 295
Sudanthi, Chander 850
Suh, G. Edward 421
Sun, Jin 529
Sutaria, Ketul 139
Suto, Gyuszi 471
Suzanne, Lesecq 994
Sylvester, Dennis 406, 1037, 1143
Sze, Cliff 465, 768, 774
Taurino, Irene 6
Taylor, Michael B. 1131
Tehranipoor, Mohammad 703
Teich, Jurgen 205
Temam, Olivier 1260
Tovinakere, Vivek D. 48
Tran, Trung N 656
Tretter, Andreas 1262
Tripakis, Stavros 656
Tripunitara, Mahesh V. 697
Tsai, Tsung-Chan 376
Tsay, Ren-Song 127
Tuzzio, Nicholas 703
Ukhov, Ivan 197
Urdahl, Joakim 334
Vassiliev, Artem 623
Vasudevan, Shobha 808, 1018
Velamala, Jyothi Bhaskarr 139
Veneris, Andreas 947
Venkataramani, Swagath 796
Venkateswaran, Natesan 1067
Vetterli, Martin 636
Vincenzi, Alessandro 636
Vinco, Sara 115
Viswanathan, Natarajan 465, 768, 774
Visweswariah, Chandu 1061, 1067
Vo, Huy 1212
Vrudhula, Sarma 1280
Vygen, Jens 459
Wachs, Megan 623, 783
Walter, Fabio Leandro 866
Wang, Cheng-Yuan Michael 453
Wang, Chundong 235
Wang, Fa 176
Wang, Guoqiang 656
Wang, Hongfei 567
Wang, Huandong 876
Wang, Jue 253
Wang, Qing 289
Wang, Yanzhi 516, 522
Wang, Yi 214
Wang, Yu 1125, 1187
Ward, Samuel 756
Wedler, Markus 334
Wei, Gu-Yeon 277
Wei, Sheng 90

AUTHOR INDEX

Wei, Yaoguang 768, 774
Welp, Tobias...................................814
Wen, Wujie....................................1187
Wentzlaff, David 265
Wille, Robert 36
Willemin, J...................................1049
Woh, Mark 980
Wong, Martin D. F..........................591
Wong, Ngai 289
Wong, Weng-Fai 235
Woo, Dong Hyuk.............................888
Wu, Hsin-I 127
Wu, Qing......................................498
Xiao, Yang 579
Xie, Fei.......................................344
Xie, Qing 522
Xie, Yuan 243, 253, 1187
Xiong, Jinjun 1067
Xu, Cong 243
Xu, Qiang.....................................555
Xue, Chun Jason 1000
Yakoushkin, Sergey 121
Yang, Chia-Lin...............................453
Yang, Fan.................................295, 1169
Yang, Hao-Yu 1012
Yang, Huazhong.............................1125
Yang, Jun.....................................907
Yang, Qing 169
Yao, Shi-Chune 642
Ye, Fangming 1024
Yeric, Greg 283
Yoo, Sungjoo 888, 897
Yoon, Dongmin...............................1037
York, Johnathan 790
Youseff, Lamia 265
Yu, Xiaochun.................................567
Yu, Yen-Ting 1163
Yuan, Feng...................................555
Yue, Siyu 516
Yuen, Kendrick 1160
Zang, Binyu 106
Zatt, Bruno 866
Zeng, Xuan 295, 1169
Zhang, Chenxi 1125
Zhang, Guangfei.............................876
Zhang, Hongbo...............................591
Zhang, Peng..................................1229
Zhang, Weihua 106
Zhang, Xiaorong 169
Zhang, Xuehui 703
Zhang, Yang 289
Zhang, Yanheng 597
Zhang, Yaojun 1187
Zhang, Youtao................................907
Zhang, Yuanrui 834

Zhao, Bo 907
Zhao, Hui 834
Zhao, Mengying...............................1000
Zhao, Xin.....................................157
Zhao, Xueqian 1119
Zheng, Jian 1000
Zhou, Huapeng 642
Zhou, Keyong.................................106
Zhou, Nancy Y................................465
Zhu, Jian-Gang...............................486
Ziv, Avi.......................................955
Zolotov, Vladimir 1061, 1067
Zou, Yi 1229
Zuluaga, Marcela 1241

CURRAN ASSOCIATES INC.
proceedings
.com

9781450311991

2012 49th ACM/EDAC/IEEE Design Automation Conference (DAC 2012)

San Francisco, California, USA
3-7 June 2012

IEEE Catalog Number: CFP12DAC-POD
ISBN: 978-1-45031-199-1

2012 49th ACM/EDAC/IEEE Design Automation Conference

(DAC 2012)

San Francisco, California, USA
3-7 June 2012

Pages 678-1304

IEEE Catalog Number: CFP12DAC-PRT
ISBN: 978-1-4503-1199-1

Copyright © 2012, Association for Computing Machinery, Inc
All Rights Reserved

***This publication is a representation of what appears in the IEEE
Digital Libraries. Some format issues inherent in the e-media version may
also appear in this print version.*

IEEE Catalog Number: CFP12DAC-PRT
ISBN 13:   978-1-4503-1199-1
ISSN: 0738-100X

Additional Copies of This Publication Are Available From:

Curran Associates, Inc
57 Morehouse Lane
Red Hook, NY 12571 USA
Phone: (845) 758-0400
Fax: (845) 758-2633
E-mail: curran@proceedings.com
Web: www.proceedings.com

2012 Table of Contents

General Chair's Message	i
Proceedings of the 49th Automation Conference®	iii
Committees	iv
Executive Committee	iv
Technical Program Committee	vi
Panel Committees	ix
Industry Liaison Committee	x
Strategy Committee	xi
User Track Committee	xii
PR/Marketing Committee	xiii
Best Paper Award Committee	xiv
Special Session Organizers	xiv
Technical Panel Organizers	xiv
Pavilion Panel Contributors	xv
Tuesday Keynote Address	xvi
Wednesday Keynote Address	xvi
Thursday Keynote	xvii
Perspective Paper Abstracts	xviii
Technical Panel Abstracts	xix
Awards	xxii
Marie R. Pistilli Women in EDA Achievement Award	xxii
The P. O. Pistilli Undergraduate Scholarships for Advancement in Computer Science and Electrical Engineering	xxii
A. Richard Newton Graduate Scholarships	xxii
ACM/IEEE A. Richard Newton Technical Impact Award in Electronic Design Automation	xxii
2011 Phil Kaufman Award for Distinguished Contributions to EDA	xxii
IEEE CEDA Outstanding Service Contribution	xxii
Donald O. Pederson Best Paper Award for the IEEE Transaction on CAD	xxiii
SIGDA Outstanding New Faculty Award	xxiii
ACM/SIGDA Outstanding Ph.D. Dissertation Award	xxiii
IEEE Fellow	xxiii
IEEE Fellow	xxiii
IEEE Fellow	xxiii
49th DAC Best Paper Candidates	xxiii
Reviewers	xxv
Author Index	

2012 Papers

Session 2: E-Health: A Killer Application for Electronic Devices?

Chair: Rajesh Gupta (*Univ. of California at San Diego*)

Biomedical Electronics Serving as Physical Environmental and Emotional Watchdogs

| 2.1 | Rudy Lauwereins (*IMEC*) | 1 |

2.2 **Integrated Biosensors for Personalized Medicine** 6

Giovanni De Micheli, Cristina Boero, Camilla Baj-Rossi, Irene Taurino, Sandro Carrara (*Ecole Polytechnique Fédérale de Lausanne*)

2.3 **Design Challenges for Secure Implantable Medical Devices** 12

Shane Clark, Ben Ransford (*Univ. of Massachusetts, Amherst*); Wayne Burleson, Kevin Fu (*Univ. of Massachusetts, Amherst*)

Session 3: Design Automation for Things Wet, Small, Spooky, and Tamable

Chair: Tsung-Yi Ho (*National Cheng Kung Univ.*)

3.1 **Design of Pin-Constrained General-Purpose Digital Microfluidic Biochips** 18

Yan Luo, Krishnendu Chakrabarty (*Duke Univ.*)

3.2 **Path Scheduling on Digital Microfluidic Biochips** 26

Daniel Grissom, Philip Brisk (*Univ. of California, Riverside*)

3.3 **Realizing Reversible Circuits Using a New Class of Quantum Gates** 36

Robert Wille (*Univ. of Bremen*); D. Michael Miller, Zahra Sasanian (*Univ. of Victoria*)

3.4 **Physical Synthesis onto a Sea-of-Tiles with Double-Gate Silicon Nanowire Transistors** 42

Shashikanth Bobba, Michele De Marchi, Yusuf Leblebici, Giovanni De Micheli (*Ecole Polytechnique Fédérale de Lausanne*)

Session 4: Be Efficient: Low-Power Design Techniques

Chair: Hamid Mahmoodi (*San Francisco State Univ.*)

4.1 **A Semiempirical Model for Wakeup Time Estimation in Power-Gated Logic Clusters** 48

Vivek D. Tovinakere, Olivier Sentieys, Steven Derrien (*Univ. de Rennes 1*)

4.2 **Cost-Effective Power Delivery to Support Per-Core Voltage Domains for Power-Constrained Processors** 56

Michael J. Schulte (*Advanced Micro Devices, Inc.*); Abhishek A. Sinkar, Hamid Reza Ghasemi, Nam Sung Kim (*Univ. of Wisconsin, Madison*)

4.3 **A Hybrid and Adaptive Model for Predicting Register File and SRAM Power Using a Reference Design** 62

Eric Donkoh, Alicia Lowery, Emily Shriver (*Intel Corp.*)

4.4 Coding-Based Energy Minimization for Phase Change Memory   68

Azalia Mirhoseini (*Rice Univ.*); Miodrag Potkonjak (*Univ. of California, Los Angeles*); Farinaz Koushanfar (*Rice Univ.*)

Session 5: Design and Data Security: Is It Even Possible?
Chair: Mohammad Tehranipoor (*Univ. of Connecticut*)

5.1 A Code Morphing Methodology to Automate Power Analysis Countermeasures   77

Giovanni Agosta, Alessandro Barenghi, Gerardo Pelosi (*Politecnico di Milano*)

5.2 Security Analysis of Logic Obfuscation   83

Youngok Pino (*Air Force Research Lab*); Ozgur Sinanoglu (*New York Univ.*); Jeyavijayan Rajendran, Ramesh Karri (*Polytechnic Institute of New York Univ.*)

5.3 Hardware Trojan Horse Benchmark via Optimal Creation and Placement of Malicious Circuitry   90

Sheng Wei (*Univ. of California, Los Angeles*); Kai Li, Farinaz Koushanfar (*Rice Univ.*); Miodrag Potkonjak (*Univ. of California, Los Angeles*)

5.4 On Improving the Uniqueness of Silicon-Based Physically Unclonable Functions via Optical Proximity Correction   96

Domenic Forte, Ankur Srivastava (*Univ. of Maryland*)

Session 6: System Simulation: The Need for Speed!
Chair: Gunar Schirner (*Northeastern Univ.*)

6.1 Transformer: A Functional-Driven Cycle-Accurate Multicore Simulator   106

Qinghao Min, Weihua Zhang, Binyu Zang (*Fudan Univ.*); Jian Li (*IBM Corp.*); Haibo Chen (*Shanghai Jiao Tong Univ.*); Zhenman Fang, Keyong Zhou, Yi Lu, Yibin Hu (*Fudan Univ.*)

6.2 SAGA: SystemC Acceleration on GPU Architectures   115

Sara Vinco (*Univ. of Verona*); Debapriya Chatterjee, Valeria Bertacco (*Univ. of Michigan*); Franco Fummi (*Univ. of Verona*)

6.3 Synchronization for Hybrid MPSoC Full-System Simulation   121

Juan Eusse, Gerd Ascheid, Rainer Leupers, Jovana Jovic, Luis Gabriel Murillo, Sergey Yakoushkin (*RWTH Aachen Univ.*)

6.4 **A Non-Intrusive Timing Synchronization Interface for Hardware-Assisted**   127
HW/SW Co-Simulation

> Yu-Hung Huang, Yi-Shan Lu, Hsin-I Wu, Ren-Song Tsay (*National Tsing Hua Univ.*)

Session 8: Can EDA Combat the Rise of Electronic Counterfeiting?

Chair: Miodrag Potkonjak (*Univ. of California, Los Angeles*)

8.1 **Can EDA Combat the Rise of Electronic Counterfeiting?**   133

> Carl McCants (*Defense Advanced Research Projects Agency*); William Bryson (*Analytical Solutions, Inc.*); Matthew Sale (*U.S. Naval Surface Warfare Center*); Saverio Fazzari (*Booz Allen Hamilton, Inc.*); Farinaz Koushanfar (*Rice Univ.*); Miodrag Potkonjak (*Univ. of California, Los Angeles*); Peilin Song (*IBM Research*)

Session 9: Reliability: From Atoms to 3-D

Chair: Angan Das (*Intel Corp.*)

9.1 **Physics Matters: Statistical Aging Prediction under Trapping/Detrapping**   139

> Jyothi Bhaskarr Velamala, Ketul Sutaria (*Arizona State Univ.*); Takashi Sato (*Kyoto Univ.*); Yu Cao (*Arizona State Univ.*)

9.2 **Library-Aware Resonant Clock Synthesis (LARCS)**   145

> Xuchu Hu (*Cadence Design Systems, Inc., Univ. of California, Santa Cruz*); Walter Condley, Matthew Guthaus (*Univ. of California, Santa Cruz*)

9.3 **Incremental Power Grid Verification**   151

> Farid N. Najm, Abhishek (*Univ. of Toronto*)

9.4 **Analysis of DC Current Crowding in Through-Silicon-Vias and its Impact on**   157
Power Integrity in 3-D ICs

> Sung Kyu Lim, Xin Zhao (*Georgia Institute of Technology*); Michael Scheuermann (*IBM T.J. Watson Research Ctr.*)

Session 10: EDA for Emerging Applications at the Kilometer, Meter, Micron, and Nanometer Scales

Chair: Sai-Wang (Rocco) Tam (*Marvell Semiconductor, Inc.*)

10.1 **Tracking Appliance Usage Information in Residential Settings Using**   163
Off-the-Shelf Low-Frequency Meters

> Deokwoo Jung (*Advanced Digital Sciences Center*); Andreas Savvides (*Yale Univ.*); Athanasios Bamis (*Univ. of Connecticut*)

10.2 **Implementing an FPGA System for Real-Time Intent Recognition for**   169
Prosthetic Legs

Xiaorong Zhang, He Huang, Qing Yang (*Univ. of Rhode Island*)

10.3   Statistical Design and Optimization for Adaptive Post-silicon Tuning of   176
MEMS Filters

Fa Wang, Gary Fedder, Larry Pileggi, Tamal Mukherjee, Jonathan Rotner, Xin
Li, Gokce Keskin, Andrew Phelps (*Carnegie Mellon Univ.*)

10.4   Generic Low-Cost Characterization of VTH and Mobility Variations in LTPS   182
TFTs for Non-Uniformity Calibration of Active-Matrix OLED Displays

Reza Chaji, Javid Jaffari (*IGNIS Innovations, Inc.*)

Session 11: Facing Dependability: System-Level Solutions and Cybercar Challenges
Chair: Hans-Joachim Wunderlich (*Univ. of Stuttgart*)

11.1   Towards Fault-Tolerant Embedded Systems with Imperfect Fault Detection   188

Jia Huang, Kai Huang, Andreas Raabe, Christian Buckl (*fortiss GmbH*); Alois
Knoll (*Technische Univ. München*)

11.2   Steady-State Dynamic Temperature Analysis and Reliability Optimization   197
for Embedded Multiprocessor Systems

Ivan Ukhov, Zebo Peng, Min Bao, Petru Eles (*Linköping Univ.*)

11.3   Considering Diagnosis Functionality during Automatic System-Level   205
Design of Automotive Networks

Michael Eberl, Michael Glass, Jürgen Teich (*Univ. of Erlangen-Nuremberg*);
Ulrich Abelein (*Audi AG*)

11.4   Meta-Cure: A Reliability Enhancement Strategy for Metadata in NAND   214
Flash Memory Storage Systems

Zili Shao, Yi Wang (*The Hong Kong Polytechnic Univ.*); Luis Angel Bathen,
Nikil Dutt (*Univ. of California, Irvine*)

11.5   EDA for Secure and Dependable Cybercars: Challenges and Opportunities   220

Hervé Seudié, Ahmad-Reza Sadeghi (*Fraunhofer SIT, and Intel-TU Darmstadt
Security Institute, Germany*); Farinaz Koushanfar (*Rice University*)

Session 12: Volatile or Non-Volatile? That's the Question
Chair: Tei-Wei Kuo (*National Taiwan Univ.*)

12.1   Software Controlled Cell Bit-Density to Improve NAND Flash Lifetime   229

Xavier Jimenez, David Novo, Paolo Ienne (*Ecole Polytechnique Fédérale de
Lausanne*)

Observational Wear Leveling: An Efficient Algorithm for Flash Memory Management

12.2   Chundong Wang, Weng-Fai Wong (*National Univ. of Singapore*)   235

12.3   **Cache Revive: Architecting Volatile STT-RAM Caches for Enhanced**   243
Performance in CMPs

Yuan Xie, Chita R. Das, Vijaykrishnan Narayanan, Cong Xu (*Pennsylvania State Univ.*); Asit K. Mishra (*Intel Corp.*); Adwait Jog (*Pennsylvania State Univ.*); Ravishankar Iyer (*Intel Corp.*)

12.4   **Point and Discard: A Hard-Error-Tolerant Architecture for Non-Volatile Last**   253
Level Caches

Jue Wang, Xiangyu Dong, Yuan Xie (*Pennsylvania State Univ.*)

Session 14: Self-Aware and Adaptive Technologies: The Future of Computing Systems?
Chair: Xiaoyun Zhu (*VMware, Inc.*)

14.1   **Self-Aware Computing in the Angstrom Processor**   259

Martina Maggio (*Lund Univ.*); Anantha Chandrakasan, Anant Agarwal, Yildiz Sinangil, Mahmut Sinangil, Srini Devadas, Eric Lau, George Kurian (*Massachusetts Institute of Technology*); Jim Holt (*Massachusetts Institute of Technology, Freescale Semiconductor, Inc.*); Henry Hoffman, Sabrina Neuman, Jason Miller (*Massachusetts Institute of Technology*)

14.2   **The Case for Elastic Operating System Services in fos**   265

Charles Gruenwald (*Massachusetts Institute of Technology*); David Wentzlaff (*Princeton Univ.*); Harshad Kasture, Nathan Beckmann (*Massachusetts Institute of Technology*); Lamia Youseff (*Google, Inc.*); Anant Agarwal (*Massachusetts Institute of Technology*)

14.3   **A Compiler and Runtime for Heterogeneous Computing**   271

Joshua Auerbach, David Bacon, Ioana Burcea, Perry Cheng, Stephen Fink, Rodric Rabbah, Sunil Shukla (*IBM T.J. Watson Research Ctr.*)

14.4   **The Helix Project: Overview and Directions**   277

Gu-Yeon Wei, David Brooks, Glenn Holloway, Simone Campanoni (*Harvard Univ.*); Timothy Jones (*Univ. of Cambridge*)

Session 15: Why Model? Because Reality is Complicated Enough!
Chair: Ibrahim Elfadel (*Masdar Institute of Science and Technology*)

15.1   **Exploring Sub-20nm FinFET Design with Predictive Technology Models**   283

Saurabh Sinha, Greg Yeric, Vikas Chandra, Brian Cline (*ARM, Inc.*); Yu Cao (*Arizona State Univ.*)

15.2 Fast Nonlinear Model Order Reduction via Associated Transforms of High-Order Volterra Transfer Functions   289

Neric Fong, Ngai Wong, Qing Wang, Haotian Liu, Yang Zhang (*The Univ. of Hong Kong*)

15.3 AMOR: An Efficient Aggregating Based Model Order Reduction Method for Many-Terminal Interconnect Circuits   295

Fan Yang, Xuan Zeng, Yangfeng Su (*Fudan Univ.*)

15.4 BLAST: Efficient Computation of Nonlinear Delay Sensitivities in Electronic and Biological Networks using Barycentric Lagrange Enabled Transient Adjoint Analysis   301

Arie Meir, Jaijeet Roychowdhurry (*Univ. of California, Berkeley*)

15.5 DAE2FSM: Automatic Generation of Accurate Discrete-Time Logical Abstractions for Continuous-Time Circuit Dynamics   311

Karthik Aadithya, Jaijeet Roychowdhury (*Univ. of California, Berkeley*)

15.6 Chip/Package Co-Analysis of Thermo-Mechanical Stress and Reliability in TSV-based 3-D ICs   317

Moongon Jung (*Georgia Institute of Technology*); David Pan (*Univ. of Texas, Austin*); Sung Kyu Lim (*Georgia Institute of Technology*)

Session 16: Is Formal Verification Ready for the System Level?

Chair: Erik Seligman (*Intel Corp.*)

16.1 Symbolic Model Checking on SystemC Designs   327

Chiao Hsieh, Yen-Sheng Ho, Chun-Nan Chou, Chung-Yang (Ric) Huang (*National Taiwan Univ.*)

16.2 System Verification of Concurrent RTL Modules by Compositional Path Predicate Abstraction   334

Joakim Urdahl, Dominik Stoffel, Markus Wedler, Wolfgang Kunz (*Univ. of Kaiserslautern*)

16.3 Equivalence Checking for Behaviorally Synthesized Pipelines   344

Kecheng Hao (*Portland State Univ.*); Sandip Ray (*Univ. of Texas, Austin*); Fei Xie (*Portland State Univ.*)

16.4 Proving Correctness of Regular Expression Accelerators   350

Christoph Hagleitner, Mitra Purandare, Kubilay Atasu (*IBM Research - Zurich*)

Sciduction: Combining Induction, Deduction, and Structure for Verification and

16.5 **Synthesis** 356

      Sanjit Seshia (*Univ. of California, Berkeley*)

Session 17: NoCs Next Top Model: From System-Level to Prototype
Chair: Fabien Clermidy (*CEA-LETI*)

17.1 **Cost-Efficient Buffer Sizing in Shared-Memory 3-D MPSoCs using Wide I/O Interfaces** 366

      Abbas Sheibanyrad (*TIMA Laboratory/CNRS*); Frédéric Pétrot, Sahar Foroutan (*TIMA Laboratory, Grenoble Institute of Technology*)

17.2 **Attackboard: A Novel Dependency-Aware Traffic Generator for Exploring NoC Design Space** 376

      Yoshi Shih-Chieh Huang, Yu-Chi Chang, Tsung-Chan Tsai, Yuan-Ying Chang, Chung-Ta King (*National Tsing Hua Univ.*)

17.3 **Towards Graceful Aging Degradation in NoCs Through an Adaptive Routing Algorithum** 382

      Sanghamitra Roy, Kshitij Bhardwaj, Koushik Chakraborty (*Utah State Univ.*)

17.4 **Explicit Modeling of Control and Data for Improved NoC Router Estimation** 392

      Siddhartha Nath, Bill Lin, Andrew B. Kahng (*Univ. of California at San Diego*)

17.5 **Approaching the Theoretical Limits of a Mesh NoC with a 16-Node Chip Prototype in 45nm SOI** 398

      Sunghyun Park, Tushar Krishna, Chia-Hsin O. Chen, Bhavya Daya, Li-Shiuan Peh, Anantha P. Chandrakasan (*Massachusetts Institute of Technology*)

17.6 **High Radix Self-Arbitrating Switch Fabric with Multiple Arbitration Schemes and Quality of Service** 406

      Trevor Mudge, Dennis Sylvester, Ronald Dreslinski, Reetuparna Das, Sudhir Satpathy, David Blaauw (*Univ. of Michigan*)

Session 18: Timing Analysis and Software-Controlled Memory: Are We Safe?
Chair: Frank Slomka (*Univ. of Ulm*)

18.1 **WCET-Centric Partial Instruction Cache Locking** 412

      Huping Ding (*National Univ. of Singapore*); Yun Liang (*Advanced Digital Sciences Center*); Tulika Mitra (*National Univ. of Singapore*)

18.2 **Worst-Case Execution Time Analysis for Parallel Run-Time Monitoring** 421

      Daniel Lo, G. Edward Suh (*Cornell Univ.*)

18.3 **Conforming the Runtime Inputs for Hard Real-Time Embedded Systems** 430

Kai Huang, Gang Chen, Christian Buckl (*fortiss GmbH*); Alois Knoll (*Technische Univ. München*)

18.4 **STM Concurrency Control for Embedded Real-Time Software with Tighter Time Bounds** 437

Mohammed El-Shambakey, Binoy Ravindran (*Virginia Polytechnic Institute and State Univ.*)

18.5 **HaVOC: A Hybrid-Memory-Aware Virtualization Layer for On-Chip Distributed ScratchPad and Non-Volatile Memories** 447

Nikil Dutt, Luis Angel Bathen (*Univ. of California, Irvine*)

18.6 **Age-Based PCM Wear Leveling with Nearly Zero Search Cost** 453

Chi-Hao Chen (*National Taiwan Univ.*); Pi-Cheng Hsiu (*Academia Sinica*); Tei-Wei Kuo (*National Taiwan Univ., Academia Sinica*); Chia-Lin Yang (*National Taiwan Univ.*); Cheng-Yuan Michael Wang (*Macronix International Co., Ltd.*)

Session 20: Routing-Driven Design Closure
Chair: Shankar Krishnamoorthy (*Mentor Graphics Corp.*)

20.1 **Algorithms and Data Structures for Fast and Good VLSI Routing** 459

Christian Schulte, Jens Vygen, Christian Panten, Tim Nieberg, Dirk Mueller, Michael Gester (*Univ. of Bonn*)

20.2 **Guiding a Physical Design Closure System to Produce Easier-to-Route Designs with More Predictable Timing** 465

Charles Alpert (*IBM Corp.*); Gi-Joon Nam (*IBM Research - Austin*); Natarajan Viswanathan (*IBM Systems and Technology Group*); Cliff Sze (*IBM Research - Austin*); Nancy Zhou (*IBM Systems and Technology Group*); Zhuo Li (*IBM Research - Austin*)

20.3 **Rule Agnostic Routing by Using Design Fabrics** 471

Gyuszi Suto (*Intel Corp.*)

Session 21: Storing, Computing, and Storing While Computing: The New Face of Non-Volatility in Systems
Chair: Charles Augustine (*Intel Corp.*)

21.1 **Making Non-Volatile Nanomagnet Logic Non-Volatile** 476

Michael Niemier, Peng Li, Gary Bernstein, Vijay Karthik Sankar, Wolfgang Porod, Xiaobo Sharon Hu, Steve Kurtz, Aaron Dingler, Gyorgy Csaba, Joseph Nahas (*Univ. of Notre Dame*)

21.2 mLogic: Ultra-Low Voltage Non-Volatile Logic Circuits Using STT-MTJ Devices ... 486

Daniel Morris, David Bromberg, Jian-Gang (Jimmy) Zhu, Larry Pileggi (*Carnegie Mellon Univ.*)

21.3 Future Cache Design using STT MRAMs for Improved Energy Efficiency: Devices, Circuits and Architecture ... 492

Sumeet Kumar Gupta, Kaushik Roy, Niladri Narayan Mojumder, Sang Phill Park, Anand Raghunathan (*Purdue Univ.*)

21.4 Hardware Realization of BSB Recall Function with Memristor Crossbar Arrays ... 498

Miao Hu, Hai Li (*Polytechnic Institute of New York Univ.*); Qing Wu, Garrett S. Rose (*Air Force Research Lab*)

Session 22: You Can Count on Me: Why it's OK to be Imprecise or Unreliable
Chair: Qinru Qiu (*Syracuse Univ.*)

22.1 A Methodology for Energy-Quality Tradeoff Using Imprecise Hardware ... 504

Gabriel Robins, Jiawei Huang, John Lach (*Univ. of Virginia*)

22.2 On the Exploitation of the Inherent Error Resilience of Wireless Systems under Unreliable Silicon ... 510

Andreas Burg (*Ecole Polytechnique Fédérale de Lausanne*); Christian Benkeser, Christoph Roth (*Eidgenössische Technische Hochschule Zürich*); Georgios Karakonstantis (*Ecole Polytechnique Fédérale de Lausanne*)

22.3 Near-Optimal, Dynamic Module Reconfiguration in a Photovoltaic System to Combat Partial Shading Effects ... 516

Xue Lin, Yanzhi Wang, Siyu Yue (*Univ. of Southern California*); Donghwa Shin, Naehyuck Chang (*Seoul National Univ.*); Massoud Pedram (*Univ. of Southern California, Los Angeles*)

22.4 Networked Architecture for Hybrid Electrical Energy Storage Systems ... 522

Qing Xie, Massoud Pedram, Yanzhi Wang (*Univ. of Southern California*); Sangyoung Park, Younghyun Kim, Naehyuck Chang (*Seoul National Univ.*)

Session 23: Optimization to the Rescue of Analog
Chair: Trent McConaghy (*Solido Design Automation, Inc.*)

23.1 A New Uncertainty Budgeting Based Method for Robust Analog/Mixed-Signal Design ... 529

Jin Sun (*Orora Design Technologies, Inc.*); Priyank Gupta (*Cirrus Logic, Inc.*); Janet Roveda (*Univ. of Arizona*)

| 23.2 | **Variability-Aware, Discrete Optimization for Analog Circuits** | 536 |

Seobin Jung, Yunju Choi, Jaeha Kim (*Seoul National Univ.*)

| 23.3 | **Efficient Multi-Objective Synthesis for Microwave Components Based on Computational Intelligence Techniques** | 542 |

Soheil Radiom, Guy A. E. Vandenbosch, Hadi Aliakbarian, Bo Liu, Georges Gielen (*Katholieke Univ. Leuven*)

| 23.4 | **Non-Uniform Multilevel Analog Routing with Matching Constraints** | 549 |

Hung-Chih Ou, Hsing-Chih Chang Chien, Yao-Wen Chang (*National Taiwan Univ.*)

Session 24: Xterminating Bugs

Chair: Sharad Kumar (*Freescale Semiconductor, Inc.*)

| 24.1 | **X-Tracer: A Reconfigurable X-Tolerant Trace Compressor for Silicon Debug** | 555 |

Feng Yuan, Xiao Liu, Qiang Xu (*The Chinese Univ. of Hong Kong*)

| 24.2 | **Quick Detection of Difficult Bugs for Effective Post-Silicon Validation** | 561 |

Farzan Fallah (*Stanford Univ.*); Nagib Hakim (*Intel Corp.*); Ted Hong, David Lin, Subhasish Mitra (*Stanford Univ.*)

| 24.3 | **Test Data Volume Optimization for Diagnosis** | 567 |

Hongfei Wang, Osei Poku, Xiaochun Yu, Sizhe Liu, Ibrahima Komara, Shawn Blanton (*Carnegie Mellon Univ.*)

| 24.4 | **Invariance-Based Concurrent Error Detection for Advanced Encryption Standard** | 573 |

Xiaofei Guo, Ramesh Karri (*Polytechnic Institute of New York Univ.*)

Session 26: Brain-Inspired Autonomous Computing and Modeling

Chair: Yiran Chen (*Univ. of Pittsburgh*)

| 26.2 | **Accelerating Neuromorphic Vision Algorithms for Recognition** | 579 |

Matthew Cotter (*Pennsylvania State Univ.*); Chaitali Chakrabarti (*Arizona State Univ.*); Vijaykrishnan Narayanan, Michael DeBole, Ahmed Al Maashri, Nandhini Chandramoorthy, Yang Xiao (*Pennsylvania State Univ.*)

| 26.3 | **Statistical Memristor Modeling and Case Study in Neuromorphic Computing** | 585 |

Robinson Pino (*Air Force Research Lab*); Hai Li (*Polytechnic Institute of New York Univ.*); Yiran Chen (*Univ. of Pittsburgh*); Miao Hu (*Polytechnic Institute of New York Univ.*); Beiye Liu (*Univ. of Pittsburgh*)

Session 27: Design, the Next Generation: From Routing to Capturing Design Expertise
Chair: Charles Chiang (*Synopsys, Inc.*)

27.1   **Triple Patterning Aware Routing and its Comparison with Double Patterning Aware Routing in 14nm Technology**   591

Martin D. F. Wong, Qiang Ma, Hongbo Zhang (*Univ. of Illinois at Urbana-Champaign*)

27.2   **GDRouter: Interleaved Global Routing and Detailed Routing for Ultimate Routability**   597

Yanheng Zhang (*Cadence Design Systems, Inc.*); Chris Chu (*Iowa State Univ.*)

27.3   **Standard Cell Routing via Boolean Satisfiability**   603

Nikolai Ryzhenko, Steven Burns (*Intel Corp.*)

27.4   **An Efficient Algorithm for Multi-Layer Obstacle-Avoiding Rectilinear Steiner Tree Construction**   613

Chih-Hung Liu, I-Che Chen (*Academia Sinica*); Der-Tsai Lee (*National Chung-Hsing Univ.*)

27.5   **Avoiding Game Over: Bringing Design to the Next Level**   623

Sameh Galal, Stephen Richardson, Artem Vassilliev, Sabarish Sankaranarayanan, Mark Horowitz, Andrew Danowitz, Megan Wachs, Ofer Shacham, John Brunhaver, Wajahat Qadeer (*Stanford Univ.*)

Session 28: Staying Cool: Modeling Thermal Effects in 3-D and Multicore
Chair: Dhireesha Kudithipudi (*Rochester Institute of Technology*)

28.1   **PowerField: A Transient Temperature-to-Power Technique based on Markov Random Field Theory**   630

Seungwook Paek (*KAIST*); Seok-Hwan Moon (*Electronics and Telecommunications Research Institute*); Wongyu Shin, Jaehyeong Sim, Lee-Sup Kim (*KAIST*)

28.2   **EigenMaps: Algorithms for Optimal Thermal Maps Extraction and Sensor Placement on Multicore Processors**   636

Juri Ranieri, Martin Vetterli, David Atienza, Alessandro Vincenzi, Amina Chebira (*Ecole Polytechnique Fédérale de Lausanne*)

28.3   **An Information-theoretic Framework for Optimal Temperature Sensor Allocation and Full-chip Thermal Monitoring**   642

Huapeng Zhou, Xin Li (*Carnegie Mellon Univ.*); Chen-Yong Cher, Eren Kursun, Haifeng Qian (*IBM T.J. Watson Research Ctr.*); Shi-Chune Yao (*Carnegie*

Mellon Univ.)

28.4   Optimizing Energy Efficiency of 3-D Multicore Systems with Stacked DRAM   648
under Power and Thermal Constraints

Ayse Coskun, Jie Meng, Katsutoshi Kawakami (*Boston Univ.*)

Session 29: SOS: Specification, Optimization, and Synthesis in System-Level Design

Chair: Brett Meyer (*McGill Univ.*)

29.1   Static Dataflow with Access Patterns: Semantics and Analysis   656

Arkadeb Ghosal, Rhishikesh Limaye, Kaushik Ravindran (*National Instruments Corp.*); Stavros Tripakis (*Univ. of California, Berkeley*); Ankita Prasad, Guoqiang Wang, Trung N. Tran, Hugo A. Andrade (*National Instruments Corp.*)

29.2   Executing Synchronous Dataflow Graphs on a SPM-Based Multicore   664
Architecture

Hyunok Oh (*Hanyang Univ.*); Sungchan Kim (*Chonbuk National Univ.*); Junchul Choi, Soonhoi Ha (*Seoul National Univ.*)

29.3   System-Level Synthesis of Memory Architecture for Stream Processing   672
Sub-Systems of a MPSoC

Glenn Leary, Weijia Che, Karam S. Chatha (*Arizona State Univ.*)

29.4   Courteous Cache Sharing: Being Nice to Others in Capacity Management   678

Akbar Sharifi, Shekhar Srikantaiah, Mahmut Kandemir, Mary Jane Irwin (*Pennsylvania State Univ.*)

Session 30: Future of IC Reliability

Chair: Alesandro Pinto (*United Technologies Research Center*)

30.1   A Hybrid Approach to Cyber-Physical Systems Verification   688

Samarjit Chakraborty (*Technische Univ. München*); Anuradha Annaswamy (*Massachusetts Institute of Technology*); Lothar Thiele (*Eidgenössische Technische Hochschule Zürich*); Dip Goswami (*Technische Univ. München*); Pratyush Kumar, Kai Lampka (*Eidgenössische Technische Hochschule Zürich*)

30.2   Reliable Computing with Ultra-Reduced Instruction Set Co-Processors   697

Aravindkumar Rajendiran, Sundaram Ananthanarayanan, Hiren Patel, Mahesh Tripunitara, Siddharth Garg (*Univ. of Waterloo*)

30.3   Identification of Recovered ICs using Fingerprints from a Light-Weight   703
On-Chip Sensor

Xuehui Zhang, Nicholas Tuzzio, Mohammad Tehranipoor (*Univ. of*

Connecticut)

30.4   Confidentiality Preserving Integer Programming for Global Routing   709

Hamid Shojaei, Azadeh Davoodi, Parameswaran Ramanathan (*Univ. of Wisconsin*)

Session 32: Breaking out of EDA: How to Apply EDA Techniques to Broader Applications
Chair: Jason Cong (*Univ. of California, Los Angeles*)

32.1   Design Tools for Artificial Nervous Systems   717

Louis K. Scheffer (*Howard Hughes Medical Institute*)

32.2   Dynamic River Network Simulation at Large Scale   723

Frank Liu (*IBM Research - Austin*); Ben R. Hodges (*Univ. of Texas, Austin*)

32.3   Humans for EDA and EDA for Humans   729

Valeria Bertacco (*Univ. of Michigan*)

32.4   Application of Logic Synthesis to the Understanding and Cure of Genetic Diseases   734

Pey-Chang Kent Lin, Sunil Khatri (*Texas A&M Univ.*)

Session 33: The Right Placement at the Right Timing
Chair: Saurabh Adya (*Magma Design Automation, Inc.*)

33.1   Exploiting Die-to-Die Thermal Coupling in 3D IC Placement   741

Krit Athikulwongse, Mohit Pathak, Sung Kyu Lim (*Georgia Institute of Technology*)

33.2   ComPLx: A Competitive Primal-Dual Lagrange Optimization for Global Placement   747

Myung-Chul Kim, Igor Markov (*Univ. of Michigan*)

33.3   PADE: A High-Performance Placer with Automatic Datapath Extraction and Evaluation through High-Dimensional Data Learning   756

Samuel Ward, Duo Ding, David Pan (*Univ. of Texas, Austin*)

33.4   Structure-Aware Placement for Datapath Intensive Circuit Designs   762

Sheng Chou, Meng-Kai Hsu, Yao-Wen Chang (*National Taiwan Univ.*)

33.5   GLARE: Global and Local Wiring Aware Routability Evaluation   768

Charles J. Alpert (*IBM Austin Research Lab*); Sachin S. Sapatnekar (*University of Minnesota*); Douglas Keller, Gustavo E. Tellez, Lakshmi Reddy

(*IBM Systems and Technology Group*); Zhuo Li (*IBM Austin Research Lab*); Natarajan Viswanathan (*IBM Systems and Technology Group*); Cliff Sze (*IBM Austin Research Lab*); Yaoguang Wei (*University of Minnesota*); Andrew D. Huber (*IBM Systems and Technology Group*)

33.6   The DAC 2012 Routability-Driven Placement Contest and Benchmark Suite   774

Natarajan Viswanathan, Charles Alpert, Cliff Sze, Zhuo Li, Yaoguang Wei (*IBM Corp.*)

Session 34: Global Views of Synthesis: Broadening the Scope
Chair: Herman Schmit (*Altera Corp.*)

34.1   Removing Overhead from High-Level Interfaces   783

Megan Wachs, Mark Horowitz, Kyle Kelley, Stephen Richardson, John Stevenson (*Stanford Univ.*)

34.2   On the Asymptotic Costs of Multiplexer-Based Reconfigurability   790

Johnathan York, Derek Chiou (*Univ. of Texas, Austin*)

34.3   SALSA: Systematic Logic Synthesis of Approximate Circuits   796

Swagath Venkataramani, Amit Sabne, Vivek Kozhikkottu, Kaushik Roy, Anand Raghunathan (*Purdue Univ.*)

34.4   Timing ECO Optimization Using Metal-Configurable Gate-Array Spare Cells   802

Iris Hui-Ru Jiang (*National Chiao Tung Univ.*); Yao-Wen Chang, Hua-Yu Chang (*National Taiwan Univ.*)

34.5   Early Prediction of NBTI Effects Using RTL Source Code Analysis   808

Kenneth Butler (*Texas Instruments, Inc.*); Heesoo Kim, Shobha Vasudevan, Jayanand Asok Kumar (*Univ. of Illinois at Urbana-Champaign*)

34.6   Generalized SAT-Sweeping for Post-Mapping Optimization   814

Tobias Welp (*Univ. of California, Berkeley*); Smita Krishnaswamy (*Columbia Univ.*); Andreas Kuehlmann (*Coverity, Inc.*)

Session 35: Adaptive Computing: When, Where, Why, How?
Chair: Philip Brisk (*Univ. of California, Riverside*)

35.1   Accuracy-Configurable Adder for Approximate Arithmetic Designs   820

Andrew B. Kahng, Seokhyeong Kang (*Univ. of California at San Diego*)

35.2   **Recovery-Based Design for Variation-Tolerant SoCs**   826

Anand Raghunathan (*Purdue Univ.*); Sujit Dey (*Univ. of California at San Diego*); Vivek Kozhikkottu (*Purdue Univ.*)

35.3   **A Hybrid NoC Design for Cache Coherence Optimization for Chip Multiprocessors**   834

Ohyoung Jang, Wei Ding, Yuanrui Zhang, Mahmut Kandemir, Mary Jane Irwin, Hui Zhao (*Pennsylvania State Univ.*)

35.4   **Architecture Support for Accelerator-Rich CMPs**   843

Jason Cong, Mohammad Ali Ghodrat, Michael Gill, Beayna Grigorian, Glenn Reinman (*Univ. of California, Los Angeles*)

35.5   **A QoS-Aware Memory Controller for Dynamically Balancing GPU and CPU Bandwidth Use in an MPSoC**   850

Min Kyu Jeong (*Univ. of Texas, Austin*); Nigel Paver (*ARM, Inc.*); Mattan Erez (*Univ. of Texas, Austin*); Chander Sudanthi (*ARM, Inc.*)

35.6   **Metronome: Operating System Level Performance Management via Self-Adaptive Computing**   856

Filippo Sironi, Davide Basilio Bartolini (*Politecnico di Milano*); Simone Campanoni (*Harvard Univ.*); Fabio Cancaré (*Politecnico di Milano*); Henry Hoffmann (*Massachusetts Institute of Technology*); Donatella Sciuto, Marco Santambrogio (*Politecnico di Milano*)

Session 36: Yin and Yang of Memories: The Power-Performance Trade-Off

Chair: Yiran Chen (*Univ. of Pittsburgh*)

36.1   **Adaptive Power Management of On-Chip Video Memory for Multiview Video Coding**   866

Muhammad Shafique, Joerg Henkel (*Karlsruhe Institute of Technology*); Sergio Bampi (*Univ. Federal do Rio Grande do Sul*); Bruno Zatt (*Karlsruhe Institute of Technology*); Fábio Leandro Walter (*Univ. Federal do Rio Grande do Sul*)

36.2   **Heterogeneous Multi-Channel: Fine-Grained DRAM Control for Both System Performance and Power Efficiency**   876

Guangfei Zhang (*Institute of Computing Tech.*); Huandong Wang (*Loongson Technology Corp., Ltd*); Xinke Chen (*Institute of Computing Tech.*); Shuai Huang (*Loongson Technology Corp., Ltd.*); Peng Li (*Institute of Computing Tech.*)

36.3   **Joint Management of RAM and Flash Memory with Access Pattern Considerations**   882

Po-Chun Huang (*National Taiwan Univ.*); Yuan-Hao Chang (*Academia Sinica*); Tei-Wei Kuo (*National Taiwan Univ., Academia Sinica*)

36.4 Hybrid DRAM/PRAM-Based Main Memory for Single-Chip CPU/GPU   888

Dongki Kim, Sunggu Lee, Sungjoo Yoo (*Pohang Univ. of Science and Technology*); Dong Hyuk Woo, DaeHyun Kim (*Intel Corp.*); Sungkwang Lee (*Pohang Univ. of Science and Technology*); Jaewoong Chung (*Intel Corp.*)

36.5 Write Performance Improvement by Hiding R Drift Latency in Phase-Change RAM   897

Youngsik Kim, Sungjoo Yoo, Sunggu Lee (*Pohang Univ. of Science and Technology*)

36.6 Constructing Large and Fast Multi-Level Cell STT-MRAM Based Cache for Embedded Processors   907

Lei Jiang, Jun Yang, Bo Zhao, Youtao Zhang (*Univ. of Pittsburgh*)

Session 38: Probabilistic Embedded Computing
Chair: Vincent Mooney (*Georgia Institute of Technology*)

38.1 Incorrect Systems: It's not the Problem It's the Solution.   913

Christoph M. Kirsch, Hannes Payer (*University of Salzburg*)

38.2 On Software Design for Stochastic Processors   918

Joseph Sloan, John Sartori, Rakesh Kumar (*Univ. of Illinois at Urbana-Champaign*)

38.3 What to Do About the End of Moore's Law, Probably!   924

Krishna Palem (*Nanyang Technological Univ., Rice Univ.*); Avinash Lingamneni (*Rice Univ.*)

38.4 Obtaining and Reasoning About Good Enough Software   930

Martin Rinard (*Massachusetts Institute of Technology*)

Session 39: Simulation-Based Verification: New Ways to Harness the Workhorse
Chair: Kerstin Eder (*Univ. of Bristol*)

39.1 Improving Gate-level Simulation Accuracy when Unknowns Exist   936

Chris Browy, Kai-Hui Chang (*Avery Design Systems, Inc.*)

39.2 **Automated Feature Localization for Hardware Designs Using Coverage Metrics** 941

Goerschwin Fey (*German Aerospace Center*); Jan Malburg, Alexander Finder (*Univ. of Bremen*)

39.3 **Path Directed Abstraction and Refinement in SAT-Based Design Debugging** 947

Brian Keng, Andreas Veneris (*Univ. of Toronto*)

39.4 **Checking Architectural Outputs Instruction-By-Instruction on Acceleration Platforms** 955

Debapriya Chatterjee (*Univ. of Michigan*); Anatoly Koyfman, Ronny Morad, Avi Ziv (*IBM Haifa Research Lab.*); Valeria Bertacco (*Univ. of Michigan*)

Session 40: Ultra-Low Power Using Subthreshold and Nearthreshold Operation
Chair: Mahadev Nemani (*Intel Corp.*)

40.1 **Standard Cell Sizing for Subthreshold Operation** 962

Bo Liu, Jose Pineda de Gyvez (*Technische Univ. Eindhoven*); Jos Huisken, Maryam Ashouei (*Holst Centre*)

40.2 **Decoupling Capacitor Design Strategy for Minimizing Supply Noise of Ultra-Low Voltage Circuits** 968

Mingoo Seok (*Columbia Univ.*)

40.3 **Regaining Throughput Using Completion Detection for Error-Resilient Near-Threshold Logic** 974

Joseph Crop, Robert Pawlowski, Patrick Chiang (*Oregon State Univ.*)

40.4 **Process Variation in Near-Threshold Wide SIMD Architectures** 980

Chaitali Chakrabarti (*Arizona State Univ.*); Trevor Mudge, Scott Mahlke, Yongjun Park, Mark Woh, Ronald Dreslinski, Sangwon Seo, David Blaauw (*Univ. of Michigan*)

Session 41: Top Picks of Run-Time Power Management Techniques
Chair: Jian-Jia Chen (*Karlsruhe Institute of Technology*)

41.1 **Run-Time Power-Down Strategies for Real-Time SDRAM Memory Controllers** 988

Karthik Chandrasekar (*Delft Univ. of Technology*); Benny Akesson, Kees Goossens (*Technische Univ. Eindhoven*)

41.2 **Embedding Statistical Tests for On-Chip Dynamic Voltage and Temperature Monitoring** 994

Lionel Vincent (*CEA-LETI Minatec*); Philippe Maurine (*Univ. Montpellier 2*); Suzanne Lesecq, Edith Beigne (*CEA-LETI Minatec*)

41.3   Quality-Retaining OLED Dynamic Voltage Scaling for Video Streaming Applications on Mobile Devices   1000

Chun Jason Xue (*City Univ. of Hong Kong*); Yiran Chen (*Univ. of Pittsburgh*); Mengying Zhao (*City Univ. of Hong Kong*); Xiang Chen, Jian Zeng (*Univ. of Pittsburgh*)

41.4   Traffic-Aware Power Optimization for Network Applications on Multicore Servers   1006

Laxmi Bhuyan, Raymond Klefstad, Jilong Kuang (*Univ. of California, Riverside*)

Session 42: The Dark Side of Test
Chair: Shreyas Sen (*Intel Corp.*)

42.1   Alternate Hammering Test for Application-Specific DRAMs and an Industrial Case Study   1012

Rei-Fu Huang (*MediaTek, Inc.*); Hao-Yu Yang, Mango C.-T. Chao (*National Chiao Tung Univ.*); Shih-Chin Lin (*United Microelectronics Corp.*)

42.2   Goal-Oriented Stimulus Generation for Analog Circuits   1018

Jayanand Asok Kumar, Shobha Vasudevan, Seyed Nematollah Ahmadyan (*Univ. of Illinois at Urbana-Champaign*)

42.3   TSV Open Defects in 3D Integrated Circuits: Characterization, Test, and Optimal Spare Allocation   1024

Krishnendu Chakrabarty, Fangming Ye (*Duke Univ.*)

42.4   Small Delay Testing for TSVs in 3-D ICs   1031

Yu-Hsiang Lin, Shi-Yu Huang (*National Tsing Hua Univ.*); Kun-Han Tsai, Wu-Tung Cheng, Stephen Sunter (*Mentor Graphics Corp.*); Yung-Fa Chou, Ding-Ming Kwai (*Industrial Technology Research Institute*)

Session 44: Design Challenges and EDA Solutions for Wireless Sensor Networks
Chair: Roman Hermida (*Complutense Univ.*)

44.1   Circuit and System Design Guidelines for Ultra-Low Power Processing   1037

Dongmin Yoon, David Blaauw, Yejoong Kim, Yoonmyung Lee, Dennis Sylvester (*Univ. of Michigan*)

44.2   Design Exploration of Energy-Performance Trade-Offs for Wireless Sensor Networks   1043

Ivan Beretta (*Ecole Polytechnique Fédérale de Lausanne*); Francisco Rincon (*Univ. Complutense Madrid*); Nadia Khaled (*Nestlé Research Center*); Paolo Grassi (*Politecnico di Milano*); Vincenzo Rana, David Atienza (*Ecole Polytechnique Fédérale de Lausanne*)

44.3   Energy Harvesting and Power Management for Autonomous Sensor Nodes   1049

Jerome Willemin (*CEA-LETI*); Christian Piguet (*Centre Suisse d'Electronique et Microtechnique SA*); Edith Beigné, Jean-Frederic Christmann, Cyril Condemine (*CEA-LETI*)

Session 45: Surviving Timing Challenges in Nanometer Designs
Chair: Florentin Dartu (*Synopsys, Inc.*)

45.1   Functional Timing Analysis Made Fast and General   1055

Jie-Hong Roland Jiang, Yi-Ting Chung (*National Taiwan Univ.*)

45.2   Timing Analysis with Nonseparable Statistical and Deterministic Variations   1061

Jeffrey Hemmett, Natesan Venkateswaran, Jeremy Leitzen (*IBM Systems and Technology Group*); Jinjun Xiong (*IBM T.J. Watson Research Ctr.*); Eric Foreman (*IBM Corp.*); Debjit Sinha (*IBM Systems and Technology Group*); Vladimir Zolotov (*IBM T.J. Watson Research Ctr.*); Chandu Visweswariah (*IBM Systems and Technology Group*)

45.3   Reversible Statistical Max/Min Operation: Concept and Applications to Timing   1067

Debjit Sinha, Natesan Venkateswaran (*IBM Systems and Technology Group*); Vladimir Zolotov (*IBM T.J. Watson Research Ctr.*); Jinjun Xiong (*IBM T.J. Watson Research Ctr.*); Chandu Visweswariah (*IBM Systems and Technology Group*)

45.4   Predicting Timing Violations Through Instruction-Level Path Sensitization Analysis   1074

Sanghamitra Roy, Koushik Chakraborty (*Utah State Univ.*)

Session 46: Special Delivery: Challenges in Packaging
Chair: Tan Yan (*Synopsys, Inc.*)

46.1   A Chip-Package-Board Co-Design Methodology   1082

Hsu-Chieh Lee, Yao-Wen Chang (*National Taiwan Univ.*)

46.2   Obstacle-Avoiding Free-assignment Routing for Flip-Chip Designs   1088

I-Jye Lin, Chin-Fang Shen, Chen-Feng Chang (*Synopsys, Inc.*); Yao-Wen Chang, Yuan-Kai Ho, Hsu-Chieh Lee, Po-Wei Lee (*National Taiwan Univ.*)

46.3   Clock Tree Synthesis with Methodology of Re-Use in 3-D IC   1094

TingTing Hwang, Fu-Wei Chen (*National Tsing Hua Univ.*)

46.4   Can Pin Access Limit the Footprint Scaling?   1100

Xiang Qiu, Malgorzata Marek-Sadowska (*Univ. of California, Santa Barbara*)

Session 47: Renovate Analog and Mixed-Signal Circuit Simulations

Chair: Chenjie Gu (*Intel Corp.*)

47.1   Yield Estimation via Multi-Cones   1107

Rouwaida Kanj (*American Univ. of Beirut*); Rajiv Joshi (*IBM T.J. Watson Research Ctr.*); Zhuo Li, Jerry Hayes (*IBM Research - Austin*); Sani Nassif (*IBM Research - Austin*)

47.2   Efficient Trimmed-Sample Monte Carlo Methodology and Yield-Aware Design Flow for Analog Circuits   1113

Wei-Yi Hu, Yi-Kan Cheng, Chin-Cheng Kuo, Yi-Hung Chen, Jui-Feng Kuan (*Taiwan Semiconductor Manufacturing Co., Ltd.*)

47.3   Towards Efficient SPICE-Accurate Nonlinear Circuit Simulation with On-the-Fly Support-Circuit Preconditioners   1119

Xueqian Zhao, Zhuo Feng (*Michigan Technological Univ.*)

47.4   Sparse LU Factorization for Parallel Circuit Simulation on GPU   1125

Ling Ren, Xiaoming Chen, Yu Wang, Chenxi Zhang, Huazhong Yang (*Tsinghua Univ.*)

Session 48: Heterogenous Platforms: Challenges and Opportunities

Chair: Norbert Wehn (*Univ. of Kaiserslautern*)

48.1   Is Dark Silicon Useful? Harnessing the Four Horsemen of the Coming Dark Silicon Apocalypse   1131

Michael Taylor (*Univ. of California at San Diego*)

48.2   Platform 2012 - A Many-Core Computing Accelerator for Embedded SoCs: Performance Evaluation of Visual Analytics Applications   1137

Luca Benini (*Univ. di Bologna, STMicrolectronics*); Denis Dutoit, Fabien Clermidy (*STMicroelectronics, CEA-LETI*); Germain Haugou, Thierry Lepley, Bruno Jego, Diego Melpignano, Eric Flamand (*STMicroelectronics*)

Session 50: Hot Chips Running Cool - Energy Efficient Near-Threshold Computing and its Barriers

Chair: David Brooks (*Harvard Univ.*)

50.1 **Assessing the Performance Limits of Parallelized Near-Threshold Computing** 1143

Kory Sewell, Trevor Mudge, David Blaauw, Dennis Sylvester, Nathaniel Pinckney, Ronald Dreslinski, David Fick (*Univ. of Michigan*)

50.2 **Near-Threshold Voltage (NTV) Design - Opportunities and Challenges** 1149

Himanshu Kaul, Mark Anders, Steven Hsu, Amit Agarwal, Ram Krishnamurthy, Shekhar Borkar (*Intel Corp.*)

50.3 **Near-Threshold Operation for Power-Efficient Computing? It Depends** 1155

Leland Chang, Wilfried Haensch (*IBM T.J. Watson Research Ctr.*)

50.4 **Not so Fast my Friend: Is Near-Threshold Computing the Answer for Power Reduction of Wireless Devices?** 1160

Matt Severson, Kendrick Yuen, Yang Du (*Qualcomm, Inc.*)

Session 51: Yielding in an Uncertain World
Chair: Rob Aitken (*ARM, Inc.*)

51.1 **Accurate Process-Hotspot Detection Using Critical Design Rule Extraction** 1163

Yen-Ting Yu (*National Chiao Tung Univ.*); Ya-Chung Chan (*Mstar Semiconductor*); Subarna Sinha (*Stanford Univ.*); Iris Hui-Ru Jiang (*National Chiao Tung Univ.*); Charles Chiang (*Synopsys, Inc.*)

51.2 **Improved Tangent Space-Based Distance Metric for Accurate Lithographic Hotspot Classification** 1169

Xuan Zeng, Jing Guo, Fan Yang (*Fudan Univ.*); Subarna Sinha (*Stanford Univ.*); Charles Chiang (*Synopsys, Inc.*)

51.3 **Simultaneous Flare Level and Flare Variation Minimization with Dummification in EUVL** 1175

Shao-Yun Fang, Yao-Wen Chang (*National Taiwan Univ.*)

51.4 **A Novel Layout Decomposition Algorithm for Triple Patterning Lithography** 1181

Shao-Yun Fang, Yao-Wen Chang, Wei-Yu Chen (*National Taiwan Univ.*)

51.5 **PS3-RAM: A Fast Portable and Scalable Statistical STT-RAM Reliability Analysis Method** 1187

Wujie Wen, YaoJun Zhang, Yiran Chen (*Univ. of Pittsburgh*); Yu Wang (*Tsinghua Univ.*); Yuan Xie (*Pennsylvania State Univ.*)

51.6 **Exploiting Narrow-Width Values for Process Variation-Tolerant 3-D Microprocessors** 1193

Sung Woo Chung, Joonho Kong (*Korea Univ.*)

Session 52: High-Level Synthesis is Not Just About Translation!

Chair: Satnam Singh (*Google, Inc.*)

52.1 **Hardware Synthesis of Recursive Functions through Partial Stream Rewriting**   1203

Christian Haubelt, Lars Middendorf (*Univ. of Rostock*); Christophe Bobda (*Univ. of Arkansas*)

52.2 **Chisel: Constructing Hardware in a Scala Embedded Language**   1212

Jonathan Bachrach, Huy Vo, Brian Richards, Yunsup Lee, Andrew Waterman, Rimas Avizienis, John Wawrzynek, Krste Asanovic (*Univ. of California, Berkeley*)

52.3 **Specification and Synthesis of Hardware Checkpointing and Rollback Mechanisms**   1222

Carven Chan, Sharad Malik, Divjyot Sethi, Daniel Schwartz-Narbonne (*Princeton Univ.*)

52.4 **Optimizing Memory Hierarchy Allocation with Loop Transformations for High-Level Synthesis**   1229

Jason Cong, Peng Zhang, Yi Zou (*Univ. of California, Los Angeles*)

52.5 **A Metric for Layout-Friendly Microarchitecture Optimization in High-Level Synthesis**   1235

Jason Cong, Bin Liu (*Univ. of California, Los Angeles*)

52.6 **Computer Generation of Streaming Sorting Networks**   1241

Marcela Zuluaga (*Eidgenössische Technische Hochschule Zürich*); Peter Milder (*Carnegie Mellon Univ.*); Markus Püschel (*Eidgenössische Technische Hochschule Zürich*)

Session 53: Wild And Crazy Ideas

Chair: Farinaz Koushanfar (*Rice Univ.*)

53.1 **CrowdMine: Towards Crowdsourced Human-Assisted Verification**   1250

Wenchao Li, Sanjit A. Seshia (*Univ. of California, Berkeley*); Somesh Jha (*Univ. of Wisconsin, Madison*)

53.2 **Extracting Design Information from Natural Language Specifications**   1252

Ian G. Harris (*Univ. of California, Irvine*)

Material Implication in CMOS: A New Kind of Logic

53.3 Elkim Roa (*Purdue Univ.*); Wu-Hsin Chen (*Purdue University*); Byunghoo   1254
Jung (*Purdue Univ.*)

53.4 **Boolean Satisfiability Using Noise-Based Logic**   1256
Pey-Chang Kent Lin, Ayan Mandal, Sunil Khatri (*Texas A&M Univ.*)

53.5 **Cognitive Computing with Spin-Based Neural Networks**   1258
Georgios Panagopoulos, Kaushik Roy, Mrigank Sharad (*Purdue Univ.*);
Charles Augustine (*Intel Corp.*)

53.6 **Capacitance of TSVs in 3-D Stacked Chips a Problem? Not for**   1260
Neuromorphic Systems!
Antoine Joubert (*CEA-LETI Minatec*); Marc Duranton (*CEA-LIST*); Bilel
Belhadj (*CEA-LETI Minatec*); Olivier Temam (*INRIA*); Rodolphe Héliot
(*CEA-LETI Minatec*)

Session 54: Optimizing Embedded Software for High Performance and Reliability
Chair: Rodric Rabbah (*IBM Research*)

54.1 **Communication-Aware Mapping of KPN Applications onto Heterogeneous**   1262
MPSoCs
Jeronimo Castrillon, Andreas Tretter, Rainer Leupers, Gerd Ascheid (*RWTH
Aachen Univ.*)

54.2 **Unrolling and Retiming of Stream Applications onto Embedded Multicore**   1268
Processors
Weijia Che, Karam Chatha (*Arizona State Univ.*)

54.3 **Exploiting Spatiotemporal and Device Contexts for Energy-Efficient**   1274
Mobile Embedded Systems
Chris Ohlsen, Sudeep Pasricha, Charles Anderson, Brad Donohoo (*Colorado
State Univ.*)

54.4 **EPIMap: Using Epimorphism to Map Applications on CGRAs**   1280
Mahdi Hamzeh, Aviral Shrivastava, Sarma Vrudhula (*Univ. of California, Los
Angeles*)

54.5 **Instruction Scheduling for Reliability-Aware Compilation**   1288
Semeen Rehman, Muhammad Shafique, Joerg Henkel (*Karlsruhe Institute of
Technology*)

54.6 **Compiling for Energy Effciency on Timing Speculative Processors**   1297
Rakesh Kumar, John Sartori (*Univ. of Illinois at Urbana-Champaign*)

Courteous Cache Sharing: Being Nice to Others in Capacity Management*

Akbar Sharifi, Shekhar Srikantaiah, Mahmut Kandemir and Mary Jane Irwin
Department of CSE
The Pennsylvania State University
University Park, PA 16802, USA
{akbar, srikanta, kandemir, mji}@cse.psu.edu

ABSTRACT

This paper proposes a cache management scheme for multiprogrammed, multithreaded applications, with the objective of obtaining maximum performance for both individual applications and the multithreaded workload mix. In this scheme, each individual application's performance is improved by increasing the priority of its slowest thread, while the overall system performance is improved by ensuring that each individual application's performance benefit does not come at the cost of a significant degradation to other application's threads that are sharing the same cache. Averaged over six workloads, our shared cache management scheme improves the performance of the combination of applications by 18%. These improvements across applications in each mix are also fair, as indicated by average fair speedup improvements of 10% across the threads of each application (averaged over all the workloads).

Categories and Subject Descriptors

B.3.2 [**Memory Structures**]: Design Styles; C.4 [**Performance of Systems**]: Design Studies

General Terms

Management, Design, Performance, Experimentation

Keywords

Shared Cache Management, Multithreaded Applications

1. INTRODUCTION

Most recent multicore processors [15, 12, 10] share caches among the cores. Memory requests made by different cores interfere with each other in different levels of the cache, causing the eviction of each core's data, which otherwise would have good hit rates. As we move towards many-core systems, interference in shared cache continues to increase, making shared cache management an important issue in increasing

---

*This work is supported in part by NSF grants 1147388, 1152479, 1017882, 0963839, 0811687, and a grant from Microsoft.

---

overall system performance. Further, to effectively make use of these many-core systems, application programmers and compiler writers alike are trying their best to exploit the increased potential of the higher number of cores by having more multithreaded workloads. Given the limited ability of current compilers to extract a large fraction of the available parallelism, we envision a typical workload on future multicore processors to consist of a dynamic and diverse range of applications that exhibit varying degrees of thread-level parallelism. This brings managing shared caches for multiprogrammed, multithreaded workloads to the limelight.

Cache sharing can bring benefits, if the threads are scheduled well (by employing schemes such as the ones presented in [1, 20, 24]) for a certain class of applications with extensive data sharing. However, for the other applications with a lesser degree of data sharing, interferences due to shared caches can be a more pressing problem than the advantages that cache sharing brings. A flurry of recent research [22, 18, 2, 19, 8, 7] has addressed shared cache management by space partitioning and/or modifying the cache replacement policies. However, the focus on shared cache management schemes for multithreaded workloads has been rather modest [16]. To the best of our knowledge, there has been no prior work on shared cache management in the presence of multiple, concurrently-executing multithreaded applications. A key characteristic of multithreaded workloads is that, if requests from different threads of different applications are not properly prioritized in usage of the shared cache space, some application threads can suffer additional misses, and consequently, overall system throughput can degrade.

In a multithreaded workload, since various applications with different numbers of threads may execute and terminate at arbitrary times during execution, threads might be mapped to the cores of a multicore in different ways. Consequently, as threads from the same application as well as co-runner applications can share the same cache space, performance variations across threads of a given application are possible. As an example, let us assume that our multicore has eight cores in two sockets, each core having a private L1 cache and each socket having a last level L2 cache shared by the four cores in that socket, as illustrated in Figure 1(a). Suppose that three multithreaded applications D, U and G are executed on this multicore with four, two and two threads, respectively. In this case, a good mapping strategy is to map the threads of D to P0-P3 and the threads of U and G to P4-P7 (see Figure 1(a)). This mapping gives the threads that belong to D the opportunity to exploit interthread data sharing in the L2 cache space, and also isolates

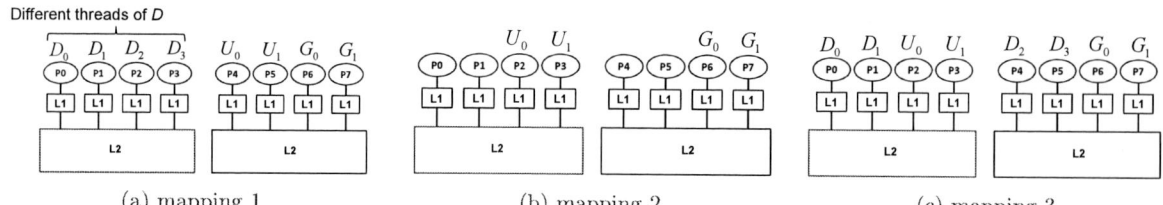

Figure 1: Different mappings of the threads that belong to three multithreaded applications running on an eight-core multicore. P0 through P3 denote cores, and D0-3, U0-1 and G0-1 are *D*'s, *U*'s and *G*'s threads, respectively.

this application from the other two. However, this mapping may not always be possible when different applications have different entry/exit times. For example, if U and G arrive earlier than D for execution and their threads are mapped to P2-P3 and P6-P7 respectively (Figure 1(b)), there are two options for D. The first one is to migrate the threads of applications U or G to the cores sharing a last level cache and map the threads of D to the other cores sharing a last level cache (as in Figure 1(a)). However, thread migration can impose significant performance overheads, since the migrated threads suffer from loosing their cache locality. The other option would be mapping the threads of the incoming application D to the available free cores, as shown in Figure 1(c).

Our goal in this paper is to design a configurable shared cache management scheme that provides high system throughput without requiring significant coordination between multiple sockets. To this end, targeting multiprogrammed, multithreaded workloads, we develop a fundamentally new approach to shared cache management, called *CCS (Courteous Cache Sharing)*, with the following key contributions: (a) CCS attempts to balance the execution speeds of different threads that belong to the same *data-parallel* multithreaded application in a multiprogrammed workload. This generally results in *de-prioritizing* the faster threads of an application, which in turn generates a *slack* that can be used to expedite threads of other applications sharing the same cache. (b) CCS uses the newly created slack to improve overall system performance by improving the performance of applications that can benefit from the additional slack. (c) We propose a simple mechanism to manage the priority of each running thread in using the shared L2 cache space.

Averaged over six workloads, our shared cache management scheme improves the performance of the combination of applications by 18%. These improvements across applications in each mix are also fair, as indicated by average fair speedup improvements of 10% across the threads of each application (averaged over all the workloads).

2. OUR PROPOSED SCHEME

We first describe a simple, yet novel prioritization scheme we employ to control the performance of each running thread by means of assigning different priorities to different cores (application threads[1]) in evicting cache lines. After that, we describe the details of our CCS approach.

2.1 Prioritization Scheme

Cache partitioning has been proposed as a mechanism to prioritize applications by allocating different capacities of the shared cache space to co-runner applications [18, 19, 21]. For

---

[1]We consider different one-to-one mappings of threads to cores.

instance, in the utility-based partitioning scheme [18], different numbers of cache ways in a set are allocated to the cores sharing a set-associative cache. There are at least three significant issues in adopting a cache partitioning scheme like [18] or [21] for multiprogrammed workloads of multithreaded applications: (1) The prior cache partitioning schemes do not consider the variations in execution speeds of different threads of an application, that may lead to the overall performance degradation. (2) To remedy the thread-obliviousness of existing cache partitioning schemes, one would need detailed performance models for each thread of each application. Each of these performance models further needs to maintain a predicted performance curve for different numbers of allocated cache ways. The area overheads of a hardware implementation of such a scheme would be prohibitively expensive in practice. (3) In general, the reaction time of cache partitioning is very slow.

To alleviate these problems, in CCS, we employ a *prioritization mechanism* which is able to manage the shared cache space with finer granularity and less overheads. Our prioritization scheme requires a simple change in the LRU eviction priority. In a normal LRU algorithm, if a cache miss occurs, the cache line which is accessed least recently is evicted from the set indexed by the memory access request. In our prioritization scheme, on the other hand, each core is assigned a positive integer number as its *priority (weight)*. First, based on the weights assigned to different cores, a core is selected as the *victim* from among the ones having cache lines in the corresponding set, and then, the LRU cache line is selected from the lines that belong to this victim core. Algorithm 1 gives the high level view of our victim core selection strategy.

2.2 CCS Architecture

This section explains the main contribution of this work, a novel cache space management scheme targeting multiprogrammed, multithreaded workloads that consist of data-parallel applications. The key insight in CCS is that, the overall performance of a data-parallel multithreaded application is determined by its slowest thread, and therefore, we can afford to *de-prioritize* the faster threads of one application, while *prioritizing* the slower threads of another application sharing the same cache. In this section, we discuss CCS for a sample architecture with eight cores, in which a set of four cores in a socket share a common L2 cache space. Note that CCS works similarly for the architectures with different number of cores and different on-chip cache topologies.

Figure 2 illustrates the high-level view of CCS. During the course of execution, CCS monitors the measured IPCs (our performance metric) of the threads that belong to each running application and, based on that, adjusts the weight value of each core (*eviction weights* of the cores). As can be observed from Figure 2, the Arbiter component takes the measured IPCs of the cores as input and determines a *rank*

679

Algorithm 1 Victim Core Selection
___
Input: The index of the set in the cache: set_num
Output: Victim Core: $core$
1: $cache\_lines = cacheLines(set\_num)$
2: $cores = \emptyset$
3: $sum = 0$
4: **for** each $cache\_line$ in $cache\_lines$ **do**
5:    $cores.add(cache\_line.core)$
6: **end for**
7: **for** each $core$ in $cores$ **do**
8:    $sum = sum + core.weight$
9: **end for**
10: **for** each $core$ in $cores$ **do**
11:    $probablity[core.num] = \frac{core.weight}{sum}$
12: **end for**
13: $random = generateRandom(0, 1)$
14: $partial\_sum = 0$
15: **for** each $core$ in $cores$ **do**
16:    **if** $partial\_sum \geq random$ **then**
17:       return $core$
18:    **else**
19:       $partial\_sum = partial\_sum+$
20:       $probability[core.num]$
21:    **end if**
22: **end for**
___

value for each of the cores. These rank values (which will be used to compute the *weights*) are obtained as follows:

$$rank_i = \frac{IPC_i - MIN}{MIN}, \qquad (1)$$

where $rank_i$ and IPC_i are, respectively, the rank value and the measured IPC of core i. Assuming A is the application to which the thread running on core i belongs, MIN is the minimum IPC of the A's threads. In other words, *rank values capture how fast each thread is, as compared to the slowest thread of the same multithreaded application.* If the rank value for core i is zero, it indicates that the thread mapped to this core has the lowest IPC across all threads that belong to the same application. Algorithm 2 shows how the rank of each core is computed based on the mappings of the running threads. In this algorithm, $mapping[i]$ indicates the application that the thread mapped to core i belongs to.

Algorithm 2 Operation of the Arbiter Component
___
Input: The measured IPCs $IPC[1 \cdots n]$
Input: Thread mapping $mapping[\ ]$
Output: Ranks of the cores $R[1 \cdots n]$
1: **for** $i = 1$ to n **do**
2:    $app = mapping[i]$
3:    $min = \inf$
4:    **for** $j = 1$ to n **do**
5:       **if** $(IPC[i] < min)$ and $(mapping[j]==mapping[i])$ **then**
6:          $min = IPC[i]$
7:       **end if**
8:    **end for**
9:    $R[i] = \frac{IPC[i]-min}{min}$
10: **end for**
___

In CCS, other than the Arbiter module, an L2 Manager is associated with each shared L2 cache in the system. An L2 Manager takes the ranks of the threads running on the connected cores as input, and adjusts the eviction weights of the running threads accordingly. For instance, in Figure 2, the ranks of the threads mapped to P0-P3 and produced by the Arbiter constitute the inputs to the L2-1 Manager. Based on these ranks, at the kth time interval, the L2-1 Manager increases the eviction weight of the thread (threads) with the highest rank value and reduces the eviction weight of the thread (threads) with the lowest rank value by the amount of δ as shown below:

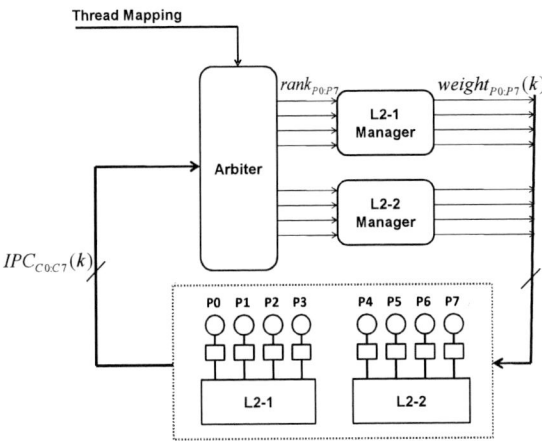

Figure 2: High-level view of CCS. The Arbiter component takes the measured IPCs of the cores as input and, at each time interval, determines a "rank value" for each of the cores in the system. Each L2 Manager takes the ranks of the threads running on its cores as input, and adjusts the eviction weights of the running threads.

$$w_i(k) = w_i(k-1) + \begin{cases} \delta, & \text{if } i \,\epsilon\, \text{MAX} \\ -\delta, & \text{if } i \,\epsilon\, \text{MIN}, \end{cases} \qquad (2)$$

where $w_i(k)$ is the eviction weight of core i at the kth time interval, and MIN and MAX are the sets of minimum and maximum rank values, respectively, received by the L2 Managers. That is, eviction weights at the kth interval are computed using the rank values as well as the eviction weights used at the $(k-1)$th interval.

Algorithm 3 shows how the eviction weights are adjusted by the L2 Managers. Note that, if an eviction weight reaches 1, since we assume that eviction weights are positive, instead of reducing this weight, the weights for the other cores connected to the same L2 cache are increased (as can be observed from lines 21 through 25).

CCS first classifies the co-runner threads into slow cores and fast ones (responsibility of the Arbiter) and then, reduces the eviction weights of the bottleneck threads (or the threads that will be likely bottleneck) to enhance their performance (responsibility of the L2 Managers). At the same time, it increases the weights of the fast threads (cores), so that they are no faster than the new slowest thread. To illustrate this, let us assume that three applications X (with four threads), Y (with two threads) and Z (with two threads) are executing on an eight-core architecture shown in Figure 1(c) (see the first two columns of Table 1 for thread-to-core mappings). Suppose that the measured IPCs at the kth time interval are as shown in the third column of Table 1. Firstly, the thread ranks (in the order of their individual speeds) are determined by the Arbiter. As can be seen in Table 1, the slowest thread of application X is X_1 and, consequently, the ranks of application X's threads are obtained based on the X_1's IPC (see the forth column of Table 1). After the rank is determined by the Arbiter, the L2 Managers select the candidate threads to increase (decrease) their eviction weights. In our example, the L2-1 Manager increases the weight of P0 (the core that X_0 is running on) and reduces the weight of

680

Algorithm 3 Operation of an L2 Manager

Input: Ranks of the cores $R[m \cdots m']$
Output: Eviction weights of the cores at time k $W_k[1 \cdots n]$
1: **for** $i = m$ to m' **do**
2:    $rank = R[i]$
3:    $increaseFlag = 1$
4:    $decreaseFlag = 1$
5:    **for** $j = m$ to m' **do**
6:      **if** $R[j] < rank$ **then**
7:        $decreaseFlag = 0$
8:      **end if**
9:      **if** $R[j] > rank$ **then**
10:       $increaseFlag = 0$
11:     **end if**
12:    **end for**
13:    $W_k[i] = W_{k-1}[i]$
14:    **if** $(increaseFlag == 1)$ and $(decreaseFlag == 0)$ **then**
15:      $W_k[i] = W_{k-1}[i] + \delta$
16:    **end if**
17:    **if** $(increaseFlag == 0)$ and $(decreaseFlag == 1)$ **then**
18:      $W_k[i] = W_{k-1}[i] - \delta$
19:    **end if**
20:    **if** $(W_k[i] < 1)$ **then**
21:      **for** $c = m$ to m' **do**
22:        **if** $c \neq i$ **then**
23:          $W_k[c] = W_k[c] + \frac{1 - W_k[i]}{m' - m}$
24:        **end if**
25:      **end for**
26:    **end if**
27: **end for**

Threads	Cores	Normalized IPCs	Ranks	Weights Before	Weights After
X_0	P0	0.8	0.66	35	37
X_1	P1	0.48	0	15	13
Y_0	P2	1.0	0.5	30	30
Y_1	P3	0.67	0.33	25	25
X_2	P4	1.0	0.1	30	30
X_3	P5	0.9	0	25	23
Z_0	P6	1.0	0.14	35	37
Z_1	P7	0.87	0	20	18

Table 1: An example that involves three applications (X, Y and Z) and 8 cores.

P1 (which has the least rank) by δ ($\delta = 2$ in this example). Similarly, the L2-2 Manager increases the weight of P6 and reduces the eviction weights of P5 and P7, with the goal of accelerating the slow threads and slowing down the fast ones dynamically to improve the overall system performance.

3. EXPERIMENTS

In this section, we present an experimental evaluation of our proposed approach for different cases.

Setup and workloads. We use SIMICS [13] as our simulation framework which is a full-system simulator that allows simulation of multicores. Table 2 gives the baseline configuration we use for our experimental evaluation.

Processors	(16 or 8) processors with private L1 data and instruction caches
Processor Model	4-way issue superscalar
Private L1 D-Caches	Direct mapped, 32KB, 64 bytes block size, 3 cycle access latency
Private L1 I-Caches	Direct mapped, 32KB, 64 bytes block size, 3 cycle access latency
Shared L2 Cache	64 bytes block size, 10 cycle access latency
Memory	4GB, 200 cycle off-chip access latency
Control Enforcement Interval	200 Million cycles
δ	2

Table 2: Baseline configuration.

Workload	Applications
Mix-1	*mgrid, bt, ocean*
Mix-2	*swim, ft, lu*
Mix-3	*swim, mg, sp*
Mix-4	*bt, sp, ocean*
Mix-5	*mgrid, lu, cg*
Mix-6	*mg, ft, cg*

Table 3: Various mixes of applications considered for our multithreaded, multiprogrammed workloads.

We performed our experiments targeting multicores with 8 and 16 cores and different workloads (application mixes). Table 3 gives the application mixes we use in our experiments. We formed our mixes using the applications from the NAS, SPLASH and SPECOMP benchmark suites. The benchmarks used are the ones that are sensitive to the variations in the L2 cache space. We formed different three-application workloads from these benchmarks, and evaluated them. In this section, we present results with six representative workloads. Note that, these applications may be executed with different numbers of threads.

After an application is initiated, based on the one-to-one mapping strategy specified, we bind its threads to available cores. To do this in Solaris 10, we use the *pbind* command.

Evaluation metrics. We use *average speedup* as our metric to evaluate the behavior of CCS. Average speedup indicates the average performance speedup achieved across the applications of a workload. Assuming equal weights, the average speedup is computed for an N-application mix as follows:

$$Average\ Speedup = \frac{\sum IPC_{scheme}(A^i)/IPC_{base}(A^i)}{N}, \quad (3)$$

$$IPC(A^i) = min(\{IPC(thread_i)|thread_i \in A^i\}), \quad (4)$$

where $IPC_{scheme}(A^i)$ is the IPC value of application A^i when CCS is employed. In our experiments, to measure the performance improvement achieved by CCS, our base case is when no scheme is used to manage shared caches ($IPC_{base}(A^i)$ is the IPC value of A^i when no scheme is employed). Note that, these values ($IPC_{scheme}(A^i)$ and $IPC_{base}(A^i)$) are determined by the slowest thread that belongs to A^i.

Average fair speedup is the other metric we employ to quantify the behavior of CCS, and can be computed as:

$$Average\_FS = \frac{FS_1 + \cdots + FS_N}{N}, \quad (5)$$

$$FS_i = \frac{N_j}{\sum_{k=1}^{j} IPC_{base}(thread_k)/IPC_{scheme}(thread_k)}, \quad (6)$$

where FS_i is the fair speedup achieved across the N_j threads that belong to application A^i. Fair speedup is also a measure of *fairness* [21].

We performed experiments with different configurations presented in supplementary section. In this section we present the result for our baseline experiment.

Experiment-1. In this experiment, we evaluate a case in which the workloads given in Table 3 are executed on a sixteen-core system. The shared cache configuration for this experiment is illustrated in Figure 3, where each L2 cache space in a socket is shared by four cores. The applications in each mix execute using 6, 6 and 4 threads in the order given by Table 3 (for instance, *mgrid*, *bt* and *ocean* in Mix-1 are executed using 6, 6 and 4 threads, respectively).

Our experiments are performed using six representative mappings that exhibit different cache sharing patterns. These

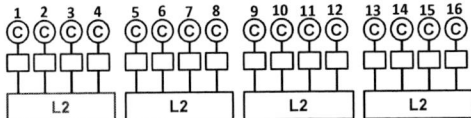

Figure 3: Experiment-1 configuration.

mappings are shown in Table 4, where X_i, Y_i and Z_i represent the threads that belong to the applications shown in Table 3 in the same order for each mix. As an example, assuming that Map-1 is used to map Mix-1 applications, four threads of *mgrid* are bound to cores 1 to 4 sharing a common L2 cache and the other two threads are bound to the cores 5 and 6 sharing an L2 cache with two threads of *bt*.

Note that, the representative mappings given in Table 4 cover different cache sharing patterns, from a mapping that offers maximum sharing across the threads from the same application (Map-1) to a mapping with the maximum L2 cache contention among the co-runner threads (Map-6). The sharing pattern for each mapping in our default configuration (16 cores and 4 L2 caches in 4 sockets) is shown in Table 4.[2]

Figure 4(a) plots the average speedups achieved in our workloads under these six different mappings. We run the workloads for 10 billion cycles and, over this period, CCS is invoked 50 times (every 200 million cycles). In this figure, the six bars presented for each mix correspond to the average speedups achieved across the applications of that mix for six different mappings. Note that the achieved performance improvements depend on the chosen mapping and also the running workload. As can be observed, CCS improves the performance of Mix-2, Mix-3, Mix-5 and Mix-6 workloads significantly through speeding up their bottleneck threads. Among the representative mappings, there is a minimum contention across the threads of different applications in Map-1, since the threads that belong to the same applications share common L2 slices in this mapping. Consequently, as can be observed from Figure 4(a) the speedups achieved when using Map-1 are not as high as those obtained with other mappings. Note that, as mentioned before, while Map-1 would be the most beneficial choice, it cannot always be done by the OS.

As one can observe in Figure 4(b), CCS also improves the performance of the individual threads of an application across different mixes in most of the cases. This is because, although CCS slows down some of the fast threads, it compensates for the performance loss through enhancing the bottleneck threads in the long run. It is further observed that the average fair speedups achieved with mixes such as Mix-4 and Mix-3 is less than the average speedups reported in Figure 4(a). As stated earlier, in computing the average speedup metric, the IPC of each application is determined by the slowest thread. CCS attempts to improve the performance of the bottleneck threads, which may in turn lead to performance degradation for the faster ones. Consequently, the average fair speedups may be less than the achieved average speedups in such cases.

There are two main factors affecting the magnitude of the savings achieved by CCS. The first one is the sensitivity of the running applications to the assigned eviction weights. For instance, performances of the applications in Mix-1 and Mix-4 are not sensitive to the cache allocation as much as

[2] $<a,b,c,d>$ shown in the third column indicates that the four L2 caches in our default configuration are shared among the threads from a, b, c and d applications.

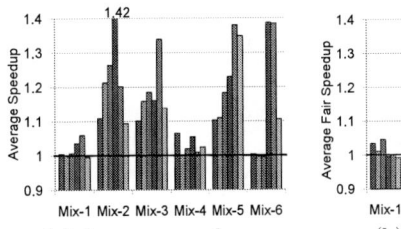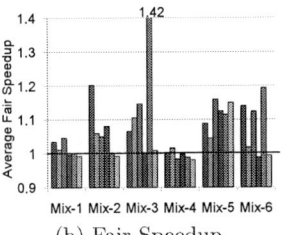

(a) Average speedup (b) Fair Speedup

Figure 4: The performance improvement achieved for the applications' threads of each mix (Experiment-1, 16 cores, 4 L2 caches). For each mix, bars correspond to the six different mappings.

the performance of the other applications and, as a result, the performance cannot be improved significantly for these mixes, as shown in Figures 4(a) and 4(b). The second factor that shapes the performance improvement is thread mapping. If the number of bottleneck threads sharing the same L2 cache increases during execution, CCS may not be able to improve the overall system performance.

Dynamic modulation. Figure 5 plots the dynamic adjustment of the eviction weights of different threads of the applications in Mix-2 when Map-2 is used in Experiment-1. As one can observe from these graphs, the eviction weights (priorities) assigned to threads of a given application are modulated dynamically to adapt to the changes dynamic behavior of application threads. Note also that, assigning the same priority values to the threads that belong to the same application, but executing in different sockets does not necessarily guarantee that the threads will run with the same speed.

Figure 6 plots the measured IPCs of the 4 threads of *lu* (in Mix-2), during execution. At each time interval, CCS adjusts the eviction weights of these threads (LU0 through LU3) based on the measured IPC values, as shown in Figure 5. For instance, as can be observed from Figure 5(c), in the 25th time epoch (marked using an oval), CCS reduces and increases the eviction weights of LU0 and LU1, respectively, since LU0 and LU1 have the lowest and highest performance among *LU*'s threads at this time (see the oval in Figure 6).

4. RELATED WORK

Shared cache partitioning has been frequently employed as a mechanism to overcome interference in shared caches among multiple single-threaded applications [22, 18, 2, 19, 11]. Qureshi and Patt [18] proposed a utility-based cache partitioning scheme where the cache-share received by an application was proportional to the utility rather than its demand. Chang and Sohi [2] studied cooperative cache partitioning (CCP), wherein they use multiple time-sharing partitions to resolve cache contention. Rafique et al [19] proposed an OS based scheme that implements a variety of policies including those for fair cache sharing and achieving differentiated instruction throughput. Several other schemes including [5, 6, 4] studied the use of shared cache management for optimizing other metrics like quality of service and fairness to multiple single-threaded applications. Cache partitioning requires extensive hardware support while the prioritization scheme used in our approach is simple to be implemented. Also, the prioritization scheme is applicable at a much finer granularity than the cache partitioning schemes proposed in the past. The key difference between alternate replacement policies [3, 14] proposed in the literature (for managing shared caches) and our prioritization scheme is that our

682

	Mappings	*Sharing Pattern*
Map-1	$< (X_0, X_1, X_2, X_3), (X_4, X_5, Y_0, Y_1), (Y_2, Y_3, Y_4, Y_5), (Z_0, Z_1, Z_2, Z_3) >$	$< 1, 2, 1, 1 >$
Map-2	$< (X_0, X_1, Y_0, Y_1), (X_2, X_3, Y_2, Y_3), (X_4, X_5, Y_4, Y_5), (Z_0, Z_1, Z_2, Z_3) >$	$< 2, 2, 2, 1 >$
Map-3	$< (X_0, Y_0, Z_0, X_1), (Y_1, Z_0, X_2, Y_2), (X_3, X_4, Y_3, Y_4), (Z_1, X_5, Y_5, Z_2) >$	$< 3, 3, 2, 3 >$
Map-4	$< (X_0, X_1, X_2, Y_0), (X_3, X_4, X_5, Z_0), (Y_1, Y_2, Y_3, Z_1), (Y_4, Y_5, Z_2, Z_3) >$	$< 2, 2, 2, 2 >$
Map-5	$< (X_0, Y_0, Y_1, Z_0), (X_1, Y_2, Z_1, Z_2), (Z_3, X_2, Y_3, Y_4), (Y_5, X_3, X_4, X_5) >$	$< 3, 3, 3, 2 >$
Map-6	$< (X_0, Y_0, X_1, Z_0), (Y_1, X_2, Z_1, Y_2), (X_3, Z_2, Y_3, Y_4), (Z_3, Y_5, X_4, X_5) >$	$< 3, 3, 3, 3 >$

Table 4: Mappings used in 16 cores experiments. X_i, Y_i and Z_i are the threads that belong to the applications shown in Table 3 in the same order for each mix.

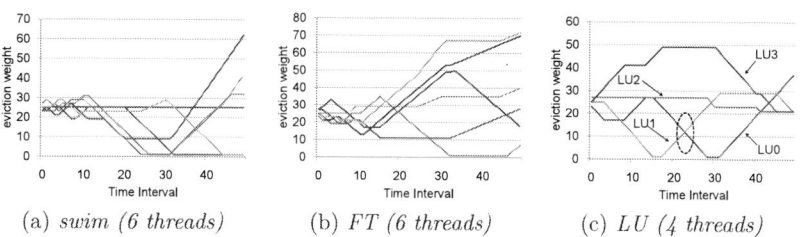

(a) *swim (6 threads)*   (b) *FT (6 threads)*   (c) *LU (4 threads)*

Figure 5: Dynamic adjustment of the eviction weights of the Mix-2's threads using Map-2 (each line corresponds to a thread).

Figure 6: The normalized measured IPCs of *LU*'s threads during execution.

prioritization scheme is able to differentiate between threads of the same application (ranking threads). Modifying insertion policies in the shared cache has also been explored as a mechanism to induce service differentiation [17, 9, 23].

Muralidhara et al [16] proposed an intra-application cache partitioning strategy that partitions the last-level shared cache among multiple threads of single multithreaded applications. This is the only prior work to our knowledge that specifically deals with cache partitioning for multithreaded applications. However, it is not directly applicable to *multiple* multithreaded applications as it does not distinguish between threads of different applications. Due to inherent differences in threads of different applications, it would not be possible to obtain fairness among them.

5. CONCLUSIONS

We introduced a novel way of managing shared caches for obtaining the best performance characteristics from multi-programmed multithreaded applications. Our proposed scheme (CCS) prioritizes the slowest thread in each multithreaded application and improves performance of each multithreaded application. We also ensure that overall system performance is improved by managing the eviction priorities of different application threads in the shared cache. We evaluated CCS on a variety of multiprogrammed multithreaded workloads on a range of systems with 8-16 cores and 2-8 L2 caches with different thread-to-core mappings for each workload. Averaged over six workloads, our shared cache management scheme improves the performance of the combination of applications by 18%.

6. REFERENCES

[1] M. Bhadauria and S. A. McKee. An approach to resource-aware co-scheduling for CMPs. In *ICS'10*.

[2] J. Chang and G. S. Sohi. Cooperative cache partitioning for chip multiprocessors. In *ICS'07*.

[3] M. Chaudhuri. Pseudo-LIFO: the foundation of a new family of replacement policies for last-level caches. In *MICRO'09*.

[4] E. Ebrahimi, et al. Fairness via source throttling: a configurable and high-performance fairness substrate for multi-core memory systems. In *ASPLOS'10*.

[5] F. Guo, et al. From chaos to QoS: case studies in CMP resource management. *SIGARCH Comput. Archit. News*, 35(1):21–30, 2007.

[6] L. R. Hsu, et al. Communist, utilitarian, and capitalist cache policies on CMPs: caches as a shared resource. In *PACT'06*.

[7] R. Iyer. CQoS: a framework for enabling QoS in shared caches of CMP platforms. In *ICS'04*.

[8] R. Iyer, et al. QoS policies and architecture for cache/memory in CMP platforms. *SIGMETRICS Perform. Eval. Rev.*, 35(1):25–36, 2007.

[9] A. Jaleel, et al. Adaptive insertion policies for managing shared caches. In *PACT'08*.

[10] J. A. Kahle, et al. Introduction to the CELL multiprocessor. *IBM J. Res. Dev.*, 49(4/5):589–604, 2005.

[11] M. Kandemir, et al. A helper thread based dynamic cache partitioning scheme for multithreaded applications. In *DAC'11*.

[12] P. Kongetira, et al. Niagara: A 32-way multithreaded sparc processor. *IEEE Micro*, 25(2):21–29, 2005.

[13] P. S. Magnusson, et al. SIMICS: A full system simulation platform. *Computer*, 35(2):50–58, 2002.

[14] R. Manikantan, et al. NUcache: An efficient multicore cache organization based on next-use distance. In *HPCA*, 2011.

[15] C. McNairy and R. Bhatia. Montecito: A dual-core, dual-thread Itanium processor. *IEEE Micro*, 25(2):10–20, 2005.

[16] S. P. Muralidhara, M. Kandemir, and P. Raghavan. Intra-application shared cache partitioning for multithreaded applications. In *PPoPP'10*.

[17] M. K. Qureshi, et al. Adaptive insertion policies for high performance caching. In *ISCA'07*.

[18] M. K. Qureshi and Y. N. Patt. Utility-based cache partitioning: A low-overhead, high-performance, runtime mechanism to partition shared caches. In *MICRO'06*.

[19] N. Rafique, et al. Architectural support for operating system-driven CMP cache management. In *PACT'06*.

[20] D. Sanchez, et al. Flexible architectural support for fine-grain scheduling. In *ASPLOS'10*.

[21] S. Srikantaiah, et al. SHARP control: controlled shared cache management in chip multiprocessors. In *MICRO'09*.

[22] G. E. Suh, et al. Dynamic partitioning of shared cache memory. *J. Supercomput.*, 28(1):7–26, 2004.

[23] Y. Xie and G. H. Loh. Pipp: promotion/insertion pseudo-partitioning of multi-core shared caches. In *ISCA'09*.

[24] S. Zhuravlev, et al. Addressing shared resource contention in multicore processors via scheduling. In *ASPLOS'10*.

SUPPLEMENTARY MATERIAL

S1. MOTIVATIONAL EXAMPLE

We go over an experiment to illustrate how sharing in the last level cache space can lead to performance variations across the threads that belong to the same multithreaded application. This experiment also demonstrates how the overall system performance can be improved by exploiting these variations. First, consider an eight-core system with each core having a private L1 cache assuming that no L2 cache space is shared among the cores (there is no L2 cache in the system). We ran three multithreaded applications, *mgrid* (from the SPECOMP benchmark suite), *lu* and *cg* (from the NAS benchmark suite) using four, two and two threads, respectively. Each thread is pinned to a core (our simulation configuration is described later in Section 3) and all threads are running simultaneously. Figure 7 plots the normalized IPC[3] values of different threads of the running applications. As can be observed, variations among the IPCs of the threads that belong to the same application are very low (below 2%). In other words, in a fixed interval of time, the number of instructions executed by the different threads of the same application are similar. This implies that the threads that belong to the same application perform similar amounts of "work" during the course of execution.

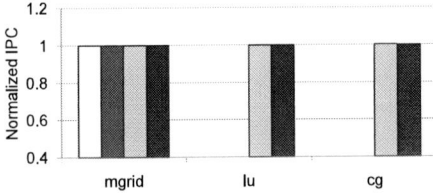

Figure 7: The measured IPCs for three different applications. For each application, bars correspond to different threads. The IPC values are normalized to the maximum IPC achieved by any thread that belongs to the same application.

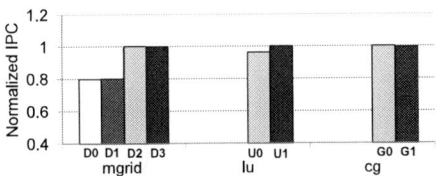

Figure 8: The measured IPCs of the threads of three co-runner applications on an 8 core multicore. D0, D1, D2 and D3 are *mgrid*'s threads; U0 and U1 are *lu*'s threads and G0 and G1 are *cg*'s threads. Each bar is normalized to the maximum IPC achieved by any thread of the same application.

Figure 8 plots the measured IPCs of the threads that belong to our three co-runner applications. The thread mapping and the multicore cache configuration are assumed to be similar to Figure 1(c) (i.e, D0, D1, U0 and U1 sharing a common L2 cache and D2, D3, G0 and G1 sharing another L2 cache). As can be observed from Figure 8, the paces of the four threads of *mgrid* (D0-D3) are not all the same. This is because, in this case, U0 and U1 are more

---

[3]We use IPC as our measurement metric in estimating and comparing the execution speed of the threads that belong to the same multithreaded application.

---

memory intensive than G0 and G1, and consequently, D2 and D3 can take better advantage of the shared L2 space. Since the performance and the execution speed of a data-parallel multithreaded application is mainly dictated by its slowest thread, the fact that D2 and D3 are very fast is not particularly beneficial to this application as far as its overall performance is concerned. To summarize, while different threads of a data-parallel application exhibit similar behavior when executing in isolation, they can exhibit different behaviors in an environment where they share a common cache space with threads of different applications; that is, *different threads of a multithreaded application can experience different magnitudes of destructive interferences from their co-runners.*

There are at least two ways in which this situation can be improved: (i) The performance of *cg* can be enhanced by giving lower priorities to D2 and D3 (as long as either D2 or D3 is not the slowest thread in the application) and correspondingly higher priorities to G0 and G1 in using the shared L2 cache. (ii) The performance of *mgrid* can be enhanced by assigning lower priorities to U0 and U1 (as long as the contribution of resultant improved performance of *mgrid* supersedes the degradation in *lu*'s performance) and correspondingly higher priorities to D0 and D1 in using the shared L2 cache. In general however, in a large multicore with many concurrently-running multithreaded applications, finding the right solution may not be at all easy.

S2. EXTENDED EXPERIMENTS

Experimental configurations. We performed experiments with different configurations listed in Table 5 and shown in Figure 9. As can be seen, these configurations differ from one another in terms of the number of cores, number of caches, number of application threads, or the value of a specific experimental parameter. We already presented the results for Experiments-1.

Experiment-2. In this experiment, we vary the number of cores attached to each L2 cache in a 16-core system (while keeping the per core cache capacity the same as in Experiment-1). Figure 9(b) illustrates a multicore with 16 cores in which every two cores sharing an L2 cache space in a socket (1MB, 16-way set associative). To employ CCS, eight L2 Managers are associated with the L2 caches (see Figure 2).

Figures 10(a) and 11(a) plot the average speedup and average fair speedup values achieved for our application mixes using the mappings given in Table 4 (as before, for a given mix, each bar corresponds to a different mapping). In this experiment, CCS has more flexibility, since the probability with which the bottleneck threads exercise the same L2 cache is reduced. Consequently, the speedups achieved in this experiment are generally higher than those reported in Experiment-1.

Figures 10(b) and 11(b) give, respectively, the average speedup and average fair speedup values, when running our mixes on a multicore with 16 cores and two shared L2 caches (as shown in Figure 9(c)). Note that, as the number of cores sharing an L2 cache increases, the performance improvement achieved by CCS reduces due to the increased conflicts across the threads of different applications. In other words, over the course of execution, the likelihood that the bottleneck threads from different applications share a common cache increases, which, in turn, affects the effectiveness of CCS.

Experiment-3. As noted earlier, in our baseline configura-

Experiment Description	Core Count	L2 Count	Number of Cores per L2	L2 Cache Capacity per Core	Number of Applications	Thread Counts for Applications
Experiment-1 (Baseline experiment)	16	4	4	512KB	3	$< 6, 6, 4 >$
Experiment-2 (L2 count sensitivity analysis)	16	2,8	8,2	512KB	3	$< 6, 6, 4 >$
Experiment-3 (Cache capacity sensitivity analysis)	16	4	4	1MB,256KB	3	$< 6, 6, 4 >$
Experiment-4 (Core count sensitivity analysis)	8	2	4	512KB	3	$< 4, 2, 2 >$
Experiment-5 (Application count sensitivity analysis)	16	4	4	512KB	2,4	$<8,8>,<4,4,4,4>$
Experiment-6 (Control enforcement interval sensitivity)	16	4	4	512KB	3	$< 6, 6, 4 >$
Experiment-7 (δ sensitivity)	16	4	4	512KB	3	$< 6, 6, 4 >$

Table 5: Our experimental configurations.

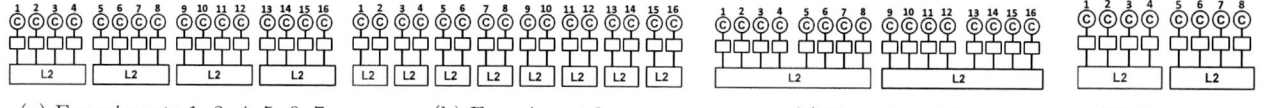

(a) Experiments-1, 3, 4, 5, 6, 7   (b) Experiment-2   (c) Experiment-2   (d) Experiment-4

Figure 9: Different multicore configurations tested in our experiments.

tion, the size of the L2 caches is 512KB per core (i.e, if four cores are attached to an L2 cache, its size is set to be 2MB). In this experiment, we vary the L2 cache capacity per core to 1MB and 256KB. Figures 12 and 13 give the average speedup and fair speedup values for these cases. As can be observed, CCS improves the overall performance significantly in several cases. Note that the results reported in these figures are comparable as CCS is able to enhance the performance of the workloads executed on the multicores with different cache capacities. When we run our workloads on a multicore with large shared caches, if the cache access contention among the co-runner threads is low, we may achieve better performance improvement by reducing the cache capacity (as in the cases of Mix-3 when using Map-2 and Map-4), since the running applications take better advantage of our shared cache management mechanism.

Experiment-4. We also evaluated CCS in a multicore with smaller number of cores (8 cores) sharing two L2 caches in two sockets (see Figure 9(d)). Figures 14(a) and 14(b) plot the average speedups and average fair speedups achieved for our mixes under three different mappings. These mappings are shown in Table 6. We see that, in Mix-1, the performance improvement achieved with Map-3 is about 10%, whereas we cannot achieve any improvement when Map-1 or Map-2 is used. This is because X_0 ($mgrid_0$) is the bottleneck thread of $mgrid$ and is not compatible with X_2 ($mgrid_2$) and Y_1 (bt_1) as far as sharing L2 capacity is concerned under Map-3; our proposed scheme has the opportunity to speedup X_0 ($mgrid_0$) and subsequently improve overall performance. The overall system performance is improved significantly in Mix-2 when Map-2 is used since, in this case, $Z_1(lu_1)$ is the bottleneck thread of lu, and when it shares the L2 cache with X_2 or X_3 ($swim_2$ or $swim_3$), its performance can be improved without causing any performance loss for $swim$ (note that X_0 and X_1 ($swim_0$ and $swim_1$) are the bottleneck threads for $swim$).

Experiment-5. We next varied the number of co-runner applications. Figures 15(a) and 15(b) show the achieved average speedups when two ($swim$ and mg, each with 8 threads) and four applications ($swim$, mg, sp and $mgrid$, each with four threads) run on a machine with 16 cores and two L2 caches using six different mappings. As can be observed, the overall performance is improved by about 7% on average. One can observe that by adding or removing one application in Mix-3, a lower performance improvement is achieved (compare the speedups presented in these figures with those reported in Figure 4(a) for Mix-3). This is because, the

magnitude of the performance improvement, as mentioned earlier, depends on the characteristics and the mapping of the bottleneck threads of the co-runner applications and not the number of them.

Map-1	$< (X_0, X_1, X_2, X_3), (Y_0, Y_1, Z_0, Z_1) >$
Map-2	$< (X_0, X_1, Y_0, Y_1), (X_2, X_3, Z_0, Z_1) >$
Map-3	$< (X_0, X_1, Y_0, Z_0), (X_2, X_3, Y_1, Z_1) >$

Table 6: Three mappings used in the experiments with an 8-core system.

Experiment-6. One of the parameters that affects the performance improvements achieved by CCS is the frequency at which the control decisions are made and enforced. In the previous experiments, this frequency, called the *enforcement interval*, is set to be 200 million cycles. If CCS is invoked more frequently, the behaviors of the running threads are monitored at higher frequency and consequently it may result in achieving better performance improvement. However, increasing the enforcement frequency means more overhead. Figures 16 and 17 plot the average and fair speedup values achieved by reducing and increasing the length of the enforcement intervals (100 million and 400 million cycles), respectively, while other parameters are similar to Experiment-1. As can be observed, the speedups achieved in Figure 16 (the experiment with smaller intervals) are generally higher than those reported in Figure 17.

Experiment-7. As explained before, the δ value (used in Equation 2) is the granularity at which the weights associated with the cores are varied in CCS. The default value of this parameter used so far in our experimental evaluation was 2 (as stated in Table 2), which corresponds to 10% of the core weights' initial values. To study the impact of this value on the CCS performance, we performed an experiment with four different δ values (1, 2, 3 and 4). Figures 18(a) and 18(b) plot the average and fair speedup values achieved for Mix-5 with our six mappings and using different δ values, while other experimental parameters are the same as Experiment-1. As one can observe, in some cases (e.g., Map-5 and 6), further performance improvements are achieved when the δ value is increased from 2 to 3. This is because, the core weights can converge to the optimal values faster in these cases through assigning a larger value to δ. However, increasing the δ value does not always result in better performance improvement (e.g, Map-5, $\delta = 3$ and 4), since the core weights may oscillate around the optimal values and not reach the optimal ones by having a larger δ.

Incorporating QoS Constraints. In a multicore system,

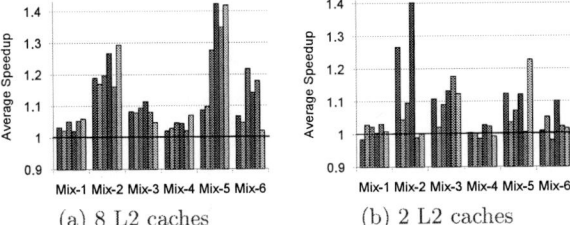

(a) 8 L2 caches	(b) 2 L2 caches

Figure 10: The average speedup values achieved for L2 count sensitivity analysis (Experiment-2).

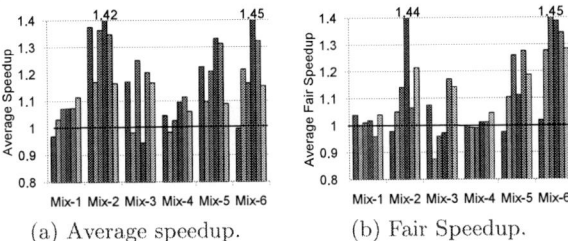

(a) Average speedup.	(b) Fair Speedup.

Figure 12: Cache capacity sensitivity analysis (Experiment-3, 1MB/core).

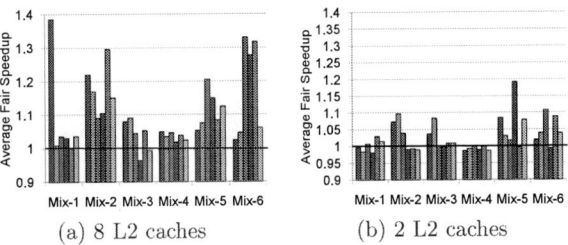

(a) 8 L2 caches	(b) 2 L2 caches

Figure 11: The fair speedup values achieved for L2 count sensitivity analysis (Experiment-2).

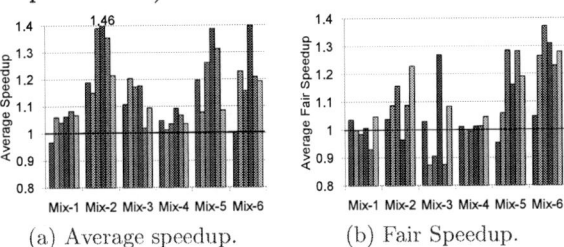

(a) Average speedup.	(b) Fair Speedup.

Figure 13: Cache capacity sensitivity analysis (Experiment-3, 256KB/core).

co-runner applications may have different *quality-of-service* (QoS) requirements. Our proposed CCS can be used in such scenarios as well. Specifically, to accommodate different QoS requirements of applications, we can tune the rank assignment process in a QoS-aware fashion. In this QoS-aware rank assignment, performance targets of different co-runner applications are compared against the current performance, and the difference between them is used during the rank assignment phase to satisfy all specified QoS targets. In QoS-aware CCS, we assume that each application A^j may have a QoS target t_j, which represents the maximum percentage performance degradation allowed with respect to an isolated execution (i.e., when the application is run alone). We can set a reference IPC value for each application based on t_j. Algorithm 4 gives the modified version of the Arbiter algorithm that supports QoS guarantees. As can be observed in line 4, the rank of the threads whose current performances are lower than the specified goals are set to -1, and as a result, the L2 Managers would select these threads for eviction weight reduction.

Algorithm 4 QoS-Aware Arbiter

Input: The measured IPCs $IPC[1 \cdots n]$
Input: Thread mapping $mapping[\ ]$
Input: QoS values $QoS[\ ]$
Output: Ranks of the cores $R[1 \cdots n]$
1: **for** $i = 1$ to n **do**
2:    **if** $(IPC[i] < QoS[mapping[i]])$ **then**
3:       $R[i] = -1$
4:    **else**
5:       $app = mapping[i]$
6:       $min = \inf$
7:       **for** $j = 1$ to n **do**
8:          **if** $(IPC[i] < min)$ and $(mapping[j]==mapping[i])$ **then**
9:             $min = IPC[i]$
10:          **end if**
11:       **end for**
12:       $R[i] = \frac{IPC[i]-min}{min}$
13:    **end if**
14: **end for**

We performed an experiment to evaluate this QoS-aware version of CCS. Figure 19 plots the normalized IPC values of the threads of different applications in Mix-2 when they are executed on our baseline multicore configuration (Map-5 is used for application mapping). The performance targets are highlighted using solid lines in this figure. As can be seen, all the performance targets are achieved by the QoS-aware CCS, whereas application B's target is not reached for two of the threads when the QoS support is not activated.

S3. IMPLEMENTATION ISSUES AND OVER-HEADS

CCS is implemented by the OS with hardware support. At each interval, the OS runs our algorithm based on the threads' mappings and the IPC values read from the hardware counters. After that, the eviction weights of the cores are determined (as explained above) and sent to the modified LRU algorithm implemented in hardware. If a cache miss occurs, the modified LRU algorithm first determines the victim core, and evicts the least recently used L2 cache block from the blocks that belong to that victim core. For an m-core machine, a $\lceil lg_2(m) \rceil$-bit tag is associated with each cache block indicating the core it belongs to.

The OS runs CCS implemented in software (both Arbiter and L2 Managers) to compute the weight of each core based on the current IPC values. The CCS algorithm for a multi-core with 16 cores takes about 6K cycles to run, which, on a 2.1GHz machine, represents a negligible overhead. Note that the value of the weight of each core is not increased beyond the saturation points. However, the weights of the cores attached to the same L2 cache can be scaled down by the minimum weight value to avoid saturation. It is also to be noted that this overhead is not incurred once in every replacement, but only once in every enforcement interval (e.g., every 200 million cycles in our experiment), when the eviction priorities of cores are recomputed.

In CCS, a pseudo-LRU tree is maintained for each core

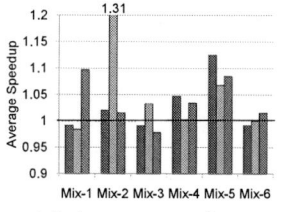
(a) Average speedup.

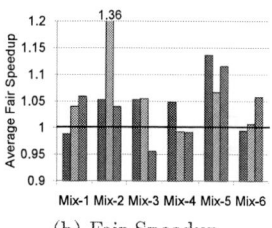
(b) Fair Speedup

Figure 14: The average and fair speedup values achieved for core count sensitivity analysis (Experiment-4).

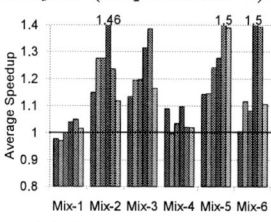
(a) Average speedup.

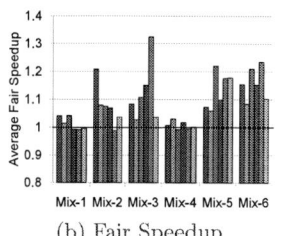
(b) Fair Speedup

Figure 16: The average and fair speedup values achieved for enforcement interval sensitivity analysis (Experiment-6). The control intervals are 100 million cycles.

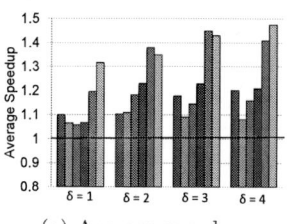
(a) Average speedup

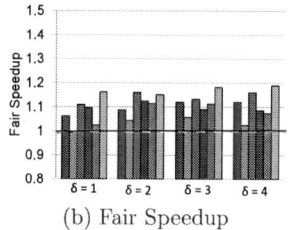
(b) Fair Speedup

Figure 18: The average and fair speedup values achieved with δ sensitivity analysis (Experiment-7) (Mix-5 with six different mappings and four δ values).

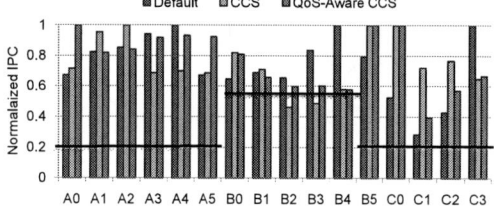

Figure 19: Achieving QoS targets. The solid lines indicate the specified QoS targets for applications.

instead of a global one. Firstly, the victim core is selected based on the cores' weights, and then the LRU cache block is determined based on the victim core's pseudo-LRU tree. Therefore, choosing the victim core (as given in Algorithm 1) is the only overhead imposed by CCS in the LRU algorithm.

S4. DISCUSSION

One can broadly divide multithreaded applications into these classes:

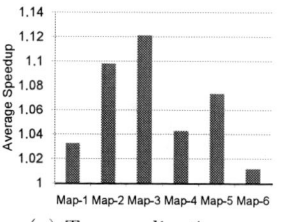
(a) Two applications.

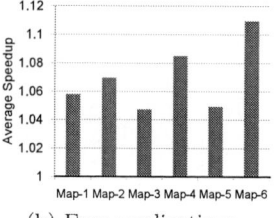
(b) Four applications.

Figure 15: The average speedup values achieved for application count sensitivity analysis (Experiment-5).

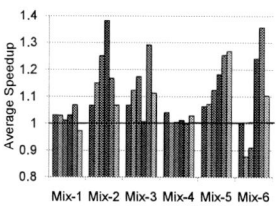
(a) Two applications.

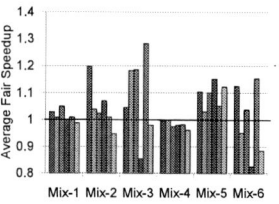
(b) Four applications.

Figure 17: The average and fair speedup values achieved for enforcement interval sensitivity analysis (Experiment-6). The control intervals are 400 million cycles.

Data parallelism. In our work, we focus on OPENMP based, data-parallel applications where the threads that belong to the same application perform very similar tasks on different sets of data. In these types of applications, the thread with the lowest IPC will be the bottleneck and other threads will have to wait for this thread to join at synchronization barriers. CCS balances the execution speeds of the co-runner threads based on the measured IPC values.

Pipeline parallelism. In applications with this parallelization model, different threads execute different functions and work as a pipeline. In this pipeline architecture, thread pools are dedicated to each parallelized stage. CCS can be employed in these applications through balancing the threads dedicated to the pipeline stages. Further, for the threads that belong to different stages, a synergy of our scheme can be established with the load balancing scheme in place, so that our scheme can be informed about the expected load on each thread. Once the relative amount of work that each thread has to do is known, our scheme can determine the existing slack and work towards the same objective.

Master-slave model. In this category of parallel multithreaded applications, typically one of the threads works as the master thread and controls the data distribution among the other co-runner threads. Consequently, the measured performance (IPC) of the master thread differs from the others and needs to be excluded from the CCS algorithm.

This paper focuses on evaluating our approach in data parallel applications. To use our approach in pipeline and master-save parallelism, some enhancements are needed. Specifically, in the pipeline parallelism based applications the threads that belong to the same pipeline stage need to be detected. For the master-slave model, master threads should be identified by the OS and not included in CCS.

A Hybrid Approach to Cyber-Physical Systems Verification

Pratyush Kumar, Dip Goswami[1], Samarjit Chakraborty[1],
Anuradha Annaswamy[2], Kai Lampka[3], Lothar Thiele

Computer Engineering and Networks Laboratory, ETH Zurich
[1]Institute of Real-Time Computer Systems, TU Munich
[2]Department of Mechanical Engineering, MIT
[3]Department of Information Technology, Uppsala University

ABSTRACT

We propose a performance verification technique for cyber-physical systems that consist of multiple control loops implemented on a distributed architecture. The architectures we consider are fairly generic and arise in domains such as automotive and industrial automation; they are multiple processors or electronic control units (ECUs) communicating over buses like FlexRay and CAN. Current practice involves analyzing the architecture to estimate *worst-case* end-to-end message delays and using these delays to design the control applications. This involves a significant amount of pessimism since the worst-case delays often occur very rarely. We show how to combine functional analysis techniques with model checking in order to derive a *delay-frequency interface* that quantifies the interleavings between messages with worst-case delays and those with smaller delays. In other words, we bound the frequency with which control messages might suffer the worst-case delay. We show that such a *delay-frequency interface* enables us to verify much tigher control performance properties compared to what would be possible with only worst-case delay bounds.

Categories and Subject Descriptors

C.3 [**Special-Purpose and Application-Based Systems**]: [Real-time and embedded systems]; D.2.4 [**Software/Program Verification**]: [Formal methods]; D.4.5 [**Reliability**]: [Fault tolerance]

General Terms

Design, Theory, Verification

Keywords

Cyber-Physical Systems, Frequency-Delay Metric, Stability, Real-Time Calculus, Timed-Automata

1. INTRODUCTION

Cyber-physical systems involve a tight interaction between the cyber (computational) and the physical components of

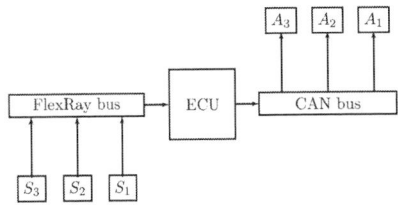

Figure 1: Architecture considered for example

an embedded system. In this paper we study cyber-physical systems in which multiple control applications are implemented in a distributed fashion on a set of electronic control units (ECUs) that communicate over buses like CAN and FlexRay (see Figure 1). Such a setup is fairly generic and can be found in automotive systems, industrial automation systems, large-scale robotic systems and in avionics.

Traditionally, the control theory community has largely ignored the implementation aspects (both software and architecture) of controllers and has focused on mathematical models, their analysis, and high-level simulation. In this process, a number of simplifying assumptions have been made when designing controllers. They include neglecting the computation times when evaluating control laws and neglecting the control message communication times. The implementation of the high-level control laws while taking into account implementation issues that have been ignored while designing these laws, mostly fell within the domain of embedded systems and software design. Several research threads within the embedded systems community have tried to systematically address these issues. They include the development of the synchronous language paradigm [4], time-triggered languages like Giotto [7] and other related formalisms such as PTIDES [16]. In parallel to these efforts, as the implementation platforms became more complex and the *semantic* gap between high-level control models and their implementations widens, the control theory community started investigating what is referred to as *networked control systems* (NCS) [13, 14], where the focus has been on distributed controllers communicating over a wireless network. Research with NCS takes into account issues such as delays suffered by control messages, message loss and jitter, and factors these issues into controller design.

In the case of distributed controller design, the current practice is to estimate worst case end-to-end message delays and design control laws that account for these delays. Within the NCS domain, statistical properties of these delays have also been taken into account while doing controller design. The problem in such cases boil down to analyzing the stability and performance of systems consisting of several switching subsystems, each designed for specific message delay values. Tools used within the control systems community to analyze such switching behavior include Multiple Lyapunov Functions [6], the Average Dwell time [15] approach, or common quadratic Lyapunov functions [9]. In all of these cases, the switching between multiple subsystems is assumed to be arbitrary and the behavior of the system is characterized by the *worst-case* switching pattern.

Our contribution: In this paper we observe that often the worst-case message delays occur very rarely. Further, in contrast to wireless networks, the architectures we study (multiple ECUs communicating over FlexRay or CAN networks and scheduled using deterministic scheduling policies) offer substantial *structure* to rule out certain delay (and hence switching) patterns. In other words, we can bound the occurrences of messages with worst-case delays, or quantify the interleavings between messages with worst-case delays and those with smaller delays. In this paper we refer to this as the *delay-frequency interface* between the underlying architecture and the control applications implemented on this architecture. We show that such *delay-frequency interfaces* may be used to verify much tighter control performance properties in comparison to using the worst-case delay alone (i.e., without the frequency with which the delay occurs).

Exploiting the above observation involves two technical challenges. (i) How to compute the *delay-frequency interface* for a distributed architecture? and (ii) How to use such an interface to give tighter control performance guarantees? Towards (i), we use a hybrid combination of (a) a functional approach to analyzing the timing properties of an architecture with (b) model checking. The functional approach is based on *real-time calculus* [12], where functions over time interval lengths are used to provide upper and lower bounds on the *service* provided by a resource (such as an ECU or a bus) or bounds on the arrival patterns of messages to be processed or transmitted by a resource. These bounds are then analyzed to estimate worst-case delays or backlogs suffered by message streams. For such analysis we have used the Matlab-based `Modular Performance Analysis` toolbox (available from `www.mpa.ethz.ch`). While the worst-case delays computed by this method consider *all* possible service and message arrival patterns that are specified by the upper and lower bounds/functions, we rule out certain architecture-specific service and arrival patterns through model checking (using UPPAAL [3]) and obtain our required *delay-frequency interface*, which is a significantly richer interface than the worst-case delay alone. Towards (ii), we utilize an analysis technique based on the existence of a common quadratic Lyapunov function [9]. As already mentioned, the main design task from control side is to guarantee control stability and performance in the presence of switching. Knowledge of the delay frequency metric essentially restricts the set of possible switching patterns. We address the stability and performance related aspects of the control loops for this restricted set of switching behavior (coming from the knowledge of delay frequency metric) by showing the existence of common Lyapunov functions among various switching subsystems as a proof of stability.

Related Works: Lately, there has been a considerable amount of activity in the area of *control/architecture* co-design. In particular, Alur *et al.* quantified the semantic gap between the control models and their implementations in [2]. In [11], an integrated co-design of control and architecture has been proposed. However, none of these and other efforts

utilized the delay properties of the underlying architecture to bound control performance. Similarly, while stochastic delay distributions have been used in the NCS domain, to the best of our knowledge, this work is the first effort to use a deterministic distribution on control message delay values in order to give tighter control performance guarantees.

The rest of the paper is structured as follows. In Section 2, we discuss a motivating example that highlight the limitation of the existing approaches. In Section 3, we describe the delay-based feedback control strategy and we then introduce the idea of delay frequency metric in light of the proposed control strategy. Section 4 illustrates how a delay frequency metric can be verified for a given architecture. The delay frequency metric is richer timing interface that we argue for. In Section 5, we demonstrate how we can apply the knowledge of the proposed timing interface to tighten the control design considering the example shown in Section 2. Several of the computations and explanations have been included in Appendices A-F.

2. MOTIVATION

In this section, we describe a motivational example of a control application that is implemented over a distributed architecture, which is retained throughout this paper. The fundamental objective of any control application is to regulate the behavior of dynamical systems (or plants) which are more commonly referred to as plant dynamics. The behavior of such dynamical systems (feedback signals) is measured using sensors, and the control algorithm (running on a processing unit) decides the necessary actuation signal which is then realized by the actuator. Hence, the feedback loop is closed over a set of spatially separated sensors and actuators communicating via a bus system and processors in distributed cyber-physical systems. In this paper, we consider an example architecture from the automotive domain. However, the analysis presented is relevant to a variety of other settings too.

The considered example is representative of three characteristic properties of today's cyber-physical systems, such as those found in modern cars or in industrial automation applications. First, it depicts how the integrated nature of the system brings together several applications, which have to be analyzed simultaneously, even though the analysis challenges may be modular. Second, it highlights the distributed nature of the system, whereby the control feedback loop goes through several components: sensors, buses, ECUs and finally actuators. Third, it characterizes the heterogeneity of the system architecture, wherein several components such as specialized buses are used.

2.1 System architecture

As a representative system, we consider the example shown in Figure 1. It consists of three sensors S_1, S_2 and S_3 that are connected to three Electronic Control Units (ECUs) (not explicitly shown in the figure). The sensor readings from these ECUs are transmitted via a FlexRay bus to the ECU shown in Figure 1. Subsequently, the sensor readings are computed upon in the ECU and the generated output is transmitted over a Controller Area Network (CAN) bus to actuators A_1, A_2 and A_3 (which are connected to three other ECUs). Each sensor/actuator (S_i, A_i) pair forms a control loop which regulates the behavior of an independent plant P_i. Thus, three sensor/actuator pairs regulate three independent plants denoted as P_1, P_2 and P_3. In this paper, we study various aspects of stability and performance of the plant P_2.

The timing properties of the above system are character-

Plant	τ^{FR}	τ^{ECU}	τ^{CAN}
1	40	50	30
2	20	15	10
3	20	15	20

Table 1: Worst-case execution times of different tasks (in ms)

ized by three aspects: (a) how often the sensors are read, (b) what is the time required to process the feedback in the different components, and (c) what are the scheduling parameters in the different components. We now describe these three aspects.

We categorize the tasks in the above architecture into three classes. First, the task τ_i^{FR} responsible for sending the sensor reading from S_i via the FlexRay bus. Second, the task τ_i^{ECU} responsible for processing the sensor signal sent by τ_i^{FR}, computing the control signal and sending the control signal via the CAN bus. Third, the task τ_i^{CAN} is responsible to actuate A_i based on the control signal sent by τ_i^{ECU}. The worst-case execution times of these tasks on their respective components are shown in Table 1.

Essentially, it is assumed that the sensor S_i and the actuator A_i are attached to plant P_i. Based on the reading from S_i and the actuation from A_i, the plant P_i is regulated. We assume that P_2 and P_3 are time-triggered with period 100ms. That is, τ_i^{FR}, τ_i^{ECU} and τ_i^{CAN} are triggered periodically with periods 100ms for $i = \{2, 3\}$. For P_1, τ_1^{FR}, τ_1^{ECU} and τ_1^{CAN} are triggered with a period of 100ms and maximum jitter of 50ms.

The FlexRay bus is used by τ_i^{FR} to send the sensor reading to the ECU. The FlexRay bus is divided into communication cycles which consists of the time-triggered static and the event-triggered dynamic segments. The static segment is divided into multiple *slots* of equal length. On the other hand, the bus access in the dynamic segment is arbitrated in a fixed-priority fashion. In this example, we consider the FlexRay bus with static and dynamic segments of length 40 ms each, and a period or cycle length of 80ms. The task τ_1^{FR} transmits message over the static segment. The messages from τ_2^{FR} and τ_3^{FR} are transmitted over the dynamic segment with increasing order of priority. On the ECU, a hierarchical scheduler is used. τ_1^{ECU} is scheduled as a high-priority task. The other two tasks are scheduled within a low-priority task which executes two constant bandwidth servers [1] with utilization 50% each, to serve τ_2^{ECU} and τ_3^{ECU}. Finally, the CAN bus schedules the messages with priorities in decreasing order, with messages from τ_1^{CAN} and τ_3^{CAN} having highest and lowest priorities respectively.

2.2 Control applications

In the context of the architecture described in the previous section, we focus on the feedback loop for plant P_2. As a plant dynamics P_2, we consider a common discrete-time linear time-invariant (LTI) system of the form,

$$x[k+1] = Ax[k] + Bu[k] \qquad (1)$$

$$A = \begin{bmatrix} 0.0 & 1 \\ 0.9 & 0.2 \end{bmatrix}, B = \begin{bmatrix} 0 \\ 1 \end{bmatrix}. \qquad (2)$$

where $x[k]$ is the vector of *state variables* and $u[k]$ is the *control input* to the system which needs to be designed. We consider a *regulation* problem which essentially means the design of $u[k]$ such that $x[k] \to 0$ from any initial condition with k. In this work, we consider static state-feedback control strategy for designing $u[k]$. In view of the above system,

we make the following observations:

- The sensor S_2 reads all the states $x[k]$, i.e., the states are *measurable*. The measurement is done periodically with constant period of $p = 100$ms. Similarly, the actuation is periodical with $p = 100$ms using A_2.

- The time when S_2 reads feedback signal $x[k]$ to the point when A_2 is actuated utilizing that signal is called the *sensor-to-actuator* delay d. The delay d consists of the transmission/waiting time of the sensor signal, processing time of the sensor signal and the transmission/waiting time of the actuator signal. Clearly, d is not constant across jobs.

- With $u[k] = 0$, the resulting plant is open-loop and the behavior of the open-loop plant is dictated by properties of the system matrix A. In this case, the absolute value of the maximum eigenvalue of A is outside the unit circle, i.e., $|\lambda_{max}(A)| > 1$. That is, the open-loop plant is unstable.

- In the presence of a *sensor-to-actuator* delay d, the task is to design static state-feedback $u[k] = Kx[k - \lceil \frac{d}{p} \rceil]$) such that $x[k] \to 0$ with $k \to \infty$ where K is the state-feedback gain.

- With above control input, we denote the closed-loop system matrix by A_{cl}. We measure the quality of control using the stability margin, i.e., $Q = (1 - |\lambda_{max}(A_{cl})|)$. A larger Q indicates a higher stability margin.

As a design criterion, we are required to guarantee a stability margin $Q \geq 0.15$ in our example.

2.3 Design Motivation

Towards designing a regulator for the control loop under consideration, the easiest way is to design the system using the worst-case sensor-to-actuator (or end-to-end) delay. There are various well-known approaches in the real-time systems literature for computing such end-to-end delays. As mentioned before, we have chosen the *Modular Performance Analysis (MPA)* toolbox because it provides several advantages like compositionality. We modelled the timing properties and scheduling parameters within this tool and obtained the worst-case sensor-to-actuator delay $d_{max} = 320$ms. The detailed analysis is presented in Appendix A.

The above analysis implies that with periodic actuation of A_2 with $p = 100$ms, the worst-case sensor-to-actuator delay is $\lceil \frac{d_{max}}{p} \rceil = 4$ samples. Then, the obvious design possibility is to set $u[k] = Kx[k-4]$, where every feedback signal delayed by 4 samples. There are several standard approaches in the control theory literature for designing K with such delayed feedback signals. With static state-feedback control scheme [10], the best stability margin of $Q = 0.1455$ is obtained with $K = [-0.14 \ -0.05]$.

The computed stability margin does not meet the design criteria ($Q \geq 0.15$), and hence a design revision is deemed necessary. Such a design revision may need an upgrade of the network or the processing architecture or the removal or redesign of some of the other control applications in the system. However, it may be possible that the existing system itself can provide the required higher level of performance, which is not exposed by our analysis technique. In other words, the traditional approach to utilize worst-case end-to-end delay as an *interface* can lead to an *analysis gap*. Indeed, the existence of such a gap can be easily argued for. Not every feedback sample will suffer the worst-case sensor-to-actuator delay d_{max}. A large number of feedback messages would suffer a delay no more than some smaller threshold delay, say d_{th}. This smaller delay can enable the guarantee for a better stability margin. This is the motivation for our work, i.e., deriving and demonstrating a richer interface between of the cyber and physical aspects of the system to enable tighter analysis.

To this end, we need to have (a) a controller design that exploits a guaranteed interleaving between messages with $d > d_{th}$ and those with $d < d_{th}$, (b) a metric that formally captures the delay variation, and (c) a computation technique to verify that this metric is indeed satisfied for the considered architecture. In the remainder of this paper, we will discuss these issues and indicate how they can be used to tighten the results for the stability margin of P_2.

3. DELAY-BASED FEEDBACK CONTROL

In this section, we will propose a controller feedback technique that will enable us to exploit the time-varying sensor-to-actuator delays to compute tighter stability margins. In addition, this will help us to define a richer interface between the cyber and the physical aspects of the system.

As discussed, not all feedback messages will suffer the largest delay d_{max}. There can exist a threshold delay, say d_{th}, such that *most* of the feedback messages suffer a delay $d \leq d_{th}$. With this observation, we propose the following control strategy: (a) we design a stabilizing controller with $u[k] = Kx[k - \left\lceil \frac{d_{th}}{p} \right\rceil]$, (b) actuator A_2 applies the control input $u[k] = Kx[k - \left\lceil \frac{d_{th}}{p} \right\rceil]$ only when delay is d_{th} or less and we denote the resulting closed-loop system by A_{cl}, and (c) whenever sensor-to-actuator delay $d > d_{th}$, we do not use that feedback signal and let the system run in open-loop, i.e., $u[k] = 0$ and the resulting system matrix is A. The control scheme is thus given as

$$u[k] = Kx[k - \left\lceil \frac{d_{th}}{p} \right\rceil], \qquad \forall\, d \leq d_{th}$$
$$= 0, \qquad \forall\, d > d_{th} \qquad (3)$$

The above control scheme leads to a switched system with two subsystems: (b) when feedback message suffers $d \leq d_{th}$ and the corresponding system is represented by A_{cl}, (a) when feedback message suffers a delay $d_{th} < d \leq d_{max}$ and the corresponding system is represented by A. For brevity of notation, we refer feedback messages of the first kind as "valid" samples while the messages of the second kind are called "invalid" samples.

There are standard techniques in the control theory literature to analyze the stability of the switched systems. However, most of these approaches assume worst-case or arbitrary switching behaviors. In our case, using our improved real-time systems analysis, we can rule out certain switching patterns. In other words, we can identify a restricted subset of switching possibilities, which can be finite for many practical design purposes. The switched system with such known switching possibilities are then analyzed using standard tools such as Average Dwell Time (ADT) [15] or Common Quadratic Lyapunov Function (CQLF) [9]. For this purpose, we propose the following metric as a richer class of interface.

DEFINITION 1 (DELAY FREQUENCY METRIC (d_{th}, n)). *If every feedback message with delay larger than d_{th} is followed by at least n feedback messages with delay no more than d_{th}, the* delay frequency metric *is said to be* (d_{th}, n).

Using the notion of valid and invalid samples, the above

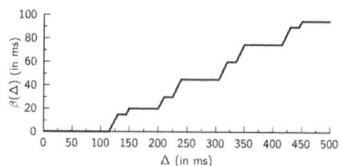

Figure 2: Service curve provided by ECU to τ_2^{ECU}

definition says that, every invalid sample is followed by at least n valid samples.

In the above context, the paper has two contributions – (a) From the real-time systems perspective, we *quantify* the interleaving between the number of messages with delay $d \leq d_{th}$ and $d_{th} < d \leq d_{max}$. (b) From the control system perspective, we utilize this more expressive *delay frequency metric* to obtain tighter bounds on the control system design. As already mentioned before, in the control systems literature such rich interfaces between the architecture and the controller has not been explored before.

In Section 4, we illustrated how a certain delay frequency metric can be verified for a given architecture. Next, we utilize the knowledge of the delay frequency metric in analyzing control stability and performance for a given architecture. We demonstrated the control theoretic analysis in Section 5 with the control loop P_2 in Figure 1.

4. VERIFYING DELAY-FREQUENCY

sIn the earlier section, we used the MPA toolbox to compute the worst-case sensor-to-actuator delay suffered by the feedback messages of P_2. However, this interface does not tightly interface the cyber and physical aspects of the system. We proposed the *delay frequency metric* along with a delay-based feedback system to allow for a tighter coupling. This section is devoted to computational techniques that can verify a given delay frequency metric for a given architectural setup. To this end, we use a hybrid approach combining analytical techniques of Real-Time Calculus [12] and state-based methods of the model checker UPPAAL [3].

4.1 Service curves

Service curves are used in Network Calculus [5] and Real-Time Calculus [12] to specify bounds on the available resource to a stream of tasks, in the interval domain. More concretely, $\beta(\Delta)$ denotes the *minimum* amount of execution time available to a stream of tasks within *any* busy window of length Δ. Techniques are known to compute the service curve for a wide class of resource arbitration policies. Examples include TDMA, fixed priority, EDF, round robin and servers. Within the scope of this paper, we consider the service curve as a given quantity and do not discuss its computation. As an example, the service curve available to τ_2^{ECU} on the ECU can be shown to be as in Figure 2.

When dealing exclusively with end-to-end timing properties, the service curves of multiple components in series can be *convolved* into a single service curve. For instance, in our example, the feedback message must go through the FlexRay bus, the ECU and the CAN bus. Each of these components is characterized by its own service curve. We can represent the timing properties of these components with a single service curve given as

$$\beta^{\text{end-to-end}} = \beta^{\text{FR}} \otimes \beta^{\text{ECU}} \otimes \beta^{\text{CAN}}, \qquad (4)$$

where $\otimes$ is the convolution operation as defined in Real-Time Calculus [12]. For the scope of this work, we may abstract the timing properties of the system, with respect to feedback messages of P_2, with a single service curve $\tilde{\beta}^{\text{end-to-end}}$.

4.2 Modelling a resource with UPPAAL

As discussed above, a service curve bounds the timing behavior exhibited by a resource. In other words, it characterizes a *set*, say **S**, of possible timed behaviors of the resource. If this set can be translated into the set of possible behaviors of a timed automata, we can model the service curve as a component in a model checking tool such as UPPAAL. Such modelling can enable us to verify richer timing properties on the system such as the delay frequency metric.

A generic service curve represents a very large set of constraints: for every interval of a busy interval, a constraint must be satisfied. Translating such constraints into an UPPAAL component can lead to a complicated automaton, with intractable verification time. We therefore propose a conservative approximation of the service curve, whereby it is represented using a series of *periodic* service curves.

Consider an approximate service curve $\tilde{\beta}$ satisfying

$$\tilde{\beta}(\Delta) \leq \beta(\Delta), \forall \, \Delta \geq 0. \tag{5}$$

Let the set of possible timing properties exhibited by a resource with a service curve $\tilde{\beta}$ be denoted as $\tilde{\mathbf{S}}$. Then it follows that

$$\mathbf{S} \subseteq \tilde{\mathbf{S}} \tag{6}$$

The above relation implies that, if we verify some property to be always true with the service curve $\tilde{\beta}$, then the property is also always true for the service curve β. In this sense, the approximation $\tilde{\beta}$ is said to be conservative. Now consider a specific approximation $\tilde{\beta}$ given as

$$\tilde{\beta} = \max(0, \beta_1(o_1, P_1, Q_1), \beta_2(o_2, P_2, Q_2), \ldots) \tag{7}$$

where $\beta_i(o_i, P_i, Q_i)$ is a periodic service curve defined as

$$\beta_i(\Delta) = \left(\left\lfloor \frac{\Delta - o_i}{P_i} \right\rfloor + 1 \right) \times Q_i \tag{8}$$

As an illustration, the service curve available to τ_2^{ECU} (Figure 2) is approximated using periodic curves in Figure 4) in Appendix B. Periodic service curves can be easily modelled in separate UPPAAL components. These components can then be interfaced together to model a resource with service curve $\tilde{\beta}$ following (7). Such an interfaced component is shown in Appendix C.

4.3 Other components

Along, with the automaton that represents the behavior of the resource, we need an automaton to represent the generation of a stream of tasks (e.g., corresponding to the control task that needs to be executed for each sampled sensor input). Computing automatons to represent generation of a wide variety of streams has been discussed in [8].

Additionally, we need an observer automaton to encode the verification of the delay frequency metric. Let the delay frequency metric that we want to verify be given as (d_{avg}, n). Recall that this means that every invalid sample is followed by at least n valid samples. This can be easily verified using a UPPAAL observer component that (a) tracks the delay suffered by the feedback messages, (b) classifies messages as valid and invalid samples, and (c) verifies the delay frequency metric.

To conclude this section, recall that we were interested in verifying timing properties which are richer than the worst-case delay. To this end, we used the analytical formulation

of service curves to represent the architecture in a modular fashion, used conservative approximation techniques to enable its efficient representation using the model checking tool, and then used an automata for the generation of the stream of tasks and an observer to verify properties such as the delay frequency metric. The observer automaton discussed here is illustrated in Appendix C.

5. THE CASE STUDY: REVISITED

In this section, we will revisit the example discussed in Section 2 (see Figure 1). Equipped with techniques from the previous two sections, we aim to demonstrate that we can tighten the computation of the stability margin for the considered architecture and the control loop P_2. In the following, we illustrate: (a) computation of delay frequency metric for the given architecture, (b) the control performance bound that we obtain without the knowledge of the delay frequency metric, and (c) the control performance bound that we obtain utilizing the knowledge of the delay frequency metric.

(a) Computing the delay frequency metric: With the worst-case delay of $d_{max} = 320$ms , $\left\lceil \frac{d_{max}}{p} \right\rceil = 4$. That is, working with the worst-case delay means setting $d_{th} = 400$ms. As demonstrated in Section 2, the control loop fails to meet design criterion ($Q \geq 0.15$). To improve upon this design, we set d_{th} to 300ms. For this value of d_{th}, we verify the delay frequency metrics for different values of n, using the UPPAAL model checker as described in the previous section. The largest value of n, in our example, for which the delay frequency metric is guaranteed by the considered architecture is 2. In other words, the architecture guarantees a delay frequency metric of (300 ms, 2). Note that this is a richer timing interface than specifying only the worst-case delay.

Coming back to the example of control loop P_2, $\left\lceil \frac{d_{th}}{p} \right\rceil = 3$ ($d_{th} = 300$ms, $p = 100$ms). Hence, we choose $u[k] = Kx[k-3]$ and $K = [0 \ -0.28]$. We apply the control scheme described in Section 3. When $d \leq 3p$, we apply $u[k] = Kx[k-3]$ with $K = [0 \ -0.28]$. The resulting closed-loop system A_{cl} is shown in Appendix D. With $d > 3p$, we run the system in open-loop and the resulting system matrix is A. The stability of this switched system depends on the interleaving between valid and invalid samples. The tighter design of the architecture needs to allow more frequent invalid samples. On the other hand, the occurrence of invalid samples causes degradation in stability margin and can potentially destabilize the system. The frequency of permissible invalid samples can be computed utilizing the traditional analysis tools such as Average Dwell Time (ADT) and Common Quadratic Lyapunov Function (CQLF).

(b) Design without the knowledge of delay frequency metric: In this case, we rely on ADT-based analysis. For ADT-based analysis, we consider switching between two subsystems A_{cl} when the delay is no more than $d_{th} = 300$ms and A when the delay exceeds 300ms. We illustrated the analysis in Appendix E. It clearly shows that only 1 invalid sample is allowed in any window of 25 samples for stability (i.e., for ensuring $Q > 0$). Hence, this does not meet the delay frequency metric computed earlier. Thus, this analysis which does not utilize the knowledge of delay frequency metric fails to guarantee desired stability margin. It should be noted that the CQLF-based analysis is not applicable to the systems with unstable subsystems. Hence, CQLF-based analysis is not suitable in this case as the open-loop system

692

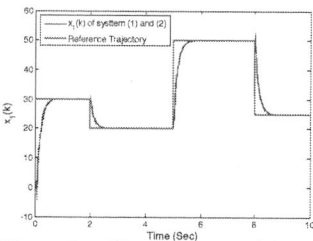

Figure 3: Plot of $x_1(k)$ of system (2) with our proposed control scheme.

A is unstable.

(c) Design exploiting the knowledge of delay frequency metric: For the delay frequency metric with $n = 2$, every invalid sample is followed by at least two valid samples when P_2 is implemented over the architecture under consideration. Hence, the overall system can be represented as follows,

$$x[k + 2 + n_i + n_j] = AA_{cl}^{n_i} \times AA_{cl}^{n_j} x[k], \quad (9)$$

where $n_i \neq n_j$ and $n_{i,j} \geq 2$. Hence, the resulting system has switching subsystems of the form $AA_{cl}^{n_i}$ with $n_i \geq 2$. For such a switched system, the stability could not be shown utilizing ADT-based analysis for the subsystems given by $AA_{cl}^{n_i}$ with $n_i \geq 2$. Now, we reply on Common Quadratic Lyapunov Function (CQLF) approach [9] for the stability analysis. The analysis presented in Appendix F shows that the above switched system is stable. The stability margin that we obtain for the given delay frequency metric ranges from $Q = 0.1812$ (when every two valid samples are followed by one invalid sample) to $Q = 0.2462$ (when no invalid sample occurs). Therefore, we could guarantee to provide a desired stability margin of $Q \geq 0.15$.

Fig. 3 shows the plot of state $x_1(k)$ of the control application (2) of control loop P_2 with the proposed control strategy. In this experiment, we have randomly dropped samples such that one invalid sample is followed by minimum two valid sample, i.e., according to the results coming from the above analysis. We have considered a tracking problem, i.e., ensuring that the state $x_1[k]$ follows the trajectory indicated by the red line. Towards this, we used an additional feedforward component (for changing the reference from *zero* to some other value) in the control input keeping the feedback component as described above. Fig. 3 shows that tracking can be done with the derived control scheme.

Hence, the system does indeed meet the design criterion on the stability margin of P_2. We have thus, highlighted how we can combine the delay-based feedback technique with the delay frequency metric to provide a tighter analysis of the cyber and physical aspects of the system.

6. CONCLUDING REMARKS

In this paper we quantify the interleavings between control messages experiencing different delay values. Further, we use this information to provide tight bounds on control performance. This a significant improvement over the case where controllers are designed only on the basis on worst-case delay values. Our proposed approach involves a computationally expensive model checking procedure and an exhaustive search over a parameter space in order to compute the delay frequency interface. As a part of future work, we will study the scalability of this procedure and make this process more efficient.

Acknowledgement This work is supported by European Community FP7 grant 248776.

7. REFERENCES

[1] L. Abeni and G. C. Buttazzo. Integrating multimedia applications in hard real-time systems. In *IEEE Real-Time Systems Symposium*, pages 4–13, 1998.

[2] R. Alur and G. Weissr. Regular specifications of resource requirements for embedded control software. In *IEEE Real-Time and Embedded Technology and Applications Symposium*, 2008.

[3] J. Bengtsson, K. G. Larsen, F. Larsson, P. Pettersson, and W. Yi. UPPAAL - A tool suite for automatic verification of real-time systems. In *Proceedings of the DIMACS/SYCON Workshop on Hybrid Systems III: Verification and Control*, 1995.

[4] G. Berry and G. Gonthier. The esterel synchronous programming language: Design, semantics, implementation. *Sci. Comput. Program.*, 19(2):87–152, 1992.

[5] J.-Y. L. Boudec and P. Thiran. *Network Calculus: A Theory of Deterministic Queuing Systems for the Internet.* Lecture Notes in Computer Science, Vol 2050. Springer, 2001.

[6] M. S. Branicky. Multiple Lyapunov functions and other analysis tools for switched and hybrid systems. *IEEE Transactions on Automatic Control*, 43(4):475 – 482, 1998.

[7] T. A. Henzinger, B. Horowitz, and C. M. Kirsch. Giotto: a time-triggered language for embedded programming. *Proceedings of the IEEE*, 91(1):84–99, 2003.

[8] K. Lampka, S. Perathoner, and L. Thiele. Analytic real-time analysis and timed automata: a hybrid method for analyzing embedded real-time systems. In *IEEE International conference on Embedded software (EMSOFT)*, 2009.

[9] O. Mason and R. Shorten. On common quadratic Lyapunov functions for stable discrete-time LTI systems. *IMA Journal of Applied Mathematics*, 69(3):271–283, 2002.

[10] W. Rugh. *Linear Systems Theory*. Prentice-Hall, N.J., 1996.

[11] S. Samii, A. Cervin, P. Eles, and Z. Peng. Integrated scheduling and synthesis of control applications on distributed embedded systems. In *Design Automation and Test in Europe (DATE)*, 2009.

[12] L. Thiele, S. Chakraborty, and M. Naedele. Real-time calculus for scheduling hard real-time systems. In *IEEE International Symposium on Circuits and Systems*, 2000.

[13] G. C. Walsh, H. Ye, and L. G. Bushnell. Stability analysis of networked control systemss. *IEEE Trans. on Control System Technology*, 10(3):438–446, 2002.

[14] M. Yongguang, C. Wenying, and L. Guangxiao. Compensation of networked control systems with time-delay and data packet losses. In *Chinese Control and Decision Conference*, 2009.

[15] G. Zhai, B. Ho, K. Yasuda, and A. N. Michel. Qualitative analysis of discrete-time switched systems. In *American Control Conference (ACC)*, 2002.

[16] J. Zou, S. Matic, E. A. Lee, T. H. Feng, and P. Derler. Execution strategies for PTIDES, a programming model for distributed embedded systems. In *IEEE Real-Time and Embedded Technology and Applications Symposium (RTAS)*, 2009.

APPENDIX

A. WORST-CASE DELAY

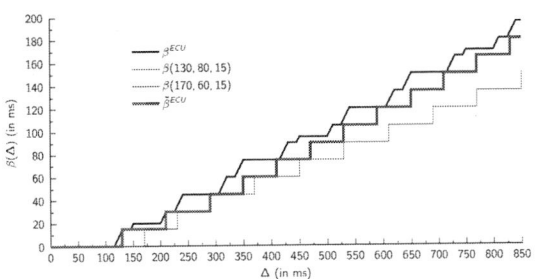

Figure 4: Approximation of a service curve using periodic service curves

Here, we will describe computation of the worst-case sensor-to-actuator delay for the feedback messages of plant P_2 on the architecture shown in Figure 1. We use the MPA toolbox to model the architectural units and the other tasks. Using this we compute the service curves provided by respective units to tasks τ_2^{FR}, τ_2^{ECU} and τ_2^{CAN}. These are shown in Figure 5. Then we use the toolbox to compute the convolution of the service curves. The worst-case end-to-end delay is then given by the horizontal distance between the convolved service curve and the periodic curve of the sensor input. This is also shown in Figure 5. The worst-case delay is obtained to be 320ms.

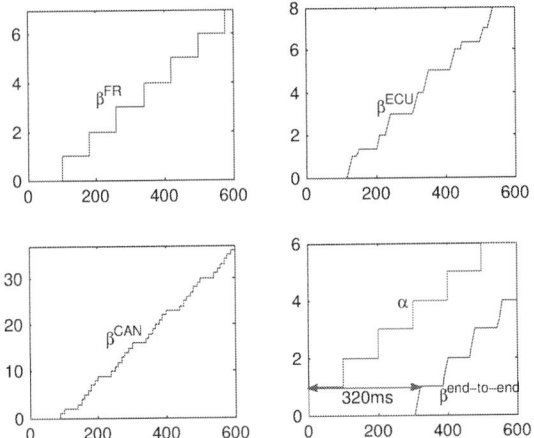

Figure 5: Computation of the worst-case end-to-end delay. Note that the y-axis shows service curve in units of messages, while the x-axis shows time in ms

B. APPROXIMATION OF SERVICE CURVES

We will describe how we can use the approximation step of (7) to conservatively and compactly represent a generic service curve. Consider the service curve of the ECU as provided to task τ_2^{ECU}, as shown in Figure 2. Clearly, the service curve is not a periodic service curve and representing it in a UPPAAL component can be very complicated. We can however, conservatively, represent the service curve by

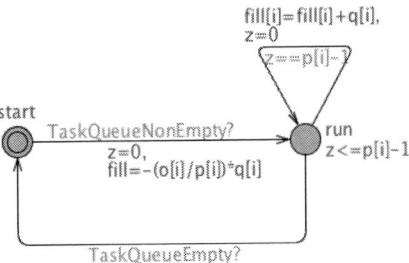

Figure 6: UPPAAL component for the periodic service curve

the following approximate service curve

$$\tilde{\beta}_2^{\text{ECU}} = \max(0, \beta(130, 80, 15), \beta(170, 60, 15)) \qquad (10)$$

where β is defined as in (8). This approximation is graphically illustrated in Figure 4. As can be noted, the approximation closely follows the original service curve while being much more amenable to representation using a model checker.

C. UPPAAL MODELS

In this section, we will illustrate the UPPAAL models of the components used in the verification of delay frequency metric.

C.1 Service curve automaton

Recall that we proposed the approximation of a generic service curve by a set of periodic service curves. The automaton shown in Figure 6 represents the working of a periodic service curve $\beta(o_i, P_i, Q_i)$. It maintains a variable fill_i which logs how the resource can be used. The initial value of fill_i is set to $\dfrac{-o_i}{P_i}Q_i$. The resource *should* serve any pending task if the variable fill_i is positive. If the task is indeed served, the fill_i level is decreased by the amount of execution provided to the task. During busy intervals, the fill level is periodically increased by Q_i every P_i units of time. When the task queue is empty, the fill level is reset to the initial value.

Several such periodic service curves can be easily combined to represent a single service defined as in (5). This is done by implicitly maintaining a single fill given as

$$\text{fill} = \max_i(\text{fill}_i). \qquad (11)$$

The resource serves any pending task if any of the fill values is positive. Executing a task will decrease each fill_i value, independently. This is abstractly depicted in the resource automaton shown in Figure 7. Here the automaton, in the location **check** decides if pending tasks are to be served at all. If any of the fill_i values is positive, the pending task is served for one time unit and the fill_i values are all decremented by 1.

Note two aspects of this system. Firstly, the resource is allowed to execute a task even if the all fill_i values are negative. This demonstrates the non-determinism in the definition of the service curve. Secondly, note that the above automatons actually represent constraints on every interval of time, but has a very efficient and compact UPPAAL component. This enables verification of properties such as the delay frequency metric.

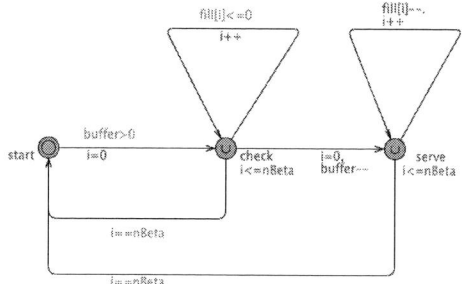

Figure 7: UPPAAL component for the resource obtained as the combination of multiple periodic curves

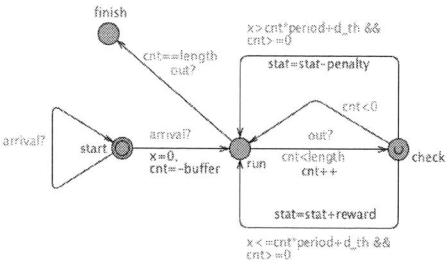

Figure 8: UPPAAL component for the observer

C.2 Observer automaton

The observer automaton is shown in Figure 8. The automaton depends on two synchronization channels `arrival` and `out` which denote the arrival of a new task and the completion of a task, respectively. The observer chooses to observe from some particular task arrival, non-deterministically. It first waits for already buffered tasks to be completed (while `cnt` is negative). Then, it begins to check whether tasks have a delay larger or smaller than d_{th}. For every invalid sample it decrements `stat` by `penalty` and for every valid sample it increments `stat` by `reward`. The variable `stat` is initialized to `penalty`. Then, the query asserting if `stat` is always non-negative, asserts whether the system guarantees a delay frequency metric $(d_{th}, \frac{penalty}{reward})$.

D. COMPUTATION OF A_{CL}

In this section, we show the computation of closed-loop system matrix A_{cl} with $u[k] = Kx[k-3]$ for the control loop P_2 with $\left\lceil \frac{d_{th}}{p} \right\rceil = 3$. We define the new system states,

$$
\begin{aligned}
z_1[k] &= x[k-3], \\
z_2[k] &= x[k-2], \\
z_3[k] &= x[k-1], \\
z_4[k] &= x[k].
\end{aligned} \tag{12}
$$

Hence, we have the following system with $u[k] = Kx[k-3] =$

$Kz_1[k]$,

$$
\begin{aligned}
z_1[k+1] &= z_2[k], \\
z_2[k+1] &= z_3[k], \\
z_3[k+1] &= z_4[k], \\
z_4[k+1] &= Az_4[k] + BKz_1[k].
\end{aligned} \tag{13}
$$

The new system states are $z[k] = [z_1[k] \ z_3[k] \ z_3[k] \ z_4[k]]^T$ and closed-loop system is,

$$
z[k+1] = A_{cl}z[k], \tag{14}
$$

where

$$
A_{cl} = \begin{bmatrix} 0_{2\times2} & I_{2\times2} & 0_{2\times2} & 0_{2\times2} \\ 0_{2\times2} & 0_{2\times2} & I_{2\times2} & 0_{2\times2} \\ 0_{2\times2} & 0_{2\times2} & 0_{2\times2} & I_{2\times2} \\ BK & 0_{2\times2} & 0_{2\times2} & A \end{bmatrix}. \tag{15}
$$

To maintain equal dimensionality for later matrix operations, the open-loop matrix A is computed by putting $K = 0$ in (15).

E. ADT APPROACH

We have two subsystems: A_{cl} and A. A_{cl} is the stabilizing subsystem using control input $u[k] = Kx[k - \left\lceil \frac{d_{th}}{p} \right\rceil]$ (computation of A_{cl} is shown in Appendix D) whereas A is the open-loop system with $u[k] = 0$. The switching between A_{cl} and A depends on the occurrence of valid and invalid samples. With arbitrary oder of valid and invalid samples, we get,

$$
x[k] = A_{cl}^{n_1} A^{n_2} A_{cl}^{n_3} A^{n_4} A_{cl}^{n_5} A^{n_6} ...x[0], \tag{16}
$$

where n_i are integers such that $\sum_i n_i = k$ (integer). Using the analysis coming from Average Dwell Time (ADT) [15], one can answer questions like what combination of n_i guarantees stability of system (16). We follow similar technique to estimate how frequently invalid samples are allowed to occur in the context of our setting.

Considering the example control loop P_2, $\left\lceil \frac{d_{th}}{p} \right\rceil = 3$, $u[k] = Kx[k-3]$ and $K = [0 \ -0.28]$, we know that the following inequality holds true,

$$
\left\| A_{cl}^k \right\| \leq 23.5 \times 0.7538^k. \tag{17}
$$

Similar property can be derived for the open-loop system,

$$
\left\| A^k \right\| \leq 1.6 \times 1.0539^k. \tag{18}
$$

To find how many valid samples must occur for every occurrence of invalid sample, system (16) becomes,

$$
x[k] = A_{cl}^{n_1} A A_{cl}^{n_2} x[0]. \tag{19}
$$

For the stability of the system, we need to make sure that $\|x[k]\| < \|x[0]\|$ for any k and $n = n_1 + n_2$. That is, the following must be true for assuring stability of (19),

$$
\begin{aligned}
\|A_{cl}^{n_1} A A_{cl}^{n_2}\| &\leq 1, \\
23.5^2 \times 1.6 \times 0.7538^n \times 1.0539 &< 1, \\
n &> 24.2.
\end{aligned} \tag{20}
$$

Hence, we conclude that ADT approach allows a 1 invalid sample within a window of any 25 samples.

F. CQLF APPROACH

We illustrate the stability analysis of the system (9). We perform the analysis considering that the control loop P_2 is implemented over an architecture described in Section 2.1 and the delay frequency metric to be $(300\ ms, 2)$. At the valid samples, the closed-loop system is A_{cl} as shown in Appendix D whereas the corresponding system for invalid samples is A as per eq. (2). From (9), we have a system which switches among various subsystems A_j,

$$A_j = AA_{cl}^{n_i}, \forall n_i \geq 2. \tag{21}$$

We resort to the concept of strong CQLF [9] to investigate the stability of the switched system (9). Towards this, we present the following standard theorems of LTI systems which are utilized in our analysis.

THEOREM F.1. (discrete-time Lyapunov equation [10]) *Let* $A \in \mathbb{R}^{n \times n}$. *If there exists* $P = P^T > 0$, $Q = Q^T > 0$ *satisfying* $A^T P A - P = -Q$, *then all eigenvalues of matrix A are inside the unit circle (or the system is stable in our context).*

THEOREM F.2. (continuous-time Lyapunov equation [10]) *Let* $A \in \mathbb{R}^{n \times n}$. *If there exists* $P = P^T > 0$, $Q = Q^T > 0$ *satisfying* $A^T P + PA = -Q$, *then matrix A is Hurwitz (or the system is stable in our context).*

LEMMA F.1. (Cayley transform [9]) *Let* $A \in \mathbb{R}^{n \times n}$, $P = P^T > 0$, $Q = Q^T > 0$. *Let P be the solution of* $A^T P A - P = -Q$. *Consider the matrix,*

$$C(A) = (A - I)(A + I)^{-1}. \tag{22}$$

Then P is also a solution of the continuous-time Lyapunov equation $C(A)^T P + PC(A) = -Q'$ *with* $Q' = 2(A+I)^{-T}Q(A+I)^{-1}$.

LEMMA F.2. ([9]) *Let A_1, A_2 be the system matrices of two stable discrete-time LTI systems with strong CQLF given by* $V(z) = z^T P z$, *i.e.,*

$$A_1^T P A_1 - P = -Q_1 < 0,$$
$$A_2^T P A_2 - P = -Q_2 < 0.$$

Then the two matrices $C(A_1)C(A_2)$ and $C(A_1)C(A_2)^{-1}$ have no real negative eigenvalues.

LEMMA F.3. *Let A_1, A_2 be the system matrices of two stable discrete-time LTI systems with strong CQLF given by* $V(z) = z^T P z$, *i.e.,*

$$A_1^T P A_1 - P = -Q_1 < 0,$$
$$A_2^T P A_2 - P = -Q_2 < 0.$$

Then the switching between the two LTI systems is asymptotically stable.

PROOF. We have two discrete-time LTI systems (23) and (24). The switching between them is modeled as a new LTI system shown in (25).

$$z[k + 1] = A_1 z[k], \tag{23}$$
$$z[k + 1] = A_2 z[k], \tag{24}$$
$$z[k + 2] = A_2 A_1 z[k]. \tag{25}$$

For the stability of (25), $A_2 A_1$ should have all eigenvalues inside the unit circle, i.e., $A_1^T A_2^T P A_2 A_1 - P < 0$ (utilizing Theorem F.1). Towards this,

$$A_1^T A_2^T P A_2 A_1 - P$$
$$= A_1^T (P - Q_2) A_1 - P$$
$$= A_1^T P A_1 - P - A_1^T Q_2 A_1$$
$$= -Q_1 - A_1^T Q_2 A_1 < 0$$

Hence, the switching between (23) and (24) is stable and the proof is complete. $\square$

Hence, there exists a strong CQLF between any two given subsystems A_i and A_j (hence, switching is stable) if $C(A_i)C(A_j)$ and $C(A_i)C(A_j)^{-1}$ do not have any real negative eigenvalue. Thus, it is possible to verify switching stability of (9) for the class of subsystems of form (21) for up to any n_i of design interest. For the control loop P_2 under consideration, the delay frequency metric $(300\ ms, 2)$ results in stable switching for all $n_i \geq 2$.

Reliable Computing with Ultra-Reduced Instruction Set Co-processors

Aravindkumar Rajendiran
University of Waterloo
a2rajend@uwaterloo.ca

Sundaram Ananthanarayanan
University of Waterloo
sundaram@uwaterloo.ca

Hiren D. Patel
University of Waterloo
hdpatel@uwaterloo.ca

Mahesh V. Tripunitara
University of Waterloo
tripunit@uwaterloo.ca

Siddharth Garg
University of Waterloo
siddharth.garg@uwaterloo.ca

ABSTRACT

This work presents a method to reliably perform computations in the presence of hard faults arising from aggressive technology scaling, and design defects from human error. Our method is based on an observation that a single Turing-complete instruction can mirror the semantics of any other instruction. One such instruction is the subleq instruction, which has been used for instructional purposes in the past. We find that the scope for using such a Turing-complete instruction is far greater, and in this paper, we present its applicability to fault tolerance. In particular, we extend a MIPS processor with a co-processor (called ultra-reduced instruction set co-processor – URISC) that implements the subleq instruction. We use the URISC to execute sequences of subleq that are semantically equivalent to the faulty instructions. We formally prove this, and implement the translations in the back-end of the LLVM compiler. We generate binaries for our hardware prototype called MIPS-URISC, which we synthesize and execute on an Altera FPGA. Our experiments indicate the performance and area overheads, and the efficacy of the proposed approach.

Categories and Subject Descriptors

B.8.1 [**Hardware**]: Performance and Reliability—*Reliability, Testing, and Fault-Tolerance*

General Terms

Reliability

Keywords

Microprocessor reliability, Turing-complete ISA

1. INTRODUCTION

Aggressive technology scaling in advanced CMOS manufacturing technology provides designers with an abundance of transistors on a single die. There is a consensus in the computing community that one way to obtain performance from these transistors is to design multicore processors [4]. The trend is towards building simple processors with shallow in-order pipelines such that the resultant multicore design remains within the constrained power budgets. Examples of such multicores are the SPEs in the Cell processor, and Sun's Niagara. While each processor is simple, the combination of many such processors on the same die has the potential to offer high throughputs. However, an impediment to such performance gains is the increased susceptibility to hard faults caused by aggressive technology scaling. Hard faults are permanent faults on a die such as a bit stuck at a single logical value. Such faults reduce the yield because the faulty core must either be discarded or disabled. Therefore, one of the major challenges in multicore designs is to discover cost-effective dependability [4] solutions for reliable computing in the presence of high defect rate technologies.

There are several research efforts that address this challenge through techniques such as redundant execution [1, 12], dynamic verification [2], and software-based recompilation [9]. Hard redundancy techniques use hardware threads or additional cores to detect faults, and recover from them using checkpointing techniques. These approaches, however, incur high hardware costs, because they consume double the resources in terms of threads and cores. Dynamic verification techniques, on the other hand, use low-cost hardware checkers to verify the validity of specific invariants that must be true during fault-free executions. DIVA [2] is an example of this approach, which extends the hardware with low-cost checkers that verify the invariants. However, the DIVA approach cannot handle faults that affect the decoding logic. A purely software-based approach called detouring [9] recompiles faulty instructions using instructions that are known to operate correctly. In fact, recent work has shown that up to 26% of the instructions in x86 ISA have no or only partial equivalence, i.e., they *cannot* be fully encoded using combinations of other instructions [3]. Rehman et al. [14] propose instruction compilation techniques to improve reliability, which is orthogonal to our approach.

In our efforts to address these issues, we argue for an approach that combines software and hardware techniques to address the challenge of reliable computing in the presence of hard faults. The novelty of our approach is based on the observation that we can represent the semantics of any faulty instruction with a single non-faulty instruction called subleq [7]. This is possible because the subleq instruction is Turing-complete. This knowledge in the past was used as a teaching tool for undergraduate computer architecture [7]. However, we identify its applicability for providing fault tolerance under high defect rate scenarios.

Furthermore, our technique can tolerate design errors as well. In doing so, we extend the TigerMIPS processor, which is a five-stage in-order pipeline implementation of the MIPS instruction-set architecture (ISA) with a low-cost co-processor that we call the ultra-reduced instruction set co-processor (URISC). Notice that the TigerMIPS is a representative processor to be used in multicore designs. The URISC implements the subleq instruction. The hardware design of the URISC separates the decoder, and execution unit from the TigerMIPS. The low complexity and area of the URISC core allows for (i) the use of formal verification techniques to protect against design bugs, and (ii) conservative physical design techniques to ensure sufficient timing guardbands and minimize layout hotspots. However, for this prototype, we assume that memories, instructions to access memories, and instruction fetch logic are also hardened.

Our fault model assumes that the subset of faulty instructions resulting from hard faults is known. For each instruction in this subset, we produce a sequence of subleq instructions that are semantically equivalent to the faulty instruction. This allows us to recompile any program given its subset of faulty instructions to use the URISC for just those faulty instructions. However, the process of discovering the sequence of subleq instructions is non-trivial, and it is imperative for us to prove its equivalence. Consequently, we formally prove that our sequences of subleq instructions are semantically equivalent to the original faulty instruction. We implement these in the back-end of the LLVM compiler to automatically generate the sequence of subleq for the faulty instructions. As of now, our prototype supports a subset of MIPS ISA instructions that can be faulty, but we capture instructions that perform arithmetic, logic and control transfer operations. Our experiments with benchmarks from Mibench [5] show the applicability of this solution, and its impact on performance. Notice that the penalty is paid only when there exist faulty instructions.

2. RELATED WORK

Gizopoulos et al. [4] present an elegant overview of recent techniques for detecting and recovering from faults. The detection techniques include redundant execution [1, 12], and dynamic verification [8]. Some of the recovery techniques include detouring [9], and reconfiguration [13]. Redundancy techniques use hardware threads or additional cores to detect faults. Once detected, checkpoint schemes assist in recovering from the faults. Another interesting approach by Powell et al. [13] advocates architectural core salvaging. This is based on the observation that a processor can execute every instruction in the ISA by using correctly operating components from other cores on the same die. They offer a hardware solution that migrates threads to other cores when the current core cannot execute an operation, but another one can. Dynamic verification techniques [8] promote using low-cost hardware checkers only to verify the validity of invariants. Upon detecting a fault the appropriate recovery mechanism is used.

Meixner and Sorin [9] present a solution to circumvent hard faults in simple cores through detouring. Detouring requires knowing the hard faults beforehand, and this is used to discover instructions that are faulty. These faulty instructions are translated into sequences of other non-faulty instructions. Notice that detouring has the disadvantage that it is not possible to recover from hard faults that render components completely unusable. For example, if the branching logic cannot be used, then detouring cannot recover from the fault. By providing low-cost hardware redundancy with the URISC, we are able to continue executing correctly under this situation. Our approach is complementary to detouring, which allows us to combine the use of operational instructions on the Tiger-MIPS, and only use the URISC when the faulty instruction cannot be represented with existing instructions from the TigerMIPS.

3. SEMANTICS AND SOUNDNESS

We have developed a theory of soundness based on semantic equivalence to express whether a sequence of instructions correctly captures another sequence of instructions. We did so to validate our realization of instructions using the subleq instruction. We provide the intuition behind our theory in this section.

We call the realization of an instruction or a sequence of instructions using another sequence of instructions a simulation. We are interested in simulations that use the subleq instruction only. What we seek in the notion of a sound simulation is semantic equivalence. In our context, an instruction is a two-tuple that comprises a name and a sequence of operands. A semantics specifies the manner in which an execution of the instruction changes the values associated with the operands. We can partition the set of operands into read-only and read-write. Only the latter's values may change as a consequence of executing the instruction.

We write the semantics of a set of operands as a set of two-tuples. Each tuple associates an operand with a function that indicates the manner in which the operand changes as a consequence of executing the instruction. The function may of course take as arguments all the operands to the instruction. We denote the value of an operand at the time of invocation of the instruction using $[\cdot]$.

For example, consider the instruction add a, b, c. In the instruction, a is the only operand that is read-write; b and c are read-only. The semantics of the instruction is $\{\langle a, f_a \rangle, \langle b, f_b \rangle, \langle c, f_c \rangle\}$, where f_a is $[b] + [c]$, f_b is $[b]$ and f_c is $[c]$. The semantics of the subleq a, b, c instruction is $\{\langle a, f_a \rangle, \langle b, f_b \rangle, \langle pc, f_{pc} \rangle\}$, where, f_a is $[a]$, f_b is $[b] - [a]$ and f_{pc} is c (an immediate value) if $[b] \leq [a]$, and $[pc] + 1$ otherwise.

Given the above characterization of the semantics of a single instruction, we can generalize it to the semantics of a sequence of instructions in a natural way. We also need to introduce the notion of an instance of an instruction. The reason is that in practice, an operand of an instruction is bound to either a register or an immediate value. Some operands may be bound to the same register.

We use $\{\cdot\}$ to distinguish an instance of an instruction from the instruction, and $\mapsto$ to indicate a binding. For example, $\{$add $r, r, 5\}$ is an instance of the add a, b, c instruction. In the instance, $a \mapsto r$, $b \mapsto r$ and $c \mapsto 5$ expresses the binding of operands to registers and immediate values. Of course, given an instruction, only some bindings may be valid. For example, for the add instruction, the first operand must be bound to a register.

We can now adapt our notion of semantics for instances of instructions. The semantics of an instance associates a register that is bound to some operand with a function that specifies how its value changes with the execution of the instance of the instruction. We can then generalize this to the semantics of an instance of a sequence of instructions. This generalization is done recursively, by applying the functions from the semantics of the next instruction to the results of the previous sub-sequence of instructions.

It is important to capture this notion of binding operands with registers and immediate values in the development of our notion of soundness of a simulation. Our notion of soundness is the following. Given two sequences of instructions, I and J, we say that J is a sound simulation of I if and only if the following are both true.

1. Every valid binding of operands for I is a valid binding for J, and,

2. For every valid binding of I (and therefore J, by (1) above),

the semantics of the instance of J is a superset of the semantics of the instance of I.

Our characterization of soundness of simulations above expresses the importance of considering every instance (valid binding) of a sequence of instructions in its simulation. That is, we do not only deal with the semantics of the abstract instruction, but with every possible instance.

Example Let I be the single instruction $i = $ "add d, s_1, s_2." Let J be the following sequence of subleq instructions. The register npc contains the value of the next program counter.

$$j_1: \quad \text{subleqi } t, t, 1$$
$$j_2: \quad \text{subleqi } s_1, t, 1$$
$$j_3: \quad \text{subleqi } s_2, t, 1$$
$$j_4: \quad \text{subleqi } d, d, 1$$
$$j_5: \quad \text{subleqi } t, d, 1$$

PROPOSITION 1. *J is a sound simulation of I.*

PROOF. A valid binding of add requires d to be bound to a register. The operands s_1 and s_2 may be bound either to registers or to immediate values. We observe that all such bindings are valid for J — only the operands t and d are read-write in J.

We need to consider the following four cases. (1) $d \mapsto r$, $s_1 \not\mapsto r$ and $s_2 \not\mapsto r$, for some register r. (2) $d \mapsto r$, $s_1 \mapsto r$, $s_2 \not\mapsto r$. (3) $d \mapsto r$, $s_2 \mapsto r$, $s_1 \not\mapsto r$. (4) $d \mapsto r$, $s_1 \mapsto r$, $s_2 \mapsto r$.

We first consider Case (1). Let $d \mapsto r$, $s_1 \mapsto a$, $s_2 \mapsto b$, where r is a register, but a and b may be registers or immediate values, and it is possible that $a = b$. The following proof is for the subcase that $a \neq b$, and a and b are registers. The proof applies with straightforward modifications for the other sub-cases.

The semantics of the instance of I is $\{\langle r, [a]+[b]\rangle, \langle a, [a]\rangle, \langle b, [b]\rangle\}$. We now recursively determine the semantics of the corresponding instance of J. For each subsequence, we consider its extended semantics. We observe that the registers to which the instance of J refers are $\{r, a, b, t, pc\}$.

We denote as J_k the instance of sub-sequence of instructions from J, $\langle j_1, \ldots, j_k \rangle$. The extended semantics of J_1 is $\{\langle r, [r]\rangle, \langle a, [a]\rangle, \langle b, [b]\rangle, \langle t, 0\rangle, \ldots\}$. The "..." in the semantics are for the tuples associated with the other registers in $R(J) - R(I)$. Continuing, for $F(J_i)$, which represents the semantics of J_i,

$$F(J_2) = \{\langle r, [r]\rangle, \langle a, [a]\rangle, \langle b, [b]\rangle \langle t, -[a]\rangle, \ldots\}$$
$$F(J_3) = \{\langle r, [r]\rangle, \langle a, [a]\rangle \langle b, [b]\rangle \langle t, -[a] - [b]\rangle, \ldots\}$$
$$F(J_4) = \{\langle r, 0\rangle, \langle a, [a]\rangle \langle b, [b]\rangle \langle t, -[a] - [b]\rangle, \ldots\}$$
$$F(J_5) = F(J) = \{\langle r, [a] + [b]\rangle, \langle a, [a]\rangle \langle b, [b]\rangle \langle t, -[a] - [b]\rangle, \ldots\}$$

We observe that the semantics of the instance of J, $F(J) \supseteq F(I)$, the semantics of the corresponding instance of I. We need to consider also the cases that (2)–(4). The proofs for these cases are similar to the proof for Case (1) above. We show Case (4) below.

$$F(I) = \{\langle r, 2[r]\rangle\}$$
$$F(J_1) = \{\langle r, [r]\rangle, \langle t, 0\rangle, \ldots\}$$
$$F(J_2) = \{\langle r, [r]\rangle, \langle t, -[r]\rangle, \ldots\}$$
$$F(J_3) = \{\langle r, [r]\rangle, \langle t, -2[r]\rangle, \ldots\}$$
$$F(J_4) = \{\langle r, 0\rangle, \langle t, -2[r]\rangle, \ldots\}$$
$$F(J_5) = F(J) = \{\langle r, 2[r]\rangle, \langle t, -2[r]\rangle, \ldots\} \supseteq F(I)$$

□

4. MIPS-URISC ARCHITECTURE

The MIPS-URISC architecture combines a five-stage pipelined implementation of the MIPS ISA known as the TigerMIPS [11] with the URISC. In doing this, we extend the MIPS ISA with instructions that facilitate transfer of register contents between the TigerMIPS and URISC register files in addition to the subleq instructions. Notice that we provide a RISC-based encoding and implementation of the subleq instructions as opposed to the original proposal that was CISC-based [7].

4.1 ISA Extensions to support URISC

We extend the MIPS ISA with the five instructions shown in Table 1. These instructions provide a programming interface for the URISC. We use co-processor instructions from the MIPS ISA to encode URISC instructions. Note that Rs, and Rt denote source registers, and Rd denotes the destination register. Immediate identifies that one of the source operands is an immediate value, which is encoded in the instruction itself. Bits I_{31-25} indicate that the instructions are for the co-processor in the MIPS ISA, and I_{25-21} are the OP-codes for each of the instructions.

Inst.	I_{31-26}	I_{25-21}	I_{20-16}	I_{15-11}	I_{10-6}	I_{5-0}
subleq	0010 00	11000	Rs	Rd	Rt	00 0000
subleqi	0010 00	11100	Rs	Rd	Immediate	
mfu	0010 00	10000	Rd	Rs	000 0000 0000	
mtu	0010 00	10100	Rs	Rd	000 0000 0000	
mtui	0010 00	10101	Rd	Immediate		

Table 1: ISA extensions for the TigerMIPS processor.

The URISC supports three types of instructions. The first type constitutes the subleq instructions. There are two variants: one where the third operand comes from a register, and one where the third operand is an immediate value. Note that the original definition of subleq [7] proposes a CISC instruction where all three operands are memory operands. Instead, we design subleq instructions as two register-operand RISC instructions as shown in Table 1. We also rely on the load and store operations of the MIPS ISA to retrieve contents from the memory to the MIPS registers. However, it is possible to extend the URISC to have its own fault-free load and store instructions. The contents of these MIPS registers are then transferred to the register file of the URISC by using the mtu (move to URISC) instructions. Hence, the source register operand (Rs) points to a register in the MIPS register file, and the destination register operand (Rd) points to a register in the MIPS-URISC. We provide the immediate variant of the move to URISC instruction (mtui) to allow directly loading an eleven-bit immediate value into a register of the MIPS-URISC. The mfu (move from URISC) instruction transfers the contents of the source register operand (Rs) pointing to a register in the URISC to the destination register (Rd) in the MIPS register file. These move instructions allow us to transfer contents between the MIPS processor and the URISC.

4.2 MIPS-URISC Hardware Design

The MIPS-URISC has three parts: the MIPS ISA implementation via the TigerMIPS processor, the URISC implementation, and their integration. We assume that memories and registers are hardened using techniques such as ECC [4].

4.2.1 TigerMIPS Design

The TigerMIPS is a five stage implementation of the MIPS ISA [11]. The five stages are Fetch, Decode, Execute, Memory and Writeback. We show this in Figure 1. The Fetch stage retrieves an

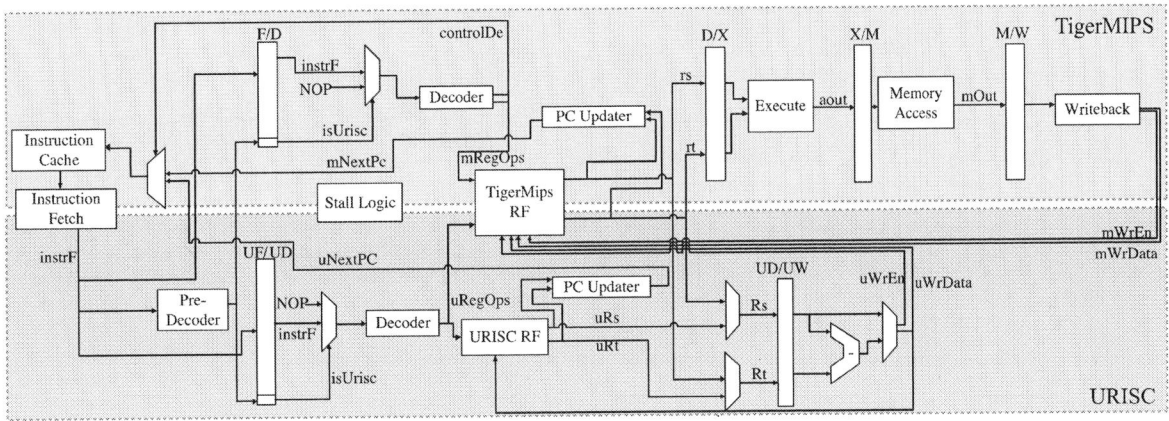

Figure 1: Hardware design for MIPS-URISC.

instruction from the instruction cache indexed using the contents of the program counter (PC). The fetched instruction is stored in a pipeline register denoted F/D. The F/D register provides the instruction as an input to the Decode stage. This is where the instruction gets decoded, and the source operands are supplied to the register file. The PC update unit implements an optimization to reduce the penalty on branches. It updates the PC with the target address given the instruction is a branch. The Decode stage produces the instrEx opcode, the values for the source operands, and corresponding control signals, and stores them in the following pipeline register D/X. The Execute stage performs the operation indicated by the instrEx opcode. The TigerMIPS has a standard arithmetic logic unit (ALU), a shifter, a multiplier, and a divider. Results of the Execute stage are fed into pipeline register X/M. In the Memory stage, a memory instruction performs its operand and stores the result in the M/W pipeline register. For non-memory instructions, there is no action performed in this stage. The Writeback stage updates the register file with the result of the instruction's execution to the destination operand. Note that there is a forwarding unit that performs operand forwarding, and a stall logic unit that does the hazard checking for dependencies between instructions in the pipeline.

4.2.2 URISC Design

The URISC implements a RISC-based implementation of the subleq instruction in three stages as shown in Figure 1. The datapath uses a decoder, a small register file, a subtractor, a unit to update the PC, pipeline registers, and several multiplexors to select the appropriate data to be passed to the next hardware components. There are three stages to the URISC: Fetch (UF), Decode (UD), and ExecuteWriteBack (UW). The Fetch stage in URISC is shared with the Fetch stage in TigerMIPS. The Decode stage receives the instruction from the UF/UD pipeline register, and identifies the instruction. The URISC supports only those indicated in Table 1. Since these instructions can have one to two source operands, we use a small register file with ten registers that holds the registers of the MIPS-URISC. The register file can either read from two operands in one clock cycle, or write to one register in one cycle. We use this register file design because it is synonymous to that of the Tiger-MIPS. The decoder provides the addresses for reading the registers from URISC and TigerMIPS register file, and the output is fed into the UD/UW pipeline register and the PC Updater unit. The PC Updater unit for URISC contains a subtract and compare unit. It sub-

tracts the input and compares the result to check if it less than or equal to zero. If it is, uNextPC is updated to the new target address. The target address can be computed by using the sign-extended immediate offset of the current PC or an absolute target address stored in a register. This is determined by the instruction. For subleq instructions, the UW stage uses a subtractor to produce the subtracted result on writeRegDataWB. The PC Updater unit of the URISC also computes the target address for the immediate variant of the PC-relative subleq instruction. Notice that the result of the UW stage is fed into the write port of the URISC register file. This updates the contents of the destination register with the computed value.

Our implementation places the PC Updater unit in UD stage to reduce penalty incurred from branching. If we update the PC using the output of the UW stage, we must flush the whole URISC pipeline, which results in a penalty of two clock cycles. However, by performing this in the UD stage, the penalty reduces to one clock cycle. Note that this version of the URISC does not have direct access to any of the memories. This is a design decision, and alternatives are possible. Our implementation of the URISC assumes that memory operations can be carried out using the TigerMIPS, and mtu instructions can transfer from the TigerMIPS register file to the URISC register file. There is a three cycle latency for executing an instruction in the URISC. We implement a forwarding unit such that read-after-write (RAW) dependencies between any URISC instructions are avoided. The design of the URISC lends itself to an implementation where there are no stalls caused by data dependencies. Since the URISC cannot directly access memories, there are also no stalls resulting from cache misses. Hence, the URISC does not have any stall logic for in-flight instructions in the URISC.

4.2.3 Integrating the TigerMIPS with the URISC

Figure 1 combines the TigerMIPS processor with the URISC. Notice that URISC shares the Fetch stage with the TigerMIPS pipeline, while other stages are all distinct. However, in the UF stage, we introduce a small pre-decoder to identify any instruction that requires the URISC. Along with the fetched instruction, we store the result of the pre-decoder in the respective pipeline registers that are input for the Decode stages (F/D and UF/UD). Both decode stages inspect the result of the pre-decoder, and if the fetched instruction is meant for the URISC, then we begin inserting no-operations (nops) in the TigerMIPS pipeline. However, if the instruction is not for the URISC, then it is executed on the TigerMIPS, but we must insert nops in the URISC. When there is a URISC instruction after a

700

standard MIPS instruction, we stall the URISC instruction until the MIPS instruction completes its write-back. Similarly, if a URISC instruction is under execution followed by a MIPS instruction, we stall the MIPS instruction until the completion of the URISC instruction.

Signal	Description
urWrEn	Controlled only by the URISC's D and WB stage. If high, URISC wants to write urWrData to the register file. Else, URISC can only read the two operands, urRdOpA and urRdOpB
urWrData	Data to be written to the register file
urRdOpA	Register number for reading first operand
urRdOpB	Register number for reading second operand
mWrEn	Controlled only by the TigerMIPS's D and WB stage. If high, TigerMIPS processor wants to write mWrData to the register file. Else, TigerMIPS can only read the two operands, mRdOpA and mRdOpB
mWrData	Data to be written to the register file
mRdOpA	Register number for reading first operand
mRdOpB	Register number for reading second operand

Table 2: Subset of input signals to register files.

The mtu and mfu instructions provide a mechanism to transfer contents between the two register files. Table 2 describes the signals that we use as input by each of the register files. There are two operand outputs from the register file for the TigerMIPS, and three operands for the URISC register file. We ensure that both enable signals are never driven high at the same time. The mtu only accesses the TigerMIPS register file. The URISC in UD drives the urWrEn signal low, and supplies the operand to read using mRdOpA. The mWrEn signal is driven low because the TigerMIPS Decode stage has a nop in it. The TigerMIPS register file then outputs the data directly to the UD/UW register. At the UW stage, we drive the urWrEn signal high to write the result to the URISC register file through urWrData. Reading from the URISC register file for the mfu instruction is straightforward. We drive the urWrEn signal low, and provide the urRdOpA. We also set the mWrEn signal low. To store the contents of the URISC register to the TigerMIPS register file, we use the UW stage.

5. MIPS-URISC COMPILER TOOLCHAIN

To program the MIPS-URISC processes, we provide a compiler toolchain based on the LLVM [6] compiler framework. In this toolchain, we extend the code generation back-end of LLVM to perform the necessary translation to subleq instructions for a given set of faulty instructions. Currently, the compiler generates a sequence of subleq instructions for only a subset of the MIPS ISA instructions. These instructions are ADD, ADDUI, SUB, BEQ, J, ANDI, OR, SLL, MULT, DIV. We provide a configuration file to the compiler that indicates the faulty instructions to replace with the corresponding subleq instructions. The toolchain accepts source programs written in C and C++, and a sequence of MIPS instructions that are known to be faulty. It parses the source files, and constructs an intermediate representation. The code generation traverses through the intermediate representation, and generates MIPS assembly instructions for the target platform. During the code generation, we identify all MIPS instructions that are known to be faulty instructions, and we replace them with an equivalent sequence of subleq instructions. We produce an assembly program that consists of a mix of MIPS ISA and the subleq instructions. This assembly program undergoes another translation where we encode every subleq instruction using the .word assembler directive. This is to allow TigerMIPS's GCC assembler to accept the extended MIPS-URISC

instruction without any alterations to GCC's assembler. We execute the resulting binary on the MIPS-URISC architecture.

6. ILLUSTRATIVE EXAMPLE: DESIGN ERROR IN MIPS R4400PC/SC ERRATA

We use a simple example to illustrate that URISC can be used to correct design errors as well. We refer the reader to the MIPS R4400 errata [10] that identifies a problem with their jump instruction. The jump instructions allow branching to an address that is within 256MB memory region of the current PC. The low-order 28 bits of the target address is calculated by left shifting the immediate by two . Then, the remaining higher-order bits of the target address are extracted from the higher-order bits of the next PC. However, if the next PC is in the next 256MB memory region, then the resulting target address is incorrect. We illustrate this issue with the sequence of instructions below.

```
0x0fff fff0    J    imm1
0x0fff fff4    J    imm2
0x0fff fff8    J    imm3
0x0fff fffc    J    imm4
----256 MB End Of PC Region-----
0x1000 0000
```

Suppose that the immediate target (imm4) is 0x3FF1010. The left-shifted value of this is then 0x0FFC4040. Now, the higher-order bits of the next PC, which is 0x10000000 are used resulting in a target address of 0x1FFC4040. This is clearly incorrect. Notice, that this is a design error where the next instruction's PC is used for computing the higher-order bits of the next instruction's PC.

We can use subleq instructions in the URISC core to rectify this design error. In order to do this, we replace the jump instructions with a sequence of subleq instructions. However, for brevity, we use subroutine calls for basic operations such as a logical AND and a logical OR, which we have already encoded using subleq instructions. We assume that register u1 holds the immediate value, and PC holds the program counter.

```
subleqi    u1,s,0      [s = -imm]
subleqi    u1,s,0      [s=-2*imm]
subleqi    r,r,0       [r=0]
subleqi    s,r,0       [r=-s;r=2*imm]
subleqi    s,r,0       [r=r-s;2*imm-(-2*imm);r=4*imm]

subleqi    s,s,0       [s=0]
subleqi    PC,s,0      [s=-PC]
subleqi    t,t,0       [t=0]
subleqi    s,t,0       [t=PC]

andsubleq(p,f,t)       [args t and f as arguments;
                        the result gets stored in p;
                        p=t&f; p = PC & 0xF000 0000]

orsubleq(s,r,p)        [call subroutine or;
                        the result is stored in s;
                        s=(PC&0xF000 0000)|(imm<<2)]
subleq     t,t,s       [branch to absolute address
                        pointed by s];
```

By using the subleq instructions, we can overcome the shortcomings of the jump instruction. Firstly, since there is a variant of the subleq instruction that has a register as its third operand, we can jump to a 32 bit target address by populating that register. Secondly, since the subleq instruction does not rely on the next PC to get the higher-order 32 bits, there is no case where it would jump to the incorrect target address. This is because, the higher-order bits of the target address are calculated from the address of the instruction itself. This illustrative example shows that the URISC can be used to resolve design errors in addition to errors resulting from hard faults.

7. EXPERIMENTAL EVALUATION

We synthesized the MIPS-URISC on an Altera FPGA, and confirmed that the URISC addition does not impact the clock frequency (approximately 27Mhz) of the TigerMIPS. The number of lookup tables (LUT) used increased by approximately 30% when compared to the TigerMIPS. However, we note that this area increase does not account for the nine embedded multipliers in the Tiger-MIPS that do not make use of any LUT resources. We therefore expect the actual overhead to be even lower.

We take a subset of the Mibench [5] benchmarks and some in-house benchmarks, and select a few combinations of faulty instructions. Our compiler recodes these using subleq instructions, and we execute the resulting program on the MIPS-URISC. We record the execution times of these benchmarks, and show the overhead with respect to latency when using the subleq instructions.

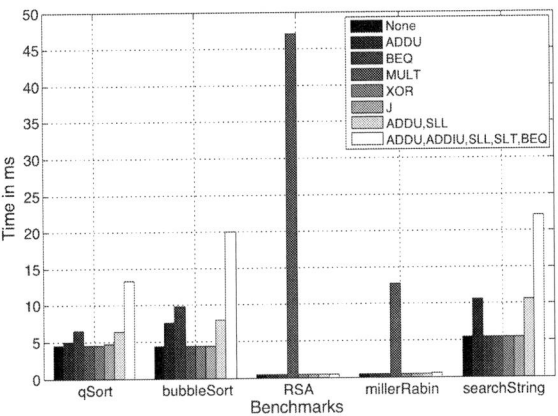

Figure 2: Performance overhead results for benchmarks.

The legend in Figure 7 describes the instructions we deemed faulty for our benchmarks. None indicates the baseline where we execute the benchmark with no faulty instructions. The other bars show the execution times with certain instructions identifies as being faulty, and recoded as subleq instructions. As expected, the dynamic instruction count of the faulty instruction, and the type of the faulty instruction has a considerable impact on the performance. However, this is the performance-correctness trade-off that we allow designers to make. We note that in some cases the performance overheard can be quite significant, for example while executing the RSA benchmark with a faulty MULT instruction. However, for security-critical functionality like RSA, this performance overhead might be an acceptable trade-off.

8. CONCLUSION

This work presents a method to continue performing computation in the presence of hard faults arising from aggressive technology scaling, and design defects from human error. We identify the usefulness of a Turing-complete instruction called subleq for tolerating hard faults, and show its applicability with a prototype that

we call MIPS-URISC. This requires recoding the faulty instructions with a sequence of subleq instructions, which we formally prove to be semantically equivalent to the faulty instructions. We implement our MIPS-URISC architecture as a prototype on an Altera FPGA, and we provide an LLVM-based compiler to generate the subleq instructions for the instructions identified as faulty.

9. REFERENCES

[1] N. Aggarwal, P. Ranganathan, N. P. Jouppi, and J. E. Smith. Configurable isolation: building high availability systems with commodity multi-core processors. volume 35, pages 470–481, New York, NY, USA, June 2007. ACM.

[2] T. Austin. DIVA: A reliable substrate for deep submicron microarchitecture design. In *proceedings of the 32nd IEEE International Symposium on Microarchitecture (MICRO)*, pages 196–207, 1999.

[3] N. Foutris, D. Gizopoulos, M. Psarakis, X. Vera, and A. Gonzalez. Accelerating microprocessor silicon validation by exposing isa diversity. In *proceedings of the 44th IEEE International Symposium on Microarchitecture (MICRO)*, pages 386–397, 2011.

[4] D. Gizopoulos, M. Psarakis, S. Adve, P. Ramachandran, S. Hari, D. Sorin, A. Meixner, A. Biswas, and X. Vera. Architectures for online error detection and recovery in multicore processors. In *proceedings of IEEE Design, Automation and Test in Europe (DATE)*, pages 1–6, 2011.

[5] M. Guthaus, J. Ringenberg, T. Austin, T. Mudge, and R. Brown. http://www.eecs.umich.edu/mibench/.

[6] LLVM Team. http://llvm.org/.

[7] F. Mavaddat and B. Parhami. URISC: the ultimate reduced instruction set computer. *International Journal of Electrical Engineering Education*, 25:327–34, 1988.

[8] A. Meixner, M. Bauer, and D. Sorin. Argus: Low-cost, comprehensive error detection in simple cores. In *proceedings of the 40th Annual IEEE International Symposium onMicroarchitecture (MICRO)*, pages 210–222, 2007.

[9] A. Meixner and D. Sorin. Detouring: Translating software to circumvent hard faults in simple cores. In *proceedings of IEEE International Conference on Dependable Systems and Networks (DSN)*, pages 80–89, 2008.

[10] MIPS. http://www.mips.com/media/files/archives/R4400PC_SCErrata,ProcessorRevision1.0.pdf.

[11] S. Moore and G. Chadwick. http://www.cl.cam.ac.uk/teaching/0910/ECAD+Arch/mips.html.

[12] S. Mukherjee, M. Kontz, and S. Reinhardt. Detailed design and evaluation of redundant multi-threading alternatives. In *proceedings of the 29th International Symposium on Computer Architecture (ISCA)*, pages 99–110. ACM, 2002.

[13] M. D. Powell, A. Biswas, S. Gupta, and S. S. Mukherjee. Architectural core salvaging in a multi-core processor for hard-error tolerance. In *proceedings of the 36th International Symposium on Computer Architecture (ISCA)*, ISCA '09, pages 93–104, New York, NY, USA, 2009. ACM.

[14] S. Rehman, M. Shafique, F. Kriebel, and J. Henkel. Reliable software for unreliable hardware: embedded code generation aiming at reliability. In *proceedings of the 7th IEEE/ACM/IFIP International Conference on Hardware/software Codesign and System Synthesis (CODES+ISSS)*, pages 237–246, New York, NY, USA, 2011. ACM.

Identification of Recovered ICs using Fingerprints from a Light-Weight On-Chip Sensor

Xuehui Zhang, Nicholas Tuzzio and Mohammad Tehranipoor
ECE Department, University of Connecticut
xuehui.zhang, nicholas.tuzzio, tehrani@engr.uconn.edu

ABSTRACT

The counterfeiting and recycling of integrated circuits (ICs) have become major problems in recent years, potentially impacting the security of electronic systems bound for military, financial, or other critical applications. With identical functionality and packaging, it is extremely difficult to distinguish recovered ICs from unused ICs. A technique is proposed to distinguish used ICs from the unused ones using a fingerprint generated by a light-weight on-chip sensor. Using statistical data analysis, process and temperature variations' effects on the sensors can be separated from aging experienced by the sensors in the ICs when used in the field. Simulation results, featuring the sensor using 90nm technology, and silicon results from 90nm test chips demonstrate the effectiveness of this technique for identification of recovered ICs.

Categories and Subject Descriptors

B.8.0 [**Performance and Reliability**]: General

General Terms

Security

Keywords

Counterfeiting, Recovered ICs, Hardware security, and Circuit aging

1. INTRODUCTION

The counterfeiting of integrated circuits (ICs) has been on the rise, potentially impacting the security of a wide variety of electronic systems. A counterfeit component is defined as an electronic part that is not genuine because it [1]:

- is an unauthorized copy;
- does not conform to original component manufacturers design, model, and/or performance;
- is not produced by the original component manufacturers or is produced by unauthorized contractors;
- is an off-specification, defective, or used original component manufacturers product sold as "new" or working;
- has incorrect or false markings and/or documentation.

The Office of Technology Evaluation, part of the U.S. Department of Commerce, reported over 10,000 incidents involving the re-sale of used or defective ICs from 2005 to 2008 alone which is much more than other types of counterfeits [1]. Business Week published an investigative report in 2008 that traced recovered ICs found in U.S. military supplies back to their sources [3]. It is reported in [2] that used or defective products being sold as new or working account for 80 to 90% of all counterfeits being sold worldwide. With such estimate on the percentage of recovered ICs being sold, and the numbers relating to semiconductor sales and counterfeiting in general presented in [7], it could be possible that the intentional sale of used or defective chips in the semiconductor market could have accounted for between $9 billion and $15 billion USD of all semiconductor sales in 2008 alone; the trends shown in [1] suggest that this number is only going to increase over time.

These used or defective ICs enter the market when electronic "recyclers" divert scrapped circuit boards away from their designated place of disposal for the purposes of removing and reselling the ICs on those boards. The recycling process involves removing ICs from board or even dies in the ICs. The security issues associated with these ICs are: (1) a used IC can act as a ticking time bomb [4] since it does not meet the specification of the unused (fresh) ICs; (2) an adversary can include additional die on top of the recovered die carrying a back-door attack, sabotaging circuit functionality under certain conditions, or causing denial of service [5]. Note that in this paper, the terms *recovered IC* and *recovered die* are used interchangeably; these are the ICs/dies which have been removed from their original boards for the purpose of illicit re-sale. It is vital that we prevent these recovered ICs from entering critical infrastructures, aerospace, medical, and defense supply chains.

These recovered ICs can be classified into two categories: partially recovered ICs and fully recovered ICs. Partially recovered ICs will have the same external appearance as the IC they are meant to mimic, but do not contain the correct die internally—they were removed from their original board and remarked as a different IC. As such, decaping of randomly selected chips and careful inspection are effective at detecting partially recovered ICs. The more difficult class of recovered IC to detect would be the fully recovered ICs. These ICs have the original appearance, functionality, and markings as the devices they are meant to mimic, but they have been used for a period of time before they were re-sold. Even the best visual inspection techniques will have a difficult time identifying these fully recovered ICs with certainty [6]. Additionally, because fully recovered ICs contain the original, correct die internally, decaping technologies will provide no assistance in their detection. It is vital to develop new techniques to measure these ICs' specifications and compare them against the unused ones.

1.1 Previous Work

Physical unclonable functions (PUFs) implement challenge and response authentication for IC identification [9] [10] [11] [12] [13]. For each physical stimulus, the circuit will react in an unpredictable way due to the complex interaction of the stimulus with the physical structure of the PUF and the inherent process variations. As the physical variations for each IC are unique, a distinct ID can be obtained for each IC through the PUF. Techniques to protect ICs against counterfeiting via active and passive authentication and identification (also known as hardware metering) have been proposed in [14] [15] [16]. Metering techniques ensure that over production of integrated circuits will be prohibited. The above approaches are effective at authenticating ICs but not at identifying recovered ICs since they are expected to have the same IDs as the unused ICs.

Computer-aided design and reliability research community has also seen an extensive research on analyzing the aging of integrated circuits. In particular, ring oscillator based reliability analysis has become a common practice. For instance, a silicon odometer has been proposed to monitor different types of aging effects in [17] [18]; however, the objective was to improve the reliability of ICs, not to identify the recovered ICs. Such sensors will be ineffective if they were to be used in detecting recovered ICs due to the presence of process and environmental variations. We believe that no existing techniques are able to effectively address the IC/die recovery problem, and to the best of our knowledge this is the first paper to propose techniques to detect recovered ICs.

1.2 Contributions and Paper Organization

The major difference between fully recovered ICs and unused ICs is that fully recovered ICs have already experienced aging, as they were removed from their original boards and re-sold in the market. Aging effects, such as negative/positive bias temperature instability (NBTI/PBTI) and hot carrier injection (HCI), would have had an impact on the performance of the fully recovered ICs due to the change in the threshold voltage. In this paper, we propose a novel fingerprinting technique using a light-weight sensor based on ring oscillators, called combating die recovery (CDR) sensor, to help detection of recovered ICs.

Our CDR sensor is composed of a reference ring oscillator (Reference RO) and a stressed ring oscillator (Stressed RO) which is conceptually similar to [17] [18]. However, the Stressed RO is designed to age at a very high rate by using high threshold voltage (HVT) gates (to expedite aging so that ICs used even for a very short period of time can be identified) while the Reference RO is gated off from the power supply during chip operation, so that it experiences no stress. The frequency difference between the two ROs could denote the usage level of the chip under test (CUT) when compared against the fingerprints generated from fresh ICs; the larger the difference is, the longer the CUT has been used, and with a higher probability the CUT could be a fully recovered IC. With close placement of the two ROs in the CDR sensor, the impact of intra-die process variations could be minimized. Data analysis can effectively distinguish the frequency differences caused by aging from those of temperature and inter-die process variations, and then identify fully recovered ICs, which is demonstrated by our simulation and silicon results. In addition, partially recovered ICs would not report frequencies from the ROs since they were recovered from totally different ICs that most likely do not contain the CDR sensor. Thus, these partially recovered ICs could also be easily detected by our technique. The proposed CDR sensor presents a negligible area overhead, imposes no constraint on circuit layout, and is resilient to removal and tampering attacks. The three working modes of the CDR sensor proposed in the paper ensure that the Reference RO cannot be gated on alone, thus the frequency difference between the two ring oscillators cannot be changed to mask detection.

The rest of the paper is organized as follows. Section 2 outlines the necessary background and analyzes the impact of aging on different circuit elements and ring oscillators. Section 3 presents the CDR sensor architecture, and the measurement flow using CDR sensor for identifying recovered ICs is described in Section 4. Simulation results as well as silicon results from our 90nm test chip are presented in Section 5. Finally, our concluding remarks and future work are given in Section 6.

2. AGING ANALYSIS

In this section, we will briefly describe aging phenomenon in integrated circuits and present their impact on different circuit components and ring oscillators, which will be used in our CDR sensor.

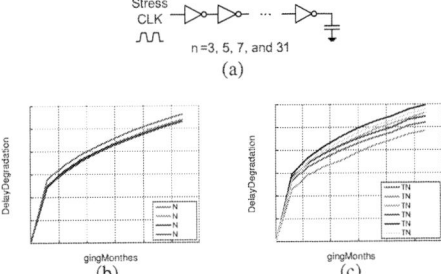

Figure 1: (a) Inverter chain structure, (b) Degradation of inverter chains with different lengths (stage count), and (c) Degradation of a 3-inverter chain with different inverter types.

When the chip operates in functional mode, the transistors age due mainly to NBTI and HCI. The aging effects of NBTI and HCI could cause parametric shifts and circuit failures, as demonstrated by reliability models [19] [21] [22]. NBTI can increase the absolute value of the PMOS threshold voltage, resulting in reduced transistor current and increased gate delay. HCI can create traps at the silicon substrate/gate dielectric interface, and can create dielectric bulk traps, and therefore impacts device parameters including threshold voltage. Since recovered ICs have been impacted by these aging effects when used in the field, the circuit parameters of recovered ICs would be different from those of fresh ICs. If a *fast-aging sensor* was embedded into the circuit to help detect its aging period, then recovered ICs could be identified.

In order to verify the effects of aging on a circuit's performance, several different inverter chains were simulated using Synopsys 90nm technology [20]. The delay of these inverter chains will represent the circuit's performance. The simulation was conducted using HSP-ICE MOSRA (Synopsys' reliability analysis tool) with combined NBTI and HCI aging effects at $25°C$. Figure 1(a) shows the basic structure of the inverter chains with the same capacitive load and the same stress coming from a 500MHz clock. These chains are composed of 3, 7, 15, and 31 standard threshold voltage (SVT) inverters. Figure 1(b) presents the delay degradation of inverter chains under clock stress for up to 27 months with no interupt. From the figure, we can see that the number of inverters does not have a significant impact on the degradation of these chains since they receive the same stress, and each inverter's speed degrades at the same rate. Aging effects are also dependent on device's threshold voltage. There are three different threshold voltage models in the Synopsys 90nm technology: SVT, HVT, and low threshold voltage (LVT). The 3-inverter chains were simulated using these threshold voltages and two different size inverters (INVX1 and INVX32). Figure 1(c) shows that the chain with the HVT inverters experiences more degradation than the chains with SVT or LVT inverters. The INVX1 inverter chain has a larger degradation than the INVX32 inverter chain.

NAND and buffer (BUF) gate chains with HVT were also simulated at $25°C$ with a 500MHz clock stress. The basic structure of these chains is the same as the inverter chains. A NAND gate will function as an inverter when its two inputs are connected together. Figure 2 shows the simulation results. From the figure, we can see that the gate type does not impact the aging speed significantly. However, the inverter chain ages slightly faster than the others while NAND gate chain and BUF chain age at almost the same speed. The difference in the amount of aging depends on the structure of gates. Therefore, inverters (INVX1) with HVT will be used to create the ring oscillators used to detect recovered ICs in our simulation analysis.

Figure 3(a) shows the frequency degradation of a 5-stage ring oscillator with HVT inverters after aging for 25 months. The frequency of the RO in a recovered IC will be smaller than in a fresh IC. If there are no environmental or process variations, we could easily identify recovered ICs by measuring the frequency of the RO

704

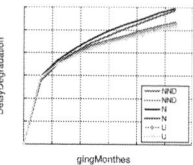

Figure 2: Delay degradation of NAND, BUF, and INV chains.

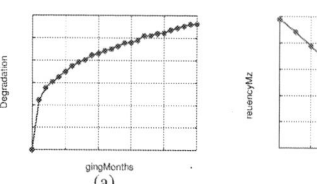

(a)   (b)

Figure 3: (a) Frequency degradation of a 5-stage RO, and (b) Frequency of a 5-stage RO decreases with increasing temperature.

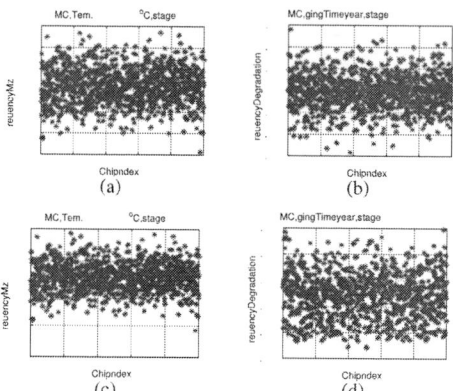

(a)   (b)

(c)   (d)

Figure 4: (a) Frequency of a 5-stage RO varying with process variations, (b) Frequency degradation of a 5-stage RO aging for one year varying with process variations, (c) Frequency of a 21-stage RO varying with process variations, and (d) Frequency degradation of a 21-stage RO varying with process variations.

embedded in the circuit. However, variations have a significant impact on the frequency of ROs. Figure 3(b) shows that the frequency of the 5-stage RO will decrease as we increase the temperature, and that the frequency variation could be very large. Note that increasing temperature can also increase the degradation of the circuit.

The 1000 Monte Carlo (MC) simulation results of the 5-stage RO are shown in Figure 4(a), at a temperature of $25°C$ with 2% Tox, 5% Vth, and 5% L inter-die variation and 1% Tox, 5% Vth, and 5% L intra-die variations. We can see that the frequency of the RO can vary as much as 20% under process variations. In addition, process variations impact the aging rate of the RO, as shown in Figure 4(b). The frequency degradation of the 1000 chips varies around 8% (7.4%-8.6%) for one year of aging. This frequency shift caused by the aging effects in recovered ICs can help separate them from those caused by process variations in fresh ICs if we are to try to use ROs to detect recovered ICs.

With a fixed stress, the number of inverters does not have a significant impact on an inverter chains' delay degradation. However, the frequency of an RO is related to the number of inverters, $f = \frac{1}{2*n*t_d}$, where n is number of stages in the RO and t_d is the delay of an inverter. Figure 4(c) shows the frequency shift of a 21-stage RO with HVT inverters. The frequency degradation is shown in Figures 4(d). Comparing the frequency degradation of the 5-stage and 21-stage ROs, we can see that the 5-stage RO experiences slightly more degradation since its oscillation frequency is higher than the 21-stage RO. However, a 5-stage RO may require a very fast counter which might be difficult to design to timing close. We will discuss this in detail in Section 5.

3. CDR SENSOR

Our main objectives in designing the CDR sensor are: (*i*) the sensor must age at a very high rate to help detect ICs used even for very short period of time, (*ii*) the sensor must experience no aging during manufacturing test, (*iii*) the impact of process variations and temperature on CDR sensor must be minimized, (*iv*) the sensor must be resilient to attacks, and finally (*v*) the measurement process must be done using a low-cost equipment and be very fast and easy.

As mentioned earlier, aging effects could slow down the frequency of the RO embedded into ICs. With an embedded RO, these recovered ICs could be identified based on its frequency, which will be smaller than that of a fresh IC. However, there are many parameters impacting the frequency of an RO, such as temperature and process variations. Our CDR sensor uses a Reference RO and a Stressed RO to separate the aging effects from process/environmental variations. Figure 5 shows the structure of our CDR sensor, which is composed of a control module, a Reference RO, a Stressed RO, a MUX, a timer, and a counter. The counter measures the cycle count of the two ROs during a time period, which is controlled by the timer.

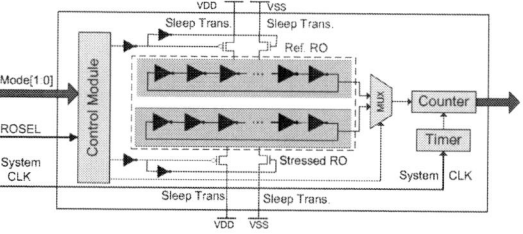

Figure 5: The structure of the CDR sensor.

System clock is used in the timer to minimize the measurement period variations due to circuit aging. The MUX selects which RO is going to be measured, and is controlled by the *ROSEL* signal. The Reference and Stressed ROs are identical, composed of HVT components. The inverters in Figure 5 could be replaced by any other types of gates (NAND, NOR, etc) only if they can construct a RO. It will not change the effectiveness of the CDR sensor significantly according to the analysis in Section 2. We use smaller-stage ROs in our CDR sensor considering the counter's measurement speed limits given a technology. For example, in our 90nm technology, a 16-bit counter can operate under frequency of up to $1GHz$; an inverter-based RO of at least 21 stages is required.

Sleep transistors are used to connect the ROs to the power supply in the CDR sensor; PMOS sleep transistors control the connection between VDD and the inverters and NMOS sleep transistors control the connection between VSS and the inverters. Both the Reference RO and the Stressed RO work in three modes, controlled by the *Mode* signal: (*i*) when the IC is in manufacturing test mode, the Reference RO and Stressed RO will be disconnected from the power supply and experience no aging. This mode only lasts a short time, depending on the test procedures of the IC. (*ii*) when the IC is in normal functional mode, the Reference RO will be disconnected from VDD and VSS but the Stressed RO will be gated on and will age. The frequency of the Stressed RO will become smaller while the Reference RO will not change. ICs will spend most of their time in this mode. (*iii*) when the IC is in authentication mode (i.e., when an IC is taken from market and its authenticity is to be verified), both the Reference RO and Stressed RO will be gated on by connecting to the power supply. The timer and counter will be enabled to measure ROs' cycle count and *ROSEL* signal will select which RO to measure. The rest of the functionality of the IC would be turned off by *Model* signals and the authentication process takes a very short period of time. The three modes of operation ensure that (*i*) the frequency difference between the Reference RO and Stressed

RO will be larger over time since the Reference RO cannot be gated on alone, and (*ii*) it is extremely difficult for adversaries to force the CDR sensor to operate in authentication mode when it is supposed to be in its normal functional mode, which would eliminate the aging difference. The only method to do that would be to modify the original CDR sensor module, which is impossible during simple recycling process.

The inverters of the Reference RO and the Stressed RO are placed physically next to each other, as Figure 5 shows, designed as a single small module. The process and environmental variations between them should be very small. Therefore, for a fresh IC, the frequency difference between the Reference RO and the Stressed RO would be within a certain small range. In a recovered IC, the Stressed RO will have suffered aging from its own oscillation since the chip has been working in normal functional mode for a long time. However, the Reference RO will not have experienced as much aging since it was gated off. The frequency difference between the Reference RO and the Stressed RO will grow larger as the chip operates longer, which is demonstrated by our simulation and silicon results. If the frequency difference is outside of the fresh ICs' frequency difference range considering process variations, we can conclude with high confidence that the CUT was recovered from used boards.

The area overhead of our CDR sensor is negligible when compared to the millions of gates in modern ICs. With a 16-bit counter, the area overhead on the ISCAS'89 benchmark s38417, a DES implementation, and an implementation of the 8051 microprocessor is 0.16%, 0.09%, and 0.006%, respectively. Power consumption is also limited to that consumed by the Stressed RO in the CDR sensor. Furthermore, this CDR sensor is resilient to removal and tampering attacks. It is inherently difficult for the recycler to remove the sensor, due to the expected measurement results from the two ROs. This feature of the CDR sensor helps detect partially recovered ICs. In addition, one cannot intentionally age the Reference RO to mask the difference between the ROs in the CDR sensor, since Reference RO cannot be gated on alone. However, one can argue that attackers with unlimited resources may be able to remove the chip package, modify the original design, and tamper the CDR sensor. For such ICs where additional security is required, alterations could be made to the CDR sensor to prevent these kinds of attacks. The CDR sensor could be obfuscated inside the IC by multiplexing functional gates. This modification would make it more difficult for an attacker to analyze the IC, making it more difficult to tamper with the sensor or modify it in any way. Additional modifications for further security may be possible as well.

4. MEASUREMENT FLOW

Figure 6 shows the measurement flow for identifying recovered ICs. First, a certain number of random, fresh ICs are used as sample chips to generate a fingerprint. The samples can come from the same or from different wafers and lots. The larger this sample is, the more process variation space will be covered, reducing the probability that fresh ICs with large process variations will be identified as recovered ICs. 1000 sample chips are tested in our simulation. In authentication mode, the Reference RO and Stressed RO's frequency is measured. The measurement environment should keep the temperature stable with as little variation as possible. However, we acknowledge that temperature variation should not impact the identification results significantly, since the Reference RO and Stressed RO will experience the same environmental temperature.

Once the sample chips have been measured, the frequency difference between the Reference RO and Stressed RO would be calculated, with $F_{diff} = F_{ref} - F_{str}$, where F_{ref} is frequency of the Reference RO and F_{str} is frequency of the Stressed RO. With 1000 sample chips, the range of F_{diff} will be determined using distribution analysis, creating a fingerprint for fresh ICs. If F_{diff} of the CUT is out of the range of the fresh ICs' fingerprint, there is a high probability that the CUT is a recovered IC. Otherwise, the CUT is assumed to be a fresh IC. The longer the CUT has been used, the more aging

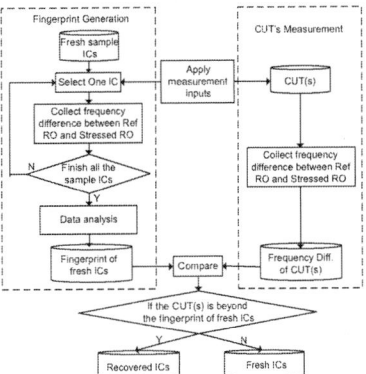

Figure 6: Measurement flow using CDR sensor for identifying recovered ICs.

effects it will have experienced, making it easier to identify. The entire measurement procedure for each CUT should take only a very short amount of time (less than 30 seconds).

5. RESULTS AND ANALYSIS

5.1 Simulation Results

In order to verify the effectiveness of the CDR sensor, we implemented and simulated it using 90nm technology [20]. HSPICE MOSRA from Synopsys is used to simulate and measure the impact of aging on the CDR sensor. The nominal supply voltage is 1.2V. During simulation, in the stress phase, the Reference RO was gated off and the Stressed RO was gated on, experiencing NBTI and HCI aging. The stress for the Stressed RO comes from its own oscillation. In the authentication phase, the Reference RO and Stressed RO were both gated on and measured one by one, selected by the *ROSEL* signal. The measurement time was set up in the timer as $100\mu s$ in our simulation. Since the clock of the counter in the CDR sensor is from the RO, the cycle count of each RO is given by the counter. The frequency of RO is equal to the cycle count divided by measurement time. The following simulation analysis is based on inverter ring oscillators.

Stage Analysis: CDR sensors with 21-stage and 51-stage ROs were simulated at $25°C$ with 2% Tox, 5% Vth, and 5% L inter-die and 1% Tox, 5% Vth, and 5% L intra-die process variations (PV0 in Table 1). 1000 chips were generated using Monte Carlo simulation by HSPICE and the total aging time was set at 24 months with a one month step.

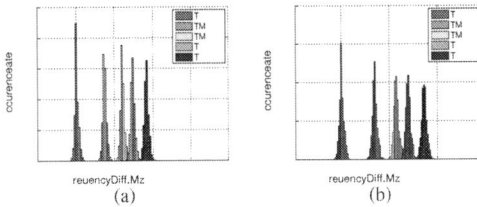

Figure 7: Frequency difference distribution of CDR sensor with PV0 using (a) 21-stage ROs, and (b) 51-stage ROs.

Figure 7(a) shows the frequency difference F_{diff} range between the 21-stage Reference RO and Stressed RO, where, in the legend, *AT* denotes aging time, *M* represents month, and *Y* represents years. From the figure, we can see that the frequency difference in fresh ICs ($AT = 0$) could be larger or smaller than 0, which is dependent on the process variations between the two ROs. In addition, the process variations of the CUTs were different from that of the 1000 sample fresh ICs, but the frequency differences still followed an identical distribution. The range of frequency differences in the fresh sample ICs is used as the fingerprint. After being used for

Table 1: Process variations.

	Inter-die			Intra-die		
	Vth	L	Tox	Vth	L	Tox
PV0	5%	5%	2%	5%	5%	1%
PV1	8%	8%	3%	7%	7%	2%
PV2	20%	20%	6%	10%	10%	4%

Table 2: Structure of CDR sensors in the test chip.

	ROs in CDR sensors			
	Reference RO	Stressed RO	RO Structure	Threshold Voltage
CDR1	R_RO1	S_RO1	1 NAND + 200 BUFs	SVT
CDR2	R_RO2	S_RO2	1 NAND + 200 BUFs	HVT
CDR3	R_RO3	S_RO3	201 NANDs	HVT

one month, the Stressed RO suffered from aging effects and its frequency became smaller. The smallest frequency difference between the Reference RO and the Stressed RO was larger than the largest frequency difference present in the fresh IC set. Therefore, the recovered IC detection rate for ICs aged for one month or longer is 100%. At 6 months, 1 year, and 2 years, the frequency difference between the Reference RO and the Stressed RO becomes larger and larger. The variation of the frequency difference becomes larger as well. This is because the aging rate is different from chip to chip due to process variations; some ICs aged faster and some others aged slower.

CDR sensors with 51-stage ROs were also implemented using the same temperature and the same process variations. Figure 7(b) shows the simulation results. Comparing Figure 7(a) and Figure 7(b), we observe that the frequency difference between aged and fresh ICs is smaller when we use the larger-stage ROs. However, the frequency difference variation becomes smaller as well, which means that the CDR sensor could still detect fully recovered ICs that had been used for one month with a 100% detection rate. If the CDR sensor uses large-stage ROs, it may impact the absolute value of the frequency difference between the Reference RO and the Stressed RO, but the detection rate will not be impacted significantly. For different technologies, the stage count of the ROs could be adjusted based on the speed of the counter. In the following, we use CDR sensors with 21-stage ROs according to our 90nm technology for further analysis.

Process Variations and Temperature Analysis: The effectiveness of our CDR sensor is partly dependent on the variations between the Reference RO and the Stressed RO. With lower rates of variation, the CDR sensor could identify fully recovered ICs that aged for shorter period of time. However, the variations between the Reference RO and the Stressed RO are determined by intra-die process variations. The smaller the intra-die variations, the more effective the CDR sensor will be. Table 1 shows the different process variation rates to analyze their impact on detection. Moving from PV0 to PV2, inter-die and intra-die variations both become larger. CDR sensors with 21-stage ROs were simulated at $25°C$ using these process variation rates.

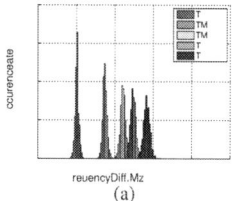 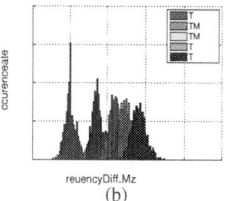

(a) (b)

Figure 8: Frequency difference distribution of CDR sensor with 21-stage ROs with (a) PV1 and (b) PV2.

By designing the sensor as a small module (hard macro), the Reference RO and the Stressed RO are placed physically close and the variations between them will be minimal. The simulation results of 1000 chips with PV1 and PV2 are shown in Figure 8(a) and Figure 8(b), respectively. Comparing Figure 7(a), Figure 8(a), and Figure 8(b), we can see that the variation of the frequency differences between the Reference RO and the Stressed RO in fresh ICs becomes larger with larger process variations. For the 1000 ICs with PV2, the detection rate of recovered ICs aged for one month is 95.2%. However, for recovered ICs that aged for six months, the detection rate is

100% again. The CDR sensor identifies shorter-aged recovered ICs with smaller intra-die process variations as in PV0, PV1, and PV2.

The 1000 circuits generated using Monte Carlo were also simulated with both process and temperature variations. Figure 9(a) shows the frequency difference occurence rate between the 21-stage Reference and Stressed ROs with process variations PV1 (shown in Table 1) and temperature variations of $±10°C$ around room temperature. Figure 9(b) shows the simulation results with process variations PV2 and temperature variations of $±20°C$ around room temperature. The results in Figure 9(a) and Figure 8(a) are from chips with the same process variations but different temperature variations. We can see that the frequency difference variations in Figure 9(a) are slightly larger than those in Figure 8(a) due to temperature variations. The same conclusion can be made by comparing Figure 9(b) and Figure 8(b). For the 1000 chips with PV2 and $±20°C$ temperature variations, the detection rate of recovered ICs aged for one months is 92.3% but it is still 100% for recovered ICs aged for six months, demonstrating that our CDR sensor is effective even with large process and temperature variations. Note that we do not expect such a large variation in temperature and process in practice when authenticating a CUT. The temperature difference and process variations between the two ROs in CDR sensor will be negligible since they are placed physically near each other.

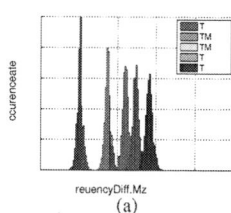 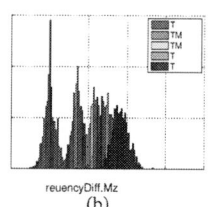

(a) (b)

Figure 9: Frequency difference distribution of CDR sensor with (a) PV1 and $±10°C$ and (b) PV2 and $±20°C$.

5.2 Silicon Results

Our CDR sensor is also verified through analysis of test chips fabricated using a 90nm technology. The test chip was originally designed to verify the effects of aging on the frequency of ROs. In this work, we use it to demonstrate the effectiveness of our CDR sensor. Since most functionality in the test chip was designed for measuring different aging effects, here, we will not describe the entire test chip's structure in detail. In total, there are 96 delay chains in the chip which can work in ring oscillator mode by controlling different input signals. Six of these ring oscillators were selected to construct three CDR sensors as shown in Table 2.

- CDR1 contains two identical ROs (R_RO1 and S_RO1) with one SVT NAND gate and 200 SVT BUFs;
- CDR2 is composed of two identical ROs (R_RO2 and S_RO2) with one HVT NAND gate and 200 HVT BUFs
- CDR3 includes ROs (R_RO3 and S_RO3) with 201 HVT NAND gates.

where R_RO1, R_RO2, and R_RO3 are Reference ROs while S_RO1, S_RO2, and S_RO3 are Stressed ROs, respectively.

Comparing ROs included in the test chip with those used for HSPICE simulation, there are two main differences: (1) the stage of ROs in the test chip is 201 while the stage of ROs used in Monte Carlo simulation is much smaller (e.g. 21). The much larger number of stages in test chip was used to make the measurement and

707

observation possible with low-end oscilloscopes. (2) the gates in ROs in the test chip are complex gates (BUFs, NANDs, etc.) while inverter-based ROs were used in simulation. That is because we aim at analyzing the impact of aging on different types of gates in test chip. However, according to our analysis in Sections 2 and 5.1, the number of stages and gate type of ROs do not present a significant impact on the effectiveness of the CDR sensor.

Currently, we only have 15 test chips in our lab and all of them are used in this experiment to present the impact of process variations and aging. To replicate the CDR sensor's stressed mode, S_RO1, S_RO2, and S_RO3 were enabled and experienced accelerated aging for 80 hours at $135°C$ with an elevated supply voltage (1.8V instead of 1.2V). The reason we used accelerated aging is that it takes a long time (usually weeks/months) to observe aging effects under normal conditions. The remaining three ROs were gated off and experienced no aging. In authentication mode, all of the ROs were enabled and the temperature was brought back to room temperature (around $25°C$). With the 15 fresh test chips, the average frequency of ROs is about $7.5Mhz$. Figure 10 shows the experimental results of the three CDR sensors over the test chips. The red bars in the figure show the frequency difference between Reference RO and Stressed RO in each CDR sensor at time zero (fresh/unused ICs). Similarly, the yellow bars are the frequency difference between the two ROs after 80 hours of aging.

Since a much larger number of stages are used in these sensors compared to those used in our simulations, the mean frequency of the ROs in test chip and the frequency difference values are very much different from that in simulations. However, even with 201 gates in these ROs, the detection rates of recovered ICs that aged 80 hours using $CDR1$, $CDR2$, and $CDR3$ are all still 100%, which demonstrates that the RO stage count in CDR sensor does not have a significant impact on the sensor's effectiveness in detecting recovered ICs. According to our detailed results, the average frequency degradation of the stressed ROs in $CDR1$, $CDR2$ and $CDR3$ (shown in Figure 10) is 3.2%, 4.0%, and 3.8%, respectively. Comparing Figure 10(a) and Figure 10(b), we can see that the frequency difference gap between fresh chips and aged chips in $CDR2$ is larger than that in $CDR1$. This is due to the fact that CDR sensors with HVT gates ($CDR2$) will be more effective than those with SVT gates ($CDR1$), which is also demonstrated in Figure 1(c) through simulation results. Comparing detection rates in Figure 10(b) using $CDR2$ (composed of HVT buffers) and Figure 10(c) using $CDR3$ (composed of HVT NAND gates), we can see that the gates used in the RO can slightly change the effectiveness of CDR sensor but not significantly.

Note that the ROs in the CDR sensors in the test chip were not placed as close as they were supposed to. For instance, the results at time zero show that for $CDR1$ and $CDR2$, the R_ROs are faster than S_ROs in most cases while this is not the case for $CDR3$. This could be because of the spatial variations that exist between the ROs not placed near each other, which made some ROs faster than others. *For a CDR sensor to be the most effective, it is recommended to place both ROs in a single localized module to reduce the variation between them.* Limited by the amount and structure of the test chips, we cannot perform the same analysis with silicon data as we did with the Monte Carlo simulations. however, the silicon results from these test chips demonstrate the effectiveness of our CDR sensor.

6. CONCLUSIONS AND FUTURE WORK

In this paper, we have presented the concept of IC/die recovery problem and proposed a technique using a light-weight on-chip sensor to detect recovered ICs. The fingerprint generated by the frequency difference between the Reference RO and the Stressed RO in the CDR sensor makes identification of fully recovered ICs easily possible. Simulation results using different process and temperature variations demonstrated its effectiveness. The silicon results further demonstrated that our CDR sensor can detect recovered ICs even used in the field for a very short period of time. In addition, our

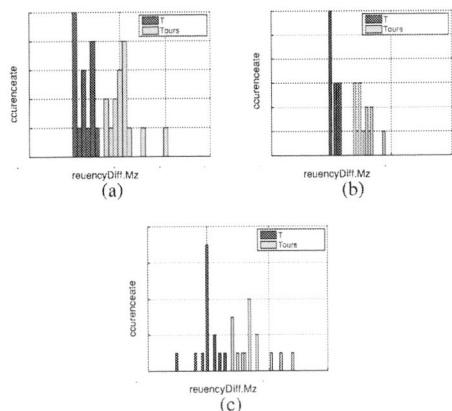

Figure 10: Frequency difference distribution in (a) $CDR1$, (b) $CDR2$, and (c) $CDR3$.

future work includes (i) using multiple CDR sensors to further improve detection resolution and capability, (ii) obfuscating the CDR sensor for further improvement of the security against tampering, and (iii) using other circuit parameters such as path-delay, leakage current, and switching power to detect recovered ICs.

7. ACKNOWLEDGEMENT

The authors also would like to thank LeRoy Winemberg of Freescale for providing the test chips for reliability analysis.

8. REFERENCES

[1] "Defense Industrial Base Assessment: Counterfeit Electronics," Bureau of Industry and Security, U.S. Department of Commence, *http://www.bis.doc.gov/defenseindustrialbaseprograms/osies/defmarketresearchrpts/final_counterfeit_electronics_report.pdf*, 2010.
[2] L. W. Kessler and T. Sharpe, "Faked Parts Detection," *http://www.circuitsassembly.com/cms/component/content/article/159/9937-smt*, 2010.
[3] Business Week, "Dangerous Fakes," *http://www.businessweek.com/magazine/content/08_41/b4103034193886.htm*, 2008.
[4] Military Times, "Officials: Fake Electronics Ticking Time Bombs," *http://www.militarytimes.com/news/2011/11/ap-fake-electronics-ticking-time-bomb-110811/*, 2011.
[5] Tezzaron Semiconductor, "3D-ICs and Integrated Circuit Security," *http://www.tezzaron.com/about/papers/3D-ICs_and_Integrated_Circuit_Security.pdf*, 2008.
[6] *http://www.combatcounterfeits.com/gallery.htm*
[7] J. Stradley and D. Karraker, "The Electronic Part Supply Chain and Risks of Counterfeit Parts in Defense Applications," *IEEE Transactions on Components and Packaging Technologies*, pp.703-705, Sept. 2006.
[8] M. Tehranipoor, and C. Wang "Introduction to Hardware Security and Trust," Springer, New York, USA. 2011.
[9] K. Lofstrom, W. R. Daasch, and D. Taylor, "IC Identification Circuit Using Device Mismatch," in *Proc. ISSCC*, pp. 370-371, 2000.
[10] R. Pappu, "Physical One-way Functions," *Phd thesis, MIT*, 2001.
[11] G. Suh and S. Devadas, "Physical Unclonable Functions for Device Authentication and Secret Key Generation," in *Proc. DAC*, pp. 9-14, 2007.
[12] E. Ozturk, G. Hammouri, and B. Sunar, "Physical Unclonable Function with Tristate Buffers," in *Proc. ISCAS*, pp. 3194-3197, 2008.
[13] A. Maiti and P. Schaumont, "Improved Ring Oscillator PUF: An FPGA-Friendly Secure Primitive," *IACR Journal of Cryptology, special issue on Secure Hardware*, 2011.
[14] F. Koushanfar "Hardware Metering: A Survey," *http://aceslab.org/sites/default/files/A5-fk-metering.pdf*
[15] J. Roy, F. Koushanfar, and I. Markov, "EPIC: Ending Piracy of Integrated Circuits," in proc. *DATE08*, pp. 1069-1074, 2008.
[16] A. Baumgarten, A. Tyagi, and J. Zambreno, "Preventing IC Piracy Using Reconfigurable Logic Barriers," *IEEE Design & Test of Computers*, 2010.
[17] T. Kim, R. Persaud, and C. H. Kim, "Silicon Odometer: An On-Chip Reliability Monitor for Measuring Frequency Degradation of Digital Circuits," *IEEE Journal of Solid-State Circuits*, pp. 974-880, 2008
[18] J. Keane, X. Wang, D. Persaud, and C.H. Kim, "An All-In-One Silicon Odometer for Separately Monitoring HCI, BTI, and TDDB," *IEEE Journal of Solid-State Circuits*, pp. 817-829, 2010
[19] S. Mahapatra, D. Saha, D. Varghese, and P. B. Kumar, "On the Generation and Recovery of Interface Traps in MOSFETs Subjected to NBTI, FN, and HCI Stress," *IEEE Trans. on Electron Devices*, vol. 53, no. 7, pp. 1583-1592, 2006.
[20] "http://www.synopsys.com/Community/UniversityProgram/Pages/Library.aspx".
[21] K. Uwasawa, T. Yamamoto, and T. Mogami, "A New Degradation Mode of Scaled P+ Polysilicon Gate P-MOSFETs Induced by Bias Temperature Instability," in *Proc. Int. Electron Devices Meeting*, pp. 871-874, 1995.
[22] P. Heremans, R. Bellens, G. Groeseneken, and H. E. Maes, "Consistent Model for the Hot Carrier Degradation in N-Channel and P-Channel MOSFETs," *IEEE Trans. Electron Devices*, vol. 35, no. 12, pp. 2194-2209, 1988.
[23] H. Luo, Y. Wang, K. He, R. Luo, H. Yang, and Y. Xie "Modeling of PMOS NBTI Effect Considering Temperature Variation," in *Proc. ISQED*, pp. 139-144, 2007.
[24] R. Vattikonda, W. Wang, Y. Cao, "Modeling and Minimization of PMOS NBTI Effect for Robust Nanometer Design," in *Proc. DAC*, pp.1047-1052, 2006.

Confidentiality Preserving Integer Programming for Global Routing

Hamid Shojaei, Azadeh Davoodi, Parmeswaran Ramanathan
Department of Electrical and Computer Engineering
University of Wisconsin at Madison, USA
Email: adavoodi@wisc.edu

ABSTRACT

Cloud computing for EDA requires a client to send problem instances containing confidential design information to an untrusted distributed network. To preserve the design information in such a framework, this work focuses on obfuscating the global routing problem modeled as an Integer Linear Program (ILP) for large industry benchmarks. Multiple transformations are introduced in a proposed framework in which the client *masks* the ILP instance before it is sent to the cloud. The cloud solves the masked instance and the client unmasks the generated solution. No approximations are involved in this process. The masked instance is shown to be substantially more immune to various introduced attacks. Otherwise layout statistics and even detailed connectivity information can easily be deciphered. When applying the transformation, the increase in immunity can be traded off with the induced runtime overhead.

Categories & Subject Descriptors
B.7.2. [Integrated Circuits]: Design Aids

General Terms Algorithms, Design

Keywords Security, Integer Programming, Global Routing

1. INTRODUCTION

The growing computational complexity of the Electronic Design Automation (EDA) tasks with the design complexity is resulting in explosion in demand for computing resources [2]. Cloud computing is emerging as a new window of opportunity for EDA. It can provide customers instant and affordable access to powerful and distributed networks of machines running complex EDA procedures. The computational resources are powerful to match even the highest demand peaks while there can exist instant elasticity to stop paying for these resources when the peak demands are over.

Despite these attractive features, a major obstacle for EDA vendors is addressing the security challenges of this computing substrate [1], especially the challenge of transferring and solving a large problem instance in the cloud which

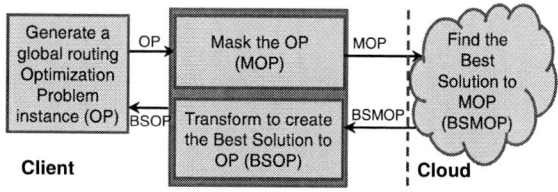

Figure 1: Overview of our framework.

contains confidential design information.

This work introduces several obfuscation techniques to mask confidential layout information in the Integer Linear Program (ILP) representing the global routing problem.

In a proposed framework shown in Figure 1, the client first creates a *masked* ILP instance before sending it to the cloud. The cloud identifies the best solution to the masked ILP and forwards the solution to the client. The solution from the cloud is unmasked at the client side. No approximations are involved in the entire process and the quality of the solution remains intact as if there was no masking.

We show the masked ILP is substantially more immune to various attacks which are introduced in this work including extracting high-level layout statistics such as the number of nets, their wirelengths, and the routing capacity of each layer. An attack is also introduced to extract detailed connectivity information from the congested areas on the layout.

The global routing constraint matrix is quite sparse for industry benchmarks. Our obfuscation techniques reduce the sparsity in the matrix which increases the immunity to attacks. As a result, more computation is necessary to solve the masked problem. However, the client can control the degree of reduced sparsity and thus explore the trade-off between the amount of immunity to attacks with the induced runtime overhead to solve the masked problem.

The summary of our contributions are listed below:

- A framework and the associated techniques through which the client masks an ILP model of global routing before sending it to cloud and unmasks the generated solution without any approximations.

- Several attacks which can decipher layout and connectivity information from the unmasked problem.

- Ability to explore the trade off between the increase in immunity and the corresponding runtime increase.

In the remainder of this paper, Section 2 describes the related works. Section 3 gives an overview of the ILP model.

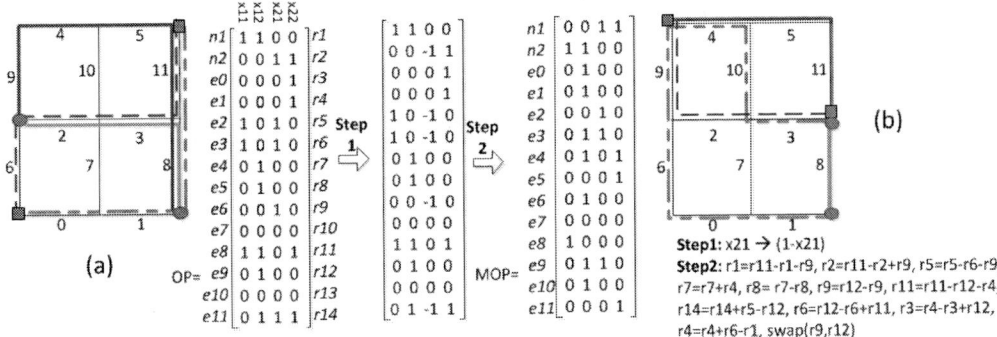

Figure 2: Example showing (a) the optimization model (OP), and (b) creating a masked instance (MOP).

Section 4 describes our obfuscation techniques. Several attacks are introduced in Section 5. Simulation results are presented in Section 6 followed by conclusions.

2. RELATED WORKS

Integer linear programming has been used in a number of recent approaches to global routing including [3], [9], [10], [13], and [12]. Among these approaches, [13] and its parallel version [12] are heavily ILP-centric and demonstrate significant improvement in the solution quality is attainable, compared to the other competitive procedures.

Related works also exist in watermarking integer linear programming solutions. Specifically, the works [11], [8] study obfuscation techniques for the ILP models representing a few EDA problems such as circuit satisfiability and scheduling of a Data Flow Graph in high-level synthesis. The procedures offered in these works are generic for any ILP and demonstrated on small-sized instances.

The work [5] considers obfuscating linear programs via transformations. The assumption is that the linear program is solved by a distributed solver composed of multiple collaborating parties where each may operate on a portion of the problem data. Other works based on secure multiparty linear programming solvers include [4] which offers a distributed Simplex algorithm and [6].

In contrast to these previous works, our work is shown for *Integer* linear programs. We introduce practical and efficient procedures which are specifically designed for a standard ILP model of global routing and are shown to be applicable to large industry problem instances. Our procedures are compatible with the above works and their hybrid combination can be investigated as part of the future research.

3. OVERVIEW OF OUR FRAMEWORK

Integer Linear Program Model: In a mathematical description of global routing, we are given a grid-graph $G = (V, E)$ and a set of multi-terminal nets denoted by $\mathcal{N} = \{T_1, T_2, \ldots, T_N\}$, (with $T_i \subset V$). Edge e in graph G is associated with capacity u_e and weight c_e $\forall e \in E$. Denote by $\mathcal{T}(T_i)$ a collection of Steiner trees, each as one *candidate* to connect the terminals in net T_i. Let the parameter $a_{te} = 1$ if Steiner tree t contains edge $e \in E$, $a_{te} = 0$ otherwise. Define the binary decision variable x_{it} that is equal to 1 if and only if net T_i is routed with tree $t \in \mathcal{T}(T_i)$. An integer linear program for the global routing problem can be written as

$$\min_x \sum_{i=1}^{N} \sum_{t \in \mathcal{T}(T_i)} c_{it} x_{it} \qquad \text{(OP)}$$

$$\begin{cases} \sum_{t \in \mathcal{T}(T_i)} x_{it} = 1 & \forall i = 1, \ldots, N, \\ \sum_{i=1}^{N} \sum_{t \in \mathcal{T}(T_i)} a_{te} x_{it} \leq u_e + \acute{o}_e & \forall e \in E, \\ x_{it} = \{0, 1\} & \forall i = 1, \ldots, N, \forall t \in \mathcal{T}(T_i). \end{cases}$$

The first set of equations in (OP), henceforth called the *route selection constraints*, enforces the routing of each net. The second set of equations expresses the *edge capacity constraints*. The utilization of edge e defined as the number of routes that contain edge e is expressed by $\sum_{i=1}^{N} \sum_{t \in \mathcal{T}(T_i)} a_{te} x_{it}$. It should be bounded by the summation of capacity u_e and a constant parameter $\acute{o}_e \geq 0$ which is set to ensure the formulation always has a feasible solution. The parameter c_{it} is the cost of candidate tree t for net T_i which is also referred to as its wire length and computed by $c_{it} = \sum_{e \ni t} c_e$, where the notation $e \ni t$ denotes that edge $e \in E$ is contained in route $t \in \mathcal{T}(T_i)$. The objective is thus minimization of the total wire length of the routed nets.

Figure 2(a) shows a global routing instance with two nets, each with two candidate trees. In the constraint matrix corresponding to formulation (OP), the first two rows (denoted by n_1 and n_2) correspond to the route selection equality. The remaining rows correspond to the edge capacity constraints. Each column corresponds to a candidate tree and contains a 1 in a row if the corresponding edge is contained in the tree.

To create an ILP instance, first, the set of candidate trees $\mathcal{T}(T_i)$ $\forall T_i \subset V$ is generated using a standard rip-up and re-route process from [9]. Using these candidate trees, the parameter $\acute{o}_e$ $\forall e \in E$ is specified which completes the description of the formulation. The details of candidate tree generation and setting this parameter are explained in Appendix 9. The procedure guarantees that a feasible solution exists for the formulation and identifies such a solution.

As shown in Figure 1 we assume the client adopts the following strategy to preserve the confidential information that may be extracted from the ILP instance: 1) Formulates the global routing instance as an optimization problem according to formulation (OP); 2) Transforms OP into a *masked optimization problem* (MOP); 3) Forwards the MOP and requests a solution from the cloud; 4) Cloud finds the best solution to the MOP (BSMOP) and returns it to the client; and 5) Client transforms BSMOP to obtain the best solution to the OP (BSOP). The example in Figure 2 shows creation of MOP from OP using a set of transformations and how the MOP may be interpreted as new problem instance.

710

4. PROPOSED SCHEME

4.1 Threat Model

We assume an honest-but-curious threat model from the cloud. This model has the following implications. First, since the cloud is honest, it will correctly perform all its computations. It also faithfully follows the specified information exchange protocol with the client. Second, since the cloud is curious, it will strive to derive as much information as possible based on all the data provided by the client and/or its own general knowledge of the problem area. Specifically, we assume the cloud knows the typical structure of OP used by the client to solve the global routing instance. Moreover, the cloud has unfettered access to MOP and BSMOP and it carries out attacks on them to derive as much salient information as possible.

4.2 ILP Masking Strategy

The original ILP formulation (OP) can be written as:

$$\min c^T x \qquad \text{(OP0)}$$

$$\begin{cases} A_E \cdot x = b_E \\ A_I \cdot x \leq b_I \\ \forall\, i,\ x_i \in \{0,1\} \end{cases}$$

where x_i is the ith element of vector x.

The client performs a sequence of masking transformations on (OP0). These transformations do not affect the correctness of the solution.

First, the client adds surplus variables to convert the inequalities to equalities.

$$\min c^T x + M^T s \qquad \text{(MOP1)}$$

$$\begin{cases} A_E \cdot x = b_E \\ A_I \cdot x + s = b_I \\ \forall\, i,\ x_i \in \{0,1\} \quad s \geq 0. \end{cases}$$

In (MOP1), M is a carefully chosen vector. The proposed heuristic for selecting M is described later in this section.

Then, it substitutes the variables in x with a new variable z such that $x = R + Qz$ with the following conditions on vector R and diagonal matrix Q: (i) for all i, $q_{ii} \in \{-1, 1\}$, where q_{ii} is the ith diagonal element of Q, and (ii) for all i, $r_i = 1$ if $q_{ii} = -1$ and $r_i = 0$ otherwise, where r_i is the ith element of vector R. Informally, due to the conditions on R and Q, this transformation does the following.

$$z_i = \begin{cases} x_i & \text{if } q_{ii} = 0 \\ 1 - x_i & \text{otherwise.} \end{cases}$$

In other words, some variables in x are replaced by their complements. After this transformation, (MOP1) can be rewritten as follows.

$$\min c^T Qz + M^T s \qquad \text{(MOP2)}$$

$$\begin{cases} A_E \cdot Q \cdot z = b_E - A_E \cdot R \\ A_I \cdot Q \cdot z + s = b_I - A_I \cdot R \\ \forall\, i,\ z_i \in \{0,1\} \quad s \geq 0. \end{cases}$$

Next, the client applies several row and column transformations to the constraint matrix along with corresponding transformation to the right-hand side (RHS) vector. These transformations are equivalent to pre-multiplying both sides of the equations in (MOP2) using non-singular matrices P_E and P_I. That is, (MOP2) can be rewritten as

$$\min c^T Qz + M^T s \qquad \text{(MOP3)}$$

$$\begin{cases} P_E \cdot A_E \cdot Q \cdot z = P_E(b_E - A_E \cdot R) \\ P_I \cdot A_I \cdot Q \cdot z + P_I \cdot s = P_I(b_I - A_I \cdot R) \\ \forall\, i,\ z_i \in \{0,1\} \quad s \geq 0. \end{cases}$$

P_E and P_I tend to increase the sparsity of the constraint matrix, which in turn, increases the time required to solve the optimization problem. Hence, these matrices P_E and P_I must be chosen with some care. A heuristic for selecting P_E and P_I is described later in this section.

Third, the client hides the cost matrix c by adding the constraints to the objective.

$$\begin{aligned} \min\ & c^T Qz + M^T s \\ & + N_E^T(P_E A_E Qz - P_E(b_E - A_E R)) \\ & + N_I^T(P_I A_I Qz + P_I s - P_I(b_I - A_I R)) \end{aligned} \qquad \text{(MOP4)}$$

$$\begin{cases} P_E \cdot A_E \cdot Q \cdot z = P_E(b_E - A_E \cdot R) \\ P_I \cdot A_I \cdot Q \cdot z + P_I \cdot s = P_I(b_I - A_I \cdot R) \\ \forall\, i,\ z_i \in \{0,1\} \quad s \geq 0. \end{cases}$$

In the above formulation, N_E and N_I are appropriately dimensioned vectors with elements comprised of 0s and 1s.

4.3 Correctness of the Masking Strategy

To show that the above transformations do not change the optimal solution, we must show that any feasible (i.e., with respect to the constraints) and optimal (i.e., feasible solution with minimum cost) to (OP0) corresponds to a feasible and optimal solution to (MOP4). This result follows from the following two lemmas.

LEMMA 1. *If a solution x is feasible in (OP0), then the solution $z = Q^{-1}(x - R)$ and $s = b_I - A_I x$ is feasible in (MOP4). Conversely, if a solution $\begin{bmatrix} z & s \end{bmatrix}^T$ is feasible in (MOP4), then $x = R + Qz$ is feasible in (OP0).*

Proof: Not included due to the page limit.

LEMMA 2. *If a feasible solution $\begin{bmatrix} z & s \end{bmatrix}^T$ minimizes the constraints in (MOP4), then $x = R + Qz$ minimizes the constraints in (OP0).*

Proof: Not included due to page limit.

4.4 Selection of M, N_E, N_I, Q, R, P_E, and P_I

The selection of M, N_E, N_I, Q, R, P_E, and P_I does not affect the correctness. These matrices affect the sparsity of the constraint matrix, which in turn, increases the time required to solve the problem. We now describe our heuristics for selection of these matrices.

Selection of M: The inclusion of surplus variables s in the objective can affect the quality of routing solution. To ensure that the quality of the routing solution obtained from (MOP4) is the same as that from (OP0), we select $M = -N_I^T P_I$ so that the surplus variables are eliminated from the objective.

Selection of N_E and N_I: N_E and N_I mask the cost vector in the optimization. In our heuristic, N_E and N_I are comprised of 0s and 1s. A 1 means that the corresponding constraint is added to objective. It is sufficient to add a few constraints to the objective to mask the cost values. Therefore, the elements in the N_E and N_I can be randomly chosen to be 0 or 1 with a small non-zero probability of selecting a 1.

711

Selection of Q and R: As mentioned above, Q is a diagonal matrix and each element is either $+1$ or -1. Given Q, R is unique. In our implementation, we randomly choose the diagonal elements of Q to be either $+1$ or -1 with 20% probability of selecting -1.

Selection of P_E and P_I: Note that, pre-multiplying the constraints with either P_E or P_I is equivalent to replacing each constraint by a linear combination of all the constraints. If P_E and P_I are invertible, then the new constraints remain linearly independent. In our implementation, we do not explicitly generate P_E and P_I. Instead, we use the following heuristic to replace the constraints with linear combination of other constraints.

We first choose a design parameter F whose value lies in the range 0 to the total number of variables, $|x| + |s|$. We consider the constraint matrix one column at a time starting from the leftmost column. In each column we find the rows with non-zero entries and replace each one of those corresponding constraints by a random linear combination of the associated constraints. The weights in each linear combination are restricted to be either $+1$ or -1. We proceed in this fashion until F columns have been considered. We can tune F to tradeoff the ability to mask and the deleterious effect on runtime. In particular, larger F results in more immunity to attack but also larger runtime. In the evaluation results, we show that limiting F to less than 20% of the total number of variables is sufficient to get almost full immunity from the attacks considered.

5. ATTACK ANALYSIS

5.1 Routing Instance Statistics

The cloud is interested in estimating salient statistics of the global routing instance such as: (i) *net-oriented statistics* such as number of nets in the instance, (ii) *edge-oriented statistics* such as minimum, maximum, mean, and distribution of edge capacities, and (iii) *route-oriented statistics* such as minimum, maximum, mean, and distribution of wirelengths.

These statistical information can be easily retrieved if the cloud can distinguish the route selection constraints from the edge capacity constraints. For example, in (OP0), there is one route selection constraint for every net in the instance. Therefore, if the cloud can identify all the route selection constraints, it can easily determine the total number of nets in the given instance. Similarly, the right-hand side of edge capacity constraints are the edge capacities. Therefore, if the cloud can identify all the edge capacity constraints, it can derive any edge capacity statistic of interest.

Attack strategies to distinguish route selection constraints from edge capacity constraints are as follows. Most of these attacks work well in (OP0). They work well in some of the masked problems (MOP1)–(MOP3). None of them work well in (MOP4).

1. *Observation:* In (OP0), route selection constraints are equalities whereas edge capacity constraints are inequalities.

   Attack1: All equalities are considered to be route selection constraints and all inequalities are considered to be edge capacity constraints.

2. *Observation:* In (OP0), right hand side (RHS) of route selection constraints is always a 1.

Attack2: Classify all constraints with a right-hand side of 1 as route selection constraints and the others as edge capacity constraints.

3. *Observation:* The surplus variable in (MOP1)–(MOP4) must be non-binary but all other variables are binary.

   Attack3: Since the surplus variable are usually introduced to convert inequalities to equalities, classify constraints with a surplus variable as edge capacity constraints.

Two attack strategies to recover route-oriented statistics are as follows.

1. *Observation:* In the edge capacity constraints, there is a non-zero coefficient for every route which makes use of that edge.

   Attack4: The total number of non-zero entries in the route-related columns (i.e., columns corresponding to 0-1 variables) is an estimate of the route's wirelength. Hence, compute the route statistics by counting the non-zero entries in each 0-1 variable corresponding column of the constraint matrix.

2. *Observation:* The cost coefficient associated with a particular route is usually a characterization of the cost of the route, which often corresponds to the route's wirelength.

   Attack5: Use cost vector statistics to estimate route statistics.

5.2 Routing Topology Attacks

The cloud is clearly very interested in retrieving the topology of the routing trees in the given instance. If a cloud is able to correctly reconstruct this topology, then the cloud will know all the nets and the selected global route for each net. In this subsection, we describe an attack strategy to reconstruct the routing topology. This attack involves formulating and solving an ILP constructed using the constraint and solution information available to the cloud.

From the constraint, the cloud first performs one or more of the net-directed and edge-directed attacks to identify the edge constraints and construct $\hat{A}_I$, an estimate of the A_I matrix. Let a_{ij} denote the element in the ith row and j^{th} column of $\hat{A}_I$. Let $\hat{E}$ denote the total number of rows in the $\hat{A}_I$ matrix, i.e., $\hat{E}$ is the cloud's estimate of the total number of edges in grid routing graph. Let $I_{a_{il}>0}$ denote an indicator function which is 1 if $a_{il} > 0$ and 0 otherwise. The cloud then constructs a new $\hat{E} \times \hat{E}$ edge adjacency matrix (denoted by W), with w_{ij}, the element in the ith row and j^{th} column of W, is chosen as follows.

$$w_{ij} = \sum_{l=1}^{|x|+|s|} I_{a_{il}>0} \cdot I_{a_{jl}>0}.$$

Informally, if a_{il} and a_{jl} are both greater than 0, it means that edges i and j are both part of the routing tree l. If the same is true for many more routing trees, then it is highly likely that edges i and j are physically close (or even neighbors) in the grid routing graph. For example, in Figure 2(a), observe that the edges labeled $e2$ and $e3$ are part of two routing trees, and hence they are likely to be adjacent to each other. Similar conclusions can be made about edges $e3$ and $e8$, $e3$ and $e11$, $e2$ and $e8$, and $e2$ and $e11$. Clearly, not all

Table 1: Attacks on extracting routing instance statistics.

Case	Number of Nets		Edge Capacity			Wirelength			Number of Trees		
	#equalities	#1s in RHS	ave	min	max	ave	min	max	ave	min	max
(OP0)	35344	35344	64	38	70	106	1	2678	4	1	22
(MOP1)	112188	35344	64	38	70	106	1	2678	4	1	22
(MOP2)	112188	13799	3	-370	165	106	1	2678	3	2	18
(MOP3)-10%	112188	6160	27	-25	3564	51	2	2654	7	3	32
(MOP3)-20%	112188	3234	46	-34	3785	23	2	3719	17	6	59
(MOP4)-10%	112188	6160	27	-25	3564	51	2	2654	7	3	32
(MOP4)-20%	112188	3234	46	-34	3785	23	2	3719	17	6	59

of these conclusions are correct, but some of the conclusions are correct. The goal for the cloud is to infer as much correct information as possible. To achieve this, the cloud solves the following ILP.

$$\max_{y} \sum_{i=1}^{\hat{E}} \sum_{j=i+1}^{\hat{E}} y_{ij} w_{ij} \qquad \text{(IP-VUL)}$$

$$\left\{ \begin{array}{ll} \sum_{j=1}^{\hat{E}} y_{ij} \leq 6 & \forall i = 1, \ldots, \hat{E}, \\ y_{ij} = y_{ji} & \forall i,j = 1, \ldots, \hat{E}, \\ y_{ij} = \{0,1\} & \forall i,j = 1, \ldots, \hat{E}. \end{array} \right.$$

Define a 0-1 variable y_{ij} whose value is 1 if the cloud estimates that the edges i and j share a common vertex in the grid graph. Otherwise y_{ij} is 0. Due to the reasoning in the previous paragraph, higher the value of w_{ij}, the higher the likelihood of edge i and j sharing a common vertex, and therefore y_{ij} should be assigned as 1 (see the objective in (IP-VUL)). There are two types of constraints. The first constraint follows from the observation that the maximum number of neighbors for each edge should be bounded depending on grid graph. For a grid graph with only horizontal and vertical edges this bound is 6, albeit some edges in the border should have fewer neighbors. Since one cannot determine which edges are on the border, a bound of 6 is imposed for all edges. The second constraint expresses the requirement that if edge i is adjacent to edge j, then edge j should also be adjacent to edge i.

6. SIMULATION RESULTS

We wrote a C++ program to create the ILP corresponding to formulation (OP) for various ISPD 2007 2D benchmark instances. The ILP was created as described in Section 3.

The experiments in this section aim to demonstrate the quality of our obfuscation techniques for immunity to various attacks and the associated tradeoffs. We note that solving the generated ILP (with or without masking) allows creating high quality global routing solutions compared to competitive procedures. (See Section 10.1 for more details.)

Extraction of Routing Instance Statistics: In this experiment, we considered benchmark instance adaptec1 and show in detail the impact of different levels of obfuscation to extract statistics from this routing instance. We consider five ILP cases, each corresponding to one row in the Table as explained in Section 4. (OP0) is the original ILP without masking and considered as the reference case for comparison. (MOP1) is the ILP after translating all the inequality constraints into equalities. (MOP2) is after further replacing 20% of the x_{it} variables with 1-x_{it}, which was done according to random selection in our experiment. (MOP3) is (MOP2) after applying our masking heuristic explained in Section 4.4. We consider two cases (MOP3)-10% and (MOP3)-20%

in which the value of parameter F in the heuristic is set to 10% and 20% of the number of columns, respectively. The columns in Table 1 correspond to different attacks. Note that, for all the considered routing instance attacks, (MOP3) is equivalent to (MOP4). Hence, the results (MOP3)-10% is identical to that of (MOP4)-10%. Similarly for the case of 20%. Therefore, we only focus of (MOP3).

Consider extracting the number of nets. Two possible attacks are done. First is counting the number of equality constraints which represents the number of route selection constraints in the ILP. Here, (OP0) reveals the number of nets which is 35344 in this instance. However, the remaining cases have transformed all inequalities to equalities and are immune to this attack.

To extract the number of nets, the second attack is by counting the number of 1s in the right-hand-side (RHS) vector in the ILP. Since the edge capacity values are typically higher than 1, each 1 entry in RHS may identify a route selection constraint. Here, both (OP0) and (MOP1) reveal the number of nets exactly. However, (MOP2) applies variable transformation to 20% of the candidate tree variables and as a result, the RHS values may change. The estimated number of nets using this method drops to 13799. (MOP3)-10% and (MOP3)-20% both further reduce the estimated number of nets to 6160 and 3234, respectively. This is because our heuristic further impacts the values on the RHS. (MOP3)-20% applies a higher number of transformations which makes the estimated number of nets more inaccurate.

The next attack is on extracting the average, minimum and maximum edge capacities as shown in columns 4 to 6. We make this attack more challenging by assuming that the attacker knows the net constraints. So the average, minimum and maximum edge capacities are computed after excluding the entries in the RHS corresponding to the net constraints in *all* the cases. Since (OP0) and (MOP1) do not change the RHS vector, they do not provide any immunity to this attack; the average, minimum, and maximum capacities are indeed 68, 38, and 70 for this benchmark. The remaining cases all provide immunity to this attack.

The next attack is on computing the wirelength of the candidate trees. In this attack, the number of non-zero and positive entries are counted in each column to compute the wirelength corresponding to the tree representing that column. From (OP0) the average, minimum and maximum wirelength of the trees are accurately found to be 106, 1 and 2678 units. (MOP1) and (MOP2) also reveal this information because they do not change the number of non-zero entries in each column. ((MOP2) swaps some +1 entries with -1). The (MOP3) cases are immune to this attack.

The last attack extracts the average, minimum and maximum number of candidate trees generated for each net. In this attack, the number of non-zero entries for each route selection constraint is counted. To make this attack more

Table 2: Comparison in the routing topology attack

Bench	#Edges	(OP0)		(MOP3)-10%		(MOP3)-20%	
		%RST	%VUL	%RST	%VUL	%RST	%VUL
adaptec1	76219	63.49	78.01	14.25	16.01	0.67	4.00
adaptec2	36737	57.61	85.17	9.87	12.20	0.00	0.00
adaptec3	188288	65.32	77.54	15.98	18.06	0.00	2.54
adaptec4	72111	74.67	92.43	19.04	21.09	0.47	9.95
adaptec5	118226	61.14	73.23	22.79	25.10	0.69	12.01
newblue1	45340	67.42	91.67	6.49	10.00	1.44	2.78
newblue2	26607	53.99	72.90	5.95	9.73	1.30	2.94
newblue3	69305	48.72	66.10	20.01	24.32	0.71	7.31
average	79104	61.55	79.63	14.30	17.06	0.66	5.19

Table 3: Tradeoff analysis.

Bench	(OP0) time	(MOP2)		(MOP3)-10%		(MOP3)-20%	
		%NZ	%OV	%NZ	%OV	%NZ	%OV
adaptec1	58	0	3.45	55.47	9.38	57.31	24.14
adaptec2	35	0	-14.29	47.93	20.45	51.22	57.14
adaptec3	92	0	7.61	39.91	17.12	43.27	64.13
adaptec4	26	0	23.08	32.18	29.73	65.61	207.69
adaptec5	109	0	-8.26	54.49	31.01	55.77	88.07
newblue1	169	0	-9.47	88.21	17.16	88.27	28.40
newblue2	28	0	10.71	18.00	24.32	79.38	67.86
newblue3	264	0	7.58	66.71	12.00	67.65	27.65
average	98	0	2.55	50.36	20.15	63.56	70.64

challenging, we assume the route selection constraints are accurately identified in *all* the cases. Here (OP0) and (MOP1) do not provide any immunity. (MOP2) provides limited immunity since only some of the route variables x_i are replaced by $1-x_i$ and some of the constraints will remain unchanged. As a result, the average, minimum and maximum values corresponding to (MOP2) are closer to the actual values shown in (OP0) and (MOP1). Both (MOP3)-10% and OP-20% provide immunity to this attack.

Routing Topology Attacks: As explained in Section 5.2, it may not be possible to fully retrieve the routing topology. To assess the effectiveness of this attack, we define a measure called %VUL as follows.

$$\%\text{VUL} = (1 - \frac{\sum_{i=1}^{|E|} \sum_{j=1}^{|E|} |v_{ij} - \hat{v}_{ij}|}{|E|^2}) \times 100 \qquad \text{(VUL)}$$

In this measure, v_{ij} is the ground truth derived from the correct routing topology reflecting whether or not edges i and j are adjacent to each other and $\hat{v}_{ij}$ is the estimated relationships between edges i and j based on the solution of (IP-VUL). Specifically, %VUL measures the percentage of edges correctly estimated by the routing topology attack.

Results are shown in Table 2. Column 2 shows the number of edges for each benchmark. We aim retrieving the *congested* edges with a utilization of 70% or higher in this attack. The number of these *congested* edges is reported in column 2 for each benchmark. For each case, we report the number of restored edges (%RST) as well as the introduced vulnerability metric given by equation (VUL). (OP0) has a vulnerability of on average 79% and over 61% of the congested edges can be retrieved in this experiment. The vulnerability is reduced to 17.06% and 5.19% in (MOP3)-10% and (MOP3)-20% respectively. The retrieved edges also significantly drop to 14.30% and 0.66% in these two cases.

Tradeoff Analysis: Table 3 shows the tradeoff analysis between the induced runtime overhead and decrease in vulnerability. We report the runtime of (OP0) in minutes in column 2. For (MOP2), (MOP3)-10% and (MOP3)-20% we report the percentage increase in the number of non-zero entries (%NZ) in their corresponding ILPs which directly relates to higher immunity to attacks as illustrated in the previous experiments. We also report the percentage runtime overhead when solving the ILP (%OV) for each case.

On average (MOP3)-10% results in 20.15% increase in runtime with 50.35% increase in the number of non-zero entries. The corresponding vulnerabilities for each case was shown in Table 2. Note, solving all the above cases results in the same solution quality, as discussed in Section 4 which we also verified by experiment.

7. CONCLUSIONS

Cloud computing has the potential to provide customers instant and affordable access to powerful and distributed computational resources for running complex EDA algorithms. However, EDA vendors are at present reluctant to make use of cloud computers due to a major concern about revealing confidential design information. In this paper, we address this confidentiality challenge for the global routing problem. We proposed obfuscation techniques to transform an ILP formulation of the global routing problem in such a way that: (i) the transformed formulation is immune to a larger number of confidentiality attacks, and (ii) there is no loss of quality of the routing solution. We presented attack strategies and evaluated their effectiveness on the ISPD 2007 global routing benchmarks. We showed that the proposed techniques dramatically increase the immunity to attacks, but also increase the time required to solve the problem. We discussed a technique to tradeoff the increase in immunity to the increase in runtime and illustrated this tradeoff using one of the ISPD 2007 benchmark.

8. REFERENCES

[1] Top Threats to Cloud Computing, Cloud Security Alliance, 2010, https://cloudsecurityalliance.org/topthreats/csathreats.v1.0.pdf.

[2] What Cloud Computing Offers the Electronic Design Community, EETimes, June 2011, https://eetimes.com/discussion/other/4217127/.

[3] M. Cho, K. Lu, K. Yuan, and D. Z. Pan. BoxRouter 2.0: A hybrid and robust global router with layer assignment for routability. *ACM TODAES*, 14(2), 2009.

[4] R. Deitos and F. Kerschbaum. Improving practical performance on secure and private collaborative linear programming. In *DEXA Workshops*, pp. 122–126, 2009.

[5] J. Dreier and F. Kerschbaum. Practical secure and efficient multiparty linear programming based on problem transformation. *IACR Cryptology ePrint Archive*, p. 108, 2011.

[6] Y. Hong, J. Vaidya, and Haibing Lu. Efficient distributed linear programming with limited disclosure. In *DBSec*, pp. 170–185, 2011.

[7] IBM ILOG CPLEX V12.0, *User's Manual for CPLEX*, 2009.

[8] S. Megerian, M. Drinic, and M. Potkonjak. Watermarking integer linear programming solutions. In *DAC*, pp. 8–13, 2002.

[9] H. Shojaei, A. Davoodi, and J. Linderoth. Congestion analysis for global routing via Integer Programming. In *ICCAD*, pp. 256–262, 2011.

[10] Tamás Terlaky, Anthony Vannelli, and Hu Zhang. On routing in vlsi design and communication networks. *Discrete Applied Mathematics*, 156(11):2178–2194, 2008.

[11] J. L. Wong, G. Qu, and M. Potkonjak. Optimization-intensive watermarking techniques for decision problems. *IEEE TCAD*, 23(1):119–127, 2004.

[12] T.-H. Wu, A. Davoodi, and J. T. Linderoth. A parallel integer programming technique to global routing. In *DAC*, pp. 194–199, 2010.

[13] T.-H. Wu, A. Davoodi, and J. T. Linderoth. GRIP: Global Routing via Integer Programming. *IEEE TCAD*, 30(1):72–84, 2011.

[14] Y. Xu, Y. Zhang, and C. Chu. Fastroute 4.0: global router with efficient via minimization. In *ASPDAC*, pp. 576–581, 2009.

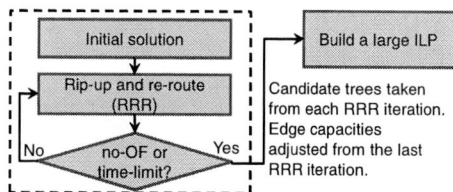

Figure 3: Creating an ILP instance.

9. CANDIDATE TREE GENERATION

To generate candidate trees, we modify a standard and iterative "rip-up and reroute" (RRR) procedure which is used by the majority of recent academic global routers. The input to the first RRR iteration is an initial tree per net, created typically when each net is routed using its shortest wire length. As a result, an edge e may be utilized above its capacity u_e. The amount of exceeding the usage on e is denoted by overflow parameter o_e. The total overflow is denoted by $\sum_{\forall e \in E} o_e$.

At each RRR iteration, the input trees of the nets that contain overflow edges are perturbed to create new trees which decrease the total overflow. This is done by visiting the nets and de-touring each tree that passes from the edges with overflow to other edges that have not been utilized to their capacities or contain lower overflow. As a result, the corresponding wirelength of a tree typically increases after de-touring. The output of each RRR iteration is a new routing solution in which the de-toured trees replace their corresponding trees given as input of that iteration. The input to the subsequent iterations of RRR is the output of the previous iteration. The process terminates either if the total overflow of the generated solution is 0 or when a time-bound has been reached. Figure 3 illustrates the process.

This work uses the RRR framework described in [9] where the time-bound to terminate RRR is set to 15 minutes. The reader is referred to [9] for details of RRR procedure. For each net T_i, the set of all its trees which are generated during the above process composes its set of candidate trees $\mathcal{T}(T_i)$. Note, each candidate tree of a net is generated from a different RRR iteration.

To create an instance of formulation (OP), the candidate trees are first generated using the described procedure. The parameter $\acute{o}_e$ used to adjust the edge capacity u_e is set to be equal to the overflow o_e corresponding to the last iteration of RRR. Note, the candidate trees generated by the last RRR iteration always serve as a feasible solution to formulation (OP). However after solving the formulation to optimality, it is possible to create a better solution with significantly lower wirelength without exceeding the adjusted edge capacities.

To control the size of the formed ILP, a certain number of nets can be fixed a-priori and the ILP is then formed for the remaining ones. This is necessary in our experiments because of the memory limitation for the large-sized benchmarks considered in this work. Specifically, the ILP is formed for the top $\kappa\%$ of the nets which have the maximum number of candidate trees generated for them. More candidate trees generated for a net indicates that more effort was spent within the RRR iterations to remove overflow on these nets so such a net is a better option to be included in the ILP. Each remaining net is fixed to the candidate tree generated by the last RRR iterations. The edge capacities u_e is adjusted accordingly to account for the utilization taken by the fixed nets. We assume $\kappa=20\%$ in our implementation.

Table 4: Comparison of solution quality.

Bench	Fastroute OF	Fastroute WL	ours (w/o cloud) OF	ours (w/o cloud) WL	ours (w/ cloud) OF	ours (w/ cloud) WL
adaptec1	0	43.75	0	44.07	0	42.03
adaptec2	0	40.44	0	40.94	0	39.62
adaptec3	0	110.43	0	110.50	0	109.04
adaptec4	0	102.47	0	103.09	0	101.96
adaptec5	0	124.03	0	123.13	0	119.43
newblue1	448	32.60	486	32.95	360	32.52
newblue2	0	56.90	0	57.58	0	55.93
newblue3	37018	89.56	37308	88.04	32042	87.34
average	4683	75.02	4724	75.04	4050	73.48

10. ADDITIONAL SIMULATION RESULTS

10.1 Comparison of Solution Quality

An ILP instance is first generated for each of the ISPD 2007 2D benchmarks and its solution quality in terms of wirelength (WL) and total generated overflow (OF) is reported. Please refer to Section 3 for description of the ILP and Section 9 for description of candidate tree generation using a rip-up and reroute framework and OF computation.

Candidate tree generation is ran with an imposed runtime limit of 15 minutes. Each ILP instance is then solved to optimality using CPLEX 12.0 [7] in multi-threading mode. All experiments ran on a four-core machine with a 2.8GHz Intel CPU and 12GB of memory.

Table 4 shows the quality of the generated solutions compared to a competitive academic global router FastRoute 4.1 (FR) [14][1]. The cost of a via associated with each wire bend is set to 1 unit to allow fair comparison with FR.

For our tool, we also report the solution quality after candidate tree generation which corresponds to the last iteration of rip-up and reroute. It reflects the best attainable solution before solving the ILP and is designated by the case (w/o cloud) in Table 4. The solution after solving the ILP using multi-threaded solver is designated by (w/ cloud). No transformation is used in this experiment to mask the ILP.

As shown in Table 4, our (w/o cloud) solution is worse than FR in terms of WL and OF. However, after solving the ILP, the (w/ cloud) solution is significantly better in terms of WL. For two benchmarks with non-zero overflow, the obtained OF is also significantly lower than FR.

The runtime of solving the ILP to optimality in our experiments is reported in Table 3. It is on average 78 minutes[2]. We note these runtimes includes the 15 minutes pre-processing to create the candidate trees. Also, these runtimes corresponds to using four cores. In the presence of more cores, the multi-threaded solver can further reduce this runtime by utilizing more cores. The tool FR has a lower runtime for each benchmark. However, in running FR, no runtime limit is used and FR terminates after concluding that no additional improvements are attainable within its internal rip-up and reroute procedure.

From this experiment we conclude that our ILP-based router allows generating high solution quality with a reasonable runtime that can be further decreased in an actual cloud network when many cores are available.

---

[1]The FR binary was obtained from the authors to conduct this experiment.

[2]The average runtime excludes `newblue3` because it is artificially made difficult by including a net with over 20K terminals.

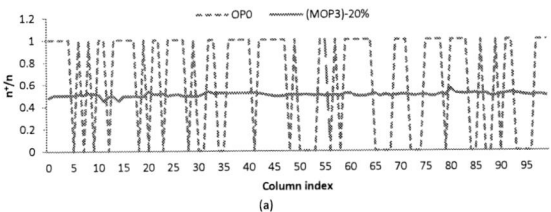

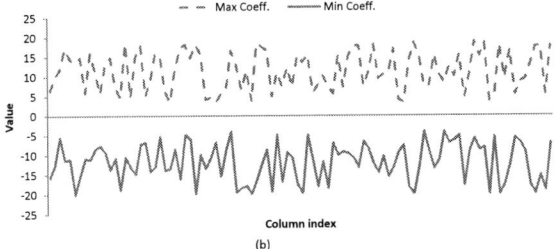

Figure 4: Detailed analysis for 100 random columns.

10.2 Detailed analysis for One Benchmark

In this section we provide detailed analysis for one benchmark to show the features and effectiveness of our methods.

For the first experiment, we randomly select 100 columns from the constraint matrix corresponding to benchmark instance `adaptec1` and study the values in these columns. Figure 4(a) for each column shows the number of positive coefficients over the number of non-zero values. We compare the original ILP problem (OP0) and (MOP3)-20% in which the value of parameter F in the heuristic is set to 20% of the number of columns. For each column we count the number of positive coefficients and divide it by the total number of non-zero coefficients. As figure shows in the original problem this value is always 0 or 1; that is all the non-zero coefficients in a column have equal signs. In (MOP3)-20%, however, almost half of the non-zero values in each columns are positive and the others are negative. The balance between number of positive and negative coefficients makes it difficult to distinguish the constraints and thereby makes the ILP instance immune to different attacks.

An obvious way to keep balance between positive and negative coefficients in each column is to replace approximately half of the constraints by an equivalent set of constraints obtained by multiplying them with -1, i.e., flip their signs. Unfortunately, this simple masking strategy is vulnerable to an equally simple attack. In particular, the cloud can identify the rows in the constraint matrix with all negative coefficients and recover the original set of constraints by multiplying them with -1. Our masking strategy is, however, not vulnerable to such attacks. In (MOP3), there is a balance between positive and negative coefficients not only in each column but also in each row.

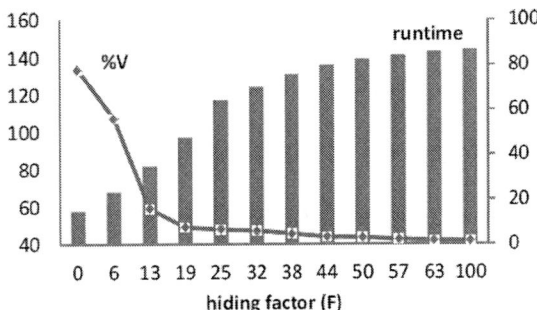

Figure 5: Tradeoff between the hiding factor vs. runtime and the vulnerability metric.

Furthermore, as shown in Figure 4(b), each column and each row has a large range of possible values. Specifically, Figure 4(b) shows the maximum and minimum coefficient values for the 100 randomly selected columns. For instance the maximum coefficient in the first column is $+6$ and the minimum value is -11. The results show that different coefficients in a single column have a range of values from negative numbers to positive values. Interestingly this range drastically changes from one column to another which shows the strength of our masking heuristic. While in the original problem all the coefficients are 0 or 1, our approach not only keeps balance between positive and negative numbers in a column, but also maps the values into a vast range of values.

An important strength of our heuristic is the possibility to bound the time needed to solve the masked ILP instance. The value of parameter F that is the number of columns considered for transformation in our masking heuristic determines the sparsity of the constraint matrix and consequently the time required for solving the masked ILP. While considering more columns for masking increases immunity to attacks, it increases the number of non-zero entries in the constraint matrix. Consequently we need more time to solve the masked instance.

We ran another experiment to study this tradeoff. For benchmark instance `adaptec1` we considered different values for parameter F. For each case we first compute the percentage of increase in the number of non-zero values in the constraint matrix. Then we run the masked instance after each transformation and compute the time required for solving the instance. Figure 5 shows the tradeoff between runtime and number of considered columns (F). The figure shows that the runtime first increases linearly with the increase in the number of considered columns. The vulnerability metric (%V) is also shown in the figure. As can be seen with slight increase in F, the rate of decrease in V is significant for small values of F. However, for F higher than 20%, V decreases at a much slower rate.

716

Design Tools for Artificial Nervous Systems

Louis K. Scheffer
Janelia Farm Research Campus
Howard Hughes Medical Institute
Ashburn, VA 20147
SchefferL@janelia.org

ABSTRACT

Electronic and biological systems both perform complex information processing, but they use very different techniques. Though electronics has the advantage in raw speed, biological systems have the edge in many other areas. They can be produced, and indeed self-reproduce, without expensive and finicky factories. They are tolerant of manufacturing defects, and learn and adapt for better performance. In many cases they can self-repair damage.

These advantages suggest that biological systems might be useful in a wide variety of tasks involving information processing. So far, all attempts to use the nervous system of a living organism for information processing have involved selective breeding of existing organisms. This approach, largely independent of the details of internal operation, is used since we do not yet understand how neural systems work, nor exactly how they are constructed. However, as our knowledge increases, the day will come when we can envision useful nervous systems and design them based upon what we want them to do, as opposed to variations on what has been already built. We will then need tools, corresponding to our Electronic Design Automation tools, to help with the design. This paper is concerned with what such tools might look like.

Categories and Subject Descriptors

B.6 [**Logic Design**]: Design Aids; J.6 [**Computer Aided Engineering**]: Computer Aided Design

General Terms

Algorithms, Design

Keywords

Design Automation, Biological systems, Neurons

1. INTRODUCTION

Biological nervous systems and conventional electronics have in theory exactly the same expressive power, according to the widely-accepted *Church-Turing thesis.*[1] However, from an engineering perspective the differences are stark. Electronics excels at high speed operation and precise timing. Biological systems can be produced, and self-reproduce, without expensive factories and in less than pristine conditions. They are tolerant of manufacturing defects, and learn and adapt for better performance. In many cases they can self-repair damage. These would be wonderful attributes to be able to incorporate in engineered systems.

However, all attempts thus far to use the nervous system of a living organism for information processing have been limited to selective breeding of existing organisms. Hunting animals, seeing eye dogs, and other useful animals are at most relatively minor variants of existing organisms. These were derived by methods that are independent of the details of internal operation because we do not yet know how such systems work, nor do we have the tools to design them.

There are at least three areas where the explicit design of nervous system features would be very helpful, and tools will be required. The first is to design direct interfaces from outside technologies to the nervous systems. Such interfaces are used in applications such as cochlear implants, and enhanced versions are common in science fiction (often called *jacking in*[2]). However, the existing interfaces have limited functionality and performance[3], at least partially because they are using the existing nervous system as-is. Presumably much better interfaces could be built if both sides were optimized for the interaction. This might involve simple steps such as making sure the relevant nerves are physically accessible, or complex operations such coding/decoding or multiplexing. In any event new design tools will be required.

The next tool opportunity is the composition of existing nervous systems. A huge number of specialized capabilities have evolved within the animal kingdom. There are animals that can sense directly magnetic[4] and electric[5] fields, animals that navigate by echo-location, creatures that sense more the visual portion of the spectrum, animals that manipulate with flexible appendages (octopuses and elephants) and so on. Each of these these capabilities requires both sensors/manipulators and supporting neural computation.

From an engineering point of view, combining the specialized traits of different organisms could be very helpful. For example, a small animal designed for use in rescue operations, might exhibit the friendliness and trainability of a dog, but the echo-location of a bat. If the ethical problems can be surmounted (or ignored), the combinations of human and other attributes could be very helpful. A human technician might appreciate the abilities to sense magnetic and electric fields with their fingertips, or many other capabilities limited only by imagination. The ability to combine features from different nervous systems will need detailed analysis and a way to resolve the many conflicts that are sure to occur. Automatic tools will surely be required.

A yet further step would be the ability to design a nervous system from scratch. This seems potentially useful since all nervous systems so far have been subject to the constraints, and happenstance, of evolution. For example, there are no higher nervous systems with tri-lateral symmetry, despite the potential usefulness of such configurations. A creature with three ears could localize sound well in both azimuth and elevation, for example. A creature with three eyes could keep track of what is behind, without losing binocular depth perception in front. Changes such as these are unlikely to evolve naturally or even by directed evolution. Explicit design will be needed, and will need tools.

1.1 How biological nervous systems are built

How a nervous system develops is far from completely understood. However, a number of the basic principles are known. Chemical gradients help define the geometrical space for construction of major structures. A sequence of similar neurons forms from a neuroblast, a type of stem cell for neurons. Growing branches of neurons, the *growth cones*, are guided by chemical cues both positive and negative. As a nervous system develops, different lineages of cells excrete, and respond to different chemical clues.[6] Some follow gradients, and others specify which cells to adhere to, and which to avoid.[7] As a nerve cell extends, the same chemical cue can cause differing effects. For example, when a nerve bundle is under construction, the surface proteins cause the nerves to stick together. However, when they reach their target, a molecular switch inside each neuron changes state, and the same surface proteins now mediate avoidance. This causes the bundle to exfoliate and enervate a larger field once it has reached its target. The operation[8] and the mechanisms[9] of this molecular guidance are a very active field of research.

Once a nerve cell reaches the right general area, connections can be made and modified by the relative activities of neurons. These changes are thought to underlie learning and adaptation in animals. Furthermore, unlike chips, biological networks are often active during their construction. This activity may be self-generated, or externally driven, or both.[10] External activity is known to be required in the development of many structures such as the visual systems of animals, where a lack of visual stimulus during development results in defects in the adult visual processing.[11] In flies, however, the visual system develops normally even in the dark, so either the system is hardwired without activity clues, or the self-generated activity suffices.[12]

These differences can be illustrated by a comparison of how a chip and a biological might be constructed, given the same underlying architecture. For this example we will use a small, dual frequency oscillator, defined by either a 3 or 7 stage ring oscillator, as selected by a mux. (This is not the way biological builds oscillators[13], or *central pattern generators*, but used only for illustration.) The electronic design might start with a hierarchical design as shown in Fig 1(a). This is flattened to a two layer hierarchy for use in place-

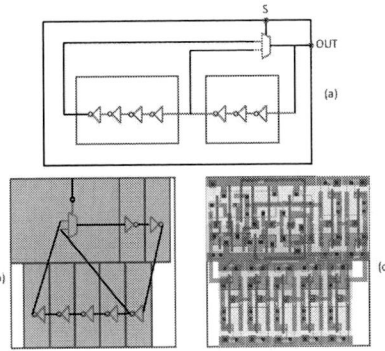

Figure 1: Sample circuit and the traditional EE hierarchies. (a) shows the design hierarchy, (b) the place and route hierarchy, and (c) the manufacturing (absence of) hierarchy. Cell layouts copyright vlsitechnology.org.

ment and routing, Fig 1(b). Then the design is completely flattened into masks for use in manufacturing, Fig. 1(c). The final design depends very little on the design hierarchy used by the original designer.

The biological system proceeds very differently, as shown by the hypothetical example of Fig. 2(a). This example is not taken from any biological creature, and is only intended to demonstrate the different growth techniques typical to biology. Biological construction starts with single cell, the egg. Through a series of cell divisions, and cell differentiations, then entire organism is grown. Each cell divides into exactly two daughter cells, though each daughter may develop differently (have a different fate). In EE terms, the entire hierarchy is constructed from a single root node, and each node in the tree is either a leaf node, or has exactly two sub-nodes.

In the example, the 'stem cell' of the oscillator might first divide into two daughter cells. In this case, one will become the MUX, and the other will give rise to all the inverter stages. Following a typical biological paradigm, each inverter stem cell divides into a daughter cell and another stem cell, and does this four times. In each case the inverter daughter is the portion in the -X direction, and the stem cell portion in the +X direction. Each daughter cell then divides into two inverter cells. Now the cells must wire themselves up. The inverter chain is easy; each output (axon) is programmed to extend along the +X gradient until it runs into a cell with the same surface proteins as itself, then forms an inhibitory synapse. The MUX cell output does the same, except it 'looks' for an inverter cell target. The MUX inputs are more complex. One will grow in the +X direction until it reaches the point where the +X chemical and the -X chemical have similar concentrations. It then makes a connection to a cell whose activity is anti-correlated with its own MUX cell output. (An important property of neurons is that inputs can tell when the cell output has been triggered through the process of *back propagation*.[14][1]) The other in-

---

[1]Not to be confused with the similarly named *back propagation* algorithm[15] for tuning artifical neural nets.

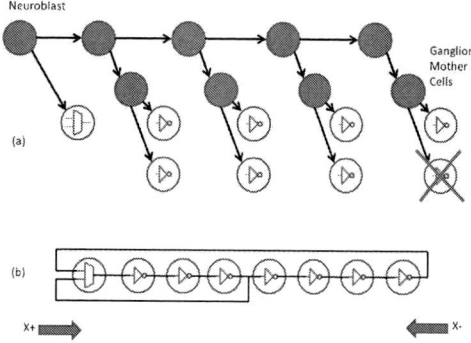

Figure 2: (a) Possible biological hierarchy that might be used to construct the same system, designed to be similar to the real biological hierarchy from [7]. The cell on the lower left is constructed, but then dies. This keeps the hierarchy simpler at the cost of extra construction effort. (b) Layout of constructed cells and dendrites.

put does the same thing until it reaches a low concentration of the -X gradient, then also connects to an anti-correlated signal. Both inputs also look for a surface protein indicating a 'select' signal, then one forms an inhibitory synapse at this point and the other an excitatory one. The logic behind this will be discussed in the section on logic synthesis.

2. OVERVIEW OF TOOLS

What design tools will be required for the three tasks specified above? At least two simulation programs will be required - one that models how a nervous system develops, and one that models how it works in operation. The analogous programs in the EE world would perhaps be SUPREM[16] for modeling the manufacturing process, and SPICE[17] for modeling a network once it has been constructed. Next, if large systems are attempted, the equivalent of logic or behavioral synthesis will be needed. Once the desired system has been determined, the appropriate set of chemical and genetic cues to cause the system to self-assemble must be determined. This is the equivalent of place and route in traditional EDA. Finally the system must be debugged and tested, the equivalent of Design For Test.

2.1 Simulation

The simulation of neuron networks once they are complete is a comparatively mature field, with accepted programs such as GENESIS[18], NEURON[19] and others[20], as covered in a review of eight such programs. However, these programs do not cover the physics of long range interneuron chemical cues (neuromodulators), nor do they model how these networks are grown. Furthermore, when behavior of the organism is the desired goal, they will need to be integrated with mechanical and kinetic models of the organism's body.

Neural development simulators are a relatively recent development. There are two major parts to this problem - determining the cell type and fate, then determining how it will grow in its environment. Determining cell fate as a

function of cell ancestry and chemical environment is an extremely challenging simulation problem in itself, since it is dependent on a complex web of chemical and genetic interactions within the cell proper. Simulation of this is recent and quite restricted.[21]

For the second part of this task, there are now programs such as NETMORPH[22] and CX3D[23] that simulate the size and shape of neurons as they grow. In particular, these programs look at known effects such as physical competition for space, and gradients and diffusion of chemical cues. However, these programs only model part of the process of neural system growth. Neuroblasts and differentiation can be modeled, but need explicit coding. Likewise, chemical and gradient guidance of each cell type are explicitly coded, and cell type, fate, and types of descendants do not arise from the expression patterns of genes and proteins. Electrical signals and their effects on synapse creation and deletion are not yet simulated. Even so, these simulators can create some simple layered structures, but not yet curved and convolved surfaces typical of nervous systems. Attempts to use these are summarized for the growth of neuronal structures *in vitro*.[24]

2.2 Synthesis of logic function

Once a designer has in mind a particular logical function, how can neurons be used to carry it out? This promises to be an extremely interesting field, particularly since the focus of much contemporary neuroscience is to figure out how the brain does the computations it does. Existing biological results are clearly impressive, with (for example) real-time speech recognition built from relatively slow components with millisecond response times.

Although all the details are not known, it is clear the brain does calculations in a very different style than conventional digital logic. The basic operation involves integrating the currents from (normally many) synapses, and then processing them through various non-linear effects. This can be a simple threshold, above which the neuron in turn sends out its own neurotransmitters, or the neuron can be driven by positive feedback into generating output spikes. In most of the brain, of almost all creatures studied, the logical depth is very low but the fanouts very high. There may be many significant non-linearities within a single neuron; the summation of signals within a single arbor is non-linear[25], and in some cases there can be a great deal of local processing in the arbors of a single nerve cell.[26] Processing is mostly asynchronous, with no global clock.

Some portions of logic synthesis into neuron-like networks have been studied. In particular, there is significant research into how to build logic out of sum-and-threshold operations, called *linear threshold gates* or *threshold logic gates*. Methods include direct synthesis of threshold gates[27] and post-processing of results from conventional logic synthesis.[28][29] In addition, researchers have considered explicit synthesis of higher level functions such as addition[30], comparison[31], and division.[32] There have been at least some studies of the tradeoffs between logic depth and fan-in (or fan-out)[33][34] though some concentrate on functions such as parity that are known to be difficult.[35] However, synthesis that can take advantage of the full capability of neurons will probably need to await better understanding of how the brain performs its calculations.

As a simple example, consider how biology might imple-

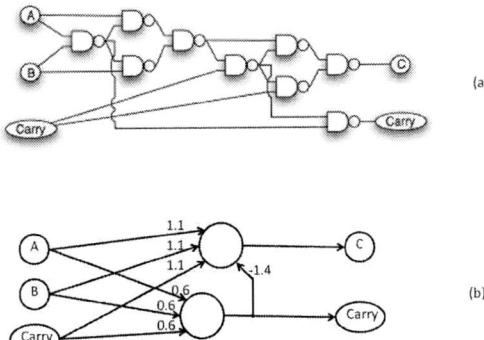

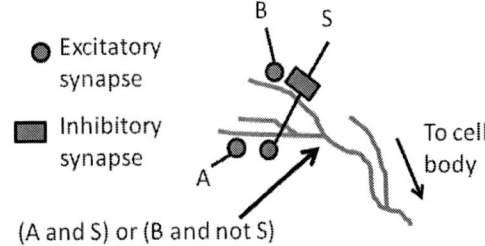

Figure 3: (a) Traditional binary digital adder constructed from 2 input NAND gates. It has a relatively large logical depth and low fanin/fanout. Credit:Wikipedia (b)Adder constructed with Linear Sum Threshold logic, with a threshold of 1. It has many fewer gates, much less logical depth, and greater fan-in.

Figure 4: Example using dendritic computation to perform the MUX function. This is modelled directly from Fig 2 of [36], with the exception that neuron S makes two different synapse types to two dendritic branches of the same neuron.

ment a full adder, assuming linear-threshold operation, compared to traditional EE boolean gates. A full adder has three inputs, a carry in and two data bits, and two outputs, a sum and carry. The sum is the XOR of all three inputs, and the carry out will be true if the sum is at least two. Using traditional EE logic, this requires 9 gates with a large logical depth, as a shown in Fig. 3 (a).

However, using linear-threshold gates this can be implemented with only two neurons. The carry out will be true provided at least 2 of the three outputs are true, since any two sum to over the threshold of 1.0. The sum neuron is slightly trickier. Its job is to produce an output if exactly one or 3 inputs are true, but not if two are true. In this case, all three data inputs have strength 1.1, but there is an inhibitory input from the carry of strength -1.4. If one input is true, the excitation of 1.1 will fire the neuron. If two inputs are true, the the carry will also be true, so the total input will be 0.8 so the neuron will not fire. If all three neurons are true, then the inputs sum to 1.9 . This overcomes the inhibition and the output returns to 1.

For a linear threshold gate, the input branches (dendrites) form a passive collection function, followed by a single nonlinearity which is typically assigned to the main body of the cell. For some neurons this simple arrangement, where the whole neuron is roughly iso-potential, is not a bad approximation. However, it is also possible (even likely) that significant processing can occur in the neural fibers themselves. Possible dendritic computations include logical operations, coincidence detection, non-linear summation, low-pass filtering, attenuation, and amplification.[36] Logical operations can be implemented by dendrites since (under certain distributions of positive feedback ion channels) any point in any branch can potentially start a wave of self propagating positive feedback that, soliton-like, will propagate throughout the whole neuron. An example might be the MUX cell shown in Fig 4.

A circuit such as this might be seen as non-biological, since it violates Dale's principle, which states that each neuron type uses only one neurotransmitter. However, Dale's

principle is known to be inexact[37] – there are known cases of two different neurotransmitters. In most cases both are inhibitory[38] or excitatory.[39] When there is evidence for both inhibitory and excitatory release from the same cell, they target different cells.[40] To the author's knowledge no case of two different sign synapses to the same target cell. This is a potential example of an arrangement that is useful and biologically possible, but apparently has never evolved.

These examples shows several features that may be important in synthesis of networks of neurons - nonlinearities giving an all or nothing response, differing size of input effects, use of inhibitory as well as excitatory inputs, computation in what is nominally the wiring, and the strong interaction of physical and logical operation (the MUX neuron above will not work if the two branches are tightly coupled, as they would be if the dendrites are too big.) Since many of these effects are at least partially analog, various DFM and design centering techniques developed for analog circuits may need to be applied even when digital systems are synthesized. This will be especially true for nervous systems that must run reliably over a significant temperature range, as those of cold-blooded animals do.

Furthermore, synthesis techniques for biology will need to pay close attention to the hierarchy of a design. In chips, construction is done through masks, which are a flattened version of any hierarchy that was used. Therefore both the logical and physical design hierarchy are largely irrelevant to the cost, speed, and quality of production. In biological systems, by contrast, two different and good hierarchies are needed for the construction phase. The first is the binary tree of cell creation, which needs to create all the cells of the correct type, in the correct place. The second is the pattern of expression of surface proteins and the guidance mechanisms, as used to specify the wiring between various branches of the cell creation tree. This too must be hierarchical, if for no other reason than there is not enough information in the DNA to specify each connection explicitly. Therefore, in biology a good hierarchy with a higher part count may be preferred to a less structured design with fewer components, or the final configuration may be developed by over-creation of a simpler hierarchy followed by pre-programmed cell death. It is also likely that a tight coupling between the synthesis and implementation phases will be needed.

These features imply that very different algorithms will

need to be employed in biological logic synthesis. In addition, the synthesis step is the stage where adaptation, learning, and robustness must be engineered into the system. Incorporating these features must await better understanding of how they work in existing biological systems, but will surely require novel tools.

2.3 Implementation

This task could be briefly described as follow: given a network of neurons to be built, propose an ordering of cell growth, cell surface molecules, and gradient processing that will generate the desired network as a result. This is analogous to the place and route steps of EE chip design flows.

Both the cell growth and the surface protein descriptions will need to be strongly hierarchical. A starting hierarchy may be available from the synthesis phase above, particularly if the synthesis is aware of placement and construction restrictions. Alternatively, the hierarchies may need to be discovered, perhaps through a process of recursive partitioning. In particular, combining features from two or more organisms will surely require refactoring both hierarchies.

Wire length optimization is the driving force behind EE place and route, since shorter wires are faster, take up less area, and are higher yield. Likewise, wiring economy has been shown to be a driving principle in worms,[41] and the optic lobe of flies,[42] likely for the exactly same reasons. Another similarity is that the same general hierarchy of wiring resources is present. Short wires will be fast since their electrical time constants are small. Longer wires can be made faster, up to a point, by making them wider - as the wire, axon, or dendrite becomes wider, resistance goes down faster than capacitance goes up. However, in both cases, when the wire becomes long enough, it is preferable to regenerate the signal periodically to keep the overall delay linear with distance. This is accomplished by *repeater insertion*[43] in EE and *nodes of Ranvier* and myelinated axons[44] in biology.

These strong similarities indicate the basic algorithms developed for global and detailed placement may be very relevant to biological design. However, there are also significant differences. Timing driven design will be different since timing and logic function are not decoupled as they are in most silicon chips. Also, the biological networks are often composed of relatively thick layered assemblies folded in complex curves into a space filling structure. The closest equivalents in the EE toolset would probably by 3-D floorplanning, followed by a folded 2-D placement. The general approach of creating a 3D placement by folding a 2-D result[45] may be relevant here.

Another similar but related task is leaving enough room for needed overhead not explicitly included in the logical design. In EE, this often includes power supplies, clock distribution, and test circuitry. In biological systems, this will include vessels for blood and air, blood-brain barriers, glia to support neural cells, and so on. New detailed design steps will be needed to include these features, and new heuristics needed to leave enough room earlier in the design process.

2.4 Debugging

Like any other complex system, human-designed nervous networks will doubtless require debugging once built. In general this requires *observability*, the ability to observe internal nodes of the system, and *controllability*, the ability to control signals that are not direct inputs. Since these systems are largely asynchronous, and since logic and memory are intermixed, traditional EE solutions such as scan design will not apply. New techniques, and tools, will be required.

Observability, in general, may be possible as an outcome of brain-machine interfaces as discussed above. Optical readout of neural state, though ion concentration[46] or voltage directly[47], is another possibility.

Controllability might be provided by designing it in, as in EE, with the corresponding overheads of space and speed. Alternatively, neural researchers today use various biological side channels to switch neurons on and off. In cold-blooded animals, temperature-sensitive ion channels can change neural function by changing the animal's environment.[48] This does not work in warm-blooded creatures, but an alternative is switching induced by small molecules that travel through the blood.[49] Light triggered modulators are also in widespread use, where a tiny fiber-optic cable can deliver light to a small portion of the brain, disabling (or triggering) subsets of neurons.[50] Most likely at least some of these techniques can be used to switch the nervous system into "debugging" mode, but whichever is chosen will need tool support for both implementation and interpretation of results.

3. CONCLUSIONS

In the (relatively) near future, it should be possible to design and specify interfaces to nervous systems, to mix, match, and combine features of existing nervous systems, and even implement biological nervous systems from scratch. Various tools will be required to do this, and are in various states of (im)maturity. Simulation of completed nervous systems is in the best shape, though it needs numerous extensions. Logic synthesis has a start, but cannot yet take advantage of many important aspects of biological systems. The physical design and implementation phase has yet to be built, as has design for test and debugging of biological systems. All of these provide excellent opportunities for the CAD research of tomorrow.

4. ACKNOWLEDGMENTS

The author would like to thank his colleagues at the Janelia Farm campus for putting up with his endless questions about how this biology stuff really works.

5. REFERENCES

[1] L.A. Rubel. Digital simulation of analog computation and Church's thesis. *Journal of Symbolic Logic*, pages 1011–1017, 1989.

[2] W. Gibson. *Neuromancer*. Ace Trade, 2000.

[3] D.R. Moore and R.V. Shannon. Beyond cochlear implants: awakening the deafened brain. *Nature neuroscience*, 12(6):686–691, 2009.

[4] W. Wiltschko and R. Wiltschko. Magnetic orientation and magnetoreception in birds and other animals. *Journal of Comparative Physiology A: Neuroethology, Sensory, Neural, and Behavioral Physiology*, 191(8):675–693, 2005.

[5] T.H. Bullock. *Electroreception*, volume 21. Springer Verlag, 2005.

[6] H. Song and M. Poo. The cell biology of neuronal navigation. *Nature cell biology*, 3(3):E81–E88, 2001.

[7] CS Goodman, M.J. Bastiani, et al. How embryonic nerve cells recognize one another. *Scientific American*, 251(6):58, 1984.

[8] DV Van Vactor and L.J. Lorenz. Neural development: The semantics of axon guidance. *Current biology: CB*, 9(6):R201, 1999.

721

[9] B.J. Dickson. Molecular mechanisms of axon guidance. *Science*, 298(5600):1959, 2002.

[10] L.I. Zhang and M. Poo. Electrical activity and development of neural circuits. *Nature Neuroscience*, 4:1207–1214, 2001.

[11] T.N. Wiesel, D.H. Hubel, et al. Effects of visual deprivation on morphology and physiology of cells in the cat's lateral geniculate body. *J Neurophysiol*, 26(978):6, 1963.

[12] P.R. Hiesinger, R.G. Zhai, Y. Zhou, T.W. Koh, S.Q. Mehta, K.L. Schulze, Y. Cao, P. Verstreken, T.R. Clandinin, et al. Activity-independent prespecification of synaptic partners in the visual map of *Drosophila*. *Current biology*, 16(18):1835–1843, 2006.

[13] E. Marder and D. Bucher. Central pattern generators and the control of rhythmic movements. *Current biology*, 11(23):R986–R996, 2001.

[14] G.J. Stuart, B. Sakmann, et al. Active propagation of somatic action potentials into neocortical pyramidal cell dendrites. *Nature*, 367(6458):69–72, 1994.

[15] S. Russell and P. Norvig. *Artifical Intelligence, A modern approach*. Prentice-Hall, 1995.

[16] C.P. Ho, J.D. Plummer, S.E. Hansen, and RW Dutton. VLSI process modeling - SUPREM III. *Electron Devices, IEEE Transactions on*, 30(11):1438–1453, 1983.

[17] L.W. Nagel and D.O. Pederson. SPICE: simulation program with integrated circuit emphasis. 1973.

[18] J.M. Bower, D. Beeman, and A.M. Wylde. *The book of GENESIS: exploring realistic neural models with the GEneral NEural SImulation System*. Telos New York, 1995.

[19] M.L. Hines and N.T. Carnevale. The NEURON simulation environment. *Neural computation*, 9(6):1179–1209, 1997.

[20] J. Aćimović, T. Mäki-Marttunen, R. Havela, H. Teppola, and ML Linne. Modeling of neuronal growth *in vitro*: Comparison of simulation tools NETMORPH and CX3D. *EURASIP Journal on Bioinformatics and Systems Biology*, 2011(1):616382, 2011.

[21] C. Li, M. Nagasaki, K. Ueno, and S. Miyano. Simulation-based model checking approach to cell fate specification during *Caenorhabditis elegans* vulval development by hybrid functional Petri net with extension. *BMC systems biology*, 3(1):42, 2009.

[22] R.A. Koene, B. Tijms, P. van Hees, F. Postma, A. de Ridder, G.J.A. Ramakers, J. van Pelt, and A. van Ooyen. NETMORPH: a framework for the stochastic generation of large scale neuronal networks with realistic neuron morphologies. *Neuroinformatics*, 7(3):195–210, 2009.

[23] F. Zubler and R. Douglas. A framework for modeling the growth and development of neurons and networks. *Frontiers in computational neuroscience*, 3, 2009.

[24] R. Brette, M. Rudolph, T. Carnevale, M. Hines, D. Beeman, J.M. Bower, M. Diesmann, A. Morrison, P.H. Goodman, F.C. Harris, et al. Simulation of networks of spiking neurons: a review of tools and strategies. *Journal of computational neuroscience*, 23(3):349–398, 2007.

[25] C. Koch, I. Segev, et al. The role of single neurons in information processing. *Nature Neuroscience*, 3:1171–1177, 2000.

[26] T. Euler and W. Denk. Dendritic processing. *Current opinion in neurobiology*, 11(4):415–422, 2001.

[27] JE Hopcroft and RL Mattson. Synthesis of minimal threshold logic networks. *Electronic Computers, IEEE Transactions on*, (4):552–560, 1965.

[28] M.J. Avedillo and J.M. Quintana. A threshold logic synthesis tool for RTD circuits. In *Digital System Design, 2004. DSD 2004. Euromicro Symposium on*, pages 624–627. IEEE, 2004.

[29] R. Zhang, P. Gupta, L. Zhong, and N.K. Jha. Synthesis and optimization of threshold logic networks with application to nanotechnologies. In *Design, Automation and Test in Europe Conference and Exhibition, 2004. Proceedings*, volume 2, pages 904–909. IEEE, 2004.

[30] S. Vassilladis, S. Contofana, and K. Bertels. 2-1 addition and related arithmetic operations with threshold logic. *Computers, IEEE Transactions on*, 45(9):1062–1067, 1996.

[31] N. Alon and J. Bruck. Explicit constructions of depth-2 majority circuits for comparison and addition. *SIAM Journal on Discrete Mathematics*, 7(1):1–8, 1994.

[32] K.Y. Siu, J. Bruck, T. Kailath, and T. Hofmeister. Depth efficient neural networks for division and related problems. *Information Theory, IEEE Transactions on*, 39(3):946–956, 1993.

[33] K.Y. Siu, V.P. Roychowdhury, and T. Kailath. Depth-size tradeoffs for neural computation. *Computers, IEEE Transactions on*, 40(12):1402–1412, 1991.

[34] V. Beiu. When constants are important. Technical report, Los Alamos National Lab., NM (United States), 1997.

[35] R. Impagliazzo, R. Paturi, and M.E. Saks. Size–depth tradeoffs for threshold circuits. *SIAM Journal on Computing*, 26:693, 1997.

[36] M. London and M. Häusser. Dendritic computation. *Annu. Rev. Neurosci.*, 28:503–532, 2005.

[37] N.N. Osborne. Is Dale's principle valid? *Trends in Neurosciences*, 2:73–75, 1979.

[38] P. Jonas, J. Bischofberger, and J. Sandkühler. Corelease of two fast neurotransmitters at a central synapse. *Science*, 281(5375):419, 1998.

[39] D. Sulzer and S. Rayport. Dale's principle and glutamate corelease from ventral midbrain dopamine neurons. *Amino Acids*, 19(1):45–52, 2000.

[40] C.B. Duarte, P.F. Santos, and A.P. Carvalho. Corelease of two functionally opposite neurotransmitters by retinal amacrine cells: experimental evidence and functional significance. *Journal of neuroscience research*, 58(4):475–479, 1999.

[41] A. Pérez-Escudero and G.G. de Polavieja. Optimally wired subnetwork determines neuroanatomy of *Caenorhabditis elegans*. *Proceedings of the National Academy of Sciences*, 104(43):17180, 2007.

[42] M. Rivera-Alba, S.N. Vitaladevuni, Y. Mischenko, Z. Lu, S. Takemura, L. Scheffer, I.A. Meinertzhagen, D.B. Chklovskii, and G.G. de Polavieja. Wiring economy and volume exclusion determine neuronal placement in the *Drosophila* brain. *Current Biology*, 2011.

[43] A.B. Kahng, S. Muddu, E. Sarto, and R. Sharma. Interconnect tuning strategies for high-performance ICs. In *Design, Automation and Test in Europe, 1998., Proceedings*, pages 471–478. IEEE, 1998.

[44] AF Huxley and R. Stämpeli. Evidence for saltatory conduction in peripheral myelinated nerve fibres. *The Journal of physiology*, 108(3):315–339, 1949.

[45] J. Cong, G. Luo, J. Wei, and Y. Zhang. Thermal-aware 3D IC placement via transformation. In *Asia and South Pacific Design Automation Conference, 2007.*, pages 780–785. IEEE, 2007.

[46] W. Göbel and F. Helmchen. *In vivo* calcium imaging of neural network function. *Physiology (Bethesda, Md.)*, 22:358, 2007.

[47] D.S. Peterka, H. Takahashi, and R. Yuste. Imaging voltage in neurons. *Neuron*, 69(1):9–21, 2011.

[48] T. Kitamoto. Conditional modification of behavior in *Drosophila* by targeted expression of a temperature-sensitive shibire allele in defined neurons. *Journal of neurobiology*, 47(2):81–92, 2001.

[49] A.Y. Karpova, D.G.R. Tervo, N.W. Gray, and K. Svoboda. Rapid and reversible chemical inactivation of synaptic transmission in genetically targeted neurons. *Neuron*, 48(5):727–735, 2005.

[50] X. Han and E.S. Boyden. Multiple-color optical activation, silencing, and desynchronization of neural activity, with single-spike temporal resolution. *PLoS One*, 2(3):e299, 2007.

Dynamic River Network Simulation at Large Scale

Frank Liu
IBM Research Austin
Austin, TX
frankliu@us.ibm.com

Ben R. Hodges
Center for Research for Water Resources
University of Texas at Austin
hodges@mail.utexas.edu

ABSTRACT

Fully dynamic modeling of large scale river networks is still a challenge. In this paper we describe SPRINT, an interdisciplinary collaborative effort between computer engineering and hydroscience to address the computational aspect of this challenge. Although algorithmic details differ, SPRINT draws many design considerations from SPICE, one of the most fundamental EDA tools. Experimental results demonstrate that SPRINT is capable of simulating large river basins at over 100× faster than real time.

Categories and Subject Descriptors

G.1 [**Numerical Analysis**]: Applications; I.6 [**Computing Methodologies**]: Simulation and Modeling; J.2 [**Computer Applications**]: Physical Sciences and Engineering

General Terms

Algorithms

Keywords

Dynamic river network simulation, Saint-Venant Equations, SPICE

1. INTRODUCTION

EDA tools are essential for designing today's VLSI semiconductor products. One of the earliest and most fundamental EDA tools is SPICE[1][2]. By working in conjunction with compact device models[3][4], SPICE provides a virtual environment so that designers can rapidly assess the function correctness, performance as well as power consumption. This "virtual prototyping" environment not only ensures the correctness of the complex design before they are committed to the lengthy and expensive manufacturing processes, but also provides the designers directions to optimize the design in order to meet the specifications.

Mathematically the function of SPICE can be described as a DAE (differential algebraic equation) solver for electronic circuits. The circuit consists of nonlinear components (e.g., BJT or MOS transistors) and linear components (e.g., resistors, capacitors or inductors). The circuit behavior is described by governing laws: KCL (Kirchhoff's Current Law), KVL (Kirchhoff's Voltage Law) and BCR (Branch Constituent Relations). The two Kirchhoff's circuit laws (KCL and KVL) are derived from Maxwell's equations. KCL states that the algebraic sum of currents flowing into any node within a circuit is zero. KVL states that the algebraic sum of voltage drops along any branch loop is zero. BCR describes the behavior of a circuit element with respect to its branch voltages and currents. For example, the BCR of a resistor is basically the Ohm's Law. The BCR of multiterminal MOSFET devices is described by a set of complex equations within the compact model. From a given circuit specification, by applying KCL, KVL and BCR, SPICE constructs a system of nonlinear differential-algebraic equations by using either sparse tableau[5], or modified nodal analysis[6]. The voltages of the circuit nodes as well as branch currents of certain devices (e.g., of inductors) are the canonical unknowns which need to be solved. For linear circuits (with only passive devices), a linear matrix solver is all required to solve the circuits. When nonlinear devices (e.g., MOS transistors) are present in the circuit, modified Newton-Raphson's method and numerical integration methods (e.g., Gear's methods[7]) are applied to solve the associated nonlinear DAE. Three types of analyses are available in the original version of SPICE: steady-state (DC), small-signal AC, and transient analysis (TR) with time-varying input excitations.

The first version of SPICE was released to public in 1971, well before Open Source Software became a big movement. Over the years, many revisions were made, including SPICE3, which is the first version written in C, instead of FORTRAN as in earlier versions[8]. Today SPICE is still part of the core curriculum on circuit simulation in universities across the globe[9][10][11]. Even over forty years after its inception, modern versions of SPICE are still an important part of the product portfolio of many EDA software companies and are generating multi-million dollar annual revenues[12][13][14].

The success of SPICE has inspired many other simulation projects; some of them are well beyond the traditional boundary of electrical engineering. For example, SUGAR[15] is an open source simulation tool for micro-electromechanical systems (MEMS). Another example is Bio-SPICE, which is an open source framework and software toolset for Systems

Biology[16]. In this paper, we describe a simulation software package which is intended to perform dynamic simulation of large scale river networks.

Water is essential for the life forms as we know today. Only 3% percent of earth's water is in the form of fresh water, mainly locked in polar icecaps and continental glaciers. River networks are the fastest distribution conduits through the landscape for fresh water, and hence play vital roles in municipal water supply, low-cost bulk transport of grain, agricultural irrigation, wildlife habitat, and recreation. River networks also play a key role in flooding, either carrying away potential flood waters or causing inundation when the carrying capacity is exceeded. Floods cause devastating loss of lives and economic harm throughout the world. For example the Thailand floods in 2011 not only caused hundreds of casualties and billion dollars of direct property damage to the local communities, but also severely disrupted the electronics supply chain with implications across the globe.

Perhaps the most commonly used software for river modeling is HEC-RAS, developed by US Army Corps of Engineers[17]. It primarily serves as a desktop tool for hydraulic engineers conducting flow analysis and flood-plain assessments over main stem rivers or local catchments. What has not been previously available is an efficient scalable river network simulator, which capable of simulating not only the main river stems, but also thousands of kilometers of upstream tributaries. This paper presents the initial development of just such a simulator.

The present work is the result of close inter-disciplinary collaboration between computer engineering and civil engineering hydroscience. Although the target applications and detailed algorithms differ, the design philosophy and many design considerations in our simulator are strongly influenced by SPICE. The result is an efficient river network simulator which is capable of simulating large river basins at over 100× of real time. With the recognition to the influence of SPICE, we name our software SPRINT (Simulation Program for RIver NeTworks). For the remaining portion of this paper, we present the first-principle physics model of river networks in Section 2; the simulation and implementation of SPRINT in Section 3, followed by some experimental results in Section 4.

2. MODELING RIVER NETWORKS

To fully describe the intricacies of water movement within a river channel requires the 3D Navier-Stokes and continuity equations. Much as in electrical circuit analyses where Maxell's equations (the fundamental physical laws with high computational costs) are replaced with simpler models with lower computational costs, in river networks a simpler model is commonly used in place of the Navier-Stokes equations. This model is named after a French mathematician, Adhémar Jean Claude Barré de Saint-Venant, and is usually called the Saint-Venant equations (although is sometimes seen as the St. Venant equations).

Several approximations are made to derive and apply a model based on the Saint-Venant equations. Some of these are:

- River flow is approximately 1-dimensional;
- River bed slope is relatively small;
- Effects of river bed friction and turbulence can be mod-

eled as resistance terms.

The 1D approximation allows us to integrate 3D Navier-Stokes and continuity equations over a river cross-section without knowledge of the exact velocity distribution. The second approximation is part of a "hydrostatic approximation" that removes smaller-scale dynamic pressure effects. The third approximation allows simplified empirical equations to model resistance, which we will discuss later. Although these approximations are generally valid for large-scale rivers analyses, there are places (such as steep mountain streams) where departures from these idealized flows may be significant.

2.1 Continuity Equation

The continuity equation describes the mass conservation in river flow. To illustrate this, consider the cross section of a river segment in Fig. 1. For a control volume highlighted in

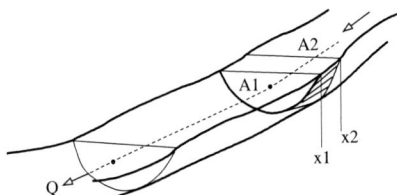

Figure 1: Illustration of a river channel cross section.

the diagram (between x_1 and x_2), the mass of water flowing into the control volume between time instances t_1 and t_2 can be written as:

$$\int_{t_1}^{t_2} \left[(\rho v A)_{x_1} - (\rho v A)_{x_2} \right] dt \tag{1}$$

where ρ is the water density, v is the flow velocity. x_1 and x_2 defines the control volume along the channel, and the cross-sectional area A is a variable that is a function of the water surface elevation and riverbed geometry.

In the same time window, the change of storage within the control volume can be written as:

$$\int_{x_1}^{x_2} \left[(\rho A)_{t_2} - (\rho A)_{t_1} \right] dx \tag{2}$$

Mass conservation dictates that the algebraic sum of the water flowing into the control volume equals the accumulation of the water in the control volume:

$$\int_{x_1}^{x_2} \left[(A)_{t_2} - (A)_{t_1} \right] dx + \int_{t_1}^{t_2} \left[(Q)_{x_1} - (Q)_{x_2} \right] dt = 0 \tag{3}$$

where $Q = vA$ is the flow rate. For fresh water in a river, variations of density ρ with temperature have negligible effects on flow dynamics, hence we can drop it from both sides of the equation.

To derive the differential form of the mass equation, we follow the Taylor expansion:

$$(A)_{t_1} = (A)_{t_2} + \frac{\partial A}{\partial t} \Delta t + \frac{\partial^2 A}{\partial t^2} \frac{\Delta t^2}{2} + \cdots \tag{4}$$

and

$$(Q)_{t_1} = (Q)_{t_2} + \frac{\partial Q}{\partial t} \Delta t + \frac{\partial^2 Q}{\partial t^2} \frac{\Delta t^2}{2} + \cdots \tag{5}$$

where $\Delta t = t_1 - t_2$.

Retaining only the first two terms in the approximations in Eqn. (4) and (5), then in the limit as $\Delta t \to 0$, we have:

$$\lim_{t_2 \to t_1} \int_{x_1}^{x_2} [(A)_{t_2} - (A)_{t_1}] \, dx = \int_{x_1}^{x_2} \int_{t_1}^{t_2} \frac{\partial A}{\partial t} \, dt \, dx \qquad (6)$$

and similarly:

$$\lim_{x_2 \to x_1} \int_{t_1}^{t_2} [(Q)_{x_2} - (Q)_{x_1}] \, dt = \int_{t_1}^{t_2} \int_{x_1}^{x_2} \frac{\partial Q}{\partial x} \, dx \, dt \qquad (7)$$

Hence the mass conservation in Eqn. (3) becomes:

$$\int_{x_1}^{x_2} \int_{t_1}^{t_2} \left[\frac{\partial A}{\partial t} + \frac{\partial Q}{\partial x} \right] dx \, dt = 0 \qquad (8)$$

or the commonly known continuity equation in differential form:

$$\frac{\partial A}{\partial t} + \frac{\partial Q}{\partial x} = 0 \qquad (9)$$

When there is lateral inflow to the channel, the continuity equation becomes:

$$\frac{\partial A}{\partial t} + \frac{\partial Q}{\partial x} = q_l \qquad (10)$$

where q_l is the lateral inflow per unit length along the channel.

2.2 Dynamic Equation

Although continuity can be used with empirical models for simplified representation of river networks, the correct dynamical solution of the Saint-Venant equations requires momentum conservation. The flows into/out of a control volume carry momentum, which is also affected by four forces on the control volume: the water pressure upstream and downstream of the control volume, the pressure from the river banks, gravity, and frictional force developed from the river bed and internal fluid shear. The relationship between forces and momentum is governed by Newton's second law of motion, which reduces to the dynamic Saint-Venant equation for 1D flow. It can be presented as:

$$\frac{\partial Q}{\partial t} + \frac{\partial}{\partial x} \left(\frac{Q^2}{A} \right) + gA \frac{\partial h}{\partial x} = gA(S_0 - S_f) \qquad (11)$$

where g is gravity, h is the depth. S_0 is the slope of the river bottom. S_f is the friction slope which we will discuss in the next subsection. Note that unlike the continuity equation in Eqn. (10), the dynamic equation has multiple nonlinearities, with variables Q, A S_f and h. The derivation for Eqn. (11) can be found in many hydraulics books, e.g. [18].

2.3 Friction Term

The frictional forces that extract momentum and energy from the flow are represented by empirical resistance models. A commonly-used model (adopted herein) is the Chézy-Manning formula, which was introduced by an Irish engineer Robert Manning[19]:

$$v = \frac{Q}{A} = \frac{1}{n} R_h^{2/3} S_f^{1/2} \qquad (12)$$

where R_h is the hydraulic radius, which is the ratio of the wetted cross section area (A) and the "wetted perimeter" (i.e. the distance measured along the river bottom across

the river). n is known as "Manning's n," serving as an empirical friction coefficient that may have a wide range of uncertainty, so serves as a calibration parameter.

2.4 Saint-Venant Equation

To summarize the behavior of a river segment is modeled by nonlinear Saint-Venant equations:

$$\begin{cases} \frac{\partial A}{\partial t} + \frac{\partial Q}{\partial x} & = q_l \\ \frac{\partial Q}{\partial t} + \frac{\partial}{\partial x}(\frac{Q^2}{A}) + gA\frac{\partial h}{\partial x} & = gA(S_0 - S_f) \end{cases} \qquad (13)$$

where the nonlinear friction slope is described by the Chézy-Manning equation:

$$S_f = n^2 \frac{Q^2}{A^2} \frac{1}{R_h^{4/3}} \qquad (14)$$

There are two independent variables in Saint-Venant equations, the flow rate Q and the wetted area A. Once these two quantities are known, the other quantities, e.g., water depth h, flow velocity v, are dependent functions of Q, A and river cross-section geometry (which may have substantial variability throughout a network).

A key difference between electric circuits and river networks is in the upstream propagation of information in the latter. Although mass flows downstream in a river, information of downstream events (e.g. a collapsed bridge that obstructs the flow) propagates upstream as a rising/falling water depth (changing A). Thus, the directionality of information propagation is both upstream and downstream through the network. Thus, the Saint-Venant equations must be solved as a distributed model and cannot be replaced with an equivalent lumped model as used in electronic circuits.

3. SIMULATION AND IMPLEMENTATION

The Saint-Venant equations in Eqn. (13) are a coupled set of nonlinear, time-varying, partial differential equations that are solvable by either explicit or implicit numerical methods. Over the years, there have been plethora of explicit methods proposed. For example, the leap-frog method in [20], the popular Lax-Wendroff method[21] in [22], and an modified Lax-Friedrichs method in [23]. The biggest advantage of explicit methods is that there is no need to construct and solve the nonlinear matrix otherwise required for implicit solution of the Saint Venant equations. However, this benefit comes at the cost of limited step size. Explicit methods have to meet the Courant-Friedrichs-Lewy (CFL) condition[24] in every element of a network; a constraint that can result in time steps on the order of seconds in a river simulation.

On the other hand, implicit methods are not stability limited by the CFL condition. A moderate step size (CFL < 10) is still preferred for accuracy, but localized high CFL conditions are tolerable. Many practical hydrology software packages use implicit methods[18] due to their robustness. The most commonly used implicit methods to solve Saint Venant equations are the method on staggered grid in [25], and the method on collocated grid in [26]. We focus on the four-points scheme from [26].

3.1 Four-Point Scheme

The four point scheme approximates spatial and temporal differences by using the average of four neighboring points in the $x - t$ plane[26]. It is generally over-damped and works very well for slow moving flow conditions.

725

In a simplified version of the four-point scheme, a time difference is approximated by:

$$\frac{\partial}{\partial t}f(x,t) \simeq \frac{1}{2\Delta t}(f_{j+1}^{n+1} - f_{j+1}^{n} + f_{j}^{n+1} - f_{j}^{n}) \quad (15)$$

where the subscript j represents spatial indices and the superscript n represents temporal indices.

The spatial difference is approximated by:

$$\frac{\partial}{\partial x}f(x,t) \simeq \frac{1}{\Delta x}(f_{j+1}^{n+1} - f_{j}^{n+1}) \quad (16)$$

and we take the average of two adjacent points to approximate the function value itself.

$$f(x,t) \simeq \frac{1}{2}(f_{j+1}^{n+1} + f_{j}^{n+1}) \quad (17)$$

Applying these discretization formulae, we can discretize Saint-Venant equations in Eqn. (13). At each time point $n+1$, the discretized continuity and dynamic equation specify a nonlinear relationship between unknowns (Q and A) at time point $n+1$, as well as at time point n.

3.2 Nonlinear Equation Solver

One of the most popular method to solve nonlinear equations is the Newton-Raphson's method. To compute the solution to a nonlinear problem $\mathbf{F}(\mathbf{x}) = \mathbf{0}$, the method iteratively improve solution $\mathbf{x}$ from a starting point $\mathbf{x}_0$ by using the local gradients:

$$\mathbf{x}_{k+1} = \mathbf{x}_k - \alpha \cdot (\frac{\partial \mathbf{F}}{\partial \mathbf{x}})^{-1} \cdot \mathbf{F} \quad (18)$$

where α is a damping factor.

Although Newton-Raphson's method is easy to implement, some cautions need to be exercised. One issue is the convergence. Since Newton-Raphson's method is a local search algorithm, it may fail to converge. In transient simulations this is rarely an issue because the solution at time $n+1$ is close to the solution at time n, which can be used as the starting point. However, for DC solve, convergence strongly depends on the selection of initial starting points; hence convergence could fail. In SPICE, this problem is partially solved by using a bounding algorithm to limit the operating range change of nonlinear devices. For difficult circuits, more advanced homotopy methods are still required circuits[27]. For river flow simulation, again this issue mainly arises for a steady-state solution with limited initialization information. We address this issue by first solving the Saint-Venant equations approximately, so that we can place a reasonable starting point for iteration. Secondly we implement a bounding algorithm based on fluid dynamics principles to limit the solution range. The combination of the two approaches provides good results.

Another issue with Newton-Raphson's method is that the functions have to be C^1 (i.e., the functions should be continuous themselves and also have continuous first-order derivatives with respect to unknowns). Depending on the complexity of the river cross-section shape, this is not always the case. We address this issue by approximating the river channel data into a C^1 function template. The approach not only solves the function continuity issues, but also help us to achieve considerable speed-up in the simulation.

As shown in Eqn. (18), Newton-Raphson's method requires the factorization of the Jacobian matrix $\frac{\partial \mathbf{F}}{\partial \mathbf{x}}$. The size of the matrix is $2m$ where m is the number of the computational nodes, which can be quite large for a continental-scale

river network. However, as we have observed in SPICE, the runtime complexity of factorizing a sparse matrix is usually $O(N^{1.2})$. In other words, it is only slightly super linear, versus the cubic complexity of factorizing dense matrix. It can be shown that the stencil of the Jacobian matrix in Saint-Venant equations using the four-point discretization scheme is only 4. Given the continuing rapid improvement of computer hardware and the efficiency of today's sparse linear matrix packages such as [28][29], the capacity required and the runtime for the factorization of the Jacobian matrix is not a concern.

3.3 General Simulation Flow

To illustrate the simulation flow, we use the symbolic three-branch river network in Fig. 2. The river network is first partitioned into computational nodes indicated by red dots. At each computational node, the river cross section description is used as input, along with the friction coefficients (Manning's n) and the bottom slope S_0. Next the computational nodes are connected from upstream to downstream following discretized Saint-Venant equations.

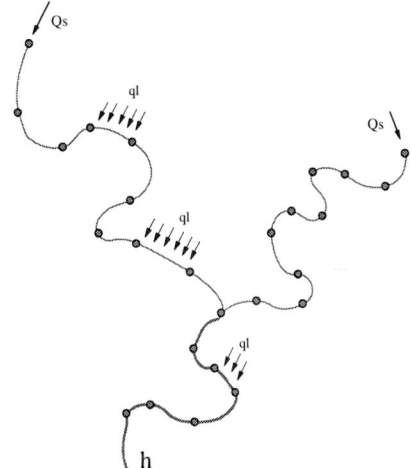

Figure 2: Illustration of computational nodes on a river network

At the junction point of the three branches, two linear relationships are specified. The first equation is mass conversation: the sum of the flow rate of two tributaries should equal the flow rate the downstream stem. Another linear relationship is used to specify the relative contributions of two upstream branches to the downstream branch.

Boundary condition flow rates are the forcing input Q_s at two upstream points, as well as the possible lateral inflows along the channels. A downstream boundary condition in the form of the depth at the most downstream node is applied. This value is typically the depth where a river enters a lake or ocean.

These procedures are used to build an entire river network with any number of tributary branchings. Once the network is constructed, we first perform steady-state solution with fixed forcing terms. We then apply the time-varying forcing terms to compute unsteady solutions. The solution provides the depth and flow velocity at every computational node for every time point in the simulation.

3.4 SPINT Implementation

SPRINT is implemented in C++ in a modular design approach. Besides the nonlinear and matrix solution methods described in the previous subsections, it also has many functionalities to enhance runtime performance and robustness, as well as other facilities such as model pre-processing and topological checking of the river networks. To facility the deployment in a cloud-based simulation environment, SPRINT also has an http front-end which enables remote simulation.

4. EXPERIMENTAL RESULTS

In this section, we present some experimental results of SPRINT.

4.1 Comparison with Analytical Solutions

In this experiment, a river segment with rectangular cross section is used. Due to its simplicity, the steady-state solutions can be calculated analytically, which are compared with the steady-state solutions computed by SPRINT. The depth at each node is plotted in Fig. 3. The difference is negligible at the resolution of the graph.

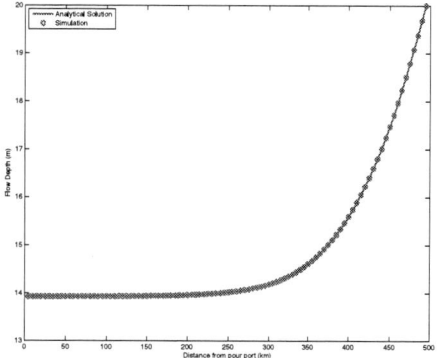

Figure 3: Comparison between SPRINT output and analytical solutions. Y-axis represents depth.

4.2 A Small Creek in Central Texas

This experiment is a creek in central Texas. It has three branches with the total length of approximately 7.2 miles. The simulation takes merely a few seconds on a common desktop computer for a 5-day event. Fig. 4 shows the simulated results at a computational node where observed gauge data are available. Note that we haven't gone through detailed calibration process so the simulated results will not match the observed depth data exactly. However, all the key waveform characteristics of the observed data are captured by the simulation results.

4.3 A River Basin in Central Texas

This experiment is a relatively large river basin in central Texas. It consists of over 3, 500 river branches with the total length of over 9, 000 miles. The whole river network is modeled by over 110K nodes and the simulation is performed on a regular desktop computer. It takes about an hour to simulate a 12-day event, which translates into about 100×

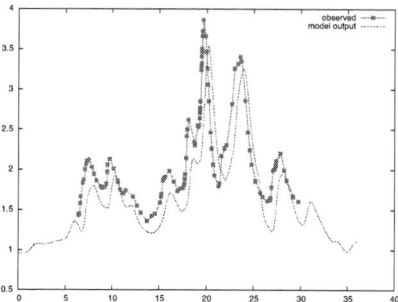

Figure 4: Comparison between SPRINT output and observed depth data. Note no calibration was performed.

speedup of real-time. Three snapshots of the simulated results are presented in Fig. 5. From the top figure, one can see the approximated diagram of the river networks. Note that many small tributaries are not included in the figure to avoid over crowding the graph. The width of each segment approximates the flow rate, while the color represents relative depths.

5. FINAL REMARKS

SPRINT represents an inter-discipline collaboration between computer engineering and civil engineering hydroscience. Although its algorithms cannot be directly applied, SPICE has been influential on the design of SPRINT, particularly in terms of design principle and design philosophy. Experimental results show that our tool can achieve considerable capacity and performance. We hope our effort will be beneficial to hydrology and hydraulic modeling community, just as SPICE has long lasting impact on the semiconductor industry.

6. REFERENCES

[1] L. W. Nagel and R. A. Rohrer, "Computer analysis of nonlinear circuits, excluding radiation," *IEEE Journal of Solid-State Circuits*, vol. 6, pp. 166–182, Aug. 1971.

[2] L. W. Nagel and D. O. Pederson, *SPICE: Simulation Program with Integrated Circuit Emphasis*. University of California, Berkeley, 1973.

[3] H. K. Gummel and H. C. Poon, "An integral charge control model of bipolar transistors," *Bell System Technology Journal*, pp. 827–852, 1970.

[4] B. J. Sheu, D. L. Scharfetter, P.-K. Ko, and M.-C. Jeng, "BSIM: Berkeley short-channel IGFET model for MOS transistors," *IEEE Journal of Solid-State Circuits*, vol. 22, pp. 558–566, Aug. 1987.

[5] G. D. Hachtel, R. K. Brayton, and F. G. Gustavson, "The sparse tableau approach to network analysis and design," *IEEE Transactions on Circuit Theory*, vol. 18, no. 1, pp. 101–113, 1971.

[6] C.-W. Ho, A. E. Ruehli, and P. A. Brennan, "The modified nodal approach to network analysis," *IEEE Transactions on Circuits and Systems*, vol. 22, no. 6, pp. 504–509, 1975.

[7] C. W. Gear, *Numerical initial value problems in ordinary differential equations*. Englewood Cliffs, NJ: Prentice-Hall, 1971.

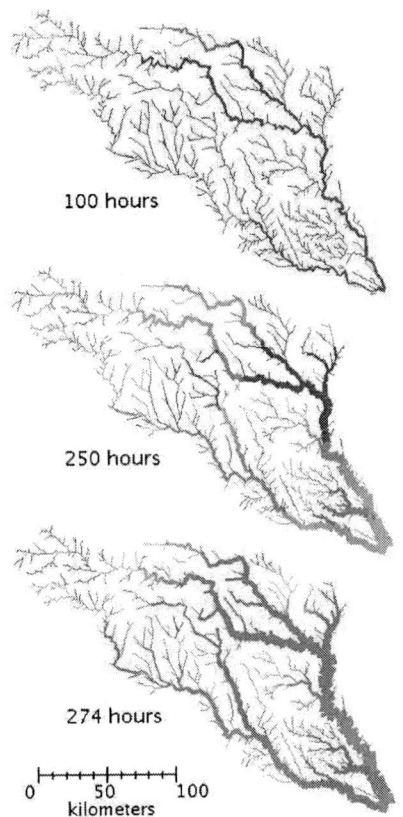

Figure 5: A river basin in central Texas. Three snapshots show the simulation output at different time points. The width represents the relative flow rate in the river branches. The color represents the relative depth.

[8] L. W. Nagel. Personal Communication, 2011.

[9] L. T. Pillage, R. A. Rohrer, and C. Visweswariah, *Electronic circuit and system simulation methods.* McGraw-Hill, 1995.

[10] J. Ogrodzki, *Circuit simulation methods and algorithms.* CRC Press, 1994.

[11] F. N. Najm, *Circuit simulation.* Wiley, 2010.

[12] Cadence Corporation, "Virtuoso spectre circuit simulator." http://www.cadence.com, 2012.

[13] Mentor Graphics Corporation, "Eldo classic." http://www.mentor.com, 2012.

[14] Synopsys Corporation, "HSPICE." http://www.synopsys.com, 2012.

[15] J. V. Clark, N. Zhou, and K. S. J. Pister, "MEMS simulation using SUGAR v0.5," in *Proc. Solid-State Sensors and Actuators Workshop*, pp. 191–196, 1998.

[16] "Bio-SPICE." http://www.biospice.org, 2012.

[17] US Army Corps of Engineers. http://www.hec.usace.army.mil/software/hec-ras/, 2011.

[18] J. A. Cunge, F. M. Holly, and A. Verwey, *Practical aspects of computational river hydraulics.* Boston MA: Pitman Publishing Ltd, 1980.

[19] R. Manning, "On the flow of water in open channels and pipes," *Transaction of the Institute of Civil Engineers, Ireland*, vol. 20, pp. 161–209, 1891.

[20] V. I. Koren and L. S. Kuchment, "Numerical integration of de Saint Venant equations with explicit schemes when computing the unsteady flow in rivers," *Trudy G.N.I.*, no. 8, 1967.

[21] P. D. Lax and B. Wendroff, "Systems of conservation laws," *Comm. on Pure and Applied Math.*, vol. 13, pp. 217–237, 1960.

[22] D. D. Houghton and A. Kashahara, "Nonlinear shallow fluid flow over an isolated ridge," *Comm. on Pure and Applied Math.*, vol. 21, 1968.

[23] J. Burguete and P. García-Navarro, "Improving simple explicit methods for unsteady open channel and river flow," *International journal for numerical methods in fluids*, vol. 45, no. 2, pp. 125–156, 2004.

[24] R. Courant, K. Friedrichs, and H. Lewy, "Über die partiellen Differenzengleichungen der mathematischen Physik," *Mathematische Annalen*, vol. 1, no. 100, pp. 32–74, 1928.

[25] M. B. Abbott and F. Ionescu, "On the numerical computation of near horizontal flows," *Journal of Hydrolic Research*, vol. 5, no. 2, pp. 97–117, 1967.

[26] A. Preissmann, "Propagation des intumescences dans les cannaux et rivièrrs," in *Congrès de l'Assoc Française de Calcul*, pp. 433–442, 1961.

[27] J. Roychowdhury and R. Melville, "Delivering global DC convergence for large mixed-signal circuits via homotopy/continuation methods," *IEEE Transactions on Computer-Aided Design of Integrated Circuits and Systems*, vol. 25, no. 1, pp. 66–78, 2006.

[28] I. S. Duff, "MA28: a set of Fortran subroutines for sparse unsymmetric linear equations," *Harwell Report AERE-R 8730, Atomic Energy Research Establishment Report*, 1980.

[29] T. A. Davids, "Algorithm 832: UMFPACK v4.3 – an unsymmetric-pattern multifrontal method," *ACM Trans Mathematical Software*, vol. 30, pp. 196–199, June 2004.

Humans for EDA and EDA for Humans

Valeria Bertacco

Department of Computer Science and Engineering, University of Michigan
Bob and Betty Beyster Building, 2260 Hayward Avenue, Ann Arbor, MI 48109

valeria@umich.edu

ABSTRACT

Two misconceptions have been plaguing the electronic design automation (EDA) industry for decades: i) EDA solutions scale to larger complexities at an insufficient rate to keep pace with improvements in silicon designs; and ii) since EDA applications target silicon chip developments, the growth of EDA as an industry is bounded by the growth of the semiconductor industry.

With this paper we address these misconceptions and we argue that they can both be overcome. To this end, we overview a number of initial studies highlighting possible directions that EDA can pursue to (i) break off from its traditional ways of scaling solutions and applications to larger complexity, that is, by developing better heuristics for its complex algorithms. (ii) We also discuss alternative domains where EDA technology can be applied, beyond that of silicon design, so that the semiconductor industry is no longer the limit of EDA growth.

Categories and Subject Descriptors

H.4 [**Information Systems Applications**]: Miscellaneous; B.6.3 [**Logic Design**]: Design Aids—*automatic synthesis, optimization, verification*

General Terms

Algorithms, Design, Verification

Keywords

EDA, Human Computing, Social Networks, Satisfiability

1. INTRODUCTION

The Electronic Design Automation (EDA) industry has been developing solutions to support silicon design for over 40 years. During this time, the scale and complexity of the problems that the industry could solve have made great strides: from supporting the development of the few-thousands-transistors chips of the 70's to that of the billion-transistors chips of the present day. However, EDA has been plagued for a long time (at least two decades) by a few misconceptions:

1. **The design complexity that EDA solutions can tackle is scaling at an increasingly slow and incremental pace.** This perception is mainly motivated by the increasing costs of silicon design developments. However, modern designs do not just have a larger scale than their previous generation counterparts. They also present a vast number of challenges that were not concerning just a few years ago, from parasitic effects, to clocking complexity, to hard-to-validate highly concurrent execution over complex communication protocols. Even if EDA is lagging in addressing and finding systematic solutions for these issues, over the past few decades it has delivered one of the best scaling trends ever observed in computing. Because of the needs outlined, however, we still ask whether we can do better (and the semiconductor industry demands it).

2. **The growth of EDA as an industry is limited by the semiconductor industry growth.** Because EDA is dedicated to developing tools and algorithmic solutions for the semiconductor industry, its market size can only grow if and at the rate that the electronics industry demand allows it. A common metric to evaluate this demand is to track the number of new ASIC design starts each year, which has suffered a persistent downtrend during the past decade, due to the steep increase in development and manufacturing costs of nanoscale-technology silicon chips [15, 20]. What we investigate in this work, though, is whether there are opportunities for EDA to serve other industries beside semiconductors, and thus unlock the potential for new growth.

With this work we explore some answers to the challenges discussed above. To address the first question, we investigate initial solutions that have the potential to break off the traditional practice of improving the scalability of EDA algorithms by relying on the performance improvements of the underlying computing hosts or by tuning and improving the algorithmic search heuristics in performance and quality. What we propose is to leverage human-computing and crowdsourcing solutions to EDA problems to develop completely different approaches that break off from our traditional way of thinking. Moreover, we want to learn from the way humans attack the complexity of these problems, and derive from it new approaches to developing automated heuristics.

To address the second challenge we look at possible additional domains to which EDA solutions can be applied. In particular, we consider the fast growing market of human-related and human-serving applications, such as social networks. Social networks are quickly expanding to provide a large number of services to humans. Their members are quickly growing in number, with Facebook leading at an estimated 845 millions registered members in 2011 [19], and more than 12 distinct social networks exceeding the

100,000,000 members level [27]. If EDA could provide services to these communities, or to companies that provide services to these communities, its applications would serve a much larger market than they are today.

The Design Automation Conference has witnessed a handful of studies in the past few years that have proposed alternative applications for the EDA field, for instance, by applying EDA solutions to the page ranking problem in web searches [12], and to social networks [7] (as we will discuss below).

2. HUMANS FOR EDA

Human computing is an approach to computing that has gained increasing interest in the past decade. In human computing, a computational process is carried out by outsourcing certain or all steps to humans. This computational model is effective when the tasks assigned to humans are those that are notoriously challenging for computers but straightforward and efficient for humans; examples of such tasks include image recognition, translation, *etc*. The term "crowdsourcing" is used when the process involves networks of individuals to achieve its goals. Humans are motivated to take part to the process by a wide range of "rewards", ranging from monetary compensation, to enjoyment in game-like processes, to gaining online reputation, *etc*.

The goal in using human computing to solve EDA problems is to break off from traditional approaches for algorithm scaling. The hope is to find completely new approaches to solving a problem by leveraging skills that are typical of humans, or by developing new solutions based on the way humans attack the problem. The main challenge in pursuing this venue is that the complexity of the problems instances in EDA is often beyond what can be managed by humans.

An example of a human computing solution for EDA is FunSAT [6]. FunSAT is a visual puzzle game recently developed in our research group. The game presents the player with visual puzzles derived from instances of the SATisfiability problem [8]. The goal is for human players to leverage their unique visual reasoning and pattern recognition abilities to solve the puzzle, that is, find a satisfiable assignment for the SAT instance.

2.1 FunSAT single-player

FunSat has some similarity with a handheld electronic game, called "Lights Out", a logic puzzle where the player manipulates a grid of buttons, some lit up and some off: every move lights up some buttons and turns off some others, while the player strives to organize his moves so to switch off all the lights [1].

In FunSAT, the game's board includes (i) rectangular control buttons representing the Boolean variables in the instance and (ii) circles of varying size representing clauses. Users can click on variables to cycle through their Boolean value assignment (true, false, not assigned) while the game shows which clauses are satisfied (green), falsified (red) or still unresolved (gray). In addition, the game represents clauses with circles of size proportional to the number of literals they include, so that players can visualize the size of each clause, and intuitively gain a sense of how easy or difficult it is to satisfy each of them (larger clauses include more literals and thus present more opportunities to be satisfied). FunSAT also uses varying gray color intensity for unresolved clauses to indicate how many unassigned literals are left in the clause; that way players can prioritize their attention in addressing unresolved clauses: the darker an unresolved clause, the most critical the situation, since there may be just one unassigned literal remaining. As a result, small, dark gray circles require top priority attention by a user: they correspond to clauses comprising just a few literals, most of which

have already been assigned.

Finally, players can zoom in and out in the game board to approach the problem by region, and can hover over a clause to see all the variables it depends on. Figure 1 shows a screenshot of the game. At the beginning, all clauses are unresolved (grey), they become green when a partial assignment satisfies them, and red if they are falsified. To leverage the human ability of spatial perception and area, we lay out clauses in a grid. Variables surround the "clause grid". Players advance in the game by levels, after solving an instance they are offered a more complex one in the next level.

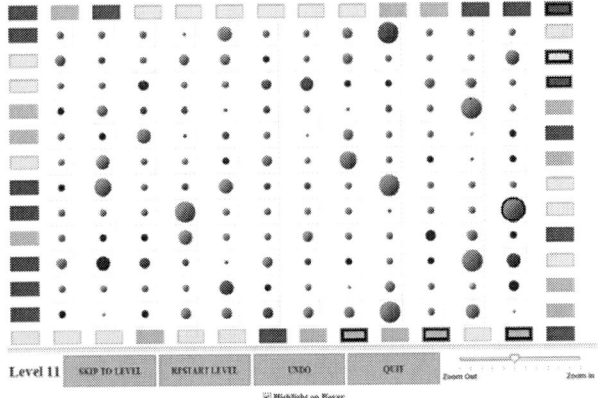

Figure 1: Screenshot of FunSAT - single-player. Circles represent clauses; rectangular buttons are Boolean variables controls. Assignments are applied by clicking a button; switching the corresponding variable from unassigned (grey), to blue (true), to yellow (false). Hovering over a clause highlights all the variables it depends on.

Through the game, human players can approach the challenge of solving SAT instances in a very unique way, compared to algorithmic solvers, using such skills as intuition and visual perception. As they click on different variables, they observe the visual impact on the grid of clauses and can progressively and intuitively tune their selection towards assignments that lead to a large fraction of satisfied clauses (visually perceivable by a higher fraction of green circles), until the entire grid is green, indicating that the instance has been solved. Classic SAT solving techniques, such as random restarts[16] and backtracking, are also naturally included in the game strategy, but with a "human twist". For instance, backtracking is naturally used by players when they feel that they are at a dead-end corner of the search, and simply change the assignment of a few variables to move away from the situation. Learning also occurs when players identify color and shape patterns that are generated by their selection and use this visual learning in developing their game strategy. This game appeared online in 2009 [4] and it is implemented in Java.

2.2 Crowdsourcing FunSAT

Scaling FunSAT to large instances presents a challenge, as it is limited by screen real estate, human patience and humans' ability to deal with complex problems. To address this aspect, we have recently developed a massively multi-player online version of this game [13]. To still leverage the visual intuition skills of humans we developed a new representation of the problem: clauses from a SAT instance are now presented in polar coordinates on a circle partitioned into one or more concentric sections. The controlling variables are placed along the outer perimeter of the circle (see Figure 2). This representation is much more compact for large instances, which typically have a large clauses-to-variables ratio. In addition, the game provides capabilities for zooming in and out conic sections of the circle, so that each player can visualize only a portion of the instance. Finally, a summary representation of the instance is provided in the fashion of a "world map" on the left side of the main game board, so that users can keep an eye on the overall situation at all times. This structure, along with the overall game controls is represented in Figure 2.

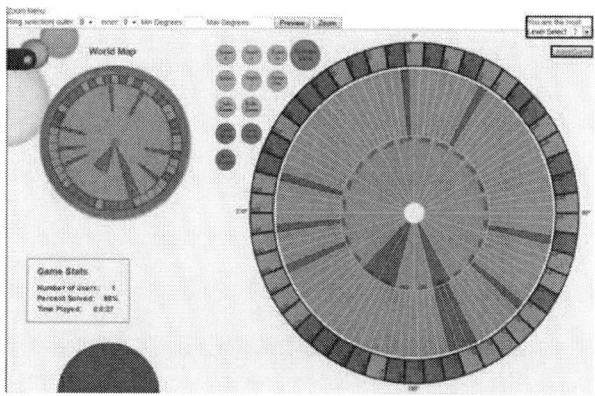

Figure 2: A FunSAT multi-player board. A game board is shown on the right with an assignment of variables (control buttons at the perimeter of the circle) that leaves 13 clauses falsified. Clauses are placed along two concentric inner circles. The smaller board on the left is a world map, useful when a player is zooming into a small portion of the game board. In between are several controls to zoom and rotate the board, create a random initial assignment and undo/redo moves.

The game setup lends itself to a wide range of multi-player interaction strategies. The one implemented in the first released version of the game is a *collaborative strategy*, where each user is assigned and controls a fraction of the instance's variables and players are meant to interact with each other to coordinate their assignments and satisfy the instance. The game is organized in game rooms; within each room, a game leader controls when to start a new level and/or end the game; incoming players can choose to start a new game room or join an existing one. The game room management and the coordination of moves among players is coordinated by a central server, which can be setup by any entity that wants to provide the online game to its community. To communicate, users can use the chat window in the game's GUI, or other chat and/or voice interaction mediums, such as Ventrilo [9].

Figure 3 is a snapshot of an advanced level of FunSAT multi-player, where clauses are laid out on several concentric circles (as shown in the world map) and one of the players is studying a conic section of the problem. Players can observe the dependency between clauses and variables by highlighting a clause or a variable. We tested this setup several times within our research group – in our case players could talk live to each other during the game – and found that collaboration enabled them to solve complex instances, of much larger size than those in the single-player version.

Other game strategies that we considered were *antagonistic strategies*, where the most effective players could over time take control

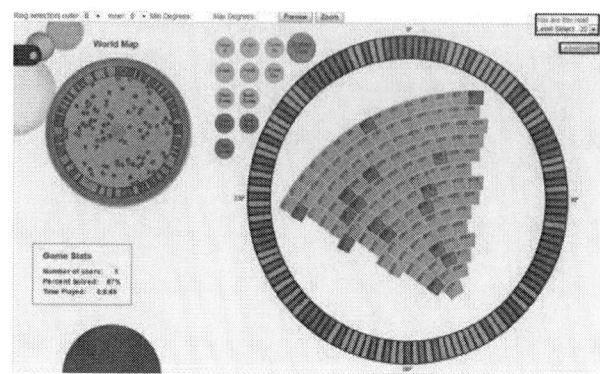

Figure 3: FunSAT multi-player, advanced game level. One the players in this game is studying a conic section of the board to evaluate if she should modify the assignment of the variables she controls or if she should negotiate an assignment modification with another player.

over other player's variables. This approach would provide positive feedback and motivate players not only with the completion of a game level, but also by increasing a player's variable control while he is solving a level.

The game is implemented in HTML5 [11] with a Python[17] back-end and requires a central server to synchronize and distribute moves across multiple players. One such central server is available at the University of Michigan and serves the players that connect to the game's website at our institution.

3. EDA FOR HUMANS

The second misconception that we want to address in this work is that EDA as an industry is limited by the market's demand of the semiconductor industry. In contrast, over the past decades, EDA has developed highly scalable and effective solutions to a number of problems that arise in many other domains. As discussed in the introduction, there has been a persistent downtrend of new ASIC design starts over the past decade: while this is a major source of concern in the industry, the shrinking size of the traditional EDA market could encourage companies to broaden their horizons and apply some of their solutions to alternative domains.

An area that has been growing at a very fast pace in the last decade is that of applications that provide services to humans: their social needs, their knowledge, their connection and communication needs. There are more than 100 social networks available online today, several of which have more than 100,000,000 registered members. Studying the characteristics and the connectivity of these networks would be a natural application for EDA's formal verification solutions. Indeed, the typical membership size of a large social network is approximately equivalent to the number of states in a Finite State Machine (FSM) representing a digital system with tens to a few hundreds storage bits: a system with 100,000,000 states requires at least 27 bits. This is a size that is commonly tackled by formal verification tools.

It seems that formal verification tools would be apt to analyzing graphs representing relational aspects of a social network group and extract a wide range of collective information: how strongly connected the group is, or which are the closely knit subgroups, whether two members could indirectly communicate with each other, *etc.* Other tasks with a more direct commercial aspect to them include searching the graph for users with similar interests, activities or background (that could be represented by edges to specific types of nodes) for the purpose of job hunting, community organizing or even match making. A similar analysis could be carried out across multiple social networks to gather more information about a member than what can be extracted from a single network and/or validate his/her identity by comparison. Studying groups of members instead of individuals could benefit targeted advertising. For instance, a high connectivity to other network members is a good indicator that the member is an influential individual in the community and would be an ideal candidate to promote a new service or product. Min-cut algorithms [3], very common in place & route solutions, are valuable in this context to identify a subgroup with relevant characteristics for the desired type of advertisement.

We recently pursued a study in this space, proposing to leverage model checking tools to analyze social networks [7]. Model checking is an approach to formal verification of digital designs that has been gaining increasing momentum and scaling to complex designs, so that today it is possible to model check complex properties over fairly involved design units. The design under study is modeled as a FSM, where each state of the design is represented by one vertex of the graph, while edges represent valid transitions between nodes. To analyze social networks, we can construct similar graphs, with vertices representing members and edges representing relevant relational properties. We performed a first evaluation of this alternative EDA application by studying the relational properties of the customer base of a Twitter [22] application startup [21] (that is, the customers are a subset of Twitter's members). We described the aspects of the group that we wanted to check as formal properties described in Computational Tree Logic (CTL); we then verified them using NuSMV [5]. Our study considered a portion of the network including only approximately 50,000 members and was able to prove or disprove the properties in a matter of seconds.

After this initial study, we strove to push the scale of our analysis further. We considered a larger Twitter subgroup of approximately 40,000,000 members [14] as the target for model checking. The main issue we encountered was in dealing with the explicit representation of the graph, that is, file size (25GB) and memory access delays. To reduce the size of the representation we pre-analyzed the graph to extract cycles (subgraphs where each user can communicate to all other users, possibly through third parties) and abstracted them away. We also needed to explore some implementation optimizations to minimize the disk seek time. This allowed us to verify some of the simpler formal properties in this large subgraph. We are currently considering the application of min-cut to aggressively abstract the graph and then apply model checking iteratively at an increasingly fine granularity, so to only expand the relevant portions within the social network graph.

4. RELATED WORK

Amazon made available one of the first human computing solutions with its Mechanical Turk [2], inspired by the "Turk", a fake chess-playing machine from the 18th century. The Turk was built to appear as an automaton that could play chess, while in reality it hid a human chess master inside who would control the machine. Similarly, Amazon's Mechanical Turk leverages a pool of human users to solve problems of disparate nature. The Mechanical Turk consists of task listings, ranging from categorizing products to writing articles, and offering a wage to anyone who completes a task: It appears to be an effective way to motivate a large group of people to perform menial, yet essentially human tasks.

Recently, a number of online human computing processes have also gained popularity, including the ESP game [24], re-captchas

[23] and duolingo [26]. All these solutions leverage humans' desire to play fun games that use their skills in image recognition and language translation, tasks that are extremely challenging for computers. The ESP game [24] is an internet game asking players to tag images: by collecting several tags from distinct individuals for each picture, the game can provide high quality tags for a large number of pictures available online, and thus greatly boosts the quality of image searches. Re-captchas [23] leverage humans to type words from scan images of books where character recognition software has failed. Finally, duolingo [26] uses human computation to translate documents between languages. To provide an estimate of the amount of human computation cycles available, von Ahn [25] reports that in 2003 humans have collectively played 9 billions hours of computer solitaire.

Complementary to these efforts, recent research has attempted to infer the computational model of the human brain [10], and has found that this model can deliver notable advantages for certain tasks (such as image recognition) over traditional computation. In addition, a recent work [18] proposes to leverage functional magnetic resonance imaging (fMRI) techniques to observe the brain activity in digital designers while at work. By studying the observed neural patterns, the authors hope to boost designers' productivity by developing better learning techniques to support their training and by selecting the most promising talent.

On the front of pursuing alternative venues and applications for EDA's solutions and algorithms, a few works have appeared in recent years at the Design Automation Conference. As an example, [12] proposes to apply the algorithmic solutions developed for parasitic extraction in silicon designs to page ranking in web searches.

5. CONCLUSIONS

In this paper we outlined approaches to overcome two classic misconceptions in EDA: (i) that scaling in EDA has been improving at a slower pace in recent years, insufficient to tackle the demands of the semiconductor industry; and that (ii) the application of EDA's solutions is limited by the needs of silicon design developments. We have shown that bringing humans into the equation has the potential to overcome both these issues.

On the first front, human computing can provide and inspire new ways to solve difficult algorithmic challenges and break the scalability barrier. Along this direction we overviewed an example solution that employs humans to solve SAT problem instances by presenting them as a game. On the second one, EDA's solutions and tools appear to be well positioned to solve large scale challenges in domains beyond semiconductors, such as those that directly benefit humans as a group. Specifically, we discussed the application of formal verification techniques to the study of social networks.

Overall, both these research directions have just began to attract the attention of a handful of EDA researchers, and we believe that much more can be gained by pursuing them further.

Acknowledgements

The author would like to thank her student Andrew DeOrio who led much of the effort in devising the solutions outlined in this paper, namely FunSAT [6], both single-player and multi-player, and the application of EDA to social networks [7]. She would also like to thank Erica Christenssen and John Krzemien for developing high quality online implementations of the FunSAT game, while overcoming the numerous issues and hurdles that are always encountered between an idea and its publicly available implementation. Finally, Joshua Lim worked on analyzing a large Twitter membership database to validate a number of formal properties.

6. REFERENCES

[1] Addicting Games. Lights out. www.addictinggames.com/puzzle-games/lightsout.jsp.

[2] Amazon. Mechanical turk, 2008. www.mturk.com.

[3] C. Chekuri, A. Goldberg, D. Karger, M. Levine, and C. Stein. Experimental study of minimum cut algorithms. In *Proc. of Symposium on Discrete Algorithms (SODA)*, pages 324–333, Jan. 1997.

[4] E. Christenssen, A. DeOrio, and V. Bertacco. FunSAT - single-player, 2009. www.funsat-single.eecs.umich.edu.

[5] A. Cimatti, E. M. Clarke, E. Giunchiglia, F. Giunchiglia, M. Pistore, M. Roveri, R. Sebastiani, and A. Tacchella. NuSMV version 2: An opensource tool for symbolic model checking. In *Proc. International Conference on Computer Aided Verification (CAV)*, pages 359–364, July 2002.

[6] A. DeOrio and V. Bertacco. Human computing for EDA. In *Proc. of the Design Automation Conference (DAC)*, pages 621–622, June 2009.

[7] A. DeOrio and V. Bertacco. Electronic design automation for social networks. In *Proc. of the Design Automation Conference (DAC)*, pages 621–622, June 2010.

[8] N. Een and N. Sörensson. An extensible SAT-solver. In *Proc. of the International Conference on the Theory and Applications of Satisfiability Testing (SAT)*, pages 502–518, May 2003.

[9] Flagship Industries, Inc. Ventrilo. www.ventrilo.com.

[10] J. Hawkins and S. Blakeslee. *On Intelligence*. Times Books, 2004.

[11] HTML Working Group. *HTML 5 Reference*. World Wide Web Consortium (W3C), Mar. 2009.

[12] V. Jandhyala. Physics-based, field-theoretic design automation tools for social networks and web search. In *Proc. of the Design Automation Conference (DAC)*, pages 280–281, June 2011.

[13] J. Krzemien, A. DeOrio, and V. Bertacco. FunSAT - multi-player, 2011. www.funsat.eecs.umich.edu.

[14] H. Kwak, C. Lee, H. Park, and S. Moon. What is Twitter, a social network or a news media? In *Proc. of the International Conference on World Wide Web (WWW)*, pages 591–600, Apr. 2010.

[15] B. Lewis and G. Ramamoorthy. Market trends: Worldwide, ASIC and ASSP design starts continue declining trend, 2012. *Gartner Research*, Feb. 2012.

[16] M. Moskewicz, C. Madigan, Y. Zhao, L. Zhang, and S. Malik. Chaff: Engineering an efficient SAT solver. In *Proc. of the Design Automation Conference (DAC)*, pages 530–535, June 2001.

[17] Phyton Software Foundation. Phyton programming language. www.phyton.org.

[18] M. Potkonjak and F. Koushanfar. (Bio)-behavioral CAD. In *Proc. of the Design Automation Conference (DAC)*, pages 351–352, June 2008.

[19] E. Protalinski. Facebook has over 845 million users. *ZDNet*, Feb. 2012.

[20] K. Shuler. The three consequences of fewer design starts. *System-Level Design (SLD)*, Mar. 2012.

[21] Twilk Inc., 2012. twilk.com.

[22] Twitter Inc., 2012. twitter.com.

[23] L. von Ahn. Games with a purpose. *IEEE Computer*, 39(6):92–94, June 2006.

[24] L. von Ahn and L. Dabbish. Labeling images with a computer game. In *Proc. Conference on Human Factors in Computing Systems (CHI)*, pages 319–326, Apr. 2004.

[25] L. von Han. Human computation. *Google Tech Talks*, July 2006.

[26] L. von Han. Massive-scale online collaboration. *TEDxCMU*, Apr. 2011.

[27] Wikipedia. List of social networking websites, 2011.

Application of Logic Synthesis to the Understanding and Cure of Genetic Diseases

Pey-Chang Kent Lin, Sunil P Khatri
Department of ECE, Texas A&M University, College Station TX 77843

ABSTRACT

In the quest to understand and cure genetic diseases such as cancer, the fundamental approach being taken is undergoing a gradual change. It is becoming more acceptable to view these diseases as an engineering problem, and systems engineering approaches are becoming more accepted as a means to tackle genetic diseases. In this light, we believe that logic synthesis techniques can play a very important role. Several techniques from the field of logic synthesis can be adapted to assist in the arguably huge effort of modeling and controlling such diseases. The set of genes that control a particular genetic disease can be modeled as a Finite State Machine (FSM) called the Gene Regulatory Network (GRN). Important problems include (i) inferring the GRN from observed gene expression data from patients and (ii) assuming that such a GRN exists, determining the "best" set of drugs so that the disease is "maximally" cured. In this paper, we report initial results on the application of logic synthesis techniques that we have developed to address both these problems. In the first technique, we present Boolean Satisfiability (SAT) based approaches to infer the logical support of each gene that regulates melanoma, using gene expression data from patients of the disease. From the output of such a tool, biologists can construct targeted experiments to understand the logic functions that regulate a particular gene. The second technique assumes that the GRN is known, and uses a weighted partial Max-SAT formulation to find the set of drugs with the least side-effects, that steer the GRN state towards one that is closest to that of a healthy individual, in the context of colon cancer. Our group is currently exploring the application of several other logic techniques to a variety of related problems in this domain.

Categories and Subject Descriptors: J.3 [**Computer Applications**]: Biology and genetics

General Terms: Algorithms, Theory

Keywords: Genomics, Gene Regulation, Logic

1. Introduction

In this section, we will present an engineering-centric view of the biological organism, with an overview of some of the relevant terminology and domain information.

The genome of a biological organism consists of a large number of base pairs (consisting of the letters A, C, G or T), arranged in a double helix sequence. This sequence uniquely identifies the individual organism. The genome fits within the nucleus of most cells of the organism. Individual "instances" of the same species have small variations in their genome, resulting in their unique "personality", although all individuals of the same species share significant commonalities in their genome (which result in the "human-ness" of the human being, for instance). The entire genome of several organisms has been *sequenced*, yielding the entire sequence of base pairs for that

organism. The human genome consists of about 3.2 billion letters in all.

The knowledge of the genome of any organism is the starting point in understanding genetic diseases such as cancer. The majority of the genome is uninteresting from a biological standpoint, while specific regions of the genome are crucial to the health of the organism. Fragments (sub-strings) of this genome are referred to as genes, and control the vital functions of the organism. The human organism is estimated to have about 30,000 genes. The organism is a complex control system in which proteins and RNA produced by genes and their products interact with and regulate the activity of other genes [1].

Based on the gene products of one or more genes in a set G_i, a gene g_i can become *repressed* (unable to produce its gene products) or *promoted or activated* (capable of producing its gene products). In this case, G_i is said to be the *predictor* of gene g_i. A *predictor* for a target gene g_i is the collection of genes directly participating in the regulation of gene g_i. As such, the predictor does not consider the type of regulation (repression versus activation), and is analogous to the *support* of a function in logic synthesis. Each gene has a single predictor (which is a set of genes) and the *predictor set* is the set consisting of predictors of each gene.

Typically, a small set of genes G are responsible for a specific biological function (or disease) in an organism. If we know the predictors G_i of each gene $g_i \in G$, and if we know the logic function of each $g_i(G_i)$, then we have the *Gene Regulatory Network* (GRN) [1] for that biological function (or disease). The GRN is essentially an FSM in logic-speak. Although the actual biological behavior of each gene has a continuous expression, the Boolean network model [2] has become popular for representing the GRN, since genes are observed to have a switch-like expression. As a result, much of the work in the literature uses Boolean networks to model GRNs.

From our knowledge of biological systems, we observe that over time, cellular processes converge to sequences of stable *attractor* states. This steady-state sequence of states is referred to as an *attractor cycle*. Some of these attractor states represent normal cellular phenomena in biology (i.e. cell cycle and division), while other attractor states are consistent with disease (i.e. metastasis of cancer).

The genome of an organism mutates, causing changes in the biological behavior of the organism over its life. Mutations are caused by several environmental factors (exposure to radiation, drugs etc.). Once a genome is mutated, then the GRN behavior may deviate from that of a healthy individual. Cancer and gene-related diseases are often the result of a failure in the signaling, leading to incorrect gene regulation and its associated functions. With the GRN-based view of the biological functioning of the organism, we can potentially target drugs which modify genetic expression of specific gene(s), thereby curing the disease. This is a promising way to treat genetic diseases, and can yield the possibility of "personalized medicine" – targeted and specific disease prevention and treatment based on an individual's genetic information [3, 4].

The task of inferring a GRN is an arduous one. Because of the complex interactions of the genes, it is hard if not impossible to construct a single biological experiment that will yield the complete GRN. Instead, several steps are employed. First, from gene expression data, biologists statistically observe that a certain subset of genes G are involved in the growth and spread of a genetic disease. Now, multiple samples of the gene expression of the genes in G (for both diseased and healthy individuals) are taken. These can also be in the form of *time course* data (where expression of the genes in G are taken for the same individual, over a sufficiently long duration). Time course data is generally not readily available, however. From the gene expression data, logic techniques can be utilized to i) find the support or predictors for each gene $g_i \in G$, and ii) infer the GRN for the disease. To validate the GRN obtained in this manner, biologists can perform targeted experiments to verify specific gene interactions within the GRN. Often, portions of the GRN are known, from targeted experiments that have already been conducted by biologists in the past. By curating the results of several such (often independently conducted) experiments, some GRNs have been inferred with reasonable confidence.

Once the GRN for a genetic disease is known, and the specific effect of candidate drugs on particular genes is known, a problem of interest is to find the best set of drugs which correct the GRN behavior of a diseased individual. This problem can be cast as an instance of modified automatic test pattern generation (ATPG).

This paper discusses approaches from the field of logic synthesis, applied to two specific problems.

- The first is the problem of inference of gene predictors in the GRN. Accurate predictors are necessary for constructing the GRN model and to enable targeted biological experiments that attempt to validate or control the regulation process. We report the results from a SAT-based algorithm to determine the gene predictor set from steady state gene expression data (attractor states). Using the attractor states as input, the states are ordered into attractor cycles. For each attractor cycle ordering, all possible predictors are enumerated and a conjunctive normal form (CNF) expression is generated which encodes these predictors and their biological constraints. Only certain combinations of predictors may form a valid predictor set, due to biological constraints. Each CNF is solved using a SAT solver to find candidate predictor sets. Statistical analysis of the resulting predictor sets selects the most likely predictor set of the GRN, corresponding to the attractor data. We demonstrate our algorithm on attractor state data from a melanoma study [5] and present our predictor set results.

- Cancer and other gene related diseases can be abstracted as faults in the regulatory function of a GRN. For effective cancer treatment, it is imperative to identify faults and select appropriate drugs to treat the faults. We present an extensible Max-SAT based automatic test pattern generation (ATPG) algorithm for cancer therapy. This ATPG algorithm utilizes the stuck-at fault model for representing signaling faults. A weighted partial Max-SAT formulation is used to enable efficient selection of the most effective drug. Our experiments are conducted on the colon cancer network, and include results on the identification of testable faults, optimal drug selection for single/multiple known faults, and optimal drug selection for overall fault coverage.

The remainder of this paper is organized as follows. Section 2 describes previous work in this arena. Section 3 presents the algorithms to perform predictor set inference, while Section 4 discusses methods to find the best set of drugs to maximally cure a genetic disease. Concluding comments are made in Section 5.

2. Previous Work

Several models have been proposed for modeling the GRN such as Markov Chains [6, 7], Coupled ODEs (ordinary differential equations), Boolean Networks [2, 8], Continuous Networks [9], and Stochastic Gene Networks [10]. We utilize the Boolean Network (BN) model that was proposed by Kauffman in 1969 [2]. In a Boolean Network, the expression activity of a gene is represented as a binary value, where 1 indicates the gene is ON (active) and producing gene-products, while 0 indicates it is OFF. Such a model cannot capture the continuous and stochastic biochemical properties of protein and RNA production. However, genes can typically be modeled as ON or OFF in any particular biochemical pathway [11]. This allows a rich set of logic synthesis algorithms to be brought to bear to the problems of GRN modeling and control.

In the context of predictor set inference, [12, 13] use dynamic Bayesian networks and probabilistic Boolean networks (PBNs). The GRN is then inferred from this data, using methods traditionally based on probabilistic transition models [14, 15]. The method proposed considers gene prediction using multinomial probit regression with Bayesian variable selection. Genes are selected which satisfy multiple regression equations, of which the strongest genes are used to construct the predictor set. The target gene is predicted based on the strongest genes, using the coefficient of determination to measure predictor accuracy.

Another method proposed by [16] also assumes a PBN model. A partial state transition table is constructed based on available attractor state data. From this state transition table, predictors with 3 or less regulating genes are selected for each target gene. All unknown values in the table are randomly set. The Boolean network is simulated for several iterations using different starting states, observing whether the states eventually transition to an attractor cycle. If the simulation successfully transitions to an attractor cycle, the selected predictors are considered as a valid predictor set. This process is repeated, to build a collection of Boolean Networks which are combined to form a Probabilistic Boolean Network (PBN).

Our larger goal is to find a small number of *deterministic* GRNs, rather than a PBN. Such a set of GRNs may be expressed as a family of Boolean relations. Towards this goal, we need to first find ways to accurately find the predictor set. Philosophically, our aim is to invest effort into accurate predictor set determination, so that the results can be used to find high quality deterministic GRNs.

In the context of optimal drug determination for cancer, the authors of [17] proposed modeling cancer as faults in the signaling network and applied fault analysis for drug intervention. Cancer is a disease that arises from fault(s) in the network leading to loss of cell cycle control and uncontrolled cell proliferation. Therapy involves both identification of the fault and a suitable drug combination to target the fault. This paper focused on the colon cancer growth factor (GF) signaling pathways, which are often associated with proliferation of cancer.

The method proposed in [17] is an ATPG technique in principle. Our approach is similar to [17] in that it uses the BN and models cancer as faults in the network. However, the differences are several. Instead of explicit enumeration of the BN, we use an extensible, implicit SAT-based ATPG approach to efficiently model and identify faults, and perform drug selection. Further, unlike [17], we include weighted clauses for outputs and drugs in the SAT formulation. Using this, the algorithm can implicitly and efficiently determine the drug combination which is maximally effective. Finally, our approach can handle multiple faults easily. The runtimes of our approach are typically much less than a second per set of faults.

In the past, ATPG has been extensively studied in research and industry. One such technique is the SAT-based ATPG [18, 19, 20] which translates the testing condition into a SAT instance that retains the circuit structure. A test for the fault can then be found by invoking a SAT solver. In the context of cancer therapy, we extend the SAT based approach to handle drug selection under the influence of multiple faults.

3. Predictor Set Inference

3.1 Definitions

DEFINITION 1. *A* **predictor** $f_i = \{g_j, g_k, \cdots\}$ *lists the set* $\{g_j, g_k, \cdots\}$ *of genes which regulate the activity of gene* g_i.

DEFINITION 2. *The* **predictor set** *is the complete set of predictors* $\{f_1, f_2, \cdots, f_n\}$ *for the GRN with n genes* $g_1, g_2, \cdots, g_n$.

3.2 Problem Formulation

Given gene expression data (a set of unordered attractor states) as input, we would like to determine the best predictor set. We first present an outline of our SAT-based algorithm, and then explain the steps through a simple example.

The algorithm has three main steps.

735

Present state			Next state		
x_1	x_2	x_3	y_1	y_2	y_3
0	1	0	1	1	0
1	1	0	0	1	0
1	1	1	1	1	1

Table 1: Example state transition table

- First, attractor states are ordered into attractor cycles in all possible ways. For each possible ordering of attractor states into attractor cycles, all possible predictors are found and a CNF is generated, encoding valid predictor sets.

- Second, the CNF is solved for All-SAT, recording all satisfying cubes. Each cube corresponds to a predictor set. The first two steps are repeated for all attractor cycle orderings.

- Finally, statistical analysis on the SAT results determines the most frequent (likely) predictor set for the GRN.

To illustrate the SAT-based algorithm, we apply it to a simple example with three genes (g_1, g_2, g_3) and gene expression data with three lines ($010, 110, 111$). The present state of these genes is represented by the variables $< x_1, x_2, x_3 >$ and the next state is represented by the variables $< y_1, y_2, y_3 >$. We assume each line was measured in steady state and therefore is an attractor state.

Step 1: We order (or arrange) the attractor states into attractor cycles for which there are six possibilities. One ordering is with each attractor state transitioning to itself with a self-edge, resulting in three singleton attractor cycles. Two possible orderings result when all three attractor states form a single attractor cycle of length three. The last three possible orderings have two attractor cycles, one cycle with length two and the other cycle of length one. We focus our example on an ordering with two attractor cycles, as shown in Table 1.

For each valid attractor cycle ordering, a *partial state transition table* is constructed, containing the attractor states. Table 1 shows the partial state transition table for the example attractor cycle ordering. To find all valid predictors of a gene, each next state column is checked against all combinations of the present state columns. For example, let us explore gene g_2 and g_3 as a candidate predictor for gene g_1. For gene g_1, the next state bit is y_1, while for gene g_2 and g_3, the present state bits are x_2 and x_3. In the first two rows of Table1, $< x_2, x_3 >= 10$. However, in row 1, $y_1 = 1$, while in row 2, $y_1 = 0$, which forms a contradiction (since the same input cannot result in different outputs). Therefore, gene g_1 cannot be predicted by genes g_2 and g_3.

Now, consider genes g_1 and g_3 as a candidate predictor for gene g_1. Since there is no contradiction, the combination is logically valid. Thus one possible predictor for gene g_1 is $f_1 = \{x_1, x_3\}$. All valid predictors with P (user-defined) or less inputs are exhaustively searched and recorded for CNF formulation (which is done in the next step). In our example, gene g_1 has 2 possible predictors $\{x_1, x_3\}$, $\{x_1, x_2, x_3\}$ which we label v_1^1, v_2^1 respectively. We assume that a gene cannot self-regulate, so $\{x_1\}$ by itself is not a valid predictor.

Step 2: After all predictors are found for each gene, we generate the SAT formula which encodes logically valid predictor sets. The jth predictor for gene i is assigned a variable v_j^i. Gene g_1 in our example will have two predictor variables $v_1^1 \equiv \{x_1, x_3\}$, $v_2^1 \equiv \{x_1, x_2, x_3\}$. Gene g_2 and g_3 will have their own corresponding predictor variables $v_1^2 \equiv \{x_1, x_2\}$, $v_2^2 \equiv \{x_1, x_3\}$, $v_3^2 \equiv \{x_2, x_3\}$, $v_4^2 \equiv \{x_1, x_2, x_3\}$ and $v_1^3 \equiv \{x_1, x_3\}$, $v_2^3 \equiv \{x_2, x_3\}$, $v_3^3 \equiv \{x_1, x_2, x_3\}$ respectively. There are three constraints that we incorporate while constructing the CNF that encodes valid predictor sets. The conjunction of these constraints forms our final CNF.

1. The first constraint (S_1) is that all genes in the GRN must have a predictor. In other words, we assume that all genes are highly correlated and are "participating" in the GRN. For gene i, all of its associated predictor variables are written in a single clause $c_i^1 = (v_1^i + \cdots + v_j^i)$. In our example, for g_1, $c_1^1 = (v_1^1 + v_2^1)$. For g_2 and g_3, we have $c_2^1 = (v_1^2 + v_2^2 + v_3^2 + v_4^2)$ and $c_3^1 = (v_1^3 + v_2^3 + v_3^3)$ respectively.

To satisfy any c_i^1 clause, at least one predictor in the clause must be chosen. To ensure that at least one predictor is chosen for all genes, we write the conjunction of all c_i^1 clauses. $S_1 = c_1^1 \cdot c_2^1 \cdot c_3^1$

	PIRIN x_1	S100P x_2	RET1 x_3	MART1 x_4	HADHB x_5	STC2 x_6	WNT5A x_7
BAD	0	0	0	0	0	1	1
	0	0	0	1	1	1	1
	1	0	1	0	0	0	1
GOOD	0	1	0	0	0	0	0
	0	1	1	1	0	0	0
	1	0	1	1	1	0	0
	1	1	0	1	1	0	0

Table 2: Attractors for Melanoma Network

2. The second constraint (S_2) specifies that for each gene, exactly one predictor is chosen. The assumption is that a gene cannot have multiple predictors. To formulate the clauses c_i^2 for gene i, smaller clauses are formed from all pairs of combinations of its predictors $v_{1 \ldots j}^i$. In each of these clauses of pairs of variables, both predictor variables are complemented. Hence, $c_1^2 = (\overline{v_1^1} + \overline{v_2^1})$, $c_2^2 = (\overline{v_1^2} + \overline{v_2^2}) \cdot (\overline{v_1^2} + \overline{v_3^2}) \cdot (\overline{v_1^2} + \overline{v_4^2}) \cdot (\overline{v_2^2} + \overline{v_3^2}) \cdot (\overline{v_2^2} + \overline{v_4^2}) \cdot (\overline{v_3^2} + \overline{v_4^2})$, and $c_3^2 = (\overline{v_1^3} + \overline{v_2^3}) \cdot (\overline{v_1^3} + \overline{v_3^3}) \cdot (\overline{v_2^3} + \overline{v_3^3})$ Any selection of two or more predictors for gene i will result in the clauses of c_i^2 becoming unsatisfiable. The c_i^1 clause ensures that at least one predictor will be chosen for gene i, and c_i^2 forces the selection of exactly one predictor for gene i. The conjunction of all c_i^2 clauses forms the constraint S_2, which forces SAT to choose only one predictor per gene. $S_2 = c_1^2 \cdot c_2^2 \cdot c_3^2$

3. The last constraint (S_3) requires that each gene must be used as a predictor for at least one other gene in the predictor set. A gene that is not used in any predictor does not perform any regulation function and could be removed from the GRN. S_3 ensures that this does not occur. To ensure that gene g_i is used in at least one predictor, we form clauses c_i^3 which include all predictors that use gene g_i as input. To specify that gene g_i must be used, we also include a single variable clause (x_i) to c_i^3. For gene g_1, $c_1^3 = (x_1) \cdot (\overline{x_1} + v_1^1 + v_2^1 + v_1^2 + v_2^2 + v_4^2 + v_1^3 + v_3^3)$. $c_2^3 = (x_2) \cdot (\overline{x_2} + v_2^1 + v_1^2 + v_3^2 + v_4^2 + v_2^3 + v_3^3)$ and $c_3^3 = (x_3) \cdot (\overline{x_3} + v_1^1 + v_2^1 + v_2^2 + v_3^2 + v_4^2 + v_1^3 + v_2^3 + v_3^3)$ for gene g_2 and g_3 respectively. To satisfy these clauses, x_i and at least one other predictor variable in the second clause of c_i^3 must be selected. Finally, $S_3 = c_1^3 \cdot c_2^3 \cdot c_3^3$

The final SAT formula S as a conjunction of the S_i formulas.

$$S = S_1 \cdot S_2 \cdot S_3$$

Step 3: The SAT solver performs an All-SAT on S. The satisfying cubes (each cube encodes a candidate predictor set) from the All-SAT output are collected. The process is repeated for the remaining attractor cycle orderings. From the results, we find the most likely predictors based on the frequency of occurrence of the predictors across all orderings. Three methods are used to analyze the statistical results, which will be described in Section 3.3.

In general, the above algorithm can be applied to input data for N genes and A attractor states. The total number of attractor state orderings is $A!$. For each ordering, there can be up to $O(N^3)$ predictors per gene. The SAT search space per ordering is on the order of $O(2^{N^3})$, resulting in overall complexity of $O(A!2^{N^3})$. Typically, the number of attractor states recorded through gene expression measurements is small. As such, $A!$ is thus much smaller than $2^{(N^3)}$, so the runtime complexity is dominated by the All-SAT operation. For pragmatic reasons, our algorithm stops each All-SAT after T minutes (or C cubes), where T or C is defined by the user.

3.3 Experimental Results

To evaluate our SAT-based algorithm for inferring gene predictors, the algorithm was tested on gene-expression data from a melanoma study done by Bittner and Weeraratna [5]. In the melanoma study, it was observed that an abundance of RNA (expression) for gene WNT5A was associated with a high metastasis of melanoma. The study measured 587 genes with 31 gene expression patterns (lines). Seven genes are believed to be closely knit: PIRIN, S100P, RET1, MART1, HADHB, STC2, and WNT5A. There are 18 distinct patterns, which were reduced to seven using Hamming-distance of one, in Table 2. These seven lines form the attractor states which are the input to our algorithm.

For the experiments, we assume two additional specifications. First, we divide attractor states into good and bad states, based on the presence of WNT5A. We allow good attractor states to cycle only to other

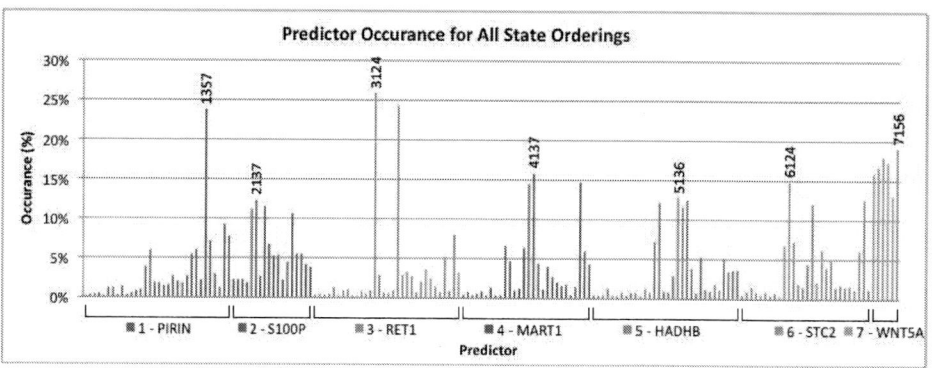

Figure 1: Method A: Predictor Occurrence for all Valid Attractor Cycle Orderings (First Iteration: No Predictor Selected)

good attractor states, and bad attractor states can only cycle to other bad attractor states. Second, we limit the maximum attractor cycle length L to 3, and the maximum number of predictor inputs P to 3, because long attractor cycles and large predictor inputs are highly complex and less likely to occur in biological systems [21, 1].

Our algorithm utilizes a modified open-source and highly efficient exact SAT-solver called MiniSAT v1.14 [22, 23]. All-SAT operations were limited to a 30 minute time-out. On average, each All-SAT run yielded 10K satisfying cubes in this duration. Our algorithm was implemented and run on a Pentium 4 Linux machine with 4GB RAM. MiniSAT was originally designed to find a single satisfying assignment. We modified MiniSAT to perform All-SAT. We further modified MiniSAT to always randomly select decision variables during the solving process to increase the activity of all variables. With the second modification, we find that a reduced runtime of 30 minutes is sufficient to achieve an average of $\leq 5\%$ difference in the predictors' occurrence frequency compared to the full All-SAT results (without the modification we obtain an 8% difference).

The following presents our results after collection of All-SAT results from all valid attractor cycle orderings. In Figure 1, we display a histogram of all logically valid predictors and their frequency of occurrence, across all attractor orderings. In the sequel, a predictor label of 2367 means that gene g_2 is predicted by genes g_3, g_6, and g_7. From this chart, we can observe that certain predictors occur with significantly higher frequency than others. For example with gene g_1, the predictor $\{x_3, x_5, x_7\}$ (PIRIN predicted by RET1, HADHB, WNT5A) occurs with much higher frequency than all other predictors for gene g_1. This indicates that this predictor is most likely to be present in the final predictor set.

From this data, we propose three methods (A, B, AB) for selecting the predictor set. In **method A**, a predictor histogram is created as in Figure 1. From the histogram, for each gene g_i, we find its predictor p_j^i such that p_j^i is the most frequently occurring predictor of gene g_i and the *resolution ratio* R_i of this predictor (defined as the ratio of the occurrence frequency of p_j^i to the occurrence frequency of the next most frequently occurring predictor of gene g_i) is maximum. Among all genes, we choose the one with the highest resolution ratio, and select its most frequently occurring predictor as its final predictor. After selecting this final predictor, we regenerate the histogram, discarding any candidate predictor sets that do not contain the final predictor(s) that have been selected in previous steps. The process repeats until all genes have a single final predictor. The set of final predictors of all genes forms the predictor set. The advantage of method A is that at every iteration, we select real predictors that have a high overall occurrence in the solution. However the method may have problems selecting final predictors if the resolution ratio is low (i.e. when the frequencies of occurrence of the predictors are nearly identical).

As an alternative, **method B** is proposed, to determine for each gene i, how likely it is that gene g_i will predict the other genes in the GRN. In other words, we ask what is the occurrence frequency of x_i in the predictors of f_j. Table 3 shows in entry (i, j) how frequently a gene g_i is used to predict a gene g_j. This table is populated by

	f_1	f_2	f_3	f_4	f_5	f_6	f_7
x_1		0.59	0.68	0.57	0.69	0.60	1.00
x_2	0.24		0.41	0.29	0.33	0.49	0.51
x_3	0.65	0.48		0.76	0.58	0.56	0.17
x_4	0.39	0.40	0.78		0.54	0.44	0.29
x_5	0.56	0.30	0.27	0.44		0.39	0.36
x_6	0.42	0.54	0.52	0.41	0.44		0.67
x_7	0.64	0.63	0.24	0.48	0.32	0.45	

Table 3: Method B: Gene occurrence for all predictors (first iteration)

summing the occurrence frequency of all predictors of g_j that have gene g_i as one of their inputs. As such, any entry can be ≥ 1, and is a measure of the usefulness of g_i as a predictor for g_j. The predictor of g_j is determined by finding, for each column j of Table 3, the three largest entries and adding their values. Suppose we call this sum s_j (the resolution score of column j). We compute the resolution score for all columns and select the final predictor for the column with the highest resolution score. This final predictor is formed by listing the 3 input genes that correspond to the 3 entries that were used to compute the highest resolution score. Similar to method A, we reiterate the process by regenerating the table after discarding all predictor sets that do not contain predictors that were selected in previous steps. Method B has the advantage of being more robust when no *single* predictor has a significantly higher occurrence frequency than others. However, there is no guarantee that the predictor selected by method B is a valid predictor. If this happens, we select the column with the next highest resolution score.

In our experiments, we also use a hybrid **method AB** which works in the following manner. Both methods A and B are used to select their best predictor. If both methods produce the same predictor f_i, we select this predictor as a final predictor. If not, we list the best predictors for each gene, for both methods. If multiple predictors match for both methods, we choose the final predictor as the one with the highest weighted sum of the resolution ratio and resolution score. The resolution ratio is weighted by 0.3 and the resolution score is weighted by 0.7. The weighting factor for the resolution ratio is lower since the resolution ratio values of any gene are often close to 1. In such a situation, we would like to favor method B. If no predictor is produced by the previous step, we look at the top five predictors of method A for each gene and calculate the weighted sum of their resolution ratio and resolution score. The predictor with the highest weighted sum is selected as the final predictor. The process is reiterated, regenerating the histogram and table at each step, discarding any predictor sets that do not contain any of the previously selected final predictors. With this combined approach, we are able to select predictors with a higher degree of confidence and robustness.

We process our All-SAT data from melanoma attractor data of [5] using methods A, B, and AB. Results are shown in Table 4 and shows what predictor was selected for each gene and the accompanying resolution ratio, resolution score, or weighted sum.

From the results, we can draw several conclusions:

• The iterative steps in regenerating the histogram (or table) retain only cubes (predictor sets) that contain previously selected final pre-

		PIRIN x_1	S100P x_2	RET1 x_3	MART1 x_4	HADHB x_5	STC2 x_6	WNT5A x_7
A	Pred. set	1357	2137	3146	4357	5124	6124	7124
	Res. ratio	2.57	1.41	1.34	1.30	1.41	1.66	1.31
B	Pred. set	1357	2137	3146	4137	5134	6137	7126
	Res. score	1.78	1.77	1.84	1.97	1.99	1.98	2.56
AB	Pred. set	1357	2367	3146	4137	5137	6357	7124
	Wt. sum	2.06	1.57	1.75	1.61	1.45	1.39	1.88

Table 4: Predictor set selection

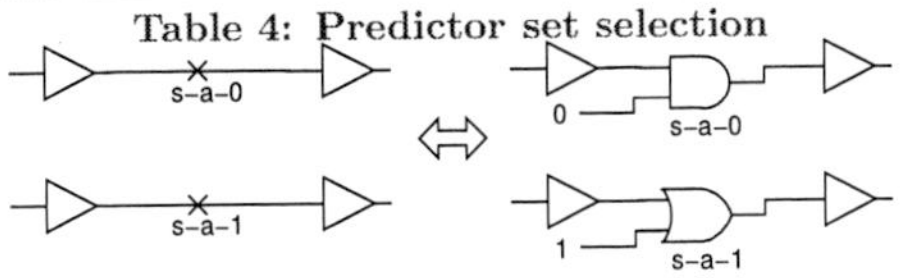

Figure 2: Fault Modeling and Injection

dictors. Hence the final predictor set from *each* method is a valid satisfying cube of the SAT formula S.

- The final predictor set is present in a select number of attractor cycle orderings. For example, the final predictor set selected by methods A, B, and AB are found in respectively 8, 4, and 6 attractor cycle orderings out of the total 5040 possible orderings. Hence the algorithm will enable us to generate a few deterministic GRNs. In practice, the entire procedure can be re-run with these new attractor cycle orderings, until a single attractor cycle ordering survives. Biologists can then design time course experiments for the disease, to verify that this ordering is indeed valid, thereby validating the predictor sets we obtain.

- Some predictors are common among the predictor sets between the three methods. For example, all three methods select $f_1 = \{g_3, g_5, g_7\}$ (PIRIN predicted by RET1, HADHB, WNT5A) as well as $f_3 = \{g_1, g_4, g_6\}$. We can conclude this predictor is highly likely to be a final predictor in the GRN. Also, a majority of the predictors selected by the three method share common input genes. For example, the predictor selected by all methods for gene g_2 (S100P) contain 2 common genes $\{g_3, g_7\}$ (RET1, WNT5A), indicating these 2 genes are likely to be contained in the final predictor of f_2. Similarly f_7 has 2 common genes g_1 and g_2 for all methods.

- Using the above results, biologists can target their research on gene regulation and control, focusing on the gene relationships determined by the predictor set results.

4. Optimal Drug Selection

4.1 Background

In an IC, the difference between a defect and a fault can be explained as imperfections in the hardware and function, respectively. In genomics, examples of biological defects can include mutations in the gene activation site, malformation of the protein folding, and problems in the gene product transport. Likewise, an example of a biological fault is a modification of the logical function of a gene, producing the incorrect output.

We use the stuck-at fault model for the GRN [17] and employ SAT-based automatic test pattern generation (ATPG) techniques to determine a drug vector (set of drugs) to maximally rectify the fault. The GRN is modeled as an interconnection of Boolean gates.

A stuck-at-0 fault is modeled by inserting a two-input AND gate at the fault site as shown in Figure 2. The side input of the gate is driven by a signal which is set to 1 to simulate a fault-free site, or set to 0 to inject the s-a-0 fault. Similarly, the circuit with a s-a-1 fault is modeled by inserting an OR gate at the site. The side input of this OR gate is set to 0 to simulate a fault-free site, or set to 1 to inject the s-a-1 fault. These gates are inserted at every net (wire), allowing the simulator to inject faults at any site. Note that drugs are modeled the same as stuck-at faults, wherein a drug that inhibits a gene is modeled as a s-a-0 "fault", while a drug that activates a gene is modeled as s-a-1 "fault". The gates for drug injection are inserted at the nets of the genes that they target.

We utilize a Weighted partial Max-SAT (WPMS) solver, in which each clause in the CNF is identified as a hard clause or a soft clause.

Each soft clause is associated with a weight. The problem then is to identify an assignment that satisfies *all* hard clauses while maximizing the total weight of the satisfied soft clauses.

When all the s-a-0 (s-a-1) variables are set to 1 (0), the CNF formula S of the biological network describes the good (fault-free) circuit behavior. The faulty circuit is a copy of the fault-free circuit, with faults (s-a-0 or s-a-1 variables) injected at the gates affected by faults.

4.2 Model implementation

We evaluate the WPMS-based ATPG methods on the GRN that models growth factor (GF) pathways for colon cancer [17]. In multicellular organisms, cell growth and replication is tightly controlled by the cell cycle control. This system receives signals from other cells which are used to decide whether the cell should grow. A failure in these signals can lead to unwanted or unregulated cell growth, leading to cancer. These signaling pathways are well studied, and several drugs have been developed to target different pathways for cancer therapy.

We begin with a BN model of the colon cancer GF pathways as derived in [17]. In this model, pathways are converted to an equivalent BN logic gate. Each interconnection (net) between logic gates is then assigned a numerical label.

As stated earlier, defects in the GRN are represented as stuck-at faults that permanently set a signal net to 1 or 0. At each net, the logic gates for injecting a s-a-0 or s-a-1 are inserted. If there is a drug that targets the net, the appropriate logic gates are also inserted. The conversion of the faults and drug locations to a logic netlist is shown in Figure 3. The final circuit is then converted to a CNF for further analysis.

In the results, stuck-at faults are referred by the net numbers that are affected (i.e. net 7 s-a-0, means that the signal corresponding to net 7 is stuck-at 0). The network has 5 primary input (PI) signals and 7 primary output (PO) signals. The PIs will be defined as a 5-bit binary vector $X = [EGF, HBEGF, IGF, NRG1, PTEN]$, while the POs will be defined as a 7-bit binary vector $Z = [FOS - JUN, SP1, SRF - ELK1, SRF - ELK4, BCL2, BCL2L1, CCND1]$. In all tests, the PIs are fixed to $X = 00001$ as this input leads to the non-proliferative output in the fault-free case.

For this network, six drugs are available, defined as a 6-bit vector. Each bit corresponds to a drug, such that a value of 1 on the ith bit indicates that drug i is selected, and a value of 0 indicates that drug i is not selected. The drug vector is D=[lapatinib, AG825, AG1024, U0126, LY249002, Temsirolimus].

Our algorithms were implemented using an open-source weighted partial Max-SAT solver called Maxsatz [24, 25]. In all examples listed in this section, the WPMS runtime was significantly less than 1 second per CNF.

4.3 Fault Modeling and Drug Selection

Case 1. Single stuck-at fault identification: In this method, we find all single stuck-at faults which are irredundant, as well as the faulty outputs that they generate. To proceed with this method, we first simulate the original circuit to determine the correct fault-free output. The circuit is "simulated" using our SAT formulation in the fault-free and drug-free mode for a specified primary input value, and the resulting primary output value for the true response is saved as Z^0.

The next step is to find all faults which are irredundant. To avoid an exhaustive search on all single stuck-at faults, we perform an All-SAT on the circuit S where we constrain the output to be not Z^0. Assuming n output signals, this constraint is formed as the clause $C^1 = (\overline{Z_0^0} + \overline{Z_1^0} + \cdots \overline{Z_n^0})$, where Z_i^0 is the variable corresponding to the ith output bit. Furthermore, we also add a constraint to S that the circuit contains only one fault that is injected at a time. This second constraint C^2 is formed by writing clauses of all pairwise combinations of faults, where k is the number of stuck-at faults and f_i is the ith fault.
$$C^2 = (\overline{f_1} + \overline{f_2}) \cdot (\overline{f_1} + \overline{f_3}) \cdots (\overline{f_{k-1}} + \overline{f_k})$$
We now form a new CNF $S^1 = S \cdot C^1 \cdot C^2$. The resulting All-SAT on S^1 is a list of all irredundant single stuck-at faults and their faulty

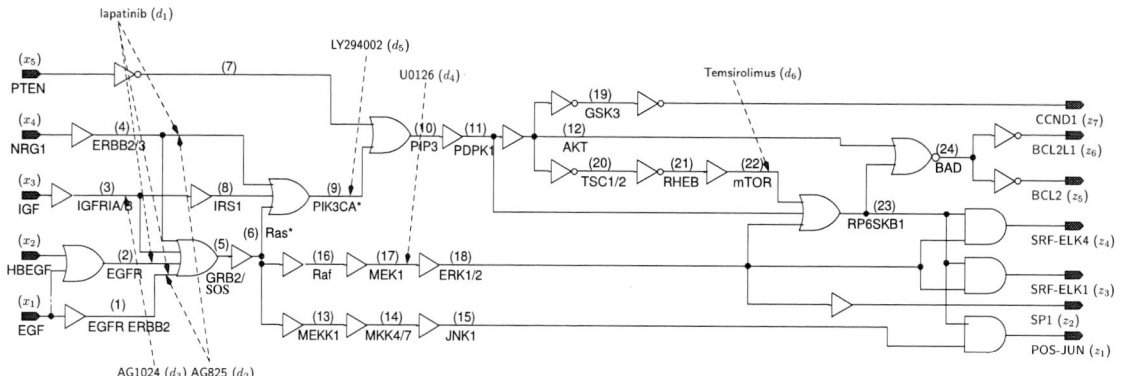

Figure 3: Logic Circuit Stuck-at Fault Model for Colon Cancer GF Signaling Pathways

Net	s-a	Faulty PO	Best PO	Drug Vector	Score
1	1	1111111	0000000	010000	85
2	1	1111111	0000000	100000	85
3	1	1111111	0000000	001000	85
4	1	1111111	0000000	010000	85
5	1	1111111	0000000	000110	84
6	1	0000111	0000000	000110	84
7	1	0000111	0000111	000000	56
8	1	1111111	0000000	000010	85
9	1	0000111	0000000	000010	85
10	1	0000111	0000111	000000	56
11	1	0000111	0000111	000000	56
12	1	0000111	0000111	000000	56
16	1	0111110	0000000	000100	85
17	1	0111110	0000000	000100	85
18	1	0111110	0111110	000000	36
19	0	0000001	0000001	000000	76
20	0	0000110	0000000	000001	85
21	1	0000110	0000000	000001	85
22	1	0000110	0000000	000001	85
23	1	0000110	0000110	000000	66
24	0	0000110	0000110	000000	66

Table 5: Drug Selection for Single stuck-at Faults

output. These faults are flagged for drug simulation using any of the next three cases.

Results: In the single stuck-at fault model, each net was simulated for s-a-0 and s-a-1 with no drugs, and results compared with the fault-free circuit. For fault-free circuit with $X = 00001$, the output vector is $Z^0 = 0000000$. All single irredundant stuck-at faults, which have an output different from the fault-free circuit, are recorded and shown in Table 5.

In this table, the first three columns show the affected net, the stuck-at value, and the faulty output, respectively. We observe that nets 13, 14, and 15 are not listed. The presence of a fault (s-a-0 or s-a-1) on these nets does not generate an incorrect PO, and as such, these are redundant faults.

The results from this case can also be used immediately in several ways. For example, this method classifies for each single stuck-at fault whether it is redundant or irredundant. That is, any fault which is redundant does not produce an incorrect output, and can be ignored from a therapy standpoint. Also, the faulty output from the stuck-at model can be compared to a previously measured output from expression data for a patient, in order to identify which genes are potentially faulty, allowing for targeted therapy. This information can be also used to target genes for potential drug development, while avoiding genes that are redundant.

Case 2. Fault rectification with fewest drugs: In the presence of a particular fault, the problem is determining whether a selection of drugs can rectify the circuit, i.e. change the faulty output to the correct output. If this is not possible, we want to obtain the "best" output ("closest" to the correct output in a Hamming-distance sense), by using drugs. To do this, we guide the WPMS solver by assigning weights to the output states. For example, in the colon cancer GF network used in our experiments, the fault-free output Z^0 is assigned the highest weight (80) and remaining output states are assigned decreasing weights (70, 60, 50, etc.) based on increasing Hamming distance (1, 2, 3, etc.) from the fault-free output. We

assume that faulty states that have a larger Hamming-distance have a more pronounced cancer proliferative effect.

Additionally, the selection of drugs to achieve the best output should use the least number of drugs to minimize the side-effects on the patient. To incorporate this in the WPMS solver, each drug that is *not* selected is given a weight of 1. The colon cancer GF network example has 6 drugs, thus if no drugs are selected, then the cumulative drug weight is 6. Likewise, if all drugs are selected, the drug weight is 0. With six drugs, the maximum score is therefore $80 + 6 = 86$.

Note that the output and drug weights are assigned in such a way as to avoid the situation where a less desirable output (with few drugs) is chosen over a more desirable output with more drugs. We assume that from a clinical standpoint, the priority is to first produce the best possible output, and secondarily to use the fewest drugs required for that output.

All faulty circuits with irredundant faults from Case 1 are augmented with the output and drug weights and simulated using WPMS. The WPMS solver will implicitly and deterministically find the assignment of drugs that achieves the best possible output and with the fewest drugs. The output values, selected drugs, and highest weight of the fault+drug circuits are recorded and compared with the drug-free circuits.

In general, several stuck-at faults can be simultaneously present in the circuit. A circuit with n lines can have $3^n - 1$ possible stuck line combinations. In our implementation, multiple stuck-at faults can easily be modeled for rectification, by injecting one or more faults on the corresponding lines.

Results: All irredundant faults from the results of Case 1 are simulated with drugs. Table 5 shows for each irredundant stuck-at fault, the best achievable output (Column 4), the drug vector to achieve this output (Column 5), and the weight score (Column 6).

We observe that for many faults, there exists a drug vector that can completely rectify the fault, and produce a fault-free circuit. Additionally, the corresponding reported drug vector is minimal in the number of drugs used (by construction), which is a desirable feature in therapy. We also determine that faults on nets 7, 10-12, 18, 19, 23, and 24 are untestable, as no combination of drugs can produce a change in the output. This can be explained as there are no drugs on the fan-out of these genes to rectify the fault. This would indicate that such genes are good targets for drug development.

To demonstrate the adaptability of our algorithm, we test it on a few examples of multiple stuck-at faults. Table 6 shows for a circuit with multiple stuck-at faults, the best drug selection for fault rectification (when possible). The columns of Table 6 have the same meaning as in Table 5.

Case 3. Fault rectification with minimal drug cost: When selecting drugs, there may be multiple drug combinations that may rectify a fault, where each drug has a different cost. In Case 2, all drugs are equal in terms of their cost. However, there may be a situation where we would want to differentiate the drugs using a cost function based on characteristics such as price, number of

Net	s-a	Faulty PO	Best PO	Drug Vector	Score
1,21	1,1	1111111	0000000	010001	84
4,9	1,1	1111111	0000000	000001	85
5,19	1,0	1111111	0000001	000110	74
6,8	1,1	1111111	0000000	000110	84
7,20	1,1	0000111	0000111	000000	56
8,21	1,0	0000111	0000000	000010	85
13,16	1,1	1111110	0000000	000100	85
1,3,6	1,0,1	1111111	0000000	000110	84
2,14,20	1,1,0	1111111	0000000	100001	84
4,7,17	1,1,1	1111111	0000111	010100	54
4,12,23	1,1,1	1111111	0000111	010000	55
8,9,11	1,1,1	0000111	0000111	000000	56
8,9,21	1,1,0	0000111	0000000	000010	85
12,18,20	0,0,0	0000110	0000000	000000	85
15,17,21	0,0,1	0000110	0000000	000001	85

Table 6: Drug Selection for Multiple Stuck-at Faults

side-effects, or ease of availability. For example, two drugs with few side-effects may be more desirable than one drug with many side-effects, if both drug selections produce the same output. As such, given a particular faulty circuit and desired output, the problem is determining a selection of drugs with lowest total cost.

Each drug that is not selected is given a weight proportional to its cost. In our example, we use the number of side-effects as the drug's cost. All faulty circuits with detectable faults from Case 2 are modified with the new drug weights. In addition, the output of the circuit is fixed to the best output as determined in Case 2. These circuits are then solved using WPMS to obtain the selected drugs with lowest cost.

Results: We use the number of side-effects as the drug's cost. Drugs AG825, lapatinib, Temsirolimus are assigned weights of 10, 15, and 35, respectively, which correspond to their approximate number of side-effects [26, 27]. However, drugs AG1024, U0126, and LY294002 have yet to under go clinical trial and the number of side-effects is unknown. As such, these drugs are assigned a weight 20, which is an average of the 3 previous weights.

In this colon cancer GF example, Case 3 simulation provides the same results as in Case 2. This is due to a lack of drugs that share paths in the circuit. In fact, for almost every irredundant fault, the best output state can only be achieved through a single drug.

Case 4. Determining therapy with fewest drugs and best coverage: From Case 2, we identify the drug selection that best rectifies a *specific* fault. However, in drug therapy, the fault location may be unknown. In this situation, a drug selection that rectifies all faults (or as many faults as possible) with the fewest drugs, is desirable.

For each faulty circuit (with a single fault), we find all combinations of 1, 2, and 3 drugs that yield the best output from Case 2. This is done by performing a WPMS All-SAT to find *all* satisfying drug selections with drug weight greater than or equal to $d-3$, where d is the total number of drugs. Each drug selection (or vector) is analyzed to see how many testable faults are rectified or covered by it. The drug vector with the highest coverage and fewest drugs is recorded as a best candidate for therapy.

Results: Using the results from Case 2, we observe that the GF network for colon cancer has 13 testable faults. For these 13 faults, we perform an All-SAT to find the top three scoring drug combinations yielding the best output. All drug combinations are analyzed across all single faults and presented in Table 7 showing drug vector, count of faults rectified, and fault coverage. Drug vectors are ordered in increasing number of drugs selected. Bold rows indicate drug vectors with the highest coverage for a given number of drugs used.

From these results, we observe that with only 1 drug selected, the best coverage is only 23% of faults using lapatinib (d_1) or Temsirolimus (d_6). When allowing for 2 drugs, coverage increases to 77% using the drug combination of U0126 (d_4) and LY294002 (d_5). Finally, we achieve 100% coverage of all testable faults when using the 3 drug combination of U0126 (d_4), LY294002 (d_5), and Temsirolimus (d_6). When the single stuck-at fault location is unknown, these selected drug combinations will be the most effective for therapy and for preventing the proliferation of cancer.

5. Conclusions

Engineering approaches are gradually becoming more accepted as a means to tackle genetic diseases. We show how several techniques

Drug Vec.	Count	Coverage	Drug Vec.	Count	Coverage
000001	**3**	**23%**	**000111**	**13**	**100%**
000010	2	15%	001011	6	46%
000100	2	15%	001101	6	46%
001000	1	8%	001110	10	77%
010000	2	15%	010011	7	54%
100000	**3**	**23%**	010101	7	54%
000011	5	38%	010110	10	77%
000101	3	23%	011001	6	46%
000110	**10**	**77%**	011010	5	38%
001001	4	31%	011100	5	38%
001010	3	23%	100011	8	62%
001100	3	23%	100101	8	62%
010001	5	38%	100110	10	77%
010010	4	31%	101001	7	54%
010100	4	31%	101010	6	46%
011000	3	23%	101100	6	46%
100001	6	46%	110001	6	46%
100010	5	38%	110010	5	38%
100100	5	38%	110100	5	38%
101000	4	31%	111000	4	31%
110000	3	23%			

Table 7: Drug Selection Count and Fault Coverage

from the field of logic synthesis can be used to model and control cancer. In particular, this paper presents SAT based approaches to help infer the predictor sets for gene regulatory networks (GRNs), and also to determine the "best" set of drugs so that the disease is "maximally" cured. The outputs from these algorithms can be used by medical practitioners to determine an optimal drug therapy, by drug developers to target drugs for specific genes, and by biologists to design experiments to extract specific gene interactions. Results from melanoma and colon cancer are presented.

6. References

[1] N. Guelzim et al., "Topological and causal structure of the yeast transcriptional regulatory network," *Nature Genetics*, vol. 31, pp. 60–63, 2002.

[2] S. A. Kauffman, "Metabolic stability and epigenesis in randomly constructed genetic nets," *Journal of Theoretical Biology*, vol. 22, no. 3, pp. 437 – 467, 1969.

[3] W. Burke and B. M. Psaty, "Personalized Medicine in the Era of Genomics," *JAMA*, vol. 298, no. 14, pp. 1682–1684, 2007.

[4] M. Teutsch et al., "The evaluation of genomic applications in practice and prevention (EGAPP) initiative: methods of the EGAPP working group," *Genetics in Medicine*, vol. 11, no. 1, pp. 3–14, 2009.

[5] M. Bittner et al., "Molecular classification of cutaneous malignant melanoma by gene expression profiling," *Nature*, vol. 406, no. 3, pp. 536–540, 2000.

[6] S. Kim, H. Li, E. R. Dougherty, N. Cao, Y. Chen, M. Bittner, and E. B. Suh, "Can Markov chain models mimic biological regulation?," *Journal of Biological Systems*, vol. 10, no. 4, pp. 337–357, 2002.

[7] G. Vahedi, B. Faryabi, J.-F. Chamberland, A. Datta, and E. Dougherty, "Intervention in gene regulatory networks via a stationary mean-first-passage-time control policy," *Biomedical Engineering, IEEE Transactions on*, vol. 55, pp. 2319 –2331, oct. 2008.

[8] I. Shmulevich and E. R. Dougherty, *Probabilistic Boolean Networks: The Modeling and Control of Gene Regulatory Networks*. Philadelphia, PA: SIAM – Society for Industrial and Applied Mathematics, 2009.

[9] N. Geard and J. Wiles, "A gene network model for developing cell lineages," *Artif. Life*, vol. 11, no. 3, pp. 249–268, 2005.

[10] A. Arkin, J. Ross, and H. H. McAdams, "Stochastic kinetic analysis of developmental pathway bifurcation in phage lambda-infected escherichia coli cells," *Genetics*, vol. 149, pp. 1633–1648, 1998.

[11] F. Jacob and J. Monod, "Genetic regulatory mechanisms in the synthesis of proteins," *Journal of Molecular Biology*, vol. 3, no. 3, pp. 318–356, 1961.

[12] X. Zhou, X. Wang, and E. R. Dougherty, "Gene prediction using multinomial probit regression with Bayesian gene selection," *EURASIP Journal on Applied Signal Processing*, pp. 115–124, 2004.

[13] W. Zhou, E. Serpedin, and E. R. Dougherty, "Inferring gene regulatory networks from time series data using the minimum description length principle," *Bioinformatics*, vol. 17, pp. 2129–2135, 2006.

[14] E. R. Dougherty, S. Kim, and Y. Chen, "Coefficient of determination in nonlinear signal processing," *Signal Processing*, vol. 80, no. 10, pp. 2219 – 2235, 2000.

[15] W. Zhao, E. Serpedin, and E. R. Dougherty, "Inferring connectivity of genetic regulatory networks using information-theoretic criteria," *IEEE/ACM Trans. Comput. Biol. Bioinformatics*, vol. 5, no. 2, pp. 262–274, 2008.

[16] R. Pal, I. Ivanov, A. Datta, M. L. Bittner, and E. R. Dougherty, "Generating Boolean networks with a prescribed attractor structure," *Bioinformatics*, vol. 21, no. 21, pp. 4021–4025, 2005.

[17] R. Layek, A. Datta, M. Bittner, and E. Dougherty, "Cancer therapy design based on pathway logic," *Bioinformatics*, vol. 27, no. 4, pp. 548–555, 2011.

[18] T. Larrabee, "Efficient Generation of Test Patterns Using Boolean Difference," in *Proc. of the Intl. Test Conf.*, pp. 795–801, 1989.

[19] P. Stephan, R. Brayton, and A. Sangiovanni-Vincentelli, "Combinational test generation using satisfiability," *Computer-Aided Design of Integrated Circuits and Systems, IEEE Transactions on*, vol. 15, pp. 1167–1176, sep 1996.

[20] N. Saluja, K. Gulati, and S. Khatri, "SAT-based ATPG using multilevel compatible don't-cares," *ACM Trans. Des. Autom. Electron. Syst.*, vol. 13, pp. 24:1–24:18, April 2008.

[21] S. A. Kauffman. *The Origins of Order: Self-Organization and Selection in Evolution*. Oxford University Press, USA, 1 ed., June 1993.

[22] N. Een and N. Sorensson, *An Extensible SAT-solver*. Lecture Notes in Computer Science, Springer Berlin / Heidelberg, 2004.

[23] "Minisat." http://minisat.se/.

[24] "Maxsatz." http://home.mis.u-picardie.fr/~cli/EnglishPage.html.

[25] C. Li, F. Manya, N. Mohamedou, and J. Planes, "Exploiting cycle structures in Max-SAT," in *Theory and Applications of Satisfiability Testing - SAT 2009* (O. Kullmann, ed.), vol. 5584 of *Lecture Notes in Computer Science*, pp. 467–480, Springer Berlin / Heidelberg, 2009.

[26] "Santa Cruz Biotechnology, Inc." http://www.scbt.com/.

[27] "PubMed Health Home." http://www.ncbi.nim.nih.gov/pubmedhealth/.

Exploiting Die-to-Die Thermal Coupling in 3D IC Placement*

Krit Athikulwongse, Mohit Pathak, and Sung Kyu Lim
School of Electrical and Computer Engineering
Georgia Institute of Technology, Atlanta, GA, USA
{krit,mohitp,limsk}@ece.gatech.edu

ABSTRACT

In this paper, we propose two methods used in 3D IC placement that effectively exploit the die-to-die thermal coupling in the stack. First, TSVs are spread on each die to reduce the local power density and vertically aligned across dies simultaneously to increase thermal conductivity to the heatsink. Second, we move high-power logic cells to the location that has higher conductivity to the heatsink while moving TSVs in the upper dies so that high-power cells are vertically overlapping below the TSVs. These methods are employed in a force-directed 3D placement successfully and outperform several state-of-the-art placers published in recent literature.

Categories and Subject Descriptors

B.7.2 [**Integrated Circuits**]: Design Aids—*Placement and routing*

General Terms

Algorithms, Design, Reliability

Keywords

3D IC, TSV, Temperature

1. INTRODUCTION

Stacking thinned dies in 3D ICs results in increasing power density, thus rising temperature, which leads to other reliability problems, such as electromigration and negative-bias-temperature instability. Because of low thermal conductivity, polymer adhesive exacerbates the problem. Moreover, if the thinned dies are silicon on insulator, an extremely high temperature can be expected. Heat must be removed from the die quickly; otherwise, reliability problems may arise.

A few recent works on temperature-aware placement for 3D ICs have been published. In [2], a force-directed approach was proposed for 3D thermal placement; however, it did not include through-silicon vias (TSVs), which are commonly found in 3D ICs. In [3],

---

*This research is funded by SRC ICSS Task 1836.075 and SRC CADTS Task 2239.001.

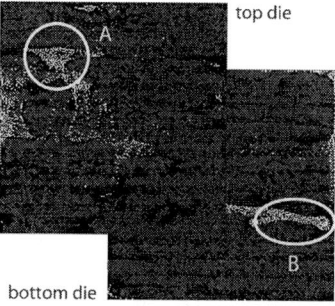
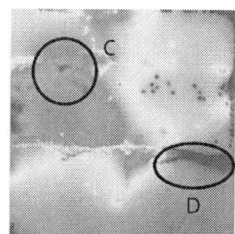

Figure 1: Die-to-die heat coupling from TSVs. TSVs are shown in white. The top die is closer to heatsink. The cold spot C is caused by the TSVs in spot A on the same die. The hot spot D is caused by the TSVs in spot B from the bottom die.

a partitioning-based approach was proposed for 3D thermal placement. The work considered the impact of parasitic resistance and capacitance of signal TSVs on power, but failed to include thermal properties of TSVs. Failing again to acknowledge TSV area, it also reported unreasonably large numbers of TSVs even for small circuits. The work in [1] considered TSV thermal properties; however, it assumed that adhesive is an ideal insulator. In reality, heat can still flow through (silicon and) adhesive because of its thinness. Based on the assumption, the work balanced only the number of TSVs in a bin to heat dissipated from cells in the same bin and bins vertically below.

The contributions of this work are as follows: (i) We propose two effective heuristics, namely TSV spread and alignment method (TSA) and thermal coupling-aware placement (CA), that exploit the die-to-die thermal coupling in 3D ICs in force-directed temperature-aware placement. We present new forces, and discuss how to manage them to obtain high quality placements. (ii) We perform extensive experiments to show the trade-off among wirelength, delay, power, and temperature results obtained from GDSII layouts. (iii) Our placers outperform several state-of-the-art placers published in recent literature [2, 5, 3, 1, 4].

2. MOTIVATION

In a 3D IC layout, logic gates cannot overlap with TSVs. Area occupied by TSVs becomes "power whitespace" because no power is consumed and thus no heat is generated. In addition, TSVs conduct majority of heat through polymer adhesive between dies toward the heatsink as shown in Fig. 1. In the figure, the hotspot D on the top metal layer of the top die is caused by the TSVs in spot B from the bottom die. Heat flows through TSVs so intensely that its effect still remains on the top die. Thus, the temperature distribution of the top die results from the combination of power profile of the top die and heat flowing from the bottom die through TSVs. Our TSV spread and alignment method presented in this paper exploits these thermal properties of TSVs by distributing TSVs evenly to reduce power density in local power hotspots and vertically aligning TSVs of adjacent dies to establish direct paths to the heatsink.

Using Ansys FLUENT, we simulate a part of bulk silicon with and without TSVs (and their related structures, e.g., landing pad and liner). We fix the temperature on the top side of the models, apply constant power density on the bottom side, and obtain the temperature distribution. The simulation results indicate that heat flowing through a TSV increases temperature far less than the same amount of heat flowing through bulk silicon and adhesive. We also observe that the temperature slowly increases in bulk silicon with TSVs. On the other hand, in bulk silicon without any TSV, low thermal conductivity of bonding adhesive results in steep temperature rise at first, but temperature does not rise as much inside the silicon. We compute the average thermal conductivity of bulk silicon with and without TSVs, and use them to guide our thermal coupling-aware placer presented in this paper.

3. GLOBAL PLACEMENT ALGORITHMS

We extend the force-directed placer [7] in two ways to perform thermal optimization in 3D ICs. In the first algorithm, we laterally spread TSVs in each die to form even thermal conductivity while perturbing TSV position to increase vertical overlap among TSVs across the dies in 3D stack. In the second algorithm, the logic cells on each die are positioned by using thermal conductivity-based force while TSVs are positioned by using power density-based force.

3.1 Design Flow

Fig. 2 shows the overall flow of our placement, where the position of cells and TSVs is determined simultaneously. Given a netlist, we partition cells into dies if the partition is not also given. Then, we insert the minimum number of TSVs required to connect cells on different dies. Once this die partitioning is fixed, we do not move cells across dies during placement. The reason is that changing cell partition results in change in the number of TSVs, and this change causes the complexity of problem to become unmanageable. Next, we minimize wirelength to obtain initial placement, which may contain high overlap among cells and TSVs. In the main loop to resolve the overlap, we use TSV density and TSV position to compute target point for TSVs in the first algorithm. In the second algorithm, we periodically perform 3D power analysis based on current cell and TSV position. Then, we use the cell power, TSV density, and average thermal conductivity of bulk silicon obtained from the simulation results in Section 2 to compute target points for cells and TSVs to move towards. After updating force equations and solving them, we update the position of cells and TSVs. This loop continues until the overlap is sufficiently reduced.

3.2 Force-directed 3D Placement

In a quadratic placement [7], quadratic wirelength Γ_x and Γ_y along x- and y-axis are separately minimized to obtain the placement result. Treated Γ_x as spring energy, its derivative can be regarded as net force $\mathbf{f}_x^{\text{net}}$. By setting $\mathbf{f}_x^{\text{net}}$ to zero, the minimum Γ_x and the corresponding placement are found; however, cells may overlap in few small areas. Hold force $\mathbf{f}_x^{\text{hold}}$ prevents $\mathbf{f}_x^{\text{net}}$ from pulling cells back to the initial placement. In addition, density-based force $\mathbf{f}_x^{\text{den}}$ reduces the overlap by spreading cells in high density region.

To extend [7] for 3D ICs, cells are not moved across dies during

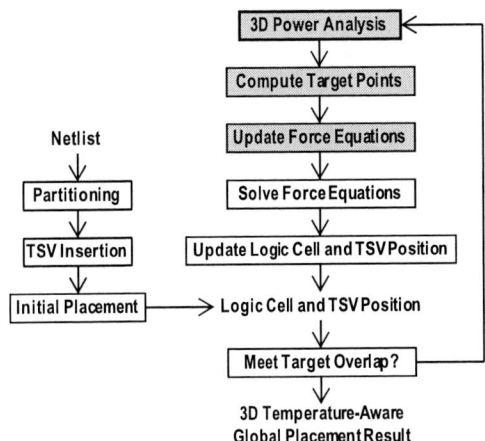

Figure 2: Design flow for our 3D IC global placement.

placement in [4] because they are already assigned into dies by the partitioner. In addition, $\mathbf{f}_x^{\text{den}}$ is computed die-by-die based on the placement density D_d of each die d, which is defined as

$$D_d(x,y) = D_d^{\text{cell}}(x,y) - D_d^{\text{die}}(x,y), \qquad (1)$$

where D_d^{cell} is the cell density on die d, and D_d^{die} is the die capacity scaled to match the total cell area on the die. Then, the placement potential Φ_d is computed by solving Poisson's equation

$$\Delta\Phi_d(x,y) = -D_d(x,y). \qquad (2)$$

The target point $\mathring{x}_i^{\text{d}}$ to connect density-based spring of cell i is computed by

$$\mathring{x}_i^{\text{d}} = x_i' - \frac{\partial}{\partial x}\Phi_d(x,y)\bigg|_{(x_i',y_i')}, \qquad (3)$$

where x_i' is the x-position of cell i on die d from the last iteration. Lastly, for each placement iteration, the placement result can be obtained by setting total force $\mathbf{f}_x$ to zero, and solve

$$\mathbf{f}_x = \mathbf{f}_x^{\text{net}} + \mathbf{f}_x^{\text{hold}} + \mathbf{f}_x^{\text{den}} = \mathbf{0}. \qquad (4)$$

3.3 TSV Spread and Alignment

In this algorithm, we exploit one of thermal properties of TSVs to help alleviate thermal problems as shown in Fig. 3(a). TSVs occupy placement area, but do not dissipate power. The existence of TSVs among cells with high power dissipation reduces local dissipated power density, which in-turn helps reduce local temperature. Therefore, spreading TSVs evenly on each die should help reduce intra-die thermal variation in 3D ICs. We propose this algorithm because it is simple yet effective. It can be viewed as a method to mimic uniform TSV position. Instead of moving TSVs based on the placement density computed from both TSV and cell area, we move TSVs based on TSV density only. In other words, we compute D_d^{cell} in Equation (1) from TSV area only, and scale D_d^{die} to match the total TSV area on the die.

In addition to TSV spread, we exploit another thermal property of TSVs to help alleviate thermal problems as shown in Fig. 3(b). TSVs conduct majority of heat through polymer adhesive between dies, causing local hot spots on the adjacent die between the TSVs and heatsink. Therefore, aligning TSVs on each die to TSVs on the adjacent die should help prevent this kind of hot spots, and direct the heat toward the heatsink quickly, resulting in overall temperature decrease. To align TSVs during global placement, we introduce an additional force for TSVs, alignment force denoted

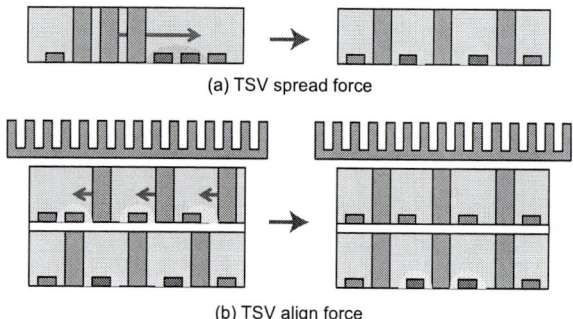

(a) TSV spread force

(b) TSV align force

Figure 3: TSV spread and TSV align forces.

$\mathbf{f}_x^{\mathrm{align}}$, into Equation (4). This force can be represented by alignment springs connected to TSVs, and defined as

$$\mathbf{f}_x^{\mathrm{align}} = \mathring{\mathbf{C}}_x^{\mathrm{a}}(\mathbf{x} - \mathring{\mathbf{x}}^{\mathrm{a}}), \qquad (5)$$

where vector $\mathring{\mathbf{x}}^{\mathrm{a}}$ represents the x-position of target points to connect alignment springs to TSVs, and diagonal matrix $\mathring{\mathbf{C}}_x^{\mathrm{a}}$ collects spring constants $\mathring{w}_{x,i}^{\mathrm{a}}$ of the alignment spring connected to TSV i.

We apply alignment force to TSV i only when its closest TSV j on the adjacent die farther from the heatsink is within a certain range so that we do not excessively increase wirelength. The range is set to the size of TSV because of the high probability of aligning the TSVs in few iterations. We balance $\mathbf{f}_x^{\mathrm{align}}$ against other forces by setting $\mathring{w}_{x,i}^{\mathrm{a}}$ to density-based spring constant $\mathring{w}_{x,i}^{\mathrm{d}}$ of $\mathbf{f}_x^{\mathrm{den}}$ and setting alignment target point $\mathring{x}_i^{\mathrm{a}}$ to x_j', the x-position from last iteration of TSV j (on the adjacent die farther from heatsink) closest to TSV i. This method naturally balances $\mathbf{f}_x^{\mathrm{align}}$ against $\mathbf{f}_x^{\mathrm{den}}$.

The intuition is that because of the high cell overlap in the early placement iterations, the target point $\mathring{x}_i^{\mathrm{d}}$ is farther away from TSV i than the alignment target point $\mathring{x}_i^{\mathrm{a}}$. Thus, $\mathbf{f}_x^{\mathrm{den}}$ dominates. When cells are evenly distributed in the late iterations of placement, $\mathring{x}_i^{\mathrm{d}}$ is closer to TSV i. Then, $\mathbf{f}_x^{\mathrm{den}}$ becomes weaker, and $\mathbf{f}_x^{\mathrm{align}}$ affects the TSV position more.

3.4 Thermal Coupling-aware Placement

In this algorithm, we consider the die-to-die thermal coupling during placement. The basic approach is to introduce two new forces, the first that moves cells and the second that moves TSVs, both in an attempt to place high-power cells closer to the TSV-to-heatsink path. Since the heat dissipated by a cell must flow toward heatsink, we place cells based on their power density and the effective thermal conductivity computed using the same die *and the dies above*. In addition, since TSV conducts heat without raising temperature too much, we place TSVs based on the total power density of the same die *and the dies below*.

Our basic approach is that the area with high power density and low thermal conductivity leads to high temperature. Thus, the temperature at a certain position depends on the difference (or imbalance) between power density and thermal conductivity. The force that moves cells (TSVs) on a die also changes the power density (thermal conductivity) distribution of the die. Our goal is to use these forces to balance the power density and the thermal conductivity at each position on the die. The force in an area with high difference should be stronger than the force in an area with low difference. The strength of a spring force depends on the distance to the connection point, so we set the strength based on this difference. Based on this concept, we first build a map of the difference, and smooth the map in an iterative fashion.

3.4.1 For Cell Movement

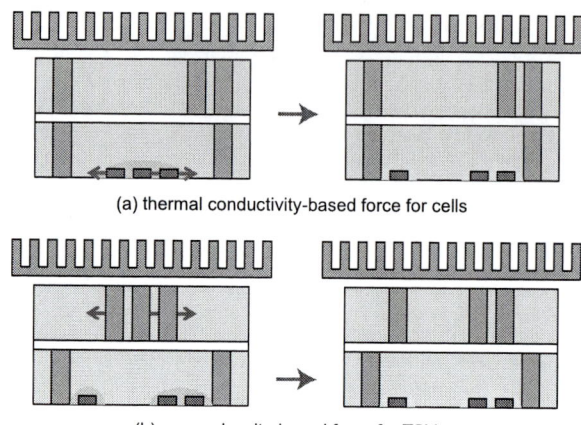

(a) thermal conductivity-based force for cells

(b) power density-based force for TSVs

Figure 4: Thermal conductivity-based vs power density-based forces.

We introduce the thermal conductivity-based force $\mathbf{f}_x^{\mathrm{cond}}$ as illustrated in Fig. 4(a). It moves high-power cells toward the position with high thermal conductivity to heatsink, and is defined as

$$\mathbf{f}_x^{\mathrm{cond}} = \mathring{\mathbf{C}}_x^{\mathrm{c}}(\mathbf{x} - \mathring{\mathbf{x}}^{\mathrm{c}}), \qquad (6)$$

where the vector $\mathring{\mathbf{x}}^{\mathrm{c}}$ represents the x-position of target points to connect thermal conductivity-based springs to cells, and the diagonal matrix $\mathring{\mathbf{C}}_x^{\mathrm{c}}$ contains spring constants $\mathring{w}_{x,i}^{\mathrm{c}}$ of the spring connected to cell i.

We compute $\mathbf{f}_x^{\mathrm{cond}}$ die-by-die by balancing the cell power density P_d^{cell} of each die d against its effective thermal conductivity to heatsink, denoted K_d^{sink}. Under the demand-supply system of the force-directed framework in [7], P_d^{cell} and K_d^{sink} represent the demand and supply to remove the heat from die d in the 3D stack. We define the thermal conductivity-based balance factor B_d^{cond} for die d as (see Fig. 5)

$$B_d^{\mathrm{cond}}(x,y) = P_d^{\mathrm{cell}}(x,y) - s_d^{\mathrm{cond}} \cdot K_d^{\mathrm{sink}}(x,y), \qquad (7)$$

where s_d^{cond} is a scaling factor to match K_d^{sink} to P_d^{cell} across the die. We use s_d^{cond} to balance the total supply (K_d^{sink}) and the total demand (P_d^{cell}), and compute it by

$$s_d^{\mathrm{cond}} = \frac{\int\int P_d^{\mathrm{cell}}(x,y)\,dx\,dy}{\int\int K_d^{\mathrm{sink}}(x,y)\,dx\,dy}. \qquad (8)$$

Here, K_d^{sink} is computed as

$$K_d^{\mathrm{sink}}(x,y) = \frac{1}{\sum_{j=d}^{N_{\mathrm{die}}} \frac{1}{K_j^{\mathrm{die}(x,y)}}}, \qquad (9)$$

where K_j^{die} is the thermal conductivity of die j, and die N_{die} is the die closest to the heatsink (see Fig. 6). Here, $K_{N_{\mathrm{die}}}^{\mathrm{die}}$ includes the thermal conductivity of thick substrate and heatsink, and K_j^{die} is computed based on the TSV density at each position on the die and the average thermal conductivity of bulk silicon with and without TSVs, obtained from the simulation results in Section 2.

The potential Φ_d^{cond} for B_d^{cond} is computed by solving Poisson's equation

$$\Delta\Phi_d^{\mathrm{cond}}(x,y) = -B_d^{\mathrm{cond}}(x,y). \qquad (10)$$

The target point $\mathring{x}_i^{\mathrm{c}}$ of cell i is computed by

$$\mathring{x}_i^{\mathrm{c}} = x_i' - \frac{\partial}{\partial x}\Phi_d^{\mathrm{cond}}(x,y)\Big|_{(x_i',y_i')}, \qquad (11)$$

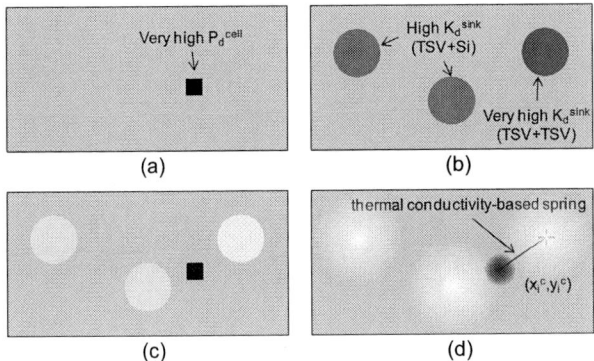

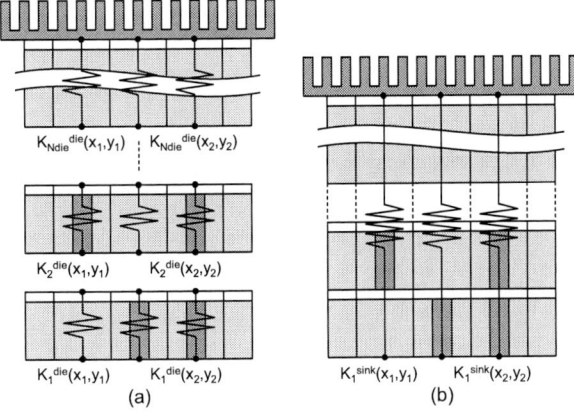

Figure 5: Illustration of B_d^{cond}. (a) P_d^{cell}, (b) $s_d^{\text{cond}} \cdot K_d^{\text{sink}}$, (c) B_d^{cond}, (d) potential for B_d^{cond} after solving Poisson's equation.

Figure 6: Computation of K_d^{sink}. (a) K_j^{die}, (b) K_1^{sink}.

where x_i' is the x-position of cell i on die d from the last iteration. We set spring constant $\mathring{w}_{\text{x},i}^{\text{c}}$ for cell i based on cell power and the total cell power by

$$\mathring{w}_{\text{x},i}^{\text{c}} = p_i / \sum_{\forall j} p_j, \qquad (12)$$

where p_i is the power of cell i, and j is a cell on die d. Therefore, a high-power cell is connected to a strong thermal conductivity-based spring.

3.4.2 For TSV Movement

We introduce power density-based force $\mathbf{f}_\text{x}^{\text{pow}}$ as illustrated in Fig. 4(b). It moves TSVs toward the position with high cell power density on the same die and the dies below. We define $\mathbf{f}_\text{x}^{\text{pow}}$ as

$$\mathbf{f}_\text{x}^{\text{pow}} = \mathring{\mathbf{C}}_\text{x}^{\text{p}}(\mathbf{x} - \mathring{\mathbf{x}}^{\text{p}}), \qquad (13)$$

where the vector $\mathring{\mathbf{x}}^{\text{p}}$ represents the x-position of target points to connect power density-based springs to TSVs, and the diagonal matrix $\mathring{\mathbf{C}}_\text{x}^{\text{p}}$ contains spring constants $\mathring{w}_{\text{x},i}^{\text{p}}$ of the spring connected to TSV i.

We compute $\mathbf{f}_\text{x}^{\text{pow}}$ die-by-die by balancing the thermal conductivity K_d^{die} of each die d against the total power density $\sum P_j^{\text{cell}}$ that flows through the die toward heatsink. Under the demand-supply system of the force-directed framework in [7], K_d^{die} and $\sum P_j^{\text{cell}}$ represent the demand and supply to conduct heat from the same die and dies below to heatsink. We define the power density-based balance factor B_d^{pow} for die d as

$$B_d^{\text{pow}}(x,y) = K_d^{\text{die}}(x,y) - s_d^{\text{pow}} \cdot \sum_{j=1}^{d} P_j^{\text{cell}}(x,y), \qquad (14)$$

where s_d^{pow} is a scaling factor to match $\sum P_j^{\text{cell}}$ to K_d^{die} across the die. We use s_d^{pow} to balance the total supply ($\sum P_j^{\text{cell}}$) and the total demand (K_d^{die}), and compute it by

$$s_d^{\text{pow}} = \frac{\int\int K_d^{\text{die}}(x,y)\,dx\,dy}{\int\int \sum_{j=1}^{d} P_j^{\text{cell}}(x,y)\,dx\,dy}. \qquad (15)$$

The potential Φ_d^{pow} for B_d^{pow} is computed by solving Poisson's equation

$$\Delta\Phi_d^{\text{pow}}(x,y) = -B_d^{\text{pow}}(x,y). \qquad (16)$$

The target point $\mathring{x}_i^{\text{p}}$ of TSV i is computed by

$$\mathring{x}_i^{\text{p}} = x_i' - \left.\frac{\partial}{\partial x}\Phi_d^{\text{pow}}(x,y)\right|_{(x_i',y_i')}, \qquad (17)$$

where x_i' is the x-position of TSV i on die d from the last iteration. We set spring constant $\mathring{w}_{\text{x},i}^{\text{p}}$ to $1/N_d^{\text{TSV}}$, where N_d^{TSV} is the total number of TSVs on die d. Therefore, the power density-based spring for each TSVs has the same strength.

3.4.3 Balancing the Forces

We balance the new forces against $\mathbf{f}_\text{x}^{\text{den}}$ because $\mathbf{f}_\text{x}^{\text{den}}$ is the main force that moves cells and TSVs. First, we scale the new forces so that they have the same magnitude as $\mathbf{f}_\text{x}^{\text{den}}$. Then, we apply weighting constants to $\mathbf{f}_\text{x}^{\text{den}}$, $\mathbf{f}_\text{x}^{\text{cond}}$, and $\mathbf{f}_\text{x}^{\text{pow}}$ so that we can control their contribution to the total force.

First, to scale $\mathbf{f}_\text{x}^{\text{cond}}$ to $\mathbf{f}_\text{x}^{\text{den}}$, we normalize P_d^{cell}, the demand for B_d^{cond} in Equation (7), to D_d^{cell} by a scaling factor s_d^{PD} defined as

$$s_d^{\text{PD}} = \frac{\int\int D_d^{\text{cell}}(x,y)\,dx\,dy}{\int\int P_d^{\text{cell}}(x,y)\,dx\,dy}. \qquad (18)$$

Then, we replace P_d^{cell} in Equation (7) and Equation (8) by $s_d^{\text{PD}} \cdot P_d^{\text{cell}}$. Second, to scale $\mathbf{f}_\text{x}^{\text{pow}}$ to $\mathbf{f}_\text{x}^{\text{den}}$, we normalize K_d^{die}, the demand for B_d^{pow} in Equation (14), to D_d^{cell} by a scaling factor s_d^{KD} defined as

$$s_d^{\text{KD}} = \frac{\int\int D_d^{\text{cell}}(x,y)\,dx\,dy}{\int\int K_d^{\text{die}}(x,y)\,dx\,dy}. \qquad (19)$$

Then, we replace K_d^{die} in Equation (14) and Equation (15) by $s_d^{\text{KD}} \cdot K_d^{\text{die}}$.

We scale both $\mathbf{f}_\text{x}^{\text{cond}}$ and $\mathbf{f}_\text{x}^{\text{pow}}$ to $\mathbf{f}_\text{x}^{\text{den}}$ based on D_d^{cell}, not on the gradient of Φ_d because of the stability issue. After normalizing P_d^{cell} and K_d^{die} to D_d^{cell} as shown in Equation (18) and Equation (19), the magnitude of B_d^{cond} and B_d^{pow} and gradient of their potential are properly normalized. At an equilibrium, a small magnitude of the gradients results in a small magnitude of $\mathbf{f}_\text{x}^{\text{cond}}$ and $\mathbf{f}_\text{x}^{\text{pow}}$. If we scale $\mathbf{f}_\text{x}^{\text{cond}}$ and $\mathbf{f}_\text{x}^{\text{pow}}$ to $\mathbf{f}_\text{x}^{\text{den}}$ based on the gradient of Φ_d instead, the magnitude of the gradient of potential of B_d^{cond} and B_d^{pow} would be exaggerated after the normalization, which in turn causes instability.

In summary, $\mathbf{f}_\text{x}^{\text{cond}}$ moves cells in such a way that high power density flows through the paths with high thermal conductivity to heatsink. In addition, $\mathbf{f}_\text{x}^{\text{pow}}$ moves TSVs in such a way that each TSV establishes a heat path for the high-power cells in the same die and the dies below. Our overall force equation is as follows:

$$\mathbf{f}_\text{x} = \mathbf{f}_\text{x}^{\text{net}} + \mathbf{f}_\text{x}^{\text{hold}} + (1-\alpha)\mathbf{f}_\text{x}^{\text{den}} + \alpha(\mathbf{f}_\text{x}^{\text{cond}} + \mathbf{f}_\text{x}^{\text{pow}}) = \mathbf{0}. \quad (20)$$

By increasing α, the forces $\mathbf{f}_\text{x}^{\text{cond}}$ and $\mathbf{f}_\text{x}^{\text{pow}}$ dominate the movement of cells and TSVs for more thermal optimization.

Table 1: Benchmark circuits.

Ckt.	#Gates	#TSVs	Util.	Footprt (mm²)	Profile
ckt1	119,040	5,725	0.66	0.50×0.50	Data encryption
ckt2	191,420	24,540	0.63	0.90×0.90	Graphic accelerator
ckt3	280,933	17,362	0.49	0.98×0.98	Video compression
ckt4	383,329	17,436	0.53	1.04×1.04	Signal processing
ckt5	644,357	15,024	0.53	1.16×1.16	Image encoder

4. EXPERIMENTAL RESULTS

We use 45-nm technology from FreePDK45 for our experiments. TSV diameter is $5\,\mu m$, and the landing pad width is $7\,\mu m$. TSV liner thickness is 250 nm. We use copper TSVs with SiO₂ liner and $2.6\text{-}\mu m$-thick benzocyclobutene bonding adhesive for our experiments. Each die in the 3D chip stack is thinned to $30\,\mu m$ except that the topmost die, which is attached to heatsink, retains its thickness at $530\,\mu m$. The ambient temperature on top of the heatsink is 300 K. The TSV parasitic resistance and capacitance are $0.1\,\Omega$ and $125\,fF$, respectively. We base all our experiments on 4-die chip stacks.

We use IWLS 2005 benchmarks and several industrial circuits from OpenCores. We synthesize the circuits using Synopsys Design Compiler to obtain gate-level netlist, and use the target clock period of each circuit when performing all analyses. The benchmark characteristics are listed in Table 1. The numbers of TSVs are based on partitioning results from our own implementation of [3]. We use the same die partitioning results for all algorithms for fair comparison in Section 4.2. Because [3] does not consider TSV area, it inserts high number of TSVs, resulting in low placement utilization.

We do not optimize the circuits after placement because buffers and sized gates can change power profile, thus affecting temperature. The results reported in this paper are from commercial tools. We use Cadence Encounter to route the layouts, Synopsys Prime-Time to analyze timing and power, and Ansys FLUENT to analyze temperature. We report all our temperature results in terms of the increase from the ambient temperature measured at the top of the heatsink.

4.1 Impact of TSV Density Uniformity

In this experiment, we show how TSV density uniformity impacts thermal profile. Our two baseline 3D placements are wirelength-driven placement with uniform TSV position [4] and wirelength-driven placement with non-uniform TSV position [4]. First, we obtain both baseline placements using our own implementation of [4]. Then, we perform power and thermal analyses on both placement results. The routed wirelength, longest path delay, and power are shown in Table 2, and temperatures are shown in Table 3. Although the placement with non-uniform TSV position has shorter wirelength, better timing, and lower power than the placement with uniform TSV position, its temperature, especially the thermal variation, is worse. Both the non-uniform power density and the non-uniform thermal conductivity, caused by the non-uniform distribution of TSVs in the 3D chip stack, contribute to the problem. In the placement with non-uniform TSV position, we observe that the area with high TSV density has low power density and low temperature, vice versa. These two opposite trends are responsible for high thermal variation.

4.2 Comparison with State-of-the-Art

We compare our temperature-aware global placement algorithms with the following recent state-of-the-art temperature-aware placers:[1]

---

[1] This task is challenging due to the discrepancy among the settings and assumptions made in each work. However, we made our best effort to provide

Table 2: Routed wirelength, longest path delay, and power of placements with uniform [4] and non-uniform [4] TSV position.

	Uniform			Non-uniform		
Ckt.	rWL (m)	D_{max} (ns)	P (W)	rWL (m)	D_{max} (ns)	P (W)
ckt1	3.897	5.320	0.752	3.014	4.836	0.728
ckt2	11.718	16.510	2.661	7.744	13.694	2.463
ckt3	13.532	8.814	2.353	9.326	6.535	2.288
ckt4	19.355	20.788	2.710	12.457	12.515	2.640
ckt5	22.708	19.772	3.209	18.711	13.798	3.122
ratio	1.405	1.350	1.039	1.000	1.000	1.000

Table 3: Temperature (°C) of placements with uniform [4] and non-uniform [4] TSV position. ($\Delta T_{ja} = T_{ja,max} - T_{ja,min}$)

	Uniform			Non-uniform		
Ckt.	$T_{ja,max}$	ΔT_{ja}	$T_{ja,ave}$	$T_{ja,max}$	ΔT_{ja}	$T_{ja,ave}$
ckt1	71.55	17.60	64.50	74.13	18.33	63.98
ckt2	101.14	47.14	69.41	94.41	50.19	64.78
ckt3	70.38	31.01	55.06	80.09	42.81	55.48
ckt4	64.91	18.76	54.32	75.98	38.01	55.16
ckt5	66.77	35.40	53.13	75.24	39.32	54.50
ratio	1.000	1.000	1.000	1.081	1.325	0.995

[2] (force-directed placer): In this work, thermal analysis is performed at the beginning of every global placement iteration. The thermal gradient obtained from the analysis is used to compute repulsive force, which moves logic cells from high-temperature area toward low-temperature area. We implement our own version of this work by calling Ansys FLUENT from inside our placer, and combining scaled thermal gradient into density-based force $\mathbf{f}_x^{den}$.

[5] (force-directed placer): Instead of moving logic cells based on placement area density, it moves logic cells based on placement power density. Therefore, logic cells are spread according to their power dissipation, and logic cells with high power dissipation occupy more space than logic cells with low power dissipation, leading to uniform power density and thermal profile across the die. We implement our own version of this work.

[3] (partitioning-based placer): In this work, logic cells are partitioned into placement area and different dies based on the switching activity and parasitic capacitance of connecting wires and TSVs. We perform global routing to determine the position of TSVs as proposed in [6] after performing global placement using our own implementation of [3].

[1] (analytical placer): We implement this method by balancing the power density combined across dies in vertical direction against the TSV density and solving the density for potential function. The gradient of potential is used to compute a force to move cells and TSVs to maintain the balance. The force is added to $\mathbf{f}_x^{den}$ with a user-defined parameter β to provide temperature-wirelength trade-off similar to the work.

Table 4 shows the routed wirelength, delay, power, and temperature comparison based on the GDSII layouts we build using these placers. The wirelength, delay, and power values are normalized to the wirelength-driven non-uniform TSV placement [4] shown in Table 2. The temperature values are normalized to the wirelength-driven uniform TSV placement [4] shown in Table 3. Recall that non-uniform placer achieves high-quality wirelength, delay, and power results while uniform placer leads to high-quality temperature values.

First, we observe that [2] produces comparable wirelength, delay, and power results to non-uniform TSV placer [4]. In case of temperature, [2] obtains worse result compared with uniform TSV placer [4]. We tried increasing the magnitude of thermal-gradient-

---

fair and meaningful comparison.

Table 4: Comparison with state-of-the-art temperature-aware placers [2, 5, 3, 1, 4]. Our placers are TSA (TSV spread and alignment) and CA (Coupling-aware placement). The routed wirelength, delay, and power values are normalized to the non-uniform TSV placement [4] shown in Table 2. The temperature values are normalized to the uniform TSV placement [4] shown in Table 3.

Ckt.	routed wirelength (m)						longest path delay (ns)						power consumption (W)					
	[2]	[5]	[3]	[1]	TSA	CA	[2]	[5]	[3]	[1]	TSA	CA	[2]	[5]	[3]	[1]	TSA	CA
ckt1	3.046	3.109	3.784	3.240	3.250	3.133	4.935	4.796	5.128	5.067	4.786	4.871	0.729	0.734	0.776	0.736	0.736	0.732
ckt2	7.740	8.780	14.924	8.349	7.892	8.314	13.679	15.004	15.231	14.416	13.588	14.785	2.463	2.548	2.564	2.521	2.487	2.523
ckt3	9.347	10.544	16.028	10.706	10.355	10.261	6.567	6.797	7.865	7.276	6.530	6.906	2.290	2.331	2.351	2.318	2.306	2.321
ckt4	12.480	13.902	19.871	15.234	14.901	14.545	12.518	12.695	16.158	13.609	13.695	13.113	2.640	2.671	2.737	2.682	2.672	2.675
ckt5	18.869	21.482	27.649	20.125	19.845	19.994	13.931	14.664	14.427	13.674	13.799	14.664	3.127	3.194	3.255	3.166	3.130	3.156
ratio	**1.005**	1.112	1.595	1.120	1.093	1.090	**1.007**	1.066	1.160	1.058	1.015	1.051	**1.001**	1.019	1.043	1.015	1.009	1.014

Ckt.	max junc.-to-amb. temp, $T_{ja,max}$ (°C)						temp difference, $T_{ja,max} - T_{ja,min}$ (°C)						average temp, $T_{ja,ave}$ (°C)					
	[2]	[5]	[3]	[1]	TSA	CA	[2]	[5]	[3]	[1]	TSA	CA	[2]	[5]	[3]	[1]	TSA	CA
ckt1	72.48	73.12	82.86	70.69	70.85	70.41	16.29	14.94	28.12	14.69	15.55	14.16	63.80	63.70	69.52	63.32	63.27	63.35
ckt2	91.70	74.21	101.00	76.89	100.19	73.05	46.96	15.15	51.16	22.39	53.87	17.15	64.81	66.84	69.36	66.07	65.14	66.14
ckt3	77.74	64.39	69.80	66.34	72.41	65.60	39.89	19.68	28.69	23.82	33.65	22.97	55.41	55.49	55.97	54.53	54.14	55.08
ckt4	73.79	62.43	80.11	60.14	65.50	59.31	35.46	16.69	39.76	15.87	21.83	14.27	55.07	54.35	60.42	53.91	53.63	53.85
ckt5	74.86	79.22	76.25	61.95	64.45	61.60	38.08	36.39	38.02	23.77	33.07	24.53	54.51	55.08	57.97	53.22	51.91	52.90
ratio	1.056	0.964	1.105	0.909	0.997	**0.895**	1.235	0.744	1.360	0.719	1.042	**0.673**	0.994	0.999	1.059	0.984	**0.973**	0.984

based force, and found large increase in wirelength without much additional temperature improvement. Moving cells out of a high-temperature area on a die may not reduce temperature if the high temperature is a result from thermal coupling with other dies. Also, without considering TSV thermal properties during thermal analysis, the thermal gradient does not capture the impact of TSVs on temperature accurately, thereby misguiding the placement. Second, we see that [5] obtains wirelength and delay results that are significantly worse than non-uniform TSV placer. This is mainly because it moves logic cells based only on power density. However, this move helps reduce maximum temperature and thermal variation inside the 3D chip stack significantly. Although it attempts to spread power over placement area, we observe that TSVs obstruct this effort frequently.

Third, the routed wirelength and delay of results from [3] are worse than all other placers. The main reason is that [3] does not consider TSV area during placement. Thus, the TSVs inserted during routing affects the placement quality significantly. The maximum temperature, thermal variation, and average temperature are also worse than uniform TSV placer. The router tends to insert TSVs in the middle of the die to minimize wirelength, leaving low thermal conductivity at chip corners, thus high temperature. Fourth, although the wirelength of result from [1] is worse than other placers, temperature improvement is among the best. Because the algorithm considers the impact of TSV on chip area and temperature, it utilizes TSVs more effectively to help improve temperature results.

Fifth, we observe that our TSV spread and alignment method (TSA) achieves comparable delay and power results at the cost of wirelength degradation compared with non-uniform placer. In case of temperature, TSA obtains better average temperature than uniform TSV and comparable maximum temperature and temperature difference. But, the wirelength of TSA method is significantly better than that of uniform TSV placer. These results show that our TSA method is better in reducing wirelength while optimizing temperature compared with uniform TSV placer.

Lastly, our thermal coupling-aware placement (CA) achieves the best temperature results among all placers [2, 5, 3, 1], including uniform TSV placer [4]. In particular, our CA method outperforms uniform TSV placer by 10% and 33% in terms of maximum temperature and temperature difference. CA obtains 9% worse wirelength and 5% worse delay results compared with non-uniform TSV placer, but CA is among the best in terms of wirelength and delay among other placers [2, 5, 3, 1]. The power overhead is negligible. The TSVs in the placement by our CA method are not spread as evenly as our TSA placer and uniform TSV placer, but they are

spread only sufficiently to help remove heat from the dies in the stack while maintaining high-quality wirelength. In addition, we observe that high-power logic cells are also placed effectively to dissipate heat using the nearby TSVs that are vertically aligned all the way to the heatsink.

The runtime of all placement algorithms is roughly in the same magnitude. Except for our TSA method, all other placement algorithms require power simulation (and thermal simulation in the case of [2]), resulting in larger runtime than [4].

5. CONCLUSIONS

In this paper we showed that temperature-aware placers must consider TSV thermal properties and die-to-die thermal coupling during placement. We presented two temperature-aware placement algorithms for 3D ICs. TSVs are spread and aligned in the first algorithm. In the second algorithm, logic cells are moved based on the thermal conductivity to the heatsink, and TSVs are moved based on the power density of the neighboring dies. Experimental results show that our placers achieve the best temperature results among all placers used in our comparison.

6. REFERENCES

[1] J. Cong, G. Luo, and Y. Shi. Thermal-aware cell and through-silicon-via co-placement for 3D ICs. In *Proc. ACM Design Automation Conf.*, pages 670–675, San Diego, CA, Jun. 5–9 2011.

[2] B. Goplen and S. Sapatnekar. Efficient thermal placement of standard cells in 3D ICs using a force directed approach. In *Proc. IEEE Int. Conf. on Computer-Aided Design*, pages 86–89, San Jose, CA, Nov. 9–13 2003.

[3] B. Goplen and S. Sapatnekar. Placement of 3D ICs with thermal and interlayer via considerations. In *Proc. ACM Design Automation Conf.*, pages 626–631, San Diego, CA, June 4–8 2007.

[4] D. H. Kim, K. Athikulwongse, and S. K. Lim. A study of through-silicon-via impact on the 3D stacked IC layout. In *Proc. IEEE Int. Conf. on Computer-Aided Design*, pages 674–680, San Jose, CA, Nov. 2–5 2009.

[5] B. Obermeier and F. M. Johannes. Temperature-aware global placement. In *Proc. Asia and South Pacific Design Automation Conf.*, pages 143–148, Yokohama, Japan, Jan. 27–30 2004.

[6] M. Pathak, Y.-J. Lee, T. Moon, and S. K. Lim. Through-silicon-via management during 3D physical design: When to add and how many? In *Proc. IEEE Int. Conf. on Computer-Aided Design*, pages 387–394, San Jose, CA, Nov. 7–11 2010.

[7] P. Spindler, U. Schlichtmann, and F. M. Johannes. Kraftwerk2–A fast force-directed quadratic placement approach using an accurate net model. *IEEE Trans. on Computer-Aided Design of Integrated Circuits and Systems*, 27(8):1398–1411, Aug. 2008.

ComPLx: A Competitive Primal-dual Lagrange Optimization for Global Placement

Myung-Chul Kim and Igor L. Markov
University of Michigan, EECS Department, Ann Arbor, MI 48109-2121
mckima@umich.edu, imarkov@eecs.umich.edu

ABSTRACT

We develop a projected-subgradient primal-dual Lagrange optimization for global placement, that can be instantiated with a variety of interconnect models. It decomposes the original non-convex problem into "more convex" sub-problems. It generalizes the recent SimPL, SimPLR and Ripple algorithms and extends them. Empirically, ComPLx outperforms all published placers in runtime and performance on ISPD 2005 and 2006 benchmarks.

Categories and Subject Descriptors

B.7.2 [**Hardware, Integrated Circuits**]: Design Aids—
Placement and routing

General Terms

Algorithms, Design, Performance

Keywords

Algorithms, optimization, physical design, placement

1. INTRODUCTION

The success of global placement determines all aspects of modern IC layout and physical synthesis [5] because it controls the amount of interconnect, which increasingly dominates on-chip resources and circuit performance [22]. However, the diverse algorithmic challenges posed by global placement and its complexity continue to surprise researchers [26]. Current algorithms still lag behind manual layout on circuits with structured components [34], do not always scale to extremely large circuits and are inconsistent in their handling of objective functions and various constraints. Analysis and comparisons of placement algorithms have been mostly empirical [26], with little formal justification.

A recent approach to global placement promises to support a variety of discrete and continuous constraints and was extended to handle routability-driven placement. Represented by the SimPL [23] and SimPLR [24] algorithms,

this approach consistently outperforms previous state of the art in speed and solution quality, is amenable to thread-level and instruction-level parallelism, requires only a modest amount of code, and was successfully re-implemented by independent researchers (Ripple [18]). SimPL was extended to power-aware placement with integrated clock-network synthesis in [25]. However, a convincing mathematical foundation for this empirical success was lacking. While the SimPL approach is based on quadratic placement, the significance of this connection has remained unclear *vis-à-vis* techniques based on the log-sum-exp interconnect model [29].

Our contributions can be summarized as follows

- A projected subgradient primal-dual Lagrange optimization (ComPLx) for global placement compatible with a variety of interconnect models, including linearized quadratic, log-sum-exp, etc.

- Convergence analysis and ensuing enhancements.

- Casting existing algorithms SimPL [23], SimPLR [24] and Ripple [18] as special cases of ComPLx. In particular, ComPLx inherits their competitiveness and lends them mathematical substantiation.

- Algorithmic extensions for mixed-size, as well as timing and power-driven placement.

- Empirical validation of the theoretical framework underlying ComPLx. On ISPD 2005 benchmarks, ComPLx is 10% faster than FastPlace [32] (including detailed placement runtime). It outperforms SimPL and RQL (the best published placers) by 1%. On ISPD 2006 benchmarks, ComPLx outperforms the leading placer RQL [33] by 1% in terms of scaled HPWL while running 2.5× faster.

In the remainder of the paper, Section 2 reviews necessary background. Section 3 introduces our primal-dual Lagrangian relaxation ComPLx, whose convergence is discussed in Section 4. Section 5 points out that the SimPL, SimPLR and Ripple algorithms are special cases of ComPLx. It then extends ComPLx to mixed-size and timing-driven placement. Section 6 presents empirical studies with improvements over these algorithms. Conclusions are given in Section 7. Comparisons to other primal-dual Lagrange optimizations are discussed in Section S4.

2. BACKGROUND

Global placement [22] of a netlist $\mathcal{N} = (E, V)$ with nets E and n nodes (cells) V seeks a set of planar node locations $(\vec{x}, \vec{y}) \in [x_{min}, x_{max}]^n \times [y_{min}, y_{max}]^n$ that minimize the weighted Half-Perimeter WireLength (wHPWL). For locations $\vec{x} = \{x_i\}$, $\vec{y} = \{y_i\}$ and net weights $\vec{w} = \{w_i\}$, wHPWL$_\mathcal{N}(\vec{x}, \vec{y})$= wHPWL$_\mathcal{N}(\vec{x})$+wHPWL$_\mathcal{N}(\vec{y})$, where

$$wHPWL_\mathcal{N}(\vec{x}) = \Sigma_{e \subset E} w_e \left[\max_{i \in e} x_i - \min_{i \in e} x_i\right] \qquad (1)$$

This piecewise-linear function lends itself to linear programming (LP) and min-cost max-flows, but these techniques have been successful only for smaller netlists. In large-scale placement, HPWL is approximated by convex twice-differentiable functions $\Phi(\vec{x}, \vec{y})$ and optimized numerically by linear or nonlinear Conjugate Gradient.

Quadratic approximations are used in many placers

$$\Phi_Q(\vec{x}, \vec{y}) = \vec{x}^T Q_x \vec{x} + \vec{f}_x \vec{x} + \vec{y}^T Q_y \vec{y} + \vec{f}_y \vec{y} \qquad (2)$$

with matrices Q_x, Q_y derived from the netlist and vectors $\vec{f}_x, \vec{f}_y$ that reflect connections to fixed objects. When sufficiently many nodes in a connected netlist are fixed, Φ_Q is strictly convex and can be optimized quickly. To approximate the HPWL by quadratic functions, one uses a *linearization* technique [30], adjusting the approximations at every global placement iteration. In particular, single-edge terms of the form $w_{ij}(x_i - x_j)^2$ are changed to $\frac{w_{ij}(x_i-x_j)^2}{|x_i'-x_j'|+\varepsilon}$ where the primed values are constants based on *the result of the last iteration* (a.k.a. the last *iterate*). Multipin nets are decomposed into sets of edges using stars, cliques or the Bound2Bound model [31]. Other differentiable approximations to the HPWL objective are outlined in supplementary material (Section S1).

Constraints in placement include *legality, target utilization, routability, resource-type* constraints, etc.

$$(\vec{x}, \vec{y}) \in \mathcal{C}$$

which prohibit multiple pairs (x_i, y_i) from concentrating in small regions. The demands for physical on-chip resources (gate area or number of routes in a region) must not exceed available supplies/design constraint (area for placing logic gates, target utilization, number of routing tracks) [22]. This is typically expressed by inequalities, e.g., allowing at most $C_{j,k}$ placeable objects in grid-cell (j, k). These inequalities are easy to satisfy when no optimization is performed. Unlike Φ, the constraints are noncovex, as illustrated by constraints on locations of two non-overlapping rectangles. Another type of constraints — routability of modern IC layouts — is NP-hard to evaluate with sufficient accuracy [22]. Some layout regions may be blocked by fixed obstacles and unavailable to (x_i, y_i), leading to discrete choices, such as placing an object on one side of an obstacle. This inhibits smooth convex optimization and, historically, motivated specialized global-placement techniques tailored to variant objective functions and constraints [26].

3. A PRIMAL-DUAL LAGRANGE METHOD

We propose a general method for handling constraints in global placement with a variety of possible interconnect models, and show how to decompose the original non-convex problem into "more convex" sub-problems.

A Lagrangian relaxation of global placement can be constructed if constraints are specified as *equalities* $\Pi(\vec{x}, \vec{y}) = 0$. Since supply-demand *inequalities* are usually given instead, the more general Karush-Kuhn-Tucker conditions may at

first seem more relevant.[1] However, working with supply-demand inequalities directly is difficult because they are specified algorithmically, not as closed-form expressions in $(\vec{x}, \vec{y})$. Without derivatives, one resorts to *subgradient optimization* [6], while the nature of the constraints calls for approximation. Placement techniques based on non-convex optimization [20, 9, 12] fit demand distribution to smooth functions using *kernel-density estimation*, and this facilitates gradient estimation. Each such step is laborious, and many steps may be required because, after moving in the gradient direction, one may need to "make turns" (as illustrated by moving around a rectangular obstacle). The reliance on local subgradient information in [20, 12] is common in analytical placement and with possible exceptions of Kraftwerk [31] and mPL6 [9] which estimate subgradients by solving second-order linear elliptic PDEs with global supply-demand information. Solutions of these PDEs can be written as convolutions of the density function with a fixed Green's function $G(s, t)$ (dependent on boundary conditions), which sometimes vanishes away from $s = t$. Further, local subgradient computations leave undefined the trade-off between demand-distribution subgradients and the gradients of the objective function. This *force modulation* problem was articulated in [33], but addressed there with *ad hoc* thresholding.

In contrast to other methods, our subgradients point to a closest $\mathcal{C}$-feasible solution, and their magnitude is modulated by respective distance. Thus, we define $\Pi_\mathcal{C}(\vec{x}, \vec{y})$ as the L_1-distance from $(\vec{x}, \vec{y})$ to a closest $\mathcal{C}$-feasible solution.

$$\Pi_\mathcal{C}(\vec{x}, \vec{y}) = \min_{(\vec{x}_*, \vec{y}_*) \in \mathcal{C}} ||(\vec{x}, \vec{y}) - (\vec{x}_*, \vec{y}_*)||_1$$

$$= \min_{(\vec{x}_*, \vec{y}_*)} \left(||\vec{x} - \vec{x}_*||_1 + ||\vec{y} - \vec{y}_*||_1 \right) \qquad (3)$$

Clearly, $\Pi_\mathcal{C}(\vec{x}, \vec{y}) = 0 \Leftrightarrow (\vec{x}, \vec{y}) \in \mathcal{C}$. Therefore, in addition to primary variables $(\vec{x}, \vec{y})$, we introduce one dual variable (multiplier) $\lambda \geq 0$, and establish the following Lagrangian

$$\mathcal{L}_{\Phi,\mathcal{C}}(\vec{x}, \vec{y}, \lambda) = \Phi(\vec{x}, \vec{y}) + \lambda \, \Pi(\vec{x}, \vec{y}) \qquad (4)$$

We use L_1-norms so that costs and penalties are expressed in meters and can be compared. Hence, λ is dimensionless.

Primal-dual Lagrangian relaxation [3] alternates minimization over the primal variables with maximization over the dual variable(s). $\min \Phi(\vec{x}, \vec{y})$ subject to $(\vec{x}, \vec{y}) \in \mathcal{C}$ can be found by *sequential unconstrained optimization*

$$\max_\lambda \min_{(\vec{x}, \vec{y})} \mathcal{L}_{\Phi,\mathcal{C}}(\vec{x}, \vec{y}, \lambda) \qquad (5)$$

Starting with $\lambda_0 = 0$, the first primal iterate is produced by minimization of $\Phi(\vec{x}, \vec{y})$ (using quadratic optimization or non-linear Conjugate Gradient, depending on the function). At subsequent iterations, primal optimization must also account for the penalty term. A straightforward argument by contradiction shows that for $\lambda_k < \lambda_{k+1}$

$$\min_{(\vec{x}, \vec{y})} \mathcal{L}_{\Phi,\mathcal{C}}(\vec{x}, \vec{y}, \lambda_k) \leq \min_{(\vec{x}, \vec{y})} \mathcal{L}_{\Phi,\mathcal{C}}(\vec{x}, \vec{y}, \lambda_{k+1}) \qquad (6)$$

As λ increases, so does the sensitivity of $\mathcal{L}_{\Phi,\mathcal{C}}$ to Π. Therefore, the minimization of $\mathcal{L}_{\Phi,\mathcal{C}}$ affects the Π term more, and this term decreases. However, since the minimized value of

---

[1] Inequalities can also be converted into equations by adding slack variables, but we avoid this common technique, to limit computational complexity.

$\mathcal{L}_{\Phi,\mathcal{C}}$ increases (per Formula 6), Φ must increase. Eventually, $(\vec{x},\vec{y})$ become $\mathcal{C}$-feasible (or very close to), making the Lagrangian insensitive to λ and indicating that an optimum is near. The following weak duality bounds hold for any $\mathcal{C}$-feasible solution $(\vec{x}^{\,\circ},\vec{y}^{\,\circ})$ and any iterate $(\vec{x},\vec{y})$ after primal optimization.

$$\Phi(\vec{x},\vec{y}) \leq \mathcal{L}_{\Phi,\mathcal{C}}(\vec{x},\vec{y},\lambda) \leq \mathcal{L}_{\Phi,\mathcal{C}}(\vec{x}^{\,\circ},\vec{y}^{\,\circ},\lambda) = \Phi(\vec{x}^{\,\circ},\vec{y}^{\,\circ}) \quad (7)$$

The first $\leq$ is due to $\lambda \geq 0$ in Formula 4. The second $\leq$ is due to $(\vec{x},\vec{y})$ being argmin from Formula 5 and the third $=$ is due to $\Pi(\vec{x}^{\,\circ},\vec{y}^{\,\circ}) = 0$ ($\mathcal{C}$-feasible). The second inequality is strict unless $(\vec{x},\vec{y})$ is $\mathcal{C}$-infeasible, hence $\Phi(\vec{x},\vec{y}) < \Phi(\vec{x}^{\,\circ},\vec{y}^{\,\circ})$. The *duality gap* is

$$\Delta_\Phi = \Phi(\vec{x}^{\,\circ},\vec{y}^{\,\circ}) - \Phi(\vec{x},\vec{y}) \quad (8)$$

minimized over best available *primal feasible* $(\vec{x}^{\,\circ},\vec{y}^{\,\circ})$ and $(\vec{x},\vec{y})$ at a given point during optimization.

Approximating the penalty term allows us to replace the nonconvex Lagrangian by a convex one. Here we use the *feasibility projection*

$$P_{\mathcal{C}}(\vec{x},\vec{y}) = \operatorname{argmin}_{(\vec{x}_*,\vec{y}_*) \in \mathcal{C}} ||(\vec{x},\vec{y}) - (\vec{x}_*,\vec{y}_*)||_1 \quad (9)$$

that finds a closest $\mathcal{C}$-feasible approximation (performs *pseudo-legalization*) of $(\vec{x},\vec{y})$.[2] Since $P_{\mathcal{C}}(\vec{x}',\vec{y}')$ is $\mathcal{C}$-feasible, $\Phi(\vec{x},\vec{y}) \leq \Phi\big(P_{\mathcal{C}}(\vec{x}',\vec{y}')\big)$ by Inequalities 7. Given that Φ is continuous, $||(\vec{x},\vec{y}) - P_{\mathcal{C}}(\vec{x},\vec{y})||_1 \to 0$ would necessitate $\Phi\big(P_{\mathcal{C}}(\vec{x},\vec{y})\big) - \Phi(\vec{x},\vec{y}) \to 0$. Hence, $\Phi\big(P_{\mathcal{C}}(\vec{x},\vec{y})\big)$ must generally decrease, providing upper bounds on final placement cost.

After finding $\mathcal{C}$-feasible *anchor locations* $(\vec{x}^{\,\circ},\vec{y}^{\,\circ}) = P_{\mathcal{C}}(\vec{x},\vec{y})$, we establish the simplified Lagrangian

$$\mathcal{L}_\Phi^\circ(\vec{x},\vec{y},\lambda) = \Phi(\vec{x},\vec{y}) + \lambda ||(\vec{x},\vec{y}) - (\vec{x}^{\,\circ},\vec{y}^{\,\circ})||_1 \quad (10)$$

To minimize it with respect to fixed $(\vec{x}^{\,\circ},\vec{y}^{\,\circ})$ and λ, the L_1-term can be approximated by the same type of function as Φ (see Section 5). Thus, for quadratic Φ, the optimality condition $\nabla\mathcal{L}_\Phi^\circ(\vec{x},\vec{y},\lambda) = 0$ turns into a system of linear equations. For other functional forms, such as the log-sum-exp expressions, one can minimize $\mathcal{L}_\Phi^\circ(\vec{x},\vec{y},\lambda)$ using the non-linear Conjugate Gradient method or other known alternatives.[3] In addition to being (strictly) convex, $\mathcal{L}_\Phi^\circ(\vec{x},\vec{y},\lambda)$ is usually separable into its x and y components which can be optimized independently. One can verify Inequalities 6 and 7 for $\mathcal{L}_\Phi^\circ(\vec{x},\vec{y},\lambda)$ subject to $(\vec{x}^{\,\circ},\vec{y}^{\,\circ}) = P_{\mathcal{C}}(\vec{x},\vec{y})$.

The ComPLx framework re-solves $\nabla\mathcal{L}_\Phi^\circ(\vec{x},\vec{y},\lambda) = 0$ and $P_{\mathcal{C}}(\vec{x},\vec{y}) = 0$ until convergence. The result of global placement can be read from the last iterate $(\vec{x},\vec{y})$ or the last $\mathcal{C}$-feasible iterate $(\vec{x}^{\,\circ},\vec{y}^{\,\circ})$ as discussed in Section 4.

4. CONVERGENCE ANALYSIS

It is sufficient for $P_{\mathcal{C}}$ to find a $\mathcal{C}$-feasible solution that is *reasonably close*, rather than closest, to a given $(\vec{x},\vec{y})$.[4] Such *approximate projected subgradient* methods are relatively recent in the operations-research literature [16, Section 1] but are proven to converge as long as $P_{\mathcal{C}}$ does not increase the

---

[2] One can additionally require breaking ties toward smaller values of Φ (or even some trade-off with Φ), but this does not seem necessary for practical success (Section 6).

[3] Techniques such as Newton's method that approximate the objective f by quadratic functions based on $\mathrm{Hessian}(f)$ essentially perform sequential quadratic optimization.

[4] Section S2 points out that this is a "more convex" problem.

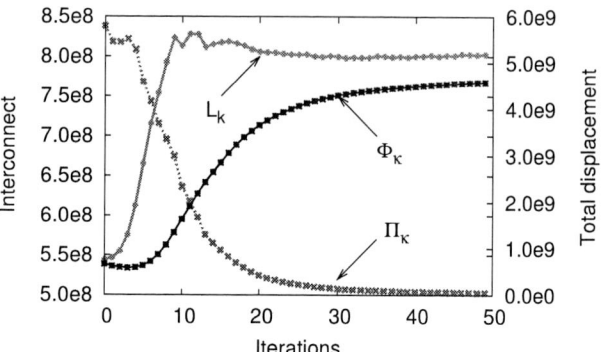

Figure 1: Progressions of $\mathcal{L}$ (the total Lagrangian), Φ (netlist interconnect), and Π (L_1-distance to legal) over ComPLx iterations on BIGBLUE4. $\mathcal{L}$ increases steeply in the early placement iterations, as λ increases. Π decreases while Φ gradually increases.

distance to the set $\mathcal{C}$ and typically reduces it during iterations [7, Sections 2 and 3]. In particular, $P_{\mathcal{C}}$ should return its input when the input is $\mathcal{C}$-feasible. Convergence can be improved if $P_{\mathcal{C}}$ exhibits reasonable *fidelity* with respect to the exact feasibility projection, and is *self-consistent*

$$||(\vec{x},\vec{y}) - P_{\mathcal{C}}(\vec{x},\vec{y})||_1 > ||(\vec{x}',\vec{y}') - P_{\mathcal{C}}(\vec{x},\vec{y})||_1 \;\Rightarrow$$
$$||(\vec{x},\vec{y}) - P_{\mathcal{C}}(\vec{x}',\vec{y}')||_1 > ||(\vec{x}',\vec{y}') - P_{\mathcal{C}}(\vec{x}',\vec{y}')||_1 \quad (11)$$

In other words, if $(\vec{x}',\vec{y}')$ is closer to $P_{\mathcal{C}}(\vec{x},\vec{y})$ than $(\vec{x},\vec{y})$, then it should also be closer to $P_{\mathcal{C}}(\vec{x}',\vec{y}')$. The ComPLx implementation of $P_{\mathcal{C}}$ reviewed in Section 5 handles both standard cells and macros. It is self-consistent through almost all iterations, as shown in Section S2. Figure 1 illustrates changes in $\mathcal{L}_k$, Φ_k, and Π_k over ComPLx iterations on BIGBLUE4. The same trends show on all other benchmarks, validating the discussion in Section 3.

Global placement iterations stop when a $\mathcal{C}$-feasible value is reached, which must happen when λ exceeds its optimal value. But the resulting solution may be far from optimal. To avoid this, we propose to improve the efficiency of the first few iterations, since the first iterates are crucial to the overall success (given that we are solving a nonconvex problem overall). The earliest non-zero value of λ must be sufficiently small so that $\Phi(\vec{x},\vec{y}) \gg \lambda\Pi(\vec{x},\vec{y})$, to make sure that $\mathcal{L}_{\Phi,\mathcal{C}}(\vec{x},\vec{y},\lambda)$ is dominated by the convex *cost* term rather than the penalty term. Hence, we initially select $\lambda_1 = \Phi/100\Pi$. This calculation is supported by the fact that Π and Φ are expressed in the same units (meters). To *avoid* premature progress, a maximum increase in λ can be imposed, say 100% per iteration.

$$\lambda_{k+1} = \min\{2\lambda_k, \lambda_k + (\Pi_{k+1}/\Pi_k)h\} \quad (12)$$

where h is a scaling constant. λ increases proportionally to Π changes to ensure that Π decreases by a sufficient amount, as Φ increases. Considering the number of iterations until λ reaches its optimal value, there is no explicit dependency on the number of variables. In practice, the maximal λ values and the iteration count do not grow with the size of the problem instance as shown in Section S3.

749

Convergence criteria can be defined in terms of $r(\vec{x},\vec{y}) = ||(\vec{x},\vec{y}) - P_\mathcal{C}(\vec{x},\vec{y})||_1$, rather than $\mathcal{L}$ — when the placement is close to $\mathcal{C}$-feasible, a detailed placer can produce optimized site-aligned legal locations. Given that pseudo-legalization $(\vec{x}^\circ,\vec{y}^\circ) = P_\mathcal{C}(\vec{x},\vec{y})$ is performed at every iteration, one can run detailed placement on $(\vec{x}^\circ,\vec{y}^\circ)$ rather than on $(\vec{x},\vec{y})$. This would allow an even more aggressive convergence criterion in terms of the duality gap $\Delta_\Phi = \Phi(\vec{x}^\circ,\vec{y}^\circ) - \Phi(\vec{x},\vec{y})$.[5] To substantiate this idea, we observe that performing detailed placement on a feasible solution $(\vec{x}^\circ,\vec{y}^\circ)$ should *not* increase costs (rather the opposite), whereas performing detailed placement on $(\vec{x},\vec{y})$ is likely to (as observed in practice). This observation upper-bounds the difference in final costs between these two scenarios by Δ_Φ.

5. SPECIAL CASES AND EXTENSIONS

We now point out that the SimPL [23], SimPLR [24] and Ripple [18] algorithms are special cases of the proposed primal-dual Lagrangian relaxation. They implement Φ as a quadratic approximation Φ_Q of HPWL, adjusted at every iteration through the linearized Bound2Bound net model [31]. Linearization is also applied to represent the L_1-norm in the penalty term Π. To model this term, each movable object is connected to its anchor location by a *pseudonet*, contributing $w_i(x_i - x_i^\circ)^2$ to the overall objective (and a similar y-term), where $w_i = \frac{\lambda}{|x_i - x_i^\circ| + \varepsilon}$ is based on the last iterate. $\varepsilon > 0$ is used to bound the denominator away from zero and make the objective function strictly convex. In SimPL and SimPLR, ε is calculated as 1.5 times row height.[6] This matches Formula 10 if the L_1-distance term is approximated by a linearized quadratic function (Section 2). SimPL, SimPLR and Ripple maintain a lower and an upper-bound placement at each iteration, and these placement satisfy conditions in Formula 7 as seen in [23, Figure 6], [24, Figure 4].

The SimPL [23], SimPLR [24] and Ripple [18] algorithms differ in how they define and implement the feasibility projection $P_\mathcal{C}$. In practice, to identify overfilled bins with respect to a *target utilization/density limit* $0 < \gamma \leq 1$ [23, Section 4], a uniform grid is superimposed over the entire layout. Then the feasibility projection seeks to satisfy the given target utilization/density limit within each grid-cell. To this end, the SimPL $P_\mathcal{C}$ first localizes the changes in $(\vec{x},\vec{y})$ to the smallest rectangular grid-cell sub-arrays that satisfy a given target utilization/density limit, and then processes each region by a top-down geometric-partitioning framework. SimPL alternates (*i*) piecewise-linear scaling in x and y directions with (*ii*) spreading locations in each dimension to even out density, while preserving the relative order (determined by sorting). This pseudo-legalization is discussed in more detail in Section S2 and can be seen as solving a convex problem in terms of (always-positive) *distances between neighboring x locations* (*y* locations). As a runtime trade-off, SimPL gradually increases the accuracy of $P_\mathcal{C}$ as the grid-cell size decreases, and we use this feature in Section 6 to show that $P_\mathcal{C}$ does not need to be implemented precisely. SimPLR and Ripple generally follow the SimPL techniques, but are concerned with routability in addition to HPWL. Therefore, they estimate congestion after placement

---

[5] As Φ is Lipschitz, $r(\vec{x},\vec{y}) \to 0$ implies $\Delta_\Phi \to 0$.

[6] In [30], a lower bound on the distance between two modules is defined as the average module width.

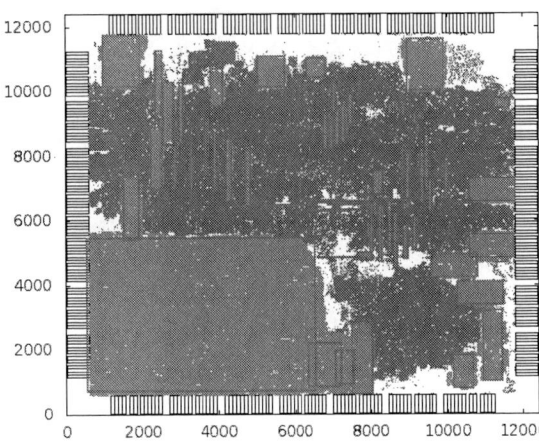

Figure 2: Macro shredding for feasibility projection $P_\mathcal{C}$ on NEWBLUE1 (an intermediate placement). Red boxes show the locations of macro cells at the centers of gravity of constituent cells (shown as green dots). Standard cells are shown as blue dots.

iterations (SimPLR calls a global router, whereas Ripple estimates congestion directly) and modify $P_\mathcal{C}$ to produce low-congestion placements. SimPLR preprocesses $P_\mathcal{C}$ by temporarily increasing the dimensions of some movable objects, so as to enhance geometric separation between them. Ripple distinguishes congestion maps for horizontal and vertical wiring, and scales minimal-sized rectangular regions differently each direction. Despite the technical differences, all these variants compute $P_\mathcal{C}$ by a series of convex optimizations. The use of feasibility projections is not only common between SimPL [23], SimPLR [24] and Ripple [18], but also distinguishes them from other placement algorithms. This is why SimPL, SimPLR and Ripple are particularly good at handling nonlinear, nonconvex layout constraints, such as numerous fixed obstacles present in modern SoC layouts.

Mixed-size placement requires careful accounting for pin offsets during quadratic optimization (since pin-offsets can be large in macros), as well as an approximate feasibility projection $P_\mathcal{C}$ which can handle macros and standard cells. We have therefore revised and extended the *macro shredding technique* from [2]. Macro cells are divided into equal-sized cells (2×2 standard-cell height), but unlike prior work, ComPLx does not connect constituent cells (shreds) with fake nets and thus does not modify the linear systems it solves. The conventional $P_\mathcal{C}$ [23] is applied to the shreds, after which the action of $P_\mathcal{C}$ on the original macro is interpolated by averaging the displacement of the shreds. Given that the conventional $P_\mathcal{C}$ [23] mostly preserves the relative placement of cells and is approximately locally isometric, the arrays of shreds are transformed into shapes similar to arrays, as seen in Figure 2. As $P_\mathcal{C}$ seeks to satisfy the given target utilization ($0 < \gamma < 1$), additional whitespace is inserted among constituent cells. Then the bounding box of projected locations of shreds outgrows the original macro cell, creating a halo around the macro, where other cells cannot be placed. To compensate, we multiply the widths and heights of constituent cells by $\sqrt{\gamma}$. Stabilizing macro positions early is important, as they greatly impact adja-

BENCHMARKS		BEST PUBLISHED as of 02/27/2011 HPWL (placer)	COMPLX					
			FINEST GRID		P_C+=FASTPLACE-DP		DEFAULT CONFIG.	
size (# of modules)			HPWL	Runtime	HPWL	Runtime	HPWL	Runtime
ADAPTEC1	211K	**77.82 (RQL**	78.95	3.86	78.39	93.52	**77.75**	**3.09**
ADAPTEC2	255K	**88.51 (RQL)**	89.81	5.23	91.09	162.07	88.76	4.31
ADAPTEC3	452K	207.67 (SimPL)	207.07	11.54	**203.70**	323.04	206.57	10.75
ADAPTEC4	496K	186.80 (SimPL)	185.13	10.90	188.37	252.40	**184.07**	9.57
BIGBLUE1	278K	94.98 (RQL)	96.15	7.22	**94.78**	140.27	95.30	7.00
BIGBLUE2	558K	**145.47 (SimPL)**	144.59	9.91	146.32	200.56	145.87	8.53
BIGBLUE3	1.10M	**323.09 (RQL)**	352.33	32.52	327.49	629.23	330.74	24.80
BIGBLUE4	2.18M	797.66 (RQL)	**787.11**	47.87	792.26	961.25	789.45	41.89
Geomean		$1.00\times$	$1.01\times$	$1.16\times$	$1.00\times$	$26.56\times$	$1.00\times$	$1.00\times$

Table 1: Legal HPWL ($\times$10e6) and total runtime (in min.) comparison on ISPD 2005 benchmarks. Each run uses a single thread on a 2.8GHz workstation. Best-published numbers are annotated with the placers that produced them – SimPL [23] or RQL [33]. mPL6 and NTUPlace3 were also considered in this comparison. We regenerated placements of SimPL without a cell-orientation optimization.

cent standard cells. To accelerate the convergence of macro cells and decrease their displacement during legalization, we extend Formulae 4 and 10 with separate, larger λ values for each macro, computed as the default λ times the ratio of the size of macro cell to the average standard-cell size. As seen in Figure 2, our mixed-size feasibility projection P_C may leave small overlaps between macros. Rather than force complete legalization, we let multiple global placement iterations (including P_C) gradually decrease these overlaps. We observe that as P_C displaces cells and macros *less*, the changes in the shapes of shredded macros also *decrease*, and this *increases* the precision of legalization for macros during P_C. Even if slight overlaps remain at the end of global placement, they can be fixed by the detailed placer without undermining the overall performance. While less sophisticated than algorithms in [10, 11, 35], our mixed-size approximate feasibility projection P_C is easy to implement and produces very good results, motivating additional studies.

Extensions for timing- and power-driven placement traditionally rely on net weights computed from activity factors and timing slacks [22, Chapter 8]. Net-weighting schemes in the literature include rigorous, provably convergent methods [8]. Since our mathematical formulation for *global placement* in Section 3 accounts for net weights in Φ, existing techniques [8] and their provable properties apply directly. An extension of SimPL with power-driven net weights is reported in [25]. However, we observe that *the impact of the feasibility projection and detailed placement* suggests revising the penalty term in the Lagrangian. Minimizing L_1-distance to $\mathcal{C}$ may leave some cells far from their legal positions, forcing P_C or the detailed placer to displace them. This may stretch out incident nets, which is undesirable for timing- and power-critical standard cells. Hence, in the simplified Lagrangian of Formula 10 we weigh the penalty term by timing/power criticality and replace

$$\lambda||(\vec{x},\vec{y}) - (\vec{x}^\circ,\vec{y}^\circ)||_1 \text{ by } \lambda(\vec{\gamma}\cdot|(\vec{x},\vec{y}) - (\vec{x}^\circ,\vec{y}^\circ)|) \quad (13)$$

where $|\ |$ represents the vector of pointwise distances and $\vec{\gamma}$ represents the vector of cell-criticalities. Initially, $\vec{\gamma}$ is populated with switching activity factors (no cells are critical). When static timing analysis, performed between placement iterations, indicates that cell i lies on a critical path (violates a timing constraint), the cell's criticality must be increased $\gamma_i = \gamma_i(1 + \delta)$, (along with the weights of critical nets in Φ).

6. EMPIRICAL VALIDATION

Our implementation of ComPLx inherits the performance and runtime advantages of SimPL [23]. Experiments ran on a 2.8GHz Intel Core-i7 860 Linux server with 8GB RAM, using one CPU core. Detailed placement was done by FastPlace-DP [28]. All settings were the same for all benchmarks.

Given that quadratic optimization in SimPL and ComPLx is optimal, we tried to improve the feasibility projection P_C. One such attempt used the finest grid during all global placement iterations. In a second attempt, we post-processed the result of P_C by the detailed placer [28] at each iteration. Our data in Table 1 show only a marginal improvement, but at a runtime cost. *Vice versa*, coarsening the grid speeds up P_C without undermining solution quality. Thus, no interconnect optimization during P_C is required. While surprising, this is consistent with the discussion in Section 4 and can be explained by the known convergence properties of Primal-dual Lagrange optimization [6, 3]. In practice this decreases the risk of incorrect implementation. On ISPD 2005 benchmarks, ComPLx outperforms SimPL, sometimes by a small amount, sometimes significantly. The similarities are not surprising because ComPLx generalizes SimPL. The improvements are due to the refined convergence criterion (Section 4) and improved scheduling of λ (λ corresponds to the pseudonet weight in [23]). ComPLx outperforms SimPL and RQL (the best published placers) by 1%. ComPLx produces best results on more benchmarks than any prior placer, while running 10% faster than FastPlace (including FastPlace-DP runtime in both cases).

Table 2 covers ISPD 2006 benchmarks, which include density constraints and movable macros, not handled by SimPL. ComPLx outperforms the best-published placer RQL [33] by 1% in terms of scaled HPWL (the official contest metric). ComPLx is about 12% faster than FastPlace (including FastPlace-DP runtime in both cases), as well as 6.88$\times$ and 8.47$\times$ faster than NTUPlace3 and mPL6, respectively. RQL is 3.1$\times$ faster than mPL6 [33], hence > 2.5$\times$ as slow as ComPLx (including detailed placement by FastPlace-DP).

The SimPL placer (generalized in this work) was extended to a routability-driven placer SimPLR in [24] with strong results on ISPD 2011 benchmarks. Section S5 demonstrates that ComPLx naturally supports region constraints (Figure 4). Section S6 illustrates timing-driven placement.

Benchmarks (Υ_{target})	NTUPL3 [12]	MPL6 [9]	RQL [33]	ComPLx
ADAPTEC5 (0.5)	451.22 (21.0)	**431.27** **(1.09)**	443.28 (9.25)	432.60 (3.09)
NEWBLUE1 (0.8)	**62.65** **(1.09)**	68.08 (0.14)	64.43 (0.34)	64.71 (0.18)
NEWBLUE2 (0.9)	205.45 (2.53)	201.85 (1.52)	199.60 (1.45)	**197.24** **(1.04)**
NEWBLUE3 (0.8)	277.87 (0.00)	284.11 (0.59)	**269.33** **(0.07)**	272.87 (0.69)
NEWBLUE4 (0.5)	306.56 (13.1)	**300.58** **(1.63)**	308.75 (15.2)	306.00 (3.00)
NEWBLUE5 (0.5)	**509.71** **(9.56)**	537.14 (1.42)	537.49 (13.6)	540.29 (2.58)
NEWBLUE6 (0.8)	520.31 (8.40)	522.54 (1.40)	515.69 (4.33)	**501.90** **(1.06)**
NEWBLUE7 (0.8)	1109.6 (5.32)	1084.4 (1.14)	1057.8 (2.57)	**1042.2** **(1.27)**
Geomean	1.01× (2.40)	1.03× (1.22)	1.01× (2.30)	**1.00×** (1.61)

Table 2: Comparison of scaled HPWL (×10e6) on ISPD 2006 benchmarks. Overflow penalties are reported in parentheses. RQL results are from [33].

7. CONCLUSIONS

We developed a global placement algorithm ComPLx based on *subgradient projected primal-dual Lagrange optimization*. In its basic form, it consists of (*i*) interconnect optimization, (*ii*) a feasibility projection P_C that represents placement constraints, (*iii*) a penalty term that includes the Lagrange multiplier λ. Our extensions for mixed-size placement handle macros through the feasibility projection P_C and establish a separate, larger λ parameter for each macro. Timing-driven extensions track separate λ for timing-critical cells and increase λ based on criticality (slack). Our baseline algorithm generalizes recent SimPL [23], SimPLR [24] and Ripple [18] algorithms and inherits their empirical success. *Vice versa*, ComPLx provides mathematical substantiation and convergence analysis for SimPL, SimPL and Ripple, suggesting improvements and algorithmic extensions.

A key difference from most prior analytical frameworks is in the spreading mechanism — rather than estimate *density gradients based on local information*,[7] we use a global feasibility projection P_C. Consequently, the handling of *region, alignment* and other types of constraints requires only the modification of the feasibility projection (Section S5). Avoiding local gradients also improves runtime (compared to APlace and NTUPlace3), and so does our avoidance of optimization by local search (compared to FastPlace and RQL). The tradeoff between spreading and interconnect optimization is controlled by Lagrange multipliers λ.

A key difference from analytical placement based on non-convex optimization [20, 9, 12] is the emphasis on decomposing the original problem into a series of convex optimizations, which enables duality and accelerates convergence. Unlike prior works limited to a single interconnect model, our technique can be used with *quadratic*, *log-sum-exp* and other models (Section S1). The closest published primal-dual Lagrangian optimizations are discussed in Section S4.

---

[7] mPL6 and Kraftwerk2 are the only competitive prior placers with a global view of supply-demand trade-offs. We are exploring theoretical comparisons to them in ongoing work.

8. REFERENCES

[1] S. N. Adya, I. L. Markov, P. G. Villarrubia, "On Whitespace and Stability in Physical Synthesis," *Integration, the VLSI Journal* vol. 39/4, 2006, pp. 340-362.

[2] S. N. Adya, I. L. Markov, "Combinatorial techniques for Mixed-size Placement," *ACM Trans. Design Autom. Electr. Syst.* 10(1), 2005, pp. 58-90.

[3] R. K. Ahuja, T. L. Magnati, J. B. Orlin, "Network Flows: Theory, Algorithms, and Applications," *Prentice Hall* 1993.

[4] C. J. Alpert, T. F. Chan, A. B. Kahng, I. L. Markov, P. Mulet, "Faster Minimization of Linear Wirelength for Global Placement," *IEEE Trans. on CAD of Integrated Circuits and Systems* 17(1), 1998, pp. 3-13.

[5] C. J. Alpert et al., "Techniques for Fast Physical Synthesis," *Proc. IEEE* 95(3), 2007, pp. 573-599.

[6] D. P. Bertsekas, "Nonlinear Programming," 2nd ed., *Athena Scientific* 1999.

[7] S. Boyd, L. Xiao, A. Mutapcic, "Subgradient Methods," Notes for EE392o, *Stanford University* 2003. http://www.stanford.edu/class/ee392o/subgrad_method.pdf

[8] T. F. Chan, J. Cong, E. Radke, "A Rigorous Framework for Convergent Net-weighting Schemes in Timing-driven Placement," *ICCAD* 2009, pp. 288-294.

[9] T. F. Chan, J. Cong, J. Shinnerl, K. Sze, M. Xie, "mPL6: Enhanced Multilevel Mixed-Size Placement," *ISPD* 2006, pp. 212-214.

[10] H.-C. Chen et al., "Constraint Graph-based Macro Placement for Modern Mixed-size Circuit Designs," *ICCAD* 2008, pp. 218-223.

[11] T.-C. Chen et al., "MP-trees: A Packing-based Macro Placement Algorithm for Mixed-size Designs," *IEEE TCAD* 27(9) 2008, pp. 1621-1634.

[12] T.-C. Chen, Z.-W. Jiang, T.-C. Hsu, H.-C. Chen, Y.-W. Chang, "NTUPlace3: An Analytical Placer for Large-Scale Mixed-Size Designs With Preplaced Blocks and Density Constraints," *IEEE TCAD* 27(7) 2008, pp.1228-1240.

[13] Y.-L. Chuang et al., "Design-hierarchy Aware Mixed-size Placement for Routability Optimization," *ICCAD* 2010, pp. 663-668.

[14] J. Cong, M. Romesis, J. Shinnerl, "Robust Mixed-Size Placement Under Tight White-Space Constraints," *ICCAD* 2005, pp. 165-172.

[15] P. E. Gill, D. P. Robinson, "A Primal-Dual Augmented Lagrangian," *Computational Optimization and Applications* 2010, DOI: 10.1007/s10589-010-9339-1.

[16] K. C. Kiwiel, T. Larsson, P. O. Lindberg, "Lagrangian Relaxation via Ballstep Subgradient Methods," *Mathematics of Operations Research* 32(3), 2007, pp. 669-686. http://mor.journal.informs.org/content/32/3/669

[17] C. Li, C.-K. Koh, "Recursive Function Smoothing of Half-Perimeter Wirelength for Analytical Placement," *ISQED* 2007, pp. 829-834.

[18] X. He, T. Huang, L. Xiao, H. Tian, G. Cui, E. F. Y Young, "Ripple: An Effective Routability-Driven Placer by Iterative Cell Movement," *ICCAD* 2011, pp. 74-79.

[19] M.-K. Hsu, Y.-W. Chang, V. Balabanov, "TSV-aware Analytical Placement for 3D IC Designs," *DAC* '11, pp. 664-669.

[20] A. B. Kahng, Q. Wang, "A Faster Implementation of APlace," *ISPD* 2006, pp. 218-220.

[21] A. A. Kennings, I. L. Markov, "Smoothening Max-terms and Analytical Minimization of Half-Perimeter Wirelength," *VLSI Design* 2002, 14(3), pp. 229-237.

[22] A. B. Kahng, J. Lienig, I. L. Markov, J. Hu, "VLSI Physical Design: from Graph Partitioning to Timing Closure," Springer 2011, 312 pages.

[23] M.-C. Kim, D.-J. Lee, I. L. Markov, "SimPL: An Effective Placement Algorithm," *IEEE TCAD* 31(1), 2012, pp. 50-60.

[24] M.-C. Kim, J. Hu, D.-J. Lee, I. L. Markov, "A SimPLR method for Routability-driven Placement" *ICCAD* 2011.

[25] D.-J. Lee, I. L. Markov, "Obstacle-Aware Clock-tree Shaping During Placement," *ISPD* 2011, pp. 123-130.

[26] G.-J. Nam, J. Cong, "Modern Circuit Placement: Best Practices and Results," *Springer* 2007.

[27] A. N. Ng et al., "Solving Hard Instances of Floorplacement," *ISPD 2006*, pp. 78-85.

[28] M. Pan, N. Viswanathan, C. Chu, "An Efficient & Effective Detailed Placement Algorithm," *ICCAD* 2005, pp. 48-55.

[29] A. E. Ruehli, P. K. Wolff, and G. Goertzel, "Analytical Power/Timing Optimization Technique for Digital Systems," *DAC* 1977, pp. 142-146.

[30] G.Sigl, K.Doll, F.Johannes,"Analytical Placement:A Linear or a Quadratic Objective Function?" *DAC'91*, pp. 427-432.

[31] P. Spindler, U. Schlichtmann, F. M. Johannes, "Kraftwerk2 - A Fast Force-Directed Quadratic Placement Approach Using an Accurate Net Model," *IEEE TCAD* 27(8) 2008, pp. 1398-1411.

[32] N. Viswanathan, M. Pan, C. Chu, "FastPlace 3.0: A Fast Multilevel Quadratic Placement Algorithm with Placement Congestion Control," *ASPDAC* 2007, pp. 135-140.

[33] N. Viswanathan et al., "RQL: Global Placement via Relaxed Quadratic Spreading and Linearization," *DAC* 2007, pp. 453-458.

[34] S. I. Ward et al., "Quantifying Academic Placer Performance on Custom Designs, " *ISPD* 2011, pp. 91-98.

[35] J. Z. Yan et al., "Handling Complexities in Modern Large-scale Mixed-size Circuit Designs," *DAC* 2009.

S1. APPROXIMATIONS OF INTERCONNECT OBJECTIVE FUNCTIONS

Alternatives to quadratic approximations to the HPWL objective (Section 2) include the β-regularization [4]

$$\sqrt{(x_i - x_j)^2 - \beta} \to |x_i - x_j|, \quad \beta \to 0,$$

the p, β-regularization for a net $e \in \mathcal{N}$, with $p \to \infty$ [21]

$$\Sigma_{i,j \in e}(|x_i - x_j|^p + \beta)^{1/p} \to \max_{i,j \in e} |x_i - x_j| = [\max_{i \in e} x_i - \min_{i \in e} x_i]$$

and the log-sum-exp technique with $\gamma \to 0$ [29]

$$\gamma \log \Sigma_{k \in e} \big(\exp(x_k/\gamma) + \exp(-x_k/\gamma) \big) \to [\max_{i \in e} x_i - \min_{i \in e} x_i]$$

Other such techniques are surveyed and compared in [17, 19]. Any one of these approximations can be used in ComPLx.

S2. ADDITIONAL DISCUSSION OF THE FEASIBILITY PROJECTION in ComPLx

The feasibility projection $P_\mathcal{C}$ is defined in this work for a variety of placement constraints $\mathcal{C}$ and illustrated for (*a*) density constraints, (*b*) region constraints. The former is related to look-ahead legalization (LAL) in the original SimPL placer [23] and routability-driven variants in SimPLR and Ripple.

Comparision to prior work. Look-ahead legalization (LAL) was earlier used for macro placement in PolarBear [14] and SCAMPI [27]. In both cases, the main issue was the feasibility of macro packing within a given fixed outline. Both algorithms use top-down min-cut partitioning to minimize interconnect and need to check if each geometric partition is feasible. The result of this check is binary — a positive result for both partitions allows top-down partitioning to proceed, while a negative result for one of partitions triggers backtracking or end-case processing. The locations of macro blocks in a feasible placement are typically not used directly, therefore the LAL algorithm in PolarBear does not seek to optimize any objective. In contrast, SimPL does not deal with movable macros, and the main

concern for LAL in SimPL is to minimize total displacement from an initial solution, which has not been considered in PolarBear and SCAMPI. Since relevant algorithms in PolarBear and SCAMPI do not work with an initial solution, they cannot be considered *feasibility projections*. In other words, $P_\mathcal{C}$ in ComPLx and LAL in SimPL differ from prior work in that they pursue different goals — finding a closest feasible solution, rather than check packing feasibility. They are used in a different context (analytic placement vs. top-down min-cut placement), employ entirely different algorithms, and their results are interpretted differently (as anchors that influence the next iteration of analytic global placement). Whereas PolarBear and SCAMPI did not anticipate primal-dual Lagrange optimization in placement, the feasibility projection in ComPLx is a key element of the proposed primal-dual Lagrange formulation.

ComPLx also differs from PolarBear and SCAMPI in how it handles movable macros. Whereas prior work seeks to ensure the feasibility of macro placements at every step and sometimes sacrifices interconnect optimization for this, ComPLx pursues a different strategy based on macro shredding — it allows for temporarily overlapping macro placements and focuses on interconnect optimization.

Implementation details and analysis of properties. As pointed out in Section 5, the feasibility projection in ComPLx *generalizes* look-ahead legalization (LAL) in SimPL, SimPLR and Ripple to deal with macros and broader placement constraints. Here we restructure LAL so as to check its convexity and self-consistency.

Whereas LAL was defined recursively in [23, Section 4], the top-level structure of $P_\mathcal{C}$ in our description is that of alternating horizontal and vertical *spreading* passes. Each pass operates over a slicing floorpan, which gets refined between the passes. Specifically, spreading occurs only inside the rooms of the floorplan. For example, at the very first iteration, there is only one room, and one-dimensional spreading evens out the density. To formalize the problem solved by one-dimensional spreading from [23, Section 4], we note that relative placements are preserved. This justifies a change of variables: initial cell locations x_i (or y_i) are sorted, and the new variables $\delta_i \geq 0$ represent distances between neighboring x (y) locations, subject to $\Sigma_i \delta_i \leq W_x$ (or W_y) for a $W_x \times W_y$ floorplan room (a convex constraint). This linear change of variables preserves the convexity of the optimization objective (L_1-distance from given locations). The density constraint requires that for (some m and) all k, $\Sigma_{i=k}^{k+m} \delta_i$ is sufficiently large (based on cell sizes). This constraint

$$\min_k \{\Sigma_{i=k}^{k+m} \delta_i - (1/\sqrt{\gamma}) \Sigma_{i=k}^{k+m} \text{width(cell}_i)\} \geq 0$$

is convex since the minimum of downward convex (linear) functions is also downward convex (the $\geq$ sign is important).

As pointed out in [23, Section 4], after one-dimensional spreading, the median location should divide cell area evenly. Since this equalizes average densities on both sides, this new median location indicates a fixed point of the spreading transform. Hence, the walls of the slicing floorplan built by alternating one-dimensional spreading steps represent fixed lines. As slicing floorplan is gradually refined, the displacement affected by later steps of $P_\mathcal{C}$ rapidly decreases.

Self-consistency (Formula 11). To establish the self-consistency of P_C it suffices to independently establish the self-consistency of each horizontal and vertical pass [23, Section 4] in each room of the floorplan (see below). While the overall algorithm described above using alternating passes differs from LAL in [23], the results produced are essentially the same. The argument for self-consistency remains valid when the algorithm is applied multiple times.

The self-consistency of our P_C seems related to convexity, but in this work we only test it empirically on ISPD 2005 and 2006 benchmarks. Since the self-consistency condition of Formula 11 is transitive, we checked it between every two consecutive ComPLx iterations. Our implementation of the approximate feasibility projection P_C was self-consistent 96.0% and inconsistent 0.6% of the time, while the sufficient condition $\|(\vec{x}, \vec{y}) - P_C(\vec{x}, \vec{y})\|_1 > \|(\vec{x}', \vec{y}') - P_C(\vec{x}, \vec{y})\|_1$ for successive iterations was not satisfied only 3.3% of the time. Thus, the approximate feasibility projection P_C used by our implementation is approximately self-consistent. The convergence plots in Figure 1 do not show any disruptions that one would expect with a seriously inconsistent P_C. Inconsistencies mostly occur in the early global placement iterations (< 5) where projected feasible placements can differ significantly between consecutive iterations.

S3. THE SCALABILITY OF COMPLX

Figure 3 plots the final values of λ (solid red line) and the number of global placement iterations of ComPLx (dotted blue line with a greater range) on ISPD 2005 and 2006 benchmarks. The number of iterations correlates with the final λ value because each iteration increases λ by a limited amount until the final value is reached. All final values in our experiments are well below 1.0, and the iteration counts do not grow systematically with the size of the input. This phenomenon is consistent with the rapid convergence for which primal-dual Lagrange optimization is known. Given that ComPLx spends near-linear time $O(n(\log n)^p)$ per iteration [23], the overall runtime is near-linear as well. In comparison, the runtime of FastPlace is estimated as $\Theta(n^{1.38})$.

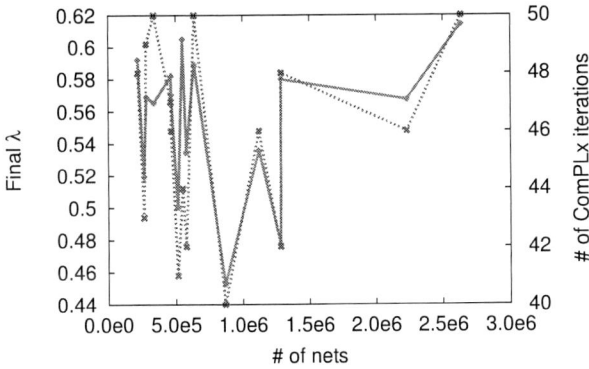

Figure 3: The final λ and total number of ComPLx iterations performed, against the number of nets.

S4. COMPARISONS TO RELATED PRIMAL-DUAL LAGRANGIAN OPTIMIZATIONS

Primal-dual optimization was used once in global placement in [4], where it was limited to explicit center-of-gravity (CoG) "spreading" constraints. These constraints appear in GORDIAN and GORDIAN-L algorithms [30], but not in modern placers — being convex and linear, they are insufficient to handle modern IC layouts (the 1997 implementation reported in [4] is not a full-fledged global placer). To deal with CoG constraints, [4] introduced slack variables, as is common in linearly-constrained primal-dual Lagrange optimization [15]. Instead, we deal with more general nonlinear, nonconvex constraints (such as fixed obstacles) by means of *approximate projected subgradient optimization*. Our primal-dual Lagrangian relaxation, in its basic form, requires only a single real-valued multiplier, making optimization very efficient. Unlike in [4, 15], we use the linear Conjugate Gradient method rather than the non-linear Newton's method.

Recent *operations research* work (unrelated to EDA) by Kiwiel et al [16] discusses *approximate subgradient projected optimization*, focusing on step-size selection and convergence analysis. Unlike prior projected subgradient methods, the Lagrangian relaxation in [16] finds both primal and dual solutions. This is also a key feature of our methods. However, [16] is not solving global placement and lacks numerous domain-specific details we described. *Vice versa*, we are not using their hallmark *ballstep strategy* that bundles multiple subgradient iterations.

S5. HANDLING REGION CONSTRAINTS IN THE FEASIBILITY PROJECTION

Chip designers often impose a region constraint on a subset of cells to express logic hierarchy and clock domains, to keep clock sinks close to clock drivers, or to assist the placer in dealing with challenging critical paths. Traditional placers convert the hard region constraints to soft constraints, which can be addressed by heavily-weighted fake nets [1, Figure 5] or modification of the objective function [13]. While ComPLx supports such techniques, it also allows for a more straightforward and robust implementation of region constraints by enforcing them as part of the feasibility projection at every global placement iteration – each cell is *snapped to* the constraining region after feasibility projection for density constraints. Figure 4 illustrates the enforcement of

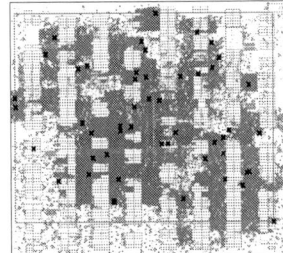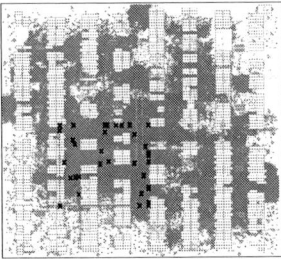

Figure 4: A hard region constraint imposed on 50 cells that were initially placed unconstrained (left). The resulting ComPLx placement (right) satisfies the constraint. HPWL drops from 143.55 to 142.70.

754

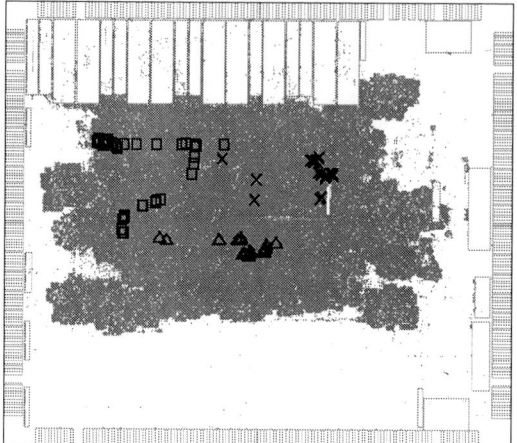

Unbiased (HPWL=94.25e6)   Net weights = 10 (HPWL=94.24e6)

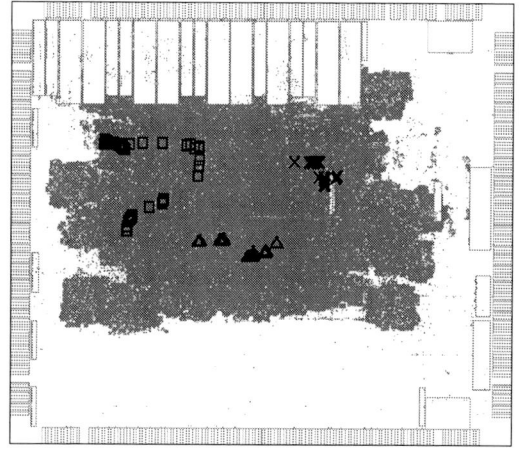
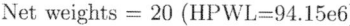
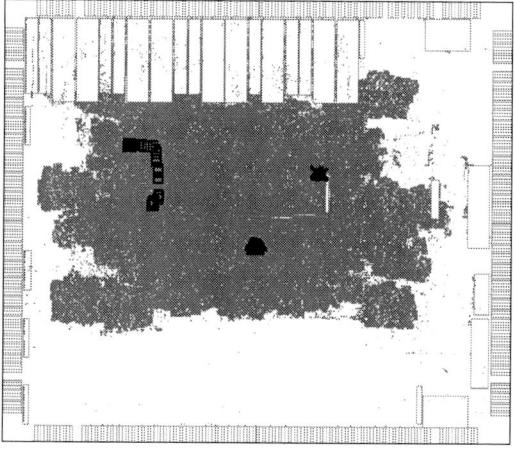

Net weights = 20 (HPWL=94.15e6)   Net weights = 40 (HPWL=94.13e6)

Figure 5: In a ComPLx placement of BIGBLUE1 **(upper left), three critical signal paths between registers are chosen. Subsequent ComPLx runs are performed with progressively larger net weights on those paths, which straightens the paths and reduces their lengths. Legal HPWL values are reported in parentheses.**

region constraints by "before" and "after" pictures. The locations from the modified feasibility projection are then used as anchors to influence the subsequent iteration of analytic global placement. Rather than degrade, HPWL actually improves — a surprising phenomenon often observed with industry placers.

S6. HANDLING TIMING-CRITICAL NETS

To demonstrate effective timing optimization, we show that timing-critical paths can be shortened and straightened by manipulating net weights without adverse effects on total HPWL. Working with the standard benchmark BIGBLUE1, we performed 30 global iterations to obtain an unbiased, stable intermediate placement that allowed us to estimate net lengths. We then selected several critical paths, increased

the weights of nets comprising these paths, and ran our placer to completion in three configurations with different net weights. Figure 5 shows that the desired outcome was achieved. With sufficiently large net weights, selected paths notably shrunk. Given that only a small fraction of net weights were modified, the overall placement and its wirelength were largely unaffected. Essential for these results was our scheduling of λ in Formula 12. While not a full-fledged demonstration of timing-driven placement, this experiment confirms that our proposed core placement algorithm is capable of controlling critical paths without tangible overhead in HPWL. Specific formulas for provably-good timing-driven net weighting can be found in [8]. As a side-effect, this experiment also demonstrates the *stability* of ComPLx to small netlist changes, which is important in the context of physical synthesis [1].

PADE: A High-Performance Placer with Automatic Datapath Extraction and Evaluation through High-Dimensional Data Learning

Samuel Ward, Duo Ding, David Z. Pan,
ECE Dept. The University of Texas at Austin, Austin, TX 78712
{wardsi}@utexas.edu, {ding,dpan}@cerc.utexas.edu

ABSTRACT

This work presents PADE, a new placement flow with automatic datapath extraction and evaluation. PADE applies novel data learning techniques to train, predict, and evaluate potential datapaths using high-dimensional data such as netlist symmetrical structures, initial placement hints and relative area. Extracted datapaths are mapped to bit-stack structures that are aligned and simultaneously placed with the random logic using SAPT [1],the SAPT, a placer built on top of SimPL [2]. Results show at least 7% average total Half-Perimeter Wire Length (HPWL) and 12% Steiner Wire Length (StWL) improvements on industrial hybrid benchmarks and at least 2% average total HPWL and 3% StWL improvements on ISPD 2005 contest benchmarks. To the best of our knowledge, this is the first attempt to link data learning, datapath extraction with evaluation, and placement and has the tremendous potential for pushing placement state-of-the-art for modern circuits which have datapath and random logics.

Categories and Subject Descriptors

B.7.2 [**Hardware, Integrated Circuits**]: Design Aids—*Placement and Routing*

General Terms

Design

Keywords

Datapath, Placement, Extraction, Physical Design

1. INTRODUCTION

Advancements in random logic placement have been impressive over the last few years with modern placers able to handle over one million placeable objects in minutes (e.g., [2]). Typically, these placers optimize the half-perimeter wire length (HPWL) objective standardized by the 2005 ISPD Placement contests [3] and is a good indicator of placement quality for random logic designs [4]. Unlike random logic, datapath logic generally is characterized by a high degree of bit-wise parallelism [5] (often called bit-stack) that modern placers have shown to be suboptimal [6]. This is partly due to the inaccuracy of the HPWL model when compared to the Steiner wire length (StWL) [1].

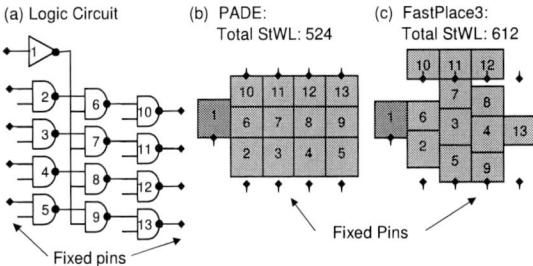

Figure 1: PADE placement example showing a 14% StWL improvement compared to FastPlace3 [7].

Figure 1 shows a toy example where modern placers are not able to handle datapaths effectively. Figure 1(a) displays the datapath circuit, where the input and output pins are fixed. Cell 1 is an inverter driving four $NAND2$ gates and there are three bit-stacks corresponding to cells: ({2, 3, 4, 5}, {6, 7, 8, 9}, and {10, 11, 12, 13}. For clarity, Fig. 1(b-c) display fixed pin locations. Fig. 1(b) displays the PADE placement solution solution. In this case, each bit-stack is tightly packed and aligned producing an StWL solution of 524. Figure 1(c) displays the placement solution from Fast-Place3 [7] where the bit-stack is not carefully aligned producing StWL of 612. In fact, with only thirteen cells, the StWL solution in 1(c) is over 14% worse than that in 1(b). In designs where there are many embedded datapaths, extracting the datapath and placing them with random logics properly has the potential for significant improvement in the overall StWL. Datapath extraction techniques in the past generally focused on functional or structural levels. Functional regularity extraction identifies logically equivalent subcircuits within a netlist that are then handled separately during placement. One example of this method is developed in [8] where a large set of templates are generated and used to search for datapath logic before placement. Another example is the hash-based approach of [9]. Structural datapath extraction techniques have focused on developing a regularity metric to represent the datapath. In [5], the datapath extraction consists of a decomposition of the netlist into a set of stages and a set of slices with one cell occurring in exactly one stage set and one slice set. The extraction algorithm expands in search-waves through the network using the regularity metric to determine the expansion direction. More recently, [10] developed a method for extracting structure within a design with the assumption that the placement distance between a pair of cells is related to the graph distance between them. Nets are weighted in a shortest path computation by assuming the distance between two cells is related to the degree of the net connecting them. Then, by extracting "corner" cells and fixing them in place, the maximum distance of the other cells can be calculated.

These previous datapath extraction algorithms, which only use functional or structure information, are not effective for modern large-scale hybrid datapath/random logic circuits with many pre-placed IP blocks: (1) the placement blockage could force the bit-stacks to be placed away from adjacent logic causing wirelegnth degradation; (2) it may give out too many or the wrong bit-stacks which would also adversely affect the overall placement quality. Compounding the problem, as shown in [11], a dedicated datapath placer often overly constrains the random logic placer. These problems lead to the general industry practice of manually designing the datapath because of the possible significant timing and wire-length improvement by careful alignment and packing of the bit-stack. However, increasing design sizes and shortening turn-around-time demand a consolidated automated datapath extraction and placement framework.

In this paper, we propose PADE, a new placement flow with automatic datapath extraction which can handle large scale designs mixed with random and datapath circuits. PADE evaluates and ranks all the first-order, important data paths, and optimizes them along with general-purpose wirelength driven placement[1]. Once extraction is complete, PADE uses the SAPT [1] placer (which extends the simPL [2] global placer) to simultaneously place datapath and random logics. The key contributions of this paper include: 1) We develop a novel high-dimensional data learning, extraction, and evaluation algorithm for datapath extraction in PADE. It considers not only logic structures, but also placement hints from initial global placement results. The flow is generic and can be applied to other state-of-the-art placers. 2) We develop an optimal algorithm for bit-stack selection (to guide data-path aware placement) using integer linear programming. 3) We show that PADE demonstrates significantly better results than previous state-of-the-art placers on both hybrid industrial designs which contain both random logics and datapaths, and even the ISPD 2005 placement benchmarks where structured datapath logics were not intended.

Section 2 outlines the overall flow and section 3 details the high-dimensional extraction data. Section 4 describes the model training and cluster evaluation and section 5 describes the binary integer programming bit-stack assignment technique. Experimental results are presented in section 6 followed by conclusions in section 7.

2. OVERALL PADE FLOW

Given a netlist $N = (V, E)$ with nodes V and nets E, placement obtains locations (x_i, y_i) for all movable nodes, such that the area of nodes within the placement boundary does not exceed the area of cell sites in that region. With $\vec{x}, \vec{y} = \{x_i, y_i\}$, HPWL is defined as: $HPWL(\vec{x}, \vec{y}) = HPWL(\vec{x}) + HPWL(\vec{y})$ where $HPWL(\vec{x}) = \sum_{e \subset E}[MAX x_i - MIN x_i]$. Modern placers often approximate HPWL by a differentiable function using the quadratic objective, defined as:

$$\Phi_G(\vec{x}, \vec{y}) = \sum_{i,j} w_{i,j}[(x_i - x_j)^2 + (y_i - y_j)^2] \qquad (1)$$

From Equation 1, (x_i, y_i) represents the coordinates of cell i, and $w_{i,j}$ represents the weight between cells i and j. A datapath netlist with p bit-stacks, each bit-stack B_k $0 < k < P$ is a disjoint set of cells $B_k \subset V$, describing the bit-wise parallelism present in the netlist. Representing the datapath as a set of cells in this manner enables implicit StWL optimization through forced alignment as presented in [1].

PADE is a new placement flow extending the SAPT [1] framework with novel autmatic datapath extraction. The flow was built

[1]The name PADE is inspired by the famous Pade approximation which is widely used in model order reduction, as they share the same principle to extract the first-order effects.

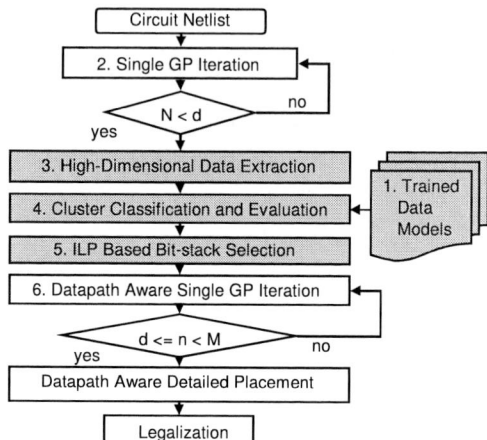

Figure 2: Overview of the PADE placement flow.

using the SimPL [2] force-directed global placer where $w_{i,j}$ is given by the Bound2Bound net model [12] and it utilizes a detailed placer similar to FastPlace3 [7].

Briefly, SAPT is a datapath aware placement flow requiring manual definition of both the bitstack and the datapath direction. For each bit-stack B_k, an alignment net is inserted, similar to a pseudo-net but remains persistent between placement iterations, connecting each cell in B_k. The alignment net is manipulated through the use of skewed weighting on $w_{i,j}$ and modified fixed-point insertion making it possible to introduce an alignment constraint to a predefined group of cells within the linear solver during global placement. A gradually increasing application of this constraint aligns each bit-stack through consecutive global placement iterations, which enables a unified placement framework that simultaneously places datapath and random logic cells without over constraining the placer. Additionally, during detailed placement the placer maintains that alignment by constraining cell movements along the bit-stacks.

The overall PADE flow is shown in Fig. 2 with the novel datapath training, extraction, evaluation and datapath bit-stack selection stages shaded. To properly handle the extraction of datapath structures, a compact and high performance knowledge base is proposed to identify datapath patterns from non-datapath logics and to evaluate the placement quality of these patterns. The knowledge base, step 1. in Fig. 2, is constructed via performing advanced data learning algorithms over a set of baseline design benchmarks after placement. By construction, it explores the placement database and captures the special characteristics that strongly correlate to datapath patterns, such as netlist connectivity automorphisms. Once these characteristics are captured and extracted, a complex decision diagram is built at a one time cost, which can later be applied to classify datapath netlists and non-datapath logics very rapidly. This knowledge base allows special treatment of the datapath logics in the placement stage without degrading the performance of the rest of the design. Additionally, these models are generic and can be applied to any circuit.

Step 2., a single global placement (GP) iteration, is one standard iteration of a force-directed placer that includes pseudo net insertion, linear system solver and fixed point generation. Let M, $0 \leq n \leq M$, be the upper bound on the global placement iterations and d an intermediate point at which time a prediction on the datapath will be made. Integrating the datapath extraction during global placement (GP) instead of before allows for enhanced prediction accuracy by taking into account physical characteristics in addition to netlist regularity measures.

Step 3 extracts the high-dimensional features from the netlist and step 4 classifies and then evaluates the datapath candidates. For all identified datapath logics, step 5 uses an ILP formulation to map cells to a bit-stack and then inserts the alignment nets. Step 6 performs datapath aware global placement with bit-stacks aligned during the following global placement iterations. The flow completes with datapath aware detailed placement and legalization as described in [1].

3. HIGH-DIMENSIONAL EXTRACTION

In this work, both graph-based and physical features are analyzed and extracted from the netlist mapping a set of parameters most critical and sensitive to datapath logics. Effective features create differentiation between random and datapath logic allowing the patterns extracted on the training set to effectively classify datapath structures in new circuits and predict the direction of the datapath. The first step in this process is to generate candidate clusters of the original netlist in which to search for datapath structures.

3.1 Seed-Based Connectivity Clustering

The connectivity based clustering stage prepares the data to analyze and extract datapath structures from. The goal is to find clusters exhibiting the structure we are looking for. Extending the seed growth method proposed in [13], the clustering method creates k clusters. It maximizes the ratio of the external to internal force of a cluster C_i, where C_i $0 \leq i < k$ indicates a group of vertices, while maintaining a maximum logic depth threshold. The external force is defined as the summation of the edge weights of nets with at least one vertex outside and one inside C_i and the internal force is defined as the summation of all internal cluster weight connections. The weight $w(u, j)$ is determined by the net model used, in this case a clique representation, where a connection c for a given edge e, $w(c) = w(e)/((|e| - 1)|e|)$. The connectivity between neighbor node u and cluster C_i is given by $conn(u, C_i) = \sum_{j \subset C_i} w(u, j)$ where suitable seed nodes are those with a large net degree.

In each subsequent pass, the neighbor node with the largest connectivity $conn$ is added to the cluster C_i while keeping the internal force of the cluster as large as possible. Once a cluster's node cardinality reaches the threshold value or the entire netlist is clustered, the high-dimensional features described in the next subsection are extracted from each cluster C_i and then the cluster is classified as a datapath or random logic cluster.

3.2 Automorphism Feature Extraction

In this section we describe the graph features used to differentiate datapath and random logic. One of the fundamental observations in this work is that datapath logic contains a high degree of graph automorphism. An automorphism of a graph, a form of symmetry, preserves the edge - vertex connectivity of the graph while mapping onto itself [2]. That is, an automorphism is a graph isomorphism from G to itself.

DEFINITION 1. *Automorphism: An automorphism of a graph $G = (V, E)$ is a permutation σ of the vertex set V, such that the pair of vertices (u, v) form an edge if and only if the pair $(\sigma(u), \sigma(v))$ also form an edge.*

Automorphism Group: The set of automorphisms of a given graph forms the automorphism group of the graph and is denoted by $Aut(G)$. The set $S \subseteq Aut(G)$ of generators for $Aut(G)$ is a set whereby combining elements of S generates every non-identity permutation in $Aut(G)$.

---

[2]Assuming reader familiarity with graph automorphisms and permutations. Please refer to [14] for further details.

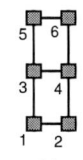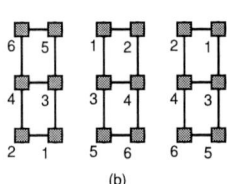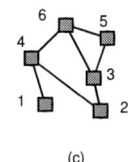

Figure 3: Graph automorphism example showing the original graph in (a) with each of the automorphisms in (b). A random netlist is shown in (c) with only trivial automorphisms.

DEFINITION 2. *Generator Set: A generator set of a group is a subset such that every element of the group can be expressed as the combination, under the group operation, of finitely many elements of the subset and their inverses.*

As an example, Fig. 3(a) displays a graph G with six labeled nodes and seven edges. The *automorphism feature* is represented with a seventeen parameter vector $(|Aut(G)|, \Delta(2:18),)$ where $|Aut(G)|$ is the cardinality of the automorphism group S_i for cluster C_i. The last sixteen parameters, $\Delta(2:18)$, are from the frequency table of the size of each automorphism.

As an example in Fig. 3, the graph G has a total of four automorphisms and two generators ($|S| = 2$, with $|Aut(G)| = 4$). The first automorphism, $G(1, 2, 3, 4, 5, 6)$, corresponds to itself and three additional automorphisms $G(2, 1, 4, 3, 6, 5)$ G flipped left-right, $G(5, 6, 3, 4, 1, 2)$ G flipped up-down, and $G(6, 5, 4, 3, 2, 1)$ G flipped left-right and up-down, displayed in Fig. 3(b). The nontrivial generator set S of G is $(1, 5)(2, 6)$ and $(1, 2)(3, 4)(5, 6)$. As this example shows, the symmetry of the graph along with the generator group provides possible bit-stack candidates including: $(1, 2)$, $(5, 6)$ or $(1, 3, 5)$, $(2, 4, 6)$.

Figure 3(c) displays a random logic netlist also with six nodes and seven edges. Unlike the clear symmetry present in Fig. 3(a), Fig. 3(c) contains no non-trivial automorphisms. In fact, this is a fundamental observation holding true for random logic netlists in general. Thus, the automorphism generators of structured logic appear very differently than the automorphism generators of random logic netlists enabling sufficient differentiation as a datapath feature.

3.3 Physical Aware Feature Extraction using Placement Hints

Graph automorphism features alone do not capture the physical nature of the placement problem, a fundamental shortcomming of prior extraction techniques. Global placement has merit in wirelength optimization, which shall be used for improved classification. Thus physical features extracted after the first few passes define the following attributes. Let a_i^c be the sum of the total cell area within cluster C_i, w_i^c be the bounding box width from the placement for C_i, h_i^c be the bounding box height and finally r_i^c be the ratio $a_i^c/(w_i^c + h_i^c)$. This physical information helps to characterize the amount of spreading and the initial cell locations for each C_i. Dense clusters indicate tightly packed logic and possibly the need for improved placement whereas sparse logic is generally less likely to improve from being passed to the datapath placer. In the next section, the training steps for building the model and the process to evaluate each cluster is described.

4. DATAPATH MODEL TRAINING AND CLUSTER EVALUATION

To classify and evaluate the datapath patterns in each cluster, we propose to combine data learning algorithms Support Vector Machine (SVM) and Neural Network (NN) to build compact and run-time efficient models as shown in Fig. 4. SVM calculates a hy-

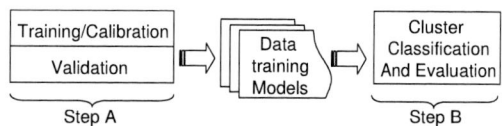

Figure 4: Major steps to build and apply the learning models

perplane boundary with maximum separation margin in-between of datapath and non-datapath. Only the critical information on the separation boundaries is preserved (the support vectors SV). All SV's are involved in the decision (score) calculation. For better quality, we combine a soft-error tolerant SVM and a special working set selection method [15]. NN works through configuring complex networks of neurons to achieve a high dimensional decision diagram-like data structure given training samples and decision hints. We employ a resilient backward propagation method based on iterative sub-gradient updates. To quantify the learning performance, we define the following two types of accuracies:

DEFINITION 3. *Datapath evaluation accuracy: the rate of correctly detected datapath (datapath-like) patterns over the total number of actual datapath structures.*

DEFINITION 4. *Non-datapath evaluation accuracy: the rate of correctly detected non-datapath (e.g., random logic) patterns over the total number of non-datapath structures processed.*

The optimization objective for both SVM and NN is to maximize the evaluation accuracies of datapath and non-datapath patterns, or equivalently, to minimize the mean square errors for both classes of pattern evaluation. This is achieved in two steps: Step A and B as shown in Fig. 4.

4.1 Training, Calibration and Validation

In Step A, we first apply data learning algorithms over a relatively small set of design patterns with known datapath information under the guidance of placement as hints. Since they are built a priori at a one time cost, the CPU run-time penalty is negligible. There are 3 major procedures involved in this step: (1) training is the process where the learning algorithms optimize both datapath and non-datapath accuracies; (2) calibration process further improves the accuracies, e.g., via properly selecting the separation threshold in (1); (3) validation process is performed over a relatively large set of known design patterns exclusive from (1) to assure the balance of learning accuracies between training data and unknown testing data, especially in Step B. These models can then be applied generically to any other designs to classify datapath clusters.

4.2 Cluster Classification and Evaluation

Once Step A is completed, in Step B the data learning models will be applied directly to classify and evaluate new unknown design patterns. As the new patterns go through the learning models, the evaluation scores could span within certain range for datapath and non-datapath patterns respectively for NN and SVM. In this step, we evaluate a pattern to be datapath like if and only if both NN and SVM evaluation scores are above certain thresholds. This helps to systematically improve the datapath evaluation accuracy without noticeable penalty in non-datapath accuracy. Usually NN and SVM have similar performance for most of binary classifications, e.g., differentiating datapath-like and non-datapath patterns. In principle, SVM guarantees the global optimum but is sensitive to data noise. NN usually has good noise-robustness, however it takes more time in the training and calibration step to reach optimal or close-to-optimal. Each C_i identified as datapath logic is passed to the bit-stack assignment in the next section.

5. BIT-STACK SELECTION WITH ILP

Once a cluster has been classified as containing structured logic, step 5 from Fig. 2 extracts the bit-stack structures from the logic clusters and passes those bit-stacks to the datapath placer. For each cluster C_i, a set of bit-stack candidates is chosen based on maximizing the total bit-stack count. This work uses the automorphism generators as the bit-stack candidates and adds in wirelength weighting to make the ILP formulation wire-length aware.

5.1 Bit-Stack Candidate List

Each generator set S_i, created during classification of each C_i, captures possible cell connections that can be used for a bit-stack assignment. Figure 3(a) provides an example for clarity. The generator group for the graph in Figure 3(a) is: $(1, 2)(3, 4)$, $(5, 6)$ and $(1, 5)(2, 6)$. Using this generator set, it is possible to create bit-stack candidates by grouping the tuple by index-0 and index-1 from each generator. For Fig. 3(a), the bit-stack candidates would be:

$$b_0 = [1:3:5] \qquad b_2 = [1:2]$$
$$b_1 = [2:4:6] \qquad b_3 = [5:6]$$

Thus, with a set of bit-stack candidates, the goal is to maximize the number of bit-stacks within the partition while maintaining mutual exclusion among the cells. This constraint maintains the requirement that a particular cell can not be assigned to multiple bit-stacks.

5.2 ILP-based Bit-Stack Selection

The bit-stack candidate selection is optimally solved using integer linear programming (ILP). A binary vector η is maximized with the linear function $\Gamma^{\mathbf{T}}(\eta)$ subject to the non-overlap constraint. Assuming there are $i = 1...n$ bit-stack candidates, let η be a binary indicator variable such that:

$$\eta_i = \begin{cases} 1 & \text{if bit-stack candidate } B_i \text{ is selected} \\ 0 & \text{otherwise} \end{cases} \qquad (2)$$

Let $\alpha_i = |B_i|$ and $\beta_i = w_i$ where w_i is equal to the Half-Perimeter Wire Length of the edges connected to each cell in bit-stack candidate B_i. The α_i term increases the value for larger bit-stack candidates (covering more cells) and the B_i term adds a penalty for larger wire-length. Then the objective function maximizes the number of bit-stack candidates selected as given in Equation 3.

$$\underset{\eta}{\text{maximize:}} \quad \Gamma = \sum_{i=0}^{n} [(\frac{\alpha_i}{\beta_i} * \eta_i)] \qquad (3)$$

$$\text{subject to:} \quad \eta_i + \eta_j \leq 1, \qquad \forall\, i, j \iff \eta_i \cap \eta_j$$
$$0 \leq i < j < n \qquad \forall\, i, j$$
$$\eta_i \in (0, 1) \qquad \forall\, i, j \qquad (4)$$

Equation 4 maintains the non-overlapping cell constraint $\eta_i \cap \eta_j = \emptyset$ between each bit-stack candidate. Though the general ILP problem is NP-Hard [16] and the solution time for the integer programming problem grows exponentially (in the worst case) with the number of integer variables, in this case the run time is negligible for two reasons: (1) The number of bit-stack candidates n from a single C_i and the number of constraints is generally very low, with n often on the order of a few hundred because we bound the size of C_i; (2) The ILP assignment only occurs when a cluster is classified as datapath logic meaning for the majority of the clusters, the ILP code does not run at all.

With the preceding steps from Fig. 2, PADE is able to quickly extract and classify datapath structures then pass them to an ILP solver to generate the bit-stacks for the logic. By making the classification and bit-stack assignment aware of physical placement information from the global placer, significant improvement in overall wirelength is possible as will be shown in the next section.

6. EXPERIMENTAL RESULTS

The first step, at a one-time cost, was training the high-dimensional models using known datapath pattern extracted from four baseline industrial hybrid circuits and the ISPD2011 Datapath Benchmark Suite [6]. Both datapath and random logic patterns were trained off the baseline circuits. Then the global placement flow was developed to extract and evaluate each C_i within the original netlist using the high-dimensional model. Clusters identified as containing datapath structures were mapped to bit-stacks using the automorphisms of the subcircuit and the ILP formulation. After the bit-stack was defined, global placement continued through completion and then detailed placement and legalization ran. All numbers reported are total wirelength results for both datapath and random logics on legal placement solutions.

PADE was implemented in C++ with g++ 4.1.2 on top of the SAPT placement flow. Running PADE without any datapath awareness results in the same wirelength results reported for SimPL because SAPT was built on the SimPL framework. Benchmark runs were performed on an Intel Xeon CPU x5570 Linux workstation running at 2.93GHz using two CPU cores. This work compared PADE against six *untrained* industrial hybrid designs and additionally on the *untrained* ISPD 2005 benchmark suite [3]. For improved experimental control, all HPWL numbers and StWL estimates were generated using coalesCgrip [17][3], every placer was run in *default mode*, and all placers were supplied a target density requirement of 1 as defined as in ISPD placement contests [3]. The tool bliss [14] was used to generate the automorphism groups for each cluster and GUROBI [18] for the lp solver. Wire-length results for the ISPD 2011 Datapath Benchmark circuits are not provided because they were used to train the high-dimensional models.

6.1 High Dimensional Learning Accuracies

We implemented and fine-tuned both SVM and NN algorithms specifically for the evaluation of datapath patterns. Then we combine both of their evaluation scores for datapath extraction. The NN accuracy is shown in Fig. 5 and SVM accuracy is shown in Fig. 6

A 2 class C-SVM algorithm is modified and configured at a one time training and calibration cost of around 3 minutes, involving: (1) training/calibrating of SVM models over some known structures with around 100 datapath and 10K non-datapath; (2) validation of the calibrated models over a relatively large set of known datapath/non-datapath structures beyond (1) with around 300 datapath and 60K non-datapath. Step(1) shows about 85% and 99.5% of datapath and non-datapath accuracy respectively, while Step(2) reaches 80% and 99% of datapath and non-datapath accuracy respectively. In the calibration and scoring process, a separation threshold of -0.9 is used for SVM models. A resilient backward propagation NN algorithm is fine-tuned within around 8 minutes using similar steps. It shows 90% and 99.9% of datapath and non-datapath accuracy in training, 87.8% and 99.6% in validation, with a separation threshold 0.05.

6.2 Wire Length Results

In the tables that follow, PADE refers to the proposed placement technique with automatic datapath extraction and evaluation. To compare the data learning and extraction effectiveness of PADE, we also implemented *Logic Based Regularity Extraction* (LBRE) based on [22]. Everything in LBRE is the same except the extraction and bit-stack assignment techniques. The logic based regularity extraction results are passed to the same datapath placer and compares the effectiveness of prior extraction techniques verses PADE. LBRE uses functional regularity to extract the datapath there-

---

[3]FastPlace3 [7] reports slightly lower HPWL than CoalesCgrip.

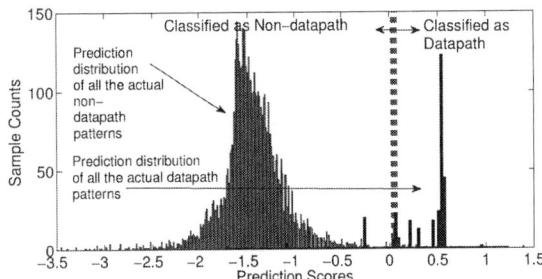

Figure 5: Validation accuracies of datapath and non-datapath by NN on relatively large data set

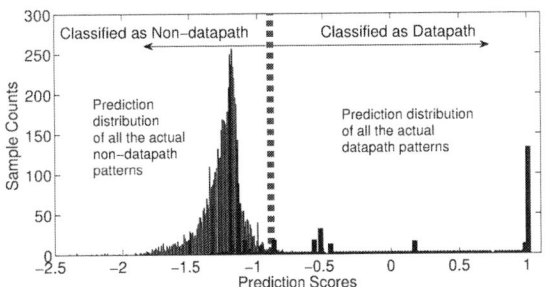

Figure 6: Validation accuracies of datapath and non-datapath by SVM on relatively large data set

fore can only be compared against the hybrid circuit designs because logical information is not provided in the ISPD 2005 benchmark circuits.

Wirelength results on six industrial hybrid circuits and the ISPD 2005 Placement Benchmarks are presented in Table 1. Each of the hybrid designs are state-of-the-art circuits containing a mixture of random and datapath logic. Though the exact ratio of datapath to random logic is not known, generally the significant majority of the logic is random. As discussed in [1], HPWL to StWL correlation can be inadequate for datapath logic. Thus, both HPWL and StWL is reported with the best StWL result in bold. In every case, PADE obtains the best StWL results. In four of the six cases, PADE also outperforms all other placers in HPWL results. The purpose of running on the ISPD 2005 placement benchmark suite is to show that the methods herein are capable of high placement quality on both random and datapath logics. Surprisingly, some datapath structure was found and on average PADE improves the HPWL 2% and StWL by 3% compared to prior academic placers. As Table 1 shows, PADE produced the best HWPL and StWL results for seven of the eight benchmarks. One notable placer missing from our comparisons is the structure aware Beacon placer [10]. Though requested, currently the placer does not work with mixed-size placement and thus direct comparison is not possible.

6.3 Runtime Comparisons

Table 2 compares the runtime of PADE against other state-of-the-art placers. For the hybrid and ISPD 2005 Benchmark circuits, FastPlace3.1 ran the fastest of all placers. For the hybrid circuits, PADE was only 19% slower than FastPlace3.1 and for the ISPD 2005 benchmarks, PADE was 32% slower than FastPlace3.1. Overall, PADE significantly outperforms CAPO10.2, mPL6, and NTU-Place3 showing speedups of 7.28x, 3.26x and 1.74x respectively on the ISPD 2005 Benchmarks. Though PADE is not the fastest, there is clearly wirelength benefit on hybrid design styles and it is possible to parallelize the clustering, evaluation and bit-stack assignment stages of the flow. Doing so would reduce runtimes to be similar with the other state-of-the-art placement algorithms.

Circuit	Capo10.5 [19]		mPL6 [20]		FastPlace3.1 [7]		NTUPlace3 [21]		SimPL [2]		LBRE		PADE	
	Total HPWL	Total StWL	Total HPWL	Total StWL	Total HPWL	Total StWL	Total HPWL	Total StWL	Total HPWL	Total StWL	Total HPWL	Total StWL	Total HPWL	Total StWL
Hybrid 1	2.39	3.12	2.27	2.89	2.06	2.61	2.04	2.67	2.19	2.75	2.32	2.89	2.05	**2.55**
Hybrid 2	1.72	2.51	1.47	2.17	1.39	2.20	1.38	2.20	1.39	2.20	1.39	2.18	1.37	**1.87**
Hybrid 3	2.68	2.75	1.89	2.41	1.81	2.28	1.77	2.25	1.77	2.25	1.79	2.26	1.71	**2.16**
Hybrid 4	2.66	3.57	3.18	4.01	2.91	3.59	2.35	3.36	2.69	3.30	2.79	3.49	2.36	**2.77**
Hybrid 5	11.36	12.90	12.76	14.41	10.88	13.27	10.59	12.30	10.57	12.22	10.56	12.22	9.71	**10.87**
Hybrid 6	7.66	9.06	9.04	10.29	7.75	9.04	9.04	10.69	6.64	7.92	6.90	8.21	6.24	**7.21**
Average	1.25	1.30	1.23	1.27	1.11	1.17	1.09	1.18	1.07	1.12	1.10	1.14	1.00	1.00
Adaptec1	88.14	97.22	77.58	86.20	79.88	88.75	81.82	91.06	78.15	87.05	-	-	76.83	**85.12**
Adaptec2	100.25	114.54	90.31	100.64	93.02	104.03	88.79	99.06	90.96	102.13	-	-	89.14	**98.92**
Adaptec3	276.80	296.22	215.88	235.06	219.78	239.70	214.83	234.52	208.81	228.32	-	-	205.32	**222.08**
Adaptec4	231.30	257.47	193.93	208.85	199.66	215.02	195.93	211.86	187.21	201.82	-	-	183.79	**196.23**
Bigblue1	110.92	127.72	97.10	108.31	94.37	**105.24**	98.41	110.02	98.64	109.94	-	-	95.86	106.98
Bigblue2	162.81	189.60	152.13	174.69	155.16	178.44	151.55	175.27	145.29	168.65	-	-	143.18	**164.33**
Bigblue3	405.40	452.91	342.50	370.70	392.72	421.31	360.66	389.39	341.55	369.61	-	-	341.72	**361.96**
Bigblue4	1016.19	1105.52	831.34	930.63	816.14	911.64	866.43	974.44	804.22	901.85	-	-	796.18	**883.82**
Average	1.21	1.22	1.03	1.04	1.06	1.07	1.05	1.06	1.02	1.03	-	-	1.00	1.00

Table 1: **Legal HPWL and StWL (x10e6) comparison on industrial hybrid designs and the ISPD 2005 Placement Benchmarks [3]. HPWL and StWL was computed using CoalesCgrip [17]. LBRE is blank for the ISPD 2005 suite because logic information is not provided for those circuits.**

	Capo10.2	mPL6	FP3.1	NTUPlace3	simPL	LBRE	PADE
hd1	7.4	1.9	1.3	1.5	1.2	2.1	1.1
hd2	8.1	2.4	1.4	1.7	2.0	2.7	1.7
hd3	8.8	4.5	1.7	2.7	2.2	4.2	1.8
hd4	8.6	4.1	2.2	3.2	2.2	6.3	2.3
hd5	25.2	10.7	6.4	9.8	5.8	12.8	7.4
hd6	45.4	22.8	4.8	9.3	4.1	14.2	7.1
Ave	**4.99**	**2.01**	**0.81**	**1.31**	**0.97**	**2.05**	**1.00**
ad1	35.7	18.3	4.7	10.0	4.2	-	5.3
ad2	42.8	19.9	2.2	9.2	4.4	-	5.6
ad3	111.9	60.3	4.4	18.6	10.7	-	12.9
ad4	110.0	58.5	9.0	19.5	18.1	-	21.4
bb1	56.6	21.8	5.4	16.2	4.5	-	5.5
bb2	107.6	64.0	9.6	32.1	17.3	-	20.4
bb3	286.0	88.4	28.3	62.5	34.8	-	40.6
bb4	543.4	172.8	58.1	141.8	62.3	-	73.0
Ave	**7.28**	**3.26**	**0.68**	**1.74**	**0.83**	-	**1.00**

Table 2: **The total runtime comparisons (sec). Runtimes on the ISPD 2005 benchmarks on LBRE are left blank because logical information is not provided by the ISPD 2005 benchmarks. (hd = hybrid, ad = adaptec, bb = bigblue, FP3.1 = FastPlace3.1)**

7. CONCLUSIONS

This work presented a new placement flow PADE with automatic datapath extraction and evaluation through high-dimensional data learning using both logical and physical information. PADE has demonstrated 7% improvements in HPWL and 12% improvements in StWL for a set of industrial hybrid circuits compared to prior placers. Even for the ISPD 2005 benchmark circuits, PADE produces 2% average improvements for HPWL and 3% improvement in StWL over prior placers. To our best knowledge, this is the first attempt that links high-dimensional data learning with placement of hybrid datapath and random logic circuits. The results are very encouraging and we believe a lot of future research can be done to further advance the state-of-the-art of modern placement.

8. ACKNOWLEDGMENTS

We would like to thank Myung-Chul Kim and Igor Markov for providing the SimPL binary and valuable feedback. This work is supported in part by IBM Faculty Award and Oracle.

9. REFERENCES

[1] S. I. Ward, M.-C. Kim, N. Viswanathan, Z. Li, C. Alpert, E. Swartzlander, , and D. Z. Pan, "Keep it straight: Teaching placement how to better handle designs with datapaths," in *Proc. ISPD*, 2012.

[2] M.-C. Kim, D.-J. Lee, and I. L. Markov, "simPL: an effective placement algorithm," in *Proc. ICCAD*, pp. 649–656, 2010.

[3] G.-J. Nam, C. J. Alpert, P. Villarrubia, B. Winter, and M. Yildiz., "ISPD 2005 placement contest benchmark suite," in *Proc. ISPD*, pp. 216–220, 2005.

[4] G.-J. Nam and J. Cong, eds., *Modern Circuit Placement: Best Practices and Results*. New York, NY: Springer, 2007.

[5] R. X. T. Nijssen and J. A. G. Jess, "Two-dimensional datapath regularity extraction," in *IFIP Workshop on Logic and Architecture Synthesis*, pp. 110–117, 1996.

[6] S. I. Ward, D. A. Papa, Z. Li, C. N. Sze, C. J. Alpert, and E. Swartzlander, "Quantifying academic placer performance on custom designs," in *Proc. ISPD*, pp. 91–98, 2011.

[7] N. Viswanathan, M. Pan, and C. Chu, "FastPlace 3.0: A fast multilevel quadratic placement algorithm with placement congestion control," in *Proceedings of ASPDAC*, pp. 135–140, 2007.

[8] A. Chowdhary, S. Kale, P. Saripella, N. Sehgal, and R. Gupta, "Extraction of functional regularity in datapath circuits," *IEEE TCAD*, vol. 18, no. 9, pp. 1279–1296, 1999.

[9] A. Rosiello, F. Ferrandi, D. Pandini, and D. Sciuto, "A hash-based approach for functional regularity extraction during logic synthesis," in *Proc. ISVLSI*, pp. 92–97, 2007.

[10] S. Ono and P. H. Madden, "On structure and suboptimality in placement," in *Proceedings of ASPDAC*, pp. 331–336, 2005.

[11] P. Ienne and A. GrieBing, "Practical experiences with standard-cell based datapath design tools," in *Proceedings of DAC*, pp. 396–401, 1998.

[12] P. Spindler, U. Schlichtmann, and F. M. Johannes, "Kraftwerk2 - a fast force-directed quadratic placement approach using an accurate net model," *IEEE TCAD*, vol. 27, no. 8, pp. 1398–1411, 2008.

[13] Q. Liu and M. Marek-Sadowska, "Pre-layout physical connectivity predictions with applications in clustering, placement and logic synthesis," in *Proc. ICCAD*, pp. 31–37, 2005.

[14] T. Junttila and P. Kaski, "Engineering an efficient canonical labeling tool for large and sparse graphs," 2007.

[15] R.-E. Fan, P.-H. Chen, and C.-J. Lin, "Working Set Selection Using Second Order Information for Training Support Vector Machines," in *Journal of Machine Learning Research*, 2005.

[16] J. E. Beasley, ed., *Advances in Linear and Integer Programming*. New York, NY: Oxford University Press, Inc., 1996.

[17] H. Shojaei, A. Davoodi, and J. Linderoth, "Congestion analysis for global routing via integer programming," in *Proc. ICCAD*, pp. 256–262, 2011.

[18] G. Optimization, "The gurobi optimizer 4.5." http://www.gurobi.com/.

[19] J. A. Roy, D. A. Papa, S. N. Adya, H. H. Chan, A. N. Ng, J. F. Lu, and I. L. Markov, "Capo: robust and scalable open-source min-cut floorplacer," in *Proc. ISPD*, pp. 224–226, 2005.

[20] T. F. Chan, J. Cong, J. R. Shinnerl, K. Sze, and M. Xie, "mPL6: enhanced multilevel mixed-size placement," in *Proc. ISPD*, pp. 212–214, 2006.

[21] T.-C. Chen, Z.-W. Jiang, T.-C. Hsu, H.-C. Chen, and Y.-W. Chang, "A high-quality mixed-size analytical placer considering preplaced blocks and density constraints," in *Proc. ICCAD*, pp. 187–192, 2006.

[22] T. Kutzschebauch and L. Stok, "Regularity driven logic synthesis," in *Proc. ICCAD*, pp. 439–446, 2000.

Structure-Aware Placement for Datapath-Intensive Circuit Designs *

Sheng Chou[1], Meng-Kai Hsu[1], and Yao-Wen Chang[1,2]

[1]Graduate Institute of Electronics Engineering, National Taiwan University, Taipei 106, Taiwan
[2]Department of Electrical Engineering, National Taiwan University, Taipei 106, Taiwan
chousheng@eda.ee.ntu.edu.tw; kaie@eda.ee.ntu.edu.tw; ywchang@cc.ee.ntu.edu.tw

ABSTRACT

Datapath is one of the most important components in high performance circuit designs, such as microprocessors, as it is used to manipulate all data. For better performance, a datapath is usually placed with high regularity and compactness. Although cell placement has been studied extensively, not much work addresses the optimization of datapaths which are often treated as big macros. In this paper, we propose a structure-aware placement algorithm that can exploit the regular structures of datapath circuits and meanwhile leverage effective techniques to achieve high quality and scalability. Our algorithm applies a nonlinear optimization for wirelength minimization and a sigmoid based density model for density control in datapath circuits. Compared with state-of-the-art works, our algorithm can achieve the best structure-aware placement results efficiently.

Categories and Subject Descriptors

B.7.2 [**Integrated Circuits**]: Design Aids [Placement and Routing]

General Terms

Algorithms, Performance

Keywords

Physical Design, Placement, Datapath

1. INTRODUCTION

Modern circuit designs, such as high-performance microprocessors and digital signal processors, can be divided into two parts: control unit and datapath. The control unit, which contains a finite state machine (FSM), determines the sequence of operations in the datapath while the datapath performs actual operations, such as addition and multiplication of two multiple-bit numbers.

Typically, datapath is implemented in bit-slice structures to manipulate multiple bits of data simultaneously [6]. The structures are also made up of several functional stages, such as registers, shifters, multiplexers, and arithmetic logic units (ALUs). For example, as shown in Figure 1, there are eight horizontal bit-slices and five vertical functional stages in the datapath circuit. Also, there are control signals for the functional stages: SEL, OP, and OE. The layouts of such regular structures are often placed in a similar fashion for better performance,

---

*This work was partially supported by IBM, SpringSoft, TSMC, and NSC of Taiwan under Grant No's. NSC 100-2221-E-002-088-MY3, NSC 99-2221-E-002-207-MY3, NSC 99-2221-E-002-210-MY3, and NSC 98-2221-E-002-119-MY3.

smaller wirelength, and higher layout density. Besides, a functional stage usually has high-degree control nets that should be aligned together. Therefore, in order to achieve high performance, it is of particular importance to consider the regular structures in datapath designs for better placement and routing results.

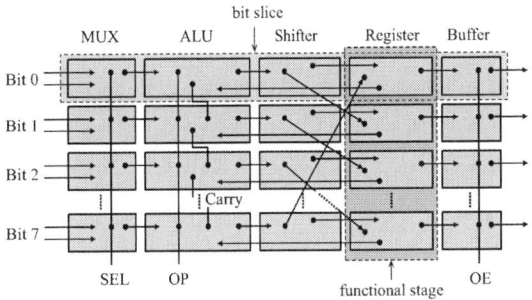

Figure 1: Example of a datapath circuit.

1.1 Previous Work

Traditionally, datapath design involved labor-intensive processes that could only be done manually. Although there were tools available to help designers construct datapath circuits, designing a datapath usually took a significant amount of time and human efforts. As a result, recent works have been proposed to automate the design process. These works can be classified into three major categories: (1) datapath synthesis, (2) regularity extraction, and (3) datapath placement. For datapath synthesis, the main objective is to preserve regularity and optimize performance [16]. Logic synthesis approaches that only consider logic minimization may lead to loss of regularity.

For regularity extraction, the objective is to identify a set of structures that repeat many times in a circuit netlist. Various techniques for regularity extraction have been proposed in the literature. Arikati et al. [4], Hirsch et al. [11], and Odawara et al. [23] proposed to choose a group of blocks of the same type as an initial reference stage, and then generate other stages by grouping neighboring blocks with similar properties, such as gate types and terminal information. Nijssen et al. [22] further extended the grouping idea to extract two-dimensional regularity in datapath circuits. Likewise, they applied the search wave method, which searches local information in a stage-by-stage manner. On the other hand, several approaches extract functional regularity from hardware description languages. In [9], given a template library, Corazao et al. performed template matching to cover a circuit.

For datapath placement, several existing works have been developed to optimize wirelength, area, routing track usage, and/or timing constraints. Most of these works formulate the datapath placement into a linear placement problem in which the main objective is to find the optimal ordering of functional stages within a datapath circuit. In [17], Luk et al. first partitioned circuits into stack (bit-slice) regions and performed placement by random swapping. Cai et al. [5] and Nakao et al. [19] presented an A*-searching-based algorithm to solve the linear placement problem. Munch et al. [18] and Kim et al. [15] solved the placement problem by integer linear programming (ILP). Yim et al. [29] proposed a hybrid approach that applies a genetic algorithm for global search and uses simulated annealing for fine tuning. In [28], Ye et al. performed quadratic placement and sliding window optimization to solve the linear placement problem. The quadratic placement finds an initial solution while the sliding window optimization refines results by local swapping.

1.2 Motivation

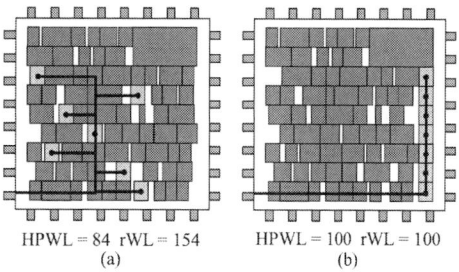

HPWL = 84 rWL = 154   HPWL = 100 rWL = 100
(a)   (b)

Figure 2: Example of the mismatch between HPWL and routed wirelength (rWL).

Placement is a very critical stage in the datapath design flow as it can make or break regularity, area, and timing constraints. In particular, placement can affect subsequent routing results considerably. To estimate routing wirelength, placement algorithms often use the half-perimeter wirelength (HPWL) model to estimate wirelenth. However, HPWL is not accurate enough particularly for high-degree nets. As shown in Figure 2, the HPWL of (a) is better than (b), but the routed wirelength of (a) is significantly worse than that of (b). Since datapath circuits usually have many high-degree control nets that connect blocks of functional stages, to remedy the insufficiency of HPWL, placement algorithms should exploit the regularity of datapath circuits to align blocks in functional stages for better routing results.

Thanks to the placement contests [20] held by the International Symposium on Physical Design (ISPD) in the past few years, modern placement algorithms have been greatly improved in terms of both quality and scalability. Academic placers can now solve large-scale problems with millions of standard cells in a short time. However, these state-of-the-art placers mainly focus on wirelength minimization in general designs and thus do not handle the regularity of datapath circuits. In addition, Ward et al. [27] recently demonstrated that modern academic placers tend to have significantly worse wirelength results for datapath designs, compared with custom solutions. Furthermore, there are few existing works that consider both regularity extraction and datapath placement. Moreover, a unified placement algorithm that considers both datapath and random logic (blocks with no regular structures) is still missing in the literature. Therefore, it is desirable to develop a new placement algorithm that can exploit the regular structures of datapath circuits and meanwhile utilize state-of-the-art techniques to achieve high quality and scalability.

1.3 Our Contributions

In this paper, we develop a high-quality structure-aware placement algorithm for datapath intensive circuit designs. We summarize the main contributions of our proposed algorithms as follows:

- A new structure-aware analytical placement algorithm is proposed. The proposed algorithm consists of three stages: (1) regularity extraction, (2) structure-aware global placement, and (3) structure-aware detailed placement. Our proposed algorithm considers the placement of both datapath and random logic.

- An effective datapath placement algorithm based on nonlinear optimization is proposed. We exploit the regular structure of datapath circuits to align functional stages of datapath circuits for better wirelength and performance. In addition, unlike the traditional datapath placement algorithms, which only consider the linear ordering of functional stages, we also consider the ordering of blocks in each functional stage. Moreover, to evenly spread blocks in datapath circuits effectively, we employ a novel sigmoid based density model for the proposed nonlinear formulation.

- Experimental results show that our proposed algorithm is effective and efficient. Compared with state-of-the-art placement algorithms, our algorithm can achieve the best structure-aware placement results efficiently. In particular, our placer generates the most competitive solutions, compared with manual designs in the literature.

The remainder of this paper is organized as follows. Section 2 gives the formulation of the structure-aware placement problem and introduces the hierarchical structure of datapath intensive circuits. Section 3 describes the overall flow of our proposed structure-aware placement algorithm. Sections 4, 5, and 6 detail the algorithm. Section 7 shows the experimental results. Finally, Section 8 concludes this paper.

2. PRELIMINARIES

We describe the structure-aware placement problem and the hierarchical structure of datapath-intensive circuits in this section.

2.1 Problem Formulation

The structure-aware placement problem can be formulated as a hypergraph $H = (V, E)$ placement problem. Let vertices $V = \{v_1, v_2, ..., v_n\}$ represent blocks and hyperedges $E = \{e_1, e_2, ..., e_n\}$ represent nets. Let x_i and y_i be the x and y coordinates of the center of block v_i, respectively. The blocks to be placed can be categorized into two types: pre-placed blocks and movable blocks. Pre-placed blocks have fixed x and y coordinates and cannot be moved. Different from traditional placement problems, pre-placed and movable blocks can be further classified into two types: (1) datapath, and (2) random logic in a datapath-intensive circuit. Section 2.2 will detail the hierarchical structure of a datapath-intensive circuit. We intend to determine the optimal positions of all movable blocks such that the target cost (e.g., wirelength) is minimized and there is no overlap among blocks. The placement problem is solved in two major steps: (1) global placement, and (2) detailed placement. Global placement distributes the blocks and finds the best position for each block while minimizing the target cost (e.g., wirelength). Detailed placement removes all overlaps and further improves the placement solution.

By dividing a placement region into a uniform non-overlapping bin grid, the global placement problem can be formulated as a constrained minimization problem as follows:

$$\min \quad W(\mathbf{x}, \mathbf{y}) \\ \text{s.t.} \quad D_b(\mathbf{x}, \mathbf{y}) \leq M_b, \quad \text{for each bin } b, \tag{1}$$

where $W(\mathbf{x}, \mathbf{y})$ is the wirelength function, $D_b(\mathbf{x}, \mathbf{y})$ is the potential function, which is the total area of movable blocks in bin b, and M_b is the maximum allowable area of movable blocks in bin b. M_b can be computed by $M_b = t_d(w_b h_b - P_b)$, where t_d is the user-specified target density value for each bin, w_b (h_b) is the width (height) of bin b, and P_b is the base potential equal to the pre-placed block area in bin b. Note that M_b is a fixed value as long as all pre-placed block positions are given and the bin size is determined.

To solve the above constrained minimization problem formulated in Equation (1), we can convert it into an unconstrained optimization problem by introducing a penalty multiplier λ. As a result, we solve a sequence of unconstrained nonlinear optimization problems of the form

$$\min \quad W(\mathbf{x}, \mathbf{y}) + \lambda \sum_b (D_b(\mathbf{x}, \mathbf{y}) - M_b)^2 \tag{2}$$

with increasing λ. The solution of the previous problem is used as the initial solution for the next one. The unconstrained problem in Equation (2) then can be solved by the conjugate gradient (CG) method. To apply the gradient search, the objective function must be smooth and differentiable everywhere. Therefore, we need to find such wirelength and density functions to be used for the above formulation.

To optimize wirelength during placement, $W(\mathbf{x}, \mathbf{y})$ is usually defined as HPWL. Since HPWL is not smooth and non-convex, several smooth wirelength approximation functions have been proposed. Both the log-sum-exp wirelength model [21] and the weighted-average wirelength model [12] give good approximations to HPWL.

To satisfy density constraints during placement, the potential function is defined as:

$$D_b(\mathbf{x}, \mathbf{y}) = \sum_{v \in V} P_x(b, v) P_y(b, v), \tag{3}$$

where P_x and P_y are the overlap functions of bin b and block v along the x and y directions, respectively. Since $D_b(\mathbf{x}, \mathbf{y})$ is neither smooth nor differentiable, smoothed density models, such as the bell-shaped function [14] can be applied. However, the bell-shaped function cannot approximate density accurately. In Section 5, we will introduce our proposed sigmoid-function-based density model, which is a more accurate smoothed approximation for the potential function above.

2.2 Hierarchical Structure of Datapath Circuits

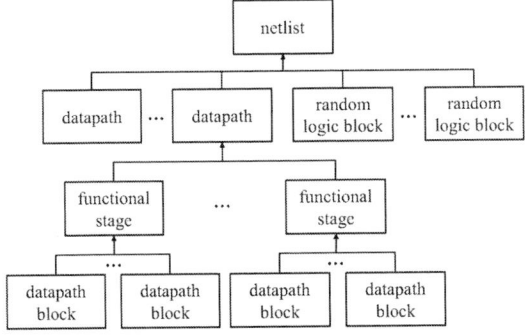

Figure 3: Hierarchical structure.

Unlike traditional circuit designs, in addition to random logic blocks, a datapath-intensive circuit might contain several datapaths. Figure 3 illustrates the hierarchical structure of a datapath-intensive circuit. As shown in Figure 3, a datapath consists of multiple functional stages, and a functional stage consists of multiple datapath blocks (e.g., ALU's, registers, and/or shifters). Since datapaths are usually intended to be placed regularly for better performance, the hierarchical structure of datapaths should be considered during placement. Regularity extraction, which finds the hierarchy of datapath circuits, will be explained in Section 4.

3. OVERALL FLOW

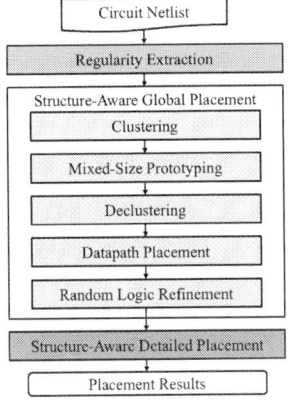

Figure 4: Overall flow.

In this paper, we propose a new structure-aware placement algorithm for datapath-intensive circuit designs. Figure 4 shows the overall flow of our proposed algorithm. Given a netlist, regularity extraction first identifies datapath blocks and the datapath hierarchy. Then, the structure-aware global placement distributes datapath blocks and random logic blocks evenly while minimizing the total wirelength. There are five steps in our structure-aware global placement: (1) clustering, (2) mixed-size prototyping, (3) declustering, (4) datapath placement, and (5) random logic refinement. Finally, the structure-aware detailed

placement places all blocks to legal placement positions and further improves the placement results. Figure 5 illustrates the main steps of our placement flow. We detail these steps in the following sections.

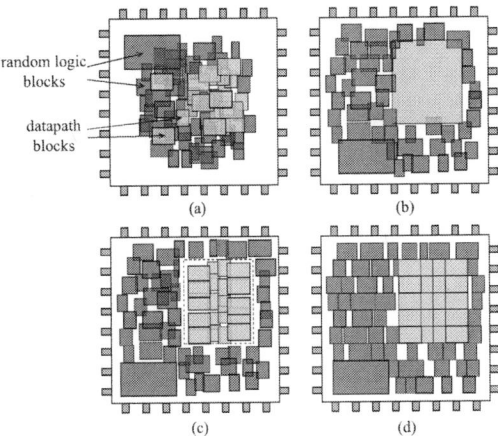

Figure 5: Main steps of our structure-aware placement flow. (a) Regularity extraction. (b) Mixed-size prototyping. (c) Datapath and random logic placement. (d) Detailed placement.

4. REGULARITY EXTRACTION

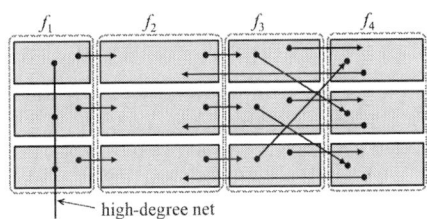

Figure 6: Example of regularity extraction.

To perform the structure-aware placement, it is necessary to obtain regularity information, which can be either specified by designers or derived by identifying the datapath hierarchy in a netlist. In this section, we introduce our regularity algorithm, which is based on the breadth-first search (BFS) algorithm. Given an initial reference stage that consists of blocks of the same type, the regularity extraction algorithm sequentially groups blocks of the same type to grow functional stages of a datapath. Assuming that f_{ref} is a reference stage with blocks $\{f_i | f_i \in f_{ref}\}$, at each time, we search all neighboring blocks that are connected to f_i's and classify these neighboring blocks according to their types. Then, the type with the number of blocks similar to the reference stage are grouped together to form new functional stages.

To identify all datapaths in a netlist, we use blocks connected by high-degree nets as initial reference stages. Note that, different types of blocks will form different initial reference stages. Our algorithm then uses a queue to record all identified stages. At first, the initial reference stages are pushed into the queue. Then, we iteratively pop a stage from the queue as a new reference stage and search neighboring blocks of the stage to form new functional stages. Once a new stage is identified, we push the new stage into the queue. The algorithm continues until there is no stage in the queue. If the number of identified stages, which indicates the regularity, exceeds a given threshold, the information of these identified stages and their corresponding datapath blocks are recorded to form a datapath. Figure 6 shows an example of identified functional stages in a datapath after regularity extraction. Note that our algorithm only identifies block memberships of stages in a datapath. Block orders in each stage and the order of

764

stages in each datapath are not determined. The placement algorithm that optimizes the positions of datapath blocks will be presented in the following sections.

5. GLOBAL PLACEMENT

In structure-aware global placement, we first cluster datapath blocks, and perform mixed-size prototyping to obtain an initial placement of datapaths and random logic blocks. Datapath placement then optimizes the positions of datapath blocks in each datapath. Finally, we refine the placement of random logic blocks to further improve the placement results.

5.1 Clustering and Mixed-Size Prototyping

In this stage, we first cluster datapath blocks as big macros (datapath macros for short). It should be noted that, datapath blocks in the same functional stages should be clustered first, then functional stages are clustered according to the datapath hierarchy identified by the regularity extraction in Section 4. The width and hight of a datapath macro are set to the total width of functional stages and the maximum height of functional stages in the datapath macro, respectively. Then, we perform mixed-size placement for datapath macros and random logic blocks by solving a nonlinear optimization problem of Equation (2) to obtain the initial placement of datapaths and random logic blocks.

5.2 Datapath Placement

After mixed-size prototyping, we decluster datapath macros, fix all random logic blocks, and place datapath blocks sequentially. To exploit the regular structure of datapaths, we apply a nonlinear optimization approach that can effectively spread blocks while utilizing regularity information of datapaths to align blocks of functional stages. In particular, we propose a novel sigmoid-function-based density model that is effective for block spreading. Unlike traditional placement problems that involve two dimensional block spreading, the horizontal spreading of functional stages and the vertical spreading of datapath blocks in each stage are independent, due to the regular structures of datapaths. As a result, we can optimize the block positions in each direction separately. Figure 7 shows our density model for the spreading. To compute the horizontal density of functional stages, we apply a set of horizontal segments in the datapath. To compute the vertical density of blocks in the functional stages, we apply multiple sets of vertical segments for each stage in the datapath.

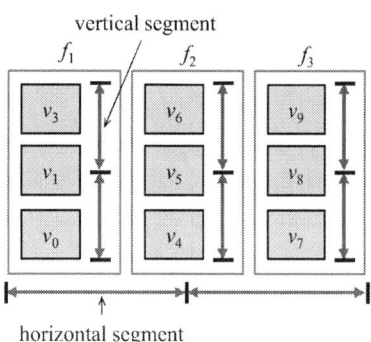

Figure 7: Density model of datapath.

5.2.1 Horizontal Optimization

We let blocks of each functional stage align vertically since datapath blocks in a functional stage usually share the same vertical control signal that goes through the stage. Therefore, in the horizontal direction, we let x-coordinates of blocks in each stage share the same variable. In this way, we can force blocks in stages to be aligned while reducing the variables for the optimization problem. To evenly distribute the stages in a datapath, the horizontal range of the datapath is divided into density segments that are used to compute the densities of a set of functional stages F in a datapath, as shown in Figure 7. We formulate

the horizontal optimization problem of datapath as follows:

$$
\begin{aligned}
\min \quad & W(\mathbf{x}_F) \\
\text{s.t.} \quad & D_s(\mathbf{x}_F) \le M_s, \quad \forall s \in S
\end{aligned}
\tag{4}
$$

where $\mathbf{x}_F$ denotes the vector of center x-coordinates of functional stages, $W(\mathbf{x}_F)$ denotes the wirelength function, S denotes the set of horizontal segments in the datapath, $D_s(\mathbf{x}_F)$ denotes the density function for segment s, and M_s denotes the expected density. Notice that the value of M_s is usually set to the width of the density segment s.

The wirelength function that computes HPWL in the x-direction is defined by

$$
W(\mathbf{x}_F) = \sum_{\text{net } e \in N} (\max_{v_i, v_j \in e} |x_i - x_j|),
\tag{5}
$$

where N is the set of nets that have at least one pin connected to the datapath. The density function is defined by

$$
D_s(\mathbf{x}_F) = \sum_{f \in F} O_x(s, f),
\tag{6}
$$

where f denotes a functional stage in F, and function $O_x(s, f)$ computes the horizontal overlap between segment s and stage f.

As mentioned in Section 2, to leverage nonlinear optimization, we have to make the wirelength and density functions smooth and differentiable. We apply the log-sum-exp wirelength model [21], which is a differentiable and effective approximation for HPWL. The log-sum-exp wirelength model in the x-coordinate is defined as:

$$
\hat{W}(\mathbf{x}_F) = \gamma \sum_{e \in N} (\log \sum_{v_k \in e} \exp(x_k/\gamma) + \log \sum_{v_k \in e} \exp(-x_k/\gamma)),
\tag{7}
$$

where γ is the parameter to control the smoothness and the approximation accuracy.

Since the density function $D_s(\mathbf{x}_F)$ is also not differentiable, we propose a sigmoid based density model to smoothly approximate the overlap function. The smoothed overlap function is defined as follows:

$$
\hat{O}_x(s, f) = \begin{cases} p(d + \frac{w_s}{2}) \cdot p(\frac{w_s}{2} - d) \cdot w_f, & \text{if } w_f \le w_s \\ p(d + \frac{w_f}{2}) \cdot p(\frac{w_f}{2} - d) \cdot w_s, & \text{if } w_f \ge w_s, \end{cases}
\tag{8}
$$

where function p is the sigmoid function, w_s is the width of segment s, w_f is the width of the functional stage f, and d is the center-to-center distance between segment s and stage f. In our implementation, we apply the quadratic sigmoid function [13], which is defined as follows:

$$
p(t) = \begin{cases} 1, & 0.5 \le \alpha t \\ 1 - 2(\alpha t - 0.5)^2, & 0 \le \alpha t \le 0.5 \\ 2(\alpha t + 0.5)^2, & -0.5 \le \alpha t \le 0 \\ 0, & \alpha t \le -0.5, \end{cases}
\tag{9}
$$

where α is the parameter for controlling the smoothness. Compared with the bell-shaped function, the sigmoid function gives a more accurate approximation to the original density overlap function. Also, the smoothness can be easily controlled by adjusting the parameter α. We illustrate the comparison of density models in Figure 8. The width of stage is 1, the width of segment is 2, and α of the function is 1.

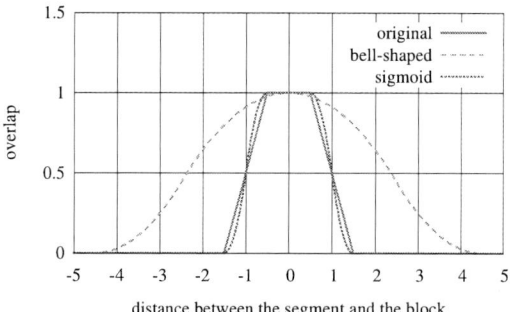

Figure 8: Comparison of density models.

To solve the nonlinear optimization problem, we apply the penalty method to transform the constrained nonlinear problem into an unconstrained problem:

$$\min \quad \hat{W}(\mathbf{x}_F) + \lambda \sum_{s \in S} (\hat{D}_s(\mathbf{x}_F) - M_s)^2 \qquad (10)$$

where functions $\hat{W}(\mathbf{x}_F)$ and $\hat{D}_s(\mathbf{x}_F)$ are for smoothed wirelength and density, respectively. In this way, we can solve the unconstrained nonlinear optimization problem by the CG method, and thus place the functional stages evenly while minimizing the wirelength in a datapath.

5.2.2 Vertical Optimization

Traditional datapath placement algorithms may not properly handle the blocks ordering in functional stages because they assume that the data flows of datapaths do not exchange between bit-slices. However, we observe that the data flows of datapaths do not necessarily go straightly through bit slices. The data can be actually exchanged in middle stages with cross connections among bit-slices. As a result, we should consider vertical block ordering and make the density even in each functional stages separately. To evenly distribute blocks, the vertical range of each stage is divided into vertical segments for computing density, as shown in Figure 7.

We formulate the vertical optimization of datapath as follows:

$$\begin{aligned} \min \quad & \sum_{f \in F} W(\mathbf{y}_f) \\ \text{s.t.} \quad & D_{s,f}(\mathbf{y}_f) \le M_{s,f}, \quad \forall s \in S_f, \forall f \in F \end{aligned} \qquad (11)$$

where $\mathbf{y}_f$ denotes the vector of center y-coordinates of blocks in stage f, $W(\mathbf{y}_f)$ denotes the wirelength function, S_f denotes the set of vertical segments in stage f, $D_{s,f}(\mathbf{y}_f)$ denotes the density function for segment s in stage f, and $M_{s,f}$ denotes the expected density in the respective segment and stage. Similar to the horizontal optimization, the wirelength function and density function is defined as follows:

$$W(\mathbf{y}_f) = \sum_{\text{net } e \in N} (\max_{v_i, v_j \in e} |y_i - y_j|) \qquad (12)$$

and

$$D_{s,f}(\mathbf{y}_f) = \sum_{v \in V_f} O_y(s, v) \qquad (13)$$

where v denotes the block, and V_f denotes the set of blocks in stage f, and function $O_y(s, v)$ computes the vertical overlap between segment s and block v.

Likewise, we apply the log-sum-exp wirelength model and our proposed sigmoid-function-based density model for the nonlinear optimization as follows:

$$\min \quad \sum_{f \in F} \hat{W}(\mathbf{y}_f) + \lambda \sum_{f \in F} \sum_{s \in S_f} (\hat{D}_{s,f}(\mathbf{y}_f) - M_{s,f})^2 \qquad (14)$$

where functions $\hat{W}(\mathbf{y}_f)$ and $\hat{D}_{s,f}(\mathbf{y}_f)$ are smoothed wirelength and density, respectively. Note that the blocks in a functional stage cannot be exchanged with the other stages. Thus, blocks contribute potential only to their respective stages. Similarly, we can solve the problem by the CG method, and thus place blocks of each functional stage with optimized wirelength.

5.3 Random Logic Refinement

After all datapaths are placed, we fix datapath blocks and refine the placement of random logic blocks to further improve the placement results. Likewise, we perform the mixed-size placement by solving the nonlinear optimization problem of Equation (2).

6. DETAILED PLACEMENT

In this section, we propose the structure-aware detailed placement to place all blocks, i.e., datapath and random logic blocks, to legal positions. The objective is to minimize the total displacements for legalization such that the quality of a given global placement result can be preserved. There are three major steps in our detailed placement: (1) datapath legalization, (2) random logic legalization, and (3) wirelength-driven refinement. We first legalize datapath blocks due to their strict structural constraints. Then, the other random logic blocks are legalized with datapath blocks being fixed. Finally, the placement results are further refined for better wirelength. We detail each step as follows.

6.1 Datapath Legalization

Since datapath circuits are usually placed regularly for better performance, smaller wirelength, and higher layout density, datapath blocks should be legalized earlier than other random logic blocks. Unlike traditional legalization techniques [10, 25] where the legal position for each block is searched and selected greedily, our proposed datapath legalization efficiently determines the positions of datapath blocks stage-by-stage by sorting the x- and y-coordinates of these blocks.

Given a global placement result of a datapath circuit, our proposed datapath legalization technique legalizes datapath blocks stage-by-stage. Since the datapath blocks in each stage is forced to be aligned during global placement as mentioned in Section 5, the horizontal positions of datapath blocks can be determined by sorting the x-coordinates of functional stages. Then, for each stage, we can sort the y-coordinates of datapath blocks to find their vertical positions. Figure 9 shows an example of our proposed datapath legalization. Given a global placement result of datapath circuits (see Figure 9 (a)), our proposed legalization technique legalizes datapath blocks horizontally and then vertically (see Figures 9 (b) and (c)). It should be noted that, it is also feasible if we perform the datapath legalization vertically and then horizontally. In our implementation, we legalize datapath blocks horizontally first.

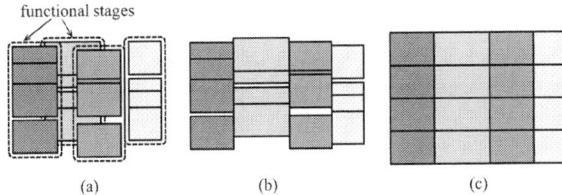

Figure 9: Illustration of the datapath legalization. (a) a given global placement result. (b) a horizontally legalized result. (c) a horizontally and vertically legalized result.

6.2 Random Logic Legalization

In the random logic legalization, we first fix datapath blocks that are legalized during the legalization. Then, traditional legalization techniques, such as Tetris [10] and Abacus [25], can be applied to legalize random logic blocks.

6.3 Wirelength-Driven Refinement

To further improve the legalized placement results, we apply traditional placement refinement techniques, such as cell matching and cell swapping [8] for wirelength optimization. Note that, we only exchange the positions of random logic blocks while keeping the regular positions of datapath blocks.

Table 1: Statistics of the placement benchmarks [3].

Circuit	#Cells	#Nets	#Pins	#Terminals
Design A	140928	174233	719904	19488
Design B	130944	148682	661340	21724

7. EXPERIMENTAL RESULTS

We conducted experiments to evaluate our proposed algorithm based on the structured placement benchmarks used in [3]. Table 1 shows the statistics of the benchmarks. We implemented our algorithm in C++ programming language, and the experiment were performed on a Linux workstation with Intel Xeon 2.4 GHz CPU and 4GB maximum allowable memory. We examined the quality of our structure-aware analytical placement algorithm by comparing with manual designs and the following state-of-the-art academic placers: APlace v1.0 [14], Capo v10.2 [24], FastPlace v3.0 [26], mPL6 v6.0 [7], and NTUplace3 v7.05.30 [8]. We used the script provided by [2] to verify the placement results, and coalesCgrip [1] to route each circuit. Note that the results of Capo, FlastPlace, and mPL6 are from [3], and the results of APlace and NTUplace3 from our experiments.

Table 2: Comparisons of placement results with state-of-the-art academic placers and the manual designs.

Placer	Design A					Design B				
	HPWL	HPWL Ratio	CPU (sec)	rWL	rWL Ratio	HPWL	HPWL Ratio	CPU (sec)	rWL	rWL Ratio
Manual	11000365	1.00	N/A	886390	1.00	8642097	1.00	N/A	881103	1.00
APlace	17837449	1.62	489	1732782	1.95	12196421	1.41	256	1642951	1.86
Capo	15945589*	1.45	1454	N/A	N/A	14381067*	1.66	1431	N/A	N/A
FastPlace	16336840*	1.49	195	N/A	N/A	N/A	N/A	N/A	N/A	N/A
mPL6	18290965*	1.66	N/A	N/A	N/A	N/A	N/A	N/A	N/A	N/A
NTUplace3	13640156	1.24	847	1392501	1.57	10829266	1.25	560	1530237	1.74
Ours	10890882	0.99	633	1005160	1.13	9028549	1.04	303	917146	1.04

* Illegal placement results where some blocks are overlapped. Routing for these illegal placement results were omitted.

Table 2 summarizes the experimental results, where HPWL, placement runtime, and routed wirelength (rWL) are compared. As shown in Table 2, our algorithm generated legal and high-quality placement results efficiently, compared with the state-of-the-art academic placers. In particular, our algorithm generates the most competitive placement solutions. Figure 10 illustrates the placement results of Design A generated by our algorithm without and with regularity considerations. Note that only partial layouts are shown due to the space limitation.

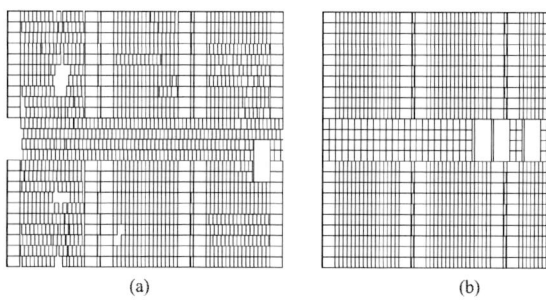

(a)   (b)

Figure 10: Placements of Design A. (a) without regularity consideration. (b) with regularity consideration.

8. CONCLUSIONS

This paper has presented a novel structure-aware placement algorithm for datapath-intensive circuit designs. Our proposed algorithm consists of three stages: (1) regularity extraction: extract regularity information of datapath circuits according to connectivity and block types, (2) structure-aware global placement: distribute datapath and random logic blocks evenly by adopting a new sigmoid-function-based density model, and (3) structure-aware detailed placement: place both datapath and random logic blocks to legal positions and further refine the placement results. Experimental results have shown that our algorithm can achieve the best structure-aware placement results efficiently, compared with state-of-the-art placement algorithms.

9. ACKNOWLEDGMENTS

We would like to thank Mr. Samuel I. Ward for providing us with the benchmarks and valuable help. We would also like to thank the authors of APlace and NTUplace3 for providing their binaries.

10. REFERENCES

[1] *coalesCgrip: A Tool for Routing Congestion Analysis.* http://homepages.cae.wisc.edu/~adavoodi/gr/cgrip.htm.

[2] *ISPD 2006 Placement Contest.* http://www.sigda.org/ispd2006/contest.html.

[3] *Structured Placement Benchmark Suite.* http://www.cerc.utexas.edu/utda/download/DP/.

[4] S. R. Arikati and R. Varadarajan. A signature based approach to regularity extraction. In *Proc. of ICCAD*, pages 542–545, 1997.

[5] H. Cai, S. Note, and H. D. Man. A data path layout assembler for high performance dsp circuits. In *Proc. of DAC*, pages 306–311, 1990.

[6] T. Chan, A. Chowdhary, B. Krishna, A. Levin, G. Meeker, and N. Sehgal. Challenges of cad development for datapath design. In *Intel Technology Journal, Q1*, 1999.

[7] T. Chan, J. Cong, J. Shinnerl, K. Sze, and M. Xie. mPL6: Enhanced multilevel mixed-size placement. In *Proc. of DAC*, pages 212–214, 2006.

[8] T.-C. Chen, Z.-W. Jiang, T.-C. Hsu, H.-C. Chen, and Y.-W. Chang. NTUplace3: An analytical placer for large-scale mixed-size designs with preplaced blocks and density constraints. *IEEE Trans. on CAD*, 27(7):1228–1240, 2008.

[9] M. R. Corazao, M. A. Khalaf, L. M. Guerra, M. Potkonjak, and J. M. Rabaey. Performance optimization using template mapping for datapath-intensive high-level synthesis. *IEEE Trans. on CAD*, 5(8):877–888, 1996.

[10] D. Hill. US patent 6,370,673: Method and system for high speed detailed placement of cells within an integrated circuit design. 2002.

[11] M. Hirsch and D. Siewiorek. Automatically extracting structure from a logical design. In *Proc. of ICCAD*, pages 456–459, 1988.

[12] M.-K. Hsu, Y.-W. Chang, and V. Balabanov. TSV-aware analytical placement for 3D IC designs. In *Proc. of DAC*, pages 664–669, 2011.

[13] M.-K. Hsu, S. Chou, T.-H. Lin, and Y.-W. Chang. Routability-driven analytical placement for mixed-size circuit designs. In *Proc. of ICCAD*, pages 80–84, 2011.

[14] A. B. Kahng and Q. Wang. Implementation and extensibility of an analytic placer. *IEEE Trans. on CAD*, 24(5):734–747, 2005.

[15] J. Kim and S. M. Kang. A timing-driven data path layout synthesis with integer programming. In *Proc. of ICCAD*, pages 716–719, 1995.

[16] T. Kutzschebauch and L. Stok. Regularity driven logic synthesis. In *Proc. of ICCAD*, pages 439–446, 2000.

[17] W. K. Luk and A. A. Dean. Multi-stack optimization for datapath chip (microprocessor) layout. In *Proc. of DAC*, pages 110–115, 1989.

[18] M. Munch, N. When, and M. Glesner. Optimum simultaneous placement and binding for bit-slice architectures. In *Proc. of ASPDAC*, pages 735–740, 1995.

[19] H. Nakao, O. Kitada, M. Hayashikoshi, K. Okazaki, and Y. Tsujihashi. A high density datapath layout generation method under path delay constraints. In *Proc. of CICC*, pages 9.5.1–9.5.5, 1993.

[20] G.-J. Nam. ISPD 2006 placement contest: Benchmark suite and results. In *Proc. of ISPD*, pages 167–167, 2006.

[21] W. C. Naylor, R. Donelly, and L. Sha. US patent 6,301,693: Non-linear optimization system and method for wire length and delay optimization for an automatic electric circuit placer. 2001.

[22] R. X. Nijssen and J. A. Jess. Two-dimensional datapath regularity extraction. In *Proc. of IFIP*, pages 110–117, 1996.

[23] G. Odawara, T. Hiraide, and O. Nishina. Partitioning and placement technique for CMOS gate arrays. *IEEE Trans. on CAD*, 6(3):355–363, 1987.

[24] J. Roy, D. Papa, A. Ng, and I. Markov. Satisfying whitespace requirements in top-down placement. In *Proc. of ISPD*, pages 206–208, 2006.

[25] P. Spindler, U. Schlichtmann, and F. M. Johannes. Abacus: Fast legalization of standard cell circuits with minimal movement. In *Proc. of ISPD*, pages 47–53, 2008.

[26] N. Viswanathan, M. Pan, and C. Chu. Fastplace 3.0: A fast multilevel quadratic placement algorithm with placement congestion control. In *Proc. of ASPDAC*, pages 135–140, 2007.

[27] S. Ward, D. A. Papa, Z. Li, C. Sze, C. Alpert, and E. Swartzlander. Quantifying academic placer performance on custom designs. In *Proc. of ISPD*, pages 91–98, 2011.

[28] T. T. Ye and G. D. Micheli. Data path placement with regularity. In *Proc. of ICCAD*, pages 264–271, 2000.

[29] J.-S. Yim and C.-M. Kyung. Datapath layout optimization using genetic algorithm and simulated annealing. In *Proc. of CDT*, pages 135–141, 1998.

GLARE: Global and Local Wiring Aware Routability Evaluation

Yaoguang Wei[1], Cliff Sze[2], Natarajan Viswanathan[3], Zhuo Li[2], Charles J. Alpert[2], Lakshmi Reddy[4],
Andrew D. Huber[4], Gustavo E. Tellez[5], Douglas Keller[4], Sachin S. Sapatnekar[1]

[1]Department of Electrical and Computer Engineering, University of Minnesota, Minneapolis, MN, USA
[2]IBM Austin Research Lab, Austin, TX, USA
[3]IBM Systems and Technology Group, Austin, TX, USA
[4]IBM Systems and Technology Group, Hopewell Junction, NY, USA
[5]IBM Systems and Technology Group, Burlington, VT, USA
Email: weiyg@umn.edu; {csze,nviswan,lizhuo,alpert,reddyl,adhuber,tellez,kellerd}@us.ibm.com; sachin@umn.edu

ABSTRACT

Industry routers are very complex and time consuming, and are becoming more so with the explosion in design rules and design for manufacturability requirements that multiply with each technology node. Global routing is just the first phase of a router and serves the dual purpose of (i) seeding the following phases of a router and (ii) evaluating whether the current design point is routable. Lately, it has become common to use a "light mode" version of the global router, similar to today's academic routers, to quickly evaluate the routability of a given placement. This use model suffers from two primary weaknesses: (i) it does not adequately model the local routing resources, while the model is important to remove opens and shorts and eliminate DRC violations, (ii) the metrics used to represent congestion are non-intuitive and often fail to pinpoint the key issues that need to be addressed. This paper presents solutions to both issues, and empirically demonstrates that incorporating the proposed solutions within a global routing based congestion analyzer yields a more accurate view of design routability.

Categories and Subject Descriptors

B.7.2 [**Hardware**]: Integrated Circuit—*Design Aids - Routing*

General Terms

Algorithms, Design, Experimentation, Measurement

Keywords

Physical design, Routing, Routability evaluation, Local wiring modeling, Congestion metric

1. INTRODUCTION

Routability has become an increasingly important and difficult issue in nanometer-scale VLSI designs, and must be addressed across the entire physical synthesis tool stack. This in turn requires fast, yet reasonably accurate techniques to identify routing-challenged regions (hot spots), for routability optimization. This work focuses on the two key components of routability evaluation: (a) the method used to analyze the congestion of a given placement or design point, and (b) the metric(s) used to score or represent the congestion.

1.1 Congestion analysis techniques

Proposed methods to perform congestion analysis include:
1. Take a design through detailed routing to determine if it is routable or not.
2. Use a probabilistic congestion estimation procedure, without performing any routing [6, 13].

3. Perform fast global routing and use its solution to perform congestion analysis [10–12].

In principle, detailed routing estimates are the most accurate, but this approach is very time-consuming and impractical during the early stages of design closure. Probabilistic methods are highly inaccurate and fail to capture the behavior of global routing, especially in modern designs with numerous IP blockages, and a large number of metal layers with varying width and spacing. Lately, the third method has become more attractive and mainstream due to the advent of fast, high-quality global routers [3, 4, 7, 14, 15].

Although global routing based congestion analysis provides a happy medium between probabilistic analysis and detailed routing, it suffers from a key drawback: local congestion, or local routing resource usage is not accounted for during global routing. Local congestion clearly consumes varying amounts of routing resources depending on factors such as design rules, size of the global routing cell and pin density. As shown in Section 5, ignoring these effects can greatly mispredict design routability. Hence, local congestion needs to be modeled; and the method should be *flexible*, for it to be adjusted in a straightforward manner from one technology to the next (as design rules are different for each technology).

1.2 Metrics to score or represent congestion

Visual inspection of congestion plots (a region-wise color-coded map marking out the hot spots with the greatest contention for wiring resources) often serves as a first order method to compare the routability of different design points. However, optimization tools and designers also require a single metric that can accurately score or represent the design congestion.

Commonly used metrics in academia and industry are:
Overflow based metrics include total overflow (TOF) and maximal overflow (MOF) that measure the excess of the routing demand over routing capacity on the edges in a global routing graph (defined in Section 2). These metrics often fail to provide a clear picture of the design routability. For example, a global routing solution with an overflow of 700 might still be routable, as the overflow could be absorbed by neighboring regions or resolved during the subsequent routing phases. Further, their lack of intuition (e.g., how good/bad is an overflow of 14, 253?) makes it difficult to quantify how much better one design point is versus another.

Net congestion based metrics [2] include[1]: (a) $ACN(x)$, the average net congestion, defined as the average congestion of the top $x\%$ congested nets, where the congestion of a net is the maximum congestion among all the global routing edges traversed by the net. (b) $WCI(y)$, the worst congestion index, defined as the number of nets with congestion greater than or equal to $y\%$. In practice, $ACN(20)$, $WCI(90)$ and $WCI(100)$ are employed. The main issue with these metrics is that they fail to differentiate between a net spanning a single congested global routing edge and one that spans multiple congested edges.

In this paper, we propose to enhance the accuracy and effectiveness of routability evaluation. Our key contributions include:

---

[1]We name the metrics differently from [2] to facilitate later references.

- A study of the inaccuracies in existing global routing based congestion analyzers, specifically due to the lack of local routing resource modeling.
- An analysis of the weaknesses in existing metrics to score or represent design congestion.
- Methods to model and incorporate the effects of local routing resource usage during global routing, the impact of which is two-fold: (a) significant improvement in the accuracy of congestion analysis, (b) better prediction of detailed routing issues such as opens and shorts.
- A new congestion metric that is more intuitive and represents the design congestion with high fidelity.
- Detailed empirical validation of our proposed techniques on advanced industrial designs.

The rest of this paper is organized as follows. Background and definitions are presented in Section 2. Section 3 presents our methods for modeling local routing congestion. Section 4 describes our new metric for routability evaluation. Empirical validation and concluding remarks are provided in Sections 5 and 6, respectively.

2. PRELIMINARIES

Typically, during global routing, the chip is tessellated into $n_r \times n_c$ grids (or *g-cells*), and the global routing graph (GRG), $G = (V, E)$, is constructed. A node in V represents a g-cell in the layout, and an edge (called a *g-edge*) in E denotes the boundary between two adjacent g-cells. An example of the GRG is shown in Figure 1.

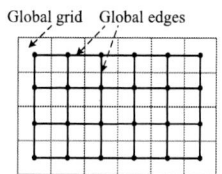

Figure 1: Global routing graph (GRG).

We now introduce some notation and terms that will be used in the remainder of this paper. For each edge e in the GRG, we define c_e as edge capacity – the total or maximal capacity of the edge, b_e as edge blockage, that needs to be discounted from c_e, and w_e as the routing demand on the edge. In global routing, c_e, b_e and w_e are generally expressed in the number of routing tracks, where a routing track is the routing resource taken by a single wire passing through an edge in the GRG. Let $o_e = \max(w_e + b_e - c_e, 0)$ be the overflow of an edge e. The total overflow of the layout is given by $\sum_{e \in E} o_e$, and the maximal overflow is given by $\max_{e \in E} o_e$. The congestion of edge e, denoted as g_e, is given by $g_e = (w_e + b_e)/c_e$.

3. MODELING OF LOCAL ROUTES

In this section, we first analyze the problems associated with existing congestion analysis methods, and then propose our model that captures the congestion due to local routes.

3.1 Limitations of existing global routing based methods

As mentioned in Section 1, global-routing based congestion analysis is now mainstream. Examples include, FastRoute [9] and NTHU-Route 2.0 [3], used as congestion analyzers within routability-driven placers IPR [10] and CRISP [11], respectively.

Global routers generally abstract the routing problem and only focus on g-cell-to-g-cell routes. In this case, they ignore the congestion due to local routes connecting the pins inside a g-cell. The problem is shown in Figure 2. The net (S, T) is a local net with two pins in a g-cell that are to the left of g-cell center b, and will definitely occupy some routing resources. However, due to the abstraction in global routers, the routing resources occupied or blocked by the local route connecting S to T are usually not modeled in congestion calculation.

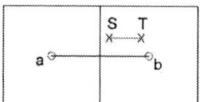

Figure 2: Local routes ignored by some global routers.

On the other hand, one of the major objectives of commercial global routers is to correlate with detailed routing. Designers expect global routers to report routing errors (usually in the form of congestion hot spots) without running detailed routing for hours or days. Therefore, industrial global routers have methods of varied sophistication to consider local routes. A simple approach is to reduce the g-cell size. Alternatively, some global routers include some form of detailed routing. However, these approaches to consider local routes greatly increase routing runtime and memory. When using global routers as congestion analyzers during physical synthesis, such routers are computationally expensive for tens or hundreds of invocations.

Our analysis in Section 5 shows that without considering local routes, a congestion analyzer significantly underestimates the congestion, and is unable to predict the problematic regions with opens and shorts in detailed routing. This motivates our work to introduce a simple and fast method to consider local routes when using a global router for congestion analysis.

3.2 Fast methods for modeling local routes

In this section, we present two methods to quickly model local routes: the first method estimates the local routes by the Steiner tree wirelength inside each g-cell, and the second method estimates the local routes based on the pin density of each g-cell.

3.2.1 Method 1: Estimation of local routes based on Steiner tree wirelength

In Figure 2, we observed that the longer the local route is, the more it would block the global routing track on that g-edge. This observation can be formulated by the following equation:

$$t_b = l_r/s_e, \qquad (1)$$

where t_b is the number of routing tracks blocked by local route (S, T), l_r is the length of the local route, and s_e is the length of the g-edge. Note that this equation just serves as the basis for our method and will be extended in later use.

One may argue that if the detailed router somehow does not route net (S, T) along g-edge (a, b) (which is possible if the router uses a detour), blocking g-edge (a, b) leads to pessimistic congestion maps. However, we usually have no way to determine how a router would connect a pin until the routing is completed. In order to perform fast congestion analysis, it is critical to model local routes in a manner that is independent of the process of routing. Moreover, it is generally a good practice for congestion analysis to be pessimistic for the sake of routing closure.

We adopt equation (1) to calculate blocked tracks on a g-edge because we can easily extend it to include local routes when there are multiple pins covered by a g-cell. Consider the case when all the pins of a net are within a g-cell. To estimate local routing, we first build a Steiner tree (alternatively, one can use minimum spanning tree) for the pins. In our experiments, we use Flute [5]. We then break each horizontal tree segment into two based on the x-coordinate of the g-cell center and apply equation (1) to calculate blocked global routing tracks for the g-edges on each side of the g-cell boundaries. Similarly each vertical Steiner tree segment can be broken using the y-coordinate of the g-cell center.

An example is shown in Figure 3, where net (A, B) is a two-pin net while (A, J) and (B, J) are the two segments of a Steiner tree. The global routing tracks blocked by net (A, B) in the horizontal direction can be calculated based on segment (A, J). Since the g-cell center b is between A and J, segment (A, J) blocks global routing tracks on g-edges (a, b) and (b, c). The blocked tracks on g-edge (a, b) can be calculated as $(x_b - x_A)/(x_b - x_a)$, where x_b denotes the x-coordinate of g-cell center b, and other notations are defined similarly. Accordingly, the blocked global routing tracks on g-edge (b, c) is $(x_J - x_b)/(x_c - x_b)$. The vertical tracks blocked by net (A, B) can be calculated similarly, based on segment (J, B). As another example, when a segment is completely on the left of (above) or right of (below) the g-cell center, such as the net (C, D) in Figure 3, the blocked tracks can be calculated as $(x_D - x_C)/(x_c - x_b)$. In this case, only the tracks on g-edge (b, c) are blocked.

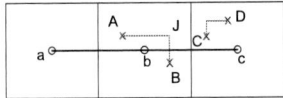

Figure 3: Local routing resource estimation for two-pin nets.

The proposed method can be easily applied to more complex Steiner trees, for example A, B, C in Figure 4. However, when a net has pins that reside in different g-cells, we must account for the synergy between global and local routes. An example is the net D, E, F, G in Figure 4. As mentioned previously, it is impractical for congestion evaluation to wait until the completion of global routing in order to consider local routes. To simplify our algorithm, we assume that all global routes connect to the center of a g-cell. In this case, we can include the g-cell center as a dummy pin when constructing the Steiner tree to model the local routes. For example, the Steiner tree connecting b, E, F, G is used to calculate the blocked global routing tracks on the four boundaries of g-cell with center b. Similarly, the Steiner tree connecting a, D is used to calculate the blocked tracks corresponding to g-cell with center a. Note that this may cause over-estimation in some cases when connection from D to E, F, G takes fewer tracks than that from a to E, F, G, e.g., D is at the bottom-left corner of the g-cell. To account for this factor, we introduce a parameter p to scale the estimated local resources, where p will be tuned empirically for each technology.

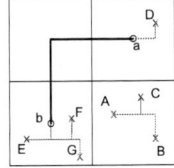

Figure 4: Local routing resources consumed by two nets.

In summary, to consider the effects of local routes in global routing, we add a pre-processing (or pre-routing) step. Specifically, we traverse all the nets, identify the pins inside each g-cell, estimate the local routes using the method presented in this section, and block the global routing tracks from the related g-edges. Local wires inside a g-cell are usually short and for pin accessibility they are typically routed in the second (M2) and third (M3) metal layers during detail routing. Hence, we only block the global routing tracks on g-edges in the M2 and M3 layers during congestion evaluation.

3.2.2 Method 2: Estimation of local routes based on pin density

Based on our experiments, Method 1 is reasonably fast and very effective in modeling congestion due to local routes. However, calculating Steiner trees for all the pins of each net in every g-cell can become a productivity bottleneck when congestion analysis is invoked hundreds of times (for example, during physical synthesis). We now propose an alternative method that is simpler and much faster than Method 1, yet equally effective. This method is based on pin density, and does not involve constructing Steiner trees to estimate the local routes. It is based on the following observations:

- Each pin is associated with a set of local wires connected to it.
- The number of pins in a g-cell is a good indicator of the number of local routes, and is a first-order estimate for routing tracks blocked by local routes within the g-cell.

Based on the above observations, we model the local routes in a g-cell by $(k \cdot n)$, where k is a technology-dependent parameter, and n is the number of pins in the g-cell. For each technology node, we empirically determine k by comparing the congestion statistics from our analyzer to those obtained from a reference industrial router. During our experiments on industry netlists, we observed that for a given technology node, k is usually similar across different designs and floorplans. This justifies the effectiveness of Method 2 while using a single k value that is technology and not design dependent.

A key benefit of using this method is that one can easily tune k for more complicated design rules at a given technology. Some design rules (such as lithography constraints) in advanced technologies, or a specific design library (smaller track), may be complicated to model through global routing. These factors may also result in significant detailed routing runtime. Tuning k to address the impacts of these issues can serve as a better guide to routability optimization during a physical synthesis flow.

As before, we use a pre-processing step to use Method 2 in a global routing based congestion evaluation tool. Specifically, we traverse all the g-cells and nets, and count the number of pins (n), inside each g-cell. Following this, we block kn global routing tracks, due to the local routes in each g-cell, on the four g-edges related to the g-cell. Similar to Method 1, we only block the global routing tracks on g-edges in the M2 and M3 layers.

Although Method 2 is empirical and much simpler than Method 1, it yields surprisingly good results. In our experiments, described in Section 5.1, Method 2 achieves a 3.6 times speedup over Method 1 with comparable accuracy.

4. METRICS FOR DESIGN CONGESTION

4.1 Limitations of current metrics

Total overflow (TOF) and maximal overflow (MOF): Naïve implementations of these metrics treat the overflow in each layer as identical; however, this is inaccurate as each layer has a different capacity. Normalizing the overflow to the layer capacity can overcome this issue, but other problems remain. The TOF metric does not capture the hot spots in the congestion map, i.e., the severity of congestion in the worst regions of the chip. MOF fares only slightly better, capturing only the maximum overflow value among all the g-edges in the routing graph. This presents a fairly incomplete picture of the congested regions in the design. Moreover, as pointed out in [2], overflow metrics fluctuate greatly, depending on design size, number of g-edges, number of routing layers, etc.

$ACN(20)$, $WCI(100)$ and $WCI(90)$: These metrics fail to differentiate between a net spanning a single congested g-edge and one that spans multiple congested g-edges.

Example: Consider two nets in the GRG: *net-A* traverses g-edges with congestion 0.50, 0.70, 0.80, 0.90 and 1.10, while *net-B* traverses g-edges with congestion 0.60, 0.80, 0.95, 1.05 and 1.10. When calculating $ACN(20)$, $WCI(100)$ and $WCI(90)$, both nets will be counted with the same congestion. However, their routability is different: clearly, *net-B* is harder to route compared to *net-A*, as it traverses more number of g-edges with higher congestion. This fact is not captured by these net congestion based metrics.

Additionally, minor design changes can cause large fluctuations in the $WCI(100)$ and $WCI(90)$ metrics.

Example: Assume a design has a g-edge e, with $c_e = 40$, $b_e = 0$ and $w_e = 39$. Assume, that we reroute a net to pass through this g-edge (say, to improve timing). Then the congestion of e becomes 100%, implying that all 40 nets crossing e now have a congestion of 100%. As a result, $WCI(100)$ will now report 40 additional congested nets, when in reality we only rerouted a single net. A similar example applies to the $WCI(90)$ metric. Such instability renders these metrics unsuitable for guiding routability optimization.

Although, $ACN(20)$ avoids large swings due to minor design changes, it suffers from the limitation of not accurately capturing design congestion (demonstrated in Section 5.5).

In addition, existing metrics improperly model the congestion along macro boundaries [1, 2], leading to an artificially high reported congestion. Referring to Figure 5, net N routes to a pin on macro block B. Due to the blockage, the congestion of edge e would be rated as being above 90%, but in practice, we find that such nets are easily routable. Including these g-edges with artificially

770

high congestion when calculating the metric introduces unnecessary noise leading to improper estimation of the routability. Note that we only suggest to exclude the edges along macro boundaries when calculating the metric *after* global routing to evaluate the routability, but the high congestion of these edges should *not* be ignored during the global routing process.

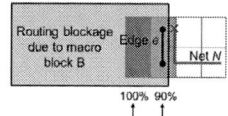

Figure 5: An example showing a net N traversing g-cells that are 90% blocked due to a routing blockage. This leads to artificially high reported congestion for edge e.

4.2 New metric for design congestion

To address the issues with existing metrics, we propose a new metric that is based on the histogram of g-edge congestion. Our metric has two features:

- It downplays the effects of g-edges with artificially high congestion due to the presence of routing blockages.
- It presents congestion as a histogram, instead of a single number.

To accurately capture the congestion, our metric, denoted as $ACE(x, y)$, computes the average congestion of the top x% congested g-edges, while ignoring g-edges that are $\geq y$% blocked. The role of the parameter y is to void counting the effects of g-edges with artificially high congestion. A typical value for y is 50, implying that all g-edges with ≥ 50% routing blockage are ignored when computing the metric. For convenience, we use $ACE(x)$ to denote $ACE(x, 50)$ in this paper.

In practice, the new metric is most useful when expressed as a vector, for different values of x, e.g., for $x \in \{0.5, 1, 2, 5, 10, 20\}$. $ACE(x)$, for a small value of x, (e.g., 0.5, 1), provides a highly local view, representing congestion in the regions with the highest contention for wiring resources (hot spots). For larger values of x, (e.g., 10, 20), it gives a broader picture of the design congestion.

5. VALIDATION AND ANALYSIS

Our proposed techniques, hereafter GLARE, are implemented within a congestion analyzer[2] that performs global routing in the spirit of MaizeRouter [8]. This section provides a detailed analysis of GLARE on advanced industrial designs listed in Table 1. Columns two and three in Table 1 give the design information and columns four and five list the technology specific pin blockage multiplier (k) and Steiner method parameter (p), respectively. These parameters are empirically tuned for each technology. Note that both k and p increase almost twice from 65nm and 45nm to 32nm. This is probably because the design rules in 32nm node become much more complex than previous technologies, and local routing takes more resources than before. All experiments were run on a 64-bit Linux server with 32 CPUs (Xeon® X7560 2.27GHz) and the common color map used for all the congestion plots in this paper is shown in Figure 6.

Table 1: Benchmark designs and the parameters used in the two methods for local routing modeling.

Designs	#Nets	Technology node	Pin blockage multiplier (k)	Steiner method parameter (p)
ckt_12	1660259	65nm	0.050	0.453
ckt_18	528500	65nm	0.050	0.453
ckt_i	350749	45nm	0.050	0.470
ckt_y	321939	45nm	0.050	0.470
ckt_s	1006029	45nm	0.050	0.470
ckt_fb	318116	32nm	0.114	1.000

Sections 5.1 through 5.4 show the impact of modeling local routing resources on routability evaluation and Section 5.5 shows the

---
[2]Our proposed techniques can also be used in global routers.

0-69%		95-99%	
70-74%		100-104%	
75-79%		105-109%	
80-84%		110-114%	
85-89%		115%+	
90-94%			

Figure 6: Common color map used for all the congestion plots in this paper.

impact of the proposed congestion metric. For the analyses presented in Sections 5.2 through 5.4, we use the following engines to evaluate the impact of our proposed techniques:

- **Reference Analyzer:** A full-blown industrial router that has a mode for performing congestion analysis with complex modeling of local wires. The reference analyzer is used to judge the quality of all results, and typically runs about 10 times slower than the GLARE based congestion analyzer.
- **Analyzer A:** A fast congestion analyzer that is based on Maize-Router [8], with the ability to perform global routing on millions of nets in less than 10 minutes.
- **GLARE:** Modification of Analyzer A, incorporating the local routing resources modeling of GLARE.

5.1 Steiner wirelength vs. pin density based modeling of local routing resources

This section compares the runtime and accuracy of the two methods to model local routing resources: Steiner tree wirelength based modeling (Method 1) and pin density based modeling (Method 2).

First, Table 2 compares the runtime of the two methods. In Table 2, "Pre-routing" gives the CPU time for modeling local routing resources, "Total" the total CPU time for congestion analysis, and "Average" the average runtime normalized to Method 2. From Table 2, Method 2 is 3.6 times faster than Method 1 when estimating local routing resources.

Table 2: Runtime comparison of Method 1 and Method 2.

Designs	Method 1 (sec)		Method 2 (sec)	
	Pre-routing	Total	Pre-routing	Total
ckt_12	28.03	401.99	9.00	384.59
ckt_18	8.72	139.50	2.17	133.12
ckt_i	5.24	46.20	1.45	39.73
ckt_y	5.64	48.63	1.93	41.65
ckt_fb	5.12	44.42	1.15	41.36
Average	3.62	1.10	1.00	1.00

Next, Figure 7 shows linear fitting results between local routing resources estimation using Method 1 and Method 2 for design ckt_fb on layer M2. The case for layer M3 is similar and omitted here. Figure 7 shows that with the right values of k and p, we can obtain very good correlation between the two methods.

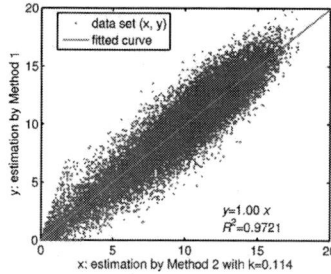

Figure 7: Linear fitting results between local routing resources estimation using Method 1 and Method 2 for design ckt_fb on layer M2.

Based on the results above and in light of its simplicity, we use Method 2 within GLARE to model and account for local routing resources for all subsequent analyses.

5.2 Improving congestion analysis accuracy

This section presents the impact of modeling local routing resources on the overall accuracy of congestion analysis.

Figure 8 shows the results of running the three analyzers on the same placement instance for design ckt_12. From Figure 8(b) we see that using a congestion analyzer with no modeling of local routing resources significantly underestimates the actual congestion. Alternatively, the plot from the GLARE based congestion analyzer (Figure 8(c)) is much closer to the one obtained from the reference analyzer[3], both in terms of the congested regions and their intensity. This result assumes significance in the context of using analyzers within congestion mitigation tools like CRISP [11], where the effectiveness of the tool is highly dependent on accurately identifying the regions of high congestion as well as their relative intensity.

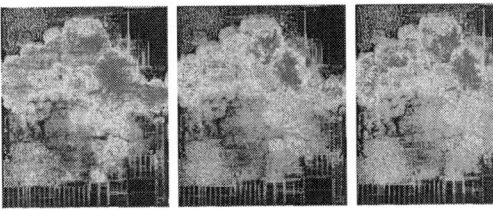

(a) Reference Analyzer   (b) Analyzer A   (c) GLARE

Figure 8: Congestion plots for ckt_12 using a g-cell size of 20 tracks (for GLARE, we set $k = 0.05$).

5.3 Better prediction of detailed routing issues

Often a design that seems routable after global routing can end up with multiple opens/shorts at the end of detailed routing. Early prediction of such issues without performing the time consuming step of detailed routing is highly beneficial as it enables designers to take appropriate measures, thereby improving overall turn-around time for design closure. This section demonstrates that our proposed techniques are able to predict detailed routing opens/shorts with higher fidelity as compared to existing methods.

As an example, we next compare the opens/shorts plots with the congestion plots from Analyzer A and GLARE for circuit ck_fb. Figure 9 shows a plot of the opens/shorts for design ckt_fb during an intermediate stage of an industrial strength detailed router. These opens/shorts indicate the problematic locations in detailed routing, which, in our experience, are usually due to high local congestion at these locations. Figure 10 shows the congestion plots generated by Analyzer A (Figure 10(a)) and the GLARE based analyzer (Figure 10(b)). Comparing Figure 9 with Figure 10(b), we see that the GLARE based analyzer clearly indicates congested hot spots, which translate to the problematic regions for detailed routing – a fact not captured by Analyzer A.

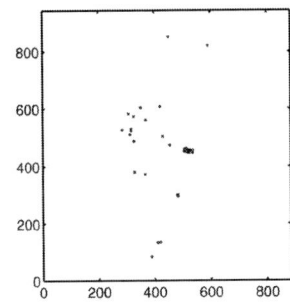

Figure 9: Opens and shorts for ckt_fb during detailed routing.

To quantitatively measure the predictability of the analyzers, we determine the ratio of the number of opens/shorts present in g-cells with global congestion greater than 85% to the total number of opens/shorts in the design. We call this as *match ratio*, and by

---

[3]It is expected that errors remain in the results generated by the GLARE-based congestion analyzer compared to the reference analyzer, since it runs much faster, and does not work as hard as the reference analyzer. However, GLARE-based congestion analyzer can generally predict the hot spots well.

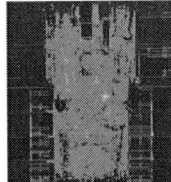 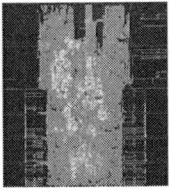

(a) Analyzer A   (b) GLARE

Figure 10: Congestion plots for ckt_fb. The GLARE based congestion analyzer predicts the problematic regions for detailed routing with higher fidelity (for GLARE, we set $k = 0.114$).

definition it measures the percentage of opens/shorts in the most congested regions of the design. We use a threshold of 85% based on prior experience that regions with high global congestion are usually problematic for detailed routing. Using this method, the *match ratios* for Analyzer A and the GLARE based analyzer are 0.20 and 0.87, respectively. Coupled with the congestion plot in Figure 10(b), this further demonstrates the effectiveness of the GLARE based congestion analyzer in predicting detailed routing opens/shorts.

5.4 Accelerating congestion analysis by using a larger g-cell

Since our method can accurately incorporate the effects of local wiring within a g-cell, it provides the freedom to increase the size of the g-cell, thereby accelerating congestion analysis. We demonstrate this by way of running the different analyzers with varying g-cell sizes as outlined below:

- Reference.20: Reference Analyzer, g-cell size = 20 tracks.
- Analyzer A.20: Analyzer A, g-cell size = 20 tracks.
- Analyzer A.80: Analyzer A, g-cell size = 80 tracks.
- GLARE.20: GLARE based analyzer, g-cell size = 20 tracks.
- GLARE.80: GLARE based analyzer, g-cell size = 80 tracks.

Figure 11 shows the results of running the different analyzers on identical placements for design ckt_12. As before, the congestion plot from the Reference Analyzer (Figure 11(a)) is considered most accurate, and used to judge the quality of all results. From Figures 11(b) and 11(c), it is apparent that the quality of the congestion analysis using Analyzer A deteriorates significantly with an increase in the g-cell size. Alternatively, Figures 11(d) and 11(e) demonstrate that the GLARE based analyzer with local routing resources modeling is still able to predict the congested hot spots with reasonable accuracy with an increase in the g-cell size. In addition, the runtime is reduced from 620 sec to 248 sec – a 60% speedup.

5.5 Comparison of routability metrics

Visual inspection of a congestion plot is widely used to quickly evaluate the routability of a design point. We now demonstrate that our new metric can capture a congestion plot with higher fidelity compared to prior metrics for routability evaluation. Consider Figure 12 displaying the congestion plots from two global routing solutions on identical placements for the design ckt_s. The corresponding values for the different congestion metrics are given in Table 3. For the new metric, the congestion is expressed as an ordered pair representing (Horizontal, Vertical) layer congestion.

(a) Solution 1   (b) Solution 2

Figure 12: Congestion plots for two routing solutions on design ckt_s.

From Table 3, the overflow-based metrics[4] indicate that Solution 1 has better congestion, while the net congestion based metrics indi-

---

[4]To counteract the drawbacks of overflow metrics discussed earlier, when

  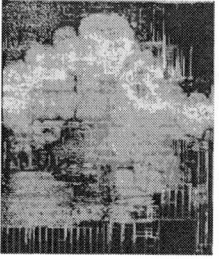 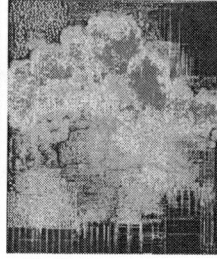 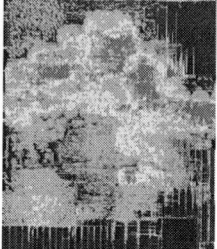

(a) Reference.20 (11287 sec) (b) Analyzer A.20 (557 sec) (c) Analyzer A.80 (207 sec) (d) GLARE.20 (620 sec) (e) GLARE.80 (248 sec)

Figure 11: Congestion plots and runtime for different analyzers with varying g-cell sizes for the design ckt_12.

cate that Solution 2 is better, as $ACN(20)$ of Solution 2 is better than that of Solution 1, even though $WCI(90)$ and $WCI(100)$ are worse.

However, a visual examination of the congestion plots indicates that they are quite similar, demonstrating the deficiencies in the existing metrics. Alternatively, our new metric correctly identifies the congestion of these two routing solutions to be similar.

Table 3: Congestion metrics for two routing solutions on design ckt_s.

Metrics		Solution 1	Solution 2
Overflow based metrics	TOF	194373	217499
	MOF	10	11
Net congestion based metrics	$ACN(20)$	84.41	77.97
	$WCI(90)$	39494	40548
	$WCI(100)$	274	276
New metric	$ACE(0.5)$	(90.47, 90.54)	(90.27, 90.46)
	$ACE(1)$	(89.23, 89.12)	(89.13, 89.10)
	$ACE(5)$	(85.93, 85.45)	(86.14, 85.49)
	$ACE(10)$	(84.16, 83.39)	(84.59, 83.51)
	$ACE(20)$	(82.48, 81.58)	(82.57, 81.53)

The significant difference in the $ACN(20)$ values for the two comparable routing solutions can be explained by Figure 13(a) which plots the distribution of the worst congestion on the nets. From Figure 13(a), Solution 1 has a considerably higher number of nets in the [78.09%, 84.41%] congestion range compared to Solution 2, leading to the difference in the $ACN(20)$ values. In practice, our experience on industry designs is that nets with congestion less than 85% are often easy to route, and considering them within the congestion metric introduces unnecessary noise during routability evaluation. In contrast, looking at Figure 13(b) which plots the distribution of the congestion on the g-edges, we observe the distributions for the two routing solutions to be quite similar above 80%. This explains why the new metric (correctly) rates the two solutions to have similar congestion.

6. CONCLUSION

Fast and accurate routability evaluation techniques are critical to address the increasingly important and difficult issue of routing closure in nanometer-scale physical synthesis. In this work, we have addressed two important aspects of routability evaluation: the accuracy of congestion estimation and a metric for evaluating the routability of a design. We have shown that ignoring the effects of local congestion can result in large errors during congestion analysis. This observation motivates our models for local congestion based on (a) the Steiner tree wirelength of the local routes and (b) the pin density. Experimental results show that the proposed modeling can improve the accuracy and fidelity of congestion analysis, and better predict detailed routing issues such as opens and shorts. It also enables designers to use larger g-cells to accelerate the process of congestion analysis, thereby speeding design closure. Furthermore, we have analyzed the limitations of existing congestion metrics including overflow, etc., and proposed a new metric based on g-edge

---

calculating overflow, the capacity is scaled down to 80% of the original, and the overflow is in unit of number of minimum-width tracks, e.g., one overflowed track on a 4X layer would be counted as four in the overflow number.

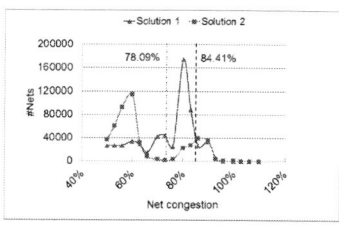

(a) Distribution of net congestion.

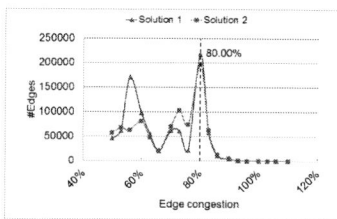

(b) Distribution of g-edge congestion.

Figure 13: Distribution of congestion for two routing solutions of ckt_s.

congestion. We have demonstrated that our new metric can represent a congestion plot with higher fidelity.

7. REFERENCES

[1] C. Alpert et al. What makes a design difficult to route. In *Proc. ISPD*, pages 7–12, 2010.

[2] C. Alpert and G. Tellez. The importance of routing congestion analysis. *DAC Knowledge Center Online Article*, 2010. http://www.dac.com/back_end+topics.aspx?article=47&topic=2.

[3] Y.-J. Chang et al. NTHU-Route 2.0: A fast and stable global router. In *Proc. ICCAD*, pages 338–343, 2008.

[4] H.-Y. Chen et al. High-performance global routing with fast overflow reduction. In *Proc. ASPDAC*, pages 582–587, 2009.

[5] C. Chu and Y.-C. Wong. Flute: Fast lookup table based rectilinear Steiner minimal tree algorithm for VLSI design. *IEEE Trans. on CAD*, 27(1):70–83, 2008.

[6] J. Lou et al. Estimating routing congestion using probabilistic analysis. *IEEE Trans. on CAD*, 21(1):32–41, 2002.

[7] C. Minsik et al. BoxRouter 2.0: Architecture and implementation of a hybrid and robust global router. In *Proc. ICCAD*, pages 503–508, 2007.

[8] M. D. Moffitt. MaizeRouter: Engineering an effective global router. *IEEE Trans. on CAD*, 27(11):2017–2026, 2008.

[9] M. Pan and C. Chu. FastRoute: A step to integrate global routing into placement. In *Proc. ICCAD*, pages 464–471, 2006.

[10] M. Pan and C. Chu. IPR: An integrated placement and routing algorithm. In *Proc. DAC*, pages 59–62, 2007.

[11] J. Roy et al. CRISP: Congestion reduction by iterated spreading during placement. In *Proc. ICCAD*, pages 357–362, 2009.

[12] H. Shojaei et al. Congestion analysis for global routing via integer programming. In *Proc. ICCAD*, pages 256–262, 2011.

[13] J. Westra et al. Probabilistic congestion prediction. In *Proc. ISPD*, pages 204–209, 2004.

[14] T.-H. Wu et al. A parallel integer programming approach to global routing. In *Proc. DAC*, pages 194–199, 2010.

[15] Y. Xu et al. FastRoute 4.0: Global router with efficient via minimization. In *Proc. ASPDAC*, pages 576–581, 2009.

The DAC 2012 Routability-Driven Placement Contest and Benchmark Suite

Natarajan Viswanathan, Charles Alpert, Cliff Sze, Zhuo Li, Yaoguang Wei
IBM Corporation, 11501 Burnet Road, Austin, TX 78758
{ nviswan, alpert, csze, lizhuo, weiyg }@us.ibm.com

ABSTRACT

Existing routability-driven placers mostly employ rudimentary and often crude congestion models that fail to account for the complexities in modern designs, e.g., the impact of non-uniform wiring stacks, layer directives, partial and/or complete routing blockages, etc. In addition, they are hampered by congestion metrics that do not accurately score or represent design congestion. This is in large part due to the non-availability of public designs depicting industrial wiring stacks and other complexities affecting design routability.

The aim of the DAC 2012 routability-driven placement contest is to address these issues, by way of the following: (a) release challenging benchmark designs that are derived from modern industrial ASICs, and contain information to perform both placement and routing, (b) present a new congestion metric, as well as an accurate congestion analysis framework to evaluate and compare the routability of various placement algorithms. We hope that a set of challenging benchmarks, along with a standard, publicly available evaluation framework will further advance research in routability-driven placement.

Categories and Subject Descriptors

B.7.2 [**Hardware, Integrated Circuits, Design Aids**]: Placement and routing

General Terms

Algorithms, Design, Experimentation, Performance

Keywords

Physical Design, Placement, Routing, Benchmarks, Congestion Analysis

1. INTRODUCTION

Design routability has emerged as one of the primary concerns during physical synthesis at advanced technology nodes [1]. In this regard, placement is perhaps one of the most powerful tools to help mitigate routing congestion during a physical synthesis flow. However, optimizing for traditional placement metrics like the total half-perimeter wire length (used in the ISPD-2005 contest [13]) or density-target based overflow (used in the ISPD-2006 contest [12]) are no longer sufficient to address this critical issue. Modern placers need to explicitly account for congestion to obtain routable placements.

Prior work (e.g., [2, 7, 16]) has incorporated some form of congestion modeling within global placement. The key issue with these techniques is that they employ rudimentary and often inaccurate congestion models like pin-density or probabilistic congestion estimation. These models fail to account for the complexities in modern designs.

To predict design congestion with reasonable accuracy, modern congestion analysis techniques need to account for:

- A complex wiring stack with anywhere from nine to twelve metal layers, along with varying width and spacing, within and across layers.

- Partial and/or complete routing blockages at various metal layers.

- Local wiring congestion due to *intra*-g-cell routes, and pin-access issues.

- Layer directives, e.g., from optimization tools.

Prior work (e.g., [10, 14, 15, 21]) has also tried to address the issue of congestion modeling by using reasonably accurate global routing based congestion analyzers – that in part handle the complex wiring constraints of modern designs. However, they are hampered by congestion metrics that often do not accurately score or represent the design congestion [18].

This paper describes the DAC 2012 routability-driven placement contest. It also presents a set of benchmarks that are representative of modern industrial designs with numerous placement and routing blockages, more metal layers, varying width and spacing across layers, etc. The primary objectives of the DAC 2012 routability-driven placement contest are:

- Release advanced benchmarks derived from modern industrial ASIC designs to advance research in the areas of placement and routing for nanometer-scale VLSI designs.

- Motivate research in the area of fast, yet reasonably accurate routing congestion analysis that can handle the complex features and constraints of modern wiring stacks.

- Motivate research to effectively integrate placement with accurate congestion analysis techniques to perform routability-driven placement.

- Provide an accurate congestion analysis framework, by releasing a set of global routing based congestion analyzers that also model local wiring or local routing resource usage.

- Present intuitive and accurate metrics to score and compare the routability of various placement algorithms.

The rest of this paper is organized as follows: Section 2 gives an overview of the new congestion metric that is used for the DAC 2012 contest. Section 3 presents the DAC 2012 contest evaluation metric. Section 4 describes the DAC 2012 benchmark suite. Finally, Section 5 provides concluding remarks.

2. CONGESTION ANALYSIS

Congestion analysis refers to the process of estimating the routability of a given placement or design point. Typically, the output of this process is a congestion or hot-spot map indicating the regions in the chip with the greatest contention for wiring resources. In a physical synthesis flow, the effectiveness of various routability-driven optimizations (e.g., placement spreading, logic re-synthesis, etc.,) are highly dependent on the speed and accuracy of congestion analysis.

2.1 Global Routing Based Congestion Analysis

With the advent of fast and high-quality global routers [5, 8, 9, 11, 19, 20], global routing based congestion analysis offers an attractive compromise between run-time and accuracy, and is gaining widespread usage. Hence, *global routing based congestion analysis was the method of choice to evaluate the routability of the placement solutions during the DAC 2012 contest.* In this regard, every design in the DAC 2012 benchmark suite (Sec. 4) contains all the relevant information to perform global routing, and/or enable realistic estimation of design congestion.

We now introduce some notations and definitions that will be used in the remainder of this paper. Typically, global routers overlay a regular grid (g-cells) on the chip image, and construct a global routing graph (Figure 1). Each node in the graph represents a g-cell in the layout, and an edge (g-edge) represents the boundary between adjacent g-cells. To represent/evaluate a global routing solution, we use the following notations, defined for each g-edge (e) in the global routing graph:

- c_e: Total or maximal capacity of edge e
- b_e: The routing blockage on edge e
- w_e: The routing demand or wire occupancy on edge e

2.2 Modeling Local Routing Resource Usage

One of the issues with existing global routing based congestion analyzers, especially the ones using academic global routers, is that they fail to consider the effect of local congestion due to *intra*-g-cell routes. To account for local routing resource usage, the congestion analyzers used for the DAC 2012 contest adopted a simple, yet effective technique based on the pin-density within each g-cell [18].

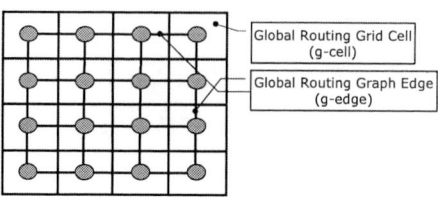

Figure 1: Global routing graph.

Briefly, before performing global routing, the pin-count, say N, within each g-cell is determined. Then, the available capacity of the g-cell (or, associated g-edges) is reduced by $N \times Pin\_Blockage\_Factor$, where $Pin\_Blockage\_Factor$ is a pre-determined technology specific parameter. Note that the capacity is only reduced on metal layers $M2$ and $M3$ to account for resource usage due to local routes.

2.3 Congestion Metrics

In addition to congestion maps, optimization tools and designers rely on a congestion metric to evaluate the routability of a particular design point. A good congestion metric should be able to accurately score or represent the design congestion as visualized by a congestion map.

2.3.1 Total Overflow and Maximal Overflow

Two commonly used congestion metrics are the total overflow (TOF) and maximal overflow (MOF). Let, $OF(e) = \max(w_e + b_e - c_e, 0)$ be the overflow of a g-edge e. Then, $TOF = \sum OF(e)$ and $MOF = \max(OF(e))$ across all the g-edges in the global routing graph.

A key issue with these metrics is that they do not provide an accurate view of the design routability. For example, looking at a given TOF value, say, 1283, a designer cannot determine if the g-edges with excess demand are all concentrated in a local region or evenly distributed across the chip image. In the former case, the design might have routability concerns, whereas, in the latter case, the design might still be routable, as the overflow could be easily absorbed by neighboring regions or resolved during subsequent routing phases. In addition, overflow based metrics can fluctuate greatly, based on the number of g-edges, metal layers, etc.

2.3.2 ACE: Average Congestion of g-cell Edges

In light of the drawbacks with existing overflow based metrics, the DAC 2012 contest adopted a new metric – the Average Congestion of g-cell Edges (ACE) [18], which is based on the histogram of g-edge congestion. Let, $Cong(e) = 100 \times (w_e + b_e)/c_e$ represent the congestion (in percentage) of a g-cell edge (or, g-edge) e. Then, the new metric, denoted as $ACE(x)$, computes the average congestion of the top $x\%$ congested g-edges. Using different values of x, say, $x \in \{0.5, 1, 2, 5, 10\}$, a set of ACE values can be computed, which together provide an accurate view of the design congestion. For example, $ACE(0.5)$ or $ACE(1)$ provide a highly local view, representing the most congested regions in the design. Whereas, $ACE(5)$ or $ACE(10)$ give a broader view of the design congestion.

2.3.3 Total Overflow versus ACE

In this section, we demonstrate the limitations of the total overflow (TOF) metric in predicting design routability, and

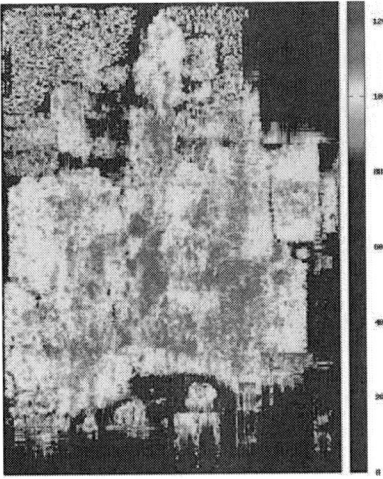
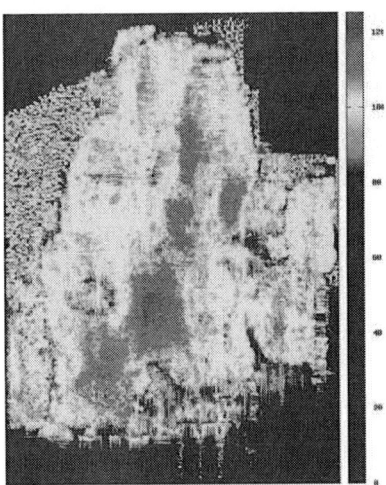

Ripple [6] with total overflow of 542786   SimPLR [10] with total overflow of 514614

Figure 2: Congestion maps and total overflow values for Ripple and SimPLR placement solutions on superblue12. The regions colored purple indicate congestion hot-spots − regions with the greatest contention for wiring resources.

Placer	$ACE(0.5)$	$ACE(1)$	$ACE(2)$	$ACE(5)$	$ACE(10)$
Ripple	126.23	123.00	120.62	114.32	109.10
SimPLR	130.89	126.34	123.17	118.97	113.49

Table 1: The $ACE(x)$ metric for Ripple and SimPLR placement solutions on superblue12, for $x \in \{0.5, 1, 2, 5, 10\}$.

the ability of the ACE metric to accurately represent the design congestion as visualized by a congestion map. For our experiments, we evaluated the placements from two of the top-performing teams (Ripple [6] and SimPLR [10]) in the ISPD-2011 routability-driven placement contest [17].

Recall, TOF was the primary evaluation metric during the ISPD-2011 contest. Let us consider the Ripple and SimPLR placements for the design *superblue12*. The TOF values as reported by the ISPD-2011 contest evaluator are: Ripple (542786) and SimPLR (514614), indicating that the SimPLR placement has better routability.

Figure 2 shows the corresponding congestion maps. From Figure 2, we make two observations: First, on this design Ripple seems to spread more than SimPLR. As a result, the congestion hot-spots for Ripple are more distributed than the hot-spots for SimPLR. Second, the regions colored purple, with congestion > 120% are much larger in the SimPLR congestion map as compared to Ripple. Based on these two observations and our experience on industry designs, we believe that the SimPLR placement is worse in terms of routability as compared to the Ripple placement. However, this is not captured by the TOF metric, highlighting its limitations.

Next, we apply the ACE metric on the same routing solutions as in Figure 2, and obtain the results shown in Table 1. From Table 1, we see that $ACE(0.5)$ and $ACE(1)$ for SimPLR are higher than that for Ripple, implying that the hot-spots in the SimPLR congestion map have higher congestion than the ones in the Ripple congestion map. More-

over, all the ACE values for SimPLR are worse than Ripple. Together, they indicate that SimPLR has worse routability compared to Ripple on this design. This correlates well with our observations from the congestion maps.

These observations indicate that ACE can represent the congestion map with higher fidelity than the TOF metric, thereby validating the choice for using it as the metric to evaluate the routability of the placement solutions during the DAC 2012 contest.

3. THE DAC 2012 CONTEST METRIC

The DAC 2012 contest used three key metrics to evaluate the quality of the placement solutions:

- The total half-perimeter wire length (HPWL) of the placement.

- The global routing based congestion, as represented by the ACE metric.

- The run-time to generate a placement solution.

3.1 Routability Metric

Let,

- Peak_Weighted_Congestion (PWC):

$$PWC = \frac{\sum(K_x \times ACE(x))}{\sum K_x}, \qquad x = 0.5, 1, 2, 5$$

where, K_x is the weight associated with $ACE(x)$.

- Routing_Congestion (RC):

$$RC = \max(100, \ PWC)$$

Based on the HPWL, and the global routing based congestion (RC), the metric to score/evaluate the routability of the placement solution is given by:

$$Routability\_Metric = HPWL \times (1 + PF \times (RC - 100))$$

776

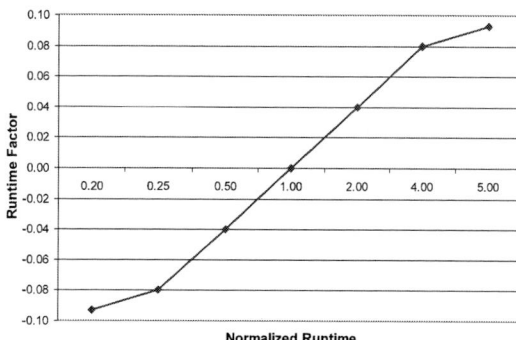

Figure 3: Run-time factor versus normalized run-time. A factor of two speed-up (Normalized Run-time = 0.5) gives a 4% wire length advantage (Run-Time Factor = −0.04).

where, PF is the penalty factor that scales the HPWL to account for routing congestion.

Let, $K_x = 1.0$, $\forall x$, and the penalty factor, $PF = 0.03$, then the interpretation of the routability metric is that for every 1% excess routing congestion ($> 100\%$), there is a 3% wire length penalty.

3.2 Run-time Factor

To encourage placer efficiency, the placement run-time was also considered in the overall contest metric. This was achieved by using a run-time factor that awards or penalizes the placers based on their run-time. For each design in the benchmark suite, the wall times for all the placers were measured. Let, *Normalized Runtime = Placer_Wall_Time / Median_Placer_Wall_Time*. Then, the run-time factor for a placer, on a particular design, is given by Figure 3. From Figure 3, a placer has a ±4% wire length advantage for a factor of two speed-up or slow-down. In addition, the maximum run-time factor is set to 10%, limiting the effect of the run-time on the overall contest metric.

3.3 Contest Metric

The overall contest metric, accounting for routing congestion and run-time, is given by the scaled wire length of the placement solution:

$$Contest\_Metric = HPWL \times (1 + PF \times (RC - 100))$$
$$\times (1 + Runtime\_Factor)$$

4. THE DAC 2012 BENCHMARK SUITE

Table 2 presents key statistics of the designs in the DAC 2012 benchmark suite. For each design, the reported statistics are:

- *Total Nodes*: The total number of nodes (movable + fixed).
- *Movable Nodes*: The number of movable nodes.
- *Terminal Nodes*: The number of fixed "terminal" nodes. There can be no overlap between the movable and terminal nodes.
- *Terminal_NI Nodes*: The number of fixed "terminal_NI" nodes. Overlap is allowed between the movable and terminal_NI nodes.

- *Total Nets*: The total number of nets.
- *Total Pins*: The total number of pins.
- *Design Util.*: The design utilization, defined as the ratio of the total area of the movable and terminal nodes to the area of the placement region.
- *Design Den.*: The design density, defined as the ratio of the total area of the movable nodes to the available free-space in the design. Where, free-space is given by the area of the placement region, less the total area of the terminal nodes.

All the designs in the benchmark suite are derived from industrial designs, and converted from the IBM internal data format to the GSRC Bookshelf placement format [3,4]. The height of all the movable and terminal nodes are an integer multiple of the standard-cell circuit row height. This allows every node to be placed aligned to the circuit rows in the design. In addition, all the macro-blocks (nodes with height > circuit row height) are fixed in-place. Hence, the only movable nodes in all the designs are the standard-cells. Movable macro-blocks greatly complicate the placement problem. In addition, they can significantly alter the congestion profile. To start with, we would like to evaluate the performance and routability of the placers when they are allowed to move only the standard-cells.

In order to preserve the integrity of the industrial designs, and enable realistic estimation of design congestion (e.g., global routing based congestion analysis), the Bookshelf placement format was augmented with some new features. These features were introduced during the ISPD-2011 placement contest [17], and are described below.

4.1 Non-rectangular Fixed Nodes

A subset of the pre-placed IP blocks in the designs are not rectangular. One such node is shown in Figure 4, where the light-red shaded (solid) box with blue outline represents a rectangular fixed node, whereas, the large gray object (hatched lines) represents a non-rectangular fixed node. To represent the non-rectangular nodes in the Bookshelf format, each non-rectangular node is fragmented into a set of single-circuit row high rectangular component shapes. This is depicted by the narrow boxes within the overall gray outline in Figure 4. A new benchmark file *circuit.shapes*, then specifies the component shape definitions for each non-rectangular node. In addition, the other files in the benchmark are modified to account for pin-offsets, node dimensions, node coordinates, etc.

4.2 RLM Pins and terminal_NI nodes

In a hierarchical design methodology, there are multiple lower level random logic macros (RLMs) that are integrated at the top level to yield the final design. To enable interconnection across different levels of the hierarchy, designs use special pins, which we denote as RLM pins. RLM pins typically reside on the upper-level metal layers associated with a particular macro. To represent such pins in the Bookshelf format, a new node class named "terminal_NI" is introduced. Each terminal_NI node contains one RLM pin. For placement, the standard-cells can be placed "below" the terminal_NI nodes without violating any legality constraints. During global routing, the pins associated with the terminal_NI nodes reside on an upper-level metal layer (e.g., M5),

Design	Total Nodes	Movable Nodes	Terminal Nodes	Terminal_NI Nodes	Total Nets	Total Pins	Design Util. (%)	Design Den. (%)
superblue19	522775	506097	286	16392	511685	1714351	78	49
superblue14	634555	567840	44743	21972	619815	2049691	72	50
superblue16	698741	680450	419	17872	697458	2280931	69	46
superblue9	846678	789064	37574	20040	833808	2898853	73	47
superblue3	919911	833370	55033	31508	898001	3110509	73	42
superblue11	954686	859771	67303	27612	935731	3071940	79	40
superblue6	1014209	919093	65316	29800	1006629	3401199	73	43
superblue2	1014029	921273	59312	33444	990899	3228345	76	28
superblue12	1293433	1278084	8953	6396	1293436	4774069	56	44
superblue7	1364958	1271887	66995	26076	1340418	4935083	76	58

Table 2: Design statistics of the DAC 2012 benchmark suite.

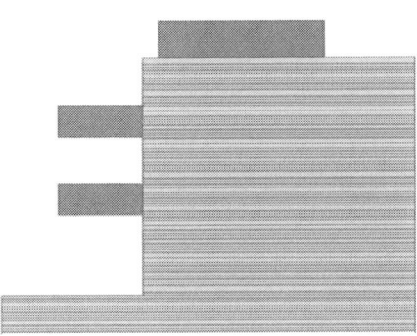

Figure 4: Non-rectangular versus rectangular fixed nodes. A light-red shaded box with blue outline represents a rectangular fixed node, whereas, the large gray object represents a non-rectangular fixed node. Note: The non-rectangular fixed node is represented as a set of rectangular component shapes (narrow boxes within the overall gray outline).

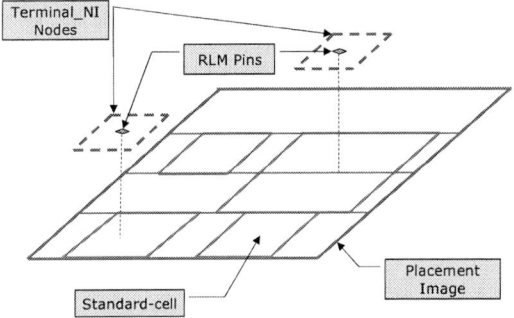

Figure 5: Terminal_NI nodes and associated RLM pins. Overlap is allowed between the movable and terminal_NI nodes during placement. The RLM pins on terminal_NI nodes reside on an upper-level metal layer (e.g., M5) during global routing.

and should be connected accordingly. Terminal_NI nodes and associated RLM pins are depicted in Figure 5.

4.3 Routing Information

To enable contestants to perform global routing, and/or other means of congestion analysis, a new file *circuit.route* is created for each design in the benchmark suite. This file has the following information:

- Global routing grid (grid size, g-cell dimensions, etc.)
- Number of routing layers.
- Maximal g-edge capacity for each routing layer.
- Wire width and spacing for each routing layer.
- Via specifications.
- Routing blockage information.
- Routing layer for RLM pins.

Figure 6 shows a sample *circuit.route* file reflecting an industrial wiring stack with nine metal layers. Layers M1–M4 have 1× width and spacing, layers M5–M7 have 2× width and spacing, and layers M8–M9 have 4× width and spacing.

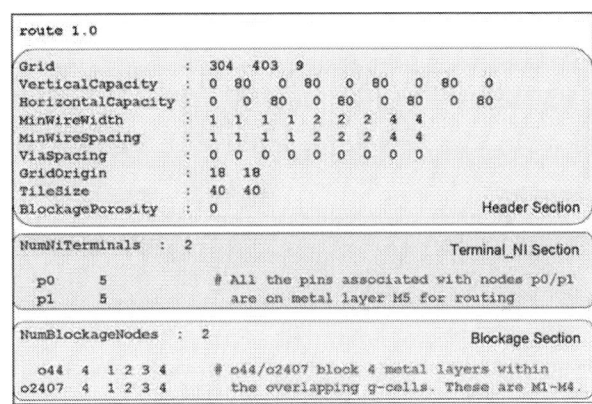

Figure 6: A sample *circuit.route* file with information to setup and perform global routing for a given placement. This example shows a layer stack with nine metal layers, along with varying width and spacing across layers.

4.4 Floorplan Layouts

Figures 7 through 9 depict the floorplans for all the designs in the benchmark suite, plotting only the fixed ("terminal") nodes. The light-red shaded boxes with blue outline represent the rectangular fixed nodes and the dark-gray shaded boxes represent the non-rectangular fixed nodes in the design. Figures 7 through 9 and Table 2 demonstrate the varying characteristics of modern ASIC designs. For example, all the fixed nodes in *superblue12* are pushed to the periphery of the placement region. As a result, the placer has a large amount of contiguous free-space in and around the center of the placement region. On the other hand, the fixed nodes in *superblue9* and *superblue7* fragment the placement region into multiple sub-regions. The design utilization of *superblue2* is greater than 75%, on account of the large fixed nodes, whereas, its design density is 28%, indicating that there is also abundant free-space to place the movable nodes. All these features offer different challenges to routability-driven placement algorithms. In addition, two of the designs *superblue12* and *superblue7* have over a million movable nodes, and are good benchmarks to test the scalability of the placers.

5. CONCLUSIONS

Routing congestion is a key issue in nanometer-scale physical synthesis flows. Numerous factors contribute to routing congestion in modern designs, e.g., increased use of embedded IPs, complicated logic structures, a high-performance layer stack that is non-uniform in nature, per-layer routing blockages, etc. As a result, routing congestion needs to be *accurately modeled* and *addressed* across the entire physical synthesis tool stack. In this regard, routability-driven placement is a highly relevant and challenging problem.

This paper describes the DAC 2012 routability-driven placement contest and benchmark suite. All designs in the DAC 2012 benchmark suite are derived from modern industrial ASICs, and contain information to perform both placement and routing. In addition, the contest provides standard metrics, as well as an accurate congestion analysis framework to evaluate and compare the routability of placement algorithms. We hope that a set of challenging benchmarks, along with a standard, publicly available evaluation framework will further advance research in placement and routing for nanometer-scale designs.

Acknowledgments

A key requirement for the contest was the availability of fast, high-quality global routers to evaluate the placement solutions. The authors would like to thank Wen-Hao Liu and Prof. Yih-Lang Li, Department of Computer Science, National Chiao Tung University, Taiwan; Jin Hu and Prof. Igor Markov, Department of Electrical Engineering and Computer Science, University of Michigan, Ann Arbor; for their time and effort to qualify their respective routers, NCTUgr and BFG-R, to be used for the DAC 2012 contest. Thanks also go to our colleagues at IBM, Bertram Bradley, Randy Darden, Gi-Joon Nam and Shyam Ramji for all the discussions and their help in releasing the benchmarks.

6. REFERENCES

[1] C. J. Alpert, Z. Li, M. D. Moffitt, G.-J. Nam, J. A. Roy, and G. Telleze. What makes a design difficult to route. In *Proc. ISPD*, pp. 7–12, 2010.

[2] U. Brenner and A. Rohe. An effective congestion-driven placement framework. *TCAD*, 22(4):387–394, 2003.

[3] A. E. Caldwell, A. B. Kahng, and I. L. Markov. Placement formats, rev. 1.2. url: *http://vlsicad.ucsd.edu/GSRC/bookshelf/Slots/Placement/plFormats.html*.

[4] A. E. Caldwell, A. B. Kahng, and I. L. Markov. Toward CAD-IP reuse: The MARCO GSRC Bookshelf of fundamental CAD algorithms. In *IEEE Design and Test*, pp. 72–81, 2002.

[5] Y.-J. Chang, Y.-T. Lee, and T.-C. Wang. NTHU-Route 2.0: A fast and stable global router. In *Proc. ICCAD*, pp. 338–343, 2008.

[6] X. He, T. Huang, L. Xiao, H. Tian, G. Cui, and E. Young. Ripple: An effective routability-driven placer by iterative cell movement. In *Proc. ICCAD*, pp. 74–79, 2011.

[7] W. Hou, H. Yu, X. Hong, Y. Cai, W. Wu, J. Gu, and W. H. Kao. A new congestion-driven placement algorithm based on cell inflation. In *Proc. ASPDAC*, pp. 723–728, 2001.

[8] C.-H. Hsu, H.-Y. Chen, and Y.-W. Chang. Multi-layer global routing considering via and wire capacities. In *Proc. ICCAD*, pp. 350–355, 2008.

[9] J. Hu, J. A. Roy, and I. L. Markov. Completing high-quality routes. In *Proc. ISPD*, pp. 35–41, 2010.

[10] M.-C. Kim, J. Hu, D.-J. Lee, and I. L. Markov. A SimPLR method for routability-driven placement. In *Proc. ICCAD*, pp. 67–73, 2011.

[11] W.-H. Liu, W.-C. Kao, Y.-L. Li, and K.-Y. Chao. Multi-threaded collision-aware global routing with bounded-length maze routing. In *Proc. DAC*, pp. 200–205, 2010.

[12] G.-J. Nam, C. J. Alpert, and P. Villarrubia. ISPD 2006 placement contest: Benchmark suite and results. In *Proc. ISPD*, pp. 167–167, 2006.

[13] G.-J. Nam, C. J. Alpert, P. Villarrubia, B. Winter, and M. Yildiz. The ISPD2005 placement contest and benchmark suite. In *Proc. ISPD*, pp. 216–220, 2005.

[14] M. Pan and C. Chu. IPR: An integrated placement and routing algorithm. In *Proc. DAC*, pp. 59–62, 2007.

[15] J. A. Roy, N. Viswanathan, G.-J. Nam, C. J. Alpert, and I. L. Markov. CRISP: Congestion reduction by iterated spreading during placement. In *Proc. ICCAD*, pp. 357–362, 2009.

[16] P. Spindler and F. M. Johannes. Fast and accurate routing demand estimation for efficient routability-driven placement. In *Proc. DATE*, pp. 1226–1231, 2007.

[17] N. Viswanathan, C. J. Alpert, C. Sze, Z. Li, G.-J. Nam, and J. A. Roy. The ISPD-2011 Routability-driven placement contest and benchmark suite. In *Proc. ISPD*, pp. 141–146, 2011.

[18] Y. Wei, C. Sze, N. Viswanathan, et al. GLARE: Global and local wiring aware routability evaluation. In *Proc. DAC*, To appear, 2012.

[19] T.-H. Wu, A. Davoodi, and J. T. Linderoth. A parallel integer programming approach to global routing. In *Proc. DAC*, pp. 194–199, 2010.

[20] Y. Xu, Y. Zhang, and C. Chu. FastRoute 4.0: Global router with efficient via minimization. In *Proc. ASPDAC*, pp. 576–581, 2009.

[21] Y. Zhang and C. Chu. CROP: Fast and effective congestion refinement of placement. In *Proc. ICCAD*, pp. 344–350, 2009.

superblue19   superblue14

superblue16   superblue9

Figure 7: Floorplan layout figures of designs superblue19, superblue14, superblue16 and superblue9. The light-red shaded boxes with blue outline represent the rectangular fixed nodes and the dark-gray shaded boxes represent the non-rectangular fixed nodes in the design.

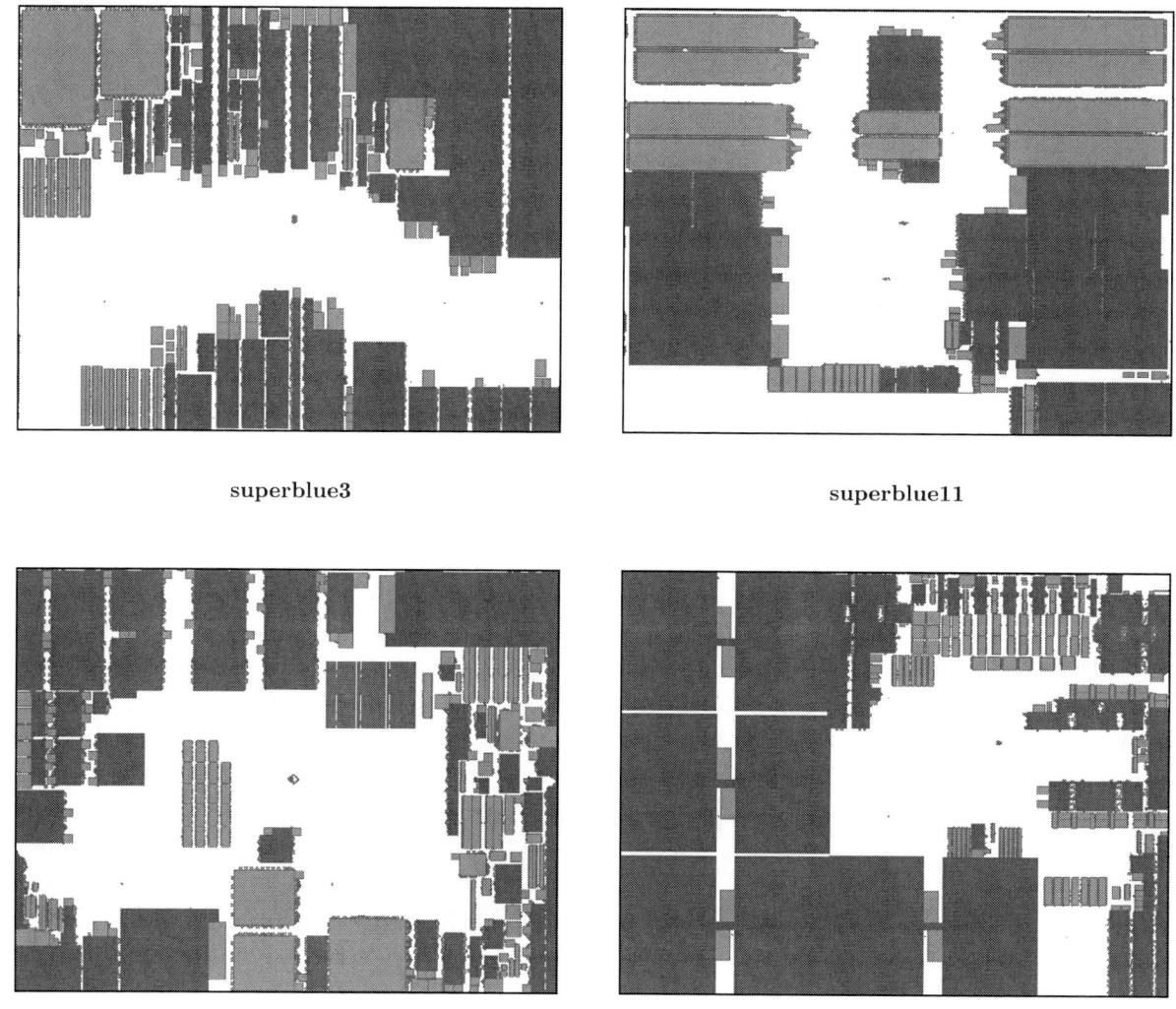

superblue3   superblue11

superblue6   superblue2

Figure 8: Floorplan layout figures of designs superblue3, superblue11, superblue6 and superblue2. The light-red shaded boxes with blue outline represent the rectangular fixed nodes and the dark-gray shaded boxes represent the non-rectangular fixed nodes in the design.

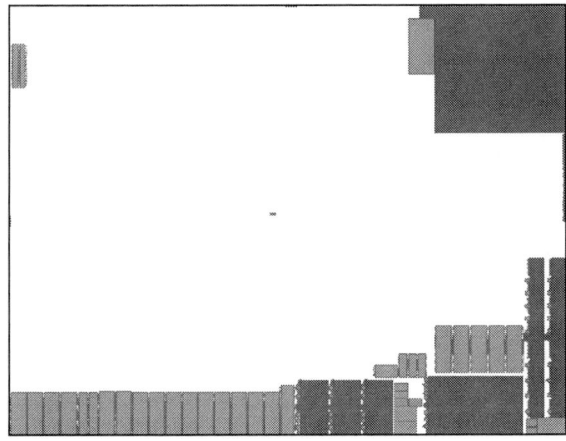 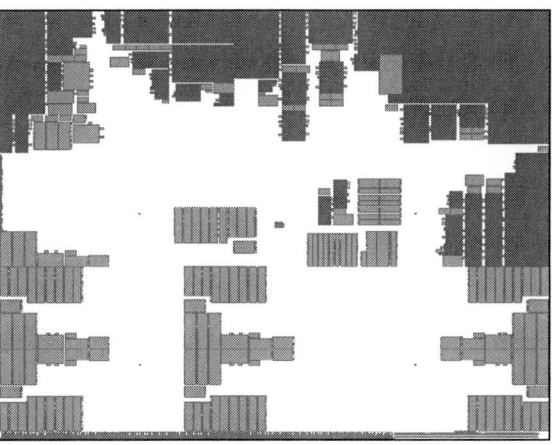

superblue12   superblue7

Figure 9: Floorplan layout figures of designs superblue12 and superblue7. The light-red shaded boxes with blue outline represent the rectangular fixed nodes and the dark-gray shaded boxes represent the non-rectangular fixed nodes in the design.

Removing Overhead From High-Level Interfaces

Kyle Kelley
kkelley@stanford.edu

Megan Wachs
wachs@stanford.edu

John Stevenson
jpeter@stanford.edu

Stephen Richardson
steveri@stanford.edu

Mark Horowitz
horowitz@stanford.edu

ABSTRACT

Hardware modules would be much easier to reuse if they supported generic flexible high-level interfaces. However, these interfaces are rarely used since they lead to timing and area overheads compared to a customized design. This paper describes a reachability analysis framework that identifies over-provisioning in instances of flexible design, and offers a technique for annotating this information so that modern synthesis tools can remove most of the overhead. Results are demonstrated on a variety of flexible structures, including functional blocks, programmable state machines, and latency-insensitive interfaces.

Categories and Subject Descriptors

B.6 [**Hardware**]: Logic Design

General Terms

Algorithms, Design, Performance

Keywords

Reachability, Synthesis, HDL, Flexibility

1. INTRODUCTION

Designing a modern day ASIC is a complex and expensive task. The exponential growth in transistors per chip afforded by Moore's Law has given rise to exponentially growing design complexity. One issue is that reuse of blocks is often difficult, since a module may contain assumptions about its environment which are true for its original use but not true in general. A common set of assumptions involves the timing of I/O signals. This rigidity of intermodule connections leads to extremely brittle systems, where it is difficult to alter one component without adversely affecting the functionality of the whole system. Adding pipeline registers to one unit can easily affect the units communicating with it. A solution to this problem would be to employ a latency-insensitive communication protocol between design components to decouple module timing from functionality. Using these higher-level protocols, the system should function if the modules maintain the right order of the messages sent on the links, and not depend on strict timing. The end result is a decoupled system that is easier to modify.

Despite these advantages, higher-level flexible VLSI design techniques are often not used in practice. Their main issue is the implementation overhead associated with flexible components. Intuitively, there is always a tradeoff between flexibility and efficiency: a module that operates correctly under more conditions inherently has more states and more logic than its customized counterpart. Since a system designer knows its timings, they want to build/use components tailored for that application and not pay for this overhead in area, energy and performance.

In theory, flexible components should not imply overheads in the implementations. If logic in the description is never needed, the synthesis system should remove it, and indeed modern synthesis tools are very good at constant propagation and dead logic elimination. Unfortunately, current logic synthesis tools are not good at identifying unused logic across sequential boundaries, so they are not able to optimize flexible interfaces properly.

This paper leverages prior work in sequential reachability analysis to create a highly-automated framework for identifying and removing unneeded overheads in flexible design instances, significantly improving the real-world viability of high-level interfaces. We show that the same analysis framework can remove overheads from a wide variety of flexible structures, suggesting it has broad applicability. Our framework easily fits into most design flows by supplementing, not replacing, commercial synthesis tools.

Examples in the next section show how overprovisioning overheads can arise. We then present an algorithm that identifies this overprovisioning, and show how this leads to significantly improved overheads.

2. DESIGN EXAMPLES

A *high-level interface* in hardware should provide a flexible, timing-independent module description analogous to *abstract functions* in software. In general, it often requires both functional flexibility within a module as well as timing flexibility at the module boundary. Aspects of these more flexible module generators are already starting to appear. Recent Verilog standards enhance the power of parameters and generate blocks to give more flexibility in the elaboration phase of compilation, while higher-level synthesis frameworks seek to generate RTL directly from higher-level language descriptions [1, 8]. Unfortunately, regardless of the method used, systems built using high-level interface descriptions tend to suffer from area/energy overheads that would not be present in equivalent customized designs, due to limitations of logic synthesis tools [6].

It is important to remember that while we are interested in creating *flexible* module descriptions to make them easier to reuse, we would like each silicon instance created from this module to be optimized for its specific environment. We do not want the instance created to be any more general than its environment requires, since that creates overhead. Section 3 describes our approach to remove this overhead; the rest of this section will set up the concrete examples we used to demonstrate the utility of these techniques.

The main example we will use in this paper is an m-x-n input-queued virtual channel router, depicted in Figure 1. The router is designed to operate on variable-length packets divided into flits. It forwards flits using cut-through flow control, and supports fanout-splitting multicast. We chose this example because, while not too complex, it demonstrates a number of ways a flexible module generator can cause overheads to occur in the implementations it creates. We created two versions of this router. The first version had fixed timing interfaces, while the second version had more flexible interfaces.

The *InputPort* module queues incoming flits per virtual channel, and holds them until ready to send across the *Fabric*. It arbitrates among virtual channels, allowing a flexible prioritization scheme.

The *RoutingTable* unit uses a flexible lookup-table to determine routing destinations from packet headers. This flexibility allows all combinations of unicast and multicast routing requests from incoming packets. It can also be uniquely programmed per *InputPort*, allowing different routing schemes from different sources.

The *Scheduler* unit arbitrates among requests, determining which inputs are granted access to the *Fabric*. Since it is designed to be flexible, it must support all combinations of unicast and/or multicast requests. To prevent system deadlock, it must first arbitrate among overlapping multicast requests so that circular wait dependencies do not occur. If the routing tables are programmed for a unicast-only system, however, this extra arbitration logic is unnecessary and thus becomes an example of logic over-provisioning.

The *Fabric* unit is a full crossbar that allows every input to route to every output. Again, depending on the specific routing table configurations, this fully-connected crossbar may be over-provisioned.

This router was initially designed using fixed timing assumptions between the modules. Figure 2(a) depicts a simplified interface and the relevant FSM control logic for the interaction of the two modules in a base design where the Scheduler is purely combinational. The *InputPort* state machine transitions between *Idle*, *Route*, and *Schedule* states. The 0-cycle latency of the *Scheduler* block is implicitly assumed in this FSM, and so a *Scheduler* with different latency characteristics will surely break this design.

To decouple these inter-module assumptions, we build control logic to account for latency behavior between blocks. Figure 2(b) depicts the modified interface. The additional *Wait* state in the *InputPort* and the extra valid bits account for a *Scheduler* with greater latency. A bypass FIFO

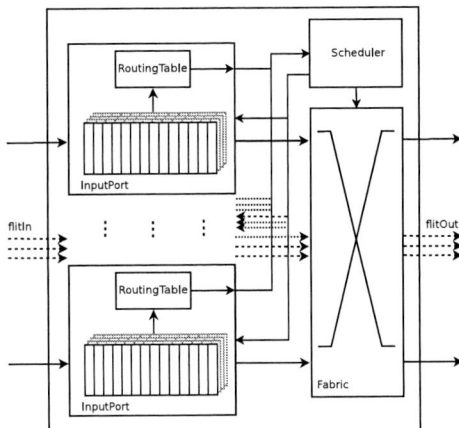

Figure 1: A flexible network router design. Note how the RoutingTable interacts with the Scheduler and, indirectly, the Fabric.

queues requests until the FSM in the *Scheduler* is ready.[1] The *InputPort* can never produce more than 1 outstanding scheduling request, so a FIFO of depth 1 is sufficient here. In general, however, an explicit backpressure mechanism would be needed to prevent overflow.

The *Scheduler* can start an allocation (dequeue FIFOs) only if all ports are *ready*. A global stall structure thus helps maintain order among the FSMs from all ports.

While our first example was chosen to be a simple block that we could use to look in depth at what structures caused the overheads in a flexible design, the second design, a memory protocol controller (MPC) for an 8-core chip-multiprocessor, serves as a realistic complex design example for our tool. The memory controller was part of a real design [5], flexible enough to support three different programming models: 1) traditional shared memory via caches, 2) streams and 3) transactions, and uses about 200K standard cells. The MPC achieves its flexibility through a series of table-based (microprogrammed) controllers. Each of these units has a superset of the functionality required to support a given memory configuration. In most memory configurations, one or more of these tables will be over-provisioned. For example, if all memories are configured in uncached modes, then all microprogram lines and state involving cache operations go unused. Likewise, in cached modes, transactional operations will never be needed.

3. REACHABILITY ANALYSIS

Reachability analysis is the process of identifying all legal states in a design. We use the term *sequential reachability* to emphasize our focus on sequential elements (flip flops) in designs. Reachability methods have been developed at many levels of abstraction, typically for formal verification of digital circuits. For example, the Murphi system uses state reachability to facilitate protocol verification [4]. Other work

---

[1] The 3 distinct *Wait* states account for different possibilities of *Scheduler* latency (0 cycles and 1+ cycles) and packet size (single-flit and multi-flit). The packet size distinction allows the network protocol to give continuing flits preference over new flits.

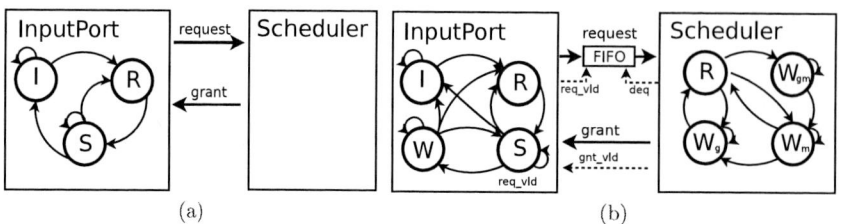

Figure 2: (a) Control logic on one interface within the Network Router. The Scheduler is purely combinational and so does not require any FSM to operate. (b) The same interface converted to be latency insensitive. Note the added FSM states and bypass-FIFO storage.

focuses more directly on reachability in gate-level netlists for formal verification of sequential circuits [9].

By their nature, unreachable states are "don't-care" conditions, and so they can be used to inform logic synthesis about additional optimizations. Our experiments have shown that modern commercial synthesis tools already do some form of reachability analysis in combinational logic, but do not propagate this information across sequential boundaries. We note there have been many prior efforts to enhance synthesis by identifying these types of optimizations. Most recently, the ABC synthesis/verification research tool utilizes a combination of simulation and SAT-sweeping to merge sequentially equivalent nodes in designs, and despite ignoring other non-equivalence node relationships, has demonstrated promising area reductions on many benchmark circuits [7]. We will use conservative approximations to capture more node relationships in our sequential reachability analysis to help eliminate waste in instances produced from more flexible module generators.

3.1 Algorithm

All sequential reachability algorithms tend to take a similar high-level approach: they start from a set of known reachable states and search for any reachable states, iterating until no new states are reachable. The reset state is a common strating point. While our underlying algorithm does not differ significantly, we include a brief discussion here for completeness. Section 3.2 presents heuristic additions, unique to our implementation, that allow the algorithm to be practical on real designs.

Our algorithm accepts a directed graph as input, and returns the reachable states for all sequential elements in the design. Generally we found it simpler to parse a gate-level netlist rather than full RTL, so we begin by doing a quick synthesis of our RTL to get a flattened gate-level netlist. The flattened gate-level netlist is then parsed into a logical directed graph. There are 3 types of nodes in our directed graph: primary inputs (PI), standard cells, and primary outputs (PO). The wires connecting these nodes form the edges of the graph. This graph data structure facilitates logical simulation as well as satisfiability (SAT) analysis.

We first isolate all combinational logic from the sequential elements (flip flops) in the graph, by cutting all edges that are outputs of a given sequential element. We connect the output of the sequential element to a new PO node and connect the original fanout of the sequential element to a new PI node. We maintain a lookup table to relate the new PO and the new PI. The result of these modifications is a

directed acyclic graph (DAG), since 1) we have severed all sequential connections and 2) combinational feedback loops are forbidden in standard cell designs.

Our algorithm requires internal register states (which correspond to some PIs) to be initialized with legal values. We rely on the fact that well-constructed designs have a global reset signal that sets the machine to a known state. We do a logic simulation of all POs, asserting the global reset PI and allowing all other PI nodes to be "don't-care", to automatically determine this initial legal state. Note that this input-to-output simulation is straightforward because we know the logical function of each node (each standard cell), and so it can be accomplished with a single pass through the graph.

By default, design inputs are assumed to reach all values. However, we can include external state constraints from the environment by additionally setting the reachable states of these PI nodes to reflect the desired constraints.

Each main loop iteration starts by seeing if the set of reached PIs has changed. If so, then we do a SAT sweep over the unreached POs, using the difference in PIs as SAT problem assumptions. Any new satisfiable states are recorded; the loop continues until no new states are found[2].

Note that the number of reachable (unreachable) states will monotonically increase (decrease) as the algorithm runs.

3.2 Conservative Heuristics

The primary concern with the algorithm is that the SAT sweeping of unreached states has exponential complexity with the number of POs, so the sweep will have difficulty completing on even modestly sized designs (those with more than 20 flop elements). We developed conservative heuristics to combat the exponential complexity, which we found to work extremely well in practice.

Imagine our problem involves sweeping 10 bits, or 1024 total sweeps. Instead of sweeping over all 10, we can break it into 2 different subgroups of 5 bits each, and sweep each subgroup independently (for a total of 64 total sweeps instead of 1024). By treating the groups independently, we can assume the design reaches the set product of states between the groups while doing a fraction of the work.

Note that any independent grouping of bits in this manner gives a legal conservative result. Since the ultimate goal is to find and remove useless logic associated with states the design cannot reach, it is perfectly okay to think some states are reachable when they are not—in fact, note that synthesis tools inherently assume *all* states are reachable. Hence, this heuristic allows us to trade off sweep time versus efficacy of

---

[2]See pseudo-code in Appendix for additional details.

logic reduction.

Since this method will assume the groups are independent, we will get the best quality of results by actually picking independent groups. Fortunately, this knowledge tends to be embedded within designs already through signal types at the RTL-level (e.g., "reg" or "logic" in Verilog), and can also be identified by instance names at the gate level (assuming the synthesis tool does not obfuscate names). While certainly not perfect, grouping based on signal names intuitively works because they come directly from the designer's intent, and typical "best practice" encourages semantically different signals to be grouped separately for improved clarity and readability.

Once we've identified the various sequential groups (SGs), we number the SGs $\{1, 2, 3, ..., g\}$, and determine the set of fan-in PIs for each. We then determine the ideal ordering of SGs that will minimize the total iterations required in the main loop. This step isn't strictly necessary but allows faster convergence of iterative maximum fixed-point solutions [2]. To do this we create a dependency graph among the SGs. The graph has g nodes, one for each group, as well as a root node that represents the original circuit inputs. The directed edges indicate dependencies, i.e., we create an edge AB if an input to SG B is driven by an output of SG A. A reverse postorder traversal of this graph gives us our ideal SG ordering. We note that this is but one approach of partitioning and traversing FSMs, and that there are a number of well-studied variations [3]. Our described method most closely resembles Cho's MBM method.

Despite using the groupings inspired by signal names, we are still likely to end up with some relatively large groups which will be difficult or impossible to sweep (recall 20+ bits becomes a challenge). For example, pipeline registers on data-paths are commonly 32 or 64 bits, and even wide decoded state registers on control-paths can be too large. Although it is easy to either ignore them (assume they reach all states) or arbitrarily divide them into smaller subgroups, both of these methods return unsatisfactory results for large control registers with many unreachable states.

The sliding window heuristic attempts to reduce the total work of these large sweeps by first eliminating many states from consideration with little effort, so that the total number of required SAT calls remains low. As an example, let's again consider a 10-bit register that has been divided into two 5-bit groups. If both groups are found to only reach 2 states, then we know the larger group can reach at *most* 4 states (their set product). We can then just do a final SAT pass over those 4 states to find the actual reachable states of the larger group. In this example, we have found the correct answer (with no approximations) using 68 SAT calls instead of 1024. If we generalize this idea beyond mutually exclusive subgroups to a series of overlapping subgroups we get a "sliding window". Each window has size w and step size s. In the example above, $w = s = 5$. Note that the efficacy of the sliding window heuristic is impacted by the order in which state bits are grouped. We observed favorable results in our designs by simply using the bit-orderings defined in the original design, but it is easy to imagine "high-effort" modes that attempt other orderings as well.

Note that the sliding window heuristic was useful in this example because the group's reachable state turned out to be sparse (4 out of 1024). If the group's potential reachable state is more densely populated, then the results obtained

from the sliding windows may not allow us to eliminate any states. In this case, we are not willing to actually sweep the full space so we abort, conservatively assuming it reaches all states. Intuitively, this is practical for our needs because we are generally interested in understanding control state in flexible designs. Since wide control registers tend to be sparse (e.g., state machines rarely have greater than 2^{20} states), this method lets us solve the groups of interest while ignoring others.

3.3 Logic Optimization

Our algorithm gives the states of sequential elements, but to use this information we need to either do our own logic optimization and mapping, or annotate those states back into the synthesis tool and leverage its combinational optimization and mapping strengths. Unfortunately, modern tools as yet provide no good way to do this annotation.[3]

To circumvent these issues, we developed a suboptimal solution that demonstrates the value of reachability information in instantiations of flexible designs while often reducing the majority of overhead. We manually instantiate pass-through decoders on the outputs of all flop groups in the gate-level netlist, and "program" the pass-through values with the determined reachable states. These pass-through decoders only let certain values appear on the outputs, treating all other conditions as don't-cares. When put through another flattened top-down synthesis flow, the tool will perform the reachability-related logic optimizations within the fanout combinational logic at the expense of the added pass-through decoder.[4] If the reductions in logic exceed the added decoder area, the synthesized design will be smaller.

An example of this method is shown in Figure 3. Figure 3(a) depicts a design with a one-hot decoder, but this one-hot reachability is lost after the flop boundary. By instantiating an additional one-hot pass-through decoder, as in Figure 3(b), we can force the synthesis tool to make the desired combinational logic optimizations (Figure 3(c)).

To force synthesis to do the desired logic optimizations we must flatten the design, obscuring the boundaries of this decoder module, which could otherwise be removed. The area result in Figure 3(c) represents an upper bound for this design, because the pass-through decoder contributes overhead.

4. RESULTS

Given the exponential nature of solving for reachability, the key question we need to address is how well our heuristic methods work on the real designs we introduced earlier. Our results are encouraging. While there are still some known limitations and scalability issues that are discussed in Section 4.2, our method has been effective at identifying and removing overhead from a number of flexible structures in real designs, often improving area by around 20%.

All of our results were obtained by running the reachability algorithm with sliding window parameters $w = 16$,

---

[3]Synopsys DesignCompiler's *set_fsm_state_vector* is intended for FSMs with clean feedback logic and often fails on larger designs; moreover it only works with one group at a time. There is planned support for a certain subclass of SystemVerilog assertions, but this isn't yet functional and doesn't allow arbitrary states.

[4]Some synthesis tools ignore decoders wider than 32 bits. We handled these these rare cases by manually injecting the key property that was proven by the reachable states.

786

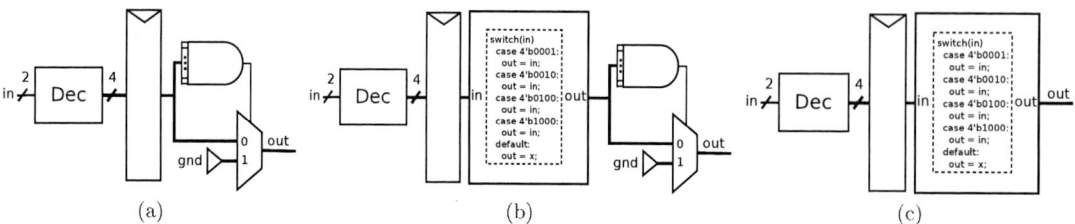

Figure 3: (a) **A sample design containing a one-hot flopped signal. Note the unneeded multiplexer logic on the output.** (b) **The design with an additional pass-through decoder, programmed to pass one-hot signals.** (c) **The design with combinational logic optimizations. The pass-through decoder still remains.**

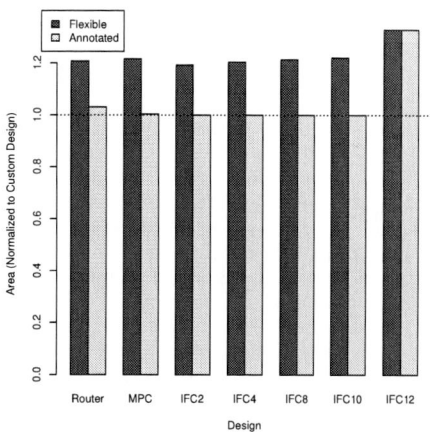

Figure 4: Synthesis results for designs with various high-level flexible structures before and after annotating reachability information using programmable decoders. IFCx represents a Router with x total ports. Results are normalized to the corresponding customized design (indicated by the horizontal dashed line).

$s = 8$, and aborting when the total possible states for a group exceeded 2^{19}.[5] Using Synopsys DesignCompiler E-2010.12-SP2 and TSMC 90nm standard cell libraries, we primarily compare synthesized areas (instead of energies or cycle time) because they provide a simple metric for understanding design complexity. Furthermore, our experience shows that for most irregular synthesized logic, area is often a good proxy for energy, and these are much smoother functions than delay.

4.1 Network Router & MPC

We performed synthesis experiments on the flexible Network Router to demonstrate how reachability analysis can remove overhead from flexible design instances. The design was configured with $m = 8$ inputs, $n = 8$ outputs, 6 virtual channels, and 72-bit flits, consuming approximately 30,000 standard logic cells (ignoring large memory queues at the inputs). The Network Router was intentionally forced into an over-provisioned case by using unicast routing tables but keeping all other blocks the same. The results at 4.2ns

---

[5]Except for IFC10, where this needed to be increased to 2^{21}.

(shown in Figure 4) indicate that the flexible Router has a 21% overhead, but the remaining overhead (after annotation with programmable decoders) was reduced to 3%.

We next turn to removing flexibility from interfaces when it is not needed. As described in Section 2, latency-insensitive interfaces contain both additional control states and bypass-FIFO storage elements. While our previous approach works for identifying unreached control states, we must modify it to identify unused bypass-FIFO elements. Intuitively, a bypass-FIFO element will be unused if its consumer is always ready to read when its producer is writing, but we cannot identify this condition by treating the producer and consumer separately. Instead, we must modify our state-grouping (after applying the name-based heuristic) to merge the producer-consumer states. This is straightforward if the designer identifies the bypass-FIFO control state (typically a counter), since the producer-consumer states will be fan-in nodes in the dependency graph.

Remember, the *Scheduler* interface enforces a global stall for all ports to ensure they a ready state. Hence, a consumer port will not only depend on its producer, but on all other producers/consumers as well, suggesting that units with more ports will have exponentially more states after the producer-consumer merging. This global stalling behavior is generally bad because the number of reachable states can quickly exceed practical values; however, it is useful for stressing our algorithm.

Figure 4 shows the area synthesis results of various-sized Network Routers at 4.0ns using a flexible interface (IFC) between *InputPort* and *Scheduler* modules. IFCx represents a router with x total ports (physical and virtual). The routers were configured using a combinational *Scheduler* (from the original custom design) so that the additional interface logic would be overhead. The areas are normalized to the corresponding custom router with no flexible interface. We are able to remove nearly all of the overhead in these examples because the bypass-FIFO becomes an unused constant and the decoder on the *Scheduler* renders the extra control states unnecessary after the tool re-optimizes the logic. Our algorithm failed to optimize IFC12 because the number of merged producer-consumer states exceeded 2^{20}.

We performed similar experiments on the MPC where it was assumed to only handle uncached memory requests at the inputs, making the original design (which also handles cached and transactional requests) over-provisioned. The MPC synthesis results are also shown in Figure 4. The extra area added by the pass-through decoders was insignificant compared to the entire design, so nearly all of the overhead

787

Table 1: Design sizes and algorithm runtimes

Design	Gates	Grps	Max	SAT	Time
Router	28512	353	$2^3/2^8$	1.9M	49.7min
MPC	209376	5166	$2^8/2^9$	2.5M	34hr
IFC2	611	35	$2^4/2^{12}$	73.3k	3.7s
IFC4	2362	69	$2^8/2^{24}$	279.4k	17.3s
IFC8	3470	137	$2^{16}/2^{48}$	868.7k	23.6min
IFC10	4802	171	$2^{20}/2^{60}$	2.6M	21hr[6]
IFC12	6973	205	$2^{24}/2^{72}$	NA	NA[7]

was recovered with our method.

4.2 Limitations and Scalability

We implemented our reachability algorithm in Python, using MiniSat-2.2 to solve SAT problems. We intended our implementation as a simple proof-of-concept to demonstrate the feasibility of our algorithm, and so it was only optimized until runtime was dominated by MiniSat calls. Table 1 presents measurements from our code on the various example designs. Runtimes were recorded on a 3GHz Core2 Duo machine with 8GB RAM.[8] Although we used a fixed set of sliding window parameters over all examples, in practice these can be tuned per-design to improve runtimes.

The runtime numbers demonstrate that the our reachability algorithm is feasible for many designs; we found the times were often comparable to that of top-down synthesis.[9] The algorithm takes longer on designs with more groups because these require more SAT calls; furthermore each SAT call takes longer on more complex designs.

The "Max" column refers to the largest sparse group that exists in each design (number of reachable states / total states), which is the most important metric for understanding the limits of our algorithm. Designs with more than 2^{20} reachable states in a sparse group are not feasible to explore, so we cannot remove any overhead associated with that group. However, since it is uncommon to find FSMs with greater than 2^{20} states, we believe this approach is practical on most high-level functionally flexible structures. This is why our algorithm scaled to the 200,000 gate MPC.

Latency-insensitive interfaces present an additional challenge because they require artificially merging a port's producer and consumer state machines together into one group. Our method of merging all groups that the FIFO-state depends on is a simple way to do this, but breaks down with global stall structures since all ports will then be viewed as dependencies, causing the merged state to quickly become too large. Instead, if additional information could correctly associate states with ports (generally difficult given a flattened gate-level netlist, but often trivial for a designer), it could merge states per-port instead of naively merging all ports, resulting in far fewer required SAT calls (similar to the grouping heuristic discussed in Section 3.2). Such user-guided partitioning solved the otherwise infeasible IFC12 design in under 5 minutes.

---

[6] 4.0min with ideal grouping (user-guided)

[7] 4.7min with ideal grouping (user-guided)

[8] The MPC example required more RAM; its runtime was measured on a 2.8GHz Opteron with 32GB RAM.

[9] In fact, the MPC example is one of the largest designs we've been able to reliably synthesize in a top-down flow.

5. CONCLUSIONS

Designing modules with flexible high-level interfaces increases reuse, and is becoming more common. However, implementation overheads often persist when these designs are synthesized with current state-of-the-art tools. This either inhibits the use of flexible techniques or requires expert designers to manually tweak design instances for specific use cases, which increases design cost and is error-prone. Although some loss of performance (clock rate, energy, or area) may be inevitable in a paradigm that emphasizes reuse through flexibility, we demonstrated that a significant portion of this overhead can be recovered with our sequential reachability analysis framework.

We believe that our reachability analysis will become an important tool for designers because (1) it provides useful information that is not leveraged in current tools, (2) it does no harm in the worst case (the analysis can be turned off, or scaled back to provide faster results), and (3) removes almost all of the overhead caused by unneeded flexibility at the module level (more than 20% in our experiments). We hope that sequential reachability analysis enables flexible module designs to better compete with customized designs in performance, so that their potential improvements in overall designer productivity become even more compelling.

6. ACKNOWLEDGEMENTS

The authors would like to acknowledge Alan Mishchenko for his helpful feedback. This work was completed with support of the Stanford PPL affiliates program, Pervasive Parallelism Lab: NVIDIA, Oracle/Sun, AMD, and NEC.

7. REFERENCES

[1] Tensilica automates architecture exploration. IEEE Review, 50(7):14, july 2004.

[2] A. V. Aho, M. S. Lam, R. Sethi, and J. D. Ullman. Compilers: Principles, Techniques, Tools. Addison-Wesley, 2nd edition, 2007.

[3] H. Cho, G. D. Hachtel, E. Macii, B. Plessier, and F. Somenzi. Algorithms for approximate fsm traversal. In Proceedings of the 30th international Design Automation Conference, DAC '93, pages 25–30, New York, NY, USA, 1993. ACM.

[4] D. L. Dill. The Murphi verification system. In Computer Aided Verification. 8th International Conference, pages 390–393. Springer-Verlag, 1996.

[5] A. Firoozshahian, A. Solomatnikov, O. Shacham, Z. Asgar, S. Richardson, C. Kozyrakis, and M. Horowitz. A Memory System Design Framework: Creating Smart Memories. In ISCA '09: Proc. 36th Annual International Symposium on Computer Architecture, 2009.

[6] K. Kelley, M. Wachs, A. Danowitz, J. P. Stevenson, S. Richardson, and M. Horowitz. Intermediate representations for controllers in chip generators. In DATE '11: Proc. Conf. on Design, Automation and Test in Europe, March 2011.

[7] A. Mishchenko, M. Case, R. Brayton, and S. Jang. Scalable and scalably-verifiable sequential synthesis. In Proc. 2008 IEEE/ACM Int'l Conf on Computer-Aided Design, ICCAD '08, pages 234–241, 2008.

[8] R. Nikhil. Bluespec system verilog: efficient, correct rtl from high level specifications. In Formal Methods and Models for Co-Design, 2004. MEMOCODE '04. Proceedings. Second ACM and IEEE International Conference on, pages 69 – 70, june 2004.

[9] D. Stoffel, M. Wedler, P. Warkentin, and W. Kunz. Structural fsm traversal. Computer-Aided Design of Integrated Circuits and Systems, IEEE Transactions on, 23(5):598 – 619, May 2004.

```
// main loop
updated = true;
while (updated) {
  updated = false;
  foreach grp in ordered-group-list {
    // inputs are outputs of other groups
    currIn = dag.getInputs(grp, reached);

    // check for any updates
    if (prevIn[grp] != currIn) {
      // we only care about new input states
      newIn = diff(currIn, prevIn[grp]);
      prevIn[grp].add(currIn);

      // unreached states are complement
      unreached = ~reached[grp];

      // do a SAT analysis on the unreached states
      // using the sliding window algorithm to avoid
      // sweeping large groups
      newReached = dag.swSAT(grp, unreached, newIn);

      // only update if new states were reached
      if (newReached.size() > 0) {
        updated = true;
        reached[grp].add(newReached);
      }
    }
  }
}
```

Figure 5: Pseudo-code of reachability algorithm main loop.

APPENDIX

A. PSEUDO-CODE

This section includes pseudo-code descriptions of our main algorithms. These were adapted from our actual implementations and should provide readers with a deeper understanding of our approach.

A.1 Main Algorithm

Figure 5 depicts the main reachability algorithm in pseudo-code. The majority of runtime is spent in the "swSAT" function.

A.2 Sliding Window

The sliding window algorithm can be formalized by assuming the full group has n bits, the window has w bits, $w \leq n$, and the step-size is s, $0 < s \leq w$. Note this will require sweeping $(1 + (n - w)/s)$ windows. At most, this implies a maximum of $2^w * (1 + (n - w)/s)$ SAT sweeps from the sliding windows. However, when $s < w$, this number of required sweeps can be reduced if unreachable states are determined in the regions of overlap. Its effectiveness on a specific problem varies with w and s. In practice, we achieved good results using $w = 14$ and $s = 8$.

Figure 6 depicts the sliding window algorithm.

```
// find the reachable states of grp in dag
// using the sliding-window algorithm
// unreached are the states to be explored
// inputs are the reachable states of inputs
function swSAT(dag, grp, unreached, inputs) {

  // set algorithm parameters
  group = dag.getGroup(grp);
  n = group.size();
  w = min(n,16);
  s = 8;
  MAX_STATES = 2**19;

  // initialize variables
  possible = group.states;
  nWindows=1+(n-w)/(s);
  num_states = 0;

  for (i=0; i < nWindows; i++) {
    // list of bits in this window
    bits_i = group.bits(i*s, i*s+w-1);
    // list of bits not in this window
    bits_i_c = group.bits() - bits_i;

    // unreached states for this window are the
    // complement of the subset of possibly reached states
    unreached_i = ~possible.subset(bits_i);

    // run SAT sweep on this window over unreached states
    reached_i = group.sweepSAT(bits_i, unreached_i, inputs);

    // retrieve reached states outside of this window
    reached_i_c = possible.subset(bits_i_c);

    // count the number of new possible states
    // (it will be the set-product)
    num_states += reached_i_c.size() * reached_i.size();

    // abort if we ever exceed our state limit
    if (num_states > MAX_STATES) {
      return null;
    }

    // keep the newly reached states
    possible.add(reached_i);
  }

  // now do a final pass over all remaining possible states
  return group.sweepSAT(group.bits(), possible, inputs);
}
```

Figure 6: Pseudo-code sliding window algorithm.

On the Asymptotic Costs of Multiplexer-based Reconfigurability

Johnathan York
The University of Texas at Austin
PO Box 8029 - F0252
Austin, TX 78713
jayork@mail.utexas.edu

Derek Chiou
The University of Texas at Austin
1 University Station C0803
Austin, TX 78712
derek@ece.utexas.edu

ABSTRACT

Existing literature documents a number of techniques for combining a set of independent datapath designs into a single datapath that is run-time configurable to the functionality of any datapath in the set. This paper explores how delay, energy and area overhead attributable to reconfigurability scales with the number of configurable functionalities, independent of the design of specific datapaths. Distinct design space regions are identified based upon common scaling properties, with implications on the design and feasible efficiency bounds of reconfigurable devices.

Categories and Subject Descriptors

C.4 [**Performance of Systems**]: Modeling Techniques

General Terms

Design,Theory

Keywords

Reconfigurable Logic, Datapath Merging

1. INTRODUCTION

Reconfigurable devices, such as Field Programmable Gate Arrays (FPGAs) and Digital Signal Processors (DSPs), are known to have substantial overhead compared to devices that cannot be reconfigured. Contemporary literature estimates that commercially-available FPGAs are 8-88X worse in area, 2-14X in delay, and 12-500X in power relative to even a standard-cell ASIC design[18]. Often worse are fetch-execute processors, which can be orders of magnitude less efficient in energy and delay than FPGA implementations[19, 24].

A middle ground that introduces flexibility into ASICs without incurring the full overhead of FPGA or processors would offer the ability to implement a limited set of applications at much higher efficiencies. Techniques to approach this middle ground have been developed for a number of application areas, under a variety of labels, including: the Datapath merging (DPM) problem, multi-mode synthesis, application-specific accelerator synthesis, Virtual Reconfigurable Architectures (VRAs), and configurable ASICs.

While extensive prior work exists on a middle ground between ASICs and general-purpose logic, most work is focused on specific applications, design problems, and/or optimization strategies. There is a literature gap at the highest levels of abstraction most useful for system-level architects. That is, there is little documented guidance on what exactly the overall design space might look like. High-level questions remain unaddressed, including: how do delay, energy, and area costs scale as the number of functionalies merged by DPM increase? How dependent are results on the specific topologies and similarities of circuits being merged? Which optimization strategies are most appropriate for a given DPM problem? This paper examines the general characteristics of the design space resulting from solving the DPM problem at a high level of abstraction, with particular attention to how overhead scales with the number of required datapath configurations.

2. BACKGROUND

As noted previously, techniques to approach a middle ground between inflexible ASICS and general-purpose programmable devices have been developed for a number of application areas, under a variety of labels. The Datapath Merging (DPM) subproblem of high-level synthesis accepts as input any number of DFGs, and produces as output a "single reconfigurable datapath", with the goal being to "design a reconfigurable datapath which incorporates all the [...] datapaths and has [the fewest] functional units and interconnections as possible" [22]. Experimental work on this subject has primarily focused on using unscheduled DFGs obtained from intermediate compiler representations of software implementations, although manual examples of the technique exist in the context of FPGA run-time reconfiguration [27]. The DPM problem has a substantial body of literature addressing algorithmic complexity [28], heuristics [2, 17], and optimization algorithms [23]. A simple example of DPM is illustrated graphically in Figure 1. Some DPM solutions may yield cyclic "false" timing paths, and these may be handled with techniques documented by Malik [21].

More recent work has applied a classic high-level synthesis argument that scheduling and binding are best jointly-optimized to the DPM problem [4, 3]. Specifically Chavet et al. [4] argue that among prior work, "four distinct ap-

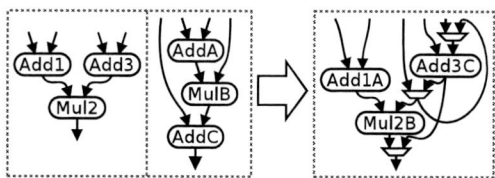

Figure 1: An simple example of the Datapath Merging problem. Here two datapaths on the left are combined into a single datapath that is configurable with the functionality of either datapath. Note the introduction of additional wires and multiplexers to support configurability.

proaches can be identified" based upon which of the steps in a conceptual high-level synthesis design are modified to be merge-aware. Chavet et al. further argue the need for a distinct "multi-mode" synthesis design flow that co-optimizes scheduling and binding, and suggest that new "scheduling, binding and register merging algorithms have to be proposed" [4]. Chiou, Bhunia and Roy [5] presented a multi-mode synthesis flow based upon a SPA-tially Chained Transformation (SPACT) in which the input DFGs are scheduled individually with estimated resource constraints, concatenated, bound, and then synthesized into HDL code.

The related application-specific accelerator synthesis problem [15, 1] has been studied to implement Application-Specific instruction set extensions for otherwise conventional microprocessors. For instance, Zuluaga and Topham [30] propose a technique that considers latency constraints during the merging process between multiple instruction set extensions. At a system level, Huang and Malik [17] discuss the DPM problem as a component of a methodology to minimize runtime reconfiguration overhead in Systems-on-a-Chip(SoC).

Addressing the problem from an angle applicable to existing FPGAs, Rullmann and Merker have developed a technique for development of virtual architectures on top of FP-GAs using datapath merging [26]. In another paper, the datapath merging technique (including a novel Ant Colony Optimization algorithm) is used to generate placement constraints to force the FPGA synthesis tool to place similar logic in similar placement between multiple designs, thereby maximizing redundant configuration bits between DFGs[25].

The Totem Project at The University of Washington has the stated goal of providing an "automatic path for the creation of custom reconfigurable hardware, targeted for use in Systems-on-a-Chip (SoCs)" [16]. This ambitious project is intended to span from high-level architecture generation, through layout of the programmable chip, ultimately including CAD suites customized for each generated architecture.

Building upon the RaPiD framework [10], Compton and Hauck [8] developed a two stage algorithm for combining multiple RaPiD netlists into an application-specific RaPiD-like structure. This paper considered area optimization as a sole metric and demonstrated that custom architectures can achieve area efficiencies of only 1.5 times a lower bound based on the minimum number of functional units able to implement each of the input netlists. These concepts were further elaborated upon in Compton's Ph.D [7], which introduced the term "configurable ASIC" for the generated architectures. Among the contributions of the dissertation is a comparison of sample configurable ASIC designs against

a traditional FPGA implementation. The comparisons with traditional FPGAs were limited to area-efficiency, but were quite favorable, with improvements ranging from 4-12X.

3. APPROACH

When examined relative to any of the input (i.e. fixed-function) datapaths, solutions of the DPM commonly rely upon the addition of multiplexers and connectivity (e.g. wires) within the datapath to introduce the required configurability. These added multiplexers and wires introduce overhead relative to the fixed-function datapath. While some of this overhead is incurred each time the device is reconfigured, the focus of this exercise is solely on overhead that is incurred post-configuration, during the operation of the datapath. That is, the focus is on overhead incurred while datapath remains a single configuration in exchange for the capability to reconfigure the datapath for other computations at a later time. To constrain the scope, we make several simplifying assumptions:

- that the configurable datapaths resulting from solving the DPM problem consist of opaque computing components present in the input datapath set, multiplexers inserted to allow configurability, and additional wires inserted to support the additional required connectivity,

- that the opaque computing components are homogenous (e.g. FPGA LUTs)[1],

- that the overhead of interest can be attributed to either multiplexer costs or the costs of the added wires,

- that no rescheduling is permissible (i.e. DPM is restricted to choosing a binding, with scheduling fixed by the datapath designs, as is common in RTL synthesis),

- that each of the input designs requires the same number of computing components,

- that the input designs have connectivity described by Rent's rule and have the same intrinsic Rent exponent, and

- that no subgraph isomorphisms are exploited in the DPM (i.e. a worst-case solution for typical DPM algorithms).

To provide insight on wiring costs independent of the peculiarities of any specific datapath, we adopt a parametrized model for wiring. The specific approach, based upon Rent's rule [20], is well known in the EDA community [6] and has proved useful in quantifying circuit characteristics in order to estimate features including wirelength distributions[14, 11] and average wirelengths [13, 29]. These results have in turn been used to estimate critical-path lengths, dynamic-power dissipation, and die areas [12]. Among the parameters used are:

- C - the number of components in a circuit (or subset thereof),

---

[1]Preliminary work suggests that a similar analysis applies to datapaths with balanced ratios of heterogenous components.

- p - the Rent exponent,
- k - the average number of terminals per component,
- α - the fraction of terminals that are inputs
- n_{cp} - the number of components in the critical path.

From these parameters, we assume wiring of the datapath designs to be merged are well-characterized by the analysis of Donath [13] and Davis et al [11]. That is, the expected total number of wires for a circuit W is given by

$$W = \alpha k C (1 - C^{p-1}). \qquad (1)$$

Moreover, the average wire length $\overline{R}$ is related to the *Rent exponent* p and the number of circuit components C, and scales as follows

$$\overline{R} \sim \begin{cases} C^{p-0.5} & \text{for } p > 0.5 \\ \log C & \text{for } p = 0.5 \\ f(p) & \text{for } p < 0.5, \end{cases} \qquad (2)$$

where $f(p)$ is an unspecified function of only p. Additionally, several parameters are defined specific to this effort:

- N - the number of datapaths (i.e. functionalities) being merged.
- β - the fraction of component input ports that need multiplexers inserted to maintain functional correctness. In the case of topological similarities, multiplexers may not be required in certain resource sharing manipulations. For a worst case bound, $\beta = 1$.
- γ - the instance-dependent fraction of dynamic energy dissipated as a result of efficiencies gained from operand isolation [9] techniques. For a worst-case bound, $\gamma = 1$.

The assumptions above suggest a bookkeeping structure with three separate sources of costs in programmable designs: the computing components themselves, multiplexers inserted to allow reconfigurability, and the wires added to connecting those multiplexers for reconfigurability. To provide insight in a manner independent of any fabrication technology, we choose to normalize the multiplexer and wiring costs relative to cost of the computing components. Therefore, with regard to any particular cost metric (e.g. area, delay, energy), one can speak of a solution to a DPM problem instance falling somewhere on the 2D plane shown in Figure 2. Towards the top and right of the space, the cost from the multiplexers and wires added for configurability dominates the cost of the opaque computing components, respectively. Similarly, towards the bottom left, the cost of computing components is dominant over the costs incurred for reconfigurability.

Based upon this accounting, we decompose the design-space into three asymptotic regions based upon the dominant source of costs. The regions are referred to as asymptotic in that their properties hold true only so far as the associated source of costs dominates the others. This decomposition offers the advantage that within an asymptotic region, costs scale distinctly with N, the number of explicit reconfiguration options. Within this framework, the remainder of this paper addresses two key questions:

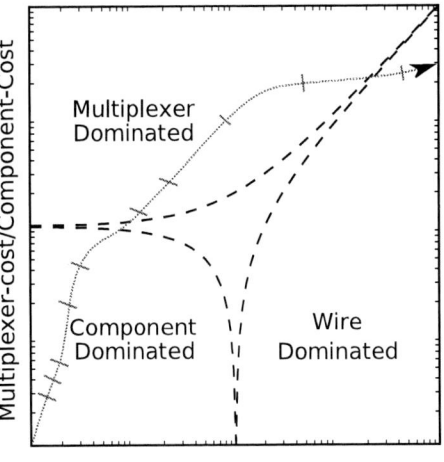

Figure 2: Depiction of the solution space implied by multiplexer-based solutions to the datapath merging (DPM) problem. A single function datapath has no configurability overhead, and would be represented by a point on the extreme bottom left. As the number of desired functionalities increase (N), the configurability overhead grows and the point traverses towards the upper right. This is illustrated graphically by the arbitrarily drawn path.

1. how do the delay, energy, and area overheads due to reconfigurability scale as the number of configurable functionalities (N) increase?

2. where do the boundaries between regions exist in terms of the number of functionalities(N) (i.e. when does the overhead due to reconfigurability become dominant)?

4. SCALING COSTS

We now address the question of how costs scale with the number of functionalities within each of the asymptotic regions. We consider the cost metrics of critical path delay, energy per operation, and area independently, using "big-O" Bachmann-Landau asymptotic notation for conciseness.

By definition, within the component-dominated asymptotic region the cost of the opaque computing components dominate the costs of multiplexers and wires added to introduce configurability. As a result, within the component dominated-region, delay, energy, and area costs are independent of the number of functionalities, and therefore scale as $O(1)$.

Within the multiplexer-dominated region, the cost of the multiplexers inserted to support configurability dominates by definition, and thus overall costs scale as do the costs of the added multiplexers. In the worst-case, it is sufficient to solve the DPM problem by inserting a N-input multiplexer at the input of every opaque computing component. That is, the N inputs of each added multiplexer are connected so as to provide the connectivity required for each of the N functionalities. Assuming a worst-case recursive implementation of 2-input multiplexers, for each component this re-

Asymptotic Region	Delay	Energy	Area
Component	$O(1)$	$O(1)$	$O(1)$
Multiplexer	$O(log(N))$	$O(N)$	$O(N)$
Wire	$O(N)$	$O(N^{\frac{3}{2}})$	Omitted

Table 1: Cost scaling with number of functionalities (N)

sults in the addition of $O(N)$ multiplexers in a configuration $O(log(N))$ deep. Therefore the critical path delay added by these multiplexers scales as $O(log(N))$. Again assuming a worst-case implementation with no operand isolation or data gating, the energy dissipated by the multiplexers switching scales with the number of multiplexers, or $O(N)$. The area required also scales with $O(N)$.

Within the wire-dominated region, the cost of wires added to support configurability dominate by definition and thus overall costs scale asymptotically as do the costs of the added wires. As noted above, in the worst-case adding N-input multiplexers for each input of each computing component is sufficient to solve the DPM problem. Each of these N inputs requires a corresponding wire in a circuit to provide the needed connectivity. In the worst case, the overall data path area scales as $O(N)$ due to the added $O(N)$ multiplexers, such that wirelength distributions would tend to scale as $O(\sqrt{N})$. If one assumes these wires are unbuffered within the datapath, critical path delay will scale then as the square of the wirelength or as $O(N)$. As the number of wires (W) scales as $O(N)$, and the wirelength distribution scales as $O(\sqrt{N})$, the dynamic energy required to charge/discharge this wire network therefore scales as $O(N^{\frac{3}{2}})$. A similar analysis for area is omitted as a straight-forward analysis is complicated by the existence of distinct resources (e.g. metal layers) for wiring that are scalable somewhat independently of the area of active resources.

These scaling derivations are summarized in Table 1. A casual inspection suggests a number of important features of the design-space implied by the DPM problem:

1. Within the component dominated region, the marginal cost of adding new functionalities to a given design is asymptotically zero. The extent of this region is of key importance and is discussed in the following section.

2. In the multiplexer and wire dominated regions, delay scales much better than energy with added functionalities. This provides a simple explanation for reports of much higher energy overheads relative to delay overheads in general purpose programmable devices (e.g. FPGAs [18]).

3. Wire-dominated reconfigurable datapaths scale more poorly than do multiplexer dominated designs. This suggests that as wiring costs become more costly relative to switching (i.e. multiplexer) costs in process technologies, rich configurability will tend to become an even more expensive design option.

5. REGION BOUNDARIES

While the prior section outlined the scaling properties within each asymptotic region, it does not necessarily follow that any particular region has a non-trivial extent. We now address the existence and extent of these asymptotic regions. The component-dominated region contains the single-functionality limit case, and therefore contains at least one trivial design point. We now attempt to predict the extent of the component dominated region in terms of the number of desired functionalities (N) by identifying when the cost of multiplexers and wires dominate the cost of the opaque computing components.

We begin by defining the delay of the configurable DPM solution as:

$$delay(programmable) = delay(fixed_{components})$$
$$+ n_{cp}ceil(log_2(N))delay(MUX2)$$
$$+ delay(fixed_{wires})\frac{\overline{R}_{programmable}}{\overline{R}_{fixed}}, \tag{3}$$

where $delay(programmable)$, $delay(fixed_{components})$, and $delay(MUX2)$ are the critical path delays of the programmable circuit, the fixed function components, and a 2 element multiplexer, respectively. The latter two terms of the sum correspond to the delay introduced by the multiplexers and the delay introduced by the wires added to support configurability. Similarly, define the dynamic energy of the configurable DPM solution as:

$$energy(programmable) =$$
$$energy(fixed_{components}) + W_{programmable}energy(MUX2)\gamma$$
$$+ energy(fixed_{wires})\gamma\frac{W_{programmable}}{W_{fixed}}\frac{\overline{R}_{programmable}}{\overline{R}_{fixed}}, \tag{4}$$

where $energy(programmable)$, $energy(fixed_{components})$, $energy(MUX2)$ are the total dynamic energy dissipated per operation in the programmable circuit, the fixed function components, and a 2 element multiplexer, respectively.

By definition, the boundary between the component-dominated asymptotic region and the multiplexer-dominated region exists when the cost of the opaque computing component cost equals the cost of the added multiplexers. We can compute the extent of the component dominated region for delay by solving the equation

$$delay(fixed_{components}) > n_{cp}ceil(log_2(N))delay(MUX2) \tag{5}$$

for N. Using the assumptions of equations 1 and 2, it can be shown that the DPM design solution lies in the component-dominated asymptotic region when

$$ceil(log_2(N)) < \frac{delay(fixed_{components})}{n_{cp}delay(MUX2)}. \tag{6}$$

Restated in prose, the total delay of the computing components is greater than the costs of the added multiplexors, provided the number of functionalities is less than two raised to the power of the delay of the average computing component expressed in units of the delay of a 2-input multiplexer. Similar derivations can be computed for energy and for the boundary with the wire dominated region. The end results of these derivations are summarized in Table 2.

Inspection of Table 2 reveals a number of interesting observations:

Boundary	Comp/Mux	Comp/Wire
Delay	$2^{\frac{delay(comp)}{n_{cp}delay(MUX2)}}$	$\left(\frac{delay(comp)}{delay(wires)}\right)^2$
Energy	$\frac{energy(comp)}{C \cdot energy(MUX2)}$	$\left(\frac{energy(comp)}{energy(wires)}\right)^{\frac{2}{3}}$
Area	$\frac{area(comp)}{C \cdot area(MUX2)}$	Omitted

Table 2: Each table entry is the number of functionalities (N) at the region boundary specificied in the first row. Thus the second column shows where component and multiplexer costs are equal and the lower row shows (N) where the component and wire costs are equal. Here n_{cp} is number of components on critical path, C is number of components in the circuit, and MUX2 is a 2-input multiplexer.

Boundary	Comp/Mux	Comp/Wire
Delay	2^{67}	10
Energy	3	6
Area	5	-

Table 3: Each table entry is the number of functionalities (N) at the region boundary specificied in the first row for an example 90nm technology node with 100 components in each functionality, and each component having the costs of a single 32-bit adder. This represents the worst-case bounds. For typical bounds, see Table 4.

1. As the cost of the computing components increase (relative to a 2-input multiplexer), the component-dominated asymptotic region grows larger. That is, the coarser-grained the reconfigurability, the more functionalities can be introduced without configurability dominating costs.

2. Considering only multiplexer delays, the number of functionalities (N) within component dominated region grows exponentially large with the delay of the computing components.

For better intuition, if one assumes typical values from 90nm standard cell technology, a computing element is a single 32-bit adder, $C = 100$ components in each merged functionality, and that average wirelength on the critical path scales as $\sqrt{area(adder) \cdot C}$, we can compute numeric values for the entries in Table 2. The resulting worst-case ($\beta = \gamma = 1$) values are shown in Table 3.

To estimate more typical values rather than worst case, we have conducted DPM experiments on a set of Digital Signal Processing cores taken from a software defined radio (SDR) platform, including 1) coordinate rotation digital computer (CORDIC) sine/cosine generator, 2) complex-value heterodyne stage (hetero), 3) Cascaded Integrating Comb (CIC) decimating filter, 4) Finite Impulse Response (FIR) filter, and 5) a Fast Fourier Transform (FFT) butterfly. To avoid overly optimal effects from topological similarities, we used a random (i.e. unoptimized) binding in the DPM process, and estimated values for α, β, γ, k, C. We found that

- only 51 of 202 possible (bus) multiplexers were inserted ($\beta = 0.25$),

- there was a 39% energy reduction via naive operand isolation ($\gamma = 0.61$),

Boundary	Comp/Mux	Comp/Wire
Delay	2^{80}	208
Energy	40	20
Area	38	-

Table 4: Each table entry is the number of functionalities (N) at the region boundary specified in the row header for an example 90nm technology node based on a case study conducted with Software Defined Radio datapaths.

- designs were dominated by 2-input, 3-terminal components ($k = 3, \alpha = 0.67$),

- there were an average of 27 components per functionality ($C = 27$), and

- computing component costs are dominated by 16-bit multipliers, driving the $\frac{delay(fixed_{components})}{n_{cp}}$ term.

Using these parameter values, we find that the expected values for the extent of the component dominated region as are shown in Table 4. Note that even in the worst case, 20 functionalities can be merged before the costs of configurability begin to dominate the costs of the computing elements. Therefore the component-dominated region is perhaps usefully large, even without exploiting topological similarities between merged datapaths, as is commonly assumed for DPM.

6. CONCLUSIONS

We have shown that by careful construction of the problem, it is possible to predict characteristics of the design space implied by the DPM problem independent of the specific datapath topologies being merged. Moreover, we have established three asymptotic regions based upon the dominant cost component that form a framework convenient for performing analysis early in the design process. This decomposition has impacts on the further study of DPM optimization algorithms. Notably, any optimization strategy should focus on the dominant cost. Current DPM optimization has primarily focused on minimizing multiplexer insertion, which is reasonable for designs in the multiplexer-dominated region. However, such optimization may be misguided for designs known to reside in the other asymptotic regions. For instance, DPM binding algorithms reminiscent of recursive partitioning placement algorithms may be more suitable in the wire dominated region. This paper lays the ground work for system-level prediction such that a designer might predict a target region early in the design process.

We have further shown both bounds and typical values for the extent of these asymptotic regions. Notably, in a conservative analysis with datapath designs from an SDR application, we predict that tens of functionalities can be merged before wiring or multiplexer costs begin to dominate the cost of the computing components. This would suggest that, with suitable design tools, a useful degree of reconfigurability can be introduced into fixed-function designs without the cost of the reconfigurability becoming substantial.

7. REFERENCES

[1] K. Atasu, C. Ozturan, G. Dundar, O. Mencer, and W. Luk. CHIPS: Custom hardware instruction

processor synthesis. *IEEE Transactions on Computer Aided Design of Integrated Circuits and Systems*, 27(3):528, 2008.

[2] R. Battiti and M. Protasi. Reactive local search for the maximum clique problem. Technical report, Algorithmica, 2001.

[3] L. Bertrand and E. Casseau. Automated multimode system design for high performance DSP applications. In *Proceedings of the 17th European Signal Processing Conference (EUSIPCO 2009)*, pages 1289–1293, 2009.

[4] C. Chavet, C. Andriamisaina, P. Coussy, E. Casseau, E. Juin, P. Urard, and E. Martin. A design flow dedicated to multi-mode architectures for DSP applications. In *Proceedings of the 2007 IEEE/ACM international conference on Computer-aided design*, pages 604–611. IEEE Press, 2007.

[5] L.-y. Chiou, S. Bhunia, and K. Roy. Synthesis of application-specific highly efficient multi-mode cores for embedded systems. *ACM Trans. Embed. Comput. Syst.*, 4(1):168–188, 2005.

[6] P. Christie and D. Stroobandt. The interpretation and application of Rent's rule. *Very Large Scale Integration (VLSI) Systems, IEEE Transactions on*, 8(6):639 –648, Dec 2000.

[7] K. Compton. *Architecture Generation of Customized Reconfigurable Hardware*. PhD thesis, Northwestern University, 2003.

[8] K. Compton and S. Hauck. Totem: Custom reconfigurable array generation. *IEEE Symposium on FPGAs for Custom Computing Machines*, 2001.

[9] A. Correale, Jr. Overview of the power minimization techniques employed in the ibm powerpc 4xx embedded controllers. ISLPED '95, pages 75–80, 1995.

[10] D. Cronquist, C. Fisher, M. Figueroa, P. Franklin, and C. Ebeling. Architecture design of reconfigurable pipelined datapaths. *20th Anniversary Conference on Advanced Research in VLSI, 1999.*, pages 23–40, 1999.

[11] J. Davis, V. De, and J. Meindl. A stochastic wire-length distribution for gigascale integration (GSI)-Part I: Derivation and validation. *IEEE Transactions on Electron Devices*, 45(3), 1998.

[12] J. Davis, V. De, and J. Meindl. A stochastic wire-length distribution for gigascale integration (GSI)-Part II: Applications to clock frequency, power dissipation, and chip size estimation. *IEEE Transactions on Electron Devices*, 45(3), 1998.

[13] W. Donath. Placement and average interconnection lengths of computer logic. *Circuits and Systems, IEEE Transactions on*, 26(4):272–277, Apr 1979.

[14] W. Donath. Wire length distribution for placements of computer logic. *IBM Journal of Research and Development*, 25(2-3):152–155, 1981.

[15] W. Geurts, F. Catthoor, S. Vernalde, and H. De Man. *Accelerator Data-Path Synthesis for High-Throughput Signal Processing Applications*. Kluwer Academic Pub, 1997.

[16] S. Hauck, K. Compton, K. Eguro, M. Holland, S. Phillips, and A. Sharma. Totem: Domain-Specific Reconfigurable Logic. *submitted to IEEE Transactions on VLSI*, 2008.

[17] Z. Huang and S. Malik. Managing dynamic reconfiguration overhead in systems-on-a-chip design using reconfigurable datapaths and optimized interconnection networks. *Design, Automation and Test in Europe Conference*, 0:0735, 2001.

[18] I. Kuon and J. Rose. Measuring the gap between FPGAs and ASICs. In *FPGA '06: Proceedings of the 2006 ACM/SIGDA 14th international symposium on Field programmable gate arrays*, pages 21–30, New York, NY, USA, 2006. ACM Press.

[19] P. Kwan and C. T. Clarke. FPGAs for improved energy efficiency in processor based systems. *Advances in Computer Systems Architecture: 10th Asia-Pacific Conference, ACSAC 2005, Singapore, October 24-26, 2005: Proceedings*, 2005.

[20] B. Landman and R. Russo. On a pin versus block relationship for partitions of logic graphs. *IEEE Transactions on Computers*, C-20:1469–1479, December 1971.

[21] S. Malik. Analysis of cyclic combinational circuits. In *IEEE/ACM International Conference on Computer-Aided Design*, pages 618 –625, Nov 1993.

[22] N. Moreano, G. Araujo, Z. Huang, and S. Malik. Datapath merging and interconnection sharing for reconfigurable architectures. In *ISSS '02: Proceedings of the 15th international symposium on System Synthesis*, pages 38–43, 2002.

[23] N. Moreano, E. Borin, C. D. Souza, and G. Araujo. Efficient datapath merging for partially reconfigurable architectures. In *IEEE Transactions on Computer Aided Design of Integrated Circuits and Systems*, pages 969–980, 2005.

[24] K. Parnell and R. Bryner. Comparing and contrasting FPGA and microprocessor system design and development. Technical report, Xilinx, 2004.

[25] M. Rullmann and R. Merker. Maximum edge matching for reconfigurable computing. In *Reconfigurable Architectures Workshop at 13th IEEE International Parallel & Distributed Processing Symposium (IPDPS 2006), Rhodes, Greece*. Citeseer, 2006.

[26] M. Rullmann, R. Merker, H. Hinkelmann, P. Zipf, and M. Glesner. An Integrated Tool Flow to Realize Runtime-Reconfigurable Applications on a New Class of Partial Multi-Context FPGAs. In *Proc. 19th Intl. Conf. on Field Programmable Logic and Appls.*, 2009.

[27] N. Shirazi, W. Luk, and P. Cheung. Automating production of run-time reconfigurable designs. *Annual IEEE Symposium on Field-Programmable Custom Computing Machines*, 0:147, 1998.

[28] C. C. d. Souza, A. M. Lima, G. Araujo, and N. B. Moreano. The datapath merging problem in reconfigurable systems: Complexity, dual bounds and heuristic evaluation. *J. Exp. Algorithmics*, 2005.

[29] D. Stroobandt. Improving Donath's technique for estimating the average interconnection length in computer logic. *ELIS Technical Report*, 1996.

[30] M. Zuluaga and N. Topham. Resource sharing in custom instruction set extensions. In *Proceedings of the 6th IEEE Symposium on Application Specific Processors.(Jun. 2008)*, 2008.

SALSA: Systematic Logic Synthesis of Approximate Circuits

Swagath Venkataramani, Amit Sabne, Vivek Kozhikkottu, Kaushik Roy and Anand Raghunathan
School of Electrical and Computer Engineering, Purdue University
{venkata0,asabne,vkozhikk,kaushik,raghunathan}@purdue.edu

ABSTRACT

Approximate computing has emerged as a new design paradigm that exploits the inherent error resilience of a wide range of application domains by allowing hardware implementations to forsake exact Boolean equivalence with algorithmic specifications. A slew of manual design techniques for approximate computing have been proposed in recent years, but very little effort has been devoted to design automation.

We propose SALSA, a **S**ystematic methodology for **A**utomatic **L**ogic **S**ynthesis of **A**pproximate circuits. Given a golden RTL specification of a circuit and a quality constraint that defines the amount of error that may be introduced in the implementation, SALSA synthesizes an approximate version of the circuit that adheres to the pre-specified quality bounds. We make two key contributions: (i) the rigorous formulation of the problem of approximate logic synthesis, enabling the generation of circuits that are correct by construction, and (ii) mapping the problem of approximate synthesis into an equivalent traditional logic synthesis problem, thereby allowing the capabilities of existing synthesis tools to be fully utilized for approximate logic synthesis. In order to achieve these benefits, SALSA encodes the quality constraints using logic functions called *Q-functions*, and captures the flexibility that they engender as *Approximation Don't Cares* (ADCs), which are used for circuit simplification using traditional don't care based optimization techniques. We have implemented SALSA using two off-the-shelf logic synthesis tools - SIS and Synopsys Design Compiler. We automatically synthesize approximate circuits ranging from arithmetic building blocks (adders, multipliers, MAC) to entire datapaths (DCT, FIR, IIR, SAD, FFT Butterfly, Euclidean distance), demonstrating scalability and significant improvements in area (1.1X to 1.85X for tight error constraints, and 1.2X to 4.75X for relaxed error constraints) and power (1.15X to 1.75X for tight error constraints, and 1.3X to 5.25X for relaxed error constraints).

Categories and Subject Descriptors

B.7.1 [**INTEGRATED CIRCUITS**]: VLSI (Very large scale integration)

General Terms

Algorithms, Design, Synthesis

Keywords

Logic Synthesis, Approximate Computing, Low Power Design, Error Resilience

1. INTRODUCTION

Error resilience can be broadly defined as the characteristic of an application to produce acceptable outputs despite its constituent computations being performed imperfectly (with errors). A plethora of emerging application domains, in both embedded and general purpose computing, exhibit this intriguing trait. For example, applications in machine learning, recognition and data mining [1] demonstrate significant *algorithmic resilience* to errors. This tolerance can be attributed to several factors like redundancies in large input data-sets; non-existence of a unique golden result; aggregating nature of the algorithms leading to errors averaging out, *etc.* [2]. *Perceptual resilience* to errors is exhibited by applications that involve a human interface. These systems could tolerate errors provided that they are not perceivable by the end user. Such applications abound in speech, video and graphics processing domains [3]. In general, most of these applications are highly compute intensive and their hardware implementations expend significant amounts of energy. By forsaking the convention of designing precise circuits and by harnessing the error tolerance provided by these application domains, significant savings in power and performance can be realized [2–8].

When it comes to the design of approximate circuits, there have been two major schools of thought. The first class of *over-scaling based approximation* methods induce timing errors in circuits by subjecting them to voltage over-scaling [8]. In contrast, the other class of *functional approximation* techniques approximate the logic functions computed by the circuits so as to reduce their implementation complexity, leading to area and energy benefits.

Initial attempts to design functionally approximate circuits focused on manual re-design of common arithmetic building blocks [9–11]. However, these techniques are confined to specific well-studied circuits such as adders and multipliers and for larger and more complex circuits, an automated synthesis procedure is indispensable.

In this work, we present SALSA, a novel systematic methodology for logic synthesis of functionally approximate circuits. Starting with an RTL description of the exact circuit and an error constraint that specifies the type and amount of error that the implementation can accommodate, SALSA automatically synthesizes a functionally approximate version of the circuit that adheres to the pre-specified error constraints. The proposed methodology rigorously reformulates the problem of Approximate Logic Synthesis (ALS) and maps it into a traditional logic synthesis problem. The problem formulation and the solution approach adopted in SALSA beget the following advantages:

- The proposed methodology provides an inherent guarantee that the specified bounds are never transgressed, thus enabling synthesis of correct-by-construction approximate circuits.

- The transformations are completely independent of the target error metric as well as the circuit considered for approximation. In essence, this decouples the synthesis procedure from the error metric, making this approach flexible and general.

- Additionally, by virtue of transforming and mapping ALS to a traditional logic synthesis problem, existing off-the-shelf logic synthesis tools could just be re-used for approximate circuit synthesis. This obviates the need for developing a custom tool for ALS, thus lowering the barrier to adoption. Further, this widens the scope of approximations that can be effected on the circuits, since the entire power of existing logic optimization algorithms can be leveraged.

We have prototyped SALSA using two different logic synthesis tools to demonstrate its generality and used it to synthesize a range of arithmetic circuits and datapaths. We demonstrate that the circuits synthesized by SALSA achieve significant reductions in area and power.

The rest of the paper is organized as follows. Section 2 overviews prior work pertaining to approximate circuit design and synthesis. The essence of the SALSA approach and the required preliminaries are elaborated in Section 3. The details of the proposed SALSA methodology are explained in Section 4. The experimental methodology and results are subsequently presented in Sections 5 and 6. Section 7 provides a brief summary and concludes the paper.

2. RELATED WORK

Previous research efforts have exploited the error resilience of applications at various levels of design abstraction. Compile time software techniques like [6], architecture level approaches like [5] and cross-layer methodologies like [7] are representative examples at higher levels of design abstraction. In this section, we provide a survey of techniques that apply approximations at the logic and transistor levels of abstraction.

A number of previous works have focused on manually approximating specific circuits like adders [9, 10] and multipliers [11] by taking advantage of their structural properties and the difference in the significance of their output bits. However, all these design techniques are confined to the specific circuits that they target. Automation becomes necessary as circuits grow functionally complex and the approximations that can be performed on them become non-intuitive.

One class of automation techniques target synthesizing circuits that trade-off accuracy for power through voltage over-scaling. Traditional synthesis optimizations result in circuits that contain a large number of near-critical paths and impede aggressive voltage scaling. To ensure a graceful degradation in the number of timing violations under over-scaling, the path delay distribution of the circuit is reshaped by increasing the slack of frequently exercised paths through cell sizing [12]. Techniques are proposed in [13] to estimate and analyze the errors caused due to such approximations.

Improvement in power and performance could be alternatively achieved by simplifying the logic functions to reduce their implementation complexity. The first automation effort in this direction focused on two-level circuits, by complementing the output for selected minterms to reduce the sum-of-products implementation [14]. For multi-level circuits, [15] proposes a scheme where a node in the circuit is assumed to have a stuck-at-fault and the circuit is simplified by propagating this redundancy. The resultant errors are then estimated using simulation and a modified automatic test pattern generation (ATPG) algorithm. This process is iterated until the pre-specified bounds are violated. A similar iterative approach is adopted in [16], however, pruning is instead carried out on paths with lowest path activation probabilities.

A common attribute of the above techniques is that they require design of a custom tool to perform the required approximations. In both cases, the quality metric is, in essence, hardwired into their synthesis procedures *i.e.*, the synthesis tools need to be substantially modified for using them with different error metrics. Also, these techniques do not perform any structural modifications to circuits but rather simplify circuits only through redundancy propagation or by pruning gates exclusive to a path. Lastly, these techniques mostly rely on simulations to testify if the approximate circuit adheres to the quality bounds.

In contrast, SALSA takes a systematic approach to approximate logic synthesis. The problem is reformulated using circuit transformations and cast in such a manner that existing logic synthesis tools could be leveraged for approximate logic synthesis. This vastly enhances the extent of approximations applied, since the full suite of techniques used in logic synthesis tools can be utilized. In addition to pruning/removing gates, this approach provides the capability to transform the functionality of circuit nodes. Also, SALSA decouples the synthesis procedure from the target error metric, which makes the approach more generic and easily adaptable. Finally, SALSA provides an inherent guarantee that the synthesized approximate circuit adheres to the pre-specified quality bounds. We believe that the above distinguishing traits make SALSA a promising approach to approximate logic synthesis.

3. SALSA: PRELIMINARIES AND APPROACH

The problem statement for approximate logic synthesis could be articulated as follows. Given the description of a logic circuit and a constraint on the errors that could be tolerated, the synthesis procedure should identify avenues for logic simplification and generate a functionally approximate version of the circuit that satisfies the pre-defined error bounds. This section describes the approach used in SALSA to accomplish this objective.

3.1 Quality Constraint Circuit

Figure 1 shows the Quality Constraint Circuit (QCC) that is used in SALSA to formulate the problem of approximate synthesis. The QCC is composed of three major blocks *viz.* the *Original circuit*, the *Approximate circuit* and the *Quality function* (Q-function). The original circuit block contains a structural description of the circuit that needs to be approximated and the error constraints that are to be satisfied are encoded into the Q-function. From the problem definition, both these blocks are available as inputs to SALSA. The task of SALSA is to synthesize the approximate circuit, so that the constraints set in the Q-function are never violated.

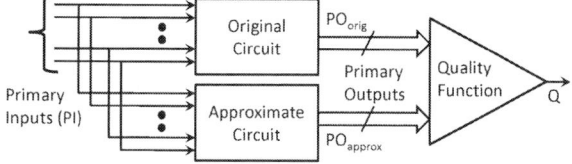

Figure 1: Quality constraint circuit

The inputs to the QCC are the primary inputs of the circuit considered for approximation. The output of the QCC is a single bit Q that indicates whether the constraints encoded into the Q-function are satisfied. The Q-function takes outputs from both the original circuit PO_{orig} and approximate circuit PO_{approx} and decides if the quality constraints are satisfied. A Q output of logic '1' means that the approximate circuit conforms to the imposed quality bounds whereas a logic '0' output indicates a transgression. Thus, the QCC determines the legitimacy of the approximate circuit. From a functional viewpoint, for the approximate circuit to be valid, we need to ensure that Q evaluates to '1' for all possible input combinations. Stated otherwise, the QCC with the synthesized approximate circuit should evaluate to a tautology. At all times during the approximate synthesis process, SALSA preserves this invariant.

3.2 Quality Function

As mentioned earlier, the Q-function takes in outputs from the original and approximate circuits and generates a single bit output indicating quality. In a circuit with M primary outputs, the Q-function maps 2M inputs into a one bit output. Thus, in SALSA, any error metric that could be expressed as a Boolean function of the original and approximate circuit output bits could be specified as the Q-function.

For our experiments, we use two different quality metrics. The first is error magnitude, where the approximate output can differ from the correct output by no more than a specified value. The other is relative error, in which the ratio of the original and approximate values is constrained to differ from 1 by at most a certain margin. The Q-functions for these quality metrics are provided in equations 1 and 2 and can be easily encoded as logic functions.

$$Q = \left(\left| PO_{orig} - PO_{approx} \right| \leq K \right) \quad ? \quad 1 : 0 \qquad (1)$$

$$Q = \left(1 - K \leq \frac{PO_{approx}}{PO_{orig}} \leq 1 + K \right) \quad ? \quad 1 : 0 \qquad (2)$$

The results obtained by applying SALSA on a wide range of circuits using these metrics are described in Section 6.

3.3 Approximation Don't Cares

We next describe the strategy used in SALSA to transform the ALS problem into a traditional logic synthesis problem. In the QCC, the primary outputs of the original and approximate circuit represent internal nodes. We know that the outputs of the approximate circuit PO_{approx} are valid provided that they do not cause the value at Q to evaluate to '0' for any input value. In other words, we could functionally modify the approximate circuit if the change can never affect the value of Q.

In multi-level logic synthesis, the Observability Don't Cares (ODCs) of a node in a logic circuit can be defined as the set of input values for which the primary outputs of the circuit remain insensitive to the node's output [17]. These input combinations can be used to simplify the node because they do not affect the primary outputs of the circuit.

Applying this concept in our scenario, finding the observability don't cares at a bit of PO_{approx} (which is an internal signal in the QCC) gives us the set of primary input values for which Q is insensitive to an output of the approximate circuit. SALSA uses this information to aid in approximating the circuit, and by virtue of their special significance these ODCs are termed as *Approximation Don't Cares* (ADCs) of the circuit.

The question that remains is how we could make use of the ADCs to approximate the circuit. We know that External Don't Cares (EXDCs) of an output in a circuit are the set of primary input combinations for which that primary output is a don't care. In our case, if the approximate circuit block is looked at in isolation, the ADCs for a given bit of PO_{approx}

could be considered as the external don't cares for that output. Therefore, by setting these input combinations (ADCs) as EXDCs of an output in the approximate circuit, we could legally simplify or (in our context) approximate the cone of logic generating that output using standard don't care based synthesis techniques [18–20].

3.4 Iterative Simplification

In the above method, it is important to note that when one output is being approximated, the functionality of all other outputs remain unaffected. This is because the ADCs, by definition, are specific to a given output bit and do not influence other outputs in any way. However, there could be avenues for approximation in the cones of logic of other output bits and hence this process of approximation should be repeated for all output bits. After each approximation, the QCC setup is updated with the latest available approximate circuit before computing the ADCs for the next output bit.

In summary, the key steps in SALSA are as follows

- Compute ODCs at an output bit of the approximate circuit to derive the set of input combinations for which Q is insensitive to that output.

- Set these ADCs as EXDCs for that output bit and simplify the circuit under this condition.

- Update the QCC with the latest available approximate circuit and iterate this process over all output bits.

The above procedure ensures that the approximations carried out never violate the specified error bounds. Also, the intermediate circuit produced after each iteration is legal and synthesis can be stopped at any point to yield a valid approximate circuit. As shown in later sections, these steps can be realized using conventional logic synthesis tools, which vastly increases the scope of optimization techniques used in ALS.

4. SALSA METHODOLOGY

This section describes the methodology that we use to realize the approach proposed in the previous section. The potential challenges in such an implementation and the speedup techniques and heuristics to overcome the same are also described.

4.1 SALSA Algorithm

Algorithm 1 SALSA

Inputs: O : Original Circuit
                   Q : Quality Function
Output : A : Approximate Circuit
Begin
Initialize $A \Leftarrow O$
for each $PO_i \in PO_{approx}$ **do**
    ## STEP 1: Obtain ADCs as f($PO_{orig} \cup PO_{approx} - PO_i$) ##
    $ADC\_PO_i \Leftarrow$ Get_ADC_PO (Q, PO_i)
    ## STEP 2: Obtain ADCs as f(PI) ##
    $ADC_i \Leftarrow$ Get_ADC (O, A, ADC_PO_i)
    ## STEP 3: Approximate PO_i using ADCs ##
    $A_i \Leftarrow$ Approx_PO (A, PO_i, ADC_i)
    Update $A \Leftarrow A_i$
end for
Return A
End

Algorithm 1 provides an overview of the steps involved in SALSA. For each output bit, SALSA computes the ADCs and uses them to approximate the logic cone that generates it. The process of finding the ADCs for a given output bit is carried out in two steps. First, the ADCs are computed as a function of other inputs to the Q-function in the QCC i.e., PO_{orig} and PO_{approx} with the exception of the output bit being processed. Next, using the original and the approximate circuits, these ADCs are expressed in terms of the primary inputs. After this, the computed ADCs are specified as EXDCs for the output bit under consideration and used to simplify its logic. The approximate circuit thus obtained is retained as the starting point for subsequent iterations. We describe below how these steps can be implemented using off-the-shelf logic synthesis tools.

STEP 1: In order to compute the ADCs of a primary output bit PO_i of the approximate circuit, we should perform ODC analysis at that node in the QCC. Finding ODCs of an internal node in a circuit, as shown in Figure 2, involves co-factoring the output with respect to the internal node and finding the set of input combinations for which both the positive and negative co-factors are equal. The resultant circuit contains a description of the ADCs of PO_i in terms of all outputs in PO_{orig} and all outputs in PO_{approx} except PO_i. We call this the ADC-PO circuit.

In this step, we have essentially extracted the information about the sensitivity of Q to the primary output of interest. This step is performed only with the Q-function and does not involve the original circuit in any way. Once we have extracted this information, the Q-function is not required any further in the algorithm.

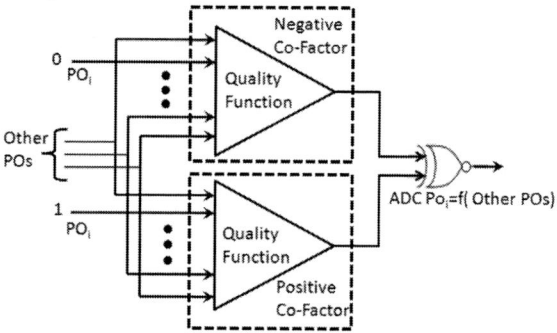

Figure 2: STEP1 - Obtaining ADCs of a primary output in terms of other original and approximate circuit outputs

STEP 2: After STEP 1, the ADCs for PO_i are available as a function of other primary outputs in the ADC-PO circuit. In this step, we express the ADCs in terms of primary inputs of the circuit. As shown in Figure 3, we connect the approximate and original circuits to the ADC-PO circuit obtained in the previous step. This concatenated circuit is simplified and the required ADCs for PO_i are thus obtained.

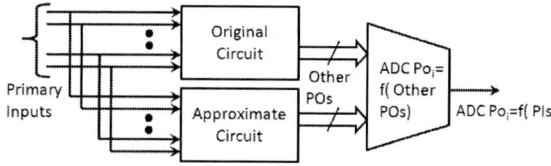

Figure 3: STEP2 - Obtaining ADCs of a primary output in terms of primary inputs

STEP 3: Given a set of ADCs for an output bit, approxi-

mating its logic can be done in a fairly straight forward manner. The computed ADCs are specified as External Don't Cares in the appropriate format required by the logic synthesis tool and conventional don't care based optimization techniques are invoked to simplify the logic cone that generates the output bit. The resultant circuit is used as the approximate circuit in the next iteration.

Thus, SALSA efficiently implements the approach described in Section 3 by reformulating the ALS problem using traditional logic synthesis operations. In each iteration of the algorithm, the synthesis tool is called thrice — once to perform each of the three steps in the algorithm.

4.2 Speedup Techniques and Other Heuristics

We next describe some optimizations that could be used to enhance the scalability of the SALSA methodology. The challenges to the above methodology could stem from two different sources - the quality function and the original circuit. If the Q-function is complex, the run-times of steps 1 and 2 of the algorithm are impacted. Also, if the original circuit has a large number of inputs or outputs, then forming the ADCs in step 2 could be a time consuming process. Speed up techniques and heuristics to overcome these challenges are discussed below.

4.2.1 Equating Un-approximated Output Bits

In SALSA, each iteration of the algorithm approximates the logic cone that generates one output bit. The hitherto unprocessed output bits should have their logic to be same as the original circuit. Therefore, while calculating the ADCs, we need not specify the entire approximate circuit, but only the logic cones that generate the output bits that have been previously approximated.

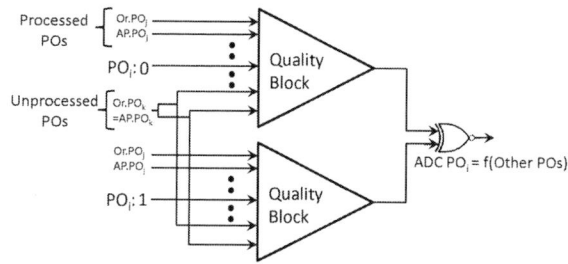

Figure 4: Equating un-approximated output bits

Using this observation, as shown in Figure 4, the unprocessed output bits of the approximate circuit are set equal to the corresponding output bits of the original circuit. This vastly simplifies the logic for ADC generation (STEP 1 and STEP 2), especially for initial output bits processed, and does not result in loss of any optimality in the approximations.

4.2.2 Quality Function Decomposition

When the number of outputs present in the original circuit is large, the complexity of the Q-function eventually grows and STEP 1 and STEP 2 of the algorithm consume significant time. A divide-and-conquer heuristic, shown in Figure 5, could be used to tackle this bottleneck. The idea is to decompose the Q-function into stages, with each stage only considering a subset of outputs from the original circuit. For the first stage, the functionality of the Q-function does not change because none of the bits have been approximated. However, for subsequent stages, the maximum error that could occur in the previously approximated bits should be considered while

799

designing the Q-function. For example, if an error magnitude based metric is used and the Q-function is decomposed in chunks of 8 bits from the LSB to MSB, then while designing the Q-function for the second set of 8 bits, the maximum error that could accrue from the lower order 8 bits should be subtracted from the actual error magnitude threshold.

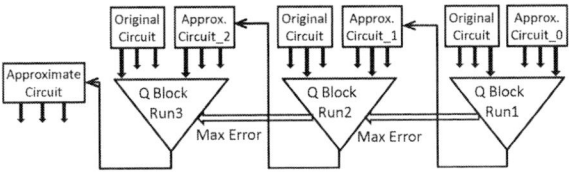

Figure 5: Quality function decomposition

This heuristic is very powerful because Q-functions of any arbitrary size can be handled by appropriately decomposing them into stages. However, we do lose some optimality in this procedure because we propagate the worst possible error across stages.

4.2.3 Exploiting Input-Output Dependencies

The previous speedup techniques were targeted at addressing the challenge of the Q-function being complex or having a large number of inputs. However, challenges may arise in finding the ADCs when the circuit to be approximated itself is large. We present two techniques to directly address this issue.

We know that, for a given output bit, not all primary inputs lie in its cone of logic. Hence, when finding the ADCs (in STEP 2) for a given output bit, we just need to define the circuit in terms of primary inputs in its transitive fan-in and generate the ADCs only in terms of these inputs. It is important to note that, this technique is exploited when generating the ADCs (STEP 2) and not when simplifying the circuit using these ADCs (STEP 3). This is because we would like to preserve the logic sharing between the output bits. In the implementation, we use the IO dependencies for ADC generation but do not extract cones of logic during logic simplification.

4.2.4 Calculating Subset of ADCs

Although the above technique is efficient in many cases, it is not effective when a primary output depends on most of the primary inputs. This scenario happens in the output MSB bits of arithmetic circuits, where the output depends on all less significant input bits. To tackle this, we resort to computing only a subset of the ADCs and use them for circuit approximation. In the implementation, we set certain dependent inputs in the cone of logic of an output to zero and then calculate its ADCs using the usual procedure. In the calculated ADC set, the condition for the dependent inputs that were set to zero is appropriately added before using them for circuit approximation. The above techniques allow us to use SALSA on larger circuits and more complex Q-functions.

5. EXPERIMENTAL METHODOLOGY

In order to demonstrate the proposed approach, we tested it on a wide range of circuits for two different error metrics *viz.* error magnitude and relative error. To demonstrate generality, the methodology was implemented using two different off-the-shelf synthesis tools, namely SIS [21] and Synopsys Design Compiler [22]. The circuits used in the experiments, listed in Table 1, range from simple arithmetic circuits to complex datapaths. The complexity of the circuits in terms of number

Table 1: Circuits used in experiments

Name	Function	Bit Width	Gate Count	I/O
RCA	Ripple Carry Adder	32	1012	64/33
KSA	Kogge Stone Adder	32	1361	64/33
CLA	Carry Look-ahead Adder	32	926	64/33
MUL	Array Multiplier	8	1055	16/16
WTM	Wallace Tree Multiplier	8	1132	16/16
MAC	Multiply and Accumulate with 32-bit accumulator	8	1910	48/33
SAD	Sum of Absolute Differences (Used in Motion Estimation)	8	1241	48/33
EU_DIST	2D-Euclidean Distance Unit (sans square root)	8	1668	32/16
BUT	Butterfly structure (Used in FFT computation)	8	496	16/18
FIR	4-tap FIR filter	8	1719	32/16
IIR	4-tap IIR filter	8	2135	56/16
DCT	8-input Discrete Cosine Transform Block	8	10817	64/72

of inputs, outputs, and gates is also listed. The circuits were approximated for a range of error values. The original and approximate circuits were mapped to the IBM 45nm technology library using Design Compiler for iso-delay and evaluated for area and power.

6. RESULTS

In this section, we present the results of experiments that evaluate the approximate circuits generated by SALSA.

Figure 6 shows the relative area (ratio of approximate circuit to original circuit) *vs.* error magnitude and relative power (ratio of approximate to original) *vs.* error magnitude plots for the benchmark circuits. The error magnitude is shown as a percentage of the maximum output value because the circuits possess different numbers of output bits and thus errors of the same magnitude have varying significance. The dynamic ranges of feasible errors are accordingly different, prompting the use of 2 different error ranges in the graphs. For 32-bit circuits like adders, MAC, and SAD, the lower X axis scale is used while other circuits, whose individual outputs have fewer bits (9 for DCT, BUT and 16 for the rest), follow the upper X axis. From the results, we see an exponential decrease (Note: X axis is in log scale) in area and power initially, which then commences to taper out as we move towards larger error values. This is explained by the fact that, in any arithmetic circuit, adjacent output bits have an exponential difference in their significance. So, for the same increase in error magnitude, the incremental potential for approximation is less as the actual value of the error increases. Also, for a given error magnitude, the cone of logic generating the LSB bits, that have exponentially lower significance compared to their MSB counterparts, have a large set of ADCs and hence have a better chance of being approximated. From Figure 6, we see that SALSA yields area savings in the range of 1.1X-1.85X for tight error constraints (less than 1%) and up to 4.75X for relaxed error constraints (upto 20%). Power benefits range from 1.15X-1.75X and 1.3X-5.25X for similar tight and relaxed error constraints respectively.

The next set of graphs, in Figure 7, show the results obtained for the relative error metric. The relative error is defined as the ratio of the approximate output to the original output. Similar trends with savings up to 1.7X in area and 1.65X in power are observed for this metric. We also observed that the ADCs derived by SALSA for the relative error metric and the error magnitude metric differed significantly. In case of the relative error metric, for small actual values of output, even a small change in the logic would prompt the output

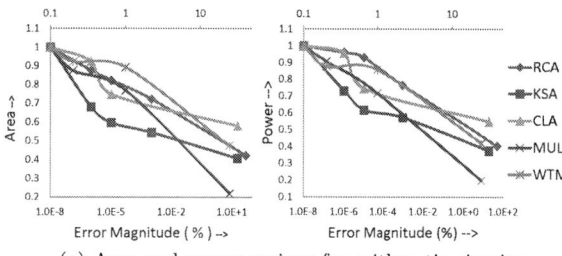

(a) Area and power savings for arithmetic circuits

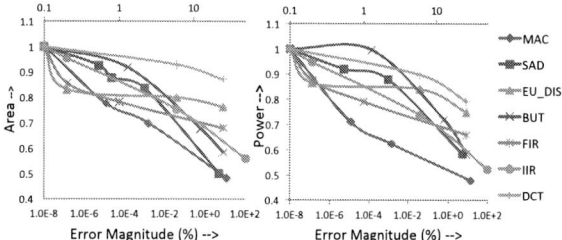

(b) Area and power savings for complex blocks and complete datapaths

Figure 6: Results for error magnitude metric

to deviate by a large percentage relative to the golden value. Moreover, the ADCs for the LSBs cannot depend on the MSB inputs. Therefore, we get a comparatively larger ADC set for the MSB output bits.

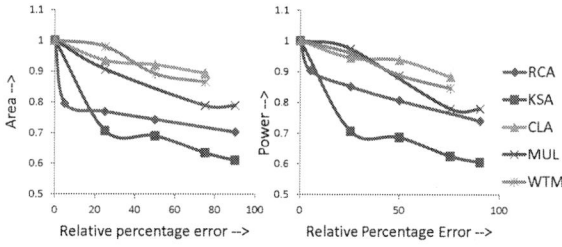

(a) Area and power savings of arithmetic circuits

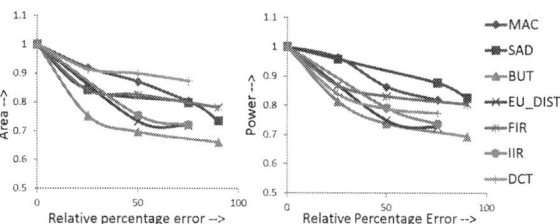

(b) Area and power savings of functional blocks and datapath modules

Figure 7: Results for relative error metric

The execution times of SALSA, on a server with an AMD Opteron 6176 (2.29 GHz) processor and 198 GB RAM, ranged from 4 minutes for smaller circuits (adders etc.) to 2.5 hours for larger datapath units (DCT).

7. CONCLUSION

Error resilient applications provide designers with a unique dimension for optimizing power consumption and area of a circuit. Paradigms like approximate computing have enabled a vast scope of implementation strategies for error resilient circuits. In our work, we have developed SALSA, a framework

to automatically synthesize approximate circuits for a given error constraint. This framework, by virtue of transforming approximate circuit synthesis into a well studied logic synthesis problem, can make use of any underlying synthesis tool to effect these approximations, thereby largely widening its scope for application. Since SALSA completely separates the notion of quality, or the error metric, from the actual synthesis procedure, it is adaptable across different error metrics. Further, it provides a guarantee that the error constraints are respected. We demonstrated the utility of SALSA by approximating various arithmetic circuits, complex blocks and entire datapaths and evaluated the benefits in terms of power and area savings.

Acknowledgment: This work was supported in part by the National Science Foundation under grant no. 1018621.

8. REFERENCES

[1] Y. K. Chen, J. Chhugani, P. Dubey, C.J. Hughes, D. Kim, S. Kumar, V.W. Lee, A.D. Nguyen, and M. Smelyanskiy. Convergence of recognition, mining, and synthesis workloads and its implications. *Proc. IEEE*, 96(5):790 –807, May 2008.

[2] S. T. Chakradhar and A. Raghunathan. Best-effort computing: Re-thinking parallel software and hardware. In *Proc. DAC*, pages 865 –870, June 2010.

[3] M.A. Breuer. Multi-media applications and imprecise computation. In *Proc. Euromicro Conf. on Digital System Design*, pages 2 – 7, Aug.- Sept. 2005.

[4] K. Palem et. al. Sustaining moore's law in embedded computing through probabilistic and approximate design: Retrospects and prospects. In *Proc. CASES*, pages 1–10, 2009.

[5] L. Leem, H. Cho, J. Bau, Q.A. Jacobson, and S. Mitra. ERSA: Error resilient system architecture for probabilistic applications. In *Proc. DATE '10*, pages 1560 –1565, Mar. 2010.

[6] H. Hoffmann et. al. Dynamic knobs for responsive power-aware computing. In *Proc. ASPLOS*, pages 199–212, 2011.

[7] V.K. Chippa, D. Mohapatra, A. Raghunathan, K. Roy, and S.T. Chakradhar. Scalable effort hardware design: Exploiting algorithmic resilience for energy efficiency. In *Proc. DAC*, pages 555 –560, June 2010.

[8] R. Hegde and N. R. Shanbhag. Energy-efficient signal processing via algorithmic noise-tolerance. In *Proc. ISLPED*, pages 30–35, 1999.

[9] V. Gupta, D. Mohapatra, S.P. Park, A. Raghunathan, and K. Roy. IMPACT: Imprecise adders for low-power approximate computing. In *Proc. ISLPED 2011*, pages 409 –414, Aug. 2011.

[10] D. Shin and S.K. Gupta. A re-design technique for datapath modules in error tolerant applications. In *Proc. ATS*, pages 431 –437, Nov. 2008.

[11] P. Kulkarni, P. Gupta, and M. Ercegovac. Trading accuracy for power with an underdesigned multiplier architecture. In *Proc. VLSI Design*, pages 346 –351, Jan. 2011.

[12] A.B. Kahng, S. Kang, R. Kumar, and J. Sartori. Slack redistribution for graceful degradation under voltage overscaling. In *Proc. ASP-DAC*, pages 825 –831, Jan. 2010.

[13] R. Venkatesan, A. Agarwal, K. Roy, and A. Raghunathan. MACACO: Modeling and analysis of circuits for approximate computing. In *Proc. ICCAD*, pages 667 –673, Nov. 2011.

[14] D. Shin and S.K. Gupta. Approximate logic synthesis for error tolerant applications. In *Proc. DATE*, pages 957–960, Mar. 2010.

[15] D. Shin and S.K. Gupta. A new circuit simplification method for error tolerant applications. In *Proc. DATE*, Mar. 2011.

[16] A. Lingamneni, C. Enz, J.-L. Nagel, K. Palem, and C. Piguet. Energy parsimonious circuit design through probabilistic pruning. In *Proc. DATE, 2011*, Mar. 2011.

[17] Giovanni De Micheli. *Synthesis and Optimization of Digital Circuits*. McGraw-Hill Higher Education, 1st edition, 1994.

[18] H. Savoj and R. K. Brayton. The use of observability and external don't cares for the simplification of multi-level networks. In *Proc. DAC*, pages 297–301, 1990.

[19] K. H. Chang, V. Bertacco, I. L. Markov, and A. Mishchenko. Logic synthesis and circuit customization using extensive external don't-cares. *ACM TODAES*, 15:26:1–26:24, June 2010.

[20] S.C. Chang and M. M. Sadowska. Perturb and simplify: optimizing circuits with external don't cares. In *Proc. ED TC*, pages 402 –406, mar 1996.

[21] E.M. Sentovich and K.J. Singh. SIS: A system for sequential circuit synthesis. Technical report, EECS, UCB, 1992.

[22] Design Compiler. Synopsys inc.

Timing ECO Optimization
Using Metal-Configurable Gate-Array Spare Cells *

Hua-Yu Chang[1], Iris Hui-Ru Jiang[2], and Yao-Wen Chang[1,3]

[1]Graduate Institute of Electronics Engineering, National Taiwan University, Taipei 10617, Taiwan
[2]Dept. of Electronics Engineering and Inst. of Electronics, National Chiao Tung University, Hsinchu 30010, Taiwan
[3]Department of Electrical Engineering, National Taiwan University, Taipei 10617, Taiwan
huayu.chang@gmail.com; huiru.jiang@gmail.com; ywchang@cc.ee.ntu.edu.tw

ABSTRACT

Due to the rapidly increasing design complexity in modern
IC designs, metal-only engineering change order (ECO) becomes inevitable to achieve design closure with a low respin
cost. Traditionally, preplaced redundant standard cells are
regarded as spare cells. However, these cells are limited
by predefined functionalities and locations, and they always
consume leakage power despite their inputs are tied off. To
overcome the inflexibility and power overhead, a new type
of spare cells, *metal-configurable gate-array spare cells*, are
considered. Therefore, in this paper, we address a new ECO
problem: Timing ECO optimization using metal-configurable
gate-array spare cells. We first study the properties for
this new ECO problem, propose a new metric, *aliveness*, to
model the capability of a spare gate array, and then develop
a timing ECO optimization framework based on aliveness,
routability, and timing satisfaction. Experimental results
show that our approach delivers superior efficiency and effectiveness.

Categories and Subject Descriptors

B.7.2 [**Integrated Circuits**]: Design Aids

General Terms

Algorithms, Design

Keywords

Engineering change order; Gate array; Mixed Integer linear
programming

1. INTRODUCTION

Due to the rapidly growing design complexity, some functional and timing failures might not be detected until late

---

*This work was partially supported by SpringSoft, TSMC, and NSC
of Taiwan under Grant No's. NSC 100-2221-E-002-088-MY3, NSC 99-
2221-E-002-207-MY3, NSC 99-2221-E-002-210-MY3, NSC 98-2221-E-
002-119-MY3, and NSC 100-2220-E-009-047.

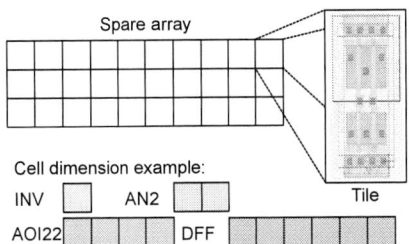

Figure 1: Metal-configurable gate-array spare cells.

design stages. To remedy these late-found failures in a short
turn-around time and with a low respin cost [1], metal-only
engineering change order (ECO) realizes incremental design
changes by only metal-layer modifications. Consequently,
metal-only ECO becomes an essential process in the modern IC design flow.

Metal-only ECO has been extensively studied in recent
literature [2, 3, 4, 5, 6, 7, 8, 9, 10, 11]. To enable metal-
only ECO, these works use pre-inserted redundant standard
cells as spare cells. To avoid gate floating, the inputs of
these redundant cells should be connected to tie cells for
ESD reliability. Once a design failure is detected, proper
spare cells are activated by rewiring their inputs and outputs. However, the weaknesses of using standard cells as
spare cells are twofold: First, these standard spare cells are
limited in functionality, quantity, and location. This inflexibility may result in unfixed failures due to a shortage of
proper spare cells. Second, these standard spare cells and
tie cells always draw leakage current. Since leakage is of particular importance not only to hand-held mobile devices but
also to standby circuit operations, the power overhead gives
a bound to the allowable number of inserted spare cells.

To overcome the inflexibility and power overhead issues, a
new type of spare cells, *metal-configurable gate-array spare
cells*, are proposed [13, 14, 15, 16, 17]. The gate-array spare
cells are developed by combining gate arrays and structured
ASICs [12]. As shown in Fig. 1, a block of gate-array spare
cells, named a *spare array*, is an array of tiles. Each tile is
composed of unwired transistors, which consume no power.
A designated functionality can be formed by configuring the
metal layers on top of several consecutive tiles. Hence, a
spare array can provide a flexible resource to realize ECO for
neighboring gates. To accommodate a sufficient and flexible
resource of spare cells for metal-only ECO, spare arrays are
scattered over the layout at the placement stage and/or filled
into empty spaces after the placement stage [17]. Further-

802

more, to facilitate timing/power characterization and mask generation, the dimension of a certain cell type is predefined. Since an irregular shape incurs performance degradation, the shape of a functional cell is typically regular. In addition, to keep the same driving strength, a functional cell generated by spare array tiles is somewhat slower and less area-efficient than a standard spare cell. Even so, if such a functional cell is close to a gate that should be fixed, this cell is definitely faster than a standard spare cell far away since interconnect dominates gate delay. On the other hand, tie cells are necessary when standard spare cells are used; their area overhead should be considered as well. Hence, the area deficiency of spare arrays is minor.

In a related work [18], Chen *et al.* propose reconfigurable decoupling capacitance (decap) cells which can be programmed as functional cells. These configurable decap cells are inserted according to IR drop consideration, and thus they are separated instead of clustered as arrays. However, a complex function may be realized by connecting several decap cells. As shown in Fig. 2(a), the internal connection among decap cells induces unwanted timing degradation. On the other hand, as shown in Fig. 2(b), selecting consecutive tiles in a spare array can realize a required function without timing degradation. Consequently, their greedy heuristic is suitable for small and local revisions. They greedily fix timing violations from the gate with the maximum loading capacitance by configuring extra decap cells as resized gates or buffers. Besides, they do not have a global view to manage all decap cells, thus leaving some timing violating paths unfixed.

Based on the above facts, using metal-configurable gate-array spare cells to accomplish ECO is practical and promising for modern IC designs. In addition, timing satisfaction is essential for ECO. Hence, in this paper, we introduce a new ECO problem: Timing ECO optimization using metal-configurable gate-array spare cells. Timing violations can be fixed by gate sizing and buffer insertion, which can be implemented by configuring spare arrays.

This new ECO problem is quite different from the conventional one. There are two issues—fragmentation and congestion—we should consider for spare arrays. A fragmented spare array is adverse to ECO because unabutted free tiles may not form the required function. On the other hand, the input and output pins of the allocated cells within a spare array should be connected to external cells, so a high pin density within a spare array and long input/output nets may induce congestion. Hence, when realizing ECO using spare arrays, we shall keep spare arrays *alive* (the free tiles should be capable of implementing as many functions as possible) and *routable* (the congestion value should be well-controlled).

To fully utilize the capability of spare arrays, we consider *aliveness*, routability, and timing satisfaction in our timing ECO optimization framework. We collect gates on timing violating paths and check if there exist gates whose timing can be improved. Most timing critical gates are extracted from these improvable gates. With a global view, we insert buffers or size the critical gates by adequate spare arrays with aliveness, routability, and timing improvement considerations. Finally, spare arrays are further packed to reduce wirelength. This procedure is repeated until no timing violations or no fixable gates can be found. We solve the spare array assignment and packing by mixed-integer linear

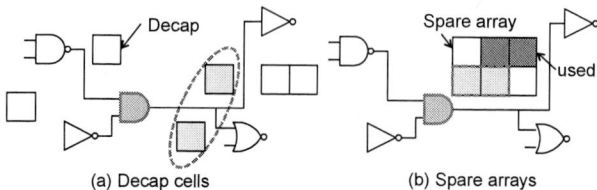

Figure 2: Reconfigurable decap cells vs. metal-configurable gate-array spare cells. Consider that the shaded AND cell is required to be resized by 2 decap cells or 2 tiles. (a) The additional wiring between decap cells induces extra delay. (b) The tiles in a spare array are adequately selected to realize the gate sizing.

programming (MILP). We summarize the features of our proposed approach as follows:

- We model the aliveness of spare arrays: To avoid fragmentation, we shall keep spare arrays alive. To achieve this goal, we propose a metric, aliveness, to model the capability of each spare array by a piece-wise linear function, which can be incorporated into our MILP.

- We adopt iterative MILP: It is impractical if we model all gates on timing violating paths into one MILP, especially for large-scale designs. In addition, the estimated timing improvement may not be accurate enough, and thus more than one MILP may be needed. Instead, we use a set of independent and small MILPs to fix timing critical gates. Experimental results show that this reduction delivers superior efficiency and effectiveness.

The remainder of this paper is organized as follows. Section 2 describes the cost metrics (including aliveness) and problem formulation. Section 3 presents our timing ECO framework using metal-configurable gate-array spare cells. Section 4 shows our experimental results. Section 5 concludes this work.

2. PROBLEM FORMULATION AND COST METRICS

In this section, we give the problem formulation and detail the cost metrics.

2.1 Problem Formulation

As mentioned in Section 1, using metal-configurable gate-array spare cells to accomplish ECO is practical and promising for modern IC designs. Different from standard spare cells, we shall avoid fragmentation and congestion when using spare arrays. Hence, we consider aliveness and routability into our problem formulation.

In addition, timing satisfaction is essential for ECO. Hence, in this paper, we focus on timing ECO optimization. Typically, timing violations are fixed by gate sizing and buffer insertion. Gate sizing is an operation that changes the driving strength of some cell on a timing violating path, while buffer insertion is an operation that inserts a buffer along a timing violating path.

The timing ECO problem is thus formulated as follows.

Problem: Given a placed design, timing violating paths, the distribution and a cell library of spare arrays, select tiles from spare arrays to perform gate sizing and/or buffer in-

803

sertion such that the timing constraint is satisfied and the aliveness and routability of spare arrays are maximized.

2.2 Cell Library of Spare Arrays

Given a spare array, a designated functionality can be formed by configuring the metal layers on top of several tiles. The cell dimension in current technology is set to multiple consecutive tiles in a single row (see Fig. 1). (An irregular shape may incur performance degradation.) The dimension, timing/power characteristics, and layout of each cell type are stored in a spare array cell library.

The timing model is based on Synopsys' Liberty library [19]. The delay and output transition of a cell depend on its input transition and output capacitance, and these values are characterized by lookup tables. The output capacitance of a cell includes its output-pin capacitance, the input-pin capacitance of its fanout gates, and the wire loading. The wire loading is proportional to the wirelength of its output net.

2.3 Aliveness

We propose a new metric, *aliveness*, to model the capability of a spare array. Given a spare array cell library which contains m different sizes of functional cells; each size of the functional cells occupies s_i tiles. Consider a spare array with k free tiles. Let z_i denote the number of cells of size s_i that are implemented by this spare array, $z_i \in N$. We have

$$s_1 z_1 + s_2 z_2 + \cdots + s_m z_m \leq k,$$
$$z_i \geq 0, \forall 1 \leq i \leq m. \tag{1}$$

The linear equation $s_1 z_1 + s_2 z_2 + \cdots + s_m z_m = k$ defines a hyperplane in the m-dimensional space. The distance between the origin to the intersection point of this hyperplane and axis z_i is k/s_i. The hyperplane and m axes form a convex polytope. Each integer point inside the polytope represents a feasible combination of cell sizes that the spare array can implement. Fig. 3 shows an example for a library with 3 different cell sizes. Since a spare array is desired to be capable of implementing as many functions as possible, we model the aliveness of a spare array by the total number of integer points within this polytope. Furthermore, we use the volume of this polytope to approximate this number. Hence, aliveness is defined as follows.

DEFINITION 1. *The aliveness of a spare array is the volume of the polytope defined by Inequality (1).*

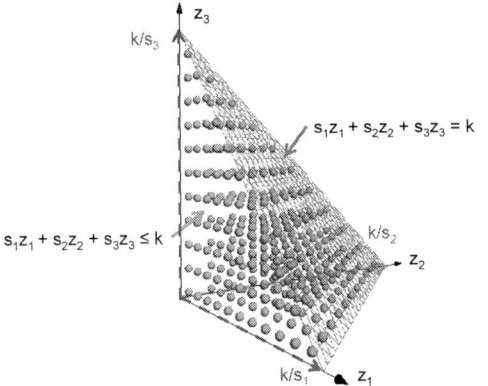

Figure 3: Aliveness. The aliveness of a spare array is modeled by the volume of the polytope defined by the hyperplane and 3 axes.

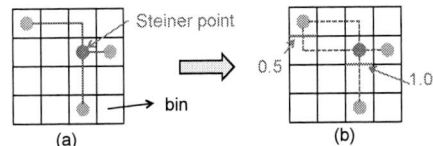

Figure 4: Routability. (a) A net is decomposed into 2-pin nets by FLUTE [20]. (b) Each 2-pin net is routed by upper-L and lower-L patterns. Each direction has 50% probability.

2.4 Routability

Congestion may induce unwanted detours, thus making the aforementioned wire loading computation inaccurate. Hence, it is necessary to consider routability during ECO. First, the layout is divided into uniform and non-overlapping bins to construct the routing grid graph. In the graph, a node represents a bin, and an edge connects each pair of adjacent bins. Each edge is associated with a routing supply, which is the number of routing tracks available for nets passing through the corresponding bin boundary. The routability of an edge is computed as the slackness between its supply and the number of occupied tracks, while the routability of a bin is thus the total routability values on the edges of its boundaries. We adopt the routing model proposed by Hsu *et al.* in [22], since it is efficient yet sufficiently accurate. Each net is first decomposed into 2-pin nets by FLUTE [20], which is a fast and accurate rectilinear Steiner minimal tree (RSMT) algorithm. Each 2-pin net is then routed by upper-L and lower-L patterns with 50% probability for each direction (see Fig. 4).

3. OUR TIMING ECO OPTIMIZATION FRAMEWORK

In this section, we detail our timing ECO framework.

3.1 Overview

In this section, we propose our timing ECO framework to achieve timing closure using metal-configurable gate-array spare cells. To fully utilize the capability of spare arrays, we consider aliveness, routability, and timing safety into our framework. It is impractical if we model all gates on timing violating paths into one mixed integer linear program (MILP), especially for large-scale designs. Therefore, it is necessary to adopt reduction techniques to reduce the number of variables and constraints. In addition, the estimated timing improvement may not be accurate enough, and thus more than one MILP may be needed. Hence, we develop iterative MILP in our framework, a set of independent and small MILPs instead of a huge MILP. Experimental results later show that our approach delivers superior efficiency and effectiveness.

Fig. 5 gives the overview of our ECO framework. First, timing analysis reports timing violating paths. Second, we collect gates on these paths and check if there exist gates whose timing can be improved by gate sizing or buffer insertion. Third, timing critical gates are extracted from these fixable gates. Fourth, we assign only extracted critical gates (instead of all gates on timing violating paths) to adequate spare arrays with aliveness, routability, and timing improvement considerations. Fifth, spare arrays are further packed

804

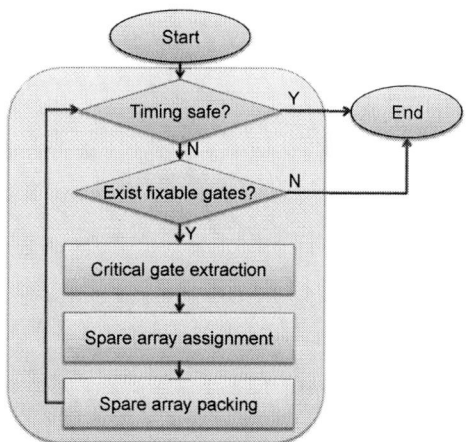

Figure 5: The overview of our timing ECO framework.

to reduce wirelength. This procedure is repeated until no timing violations or no fixable gates can be found.

3.2 Fixable and Critical Gate Extraction

To effectively reduce the problem size for MILP, we extract only timing critical gates from timing violating paths. Chang *et al.* propose a new metric of timing criticality—fixability—which can accurately identify the most timing critical gates along the timing violating paths [11]. Hence, in this paper, we extract the timing critical gates based on fixability.

First of all, we check if there exist gates on the timing violating paths whose timing can be improved by gate sizing or buffer insertion. Considering a gate and a spare array, the configuration resulting in the best timing improvement is recorded for subsequent steps. Second, we calculate the fixability for these fixable gates. Finally, timing critical gates are extracted from these fixable gates based on the method proposed in [11].

To further reduce the problem size, for each investigated gate, we consider only the spare arrays located inside its bounding polygon. The bounding polygon defined in [2] specifies a search region so that the spare arrays outside the bounding polygon are negligible.

In addition, the query of spare arrays is accelerated by using R-tree to efficiently categorize the neighboring resource.[1]

3.3 Spare Array Assignment

After extracting critical gates, we assign adequate spare arrays to perform gate sizing or buffer insertion for these gates. Since we adopt iterative MILP, we relax the timing satisfaction constraint to the objective function. Therefore, at each iteration, the objective function of spare array assignment is to maximize the aliveness, routability, and timing improvement.

The notation used in the MILP formulation for spare array assignment is listed as follows:

- G: set of critical gates.
- S: set of spare arrays, a spare array means a maximal consecutive free tiles in a single row.

---

[1]R-tree is a height-balanced tree which is widely used for indexing spatial objects such as points, rectangles, or polygons [21].

- $s_{i,j}$: required size of the sized gate or inserted buffer if gate i is assigned to spare array j, this value is obtained from critical gate extraction described in Section 3.2.
- k_j: the number of free tiles of spare array j.
- $x_{i,j}$: 0-1 variable indicating if critical gate i is assigned to spare array j. For critical gate i, only the spare arrays within its bounding polygon are considered.
- a_j: aliveness of spare array j.
- $r_{i,j}$: routability by assigning gate i to spare array j.
- r_j: routability contributed by spare array j.
- $t_{i,j}$: timing improvement by assigning gate i to spare array j.
- t_j: timing improvement contributed by spare array j.
- d_j: the pin density bound of spare array j.
- p_i: pin count of gate i.
- α, β, γ: user-specified parameters. $\alpha + \beta + \gamma = 1$.

Based on the above notation, the spare array assignment problem can be formulated as follows:

$$maximize \quad \alpha \sum_{j \in S} a_j + \beta \sum_{j \in S} r_j + \gamma \sum_{j \in S} t_j$$

$$subject\ to \quad a_j = f(k_j - \sum_{i \in G} x_{i,j} s_{i,j}), \forall j \in S, \quad (2)$$

$$r_j = \sum_{i \in G} r_{i,j}, \forall j \in S, \quad (3)$$

$$t_j = \sum_{i \in G} t_{i,j}, \forall j \in S, \quad (4)$$

$$\sum_{i \in G} x_{i,j} s_{i,j} \le k_j, \forall j \in S, \quad (5)$$

$$\sum_{j \in S} x_{i,j} = 1, \forall i \in G, \quad (6)$$

$$\sum_{i \in G} x_{i,j} p_i \le d_j, \forall j \in S, \quad (7)$$

$$x_{i,j} \in \{0, 1\}, \forall i \in G, j \in S. \quad (8)$$

The objective function is to maximize the weighted sum of aliveness, routability, and timing improvement. Equations (2), (3), and (4) define the aliveness, routability, and timing improvement of each spare array, respectively. Based on Definition 1, the aliveness of a spare array is defined by the volume of the polytope formed by Inequality (1). To incorporate the volume computation into our MILP, we use a piece-wise linear function f to approximate the volume. Since the input and output pins of the allocated cells within a spare array should be connected to external cells, long input/output nets and a high pin density within a spare array may induce congestion. The routability r_j of a spare array j is used to optimize the total congestion values induced by its related input/output nets. $r_{i,j}$ is computed by $c_{\max} - c_{i,j}$, where $c_{i,j}$ denotes the sum of congestion values of input/output nets induced by assigning gate i to spare array j, and $c_{\max} = \max_{i,j} c_{i,j}$. The congestion value of a net is the number of bin boundaries crossed based on FLUTE [20], e.g., the congestion value of the net shown in Fig. 4(a) is 6.0. The timing improvement $t_{i,j}$, relaxed from the timing satisfaction constraint, is computed assuming the resized gate or inserted buffer i is located at the center of spare array j. Constraint (5) ensures that the total sizes of allocated cells do not exceed the number of free tiles for each spare array. Constraint (6) guarantees that each critical gate is assigned to exactly one spare array. In addition to the routability optimized by the objective function, Constraint (7) limits the pin density of each spare array [22].

805

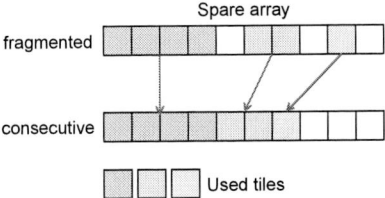

Figure 6: Fragmented vs. consecutive packing. A fragmented spare array is adverse to ECO.

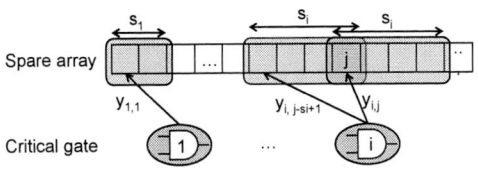

Figure 7: Spare array packing. Each tile can be occupied by at most one gate.

3.4 Spare Array Packing

After spare array assignment, each spare array is assigned with a set of critical gates to which gate sizing or buffer insertion will be applied. Spare array packing further determines the actual tiles allocated for each of these critical gates with minimum wirelength.

Since a spare array is much smaller than a bin size, different packings do not influence routability, but affect aliveness and wire loading. First of all, as shown in Fig. 6, if allocation is not in continuity, there is some gap between tiles. Consecutive allocation is better in terms of aliveness, and thus we have the following property that can be proved by an exchange argument.

THEOREM 1. *There is an optimal packing with no fragmentation.*

Based on the property, we adopt consecutive allocation for spare array packing. Furthermore, we formulate an MILP for each spare array to determine the actual allocation with wirelength minimization. For spare array $j*$, we pack the tiles for the gates whose $x_{i,j*} = 1$ after spare array assignment. The notation used in the MILP formulation for spare array packing is as follows:

- G: set of critical gates assigned to the investigated spare array.
- T: set of free tiles in the investigated spare array.
- s_i: the required size of gate i. s_i is determined by spare array assignment, $s_i = \sum_j s_{i,j} x_{i,j}$.
- $y_{i,j}$: 0-1 integer variable that denotes if gate i is assigned to tile j. When $y_{i,j} = 1$, gate i occupies tiles $j, j+1, \ldots, j+s_i - 1$.
- $w_{i,j}$: wirelength for gate i assigned to tile j. $w_{i,j}$ counts the external connections induced by input/output nets since the wirelength between tiles within the same spare array is negligible.

Based on the above notation, the spare array packing problem can be written as follows.

$$minimize \sum_{i \in G} \sum_{j \in T} w_{i,j} y_{i,j}$$

$$subject\ to \sum_{j} y_{i,j} = 1, \forall i \in G, \tag{9}$$

$$\sum_{i \in G} \sum_{k=j-s_i+1}^{j} y_{i,k} = 1, \forall j = 1, \ldots, \sum_{i \in G} s_i, \tag{10}$$

$$\sum_{i \in G} \sum_{k=j-s_i+1}^{j} y_{i,k} = 0, \forall j = \sum_{i \in G} s_i + 1, \ldots, |T|. \tag{11}$$

The objective function is to minimize the total wirelength. Constraint (9) ensures that each critical gate is assigned to exactly one position. Constraints (10) and (11) guarantee that each tile is occupied by at most one gate. Based on Theorem 1, tiles 1 to $\sum_{i \in G} s_i$ should be occupied, and tiles $\sum_{i \in G} s_i + 1$ to $|T|$ should not be occupied. When $y_{i,j} = 1$, gate i occupies tiles $j, j+1, \ldots, j+s_i - 1$, as shown in Fig. 7. Hence, tile j is occupied by gate i if $y_{i,j-s_i+1} = 1, \ldots,$ or $y_{i,j} = 1$.

4. EXPERIMENTAL RESULTS

Our algorithm was implemented in the C++ programming language on a platform with a 2.53 GHz Intel Core™2 Duo T9400 CPU and 4 GB memory. The CPLEX [23] is applied to solve the formulated MILPs. The experiments were conducted with five industrial designs, where the spare arrays are uniformly filled into the layout. Because of the flexibility of spare arrays, only 0.5% to 1% of the chip area is occupied by spare arrays. (This ratio is very low, compared with 2% to 5% for standard spare cells.) A cell in the spare array cell library incurs an average 20% timing degradation compared with the cell in the standard cell library with the same driving strength. The statistics of these circuits are summarized in Table 1, including the benchmark name (Circuit name), the number of gates in each design (Gate count), the number of available spare cells (#Spare cells), the clock period (Cycle), the number of timing violating paths (#Critical paths), the total number of gates passed by the critical paths (#Gate passed), and the total negative slack (TNS).

For fair comparison, we implemented the greedy heuristic proposed by Chen *et al.* in [18] with a small modification to handle the spare arrays. For each timing violating path, they greedily fix timing from the gate with the largest output loading. At each iteration, for the investigated gate, they find the spare array that can improve the delay of this gate best (either by gate sizing or by buffer insertion). The search region is set to the bounding box defined by this gate and its fanin/fanout gates. They then configure the selected spare array and perform the incremental STA. Because of the greedy assignment, some timing violations cannot be fixed well because the best candidates have been occupied in previous iterations.

Table 1 compares the above heuristic [18] with our approach in terms of the resulting TNS (TNS), the number of iterations (#Ite.), and the running times (Runtime). 'STA' means the runtime consumed by incremental STA, 'ILP' means the runtime used by MILP, and 'CPU' means the runtime consumed by the remaining part. For [18], '#Ite' represents the number of iterations actually applied, while for our approach, '#Ite.' represents the number of MILPs generated. In these experiments, $\alpha = \beta = \gamma$. First of all,

Table 1: Comparison between [18] and our framework.

Circuit name	Initial								Greedy [18]					Ours					
	Gate count	Spare arrays	Cycle (ns)	#Critical paths	Max #gate	#Gate passed	WNS (ns)	TNS (ns)	TNS (ns)	#Ite.	Runtime (s) CPU	STA	Total	TNS (ns)	#Ite.	Runtime(s) ILP	CPU	STA	Total
Industry1	28,927	1.0%	38	16	164	2,604	1.1	9.8	0.0	1	0.44	0.29	0.73	0.0	1	0.43	0.55	0.49	1.47
Industry2	200,504	0.5%	40	80	178	13,627	10.8	312.0	0.0	32	65.29	2.02	64.31	0.0	2	4.18	4.19	2.44	10.80
Industry3	91,107	0.5%	37	27	173	4,059	19.3	319.0	8.7	542	327.68	0.90	328.57	0.0	2	4.26	1.34	1.13	6.72
Industry4	18,932	0.1%	18	22	85	1,278	6.8	70.0	17.4	366	0.51	1.02	1.53	0.0	1	0.74	0.31	0.18	1.23
Industry5	38,011	0.5%	18	137	72	9,160	2.8	161.0	76.3	3,054	254.78	0.75	255.53	0.0	27	9.68	8.82	1.07	19.56
Ratio													16.43						1.00

we can successfully fix all timing violations because we consider aliveness during ECO optimization, while [18] fails for Industry3, Industry4, and Industry5 because their method does not have a global view to manage all spare arrays. Second, our iterative MILP is very efficient, achieving a 16.43X speedup. Fig. 8 shows the routing congestion map for Industry5. It can be seen that ECO with routability consideration indeed results in better routability, thus facilitating subsequent rewiring.

5. CONCLUSION

Traditionally, preplaced redundant standard cells are regarded as spare cells for metal-only ECO. To overcome the inflexibility and power overhead, in this paper, we introduce a new ECO problem, timing ECO optimization using a new type of spare cells, metal-configurable gate-array spare cells. We first study the properties for this new ECO problem and propose a new metric, aliveness, to model the capability of a spare gate array. We then adopt iterative MILP to solve the new ECO problem with aliveness, routability, and timing safety consideration. Experimental results show that our approach delivers superior efficiency and effectiveness.

6. REFERENCES

[1] A. Balasinski. Optimization of sub-100-nm designs for mask cost reduction. *SPIE JM3*, vol. 3 no. 2, pp. 322–331, Apr. 2004.

[2] K.-H. Ho *et al*. ECO timing optimization using spare cells and technology remapping. *IEEE TCAD*, vol. 29, no. 5, pp. 697–710, May 2010.

[3] S.-Y. Fang *et al*. Redundant-wires-aware ECO timing and mask-cost optimization. In *Proc. ICCAD*, pp. 381–386, Nov. 2010.

[4] K.-H. Ho *et al*. TRECO: dynamic technology remapping for timing engineering change orders. In *Proc. ASP-DAC*, pp. 331–336, Jan. 2010.

[5] Y.-M. Kuo *et al*. Engineering change using spare cells with constant insertion. In *Proc. ICCAD*, pp. 544–547, Nov. 2007.

[6] N. Modi and M. Marek-Sadowska. ECO-map: technology remapping for post-mask ECO using simulated annealing. In *Proc. ICCD*, pp. 652–657, Oct. 2008.

[7] I. H.-R. Jiang *et al*. Matching-based minimum-cost spare cell selection for design changes. In *Proc. DAC*, pp. 408–411, Jul. 2009.

[8] S.-L. Huang *et al*. A robust ECO engine by resource-constraint-aware technology mapping and incremental routing optimization. In *Proc. ASP-DAC*, pp. 382–387, Jan. 2011.

[9] C.-P. Lu *et al*. A metal only-ECO solver for input slew and output loading violations. In *Proc. ISPD*, pp. 191–198, Mar. 2009.

[10] H.-Y. Chang *et al*. Simultaneous functional and timing ECO. In *Proc. DAC*, pp. 140–145, Jun. 2011.

[11] H.-Y. Chang *et al*. Timing ECO optimization via Bézier curve smoothing and fixability identification. In *Proc. ICCAD*, pp. 742–746, Nov. 2011.

[12] K.-C. Wu and Y.-W. Tsai. Structured ASIC, evolution or revolution? In *Proc. ISPD*, pp. 103–106, Apr. 2004.

[13] T. Petit, STMicroelectronics. Important ECOs implementation using gate-array-like mask configurable cells. Cadence CDNLive! EMEA User Conference, May 2011.

[14] ARM Artisan Physical IP. SC12 Standard Cell Library ECO Kit, High Performance (TSMC 40nm G).

[15] L. Ciccarelli *et al*. Base cell for engineering change order (ECO) implementation. US Patent US 2010/0164547 A1, Jul. 2010.

[16] L.-C. Tien. Method for reducing layers revision in engineering change order. US Patent 7137094 B2, Nov. 2006.

[17] G. S. Tsapepas *et al*. Spare gate array cell distribution analysis. US Patent 7676776 B2, March 2010.

[18] H.-T. Chen *et al*. New spare cell design for IR drop minimization in engineering change order. In *Proc. DAC*, pp. 402–407, Jul. 2009.

[19] Liberty: The EDA library modeling standard. http://www.opensourceliberty.org/.

[20] C. Chu and Y.-C. Wong. FLUTE: Fast lookup table-based rectilinear Steiner minimal tree algorithm for VLSI Design. *IEEE TCAD*, vol. 27, no. 1, pp. 70–83, Jan. 2008.

[21] A. Guttman. R-trees: A dynamic index structure for spatial searching. In *Proc. SIGMOD*, pp. 47–57, Jun. 1984.

[22] H.-K. Hsu *et al*. Novel routability-driven analytical placement for mixed-size circuit designs. In *Proc. ICCAD*, pp. 80–84, Nov. 2011.

[23] IBM ILOG CPLEX Optimizer. http://www.ilog.com/products/cplex/.

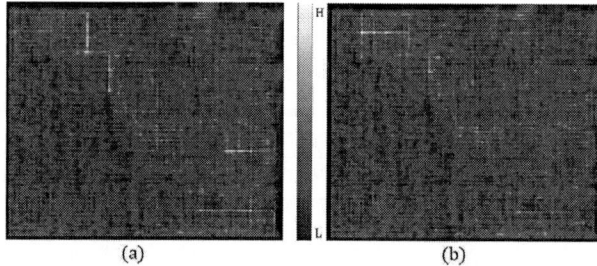

(a) (b)

Figure 8: Routing congestion map for Industry5 after timing ECO is applied. H (L) indicates high (low) congestion. (a) Without routability consideration. (b) With routability consideration.

Early Prediction of NBTI Effects Using RTL Source Code Analysis

Jayanand Asok Kumar, Kenneth M. Butler [†], Heesoo Kim and Shobha Vasudevan
University of Illinois at Urbana-Champaign   [†]Texas Instruments Inc.
{jasokku2, hkim217, shobhav}@illinois.edu   kenb@ti.com

ABSTRACT

In present day technology, the design of reliable systems must factor in temporal degradation due to aging effects such as Negative Bias Temperature Instability (NBTI). In this paper, we present a methodology to estimate delay degradation early at the Register Transfer Level (RTL). We statically analyze the RTL source code to determine signal correlations. We then determine probability distributions of RTL signals formally by using probabilistic model checking. Finally, we propagate these signal probabilities through delay macromodels and estimate the delay degradation. We demonstrate our methodology on several benchmarks RTL designs. We estimate the degradation with <10% error and up to 18.2x speedup in runtime as compared to estimation using gate-level simulations.

Categories and Subject Descriptors

B.5.2 [**Hardware**]: RTL Implementation—*Design Aids*

General Terms

Algorithms, Performance, Design, Reliability

Keywords

Aging, NBTI, Statistical Analysis, RTL, Static Analysis

1. INTRODUCTION

Negative Bias Temperature Instability (NBTI) [1] in PMOS transistors has become a significant concern in the design of reliable digital circuits. Circuit simulations show that NBTI effects over a lifetime of 10 years can degrade delay by up to 10% which may potentially violate timing constraints and hence result in a circuit failure [2] [3]. Typically, such failures are detected only while performing extensive pass/fail checks on the circuit after most of the timing closure is achieved. However, at this stage, it is often too late to revise circuit topology or architecture in order to improve circuit reliability. Therefore, it is desirable to have a methodology that can avoid such "surprises" by estimating delay degradation in earlier stages of the design flow.

There are several aging analysis techniques at the transistor level and the gate level [4] [5] [6] [7] that predict delay degradation of a circuit by computing signal probabilities of the circuit nodes. However, if such analyses are present at the Register Transfer Level (RTL), the signal probabilities that are computed would be representative of the actual usage statistics /workload of the design. As a consequence, RTL analysis provides a wider perspective of the effects of NBTI since the delay degradation of a small block is estimated in the context of a larger operating environment specified by the RTL. To the best of our knowledge, there is no technique that estimates delay degradation in RTL.

RTL aging analysis maps the effects of NBTI, an artifact of lower physical level elements, to higher level design choices. Typically, such mapping involves a loss in lower-level information and therefore, RTL estimates are distant approximations. In this paper, we present a methodology for analyzing aging in RTL. We demonstrate that, with intelligent mapping, RTL analysis can be made up to 18.2x faster (Section 5) than gate-level analysis while providing reasonably accurate estimates (<10% error). Such RTL analysis can be used as an upstream "triage" tool that trades off accuracy for speed.

In our methodology, we obtain RTL estimates of delay degradation based on the RTL signal probabilities. We compute these probabilities entirely in RTL due to their independence from the gate-level implementation. In order to map NBTI effects to RTL within specified bounds of error, we construct *macromodels* [8]. We then use these macromodels to estimate delay degradation based on the computed RTL signal probabilities.

In order to shift aging analysis to RTL, we consider delay degradation for each statement in the RTL source code. For all RTL operators appearing in a statement, we extract macromodels based on a set of heuristics. Our macromodels are subcircuits of the gate-level implementations of RTL operators. We estimate the delay degradation of an RTL operator by propagating the computed RTL signal probabilities through all the gates in the corresponding macromodel.

In order to compute RTL signal probabilities, we represent the RTL design as a Discrete Time Markov Chain (DTMC) [9]. This DTMC model is an FSM where the state transitions are labeled based on the probability distributions of the RTL input signals. We perform static analysis on the RTL source code and determine the correlations/relationships between the internal RTL signals and the RTL inputs. We then propagate the input signal probabilities through the RTL design by employing probabilistic model checking [10] on the RTL DTMC model. Probabilistic model checking explores the probabilistic executions of the RTL DTMC exhaustively and therefore, the computed RTL signal probabilities are exact.

Our methodology computes RTL signal probabilities entirely in RTL. This is significantly faster than gate-level aging analysis that propagates probabilities entirely at the gate-level. We perform gate-level analysis only when computing the degraded delays of the RTL operators by using

```
always @(posedge clk)
X1 = I1 + I2; X2 = I2 + I3; X3 = I4 + I5;
Y1 = X1 + X2; Y2 = X3;
O1 = Y1 + Y2;
```

Figure 1: An example RTL block where *I1* to *I5* are the RTL input signals and *O1* is the RTL output. *X1* to *X3* and *Y1*, *Y2* are intermediate RTL signals.

the corresponding macromodels.

We demonstrate the effectiveness of our methodology on several data-intensive RTL benchmark designs. The speedup afforded by our methodology (Section 5) comes mainly from the abstraction of behavior of datapath operators that correspond to large gate-level implementations. Although our methodology can also be applied in the context of control-intensive RTL designs, the speedup provided may be quite modest.

2. BACKGROUND CONCEPTS

An n-bit variable v can be assigned any one of 2^n possible values with associated probabilities. Collectively, these define the *probability distribution* (called PMF) of v. The *expected* value of v is the probability-weighted sum of the possible values. The *joint probability* of a set of variables is the probability with which a set of values are collectively assigned to the variables. Two variables a and b are *independent* if the probability of an assignment to a is not affected by the assignment to b. We refer to non-independent variables as *correlated variables*. PMFs that do not vary across time are said to be *stationary*.

We assume knowledge of the distribution of primary input variables. We also assume stationary PMFs for our inputs, and therefore for all variables in the system [9]. In this paper, we assume that all the primary inputs are independent. This independence assumption is frequently used in techniques [6] that employ analytical propagation of probabilities. However, our methodology is not restricted to this assumption (Section 3.2.2).

DEFINITION 1. *If a set of variables are independent, their joint probability is simply a product of their individual probabilities.*

We consider RTL designs written in Verilog HDL. We view the RTL design as a Verilog program [11] on which we can perform static analysis techniques. We consider the synthesizable subset of Verilog for our analysis. A Verilog program statement is a conditional (*if-else, case*) statement or an assignment. Every statement uses RTL operators (addition, subtraction, bitwise operators etc.). If a variable is on the right hand side of an assignment to a variable v, it is an *operand*. The set of all the operands is called $RHS(v)$.

3. OUR METHODOLOGY

In this section, we describe our methodology by using the RTL code fragment in Figure 1 as a running example. We wish to compute the degraded delay at the output $O1$. In general, the RTL may also contain sequential variables.

We consider sequential variables while computing the signal probabilities (Section 3.2.2). However, we estimate RTL delay degradation only for the datapath.

To restrict the aging analysis to RTL, we define degraded delays in terms of the RTL statements of interest. The degraded delay of a statement depends on the operation that it performs and the PMFs of the operands (elements of set $RHS(v)$). In order to find the degraded delay of a RTL block/module, we need a characterization of delay degradation for all the statements that appear in the module.

A summary of the steps in our methodology is as follows.

- For every RTL operator, we construct a gate-level circuit that can be used to estimate the degraded delay based on the PMFs of the operands. We refer to this circuit as a *macromodel* for that operator (Section 3.1).
- We compute exact PMFs for RTL operands by employing probabilistic model checking on the RTL design (Section 3.2).
- Finally, we compute the degraded delays of each RTL statement by using the PMFs of the corresponding operands and the macromodel for the corresponding operator (Section 3.3).

We now describe these steps in detail.

3.1 Modeling delay in the presence of NBTI

A macromodel for an RTL operator is a gate-level circuit with the RTL operands as inputs. We now define necessary terminology and describe our technique for constructing macromodels.

DEFINITION 2. *A logic cone of a gate input node i_g is the set of all gates and nodes of circuit C that can affect the value of i_g. A logic cone for a path P is the union of the logic cones of all the gate input nodes in P.*

DEFINITION 3. *A subcircuit of C contains the path P if all the gates and nodes in the logic cone of P are present in the subcircuit.*

DEFINITION 4. *A subcircuit of C is minimal with respect to a set of paths Π if it contains only those gates and nodes from C that are present in the logic cone of at least one path $P \in \Pi$.*

Let circuit C be a possible gate-level implementation that is obtained by synthesizing the RTL operator. The corresponding RTL operands are the inputs of C. Let Π be the set of all simple paths of C. The delay of a circuit C is defined by the delay of its slowest path $P_{crit} \in \Pi$, *i.e.* the critical path.

The circuit C can be used to accurately estimate the degraded delay of the corresponding RTL operator by computing the signal probabilities for all internal nodes [6]. The complexity of accurate probability propagation is known to grow exponentially with the size of the circuit. In order to obtain quick estimates, we consider only a smaller subcircuit of C. We use this subcircuit as the macromodel M for the RTL operator.

In order to compute the delay degradation of a path P in circuit C, we need to propagate signal probabilities from the circuit inputs to the inputs nodes of all the gates on the path [6]. However, we can also achieve this by considering a subcircuit of C which contains (Definition 3) the path P. We construct the macromodel M by extracting a subcircuit that contains a set of relevant paths in C.

Let $P_{crit} \in \Pi$ be the critical path that determines the delay of C in the absence of NBTI. Therefore, a macromodel M that contains P_{crit} can be used to estimate the delay of C in the absence of NBTI. However, in the presence of NBTI effects, the delay of all paths $\in \Pi$ degrade to different extents. Therefore, the new critical path $P'_{crit} \in \Pi$ of circuit C can potentially be different from P_{crit}. Since we intend our macromodels to provide reliable estimates of degraded delay of C, M should ideally contain the path P'_{crit}.

The delay degradation of paths depend on the signal probabilities of the internal nodes which in turn depend on the signal probabilities of the circuit inputs. These input signal probabilities are determined as a function of the design workload which is available only during runtime. Therefore, it is difficult to identify P'_{crit} apriori and include it in the macromodel M.

Let M be a subcircuit of C that contains a subset of K paths from Π, where K is a user-defined value. Consider that M does not contain P'_{crit} for the given value of K. Let $P_{approx} \in \Pi$ be the critical path of M in the presence of NBTI effects. If the degraded delay of $P_{approx} \in \Pi$ closely approximates that of P'_{crit}, we can still use M to obtain reliable estimates of the degraded delay of C. This forms the basis of our macromodeling technique.

DEFINITION 5. *For each gate-level implementation C of an RTL operator, the corresponding macromodel M is the subcircuit that is minimal (Definition 4) with respect to the set of K slowest paths in C.*

In order to construct M (Definition 5), we first determine the set of K slowest paths in C (denoted Π^K) by performing gate-level static timing analysis without considering NBTI effects. We then obtain M by pruning C and removing all gates and nodes that are not in the logic cones of any path in Π^K.

The delay degradation due to NBTI is typically less than 10%. Therefore, it is reasonable to expect that P'_{crit} is among the slowest paths of C even in the absence of NBTI effects. For small values of K, only a small fraction of the paths in Π is present in Π^K. As the value of K increases, the likelihood of $P'_{crit} \in \Pi^K$ increases. The likelihood of P_{approx} closely approximating P'_{crit} also increases with K.

Since we wish to obtain only an estimate of the degraded delay by using M, it is not necessary to choose K large enough such that M contains P'_{crit}, i.e. $P'_{crit} \in \Pi^K$. Instead, by setting a smaller value of K, we can construct M that provides estimates (using P_{approx}) within a specified range of error.

Constructing a library of macromodels

We obtain macromodels for every RTL operator and construct a library of macromodels. This macromodeling is a one-time effort and can be done offline for a given technology library. We obtain the synthesized gate-level implementations by using a library that is constructed based on the PTM 45nm model files [12].

We obtain a distinct macromodel corresponding to each possible gate-level implementation of an RTL operator. While estimating RTL delay degradation (Section 3.3), the user can guide the selection of macromodels by specifying the gate-level implementation for each RTL operator. In the absence of such guidance, our methodology will select the default macromodel for each operator.

In order to model the effects of synthesis optimizations, we consider three scenarios: Optimization in 1) logic only (technology independent), 2) delay only (technology dependent) and 3) both logic and delay. For each scenario, we perform the corresponding type of optimization on the gate-level circuit of the RTL operator. We then extract the macromodel, which we refer to as the *optimized macromodel*, from this optimized gate-level circuit. If more scenarios are considered, a richer library of macromodels can be constructed.

3.2 Computing the PMFs of RTL signals

In order to compute the degraded delay of $O1 = Y1 + Y2$ (Figure 1), we require the PMFs of $Y1$ and $Y2$. We can symbolically express $Y1$ and $Y2$ in terms of the RTL inputs as $I1 + 2*I2 + I3$ and $I4 + I5$, respectively. Once the input signal probability distributions are given, we can derive the exact PMFs of $Y1$ and $Y2$ by using these expressions.

We now describe a systematic technique to derive the PMFs for all variables of interest.

3.2.1 Source Code Static Analysis

We statically analyze the Verilog program [11] in order to symbolically express each variable of interest in terms of RTL inputs.

For an output signal v, the *signal function* $f(v)$ is the symbolic expression that includes inputs, or the "formula" that corresponds to its evaluation. The *support* of a signal function $f(v)$ is the set of all input variables in $f(v)$. Since we assume that the input PMFs are independent, the joint PMF of any subset of the set of inputs that are included in $f(v)$ are easily calculated as in Definition 1.

We statically traverse the source code for computing the signal function for each variable v of interest. For a Verilog statement where a value of v is assigned, $RHS(v)$ gives the set of variables that are included in the right hand side of the assignment statement. We recursively step back through each variable in $RHS(v)$ until the RTL inputs are reached. We define the support of v by considering the union of all RTL inputs that are reached. A detailed description of this static analysis technique can be found in [11].

In the RTL example, $RHS(Y1) = \{X1, X2\}$. Since $X1$ and $X2$ are not RTL inputs, we step the design backwards once more. We find that the supports of $Y1$ and $Y2$ are $\{I1, I2, I3\}$ and $\{I4, I5\}$, respectively. The corresponding signal functions for $Y1$ and $Y2$ are $I1 + 2*I2 + I3$ and $I4 + I5$, respectively.

3.2.2 Propagating input PMFs to RTL variables

We now describe the process of deriving the PMF of RTL variables by using the corresponding signal functions that we derive using static analysis.

The probabilistic behavior (*i.e.* the PMF) of a signal v is completely specified by the signal function $f(v)$ and the joint PMF of the support of v. We model this by using a finite DTMC, which we call an *RTL-DTMC*. We then use this RTL-DTMC in order to derive the PMF of v.

A finite *DTMC* is a finite state machine where each transition is associated with a probability. A DTMC *state* is a unique assignment of values to a set of variables called *state variables*. A transition in a DTMC is a movement from one state to another, *i.e.* an assignment of a different set of values to the state variables. The transition is labeled with the joint probability with which the state variables would be assigned their respective values in the new state.

810

For each v of our interest, we construct the RTL-DTMC by using the support of v as the state variables and their joint PMF as the transition probabilities. In this work, we assume that the support is independent. Therefore, we obtain the transition probabilities by taking the product of the individual probabilities of the variables in the support (Definition 1). In general, the support may comprise variables that are correlated. However, if their joint PMF is provided to us, we can directly use them to define the transition probabilities of the DTMC.

Each RTL-DTMC state can be viewed as corresponding to a unique input pattern that is applied to the RTL. Therefore, the probability with which the RTL-DTMC transitions to a new state is equivalent to the probability of applying the corresponding new input pattern.

Each DTMC state corresponds to a unique set of values assigned to the support of v (*i.e* the state variables). Therefore, the value assigned to v in a DTMC state can be computed by using $f(v)$. Let i be one of the possible values that v can be assigned. The probability of $v = i$ at a given time can be obtained by computing the probability of being in a DTMC state where $f(v) = i$. We repeat this for all i and compute the complete PMF of v.

In order to compute the probability of $v = i$, we tag every state in the model where $f(v) = i$ is satisfied. We would like to find the probability of being in a tagged state, at the end of N transitions of the DTMC. We employ probabilistic model checking in order to compute this probability.

If a sufficiently large N is considered (we use $N{=}10$), the probability converges to a steady value. Sequential variables in RTL introduce cycles/loops in the DTMC state graph. In such cases, N needs to be larger than the sequential depth of the design.

In the RTL example (Figure 1), the PMF of $Y1$ can be derived by using the DTMC with $I1$, $I2$ and $I3$ as the state variables. For any value i that can be assigned to $Y1$, the probability of $Y1 = i$ is equal to the probability of being in a state where $I1 + 2 * I2 + I3 = i$.

We use PRISM [10], a symbolic probabilistic model checking tool that analyzes all possible DTMC executions of length N. Therefore, the probabilities that we obtain are high in confidence as compared to those obtained using simulation-based techniques.

3.3 Computing the degraded delays in RTL

We compute the degraded delays in RTL on a statement-by-statement basis. For each RTL statement, we compute the degraded delay of the appropriate macromodel for the corresponding RTL operator. We use the PMFs of the corresponding operands and propagate them through the gates in the macromodel circuit. For each gate, the degraded delay can be analytically computed as a function of the signal probabilities of the gate inputs [6] [13]. We then employ gate-level static timing analysis using these degraded gate delays and compute the degraded delay of the macromodel.

While propagating the signal probabilities through the macromodel, we ignore the correlations among the internal gate nodes. This approximation significantly reduces the complexity of signal probability computation. For small gate-level circuits such as our macromodels, the delay degradation estimates that are obtained with this approximation (Section 5) are not far from the estimates obtained using exact probabilities [13].

The RTL signal probabilities that we compute are exact and account for all correlations amongst RTL signals (Section 3.2). We approximate the signal probabilities only at the nodes within each macromodel. In our methodology, the computation of RTL signal probabilities is separated from the computation of the signal probabilities of the internal nodes of the macromodels. Therefore, the approximation does not affect the exactness of the RTL signal probabilities. As a result, we restrict the approximation errors to the small macromodel circuits and prevent them from propagating throughout the RTL design.

Once we obtain the degraded delays for all statements in the RTL block, we compose them together and compute the degraded delay of the RTL block. We now describe the semantics for composing delays of RTL statements.

The *delay of an RTL assignment statement* is the time taken from the (rising) clock edge for the effect of the statement execution to be observed. We consider the rising edge of clock as the reference point for measuring delay.

In the RTL example (Figure 1), the values of the inputs $I1$ to $I5$ are available synchronously at the rising edge of clock. We assume that the signal arrival time T_{O1} at the output is $T_{O1} = T_{add} + max(T_{Y1}, T_{Y2})$ where T_{add} is the operator delay.

The estimates of degraded delays computed are provided as feedback to the RTL designer for revising the design. This process can be iterative until the reliability constraints are satisfied. The use of macromodels avoids the need to synthesize the design to the gate-level in each iteration.

4. SCALING OUR METHODOLOGY

Probabilistic model checking is known to encounter the problem of state-space explosion which can potentially restrict our analysis to smaller designs. In [11], a compositional reasoning approach for probabilistic model checking of RTL designs has been presented. We now briefly outline a related strategy to significantly improve the scalability of our methodology.

DEFINITION 6. *RTL signals v_1 and v_2 are independent if all the signals in the support of v_1 are independent of the signals in the support of v_2. RTL blocks B_1 and B_2 are independent of each other if there exists no signal in block B_1 that is correlated with a signal in block B_2.*

We wish to compute the PMF for the output signal v of an RTL block B. Consider that B can be structurally decomposed into a set of independent blocks B_i (Definition 6). We first compute the PMFs for the output signals v_i of these blocks by using their corresponding RTL-DTMCs (Section 3.2.2). We then "compose" the PMFs of all v_i and compute the PMF of v. Instead of constructing the RTL-DTMC for the entire block B, we now construct smaller RTL-DTMCs corresponding to B_i. This forms the basis of our compositional reasoning strategy.

We modify our static analysis algorithm, described in Section 3.2.1, in order to determine the set of independent RTL signals v_i (both sequential and combinatorial) in RTL block B. These v_i can be viewed as output signals of a set of independent RTL block B_i. Therefore, the modified static analysis algorithm decomposes the original RTL block B into a set of independent blocks B_i.

We illustrate the compositional reasoning approach by using the RTL example from Figure 1. Figure 2 shows a

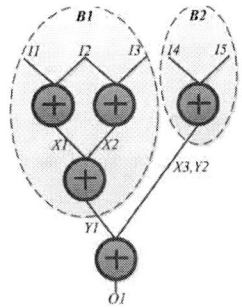

Figure 2: In the example RTL block B, $Y1$ and $Y2$ are independent signals. Therefore, B can be decomposed into two independent blocks, $B1$ and $B2$.

block diagram that depicts the decomposition of the example RTL block. We wish to compute the PMF of the output signal $O1$. Through static analysis, we determine that $O1=Y1+Y2=I1+2*I2+I3+I4+I5$. In the absence of compositional reasoning, we compute the PMF of $O1$ by using the support $\{I1, I2, I3, I4, I5\}$ as state variables of the RTL-DTMC. Consider that all the signals in the RTL example are of 10 bits each. Therefore, the RTL-DTMC of $O1$ has 2^{50} states. With compositional reasoning, we first compute the PMFs of $Y1$ and $Y2$. The RTL-DTMCs for $Y1$ and $Y2$ have 2^{30} and 2^{20} states, respectively. We then construct the RTL-DTMC for $O1$ by using $\{Y1, Y2\}$ as state variables and compute the PMF of $O1$ by using $O1=Y1+Y2$. The RTL-DTMCs for $O1$ now has only 2^{20} states.

5. EXPERIMENTAL RESULTS

We perform our experiments on a 3 GHz, 3.25 GB machine (Intel Core2 Duo CPU). We obtain the synthesized gate-level implementations by using a library that is constructed based on the PTM 45nm model files [12].

Validating the macromodels

We constructed macromodels for several gate-level MCNC benchmark circuits. We estimated the degraded delay for five sets of input patterns. Each of these patterns corresponds to a different probability distribution. The confidence of validation can be increased by using more input pattern sets. We determine the average error of our macromodel estimates as compared to those obtained by simulating the full gate-level circuit up to 10^5 cycles.

Figure 3 shows the accuracy of the macromodels that we obtain. Our macromodels contain only a subset of the paths from the original circuit. Moreover, we ignore internal correlations while propagating signal probabilities through the macromodel (Section 3.3). Both these factors may contribute to the macromodel estimation error. However, for large K (more paths included in the macromodel), we observe that the estimation error can be removed almost completely. We thus confirm that estimation error due to ignoring internal correlations is negligible. Typically, NBTI effects cause the delay to degrade by about 10%. Therefore, we wish to keep the macromodel estimation error **less than 2%**, *i.e.* much smaller than 10%.

Speedup and reliability of RTL estimation

We consider several data-intensive RTL designs from the High-Level Synthesis benchmarks suite (HLS95) [14]. We consider different gate-level implementations for adders. We

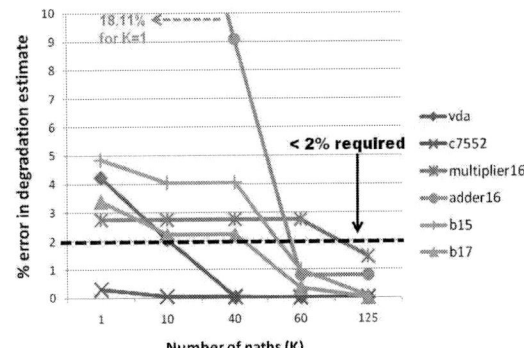

Figure 3: Accuracy of the macromodels as a function of the number of included paths K.

use the notations `_bk`, `_mc` and `_cla` to denote designs implemented with BrentKung, Manchester Carry and Carry Look Ahead adders, respectively. For deriving each macromodel, we choose K such that the error is less than 2% (Figure 3).

We assume that the RTL inputs are uniformly distributed. We obtain the PMFs of all variables in the RTL design. We start the DTMCs from a known initial state that we specify and then employ probabilistic model checking to compute the PMFs. For $N=10$ (*i.e.* 10 DTMC transitions), we find that the PMFs of all variables converge to a steady value.

We use our macromodels to estimate delay in RTL in the absence of NBTI effects (D_{RTL}) as well as the degraded delay (D'_{RTL}). We compute the degradation $\Delta D_{RTL}=D'_{RTL}-D_{RTL}$. In Table 1, we compare these RTL estimates with those obtained by using gate-level simulations, denoted by $\Delta D_{gate}=D'_{gate}-D_{gate}$. We simulate the gate-level netlist until the estimated signal probabilities converge to within 10% of their steady value (we use 10^5 simulation cycles).

In Column 2 of Table 1, we list the error in estimating D'_{RTL} (with respect to D'_{gate}). The estimation error is mainly due to the sizing mismatch between the gates in the netlists that we use for constructing macromodels and the actual netlists obtained by synthesizing the RTL designs. However, both D_{RTL} and ΔD_{RTL} are affected to the same extent by this error. Therefore, we estimate the % degradation $\frac{\Delta D_{RTL}}{D_{RTL}}$ (Column 3) and find that it is in good agreement with $\frac{\Delta D_{gate}}{D_{gate}}$ (Column 4).

In Column 5 of Table 1, we show that our runtimes are an order of magnitude lower (18.2x faster for `ellipf_cla`) than those for gate-level simulation. The time taken for computing the PMFs of RTL variables dominates the runtimes for our RTL estimation.

Table 1: Accuracy and speedup of our RTL estimates for degradation.

RTL design	$\frac{D'_{RTL}-D'_{gate}}{D'_{gate}}$	$\frac{\Delta D_{RTL}}{D_{RTL}}$	$\frac{\Delta D_{gate}}{D_{gate}}$	Speedup
filter_cla	2.13%	5.56%	5.63%	6.1x
kalman_bk	4.89%	5.41%	6.31%	11.9x
kalman_mc	4.03%	5.68%	6.42%	12.3x
fft_bk	2.10%	5.12%	5.31%	8.1x
fft_cla	2.43%	5.24%	5.31%	8.9x
ellipf_bk	8.82%	4.58%	5.52%	13.6x
ellipf_mc	8.72%	4.75%	5.78%	14.7x
ellipf_cla	9.79%	5.10%	5.23%	18.2x

Robustness to synthesis optimizations

We compare our RTL estimates against those obtained using the synthesized gate-level netlists that have undergone heavy optimizations. For each optimization scenario (Section 3.1), we obtain our RTL estimates by choosing the corresponding optimized macromodels.

We derive the optimized macromodels by considering synthesis optimizations on an operator-by-operator basis. Although this does not model global optimizations, a large number of local optimizations are captured by this approach. Therefore, our RTL estimates for % degradation faithfully follow the gate-level estimates even in the presence of optimizations (Table 2). For example, in all the RTL designs, the % degradation with Logic+Delay optimization (Column 7, Table 2) is higher than that without optimization (Column 4, Table 1). Correspondingly, our RTL estimates using the optimized macromodels (Column 6, Table 2) indicate a higher degradation compared to our RTL estimates using the unoptimized macromodels (Column 3, Table 1).

Degradation of RTL blocks in 'SLEEP' mode

During SLEEP mode, the inputs to an RTL block are held constant in order to reduce switching activity. However, based on this constant input vector, NBTI effects could be exacerbated [15]. We evaluate the effectiveness of holding all inputs constant at logic '1' during SLEEP mode.

We consider the adder block in the 32-bit ALU of an OR1200 processor. We assume that the adder is in SLEEP mode whenever its output is not used. We use RTL-DTMCs to compute the probability of being in SLEEP mode to be equal to 0.1875. Our methodology estimates that the delay degrades by 3.23%, which is an improvement from the 3.79% degradation if inputs are not held at '1'. This conforms to estimates from gate-level simulations which show that degradation reduces from 3.2% to 2.72% if inputs are held at '1' during SLEEP mode.

Our methodology estimates that the degradation of `ellipf_bk` and `ellipf_mc` is 4.44% and 2.88%, respectively, when the inputs are held at '1'. Since these are lower than the average-case degradation in Table 1, the application of all 1's during SLEEP mode is effective for these designs. However, this input worsens the degradation of `fft_bk` and `fft_cla` to 7.57% and 6.78%, respectively. We confirm all our estimates by using gate-level simulations.

6. RELATED WORK AND CONCLUSION

Commercial tools like RelXpert [16] estimate the delay degradation of the circuit by performing extensive simulations at the transistor-level. In [4] [5] [6] [7] aging effects are analyzed at the gate-level. Our methodology is more scalable since we perform analysis at a higher level of abstraction. In [5], the delay degradation of microarchitectural components is estimated by synthesizing them into gate-level netlists. Our methodology operates at a finer granularity since the degradation is estimated for each RTL statement.

In conclusion, we have presented a methodology for providing quick and accurate estimates of NBTI-induced delay degradation in RTL.

7. ACKNOWLEDGMENTS

We would like to thank Professor Sachin Sapatnekar (University of Minnesota) for his valuable comments. We would also like to thank Sanjay V. Kumar for granting us access to the source code he developed for performing gate-level aging analysis [13].

Table 2: Accuracy of our RTL estimates while considering three gate-level optimization scenarios.

Design	Logic optimization only		Delay optimization only		Logic + Delay optimization	
	$\frac{\Delta D_{RTL}}{D_{RTL}}$	$\frac{\Delta D_{gate}}{D_{gate}}$	$\frac{\Delta D_{RTL}}{D_{RTL}}$	$\frac{\Delta D_{gate}}{D_{gate}}$	$\frac{\Delta D_{RTL}}{D_{RTL}}$	$\frac{\Delta D_{gate}}{D_{gate}}$
`filter_cla`	5.42%	5.92%	6.21%	5.69%	5.67%	5.77%
`kalman_bk`	5.86%	5.97%	6.51%	6.37%	6.11%	6.52%
`kalman_mc`	5.86%	5.98%	6.52%	6.31%	6.11%	6.51%
`fft_bk`	5.86%	5.84%	6.33%	6.38%	6.08%	5.92%
`fft_cla`	5.93%	6.02%	6.41%	6.16%	6.21%	6.31%
`ellipf_bk`	5.52%	5.70%	4.51%	5.51%	5.25%	5.94%
`ellipf_mc`	5.74%	6.02%	4.96%	5.68%	6.19%	6.14%
`ellipf_cla`	5.74%	5.85%	5.31%	5.61%	5.60%	5.99%

8. REFERENCES

[1] M. A. Alam and S. Mahapatra, "A comprehensive model of PMOS NBTI degradation," *Microelectronics Reliability*, vol. 45, no. 1, pp. 71 – 81, 2005.

[2] B. C. Paul, K. Kang, H. Kufluoglu, M. A. Alam, and K. Roy, "Temporal performance degradation under NBTI: Estimation and design for improved reliability of nanoscale circuits," in *Proc. of DATE'06*, 2006, pp. 780–785.

[3] S. Borkar, "Designing reliable systems from unreliable components: The challenges of transistor variability and degradation," *IEEE Micro*, vol. 25, pp. 10–16, 2005.

[4] W. Wang, S. Yang, S. Bhardwaj, R. Vattikonda, S. Vrudhula, F. Liu, and Y. Cao, "The impact of NBTI on the performance of combinational and sequential circuits," in *Proc. of DAC'07*, 2007, pp. 364–369.

[5] M. DeBole, K. Ramakrishnan, V. Balakrishnan, W. Wang, H. Luo, Y. Wang, Y. Xie, Y. Cao, and N. Vijaykrishnan, "A framework for estimating NBTI degradation of microarchitectural components," in *Proc. of ASP-DAC'09*, 2009, pp. 455–460.

[6] S. Kumar, C. Kim, and S. Sapatnekar, "NBTI-aware synthesis of digital circuits," in *Proc. of DAC'07*, 2007, pp. 370 –375.

[7] J. Abella, X. Vera, and A. Gonzalez, "Penelope: The NBTI-aware processor," in *Proc. of MICRO'07*, 2007, pp. 85–96.

[8] S. Sambamurthy, J. Abraham, R. Tupuri, and S. Raghuram, "A robust top-down dynamic power estimation methodology for delay constrained register transfer level sequential circuits," in *Proc. of VLSID'08*, 2008, pp. 521–526.

[9] M. Mitzenmacher and E. Upfal, *Probability and Computing: Randomized Algorithms and Probabilistic Analysis*, 2005.

[10] M. Kwiatkowska, G. Norman, and D. Parker, "PRISM 2.0: A tool for probabilistic model checking," in *Proc. of QEST'04*, 2004, pp. 322–323.

[11] J. A. Kumar and S. Vasudevan, "Automatic compositional reasoning for probabilistic model checking of hardware designs," in *Proc. of QEST'10*, 2010, pp. 143–152.

[12] "Predictive technology model," Device Group at Arizona State University. Available at www.eas.asu.edu/~ptm.

[13] S. V. Kumar, "Reliability-Aware And Variation-Aware CAD Techniques," Ph.D. dissertation, University of Minnesota, 2009.

[14] P. R. Panda and N. D. Dutt, "1995 high level synthesis design repository," in *ISSS*, 1995, pp. 170–174, available at http://ftp.ics.uci.edu/pub/hlsynth/.

[15] Y. Wang, X. Chen, W. Wang, V. Balakrishnan, Y. Cao, Y. Xie, and H. Yang, "On the efficacy of input vector control to mitigate NBTI effects and leakage power," in *Proc. of ISQED'09*, 2009, pp. 19–26.

[16] "Cadence RelXpert Manual," www.cadence.com.

Generalized SAT-Sweeping for Post-Mapping Optimization

Tobias Welp[1]   Smita Krishnaswamy[2]   Andreas Kuehlmann[1,3]

[1] University of California at Berkeley, CA, USA
[2] Columbia University, NY, USA
[3] Coverity, Inc., San Francisco, CA, USA

Abstract

Modern synthesis flows apply a series of technology independent optimization steps followed by mapping algorithms which bind the optimized network to a specific technology library. As the exact solution of the mapping problem is computationally intractable, algorithms used in practice use heuristic, typically tree-based approaches. The application of these algorithms results in mapped but suboptimal networks. In this work, we present a novel, efficient, and effective optimization algorithm for mapped networks which can be considered a generalization of SAT-sweeping. Our algorithm searches for alternative, more efficient implementations of each net in the network. Candidate support nets for reimplementation are selected using simulation signatures and verified using Boolean satisfiability. We report experimental results on the quality of our algorithm obtained from an implementation of the approach using the logic synthesis system ABC.

Categories and Subject Descriptors:

B.6.3 [Logic Design]: Design Aids–*Optimization*
General Terms: Algorithms
Keywords: SAT sweeping, post mapping optimization

1 Introduction

For decades, technology independent optimization has been a field of intensive research. Early attempts have often leveraged optimization techniques known to work well during software compilation. Examples of these approaches include pattern matching and global flow optimization [1]. Later, with the emergence of efficient packages for binary decision diagrams (BDDs) [2], a large number of approaches using BDDs were proposed. A notable example is the logic restructuring technique presented in [3]. However, as the size of the industrial designs grew substantially over time, the construction of the required BDDs became infeasible, dramatically limiting the utility of these approaches in practical settings. In contrast to BDDs, the solvers for Boolean satisfiability problems (SAT-solvers) have only moderate memory requirements. Also, enormous efficiency improvements of SAT-solvers have been possible by techniques such as conflict learning, fast Boolean constraint propagation with two-literal watching, and random restarts. Efficiently implemented SAT-solvers were shown to perform well on problems typically arising in electronic design automation. There-

fore, many recent approaches to technology independent optimization successfully use SAT-solvers as backbone. A well-known example is SAT-sweeping, first published in [4].

The intensive research in the area of logic synthesis has resulted in a good understanding of logic optimization techniques. Even though recent research [5] shows that there is still significant potential for improvement, due to the enormous research advances, modern academic and commercial synthesis tools are able to synthesize designs with millions of gates yielding excellent results.

After technology independent optimization, technology mapping is used to bind the abstract logic network to a specific technology library. This mapping problem is known to be NP-hard. Hence, exact solutions can only be calculated for small designs. Instead, one usually resorts to tree-mapping algorithms, first presented in [6]. Although the application of the algorithm heuristically yields good results, the design loses quality during the technology mapping, in particular if the netlist is strongly connected. This motivates the application of optimization algorithms which are applied after mapping.

Here, we present an efficient and effective algorithm for post mapping optimization that combines several key ideas used in logic synthesis, such as simulation based heuristics, SAT-solving, node addition and removal, etc. The algorithm iterates through the network and attempts to find a better implementation of each net. Therefore, for each net, it collects a set of support nets and attempts to find a gate driven by the support nets which could be added to the network and serve as a valid and advantageous replacement for the original implementation of the net. If such a replacement is found, the algorithm adds the found gate and removes the original fanout free cone which implemented the function of the net.

2 Preliminaries

In this work, we attempt to improve a mapped circuit by finding a better implementation of a net n within the circuit. We will refer to net n as *target net* and to the potential replacement as *alternative implementation*.

In the presented algorithm, the alternative implementation is an arbitrary *monotone* gate driven by available nets in the circuit. We call a gate monotone if it fulfills the following two conditions: First, the value at the output of the gate can be calculated by iteratively applying the same operator to all fanins and optionally applying a unary operation to the result. Second, adding another fanin to the gate must impact the value of the gate in only one direction. As an example, a NAND-gate is monotone, because its value can be calculated by combining the fanins of the gates by AND-ing the fanins in an arbitrary order and finally inverting the result. Also, adding an additional fanin to a NAND-gate can only change the value at the output to one if the output is zero otherwise. If it is one not considering the additional fanin, the value of the additional fanin does not matter. On the contrary, an XOR-gate is not monotone because the value at the output of the gate can change in both directions by adding an additional fanin. Monotone gates are ample in technology libraries, e.g. in the library we used in our experimentation, 75% of the gates are monotone.

Our optimization algorithm is guided by the use of simulation results. For efficiency, we use bit-parallel simulation. In our examples, explanations, and formulas, we will denote the simulation vector associated with net n as $\phi(n)$ and we refer to the ith bit within this vector by using the notation $\phi(n)[i]$.

3 Related Work

An interesting approach for post-mapping optimization for FPGAs has been proposed in [7]. As in our work, the objective is to find an alternative, better implementation of a target net within a mapped design. The authors find the actual replacement of the target net by encoding a general configuration consisting of n look-up tables (LUTs) and the programmable routing structure as a SAT-instance. Next, the SAT-instance is solved using a Boolean satisfiability solver. If satisfiable, the emitted model by the SAT-solver encodes an alternative implementation of the target net. Unfortunately, the approach transfers only with difficulties to a setup using a library of standard cells. This is because in contrast to LUTs, different standard cells have different cost and the satisfiability solver does not discriminate between better and worse solutions. An additional problem of the approach in [7] is its poor scalability caused by the fact that the size of the SAT-instance grows exponentially with the number of inputs of the general configuration.

In SAT-sweeping [4], functionally equivalent nodes are merged. Candidates for merging are found using simulation and verified using a satisfiability solver. In [8], the approach has been strengthened so that it also takes observability don't cares into account. Even though introduced as an algorithm to be performed on AND-inverter graphs, SAT-sweeping could equally well be used for post-mapping optimization. Following this, the approach presented here can be considered a generalization of this work. As in SAT-sweeping, we use simulation to assist finding candidates for an alternative implementation of a target net. Also, the actual validity of the replacement is verified using a satisfiability solver. However, in SAT-sweeping, the target net can only be implemented using another net with equivalent functionality. In contrast, we allow the addition of arbitrary monotone gates in our work.

Node addition and removal as presented in [9] is similar to our technique as it also allows the addition of nodes to the network. However, our technique is more general as it not only allows the addition of AND-nodes, but also of nodes with other functionality and can therefore be used as an algorithm for post-mapping optimization. Also, our approach uses different techniques for finding potential alternative implementations of a target net and its verification.

In [10], the concept of entropy of simulation signatures has been introduced as a means to guide the search of nets within a circuit which are useful for correcting a design error. In our approach, we use a similar technique to evaluate the utility of support nets and their combinations for driving the alternative implementation of the target net.

4 Motivational Example

To motivate our algorithm, we consider the example given in Figure 1 which illustrates the need for an optimization algorithm applied after technology mapping. For this purpose, we assume that the circuit has been thoroughly optimized using technology independent optimization techniques. As the first step of technology mapping, the design is transformed to a subject graph consisting of NAND-gates and inverters as given in Figure 1(a). Assume that the desired technology library is the 180 nm generic library included in the IWLS benchmark set [11]. A choice of the available gates in this library along with their area requirements is illustrated in Figure 1(b). We assume that we want to apply a tree-based mapping algorithm as first introduced [6]. Therefore, the subject graph needs to be cut into a forest of trees as illustrated by the lines in Figure 1(a). Now, every tree is mapped to the technology library. An optimal mapping of the individual trees is given in Figure 1(c) and requires area 99. Note that this mapping, though optimal for the trees, is not an optimal mapping for the complete circuit. Figure 1(d) shows an optimized version of the circuit as would be found by our algorithm. The mapping requires area 89, i.e. is 10% smaller than the original mapping.

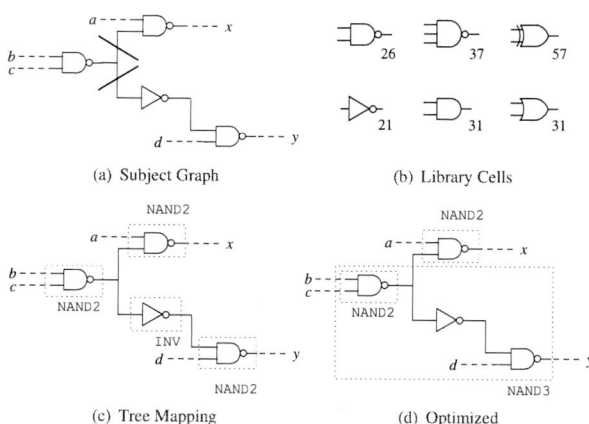

(a) Subject Graph   (b) Library Cells

(c) Tree Mapping   (d) Optimized

Figure 1: In this example, the tree mapping causes a suboptimal mapping result. After application of our algorithm, we obtain a mapping which requires less area.

5 Post-Mapping Optimization (PMO)

For a net n in the network, our algorithm attempts to find an alternative and more efficient implementation using a single monotone gate. We will describe our algorithm using a top-down methodology. We start in the following subsection with describing the top-level algorithm followed by detailed information about the individual subroutines in the remaining subsections. For ease of exposition, we will describe our algorithm for the special case that the alternative implementation of n is an AND-gate. It is straightforward to generalize the algorithm for arbitrary monotone gates.

5.1 Top-level Algorithm

The top-level algorithm is illustrated in Algorithm 1. The algorithm iterates through the network and attempts to re-implement each net n. Therefore, it collects a set S of support nets which are potential inputs of the alternative implementation of the target net n. Next, the circuit is simulated and the simulation vectors are used to streamline and guide the search for candidates of alternative implementations. For each candidate, we check if the associated replacement would be worthwhile with respect to optimization targets such as size, speed, or power consumption of the circuit. In case the replacement would be worthwhile, a SAT check is used to verify that it is actually a valid replacement. In this case, the circuit is changed, otherwise a new candidate is searched.

Algorithm 1 POST-MAPPING OPTIMIZATION

1: **for** each net n in network **do**
2:     $S = \texttt{collectSupportNets}(n,d,D)$
3:     $\texttt{randomSimulate}(S)$
4:     $S = \texttt{filterSupportNets}(S)$
5:     **if** $\texttt{checkFeasibility}(S,n)$ **then**
6:       **while** $c = \texttt{searchCandidate}(S,n)$ **do**
7:         **if** $\texttt{advantageousReplacement}(c,n)$ **then**
8:           **if** $\texttt{validReplacement}(c,n)$ **then**
9:             $\texttt{replace}(c,n)$

5.2 Collection of Support Nets

For each net n, we collect a set S of support nets, which will be the inputs for a potential alternative implementation of n. Nets which are in the transitive fanout of n must not be used as support nets as a re-implementation of n using a net which is in the transitive fanout of n could potentially yield loops in the network. Conceptually, all other nets could be used. However, for performance reasons, we want to limit the number of support nets to a small fraction of this set. Heuristically, nets in the transitive fanin of n which are only a few levels away from n are the most useful choice of support nets. Therefore, in our algorithm, we populate S with all nets in the transitive fanin of net n which are at least d levels and at most D levels away from n. The details of our implementation of the routine for the collection of support nets are given in Algorithm 2. Note that a net is added to the set of support nets regardless of if it is at least d away from n if it does not have any fanins. For instance, this assures that a primary input, which drives the gates implementing the target net would be added to S even if d is chosen to be larger than 1. Optimal concrete values for the parameters d and D depend on the circuit under synthesis. A small value for d and large value for D results in a large number of support nets. This increases the probability of finding a good alternative implementation of the target net but increases the runtime dramatically.

Algorithm 2 COLLECTSUPPORTNETS(n,d,D)

1: **if** $d \le 0$ and $D \ge 0$ **then** $S = S \cup \{n\}$
2: **if** $D > 0$ **then**
3:     **foreach** fanin f of n **do**
4:       $S = S \cup \texttt{collectSupportNets}(f,d-1,D-1)$
5:     **if** n has no fanins **then** $S = S \cup \{n\}$

5.3 Random Simulation

To guide our heuristics for candidate selection, we use simulation vectors obtained by random simulation of w words where w is a parameter of the algorithm. In general, large values for w yield more precise heuristics but result in a linear increase in runtime due to simulation cost. Given a target net n and a set of support nets S, simulation vectors can be produced by assigning random values to the support nets and simulating only the cone of the network between the support nets and the target net. However, this allowed for unjustifiable value combinations at the support nets. To avoid this effect, i.e. to take satisfiability don't cares (SDCs) into account, we assign random values to the primary inputs of the complete circuit.

Then, we simulate the network in a lazy fashion, i.e., we calculate the values for a specific net only if the value is actually needed by our heuristics. To avoid unnecessary recalculation of simulation values which have been calculated already, all intermediate simulation values are memoized using a hash-table.

The network does not need to be re-simulated until changes are made to the network. If the functions of all alternative implementations were equivalent to the functions of the corresponding target nets, the network never needed to be re-simulated. However, as described in the following subsections, we also consider observability don't cares (ODCs). As a consequence, the simulation values in the fanout of the target net might change after a replacement. Therefore, after each restructuring operation, we mark the simulation vectors in the transitive fanout of a target net as "tainted". This indicates that they must be re-simulated if they are to be used in candidate selection again.

5.4 Filtering Support Nets

In the filtering step, we try to reduce the number of support nets by filtering out nets which cannot possibly be considered as a fanin of an alternative implementation of the target net. Recall that for ease of exposition, we assume here that the added gate is an AND-gate. As a 0 in one of the fanins of an AND-gate causes the output of the AND-gate to be 0 regardless of the value of the other fanins, we can directly filter out any support nets whose simulation vector has zeros at positions where the simulation vector of the target net has a 1. As an example, consider that we have a target net n with simulation vector $\phi(n) = 1001\ldots$ and support vectors s_1, s_2, and s_3 with simulation vectors $\phi(s_1) = 1101$, $\phi(s_2) = 0111$, and $\phi(s_3) = 1011$. In this case, we can directly drop s_2 from further consideration, because $\phi(s_2)[1] = 0$ whereas $\phi(n)[1] = 1$.

The filtering heuristic can be improved by also considering ODCs: The bits where the simulation vector of the target net are not observable at any primary output do not need to be taken into account during the filtering of the support nets. The target net cannot be observed at the output if on each path between the target net and a primary output, there is at least one gate which blocks the path. A gate blocks a path when a side-input dominates the output of the gate. An example of this idea is given in Figure 2 where we assume that net c is the target net. Consider the first bit of the simulation vectors. Here, the target net cannot be observed at any output because the path through the NAND-gate is blocked by net b and the path through the NOR-gate is blocked by net d. In contrast, the target net is observable for the second bit of the simulation vectors. Even though the path through the NOR-gate is blocked by net d, the path through NAND-gate and the OR-gate is active.

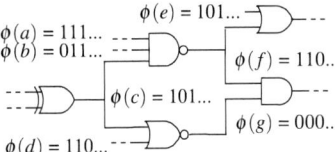

Figure 2: An example to illustrate the calculation of ODCs.

As pointed out in [12], special care has to be taken in the presence of reconvergent paths. To illustrate the problem, consider the third bit of simulation vectors. Note that at the AND-gate, net f appears to block the path through g and vice versa. However, if $\phi(c)[3] = 0$, then the value at the output of the AND-gate would change, i.e. the target net is indeed observable. We resolve this problem by using the conservative solution presented in [8]: In case a gate is blocked solely by a reconvergent signal, we assume that the path is active anyways.

5.5 Feasibility Check

The aim of the feasibility check is to avoid spending runtime on trying to find a candidate for an alternative implementation of the target net in case there is none using the set of filtered support nets. We know that there is no such candidate if there exists a simulation slice for which the value of the target net cannot be obtained even if all support nets were considered. In this case, we can bail out of the optimization attempt and do not need to spend runtime on trying to find a subset of the support nets which can serve as the fanin of an alternative implementation. As a concrete example, imagine that the alternative implementation of the target net is an AND-gate. Then the conjunction of the simulation vectors of all support nets must be equivalent to that of the target net. As in the previous subsection, we disregard bits for which the target net is not observable.

5.6 Searching Candidates for Alternative Implementations

The next step of the algorithm is trying to find a concrete alternative implementation of the target net guided by using the simulation vectors. In the case where the alternative implementation is an AND-gate, this means that we search for a subset S^* of S such that the conjunction of the simulation vectors corresponding to the nets in S^* is equivalent to the simulation vector of the target net n. Formally, such a set is described as

$$S^* = \{s_1, s_2, \ldots, s_k\} \subseteq S | \Pi_{i=1}^k \phi(s_i) = \phi(n)$$

As in the preceding subsections, we also want to consider ODCs. Therefore, we only check for equivalence of the simulation slices for which the target net is observable. Assume that $\omega(n)[i]$ is 1 if n is observable in the ith simulation slice. Then, we have

$$S^* = \{s_1, s_2, \ldots, s_k\} \subseteq S | \omega(n)\Pi_{i=1}^k \phi(s_i) = \omega(n)\phi(n)$$

In general, S^* is not unique and we prefer an S^* with small cardinality. We use the heuristic approach as coded in Algorithm 3 to find a good S^*. The algorithm uses a priority queue which is initialized with S. Next, we repetitively merge the pivot element of the priority queue with the vectors of S until we find a match. In case the priority queue gets empty, no potential alternative implementation was found and the optimization attempt of n failed, i.e. n will remain implemented in its original form.

Algorithm 3 SEARCHCANDIDATE(S, n)

1: {C is priority queue initialized with S.}
2: **while** C.notEmpty() **do**
3:     $c = C$.pop()
4:     **if** $\phi(c) = \phi(n)$ **then return** c
5:     **foreach** $s \in S$ **do** C.push(merge(s, c))
6: **return false**

As ordering criterion for the priority queue, we utilize a fitness function f which can be composed according to the emphasis of the optimization. For our experimentation, we used as fitness function the linear combination

$$f(c) = -f_d \text{dist}(\omega(n)\phi(c), \omega(n)\phi(n)) - f_c|c|$$

composed of the hamming distance between the simulation vector associated with the candidate and that of the target net (considering ODCs) and the cardinality of the candidate. We weight these values with the parameters f_d and f_c. The bigger f_d, the stronger the emphasis on quickly finding a candidate, the larger f_c, the stronger the emphasis on a small added gate.

Note that other implementations of the fitness function are possible. A promising perspective, for instance, would be to take characteristics of individual support nets into account. Imagine, for example, one specific support net had already very high fanout and has therefore already a big delay. In this case, it would be wise if adding this net would decrease the fitness of the combination.

5.7 Evaluating the Value of a Replacement

Before the costly check of whether a specific potential replacement is actually valid, we evaluate its value. If the replacement deteriorates the circuit, we will not implement it, regardless of whether a SAT-check confirms that the replacement preserves functional equivalence of the circuit or not. Hence, we only perform SAT-checks for valuable potential replacements, which saves a substantial amount of running time.

The evaluation of whether a replacement is valuable or not depends on the optimization target of the designer. In our experimentation, we attempt to optimize the size of the circuit under the constraint that timing does not deteriorate.

Therefore, we calculate the value of a replacement as follows. Firstly, we calculate the required time for the target net such that the overall delay of the circuit does not deteriorate. If the arrival time of the target net using the alternative implementation is larger than the required time, the delay of the circuit will deteriorate and we abort the optimization attempt. Otherwise, we evaluate the impact of the replacement on the area of the circuit. If the target net n is implemented using an alternative implementation, the complete fanout free cone in the transitive fanin of n can be removed. Hence, the sum of the area of each cell in this fanout free cone is the gain of the potential replacement. On the other hand, the cost of the newly added gate must be considered as the loss and is deducted from the gain yielding the value of the replacement.

5.8 Verifying the Validity of Replacements

We use a miter structure for verifying the validity of replacements. However, note that it is not necessary to instantiate the miter of the complete design. Instead, we only have to add the transitive fanin cone of the target net, its transitive fanout cone, and all additional logic which is in the transitive fanin of the transitive fanout cone of the target net. For the revised network, we only need to clone the gates in the transitive fanout cone of the target net and rewire all nets which are driven by the target net to be driven by the alternative implementation. Figure 3 illustrates the construction on an example circuit. Lastly, we constrain that at least one pair of combinational outputs must be non-equivalent. This assures that the constraint circuit is satisfiable iff the proposed alternative implementation is invalid. To use a Boolean satisfiability solver, we apply the Tseitin-transformation [13].

Note that by using the miter structure as outlined above takes SDCs and ODCs into account implicitly. If there is a difference between original and alternative implementations which either cannot be stimulated or not observed at the output, the SAT-instance will not be satisfiable, i.e. our algorithm will change the circuit.

The presented exact solution can result in very excessive runtime for large circuits. Hence, for performance reasons, it is often advantageous to consider much smaller SAT-instances which

817

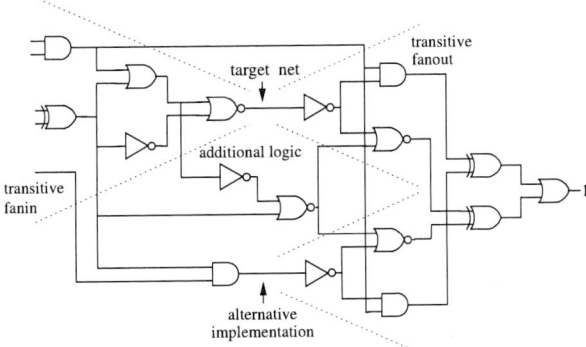

Figure 3: Illustration of the used miter-structure.

may be satisfiable if the replacement is valid but maintaining the property that it cannot be unsatisfiable if the replacement is invalid. Therefore, we add only those gates to the SAT-instance, which are at most b levels away from the target net in both fanin and fanout direction. Also, instead of constraining that at least one pair of combinational outputs must be non-equivalent, we constrain that at least one pair of the nets within the SAT-instance have non-equivalent values which are either associated with combinational outputs or are driving gates which are not added to the SAT-instance because they are more than b levels away from the target net. Evidently, for $b \to \infty$, we have the originally presented exact approach. For smaller values of b, the SAT-instances become smaller and are usually easier to solve but the probability that a valid replacement is not identified increases. Note that this solution resembles windowing as e.g. described in [14].

6 Experimental Setup and Results

For experimentation, we implemented a software prototype of our optimization algorithm into the synthesis system ABC [15], callable as an additional command *pmo*. As satisfiability solver, we used MiniSAT [16]. In its current version, our algorithm is only able to add NAND-gates for reimplementing a target net.

To generate the empirical results, we applied *pmo* to the designs of the IWLS benchmark set [11] originating from the OpenCores repository[1] which have already been synthesized, optimized, and mapped using a commercial synthesis tool. We performed our experimentation on a computer with an Intel Pentium 4 CPU with 3.4 GHz.

In the following subsection, we report the quality of results achieved by our algorithm using default values for the parameters. Next, in Subsection 6.2, we report relevant insights obtained during our experimentation with respect to parameter settings.

6.1 Quality of Results

To obtain a quantitative estimate of the performance of the presented algorithm, we applied the algorithm to the benchmarks using the following values for the parameters: The set of support vectors is composed by using all nets in the transitive fanin which are at least one level ($d = 1$) and at most two levels ($D = 2$) away from the target net. Further, the heuristics are guided using 256 simulation slices (four words of size 64 bits, $w = 4$) and the SAT window includes all gates which are at most 3 levels away from the target net ($b = 3$). We verified that all modified circuits are functional

---
[1]We report results of all benchmarks in this set with the exception of *wb_dma.v*. For this design, ABC reports a specious parsing error.

Benchmark	Area [μm^2]	ΔArea [%]	Del [ns]	ΔDel [%]	Mem [MB]	Runtime [ms]
ac97_ctrl	336447	1.51	0.48	6.65	26	10310
aes_core	613457	3.47	0.92	2.00	34	60230
des_area	139539	4.87	1.03	0.78	22	11100
des_perf	2614650	8.25	0.84	5.79	87	1680740
ethernet	1783870	0.25	1.09	0.28	46	106980
i2c	30377	1.69	0.46	4.76	21	650
mem_ctrl	353599	1.24	1.20	0.00	32	25870
pci_bridge32	542931	0.86	0.88	0.34	31	16190
pci_conf	3266	4.41	0.18	0.56	20	30
pci_spoci_ctrl	37021	2.76	0.59	0.51	21	1030
sasc	18072	1.56	0.31	7.19	21	100
simple_spi	25781	1.60	0.37	0.00	21	280
spi	94807	1.67	1.10	2.32	22	3590
ss_pcm	11512	1.78	0.28	0.00	21	70
stepper	5884	2.83	0.35	0.00	20	100
systemcaes	271593	0.97	1.55	1.97	24	13540
systemcdes	93703	4.64	0.98	1.87	21	3820
tv80	222137	1.02	1.63	2.20	27	13250
usb_funct	366064	0.67	0.89	2.41	29	11100
usb_phy	13175	3.10	0.33	0.00	21	150
vga_lcd	3245920	1.15	0.81	0.25	110	875430
wb_conmax	929706	0.44	0.88	1.62	52	40110

Table 1: Impact of *pmo* on benchmark circuits.

equivalent to the corresponding original circuits using the ABC-routine *cec*.

Table 1 summarizes the impact of *pmo* on the benchmark circuits and the cost of its execution: It reduces the sizes of the circuits by 0.5% to 8% and speeds up the longest path within the circuit by 0% to over 7%. Note that ABC estimates the delay of the circuits by using point-to-point delays as defined in the technology library. Running time and memory usage of the algorithm grow roughly linear with the size of the benchmark circuit.

6.2 Impact of Parameters

In the following, we provide insights into two aspects of the algorithm. Firstly, we analyze the impact of w, the number of simulation slices. We will conclude that a relatively small value for w is sufficient to guide our heuristics well and to keep runtime and memory costs moderate. Secondly, we investigate the impact of the size of the window considered in the SAT-instances on the efficacy and efficiency of the algorithm. We provide data showing that the presented algorithm is effective even if small SAT-instances containing only gates close to the target net are used. On the other hand, we show that big windows result in excessive runtime of the algorithm, providing strong evidence that small windows are preferable.

6.2.1 Number of Simulation Slices

Intuitively, we expect that the more simulation values are used, the more information of the circuit is available to guide our heuristics. However, the larger w, the longer simulation takes and the more storage is required to keep the simulation vectors in memory. To find a value for w which provides a good trade-off between quality and cost, we ran *pmo* for different values of w and recorded the impact of w on the quality of the heuristics and its cost. We express the quality of the heuristics in terms of the success rate of a call of the satisfiability solver where we define a SAT-call as successful if the solver returns that the instance is not satisfiable, meaning that the replacement is valid. The cost is expressed in terms of runtime and maximum memory usage of the algorithm.

Figure 4 summarizes the results of this experiment for the benchmark *spi* where Figure 4(a) shows the impact of w on the preciseness of the heuristics and Figure 4(b) the impact of w on the cost of the algorithm. The data allows the conclusion that more simulation slices lead to more precise heuristics. However, this effect

818

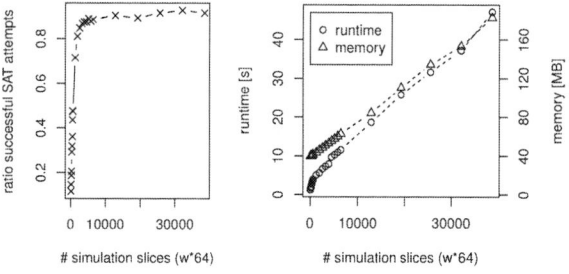

Figure 4: The quality of the heuristics increase with simulation effort. However, this effect tapers off quickly whereas the costs of calculating and storing simulation slices increases linearly.

tapers off quickly beyond approximately 5000 simulation slices. On the other side, running time and memory use of the algorithm increase roughly linear with w.

Note that we repeated the experiments with all other benchmarks and obtained qualitatively equivalent results.

6.2.2 Window Size in SAT-Instance

If parameter $b \rightarrow \infty$, the SAT-instance is unsatisfiable iff the corresponding replacement is valid. Unfortunately, in this case, the SAT-instance contains clauses for the complete fanin and fanout cones of the target net, which is often a substantial fraction of the entire circuit. For large circuits, this can cause that the overall efficiency of the algorithm suffers from excessive runtime for the SAT-calls. It is possible to avoid this by assigning a small value to b such that only those gates are added to the SAT-instance which are at most b levels away from the target net. The disadvantage of using small values for b is that they can cause that replacements are mistakenly evaluated invalid. This is because SDCs or ODCs caused by gates which are more than b levels away from the target net are not taken into consideration.

To measure the practical impact of parameter b, we applied *pmo* with different values for b to the benchmarks. The results for a choice of the benchmarks are contained in Figure 5. The results of the remaining benchmarks are similar but not displayed to improve readability. As can be seen in Plot 5(a), a large fraction of the optimization is achieved using small values for b. However, as Plot 5(b) shows, the runtimes increase exponentially with higher values for b. These results suggest the use of small values for b, such as 2 or 3.

7 Conclusions and Future Work

Heuristic algorithms for technology mapping yield suboptimal mapped designs, motivating post-mapping optimization. In this paper, we outlined an efficient and effective optimization algorithm for mapped designs which leverages multiple ideas from technology-independent optimization such as SAT-sweeping and node addition and removal. We implemented a software prototype of our algorithm into the logic synthesis system ABC. In our experimentation, we found that our algorithm is able to decrease the size of the mapped area of the circuits in the OpenCores repository of the IWLS-benchmarks by up to 8% while speeding up the delay to up to 7% even though we only considered NAND-gates as alternative implementation. The algorithm has moderate execution costs that increase roughly linear with the size of the circuit, allowing its use in industrial settings where circuits are very large.

As future work, we plan on tuning the presented algorithm for

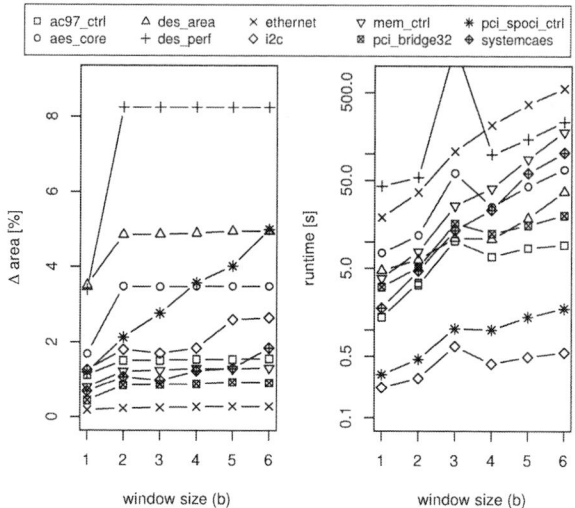

Figure 5: Small windows are sufficient to capture most of the area gain and require dramatically shorter runtime.

optimizing circuits with respect to other optimization goals such as the power consumption of the circuit.

References

[1] C. L. Berman and L. H. Trevillyan, "Global flow optimization in automatic logic design," *IEEE Trans. Computer-Aided Design*, vol. 10, pp. 557–564, May 1991.

[2] R. E. Bryant, "Graph-based algorithms for Boolean function manipulation," *IEEE Trans. Computers*, vol. 35, pp. 677–691, Aug. 1986.

[3] V. N. Kravets and P. Kudva, "Implicit enumeration of structural changes in circuit optimization.," in *Proceedings of the 41th ACM/IEEE Design Automation Conference*, pp. 526–532, 2004.

[4] A. Kuehlmann, "Dynamic transition relation simplification for bounded property checking," in *Digest Tech. Papers IEEE/ACM Int'l Conf. Computer-Aided Design*, (San Jose, California), pp. 50–57, Nov. 2004.

[5] Z. Vasicek and L. Sekanina, "A Global Postsynthesis Optimization Method for Combinational Circuits," in *Design Automation and Test in Europe*, March 2011.

[6] K. Keutzer, "DAGON: technology binding and local optimization by DAG matching," in *Proceedings of the 24th ACM/IEEE Design Automation Conference*, DAC '87, (New York, NY, USA), pp. 341–347, ACM, 1987.

[7] A. Ling, D. P. Singh, and S. D. Brown, "FPGA Technology Mapping: A Study of Optimality," in *Proceedings of the 42th ACM/IEEE Design Automation Conference*, pp. 121–126, 2005.

[8] Q. Zhu, N. Kitchen, A. Kuehlmann, and A. L. Sangiovanni-Vincentelli, "SAT Sweeping with Local Observability Don't Cares," in *Proceedings of the 43th ACM/IEEE Design Automation Conference*, pp. 202–207, 2006.

[9] Y. C. Chen and W. Chun-Yao, "Node addition and removal in the presence of don't cares," in *Proceedings of the 47th ACM/IEEE Design Automation Conference*, pp. 505–210, 2010.

[10] K.-H. Chang, I. L. Markov, and V. Bertacco, "Fixing Design Errors with Counterexamples and Resynthesis," in *Proceedings of the Asia-South Pacific Design Automation Conference*, pp. 944–949, 2007.

[11] C. Albrecht, *IWLS benchmark library*. http://www.iwls.org/iwls2005/benchmarks.html, 2005.

[12] H. Savoj and R. K. Brayton, "The use of observability and external don't cares fo the simplification of multi-level networks," in *Proceedings of the 27th ACM/IEEE Design Automation Conference*, pp. 297–301, 1990.

[13] G. Tseitin, "On the complexity of derivation in propositional calculus," vol. 8 of *Seminars in Mathematics*, Leningrad: V. A. Steklov Mathematical Institute, 1968. English translation: Studies in mathematics and mathematical logic, Part II, 1970, pp. 115-125.

[14] A. Mishchenko and R. K. Brayton, "SAT-Based Complete Don't-Care Computation for Network Optimization," in *Proceedings of the conference on Design, Automation and Test in Europe - Volume 1*, DATE '05, (Washington, DC, USA), pp. 412–417, IEEE Computer Society, 2005.

[15] Berkeley Logic Synthesis and Verification Group, "A System for Sequential Synthesis and Verification," Release 70930.

[16] N. Eén and N. Sörensson, "An Extensible SAT-Solver," in *Proc. 6th Int'l Conf. Theory & Appl. Satisfiability Testing (SAT)*, pp. 502–518, May 2003.

Accuracy-Configurable Adder for Approximate Arithmetic Designs

Andrew B. Kahng[†‡] and Seokhyeong Kang[†]

[†]ECE and [‡]CSE Departments, University of California at San Diego
abk@cs.ucsd.edu, shkang@vlsicad.ucsd.edu

ABSTRACT

Approximation can increase performance or reduce power consumption with a simplified or inaccurate circuit in application contexts where strict requirements are relaxed. For applications related to human senses, approximate arithmetic can be used to generate sufficient results rather than absolutely accurate results. Approximate design exploits a tradeoff of accuracy in computation versus performance and power. However, required accuracy varies according to applications, and 100% accurate results are still required in some situations. In this paper, we propose an *accuracy-configurable approximate (ACA) adder* for which the accuracy of results is configurable during runtime. Because of its configurability, the ACA adder can adaptively operate in both approximate (inaccurate) mode and accurate mode. The proposed adder can achieve significant throughput improvement and total power reduction over conventional adder designs. It can be used in accuracy-configurable applications, and improves the achievable tradeoff between performance/power and quality. The ACA adder achieves approximately 30% power reduction versus the conventional pipelined adder at the relaxed accuracy requirement.

Categories and Subject Descriptors

B.7.2 [**Hardware**]: INTEGRATED CIRCUITS—*Design Aids*; J.6 [**Computer Applications**]: COMPUTER-AIDED ENGINEERING

General Terms

Algorithms, Design, Performance

Keywords

Approximate Arithmetic, Error-Tolerance, Power Minimization, Accuracy-Configurable Adder

1. INTRODUCTION

Guardbands for dynamic variations severely limit performance and energy efficiency of conventional IC designs. To overcome consequences of overdesign, several recent mechanisms for variation-resilient design [4] allow timing errors and manage design reliability dynamically. Relaxing the requirement of correctness for designs may dramatically reduce costs of manufacturing, verification and test [16]. In resilient designs, errors can be corrected with redundancy techniques (*error-tolerance*), or accepted in some applications relating to human senses such as hearing and sight (*error-acceptance*). In the error-acceptance regime, approximation via a simplified or inaccurate circuit can increase performance and/or reduce power consumption.

Various approximate arithmetic designs have been previously proposed. Lu [7] introduces a faster adder which has shorter carry chains and considers only the previous k bits of input in computing a carry bit. Verma et al. [12] provide a variable latency speculative adder ($VLSA$), which is a reliable version of the Lu adder [7] with error detection and correction. Shin et al. [10] also propose a data path redesign technique for various adders which cuts the critical path in the carry chain. Zhu et al. [14] [13] propose three approximate adders – $ETAI$, $ETAII$ and $ETAIIM$. ETAI is divided into an accurate part and an inaccurate part to achieve approximate results. ETAII cuts carry propagation to speed up the adder, and ETAIIM modifies ETAII by connecting carry chains in accurate MSB parts. Kulkarni et al. [5] present a 2x2 under-designed multiplier, and use it to build large power-efficient approximate multipliers. George et al. [3] define the concept of *probabilistic CMOS (PCMOS)*, and implement efficient arithmetic using $PCMOS$. Shin et al. [11] propose a logic synthesis approach to design an approximate circuit.

The approximate designs produce almost-correct results at the given required accuracy, and obtain power reductions or performance improvements in return. In some applications, however, more accurate or totally accurate results are required *under certain conditions* – e.g., image processing in security cameras would require cleaner images after detecting a motion. In contexts where the required accuracy changes during runtime, the accuracy of results should be configurable to maximize the benefit of approximate operations. Figure 1 illustrates how power benefits can be achieved with an accuracy-configurable design. The accuracy-configurable design can adapt to changing accuracy constraints by using different modes in each situation. To our knowledge, no previous work can configure the output accuracy during runtime, and each is thus restricted (or, best-suited) to particular application contexts. In contexts where the accuracy requirement can change dynamically, the previous methods' benefits from the accuracy tradeoff are reduced since the implementation must be targeted to the maximum accuracy requirement.

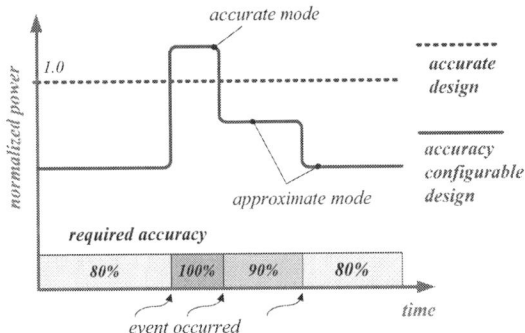

Figure 1: Power benefits from accuracy-configurable design.

In this paper, we propose an *accuracy-configurable approximate* (ACA) adder, which can configure the accuracy of results during runtime. The main contributions of our work are the following.

- The proposed ACA adder has runtime-configurable accuracy to better enable tradeoff of accuracy in computation versus performance and power.

- We provide quantitative metrics for an approximate arithmetic design. We compare the ACA adder to previous approximate adders based on these metrics.

- We demonstrate the power benefits of the ACA adder over previous approximate and conventional adder designs for accuracy-configurable applications.

The rest of the paper is organized as follows. Section 2 presents the proposed ACA adder design. Section 3 provides experimental results and analysis. Section 4 summarizes and concludes the paper.

2. ACCURACY-CONFIGURABLE ADDER

2.1 Approximate Adder Implementation

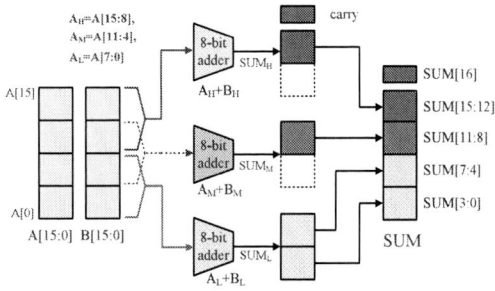

Figure 2: Proposed approximate adder – 16-bit adder case.

Previous approximate adders [7] [10] [14] have difficulty detecting and correcting errors since they are designed for error-acceptable applications with a target accuracy. However, accurate computations are still required at certain times, according to the application. VLSA [12] can provide accurate results, but has large delay and area overhead for the error detection and correction. The central contribution of our present work is to propose an approximate adder which supports both accurate and inaccurate computation with error-correction and accuracy-configuration capability. Figure 2 shows our proposed approximate circuit for the case of a 16-bit adder. In the adder, the carry chain is cut to reduce critical-path delay, and three sub-adders generate results of partial summations. With the reduced critical-path delay, high performance (by increasing the clock frequency) or low power consumption (by decreasing the operating voltage) is obtained. A middle sub-adder ($A_M + B_M$) is introduced to increase accuracy. Without the middle sub-adder (as in ETAII [13]), error occurs when the eighth carry bit is high, and for random input patterns the error rate is 50.1%. On the other hand, with the introduction of the middle sub-adder, error rate for random input patterns is reduced to 5.5%. (In the real implementation, all redundant parts (four-LSB output of $A_H + B_H$ and $A_M + B_M$ sub-adders) are optimized only for carry-generation.)

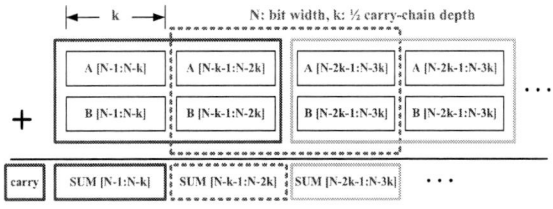

Figure 3: General implementation for the proposed adder.

We can generalize the implementation of the proposed approximate adder. Figure 3 shows the general implementation of an N-bit

adder with a parameter k, which is the bit-width of the sub-adder result. In the adder, each divided sub-module produces a k-bit result except for the last sub-module, which produces a $2k$-bit result. The approximate adder thus consists of the $(N/k - 1)$ sub-modules as described in Equation (1).

$$
\begin{aligned}
SUM[N - ik - 1 &: N - (i + 1)k] = \\
A[N - ik - 1 &: N - (i + 2)k] + \\
B[N - ik - 1 &: N - (i + 2)k], \\
&where \ i = 0, ..., N/k - 2
\end{aligned} \tag{1}
$$

In modern adder designs, such as carry-lookahead (CLA), carry-select and Kogge-Stone adders, the path depth and area are asymptotically proportional to $log_2 N$ and $N log_2 N$ respectively, where N is the bit-width of the adder [15]. Based on this, we can express delay, area and power consumption of the proposed adder in terms of the parameters N and k. The proposed ACA adder has $(N/k - 1)$ sub-adders, each of which is a $2k$-bit adder. Therefore, delay of the critical path can be expressed with Equation (2) and area can be estimated with Equation (3), where C_{delay} and C_{area} are constants for delay and area, respectively.

$$
delay = C_{delay}(log_2 k + 1) \tag{2}
$$

$$
area = C_{area}(N - 2k)(log_2 k + 1) \tag{3}
$$

$$
Power_{dyn} = C_{power}(N - 2k)(log_2 k + 1)^2 \tag{4}
$$

Power consumption of the ACA adder can be roughly estimated as follows. Dynamic power consumption with voltage scaling at a fixed frequency is proportional to $capacitance \cdot V_{dd}^2$, where the $capacitance$ is proportional to the area. Cell delay is proportional to $1/(V_{dd} - V_t)^\beta$, and V_{dd}^2 is roughly proportional to $1/(cell \ delay)$ if we assume that β is 2. Since $(cell \ delay) \times (path \ depth)$ is constant at a fixed frequency, V_{dd}^2 is proportional to the path depth, which is $log_2 k + 1$. Consequently, dynamic power with voltage scaling can be expressed using Equation (4), where C_{power} is a constant fixed for given V_{dd} for dynamic power consumption. Static power consumption of the adder can be roughly estimated as proportional to the area in Equation (3).

In our proposed adder design, the output of each sub-adder (except the last sub-adder) is incorrect when a carry input should be propagated to the results. In Figure 2, when the $carry[4]$ (carry bit from $A_L + B_L$) is '1' and $SUM_M[3 : 0]$ is $1111_{(2)}$, the output result has an error in $SUM[11 : 8]$. In the general implementation, the output result will be correct when there are no errors in all $(N/k - 1)$ sub-adders. In the ith sub-adder, errors occur when (1) the LSB part of the result ($SUM_i[k - 1 : 0]$) has all '1' values (probability $P = \frac{1}{2^k}$) and (2) the LSB part ($[k - 1 : 0]$) of the $(i + 1)^{th}$ sub-adder produces a carry bit (probability $P = \frac{1}{4} + \frac{1}{2} \cdot \frac{1}{4} + \frac{1}{2} \cdot \frac{1}{2} \cdot \frac{1}{4} + ...$). Therefore, with a random input vector, the probability of having a correct result in the proposed adder is

$$
P(N, k) = (1 - \frac{1}{2^k} \cdot \frac{2^k - 1}{2^{k+1}})^{\frac{N}{k} - 2} \tag{5}
$$

Table 1 shows the estimated results of 16-bit ACA adders with different parameter values k. With smaller k value, the minimum clock period and dynamic power can be reduced, but the pass rate (probability of having a correct result) will be decreased. The estimations come from Equations (2), (3), (4) and (5). In Section 3.3 below, we validate the above estimation with real implementations.

Table 1: Estimated minimum clock cycle, area, dynamic power and pass rate for each k value when $N = 16$ (normalized to the conventional CLA 16-bit adder).

	k=2	k=3	k=4	k=5	k=6
min. clock period	0.5	0.65	0.75	0.83	0.89
area	0.87	1.05	1.12	1.15	1.12
dynamic power	0.44	0.68	0.84	0.95	1.00
pass rate	0.554	0.829	0.942	0.982	0.995

2.2 Error Detection and Correction for Accurate Computation

As described in Section 2.1, our proposed adder is incorrect when a carry bit is propagated between sub-adders. However, the error can be detected and corrected with a small overhead. We detect an error for each sub-adder by checking the output of the sub-adder and the carry-in signal that comes from the previous sub-adder. Error detection can be implemented with several '*and*' gates. To correct the error, '1' should be added to the approximate (inaccurate) output, and the error correction can be implemented with an incrementor circuit.

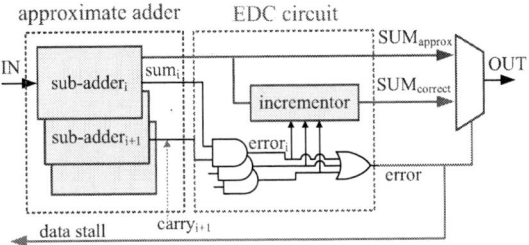

Figure 4: Error detection and correction with the approximate adder.

With these simple error detection and correction circuits, our proposed adder can be implemented to have variable latency like the previous VLSA adder [12], with a small overhead for an error detection and correction (EDC) system. Figure 4 shows an EDC system with our proposed adder. The error detection circuit ('*and*' gates) checks the carry propagation and generates an error signal. The error correction (incrementor) circuit produces an error-free output by adding compensation data, and requires an additional clock cycle. When errors are detected from input patterns, the *error* signal is activated. The error signal holds the input pattern during the error correction and chooses the error-corrected value ($SUM_{correct}$) as an output. With this approach, our approximate adder can provide accurate results at a higher clock frequency than that of conventional adders (e.g., CLA). According to the estimated results in Table 1, clock period can be reduced by 25% with 6% (= error rate) recovery-cycle overhead (16-bit ACA, $k = 4$).

2.3 Accuracy Configuration with Pipelined Architecture

When our proposed adder is combined with a pipelined architecture, we can obtain accurate results with the same throughput as a conventional adder. In the pipelined architecture, approximate additions are computed at the first pipeline stage, and error correction can be completed at the second stage. Figure 5 shows the conventional pipelined adder (above) and the approximate adder (below). The pipelined implementation of approximate adder has a structural analogy with the pipelined adder of the 2006 U.S. patent [8] in which partial summations are performed at the first stage and carry bits are added at the later stages. However, the patent is clearly directed to accurate operations, not approximate computations. In addition, we use our approximate adder (Figure 3) in the first stage. In the pipelined approach, there is no improvement of the clock frequency since the achievable clock period is the same as that of the conventional adder. However, power benefits are obtained through configuration of accuracy: in the approximate mode, the error correction stage is power-gated with foot (or, head) switches in Figure 5, and power reduction over the conventional adder design can be achieved. We compare the conventional and approximate pipelined adders in Section 3.

In the proposed adder implementation, to achieve higher performance or lower power consumption, we can reduce the carry chain depth (k) of sub-adders (see Table 1). However, when k is less than $N/4$, it is impossible to correct all errors and achieve 100% correct results within one clock cycle since the error-correction paths become critical. To achieve correct results in the pipelined implementation, the error-correction stage should be extended to mul-

tiple stages. Figure 6 shows the pipelined adder implementation ($k = N/8$ case), in which four pipeline stages are required to achieve a 100% accurate result. In the pipelined adder, each stage generates a result with different accuracy; the output accuracy increases as the number of pipeline stages increases. According to the accuracy requirement, we can turn off the later stages with a power gating technique, and we can reduce the power consumption further with the accuracy tradeoff.

Since the proposed adder supports both approximate and accurate results, it can be used in applications that require accurate results only under certain conditions. Conventional accurate designs are energy-inefficient in the error-acceptable application context, because they always compute the exact function. Previous approximate designs cannot handle a varying accuracy requirement, and this limits the benefit of the accuracy tradeoff: as noted above, the approximate function must meet the maximum accuracy threshold across all applications. Moreover, if the application requests an exact computation, additional accurate circuits must be added to the previous approximate designs. By contrast, the ACA design efficiently exploits a tradeoff between accuracy and power/performance with its runtime accuracy configurability.

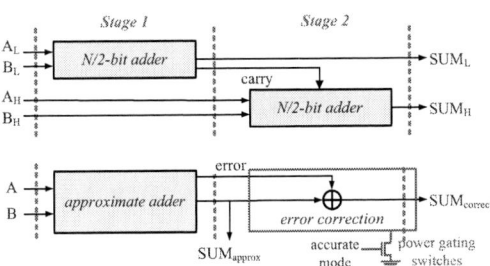

Figure 5: Pipelined adder implementation – conventional adder (above) and approximate adder (below). In approximate operation, the error correction stage is power-gated.

3. EXPERIMENTAL SETUP AND RESULTS

3.1 Experimental Setup

To test approximate designs, we have written each design in Verilog and synthesized it to a TSMC 65GP cell library with *Synopsys DesignCompiler* [17]. We then perform gate-level simulations using *Cadence NC-Sim* [18]. In the simulation, gate delay is taken from an SDF (standard delay format) file. For voltage scaling experiments, we prepare *Synopsys Liberty* (.lib) files for each voltage from 1.00V to 0.60V in 0.01V increments, using *Cadence Library Characterizer v9.1* [19]. The prepared libraries are used for SDF file generation and power estimation at each voltage. Each simulation is performed with input patterns for one million cycles. During the simulation, each output value is compared with a reference (correct) value to produce the accuracy metrics. For the input patterns, we use random data, as well as actual data from *SPEC 2006* [20] benchmarks. We extract operand data from ADD instructions in the SPEC benchmarks.

3.2 Metric for Approximate Design

To quantify errors in approximate designs, two metrics have been previously proposed [1]. *Error rate* (ER) is the percentage of cycles in which output value is different from the correct value. *Error significance* (ES) is the numerical difference between correct and output results; this quantifies the amount of error. In image/video applications, [2] uses the product of ES and ER as a metric of error tolerance. [10] introduces a criterion for acceptability: $ES \times ER \leq acceptance\ threshold$, where the acceptance threshold is specified according to the application. For the error significance (ES) metric, [14] considers only amplitude of error. This is useful for many digital signal processing (DSP) systems that process, e.g., sound and image data. However, in communication systems that mainly handle information data, the number of incorrect bits

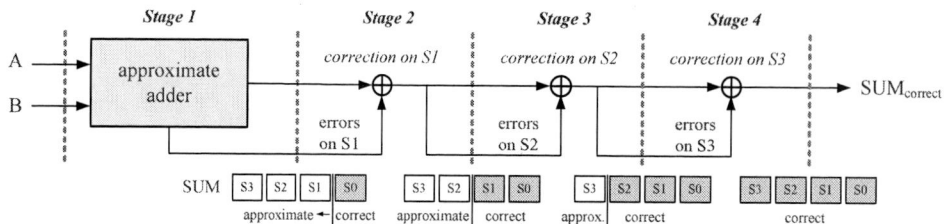

Figure 6: Accuracy-configurable implementation for pipelined adder.

(Hamming distance) is a more meaningful metric for accuracy – e.g. a *(32,28) Reed-Solomon code* can correct up to 2-byte errors. This consideration for the ES metric is required when approximate arithmetic is applied to error-tolerant systems with a redundancy technique.

Table 2 shows two accuracy metrics for amplitude data and information data. ACC_{amp} used in [14] quantifies the amplitude of errors, where R_c and R_e are the correct and obtained results, respectively. We propose another accuracy metric, ACC_{inf}, which measures error significance as Hamming distance, where B_e is the number of error bits and B_w is the bit-width of the data. For example, when the correct (reference) data is $1000\_0000_{(2)}$ and the result data is $1100\_0000_{(2)}$, accuracy with ACC_{amp} and ACC_{inf} will be $\frac{1}{2}$ and $\frac{7}{8}$, respectively. To evaluate the approximate circuits, we obtain average values of accuracy metrics ACC_{amp} and ACC_{inf} over the entire simulation to consider both ER and ES.

Table 2: Accuracy metrics for error significance (ES).

metric	definition	data type
ACC_{amp}	$1 - |R_c - R_e|/R_c$	amplitude data
ACC_{inf}	$1 - B_e/B_w$	information data

Table 3: ACA adder results with different k values.

k	2	3	4	5
min. clock period (ps)	180	190	220	230
area (um^2)	550	990	920	840
pass rate (%)	55.3	82.8	94.0	98.1
throughput improvement (%)	11.3	24.6	22.3	21.4

Table 4: Design comparison for each adder design.

	CLA	LU	ACA	ETAI	ETAIIM
area (um^2)	910	1356	923	576	678
min. clock period (ps)	280	210	200	200	260
pass rate (%)	100	99.2	94.1	10.0	97.0
ACC_{amp} (maximum)	1.000	0.998	0.997	0.999	0.999
ACC_{inf} (maximum)	1.000	0.999	0.993	0.694	0.996
area overhead for EDC	N/A	75%	28%	N/A	15%

3.3 Approximate Adder with Different Parameters

We explore the proposed adder with different parameters (k: half of carry-chain depth). Table 3 summarizes results – minimum clock period, area, error rate and throughput improvements – for each implementation of the 16-bit adder with different k values. According to the results, with smaller k, the maximum operating frequency increases, and the error rate increases as well. With higher k, the error rate is reduced significantly, but the benefit of the approximate circuit, i.e., clock period reduction, is small. In the table, throughput improvement over conventional design is calculated including error recovery overhead. From the implementations, a maximum throughput improvement is achieved when $k = 3$. If we correct erroneous results with EDC as in Figure 4, then 17.2% additional clock cycles are required for error correction. With this overhead, ACA adder can improve data throughput by 24.6% over the conventional CLA adder.

3.4 Approximate Adder Comparison

We evaluate each approximate adder with respect to the pass rate and the accuracy metrics which we have proposed. We use gate-level simulation at each possible clock period to compare five

adders: CLA, Lu's adder [7], ETAI, ETAIIM [14] and the proposed ACA adder (without error correction). In the experiment, the same carry-chain width (8-bit) is selected for the four approximate adders. In the implementation, a register (flip-flop) is inserted in each output port to detect timing errors.

Table 4 shows area, pass rate, accuracy, minimum clock period and EDC overhead for each adder design. According to the results, the ETAI adder has the smallest design area, but has a low pass rate and limited accuracy with respect to the ACC_{inf} metric. Therefore, the ETAI adder is preferred for applications which allow low accuracy in results. The ETAIIM adder shows fairly high accuracy, but does not have speed (clock period) benefit. Lu's adder shows a smaller error rate and high accuracy with respect to both ACC_{amp} and ACC_{inf} metrics. However, it requires larger area than the other designs. The proposed adder shows similar results for both metrics as Lu's adder. However, the area of the ACA adder is smaller than that of Lu's adder, and EDC is possible with small area overhead (28%). With the ACA adder, the minimum clock period can be reduced by 26% compared to the accurate CLA.

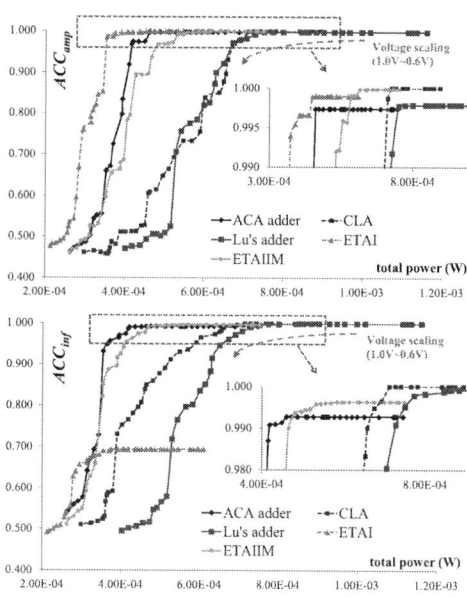

Figure 7: Accuracy (y-axis) vs. power consumption (x-axis) under fixed clock period ($0.25ns$) and scaled voltage (from $1.0V$ to $0.6V$).

Figure 7 shows a power vs. accuracy tradeoff in a voltage scaling scenario: the x-axis shows total power consumption, and the y-axis shows the accuracy (ACC_{amp}, ACC_{inf}). The power consumption and the accuracy are measured with different voltage libraries characterized using *Cadence Library Characterizer* [19]. The clock period is fixed at $0.30ns$ during the simulations. In the results, Lu's adder does not show power benefits due to its design size. ETAI shows low power consumption and high ACC_{amp} accuracy, but has low ACC_{inf} accuracy, and cannot detect and correct errors. ETAIIM shows similar characteristics to ACA in the voltage scaling case, but the adder cannot be used for a high-performance (high-frequency) design, as shown in Table 4. The results in Figure 7 imply that our proposed adder can provide a significant power

823

reduction with small accuracy penalty. When the required accuracy is 0.970 (ACC_{amp}), the ACA adder shows 37.0%, 36.4% and 15.9% total power reduction over CLA, Lu's adder and ETAIIM, respectively.

We have tested our approximate adder on a real application – a Gaussian smoothing filter used in [6]. Gaussian smoothing is performed on the input image by convolving with a matrix in the spatial domain. In the convolution, the addition operation is done with approximate 16-bit adders. Other operations, such as multiplication and division, are accurate computations. Figure 8 shows results for various approximate adders when they consume 50% of the power of accurate CLA. From the results, the ACA adder has PSNR of 24.5dB, and this suggests that image processing/filtering applications could employ our proposed adder with significant power savings and only small loss in image quality.

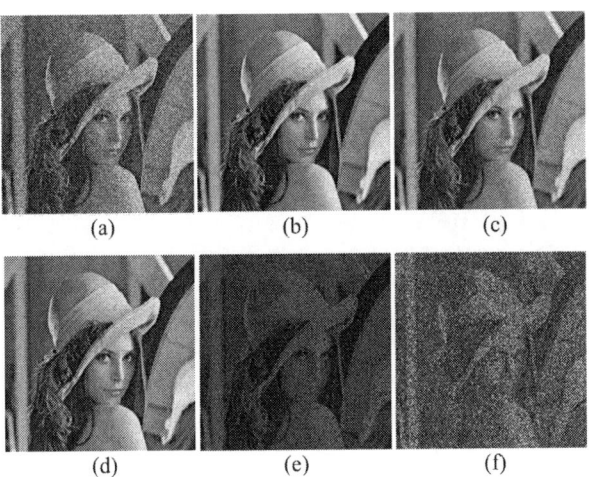

Figure 8: Image smoothing: (a) original image with noise; (b) accurate adder; (c) ACA, PSNR: 24.5 dB; (d) ETAI, PSNR: 25.3 dB; (e) ETAIIM, PSNR: 16.2 dB; (f) Lu's adder, PSNR: 11.1dB.

Table 5: Comparison between conventional and approximate (2-stage) pipelined adders at the accurate mode.

adder width (N)	conventional pipelined			approximate pipelined			
	area (um^2)	clock period (ns)	total power (mW)	k	area (um^2)	clock period (ns)	total power (mW)
8	459	0.313	0.557	2	576	0.312	0.564
16	1082	0.357	1.558	4	1171	0.358	1.669
32	2252	0.404	2.860	8	2420	0.414	2.914

Table 6: Implementation results of 32-bit ACA adder with 4-stage pipeline (power consumption of each mode and power reduction over conventional pipelined adder).

config.	power-gating	ACC_{amp} (max.)	ACC_{inf} (max.)	total power (mW)	reduction (%)
mode-1	none	1.000	1.000	5.962	-11.5%
mode-2	stage-4	0.998	0.960	4.683	12.4%
mode-3	stage-3, 4	0.991	0.925	3.691	31.0%
mode-4	stage-2, 3, 4	0.983	0.900	2.588	51.6%

3.5 Accuracy Configuration and Power Savings

When the architecture allows pipelining for addition, our proposed adder can be implemented as shown in Figure 5. We implement both the conventional pipelined adder and the approximate pipelined adder to compare the designs in terms of area, timing and power. In the implementation, registers (flip-flops) are included at each pipeline stage (before *stage-1*, between *stage-1* and *stage-2*, and after *stage-2*).

Table 5 shows the implementation results for the conventional and approximate pipelined adders. The parameter k has been selected as $N/4$ for a two-stage pipelined implementation. In the table, minimum clock period is measured at a fixed voltage ($1.0V$), and total power is measured at a fixed frequency ($2.5GHz$) with voltage scaling. In the ACA adder case, timing and power overheads from power gating cells, output MUXes, and IR drop are included. We can see that area, timing and power of both designs are similar when the ACA adder operates in the accurate mode. Total power of the approximate adder is comparable to that of the conventional adder, even though ACA has additional EDC circuits. This is because ACA has fewer registers between *stage-1* and *stage-2* than the conventional pipelined adder. (In Figure 5, the conventional adder requires registers for A_H, B_H, SUM_L and carry at the first stage. For a 16-bit adder, 25 registers ($8 + 8 + 8 + 1$) are required. On the other hand, ACA requires 18 registers (16 for SUM_{approx} and 2 for error indication).)

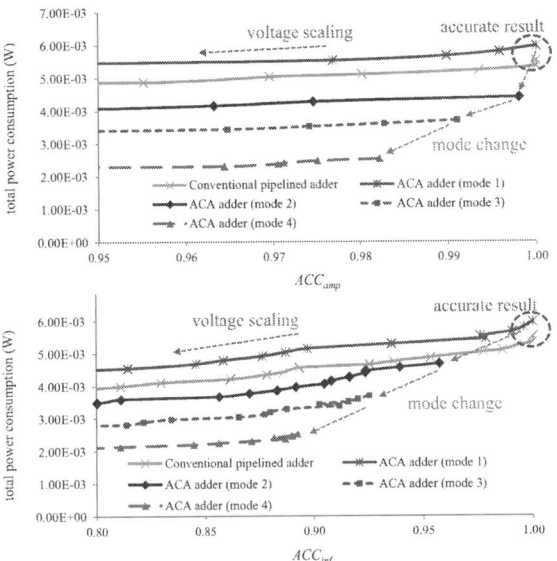

Figure 9: Accuracy metric ACC_{amp} (above) and ACC_{inf} (below) vs. power consumption for conventional pipelined adder, ACA adder in accurate mode, and ACA adder in approximate mode (4-stage, 32-bit adder).

In the pipelined architecture, the ACA adder can provide various configurable modes according to the pipeline depth. To improve the design performance, we increase the pipeline depth; the deeper pipeline reduces the path depth of the design. In the conventional pipelined adder, bit-width of the adder in each stage can be reduced to $N/\#stage$, where N is the entire bit-width and $\#stage$ is the depth (number) of the pipeline stages. In the ACA adder, we can reduce the value of parameter k with deeper pipeline depth as shown in Figure 6. To show the benefit of accuracy configuration, we have implemented a 32-bit ACA adder ($N = 32$, $k = 4$) with 4-stage pipeline, and compared it with a conventional pipelined adder with an 8-bit CLA in each stage. Table 6 shows the implemented results for the 32-bit ACA adder. For the accuracy estimation, one million cycles of random patterns are used. The ACA adder can operate in four different modes, based on the power gating of each stage. We can see that the modes show different power consumptions and different achievable accuracies. The ACA adder consumes 11.5% more power than the conventional adder in accurate mode (mode-1) due to the presence of recovery circuits. At the same time, it shows a significant power reduction in the approximate modes: 12.4%, 31.0% and 51.6% in mode-2, mode-3 and mode-4, respectively. Figure 9 shows detailed results for power consumption versus accuracy metrics in each configuration. From the results, we can see that accuracy configuration with the mode change is much more effective than with voltage scaling, in terms of the tradeoff between accuracy and power.

Table 7: Accuracy (ACC_{amp}, ACC_{inf}) results of 32-bit ACA adder for real benchmarks (SPEC 2006).

accuracy metric	benchmark	astar	bzip2	calculix	gcc	h264ref	mcf	sjeng	soplex
ACC_{amp}	mode-1	1.0000	1.0000	1.0000	1.0000	1.0000	1.0000	1.0000	1.0000
	mode-2	0.9999	1.0000	0.9999	0.9992	0.9999	0.9997	0.9998	0.9999
	mode-3	0.9993	0.9998	0.9972	0.9990	0.9990	0.9997	0.9995	0.9998
	mode-4	0.9979	0.9970	0.9958	0.9951	0.9978	0.9991	0.9981	0.9953
ACC_{inf}	mode-1	1.0000	1.0000	1.0000	1.0000	1.0000	1.0000	1.0000	1.0000
	mode-2	0.9979	1.0000	0.9978	0.9881	0.9953	0.9819	0.9897	0.9985
	mode-3	0.9949	0.9984	0.9967	0.9849	0.9897	0.9809	0.9876	0.9965
	mode-4	0.9940	0.9931	0.9910	0.9617	0.9851	0.9596	0.9787	0.9925

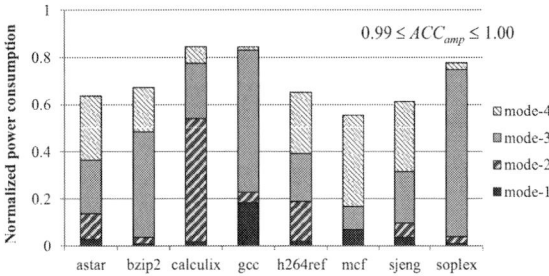

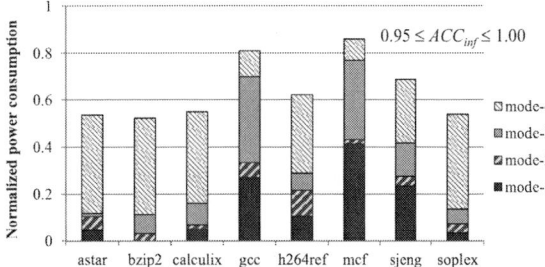

Figure 10: Normalized power consumption versus conventional pipelined design when the accuracy requirement is varied uniformly over the interval 0.99 $\leq ACC_{amp} \leq$ 1.00 and 0.95 $\leq ACC_{inf} \leq$ 1.00.

We also obtain the accuracy results in each accuracy mode with real input patterns extracted from SPEC 2006 benchmarks. Table 7 shows accuracy results of a 32-bit ACA adder with such real input patterns. The accuracy results are different for each benchmark, e.g, the measured accuracy for *bzip2* is higher than for *gcc*. Furthermore, the accuracy with real patterns is greater than with random input patterns (Table 6), most likely because addition inputs for MPU have infrequently and/or systematically changing patterns in the applications. We evaluate power reductions across accuracy requirements with the patterns from SPEC 2006 benchmarks. Figure 10 shows power reduction achieved by the ACA adder versus the conventional pipelined adder under the accuracy requirements. We assume that required accuracy is from 0.99 (0.95) to 1.0 for ACC_{amp} (ACC_{inf}), and that it varies uniformly over this range during the entire runtime. From the results, dynamic accuracy configuration achieves up to 44.5% (30.0% on average) and 47.1% (35.8% on average) power reduction over the conventional pipelined design for ACC_{amp} and ACC_{inf} metrics, respectively.

4. CONCLUSIONS

In this paper, we propose an accuracy-configurable approximate (ACA) adder for which the accuracy of results is configurable during runtime. Due to its configurability, the ACA adder can operate adaptively in both approximate (inaccurate) mode and accurate mode. To quantify the accuracy in approximate computation, we provide two metrics for amplitude data and information data. We compare the ACA adder against previous approximate adders based on the proposed metrics. The ACA adder shows high accuracy with respect to the metrics, and can provide up to 24.6% throughput improvement and 37.0% power reduction over the conventional CLA adder. The ACA adder can also be used in accuracy-configurable applications with pipelining. We demonstrate that the ACA adder can provide approximately 30% power reduction under a relaxed accuracy requirement versus the conventional pipelined adder. Finally, we show that our ACA adder can improve the achievable tradeoff between performance, power and quality for given accuracy requirements.

Our ongoing work seeks to implement accuracy-configurable designs for other arithmetic components such as multipliers, multi-input adders, etc. More broadly, our research addresses additional aspects of (runtime) accuracy-configurable systems and applications.

5. REFERENCES

[1] M. A. Breuer, "Intelligible Test Techniques to Support Error-Tolerance", *Proc. Asian Test Symp.*, 2004, pp. 386–393.

[2] I. Chong, H. Y. Cheong and A. Ortega, "New Quality Metric for Multimedia Compression Using Faulty Hardware", *Proc. International Workshop on Video Processing and Quality Metrics for Consumer Electronics*, 2006, pp. 267–272.

[3] J. George, B. Marr, B. E. S. Akgul and K. V. Palem, "Probabilistic Arithmetic and Energy Efficient Embedded Signal Processing", *Proc. CASES*, 2006, pp. 158–168.

[4] S. Ghosh and K. Roy, "Parameter Variation Tolerance and Error Resiliency: New Design Paradigm for the Nanoscale Era", *Proceedings of the IEEE* 98(10) (2010), pp. 1718–1751.

[5] P. Kulkarni, P. Gupta and M. Ercegovac, "Trading Accuracy for Power with an Underdesigned Multiplier Architecture", *Proc. IEEE/ACM International Conference on VLSI Design*, 2011, pp. 346–351.

[6] M. S. Lau, K.-V. Ling and Y.-C. Chu, "Energy-Aware Probabilistic Multiplier: Design and Analysis", *Proc. CASES*, 2009, pp. 281–290.

[7] S.-L. Lu, "Speeding Up Processing with Approximation Circuits", *IEEE Computer* 37(3) (2004) pp. 67-73.

[8] H. D. Mohammed and L. Hemmert, "Fast Pipelined Adder/Subtractor using Increment/Decrement Function with Reduced Register Utilization", *U.S. Patent* No. 7,007,059, 2006.

[9] B. J. Phillips, D. R. Kelly and B. W. Ng, "Estimating Adders for a Low Density Parity Check Decoder", *Proc. SPIE*, vol. 6313, 2006, pp. 1–9.

[10] D. Shin and S. K. Gupta, "A Re-Design Technique for Datapath Modules in Error Tolerant Applications", *Proc. Asian Test Symp.*, 2008, pp. 431–437.

[11] D. Shin and S. K. Gupta, "Approximate Logic Synthesis for Error Tolerant Applications", *Proc. DATE*, 2010, pp. 957–960.

[12] A. K. Verma, P. Brisk and P. Ienne, "Variable Latency Speculative Addition: A New Paradigm for Arithmetic Circuit Design", *Proc. DATE*, 2008, pp. 1250–1255.

[13] N. Zhu, W. Goh and K. Yeo, "An Enhanced Low-Power High-Speed Adder For Error-Tolerant Application" *Proc. Intl. Symp. on Integrated Circuits*, 2009, pp. 69–72.

[14] N. Zhu, W. Goh, W. Zhang, K. Yeo and Z. Kong, "Design of Low-Power High-Speed Truncation-Error-Tolerant Adder and Its Application in Digital Signal Processing", *IEEE Trans. on VLSI Systems* 18(8) (2010), pp. 1225–1229.

[15] M. Ziegler and M. Stan, "Optimal Logarithmic Adder Structures with a Fanout of Two for Minimizing the Area-Delay Product", *Proc. ISCAS*, 2001, pp. 657–660.

[16] *International Technology Roadmap for Semiconductors*, 2009, http://www.itrs.net .

[17] *Synopsys Design Compiler User's Manual*. http://www.synopsys.com .

[18] *NC-Sim User's Manual*. http://www.cadence.com .

[19] *Cadence LC User's Manual*. http://www.cadence.com .

[20] *Standard Performance Evaluation Corporation (SPEC) CPU2006*. http://www.spec.org/cpu2006 .

Recovery-Based Design for Variation-Tolerant SoCs*

Vivek Kozhikkottu [†], Sujit Dey[*] and Anand Raghunathan[†]
[†] School of Electrical and Computer Engineering, Purdue University
[*]School of Electrical and Computer Engineering, UC San Diego
{vkozhikk,raghunathan}@purdue.edu, dey@ece.ucsd.edu

ABSTRACT

Parameter variations have emerged as a significant threat to continued CMOS scaling in the nanometer regime. Due to increasing performance penalties associated with worst-case design, recovery based design has emerged as a promising approach for dealing with the impact of variations. Previous work has applied recovery based design at the circuit and micro-architecture levels of abstraction. In this work, we address the problem of designing variation-tolerant SoCs using the recovery based design paradigm. We demonstrate that a monolithic implementation of recovery based design fails to scale for large SoCs. We propose the concept of *recovery islands*, wherein each island consists of one or more SoC components that can recover independent of the rest of the SoC, and demonstrate how our proposal can be easily realized via minor changes to a traditional SoC design flow. We study the trade-offs involved in applying recovery based design at the system level. We demonstrate that it is critical to account for (i) the inherent diversity of the error-voltage profiles among various components in an SoC, and (ii) the impact of error recovery in a component on overall system performance. We then propose a systematic recovery-based SoC design methodology that partitions a given SoC into recovery islands and also computes the optimal operating points for each island, taking into account the various system level trade-offs involved. We evaluate our framework on three different SoC designs, an 802.11b MAC processor, an MPEG encoder and a Wireless Video Capture system and demonstrate an average of 32% energy savings over conventional designs.

Categories and Subject Descriptors

B.7.1 [**INTEGRATED CIRCUITS**]: VLSI (Very large scale integration)

General Terms

Algorithms, Design

Keywords

System-on-chip, Variation Aware Design, Variation Tolerance, Low Power Design

1. INTRODUCTION

Continued scaling of CMOS technologies has resulted in parameter variations emerging as a critical design concern. Parameter variations can be broadly classified as process variations caused due to the inherent nature of the manufacturing

---

*This material is based upon work supported in part by the National Science Foundation under Grant No. 0916117.

process and environmental variations due to fluctuations in temperature and supply voltage. These parameter variations manifest as statistical behavior in the delay and power consumption of circuits, and have traditionally been dealt with by over-design. However, with continued scaling into the nanometer regime, the gap between typical-case and worst-case design is growing too large, and the performance and energy cost of worst-case design can no longer be ignored.

To overcome the problems with worst-case design, recovery based design techniques such as Razor [1] and EDS [2] have been proposed. These techniques employ embedded error detection and recovery circuitry to help detect and recover from timing errors induced by variations. They help eliminate conservative voltage guard bands by dynamically controlling the supply voltage in response to the occurrence of timing errors. Moreover, components can be voltage "overscaled" even beyond their zero error operating points to achieve considerable energy reductions for a negligible loss in performance [1]. These recovery based design techniques have hitherto been applied only at the circuit and micro-architecture levels [3,4]. We believe that ours is the first effort to explore the application of recovery-based design in a systematic manner to entire SoCs.

1.1 Paper Overview and Contributions

In this work, we address the problem of designing variation-tolerant SoCs using the recovery based design paradigm. The significant contributions of our work are as follows:

- We demonstrate that applying recovery based design in a monolithic fashion is not scalable for large SoC designs. We propose a new design approach in which SoCs are divided into multiple recovery islands, each of which can detect and recover from errors independent of the rest of the SoC. We also demonstrate that the communication architecture serves as an ideal variable latency interface for partitioning the SoC into recovery islands.

- We study the trade-offs involved in applying recovery based design at the system level. We demonstrate that each component's distinct error-voltage characteristics as well as its impact on overall system performance need to be considered while clustering them into recovery islands and computing their operating points.

- We propose a methodology that systematically partitions a given SoC into recovery islands and also computes the optimal operating point for each island. The framework takes into account the above trade-offs, as well as the complex interactions between different islands, using an emulation based performance analysis framework.

- We apply recovery based SoC design to three different SoC designs — an 802.11b MAC processor, an MPEG encoder and a Wireless Video Capture system — and obtain an average of 32% energy savings over conventional designs.

The rest of this paper is organized as follows. Section 2 summarizes prior work on variation-aware system design. Section 3 describes the challenges involved in applying recovery based design to SoCs. Section 4 gives an overview of the proposed concept of recovery islands and the various interfaces needed to enable it. Section 5 analyzes the various system-level trade-offs involved in recovery island based SoC design with the help of an example. Section 6 describes our systematic recovery based SoC design methodology. Section 7 describes our experimental setup and presents the results obtained by applying the proposed framework to three example SoC designs.

2. RELATED WORK

In the context of SoCs, several previous efforts have demonstrated the strong potential of addressing variations at the system level. In the context of multiple voltage-frequency island based SoC design, several efforts [5, 6] have exploited the inherent flexibility of the multi-island design paradigm to mitigate the impact of variations. Techniques for analyzing the impact of process variations on system performance and power were developed in [7] and [8]. A variation tolerant on-chip communication architecture was discussed in [9]. In [10], the authors develop techniques to optimize system-level power management policies under the impact of variations. [11] proposes partitioning an SoC into fine grained body bias islands to help mitigate the impact of within-die leakage variations. However, most of these techniques only deal with manufacturing induced process variations and do not deal with workload, voltage and temperature based variations. Recovery based design, due to its dynamic and adaptive nature [1] [2], deals with all sources of variations and thus eliminates the need for conservative design margins.

Due to the various power-performance penalties associated with worst-case design, researchers have started actively developing recovery based design techniques. Razor [1] and EDS [2] propose circuit level mechanisms to detect and correct timing based errors, providing a safety net that allows the elimination of guard bands and design margins. Furthermore, these mechanisms achieve substantial energy savings by facilitating voltage overscaling, a technique of scaling the supply voltage beyond the circuit's critical operating point, resulting in timing errors. In this context, [12] and [13] have proposed using cell sizing and dual threshold voltage cells to modify the timing slack of the frequently-occurring, near-critical timing paths to facilitate further voltage overscaling, thereby achieving additional energy savings. Similarly, at the architecture level, [14] and [15] have suggested architectural modifications to reshape the error-voltage profiles of underlying micro-architectural blocks so as to increase their potential for voltage overscaling. In [3] and [4] the authors argue that fine-grained adaptive biasing and voltage interpolation based techniques can be applied to processors instrumented with recovery mechanisms to help mitigate the impact of within-die parameter variations. However, as noted earlier, these techniques focus on the circuit and micro-architecture level trade-offs involved in applying recovery based design. In this paper, we focus on identifying the key system level characteristics and trade-offs that must be taken into account for applying recovery based design in the context of SoCs.

3. MOTIVATION

In this section, we motivate the need for a new approach to recovery based design for SoCs, by outlining two major scalability concerns associated with applying recovery based techniques in a monolithic fashion. We utilize an example SoC design to help quantify these concerns. Figure 1 shows the block diagram of a Wireless Video Capture Device (WVCD) SoC consisting of ten components connected to a system bus. The SoC performs two main functions, namely, i) it encodes video frames stored in an on-chip frame buffer, and ii) it packetizes

the frames using the 802.11b protocol and sends the packets out to a wireless interface for transmission. The four important compute-intensive functions i) Checksum Computation (CRC), ii) Wired Equivalent Privacy encryption (WEP), iii) Motion Estimation (ME), and iv) DCT compression (DCT), are all implemented as hardware accelerators.

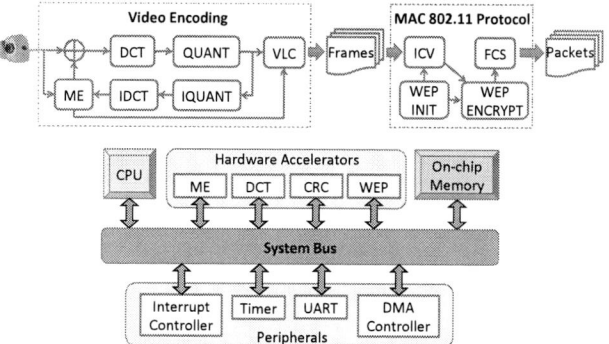

Figure 1: Wireless video capture SoC

The first major factor limiting the scalability of a monolithic recovery based scheme is the impact of within-die parameter variations [16]. Within-die variations cause components within a given instance of the SoC to have differing performance-power characteristics. Recovery based design techniques typically try to operate a component at its optimal operating voltage point so as to eliminate the conservative voltage guard bands needed to deal with variations. However, in a monolithic implementation, the operating voltage of the entire SoC would be determined by the voltage of its slowest component that has been impacted most negatively by variations. As a consequence, a large number of components would be forced to operate at sub-optimal voltages, leading to reduced energy benefits.

Figure 2 shows the mean energy savings (for 10000 WVCD SoC chips) obtained by a monolithic implementation of recovery based design, for increasing values of within-die process variations. The figure shows that for higher values of within-die variations, the energy savings attained by monolithic recovery based design decreases significantly. Moreover, increased within-die variations in other important parameters such as voltage, temperature and workload, would only exacerbate this effect. In summary, within-die parameter variations pose a severe challenge to scaling recovery based design to large SoCs.

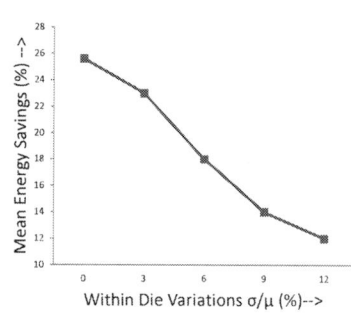

Figure 2: Mean energy savings vs. within-die variations

The second major concern affecting the scalability of monolithic recovery based design is the strict timing constraint required for performing error detection and correction. The timing constraint can be expressed as follows:

$$T_{\text{clk\_tree}} + T_{\text{delay\_sample}} + T_{\text{clk\_error}} + T_{\text{error\_agg}} < T_{\text{clk\_period}},$$
(1)

where $T_{\text{clk\_tree}}$ is the clock to flip-flop delay, $T_{\text{delay\_sample}}$ and $T_{\text{clk\_error}}$ represent the delays associated with generation of the error signal by the shadow flipflop and finally $T_{\text{error\_agg}}$ refers to the delay required for aggregating all the error signals back to gate the clock source. All the above delays must add up to less than the system's clock period ($T_{\text{clk\_period}}$) so as to successfully perform clock gating before the start of the next clock cycle, when an error is detected. However, with increasing SoC sizes and the poor scaling of interconnect delays [17], the global delay components ($T_{\text{clk\_tree}}$ and $T_{\text{error\_agg}}$) restrict the applicability of monolithic implementations of recovery based design to small systems. For example, for an operating frequency of 1 GHz at the 45nm technology node, our analysis suggests that the monolithic scheme would be feasible only for circuits of size up to $0.9 mm^2$. Thus, interconnect delays enforce a strict limit on the size of SoC designs for which monolithic recovery based design is applicable.

Due to the above mentioned limitations associated with monolithic recovery based design, we propose applying recovery based techniques to SoCs in a more fine grained manner. There are two key issues that must be addressed in order to realize this proposal. First, we need to allow components to recover independent from the rest of the SoC, while maintaining correct operation. Second, we must explore the design space of possible partitions ranging from a monolithic recovery based design on one extreme to a fine-grained partitioning where each SoC component is in its own recovery island. In doing so, it is necessary to consider the area and energy overheads associated with creation of recovery islands. We address these issues in the following sections.

4. RECOVERY ISLANDS: SCALING RECOVERY BASED DESIGNS TO SOCS

Traditionally, SoCs are partitioned into coarse-grained voltage and frequency islands. We propose partitioning these voltage-frequency islands further into more fine grained recovery islands. Each recovery island consists of one or more SoC components and must possess the following key properties i) the ability to detect and recover from errors in any of its components independent of the rest of the SoC, and ii) the dimensions of the recovery island must allow the timing constraints imposed by Equation 1 to be satisfied. One of the challenges associated with enabling recovery islands to recover independently is to find suitable points for partitioning a given SoC into islands. Creating recovery islands from arbitrary partitions of logic would require extensive re-design of the underlying components and their respective interfaces. From an external perspective, each recovery island can take a variable number of cycles to respond to a transaction, depending on whether a component in the island has encountered an error or not. We note that the system-level communication architecture (bus or network-on-chip) used in most SoCs is already designed so as to tolerate variable latencies, be it due to bus contention or a component being busy. The communication architecture thus serves as an ideal variable latency interface to partition the SoC into recovery islands. However, one cannot just directly connect the recovery islands to the interconnect fabric, as they would violate the established interface protocols. Appropriate cross recovery-island interfaces need to be designed to interface recovery islands with the rest of the system. A more detailed description of these interfaces is provided in Section A of the supplementary material.

For each recovery island, the timing critical flip-flops are in-

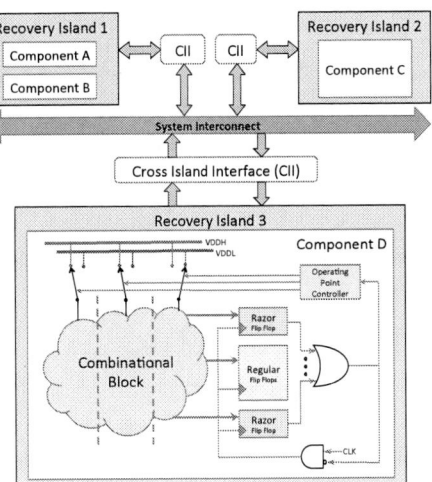

Figure 3: Recovery island based design

strumented with error detection and recovery circuitry [1, 2]. When an error is detected, the clock is gated for the next cycle to allow the correct values to be restored to all the flip-flops. All the error signals are then aggregated and fed into the operating point controller [1], which is responsible for dynamically controlling the supply voltage of the island to maintain a desired error rate. We achieve supply voltage scaling by utilizing voltage interpolation [18] so as to avoid the significant overheads associated with voltage regulators and converters. Voltage interpolation provides the ability for different groups of logic gates within a block to select between two static supply voltages VDDH and VDDL. The scheme enables dynamic modulation of a circuit's delay by choosing an appropriate combination of logic segments within a block to be connected to VDDH and VDDL, respectively.

The recovery island based design methodology incurs area and energy penalties associated with additional cross-island interfaces and operating point control mechanisms. As a consequence, partitioning the SoC at the granularity of individual components can lead to considerable energy overheads and may significantly diminish the system level energy benefits obtained by the framework. Thus, it is imperative to find an energy-optimal partitioning of the SoC into recovery islands.

5. DESIGN TRADEOFFS

In this section, we explore the various system level design trade-offs involved in recovery island based SoC design. We utilize the previously described WVCD SoC (Figure 1) for illustrating some of these trade-offs.

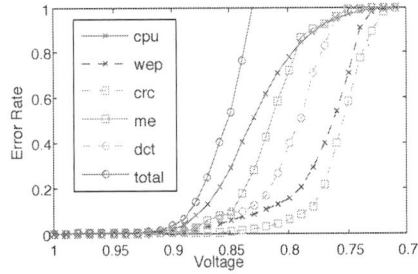

Figure 4: Error rate versus voltage profile

We first describe the component level characteristics to be considered when partitioning an SoC into recovery islands. The operating voltage of a component in recovery based design depends on its inherent error-voltage profile, which is in turn determined by various factors such as circuit structure, component size, application workload (path activation probabilities) as well as process, voltage and temperature variations. For the example WVCD SoC, Figure 4 shows the error versus voltage profiles for the largest five components. We also plot the total error versus voltage profile for the entire SoC. As the figure shows, the total system error is mostly dominated by errors in the CPU. In a scheme in which the entire SoC is treated as a single recovery island (monolithic implementation of recovery based design), each component would be operated at a voltage mostly determined by the CPU's error-voltage profile. On the other hand, if we perform recovery at a component level, each component would be operated at its own optimum voltage based on its error-voltage profile, which would lead to substantial energy savings. However this scheme would also involve excessive overheads associated with implementing recovery islands. In general, SoC's have components that tend to greatly differ in their structure, complexity, size and workload, leading to diverse error-voltage profiles across components. This diversity is further amplified because of intra-die variations in process parameters, temperature gradients across the chip, as well as local voltage fluctuations. The partitioning scheme thus needs to incorporate this inherent diversity in the error-voltage profiles among various components, and the overheads associated with recovery islands, in addition to system-level factors as discussed next.

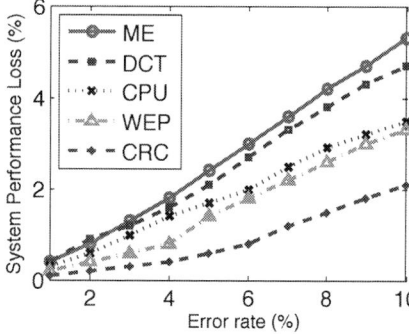

Figure 5: System performance loss versus error rate

We now motivate the need to consider system level effects while choosing optimal operating points for each recovery island. Errors in a component force it to spend clock cycles in recovery and thereby affect system performance. However, depending on how critical a component is to overall system performance, the same error rate in different components can have different effects on system performance. Also, due to complex inter-dependencies between components (*e.g.*, concurrent execution and synchronization), the system performance impact due to errors in different components need not be additive. Figure 5 plots the system performance loss versus error rate for different components of the WVCD SoC. As can be seen, errors in the ME accelerator have the greatest impact on system performance, followed by the DCT accelerator. Therefore, for a given system level performance target, a configuration in which the rest of the components (CRC, CPU, WEP) operate at higher error rates and thereby can be voltage scaled more aggressively, is more energy efficient than a configuration in which the ME and DCT components operate at higher error rates. In complex SoCs, the correlation between error rate and system performance loss can be quite varied across components and this diversity needs to be considered while selecting the optimal operating points for each SoC component. Recovery islands that consist of components that are more critical to system performance should be operated at lower error rates, whereas those that contain components with a lower impact on system performance should be voltage scaled more aggres-

sively so as to reduce overall system power. Also note that it is beneficial during partitioning to group together components that have similar impact on system performance.

6. RECOVERY BASED SOC DESIGN

In this section, we describe a systematic methodology for recovery based SoC design that considers the issues and trade-offs described in the previous section. The proposed methodology, shown in Figure 6, takes as its input the given SoC architecture, the application software, the desired performance target and component-level clustering constraints derived from the SoC floorplan. It produces as its output, the best SoC partitioning scheme along with optimized operating points for each island. The methodology consists of three main steps. In the *component characterization* step, we compute the error rate, system performance loss and energy savings for each SoC component at each possible operating voltage. The *island partitioning and optimization* step partitions the SoC appropriately into recovery islands, and computes the best operating point for each island. Finally, the *local search* step further tunes the operating points obtained in the previous step while considering the complex performance interactions between different SoC components. We elaborate upon these steps in the rest of this section.

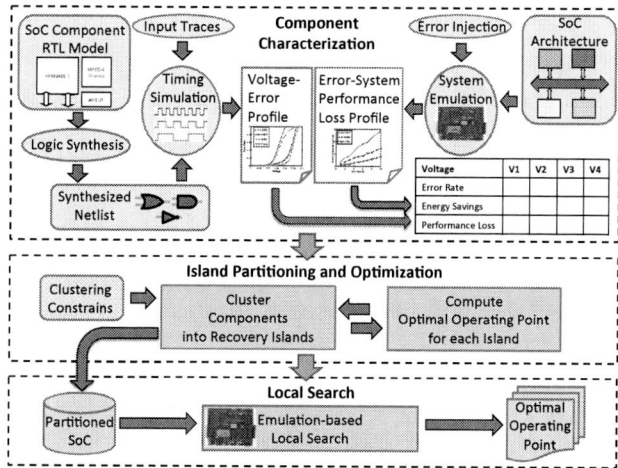

Figure 6: Recovery island based design methodology

6.1 Component Characterization

In this step, we first obtain the error-voltage profile and the error-system performance loss profile for each component. For generating the error-voltage profiles, we first capture bus level input traces for each component by performing cycle-accurate functional simulation of the SoC for representative workloads. We then use the captured traces as input vectors to perform post-synthesis simulations at different operating voltages to obtain the error-voltage profile for each component. For the error *vs.* system performance loss profile, we use an emulation based performance analysis framework. We instrument each SoC component with error injectors and circuitry that mimics error recovery, and obtain the system performance loss for increasing error rates in each component. More details on the emulation setup are provided in Supplemental Section B. We now combine both the profiles with component level energy estimates to obtain an error, system performance loss and energy tuple for each operating point (voltage).

6.2 Island Partitioning and Optimization

In this step we derive an optimized partition of the SoC into recovery islands and obtain the best operating point for each island. The number of ways an SoC consisting of N components can be partitioned into k recovery islands can be quite large (N^k). The search space involved in identifying an optimal operating point for each island further increases the design space by O^k, where O is the number of operating points. To efficiently explore this design space, we adopt an iterative procedure wherein we start off with each component in a separate partition, and iteratively apply the operating point selection and island clustering steps until we can no longer find a better partition. Consider the initial partition where each component is in a separate recovery island. We can compute the best operating point for each component by modeling it as a convex optimization problem as shown in Equation 2.

$$\underset{V_i}{\text{minimize}} \quad \sum_{i=1}^{n} E_i(V_i) \qquad \text{subject to} \quad \sum_{i=1}^{n} P_i(V_i) \leq \zeta;$$
$$V_{min} \leq V_i \leq V_{max} \quad i = 1, \ldots, n$$
$$(2)$$

In the above equation, $E_i(V_i)$ and $P_i(V_i)$ refer to the energy and system performance loss respectively of the i-th component operating at voltage V_i and ζ is the constraint on acceptable system performance loss. For this step, we make the simplifying assumption that system performance loss due to N different components is linearly additive. This is not true in general, due to effects such as communication dependencies, shared resources, system-level critical paths, *etc.* We ignore these effects in the island partitioning and optimization step to make the problem tractable, but account for them in the subsequent local search step.

Once we obtain the optimal operating points for each component, we group together the two components whose operating points are closest, if this clustering is valid based upon the floorplan derived constraints. These constraints are represented as a clustering matrix that specifies which component pairs could be grouped together, based on their proximity in the SoC's floorplan. Grouping components reduces the overheads associated with recovery islands, at the cost of forcing the grouped components to operate at the same voltage. The grouping heuristic therefore minimizes the sub-optimality in operating points. We choose the two components j and k that, when grouped together, give the best energy savings $E = E_{ri} - (E_k(V_j) - E_k(V_k))$. The first term E_{ri} refers to the energy savings due to the reduced overheads of having one less recovery island and the second term refers to the energy loss due to one of the components (in this case, k) going from a lower operating voltage V_k to a higher point V_j. We iteratively perform the island partitioning and operating point selection steps until we find no further grouping that can lead to energy savings.

6.3 Local Search

In this step, we tune the operating points for the partitioned SoC obtained from the previous phase taking into account the inter-dependencies between various SoC components. We achieve this by first performing emulation of the SoC at the operating points obtained from the previous step to compute the actual system performance loss. Next, based on whether the performance loss is larger or smaller than the given target, we increase or decrease the operating voltages for each island by one unit step and measure the resulting system performance loss. We now greedily change the operating voltage of the island with the best energy savings to performance loss ratio, and repeat this process until the specified performance target is just satisfied.

7. EXPERIMENTAL RESULTS

In this section, we first describe our experimental set up and the example SoCs used in our study. We then present the energy savings obtained by utilizing our framework on three different SoC designs.

Our experimental methodology to evaluate the proposed concepts consists of various commercial and research tools. For obtaining the error-voltage profiles, we first perform logic synthesis of each component with Synopsys Design Compiler using the IBM 45nm technology cell library. We utilize VARIUS [19] for modeling the impact of inter-die and intra-die process and temperature variations on component-level error-voltage profiles. For each of our experiments we generate 10000 chips, each of which has different intra-die variation profiles for V_{th} and L_{eff} values. To obtain the temperature distribution across the chip, we provide average power consumption values of each SoC component, along with the SoC floorplan to the HotSpot thermal modeling tool [20]. We use NANOSIM [21], a transistor level simulator, to obtain the power consumption data for each of the SoC components at different operating voltages. The memory energy consumption and access times are modeled using CACTI5.3 [22]. We use an Altera DE3 board [23] as our emulation platform for obtaining the component level error-rate versus system performance loss profiles.

We evaluate our framework on three example SoC designs, an 802.11b MAC processor, an MPEG encoder, and a Wireless Video Capture Device. The WVCD system was described in detail in Section 3. We now briefly describe the MPEG and MAC systems. MPEG encoding entails two compute-intensive operations - Motion Estimation (ME) and DCT Compression, which are implemented as hardware accelerators. The input frames to be encoded are stored in an on-chip frame buffer and an embedded processor is in charge of co-ordinating the transfer of frames between the frame buffer and the two accelerators, and also executes the remaining tasks. The 802.11 MAC processor implements the key steps of the 802.11b MAC protocol, and consists of a processor, hardware accelerators for CRC and WEP computation, and peripherals connected by a system interconnect. In order to verify the functional correctness of the various cross island interfaces as well as to accurately model the impact of errors on system performance, each SoC was partitioned into recovery islands using the proposed methodology and emulated on the DE3 platform.

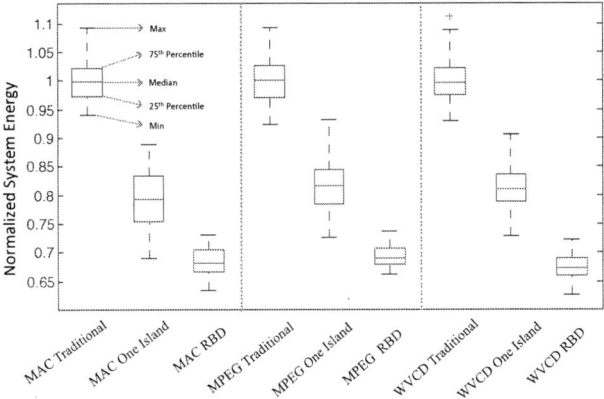

Figure 7: Energy distribution for conventional and recovery based SoC designs

Figure 7 presents a box whisker plot of the normalized energy consumption of 10000 distinct chip instances for each of the three example SoCs. For each SoC, we evaluate the energy consumption under three different design schemes. *Traditional* refers to a guard band based design scheme wherein the voltage is chosen based on timing analysis using the worst case process/temperature corner provided in the cell library.

830

One Island refers to a recovery-based design wherein the entire SoC is treated as a single recovery island, ignoring the feasibility of timing constraints described in Equation 1. Finally, *RBD* refers to the proposed recovery island based design framework. Both the *One Island* and *RBD* cases are designed for a target system performance loss of no more than 2% due to error recovery. As can be seen from the figure, the *RBD* design achieves the best energy distribution - the median of the energy distribution is reduced by 31%-33% compared to the *Traditional* design. The *One Island* design is able to eliminate the overheads associated with die-to-die variations and thereby achieves 18%-21% improvements in the median of the energy distribution. *RBD* outperforms the *One Island* case by 11%-14%. These results clearly illustrate that (i) recovery based SoC design can significantly optimize energy consumption under variations, and (ii) the proposed recovery island based SoC design framework maximally leverages the potential of recovery based design.

Figure 8 plots the percentage energy savings offered by the *RBD* scheme over the *One Island* scheme with increasing values of within die manufacturing variations. With increasing values of σ/μ, the energy savings offered by the RBD

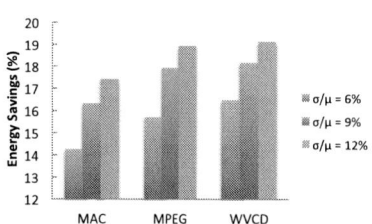

Figure 8: Energy savings sensitivity to magnitude of WID variations

scheme increases as it is able to locally reconfigure the operating voltage to each island's characteristics. Also note that the *RBD* scheme performs better for the Wireless Video Capture Device as it has a larger number of components and hence displays more diversity across components.

Figure 9 plots the percentage energy savings obtained for the WVCD SoC as a function of the number of recovery islands. As can be seen from the figure, the optimal energy savings are obtained for an SoC partitioned into three recovery islands. For larger numbers of recovery islands, the overheads associated with recovery

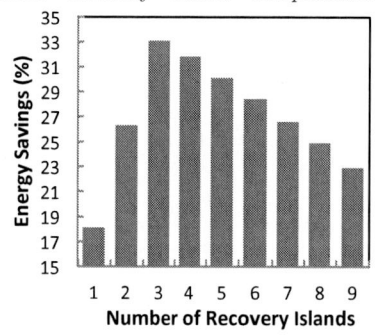

Figure 9: Energy savings sensitivity to number of recovery Islands for the WVCD system

islands begin to dominate over the potential energy savings attainable by performing recovery at a finer granularity. The above example clearly demonstrates the need for performing recovery based design at an optimal granularity and hence the partitioning methodology presented in Section 6. Table 1 details the number of components, the area overheads and the number of recovery islands in the final clustering, for each of the three example SoC designs. The area and energy overheads of the required cross-recovery island interfaces and operating point controllers were estimated by synthesizing them using the IBM 45nm library. These overheads are added to the overheads reported in [1] to estimate the total overheads of recovery based design.

In summary, we believe that our experiments clearly illustrate the potential benefits of recovery based SoC design in

Table 1: Recovery island design details

SoC Design	No. of Components	Area (%) Overhead	Best No. of Recovery Islands
MAC	8	3.2%	2
MPEG	8	4.7%	3
WVCD	10	4.1%	3

optimizing energy consumption under variations.

8. CONCLUSION

We explored the concept of recovery based design and demonstrated how one can implement such a paradigm in the context of modern SoC designs. We presented a variation aware framework for partitioning an SoC into recovery islands and also finding the optimal operating points for each island. We applied the proposed framework to three example SoCs and demonstrated substantial energy benefits over traditional guard band based design.

9. REFERENCES

[1] D. Ernst et al., "Razor: a low-power pipeline based on circuit-level timing speculation," in *Proc. MICRO*, 2003, pp. 7–18.

[2] K. Bowman, J. Tschanz, C. Wilkerson, S. Lu, T. Karnik, V. De, and S. Borkar, "Circuit techniques for dynamic variation tolerance," in *Proc. DAC*, 2009, pp. 4–7.

[3] M. Gupta, J. Rivers, P. Bose, G. Wei, and D. Brooks, "Tribeca: Design for PVT variations with local recovery and fine-grained adaptation," in *Proc.Micro*, 2009, pp. 435 –446.

[4] S. Sarangi, B. Greskamp, A. Tiwari, and J. Torrellas, "EVAL: Utilizing processors with variation-induced timing errors," in *Proc. MICRO*, 2008, pp. 423–434.

[5] U. Y. Ogras, R. Marculescu, and D. Marculescu, "Variation-adaptive feedback control for networks-on-chip with multiple clock domains," in *Proc. DAC*, 2008, pp. 614–619.

[6] S. Garg and D. Marculescu, "System-level throughput analysis for process variation aware multiple voltage-frequency island designs," *ACM TODAES*, vol. 13, no. 4, pp. 1–25, 2008.

[7] V. J. Kozhikkottu, R. Venkatesan, A. Raghunathan, and S. Dey, "VESPA: Variability emulation for System-on-Chip performance analysis," in *Proc. DATE*, 2011, pp. 2–7.

[8] S. Chandra, K. Lahiri, A. Raghunathan, and S. Dey, "Considering process variations during system-level power analysis," in *Proc. ISLPED*, 2006, pp. 342–345.

[9] S. Pasricha, Y. Park, N. Dutt, and F. J. Kurdahi, "System-level PVT variation-aware power exploration of on-chip communication architectures," *ACM TODAES*, vol. 14, no. 2, pp. 1–25, 2009.

[10] S. Chandra, K. Lahiri, A. Raghunathan, and S. Dey, "Variation-tolerant dynamic power management at the system-level," *IEEE TVLSI*, vol. 17, no. 9, pp. 1220–1232, 2009.

[11] S. Garg and D. Marculescu, "System-level mitigation of WID leakage power variability using body-bias islands," in *Proc. CODES+ISSS*, 2008, pp. 273–278.

[12] A. Kahng, S. Kang, R. Kumar, and J. Sartori, "Designing a processor from the ground up to allow voltage/reliability tradeoffs," in *Proc. HPCA*, 2010, pp. 1 –11.

[13] L. Wan and D. Chen, "DynaTune: circuit-level optimization for timing speculation considering dynamic path behavior," in *Proc. ICCAD*, 2009, pp. 172–179.

[14] B. Greskamp et al., "Blueshift: Designing processors for timing speculation from the ground up." in *Proc. HPCA*, 2009, pp. 213 –224.

[15] N. Zea, J. Sartori, B. Ahrens, and R. Kumar, "Optimal power/performance pipelining for error resilient processors," in *Proc. ICCD*, 2010, pp. 356 –363.

[16] K. Bowman, A. Alameldeen, S. Srinivasan, and C. Wilkerson, "Impact of die-to-die and within-die parameter variations on the clock frequency and throughput of multi-core processors," *IEEE TVLSI*, vol. 17, no. 12, pp. 1679 –1690, dec. 2009.

[17] ITRS, "http://www.itrs.net/links/2005itrs/home2005.htm."

[18] K. Brownell, G. Wei, and D. Brooks, "Evaluation of voltage interpolation to address process variations," in *Proc.ICCAD*, 2008, pp. 529–536.

[19] S.R. Sarangi et al., "VARIUS: A Model of Process Variation and Resulting Timing Errors for Microarchitects," *Semiconductor Manufacturing, IEEE Trans.*, vol. 21, no. 1, pp. 3 –13, 2008.

[20] Skadron, K. et al., "Temperature-aware microarchitecture," in *Proc. ISCA*, 2003, pp. 2 – 13.

[21] Nanosim, "Synopsys Inc."

[22] CACTI-5.3, "http://quid.hpl.hp.com:9081/cacti/detailed.y."

[23] ALTERA, "http://www.altera.com/."

SUPPLEMENTAL SECTION

A. CROSS ISLAND INTERFACES

In this section, we give an overview of the cross island interfaces needed to ensure correct functioning of the recovery island based designs. As noted earlier, each recovery island needs to adhere to the existing system bus protocols and hence appropriate wrappers need to be designed for each island. In this work, we implemented and tested interface wrappers (explained below) for both master and slave interfaces of a commercially available communication architecture, the Avalon Interconnect Fabric from Altera [23]. A similar procedure can be applied to design interface wrappers for any other standard communication architecture.

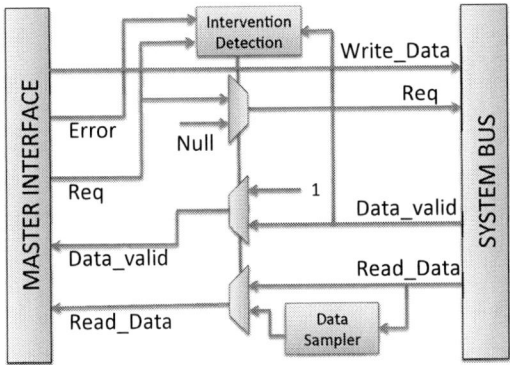

Figure 10: Cross island interface logic

Figure 10 shows an overview of the wrapper interface for a Read-Write Avalon Master. The wrapper needs to deal with two scenarios. First, it needs to ensure that a read or write request sent out by the component during an error recovery cycle is not interpreted by the communication architecture as two requests. Second, it needs to make sure that any data returned by the communication architecture during an error recovery cycle is always captured and not lost. As can be seen from the figure, the wrapper consists of two major components, the intervention detection logic and the selection and sampling logic. The intervention detection logic analyzes the error signal coming from an island, the request signals from the master and the *data_valid* signal from the bus to determine if there is a need to intervene in the current cycle. The selection logic is a set of multiplexers that perform the desired modification to the bus signals.

Consider a scenario wherein the master interface sends out a write request during a cycle in which an error occurred. The intervention detection logic should detect this scenario and deassert the write request signal in the next cycle, so that the system bus does not treat it as two distinct write requests. This functionality is achieved with the help of simple multiplexer logic. The more complicated scenario arises when the master issues a read request and the system bus responds to it during a recovery cycle. In this case, we need to ensure that the data returned is appropriately captured and is available to the master in the next cycle. This functionality is achieved with the help of sampler logic which always stores a clock delayed version of the *read_data* bus signal. The selection logic now sets the *data_valid* signal high on the next cycle and the sampled *read_data* signal is appropriately routed to the master interface.

B. EMULATION AND ERROR INJECTION FRAMEWORK

In this section, we describe in detail the emulation and error injection framework utilized to obtain the error *vs.* system performance loss profiles for each SoC component.

Figure 11 gives an overview of the proposed emulation based error injection framework. To perform the required analysis we first instrument each SoC component with cross island interfaces described in the previous section. To analyze the impact of error recovery on system performance, we mimic the error aggregation signal generated by shadow latches using a synthetic error injection module and clock gate each component using the generated error signal. The error injection circuit consists of a random number generator (LSFR) and a software programmable control register. The error signal is produced by comparing the generated random number to the threshold value programmed into the control register. Thus, the error rate in a component can be appropriately controlled by writing the required value into a threshold register through software.

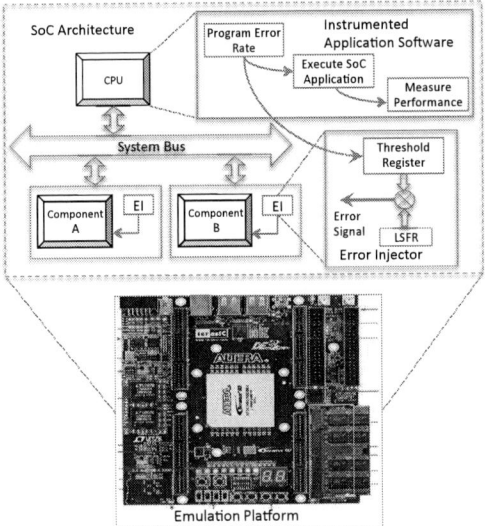

Figure 11: Emulation based error injection framework

The application program that runs on the SoC is instrumented with a software control loop that is in charge of programming the error rate for a given component, executing the application and finally measuring the overall system performance with the help of hardware performance counters. We note that the system performance metric is chosen by the system designer and can be anything ranging from throughput, latency or a pre-defined performance score over a set of benchmarks. The emulation board used for our experiments is an Altera DE3 board equipped with a Stratix III EPS3SL150 FPGA. The proposed methodology can also be applied to any state-of-the-art emulation platform.

C. DISCUSSION

For the recovery based design paradigm to be widely applicable to a large class of SoCs it needs to be compatible with current design flows. In this section, we discuss key considerations in this regard. We also explore alternative methodologies that could be incorporated into our proposed framework for differing design requirements.

Incorporating recovery based design into current design flows: A key requirement needed to utilize the proposed recovery based design paradigm is the ability to partition an SoC into multiple recovery islands. As noted in Section 4, variable latency interfaces in the system serve as ideal points around which the system may be partitioned. Most interfaces which exist in commercial SoCs such as communication channels, system buses and on-chip networks utilize latency insensitive protocols and hence can be appropriately re-designed or instrumented with interface wrappers to ensure correct functionality even when a component is unavailable during the error recovery process.

Most commercial SoCs include components which are equipped with recovery mechanisms like pipeline flushes, state machine rollback *etc.* These mechanisms are essential for correcting errors from sources such as speculative execution and soft errors. Although in this study we chose to utilize a single-cycle clock gating based recovery scheme for each island, they can instead utilize their own inbuilt recovery mechanisms for dealing with timing violations. The proposed framework is not restricted to any specific error recovery scheme and can easily be adapted to deal with multiple recovery mechanisms that can exist within an SoC.

Current SoC platforms make use of various power management schemes to dynamically adapt to an application's time varying power-performance requirements. Dynamic voltage-frequency scaling (DVFS) is one such widely used mechanism which modulates the voltage and frequency of individual SoC components/islands based on workload characteristics. The voltage interpolation scheme utilized by the framework requires two supply voltage rails –VDDH and VDDL. One possible integration scheme with DVFS would involve utilizing existing DVFS controllers to decide the VDDH and VDDL operating voltages. The recovery based design scheme's operating point controller can then perform more fine grained voltage interpolation based on the current error rate. Thus, the proposed framework can be integrated with DVFS with minimal changes to the overall design flow.

Another common practice employed in current commercial SoC design involves using IP modules procured from external vendors. These components are often non-modifiable and cannot be instrumented with the required circuitry needed to detect and recover from errors. In such a scenario, these components alone may be operated with design margins and only other system components are considered in the recovery based design process.

Alternative design methodologies: In this work, we made several choices such as floorplan driven clustering, emulation based local search for eliminating non-linearities associated with component inter-dependencies, number of operating points under consideration and a static error-voltage profile based evaluation methodology. We now analyze the various alternative choices that could be adopted for achieving different end design objectives such as improved energy benefits, reduced emulation runtime *etc.*

One such alternative involves performing cluster driven floor planning wherein various components are first clustered together without considering the delay constraints essential for correct functionality. The clustered system is then floorplanned and evaluated for delay violations. Performing partitioning prior to floorplanning could potentially help in physically grouping together components that are most suited for clustering, thereby leading to improved energy savings. However, if the current clustering configuration violates the delay requirements, the above described process would have to repeated with the next best clustering configuration. Also in some commercial design flows, floorplanning is done quite early in the design cycle and may not be flexible to changes thereafter.

In this study we considered ten distinct operating points at which each component was characterized. Thus, a k component system would require $10k$ emulation runs to derive the error *vs.* system performance loss profiles. Reducing the number of operating points proportionately decreases the total run time at the cost of increased energy consumption due to a more coarse grained search space. The emulation based local search phase could also be replaced with an appropriate analytical performance model to attain similar run time savings. However, this scheme would be applicable only for systems that have relaxed overall system performance constraints as the complex inter-component interactions cannot be completely captured by an analytical framework.

A Hybrid NoC Design for Cache Coherence Optimization for Chip Multiprocessors [*]

Hui Zhao, Ohyoung Jang, Wei Ding, Yuanrui Zhang, Mahmut Kandemir, Mary Jane Irwin

Department of Computer Science and Engineering, The Pennsylvania State University

{ hzz105, oyj5007, wzd109, yuazhang, kandemir, mji}@cse.psu.edu

ABSTRACT

On chip many-core systems, evolving from prior multi-processor systems, are considered as a promising solution to the performance scalability and power consumption problems. The long communication distance between the traditional multi-processors makes directory-based cache coherence protocols better solutions compared to bus-based snooping protocols even with the overheads from indirections. However, much smaller distances between the CMP cores enhance the reachability of buses, revitalizing the applicability of snooping protocols for cache-to-cache transfers. In this work, we propose a hybrid NoC design to provide optimized support for cache coherency. In our design, on-chip links can be dynamically configured as either point-to-point links between NoC nodes or short buses to facilitate localized snooping. By taking advantage of the best of both worlds, bus-based snooping coherency and NoC-based directory coherency, our approach brings both power and performance benefits.

Categories and Subject Descriptors

B.4.3 [**Interconnections (Subsystems)**]: Topology; C.1.2 [**Multiple Data Stream Architectures (Multiprocessors)**]: Interconnection architectures

General Terms

Design, Management, Performance

Keywords

Multi-core, NoC, Cache Coherence, Bus

1. INTRODUCTION AND MOTIVATION

As modern fabrication technologies advance into deep submicron era, chip multiprocessors (CMPs) are moving from multi-core to many-core architectures in order to fully take advantage of the increasing number of transistors available [1, 2]. Although in many aspects, many-core CMP systems

---

[*]This work is supported in part by NSF grants 1147388, 1152479, 1017882, 0963839, 0811687, and a grant from Microsoft.

exhibit similarities to their predecessors, multi-processor systems, there are two major differences between these two types of systems. First, many-core systems are more constrained by the limited on-chip memories. Second, inter-core communication latencies are greatly reduced because of the short distance between on-chip cores. The first difference brings new challenges on how to design an efficient memory system with limited capacity. The second difference opens up opportunities for reducing the memory access latencies. A very important component of memory system design is the cache coherence protocols. Cache coherence protocols should not only ensure data access consistency, but should also have low performance and energy overheads. In this direction, several techniques have been proposed, such as tree-based coherence and token coherence [3, 4, 5].

In this paper, we investigate new schemes to optimize cache coherence by taking advantage of the short communication distances in many-core CMPs. Specifically, we propose a novel network-on-chip (NoC) architecture that can configure on-chip links into high-speed snooping buses in order to support cache-to-cache data transfer. Conventional multi-processor systems employ directory-based cache coherence protocols because bus-based snooping coherence is not scalable to high core counts. Bus-based snooping is not considered a good option for many-core CMPs due to the similar scalability concerns. However, when an L1 access misses in the directory-based coherence, directories have to be accessed to obtain the sharer information, resulting in extra delays and power overheads . We observe that, when multiple cores are running threads of a same application, there exist opportunities for a core to find data sharers in its neighborhood (nearby cores). To exploit such opportunities, we propose a scheme that connects several point-to-point links of the NoC together to form short-ranged snooping buses. When a core does not find the requested data in its L1 cache, it first snoops opportunistically for a copy of the data in the L1 caches of nearby cores. Indirections to the directories are avoided if such snoops result in hits. Consequently, directories are accessed only when the snooping cannot find a copy of the data. Our proposed snooping scheme has the advantage of short latencies and low power overhead because snooping messages are transferred on buses instead of NoC routers.

Our other important observation is that the effectiveness of our local snooping scheme is closely related to the application mappings. Fixed buses cannot exploit the communication locality if the data sharers are not mapped to the cores connected by a bus. In order to decrease the dependence

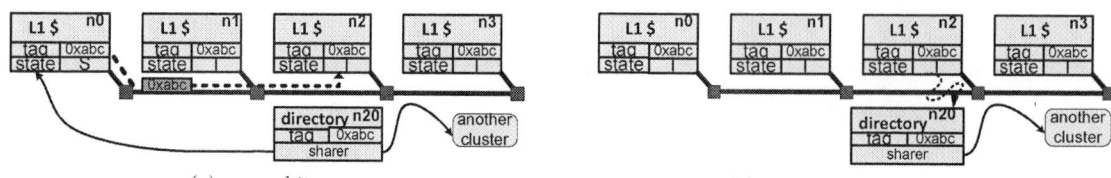

(a) snoop hit   (b) snoop miss

Figure 1: Snoop coherence in case of an L1 miss.

of snooping effectiveness on the application mappings, we propose novel schemes to dynamically build snoop buses, in an on-demand fashion, based on the locality of data sharers.

We make the following contributions in this paper:

- We propose a hybrid approach to cache coherence that employs re-configurable snooping buses to reduce the effect of application mappings on snooping effectiveness.

- We propose dynamically constructing snooping buses by connecting on-chip point-to-point links together. This technique can reduce the hardware overhead without compromising on-chip bandwidth. To the best of our knowledge, we are the first one to propose such techniques.

- We provide a detailed design of the snooping bus, including the bus arbitrator and bus switch interfaces.

- We design clustering and bus-building algorithms to group data sharing cores into local groups and build short buses to facilitate the broadcasting of snooping messages (presented in the supplemental section).

2. DESIGN OF CACHE COHERENCE

Our cache coherence proposal is built upon the principle that, before sending requests to the directory, a core running parallel programs first snoops cores in its vicinity for shared data. If a snooped core can provide the data requested, which we call a snoop hit, a cache-to-cache transfer is performed between the data requester and provider. Such cache-to-cache data transfers are made possible by connecting the point-to-point links of the NoCs to form snooping buses. Our design of cache coherence involves two mechanisms: *global directory-based protocol* and *local snooping protocol*. We optimistically group data-sharing cores that are located in close distance to build a localized snooping cluster. In such a cluster, snooping coherence is employed to facilitate cache-to-cache data transfer, without involving global (chip wide) directory. Globally, among different clusters, cache coherence is maintained through structures similar to the traditional directories.

2.1 A Walkthrough Example

Figure 1 shows the detailed behavior of our coherence protocol in the case of an L1 miss. Node 2 issues a load to access a cache line with a tag of 0xabc, but results in a miss. Instead of sending request to the directory node, it first sends a snooping query on the locally connected snoop bus. In this case, three other cores are connected by the snooping bus in the same local cluster with core 2: cores 0, 1 and 3. All other cores search their L1 cache for the tag of the requested cache line. If there is another core in the this cluster has a valid copy of the cache line, for example core 0, then core 0 puts the data on the bus. After a few cycles, core 2 can grab the data from the bus and save it to its own L1 cache, concluding a cache-to-cache data transfer. If, on the other hand, none of the cores snooped has a valid copy of the sought

cache line, then core 2 needs to send a request to directory (node 20), as in the directory-based protocols. Since our coherence protocol involves both snooping and directory-based cache coherences, we need to pay extra attention to coordinate state changes both locally and globally. we create a new state: *shared-exclusive*. This state is used to identify this locally shared and globally exclusive status. Had a core shared globally exclusive data locally, both this core and its sharers need to change to shared-exclusive state. Otherwise, its state remains exclusive. In a similar manner, modified states need to be distinguished between whether or not the only copy in the directory's view has actually been shared inside the snooping cluster. So, we also add a new state, called *shared-modified* to handle this situation.

2.2 Writes and Invalidations

If an L1 core gets a write hit, the new state is decided based on the cache line's current state. For exclusive or modified state, the operations needed are the same as in a conventional directory based protocol. In the situation that the returned state is shared (meaning there are shared data copies outside its snooping cluster), an invalidation message has to be sent to the directory to nullify all other possible sharers. If the cache line state is shared-exclusive or shared-modified, that means there are several copies of the cache line in the same snooping cluster. However, globally, there is no other copy of the data in any cores outside this core's snooping cluster. As a result, the core only needs to broadcast locally to invalidate other sharers in the same cluster before modifying the data. In this case, no involvement of the global directory is needed and an indirection to directory is avoided.

In the only remaining case where an L1 write returns a miss, our coherence protocol also tries to avoid indirections to directories. Before going to the directory to fetch the data, the requesting core first broadcasts a message to local cores in the same snooping cluster. If there is any core that has the data in shared state, we invalidate local sharers and access the directory to invalidate global sharers. However, if there are other cores with the data in Modified or Exclusive state, indicating that the only one copy of data of the whole system is in the same cluster, then there is no need to go to the directory to get this information. Instead, the core can invalidate the other copy, and change its own state to be modified. If there exist some cores in the same snooping cluster with the state of shared-exclusive or shared-modified, this means all the copies of the data line is in this cluster. Therefore, the requester can invalidate all other local sharers and change its own state to be modified and no message needs to be sent to the directory.

2.3 Organizing Local Clusters in the Directory

In the conventional MESI protocols, silent evictions are performed of exclusive and shared lines in order to avoid

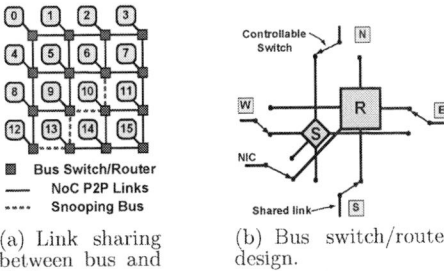

(a) Link sharing between bus and NoC.

(b) Bus switch/router design.

Figure 2: Proposed hardware design

the bandwidth overhead of notifying the directory. In our proposed protocol, when replacing an L1 line, silent evictions are still valid for shared, exclusive and shared-exclusive states. However, for the shared-modified state, we need to ensure that modified data are written to the next level cache properly. Our policy for replacing cache lines with shared-modified state is as follows: the cache first needs to snoop sharers in local cluster; if a sharer is found, then the line can be evicted silently. If there is no other cores holding the line in the same cluster, i.e., this is the only core holding the modified data in the cluster, a writeback needs to be performed just like replacing a cache line with modified state.

In our design, we assume a cluster is a group of cores that are running threads that belong to the same application. When a new application's threads are mapped to CMP cores, we create local clusters of cores running that application's threads. Such cluster structures will exist until the application finishes its execution. In traditional directory-based cache protocols, there is a bit map for each cache line associated with one directory. If a core's L1 cache has the valid data of the cache line, the corresponding bit is set to 1. In our protocols, on the other hand, since we group cores that possibly share data into clusters, each cluster only needs one representative in the directory. From the directory's point of view, the cores inside the cluster behave like just one core because all the copies of data are consistent inside a cluster. That is, either there is only core having the data inside the cluster, or all copies are exactly same.

3. CONSTRUCTING SNOOPING BUSES FROM NOC LINKS

3.1 Reusing NoC links to build snooping buses

Buses are widely used in off-chip interconnections. However, with the growing discrepancy between wire delay and gate delay, buses are not scalable to be employed in many-core systems. Consequently, for on-chip communications, packet switched network-on-chips are considered a better solution. However, the power consumptions of such NoC networks can be very high as the routers are power hungry, due to their inside components such as buffers and crossbars [20]. It has been observed that the on-chip routers can consume up to 40% of the network power [10].

Our belief is that, even if buses are not suitable to be used as global connections, there are still advantages to employ them as local connections. When applied to a subarea of the NoC network, buses have the advantages of short delays and lower energy consumptions compared to router based NoCs. If fact, there have been several previous works that

have reconsidered the application of buses to connect on-chip cores [8, 10, 11].

In this work, *we propose to use buses to support localized cache snooping.* The advantage of doing so is three fold: firstly, buses have inherent ordering property which can support cache access coherency; secondly, buses can provide simultaneous broadcasting without accessing the cores one after another in a sequential manner; and thirdly, buses can reduce transaction delays and consume less power compared to the routers. In this work, our baseline network is point-to-point link connected router-based NoC. We expand the functionality of routers by adding bus switches to them so that several links can be connected together to form a short-ranged snooping bus. At one time, the on-chip links can be used either to transfer data packets between routers or to form a short snooping bus. As shown Figure 2(a), cores 12, 13, 9, 10 and 6 are committing a snooping bus transaction on the dotted lines. At the same time, all other links in the NoC are available to transmit packets between the routers. In a conventional NoC, the links connecting routers are unidirectional, and as a result, one router can both send and receive packets in one direction at the same time. In our proposed design, when links are connected to build snooping buses, all the wires are used as bi-directional links. Consequently, the snooping buses have bandwidth doubled compared to the point-to-point links which can offset the delays incurred by increased length of the buses. By reusing the on-chip links between buses and NoC links in a time-division multiplexing manner, our design can reduce the hardware overhead without compromising the NoC bandwidth.

3.2 Bus Switch Design

In order to connect NoC links to build snooping buses, we need to extend the routers to provide bus switching functionality. Figure 2(b) shows our proposed link interconnection interface design. There are two major components: packet routing component (R) and bus switching component (S). The routing component is similar to the conventional NoC routers consisting of buffers, routing logic and a crossbar. The bus switching component is a programmable switch that can connect each segment (an NoC link) to form a bus. Links from each direction connected to our Routing/Bus Switching unit are controlled by programmable switches. As shown in Figure 2(b), links in the north and west directions are programmed to be used by a bus, whereas links in the south and east directions can serve as NoC links. When links are used as part of a bus, packets need to stay inside the router until the bus transaction is over. This is similar to the scenario in the conventional NoC routers where multiple packets are competing for the same output link. Packets failed to be granted the output link have to stay inside the buffer for the next chance. Our scheme uses the same mechanism to hold data packets when their output links are used by the bus.

3.3 Bus Arbiter Design

Another essential component of a bus design is the arbiter. The reason buses can work as a mechanism to provide transaction ordering is because the users need to take order to control a bus. Such arbitrations of bus control are implemented through bus arbiters. A typical arbitrator design is shown in Figure 3(a). Each core connected to a bus first send requests to the arbiter. The arbiter then makes

836

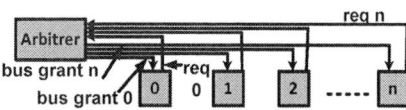
(a) centralized arbitration

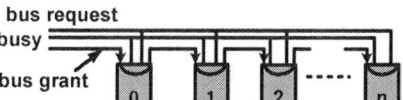
(b) decentralized arbitration

Figure 3: Traditional centralized bus arbiter design and our proposed decentralized bus arbiter design

a decision based on some priority policies to grant the bus control to only one requester. It is easy to build such an arbiter if the bus configuration is fixed. However, this type of centralized arbiters do not fit in our reconfigurable environment since the bus topology is changing from time to time, based on the applications' mappings.

We propose to employ *de-centralized* bus arbitration mechanisms [13]. As shown in Figure 3(b), there is no centralized arbiters. Instead, the arbitration logic is distributed across all bus switches and is also reconfigurable. The bus grant input indicates that the bus can be granted to a requestor. Bus busy indicates whether the bus is being used by some device and the bus request line shows if another device has made a request. Each device needs to make a request first. If the bus busy is negative, then the device negate its own bus grant out signal and wait to see if its grant IN signal is asserted. If so, the device can grab the bus and assert the bus busy signal. The bus grant out signal also needs to be asserted so other devices can have their grant in to be set in order to compete in the next bus arbitration. More detailed design description can be found in [13].

Similar to the data lines of the snooping buses, the bus arbitration control lines are segmented and can be programmed through the bus switches. The major arbitration lines are the request, busy and grant lines, as illustrated in Figure 3. Each data link is associated with a set of such segmented control lines. At the time the data links are configured to build connected buses, the corresponding control lines are connected in the same way to build a separate arbitration bus. Then, each core can send its requests across the arbitration bus in a decentralized manner. All the arbitration logics on a snooping bus work together to decide the winner of the bus control.

4. EXPERIMENTAL EVALUATION

4.1 Experimentation Setup

We evaluate our proposed techniques using a trace-driven, cycle-accurate CMP simulator that has a built in NoC network. We use GEMS [14] to generate traces from SPLASH [15] and SPEC OMP [16] benchmarks and feed the traces into our CMP simulator. Application threads are randomly mapped to CMP cores. Our baseline architecture has 64 cores organized as a 2D 8 by 8 mesh. Each NoC node consists of a core, a private write-back L1 cache and a tiled L2 bank. The default memory hierarchy is a two-level directory-based MESI cache coherence protocol. Each router has two pipeline stages with the input buffer depth of four. Our NoC employs wormhole switching and virtual-channel flow control and use the deterministic X-Y routing algorithm to route packets. Table 1 provides the main parameters of our simulation platform.

We evaluate two types of snooping bus configurations. The first one is called *fixed bus configuration*, where the cores connected by a bus is fixed, no matter how the applications are mapped to the NoC nodes. In the second configuration, called *dynamic bus configuration*, we use our clustering and

Fixed bus with length of 4

Fixed bus with length of 8

Dynamic bus with length of 7

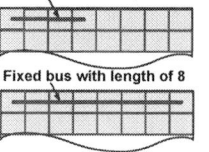

Dynamic bus with length of 4

(a) fixed bus configurations

(b) dynamic bus configurations

Figure 4: Experimented bus configurations

bus-building algorithms (described in detail in the supplemental section) to dynamically configure buses that connect cores. For each of the bus configurations we have, we experiment with different bus lengths of 4 links and 8 links respectively. In the following discussion, we refer to these bus configurations as Fix-4, Fix-8, Dynamic-4 and Dynamic-8 respectively. Figure 4 (a) shows two fixed bus configurations with length of 4 and 8 respectively. Figure 4 (b) illustrates two dynamically constructed buses proposed by our scheme: the first bus connects 5 cores running threads of application 0 and the second bus connects 3 cores running threads of application 2.

We use CACTI 6.5 [17] to estimate the delay and power values for the links. Additional loading due to multiple senders and receivers is considered when we get these parameters. We use Orion [19] to obtain the router power. Both our network and NoC structures run at a frequency of 3GHz. Table 2 gives our bus related parameter setups.

Processors	SPARC 3 GHz processor, two-way out of order, 64-entry instruction window
L1 Cache	64 KB private cache, 4-way set associative, 128B block size, 2-cycle latency, split I/D caches
L2 Cache	shared L2 cache, with 1MB banks, 16-way set associative, 128B block size, 6-cycles latency, 32 MSHRs
Memory	4GB, 260 cycle off-chip access latency
NoC	2-stage pipeline, 128 bit flits, 4 flits per packet, X-Y routing

Table 1: Baseline CMP configuration.

Parameters	link	Bus of 4 links	Bus of 8 links
Length(mm)	3.1	9.6	22.4
Delay(ns)	0.13	0.41	0.93
Dynamic Energy(pJ)	0.93	2.88	6.51
Leakage Pwr(mw)	0.03	0.09	0.21

Table 2: Energy and delay of buses and links.

4.2 Results

Impact on Memory Latency. Figure 5 plots the impact of the localized snooping on L1 load miss latencies. We observe that, on average, the Dynamic-4 configuration can reduce the load miss latency by about 10%. Dynamic-8 provides further improvements, lowering the load miss latencies by 20% on the average, with a maximum reduction of 30% in the case of *wupwise*. These improvements are achieved by avoiding unnecessary indirections to the directory, as discussed in Section 2. In most of the cases, Dynamic-8 incurs lower load miss latency. This is due to the fact that a larger number of cores are snooped by this configuration.

837

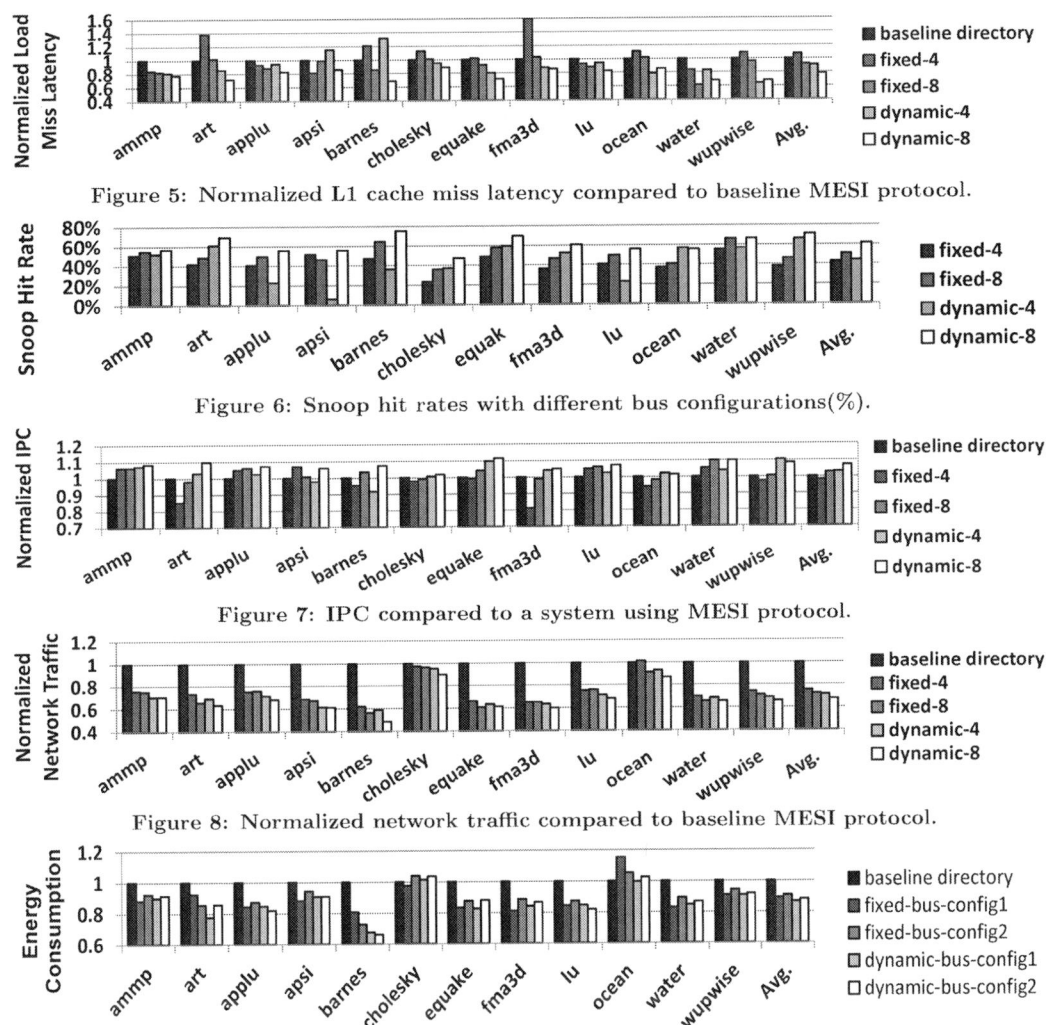

Figure 5: Normalized L1 cache miss latency compared to baseline MESI protocol.

Figure 6: Snoop hit rates with different bus configurations(%).

Figure 7: IPC compared to a system using MESI protocol.

Figure 8: Normalized network traffic compared to baseline MESI protocol.

Figure 9: Normalized network energy consumption compared to a system using MESI protocol.

Compared to the dynamically configured buses, the fixed buses experience larger miss latencies in most cases. In several cases, the snooping results for Fix-8 are even worse compared to Dynamic-4. This proves that our dynamic buses perform better than fixed buses when the application mapping is not optimal. In such situations, longer snooping distance does not necessarily guarantee more snoop hits, but instead results in increased load latency due to more cycles being spent on snooping transactions.

Snoop Hit Rates. Figure 6 shows the measured snoop hit rates of the experimented snoop bus configurations. Our dynamically configured buses can improve the snoop hit rate compared to the fixed buses by up to 50%. There are some interesting cases where our Dynamic-4 configuration has lower hit rate than the Fix-4 configurations, e.g., in *apsi*, *barnes* and *lu*. This is because our clustering and bus-building algorithms use local information to generate group of sharers to be snooped together. Also, since links are not reused between clusters, some nodes could not get connected by the snooping buses in our dynamic scheme, even if this is

possible under a fixed bus scheme. This is more pronounced when the bus length is short, as is in the Dynamic-4 in our case. However, in most cases, Dynamic-4 performs better compared to Fix-4 in terms of the snoop hit rates. Note that higher snoop hit rate does not necessarily indicate improvements in performance because longer buses also incur more delay cycles.

Performance Improvements. Figure 7 shows the overall improvement in execution time of our proposed scheme compared to the baseline MESI directory protocols. Our dynamic bus configurations deliver performance improvements of up to 12%, with only *barnes* and *apsi* suffering a slight slowdown when the bus length is short. The reason is that sometimes short dynamic bus constructions suffer from generating isolated nodes, as explained in the snoop hit rate analysis. On an average, the dynamic bus configuration improves performance by about 8%. This is significant considering the high L1 cache hit rates which limit the impact of our optimizations on the execution time.

When we compare fixed and dynamic bus configurations,

838

we can easily notice the advantages of the dynamic configurations. In the cases of *art, ocean, cholesky* and *wupwise*, both of the fixed bus configurations results in lower IPC, compared to the baseline directory protocols. On average, the Fix-4 configuration performs worse than the baseline directories. This demonstrates that it is difficult to design effective snooping schemes without considering the affect of application mappings. These results also underline the importance of our dynamic bus structure that works even if the application mapping is not ideal.

Impact on Network Traffic. As shown in Figure 8, except for *cholesky* and *ocean*, our proposed scheme can reduce the network traffic by about 20%. This explains why the performance still improves even if the snooping buses preempt the NoC links, increasing the latency required for a data packet to reach its destination. Since the snooping hits decrease the number of directory visits, the network traffic also gets reduced. Even though some packets have to be held in buffers due to conflicted output link with snooping buses, due to the distributed nature of our snooping buses, the unoccupied links in the NoC can still transmit packets in other areas of the network at the same time. In addition, an NoC network already provides high bandwidth for data transmission by packetizing and routing the data on different routes, losing some links for a few cycles does not affect the network performance in most cases.

Energy Savings. Besides the performance improvements, our proposed snooping bus scheme can significantly reduce the network energy consumption. Figure 9 plots the normalized network energy consumptions under different schemes. For example, in *barnes*, the energy savings can be as much as 37% for dynamic bus configurations. In most of our benchmarks, the proposed scheme can save energy between 15% to 20%. The only exceptions are *cholesky* and *ocean*. As we have analyzed in the performance sections, these two benchmarks have comparatively low snoop hit rates. Therefore, the overall energy consumed is even higher compared to schemes without localized snooping. For other benchmarks, even if the bus transactions consume more power, our approach still benefits by avoiding the directory indirections. Even if longer buses consume more power for snooping transactions, the difference in energy consumption is rather small. This is because, as compared to the energy hungry routers, longer buses are more efficient as far as energy consumption is concerned.

5. RELATED WORK AND CONCLUSION

Cache coherence designs to exploit the proximity of data sharers have been proposed in [6, 7]. Williams et. al. [7] propose to add direct links in four directions of NoC routers to snoop sharers in direct neighbors. However, their scheme depends on specific application mapping to work and has more hardware overhead. There have been several prior efforts on utilizing buses to optimize the network on chip designs [8, 9, 10]. The difference between their work and ours is that we propose to use buses to optimize cache coherence protocols. Reconfigurable NoC designs have been proposed in [12, 11]. [11] designed reconfigurable bus-based networks based on inter-core data sharing. Kim et.al [12] proposes to reconfigure the network in order to suit for the specific application characters.

In this paper, we proposed a novel hybrid NoC architecture that takes advantage of both snooping and directory based cache coherence protocols. We first investigated how application mappings can affect the performance of proximity snooping schemes. We then explained the design of a localized bus-based snooping cache coherence protocol. We also presented the design details of our configurable snooping buses using the on-chip links. In order to reduce the dependency of the local snooping on application mappings, we further designed two algorithms to dynamically group sharing cores into local clusters and build buses to connect those cores. Our experiment results showed that the proposed techniques not only increase system performance but they can also reduce energy consumption.

6. REFERENCES

[1] Intel. From a few cores to many: A tera-scale computing research overview.
http://download.intel.com/research/platform/terascale/terascale_overview_paper.pdf.

[2] W. J. Dally and B. Towles. Route Packets, Not Wires: On-Chip Interconnection Networks. *DAC*, 2001.

[3] N. Jerger, et.al. Virtual Tree Coherence: Leveraging Regions and In-network Multicast Trees for Scalable Cache Coherence *MICRO*, 2008.

[4] M. R.Marty and M. D.Hill. Coherence ordering for ring based chip multiprocessors. *MICRO*, 2006.

[5] K. Strauss, et.al. Uncorq: Unconstrained snoop request delivery in embedded-ring multiprocessors. *MICRO*, 2007.

[6] J. A. Brown, et.al. Proximity-Aware Directory-based Coherence for Multi-core Processor Architectures. *In Proceedings of SPAA, 19*, 2007.

[7] N. Barrow-Williams, et.al. Proximity coherence for chip multiprocessors. *In Proceedings of PACT*, 2010.

[8] R. Das, et.al. Design and Evaluation of Hierarchical On-Chip Network Topologies for next generation CMPs. *HPCA*, 2009.

[9] L. Cheng, et.al. Interconnect-Aware coherence Protocols for Chip Multiprocessors. *ISCA*, 2006.

[10] A. N. Udipi, et.al. Towards Scalable, Energy-Efficient, Bus-Based On-Chip Networks. *HPCA*, 2010.

[11] S. Akram, et.al. A Workload-Adaptive and Reconfigurable Bus Architecture for Multicore Processors. *International Journal of Reconfigurable Computing*, 2010.

[12] M. Kim, et.al. Polymorphic On-Chip Networks. *ISCA*, 2008.

[13] A. S. Tanenbaum. Computer Networks. *Prentice Hall Pub.*, 1999.

[14] M. M. K. Martin, et.al. Multifacets General Execution-driven Multiprocessor Simulator (GEMS) Toolset. *SIGARCH*, Nov. 2005.

[15] S. C. Woo, et.al.The SPLASH-2 Programs: Characterization and Methodological Considerations. *ISCA*, 1995.

[16] V. Aslot, et.al. SPEComp: A new benchmark suite for measuring parallel computer performance. *Lecture Notes in Computer Science (WOMPEI2001)*, 2001.

[17] N. Muralimanohar, et.al.Optimizing NUCA organizations and wiring alternatives for large caches with Cacti 6.0. *MICRO*, 2007.

[18] R. Mukherjee, et.al.Thermal sensor allocation and placement for reconfigurable systems. *ICCAD*, 2006.

[19] H. Wang, et.al. Orion: A Power-Performance Simulator for Interconnection Networks. *MICRO*, 2006.

[20] J. Kim, D. Park, C. Nicopoulos, N. Vijaykrishnan and C. R. Das. Design and Analysis of an NoC Architecture from Performance, Reliability and Energy Perspective. *ANCS'05*. 2005.

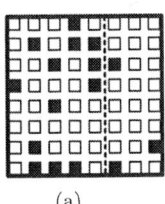 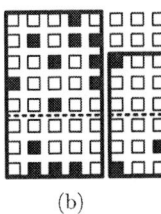 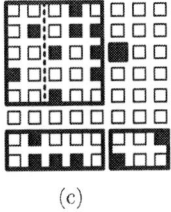 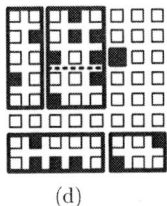 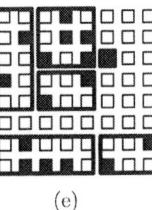

(a)   (b)   (c)   (d)   (e)

Figure 10: Stages of our bisection clustering algorithm to build localized clusters of cores on chip. There are 16 cores running one application's threads on an 8 by 8 CMP. After 4 iterations of bisections, the 16 cores are grouped into 6 clusters.

APPENDIX

A. ALGORITHMS TO BUILD SNOOPING BUSES

In this section, we present our algorithms that group data sharing cores into clusters and then build snooping buses to connect the cores inside a cluster. We assume that the information about which cores on the CMP are running threads of a certain parallel program is exposed to our approach. This information can either be retrieved at thread mapping time before application execution or be inferred from data sharing patterns collected by filters at run time. We first employ a *recursive bisection algorithm* similar to [18] to build clusters of sharers from on-chip cores. We define the terms used by our algorithms as below:

Cluster: A cluster denoted by $C(pid)$ is a rectangular region owned by the process *pid*. A cluster must contain at least one owner core in it.

Density of a cluster: The density of a cluster $C(pid)$ is the ratio of the occurrence of owner cores to the number of all cores in the cluster.

Path: A path denoted by $P_k(pid, cluster)$ is the kth path, where pid is the process id of owner and cluster owned by the same process *pid*. A path forms a snooping bus that passes connects all owner cores in the chip.

Length of a path: The length of a path is the maximum Manhattan Distance between two cores on the path.

Weight of a path: The weight of a path is defined by the number of owner cores on the path.

A.1 Clustering Algorithm

Our clustering algorithm recursively forms rectangles that contain cores running threads of a same application, which we call the owner of the cluster. Clusters owned by one application do not overlap with each other, but clusters owned by different applications may overlap. During the clustering procedure, we identify a core by its location on the chip as $N_{r,c}$, where r and c are respectively the indices of row and column of the core on the chip. The range of a cluster is represented by $\{row_b, col_b, row_e, col_e\}$, where subscripts b and e represent begin and end, respectively. Initially, each cluster encloses all cores on the chip. After the maximum length of a cluster is given as the input, the clusters are recursively tightened by dividing each cluster into two clusters.This procedure continues until no clusters have edges longer than the maximum cluster size D.

Figure 10 illustrates a clustering example of one application with 16 threads running on an 8 by 8 NoC. The maximum edge length D is three. The locations of the appli-

cation's threads are represented by black squares. A solid rectangle encloses a cluster and a dashed line represents a bisection edge. A cluster region with a dashed line is going to be divided into two clusters in the next iteration. At the initialization stage, there exists only one cluster including all cores in the chip. The cluster remains the same after tightening, for each of all four edges is touched by at least one owner core. During the first iteration, six possible bisection points are explored, three for the vertical edges and three for the horizontal edges. After that the bisection clustering algorithm selects one bisection point to divide the initial cluster into two clusters as shown by the dashed line. At the second iteration, two bisection points are found inside both the clusters considering that they all have edges longer than D. Both clusters are further divided into two clusters each. This procedure continues until no cluster has edges longer than D. Algorithm 1 describes our clustering algorithm in detail.

Algorithm 1 Bisection_Clustering.

INPUT: A 2D array of process core mapping M and its region R

INPUT: Process id *pid* and max diameter D

1: Create a empty queue R_i
2: Create a empty list R_o
3: Enqueue $Tighten(R)$ into R_i
4: **while** (R_i is not empty) **do**
5:     $rec \leftarrow dequeue(R_i)$
6:     Set Len_h by the length of horizontal edge of the rec
7:     Set Len_v by the length of vertical edge of the rec
8:     **if** $Len_h \leq D$ and $Len_v \leq D$ **then**
9:        Enqueue rec to the R_o
10:       Continue
11:    **end if**
12:    **if** $Points(rec, M) = 0$ **then**
13:       Continue
14:    **end if**
15:    $rec1, rec2 \leftarrow Bisect_Clustering(rec, M)$
16:    Enqueue $Tighten(rec1)$ to R_i
17:    Enqueue $Tighten(rec2)$ to R_i
18: **end while**

OUTPUT: A set of clusters

A.2 Bus Building Algorithm

After we have grouped data sharing cores into clusters, we use our bus building algorithm (Algorithm 2) to connect related cores to build snooping buses within a cluster. The inputs of this algorithm are the clusters found by the bi-

840

Algorithm 2 Bus Building Algorithm.

INPUT: A set of clusters CS
INPUT: Maximum length of a bus L
1: Compute density of all cluters $D(c) = \frac{\#of core of corresponding pid}{\#of all cores in the cluster}$
2: Build a graph G, where a vertex is a core and an edge is unoccupied link between two adjacent cores
3: **while** not empty(CS) **do**
4:     Pick up the densest cluster c in the CS.
5:     Set process id of the c to pid
6:     Build a list of cores N that runs the process pid
7:     Build a empty list $buses$
8:     **for** core in N **do**
9:         BuildBus($core$, G, L)
10:     **end for**
11:     Select the bus P_{opt} with maximum number of cores running the process pid
12:     Append P_{opt} to $buses$
13:     Remove edges which are belong to P_{opt} from G
14:     **if** not all cores in N belong to the P_{opt} **then**
15:         Build a core list N_r containing the cores in N that do not belong to the P_{opt}.
16:         Create a new minimum-size cluster c_r enclosing all cores in N_r.
17:         Compute density of the c_r and plug it into the SC at the appropriate position.
18:     **end if**
19: **end while**
OUTPUT: All buses built

---

section clustering algorithm and the maximum length of a snooping bus desired. The outputs are the buses that can be used for cache snooping purpose.

Algorithm 2 describes our bus building strategy in detail. First we compute the densities of the clusters found by the bisection clustering algorithm. Next, a graph G is built where vertices are the cores in the chip and edges are the links connecting two cores. Edges are removed from the graph G if they are assigned to a specific bus. The algorithm always starts with the cluster c with the maximum density. Then $BuildPath$ function build buses for the owner cores in the cluster c, and selects the path P_{opt} with largest weight. Later links belonging to the path P_{opt} are removed from the graph G to ensure the links of two different buses do not overlap. Owner cores on the path P_{opt} are marked as non-owners of the cluster c before the cluster c is tightened. If the region of the cluster c is connected by buses, it is removed from the cluster set SC. The procedure continues until the cluster set SC becomes empty.

B. IMPACT OF APPLICATION MAPPINGS ON THE EFFECTIVENESS OF SNOOPING BUSES

B.1 Mapping of parallel programs onto a CMP platform

In order to take advantage of the large number of cores available in a many-core system, programs are usually split into multiple threads that can be executed in parallel. The ideal mapping of an application's threads is to place them as close as possible to exploit the communication locality.

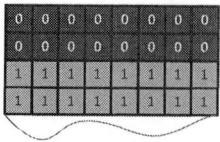 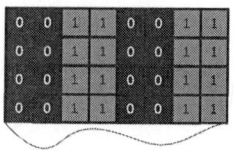

(a) Mapping 1 of applications.   (b) Mapping 2 of applications.

Figure 11: Two schemes of application mappings.

However, in reality, such mappings may not be possible. For example, we have a queue of parallel programs each can be executed in different number of threads (2, 4, and 8 etc.). Initially, when most of the cores are available, we can choose to map the threads of a same application to cores as close as possible, resulting in mappings close to the ideal ones. Since the threads of different applications take different length of time to finish, there will be holes (of available cores) generated in the CMP in the later time. A similar scenario is the dynamic memory allocation in the Operating Systems where holes of free memory are generated when previously allocated memory is returned. In such cases, even if we still try to map an application's threads to cores next to each other, the resulted mapping will not be a close-knit group, but several local clusters scattered across the CMP.

As a result, if the bus configurations are fixed as in [10] and [7], the effectiveness of the snooping will be compromised since the number of possible shares searched is not related to the bus length. In another word, even if we increase the length of snooping buses, the sharers found may not increase accordingly. This motivated our work that we build dynamic buses based on the location of cores sharing data. Because we intelligently build buses that connect sharers together using our algorithms described in section A, in our scheme, the longer buses are guaranteed to connect more data sharing cores together.

B.2 Analysis of experimental results

We experimented with two application mappings in order to further validate our motivations. Figure 11 illustrates our two types of mappings. Mapping 1 depicts an ideal mapping, where all threads of an application are assigned to cores close to each other in each row. In mapping 2, only a pair of threads of an application are placed next to each other per row. We use the two types of fixed buses to snoop sharers as illustrated in Figure 4 (a).

Figure 12 and Figure 13 plots the impact of application mappings on performances for both types of bus configurations. We observe that, when the application mapping is ideal as in mapping 1, bus configuration of fixed-4 achieves higher performance compared to fixed-8. In all of the benchmarks, the IPC of fixed-4 is higher compared to that of fixed-8. The reason is that short buses are efficient enough to find data sharers and incur lower bus delays for ideal mappings. The benchmark *equake* exhibits the most significant difference in terms of performance between these two types of bus configurations. On the other hand, even if longer buses can increase the performance compared to the baseline directory, they become overkill for mapping 1 and lead to higher snooping cost. However, if the mapping is not ideal as in mapping 2, longer buses bring more performance benefits than short buses as shown in Figure 13. Performance of fixed-4 in some benchmarks is even worse than the baseline directory, such as in *art*, *barnes* and *fma3d*. This is because

841

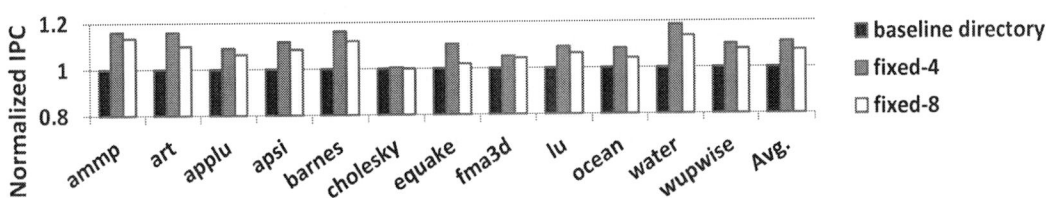

Figure 12: Normalized IPC with mapping 1.

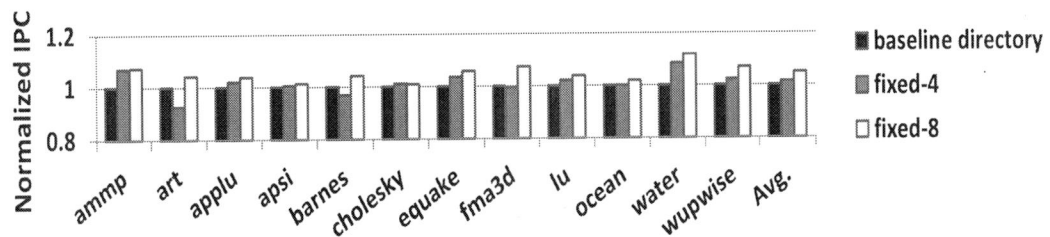

Figure 13: Normalized IPC with mapping 2.

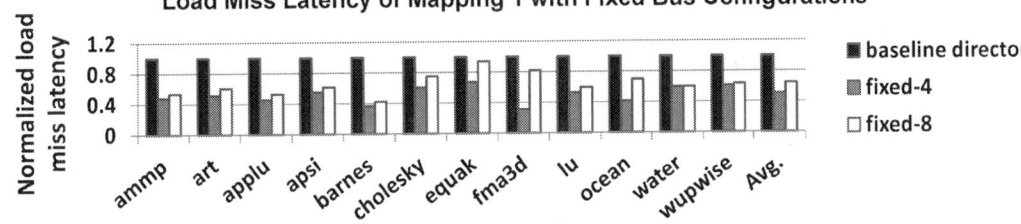

Figure 14: Load miss latency with mapping 1.

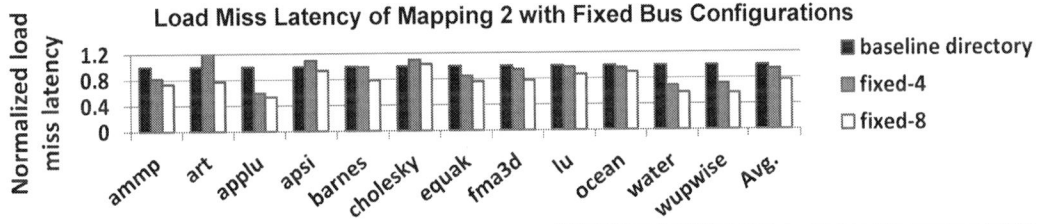

Figure 15: Load miss latency with mapping 2.

the short buses are not able to find sharers and add more delays in addition to directory indirections.

The impact of application mappings on load miss latency is shown in Figure 14 and Figure 15. We observe that if the mapping is ideal, buses of configuration fixed-4 can reduce the load miss latency by about 20% compared to fixed-8 configuration on the average. However, in mapping 2, the load miss latency of fixed-4 is 12% more than the fixed-8 buses. Even though fixed-8 buses have longer delays in bus transactions, the overall miss latency is still reduced compared to fixed-4 for mapping 2.

Our results in this section prove that the effectiveness of snooping buses not only depends on the lengths of the buses, but also is affected by application mappings. Fixed buses are

not able to adjust with flexible application mappings, and thus cannot bring guaranteed performance benefits. On the contrary, our proposed scheme can take the application mappings into consideration when constructing snooping buses. As a result, our scheme can improve the performance and reduce the snooping overhead at the same time.

Architecture Support for Accelerator-Rich CMPs

Jason Cong
UCLA, CS and Center for
Domain Specific Computing
cong@cs.ucla.edu

Mohammad Ali Ghodrat
UCLA, CS and Center for
Domain Specific Computing
ghodrat@cs.ucla.edu

Michael Gill
UCLA, CS and Center for
Domain Specific Computing
mgill@cs.ucla.edu

Beayna Grigorian
UCLA, CS and Center for
Domain Specific Computing
bgrigori@cs.ucla.edu

Glenn Reinman
UCLA, CS and Center for
Domain Specific Computing
reinman@cs.ucla.edu

ABSTRACT

This work discusses a hardware architectural support for accelerator-rich CMPs (ARC). First, we present a hardware resource management scheme for accelerator sharing. This scheme supports sharing and arbitration of multiple cores for a common set of accelerators, and it uses a hardware-based arbitration mechanism to provide feedback to cores to indicate the wait time before a particular resource becomes available. Second, we propose a light-weight interrupt system to reduce the OS overhead of handling interrupts which occur frequently in an accelerator-rich platform. Third, we propose architectural support that allows us to compose a larger *virtual* accelerator out of multiple smaller accelerators. We have also implemented a complete simulation tool-chain to verify our ARC architecture. Experimental results show significant performance (on average 51X) and energy improvement (on average 17X) compared to approaches using OS-based accelerator management.

Categories and Subject Descriptors

C.1 [**PROCESSOR ARCHITECTURES**]: C.1.3—*Heterogeneous systems*

General Terms

Design

Keywords

Chip multiprocessor, Hardware Accelerators, Accelerator Virtualization, Accelerator Sharing

1. INTRODUCTION

Power-efficiency has become one of the primary design goals in the many-core era. While ASIC/FPGA designs can provide orders of magnitude improvement in power-efficiency over general-purpose processors, they lack reusability across different application domains, and significantly increase the overall design time and cost [24]. On the other hand, general-purpose designs can amortize their cost over many application domains, but can be 1,000 to 1,000,000 times less efficient in terms of performance/power ratio

in some cases [24]. A recent industry trend to address this is the use of on-chip accelerators in many-core designs [16][25][17]. According to an ITRS prediction [2], this trend is expected to continue as accelerators become more common and present in greater numbers (close to 1500 by 2022). On-chip accelerators are application-specific implementations that provide power-efficient implementations of a particular functionality, and can range from simple tasks (i.e., a multiply accumulate operation) to tasks of more moderate complexity (i.e., an FFT or DCT) to even more complex tasks (i.e., complex encryption/decryption or video encoding/decoding algorithms). We believe that future computing servers will improve their performance and power efficiency via extensive use of accelerators.

Accelerator-rich architectures also offer a good solution to overcome the *utilization wall* as articulated in the recent study reported in [28]. It demonstrated that a 45nm chip filled with 64-bit operators would only have around 6.5% utilization (assuming a power budget of 80W). The remaining *un-utilizable* transistors are ideal candidates for accelerator implementations, as we do not expect all the accelerators to be used all the time.

We classify on-chip accelerators into two classes: 1) tightly coupled accelerators where the accelerator is a functional unit that is attached to a particular core (e.g., [17][15]); and 2) loosely coupled accelerators (e.g., [3]) where the accelerator is a distinct entity attached to the network-on-chip (NoC), which can be shared among multiple cores. This paper focuses on the efficient use of loosely coupled accelerators, which have been studied much less. These accelerators are not tied to any particular core, and can potentially be shared among all cores on-chip – but this does require some form of arbitration and scheduling.

In order to increase the utilization of accelerators, and allow application developers to take advantage of the performance and energy consumption benefits they offer, it is necessary to reduce the overhead involved in their use. This overhead currently comes in the form of interacting with the operating system (OS) that is responsible for managing accelerator resources. Another key issue in such accelerator-rich architectures is efficient management for sharing of accelerators among different cores and across different applications. Additionally, an application author who targets a platform featuring accelerators produces code that is bound to that platform, because accelerators are potentially unique to a given platform. We aim to develop an efficient architectural framework and an associated set of algorithms that minimize the overhead associated with both using accelerators and targeting a platform that extends accelerators to an application.

With these goals in mind, we propose an accelerator-rich CMP architecture framework, named ARC, with a low-overhead resource management scheme that (i) allows accelerators to be shared and virtualized in flexible ways, (ii) is minimally invasive to core designs, and (iii) is friendly for application programs to use. Our paper provides the following contributions:

- An accelerator allocation protocol to avoid OS overhead in scheduling tasks to shared, loosely coupled accelerators
- An approach to accelerator composability that allows multiple accelerators to work collaboratively as a single complex virtual accelerator that is transparent to program authors
- A fully automated simulation tool-chain to support accelerator generation and management

An early version of this work without virtualization, light-weight interrupt and visual-navigation study was presented in [9]. The rest of the paper is organized as follows. Section 2 reviews some related work. The architectural support for our proposed method is reviewed in Section 3. Section 3 also discusses the algorithms we have developed to efficiently share and virtualize accelerators. Section 4 discuss our experimental results which support our proposed methods.

2. RELATED WORK

There is a large amount of work that implements an application-specific coprocessor or accelerator through either ASIC or FPGA [4] [7]. These works mostly consider a single accelerator dedicated to a single application. Convey [1] and Nallatech [3], target reconfigurable computing in which customized accelerators are off-chip from the processors, unlike our work which target CMP architectures with on-chip accelerators. Some previous work considered on-chip integration of accelerators. Garp [14], UltraSPARC T2 [17], Intel's Larrabee [25] and IBM's WSP processor [16] are examples of this. Most of these platforms (except WSP) are tightly coupled with processor cores (or core-clusters). Our paper focuses on loosely coupled accelerators in a way where accelerators can be shared between multiple cores. OS support for accelerator sharing and scheduling is presented in [13]. In contrast, we focus on hardware support for accelerator management. To the best of our knowledge, this is the first work to address this issue.

There have also been a number of recent designs of heterogeneous architectures, like EXOCHI [29], SARC [23], and HiPPAI [26]. Similar to our work, EXOCHI's focus is on a heterogeneous non-uniform ISA. HiPPAI, like our work, eliminates system overhead involved in accessing accelerators, only it does so using a software layer (portable accelerator interface). SARC also has a core and accelerator architecture similar to our work, yet it also lacks a hardware management scheme. Unlike these works that focus on software-based methodologies, our approach fully advocates the use of hardware for managing and interfacing with accelerators.

There are some related work in accelerator virtualization, namely VEAL [6] and PPA [21]. VEAL [6] uses an architecture template for a loop accelerator and proposes a hybrid static-dynamic approach to map a given loop on that architecture. The difference between our virtualization technique and theirs is that their work limits to nested loops, while in our approach we seek any accelerator such that its composition can be described by some set of rules. PPA [21] uses an array of PEs which can be reconfigured and programmed. PPA, uses a technique called virtualized modulo scheduling which expands a given static schedule on available hardware resources. Again in this work the input is a nested loop, where in our approach this is not a limitation.

3. ARCHITECTURE SUPPORT OF ARC

In an accelerator-rich platform, one main issue is how to increase the utilization of accelerators and also how to make them reusable between multiple applications. Our approach uses several techniques, namely accelerator sharing and accelerator virtualization. In the following subsections we first discuss the motivation for our work and then show how we efficiently implement these techniques.

3.1 Motivation

In a typical heterogeneous system which uses accelerators, when a core wants to access an accelerator, it does that by using an accelerator driver (OS call) [13] [27]. Using the Simics/GEMS simulation [19] [20] platform to model a system consisting of Ultra-SPARC III-i processors running Solaris 10, we measured the delay for different system call operations. These results are shown in Table 1 (ioctl is the system call for device-specific operations). In an accelerator-rich platform, this simplistic approach becomes very inefficient, both in terms of energy and performance. The first motivation for our work (efficient sharing) is to minimize this overhead when there are many accelerators. The second motivation for our work is to increase the utilization of these accelerators by creating new or larger accelerators through composition and chaining.

Table 1: OS overhead to access accelerators(cycles)

Operation	1 Core	2 Cores	4 Cores	8 Cores	16 Cores
Open driver	214,413	256,401	266,133	308,434	316,161
ioctl (average)	703	725	781	837	885
Interrupt latency	16,383	20,361	24,022	26,572	28,572

3.2 Microarchitecture of ARC

Figure 1 shows the overall architecture of ARC which is composed of cores, accelerators, the Global Accelerator Manager (GAM), shared L2 cache banks and shared NoC routers between multiple accelerators. All of the mentioned components are connected by the NoC. Accelerator nodes include a dedicated DMA-controller (DMA-C) and scratch-pad memory (SPM) for local storage and a small translation look-aside buffer (TLB) for virtual to physical address translation. GAM is introduced to handle accelerator sharing and arbitration.

In order to interact with accelerators more efficiently, we have introduced an extension to the instruction set consisting of four instructions used specifically for interacting with accelerators. These instructions are briefly described in Table 2. A processor uses *lcacc-req* to request information about accelerator availability, consisting of pairs of accelerator identifiers and predicted wait times for each available accelerator. A processor will then use *lcacc-rsv* to request use of a specific accelerator. *lcacc-cmd* is used for interacting directly with an accelerator. When a job is completed, *lcacc-free* is used to release an accelerator to be used by another cpu. These instructions are accessible directly from user code, and do not require OS interaction. Communication with accelerators is done with the use of virtual addresses, accessing resources that are already accessible from user code. Execution of each of these instructions results in a message being sent to a device on the network, either the GAM or an accelerator. Attached to each of these messages is the thread ID of the executing thread that can be used to track requesting threads in an environment where context switches are possible.

Figure 2 shows the communication between a core, the GAM, an accelerator and the shared memory detailing the use of an accelerator by a core. The numbers on the arrows in Figure 2 show the steps taken when a core uses a single accelerator. They are described below.

1. The core requests an enumeration of all accelerators it may

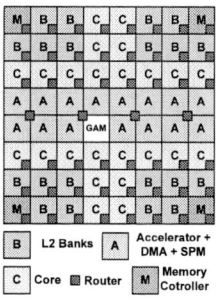

Figure 1: Overall architecture of ARC

Legend:
- B = L2 Banks
- C = Core
- A = Accelerator + DMA + SPM
- Router
- M = Memory Cotroller

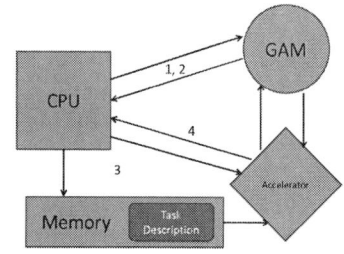

Figure 2: Communication between core, GAM, and accelerator

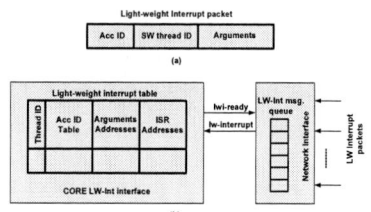

Figure 3: Light-weight interrupt support

Table 2: Instructions used to interact with accelerators.

lcacc-req x	Request information from GAM about availability of accelerators implementing functionality *x*
lcacc-rsv x y	Reserve the accelerator with ID *x* for a predicted duration *y*
lcacc-cmd accl cmd addr x y z	Send a command *cmd* to an accelerator *accl* with parameters *x*, *y*, and *z*. Performs an address translation on *addr*, sending both logical and physical address.
lcacc-free accl	Sends a message to GAM releasing accelerator *accl*.

Table 3: Instructions to handle light-weight interrupts.

lwi-reg x y z	Register service routine *y* to service interrupts arriving from accelerator *x*. LWI message packet will be written to *z*
lwi-ret	Return from an interrupt service routine.

potentially need from the GAM (*lcacc-req*). The GAM responds with a list of accelerator IDs and associated estimated wait times.

2. The core sends a sequences of reservations (*lcacc-rsv*) for specific accelerators to the GAM. The core waits for the GAM to give it permission to use these accelerators. The GAM also configures the reserved accelerators for use by the core.

3. The core writes a task description detailing the computation to be performed to the shared memory. It then sends a command to the accelerator (*lcacc-cmd*) identifying the memory address of the task description. The accelerator loads this task description, and begins working.

4. When the accelerator finishes working, it notifies the core. The core then sends a message to the GAM freeing the accelerator (*lcacc-free*).

3.3 Light-weight interrupt support

A platform that features accelerators requires a mechanism for a processor to be notified of the progress of an accelerator. In the ARC platform, we handle this issue with the use of light-weight interrupts. ARC light-weight interrupts are interrupts handled entirely as user code, and do not involve OS interaction, as this interaction can be a major source of inefficiency. Table 1 shows the cost in cycles of interacting with accelerators through a device driver and the overhead associated with OS interrupts.

There are three main sources of interrupts associated with accelerator interaction: 1) GAM responses 2) TLB misses 3) notification that the accelerator has finished working. GAM responses come either because a core sent a request or a reserve message. TLB misses occur when an accelerator fails to perform address translation with the use of its own private TLB, and requires a core's assistance in performing the lookup. Interrupts notifying the completion of work arrive when an accelerator has completed all work given to it.

Figure 3 shows the microarchitecture components added to the cores in ARC in order to support the light-weight interrupt. An interrupt is sent via an interrupt packet (shown in Figure 3-a) through the NoC to the core requested accelerator. Each interrupt packet includes the thread ID which identifies the thread which this in-

terrupt belongs to, and a set of interrupt-specific information. The main microarchitecture components added to support light-weight interrupt are listed below:

1. Interrupt controller located at the core's network interface. This is responsible for receiving the interrupt packets and queuing them until being serviced by the core.

2. Light-weight interrupt interface in the core. This is responsible for: 1) receiving the interrupt from the interrupt controller, 2) providing a software interface to setup the information needed to service the interrupt.

The interrupt controller has a queue for buffering the received interrupt packets, so they don't get lost if the core is busy handling other interrupts. Without loss of generality we assume that for each thread we can only have one level nest for interrupt. This means no other light-weight interrupt will be serviced, while servicing another light-weight interrupt. If an interrupt arrives for a thread that is currently scheduled, it is executed immediately. If the thread is not scheduled, a normal OS-based interrupt occurs.

In order to support light-weight user-level interrupts, we introduce a set of instructions to enable user code to handle interrupts. These instructions are described in Table 3. *lwi-reg* registers the interrupt handlers. *lwi-ret* returns from an interrupt handler routine. A program segment using accelerators is then designed as a series of interrupt service routines.

3.4 Invoking accelerators

In this work, we assume an accelerator will be used to process a relatively large amount of data. The initial overhead associated with acquiring permissions to use an accelerator is large enough that it should be amortized over a large amount of work. To that end, we introduce two accelerator features that explicitly deal with efficiently processing large amounts of data: (1) task descriptions to limit communication between accelerators and the controlling core, and (2) methods to handle TLB misses.

To communicate with an accelerator, a program would first write to a region of shared memory a description of the work to be performed. This description includes location of arguments, data layout, which accelerators are involved in the computation, the computation to be performed, and the order in which to perform necessary operations. This detail is included to allow accelerators to be both general, and to allow coordination of accelerators in groups to perform more complex tasks through virtualization(described in Sec-

845

tion 3.6). Evaluating the task description yields a series of steps to be performed in order, with each step consisting of a set of memory transfers and computations that can be executed concurrently. This allows accelerators to overlap computation with memory transfer within a given step. When all computations and memory transfers of a given step are completed, the accelerator moves onto the next step. In this work, we refer to these individual steps as tasks, and the structure detailing a sequence of tasks as a task description.

To further decouple the accelerator from the controlling core, each accelerator contains a small local TLB. This is required because the accelerator operates within the same virtual address space as the software thread that is using the accelerator. The accelerator relies on the controlling core to service any detected TLB misses. It does this by sending a light-weight interrupt to the controlling core when a TLB miss occurs with the address that caused the TLB miss. Handling this interrupt would involve the core executing the same TLB miss handler that is executed when the core normally encounters a miss in its own TLB. Because this is an OS action, and involves trapping to an OS handler regardless, it is not actually necessary that the original software thread that is using the accelerator be currently scheduled. If it is scheduled, the lightweight interrupt interface can be used to limit overhead associated with interrupt handling. Otherwise, the OS can be notified directly (e.g. by invoking a software interrupt or real hardware interrupt) without having to wait for or force a context switch to reschedule the controlling thread. The resolved address is then sent back to the accelerator that had encountered the TLB miss.

3.5 Sharing accelerators

When accelerators are shared among all the on-chip cores, it is possible for there to be several cores competing for the same accelerator. Even in architectures with large numbers of accelerators, there may be a limited number of one particular type of accelerator that is suddenly in high demand. In this situation, some of these cores may choose to eschew the use of the accelerator and simply execute the task to be offloaded using their own core resources. While the core is certainly less power efficient in executing this task, it may make sense for it to do so in situations where the wait time for an accelerator will eliminate any potential gains. In this paper we propose a sharing and management scheme which can dynamically determine whether the core should wait to use an accelerator or should instead choose a software path, based on an estimated waiting time. This proposed sharing and management strategy is performed by the GAM. The GAM tracks: 1) the types of available accelerators; 2) the number of accelerators of each type; 3) the jobs currently running or waiting to run on accelerators, their starting time and estimated execution time (Section 3.5.1); 4) the waiting list for each accelerator and the estimated run time for each job in the waiting list (Section 3.5.2).

3.5.1 Accelerator run-time estimation (by the core)

The execution time of a certain job on an accelerator is data-dependent. For most of our examples, we found that a low order polynomial regression model was sufficient to estimate execution time. The regression model is provided by the accelerator application programming interface (more info in [11]).

3.5.2 Wait-time estimation algorithm (by the GAM)

After receiving the reserve request message from the core, the GAM will add the requesting core's ID to the tail of the waiting list for that accelerator. The estimated waiting time can be simply derived by summing the expected execution time of all jobs in the waiting list for an accelerator. These tasks are issued in a first-

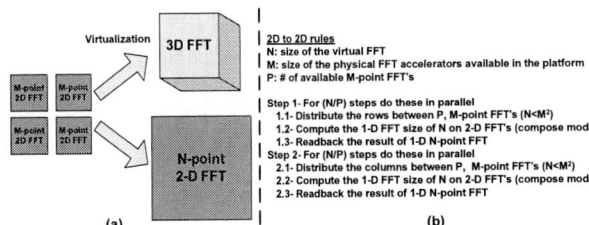

Figure 4: An example of accelerator composition

come-first-serve (FCFS) order. This is simple, and is practical for a hardware implementation.

3.6 Accelerator virtualization

A key contribution of our work is to increase the utilization of the available accelerators by either composing different accelerator types to create new types of accelerators or to compose the same type of accelerators to create a larger accelerator. In the next two sections we discuss these two techniques.

3.6.1 Accelerator chaining

In an accelerator-rich platform, there are many cases when the output of one accelerator feeds the input of another accelerator (like many streaming applications). In a traditional system, these two accelerators communicate through system memory, i.e., the controlling core reads the output of the first accelerator from its SPM, stores it to shared memory, and writes it to the second accelerator's SPM. To remove this inefficiency, two DMA-controllers can communicate and the source DMA-controller can send the content of its SPM to another DMA-controller to be written in its SPM.

3.6.2 Accelerator composition

For many types of problems, it is not practical to provide an accelerator to directly solve each possible problem instance. Additionally, it is not practical to demand that an application author target a single architecture. For this reason, we provide a set of virtual accelerators to decouple hardware design and software development. A virtual accelerator is an accelerator that is implemented as a series of calls to other physical accelerators, available in hardware (Figure 4(a)). A large library of virtual accelerators can be provided to the application author as if they were implemented in hardware. These accelerators would actually be implemented as a series of decomposition rules that break down a large problem into a number of smaller problems (Figure 4(b)), similar in style to the approach presented in [22]. These small problems would then be solved directly by hardware. These rules describe two things: 1) computation that must be performed by accelerators capable of solving sub-problem instances, and 2) how data is communicated to, from, and between these various smaller accelerators. Rules would be applied recursively to express an implementation for each virtual accelerator in terms of calls to physical accelerators.

These statically determined decomposition rules can thus be applied at run-time. Figure 5 describes the process of invoking a virtual accelerator from within the application binary. When an accelerator is called, a *lcacc-req* message is sent to the GAM for wait times for all functional units that may be required by the decomposition result. While waiting on this request, the requesting core either begins calculating the decomposition or begins fetching the data structures associated with the statically computed solution. Once the GAM responds and the requesting core has a fully decomposed problem available, the core calculates the wait time for the

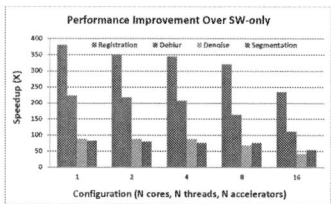

Figure 6: MI - Speedup over SW-only

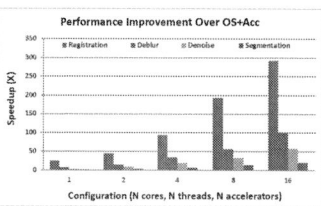

Figure 7: MI - Speedup over OS+Acc

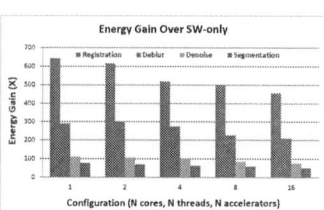

Figure 8: MI - Energy gain over SW-only

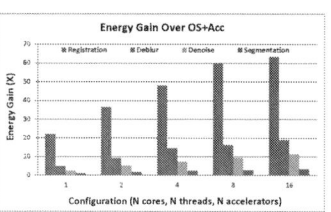

Figure 9: MI - Energy gain over OS+Acc

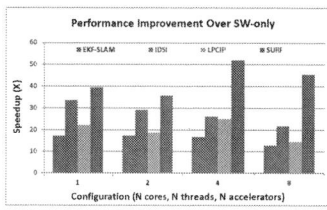

Figure 10: VN - Speedup over SW-only

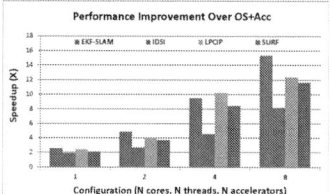

Figure 11: VN - Speedup over OS+Acc

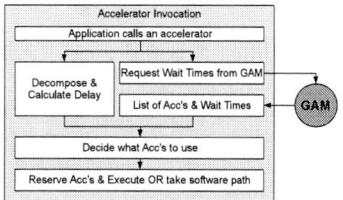

Figure 5: Accelerator composition steps

entire computation. It does this by adding the delay calculated with the use of the regression model to the largest of the delays provided by GAM. The core then executes a series of *lcacc-rsv* instructions for each required accelerator, specifying the wait time for the entire operation as the estimated duration of use of each accelerator reserved. GAM will not assign any accelerators until it can assign all accelerators requested. The core releases accelerators in the same way as it normally would. With these mechanisms, an application author can use a simple API to invoke virtual accelerators, and a hardware developer can implement accelerators based on need and available resources.

We will show more details on programming interface in ARC in Supplemental Section 8.1. More info on accelerator extraction methodology can be found in [11].

4. EXPERIMENTAL RESULTS

To illustrate the effectiveness of our ARC platform, we evaluate a number of compute intensive benchmarks, primarily from the medical imaging(MI) and computer vision and navigation(VN) domains. More information on our benchmark can be found in [11] and Supplemental Section 8.2.2. Our experiments were conducted using a heavily modified version of the Simics and GEMS [19] [20] simulation platform. More information about our simulation platform can be found in Supplemental Section 8.2. Additional experimental results not presented here can also be found in [11].

We used the following schemes for ARC evaluation:

- **Original benchmark (SW-only)**: The baseline for the experiments is the execution of these multithreaded benchmarks on a multiprocessor (one thread per processor).
- **Accelerators + OS management (OS+Acc)**: This is a system which has accelerators managed by OS drivers.

- **Accelerators + HW management (ARC)**: This is a system which features all enhancements discussed thus far, including hardware resource arbitration managed by the GAM.

We show the simulation configuration using Cc-Tt-Aa-Dd mnemonic. Here "C" is the number of cores, "T" is the number of threads, "A" is the number of replicates of each accelerator needed by a benchmark, and "D" is data size. For example, a benchmark featuring 4 cores, 2 threads, 1 replicate of each accelerator, and an argument that is 64-cubes of data would be described as 4c-2t-1a-64d. For MI benchmarks since data is cubic in form, "D" shows a cube of $D \times D \times D$ data elements for each argument. For VN benchmarks data is linear, thus "D" shows the absolute data size. Next the results for baseline speedup and energy improvement are discussed.

4.1 Speedup and energy improvements

Figures 6, 10, 8, and 12 shows the speedup and energy gain result for the ARC base configuration (Nc-Nt-Na) compared to running the software-only version of the benchmark on the same number of processors, threads, and data size. The highest speedup is for *Registration* (485X for 1c-1t-1a-32d case) and the lowest is for EKF-SLAM (13X for 16p-16t-16a case). The best energy gain is for registration with 641X improvement. On average we get 241X energy improvement over all the benchmarks and configuration. VN benchmarks are shown benefiting less from acceleration as compared to MI benchmarks due largely to data sizes selected. A study of the impact of data size on accelerator efficiency can be found in [11].

We observe a reduction in speedup as we increase the number of cores and threads. This reduction is attributed to several sources. First, we measure the time from the start of all threads, to the end of the last thread, thus the results shown are the measured time of the longest running thread. Adding more threads increases the likelihood of observing normal fluctuations in run time. Lastly, while we increase the number of cores and accelerators, we do not correspondingly increase network resources, memory bandwidth, or cache capacity. As a result, increasing the number of cores and threads resulted in additional contention for communication and memory resources. This impacted accelerated cases more than software-only cases because, while the same amount of data is accessed, the accelerated cases access this data over a much shorter time period.

Figures 7 and 11 show the speedup gain ARC achieves com-

847

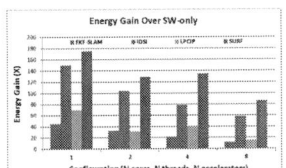
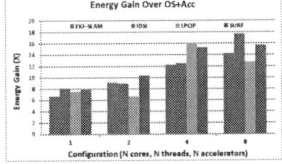
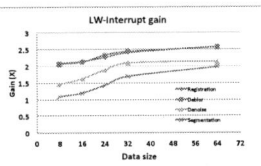
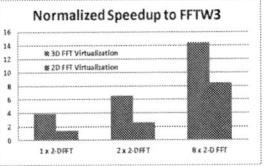

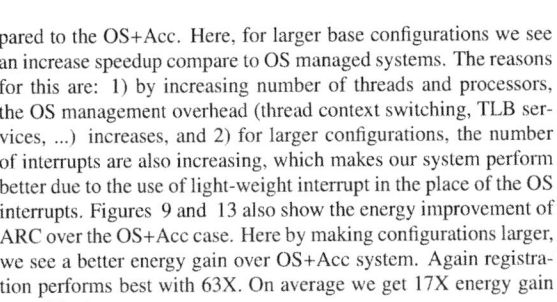

Figure 12: VN - Energy gain over SW-only

Figure 13: VN - Energy gain over OS+Acc

Figure 14: Benefit of using light weight interrupt

Figure 15: FFT virtualization (2D and 3D)

pared to the OS+Acc. Here, for larger base configurations we see an increase speedup compare to OS managed systems. The reasons for this are: 1) by increasing number of threads and processors, the OS management overhead (thread context switching, TLB services, ...) increases, and 2) for larger configurations, the number of interrupts are also increasing, which makes our system perform better due to the use of light-weight interrupt in the place of the OS interrupts. Figures 9 and 13 also show the energy improvement of ARC over the OS+Acc case. Here by making configurations larger, we see a better energy gain over OS+Acc system. Again registration performs best with 63X. On average we get 17X energy gain over OS+Acc case.

4.2 Accelerator virtualization results

Figure 15 shows the result of virtualizing a 512x512 2D FFT and a 128x128x8 3D FFT on multiple 128x128 2D FFTs. The SW case is compared to having 1, 2, and 8 copies of 128x128 2D accelerator on the chip (8 FFT is based on assigning a maximum 5% of the chip area to FFT). The SW case is the result of running FFTW3 [12]. In the best case for 3D-FFT we obtained 14.4X speedup and for 2D-FFT we obtained 8.4X speedup.

4.3 Light-weight interrupt benefit

To measure the benefit of light-weight interrupts, we examined a platform lacking light-weight interrupts to compare our ARC platform against a system that relies instead on OS handling of interrupts. Figure 14 shows the speedup measured over a platform lacking light-weight interrupts. ARC is up to 2.5X faster than an otherwise identical system that lacks light-weight interrupts. The larger the data size, the more interrupts are generated, so the benefits of ARC increases as the data size grows.

4.4 Accelerator sharing results

Run-time estimation was calculated using a simple regression model based on profiled runs. Additional details regarding this regression model can be found in [11]. Wait-time estimation was based on the accumulated run-time estimates. Our results shows that the estimated error ranges from < 1% to 6% of execution times on accelerators, which is sufficiently predictable for this to be a very practical approach.

5. CONCLUSION AND FUTURE WORK

We have discussed hardware architectural support for accelerator-rich CMPs. This was motivated by our belief that future supercomputers, especially green supercomputers, will improve their performance and power efficiency through extensive use of accelerators. First, we presented a hardware resource management scheme for sharing of accelerators and arbitration of multiple requesting cores. Second, we presented a mechanism that allows us to efficiently compose a larger virtual accelerator out of multiple smaller accelerators. Our results showed large performance and energy efficiency improvement over a software implementation, and also using OS-based accelerator management, with minimal hardware overhead

6. ACKNOWLEDGEMENTS

This research is partially supported by the Center for Domain-Specific Computing (CDSC) funded by the NSF Expedition in Computing Award CCF-0926127, GSRC under contract 2009-TJ-1984 and NSF Graduate Research Fellowship Grant # DGE-0707424.

7. REFERENCES

[1] Convey computer. http://conveycomputer.com/.

[2] ITRS 2007 system drivers. http://www.itrs.net/.

[3] Nallatech FSB - development systems. http://www.nallatech.com/Intel-Xeon-FSB-Socket-Fillers/fsb-development-systems.html.

[4] D. Bouris et al. Fast and efficient FPGA-based feature detection employing the SURF algorithm. FCCM '10, pages 3–10.

[5] A. Bui et al. Platform characterization for domain-specific computing. In ASPDAC, 2012.

[6] N. Clark, , et al. VEAL: Virtualized execution accelerator for loops. ISCA '08, pages 389–400.

[7] J. Cong et al. FPGA-based hardware acceleration of lithographic aerial image simulation. ACM Trans. Reconf. Technol. Syst., pages 1–29, 2009.

[8] J. Cong et al. Accelerating vision and navigation applications on a customizable platform. In ASAP, 2011.

[9] J. Cong et al. AXR-CMP: Architecture support in accelerator-rich CMPs. 2nd Workshop on SoC Architecture, Accelerators and Workloads, February 2011.

[10] J. Cong et al. High-level synthesis for FPGAs: From prototyping to deployment. Computer-Aided Design of Integrated Circuits and Systems, IEEE Transactions on, 30(4):473 –491, April 2011.

[11] J. Cong, M. A. Ghodrat, M. Gill, B. Grigorian, and G. Reinman. UCLA computer science department technical report #120008.

[12] M. Frigo et al. The design and implementation of FFTW3. Proc. of the IEEE, 93(2):216–231, 2005.

[13] P. Garcia et al. Kernel sharing on reconfigurable multiprocessor systems. FPT 2008, pages 225 –232.

[14] J. Hauser et al. Garp: a mips processor with a reconfigurable coprocessor. FCCM'97, pages 12 –21.

[15] W. Jiang et al. Large-scale wire-speed packet classification on FPGAs. FPGA '09, pages 219–228.

[16] C. Johnson et al. A wire-speed power™ processor: 2.3ghz 45nm soi with 16 cores and 64 threads. ISSCC'10, pages 104 –105.

[17] T. Johnson et al. An 8-core, 64-thread, 64-bit power efficient sparc soc (niagara2). ISPD '07, pages 2–2.

[18] S. Li et al. McPAT: an integrated power, area, and timing modeling framework for multicore and manycore architectures. MICRO 42, 2009.

[19] P. S. Magnusson et al. Simics: A full system simulation platform. Computer, 35:50–58, 2002.

[20] M. M. K. Martin et al. Multifacet's general execution-driven multiprocessor simulator toolset. SIGARCH Comput. Archit. News, 33, 2005.

[21] H. Park et al. Polymorphic pipeline array:a flexible multicore accelerator with virtualized execution for mobile multimedia application. MICRO, 2009.

[22] M. Puschel et al. Spiral: Code generation for dsp transforms. Proc. of the IEEE, (2):232 –275, 2005.

[23] A. Ramirez et al. The SARC architecture. Micro, IEEE, 30(5):16 –29, Sep 2010.

[24] P. Schaumont et al. Domain-specific codesign for embedded security. Computer, 36:68–74, 2003.

[25] L. Seiler et al. Larrabee: A many-core x86 arch. for visual computing. IEEE Micro, 29:10–21, 2009.

[26] P. Stillwell et al. HiPPAI: High performance portable accelerator interface for SoCs. HiPC 2009.

[27] N. Sun et al. Using the cryptographic accelerators in the ultrasparc t1 and t2 processors. Sun BluePrints Online, 2007.

[28] G. Venkatesh et al. Conservation cores: reducing the energy of mature computations. ASPLOS '10.

[29] P. H. Wang et al. EXOCHI: architecture and programming environment for a heterogeneous multi-core multithreaded system. PLDI '07.

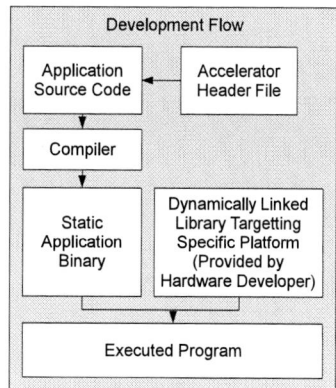

Figure 16: ARC development flow

8. SUPPLEMENTAL

8.1 Programming interface to ARC

The Application Programming Interface (API) involved in using accelerators is presented in Figure 16. For each type of accelerator, one dynamic linked library (DLL) is provided. This DLL is specific to a target platform, and provides a mapping from accelerator calls to actual invocations of physical accelerators. Calls to accelerators have their implementations dynamically linked to application code.

8.2 EVALUATION METHODOLOGY

8.2.1 Simulation tool-chain

In order to make the exploration of this topic practical, a number of supporting tools were created. These tools simplified the authoring of programs that used accelerators, and automated the process of implementing our chosen accelerators in our simulator framework. These tools were used in place of hand-written implementations and hand-adapted benchmarks to allow us to simulate systems that would have been prohibitively complex to manually author, such as those that utilized many accelerators or featured complicated inter-accelerator communication. Additionally, we believe that this is representative of what will be done in the development of future accelerator exploiting libraries, to simplify the job of programmers who would use these libraries without compromising any of the capabilities of these accelerators.

With this toolchain, generation of accelerators is only a matter of identifying a function in an application's source code to accelerate. We have automated the process of extracting these functions, compiling these modules into VHDL, and synthesizing these modules to extract timing and energy information. This process yields a module that plugs into our cycle-accurate simulation infrastructure to model this hardware unit, and coordinates the execution of this selected function in a pipelined fashion.

Once we select the functions we want to accelerate, typically encompassing the kernel of the benchmark, we procedurally generate a program segment to use these accelerators. We described communication between accelerators in a simple data-flow language that we use to generate C source code. These program segments together make up the platform specific DLL mentioned above. This code is responsible for coordinating interactions between accelerators, registering and handling interrupts, managing task descriptions and accelerator resources, and dealing with synchronization between accelerators and the CPU. Figure 17 illustrates the work flow described here. The AutoPilot [10] behavioral synthesis tool

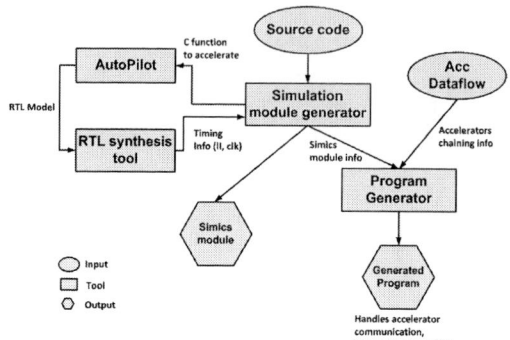

Figure 17: Process used to generate simulation structures and Accelerator using programs

is used to synthesize the C modules into ASIC.

8.2.2 Benchmarks

To illustrate the effectiveness of our ARC platform, we evaluate a number of compute intensive benchmarks from both the medical imaging domain as well as the computer vision and navigation domain. Information on medical imaging domain can be found in [5]. Information on computer vision and navigation domain can be found in [8].

8.2.3 Simulation Platform

Our experiments were conducted using a heavily modified version of the Simics and GEMS [19] [20] simulation platform. The machine we modeled was based on a multicore system consisting of a mix of Ultrasparc III processors and accelerators. In order to create a fair comparison between machines of different configurations, we maintained a fixed cache and network configuration. Our network topology was a mesh modeled on a system normally used to support 32 processors. These nodes were then configured to either be processors, accelerators, or empty sockets. We featured a per-processor split L1 cache, and a distributed L2 spread across all nodes that relied on a directory based coherence protocol. Table 4 shows the machine configurations which is modeled in simulations.

8.2.4 Area/Timing/Power Measurements

The AutoPilot [10] behavioral synthesis in combination with the Synopsys design compiler was used to synthesize the C modules into ASIC (using 32nm ASIC library from Synopsys). The timing information produced by the synthesis process was back-annotated to our accelerator modules to model cycle accurate accelerators. For computing energy we used power reports from Synopsys for accelerators and McPAT [18] for CPU power. Table 5 shows the synthesis results for the accelerators in our selected benchmarks together with the GAM and DMA-controller.

Table 4: Simics/GEMS configuration

CPU	Ultra-SPARC III-i @ 2.0GHz
Number of cores	1, 2, 4, 8, 16
Coherence protocol	MSI_MOSI_CMP_directory
L1 cache	32 KB, 4 way set-associative
L2 cache	8 MB, 8-way set-associative
Memory latency	1000 cycles
Network topology	Mesh
Operating System	Solaris10

Table 5: Synthesis results

	Deblur	Registration	Denoise	Segmentation	GAM	DMA-C
Clock(ns)	2	2	2	2	1	1
Area (μm^2)	2013228	3853095	496908	688298	12270	10071
Power (mW)	98.28	256.3	57.69	80.93	2.64	0.59

A QoS-Aware Memory Controller for Dynamically Balancing GPU and CPU Bandwidth Use in an MPSoC

Min Kyu Jeong, Mattan Erez
Dept. of Electrical and Computer Engineering,
The University of Texas at Austin
{mkjeong, mattan.erez}@mail.utexas.edu

Chander Sudanthi, Nigel Paver
ARM Inc.
{Chander.Sudanthi, Nigel.Paver}@arm.com

ABSTRACT

Diverse IP cores are integrated on a modern system-on-chip and share resources. Off-chip memory bandwidth is often the scarcest resource and requires careful allocation. Two of the most important cores, the CPU and the GPU, can both simultaneously demand high bandwidth. We demonstrate that conventional quality-of-service allocation techniques can severely constrict GPU performance by allowing the CPU to occasionally monopolize shared bandwidth. We propose to dynamically adapt the priority of CPU and GPU memory requests based on a novel mechanism that tracks progress of GPU workloads. Our evaluation shows that the proposed mechanism significantly improves GPU performance with only minimal impact on the CPU.

Categories and Subject Descriptors

C.1.3 [**Processor Architectures**]: Other Architecture Styles—*Heterogeneous (hybrid) systems*
; I.3.1 [**Computer Graphics**]: Hardware Architecture—*Graphics processors*
; C.3 [**Special-purpose and Application-based Systems**]: *Real-time and embedded systems*
; C.4 [**Performance of Systems**]: *Design studies*

General Terms

Design, Experimentation, Performance

Keywords

Graphics processor, Memory controller, Quality of service, System on chip

1. INTRODUCTION

A modern system-on-chip (SoCs) is typically composed of multiple types of *intellectual property cores* (IP cores) with different functionality. This is done because heterogeneity increases efficiency and decreases development time. All integrated cores share off-chip memory, which is often one of the most constrained resources. High end SoCs, for example, now include powerful CPU and GPU cores, which are both very demanding of the memory system. Optimally allocating the scarce memory bandwidth resource between the

CPU and GPU cores is critical yet challenging. The CPU is latency sensitive and cannot tolerate long memory latency without performance loss. The GPU, on the other hand, is designed to tolerate long latencies but requires consistent high bandwidth for periods of time to meet its real-time deadlines. Because the CPU is sensitive to latency, it is common practice to always prioritize requests from the CPU over those of the GPU. We show that such a static policy can lead to an unacceptably low frame rate for the GPU. Conversely, prioritizing GPU requests significantly degrades CPU performance.

We propose a new mechanism to solve this challenge by dynamically adjusting the memory controller's quality-of-service (QoS) policy. As is done today, we prioritize CPU requests by default and GPU requests are serviced opportunistically. When the GPU is expected to miss a deadline, however, we increase the GPU service rate by raising its priority. The key to our technique is identifying when the default policy should be adjusted. We do this by utilizing knowledge of the GPU architecture and monitor the progress of processing a frame against the frame deadline. The memory controller can than determine when a deadline is likely to be missed and boost the GPU service quality.

To the best of our knowledge, we are the first to propose and provide a detailed evaluation of adjusting the memory controller QoS policy in response to progress towards a real-time deadline. We show how the dynamic technique balances both real-time constraints and best-effort memory accesses and maintains GPU target performance with only a small impact on the CPU. We are also first to present a detailed analysis based on a combined cycle-accurate simulation of GPU and CPU cores with a detailed memory system. [1] We draw important insights into how these components interact, which contradicts current best practices.

The rest of this paper is organized as follows: Section 2 provides background on CPU and GPU architecture, memory controllers, and the commonly used QoS mechanisms. Section 3 describes our dynamic QoS mechanisms, based on our technique to monitor GPU's workload progress. We present our evaluation methodology and results in Section 4 and 5, then conclude in Section 6.

2. BACKGROUND AND RELATED WORK

This section briefly discusses the memory access and execution characteristics of CPU and GPU cores, as well as the fundamental principles and design of modern memory controllers and QoS mechanisms.

---

[1] Concurrent work by Ausavarungnirun et al. [3] also explores the memory-system implications of heterogeneous GPU/CPU processors.

2.1 CPU

Modern general purpose processors are designed mainly to maximize the performance of a single thread of execution. Single-thread performance is very sensitive to long-latency memory requests because instructions dependent on the long latency load cannot proceed until the load completes. Caching and out-of-order execution can mitigate the impact of long main memory latency. Main memory access latency, however, is much higher than what the out-of-order structure can tolerate and cache misses that go out to main memory inevitably stall the accessing thread [7]. Therefore, any increase in CPU memory access latency, such as delays introduced by contention from the GPU, decreases CPU performance.

2.2 GPU

Mobile GPU cores often utilize tile based rendering to reduce off-chip memory bandwidth consumption. The screen is subdivided into many blocks/tiles, which can be processed independently of one another (Figure 1). As the tiles are small enough, the entire pixel data of a tile can be kept in on-chip buffers while being rendered so that repeated accesses to the same pixel do not incur off-chip memory accesses. GPUs can process all vertices and fragments within the tile and multiple tiles in parallel and can therefore tolerate very long memory latencies. They still require high bandwidth and are sensitive to disruptions in available bandwidth.

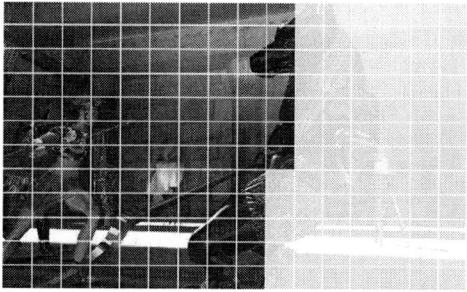

Figure 1: Tile based rendering in progress. Shaded tiles are to be rendered.

Simple scenes can be processed rapidly and generate correspondingly low memory traffic, while others may take the entire time allotted or longer, resulting in skipped frames and degraded user experience. The GPU may idle between finishing a frame and starting the next frame, because frame rate is fixed and frame render time varies, Figure 2 shows an example of processing one frame from the Taiji GPU workload [1] with two CPU cores running together. The figure shows how the GPU only requires about half the frame time to process a scene (GPU bandwidth consumption shown with dashed lines) and consumes up to 62% of total memory bandwidth when not constrained by the memory controller as discussed below. The figure also shows how the heavy bandwidth use of the GPU hurts CPU performance (CPU's instructions per cycle (IPC) shown with solid lines) compared to when the GPU is idling. The two subfigures show different CPU workloads and the same set of GPU frames. Both mcf and art are memory-intensive applications from the MinneSPEC suite [9], with art requiring somewhat higher bandwidth.

2.3 Memory Controller Basics

Modern DRAM architecture is optimized for access patterns with spatial locality. Within a DRAM chip, each access is performed at a granularity of an entire row, which is 8Kb or 16Kb in current technology [10]. To amortize the time and energy involved in activating an entire DRAM row, each DRAM bank contains a *row buffer*. Consecutive accesses to the same row (row buffer hits) can be served directly from the row buffer, saving time and energy in activating rows. In contrast, accesses to a different row (row buffer misses) need additional steps of precharging the array and activating the new rows. When the row buffer hit rate is low, DRAM can only supply a small fraction of its peak bandwidth. Therefore, modern out-of-order memory controllers schedule accesses to the same row together by prioritizing row buffer hit requests [13].

2.4 Quality of Service

An out-of-order memory scheduler increases overall bandwidth, but in a shared memory SoC, the priority scheme can starve some cores when other cores offer frequent requests with high spatial locality, like GPUs do. To prevent such unfairness, the memory controller must balance the accesses from different cores and provide QoS mechanisms. Because of the heavy competition in the SoC industry, very little information on how commercial SoCs manage shared memory bandwidth is publicly available.

Most previous literature on off-chip memory bandwidth QoS focuses on a different context than our multi-processor SoC (MPSoC), such as real-time systems and general purpose chip multiprocessors (CMP). High-end SoCs combine both real-time and best-effort components and place very

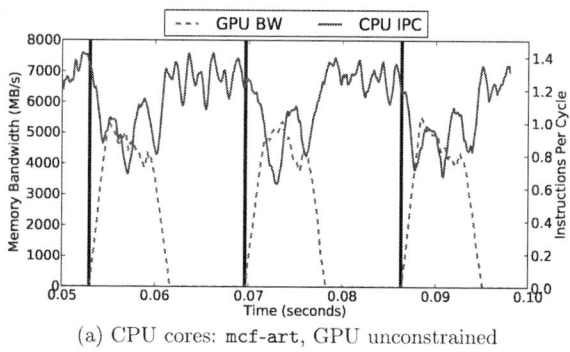

(a) CPU cores: mcf-art, GPU unconstrained

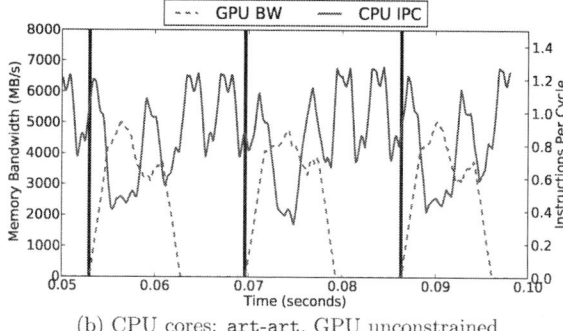

(b) CPU cores: art-art, GPU unconstrained

Figure 2: GPU activity (bandwidth consumption) and CPU performance. Vertical lines represent frame deadlines. Experimental setup discussed in Section 4.

high pressure on shared memory bandwidth. Prior work on QoS for CMPs (e.g., [11, 12]) does not consider real-time constraints, thus can lead to an unacceptable rate of missed deadlines for the GPU. Work on real-time systems, on the other hand, has focused exclusively on bounding the latency of individual requests and ensuring a minimal fraction of shared throughput [2]. This approach sacrifices effective memory scheduling in favor of guaranteed deadlines, which leads to very poor utilization of available DRAM bandwidth and requires significant over-provisioning of this scarce and expensive resource. Recent white papers [15, 16] discuss general quality-of-service techniques and recommendations and appear to describe the status quo. This status quo is that two techniques are effective when combined: regulating the number of outstanding GPU requests and prioritizing CPU requests over ones from the GPU. Note that previous academic literature does not address this particular problem of sharing bandwidth between best-effort and real-time workloads from different cores.

Restricting the number of outstanding GPU requests reduces the GPU's ability to continuously send requests to the memory system, even though the abundant parallelism associated with graphics allows many concurrent requests. The smaller the number of outstanding GPU requests, the greater the number of memory access issue slots that are available for other cores. There are several equivalent mechanisms that can be used to constrain the number outstanding GPU requests, including separate memory controller queues for each core, which may be either physical or virtual.

Guaranteeing available request queue slots is insufficient because the memory controller may still prefer to always issue GPU requests. To avoid this situation, an age-based QoS technique can be used [11]. Recent guidelines, however, suggest that CPU requests should receive higher priority to decrease the performance lost to contention-induced high latency memory accesses [15, 16]. While the priority policy is static, the GPU may take advantage of much of the available bandwidth when CPU cores do not access main memory frequently.

Prioritizing CPU requests indiscriminately, however, can hurt GPU performance significantly, when a CPU core continuously uses high memory bandwidth and a GPU workload is complex enough to mandate high memory bandwidth as well. The impact of the aggressive QoS is shown in Figure 3. The figure shows the same workload scenario as in Figure 2, but with a QoS mechanism that balances memory performance by restricting the GPU to 8 outstanding requests and prioritizing all latency-sensitive CPU requests. When compared to Figure 2, the QoS mechanisms successfully prevent CPU starvation and CPU performance is not impacted by the GPU. With this static QoS, however, the GPU suffers. When mcf and art are run together with the GPU, the GPU receives barely enough bandwidth to maintain the frame rate. With the higher bandwidth art-art CPU workload, frame deadlines are missed.

From this example and discussion, we can conclude that for the static QoS scheme to generally work for any workload scenario, the memory bandwidth needs to be over-provisioned for the worst case. Otherwise, it is possible that frames must be dropped, either while reconfiguring or while programming. A better alternative to costly over-provisioning, is to identify when the GPU should be allowed to nearly monopolize bandwidth to meet its real-time constraints, which we discuss in the next section.

3. DYNAMIC QUALITY-OF-SERVICE

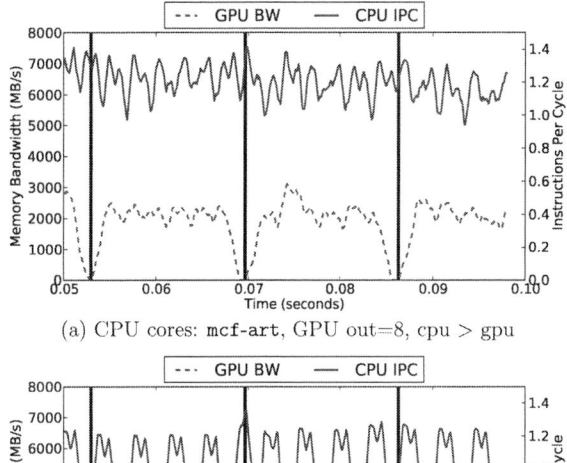

(a) CPU cores: `mcf-art`, GPU out=8, cpu > gpu

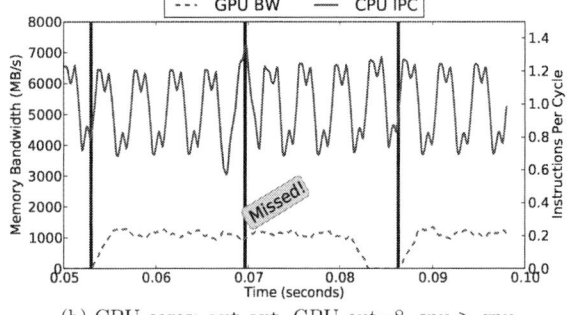

(b) CPU cores: `art-art`, GPU out=8, cpu > gpu

Figure 3: GPU activity (bandwidth consumption) and CPU performance. GPU is restricted to at most 8 outstanding memory requests and CPU requests are given higher priority. Vertical lines represent frame deadlines. Experimental setup discussed in Section 4.

Static QoS mechanisms lack the ability to adapt to the dynamic behavior of real workloads, resulting in either degraded CPU performance (Figure 2), or missed GPU deadlines (Figure 3(b)). In order to achieve high CPU performance while satisfying real-time constraints, we propose to dynamically adjust the QoS policy based on runtime workload characteristics. Ideally, CPU requests should be prioritized as long as the CPU does not compromise the GPU target frame rate. The key to achieving behavior that is near this ideal is to identify when a deadline is likely to be missed and only then adjust the QoS policy and either treat the GPU and CPU as equals, or even prioritize GPU requests. We discuss how to predict when the GPU makes insufficient progress and a heuristic to adjust priority below.

3.1 Monitoring GPU workload progress

As discussed in Section 2.2, mobile GPUs typically partition the screen into equal-sized tiles and process them in order. Each tile is processed once for all primitives that overlap it, then it is not accessed again until the following frame. We exploit this to track the progress the GPU is making in the current frame. The GPU hardware is aware of how many tiles in total it must process, the order in which tiles are processed, and what tiles are currently active. Progress is thus simply the current position within the total frame, as described by Equation 1. This information can readily be communicated from the GPU to the memory controller to affect the QoS policy.

$$FrameProgress = \frac{Number\ of\ tiles\ rendered}{Number\ of\ tiles} \quad (1)$$

Although this progress monitoring mechanism is simple, it is very effective in our system because the mobile GPU uses fine-grained tiles. With coarser tiles or non-tiled GPU architecutres, a more sophisticated estimation of workload can be used, such as those suggested by prior work in the context of coarse-grained adjustments to the GPU voltage and frequency [5, 14] [2] .

3.2 Dynamic QoS policy

To determine the QoS policy, the memory controller compares the frame progress rate, obtained above, with the *expected progress rate*. The expected progress rate can be calculated by dividing the elapsed from the beginning of the frame by the target frame time, (e.g. $16.67ms$ for 60 frames-per-second (FPS)) as shown in Equation 2. As with tracking progress, more sophisticated techniques can be used to obtain higher accuracy estimates of expected progress [5].

$$ExpectedProgress = \frac{Time\ elapsed\ in\ current\ frame}{Target\ frame\ time} \quad (2)$$

The memory controller then chooses a QoS policy based on how far the GPU is behind its expected progress point. Algorithm 1 shows an example dynamic QoS policy, which we employ in this paper and which works well in our experiments. There are two priority levels, and the CPU gets the higher priority as long as the current GPU progress rate is above the expected rate. When the progress falls behind the expected rate, GPU priority is increased to equal that of the CPU. When only 10% of the frame time remains until the deadline and if the GPU has not yet caught up to its expected point, the GPU is prioritized above the CPU in an attempt to make the frame deadline. This 10% buffer was chosen arbitrarily and can be tuned for better performance. Again, we favored a simple design that can demonstrate the benefits and importance of the dynamic approach. We leave refinements of this QoS selection algorithm to future work.

Algorithm 1 Dynamic QoS policy

> **if** $FrameProgress > ExpectedProgress$ **then**
> $CPU_{priority} = High$
> $GPU_{priority} = Low$
> **else if** $ExpectedProgress > 0.9$ **then**
> $CPU_{priority} = Low$
> $GPU_{priority} = High$
> **else**
> $CPU_{priority} = GPU_{priority}$
> **end if**

Figure 4(a) demonstrates how a static mechanism that prioritizes the CPU can lead to a missed GPU deadline. With our dynamic scheme, GPU priority is dynamically increased to enable it to meet its deadlines (Figure 4(b)). In the first frame, the GPU makes acceptable progress most of the time even with CPU priority. In the second frame,

---

[2]Prior work estimates workload at frame granularity, and does not discuss monitoring in-frame progress dynamically. An additional hardware counter is needed to keep track of in-frame progress, such as number of geometries processed. Our tile-based monitor does not need any additional hardware, as the tile bookkeeping is an integral part of the GPU's job management.

however, the GPU requires equal priority for much of the frame and higher priority towards the end of the frame to ensure the deadline is met.

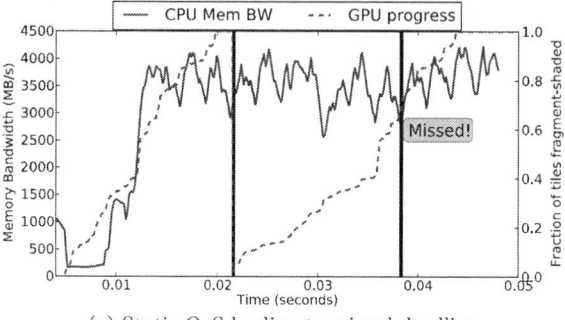

(a) Static QoS leading to missed deadline

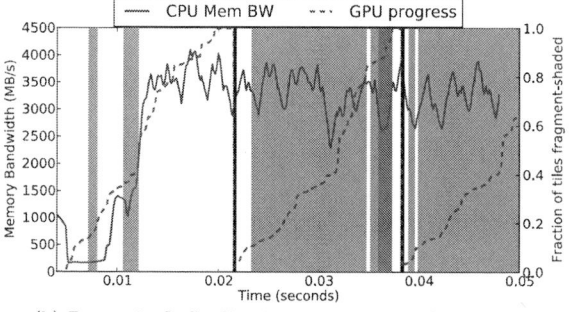

(b) Dynamic QoS adjusting priority to make deadline

Figure 4: CPU memory bandwidth consumption and GPU progress over several frames for static and dynamic QoS. Time intervals shaded in light red indicate times that GPU progress was insufficient and CPU and GPU priority are equal. Dark red shading indicates the critical periods of time that the dynamic scheme prioritizes GPU requests over CPU accesses.

4. EVALUATION METHODOLOGY

We evaluate our proposed dynamic QoS scheme using cycle level simulations. We use a combination of the gem5 system simulator [4], a proprietary next-generation GPU simulator, and the DrSim DRAM simulator [6]. The gem5 out-of-order CPU model and the GPU model share the DRAM model through the gem5 bus. DrSim models memory controllers and DRAM modules faithfully, simulating the buffering of requests, scheduling of DRAM commands, contention on shared resources (such as address/command and data buses), and all latency and timing constraints of LPDDR2 DRAM.

System configuration

The QoS schemes we simulate include uncontrolled CPU and GPU (noqos), static CPU priority over GPU (static), and our dynamic scheme (dynamic). Constraining the number of outstanding GPU requests to N (outN) is used in combination with static and dynamic.

Table 1 summarizes the parameters of our simulated systems. We believe the simulated system is representative of the next-generation high-end mobile SoC. Memory scheduling queues are large enough to guarantee room for CPU requests even when GPU was not constrained in noqos.

853

CPU	Dual-core, 1.2GHz ARM out-of-order superscalar
Caches	32KB private L1 I/D, 1MB shared L2
GPU	8 Unified shader cores, 600MHz
GPU L2	128KB shared
System bus	128-bit wide, 1GHz
Memory	FR-FCFS scheduling, open row policy
controller	64 entries read queue, 64 entries write queue
Main memory	1 channel, 1 rank / channel, 8 banks / rank
	4 x16 LPDDR2-1066 chips / rank
	8.3GB/s peak BW, All chip parameters from the
	latest Micron datasheet [10]
	XOR-interleaved bank index [17]

Table 1: Simulated system parameters

GPU workload	Source	Target FPS
taiji	3DMarkMobile ES 2.0 [1]	60
egypt	GLBenchmark [8]	60
taiji1080p	3DMarkMobile ES 2.0	30
farcry	Game	30

CPU workload	Average Mem Bandwidth
art-art	5.8GB/s
mcf-art	3.5GB/s
mcf-mcf	800MB/s

Table 2: Workloads used and their characteristics.

Workloads

Due to the slow GPU simulation speed, it is impractical to run a GPU accelerated application and other memory intensive applications on top of the full OS and GPU driver stack. Instead, we run CPU workloads on the CPU cores in parallel with graphics workloads on the GPU to approximate the memory bandwidth constrained usage scenario.

Table 2 shows the CPU and the GPU benchmarks used. We selected two SPEC CPU 2000 benchmarks with the MinneSPEC input set [9], which place significant demand on the memory system. Dual-core CPU workloads are multi-programmed to simulate three levels of CPU memory bandwidth usage. GPU workloads are post-driver output of a representative frame from each graphics benchmark; taiji and egypt are WVGA resolution and taiji1080p and farcry are 1080p. Their target performance in frames-per-sercond (FPS) was determined by measuring the execution time on GPU without CPU interference. Two workloads, taiji1080p and farcry were not able to finish in 16.67ms on our simulated system, but finished within 33.34ms. We assume that missed frames are skipped and the behavior is repetitive, so the target FPS is an integer divisor of the 60 FPS base.

5. RESULTS

In this section we present experimental results that demonstrate the effectiveness of the dynamic mechanism. We focus on challenging workloads that are constrained by the available memory bandwidth of the SoC. In such cases, it is impossible to simultaneously meet the target frame rate and service the CPU without GPU interference. We analyze the interaction between the components and show how dynamic can adapt to the changes in workload demand combination and provide the near-optimal QoS strategy, while current guidelines for static QoS fail.

To better quantify these interactions, we show the results of multiple QoS schemes for GPU and CPU performance in Figure 5 and Figure 6, respectively. The results point to two important insights, which contradict the current best practice of prioritizing the CPU and constraining the GPU.

First, restricting the number of outstanding GPU requests can cause the GPU to miss deadlines, while at the same

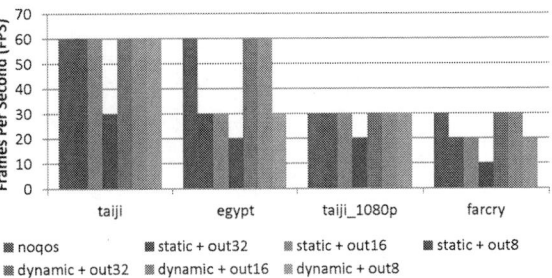

Figure 5: GPU performance in frames per second (FPS) when both CPU cores run art.

time often *degrading* CPU performance. As shown in Figure 5, restricting the GPU to 8 outstanding requests reduces the frame rate by 33% or 50%, dropping one of every three or two frames respectively. The impact on the CPU is interesting. Figure 6 shows that as long as the GPU meets its deadlines (configurations for which the GPU fails are shaded black in Figure 6) the CPU either sees little benefit from constraining the GPU, or experiences noticeable performance degradation. We found that having fewer GPU requests for the memory scheduler to choose from prevents efficient scheduling and reduces the effective memory bandwidth. Both the GPU and CPU cores suffer from the longer low effective bandwidth period.

The CPU benefits from a constrained GPU only when the GPU fails to meet its required frame rate (black bars). For example, restricting the GPU to 8 outstanding accesses leads to only a 5% degradation in CPU performance in farcry-art-art. The GPU however, only achieves 10 FPS instead of the targeted 30 FPS. This is generally unacceptable.

Second, our GPU progress-aware dynamic QoS mechanism can indeed adapt to the changes in CPU and GPU workloads and provide the performance of the best QoS setting for each workload. When bandwidth is not sufficient for the workloads (egypt and farcry in Figure 5), only noqos and dynamic meet GPU performance requirements. Even in such severely bandwidth-constrained cases, dynamic can still find opportinity to give the CPU some priority and reduces slowdown by 3.6% and 6.9% from noqos. When there is sufficient bandwidth to serve the GPU and CPU cores simultaneously, CPU performance of dynamic + out32 roughly matches the best static configuration of each workload (shown as the lowest non-black bar in Figure 6) and provides up to 9.4% reduction in slowdown from noqos.

Again, constraining the outstanding number of requests from the GPU hurts CPU performance even more for dynamic with egypt and mcf-art. It turns out that the constraint slows down GPU progress and our dynamic scheme forces the memory controller to raise GPU priority, and therefore the CPU suffers. Since mcf-art uses a moderate amount of memory bandwidth, GPU requests can be issued opportunistically even with low priority. Therefore, having many outstanding requests at the memory controller ready allows them to be scheduled efficiently and enables the GPU to make good progress. In the dynamic + out32 configuration, the memory controller doesn't raise GPU priority and the CPU gets low-latency priority accesses, yielding better performance.

6. CONCLUSION

In this paper we carefully analyzed the performance of a system that is representative of current and upcoming ad-

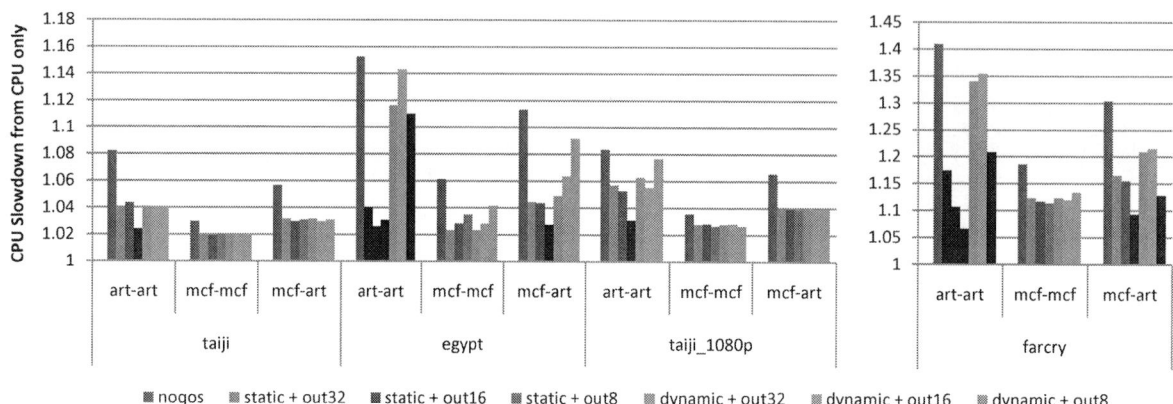

Figure 6: Slowdown of the CPU relative to its performance in an SoC with the same memory configuration but no GPU. Black bars represent configurations in which the GPU could not meet the workload's performance target.

vanced SoCs. We used a cycle-level simulator that accurately models an SoC with two CPU cores and a mobile GPU, which all share a single DRAM main memory system. By evaluating the complex interactions between these components we show that current best-practice QoS mechanisms are insufficient and often apply the wrong QoS policy. We determine that there is no single static policy that can be used to simultaneously meet the requirements of the GPU without significantly impacting the CPU cores.

We use this insight to develop a dynamic QoS scheme that maintains CPU priority when possible, but shifts priority towards the GPU if it predicts that the GPU will miss a real-time deadline. We propose a simple, yet effective, tile based frame progress tracking mechanism to enable dynamic QoS policy decisions and show that it both enables the GPU to meet its deadlines and minimizes impact on the CPU. Using our technique, we also conclude that restricting the number of outstanding GPU requests, a static QoS mechanism in use today, often degrades the performance of all cores, because it limits the ability of the memory scheduler to exploit locality.

These conclusions are important and open the way to additional research. This paper used a GPU and CPU as an example of cores with conflicting demands: latency-sensitive best-effort CPU cores, and a bandwidth-sensitive real-time GPU. These diverse requirements are common and a growing number of cores share an increasingly constrained memory system. We believe dynamic techniques, such as the one we present, are the key to enabling such future systems to meet user requirements while still efficiently utilizing scarce shared resources.

7. REFERENCES

[1] 3DMarkMobile ES 2.0.
http://www.futuremark.com/products/3dmarkmobile,
2011.

[2] B. Akesson, K. Goossens, and M. Ringhofer. Predator: A Predicatable SDRAM Memory Controller. In *Proceedings of the 5th IEEE/ACM international conference on Hardware/software codesign and system synthesis - CODES+ISSS '07*, page 251, New York, New York, USA, Sept. 2007. ACM Press.

[3] R. Ausavarungnirun, G. Loh, K. Chang, L. Subramanian, and O. Mutlu. Staged memory scheduling: Achieving high performance and scalability in heterogeneous systems. In *Proc. the 39th Ann. Int'l Symp. Computer Architecture (ISCA)*, ISCA '12, New York, NY, USA, 2012. ACM.

[4] N. Binkert, B. Beckmann, G. Black, S. K. Reinhardt, A. Saidi, A. Basu, J. Hestness, D. R. Hower, T. Krishna, S. Sardashti, R. Sen, K. Sewell, M. Shoaib, N. Vaish, M. D. Hill, and D. A. Wood. The gem5 simulator. *SIGARCH Comput. Archit. News*, 39:1–7, Aug. 2011.

[5] Y. Gu and S. Chakraborty. A Hybrid DVS Scheme for Interactive 3D Games. In *2008 IEEE Real-Time and Embedded Technology and Applications Symposium*, pages 3–12. IEEE, Apr. 2008.

[6] M. K. Jeong, D. H. Yoon, and M. Erez. DrSim: A platform for flexible DRAM system research.
http://lph.ece.utexas.edu/public/DrSim.

[7] T. Karkhanis and J. E. Smith. A day in the life of a data cache miss. In *Workshop on Memory Performance Issues*, 2002.

[8] Kishonti Informatics Ltd. GLBenchmark.
http://www.glbenchmark.com, 2011.

[9] A. J. KleinOsowski and D. J. Lilja. Minnespec: A new spec benchmark workload for simulation-based computer architecture research. *IEEE Comput. Archit. Lett.*, 1:7–, January 2002.

[10] Micron Corp. *Micron 2 Gb ×16, ×32, Mobile LPDDR2 SDRAM S4*, 2011.

[11] O. Mutlu and T. Moscibroda. Stall-time fair memory access scheduling for chip multiprocessors. In *International Symposium on Microarchitecture*, pages 146–160, 2007.

[12] K. Nesbit, N. Aggarwal, J. Laudon, and J. Smith. Fair queuing memory systems. In *Proceedings of the 39th Annual IEEE/ACM International Symposium on Microarchitecture*, pages 208–222. IEEE Computer Society, 2006.

[13] S. Rixner, W. J. Dally, U. J. Kapasi, P. R. Mattson, and J. D. Owens. Memory access scheduling. In *Proc. the 27th Ann. Int'l Symp. Computer Architecture (ISCA)*, Jun. 2000.

[14] B. Silpa, G. Krishnaiah, and P. R. Panda. Rank based dynamic voltage and frequency scaling for tiled graphics processors. In *Proceedings of the eighth IEEE/ACM/IFIP international conference on Hardware/software codesign and system synthesis*, CODES/ISSS '10, pages 3–12, New York, NY, USA, 2010. ACM.

[15] A. Stevens. Qos for high-performance and power-efficient hd multimedia. Technical report, Arm, 2010.

[16] A. Tune and A. Bruce. How to tune your SoC to avoid traffic congestion. In *DesignCon*, 2010.

[17] Z. Zhang, Z. Zhu, and X. Zhang. A permutation-based page interleaving scheme to reduce row-buffer conflicts and exploit data locality. In *Proc. the 33rd IEEE/ACM Int'l Symp. Microarchitecture (MICRO)*, Dec. 2000.

Metronome: Operating System Level Performance Management via Self–Adaptive Computing

Filippo Sironi[1,2], Davide B. Bartolini[1], Simone Campanoni[3], Fabio Cancare[1]
Henry Hoffmann[2], Donatella Sciuto[1], Marco D. Santambrogio[1,2]

[1]Politecnico di Milano, [2]Massachusetts Institute of Technology, [3]Harvard University
{sironi, bartolini}@elet.polimi.it, xan@eecs.harvard.edu, cancare@elet.polimi.it
hank@csail.mit.edu, {sciuto, santambrogio}@elet.polimi.it

ABSTRACT

In this paper, we present Metronome: a framework to enhance commodity operating systems with self-adaptive capabilities. The Metronome framework features two distinct components: Heart Rate Monitor (HRM) and Performance–Aware Fair Scheduler (PAFS). HRM is an active monitoring infrastructure implementing the observe phase of a self–adaptive computing system Observe–Decide–Act (ODA) control loop, while PAFS is an adaptation policy implementing the decide and act phases of the control loop. Metronome was designed and developed looking towards multi–core processors; therefore, its experimental evaluation has been carried on with the PARSEC 2.1 benchmark suite.

Categories and Subject Descriptors

C.4 [**Performance of Systems**]: *Measurement techniques, Performance attributes*; D.4.1 [**Operating Systems**]: Process Management—*Scheduling*; D.4.8 [**Operating Systems**]: Performance—*Measurements, Monitors*

General Terms

Design, Management, Measurement, Performance

Keywords

Self-Adaptive Computing, Operating Systems, Performance Management

1. INTRODUCTION

In the recent years, the demands in terms of computing performance, functionality, reliability, availability, and serviceability has grown exponentially, raising the overall complexity of the hardware/software execution stack [15]. Hardware developers multiply the amount of resources (i.e., cores count, memory size, etc.), making them more and more heterogeneous and posing an ever–increasing burden on both

system and application developers. This is even more evident in the embedded systems domain, where capabilities may present huge variations among different system configurations. Moreover, embedded systems might be required to operate continuously for years in possibly uncertain conditions where some of the environmental characteristics might affect the behavior of the system. As an example, consider a mobile phone: it must consume the least possible amount of battery power while operating at different signal strengths and perhaps with different signal types (i.e., GSM, EDGE, 3G, etc.). The amount of requirements and constraints is gigantic.

One approach to simplify the duty of application developers is the adoption of *autonomic* or *self–adaptive computing* [20] through *self–adaptive hardware* [19, 7] and *self–adaptive software* [21]. Self–adaptive systems (i.e., systems employing either self-adaptive hardware or software) rely on control loops to adjust their behavior to internal and environmental changes. Such systems are required to observe themselves and the environment, decide on a sequence of actions to perform, and apply them in order to optimize their operations. The process of observing, deciding, and acting is referred to as either Observe–Decide–Act (ODA) or Monitor–Analyze–Plan–Execute with Knowledge (MAPE–K) control loop.

In this paper we make the following contributions:

- We present *Metronome*, a framework for self-adaptive computing constituting an implementation of the ODA control loop to enhance commodity operating systems. The reference implementation of Metronome is publicly available as free software[1].

- We compare our active monitoring infrastructure, i.e., *Heart Rate Monitor* (HRM), with an open source, state–of–the–art solution designed over the same concept, namely Application Heartbeats [16]. We show a considerable reduction of the worst-case overhead by a factor greater than $160\times$, thanks to our efforts in considering multi–core processors–related issues such as synchronization and cache sharing.

- We present an adaptation policy, called *Performance–Aware Fair Scheduler* (PAFS), to implement the decide and act phases of the ODA control loop. PAFS is an adaptive scheduling infrastructure relying on HRM to take informed decisions and actions. We experimentally evaluated PAFS through concurrent runs of a subset of the PARSEC 2.1 benchmark suite [6].

---

[1]http://www.changegrp.org/acos/.

The remainder of this paper is organized as follows. Section 2 discusses related work. Section 3 presents the proposed solution to implement the ODA control loop to enhance a commodity operating system. Section 4 and Section 5 describe respectively HRM and PAFS. Finally, Section 6 concludes the paper.

2. RELATED WORK

At the beginning of 2000, IBM published the autonomic computing manifesto, proposing a vision [20] into which computing systems manage themselves according to user–defined goals and system–defined constraints; lately, autonomic computing systems have been referred to as adaptive or self–* computing systems. The big idea is to ease the work of system and application developers in exploiting the available amount of resources in accordance with user–defined goals and system–defined constraints. Software implementations include a language and compiler for algorithmic choice with auto–tuning capabilities [4], a framework to statistically guarantee the Quality of Service (QoS) trading performance for energy and vice versa [5], a self–tuning scheduler to guarantee the QoS for soft real–time applications [12], a framework for adaptive data structures and algorithms selection [24], and a self–adaptive synchronization library [14]. More recently, a set of more comprehensive solutions like PowerDial [18] and SEEC [17] were presented.

Examining to a greater extent the self–adaptive computing literature, the observe phase of the ODA control loop received a lot of attention and contributions. With the recent advances, processors significantly improved the capabilities of Performance Monitoring Units (PMUs). However, alongside with capabilities, the complexity of PMUs rose too, making the tasks of understanding and using them progressively more difficult [23]. Various approaches using PMUs [9, 23] are inadequate to capture user–defined goals and system–defined constraints; in addition, they are also difficult to port between platforms and operating systems. Performance and Evaluation Monitoring (PEM) [10] is an enabling technology for autonomic computing systems and Continuous Program Optimization trying to abstract and extend PMUs. On one hand, the proposed infrastructure is complete (i.e., potentially supporting multiple programming languages) and has been successfully ported on the K42 research operating system [22]. On the other hand, PEM poses a notable burden on system and application developers, who must provide both an XML specification of the events they want to track and the code to handle the event.

Application Heartbeats [16] is an active monitoring infrastructure designed and developed for self–adaptive computing. It provides application developers a way to expose user–defined performance goals and a method to signal execution progresses; both the performance goals and measures are then made available to system developers. Application Heartbeats has the unquestionable advantage of being extremely simple with respect to other solutions.

3. METHODOLOGY

Our methodology requires partitioning the components that make the system self-adaptive in three distinct classes: *applications*, *monitoring infrastructures*, and *adaptation policies*. An application is an element capable of making one or more entities of the system aware of its goals and progresses;

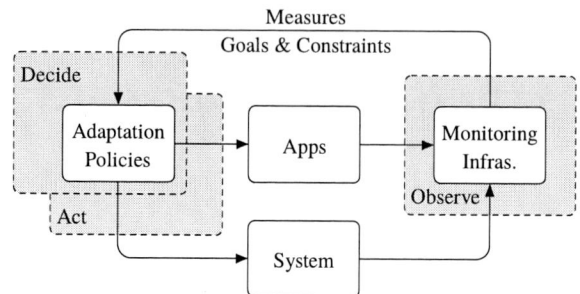

Figure 1: System architecture diagram showing the relations incurring among system, applications, monitoring infrastructures, and adaptation policies.

we refer to those applications that do not provide such information as *legacy applications*. A monitoring infrastructure is an entity equipped with sensors able to gather information from applications or from the system. Within this context, it is important to notice that goals and constraints are defined using data measurable by means of a monitoring infrastructure. An adaptation policy is an element whose purpose is to observe applications through monitors, decide on a strategy to change the overall behavior of applications or the system, and act via a set of predefined actuators, with the objective of meeting goals and satisfying constraints. A *self-optimizing application* is a special kind of application in which the roles of application and adaptation policy coexist. The three distinct classes of components cooperate to establish the ODA control loops as highlighted in Figure 1.

Two distinct roles take shape: application developers and system developers. Application developers create applications and, if required, instrument them to provide user-defined goals and execution progresses to the self-adaptive system. System developers design and implement monitoring infrastructures and adaptation policies; monitoring infrastructure can be either active (i.e., requiring to manually instrument the applications) or passive, and allow the self-adaptive system to collect as much information as possible, while the adaptation policies provide as many ways as possible to change the behavior of the self-adaptive system.

We propose Metronome, a framework capable of exploiting the availability of user-defined performance goals and measures collected through Heart Rate Monitor (HRM) to enhance the scheduling infrastructure of a commodity operating system by means of Performance–Aware Fair Scheduler (PAFS). PAFS is an adaptation policy demonstrating the applicability of the methodology over the Completely Fair Scheduler (CFS), the scheduling infrastructure of the Linux kernel. Extending a commodity operating system kernel such as Linux comes naturally, since it is widespread and natively collects most of the information to take informed decisions and actions.

4. HEART RATE MONITOR

The ideas behind Heart Rate Monitor (HRM) resemble those of the Application Heartbeats Application Programming Interface (API) and exploit the well–known idea of *heartbeat*, already used in the past for measuring performance, expressing progresses, and signaling availability [11]. Hoffmann et al. [16] proposed the Application Heartbeats

API. It is a simple yet effective interface [14, 18, 17] for application developers to express both performance goals and execution progresses and for system developers to retrieve performance measures.

The Application Heartbeats API intentionally leaves some behavior undefined so that it can be customized for the needs of particular implementations. Performance–Aware Fair Scheduler (PAFS), for example, needs to be able to access the heartbeat data from within the kernel at extremely low latency. To meet this goal, we develop a partitioned implementation of the API, in which the kernel makes shared pages available for storing heartbeat data. These pages can be accessed much more quickly from the kernel–space than the POSIX shared pages used by the reference implementation of the Application Heartbeats API [1]. Taking advantage of the opportunity to customize the implementation allows a much higher performance implementation of PAFS, and other potential kernel–space adaptation policies.

HRM is an active monitoring infrastructure integrated within Linux, which slightly revise the interface of Application Heartbeats API. HRM provides and high performance implementation supporting diverse parallelization models (i.e., multiple processes, multiple threads, or any feasible combination) and avoiding synchronization. The design of HRM makes its porting to new platforms a negligible task[2] without losing in functionality. HRM comes with a slightly modified API with respect to the Application Heartbeats API. However, it still allows application developers to easily instrument applications and system developers to build both user and kernel–space adaptation policies. The interaction model between applications and adaptation policies can be seen, similarly to PEM [10], as a producer/consumer model in which applications work as producers and adaptation policies work as consumers, with the monitoring infrastructure in the middle.

4.1 Definitions

This section provides a set of definitions to better understand the remainder of this paper.

A running instance of a program, including both its code and data, is called a *process*; a unique Process IDentifier (PID) identifies a process in Linux. A thread is a finer grained unit of execution and conceptually exists within a process, sharing both the code and the data with other sibling threads; in Linux, a unique Thread IDentifier (TID) identifies a thread. A *task* is any unit of execution, being either a process or a thread. Given these definitions, an *application* can be defined as a set of tasks pursuing a set of objectives (e.g., encoding an audio/video stream). Being a set of tasks, an application can be single–threaded, multi–threaded, multi–processed, or any combination of them and a monitoring infrastructure should account for this.

A *heartbeat* is a signal emitted by any task of an application at a certain point in the code indicating execution progresses. A *hot–spot* is a performance–relevant portion of code executed by any task of an application and usually abstracts the most time consuming portion of an application. It is useful to define the concept of *group*[3]; a group is a subset

of an application's tasks pursuing a common objective (e.g., encoding a video stream in audio/video encoder). Groups are non–intersecting subsets meaning a task belongs to only one group at a time. It is important to notice how such a constraint does not neglect the existence of multi–grouped applications (e.g., a group encoding the audio stream and a group encoding the video stream in an audio/video encoder), a case Application Heartbeats completely neglects. The concept of group is key to support diverse parallelization models, the only thing needed is to link a task, being it a process or a thread, to a group. Within HRM, a unique Group IDentifier (GID) identifies a group. Given the definitions of hot–spot and group, it is possible to define a many–to–one relations between such entities. Each of the tasks belonging to a group executes the same hot–spot, which is characterized by its *heartbeats count, performance measures,* and *performance goal*. The heartbeats count is linked to the number of times the hot–spot is executed. Performance measures are expressed in heartbeats per second and capture the concept of heart rate, which is the frequency at which tasks emit heartbeats. The performance goal is expressed as a desired heart rate range, delimited by a *minimum heart rate* and a *maximum heart rate*.

4.2 Evaluation

The implementation of HRM is an extension of Linux. In the remainder of this section, we compare HRM to the reference implementation of the Application Heartbeats API [1] looking towards efficiency. Experimental results were collected on a workstation equipped with a single Intel Core i7–870 quad–core processor running at 2.97 GHz featuring 8 MB of shared LLC (L3), 4 GB of DDR3–1066 non–ECC RAM, and a 500 GB 7200 RPM SATA2 hard disk. Advanced features such as Intel Hyper–Threading Technology, Intel Turbo Boost Technology, and Enhanced Intel Speed-Step Technology were disabled. The AMD64 version of Debian 6.0, alias "squeeze", was configured to run the Linux kernel [3] 2.6.35.13 extended with HRM.

We evaluated the overhead of the two monitoring infrastructures through a multi–threaded micro–benchmark. The micro–benchmark allows specifying the level of parallelism (i.e., the number of threads to spawn) and the amount of heartbeats to emit. Since the hot–spot of this application is a tight loop emitting heartbeats, the heart rate (i.e., the throughput) quantifies the overhead of the employed monitoring infrastructure: the higher the throughput, the lower the overhead.

Figure 2 shows the throughput emitting 1 million heartbeats varying the amount of parallelism from 1 to 8 threads. The experimental results yield evidence of how HRM outperforms the reference implementation of the Application Heartbeats API. HRM scales almost perfectly with the number of threads. As expected, the peak performance of HRM is obtained with 4 threads, which saturate the quad–core processor of the workstation. According to the experimental results, HRM poses a worst–case overhead between 1 and 2 orders of magnitude (up to a 160× factor) lower than the reference implementation of the Application Heartbeats API. We argue the advantage is due to the multi–core processor aware design accounting for issues such as synchronization and cache sharing [25].

The reference implementation of the Application Heartbeats API employs a protected shared data structure; syn-

---

[2] The design of HRM makes the porting process from Linux to other kernels (e.g., BSD kernels) a straightforward task.

[3] This definition of group does not relate to any other group currently supported in Linux nor in other UNIX–like operating systems.

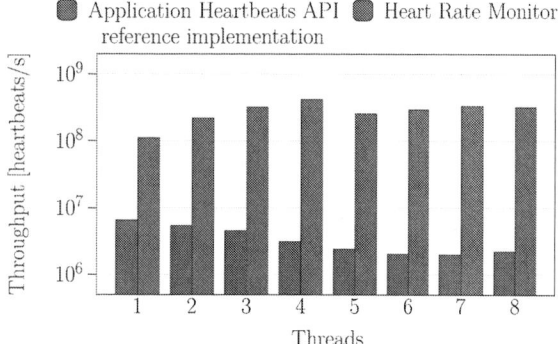

Figure 2: Average micro–benchmark throughput emitting 1M heartbeats with [1, 8] threads over 1000 executions. Higher is better.

chronization guarantees its consistency. The emission of heartbeats requires synchronization among threads, imposing their serialization, potentially compromising applications scalability and it is synchronous with the performance measures computation. Moreover, time–consuming boundary crosses between user and kernel–space occur when retrieving the wall–clock time. Conversely, HRM adopts a data structure distributed among all the threads of a group and across the user and kernel–space boundary. The emission of heartbeats reduces to an *atomic* increment of a cache line–aligned per–thread counter, while a kernel–space high–precision timer handler computes the performance measures asynchronously with a MapReduce–like model [13]. The asynchronous computation of performance measures avoids time–consuming boundary crosses.

5. PERFORMANCE–AWARE FAIR SCHEDULER

Within an operating system, the scheduling infrastructure is the component in charge of determining the allotment of the available computational resources to the running tasks. The choice of the policy or policies ruling the scheduling infrastructure can highly impact the behavior of the system, favoring either run time (i.e., throughput), latency (i.e., response time), or overall fairness (i.e., wait time). This trade-off can usually be statically tuned in commodity operating systems. Due to its high impact on the behavior of the system, the scheduling infrastructure represents a suitable component in which adaptive capabilities can be embedded, enabling it to pursue user-defined performance goals.

Performance–Aware Fair Scheduler (PAFS) is an adaptation policy extending the scheduling infrastructure of Linux; more precisely, it enhances Completely Fair Scheduler (CFS), which is one of the subclasses of the hierarchical scheduling infrastructure found in Linux. PAFS exploits the information provided by HRM (i.e., performance measures and user-defined performance goals) to introduce *performance-awareness*, a new factor that is taken into account when defining *fairness*. At first glance, PAFS can be misleadingly considered a sort of (soft) real–time scheduling infrastructure [8] with a QoS definition based on the user–defined performance goals. However, since we extended a *best–effort* scheduling infrastructure without altering all of its desirable properties (e.g., management of both legacy and non–legacy

applications, non-starvation, absence of admission control, etc.), PAFS cannot provide any guarantees on matching user–defined performance goals.

5.1 Design

The definition of fairness of CFS regards processor time; the basic idea is simple: being fair in providing processor time to tasks. When the time for tasks is out of balance (i.e., one or more tasks are not given a fair amount of time relative to others), then those out-of-balance tasks should be given time to execute. CFS maintains the amount of processor time provided to a given task in what is called the *virtual run–time*. The smaller a task's virtual run–time the higher its need for the processor[4].

All runnable tasks are sorted in a time–line implemented with a red–black tree[5] according to the key value reported in Equation (1), where i represents the i–th task, $vruntime_i$ is the virtual run–time of the i–th task, and $vruntime_{min}$ is minimum virtual run–time within the time–line.

$$vruntime_i - vruntime_{min} \qquad (1)$$

Tasks with the gravest need for processor time (i.e., lowest virtual run–time) are located toward the left side of the tree while tasks with the least need for processor time (i.e., highest virtual run–times) are located toward the right side of the tree. The scheduling infrastructure always picks up the left–most task in the time–line; the task makes use of its processor time, its virtual run–time is updated, and then it is put into the time–line again. In this way, tasks on the left side are given processor time and tend to migrate to the right side. The virtual run–time is updated as reported in Equation (2), where i represents the i–th task, $\Delta exectime_i$ is the execution time spent by the i–th task, w_i is the weight associated with the *nice* value of the i–th task, and w_0 is the weight associated with nice value 0.

$$vruntime_i = vruntime_i + \frac{\Delta exectime_i}{w_i} \times w_0 \qquad (2)$$

PAFS harnesses this infrastructure and, at each update of the virtual run–time of a non–legacy application task, it weighs the execution time using a performance–aware indicator, according to Equation (3), $g(i)$ represents the group (i.e., the HRM group) containing the i–th task, $\Pi_{g(i)}$ is the performance–aware indicator of the $g(i)$ group, $heart\_rate_{g(i)}$ is the current heart rate of the $g(i)$ group, and $\overline{heart\_rate}_{g(i)}$ is computed according to Equation (4) and represents the user–defined performance goal. In Equation (4) $min\_heart\_rate_{g(i)}$ and $max\_heart\_rate_{g(i)}$ are the lower bound and the upper bound of the desired heart rate window.

$$\Pi_{g(i)} = \frac{heart\_rate_{g(i)}}{\overline{heart\_rate}_{g(i)}} \qquad (3)$$

$$\overline{heart\_rate}_{g(i)} = \frac{min\_heart\_rate_{g(i)} + max\_heart\_rate_{g(i)}}{2} \qquad (4)$$

---

[4]The definition of fairness of CFS also accounts for sleeping tasks (e.g., tasks waiting for I/O operations); sleeping tasks receive a fair amount of processor time when they eventually need it.

[5]A red-black tree is a self-balancing binary tree where the left–most leaf has the smallest key value.

859

The performance–aware indicator is greater than 1 when the $g(i)$ group's current heart rate is over the middle of the desired heart rate window), it is bounded between 0 and 1 when the $g(i)$ group's current heart rate is below the middle of the desired heart rate window. When the $g(i)$ group is either over (i.e., $heart\_rate_{g(i)} > max\_heart\_rate_{g(i)}$) or under (i.e., $heart\_rate_{g(i)} < min\_heart\_rate_{g(i)}$) performing, the virtual run–time of tasks belonging to non–legacy applications is updated as reported in Equation (5), while it is still updated as reported in Equation (2) if the $g(i)$ group is performing within the desired heart rate window (i.e., $min\_heart\_rate_{g(i)} < heart\_rate_{g(i)} < max\_heart\_rate_{g(i)}$).

$$vruntime_i = vruntime_i + \frac{\Delta execime_i \times \Pi_{g(i)}}{w_i} \times w_0 \quad (5)$$

Weighing the execution time spent by tasks belonging to non–legacy application through the performance–aware indicator either speeds up or slows down the migration of task from the left side to the right side of the time–line according to their performance measure and performance goal. Moreover, the non–starvation property of CFS is preserved, since we impose a minimum value grater than 0 for the performance–aware indicator.

5.2 Evaluation

The implementation of PAFS is an extension of CFS of Linux. Experimental results were collected using the same workstation described in Section 4.2; the AMD64 version of Debian 6.0, alias "squeeze", was configured to run Linux 2.6.35.13 extended with both the HRM and PAFS.

We evaluated PAFS with 3 different workloads, each synthesized using two 4–threaded applications from the PARSEC 2.1 benchmark suite. The first workload (i.e., *mix 1*) comprised *facesim* and *ferret*. Facesim is a virtual reality application simulating the underlying physics of a human face and generating corresponding frames. Ferret is a content–based similarity search application. The second workload (i.e., *mix 2*) consisted of *blackscholes* and *swaptions*. Blackscholes is an application to price portfolios of options using partial differential equations. Swaptions is an application to price portfolios of options using Monte Carlo experiments. The third workload (i.e., *mix 3*) embodied *facesim* and *fluidanimate*, which is a virtual reality application simulating incompressible fluids underlying physics.

We ran each workload with CFS (i.e., legacy execution) and PAFS (i.e., non–legacy execution). Regarding the non–legacy execution, we defined reachable performance goals, advantaging either one application or the other. Table 1 reports the execution time for each application and workload; PAFS consistently ends the execution of the entire workload before CFS. Figure 3 highlights how PAFS constantly reduces the normalized mean squared error between the performance measure (i.e., the current heart rate) and the performance goal (i.e., desired heart rate window).

Figure 4 provides insights regarding the first workload and PAFS behavior. The execution time of facesim is higher than the execution time of ferret; when ferret ends its execution, the performance measure of facesim grows with both CFS (see Figure 4a) and PAFS (see Figure 4b). The weighing of the virtual runtime is not enough to constrain the performance measure of an application when there is little to no contention. This behavior explains the high values of the normalized mean squared error reported in Figure 3 for

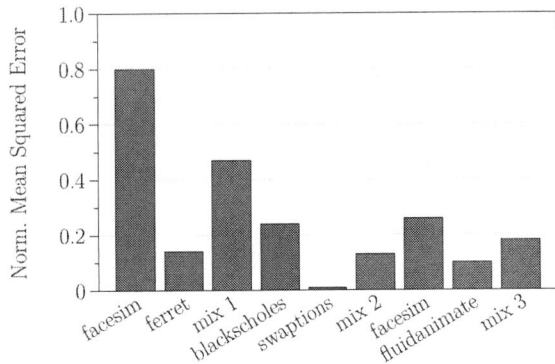

Figure 3: NMSE between the performance measure and goals of non–legacy executions with PAFS where 1.0 is the NMSE of the legacy execution with CFS. Lower is better.

both facesim and, consequently, mix 1.

6. CONCLUSIONS

This paper presented Metronome, a framework to enhance Linux with self–adaptive computing capabilities. Metronome features a monitoring infrastructure, namely HRM, and an adaptation policy, namely PAFS. HRM computes performance measures for non–legacy applications and allows them to expose user–defined performance goals. PAFS drives non–legacy applications towards meeting user–defined performance goals, modifying the scheduling infrastructure of Linux. HRM represents an improvement with respect to Application Heartbeats, an open source, state–of–the–art monitoring infrastructure, both in terms of functionality and performance. In addition, the methodology at the very base of Metronome decouples non–legacy applications from adaptation policies, which become completely transparent to them thanks to the presence of monitoring infrastructures, leaving a lot of space for further improvements.

7. REFERENCES

[1] Application Heartbeats. http://code.google.com/p/heartbeats/.

[2] Linux Programmer's Manual. http://kernel.org/doc/man-pages/.

[3] The Linux Kernel. http://www.kernel.org/.

[4] J. Ansel, C. Chan, Y. L. Wong, M. Olszewski, Q. Zhao, A. Edelman, and S. Amarasinghe. PetaBricks: A Language and Compiler for Algorithmic Choice. In *Proceedings of the 2009 ACM SIGPLAN Conference on Programming Language Design and Implementation*, 2009.

[5] W. Baek and T. M. Chilimbi. Green: A Framework for Supporting Energy–Conscious Programming using Controlled Approximation. In *Proceedings of the 2010 ACM SIGPLAN Conference on Programming Language Design and Implementation*, 2010.

[6] C. Bienia. *Benchmarking Modern Multiprocessors*. PhD thesis, Princeton University, 2011.

[7] R. Bitirgen, E. Ipek, and J. F. Martinez. Coordinated Management of Multiple Interacting Resources in Chip Multiprocessors: A Machine Learning Approach. In *Proceedings of the 41st Annual IEEE/ACM International Symposium on Microarchitecture*, 2008.

[8] S. A. Brandt, S. A. Banachowski, C. Lin, and T. Bisson. Dynamic Integrated Scheduling of Hard Real–Time, Soft Real–Time and Non–Real–Time Processes. In *24th IEEE*

Table 1: Standard mean and standard deviation of the execution time for each application and workload run with CFS and PAFS over 100 executions

Workload	Application	Completely Fair Scheduler		Performance–Aware Fair Scheduler	
		Std. Mean [s]	Std. Deviation [s]	Std. Mean [s]	Std. Deviation [s]
mix 1	facesim	**250.890**	**0.169**	**240.310**	**0.237**
	ferret	163.337	0.247	131.402	0.126
mix 2	blackscholes	92.819	0.059	**108.659**	**0.037**
	swaptions	**131.931**	**0.028**	103.431	0.017
mix 3	facesim	**237.206**	**0.673**	**227.559**	**0.679**
	fluidanimate	184.537	0.444	212.766	0.405

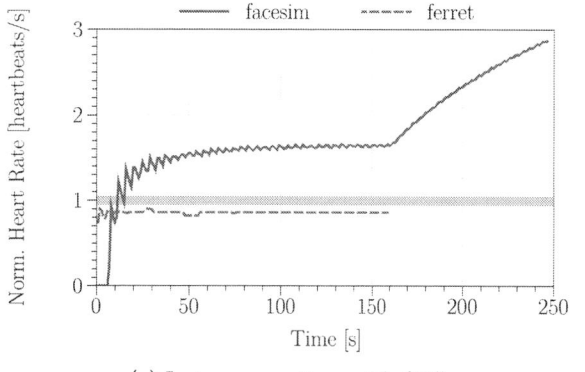

(a) Legacy executions with CFS.

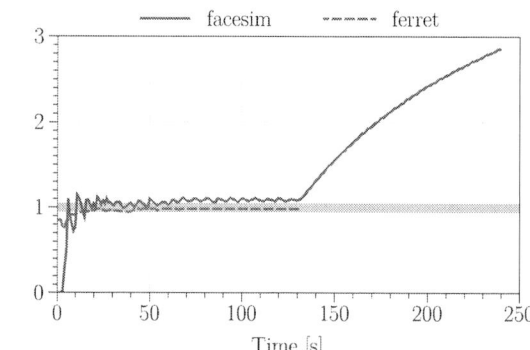

(b) Non–legacy executions with PAFS.

Figure 4: Normalized current heart rates for mix 1 where [0.95, 1.05] is the normalized performance goal (i.e., gray–shaded area). Nearer the gray–shaded area is better.

Real–Time Systems Symposium, 2003.

[9] S. Browne, J. Dongarra, N. Garner, G. Ho, and P. Mucci. A Portable Programming Interface for Performance Evaluation on Modern Processors. *International Journal of High Performance Computing Applications*, 14(3), 2000.

[10] C. Cascaval, E. Duesterwald, P. F. Sweeney, and R. W. Wisniewski. Performance and environment monitoring for continuous program optimization. *IBM Journal of Research and Development*, 50(2.3), 2006.

[11] W. Chen, S. Toueg, and M. Aguilera. On the Quality of Service of Failure Detectors. *IEEE Transactions on Computers*, 51(1), 2002.

[12] T. Cucinotta, F. Checconi, L. Abeni, and L. Palopoli. Self–tuning Schedulers for Legacy Real–Time Applications. In *Proceedings of the fifth European Conference on Computer Systems*, 2010.

[13] J. Dean and S. Ghemawa. MapReduce: Simplified Data Processing on Large Clusters. In *Proceedings of the 6th USENIX Symposium on Operating Systems Design and Implementation*, 2004.

[14] J. Eastep, D. Wingate, M. D. Santambrogio, and A. Agarwal. Smartlocks: Lock Acquisition Scheduling for Self–Aware Synchronization. In *Proceedings of the seventh International Conference on Autonomic Computing*, 2010.

[15] S. Fuller and L. Millett. Computing Performance: Game Over or Next Level? *Computer*, 44(1), 2011.

[16] H. Hoffmann, J. Eastep, M. D. Santambrogio, J. E. Miller, and A. Agarwal. Application Heartbeats: A Generic Interface for Specifying Program Performance and Goals in Autonomous Computing Environments. In *Proceedings of the seventh International Conference on Autonomic Computing*, 2010.

[17] H. Hoffmann, M. Maggio, M. D. Santambrogio, A. Leva, and A. Agarwal. SEEC: A Framework for Self–aware Management of Multicore Resources. Technical Report

MIT–CSAIL–TR–2011–016, Massachusetts Institute of Technology, Computer Science and Artificial Intelligence Laboratory, 2011.

[18] H. Hoffmann, S. Sidiroglou, M. Carbin, S. Misailovic, A. Agarwal, and M. C. Rinard. Dynamic knobs for responsive power–aware computing. In *Proceedings of the 16th International Conference on Architectural Support for Programming Languages and Operating Systems*, 2011.

[19] E. Ipek, O. Mutlu, J. F. Martínez, and R. Caruana. Self–Optimizing Memory Controllers: A Reinforcement Learning Approach. In *Proceedings of the 35th Annual International Symposium on Computer Architecture*, 2008.

[20] J. O. Kephart and D. M. Chess. The Vision of Autonomic Computing. *Computer*, 36(1), 2003.

[21] M. Salehie and L. Tahvildari. Self–Adaptive Software: Landscape and Research Challenges. *ACM Trans. Auton. Adapt. Syst.*, 4(2), 2009.

[22] D. D. Silva, O. Krieger, R. W. Wisniewski, A. Waterland, D. K. Tam, and A. Baumann. K42: An Infrastructure for Operating System Research. *SIGOPS Oper. Syst. Rev.*, 40(2), 2006.

[23] B. Sprunt. Managing The Complexity Of Performance Monitoring Hardware: The Brink Andabyss Approach. *International Journal of High Performance Computing Applications*, 20(4), 2006.

[24] N. Thomas, G. Tanase, O. Tkachyshyn, J. Perdue, N. M. Amato, and L. Rauchwerger. A Framework for Adaptive Algorithm Selection in STAPL. In *Proceedings of the Tenth ACM SIGPLAN Symposium on Principles and Practice of Parallel Programming*, 2005.

[25] E. Z. Zhang, Y. Jiang, and X. Shen. Does cache sharing on modern CMP matter to the performance of contemporary multithreaded programs? In *Proceedings of the 15th ACM SIGPLAN Symposium on Principles and Practice of Parallel Programming*, 2010.

APPENDIX

A. HEART RATE MONITOR

This section explains the design and implementation of HRM (see Appendix A.1) and shows additional experimental results gathered during the evaluation of HRM (see Appendix A.2).

A.1 Design and Implementation

HRM is an active monitoring infrastructure designed to be simple, effective, and efficient. *Simplicity* is achieved through a very compact API; *effectiveness* comes with the support of diverse parallelization models through groups and the support of both user and kernel–space adaptation policies; finally, *efficiency* is achieved thanks to the partitioned design between user and kernel–space accounting for multi-core related issues such as synchronization and cache sharing.

A.1.1 User–space

The user–space partition of HRM is implemented through a library, namely *libhrm*, exposing the API for both *producers* (i.e., applications) and *consumers* (i.e., adaptation policies). The complete API is reported in Table 2.

The API exposes a function to attach the current task to the group identified by GID as either a producer or a consumer: `hrm_attach`; this function involves multiple boundary crosses since it switches from user to kernel–space to setup data structures, memory, and timers. Once the virtual memory is mapped, the current task can set the user–defined performance goal making it system–wide available through: `hrm_set_min_heart_rate`, `hrm_set_max_heart_rate`, and `hrm_set_window_size`. These three functions respectively set the lower bound over the performance measure, the upper bound over the performance measure, and the amount of sample to use to compute the window heart rate. As highlighted in Table 2, the functions to set the user–defined performance goal are accessible to producers only, while consumers are allowed to call the functions to retrieve the user–defined performance goal: `hrm_get_min_heart_rate`, `hrm_get_max_heart_rate`, and `hrm_get_window_size`. Performance measures can be retrieved, like the user–defined performance goal, by either producers or consumers through: `hrm_get_global_heart_rate` and `hrm_get_window_heart_rate`. HRM provides both a global performance measure and a window performance measure because different applications may be concerned with either long or short–term trends. The global performance measure is supposed to catch long–term trends since it averages the whole execution of an application while the window performance measure is meant to capture short–term trends being a moving average. Short–term trends should be intended as variable–length trends since the length is controlled by the window size, which is part of the user–defined performance goal, and the timer period, which is a kernel compile time parameter controlling the update frequency for the performance measures.

The most important function exported by *libhrm* is `heartbeat`. Calls to this function are inserted within the hot–spot of an application to signal progresses in the execution; the amount of progresses can be specified through the integer parameters the function accepts.

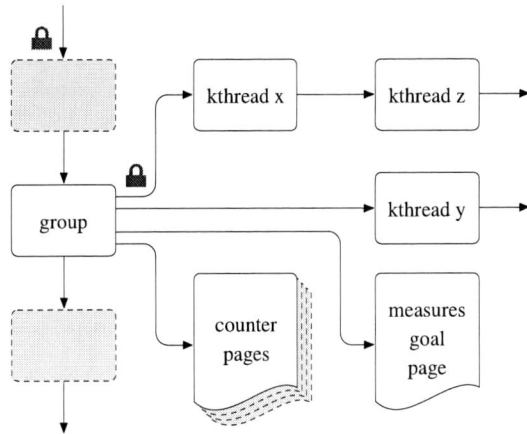

Figure 5: Data structure involving the linked–list of groups and the linked–list of producers and consumers (per group), set of pages to store per thread counters, and single page to store both the performance measures and the performance goal.

A.1.2 Kernel–space

The kernel–space partition of HRM is implemented as an extension of the Linux kernel and can be considered the core of the active monitoring infrastructure.

The basic data structure behind HRM is a linked–list of groups, shown in Figure 5. The linked–list of groups is protected by a global spinlock, which is needed to guarantee consistency when modifying the linked–list (i.e., group addition or deletion). More common operations involving a single group are not required to own the global spinlock; instead, they may be required to own the members read–write lock (read/write lock). A read/write lock was chosen instead of a spinlock because of the unbalance between read and write operations over the linked–lists of producers and consumers. Each group allocates a set of pages to store per thread counters (i.e., default 1 page up to 16 pages[6]) and a single page to store both the performance measures and the performance goal as depicted in Figure 5. As Figures 6 and 7 shows, pages are shared across the kernel and the user address spaces; moreover, since HRM supports diverse parallelization models through the concept of group, pages may be shared across multiple user address spaces since each of the user thread may belong to a different process.

Figure 6 gives a more accurate view of the layout of the pages storing the counters; as the least significant portion of the addresses highlight, each counter is aligned to the size of the cache line. Cache line alignment results in a slightly less efficient use of the available memory; however, we argue the performance improvements due to cache line alignment on multi and many–core processor and especially on multi–processor systems, which necessitate off–chip communication to maintain cache coherency, is such that more memory can be allocated, being an increasingly available resource in modern computing systems. The content of the

---

[6]The limit of 16 pages to store per thread counters is completely arbitrary and can be boosted through the Linux kernel configuration utility.

Table 2: *libhrm* API[1,2]

Function	Description
hrm_attach(int gid, bool_t consumer)	Attach the current task to the group identified by **gid** as either a producer or consumer
hrm_detach()	Detach the current task
hrm_set_min_heart_rate(uint32_t min_heart_rate)[3]	Set the minimum heart rate in the user–defined performance goal
hrm_set_max_heart_rate(uint32_t max_heart_rate)[3]	Set the maximum heart rate in the user–defined performance goal
hrm_set_window_size(size_t window_size)[3]	Set the window size in the user–defined performance goal
hrm_get_min_heart_rate(uint32_t *min_heart_rate)	Get the minimum heart rate from the user–defined performance goal
hrm_get_max_heart_rate(uint32_t *max_heart_rate)	Get the maximum heart rate from the user–defined performance goal
hrm_get_window_size(size_t *window_size)	Get the window size from the user–defined performance goal
hrm_get_global_heart_rate(uint32_t *global_heart_rate)	Get the global heart rate from the performance measure
hrm_get_window_heart_rate(uint32_t *window_heart_rate)	Get the window heart rate from the performance measure
int heartbeat(uint64_t n)[3]	Emit **n** heartbeats

[1]Every function receive an additional parameter of type **hrm_t *** pointing to the underlying data structure
[2]Every function return a value of type **int** containing either 0 or an error number
[3]Every task attached as a consumer is not allowed to call this function

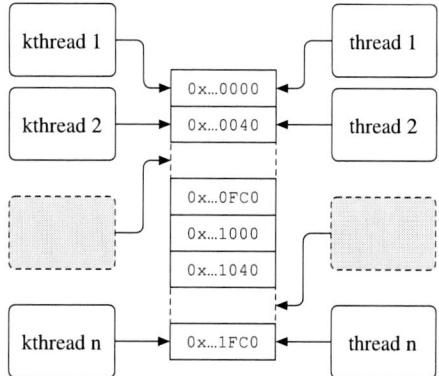

Figure 6: **Memory layout and access pattern of the pages storing per thread counters.**

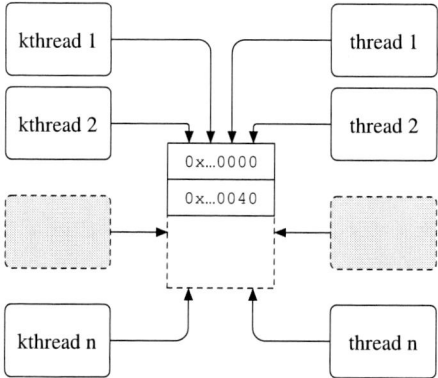

Figure 7: **Memory layout and access pattern of the page storing the performance measures and the performance goal.**

pages devoted to store the counters is the most critical to HRM since it can be concurrently accessed at a high rate by both the kernel and user threads. Distribution avoid synchronization among user threads, while heavy weight synchronizations between kernel and user–space are avoided by adopting atomic operations; hence, a function call to **heartbeat** reduces to an atomic increment of a per–thread counter. Due to cache line alignment, the number of counters is architecture–dependent; the reference implementation of HRM allocates standard sized pages whose size is 4 kB, while the size of cache lines of x86 and x86–64 processors is 64 bytes, with such parameters, each page can contain up to 64 counters.

Figure 7 shows the single page each group allocates to store both the performance measures and the performance goal; conversely to what happens with the counters, the performance measures and the performance goals are not distributed across the group. As reported in Section 4, HRM provides both a global heart rate (i.e., long–term performance measure) and a window heart rate (i.e., short–term performance measure); they are respectively computed according to Equations (6) and (7) in which g indicates the group, t is the current time stamp, t_0 is the group creation time stamp, and t_w is the time stamp at which window started. The performance measures are asynchronously

updated by the kernel in the context of a High–Resolution (HR) timer after acquiring the members read–write lock in read mode; the adoption of asynchronous updates for performance measures avoids boundary crosses to retrieve the current time stamp. The period of the HR timer can be tuned through a kernel compile time parameter.

$$ghr_g(t) = \frac{\sum_i cnt_i(t)}{t - t_0} \qquad (6)$$

$$whr_g(t) = \frac{\sum_i cnt_i(t) - cnt_i(t_w)}{t - t_w} \qquad (7)$$

The asynchronous computation of the performance measures is fundamental for providing and high performance implementation of the Application Heartbeats API. To guarantee high performance, HRM sacrifices a tiny bit of accuracy: heartbeats may be accounted with a delay which is at most equal to the period of the HR timer. This behavior fits into one of the open spot of the Application Heartbeats API. The performance goal is made up of a lower and an upper bound defining a heart rate range; moreover, the performance goal contains also the window size to compute the window heart rate.

HRM extends the **proc(5)** process information pseudo–file system (*procfs*) [2] to provide all the necessary entry

863

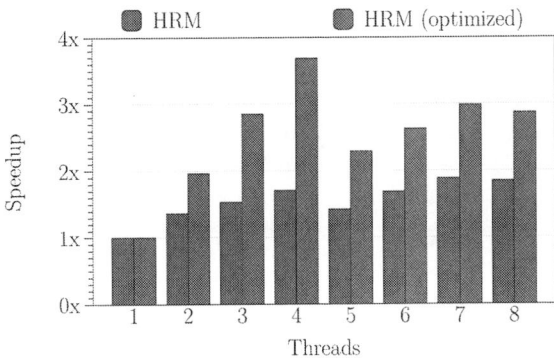

Figure 8: Speedup on the throughput of the optimized vs. non–optimized implementation of HRM with 1 to 8 threads over 1000 executions. Peak performance should be reached with 4 threads since the workstation features a quad–core processor. Higher is better.

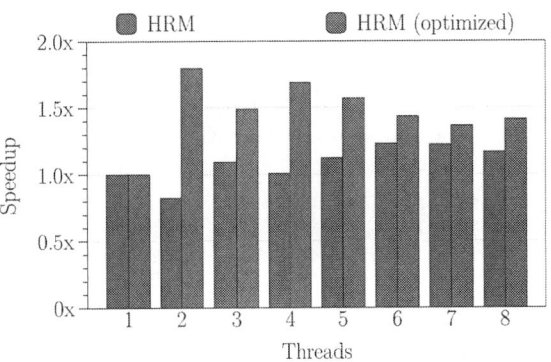

Figure 9: Speedup on the throughput of the optimized vs. non–optimized implementation of HRM with 1 to 8 threads over 1000 executions. Peak performance should be reached with 2 threads since the workstation features a dual–core processor. Higher is better.

points to attach (detach) threads to (from) a group and `mmap(2)` [2] the pages storing per thread counters and both the performance measures and the performance goal.

A.2 Evaluation

As reported in [25], cache sharing is an important factor in modern multi–core processors and, we add, in multiprocessor systems. The initial implementation of HRM did not adopt any smart page layout, allowing more than one per–thread counter to reside in a single cache line. False sharing of cache lines resulted in an unexpected contention over the memory hierarchy and suboptimal performance. Figure 8 shows the speedup thanks to page layout optimization; experimental results were collected on the same workstation and with the same procedure described in Section 4.2 using the non–optimized and the optimized implementations of HRM; note how HRM scales almost perfectly reaching 3.7× speedup with 4 threads.

The same experiment was repeated on a second workstation equipped with a single Intel Pentium D 820 dual–core processor running at 2.80 GHz featuring 1 MB of core private LLC (L2) per core, 2 GB of DDR2–800 non–ECC RAM, and a 250 GB 7200 RPM SATA hard disk. Enhanced Intel SpeedStep Technology was disabled. The AMD64 version of Debian 6.0, alias "squeeze", was configured to run the Linux kernel 2.6.35.13 extended with HRM. Due to the per–core private last–level cache, cache coherency necessitate off–chip communication through the Front–Side Bus (FSB) and the northbridge; this limitation makes the workstation more similar to a multi–processor system instead of a multi–core processor system. Figure 9 shows the speedup due to the page layout optimization when costly off–chip communication is employed to maintain cache coherency. When 2 threads are employed, the non–optimized implementation of HRM incurs in a sensible slowdown while the optimized implementation of HRM scales almost linearly (i.e., speedup near 2×) showing the real advantage of the smart page layout.

To further investigate the implications of HRM, experimental results were collected on the same workstation described in Section 4.2 using 5 out of 13 applications of the PARSEC 2.1 benchmark suite. For the non–legacy applications, HRM was setup to provide only the global heart rate and the heart rate computation period was set to 100 ms. Table 3 put forth evidence that HRM imposes low monitoring overhead on real applications, with a maximum of 1.26% for facesim.

B. PERFORMANCE–AWARE FAIR SCHEDULER

This section provides insights regarding the second and the third workloads described in Section 5.2, whose results were reported in Table 1 and Figure 3 (see Appendix B.1).

B.1 Evaluation

Figure 10 explains the very low normalized mean squared error reported in Figure 3 for the first workload. Both blackscholes and swaptions are fairly regular applications; as Figure 10b shows, their behavior is simple to anticipate and control and they can be driven towards their performance goal.

Figure 11 provides insights regarding the third workload and PAFS behavior. Facesim is not as regular as either blackscholes or swaption; in fact, the third workload is more complicated to control for PAFS, even though the presence of fluidanimate in place of ferret makes it simpler with respect to the first workload. As Figure 11b shows, the amount of available resources is greater than the amount required to satisfy the performance goals of both the application (i.e., the performance measure of fluidanimate is constantly past the upper bound of the performance goal). Although on a smaller scale, the third workload shows the same behavior of the first in which one of the application (i.e., fluidanimate) ends its execution before the other application (i.e., facesim) whose performance measure starts to increase due to the absence of resource contention.

In conclusion, these experiments show how PAFS is able to drive applications' performance towards a user–defined performance goal in presence of contention over the computational resources (i.e., in this case, processor time).

Table 3: Standard mean and standard deviation of the execution time over 100 consecutive runs of the benchmark suite. The overhead is computed using the ratio between standard mean of the non–instrumented version execution time and the standard mean of the instrumented version execution time

| Application | Legacy | | Non–Legacy | | Overhead |
	Std. Mean [s]	Std. Dev. [s]	Std. Mean [s]	Std. Dev. [s]	
blackscholes	69.745	0.178	70.541	0.205	1.14%
facesim	140.645	0.810	142.419	0.716	1.26%
ferret	114.856	0.079	114.903	0.110	0.04%
fluidanimate	103.052	0.061	103.088	0.067	0.03%
swaptions	83.989	0.128	84.220	0.124	0.27%

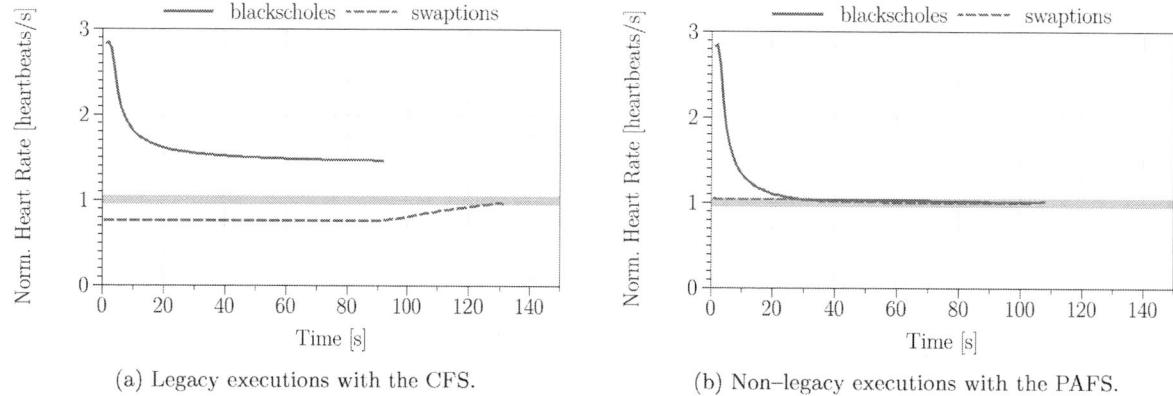

(a) Legacy executions with the CFS.   (b) Non–legacy executions with the PAFS.

Figure 10: Normalized current heart rates for mix 2 where [0.95, 1.05] is the normalized performance goal (i.e., gray–shaded area). Nearer the gray–shaded area is better.

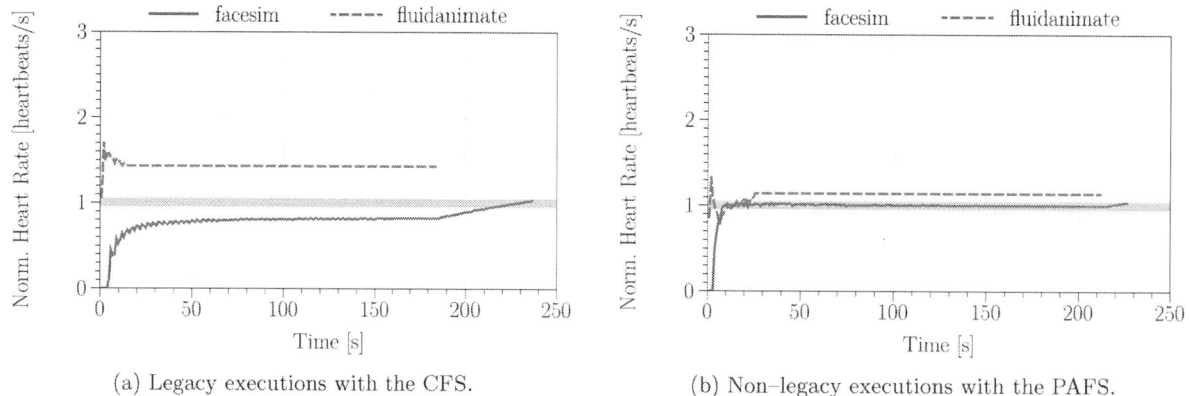

(a) Legacy executions with the CFS.   (b) Non–legacy executions with the PAFS.

Figure 11: Normalized current heart rates for mix 3 where [0.95, 1.05] is the normalized performance goal (i.e., gray–shaded area). Nearer the gray–shaded area is better.

Adaptive Power Management of On-Chip Video Memory for Multiview Video Coding

Muhammad Shafique[1], Bruno Zatt[1,2], Fabio Leandro Walter[2], Sergio Bampi[2], Jörg Henkel[1]

[1]Karlsruhe Institute of Technology (KIT), Chair for Embedded Systems, Karlsruhe, Germany
[2]Federal University of Rio Grande do Sul (UFRGS), Informatics Institute/PGMICRO, Porto Alegre, Brazil
{muhammad.shafique, bruno.zatt, henkel}@kit.edu; {bzatt, bampi}@inf.ufrgs.br

Abstract—An adaptive power management of on-chip video memory for Multiview Video Coding is presented. It leverages texture, motion and disparity properties of objects and their correlations in the 3D-neighborhood. It groups different Macroblocks of a frame and predicts the highly-probable motion/disparity search direction in order to power-gate idle memory regions. Exploited are the statistical properties of Macroblock groups to predict idle sectors. Our approach achieves on average 32% and 61% energy reduction (averaged over various video sequences) compared to state-of-the-art DSW [7] and Level C [12], respectively. The Motion/Disparity Estimation architecture with video memory and power management scheme is implemented using an ASIC flow (IBM-65nm Low-Power technology) and it processes 4-view HD1080p@33fps.

Categories and Subject Descriptors: C.3 [**Special-Purpose and Application-Based Systems**]: Real-time and embedded systems; B.3.2 [**Design Styles**]: Cache Memories; I.4.2 [**Compression (Coding)**]: Approximate Methods

General Terms: Algorithms, Design, Management

Keywords: MVC, Video Coding, Motion Estimation, Disparity Estimation, Low-Power, Power-Management, On-Chip Memory, Video Memory, Adaptivity, Power-Gating

I. INTRODUCTION AND RELATED WORK

The Multiview Video Coding (MVC) standard [2] compresses the multiview video sequences (captured using multiple cameras) to realize emerging 3D-multimedia applications (like 3D-video recording/playback) on mobile devices [3][4]. MVC provides 20%-50% improved compression compared to simulcast H.264 (i.e. independent encoding of each view) by employing multiple block-sized Motion and Disparity Estimation (ME, DE) that exploit temporal and interview correlations at the cost of significantly high complexity and energy consumption [3]. Typically, ME/DE accounts for more than 90% of the total MVC energy consumption, out of which the major energy consuming part is the (on-chip and off-chip) memory [7]. Therefore, *memory is the key focus for energy reduction in ME and DE in order to implement MVC on battery-powered devices.*

The high memory energy consumption is primarily due to the frequent access of reference pixel data used in SAD (Sum of Absolute Differences) computations during the block matching process [7]. ME and DE search the best match of a Macroblock (MB, 16x16 pixel block) in different *search directions* (i.e. neighboring reference frames in the left, right, top, and down directions). For a given search direction, the search is performed in a predefined *search window* such that the reference pixels in a search window can be used for

multiple SAD computations (see Section S1 for ME/DE overview). The search direction that contains the best matching of an MB is denoted as the *best search direction*. State-of-the-art techniques employ an on-chip memory[1] to incorporate *search window prefetching* and *data reuse* [12]-[15], search window follower [11], or asymmetric search windows [16] for reducing the off-chip memory power. These on-chip memories suffer from non negligible leakage power due to their large footprint (≈2 Mbit memory is required for a search range of ±128 and 4 search directions). Furthermore, not all parts of the search window stored in the on-chip memory are accessed because of the adaptive nature of fast ME/DE algorithms (like TZ Search [5]) and diverse texture/motion properties of MBs (see memory usage analysis in Section II.A). To address this issue, the work of [14] employs adaptive window sizing. However, this work targets a fixed Four-Step Search and does not account for DE and leakage power, which is a crucial power component.

To reduce the leakage power, advanced techniques in power-gating switch-off parts of the on-chip memory using sleep transistors with multiple sleep modes [17][18][27]. Some sleep modes are data-retentive, i.e. data loss in the memory is avoided while providing relatively little leakage savings. For power management, state-of-the-art techniques employ prediction techniques based on *either* hardware monitoring [24] *or* exploiting limited application knowledge at frame level [25]. As a result, these techniques perform power-wise inefficient due to severe miss-predictions under high variations of memory usage. The work [29][30] illustrates the feasibility of application-aware power management for power-gating idle ASIP cores in a multimedia pipelined processor. The work in [7][8][9] presents an MVC ME/DE architecture with an on-chip video memory and a dynamic search window formation algorithm. The power-gating scheme evaluates the predicted memory usage of consecutive MBs to make a power-gating decision for the idle memory sectors, but do not employ methods (like computation reordering) to increase sleep durations.

The above-discussed techniques provide limited leakage savings as they do not exploit (i) the relationship of MB properties (texture, motion, and disparity) with the distribution of ME and DE as the best search direction, (ii) best search direction and memory usage correlation in the 3D-neighborhood (i.e. spatial, temporal, and view domains). These *multiview video content characteristics* may provide a higher potential for leakage savings.

Summarizing: in order to realize ME/DE of real-time full-HD MVC with low-power consumption, *an adaptive power management scheme for on-chip memory is required* that leverages the multiview video content characteristics at various levels (search direction, frame, MB, etc.) to predict the memory requirements (number and duration of idle memory sectors) and to power-gate them in an appropriate sleep mode.

Before proceeding to our novel contribution, we present an analysis of the best search direction and memory usages during ME/DE, which provides the motivation and foundation for this work.

---

[1] An 'on-chip memory' in this paper denotes an 'on-chip video memory'.

II. MOTIVATION AND NOVEL CONTRIBUTION

A. Motivational Analysis of Motion and Disparity Estimation

Our experiments on the "Rena" test video sequence in Fig. 1 and Fig. 2 illustrate the distribution of ME and DE as the best search direction (see Section S2 for detailed analysis for other test video sequences). Fig. 1 depicts that the majority of the Macroblocks (MBs; 70%-90% cases) are encoded using ME as the best search direction. Note, in case of the first view V0, all MBs are encoded using ME, because no neighboring views are available for prediction. Similarly, in case of other views (V1-V2), for the first frame of each GOP (Group of Pictures; i.e. T0 and T8 in Fig. 2), only DE is performed. When performing a detailed analysis of various frames in view V1 at QP=27 (Fig. 2), it can be observed that background objects (low-texture, low-motion, static blocks) are mostly encoded using ME. In contrast, foreground objects (medium-high texture, high motion) are encoded using DE. Moreover, some background objects with medium-high texture may also be coded using DE (see 'curtains' in T4, Fig. 2). The decision of ME and DE also depends upon the available correlation in the temporal domain. For instance, the number of DE-coded MBs is higher in T4 compared to T5.

Hint-1: If the best search direction (ME or DE) can be correctly predicted, significant energy savings can be obtained by avoiding ME/DE over unused search directions and power-gating the sectors storing the search windows for these unused search directions. It will also lead to a reduced amount of external memory transfers and computations. *The key is to use the texture and motion/disparity properties of the MBs in the 3D-neighborhood for correct prediction of the best search direction.*

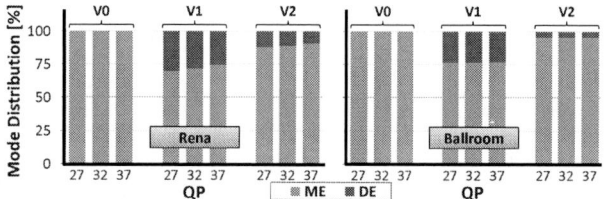

Fig. 1 (a) ME and DE distribution for three views of "Rena" and "Ballroom" test multiview video sequences

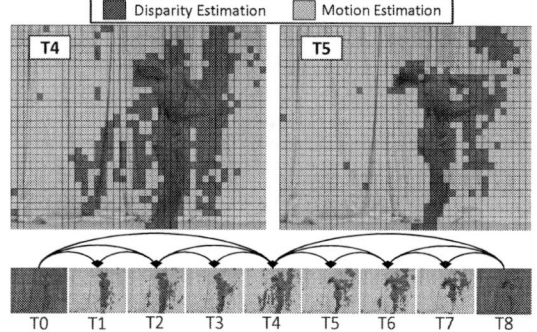

Fig. 2 ME/DE distribution in view V1 of "Rena" sequence

In state-of-the-art schemes [7]-[16], ME/DE of MBs is processed in a raster scan order. However, our experiments (Fig. 2, see also Section S2) illustrate that objects often consist of MBs that do not lie on the raster scan order. Therefore, these schemes suffer from severe variations in the memory usage as MBs of different objects typically exhibit diverse memory requirements for ME/DE. This leads to reduced sleep durations and frequent wakeups of memory sectors, thus low leakage savings. Our analysis in Sections IV.A, S2 shows that different MBs sharing similar texture and motion/disparity properties have similar memory requirements.

Hint-2: Longer sleep durations (thus higher leakage savings) can be achieved if the ME/DE processing of the MBs with similar properties

is performed together, i.e. in a *non-raster scan order. The key challenges are MB grouping and ME/DE computations reordering.*

Summarizing the analysis, **the key research challenges** for reducing the power of on-chip video memory of ME/DE are:

a) *Grouping the MBs* of a frame w.r.t. their texture, motion, and disparity properties,

b) *Adaptively predicting the best search direction* for MBs in different groups to power-gated on-chip memory sectors of unused search directions,

c) *Reordering the ME/DE processing computations* to increase the sleep durations of on-chip memory sectors,

d) Leveraging the *multiview video content characteristics* to enable a *content-driven power management* at various granularities (group-level, MB-level).

B. Overview of Our Concept and Novel Contributions

To address these challenges, a novel adaptive power management scheme is proposed for on-chip video memory that incorporates:

1. **An MB-Group Formation Scheme (Section IV.A)** that performs *texture and activity (i.e. motion and disparity) classification* for MBs considering the correlated neighboring MBs in their 3D-neighborhood (i.e. spatial, temporal, view domains). This classification is used to form groups of MBs that share similar texture and activity properties.

2. **An Adaptive ME/DE Search Direction Prediction Algorithm (Section IV.A)** that adaptively predicts the highly-probable ME/DE search direction for MBs in different groups based on the best search direction correlation in the 3D-neighborhood and their respective texture differences. For each MB in a group, our scheme power-gates the memory sectors of the unused search direction.

3. **A Content-Driven Power Management Scheme (Section IV.B)** that leverages the *multiview video content characteristics* to manage the power at multiple levels (i.e. search direction, MB-group, MB).

III. MEMORY AND POWER MODELS

Now, we describe the model of our multibank on-chip video memory [7][8] and power-gate model of [17] which is used in this work to enable power-gating of multibank memory at a fine-granularity.

Memory Model: The on-chip memory consists of N_B number of banks. Each bank $B_{k;\ k=1...NB}$ contains N_S number of sectors each having S_L number of 128-bit memory lines (see Fig. 5 for an abstract view). The size of a sector is given as $S=S_L \times 128$ bits. In order to provide parallel data access for SAD computing hardware accelerators, different rows of an MB are stored in different banks. The leakage energy is given as $E_{Leak}=T_{MEDE} \times P_{Leak}$, where the T_{MEDE} denotes the time for processing motion and disparity estimation. The miss energy is given as: $E_{Miss}=\sum_{i=1...NMiss} E_{Missi}$, where N_{Miss} is the number of misses. Such a memory model can also be realized with multiple SRAM blocks, each having multiple sub-arrays [27] or considering the SRAM model of [26].

Power-Gate Model: We assume a power-gate model with three power modes: P_{ON}, P_{DR}, and P_{OFF}. P_{ON} is the Power-ON mode. P_{DR} is the Data-Retentive (DR) low-leakage mode that preserves the data in SRAM cells. P_{OFF} is the Power-OFF mode with data loss; it requires re-fetching of data from the external memory. Fig. 3 shows the power state machine with leakage energy savings and wakeup latency/energy overhead [17]. The wakeup latency of P_{DR} is quite short compared to P_{OFF}, therefore, it is beneficial for short sleep durations (see values in Table I; Section S3). Contrarily, P_{OFF} is beneficial for long sleep durations. Multiple sleep modes facilitate different wakeup-overhead vs. leakage-saving tradeoff options. Since different collocated sectors in different banks store

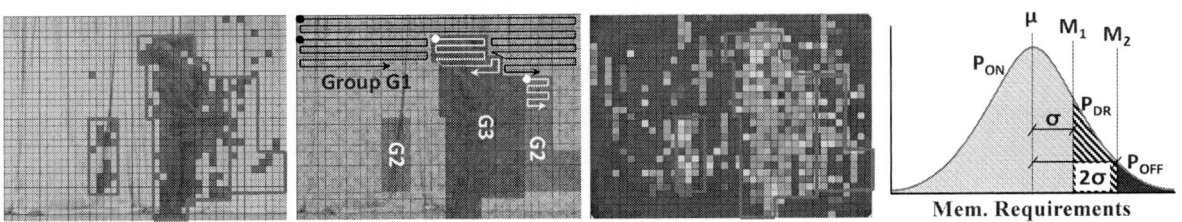

Fig. 4 "Rena" test video sequence encoded at QP=22: (a) Distribution of ME and DE; (b) Macroblock grouping w/ computation reordering; (c) Distribution of memory usage for ME; (d) Memory requirement prediction using PDFs for ON, OFF, and DR mode.

the data from the same MB, same sleep control is issued to these sectors. Power-gating at sector level enables a fine-grained power management control. Similar style of power-gating can be found in sub-array level power-gating [27] or even further fine-grained using wordline-level power gating [28].

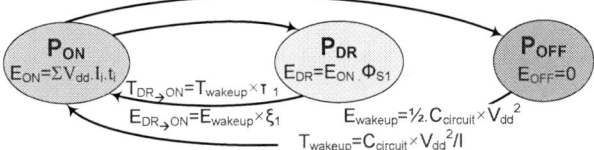

Fig. 3 Power state machine with multiple sleep modes [17]

IV. OUR ADAPTIVE POWER MANAGEMENT OF ON-CHIP VIDEO MEMORY FOR ME/DE IN MVC

Fig. 5 shows an overview of our adaptive power management scheme (novel contribution in green boxes) for an on-chip multi-bank video memory integrated with an ME/DE architecture.

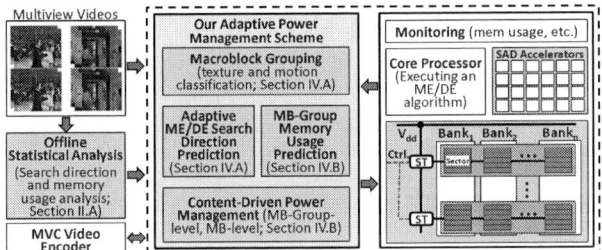

Fig. 5 ME/DE architecture with an on-chip memory and our power management scheme (novel contribution in green boxes)

Our scheme works in five phases:

i) *Macroblock (MB) Grouping*: First, the texture and activity classification of MBs is performed and MBs with similar texture and activity properties are grouped together,

ii) *Predicting the highly-probable best search direction* based on the correlation in the 3D-neighborhood,

iii) *Predicting the memory usage of MB-groups* using a statistical analysis of the memory usage of different groups and memory usage correlation of same groups in 3D-neighborhood,

iv) *Power-gating the unused memory sectors* in appropriate sleep modes based on the predicted memory requirements,

v) *Computation-reordering and fine-tuning the power modes of different sectors at MB level*: Since all MBs in a given group exhibit similar memory requirements for ME and DE, ME/DE processing computations of MBs are reordered in order to increase the sleep durations of on-chip memory sectors. Computation reordering is performed within a group, where the next MB for ME/DE processing is selected by evaluating its texture difference w.r.t. to the currently processed MB.

A. Macroblock Grouping and Search Direction Prediction

As discussed in Section II.A, in a conventional raster scan coding order, ME/DE is performed for all MBs in a row-wise fashion.

Each row typically has MBs from different objects that typically span over many MBs both horizontally and vertically (see Rena dancing picture in Fig. 2 and Fig. 4). Since MBs from different objects exhibit distinct memory usage properties, that results in memory usage variations that lead to short sleep durations and frequent ON and OFF switching of the unused memory sectors. To avoid this, our scheme aggregates different MBs that share similar texture and activity (i.e. motion and disparity) properties in so-called *MB-groups* (see an example in Fig. 4a).

Fig. 6 shows the algorithm for MB grouping. The input is the frame $F_{(T,V)}$, where T denotes the temporal location of the frame in view V. Other inputs are variance of the MB (σ, Eq. 2) as the light-weight texture approximation and texture difference (ξ, Eq. 3) w.r.t. the neighboring MBs in the 3D-neighborhood (i.e. spatial, temporal, and view domains). There are 4 spatial, 18 temporal, and 18 view neighboring MBs (see Fig. 15 in Section S1). First, the texture classification of the current MB is performed as *low-texture* (L), *medium-texture* (M), and *high-texture* (H); line 5. Afterwards, the matching neighbors (i.e. MBs in the 3D-neighborhood having similar texture properties as of the current MB) are found (lines 6-7). Since the MBs with similar texture properties most-probably belong to the same object, these MBs share the motion/disparity properties, i.e. so-called *activity*. Therefore, the activity of the current MB is predicted as *low-motion* (L), *medium-motion* (M), and *high-motion* (H) from the average activity of the matching neighbors (lines 8-9). Based on the texture and activity classification, an MB is assigned to a group, such that all the MBs in that group exhibit similar texture and activity properties (lines 10-11). The output is the composition of all three groups and the set of matching neighbors (line 13).

---

1.   **groupMBs**(*Input: Frame* $F_{(T,V)}$, *Variance* σ, *Texture Difference* ξ;
      Output: MB-Group **G**, *Matching Neighbors* N_{match})
2. $G \leftarrow \varnothing$; $N_{match} \leftarrow \varnothing$;
3. $\forall mb \in F_{(T,V)}$ {
4. $N_{MBMatch} \leftarrow \varnothing$;
5. $T := (\sigma_{MB} \leq \tau_{\sigma 1})?L:(\sigma_{MB} > \tau_{\sigma 2})?H:M$;
6. $N \leftarrow mb.getNeighbors(\;)$; // *see Fig. 15 in Section S1*
7. $\forall n \in N$ if$(\xi_n < \tau_\xi)$ $N_{MBMatch} \leftarrow N_{MBMatch} \cup n$;
8. $M := \sum_{\forall n \in N_{MBMatch}} (|v_X| + |v_Y|)_n / size(N_{MBMatch})$;
9. $\beta := (M_{MB} \leq \tau_{m1})?L:(M_{MB} > \tau_{m2})?H:M$;
10. $G_{MB} := (T=L \& \beta=L)?G_1:(T=H \& \beta=H)?G_3:G_2$;
11. G.store(mb, G_{MB});
12. N_{match}.store(mb, $N_{MBMatch}$);   }
13.   return(G, N_{match});

Fig. 6 Pseudo-code for macroblock grouping

The thresholds ($\tau_{\sigma 1}$, $\tau_{\sigma 2}$, τ_ξ, τ_{m1}, τ_{m2}) are obtained using the statistical distribution analysis of texture and activity properties of MBs of numerous background and foreground objects in various test video sequences (like "Rena", "Ballroom", "Vassar", etc.) [1]. Highly-probable value of these thresholds are obtained as "$\mu+3\times\sigma$" (μ denotes the mean, σ denotes the standard deviation) using the probability density functions (PDF) following a Gaussian distribution; Eq. 1.

$$F(\mu_k+3\sigma_k; \mu_k, \sigma_k^2) - F(0; \mu_k, \sigma_k^2)_{k=[\text{variance, motion/disparity vectors}]} \approx 0.99 \quad (1)$$

$$\sigma_{MB} = \sum_{i=1}^{16} \sum_{j=1}^{16} \left(\rho_{(i,j)} - \rho_{Avg}\right)^2 / 256 \quad (2)$$

$$\xi_n = |\sigma_{CurrMB} - \sigma_n| / \sigma_{CurrMB} \quad (3)$$

MBs in a group share the best prediction direction due to their correlation as they most-probably belong to the same object. Fig. 4 illustrates an example scenario, where for the MBs of the dancing girl (group G3), DE is selected as the best search direction. In contrast, for the MBs of the background curtains (group G1), ME is selected as the best search direction. Therefore, grouping MBs also provides a potential for search direction prediction for the complete group (as we will discuss using Fig. 7). Note, in case of group G2, the decision becomes challenging as in case of medium-texture nature with slow-medium motion, the best match can be found using ME or DE. Therefore, in case of group G2, our scheme adaptively selects the highly-probable search direction depending upon 3D-neighborhood.

Adaptive Search Direction Prediction: Fig. 7 shows the algorithm for adaptively predicting the highly-probable best search direction for three groups (Fig. 6). As discussed in Section II.A and Fig. 4, background/low-textured MBs with low-motion (i.e. MBs in the group G1) are typically encoded using ME, and MBs with high-texture and high-motion (i.e. MBs in the group G3) are encoded using DE. Therefore, our algorithm predicts ME and DE as the best search directions for groups G1 and G3, respectively (lines 2-3). The decision about the MBs in the group G2 is made adaptively by taking into consideration the best search directions of the matching neighboring MBs (lines 4-12). If there are sufficient number of matching neighbors (for a high confidence of prediction), a prediction is performed considering the texture difference of the matching neighbors (lines 6-9). A cost $cost_{ME}$ is computed by accumulating the inverse of texture differences for all neighbors with ME as the best search direction (line 8). Similarly, $cost_{DE}$ is computed (line 9). If $cost_{ME}$ is greater than or equal to $cost_{DE}$, ME is predicted as the best search direction, otherwise, DE is selected (lines 10-11). In case of insufficient correlation in the 3D-neighborhood, ME is predicted as the best search direction (line 12). Finally, the best search direction D_{Best} is returned (line 13).

---

1.   **searchDirectionPrediction**(*Input: MB-Group* **G**, *Matching Neighbors* **N**$_{\text{match}}$, *Texture Difference* ξ; *Output: Best Search Directions* **D**$_{\text{Best}}$)
2.   if (G_1) $\forall mb \in G_1$ mb.$D_{Best} := ME$;
3.   if (G_3) $\forall mb \in G_3$ mb.$D_{Best} := DE$;
4.   if (G_2) {   *// adaptively select ME or DE for MBs in group G2*
5. $\forall mb \in G_2$
6.   if $(size(mb.N_{match}) > \tau_{Match})$ {
7. $\forall n \in mb.N_{match}$
8.   if $(n.D_{Best}=ME)$ cost$_{ME} := $cost$_{ME} + (1 / \xi_n)$;
9.   else cost$_{DE} := $cost$_{DE} + (1 / \xi_n)$;
10.   predDir := (cost$_{ME} \geq$ cost$_{DE}$)? ME : DE;
11.   mb.D_{Best} := predDir;   }
12.   else   mb.D_{Best} := ME;   }
13. return D_{Best};

Fig. 7 Pseudo-code for adaptive search direction prediction

For each MB, only one motion or disparity search in the selected search direction is performed. It leads to significant energy savings by avoiding external memory transfers and excessive computations. Furthermore, the sectors storing the search windows for the unused search directions are power-gated to reduce the leakage, which provides further energy savings (see Fig. 8).

Note, Fig. 4a shows that in case of group G1 there are a few MBs that have DE as the best search direction. However, our scheme predicts ME as the best search direction for the group G1, so it might incur some video quality loss. Furthermore, a miss-prediction may also results in quality loss. Experiments in Section V, S3 show that this loss is visually imperceptible.

B. Video Content-Driven Power Management

Once the MB-groups are formed, the challenge is to accurately predict the memory usage requirements of an MB-group. The key is to leverage the multiview video content properties and the offline-statistical analysis of memory usage of different groups.

Step-1: Memory Usage Prediction of MB-Groups: Fig. 4c shows the memory usage of different groups, where the memory usage of G1 is much lower than in other groups. Our scheme computes two different highly-probable memory requirement predictions (M_1 and M_2) from the probability density function (PDF obtained through an offline-analysis over various test video sequences, see details in Section S2). The M_1 amount of memory is kept in P_{ON} mode as the probability of using these memory sectors is high. The memory requirement M_2-M_1 is kept in the P_{DR} mode, as others MBs of the same group may use this data and the wakeup overhead is minimal to avoid delay. Fig. 4d shows an abstract representation of obtaining these predictions. *The memory requirements [M_1, M_2] of an MB-group can also be predicted with a high accuracy from the memory usages of the same MB-group in the neighboring frames or even views.* (see experimental evidence in Section S2). These predicted memory requirement values are then forwarded to the power-management scheme to determine the number and mode of gated sectors.

Step-2: MB-Group-Level Power Management: Fig. 8 presents the algorithm of our content-driven power-management. First the MB grouping is performed (line 2). Afterwards, each group is sequentially processed, i.e. ME/DE of the MBs from the group G1 is processed first followed by MBs from groups G2 and G3, respectively. It demonstrates the first reordering of the ME/DE computations, as MBs are now processed in a *non-raster scan order* (lines 3-28). The second reordering occurs when processing MBs within a group (lines 23-27).

First, for each group, the best search direction is predicted and the memory sectors of the unused search directions are power-gated in power state P_{OFF} (line 4-5), as they will not be used during the complete ME/DE of this MB. Afterwards, the highly-probable memory usage is predicted from the PDF obtained by the offline statistical analysis (line 6); as also shown in Fig. 4. Based on this predicted memory usage, number of sectors that are candidate for power-gating in different power modes (P_{ON}, P_{OFF}, P_{DR}) are computed (lines 7-8). To cope with the potential misprediction, the correlation of the monitored memory usage of similar MB-group in the 3D-neighborhood is exploited (line 9). For G1 and G3, average memory usages of the same group in the temporal neighbors (Frame$_{Left}$, Frame$_{Right}$) and in the disparity neighbors (Frame$_{Top}$, Frame$_{Down}$) are considered, respectively. For G2, the average of all the four neighbors is computed. The candidate sectors for power-gating in P_{ON} and P_{DR} power modes are determined considering this correlated memory usage (line 10). PDF-based and neighborhood-based predicted memory usages are averaged to obtain the number of sectors that are candidate for power-gating in P_{ON}, P_{OFF}, and P_{DR} power modes (lines 11-12).

To amortize the wakeup energy overhead, our scheme predicts the leakage energy benefit of gating sectors in different power modes. For this, first the sleep duration is predicted as the predicted ME/DE processing time of all the MBs in the group (line 13). The ME/DE of an MB is predicted as the average of the ME/DE processing time of all the matching neighbors in the 3D-neighborhood. Afterwards, the leakage savings are compared with the wakeup energy overhead and the sectors are set in their respective power modes (lines 14-17). In case of P_{OFF}, additionally $E_{MissGroup}$ is considered as P_{OFF} results in the loss of data in memory sectors and require a re-fetching (line 16).

1.	**ContentDrivenPM**(*Input: Frame* $F_{(T,V)}$)	
2.	$(G, N_{match}) \leftarrow groupMBs(F_{(T,V)}, \sigma, \xi);$	// Fig. 6
3.	$\forall g \in G \{$	
4.	$D_{Best} \leftarrow searchDirectionPrediction(g, N_{match}, \xi);$	// Fig. 7
5.	if$(G_1$ or $G_3)$ PowerGate(D / D_{Best}, P_{OFF}, $S_{(D/D_{Best})}$);	
6.	$[M_1, M_2] \leftarrow MemUsagePDF(g);$	// Fig. 4
7.	$S_{OFF(PDF)} := \lfloor (S - M_2)/S_{Sector} \rfloor;$ $S_{DR(PDF)} := \lfloor (M_2 - M_1)/S_{Sector} \rfloor;$	
8.	$S_{ON(PDF)} := S/S_{Sector} - (S_{OFF} + S_{DR});$	
9.	$M_{Nbs} \leftarrow AVG_{\forall d \in N \mid N=[Left,Right,Top,Down]} d.Group(g).getMemUsage();$	
10.	$S_{ON(Nbs)} := \lceil M_{Nbs}/S_{Sector} \rceil;$ $S_{DR(Nbs)} := AVG_{\forall mb \in g} mb.S_{DR};$	
11.	$S_{ON} := (S_{ON(Nbs)} + S_{ON(PDF)}) / 2;$ $S_{DR} := (S_{DR(Nbs)} + S_{DR(PDF)}) / 2;$	
12.	$S_{OFF} := (S - S_{ON} - S_{DR});$	
13.	$\forall mb \in g$ $mb.T_{pred} := AVG_{\forall n \in Nmatch} n.T_{MEDE};$	
14.	if$\left(\left(\sum_{\forall mb \in g} mb.T_{pred} \times P_{Leak(SDR)} \right) > E_{wakeup(DR \rightarrow ON)} \right)$	
15.	PowerGate(g, P_{DR}, S_{DR});	
16.	if$\left(\left(\sum_{\forall mb \in g} mb.T_{pred} \times P_{Leak(OFF)} \right) > (E_{wakeup(OFF \rightarrow ON)} + E_{MissGroup}) \right)$	
17.	PowerGate(g, P_{OFF}, S_{OFF});	
18.	else PowerGate(g, P_{DR}, S_{DR});	
19.	PowerON(g, S_{ON});	
20.	$g' \leftarrow g;$ $mb' \leftarrow \varnothing;$ $mb \leftarrow g.getFirstMB();$	
21.	while$(g' \neq \varnothing)$ {	
22.	if(G_2) PowerGate(D / $mb.D_{Best}$, P_{OFF}, $S_{(D/mb.D_{Best})}$);	
23.	$mb' \leftarrow g'.getCorrelatedMB(mb, mb_{\forall n \in N \mid N=[Left,Right,Top,Down]});$	
24.	if$(mb' = \varnothing)$ $mb' \leftarrow g'.getNextMB(mb);$	
25.	MBLevelPM(mb', S_{ON}, S_{OFF}, S_{DR});	
26.	$[E_{MEDE}, E_{Miss}, E_{Leak}, M_{mb}] \leftarrow performSearch(mb', D_{Best});$	
27.	$g' \leftarrow g'/mb';$ }	
28.	}	

Fig. 8 Pseudo-code for our content-driven power management

Step-3: Computation Reordering: In the next step, MBs of the group are processed one-by-one (lines 21-27, Fig. 8). As discussed earlier in Section I, processing ME/DE of MBs in a raster scan order results in frequent sleep and wakeup fluctuations, as MBs in a row may belong to different objects. Since an object typically spans over MBs of different rows (see Fig. 4, Section S2), sleep durations of the unused sectors can be lengthened (thus increasing the potential to put them in P_{OFF} mode) by processing MBs group by group, as MBs of the same group exhibit similar memory requirements. This will reduce the sleep-wakeup fluctuations and lead to relatively higher leakage savings. Fig. 4 shows that MB-groups can be of non-rectangular shape and the ME/DE processing order of MBs in different groups is non-raster scan order; see Fig. 4b for a possible ME/DE processing order of MBs in group G1. To avoid sleep-wakeup fluctuations at fine-granularity, even the computations inside the MB-groups are reordered. Inside the group, the next MB for ME/DE processing is selected by evaluating its texture difference w.r.t. to the current MB, such that consecutively executing MBs exhibit similar memory requirements. The algorithm in Fig. 8 first determines a correlated MB in the spatial neighborhood (Left, Right, Top, Down); line 23. Then MB-level fine-grained power-management (see Fig. 9) is performed; line 25. Afterwards, the ME/DE is performed based on the decision of the best search direction and E_{MEDE}, E_{Miss}, E_{Leak}, M_{mb} are monitored as the ME/DE processing energy, miss energy, leakage energy, and actual memory usage, respectively (line 26).

Step-4: Macroblock-Level Power Management: Fig. 9 shows the algorithm for MB-level power management. First the memory requirements of the MB are predicted from the matching neighbors (line 2) and the number of required memory sectors is computed (line 3). In case the number of required sectors is equal to the number of ON sectors, power modes of different sectors are not changed (line 5). In case the required memory is less than the P_{ON} memory, the difference is put into data-retentive sleep mode P_{DR}

(lines 6, 8). Otherwise, more sectors are powered-on from P_{DR} mode to P_{ON} mode (lines 7, 8).

1.	**MBLevelPM**(*Macroblock* mb, *Group-Level number of memory sectors in different power modes* S_{ON}, S_{OFF}, S_{DR})
2.	$M_{pred} := AVG_{\forall n \in mb} N_{match} n.getMemUsage();$
3.	$S_{MB} := \lceil M_{pred}/S_{sector} \rceil;$
4.	$\Delta S := S_{ON} - S_{MB};$
5.	if$(\Delta S == 0)$ return;
6.	else if$(\Delta S > 0)$ $S_{ONmb} := S_{ON} - \Delta S;$ $S_{DRmb} := S_{DR} + \Delta S;$
7.	else $S_{ONmb} := S_{ON} + \Delta S;$ $S_{DRmb} := S_{DR} - \Delta S;$
8.	PowerGate(g, P_{DR}, S_{DRmb}); PowerON(g, S_{ONmb});

Fig. 9 Pseudo-code for macroblock-level power management

V. RESULTS AND EVALUATION

For energy and quality comparison, several multiview video sequences with different resolutions are used; VGA (480x640; "Ballroom", "Exit", "Flamenco2", and "Vassar") and XGA (1024x768; "Breakdancers" and "Ballet") [1]. "Rena" is a part of the training set, so we do not employ it for the evaluation to avoid biasing effects. Further test conditions are: *TZ Search* ME/DE algorithm, 193x193 search window, QP={22,27,32,37}. Note, the energy results include the overhead of our scheme.

A. Comparison to State-of-the-Art

We compare the on-chip memory energy savings of our adaptive power-management scheme with state-of-the-art memory energy reduction techniques like Level-C and Level-C+ [12] (search window-data reuse) and a memory power management scheme with dynamic search windows (DSW) [7]. For fairness of comparison same test videos and QP set are used for all schemes. Fig. 10 shows the on-chip energy consumption normalized to Level-C+ that presents the highest energy consumption among all schemes. Level-C+ and Level-C incur significant on-chip memory energy due to their large-sized search window that is active all the time, i.e. not exploiting the idle periods of memory to save power. Compared to Level-C and Level-C+, our scheme provides on average 61% and 67% on-chip memory energy reduction, respectively. Compared to the DSW scheme [7] that employs power-gating of memory sectors based on the memory requirements of consecutive MBs, our scheme provides on average 32% higher energy savings. Energy reduction is achieved by (i) increasing sleep duration using computation reordering, (ii) power-gating memory sectors due best search direction prediction, and (iii) leveraging the video content knowledge for a multi-level power management. On average, 51% of the sectors are in P_{OFF} mode while 9.5% are in P_{DR} mode (see further details in Section S3). Our experiments show that high motion/texture allow relatively less energy savings because more data from search area is accessed and less sectors are gated.

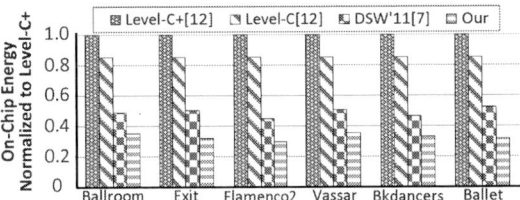

Fig. 10 On-chip memory energy savings comparison

B. Overhead: Mispredictions and Memory Misses

Fig. 11a shows that our scheme predicts the best search direction with an accuracy of 87% for high-activity sequences and 94% for low-activity sequences. This incurs a video quality loss of average 0.054 dB BD-PSNR (Bjøntegaard Delta PSNR) with an average increase of 1.86% BD-BR (Bjøntegaard Delta Bitrate), compared to the exhaustive search of JMVC 6.0 [2]. However, this loss is visually imperceptible (see Fig. 21, Section S3). Due to the predictive nature, our scheme incurs on average 8.5% on-chip memory misses compared to when storing the complete search window Fig. 11b. However, including all latency overhead due to misprediction and the power-management decision logic, our scheme still provides a minimum throughput of 33fps (see Fig. 12).

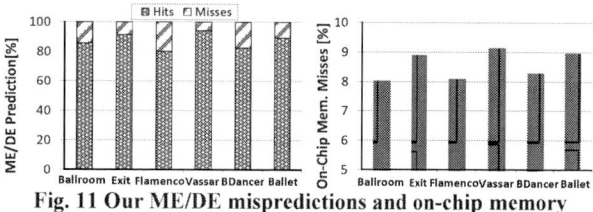

Fig. 11 Our ME/DE mispredictions and on-chip memory misses

C. Hardware Implementation

The hardware prototype is implemented using an ASIC flow using the Cadence tool chain for standard-cell synthesis with an IBM 65nm Low-Power technology. The IC layout and comparison table is shown in Fig. 12. The designed architecture employs 64x4-sample SAD operators and 21 SAD trees fed by the 16 on-chip memory banks. Compared to the state-of-the-art our architecture reduces the on-chip energy by 76% and 25% when compared to [13] and [7], respectively. Note, the work of [13] is implemented in 90nm technology. Considering a 30% power reduction (in case of SRAMs) when moving from 90nm to 65 nm technology node [6], our proposed scheme still provides >60% reduction in the on-chip energy. The provided throughput is capable of providing real-time HD1080p ME and DE at 33fps. The performance increase in relation to [7] is mainly due to the complexity reduction resulting from the search direction prediction. Note the 8x increase in the number of on-chip bits in comparison to [13] is due to the different-sized search windows. Our scheme supports 193x193 search windows (which are mandatory for DE to provide good video quality), while the architecture of [13] supports 33x33 search windows.

	Tsung'09 [13]	DSW'11 [7]	Our
Technology	TSMC 90nm Low Power LowK Cu	ST 65nm LP 7 metal layer	ST 65nm LP 7 metal layer
Gate Count	230k	102k	104k
SRAM	64 Kbits	512 Kbits	512 Kbits
Frequency	300 MHz	300 MHz	300 MHz
Power	265mW, 1.2v	74mW, 1.0v	63mW, 1.0v
Throughput (Resolution, Frame Rate)	4-views 720p @30fps	4-views HD1080p @30fps	4-views HD1080p @33fps

SAD Units:	Sum of Absolute Differences Operators
ME/DE Ctrl:	Motion/Disparity Estimation Control
AGU:	Address Generation Unit
APM	Adaptive Power Management

Fig. 12 (a) Chip Layout, (b) Hardware results comparison

VI. CONCLUSIONS

We propose a novel adaptive power management scheme for on-chip video memory targeting MVC. It leverages the multiview video content knowledge and computation reordering to achieve high energy savings with an imperceptible video quality loss. Key enabling attributes are MB-grouping based on texture and activity classification, best search direction prediction, and a video content-driven multi-level power management policy. Our scheme achieves on average 32%-61% on-chip energy reduction compared to state-of-the-art [7][12]. We demonstrate the potential of leveraging the multiview video properties for low-power MVC realization on battery-powered devices.

REFERENCES

[1] Y. Su, A. Vetro, A. Smolic, "Common Test Conditions for Multiview Video Coding", ISO/IEC JTC1/SC29/WG11 and ITU-T SG16 Q.6, Doc. JVT-T207, July 2006.

[2] JMVC 6.0, garcon.ient.rwthaachen.de, Sep. 2009; Joint Draft 8.0 on Multiview video coding, JVT-AB204, 2008.

[3] P. Merkle et al., " Efficient Prediction Structures for Multiview Video Coding" IEEE TCSVT, vol.17, no.11, pp. 1461- 1473, 2007.

[4] Lynx: http://www.sharp.co.jp/products/sh03c/index.html

[5] J. Yang et al., "Multiview video coding based on rectified epipolar lines", International CICSP, pp. 1-5, 2009.

[6] Cypress Seminconductor Corp., "Advantages of 65 nm Technology over 90 nm Technology QDR® Family of SRAMs", 2010.

[7] B. Zatt, M, Shafique, F. Sampaio, L. Agostini, S. Bampi, J. Henkel, "Run-time adaptive energy-aware motion and disparity estimation in multiview video coding", IEEE DAC, pp. 1026-1031, 2011.

[8] B. Zatt, M, Shafique, S. Bampi, J. Henkel, "A Low-Power Memory Architecture with Application-Aware Power Management for Motion & Disparity Estimation in Multiview Video Coding", IEEE ICCAD, pp. 40-47, 2011.

[9] B. Zatt, M, Shafique, S. Bampi, J. Henkel, "Multi-Level Pipelined Parallel Hardware Architecture for High Throughput Motion and Disparity Estimation in Multiview Video Coding", IEEE DATE, pp. 1448-1453, 2011.

[10] M, Shafique, B. Zatt, J. Henkel, "A Complexity Reduction Scheme with Adaptive Search Direction and Mode Elimination for Multiview Video Coding", Picture Coding Symposium, 2012.

[11] S. Saponara, L. Fanucci, "Data-adaptive motion estimation algorithm and VLSI architecture design for low-power video systems", IEE Comp. & Digital Tech., vol.151, no.1, pp. 51- 59, 2004.

[12] C.-Y. Chen et al., "Level C+ data reuse scheme for motion estimation with corresponding coding orders", IEEE TCSVT, vol.16, no.4, pp. 553- 558, 2006.

[13] P.-K. Tsung et al., "Cache-based integer motion/disparity estimation for quad-HD H.264/AVC and HD multiview video coding", IEEE ICASSP, pp.2013-2016, 2009.

[14] C.-Y. Tsai et al., "Low power cache algorithm and architecture design for fast motion estimation in H.264/AVC encoder system", IEEE ICASSP, vol. 2, pp. II-97-II-100, 2007.

[15] H. Shim, C.-M. Kyung, "Selective search area reuse algorithm for low external memory access motion estimation", IEEE TCSVT, vol.19, no.7, pp.1044-1050, 2009.

[16] X. Xu, Y. He, "Fast disparity motion estimation in MVC based on range prediction", IEEE ICIP, pp.2000-2003, 2008.

[17] H. Singh et al., "Enhanced leakage reduction techniques using intermediate strength power gating", IEEE TVLSI, vol. 15, no. 11, pp. 1215-1224, 2007.

[18] S. Roy, N. Ranganathan, S. Katkoori, "State-retentive power gating of register files in multi-core processors featuring multithreaded in-order cores", IEEE Transaction on Computers, 2010.

[19] L. Shen et al., "View-adaptive motion estimation and disparity estimation for low complexity multiview video coding", IEEE TCSVT, vol.20, no.6, pp.925-930, 2010.

[20] H.-C. Chang et al., "A dynamic quality-adjustable H.264 video encoder for power-aware video applications", IEEE TCSVT, vol.19, no.12, pp.1739-1754, Dec. 2009.

[21] S.-H. Wang, S.-H. Tai, T. Chiang, "A low-power and bandwidth-efficient motion estimation IP core design using binary search", IEEE TCSVT, vol.19, no.5, pp.760-765, 2009.

[22] T. Tuan et al., "A 90nm Low-power FPGA for battery-powered applications", ACM/SIGDA FPL, pp. 3-11, 2006.

[23] X. Xu, Y. He, "Fast disparity motion estimation in MVC based on range prediction", IEEE ICIP, pp.2000-2003, 2008.

[24] S. Mondal, S.O. Memik, "Fine-grain leakage optimization in SRAM based FPGAs", IEEE GLSVLSI, pp. 238-243, 2005.

[25] X. Liu, P. J. Shenoy, and M. D. Corner, "Chameleon: application-level power management," IEEE TMC., vol. 7, no. 8, pp. 995–1010, 2008.

[26] G. Fukano et al., "A 65nm 1Mb SRAM Macro with Dynamic Voltage Scaling in Dual Power Supply Scheme for Low Power SoCs", NVSMW/ICMTD. pp.97-98, 2008.

[27] M. Khellah et al. "A 4.2GHz 0.3mm2 256kb Dual-V/sub cc/ SRAM Building Block in 65nm CMOS", IEEE ISSCC, pp.2572-2581, 2006.

[28] G. Gerosa et al., "A Sub-2 W Low Power IA Processor for Mobile Internet Devices in 45 nm High-k Metal Gate CMOS", IEEE ISSCC,73-82, 2009.

[29] H. Javaid, M, Shafique, S. Parameswaren, J. Henkel, "Low-power adaptive pipelined MPSoCs for multimedia: an H.264 video encoder case study", IEEE DAC, pp. 1032-1037, 2011.

[30] H. Javaid, M, Shafique, J. Henkel, S. Parameswaren, "System-Level Application-Aware Dynamic Power Management in Adaptive Pipelined MPSoCs for Multimedia", IEEE ICCAD, pp. 616-623, 2011.

[31] Supplementary Material

S1. Motion and Disparity Estimation in MVC

MVC exploits the redundancies available in temporal and inter-view domains using multiple block-sized Motion Estimation (ME) and Disparity Estimation (DE), respectively. The ME/DE search is performed in previously encoded frames (i.e. reference frames) for finding a block that best matches the currently encoded Macroblock (MB) given a similarity criterion (like Sum of Absolute Differences, SAD). ME searches in temporal neighboring reference frames, while DE searches in frames of the neighboring views (see Fig. 13). Note, a *search direction* refers to the relative position of a reference frame with respect to the current frame. According to the MVC standard, multiple reference frames may be used to additionally improve the coding efficiency. However, in this work, we consider one reference frame per search direction, i.e. one forward and one backward reference in the temporal domain plus one forward and one backward reference in the view domain (if available). The search is performed by comparing a set of candidate blocks (selected depending upon given search patterns) inside a predefined search window (see Fig. 13) in order to find the best matching block.

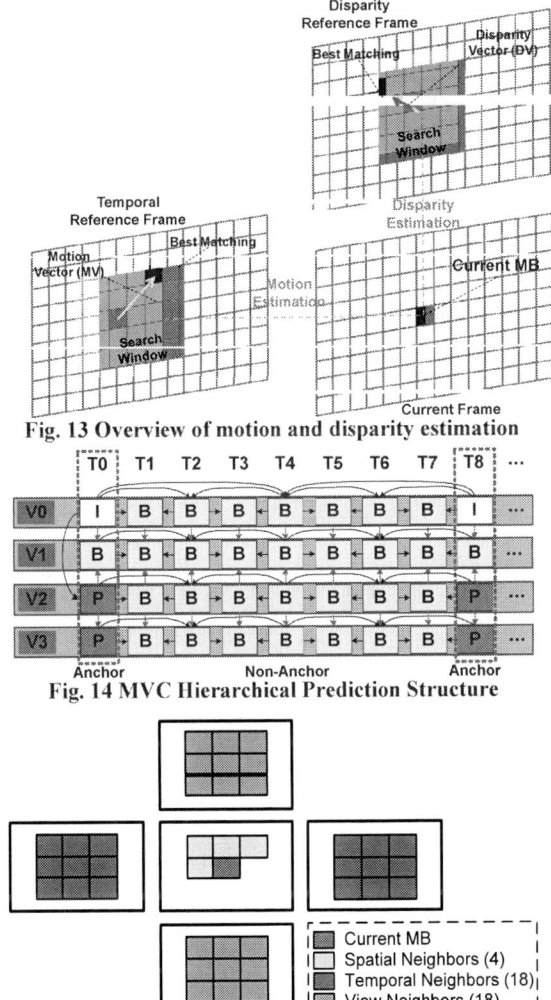

Fig. 13 Overview of motion and disparity estimation

Fig. 14 MVC Hierarchical Prediction Structure

Fig. 15 Neighboring MBs in the 3D-neighborhood

Once the best matching block is found, a Motion or Disparity Vector (MV, DV) is determined in order to represent the displace-ment between the current MB position and the best matching block position. Note, although ME and DE are conceptually similar, their search behavior and consequently the computational requirements, memory access pattern, and vector properties are distinct (see discussion in Section II.A).

Fig. 14 illustrates the MVC prediction structure and coding sequence. Fig. 15 shows the neighboring MBs in spatial, temporal, and view domains (i.e. 3D-neighborhood).

S2. Detailed Analysis of Multiview Videos

A fast ME/DE TZ Search [5] algorithm is deployed for this analysis in order to represent a real-world scenario. Fast ME/DE algorithms are based on multiple search stages and patterns. These algorithms evaluate different number of search candidates for different MBs, thus exhibit highly-varying memory usage profile.

A. Motion and Disparity Estimation Distribution

This section reinforces our analysis of ME/DE search direction distribution (presented in Section II.A) by evaluating for different video sequences with diverse motion/disparity and texture properties. The distribution in Fig. 16 and Fig. 17 illustrates that most of the MBs (typically from the background objects with low-texture, low-motion, static blocks) are encoded using ME. While the MBs of foreground objects and object borders (with medium to high texture, high motion) are encoded using DE.

It is noteworthy in Fig. 16 that the view V1 exhibits a higher number of DE encoded MBs compared to the other views. This is due to the fact that V1 has two references views available that increases the possibility to find a good match. The view V0 has no reference view available and consequently, all MBs are encoded using ME.

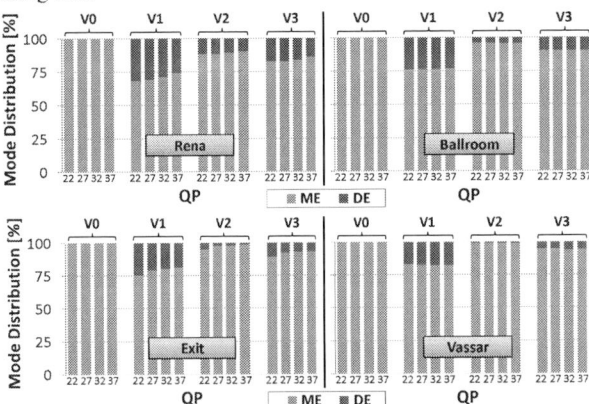

Fig. 16 (a) ME and DE distribution for four views of "Rena", "Ballroom", "Exit" and "Vassar" test video sequences

The decision of ME and DE (as the best search direction) also depends upon the correlation available in the temporal domain. For instance, the number of DE-coded MBs is higher in T4 compared to T5 since T4 is farther to the temporal references.

Our memory usage analysis in Fig. 18 shows that the pattern of memory usage in ME is less scattered compared to that in DE, especially in case of low-motion sequences with smaller objects like "Ballroom". The probability density function (PDF) in Fig. 18 shows that the distribution patterns of three groups are quite diverse. The PDF of the group G1 is quite centered in a low range (8-15Kpixels), while the PDF of the group G3 is quite dispersed over a big range (10-35Kpixels). Moreover, there is a minimal overlap between the PDFs different groups, which hints towards distinct memory predictions using PDFs. Therefore, based on this PDF analysis, our scheme computes two different highly-probable memory requirement predictions (M_1 and M_2) from the PDF

872

(obtained through an offline-analysis over various test video sequences) considering a Gaussian distribution. M_1 is obtained with a probability of 0.84 $[(F(\mu+\sigma;\ \mu,\ \sigma^2) - F(0;\ \mu,\ \sigma^2)]$ and M_2 is obtained with a probability of 0.975 $[(F(\mu+2\sigma;\ \mu,\ \sigma^2) - F(0;\ \mu,\ \sigma^2)]$. μ and σ are the mean and standard deviation, respectively. The M_1 amount of memory is kept in P_{ON} mode as the probability of using these memory sectors is high. The memory requirement M_2-M_1 is kept in the P_{DR} mode, as others MBs of the same group may use this data and the wakeup overhead is minimal to avoid delay.

Furthermore, neighborhood correlation of memory usage can also be exploited to predict memory usage of a given MB-group, because MBs of the same groups typically contain same object, thus exhibiting similar memory requirements for ME and DE. Fig. 19 shows that there is an extensive correlation between the neighboring frames T1 → T2 → T3. Therefore, *memory requirements [M₁, M₂] of an MB-group can also be predicted with a high accuracy from the memory usages of the same MB-group in the neighboring frames or even views*. Similar observation can be made from the memory requirements correlation shown in Fig. 20. The regions that require more memory are located in the same region for different instants of time. It shows that it is possible to infer the memory behavior based on the neighborhood knowledge. Similar observation can be made for view neighbors.

These predicted memory requirement values are then forwarded to the power-management scheme to determine the number and mode of gated sectors (see details in Section IV.B).

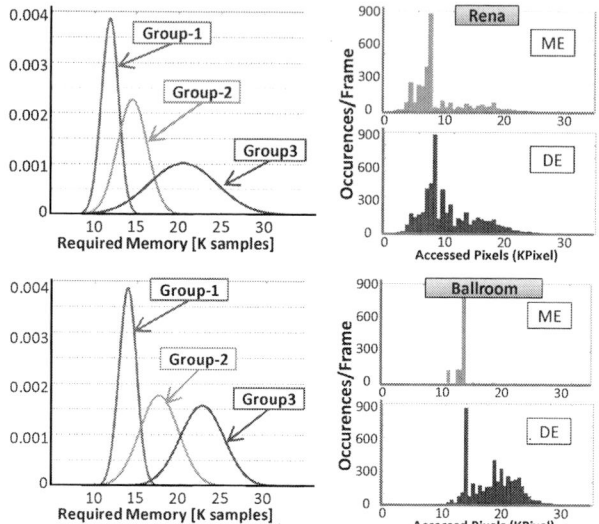

Fig. 18 (a) Probability density function (PDF) for the memory usage requirements of different groups for various test video sequences; (b) Histograms of memory usage during ME and DE processes for "Rena" and "Ballroom" sequences

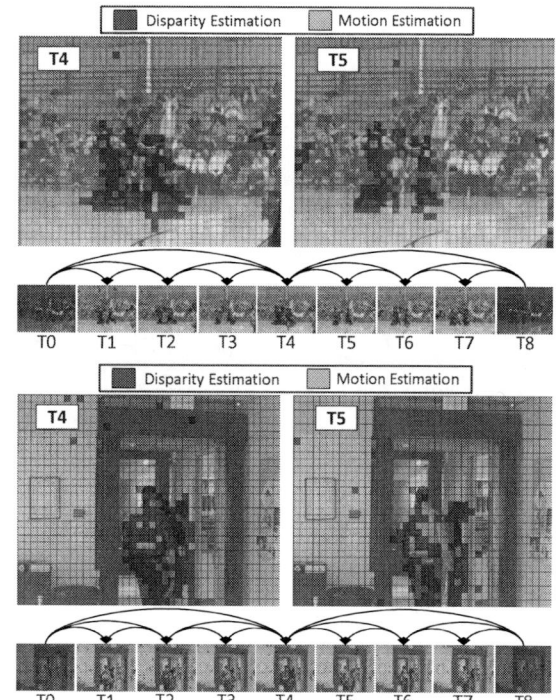

Fig. 17 ME/DE distribution for different frames in the view V1 of the "Rena" test multiview video sequence

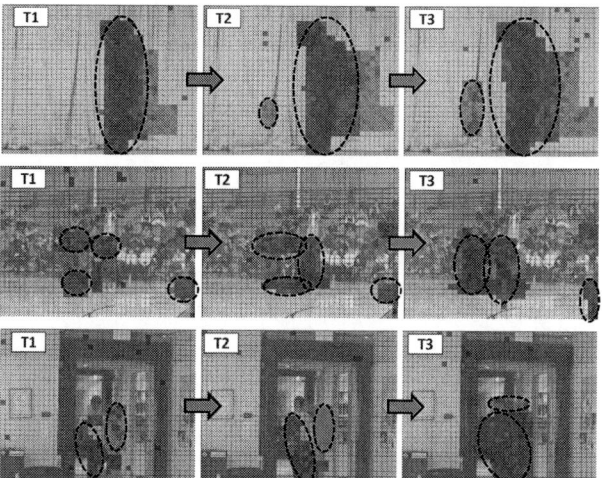

Fig. 19 MB-group correlation in different neighboring frames

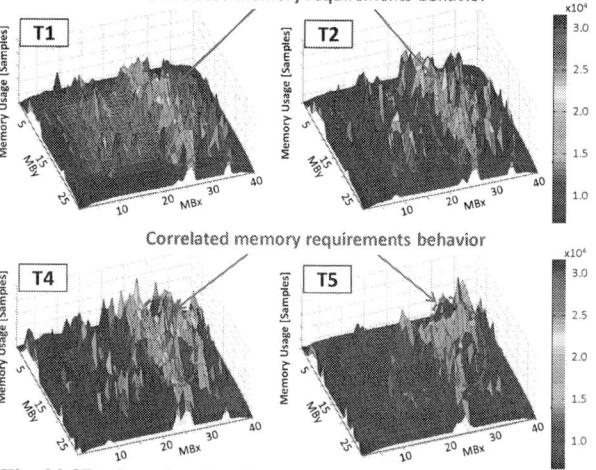

Fig. 20 3D-plots showing the correlation in the memory usage of MBs in the same frame and its temporal neighbors

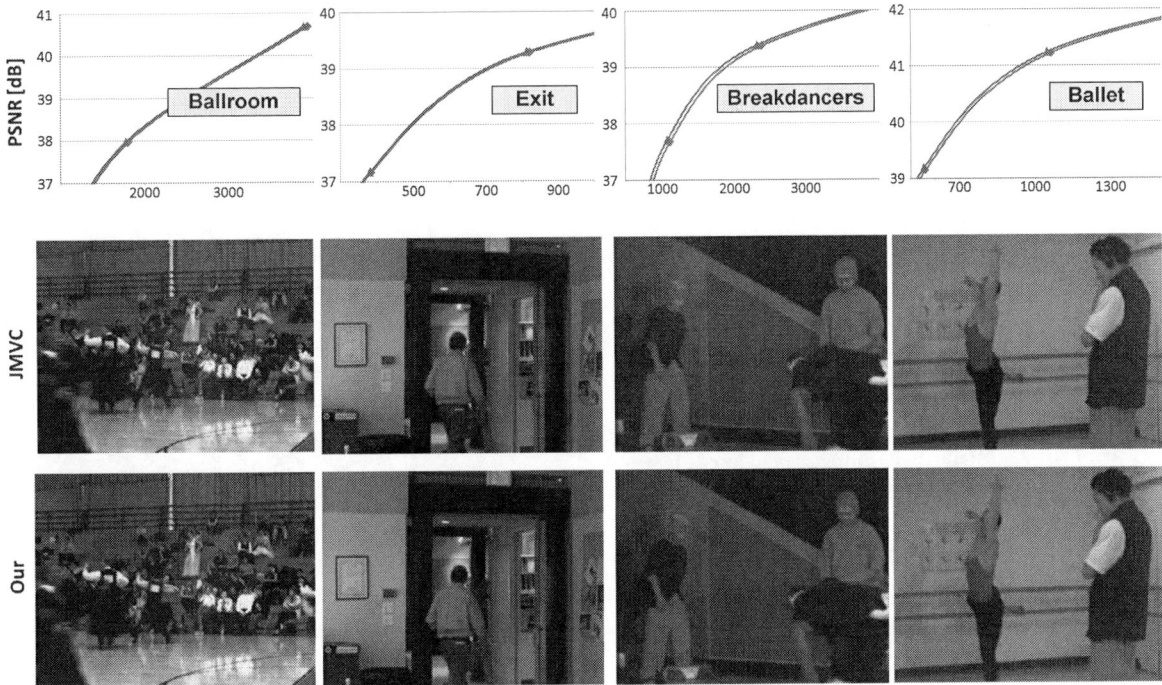

Fig. 21 Comparing the objective video quality (rate-distortion curves) and subjective video quality (pictures) of our scheme with the exhaustive ME/DE search of JMVC 6.0 [2]

S3. Additional Detailed Results

The on-chip memory power reduction is achieved by applying the computing reordering (that increases the number and sleep durations of idle memory sectors) and power management at different levels (MB-groups, MBs, etc.). The power state machine parameters are provided in Table I, based on the model of [17] (see Section III for power model details). Fig. 22 shows that on average 51% of the sectors are on P_{OFF} mode (up to 63%) while 9.5% are in P_{DR} mode (up to 15%). These results highly depend on the accuracy of MB-level memory requirements prediction. Fig. 24 presents the comparison between our application-driven memory requirements predictor and traditional history-based median predictor. Note, our proposed predictor reacts better and faster to the sudden variations of memory requirements. The high prediction accuracy is achieved by taking into consideration the correlation on the 3D-neighborhood along with texture and activity properties of different MBs, frames, and views.

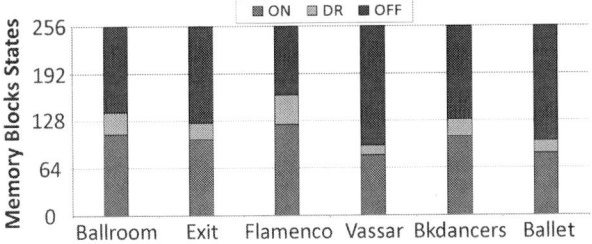

Fig. 22 Power modes distribution of the on-chip video memory

Compared to search window-based schemes (like in [12]), our approach requires much less external memory access since only a part of the search window is prefetched. Fig. 23 shows that our approach reduces the off-chip energy by 89% and 95% (on average) compared to Level-C and Level-C+ [12], respectively. Due to the computation reordering, our scheme reduces on average 15% of external memory access compared to our previous work of [7].

Table I: Power state machine parameters

Sleep Mode	Leakage Energy	Wakeup Energy	Wakeup Latency
P_{ON}	1	0	0
P_{DR}	0.3	0.35	0.3
P_{OFF}	0	1	1 + refetching

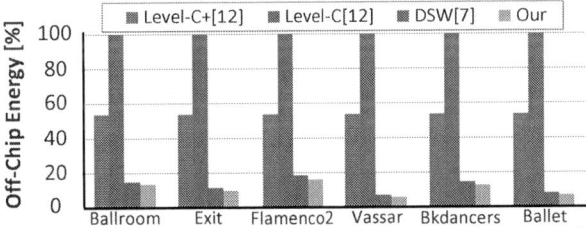

Fig. 23 Off-chip memory energy savings compared to state-of-the-art search window prefetching techniques

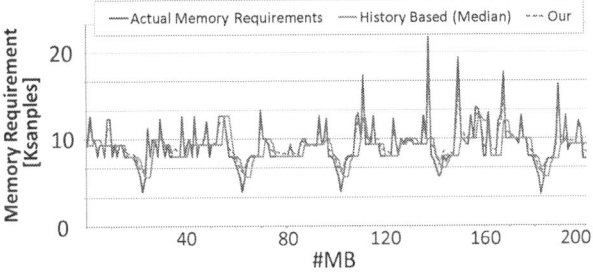

Fig. 24 comparing the accuracy of our application-driven memory requirement predictor with the history-based median predictor at MB-level

The detailed video quality results are shown in Fig. 21 and Fig. 25. The objective video quality (rate-distortion curves) and subjective video quality (decoded frames) results in Fig. 21 illustrate that our scheme achieves video quality close to that of the exhaustive ME/DE search of the JMVC 6.0 reference software [2]. Note: this implementation in JMVC 6.0 reference software [2] is provided to benchmark for the video quality

Fig. 25 and Table II show the BD-PSNR (Bjøntegaard Delta PSNR) and BD-BR (Bjøntegaard Delta Bitrate) comparison of our scheme with the exhaustive search. Fig. 25 shows that our scheme incurs a negligible loss in the rate-distortion. On average, 0.054 dB BD-PSNR drop (worst case 0.095dB) and 1.86% BD-BR increase (worst case 4.16%) are observed that lead to an imperceptible visual loss (see Fig. 21).

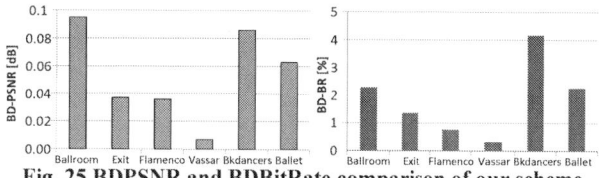

Fig. 25 BDPSNR and BDBitRate comparison of our scheme with the exhaustive search of JMVC

Table II: PSNR and Bitrate results; BD-Comparison

Multiview Video Sequences	JMVC		Our		Video Quality Loss	
	PSNR [dB]	BR [%]	PSNR [dB]	BR [%]	BDPSNR [dB]	BDBR [%]
Ballroom	1776,974	36,492	1807,220	36,485	-0,095	2,284
Exit	941,603	38,116	949,847	38,101	-0,037	1,372
Flamenco2	1910,379	38,473	1919,693	38,463	-0,036	0,759
Vassar	1176,496	36,530	1179,457	36,521	-0,007	0,318
Breakdancers	2706,915	38,472	2766,592	38,441	-0,086	4,168
Ballet	1039,264	39,949	1053,252	39,919	-0,063	2,260
Average	1591,938	38,005	1612,676	37,988	-0,054	1,860

S4. Details of the Hardware Architecture

The complete ME/DE architecture with on-chip video memory and our adaptive power management scheme is implemented (synthesis, place & route) using ASIC flow for an IBM-65nm Low-Power technology. Fig. 26 shows the schematics of the ME/DE top module architecture, the control unit, SAD calculator, and on-chip memory unit.

Control Unit performs the 3D-neighborhood evaluation, image properties extraction, memory requirements prediction, adaptive power management, and search algorithm control.

SAD Calculator is composed of 64x4samples SAD operators, 21 SAD trees and one SAD comparator.

On-Chip Memory consist of 16 memory banks, a register file to contain the tags of the data available in the on-chip memory, and an Address Generating Unit (AGU).

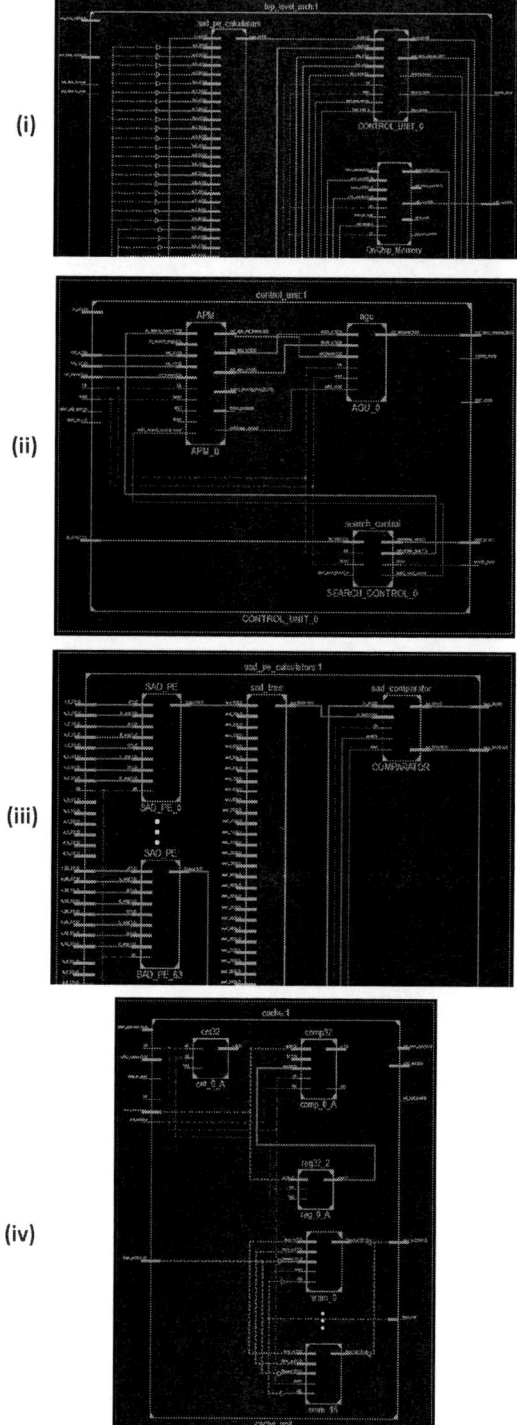

Fig. 26 Circuit Diagrams for (i) Top Level ME/DE Architecture; (ii) ME/DE Control Unit; (iii) ME/DE SAD Calculator; (iv) ME/DE On-Chip Memory Unit

Heterogeneous Multi-Channel: Fine-Grained DRAM Control for Both System Performance and Power Efficiency[*]

Guangfei Zhang, Huandong Wang, Xinke Chen, Shuai Huang, Peng Li

Loongson Corporation, Beijing 100190, China

Graduate University of Chinese Academy of Sciences, Beijing 100049, China

State Key Laboratory of Computer Architecture, ICT, CAS, Beijing 100190, China

{zhangguangfei, chenxinke, lipeng-cpu}@ict.ac.cn, {wanghuandong, huangshuai}@loongson.cn

ABSTRACT

We propose a novel architecture of memory controller, called HMC (Heterogeneous Multi-Channel), as an improvement to the previous homogeneous multi-channel memory controller. HMC groups physical DRAM devices into logical sub-ranks with different data bus width, and controls them simultaneously. Employing new proposed memory access algorithm, HMC manages the number of devices involved in a single memory access flexibly, and achieves the best performance/power efficiency. Using four-core multiprogramming workloads, our experimental results show that HMC improves system performance by 27.6% with 24.2% reduction in DRAM power consumption on average.

Categories and Subject Descriptors: B.3.1 [Memory Structures]: Semiconductor Memories-*Dynamic memory (DRAM)*

General Terms: Design, Performance, Experimentation

Keywords: Chip multiprocessor, DRAM, Memory access controller, Power efficiency

1. INTRODUCTION

Multiple threads run simultaneously in chip multiprocessors (CMP) which are integrated in a single chip. A CMP system consists of multiple independent processing cores that can access the memory system concurrently. The organization of CMP systems has benefits in terms of power-efficiency, scalability, and system throughput compared to single-core systems [1]. However, the resource management problem posed by the sharing of last level cache, interconnection and DRAM system is significant in designing CMP platforms. The demands of the working-set size from increasingly complex applications has led to the dramatically increase of memory bandwidth and capacity. Therefore memory system contributes significantly to the power consumption of the overall system, and the memory power consumption has approached to that of processors [2].

This study is focused on improving bandwidth utilization, reducing average latency and achieving high power efficiency of DRAM systems. Nowadays, with the increasing demand for high memory bandwidth from multi-core processors, the data bus

---

[*]This work is supported by the National Sci&Tech Major Project (No.2009ZX01028-002-003, 2009ZX01029-001-003), National Natural Science Foundation (No.60921002, 61003064, 61050002, 61070025, 61100163, 61133004, 61173001) of China.

---

frequency of a DRAM system has increased beyond 400MHz besides the use of double data transfer. For example, a DDR3-1600 DRAM system runs at 800MHz and can transfer a 64-byte cache block in 5ns. A typical memory device may provide 4-bit (X4), 8-bit (X8) or 16-bit (X16) data output. A DRAM rank groups several DRAM devices together to serve requests from the memory controller. A JEDEC standardized DRAM system is organized by ranks that operate on a 64-bit data bus. The use of 64-bit rank is good for reducing the data transfer time and thus memory idle latency, but it is a limit for DRAM system to allow meaningful trade-offs between performance and power [3].

We propose a novel architecture of memory access controller called HMC (Heterogeneous Multi-Channel) to balance the performance and power consumption of the DRAM system. It adopts configurable memory access channels to manage the DRAM devices independently. The DRAM devices, which are grouped into 64-bit ranks by tradition, can be grouped into 64-bit ranks, 32-bit ranks, and 16-bit ranks logically. In this case, all the DRAM devices share the same command and address bus, but different data buses of the DRAM devices are grouped together to form data buses with different width. In other words, different DRAM commands can be processed serially, while different data streams corresponding to different DRAM commands can be processed in parallel.

By controlling DRAM devices concurrently, HMC has four merits crucial for system performance and power savings. Firstly, multiple banks can be opened simultaneously, which dramatically decreases bank conflicts and promotes DRAM data bus utilization. Secondly, all the channels share the same DRAM command bus in sequence, which can largely increase DRAM command bus utilization. Furthermore, the DRAM operation power is reduced to a large extent since fewer DRAM devices are involved in every DRAM command. Because of the reduction of bank conflicts, there are fewer DRAM commands are issued to DRAM devices, which also saves the power consumption of the DRAM system. More importantly, different data streams corresponding to different DRAM commands can flow from or to DRAM devices concurrently. The memory queuing latency, which accounts for most of the time that a memory request stays in the memory access controller, can be greatly reduced.

We make the following contributions in this paper:

- We propose a new architecture of memory access controller that can achieve fine-grained control of DRAM devices for both system performance and power efficiency.
- We introduce a novel memory access scheduling algorithm that is good for reducing the memory request queuing time and improving data bus parallelism, which obtains high DRAM data bus utilization.
- We qualitatively and quantitatively evaluate our new memory access controller with extensive comparisons to homogeneous multi-channel memory access controllers in terms of system performance and power consumption. Our

experimental results show that HMC provides the best tradeoff between high-performance and power saving.

The remainder of this paper is structured as follows. Section 2 describes the background of DRAM system and related work. Section 3 presents the proposed Heterogeneous Multi-Channel memory access controller. Sections 4 and 5 show our evaluation methodology and experimental results of HMC, respectively. Finally, we conclude this paper in Section 6.

2. BACKGROUND AND RELATED WORK

2.1 DRAM System and Power Model

Modern DRAM components are memories with 3D structure of banks, rows, and columns. Data in a DRAM bank can only be accessed from the bank's row buffer which essentially serves as a buffer for the last accessed memory row in that bank. There are three major steps required when accessing a data element [4]:

- Precharge the bank's bit lines and write the page in the bank's row buffer back.
- Active the target row and read the page from activated row to row buffer.
- Do write or read operation in the row buffer.

Subsequent accesses to the same row can be performed by just accessing the row buffer, which is called a *row hit*. Otherwise, it is called a *row conflict*. To maximize memory bandwidth, a DRAM device consists of multiple banks so that DRAM commands to different banks can be served concurrently. The notion of servicing multiple requests in parallel in different DRAM banks is called *DRAM bank-level parallelism* (BLP). Write accesses always interfere with read accesses in modern DRAM systems, causing idle cycles on the DRAM data bus, which is called *read/write turnaround*.

TABLE I DDR3-1600 SDRAM DEVICE CONFIGURATION

Data Bus Width	16 Bits	Speed Grade	-125
DRAM Density	2Gb	Precharge PD Exit Mode	Fast
Maximum Vcc	1.575 V	Minimum Vcc	1.425 V
tRRD	7.5 ns	tRC	48.75 ns
tRAS	35 ns	tRFC	160 ns
tREFI	7.8 μs	tFAW	40ns

TABLE II DDR3-1600 POWER MODEL

Power Categories	Power State	Power (mW)
Background Power	Precharge power-down power	3.4
	Idle standby power	14.4
	Active power-down power	9.0
	Active standby power	36.0
	Power to complete refreshes	5.0
Operation Power	ACT/PRE commands power	307.8
Read/Write Power	DRAM write power	73.9
	DRAM read power	144.6
I/O Power	DQ output power	44.4
	Termination power for writes	56.2
	Termination power for reads from another rank	60.7
	Termination power for writes from another rank	24.7

The power consumption of a DRAM device can be classified into four categories: background power, operation power, read/write power, and I/O power. Table I and Table II summarize a representative DRAM configuration [5] and its corresponding power model [6] used in this study. The power model is calculated based on the voltage and current values of DDR3-1600 data sheet [7]. Background power is the power that a DRAM device has to consume all the time with or without operations. When the device performs activation or precharge operations, it consumes the operation power. The read/write power is consumed when the device reads or writes data during column accesses. The I/O power is spent on bus transactions and terminations. To reduce the background power, an idle DRAM device can be put into a low-power mode, such as standby, power-down or self-refresh. In general, a low-power mode that consumes less power has longer exit latency back to the active mode to serve incoming requests [8].

2.2 Sub-ranked Memory System

A group of DRAM devices operates together to serve requests from the memory controller. All the DRAM devices share the same command and address bus, but data buses of different DRAM devices are grouped together to form a wider, monolithic data bus. As shown in Figure 1(a), four X16 DRAM devices form a X64 DARM rank, and these DRAM devices consume the same power while present the same bandwidth/latency characteristics.

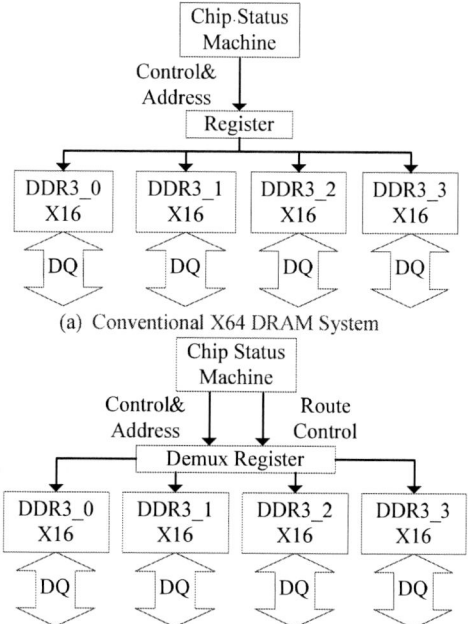

(a) Conventional X64 DRAM System

(b) Sub-ranked X16 DRAM System

Figure 1. Comparison of a conventional memory system and a sub-ranked memory system.

Recently, there are several works that have been proposed to balance the DRAM system performance and efficiency by controlling individual DRAM devices within a rank independently. Micro-threading [9] and multi-threading [10] permits several small accesses to take the place of a single large access, and use additional chip select signals to direct the DRAM commands to independent DRAM devices. SG-DIMM [11] allows access to physical memory by 8-byte words instead of by 64-byte cache lines. Mini-rank [12] adds MRB (Mini-rank Buffer) on the DRAM chip. MRB receives DRAM chip control signals and translates them into independent DRAM device control signals. MC-DIMM [13] uses a demux register, which is similar to MRB, to group multiple DRAM devices into a virtual memory device. For example, as shown in Figure 1(b), four X16 DRAM devices can be controlled independently, and form four X16 DARM ranks. AGMS extends the work of MC-DIMM, and proposes adaptive granularity memory systems. AGMS [14] can process both coarse- and fine-grained requests issued to memory system. All these previous works are called sub-ranked memory system.

All the previous works focus on the modification to the interface of the DRAM chip, and the design space has not been fully explored. HMC improves the architecture of conventional memory access controller, and proposes the corresponding memory access algorithm. The combination of the architecture and memory access algorithm achieves the best balance of performance and power consumption.

3. HETEROGENEOUS MULTI-CHANNEL MEMORY ACCESS CONTROLLER

3.1 Design and Implementation

As is shown in Figure 2, our experimental DRAM system is composed of four X16 DRAM devices, and each DRAM device has its independent chip select signal. HMC configures the DRAM system into a heterogeneous memory system, and divides the DRAM system into three types of channels: one 64-bit channel (X64), two 32-bit channels (X32_0 and X32_1) and four 16-bit channels (X16_0, X16_1, X16_2 and X16_3). Each channel has 8 independent banks. Through software configuration, the DRAM system can also be configured into a homogeneous memory system, in which only one type of channel exists, such as one 64-bit, two 32-bit or four 16-bit homogeneous memory channels.

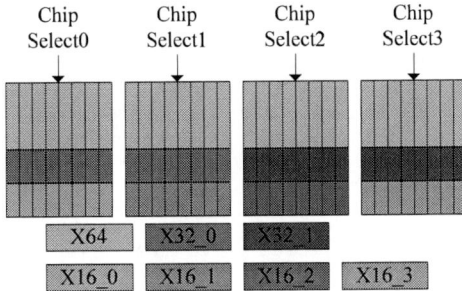

Figure 2. Main Idea of HMC

We implement HMC in our memory access controller, and Figure 3 shows the diagram of HMC. For a single memory request, there are mainly four stages:

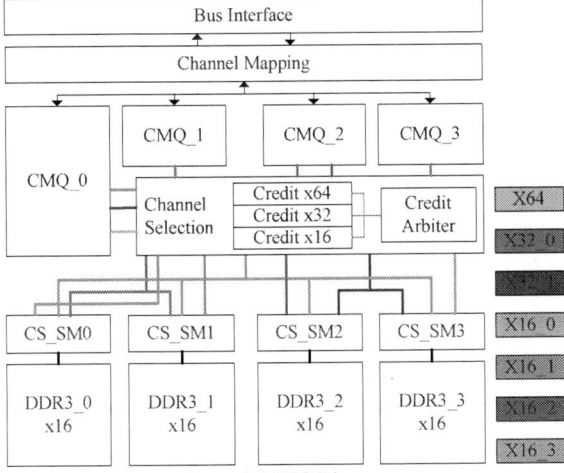

Figure 3. HMC Diagram

Stage 1: **Channel Mapping**. There are three functional modules in Channel Mapping Stage:

i.   Translate the target address of the memory request into CS (Chip Select), RA (Row Address), BA (Bank Address) and CA (Column Address).

ii.   Decide which channel the memory request should be routed to. HMC divides the address space into several address windows, and each address window has its independent channel width configuration. For example, as is shown in Figure 2, we configure the address space into seven address windows. Different memory requests can be mapped to different channels based on its address information, and different channels target to different groups of DRAM devices.

iii.   Decide which CMQ (Command Queue) the memory request should be routed to. We implement four CMQs in HMC, and different CMQs control different DRAM devices independently. In our implementation, CMQ_0 is responsible for the control of channel X64, X32_0 and X16_0; CMQ_1 is responsible for the control of channel X16_1; CMQ_2 is responsible for the control of channel X32_1 and X16_2. CMQ_3 is responsible for the control of channel X16_3.

Stage 2: **Command Queue Placement**. The memory request is placed into the command queue based on the Channel Grouping and FR-FCFS (First-Ready First-Come-First-Serve) policies, which will be introduced in Section 3.2.

Stage 3: **Channel Selection**. Channel Selection Stage use the Channel Selection policy which will be introduced in Section 3.2 to select a memory request from the four CMQs, and issues the memory request to the corresponding CS_SMs (Chip Select Status Machines).

Stage 4: **DRAM Device Access**. The memory request accesses the DRAM devices and returns the memory access results to the Bus Interface.

3.2 Memory Scheduling Algorithm

HMC is designed to provide fine-grained control of DRAM devices, and the memory scheduling algorithm employed in HMC must be adjusted accordingly. The data bus parallelism of DRAM devices is crucial to the performance of HMC. But if the memory requests with different channel widths are not scheduled appropriately, the parallelism of data buses will be destroyed. We introduce the following policies to achieve the maximum parallelism:

Policy 1: Channel Interleaving. In order to balance the memory access requests routed to different channels, we use the lower order address bits to select channels. For example, address bit [5:4] is used to select an X16 channel among X16_0, X16_1, X16_2 or X16_3, and address bit [5] is used to select an X32 channel between X32_0 and X32_1.

Policy 2: Channel Grouping. In a command queue, memory requests with the same channel width are grouped together. As is shown in Figure 4(a), memory access requests arrive in command queues randomly. Figure 4(b) shows the result of channel grouping, and there are three channel groups in CMQ_0. Each channel group will be scheduled as an independent unit.

Policy 3: FR-FCFS. HMC employs the FR-FCFS scheduling policy to schedule the memory requests in the same channel group. FR-FCFS prioritizes ready DRAM commands from row-hit requests over others and older requests over younger ones.

Policy 4: Channel Selection. As is shown in Figure 4(b), different channel groups from different CMQs will access the DRAM devices concurrently. To achieve the maximum utilization of DRAM data buses, we introduce channel selection policy into HMC. Channel selection policy selects channel groups with the same channel width from different CMQs, and issues the memory request at the front of each group to DRAM devices in sequence.

As is shown in Figure 4(c), narrow DRAM data buses transfer data concurrently and the data bus utilization can be dramatically improved.

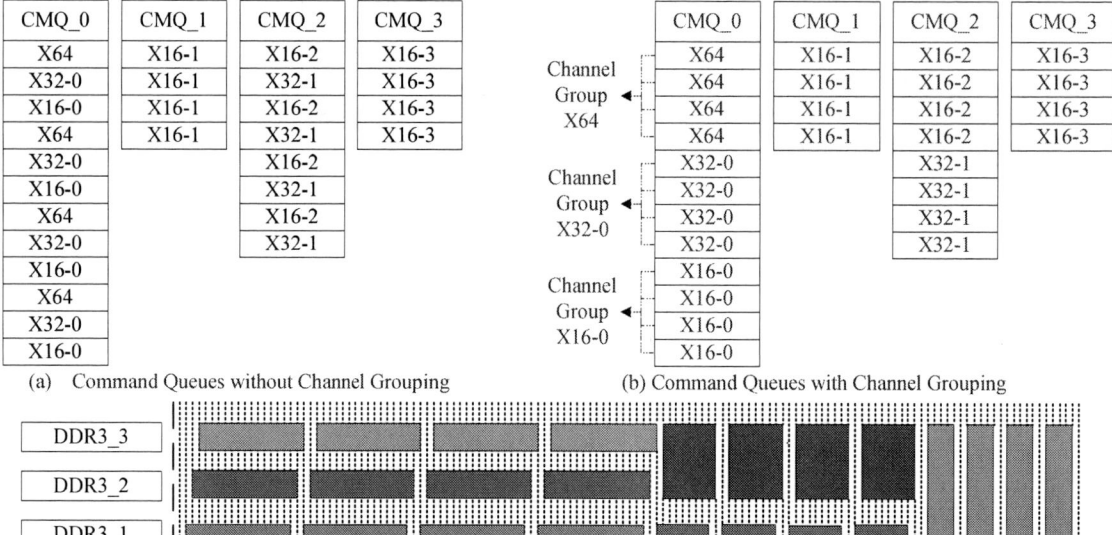

(a) Command Queues without Channel Grouping   (b) Command Queues with Channel Grouping

(c)   Result of Memory Access Scheduling

Figure 4. Example of Memory Scheduling Algorithm. X16 channel, X32 channel and X64 channel need 16, 8 and 4 cycles to transfer the same amount of data (64 bytes in this example), respectively. Different DRAM commands are issued to the corresponding DRAM devices in sequence. In this example, the relative priorities of X16, X32 and X64 channels decline according to the order.

4. METHODOLOGY

4.1 Simulation Environment

We implement HMC in RTL (Register Transfer Level) using Verilog hardware description language on a four-core multiprocessor system, which is a highly modular and hierarchical shared-memory multiprocessing design based multi-core processor, and the detailed configuration of the processor is summarized in Table III. Our baseline has an on-chip memory controller attached to a DDR3-1600 DRAM system, and the detailed configuration and its power model are introduced in Table I and Table II. We simulate the whole system on EVE Zebu platform. Zebu is an extremely high capacity system emulator that supports up to 30MHz emulation frequency, requires no RTL modifications and has up to 1 billion ASIC gates.

TABLE III SYSTEM CONFIGURATION

Technology	32 nm	
Chip Size	300 mm²	
Clock Frequency	2.5 GHz	
Power Dissipation	40 W	
Transistor Count	582.6 M	
Micro Architecture	4 cores, 4-issue out-of-order superscalar, 13-stage dynamic pipeline of each	
L1 cache of each core	64 K 4-way set associative I-Cache and 32 K 4-way set associative D-Cache	
L2 Cache	8 M 4-way set associative	
Memory Controller	Rank	2 ranks
	Depth of Four Command Queues	32/16/16/16
	Narrow Channel Number	7 channel

4.2 Workload

In this paper, we use SPEC CPU2006 to form four-core multiprogramming workloads for evaluation. SPEC CPU2006 benchmark suite [15] is the best available source of heterogeneous applications. Each benchmark was compiled using Gcc 4.3.2 with –O3 optimizations. Based on each benchmark's memory access characteristic, we classify the benchmarks into four categories:

1. Memory intensive and memory friendly (MIMF).
2. Memory intensive and memory non-friendly (MIMNF).
3. Memory non-intensive and memory friendly (MNIMF).
4. Memory non-intensive and memory non-friendly (MNIMNF).

We refer to a benchmark as memory intensive if its LLC MPKI (Last Level Cache Misses Per 1000 Instructions) is greater than 2, otherwise the benchmark can be regarded as memory non-intensive. We refer to a benchmark as memory friendly if its RHR (Row Hit Rate) is greater than 60%, otherwise the benchmark can be regarded as memory non-friendly [16]. We evaluate 10 different combinations of SPEC CPU2006 benchmarks running on a four-core CMP.

Table IV shows the specific classifications and combinations of these benchmarks. MNIMF and MNIMNF type of benchmarks are latency sensitive, and we map them into X64 channel. MIMF type of benchmarks is bandwidth sensitive, and we map it into X32 channels. MIMNF type of benchmarks is bandwidth sensitive, and can cause severe bank conflicts with its and other benchmarks' memory access requests, so we map it into X16 channels.

We measure overall system throughput using the weighted speedup metric, defined as the sum of relative IPC performances of each thread in the evaluated workload [17]:

$$\text{Weighted Speedup} = \sum_{i=0}^{M-1} \frac{\text{IPC}_i^{shared}}{\text{IPC}_i^{alone}}$$

TABLE IV SPEC CPU2006 BENCHMARK CLASSIFICATIONS AND COMBINATIONS

Com	MIMNF	MIMF	MNIMF	MNIMNF	Com	MIMNF	MIMF	MNIMF	MNIMNF
Com1	403.gcc	401.bzip2	444.namd	400.perlbench	Com6	481.wrf	473.astar	434.zeusmp	400.perlbench
Com2	429.mcf	433.milc	447.dealII	445.gobmk	Com7	444.namd	445.gobmk	450.soplex	483.xalancbmk
Com3	456.hmmer	458.sjeng	453.povray	462.libquantum	Com8	458.sjeng	447.dealII	482.sphinx3	436.cactusADM
Com4	470.lbm	416.games	454.calculix	464.h264.ref	Com9	401.bzip2	416.games	453.povray	437.leslie3d
Com5	465.tonto	410.bwaves	471.omnetpp	435.gromacs	Com10	433.milc	454.calculix	435.gromacs	459.GemsFDTD

5. EXPERIMENTAL RESULTS

In this paper, we will compare Heterogeneous Multi-Channel memory access controller (HMC) with 64-bit, 32-bit and 16-bit Homogeneous Multi-Channel memory access controller (HMC64, HMC32 and HMC16) in terms of bandwidth, latency, system throughput and power consumption.

5.1 System Performance Improvement

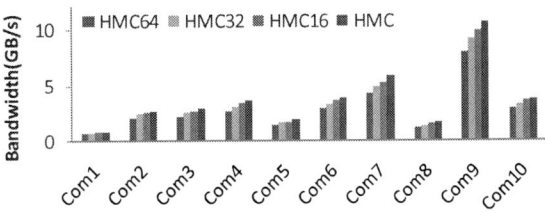

Figure 5. Bandwidth Results of Benchmarks Combinations

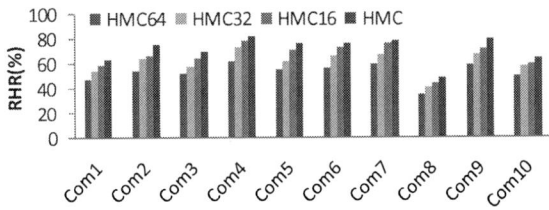

Figure 6. RHR Results of Benchmarks Combinations

Figure 5 shows the average bandwidth of benchmark combinations provided by HMC64, HMC32, HMC16 and HMC. Compared with HMC64, the average bandwidth of HMC32, HMC16 and HMC are improved by 14.7%, 24.7% and 34.6%, respectively. The bandwidth improvements of HMC16, HMC32 and HMC are mainly due to the following four reasons:

1. **High row hit rate.** As is shown in Figure 6, compared with HMC64, the row hit rates of HMC32 and HMC16 are improved by 9.47% and 13.2%, respectively. By controlling DRAM devices independently, multiple banks will be opened concurrently, which decreases bank conflicts and promotes DRAM data bus utilization dramatically.
2. **High bank level parallelism.** The DRAM parameters, such as tRRD (ACTIVE to ACTIVE command period) and tFAW (During the time period of tFAW, only four ACTIVE commands can be issued to a rank), and the terrible bank conflicts prevent BLP to be fully explored. By separating the control of DRAM devices, more sub-banks can be accessed concurrently, which reduces the restrictions of DRAM parameters on BLP and improves RHR.
3. **Hidden delay of read/write turnaround.** Conventional 64-bit DRAM data bus is partitioned into several narrow buses. Because multiple DRAM data buses can be managed separately, while a narrow data bus is pending for read/write turnaround, other narrow data buses can be accessed. Then the delay of DRAM data

bus turnaround can be largely hidden.

4. **Fully utilized DRAM command bus.** All the channels share the same DRAM command bus in sequence, which can increase DRAM command bus utilization largely.

HMC improves DRAM bandwidth more than HMC32 and HMC16, which is because of the following two reasons:

1. The channel interleaving policy implemented in HMC improves the parallelism of different channels.
2. Channel grouping policy makes memory requests with the same channel width be grouped into channel groups. Scheduling memory requests based on channel groups can throttle the memory requests that do not have the same channel width as the channel group being processed. Memory requests with the same channel width usually target for the close address space, and have high RHR. Then FR-FCFS policies schedule memory request within the same channel group to maximize RHR and BLP. As a result, the DRAM data bus utilization is further improved.

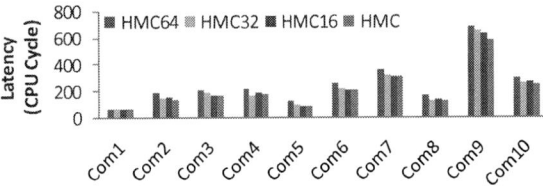

Figure 7. Latency Results of Benchmarks Combinations

Figure 7 shows the average latency of benchmark combinations provided by HMC64, HMC32, HMC16 and HMC. Compared with HMC64, the average latency of benchmark combinations provided by HMC32, HMC16 and HMC are reduced by 14.5%, 15.3% and 20.9%, respectively. To transfer a 64-byte cacheline, a 64-bit, 32-bit, and 16-bit DDR3-1600 rank will cost 4, 8, and 16 DRAM cycles, respectively. Typically, the average latency of a memory access is hundreds of DRAM cycles. The data transferring time between memory access controller and DRAM devices occupies a small part of the memory access latency, and most of the memory access latency is spent on the command queues of the memory access controller. HMC controls DRAM devices independently, and different data streams corresponding to different DRAM commands can flow from or to DRAM devices concurrently. The memory queuing delay, which accounts for most of the time that a memory access stays in the memory access controller, can be greatly reduced.

From a memory intensity perspective, threads can be classified into two categories: latency-sensitive and bandwidth-sensitive. Latency-sensitive threads spend most of their time on computation and average latency of the memory system is crucial to the performance of these threads. Bandwidth-sensitive threads have high LLC MPKI and average bandwidth of the memory system is crucial to the performance of these threads. HMC groups different threads into different channels based on their different memory behaviors, and prioritizes latency-sensitive channel over bandwidth-sensitive channel. Then not only the average memory access latency of all threads is decreased, but also the system performance is largely improved. As is shown in Figure 8, compared with HMC64, the average weighted speedup of benchmark combinations provided by

HMC32, HMC16 and HMC are improved by 7.6%, 16.4% and 27.6%, respectively.

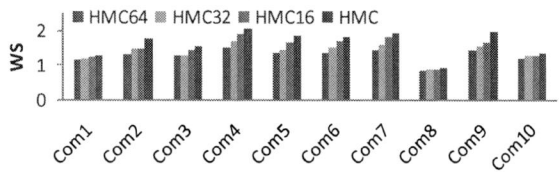

Figure 8. System Performance Results of Benchmarks Combinations

5.2 Implementation Cost and Power Saving

The design of HMC does not require any changes to the DRAM devices, but it needs some additional chip-select (CS) signals for fine-grained control of DRAM devices. There are multiple channels in HMC, and each channel has independent command queue and CS_SM, which introduces additional power dissipation and on-chip area that have to be seriously considered. We have implemented our design in a memory access controller whose maximum clock frequency is 800MHz, and have synthesized with Synopsys IC Compiler on a 32nm CMOS general purpose process library provided by STMicroelectronics Corporation. The average power consumption of memory controller is calculated by Synopsys PrimeTime Suite. The basic technology and characteristics are list in Table V.

TABLE V IMPLEMENTATION COST

Technology	32nm	Clock Frequency	800MHz
Area (μm^2)	Memory Controller		939050
	per 32-entry Command Queue		90057
	per 16-entry Command Queue		63507
	per CS_SM		862
	Physical Interface		424918

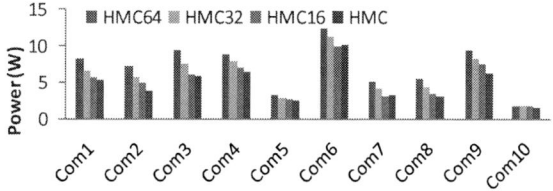

Figure 9. DRAM System Power Consumption (Lower Is Better)

Figure 9 shows the memory system's power consumption of benchmark combinations provided by HMC64, HMC32, HMC16 and HMC. The power consumption includes the power of memory controller. Compared with HMC64, the average power of HMC32, HMC16 and HMC are reduced by 13.3%, 22.5% and 24.2%, respectively. By controlling DRAM devices separately, HMC has three advantages for power savings:

1. The DRAM operation power is reduced to a large extent since fewer DRAM devices are involved in a single memory request.
2. Because of the reduction of bank conflicts, there are fewer DRAM commands are issued to DRAM devices, which also saves the power consumption of DRAM system.
3. Although HMC tries to make the independent narrow channels run in parallel, the parallelism cannot be fully explored and the idle DRAM devices can be transferred to low power mode automatically, which further decreases the power consumption. For space limitation, we don't present the experimental results of low power modes.

6. CONCLUSION

A novel memory access controller, HMC, was presented and implemented targeting for both high performance and power efficiency of the shared DRAM system. Since HMC is superior in terms of memory bandwidth, latency and power consumption, it has significant influence on the design of future multi-core embedded system.

Further improvements can be achieved when applying HMC to multimedia embedded system, which may integrate a variety of functional units, such as GPU (Graphic Processing Unit), DC (Display Controller), HDV (High Definition Video), etc. The memory access behaviors of these functional units can be quite different from each other, and there will be more space for optimization.

7. REFERENCES

[1] K. J. Nesbit, N. Aggarwal, J. Laudon, and J. E. Smith, "Fair queuing memory systems," Proceedings of MICRO, 2006.

[2] B. Diniz, D. Guedes, J. Wagner Meira, and R. Bianchini, "Limiting the power consumption of main memory," Proceedings of ISCA, 2007.

[3] H. Zheng, and Z. Zhu, "Power and Performance Trade-Offs in Contemporary DRAM System Designs for Multicore Processors," IEEE Transactions on Computers, 2010.

[4] S. Rixner, W. J. Dally, U. J. Kapasi, P. Mattson, and J. D. Owens., "Memory access Scheduling," Proceedings of ISCA, 2000.

[5] Micron Corp. Micron 1Gb x4, x8, x16, DDR3 SDRAM: MT41J256M4, MT41J128M8, and MT41J64M16, 2006.

[6] T. Vogelsang, "Understanding the Energy Consumption of Dynamic Random Access Memories," Proceedings of MICRO, 2010.

[7] Calculating memory system power for DDR3. Technical Report TN-41-01, Micron Technology, 2007.

[8] V. Delaluz, M. Kandemir, N. Vijaykrishnan, A. Sivasubramaniam, and M. J. Irwin, "DRAM energy management using software and hardware directed power mode control," Proceedings of HPCA, 2001

[9] F. A. Ware and C. Hampel, "Micro-threaded row and column operations in a DRAM core," Proceedings of UCAS, 2005.

[10] F. A. Ware and C. Hampel, "Improving power and data efficiency with threaded memory modules," Proceedings of ICCD, 2006.

[11] T. M. Brewer., "Instruction set innovations for the Convey HC-1 computer," Proceedings of MICRO, 2010.

[12] H. Zheng, J. Lin, Z. Zhang, E. Gorbatov, H. David, and Z. Zhu, "Mini-rank: Adaptive DRAM architecture for improving memory power efficiency," Proceedings of MICRO, 2008.

[13] J. H. Ahn, J. Leverich, R. Schreiber, and N. P. Jouppi, "Multicore DIMM: An energy efficient memory module with independently controlled DRAMs," IEEE Computer Architecture Letters, 8(1):5-8, 2009.

[14] D. H. Yoon, M. K. Jeong, and M. Erez, "Adaptive Granularity Memory Systems: A Tradeoff between Storage Efficiency and Throughput," Proceedings of ISCA, 2011.

[15] Standard Performance Evaluation Corporation. SPEC CPU 2006. http://www.spec.org/cpu2006/, 2006.

[16] O. Mutlu and T. Moscibroda, "Parallelism-aware batch scheduling: Enhancing both performance and fairness of shared DRAM systems," Proceedings of ISCA, 2008.

[17] S. Eyerman and L. Eeckhout, "System-level performance metrics for multiprogram workloads," Proceedings of MICRO, 2008.

Joint Management of RAM and Flash Memory with Access Pattern Considerations

Po-Chun Huang[1], Yuan-Hao Chang[2,4], and Tei-Wei Kuo[1,3,4]

[1] Department of Computer Science and Information Engineering, National Taiwan University
[2] Institute of Information Science, Academia Sinica
[3] Graduate Institute of Networking and Multimedia, National Taiwan University
[4] Research Center for Information Technology Innovation, Academia Sinica
f95070@csie.ntu.edu.tw, johnson@iis.sinica.edu.tw, ktw@csie.ntu.edu.tw

ABSTRACT

The popularity of flash memory has triggered the emerging of various products with flash memory as storage medium. More advanced architectures with better hardware resources are now explored by vendors to fit different market needs. Different from the past work, this paper proposes to consider RAM as a storage medium together with flash memory to take advantage of the characteristics of both RAM and flash memory. In particular, an adaptive management strategy is proposed with the considerations of access patterns to improve both the system performance and the system endurance. The capability of the proposed approach is evaluated by a series of experiments, for which we have very encouraging results[1].

Categories and Subject Descriptors

D.4.2 [**Operating Systems**]: Storage Management—*Secondary Storage, Storage Hierarchy*

General Terms

Design, Management

Keywords

Hybrid Storage Device, Flash Memory

1. INTRODUCTION

As flash memory becomes popular for data storage, vendors start exploring different product designs and applications with flash memory. Well-known example products are solid-state disks and flash-memory-equipped servers. Due of the explosion of application domains, it becomes important not being restricted by the common RAM-size constraint of the controllers of most flash-memory storage devices. *This work is motivated by this observation and explores the question on what to do if more RAM space is possible for flash-memory storage devices.*

A flash-memory chip usually consists of multiple planes, and each plane is composed of many blocks. Each block is

---

[1] This work was supported in part by the National Science Council under grant Nos. 98-2221-E-002-120-MY3, 100-2628-E-027-008-MY2 and 101-2220-E-001-001, and the Excellent Research Projects of National Taiwan University under grant Nos. 10R80300 and 10R80919-2.

---

of a fixed number of pages, and a block is the smallest unit for erase operations. A page is the basic unit for read/write operations, and a data update to a page is usually done by writing to a free page such that the previous page becomes invalid (referred to as *out-place updates*). Because of out-place updates, a management software layer, referred to the *Flash Translation Layer* (FTL), is usually adopted to handle the address mapping for the logical block addresses (LBA's) of read/write requests and the garbage collection of invalid pages. In the past years, a number of excellent FTL designs were proposed to provide efficient address mapping with limited RAM space and garbage collection overheads. The proposed designs could be classified as fine-grained and coarse-grained address mapping designs:

FTL [1], as one of the very first designs, is a fine-grained design and adopts a mapping table indexed by LBA's. Because of the huge RAM capacity demands from many flash-memory storage devices, various coarse-grained designs were proposed. Excellent examples are such as NAND-type FTL (NFTL) [1] and Block-Associative Sector Translation (BAST) [8], and they basically adopted a mapping table indexed by the corresponding block numbers of LBA's to save the RAM space. To explore a balance between the address mapping efficiency and the storage space utilization (/management overheads), researchers proposed hybrid FTL designs that incorporate both fine-grained and coarse-grained address mapping such that frequently used LBA's are managed by a fine-grained mechanism. Example designs are such as the Superblock [5], K-Associative Sector Translation (KAST) [2], and Adaptive FTL (AFTL) [17]. More recently, some address mapping designs that target the FTL scalability were proposed, such as Demand-based FTL (DFTL) [4] and Convertible FTL (CFTL) [16]. For such approaches, the mapping information of less frequently accessed LBA's is saved on flash memory to reduce the RAM space needs but at a tradeoff in extra flash-memory accesses and performance overheads.

Different from the past work, this paper aims at the joint management of RAM and flash memory by utilizing their individual advantages in performance and capacity. The objective is to explore effective RAM utilization if more RAM space is possible for product designs. In particular, we propose to consider RAM as a storage medium together with flash memory to take advantage of the characteristics of both RAM and flash memory. The closely related idea is to cache frequently-used data over RAM, such as Least-Recently-Used (LRU), Least-Frequently-Used (LFU), Least-Recently/Frequently-Used (LRFU) [9] and Adaptive Replacement Cache (ARC) [14], that is proven very effective as a common practice to close the performance gap between a storage device and the processor. Although the direct application of such a caching algorithm should also be fine in utilizing extra RAM space for existing FTL designs, it does not consider the needs of FTL designs and the characteristics of storage media. On the other hand, some other work such as Cold-Clean-First LRU (CCF-LRU) [10] and Block-Padding LRU (BPLRU) [7] focus on the effective design of write buffers, without much emphasis on the management of mapping information. Such concerns motivate this work in the proposing of joint management of RAM and flash memory in the improvement of the system performance and the system endurance. In particular, an adaptive management strategy is proposed with the considerations of access patterns, in which the RAM space is divided into two parts: One is to keep the most frequently accessed data and their address mapping over RAM, especially those by writes, to reduce the number of writes to flash memory and, thus, to improve both the system performance and the device lifetime. The other one is to keep the mapping information of frequently-used and flash-resident data over RAM to improve the overall performance, especially the read performance. We must point out that this work is orthogonal to the past FTL work. Note that the proposed design could also be used together with many existing FTL designs to serve as a caching module for frequently accessed data and their mapping information. The capability of the proposed approach is evaluated by a series of experiments, which show that in many configurations of RAM space, the proposed scheme outperforms LRU and CCF-LRU for up to 45% of I/O time.

The rest of this paper is organized as follows. Section 2 covers the system architecture and the motivations. Section 3 presents the joint management design of RAM and flash memory, especially a group-based structure and its manipulation strategies. In Section 4, a series of experiments is done to evaluate the design. Section 5 is the conclusion.

2. MOTIVATION

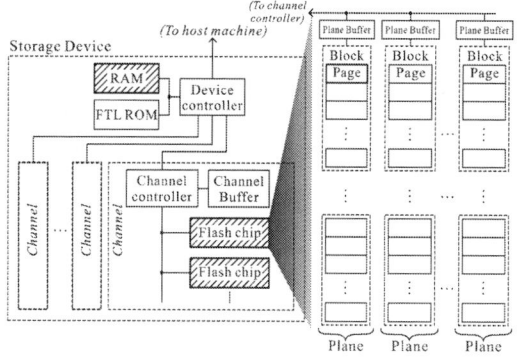

Figure 1: Flash-Memory Device Architecture

A typical flash-memory device has a RAM chip and one or more flash memory chips, as shown in Figure 1. The RAM chip is directly attached to a device controller, which executes the firmware code stored in an ROM chip (and loaded during the device initialization) to perform various management tasks for the device. The flash chips may be grouped into one or more channels, each connected through an independent bus to the device controller. The flash chips of each channel and the planes of the same flash chip could operate in parallel, but the data transmission between the device controller and each flash chip/plane would be one by one. The limitation is due to the constraints of the bus between each channel controller and the flash chips. However, there

could be an independent bus between each channel controller and the device controller for better parallel accesses.

Each plane of a flash chip contains a number of *blocks*, where a block is the unit for erase operations. Each block is of a fixed a number of *pages*, that is the unit for read and write operations. Due to the *out-place-update* property of flash memory, a page can not be overwritten unless its residing block is erased. Such a property complicates the address-translation designs because a given logical block address of an access operation must be mapped to the right page, and pages would become *invalid* when their contents are updated and written to other *free pages* (referred to as *address translation*)). Note that two pages/blocks of different planes of the same flash chip could operate in parallel only when the targeted page/block address within the residing plane is identical. The designs are further complicated because garbage collection might be triggered, when necessary, to reclaim the space of the invalid pages, and the resulted erase operations are seriously constrained by the limit on the number of erase operations over blocks (referred to as the *wear leveling* problem) [3]. Because of the cost considerations, Multi-Level Cell (MLC) flash chips have gained their momentum in recent years, where each cell contains more than one bit of data. However, MLC flash memory comes with slower performance and a worse wear leveling problem and additional constraints, such as sequential writing within a block and no support of the invalidity marking of a page, compared to Single-Level Cell (SLC) flash chips.

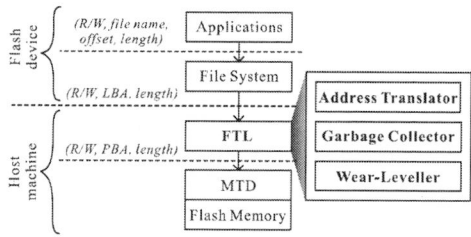

Figure 2: A Typical System Architecture

As shown in Figure 2, file systems receive services from the Flash Translation Layer (FTL) to provide uniform file services to applications and users. The FTL provides LBA address translation, garbage collection, and wear leveling. The Memory Technology Device (MTD) provides the basic services of reads, writes, and erases. The RAM chip of a flash device is usually used to store the house-keeping information for the FTL, such as address translation information. Although it is possible to steal a part of the RAM space to cache data on flash memory (and even some address translation information stored on flash memory) to improve the performance and reliability of the device, typical designs of the FTL usually do not try to utilize both the characteristics of flash memory and RAM in the FTL designs. Instead, the RAM space is only a place for house-keeping information and even caching. Even though caching over the RAM space of a flash device is a good idea, many excellent caching strategies, such as the LRU strategy, should be applied to the caching with joint considerations of the characteristics of flash memory and RAM. Furthermore, the rapid increasing of the flash-memory capacity has imposed tremendous impacts on the handling of address mapping information for data stored on flash memory.

With the above observations, we propose to consider the RAM space as a storage resource together with the flash memory space to take an advantage of the in-place update capability of the RAM and the huge capacity of the flash

883

memory chips at the same time. Different from the past work on FTL designs, an LBA of an FTL design could now be possibly mapped to some place on the RAM space to both improve the performance and the reliability of a flash device. The approach also takes access patterns/locality into considerations in the joint management of the RAM and flash-memory space, compared to approaches with separated caching and FTL designs. The proposed design also intends to have an option to be used together with existing FTL designs, such as DFTL and FAST, as a caching design for the mapping information and flash-resident data over RAM. In the following sections, we shall first propose an FTL design with joint considerations of flash-memory and RAM chips and have the option being explored in the presentation.

3. JOINT MANAGEMENT OF RAM AND FLASH MEMORY

3.1 Overview

This section presents a joint management scheme design for RAM and flash memory. The objective is to explore the characteristics of RAM in in-place updates and those of flash memory in out-place updates and wear leveling. In particular, a group-based structure and strategies are proposed to map recently-used data and writes over RAM to improve the system performance and endurance.

Section 3.2.1 proposes a group-based structure to facilitate the searching of the corresponding RAM or flash-memory locations of the LBA's of read and write requests and their data, if they are RAM-resident. Section 3.2.2 presents strategies to identify the recently-used data, where the least-recent-used (LRU) idea is approximated together with the concept of access frequencies. Random writes and reads are favored and cached as much as possible, due to the wear leveling and performance concerns. In the strategy designs, least recently used LBA's will be degraded gradually in the group-based structure and eventually removed out of the structure. Less recently used data can be managed by existing flash translation layer designs.

3.2 Access-Pattern-Adaptive Address Mapping

3.2.1 A Group-Based Caching Design for Recently-Used Data and Address Mapping

The joint management of RAM and flash memory requires a scalable data structure to not only efficiently identify RAM-residing data but also be adaptive to current access patterns with extremely low overheads and without being constrained by the flash memory capacity. We propose to deploy a RAM-resident mapping table with a two-level address mapping structure and a self-balancing binary search tree (such as red-black tree) to track LBA mapping and their recent usage patterns, as shown in Figure 3. The mapping table is designed for LBA lookup of frequently accessed data, where the address translation information of less-recently-used data are stored on flash memory.

A group-based structure is proposed to provide the mapping information of most-recently-used RAM-resident and flash-memory-resident data, referred to the G_0 and G_1 parts, respectively. The numbers of the levels in G_0 and G_1 are predetermined, depending on the RAM size (Please see the experimental section for the setting exploring). Suppose that G_0 and G_1 have N_0 and N_1 levels, respectively, and the table, thus, has $(N_0 + N_1)$ levels. The j_{th} level, denoted as $G_{0,j}$, of G_0 has 2^j entries, for $0 \le j < N_0$, and occupies the mapping-table entries from $T[2^j - 1]$ to $T[2^{(j+1)} - 2]$. The j_{th} level, denoted as $G_{1,j}$, of G_1 has $2^{(N_0+j)}$ entries, for $0 \le j < N_1$, and occupies the mapping-table entries from

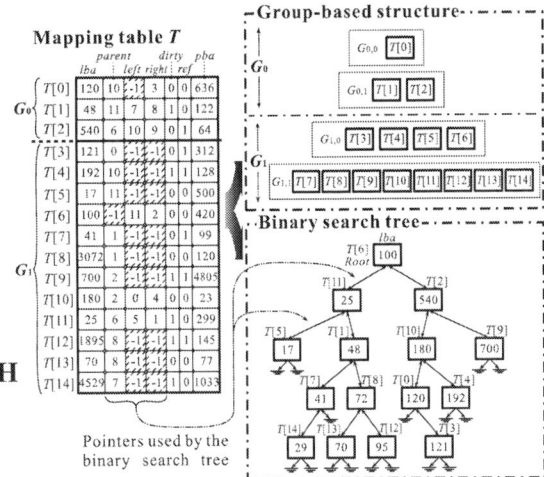

Figure 3: A Group-Based Data Structure for Data and Mapping-Information Caching

$T[2^{(N_0+j)} - 1]$ to $T[2^{(N_0+j+1)} - 2]$. The binary search tree is to facilitate the lookup of the corresponding LBA's of the mapping-table entries. Each entry of the mapping table has seven fields: Field *lba* denotes its corresponding logical block address and serves as its key in the binary search tree. Field *pba* denotes the corresponding physical address, that is either a RAM or flash-memory address, depending on where the corresponding data are. Fields *parent*, *left* and *right* denote its parent, left child and right child entries of the binary search tree, respectively. Field *ref* is a reference bit to show whether the corresponding data are accessed recently. Field *dirty* is a flag to indicate whether the corresponding data have been modified since they were saved on RAM (if they are). The manipulation procedures are in Section 3.2.2.

The design rationale is to keep the most recently used data on RAM and the mapping information of more popular flash-memory-resident data on RAM. The address translation information of less popular flash-memory-residing data can be maintained over flash memory, such as that adopted by DFTL [4]. Note that the proposed RAM-resident mapping table could also work with many fine-grained or coarse-grained address translation designs, such as NFTL [1] or DFTL, and those address mapping designs could be implemented by saving everything over flash memory or be done according to their original designs. What is needed is a mechanism in the merging of mapping information when some mapping information is evicted from the proposed RAM-resident mapping table. In order to reduce the management cost, we propose to store the flash-memory-residing address translation information in a partitioned and specific area of the flash memory. Note that the area would not be heavily updated because updates to frequently/recently used data and their mapping information would be mostly done over RAM (Please see Section 3.2.2). When a flash memory device is unmounted, the RAM-residing data and their address translation information are saved back to the flash memory. When a flash device is mounted, RAM is assumed being empty, and the RAM-residing mapping table is built up during the access processing. For the simplicity of discussion, we assume that the flash-memory-residing address translation information is saved as a huge mapping table indexed by LBA's in a partitioned area of the flash memory for the rest of this paper.

3.2.2 Identification Strategies of Recently-Used Data and Address Mapping

This section proposes a strategy in keeping the mapping information of the recently-used and the most popular data on RAM. The strategy also aims at the minimization of the write amplification and wear-leveling problems by having random writes (that are not writes of sequentially accessed data) done over RAM as much as possible. For the simplicity of discussion, we assume that sequential reads/writes are reads/writes of large sizes.

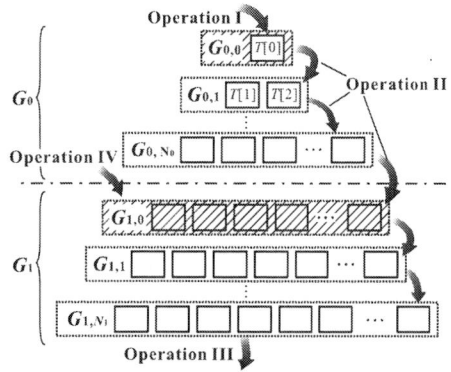

Figure 4: LBA Manipulation for Random Reads/Writes

When a random write request is received, its data and address mapping information are saved on RAM as follows: The LBA of the write request is saved on $T[0]$, $i.e.$, the only entry of $G_{0,0}$ (Operation I in Figure 4), with Fields $dirty$ and ref both being set to 1 to show that the data have been recently updated over RAM (and other fields set accordingly). (Note that when a request has multiple LBA's, each LBA and its data are processed in the following same procedure.) If the LBA is already in the group-based structure, then the corresponding entry is freed, and its corresponding data are also freed (if they are in RAM). It is assumed that each group maintains a singly-linked list of free entries with Field pba as the next pointer. If the original $T[0]$ already has an LBA, then the original $T[0]$ (or simply referred to as an LBA) is moved down to the next group, $i.e.$, Group $G_{0,1}$ (Operation II in Figure 4); otherwise, the procedure ends. If there exists a free entry in Group $G_{0,1}$, then a free entry is allocated to keep the LBA moved down from $T[0]$ (and Group $G_{0,0}$). Otherwise, one entry of Group $G_{0,1}$ is randomly picked up. If Field ref of the randomly-selected entry of Group $G_{0,1}$ is 0, then it is chosen to save the LBA moved down from Group $G_{0,0}$ because the value 0 denotes that its corresponding data are not recently used. The original LBA of the randomly-selected entry is moved down to the next group, $i.e.$, Group $G_{0,2}$. If Field ref of the randomly-selected entry is 1, then it is reset to 0 to give the data a second chance to stay in the current group. The LBA of the original $T[0]$ is moved directly down to the next group, $i.e.$, Group $G_{0,2}$. Each moving step of the above procedure might have the copying of an LBA to a randomly selected entry of a lower group, and the step ends with the finding of a free entry in the group, the triggering of a following step to move an LBA to the next lower group, or the reaching of the last group of G_1. When an LBA is moved down from a group of G_0 to the top group of G_1 (Operation IV in Figure 4), the above step repeats, except that the RAM-resident data must be written to the corresponding flash-memory page managed by the adopted flash-memory

address translation scheme such as NFTL or DFTL, if the data are dirty, and the address mapping information is updated accordingly. If the above step reaches the last group of G_1, $i.e.$, Group G_{1,N_1} (Operation III in Figure 4), then the mapping information of the to-be-moved LBA is simply discarded because the adopted flash-memory address translation scheme has the information. Note that the execution of the above procedure involves the copying of entries, and the parent, left child, and right child pointers of entries in the binary search tree must be updated accordingly. If the LBA of a write request is new to the binary search tree, then it is inserted into the binary search tree after the above procedure. A removed LBA from the last group, if it happens, is deleted from the binary search tree.

When a random read request is received, the corresponding LBA's are first looked up in the binary search tree. Each LBA is processed as follows: There are three cases: (1) If the LBA under considerations is in G_0, then the corresponding entry is copied to $T[0]$ with Field ref being set to 1 (to denote its recent access), and the entry is freed. The original contents of $T[0]$ (or simply referred to as an LBA) is moved down to the next group, $i.e.$, Group $G_{0,1}$ (Operation II in Figure 4); otherwise, the procedure ends. The procedure repeats in the same way as that of random writes. Since the initial LBA under considerations is found in G_0, the procedure will end before any group of G_1. (2) If the LBA under considerations is found in G_1, the corresponding entry is copied to either a free entry or a randomly selected entry of Group $G_{1,0}$ with Field ref being set to 1, and the corresponding entry of the LBA is freed. If there is a free entry, the procedure ends. Otherwise, the procedure repeats in the same way as that of random writes. Since the initial LBA under considerations is found in G_1, the procedure will end before any LBA is dropped from G_1. (3) If the LBA is not found in the binary search tree, the mapping information of the LBA is first found by the flash-memory address translation scheme, such as NFTL or DFTL, and copied to either a free entry or a randomly selected entry of Group $G_{1,0}$ with Field ref being set to 1. The procedure repeats as that of Case (2), except that some LBA might be dropped from G_{1,N_1}, and the initial LBA under considerations must be inserted into the binary search tree. Note that the parent, left child, and right child pointers of the binary search tree must be updated accordingly during the entry copyings.

The above procedure could be illustrated by the processing of a random write: Suppose that a random write of LBA 17 is received, where the group-based structure is as in Figure 3. LBA 17 (with its corresponding fields) is first saved on $T[0]$, and the to-be-written data are also saved on RAM. Since LBA 17 is also found in $G_{1,0}$, the corresponding entry, $i.e.$, $T[5]$, is freed. The original $T[0]$ is then copied to a randomly selected entry in $G_{0,1}$, e.g., $T[1]$, since there is no free entry in $G_{0,1}$, and Field ref of $T[1]$ is 1. The original $T[1]$ is copied to $T[5]$ with its data written to flash memory according to the flash-memory address translation scheme. Note that $T[5]$ becomes a free entry after LBA 17 is processed. If $T[2]$, instead of $T[1]$, is selected in the above process, $T[2]$ will be given a second chance, and the original contents of $T[0]$ will be moved to $T[5]$ because Field ref of $T[2]$ is 1. The parent, left child, and right child pointers of the above entries should be update accordingly.

When sequential reads or writes are received, they are processed directly by the adopted flash-memory address translation scheme. It is because sequential reads or writes often access cold data. Suppose that we adopt a native scheme with a huge mapping table indexed by LBA's in this paper, for the simplicity of discussion. When a sequential read is received, the group-based structure is first searched for its corresponding LBA's, and all of the found data in G_0 are returned. If there is any pending LBA of the read re-

885

quest, then their corresponding physical block addresses are retrieved through the table, and the data are then retrieved from flash memory. When a sequential write is received, the needed free flash-memory space is allocated for the write, and the mapping table is updated accordingly. If any LBA of a sequential write is found in the binary search tree of the group-based structure, then its corresponding entry is freed.

4. PERFORMANCE EVALUATION

4.1 Experimental Setup

In this section, the performance of the proposed group-based caching design and several state-of-art work are evaluated through trace-driven simulation. To our best knowledge, the majority of the related work focus on the design of the caching/buffering of either the address mapping information or data that are likely to be referenced in the near future, but not both. For a fair comparison, we conduct necessary extensions to two existing caching schemes, *LRU* and *CCF-LRU*, so that they could manage address mapping information and data simultaneously[2]. The extensions adds an extra LRU queue to LRU and CCF-LRU schemes, where each entry of the original LRU and CCF-LRU caching structures manages the RAM-resident data of an LBA, and each entry of the extra LRU queue stores the address mapping information of a data page on flash memory. When the original LRU or CCF-LRU caching structures remove an LBA for a newly-accessed LBA, the LBA and its address mapping information are added to the head of the extra LRU queue. (The dirty data are written to flash memory, if needed.)

To avoid moving/traversing as many entries as those of the whole list in the worst case, the proposed joint management scheme suggests a space(/time)-efficient group-based structure to inherently cluster the likely-to-be-accessed LBA's in RAM. Meanwhile, LRU and CCF-LRU have to use doubly-linked lists to organize their mapping entries, so that newly-accessed LBA's could be added to the queue head without moving a great number of mapping entries. However, a linked list requires a considerable amount of RAM space on the internal pointers that connects the neighboring entries. The size of a mapping entry with different caching and searching schemes, along with the relative space savings of the group-based structure compared to caching schemes using doubly-linked lists, are summarized as Table 1[3].

	Hash Table (Open Addressing)	Hash Table (Chaining)	Red-black Tree
Group-based	9 Bytes	13 Bytes	21 Bytes
LRU/CCF-LRU	17 Bytes	21 Bytes	29 Bytes
Space Saving	47.1%	38.1%	27.6%

Table 1: Mapping Entry Size of Various Schemes

In the simulation, DFTL, an efficient flash management scheme, is integrated with the proposed caching design and existing work. DFTL stores a page-grained mapping table on flash, where the mapping table is scattered across many mapping pages, each of which saves the address mapping information of pages with adjacent LBA's, and the address of the mapping pages are kept in RAM. Various caching schemes are then used to maintain the address mapping information of frequently-used data pages for efficient accesses.

[2]Some caching schemes such as ARC and LRFU are not compared as it is not directly feasible to extend them for the experiments. Other schemes such as BPLRU are not included, for that they are specifically designed for log-block-based FTL schemes.

[3]For the simplicity of computation, we assume that the three bit flags *ref*, *dirty* and *color* (used by the red-black tree) use one byte in total, while each remaining field requires four bytes.

As a dirty entry is removed from RAM, its address mapping information is merged with the flash-resident table of DFTL. Section 4.2 compares the overheads of the proposed scheme and those of the prior arts, based on DFTL.

For the performance evaluation, we use a server-level disk access workload from MSR Cambridge traces in SNIA IOTTA repository [13], two widely-used benchmarks (PostMark and Iozone) [6, 11], as well as a trace of common applications, such as web browsing and program installation, on a generic personal computer. The application traces are collected with the Disk Monitor utility in Windows Sysinternals Suite under Windows XP service pack 3 [12]. The disk capacity is 250GB. Basic parameters of the traces used throughout the simulation are listed as in Table 2. On the other hand, the characteristics of the NAND flash-memory chip used in the simulation are summarized in Table 3 [15].

Workload	#I/O req.	#LBA's	Write ratio
IOTTA	14235176	1378321429	1.83%
Postmark	8401557	85317907	63.82%
IOZone	428800	43885863	49.56%
Web-based apps	766404	9933456	35.54%

Table 2: Basic Parameters of the Workloads

Page size (data area/spare area)	4KBytes/224Bytes
Block size	512KBytes
Page read/write time	$25\mu sec./500\mu sec.$
Block erase time	2msec.

Table 3: Properties of Common Flash-Memory

4.2 Analysis and Experimental Results

4.2.1 A Remark of the Lifetime of an Accessed LBA

Due to the use of randomization techniques of the LBA migration between groups, we need to clarify that an LBA will not be undesirably removed soon after it was accessed. The expected "lifetime" of a accessed LBA, *i.e.*, the number of LBA accesses that a recently-accessed LBA lasts in the group-based structure, could be derived easily and is found linear to the number of entries of the group-based structure.

4.2.2 Experimental Results and Discussion

We first evaluate the performance of the group-based structure with LRU and CCF-LRU, which have been extended to enable them to manage the data and address mapping information simultaneously. Figure 5 shows the flash memory I/O overheads with the proposed group-based structure (shown as *Joint Translation Layer (JTL)*) and existing caching schemes (*LRU+* and *CCF-LRU+*) with different RAM size configurations under the four workloads. In each figure, the X-axis is the size of the total RAM space available for the maintenance of data and address mapping information, while the Y-axis is the total time spent on the flash read/write operations throughout the workloads. For a fair comparison, in all configurations, half of the RAM space is allocated for the RAM-resident data, and all three schemes are equipped with the same number of RAM-resident sectors. In the meantime, the other half of the RAM space is allocated to maintain the address mapping information that is likely to be reused soon. Due to the additional space overheads of LRU+ and CCF-LRU+, each entry of these two schemes takes 29 bytes of space, while an entry of JTL only takes 21 bytes, as red-black tree is used in all schemes.

As in Figure 5, the performance of all three schemes is enhanced as the RAM space grows, and JTL outperforms

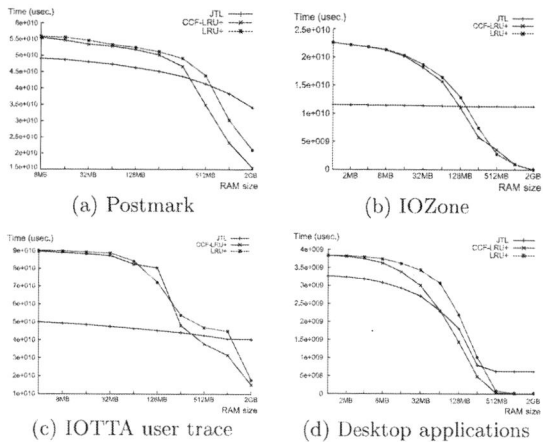

(a) Postmark   (b) IOZone

(c) IOTTA user trace   (d) Desktop applications

Figure 5: Read/Write Overheads of Four Workloads

LRU+ and CCF-LRU+ by 10%–45% of flash-memory I/O time in most of the cases, except for those with an excessive size of RAM space. This is due to JTL's idea that sequentially-accessed data are treated as cold and directly read from/written to flash memory. As the contrast, LRU+ and CCF-LRU+ try to keep all data in RAM as there is sufficient RAM space. Thus, for the cases with an excessive amount of RAM space, LRU+ and CCF-LRU+ could store all data in RAM, and yield lower overheads than JTL (those data shall still be written back to flash memory when the device is unmounted). Note that with common sizes of the on-device RAM space, such as 256MB or less, JTL and CCF-LRU+ both outperforms LRU+, as they exploit the read-write performance skew of flash memory by giving the written data with higher tendency to stay in RAM.

It is an issue that, given the total RAM size, how much RAM space should be assigned to keep data and mapping information respectively. Based on the desktop application workloads, we fix the size of overall RAM space and observe the performance impacts of various sizes of the area for the RAM-resident data. The results for the on-device RAM size (128MB) of a typical mid-end or high-end solid-state disk and those under an extreme case (2GB) are given in Figure 6. As system performance deteriorates with the decreasing of the RAM-resident data size, it could be concluded that the size of the RAM space dedicated to the storing of RAM-resident data should be set to a half of the total RAM space, so as to respect the slow write performance of flash memory.

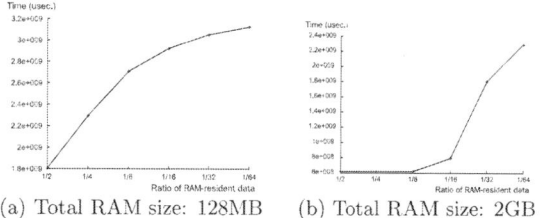

(a) Total RAM size: 128MB   (b) Total RAM size: 2GB

Figure 6: Performance w.r.t. the Ratio of Data against Mapping Information in RAM

5. CONCLUSION AND FUTURE WORK

This paper is motivated by the emerging of various flash-memory products in which more RAM space might be possible, compared to low-end flash-memory devices like SD cards. In this work, we do not intend to re-invent LRU-like caching algorithms to cache mapping information or data or to propose address mapping designs that could use extra RAM space in a more luxury way. Instead, this paper aims at the joint management of RAM and flash memory by utilizing their individual advantages in performance and capacity. In particular, a group-based structure and manipulation strategies are proposed to approximate the ideas of recently frequently used algorithms w.r.t. the characteristics of flash memory and RAM. The RAM space is divided into two parts: One is to keep the most frequently accessed data and their address mapping over RAM, especially those by writes, to reduce the number of writes to flash memory and, thus, to improve both the system performance and the device lifetime. The other one is to keep the mapping information of frequently-used and flash-resident data over RAM to improve the overall performance, especially the read performance. The capability of the proposed approach is evaluated by a series of experiments, which show that in many configurations of RAM space, the proposed scheme outperforms LRU and CCF-LRU by up to 45% of I/O time.

For future research, we shall further explore the designs of high-end flash-memory storage devices and their best configurations. As flash memory gains more and more momentum, more research is needed to exploit flash memory in the designs of a memory hierarchy with flash memory included.

6. REFERENCES

[1] A. Ban. Flash file system. *US Patent 5404485*, 1995.

[2] H. Cho, D. Shin, and Y. I. Eom. KAST: K-associative sector translation for NAND flash memory in real-time systems. In *Design, Automation Test in Europe Conference*, DATE '09, April 2009.

[3] E. Gal and S. Toledo. Algorithms and data structures for flash memories. *ACM Comput. Surv.*, 37(2), June 2005.

[4] A. Gupta, Y. Kim, and B. Urgaonkar. DFTL: a flash translation layer employing demand-based selective caching of page-level address mappings. In *International Conference on Architectural Support for Programming Languages and Operating Systems*, ASPLOS '09, 2009.

[5] J.-U. Kang, H. Jo, J.-S. Kim, and J. Lee. A superblock-based flash translation layer for NAND flash memory. In *ACM/IEEE International Conference on Embedded Software*, EMSOFT '06, 2006.

[6] J. Katcher. PostMark: A new file system benchmark. Technical Report TR3022, Network Appliance Inc., 1997.

[7] H. Kim and S. Ahn. BPLRU: a buffer management scheme for improving random writes in flash storage. In *USENIX Conference on File and Storage Technologies*, FAST'08, 2008.

[8] J. Kim, J. M. Kim, S. Noh, S. L. Min, and Y. Cho. A space-efficient flash translation layer for compactflash systems. *IEEE Trans. Consum. Electron.*, 48(2), May 2002.

[9] D. Lee, J. Choi, J.-H. Kim, S. Noh, S. L. Min, Y. Cho, and C. S. Kim. LRFU: a spectrum of policies that subsumes the least recently used and least frequently used policies. *IEEE Trans. Comput.*, 50(12), December 2001.

[10] Z. Li, P. Jin, X. Su, K. Cui, and L. Yue. CCF-LRU: a new buffer replacement algorithm for flash memory. *IEEE Trans. Consum. Electron.*, 55(3), August 2009.

[11] IOzone Filesystem Benchmark. http://www.iozone.org.

[12] Microsoft Diskmon for Windows. http://technet.microsoft.com.

[13] SNIA: The IOTTA Trace Repository. iotta.snia.org.

[14] N. Megiddo and D. Modha. Outperforming LRU with an adaptive replacement cache algorithm. *Computer*, 37(4), April 2004.

[15] Micron. Spectek FxxM72A flash-memory chip datasheet. 2010.

[16] D. Park, B. Debnath, and D. Du. CFTL: a convertible flash translation layer adaptive to data access patterns. In *ACM SIGMETRICS International Conference on Measurement and Modeling of Computer Systems*, SIGMETRICS '10, June 2010.

[17] C.-H. Wu and T.-W. Kuo. An adaptive two-level management for the flash translation layer in embedded systems. In *ACM/IEEE International Conference on Computer-Aided Design*, ICCAD '06, 2006.

Hybrid DRAM/PRAM-based Main Memory for Single-Chip CPU/GPU

Dongki Kim[†]   Sungkwang Lee[†]   Jaewoong Chung[‡]   Dae Hyun Kim[‡],

Dong Hyuk Woo[‡]   Sungjoo Yoo[†]   Sunggu Lee[†]

[†]Dept. of Electrical Engineering, POSTECH   [‡]Intel Labs, Intel Corp.

{dongki.kim, gwangyi, sungjoo.yoo, slee}@postech.ac.kr   {jaewoong.chung, daehyun.kim, donghyuk.woo}@intel.com

ABSTRACT

Single-chip CPU/GPU architecture is being adopted in high-end (embedded) systems, e.g., smartphones and tablet PCs. Main memory subsystem is expected to consist of hybrid DRAM and phase-change RAM (PRAM) due to the difficulties in DRAM scaling. In this work, we address the performance optimization of the hybrid DRAM/PRAM main memory for single chip CPU/GPU. Based on the tight requirements of low latency from CPU and the relative tolerance to long latency from GPU, DRAM is first allocated to CPU while PRAM with longer write latency is allocated to GPU. Then, in order to improve the write performance of GPU traffic, we propose (1) an in-DRAM write buffer to accommodate GPU write traffics, (2) dynamic hot data management to improve the efficiency of write buffer, (3) runtime-adaptive adjustment of write buffer size to meet the given CPU performance bound, and (4) CPU-aware DRAM access scheduling to give low latency to CPU traffics. The experiments show that the proposed method gives 1.02~44.2 times performance improvement in GPU performance with modest (negligible) CPU performance overhead (when compute-intensive CPU programs run).

Categories and Subject Descriptors: B.3.2 [Memory Structures]: *Design Styles*

General Terms: Design, Management, Performance

Keywords: Main memory subsystem, single-chip CPU/GPU, phase-change RAM

1. Introduction

Single chip CPU/GPU is becoming popular in both PC and high-end embedded systems, e.g., Intel Sandy Bridge for PC [1] and nVidia Tegra 3 for smartphone and tabletPC [2]. Such a system is expected to require more memory performance and capacity due to more advanced graphics and video applications and user programs requiring large data sets [3][4].

DRAM is expected to face a significant problem to meet the increasing demand of high performance and large capacity main memory. It is mainly because DRAM is approaching its scaling limit [5]. Recently, emerging memory technologies are actively studied to complement or replace DRAM in the near future. Phase-change RAM (PRAM) is one of the most promising candidates [6][7] since it is already in mass production [8], standardized as LPDDR2-N [9] and large size designs are available [10][11].

When PRAM is applied in the main memory, a hybrid

DRAM/PRAM configuration is considered to be the practical choice. DRAM provides low latency reads/writes and practically unlimited write endurance, which compensates for slow reads/writes and poor write endurance in PRAM. A recently released chip enables such a hybrid configuration by including both PRAM and DRAM dies in a single chip package [11].

In the hybrid DRAM/PRAM main memory, there can be two structures: hierarchical or flat. Figure 1 illustrates both structures. In the hierarchical structure, DRAM plays the role of last level cache. PRAM works as the large main memory exploiting both the benefits of larger capacity (due to better scaling) and low standby power (due to non-volatility). The flat structure can provide higher memory bandwidth due to dual channels requiring that the memory address space is partitioned into DRAM and PRAM regions.

The single chip CPU/GPU is expected to take the flat structure. It is due to two reasons. First, GPU programs require large memory bandwidth. Thus, two memory channels provided by the flat structure can give double the memory bandwidth of hierarchical structure. Second, the flat structure can give a better performance isolation to CPU programs which have tight latency requirements. In the flat structure, DRAM can be allocated to CPU data and some write-intensive GPU data while most of GPU data are allocated to PRAM. Thus, CPU can access DRAM without significant interference from GPU traffics.

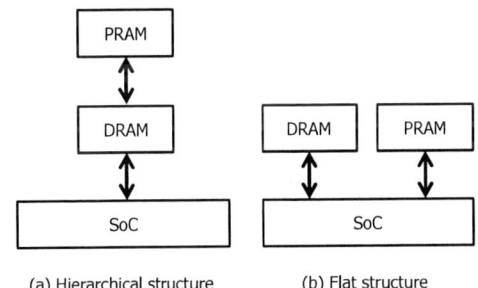

(a) Hierarchical structure   (b) Flat structure

Figure 1 Two structures of hybrid DRAM/PRAM

In our work, we aim at the high performance flat structure of hybrid DRAM/PRAM for CPU/GPU. One of the most important performance problems in the flat structure is how to assign GPU data to DRAM and PRAM. Keeping write-intensive data in DRAM will improve the write performance of GPU program compared with the case that all the GPU data are assigned to PRAM with long write latency. However, too much GPU traffic to DRAM will increase the latency of CPU requests via increased queuing delay in the memory controller and increased memory bank/channel conflicts thereby degrading CPU program performance. In order to resolve this problem, we propose the following novel ideas,

- an in-DRAM write buffer which contains write-intensive GPU data in DRAM thereby improving the write performance of GPU programs

- a hot write data management for the in-DRAM write buffer which dynamically identifies hot write data during runtime thereby improving the efficiency of in-DRAM write buffer

- a runtime-adaptive method to adjust the write buffer size to meet the given CPU performance bound

- a novel method of DRAM memory access scheduling which prioritizes latency-sensitive CPU requests in a way that CPU performance is improved while GPU performance degradation is limited.

This paper is organized as follows. Section 2 reviews related work. Section 3 gives the preliminary. Sections 4 and 5 explain our motivation and proposed methods, respectively. Section 6 reports experimental results. Section 7 concludes the paper. Appendix includes additional experimental details and results.

2.   Related Work

Recently, there have been active studies on PRAM for the main memory. In this section, we focus on performance-related studies. In [12], Qureshi et al. present a hybrid DRAM/PRAM main memory where DRAM works as the working memory and PRAM as the large background memory. In [13], Dhiman et al. present a data migration method for the flat DRAM/PRAM which identifies hot write data to PRAM by comparing the write count with the given threshold and, if the write count exceeds the threshold, they are migrated from PRAM to DRAM. This work is similar to ours in that write-intensive data are stored in DRAM. Our difference is two-fold. First, our proposed method takes into account CPU performance in both ways of considering the given performance bound of CPU program and prioritizing CPU requests in DRAM memory access scheduling. Second, our proposed method is dynamic in identifying hot write data. Thus, it can adapt to dynamically changing write behavior (e.g., time varying hot write data) in GPU programs thereby giving better performance.

Differential writes are typically adopted to reduce the number of bit updates in write operations [11][14][15]. In the differential writes, the existing PRAM data are compared with the new write data. Then, only the updated bits are stored in PRAM. PRAM has typically an internal write bandwidth constraint to meet the chip-level peak power constraints. For instance, the PRAM in [16] writes 16 bits at a time while the one in [11] writes 32 bits per internal write. Thus, PRAM write latency is determined by the amount of updated bits divided by the internal write data width.

In wear leveling, the performance overhead of data migration required in wear leveling can cause significant performance degradation. Especially, due to the limited write endurance, PRAM is vulnerable to malicious attacks, e.g., repeat address attacks. In [17], Qureshi et al. present a wear leveling which adapts the rate of wear leveling depending on the possibility of malicious attacks.

In order to optimize memory access scheduling for GPU applications, conventional methods for stream processors can be applied, e.g., scatter and gather [18], and thread-aware data parallel memory subsystems [19]. However, there are few GPU-specific approaches. Commercial GPU designs perform batch scheduling where requests accessing the same DRAM row are accumulated and processed in one shot thereby reducing the number of row activations and improving memory utilization [20]. Recently, in [21], Lakshminarayana and Kim present a set of memory access scheduling policies for GPU where the warp-level fairness is imposed by considering the number of retired and to-be-executed instructions per warp.

3.   PRAM Characteristics

The PRAM cell stores information based on phase change between amorphous (reset) and crystalline (set) states [7]. The write operation, i.e., the phase change requires melting the phase-change material at high temperature (about 600°C for reset and 300°C for set). Thus, the write operation requires high write latency and power consumption. Due to the heating process, the PRAM cell has write endurance limit. Thus, it is imperative to minimize writes in order to reduce the overhead and improve lifetime.

Table I compares LPDDR2 DRAM[1], LPDDR2-N PRAM [9][11], and our projections of PRAM in 28nm and 14nm technologies[2]. Read bandwidth is not shown in the table since the PRAM already gives the same bandwidth as LPDDR2 DRAM [11]. However, PRAM read latency is slightly worse than DRAM. More importantly, both current and projected PRAMs give inferior write bandwidth to DRAM. Thus, in the hybrid DRAM/PRAM main memory, DRAM needs to be utilized to store write-intensive data in order to overcome such an overhead.

Table 1 DRAM and PRAM Characteristics

	DRAM	PRAM		
Technology node		58nm	28nm	14nm
Read latency	45ns	98ns		
Write latency	45ns	490μs	130μs	40μs
Write bandwidth	1.6GB/s	2.0MB/s	7.5MB/s	24.4MB/s

There have been presented two interface architectures for PRAM. One is LPDDR2-N interface where the read path is like DRAM having (small) row buffers while the write path is like conventional non-volatile memory, e.g., NAND flash, having a program buffer [9]. In the other architectures [12][22], both read and write paths are like DRAM. Our methods can be applied to both architectures.

4.   Observation

Figure 2 shows two histograms on write data in two GPU benchmark programs, RAY and BLK which are obtained from running GPGPUsim [23] (more details of experimental setup will be given in Section 6.1). The x-axis represents data at the granularity of 64B. As shown in Figure 2 (a), in the case of RAY, only a small portion (1.9%) of data generates the majority (95.4%) of write traffic to main memory. Thus, the idea of storing write-intensive data in DRAM is expected to require small space in DRAM.

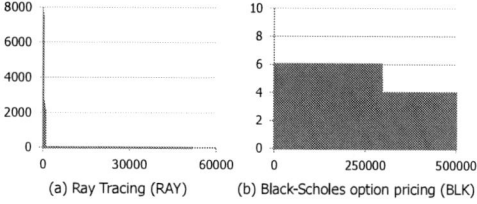

(a) Ray Tracing (RAY)   (b) Black-Scholes option pricing (BLK)

Figure 2 Write histograms (y: # writes, x: sorted addresses)

In Figure 2 (b), the program BLK shows a different write behavior with uniform writes over wide address regions. Even in such a case, we expect that, if the write behavior has temporal locality and the temporal locality can be tracked during runtime (as in our work),

---

[1] LPDDR2 DRAM, 16b, 6-6-6, 400MHz.

[2] The projection is based on the real PRAM performance data at 58nm technology in [11] and the fact that, given a chip-level peak current constraint, 2X technology scaling gives 4X improvement in PRAM write bandwidth (2X from write latency reduction and 2X from write power reduction).

our proposed in-DRAM write buffer can be effective to obtain write coalescing for GPU traffics.

Write behavior changes dynamically during runtime. Figure 3 shows the dynamically changing write behavior for the hot write data (top 16% of data in Figure 2 (a)). As the figure shows, the required write bandwidth changes over time. Too much traffic to the DRAM write buffer can increase the latency of CPU request due to increased conflicts in DRAM (banks and channel). Thus, the write buffering in DRAM needs to take into account the dynamically changing behavior in order to avoid too much conflict in DRAM accesses (during hot GPU write periods).

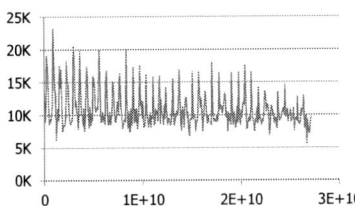

Figure 3 Dynamic write behavior (y: # writes, x: time)

5.   Proposed Methods

5.1 Basic Ideas

Figure 4 illustrates our proposed ideas. Basically, we assign CPU data on DRAM address and GPU data on PRAM address spaces in the flat DRAM/PRAM memory subsystem. It is because GPU programs tend to be latency-tolerant by exploiting thread-level parallelism to hide memory access latency [21]. However, due to the low write bandwidth in PRAM, if all the GPU data are stored in PRAM, GPU write performance is degraded significantly according to our experiments. Thus, we allow all the GPU write traffics to pass through DRAM as the dashed arrow (write request) from GPU to DRAM shows in Figure 4.

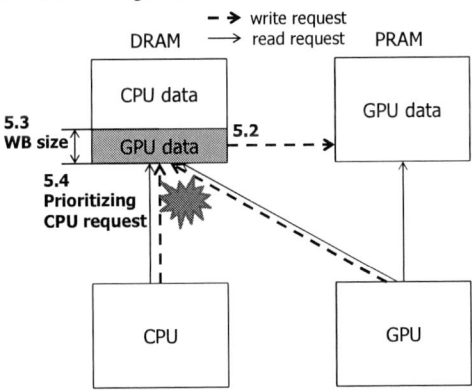

Figure 4 Basic ideas of the proposed method

In our hybrid DRAM/PRAM architecture, DRAM is divided into two regions: one for CPU data and the other for GPU data. The region for GPU data (shaded rectangle in Figure 4) corresponds to the in-DRAM write buffer. The write buffer has a maximum size. Thus, if the footprint of GPU write data is larger than the write buffer size, then there is replacement when data with a new GPU address enter the write buffer (more details in Section 5.2). The OS memory manager in CPU is aware of the write buffer area (with a maximum size) and does not perform memory allocation for the DRAM address space dedicated to the write buffer.

The GPU-induced DRAM traffics (both reads and writes from/to

the GPU data in the write buffer) incur additional memory access latency in CPU requests. Thus, we assume that a bound of CPU performance loss, in short performance bound (PB) is given (by the OS or the designer). In order to maximize the utility of write buffer (reduction in PRAM writes) while meeting the given performance bound, the write buffer is controlled in two ways. First, the size of write buffer is dynamically adjusted (Section 5.3). Second, the GPU data to be stored in the write buffer are selected to better utilize the write buffer. To do that, the list of write-intensive address is managed dynamically in order to select the most effective data for write buffering (Section 5.2). In addition, we try to reduce the latency of CPU requests even when DRAM receives additional GPU traffics (Section 5.4).

5.2 Dynamic Hot Write Data Management

Figure 5 illustrates our method of identifying write-intensive data. The memory controller keeps track of a list of recent write addresses from GPU programs which is called *write address list*. Figure 5 (a) exemplifies the list. The highest (lowest) entry represents the most (least) frequently written address. The list is divided in halves into two hot and cold regions. Each entry in the list has the original address in PRAM and an allocated address in DRAM. Note that the number of write address list entries corresponds to the write buffer size. In our experiments, each entry in the list corresponds to 64B data.

On each GPU write access, the list is updated. In Figure 5 (a), assume that address B in the hot region receives a write request. Then, its rank is promoted by one rank as the upward arrow shows while its direct high-rank neighbor (address A in this case) is demoted by one rank. In the case that an address in the cold region receives a write request, then the corresponding entry enters the hot region as the bottom entry of hot region as the arrow from F to C shows in Figure 5 (a). Correspondingly, the previous high-rank entries (from addresses C to E in this case) are demoted by one rank.

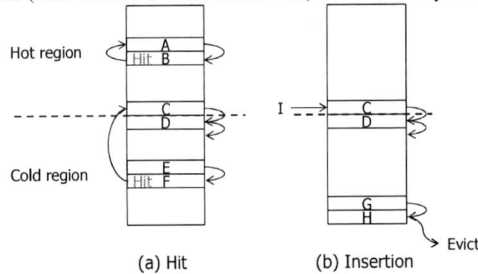

Figure 5 Hot write data identification

In the case that a new GPU address, which is not present in the write address list, receives a write request, the write address list is updated as shown in Figure 5 (b). The newly arrived address is inserted at the bottom of hot region while the lowest priority entry in the cold region is evicted from the write address list. The insertion of new address into the write address list requires storing the corresponding write data in the write buffer. The eviction from the write address list incurs a write-back of corresponding data from DRAM to PRAM.

The write address list is utilized in locating required data in the write buffer during both read and write operations. When a request arrives at the memory controller. First, the address is checked to see whether it belongs to GPU data or not (based on CPU and GPU address ranges). If it belongs to GPU data, then the write address list is consulted to see if the corresponding data are stored in the write buffer. If found in the write buffer, the corresponding data in the write buffer are accessed (read or written). If the data are not found in the write buffer, in the case of read access, the PRAM provides the data. In the case of write access, the data are written to the write buffer as previously explained in the insertion into the write address list (Figure 5 (b)).

5.3 Runtime Adjustment of Write Buffer Size

In the case that a CPU performance bound is given, the dynamic hot write data management in Section 5.2 requires a design-time step which explores the write buffer sizes and finds a suitable write buffer size to meet the given bound. In this section, we present a runtime method to meet the given CPU performance bound which does not require a design-time step. In order to meet the performance bound, we adjust the write buffer size during runtime. Figure 6 illustrates our runtime adjustment. We divide time into epochs. On each epoch, we calculate the difference between the target and current CPU performance (Section 5.3.1), and adjust the write buffer size for the next epoch (Section 5.3.2).

5.3.1 Calculating Performance Difference

On each epoch (100 thousand cycles in our experiments), we obtain both the current CPU performance (IPC_{cur}) in terms of instruction per cycle (IPC) and the estimation of CPU performance (IPC_{alone}) which might be obtained when the CPU alone accesses DRAM. In order to obtain the current performance level, IPC_{cur} ($=N_{inst}/T_{epoch}$, where N_{inst} is the number of instructions retired in the epoch and T_{epoch} epoch length in cycles), we utilize profiling capabilities available in modern CPUs [24]. IPC_{alone} is estimated as follows.

$$IPC_{alone} = N_{inst}/(T_{epoch}-AL) \qquad (1)$$

where AL represents additional memory access latency incurred by the GPU traffics to DRAM. The memory controller can calculate AL by identifying whether each CPU request is delayed by the GPU request and, if so, accumulating wait cycles.

In our experiments, we used an in-order uni-processor architecture. In case of multi-core architectures, we need to calculate IPC_{alone} for each core. To do that, AL needs to be calculated for each core. In this case, AL includes the wait cycle caused by traffics from the other CPU cores as well as GPU traffics. The evaluation of our proposed performance estimation on multi-core architectures will be our future work.

5.3.2 Adjusting Write Buffer Size

The runtime adjustment tries to keep CPU performance around the target performance as shown in Figure 6. The target performance is set to IPC_{alone} − PB. In order to meet the target performance, we utilize two limits, upper and lower limits (UL and LL) as shown in Figure 6. The limit size is set to ±1% of the target performance in our experiments.

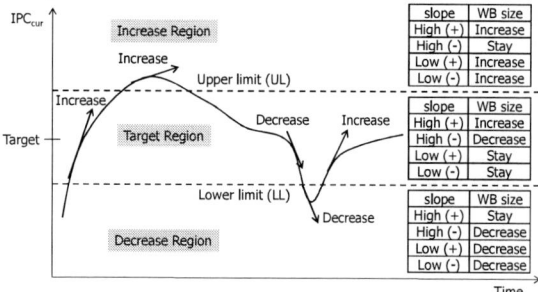

Figure 6 Runtime adjustment of write buffer size

During runtime, if the current CPU performance, IPC_{shared} becomes out of the limit, then the write buffer size can be adjusted as shown in the tables of Figure 6. If IPC_{shared} moves lower than LL, then depending on the trend of IPC change, the write buffer size is adjusted. If IPC_{shared} tends to increase rapidly (corresponding to 'High (+)' in the table in the Decrease region), then we keep the current write buffer size expecting that the performance will recover soon. In the other cases, we decrease the write buffer size to reduce GPU traffics to DRAM. The step size is 1024 entries (128 entries when the write address list size is 1024) in our experiments.

When the CPU performance is between UL and LL, if there is no abrupt change in performance, we keep the current write buffer size as the table in Target region shows in Figure 6. When the CPU performance is beyond UL, we try to increase the write buffer size to allow for more GPU traffics to DRAM in the opposite manner to the case of Decrease region.

5.4 Prioritizing CPU Requests in DRAM Access Scheduling

The idea of the proposed CPU-aware memory access scheduling (MAS) is to favor CPU requests over GPU ones in order to give low latency to CPU requests which are typically more sensitive to the increase in memory access latency than GPU ones. Figure 7 shows the flow of our CPU-aware memory access scheduling. In our work, we implement two separate request buffers, CPU and GPU request buffers in the memory controller.[3]

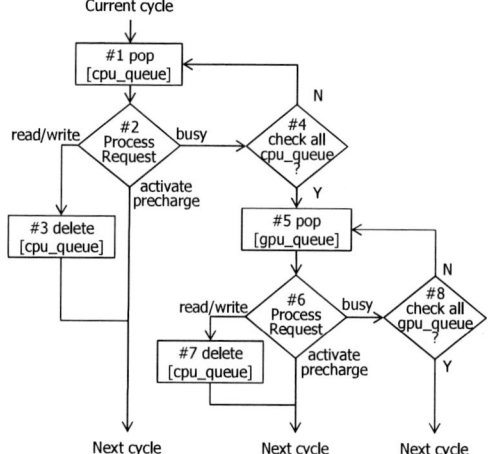

Figure 7 CPU-aware memory access scheduling

On each clock cycle, the memory controller checks the request buffers. First, the CPU request buffer is searched to prioritize CPU requests and the oldest request is selected from the CPU request buffer (step #1 in Figure 7). We try to serve the selected request according to the FR-FCFS (first-ready-first-come-first-serve) policy [25] (for more details on the FR-FCFS policy, see Appendix). If the selected request can be served, i.e., the request is ready, then a read or write command is issued according to the FR-FCFS policy (step #2, 'Process Request' in Figure 7 and, for more details, see Figure 12 in Appendix). Then, the request is deleted from the CPU request

---

[3] The request buffer can be implemented as a shared buffer or separate ones. We manage CPU and GPU requests in separate request buffers for better implementations, e.g., to meet high clock frequency.

buffer (step #3) and the current cycle finishes (arrow to Next cycle in the figure). If the selected request is not ready, i.e., the target DRAM bank is busy, then we check the CPU request buffer to see if there is any un-tried request in the current cycle (step #4). If there is any un-tried one, we follow the previous steps to serve it. If there is no more un-tried CPU request, then the GPU request buffer is searched and the selected request is served in steps #5~#8 in the same manner as in the CPU requests.

5.5 Overhead Analysis

The write address list is implemented with a cache which has, in each cache entry, two additional pointers for a linked list management. On each input address, the cache is looked up and the tag comparison gives hit or miss as explained in Section 5.2. We use 32b address. The cache-level address requires 26b (=32b-6b since the cache line size is 64B) and the additional pointer requires 13b (for maximum 8K entries in the write address list). Each entry in the list has 79b for two DRAM and PRAM addresses (=2*26b), two additional pointers (=2*13b), and one valid flag bit. In our experiments, the size of write address list ranges between 1K and 8K entries. Thus, the maximum area overhead of write address list is 79KB (=8K*79b).

6.   Experiments

6.1 Target Architecture and Applications

We utilize an architecture having one CPU and one GPU with a hybrid DRAM/PRAM memory. Table 2 shows the details of the target architecture used in our experiments. We developed an event-driven simulation environment by integrating McSim [26] (for the CPU, caches, DRAM and PRAM) and GPGPUsim [27] (for the GPU core).

Table 2 Architectural parameters

Component	Details
CPU core	x86, in-order, 800MHz
GPU core	# shaders = 2 (8 scalar processors/shader), 400MHz
L2 cache	16-way 512KB, I/D shared, 64B cache line, 800MHz
Memory controller	FR-FCFS policy [25], closed page scheme
DRAM	1Gb, 16b DDR2-800, 2KB pages, 8 banks, $t_{CL}/t_{RP}/t_{RCD}$=15ns/15ns/15ns
PRAM	1Gb, 16b LPDDR2-N-800, 16 banks, 4x32B row buffers [11], latency/energy parameters at 14nm from Table 1

In our experiments, we utilize SPEC2006 benchmarks as the CPU programs and classify them into compute-intensive and memory-intensive ones. For GPU programs, we use GPGPUsim benchmarks [27] and classify them into high spatial locality (like RAY in Figure 2 (a)) and low spatial locality (like BLK in Figure 2 (b)) programs. We use four CPU programs (two compute-intensive and two memory-intensive programs) and four GPU programs (two high spatial locality and two low spatial locality programs). Using them, we construct 16 combinations of CPU and GPU programs as shown in Table 3 in Appendix. We run the simulation for 100 million CPU instructions.

6.2 Experimental Results

In our experiments, we first compare the following four DRAM/PRAM configurations,
-    #1 NWB (no write buffer)
-    #2 SL (static list without CPU-aware MAS)
-    #3 SL+CM (static list + CPU-aware MAS)
-    #4 DL+CM (dynamic list + CPU-aware MAS)

where 'static list' represents the case that the write address list is obtained during design time and utilized during runtime to identify GPU data to store in the write buffer. Note that the static list does not change its entries during runtime. SL is similar to the method in [13] in that once PRAM data are migrated to DRAM, its location is fixed. SL can give a better performance than the one in [13] since the static list in SL is obtained from design-time profiling, thus, the list reflects average (global) behavior. The dynamic list represents the write address list proposed in Section 5.2 and changes its entries during runtime in order to adapt to dynamically changing behavior.

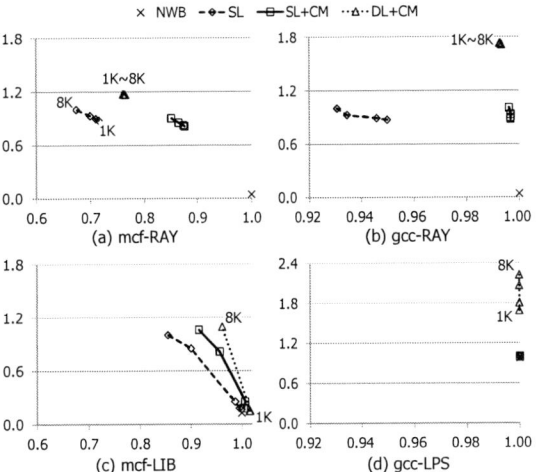

Figure 8 Performance comparisons: normalized IPCs

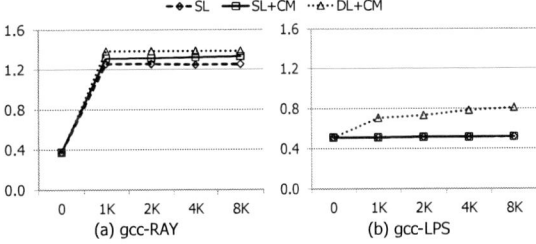

Figure 9 Performance comparisons: harmonic means

Figures 8 and 9 show performance comparisons for the four configurations (#1~#4) of DRAM/PRAM architecture while the size of write buffer varies from 1K to 8K entries. In terms of performance metric, we utilize normalized IPC (in Figure 8) and harmonic mean of IPCs (Figure 9). Figure 8 shows that compared with NWB, the write buffer improves the performance of GPU programs significantly in the four graphs. For instance, in the case of mcf-RAY (Figure 8 (a)), the write buffer with the static write address list (SL) improves the GPU performance by 24.1 times while sacrificing the CPU performance by 32.6%. DL (where the dynamic list in Section 5.2 is used) gives better overall performance than SL. Especially, DL+CM gives noticeable improvement in GPU performance. For instance, Figure 8 (a) shows that, with the write buffer of 8K entries, DL+CM gives larger improvement (28.2 times compared with 24.1 times in SL) in GPU performance with less CPU performance degradation of 23.9% (32.6% in SL). As shown in Figure 13 in Appendix, both dynamic hot write data management and CPU-aware memory access scheduling (DL+CM) give significant improvements in GPU performance, 2.0~44.2 times across 16 mixes of CPU/GPU programs, with modest degradation in CPU performance.

The cases of mcf in Figure 8 (a) and (c) and those of gcc in Figure 8 (b) and (d) show different trends since mcf is memory-intensive while gcc is not. Figure 8 (b) and (d) show that the write buffer with CPU-aware memory access scheduling improves GPU performance with negligible degradation in CPU performance. For instance, in the case of gcc-RAY (Figure 8 (b)), DL+CM with a 8K entry write buffer gives 44.2 times improvement in GPU performance with only 0.7% degradation in CPU performance.

Figure 9 shows the performance comparisons in terms of harmonic mean of CPU and GPU IPCs while the write buffer size (x axis) varies up to 8K entries. The figure shows that the proposed dynamic hot data management (DL) and CPU-aware memory access scheduling (CM) improve the harmonic mean over most cases where a static list (SL) is utilized. In Figure 14 of Appendix, the DL+CM gives inferior harmonic mean in the cases of LIB mixes. It is due to the write buffer size. Additional experiments show that the harmonic mean of DL+CM becomes comparable to or better than the others with 16K and 32K entry write buffers.

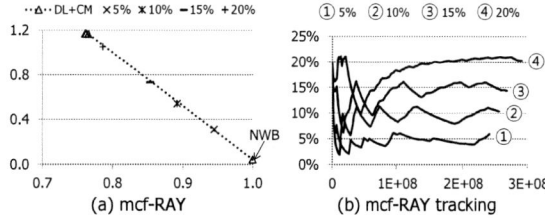

Figure 10 Evaluation of runtime methods: mcf-RAY

Figure 10 shows the effectiveness of the proposed runtime adjustment of write buffer size (Section 5.3). We use four CPU performance bounds, 5%, 10%, 15% and 20%. Figure 10 (a) shows that the four points obtained from the runtime adjustment track well the result of DL+CM. Note that, as mentioned in Section 5.3, compared with DL+CM (which requires a design-time step), the runtime adjustment does not require a design-time step. Figure 10 (b) shows the CPU performance overhead obtained by the runtime adjustment tracks well the given performance bound over time (x axis). Figure 15 in Appendix shows that when a compute-intensive program, gcc runs, the CPU performance degradation of DL+CM is less than 5% for the 8K entry write buffer. Thus, we obtain similar performance results for all the four performance bounds.

The proposed method gives performance improvement mostly by reducing writes to PRAM. Thus, it also improves PRAM lifetime by reducing PRAM writes. Figure 16 in Appendix shows that the proposed method gives significant reductions (2.1~88.6 times) in PRAM writes compared with the no write buffer cases (NWB).

7. Conclusion

In this paper, we proposed a hybrid DRAM/PRAM main memory subsystem. In order to reduce write traffics to PRAM with long write latency, an in-DRAM write buffer is utilized. In order to improve the efficiency of write buffer, hot write data are dynamically identified and stored in the write buffer. The write buffer size is dynamically adjusted to meet the given CPU performance bound. In addition, to improve CPU performance, DRAM access scheduling favors CPU requests over GPU ones. Both dynamic hot write data management and CPU-aware memory access scheduling give significant improvements (1.02~44.2 times across 16 mixes of CPU/GPU programs) in GPU performance with

modest degradation in CPU performance.

8. References

[1] Intel, Co., 2nd Generation Intel® Core™ Processor Family (Codemane Sandy Bridge), http://software.intel.com/en-us/articles/sandy-bridge.

[2] NVIDIA. Co., TEGRA 2 & TEGRA 3 SUPER CHIP PROCESSORS, http://www.nvidia.com/object/tegra-superchip.html.

[3] S. Dumas, "Mobile Memory Forum: LPDDR3 and WideIO," JEDEC Mobile Forum, June 2011.

[4] S. Keckler, et al., "GPUs and The Future of Parallel Computing," IEEE MICRO, vol. 32, issue 5, pp. 7-17, Sept/Oct. 2011.

[5] International Technology Roadmap for Semiconductors (ITRS), available at www.itrs.net.

[6] M. Abdulla, and M. Greenberg, "Will Phase Change Memory (PCM) Replace DRAM or NAND Flash?," Flash Memory Summit, Aug. 2010.

[7] Numonyx, "Phase Change Memory (PCM): A new memory technology to enable new memory usage models," available at www.numonyx.com/en-us/MemoryProducts/PCM/Pages/PCM.aspx.

[8] EE Times, Samsung to ship MCP with phase-change, http://www.eetimes.com/electronics-news/4088727/Samsung-to-ship-MCP-with-phase-change.

[9] JEDEC Standard, Low Power Double Data Rate 2 (LPDDR2), JESD209-2E, April 2011.

[10] C. Villa, et al., "A 45nm 1Gb 1.8V Phase-Change Memory," Proc. ISSCC, 2010.

[11] H. Chung, et al., "A 58nm 1.8V 1Gb PRAM with 6.4MB/s Program BW," Proc. International Solid-State Circuits Conference (ISSCC), 2011.

[12] M. K. Qureshi, V. Srinivasan, and J. A. Rivers, "Scalable High Performance Main Memory System Using Phase-Change Memory Technology," Proc. ISCA, 2009.

[13] G. Dhiman, R. Ayoub, and T. Rosing, "PDRAM: A Hybrid PRAM and DRAM Main Memory System," Proc. DAC, 2009.

[14] G. Sandre, et al., "A 90nm 4Mb Embedded Phase-Change Memory with 1.2V 12ns Read Access Time and 1MB/s Write Throughput," Proc. ISSCC, 2010.

[15] B. D. Yang, et al., "A Low Power Phase-Change Random Access Memory Using a Data-Comparison Write Scheme," Proc. ISCAS, 2007.

[16] K. Lee, et al., "A 90nm 1.8V 512Mb Diode-Switch PRAM with 266MB/s Read Throughput," IEEE J. Solid-State Circuits, vol. 43, no. 1, pp. 150-162, Jan. 2008.

[17] M. Qureshi, et al., "Practical and Secure PCM Systems by Online Detection of Malicious Write Streams," Proc. HPCA, 2011.

[18] L. Zhang, et al., "The Impulse Memory Controller," IEEE Trans. Computers, vol. 50, no. 11, Nov. 2001.

[19] J. Ahn, M. Erez, and W. J. Dally, "The Design Space of Data-Parallel Memory Systems," Proc. SC, 2006.

[20] Personal communications with Intel CPU/GPU designers, 2011.

[21] N. B. Lakshminarayana, and H. Kim, "Effect of Instruction Fetch and Memory Scheduling on GPU Performance," Workshop on Language, Compiler, and Architecture Support for GPGPU, in conjunction with HPCA/PPoPP, 2010.

[22] B. C. Lee, et al., "Architecting Phase Change Memory as a Scalable DRAM Alternative," Proc. ISCA, 2009.

[23] T. M. Aamodt, et al., "GPGPU-Sim: A Performance Simulator for Massively Multithreaded Processor Research," available at http://www.ece.ubc.ca/~aamodt/gpgpu-sim/tutorial/GPGPU-Sim-Tutorial-MICRO42.pdf

[24] Intel, Co., "Performance Analysis Guide for Intel Core i7 Processor and Intel Xeon 5500 Processors," available at http://software.intel.com/sites/products/collateral/hpc/vtune/performance_analysis_guide.pdf.

[25] S. Rixner, et al., "Memory Access Scheduling," Proc. ISCA, 2000.

[26] S. Li, et al., "McPAT: An Integrated Power, Area, and Timing Modeling Framework for Multicore and Manycore Architectures," Proc. MICRO, 2009.

[27] A. Bakhoda, et al., "Analyzing CUDA Workloads Using a Detailed GPU Simulator," Proc. ISPASS, 2009.

APPENDIX
Hybrid DRAM/PRAM-based Main Memory for Single-Chip CPU/GPU

DRAM Access Scheduling

In this subsection, we give a preliminary on the existing DRAM access scheduling policies which are the baseline of our proposed CPU-aware DRAM access scheduling in Section 5.4. Figure 11 (a) illustrates a simplified DRAM structure. DRAM consists of banks (four banks in the figure). Each bank has data array consisting of rows (each 1KB or 2KB size) and a row buffer (sense amplifier, latch, and column address decode logic). Given an address (Addr in the figure), the memory controller issues a row access command (called activation command, ACT) utilizing the row address (Row addr in the figure). After t_{RCD} (row activation delay), the entire data in the selected row is copied to the row buffer (the arrow annotated with ACT in the figure). Then, the read (RD) or write (WR) command is issued by the memory controller to access desired data in the row buffer utilizing the column address (Col addr in the figure). In case of read operation, after the column access latency (t_{CL} in read), the required data appear on the DRAM data I/O, DQ signals. If there is no more request to access the row in the row buffer, a precharge (PRE) command is issued to precharge the bank to prepare for a next row activation. The latch in the row buffer allows the precharge command to be overlapped with the last read operation.

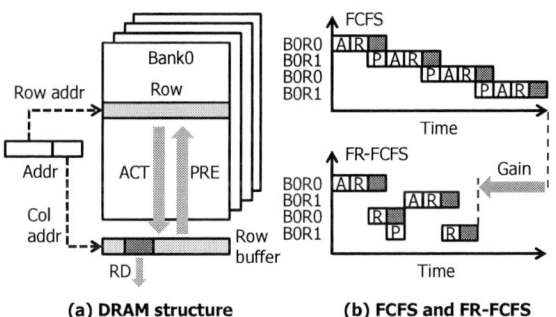

(a) DRAM structure **(b) FCFS and FR-FCFS**

Figure 11 DRAM structure and scheduling policies

Figure 11 (b) shows two representative memory access scheduling policies, first-come-first-serve (FCFS) and first-ready-first-come-first-serve (FR-FCFS) [MAS]. Assume that the memory controller received four read requests to access two rows, R0 and R1 in bank B0. For instance, R0R1 represents a read request accessing data in row R1 in bank B0. The four requests arrived in the order of B0R0, B0R1, B0R0, and B0R1. Assume that the memory controller issues memory commands (ACT, RD and PRE) to serve the four requests. In the FCFS policy, the four requests are served in the order of their arrival at the memory controller. Assuming that all the banks were initially precharged, the memory controller issues ACT (rectangle labeled with 'A') and RD (labeled with 'R') commands to serve the first request B0R0 as shown at the top of Figure 11 (b). After t_{CL}, DRAM gives data (shaded rectangle). As soon as the first data start to be read, the PRE command can be issued to prepare for serving the second request B0R1 accessing a different row from the one in

the row buffer. Note that, in Figure 11 (b), the FCFS policy requires four ACT commands to serve the four read requests.

The FR-FCFS policy allows for reordering the service of requests in order to improve memory utilization. As shown at the bottom of Figure 11 (b), after serving the first read request, instead of serving the second one, the FR-FCFS serves the third request which accesses the same row as the first one. By doing that, as shown in the figure, the memory controller has only to issue a RD command to serve the third request. After the first row R1 is precharged, the second and fourth requests are served in their arriving order. As shown in Figure 11 (b), the FR-FCFS policy can give a reduction in execution cycle by optimizing DRAM accesses (by removing two ACT commands in this case).

Memory access scheduling: FR-FCFS

```
1  //dq_busy<A:B> dq_busy from (Current cycle+A)
2  //                    to (Current cycle+B)
3  function ProcessRequest
4    if busy_count[bank] > 0
5      busy_count[bank]--
6      return busy
7    else
8      state[bank] = next_state[bank]
9      if state[bank] == precharge
10       next_state[bank] = activate
11       busy_count[bank] = tRCD
12       row_buffer[bank] = page
13       return activate
14     else if state[bank] == activate
15       if row_buffer[bank] == page
16         next_state[bank] = activate
17         if dq_busy<CL:CL+BL/2> == 0
18           dq_busy<CL:CL+BL/2> = 1
19           return read/write
20         else
21           return busy
22       else
23         next_state[bank] = precharge
24         busy_count[bank] = tRP
25         return precharge
```

Figure 12 Per-bank access scheduling for FR-FCFS

Experimental Setup

Table 3 Application sets

		GPU: high spatial locality		GPU: low spatial locality	
		RAY	LIB	BLK	LPS
CPU: memory-intensive	mcf	mcf-RAY	mcf-LIB	mcf-BLK	mcf-LPS
	lbm	lbm-RAY	lbm-LIB	lbm-BLK	lbm-LPS
CPU: compute-intensive	gcc	gcc-RAY	gcc-LIB	gcc-BLK	gcc-LPS
	omn	omn-RAY	omn-LIB	omn-BLK	omn-LPS

Experiments

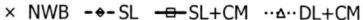

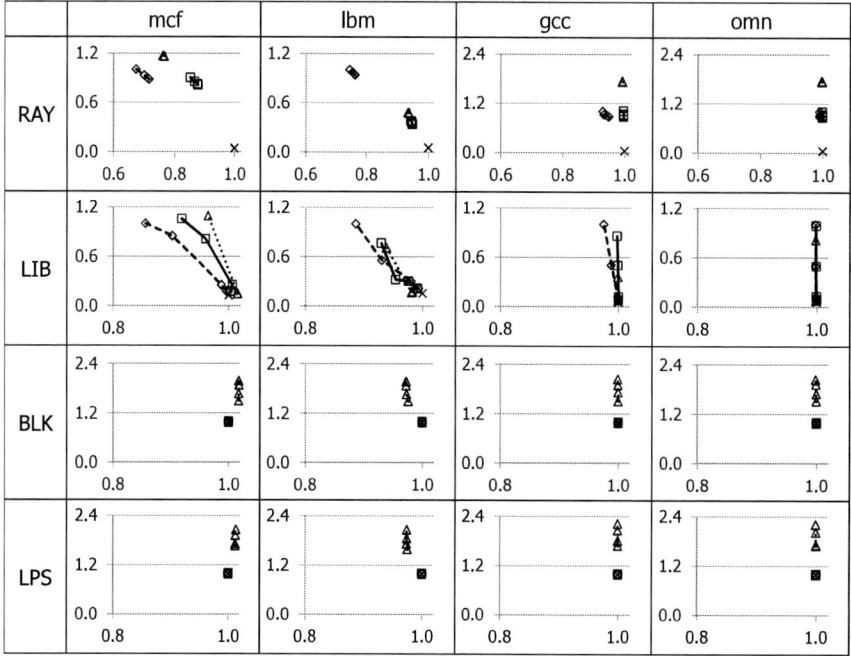

Figure 13 Performance comparisons: normalized IPCs

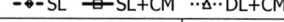

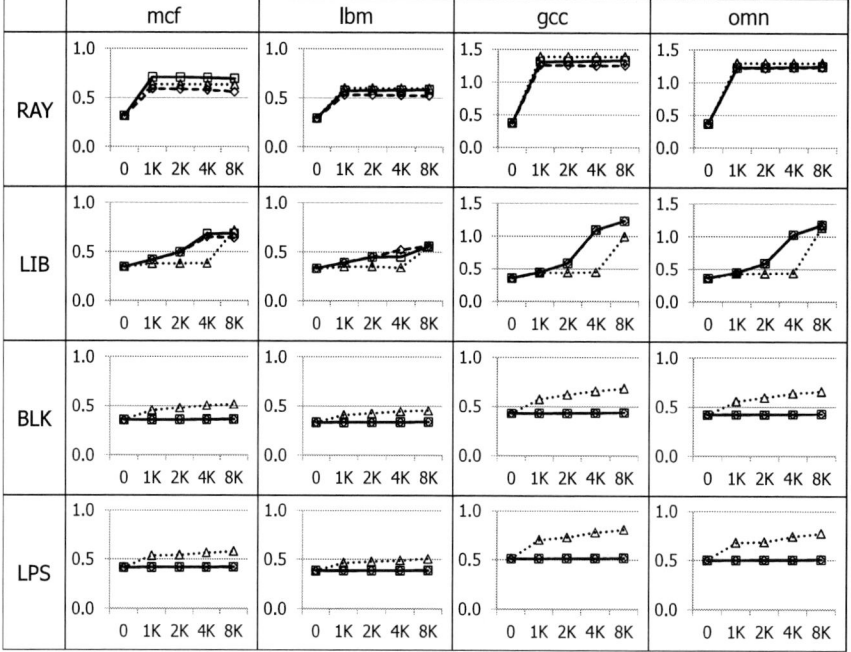

Figure 14 Performance comparisons: harmonic mean of IPCs

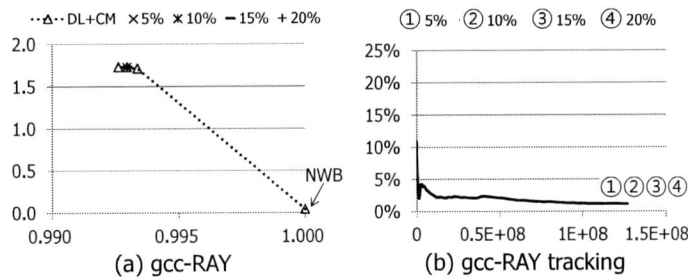

Figure 15 Evaluation of runtime methods: gcc-RAY

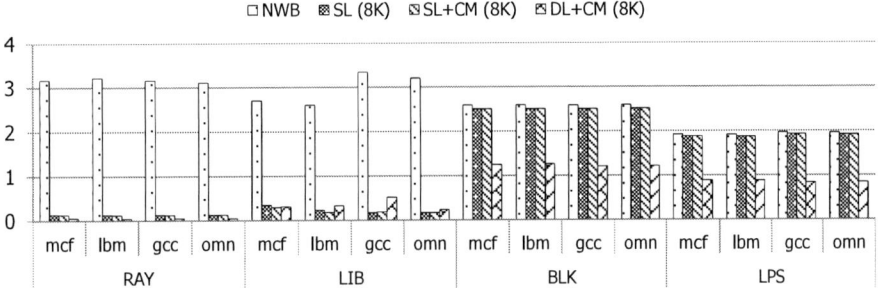

Figure 16 Comparison of PRAM writes (for the execution of the same number of CPU instructions in each combination)

Write Performance Improvement by Hiding R Drift Latency in Phase-Change RAM

Youngsik Kim, Sungjoo Yoo, and Sunggu Lee

Department of Electronic and Electrical Engineering, POSTECH

{kaengsik, sungjoo.yoo, slee}@postech.ac.kr

ABSTRACT

Phase-change RAM (PRAM) is considered to be one of the most promising candidates to complement or replace DRAM in the near future. However, it is imperative to overcome the limitations of PRAM, especially, long write latency for its widespread applications. R drift latency occupies a significant portion in PRAM write latency thereby adversely affecting system performance. In this paper, we propose a novel method called write status holding register (WSHR) to reduce the write latency due to R drift latency. The WSHR allows for non-blocking accesses to PRAM during R drift latency thereby improving system performance. Our experiments with SPEC benchmarks show that the proposed WSHR gives 53.6%~0% performance improvements in the hybrid DRAM/PRAM main memory (256MB DRAM and 14nm PRAM).
Categories and Subject Descriptors: B.3.2 [Memory Structures]*: Design Styles*

General Terms: Design, Management, Performance

Keywords: Phase-change RAM, write performance, R drift

1. Introduction

In high-end embedded systems, e.g., smart phones, the main memory is expected to require larger capacity and higher performance in the future [1]. However, DRAM is approaching its scaling limit in sub-20nm technology [2]. Among emerging memory technologies, phase-change RAM (PRAM) is considered to be one of the most promising ones since large size designs (1Gb) are already available [3][4] and it is in mass production [5].

PRAM gives the benefits of better scalability and non-volatility. However, it has limitations in terms of write endurance, write latency/power, and read latency. Recently, there have been many studies to overcome the limitations. As an industrial solution, a hybrid DRAM/PRAM [4] is proposed to mitigate the PRAM limitations. Our experiments (more details in Section 6) with the Verilog model of the industrial hybrid DRAM/PRAM [4] show that even in the hybrid DRAM/PRAM, low PRAM write performance often determines system performance especially for programs with large footprints.

A significant portion of PRAM write latency is incurred by R drift latency. R drift latency is the time duration that the resistance of PRAM cell drifts after programming [6]. In real PRAM designs, R

drift latency amounts to order of 10µs [7]. During R drift latency, the read operation to the PRAM cell under R drift latency is not allowed since it will disturb the PRAM cell thereby possibly causing a bit error. Thus, R drift latency has been included in the write latency of PRAM. According to our experiments, R drift latency affects up to 41.9% in total write latency in the 58nm technology PRAM. Moreover, the problem of R drift latency will become much more significant in the future PRAM. It is because PRAM write latency scales down as device scaling. However, R drift latency does not scale. Thus, the negative impact of R drift latency is expected to increase in future PRAM designs.

In this work, we propose a novel method which tackles the R drift latency problem in order to improve PRAM write performance, thereby enhancing system performance. The basic idea is to hide R drift latency with concurrent PRAM accesses exploiting the fact that a very small portion (maximum 0.50% in our experiments with SPEC benchmarks) of read requests access data which undergo R drift.

This paper is organized as follows. Section 2 reviews related work. Section 3 gives the preliminaries. Section 4 explains the problem and our observation. Section 5 offers the WSHR method of hiding R drift latency. Section 6 gives experimental results. Section 7 concludes the paper.

2. Related Work

Recently, there have been several studies on applying PRAM in main memory. In [8][9][10], the hybrid DRAM/PRAM is proposed where DRAM is utilized together with PRAM to compensate for the overhead of long read/write latency and high write power consumption in PRAM. Existing works to resolve the limitations of PRAM can be classified into four categories: write reduction, wear leveling, error correction and performance improvement.

In [8][11][12], partial writes are proposed to update only modified data in PRAM. To do that, the dirty information is managed, e.g., at the granularities of cache line. In [13], Yang et al. propose a differential write which compares existing PRAM data and new write data and writes only the modified bits in PRAM. In [14], Cho et al. propose an invert coding called Flip-n-Write. In this method, if the bit difference between existing PRAM and new write data is larger than the half of data width, then new write data are inverted and stored in differential writes together with the invert code (1 bit per granularity of inversion).

The lifetime of PRAM is determined by the most worn-out cell. Thus, wear leveling tries to evenly distribute writes across the PRAM cells. Rotation-based wear leveling is a representative wear leveling method where the offset of start address is rotated in a unit of rotation, e.g., cache line [8], page, super-pages [12], etc. In [15], Qureshi et al. present a rotation-based wear leveling called start gap wear leveling with randomization. Especially, address randomization is applied before wear leveling in order to distribute spatially localized write traffics. In [16], Seong et al. present security refresh to cope with malicious attacks, e.g., repeat address attacks. In security refresh, two-level address remapping, i.e., wear leveling is performed in a similar fashion to DRAM refresh. In [17], Dong et al. present process variation-aware wear leveling called wear rate leveling. In this method, hot data (frequently written data) are dynamically mapped to strong regions (in terms of process variation) thereby improving PRAM lifetime.

Error correction is required to increase PRAM lifetime. In [18], Schechter et al. present error correction pointer (ECP) which exploits the fact that PRAM bit errors are persistent. An ECP stores the information of faulty bit location and correct bit value in the spare area. During the write operation, the correct bit value corresponding to the faulty bit location is stored in the spare area and, during the read operation, the stored correct bit value is utilized to correct the bit error in the faulty bit location. In [19], Seong et al. present an error correction method called SAFER which reduces the overhead of required spare area for error correction by utilizing one bit error correction (e.g., Hamming code) and bits grouping.

In terms of system performance, read performance is more critical than write performance unless write bandwidth is a limiting factor. In [20], Qureshi et al. present write preemption which preempts the on-going write operation to serve a newly arrived read request to the same bank. In this work, preemption latency is not considered. However, in reality, a significant preemption latency is incurred due to R drift latency [6][7]. Due to the large preemption latency, write preemption is hardly applied to the write operation of small data which has a large portion of R drift latency in the write latency, which adversely affects read performance. In our method proposed in this paper, the preemption latency is minimized by hiding R drift latency in write preemption thereby contributing to read performance improvement.

3.   Preliminary: PRAM Operation

Figure 1 illustrates the PRAM internal structure based on the JEDEC LPDDR2-N standard [21]. The PRAM consists of banks. A bank consists of rows. As shown in the figure, read and write paths are separate. For a read operation, the memory controller first needs to issue two activation commands (Pre-Act and Act). The Pre-Act command gives the upper part of address bits and the Act command having the lower part initiates row activation, i.e., copying the contents of the row to one of four row buffers. Then, the memory controller issues a read (RD) command to access the row buffer data. After t_{CL} (read latency), a burst of data (whose size is called burst length, BL) appear on the IO data channel.

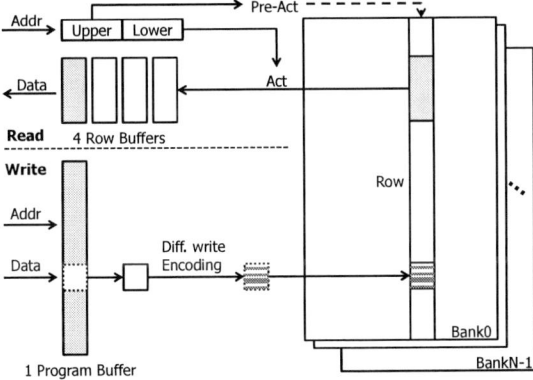

Figure 1 PRAM with the LPDDR-N interface

The write path has a program buffer as shown at the bottom of Figure 1. Write data and address are first written to the program buffer and control register, respectively. Then, internal write operation starts. As shown in Figure 1, differential writes and data encoding, e.g., Flip-N-Write [14] can be applied to reduce bit updates in PRAM.

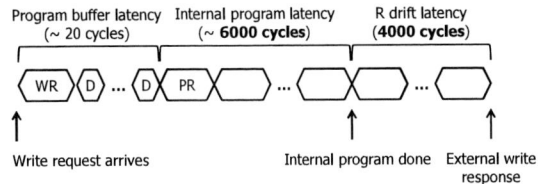

Figure 2 Write operation in PRAM

Figure 2 gives a detailed view of write operation in the case of writing 128B to PRAM assuming 28nm technology.[1] When a write request arrives at the memory controller, the write information (address and data size) is written to the control register ('WR' in the figure). Then, the write data are written to the program buffer ('D' in the figure). The internal write operation is initiated by the program command ('PR' in the figure). After the internal write delay, which is 6,000 cycles (@400MHz) in this case, the write operation is finished. However, the last written data require R drift latency which is 4,000 cycles. After R drift latency, the PRAM gives the external write response. Then, the PRAM bank can be accessed to serve next requests to the bank. As shown in Figure 2, R drift latency can occupy a significant portion of total write latency.

Given an implementation technology (e.g., 28nm), write latency is a function of bit difference (between old and new data) and internal write bandwidth in PRAM (required to meet the chip-level peak current constraint). For instance, 16b data can be written at a time in [22] and 32b data in [4]. Thus, in the case of 32b internal write width, for instance, the data write with 50b difference takes two internal 32b writes while that with 20b difference takes one internal 32b write. Given the same chip-level peak current constraint, as the technology scales twice (e.g., 58nm → 28nm), write bandwidth improves 4 times. It is because both PRAM cell write latency (contributing to 2X bandwidth improvement) and write power consumption (another 2X) are reduced linearly as technology scales.

Read latency is crucial in system performance. Thus, the PRAM design [4] has two optimizations for read latency reduction. First, read requests to currently free banks (not performing write operation) can be served while the write operation (including R drift latency) is being performed on another bank. However, there is no write parallelism between write operations. There is only one write operation at a time due to the peak current constraints. Second, the write preemption is adopted for read performance improvement as in [20]. Note that, as mentioned in Section 2, R drift latency has a strong impact on the efficiency of write preemption since R drift latency becomes the write preemption latency.

4.   Problem and Observation

Figure 3 shows the relationship between write latency (in clock cycles at 400MHz) and write data size for a 28nm PRAM.[2] As the

---

[1] Based on the write latency data at 58nm technology in [7], we estimated write latency at 28nm technology.

[2] We first obtained the relationship from the Verilog model of 58nm PRAM in [4][7]. Then, we scaled the write latency for 28nm technology. For simplicity, we assumed maximum bit updates per data size after Flip-n-Write is applied.

write data size increases, the write latency increases almost linearly. Note that the x axis is in log scale. The figure also shows that for each write data size, R drift latency occupies a constant portion (10μs in this case) of write latency. Thus, when writing small data, e.g., 64B, R drift latency occupies the majority of total write latency.

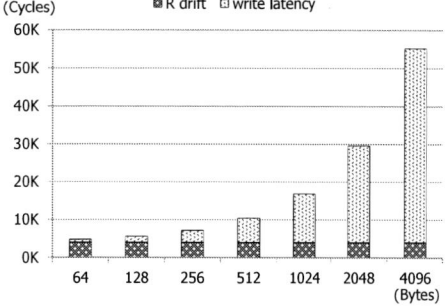

Figure 3 Write latency (y-axis) vs. write data size (x-axis)

In our work, DRAM plays the role of last level cache and PRAM is the large background memory in the hybrid DRAM/PRAM main memory. Figure 4 shows the histogram of bit difference (obtained after applying Flip-n-Write) in writing back a dirty DRAM row (4KB size) to PRAM on a DRAM/PRAM main memory (for more details, see Section 6.1). They are obtained from two programs (sjeng and milc) in SPEC 2006 benchmarks. Figure 4 shows that write data sizes are mostly less than 64B in sjeng and 192B in milc. Most programs used in our experiments show similar write behavior. Based on both Figures 3 and 4, we found that programs can suffer from the adverse effect of R drift latency since the majority of PRAM writes are for short write data which have a dominant portion of R drift latency in the write latency. In the next section, we present our idea and implementation to resolve this problem.

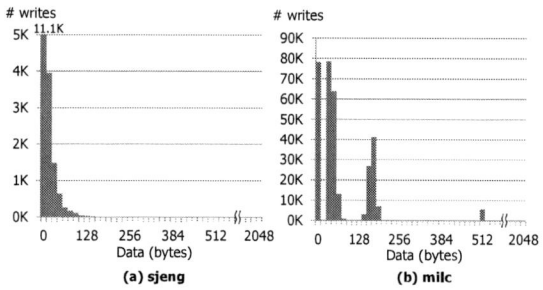

Figure 4 Write data size histogram (in Flip-n-Write)

5. PRAM Access Scheduling based on Write Status Holding Register

5.1 Basic Idea

There are three cases where the proposed idea can give performance benefits: hiding R drift latency with concurrent reads and writes, and eliminating R drift latency in the write preemption latency. Figure 5 illustrates the case of hiding R drift latency with concurrent reads. The figure gives an example of PRAM accesses to bank 0. Assume that a write operation at bank 0 finishes internally at 6000 cycle and gives an external write response at 10000 cycle after the R drift latency. Four read requests to the same bank 0 arrive during the R drift latency. In the conventional PRAM, the four read requests need to wait for R drift to finish. Thus, the four read requests are served after the R drift latency as the four blank

rectangles (labeled with 'RD') show. Note that write preemption cannot be applied to serve these read requests since they arrive during R drift latency after the internal write operation finishes.

Our idea is that, during R drift latency, the memory controller keeps the address (0x100 in this example) of write data under R drift latency (which is called incomplete write data). Then, in memory access scheduling, the memory controller delays only the read request for the incomplete write data while allowing other read requests on the bank under R drift latency to be served thereby reducing read latency. Figure 5 illustrates the effect of our idea. The first read request to address 0x200 in the same bank can be served during the R drift latency as the shaded rectangle (labeled 'RD') shows, which reduces the read latency of this request significantly, i.e., by 4000 cycles. The third and fourth read requests can also be served much earlier than in the conventional case thereby significantly reducing their read latency. However, the second read request is for the incomplete write data. Thus, its service is delayed until the R drift latency finishes.

According to our experiments, the case of reading incomplete write data rarely happens. Table 1 shows the percentage of read requests to the incomplete write data (=# of read requests to incomplete write data/# of total read requests which arrive during R drift latency) in two programs in SPEC 2000 and 2006 benchmarks. We show only the two programs since the other programs in both benchmarks do not have such a case with the DRAM size of 64MB. As the table shows, maximum 0.50% latency access the incomplete write data. Moreover, as the DRAM size increases, the percentage is reduced due to reduced write traffics to PRAM. of read requests arriving during R drift

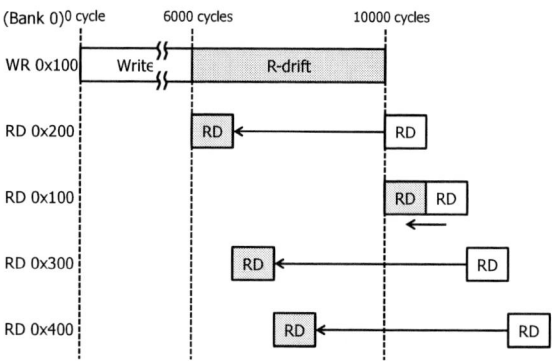

Figure 5 Hiding R drift latency

Table 1 % of read requests on write data under R drift latency

Benchmarks	DRAM cache size	
	64MB	128MB
171.swim	0.07%	0%
458.sjeng	0.50%	0%

R drift latency is a kind of blocking period in PRAM access. Our idea is to enable non-blocking accesses to PRAM. Thus, we call our idea write status holding register (WSHR), named after miss status holding register (MSHR) for non-blocking caches [23]. In the following subsections, we give our implementation details.

5.2 WSHR-based Requests Handling

The memory controller has a request buffer to store arrived read and write requests until their service is finished. Each request in the request buffer follows the flow in Figure 6 in order to be selected by the scheduler and issue their PRAM commands.

899

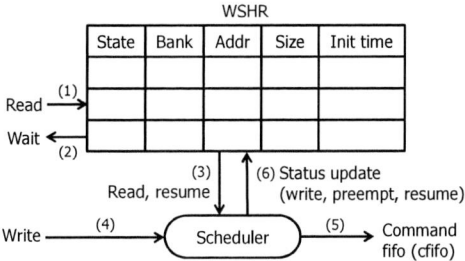

Figure 6 Request and commands flow

Each entry in the WSHR (shown at the top of Figure 6) corresponds to a program buffer. Program buffer to bank allocation depends on PRAM designs. In [4][21], there is one shared program buffer for 16 banks. Our method allows for multiple program buffers in order to enable write-under-write-preemption where a program buffer can initiate its internal write operation as soon as a previous write operation of another program buffer for a different bank is preempted (to serve a newly arrived read request to the bank). The write-under-write-preemption enables us to maximally utilize write bandwidth. Thus, there can be multiple entries in the WSHR. Each entry has a tuple, bank state, bank ID, address, data size, and initiation time as shown in Figure 6.

When a read request arrives (step 1 in Figure 6), the WSHR checks to see if the request is for the write data (including incomplete write data) in the WSHR. If the read request is for the write data (under internal writes) or incomplete write data (in R drift latency), its service is delayed until R drift latency finishes (step 2). If it is not for the write data (step 3), then it can participate in the memory access scheduling (round rectangle in the figure). The write request participates in the memory access scheduling without checking the WSHR (step 4). The scheduler selects the winner among read and write requests, and sends corresponding PRAM commands (read, program, preempt, and/or resume) to the command fifo (step 5). The details of scheduler operations will be given in Section 5.3. The command fifo (cfifo) is in charge of issuing PRAM commands based on the timing constraints, e.g., issuing an ACT command respecting bank preactive time, t_{RP} [21].[3] If the selected one is a write, preempt, and resume request, then its status is registered or updated in the WSHR (step 6).

5.3 PRAM Access Scheduling

The memory controller manages a finite state machine for each bank. Figure 7 illustrates the state machine which has three states: Ready, RD busy, and WR busy. The bank at Ready state can receive either RD or WR commands. During busy states, no command can be issued to the bank. During RD busy state, row activation is being performed to copy the contents of selected row to the row buffer (RB). During WR busy state, internal write operation is being performed writing the data in the program buffer (PB) on the selected PRAM cells.

The bank enters WR busy state in two cases that a program[4] or resume command is initiated to the bank ('program' or 'resume' on the edge from Ready to WR busy state in Figure 7). The bank exits WR busy state when the internal write operation finishes ('WR done' in Figure 7) or a preempt command is issued to the bank.

---

[3] In the LPDDR2-N specification, t_{RP} represents the timing constraint between Pre-ACT and ACT commands.

[4] We use two terms, program and write, interchangeably.

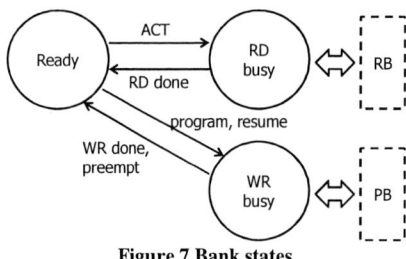

Figure 7 Bank states

Our PRAM access scheduling is based on the first-ready-first-come-first-serve (FR-FCFS) policy in [24]. Figure 8 shows the pseudo code of row activation and program initiation functions. Each bank performs the functions.[5] If there is any read request (in the request buffer of the memory controller) whose row is not yet activated (line 1 in Figure 8), then we check the bank state. If the bank state is Ready and there is an available row buffer (line 2), then the memory controller starts a row activation by sending Pre-ACT and ACT commands to the command fifo (cfifo in line 3, Figure 8). If there is a not-yet-activated read request but the current bank state is WR busy (line 5), then write preemption is tried. If there is an available row buffer (line 6), then write preemption is performed (line 7). Then, activation commands for the read request are sent to the command fifo and the bank state is set to RD busy (lines 8 and 9).

If there is no read request (whether not-yet-activated or activated one) to the bank in the request buffer, but there is a write request (including a resumption of preempted write) to the bank (line 10), then we first check to see if there is any on-going write in another bank since there can be only one on-going write operation at a time due to the peak current constraints as mentioned in Section 3. If the condition in line 11 is met, then the write operation (including write resumption) starts and the bank state is set to WR busy (lines 12 and 13). Note that the program, preempt, and resume commands require updating or registering the status in the WSHR.

```
// Each bank
1   if any not-yet-activated read request
2       if Ready state & available row buffer
3           Send Pre-ACT and ACT commands to cfifo
4           Bank state = RD busy (return to Ready after t_RCD)
5       else if WR busy state
6           if available row buffer
7               Preempt the current write and update it in WSHR
8               Send Pre-ACT and ACT commands to cfifo
9               Bank state = RD busy (return to Ready after t_RCD)
10  else if any write request and no activated read request
11      if Ready state & no on-going write in other banks
12          Send write address, data and program/resume cmd to PB
13          Bank state = WR busy
```

Figure 8 Row activation and program initiation

Figure 9 shows the pseudo code of row buffer access. If there is an activated read request which has the desired data in one of the four row buffers (line 1 in Figure 9), then we check to see if there will be any conflict in the IO data channel at $T_{cur}+t_{CL}$ when the data will appear in the data channel if a RD command is issued at the present time, T_{cur} (line 2). If the IO data channel is free at that time, then a RD command is issued to the command fifo. The state of IO data channel is set to busy for the time period of read data transfer, $[T_{cur}+t_{CL}:T_{cur}+t_{CL}+BL/2]$ where BL represents the burst length in PRAM data transfer (line 4). After the RD command is issued, the

---

[5] In reality, there are inter-bank timing constraints, e.g., t_{RRD} and t_{FAW}. We omit them for simplicity in this figure.

corresponding read request is excluded from memory access scheduling (line 5).[6] If the IO data channel is not free at $T_{cur}+t_{CL}$, then the service of the read request is tried in the next cycle (line 7).

```
1    if an activated read request
2        if IO data channel is free at Tcur+tCL
3            Send RD command to cfifo
4            IO_data_state[Tcur+tCL:Tcur+tCL+BL/2] = busy
5            Exclude the request from scheduling
6        else
7            Try in the next cycle
```

Figure 9 Row buffer access

6. Experiments

6.1 Target Architecture and Applications

We used an event-driven simulator, McSim [25] that is based on the Pin dynamic instrumentation tool [26] and includes x86 in-order cores, L1/L2 caches, and memory controller. The detailed architectural parameters of our experiments are shown in Table 2. The memory controller is in charge of both DRAM and PRAM. It supports the write preemption. The main memory subsystem has two configurations: hybrid DRAM/PRAM memory and DRAM only cases. The characteristics of DRAM and PRAM are shown in Table 3.

Table 2 Architectural parameters

Component	Details
CPU core	In-order x86 core, 2 GHz
L1 cache	4-way, 32KB/32KB I/D cache, 64B cache line, 1 cycle
L2 cache	16-way, 1MB, I/D shared, 64B cache line, 6 cycles
DRAM	16-way, 64~512MB, 4KB cache line (cache) LPDDR2-800, 4x16b, $t_{CL}=t_{RP}=t_{RCD}=6$ cycles FR-FCFS policy, open page scheme
PRAM	1GB, LPDDR2-N-800, 4x16b, $t_{CL}/t_{RP}/t_{RCD}=6/1/32$ cycles, write/R drift latency=Max.48000/4000 cycles (28nm), FR-FCFS policy, read priority, write preemption, Flip-N-Write

Table 3 DRAM and PRAM characteristics per chip

Parameters		DRAM	PRAM		
			58nm	28nm	14nm
IO bits		x16	x16		
# banks		4	16		
Row buffer size		1KB/bank	32B x 4		
Program buffer size		-	1KB x 4		
Read	Latency	45ns	98ns		
Write	Latency	45ns	490µs (1KB)	130µs (1KB)	40µs (1KB)
	Bandwidth	1.6GB/s	2.0MB/s	7.5MB/s	24.4MB/s

In our experiments, we compare the proposed WSHR with the conventional method of having R drift latency in the total write latency and write preemption latency, respectively. We vary technologies (58nm, 28nm, and 14nm) and the DRAM size (64MB~512MB) in the hybrid DRAM/PRAM subsystem. Note that we have a conservative assumption that the read latency of PRAM does not scale as technology scales though the write bandwidth scales (4X) as technology scales (2X) as mentioned in Section 3.

---

[6] After the corresponding data are received from the memory and forwarded to the core (which sent the read request to the memory controller), the read request is removed from the request buffer.

Since we used the DRAM of 64~512MB, we needed applications which incur DRAM cache misses with those DRAM sizes. Thus, after evaluating SPEC 2000 and 2006 benchmark programs, we selected 9 programs (swim form SPEC 2000, and bzip2, sjeng, astar, milc, zeusmp, cactusADM, lbm and wrf from SPEC 2006) whose working set sizes are larger than 64MB. Thus, the other programs of SPEC 2000 and 2006 show almost the same execution cycles in the hybrid DRAM/PRAM main memory as in the DRAM only case. We ran simulations for 1 billion instructions.

6.2 Experimental Results

Figure 10 shows the comparison of total execution cycles when a 256MB DRAM and a 14nm PRAM are used in the hybrid main memory. A complete set of results for the hybrid DRAM/PRAM main memory with 64MB~512MB DRAM across the three technologies can be found in Appendix. Figure 10 shows that for 9 programs the proposed WSHR gives 53.6%~0% improvements in total execution cycles compared with the conventional method. Three programs, bzip2, milc, and lbm give more than 2 times longer execution cycles in the hybrid memory than in the DRAM only case due to high DRAM cache miss rates incurred by their large footprints. The larger DRAM (512MB) reduces the execution cycles down to 1.12 times as shown in Appendix.

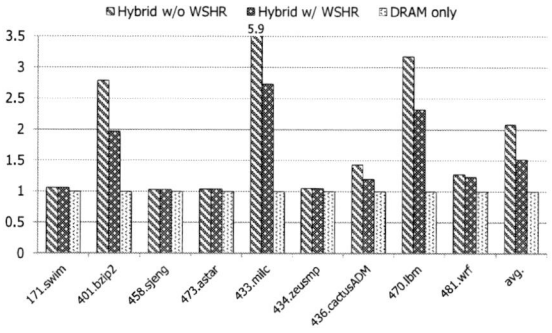

Figure 10 Comparison of total execution cycles

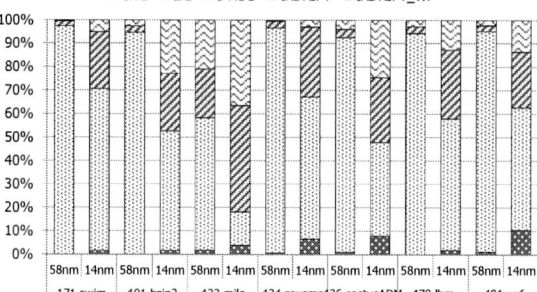

Figure 11 Write latency decomposition

Figure 11 shows the decomposition of total PRAM write latency at both 58nm and 14nm technologies for the DRAM size of 128MB in the baseline hybrid main memory.[7] The figure shows that R drift latency (sum of tRDRIFT and tRDRIFT_WP)[8] becomes significant

---

[7] In the case of 256MB or larger DRAM, there are only a few programs incurring DRAM cache misses. Thus, in the following figures, we show the results for the 128MB DRAM case to show the trends for more programs.

[8] t_{RDRIFT_WP} represents the R drift latency during write preemption.

in the total PRAM write latency at 14nm technology. Thus, the proposed WSHR can be effective to reduce the write latency to access PRAM at advanced technologies. Figure 12 confirms this. The figure shows the performance gain obtained by the proposed WSHR method (w.r.t. the baseline hybrid DRAM/PRAM) across the three technologies in the case of 128MB DRAM. As technology advances, the average performance gain tends to increase.

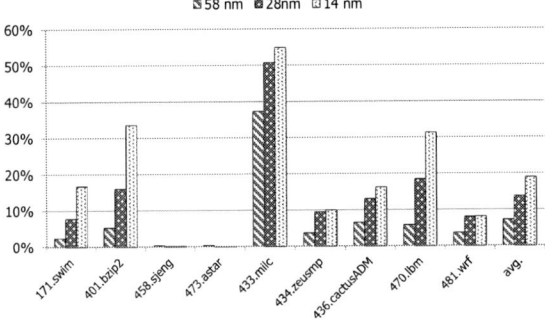

Figure 12 Performance gain vs. technologies (128MB DRAM)

Note that the system performance with the hybrid DRAM/PRAM shown in Figures 10 and 12 is conservative since, as mentioned in Section 6.1, we did not scale down the PRAM read latency at advanced technologies. Thus, with a scale down of read latency at those technologies, further performance improvement can be obtained with the hybrid DRAM/PRAM main memory.

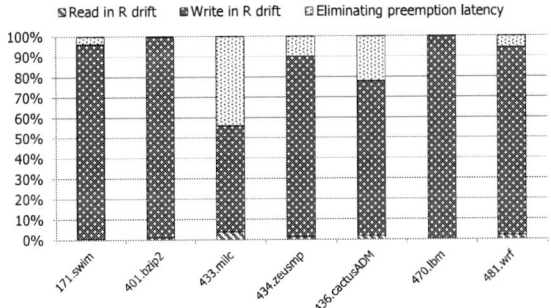

Figure 13 WSHR effect decomposition (128MB DRAM, 14nm)

Figure 13 shows the decomposition of WSHR effect on performance gain into three factors: hiding R drift latency performing concurrent reads to the same bank (as in Figure 5) and concurrent writes during R drift latency after the execution of previous write, and eliminating R drift latency in the write preemption latency. Figure 13 shows that the majority of performance gain comes from being able to consecutively execute multiple writes to different addresses without intervening R drift latency (write in R drift latency) and eliminating R drift latency in the write preemption latency. The effect of hiding R drift latency with concurrent reads is small in Figure 13 since our target architecture has a single core. We expect a more significant portion of the effect of concurrent reads in multi-core architectures.

The proposed WSHR has an area overhead. In our experiments, each entry in the WSHR requires 33 bits (2b state, 4b bank, 16b addr, 5b size, and 6b time). In the case of four program buffers, the area overhead is only 132 bits. The proposed WSHR does not affect the number of writes to PRAM since it does not change the number of internal write operations.

7. Conclusion

In this paper, we described the R drift latency problem that R drift latency dominates both the PRAM write latency of short write data and the write preemption latency. Such a long write latency incurred by R drift latency adversely affects system performance. In order to resolve the problem, we proposed the write status holding register (WSHR) for non-blocking PRAM accesses by hiding R drift latency with concurrent PRAM accesses. Our experiments show that the proposed WSHR gives 53.6%~0% performance improvements for large footprint programs in SPEC 2000 and 2006 benchmarks in the hybrid DRAM/PRAM main memory with 256MB DRAM and 14nm PRAM.

8. References

[1] S. Dumas, "Mobile Memory Forum: LPDDR3 and WideIO," JEDEC Mobile Forum, June 2011.

[2] International Technology Roadmap for Semiconductors (ITRS), available at www.itrs.net.

[3] C. Villa, et al., "A 45nm 1Gb 1.8V Phase-Change Memory," Proc. ISSCC, 2010.

[4] H. Chung, et al., "A 58nm 1.8V 1Gb PRAM with 6.4MB/s Program BW," Proc. ISSCC, 2011.

[5] EE Times, Samsung to ship MCP with phase-change, http://www.eetimes.com/electronics-news/4088727/Samsung-to-ship-MCP-with-phase-change.

[6] D. Ielmini, A. L. Lacaita, and D. Mantegazza, "Recovery and Drift Dynamics of Resistance and Threshold Voltages in Phase-Change Memories," IEEE Trans. on Electron Devices, vol. 54, no. 2, Feb. 2007.

[7] Samsung Electronics, Datasheet 1Gb(64Mx16) LPDDR2-PRAM, 2010.

[8] M. K. Qureshi, V. Srinivasan, and J. A. Rivers, "Scalable High Performance Main Memory System Using Phase-Change Memory Technology," Proc. ISCA, 2009.

[9] G. Dhiman, R. Ayoub, and T. Rosing, "PDRAM: A Hybrid PRAM and DRAM Main Memory System," Proc. DAC, 2009.

[10] H. Park, S. Yoo, and S. Lee, "Power Management of Hybrid DRAM/PRAM-based Main Memory," Proc. DAC, 2011.

[11] B. C. Lee, et al., "Architecting Phase Change Memory as a Scalable DRAM Alternative," Proc. ISCA, 2009.

[12] P. Zhou, et al., "A Durable and Energy Efficient Main Memory Using Phase Change Memory Technology," Proc. ISCA, 2009.

[13] B. D. Yang, et al., "A Low Power Phase-Change Random Access Memory Using a Data-Comparison Write Scheme," Proc. ISCAS, 2007.

[14] S. Cho and H. Lee, "Flit-N-Write: A Simple Deterministic Technique to Improve PRAM Write Performance, Energy and Endurance," Proc. MICRO, 2009.

[15] M. K. Qureshi, et al., "Enhancing Lifetime and Security of PCM-Based Main Memory with Start-Gap Wear Leveling," Proc. MICRO, 2009.

[16] N. H. Seong, et al., "Security Refresh: Prevent Malicious Wear-out and Increase Durability for Phase-Change Memory with Dynamically Randomized Address Mapping," Proc. ISCA, 2010.

[17] J. Dong, et al., "Wear Rate Leveling: Lifetime Enhancement of PRAM with Endurance Variation," Proc. DAC, 2011.

[18] S. Schechter, et al., "Use ECP, not ECC, for Hard Failures in Resistive Memories," Proc. ISCA, 2010.

[19] N. Seong, et al., "SAFER: Stuck-At-Fault Error Recovery for Memories," Proc. MICRO, 2010.

[20] M. K. Qureshi, et al., "Improving Read Performance of Phase Change Memories via Write Cancellation and Write Pausing," Proc. HPCA, 2010.

[21] JEDEC Standard, Low Power Double Data Rate 2 (LPDDR2), JESD209-2E, April 2011.

[22] K. Lee, et al., "A 90nm 1.8V 512Mb Diode-Switch PRAM with 266MB/s Read Throughput," IEEE J. Solid-State Circuits, vol. 43, no. 1, pp. 150-162, Jan. 2008.

[23] D. Kroft, "Lockup-free Instruction Fetch/Prefetch Cache Organization," Proc. 8th Annual Symposium on Computer Architecture, 1981.

[24] S. Rixner, et al., "Memory Access Scheduling," Proc. ISCA, 2000.

[25] S. Li, et al., "McPAT: An Integrated Power, Area, and Timing Modeling Framework for Multicore and Manycore Architectures," Proc. MICRO, 2009.

[26] C. Luk, et al., "Pin: Building Customized Program Analysis Tools with Dynamic Instrumentation," Proc. PLDI, June 2005.

APPENDIX
Write Performance Improvement by Hiding R Drift Latency in Phase-Change RAM

Experiments

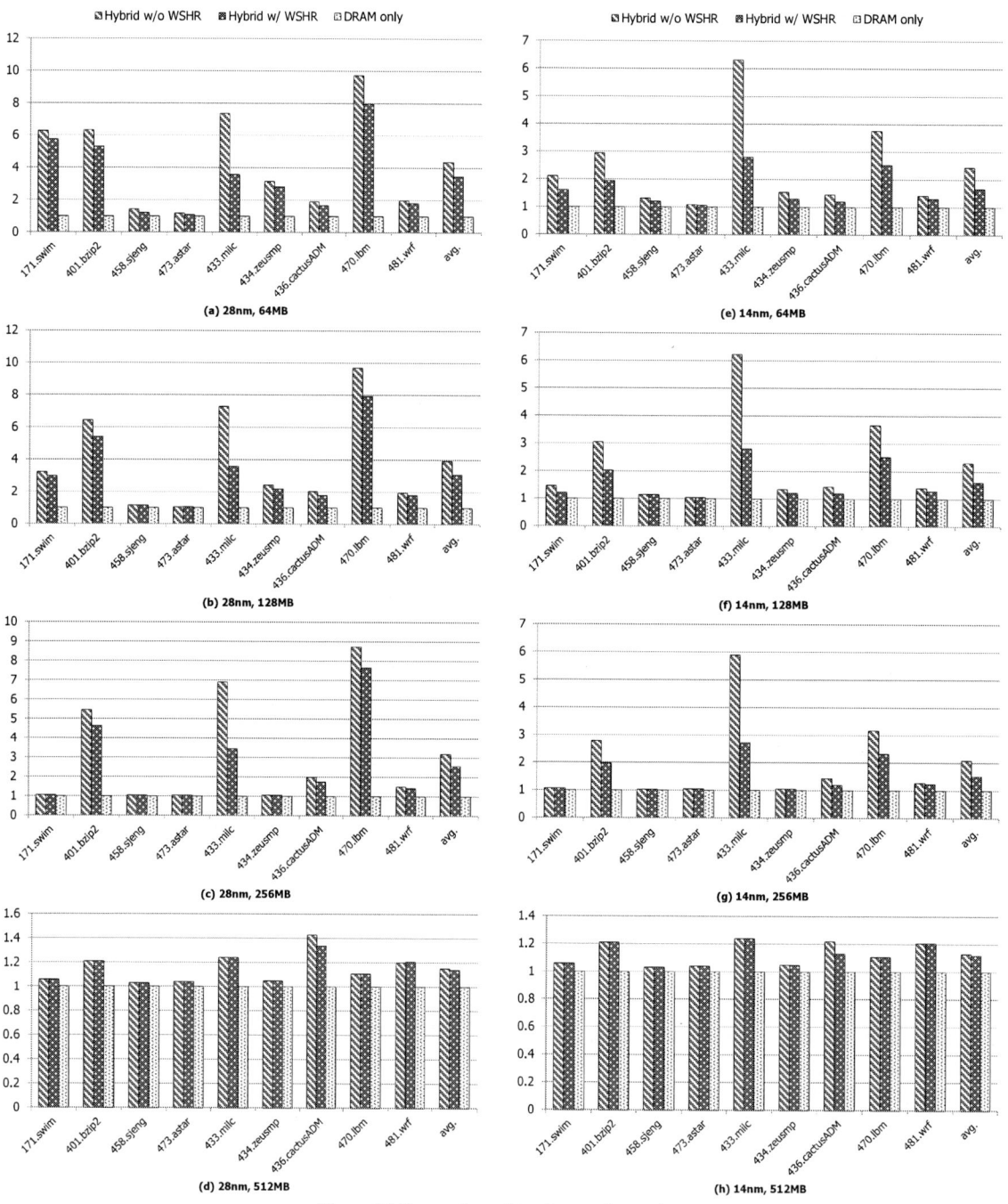

Figure 14 Comparison of total execution cycles

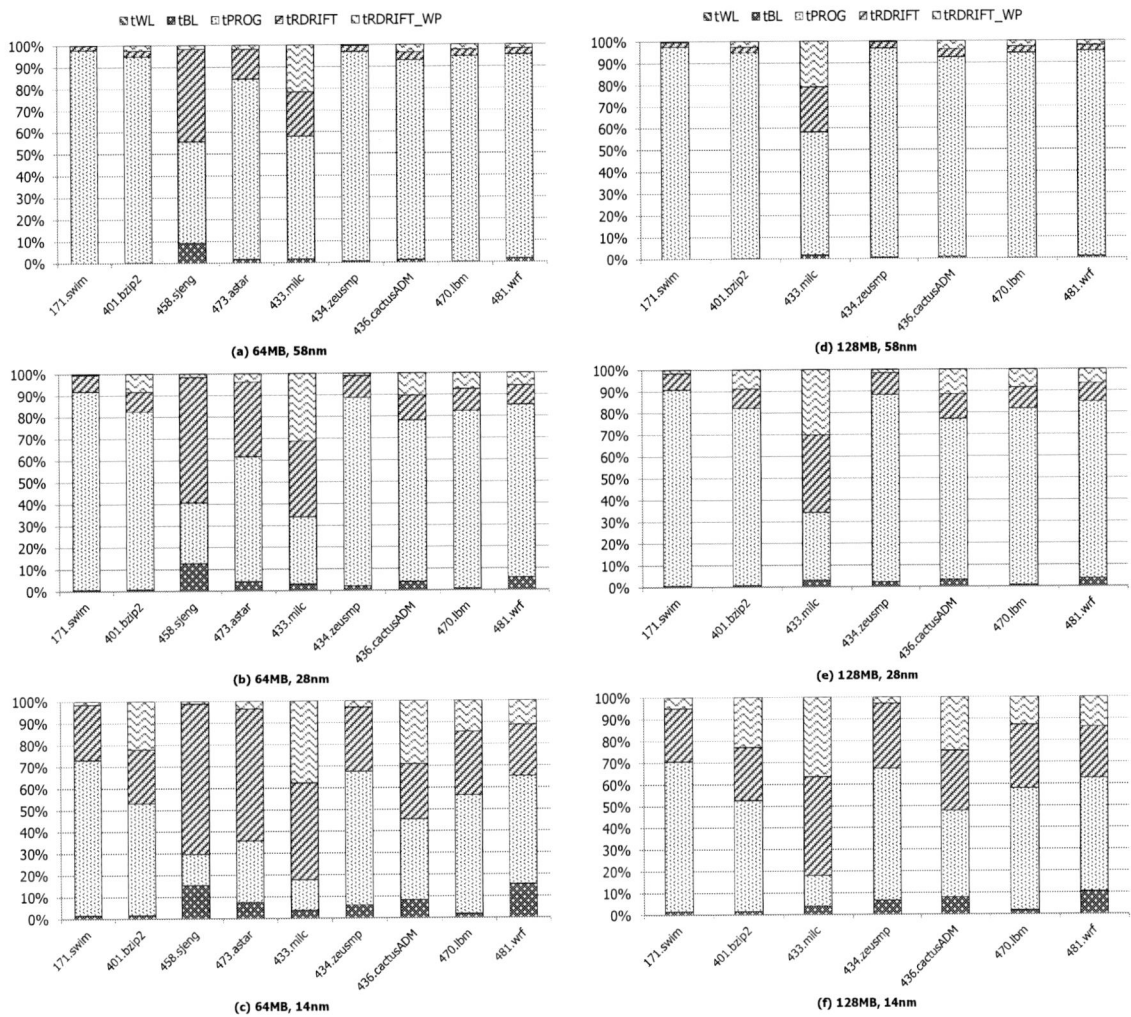

Figure 15 Write latency decomposition (obtained from the baseline hybrid main memory)

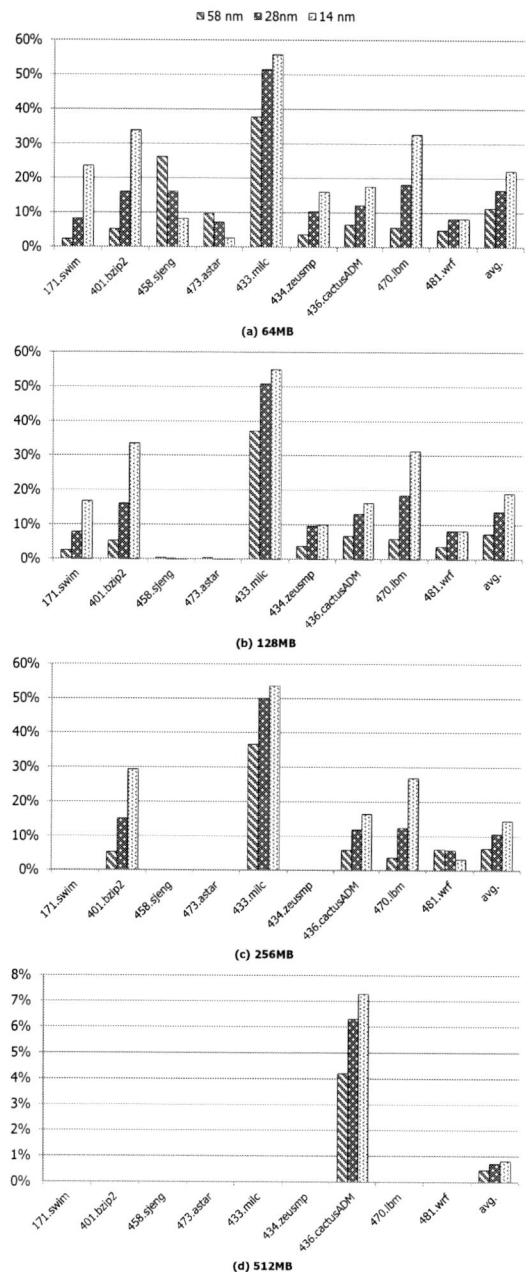

Figure 16 Total cycle improvements with different technologies

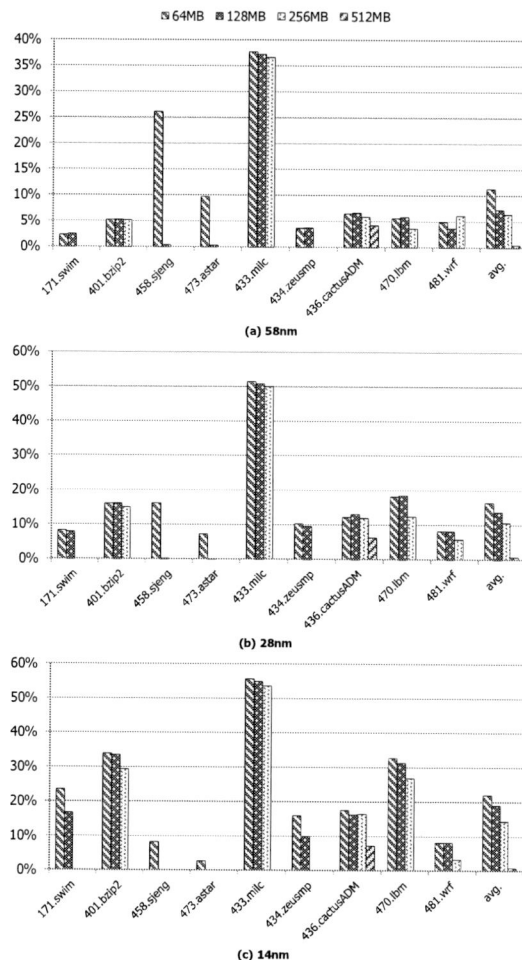

Figure 17 Total cycle improvements with different DRAM sizes

Note: Figure 16 (a) shows that two programs sjeng and astar show a different behavior that the gain of the proposed WSHR decreases as technology advances. It is because the relative portion of PRAM write latency decreases in the total execution cycle as the PRAM cell write latency scales down.

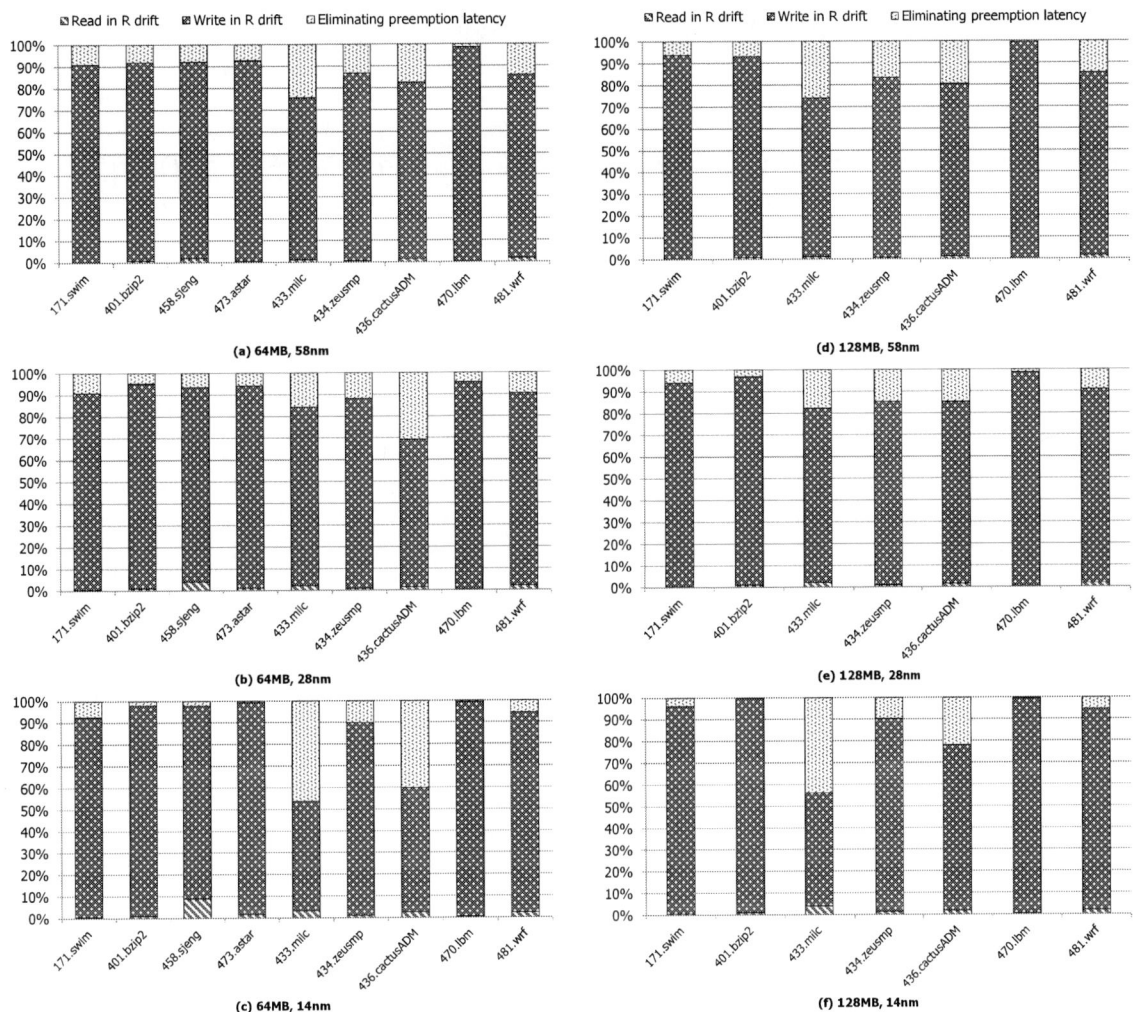

Figure 18 WSHR effect decomposition

Constructing Large and Fast Multi-Level Cell STT-MRAM based Cache for Embedded Processors [*]

Lei Jiang [†], Bo Zhao [†], Youtao Zhang [‡], Jun Yang [†]
† Electrical and Computer Engineering Department
University of Pittsburgh
Pittsburgh, PA 15261, USA
†{lej16,boz6,juy9}@pitt.edu

‡ Computer Science Department
University of Pittsburgh
Pittsburgh, PA 15260, USA
‡zhangyt@cs.pitt.edu

ABSTRACT

MLC STT-MRAM (Multi-level Cell Spin-Transfer Torque Magnetic RAM), an emerging non-volatile memory technology, has become a promising candidate to construct L2 caches for high-end embedded processors. However, the long write latency limits the effectiveness of MLC STT-MRAM based L2 caches. In this paper, we address this limitation with two novel designs: Line Pairing (LP) and Line Swapping (LS). LP forms fast cachelines by re-organizing MLC soft bits which are faster to write. LS dynamically stores frequently written data into these fast cachelines. Our experimental results show that LP and LS improve system performance by 15% and reduce energy consumption by 21%.

Categories and Subject Descriptors

B.3.2 [**Memory Structures**]: Design Styles

General Terms

Design, Performance

Keywords

MLC, STT-MRAM, LLC

1. INTRODUCTION

Modern embedded processors, in order to meet the fast growing computation and communication demands from end users, often integrate multiple cores with high computing capability and low power consumption. For example, high-end embedded processors such as nVIDIA Tegra [1] and Intel Atom [2] have been integrated in commercial tablet [3] and smartphone [4] products. Multi-core embedded processors require large on-chip caches to achieve scalable performance.

It is challenging to construct large caches that achieve both high performance and low power consumption. Many

---

[*]This work was supported in part by NSF CSR #1012070 and NSF CAREER #0747242.

high-end embedded processors have 256KB~2MB [1, 2] L2 SRAM caches. Choosing SRAM for future larger caches is less appealing due to SRAM's low density, serious leakage and reliability problems at nanoscale. Studies have been conducted to integrate emerging memory technologies, such as Phase-Change Memory (PCM) [5] and Spin-Transfer Torque magnetic RAM (STT-MRAM) [6], in the cache hierarchy. Among different memory technologies, STT-MRAM has been identified as a promising candidate for on-chip caches. STT-MRAM has no leakage from the memory cell and a comparable read speed as SRAM. Recently proposed Multi-level cell (MLC) STT-MRAM [7, 8, 9, 10] can store multiple bits per cell, reduce per-bit cost, and make STT-MRAM even more attractive for constructing large and low power onchip caches.

However, while MLC STT-MRAM improves density, it has almost doubled read and write latencies. This degrades the cache performance and diminishes the benefits gained from enlarged capacity. In this paper, we propose two novel designs, Line Pairing (LP) and Line Swapping (LS), to address the latency limitation and improve the effectiveness of MLC STT-MRAM based caches. The existing fabrication of a 2-bit MLC STT-MRAM cell includes a *hard-bit* and a *soft-bit* [7, 9]. The hard-bit is fast to read but slow to write, while the soft-bit is fast to write but slow to read. LP re-organizes two physical cachelines into a *read slow write fast* (RSWF) soft-bit cacheline and a *read fast write slow* (RFWS) hard-bit cacheline. By promoting frequently written data into soft-bit lines and frequently read data into hard-bit lines, LS greatly reduces L2 hit latency and improves the overall performance. We evaluate LP and LS using a set of different benchmark programs. Experimental results show that LP and LS can improve performance by 15% and reduce energy by 21% over the naive MLC STT-MRAM based L2 cache design on a high-end embedded processor platform.

In the rest of paper, Section 2 describes the STT-MRAM background. Section 3 elaborates the design details of LP and LS. Section 4 summarizes the experiment methodology. We report and analyze the simulation results in Section 5. We present more related work in Section 6 and conclude the paper in Section 7.

2. BACKGROUND

STT-MRAM (Spin-Transfer Torque Magnetic RAM) is a new generation of magnetic random access memory (MRAM). As shown in Figure 1(a), a single level cell (SLC) STT-MRAM uses one MTJ (Magnetic Tunnel Junctions) to store binary information [11]. A MTJ consists of two ferromagnetic layers separated by an oxide barrier layer (MgO). The magnetization direction of one ferromagnetic layer is fixed (reference layer) while the other (free layer) can be changed by injecting a current. When the magnetic fields of two layers are parallel, the MTJ resistance is low representing a logical '0'; when the magnetic fields are anti-parallel, the MTJ resistance is high indicating a logical '1'.

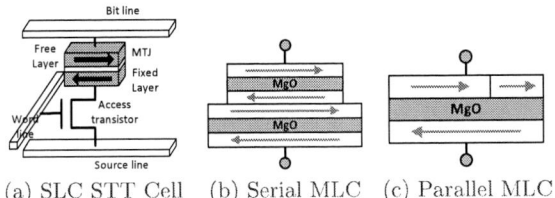

(a) SLC STT Cell   (b) Serial MLC   (c) Parallel MLC

Figure 1: STT-MRAM cell structure of SLC and 2-bit MLC cells.

2.1 STT-MRAM MLC Structure

The advances in device research [7, 9] have enabled the fabrication of multi-level cell (MLC) STT-MRAM. There are two types of MLC cells in the literature (Figure 1(b) and 1(c)). A 2-bit *serial* MLC MTJ (Figure 1(b)) possesses two vertically stacked MTJs that have different MgO and layer thicknesses. Since serial MTJ has a simple structure, it is relatively easy to design and fabricate, e.g., Hitachi has announced the successful tapeout of serial MLC MTJs [7]. However, serial MTJ requires a large critical switching current [9], which increases onchip dynamic power of caches. Therefore, Seagate proposed another type of MLC MTJ, *parallel* MLC MTJ (Figure 1(c)), which utilizes a single MgO MTJ whose free layer has two fields [9]. Parallel MLC STT-MRAM has been adopted in onchip caches [8, 10]. In this paper, we assume parallel MLC STT-MRAM is used.

For both types of MLC cells, the two magnetic fields are switched at different spin-polarized currents. Their combinations form multiple resistance levels to represent multi-bit values. The field that requires large current to switch is referred to as *hard-bit*; the field that requires smaller current to switch is referred to as *soft-bit*. For 2-bit data, the least significant bit (LSB) is the soft-bit, while the most significant bit (MSB) is the hard-bit.

Unlike SLC STT-MRAM, MLC STT-MRAM may have endurance problem. Wear leveling and error correction based techniques [10, 12] have been proposed for MLC STT-MRAM based caches.

2.2 MLC Read and Write Operations

Two-step MLC read: It requires a two-step operation to read the value stored in a 2-bit MLC STT-MRAM cell [8]. As shown in Figure 2(a), the read operation senses the resistance of a MLC MTJ and compares it to two of three resistance references. Each reference is generated by using two reference memory cells [7]. The sense amplifier first compares the cell with Ref_0 to determine the hard-bit, and then compares it with either Ref_1 or Ref_2 to determine the soft-bit. The read latency of a MLC cell is 1.5 times that of a SLC cell [8].

Two-step MLC write: Writing a 2-bit value into the cell is more complicate. In particular, two MTJs are controlled by one access transistor so that the write current

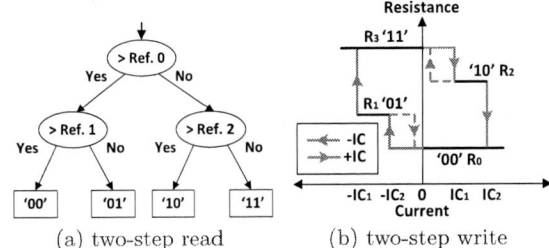

(a) two-step read   (b) two-step write

Figure 2: Accessing MLC STT-MRAM cells.

always passes through both MTJs at runtime. Since the current required to switch the hard-bit is higher than that of the soft-bit (Figure 2(b)), changing the hard-bit (MSB) always changes the soft-bit (LSB) to have the same magnetic field, i.e., either '00' or '11' is written. In the case if the LSB is not the same as MSB, a smaller write current is required in the second step to switch the soft-bit. Since the time spent in writing each bit is comparable to that of SLC, the latency of a 2-bit MLC write roughly doubles that of a SLC write. MLC cell write can terminate early in some cases: (i) if the LSB is the same as the MSB, then the second step can be skipped; (ii) if the MSB is not changed and only the LSB (soft-bit) needs to be changed, then the first step can be skipped [8]. However, when writing a cacheline (64B), it is less likely that the write can be terminated early as most likely at least one cell requires both steps.

3. PROPOSED TECHNIQUES

In this section, we elaborate the design details of *Line Pairing* (LP) and *Line Swapping* (LS). Our goal is to reduce the effective hit latency of MLC STT-MRAM caches.

3.1 Line Pairing

Given a 64B cacheline, one access to the data array activates 256 2-bit MLC cells in one bank. Each 2-bit MLC cell consists of one hard-bit and one soft-bit. The hard-bit is fast to read but slow to write while the soft-bit is fast to write but slow to read. The access latency of a cacheline is determined by finishing the read or write operations on the slow bits.

Here we use the cache parameters in Table 1 and 2 for discussion purpose. The design is applicable to caches with different cache sizes and line sizes. In this paper, we assume the two-phase cache access scheme that is widely adopted for highly associative caches, i.e., a parallel tag matching phase followed by a data array access phase (only to the matched line). Two-phase access achieves better trade-off between performance and energy consumption.

To exploit different read/write characteristics of hard- and soft- bits, we propose Line Pairing (LP) scheme to combine cells from two physical cachelines (having the same access latency) and re-arrange them to construct one **RFWS** line (*read fast and write slow line*) and one **RSWF** line (*read slow and write fast line*). As shown in Figure 3, while each physical line has both hard- and soft- bits, a logical cacheline has either all hard-bits or all soft-bits, but not mixed. The cacheline that only has soft-bits is referred to as a RSWF line; the cacheline that only uses hard-bits is referred to as a RFWS line. Within each cache set, half of cachelines are

908

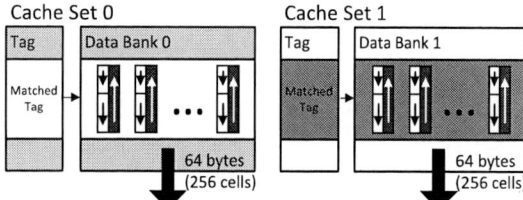

(a) each cacheline contains **both hard bits and soft bits**

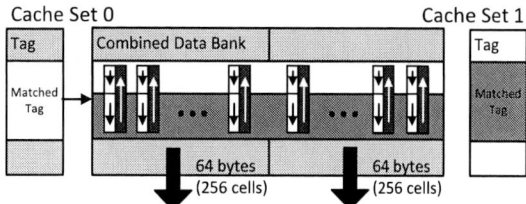

(b) after Line Pairing, each cacheline contains **either hard bits or soft bits**

Figure 3: Employ Line Pairing (LP) to reduce hit latency. (After LP, each cache set is stored across two banks but requires half of the original storage space in each bank)

RSWF while others are RFWS. A one-bit flag per cacheline is introduced in the tag array to indicate if the line is a RSWF or RFWS line. After tag access, LP knows the type of the cacheline.

By constructing RSWF lines, we can greatly speedup write operations when they fall in these lines. The latency is 19 cycles, which is only ~50% of the write latency in baseline design (Table 2). However, LP penalizes the writes to RFWS lines. As discussed in Section 2.2, the current required to write a hard-bit always destroys the soft-bit in the same MLC cell. Therefore, with LP, directly writing a hard-bit line erases the data of its co-located soft-bit line. To prevent such losses, LP first reads the soft-bit line, merges the data from both lines, and then writes both lines into the MLC cells. It takes 42 cycles(=5 cycles read + 37 cycles write) to write a hard-bit RFWS line. It is 14% slower than the baseline.

To reduce the hardware cost, LP uses the same number of port (i.e., read/write circuits) as baseline has. Since one bank has only 256 read/write circuits, to finish one cache access in one round, LP spreads the cells from one logical cacheline into two banks. As shown in Figure 3, *cache set 0* and *set 1* are stored in *data bank 0* and *bank 1* before LP. After LP, *cache set 0* stores its lines in both banks but uses one bit per cell. Activating 512 cells from two banks enables the access of one 64B cacheline in one round.

LP improves cache performance if many accesses fall in the RSWF lines. Assuming writes are evenly distributed among RSWF and RFWS lines, the average write latency is 31 cycles (= (19+42)/2 cycles), which is better than the baseline 37 cycles. However, LP activates two banks per access and thus reduces the number of parallelizable accesses. On the one hand, if the baseline issues two accesses to two banks in parallel, these accesses can finish in 37 cycles. Instead, LP has to sequentialize and finish two accesses in 38 cycles (two fast accesses), or up to 84 cycles (two slow ac-

cesses). On the other hand, LP can finish one write access in 19 cycles and thus resume following reads earlier to improve performance. We studied the overall impact and the results (as shown in Section 5) showed that L2 caches are frequent but do not have many overlaps such that LP speedups the overall performance.

3.2 Line Swapping

To maximize the benefits provided by LP, we further propose to dynamically promote write-intensive data to RSWF lines and read-intensive data to RFWS lines, referred to as line swapping (LS).

For a simple implementation of LS, a write-intensive line may be placed in a RFWS line initially; when it is identified as write intensive, it is always swapped to a RSWF line. A read-intensive line may be swapped from a RSWF line to a RFWS line as well. Clearly, too many such swaps increase the number of write operations, incur more bank contentions, and consume more energy. In addition, a line that is both read-intensive and write-intensive may keep thrashing between RSWF and RFWS lines. To achieve better swapping effectiveness and to mitigate thrashing, we introduce two counters — one swap counter Scnt and one weight counter Wcnt. Scnt is introduced to indicate if a line should be swapped: the one added to RSWF (or RFWS) line indicates if the line becomes read-intensive (or write-intensive respectively). Wcnt is introduced to prevent thrashing. A line with large weight value is less likely to be swapped.

Figure 4 illustrates the Line Swapping (LS) algorithm. Wcnt is initialized to be 1, increments when a swap happens, and saturates when it reaches M. Scnt is initialized to be Wcnt×N, and decrements when a RFWS line has a write hit, or a RSWF line has a read hit. Here M is the maximal value that the hardware counter can represent; N is the swap threshold to be studied in the experiments. When Scnt reaches zero, LS swaps either the current RSWF line with a LRU (least recently used) RFWS line, or the current RFWS line with a LRU RSWF line. If a miss happens and the victim line is a RFWS line, LS moves the LRU RSWF line here and loads the new data to the RSWF line location.

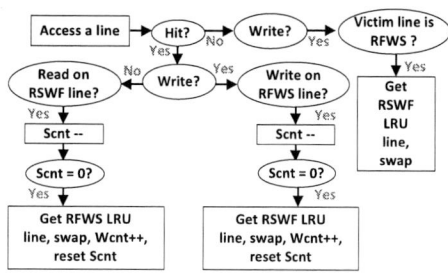

Figure 4: The line swapping (LS) algorithm.

The hardware cost is very modest. Our results showed that using a total of 6 bits per line (2 bits for Wcnt and 4 bits for Scnt) makes most accesses fall in corresponding fast lines. The storage overhead is ~1% (42 tag bits and 512 data bits). Note that supporting line swapping does not require extra buffer to hold extracted lines — the baseline has two 512-bit buffers for two banks, which are enough to hold two lines to enable LS.

4. EXPERIMENT METHODOLOGY

Simulator: We evaluated our proposed schemes using Virtutech Simics [13] simulation environment. We upgraded Simics *g-cache* cache module to simulate bank conflicts, bus contention, RFWS/RSWF MLC STT-MRAM cache lines, Line Pairing (LP) and Line Swapping (LS).

Baseline Modeling: Table 1 summarizes the core and cache configuration that we simulated in this paper. We modeled Intel Atom D525 processor [2] that has two in-order single-issue cores. Each core has separate 32KB I-L1 and 32KB D-L1 write-back caches. The L1 caches are private, and L2 cache is shared. MESI snoopy protocol is deployed to maintain the cache coherence. If a L1 miss happens, the host cache queries the other private L1 caches and waits until receiving a response. We overlapped L2 cache tag lookup with snooping. Based on the type of memory used in L2 cache, we summarized L2 cache tag lookup latency, read/write hit latency and energy consumption in Table 2.

L2 Cache Modeling: Intel Atom D525 is equipped with 1MB shared SRAM L2 cache [2]. We modeled such L2 cache area by CACTI [14] and obtained the baseline L2 cache area overhead — $5.1mm^2$. By assuming a SLC STT-MRAM cell size of $14F^2$ [15], we calculated that 5MB SLC can fit in a $5.1mm^2$ area overhead under $45nm$ technology. Since the area of a STT-MRAM cell is dominated by its access transistor, the size of MLC cell is slightly bigger than that of SLC cell [10] — 4M cells (or 8MB data) MLC STT-MRAM can fit within $5.1mm^2$ area. We adopted SLC STT-MRAM for the tag array of our MLC caches. Comparing to SRAM and MLC tag arrays, SLC tag array provides a better trade-off between area overhead and access latency. We also modeled an eDRAM L2 cache for comparison. We included a 4MB eDRAM L2 cache as the density of eDRAM is slightly larger than that of SLC STT-MRAM [5].

We used CACTI [14] to model STT-MRAM dynamic energy based on the reported results [8, 10, 16]. The write energy of SLC STT-MRAM is close to that of a soft bit for MLC STT-MRAM.

Processor	Intel Atom, 2-core, 1.8GHz, single-issue, in-order
Private L1	32KB I & 32KB D caches; 4-way, write-back SRAM-Based, 32B line, 1-cycle hit latency
Shared L2	16-way, 64B line, write-back, 4 banks **single** read/write port per bank
Main Memory	2GB, 300-cycle latency for the critical block

Table 1: Baseline Chip Configuration.

Type	Latency (cycles)	Dynamic Energy (64B) (nJ)	Leakage Power (W)
SRAM (1MB)	tag lookup: 1 data hit: 3	0.31	1.354
eDRAM (4MB)	tag lookup: 3 data hit: 5	0.51	0.396 refresh
SLC STTMRAM (5MB)	tag lookup: 2 R-hit: 3 W-hit: 19	read: 0.32 write: 1.29	0.156
MLC STTMRAM (8MB)	tag lookup: 3 R-hit: 5 W-hit: 37	read :0.32 write: 1.58	0.152
MLC STTMRAM (8MB) with LP	hard R-hit: 3 soft R-hit: 5 soft W-hit: 19 hard W-hit: 42	hard-R: 0.34 soft-R: 0.38 hard-W: 1.93 soft-W: 1.28	0.152

Table 2: L2 Cache Configuration.

Benchmarks: We compiled a set of benchmark programs

with different memory access characteristics to evaluate our proposed schemes. Since high-end embedded chips usually have GPUs [1, 2], we did not constrain our selections in multimedia benchmarks. Instead, we picked up various workloads from SPEC-CPU2006, SPEC-OMP2001 and PARSEC to study memory accesses for embedded cores. Table 3 lists the workload details.

Benchmark	Description	L1 Read MPKI	L1 Write MPKI
art	SPEC-OMP2001	106.4	40.1
bzip2	SPEC-CPU2006	463.4	90.3
gromacs	SPEC-CPU2006	10.1	4.2
hmmer	SPEC-CPU2006	21	10.79
tonto	SPEC-CPU2006	18.7	8.76
zeusmp	SPEC-CPU2006	36.3	11.4
dedup	PARSEC	14.6	8.13
ferret	PARSEC	18.7	13.6
vips	PARSEC	16.1	7.58
X264	PARSEC	35.3	4.6

Table 3: Simulated Benchmarks.

5. EXPERIMENTAL RESULTS

We compared four schemes as follows.

- `Baseline` indicates the MLC STT-MRAM cache without LP and LS.
- `LP` means the scheme using LP only.
- `LP+LS` is the scheme using both LP and LS.
- `Ideal` represents an ideal but unrealistic setting that defines the upper bound. In this setting, the read and write latencies of MLC are the same as those of SLC.

5.1 Performance Improvement

Figure 5 compares the performance using different schemes. We report the weighted speedup that is computed as follows [17].

$$Weighted\ Speedup = \sum_{thread} \frac{IPC_{tech}}{IPC_{baseline}} \quad (1)$$

On average, LP and LP+LS achieve 7% and 14.8% performance improvements, respectively, over `Baseline`. From Figure 5, LP+LS always outperforms LP indicating that dynamically promoting write-intensive data into RSWF lines and read-intensive data into RFWS lines is effective. LP+SW is only ~1.6% worse than `Ideal` indicating that LP+LS has explored most opporunties — most reads fall in RFWS lines while most writes fall in RSWF lines.

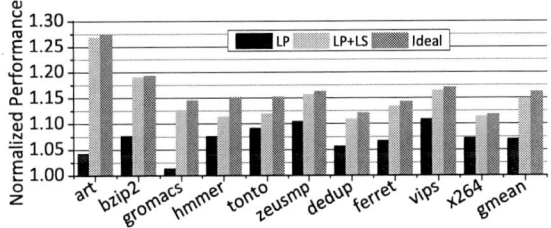

Figure 5: Performance improvement (Normalized to baseline).

We discussed in Section 3.1 that LP trades the maximal parallelizable accesses for short hit latency. From Figure 5, we observed that memory-intensive workloads with large L1 MPKIs tend to gain large performance improvements. For example, *art* has 40.1 write MPKI and achieves 27.3% performance improvement. The results suggest that for L2 cache accesses, it is more critical to speed up per access.

5.2 Energy Reduction

We next studied energy consumption of different schemes. As LP needs to activate more banks per cache access, and penalize more if writes fall into RFWS lines, our design may consume more energy. We used CACTI [14] to model the bank activation power with details reported in Table 2.

Read Energy: Since LP activates two banks for each access, the read energy increases — on average LP+LS consumes 9% more read energy, as shown in Figure 6(a).

Write Energy: Table 2 reports the required energy for two types of write operations — RFWS line write and RSWF line write. The former is more expensive due to its extra operations. Figure 6(b) shows that on average LP+LS reduces 8% write energy over baseline. With LS, most writes happen on RSWF lines such that write energy consumption is greatly reduced.

Dynamic Energy: Figure 6(c) reports that LP+LS saves 8% dynamic energy over baseline. This is because write energy dominates the dynamic energy consumption in STT-RAM based caches. This observation was also reported in recent studies [16, 18].

Total Energy: LP+LS spends less CPU cycles in executing the same number of instructions, which results in less leakage energy. Given leakage energy dominates the total energy consumption, Figure 6(d) shows that on average LP+LS reduces 12% total energy consumption.

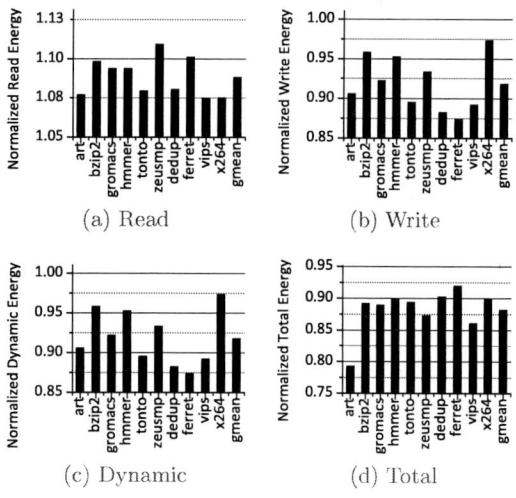

(a) Read   (b) Write

(c) Dynamic   (d) Total

Figure 6: Energy comparison (Normalized to baseline).

5.3 Line Swapping Overhead

Since LS introduces both latency and energy overheads, we prefer less swaps at runtime. The number of swaps depends on how effective we can find read-/write-intensive lines and if these lines are stable. We adjusted the swap threshold value and studied read- and write-triggered swaps independently. Read-triggered swap happens when Scnt of RSWF

line reaches zero while write-triggered swap happens when Scnt of RFWS line reaches zero.

Figure 7(a) summarizes the performance improvement when adjusting the initial value of Scnt from 1 to 5. Initializing Scnt of RFWS lines to 2 achieves the best performance, which shows that the space locality of writes is strong — a swap should be conducted if the RFWS line gets the second write hit. We then studied read-triggered swaps. Figure 7(b) shows that initializing Scnt of RSWF lines to 4 achieves the best performance improvement. This is because a swap operation has a long latency. It is not worthy to pay the overhead if a line to be swapped will be read only once or twice in the future. If the data has been read for 4 times, it is very likely for it to receive more reads. Swapping such kind of data into RFWS lines reduces both the effective read latency and the number of unnecessary swaps.

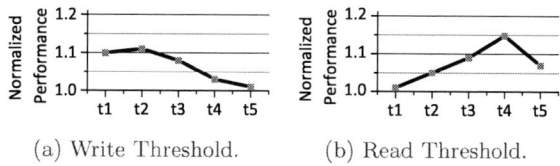

(a) Write Threshold.   (b) Read Threshold.

Figure 7: Performance geomean comparison when using different Scnt swap thresholds (Normalized to baseline).

We thus set Scnt to be 4 for RSWF lines and to be 2 for RFWS lines. We also studied Wcnt and found that incrementing Wcnt per swap effectively eliminates almost all thrashing in our tested programs. We used two bit for Wcnt and let it saturate at 3. Accordingly we need four bits for Scnt. In total, we need 6 bits per line to support LS. With this setting, Figure 8(b) reports that 97% writes hit in RSWF lines. Figure 8(a) shows the read hit rate in RFWS lines — on average, 79% reads occur in RFWS lines. The read hit rate in RFWS lines is lower because using a large Scnt threshold has more reads occurring in RSWF lines.

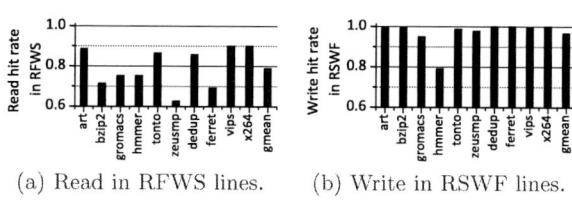

(a) Read in RFWS lines.   (b) Write in RSWF lines.

Figure 8: Access hit rate in fast cachelines.

5.4 Various Memory Technologies

The last experiment that we conducted was to compare SRAM, eDRAM, SLC STT-MRAM and MLC STT-MRAM caches, under the same die area constraint. The configuration details are in Table 2.

Performance: Figure 9 summarizes the performance comparison of L2 caches using SRAM, eDRAM, SLC STT-MRAM, MLC STT-MRAM, and MLC STT-MRAM with LP and LS. The results are similar to the findings in [5]. For most benchmark we simulated, 512KB per core SRAM cache can not hold their working sets. Too many offchip accesses significantly hurt the performance of SRAM-based caches.

911

eDRAM has similar performance as SLC STT-MRAM. With a larger capacity and similar access latency, MLC STT-MRAM with LP and LS achieves much better performance than SLC STT-MRAM.

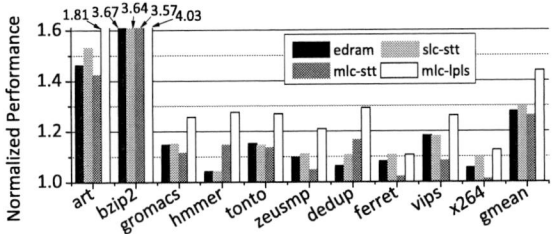

Figure 9: Performance comparison (Normalized to SRAM performance).

Total energy: Figure 10 compares the total energy when using SRAM, eDRAM, SLC STT-MRAM based caches. For all configurations, leakage energy dominates the total energy. It is not surprising that SRAM L2 cache consumes the highest total energy. Due to refreshes, eDRAM spends 23% of SRAM total energy on average. The energy consumptions on SLC and MLC STT-MRAM are similar, which are about 10% of SRAM total energy. MLC STT-MRAM with LP and LS always obtains the smallest leakage energy among all benchmarks, because of its shortest execution time. On average, MLC STT-MRAM with LP and LS consumes only 8% of the total energy of SRAM, when executing the same number of instructions.

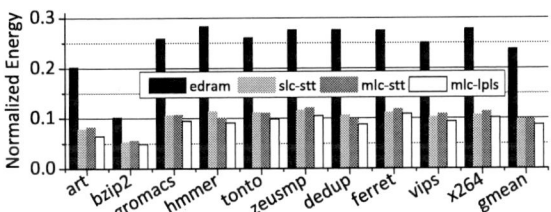

Figure 10: Total energy comparison (Normalized to SRAM total energy).

6. RELATED WORK

As a promising non-volatile memory, STT-MRAM has been proposed as SRAM or DRAM replacement for onchip caches. Previous works [5, 6, 19] presented a 3D-stacked processor architecture with onchip SLC STT-MRAM cache and compared SLC STT-MRAM to other different memory technologies for onchip caches. Zhou et al. designed early write termination to save write energy when the bit value is not changed [16].

Multi-level cell (MLC) STT-MRAM is a promising approach to improve the density of STT-MRAM [8, 7, 9, 10]. Ishigaki et al. proposed serial MLC MTJ and studied two-step read and write schemes [8]. Lou et al. presented parallel MLC MTJ and compared it to serial MLC STT-MRAM [9].

Recent studies of MLC flash devices have also found that some MLC flash pages are as fast as SLC pages while others are slow [20]. Flash manufacturers utilize FTL (flash translation layer) support to pack bits from two different pages into each MLC cell. The Line Pairing (LP) scheme proposed

in this paper shares similarity except that LP is performed at line granularity. A big difference between our design and Flash designs is that we developed Line Swapping (LS) that can greatly improve system performance and reduce the energy overhead.

7. CONCLUSIONS

Integrating a large and fast onchip L2 cache is a simple and effective way to mitigate the memory wall on high performance embedded processors. As technology scales, MLC STT-MRAM shows many advantages over SRAM, eDRAM, and SLC-MRAM in constructing onchip caches. However, MLC STT-MRAM suffers from long read and write access latencies. In this paper, we proposed two novel schemes: line pairing (LP) and line swapping (LS), to reduce average cache hit latency of MLC STT-MRAM based L2 caches. Our experimental results showed that these schemes are effective and on average achieve 15% performance improvement and 21% energy reduction over the baseline design.

8. REFERENCES

[1] nVIDIA Tegra.
http://www.nvidia.com/object/tegra-superchip.html.

[2] Intel Atom D525. http://ark.intel.com/products/49490.

[3] P. Gralla. Motorola Xoom: The Missing Manual. In *O'Reilly Media*, 2011.

[4] Fujitsu LOOX. http://solutions.us.fujitsu.com/LOOX/.

[5] X. Wu et al. Hybrid Cache Architecture with Disparate Memory Technologies. In *ISCA*, 2009.

[6] X. Dong et al. Circuit and Microarchitecture Evaluation of 3D Stacking Magnetic RAM (MRAM) as a Universal Memory Replacement. In *DAC*, 2008.

[7] T. Ishigaki et al. A Multi-level-cell Spin-transfer Torque Memory with Series-stacked Magnetotunnel Junctions. In *Symposium on VLSI Technology*, 2010.

[8] Y. Chen et al. Access Scheme of Multi-Level Cell Spin-Transfer Torque Random Access Memory and Its Optimization. In *IEEE International Midwest Symposium on Circuits and Systems*, 2010.

[9] X. Lou et al. Demonstration of multilevel cell spin transfer switching in mgo magnetic tunnel junctions. *Applied Physics Letters*, 2008.

[10] Y. Chen et al. Processor Caches with Multi-level Spin-transfer Torque RAM Cells. In *ISLPED*, 2011.

[11] M. Hosomi et al. A novel nonvolatile memory with spin torque transfer magnetization switching: Spin-ram. In *IEDM*, 2005.

[12] W. Xu et al. Improving STT MRAM Storage Density through Smaller-than-worst-case Transistor Sizing. In *DAC*, 2009.

[13] Virtutech Simics. http://www.virtutech.com.

[14] HP CACTI. http://www.hpl.hp.com/research/cacti/.

[15] S. Chung et al. Fully integrated 54nm STT-RAM with the Smallest Bit Cell Dimension for High Density Memory Application. In *IEDM*, 2010.

[16] P. Zhou et al. Energy Reduction for STT-RAM using Early Write Termination. In *ICCAD*, 2009.

[17] D. Tullsen and J. Brown. Handling Long-latency Loads in a Simultaneous Multithreading Processor. In *MICRO*, 2001.

[18] C. Smullen et al. Relaxing Non-volatility for Fast and Energy-efficient STT-RAM Caches. In *HPCA*, 2011.

[19] G. Sun et al. A Novel Architecture of the 3D Stacked MRAM L2 Cache for CMPs. In *HPCA*, 2009.

[20] L. Grupp et al. Characterizing Flash Memory: Anomalies, Observations, and Applications. In *MICRO*, 2009.

Incorrect Systems: It's not the Problem, It's the Solution*

Christoph M. Kirsch
Department of Computer Sciences
University of Salzburg
ck@cs.uni-salzburg.at

Hannes Payer
Department of Computer Sciences
University of Salzburg
hpayer@cs.uni-salzburg.at

ABSTRACT

We present an overview of state-of-the-art work in the engineering of digital systems (hardware and software) where traditional correctness requirements are relaxed, usually for higher performance and lower resource consumption but possibly also for other non-functional properties such as more robustness and less cost. The work presented here is categorized into work that involves just hardware, hardware and software, and just software. In particular, we discuss work on probabilistic and approximate design of processors, unreliable cores in asymmetric multi-core architectures, besteffort computing, stochastic processors, accuracy-aware program transformations, and relaxed concurrent data structures. As common theme we identify, at least intuitively, "metrics of correctness" in each piece of work which appear to be important for understanding the effects of relaxed correctness requirements and their relationship to performance improvements and resource consumption.

Categories and Subject Descriptors

C.0 [**Computer Systems Organization**]: General

General Terms

Algorithms, Design, Performance, Reliability

Keywords

relaxed correctness, probabilistic computing, performance, scalability, power consumption, robustness

---
*This work has been supported by the National Science Foundation (CNS1136141), the European Commission (ArtistDesign NoE on Embedded Systems Design, 214373), and the Austrian Science Fund (RiSE NFN on Rigorous Systems Engineering, S11404-N23).

1. INTRODUCTION

We acknowledge the emergence and advocate the study of relaxed, possibly quantitative approaches to describing and establishing the correctness of digital systems. The notion of hardware or systems software either computing the correct result for a given input or not has been the dominating principle in systems engineering for a long time whereas other areas of computer science such as scientific computing and machine learning as well as audio, video, and image processing, to name a few, have adopted relaxed notions of correctness early on. While the promise in special-purpose areas is typically higher performance and lower power consumption at a bounded loss in quality the effect of relaxed notions of correctness in systems may add other non-functional properties to the list such as robustness as well as production and development cost. Yet systems engineering tolerating, beware, incorrect results has up until recently played a rather secluded role.

The discrete nature of mathematics relevant in digital systems is probably a promising factor to look at for an explanation. Clearly, constructing a digital artifact and then showing that it produces results good enough for general (as opposed to special) purpose rather than results that are simply right or wrong is difficult in the presence of discrete semantics. Just modelling systems that involve both discrete and continuous concepts and then argueing about their properties is already a challenge [6]. Yet we feel that the emergence of relaxed notions of correctness in systems engineering is a sign of a maturing field taking a turn with new potential for the current and next generation of computer scientists and engineers. And it is not just about improving robustness and reducing cost but also about being able to utilize the emerging generations of systems, some increasingly parallel, using hopefully less energy so that some of the limitations of traditional design may eventually be overcome.

The purpose of this paper is to provide a brief overview of state-of-the-art work in the field of which some is related to material presented during a special session on probabilistic embedded computing at DAC 2012 organized by the first author. We also describe the key ideas behind our own work on concurrent data structures related to the topic. The common theme is here to identify "metrics of correctness" in each example and discuss their properties and possibly ways to obtain quantities in them. Note that this is an extremely short list of work far from being complete or even representative. The material cited here should only be seen as a hopefully reasonable starting point for finding other related work not cited here.

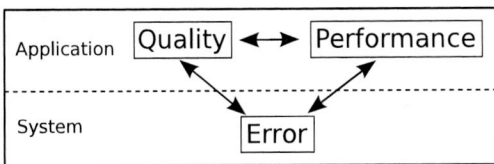

Figure 1: "Metrics of Correctness"

We work with three types of metrics for quantifying quality and performance of computation as well as error degree of systems whose design has been relaxed such that they may make mistakes, as shown in Figure 1. An error metric quantifies the degree of errors introduced by a relaxed system. A quality metric quantifies the computational quality produced by an application running on a relaxed system. We say that an application is error-tolerant on a relaxed system if the quality it produces increases whenever the system makes less errors, i.e., if the degree of errors in the computed results is proportional to the degree of errors introduced by the system. Relevant performance metrics are here execution time and power consumption. We say that an application is error-scalable if its performance increases whenever the system may make more mistakes.

2. HARDWARE

An important source of complexity in hardware design is reliability of computation, e.g. in arithmetic units and mechanisms such as hardware-based error detection and correction. Unreliable hardware design allows to reduce that complexity potentially providing benefits such as lower production cost, lower test and verification cost, smaller form factors, higher performance, and lower power consumption.

So-called probabilistic design as well as approximate design are two unreliable hardware design principles [11, 12] for trading-off reliability of computation and power consumption [16]. The idea of probabilistic design is to develop hardware that produces an output value for a given input value with a certain probability. Experimental data obtained on actual hardware shows that there is a monotone relationship between the probability of correct computation and power consumption [4], suggesting the probability of correct computation as quality metric. Approximate design results in hardware that deterministically produces, for a given input value, an output value that may, however, be incorrect. Similarly, experimental data obtained on actual hardware shows that there is a monotone relationship between arithmetic error and power consumption [5]. Here, the arithmetic error is an obvious candidate for a quality metric.

Unreliable hardware may also help increase hardware parallelism since unreliable hardware may require significantly less space than reliable hardware. An asymmetric multi-core architecture where a small number of reliable cores is combined with a large number of unreliable cores is an example of a design with a higher degree of parallelism than a conventional design of the same size [10]. Intuitively, the number of unreliable cores may be useful as error metric, and inversely even as quality metric if the quality of computations on unreliable cores deteriorates monotonically with the number of unreliable cores.

3. HARDWARE-SOFTWARE

The correctness of conventional software typically relies on hardware that returns deterministic output values for any given input values. Best-effort computing [3, 1] is a system design methodology for taking advantage of combinations of unreliable hardware and error-tolerant software to gain higher performance and lower power consumption. Errors introduced by unreliable hardware may be tolerated by certain types of software or handled by higher-level software layers. The challenge here is to divide applications into parts that tolerate errors and parts that do not. Such applications can also take advantage of the previously mentioned asymmetric multi-core architectures where application parts that tolerate errors run on unreliable cores [3, 10].

Stochastic processors [15, 19] produce so-called stochastically correct values, as with probabilistic design, through simplified hardware design, which may again enable higher performance and lower power consumption. Higher-level software layers may handle incorrect values either by tolerating the error or by detecting and correcting the error [21].

Control divergence in control-flow graphs makes branch prediction difficult and unreliable and may thus decrease performance. Branch herding performed by hardware or software reduces control divergence by forcing threads to take only a subset of all possible paths through the control-flow graph which may result in higher performance [21]. The degree of branch herding is an error metric whose inverse may also serve as quality metric if eliminating any branch always maintains or increases the error of the output. In this case, the application tolerates branch herding.

Detecting and correcting errors may incur high overhead which may eliminate the performance gains of stochastic processors. Therefore, it might be beneficial to detect just certain types or certain numbers of errors and correct them to stay within given error boundaries [21]. Here an error metric may be the number of errors that get detected and to which degree they get corrected. Again, the inverse may serve as quality metric if the actual error of the output is monotone in the error metric.

A key enabler of stochastic processors may be automatic transformation tools [19, 21] that generate error-tolerant versions out of regular applications, i.e., a transformed application may produce results at a quality that increases, at least within certain boundaries, whenever stochastic processors make less errors.

4. SOFTWARE

Relaxing software specifications may result in higher performance and lower power consumption, and may even increase reliability and robustness of software [18]. We distinguish relaxation techniques based on program transformations and concurrent data structure design.

4.1 Program Transformations

For certain types of applications, accuracy-aware program transformations [22] may generate error-tolerant code that may perform better than the original code. Here are three examples.

Substitution transformations [22] replace parts of a program with code that computes approximations of the output computed by the original parts but with less computational overhead. The approximate versions of the code are given

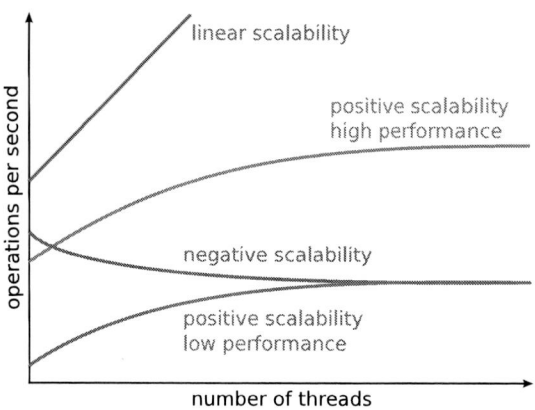

Figure 2: Concurrent data structure performance and scalability in number of data structure operations per second with an increasing number of threads sharing the same data structure in an exemplified benchmark scenario

and relaxed to different degrees in terms of some error metric. The program is error-tolerant if the approximations are compositional in terms of some quality metric.

Sampling transformations [22] work with code that computes output from a given set of elements. The transformed code performs the same computation as the original code but only on a subset of the elements obtained by some sampling policy. The code is error-tolerant if the quality of the result improves with larger subsets.

Loop perforation [20] is another type of program transformation which transforms a loop into a new loop where only a subset of the original loop iterations is performed. Decreasing the subset size of loop iterations decreases the runtime of executing the loop but also increases the error of the loop output.

4.2 Concurrent Data Structures

Concurrent data structures require synchronization to implement the exact specification of their sequential counterpart in a concurrent environment. However, synchronization operations may incur high overhead and prevent program code from executing in parallel. We discuss two approaches which may decrease synchronization overhead and increase parallelism, either by reducing contention on synchronization bottlenecks or by eliminating synchronization operations entirely, at the expense of adherence to exact data structure semantics. As a result, relaxed versions of concurrent data structures emerge which may perform and scale better on increasingly parallel hardware, and still be tolerated by certain applications.

In terms of performance and scalability, the goal is to achieve throughput in terms of number of data structure operations per second that is higher than of conventional designs (high performance) and grows with the number of concurrent units such as threads, for example, sharing the same data structure, to more threads than with conventional designs (positive scalability), as shown in Figure 2. Both, negative scalability even with high performance for low numbers of threads and positive scalability but with low performance are undesirable.

Reducing Contention on Synchronization Bottlenecks.

Our own work is on relaxing the semantics of concurrent data structures by reducing contention on synchronization bottlenecks. We achieve that by relaxing the sequential specification of a concurrent data structure. Consider, for example, a regular first-in-first-out (FIFO) queue where elements are enqueued at the queue tail and dequeued at the queue head. The problem with this specification is that it leaves little room for optimization in concurrent implementations [13, 7] so that scalability in the presence of high contention on the queue may still be limited to relatively low numbers of threads [9].

Instead of maintaining the original specification of a FIFO queue, we propose to relax the specification to what we call a k-FIFO queue with $k \geq 0$, which may dequeue elements out of FIFO order up to k [8, 9]. Retrieving the oldest element from the queue may require up to $k + 1$ dequeue operations (bounded lateness), which may return elements not younger than the $k + 1$ oldest elements in the queue (bounded age) or nothing even if there are elements in the queue.

A k-FIFO queue is starvation-free for finite k where $k + 1$ is what we call the worst-case semantical deviation (WCSD) of the queue from a regular FIFO queue. The WCSD bounds the actual semantical deviation (ASD) of a k-FIFO queue from a regular FIFO queue when applied to a given workload. Intuitively, the ASD keeps track of the number of dequeue operations necessary to return oldest elements and the age of dequeued elements.

Here, the error metric is semantical deviation: an implementation of a k-FIFO queue is correct if ASD $\leq$ WCSD for all workloads. Since semantical deviation is monotone, increasing k means more room for performance and scalability improvements. However, it is also important to consider which properties of an implementation determine k, i.e., whether k is configurable, depends on the workload, or is even probabilistic. For example, there are concurrent algorithms that implement k-FIFO queues whose WCSD is determined by configurable constants independent from any workload [2].

We took an entirely different approach called Scal queues based on distributed data structures and load balancing by creating p copies of a standard, non-blocking FIFO queue [13] and then, upon each queue operation, selecting one out of the p so-called partial queues for performing the actual operation, independently of and concurrently to any other operations that might hit the other $p - 1$ partial queues [8, 9]. Thus the load balancing algorithm for selecting partial queues and p itself determine the WCSD of the resulting Scal queue. For example, selecting partial queues in a round-robin fashion for enqueueing, and independently for dequeueing, using two atomic indexes, limits the WCSD to p since the maximum imbalance of the partial queues cannot become larger than p. However, performance and scalability truly improves only if selection and actual partial queue operation are done non-atomically, increasing the WCSD to a workload-dependent p times the number of threads in the system [8, 9]. Even more performance and better scalability are possible if selection is done randomly (to avoid any explicit synchronization in selection [8, 9]) and hierarchically (to exploit memory hierarchies [9]). The resulting WCSD may then only be bounded probabilistically.

Another interesting aspect of semantical deviation is the problem of measuring it. Clearly, improved performance and scalability comes at the expense of increased semantical deviation, which may or may not be tolerated by applications using queues. But how bad is it? Well, ASD cannot be measured directly, at least as long as individual machine instructions cannot be time-stamped without introducing significant overhead. Instead, we obtain at runtime with low overhead so-called concurrent histories, which are sequences of time-stamped invocation and response events of the queue operations. A concurrent history represents the set of sequential histories—sequences of queue operations—that preserve precedence, i.e., if the response event of an operation A is before the invocation event of an operation B then A occurs before B in any of the sequential histories. One of these sequential histories actually took place with ASD as its semantical deviation. However, given a concurrent history, possibly containing millions of events, we are only able to compute offline the semantical deviation of one of the sequential histories with minimal semantical deviation, i.e., the lower bound on ASD, which nevertheless enables interesting relative comparisons of semantical deviation. In particular, we show that some Scal queues outperform and outscale existing implementations at the expense of moderately increased lower bounds on ASD [8, 9]. Computing upper bounds is more difficult and remains future work since it seems to involve enumerating possibly all precedence-preserving permutations of queue operations in a concurrent history.

Eliminating Synchronization Bottlenecks.

Synchronization bottlenecks can be eliminated by eliminating the corresponding synchronization operations. This approach may lead to race conditions of which some may result in effects such as data duplication or loss which may nevertheless still be tolerated by certain applications.

Idempotent work-stealing queues are distributed queues, one per thread, where a thread may either dequeue an element from its local queue without synchronization or dequeue (steal) an element from the queue of another thread with synchronization [14]. Queue elements may be returned multiple times instead of just once because of races between unsynchronized local dequeue and synchronized global steal operations. Intuitively, the number of races or, even more accurate, the amount of element duplication may be useful as error metric and inversely even as quality metric if the quality of computation deteriorates monotonically with element duplication. Moreover, error and quality metric may also be related to the number of involved threads since a larger number of threads increases the probability of races.

Another example of in fact full elimination of synchronization is a space-subdivision tree construction algorithm that does not use any synchronization operations and yet provides a well-defined and consistent tree state that may be good enough for some applications [17]. The race conditions that may occur result in subtree losses which reduces the amount of data held by the tree. Similar to the idempotent work-stealing queues, the number of races or, again even more accurate, the amount of subtree loss may serve as a useful error metric and inversely as a quality metric if the quality of computations deteriorates monotonically with subtree loss. Again, both error and quality metric may also be monotone in the number of involved threads since more threads make races more likely.

5. CONCLUSIONS

Relaxed and possibly quantitative notions of correctness are likely to play an increasingly important role in systems engineering. We believe that one of the key challenges is identifying "metrics of correctness" such that the effects of more errors gracefully, as opposed to abruptly, degrade the quality of a system and yet translate into higher performance and lower resource consumption. The idea seems to apply in virtually any area of systems engineering. Go for it!

6. REFERENCES

[1] Designing chips without guarantees. *IEEE Design and Test of Computers*, 27:60–67, 2010.

[2] Y. Afek, G. Korland, and E. Yanovsky. Quasi-linearizability: Relaxed consistency for improved concurrency. In *Proc. Conference on Principles of Distributed Systems (OPODIS)*, pages 395–410. Springer, 2010.

[3] S. Chakradhar and A. Raghunathan. Best-effort computing: re-thinking parallel software and hardware. In *Proc. Design Automation Conference (DAC)*, pages 865–870. ACM, 2010.

[4] L. Chakrapani, P. Korkmaz, B. Akgul, and K. Palem. Probabilistic system-on-a-chip architectures. *Transactions on Design Automation of Electronic Systems (TODAES)*, 12(3):29:1–29:28, May 2008.

[5] L. Chakrapani, K. Muntimadugu, A. Lingamneni, J. George, and K. Palem. Highly energy and performance efficient embedded computing through approximately correct arithmetic: a mathematical foundation and preliminary experimental validation. In *Proc. Conference on Compilers, Architectures and Synthesis for Embedded Systems (CASES)*, pages 187–196. ACM, 2008.

[6] T. A. Henzinger. The theory of hybrid automata. In *Proc. Symposium on Logic in Computer Science (LICS)*, pages 278–292. IEEE, 1996.

[7] D. H. I. Incze, N. Shavit, and M. Tzafrir. Flat combining and the synchronization-parallelism tradeoff. In *Proc. Symposium on Parallelism in Algorithms and Architectures (SPAA)*, pages 355–364. ACM, 2010.

[8] C. Kirsch, H. Payer, H. Röck, and A. Sokolova. Brief announcement: Scalability versus semantics of concurrent FIFO queues. In *Proc. Symposium on Principles of Distributed Computing (PODC)*, pages 331–332. ACM, 2011.

[9] C. Kirsch, H. Payer, H. Röck, and A. Sokolova. Performance, scalability, and semantics of concurrent FIFO queues. Technical Report 2011-03, Department of Computer Sciences, University of Salzburg, September 2011.

[10] L. Leem, C. Hyungmin, J. Bau, Q. Jacobson, and S. Mitra. Ersa: Error resilient system architecture for probabilistic applications. In *Proc. Design, Automation Test in Europe Conference Exhibition (DATE)*, pages 1560 –1565, 2010.

[11] A. Lingamneni, K. Muntimadugu, C. Enz, R. Karp, K. Palem, and C. Piguet. Algorithmic methodologies for ultra-efficient inexact architectures for sustaining technology scaling. In *Proc. Computing Frontiers (CF)*. ACM, 2012.

[12] A. Lingamneni and K. Palem. What to do about the end of Moore's law, probably! In *Proc. Design Automation Conference (DAC)*. ACM, 2012.

[13] M. Michael and M. Scott. Simple, fast, and practical non-blocking and blocking concurrent queue algorithms. In *Proc. Symposium on Principles of Distributed Computing (PODC)*, pages 267–275. ACM, 1996.

[14] M. Michael, M. Vechev, and V. Saraswat. Idempotent work stealing. In *Proc. Symposium on Principles and Practice of Parallel Programming (PPoPP)*, pages 45–54. ACM, 2009.

[15] S. Narayanan, J. Sartori, R. Kumar, and D. Jones. Scalable stochastic processors. In *Proc. Conference on Design, Automation and Test in Europe (DATE)*, pages 335–338. European Design and Automation Association, 2010.

[16] K. Palem, L. Chakrapani, Z. Kedem, A. Lingamneni, and K. Muntimadugu. Sustaining moore's law in embedded computing through probabilistic and approximate design: Retrospects and prospects. In *Proc. Conference on Compilers, Architecture, and Synthesis for Embedded Systems (CASES)*, pages 1–10. ACM, 2009.

[17] M. Rinard. A lossy, synchronization-free, race-full, but still acceptably accurate parallel space-subdivision tree construction algorithm. Technical Report 2012-03-005, MIT-CSAIL, February 2012.

[18] M. Rinard. Obtaining and reasoning about good enough software. In *Proc. Design Automation Conference (DAC)*. ACM, 2012.

[19] J. Sartori, J. Sloan, and R. Kumar. Stochastic computing: embracing errors in architectureand design of processors and applications. In *Proc. Conference on Compilers, Architectures and Synthesis for Embedded Systems (CASES)*, pages 135–144. ACM, 2011.

[20] S. Sidiroglou-Douskos, S. Misailovic, H. Hoffmann, and M. Rinard. Managing performance vs. accuracy trade-offs with loop perforation. In *Proc. Symposium and Conference on Foundations of Software Engineering (ESEC/FSE)*, pages 124–134. ACM, 2011.

[21] J. Sloan, J. Sartori, and R. Kumar. On software design for stochastic processors. In *Proc. Design Automation Conference (DAC)*. ACM, 2012.

[22] Z. Zhu, S. Misailovic, J. Kelner, and M. Rinard. Randomized accuracy-aware program transformations for efficient approximate computations. In *Proc. Symposium on Principles of Programming Languages (POPL)*, pages 441–454. ACM, 2012.

On Software Design for Stochastic Processors

Joseph Sloan, John Sartori, Rakesh Kumar
University of Illinois,
Urbana-Champaign
jsloan,sartori2,rakeshk@illinois.edu

ABSTRACT

Much recent research [8, 6, 7] suggests significant power and energy benefits of relaxing correctness constraints in future processors. Such processors with relaxed constraints have often been referred to as stochastic processors [10, 15, 11]. In this paper we present three approaches for building applications for such processors. The first approach relies on relaxing the correctness of the application based upon an analysis of application characteristics. The second approach relies upon detecting and then correcting faults within the application as they arise. The third approach transforms applications into more error tolerant forms. In this paper, we show how these techniques that enhance or exploit the error tolerance of applications can yield significant power and energy benefits when computed on stochastic processors.

Categories and Subject Descriptors

D.2.10 [**Software Engineering**]: Design—*Methodologies, Representation*; D.2.4 [**Software Engineering**]: Software/Program Verification—*Reliability*

General Terms

Algorithms, Design, Reliability

Keywords

Application Error Tolerance, Stochastic Processors, ABFT

1. INTRODUCTION

With growing challenges to sustaining Moore's law with process scaling alone, stochastic computing [4, 13, 14] is being investigated aggressively as a possible path for realizing future high performance and low power technologies. The key to realizing the stochastic computing vision will rest with the ability of researchers and developers to design efficient techniques which both enhance and exploit the natural error tolerance of applications. Due to growing fault rates and diverse failure behaviors, software techniques will be especially important for providing low power and low energy stochastic computation. In this paper, we discus three approaches that can be used to design robust software on stochastic processors.

The first approach involves relaxing application correctness constraints. Many applications have inherent algorithmic and cognitive error tolerance. For such applications, significant performance and energy benefits can be obtained by selectively allowing errors. As an example, we consider GPU applications where control

divergence [3] incurs large overheads. By selectively eliminating control divergence through a technique called *branch herding* (i.e., forcing all threads to take the same control path for certain branches), we show that performance can be improved significantly while maintaining acceptable output quality acceptable for many applications.

For the second approach, faults are handled in a more traditional manner by detecting and then correcting or recovering from the faults. Unfortunately, the overheads of many typical fault detectors (e.g. dual modular redundancy (DMR)) are simply too large for these techniques to be utilized as a viable solution for stochastic computing. For this reason, techniques for low-overhead algorithmic fault detection and correction are extremely important. We present one example of low-overhead Algorithmic Based Fault Tolerance (ABFT) [5] for an important class of algorithms (sparse linear algebra) that will be used in many future applications. These techniques exploit both inherent properties of the data, as well as inherent fault tolerance characteristics of common iterative linear solvers. As error rates increase, however, the overheads incurred from frequent recovery events can make a detection-only-based fault tolerance approach (even an algorithmic one) infeasible. Frequently, applications may not be concerned with correcting all errors exactly, but instead with simply reducing the amount of noise below a certain threshold. Below that threshold, applications can still make efficient forward progress by naturally tolerating noise. In this paper, we present an approach for algorithmic fault correction by approximately correcting errors that occur in Matrix-Vector operations of an iterative linear solver.

The third approach aims to take any arbitrary application and transform it into a more robust form capable of efficiently running on stochastic processors. We describe one such approach for application transformation for robustness that utilizes a numerical optimization framework to naturally tolerate errors. By converting applications to numerical optimization problems with minima corresponding exactly to the original programs' output, we can use robust solvers to efficiently compute the original programs' deterministic output.

Section 2 describes the approach of application relaxation by using branch herding. Sections 3 and 4 describe the approaches for algorithmic detection and algorithmic correction, respectively, with examples involving linear algebra-based applications. Finally, Section 5 introduces a general approach for transforming applications for increased robustness, which we call *Application Robustification*. Section 6 concludes.

2. RELAXING CORRECTNESS

One approach to building software for stochastic processors involves relaxing the correctness constraints of the applications to improve performance or power characteristics [1]. We focus on GPU applications with frequent control divergence [3]. GPUs utilize SIMD-based architectures where multiple execution pipelines are designed to run in lockstep. Applications incur large performance overheads for synchronization when threads running on different execution pipelines encounter control divergence. However, many GPU applications also exhibit typical fault tolerance characteristics commonly seen in data-parallel applications with

```
while (--i && (xx + yy < T(4.0))) {
    y = x * y * T(2.0) + yC;
    x = xx - yy + xC;
    yy = y * y;
    xx = x * x;
} return i;
```

Figure 1: The main computation loop for Mandelbrot. The loop is unrolled 20 times in the actual application kernel.

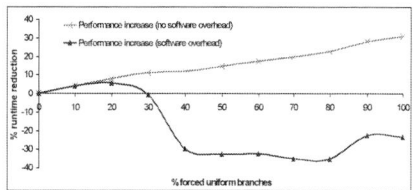

Figure 2: The performance of Mandelbrot can be increased by herding more branches. However, if software overhead is added to ensure branch uniformity, increasing the number of affected branches increases overhead and can even result in degraded performance.

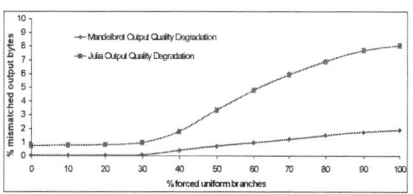

Figure 3: While eliminating control divergence can increase performance, blindly herding can result in degraded output quality.

inherent redundancy across time and space. So by carefully eliminating control divergence, we may achieve significant performance benefits.

Figure 1 shows an example of a kernel that exhibits control divergence, called Mandelbrot [18]. Control divergence arises in Mandelbrot because the number of iterations required to determine whether a particular is in the Mandelbrot (or Julia) set varies based on the point's location, especially in image regions near the set boundary, where some threads execute many iterations while others finish quickly.

The effect of control divergence on performance can be significant. Figure 2 shows the potential performance increase (runtime reduction) if control divergence is eliminated for a fraction of the static branches in Mandelbrot (from 0% to 100% of branches). The branches are chosen uniformly randomly when the fraction is less than 100%. Control divergence is eliminated by changing the source code to vote within a warp on the condition of a branch and forcing all threads in the warp to take the same (majority) direction at the branch. We call the technique *branch herding*. Branch herding can be implemented relatively efficiently in software, using the CUDA intrinsics ballot (_ballot) and population count (_popc). The ballot intrinsic is a warp vote function that combines predicates computed by each thread in a warp and sets the N_{th} bit in a 32-bit integer if the predicate evaluates to non-zero for the N_{th} thread in the warp. The ballot result is broadcasted to a destination register for each thread in the warp. We use the population count intrinsic to count the number of set bits in the ballot result. Branch herding can also be implemented easily in hardware by adding a 32-bit majority logic block.

While only 10% of dynamic instructions in Mandelbrot are branches, and less than 1% of branches diverge, performance can potentially be increased by 31% by eliminating control divergence. As the *no software overhead* performance series in Figure 2 demonstrates, performance increases for Mandelbrot as control divergence is eliminated for more branches. Figure 3 shows that the quality of the Mandelbrot output set degrades by less than 2%, even when divergence has been eliminated for all static branches. This shows

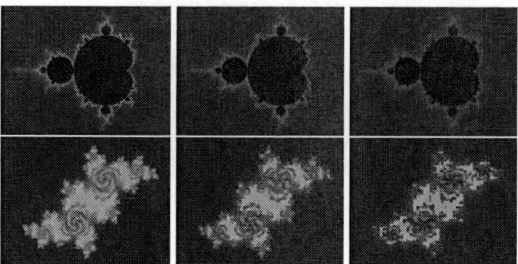

Figure 4: Progression of Mandelbrot (top) and Julia (bottom) images from 20% to 100% forced branch uniformity in 40% intervals.

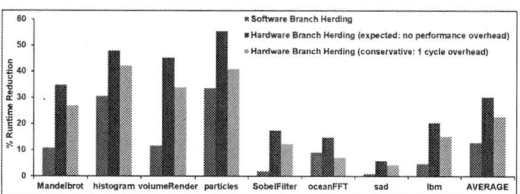

Figure 5: Performance for software and hardware branch herding.

that for certain error-tolerant applications, it may be possible to get significant performance benefits by relaxing correctness constraints related to control divergence for minimal output quality degradation.

Figure 4 shows the quality of the Mandelbrot and Julia output sets as the percentage of herded branches increases from 20% to 100% in increments of 40%.

A quick look at the last Julia output set, however, also suggests that an indiscriminate selection of branches for herding may result in significant output quality degradation for several applications. Therefore, any implementation of branch herding needs to carefully select the branches to target.

One policy for determining which branches are profitable for herding is to use profiling information in a feedback loop. After instrumenting a branch (or some fraction of the candidate branches) for herding (which can be done automatically by the compiler or manually by the programmer), the code is re-compiled and profiled to measure performance and output quality. If performance increases and output quality degradation remains below the acceptable threshold specified by the programmer, the branches are accepted for herding, else they are reverted to their original state. In terms of output quality, we find that in several cases, outputs may be considered acceptable even when herding is used for all the branches in a kernel function. In this case, herding can simply be switched with a compiler flag.

Experimental performance results for branch herding are shown in Figure 5. Software branch herding performance and output quality are measured directly at runtime (i.e., native execution on NVIDIA GeForce GTX 480.). The hardware branch herding technique in Figure 5 assumes some simple logic that eliminates the software overhead of herding. Hardware branch herding increases performance by 30% on average and up to 55% for individual applications. The software branch herding implementation achieves 13% performance benefits, on average.

Since branch herding exploits error tolerance to eliminate divergence, it may result in output quality degradation. Table 1 compares output quality degradation for the benchmarks with and without branch herding. Overall, branch herding does not result in much additional output quality degradation, while providing fairly substantial performance benefits.

3. ALGORITHMIC FAULT DETECTION

The second approach relies on detecting and then recovering from faults. In context of stochastic processors, we focus on techniques for algorithmic fault detection which exploit application er-

Table 1: Output Quality Degradation (%) for Branch Herding compared to Original

% Mismatch	Mandelbrot	histogram	volumeRender	particles	SobelFilter	oceanFFT	sad	lbm
Original	0.03	0.00	6.72	18.24	0.00	0.03	0.00	6.7E-7
Branch Herding	1.87	5.82	7.61	18.24	6.00	0.03	0.42	5.6E-5

ror tolerance as only those errors that adversely affect application output are considered . Below we discuss one example application class (sparse linear algebra) which exhibits inherent fault tolerance that algorithmic techniques can exploit.

Sparse linear algebra problems frequently have well defined structures. Common examples of structure are diagonal, banded diagonal, and block diagonal matrices. For example, *qpband* (Figure 6), which represents a canonical indefinite optimization problem, illustrates a typical banded diagonal structure (the nonzero pattern is on the left). Similarly, the matrix *msc00726* (Figure 7), representing a structural engineering problems from the Boeing test matrix group [2], also contain banded diagonal type structures.

Such structures in sparse problems commonly translate into uniform distributions of the column sums, which are directly used in algorithmic checks [17, 5] ($c^T(Ax) = (c^T A)x$ where $c = \bar{1}$). These matrices with well-defined distributions of column sums present an opportunity to sample only a fraction of the columns, which gives up a small degree of coverage (some errors may be missed) for a significant reduction in overhead. We call the technique using a random sampling and a sampling based upon clustering, Approximate Random and Approximate Clustering respectively. These checks are especially valuable since they exploit the fact that some errors can be tolerated by the application itself (e.g. iterative methods that converge to more accurate solutions).

Another opportunity for reducing algorithmic detection overhead for sparse linear algebra applications is that many such applications typically use the same matrix as part of many individual operations. For example, iterative solvers for linear systems ($Ax = b$) use MVM multiple times over thousands of iterations. This property of frequent data reuse makes it possible to analyze the structure of a given matrix or precondition the matrix to have a more amenable form for low overhead algorithmic fault detection, thus amortizing the setup cost by using lower overhead checks for subsequent MVM operations.

Identity conditioning (IC) transforms the high variance column sum distribution of the original matrix (A) into a more uniform set of values by using a check vector tailored to the given problem, instead of the traditional checksum: $c = \bar{1}$. IC finds such a tailored check vector by solving the system:

$$c^T A = \bar{1}^T \quad \text{(identity equation)}$$

The effect of A and the variance of the column sums can be minimized by then using:

$$c^T y = (c^T A)x = \bar{1}^T x = \sum x \quad \text{(IC)}$$

This makes the problem directly amenable to low-cost sampling as the variance in A now has a smaller effect on the product $c^T A$, making the sampling in AR and AC more representative than when sampling the check vector $c = \bar{1}$.

While Identity Conditioning eliminates the influence of A on the check, additional conditioning can also eliminate the influence of x. The *Null Conditioning* (NC) algorithm finds a check vector in the null space of the matrix A, solving the equation

$$c^T y = (c^T A)x = 0 \quad \text{(NC)}$$

This significantly reduces the runtime overhead of the check, since the right side of the check requires no additional computation (e.g. the sum equals zero) and the memory locality is improved since the input is no longer read in the check.

Finding a vector in (or near) the null space of A is done by computing its smallest singular value using singular value decomposition (SVD). The accuracy of fault detection for NC depends on the size of the problem's smallest singular value.

To evaluate the effectiveness of these detection techniques, we compare each technique when applied to a single MVM operation over a set 100 problems from the University of Florida Sparse

Matrix Collection [2]. The analysis includes the overhead incurred during the execution of the MVM operation and excludes the set-up cost, such as clustering and conditioning. The utility of our fault detection algorithms depends on both their detection accuracy and performance overhead.

Figure 8 presents the results. The three columns on the left-hand side correspond to the three full algorithms we're evaluating: the traditional dense check, the Oracle algorithm, and a Decision Tree based-algorithm (picks best technique and parameters based on learned parameters from matrix characteristics). The eight columns on the right-hand side correspond to each base technique, highlighting their individual capabilities. The four techniques on the far right are combinations of the others (e.g. ICAR is IC + AR, while NCAR is NC + AR). For a given technique and input problem, we choose the configuration parameters (detection threshold, sampling rate, conditioning quality) that minimize its overhead while meeting the F-Score bound. The bars (left vertical axis) show the fraction of problems on which each detection technique achieved the target F-Score. The empty circles show each technique's overhead on each of the problems meeting the F-Score target, and the red filled circle within each column is the detector's average overhead across all of these problems. Finally, the lines within each column indicate the range of overheads within $\pm$ one standard deviation of the average as well as the minimum and maximum overheads.

In general, the results show that the traditional dense check has an average overhead of 32%, ranging from 5% for denser problems to 80% for larger sparse problems with poor locality. While in contrast, the overhead of AR was 16% on average, over the same set of sparse problems (i.e. 16% lower than the traditional dense check).

Figure 8 illustrates a scenario where the dense check is significantly brittle, meeting the F-Score target with only 10% of the problems.In contrast, the Oracle can combine checks to cover 94% of the problems and the Decision Tree succeeds with 77%. This is because faults directly in the check are more likely to occur with the traditional dense check, which performs more operations than the proposed techniques.

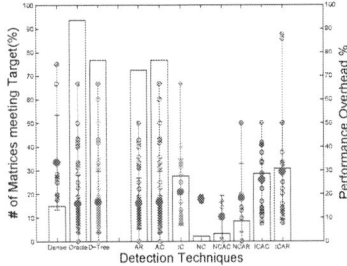

Figure 8: Right axis: Runtime overhead of each technique. Left axis: Number of problems meeting F-Score target. F-Score target=0.9, Fault Rate=1e-6, FaultModel=1

We also evaluate the fault detection techniques in the context of CG and IR sparse linear solvers. Errors affect linear solvers in two ways. First, since iterative algorithms converge from a poor solution to an accurate one, undetected errors are likely to slow down the algorithm's convergence or even cause it to diverge. Further, detected errors are managed using the classic rollback-restart technique, where the application is rolled back to some prior point in its execution and its execution is resumed. Our experiments use the simplest variant of this technique where the solver rolls back to the start of the current iteration every detected fault. In our experiments, each solver is executed until it reaches an error residual of 1e-6, meaning that if errors are detected, they may restart

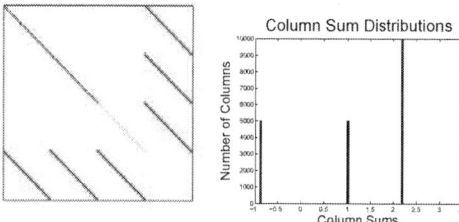
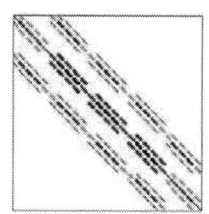

Figure 6: qpband (Variance $= 1.6071$) The matrix has a well defined and low variance ($< 1e3$) column sum distribution and is a good candidate for both Approximate Random and Approximate Clustering.

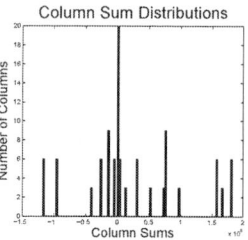

Figure 7: msc00726 (Variance $= 9.4724e14$) The matrix has high variance ($> 1e3$) column sums. This matrix is a good candidate for clustering given the finite sets of unique values shown above.

many iterations multiple times before they reach this goal. Figures 9 make this comparison between the execution time of the solver implementations employing a sparse check (on the x-axis) and the corresponding implementation employing the traditional dense check, via a sequence of graphs. The difference is measured as

$$overhead = \frac{Time_{sparse\_check} - Time_{dense\_check}}{Time_{dense\_check}}$$

which means that a difference of -50% corresponds to the linear solver executing twice as fast with the sparse detector than with the traditional dense detector. The overall difference in execution times of the linear solvers is consdiered in Figure 9.

Each detector and linear solver combination is evaluated on 5 different linear problems (separate sets for basic and preconditioned solvers). The average overhead over these matrices, for each detector, is shown as a red filled circle. Red lines correspond to the standard deviation and the max/min are shown using blue lines. The set of problems, for use with the basic solvers, were chosen randomly from those used in the MVM experiments.

The results also show that in the context of linear solvers the dense checks can have fairly large performance overheads (30-50%). For CG, the sparse check based implementation spent 17% less time in MVM operations on average than the traditional dense check-based implementations. This corresponds to a total execution time that is 9% lower on average. For IR, the sparse check based implementations spent 10% less time in MVM operations than the dense check-based implementations on average. This corresponds to 5% lower total execution time on average.

The results show that the impact of larger setup overheads for some of the techniques (e.g. clustering and preconditioning), in the context of both the IR and CG, is fairly negligible ($< 0.01\%$), since the amount of reuse is high. We observed that the absolute amount of reuse in the context of CG is dependent on the conditioning of the problem which impacts the number of iterations required to reach the desired solution. The error rate can also have an impact on the number of iterations and hence the amount of reuse within the algorithm

Upon analyzing the performance of the techniques in the different scenarios shown in Figure 9, we observed that the overall overhead can be reduced by 5%-20% by configuring the techniques to minimize the overhead from missed faults and false positives.

CG, and IR are two real application contexts that demonstrate that the sparse techniques are frequently able to exploit structure and reuse in sparse problems to reduce the overall overhead of algorithmic fault tolerance compared to the traditional dense checks. More details of our work on algorithmic fault detection for sparse linear algebra applications can be found in [17].

4. ALGORITHMIC FAULT CORRECTION

Previous algorithmic techniques for correcting errors [5] are primarily limited to scenarios involving rare error events, because of high overheads and the inability to make forward progress. Many applications contain inherent fault tolerance however, and are not always concerned with correcting all errors exactly. Therefore we can frequently use techniques that only approximately cor-

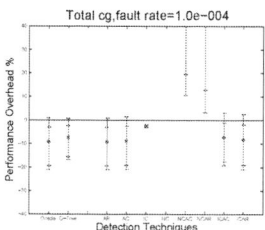
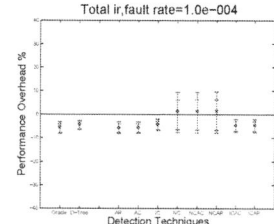

Figure 9: Percent difference between the execution time of the sparse techniques vs. dense check applied to CG & IR. Each column shows the total execution time overhead.

rect errors, ensuring that the aggregate effect of errors on the application's correctness and performance is bounded.

The general problem formulation of Algorithmic Fault Correction is, therefore: Given an application with an unknown correct output y, ensure that the application, even in the presence of faults produces an output y^* within a certain threshold of y.

As an example, in the context of Linear Algebra, we consider an MVM operation ($v = Au$) with k faulty entries in the output vector(v'). The traditional approach would explicitly detect and correct each of the k faults. In reality, the application may only care about approximately correcting the error ($e = v' - v$), and improving the accuracy (i.e. RMS $\|v' - v\|^2$). Therefore an algorithmic correction technique for the MV product could involve subtracting the projection of the error onto a code space. The partially corrected MV product(v'') in then found by:

$$v'' = v' - \frac{(c^T e)c}{\|c\|^2} \qquad (1)$$

One of the primary advantages of this particular approach, is that this type of approximate correction is guaranteed to always improve accuracy:

$$\|v'' - v\|^2 = \|v' - v\|^2 - \frac{(c^T e)^2}{\|c\|^2}$$

$$\|v'' - v\|^2 \leq \|v' - v\|^2 \qquad (2)$$

The above algorithmic correction can be easily adapted to account for the most important faults in terms of performance and accuracy. The developer has significant flexibility in the amount and types of codes chosen for the correction, depending on the accuracy targets which are desired.

5. APPLICATION TRANSFORMATIONS FOR ROBUSTNESS

The approaches described in prior sections are application-specific. Having the ability to transform arbitrary programs into more error tolerant forms is also an important consideration for stochastic processors. We call this *Application Robustification* [16]. In this section, we describe one approach for taking an arbitrary application and converting it into a more error tolerant form by reformulating it as a numeric optimization problem. We express the applications as constrained optimization problems, mechanically convert these to an unconstrained exact penalty form, and then

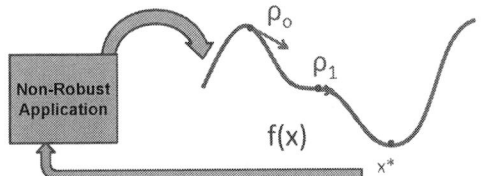

Figure 10: Application Robustification involves converting an application to an unconstrained optimization problem, where the minimum corresponds to the output of the original non-robust application.

solve them using gradient descent and conjugate gradient algorithms. This approach is quite generic, since linear programming, which is P-complete, can be implemented this way.

Let the vector x^* denote the (unknown) solution to our problem. To devise a robust algorithm, we construct a cost function f whose minimum is attained at x^*. Solving the problem then amounts to minimizing f. The main challenges, as illustrated in Figure 10:

- How to construct f without knowing the actual value of x^* a priori?
- How to choose an optimization engine that converges quickly and tolerates CPU noise?

For some applications, the natural conversion is to a general constrained variational form

$$\min_{x \in \mathbf{R}^d} f(x) \text{ s.t. } g(x) \le 0, h(x) = 0 \qquad (3)$$

for some functions f, g, and h. Commonly, the transformation of a given problem into its general variational form (3) is often immediate from the definition of the problem. We provide several illustrative examples below.

Least Squares Given a matrix A and a column vector b, a fundamental problem in many problems is to find a vector x that minimizes $\|Ax - b\|^2$ This problem is typically implemented on current CPUs via the SVD or the QR decomposition of A. The robust formulation of this problem is constructed by minimizing the quadratic function: $f(x) = \|Ax - b\|^2 = x^\top A^\top A x - 2b^\top x + b^\top b$. is The Least Squares problem is commonly thought of as a more intrinsically robust application, due to the continuous nature of much of it computation. We'll now consider the formulation of a problem not typically seen as fault tolerant, sorting.

Sorting: To sort an array of numbers on current CPUs, one often employs recursive algorithms like QUICKSORT or MERGE-SORT. Sorting can be recast as an optimization over the set of permutations. Among all permutations of the entries of an array $u \in \mathbf{R}^n$, the one that sorts it in ascending order also maximizes the dot product between the permuted u and the array $v = [1 \ldots n]^\top$ In matrix notation, for an $n \times n$ permutation matrix X, Xu is the sorted array u if X maximizes the linear cost $v^\top X u$. Since permutation matrices are the extreme points of the set of doubly stochastic matrices, which is polyhedral, such an X can be found by solving the linear program

$$\max_{X \in \mathbf{R}^{n \times n}} v^\top X u \quad \text{s.t. } X_{ij} \ge 0, \sum_i X_{ij} \le 1, \sum_j X_{ij} \le 1. \qquad (4)$$

Note that sorting is traditionally not thought of as an application that is error tolerant. Our methodology produces a potentially error tolerant implementation of sorting.

Bipartite Graph Matching The maximum weight bipartite graph matching problem can also be solved with a linear program, similar to sorting but with a more generalized objective function. Typical implementations are again not considered error tolerant (e.g. Hungarian algorithm). Our methodology however produces a potentially error tolerant implementation of Bipartite Graph Matching.

Once we have converted the programs (both those that require precisely correct outputs –fragile applications, as well as those that

do not–intrinsically robust applications) into optimization formulations,the best solver to compute the programs output can now be determined.

Under mild conditions, as long as step sizes are chosen carefully, gradient descent converges to a local optimum of the cost function even when the gradient is known only approximately. For this reason, we rely on gradient descent as the primary optimization engine to construct algorithms that tolerate noise in the CPU's numerical units. To minimize a cost function $f : \mathbf{R}^d \to \mathbf{R}$, gradient descent generates a sequence of steps $x^1 \ldots x^i \in \mathbf{R}^d$ via the iteration

$$x^i \leftarrow x^{i-1} + \lambda^i \nabla f(x^{i-1}), \qquad (5)$$

starting with a given initial iterate $x0 \in \mathbf{R}^d$. The vector $\nabla f(x^{i-1})$ is a subgradient of f at x^{i-1}, and the positive scalar λ^i is a step size that may vary from iteration to iteration. The goal is for the sequence of iterates to converge to a local optimizer, x^*, of f. The bulk of the computation in gradient descent is in computing the gradient ∇f. The suitability of gradient descent for processors with reduced guardbands is due to the fact that under various assumptions of local convexity on f, x^i is known to approach the true optimum as iterations progress [12]. As long as the ∇f is unbiased, gradient descent can eventually extract a solution with arbitrarily high accuracy. [16]

We rely on an exact penalty method to convert constrained problems, such as (3) into unconstrained problems that can be solved by gradient descent:

$$f(x) + \mu \sum_i |h_i(x)| + \mu \sum_j [g_j(x)]_+ . \qquad (6)$$

The operator $[\cdot]_+ = \max(0, \cdot)$ returns its argument if it is positive, and zero otherwise. A similar result for quadratic exact penalty functions of the form $f(x) + \mu \sum_i h_i 2(x) + \mu \sum_j [g_j(x)]_+^2$ also hold.

For example, the linear program for Sorting (4) can be converted into a corresponding unconstrained problem by using an exact quadratic penalty function:

$$f(X) = -v^\top X u + \lambda_1 \sum_{ij} [X_{ij}]_+^2 + \lambda_2 \sum_i \left[\sum_j X_{ij} - 1 \right]_+^2$$

$$+ \lambda_2 \sum_j \left[\sum_i X_{ij} - 1 \right]_+^2$$

where λ_1 and λ_2 are suitably large constants. $\qquad (7)$

While we use gradient descent as a search strategy for most of our kernels, some kernels may warrant the use of other search strategies. For example,the conjugate gradient (CG) method can be used for well conditioned (quadratic objective functions) to obtain very fast convergence with large problems.

To evaluate the robust versions of the above algorithms, we built an FPGA-based framework with support for controlled fault injection [16]. To calculate the energy benets from application robustifcation, we also used circuit-level simulations to calculate the relationship between voltage and error rate for the FPU.

To explore the feasibility of the proposed approach to provide robustness and energy benefits, we evaluated stochastic gradient descent (SGD) on the problems for Bipartite Graph Matching and Sorting across a wide range of fault rates.

The metric used to describe the quality of output is different for each benchmark. For Sorting, the y-axis represents the percentage of outputs where the entire array is sorted correctly (any undetermined entries (NaNs), wrongly sorted number, etc., is considered a failure). For Bipartite Graph Matching, the y-axis represents the percentage of outputs where all the edges are accurately chosen.

We chose small problem sizes for our evaluations due to low FPGA-based simulation speeds and the need to manually orchestrate each experiment (e.g., identify coefficients, parameters, etc.). For sorting, array size is 5 elements. Bipartite Graph Matching is performed for a graph with 11 nodes and 30 edges. State of the art deterministic applications are used for each of the application baselines (i.e. the C++ Standard Template Library (STL) and Hungarian Algorithm)

Examining the results, we see that we are able to achieve high quality results for both the fragile and the intrinsically robust applications. Sorting (Figure 11) performs poorly with linear step size scaling, but with sqrt step size scaling is able to achieve 100% accuracy even with large fault rates.

Bipartite Graph Matching (Figure 12) using 10000 iterations of SGD showed little performance degradation with increasing fault rates. However, using a combination of preconditioning, step sizing, and annealing techniques with gradient descent showed that 100% accuracies were also obtainable across even the largest fault rates.

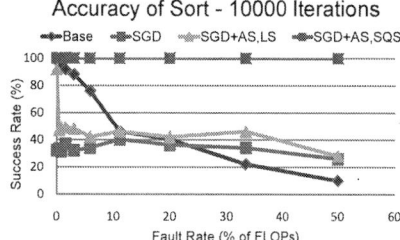

Figure 11: Success rate for different implementations of Sorting as a function of fault rate

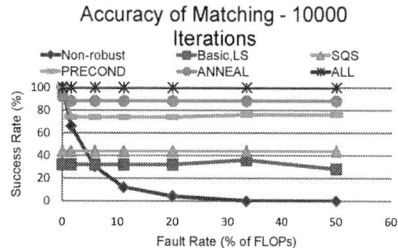

Figure 12: Success rate for different implementations of Bipartite Graph Matching as a function of fault rate

While stochastic gradient descent-based techniques provide high robustness, it often comes at the expense of significantly increased runtime due to the large number iterations required for convergence. For some applications where the transformed implementation has complexity per iteration less than the original applications complexity, it may be possible to show energy benefits by voltage overscaling a processor and letting the processor have errors. For other applications where the complexity per iteration of the optimization form is equal or greater than the original implementation complexity, it may be more difficult to show energy benefits.

Figure 13 shows an example of a problem where the complexity of the transformed implementation is lower. In this Figure, the y-axis shows the normalized energy results for the FPU for a Least Squares problem ($Ax = B$) assuming a voltage overscaled processor (non-zero error rates). The quadratic nature of the problem allows for CG to converge in fewer iterations (compared to gradient descent), while also naturally tolerating certain types of errors. Additionally, by using a specialized accelerator for linear operations [9], more applications such as Graph Matching can achieve energy benefits by using stochastic processors. For some problems, such as Sorting, it will be remain difficult however, even with a accelerators, to achieve energy benefits with voltage/scaling type models.

6. CONCLUSIONS

Harnessing the power of stochastic computing systems depends heavily on the ability of researchers and developers to design efficient algorithms and techniques which both enhance and exploit the error tolerance of applications. This paper presented three approaches for building applications for stochastic processors. This included: relaxing the correctness of applications, algorithmic detection/correction as faults arise, and application transformation

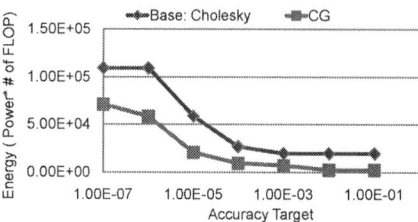

Figure 13: Energy for a CG-Based implementation of Least Squares

for robustness. In this paper, we show how these techniques that enhance or exploit the error tolerance of applications can yield significant power and energy benefits when computed on stochastic processors.

7. REFERENCES

[1] M. Carbin, D. Kim, S. Misailovic, and M. C. Rinard, editors. *the 33rd ACM SIGPLAN Conference on Programming Language Design and Implementation(PLDI), Beijing.* ACM, 2012.

[2] Timothy A. Davis. University of florida sparse matrix collection. *NA Digest*, 92, 1994.

[3] W. Fung, I. Sham, G. Yuan, and T. Aamodt. Dynamic warp formation and scheduling for efficient gpu control flow. In *MICRO*, pages 407–420, 2007.

[4] R. Hegde and N.R. Shanbhag. Energy-efficient signal processing via algorithmic noise-tolerance. In *Low Power Electronics and Design, 1999. Proceedings. 1999 International Symposium on*, pages 30 – 35, 1999.

[5] Kuang-Hua Huang and J.A. Abraham. Algorithm-based fault tolerance for matrix operations. *Computers, IEEE Transactions on*, C-33(6):518 –528, 1984.

[6] A. Kahng, S. Kang, R. Kumar, and J. Sartori. Designing a processor from the ground up to allow voltage/reliability tradeoffs. In *IEEE International Symposium on High-Performance Computer Architecture(HPCA)*, 2010.

[7] A. Kahng, S. Kang, R. Kumar, and J. Sartori. Recovery-driven design: A methodology for power minimization for error tolerant processor modules. In *the 47th Design Automation Conference (DAC)*, June 2010.

[8] A. Kahng, S. Kang, R. Kumar, and J. Sartori. Slack redistribution for graceful degradation under voltage overscaling. In *Asia and South Pacific Design and Automation Conference (ASPDAC)*, January 2010.

[9] D. Kesler, B. Deka, and R. Kumar. A hardware acceleration technique for gradient descent and conjugate gradient. In *Application Specific Processors (SASP), 2011 IEEE 9th Symposium on*, june 2011.

[10] R. Kumar. Stochastic processors. In *NSF Workshop on Science of Power Management*, March 2009.

[11] S. Narayanan, J. Sartori, R. Kumar, and D.L. Jones. Scalable stochastic processors. In *Design, Automation Test in Europe Conference Exhibition (DATE)*, 2010.

[12] A Nemirovski, A Juditsky, G Lan, and A Shapiro. Robust stochastic approximation approach to stochastic programming. *SIAM Journal on Optimization*, 19(4), 2009.

[13] J. Sartori and R. Kumar. Architecting processors to allow voltage/reliability tradeoffs. In *CASES*, 2011.

[14] J. Sartori and R. Kumar. Compiling for energy efficiency on timing speculative processors. In *the 49th Design Automation Conference(DAC)*, June 2012.

[15] N. Shanbhag, R. Abdallah amd R. Kumar, and D. Jones. Stochastic computation. In *the 47th Design Automation Conference(DAC)*, June 2010.

[16] J. Sloan, D. Kesler, R. Kumar, and A. Rahimi. A numerical optimization-based methodology for application robustification: Transforming applications for error tolerance. In *Dependable Systems and Networks (DSN), 2010*, June 2010.

[17] J. Sloan, R. Kumar, G. Bronevetsky, and T. Kolev. Algorithmic approaches to low overhead fault detection for sparse linear algebra. In *Dependable Systems and Networks (DSN), 2012*, 2012-july 1 2012.

[18] Wikipedia. Mandelbrot set, 2011. http: //en.wikipedia.org/wiki/Mandelbrot_set.

What to Do About the End of Moore's Law, Probably!

Krishna Palem
NTU-Rice Institute of Sustainable and Applied
Infodynamics, Nanyang Technological University
639798, Singapore
Department of CS & ECE, Rice University,
Houston, TX 77005, USA
kvp1@rice.edu

Avinash Lingamneni
Department of ECE, Rice University
Houston, TX 77005, USA
Wireless and Integration Systems, CSEM SA,
2000 Neuchatel, Switzerland
avinash.l@rice.edu

Categories and Subject Descriptors

B.8.0 [**Hardware**]: Performance and Reliability—*General*

General Terms

Algorithms, Design, Reliability

Keywords

Co-design, EDA, Energy-Accuracy Tradeoff, Moore's Law, Inexact Circuit Design, Probabilistic CMOS

1. RELIABLE COMPUTING IN THE BEGINNING

Computers process *bits* of information. A bit can take a value of 0 or 1, and computers process these bits through some physical mechanism. In the early days of electronic computers, this was done by electromechanical relays [28] which were soon replaced by vacuum tubes [6]. From the very beginning, these devices and the computers they were used to build were affected by concerns of reliability. For example, in a relatively recent interview with Presper Eckert [1] who co-designed ENIAC, widely believed to be the first electronic computer built, he notes: "we had a tube fail about every two days, and we could locate the problem within 15 minutes."

The computer pioneer John von Neumann who had worked with Eckert and his colleague John Mauchly clearly understood the importance of reliability [38]. It is worth spending some time reviewing this relatively old history and issues from this period, both as a curiosity, but notably since there are important lessons to be learnt from his historically significant lectures delivered at Caltech [38]. Back in 1952, he notes, *"The subject matter, as the title suggests is error in logics, or in the physical implementation of logics–in automata synthesis. Error is viewed therefore, not as an extraneous and misdirected or a misdirecting accident, but as an essential part of the process under consideration ...."* von Neumann is therefore concerned with errors that occur *intrinsically* in implementations of logics, and today, this would mean VLSI and ULSI circuits built out of CMOS transistors. Surprisingly, sixty years after von Neumann's lectures, we are again grappling with the same issue in the modern

era, since many now claim that the laws of physics dictating the exponentially improving benefits of *Moore's Law* will end in the next 10 to 20 years. So, after six decades from the time von Neumann gave his lectures, we have returned to the same concerns again.

2. A WALK THROUGH HISTORY STARTING WITH VON NEUMANN'S SEMINAL LECTURES

More than half a century ago, as each new generation of technology emerged, the issue of failures and reliability invariably reared its ugly head. We believe that the awareness by working with engineers who built the early vacuum tube based computers motivated von Neumann and his collaborators to use *probability* to model error or failures, through abstract models of "hardware". Since these models were based on Turing's now classical paper and were presented in the McCulloch-Pitts style, they had a cybernetic flavor, but nevertheless captured the essence of a state machine or automaton widely used to capture hardware behaviors today. We must remember that modern automata theory was nascent at this time and so, some of von Neumann's constructions might seem cumbersome. Nevertheless, we think the insights were incisive and conclusive. For example, the model in his lectures is essentially an abstract form of a modern computer, represented by a combinational logic component, and a communication component corresponding to wires or interconnect including a mechanism for encoding the state of a computation. In this model and with an element of error introduced into the elements, he considers ways for realizing, in modern terms, correct computational systems given that individual elements are vulnerable to failure.

Four years later, Moore and Shannon [26] in a sense reworked and extended the work presented in these lectures by introducing a model that is based on *switches*, which are the ubiquitous building blocks of computing systems even today. The particular model they used is based on Shannon's celebrated masters thesis, *A Symbolic Analysis of Relay and Switching Circuits*, from 1937. In this thesis, Shannon used the switch to abstractly represent an electro-mechanical relay to be used as a basis for building digital circuits. Once again, probabilities were used to model the correct or incorrect functioning of a switch. Their focus was on the important question of correcting errors introduced by potentially faulty switches and determining the cost of such correction in realizing digital circuits.

Since these early concerns, close to five decades passed with the spectacular success of the transistor, their integration at a very large scale leading to VLSI, the revolutionary invention of the microprocessor and the historically unprecedented march of ever-decreasing transistor feature sizes prophesied by Gordon Moore [27]. As physicist Gell-Mann notes in a recent interview [7] where he also mentions his role with K. A. Bruckner in the work leading to von Neumann's celebrated lecture, concerns of reliability became insignificant with the remarkable reliability achieved with transistors replacing vacuum tubes as switches. However, by the 1990s, the call for approaches to help sustain Moore's law as CMOS transistor feature sizes approached the nanoscale dimensions, were becoming increasingly strident. Scholarly articles started appearing with daunting titles such as *"End of Moore's law: thermal (noise) death of integration in micro and nano electronics"* [16]. So, in keeping with what seems to be history's penchant for repeating itself, the need for realizing reliable computing architectures from unreliable switches resurfaced again close to five decades after von Neumann's lectures, this time in the modern context of CMOS transistor based switching devices [14].

3. CROSSING OVER TO THE DARK SIDE BY USING UNRELIABLE ELEMENTS *UNRELIABLY*

Around 2002, one of us asked the following question: what if we consider building unreliable circuits and computing blocks from unreliable elements, rather than striving to build reliable switches, circuits and computing hardware from potentially unreliable components? We could potentially have a richer domain of switches to draw upon and therefore be much less constrained, than having to strive for reliability all the time. Fifty years after von Neumann's lectures, this idea was shown to be viable using a variety of mathematical models [31] including a *random access machine* and those representing circuits as switches whose probabilities determine error [32]. Of course, why would anyone want to build unreliable circuits and computing elements from them, that do not compute correct answers? While this goal might seem counterintuitive, it turns out that there is an entirely different resource associated with the physical implementations of switches, that their probabilistic and erroneous behaviors are intimately tied to. This has to do with the *energy* consumed by the physical implementation of a switch. Notably, the work from [32] characterized, and as far as we can determine for the first time, the amount of energy savings in the physical implementation of switches as we vary the correctness or error.

Hand in hand with the concerns about being able to sustain Moore's law, by the late 1990s, there was a rapidly increasing concern, if not alarm, about the amount of energy consumed by computing systems. Therefore, implementing probabilistic switches designed to be erroneous as a basis for realizing energy savings and building computing systems from them *for doing useful work*, seemed very attractive. One of us became a "heretic" [12] as a result of starting an active project and building a group to work on this idea, starting in 2001-2002 time period. This effort received significant stimulation with support from DARPA through a seedling grant in 2002. A summary of our work over the next decade caricatured against the historical legacy of working with erroneous computing elements is the primary subject of this paper.

3.1 The Energy-error Relationship from a Thermodynamic Perspective

That error and the energy consumed by a switch are related is not entirely a surprising fact. von Neumann in fact came close to this connection when he remarks in his lecture that "Our present treatment of error is unsatisfactory and ad-hoc. It is the author's conviction, voiced over many years, that error should be treated by thermodynamical methods and be the subject of a thermodynamical theory, as information has been, by the work of L. Szilard and C. E. Shannon (Cf. 5.2). [38]" This seemingly passing comment is actually

central to the subject matter of this paper, where we are concerned with energy efficiencies, realized through using unreliable switches.

To understand its full import, a short digression first. We note that physical implementations of switches using electrical means, whether they are built using vacuum tubes or transistors, consume energy when they perform the switching function. Thus, they are governed by the laws of classical physics in general, and thermodynamics in particular. By the time von Neumann delivered his lectures, classical thermodynamics had a clear statistical foundation and interpretation following the seminal work of Maxwell, Boltzmann and Gibbs [2, 8]. In particular, Szilard's work [36] that von Neumann refers to is an important part of this development. One of the most important debates in classical physics which lasted over sixty years has to with the celebrated second law of thermodynamics. Through a very clever construction, Szilard created a single object which has since come to be known as Szilard's engine, for trying to analyze and understand the validity of this law. Loosely speaking, Szilard's engine is a physical structure which we can think of as a cylinder delineated into two halves which are separated by a (weightless) trap door. For simplicity, we can think of a single molecule (of some gas) in motion in the cylinder. An external agent can, by raising and lowering the trapdoor, trap the molecule in either half of the cylinder. This agent also has the power to observe and *record* the half of the cylinder in which the molecule is trapped.

While it is not widely known, this construct had and continues to have a powerful influence on the way we reason about information, and its relationship to physical implementations, especially as they relate to issues of energy consumption. It is the first device that we know of which can be in one of two *states*, and which can be *switched* to induce a change of state by raising and lowering the trapdoor. It is therefore, the earliest instance of an abstract construction, as well a physical implementation, of a bistable switch. In this sense, it is also the first known link between a physical object and a method for producing abstract information through switching, which in this instance, is recorded through the actions of the external agent. The record can be thought of as a precursor of our modern digital memory in technological terms. Thus, each act of raising and lowering the trapdoor represents a switching step and produces a bit of *information* which is recorded, encoding the *state* of the switch. With this interpretation, Szilard's construct can be viewed as an engine which, through a switching step, produces a single bit of information. Since the state of this device is determined by the location of the gas molecule, the laws of statistical thermodynamics naturally apply. Therefore, it provides a very natural and intrinsic statistical basis for relating the energy consumed to the information being produced or *computed* through switches.

Returning to von Neumann's comment, the "unsatisfactory" aspect, we believe, has to do with the fact that in the approach he described, the probability of error is modeled synthetically in a manner that is extrinsic to the implementation. The probability of a switch performing its activity correctly is simply a numerical value associated with it as a number which is not derived from the inherent, and therefore intrinsic, physical implementation of a switch. We believe this was intentional of course since it allows reasoning in purely abstract terms without being encumbered by the physical details as exemplified and exploited by Moore and Shannon subsequently. As a result, the relationship between the probabilistic error as it varies, and the associated thermodynamic or energy cost, was not a part of the development. An exception of course is the remarkable body of work leading to Landauer's [19] historic insight on the minimum amount of energy needed to perform the act of recording the bit of information by the agent, during a single switching step of Szilard's engine. However, to use our terminology, it is very important to observe that Landauer is concerned about producing a bit correctly and does not concern himself with the energy associated with probabilistic switches with varying probabilities of correctness. So, mathematical models of probabilistically correct switching which do not have a relationship with energy consumption is what we had on the one hand. On the other hand, through Landauer's work, we have a connection between the energy cost as it relates to a switching step for producing one bit of information and recording it correctly without error.

3.2 Connecting Switches Back to the Physical Reality

What if were to go back to Szilard's roots and rather than his focus on understanding the profound nature of the second law of thermodynamics, instead look at his engine as a technical and perhaps even as a technological construct. By doing this, we could extend his engine to a model of a switch which can be switching correctly with some probability of correctness p and relate this probability very naturally to the associated energy consumption. As a result, we now have a switching device which can produce a *bit* of information through some physical medium wherein, the probability of it being correct is the parameter p, and the greater its value, the greater the energy consumed. The first attribute of our switch is the probability of correctness p associated with each bit of information being produced by it, whereas the second attribute is the associated energy cost.

In 2003, this was done through a probabilistic switch that captures these two attributes simultaneously, and was the subject of our work referred to earlier [32]. In spirit and philosophy, it went against the direction of abstracting away the physical attributes through clean models as von Neumann and Moore-Shannon set out to do, and connected probabilistic error during switching back to the physics. In our experience, this formulation of a probabilistic energy aware switch has proven to be a very useful foundational construct to understand and reason about *potentially unreliable* and energy-efficient hardware.

However, in order to really use this idea, CMOS implementations had to be considered. By 2004, we could show that probabilistically correct CMOS switches, which have since been referred to as PCMOS (switches), did exhibit this trade-off (summarized in [15]. In fact, the trade-off between p and the associated energy was modeled mathematically and also measured physically. It turned out to be much more favorable than we had anticipated in the following sense. As the probability of a PCMOS switch being correct approached 1, the energy consumed grew extremely rapidly. In other words, a small decrease in the probability of correctness (below 1) can be traded for a significant drop in the energy consumed, illustrated in Figure 1 during one switching step involving an XOR gate. For example, the point A in that figure has a p of 0.988 with a corresponding energy consumption of 20 fJ while point B requires only 7.5 fJ with a p of 0.975 per switching step indicating a ~3X reduction in energy consumption for a 2.3% decrease in the probability of correctness. Over the course of the next two years, we fabricated working devices and demonstrated that indeed, measured behaviors matching those predicted by the models and simulations could be realized [17].

4. THE LOGIC OF HARDWARE DESIGN AND A *PROBABILISTIC BOOLEAN LOGIC*

The behavior of probabilistic switches and the way they can be combined to design electrical circuits is based on Boole's strikingly contemporary formulation of logic [3] going back to the nineteenth century! However, if we set out

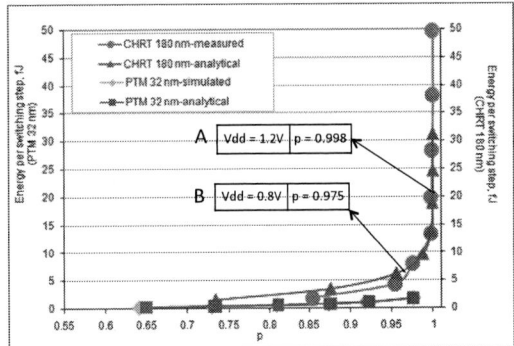

Figure 1: Measured and modeled energy-probability of correctness relationship for an XOR gate

Input			Truth Value
x	y	z	
0	0	0	0
0	0	1	0
0	1	0	0
0	1	1	1
1	0	0	0
1	0	1	1
1	1	0	1
1	1	1	1

(a)

Input			Probabilities	
x	y	z	Truth Value = 1	Truth Value = 0
0	0	0	¼	¾
0	0	1	¼	¾
0	1	0	¼	¾
0	1	1	¾	¼
1	0	0	¼	¾
1	0	1	1	0
1	1	0	1	0
1	1	1	1	0

(b)

Figure 2: (a) Conventional truth table for the carry logic function of a full adder $(((x \wedge y) \vee (x \wedge z)) \vee (y \wedge z))$ built from reliable hardware; (b) A probabilistic Boolean truth table for $(((x \wedge_1 y) \vee_1 (x \wedge_1 z)) \vee_1 (y \wedge_{3/4} z))$ for the full adder carry bit built from unreliable hardware [4].

to design hardware that is erroneous, Boolean logic has to be extended to admit incorrect outcomes. Specifically, we needed to develop a framework for modeling erroneous gates and furthermore, *extensions of Boole's rules* for combining their outputs.

In our *probabilistic boolean logic* (PBL) [4] that we published in 2008, the operators have the probability parameter p tied to them inextricably as shown in Figure 2 (b). Thus, each probabilistic Boolean operator, say $\wedge_p$ denoting a probabilistic AND, has an explicitly associated probability of correctness parameter p, which relates it simultaneously to its energy cost. Alternately, we can think of each of our operators $\wedge_p$, $\vee_p$ and others as having a thermodynamic interpretation, in addition to capturing correctness, through p.

4.1 From Gates and Logic to Applications

The next obvious question was to see how best to use PC-MOS hardware in mainstream computing. We considered applications that embodied probabilities naturally such as decision systems and pattern recognition based on bayesian and random neural networks respectively, as well as cryptographic applications. Using a system-on-a-chip type architecture, we could show that significant energy and speed gains were simultaneously possible through using PCMOS hardware [18]. Furthermore, it occurred to us that as we interact with information in our every day lives, we are, increasingly consuming it through our senses. So, the results of computing can be erroneous, albeit *perceptually* acceptable, if not indistinguishable from those produced using correct and less efficient hardware which we did for embedded signal processing. As far as we can tell, this is the first foray of using hardware to design a system that embodies error, and significantly, there is no intention or effort to correct it.

5. FROM PROBABILISTIC TO *INEXACT* CIRCUITS AND COMPUTING

Thanks to the greatly successful effort that continues to sustain Moore's law, transistors and VLSI circuits built from them are not very unreliable even today, probabilistically or otherwise. So, we were faced with the question as to how to use existing reliable VLSI technologies to be able to design circuits and architectures, and achieve gains in energy, and possibly in other physical attributes such as area and speed by exploiting this principle we have been developing.

For terminological clarity, let us refer to approaches that seek to garner resource gains by trading accuracy at the hardware level to belong to the *inexact design* class. We have shown the trajectory of inexact design in Figure 3 where we summarize the chronology of work starting with our departure in 2003 into the branch of *inexact* or unreliable yet useful computing. Traditionally, the term *reliable* has always been used to refer to component switches and circuits or systems built by composing them, when they have an associated probability of correctness of 1, or extremely close to one in practical settings; they are considered to be unreliable otherwise. As a result, realizing or synthesizing reliable systems from unreliable switches entails boosting the probability of correctness by compensatory error correction mechanisms. In contrast, as shown in Figure 3, since our work in 2003, the field has split from the legacy of von Neumann's lecture by explicitly seeking to build *unreliable* yet useful computing systems from unreliable components, *without* compensating through error correction. This departure from traditional approaches to designing reliable systems had led to the emergence of the area of inexact design.

At this point, we will digress from inexact design to identify two very innovative ideas with great potential for utility and impact that appeared between 1995-2001 that aimed to realize *reliable* low-cost DSP primitives. Sometimes, we have found references in literature that indicate a close affinity of these ideas to our work. The first idea of *approximate signal processing* [13] was inspired by the approaches to designing resource-constrained (typically execution time) systems in the areas of Artificial Intelligence and Real-Time Systems. The second involved *algorithmic noise tolerance* [9] techniques which allow circuits to be unreliable but combining techniques from signal processing, correct the resulting errors. Another similar approach that appeared at the same time as the inception of our work is the RAZOR effort from University of Michigan [5]. We observe these approaches are fundamentally different from our work described so far and do not follow our split from the legacy embodied in von Neumann's lectures as shown in Figure 3 since they eventually realize reliable computations.

Returning to inexact design, several research efforts have been launched by us as well as others to investigate techniques that would help realize inexact circuits starting with *mostly* reliable components of current CMOS technologies. Again, we refer the reader to Figure 3 where we have listed some of the highlights from this chronology along the inexact design branch. Most of the initial approaches to induce erroneous behavior in correctly functioning hardware involved varied forms of voltage overscaling. Voltage overscaled circuits were used as the basis for realizing cost-accuracy trade-offs for a wide variety of applications such as datapath circuits [24, 15] through Biased Voltage Overscaling (BiVOS), Motion Estimation, Discrete Cosine Transform [29], and

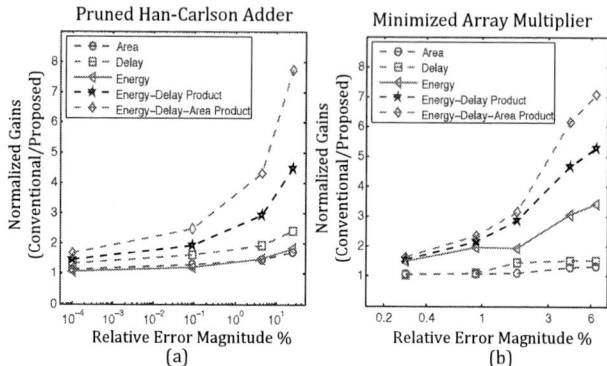

Figure 4: (a) Results of the Probabilistic Pruning technique on 64-bit Han-Carlson Adder; (b) Results of Probabilistic Logic Minimization technique on 16-bit array multiplier [23]

Image Processing. More recent and promising efforts applied voltage scaling at the granularity of processor modules. These approaches are popularly referred to as *stochastic computation* [11].

Canonically, voltage overscaling provides a fine-grained approach to enable energy-accuracy trade-offs, but have associated overheads due to a need for level-shifters, metastability resolutions circuits, and the routing of multiple voltage lines. The practical realizations only implement a subset of well characterized voltage levels owing to these overheads, and also due to the inevitable power supply fluctuations. While, these techniques do provide more dynamic control and have the potential for quadratic energy savings, the associated overheads are seldom amortized at fine-grain levels. Thus, more recent techniques have focused on the architecture as well as the logic layer for realizing circuits with *zero hardware overheads* while gleaning significant savings across the energy, delay and area dimensions simultaneously.

Continuing, in this context, we have proposed two novel techniques at the architectural and logic-layers which we have referred to as *probabilistic pruning* [20] and *probabilistic logic minimization* [21]. While the former is used to systematically *prune* or delete components and their associated wires along the paths of the circuit that have a lower significance or a lower probability of being active during circuit operation or both, the latter transforms logic functions to lower cost variants by manipulating the corresponding Boolean function through "bit-flips" (1 → 0 or vice-versa) for some of the input vector combinations that are not significant. Through these approaches, we have been able to demonstrate cumulative savings of a multiplicative factor of 8 in the energy-delay-area product (EDAP) for critical datapath elements such as adders and multipliers, as shown in Figure 4. The proposed pruning technique has been applied to a variety of standard 64-bit adder designs and a prototype chip has been fabricated using TSMC 180nm (low power) technology. A photograph of the 86-pin fabricated chip implementing our pruned adders along with its testing framework is shown in Figure 5 [22].

6. FROM INEXACT CIRCUITS TO DESIGN AUTOMATION

While holding great promise, the ability of the proposed inexact design techniques to influence the broader milieu of computing is limited as they are mostly based on ad-hoc hand designs and did not consider algorithmically well-characterized *automated* design methodologies. Also, existing design approaches were limited to particular layers of abstraction such as physical, architectural and algorithmic

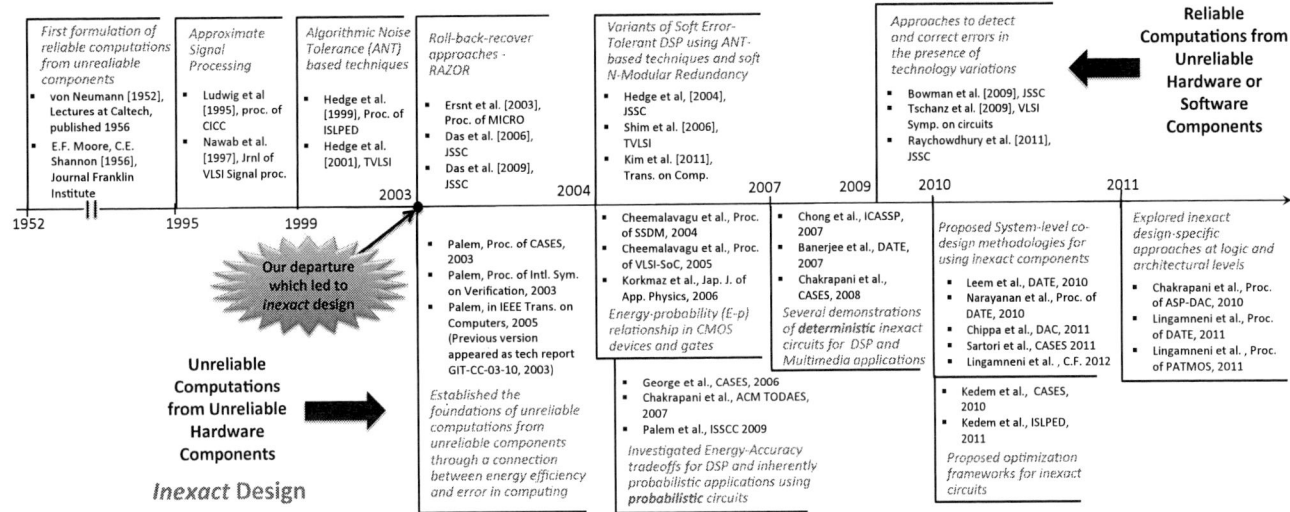

Figure 3: Timeline of important papers and innovations that shaped the domain of *inexact* circuit design

Figure 5: (a) A die photograph of the fabricated prototype chip; (b) The prototype chip integrated into the *icy*board test platform [22]

or more broadly software. However, it is well-known that significant gains can be achieved by optimizing across the layers. This is in stark contrast to conventional computing systems and hardware design wherein the hardware is expected to operate correctly all the time. In this context, automatic algorithmic methods are used widely across the layers of abstraction to achieve gains, to great effect [25]. To respond to this need, we presented an algorithmically well-founded cross-layer co-design framework (CCF) for automatically designing *inexact* hardware in the form of datapath elements that could significantly reduce the time needed for design space exploration [22]. We believe that this is an important first step in injecting the twin ideas of EDA and co-design into the milieu of inexactness.

7. FUTURE DIRECTIONS AND REMARKS

When we were invited to submit a paper on the genesis of inexact design viewed against the backdrop of probabilistic methods in computing, we considered three possible approaches. The first was to present a survey of the field. An alternative was to discuss our own technical work in some depth. Upon consultation, we concluded that both of these alternatives could be easily done by compiling and perusing existing publications. Therefore, we decided to write this paper by not intending it to serve either of these purposes. However, what seemed to be missing and therefore of some value as a contribution to the scholarship of this emerging field, was in connecting it to its rich legacy spanning the last century. This seemed a worthwhile endeavor, especially given the powerful ideas that served as a basis. In doing

so, we were guided by Longfellow's inspirational comments from his poem *'A Psalm of Life'* where he eloquently and beautifully characterizes the legacy of great ideas and the influence of people behind them, in that their lives : "..remind us, We can make our lives sublime. And, departing, leave behind us Footprints on the sands of time." Given the fast and furious pace at which we have been progressing in the CMOS world, we felt that to pause and reflect on the history behind topics in VLSI design and its automation wherein the "probability" is increasingly being used, would be a worthwhile effort. We hope to have done some justice to this goal within the restrictions of space that this paper has afforded us. A more detailed and complete version of this paper is to appear in a forthcoming special issue of the ACM Transactions on Embedded Computing Systems [30], and the reader is referred to this forthcoming paper for a more comprehensive overview and the many rich and exciting prospects for future work. Also, the deeper physical implications of CMOS devices and their switching has been analyzed comprehensively, and with great clarity, by Meindl [10]. The reader is referred to this paper for a deeper understanding of the issues underlying our discussion, with the context of reliable CMOS switches. Finally, probabilistic methods have been a topic of great import and utility in the software and algorithmic domains. Starting with Monte Carlo simulations [37] through the concepts of randomized algorithms [34, 35] to average case analysis [33], the fruitful use of probabilistic methods has deep roots. In contrast, the concepts and ideas that we have discussed in this paper represent a novel foray of probabilistic ideas and the frameworks of inexact computing that they have inspired, in the domain of hardware and its design.

8. ACKNOWLEDGMENTS

Parts of this work were supported in part by the US Defense Advanced Research Projects Agency (DARPA) under seedling contract number F30602-02-2-0124, the DARPA ACIP program under contract FA8650-04-C-7126 through a subcontract from USC-ISI, and an award from the Intel Corporation. We also wish to thank the Moore distinguished faculty fellow program at the California Institute of Technology, the Canon distinguished professorship program of the Nanyang Technological University (NTU) at Singapore, the NTU-Rice Institute for Sustainable and Applied Infodynamics and CSEM SA at Switzerland, which enabled pursuing this work in part. We gratefully acknowledge the contributions of our colleagues who are co-authors in our work which we have surveyed through citations. And last but not least, we thank the organizers of the special session at DAC 2012, Christoph Kirsch and Vincent Mooney, for inviting us to share our perspective and present this paper.

9. REFERENCES

[1] Alexander Randall 5th. A lost interview with ENIAC co-inventor J. Presper Eckert. Computer World, Retrieved 2011-04-25.

[2] L. Boltzmann and S. Brush. *Lectures on Gas Theory, English translation by S.G. Brush.* Dover Publications, 1995.

[3] G. Boole. The mathematical analysis of logic: Being an essay towards a calculus of deductive reasoning. 1847.

[4] L. N. B. Chakrapani and K. V. Palem. A probabilistic boolean logic and its meaning. Technical Report TR08-05, Rice University, Department of Computer Science, Jun 2008.

[5] D. Ernst et al. Razor: A low-power pipeline based on circuit-level timing speculation. *in proc. of MICRO*, pages 7–18, Oct. 2003.

[6] J. A. Fleming. Intrument for converting alternating electric currents into continuous currents. US Patent 803684, Nov 1905.

[7] M. Gell-Mann. Trying to make a reliable computer out of unreliable parts. *http://www.webofstories.com/play/10585?o=MS.*

[8] J. Gibbs. *Elementary Principles in Statistical Mechanics.* Scribner, New York, 1902.

[9] R. Hegde and N. R. Shanbhag. Energy-efficient signal processing via algorithmic noise-tolerance. *In Proc. Int. Symp. on Low Power Electronics and Design*, pages 30–35, 1999.

[10] J. D. Meindl et al. Nanoelectronics in retrospect, prospect and principle. *in proc. of ISSCC*, pages 31–35, 2010.

[11] John Sartori et al. Stochastic computing: Embracing errors in architecture and design of processors and applications. *in proc. of CASES*, 2011.

[12] E. Jonietz. Probabilistic chips. *Technology Review, Published by MIT*, *http://www.technologyreview.com/energy/20246/*, 2008.

[13] J.T. Ludwig et al. Low power filtering using approximate processing for DSP applications. *in proc. of CICC*, pages 185–188, 1995.

[14] K. Nikolic et al. Architectures for reliable computing with unreliable nanodevices. *in the proc. of IEEE conf. on Nanotechnology*, pages 254–259, 2001.

[15] K. V. Palem et al. Sustaining moore's law in embedded computing through probabilistic and approximate design: retrospects and prospects. In *in proc. of CASES*, pages 1–10, 2009.

[16] L. B. Kish. End of Moore's law: Thermal (noise) death of integration in micro and nano electronics. *Physics Letters A*, 305:144–149, 2002.

[17] P. Korkmaz. *Probabilistic CMOS (PCMOS) in the Nanoelectronics Regime.* PhD thesis, Georgia Institute of Technology, 2007.

[18] L. N. B. Chakrapani et al. Probabilistic system-on-a-chip architectures. *in ACM Trans. on Design Automation of Elec. Sys*, 12(3):1–28, 2007.

[19] R. Landauer. Irreversibility and heat generation in the computing process. *IBM J. Research and Development*, 3:183–191, July 1961.

[20] A. Lingamneni et al. Energy parsimonious circuit design through probabilistic pruning. *in proc. of DATE*, pages 764–769, Mar 2011.

[21] A. Lingamneni et al. Parsimonious circuit design for error-tolerant applications through probabilistic logic minimization. *in the proc. of the PATMOS*, pages 204–213, 2011.

[22] A. Lingamneni et al. Algorithmic methodologies for ultra-efficient inexact architectures for sustaining technology scaling. *in the ACM International Conference on Computing Frontiers*, May 2012.

[23] A. Lingamneni et al. Synthesizing parsimonious inexact circuits through probabilistic design techniques. *in the ACM Trans. on Embedded Computing Systems (spl. issue on Probabilistic Embedded Computing)*, 2012.

[24] L.N.B. Chakrapani et al. Highly energy and performance efficient embedded computing through approximately correct arithmetic: A mathematical foundation and preliminary experimental validation. In *proc. of IEEE/ACM CASES*, pages 187–196, 2008.

[25] G. D. Micheli. *Synthesis and Optimization of Digital Circuits.* McGraw-Hill, 1994.

[26] E. Moore and C. Shannon. Reliable circuits using less reliable relays I. *Journal Franklin Institute*, 262:191–208, Sept 1956.

[27] G. E. Moore. Cramming more components onto integrated circuits. *Electronics Magazine*, 38(8), 1965.

[28] S. F. B. Morse. Improvement in the mode of communicating information by signals by the application of electro-magnetism. US Patent 1,647, June 1840.

[29] N Banerjee et al. Process variation tolerant low power DCT architecture. In *Design, Automation and Test in Europe Conference*, Apr 2007.

[30] K. Palem and A. Lingamneni. Computing unreliably from unreliable elements: Inexact computing in perspective. *in the ACM Trans. on Embedded Computing Systems (spl. issue on Probabilistic Embedded Computing)*, 2012.

[31] K. V. Palem. Energy aware algorithm design via probabilistic computing: From algorithms and models to Moore's law and novel (semiconductor) devices. In *proc. of CASES*, pages 113 – 116, 2003.

[32] K. V. Palem. Energy aware computing through probabilistic switching: A study of limits. *IEEE Transactions on Computers*, 54(9):1123–1137, 2005; (abridged form appeared as – K.V. Palem, Energy Aware Computing through Randomized Switching, *Technical Report GIT-CC-03-16, Georgia Inst. of Technology*, May 2003).

[33] R. M. Karp et al. Average case analysis of a heuristic for the assignment problem. *Mathematics of Operations Research*, 19(3):513–522, Aug 1994.

[34] M. O. Rabin. Probabilistic automata. *Information and Control*, 6:230–245, 1963.

[35] R. Solovay and V. Strassen. A fast monte-carlo test for primality. *SIAM Journal on Computing*, pages 84–85, 1977.

[36] L. Szilard. Reduction in entropy of a thermodynamic system caused by the interference of intelligent beings. *Z. Physik*, 53:840–856, 1929.

[37] S. Ulam, R. D. Richtmyer, and J. von Neumann. Statistical methods in neutron diffusion. *Los Alamos Scientific Laboratory report LAMS–551*, 1947.

[38] J. Von Neumann. Probabilistic logics and the synthesis of reliable organisms from unreliable components. In *Automata Studies (C.E. Shannon and J. McCarthy eds.)*, Priceton Univ. Press, Princeton, N.J., 1956.

Obtaining and Reasoning About Good Enough Software

Martin Rinard

MIT EECS, MIT CSAIL

rinard@csail.mit.edu

Abstract

Software systems often exhibit a surprising flexibility in the range of execution paths they can take to produce an acceptable result. This flexibility enables new techniques that augment systems with the ability to productively tolerate a wide range of errors. We show how to exploit this flexibility to obtain transformations that improve reliability and robustness or trade off accuracy in return for increased performance or decreased power consumption. We discuss how to use empirical, probabilistic, and statistical reasoning to understand why these techniques work.

Categories and Subject Descriptors D.2.5 [*Testing and Debugging*]: Error Handling and Recovery

General Terms Reliability, Security, Verification

Keywords Recovery, Fault, Error

1. Introduction

A primary goal of many software development projects is to produce a system that is as close to correct as possible (in the sense that it contains as few errors as possible). In support of this goal, the programming languages and software engineering communities have invested significant time and effort developing techniques to either ensure the absence of errors in the system or to detect errors before the system is deployed (which the developers would then presumably correct before deployment).

In this position paper we present an alternate perspective. Instead of viewing systems as correct or incorrect, instead of viewing actions that the system takes as correct actions or errors, we instead propose to take a broader, more general perspective. This perspective focuses on systems and actions as acceptable or unacceptable. Unless an action causes the system to behave in an unacceptable way, we may see no need to classify the action as an error or the system as incorrect. And even if the action does cause the system to behave unacceptably, it is often possible to apply a simple modification that (while not eliminating the error) rehabilitates the action to have an acceptably benign effect on the overall behavior of the system.

1.1 Good Enough Software

So while we may not have correct software, or even software that a traditional software engineer would call good, we can obtain *good enough* software. And good enough software can be far better than software that aspires (and inevitably fails) to be correct when one considers broader aspects such as development cost, performance, robustness, reliability, and fault tolerance.

This perspective makes new techniques, optimizations, and approaches available to us. Freed from the burden of developing correct systems, we can instead focus on developing systems that best satisfy a range of desirable properties. We can appropriately invest engineering effort where it is most effectively deployed — if certain kinds of correctness are not directly relevant, we have the freedom to invest only as much engineering effort as necessary to produce an acceptable, good enough, but not necessarily correct system.

1.2 Obtaining Good Enough Software

With this perspective, we can use acceptably incorrect components with no modifications whatsoever. Given an unacceptably incorrect component, we can apply simple transformations that rehabilitate the incorrectness to give us an acceptably incorrect component. Given an overly engineered or rigid correct component, we can apply transformations that relax the correctness to obtain other benefits such as robustness, reliability, performance, or reduced resource consumption. Examples of such transformations include the following:

- **Precondition Expansion:** Many components execute correctly only if their inputs satisfy certain preconditions. In some systems it may be desirable to use the component with inputs that violate the preconditions. Precondition expansion transforms the component so that it can survive any otherwise fatal errors that might occur when given an input that does not conform to the preconditions. Examples of such transformations include infinite loop termination [3] and failure-oblivious computing [20]. Such techniques enable the component to generate (ideally acceptable) outputs even for inputs that do not satisfy the precondition. They also enable the component to survive to process additional inputs.

- **Input Rectification:** Instead of modifying the component to process inputs that violate its precondition, *input rectification* instead modifies inputs so that they satisy the precondition [8]. In many cases, input rectification preserves most or even all of the useful information in the input, nullifies otherwise fatal vulnerabilities, and enables the component to produce acceptable outputs even for otherwise problematic inputs.

- **Discarding Computation:** Task skipping [16, 17], loop perforation [10, 22], and reduction sampling [23] discard computations. When appropriately applied, the result is a significant reduction in the amount of computational resources (time or energy) required to complete the computation combined with an acceptably small change in the output that the computation produces.

- **Removing Functionality:** Many systems build on general-purpose software bases that provide more functionality than the system requires. This excess functionality can make the system vulnerable to security attacks and prone to exhibit irrelevant behaviors. Indeed, acceptable functionality may be available with a fraction of the code in the original implementation [18]. Automatically eliminating excess functionality can shrink the size of the code base and eliminate undesirable unanticipated behaviors.

- **Race-Full Parallelization:** Data races are often seen as unacceptable behavior [1]. The facts show, however, that many acceptable parallelizations have data races [9, 15]. Advantages of considering computations with data races include the elimination of synchronization overhead [15] and compilers that can automatically parallelize a much broader range of computations [9].

- **Data Structure Repair:** If a system's data structures violate key consistency properties, a system can produce unacceptable outputs or crash. Data structure repair detects and repairs corrupted data structures [4–6]. In many cases this technique can rehabilitate the error that originally caused the corruption, enabling the system to generate acceptable output for the input that caused the corruption and (in many cases more importantly) continue to execute to successfully provide service to its clients.

1.3 Critical and Forgiving Regions and Developer Observation Bias

In our experience, many developers are surprised that the techniques we outline above can improve software systems — the perception is that the system must walk a narrow path to execute correctly and that any deviation from this path is likely to cause the system to fail. Our experimental results show that, at least for the benchmark systems that we use in our experiments, this perception is simply false.

So why do some developers have this incorrect perception? Our hypothesis is that observation bias is a large part of the reason. Our results indicate that systems usually have *critical* regions, which must be close to correct for the system to operate acceptably, and *forgiving* regions, which can tolerate significant changes [2, 10, 16, 17, 21, 22]. Every developer has encountered a memorable situation (typically associated with debugging) in which a small change to the system caused large changes to its behavior. Our hypothesis is that these memorable events often involve errors in the critical regions of the system. Errors in forgiving regions, to the extent that developers notice them at all, may have much less memorable consequences. This experience may bias the perceptions of some developers and impair their ability to conceive of, understand, and realize the significant benefits that are available from the techniques outlined above.

1.4 Reasoning About Good Enough Software

So how might we help such developers obtain a more balanced understanding of the systems that they develop? And how might we develop better explanations for the reasons why systems exhibit these surprising characteristics?

We present several different reasoning approaches that can help explain results that we have observed in existing systems and predict outcomes across a broader range of systems. We consider different transformations in turn and reason about the interaction of the system with each of these transformations. Depending on the transformation, the system, and the usage context, different reasoning approaches may be appropriate.

2. Empirical Reasoning

With empirical reasoning, we apply the transformation, then use executions on representative inputs to explore the effect of the transformation. This is essentially a form of software testing, which is currently the dominant way to validate software systems. An advantage of this approach is its universality — it is possible to apply it to virtually any transformation and any system.

As with any reasoning approach based on empirical observations, a question that can arise is how to generalize the reasoning to other systems and other inputs. Our initial approach analyzes the implementation of the system to understand why the transformation produces an acceptable system. This analysis often enables us to recognize general system properties that make the system interact well with the transformation. When given a new system or input, we can then analyze the system (potentially in the context if the input) to understand if it satisfies these properties.

2.1 Failure-Oblivious Computing

Failure-oblivious computing is a technique that renders systems oblivious to memory access errors such as out of bounds accesses or null pointer dereferences [20]. We have explored two techniques for out of bounds writes: discarding the write and modulo writes (in which each out of bounds access wraps back around to write a location in the accessed data block). We have also explored two similar techniques for out of bounds reads: manufactured values (which makes up values for out of bounds reads) and modulo reads (in which the out of bounds access wraps back around to read a location in the accessed data block). Our empirical results indicate that, for a range of applications, failure-oblivious computing can eliminate security vulnerabilities and enable applications to survive otherwise fatal memory accessing errors.

Failure-oblivious computing works well for applications with short error propagation distances. In many servers, for example, the computations that process each request are largely independent. Failure-oblivious computing can eliminate data structure corruption and prevent the server from crashing when it encounters an out of bounds access or null pointer dereference. The server can then survive to successfully process subsequent requests. Modulo accesses can be effective in ensuring that the server observes values that conform to the data structure consistency constraints even for out of bounds accesses. Consider, for example, out of bounds accesses to an array of structures. Modulo reads redirect the accesses back into the array to observe an existing structure that will typically conform to the consistency constraints.

We anticipate that failure-oblivious computing will also work well with self-stabilizing computations, which, as long as they survive and continue to execute, eventually discard the effect of any errors or perturbations [4–6].

Finally, we anticipate that failure-oblivious computing may be appropriate in any situation with a need for continued execution. For example, it may be critical in ensuring the continued execution of systems that control unstable physical phenomema. To cite one example, simply ignoring arithmetic overflow and using whatever value was produced would have eliminated the cause of the Ariane 5 launch failure [7].

2.2 Boundless Memory Blocks

Boundless memory blocks store out of bounds writes in a hash table for retrieval when the system generates a corresponding out of bounds read [19]. With this technique, each memory block is conceptually unbounded, with its initial range implemented efficiently with a contiguous block of memory. Boundless memory blocks work well when the developer has produced a program that is mostly correct but produces data block sizes that are smaller than some executions require.

More generally, systems often have multiple interacting aspects, each of which must operate acceptably for the system as a whole to operate acceptably. Because of the redundancy between aspects, it may be possible to use the behavior of one aspect to adjust another aspect to become more correct. It is possible, for example, to examine the array accessing patterns of applications to find out of bounds accesses that expose errors in the computation of the required array size [14]. Using the offset of the out of bounds index to compute a new, larger, array size may (but, unlike boundless memory blocks, is not guaranteed to) eliminate the out of bounds accesses.

2.3 Data Structure Repair

Data structure repair finds data structures that violate key consistency constraints, then modifies the data structures to eliminate the inconsistency [4–6]. Note that there is no guarantee that the repair will create the data structure that a (hypothetical) correct execution would have produced — the error that caused the inconsistency may have destroyed information required to obtain this data structure, or the repair algorithm may be unable to determine which of several alternative consistent data structures the correct execution would have generated. Nevertheless, the results show that data structure repair can restore acceptable, if not perfect, execution and enable the system to continue to execute productively. A key aspect of this technique (like failure-oblivious computing) is that it ensures consistent (even though perhaps not perfect) data structures, prevents the system from crashing, and enables the system to continue to provide service. The general pattern that repeatedly emerges is that continued execution with consistent data structures, regardless of the specific mechanism used to obtain this continued execution, typically delivers acceptable results.

2.4 Cyclic Memory Allocation

Cyclic memory allocation eliminates memory leaks by allocating a fixed-size buffer, then cyclically allocating memory out of that buffer [13]. With this technique, it is possible to allocate two objects into the same slot in the buffer, in effect overlaying live data. Maintaining a separate buffer for each different allocation site tends to ensure that each individual object in the buffer preserves the basic consistency constraints for that object (but not necessarily consistency constraints that involve linked relationships between objects). An examination of the behavior of the systems after overlaying indicates that this form of consistency facilitates continued acceptable execution.

Our results show that, when cyclic memory allocation overlays live objects, the system may lose some functionality, but typically continues to execute acceptably for many inputs. An analysis of the system also indicates that preserving basic object integrity constraints in the face of overlaid live data (by maintaining separate buffers for different allocation sites) facilitates this acceptable continued execution.

Cyclic memory allocation is conceptually similar to failure-oblivious computing and data structure repair in that it is designed to rehabilitate otherwise fatal errors to keep the system executing acceptably although not necessarily perfectly. it differs in that the threat to the application is different. Instead of a single error that kills the application immediately (like a heart attack), memory leaks are a form of unbounded resource consumption that (more like cancer) eventually monopolizes all of the resources that the system needs to survive.

2.5 Infinite Loop Termination

Infinite loops can cause systems to become unresponsive. Infinite loop termination techniques detect (in some cases only likely) infinite loops, then exit the loop [3, 21]. One approach compares states before and after loop iterations to detect repeated states [3]. Another approach learns how many iterations loops typically execute, then terminates loops after they exceed this number of iterations by some conservative factor [21]. After exiting the infinite loop, the system can then proceed on to perform the rest of the computation required to generate the anticipated output. Our results show that this continued execution typically produces a better outcome than the alternative (terminating the program).

Like cyclic memory allocation, infinite loop termination eliminates an (effectively fatal) unbounded resource consumption problem — cyclic memory allocation eliminates the fatal monopolization of memory; infinite loop termination eliminates the fatal monopolization of the program counter (which must typically be shared between different parts of the system for the system to produce acceptable outputs).

2.6 Immortal Systems

It is possible to combine failure-oblivious computing, cyclic memory allocation, and infinite loop termination to obtain a conceptually (at the software level) immortal system. Specifically, the system will keep executing, will not exhaust memory, and will not become stuck in an infinite loop (of course, we provide a way for the developer to specify that a specific loop, for example the main control loop of the system, should never terminate). The acceptability of the results that such an immortal system will produce will vary depending on the system and the context in which it is used. However, our results show that, when augmented with such techniques, systems often have a surprising ability to produce acceptable outcomes even in the fact of otherwise fatal errors.

2.7 Injected Errors

Given the success of these techniques in enabling systems to tolerate otherwise fatal errors, a natural question to ask is How many errors can the system contain and still execute acceptably? We explored this question by injecting errors into the source code of the system, then using various techniques to ensure that the system executes through the errors [21].

Our specific error injection mechanism changed loop termination conditions to simulate off by one errors. Our results indicate that software systems with these injected off by one errors often execute acceptably even when the errors visibly perturb the execution.

2.8 Task Skipping and Loop Perforation

Inspired by our success in enabling programs to tolerate off by one errors, we next explored transformations designed to increase robustness and performance. Two transformations include skipping tasks in parallel programs [16, 17] and skipping iterations of time-consuming loops [10, 22]. The motivation is to discard pieces of computation that contain errors (thereby preserving the integrity of the system and enabling it to survive the error) or to reduce the amount of computational resources required to obtain the result. Our results show that this technique can deliver significant improvements in robustness and performance at the cost of small changes in the result that the system produces.

2.9 Critical and Forgiving Code and Data

One of the results of this research was the distinction between critical and forgiving code and data. Our results indicated that systems typically contain some components that must be essentially perfect for the system to execute acceptably. Other components can, if appropriately augmented with techniques such as failure-oblivious computing that enable the system to execute through errors, tolerate significant imperfection or transformations that significantly change what the component does [2, 10, 16, 17, 21, 22].

2.10 Developer Observation Bias

In our experience many software developers view systems as walking a single narrow correct execution path, with any deviation from this path causing the system to execute incorrectly. This belief has produced software development approaches that focus on bringing systems as close to perfect as possible — after all, if the slightest deviation from correct execution is unacceptable, anything less than perfection is simply pointless. Another counterproductive consequence of this belief is underinvestment in techniques that enable systems to tolerate errors — after all, if only the correct execution is acceptable, techniques that attempt to rehabilitate systems when they diverge from the correct path are irrelevant.

Our experimental results show that this understanding of software systems is simply incorrect — our results demonstrate, time and again, that software systems, when appropriately transformed to better tolerate unanticipated errors, exhibit remarkable flexibility in generating acceptable results across a large range of behaviors.

So why do some software professionals believe something that is simply wrong? Observation bias may account for part of this misconception. Every developer has encountered situations in which a very small change to the source code of the software system has a huge impact on the overall behavior. Developers may be (mistakenly) generalizing from this experience to conclude that *any* small change will make a large difference. The concept of critical and forgiving regions may be particularly important here [2, 10, 16, 17, 21, 22]. Our results indicate that programs tend to have critical regions which must be perfect (or close to perfect) for the system to execute acceptably. It is our hypothesis that errors in these critical regions shape some developers' beliefs about the need for perfection in software systems — errors in forgiving regions typically have less memorable effects and may even go largely unnoticed.

3. Probabilistic Reasoning

Probabilistic reasoning models uncertainty about various aspects of the system and its execution (for example, the values of input variables or the local effect of certain transformations), then reasons how this uncertainty may affect the execution of the system and the results that it produces. In our research we have focused on obtaining probabilistic bounds of the form $\Pr(e > b) < p$, where e is a measure of the inaccuracy of the transformed computation, b is a bound on the inaccuracy, and p is an upper bound on the probability with which the inaccuracy e exceeds the inaccuracy bound b [12, 23]. The analyzed transformations include loop perforation [12] and the combination of approximate function substitution (using less accurate but more efficient implementations of functions) and reduction sampling (approximating a reduction using only a subset of the inputs to the reduction) [23].

An advantage of probabilistic reasoning is that it provides guarantees that are quantified over all inputs and all executions. This universal quantification is important because the guarantee characterizes all system behaviors, not just those exposed via representative inputs.

4. Statistical Reasoning

Our statistical reasoning uses families of mathematical objects to model aspects of the transformed system. We then use observations from representative executions to select a specific mathematical object, or, more generally, a set of mathematical objects, that characterize the transformed computation. For example, we use multiple linear regression to model the effect of task skipping on the accuracy of the result that the system produces [16, 17]. Starting with observations from representative executions, the regression algorithm computes linear coefficients to obtain a single linear model for the effect of task skipping on the system. We have also used statistical approaches to select appropriate probability distributions to model the values that perforated loops manipulate [11]. With these probability distributions, we then use probabilistic reasoning to model the effect of loop perforation.

As these examples illustrate, our statistical approaches combine elements of both the probabilistic approach (they produce probabilistic models of the transformed system) and the empirical approach (they rely on observations from representative executions to select the final model).

5. Future Directions

At this point we have accumulated significant empirical evidence that systems have a significant degree of flexibility in the computation they execute to produce an acceptable result. Systems typically have both critical parts, which exhibit little flexibility to vary their execution, and forgiving parts, which exhibit substantial flexiblity. Our results show that empirical test executions on transformed programs can effectively separate critical and forgiving regions [2, 10, 16, 17, 22]. These results, along with our manual analysis of the behavior of the transformed systems, also show that empirical techniques can identify transformations that are appropriate for all inputs and not just those inputs used in the representative executions used to evaluate the effect of the transformations.

The next step is to develop more general explanations for these phenomena. We have already obtained the first results in this area, which use probabilistic and statistical reasoning to model systems which exhibit these phenomena. But these results, as important as they may be, only explain a few classes of behaviors. As this new approach to program analysis and transformation continues to evolve, we anticipate the development of increasingly sophisticated techniques that explain ever broader ranges of techniques. And we also anticipate the development of new and more powerful techniques for productively tolerating errors and optimizing various aspects of (not necesssarily perfect) systems. While the resulting systems may not be correct or even good, they will be better than correct and better than good — they will be good enough. An exciting and interesting time to be working in this area!

References

[1] H. Boehm and S. Adve. You don't know jack about shared variables or memory models. *Commun. ACM*, 55(2), 2012.

[2] M. Carbin and M. C. Rinard. Automatically identifying critical input regions and code in applications. In *ISSTA*, pages 37–48, 2010.

[3] M. Carbin, S. Misailovic, M. Kling, and M. C. Rinard. Detecting and escaping infinite loops with jolt. In *ECOOP*, pages 609–633, 2011.

[4] B. Demsky and M. C. Rinard. Automatic detection and repair of errors in data structures. In *OOPSLA*, pages 78–95, 2003.

[5] B. Demsky and M. C. Rinard. Data structure repair using goal-directed reasoning. In *ICSE*, pages 176–185, 2005.

[6] B. Demsky and M. C. Rinard. Goal-directed reasoning for specification-based data structure repair. *IEEE Trans. Software Eng.*, 32(12):931–951, 2006.

[7] J. L. Lions. Ariane 5 flight 501 failure report by the inquiry board, July 1996. URL http://www.di.unito.it/ damiani/ariane5rep.html.

[8] F. Long, V. Ganesh, M. Carbin, S. Sidiroglou, and M. Rinard. Automatic input rectification. In *ICSE*, 2012.

[9] S. Misailovic, D. Kim, and M. Rinard. Parallelizing sequential programs with statistical accuracy tests. Technical Report MIT-CSAIL-TR-2010-038, MIT, 2010.

[10] S. Misailovic, S. Sidiroglou, H. Hoffmann, and M. C. Rinard. Quality of service profiling. In *ICSE (1)*, pages 25–34, 2010.

[11] S. Misailovic, D. Roy, and M. Rinard. Probabilistic and statistical analysis of perforated patterns. Technical Report MIT-CSAIL-TR-2011-003, MIT, 2011.

[12] S. Misailovic, D. M. Roy, and M. C. Rinard. Probabilistically accurate program transformations. In *SAS*, pages 316–333, 2011.

[13] H. H. Nguyen and M. C. Rinard. Detecting and eliminating memory leaks using cyclic memory allocation. In *ISMM*, pages 15–30, 2007.

[14] G. Novark, E. Berger, and B. Zorn. Exterminator: Automatically correcting memory errors with high probability. In *PLDI*, 2007.

[15] M. Rinard. A lossy, synchronization-free, race-full, but still acceptably accurate parallel space-subdivision tree construction algorithm. Technical Report MIT-CSAIL-TR-2012-005, MIT, 2012.

[16] M. C. Rinard. Probabilistic accuracy bounds for fault-tolerant computations that discard tasks. In *ICS*, pages 324–334, 2006.

[17] M. C. Rinard. Using early phase termination to eliminate load imbalances at barrier synchronization points. In *OOPSLA*, pages 369–386, 2007.

[18] M. C. Rinard. Living in the comfort zone. In *OOPSLA*, pages 611–622, 2007.

[19] M. C. Rinard, C. Cadar, D. Dumitran, D. M. Roy, and T. Leu. A dynamic technique for eliminating buffer overflow vulnerabilities (and other memory errors). In *ACSAC*, pages 82–90, 2004.

[20] M. C. Rinard, C. Cadar, D. Dumitran, D. M. Roy, T. Leu, and W. S. Beebee. Enhancing server availability and security through failure-oblivious computing. In *OSDI*, pages 303–316, 2004.

[21] M. C. Rinard, C. Cadar, and H. H. Nguyen. Exploring the acceptability envelope. In *OOPSLA Companion*, pages 21–30, 2005.

[22] S. Sidiroglou-Douskos, S. Misailovic, H. Hoffmann, and M. C. Rinard. Managing performance vs. accuracy trade-offs with loop perforation. In *SIGSOFT FSE*, pages 124–134, 2011.

[23] Z. A. Zhu, S. Misailovic, J. A. Kelner, and M. C. Rinard. Randomized accuracy-aware program transformations for efficient approximate computations. In *POPL*, pages 441–454, 2012.

Improving Gate-level Simulation Accuracy when Unknowns Exist

Kai-Hui Chang and Chris Browy

Avery Design Systems, Inc., Andover, MA, USA
changkh@avery-design.com, cbrowy@avery-design.com

ABSTRACT

Unknown values (Xs) may exist in a design due to uninitialized registers or blocks that are powered down. Due to X-pessimism in gate-level logic simulation, such Xs cannot be handled correctly, producing false Xs that result in inaccurate simulation values. To improve gate-level simulation accuracy when Xs exist, we first trace the fan-in cone of Xs to check whether they are real. For the Xs that are not real, we extract small sub-circuits responsible for creating the false Xs. We then generate auxiliary code to repair gate-level simulation by replacing the Xs with the correct values. Our experimental results on commercial designs show that the proposed methods are both effective and efficient.

Categories and Subject Descriptors

B.6.3 [**Logic Design**]: Design Aids—*Simulation*

General Terms

Algorithms, Design, Verification

Keywords

X-pessimism, Gate-level logic simulation, Formal methods

1. INTRODUCTION

Gate-level logic simulation is one of the most commonly-used methods for verifying the correctness of design netlists. Even though equivalence checking between Register Transfer Level (RTL) code and gate-level netlists replaced some of the gate-level simulation tasks, the use of physical synthesis optimizations as well as Engineering Change Order (ECO) modifications created new needs to perform gate-level simulation. To utilize gate-level simulation for verification, stimuli are applied to the inputs of the netlist, and the simulation results are compared with a golden model or certain pre-defined checkers for correctness.

Gate-level logic simulation mimics the digital behavior of netlists and handles Boolean (0/1) values well. When unknown values (Xs) exist, however, it can no longer produce correct results due to X-pessimism. A simple example to illustrate the problem is shown in Figure 1. In the example, the output of gate $g6$ should be 0, but logic simulation generates an X. Such inaccuracy has a ripple effect and can produce numerous false Xs, rendering gate-level simulation useless. This problem is becoming severe due to physi-

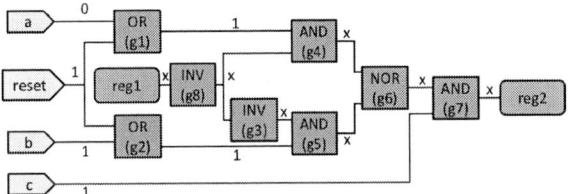

Figure 1: X-pessimism example. The output of $g6$ should be 0, but logic simulation produces X.

cal optimizations and low-power requirements that allow more and more Xs to reside in the design after reset.

One simple solution to address such simulation problems is to replace Xs in registers with random values, such as the work by Hira *et al.* [5]. Such an approach eliminates X problems by converting the Xs into non-X values. Since Xs no longer exist in the design, logic simulation can produce correct simulation results. However, each deposited value only represents one of the two possible values that the X can have, and randomly choosing one of them for simulation can cause bugs to escape verification. For example, if a bug only manifests itself when twenty unresettable registers happen to be 1 after power up, then the probability for the random-deposit approach to detect the problem is less than one in a million. To properly fix gate-level simulation without masking any bugs, Chang *et al.* [1] proposed a formal-based methodology to find false Xs during simulation. Although effective in finding false Xs, their solution to fix the problem is not generic. More specifically, their solution replaces the false Xs in registers with the correct non-X values at the simulation time when the formal analysis was applied. This solution clears false Xs at the particular time, but it does not resolve subsequent false Xs even if the conditions that produced the false Xs are identical.

In this paper we propose a simple yet effective solution to improve gate-level simulation accuracy when Xs exist[1]. Our solution first identifies false Xs when simulating a given input trace. It then analyzes the combinational fan-in cones of such false Xs to find small sub-circuits responsible for the false Xs. Finally, it generates auxiliary-code to eliminate those false Xs. To achieve this goal, we propose a novel methodology and several innovative algorithms based on logic simulation and formal analysis. By utilizing logic simulation results in our analysis, we can considerably reduce the search space of formal engines, making our methods scalable and efficient. Our empirical results show that we can analyze a multi-million gate industrial design in two hours, and the generated fixes successfully eliminated the identified false Xs. The techniques proposed in this paper are currently in commercial production use, which further demonstrates their usefulness in solving real industrial problems.

---

[1]The proposed solution is currently patent pending.

The rest of the paper is organized as follows. Section 2 provides a brief overview of related work. Section 3 presents our analysis of the gate-level X-pessimism problem. Our simulation repair solution is described in Section 4. Section 5 shows our experimental results, and Section 6 concludes this paper.

2. RELATED WORK

Most work that addresses the X problem in logic simulation focuses on the RTL because X-optimism in RTL logic simulation can easily mask bugs, whereas the X-pessimism problem at the gate level corrupts simulation but does not mask bugs. To this end, the paper by Piper *et al.* [8] provided comprehensive background on how Xs are generated in industrial designs. They also surveyed several engineers to find out what X problems they have and what solutions they need. However, even though the paper claimed that they provided a complete solution for X-optimism and X-pessimism problems, the proposed solution lacks technical details and is only tested on a small OpenCores design. Therefore, the effectiveness of their solution on industrial designs is questionable. Unlike Piper's work, Chou *et al.* [2] and Chang *et al.* [1] reported their results using real industrial case studies, suggesting that their solutions are more practical than Piper's work. More specifically, Chou *et al.* [2, 3] proposed to use formal methods to find Xs that are masked by X-optimism. They also proposed a methodology that is effective in reducing the amount of analysis that engineers need to perform. Chang *et al.* [1] extended Chou's work to find false Xs produced by X-pessimism at the gate level. Chang's technique successfully analyzed two industrial designs in a few hours. However, their analysis only provides correct X status at the checkpoint and cannot eliminate false Xs that occur later in the simulation trace. The solution provided in this work addresses this limitation.

Haufe *et al.* [4] proposed to change the RTL coding style to reduce X-optimism so that RTL simulation results are closer to the gate level when Xs exist. However, such work is typically not applicable to fixing gate-level simulation because it focuses on finding problems caused by X-optimism instead of X-pessimism.

Petlin [7] proposed a technique to find X sources as well as the locations where such Xs can be trapped. However, he did not provide solutions on how to use the analysis to improve gate-level logic simulation accuracy. In addition, his method analyzes all possible paths for X-propagation, while in this work we only analyze the paths that caused the false Xs. The latter has significant performance advantage because the search space is much smaller.

In industry, engineers also have ad hoc solutions for X-pessimism problems. For example, one company developed a sophisticated Perl script to recognize multiplexers and other gate-level structures that can create X-pessimism problems. However, the recognition was based on structure templates and could easily miss new constructs generated by physical synthesis tools. Maintaining the accuracy of the script eventually became a major challenge for the engineers, and they decided that a solution based on logic instead of structural analysis was required to properly address the X-pessimism problem in gate-level logic simulation.

3. ANALYSIS OF THE GATE-LEVEL X-PESSIMISM PROBLEM

Unlike RTL simulation where Xs may be injected to represent don't-cares or erroneous conditions, at the gate level Xs are typically from uninitialized registers or power-down blocks. Such Xs tend to disappear when simulation proceeds because known values will be written to those registers during the reset or power-up sequences. Given that specific sequences are required to reset a de-

sign or power up a block, typically only a small number of short traces need to be analyzed for X-pessimism problems. Once the false Xs are eliminated in those reset or power-up sequences, gate-level simulation should be clean afterward. In this work we assume that a trace is given, and the purpose of our flow is to generate fixes to eliminate the false Xs so that no false X can be latched into any register when simulating the trace. If there is more than one reset or power-up sequence, all the sequences need to be analyzed. However, since the fix generated by one trace can be applied to other traces, we expect the number of generated fixes to reduce when each additional trace is analyzed. This is not the case for other methods such as [1], where the fix for one trace cannot be applied to another one.

If a netlist is modeled using basic gate types such as AND, OR, XOR, INV, etc., then gate-level simulation has a special characteristic that simulating the combinational logic can only produce X-pessimism problems and cannot produce X-optimism problems. The reason is that for each gate, all its inputs will be evaluated during simulation, and the output of the gate is determined pessimistically — it produces a known value only if the Xs on the inputs are guaranteed not to propagate to the output. As a result, the non-X values in gate-level simulation are always correct, while the Xs may be false. In our algorithm we utilize this characteristic to reduce formal analysis, which can provide considerable performance gain compared with pure formal methods. Since combinational cells in cell libraries are typically composed of such basic gates, most designs possess this characteristic.

4. OUR SIMULATION REPAIR SOLUTION

In this section we present our methodology for correcting X-pessimism problems in gate-level simulation and the algorithms for implementing the methodology. In the following discussions we assume the Xs are eliminated by combinational logic. In Section 4.5 we will describe how our analysis can be extended to consider sequential elements.

4.1 Overall Methodology

The inputs to our methodology are a trace as input stimuli, a gate-level netlist, and a set of time points (called checkpoints) that the Xs should be checked to determine whether they are false or not. The output is auxiliary-code that when simulated with the trace, the false Xs at the checkpoints will be replaced with the correct values. If each cycle has a checkpoint, then it can be guaranteed that no false Xs will be latched into registers based on the current simulation values.

Our methodology works as follows: (1) at each checkpoint we check Xs in register data inputs (typically denoted as "d") to determine if they are false; (2) for each false X, we trace the fan-in cone of the register input to find a portion of the cone, called a sub-circuit, whose inputs have real Xs and whose output is a false X; (3) we generate auxiliary-code based on the sub-circuit to eliminate such Xs; and (4) the original trace is resimulated with the auxiliary-code and the code eliminates false Xs. This methodology allows gate-level simulation to produce correct results.

To support step (1), (2), and (3) of the methodology, we develop novel methods and algorithms, which will be discussed in the rest of the section.

4.2 Identifying False Xs

The algorithm for identifying whether an X is false is shown in Figure 2. The input to the algorithm is a register data input, d, that has X in logic simulation. The algorithm returns whether the X is false. The fan-in cone of d is also returned in *subckt*.

937

```
function checkX(input d, output subckt);
1   pi_frontier ← d;
2   while (pi_frontier not empty)
3     var ← pi_frontier.pop();
4     gate ← var.get_fanin_gate();
5     subckt ← subckt ∪ gate;
6     foreach input ∈ gate.get_inputs()
7       if (input.value = x &&
          input ∉ {design inputs, register outputs})
8         pi_frontier ← pi_frontier ∪ input;
9   return proveX(subckt);
```

Figure 2: Algorithm for checking whether an X is false. The sub-circuit responsible for producing the X is also returned.

In line 1 of the algorithm, *d* is inserted into *pi_frontier*, which is a set of inputs to the fan-in cone logic collected in *subckt* so far. We then expand the fan-in cone by popping a variable, *var*, from *pi_frontier* (line 3) and get the gate, *gate*, that fans out to the variable in line 4. The gate is then added to *subckt* in line 5. In line 6 we check the inputs of *gate* and add an input to *pi_frontier* if (1) the input has an X value in logic simulation, and (2) the input is not a primary input or a register output. The latter condition stops fan-in extraction at register boundaries, making our analysis combinational. Line 9 calls function *proveX* to prove whether the X in *subckt*'s output, *d*, is real.

Function *proveX* is implemented as follows. It first builds a Boolean function from *subckt*. For each input of *subckt*, if its corresponding variable has an X value in logic simulation, we make it an input of the function; otherwise, we assign the non-X value to the input and propagate the constant into the logic function. We then use a formal solver to check whether the output can have different values. In our implementation, we first use random simulation to calculate several values of the Boolean function. If the values can be different, the X is real, and *proveX* returns true. If all the values are identical, we form a SAT instance from the Boolean function and constrain the output of the function to the opposite value from simulation. We then use a SAT solver to solve the instance. If SAT found a solution, the X is real and *proveX* returns true. Otherwise, *proveX* returns false.

Since there is no X-optimism in gate-level simulation, all non-X values are correct, allowing us to use them directly in *checkX* without the risk of masking any real X problems. Therefore, only Xs need to become inputs to the Boolean function. This Boolean function is typically much smaller than the complete fan-in cone of *d*, which allows us to prove false Xs efficiently.

4.3 Minimizing X-eliminating Sub-circuits

In the previous algorithm, when *proveX* returns false, the returned sub-circuit (*subckt*) produces a false X on its output. To repair logic simulation, we can monitor the simulated values on the inputs of *subckt* and eliminate the X on its output when the condition matches. Nonetheless, *subckt* may be unnecessarily large because its inputs are either primary inputs or register outputs, but the logic that produces the false X may only be a small portion of *subckt*. If we can identify this portion, the generated fix to repair logic simulation will be much more compact. In addition, the fix can potentially eliminate multiple false Xs that have overlapping fan-in cones, reducing the number of fixes that need to be generated. To achieve this goal, we propose two new algorithms. The first one reduces the sub-circuit from its output by tracing the X towards its inputs, while the second one proceeds from the inputs towards the output.

The first algorithm is called *ckt_minimize*1 and is shown in Figure 3. The input to the algorithm is *subckt* from *checkX*, and the output is a new sub-circuit, called *subckt_n*, that is a subset of the original sub-circuit and still produces false Xs.

```
function ckt_minimize1(input subckt, output subckt_n);
1    new_po ← subckt.get_output();
2    subckt_n ← subckt;
3    do
4      gate ← new_po.get_fanin_gate();
5      c_po ← new_po;
6      foreach input ∈ gate.get_inputs()
7        subckt_input ← fanin cone of input in subckt;
8        if (input.value = x &&
           proveX(subckt_input) = false)
9          new_po ← input;
10         subckt_n ← subckt_input;
11     if (c_po = new_po)
12       break;
13   return;
```

Figure 3: Algorithm for tracing the fan-in of false Xs to reduce the X-elimination sub-circuit from the output.

The algorithm starts from the original output of *subckt* and traces the Xs in its fan-in cone until a real X is reached. The last false X then becomes the new output of the sub-circuit, *subckt_n*, that also eliminates false Xs. Note that during the tracing, if there is more than one input that has a false X for a given gate, our algorithm always picks the first one (lines 6-8). When this happens, more than one iteration may be necessary to eliminate all the false Xs because the X-eliminating sub-circuit now only eliminates the X on the chosen input. To handle this situation, after a fix is found, we replace the X in the fixed variable with its non-X value and perform logic simulation on the original fan-in cone of *d*. We then check if *d* is X. If it is, the same repair analysis is performed again. This process should repeat until *d* is no longer X. At this point, the false X at *d* is successfully repaired by the fixes generated for its fan-in cone. A simple example to illustrate the algorithm is shown in Figure 4.

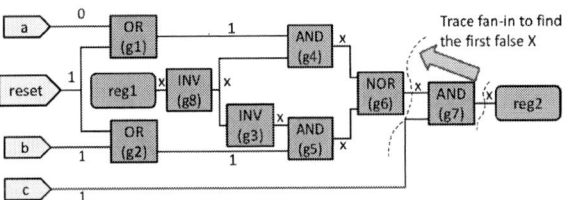

Figure 4: A simple example to illustrate the first step of reduction: tracing the fan-in cone of the false X to find the first appearance of false Xs. This analysis reduces the X-eliminating sub-circuit from its output.

Algorithm *ckt_minimize*1 reduces the sub-circuit from its output. To further minimize the sub-circuit, we propose another algorithm named *ckt_minimize*2. The algorithm is shown in Figure 5, and it moves the input frontier of the sub-circuit towards the output. The input to the algorithm is *subckt*, which is the sub-circuit (*subckt_n*) returned by *ckt_minimize*1 shown above. The output is a new sub-circuit saved in *subckt_n*.

In line 1 of the algorithm, we copy *subckt* to *subckt_n*. In lines 4-10 of the algorithm, we iteratively remove each gate that connects to the primary inputs of *subckt_n* and then check whether the output

938

```
function ckt_minimize2(input subckt, output subckt_n);
 1    subckt_n ← subckt;
 2    do
 3      changed= false;
 4      foreach gate connected to subckt_n.get_inputs();
 5        subckt_n ← subckt_n \ gate;
 6        if (proveX(subckt_n) = false)
 7          changed = true;
 8          break;
 9        else
10          subckt_n ← subckt_n ∪ gate;
11    while (changed);
12    return;
```

Figure 5: Algorithm for minimizing an X-elimination sub-circuit by moving its primary inputs towards its output.

is still a false X. If the X is still false, we keep the change (lines 6-8). Otherwise, we add the gate back (line 10). This process repeats until no further gates can be removed (the do-while loop from line 2 to line 11). At this point, we have moved the inputs of the sub-circuit as close to its output as possible. An example to illustrate the procedure is shown in figure 6.

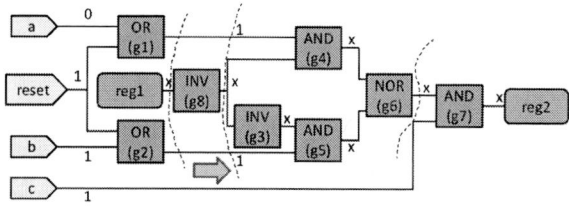

Reduce the fan-in cone for more compact fix

Figure 6: The second step of reduction: moving primary inputs of the sub-circuit towards its output.

4.4 Generating Simulation Repair Code

The sub-circuit ($subckt_n$) produced in the algorithms shown in Section 4.3 can be used to generate auxiliary-code to repair gate-level logic simulation. The algorithm to generate the repair code works as follows.

1. Traverse the inputs of $subckt_n$ to collect the condition for the false X to occur based on their logic simulation values. For example, if variable $var1$ is 1 and $var2$ is X in logic simulation, the condition will be "$var1 === 1'b1$ && $var2 === 1'bx$" when expressed in the Verilog language.

2. Generate code to replace the X on the output of the sub-circuit with the non-X value when the condition matches. The non-X value can be derived by assigning random values to $subckt_n$'s inputs and then check its output. Alternatively, the constant value proven by $proveX$ can be retained and used here. When the condition does not match, such value-overwrite should be disabled. In the Verilog and System-Verilog hardware description languages, commands "force" and "release" can be used.

The fix for the example in Figure 1 is shown in Figure 7. The generated code can correct all false Xs that match the condition even if the Xs are not within the analyzed period of the trace or are in a different trace.

The reason why the generated code can correctly eliminate the false Xs is because the constructed sub-circuits follow the seman-

```
always @(g1.o or g2.o or g8.o)
if (g8.o === 1'bx && g1.o === 1'b1 && g2.o === 1'b1)
    force g6.o = 1'b0;
else
    release g6.o;
```

Figure 7: Code generated for correcting the X-pessimism problem in Figure 1 using the Verilog language. In the example, the output port of a combinational gate is named "o".

tics of X and non-X values in logic simulation. In logic simulation, X can be either 0 or 1, which is consistent with our construction that the variable is an input to the sub-circuit. For non-X values, we propagate them into the sub-circuit by stopping at non-X boundaries when constructing the Boolean function for X-proving, and this is also consistent with the semantics of logic simulation. Therefore, the sub-circuit faithfully captures the behavior of the netlist, and this allows us to replace the output of the sub-circuit with the correct value during gate-level simulation without creating or masking any problem.

4.5 Analysis of Our Solution

If every cycle is a checkpoint, our methodology is guaranteed to repair all false Xs for a given trace because the number of false Xs produced when simulating the trace is limited, and our procedure repairs at least one false X in each iteration. Therefore, the repair process will eventually end. If not all cycles are checked, however, false Xs in the unchecked cycles can be missed. Nonetheless, checking every cycle can be time-consuming, making the selection of checkpoints important for the completeness and the efficiency of our analysis. Designing an effective heuristic for checkpoint selection is our on-going work.

Our analysis traces the fan-in cone of a register's data input and stops at primary inputs or registers. Therefore, the fixes we found are combinational. To repair false Xs that can be eliminated when sequential elements are involved, we can trace across register boundaries by backward time-frame expansion. In this way, false Xs that can be eliminated within N cycles can be found by expanding N time frames. To generate repair code, additional registers need to be used to keep track of the values of the involved registers up to N cycles before the X-elimination point. In practice, we have analyzed tens of netlists from several different companies and have only seen one case where sequential analysis was necessary. Given that sequential analysis is much more time-consuming than combinational analysis and is rarely used, we currently have not implemented this analysis mode.

Our algorithm is more efficient than other methods such as [1] because we simplify our analysis based on the fact that all non-X values in gate-level simulation are correct. As a result, the sub-circuit that needs to be proved only includes the logic along the paths that are Xs in logic simulation. By analyzing a smaller logic cone, the false Xs can be proved more efficiently.

When designing our solution for X-pessimism problems, we have considered three other approaches and found that they are impractical. The first approach is to abstract the gates to higher-level structures and then eliminate false Xs for those structures. For example, if multiplexers can be recognized, the false Xs on their outputs can be easily eliminated. However, we found that most netlists in industry have undergone various physical synthesis optimizations, making accurate abstraction extremely difficult. The second approach is to carry RTL information over to the gate level so that logic structures that cause X-pessimism problems can be known in advance. However, this is difficult without the support from synthesis tools. In addition, X-pessimism problems can be caused by physical synthesis tools and ECO changes, limiting the X-pessimism problems that can be found by this approach. The third approach is to use pure-formal methods to enumerate all possible X-pessimism conditions and generate fixes before gate-level simulation is performed. However, we found that this approach can easily generate a huge number of fixes. For example, for a two-input multiplexer, at many as six fixes can be generated depending on how the multiplexer is implemented. However, most of the fixes are useless because the X-pessimism conditions never appear in real traces, and the large volume of fixes can significantly slow down gate-level simulation. Therefore, this approach is also impractical for industrial designs.

5. EXPERIMENTAL RESULTS

We implemented our methods on a proprietary simulator [9] and the ABC package from UC Berkeley [10]. We applied our methods to a multi-million gate commercial design to repair false Xs in its reset sequence. The reset sequence was 74 cycles long, and the checkpoint was set to the last cycle. Due to the non-disclosure agreement with the company, we could not describe the benchmark in more detail.

The work by Chang *et al.* [1] is one of the state-of-the-art methods to solve X-pessimism problems. In order to compare our results with Chang's, we used the same partitioned flow described in [1] even though such a step is not necessary due to the scalability of our methods. 202 partitions were produced by the partitioner, and 140760 register inputs were checked for false Xs. Our method used 1h36m to identify 125 false Xs, and it produced 145 fixes. Most of the fixes involved only three wires in their conditions, and a few of them involved more than 10 wires. Those involving more than 10 wires were mostly in arithmetic blocks. We also produced fixes whose conditions are all Xs. Further analysis showed that the wires being repaired were constant under all possible conditions. In other words, they were stuck-at-0 or stuck-at-1. To repair such false Xs, we generated force statements without any condition. We have rerun logic simulation with the fixes and confirmed that all the false Xs have been eliminated. We did not observe simulation speed degradation in this benchmark because the fixes only contained 145 forces for this multi-million gate design and the fixes were mostly inactive throughout simulation.

We also applied Chang's methods on the same design for comparison. Runtime of Chang's flow was 4h46m, and it identified 201 false Xs. The false Xs we found were a subset of those found by Chang's work. The reason is that we were performing combinational analysis, while Chang's method performed full-fledged sequential analysis. There were Xs eliminated in earlier cycles and then propagated to a register as a false X. For example, a register whose fan-in cone is a buffer may have latched a false X from its source register. In this case, Chang's work found the X to be false, while our analysis showed that the buffer was not able to eliminate the X. However, since the repair code we produced may eliminate the false Xs at earlier cycles if the conditions match, we actually eliminated 140 false Xs out of the 201 Xs proven to be false by Chang's flow, 15 more than the 125 false Xs identified in our flow. This result shows one advantage of this work compared with Chang's work: our fixes can repair unseen problems as long as the conditions match, while Chang's work can eliminate false Xs correctly at the checkpoint but cannot eliminate false Xs elsewhere during simulation.

To further illustrate the value of our methods, we provide two case studies from our industrial partners. In the first case, Xs kept reappearing during simulation due to power-related operations. Because the problematic block was from an IP vendor, no RTL was available for comparison, forcing the engineers to analyze the gate-level netlist directly. They spent a month tracing the problem and finally relied on a formal tool to prove that the Xs were harmless. Our methods analyzed the design in 1h30m and the generated fixes successfully repaired gate-level simulation. In the second case, our methods were used as a debugging tool to analyze gate-level X problems. In this use model, the X-proving step tells the engineer whether the X needs to be analyzed. If the X is false, the generated fix not only repairs simulation but also explains how the false X is generated. In one real case, a designer spent almost three hours to trace an X problem. Our solution took just a few minutes to prove that the X is false and produced a fix that involved only 7 gates. By inspecting the generated repair code, the engineer easily found that the inverters inserted by physical synthesis tools caused the false X. Without the rigorous formal analysis and repair minimization steps, it would be difficult to identify the root cause of the problem.

6. CONCLUSION

In this work we proposed a new technique for improving gate-level logic simulation accuracy when unknowns (Xs) exist. Our method uses simulation results to reduce the scope of analysis. We then apply formal techniques to rigorously prove whether the Xs are false. For false Xs, we identify a small portion of design logic responsible for creating the false Xs and then generate code for correcting the simulation results. Unlike existing solutions that repair X problems by depositing random values, our solution does mask real problems due to its rigorous formal analysis step and the robust repair method. The tool is in commercial production use, suggesting its effectiveness in solving industrial problems. We are currently integrating our methods with commercial simulators for easier adoption of our simulation-repair solution.

7. REFERENCES

[1] K.-H. Chang, H.-Z. Chou, H. Yu, D. Dobbyn and S.-Y. Kuo, "Handling Nondeterminism in Logic Simulation So That Your Waveform Can Be Trusted Again", *IEEE D&T*, DOI:10.1109/MDT.2011.75

[2] H.-Z. Chou, H. Yu, K.-H. Chang, D.Dobbyn and S.-Y. Kuo, "Finding Reset Nondeterminism in RTL Designs – Scalable X-Analysis Methodology and Case Study", *DATE, 2010*, pp. 1494-1499.

[3] H. Z. Chou, K. H. Chang, and S. Y. Kuo, "Handling Don't-Care Conditions in High-Level Synthesis and Application for Reducing Initialized Registers," *DAC, 2009*, pp. 412-415.

[4] C. Haufe and F. Rogin, "Ad-Hoc Translations to Close Verilog Semantics Gap," *workshop on DDECS, 2008*, pp. 1-6.

[5] K. Hira and N. A. Panchal, "Random Initialization of Latches in an Integrated Circuit Design for Simulation", *US Patent Application 2010/0017187 A1*

[6] A. Mishchenko, S. Chatterjee, R. Brayton, "DAG-Aware AIG Rewriting, A Fresh Look at Combinational Logic Synthesis", *DAC, 2006*, pp. 532-535.

[7] O. A. Petlin, "Verification Systems and Methods", *US Patent Application 2010/0313175 A1*

[8] L. Piper and V. Vimjam, "X-Propagation Woes: Masking Bugs at RTL and Unnecessary Debug at the Netlist", *DVCon, 2012*, session 5.3.

[9] Avery Design Systems Inc., http://www.avery-design.com

[10] Berkeley Logic Synthesis and Verification Group, ABC: A System for Sequential Synthesis and Verification, http://www.eecs.berkeley.edu/~alanmi/abc/abc.htm

Automated Feature Localization for Hardware Designs Using Coverage Metrics.

Jan Malburg*
malburg@informatik.uni-bremen.de

Alexander Finder*
final@informatik.uni-bremen.de

Görschwin Fey*†
Goerschwin.Fey@dlr.de

*University of Bremen
28359 Bremen, Germany

†German Aerospace Center
28359 Bremen, Germany

ABSTRACT

Due to the increasing complexity modern System on Chip designs are developed by large design teams. In addition, existing design blocks are re-used such that the knowledge about these parts of the design entirely depends on the quality of the documentation. For a single designer it is almost impossible to have detailed knowledge about all blocks and their interaction.

We introduce a simulation-based automation technique to support design understanding. Based on use cases provided by the designer and on their coverage information, the proposed technique identifies parts of the source code that are relevant for a certain functional feature. In two case studies the technique is shown to be at least as exact as reading the documentation with two important advantages: the automated approach is fast and more precise than the existing documentation for the inspected designs.

Categories and Subject Descriptors

B.7.2 [**Integrated Circuits**]: Design Aids

General Terms

Design, Documentation, Experimentation

Keywords

Feature Localization, Design Understanding, Simulation

1. INTRODUCTION

Modern chip designs, especially Systems on Chip, grow with respect to their transistor count as well as their supported features. Such chips are developed by large design teams consisting of hundreds of people [9] and are far beyond the point where a single designer knows every detail about the design. Furthermore, chips are assembled of design blocks from different sources. A design block is a part of a chip which provides a defined functionality. The functionality ranges from very complex, for example complete

---

*This work was supported in part by the German Research Foundation (DFG, grant no. FE 797/6-1)

CPU-cores, to simple encoder and decoder blocks. Typically, complex design blocks are assembled from several less complex design blocks. A design block could be a new block developed especially for the new chip, or it might be a block already used in previous designs, or even a third-party block. All of those design blocks in a chip are responsible for one or more different features. Some of the features might be realized by combining the functionality of several different blocks.

A feature is a distinguishing characteristic of a design. A functional feature defines the expected output of the system under specific input. Other types of features are for example robustness, defining the amount of errors which can occur before a result becomes incorrect, or performance, limiting the time until a design has to return the expected output. In the following only functional features are considered and for simplicity called features.

For design improvement, design extension, and bug fixing a developer has to understand the design. These tasks become even more important, since the amount of re-used design blocks is continuously increasing in future [9]. In order to understand a design it is mandatory to know where in the design which feature is implemented. This does not only mean to know the block providing a feature, but where exactly in the block the feature is implemented. In general, it is unlikely that a developer has this knowledge. Purely manual inspection of the *Hardware Description Language* (HDL) code is a laborious, and therefore cost intensive task and the developer still might miss relevant parts of the implementation. Therefore, it is desirable to have tools which help the developer to find the relevant code for a feature and to understand the design.

In this paper we present a new approach for locating parts of the HDL code which are relevant for a functional feature. The proposed technique uses a dynamic approach relating coverage information gathered by simulation to features executed by use cases. This technique is basically usable with all HDLs and representations given the design can be simulated and coverage can be measured, e.g. Register-Transfer-Level or Transaction-Level descriptions. We currently support Verilog in our prototype. Results are presented to the user by coloring the source code. The contributions of this paper are:

- a feature localization technique for hardware designs,
- a unified notation to compare existing coloring schemes,
- a new coloring scheme,
- the use of toggle coverage for feature localization,
- an adaptive ranking of source files according to their likelihood to be related to a feature,
- a comparison of orthogonal features to improve design understanding,

Experiments showed that, our approach even provides good results when applied to designs with poorly separated features.

The remainder of this paper is organized as follows: Section 2 gives an overview of related work. Notations are introduced in Section 3. In Section 4 we present our technique. Section 5 describes the application of our prototype to two open source designs. Section 6 concludes the paper and discusses results.

2. RELATED WORK

Techniques for design understanding of HDL descriptions concentrated on inferring specifications from traces [7, 4, 11] or merging partial specifications to more abstract ones [13]. Another technique is program slicing [3], which differs from feature localization as it answers the questions which parts of the code can affect or be affected by a signal. Instead feature localization answers the question which parts of the code are responsible for creating a defined output under certain input. So far, no technique has been published about feature localization in HDL descriptions.

Feature localization for software designs is an active research area. Often the statement coverage of runs using a wanted feature is compared to the coverage of runs not using this feature [16]. Simple approaches only consider program statements which are covered by runs using a wanted feature, but are not covered by runs that do not use the feature [15]. More advanced approaches use more fine-grained categorizations, where the statements are classified based on the relation of runs using (not using) a certain feature [5].

Techniques for bug localization in software using coverage information are similar to coverage-based feature localization. A well-known tool in this context is *Tarantula* [10]. *Tarantula* is a visualization tool for bug localization which colors statements depending on their suspiciousness of causing a bug. The suspiciousness is computed by comparing the percentage of failing runs which execute a statement to non-failing runs executing the statement. Abreu et. al. [1] showed that using the *Ochiai coefficient*, a similarity coefficient used in molecular biology, yields better results for computing the suspiciousness of a statement compared to the *Tarantula* formula. Later Santelices et. al. [12] presented an approach to relate branch coverage and definition-use coverage to statements. They showed that using the average of several different coverage criteria to compute the final suspiciousness creates better results than each coverage criterion on its own.

The approach presented in this paper is based on coverage information gathered by simulating the design under test. To this extend it is similar to the approaches described above. However, all previous approaches are used for software, while we consider hardware systems. There are several differences between software and hardware. For hardware there exist different coverage metrics, like toggle coverage, which we also use for feature localization. Moreover, in HDL descriptions of a design, there is several code which continuously is executed without being called from any other function, for example *always-blocks* and *assign-statements* in Verilog [8]. Finally, hardware designs are inherently parallel.

3. PRELIMINARIES

In this section we provide some basic definitions and introduce some terminology required for the rest of the paper.

Let D be the design under test. A *use case* u for D is given by a sequence $u = (i_1, i_2, ..i_m)$ of input values $i_j, j = 1, ..., m$ for D. A use case may either be directly defined by the user, or a test case from the test bench of D may be considered as use case. A *run* r is the simulation of D applying a use case. The set $R = \{r_1, r_2, r_3, ..., r_n\}$ is the set of all runs. A *coverage metric* C with respect to D is a set of conditions over elements in R. A *coverage item* $c \in C$ is a single condition. The form of these conditions is defined by C. A feature f is a distinguishing characteristic of D, defining the expected output of D under specific

input. The set $F = \{f_1, f_2, f_3, .., f_k\}$ is the set of all features supported by D. The user defines, if a run r *uses* a feature f. A feature f is implemented by a set of coverage items C_f. The goal of feature localization is to determine C_f.

Two typical coverage metrics used in hardware design are statement coverage and toggle coverage [14]. Toggle coverage is a coverage metric especially for hardware design. Statement coverage is also used in software design [10]. In case of statement coverage C_s for each statement s contained in the HDL code of D, there exists exactly one condition c_s, where c_s has the form "r executes s". In case of toggle coverage C_t for each wire and each register t there exist exactly two conditions c_{t_1} and c_{t_2}, where c_{t_1} is of the form "r switches t from 0 to 1" and c_{t_2} is of the form "r switches t from 1 to 0". The set $C = C_s \cup C_t \cup ...$ is the union of all coverage metrics with respect to D. A run $r \in R$ covers c, if r fulfills c. Standard coverage tools determine the following sets:

1. Coverage items covered by r:
$$coveredBy(r) = \{c \in C | r \text{ covers } c\}$$

2. Coverage items not covered by r:
$$uncoveredBy(r) = C \backslash coveredBy(r)$$

3. Coverage items covered by R_s:
$$coveredBySet(R_s) = \bigcup_{r \in R_s} coveredBy(r)$$

4. Coverage items not covered by R_s:
$$uncoveredBySet(R_s) = C \backslash coveredBySet(R_s)$$

5. Runs covering c:
$$hit(c) = \{r \in R | r \text{ covers } c\}$$

6. Runs not covering c:
$$miss(c) = R \backslash hit(c)$$
with $r \in R$, $R_s \subseteq R$, and $c \in C$.

4. LOCATING FEATURES

In this section we will present our approach for feature localization in hardware designs. The main idea of feature localization using coverage metrics is to compare the coverage of runs which use a certain feature with those not using this feature. Therefore, several different runs of the system under test are required. An underlying assumption for our approach is, that for a developer it is easier to decide if a run is related to a feature than deciding if a coverage item is related to a feature. Which feature $f \in F$ is used by a run must either be specified by a developer or taken from the test bench documentation: Based on the user input and the coverage informa-

7. Runs using f:
$$use(f) = \{r \in R | r \text{ uses } f\}$$

8. Runs not using f:
$$notuse(f) = R \backslash use(f)$$

tion the relation between features and coverage items is computed: Intuitively, a coverage item c is likely related to a feature f, if c is

9. Runs covering c and using f:
$$pass(c, f) = hit(c) \cap use(f)$$

10. Runs covering c and not using f:
$$fail(c, f) = hit(c) \cap notuse(f)$$

covered whenever f is used but never covered when f is not used, or formally: $(pass(c, f) \equiv use(f)) \wedge (fail(c, f) \equiv \emptyset)$. Still the difference in the coverage may have other reasons. For some coverage item c it might be possible that $c \notin setCoveredBy(use(f))$, even though c is related to the implementation of f. For instance, if c is related to a special case of f. Having only small differences between the runs, which use a feature and which do not, as well as having runs which use as few other features as possible often improves the result [6]. Next we will present three coverage based heuristics for computing the likelihood of a coverage item to be related to a feature.

942

4.1 Coloring heuristics

For feature localization, there exist several heuristics to relate the source code parts to a feature [5, 16]. For evaluating which heuristics are best for the localization of features in hardware designs, three heuristics from literature have been adapted for our technique. To present the results to the user we use color coding. This way of presentation is inspired by the *Tarantula* tool [10].

In [5] a categorization for feature localization is described. This categorization is defined over a set of computational units. Based on how fine-grained the partition should be, a computational unit can be for example a source code statement, a basic block, or a function. For our approach we define the categorization over the set of coverage items $\mathcal{C}$. This categorization (Cat) partitions the coverage items with respect to a certain feature into five groups defined as: In addition to the presented categorization, two coloring

1. Coverage items covered if and only if f is used:
$$specific(f) = \{c \in \mathcal{C}|\ (pass(c,f) \equiv use(f)) \wedge (fail(c,f) \equiv \emptyset)\}$$

2. Coverage items sometimes covered when f is used and never when f is not used:
$$conditional(f) = \{c \in \mathcal{C}|\ 0 < |pass(c,f)| < |use(f)| \wedge (fail(c,f) \equiv \emptyset)\}$$

3. Coverage items always covered when f is used and at least once when f is not used:
$$relevant(f) = \{c \in \mathcal{C}|\ (pass(c,f) \equiv use(f)) \wedge (fail(c,f) \neq \emptyset)\}$$

4. Coverage items sometimes covered when f is used and at least once when f is not used:
$$shared(f) = \{c \in \mathcal{C}|\ 0 < |pass(c,f)| < |use(f)| \wedge (0 < |fail(c,f)|)\}$$

5. Coverage items never covered when f is used:
$$irrelevant(f) = \{c \in \mathcal{C}|\ pass(c,f) \equiv \emptyset\}$$

schemes from bug localization in software are adapted for our approach. The first scheme extends the two-dimensional *Tarantula* scheme [10] to differentiate multiple features. One dimension is the likelihood $like_T$ of a coverage item c to be related to feature f:

$$like_T(c,f) = \begin{cases} \frac{passed(c,f)}{passed(c,f)+failed(c,f)} & \text{if } hit(c) \neq \emptyset \\ 0 & \text{otherwise} \end{cases}$$

with $passed(c,f) = \frac{|pass(c,f)|}{|use(f)|}$ and $failed(c,f) = \frac{|fail(c,f)|}{|notuse(f)|}$. Our formula is a generalization of the original formula. The value of the original formula can be computed by fixing the feature f to "does not pass the test case". For our approach, the hue of a coverage item is defined by its likelihood. The hue reaches from green ($like_T = 1$) over yellow ($like_T = 0.5$) to red ($like_T = 0$). The other dimension of the *Tarantula* scheme estimates the confidence con towards the likelihood value of c. The confidence is defined as:

$$con(c,f) = max(passed(c,f), failed(c,f))$$

The confidence is visualized as the brightness in which c is colored. The brightness of c is linear to its confidence. The highest confidence ($con = 1$) is colored brightest and the lowest confidence ($con = 0$) is colored darkest.

The other coloring scheme adapted from bug localization in software uses the *Ochiai coefficient* for computing the likelihood. Compared to the *Tarantula* scheme, the *Ochiai* coloring scheme yields better results in case of bug localization in software [2]. Again we generalize the formula by adding a parameter for the wanted feature f, such that we compute the likelihood $like_O$ of a coverage item c to be related to f by:

$$like_O(c,f) = \begin{cases} \frac{|pass(c,f)|}{\sqrt{|use(f)| * |hit(c)|}} & \text{if } hit(c) \neq \emptyset \\ 0 & \text{otherwise} \end{cases}$$

The computation of the confidence is identical to the *Tarantula* scheme.

4.2 Comparison

Early experiments with Cat have shown that for hardware designs a large portion of the coverage items is categorized as *relevant* even if it has nothing to do with the feature. This is due to the fact that there is much code which is always executed, like *always-blocks* and *assign-statements*. For overcoming these problems we propose an extension of Cat. This extension Cat_{ext} introduces the category *common* and redefines *relevant* as: The categorizations

2a. Coverage items always covered:
$$common(f) = \{c \in \mathcal{C}|\ \forall r \in R, c \in coveredBy(r)\}$$

2b. Coverage items always covered when f is used and sometimes covered when f is not used:
$$relevant_{ext}(f) = \{c \in \mathcal{C}|\ (pass(c,f) \equiv use(f)) \wedge (0 < |fail(c,f)| < |notuse(f)|)\}$$

and the *Tarantula* scheme are related to each other. Categorization Cat_{ext} subsumes Cat and the *Tarantula* scheme subsumes the categorizations. By partitioning the *Tarantula* scheme in different classes the categorizations can be computed. Table 1 shows the relation between the three schemes and describes which color is used for which category. The *Ochiai* scheme cannot be related to the other coloring schemes, because this scheme also considers the total number of runs covering c.

As an advantage the *Tarantula* and the *Ochiai* scheme provide a continuous range preventing runs with very high or very low coverage to have a disproportionately strong effect on the result. Both schemes also identify coverage items that are not covered if and only if a certain feature is used. The other two do not distinguish between not covered while using a feature f, and items never covered.

4.3 Feature comparison

Often there are sets of features for which at a point in time at most one feature in the set can be used. Such features are called *orthogonal*. The user can define features as *orthogonal* to each other. An extension unique to our approach is the comparison of two *orthogonal* features. This allows the user to see which are the parts where the features differ from each other and therefore gives additional insight to the implementation of the features and the design as a whole. This comparison is defined over the likelihood and the confidence of the features and therefore only usable for the *Tarantula* and *Ochiai* coloring schemes. The comparison value $comp$ computes how likely a coverage item c is covered by one feature f_b but not by a feature f_c orthogonal to f_b. The comparison value is defined as:

$$comp(c, f_b, f_c) = \frac{(1 + (like(c, f_b) - like(c, f_c))}{2}$$

with $like \in \{like_T, like_O\}$ defining which coloring scheme is used for the comparison. The feature $f_b \in F$ is the feature which we want to inspect and $f_c \in F$ is the orthogonal feature which we want to compare with f_b. The mapping of the comparison value to hue is equivalent to mapping the likelihood to the hue. For the brightness, the maximum of the confidences is used:

$$brightness(c, f_b, f_c) = max(con(c, f_b), con(c, f_c))$$

4.4 File ranking

Another extension particular to our approach provides additional guidance for feature localization by ranking the different files based on their likelihood to be related to a feature. As the mapping of signals to files is a non-trivial task, our current implementation only considers statement coverage for the ranking. The ranking works

Table 1: Relation between categorizations and Tarantula coloring scheme; colors encoding categories

Cat		Cat_{ext}		Tarantula
Category	Color	Category	Color	equivalent class
specific	bright green	*specific*	bright green	$like_T = 1 \wedge con = 1$
relevant	yellow green	*common*	bright yellow	$like_T = 0.5 \wedge con = 1$
		relevant$_{ext}$	yellow green	$0.5 < like_T < 1 \wedge con = 1$
conditional	dark green	*conditional*	dark green	$like_T = 1 \wedge con < 1$
shared	dark yellow	*shared*	dark yellow	$(con = 1 \wedge 0 < like_T < 0.5) \vee$ $(0 < con < 1 \wedge like_T < 1)$
irrelevant	dark grey	*irrelevant*	dark grey	$like_T = 0$

Table 2: Overview of the designs used in the case study

Design	LOC	Files	use cases	features	time
double_fpu_verilog	2555	7	144	8	22.4 sec
SD/MMC Controller	3840	17	5	5	2.8 sec

as follows: Initially the user has to choose a threshold for the computation. In case of the categorization this is a category and in case of the *Tarantula* and *Ochiai* scheme this is a minimum value for the likelihood. Then starting with the highest value (specific or likelihood of 1, respectively) as the upper bound and the lower bound, all files having statements within these bounds are considered and then ordered based on the percentage of statements within these bounds. Those files are added to the ranking in this order. The lower bound is reduced until more files are found or the given threshold is reached. In case a new file is found, it is added to the ranking. If several files are found at the same time, they are added to the ranking ordered by the percentage of statements within the bounds.

5. CASE STUDIES

For testing our approach we have implemented a prototype. This prototype uses *ModelSim*, to compute the coverage of the different use cases. The current version of our prototype supports only Verilog, but this is only a technical limitation of our prototype. Adding the support for additional HDLs requires only adding an additional parser, which can translate the hierarchical signal name to the local signal names in each source code file. In the current version our implementation supports statement and toggle coverage, where statement coverage is represented by coloring the corresponding lines. Toggle coverage is represented by overlining, in case of a toggle from 0 to 1, or underlining, in case of a toggle from 1 to 0, the corresponding signals.

In order to evaluate our approach, we considered designs that have to fulfill the following requirements: they provide several different features, they are written in Verilog, the designs and the corresponding test benches run in *ModelSim*, and they have a well commented test bench either distinguishing the different features or allowing to easily use the test bench as template for use cases. Two designs from the website *OpenCores.org*, fulfilling these requirements, have been chosen. We conducted our case study as follows:

1. we looked for a design, unknown to us, which provides several features and including a test bench testing those features,
2. we analyzed the design using our prototype,
3. we wrote down all our findings,
4. finally, we checked our findings against the documentation.

Note, as we only used designs which were originally unknown to us, the only information we had about the designs were their descriptions at *OpenCores.org* and the structure of their test benches. Table 2 gives a brief overview of the designs used for the case studies. The column *Design* contains the title under which the designs are listed at *OpenCores.org*. Lines of code (*LOC*) is the number of all non-comment and non-empty lines of the design. In column *time* the time required to compute and present the heuristics

is shown. Compared to the time for simulation and coverage gathering, which takes 30 minutes for double_fpu_verilog and 18 seconds for SD/MMC Controller, the computation of the heuristic is rather fast, making the simulation the main limitation of our technique. In many cases this coverage information will already be computed during the validation of the design. In our studies, gathering the coverage information has not increased the time needed for the simulation, i.e. the computational overhead of our approach is negligible.

5.1 Case Study: double_fpu_verilog

This case study considers a double precision FPU which requires 20 (addition) to 71 (division) clock cycles per operation. The supported features are four arithmetic operations:

- addition
- subtraction
- multiplication
- division

and four rounding modes:

- round to nearest even
- round to zero
- round to +INF
- round to -INF

For each combination of operation and rounding mode, there exist nine use cases. The documentation consists of a pdf-file with twelve pages and very few source-code comments.

There is a huge difference between the difficulty to localize arithmetic operations and to localize rounding modes. For the arithmetic operations, the statement-coverage-based coloring schemes provide several locations related to the feature. But still 56% of the statements are executed for all use cases. For these statements, statement coverage cannot help to decide whether they are part of the feature or not. The information provided to the user based on statement coverage is very similiar for all coloring schemes such that no qualitative difference can be found between them.

When in addition considering toggle coverage, it is easy to partition the statements always executed in statements that use toggling signals and statements using not toggling signals. Since statements that operate on constant values are unlikely to be part of the computation, they can be filtered out. When considering toggle coverage all coloring schemes can support the user by locating features. The *Tarantula* coloring scheme provides the strongest contrast and therefore shows the difference in toggle coverage very clearly. The *Ochiai* scheme also provides the information clearly, but with less contrast, making it harder to recognize. These two coloring schemes show whether a coverage item is not covered if and only if a feature is executed. When relating this information to toggle coverage, this translates to a given register or wire staying constant if and only if a given feature is used. This information helps to understand a feature, as already assumed in Section 4.2. As the two categorization schemes do not provide this information, it is not possible to recognize which signals are changing and which are not using them.

Figure 1 gives two examples how the FPU design is presented to

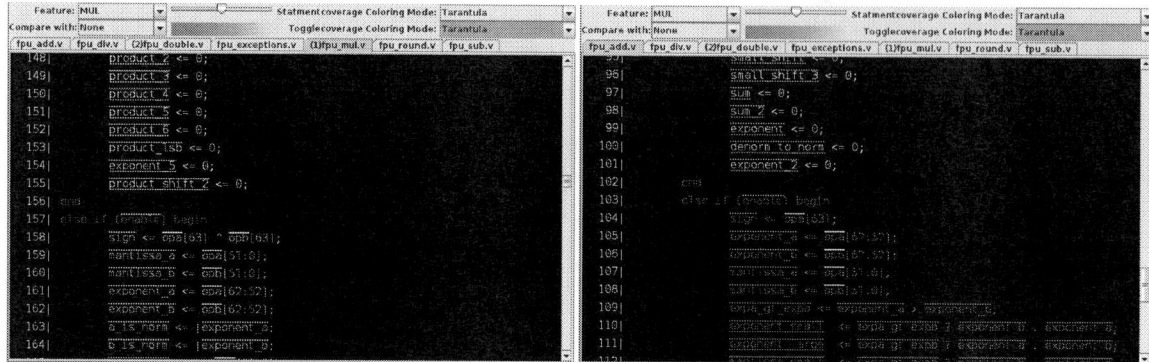

Figure 1: Screenshots of our prototype inspecting a part of the design belonging to a feature, as claimed by the documentation (left), and a part of the design which does not (right)

Table 3: The file ranking for the arithmetic operations of the double_fpu_verilog design compared to the documentation.

Feature	*Tarantula* scheme	documentation
Addition	fpu_sub fpu_add	fpu_add fpu_sub
Substraction	fpu_double fpu_sub fpu_add	fpu_sub fpu_add
Multiplication	fpu_mul	fpu_mul
Division	fpu_div	fpu_div

the user, and how clearly the design is partitioned in case of an arithmetic operation (multiplication). The example shows that statement coverage provides a clear distinction for some parts of the design, but also that the toggle coverage provides additional information to further distinguish statements always executed (yellow statements). More difficult is the localization of the rounding features. Based on statement coverage there is no difference between the rounding modes, forcing the user to completely rely on toggle coverage. Even for toggle coverage there is only very little difference. In case of round to nearest even only the *Tarantula* or the *Ochiai* scheme show a difference, still for identifying the feature it is necessary to use the feature comparison functionality of our approach. Altogether in case of the rounding modes the feature localization results in 2-4 statements corresponding to each rounding mode. The comparison of our findings with the documentation shows the benefits of our approach. First the documentation only describes in which module a feature is implemented, and all the positions found with our approach are placed in the corresponding module. Therefore, we are able to get at least as good results as someone reading the documentation of the design. Also there are special cases for addition and subtraction based on the signs of the operands. An addition could be executed by the subtraction unit and vice versa. The documentation does not include this information in the description of the two operations, but in the description of the design hierarchy. By this, someone only reading the operation descriptions would miss this peculiarity. Additionally, our approach determines the signal that defines which variant is used. This information does not even exist in the documentation. After inspecting the corresponding code for the rounding modes we are confident that the lines marked by the prototype in fact are the main parts implementing the rounding features. Again this is information not included in the documentation. The result of the comparison of the file ranking and the documentation is shown in Table 3. Only the file ranking for the *Tarantula* scheme is shown because this scheme yields the best results. As the file ranking currently only considers statement coverage only the arithmetic operations are shown. For the rounding mode all

files have been included for each rounding mode. Similarly to the approach in [12] only the ranked files are shown until the point where all files are included which the documentation claims to belong to the feature. Except of for subtraction these are exactly those file which the documentation relates to the feature. In case of subtraction also the top-module is included as it contains some statements executed if and only if subtraction is used.

In conclusion, the *Tarantula* coloring scheme has provided the best results and statement coverage gives a first overview. Toggle coverage allows to differentiate those statements which are always executed (yellow statements). The arithmetic operations were practically found at the first glance, and except for the round to nearest even all features were found faster than by looking at the documentation. In addition our prototype yields more information about the design than the documentation does. The file ranking feature was very useful in several cases.

5.2 Case Study: SD/MMC Controller

The design of a controller chip for SD/MMC cards for up to 2GB, is used in this case study. The controller is accessed through a Wishbone-slave-interface. The test bench of the controller includes a Wishbone simulator and an SD-card simulator used for testing. The test bench defines five different features:

- Register access
- SPI bus access
- SD init
- SD write
- SD read

The test bench consists of one test case, but clearly defines when which feature is used, such that we used this distinction to measure the different coverages for the corresponding executions.

The documentation of this design consists of two pdf-files, one consisting of 23 pages and the other one consisting of 17 pages. Additionally, there are several source code comments.

In contrast to the first case study, in the SD/MMC Controller all features are equally easy to find. They are less easy to spot than the arithmetic operations in the first case study, but far easier than the rounding modes. When comparing the different coloring schemes, we observed that the run for the Register access feature covers very few coverage items, resulting in the effect that the categorization based schemes mark the coverage items which are covered by all the other runs as *relevant* or *indispensable*, respectively. This practically rendered the categorization schemes useless. This is very similar to the effect which motivated us to introduce Cat_{ext}. However, this causes no problem for the schemes with continuous range (*Tarantula* and *Ochiai*) as the computed likelihood is only minimally affected, both schemes showed good results, with no

Table 4: The file ranking for the SD/MMC Controller design compared to the documentation.

Feature	*Tarantula* scheme	documentation	
	Ranking	Belongs	Possible
Register access	ctrlStsRegBl[2]		
SD init	initSD spiTxRxData spiCtrl	initSD spiCtrl	sendCMD
SD read	readWriteSDBlock spiMasterWishBoneBl[2] spiCtrl sm_RxFifoBl[1]	readWriteSDBlock spiCtrl	sendCMD
SD write	readWriteSDBlock sm_TxFifoBl[1] spiMasterWishBoneBl[2] spiCtrl	readWriteSDBlock spiCtrl	sendCMD
SPI bus access	spiCtrl spiTxRxData ctrlStsRegBl[2] readWriteSPIWireData	readWriteSPIWireData	spiCtrl spiTxRxData

[1] File that is not documented or the documentation does not relate it to any feature
[2] File that the documentation claims to belong to the Wishbone-interface and therefore is commonly used for all features

visible differences between each other. As there are no features which are clearly *orthogonal* the comparison function of our technique was not used.

Again, after we finished our inspection we checked the documentation to find out where which feature was implemented. The pdf-files of the documentation did not help because they only explain how to use the design. However, most of the source code files have a description explaining their purpose. In many cases this description can directly be related to a feature. However, there are some files without any description, e.g. sm_RxFifoBl.v, or files where the description could not be related to any feature, e.g. sm_fifoRTL.v. Additionally, the design is accessed through a Wishbone-interface. The Wishbone-interface identifies the commands and forwards them to the corresponding modules. Therefore, we consider the files related to the Wishbone-interface as commonly used by all features. Table 4 compares our findings with the claims of the documentation. The files are ordered based on their ranking. If the documentation clearly relates a file to a feature this file is listed in column *Belongs* and those files where the documentation is unclear, listed in column *Possible*. In case of Register access only the files in the ranking with a threshold of 1.0 are shown. In all other cases all files are shown until the point where all files from *Belongs* are included. Over all features only one file is included by our approach which the documentation relates to another functional behavior.

This case study shows clearly that the three categorization based coloring heuristics are inaccurate when being faced with a single use case yielding very low coverage. The *Tarantula* and the *Ochiai* scheme provide equally good results as there are no visible differences between both schemes. Again, our approach gives at least as good information as the documentation. Moreover, the approach often provides additional information for feature localization. Therefore techniques for feature localization, like the one presented in this paper, are needed for design understanding.

6. CONCLUSION

We described an approach for feature localization in hardware designs. Our approach uses coverage information gathered by simulation to relate different coverage items to different features. Our prototype supports statement coverage and toggle coverage. Four different coloring schemes to present the results have been imple-

mented. Two categorize the coverage items into different groups. The other two schemes compute the likelihood of a coverage item to be related to a feature and the confidence in this likelihood. These values are then presented as the hue and the brightness of the coverage items. We also introduced a heuristic to rank file based on their likelihood of being related to a feature, allowing to guide the user faster to the corresponding code. Additionally we introduced a comparison for *orthogonal* features to improve design understanding.

The case studies emphasize the strength of our approach. They also showed that the *Tarantula* scheme performs best and that coverage metrics typically used for feature localization in software systems are not sufficient for feature localization in hardware designs. Therefore, hardware specific coverage metrics must be used as well. Additionally, the two case studies showed that the main advantages of the *Tarantula* and *Ochiai* scheme are their continuous range and their notion of not covering a coverage item if a certain feature is used. Altogether our approach often yields more information about the implementation of the features than the documentation, even in difficult cases.

7. REFERENCES

[1] R. Abreu, P. Zoeteweij, and A. J. C. van Gemund. An evaluation of similarity coefficients for software fault localization. In *Pacific Rim International Symposium on Dependable Computing*, pages 39 –46, 2006.

[2] R. Abreu, P. Zoeteweij, and A. J. C. van Gemund. On the accuracy of spectrum-based fault localization. In *Testing: Academic and Industrial Conference Practice and Research Techniques - MUTATION*, pages 89 –98, 2007.

[3] E. Clarke, M. Fujita, S. Rajan, T. Reps, S. Shankar, and T. Teitelbaum. Program slicing of hardware description languages. In *Correct Hardware Design and Verification Methods*, volume 1703 of *Lecture Notes in Computer Science*, pages 72–72. 1999.

[4] A. DeOrio, A. Bauserman, V. Bertacco, and B. Isaksen. Inferno: Streamlining verification with inferred semantics. *IEEE Transactions on Computer-Aided Design of Integrated Circuits and Systems*, 28(5):728 –741, 2009.

[5] T. Eisenbarth, R. Koschke, and D. Simon. Locating features in source code. *IEEE Transactions on Software Engineering*, 29:210–224, 2003.

[6] A. Fantozzi. Locating Features in Vim: A Software Reconnaissance Case Study. Technical report, 2002.

[7] G. Fey and R. Drechsler. Improving simulation-based verification by means of formal methods. In *Asia and South Pacific Design Automation Conference*, pages 640–643, 2004.

[8] IEEE 1364 Working Group. IEEE Standard for Verilog Hardware Description Language. *IEEE Std 1364-2005 (Revision of IEEE Std 1364-2001)*, 2006.

[9] ITRS Working Group. International technology roadmap for semiconductors 2009 update system drivers, 2009.

[10] J. A. Jones, M. J. Harrold, and J. T. Stasko. Visualization for fault localization. In *Proceedings of the Workshop on Software Visualization*, pages 71 –75, 2001.

[11] W. Li, A. Forin, and S. A. Seshia. Scalable specification mining for verification and diagnosis. In *Design Automation Conference*, pages 755 –760, 2010.

[12] R. Santelices, J. A. Jones, Y. Yu, and M. J. Harrold. Lightweight fault-localization using multiple coverage types. In *International Conference on Software Engineering*, pages 56–66, 2009.

[13] A. Sinha, P. Dasgupta, B. Pal, S. Das, P. Basu, and P. P. Chakrabarti. Design intent coverage revisited. *ACM Transactions on Design Automation of Electronic Systems*, 14:9:1–9:32, 2009.

[14] S. Tasiran and K. Keutzer. Coverage metrics for functional validation of hardware designs. *IEEE Design Test of Computers*, 18(4):36 –45, 2001.

[15] N. Wilde and C. Casey. Early field experience with the software reconnaissance technique for program comprehension. In *Working Conference on Reverse Engineering*, pages 270 –276, 1996.

[16] N. Wilde and M. C. Scully. Software reconnaissance: Mapping program features to code. *Journal of Software Maintenance: Research and Practice*, 7(1):49–62, 1995.

Path Directed Abstraction and Refinement in SAT-Based Design Debugging

Brian Keng
University of Toronto
ECE Department, Toronto, Canada
briank@eecg.toronto.edu

Andreas Veneris
University of Toronto
ECE & CS Department, Toronto, Canada
veneris@eecg.toronto.edu

ABSTRACT

The past decade has seen a disproportionate amount of resources dedicated towards verification as compared to actual design. It is reported that one third of this overhead is due to the resource-intensive task of manual debugging. To relieve this burden, this work introduces the novel concept of path directed debugging within a window-based abstraction/refinement framework. The algorithm divides the error trace into non-overlapping time-windows where each window is analyzed separately. Subsequent windows are replaced with abstracted over-approximations derived from failing paths in the time domain. Using this abstracted model, each solution found is processed through an additional verification step that removes spurious solutions and simultaneously refines the problem. This paper also develops the theory that shows that the proposed approach is complete, a fact that mitigates the incompleteness inherent in past time-window based debugging methods. Experimental results on industrial designs with long error traces show a 55% decrease in peak memory usage resulting in 78% more instances being solved when compared to previous work.

Categories and Subject Descriptors

B.5.2 [**Design Aids**]: Verification; B.6.2 [**Design Aids**]: Verification

General Terms

Algorithms, Verification

Keywords

debug, diagnosis

1. INTRODUCTION

Computer-Aided Design (CAD) tools are continuously improving their scalability and efficiency to mitigate the high cost associated with the design of modern VLSI systems. In the past decade, the effort to verify these systems has increased disproportionately [4], a trend coined as the *verification gap*. This is also confirmed by the 3:1 ratio between the number of verification engineers and designers. To make matters worse, this trend has been projected to increase almost seven-fold by 2015 [8]. The resource-intensive task of *debugging* is a significant component of this gap. Technical road-maps and market studies indicate that once a design fails verification, fixing it can take up to 32% of the total verification effort [4].

Automated design debugging techniques [5, 11] aim to increase debugging efficiency by localizing the root-cause of the error. However, the tremendous growth in modern design size (>5M gates per typical design block) and error trace lengths (tens of thousand of cycles) can limit their applicability. The primary cause for this limitation is their use of *time-frame expansion* [11] to model the problem, where the combinational part of the circuit is replicated for the length of the error trace. As such, industrial design sizes coupled with excessive error trace lengths inevitably lead to memory explosion issues.

Recent research to reduce this massive memory footprint focuses on these two pertinent complexity factors (*i.e.*, design size and trace length). Abstraction and refinement techniques [7, 10] have shown to greatly reduce the effective design size by iteratively refining the abstract design until a minimal set of necessary components are determined. Orthogonally, *time-windowing* techniques [6, 13] use a sliding window of consecutive clock cycles to analyze the problem. In this fashion, they model only a segment of the entire error trace and they use different methods to approximate the remaining non-modeled parts. These methods greatly reduce the problem size but their use of approximations leads to many spurious solutions negating the benefit of localization.

In this work, we present a novel design debugging algorithm utilizing time-windows in an abstraction and refinement Boolean Satisfiability (SAT) based framework. The algorithm is also shown to be complete in contrast to previous approximate time-windowing methods that provide no means of refinement [6]. The process begins by dividing the error trace into non-overlapping time-windows where each window is analyzed separately. Just like other window-based methods, only one time-window is explicitly modeled while subsequent time-frames are over-approximated using the abstraction.

The key novelty to our work is that the proposed abstraction initially consists of paths in the time-frame expanded circuit that directly lead to the observed failure during simulation. As before, during each iteration, solutions to this abstract problem may either be valid or spurious. Therefore, an additional verification step is necessary for each solution. This step works by iteratively propagating solutions forward to successive time-windows. If the solution is not spurious then successive time-windows will be satisfiable. Otherwise, they will return UNSAT and additional paths are extracted from the resulting conflict to refine the abstract problem. The net result is a dramatically reduced memory footprint that – as theoretically proven in this paper – mitigates the incompleteness issues of past time-windowing methods.

Experimental results on large industrial designs with long error traces from OpenCores [9] and from our industrial partners illustrate the benefits of this work. Across all instances, the proposed approach solves 78% more cases when compared to a previous time-windowing technique. This results in an average 55% reduction in peak-memory while being able to debug traces that are 31% longer on average. The abstraction is also shown to be very efficient in reducing problem size using only an average of 4.3% of the clauses needed compared to explicit modeling of the problem.

947

The remaining sections of this paper proceed as follows. Section 2 presents background material. Section 3 and Section 4 describe the proposed approach and extensions, respectively. Section 5 presents experimental results and Section 6 concludes this work.

2. PRELIMINARIES

2.1 SAT-based Design Debugging

Automated design debugging techniques aim to find error locations, known as *suspects*, that can explain an observed failure during verification [5] given an error trace (*i.e.*, *counter-example*) and a user-defined number of errors. SAT-based design debugging [11] encodes this problem as a SAT instance where the solutions correspond to the set of all possible suspects for the given number of errors.

This SAT instance is constructed in several steps. First, the combinational part of the circuit (T) is enhanced with an error model for each location in the design, denoted as T_{en}. Each error model has an associated *suspect variable* (e_i) that when active (*i.e.*, $e_i = 1$), disconnects the location's fan-out from its fan-in, allowing it to be free. Next, the enhanced combinational circuit is replicated as a time-frame expanded model for the length of the error trace. Observe, the suspect variables are not replicated since they represent the same location regardless of time-frame. On the unrolled model, the initial state (S^p), vector of inputs (X^i), and vector of *expected* or correct outputs (Y^i) are constrained for each clock cycle i to model the error trace. Finally, the number of active suspect variables are constrained $(\Phi(N))$ to exactly N to denote the search for N simultaneous errors. A *debugging solution* to this instance consists of the set of N active suspect variables which, once found, are blocked in order to find all other solutions. The following equation models a window of an error trace with width w from cycle p to $p + w - 1$.

$$Debug_p^{p+w-1}(N) = S^p \wedge \Phi(N) \wedge \bigwedge_{i=p}^{p+w-1} X^i \wedge Y^i \wedge T_{en}^i \quad (1)$$

Solutions to this equation correspond to suspects that can explain the observed error. Note that the observed failure must occur within cycles p to $p + w - 1$ to generate a mismatch between the erroneous circuit behavior and expected outputs. If this is not the case, then the equation is trivially satisfiable at $N = 0$ indicating that the failure cannot be observed in the modeled window and additional cycles must be added.

2.2 Time Diagnosis

An alternate formulation of a debugging instance, known as *time diagnosis* [12], can be generated by utilizing a variation of Eq. 1. The key difference with the one from Section 2.1 is that suspect variables for a location are shared only within a fixed time-window of clock cycles and not over the entire error trace. In other words, this time-window can model the excitation of a suspect within the given set of clock cycles, but it cannot model it across different time-windows. It has been shown [12] that this trade-off can dramatically reduce the problem complexity in an effort to model real-life bugs with long error traces. Due to its relation to the work presented here, we introduce it in some more detail.

To model a debugging instance for a single time-window Eq. 1 can be used. However, as explained earlier, if the observed error is not within this time-window the instance will be trivially satisfiable. To overcome this issue, one can model the circuit behavior for subsequent time-frames in order for the observed failure to be included. The modeling of subsequent times-frames is shown in the next equation for

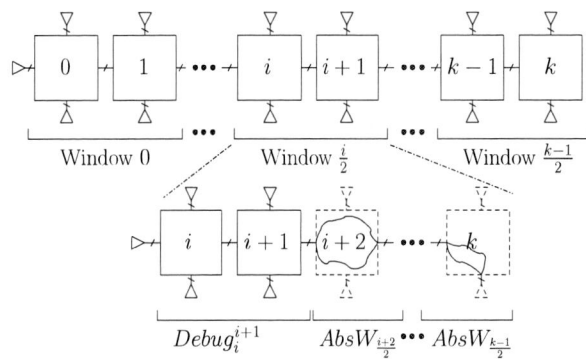

Figure 1: Path Directed Abstraction and Refinement

a *width* of w time-frames starting from cycle $p \cdot w$:

$$W_p = \bigwedge_{i=p \cdot w}^{(p+1) \cdot w - 1} X^i \wedge Y^i \wedge T^i \quad (2)$$

Note that if $N = 0$ is set in Eq. 1, it simplifies to a time-frame shifted version of Eq. 2 (with the addition of the initial state constraints S^p) because it eliminates all the additional circuitry that relates to the error models.

By combining Eq. 1 and 2, one can overcome the issue of debugging a time-window that does not contain the observed failure. This formulation is shown with respect to a set of time-windows from p to q with a width of w:

$$TimeDebug_p^q(N) = Debug_{p \cdot w}^{(p+1) \cdot w - 1}(N) \wedge \bigwedge_{i=p+1}^{q} W_i \quad (3)$$

For the interested reader, a detailed example of time diagnosis is provided in the supplemental material in Section S1.1.

3. PATH DIRECTED ABSTRACTION AND REFINEMENT

Path directed abstraction and refinement is a methodology built around the debugging model presented in Eq. 3. It also provides a complete technique to iteratively analyze long industrial traces. It begins by dividing an error trace into multiple non-overlapping time-windows. Each window is iteratively analyzed starting from the last time-window in the error trace that contains the observed failure. This "sliding window" of time-frames iteratively moves towards earlier cycles of the trace searching for suspects without the need to explicitly model later windows. To properly constrain the observed failure, each of these subsequent windows are abstracted to a smaller set of constraints. Intuitively, these constraints are modeled with information obtained by structural circuit paths in the time-frame expanded circuit that propagate to the observed failure on the primary output(s).

This key idea is shown in Figure 1 with an error trace containing $k + 1$ time-frames. In this example, the trace is divided up into $\frac{k+1}{w}$ non-overlapping time-windows with $w = 2$. The abstract time diagnosis instance for time-window $\frac{i}{w}$ is shown in greater detail, where the first w time-frames are modeled explicitly with the use of Eq. 1. In the proposed method, the subsequent time-frames (past cycle $i + 1$) are not modeled explicitly. Instead, they are abstracted and replaced with information obtained by paths in the time-frame expanded circuit such that the original observed failure can still be modeled. This abstraction over-approximates the subsequent time-frames similar to previous time-windowing techniques [6]. However, the key benefit

948

is that it provides a compact abstraction combined with an efficient refinement strategy, which is not available in [6], to remove spurious solutions that do not occur in the concrete problem.

These spurious solutions are removed by a refinement step that propagates the associated satisfying assignments forward to each successive explicitly modeled time-window using a separate SAT instance. If all successive windows yield a SAT result, then the solution is valid. Otherwise, a conflict is generated indicating that the current abstraction needs to be refined. The refinement step adds back all clauses involved in the conflict, intuitively adding more paths to the original abstraction. This process is repeated until the refined abstract problem is unsatisfiable. In this case, the current time-window is finished with complete results and the algorithm proceeds to analyze the next one.

The following subsections describe the formulation, theoretical results and pseudo-code for the path-directed abstraction and refinement algorithm in much greater detail. Due to space requirements, formal proofs of the lemmas and theorems as well as a descriptive example can be found in the supplemental material section of this paper.

3.1 Path-based Abstraction

The path-based abstraction approximates Eq. 3 by replacing consecutive windows of time-frames (W_i) with an abstract version, denoted by $AbsW_i$. This is initially formulated by adding paths in the time-frame expanded circuit that are directly involved in the observed failure. Practically, this is a set of clauses extracted from the SAT-solver conflict graph generated while modeling the observed failure. The rest of this subsection describes this process in greater detail.

Consider the first abstract time-window occurring at the end of the trace ($AbsW_q$) for cycles $q \cdot w$ to $(q+1) \cdot w - 1$. This abstract window is first needed for use in the next iteration once we have finished analyzing window $p = q$ using Eq. 3. Notice that in this case by setting $N = 0$ in Eq. 3, it simplifies to W_q with the addition of the initial state constraints. This precisely models the circuit behavior for cycles $q \cdot w$ to $(q+1) \cdot w - 1$ that led to the observed failure. The resulting conflict graph from the SAT-solver will contain clauses that are directly involved in the observed failure at the primary outputs. These clauses, excluding the initial state constraints, form the initial abstraction $AbsW_q$.

In general, the abstraction is formed by maintaining a set of clauses for each time-window that has been previously analyzed. When a time-window has completed analysis, we set $N = 0$ for that instance and clauses within that window are extracted. This is shown in the following equation:

$$AbsW_i = Conflict_i(AbsDebug_i^q(N = 0)) \qquad (4)$$

Here, $Conflict_i$ denotes all clauses with variables within time-frames i to $i + w - 1$ involved in the conflict of the input formula. $AbsDebug_i^q(N)$ denotes the abstract debugging instance for the previously analyzed time-window. In the base case for $i = q$, this simplifies to Eq. 1.

Using Eq. 4, the initial abstract debugging problem can be created, shown in the following formula:

$$AbsDebug_p^q(N) = Debug_{p \cdot w}^{(p+1) \cdot w - 1}(N) \wedge \bigwedge_{i=p+1}^{q} AbsW_i \quad (5)$$

This formula mirrors Eq. 3 except it uses the abstract windows for later time-frames instead of explicitly modeling them. It results in a greatly reduced memory footprint because the abstract windows typically are much smaller than the explicit model. In addition, since each abstract window ($AbsW_i$) contains either clauses that were in the explicit model of that window (W_i) or implied by it, Eq. 5 is in fact

an over-approximation of Eq. 3 as stated in the next lemma.

Lemma 1 *Let $\mathcal{E}$ be a set of N active suspect variables found in a satisfying assignment to $TimeDebug_p^q(N)$, then there is a satisfying assignment to $AbsDebug_p^q(N)$ that will also contain the active suspect variables from $\mathcal{E}$.*

Lemma 1 implies that the initial abstract instance will return a superset of debugging solutions when compared to explicit modeling. To filter out spurious solutions, each one of them will need to be verified by propagating its satisfying assignment forward to each subsequent concretely modeled time-window. If a solution is indeed spurious, the resulting conflict will act as a refinement step, a process discussed in the following subsection.

3.2 Path Directed Refinement

Due to the use of abstract windows in Eq. 5, a satisfying assignment, $\mathcal{A}$, may not correspond to one in the concrete instance and therefore may be spurious. A straightforward method to verify $\mathcal{A}$ would be to apply it to the entire concrete instance in Eq. 3. However, this would negate the benefit of a reduced memory footprint since the instance would involve explicit modeling of all time-frames. Instead, by utilizing the previous abstract windows, it is possible to create several smaller SAT instances that can equivalently verify $\mathcal{A}$.

In order to propagate $\mathcal{A}$ forward, only a subset of assignments are actually needed. Notice that in Eq. 5, time-frames from $p \cdot w$ to $(p+1) \cdot w - 1$ are modeled explicitly, while the remaining are not. This means that an assignment for the first $p \cdot w$ to $(p+1) \cdot w - 1$ time-frames on the abstract model should also work for the respective frames on the concrete model in Equation 3. However, the only way for these concrete time-frames to affect forward time-frames are through the state variables at time-frame $(p + 1) \cdot w$. If these state assignments are used to constrain subsequent concretely modeled time-frames, then we can iteratively propagate the effect of the original assignment forward to each concretely modeled time-window. We denote this subset of assignments of $\mathcal{A}$ on state variables for time-frame i by $cube^i$.

To accomplish this propagation, we create multiple instances each of which models precisely w concrete time-frames and uses the abstract windows for the others. This construction is represented by the following equation:

$$Prop_r^q = cube^{r \cdot w} \wedge W_r \wedge \bigwedge_{i=r+1}^{q} AbsW_i \qquad (6)$$

The state assignment $cube^{(p+1) \cdot w}$, extracted from $\mathcal{A}$, is propagated using the above equation to generate $Prop_{p+1}^q$. If this results in SAT, another cube, $cube^{(p+2) \cdot w}$, will be extracted and propagated to the next instance, $Prop_{p+2}^q$, and so on, until all time-frames have been verified. If all subsequent instances result in SAT, then the original abstract satisfying assignment can be extended to one in the the concrete model. This is stated more precisely in the following lemma.

Lemma 2 *Let $\mathcal{E}$ be a set of N active suspect variables found in a satisfying assignment of $AbsDebug_p^q(N)$. If $AbsDebug_p^q(N)$, $Prop_{p+1}^q$, ..., $Prop_q^q$ are SAT, then there is a satisfying assignment to $TimeDebug_p^q(N)$ that will also contain the active suspect variables from $\mathcal{E}$.*

After a set of active suspect variables is found, they are blocked so that the next solution can be found for $AbsDebug_p^q$. Notice that the propagation generates a separate instance for every w time-frames resulting in exactly w concretely modeled time-frames for any given instance. This key point ensures that the memory footprint of the entire process is kept to a minimum.

949

On the other hand, if $Prop_r^q$ is UNSAT, this means that the original assignment $\mathcal{A}$ cannot be extended to the concrete model. In this case, the abstract window was not sufficiently refined and additional clauses must be added. Similar to the initial abstraction, we extract clauses in the SAT-solver conflict graph and add them to the abstract window. Intuitively, this indicates that additional paths are required. This is shown in the following equation:

$$\hat{AbsW}_r = AbsW_r \cup Conflict_r(Prop_r^q) \quad (7)$$

Notice that the refined abstract window, $\hat{AbsW}_r$, still is an over-approximation of the concrete window because it only contains clauses that are either a subset, or implied by, the concrete window it models.

With the refined abstract window $\hat{AbsW}_r$, we can similarly refine all previous abstract windows. This can be accomplished by re-creating Eq. 6 with $j < r$ and the current refined abstract window $(\hat{AbsW}_r)$. Since $\mathcal{A}$ is known to be invalid, each one of these instances will be UNSAT because they are just an extension of the assignment $\mathcal{A}$. After the abstract windows have been updated, the refined formula defined by updating Eq. 5 can be solved for all solutions again. These solutions similarly can be either be confirmed or used to refine the abstract time-windows until the refined instance, $AbsDebug_p^q$, is unsatisfiable.

Once all solutions are found, $\hat{AbsDebug}_p^q$ is UNSAT indicating that debugging has been completed on this time-window. The next theorem confirms the completeness of our approach, that is, the set of solutions found in the abstract model equals to the one found for the concrete model.

Theorem 1 *Let $sols_{abs}$ be the set of confirmed debugging solutions returned by iteratively debugging and refining $\hat{AbsDebug}_p^q$ and let $sols_{time}$ be the set of debugging solutions returned by $TimeDebug_p^q$. If the final refined abstract instance $\hat{AbsDebug}_p^q$ is UNSAT by blocking all solutions in $sols_{abs}$, then $sols_{abs} = sols_{time}$.*

3.3 Overall Algorithm

Algorithm 1 presents pseudo-code for the path directed abstraction and refinement algorithm for time-windows from p to q. The main loop from lines 3-13 iterates through each time-window analyzing them separately. Line 4 constructs the initial abstract instance using Eq. 5 with any previously generated abstract windows $(AbsW_i)$. In the first iteration, this equation simplifies to Eq. 1 where the very last time-window is analyzed. The inner WHILE loop (lines 5-11) finds all satisfying assignments and confirms each one by passing the result to the VERIFY procedure. If the assignment is not confirmed, then the procedure refines the relevant abstract windows. Once all the current assignments have been verified, the refined abstract problem is reconstructed (line 10), blocking any solutions that were confirmed from being found again. When the algorithm has finished analyzing the current time-window, it exits the WHILE loop and generates the initial abstraction for the current window on line 12 for use in the next time-window.

The pseudo-code for the VERIFY procedure is also shown in Algorithm 1. The procedure begins by extracting the state cube for window $p+1$ from the assignment $\mathcal{A}$ on line 17. The outer FOR loop (lines 18-28) propagates the cube forward to subsequent windows using Eq. 6 (line 19). If any of the subsequent time-windows are UNSAT (line 20), then the abstract windows are refined by iterating backwards through the windows (lines 21-24). Each backward iteration reconstructs Eq. 6 with the refined abstract windows on line 22. The refinement step (line 23) extracts clauses from each of these UNSAT instances to update the abstract window. The procedure either returns the confirmed solution (line 29) or will return an empty solution otherwise (line 25).

Algorithm 1 Path Directed Abstraction and Refinement

```
 1: procedure PATHDEBUG
 2:     sols ← ∅
 3:     for p ∈ q, q − 1, . . . , 1, 0 do
 4:         inst ← AbsDebug_p^q∧BLOCK(sols)
 5:         while inst is SAT do
 6:             Assignments ←SOLVEALL(inst)
 7:             for A ∈ Assignments do
 8:                 sols ← sols ∪ VERIFY(A, p, q)
 9:             end for
10:             inst ← AbsDebug_p^q∧BLOCK(sols)
11:         end while
12:         AbsW_p ← Conflict_p(inst)
13:     end for
14:     return sols
15: end procedure
16: procedure VERIFY(A, p, q)
17:     cube^{(p+1)·w} ←EXTRACTCUBE(A, p + 1)
18:     for i ∈ p + 1, p + 2, . . . , q − 1, q do
19:         inst ← Prop_i^q
20:         if inst is UNSAT then
21:             for j ∈ i, i − 1, . . . , p + 2, p + 1 do
22:                 inst ← Prop_j^q
23:                 AbsW_j ← AbsW_j ∪ Conflict_j(inst)
24:             end for
25:             return ∅
26:         end if
27:         cube^{(i+1)·w} ←EXTRACTCUBE(inst, i + 1)
28:     end for
29:     return EXTRACTSUSPECT(A)
30: end procedure
```

---

4. PRACTICAL CONSIDERATIONS

This section presents implementation details that provide significant performance gains to Algorithm 1.

4.1 Leveraging the SAT-solver

There are two main efficiency improvements that leverage the capabilities of modern SAT-solvers. The first improvement involves extensive use of unit assumptions and incremental SAT capabilities of modern solvers [1, 3]. This also requires a minor modification to Algorithm 1. During the VERIFY procedure instead of verifying one assignment at a time, multiple assignments can be verified using one instance of $Prop_i^q$. This is accomplished by setting each cube, $cube^{i·w}$, as unit assumptions and re-solving the instance using incremental SAT. This greatly reduces the overhead of re-solving the same instance (with the exception of the state cube) multiple times. The second improvement makes merging clauses practical by using *clause subsumption* [1]. Since this step is performed each time an abstract window is refined, one can reduce its overhead by using efficient implementations available in many modern solvers.

4.2 Improved Refinement Techniques

Although it will be experimentally shown that the path-based abstraction in Algorithm 1 is memory efficient, in certain cases it may require an excessive number of refinement iterations. We briefly present three additional techniques to manage this issue.

The first technique aims to speed up refinement by attempting to find additional state cubes that may cause a conflict. Multiple additional state cubes can be generated from the original cube ($cube^{i·w}$) by simply inverting a subset of its bits. These additional cubes are applied to Eq. 6 on line 22 of Algorithm 1 using incremental SAT and unit assumptions as described above. If the resulting instance is UNSAT, clauses can be extracted from the conflict-graph accelerating refinement in an efficient manner.

The next technique re-uses clauses from previous abstract windows. When an initial abstraction for $AbsW_p$ is created on line 12 of Algorithm 1, one can add the equivalent clauses from the previous abstract window, $AbsW_{p+1}$. Note that not all clauses will be valid in the current window due to the differences in the error trace between windows. However, it is efficient to verify if the clause is valid by propagating the negation of the clause using the SAT-solver and the concrete model. The rationale behind this technique is that the same circuit paths may be needed in future abstract windows, so there is no need to duplicate the work by refining them repeatedly.

The last technique handles the worse-case scenario where refinement may require an excessive number of iterations. These cases typically occur when trying to verify certain suspects that require a large number of circuit paths. This exposes itself on line 5 of Algorithm 1 where the WHILE loop takes an excessive number of iterations due to the same suspect appearing in the satisfying assignments. The solution to this problem is to skip these suspects by blocking them if they appear as a solution on line 6 of Algorithm 1 more than *skip* times for a given time-window. *skip* is a user-defined parameter that allows the user to trade-off confirming these suspects for a speed-up in run-time. Practically, the number of suspects skipped is small that it does not have a major impact on quality of results while providing large benefits in run-time.

5. EXPERIMENTS

This section presents the experimental results for the proposed path directed abstraction and refinement algorithm. All experiments are run on a single core of a Intel Core i5 3.1 GHz quad-core workstation with 8 GB of RAM and a timeout of 7200 seconds. Algorithm 1 with the improvements described in Section 4 are implemented and compared against the time-windowing technique in [6] with a time-diagnosis framework. This algorithm named *window expansion* is chosen because it is the only other time-windowing technique that can return exact results through its use of explicit modeling. This is not the case with other existing time-windowing techniques that use approximations with no means of refinement leading to spurious solutions. This time-windowing technique analyzes time-windows starting from the end of the trace. However, it does not approximate suffix time-frames and instead it models them explicitly as in Eq. 3. Note, comparisons to previous abstraction and refinement techniques [7,10] for debugging are omitted because these techniques are orthogonal to the time-windowing techniques presented here. Minisat [3] is used to solve all SAT instances.

The proposed algorithm is exercised on industrial Verilog RTL designs from OpenCores [9] and two commercial designs (fxu, comm) provided by our industrial partners. Each debugging instance is created by randomly selecting a line in the RTL and inserting a typical industrial RTL error (wrong state transition, incorrect operator, erroneous module instantiation, etc.). Such errors typically map to multiple gate-level errors. Next, the buggy design is simulated with its accompanying testbench to expose the failure and record the error trace. After this step, all designs with the inserted error are run through a cone of influence reduction [2] to remove logic in the circuit that is not involved in the debugging problem. This combined with the inserted errors may result in different effective circuit sizes for instances created using the same design. Finally, both algorithms are run for the length of the error trace with error cardinality $N = 1$, time-window size $w = 10$, and $skip = 10$ for the proposed algorithm. The window size w is chosen so that *window expansion* could be run for at least one iteration on

the biggest design. The *skip* parameter is chosen to balance the performance and number of skipped solutions over all instances. The solutions returned from both these algorithms correspond to RTL structures such as if statements, assign statements, instantiations, etc.

The results of the experiments are shown in Table 1. Each row of the table shows results for a separate debugging instance with a different bug. Each instance is labeled by appending the design name with a number to indicate different bugs. Two sets of experiments are run, one for the *window expansion* [6] algorithm, and the other the proposed technique from Algorithm 1.

The first five columns of Table 1 show the instance name, number of gates, number of cycles in the error trace, number of potential suspect locations, and base memory footprint of the design. The base memory footprint is common among both algorithms and includes the memory used to parse the RTL, load the error trace and setup the common circuit data structures. The next four columns show the results for *window expansion* which include the run-time in seconds, peak-memory used for the problem over and above the base memory, the number of solutions returned by the algorithm, and the total number of frames analyzed. The following five columns show results for the proposed method which include run-time, peak-memory for the problem, number of solutions, total number of frames analyzed, and number of skipped solutions. A **TO** (**MO**) entry in the table is used to refer to a time-out (memory-out) condition for that experiment. If a **TO** or **MO** condition occurs, partial results for the corresponding experiments are shown in the table.

The benefit of the proposed technique is apparent by the significant reduction in peak-memory when compared to *window expansion*. The proposed algorithm is able to complete analysis on the entire trace for 78% more instances compared to previous work primarily due to **MO** conditions by the previous technique. This translates to 16 instances for the proposed method versus 9 for *window expansion*. Moreover for instances of Algorithm 1 that ran to completion, the proposed algorithm showed a 55% average reduction in peak problem memory while being able to debug traces that were 31% longer on average. This shows the efficacy of the path-based abstraction to dramatically reduce the memory footprint of large industrial problems that were previously too large to debug.

The run-time results for Algorithm 1 show mixed results compared with *window expansion*. In some cases such as div64bits_2, it can perform better. This is primarily because the error propagation path is relatively narrow so very few refinement iterations are necessary. In other cases where both algorithms finished, the run-time can be greater. This is most apparent in mips_sys_1/2 where the number of paths that needed to be refined was too large causing a time-out. Despite this increase in run-time, it is only comparing instances that finished for both algorithms. Admittedly, the central benefit is that it can solve instances that are too large to fit into memory compared to previous methods.

The number of confirmed solutions returned between both algorithms are approximately the same. In some cases such as comm_1, the proposed algorithm was able to find additional suspects that the previous method could not due to a **MO** condition. In other cases, the use of skip suspects reduced the number of confirmed solutions that the path directed abstraction returned. However, the number of unconfirmed suspects is relatively small with 0.36% of total suspects skipped on average. Additionally, the absolute numbers of skip suspects range from 0 for most cases to at most 20, which practically does not decrease the effectiveness of the results for human analysis.

Figure 2 shows a more detailed analysis of the proposed abstraction. This graph shows the relative size of the abstraction in terms of clauses compared with explicit model-

Table 1: Path Directed Abstraction and Refinement Experiments

Instance Info					Window Expansion [6]				Path Directed Abstraction				
instance name	# gates (k)	# total cyc	# total susp	base mem (MB)	time (s)	prob mem (MB)	# sols	# debug cyc	time (s)	prob mem (MB)	# sols	# debug cyc	# skip
ac97_ctrl_1	22.9	1000	1380	278	2641	**MO**	46	610	1886	2478	46	1000	0
ac97_ctrl_2	22.9	1000	1388	278	4077	**MO**	122	550	4023	2571	123	1000	0
comm_1	164.4	150	20155	1453	899	**MO**	72	70	7163	4925	74	150	2
comm_2	164.4	120	20155	1453	1018	**MO**	73	70	4803	3879	75	120	2
div64bits_1	59.7	110	197	347	390	3647	38	110	1929	1122	36	110	8
div64bits_2	57.8	100	198	347	374	3193	21	100	69	873	21	100	0
fdct_1	239.5	80	4662	1302	2450	**MO**	78	50	6673	3611	80	80	3
fdct_2	243.2	70	4661	1302	3304	**MO**	69	50	6414	3154	61	70	8
fpu_1	72.5	240	1982	388	702	**MO**	35	220	312	2024	37	240	0
fpu_2	71.5	40	1981	385	35	1583	5	40	26	674	5	40	0
fxu_1	447.3	100	33087	1872	758	**MO**	174	20	939	**MO**	174	50	0
fxu_2	447.3	100	33087	1872	266	**MO**	37	20	326	**MO**	37	20	0
mips_sys_1	50.3	250	2636	372	2073	7105	123	250	**TO**	2057	104	210	12
mips_sys_2	50.1	200	2636	434	3983	5322	260	200	**TO**	1578	251	140	5
rsdecoder_1	14.5	80	2040	562	5187	1311	81	80	6816	680	78	80	12
rsdecoder_2	14.5	110	2040	489	489	627	384	110	2471	579	384	110	0
usb_funct_1	35.8	410	4061	328	3413	**MO**	105	410	306	1684	105	410	0
usb_funct_2	35.8	690	4061	328	3779	**MO**	106	420	2542	2950	106	690	0
vga_1	48.8	100	1731	624	4955	3545	20	100	3447	1167	20	100	9
vga_2	20.7	150	1729	624	720	1911	46	150	4138	818	46	150	20

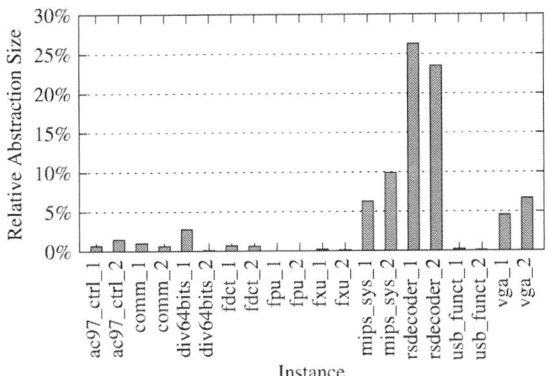

Figure 2: Relative Abstraction Size

ing of those same time-frames. On average, the abstraction is 4.3% the size of the explicit model. Most cases are well below the average indicating that most paths are not needed to solve the debugging instance. However, in some cases such as rsdecoder_1 the number of clauses is higher at 26%. This translates into less benefit in terms of memory reduction but still a small fraction of the explicit model.

6. CONCLUSION

In this work, a novel path directed abstraction and refinement algorithm is proposed. The algorithm divides the error trace into non-overlapping time-windows where each window is analyzed separately. Subsequent windows are replaced with abstracted over-approximations derived from failing paths in the time domain. Using this abstracted model, each solution found is processed through an additional verification step that removes spurious solutions and simultaneously refines the problem. Experimental results on industrial designs with long error traces show a dramatic decrease in peak memory usage when compared with previous work.

7. REFERENCES

[1] A. Biere, M. Heule, H. van Maaren, and T. Walsh, editors. *Handbook of Satisfiability*, volume 185 of *Frontiers in Artificial Intelligence and Applications.* IOS Press, 2009.

[2] E. Clarke, O. Grumberg, and D. Peled. *Model Checking.* MIT Press, 1999.

[3] N. Eén and N. Sörensson. An extensible SAT-solver. In *Int'l Conf. on Theory and Applications of Satisfiability Testing*, pages 502–518, 2003.

[4] H. Foster. From Volume to Velocity: The Transforming Landscape in Function Verification. In *Design Verification Conference*, 2011.

[5] S. Huang and K. Cheng. *Formal Equivalence Checking and Design Debugging.* Kluwer Academic Publisher, 1998.

[6] B. Keng, S. Safarpour, and A. Veneris. Bounded Model Debugging. *IEEE Trans. on CAD*, 29:1790–1803, November 2010.

[7] B. Keng and A. Veneris. Managing complexity in design debugging with sequential abstraction and refinement. In *ASP Design Automation Conf.*, pages 479–484, 2011.

[8] D. McGrath. De Geus touts new products, says ICs will rebound. *EE Times*, March 2009.

[9] OpenCores.org. http://www.opencores.org, 2007.

[10] S. Safarpour and A. Veneris. Automated design debugging with abstraction and refinement. *IEEE Trans. on CAD*, 28(10):1597–1608, 2009.

[11] A. Smith, A. Veneris, M. F. Ali, and A. Viglas. Fault diagnosis and logic debugging using Boolean satisfiability. *IEEE Trans. on CAD*, 24(10):1606–1621, 2005.

[12] Y.-S. Yang, A. Veneris, and N. Nicolici. Automating data analysis and acquisition setup in a silicon debug environment. *IEEE Trans. on VLSI Systems*, PP(99):1–14, 2011.

[13] C. S. Zhu, G. Weissenbacher, and S. Malik. Post-Silicon Fault Localisation Using Maximum Satisfiability and Backbones. In *Formal Methods in CAD*, 2011.

SUPPLEMENTAL MATERIAL

S1. DETAILED EXAMPLES

S1.1 Time Diagnosis Example

This example illustrates time-diagnosis described in Section 2.2. Figure 3(a) visualizes Eq. 3 when $p = q = 1$, $w = 1$ and $N = 1$. The circuit under debug has three internal gates (g_1, g_2, g_3), one input (x_0), one output (y_0), and three state-variables (s_0, s_1, s_2). In this circuit, g_1 and g_2 form module A and g_3 forms module B. The error models are denoted by $\otimes$. There are two suspect variables, e_a, e_b corresponding to module A and B respectively. The suspect variable e_a, corresponds to the two outputs of g_1 and g_2, similarly for suspect variable e_b and g_3. Error trace values for the initial state, input and expected outputs are shown directly in the figure.

When the instance in Figure 3(a) is passed to a SAT-solver, one solution is returned where $e_b = 1$ corresponding to module B being returned as a suspect. When $\overline{e_b}$ is added as a unit clause to block it from appearing again, the instance is UNSAT indicating that no more solutions are found.

Figure 3(b) shows a visualization of Eq. 3 with the next time-window where $p = 0, q = 1, w = 1$ and $N = 1$. In this instance, the suspect variables now only correspond to modules in the earliest time-frame. When sent to the SAT-solver, one solution is returned where $e_a = 1$. After blocking it as a solution, the instance is UNSAT indicating that all solutions have been found.

S1.2 Path Directed Abstraction and Refinement Example

This example shows how Algorithm 1 works by using the circuit and error trace from Section S1.1. The first iteration of Algorithm 1 is identical to the instance generated from Figure 3(a). Once this instance returns UNSAT, the algorithm sets $N = 0$, sends it to the SAT-solver and extracts clauses involved in the resulting conflict graph to generate $AbsW_1$.

In the next iteration of the main loop, it generates an abstract debugging problem using Eq. 5 with $p = 0, q = 1, w = 1$ and $N = 1$. A visualization of this instance is shown in Figure 4(a). Here, most of circuitry from time-frame 1 has been abstracted leaving only the path directly involved in the conflict from the SAT-solver. Note this path is not unique and another conflict graph could have generated a different path.

When given to the SAT-solver, this instance returns one solution where $e_a = 1$ since e_b was already blocked from the previous time-window. The associated cube with this satisfying assignment is $cube^1 = \overline{s_0^1} \wedge s_1^1 \wedge s_2^1$. This cube is verified by generating a new instance using Eq. 6 with $r = q = 1$, it is shown in Figure 4(b) and one can show that it is UNSAT. The clauses from the conflict graph are extracted and the abstract window $AbsW_1$ is refined.

Figure 4(c) shows the refined abstract debugging problem $Abs\hat{D}ebug_0^1(N = 1)$. This instance again finds a solution with $e_a = 1$ and a new cube, $\hat{cube}^1 = s_0^1 \wedge s_1^1 \wedge s_2^1$. This cube is verified using Eq. 6 which returns SAT. Thus $e_a = 1$ is confirmed as a solution matching the results from Section S1.1. Once this solution is blocked, the refined instance returns UNSAT indicating that this time-window has completed analysis.

S2. FORMAL PROOFS

Lemma 1 *Let $\mathcal{E}$ be a set of N active suspect variables found in a satisfying assignment to $TimeDebug_p^q(N)$, then there is a satisfying assignment to $AbsDebug_p^q(N)$ that will also contain the active suspect variables from $\mathcal{E}$.*

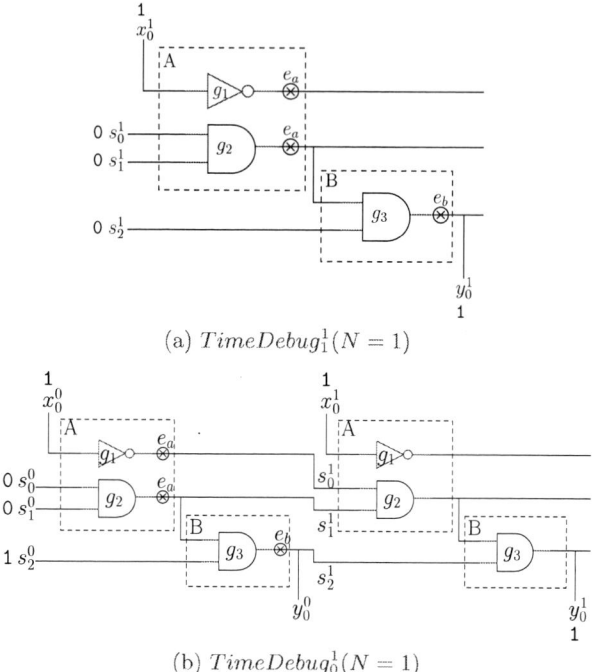

(a) $TimeDebug_1^1(N = 1)$

(b) $TimeDebug_0^1(N = 1)$

Figure 3: Time Diagnosis Example

PROOF. Let $\mathcal{A}$ be the satisfying assignment to $TimeDebug_p^q(N)$. We will show that each of the sub-formulas of $AbsDebug_p^q(N)$ is satisfiable under $\mathcal{A}$. $Debug_{p \cdot w}^{(p+1) \cdot w - 1}(N)$ is SAT under $\mathcal{A}$ because it is a subset of the clauses in $TimeDebug_p^q(N)$. Each $AbsW_i$ is derived from extracting clauses from a SAT-solver conflict graph that are within the same time-frames modeled by W_i. This means that these clauses are either a subset of W_i or implied by the overall problem. Therefore, if all W_i are SAT under $\mathcal{A}$, so are all $AbsW_i$. Since each component of is SAT under $\mathcal{A}$ so is $AbsDebug_p^q(N)$ with active suspect variables $\mathcal{E}$. $\square$

Lemma 2 *Let $\mathcal{E}$ be a set of N active suspect variables found in a satisfying assignment of $AbsDebug_p^q(N)$. If $AbsDebug_p^q(N)$, $Prop_{p+1}^q$, ..., $Prop_q^q$ are SAT, then there is a satisfying assignment to $TimeDebug_p^q(N)$ that will also contain the active suspect variables from $\mathcal{E}$.*

PROOF. Let $\mathcal{A}_p, \mathcal{A}_{p+1}, \ldots, \mathcal{A}_q$ be the satisfying assignments to $AbsDebug_p^q(N)$, $Prop_{p+1}^q$, ..., $Prop_q^q$, respectively. We show how to construct an assignment $\mathcal{A}$ to $TimeDebug_p^q(N)$ from $\mathcal{A}_p, \mathcal{A}_{p+1}, \ldots, \mathcal{A}_q$.

From $\mathcal{A}_p$, we add all assignments to variables involved in the subformula $Debug_{p \cdot w}^{(p+1) \cdot w - 1}(N)$ to $\mathcal{A}$. From each of the subsequent $\mathcal{A}_r$, we add assignments to all variables of W_r from the respective instance $Prop_r^q$. Notice the overlapping variables that were added to $\mathcal{A}$ are precisely the state variable from the cubes $cube^{p \cdot w}$, $cube^{(p+1) \cdot w}$, ..., $cube^{q \cdot w}$. But from the formulation of $Prop_{r-1}^q$, each cube, $cube^{r \cdot w}$, is generated from an instance that is constrained by the previous cube, $cube^{(r-1) \cdot w}$, except for the first one derived from the assignment $\mathcal{A}_p$. This means that there is no conflict between these overlapping state variables because each one is implied by the previous one in the sequence which originates from $\mathcal{A}_p$.

The final constructed assignment $\mathcal{A}$ composes a full assignment to $TimeDebug_p^q(N)$ since it contains assignments to all variables in each component of the formula. Moreover,

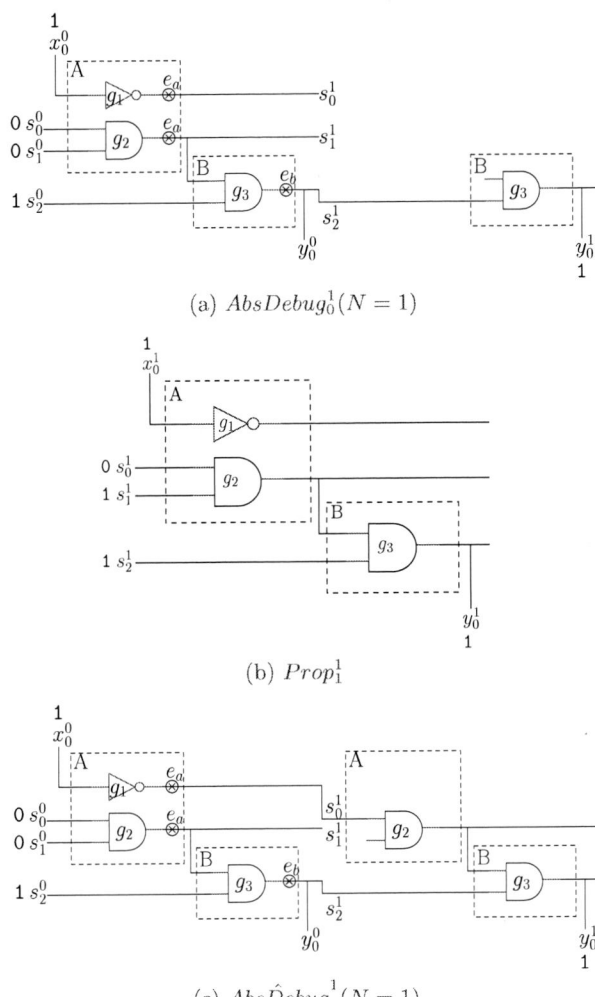

(a) $AbsDebug_0^1(N = 1)$

(b) $Prop_1^1$

(c) $Abs\hat{D}ebug_0^1(N = 1)$

Figure 4: Path Directed Abstraction and Refinement Example

$\mathcal{A}$ is a valid satisfying assignment to each of the components of $TimeDebug_p^q(N)$. We can see this because we constructed $\mathcal{A}$ by extracting exactly the assignments involved in the concrete parts of the abstract formulas which are identical to the concrete model. These concrete parts compose exactly the clauses of $TimeDebug_p^q(N)$, so it too is SAT under $\mathcal{A}$ with active suspect variables $\mathcal{E}$. $\square$

Theorem 1 *Let $sols_{abs}$ be the set of confirmed debugging solutions returned by iteratively debugging and refining $Abs\hat{D}ebug_p^q$ and let $sols_{time}$ be the set of debugging solutions returned by $TimeDebug_p^q$. If the final refined abstract instance $Abs\hat{D}ebug_p^q$ is UNSAT by blocking all solutions in $sols_{abs}$, then $sols_{abs} = sols_{time}$.*

PROOF. From Lemma 1, we know that $sols_{abs} \supseteq sols_{time}$. From Lemma 2, we also know that any set of debugging solutions $\mathcal{E}$ found and verified by $Abs\hat{D}ebug_p^q$ is also valid for $TimeDebug_p^q$ i.e., $sols_{abs} \subseteq sols_{time}$. Therefore, $sols_{abs}$ is both a superset and subset of $sols_{time}$, so $sols_{abs} = sols_{time}$. $\square$

Checking Architectural Outputs Instruction-By-Instruction on Acceleration Platforms

Debapriya Chatterjee†, Anatoly Koyfman‡, Ronny Morad‡, Avi Ziv‡ and Valeria Bertacco†

†University of Michigan
Ann Arbor, MI
{dchatt,valeria}@umich.edu

‡IBM Research Lab
Haifa, Israel
{anatoly,morad,aziv}@il.ibm.com

ABSTRACT

Simulation-based verification is an integral part of a modern microprocessor's design effort. Commonly, several checking techniques are deployed alongside the simulator to detect and localize each functional bug manifestation. Among these, a widespread technique entails comparing a microprocessor design's outputs with a golden model at the architectural granularity, instruction-by-instruction. However, due to exponential growth in design complexity, the performance of software-based simulation falls far short of achieving an acceptable level of coverage, which typically requires billions of simulation cycles. Hence, verification engineers rely on simulation acceleration platforms. Unfortunately, the intrinsic characteristics of these platforms make the adoption of the checking solutions mentioned above a challenging goal: for instance, the lockstep execution of a software checker together with the design's simulation is no longer feasible.

To address this challenge we propose an innovative solution for instruction-by-instruction (IBI) checking tailored to acceleration platforms. We provide novel design techniques to decouple event tracing from checking by including specialized tracing logic and by adding a post-simulation checking phase. Note that simulation performance in acceleration platforms degrades when increasing the number of signals that are traced; hence, it is imperative to generate a compact summary of the information required for checking, collecting and tracing only a few bits of information per cycle.

Categories and Subject Descriptors

B.6.3 [**Logic Design**]: Design Aids—*Verification*

General Terms

Design,Verification

Keywords

Simulation Acceleration, Checking, Checking on Acceleration

1. INTRODUCTION

Verification remains one of the most challenging and time consuming activities in the modern microprocessor design process. Shrinking transistor sizes have enabled a massive increase of microarchitectural complexity in microprocessors over the past decades. As a result, the verification effort needed for these designs has also increased tremendously. Simulation-based validation is still the workhorse of verification in the industry: a large collection of test regression suites are simulated on different models (architectural, RTL-level, structural) of the processor under verification, to check whether the design adheres to the original specification. To attain an acceptable degree of functional verification coverage for a modern microprocessor, billions of simulation cycles are executed

on each new revision of the processor in development. Clearly, the success of simulation-based validation is closely tied to simulation performance. Unfortunately, the performance of software simulation tools on complex designs, such as microprocessors, falls far short (1-10 cycles per second) of what is required to complete validation in a reasonable amount of time. Hence many functional verification teams in the industry rely on acceleration and prototyping platforms to meet their verification performance needs. However, even if simulation performance is much higher on these platforms, checking the functional correctness of a complex system, such as a processor design, presents many new challenges:

Mapping checkers to acceleration platforms: Many software-based checkers designed for microprocessor validation are expected to execute in lock-step with the RTL or gate-level simulation of the processor model. However, acceleration platforms can only simulate synthesized logic descriptions. Hence, if a software checker is sufficiently complex that it cannot be easily mapped into hardware, the checking solution cannot be brought onto the acceleration platform. Lock-step execution of software checkers in the host is infeasible since it would require frequent transfers of values from the platform and thus hinder performance unacceptably.

Acceleration performance: The performance of acceleration degrades when increasing the number of recorded/monitored signals or events. This effect is present even in software simulation-based validation; however, it is much more prominent in acceleration. Indeed, tracking a large number of signals in acceleration can degrade its performance to the point of cancelling its benefits over software simulation (as discussed in Section 4). Given an accelerator architecture and a design mapped to it, it is possible to estimate the slowdown incurred from the number of bits being traced. Thus, checkers that require to monitor a large number of signals cannot be adopted in acceleration in a straightforward fashion.

In the wake of these challenges there is a growing need of innovation to transform traditional microprocessor checking methodologies so that they fit in the constraints of accelerator/prototyping platforms while still delivering comparable verification quality and accuracy as their software simulation-based counterpart.

In this work, we present an architectural checking solution for microprocessor cores on acceleration platforms. Our solution performs what we call "instruction-by-instruction" (IBI) checking, that is, it verifies the outcome of each instruction completed in the accelerated simulation by comparing it with an architectural golden model. We achieve our goal by applying a number of major transformations to a baseline software simulation-based validation methodology. We solve the challenge of mapping complex logic to the platform by decoupling the recording of events from their checking. We also address the challenge of poor performance due to large data tracking by computing a summary of the information required by the checker before transferring the data off-platform. The proposed solution retains almost all the capabilities of its software counterpart but does not compromise the performance of acceleration. We successfully deployed and evaluated this solution in the validation of an upcoming IBM POWER processor.

2. RELATED WORK

Simulation accelerators and emulation platforms have been traditionally used to boost the productivity of the microprocessor validation effort [8, 12], and they play an even more critical role today, in light of the increased complexity of these designs. However, acceleration-based flows usually have a coarse checking granularity, that is, they can label a test as passed or failed after its completion but, in case of failure, no additional information is available related to the time/location of the bug manifestation. Comparing architectural state between a purely software-simulated design model and a golden architectural software model at instruction boundaries, or at other synchronizing boundaries, has also been a commonly deployed method for microprocessor validation[14, 5]. The key reason why this methodology has not yet been considered for acceleration, with the golden model running in software on a host platform, is that connecting these two components (golden model and accelerated design) is both difficult (due to lack of debugging support) and detrimental to performance [5]. Obtaining scan values from a silicon prototype and comparing them to a RTL golden model to detect divergence analysis during post-silicon debug has been proposed in [4]; however, this solution is only used to diagnose electrical faults.

More recent silicon-debug solutions, such as IFRA [11], introduce additional logic into the design to trace the flow of an instruction through various microarchitectural blocks and use this information with a post-simulation analysis tool to locate the manifestation of a possible design bug. Though our solution has a similar organization, *i.e.*, decoupled tracing and checking components, we are interested in the manifestation of a failure in the architectural state. Moreover, IFRA cannot detect divergence of the processor execution from the ideal model on its own, in fact it relies on post-triggers for this information. Our solution is focused on detecting the first point of divergence in the architectural state, hence it solves an orthogonal problem. Certain runtime verification techniques such as DIVA [2], introduce a lightweight companion processor to check the architectural state of the main processor, but these solutions operate at runtime, past design debug. Standards for hardware support for a variety of trace and control instrumentation for system debug has been proposed for the embedded domain [10], though they are not meant to be used for debugging the processor core design.

3. IBI BACKGROUND

Instruction by instruction (IBI) checking, or golden model based validation, is a well known checking technique that has been used in processor verification for many years [9, 13]. IBI compares the architectural events produced by each executed instruction with those required by the processor specification. This technique provides a simple way to distinguish deviations from the desired behavior. It does not depend on the internal implementation of the processor, and can be used with any microarchitecture implementing the same instruction set. An additional benefit of this approach is the relative ease of debugging: the corresponding checker recognizes the exact spot of the deviation in time and thus it enables the time localization of the problem.

A typical IBI checking methodology works as follows. A test generator (*e.g.* [7, 1]) produces a test program containing the results expected by the processor specification after each instruction (the expected results). These results are usually obtained using a software that can calculate the expected results after each instruction, known as a golden model. Then the checker environment compares these results to the ones produced by the processor simulator for the same test program [9, 13]. The checker environment needs to identify when an instruction execution completes and what resources were modified because of the instruction execution. It

also needs to account for the behavior that cannot be predicted by the golden model (*e.g.* external interrupts), or are not fully defined by the specification (*e.g.* values of some registers become "undefined" when exceptions occur).

4. ACCELERATION BACKGROUND

To boost the performance of simulation, a number of platforms have recently attracted interest as alternatives to software-based simulation: acceleration [6], emulation / proto-typing platforms [3] and post-silicon validation [11]. Hardware-accelerated simulation platforms are composed of large arrays of customized ASIC processors, specifically designed to simulate logic gates concurrently. To target these platforms, the design under verification (DUV) must be synthesized into a structural netlist, and then the corresponding logic gates are mapped to the execution substrate. Acceleration platforms have limited logic capacity, and even within their capacity limit, they may experience a performance penalty for large designs and depth of simulation. This logic capacity limit prohibits the mapping of any arbitrary checking solution into equivalent hardware and simulating it alongside the design. Thus, only checkers that result in low logic overhead can be tolerated. In our case, we evaluated our solution on a single core of an upcoming POWER processor, which in itself fit within the capacity limit of the accelerator used in our evaluation. The limit was also maintained after the addition of the logic required by our technique.

Acceleration platforms usually allow the collection and transfer of waveforms for debugging purposes [6], but the transfer slows down the simulation, eroding the key benefit of acceleration. In general, the more signals are observed and transferred, the lower the acceleration performance. However, the precise relation between acceleration performance impact and signals traced depends on the architecture of the accelerator. Reducing the number of recorded signals per cycle (thus the trace data generation rate) is extremely important for a successful checking solution for acceleration platforms. Emulation platforms have very similar trade-offs, except that the hardware acceleration fabric consists of programmable look-up tables (FPGAs). Hence, our solution could be adapted to emulation.

5. IBI FOR ACCELERATION PLATFORMS

In this section we present our instruction-by-instruction checker solution for acceleration platforms. Our technique enables this validation methodology on fast accelerated simulations, thus boosting the amount of simulation cycles that can be checked within a given amount of time. In our solution, we run the same test on the processor model simulated in the acceleration platform and on the golden model running on the off-platform host, and then compare results. To make the comparison possible, we need to collect relevant information about the retired instructions and architectural resources modified from the acceleration platform, and transfer it off-platform. The actual comparison is then performed by a dedicated software checker, capable of running the golden model on the same test and compare the two sets of results. As mentioned earlier, the acceleration advantage decreases when increasing the amount of recorded information and the size of simulated logic. Hence, one of our design goals is to record as little information as possible and incur as little hardware overhead as possible, all while delivering accurate bug detection capabilities.

Based on the observations above, our solution comprises the following two components: i) a dedicated, on-platform, logic block to record a compact summary of architectural events and ii) an off-platform software checker module that considers the recorded data and analyzes it in light of a golden model output. This decoupled approach enables us to get around one of the fundamental challenges discussed previously, minimizing on-platform logic

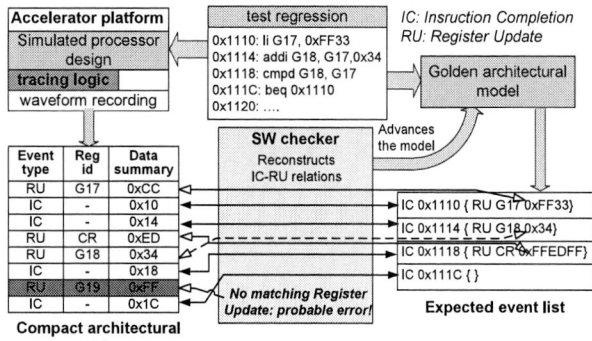

Figure 1: Overview of our solution to provide IBI checking on acceleration platforms. The Figure illustrates the test running on the platform (left) and on the off-platform software (right). The bottom left table shows an example of data transferred off-platform.

overhead. However, it also imposes a substantial redesign of the checking approach. We will check instruction completions and registers only (similar to many other IBI solutions) because memory behavior is very difficult to trace and predict in modern architectures. To achieve this we will record two types of events on the acceleration platform - instruction retirements and register updates. We do not focus on memory behavior, as it is common for other golden model solutions, since that requires specialized solutions beyond the scope of this work. We then compress the collected information on-platform to minimize the amount of data transferred. As a result, we must only record and transfer a few bits per cycle, thus maintaining the acceleration performance advantage. The on-platform tracing logic is simulated along with the processor in the acceleration platform. To minimize data recording, we do not track information that ties registers to a specific instruction; instead, we rely on the off-platform software, to reconstruct these connections based on the information recorded. Figure 1 presents an overview of our solution showing the components on the accelerator and on the off-platform software. It also outlines the type of data that is traced and transferred.

5.1 Hardware-based data tracing

From a high level standpoint the collection of information for our purposes appears to be straightforward; however, when applied to an industry processor, many aspects become challenging. The processor in question is a modern, server class, superscalar out-of-order processor with simultaneous multi-threading allowing 8 simultaneous threads per core. Hence, each architectural event is a complex combination of several microarchitectural events. To correctly identify and log individual architectural events, we need a number of microarchitectural monitor points, mapped together with the design onto the accelerator. The main architectural events to be collected for our purposes can be grouped into the following 3 major classes:

Instruction completion: Since the underlying processor is out-of-order, we can only obtain a finalized instruction retirement event when an instruction is committed. This information is gathered from the group completion table of the processor design, where instruction completion events are built from micro-operation completion information.

General purpose register activity: This group of registers includes integer general purpose registers (GPR), floating point registers and vector registers (VR). Accessing update events and values incurs an additional layer of indirection due to register renaming deployed in out-of-order microarchitectures.

Special purpose register activity: Special purpose registers (SPR),

such as several status registers, are easier to handle, since they are directly mapped and have explicit signals that identify a write to a special purpose register. We chose to collect information on a subset of special purpose registers that are either part of or closely related to the architectural state.

Note that we record all instruction completion events and all update events on the monitored registers. However, we perform lossy compression on the data associated with each event, *i.e.* completed instruction addresses or values written to a register, to reduce the number of bits recorded on the acceleration platform.

5.2 Off-platform software checker

As discussed in previous sections, our instruction-by-instruction checker strives to identify all discrepancies between the simulated processor behavior and its golden model. A processor's architectural state is defined by the values of the architectural registers (including general purpose registers, certain special purpose registers that affect execution flow and program counter) and the contents of memory. We assume that events that are not captured by the golden model (such as memory updates due to shared memory) do not appear in the test case. Thus, our single core processor model can be considered to be executing correctly, as long as program flow and architectural state are identical to that of the golden model. Hence, tracking the completion of instructions (program flow) and any modification to architectural registers is sufficient to check the correctness of execution. We encountered two key challenges in developing the off-platform checker, discussed below:

Reconstruction of instruction flow: A significant problem we had to address was the lack of close time correlation between an instruction retirement and its register events. This information cannot be reconstructed simply from the acceleration trace. Thus, in our solution we maintain a list of all registers that should have been modified by a completed instruction. We expect that for each such register, the first modification report that appears after the completed instruction will contain the correct value, and this report will appear within a bounded number of cycles. This solution is based on the assumption that registers are modified only after the corresponding instruction completes, and all associated register modifications are reported within a bounded number of cycles. However, we also had to consider the case where a register update is received before its corresponding instruction completion: in this case we must search for a matching event from the golden model over a few instructions downstream. If we do not find the matching event within a few instructions, we flag an error. We have run experiments to compare the results reported by a state-of-the-art software-based IBI checker to the results reported by our solution. We learned that the only difference lies in identifying which instruction is the root of the execution path deviation from the golden model execution (when such deviation exists). Our checker may report an instruction that is close to the actual deviating instruction (usually the next instruction), which we found satisfactory for effective debugging.

Handling interrupts for checking purposes: External interrupts and other non-deterministic events are not predictable by the golden architectural model; however, they are still included in the acceleration traces. External interrupts can still be identified from the address of the corresponding interrupt handler and specific values of the related control registers. Our solution mimics the effect of the interrupt routine by modifying the associated status registers and other architectural resources in the golden model and then it resynchronizes the model with the trace.

6. DATA COMPRESSION

As discussed in Section 5.2, a central goal of our work is to keep the amount of data recorded per cycle at a bare minimum, to maintain the performance advantage of acceleration, while still providing acceptable detection accuracy. To this end, we compress the data associated with each event, such as register update values and addresses of completed instructions. A lossy compression scheme, such as a checksum is ideal for this purpose, since we are only interested in identifying value deviations. So, as long as a different value produces a different checksum with high likelihood, it serves the purpose. Moreover, another important aspect in the development of our solution, is that the additional hardware required to implement the compression scheme should have minimal logic overhead and minimal logic depth. Hence, a compression scheme that involves little additional logic and does not add substantial delay to the critical path is favored over a more complex scheme.

6.1 Register update values

Value discrepancies in register updates can often be discerned using a checksum over a small subset of the bits, without requiring a complete value comparison. We strive to use only a few (say, less than 8) bits of encoded information for each register value field (32 bit / 64 bit). The basic idea is to compute a checksum from the value generated by the simulated hardware and perform the same operation on the value generated by the reference model for each register update in the software checker. For the sake of our checker solution, a checksum match is considered a valid register update. Since all checksum schemes are a hash function from a set of size 2^{64} (for 64 bit registers) to a set of size 2^c, where c is a small value, some amount of aliasing is unavoidable. However, we found that blocked parity schemes, presented below, provide sufficient accuracy in practice for the typical error scenarios that we encountered.

Blocked parity schemes partition the data vector into several distinct blocks and then compute single bit checksums for each block. The concatenation of these bits provides the final checksum. This approach is guaranteed to detect any bit value difference, as long as the number of single bit errors within each block is odd. A benefit of this approach is that its computation is extremely low cost in hardware, simply requiring a few XOR gates. However, this approach is ineffective for scenarios where errors manifest with an even number of localized bit-flips, which may occur all within one, or a few, blocks. To address this situation we build blocks on non-contiguous bits, scattering the bits over the checksum blocks. With this technique, an error affecting a few contiguous bits has a much higher chance of detection. The experimental evidence presented in Section 8 supports this intuition.

6.2 Retired instruction addresses

The data associated with each retired instruction is the address of the committed instruction. To compress this values we use a very simple scheme, recording only the last few bits of the address. Even though this scheme is prone to aliasing, it works very well in practice. Indeed, it allows us to identify an execution divergence from the golden model fairly precisely, since the probability of execution starting at an aliased address leading to the same sequence of register updates as the correct execution is extremely low.

7. ON-PLATFORM TRACING UNIT

As discussed earlier, there are several types of data collected on the acceleration platform originating in different regions of the design at a variable rate. To manage this flow of data, we developed a novel unified scheme to collect and organize it for on-platform storage, before it can be transferred off-platform. To this end, we first need a mechanism to identify which registers are updated on a particular cycle or which instruction groups have completed, so that we only record new values for the relevant registers/addresses. Second, we need a mechanism to present this data in a structured fashion, so that it can be recorded efficiently by the acceleration platform's data logging mechanism. We note that, although the

maximum number of simultaneous events in a clock cycle can be quite high, the average number of events per cycle is fairly small. Hence, a recording mechanism that can handle transient peaks in the number of events and can present data at a constant rate to the platform's debug support unit would be ideal. A possible solution to this second requirement is a first-in first-out buffer that allows the storing of up to a few entries at a time and it is drained at a constant rate. This section discusses how we achieved these requirements.

7.1 Select and encode logic

The first task of the tracing unit focuses on selecting and encoding different types of events as they are flagged during a clock cycle. In the platform there are a number of data lines and corresponding valid lines coming from different parts of the processor and corresponding to different special purpose registers or instruction completion events that we want to track. Our goal is to be able to store the relevant data at each cycle (as signaled by the corresponding valid lines) while also tracking the correct source for the data. By doing so the off-platform software is able to reconstruct the sequence of events to be checked against the golden model.

The goal of the select and encode logic unit can be formally expressed as follows: given a collection of N signal lines, presented as an ordered list, up to any M lines among those can request data logging on any given clock cycle. The task of this unit is to identify and encode the position in the list of the M lines in preparation for storing them along with the data itself. Ultimately, these positions will be used to identify the source of the corresponding data value. This problem is also known as the "detect and encode all ones" problem: one straightforward solution would be to use a chain of priority encoders: the first encoder is responsible for the highest order position, which is then masked and the entire vector of N lines is passed down to the next encoder. While simple, this solution creates a deep combinational logic block, which could hamper the performance of acceleration.

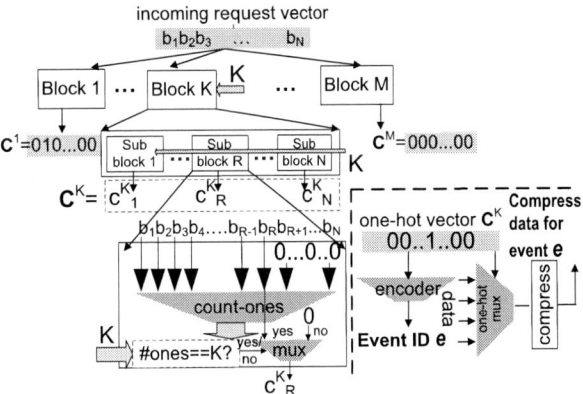

Figure 2: **Detector block to identify the source of data** to be logged in a given clock cycle of simulation acceleration.

Our goal in developing this unit is to develop a design that is most suited for acceleration platforms, even if it may entail a non-minimal area footprint in silicon. To this end, we devised an alternative solution, that has a much smaller logic depth. Our solution uses a parallel detection scheme, where each detection block is responsible for generating a one-hot encoded vector corresponding to the line position for which the block is responsible, if that line has data available. If no logging data is generated from that line during a cycle, the block should simply output a vector of zeros. Figure 2 illustrates our solution: we use M detection blocks, since we have at most M lines generating data within one cycle. Each block receives in input a value K, and generates a one-hot encoded vector where the 1-bit is in the position of the K-th line producing data in that cycle. For instance, if during a cycle lines 4, 7 and 11 produce data to be logged, then block 1 should have a one in position 4, block 2 should have a one in position 7 and block 3 should have a one in position 11.

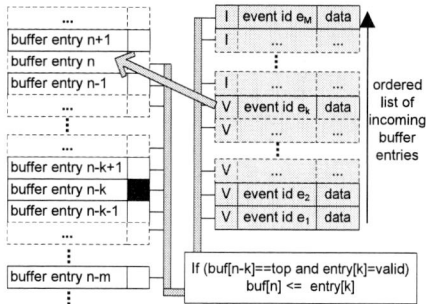

Figure 3: **Trace buffer writing unit**. Each buffer entry is associated with a writing unit. Each unit determines which data logged in the cycle should be stored in the position for which it is responsible.

7.2 Trace buffer

Once the relevant data has been selected and encoded for logging, we need a hardware block to record the architectural events. To this end we use a trace buffer that must be capable of handling up to M entries in each clock cycle, while allowing a constant R entries to be read. Such a buffer is typically realized via a circular buffer with read and write pointers. However, multiplexors are needed to realize these pointers. Unfortunately, they also increase the logic depth of the design, particularly when the number of buffer entries is large. Hence, we adopted an alternative design, where the buffer is implemented as a shift-buffer, so that the constant number of read operations in each simulation cycle corresponds to a constant number of shifts. A bit is associated with each entry to indicate the first free entry, and independent write units are associated with each buffer entry. Each write unit has access to its corresponding entry and the M preceding ones, and it determines what to write in its entry based on the number of write operations to be completed in the cycle. This design is shown in Figure 3: the implementation is parallel and logic depth is kept minimal.

8. OPTIMIZING DESIGN PARAMETERS

In this section we discuss a number of analyses that we conducted to optimize our checker design. The most influential parameter for the performance of our checker is the number of bits traced per cycle. This parameter is determined by the product of two other parameters: (i) the number of trace buffer entries being drained per clock cycle and (ii) the number of bits per entry in the trace buffer. The first one has to be equal or greater than the average rate of traced events generated by the processor, since we are using a finite size buffer. The other is determined by the number of bits required to describe the event type (which is a fixed value), along with the number of data-bits associated with each entry (a variable value). Below we discuss how we computed the near ideal value for each of these components.

Trace buffer size: Since the number of write operations to the buffer varies from cycle to cycle, while draining rate remains constant, we need to ensure that the average draining rate is higher than the average generation rate. Even then, bursts of generation may create backlogs in the buffer. In Figure 4 we show how different buffer draining rates affect instantaneous buffer occupancy. We derived a queuing theory-based estimate for our buffer size, which

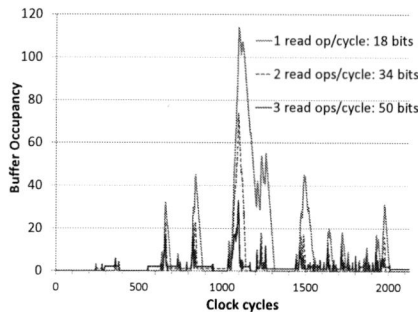

Figure 4: **Buffer occupancy peaks at varying rates of draining**.

ensures a very low probability of overflow, while using the lowest possible draining rate. Assuming a pessimistic generation rate, a draining rate of 3 entries/cycle was found to be sufficient and a corresponding buffer size of 512 is adequate.

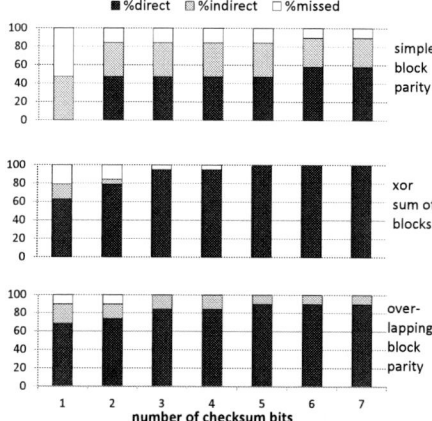

Figure 5: **Detection accuracy of a range of checksum schemes**. Register value discrepancies can be either detected at register update (direct), or in downstream computation (indirect), or missed.

Checksum width: The number of bits recorded per buffer entry is dictated by the number of data bits. We want to store a minimal number of bits in the checksum, while still detecting value discrepancies caused by a functional bug. Hence, we investigated the detection accuracies of several blocked parity schemes, as described in Section 6.1, over buggy traces diverging on a register value update.

To this end, we varied the number of checksum bits from 1 to 7, while the original register values are 64-bits wide. We studied three different checksum schemes as reported in Figure 5, and estimated the minimum checksum bit width required to detect typical value discrepancies. The schemes we evaluated are: (i) Simple blocked parity, where a single parity bit is computed from each portion of register data and appended to the final checksum. (ii) XOR sum of blocks, where the checksum is obtained by applying bitwise XOR to all same size sub-blocks of the register value. (iii) Overlapping block parity, similar to (i), but with overlapping partitions. The sample size for this study was 500 traces with register value corruptions similar to those of actual buggy traces. From Figure 5, it can be gathered that typical discrepancies can be detected with as little as 5 bits of XOR sum of blocks. This allows storing only 8 (event-type ID, one out of 256) +3 (thread ID) +5 (checksum)=16 bits in each trace buffer entry. In addition to this we have 1-bit indicating that the values at the buffer output are valid and another 1-bit to indicate buffer overflow. These 2 bits are constant overhead irrespective of number of entries read.

959

9. EXPERIMENTAL RESULTS

Our solution was implemented for an upcoming POWER processor core design on the AWAN accelerator [6] platform. We evaluated the capability of our solution to detect bugs as well as its performance. The IBM SixthSense tool-chain was used to design and synthesize the hardware blocks for our solution. The processor core netlist consisted of a few million logic gates, and the resulting logic overhead was within 20%.

9.1 Bug detection capability

Any discrepancy of the processor's behavior from the golden architectural model due to a probable functional bug is detected as one of the following situations (symptoms) by our IBI checker:

1. Register value mismatch: Updated value of a register does not match with predicted value from golden model;
2. Unexpected register update: An architectural register update event takes place in the design but not in the golden model;
3. Unaccounted register update: A register update event takes place in the golden model but does not occur in the design;
4. Wrong instruction: The instruction address of an executed instruction is in disagreement with the golden model;

We obtained a set of 145 architectural event traces that exposed actual functional bugs. These 145 constituted the entire set of buggy traces that we had access to. To evaluate the bug detection capability of our checker, we ran the same traces on our off-platform software checker to determine if our accelerator-based checker could also detect the occurrence of the bugs. All 145 test-cases exposed a bug in our setup; in addition the symptoms reported matched those of the software-based golden model solution. We report in Table 1 the distribution of the bugs detected according to the type of symptom flagged by our checker. As it can be noted, a large portion of the issues are due to unaccounted/unexpected register updates. All these problems were detected within 5 instructions from the first point of golden model/accelerator divergence.

Symptom	#occurences
Register value mismatch	21
Unexpected register update	30
Unaccounted register update	89
Wrong instruction	5

Table 1: **Distribution of bugs detected by our solution**.

Since we do not compress the information regarding which architectural register (among the monitored subset) is updated, we detect all discrepancies that are not affected by checksum aliasing. However, even in this latter case, often the program flow diverges substantially due to the bug, and we can still flag the issue a few instructions downstream.

9.2 Tracing overhead

The amount of logic added for on-platform tracing purposes may impact the performance of the simulation. However, this is only the case if the overall logic size mapped to the platform (design + checkers) exceeds a certain threshold, dependent on the accelerator's characteristics. When abiding this threshold, the performance degradation due to the tracing logic comes from two sources (i) additional logic to simulate (ii) signal recording time.

To evaluate these effects, we measured the simulation acceleration performance of the POWER core design in several situations. The stimuli used for this study were regression tests lasting few million cycles. First, we run a baseline design with no tracing logic. Then we added the tracing logic, but without observing the trace buffer output. Then we also enabled tracing for the typical case,

that is, 3 buffer entries are read per cycle, amounting to 50 bits of recorded information per cycle. Finally, we considered an extreme situation where 10 buffer entries are read per cycle, for a total of 162 bits. Figure 6 summarizes our findings, normalized to the simulation performance (between 10-100 kHz) of the baseline design with no tracing logic.

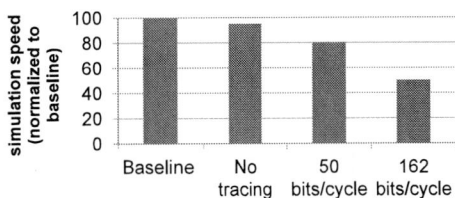

Figure 6: **Impact of tracing logic** on acceleration performance.

From Figure 6 we gather that our solution introduces only a 5% slowdown due to the tracing logic alone, and another 15% due to data logging. Even the extreme situation causes no more than a 50% slowdown in acceleration performance, a value still order of magnitudes better than software-based simulation.

10. CONCLUSIONS

In this work we presented a novel microprocessor design checking solution that provides architectural checking against a golden model for simulation acceleration. Our solution provides the same bug detection quality as its software-based counterpart. It enables architectural validation on acceleration platforms with negligible accuracy loss and moderate performance loss of approximately 20%.

Acknowledgments

This work was developed with partial support from the Air Force Office of Scientific Research, Grant #FA9550-09-1-0164. The authors would also like to thank Ariel Birnbaum, Brian Mestan, Chris Abernathy and Kenneth Ward from IBM.

11. REFERENCES

[1] A. Adir, E. Almog, L. Fournier, E. Marcus, M. Rimon, M. Vinov, and A. Ziv. Genesys-Pro: Innovations in test program generation for functional processor verification. *IEEE Design & Test*, 21(2), 2004.

[2] T. M. Austin. DIVA: A reliable substrate for deep submicron microarchitecture design. In *Proc. MICRO*, 1999.

[3] M. Boulé, J.-S. Chenard, and Z. Zilic. Adding debug enhancements to assertion checkers for hardware emulation and silicon debug. In *Proc. ICCD*, 2006.

[4] O. Caty, P. Dahlgren, and I. Bayraktaroglu. Microprocessor silicon debug based on failure propagation tracing. In *IEEE Transactions on Computers*, 2005.

[5] Y.-S. Chang, S. Lee, I.-C. Park, and C.-M. Kyung. Verification of a microprocessor using real world applications. In *Proc. DAC*, 1999.

[6] J. Darringer, E. Davidson, D. Hathaway, B. Koenemann, M. Lavin, J. Morrell, K. Rahmat, W. Roesner, E. Schanzenbach, G. Tellez, and L. Trevillyan. EDA in IBM: past, present, and future. *IEEE Transactions on Computer-Aided Design of Integrated Circuits and Systems*, 19(12), 2000.

[7] L. Fournier, Y. Arbetman, and M. Levinger. Functional verification methodology for microprocessors using the Genesys test-program generator. In *Proc. DATE*, pages 434–441, March 1999.

[8] G. Ganapathy, R. Narayan, C. Jorden, M. Wang, and J. Nishimura. Hardware emulation for functional verification of K5. In *Proc. DAC*, 1996.

[9] J. M. Ludden et al. Functional verification of the POWER4 microprocessor and POWER4 multiprocessor systems. *IBM Journal of Research and Development*, 46(1):53–76, 2002.

[10] A. Mayer, H. Siebert, and K. McDonald-Maier. Boosting debugging support for complex systems on chip. *Computer*, 40(4), 2007.

[11] S.-B. Park, T. Hong, and S. Mitra. Post-silicon bug localization in processors using instruction footprint recording and analysis (IFRA). *IEEE Trans. on CAD*, 28(10), 2009.

[12] V. Popescu and B. McNamara. Innovative verification strategy reduces design cycle time for high-end SPARC processor. In *Proc. DAC*, 1996.

[13] D. W. Victor et al. Functional verification of the POWER5 microprocessor and POWER5 multiprocessor systems. *IBM Journal of Research and Development*, 49(4), 2005.

[14] J.-S. Yim, Y.-H. Hwang, C.-J. Park, H. Choi, W.-S. Yang, H.-S. Oh, I.-C. Park, and C.-M. Kyung. A C-based RTL design verification methodology for complex microprocessor. In *Proc. DAC*, 1997.

Supplementary material

S1: Insights on micro-architectural to architectural event translation

Instruction completion: To properly sample the information required for each instruction at its correct completion time, we leverage the fact that instruction retirement is performed in the program order. This remains true even when the processor has an out-of-order microarchitecture. This observation is utilized to create instruction completion events from the retirement of an instruction at the reorder buffer. However, there is another layer of complexity associated with identifying instruction completion events. Even though the processor has a RISC architecture, for performance reasons each instruction is sub-divided into micro-operations. At the retirement stage, the actual visible events are micro-operation retirements. Hence, to generate retirement events at the architectural level, micro-ops completions are collapsed together based on the directives of the instruction dispatch table. The associated instruction address is also gathered from the instruction dispatch table.

General purpose registers: All general purpose architectural registers are dynamically mapped to registers in the physical register file; hence, a map entry is associated to each register update event, and it provides the architectural index along with the tag that indexes the written value in the physical register file. From this information, we collect the most recent written value to an architectural register on the notification of a write event. This is achieved using a significant amount of decoder and multiplexing logic.

S2: Select and encode logic - insights

As mentioned in Section 7.1, a straightforward solution to this problem entails using a cascade of priority encoders. However, combinational implementations of this approach would lead to an unacceptable logic depth, while sequential solutions would require more than one clock cycle to compute the output value.

Though the cascade solution and some slight variations of it (optimizing the detection part but retaining the cascaded logic structure) describes a correct logic implementation, the problem that remains is the enormous logic depth required for implementing even a small cascade depth. This would lead to forming a very long critical path in the logic and would severely limit simulation performance in the accelerator. Solutions that employ sequential logic to get around the cascading problem exist but we can not use this solution since we are not assured to have a fixed number of idle cycles where absolutely no request arrives, we need to finish detect and encode in one cycle only.

In contrast, our solution, shown in Figure 2, is combinational with a small logic depth. In the Figure, we use M concurrent detection blocks, each receiving in input a distinct value K. Each block outputs a one-hot vector with a 1 in position of the K-th input line producing data in that cycle. Blocks for which the input value K is larger then number of data lines producing data will simply output a zero vector.

To achieve this functionality, each block is organized into N sub-blocks in order, one for each of the incoming data lines. The input value K is passed along to all sub-blocks. The one sub-block connected to the line with the K-th data value must output a 1 is there is a new value on the line, and a 0 otherwise. All other sub-blocks simply output 0. In order for a sub-block to determine if it is connected to the line with the K-th data value, it counts the number of lines providing data values in line indices lower than its own

order position. The only logic required to implement this solution includes a tree of few-bits adders and a comparator of only $\log_2 M$ bits for each sub-block.

S3: Trace buffer size selection

To determine an adequate size for the trace buffer, we applied the following reasoning. Let's call I the average number of instructions completed per clock cycle (*i.e.* IPC) and W the average number of register updates per completed instruction (including SPR updates). Then, the average number of entries generated per cycle is: $G = I \times (1 + W)$. The draining rate of the tracebuffer (D) has to be greater or equal than the generation rate to avoid overflow, *i.e.*, $D \geq G$. In addition, we need to take burst into account: if the average of G over a window of T clock cycles is $G_T > D$, then a backlog of $(G_T - D) \times T$ entries has been created, which the buffer must accommodate. As it can be noted, the buffer always drains at steady state; however, its instantaneous behavior maybe adversely affected by small draining rates. Our ideal situation is to have a minimal draining rate and experience overflow rarely.

To determine the probability of overflow we use a theoretical model from queuing theory: our problem can be modeled as a variant of a M/M/1 queuing model, where the average number of writes per cycle is the arrival rate λ and the service rate μ is the number of entries drained per clock cycle. This model holds since the number of events reported per clock cycle can be considered as a Poisson arrival process with mean rate $\lambda = G$, whereas the average rate of departure is $\mu = D$. Now, the probability of queue occupancy q exceeding a given value Q, $P(q > Q)$ is given by $(\frac{\lambda}{\mu})^{Q+1}$. M/D/1 is a more accurate model since our draining rate is constant; however, since there is no closed form expression for this case, we conservatively approximated it with M/M/1. We can plot this probability as a function of buffer size, for the worst case generation rate. To obtain the worst case generation rate we used an artificial regression that dispatches a large number of independent ADD instructions for each of the 8 threads simultaneously. This forces the processor to operate at high throughput and approach the theoretical maximum of the average rate of event generation.

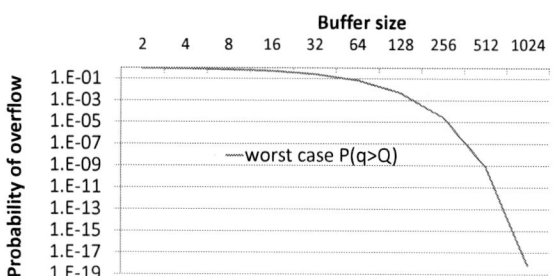

Figure 7: **Overflow probability for finite buffer size assuming M/M/1 queuing model.**

We observed an average instruction completion rate I of 1.2 per clock cycle, along with an average rate of register updates per instruction W of 1.0. We pessimistically assumed 0.4 SPR writes per instruction completion, leading to a total generation rate of $1.2 \times (1 + (1.0 + 0.4)) = 2.88$, which is then rounded to the closest larger integer, leading to a buffer draining rate of 3. With these parameters, we plotted the overflow probability in Figure 7 for different buffer sizes and concluded that a buffer size of 512 corresponds to a minimal probability of overflow.

Standard Cell Sizing for Subthreshold Operation

Bo Liu[1,2], Maryam Ashouei[2], Jos Huisken[2], and Jose Pineda de Gyvez[1]

[1] Department of Electrical Engineering, Technische Univ. Eindhoven, Eindhoven, NL

[2] Holst Centre/imec-nl, Eindhoven, NL

{B.Liu, J.Pineda.de.Gyvez}@tue.nl, {Maryam.Ashouei, Jos.Huisken}@imec-nl.nl

ABSTRACT

Process variability severely impacts the performance of circuits operating in the subthreshold domain. Among other reasons, this mainly stems from the fact that subthreshold current follows a widely spread Log-Normal distribution. In this paper we introduce a new transistor sizing methodology for standard cells. Our premise relies on balancing the N and P network currents based on statistical formulations. Our approach renders more robust cells. We observe up to 57% better performance and 69% lower energy consumption on a set of ISCAS circuits when they are synthesized with our library as opposed to a commercial library in a CMOS 90nm technology.

Categories and Subject Descriptors

B.7.1 [**Hardware**]: Integrated Circuits-Types and Design Styles

General Terms

Algorithms, Design, Reliability, Theory.

Keywords

Process variation, standard cell library, subthreshold design, transistor sizing

1. INTRODUCTION

Low voltage digital design, especially near/subthreshold design, is becoming more popular in application domains where performance is not the primary concern. More and more systems with low performance requirements are operated from a near/subthreshold supply voltage in order to save power [1]. However, because of the small gate voltage drive of the transistors operating in the subthreshold domain, standard logic cells tend to be very sensitive to process variations. Commercial cell libraries are designed and characterized for super-threshold voltage operations. Without any optimization, most cells of such conventional libraries will not have robust operations in the presence of process variability at a low operating voltage. We will show that, if the conventional sizing methodology is used for subthreshold sizing, we end up with a library that has much smaller drive currents, and more sensitive to process variations.

Several relevant research results have been presented about subthreshold sizing. In [1], the authors calculate the optimum supply voltage to minimize energy for a given frequency in subthreshold operation. It is also claimed that, theoretically, minimum sized cells are optimal for energy reduction. We show in this paper that under speed constraints and when process variability is taken into account this is not the case. The concept

of subthreshold logical effort for complex gate sizing is presented in [2]. Particularly interesting is a closed form current equation is derived for stacked transistors in relation to other transistors in the same stack. Compared to [1] and [2], our sizing focus is to narrow the current/delay distribution spread and to increase the performance through a new balancing theory that slows down fast transistors and vice-versa. In [3] the transistor reverse short channel effect (RCSE) is used for device sizing optimization. Essentially, the channel length is increased to have an optimal threshold voltage which makes the transistors have higher current, be less sensitive to random variations, and have smaller area. With higher current and less capacitance, the delay and power are both reduced. Moreover, in [3], the channel lengths of the nMOS and pMOS are increased to achieve the maximum currents for both nMOS and pMOS transistors. Unlike [3], our optimization does not always lead to the maximum active current for both the nMOS and pMOS transistors. Only the transistors on slower timing arc are allowed to be upsized, the ones on faster timing arc are down sized to save area.

Overall, in our work, we introduce a new statistical formulation to size standard cells. The differences of our work from other sizing methods are that we treat the threshold voltage variation as one of the statistical parameters in the current/delay equation and we optimize cells to have balanced current/delay distributions. Our sizing approach is derived from the observation that the transistor's current distribution in the subthreshold regime follows a Log-Normal spreading, whereas conventional sizing treats the transistor's current as a Normal distribution, which is only true for the super-threshold region. Considering the above mentioned fact and the observation that process variability can be mapped onto threshold voltage variability with a first order approximation, we develop a methodology for robust standard cell design. Our results indicate that it is possible to obtain standard cells which have smaller area, stronger drive current, and that are more robust when compared to the cells from a commercial 90nm library operating in the subthreshold regime.

2. TRANSISTOR IN SUBTHRESHOLD

Transistors operating in the subthreshold regime obey an exponential dependence on the gate drive voltage [3]:

$$I = \mu C \frac{W}{L} e^{1.8} U^2 e^{\frac{V_{gs}-V_{th}}{nU}} (1 - e^{\frac{-V_{ds}}{U}}) \qquad (1)$$

where μ is the mobility, C is the oxide capacitance, n the subthreshold slope factor, and U is the thermal voltage. V_{gs} is the gate to source voltage. V_{ds} is the drain to source voltage. V_{th} is the threshold voltage.

Fig. 1 shows the distributions of V_{th} and current in the subthreshold regime of a single nMOS transistor (W=0.3um, L=0.1um) in a 90nm CMOS technology, working from a supply voltage of 0.3V at 25°C. The results are obtained from 3000 Monte Carlo simulations with the foundry model (in the remaining of the paper we use the same commercial CMOS 90nm technology as our reference, and same foundry model for the Monte Carlo simulations). From these simulations it is evident

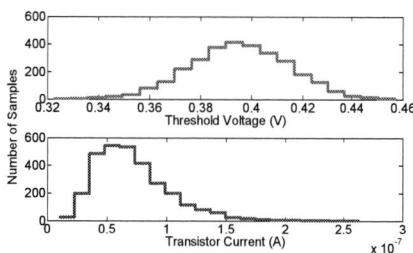

Figure 1. V_{th} and subthreshold current distribution

that the V_{th} distribution obeys a Normal distribution, and that the current obeys a Log-Normal distribution. Therefore the subthreshold current can be formulated statistically as [4, 5]

$$E[I] = \mu C \frac{W}{L} e^{1.8} U^2 e^{\frac{V_{gs}-E[V_{th}]}{nU} + \frac{Std^2[V_{th}]}{2(nU)^2}} (1 - e^{\frac{-V_{ds}}{U}})$$

$$Std^2[I] = (e^{\frac{Std^2[V_{th}]}{(nU)^2}} - 1)(E[I])^2$$

(2)

where $E[]$ stands for the mean value and $Std[]$ stands for the standard deviation. In this model we regard $E[V_{th}]$ and $Std[V_{th}]$ as given technology parameters (typically found in process technology package manuals).

Before we move on to our sizing methods, it is useful to understand the impact of short channel and narrow width effects for transistors operating in subthreshold. It is known that bigger transistors have less mismatch problems [6]. Consequently, transistors' width is conventionally upsized to reduce the impact of mismatch. We will illustrate that in subthreshold this is not always necessary. Without loss of generality, consider now an nMOS transistor operating at 0.3V and at 1.2V to investigate the spread of V_{th}, and active current I_{on} w.r.t. transistor width and length dimensions. The results are shown in Fig. 2.

In Fig. 2-a, we compare threshold variation trends at 0.3V and at 1.2V for different widths and lengths. One can see that, when the length and width increases, the standard deviation of V_{th} decreases. RSCE can also be observed in Fig. 2-a. The arrow indicates that transistors A and B have the same V_{th} variation but A is smaller than B. To minimize the threshold variation it is more beneficial to increase the length than to increase the width. This is because the resulting transistor area is smaller, besides that a smaller width positively affects the gate capacitance. Also observe that the gate-source capacitance grows linearly with transistor width, so this eventually translates into higher power consumption as a higher I_{on} is needed to drive the capacitance. Furthermore, wider transistors have a bigger I_{off} [7], which may be comparable to the I_{on} in subthreshold. When the transistor is not carefully upsized, it is easy to observe that the I_{on}/I_{off} ratio is not big enough to drive the increased gate capacitance.

In Fig. 2-b we show the differences between I_{on} and I_{off} for different areas at 0.3V TT corner. Note that the I_{off} of big

transistors can be greater than the I_{on} of the small transistors. Unless we size every transistor big enough to make sure that the maximum value of I_{off} does not exceed the minimum value of I_{on} within one circuit, we will have malfunction transistors in the circuit.

Consider now a simple inverter with FO4 loading. We show in Fig. 2-c a delay comparison when the inverter is operated at 0.3V and at 1.2V. We can see that increasing the transistor's width will obviously speed up the inverter, but increasing the length will also speed up the inverter at 0.3V. Also, as the arrow indicates, at 0.3V, transistors A and B have the same speed but A has smaller area which means when sizing a cell, including length tuning is more area efficient as compared to width tuning only.

In Fig. 2-d the RSCE effect at 0.3V is illustrated. As channel length is increased, $E[V_{th}]$ decreases and the $E[I]$ increases to reach a maximum value.

From Fig. 2, we can conclude that, upsizing will reduce the impact of variation and increase the speed of the transistors in subthreshold. But we do not necessarily need big transistors when considering the ratio I_{on}/I_{off} and the standard deviation of V_{th}. When upsizing transistors, increasing the length will provide transistors with smaller area and faster speed when compared to width tuning only. With the above understanding of the behavior of the transistors in subthreshold, we propose two sizing methods. One is a DC based current sizing method, which utilizes the reverse short channel effect and a new balancing theory to have equal pull-up and pull-down currents for different transistors. The second approach takes into account transition delays of different input patterns. In this case the worst rise and fall transition times are balanced and compensated by the best rise and fall time delay. In the next sections, we will show our sizing methods in detail.

3. SIZING FOR COMBINATIONAL LOGIC

3.1 Transistor Sizing for Balanced Current Distributions

In traditional CMOS design, the transistor geometry ratio (W/L) of the pull-up pMOS network to the pull-down nMOS network is carefully tuned to compensate for the difference between the mobility of electrons and holes. This ratio is derived with the objective of balancing the rise/fall-time delays and minimizing the propagation delay of a cascaded inverter chain.

In our sizing methodology, the ratio of the pull-up to pull-down transistors is determined by the balance between the current distributions of the pMOS and nMOS transistors. The difference w.r.t the conventional sizing approach is that the current spread caused by the V_{th} variation is taken into account. Based on this observation, the mean currents of the pMOS and nMOS networks should be equal, $i.e.$ $E[I_n] = E[I_p]$. From this we can derive:

$$\frac{W_n L_p}{W_p L_n} = \alpha e^{\frac{E[V_{thn}]-E[V_{thp}]}{nU}} e^{\frac{Std^2[V_{thp}]-Std^2[V_{thn}]}{2(nU)^2}}$$

(3)

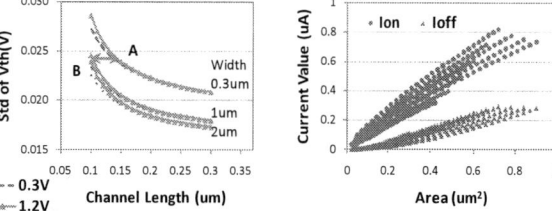

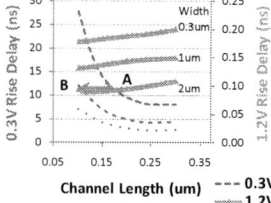

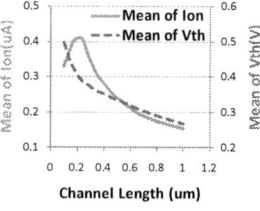

Figure 2-a. V_{th} variation comparisons b. I_{on} and I_{off} comparisons   c. Delay comparisons   d. RCSE at the same area 0.3um²

where $\alpha = \mu_p C_p / \mu_n C_n$ is a technology parameter defined by the mobility and oxide capacitance of the nMOS and pMOS transistors. α is also used as the conventional sizing factor. Given the V_{th} mean and variance values, Eq. (3) serves as our current balancing equation. The nMOS and pMOS current distributions can be closely matched based on Eq. (3).

Using Eq. (3), one can either maximize cell current with area constraints (say, maximum area K) or minimize cell area with current constraints (say minimum current J) as follows

$$
\begin{aligned}
Maximize \quad & E[I] \\
Subject\ to \quad & W_n L_n + W_p L_p \leq K \\
& E[I_n] = E[I_p]
\end{aligned}
\tag{4}
$$

or

$$
\begin{aligned}
Minimize \quad & W_n L_n + W_p L_p \\
Subject\ to \quad & E[I] \geq J \\
& E[I_n] = E[I_p]
\end{aligned}
\tag{5}
$$

Note that the area under the channel is used in (4) as an indication of the total cell area. During the optimization process, the space of transistor widths and lengths is divided into many intervals with a step of 0.01um. For each interval (i) with $[W_{i\text{-}min}, W_{i\_max}]$ and $[L_{i\text{-}min}, L_{i\_max}]$, the V_{th} has a normal distribution with $E[V_{thn,p}](i)$ and $Std[V_{thn,p}](i)$ parameters. For each interval i, we find (W_n, L_n, W_p, L_p) that satisfies Eq. (3) using an exhaustive search. Note that there might be multiple W, L's that satisfy Eq. (3). We consider all of them in the next step of the algorithm in which we found the sizes that also satisfy the area or timing constraints specified by Eq. (4) or Eq. (5) respectively. This is shown in Algorithm I.

Algorithm I:
$M \leftarrow \{\}$
For each *interval(i) of range of widths and lengths*
  Interval(i) $\leftarrow$ *$E[V_{thn,p}](i)$ and $Std[V_{thn,p}](i)$*
  $\left(\frac{W_n L_p}{W_p L_n}\right)_i \leftarrow$ Solve Eq. (3)
  $M \leftarrow M \cup \{ W_{n,i}, L_{n,i}, W_{p,i}, L_{p,I} \}$
End
Find (W_n, L_n, W_p, L_p) in M that satisfy Eq. (4) or Eq. (5)

As an example let us see the results of applying our optimization methodology to a simple inverter. We use an inverter from a CMOS 90nm commercial library to set the area and current constraints for Eqs. (4) and (5). Our Monte-Carlo simulation results are shown in Table I. Note that the area of the optimized cell is 1.5x smaller than the one of the library inverter when the mean current is the same. On the other hand, with equal area, the optimized inverter has 1.67x higher mean current when compared to the commercial library cell. The optimized inverter has 10% less variation than the library cell.

3.1.1 Transistor Sizing for Stack Topologies
Let us broaden the previous concept to series-connected transistors. The magnitude of the current flowing through the transistor stack depends on the number of the transistors and the

TABLE I INVERTER COMPARISON

	Same Mean Current		Same Area	
	Our Library	Commercial Library	Our Library	Commercial Library
Sizes(W/L) (um)	n:0.4/0.2 p:0.41/0.26	n:1.2/0.1 p:1.6/0.1	n:0.3/0.2 p:0.31/0.26	n:0.6/0.1 p:0.8/0.1
Area(um²)	0.19	0.28	0.14	0.14
E[I] (nA)	69.30	69.68	63.89	38.43
Std[I]/E[I]	32.12%	41.69%	31.27%	41.57%
100%Yield	27.2 ns	44.6 ns	27.4 ns	64.2 ns

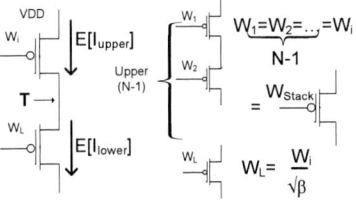

Figure 3. pMOS stack schematic

size of each transistor. Making the current distributions of any two consecutive transistors, $E[I_{upper}]$ and $E[I_{lower}]$ equal, one can solve the voltage at the common node T as in [2, 8], see Fig. 3.

Let us consider now a stack of N pMOS transistors lexicographically enumerated in descending order as a function of their proximity to the power supply V_{DD}. (By analogy consider a stack of nMOS transistors enumerated as a function of their proximity to Ground.) Simulation results show that the upper $(N-1)$ pMOS transistors (lower $(N-1)$ nMOS transistors) have a similar impact on the current behavior of the stack. Therefore, we consider these $(N-1)$ transistors to have equal sizes. Using the results of [2, 8] to calculate the equivalent transistor width of the stack, W_{Stack} (To simplify the equation length is not included), we calculate the mean current of N transistors in a stack as follows

$$
E[I_{stack}] = K_E W_{Stack} e^{\frac{V_{dd} - E[V_{th}]}{nU} + \frac{Std^2[V_{th}]}{2(nU)^2}}
$$
$$
W_{Stack} = \frac{\beta W_L}{\left(1 + \beta W_L \left(\sum_1^{N-1} \frac{1}{W_i}\right)\right)}
\tag{6}
$$
$$
\beta = e^{\frac{-\lambda V_{dd}}{nU}}
$$

where K_E is a technology fitting parameter and λ is the DIBL effect coefficient [2]. To simplify the calculation of the equivalent transistor size of the stack, the length of each transistor in the stack is held fixed. Let the width of all $(N-1)$ transistors be W_i and the width of the remaining transistor be $W_i/\sqrt{\beta}$, see Fig. 3. We denote the width of the equivalent transistor as to W_{Stack}. The same procedure holds for nMOS transistors.

The variance of the stack is determined by the variance of each transistor in the stack. Since each transistor has the same impact on the total variance, the stack variance is the sum of the variances of each transistor divided by the square of the number of transistors in the stack.

$$
Std^2[I_{stack}] = \frac{1}{N^2}\left(Std^2[I_L] + \sum_{i=1}^{N-1} Std^2[I_i]\right)
$$
$$
Std^2[I_{stack}] = \left(\frac{\beta\left(\sum_1^{N-1} W_i\right) + W_L}{K_{Std} N^2 W_{stack}}\right) (e^{\frac{Std^2[V_{th}]}{(nU)^2}} - 1) E^2[I_{stack}]
\tag{7}
$$

where K_{Std} is also a technology dependent fitting parameter. With Eqs. (6) and (7), one can easily derive the optimal stack width ratio for the stack's maximum current or minimum current spread. To achieve maximum current, the lower pMOS (upper nMOS) transistor needs to be sized $1/\sqrt{\beta}$ times smaller w.r.t. to upper pMOS (lower nMOS) transistors. The variation of the current stack can be written as

$$
\frac{Std[I_{stack}]}{E[I_{stack}]} \propto \sqrt{\frac{\left(\beta\sqrt{\beta}(N-1) + 1\right)\left(1 + \sqrt{\beta}(N-1)\right)}{K_{Std} N^2 \beta}}
\tag{8}
$$

Eq. (8) helps us understand how many transistors can be stacked within current variation and area constraints. Ultimately, this is a

TABLE II CURRENT VARIATION IN SERIES-CONNECTED TRANSISTORS

Number of transistors in series	Simulation results		Normalized $Std[I]/E[I]$	Calculation from (8)
	$E[I]$ (A)	$Std[I]/E[I]$		
$2\times1.50um$	1.45E-07	59.98%	1	1
$3\times1.00um$	7.29E-08	67.35%	1.123	1.107
$4\times0.75um$	3.59E-08	70.42%	1.174	1.161
$5\times0.60um$	2.43E-08	73.55%	1.226	1.192

very important criterion for robust operation. We ran 3000 Monte-Carlo simulations for 2, 3, 4 and 5 nMOS transistors in a stack working at 0.3V and at 25°C(unless mentioned all the Monte-Carlo simulations are at 0.3V and at 25°C). As the length of each transistor is made equal, the total width for each simulation set-up is set to 3um. In Table II it is shown that Eq. (8) predicts correctly the trend of the variation. The mismatch between the calculation and the simulation values is because in our model V_{th} variation is treated as a given technology dependent parameter (source bulk modulation is not taken into account).

3.1.2 Transistor Sizing for Parallel Topologies

For parallel sizing, we need to calculate the mean and variance of the total current of N transistors connected in parallel. The resulting current is the sum of the Log-Normal distributions. The sum of Log-normal distributions with the same variance can be approximated by one Log-Normal distribution [9]. However, the sum will not represent the actual current of the uncorrelated parallel transistors. Hence, a correlation factor ρ_p for V_{th} needs to be introduced. This correlation factor was not needed in series-connected transistors because in that case the source-bulk modulation overshadows the correlation. The mean and variance of the current of N identical parallel transistors is

$$E[I_{para}] = NK_{Ep}\frac{W_{one}}{L}e^{\frac{V_{gs}-E[V_{th}]}{nU}+\frac{Std^2[V_{th}]}{2(nU)^2}+\frac{N^2}{\rho_p}}$$

$$K_{Ep} = \mu Ce^{1.8}U^2(1-e^{\frac{-V_{ds}}{U}})$$ (9)

$$\rho_p \propto Std^2[V_{th}]$$

$$Std^2[I_{para}] = (e^{\frac{Std^2[V_{th}]}{(nU)^2}+\frac{2N}{\rho_p}}-1)(E[I_{para}])^2/N^2$$

where W_{one} is the width of one single transistor. The equivalent width for parallel transistors can be calculated from Eq. (9):

$$W_{Para} = \gamma(N)W_{one}$$
$$\gamma(N) = Ne^{\frac{N^2}{\rho_p}}$$ (10)

Hence the width of a single transistor, which has the same mean current as the one of N transistors in parallel, is $\gamma(N)$ times the width of the transistors in parallel.

We ran 3000 Monte-Carlo simulations for 1 to 6 nMOS transistors in parallel, with a total width of 1.2um. The simulation and calculation results are shown in Table III. It is worth observing

TABLE III MEAN CURRENT OF PARALLEL CONNECTED TRANSISTORS

Number of parallel transistors	Simulated $E[I]$ (A)	Normalized $E[I]$	Calculation from (9)
$1\times1.20um$	3.54E-07	1.00	1.00
$2\times0.60um$	3.71E-07	1.05	1.15
$3\times0.40um$	4.13E-07	1.17	1.33
$4\times0.30um$	5.09E-07	1.44	1.54
$5\times0.24um$	6.29E-07	1.78	1.77
$6\times0.20um$	7.64E-07	2.16	2.04

TABLE IV CURRENT FLOW DISTRIBUTION OF A 2-INPUT NAND CELL

Output	Input Pattern	Current at CL
$1\rightarrow0$	$00\rightarrow11$	$I_{onStack}-2I_{offP}$ [a]
$1\rightarrow0$	$01\rightarrow11$	$I_{onStack}-2I_{offP}$ [b]
$1\rightarrow0$	$10\rightarrow11$	$I_{onStack}-2I_{offP}$ [b]
$0\rightarrow1$	$11\rightarrow00$	$2I_{onP}-I_{offStack}$
$0\rightarrow1$	$11\rightarrow01$	$I_{onP}-I_{offStack}$
$0\rightarrow1$	$11\rightarrow10$	$I_{onP}-I_{offStack}$

a. Stack referes to the nMOS stack, p denotes the pMOS

b. Up and Low denotes the upper and lower nMOS

these results in more detail. Namely, the joint correlated Log-Normal distribution indicates that the mean current is bigger than that of the uncorrelated sum of individual transistor currents [10].

3.2 Transistor Sizing for Balanced Transition Distributions

In this method, the best-case and worst-case transition paths determined by different input patterns are fed into our balacing process. The area of transistors on the best-case transition path are reduced and instead the transistors on the worst-case transition paths are upsized, and the balance between the pull up and pull down paths is considered. So we call it transistion balancing sizing method. Without loss of generality, consider a 2-input NAND cell loaded with a fixed load C_L. Table IV enumerates all possible switching conditions.

One can see that the worst pull-down delay occurs when the input vector changes from 00 to 11, while the worst pull-up delay occurs when the input vector changes from 11 to 10 (or 01). Thus, for a 2-input NAND cell we need to balance the current of one pMOS transistor with the current of the nMOS stack as the worst-case balancing criteria. In this method, the best-case paths are weakend to compensate the worst-case paths, such that the worst-case paths are optimized without area overhead.

We can derive the sizing criterion with the help of our stack and parallel sizing approaches by balancing one active pMOS transistor with the nMOS stack. The sizing ratio is

$$\frac{W_{stack}}{W_p} = \alpha e^{\frac{E[V_{thn}]-E[V_{thp}]}{nU}}e^{\frac{Std^2[V_{thp}]-Std^2[V_{thn}]}{2(nU)^2}}$$ (11)

Monte-Carlo simulations have been done for a commercial library NAND cell and the optimized NAND cell. The results are shown in Fig. 4. Cumulative distribution function (CDF) curves are presented for both the best and the worst case (determined by the input patterns) rise delays. The optimization procedure reduces the highest current to compensate the lowest current. Therefore, the best case delay is slightly increased compared to the cell from the library, while the worst case delay is reduced. Here, we see the advantages of using the statistical distribution of the current for

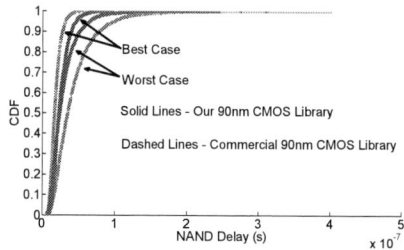

Figure 4. NAND simulation results (best and worst cases)

965

sizing purposes, since one can optimize the worst case current without any area penalty. The worst delay decreases from 43.87ns to 29.95ns, the standard deviation decreases from 31.08ns to 19.03ns, and the variation (standard deviation/mean) decreases from 70.85% to 63.54%.

3.3 Complex Cell Sizing Example

Complex cells can be sized by finding equivalent transistor sizes from reducing stack and parallel arrangements to their equivalent reference transistors.

Without loss of generality, we take a complex cell as the one depicted in the left part of Fig. 5 to explain how the cell is balanced. Both our sizing methods require path identification. To calculate the equivalent transistors we need to calculate the sizing ratio for the N/P network. In the right part of Fig. 5 the nMOS network is used as an example to show how the sizing ratio is determined by Algorithm II.

Algorithm II:
If n transistors in *Parallel*
Then *Size* of parallel transistors:
$$W_1 = W_2 = \cdots = W_n$$
Parallel Equivalent Size: *Eq. (10)*
If m transistors in *Series*
Then *Size* of transistors in stack:
$$W_{L1} = W_{L2} = \cdots W_{L(m-1)} = \sqrt{\beta} W_U$$
Stack Equivalent Size: *Eq. (6)*

*U means next to output node; L means away from output node.

With the described sizing factors of Fig. 5, we can calculate the proper equivalent transistor sizes required for the balancing optimization of Eqs. (4) and (5). The optimization results determine the aspect ratio between the equivalent nMOS and pMOS transistors. After the sizing optimization, the size of the equivalent transistors can be calculated back to each transistor in the network, obtaining an optimized cell. Monte-Carlo simulation results are shown in Fig. 6. The worst case (determined by the input patterns) rise delay CDF curves show that both methods provide faster cells with narrower distribution than the cell from the commercial library. Also, the transition balancing sizing method has more improvement when compared to the current balancing sizing method.

4. FLIP-FLOP DESIGN

Flip-flops contain feedback logic that are needed in sequential logic. The small inverter in the feedback loop of the slave may not be able to retain the data if not sized correctly. Therefore, proper sizing is needed. In this work, the flip-flop shown in Fig. 7 is chosen for comparison purpose only.

The flip-flop sizing approach is similar to our combinational logic sizing methodology. In transmission gate structures the nMOS and pMOS transistors are balanced to have equal current flow through the source and drain. The single inverter is balanced to

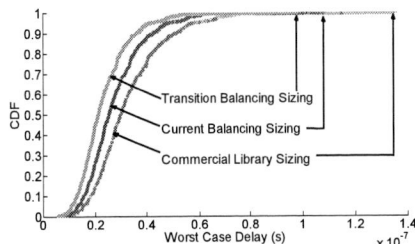

Figure 6. Complex gates simulation result (worst cases)

have equal rise and fall delays. The delay is calculated based on a differential equation [11]

$$V'_{out}(t) C_{Load} = \mu C \frac{W}{L} e^{\frac{V_{gs}-V_{th}}{nU}} \left(1 - e^{\frac{V_{ds}}{U}}\right) dt \qquad (12)$$

Solving the differential equation (12) and calculating the mean delay for pMOS and nMOS transistors results in

$$E[Delay] = \frac{UF\left(\frac{V_{dd}}{U}\right) C_{Load} e^{\frac{-V_{gs}+E[V_{th}]}{nU} + \frac{Std^2[V_{th}]}{2(nU)^2}}}{\mu C \frac{W}{L}}$$

$$F\left(\frac{V_{dd}}{U}\right) = \log\left(\frac{1 - e^{V_{dd}/U}}{1 - e^{V_{dd}/2U}}\right) \qquad (13)$$

One can also optimize this standard cell from a delay perspective using Eq. (13). For comparison purposes, the commercial library area constraint is applied to our balancing sizing strategy. The Clock to Q delay distributions are shown in Fig. 7. It is shown that after sizing, the mean value of the Clock to Q delay decreased from 63.07ns to 35.87ns, the standard deviation decreased from 32.80ns to 14.44ns, and the variation (deviation/mean) decreased from 52.01% to 40.26%.

5. BENCHMARKING

With the described methodology, we characterized our 90nm standard cell library at the slow process (SS) corner and typical process (TT) corner at 25°C and at 0.3V. The library contains 144 commonly used cells including inverters, logic gates and flip-flops. For the purpose of benchmarking, the ISCAS benchmark circuits were synthesized using Cadence RC Compiler with constraints of minimum area and leakage power. We evaluated timing and power w.r.t. the commercial library and the work in [2] and [3], note that in [3] 130nm technology is used.

In Table V, we compare our library w.r.t. the commercial 90nm library. We can see that for all shown circuits in the slow (SS) corner, our library has up to 57% and 69% improvement in timing and energy, respectively. In the typical (TT) corner the values are 55% and 74%. In general, the synthesis with our library yields a

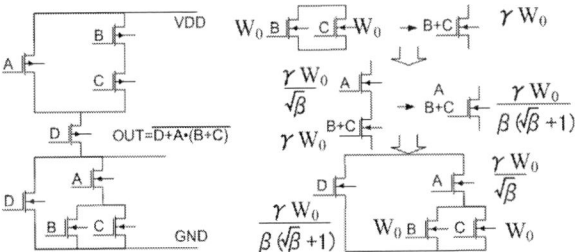

Figure 5. Complex cell sizing example

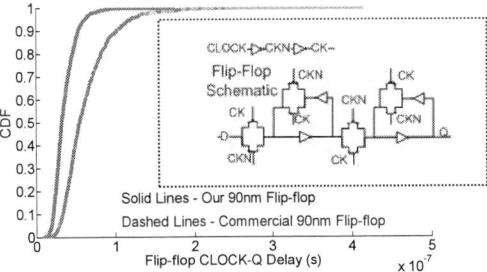

Figure 7. Flip-flop delay distributions comparison

TABLE V Iscas Benchmark

| | | Commercial Library | | | | Our Work | | | | Improvement Comparison % | | | | |
| | | Timing (ns) | Power (uw) | | Gate Count | Timing (ns) | Power (uw) | | Gate Count | Our Work | | [2] | [3]* | |
			Leakage	Dynamic			Leakage	Dynamic		Timing	Energy	Timing	Timing	Energy
C6288	SS	199.32	9.86	137.31	1592	113.45	11.28	69.24	1356	43.1	68.8	12.6	10.1	29.8
	TT	34.86	16.48	274.23		20.69	20.76	150.24		40.6	65.1			
C3450	SS	138.75	8.79	112.28	1423	63.31	15.32	65.12	1320	54.4	69.7	4.4	Not available	
	TT	30.78	15.74	217.54		14.65	27.48	131.28		52.4	67.6			
C1355	SS	57.21	2.24	34.25	286	36.32	3.03	17.31	217	36.5	64.6	13.5	10.4	41.2
	TT	13.44	4.12	68.50		7.14	5.19	30.06		46.9	74.2			
74283	SS	220.91	0.85	7.76	44	122.35	0.97	4.42	27	44.6	65.3	5.3	10.4	22.7
	TT	41.19	1.40	13.89		22.32	1.87	7.72		45.8	66.0			
74L85	SS	272.74	1.71	14.27	60	140.28	2.84	8.92	43	48.6	62.1	6.6	9.1	37.8
	TT	55.48	3.12	27.54		30.01	3.65	15.28		45.9	66.6			
74182	SS	167.15	0.77	6.29	22	72.25	1.27	4.58	18	56.8	64.2	33.1	7.8	12.4
	TT	31.43	1.39	12.41		14.32	1.95	8.21		54.4	66.5			

*note that the result in [3] is at 0.2V and the rest is at 0.3V

lower gate count number. This is because with our library RC Compiler tends to use more complex cells. This can be explained by better complex gates sizing using the proposed methodology. The results shows that our balancing theory is effective for delay and power savings without area penalty.

In Table V we also show the comparison w.r.t. the work of [2] and [3]. Due to lack of information of [2] and [3], we can not rerun the synthesis to compare directly with our work, only the ISCAS benchmark improvement data (w.r.t. their own reference library at their own technology) is compared. The differences between the approach of [2] and ours is that V_{th} variation is considered. Also, our transition balancing sizing allows us to compensate the worst case delay by increasing the best case delay (determined by the input patterns). This allows us to match the rise and fall delays. In [3], the channel length is increased to decrease V_{th}, leading to higher current through the transistors. Because the channel length is increased to minimize V_{th}, the width is decreased such that the gate capacitance is not increased substantially. Therefore, the delay can be optimized. As mentioned in Section III-A, our optimization interest is not maximizing the current, but balancing the rise and fall transitions. Observe that our library has on average 3.3x and 4.8x better timing than the work in [2] and [3], respectively, and has 1.6x better energy improvement when compared to [3].

6. CONCLUSION

We presented an analytical framework for robust subthreshold cell sizing. Our approach renders cells that have a narrower delay distribution as well as better active current specifications. Our sizing methodology relies on the statistical variations observed in the current distribution of a transistor when it is operated in the subthreshold regime. Compared with [2, 3] and a commercial 90nm library, our cells achieved up to 54% and 57% timing improvement and 74% and 66% energy improvement for ISCAS circuits and without area penalty. We have proven the effectivenes of our methodology for a 90nm technology. In the future work, we will verify our sizing methodology in smaller CMOS technology nodes.

7. REFERENCES

[1] Calhoun, B.H., Wang, A., Chandrakasan, A. 2005, "Modeling and sizing for minimum energy operation in subthreshold circuits", IEEE J. of Solid-State Circuits, pp. 1778 - 1786, 2005.

[2] Keane, J., Hanyong Eom, Tae-Hyoung Kim, Sapatnekar, S., Kim, C. 2006, "Subthreshold logical effort: a systematic framework for optimal subthreshold device sizing", 43rd ACM/IEEE Design Automation Conference, pp. 425 - 428, 2006.

[3] Tae-Hyoung Kim, Hanyong Eom, Keane, J., Kim, C. 2006, "Utilizing reverse short channel effect for optimal subthreshold circuit design.", Int. Symp. Low Power Electronics and Design, pp. 127 - 130, 2006.

[4] E. L. Crow and K. Shimizu, 1988, "Lognormal distributions: Theory and applications", CRC, 1988.

[5] B. Zhai, S. Hanson, D. Blaauw, and D. Sylvester, 2005, "Analysis and mitigation of variability in subthreshold design", Int. Symp. Low Power Electronics and Design, pp. 20–25, 2005.

[6] M. Pelgrom, A. Duinmaijer, A. Welbers, and A. P. G. Welbers, 1989, "Matching properties of MOS transistors", IEEE J. Solid-State Circuits, vol. 24, pp. 1433–1440, 1989.

[7] J. Kwong and A. P. Chandrakasan, 2006, "Variation-driven device sizing for minimum energy subthreshold circuits", Int. Symp. Low Power Electronics and Design. pp. 8–13, 2006.

[8] H. Al-Hertani, D. Al-Khalili, and C. Rozon, 2007, "A new subthreshold leakage model for NMOS transistor stacks", IEEE Northeast Workshop on Circuits and Systems, Montreal, QC, Canada, pp. 972-975, 2007.

[9] L. Fenton, 1960, "The sum of log-normal probability distributions in scatter Transmission Systems", IRE Transactions onCommunications Systems, vol. 8, no. 1, pp. 57-67, 1960.

[10] Y. Pu, J. Pineda de Gyvez, H. Corporaal, and Y. Ha, 2010, "An ultra low energy multi-standard JPEG co-processor in 65nm CMOS with sub/near threshold supply voltage", IEEE J. Solid-State Circuits, vol. 45, no. 3, pp. 668–680, 2010.

[11] N. Lotze, J. Goppert, and Y. Manoli, 2010, "Timing modeling for digital sub- threshold circuits," in Design, Automation & Test in Europe Conference, pp. 299-302, 2010.

Decoupling Capacitor Design Strategy
for Minimizing Supply Noise of Ultra Low Voltage Circuits

Mingoo Seok

Department of Electrical Engineering, Columbia University

mgseok@ee.columbia.edu

ABSTRACT

Supply noise is a critical problem for the robust operation of integrated circuits at ultra low voltage regimes. Although decoupling capacitance is a traditional solution, the reduction of gate capacitance at subthreshold voltage can cause area overhead. In this paper, we propose a decoupling capacitor design strategy to reduce area overhead. The strategy consists of two parts: 1) enhancing gate capacitance through circuit optimizations and 2) using remote decoupling capacitors. Remote decoupling capacitors, which can be placed far from the block to compensate, can minimize the area overhead of the capacitance-enhancing optimizations. They also exploit less utilizable silicon area. The proposed strategy improves the capacitance density by 6.1× without extra process steps, compared to the conventional approach. The gained robustness may be traded off for higher energy efficiency.

Categories and Subject Descriptors
B.7 [**Integrated Circuits**]: General
General Terms: Design
Keywords: ultra-low power, subthreshold operation, ultra low voltage operation, supply noise, decoupling capacitor

1. Introduction

Recently ultra low voltage (ULV) operation, where the supply voltage of metal oxide semiconductor field effect transistors (MOSFET) is scaled down to near or below transistor threshold voltage (V_{th}), has gained a significant amount of attention due to the large (10-20×) energy savings [10][11]. ULV operation can benefit a variety of energy-constrained systems for increasing battery lifetime and reducing the volume of power sources. The applications range from implantable medical devices, to infrastructure and environment monitoring systems, to active radio frequency identification (RFID) tags.

However, integrated circuits (IC) operating at ULV regimes become more sensitive to various sources of noise. Especially, noise on the supply voltage should be minimized since it can cause robustness and performance degradations, which can affect the minimum functional supply voltage (V_{min}) for both logic and memory circuits.

For mitigating supply noise, traditionally, decoupling capa-

ous current, reducing fluctuations of supply voltage. Since a large amount of decap, e.g. 10× the switching capacitance, is typically needed [1], the gate capacitors of MOSFETs are often used to maximize capacitance density and thus reduce area overhead.

However, the gate capacitance of MOSFETs reduces as we scale down supply voltage near or below V_{th} since the inversion layers of MOSFETs are not completely formed at these voltage regimes [2]. Figure 1 shows that the capacitance of decaps, which are provided in a 65nm standard-cell library, reduces by up to 2.5× at lower supply voltages, which can incur a significant amount of area overhead.

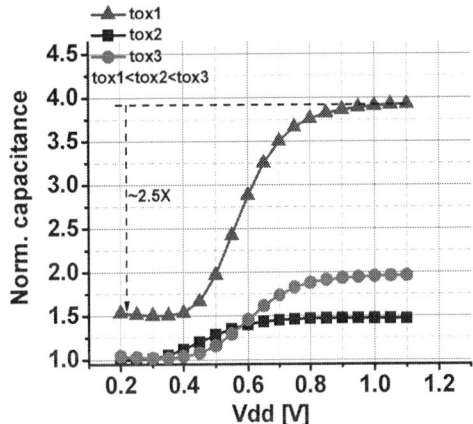

Figure 1. Degradation of capacitance of every standard-cell decap available in an industry 65nm CMOS.

Therefore, in this paper, we investigate the strategy of designing decaps at ULV regimes to reduce area overhead without compromising decap performance. The proposed strategy consists of two parts: 1) improving capacitance density through circuit and layout optimizations and 2) utilizing remote decaps. Remote decaps, which can be placed far from the blocks to compensate, can amortize the area overheads associated with some of the circuit optimizations of enhancing capacitance density. It can also exploit small and less-utilizable silicon area in a floor plan. In the modern design practice, this area is usually filled with decap fillers for compensating the blocks around it. However, the remote decap in unused space can be used to compensate the block placed further away, providing more flexible design.

The paper is organized as follows. In Section 2, we identify one of the critical sources of supply noise at ULV regimes: supply noise caused by power gating switches (PGSs). It will be used as an example problem to mitigate throughout the paper. We also investigate the required amount of decaps at

ULV regimes. Then, in Section 3, we propose several circuit-level techniques to improve gate capacitance density. In Section 4, we investigate the feasibility of using remote decaps in ULV regimes. SPICE simulations confirm that remote decap can compensate the supply noise induced by PGSs even if it is connected to noise source (i.e. gates) through a highly resistive (1k-10kΩ) path. In Section 5, we propose our decap design strategy and compare it to the conventional one. The proposed remote on-chip decaps achieve 6.1-8.2× higher density capacitance , compared to the conventional design practices. Finally, the paper concludes in Section 6.

2. Decap at ULV regimes

In this section, we investigate one of the important sources of supply noise at ULV design: the supply noise from PGSs. We also study the required amount of decaps for mitigating the performance degradation from this kind of supply noise.

2.1 Supply noise at ULV regimes

Dynamic IR drops induced by PGSs are one of the problematic sources of supply noise at ULV regimes. As shown in Figure 2, typically NFET (or PFET) PGSs are employed between virtual ground (or virtual supply) and the real ground (or real supply) to reduce power consumption during standby modes. For energy-constrained applications, these PGSs are often aggressively down-sized for maximizing standby power savings [5]. The small PGSs can exacerbate supply noise as a large amount of impedance is added in series to power grids.

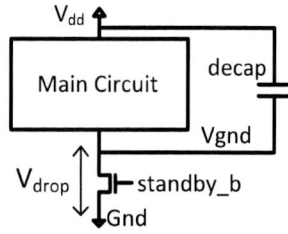

Figure 2. Main circuit with power gating switches.

For investigating the impact of this supply noise, we run SPICE simulations, using a 16b pipelined multiplier with a PGS designed in an industrial 65nm low power (LP) CMOS technology. The supply voltage is set at 0.35V, which is near the energy-optimal point [12]. As shown in Figure 3, more aggressively sized PGSs cause a larger amount of performance degradations. The width of PGSs (W_{PGS}) on the x-axis of Figure 3 is shown as the percentage of the total NFET width used in the main circuits, i.e. the multiplier. When W_{PGS} is downsized to 0.1% of the total NFET width, the delay increases by ~2.2×, compared to the case without using PGSs. The peak voltage drop in this configuration is ~63mV (V_{drop}/V_{dd}=18%). At 1.0V, the drop is ~260mV. (V_{drop}/V_{dd}=26%). Note that the relative magnitude of supply noise (V_{drop}/V_{dd}) is smaller at ULV regimes. However, the exponential sensitivity of delay to supply voltage incurs a larger degradation at ULV regimes. In addition, supply noise can hurt circuit robustness, in particular for SRAM arrays employing PGSs [3][4].

One simple remedy for mitigating supply noise is to upsize PGSs. However, this approach can increase the standby power

consumption. As shown in Figure 3, upsizing the PGSs from 0.1% to 10% can mitigate the dynamic IR drops over PGSs while increasing the standby power consumption by 100×. Given the importance of the standby power consumption in ULP systems [4], it is unlikely to be an attractive option.

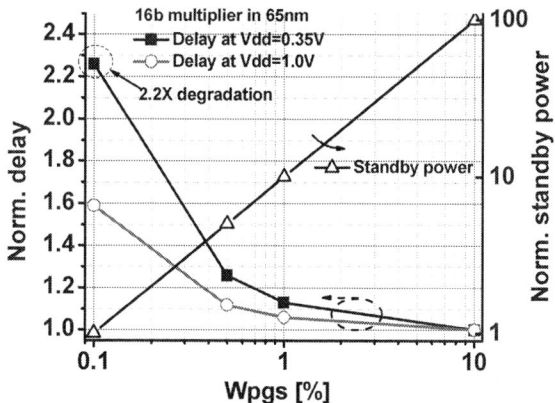

Figure 3. A large dynamic IR drop over aggressively-sized PGSs. The size of PGSs are normalized to the total NFET width used in the main circuits

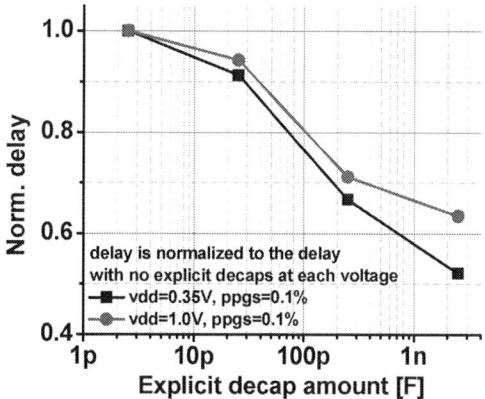

Figure 4. The required amount of decaps.

A higher voltage, if available, can be used as the control voltage (the standby_b signal in Figure 2) to substantially reduce the resistance across PGSs. However, this is not a fundamental solution since designers are tempted to use a PGS which is downsized by the same factor by which the resistance reduces [5]. Therefore, the supply noise problem still exists if lower standby power is sought. In fact, the supply noise and standby power consumption pose a fundamental tradeoff relationship in PGS designs.

2.2 Required amount of decaps at ULV regimes

Decaps can be a solution to mitigate the supply noise without compromising standby power consumption. Since decaps can reduce the amount of peak current flowing through PGSs, it can reduce the supply noise. Although decaps may consume gate leakage at high supply voltage, its impact is negligible at ULV regimes.

A large amount of decap is needed due to the exponential impact of supply voltage on circuit delay. Figure 4 shows the SPICE simulation results from a multiplier at subthreshold and super-threshold supply voltages (0.35V and 1.0V). At V_{dd}=0.35V, the multiplier needs more decaps for the same amount of delay reduction, which can incur area overhead for decaps at ULV regimes.

3. Gate capacitance density enhancement

Although decaps are critical to mitigate supply noise, the gate capacitance density decreases by ~2.5× at ULV regimes as shown in Figure 1. This can incur an area overhead for implementing on-chip decaps. Therefore, in this section, we investigate the circuit and layout techniques to improve gate capacitance.

3.1 Impact of transistor threshold voltage

The primary reason for the decrease of capacitance density is that the channel of MOSFETs is not fully inverted at ULV regimes. Therefore, we need techniques to help forming inversion layers even at ULV regimes for improving the capacitance density.

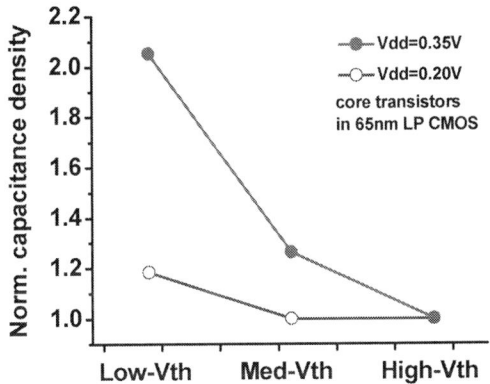

Figure 5. Impact of transistor V_{th} on capacitance density.

One straightforward method is to use low V_{th} MOSFETs for decaps. As shown in Figure 5, decaps with lower V_{th} exhibits higher capacitance density at ULV regimes. At 0.35V, moving from high-V_{th} to low-V_{th} results in 2× increase in capacitance density for decaps. Note that V_{th} does not change much between 0.35V and 0.5V due to the already reduced short channel effects (SCE). Although the decaps with lower V_{th} transistors exhibits better capacitance density at ULV regimes, the capacitance density still reduces with even lower supply voltage; at V_{dd}=0.2V, it become about the half of the largest available gate capacitance (Figure 5).

In addition, we select only the transistor type with higher capacitance density for use in decaps. The capacitance density is heavily influenced by the different NFET and PFET V_{th}s, for devices of the same geometry. In this technology, NFET has ~50% higher capacitance density than PFET due to the lower V_{th}, making NFET-only decap an attractive option.

3.2 Enhancements through Sizing

Reverse short channel effect (RSCE) is the phenomenon where the V_{th} of MOSFETs decreases with longer channel length at ULV regimes. RSCE is a byproduct of HALO doping for mitigating SCE at super-threshold voltages [7][14]. At nominal supply voltage, due to the SCE such as drain-induced barrier lowering (DIBL), RSCE is not pronounced. However, RSCE becomes a larger contributor on V_{th} change since SCE becomes negligible at ULV regimes.

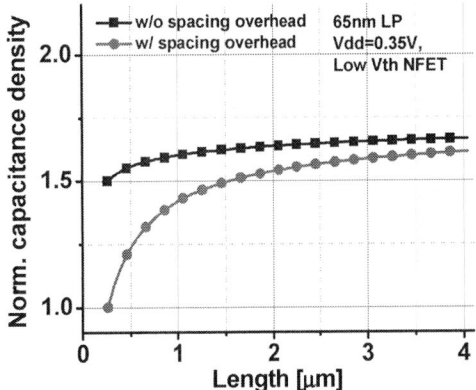

Figure 6. Capacitance improvement with longer channel length at ULV regimes.

Therefore, we can use longer channel MOSFETs to improve the capacitance density, as shown in Figure 6. The improvement becomes even larger when considering the non-proportional relationship between gate and decap cell area. Therefore, it is almost always preferable to use the longest allowed transistors for higher capacitance density in ULV decaps.

We may also want to consider inverse narrow width effects (INWE) [8], to reduce V_{th} and improve capacitance density. However, this technique is less feasible since the capacitance improvement is not often sufficient to overcome the area overhead from narrow-width fingers. SPICE simulations show that using wider width transistors can be a better option for decap design in this technology, instead of using many narrow fingers.

Some of the layout-dependant stresses can be exploited to reduce V_{th} [15]. Note that not all stress techniques modulate V_{th}; some of them only affect channel mobility [16]. Also, careful layouts to avoid well proximity effect (WPE) can reduce V_{th}. However, some of the techniques may become impractical since their gain in capacitance density does not offset the associated area overhead.

3.3 Higher overhead techniques: accumulation mode capacitors and body biasing

Accumulation-mode gate capacitors such as n+ poly over n-well can be an attractive option for ULV decaps. This capacitor has a conduction channel even with a fairly low gate voltage [9], which helps maintaining a good amount of capacitance. However, as the n-well of this device needs to be biased to ground voltage, it may suffer from substrate noise. Therefore, a guard ring with p-substrate ties is often deployed around the capacitor, which may cause area penalty.

In addition, forward body bias may be used to improve capacitance density for inversion mode decaps. As shown in Figure 7, the capacitance density increases roughly linearly with forward body bias. The forward body bias can be increased up to ~0.5V until parasitic diodes in well interface are turned on. However a technology-specific evaluation is needed as it may incur a considerable area penalty; triple well is needed for NFET body biasing. PFET biasing needs separate n-wells which have to be spaced out from other n-wells at a different potential.

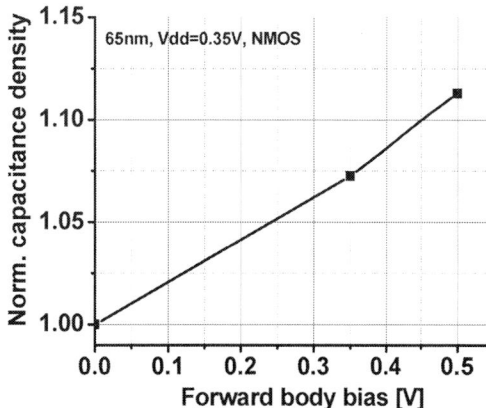

Figure 7. Forward body bias to improve capacitance.

4. Remote decoupling capacitors
4.1 Distributed and remote decaps

In the modern standard-cell-based automatic place and route (APR) flow, decaps can be designed in standard-cell form factors and placed between gates. The low resistance from decaps to gates enables a fast charge transfer, making them more effective to mitigate supply noise.

On the other hand, we can also implement larger size decaps outside the design. These decaps, which we called remote decaps, can provide a higher capacitance density since they are not restricted by the layout and routing limitations of standard-cells. For example, metal-to-metal capacitors using higher level interconnect layers can be easily stacked upon gate capacitors for improving capacitance density, while the same approach can be infeasible for distributed decaps as it can cause serious routing congestion. Also, remote decaps are more suitable for using some of the optimization techniques discussed in Section 3.3 since the associated overhead can be easily amortized.

Despite the benefits, remote decaps have a longer path to noise sources, which may limit the performance. This limitation mandates using both distributed and remote decaps for super-threshold supply voltage operations. Also, a low resistance interconnect should be used between remote decaps and the target main circuits.

4.2 Feasibility of remote decaps

In this subsection, we investigate the feasibility of remote decaps at ULV regimes for mitigating PGS-induced supply noise. In particular we study the impact of resistance between decaps and main circuits, i.e. a multiplier, in Figure 1. As shown in Figure 8, at super-threshold supply voltage (1V), a relatively small resistance (10Ω) between decaps and main circuits can degrade the performance of the 250pF decap. The decap becomes totally disabled when the resistance reaches ~10kΩ. In other words, decaps cannot mitigate supply noise at all and the circuits operate at the same frequency as the circuits without decaps.

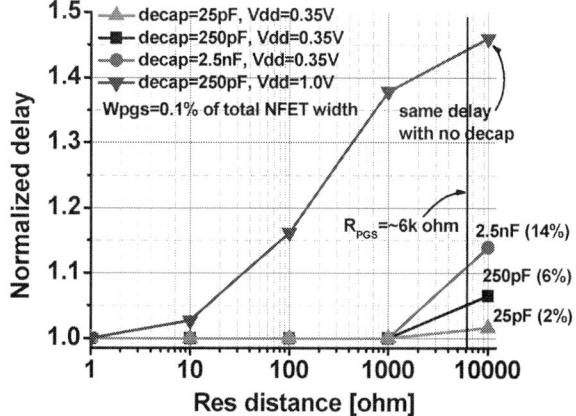

Figure 8. Impact of resistance distance of decaps.

However, at ULV regimes, the remote decap still provides a comparable amount of compensation for the supply noise even through a highly resistive path. As shown in Figure 8, the performance of 250pF decap is not compromised with 1kΩ resistance. The circuits operate only 6% slowly even with 10kΩ resistance. Note that the same amount of resistance completely disables the function of decaps at V_{dd}=1.0V. The 10kΩ resistance is equivalent to a >5000μm long metal trace when using minimum-width and thin layer metal in this technology (65nm LP), confirming the feasibility of using remote decaps in ULV regimes.

The higher resilience of decaps against resistance at ULV regimes comes from two factors: slow circuits and large impedance of PGSs. In other words, ULV circuits have a dominant portion of noise in lower frequency, which falls within the corner frequency of remote decap. Interconnect resistance is negligible due to the large series resistance of power gating switches.

We model the delay penalty of main circuits in EQ1 and EQ2 along with Figure 9. The penalty is proportional to current change and equivalent impedance at virtual ground (Z_{VG}). The 'func' in EQ1 can be exponential in ULV regimes or alpha-power at nominal supply voltage [17], based on the governing driving current type. In EQ2, Z_{VG} can be modeled as the parallel sum of three impedance sources; the lumped impedance of MOSFETs in main circuits (Z_{gate}), the lumped impedance of PGSs (Z_{PGS}), and the impedance of decaps (C_{decap}) with the resistance (R_{dist}). The w is the frequency of interest.

971

R_{dist} can be ignored when both Z_{PGS} and $1/jwC$ are substantially larger than R_{dist}. Note that in order to maintain the virtual ground to be close to real ground, Z_{gate} should be larger than Z_{PGS} and thus can be ignored in calculating Z_{VG}. At ULV regimes, Z_{PGS} (W_{PGS}=0.1%) become significantly large; the Z_{PGS} in this example is ~60kΩ. Due to this large impedance, a considerably large (up to 1kΩ) R_{dist} can be ignored. However, at nominal V_{dd}, the Z_{PGS} of the same PGS is only ~100Ω, limiting the value of R_{dist} to grow without affecting Z_{VG}.

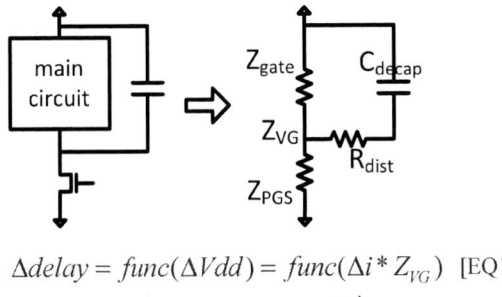

$$\Delta delay = func(\Delta Vdd) = func(\Delta i * Z_{VG}) \quad [EQ1]$$

$$Z_{VG} = (Z_{gate} \| Z_{PGS} \| (R_{dist} + \frac{1}{jwC_{decap}}) \quad [EQ2]$$

Figure 9. Analysis of performance of remote decaps.

The decap impedance ($1/jwC_{decap}$) is also substantially large at ULV regimes because the frequency of noise decreases. Based on the FO4 delay scaling, it increases by >3 orders of magnitude at the same amount of C_{decap}. This also implies that smaller amount of decaps have a higher resilience against R_{dist}. As shown in Figure 8, with 10kΩ resistance, the design with 25pF has only <2% performance degradation while the design with 2.5nF operates at ~14% slower clock frequency.

5. Decap design strategies at ULV regimes

Given the feasibility of remote decaps at ULV regimes, we propose a strategy which utilizes the remote decaps with the capacitance enhancing techniques. The remote decaps can amortize the area overhead associated some of the techniques (such as using accumulation-mode decaps). They can be placed in un-used areas and connected to blocks through low overhead routing (e.g. a minimum-width thin-metal interconnect). Moreover, designers may want to place no distributed decaps inside designs and use only remote decaps, which can help saving silicon area as well as improving performance and energy efficiency from shorter interconnects.

Figure 10 shows the cumulative improvements of capacitance density through each optimization technique. The baseline is the decap in the industrial standard cell library targeting for super-threshold operation, which has both NFET and PFET capacitors in a cell. Among the series of techniques, the first five techniques can be used even in the standard cell layout with little overhead. The optimal V_{th} selection, denoted as Vth in Figure 10, boosts >2× capacitance density. We also apply length optimization (denoted as length in Figure 10), NFET-only design (NFET in Figure 10), and consequent area savings (spacing1, spacing2 in Figure 10), which collectively provide an additional 1.85× improvement in capacitance density. All the techniques that can be used in standard cell layouts yield a 3.7× improvement, compared to the baseline design.

We can design decaps with higher density by removing standard cell layout limitations created by fixed row heights. Figure 11 shows an example decap layout, where a larger length and width can be chosen. One constraint is the maximum spacing between the p-substrate ties, which is denoted as L_{crit} and is technology-dependent. With the constraints in this technology, the area efficiency (AE), which is defined as oxide area over total area (EQ3), is estimated as ~91%. This new layout, denoted as non-stdcell in Figure 10, improves the capacitance density by 38%.

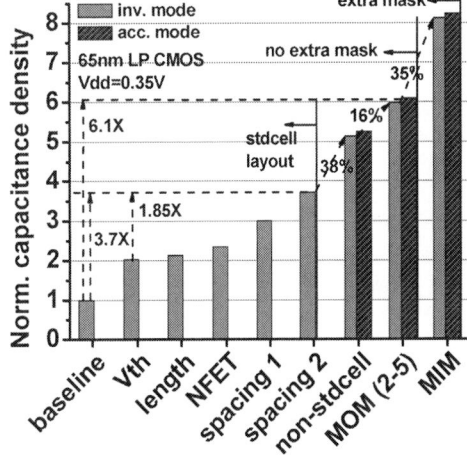

Figure 10. Improvement of the capacitance density of on-chip decaps through multiple and cumulative optimizations.

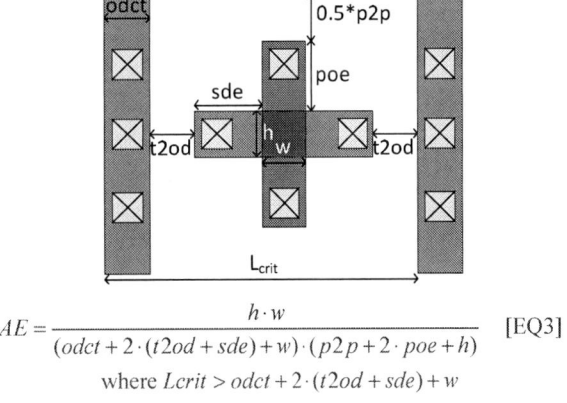

$$AE = \frac{h \cdot w}{(odct + 2 \cdot (t2od + sde) + w) \cdot (p2p + 2 \cdot poe + h)} \quad [EQ3]$$

where $Lcrit > odct + 2 \cdot (t2od + sde) + w$

Figure 11. An example decap design without adhering to standard cell layout

With the freedom in layout, accumulation-mode capacitors with n+ poly on n-well structure can be used for decap design. Due to the n-well and guard ring for mitigating substrate noise, it can have a lower area efficiency of ~87%. However, the effective capacitance density is slightly better than inversion-mode capacitors due to higher capacitance per oxide area. In addition, the accumulation-mode capacitors maintain almost the same capacitance density (~6% degradation) at ~200mV while inversion-mode decaps exhibit 42% lower density compared to V_{dd}=0.35V. This makes accumulation-mode decaps an attractive option for designing remote decaps at ULV regimes.

Additionally, multi-layer metal-oxide-metal capacitor can be stacked on top of the optimized gate capacitors for boosting capacitor density. We add a metal-oxide-metal capacitor using layer 2 to 5 to the accumulation mode capacitor, which improves capacitance density by 16% (denoted as MOM(2-5) in Figure 10). Finally, another metal-based capacitor, metal-insulator-metal (MIM) capacitor can be stacked at the cost of extra masks. MIM capacitors provide extra 35% improvement in capacitance density. Overall, these design strategies can save silicon area by improving capacitance density by 6.1× without an extra mask, and 8.2× with an extra mask at V_{dd}=0.35V.

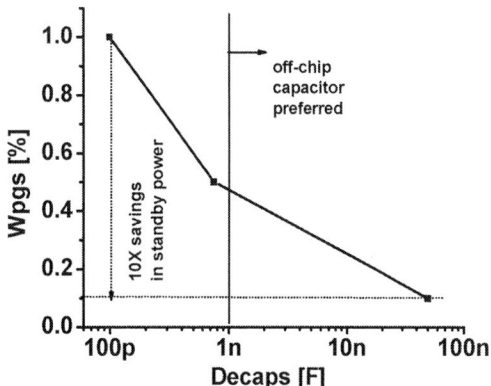

Figure 12. Standby power savings with remote decaps at iso-performance.

The improved decap can be used for mitigating the supply noise induced by PGSs. The reduced supply noise enables the use of even smaller PGSs, which in turn can achieve a significant amount of standby power reduction. Figure 12 shows the allowed size of PGSs at different values of decaps. At the design points in the line, the main circuits operate at the same clock frequency. Although an on-chip decap larger than 1nF may not be feasible due to large silicon area, we can easily use off-chip capacitors since decaps at ULV regimes are more tolerant to the resistance of the path.

6. Conclusions

In this paper, we investigate the design strategy for decaps at ULV regimes in order to mitigate the degradation of gate decoupling capacitance at lower supply voltage. The proposed decap design strategy of using capacitance-enhanced remote decaps reduces the area overhead by 6.1-8.2×, compared to conventional design approaches. When used in a design with PGSs, it can reduce standby power consumption through improved robustness against supply noise.

Acknowledgement – the author appreciate the valuable discussion with Gregory K. Chen at Intel.

References

[1] P. Larsson, "Parasitic resistance in an MOS transistor used as on-chip decoupling capacitance," *IEEE Journal of Solid-State Circuits*, Vol. 32, no.4, pp.574-576, Apr, 1997

[2] D. Vasileska, D. K. Schroder, D. K. Ferry, "Scaled silicon MOSFET's: degradation of the total gate capacitance," *IEEE Transaction on Electron Devices*, vol.44, no.4, Apr 1997

[3] E. Seevinck et al., "Static noise margin analysis of MOS SRAM cells," *IEEE Journal of Solid-State Circuits*, vol. sc-22, no.5, pp.748-754, Oct. 1987

[4] M. Seok, et al., "Analysis and optimization of sleep mode in subthreshold circuit design," *ACM/IEEE Design Automation Conference*, 2007

[5] G. Chen, et al, "Millimeter-scale nearly perpetual sensor system with stacked battery and solar cells," *IEEE International. Solid-State Circuits Conference*, pp.288-289, 2010

[6] J. Kil, et al, "A high-speed variation-tolerant interconnect technique for sub-threshold circuits using capacitive boosting," *IEEE International Symposium on Low Power Electronics and Design*, pp.67-72, Aug, 2006

[7] C.-Y. Lu, J.M. Sung, "Reverse short-channel effects on threshold voltage in submicrometer scaled devices," *IEEE Electron Device Letters*, vol.10, issue.10, pp.446-448, Oct. 1989

[8] L.A. Akers, "The inverse narrow width effect," *IEEE Electron Device Letters*, vol.7, issue.7, pp.419-421, Jul. 1986

[9] T. Soorapanth, et al., "Analysis and optimization of accumulation-mode varactor for RF ICs," *IEEE Symposium on VLSI circuits*, 1998

[10] B. Zhai, et al, "Theoretical and practical limits on dynamic voltage scaling", *ACM/IEEE Design Automation Conference*, 2004

[11] B. Calhoun, et al, "Characterizing and modeling minimum energy operation for subthreshold circuits", *IEEE International Symposium on Low Power Electronics and Design*, 2004

[12] D. Bol, D. Flandre, J.-D. Legat, "Technology flavor selection and adaptive techniques for timing-constrained 45nm subthreshold circuits," *IEEE International Symposium on Low Power Electronics and Design*, 2009

[13] A. Agarwal, H. Li, K. Roy, "DRG-cache: a data retention gated-ground cache for low power," *ACM/IEEE Design Automation Conference*, 2002

[14] B.C. Paul, "Device optimization for digital subthreshold logic operation, "*IEEE Transaction on Electron Devices*, vol.52, no.2, pp.237-247, Feb. 2005.

[15] A. Kahng, P. Sharma, R. O. Topaloglu, "Exploiting STI stress for performance," *IEEE International Conference on Computer-aided Design*, 2007

[16] V. Joshi, B. Cline, D. Sylvester, D. Blaauw, K. Agarwal, "Stress aware layout optimization," *IEEE International Symposium on Physical Design*, 2008

[17] T. Sakurai, A.R. Newton, "Alpha-power law MOSFET model and its applications to CMOS inverter delay and other formulas," *IEEE Journal of Solid-State Circuits*, vol. 25, issue 2, pp.584-594, 1990

[18] P. Andreani, S. Mattison, "On the use of MOS varactors in RF VCO's," *IEEE Journal of Solid-State Circuits*, vol.35 no.6, pp.905-910, Jun 2000

Regaining Throughput Using Completion Detection for Error-Resilient, Near-Threshold Logic

Joseph Crop
Oregon State University
Corvallis, OR, USA
cropj@eecs.orst.edu

Robert Pawlowski
Oregon State University
Corvallis, OR, USA
pawlowsr@eecs.orst.edu

Patrick Chiang
Oregon State University
Corvallis, OR, USA
pchiang@eecs.orst.edu

ABSTRACT

Operating in the near-threshold regime can result in significant energy savings. Unfortunately, the increased timing variation prevents conventional error-detection techniques from properly functioning. This paper introduces two circuit-level timing error detection techniques that aim to increase throughput while operating in the near-threshold voltage regime: current-sensing completion detection and transition-aware completion detection. Each method allows any digital circuit to operate at speeds not limited by the worst-case critical path. Throughput improvements and energy savings are reported for implementations on a 16-bit adder.

Categories and Subject Descriptors

B.7.1 [**Integrated Circuits**]: Types and Design Styles— *VLSI – very large scale integration*

General Terms

Design, Performance, Reliability

Keywords

Low Power, Variation Tolerance, Completion Detection, Error Detection

1. INTRODUCTION

Emerging mobile and embedded applications are becoming critically constrained by power consumption. Dynamic-voltage and/or frequency scaling (DVFS), as well as other methods (clock gating and power gating) are all widely proposed methods for reducing power consumption. While V_{dd} scaling has been shown to be one of the most effective ways to improve power, the unpredictable increase in timing delay spread prevents further reductions of the supply voltage into the near-threshold regime.

In this paper, we propose two new ideas for ensuring timing-resilient circuit operation at a lower supply voltage.

We will begin by discussing near-threshold supply voltage (NTV) operation, discussing both its benefits and drawbacks. Conventional methods for circuit-level timing error detection will then be covered, and their limitations in the near-threshold domain will be discussed. Next, our two methods will be presented, illustrating their effectiveness in the near-threshold voltage regime. Finally, our experimental simulation setup will be presented and discussed, comparing our results with previous works and illustrating the potential throughput advantages of our proposed work.

2. NEAR-THRESHOLD OPERATION

One of the most popular methods to reduce power consumption is to aggressively lower the supply voltage into the sub-threshold or near-threshold voltage (NTV) region. With NTV operation, the supply voltage is lowered to just above the threshold voltage of the transistors. This has previously shown to lower energy by ~5-10X while decreasing the operating frequency by ~10X. In [4] it is shown that this region of operation provides the best energy savings without introducing significantly long delays, when compared to sub-threshold operation. Unfortunately, NTV operation comes with significant challenges. For example, because the supply is lowered, large V_{dd} noise events or temperature shifts can affect performance.

However, the primary challenge with NTV designs is process variations that exacerbate timing variations, where delay spread may increase from ~30% (super-threshold) to 400% (near-threshold) [4, 10]. Some sources of static process variations include line edge roughness, random dopant fluctuations, well-proximity effects, and gate-length variations. Such variations are expected to increase with continued process scaling, resulting in a projected wider distribution between min-max timing delays.

3. CONVENTIONAL TECHNIQUES

There are several methods that have been proposed to combat timing uncertainties. These generally fall into two categories, pre-fabrication techniques such as delay margining, and post-fabrication techniques such as in-situ error detection.

3.1 Delay Margining

Traditionally, digital circuits are engineered to satisfy a particular worst-case timing constraint. This timing guard band ensures 'no-error' operation across a wide range of possible conditions, such as process variations, supply noise droops, soft error, and temperature shifts. These circuits are typically verified across multiple process corners to ensure all outputs will be correct at the required synthesized frequency.

Unfortunately, in the NTV regime, it is extremely difficult to insure that these timing constraints will be met in the presence of large process variations. Characterizing an entire processor and its worst-case delay path under Monte Carlo is usually out of the question. In [6], a Monte Carlo approximation is proposed to ensure correct timing under NTV variations. These pre-silicon timing characterization methods, however, may still be much different than the actual silicon that is fabricated. For example, the 0.6V DSP in [6] was pre-silicon characterized at $14MHz$ after variations and aging margins, but after post-silicon fabrication, was able to successfully operate at $43.4MHz$. With continued process scaling, guaranteeing post-fabricated operation that correlates well with pre-silicon simulation is becoming more challenging, especially for NTV operation across multiple dies and wafers.

3.2 Error Detection

A different solution to combating variations is in-situ error detection. After digital circuits are synthesized, an error detection method is employed that detects if timing constraints are met during normal chip operation. Therefore, the design is not required to undergo simulation of every possible variation-induced timing error that might occur under NTV.

The most common error detection method is Razor [5], or related timing speculative approaches [1]. With Razor, each datapath flip-flop is modified similar to the circuit shown in Figure 1. Each flop-flop is compared with the value stored in a 'shadow' latch that is sampled by a delayed clock, typically around 20% after the main clock edge. If the outputs have not settled to their final result before the main clock edge, they will be caught in this delayed latch afterward. After XOR comparison, if the two values sampled by the main latch and the delayed latch are different, an error is flagged.

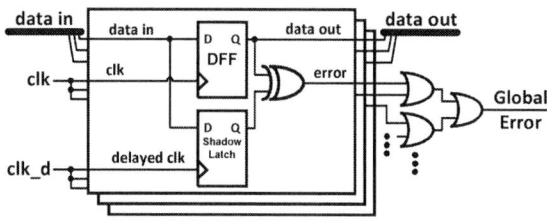

Figure 1: Razor flip-flops with global error flag

These errors are then managed by an error recovery mechanism that ensures all instructions are executed correctly after detection of an error. There are many different recovery methods, with different recovery speeds and energy overheads [3]. The clearest advantage of Razor is the potential for achieving approximately a 20% speed improvement. Using this timing speculative method allows for operation faster than the worst-case STA, improving both throughput and energy efficiency.

However, Razor circuits can exhibit several disadvantages in the NTV regime. First, the addition of an extra set of latches to the datapath increases energy, significantly impacting any energy improvements due to increased clock

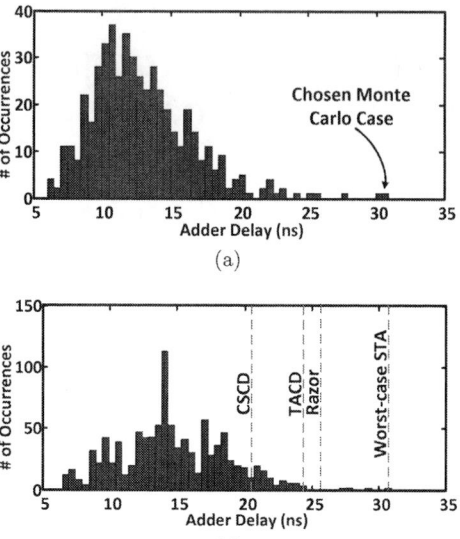

Figure 2: histograms of (a) Monte Carlo chip-to-chip delay of the STA and (b) delay of changing FIR filter data on a 16-bit adder with error-detection speeds marked.

speed. Second, in order to guarantee that the 20% window can correctly catch all errors, minimum-delay buffers must be added to all min-delay fast paths in the logic, in order to eliminate race conditions that prevent Razor from operating correctly. These buffers add to the energy overhead of Razor as well. Hence, while it is common to apply Razor and min-delay path insertion into only a subset of the logic paths at super-threshold, at near-threshold, this may not guarantee correct operation due to the unknown delay distribution. The final disadvantage with Razor, specifically in the NTV regime, is its inability to improve throughput beyond approximately 20%, limited by the min-path race conditions. [1]

4. VARIATION STUDY

In order to explore the effects of process variations in the NTV regime, a 16-bit Carry-Lookahead (CLA) adder was synthesized in a 90nm CMOS process. Figure 2a shows the histogram of a 500-point Monte-Carlo simulation of the CLA adder, where the inputs are the worst-case static timing analysis (STA) vectors determined by the synthesizer. The figure shows a large standard deviation in delay while operating at a NTV voltage of 500mV. In addition to process variation-induced timing uncertainty is input vector variation. Figure 2b shows the simulated delays of the same adder at one particular Monte-Carlo case, where the input vectors are supplied from the outputs of a FIR filter. Further, these vector-to-vector timing variations worsen as the circuits operate deeper in the NTV regime.

Figure 3 shows the potential speedup that can be ideally achieved with the ability to detect all timing errors. For this simulation, we chose the worst-performing Monte Carlo case

---

[1]Typical maximum delay windows for Razor are around 20%-30%. Unfortunately, NTV operation exhibits unpredictable delays that may push this limit much lower.

975

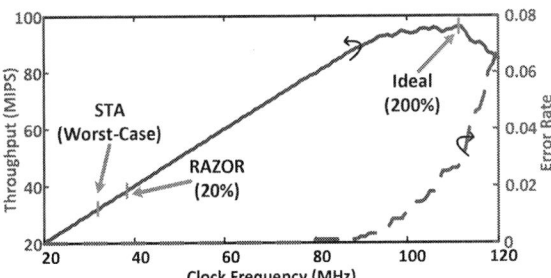

Figure 3: Potential throughput improvement with ideal error detection

from the 500 that were simulated, resulting in a worst-case clock speed of $32MHz$ (as opposed to a best-case speed of $166MHz$). Next, 1000 add-vectors from a low-pass FIR filter using electroencephalography (EEG) data were extracted from a Matlab simulation, and then simulated with a 16-bit carry-lookahead adder operating in NTV. Using either the micro-rollback or counterflow pipeline error-recovery methods [3], assuming that an ideal error detection method exists (one that can perfectly detect all errors at any clock speed), the potential speedup can be as much as 200%.

These HSPICE simulation results suggest that circuits operating in the NTV regime cannot afford to be margined for the worst-case while still ensuring error-free operation with predictable yield, in regards to both throughput and energy-efficiency.

5. PROPOSED ERROR DETECTION METHODS

The following section introduces two completion-based error detection techniques that can improve throughput beyond the limitations of conventional Razor circuits. The first, Transition Aware Completion Detection (TACD), is a fully synthesizable method similar to Razor circuits. The second, known as Current Sensing Completion Detection (CSCD), is an analog approach that uses a current sensor to monitor the supply droop to detect errors.

5.1 Transition Detecting

Conventional Razor circuits detect errors due to output changes after the clock edge. The proposed TACD method detects errors based not on output value correctness but on output value transitions, and not after computation completion but during the computation.

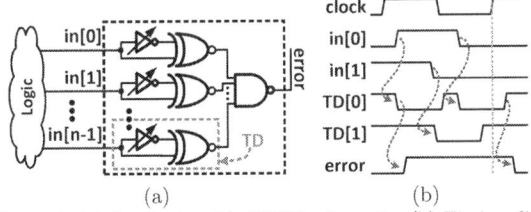

Figure 4: (a) Synthesize-able TACD schematic, (b) Timing diagram of TACD.

TACD is comprised of a variable-delay inverter chain with

a single XNOR gate per output wire (termed Transition Detectors (TDs)), as illustrated in Figure 4a. The output of each XNOR is NANDed globally to produce an error/done signal for each pipeline stage. As switching activity is present for each output, the XNORs will transition low for the duration of the inverters' delay, indicating that operation has not completed. The timing diagram for this operation can be seen in Figure 4b. Note that the inverter delays need to be calibrated, depending on the amount of switching activity – not for a single output but for all the outputs combined (i.e. just long enough until another output toggles). The total inverter delay is summarized in Equation 1:

$$d_{inv} = d_{NAND\ tree} + \Delta_{toggle\ max} + d_{margin} \qquad (1)$$

where $d_{NAND\ tree}$ is the worst-case delay of the NAND tree, $\Delta_{toggle\ max}$ is the worst-case delay time between any two output transitions, and d_{margin} is an added margin to account for dynamic variations. The synthesizer can be used to determine the optimal value for d_{inv}. In the case of this work, since the adder implemented was a simple carry-lookahead architecture, it is easily determined that $\Delta_{toggle\ max}$ results in the longest carry propagation path. In the presence of a synthesized circuit that generates glitches, the equation becomes slightly more complicated in that any given Δ_{toggle} must be less than the time between either another glitch or a legitimate output toggle. Glitches can be removed in some cases by adding a prime implicant. However, this requirement may be hard to ensure with more complicated logic.

In order to improve the performance of TACD, TDs can be added at strategic points within the logic. In the case of a CLA adder, a TD can be connected to the carry-out of each lookahead unit. This allows each inverter delay to be smaller, requiring less area for each TD. Even though there are more TDs, the total area increase is minimal for relatively complex digital logic blocks.

5.1.1 Detector Resiliency

Because the TACD error-detector will be used in the same highly-variable NTV environment as the combinational logic, it must be designed for error-free operation. Because the detector uses only simple logic parts, it can be easily tuned for NTV operation, and the detector as a whole can continue to operate in the presence of variations. d_{inv} can be determined using simple delay tests, and can therefore be easily tuned using off-line calibration after post-silicon fabrication. For example, off-line delay-path tuning was previously proposed with tunable replica circuits (TRCs) [2].

5.2 Current Sensing Completion Detection

CSCD [9] consists of an analog sensor that senses the current flowing through a group of combinational logic via a resistor or power-gate transistor. When a system begins to compute on new input vectors, the logic's current consumption increases. When this current consumption abates to a steady state, the computation has completed. While these types of circuits have been evaluated for asynchronous operation, to the best of the authors' knowledge, they have not been applied specifically to synchronous systems operating in NTV where timing variations are a critical concern.

The proposed CSCD method consists of a clocked, offset-programmable, dynamic sense-amplifier that measures the

976

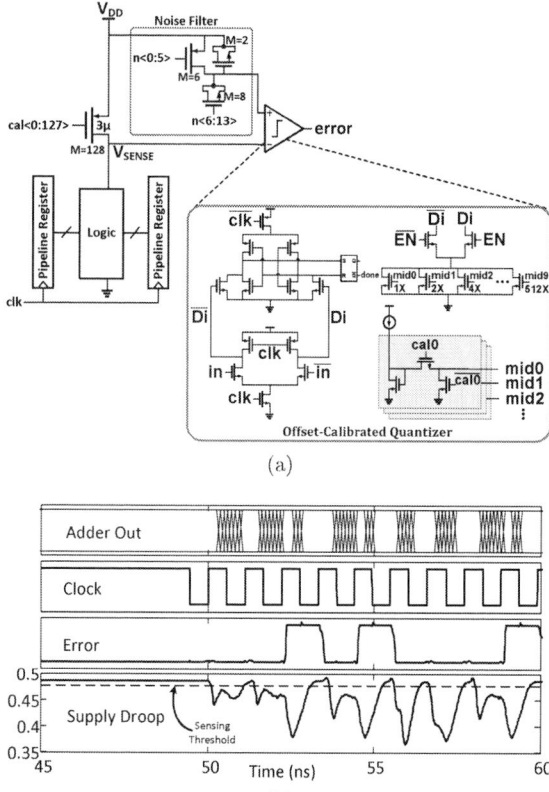

(a)

(b)

Figure 5: (a) CSCD schematic, (b) CSCD timing diagram.

voltage droop across a large PMOS power-gate transistor (Figure 5a). As power gates are becoming more common in modern digital designs [8], it is assumed that they do not add significant area overhead for this work. Because the current consumption only needs to be measured at the clock edge, there is no need to use a continuously-monitoring sensor, like that in asynchronous versions of CSCD [9]. Figure 5b shows a sample SPICE waveform of the virtual supply droop and resulting error detection at the clock edge.

5.2.1 Sensor Resiliency

The CSCD sensor will operate in conditions that are more harsh than typical super-threshold operation. These conditions include process variations, slow NTV operation, temperature variations, small virtual supply droops (affecting minimum input sensitivity), and supply noise.

In order to combat the exacerbated process variations that occur in the NTV regime, a well-known offset calibration scheme in the form of current steering is chosen [7], as shown in Figure 5a. A current-steering DAC along with a simple one-time calibration procedure is used to set the residual offset below 5mV under most extreme variations, including near-threshold operation. To perform the calibration, one tail of the sense-amplifier is chosen and the calibration bits of the current DAC are incremented once for each calibration cycle. Once the sensor reports the error signal the calibration is subtracted to set the sensor threshold just below the settling voltage of the supply. This calibration scheme can

be extended to combat slow-changing variations like temperature by performing live in-situ calibrations periodically.

Figure 6a shows a histogram of the calibrated offset using 8-bits of calibration of the CSCD sensor. Only 2 of the 100 cases have large offsets above 5mV. Larger offsets can be compensated for by increasing the dynamic range of the reference currents into the quantizer.

The CSCD sensor must make a quantization before a new set of data is clocked into the pipeline stage. This is analogous to the min-path race condition problem that exists for Razor-based systems. Figure 6b shows a plot of the sensor conversion speed relative to flip-flop D-Q delay across 100 Monte Carlo points. The majority of cases result in faster conversion speed than the D-Q delay. If timing is a concern, small delay buffers can be added between the sensor and the flip-flop clocks, or a more robust sample-and-hold can be added at the input to the quantizer.

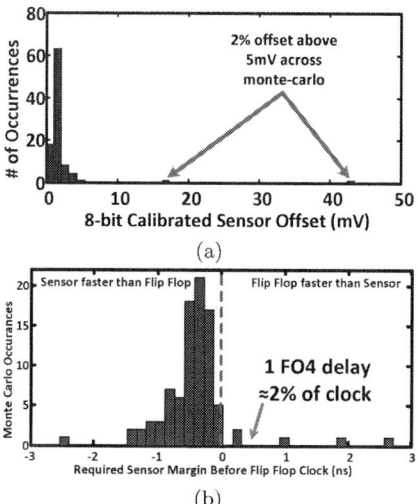

(a)

(b)

Figure 6: (a) Offset calibration of CSCD across 100 Monte Carlo simulations, (b) Speed of sensor stays relatively fast in NTV regime.

One important concern with all CSCD methods is sensing margin. The voltage droop on the virtual supply needs to be large enough to allow the detection of computational errors, but small enough to mitigate a negative impact on performance, due to the large voltage drop on the virtual supply. Figure 7 plots both logic speed and sensing margin versus power gate size. It can be seen that this voltage drop can be quite large (40mV) without negatively impacting speed. For this work, a $100\mu m$ power gate was chosen not only for its droop and speed characteristics, but also due to its smaller impact on area, compared with a larger power gate that provides a minimal speed improvement. The power-gating transistor is parallelized and digitally controlled (30 parallel header-PMOS transistors), thus allowing for programmable amounts of voltage droop across process skews. It is also important to note that this method is strictly limited to supply voltages above or near the NTV operating point, as it will only work when the switching current is discernible from the leakage current. This limits this sensor to process nodes with lower leakage and possibly higher operating voltages than the optimal NTV voltage.

Another major concern for this type of circuit is noise.

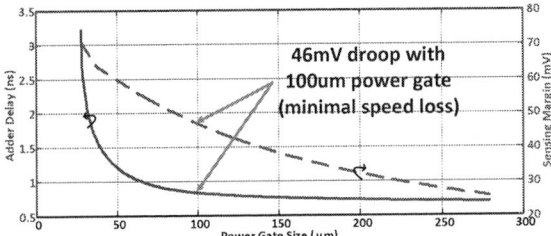

Figure 7: Power gate sizing has very minimal effect speed while maintaining a reasonable sensing margin.

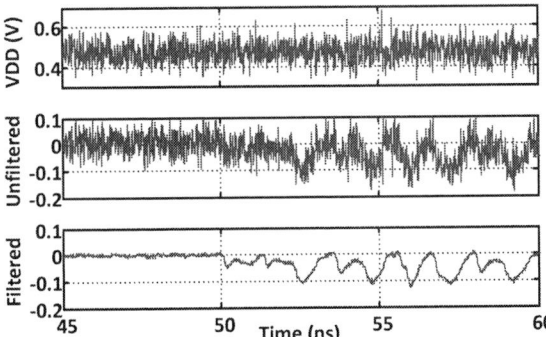

Figure 8: With 200mVpp of supply noise input referred noise is reduced to less than 5mVpp after calibrating CSCD sensor's RC noise filter.

Measuring a small voltage drop across a header transistor can be extremely difficult, especially with supply noise and other sources of noise. Hence, a proposed differential configuration of the sensor can cancel common-mode noise at the inputs, assuming both inputs experience the same noise filtering. In order to make sure the two differentially inputs are correctly correlated with any power-supply noise, a replica RC-matching circuit was designed. Shown in Figure 5a, the circuit consists of digitally-controlled resistances and capacitances that can be tuned post fabrication to match the RC characteristics of the power gate and logic. Simulations show successful power-supply noise reduction of 20x (from 200mVpp to under 5mVpp) at the differential inputs after proper digital calibration (Figure 8).

6. RESULTS

In order to quantify the robustness of the two NTV error detection methods, simulations were performed at a near-threshold voltage of 500mV, comparing Razor, TACD, and CSCD. Using the worst-case static variations utilized in section 4, HSPICE simulations were carried out on all three error-detection methods.

The simulations were designed to find the fastest clock speed at which an error could be detected. Hence, the 16-bit CLA adder was simulated across Monte Carlo process variations, on 1000 input vectors extracted from the EEG FIR filter. To ensure simulation coherency between the three different error-detection methods, the outputs and current consumption of the adder were first extracted and then used as input stimuli for separate simulations of each error detector. Each error-detection method was simulated to find its

optimal operating speed, given the simulated delays of the 1000 vectors.

6.1 Razor Results

In the case of Razor, the fastest clock speed can only increase 20% faster than the STA, whereas TACD and CSCD can be clocked much faster. For Razor, given the limited input data simulated, no errors were generated because all delays were 20% faster than the worst case. This implies that Razor does yield a throughput of 20%, but it is clear that it could benefit from an even faster clock speed. Hence, the choice of these 1000 input vectors may not have stressed the worst-case logic delays, which sets the delay of the Razor clock, and therefore the best possible clock speedup.

6.2 TACD Results

Because TACD essentially lengthens the datapath by adding inverter delays before the error signal, its throughput exceeds Razor only marginally. Figure 9 shows the simulated throughput using TACD. Because of the finite delay of the TDs, many residual error signals are flagged, resulting in a 29% improvement in throughput with TACD, after considering the error rollback delay overhead. Note that TACD does not require a min/max logic delay guarantee within 20%, as required with Razor, but does require an initial off-line calibration procedure for calibrating the inverter buffer delays.

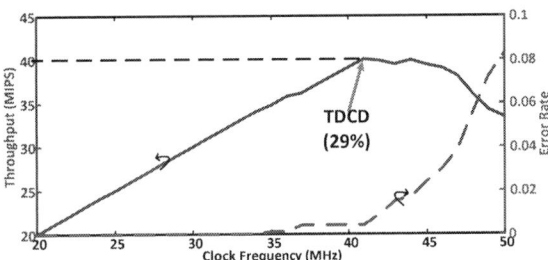

Figure 9: Simulated throughput of TACD.

6.3 CSCD Results

Simulated throughput for CSCD in the NTV regime shows significant improvements over both Razor and TACD. Since CSCD does not add any delays to the datapath, its throughput nearly triples Razor's average performance improvement (56%). As shown in Figure 10, the throughput saturates due to errors generated by the finite settling time of the virtual supply droop and small delay increase associated with the droop, as seen in Figure 7.

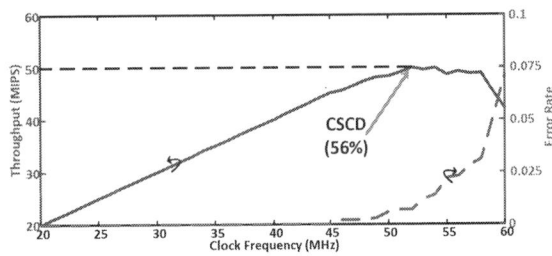

Figure 10: Simulated throughput of CSCD

Table 1: Comparison of error-detection methods.

	No Error Detection	Razor	TACD	CSCD
Average Energy	192.5fJ/comp	976.5fJ/comp	910.3fJ/comp	195.9fJ/comp
Throughput (% Increase)	32MIPS (0%)	38.4MIPS (20%)	41.2MIPS (29%)	49.9MIPS (56%)
Area	15495μm^2	24080μm^2	15796μm^2	16005μm^2
Complexity	NA	Medium	Medium	High
NTV Variation Adaptability	None	None	Tunable TDs	Robust Calibration

6.4 Energy, Area and Complexity

Along with the throughput improvements, Table 1 compares the energy and area overheads of the error-detection methods. Error recovery methods [3] are neglected because their overhead is independent of the detection method used. At 976.5fJ, Razor's energy consumption is dominated by the energy of the shadow latch. The minimum-delay buffer insertion contributes 21 percent to the overall energy increase.

Exhibiting a significantly less area footprint to Razor (mainly due to the large overhead of Razor's inserted min-delay buffers), TACD consumes energy similar to Razor.

CSCD, with a capacitance switching equivalent to one large logic gate, consumes the least amount of energy. Furthermore, because it is clocked only once per cycle, CSCD's dynamic contribution to the energy is much smaller than the other two methods. However, it does require more area than TACD, due to the large offset and process calibration required to ensure correct NTV operation.

One other key factor contrasting all three of these designs is complexity. First, it has not been proven that the minimum-delay buffer insertion required for Razor's operation will scale correctly for NTV operation. Therefore, although any synthesizable form of Razor may be relatively simple to implement, Razor circuits are difficult to guarantee error-free operation across instances of extreme variations. TACD, being fully synthesizable, is easy to implement with logic. However, improving its performance using architecture-dependent techniques and post-fabrication tuning of the transition detectors increases TACD's implementation complexity. This is especially true for ensuring glitch-free operation of the error signal across process corners. CSCD, exhibiting the best throughput improvements, is also the most complex to implement. Designing and adding the analog sensor to a digital circuit will be challenging, such as the post-fabrication calibration required for proper operation across process and supply voltage variations.

7. CONCLUSION

This paper introduces two new approaches to circuit-level timing error detection. Transition-Aware Completion Detection (TACD) observes the activity of the outputs of the combinational logic using an XNOR gate and a variable-delay inverter chain, which is calibrated based upon the amount of switching activity that exists in the logic. The technique of Current-Sensing Completion Detection (CSCD) to the NTV domain was also introduced. CSCD consists of a current sensor that bases its completion/error signal on the current consumption profile of combinational logic across a power gate. These methods were compared to the well known Razor error-detection technique operating in near-threshold. Comprehensive HSPICE simulations show that both TACD and CSCD outperform Razor in throughput, area, and energy and provide a good basis for future work

in NTV error detection.

8. REFERENCES

[1] K. Bowman, J. Tschanz, N. S. Kim, J. Lee, C. Wilkerson, S.-L. Lu, T. Karnik, and V. De. Energy-efficient and metastability-immune resilient circuits for dynamic variation tolerance. *Solid-State Circuits, IEEE Journal of*, 44(1):49 –63, jan. 2009.

[2] K. Bowman et. al. A 45 nm resilient microprocessor core for dynamic variation tolerance. *Solid-State Circuits, IEEE Journal of*, 46(1):194 –208, jan. 2011.

[3] J. Crop, E. Krimer, N. Moezzi-Madani, R. Pawlowski, T. Ruggeri, P. Chiang, and M. Erez. Error Detection and Recovery Techniques for Variation-Aware CMOS Computing: A Comprehensive Review. *Journal of Low Power Electronics and Applications*, 1(3):334–356, 2011.

[4] R. G. Dreslinski, M. Wieckowski, D. Blaauw, D. Sylvester, and T. Mudge. Near-Threshold Computing: Reclaiming MooreâĂŹs Law Through Energy Efficient Integrated Circuits. *Proceedings of the IEEE*, 98(2):253–266, 2010.

[5] D. Ernst, S. Das, S. Pant, R. Rao, C. Ziesler, D. Blaauw, T. Austin, K. Flautner, and T. Mudge. *Razor: a low-power pipeline based on circuit-level timing speculation*. IEEE Comput. Soc.

[6] G. Gammie, N. Ickes, M. Sinangil, R. Rithe, J. Gu, A. Wang, H. Mair, S. Datla, B. Rong, S. Honnavara-Prasad, and Others. A 28nm 0.6 V low-power DSP for mobile applications. In *Solid-State Circuits Conference Digest of Technical Papers (ISSCC), 2011 IEEE International*, number March 2010, pages 132–134. IEEE, 2011.

[7] K. Hu, T. Jiang, J. Wang, F. O'Mahony, and P. Chiang. A 0.6 mw/gb/s, 6.4–7.2 gb/s serial link receiver using local injection-locked ring oscillators in 90 nm cmos. *Solid-State Circuits, IEEE Journal of*, 45(4):899 –908, april 2010.

[8] M. Keating, D. Flynn, R. Aitken, and K. Shi. *Low power methodology manual: for system-on-chip design*. Springer Verlag, 2007.

[9] L. Nagy and V. Stopjakova. Current Sensing Completion Detection in deep sub-micron technologies. In *Design and Diagnostics of Electronic Circuits and Systems (DDECS), 2010 IEEE 13th International Symposium on*, pages 145–148. IEEE, 2010.

[10] R. Pawlowski, E. Krimer, J. Crop, J. Postman, N. Moezzi-Madani, M. Erez, and P. Chiang. A 530mv 10-lane simd processor with variation resiliency in 45nm soi. In *Solid-State Circuits Conference Digest of Technical Papers (ISSCC), 2012 IEEE International*, pages 492 –494, feb. 2012.

Process Variation in Near-Threshold Wide SIMD Architectures

Sangwon Seo[*1], Ronald G. Dreslinski[1], Mark Woh[1], Yongjun Park[1],
Chaitali Charkrabari[2], Scott Mahlke[1], David Blaauw[1], Trevor Mudge[1]

[1]University of Michigan, Ann Arbor, MI 48109   [2]Arizona State University, Tempe, AZ 85287

ABSTRACT

Near-threshold operation has emerged as a competitive approach for energy-efficient architecture design. In particular, a combination of near-threshold circuit techniques and parallel SIMD computations achieves excellent energy efficiency for easy-to-parallelize applications. However, near-threshold operations suffer from delay variations due to increased process variability. This is exacerbated in wide SIMD architectures where the number of critical paths are multiplied by the SIMD width. This paper provides a systematic in-depth study of delay variations in near-threshold operations and shows that simple techniques such as structural duplication and supply voltage/frequency margining are sufficient to mitigate the timing variation problems in wide SIMD architectures at the cost of marginal area and power overhead.

Categories and Subject Descriptors

C.1.2 [**Processor Architectures**]; C.1.4 [**Parallel Architectures**]; C.4 [**Performance of Systems**]

General Terms

Design, Experimentation, Reliability

Keywords

Near-threshold Computing, Wide SIMD, Process Variation

1. INTRODUCTION

An attractive approach for energy-efficient system design is the combination of near-threshold operation [1] for reduced energy consumption and wide SIMD (Single Instruction Multiple Data) architectures to improve parallel performance. This approach is particularly suited for hand-held devices running signal processing algorithms for high throughput applications. However, near-threshold designs are impacted greater by process variations than traditional designs, because the on-current (I_{on}) in the near-threshold voltage region is highly sensitive to variations in threshold voltage, V_{th}. Increased process variations in advanced technology nodes further

---

[*]Currently at Qualcomm Incorporated, San Diego, CA

---

exacerbates the problem, providing many challenges for process engineers and circuit designers [2]. These variation-induced timing errors are much more critical in wide SIMD architectures for two reasons. First, the probability that all SIMD datapaths are error-free decreases when variations are severe, because the number of critical paths are multiplied by the SIMD width. Recent work also shows that there is a significant performance drop in SIMD architectures as single-stage-error probabilities increase [3]. Second, commonly used error-tolerating methods such as pipeline stalling or re-execution result in greater performance and power penalties due to problems in one lane impacting all other lanes. To tolerate variation-induced timing errors in near-threshold operations, complex architectural enhancements have been considered. For example, Synctium [3] proposed decoupled parallel SIMD pipelines and pipeline weaving using decoupling queues and micro-barriers.

In this paper, we investigate the effect of process variations in wide SIMD architectures operating at near-threshold voltages. Delay variations in the near-threshold regime are first analyzed for present and future technology nodes (90nm, 45nm, 32nm, and 22nm). Our study shows that delay variations in near-threshold operations have been over-estimated in the past. In 90nm technology, although delay variation ($3\sigma/\mu$) at 0.5V in a single gate increases by $\sim$2.5x compared to that at 1V, the variation decreases in a chain of gates. For instance, the variation is only $\sim$1.5x for a chain of 50 gates. This is an example of mean-value theorem where the uncorrelated variations are averaged out over the chain. Working against this effect is the fact that the datapath is a wide SIMD machine, thus increasing the number of these critical paths. Nevertheless, the corresponding performance degradation for such wide systems in 90nm technology is less than 5%. Therefore, simple techniques are sufficient to tolerate and mitigate the timing variation problems. Three techniques are explored in this work: 1) structural duplication to replace underperforming modules, 2) voltage margining to reduce both average delay and its variation, and 3) frequency margining to increase delay margins. The analysis shows a combination of these simple techniques can effectively reduce variation-induced timing errors in wide SIMD architectures such as Diet SODA [4] with marginal area and power overhead.

The rest of the paper is organized as follows. Section 2 introduces near-threshold operation. Section 3 discusses variation issues at circuit- and architecture-levels. Section 4 explores techniques to tolerate and mitigate the variation-induced timing errors. Section 5 discusses the related work and Section 6 concludes the paper.

2. NEAR-THRESHOLD OPERATION

There are three regions of operating voltage: super-threshold, near-threshold and sub-threshold (See Figure 9 in Appendix A). In the super-threshold region ($V_{dd} > V_{th}$), energy is highly sensitive to V_{dd} due to the quadratic scaling of switching energy with V_{dd}.

Hence, voltage scaling down to the near-threshold region ($V_{dd} \sim V_{th}$) yields an energy reduction on the order of 10x at the expense of approximately 10x performance degradation. However, the dependence of energy on V_{dd} becomes more complex as voltage is scaled below Vth. In the sub-threshold regime ($V_{dd} < V_{th}$), circuit delay increases exponentially with V_{dd}, causing leakage energy (the product of leakage current, V_{dd}, and delay) to increase in a near-exponential fashion. This rise in leakage energy eventually dominates any reduction in switching energy, creating an energy minimum.

Although the energy minimum is achieved in the sub-threshold region, the performance improves by 50~100x when V_{dd} is scaled from the sub-threshold regime to the near-threshold regime while the energy increases by only 2x. Therefore, near-threshold operations achieve a good balance between performance and energy. The near-threshold region offers an opportunity for applications that require high processing power with high energy efficiency. Furthermore, data parallel architectures like SIMD can be used to compensate for the reduced performance when operating in the near-threshold regime for DLP (Data Level Parallelism)-intensive applications.

3. VARIATIONS IN NEAR-THRESHOLD OPERATION

As described in Section 2, near-threshold designs significantly reduce energy consumption. However, I_{on} is highly sensitive to variations in V_{th}, resulting in delay variations which diminish the advantage of near-threshold operations. RDFs (Random Dopant Fluctuations) are known to be the dominant factor of I_{on} variations in near-threshold operation [5]. In addition, LER (Line Edge Roughness) is a significant factor for advanced technology nodes. To evaluate the effect of cross chip variations in the near-threshold voltage regime, Monte Carlo simulations with Hspice are performed for 90nm/45nm commercially used GP (General Purpose) models and 32nm/22nm PTM (Predictive Technology Model [6]) HP (High Performance) models. Two dominant variation sources, V_{th} and LER, are represented by normal distributions and inserted into the 32nm/22nm PTM HP models.

In this section, we examine how much delay variations occur in the near-threshold voltage region at two levels: (A) circuit-level and (B) architecture-level.

3.1 Circuit-level Variations

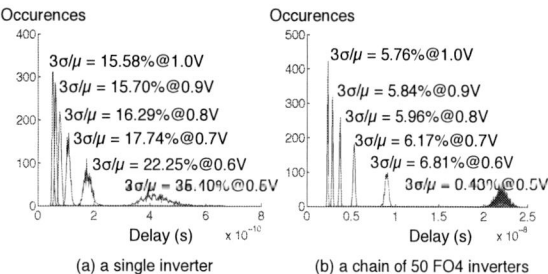

(a) a single inverter   (b) a chain of 50 FO4 inverters

Figure 1: Delay distributions of (a) a single inverter and (b) a chain of 50 FO4 inverters with different supply voltages (0.5V, 0.6V, 0.7V, 0.8V, 0.9V and 1.0V) using 90nm GP technology. A thousand samples are simulated for each supply voltage.

Figure 1 shows that the delay distributions of a single inverter and a chain of 50 FO4 (Fan-out of 4) inverters using 90nm GP models. The delay variation ($3\sigma/\mu$) of a single inverter signifi-

cantly increases as V_{dd} reduces; for example, $3\sigma/\mu$ increases from 15.58%@1.0V to 35.49%@0.5V. Although the delay variations in near-threshold voltage region cause large performance degradation on a single gate, the uncorrelated random within-die variations average out over a long chain of gates as shown in Figure 1(b). The delay variation ($3\sigma/\mu$) of a chain of 50 FO4 inverters is only 9.43% @0.5V compared to that of a single inverter (35.49%@0.5V). Thus the delay variation is not significant for medium to long chains and is expected to not be significant for datapath components. A similar observation was made in [7] which showed only 8.4%@0.5V delay variation for a 64-bit Kogge-Stone adder. Therefore, part of the delay variation problem can be alleviated by implementing longer logic chains [5].

Although delay variations reduce as a chain length (N) increases, additional study shows the amount of reduction ($\frac{\Delta 3\sigma/\mu}{\Delta N}$) decreases with N (see Figure 11 in Appendix C). Therefore, implementing the logic with a very long chain of gates will not solve all the timing variation problems. In addition, technology scaling exacerbates the delay variations [2]; for example, technology scaling from 90nm to 22nm increases delay variation of a chain of 50 FO4 inverters by ~2.5x when operating at 0.55V.

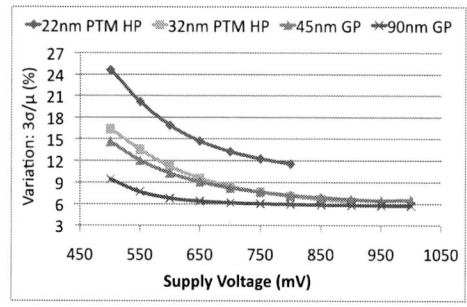

Figure 2: Delay variations ($3\sigma/\mu$) (%) of a chain of 50 FO4 inverters vs. supply voltage (V_{dd}) using four technology models (90nm GP, 45nm GP, 32nm PTM HP, and 22nm PTM HP). A thousand samples for each data point are simulated.

Figure 2 shows the delay variations of a chain of 50 FO4 inverters as a function of V_{dd}. The 22nm PTM HP and 32nm PTM HP models are simulated up to their nominal voltages—800mV and 900mV respectively. As V_{dd} decreases, the delay variations exponentially increases. This trend exacerbates with technology scaling; for example, the increase in delay variation ($3\sigma/\mu$) from 1V to 0.5V is only ~4% in 90nm technology, which is very small compared to ~14% increase in 22nm technology (from 11%@0.8V to 25%@0.5V). This is because LER causes relatively high variations on devices in advanced technology nodes [8]. Advances in lithography like double patterning and immersion are likely to reduce the effect of LER. In addition, strict design rules and new manufacturing processes such as the use of metal-gates with high-k material or silicon-on-insulator (SOI) are expected to help limit the variability. However, in this paper, delay variations presented in Figure 2 are used to analyze variation effects on wide SIMD architectures.

3.2 Architecture-level Variations

To examine the variation effects of near-threshold operations in parallel computing, a 128-wide SIMD architecture, Diet SODA [4], is studied in this paper. A brief description of Diet SODA is included in Appendix B. We focus on the 128-wide SIMD pipeline.

To expedite the study of variation effects in this wide SIMD architecture, several reasonable simplifications were made in this study. First, a chain of 50 FO4 inverters is used to emulate a criti-

981

cal path of the SIMD datapath because they are similar in terms of average delay and variation at all voltages, not just at near-threshold voltages. We chose a chain configuration because it is a standard practice in circuit-level analysis. Second, a hundred critical paths are assumed to exist in one SIMD lane because of two reasons: 1) a generated synthesis report for Diet SODA [4] shows ~50 critical paths in each SIMD lane; 2) another 50 near-critical paths are also considered because they could become critical due to increased variations in the near-threshold regime. Third, we used the following two properties: 1) the delay of one SIMD lane (1-wide) is determined by the slowest critical path in the lane; 2) the delay of an N-wide SIMD datapath is determined by the slowest of the N SIMD lanes in simulations.

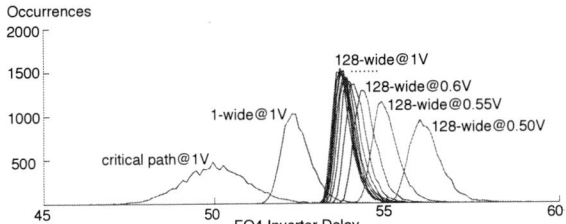

Figure 3: Delay distributions for a critical path (a chain of 50 FO4 inverters) at V_{dd}=1V, one SIMD lane at V_{dd}=1V, and 128-wide SIMD datapath at near-threshold supply voltages from 0.5V to 1V. 90nm GP model is used and a 10,000 samples are simulated.

Figure 3 shows the delay distributions for a critical path (a chain of 50 FO4 inverters), one SIMD lane (1-wide system) operating at 1V, and 128-wide systems operating at near-threshold supply voltages. The delay unit on the x-axis is FO4 inverter delay which is different from absolute delay (in *ns*) used in Figure 1. For example, the delay of a chain of 50 FO4 inverters operating at 0.5V is 22.05ns (= 50 FO4 delay@0.5V); on the other hand, that at 0.6V is 8.99ns (= 50 FO4 delay@0.6V). In this paper, FO4 delay is used to measure variation effects in the near-threshold voltage region.

As can be seen in Figure 3, the delay distribution of a 1-wide SIMD datapath@1V is shifted to the right compared to that of one critical path@1V because the delay of a 1-wide system is determined by the maximum delay of a hundred critical paths. The same reasoning can be made to explain the shift in the delay distribution from 1-wide@1V to 128-wide@1V. The 128-wide SIMD datapath is slower than the 1-wide SIMD datapath because the possibility of having slow critical paths increases. Another characteristic is that the delay distributions of 128-wide systems operating at low supply voltages drift to the right. This shift is because the delay distribution of a critical path at near-threshold voltages has a wider spread than that at nominal voltage.

In order to evaluate performance degradation due to near-threshold voltage operations, we compare the 99% point of FO4 chip delay ($fo4chipD$) distributions. The performance degradation of a 128-wide SIMD architecture operating at near-threshold voltage (NTV) region compared with the performance at nominal voltage (or full voltage, FV) is given by $\frac{fo4chipD@NTV - fo4chipD@FV}{fo4chipD@FV}$. Figure 4 shows the performance drop as a function of supply voltage in four technology nodes. As expected, the performance drop increases as the supply voltage decreases. For example, in 90nm GP model, the performance drop at 0.5V, 0.55V, and 0.6V is ~5%, ~2.5%, and ~1.5% respectively compared to 1V operation. In addition, the increase in performance degradation of lower technology nodes is much higher. For example, the performance drop at 0.5V climbs to

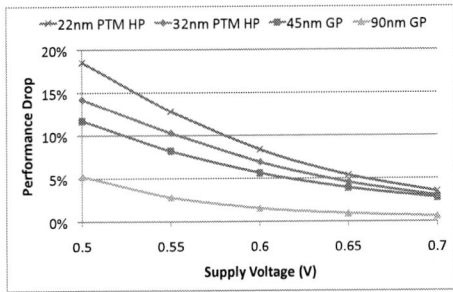

Figure 4: Performance drop (%) in the near-threshold voltage region for a 128-wide SIMD architecture. 90nm/45nm GP and 32nm/22nm PTM HP models are used.

~18% in 22nm PTM HP model.

This analysis shows that delay variations in wide-SIMD architectures is not that large. It is only ~5%@0.5V in 90nm GP and increases to ~20% for 22nm PTM HP model. It is very likely that the variations will be lower in 22nm real silicon. Thus complex architectural enhancements are not needed to handle these delay variations. In fact, simple techniques are sufficient to handle the variation-induced delay variations in wide SIMD architectures, as will be described in the following section.

4. TECHNIQUES TO CONTROL EFFECT OF VARIATIONS

There are two mechanisms to tolerate variation-induced timing errors in a scalar pipeline: 1) flushing the pipeline and re-executing a instruction with relaxed timing or 2) waiting one more clock cycle for the pipeline to generate the correct output. However, applying these approaches to wide SIMD architectures is problematic because the power penalty of the flush-rollback process in the SIMD pipeline is much larger than that of a scalar pipeline. For example, an error encountered in one SIMD lane would cause the other SIMD lanes to stall, flush, and execute the same operations again. Recent work also shows that there is a significant performance drop in SIMD architectures as single-stage-error probabilities increase [3]. To prevent variation-induced timing errors in near-threshold operation, we analyzed the effect of three techniques: 1) structural duplication, 2) voltage margining, and 3) frequency margining.

4.1 Structural Duplication

Structural duplication is a well-known technique for extending reliability. Redundant micro-architectural structures are added to the processor and designated as spares [9]. When some architectural modules fail in time, the spare structures replace the failed ones to extend lifetime reliability. This structural duplication idea can be used to handle slow SIMD lanes that fail to operate within a given clock period. If the faulty SIMD lanes can be identified at test time, the spare SIMD lanes can be used to replace them.

We studied a 128-wide SIMD architecture and analyzed how many SIMD functional unit duplications (α spares) are required to tolerate variation-induced timing errors while running in the near-threshold voltage regime. Monte Carlo simulations were performed to generate FO4 delay distribution curves for the duplicated systems as shown in Figure 5.

The delay distribution of a 128-wide SIMD system operating at 1V (*128-wide*@1V) is used as the baseline and the delay distribution of *128-wide+α-spares*@ 0.55V is used to demonstrate the effect of SIMD functional unit duplications. For example, the dis-

982

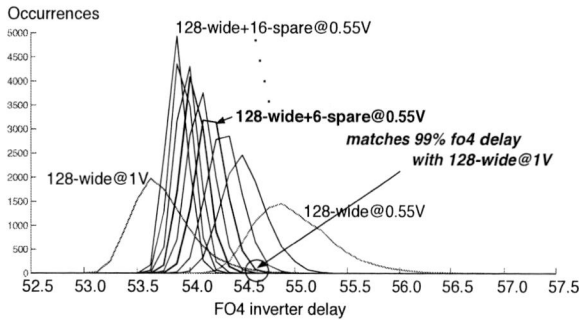

Figure 5: Delay distributions for SIMD duplicated systems (128-wide + α-spares) using 90nm GP model. For each curve, 10,000 samples are simulated.

tribution curve of *128-wide+6-spares*@0.55V is essentially the distribution of 128 *good* SIMD datapaths out of 134 (128+6) SIMD datapaths; i.e. six slowest SIMD datapaths are dropped to generate this delay distribution. As can be seen, extra SIMD datapaths help shift delay distributions to the left and make the spread smaller.

We match the 99% FO4 delay point of the duplicated systems operating at near-threshold voltages with that of the baseline architecture (128-wide) operating at nominal voltage to obtain the required number of additional SIMD spares. This experiment is repeated for four technology nodes (90nm, 45nm, 32nm, and 22nm), and the number of spares and corresponding area and power overhead at each supply voltage are presented in Table 1. We see that as supply voltage reduces, the number of SIMD spares exponentially increases to tolerate effect of delay variations. For example, in 90nm technology node, the number of spares increases from two spares for 0.6V to six spares for 0.55V and 28 spares for 0.5V. This is because, as shown in Figure 5, adding more spare units shifts the chip delay distribution to the left, but makes it tighter. For lower technology nodes, delay variations are larger and excessive number of spares is required to match the 99% FO4 delay point of the baseline architecture.

The additional SIMD functional unit (FU) spares are used to replace underperforming ones that are identified at test time. The faulty SIMD FUs can be power-gated because they are not used at run time. Therefore, the power overhead of the structural duplication scheme is limited only to enlarged routing, thus leading to minimal impact on power consumption. However, the increased SIMD width also requires a wider shuffle network operating at nominal voltage whose power consumption cannot be ignored. Thus, for low voltages (∼0.50V) where the variation-induced timing errors are severe, the structural duplication scheme has a large overhead.

Based on the analysis in Table 1, the number of additional SIMD spares can be determined. However, how to place the spares is another interesting design choice in wide SIMD architectures. We investigate two placement methods: global sparing and local sparing. The local sparing scheme groups SIMD functional units into clusters and places a spare for each cluster while the global sparing scheme places all the spares together. Recently proposed Synctium [3] suggests a local sparing method such as assigning one spare per every cluster of four SIMD lanes; here, the spare substitutes any one of four faulty SIMD lanes. Although the local redundancy overcomes complex re-routing problems, this local sparing method does not work when there are more than one faulty SIMD lanes in a cluster. On the other hand, a global sparing method is capable of dealing with any bursty failures in adjacent SIMD lanes because spares are not assigned to specific clusters. To avoid com-

plex re-routing that is required of most global sparing schemes, the XRAM crossbar [10] is used. It exploits the circuit topology of SRAM cells by holding shuffle configurations at crossing points of the cells and is both area- and power-efficient. An application of this scheme is illustrated in Appendix D.

4.2 Voltage Margining

As supply voltage (V_{dd}) decreases, the delay of a chain of 50 FO4 inverters exponentially increases. Therefore, a small increase in supply voltage in the near-threshold voltage region can help compensate for variation-induced timing errors without increasing the clock period.

To gauge how much extra supply voltage is required, we first generated the FO4 chip delays ($fo4chipD$) and the corresponding absolute chip delays ($chipD$ in *ns*) of a 128-wide SIMD architecture operating at near-threshold voltages (NTVs). Then, the $chipD@NTV$ is scaled based on the ratio of $fo4chipD@FV$ and $fo4chipD@NTV$. The normalized $chipD@NTV$ is used as the baseline *target delay* for the architecture operating at near-threshold voltage to achieve the same level of variations at nominal voltage. Next, we increase supply voltage at a fine grain to find required voltage margin (V_M) that makes $chipD@(NTV+V_M)$ less than the *target delay*. Figure 6 illustrates how voltage margin is obtained for a 128-wide SIMD datapath operating at 600mV for a specific *target delay*. Delay distributions of a 128-wide SIMD architecture operating at 600mV, 605mV, 610mV, 615mV, and 620mV are generated. In addition, delay distributions of 128-wide+α-spare SIMD duplicated systems operating at 600mV are also shown in the figure. As can be seen, the $chipD$ (99% point of delay distribution) of a 128-wide SIMD architecture operating at ∼615mV is less than *target delay*. Therefore, ∼15mV is the voltage margin at design time that is required for a 128-wide SIMD architecture operating at 600mV to tolerate its delay variation.

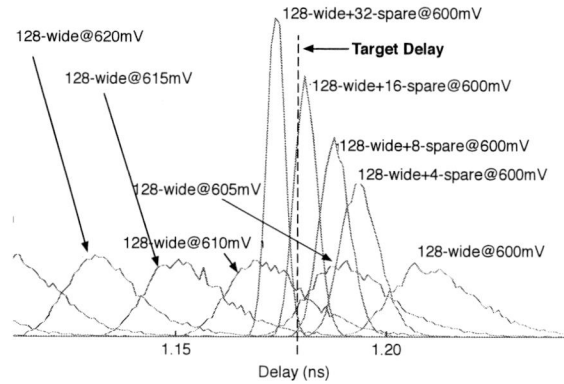

Figure 6: Delay distributions of 128-wide SIMD architecture operating at 600mV, 605mV, 610mV, 615mV and 620mV. For comparison, delay distributions of 128-wide+α-spare SIMD duplicated systems operating at 600mV are also presented. A 10,000 samples for each curve are simulated with 45nm GP model.

Table 2 lists supply voltages (V_{dd}), voltage margins and the corresponding power overhead for four different technology nodes. Although a very small increase in supply voltage is sufficient for the 90nm technology node, lower technology nodes require much larger supply voltage margins. For example, in 90nm technology, at V_{dd}=500mV, the supply voltage has to be increased by 5.78mV to ∼506mV, but this jumps to ∼520mV in 45nm technology.

983

	90nm			45nm			32nm			22nm		
Vdd	spares	area ovhd.	power ovhd.	spares	area ovhd.	power ovhd.	spares	area ovhd.	power ovhd.	spares	area ovhd.	power ovhd.
0.50V	28	12.1%	4.6%	>128	> 57.8%	> 25.0%	>128	> 57.8%	> 25.0%	>128	> 57.8%	> 25.0%
0.55V	6	2.6%	1.0%	84	37.2%	15.3%	>128	> 57.8%	> 25.0%	80	35.3%	14.5%
0.60V	2	0.9%	0.3%	26	11.2%	4.3%	48	20.9%	8.2%	22	9.5%	3.6%
0.65V	1	0.4%	0.2%	10	4.3%	1.6%	12	5.1%	1.9%	7	3.0%	1.1%
0.70V	1	0.4%	0.2%	4	1.7%	0.6%	6	2.6%	1.0%	3	1.3%	0.5%

Table 1: The required number of spares and corresponding area and power overhead of structural duplication scheme for four technology nodes. The area and power numbers are based on Diet SODA [4].

	90nm		45nm		32nm		22nm	
Vdd	Vdd margin	power ovhd.	Vdd margin	power ovhd.	Vdd margin	power ovhd.	Vdd margin	power ovhd.
0.50V	5.8 mV	1.0%	19.6 mV	3.3%	12.1 mV	2.0%	16.4 mV	2.8%
0.55V	4.1 mV	0.6%	18.2 mV	2.8%	11.1 mV	1.7%	17.6 mV	2.7%
0.60V	2.9 mV	0.4%	16.2 mV	2.3%	10.4 mV	1.5%	11.1 mV	1.6%
0.65V	2.2 mV	0.3%	14.0 mV	1.8%	8.9 mV	1.1%	11.5 mV	1.5%
0.70V	1.7 mV	0.2%	12.8 mV	1.5%	7.7 mV	0.9%	9.6 mV	1.1%

Table 2: Required voltage margin (V_M) to tolerate variation-induced timing errors for a 128-wide SIMD architecture operating at near-threshold voltages and corresponding power overhead for four technology nodes. The final supply voltage should be $V_{dd} + V_M$. The power overhead is based on Diet SODA [4].

This extra supply voltage margin applies to all modules operating in the near-threshold voltage domain and thus incurs more power consumption than structural duplication methods for low variations. However, as variation increases, the voltage margining method offers a more power-efficient solution than the structural duplication scheme.

4.3 Frequency Margining

To avoid variation-induced timing errors, the clock period can be increased when there is a very loose realtime constraint so that the increased clock period can still make the timing requirements. However, as we move to advanced technology nodes, required delay margins reach almost 20% (details in Table 4 in Appendix E), which makes the frequency margining scheme inappropriate for handling variation-induced timing errors. In addition, the clock frequency of near-threshold SIMD datapath is closely related to that of a memory system; for example, the SIMD datapath clock period (T_{clk}@NTV) has to be multiples of the memory clock period (T_{clk}@FV) to avoid complex synchronization between two subsystems. Therefore, frequency margining can only be supported after careful consideration of the underlying architecture.

4.4 Comparisons Between Variation-Tolerant Techniques

In this section, the power overhead of structural duplication and voltage margining is compared and summarized (see Figure 7). To achieve iso-throughput performance, frequency margining is not considered here.

Structural duplication scheme outperforms voltage margining sc-heme in high near-threshold voltage regions (0.6V~0.7V) where variations are very low. However, as technology scales and supply voltage decreases, the voltage margining scheme starts to outperform the structural duplication scheme. This is because a slight increase in supply voltage exponentially reduces delay. Figure 7 serves as a guideline in which variation-tolerating scheme must be selected for each supply voltage. For example, in 45nm technology node, when V_{dd}=0.6V, duplication method incurs ~4% power overhead compared to ~2% overhead of voltage margining scheme; therefore voltage margining is the preferred choice.

Although voltage margining offers a better solution than structural duplication for lower technology nodes as V_{dd} decreases, the structural duplication scheme still can significantly help manage variation-induced timing errors. Figure 8 shows chip delays for a 128-wide SIMD architecture operating at 600mV, 605mV, 610mV, 615mV and 620mV using 45nm GP model. Target chip delay is

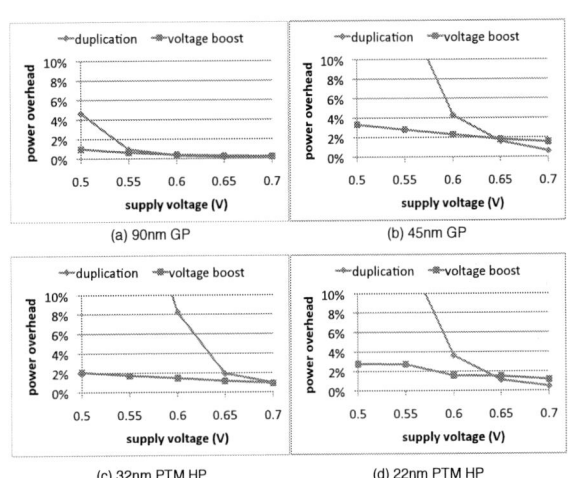

Figure 7: Power overhead comparison between structural duplication and voltage margining schemes for four technology nodes: (a) 90nm GP, (b) 45nm GP, (c) 32nm PTM HP, and (d) 22nm PTM HP

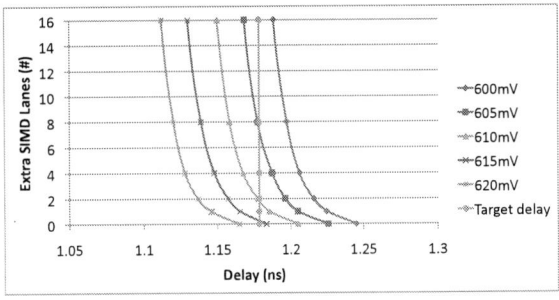

Figure 8: Chip delays for a 128-wide SIMD datapath operating at from 600mV to 620mV. Target delay is a design constraint for the 128-wide near-threshold system operating at 600mV. 45nm GP model is used.

calculated as described in Section 4.2. Based on this figure, the target chip delay can be achieved by having 1) two additional SIMD lanes with 10mV voltage margin or 2) eight additional SIMD lanes with 5mV voltage margin.

Table 3 summarizes several design choices and the corresponding power overhead. As can be seen, a combination of two additional SIMD lanes and 10mV voltage margin achieves minimal power overhead (1.72%) compared to only structural duplication (4.28%) or only voltage margining (2.39%). Therefore, a combination of voltage margining and structural duplication can effectively tolerate and mitigate timing variation problems for lower technology nodes.

duplications	voltage margin	power overhead
26	0 mV	4.3 %
8	5 mV	2.0 %
2	10 mV	1.7 %
1	15 mV	2.3 %
0	17 mV	2.4 %

Table 3: Design choices for a 128-wide@600mV system in 45nm technology node. Combinations of structural duplication and voltage margining are presented with corresponding power overhead.

5. RELATED WORK

There has been a large interest in sub-threshold designs, resulting in a wide range of working processors for ultra low power applications. Examples include Subliminal [11], Phoenix processors [12], and the 180mV FFT processor [13]. However, to improve processing throughput significantly while marginally affecting the high energy efficiency, near-threshold operations are proposed. In addition, near-threshold operation also combines with parallel computing platforms in a synergistic manner. Zhai et al. show that exploiting near-threshold techniques achieves substantial energy savings in chip multi-processing [14] and Kaul et al. present 494 GOPS/W SIMD vector processing accelerators operating at 300mV [15].

Although these sub-threshold and near-threshold techniques offer great energy efficiency, variability has become a serious concern for operating at extremely low voltages. Variation-aware architectures are implemented using circuit techniques such as clock/power gating and dynamic voltage-frequency scaling [16], and fine-grained power management using both dual-supply voltage and power gating [15]. EVAL [17] provides a framework to show how several techniques such as ABB (Adaptive Body Biasing) / ASV (Adaptive Supply Voltage), FU (Functional Unit) replication, and issue-queue resizing can trade off variation-induced errors for power and performance. However, little analysis has been performed to investigate the impact of process variability on large parallel architectures such as a SIMD machine. Recently, Synctium [3] studied the variation issues in near-threshold SIMD architectures and proposed decoupled parallel SIMD pipelines and pipeline weaving using decoupling queues and microbarriers to tolerate variation-induced timing errors. Our work differs in that we first provide a detailed analysis of variation impact on wide SIMD architectures for different technology nodes, and show past studies have over-estimated the effect of delay variations in near-threshold operations. We also propose simple techniques to handle delay variations in multiple technology nodes and present a variation-aware wide SIMD architecture that effectively tolerates the timing variability problems in 90nm technology by exploiting simple SIMD functional unit duplications connected via an XRAM crossbar.

6. CONCLUSIONS

Near-threshold operation enables a more energy-efficient architecture. In particular, a combination of near-threshold circuit techniques and parallel SIMD computations has the capability of providing high energy efficiency with high-throughput performance.

Although near-threshold techniques offer new promising architectural design options, they suffer from large delay variations due to increased process variability. In this work we provide a systematic study of variation issues of near-threshold wide SIMD architectures and show that the variation-induced timing errors in wide SIMD architectures are fairly small, and can be allayed with combinations of three simple techniques: structural duplication, voltage margining and frequency margining. In 90nm technology node, we show the variation-induced timing errors in wide SIMD architectures can be handled by only structural duplications. However, for lower technology nodes, use of only structural duplication is not as efficient; rather a combination of structural duplication and voltage margining results in a solution with the lowest power overhead.

7. ACKNOWLEDGMENTS

This research is supported in part by the National Science Foundation grants CSR-091699, CNS-0910851 and ARM. Thanks also to Yoonmyung Lee and Mingoo Seok for their help and feedback.

8. REFERENCES

[1] R. Dreslinski et al. Near-Threshold Computing: Reclaiming Moore's Law Through Energy Efficient Integrated Circuits. *Proceedings of the IEEE*, vol. 98, no. 2, pages 253–266, Feb. 2010.

[2] K. Bernstein et al. High-performance CMOS variability in the 65nm regime and beyond. *IBM Journal of Research and Development*, vol. 50, no. 4.5, pages 433–449, 2006.

[3] E. Krimer et al. Synctium: a near-threshold stream processor for energyconstrained parallel applications. *IEEE Computer Architecture Letters*, pages 21–24, 2010.

[4] S. Seo et al. Diet SODA: A Power-Efficient Processor for Digital Cameras. *Proceedings of the 16th ACM/IEEE International Symposium on Low Power Electronics and Design*, pages 79–84, 2010.

[5] B. Zhai et al. Analysis and mitigation of variability in subthreshold design. *Proceedings of the 2005 International Symposium on Low Power Electronics and Design*, pages 20–25, 2005.

[6] Nanoscale Integration and Modeling (NIMO) Group. Predictive technology model (PTM). *[online] http://www.eas.asu.edu/~ptm/*

[7] N. Drego et al. All-digital circuits for measurement of spatial variation in digital circuits. *IEEE Journal of Solid-State Circuits* vol. 45, no. 3, pages 640–651, Mar. 2010.

[8] Y. Ye et al. Statistical Modeling and Simulation of Threshold Variation Under Random Dopant Fluctuations and Line-Edge Roughness. *IEEE Transactions on Very Large Scale Integration (VLSI) Systems*, no. 99, pages 1–10, 2010.

[9] J. Srinivasan et al. Exploiting structural duplication for lifetime reliability enhancement. *Proceedings of the 32nd annual international symposium on Computer Architecture*, pages 520–531, 2005.

[10] S. Satpathy et al. A 1.07 Tbit/s 128x128 Swizzle Network for SIMD Processors. *IEEE Symposium on VLSI Circuits*, June 2010.

[11] B. Zhai et al. Energy-efficient subthreshold processor design. *IEEE Transactions on Very Large Scale Integration (VLSI) Systems*, vol. 17, no. 8, pages 1127–1137, Aug. 2009.

[12] M. Seok et al. The Phoenix processor: A 30pw platform for sensor applications. *IEEE Symposium on VLSI Circuits*, pages 188–189, June 2008.

[13] A. Wang and A. Chandrakasan. A 180mV FFT processor using subthreshold circuit techniques. *IEEE International Solid-State Circuits Conference, Digest of Technical Papers*, pages 292–529, 2004.

[14] B. Zhai et al. Energy efficient near-threshold chip multi-processing. *Proceedings of the 2007 International Symposium on Low-Power Electronics Design*, pages 32–37, 2007.

[15] H. Kaul et al. A 300mV 494GOPS/W reconfigurable dual-supply 4-way SIMD vector processing accelerator in 45nm CMOS. *IEEE Journal of Solid-State Circuits*, vol. 45, no. 1, pages 95–102, 2010.

[16] S. Dighe et al. Within-die variation-aware dynamic-voltage-frequency scaling core mapping and thread hopping for an 80-core processor. *IEEE International Solid-State Circuits Conference Digest of Technical Papers*, pages 174–175, Feb. 2010.

[17] S. Sarangi et al. Eval: Utilizing processors with variation-induced timing errors. *Proceedings of the 41st Annual International Symposium on Microarchitecture*, pages 423–434, Dec. 2008.

APPENDIX

A. NEAR-THRESHOLD OPERATION

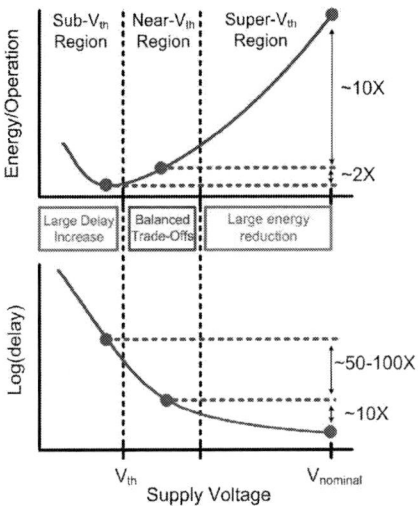

Figure 9: The energy and delay associated at each supply voltage point is presented for three regions of operation: super-threshold ($V_{dd} > V_{th}$), near-threshold ($V_{dd} \sim V_{th}$) and sub-threshold ($V_{dd} < V_{th}$).

Figure 9 defines three regions of operations, namely, super-threshold, near-threshold and sub-threshold. Voltage scaling down to the near-threshold region from the super-threshold region yields an energy reduction on the order of 10x at the expense of approximately 10x performance degradation. Although the energy minimum is achieved in the sub-threshold region, the performance improves by 50~100x when V_{dd} is scaled from the sub-threshold regime to the near-threshold regime while the energy increases by only 2x. Therefore, near-threshold operations achieve a good balance between performance and energy.

B. A NEAR-THRESHOLD WIDE SIMD ARCHITECTURE: DIET SODA

Figure 10 shows the architectural details of a single processing element (PE) of a wide SIMD architecture, Diet SODA [4]. The PE consists of 1) 64 KB multi-banked SIMD memory, 2) 4 KB scalar memory, 3) SIMD data prefetcher, 4) SIMD pipeline for vector operations, 5) scalar pipelines for sequential operations, and 6) 4-wide address generation unit (AGU) pipeline for providing local memory addresses for four memory banks. The PE operates in two different voltages: full voltage and near-threshold voltage. Memory-related modules (1, 2, 3, 5a, and 6 in Figure 3) operate at full voltage because of data retention issues in the near-threshold voltage regime while SIMD datapath (4 and 5b in Figure 3) can operate at near-threshold voltage to lower power consumption.

The multi-banked SIMD memory system consists of four memory banks; each bank is 32-wide 16-bit 256-entries (16KB). The SIMD data prefetcher coordinates with 128-wide buffer and 128x128 XRAM crossbar to support complex alignment operations such as two-dimensional data access that are widely used in multimedia algorithms. The four AGU pipelines are dedicated to the four SIMD memory banks and SIMD data prefetcher to handle memory address calculations. The SIMD pipeline consists of a 128-wide 16-bit 32-entry SIMD register file (RF), 128 functional units (FUs), a

128 x 128 XRAM crossbar (SIMD shuffle network (SSN)), and a multi-output adder tree. There are two scalar pipelines, one in each voltage domain; both pipelines consist of one 16-bit datapath and are used to perform sequential algorithms in addition to coordinating the SIMD datapath.

C. DELAY VARIATION VS. LOGIC CHAIN LENGTH

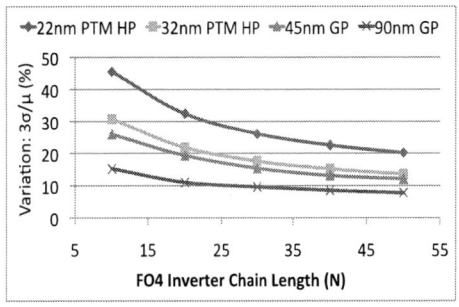

Figure 11: Delay variations ($3\sigma/\mu$) (%) at 0.55V of a chain of FO4 inverters vs. chain length (N) using four technology models (90nm GP, 45nm GP, 32nm PTM HP, and 22nm PTM HP). A thousand samples are simulated for each data point.

Figure 11 shows the delay variations ($3\sigma/\mu$) (%) at 0.55V as a function of chain length (N) of FO4 inverters at four technology nodes (90nm, 45nm, 32nm and 22nm). The amount of reductions, $\frac{\Delta 3\sigma/\mu}{\Delta N}$, decreases with N and therefore implementing the logic with a long chain of gates will not solve the timing variation problem.

D. PLACEMENT METHOD: GLOBAL VS. LOCAL

Figure 12(a) shows how local functional unit (FU) spares work. Here, functional unit spare (FU-S-0) is used as a spare for a cluster consisting of FU-0, FU-1, FU-2, and FU-3. If multiple timing errors occur in this cluster, FU-S-0 cannot replace all the failing FUs. In that case, either the entire system must slow down or waste energy by increasing the voltage to meet timing constraints. On the other hand, a global sparing method is capable of dealing with bursty FU failures because spares are not assigned to specific clusters.

Although global sparing effectively solves timing variability issues, it requires complex re-routing. Satpathy et al. recently proposed an area- and power-efficient XRAM crossbar [10], which exploits the circuit topology of SRAM cells and stores shuffle configurations at crossing points of the cells to improve performance while reducing area, power and routing congestions. We make use of the XRAM crossbar to effectively support bypassing under-performing SIMD lanes. Figure 12(c) shows that how an XRAM crossbar bypasses faulty SIMD FU-2 & FU-3 and fully utilizes the remaining eight SIMD functional units based on the configuration registers stored in the XRAM crossbar shown in Figure 12(b).

E. FREQUENCY MARGINING

Table 4 presents desired clock period (T_{clk}), variation-aware clock period (T_{va-clk}), and corresponding performance degradation for several near-threshold voltages. For advanced technology nodes, frequency margining is not a usable option.

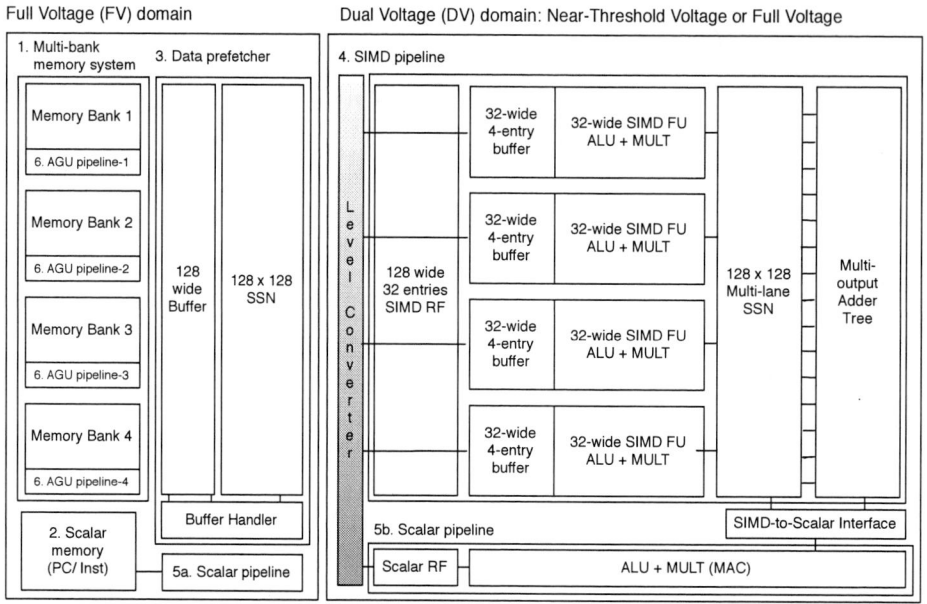

Figure 10: Processing element (PE) of a wide SIMD architecture. The PE contains two different voltage domains: full voltage (FV) and dual voltage (DV). DV domain operates at either full or near-threshold supply voltage. The PE consists of 1) multi-banked SIMD memory; 2) scalar memory; 3) SIMD data prefetcher, 4) SIMD pipeline, 5a) scalar pipeline in FV domain, 5b) scalar pipeline in DV domain, and 6) 4-wide address generation unit (AGU) pipelines.

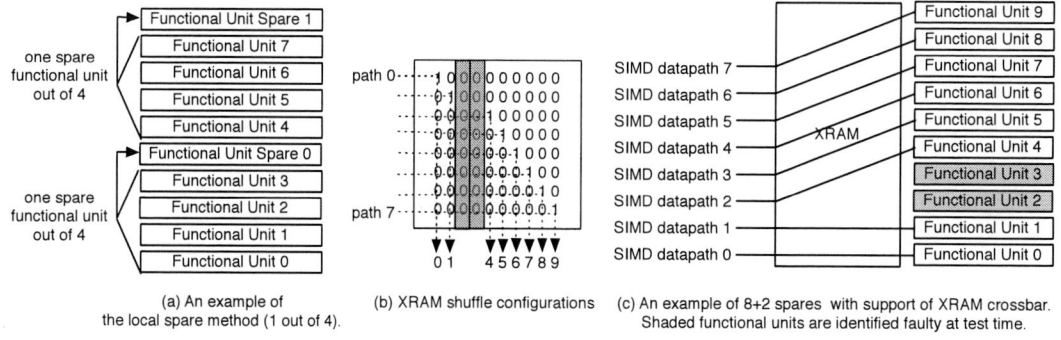

Figure 12: (a) Local sparing method. An example of *1 out of 4*. (b) XRAM shuffle configuration to bypass faulty SIMD lanes. (c) Global sparing method. An example of 10 functional units (8 + 2 spares) with support of XRAM crossbar. Shaded SIMD functional units are identified as faulty ones at test time.

	90nm			45nm			32nm			22nm		
Vdd	Tclk(ns)	Tva-clk(ns)	perf. drop	Tclk(ns)	Tva-clk(ns)	perf. drop	Tclk(ns)	Tva-clk(ns)	perf. drop	Tclk(ns)	Tva-clk(ns)	perf. drop
0.50V	24.0	25.3	5.2%	1.9	2.1	11.7%	5.5	6.3	14.2%	3.1	3.7	18.5%
0.55V	14.3	14.7	2.8%	1.4	1.6	8.2%	2.9	3.2	10.3%	1.8	2.0	12.8%
0.60V	9.8	9.9	1.5%	1.2	1.2	5.6%	1.8	1.9	6.9%	1.3	1.4	8.4%
0.65V	7.3	7.4	0.9%	1.0	1.0	3.9%	1.3	1.3	4.5%	0.9	0.9	5.4%
0.70V	5.8	5.8	0.6%	0.9	0.9	2.7%	1.0	1.0	3.0%	0.7	0.7	3.5%

Table 4: Designed clock period (T_{clk}), variation-aware clock period (T_{va-clk}), and corresponding performance degradation at near-threshold voltages for four technology nodes. The power overhead is based on Diet SODA [4]. With technology scaling, frequency margining becomes infeasible solution.

Run-Time Power-Down Strategies for Real-Time SDRAM Memory Controllers

Karthik Chandrasekar[1], Benny Akesson[2], Kees Goossens[2]

[1]Computer Engineering, TU Delft, The Netherlands
[2]Electronic Systems, TU Eindhoven, The Netherlands
[1]k.chandrasekar@tudelft.nl
[2]{k.b.akesson, k.g.w.goossens}@tue.nl

ABSTRACT

Powering down SDRAMs at run-time reduces memory energy consumption significantly, but often at the cost of performance. If employed speculatively with real-time memory controllers, power-down mechanisms could impact both the guaranteed bandwidth and the memory latency bounds. This calls for power-down strategies that can hide or bound the performance loss, making run-time memory power-down feasible for real-time applications.

In this paper, we propose two such strategies that reduce memory energy consumption and yet guarantee real-time memory performance. One provides significant energy savings without impacting the guaranteed bandwidth and latency bounds. The other provides higher energy savings with marginally increased latency bounds, while still preserving the guaranteed bandwidth provided by real-time memory controllers. We also present an algorithm to select the most energy-efficient power-down mode at run-time. We experimentally evaluate the two strategies at run-time by executing four media applications concurrently on a real-time MPSoC platform and show memory energy savings of 42.1% and 51.3% for the two strategies, respectively.

Categories and Subject Descriptors

C.3 [**Special-Purpose and Application-Based Systems**]: Real-time and embedded systems; B.8.2 [**Performance and Reliability**]: Performance Analysis and Design Aids

General Terms

Performance, Design, Algorithms

Keywords

SDRAM, Power-Down, Real-Time, Memory Controller

1. INTRODUCTION

Increasing performance demands of modern MPSoCs often reflect poorly in overall system energy consumption. SDRAMs in particular, contribute considerably to the system energy consumption [1] and have the option of powering down [2] at run-time to save energy. However, these power-down mechanisms come at the cost of performance (bandwidth and latency), due to their power-up latencies [10].

Applications with real-time requirements demand worst-case performance guarantees from every component in the system, including the SDRAMs, where these guarantees are at the memory transaction level. Real-time SDRAM controllers provide such guarantees to a memory requester, such as a processor, in terms of a minimum guaranteed bandwidth and/or a maximum latency bound for memory accesses. Real-time SDRAM controllers, such as [3–8], employ predictable *memory arbiters*, such as Round-Robin or Time Division Multiplexing, to schedule memory accesses from different requesters and to provide performance guarantees. If they speculatively employ power-down mechanisms at run-time when the memory is idle, it can affect both the latency and the bandwidth guarantees provided, due to the power-up latencies [10]. Hence, they do not support run-time power-down. However, to design efficient future real-time systems [9], it is essential to reduce memory power consumption while satisfying performance requirements.

This paper proposes *two run-time power-down strategies* that reduce memory power consumption, while preserving the original bandwidth guarantees and also providing memory access latency bounds to guarantee real-time behavior. The first strategy provides significant energy savings without impacting either the maximum latency bounds or the minimum guaranteed bandwidth. The second strategy provides higher energy savings with marginally increased bounds on the memory latency, while still preserving the original guaranteed bandwidth provided by the real-time memory controller. Both these strategies can be employed with any of the real-time memory controllers presented in [3–8].

SDRAMs support different power-down modes viz., *fast exit* and *slow exit*. The former has a short power-up latency and saves some power, while the latter has a longer power-up latency, but saves more power. This paper also proposes an *algorithm to select the most energy-efficient power-down mode at run-time* based on the memory state and the power-down duration, for both the power-down strategies.

We experimentally evaluate the two proposed strategies by concurrently executing four media applications on an MPSoC platform using a real-time SDRAM memory controller. We show around 42.1% memory energy savings using the first power-down strategy and around 51.3% using the second strategy. The second strategy almost reaches the theoretical maximum of 51.4% memory energy savings using power-down modes, but slightly increases the execution time of the applications by 0.25%, due to the marginal increase in latency bounds. We also compare these two strategies against a speculative power-down policy on energy savings and impact on performance guarantees.

The remainder of this paper is organized as follows: Section 2 describes the related work on real-time SDRAM controllers and memory power optimization strategies. Section 3 gives the background on SDRAMs and introduces real-time memory arbitration. Section 4 presents the proposed power-down strategies, followed by deriving the impact of these strategies and a speculative power-down policy on performance guarantees in Section 5. Section 6 describes an algorithm to select the most energy-efficient power-down mode at run-time. In Section 7, we experimentally evaluate our solutions using four media applications and compare against the speculative and theoretical-best power-down options. Section 8 concludes the paper by highlighting the contributions of this work.

2. RELATED WORK

Real-time SDRAM memory controllers like [3–8] employ predictable arbiters, such as Round-Robin or TDM, and provide latency and/or bandwidth (rate) guarantees by bounding the temporal interference between requesters.

[4] employs Round-Robin arbitration and provides upper bounds on delays for different memory accesses. Similarly, [6] employs Round-Robin arbitration and uses worst-case response time to bound memory access latency. [3] adopts a budget-based static-priority arbitration and provides bounds on latency and guarantees a minimum bandwidth for every memory requester. It also supports Round-Robin or TDM arbiters. [7] uses TDM arbitration and provides bandwidth guarantees and a worst-case execution time for memory accesses. In [5], weighted Round-Robin arbitration is used to provide both bandwidth guarantees and latency bounds. [8] uses static scheduling and provides predictable memory accesses. However, none of these real-time memory controllers support power-down at run-time, due to the impact of power-up latencies on performance guarantees.

When it comes to work on SDRAM memory power minimization, there exists no generic run-time SDRAM power-down solution for real-time systems. For instance, [15] proposed to reduce idle power consumption by using a compiler-directed selective power-down and a hardware-assisted run-time power-down. However, the former is not suitable for run-time use and the latter can incur large performance penalties due to mis-predictions of future idleness. [16] proposed history-based scheduling and an adaptive memory throttling mechanism to allow memory to remain in the idle mode for longer periods of time to employ power-down longer. However, these methods also incur performance penalties and cannot be used for real-time applications.

In short, real-time memory controllers do not currently support power-down mechanisms, and existing power-saving solutions are not applicable at run-time and cannot be used with real-time memory controllers. This paper bridges this gap and provides run-time power-down strategies for real-time SDRAM memory controllers.

3. BACKGROUND

This section discusses SDRAM organization, operation and power-down options. This is followed by an introduction to predictable arbiters and how they guarantee latency and bandwidth (rate) in a real-time memory controller.

3.1 SDRAM Essentials

SDRAMs are organized as a set of memory banks that include memory elements arranged in rows and columns. A row buffer also resides in every bank to store contents of the currently accessed memory row. The banks in an SDRAM operate in a parallel and pipelined fashion, although only one bank can perform an I/O operation (data transfer) at a particular instance in time, due to the shared data bus. To read contents from the memory, an *Activate* command is first issued by the memory controller to the SDRAM, which opens the requested row and brings data from the SDRAM cells into the row buffer. Then, any number of *Read* or *Write* commands can be issued to read out or write into specific columns of data in the row buffer. Subsequently, a *Precharge* command is issued and the contents of the row buffer are stored back into the corresponding memory row. Reads and writes can also be issued with an *auto-precharge flag* to au-

tomatically precharge as soon as the request completes. If any row is active, the memory is said to be in the *active* state, else it is in the *precharged* state. Switching between a read and a write transaction, or vice versa, takes a few clock cycles. Further, to retain data in the memory, all rows in the SDRAM need to be refreshed at regular intervals, which is done by issuing a *Refresh* command. The number of words of data transferred in a single read/write command is called a burst, and its size is given by the Burst Length (BL) (usually 8 words for DDR3). A memory controller may serve a single request by issuing a number of read/write bursts per bank (defined by the Burst Count (BC) parameter) and interleaving over more than one banks (given by the degree of Bank Interleaving (BI)). The BL, BC and BI parameters determine the data *access granularity* with which the memory controller accesses the memory and have a large impact on performance and power consumption [18].

3.2 Power-Down Options in SDRAMs

It is possible to power down the SDRAM memory at run-time to reduce power consumption if it is not in use. SDRAMs support different power-down modes, such as *fast exit* and *slow exit* [10, 11]. The former has shorter power-up latency but saves less power, while the latter has longer power-up latency, but saves more power. These power-down modes can be entered either in the active or precharged state, based on certain timing constraints and the type of memory being used. For DDR2 memories, an active power-down can be used in the fast or slow exit mode, but only in the slow exit mode for DDR3. Conversely, a precharged power-down can be used in the fast or slow exit mode for DDR3, but only in the slow exit mode for DDR2.

When transitioning into and out of these power-down (PD) modes, certain timing constraints must be respected, as shown in Figure 1. In the figure, the t_{TRANS} parameter gives the *transition in* timing constraint before the memory can switch to a power-down mode after a read/write command is issued. The t_{PD} parameter gives the *power-down time*, which may vary from a minimum of $tCKE$ (Clock Enable pulse width) to a maximum of $9 \times tREFI$ (refresh interval). The t_{PUP} parameter gives the *power-up* timing constraint before the next command can be issued. This t_{PUP} parameter for the fast exit mode is given by t_{XP} for DDR3 and t_{XARD} for DDR2 memories. For the slow exit mode, the same power-up timing constraints are applicable if the next issued command is an ACT/PRE/REF. However, before a Read/Write command is issued after powering-up, the timing constraints of t_{XPDLL} for DDR3 and t_{XARDS} for DDR2 must be satisfied, for the DLL to be activated before issuing these commands.

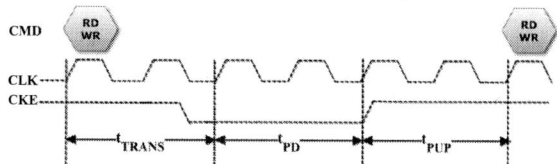

Figure 1: Power-Down Transitions

3.3 Arbiters and Latency-Rate Servers

Real-time SDRAM memory controllers employ predictable arbiters to provide latency and bandwidth (rate) guarantees to applications (requesters) accessing the memory. These arbiters use scheduling algorithms like Round-Robin and TDM, and can be analyzed using the Latency-Rate ($\mathcal{LR}$) server model [12], to characterize their performance guarantees. These guarantees are provided to a requester in terms of a *minimum rate of service* (ρ) and a *maximum initial service latency* (Θ), whenever it is *busy* (requesting a higher rate of service on average than allocated to it). Figure 2 depicts this rate guarantee and the initial service latency

989

bound provided by $\mathcal{LR}$ arbiters. As shown in the figure, a *busy period* for a requester corresponds to a time interval when its *requested service rate* is above the *busy line*, else, it is considered to be *not busy*. In the figure, the *allocated service line* indicates the minimum rate guarantee (ρ) given to a requester. ρ corresponds to the fraction of the net memory bandwidth that is provided as the bandwidth guarantee (β) to that requester. The maximum initial service latency (Θ) gives the maximum duration a requester has to wait after its arrival, to start getting served by the memory at the guaranteed rate (ρ). As can be noticed in the figure, the actual *provided service* may be higher than the allocated rate, if the system has the capacity to support it.

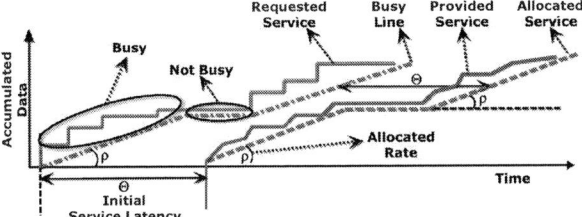

Figure 2: Latency-Rate Server

In short, a Latency-Rate ($\mathcal{LR}$) arbiter provides a busy requester, a guaranteed bandwidth β in the form of a guaranteed rate of service ρ after an initial service latency (Θ). These guarantees can be used for formal verification of an application's real-time behavior.

3.4 $\mathcal{LR}$ Arbiters and Memory Controller Guarantees

This section describes how latency and bandwidth (rate) guarantees are derived for a real-time SDRAM controller.

The initial service latency bound (Θ) of a requester can be intuitively seen as the duration between the time of arrival of a request at the arbiter and the time at which the request is accepted by the memory for service. It is given by the sum of the service time of the request currently being served and that of other interfering requesters, including refreshes (if any), as discussed later in Section 5. The service time for any given request is defined as the *service cycle length* (SCL) of that request. This can be highly variable depending on whether the request is a read or a write and if there is switching time involved (from read to write or vice versa) between the last and the next request. Hence, we use the longest SCL denoted by *maximum service cycle length* (*max_SCL*) to derive a conservative worst-case initial service latency bound Θ (as will be shown in Section 5.1).

As stated before, once the request is accepted for service by the memory after Θ, it is guaranteed a minimum rate of service, ρ. This rate of service (ρ) defines the bandwidth guarantee (β), based on the net memory bandwidth (*net_BW*). This net memory bandwidth is predominantly defined by the request size and the *max_SCL* for the particular request size. The size of a request can be defined as a multiple of the *access granularity* parameter (described in Section 3.1), which is the minimum size of data accessed by the memory controller. For efficient memory access, the access granularity should be of the same size as the request size, although it can be smaller. In this work, we assume all requests to be of one size, and the access granularity to be of the same size as the requests, for efficient memory access and simplicity of analysis. The *max_SCL* for the given access granularity is then used to compute the net memory bandwidth (*net_BW*) and along with ρ, is used to provide the bandwidth guarantee β (also shown in Section 5.1).

An $\mathcal{LR}$ arbiter employs a *scheduling interval* parameter, to schedule different requesters to memory. This parameter gives the duration after which, in every service cycle, a subsequent requester is selected to be scheduled at the end of the current request. This scheduling interval is statically defined as the minimum service cycle length among all requests (*min_SCL*), since this is the minimum period after which, the next requester could be scheduled.

As an example, consider 64-byte requests from a real-time memory controller accessing a 1Gb DDR3-800 memory with a BC of 4 interleaving over 1 bank. The SCLs of read and write requests (including any switching) corresponding to the 64-byte access granularity are shown in Figure 3. As can be noticed, the SCLs vary depending on the request type (read/write). The shortest SCL (*min_SCL*) is 26 clock cycles for a read transaction and this defines the *scheduling interval* for all service cycles and the length of an idle service cycle. The longest SCL *max_SCL* is 37 cycles and it includes write SCL (t_{WR}) and read to write switching time (t_{RTW}).

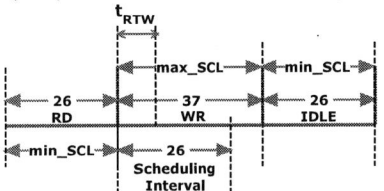

Figure 3: Scheduling Interval & SCLs

4. REAL-TIME POWER-DOWN STRATEGIES

Existing real-time SDRAM controllers employ predictable $\mathcal{LR}$ arbiters, such as Round-Robin and TDM, to provide latency and/or bandwidth guarantees, but do not address power optimization. In this section, we propose two runtime power-down strategies for such memory controllers; one a conservative latency-bandwidth-neutral strategy and the other an aggressive bandwidth-neutral strategy. These strategies can be employed whenever the memory is idle to reduce memory power consumption, while preserving the original guaranteed bandwidth, and providing bounds on memory latencies. The analysis here is restricted to DDR3 memories, although it is easily adaptable for DDR2 as well.

4.1 Conservative Latency-Bandwidth-Neutral Strategy

The first strategy involves triggering a special *power-off* request whenever an arbiter service cycle is idle. This power-off request is designed to power down the memory and power it back up within the scheduling interval, thus hiding the power-up transition latencies within the idle service cycle. This ensures that the scheduling of memory access requests is not disturbed and the power-down mechanism is effectively hidden from the requesters. This latency-bandwidth-neutral strategy provides significant energy savings and preserves both the guaranteed initial service latency bounds and the bandwidth, as is shown later in Section 7.

4.2 Aggressive Bandwidth-Neutral Strategy

The second strategy is more aggressive, since it checks for new requests before powering up the memory. It involves issuing a *power-down* request when there are no pending requests at the arbiter, and snooping the arbiter inputs for new requests before the end of the current idle service cycle. If there are any new pending requests, a *power-up* request is issued to the memory to power it up by the end of the idle service cycle (thus maintaining scheduling interval). To implement this strategy, we introduce a *Snooping Point* at a pre-defined time instance, before the end of the scheduling interval, as shown in Figure 4. This snooping point can be derived by subtracting the worst-case power-up time (t_{PUP_max}) (given by Equation (1)) from the scheduling interval (given by Equation (2)).

$$t_{PUP\_max} = max(t_{XP}, t_{XPDLL} - t_{RCD}) \qquad (1)$$

$$t_{SNOOP} = t_{SCHED\_INTERVAL} - t_{PUP\_max} \qquad (2)$$

This strategy assures that the memory powers-up in time and following request is scheduled on-time, as in the conservative case, if it arrives before this snooping point.

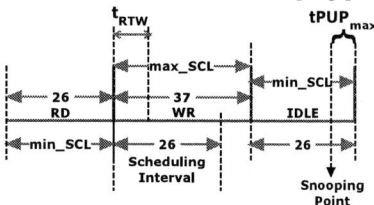

Figure 4: Snooping Point in Aggressive Power-Down

As can be noticed in Equation (1), t_{PUP_max} considers the minimum timing constraint between an ACT and a RD/WR command (t_{RCD}) [10], besides the fast exit (t_{XP}) and slow exit (t_{XPDLL}) power-up timing constraints. The rationale behind it is as follows: In DDR3 memories, the slow exit power-down can be employed only in the precharged state, which implies that every read/write transaction has to end with a precharge and begin with an ACT command. The power-up timing constraint before issuing an ACT after a slow-exit power-down is given by t_{XP}, which is shown as the first constraint in Equation (1). For efficient memory access, the first RD/WR command is scheduled immediately after t_{RCD} is satisfied, after an ACT command is issued. However, since a RD/WR can be issued only after a duration of t_{XPDLL}, after the memory begins to power-up, the corresponding ACT in the transaction can be issued after $max(t_{XP}, t_{XPDLL} - t_{RCD})$ is satisfied.

Now consider the scenario, when the next request arrives after the snooping point but before the end of scheduling interval. In this case, no power-up will be issued and the memory continues in the power-down mode. This results in the next request missing a service cycle and getting scheduled in the following service cycle, if no other interfering requesters show up. However, as will be shown in Section 5, this only increases the latency bounds (Θ) by the power-up transition time (in the worst-case) and does not impact the maximum service cycle length (max_SCL) and therefore the bandwidth guarantee (β). In short, this strategy is bandwidth-neutral and provides marginally increased latency bounds, thus guaranteeing real-time memory performance. The advantage of this strategy is that all contiguous idle periods are combined into one large idle period, thereby avoiding frequent powering-up of memory (every idle service cycle), as in the conservative strategy, to save more energy.

5. IMPACT ON LATENCY-BANDWIDTH BOUNDS

In this section, we first derive the initial service latency bounds and bandwidth guarantees offered by real-time memory controllers. These guarantees are conservative and simpler than those presented in [14], for ease of understanding. We then analyze the impact of the conservative, aggressive and speculative power-down strategies on these bounds.

5.1 Latency and Bandwidth Guarantees

As stated in Section 3.4, the bandwidth guarantee (β) depends on net memory bandwidth (net_BW), which can be derived using the max_SCL for a given access granularity. The worst-case bound for max_SCL can be computed as the maximum of: (1) the service time of a read (t_{RD}) and the time to switch to a read after a write (t_{WTR}), and (2) the service time of a write (t_{WR}) and the time to switch to a write after a read (t_{RTW}) (given by Equation (3)).

$$max\_SCL = max(t_{WTR} + t_{RD}, t_{RTW} + t_{WR}) \qquad (3)$$

To compute the net memory bandwidth, let us assume that a requester is busy throughout a refresh interval (t_{REFI}), when it is interrupted by a refresh. For a worst-case estimate, we consider every service cycle during t_{REFI} to be as

long as max_SCL. Hence, the total number of service cycles (num_SCL) in t_{REFI} is given by Equation (4), where t_{Ref} is the length of a single refresh request in the refresh interval.

$$num\_SCL = \lfloor (t_{REFI} - t_{Ref})/max\_SCL \rfloor \qquad (4)$$

The total data transfer during this period, assuming an access granularity of G is $num\_SCL \times G$. Hence, the next memory bandwidth is given by Equation (5).

$$net\_BW = num\_SCL \times G/tREFI \qquad (5)$$

Consider a use-case with 'x' requesters accessing the memory through a real-time memory controller using Round-Robin arbitration. Each requester is guaranteed a service rate of $\rho = $ '1/x' in the form of *1 out of x Round-Robin time slots*, if it is *busy*. Hence, the minimum bandwidth guaranteed to each requester (β), is given by Equation (6).

$$\beta = net\_BW \times \rho \qquad (6)$$

Next, we derive the initial service latency (Θ) for a requester (in Equation (7)), considering the max_SCL from Equation (3)), in the presence of 'x' interfering requesters. We also consider a refresh request length t_{tRef} for any interference from a refresh during the busy period. In addition, Θ would also include the SCL of the currently scheduled request and a waiting period t_{wait} equal to the difference between the max_SCL and the scheduling interval (min_SCL).

$$\Theta = t_{wait} + t_{Ref} + (max\_SCL \times (x + 1)) \qquad (7)$$

In Equation (7), the '1' corresponds to service cycle of the currently scheduled request. The t_{wait} period corresponds to the difference between the time of scheduling a request and the end of the current SCL (max_SCL in the worst-case).

These conservative guarantees are applicable to most real-time memory controllers discussed in this paper.

5.2 Impact of Conservative and Aggressive Power-Down Strategies

To explain the implications of both the conservative and aggressive strategies, we illustrate a use-case with four requesters (R1, R2, R3 and R4) connected to a Round-Robin arbiter, as shown in Figure 5. Each requester is provided with a initial service latency bound (Θ) and a bandwidth guarantee (β) based on a rate guarantee (ρ).

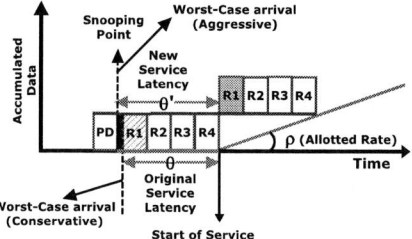

Figure 5: Impact on Latency and Rate Guarantees

We first re-visit the condition for deriving the worst-case initial service latency bound for requester R1 (given by Θ). Consider that R1 arrives at the arbiter one clock cycle after the end of the current scheduling interval, depicted in Figure 5 as worst-case arrival. In this case, it would miss out on being scheduled in its first service slot (indicated by the striped slot). It would eventually get serviced after waiting four service cycles, at the next allotted service slot (indicated by the shaded slot), with a guaranteed rate of service ρ at a guaranteed bandwidth β. Hence, its worst-case latency bound, Θ, includes four max_SCLs, apart from the t_{Ref} and t_{wait} discussed in Equation (7).

In the case of conservative power-down, the SCLs are not modified, since power-up is always completed within the scheduling interval, which is given by the shortest service cycle. Hence, max_SCL remains unaltered and the bandwidth guarantee (β) does not change. The initial service

991

latency bound is also maintained as is, since any request arriving before the end of scheduling interval in the idle service cycle is scheduled as before, when no power-down was used.

In the case of aggressive power-down, a power-down (PD in Figure 5) was issued during the idle period, preceding R1's request for service. If R1 arrives before the snooping point, it is scheduled during its first available service slot (striped slot). If R1 arrives after the scheduling interval, as discussed above, it is scheduled after waiting four service cycles. However, if R1 arrives after the snooping point and before the end of scheduling interval, it also misses out on its first service slot, since the next slot is already scheduled to be in power-down. But, after waiting over the next four service cycles and any memory-generated refresh, R1 would get serviced at the next allotted service slot (indicated by the shaded slot), with the guaranteed rate of service ρ at a guaranteed bandwidth β. This is the same, as the case when R1 arrives one clock cycle after the scheduling interval, as discussed above. The only difference in this scenario is that, the worst-case initial service latency bound Θ would increase marginally by the worst-case power-up transition time t_{PUP_max}, as shown in Equation (8).

$$\Theta' = \Theta + t_{PUP\_max} \qquad (8)$$

Since powering-up of memory is not allowed beyond the snooping point, the power-up is always completed by the end of scheduling interval in the idle service cycle. Hence, the worst-case bound on max_SCL (shown in Equation (3)) is not affected by the power-up and hence, the net memory bandwidth (net_BW) and the bandwidth-guarantee (β) do not change. To quantify the increase in Θ (shown in Equation (8)), we consider service cycle lengths from the illustration in Figure 3 with request size of 64 bytes. The original Θ for requester R1 in the presence of three interfering requesters (R2, R3 and R4) is derived as 203 clock cycles (cc) using Equation (7), where t_{Ref} is 44 cc for 1Gb DDR3-800. The increased Θ' is calculated as 208 cc ($t_{PUP\_max} = 5$ cc), thus, showing marginal increase in latency bounds (2.4%).

In conclusion, the initial service latency hit of t_{PUP_max} is observed only once per busy period and only by the requester waking up the memory from power-down. Also, there is no impact on the requester's bandwidth guarantee.

5.3 Impact of Speculative Strategies

In this subsection, we derive the impact of speculative power-down policies on latency and bandwidth guarantees.

A *speculative power-down strategy* can be defined as one that powers-down the memory whenever it is idle and *allows it to power-up even after the snooping point in the idle service cycle*. In the worst-case, a request may arrive at the last clock cycle of the idle service cycle and hence, the power-up transition time (t_{PUP_max}) gets added to the SCL of the following request, which may originally have been max_SCL in length. This impact on max_SCL as a result of a speculative power-up, is shown in Equation (9). This reduces the net memory bandwidth (net_BW) and thereby, the bandwidth guarantee (β) provided by the memory controller, as shown in Equations (4), (5) and (6). It also increases Θ in the presence of 'x' interfering requesters, by $t_{PUP\_max} \times (x + 1)$.

$$max\_SCL' = max(t_{PUP\_max} + t_{RD}, t_{PUP\_max} + t_{WR}, max\_SCL) \qquad (9)$$

Using the SCLs from Figure 3, max_SCL increases to 42 cc from 37 cc (by 13.5%). As a result, the service latency bound increases to 228 cycles, showing a larger increase (around 12.3% using Equation (7)) compared to the aggressive strategy. Most importantly, the net memory bandwidth reduces from 681 MB/s to around 599 MB/s and the bandwidth guarantee (β) reduces from around 170.27 MB/s to 149.72 MB/s (around 12.1% using Equation (5)), which is unacceptable for real-time memory controllers, since it results in

an inefficient use of the already scarce memory bandwidth.

Moreover, the bandwidth and latency impact of the speculative policy depends on the number of requesters accessing the memory and gets worse with an increase in the same.

6. POWER-DOWN MODE SELECTION

In this section, we present a power-down mode selection algorithm that determines the most appropriate mode of power-down (fast exit or slow exit) based on the state of the memory and idle service cycle length. Using the power-down equations presented in [13], and the current and voltage numbers from SDRAM datasheets [17], the algorithm evaluates the different power-down modes (fast exit or slow exit, active or precharged) and selects the best power-down mode with the least energy consumption.

To employ Algorithm 1, we derive a 'power-off' (t_{Off}) period for the entire power-down request including the transitions in and out of the power-down mode ($t_{TRANS} + t_{PD} + t_{PUP}$ in Figure 1), equal to the idle service cycle length (min_SCL). We then forward this information to the algorithm, along with the memory state information (precharged or active), which then selects the most energy-efficient power-down mode for the given system configuration. If there can be no energy savings with any of the power-down modes, the algorithm opts for no power-down (No_PD).

Algorithm 1 Power-Down Mode Selection

Require: mode_select(t_{Off}, mem_state)
1: **if** $t_{Off} > t_{CKE} + t_{XPDLL} - t_{RCD}$ **then**
2:   {Comment: Minimum PRE Slow Exit Duration}
3:   **if** mem_state == PRE **then**
4:     Mode ←Min_Mode(E(t_{Off}, S_PRE), E(t_{Off}, F_PRE))
5:   **else**
6:     Mode ← F_ACT
7:   **end if**
8: **else if** $t_{Off} > t_{CKE} + t_{XP}$ **then**
9:   {Comment: Mimimum ACT/PRE Fast Exit Duration}
10:   **if** mem_state == PRE **then**
11:     Mode ←Min_Mode(E(t_{Off}, F_PRE), E(t_{Off}, No_PD))
12:   **else**
13:     Mode ← F_ACT
14:   **end if**
15: **else**
16:   Mode ← No_PD
17: **end if**
18: **return** (Mode)

It should be noted that the algorithm presented here is for DDR3 memories. For DDR2, the appropriate power-down modes and timings described in [11], must be used.

7. EXPERIMENTS AND RESULTS

In our experiments, we employ a 1Gb Micron DDR3-800 [17] memory and four common media applications: (1) H.263 encoder, (2) EPIC Encoder, (3) JPEG Encoder and (4) MPEG2 Decoder. Using these applications, we evaluate the two proposed real-time power-down strategies and a speculative power-down policy with respect to energy savings, average execution time, initial service latency bounds and bandwidth guarantees. We also derive the theoretical best-case energy savings when using power-down by performing memory trace post-processing. Once the trace is obtained, we manually insert a power-down request at the start of every idle period and power-up the memory in-time for the next transaction to be served, in order to avoid any impact on execution times and performance guarantees.

7.1 System and Experiments Setup

We executed the four test applications independently on the Simplescalar simulator [20] with a 16KB L1 D-cache, 16KB L1 I-cache, 128KB shared L2 cache and 64-byte cache line configuration. We filtered out the L2 cache misses, and obtained a trace of the transactions meant for the SDRAM memory. To simulate four requesters in our experimental setup (similar to the illustration in Section 5.2), we employed the traces from these four applications on four trace players in a SystemC model of our real-time MPSoC platform [19]. We forwarded these transactions to our real-time SDRAM memory controller [3], fitted with a Round-Robin arbiter. To obtain 64-byte access granularity for DDR3-800, we used a BL of 8 words, each 2 bytes long, with a BC of 4, interleaving over 1 bank, for most power-efficient memory accesses [13]. For all our power analysis, we employed our open-source DRAM energy estimation tool [21] based on the power model presented in [13], and the current and voltage numbers from SDRAM datasheets [17].

7.2 Results and Analysis

In our first experiment, we analyze the impact of the different power-down policies on total memory energy consumption when executing the four application traces concurrently. In doing so, we also observe the average-case impact on total execution time, due to these power-down policies. We compare the conservative (CPD) and aggressive (APD) power-down strategies against the theoretical best-case power-down, no power-down (No PD) and the speculative power-down (Spec PD) policies, as depicted in the graph in Figure 6. We observed that the conservative strategy saves around 42.1% of total memory energy, compared to using no power-down, without impacting the execution times. We also observed that the aggressive power-down strategy saves 51.3% of the total memory energy (very close to the theoretical best-case of 51.4%), at a marginal increase of 0.25% (approximately 510 μs) in the total execution time. The speculative power-down policy saves around 51.1% of the memory energy consumption, but at an increase of about 1.32% (approximately 2640 μs) in the total execution time.

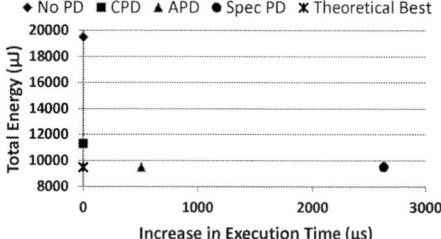

Figure 6: Total Energy & Penalties using Strategies

In our second experiment, we analyze the difference in the energy savings, and the bandwidth guarantee (β) and initial service latency bound (Θ) of the different policies, as depicted in the graph in Figure 7. The data labels indicate the energy consumption, and the bandwidth and latency guarantees of the different policies.

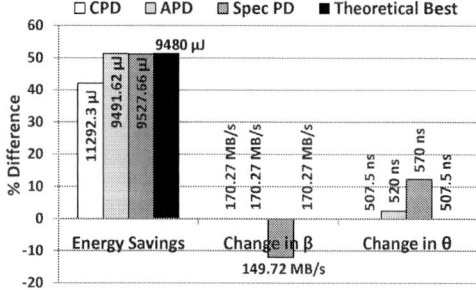

Figure 7: Energy Savings & Latency-Rate Impact

As can be observed, the aggressive and speculative power-down strategies save almost as much energy as the theoretical best power-down solution. However, in the case of the aggressive power-down strategy, the service latency bound

(Θ) increases only marginally by around 2.4% to 520 ns, whereas it increases by around 12.3% to 570 ns in the case of the speculative power-down policy. Moreover, the bandwidth guarantee (β) of the speculative power-down policy takes a large of around 12.1% and reduces to 149.72 MB/s from the initial 170.27 MB/s, while it is maintained by the aggressive power-down policy at 170.27 MB/s. This decrease in bandwidth-guarantee occurs for any speculative power-down policy that decides to power-up the memory after the snooping point, since it increases max_SCL by 13.5% and decreases net_BW by 12.1%, as shown in Section 5.3. Both the proposed bandwidth-neutral strategies can be employed at run-time by any real-time SDRAM memory controller.

8. CONCLUSION

This paper presented two run-time power-down strategies that reduced SDRAM memory energy consumption and yet guaranteed real-time memory performance. The conservative strategy provided significant energy savings of around 42.1% when running traces from four media applications, without impacting the guaranteed latency bound and bandwidth. The aggressive strategy provided higher energy savings of around 51.3% (close to the theoretical best of 51.4%) and only marginally increased the latency bounds by 2.4%, while still preserving the original guaranteed bandwidth. This paper also showed that a speculative power-down policy cannot do any better than the aggressive strategy in terms of energy savings and would increase the latency bounds by 12.3% and reduce the guaranteed bandwidth by 12.1%, which is unacceptable for real-time memory controllers. Finally, this paper also presented an algorithm to select the most energy-efficient power-down mode at run-time for both the strategies, thereby providing a complete power-down solution for all real-time memory controllers using LR arbiters.

9. REFERENCES

[1] O. Vargas, *Achieve minimum power consumption in mobile memory subsystems*, EE Times Asia, March 2006.
[2] B. Jacob et al., *Memory Systems: Cache, DRAM, Disk*, Morgan Kaufmann Publishers, 2007.
[3] B. Akesson et al., *Architectures and Modeling of Predictable Memory Controllers for Improved System Integration*, In Proc. DATE 2011.
[4] M. Paolieri et al., *An Analyzable Memory Controller for Hard Real-Time CMPs*, IEEE Embd. Sys. Letters, Vol.1, No.4, 2009.
[5] A. Burchard et al., *A Real-Time Streaming Memory Controller*, In Proc. DATE 2005.
[6] S.A. Edwards et al., *A Disruptive Computer Design Idea: Architectures with Repeatable Timing*, In Proc. ICCD 2009.
[7] C. Pitter, *Time-predictable memory arbitration for a Java chip-multiprocessor*, In Proc. JTRES 2008.
[8] J. Reineke et al., *PRET DRAM Controller: On the Virtue of Privatization*, In Proc. CODES+ISSS 2011.
[9] ITRS, *http://www.itrs.net*
[10] JEDEC, *DDR3 SDRAM Standard*, JESD79-3E, 2010.
[11] JEDEC, *DDR2 SDRAM Standard*, JESD79-2F, 2009.
[12] D. Stiliadis et al., *Latency-rate servers: a general model for analysis of traffic scheduling algorithms*, IEEE Trans. on Netw., Vol. 6, No. 5, 1998.
[13] K. Chandrasekar et al., *Improved Power Modeling of DDR SDRAMs*, In Proc. DSD 2011.
[14] B. Akesson et al., *Memory Controllers for Real-Time Embedded Systems*, Springer, 2011.
[15] V. Delaluz et al., *Hardware and Software Techniques for Controlling DRAM Power Modes*, IEEE Trans. on Comp., Vol.50, No.11, 2001.
[16] I. Hur et al., *A comprehensive approach to DRAM power management*, In Proc. HPCA 2008.
[17] Micron, *1Gb: X4, X8, X16 DDR3 SDRAM*, Rev. 2010.
[18] S. Goossens et al., *Memory-Map Selection for Firm Real-Time Memory Controllers*, In Proc. DATE 2012.
[19] A. Hansson et al., *CoMPSoC: A template for composable and predictable multi-processor system on chips*, ACM TODAES, Vol. 14, No. 1, 2009.
[20] D. Burger et al., *The SimpleScalar tool set, Version 2.0*, ACM SIGARCH Comp. Arch. News, Vol. 25, No. 3, 1997.
[21] K. Chandrasekar et al., *DRAMPower: Open-source DRAM power & energy estimation tool*, www.es.ele.tue.nl/drampower

Embedding Statistical Tests for On-Chip Dynamic Voltage and Temperature Monitoring

Vincent Lionel
CEA-LETI, MINATEC Campus
17 rue des Martyrs,
38054 Grenoble Cedex 9, France

lionel.vincent@cea.fr

Maurine Philippe
LIRMM, Université Montpellier 2
161 rue Ada,
34095 Montpellier, Cedex 5, France

philippe.maurine@lirmm.fr

Lesecq Suzanne, Beigné Edith
CEA-LETI, MINATEC Campus
17 rue des Martyrs,
38054 Grenoble Cedex 9, France

{firstname.lastname}@cea.fr

ABSTRACT

All mobile applications require high performances with very long battery life. The speed and power consumption trade-off clearly appears as a prominent challenge to optimize the overall energy efficiency. In MultiProcessor System-On-Chip architectures, the trade-off is usually achieved by dynamically adapting the supply voltage and the operating frequency of a processor cluster or of each processor at fine grain. This requires monitoring accurately, on-chip and at runtime, the supply voltage and temperature across the die. Within this context, this paper introduces a method to estimate, from on-chip measurements, using embedded statistical tests, the supply voltage and temperature of small die area using low-cost digital sensors featuring a set of ring oscillators solely. The results obtained, considering a 32nm process, demonstrate the efficiency of the proposed method. Indeed, voltage and temperature measurement errors are kept, in average, below 5mV and 7°C, respectively.

ACM Categories and Subject Descriptors
B.7.1 Types and Design Styles
General Terms: Measurement, Performance, Reliability
Keywords: Hypothesis testing, GALS, PVT sensor, ring oscillator, Multiprobe, variability, supply voltage and temperature estimation, energy efficiency

1. INTRODUCTION

Modern embedded platforms must provide ever increasing computational capabilities under tight power consumption constraints. Single processor architectures are not anymore able to provide the computational performances required under ultra-low power consumption. MultiProcessor Systems-on-Chip (MPSoC) recently appear as a promising alternative, especially with the downscaling of CMOS technology.

However, this downscaling together with the supply voltage decrease enhances side effects such as performance variability between similar platforms. Actually, several variability sources can be distinguished within a chip. One of them is related to the manufacturing variations (Process, P) with consequences for the whole platform (inter-die) as well as for different areas of the circuit (intra-die) [6][21]. Each platform and even each part of it has

therefore unique characteristics. The process variations can be considered static as we will not mention in this paper transistor ageing.

The platform is also affected by temperature (T) and supply voltage (V) induced dynamic variations. Because these dynamic variations affect the whole platform, as they are related to environmental operating conditions, they can be considered as global. However, they are also local as they depend on the instantaneous computational activity of the MPSoC [1][12][17] and on its topology [15][8][16].

As V and T variations are local and dynamic, the MPSoC architectures must be adaptive [18] to operate at the best possible compromise between performances and power consumption while increasing the parametric yield.

Globally Asynchronous Locally Synchronous (GALS) architectures [14] enable to mitigate variability consequences because of their island organization [4]. Indeed, each island may be designed to have its power independency so as to allow a smart and local management of its performances. In such a design scenario, each island is a power/frequency domain featuring a local control loop to dynamically adapt [5] the supply voltage, the clock frequency and other parameters [13].

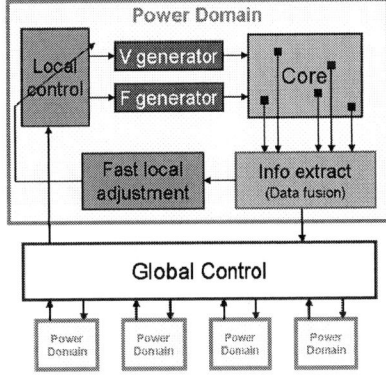

Figure 1: Architecture of an adaptive system

The closed-loop local controller shown Figure 1 can be implemented in hardware or software. In both cases, it computes the control values applied to the voltage and frequency actuators according to temporary performance targets and functionality guard bands [9][20]. Note that these control values depend on the computational load but also on the instantaneous supply voltage V and the temperature T values of the Voltage-Frequency Island (VFI). The instantaneous state {V,T} of all VFI must be monitored using local sensors that may also measure different process parameters and/or performance figures of merit.

Various integrated sensors have been proposed in the literature. Most of them are "specialized" i.e. they provide an accurate and absolute measure of a specific variable (e.g. V) regardless of any process parameter values and others operating conditions (e.g. T, substrate bias). As an example, [7] and [3] measure T and V with an accuracy of a few degrees and mV respectively. One of the main difficulties while designing such sensors, is to obtain compact sensors with (a) a high sensitivity to a single parameter (e.g. V) and (b) a negligible sensitivity to other parameters including process ones. These sensors [13][7][11] are mainly differential analogue blocks monitoring the evolution of V or T according to a given reference point. Thus each sensor embeds an Analogue-to-Digital Converter (ADC) to provide valuable data to digital blocks computing the control values to be applied to the actuators. As a result, these specialized sensors occupy a too large silicon area, to be duplicated in each VFI of a complex GALS and can not be used for a local variations sensing.

Contrarily to specialized sensors, compact and general purpose sensors can be designed with one or several Ring Oscillators (RO) associated with a counter acting as a compact ADC to limit the area overhead [10][19][22]. The obtained information is not absolute but relative to different parameters evolution. However, the binary code provided by such sensors depends on V and T, but also on many process parameters. Therefore, numerous RO measurements must be processed by a digital block (data fusion block in Figure 1) shared by several sensors, to extract the effective voltage V and T.

Within this context, the main contribution of this paper is to propose a method to extract from several RO frequency measurements the {V,T} state of a VFI. The RO outputs are processed with M non-parametric hypothesis testing procedures, making use of M models. The results of the M tests are then aggregated to infer the {V,T} state.

The rest of the paper is organized as follows. The principle of the method proposed is given in Section 2. Then the data fusion method is developed in Section 3. The validity of the technique proposed is exhibited in Section 4 that presents its efficiency and the computational cost. Then possible actions are given in Section 5 to improve the method efficiency. Finally Section 0 provides concluding remarks and future works.

2. CONTEXT AND PRINCIPLE OF THE METHOD

The digital output of a general purpose sensor provides information (F) which depends on the current PVT conditions. In case of a RO-based sensor, this information is a frequency value that depends on the supply voltage V, the temperature T and several process parameters P. Actually, this frequency is a complex non-linear function of all these parameters. Therefore, it is obviously impossible to find the values of V and T from a single frequency measurement. Thus a set of N measurements from N different sensors is required to estimate the values of V and T. Solving the set of N equations appears mathematically infeasible because the models are unavailable. The resulting question is how can we estimate, efficiently, accurately and on-chip, the values of V and T from N probably noisy measurements.

To tackle this issue a solution based on (a) statistical tools such as hypothesis tests and (b) a calibration phase of the Integrated Cir-

cuit (IC) at the end of fabrication, i.e. during test and characterization steps is proposed in the present paper.

Figure 2 illustrates the method. A set of localized sensors, as the Multiprobe [22], returns information which depends on the process P (assumed constant) and on the dynamic parameters V and T.

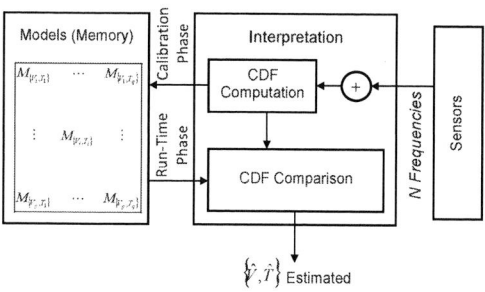

Figure 2: Principle of the method

During the calibration stage, the circuit is stimulated to collect a set of N reference frequencies for each state {V,T}. Then, in a second step and for each {V,T} state, the *interpretation* block computes the sum of all pairs of frequencies. This provides $N \cdot (N-1)/2$ values rather than N that allow building for each {V,T} a sufficiently rich Cumulative Distribution Function (CDF) representative of the sensor behavior under this given {V,T} condition. The CDFs are finally memorized on-chip as response models, $M_{\{V,T\}}$. Note that different solutions can be adopted to combine pairs of frequencies. This point will be discussed in Section 5.2. However, one must remember that it is a key step to obtain a sufficient amount of data from a reduced set of on-chip measurements. To sum up, during this calibration stage, a set of fingerprints is built and each fingerprint takes, in the present proposal, the form of a CDF.

At runtime, the current fingerprint is first collected so as to estimate the current {V,T} state of the IC. Thus, N frequencies are collected from the sensor used during the calibration stage. Then, these N measurements are combined by the *interpretation* block to build the current CDF $E_{\{V,T\}}$. Finally, a statistical procedure is applied to identify which $M_{\{V,T\}}$, among all available response models, matches with the current CDF $E_{\{V,T\}}$.

Several statistical tests are available in the literature such as the Kolmogorov-Smirnov, the Cramer-von Mises and the Student ones. Some of them are parametric, i.e. they compare experimental data with a theoretical distribution (e.g. the Normal distribution). Other tests do not consider any a priori assumption about the distribution shape. They are usually defined as non-parametric tests. The P parameters being unknown, assume herein that the distribution shape is unknown. Therefore consider only non-parametric tests to find the $M_{\{V,T\}}$, matching with $E_{\{V,T\}}$. In the sequel, only the Kolmogorov-Smirnov (KS) and Cramer-von Mises (CvM) tests are considered as they are well suited for hardware or software integration because of their simplicity.

3. PARTICULAR IMPLEMENTATION OF THE PROPOSED METHOD

A particular implementation of the proposed method is now described. In this implementation the measurements were provided by the Multiprobe sensor [22]. This sensor has been designed to have a silicon footprint as small as possible (450µm² in 32nm STMicroelectronics technology), and to provide several data related to the current state of the IC. Actually, the Multiprobe features a set of 7 localized ring oscillators. Thus, it provides up to 7 digitalized frequencies.

In the calibration stage, the 7 frequencies are first combined to get 21 values and built all $M_{\{V,T\}}$ CDF made of 21 bins. At run time, $E_{\{V,T\}}$ is built and the Kolmogorov-Smirnov (KS) test is applied. This non-parametric test checks the correctness of the hypothesis H_0 according to which a given $M_{\{V,T\}}$ and $E_{\{V,T\}}$ are issuing from the same distribution law. The decision criterion of this test is the maximum difference Δ_{max} between the two tested CDF:

$$\Delta_{max} = \underset{f}{Sup}\left| M_{\{V,T\}}(f) - E_{\{V,T\}}(f) \right| \quad (1)$$

where $M_{\{V,T\}}(f)$ and $E_{\{V,T\}}(f)$ are the probabilities to obtain sum of two frequencies inferior to f. Then, the probability $pvalue_{\{V,T\}}$ that $M_{\{V,T\}}$ and $E_{\{V,T\}}$ come from the same distribution law can be computed:

$$pvalue(y) = 2\sum_{k=1}^{+\infty}(-1)^{k+1}e^{\left(-2k^2y^2\right)} \quad (2)$$
$$with \quad y = \sqrt{n}\cdot\Delta_{max}$$

To estimate the circuit current state, it could be sufficient to find the model $M_{\{V,T\}}$ obtaining the highest $pvalue_{\{V,T\}}$. However only a limited number of $\{V,T\}$ cases were considered to create the catalogue of fingerprints $M_{\{V,T\}}$ during the calibration stage. Thus, it is highly probable that the current $\{V,T\}$ is not referenced in the model table. As a consequence, many neighboring $M_{\{V,T\}}$ conditions score a high $pvalue_{\{V,T\}}$. Therefore, a threshold $TH \in [0,1]$ is chosen to retain only the $M_{\{V,T\}}$ with a $pvalue_{\{V,T\}}$ greater than $TH \times pvalue_{max}$, with $pvalue_{max}$ the maximum $pvalue_{\{V,T\}}$ obtained for all tests. As a consequence, only $100 \times (1-TH)$ % of the $M_{\{V,T\}}$ are kept. Note that the influence of TH value on the efficiency of the method proposed is discussed in Section 5.3.

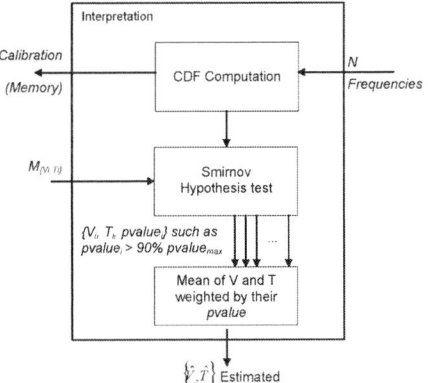

Figure 3: Interpretation block

In the present example, TH was fixed to 0.9 to select only 10% of the $M_{\{V,T\}}$. The current operating conditions are estimated thanks to the weighted means:

$$\hat{V} = \frac{\sum\limits_{(k,l)\in PKS} pvalue_{\{V_k,T_l\}}V_k}{\sum\limits_{(k,l)\in PKS} pvalue_{\{V_k,T_l\}}} \qquad \hat{T} = \frac{\sum\limits_{(k,l)\in PKS} pvalue_{\{V_k,T_l\}}T_k}{\sum\limits_{(k,l)\in PKS} pvalue_{\{V_k,T_l\}}} \quad (3)$$

where PKS is the set of $\{V,T\}$ conditions that have a $pvalue_{\{V,T\}}$ higher than $0.9 \times pvalue_{max}$.

The procedure described above is represented on Figure 3 which

summarizes the hardware or software interpretation block that must be embedded on chip for a fast and efficient adaptability. As can be seen, the frequencies F_i provided by the sensor are first summed 2 by 2, to build the CDF $E_{\{V,T\}}$. Then the $pvalue_{\{V,T\}}$ are computed. Finally an estimate of the current supply voltage $\hat{V}$ and temperature $\hat{T}$ are obtained.

4. APPLICATION AND EVALUATION OF THE METHOD

Some experiments have been performed to highlight the powerfulness of the method proposed and to quantify its efficiency. The Multiprobe behavior is simulated with Eldo for 32nm CMOS technology, in many operating conditions. V and T were increased respectively from 0.5V to 1.3V by steps of 0.01V and from -40°C to 120°C with steps of 10°C. 1377 response models $M_{\{V,T\}}$ for the Mutliprobe were thus built. Then, the estimation method, coded in Matlab and C languages, has been applied considering intermediate $\{V,T\}$ conditions. Note that all frequencies provided during the simulations were represented by 32bit words, during the procedure to take into account the counter side-effects.

4.1 Analysis of a particular $\{V, T\}$ state

Figure 4 illustrates for a simple example the method efficiency. The main objective is to verify that the estimate is as close as possible to the current state $\{V,T\}=\{0.83V, 12°C\}$. The black dots correspond to $\{V,T\}$ points that passed the KS tests, i.e. their $pvalue_{\{V,T\}}$ are higher than $0.9 \times pvalue_{max}$. The circle represents the current state $\{V,T\}$ while the "+" sign is the state $\{\hat{V}, \hat{T}\}=\{0.831V, 11.7°C\}$ estimated with the method proposed. The errors are respectively of 0.01V and 0.3°C.

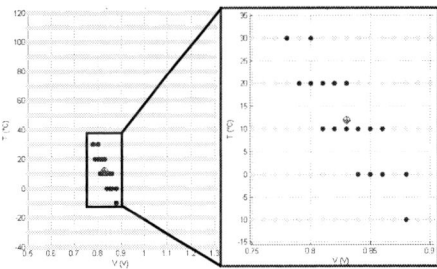

Figure 4: Result for the case $\{V,T\}=\{0.83V,12°C\}$

4.2 Method performances analysis for various $\{V, T\}$ states

The estimation error for many simulated $\{V,T\}$ cases is now computed to quantify fairly the method accuracy. Note that most of the different current $\{V,T\}$ states are not in the set of models.

Table 1 presents the results obtained for 51681 $\{V,T\}$ estimation procedures performed in simulation. During this experiment, V was increased from 0.5V to 1.3V by steps of 5mV and T scaled from -40°C to 120°C by steps of 0.5°C. $<|\varepsilon_V|>$ and $<|\varepsilon_T|>$ are the mean absolute estimation errors in the supply voltage and temperature respectively. $\sigma_{\varepsilon V}$ et $\sigma_{\varepsilon T}$ are the associated standard deviations (STD). ΔV and ΔT are the increments of V and T considered during the calibration stage to elaborate the response models $M_{\{V,T\}}$. As shown in the third row, with a large set of models ($\Delta V=10mV$, $\Delta T=10°C$), the mean estimation errors are lower than ΔV and ΔT considered during the calibration stage.

Table 1: Mean and STD estimation errors achieved with the Kolmogorov-Smirnov test

$<\|\varepsilon_V\|>$	$\sigma_{\varepsilon V}$	$<\|\varepsilon_T\|>$	$\sigma_{\varepsilon T}$
5 mV	9 mV	7.5 °C	12.6°C
$<\|\varepsilon_V\|>/\Delta V$	$\sigma_{\varepsilon V}/\Delta V$	$<\|\varepsilon_T\|>/\Delta T$	$\sigma_{\varepsilon T}/\Delta T$
0.50	0.90	0.77	1.28

4.3 Influence of the model tables

The results in Table 1 validate the approach proposed. However one may wonder if the method remains valid with a smaller set of models of the Multiprobe behavior, i.e. while considering larger values of ΔV and ΔT during the calibration stage. Obviously the smaller these values are, the more precise the sensor modeling is, at the cost of a longer and more expensive calibration stage.

To answer this question, the previous procedure is performed considering wider grids of response models. The ΔV increment between two models $M_{\{V,T\}}$ has been successively fixed to 20mV, 50mV and 100mV. During this experiment, ΔT was kept at 10°C.

Table 2: Mean and STD estimation errors achieved with the Kolmogorov-Smirnov test for different ΔV and ΔT=10°C

ΔV	10 mV	20 mV	50 mV	100mV
$<\|\varepsilon_V\|>$	5mV	7 mV	17 mV	28mV
$<\|\varepsilon_T\|>$	7.5°C	11.2 °C	26°C	46°C
$<\|\varepsilon_V\|>/\Delta V$	0.5	0.35	0.34	0.28
$<\|\varepsilon_T\|>/\Delta T$	0.75	1.12	2.6	4.6

Table 2 shows that the estimation errors on V remain low even for large values of ΔV. However, the estimation errors on T quickly increase with ΔV to become unacceptable for ΔV>50mV. This clearly highlights that accurately estimating T requires a really accurate modeling of the Multiprobe behavior with V. Actually, the impact of V on the frequency of the ROs is much more important than the temperature one.

Table 3: Mean and STD estimation errors achieved with the Kolmogorov-Smirnov test for different ΔT and ΔV=10mV

ΔT	10°C	20°C	40°C	80°C
$<\|\varepsilon_V\|>$	5mV	7 mV	10 mV	15 mV
$<\|\varepsilon_T\|>$	7.5°C	10.4°C	15.4°C	27.8°C
$<\|\varepsilon_V\|>/\Delta V$	0.5	0.7	1	1.5
$<\|\varepsilon_T\|>/\Delta T$	0.75	0.53	0.39	0.35

Experiments have been also conducted with a fixed value of ΔV (equal to 10 mV) and for different ΔT. Results are summarized in Table 3. As expected from Table 2, increasing significantly ΔV does not affect severely the errors on $\hat{V}$ and $\hat{T}$. Indeed, the choice ΔV=10mV and ΔT=40°C to build the set of models (calibration stage) is sufficient to estimate with the Multiprobe designed in a 32nm process the current supply voltage and temperature values with accuracy of ±10mV and ±15°C. This corresponds to a set of 405 CDF models to cover the supply voltage and temperature ranges of [0.5,1.3]V and [-40,120]°C respectively. Note that only 205 models are sufficient to cover the voltage range [0.9,1.3]V with the same temperature range.

4.4 Computational and memory costs

In order to verify if this method can be used efficiently at runtime, it has been implemented in software. The platform available is a simulator of a Multiprocessor architecture based on a cluster of 8 STMicroelectronics xP70 processors implemented in 32 nm. Each processor is a VFI and is instrumented with 8 Multiprobes in order to dynamically adapt the Voltage/Frequency (VF) point to reach an optimum energetic and functional point. One of the cluster processor is not instrumented and is used to implement the KS test.

The aim of this experiment is to quantify the computational cost in terms of clock cycles. The C code of the method implemented on the simulator needs 1150 cycles by KS test. With a 680MHz processor the KS test with a model $M_{\{V,T\}}$ and a measurement $E_{\{V,T\}}$ can by performed in 1.7µs.

The maximum dynamics for V and T that can be caught is partially limited by the time the method needs to estimate the state $\{\hat{V},\hat{T}\}$. Environmental variations have very different timing dynamics [23][17][1] covering a wide frequency range. The computational cost is therefore a limiting factor of the method as it is a significant part of the total timing budget for the control loop. Obviously the user has to find the accuracy/speed trade-off according to its application. Hardware KS-test accelerator, as well as smart representation of CDF in memory are under investigation to reduce costs.

5. IMPROVEMENT

In Section 4 the influence of the granularity of the response models set on the estimation accuracy has been pointed out. Furthermore it has also be noticed that this model set size has a direct impact on the cpu time (number of cycles) required to perform an estimation. Therefore, there is a trade-off between the accuracy and the cpu-time. This section presents possible solutions to achieve such a trade-off.

5.1 Reduction of the number of models

As explained above, the estimation of state {V,T} is performed via the comparison between $E_{\{V,T\}}$ and a set of models $M_{\{V,T\}}$ elaborated during the calibration stage and stored in the Response Model Table (RMT). It has been shown in Section 4 that the reduction of the RMT limits the accuracy of the estimates, especially with regards to the temperature T.

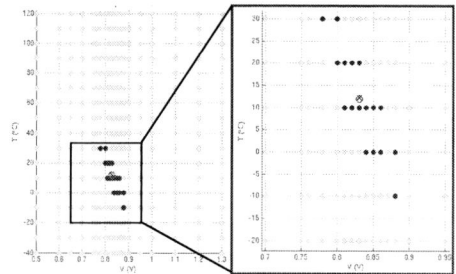

Figure 5: Iterative windowing procedure to reduce the cpu-time spent to obtain a {V, T} estimate

To reduce the cpu-time spent to estimate {V,T} while keeping the accuracy required by the user, the basic idea is to estimate the state with a crude description of the state space (i.e. few models; large ΔV and ΔT). As a consequence, a narrower {V,T} area of interest is determined. Then a zoom is performed in the area of interest with smaller ΔV and ΔT. Actually, successive analyses are iteratively performed with smaller and smaller ΔV and ΔT on the area of interest. This iterative process is carried out up to the finest RMT available (the one elaborated during the calibration stage) considering narrower and narrower {V,T} windows of interest.

Figure 5 illustrates this procedure. Note that it preserves a high accuracy while reducing by 56%, in average the cpu-time spent to obtain a {V,T} estimate. In this figure light grey points correspond to {V,T} conditions effectively tested. Black ones are those passing the final KS test. The circle and "+" sign are respectively the current state and its estimate computed with (3). Table 4 reports the mean errors and their related standard deviations obtained when considering a really fine grain RMT (i.e. a large set of models), and the ones obtained using the windowing procedure. As shown the errors are kept roughly constant while the number of KS tests performed is drastically reduced.

Table 4: Results comparison exhaustive vs iterative method

	Method of Section 4.2	Iterative method
$<|\varepsilon_V|>$	5mV	5mV
$\sigma_{\varepsilon V}$	9mV	9mV
$<|\varepsilon_T|>$	7.5°C	7.7°C
$\sigma_{\varepsilon T}$	12.6°C	12.8°C
# of effectively tested $M_{\{V,T\}}$	1377	605

5.2 Combinations of frequency measurements

As mentioned in Section 2, a sufficient number of data from a reduced set of frequency measurements is obtained by their combination. Up to now, measurements have been combined by the simple bitwise sum on-chip operation. However, there are several basic operations to combine pairs of frequencies. Among them, one may consider the subtraction but also the product. These two alternative solutions have been tested. Table 5 provides a comparison of the estimation errors obtained using the product and the subtraction instead of the sum.

Table 5: Comparison of mean and STD estimation errors for different combinations of pairs of frequencies

	Sum	Subtract	Product
$<|\varepsilon_V|>$	5 mV	15 mV	7 mV
$\sigma_{\varepsilon V}$	9 mV	20 mV	13 mV
$<|\varepsilon_T|>$	7.7°C	12.4°C	11.7°C
$\sigma_{\varepsilon T}$	12.8°C	17.2°C	19.2°C
Integrability	++	+	-

As can be seen, the sum of frequencies provides the smallest errors and standard deviations. This result is counter-intuitive as one might figure the best results from the subtract solution which removes common behaviors. Actually, the discrepancy between measurements has to be maximized in order to improve the estimation capability of the method proposed. Using the sum pre-treatment method amplifies the influence of V and T on the measurements. Note that the frequencies could be combined by 3, 4 or more to obtain larger sample of data. But others experiments have shown that the results accuracy is better for combination by 2.

However using the sum pre-treatment method provides worse results for some {V,T} conditions, when all the ROs have almost the same behavior, i.e. they oscillate almost at the same frequency. These few cases can be detected by the measurement of the $pvalue_{\{V,T\}}$. This latter is an indicator of the confidence one can set in the result. Thus when the $pvalue_{\{V,T\}}$ is "low", the estimation procedure might be run again with a different pre-treatment method, e.g. using the subtract of frequencies pairwise.

Table 6: Gain in result quality using low $pvalue_{\{V,T\}}$ detection

	Sum	$pvalue_{\{V,T\}}$ detection	Improvement
$<|\varepsilon_V|>$	5 mV	5 mV	0%
$\sigma_{\varepsilon V}$	9 mV	7 mV	22%
$<|\varepsilon_T|>$	7.7°C	6.8°C	11%
$\sigma_{\varepsilon T}$	12.8°C	9.9°C	22%
# of tests	605	609	-0.8%

This new strategy allows taking into account the qualitative aspect of the result during the estimation. Table 6 illustrates the results improvement for a number of test increased of 0.8% on average.

5.3 Threshold impact

The *threshold TH*, has been arbitrarily fixed in the previous experiments to 0.9 in order to retain for the final computations (3) of $\{\hat{V}, \hat{T}\}$, only the $M_{\{V,T\}}$ with the 10% highest *pvalues*. This value was chosen heuristically according to the following engineering intuitions:

- when a too small value of TH is chosen, the selected $M_{\{V,T\}}$ are really far from the current $E_{\{V,T\}}$ which leads to large estimation errors;

- on the contrary, if a too large value of TH is fixed, the number of $M_{\{V,T\}}$ considered during the estimate computation (3) could be too small to fully benefit from the weighted sum.

In order to analyze the influence of the TH value, the evolution of the estimation errors is analyzed wrt the TH value. The procedure described in Section 4.2 is performed for TH values ranging from 0.5 to 0.99. Table 7 reports the maximum normalized (noted TH/0.9) variation of identification errors. As can be seen, the estimation errors are quite insensitive to TH. This constitutes a main practical advantage of the proposed method.

Table 7: Maximum results variation due to *threshold*

	$<|\varepsilon_V|>_{TH/0.9}$	$\sigma_{\varepsilon V \, TH/0.9}$	$<|\varepsilon_T|>_{TH/0.9}$	$\sigma_{\varepsilon T \, TH/0.9}$
Max variation	3.5%	2.2%	3.8%	1.4%

5.4 Other hypothesis test

Here, the KS test was chosen because it is well suited for hardware or software integration. However, it could be replaced by any non-parametric test available in statistical toolboxes. Among them, one may choose the Cramer-von Mises (CvM) [2] test known to be more efficient in most cases than the KS test. The statistic of the CvM test is the difference between the integral of the two CDF.

Table 8: KS vs CvM tests

	KS	CvM
$<|\varepsilon_V|>$	5 mV	4 mV
$\sigma_{\varepsilon V}$	7 mV	6 mV
$<|\varepsilon_T|>$	6.8°C	6.3°C
$\sigma_{\varepsilon T}$	9.9°C	9.1°C

Table 8 shows the results for the same experiment as Section 5.2, obtained by replacing the KS test by the CvM one. The CvM test provides slightly better results although its powerfulness has never been theoretically demonstrated.

6. CONCLUSION

A new method has been proposed, based on hypothesis tests, to dynamically monitor, on-chip, the environmental parameters of a digital block. Its application to the interpretation of the data provided by a low cost and fully digital sensor, gives many evidence of its efficiency and accuracy. Therefore it appears possible to monitor, using low cost and fully digital sensors, the environmental variables in each power frequency domain with accuracy and high reactivity, in order to dynamically track the optimal energetic point under performance constraints. The definition of interesting local control policies of the voltage and frequency actuators of power/frequency domains of a GALS is undergoing.

7. ACKNOWLEDGMENTS

This work has been done in the context of the Platform 2012 program, a joint project between STMicroElectronics, CEA and the UE (ENIAC Joint Undertaking) in the MODERN project (ENIAC-120003).

8. REFERENCES

[1] J. Altet, A. Rubio, E. Schaub, S. Dilhaire, and W. Claeys. Thermal coupling in integrated circuits: application to thermal testing. *Solid-State Circuits, IEEE Journal of,* 36:81–91, 2001.

[2] T. W. Anderson. On the Distribution of the Two-Sample Cramer-von Mises Criterion. *The Annals of Mathematical Statistics,* 33(3):1148–1159, 1962.

[3] H. Aoki, M. Ikeda, and K. Asada. On-chip voltage noise monitor for measuring voltage bounce in power supply lines using a digital tester. *Microelectronic Test Structures, 2000. ICMTS 2000. Proceedings of the 2000 International Conference on,* pages 112–117, 2000.

[4] E. Beigne, F. Clermidy, H. Lhermet, S. Miermont, Y. Thonnart, X.-T. Tran, A. Valentian, D. Varreau, P. Vivet, X. Popon, and H. Lebreton. An Asynchronous Power Aware and Adaptive NoC Based Circuit. *Solid-State Circuits, IEEE Journal of,* 44:1167–1177, 2009.

[5] E. Beigne and P. Vivet. An innovative local adaptive voltage scaling architecture for on-chip variability compensation. *New Circuits and Systems Conference (NEWCAS), 2011 IEEE 9th International,* pages 510–513, 2011.

[6] J. Cain and C. Spanos, "Electrical linewidth metrology for systematic CD variation characterization and causal analysis," *Metrology, Inspection, and Process Control for Microlithography XVII, Proceedings of SPIE,* vol. 5038, pp. 350–361, 2003.

[7] P. Chen, C.-C. Chen, C.-C. Tsai, and W.-F. Lu. A time-to-digital-converter-based CMOS smart temperature sensor. *Solid-State Circuits, IEEE Journal of,* 40:1642–1648, 2005.

[8] J.-H. Chien, C.-L. Lung, C.-C. Hsu, Y.-F. Chou, and D.-M. Kwai. Floorplanning 1024 cores in a 3D-stacked network on-chip with thermal-aware redistribution. *Thermal and Thermomechanical Phenomena in Electronic Systems (ITherm), 2010 12th IEEE Intersociety Conference on,* pages 1–6, 2010.

[9] S. Das, C. Tokunaga, S. Pant, W.-H. Ma, S. Kalaiselvan, K. Lai, D. Bull, and D. Blaauw. RazorII: In Situ Error Detection and Correction for PVT and SER Tolerance. *Solid-State Circuits, IEEE Journal of,* 44:32–48, 2009.

[10] B. Datta and W. Burleson. Low-power and robust on-chip thermal sensing using differential ring oscillators. *Circuits and Systems, 2007. MWSCAS 2007. 50th Midwest Symposium on,* pages 29–32, 2007.

[11] I. Filanovsky and S. T. Lim. Temperature sensor applications of diode-connected MOS transistors. *Circuits and Systems, 2002. ISCAS 2002. IEEE International Symposium on,* 2:II–149, 2002.

[12] H. F. Hamann, A. Weger, J. A. Lacey, Z. Hu, P. Bose, E. Cohen, and J. Wakil. Hotspot-Limited Microprocessors: Direct Temperature and Power Distribution Measurements. *Solid-State Circuits, IEEE Journal of,* 42:56–65, 2007.

[13] K. Kang, S.-P. Park, K. Kim, and K. Roy. On-Chip Variability Sensor Using Phase-Locked Loop for Detecting and Correcting Parametric Timing Failures. *Very Large Scale Integration (VLSI) Systems, IEEE Transactions on,* 18:270–280, 2010.

[14] M. Krstic, E. Grass, F. Gurkaynak, and P. Vivet. Globally Asynchronous, Locally Synchronous Circuits: Overview and Outlook. *Design & Test of Computers, IEEE,* 24:430–441, 2007.

[15] J. Kung, I. Han, S. Sapatnekar, and Y. Shin. Thermal signature: A simple yet accurate thermal index for floorplan optimization. *Design Automation Conference (DAC), 2011 48th,* pages 108–113, 2011.

[16] S. Logan and M. R. Guthaus. Fast thermal-aware floorplanning using white-space optimization. *Very Large Scale Integration (VLSI-SoC), 2009 17th IFIP International Conference on,* pages 65–70, 2009.

[17] A. Muhtaroglu, G. Taylor, and T. Rahal-Arabi. On-die droop detector for analog sensing of power supply noise. *Solid-State Circuits, IEEE Journal of,* 39:651–660, 2004.

[18] M. Nakai, S. Akui, K. Seno, T. Meguro, T. Seki, T. Kondo, A. Hashiguchi, H. Kawahara, K. Kumano, and M. Shimura. Dynamic Voltage and Frequency Management for a Low-Power Embedded Microprocessor. *Solid-State Circuits, IEEE Journal of,* 40:28–35, 2005.

[19] G. Quenot, N. Paris, and B. Zavidovique. A temperature and voltage measurement cell for VLSI circuits. *Euro ASIC '91,* pages 334–338, 1991.

[20] B. Rebaud, M. Belleville, E. Beigne, M. Robert, P. Maurine, and N. Azemard. On-chip timing slack monitoring. *Very Large Scale Integration (VLSI-SoC), 2009 17th IFIP International Conference on,* pages 89–94, 2009.

[21] D. Sylvester, K. Agarwal, and S. Shah, "Variability in nanometer CMOS: Impact, analysis, and minimization," *Integration, the VLSI Journal,* vol. 41, pp. 319–339, 2008.

[22] L. Vincent, E. Beigné, L. Alacoque, S. Lesecq, C. Bour and P. Maurine, "A Fully Integrated 32 nm MultiProbe for Dynamic PVT Measurements within Complex Digital SoC," *2nd European Workshop on CMOS Variability, VARI'11,* 2011.

[23] R. Zheng, J. Velamala, V. Reddy, V. Balakrishnan, E. Mintarno, S. Mitra, S. Krishnan, and Y. Cao, "Circuit aging prediction for low-power operation," *Custom Integrated Circuits Conference, 2009. CICC '09. IEEE,* pp. 427–430, 2009.

Quality-retaining OLED Dynamic Voltage Scaling for Video Streaming Applications on Mobile Devices

Xiang Chen[1], Jian Zheng[3], Yiran Chen[5]
Department of Electrical and Computer Engineering
University of Pittsburgh
Pittsburgh, PA, USA 15261
{xic33[1], jiz69[3], yic52[5]}@pitt.edu

Mengying Zhao[2], Chun Jason Xue[4]
Department of Computer Science
City University of Hong Kong
Hong Kong, China 999077
mengyzhao2@student.cityu.edu.hk[2], jasonxue@cityu.edu.hk[4]

ABSTRACT

This paper developed a dynamic voltage scaling (DVS) technique for the power management of the OLED display on mobile devices in video streaming applications. An optimal voltage control scheme is proposed under input constraints. Fine-grained DVS technique is applied to maximize the power saving by leveraging the locality of the display content. The display quality is retained by monitoring structural-similarity-index (SSIM) during the optimization, subject to the hardware constraints like voltage regulator response time. Simulation results on four typical video test benchmarks show that the proposed technique saves 19.05%~49.05% OLED power on average while maintaining a high display quality (SSIM > 0.98) all the time. The power saving efficiency of the proposed technique varies at different display resolutions, refresh rates, and display contents.

Categories and Subject Descriptors

C.3 **[Special-Purpose and Application-Based Systems]**: Real-Time and Embedded Systems

General Terms

Design, Management, Performance

Keywords

OLED, dynamic voltage scaling, video streaming, mobile device

1. INTRODUCTION

The explosive growths of internet services and wireless net-work motivate the increased investment on research and development of mobile multimedia devices. Smart phones, tablets, and laptops have become the main horsepower of semiconductor industry. As a major component of mobile devices, display plays an important role in the applications of communication, entertainment and business. The invention of portable display technology, e.g. liquid crystal display (LCD), makes an efficient and informational human-machine interaction possible.

Organic light emitting diode (OLED) is one of the most promising display technologies to replace LCD. In an OLED pixel, a set of organic thin films (i.e., three corresponding to the RGB colors), is sandwiched between two conductors. When a current is applied, fluorescent light is emitted from the organic thin films. Different colors can be generated by the organic thin films with different dopants. Compared to LCD, the advantages of OLED are: 1) because OLED pixels can emit light by themselves, the power efficiency is significantly improved. The power consumption of an OLED panel is generally around 60% of a LCD panel with the same size [1]; 2) OLED has much brighter colors and higher contrast ratio, and its displaying quality is not constrained by view angles or temperature [2]; 3) OLED panels can be built on a flexible and transparent substrate [3]. Also, the latest research has demonstrated a sufficiently long device lifetime, which was once a major concern in OLED applications [4].

However, display even with OLED is still the biggest power consuming component in mobile devices. To reduce OLED display power consumption, some optimization methods are developed. Despite traditional user-behavior level display power management techniques i.e., completely or partially turning off displays, many OLED power management techniques focus on controlling pixel color composition, i.e., remapping the pixel color to the one with lower power consumption. However these techniques can substantially influence viewer experiences [5]. Thus, recently, Shin *et al.* proposed a dynamic voltage scaling (DVS) technique to adaptively adjust the supply voltage to the OLED panel based on the image content [6]. However, it is still unclear about the effectiveness of the OLED DVS technique in video streaming applications, in which, the optimization complexity may significantly increase due to the exploration of the optimal solution for both spatial and temporal display qualities.

In this work, we propose an OLED DVS technique for video streaming applications. Compared to the existing works on the power management techniques on LCD or OLED displays, the main contributions in this paper are:

•Proposed a fine-grained DVS technique to minimize the OLED power consumption in video streaming applications under realistic voltage regulator constraints;

•Proposed an SSIM-based dynamic image quality assessment index to quantitatively measure the spatial and temporal quality degradation of a video stream;

•Explored the effectiveness and the cost of pixel color repairing technique to further improve the power consumption and video quality in our proposed OLED DVS technique.

Experimental results on four test benches show that the proposed technique can reduce the OLED display power consumption by 19.05%~49.05% while maintaining a high display quality. The measured SSIM constantly keeps at 0.98 or higher.

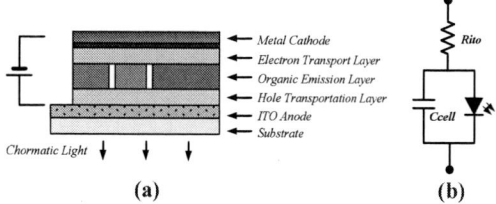

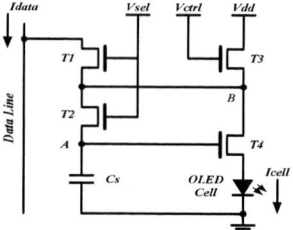

Figure 1: OLED Structure. (a) Equivalent schematic of an OLED cell. (b) Structure of a RGB color tunable OLED.

Figure 2: Schematic of DVS-friendly AMOLED Driver.

2. PRELIMINARY

2.1 OLED Cell Structure and Modeling

A chromatic color tunable OLED pixel is composed of three basic cells with RGB color space, i.e., red, green and blue. Fig. 1(a) shows a popular OLED pixel structure where the three basic cells are built on the same substrate and aligned side-by-side. Each cell is driven by an independent driver though all drivers share the same power supply. Since the emitting efficiency of the cells varies with different colors, the sizes of the cells are skewed to ensure balanced color composition [7]. The equivalent schematic of an OLED cell is shown in Fig. 1(b), including a capacitor, an ESR (electrical series resistor), and a light emitting diode. The current-voltage (I-V) characteristic of an OLED cell is determined by dopant types and cell sizes [7].

2.2 DVS-friendly Driver

In [8], a dynamic voltage scaling (DVS) friendly AMOLED driver was proposed, as shown in Fig. 2. When V_{select} is pulled down, I_{data} is cut off from the OLED cell. V_{ctrl} keeps at HIGH to turn on T_3, which keeps the OLED cell working in emission period. I_{cell} is unchanged ($\approx \square I_{data}$) since the bias condition of T_4 is maintained by C_s. $\square$ is the current mirror ratio. When V_{dd} scales, I_{cell} keeps the same as $\square I_{data}$ as long as the V_{dd} is sufficiently high. In other words, OLED color distortion occurs only when the V_{dd} is too low to supply $I_{cell} = \square I_{data}$. Fig. 3 depicts the relationship between I_{cell} and the programmed RGB levels of a DVS-friendly AMOLED driver at different V_{dd}'s [8].

2.3 OLED Power Calculation

In display techniques, the power saving is based on the light emitting efficiency, which in an OLED cell is measured by the ratio between its luminance and power consumption, i.e., $\eta = L/P$. Here power consumption $P \approx I_{cell} \cdot V_{dd}$. For an OLED pixel based on RGB color space, the total power consumption is:

$$P_{pixel} = P_{cell}(R) + P_{cell}(G) + P_{cell}(B). \tag{1}$$

Here $P_{cell}(R)$, $P_{cell}(G)$ and $P_{cell}(B)$ are the power consumptions of red, green, and blue cells, respectively. Because I_{cell} almost linearly determines the cell luminance, the OLED cell power consumption can be reduced by scaling V_{dd} without incurring any display distortions as long as a sufficient I_{cell} can be supplied. The light emitting efficiency of OLED cells varies by colors, active areas, programing methods and V_{dd}'s. In our work, we adopt the power model of a real display module, μOLED-32028-PMD3T [9]. The power consumption of a single red or green cell is 4□W at full luminance while a blue cell consumes 8□W power.

3. CHALLENGES IN OLED DVS

3.1 Multi-Level Voltage Scaling

In a conventional OLED panel design, a static supply voltage that is high enough to sustain the maximum luminance of OLED pixels is selected. However, the maximum luminance is seldom reached during the normal OLED operations. Some images allow the OLED supply voltage to be safely scaled with or without the minimized display quality degradations. In most common cases, the OLED pixels in some regions of the OLED panel usually do not reach their maximum luminance, i.e., a dark background of video games. The supply voltage to those regions may be scaled separately for power saving without incurring local color distortions, as proposed in [8]. We refer to the voltage level selection for different display regions as spatial supply voltage optimization (SSVP).

Since the optimal V_{dd} highly relies on the color profile of the image, the optimal DVS solutions change from scene to scene. The hardware cost and the power consumption required to search the optimal solutions can be considerably high. We refer to the voltage level selection associated with the video frame transitions as temporal supply voltage optimization (TSVP). We note that the changing rate of supply voltage may be subject to the constraints of human-vision perception and hardware implementations, i.e., the response time of voltage regulator.

3.2 Color Distortion and Compensation

As we discussed in Section 2.3, color distortions may occur when a lower-than-expected driving current is applied to OLED cells. In DVS technique, this issue happens when the OLED cell driving current ($I_{cell,sca}$) under the scaled V_{dd} is lower than the OLED cell driving current ($I_{cell,reg}$) under the normal V_{dd}. Initial input signals, which determine the relative OLED driving current, may be increased to compensate the luminance loss due to DVS. One example is to build a lookup table in driver system and remap the RGB level to a new one under the new V_{dd} [10]. Another example is to directly modify RGB information in the display RAMs [11]. In LCD technology, color compensation technique incurs huge cost because all LCD pixels on the panel must be compensated according to the BLU luminance loss. In OLED technology, we only need to compensate the pixels of which the driving currents change under the scaled voltage, as shown in Fig. 3. The lower V_{dd} applied, the more pixels require color compensation due to the reduction of the maximum driving current the OLED cell driver can supply. With color compensation, more aggressively voltage scaling can be achieved to further reduce power consumption.

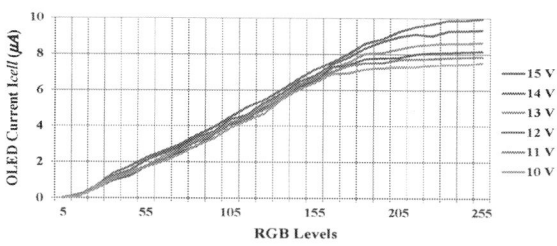

Figure 3: Relationship between I_{cell} and RGB levels of the DVS-friendly AMOELD Driver at different V_{dd}'s [8].

Table 1: Regulating Response Failure Rate (RRFR)

Refresh Rate	Movie	Cartoon	Game	News
30fps	7.82%	4.45%	17.67%	0.36%
60fps	16.73%	10.09%	36.32%	1.11%

3.3 HVS-Aware Video Quality Assessment

Different video quality assessment (VQA) methods have been investigated in the research on human visual systems (HVS) [12]. For instance, structural similarity index (SSIM) was proposed to quantitatively measure the image distortions by comparing the similarity between two images: For a color image, an image matrix is created as MSSIM = [R:G:B], where R, G, B are the matrices of the luminance of the red, green and blue cells of every pixel on the OLED panel. The SSIM of the new image X at the scaled V_{dd} w.r.t. the original image Y at the normal V_{dd} is calculated as [13]:

$$SSIM = \frac{(2\mu_x\mu_y + C_1)(2\sigma_{xy} + C_2)}{(\mu_x^2 + \mu_y^2 + C_1)(\sigma_x^2 + \sigma_y^2 + C_2)}. \quad (2)$$

Here $\Box_x$ and $\Box_y$ are the average of the image matrices of the new image X and the original image Y, respectively; $\Box_x$ and $\Box_y$ are the variances of each image matrices, respectively; $\Box_{xy}$ is covariance. C_1 and C_2 equal 2.55 and 7.56 for RGB color space, respectively. SSIM value denotes the image quality: 1~0.98, 0.98~0.96, and 0.96~0.94 means high quality, medium quality, and low quality, respectively. Below 0.94 means unacceptable. In the proposed OLED DVS technology, SSIM is used during both SSVP and TSVP to retain the video quality.

3.4 Voltage Regulator Response Time

As aforementioned, our OLED DVS technique requires real-time supply voltage adjustment based on the display content. In OLED panel designs, the ratio between the time durations spent on programming the OLED cell driving currents and driving the OLED cells is about 1:10 [14]. If we keep the OLED cell driving currents constant during the display time to ensure a steady display quality, the time left for adjusting the output of the voltage regulators is only about 3ms for a 30fps video stream (or 1.5ms for 60fps), including multiple steps such as process enabling, programing, disabling and the delays between two steps. Also, the adjusting rates of voltage regulators are asymmetric in increasing and decreasing the output voltage: For TPS61060 [15], a multi-step dynamic voltage regulator up to 15V, maximum output change at rising and failing outputs can be 2V and 0.5V, respectively, between the two frames at 30fps. For 60fps, the corresponding maximum changes decrease to 0.8V and 0.2V.

We define regulating response failure rate (RRFR) to denote the probability that a voltage regulator fails to adjust the voltage to the required level of a video frame. Table.1 shows the RRFR of four test benches in which the voltage maps for the display regions of every frame is the optimum one that is directly derived from SSVP. The test benches with frequent scene transitions (like Game) and large supply voltage gaps between two scenes (like Movie) have higher RRFR. In Section 4.2 we will give the details on how the proposed DVS technique addresses RRFR issue.

4. VIDEO STREAM DVS TECHNIQUE

In this paper, we proposed an optimal OLED DVS technique, the optimization is conducted in two different steps as spatial supply voltage optimization (SSVP) and temporal supply voltage optimization (TSVP). In SSVP, a fine-grained voltage scaling is applied to every frame. The OLED panel is defined into multiple regions, each of which has its own voltage regulator. The optimal voltage maps of every video frames are then generated by optimizing the voltage of each region based on SSIM. In the second step, TSVP, based on the analysis on frame-to-frame display content transition, the voltage maps of every frame are further adjusted to meet the voltage regulator response time constraints and the video quality requirement simultaneously. Color compensation may be applied to further improve the power efficiency and video quality of the OLED panel.

4.1 Spatial Supply Voltage Optimization

In SSVP step, display quality is controlled at both region and frame levels. We note that the satisfaction for SSIM in regions cannot definitely guarantee the satisfaction at frame level [8]. Without concerning the whole frame's SSIM, the sharp edge between two regions can be easily observed. Thus, a second voltage adjustment must be applied to ensure the SSIM of the whole image above the threshold. According to the characters of algorithm in [8], each region's voltage to be scaled will be firstly predicted with "sacrificed luminance ratio (S.R.)", which is introduced to measure this DVS-induced luminance loss by calculating the ratio between the sacrificed luminance at the scaled V_{dd} and the luminance of the whole OLED panel at the normal V_{dd}. S.R. is used to initially direct SSVP, the voltage of each region is scaled to the level that ensures a pre-determined luminance loss, say, S.R. = 0.1. Then we calculate the SIMM of every region and the whole frame. If the SSIM of the whole frame is higher than the quality requirement (say, 0.98), the voltage of the region with the highest SSIM will be decreased; otherwise, the voltage of the region with the lowest SSIM will be increased. Our experiments found that a S.R. of 0.1 gives us the best optimization result and run time cost for all test benches [8].

4.2 Temporal Supply Voltage Optimization

In this section, we will first give formal formulation of the temporal supply voltage optimization problem (TSVP). Then we will present a polynomial time optimal algorithm.

Problem Formulations

Assume the size of frame set is n, each frame has m partitions and there are x number of voltage levels to choose from. We use f_i to represent the i_{th} frame in the frame set, p_{i_j} for the j_{th} partition of f_i, and $\overline{v_k}$ for the k_{th} voltage level.

We further assume, for each voltage level $\overline{v_k}$, the corresponding variable $x_{i\_j\_k}$ stands for whether it's picked by partition p_{i_j}.

$$x_{i\_j\_k} = \begin{cases} 1, \text{ if partition } p_{i\_j} \text{ chooses voltage level } \overline{v_k} \\ 0, \text{ otherwise} \end{cases} \quad (3)$$

And for each partition p_{i_j}, $i \in \{1,...,n\}$, $j \in \{1,...,m\}$,

$$\sum_{k=1}^{x} x_{i\_j\_k} = 1 \quad (4)$$

so v_{i_j}, the voltage level of partition p_{i_j}, is represented by:

$$v_{i\_j} = \sum_{k=1}^{x} x_{i\_j\_k} \overline{v_k} \quad (5)$$

Variable set $y_{i\_j\_k}$ stands for voltage difference between v_{i+1} and v_i. The value of $v_{i-1} - v_i$ is limited to set Var, whose size is q, due to the voltage change delay constraint. Each element $\overline{var_k}$ in Var is corresponding to a variable $y_{i\_j\_k}$, which is used to indicate its effectiveness.

$$y_{i\_j\_k} = \begin{cases} 1, \text{ if } v_{i+1\_j} - v_{i\_j} = \overline{var_k} \\ 0, \text{ otherwise} \end{cases} \quad (6)$$

1002

For all adjacent partition pair $(p_{i\_j},\ p_{i+1\_j})$, $i \in \{1,...,n-1\}$, $j \in \{1,...,m\}$, the following guarantees the voltage delay constraint.

$$\sum_{k=1}^{q} y_{i\_j\_k} = 1 \tag{7}$$

$$v_{i+1\_j} - v_{i\_j} = \sum_{k=1}^{q} y_{i\_j\_k} \overline{\text{var}_k} \tag{8}$$

At the same time, frame image f_i, $i \in \{1,...,n\}$, should be visually acceptable:

$$SSIM_i \geq SSIM_{req} \tag{9}$$

where the $SSIM_i$ describes the display quality of f_i and $SSIM_{req}$ is the lowest requirement.

Temporal Supply Voltage Optimization (TSVP) Objective: According to the predefined variables and constraints, we want to minimize the total power of the OLED display. Given the width w and height h, the power of partition p_{i_j} is calculated by adding all the pixel power:

$$power_{i\_j} = \sum_{s=1}^{w} \sum_{t=1}^{h} v_{i\_j} c_{s\_t} \tag{10}$$

where c_{s_t} is the current going through pixel $(s,\ t)$ in p_{i_j}, which is estimated by RGB information in voltage level v_{i_j}.

The objective is

$$Min \sum_{i=1}^{n} \sum_{j=1}^{m} power_{i\_j} \tag{11}$$

Theoretically, this optimization problem can be solved by ILP solver to achieve the best voltage decision for each partition with the VQA requirement. However, ILP solution is too time-consuming and not scalable. In this paper, we have identified an optimal algorithm which can solve the TSVP problem in polynomial time. In the following, we present the polynomial time optimal algorithm.

Optimal Algorithm

After SSVP, the local optimal voltage of every partition is obtained. The following optimal algorithm can be applied to adjust the voltage map to meet the temporal VQA and voltage regulating constraints.

The result from SSVP is set to be the low bound of voltage levels as it guarantees the image quality of each frame. The same or higher voltage levels will always satisfy the display quality based on SSVP. The larger voltage gap between corresponding partitions of adjacent frames, the more time we need to regulate the voltage. So the voltage gap is considered instead of responding time. Intuitively, the algorithm needs to decide which partition should raise its voltage level to meet the regulator constraints.

The inputs of TSVP are:
  (1) Frame set $F=\{f_i \mid i=1,...,n\}$;
  (2) Voltage set $\bar{V}=\{v_k \mid k=1,...,x\}$;
  (3) Initial voltage level set from SSVP. $V=\{v_{i\_j} \mid i=1,...,n, j=1,...,m\}$;
  (4) Voltage-change set Var.

The objective is to adjust the voltage levels which violate the voltage change constraints and achieve an optimal voltage set $V'=\{v_{i\_j}' \mid i=1,...,n, j=1,...,m\}$.

The updating method is described in Algorithm 2. If the voltage levels of one adjacent partition pair violate the constraints, we raise the lower voltage level to reduce the difference. It's raised one level each time as we would like to keep it as low as possible to save power. After each improvement, the neighbors of the adjusted partition are checked to make sure both are eligible with this update, any gap larger than the regulator constraints caused by this update is handled by recursively calling the *Adjust_Voltage* function. This procedure is executed for each partition pair, like the expression in Algorithm 1, to eliminate all the constraint violations.

Algorithm 1 *Optimizer*

  Input: F, $\bar{V}$, V, Var.
  Output: V'.
  For each pair of adjacent partitions (p_{i_j}, p_{i+1_j}) **do**
    Adjust Voltage (p_{i_j}, p_{i+1_j});
  Return V'

Algorithm 2 *Adjust_Voltage* (p_{i_j}, p_{i+1_j})

  if $(v_{i+1\_j} - v_{i\_j}) \notin Var$ **then**
    if $v_{i+1\_j} > v_{i\_j}$ **then**
      Raise v_{i_j} by one level;
        Adjust_Voltage (p_{i_j}, p_{i+1_j});
        if there exist one frame f_{i-1} before f_i **then**
          Adjust_Voltage (p_{i-1_j}, p_{i_j});
        endif
    else
      Raise v_{i+1_j} by one level;
      Adjust_Voltage (p_{i_j}, p_{i+1_j})
      if there exist one frame f_{i+2} after f_{i+1} **then**
        Adjust_Voltage (p_{i+1_j}, p_{i+2_j});
      endif
    endif
  endif

In this algorithm, V is given as the lower bound of V', describing satisfying image quality for each partition, which means we could not lower voltages to below V.

To prove the efficiency of this algorithm, let sequence $S= (v_i, v_{i+1}, ... , v_{j-1}, v_j)$ be a subsequence of output $(v_0, v_1, ... , v_{n-1}, v_n)$ with continuous voltage levels have been regulated, which means improved. For convenience, we use δ ($\delta > 0$) to represent the allowed voltage maximum gap of regulator.

1. If $|S|=1$, then v_i is improved due to $v_{i-1}-v_i \geq \delta$ or $v_{i+1}-v_i \geq \delta$, where v_{i+1} and v_{i-1} are already smallest values, so the proposed method raises v_i least to respect the constraint and can't be lower.

2. If $|S|>1$, considering v_i is raised with two possibilities:
  (1) $v_{i-1}-v_i \geq \delta$, where v_{i-1} makes this improvement least and necessary;
  (2) $v_{i+1}-v_i \geq \delta$, in which case v_{i+1} is regulated only if $v_{i+2} - v_{i+1} \geq \delta$ because it has no chance to increase due to lower voltage v_i, similarly, we will have $v_i < v_{i+1} < ... < v_{j-1} < v_j < v_{j+1}$. Consequently v_j could not be lower because it is on the appropriate level the optimal method sets to catch up with v_{j+1}. So as v_j, v_{j-1},..., v_{i+1} and v_i.

In a conclusion, the first voltage of v_i could not be downgraded. Update S with $(v_{i+1},..., v_{j-1}, v_j)$ and repeat the procedure from the beginning. In this way, we can prove that all the voltage levels are already the lowest and could not be reduced any further.

As there are a total of $n \times m$ partitions and for each partition there are at most p voltage levels to regulate, so the proposed optimal algorithm has the time complexity of $O(n \times m \times p)$. This polynomial time optimal algorithm works efficiently and can handle with 900 frames in seconds, and the exact time differs with various test

1003

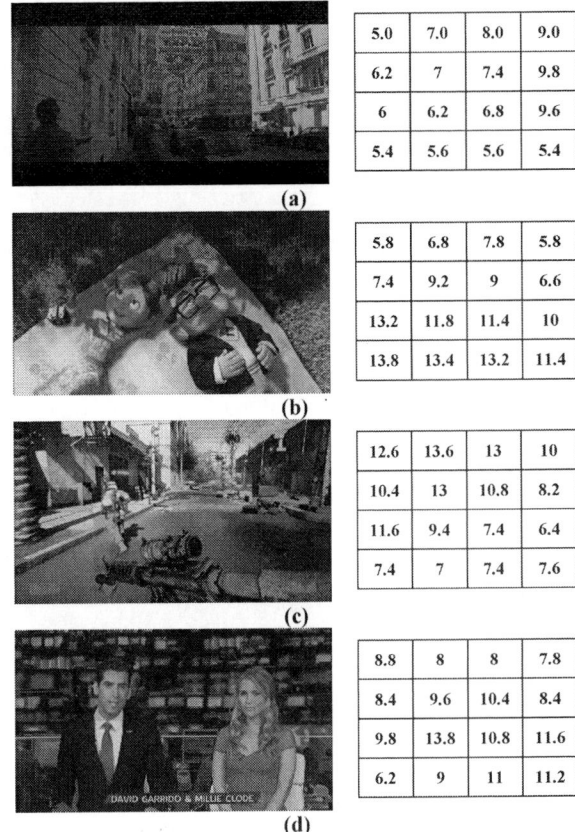

Figure 4: Typical single Frame SSVP results with voltage maps. (a) Movie; (b) Cartoon; (c) Game; (d) News Report

benches. For longer streams, we could partition the video based on scene transitions, which can be determined by many methods

4.3 Color Compensation Mechanism

In the proposed DVS technique, the OLED pixels under a scaled V_{dd} can be divided into three groups: 1) the pixels do not need color compensations ("none"), or $I_{cell,sca} = I_{cell,reg}$; 2) the pixels can be fully compensated ("full"), or $I_{cell,reg} < max(I_{cell,sca})$. $max(I_{cell,sca})$ is high output of the OLED cell driver can achieve at the scaled V_{dd}; and 3) the pixels can be partially compensated ("partial"), or $I_{cell,reg} > max(I_{cell,sca})$. Since the original luminance of the "partial" OLED cells is beyond the highest luminance of the OLED cells can have under the scaled V_{dd}, the luminance loss become inevitable even color compensation is applied.

5. EXPERIMENTS AND DISCUSSIONS

5.1 Evaluation of SSVP

Fig. 4 shows the frame partition schemes of all test benches. Each frame is divided into 16 regions with independent voltage regulator. The typical voltage maps an SSIM = 0.98 are also shown in Fig. 4. We made the following observations:

1) In test bench "Movie", the black borders at the top and bottom are commonly used to adapt the resolutions. The voltages of these regions can be safely reduced to 5.0V. The voltage of the right-top regions displaying white buildings, however, cannot be scaled lower than 9.0V; similarly, in "News Report", the background with black and blue colors is static and dim. The voltages of the background regions can be scaled significantly lower than that of the other regions.

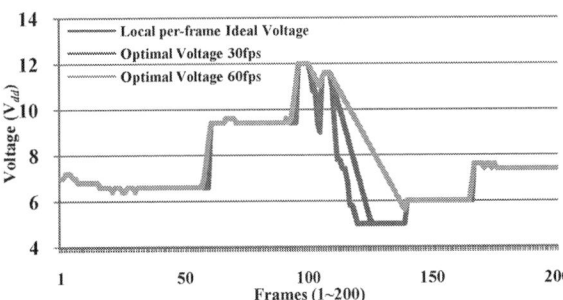

Figure 5: Voltage Regulating Histogram with Optimal Optimization

2) In "Cartoon" and "Game" test benches, the applications of CG (computer graphics) technique created complex display contents with high contrast ratios. The voltage differences between the adjacent regions can be large. The voltage level of a region can be scaled heavily depends on the major color components displayed in the region.

3) We note that the fine-grained OLED DVS technique demonstrates overwhelming power benefits compared to the global OLED DVS technique. For example, the minimum voltages for the above four images to ensure a SSIM = 0.98 with global DVS technique are 8.8V, 11.2V, 11.6V, and 10.4V, respectively, which are 7.4%, 8.5%, 10.9%, 4.3% higher than that the fine-grained DVS technique can achieve. It is why we use fine-grained OLED DVS technique as the baseline technology in SSVP.

5.2 Evaluation of TSVP

To evaluate the TSVP performance, we first verify the voltage optimization mechanism based on per-frame ideal voltage setting. Ideally, when the voltage regulator's response is fast enough, all the voltage gaps can be met between adjacent frames, and the most aggressive power saving ratio can be achieved. However, due to practical product's narrow regulating range, the regulating response failures exist. Optimal algorithm is applied to repair all regulating response failures, which can cause unstable image quality and OLED life spam damage, by increasing the lower voltage to adjust to the voltage gap of the same region in two adjacent frames. For example one area's voltage regulating from test bench of Movie is shown in Fig. 6. It can be seen that, around 100 frames, huge voltage gaps take place. For the ideal case, the V_{dd} only changes for the optimal power saving purpose and performs the fattest changing speed. However, when it comes to 30fps, due to limited regulating time, the voltage regulator has to

Table 2: Power Savings with Different Resolutions and Refreshing Rates

Resolution	640×360		
Opt. Condition	Ideal	30fps	60fps
Movie	48.65%	46.82%	40.30%
Cartoon	30.79%	28.98%	26.54%
Game	35.78%	24.14%	19.05%
News	34.27%	34.12%	33.86%
Resolution	1280×720		
Opt. Condition	Ideal	30fps	60fps
Movie	52.18%	49.05%	43.23%
Cartoon	31.76%	29.14%	27.17%
Game	36.15%	25.35%	20.50%
News	37.12%	36.56%	36.34%

adjust the voltage over extra frames with narrow steps to meet the voltage after gap, especially in the decreasing process where the regulator has the poorest performance. And when it comes to 60fps, the slow voltage regulating process is more obvious.

Table 2 shows the energy savings of each test bench when the optimal DVS technique is applied, compared to the energy of the OLED panel working at the normal V_{dd}. Each frame's quality is maintained at SSIM = 0.98. From Table 2, it can be seen that, the power saving ratio vary from benchmarks, due to the scene transient rate and luminance distribution as we analyzed before. However, within each benchmark, when fresh rate increases from 30fps to 60fps, the maximum output change of voltage regulators between two frames significantly reduces. The increased RRFR makes the optimal algorithm keep the voltage gap of the same region in two adjacent frames within a smaller range. It results in a lower power saving than 30fps. For example, in "Game" test bench which has the most frequent scene transitions, The RRFR to be repaired increases from 17.67% to 36.32% when the refresh rate rises from 30fps to 60fps. And the power saving also reduces from 24.14% to 19.05% and 25.35% to 20.50% accordingly. The optimal algorithm in the above simulations target 0% RRFR in the optimization process to ensure no frame will have an SSIM less than 0.98. However, if we allow a certain RRFR exist after the optimization, we should be able to achieve higher energy saving.

Also, our results show that the energy saving is insensitive to the display resolution; generally, with the same display content, our power saving method will achieve aggressive performance, which will definitely benefit from the display development trend and the details will be explained in later color compensation part.

5.3 Color Compensation

The cost of color compensation can be measured by the ratio between the numbers of the pixels need to be applied with color compensation and the total pixels across all frames. Fig. 6 shows the color compensation ratio of all test benches with different OLED panel configurations, including all compensation necessary and full compensation successfully achieved. In general, color compensation cost grows with the display resolution. Although the display contents are the same with different resolutions, the maximum luminance can be even bigger in the original high resolution video stream, which performs more luminance transition levels, higher contrast ratio and more details in unit display area. When a smaller resolution stream is recoded (generally converted to portable devices) from the original, some luminance are reduce to adapt less grained pixels, hence the all

compensation ratio is less in 640×360. However, fortunately, with DVS-friendly driver, most of higher luminance pixels existed in the bigger resolution can be fully compensated. Hence the ratio between the "full compensation" pixels over the total pixels in the high resolution case is also higher than that in low resolution case. For the four test benches, the ratio of full compensation over all compensation is 53.5%~73.5% in 640×360 and rise up to 58.1%~75.3% in 1280×720. Thus, with higher resolution, the proposed method can achieve better compensation performance. In Fig. 6, compensation different also happens in different refreshing ratio. In 60fps case, the higher working voltage due to the limited voltage regulator response time makes the full color compensation cost higher than the 30fps. It indicates less number of pixels suffer from the color distortions.

6. CONCLUSION

This paper proposed an optimal DVS optimization method to manage the power consumptions of OLED panels in video stream applications. Two optimization steps – spatial supply voltage optimization and temporal supply voltage optimization guarantee the real-time video quality while receiving the minimal energy consumption. The experimental results on four typical test benches show that comparing to conventional global DVS solution, our technique saves 19.05%~49.05% OLED power on average while maintaining a high display quality (SSIM > 0.98).

7. REFERENCE

[1] "Samsung, LG in Legal Fight over Brain Drain", *The Korea Times*, Jun. 2010.

[2] W. Graupner, C. M. Heller, A. P. Ghosh, W. E. Howard, "High-resolution Color Organic Light-emitting Diode Micro-display Fabrication Method", in Proc. of *Int'l Society for Optical Engineering*, 4207, Nov. 2000.

[3] G. Gustafsson, Y. Cao, G. M. Treacy, F. Klavetter, N. Colaneri, A. J. Heeger, "Flexible Light-emitting Diodes Made from Soluble Conducting Polymers," *Nature*, vol. 357, pp. 477 – 479, Jun. 1992.

[4] Universal PHOLED Technology and Materials, Product Data, Universal Display Corporation, 2011.

[5] D. Mian, L. Zhong, "Chameleon: A Color-adaptive Web Browser for Mobile OLED Displays," *ACM/USENIX Int. Conf. Mobile Sys .Appl.*, Jun. 2011.

[6] D. Shin, Y. Kim, N. Chang, M. Pedram, "Dynamic Voltage Scaling of OLED Displays," *Des. Auto. Conf. (DAC)*, Jun. 2011.

[7] J. Jacobs, D. Hente, E. W. Schmidt, "Drivers for OLEDs", *Industry Appl. Society Annual Meeting*, pp. 1147-1152, 2007.

[8] X. Chen, J. Zeng, W. Zhang, H, Li, Y. Chen, "Fine-grained Dynamic Voltage Scaling on OLED Display" in Proc. of *Asia and South Pacific Des. Auto. Conf. (ASPDAC)*, 2012.

[9] M. Dong, K. Choi, L. Zhong, "Power Modeling of Graphical User Interfaces on OLED Displays," in Proc. of *ACM/IEEE Des. Auto. Conf. (DAC)*, Jul. 2009.

[10] S. Lee, T Kim, J. Choi, "A Color Correction System using a Color Compensation Chart," *Int'l Conf. on Hybrid Info. Tech. (ICHIT)*, vol.1, pp.409-416, Nov. 2006.

[11] J. Zhong, S. Yao, J. Xu, "Pixel Cross-talk Compensation for CMOS Image Sensors," *Intelligent Info. Tech. Appl. (IITA)*, vol.2, pp.165-168, Nov. 2009.

[12] A. Iranli, W. Lee, M. Pedram, "HVS-Aware Dynamic Backlight Scaling in TFT LCD's," *IEEE Trans. on VLSI Systems*, vol. 14, no. 10, pp. 1103-1116, Oct. 2006.

[13] Z. Wang, A. C. Bovik, H. R. Sheikh, E. P. Simoncelli, "The SSIM Index for Image Quality Assessment," http://ece.uwaterloo.ca/ ~z70wang/research/ssim/.

[14] Y. He, R. Hattori, J. Kanicki, "Improved a-Si:H TFT Pixel Electrode Circuits for Active-Matrix Organic Light Emitting Displays," *IEEE Trans. on Electron Devices*, vol.48, no.7, php.1322-1325, Jul. 2001.

[15] TPS61060/61/62: Single-Wire Digital Brightness Control, Product Data, Texas Instruments Corporation, 2006.

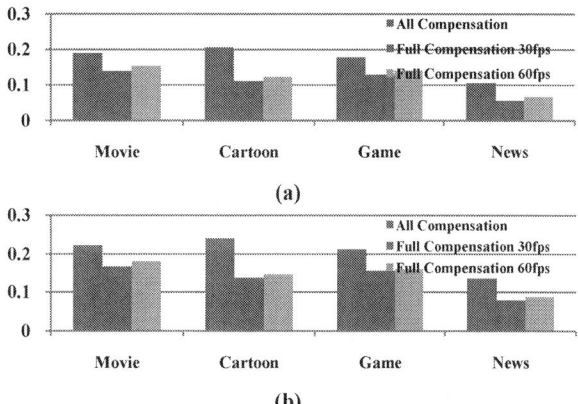

Figure 6: Color compensation ratios of all test benches with different resolution (a) 640×360; (b) 1280×720

Traffic-Aware Power Optimization for Network Applications on Multicore Servers *

Jilong Kuang, Laxmi Bhuyan and Raymond Klefstad
Computer Science & Engineering Department
Unviersity of California, Riverside
900 University Ave, Riverside, CA 92521, USA
{jkuang, bhuyan, klefstad}@cs.ucr.edu

ABSTRACT

In this paper, we design, implement, and evaluate a traffic-aware and power-efficient multicore server system by translating incoming traffic rate to appropriate system operating level, which is then translated to optimal per-core frequency configuration. According to the varying traffic rate, the system can adjust the number of active cores and per-core frequency "on-the-fly" via the use of per-core DVFS, power gating, and power migration techniques based on our new power model which considers both dynamic and static power consumption of all cores. Results on an AMD machine with two Quad-Core Opteron 2350 processors for six real network applications chosen from NetBench [19] show that our scheme reduces power consumption by an average of 41.0% compared to running with full capacity without any reduction in throughput. It also consumes less power than three other approaches, chip-wide DVFS [22], power gating [17], and chip-wide DVFS + power gating [15], by 35.2%, 24.3%, and 10.5% respectively.

Categories and Subject Descriptors

C.4 [**Computer Systems Organization**]: Performance of Systems—*Performance attributes, Design studies*

General Terms

Design, Performance

Keywords

Packet processing, power efficiency, multicore architecture

1. INTRODUCTION

Explosive growth of Internet high-traffic applications, such as video streaming, cloud computing and file sharing, requires orders-of-magnitude increase in system throughput. Affordable multicore servers, such as Cavium's OCTEON [2], Cisco's AON [3], and IBM's BladeCenter [6], can now meet this throughput demand. Along with increased throughput, however, comes significantly increased power consumption [8]. Collectively, millions of servers in the global network consume a great deal of power [12]. And chip manufacturers continue to increase both the number of cores and their frequencies, substantially increasing both dynamic and static power consumption.

At the hardware level, there are two main techniques to reduce power consumption. The first technique, Dynamic Voltage and Frequency Scaling (DVFS), which is widely used, reduces or increases processor voltage/frequency just enough to meet performance requirements. DVFS can be either chip-wide, where the entire chip is scaled as one unit (e.g., Intel's Foxton technology [18]), or per-core, where individual cores on the chip can be scaled at different rates (e.g., AMD's Opteron processor [1]). The second

---

*This work was supported by National Science Foundation grants CNS-0832108 and CSR-0912850.

hardware-level technique, called power gating, minimizes leakage current when a core is inactive by powering it down almost completely. Power gating has been introduced only recently by major chip manufacturers (e.g., Intel's Nehalem [14]).

To determine when to reduce power to the core, various application run-time characteristics are exploited, such as program phase analysis [10], degree of parallelism [15], and time slack detection [20]. We see great potential power-saving opportunity, however, in an additional aspect: *network traffic*. Computing power needs fluctuate dramatically with the large fluctuations in network traffic. For example, Figure 1 shows real-time network traffic in a typical day monitored by Equinix data center [4] at San Jose, CA. The traffic rate varied from 320K packets/s to 720K packets/s. Power consumption could be greatly reduced when traffic is low.

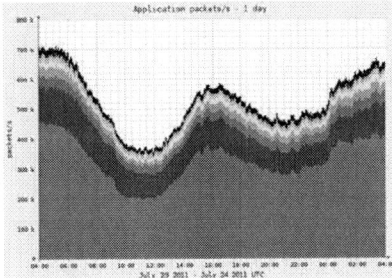

Figure 1: Traffic variation versus time in 24 hours. Different colors represent the breakdown of different packet types.

Existing studies that consider traffic variation, however, are limited in the following two ways:

1. **Dynamic Power Only**. They assume that dynamic power dominates total power consumption, and that static power can be ignored [11, 17, 22]. However, static power has increased dramatically with increases in device speed and chip density. According to a projection by the International Technology Roadmap for Semiconductors, leakage power increases its dominance of total power consumption as semiconductors progress toward 32nm [7].

2. **Single Dimensional**. Traffic-aware studies focus either on single-core platforms and chip-wide DVFS [22], or adopt power gating only [11, 17]. Thus, these approaches cannot be applied to multicore systems that support both per-core DVFS and power gating.

Using a combination of per-core DVFS and power gating can potentially minimize power consumption when network traffic is low. However, with this approach, cores perform different amounts of work. Not all cores actively run all the time, and each core may run at a different frequency. Some cores may then be stressed more than others, and overworked cores will generate excess heat, increasing static power consumption exponentially with temperature [23]. It is therefore advisable to migrate active cores periodically for lower peak core temperature and less static power consumption. A software approach called *power migration* can be used to achieve thermal load balancing across the cores. Locations of more- and less-active cores can be dynamically changed according to some policy while keeping the same system operating level. Given the same amount of heat generation depending on the number of active cores and core frequency, power migration can

1006

redistribute the generated heat in space and time to reduce peak core temperature and improve thermal uniformity.

This paper describes a power-efficient multicore server system for network applications which dynamically adjusts system operating level and per-core frequency configuration based on incoming traffic rate. Our on-line algorithm optimizes a novel power model that considers both dynamic and static power. The dynamic per-core frequency configuration is achieved through a combination of per-core DVFS, power gating, and power migration.

We first derive a formula to translate traffic arrival rate to required cumulative core frequency. Then, based on our power model, we derive the optimal system operating level to maintain sufficient system throughput for the current traffic while using minimal dynamic power. Lastly, because each core may be configured at a different operating frequency, we migrate active cores in the system periodically to achieve thermal balancing and reduce peak core temperature. To the best of our knowledge, we are the first to target power optimization considering both dynamic and static power for network applications running on multicore servers.

To verify our design, we implement our approach on a multicore server system with varying traffic loads, running six real network applications from NetBench [19]. Our approach reduces power consumption by an average of 41.0% compared to running with full capacity without any reduction in throughput. Our approach also outperformed three other approaches with negligible overhead: chip-wide DVFS [22], power gating [17], and a hybrid combination of chip-wide DVFS and power gating [15].

The rest of this paper is organized as follows: Section 2 presents our system design which includes the traffic-aware power optimization scheme in a three-step approach. Section 3 presents our implementation and performance evaluation. Finally, Section 4 concludes this paper.

2. TRAFFIC-AWARE POWER OPTIMIZATION

2.1 System Design

The typical application supported by this work runs on a multicore server and processes a stream of network requests. Figure 2 shows the system overview, where incoming packets from the network are first stored in a global FIFO queue and then scheduled to proper cores for packet processing. The core component is system manager, which consists of four functional modules: traffic monitoring, power managing, core configuring, and task scheduling.

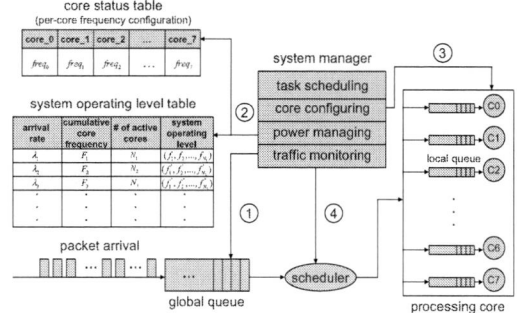

Figure 2: Overview of the traffic-aware system.

1. **Traffic monitoring module** tracks packet inter-arrival times to obtain the packet arrival rate and detect the rate change point whenever the traffic rate varies. We monitor the traffic and detect the rate change point by using the sampling technique based on maximum likelihood ratio [21, 22], which is particularly useful for server/router deployment where traffic changes can be predicted only based on prior information.

2. **Power managing module** manages two runtime tables, a system operating level table, which caches optimal system operating level for a given traffic arrival rate, and a core status table, which tracks the actual per-core frequency configuration. The system operating level is represented by tuple $(f_1, f_2, ..., f_N)$ throughout this paper, where $f_1 \geq f_2 \geq ... \geq f_N$. This tuple indicates that N cores are active running and the ith core has the frequency f_i $(1 \leq i \leq N)$. Given a certain arrival rate, the power managing module appropriately derives the optimal system operating level based on our dynamic power optimization scheme. In addition, it also initializes the core status table at each traffic rate change point, and periodically updates the core status table to enable

power migration for active cores between two consecutive traffic rate change points based on our static power optimization scheme.

3. **Core configuring module** adjusts the frequency level of each core based on the information from the core status table managed by the power managing module. At each configuration point, it applies power gating to cores labeled as inactive as soon as their local queues become empty, and applies per-core DVFS for cores labeled as active and adjusts their frequencies according to their respective configurations chosen from one of the five frequency levels, namely 1GHz, 1.2GHz, 1.4GHz, 1.7GHz and 2GHz. The core configuring module is critical in the system because it is where the three applied power techniques are actually enabled.

4. **Task scheduling module** appropriately schedules packets in the global queue to active cores in our system. When the per-core frequency configuration updates, the scheduler stops sending packets to power-gated cores. For active cores with different core frequencies, the scheduler distributes the workload, which is determined by the number of packets, per-packet size, and application type, in proportion to the core frequency. This approach achieves weighted load balancing across cores under varying traffic rate and avoids loss of system throughput due to the change of system operating level and update of per-core frequency configuration.

As we focus on the power managing module in this paper, we propose a three-step approach as shown in Figure 3 to solve the power optimization problem.

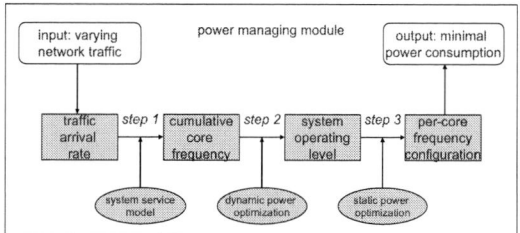

Figure 3: A three-step power optimization scheme.

2.2 Step 1: System Service Model

The system service model translates the traffic arrival rate to required cumulative core frequency in the multicore system. As traffic rate varies, an ideal cumulative core frequency should be just sufficient to satisfy the demand without over-provisioning.

First, we let service rate equal to arrival rate, because 1) to guarantee a stabilized system without packet overflow, service rate should be no less than arrival rate, and 2) to avoid over-provisioning and achieve power efficiency, service rate should be no greater than arrival rate.

Second, [9, 22] have shown that the service rate is linearly proportional to the CPU frequency for a single-core system. However, because we target multicore architectures where different cores may run at different frequencies, it is necessary to re-think and justify the relationship between the service rate and cumulative core frequency. We, therefore, conduct two empirical trace-driven studies with 6 chosen network applications from NetBench on our multicore machine to help establish the relationship.

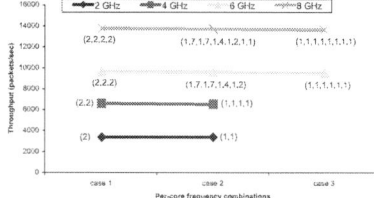

Figure 4: Throughput versus frequency combinations.

The first study examines the effect of various per-core frequency combinations versus throughput given the same cumulative core frequency. Figure 4 shows the results of the URL application when we vary the cumulative core frequency from 2GHz to 8GHz. For each cumulative core frequency, we change the per-core frequency combinations. From this figure, we observe that the throughput (service rate) only depends on cumulative core frequency, regardless of per-core frequency combinations. This is because when incoming packets are processed on multiple cores with different service rates, we can equivalently treat this multicore server as a single-core system with the aggregated service rate equal to the sum of per-core service rate.

The second study builds the relationship between the service rate and cumulative core frequency in our system. We vary the cumulative core frequency from the minimum (1GHz) to the maximum (16GHz) and record the system throughput. The results show that for our multicore server, the throughput (service rate) is also linearly proportional to the cumulative core frequency. Figure 5 illustrates both the experiment result and the fitted line for the URL application (Other applications have the similar results with different parameters and coefficients). Therefore, our system service model for the URL application is given by the linear function in Equation 1, where X represents cumulative core frequency and Y represents the service rate, or the arrival rate in our case.

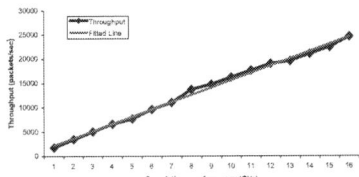

Figure 5: Throughput versus cumulative core frequency.

$$Y = 1496 \cdot X + 628 \tag{1}$$

2.3 Step 2: Dynamic Power Optimization

The dynamic power optimization scheme takes cumulative core frequency as input and produces the optimal system operating level as output. More specifically, it answers the following two questions: Q1) what is the theoretically optimal number of active cores from our power model? Q2) what is the frequency assignment for active cores considering the discrete frequency levels?

2.3.1 Power Model

Consider a network application running on a core at voltage v and frequency f. The dynamic power consumption is given by: $P_{dynamic} = K_a \cdot f \cdot v^2$, where K_a is a task/core dependent factor determined by the switched capacitance. Besides, the frequency f is almost linearly related to the voltage v : $f = K_b \cdot (v - v_t)^2/v$, where v_t is the threshold voltage and K_b is a constant. For a sufficiently small threshold voltage, the frequency is approximated to $K_b \cdot v$. Therefore, we assume the dynamic power consumption is cubic to the frequency as shown in Equation 2, where $K = K_a/K_b^2$.

$$P_{dynamic} = K \cdot f^3 \tag{2}$$

With respect to static power, we assume that for power-gated cores, they consume zero static power. For active cores, static power consumption is exponential to core temperature [23]. However, as this step focuses on dynamic power optimization, we ignore the temperature effect and assume the static power of each active core is constant, P_s. Detailed discussion for static power consumption is given later because it is related to thermal balancing and power migration. Thus, the total power consumption of an active core is given by Equation 3:

$$P_{core} = P_{static} + P_{dynamic} = P_s + K \cdot f^3 \tag{3}$$

In a multicore system with N active cores, suppose f_i is the frequency on core i and $P(f_1, f_2, ..., f_N)$ is the total system power consumption as a function of system operating level denoted as $(f_1, f_2, ..., f_N)$. We have the following:

$$P(f_1, f_2, ..., f_N) = N \cdot P_s + K \cdot (f_1^3 + f_2^3 + ... + f_N^3) \tag{4}$$

In the following, we focus on answering Q1 in a quantitative approach under the assumption that the core frequency is continuous. Later on, to answer Q2, we will relax this constraint in practical scenario with discrete frequency levels.

Suppose we have x active cores to handle cumulative core frequency F. As the dynamic power is proportional to the cube of core frequency, we know that when every active core is running at the same frequency of F/x, the total dynamic power consumption reaches minimum. Thus, from Equation 4, we can derive the total power consumption as follows.

$$P = P(f_1, f_2, ..., f_x) = x \cdot P_s + K \cdot (f_1^3 + f_2^3 + ... + f_x^3) \tag{5}$$

$$\geq x \cdot P_s + x \cdot K \cdot (F/x)^3 = x \cdot P_s + \frac{K \cdot F^3}{x^2}$$

This function $(P = x \cdot P_s + \frac{K \cdot F^3}{x^2})$ is a unimodal function and has a global minimum as illustrated in an example in Figure 6. This curve is drawn for the URL application when we set $P_s = 5.8$, $K = 1.6$ and $F = 3$. More details about the parameters can be found in Section 3. It shows that starting from a single active core $(x = 1)$, increasing the number of active cores will reduce the total power consumption while satisfying the cumulative core frequency requirement, until the number of active cores increases past a certain threshold value. We call this value x^*, which is the optimal number of active cores that strikes a good balance between static and dynamic power. In fact, from the classic algebra inequality as shown in Equation 6, we can easily solve the problem.

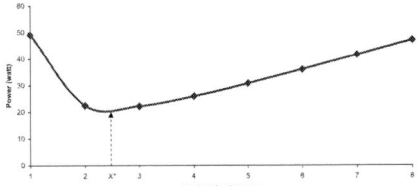

Figure 6: Power consumption as the number of active cores varies given the same cumulative core frequency.

$$\frac{a + b + c}{3} \geq \sqrt[3]{a \cdot b \cdot c} \tag{6}$$

when a=b=c, left side reaches minimum.

$$P = x \cdot P_s + \frac{K \cdot F^3}{x^2} = \frac{x \cdot P_s}{2} + \frac{x \cdot P_s}{2} + \frac{K \cdot F^3}{x^2} \tag{7}$$

$$\geq 3 \cdot \sqrt[3]{\frac{P_s^2 \cdot K \cdot F^3}{4}}$$

In addition, the minimal power consumption is achieved if and only if Equation 8 is satisfied.

$$\frac{x \cdot P_s}{2} = \frac{K \cdot F^3}{x^2} \Rightarrow x^* = \sqrt[3]{\frac{2K \cdot F^3}{P_s}} \tag{8}$$

2.3.2 Frequency Assignment

After obtaining the optimal number of active cores, we address Q2, the frequency assignment problem to appropriately assign frequency to each active core considering the discrete frequency levels. We propose two rules to guide the frequency assignment [1].

- *Rule 1: Always provide the minimal cumulative core frequency that satisfies the traffic demand.*

- *Rule 2: For a given cumulative core frequency, the per-core frequency combination with the least standard deviation consumes the least power.*

To demonstrate the two rules, we carry out two empirical studies with the same settings as in Section 2.2. In the first study, we vary the cumulative core frequency from 2GHz to 8GHz. For a given cumulative core frequency, we vary the per-core frequency combinations and record the net power consumption (load power minus idle power). Figure 7 shows the results for the URL application. From this figure, we observe that power consumption varies substantially, as much as 114% when comparing $(2, 2)$ to $(1, 1, 1, 1)$, with different per-core frequency combinations, which indicates that a proper frequency assignment is very necessary for multicore servers supporting per-core DVFS and power gating.

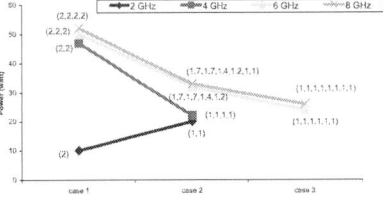

Figure 7: Power versus per-core frequency combinations.

In the second study, we take two cores and change the frequency in all possible combinations and record the power consumption.

---

[1] Algorithm pseudocode is omitted due to space limit. Interested readers can refer to [13] for details.

Figure 8 shows the 3D plot for the results with the URL application, where each black point represents a per-core frequency combination and its corresponding power consumption. In addition, based on Figure 8, we also plot Figure 9 illustrating the power consumption versus cumulative core frequency. When a certain cumulative core frequency corresponds to multiple power consumptions, we take the minimal one. From these two figures, we notice: 1) If we only consider the minimal power consumption for a given cumulative core frequency as shown in Figure 9, we see higher cumulative core frequency corresponds to higher power consumption. 2) For the same cumulative core frequency, the more evenly-distributed per-core frequency combination results in less power consumption as shown in Figure 8. For example, point $(1.7, 1.7)$ is lower than point $(2, 1.4)$ and point $(1.2, 1.2)$ is lower than point $(1.4, 1)$, although they have the same cumulative core frequency in both cases. In summary, this exhaustive study empirically validates our two rules.

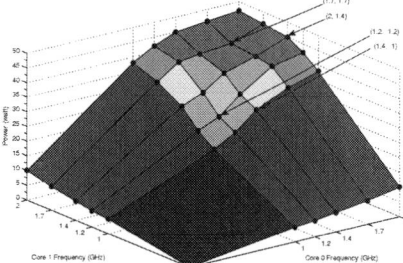

Figure 8: Power versus two-core frequency combinations.

Figure 9: Power versus cumulative core frequency (two cores).

2.4 Step 3: Static Power Optimization

The static power optimization scheme takes system operating level as input, which virtually contains an array of core frequencies optimized for a given traffic rate, and produces as output the actual per-core frequency configuration that is dynamically updated for power migration. This step focuses on the power migration design for active cores to achieve thermal balancing and reduce peak core temperature that effectively reduces static power consumption.

2.4.1 Design Overview

While keeping the system operating level constant, we appropriately vary the physical location for active cores so that thermal balancing is achieved across all cores. If the time interval between two migration points is small enough, we can minimize peak core temperature and effectively reduce static power consumption. Figure 10 illustrates the overview of our power migration scheme with varying network traffic rate. At each traffic change point (T_i), we apply the dynamic power optimization scheme to obtain the optimal system operating level. Because the number of active cores may be less than the total number of cores, and per-core frequency is heterogeneous, we periodically redistribute the power dissipation at each migration point (t_i) among all cores. In Figure 10, color squares represent active cores with different frequency levels (the darker the color, the higher the frequency), whereas white squares represent power-gated cores. In this example, the migration process happens among core pairs (C1, C2), (C6, C4), and (C5, C3), where the highest frequency cores C1, C6 and C5 are swapped with the lowest frequency cores C2, C4 and C3.

2.4.2 Migration Policy

The policy of our power migration consists of both long-term update and short-term update of core status table. The long-term update refers to the initialization of core status table at each traffic change point, which is in the order of minutes based on network

traffic studies in [11, 17, 22]. The short-term update refers to the periodic update of core status table at each migration point between two consecutive traffic change points. Considering core thermal behavior, packet processing time and system reconfiguration overhead, we find an update frequency of 1 second to be a good value for short-term update in our system.

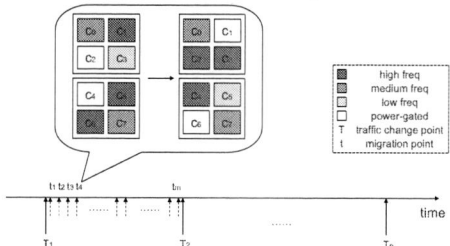

Figure 10: Illustration of power migration for active cores.

Long-term update: The long-term update should be based on previous history in the core status table for thermal balancing. At the traffic change point, because the system operating level will change in terms of the number of active cores and core frequency, we want every core to have even power dissipation over a period of time. Therefore, our long-term update can be described as follows:

General policy: Given the current system operating level ($f_1, f_2, ..., f_N$), we first sort the per-core frequency configuration in the previous core status table according to the frequency level from the lowest to highest. Then, we assign frequency f_1 to the first core in that list and frequency f_2 to the second core in that list and so on. For cores that are not assigned a frequency level, we leave them to be power-gated.

We have one exception for the above-mentioned long-term update. As we target servers with multicore processors, we should use as few processors as possible while satisfying traffic demand. Thus, when all active cores can fit into one processor, we should always use only one processor. Considering the general policy, we add the following exception rule:

Exception: If all active cores can fit into one processor, we choose the processor which contains the core that is assigned the frequency f_1.

Short-term update: The short-term update aims to achieve thermal balancing across all cores in the system through power migration. We argue that the migration policy has to be temperature-aware to guarantee thermal balancing during two consecutive short-term updates. Because core frequency is directly related to core temperature and per-core frequency is easy to obtain, we propose a frequency-aware migration policy to guide the short-term update as follows:

General policy: We sort the per-core frequency configuration in the current core status table according to the frequency level from the lowest to highest. Then, we swap the frequency between the first core in the list and the last core, and between the second core and the last but one core, and so on.

This strategy lets the power dissipation be evenly distributed across all cores during two consecutive short-term updates; thus overall thermal balancing will be achieved as expected. In the exceptional case where only one processor is used, we apply the following rule:

Exception: If all active cores can fit into one processor, we switch the active processor at every migration point and copy the same per-core frequency configuration within a processor from one to the other. However, at every other migration point, we update the per-core frequency configuration within a processor following the short-term update general policy.

This exception rule ensures we will keep the frequency assignment to one processor when it is possible. Using the regular short-term policy without this exception will likely split the frequency assignment across multiple processors.

3. EXPERIMENTAL EVALUATION

3.1 Experiment Setup

We implement our scheme along with three other schemes on an AMD server with two Quad-Core Opteron 2350 processors. For power measurement, we use a power analyzer (model EXTECH 380801 [5]) to obtain the real-time whole system power. We use the net power consumed exclusively by network applications as the metric for fair comparison. Net power is obtained by subtracting idle power from load power.

In our experiment, per-core DVFS is achieved by setting the core frequency to one of the five predefined frequency levels: 1GHz, 1.2GHz, 1.4GHz, 1.7GHz and 2GHz. We rely on the Linux kernel CPUfreq subsystem to implement the frequency scaling. Power gating is achieved by removing cores from active working set based on kernel's built-in CPU "hotplug" support, which mimics precisely the behavior of power gating [16]. Task scheduling module achieves power migration by dynamically scheduling incoming packets to active cores.

We parallelize six network applications from NetBench [19] (as listed in Table 1) and execute them in a multi-threaded fashion with packet-level parallelism. To guarantee each active core is running a thread, we enforce thread-to-core binding by setting thread affinity. We select two applications from each category (i.e., Micro-level, IP-level and Application-level). The packet trace is from NetBench with 10,000 packets, which are repeatedly processed in our experiment. The packet size ranges from 40 bytes to 1500 bytes with an average of 723 bytes. The routing table size for TL, Route and DRR is 128, and we use the small_input file for URL.

Table 1: Six network applications from NetBench.

Name	Functionality	Category
CRC	CRC-32 checksum calculation	Micro level
TL	Radix-tree table lookup routine	Micro level
Route	IPv4 routing based on radix	IP level
DRR	Deficit-round robin scheduling	IP level
URL	URL-based switching	Application level
MD5	Message digest algorithm	Application level

Table 2: Application-specific parameters.

App.	System Service Model	Latency (μs)	K
CRC	$Y = 81109 \cdot X + 26662$	0.008·size+0.3	1.5
TL	$Y = 571389 \cdot X + 328743$	0.8	1.4
Route	$Y = 253707 \cdot X + 87975$	1.8	1.4
DRR	$Y = 74965 \cdot X + 54945$	5.5	1.4
URL	$Y = 1496 \cdot X + 628$	0.131·size+73.2	1.6
MD5	$Y = 76016 \cdot X + 35024$	0.005·size+3.2	1.8

Table 2 shows application-specific parameters. We profile each application and obtain their system service model, where X represents the cumulative core frequency [2] and Y represents the service rate (packets/sec), equivalent to the arrival rate. To quantify the workload for weighted load balancing scheduling, we also obtain the per-packet latency for each application when running on a single core with 2GHz frequency. It is worth noting that our method also applies to non-linear applications, as long as we can model and translate the traffic arrival rate to cumulative core frequency (step 1 in Figure 3). This is because step 2 and step 3 are solely based on the result of step 1. We derive the dynamic power parameter K for each application based on Equation 2 and Equation 3 by substituting known frequency and measured power consumption. In addition, to calculate the static power P_s, we refer to manual specification, and use $V_{dd} \in (1.06V, 1.35V)$ and $I_{leak} \in (4.2A, 5.3A)$ as the 65nm technology parameters [16]. Hence, we take the average of 5.8W as the input for our power model.

To achieve traffic variation, we experiment with both synthetic and real-world workloads. For synthetic workload, we set the required cumulative core frequency (F) for incoming traffic to be one of the following five cases (as shown in Table 3). For real-world workload, we take the 24-hour traffic as shown in Figure 1. We consider the total volume as the arrival traffic for packet processing, and sample 24 different average traffic rates at each hour to obtain the required cumulative core frequency. Without loss of generality, the cumulative core frequency is then scaled according to our system capacity from 1GHz to 16GHz. For both workloads, we change the traffic rate every minute and set the power migration frequency to be 1 second.

Table 3: Synthetic workload for different traffic rate.

Traffic	extra low	low	medium	high	extra high
F	1GHz	4GHz	8GHz	12GHz	16GHz

In the experiment, we first compare our scheme to a traffic-unaware native system without power management. In addition, we compare our scheme with three other traffic-aware schemes, i.e., PG [17], which turns off cores when traffic is light using power gating, C-DVFS [22], which assumes a unified frequency adjustment across all cores using chip-wide DVFS, and C-Hybrid [15], which combines both chip-wide DVFS and power gating.

[2]In our experiment, X is between 1GHz and 16GHz.

3.2 Power Savings

Figure 11 shows power savings percentage for our scheme under different synthetic workloads compared to a native system. We observe that our scheme can achieve power savings in four out of five rates ranging from 18.0% for the CRC application in high traffic rate, to as high as 90.0% for the URL application in extra low traffic rate. The only exception is the extra high traffic rate, where all the cores must be running at the maximal 2GHz. Overall, our scheme reduces an average of 41.0% power consumption for the six applications and their five different workloads. In addition, we find our scheme especially useful when the traffic is light, e.g., in medium, low and extra low cases. This is because under light load we have more potential to apply per-core DVFS, power gating and power migration to achieve power savings.

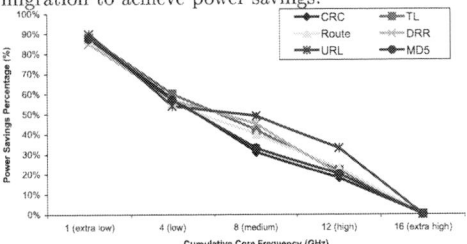

Figure 11: Power savings under different workloads.

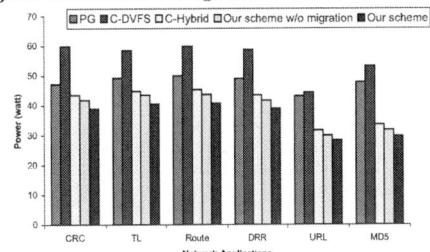

Figure 12: Power consumption with three other schemes.

Figure 12 shows the average power consumption for different applications comparing our scheme with three other schemes for the five different synthetic workloads. We observe that our scheme performs the best across all applications with an average of 35.2% power savings over C-DVFS, 24.3% over PG and 10.5% over C-Hybrid. C-DVFS performs the worst due to significant over-provisioning and excessive static power consumption, as it always keeps all the cores actively running. PG improves upon C-DVFS by turning off unnecessary cores to mitigate over-provisioning and save static power. However, without frequency scaling, it still suffers from excessive power consumption during extra low traffic. C-Hybrid outperforms both C-DVFS and PG due to its more flexible power management scheme using both chip-wide DVFS and power gating. But, C-Hybrid fails to achieve the best power savings because it does not consider static power or support more advanced per-core DVFS and power migration. Our scheme outwins all other schemes by providing the optimal system operating level and dynamically changing the per-core frequency configuration.

In addition, to emphasize the importance of power migration, we also experiment with our scheme without migration. Our scheme with power migration achieves an additional 2.5W reduction of power on average over our scheme without power migration. This highlights the advantage of including power migration in our power optimization scheme. In particular, when the traffic rate is extra low, low and medium, power migration can significantly reduce peak core temperature and hence effectively reduce static power.

3.3 Energy Savings

Figure 13 shows normalized energy consumption compared to a native system for different schemes using the real-world workload (24-hour traffic in Figure 1). We observe that all four schemes can achieve energy savings, ranging from the least energy consumption of 0.45 for the URL application in our scheme, to the most energy consumption of 0.71 for the Route and DRR application in C-DVFS scheme. However, upon averaging out all six applications over the 24-hour period, we still find our scheme outperforms PG, C-DVFS and C-Hybrid by 22.0%, 19.1% and 8.4%, respectively. The poor performance of PG and C-DVFS is due to the following two reasons: 1) PG always lets the cores run at full speed without frequency scaling, and 2) C-DVFS always has all 8 cores actively running without power gating. Compared to PG and C-DVFS, C-Hybrid improves the energy performance by combining both chip-wide DVFS and power gating. However, because C-Hybrid is unable to provide the optimal system operating level and ignores static power, it fails to achieve the best energy savings.

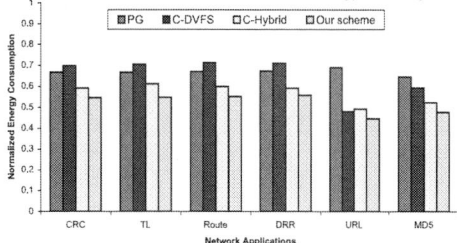

Figure 13: Normalized energy with three other schemes.

3.4 Reconfiguration Overhead

First, we individually measure the overhead for DVFS and power gating. For DVFS, it takes 0.008 seconds to change the per-core frequency level. For power gating, it takes 0.11 seconds to turn off a core and 0.08 seconds to turn on a core. In addition, we notice that in power gating, turning off a core does not add overhead as that power-gated core will be inactive in the next second. Also, not every core changes status every second. Therefore, we measure the average per-core reconfiguration overhead over the 24-hour traffic periodic at each hour as shown in Figure 14. This figure shows the result for TL, Route and DRR, which have the same system operating level with the same K value. The other three applications, CRC, URL and MD5 have very similar performance. Every second, we count the invoked number of DVFS and power gating for all the cores and divide the aggregated total overhead by 8. From this figure, we observe that the overhead ranges between 0.2% and 3.3% with an average of 1.7%, which is negligible. It is also easy to see that during low traffic hours (i.e., 8:00-14:00 and 17:00-23:00), the overhead is higher due to more frequent power migrations.

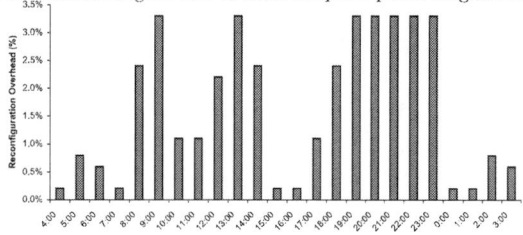

Figure 14: Overhead versus time in our scheme.

3.5 Thermal Behavior

Finally, to demonstrate the effectiveness of power migration in reducing peak core temperature, we use IPMItool utility to read processor thermal sensor and obtain the temperature for each processor every second. Figure 15 shows the maximal temperature increase at extra low (XL), low (L) and medium (M) traffic rate, where power migration is playing a significant role. Since all the starting temperatures are the same, we can see our scheme has the minimal peak core temperature in all cases. In this figure, DVFS represents DVFS-based schemes, including both C-DVFS and C-Hybrid, as they have the same thermal behavior.

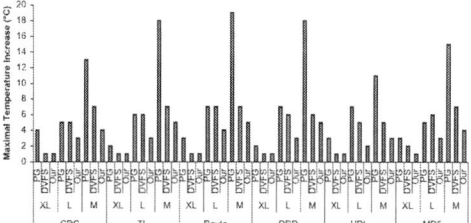

Figure 15: Temperature under different workloads.

More specifically, we see that PG causes the highest temperature increase (up to 19°C for the Route application in medium traffic

rate) because it lets all the cores run at the maximal frequency all the time. DVFS-based schemes, on the other hand, achieve better thermal behavior with frequency scaling, especially in extra low traffic rate. However, it still suffers from 1°C to 3°C higher peak core temperature compared to our scheme when the traffic rate is low or medium, as it always stresses the same active cores. We observe that our scheme on average reduces peak core temperature by 6°C compared to PG in all traffic rates and by 2°C compared to DVFS in low and medium traffic rate. This observation clearly shows that our scheme is able to achieve thermal balancing and keep a lower peak core temperature through power migration.

4. CONCLUSION

We design, implement, and evaluate a traffic-aware and power-efficient multicore server system that appropriately changes the system operating level according to varying traffic rate and dynamically adjusts the per-core frequency configuration using a combination of per-core DVFS, power gating, and power migration techniques to minimize power consumption. Our experimental results show that on an average our system saves 41.0% power compared to a native system. It also consumes less power than three other approaches, C-DVFS [22], PG [17], and C-Hybrid [15], by 35.2%, 24.3%, and 10.5% respectively.

5. REFERENCES

[1] AMD Opteron Processor. http://www.amd.com/opteron.
[2] Cavium OCTEON Processor Family. http://www.caviumnetworks.com/OCTEON_MIPS64.html.
[3] Cisco AON Technology. http://www.cisco.com/en/US/products/ps6692/Products_Sub_Category_Home.html.
[4] Equinix-sanjose. http://www.caida.org/data/monitors/passive-equinix-sanjose.xml.
[5] EXTECH Power Analyzer. http://www.extech.com/instruments.
[6] IBM BladeCenter System. http://www-03.ibm.com/systems/bladecenter/.
[7] International Technology Roadmap for Semiconductors. http://public.itrs.net.
[8] K. Greene. Data centers' growing power demands. *MIT Technology Review*, 2007.
[9] C. Hughes, J. Srinivasan, and S. Adve. Saving energy with architectural and frequency adaptations for multimedia applications. In *Proc. of Micro '01*, 2001.
[10] C. Isci, G. Contreras, and M. Martonosi. Live, runtime phase monitoring and prediction on real systems with application to dynamic power management. In *Proc. of Micro '06*, 2006.
[11] R. Kokku, U. B. Shevade, N. S. Shah, M. Dahlin, and H. M. Vin. Energy-efficient packet processing. *UT-Austin Technical Report TR04-04*, 2004.
[12] J. Koomey. Estimating total power consumption by servers in the us and the world. *Analytics Press*, 2007.
[13] J. Kuang, D. Guo, and L. Bhuyan. Power optimization for multimedia transcoding on multicore servers. In *Proc. of ANCS '10*, 2010.
[14] R. Kumar and G. Hinton. A family of 45nm ia processors. In *Proc. of ISSCC '09*, 2009.
[15] J. Lee and N. S. Kim. Optimizing throughput of power- and thermal-constrained multicore processors using dvfs and per-core power-gating. In *Proc. of DAC '09*, 2009.
[16] J. Leverich, M. Monchiero, V. Talwar, P. Ranganathan, and C. Kozyrakis. Power management of datacenter workloads using per-core power gating. *HP Labs Technical Report HPL-2009-326*, 2009.
[17] Y. Luo, J. Yu, J. Yang, and L. Bhuyan. Conserving network processor power consumption by exploiting traffic variability. *ACM TACO*, 2007.
[18] R. McGowen, C. A. Poirier, C. Bostak, J. Ignowski, M. Millican, W. H. Parks, and S. Naffziger. Power and temperature control on a 90-nm itanium family processor. *Journal of Solid-State Circuits*, 2006.
[19] G. Memik, W. H. Mangione-Smith, and W. Hu. Netbench: A benchmarking suite for network processors. In *Proc. of ICCAD '01*, 2001.
[20] R. Mishra, N. Rastogi, and D. Zhu. Energy aware scheduling for distributed real-time systems. In *Proc. of IPDPS '03*, 2003.
[21] J. W. Pratt. F. y. edgeworth and r. a. fisher on the efficiency of maximum likelihood estimation. *The Annals of Statistics*, 1976.
[22] T. Simunic, L. Benini, A. Acquaviva, P. Glynn, and G. D. Micheli. Dynamic voltage scaling for portable systems. In *Proc. of DAC '01*, 2001.
[23] K. Skadron, M. Stan, K. Sankaranarayanan, W. Huang, S. Velusamy, and D. Tarjan. Temperature-aware microarchitecture: Modeling and implementation. *ACM TACO*, 2004.

Alternate Hammering Test for Application-Specific DRAMs and an Industrial Case Study

Rei-Fu Huang

MediaTek Inc.,
Hsinchu, Taiwan

rf.huang@mediatek.com

Hao-Yu Yang, Mango C.-T. Chao

Dept. of Electronics Engineering
& Institute of Electronics,
National Chiao Tung University,
Hsinchu, Taiwan

(max0327.eecs94@, mango@faculty).nctu.edu.tw

Shih-Chin Lin

United Micro-
electronics Corp.,
Hsinchu, Taiwan

Shih_Chin_Lin@umc.com

Abstract

This paper presents a novel memory test algorithm, named alternate hammering test, to detect the pairwise word-line hammering faults for application-specific DRAMs. Unlike previous hammering tests, which require excessively long test time, the alternate hammering test is designed scalable to industrial DRAM arrays by considering the array layout for potential fault sites and the highest DRAM-access frequency in real system applications. The effectiveness and efficiency of the proposed alternate hammering test are validated through the test application to an eDRAM macro embedded in a storage-application SoC.

Categories and Subject Descriptors

B.3.4 [**Memory Structures**]: Reliability, Testing, and Fault-Tolerance—*test generation*

General Terms

Design

Keywords

embedded-DRAM, hammering test

1. INTRODUCTION

DRAM has the advantage of structure simplicity, high density, and low-power consumption, and has been widely used in the commodity-memory market since its invention. Due to higher capacity of DRAMs as compared to SRAMs, the test time of DRAMs has always been one of the most critical factors to the overall cost of a DRAM product line. In practice, DRAM companies often price their DRAM chips differently according to the length of the applied test. As a result, DRAM companies need to identify the faults or physical defects that really occur to the adopted memory design and process technology, and develop an effective but economic test algorithm to detect them.

Compared to the research publications on advancing the DRAM process technologies, such as trench [1] [2] [3], metal-insulator-metal [4] [5] [6], or embedded DRAM (eDRAM) [7] [8] [9], the publications on the DRAM fault models or test algorithms are relatively limited during the past decade. [10] and [11] shared their industrial experience on constructing DRAM test and testing eDRAM, respectively, but did not discuss the fault models covered by their test methods. [12] summarized the result of applying various types of test algorithms and stress conditions to 50 industrial DRAM SIMMs. [13] listed the potential functional fault models of DRAMs and proposed their corresponding tests. [14] discussed the targeted fault models for eDRAMs. None of the previous works reported the physical defects that results in the targeted faults in practice except [15], which discussed the physical defect, simulation model, and test reduction for the strap problem in DRAMs.

Several hammering test algorithms were presented by [12] [13] [14] and can be used to detect a weak data retention fault linked with a coupling fault, partial faults, or word-line coupling fault in DRAMs or eDRAMs. After repeatedly reading/writing (hammering) an adjacent aggressor row or column which couples the victim cell, the value stored in the victim cell may be leaked out before the next refresh. However, as reported in the conclusion of [12], those hammering tests take excessively long test time to repeatedly access each word-line for a refresh period and hence are not economically practical for a DRAM production line. Also, repeatedly accessing a word-line for the entire refresh period may over-test the linked retention fault since a word-line may not be accessed that frequently in the real applications.

In this paper, we would like to share our experience on applying a hammering test to a 65nm eDRAM macro, which is embedded in a storage-application SoC. First, we will introduce a specific fault model for the targeted eDRAM technology, named as the *pairwise word-line hammering fault*, which requires a hammering test to detect. The physical root cause and a silicon instance of this targeted fault will also be provided. Second, we will propose a new scalable hammering test, named *alternate hammering test*, which can detect the targeted hammering fault with acceptable test time by considering the physical layout location of a possible fault. Next, we will calculate the number of hammering times that should be applied to each word-line based on the system application of the SoC, such that the eDRAM macro will not be over-tested or under-tested. Also, we will further introduce a WAT test key that can monitor the potential occurrence of the targeted fault. Last, we will show the experimental result of applying the proposed alternate hammering test to defect-impacted wafers and demonstrate the effectiveness and efficiency of the proposed alternate hammering test.

2. BACKGROUND

2.1 Overview of the eDRAM Macro in Use

The targeted eDRAM macro of this paper is a 16Mb eDRAM macro with ECC (error correction code), embedded in a storage-application SoC. This eDRAM macro utilizes deep trench capacitors and is implemented in a UMC 65nm low-leakage logic process. The word size on the interface of this eDRAM macro is 32 bits. Due to the use of ECC, 6 more additional bits are added to the physical array for each word. The area of the eDRAM macro is around 4 mm^2, which contains two symmetric eDRAM arrays. Each array contains 128 banks, and each bank contains 64 word-lines and its own local sense amplifier. Each word-line on each array is connected to 64 half-words. This eDRAM macro runs at 100 MHz and its bandwidth is 3.125 Gb/s (32 bits x 100 MHz). The refresh period is 4ms.

Note that the layout of this eDRAM macro utilizes distributed folding, address decoder scrambling, and bit-line twisting for the optimization of area and performance. Also, eDRAM utilizes differential sensing to read the stored value of a cell while the cell's capacitor is only connected to either one of the corresponding bit-line and bit-line-bar. In this eDRAM macro, half of the capacitors are connected to the bit-line, and the other half to the bit-line-bar. As a result, the logic address order of this eDRAM macro is not its physical address order; and the logic value of a cell may not be its physical value (the value stored in the capacitor). In the following content of this paper, the address and value are all referred to the physical address and physical value. When designing our memory BIST, a scrambling table is used to re-map the physical address and the physical value back to its logic address and logic value.

Figure 1 shows the layout up to the metal 1 layer for a small portion of our eDRAM array, where the corresponding word-lines (WL) and bit-line pairs (BL and BLB) are labeled on the sides, respectively. Figure 2 illustrates the cross-section view from label-A to label-F along the BLB_i in Figure 1. Label-A to D represents the cell accessed by WL_i and the bit-line pair i. Label-A points to the shared contact connecting to the metal of BLB_i. Label-B points to the poly gate of the cell's pass transistor. Label-C points to the contact connecting the pass transistor with the cell's capacitor. Label-D points to the poly of WL_{i+1} passing through the top of the capacitor. Label-E and Label-F point to the passing poly and the contact of the cell accessed by WL_{i+2} and the bit-line pair (i), which is symmetric to the layout of the above cell. Note that the paired bit-line and bit-line-bar may switch with each other several times along the columns due to the use of bit-line twisting, which is done on the upper metal layers and hence cannot be observed in Figure 1.

2.2 Hammering Tests and Detectable Faults in Previous Works

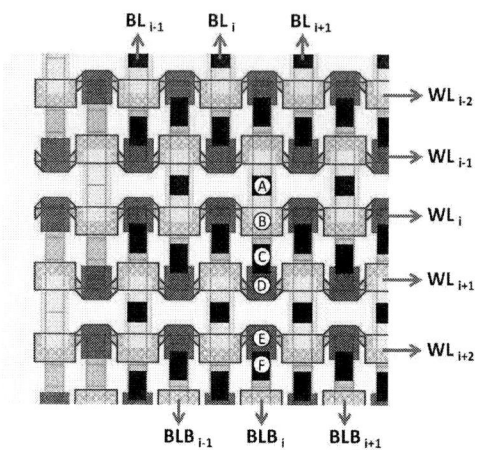

Figure 1: Layout of the targeted eDRAM array.

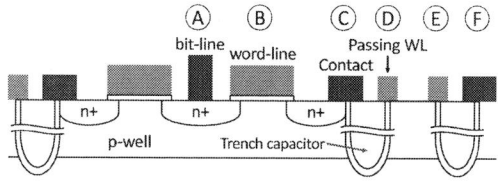

Figure 2: Cross-section view along a column in Figure 1.

A hammering test is referred to a memory test algorithm that can trigger certain specific faults by repeatedly accessing a cell and then detect them. [12] applied several hammering tests, called disturb tests, to detect a weak data retention fault linked with a coupling fault. As a result, a disturb test has to repeatedly access each word-line for its refresh period and requires excessively long test time. Figure 3 shows the shortest disturb test applied by [12] and the complexity of its test time.

$\{\uparrow (w0);_R \uparrow_{x=0}^{R-1} (T_{REF}(_C\uparrow_{r=x+1} (w1)); Ref;_C \uparrow_{r=x+1} (w0);_C \uparrow_{r=x} (r0));$
$\uparrow (w1);_R \uparrow_{x=0}^{R-1} (T_{REF}(_C\uparrow_{r=x+1} (w0)); Ref;_C \uparrow_{r=x+1} (w1);_C \uparrow_{r=x} (r1))\}$
Time complexity : $8n + t$
n : number of words
$t = 2 *$ number of wordlines $* T_{REF}$
T_{REF} : refresh period

Figure 3: The shortest hammering (disturb) test used in [12].

[13] proposed another hammering test, named March H2C, to detect the *partial faults*. A partial fault is defined as a functional fault model which can only be sensitized during its initialization or activation by repeating a specific operation. Such a fault model does not tie to a data retention fault and hence requires shorter test time. However, March H2C does not consider the impact of the auto-refresh mechanism of DRAM, which may reset the accumulated faulty effect and result in test escape. Also, March H2C did not mention how many hammering times are sufficient to detect the targeted partial faults, which makes the effectiveness and the occasion to apply March H2C unclear.

[14] reported another fault requiring a hammering test to detect, i.e., the word-line coupling fault. When a word-line coupling fault occurs, turning on one word-line may slightly turn on its adjacent word-line due to an unexpected large coupling capacitance in between and hence may leak the stored value on the adjacent word-line. A word-line coupling is more likely to occur in eDRAMs since a word-line in eDRAMs is usually longer than that in DRAMs (due to the extra metal layers in eDRAMs that can be used to parallel-connect the word-line and reduce the word-line resistance [14]). However, such a fault can be prevented by applying a negative voltage to the unselected word-lines, which is exactly the case in our targeted eDRAM macro. Also, [14] did not provide the sufficient number of the hammering times for detecting such a fault either.

3. PAIRWISE WORD-LINE HAMMERING FAULT

3.1 Fault Behavior and Physical Root Cause

A pairwise word-line hammering fault is caused by an unexpected bridging between two adjacent contacts. Figure 4 (a) shows an example, where x and y represent the two bridging contacts (as well as the bridged cells). Note that the two bridging contact are placed almost on the same horizontal line in the layout, but x and y are accessed by different word-lines (WL_{i+1} and WL_i respectively). Figure 4 (b) illustrates the corresponding schematic.

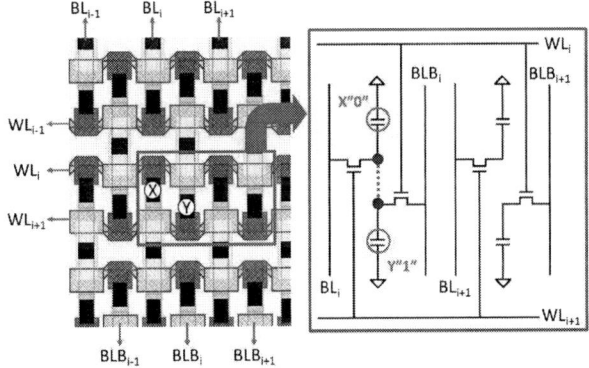

Figure 4: Root cause of a pairwise word-line hammering fault.

Once such a defect occurs and the two bridged cells store different values, the cell with value 1 may lose some of its stored charges to the cell with value 0. However, this lost of charges for the cell with value 1 may not be significant enough due to the large resistance of the bridging, such that the value 1 can still be successfully read out before the next refresh. Also, once the a cell is read out, the value will be automatically write back, meaning that its faulty effect cannot be accumulated. Therefore, such a fault cannot be detected by simply reading or writing the cell, which may escape a general DRAM test and become a source of DPPM.

3.2 Hammering Test for Fault Detection

In order to detect the bridging between x and y as shown in the above example, we first need to write 0 into x and

write 1 into y. Next, we start hammering x by applying consecutive read operations to it. The impact of each read operation can be divided into three stages as shown in Figure 5. In the first stage, WL_{i+1} turns on and the value 0 of x is sensed through its charge sharing with BL_i, which at the same time creates a current from y to BL_i and some charges stored in y leak through BL_i as well. In the second stage, value 0 is written back to x after the sense amplifier differentiates the stored value, which creates a larger current from y to BL_i than the first stage. In the third stage, WL_{i+1} turns off after value 0 is successfully written back to x, and the charges stored in y only slightly leak to BLB_i through the leakage current of y's pass transistor or leak to x through the slow charge sharing. After several read operations, the charges stored in y will be lowered to a level such that the stored value 1 cannot be successfully read out.

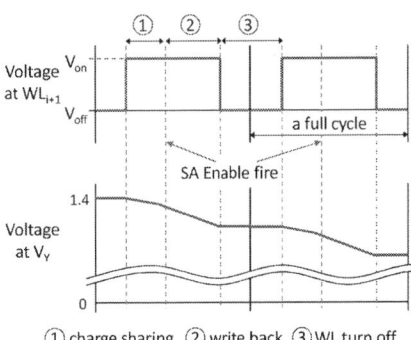

Figure 5: Impact of applying consecutive read operations on the bridged cell with value 1.

In fact, detecting this bridging defect between x and y not necessarily needs to perform consecutive read operations directly to x. As long as any word at WL_{i+1} is read, all cells at WL_{i+1} (including x) will also be read (but not selected by the DO multiplexor) and automatically written back. Thus, the same effect can be generated. In other words, all the contact-bridging defects between WL_{i+1} and WL_i can be simultaneously detect when hammering any one cell at WL_{i+1}. It means that only one hammering test needs to be applied to each word-line. Also, in the above example, if we switch the values between x and y and then repeatedly reading y, the same effect can be generated as hammering x.

In addition, based on the array layout, the contact at WL_i can only bridge the contact at one adjacent word-line WL_{i+1}, not the other adjacent word-line WL_{i-1}. In other words, such a bridging defect can only occur on separate pairs of word-lines. It means that the hammering test should be applied to each pair of word-line, not each two adjacent word-lines, which can cut the test length into half.

Note that we have to use repeated read operations instead of repeated write operations to hammer the targeted word-line and create the worst-case charge lose for the bridged cell with value 1. This is because a read operation contains a write-back mechanism after the charge sharing and in turn requires longer word-line pulse, which can result in larger leakage to the cell.

3.3 Physical Probing to a Fault Site

Figure 6 shows a nanoprobe SEM image of a defective

1014

eDRAM macro. On the left of the figure shows the mapping of the SEM image to the corresponding array layout as labeled from 1 to 8. Label-1 and label-2 are two bridged contacts, which was diagnosed by our proposed hammering algorithm. We tried to confirm the existence of this defect through physical probing.

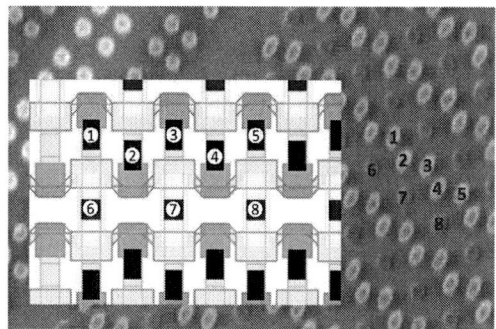

Figure 6: SEM image and corresponding layout for nanoprobe.

Figure 7 shows the measurement result, where we applied different voltages to different pairs of adjacent contacts and measured the corresponding current in between. The red line represents the measured current for two bridging adjacent contacts (label-1 and label-2). The blue line represents the measured current for two defect-free adjacent contacts. As the result shows, the measured current between two defect-free contacts stays at 0 until the voltage is raised to 3.7V. However, for the two defective contacts (label-1 and label-2), their current starts to increase when the voltage is more than 1V, which proves the existence of the bridging defect between the two contacts.

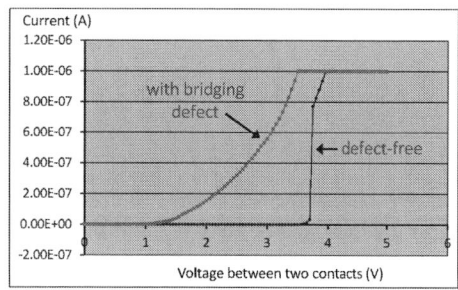

Figure 7: Current between two adjacent contacts w.r.t. different voltages.

4. PROPOSED ALTERNATE HAMMERING TEST

In this section, we would like to present the proposed *alternate hammering test* in detail and how it can effectively and efficiently detect the pairwise word-line hammering faults. Following are the main ideas behind the alternate hammering test. First, since the effect of a pairwise word-line hammering fault is related to the leakage of the stored charges of a DRAM cell, the detection of this fault should be subject to the adopted refresh mechanism, which can periodically recover the weak stored value. Therefore, the alternate hammering test will hammer (repeatedly read) a corresponding

word-line and check the correctness of the stored value for the word-line after the next refresh, such that each hammered cell has a complete refresh period to leak. Second, unlike previous hammering tests, such as the disturb tests, the alternate hammering test only hammers a word-line for a pre-specified number of times, not the entire refresh period. The number of hammering times (repeated read operations) is the maximum number of the DRAM accesses that may happen to the system application during a refresh period. Third, the alternate hammering test hammers only one word-line of each word-line pair, not all word-lines as in the disturb tests, which is why the name "alternate" is used. Also, the background of both the hammered and not-hammered word-lines need to be properly set as introduced in Section 3.2. Fourth, the alternate hammering test divides the word-line pairs into several *hammered blocks* and performs the test based on each block. For a hammered block, all word-line pairs are hammered in order within a refresh period. The size of a hammered block is determined by the number of hammering times to each word-line and the refresh period, which will be introduced later.

4.1 Test Algorithm

Figure 8 shows the detailed test algorithm of the alternate hammering test. For each block, we will first prepare the data background for all word-line pairs in the block by setting even word-lines to 0 and odd word-lines to 1 (line 2-4). Second, refresh all odd word-lines (line 5-6). Third, hammering each even word-line by reading the first word at the word-line for H times (line 7-8). Fourth, wait some cycles (w_1 or w_2 cycles) until the refresh period ends (line 9-12). Next, refresh all odd word-lines again (line 13-14). Last, read each word at odd word-lines and check whether all 1s can be successfully read out. Figure 9 illustrates the time line of applying the alternate hammering test to a hammered block b. Note a word-line is only hammered for H times and remains idle for the rest time of the refresh period. In addition, all hammered blocks contain the same number of word-line pairs except the last hammered cycle. Therefore, the final waiting time of the last block (w_2) is different from all the others (w_1).

4.2 Number of Hammering Times to a Word-line

In our system application, a word-line of the eDRAM macro will not be repeatedly accessed for the entire refresh period since the eDRAM macro functions as the main memory. The processor usually get data from cache and only access the eDRAM macro when the cache data is invalid. To determine the most frequent DRAM accesses within a refresh period, we applied different representative benchmarks to the system and observed the real-time access of the eDRAM macro through a diagnosis-mode pin, which directly connects to the eDRAM enable signal (or the so-call chip-enable signal). As a result, by preserving this diagnosis-mode pin, we can avoid the time-consuming simulation and directly collect data from a manufactured chip, which greatly speed up the whole analysis process. Among the data collected from all the benchmarks, the number of the most frequent accesses to eDRAM in a refresh period (4ms) is a little bit less than 8000. Thus, when applying the alternate hammering test to this eDRAM macro, we use 8000 hammering times as a conservative bound.

1015

Test procedure : Alternate Hammering Test
#define P = the # of the WL-pairs in each hammering block
#define H = the max # of system accesses to DRAM during a refresh period
1   for each hammering block b
2   for each pair i // data prepare
3   write 0 to words at WL_{b*P+2i} (even WL)
4   write 1 to words at $WL_{b*P+2i+1}$ (odd WL)
5   for each pair i // refresh odd WL
6   Refresh $WL_{b*P+2i+1}$ (odd WL)
7   for each pair i // hammering even WL
8   Read any word at WL_{b*P+2i} (even WL) for H times
9   if b is not last block   // wait till refresh
10   wait w_1 cycles
11   else
12   wait w_2 cycles
13   for each pair i // refresh odd WL
14   Refresh $WL_{b*P+2i+1}$ (odd WL)
15   for each pair i // read odd WL
16   Read each word at $WL_{b*P+2i+1}$ (odd WL) and compare

Figure 8: Detail steps of the alternate hammering algorithm.

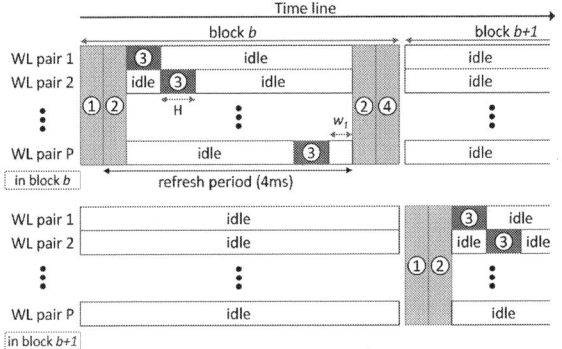

① data prepare ② refresh odd WL ③ hammer even WL ④ read odd WL

Figure 9: Time-line for hammering a block of word-lines in the alternate hammering test.

4.3 Size of a Hammered Block

The size of a hammered block is determined by the maximum number of word-line pairs that can be hammered in a refresh period, denoted T_{ref}. As the time line shown in Figure 9, the refresh period first starts with refreshing all odd word-lines in the block, followed by hammering the even word-lines. Thus, the maximum number of word-line pairs in a hammered block, denoted by P, can be calculated by solving Equation 1, where Fq represents the operating frequency.

$$P = (T_{ref} \times Fq - P)/H \qquad (1)$$

The total number of hammered blocks (denoted by B) is equal to the total number of word-line pairs in the memory divided by P. Take our target eDRAM as an example, $Tref$, Fq, and H are 4ms, 100MHz, and 8000. The resulting P and B are 49 (floor function) and 84 (ceiling function).

4.4 Test Time

For each hammered block, we need to spend one refresh period plus the time accessing all word-line pairs for 3 times (two for preparing data background and one for reading odd word-lines) and refreshing odd word-lines, as shown in Figure 9. Thus, the test time of the alternate hammering test (denoted as T_{test}) can be calculated by Equation 2, where N_{WL} and W denotes the total number of word-lines and the number of words per word-line, respectively.

$$T_{test} = \frac{N_{WL}}{2P} \times (T_{ref} + 3 \times \frac{P \times W}{Fq} + \frac{P}{Fq}) \qquad (2)$$

5. WAT TEST KEY

In this section, we would like to present the test key that can be used for monitoring the potential occurrence of the pairwise word-line hammering faults. The test key is put inside the scribe line and will be measured during WAT (wafer acceptance test). The idea of this test key is to use a similar layout of the real eDRAM array, where (1) all bit-lines, all bit-line-bars, and all word-lines are connected, respectively, and (2) the size of the layout fits the scribe line. As a result, the test key requires only three IO pads for the measurement (one for each of BL, BLB, and WL). Figure 10(a) first illustrates the idea of the proposed test key. Figure 10(b) shows part of the masks containing capacitors and the first few metal layers, where all bit-lines are connected together on the bottom.

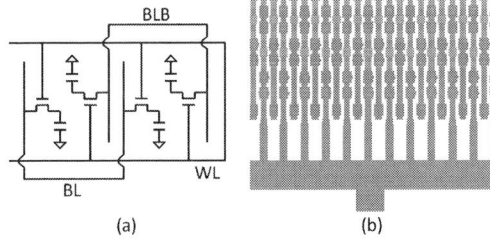

Figure 10: Idea and part of the mask layers for the proposed test key.

When measuring the test key, WL is set to V_{on}, BL is set to 0, and BLB is set to a perfect value 1's voltage stored in a capacitor, which is 1.4V for our eDRAM macro. Then we measure the current between BL and BLB, which should be close to 0 for a fault free case, to determine whether any significant bridging exists between two adjacent contacts. In other words, we need to first define the minimum cut-off current that may cause a pairwise word-line hammering fault for the measured current between BL and BLB. Note that this cut-off current needs to be conservative enough since we would rather over-test than under-test in this situation. Following is the derivation of this minimum cut-off current.

First, a storage capacitor in this eDRAM technology is 7fF. Next, we found by simulation that the sense amplifier cannot successfully sense a value 1 when the stored voltage is lower than 1.1V. So the quantity of charges that need to be leaked for failing a read-1 operation is 7fF*(1.4V-1.1V). i.e., 2.1fC. The quantity of leaked charges is determined by the current passing through the bridging, denoted as I_{short}, and the leakage current of the cell's pass transistor, denoted as I_{leak}. The leaked charges through I_{short} is also determined by the time that the word-line is turned on. For a read

1016

operation, the word-line is turned on for 6.5ns. During the alternate hammering test, a word-line is repeatedly read for 8000 times and hence a word-line is turned on for 8000*6.5ns long within a refresh period. Note that, in this eDRAM technology, the pass transistor is a long-channel IO device and the word-line turn-off voltage (V_{off}) is -0.5V, such that I_{leak} can be 120fA at $85^\circ C$, which is much lower than the logic device. To result in a conservative estimation of I_{short}, we double the time for a refresh period to 8ms in the analysis. As a result, the minimum I_{short} that may cause a failing read-1 operation can be calculated as follows.

$$2.1fC \leq I_{leak} * 2 * 4ms + I_{short} * 8000 * 6.5ns$$
$$I_{short} \geq 21.92pA$$

Another conservative estimation used in the above analysis is that the voltage at the storage capacitor will gradually decrease along the time, not constantly staying at 1.4V as our measurement. In other words, the actual leaked charges in the real application is less than our measurement. Therefore, when the measured current between BL and BLB is larger than 21.92pA, we consider the wafer as defect-impacted.

6. EXPERIMENTAL RESULTS

We have applied the proposed alternate hammering test to different silicon versions of the storage-application SoCs (all implemented by a UMC 65nm low-leakage process), and the pairwise word-line hammering fault was first identified in an early silicon version. Figure 11(a) and Figure 11(b) show the pass/fail wafer map without and with applying the alternate hammering test for a highly impacted 12-inch wafer of the early silicon version. The failed dies are highlighted by red in the wafer maps. Note that a general eDRAM test without any hammering test was first applied to the wafer for both Figure 11(a) and Figure 11(b).

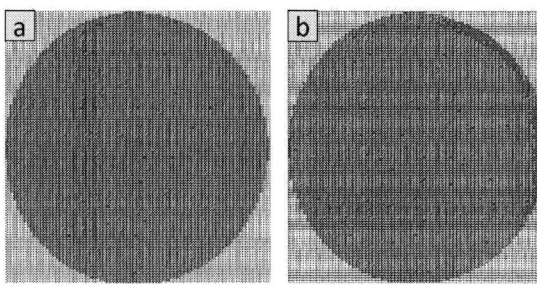

Figure 11: Pass/fail wafer maps (a) without and (b) with applying the alternate hammering test.

As the result shows, the proposed alternate hammering test can indeed detect certain defects that cannot be detected by a general eDRAM test without a hammering test. The yield difference in this highly impact wafer is around 7%. The general yield difference with and without applying the alternate hammering test ranges from 1% to 7% for the impacted wafers of this early silicon version. The average yield loss due to the pairwise word-line hammering fault is less than 1% for the latest silicon versions. Note that we have further confirmed the root causes of some sampled failing dies through the nanoprobe as shown in Section 3.3

and found that the yield loss does result from the pairwise word-line hammering faults. In addition, the test time of applying the alternate hammering test to this eDRAM macro is 343.94ms, which is still a little bit longer than general non-hammering test but within an acceptable range. Compared to the shortest hammering test used by [12] (as shown in Figure 3), which takes 65.58s on this eDRAM macro, the proposed alternate hammering test is 190X faster and hence more scalable in practice.

7. CONCLUSION

In this paper, we have first introduced the pairwise word-line hammering fault and discussed its fault behavior and physical root cause. Next, we have proposed the alternate hammering test to effectively and efficiently detect the pairwise word-line hammering fault by considering the layout information and the system application. A new test key was also presented to monitor the occurrence of the faults. Last, we have shown the silicon result of applying the alternate hammering test to an eDRAM macro used in storage-application SoC and demonstrated the effectiveness and efficiency of the proposed alternate hammering test.

8. REFERENCES

[1] T. Tran, et al., "A 58nm Trench DRAM Technology," IEEE International Electron Devices Meeting, 2006.

[2] C.W. Teng, Y. Okumoto, J. Liu, Ih-Chin Chen, K. Yuhara, and Y. Yoneoka, "A 64 Mbit DRAM trench capacitor cell with field-plate isolation," International Symposium on VLSI Technology, Systems, and Applications, 1991.

[3] J. Lutzen, et al., "Integration of capacitor for sub-100-nm DRAM trench technology," Symposium on VLSI Technology, 2002.

[4] M.A. Pawlak, et al., "Enabling 3X nm DRAM: Record low leakage 0.4 nm EOT MIM capacitors with novel stack engineering," IEEE International Electron Devices Meeting, 2010.

[5] N. Mise, et al., "Scalability of TiN/HfAlO/TiN MIM DRAM capacitor to 0.7-nm-EOT and beyond," IEEE International Electron Devices Meeting, 2009.

[6] N. Menou, et al., "0.5 nm EOT low leakage ALD SrTiO3 on TiN MIM capacitors for DRAM applications," IEEE International Electron Devices Meeting, 2008.

[7] G. Wang, et al., "A 0.127 μm^2 High Performance 65nm SOI Based embedded DRAM for on-Processor Applications," International Electron Devices Meeting, 2006.

[8] J. Barth, et al., "A 45 nm SOI Embedded DRAM Macro for the POWER? Processor 32 MByte On-Chip L3 Cache," IEEE J. Solid-State Circuits, vol. 46, no. 1, pp. 64 - 75, Jan. 2011.

[9] A. Berthelot, et al., "Highly Reliable TiN/ZrO2/TiN 3D Stacked Capacitors for 45 nm Embedded DRAM Technologies," Solid-State Device Research Conference, 2006.

[10] J. Vollrath, "Tutorial: Synchronous Dynamic Memory Test Construction - A Field Approach," IEEE International Workshop on Memory Technology, Design and Testing, pp. 59-64, 2000.

[11] R. McConnell, U. Moller, D. Richter, "How we test Siemens Embedded DRAM Cores," IEEE International Test Conference, pp. 1120-1125, 1998.

[12] A.J. van de Goor, A. Paalvast, "Industrial Evaluation of DRAM SIMM Tests," IEEE International Test Conference, pp. 426-435, 2000.

[13] Z. Al-Ars, S. Hamdioui, A.J. van de Goor, G. Gaydadjiev, J. Vollrath, "DRAM-Specific Space of Memory Tests," IEEE International Test Conference, pp. 1-10, 2006.

[14] M. C.-T. Chao, H. Yang, R. Huang, S. Lin, C. Chin, "Fault Models for Embedded-DRAM Macros," ACM/IEEE Design Automation Conference, pp. 714-719, 2009.

[15] Z. Al-Ars, S. Hamdioui, A.J. van de Goor, G. Mueller, "Defect Oriented Testing of the Strap Problem Under Process Variations in DRAMs," IEEE International Test Conference, pp. 1-10, 2008.

Goal-Oriented Stimulus Generation for Analog Circuits

Seyed Nematollah Ahmadyan,   Jayanand Asok Kumar,   Shobha Vasudevan
Coordinated Science Lab, Electrical and Computer Engineering Department
University of Illinois at Urbana-Champaign
{ahmadya2, jasokku2, shobhav}@illinois.edu

ABSTRACT

We present a methodology to generate goal-oriented test cases for verifying nonlinear analog circuits. We use a learning-based approach to identify the goal regions in circuit's state space. We use the information that we learn to guide the growth of Rapidly-exploring Random Trees (RRTs) towards these goal regions. Compared to previous approaches for test generation, our methodology generates several test cases of the circuit that are more concentrated in the relevant operating regions. We demonstrate the effectiveness of our approach on typical case studies. We show that our methodology can be used to generate test cases for undesirable behavior that was previously hard to detect.

Categories and Subject Descriptors

B.7.2 [**Integrated Circuits**]: Design Aids—*Verification*;
B.7.3 [**Integrated Circuits**]: Reliability and Testing—*Test generation*

General Terms

Design, Algorithms, Verification

Keywords

Pre-Si testing, Rapidly-exploring random trees

1. INTRODUCTION

Verifying non-linear analog circuits is a major challenge and is an ongoing topic of intensive research. Analog verification can be broadly classified as being based on either formal methods[14][2] or simulation-based methods[8][17][10]. Formal verification methods exhaustively analyze all possible behaviors of the circuit. However, for complex circuits, such exhaustive analysis is not feasible. Therefore, simulation-based verification is a more widespread approach for verifying complex analog circuits [3].

Simulation-based verification is performed using several test cases[1] for the circuit. Each test case is a sequence of

---

[1] In this work, we use the term test in the context of pre-silicon analog verification, not manufacturing level testing/chip measurement data.

values that is applied to the circuit inputs. For each test case, the circuit is simulated by applying the corresponding sequence of values to the inputs. The behavior induced by these sequences is then analyzed. If an erroneous or illegal set of circuit states states (*i.e.*, a"bad" region) is known, it is desirable to check whether there is a legal sequence of input values that takes the circuit from an initial state to the bad region. Simulation-based verification is non-exhaustive and therefore, checks only a subset of all possible behaviors. The quality of simulation-based verification is determined by the choice of test cases.

We present a methodology to efficiently generate goal-oriented test cases for simulation-based analog verification. We use an algorithm called Rapidly exploring Random Tree (RRT) algorithm[12]. An RRT is a tree data structure that can be grown in the state space of a system. If there is a feasible path in the state space, the RRT algorithm will eventually find it[12]. Recently, in [15][8][10], RRTs were used to generate test cases for analog verification. However, these test cases were intended to obtain maximum coverage of the state space. Therefore, the RRTs were grown omnidirectionally by uniformly sampling the state space. In our methodology, we achieve our objective by modifying the classical RRT algorithm to grow towards a goal region in the state space.

We use a learning-based approach to guide the RRT growth towards a goal region. We first sample the circuit to learn the statistics regarding frequently occurring regions in the state space. Instead of uniformly sampling the state space, we use this information to generate samples biased towards such *goal regions*. Consequently, the samples that we generate guide the RRT to grow towards the goal regions in the state space. The test cases that we generate are more concentrated in the goal regions, compared to those obtained using the classical RRT algorithm. Our approach can be applied to linear and nonlinear time-invariant analog circuits.

Our algorithm for guiding the RRT towards a goal region is as follows. In the learning phase, we sample the circuit to identify the goal regions in the state space. We then partition the entire state space into a network of equally-sized grids. We employ a *grid-based clustering* scheme group together grids based on their relative relevance to the goal region. In the exploring phase, we use these grid-based clusters to generate samples of the state space that are biased towards the goal region. This biased generation of samples guides the RRT to grow towards the goal region. We generate test cases based on the final RRT data structure. Due to the guided growth, the RRT will converge rapidly on the goal regions thereby giving more focused test cases.

In classical RRT algorithms, the RRT growth is optimized at every step without considering the larger context of state space. Such locally optimal solutions are frequently not efficient in arriving at the global goal. Therefore, the growth of RRTs in these algorithms is locally optimal at best. Several techniques have tried to address this issue either by computing the discrepancy or disparity of sampling[8][7] or by changing the variance of the sampling distribution [10], to guide the direction of the growth to where it would encounter more samples. However, the guiding process used by these techniques are very resource-intensive and slow.

Our main contribution is as follows. We present a learning-based approach to guide the RRTs to grow towards a goal region. We apply this approach as an efficient methodology for generating goal-oriented test cases for analog verification. Our approach is very efficient since it does not waste any effort in exploring the irrelevant regions of the state space.

We demonstrate the effectiveness of our methodology on a tunnel diode circuit[9] and a Josephson junction circuit. In both circuits, we show that our methodology yields test cases that are more concentrated in the goal regions as compared to those obtained using previous RRT-based test generation methods. For the Josephson junction circuit, we easily obtain several test cases that drive the circuit into undesirable states. This undesirable behavior is known to to be hard to detect using conventional test generation methods.

The rest of this paper is organized as follows. In Section 2, we briefly describe some background material on non-linear systems and RRTs. In Section 3, we describes all the steps in both the learning phase and the exploring phase of our algorithm for goal-oriented test generation. In Section 4, we demonstrate the effectiveness of our algorithm on two analog circuits. Finally, in Section 6, we present related work and conclude the paper.

2. PRELIMINARIES

We define the terminology that we use for modeling non-linear systems. We also present some background on the *Rapidly-exploring Random Tree* (RRT) algorithm[11].

2.1 Models for non-linear systems

A non-linear system can be modeled as a set of differential algebraic equations (DAEs) by applying modified nodal analysis (MNA) to the circuit netlist, modeled as:

$$f(\mathbf{x}(t), \dot{\mathbf{x}}(t), \mathbf{u}(t)) = 0 \qquad (1)$$

where $t \in [0, \infty)$. Let $\mathbb{S} \subseteq \mathbb{R}^n$ denote the continuous state space of the system. Let $\mathbb{U} \subseteq \mathbb{R}^m$ denote the input space of the system. $\mathbf{x}$ denotes the state variables and $\mathbf{u}$ denotes the input variables of the system. $\mathbf{x}(t)$ denotes the state of the system at time t. The initial state of the system is $\mathbf{x}(0)$.

DEFINITION 1. *A trajectory of the system in the time interval $[t_1 \ t_2]$ is the path taken by the system from state $\mathbf{x}(t_1)$ to state $\mathbf{x}(t_2)$.*

For a given state $\mathbf{x}(t_1)$ and input $\mathbf{u}(t_1)$, the differential constraints in Equation 1 determine the trajectory of the system in the interval $t \in [t_1 \ t_2]$. for some initial state $\mathbf{x}(t_1)$ at time $t = t_1$, a state trajectory derived from action trajectory is defined by

$$\mathbf{x}(t) = \mathbf{x}(t_1) + \sum_{t_1}^{t} f(\mathbf{x}(t'), \mathbf{u}(t')) \Delta t' \qquad (2)$$

In practical applications, the trajectory of non-linear systems can be computed by using a numerical ODE solver. For example, in analog circuits, the trajectory can be obtained by simulating the circuit with a SPICE program.

2.2 Rapidly-Exploring Random Trees

We briefly describe the RRT algorithm presented in [11]. The core of the RRT algorithm is the evolution of a tree data structure. The tree is initialized by fixing its root at a specified point in the state space $\mathbb{S}$. The tree is then *grown* incrementally by adding an edge between an existing node and a new point (*i.e.*, state) selected from the state space. The selection of these new points determines the manner in which the tree grows in the state space. Typically, these new points are selected at random by uniformly sampling the state space.

Algorithm 1 RRT Algorithm using uniform sampling

$\mathbb{G}$.init $(\mathbf{x}(0))$:
for $i \leftarrow$ **to** K **do**
    $q_{sample} = \text{UniformSampling}(\mathbb{S})$:
    $q_{near} = \text{FindNearestNodeInTree}(\mathbb{S}, q_{sample})$;
    $q_{new} = \text{FindOptimumTrajectory}(q_{near}, q_{sample})$;
    $\mathbb{G}$.expand(q_{new});
end

Let $\mathbb{G}$ be the RRT data structure. Each node of G corresponds to a state in $\mathbb{S}$, *i.e.* a unique set of values assigned to the state variables $\mathbf{x}$. Each edge represent a trajectory (Definition 1) of the system from initial condition $\mathbf{x}$ for a given assignment of values to the input variables $\mathbf{u}$.

Algorithm 1 describes the growth of the tree $\mathbb{G}$ in the classic RRT algorithm [11]. For every new generated state q_{sample}, the RRT algorithm will find a closest state, q_{near}, and will determine which trajectory (Definition 1) for any $u \in U$ will brings node q_{near} closer to the sampled state. The closest state is determined based on Euclidean distance between two states. The RRT does this by *shooting* different trajectories and selecting the optimum one [8][10]. From initial state q_{near} algorithm will randomly sample the input space $\mathbb{U}$ and generate the corresponding trajectory by simulating the circuit (*shooting*) for a small time. The algorithm will then select the trajectory that would result in the closest final state (based on Euclidean distance) to the q_{sample} as the optimal trajectory. When the trajectory is determined, the tree will be expanded from q_{near} towards q_{new} which is in the trajectory for duration Δt and the edge e_{new} is added to the tree. The algorithm would store the state q, time t and the trajectory u for each node in the tree so later a test trace can be reproduced using these information. Figure 1 shows the growth of the RRT tree toward a given sample node. The RRT algorithm will terminate after a fixed number of iteration K.

The RRT grows rapidly and quickly visits unexplored regions of the state space[12]. Moreover, as the number of samples approaches infinity, the RRT provably covers the entire state space[11][16]. For these two reasons, the RRT algorithm is ideal for test generation methodologies.

2.3 Test cases for analog circuits

In the RRT $\mathbb{G}$, the root is fixed at $\mathbf{x}(0)$. Each path in $\mathbb{G}$

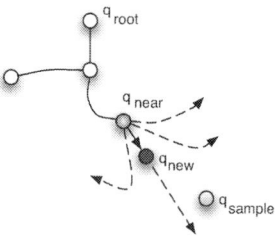

Figure 1: Growth of RRT by adding a new node sampled from the state space.

from the root to a given node represents a trajectory of the circuit. Since the tree is grown according to the constraints in Equation 1, this trajectory represent a feasible behavior of the circuit. For a given initial state $\mathbf{x}(0)$, a test case for an analog circuit consists of a sequence of input values and the time intervals for which each of these input values is applied.

Each edge in the tree corresponds to a value assigned to the input variables $\mathbf{u}$. Therefore, by traversing the edges along a path in $\mathbb{G}$, a test case can be extracted. We consider a test case to be *goal-oriented*, if its application drives a circuit from an initial state $x(0)$ to a state that lies in the relevant region of the circuit's state space.

3. GOAL-ORIENTED TEST GENERATION: OUR ALGORITHM

We describe our RRT-based algorithm (Algorithm 2) for generating goal-oriented test cases for analog circuits.

Our algorithm consists of a learning phase and an exploring phase. In the learning phase, we identify the regions in the state space that are relevant to the circuit. We define these regions as the *goal* towards which the RRT must be grown. In the exploring phase, we enhance the classical RRT-based test generation approach (Section 2.2) by guiding the growth of the RRT towards this goal. We achieve this by biasing the generation of samples used for growing the RRT. The test casesthat we generate are concentrated in the relevant regions of circuit operation.

The inputs to the algorithm are the state space $\mathbb{S}$, the input space $\mathbb{U}$ and the initial condition $\mathbf{x}(0)$ of the circuit. The outputs of the algorithm are the RRT data structure and a batch of test cases that drive the circuit from the given initial conditions $(\mathbf{x}(0))$ to the goal region in $\mathbb{S}$.

The steps in our algorithm are as follows:

1. **Identifying goal regions:** We randomly generate several learning sample traces of the system to identify the relevant regions (*i.e.*, the goal) in the state space.

2. **Grid-based clustering:** We partition the entire state-space into sets of equal-sized grids. We cluster the grids together based on the number of samples in each grid.

3. **Guiding the growth of RRT:** We use the grid-based clusters to generate samples that guide the RRT growth towards the goal region.

4. **Extracting tests from RRT:** Finally, we generate test cases for the circuit by analyzing the RRT data structure. Each test case can be used to drive the circuit from a given initial state to the goal region that we identified in the state space.

Algorithm 2 Goal-oriented RRT algorithm using biased sampling

---

$\mathbb{N}$ = Generate grid network $(\mathbb{S})$;
Filtering $(\mathbb{N})$;
$\mathbb{L}$ = Grid-based clustering $(\mathbb{N})$;
SamplesList = BiasedSampling $(\mathbb{L})$;
$\mathbb{G}$.init $(\mathbf{x}(0))$;
for $i \leftarrow$ *to* K **do**
   q_{goal} = SamplesList (i) ;
   q_{near} = FindNearestNodeInTree $(\mathbb{S}, q_{goal})$;
   q_{new} = FindOptimumTrajectory (q_{near}, q_{goal}) ;
   $\mathbb{G}$.expand (q_{new});
end

3.1 Identifying the goal regions

We sample the circuit by performing a limited number of transient simulations. In each simulation, we randomly choose the initial point as well as the duration of the simulation. At the termination of each simulation, we record the state of the circuit. We refer to this as *learning samples*.

For typical circuits, most simulations terminate in a small region in the state space. Therefore, we use the concentration of the termination states as a measure of the importance of a region. Figure 2 depicts the concentration of learning samples for a tunnel diode circuit[9]. The two dense regions correspond to the stable equilibrium points which are important for analyzing the circuit. We shall describe the tunnel diode circuit in more detail in Section 4.1.

3.2 Grid-based clustering

We partition the entire state space $\mathbb{S}$ into a network $\mathbb{N}$ comprising of equally-sized grids. For each grid in $\mathbb{N}$, we determine the number of the learning samples containted in the grid by sampling the circuit in Section 3.1. We then group the grids into M clusters, where M is a user-defined value, based the number of the learning samples in each grid.

We identify the grids containing the maximum and minimum number of learning samples. Let MAX and MIN represent the maximum and minimum number of learning samples in a grid, respectively. We divide the range $MAX-MIN$ into M equal intervals of length $\Delta = \frac{MAX-MIN}{M}$. Then, we construct a list $\mathbb{L}$ with M elements, L_1 to L_M. The ith element in the list points to all grids for which the number of learning samples lie in the ith interval $[(i-1)\Delta \ i\Delta]$. Therefore, each element in $\mathbb{L}$ corresponds to a grid-based cluster. We construct the list by traversing the grid network $\mathbb{N}$ and assigning each grid to the corresponding element (*i.e.*, cluster) in $\mathbb{L}$. Since the Mth cluster L_M has the highest number of learning samples, we consider L_M to be a cluster of grids in the region that are most relevant to the circuit (Section 3.1). Therefore, we define L_M to be the goal towards which the RRT must be grown.

Optimization for clustering:.

We perform an optimization to reduce the number of samples that need to be generated (Section 3.1) for accurate grid-based clustering. In this work, we consider systems that exhibit continuity in both space and time (Section 2.1). Therefore, if a region in the state space is relevant, it is reasonable to expect that the neighboring regions (in terms of Euclidean distance) are also relevant. We use this as the

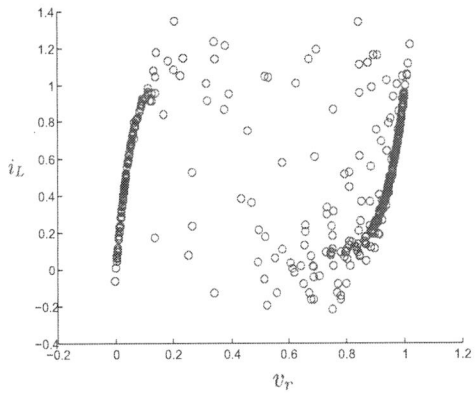

Figure 2: Concentration of learning samples indicating the relevant regions of a tunnel diode circuit (Section 4.1).

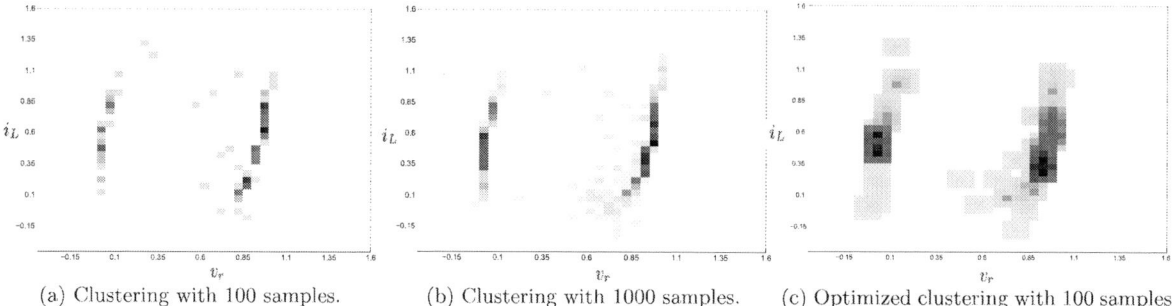

(a) Clustering with 100 samples.   (b) Clustering with 1000 samples.   (c) Optimized clustering with 100 samples.

Figure 3: Grid-based clustering using the samples generated for the tunnel diode circuit (Figure 2). The darker regions represent grids with a higher concentration of samples.

basis for our optimization.

For each grid in the network $\mathbb{N}$, we compute the number of learning samples that have been terminated in that grid. Then, we average this number with the number of learning samples in the grids that are immediately adjacent to it and have a higher number of samples in them. In effect, we are performing a *filtering* operation on the samples generated in Section 3.1. We perform the grid-based clustering using the filtered samples. Filtering reduces the statistical discrepancy in the sampling process. As a result, we are able to quickly identify the broad regions of relevance in the state space.

Figure 3 depicts the grid-based clustering using the samples generated by simulating the tunnel diode circuit (Figure 2). Clustering using 1000 samples (Figure 3a) is more accurate than clustering using only 100 samples (Figure 3b). However, with our filtering optimization, we obtain reasonably accurate clusters using only 100 samples (Figure 3c).

3.3 Guiding the growth of RRT

The RRT algorithm that we use is mostly similar to classic RRT algorithm described in Algorithm 1 (Section 2.2). In the classical RRT algorithm, the RRT is grown by uniformly sampling the state space. However, since we wish to guide the RRT to grow towards a goal region, we bias the generation of samples accordingly.

For generating samples, we start with cluster L_1 and we will generate n_1 samples that are normally (or uniformly) distributed in L_1 space. For generating each samples, we select a random grid in L_1 and we will bias the generator towards that grid and will generate a random number inside that grid (the random number distribution can either be normal or uniform, but we got better results with normal distribution). After we have covered cluster L_1, we increment number of samples n accordingly and will move to cluster L_2 and we will generate n_2 samples. In the end we would have $\sum_{i=1}^{M} n_i = K$ samples where K is the total number of generated samples.

After generating samples, we use these goal-oriented and prioritized samples to guide the tree towards goal regions. Similar to classic RRT algorithm, for each goal sample in *SampleList* we select the nearest node in the tree and will find the optimum trajectory from that node to the goal sample. Then we expand the tree by Δt. Since the samples are more concentrated in the goal regions, the tree will also converge toward the goal region and will generate a lot of concentrated traces from the root of the tree toward the samples in the goal regions.

3.4 Extracting tests from RRT

To generate a test case we select the desired circuit state as our target and choose the appropriate target node (q_{target}) in the tree that is inside our target regions. We extract a (unique) path between the target node and a root of the tree. By traversing this path in reverse, we can generate the test case that would take us from the initial state (q_{root}) to our desired state (q_{target}). An example counter example is shown in Figure 8c in Section 4.2.

3.5 Complexity analysis

An RRT can be optimally constructed using *KD-Tree* data structure[11]. Therefore the algorithm has a computational complexity of $\mathcal{O}(dk \times logk)$ where k is the number of points and d is the number of variable dimensions. But there is a constant that increase exponentially with the dimension d, therefore in practice, this data structure is only useful for problems of up to about 20 dimensions. After this, the performance usually degrades too much [11].

In previous guided RRT algorithms that use incremental sampling, sample generation is iterative and should be performed at each step. This process can be (and usually is) costly. Therefore researchers proposed using bounded values to approximate the optimum direction of the tree[8]. In our approach, we perform a pre-process for identification of relevant regions and grid-based clustering, In our learning phase, we use l samples. The number of learning samples is limited and fixed in advance ($l << k$), Hence it does not impose any computational complexity on the algorithm. The order of generating samples from the clusters is $\mathcal{O}(k)$; Therefore the complexity of our algorithm is

$$\mathcal{O}(l) + \mathcal{O}(k) + \mathcal{O}(dk \times logk) = \mathcal{O}(dk \times logk) \quad (3)$$

4. EXPERIMENTAL RESULTS

We have implemented our algorithm in *MATLAB* to evaluate its accuracy and efficiency. Here, we show the results for two non-linear systems. The first case-study is a simple analog circuits that we use as a proof of concept. The seconds case-study is a Josephson junction circuits that we intentionally designed to show how we use our technique to generate counter examples. In both cases, we set $\Delta t = 0.05$ and executed the RRT algorithm for 3000 samples for randomized algorithm and 3000 total samples for goal oriented algorithm (2700 samples for growing the RRT plus 300 learning sample).

4.1 Tunnel Diode

We use a tunnel diode circuit [9] to show how our technique works under parameter and input variation. The circuit's behavior is specified by the following differential equations:

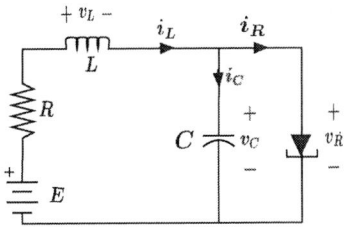

Figure 4: Tunnel-diode circuit

$$\dot{i_L} = \frac{1}{C}(i_L - h(v_r)) \tag{4}$$

$$\dot{v_r} = \frac{1}{L}(-i_d - R \times i_L + E) \tag{5}$$

i_L is the current through the inductor L and v_r is the voltage across the tunnel diode. where as $E = 1.2V, R = 1.5k\Omega, C = 2pF, L = 5\mu H$. Diode's behavior is governed by a nonlinear current-voltage relation $h(x) = 17.76x - 103.79x^2 + 229.62x^3 - 226.31x^4 + 83.72x^5$. We analyze the circuit under disturbance input variables for for the current throughout the diode (modeled as $I_d = h(V_d) + \Delta p_1$) and voltage variation from the DC source (modeled as $E = U_0 + \Delta p_2$. The range of the disturbances are $[-0.1, 0.1]$. The system has three equilibrium points at $(0.063, 0.758)$, $(0.285, 0.61)$ and $(0.884, 0.21)$. Stable equilibrium regions are identified from the learning step of the algorithm. We select unstable equilibrium point $(0.285, 0.61)$ as the root of the RRT.

Figure 6a and 6b shows the result of the analysis using both randomized RRT algorithm and our proposed algorithm. As shown in the figure, in comparison to the randomized RRT algorithm we generate more traces ending in the target regions (in this case, regions around two stable equilibrium points of the tunnel diode circuit); whereas the randomized algorithm would waste a lot of samples to find its path toward equilibrium points.

4.2 Josephson junction

Josephson junction circuit is shown in Fig. 7. The Josephson junction is a time-invariant nonlinear inductor governed by equation 6.

$$i_L = I_0 \times sin(k\Phi_L) \tag{6}$$

Therefore the differential system for circuit of Fig. 7 is

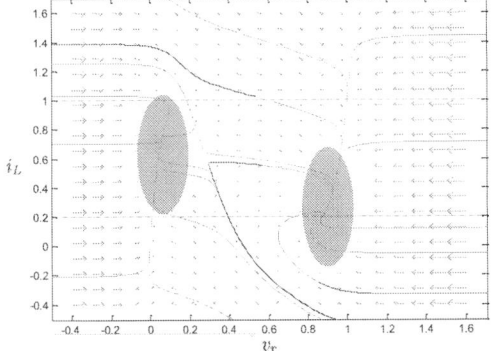

Figure 5: Tunnel diode phase portrait

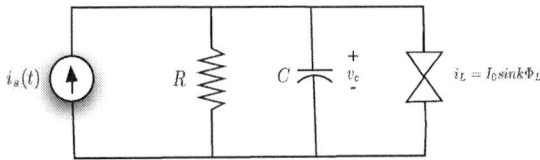

Figure 7: Josephson junction circuit

$$\dot{v_c} = \frac{1}{C}(\frac{1}{R}v_c - I_0 sin(k\Phi_L) + i_s(t)) \tag{7}$$

$$\dot{\Phi_L} = v_c \tag{8}$$

Where $I_0 = 1, R = 4$ and $C = 1$. The inputs of the circuit are the current source ($i_s(t)$) and the variation in $\Phi_L(\Delta_\Phi)$, which both are within range of $[-0.1, 0.1]$. Figure 8 shows the result of goal-oriented RRT algorithm versus the randomized algorithm. The initial state of the circuit was chosen at point (-1,3). For a circuit with no input and variation (i.e. $i_s(t) = 0, \Delta_\Phi = 0$) this initial condition will end up at equilibrium state at (0, 6.3). However since we identified regions around the state (0,0) as our goal regions, the algorithm will guide the tree toward the state (0,0) and will generate many traces toward this point. As shown in Fig. 8, in goal-oriented RRT the algorithm does not waste any samples in the irrelevant regions and will quickly converge directly towards the center of the space at (0,0). On the other hand, in randomized sampling, the algorithm will spend a lot of it samples in uninteresting regions at eventually will not converge toward the center.

5. RELATED WORK AND CONCLUSION

Recently, a lot of research has been done on formal verification of safety of analog circuits using reachability analysis [3]. Reachability analysis of nonlinear analog circuits is an intractable and challenging problem. Various techniques has been proposed to solve this problem via symbolic methods [1] or numerical techniques [14][2]. Another approach to validate the system is discovering safety violation (falsification) by generating a set of test inputs and simulating the system [17][10]. These randomized approaches (including Rapidly-exploring Random Trees) are promising for validating circuits.

The basic RRT algorithm described in this paper was derived from RRT technique in robotic motion planning literatures [11][6][18][4] [5]. Guiding the RRT algorithm in state space is one of the fundamental issues concerning the algorithm efficiency. Dang *et al.* use RRT with coverage measurement to improve a coverage quality and uses them to verify analog and hybrid systems [8][16][7][15]. A similar technique in robotic motion planning is sampling domain control [19]. [13] proposed a dynamic domain method for biasing the exploration by reducing the dispersion in an incremental approach. Another similar method for defining an adaptive biased sampling was proposed in [10]. These algorithm have no knowledge of the system's state space; hence the guidance and growth of the RRTs is only locally optimal. Therefore, they are not very efficient at generating traces toward the goal regions. We addressed these issue by utilizing a learning-based technique to identify regions of interest in the circuit.

In conclusion, we have presented a novel algorithm for test generation in nonlinear analog circuits. We use RRTs and guide them towards the relevant regions in the state space. As a result, the test cases that we extract from the RRTs

1022

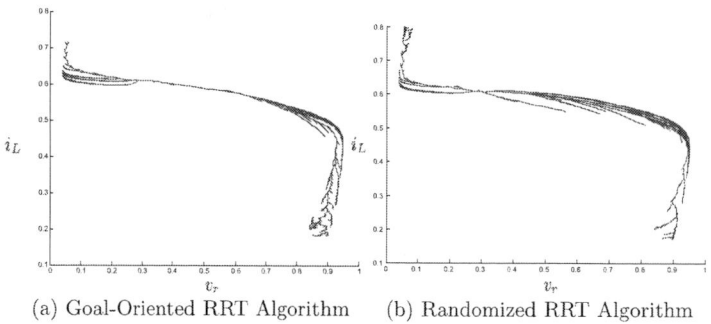

(a) Goal-Oriented RRT Algorithm   (b) Randomized RRT Algorithm

Figure 6: Tunnel diode results for randomized and goal-oriented RRT algorithm, while the classic RRT algorithm will generate a lot of samples to find its path towards two stable equilibrium points, the goal-oriented RRT algorithm will rapidly converge and will generate more traces in relevant regions.

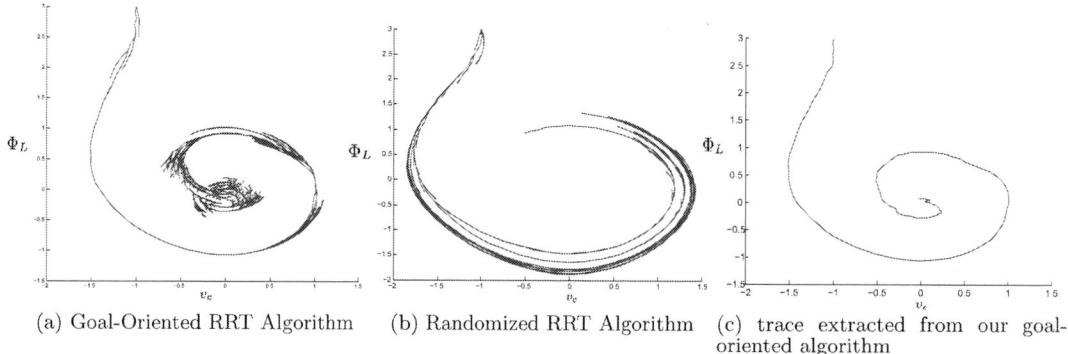

(a) Goal-Oriented RRT Algorithm   (b) Randomized RRT Algorithm   (c) trace extracted from our goal-oriented algorithm

Figure 8: Josephson junction circuit result illustrating a trace for a counter example. For the same number of samples, the goal-oriented algorithm will converge faster and provides a more coverage of the region around equilibrium point (0,0). For generating a test, a trace has been extracted using our goal-oriented algorithm.

are concentrated in the relevant regions of operation.

6. REFERENCES

[1] R. Alur, T. Henzinger, and P.-H. Ho. Automatic symbolic verification of embedded systems. In *Real-Time Systems Symposium, 1993.*, pages 2 –11, dec 1993.

[2] E. Asarin, T. Dang, and O. Maler. d/dt: a verification tool for hybrid systems. In *IEEE Conference on Decision and Control, 2001.*, volume 3, pages 2893 –2898 vol.3, 2001.

[3] E. Barke, D. Grabowski, H. Graeb, L. Hedrich, S. Heinen, R. Popp, S. Steinhorst, and Y. Wang. Formal approaches to analog circuit verification. In *Design, Automation Test in Europe Conference, 2009.*, pages 724 –729, april 2009.

[4] M. S. Branicky, M. M. Curtiss, J. A. Levine, and S. B. Morgan. Rrts for nonlinear, discrete, and hybrid planning and control. In *in IEEE Conf. on Decision and Control*, pages 9–12, 2003.

[5] M. S. Branicky, M. M. Curtiss, J. A. Levine, and S. B. Morgan. Sampling based reahability algorithms for control and verification of complex systems. In *Thirteen Yale workshop on Adaptive and Learning Systems, New Haven, CT*, 2005.

[6] P. Cheng. *Sampling-based motion planning with differential constraints.* Ph.D. dissertation, University of Illinois at Urbana-Champaign, Urbana, IL, 2006.

[7] T. Dang and T. Nahhal. Using disparity to enhance test generation for hybrid systems. In *Proceedings of the 20th IFIP TC 6/WG 6.1 international conference on Testing of Software and Communicating Systems: 8th International Workshop*, TestCom '08 / FATES '08, pages 54–69, 2008.

[8] T. Dang and T. Nahhal. Coverage-guided test generation for continuous andÂåhybrid systems. *Formal Methods in System Design*, 34:183–213, 2009. 10.1007/s10703-009-0066-0.

[9] H. K. Khalil. *Nonlinear Systems (3rd Edition).* Prentice Hall, 2001.

[10] E. J. Kim, J. Adaptive sample bias for rapidly-exploring random trees with applications to test generation. *American Control Conference, 2005*, 2005.

[11] S. M. LaValle. *Planning Algorithms.* Cambridge University Press, Cambridge, U.K., 2006.

[12] S. M. Lavalle, J. J. Kuffner, and Jr. Rapidly-exploring random trees: Progress and prospects. In *Algorithmic and Computational Robotics: New Directions*, pages 293–308, 2000.

[13] S. Lindemann and S. LaValle. Incrementally reducing dispersion by increasing voronoi bias in rrts. In *IEEE International Conference on Robotics and Automation, 2004.*, volume 4, pages 3251 – 3257 Vol.4, 26-may 1, 2004.

[14] I. Mitchell and C. Tomlin. Level set methods for computation in hybrid systems. In *Proceedings of the Third International Workshop on Hybrid Systems: Computation and Control*, HSCC '00, pages 310–323, London, UK, 2000. Springer-Verlag.

[15] T. Nahhal and T. Dang. Guided randomized simulation. In A. Bemporad, A. Bicchi, and G. Buttazzo, editors, *Hybrid Systems: Computation and Control*, volume 4416 of *Lecture Notes in Computer Science*, pages 731–735. Springer Berlin / Heidelberg.

[16] T. Nahhal and T. Dang. Test coverage for continuous and hybrid systems. In W. Damm and H. Hermanns, editors, *Computer Aided Verification*, volume 4590 of *Lecture Notes in Computer Science*, pages 449–462. Springer Berlin / Heidelberg.

[17] E. Plaku, L. E. Kavraki, and M. Y. Vardi. Hybrid systems: From verification to falsification by combining motion planning and discrete search. *Formal Methods in System Design*, 34:157–182, 2009.

[18] A. Shkolnik, M. Walter, and R. Tedrake. Reachability-guided sampling for planning under differential constraints. In *IEEE International Conference on Robotics and Automation*, pages 2859 –2865, may 2009.

[19] A. Yershova, L. Jaillet, T. Simeon, and S. M. LaValle. Dynamic-domain rrts: Efficient exploration by controlling the sampling domain. In *IEEE International conference on robotics and automation*, pages 3867–3872, 2005.

TSV Open Defects in 3D Integrated Circuits: Characterization, Test, and Optimal Spare Allocation*

Fangming Ye and Krishnendu Chakrabarty
Department of Electrical and Computer Engineering
Duke University, Durham, NC 27708, USA

ABSTRACT

Three-dimensional integration based on die/wafer stacking and through-silicon-vias (TSVs) promises to overcome interconnect bottlenecks for nanoscale integrated circuits (ICs). However, TSVs are prone to defects such as shorts and opens that affect circuit operation in stacked ICs. We analyze the impact of open defects on TSVs and describe techniques for screening such defects. The proposed characterization technique estimates the additional delay introduced due to a resistive open defect as well as due to re-routing based on spare TSVs. We also present an optimization method based on integer linear programming (ILP) that allocates spares to functional TSVs such that the spare for a functional TSV is neither too close to a functional TSV (to avoid the case of both functional and spare TSV being defective) nor too far to ensure that the additional delay due to rerouting is below an upper limit. Results are presented using Hspice simulations based on a 45 nm predictive technology model, recently published data on TSV parasitics, and a commercial ILP solver.

Categories and Subject Descriptors: B.7.2 [Integrated Circuits]: Design Aids

General Terms: Design, Algorithms

Keywords: 3D-ICs, ILP, TSV redundancy

1. INTRODUCTION

Semiconductor technology scaling has followed Moore's law over the last 40 years. However, due to the relentless decrease in feature sizes and increase in chip complexity, long interconnects dominate circuit delay and power consumption today [1]. Three-dimensional (3-D) integration offers a promising solution for overcoming interconnect challenges in nanoscale integrated circuits (ICs) [2]. 3D-ICs offer the added benefit of smaller footprint, fewer metal layers (and fewer masks for fabrication), and potentially lower cost [3]. Since 3D-ICs can scale 'up' with multiple layers, higher packing density can be achieved and heterogeneous integrated circuits become feasible [4].

A number of 3-D integration methods have been proposed and several of them have been demonstrated through manufactured prototypes [5]. In this paper, we focus on through-silicon-via (TSV) vertical interconnects, since they provide

---

*This work was supported in part by the National Science Foundation (NSF) under grant no. CCF-1017391, by the Semiconductor Research Corporation (SRC) under contract no. 2118.001, and by a research grant from Intel Corporation.

---

the highest interconnect density compared to other 3-D integration technologies. Multiple device layers can be bonded together through wafer or die stacking, with neighboring layers connected using TSVs and microbumps.

Testing is a major concern in 3-D integration and the related problems of quality assurance, yield, and defect tolerance must be addressed before 3-D technology can be ready for mainstream adoption. Since TSV manufacturing processes are still evolving, TSV defects and TSV yield have been highlighted in the literature as potential problems [6]. Since it is especially challenging to test TSVs at a pre-bond stage due to the much smaller TSV and microbump dimensions relatively to the sizes of probe needles in today's wafer probes, TSVs must also be tested and characterized after stacking, i.e., at a post-bond stage. Even if pre-bond TSV testing becomes feasible through advances in technology or innovative test solutions, post-bond TSV testing is desirable because the stacking process can introduce additional defects due to high temperature, high pressure, and additional processing steps involved in bonding. It is also desirable that we can repair faulty TSVs or bypass faulty TSVs using spare TSVs and re-routing of signals, but with minimum impact on interconnect delay.

Due to the importance of the TSV testing problem, several papers have been published in recent years on this topic and on design-for-testability innovations for testing TSVs in 3D-ICs. A test and repair method based on signal restoration was presented in [7]. A TSV short (e.g., due to a pinhole defect) forms a low resistance path between the TSV and ground, and results in partial or complete degradation of signal quality between two dies. While the approach presented in [7] is effective for detecting shorts and recovering from resistive shorts that cause a limited amount of signal degradation, it is not applicable to open defects. Open defects (especially resistive opens) have been identified as a serious concern for TSVs [8]. Underlying reasons for TSV opens include incomplete fills (pre-bond) or misalignment during bonding (post-bond). These defects lead to an increase in the TSV resistance. While open defects have been extensively studied for CMOS ICs [9], they have not received much attention for 3D-ICs, and since TSVs are different from on-die interconnects in terms of dimensions, aspect ratios, and sometimes also material [6], existing open defect models and test techniques for interconnects cannot be directly applied to vertical vias. Recent work has also addressed delay characterization of TSVs using oscillation test applied to a pair of TSVs [10], but this approach is intended for the evaluation of process variations: it does not provide any capabilities for targeting TSV open defects.

In this paper, we address open defects (including resistive-open defects) in TSVs. Resistive-open defects cause unintended signal propagation delays, while open defects with high resistance lead to a floating end for a TSV: both types of defects adversely affect chip functionality. We model TSVs in 3D-ICs using the Elmore delay model and analyze the delay for defect-free TSVs and TSVs with defects. The analytical model for characterization is compared to Hspice simulations and we show there is good correlation between these two methods for determining TSV interconnect delay in the presence of defects. We also consider the problem of post-bond testing of TSVs for open defects with only a small amount of additional hardware.

To facilitate the repair of faulty TSVs, we consider the problem of allocating spare TSVs to functional TSVs. Recently published work on grouping spares and functional TSVs provides an technique for trading-off yield with hardware overhead [11], but it does not consider the problem of minimizing the additional interconnect length due to rerouting. The goal of our work is to designate for each functional TSV (T_i), one or more spare TSVs (T_i^S) with the constraints that: (i) each spare T_1^S is located at a distance greater than or equal to a minimum specified distance from T_i to minimize the likelihood that both T_i^S and T_i are faulty due to the same defect cause; (ii) T_1^S is located at a distance less than or equal to a maximum specified distance beyond which the added interconnect delay becomes unacceptable. Spare allocation is optimized using integer linear programming (ILP) to minimize the additional wire length required when T_i is replaced by T_i^S. We assume that on-chip reconfiguration structures, such as fuses, are available for re-routing, therefore based on the spare allocation carried out during design and depending on the level of built-in self-test available, repair can also be carried in the field using fuses after the 3D-IC is assembled, packaged, and shipped [12]. To the best of our knowledge, this work is the first attempt to characterize and test TSVs for open defects, and incorporate an optimized spare allocation scheme for repair based on interconnect re-routing. In this work, each functional TSV is assigned to more than one spare TSV, in order to achieve a higher yield.

The rest of paper is organized as follows. Section 2 presents an analysis of TSV resistive-open defects in terms of their impact on TSV delay. Hspice simulations using a 45 nm predictive technology model [13] are also presented and correlated with analytical results. In Section 3, a test structure is described for detecting resistive-open defects. Section 4 provides the delay-aware routing ILP model and simulation results for TSV repair. Section 5 concludes the paper.

2. TSV OPEN DEFECTS AND INTERCONNECT ANALYSIS

In this section, we model TSV open defects, analyze their impact on TSV delay, and present Hspice simulation results to evaluate the accuracy of the analytical model. We also evaluate the additional delay introduced by re-routing to avoid fault TSVs.

2.1 Modeling of Open Defects in TSVs

We first carry out an analysis of open defects in TSVs using the Elmore delay model [10, 14], in view of the high fidelity of Elmore delay model with respect to the actual circuit delay [15]. A resistive-open (with an open defect as a limiting case with very high resistance) can be modeled as an additional resistance in the TSV. Figure 1 illustrates the TSV resistive-open defect model.

We consider TSV resistance and capacitance values for TSVs that have been published recently in the literature

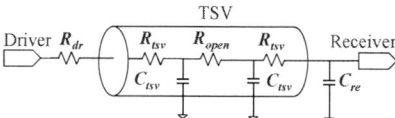

Figure 1: Circuit model of a TSV resistive-open defect.

Table 1: Parameter values [13, 18].

Parameters Used For Delay Calculations	Symbols	Values
Resistance of intermediate metal layers	R_{inter}	3.31 $\Omega/\mu m$
Capacitance of intermediate metal layers	C_{inter}	0.171 fF/μm
Output resistance of a 20-stage driving buffer	R_{dr}	0.86 kΩ
Input capacitance of a buffer (receiving buffer)	C_{re}	1.55 fF
TSV resistance (including bump and interface resistance)	R_{tsv}	100 Ω

[16]. For device parameters of transistors and parasitic values of on-die interconnects, we use the 45 nm predictive technology model (PTM) that is available in the public domain [13]. The various parameters are listed in Table 1.

In this work, we consider device and interconnect parameters for digital circuits only; we do not consider RF circuits, even though the latter are important for 3-D integration. The substrate doping concentration is much higher in RF circuits and the high doping concentration leads to lower substrate capacitance and higher coupling capacitance. Nevertheless the resistance values are the same in both types of circuits [17]. Here, we rely on published data and use the representative value $C_{tsv} = 20$ fF for a TSV of diameter 5 μm, oxide thickness 120 nm, and dopant concentration $N_a = 2e^{15}/cm^3$. For delay calculations, we also assume a TSV of height 20 μm, which is typical for 3-D stacks that have been demonstrated and as listed in ITRS 2009 [1]. Thus, the TSV material resistance is only 280 mΩ, which is smaller than that for a very short wire. On the other hand, the contact resistance is strongly dependant on TSV manufacturing and die bonding. Here, we use 100 Ω as the baseline TSV resistance, which is the sum of the material resistance and contact resistance [18].

2.2 Delay Analysis of Resistive-Open Defects

For delay analysis, we apply a signal at the driving end of the TSV (Die 1) and consider the waveform at the receiving end (Die 2). The input signal can be a step, pulse or impulse input. In this paper, we apply a pulse signal at the driver and capture the output at the receiver after signal propagation through the TSV. For circuit-delay analysis, the 50%-to-50% propagation delay, referred to as $D_{50\%}$ here, is commonly used as a delay metric [14]. We can analytically determine $D_{50\%}$ and the corresponding delay under a TSV open defect, denoted by $D_{open-50\%}$, using the Elmore delay model. The Elmore delay model is inaccurate for "near-end" nodes in RC trees that have many capacitive branches because it ignores resistive shielding [19]. However, the circuit models in this work consist of only a few branches and we are not calculating the delay at any near-end nodes. The Elmore delay model is known to be accurate for the "far-end" nodes that are being studied in this work [15], regardless of the power supply noise [20].

The TSV interconnect delay in the presence of the open defect can be calculated using the circuit model of Figure 1 as follows:

$$D_{open-50\%} = 0.69 \cdot (\frac{1}{2}(R_{tsv} + R_{open}) \cdot C_{tsv} + R_{dr} \cdot C_{tsv} + (R_{dr} + R_{tsv} + R_{open}) \cdot C_{re}) \quad (1)$$

By substituting the values of the parameters from Table 1 into (1), we get the following results. First we note that $D_{open-50\%} = 7.97 R_{open} + 13.58$, where R_{open} is in Ω and $D_{open-50\%}$ is in ps. For a fault-free TSV, $R_{open} = 0$, hence $D_{50\%} = 13.58$ ps. If $R_{open} = 1$ kΩ then $D_{open-50\%} = 21.55$ ps, which is not much different from $D_{50\%}$. For a larger

1025

Table 2: Comparison between analytical and simulation results for TSV delay under open defects.

Value of Resistance R_{open} (kΩ)	Simulated Delay Value $D_{open-50\%}$ (ps)	Analytical Delay Value $D_{open-50\%}$ (ps)
0 (no defect)	12.0	13.58
1	19.5	21.5
10	81.0	93.3
50	371	412.1
100	710	810.5

value of the resistance, e.g., $R_{open} = 10$ kΩ, $D_{open-50\%} = 93.3$ ps, which is considerably different from the TSV delay in the fault-free case.

We also simulated the circuit model of Figure 1 using Hspice. The TSV delays for various values of $D_{open-50\%}$ are shown in Table 2.

We note that the simulated delay values reported by Hspice are close to the analytical values obtained using Equation (1). In order to accurately match the simulated delay to a closed-form formula, we use linear regression and derive the following delay equation: $D_{open-50\%} = 7R_{open} + 12$, where $D_{open-50\%}$ is in ps and R_{open} is in Ω.

We have shown that the TSV delay has a linear relationship with the size of the open defect. Therefore, we partition the range of TSV resistances into two categories: (i) benign TSV defects and (ii) resistive-open TSV defect. Since no work has been published thus far on any feasible signal restoration/enhancement structure for resistive-open TSV defects, we propose to replace a faulty TSV with a spare such that the additional routing overhead is minimum.

We evaluate TSV open defects based on the voltage degradation at the receiver (Die 2) when a high voltage (1.2 V) is transmitted from the driver side (Die 1). The TSV voltage on the receiver side is used to detect a defect in a TSV (Section 3) and enable re-routing (Section 4). We next consider the problem of determining the additional delay contributed by re-routing using a spare TSV.

2.3 Interconnect Delay Overhead Calculation

We consider two methods to determine the delay overhead.

2.3.1 TSV Connected Directly to the Interconnect

Figure 2 shows the circuit model for the repair method. Multiplexer and fuses required for switching are not shown. R_{int} refers to the wire resistance, obtained by multiplying R_{inter} (Table 1) with the wire length. Using Equation (1), Figure 2 and the data listed in Table 1, we calculate the

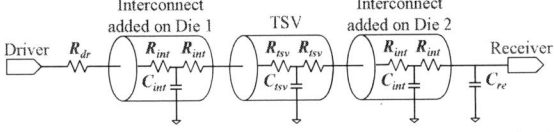

Figure 2: Circuit model used to evaluate the method that directly connects a spare TSV to on-die interconnect.

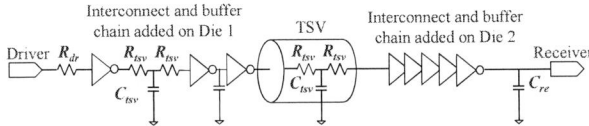

Figure 3: Circuit model used to evaluate method that connects a spare TSV to on-die interconnect using buffers.

Table 3: Delay due to re-routing (direct connection).

Interconnect length (μm)	Simulated delay value (ps)	Analytically computed delay value (ps)
0	12	13.58
10	14.9	16.33
100	35.8	48.13
500	287	342.33
1000	1002	1061.1

Table 4: Delay due to re-routing (buffered connection).

Interconnect length (μm)	Delay $D_{50\%}$ (ps)	Delay without repeater $D_{50\%}$ (ps)
100	32.8	32.8
500	179.2	160.4
1000	368	496
2000	736	1964

additional delay D_r in ps due to re-routing as follows: $D_r = 0.78L^2 + 267.5L + 13.58$, where D_r is in ps and L (in μm) is the added interconnect length for each die. Note that the use of the spare TSV impacts interconnect length in the same manner for each die. We also note that the additional delay increases quadratically as L. The analytically computed delay values and delay values obtained by Hspice simulations are shown in Table 3. Depending on the additional delay that can be tolerated, the results in Table 3 provide guidelines on the extent to which re-routing is acceptable and influence the choice a key parameter value (upper limit on distance) in the ILP-based optimization method described in Section 4.

2.3.2 TSV Connected to the Interconnect With Evenly-Distributed Buffer Insertion

In this re-routing method, we add buffers to the on-die interconnect to reduce delay due to the additional wire length. Figure 3 in the appendix illustrates the circuit model for this type of TSV re-routing. Note that the buffer chain is expanded in the figure for Die 1 but shown in symbolic form for Die 2 to save space.

As shown in Table 3, the delay D_r grows quadratically with L. To reduce this delay, we consider the uniform insertion of buffers (repeaters) at distances of 100 μm. For such a design, we know that the delay $D_{repeaters}$ grow linearly with L [21]. We ran Hspice simulations to obtain delay values due to re-routing for various values of wire length. The results are shown in Table 4. As in the case of Table 3, these results allow us to set limits on the additional wire length that can be accepted for re-routing using spare TSVs.

3. TEST CIRCUIT FOR TSV RESISTIVE-OPEN DEFECTS

In this section, we describe a method to detect resistive-open defect in TSVs. Figure 4 shows the block diagram of the proposed test structure based on [7]. Note that this block diagram also includes the spare TSV to be used for repair. It consists of a selector circuit (on Die 2) and a voltage divider (across Die 1 and Die 2). The voltage V_{tsv} depends on the magnitude of the open resistance. The voltage V_{tsv} is then compared to a reference voltage V_{ref} and sampled to a scan register connected to a comparator. The reference voltage is selected so that it represents the voltage V_{tsv} for a benign TSV (e.g, $V_{ref} > 0.8V_{DD}$). The output of the comparator is stored in the scan register and finally scanned out for locating a faulty TSV, and for subsequent repair. If

1026

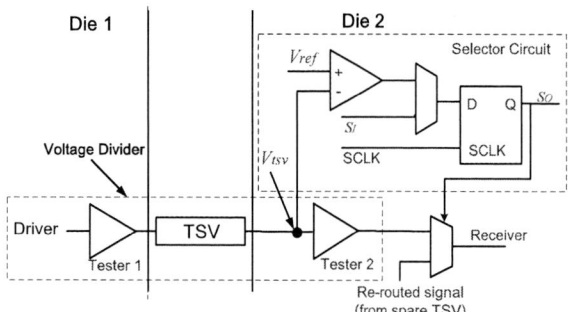

Figure 4: Basic TSV test structure.

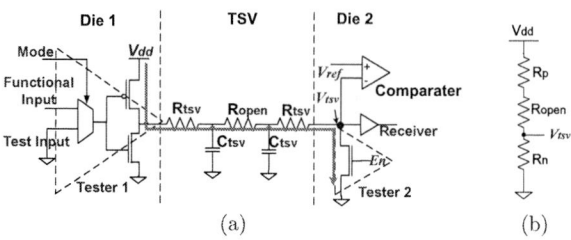

Figure 5: (a) Voltage divider for generating V_{tsv} to detect a TSV resistive-open defect, (b) Simplified circuit model.

the comparator indicates that the TSV is faulty, i.e., $V_{tsv} < V_{ref}$, the output S_O will select the re-route signal, which comes from the TSV allocated to this functional TSV.

Figure 5(a) shows the details of the voltage divider test circuit that is used to generate the V_{tsv} voltage. The circuit consists of "Tester 1", the TSV, and "Tester 2". The voltage divider operates in two modes. In the normal functional mode, the multiplexer on Die 1 (Tester 1) selects the driver input, the enable signal $En = 0$ (Tester 2), and the TSV drives the receiver in Die 2. The inverter in Tester 1 also serves as the driver for the TSV in functional mode. In the test mode for resistive-open defects, the multiplexer in Tester 1 selects "Test Input", which is hardwired to logic zero. The enable signal in Tester 2 is asserted, i.e., $En = 1$. We now have a voltage divider circuit, which is shown in simplified form in Figure 5(b). Thus the voltage V_{tsv}, which is fed to the comparator, depends on the values of the resistances R_p, R_{open} and R_n, where R_r (R_n) is the on resistance of the pMOS (nMOS) transistor in Tester 1 (Tester 2). The resistance R_{open} depends on the magnitude of the resistive-open defect. For a defect-free TSV, $R_{open} = 0$. Suppose the voltage V_{tsv} must be designed to drop below $0.8V_{DD}$ in this case so that the comparator can indicate a fault. We can ensure this by selecting R_n (i.e., sizing the nMOS transistor in Tester 2) on the basis of the following relationship:

$$V_{tsv} = \frac{R_n}{(R_p + R_{open} + R_n)} \cdot V_{DD}$$
$$< 0.8V_{DD}$$

Therefore, for $V_{DD} = 1.2$ V, we get $R_n < 4(R_p + R_{open})$. For $R_{open} = 1\text{k}\Omega$, we get $R_n < 4R_p + 4\text{k}\Omega$. This bound on R_n satisfies the requirement for larger values of R_{open}. For example, to detect $R_{open} = 10$ kΩ, we need $R_n < 4R_p + 40$ kΩ. Therefore, we can size the nMOS transistor in Tester 2 on the basis of the smallest open defect resistance being targeted. Once the Tester 2 circuit is designed and implemented for a target value of R_{open}, larger values of R_{open}, which cause longer TSV interconnect delay (Section 2), will lead to lower values of V_{tsv}, making fault detection by the comparator easier. As illustrated in Figure 9 in the appendix, V_{tsv} decreases as R_{open} increases. The least value of R_{open} that can be detected is 10 kΩ, since this value of R_{open} corresponds to $V_{tsv} = 0.96$ V $= 0.8 \ V_{DD}$.

In order to determine the appropriate sizing of the nMOS transistor in Tester 2, we compute the value of R_n using $R_p = 0.86$ kΩ (R_{dr}) from Table 1. For $R_{open} = 1$ kΩ as the smallest defect resistance being targeted, we get $R_n = 6.44$ kΩ. For a minimum-size nMOS transistor in 45nm PTM technology, $R_n = 14.5$ kΩ [13]. Therefore, a ratio of W/L = 3/1 is sufficient to ensure fault detection for $R_{open} = 1$ kΩ and higher values.

4. OPTIMIZATION FOR RE-ROUTING BASED ON SPARE TSVS

In this section, we present an ILP-based optimization model to minimize the wire delay due to the additional interconnect introduced when a spare TSV is used to replace a faulty TSV after testing and diagnosis. We address the problem of allocating spare TSVs to functional TSVs, given the locations of all these TSVs in the die floorplan. We assume that the TSVs, including spares, are placed on the die based on design and functional considerations and available chip area. The number of spare TSVs is determined a priori based on the acceptable hardware overhead, and provided as input to the ILP model. The overall design, fabrication, test and repair flow is illustrated in Figure 6. Note that such an optimization model can be applied to a TSV affected by any type of defect that impairs its functionality.

4.1 ILP model

Let S_1 be the set of functional TSVs and let S_2 be the set of available spare TSVs. The optimization goal is to derive a mapping function $\Phi: S_1 \to S_2$, whereby each functional TSV is mapped to a spare TSV (T_i^S), such that the distance between T_i and T_i^S lies between a lower limit L_{min} and an upper limit L_{max}. The above limits are determined by the need to eliminate "common-mode failures" that can affect both T_i and T_i^S (hence L_{min}) and minimize the additional wire length due to re-routing (hence L_{max}). As a generalization, we can also consider that the interconnect distance limits may vary for each TSV due to the chip layout, local

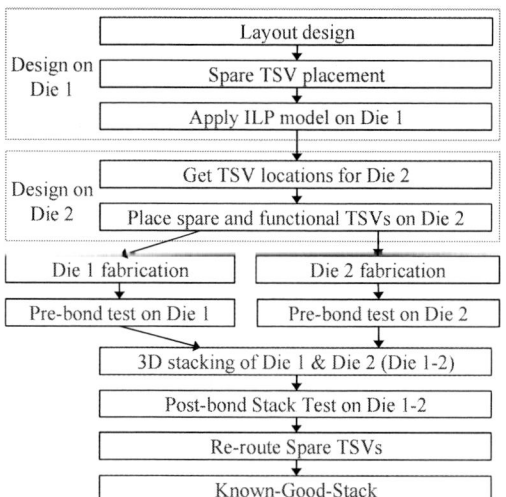

Figure 6: 3-D design, fabrication, test and repair flow assumed in this paper.

voltage droops or temperature impact on interconnect delay [22]. Therefore, each functional TSV T_i can also be associated with its own limits L_{max}^i and L_{min}^i, respectively. Let L_{ij} be the routing distance between functional TSV i and spare TSV j. We note that $L_{min}^i \leq L_{ij} \leq L_{max}^i$.

Note that not every element of S_2 needs to lie in the range of the mapping function Φ, i.e., be allocated to a functional TSV. Such a situation (of not all spares being allocated) can arise if strict limits are placed on the additional wire length that can be tolerated after repair, e.g., based on the data in Tables 3 and 4. In fact, the TSVs that are not allocated to any functional TSVs by the optimization process do not have to be fabricated for use in a subsequent repair step.

Let x_{ij} be a binary variable defined as follows: $x_{ij} = 1$ if TSV i is mapped to spare TSV j, and $x_{ij} = 0$ otherwise. Once a spare TSV is used to repair one of its corresponding defective functional TSVs, an alternative spare TSV is needed to ensure that the rest of functional TSVs can be repaired. Every functional TSV is therefore mapped to one or more spare TSVs. Thus we must satisfy the constraint $\sum_{j=1}^{|S_2|} x_{ij} = k$ for all values of i, where $k \geq 1$ is a user-defined "grouping ratio" parameter. The grouping ratio of spare to functional TSVs has been studied in [11, 23].

Next we place on upper limit N^* on the number of functional TSVs that can be mapped to a spare TSV, as follows. This constraint increases the likelihood of sucessful repair when multiple TSVs are faulty. We have the constraint $\sum_{i=1}^{|S_1|} x_{ij} \leq N^*$ for all values of j.

Finally we state the objective function of our mathematical programming model. The total additional wire length L^* when a faulty TSV is replaced by a spare TSV is given by:

$$L^* = \max_i \left\{ \sum_{j=1}^{|S_2|} L_{ij} x_{ij} \right\} \qquad (2)$$

Equation (2) gives rise to the objective function:

$$\text{Minimize } L^*. \qquad (3)$$

The above minmax objective can be linearized using standard techniques to yield an ILP model. The complete ILP model is shown in Figure 7.

Minimize L^* subject to:

1. $L^* \geq \sum_{j=1}^{|S_2|} L_{ij} x_{ij}, \ \forall i$

2. $\sum_{i=1}^{|S_1|} x_{ij} \leq N^*, \ \forall j$

3. $\sum_{j=1}^{|S_2|} x_{ij} = k, \ \forall i$

4. $L_{min}^i \leq L_{ij} \leq L_{max}^i, \ \forall i,j$

Figure 7: ILP model for spare TSV allocation.

4.2 Experiments and Results

We used a commercial ILP tool, FICO^{TM} Xpress [24], on a Linux server with 64 GB of RAM and two quad-core Intel Xeon processors running at 2.53 GHz each. As an illustration, we first consider an example of 30 functional TSVs and 10 spare TSVs, placed on an 0.8 mm × 0.8 mm die, as shown in Figure 8. We let $L_{min} = 50\ \mu m$ and let $L_{max} = 600\ \mu m$, and set $N^* = 10$. Note that $k = 1$ in Figure 8 and $k = 2$ in Figure 10. (The latter is included in the appendix.) The optimal allocation of spares to functional TSVs to minimize

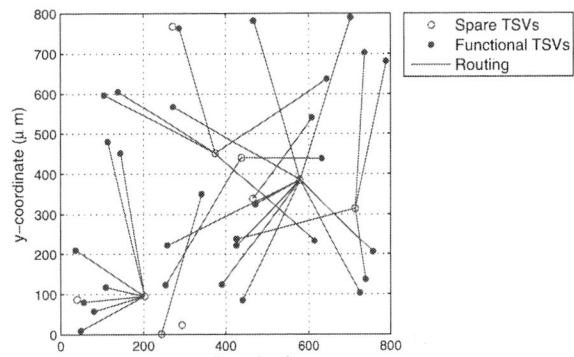

Figure 8: Illustration of optimal spare allocation (grouping ratio $k = 1$).

the additional interconnect length for re-routing due to a faulty TSV, determined using the ILP model of Figure 7, is also shown. Only seven out of the 10 designated spare TSVs are allocated for repair (thus the three remaining TSVs are not necessary). The minimized additional wire length for TSV repair is 424.4 mm ($k = 1$) and 435.6 mm ($k = 2$).

Next, we consider a medium-sized die of 5 mm on the side and a smaller die of 2 mm on the side. These die sizes are chosen because the popular (and representative) microprocessor chip Intel Core^{TM} 2 Duo E6700 has a die of 12 mm on the side [25], and 3-D integration is expected to reduce die size [26]. We consider three test cases for this die—5,000, 10,000, and 20,000 functional TSVs. The last test case is considered only for the larger die. While larger numbers of TSVs on a die have been predicted [27], we expect TSV repair to be local, using only spare TSVs in the vicinity, due to wire length constraints. Hence larger numbers of TSVs do not have to be considered for evaluating ILP-based spare allocation. For each set of parameter values, we repeat the experiment for various random placements of TSVs on the die.

For each test case, we consider three subcases, whereby the number of spare TSVs is 5%, 10% and 20% of the number of functional TSVs, i.e., 250, 500 and 1,000 spare TSVs for 5,000 functional TSVs, and 500, 1,000 and 2,000 spare TSVs for 10,000 functional TSVs. We set $L_{max} = 250\ \mu m$ for the smaller die and $L_{max} = 500\ \mu m$ for the larger die. The parameter L_{min} is set to 10 μm for both test cases, based on defect distribution data reported in [28]. The parameter N^* is set to 50.

As a baseline method, we consider the random allocation of spares to functional TSVs for the same test cases and parameter values. In Table 5, we report the worst-case and average-case re-routing wire length for the proposed method ($k = 1$) and the baseline method for the larger die. We find that the worst-case wire length is up to 50.7% less and the average-case wire length is up to 50.5% less for the proposed method. Table 6 shows results for the smaller die for $k = 1$. Once again, we obtain a significant reduction in worst-case and average-case wire length, up to 60.4% and 60.3%, respectively. The result for $k = 2$ are presented in the appendix. The CPU time for ILP was less than 6 hours for the largest test case and only a few minutes for the smallest test case.

5. CONCLUSIONS

TSVs are prone to defects such as shorts and opens that affect circuit operation in stacked 3D-ICs. We have presented a rigorous analysis of the delay impact of open defects

Table 5: ILP-based optimization results (large die, grouping ratio $k = 1$).

No. of functional TSVs	No. of spare TSVs	Proposed method	Baseline method	Interconnect length reduction (worst case–average case)
		Min additional wire length (μm) (worst case–average case)	Min additional wire length (μm) (worst case–average case)	
5000	250	435.460–285.755	499.930–328.992	12.9%–13.1%
	500	327.674–215.673	499.842–328.517	34.4%–34.3%
	1000	271.046–180.720	499.938–330.898	45.7%–45.4%
10000	500	392.298–258.937	499.985–331.750	21.5%–21.9%
	1000	309.013–204.184	499.925–330.540	38.2%–38.2%
	2000	246.546–163.924	499.938–330.952	50.7%–50.5%
20000	1000	291.414–192.209	499.938–332.051	41.7%–42.1%
	2000	247.542–164.145	499.925–331.346	50.5%–50.5%

Table 6: ILP-based optimization results (small die, grouping ratio $k = 1$).

No. of functional TSVs	No. of spare TSVs	Proposed method	Baseline method	Interconnect length reduction (worst case–average case)
		Min additional wire length (μm) (worst case–average case)	Min additional wire length (μm) (worst case–average case)	
5000	250	207.328–137.281	249.900–165.005	17.0%–16.8%
	500	172.351–113.491	249.898–164.376	31.0%–30.9%
10000	500	173.833–114.882	249.954–165.517	30.5%–30.6%
	1000	99.040–65.830	249.930–165.653	60.4%–60.3%

on TSVs in 3-D integrated circuits. We have presented a new method for detecting such defects using a simple test structure. The proposed test and characterization technique estimates the additional delay introduced due a resistive open defect as well as due to re-routing based on spare TSVs. To recover from open defects and increase the yield for TSV stacks, we have presented an optimization method based on integer linear programming (ILP) that allocates spares to functional TSVs. The optimization is carried out under constraints that the spare for a functional TSV is neither too close to a functional TSV (to avoid the case of both functional and spare TSV being defective) nor too far to ensure that the additional delay due to rerouting is below an upper limit. Results have been presented using Hspice simulations based on a 45 nm predictive technology model, recently published data on TSV parasitics, and a commercial ILP solver. As part of ongoing work, we are extending the test, characterization, and repair method to handle resistive opens and resistive shorts in a unified framework.

6. REFERENCES

[1] International Technology Roadmap for Semiconductors 2009 (ITRS'09).

[2] R. Weerasekera *et al.*, "Extending systems-on-chip to the third dimension: performance, cost and technological trade-offs," in *Proc. International Conference on Computer-Aided Design (ICCAD)*, pp. 212–219, Nov. 2007.

[3] Y. Xie *et al.*, "Design space exploration for 3D architectures," *ACM Journal on Emerging Technologies in Computing Systems*, vol. 2, pp. 65–103, 2006.

[4] T.-Y. Chiang *et al.*, "Thermal analysis of heterogeneous 3D ICs with various integration scenarios," in *Proc. IEEE International Electron Devices Meeting (IEDM)*, pp. 31.2.1–31.2.4, Dec. 2001.

[5] P. Garrou, C. Bower, and P. Ramm, *Handbook of 3D Integration: Technology and Applications of 3D Integrated Circuits.* Wiley-VCH, Oct. 2008.

[6] H.-H. S. Lee and K. Chakrabarty, "Test challenges for 3D integrated circuits," *IEEE Design & Test of Computers*, vol. 26, no. 5, pp. 26–35, 2009.

[7] M. Cho *et al.*, "Pre-bond and post-bond test and signal recovery structure to characterize and repair TSV defect induced signal degradation in 3-D system," *IEEE Trans. Components, Packaging and Manufacturing Technology*, vol. 1, pp. 1718–1727, Nov. 2011.

[8] P.-Y. Chen, C.-W. Wu, and D.-M. Kwai, "On-chip TSV testing for 3D IC before bonding using sense amplification," in *Proc. IEEE Asian Test Symposium (ATS)*, pp. 450–455, Nov. 2009.

[9] D. Arumi, R. Rodriguez-Montanes, and J. Figueras, "Experimental characterization of CMOS interconnect open defects," *IEEE Trans. CAD*, vol. 27, pp. 123–136, Jan. 2008.

[10] J.-W. You *et al.*, "Performance characterization of TSV in 3D IC via sensitivity analysis," in *Proc. IEEE Asian Test Symposium (ATS)*, pp. 389–394, Dec. 2010.

[11] Y. Zhao, S. Khursheed, and B. M. Al-Hashimi, "Cost-effective TSV grouping for yield improvement of 3D-ICs," in *Proc. IEEE Asian Test Symposium (ATS)*, pp. 201–206, Nov. 2011.

[12] A.-C. Hsieh *et al.*, "TSV redundancy: architecture and design issues in 3D IC," in *Proc. Design, Automation, and Test in Europe (DATE)*, pp. 166–171, Mar. 2010.

[13] Predictive Technology Model (PTM), Available on http://ptm.asu.edu/.

[14] D. Hodges, H. Jackson, and R. Saleh, *Analysis and Design of Digital Integrated Circuits.* McGraw-hill, July 2003.

[15] K. D. Boese *et al.*, "Fidelity and near-optimality of Elmore-based routing constructions," in *Proc. International Conference on Computer Design (ICCD)*, pp. 81–84, Oct. 1993.

[16] G. Katti *et al.*, "Electrical modeling and characterization of through silicon via for three-dimensional ICs," *IEEE Trans. Electron Devices*, vol. 57, no. 1, pp. 256–262, 2010.

[17] R. Weerasekera *et al.*, "On signalling over through-silicon-via (TSV) interconnects in 3-D integrated circuits," in *Proc. Design, Automation, and Test in Europe (DATE)*, pp. 1325–1328, Mar. 2010.

[18] D. H. Kim and S. K. Lim, "Through-silicon-via-aware delay and power prediction model for buffered interconnects in 3D ICs," in *Proc. ACM/IEEE International workshop on System level interconnect prediction*, pp. 25–32, 2010.

[19] C. J. Alpert, A. Devgan, and C. V. Kashyap, "RC delay metrics for performance optimization," *IEEE Trans. CAD*, vol. 20, pp. 571–582, May 2001.

[20] M. Saint-Laurent and M. Swaminahant, "A model for power-supply noise injection in long interconnects," in *Proc. IEEE International Interconnect Technology Conference (IITC)*, pp. 113–115, June 2004.

[21] V. Adler and E. G. Friedman, "Repeater design to reduce delay and power in resistive interconnect," in *IEEE Trans. Circuits and Systems II*, vol. 45, pp. 607–616, May 1998.

[22] R. H. Havemann and J. A. Hutchby, "High-performance interconnects: an integration overview," *Proc. of the IEEE*, vol. 89, pp. 586–601, May 2001.

[23] L. Jiang, Q. Xu, and B. Eklow, "On effective TSV repair for 3D-stacked ICs," in *Proc. Design, Automation, and Test in Europe (DATE)*, Mar. 2012.

[24] FICO™ Xpress Opitmization Suite, http://www.fico.com/en/Products/DMTools/.

[25] Intel Core™ Duo Processor E6700, http://ark.intel.com/products/27251/Intel-Core2-Duo-Processor-E6700-(4M-Cache-2_66-GHz-1066-MHz-FSB).

[26] A. Jain, "Thermal characteristics of multi-die, three-dimensional integrated circuits with unequally sized die," in *Proc. IEEE Intersociety Conference on Thermal and Thermomechanical Phenomena in Electronic Systems*, June 2010.

[27] C. Bermond *et al.*, "High frequency characterization and modeling of high density TSV in 3D integrated circuits," in *IEEE Workshop on Signal Propagation on Interconnects (SPI)*, May 2009.

[28] R. Glang, "Defect size distribution in VLSI chips," *IEEE Trans. Semiconductor Manufacturing*, vol. 4, pp. 265–269, Nov. 1991.

APPENDIX

We provide additional details related to models and findings reported in Sections 2-4.

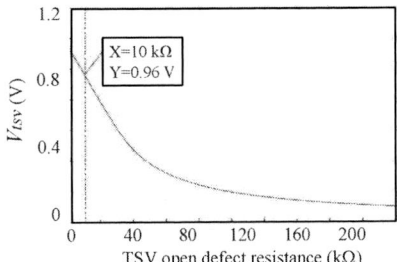

Figure 9: Simulation of the TSV test structure of Figure 4.

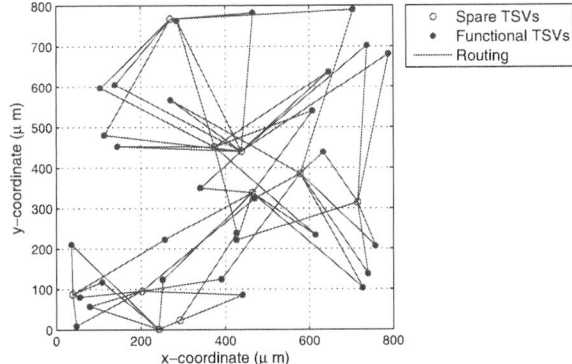

Figure 10: Illustration of optimal spare allocation (grouping ratio $k = 2$).

Table 7: ILP-based optimization results (large die, grouping ratio $k = 2$).

No. of functional TSVs	No. of spare TSVs	Proposed method Min additional wire length (μm) (worst case– average case)	Baseline method Min additional wire length (μm) (worst case– average case)	Interconnect length reduction (worst case– average case)
5000	250	NF[1]	NF[1]	NF[1]
	500	460.570–304.023	499.833–329.813	7.85%–7.81%
	1000	300.719–200.68	499.917–329.488	39.8%–39.1%
10000	500	NF[1]	NF[1]	NF[1]
	1000	328.288–215.881	499.925–329.572	34.3%–34.5%
	2000	255.494–178.713	499.988–330.306	48.8%–45.9%
20000	1000	NF[1]	NF[1]	NF[1]
	2000	249.236–171.589	499.869–328.924	50.1%–47.9%

[1]No feasible solution found for given parameter values.

Table 8: ILP-based optimization results (small die, grouping ratio $k = 2$).

No. of functional TSVs	No. of spare TSVs	Proposed method Min additional wire length (μm) (worst case– average case)	Baseline method Min additional wire length (μm) (worst case– average case)	Interconnect length reduction (worst case– average case)
5000	250	220.952–141.591	249.000-164.203	11.3%–13.7%
	500	237.794–154.501	249.899–165.027	4.8%–6.3%
10000	500	199.620–121.928	249.923–164.856	20.1%–26.0%
	1000	119.231–79.161	249.754–165.528	52.2%–52.2%

Small Delay Testing for TSVs in 3-D ICs

Shi-Yu Huang   **Yu-Hsiang Lin**

Electrical Engineering Department, National Tsing Hua University, Taiwan

Kun-Han (Hans) Tsai   **Wu-Tung Cheng**   **Stephen Sunter**

Silicon Test Solutions, Mentor Graphics

Yung-Fa Chou   **Ding-Ming Kwai**

Information and Communications Research Labs, Industrial Technology Research Institute, Taiwan

Abstract—In this work, we present a robust small delay test scheme for through-silicon vias (TSVs) in a 3D IC. By changing the output inverter's threshold of a TSV in a testable oscillation ring structure, we can approximate the propagation delay across that TSV, and thereby detecting a small delay fault. SPICE simulation reveals that this Variable Output Thresholding (VOT) technique is still effective even when there is significant process variation in detecting a slow TSV with some resistive open defect that may escape the traditional at-speed test.

Categories and Subject Descriptors:
   B.8.1 [Reliability, Testing, and Fault-Tolerance]

General Terms
Measurement, Reliability

Keywords
TSV Testing, Small Delay Testing, Design for Testability, 3D IC

I. INTRODUCTION

Since a 3D IC uses lots of TSVs to connect several dies, the functions of the chip may fail if some of TSVs have defects. To increase the yield of a 3D IC, all kinds of techniques are needed, such as performance characterization, fault-tolerant design, testing, debugging, redundancy analysis and repair, etc.

Delays across the TSVs are relatively less predictable in a 3D IC than the intra-die interconnects (due to the potential mechanical stress of wafer thinning process and the die-stacking process). As reported in [9], 3D IC could suffer from non-negligible yield loss due to the malfunctioning of the TSVs. From a performance point of view, there is an essential need to predict the actual delay across every TSV in a 3D IC even in the mass-production stage for two main reasons – i.e., (1) delay testing, and (2) fault diagnosis or silicon debugging.

In general, a TSV is nothing but an interconnecting wire, only with often larger capacitance. An in-depth investigation of the interconnect delay model can be found in [10]. A design for direct delay measurement was ever proposed in [11], and the oscillation ring concept was applied to characterize the delay of a cell in a closed-loop manner. Throughout this paper, we will use RO to denote a *ring oscillator* and will use this abbreviation frequently. Basically, RO consists of a cyclic circuit path of negative polarity and will thus autonomously oscillate with a certain period depending on twice the propagation delay along

the ring. In [4], RO is further utilized to characterize the capacitance of some transmission line. In [1] and [3], a cell delay inside an RO can be obtained by some post-measurement mathematical analysis. Also, the RO can be used to test the stuck-at, open and delay faults associated with the interconnecting wires in a 2D IC, as demonstrated in [7][8][12]. However, these schemes need some modification before they can be applied to the TSVs in a 3D IC.

Recently, various built-in techniques tailored for testing the TSV have been actively explored. At the pre-bond phase (i.e., when bare dies are not stacked yet and the TSVs have not been completely formed), a test method was proposed in [2] to decide if there is a delay fault associated with a target TSV via a charging/discharging process that converts the intrinsic parasitic capacitance of a TSV into a measureable delay time for ease of fault detection. Another method has also been proposed to perform post-bond test of the TSVs through a scan chain based test infrastructure [6]. Although these methods may perform well for detecting if a TSV is failing or not, they may not be fully adequate when it comes to the detection of *small delay fault* in a TSV, e.g., detecting a 100ps extra delay that added to a normally 300ps TSV.

The oscillation ring concept with a so-called *input sensitivity* analysis technique proposed in [13] (which will be referred to as *IS analysis* in the sequel) provides more insight into the delay of a TSV in that it can approximate the amount of capacitance of every TSV. *Still, it has one drawback — it does not scale well to the condition when a TSV under measurement is faulty with significant resistance.* It is noteworthy that the fault-free resistance of a TSV is only in the range of several mΩ, but a faulty TSV could be highly resistive when it has some resistive open fault. *Under such a condition, the IS analysis will report a misleading result on the TSV delay for not being able to take into account of the effect of the resistance.* In light of this, we need some other scheme.

In this paper, we propose such an on-chip measurement technique that can be used to *approximate the propagation delay across a TSV* even when the TSV under measurement is resistively open. The technique to be proposed, referred to as *Variable Output Thresholding*, or *VOT* for short, when used in conjunction with the *IS analysis* can jointly shed more light on the characteristics of a TSV under measurement, such as delay, resistance, and capacitance. The *VOT* scheme is implemented on an RO architecture consisting of two TSVs and some logic gates. **The major feature that differentiates it from all the previous RO-based schemes is that we equip each TSV's output inverter with *variable thresholds*.** In other words, sometimes a TSV's output inverter behaves like a normal CMOS logic inverter, while at some other time its 0→1 and

$1 \rightarrow 0$ thresholds are made different, by switching it to a Schmitt-Trigger inverter as will be detailed later. The most important concept of this work can be stated as follows - *By changing the TSV's output inverter from a normal one to a Schmidt-Trigger, the oscillation period of the clock signal generated by the RO will change accordingly, with the difference reflecting the propagation delay across the TSV.*

The rest of this paper is organized as follows. Section 2 provides the background information, including the basic electrical model of a TSV, and a brief review of the *IS analysis* [14], which forms the basis of our work. Section 3 proposes the new architecture supporting the *VOT* technique and its operations. Section 4 presents the experimental results, and Section 5 concludes.

II. TSV DELAY MODEL

A. TSV Modeling

For a TSV that has the following physical parameters as $\{d = 30\ \mu m,\ t = 1\ \mu m,\ \text{and}\ l = 75\ \mu m\}$ [14], the resistance (R_{TSV}) and capacitance (C_{TSV}) will be approximately 2 mΩ and 242 fF, respectively, based on the analytical models presented in [5]. Then, the fundamental time constant of this wire (which is R_{TSV} multiplied by C_{TSV}) will be about 0.484×10^{-3} ps. This is an extremely small value as compared with the intrinsic gate delay (which is in the range of tens to hundreds of picoseconds in a typical 0.18 μm process). Unfortunately, the capacitance of a TSV is much larger than that of a usual wire (which is in the range of just a few femto-farads). As a result, the delay a TSV actually causes is not on the wire itself, but on its driving gate. Based on this observation, we define the *TSV delay (which is basically a quantity we wish to decide if there is too much delay)* as the overall propagation delay across not just the TSV structure but also its driver cell.

B. RLCG-Based Delay Model of a TSV

A TSV can be viewed as a transmission line that takes into account the effects of the inductance and the conductance to ground, as shown in Fig. 1(a). This figure also shows the typical values of the R, L, C, and G as $\{R_{TSV} = 2\ m\Omega,\ L_{TSV} = 1\ pH,\ C_{TSV} = 242\ fF,\ G_{TSV} = 1\ nS\}$. It is notable that the effects of the inductance L and the conductance G may not be negligible when the TSV is used for transmitting high-frequency analog signal. However, as a channel of digital signal, these two effects can be mostly neglected as will be illustrated below.

To verify this point, we perform the SPICE simulation using the RLCG model, with the inductance increased by 5000X from its typical value to 5000pH. The simulation results are shown in Fig. 1(b). We apply a $0 \rightarrow 1$ transition at the input of a TSV driver (i.e., node A). There will be a $1 \rightarrow 0$ transition observed at the endpoint of the TSV (i.e., node B). The output transition at node B exhibits some low-amplitude ripples. But overall, the propagation delay of the signal remains almost unchanged.

Generally speaking, the capacitance (denoted as C_{TSV}) will dominate the TSV delay when it is relatively fault free [13]. However, the resistance (denoted as R_{TSV}) in a faulty TSV

could have some impact if its value is large enough to be comparable to the effective on-resistance of the driver (which is usually in the range of several hundred Ω to several kΩ).

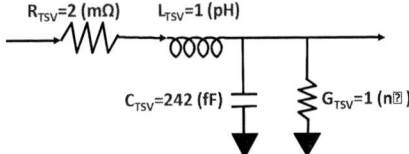

(a) An RLCG-based transmission line model for a TSV

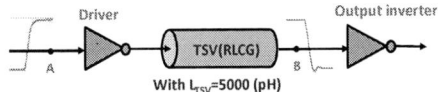

(b) Simulation of signal across a TSV with exaggeratively big L_{TSV}.

Fig. 1: A transmission line based circuit model for a TSV.

C. Review of Input-Sensitivity Analysis

In this subsection, we review a most related previous work proposed in [14] – referred to input-sensitivity analysis (IS) herein. It aimed to predict the capacitance of each TSV under monitoring.

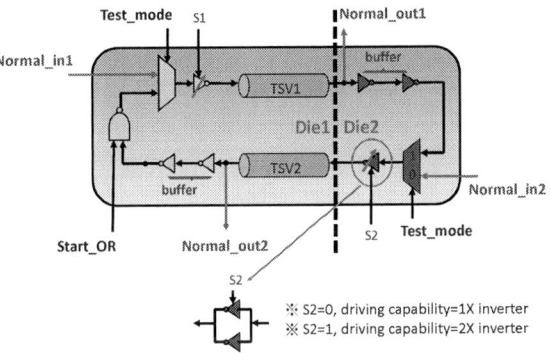

Fig. 2: The architecture for a TSV pair in [14].

As shown in Fig. 2, two TSVs are paired up to form an RO. The resulting period of the RO is twice the accumulated propagation delay along the ring. Observing the oscillation period of this RO can reveal some delay information. The RO can be dynamically set to one of three different configurations by tuning the strengths of the input derivers of the two TSVs – i.e., (1) *normal configuration*, (2) *TSV1-driver-reduced configuration*, and (3) *TSV2-driver-reduced configuration*. During the test mode, the three RO configurations are turned on in sequence one by one, and each oscillation period is measured. A procedure is followed to map the measurement results to the corresponding capacitance and the propagation delay across each of the two TSVs. In summary, the idea of the IS analysis is that – the delay of a particular TSV can be reflected in the change of the clock period of the RO when we perturb the driving strength of the input buffer of that TSV, regardless of the periphery circuit in the oscillation ring.

Although the *IS analysis* can capture the TSV capacitance effect, it has a drawback in not being able to factor in the TSV resistance effect when the TSV is highly resistive. This is a drawback that may not seem so detrimental if one only attempts to use it for profiling the delays across fault-free TSVs. But it is indeed a weakness that needs to be strengthened if one attempts to measure the delay across a faulty TSV (e.g., one with a resistive open fault). This phenomenon is illustrated in Fig. 3. As we sweep the capacitance of a TSV in an RO from a small value to a large value, we can see that a proportional growth in the measureable quantity, denoted here as ΔT_{IS} [14], as shown in Fig. 3(a). On the other hand, as we sweep the resistance of a TSV in an RO from a small value to a large value, we cannot see any proportional reflection in the measureable quantity, as shown in Fig. 3(b). This phenomenon indicates that we need some other complementary scheme, as the *VOT analysis* to be proposed.

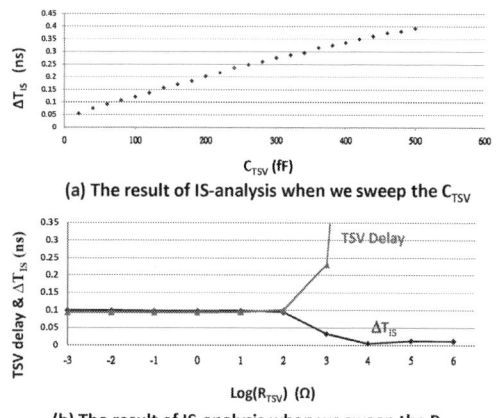

(a) The result of IS-analysis when we sweep the C_{TSV}

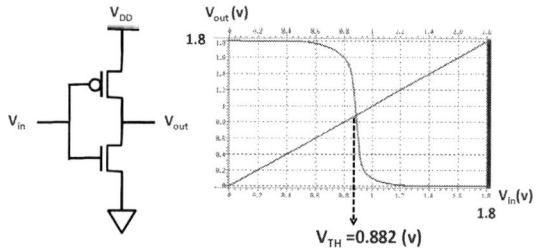

(b) The result of IS-analysis when we sweep the R_{TSV}

Fig. 3: The responses of IS Analysis [14] to varying TSV capacitance and resistance.

III. METHODOLOGY OF VARIABLE OUTPUT THRESHOLDING (VOT)

In this section, we introduce the architecture needed for VOT analysis, and then describe the overall delay measurement procedure. To avoid possible confusion in our discussion, we define the threshold voltages of some logic cells more formally.

A. Thresholds of Inverters

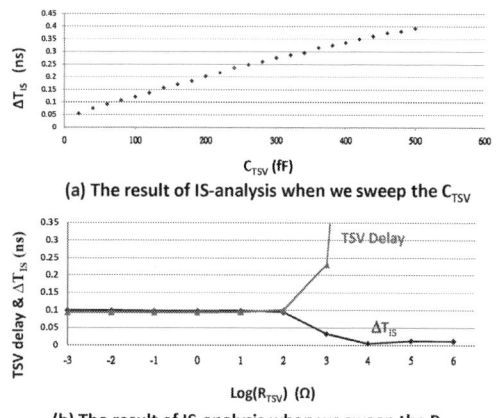

Fig. 4: Threshold voltage of a normal CMOS inverter.

Definition 1: (Threshold of an inverter) The threshold voltage of a normal CMOS inverter in this paper refers to the input voltage that causes the output to change the state (from '0' to '1' or '1' to '0'). In general, this voltage is defined by intersecting the VTC (Voltage Transfer Curve) of an inverter with a 45^0 straight line through the origin point, as shown in Fig. 4.

Definition 2: (Thresholds of a Schmitt-Trigger inverter) A Schmitt-Trigger inverter (or simply called **ST inverter**) as shown in Fig. 5 is known to have a *hysteresis property*, meaning that a stronger-than-normal input voltage is needed in order to flip the output state (which has a larger inertia). In other words, we need a much higher voltage to change the output of a ST inverter from '1' to '0', and much lower voltage to change the output from '0' to '1'. This property implies two separate thresholds — (1) the high-to-low threshold, denoted as $V_{TH(1->0)}$, and (2) the low-to-high threshold, denoted as $V_{TH(0->1)}$. More precisely, the high-to-low threshold $V_{TH(1->0)}$ is the input voltage beyond which one can change the output state from '1' to '0', while low-to-high threshold $V_{TH(0->1)}$ is the input voltage below which one can change the output state from '0' to '1'. The hysteresis property of an ST inverter is useful in the measurement of the propagation delay across a TSV as to be revealed later.

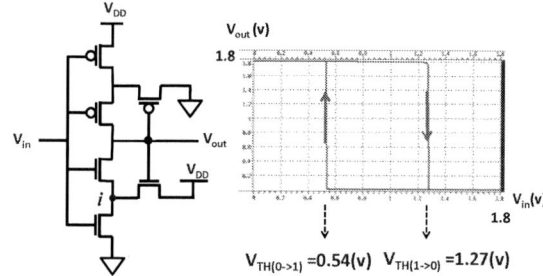

Fig. 5: Threshold voltages of a Schmitt-Trigger (ST) inverter.

Fig. 5 shows a typical ST inverter. Both the pull-up and pull-down paths are now composed of two transistors in cascade. The hysteresis property is generated by slightly dragging the voltage of the middle points of the pull-up and pull-down paths, to take advantage of the body effect of MOS transistors. To explain the detail, let us consider a scenario when the output state is logic '1'. We increase the input voltage gradually until the output flips to '0' to derive the high-to-low threshold $V_{TH(1->0)}$, which is 1.27V in this example. Before the flipping, the pass transistor is conducted, forcing the middle point of the pull-down path, i.e., node i, to a relatively high voltage. This situation will increase the threshold voltage of the nMOS transistor due to the body effect and thus make the pull-down path of the ST inverter harder to be activated, unless when the input voltage is high enough to overcome the body effect. A similar scenario applies when the output state is originally '0'. In that case, we need to apply fairly low voltage to overcome the body effect of the pMOS transistor in the pull-up path, and thus leading to a small low-to-high threshold $V_{TH(0->1)}$, which is 0.54V in this example.

B. DfT Architecture for VOT Analysis

As illustrated in Fig. 6, a TSV pair is now configured into an RO, similar to the *IS analysis* (with some peripheral circuitry). To support the *VOT* analysis, we change the output inverter of each TSV with variable thresholds, as defined below.

Definition 3: (VOT Inverter) A Variable-Output-Threshold (VOT) inverter is the combination of a normal CMOS inverter and a ST inverter, with the function depending on the value of a control signal, say Z. In other words, a VOT inverter operates in two different modes – (1) *normal mode*, and (2) Schmitt-Trigger (ST) *mode*.

If (Z==0), VOT inverter = normal CMOS inverter;

Else if (Z==1), VOT inverter = Schmitt-Trigger inverter; ∎

There are two VOT inverters in a test unit composed of two TSVs (as shown in Fig. 6), one for each TSV. Their controlling signals are denoted as Z_1 (for the VOT inverter of the TSV1) and Z_2 (for the VOT inverter of the TSV2). Based on the value combination of $<Z_1, Z_2>$, we define three configurations in our VOT analysis, as summarized in Fig. 7.

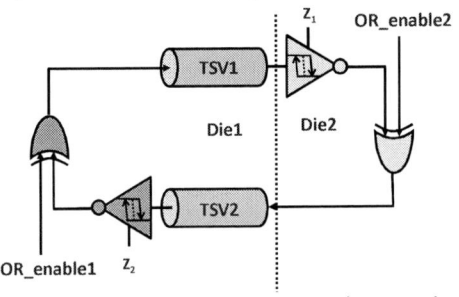

Fig. 6: A simplified architecture of a TSV pair supporting *VOT (Variable Output Threshold) analysis.*

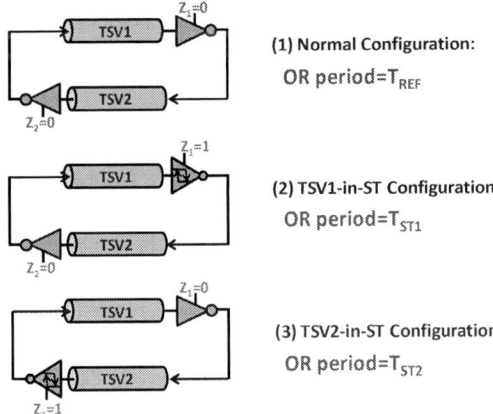

Fig. 7: Summary of three configurations of *VOT analysis.*

(1) *Normal Configuration*: $<Z_1, Z_2> = <0, 0>$, i.e., both VOT inverters are in their normal modes. The clock period of the RO is denoted as T_{REF}.

(2) *TSV1-in-ST Configuration*: $<Z_1, Z_2> = <1, 0>$, i.e., the VOT inverter associated with TSV1 is in the ST mode, while that of TSV2 is in its normal mode. The clock period of the resulting RO is denoted as T_{ST1}.

(3) *TSV2-in-ST Configuration*: $<Z_1, Z_2> = <0, 1>$, i.e., the VOT inverter associated with TSV2 is in the ST mode, while that of TSV1 is in its normal mode. The clock period of the resulting RO is denoted as T_{ST2}.

C. Implementation of a Variable-Threshold Inverter

A VOT (Variable-Output-Threshold) inverter can be realized by a schematic shown in Fig. 8. Fig. 8(a) is the overall schematic, Fig. 8(b) is the equivalent circuit when the control signal $Z = 0$ (or when it degenerates to a normal inverter), and Fig. 8(c) is the equivalent circuit when the control signal $Z = 1$ (or when it degenerates to an ST inverter). It is notable that the two ST thresholds (i.e., $V_{TH(1->0)}$ and $V_{TH(0->1)}$) can be set by tuning the transistor sizes.

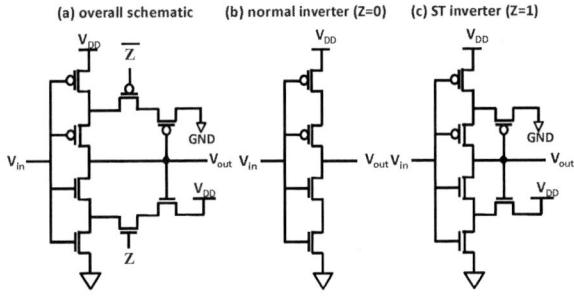

Fig. 8: An implementation of the VOT inverter which can switch its configuration between a normal inverter and a ST inverter.

D. Overall Flow

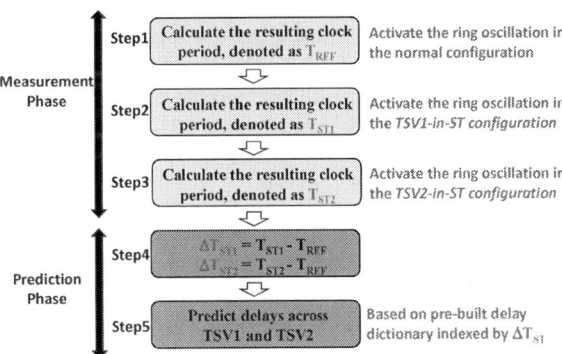

Fig. 9: Overall flow for TSV delay prediction.

As shown in Fig. 9, the entire flow can be further divided into two phases: (1) the measurement phase and (2) the prediction phase. There are three major steps in the measurement phase for each test unit (composed of a TSV pair). In step 1, we activate the RO in the *normal configuration*. Calculate the resulting clock period, denoted as T_{REF}. In step 2, we activate the RO in the *TSV1-in-ST configuration*. Calculate the resulting clock period, denoted as T_{ST1}. In step 3, we activate the RO in the *TSV2-in-ST configuration*. Calculate the resulting clock period, denoted as T_{ST2}. In the prediction phase, we first do the following computations:

1034

$$\Delta T_{ST1} = T_{ST1} - T_{REF}$$

$$\Delta T_{ST2} = T_{ST2} - T_{REF}$$

The derived two timing parameters ΔT_{ST1} and ΔT_{ST2} can be regarded as the *timing signatures* of the two TSVs (ΔT_{ST1} for TSV1 and ΔT_{ST2} for TSV2), considering that they are directly linked to the TSV delays. The rationale behind this linking can be explained as follows:

A larger TSV delay (due to excessive R, excessive C or both) often induces a larger rise/fall time of the signal waveform observed at the endpoint of the TSV. After the VOT inverter, the amount of the rise/fall time will be translated into a difference of the propagation delays across the VOT inverter between the normal configuration and the TSV-in-ST configuration.

It is notable that the delays due to the peripheral circuitry in the RO (such as the multiplexers and the triggering XOR gate) have been canceled out, as we perform the deduction in calculating $\Delta T_{ST} = (T_{ST} - T_{REF})$, and thus ΔT_{ST} will reflect only the delay across the TSV.

IV. EXPERIMENTAL RESULTS

Example 1: We assume that $\{R_{TSV1} = 10\Omega, C_{TSV1} = 400\text{fF}\}$ and $\{R_{TSV2} = 1k\Omega, C_{TSV2} = 800\text{fF}\}$. That is, TSV1 is normal while TSV2 has a resistive open fault that makes the resistance rise by 100 times. At the same time, we assume that the capacitance of TSV2 is also larger at 800fF. We use these two sets of RC values to replace the two TSVs in the RO model and perform the SPICE simulation using a 0.18μm process. Fig. 10(a) shows the waveforms of signals right at the endpoints of the two TSVs, denoted as B_1 and B_2, respectively. It can be seen that the waveform of B_1 is relatively normal, whereas the waveform of B_2 rises and falls much more slowly due to the joint effect of larger R and larger C in TSV2. Fig. 10(b) shows the simulated waveforms under the two ST configurations. The clock periods under both the *TSV1-in-ST configuration* and the *TSV2-in-ST configuration* are larger than their reference clock period under the normal configuration — implying that using an ST inverter as the output driver of a TSV does expand the clock period. In particular, the amount of expansion in the clock period is much larger in the *TSV2-in-ST configuration* than in the *TSV1-in-ST configuration*, since the TSV2 delay is much larger than TSV1 delay. As shown in Fig. 10(c), we can derive the following data (which are assumed to be measured from the IC during the testing process): $\{T_{REF} = 4.42\text{ns}, T_{ST1} = 5.05\text{ns}, \text{and } T_{ST2} = 6.49\text{ns}\}$. After simple calculation, we arrive at $\{\Delta T_{ST1} = 0.63\text{ns} \text{ and } \Delta T_{ST1} = 2.07\text{ns}\}$.

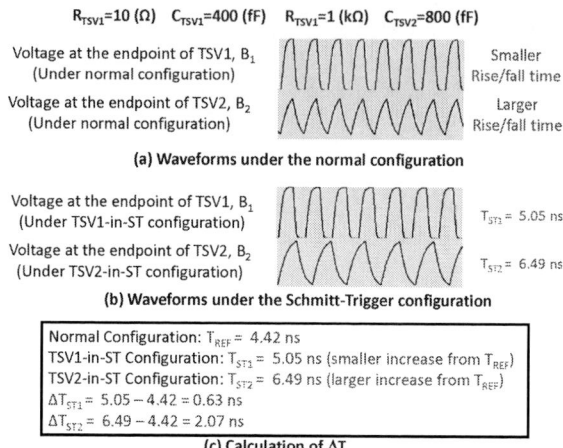

(a) Waveforms under the normal configuration

(b) Waveforms under the Schmitt-Trigger configuration

Normal Configuration: $T_{REF} = 4.42$ ns
TSV1-in-ST Configuration: $T_{ST1} = 5.05$ ns (smaller increase from T_{REF})
TSV2-in-ST Configuration: $T_{ST2} = 6.49$ ns (larger increase from T_{REF})
$\Delta T_{ST1} = 5.05 - 4.42 = 0.63$ ns
$\Delta T_{ST2} = 6.49 - 4.42 = 2.07$ ns

(c) Calculation of ΔT_{ST}

Fig. 10: An example illustrating the correspondence between the TSV delay, the signal rise/fall time, and ΔT_{ST}.

Fig. 11 shows the derived ΔT_{ST} versus the TSV delay for one TSV with the resistance gradually increasing from 10Ω to $10k\Omega$ in a step-size of 10Ω. During the sweeping, the TSV delay rises from its fault-free value of 363ps up to more than 1800ps. It can be seen that there is a strong linear relationship between the measurable quantity ΔT_{ST} and the TSV delay. The average absolute error percentage is 1.6% between the linear regression model and the samples, while the maximum error is 15.9%.

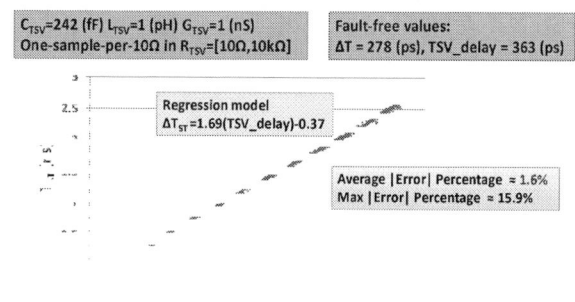

Fig. 11: The relationship of measurable quantity ΔT_{ST} versus the TSV delay, where ($\Delta T_{ST} = T_{ST} - T_{REF}$).

Fig. 12 shows the results that take into account the process variation in a CMOS 90nm process technology. For each of the TSV model (with a faulty R_{TSV} value) we perform 10,000 times of Monte-Carlo simulation to reflect the potential *variation of the circuit elements* in our VOT architecture. After applying the linear regression, we can calculate that the average absolute error is 5.85%, while the maximum error is 29.28%. Since the number of samples experimented for each TSV model has been 10,000, we can interpret this result in a statistical sense that the confidence level is 99.99% that the TSV delay is within the 29.28% error bound of the estimated value.

1035

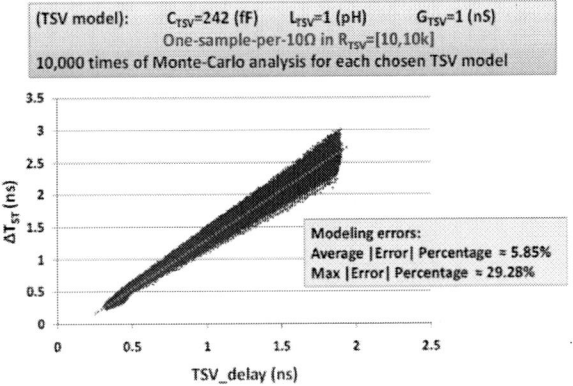

Fig. 12: The relationship of the measurable quantity ΔT_{ST} versus the TSV delay, taking into account the process variation in a 90nm CMOS process technology.

The overall test flow for *small delay fault detection* can be depicted as follows. For a test unit, we derive the ΔT_{ST1} and ΔT_{ST2} for the two TSVs by our VOT analysis, respectively. Then, we use the regression model that we have derived in Fig. 12 under a specific process technology to serve as a ΔT_{ST}-*to-TSV-delay dictionary* to derive the *predicted delay values* for TSV1 and TSV2. For each of these two values, we multiply it by (1-29.28%) and (1+29.28%) to derive its statistical lower-bound and upper-bound, respectively, and denote them for example by (LB-Delay, and UB-delay). For a given delay test threshold of a TSV, then we can report one of the following three test result:

- (1) If (delay test threshold > UB-delay), then it is a pass.

- (2) If (delay test threshold < LB-delay), then it is a fail.

- (3) Otherwise, we can report it as a conditional fail, depending on the system-level test requirement and decision.

V. CONCLUSION

Knowing the delay across each TSV in a 3D IC is often a plus to assist either the silicon debugging during the yield ramp-up stage or to catch small delay faults that may escape the at-speed test during the mass-production stage. We propose in this paper a VOT (Variable Output Thresholding) technique that can be used to gauge the delay across a TSV even when it has serious resistive open defect. SPICE simulation validates that there indeed exists a linear relationship between the TSV delay to be quantified and the quantity that can be measured (denoted as ΔT_{ST} by the proposed scheme). With 10,000 times of Monte-Carlo simulation, the average absolute error is 5.85% and maximum is 29.28%. It implies that this scheme is able to detect a small TSV delay that causes more than 29.28% deviation from its original fault-free value with a confidence level of 99.99%.

ACKNOWLEDGMENTS

This work was supported in part by National Science Council (NSC) of Taiwan under grant NSC-99-2220-E-007-009, Mentor Graphics, and ITRI. The authors are also grateful to CIC (Chip Implementation Center) of Taiwan, for the provision of the commercial EDA tools used in our experiments.

REFERENCES

[1] A. Bassi, A. Veggetti, L. Croce, and A. Bogliolo, "Measuring the Effects of Process Variations on Circuit Performance by Means of Digitally Controllable Ring Oscillators," *Proc. of Int'l Conf. Microelectronic Test Structures*, pp. 214–217, Mar. 2003.

[2] P.-Y. Chen, C.-W. Wu, and D.-M. Kwai, "On-chip TSV Testing for 3D IC Before Bonding using Sense Amplification," *Proc. of IEEE Asian Test Symp.*, pp. 450–455, Nov. 2009.

[3] B. P. Das, B Amrutur, H. S. Jamadagni, N. V. Arvind, and V. Visvanathan, "Within-Die Gate Delay Variability Measurement using Re-configurable Ring Oscillator," *Proc. of IEEE Custom Integrated Circuits Conf. (CICC)*, pp.133–136, Sept. 2008.

[4] L. S. Dutta and T. Hillmann-Ruge, "Application of Ring Oscillators to Characterize Transmission Lines in VLSI Circuits," *IEEE Trans. Components, Packaging, Manufacturing Tech., Part. B.*, vol. 18, no. 4, pp. 651–657, Nov. 1995.

[5] G. Katti, M. Stucchi, K. De Meyer, and W. Dehaene, "Electrical Modeling and Characterization of Through Silicon Via for Three-Dimensional ICs," *IEEE Trans. Electron Devices*, vol. 57, no. 1, pp. 256-262, Jan. 2010.

[6] I. Loi, S. Mitra, T. H. Lee, S. Fujita, and L. Benini. "A Low-Overhead Fault Tolerance Scheme for TSV-Based 3D Network on Chip Links," *Proc. of Int'l Conf. Computer-Aided Design*, pp. 598–602, Nov. 2008.

[7] K. S.-M. Li, C. L. Lee, C. Su, and J. E. Chen, "Oscillation Ring based Interconnect Test Scheme for SoC," *Proc. of IEEE Asia South Pacific Design Automation Conf. (ASP-DAC)*, pp. 184–187, 2005.

[8] K. S.-M. Li, C.-L. Lee, C.-C. Su, Y.-M. Chang, and J.-E Chen, "IEEE Standard 1500 Compatible Sscillation Ring Test Methodology for Interconnect Delay and Crosstalk Detection," *J. of Electronic Testing: Theory and Applications*, vol. 23, no. 4, pp. 341–355, Aug. 2007.

[9] E. J. Marinissen and Y. Zorian, "Testing 3D Chips Containing Through-Silicon Vias," *Proc. of Int'l Test Conf.*, pp.1-11, 2009.

[10] P. R. O'Brien and T. L. Savarino, "Modeling the Driving-Point Characteristic of Resistive Interconnect for Accurate Delay Estimation," *Proc. of Design Automation Conf.*, pp. 512–515, Nov. 1989.

[11] C. C. Su, Y. T. Chen, M. J. Huang, G. N. Chen, and C. L. Lee, "All Digital Built-In Delay and Crosstalk Measurement for On-Chip Buses," *Proc. of Design, Automation & Test in Europe Conf. (DATE)*, pp. 527–531, March 2000.

[12] W. C. Wu, C.-L. Lee, M. S. Wu, J. E. Chen, and M. Abadir, "Oscillation Ring Delay Test for High Performance Microprocessor", *J. of Electronic Testing: Theory and Applications (JETTA)*, vol. 16, no. 1–2, pp. 147–155, 2000.

[13] C. Xu, H. Li, R. Suaya, and K. Banerjee, "Compact AC Modeling and Analysis of Cu, W, and CNT Based Through-Silicion Vias (TSVs) in 3D ICs," *Tech. Digest of Int'l Electron Devices Meeting (IEDM)*, pp. 521-524, (2009).

[14] J.-W. You, S.-Y. Huang, D.-M. Kwai, Y.-F. Chou, and C.-W. Wu, "Performance Characterization of TSV in 3D IC via Sensitivity Analysis," *Proc. of Asian Test Symp. (ATS)*, pp. 389–394, Dec. 2010.

Circuit and System Design Guidelines
for Ultra-Low Power Sensor Nodes

Yoonmyung Lee, Yejoong Kim, Dongmin Yoon, David Blaauw, Dennis Sylvester

University of Michigan, Ann Arbor

{sori, yejoong, dmyoon, blaauw, dmcs}@umich.edu

ABSTRACT

Designing an ultra-low power sensor node requires careful consideration of the system-level energy budget. Depending on applications, various components can dominate total energy. In this paper, we review three different system energy budget scenarios where any of the microprocessor, memory, and timer of a sensor node can dominate the energy budget. The design space and corresponding trade-offs for these three components are explored to suggest guidelines for the design of ultra-low power sensor nodes.

Categories and Subject Descriptors

B.7.1 [Types and Design Styles]

General Terms

Design

Keywords

Ultra-low Power, Wireless Sensor Node, Low Voltage

1. INTRODUCTION

Sensor nodes collect useful data from their environment in a distributed fashion and process the collected raw data to extract essentials to be transferred to user or central data collection point. Recently, with advances in circuit design, packaging and battery technologies, tiny sensor nodes whose volume is as small as a few cubic-millimeters are demonstrated [1].

Figure 1 shows key building blocks of ultra-low power sensor nodes. Sensor nodes typically consist of sensors for measurement, a microprocessor for data processing, a memory for data and execution code storage, a radio for data transfer, a timer for wake up and communication control, and a power management unit. Although sensor nodes commonly includes these function units, each function unit's frequency or length of activation can largely vary from applications to applications which makes the system energy budget varies significantly. Other function units, such as wake up receiver, energy harvesting unit can be added for more complete system, but scope of discussion in this paper will be limited to the key function units illustrated in Figure 1 for simplicity.

Table 1 shows an example energy/power consumption of function blocks in state-of-the-art literatures [1]. Table 2 shows usage models of these function blocks in three different scenarios. Scenario 1 represents a surveillance application. A sensor node wakes up every 5 seconds and takes an image of a monitored object. The image is then processed with microprocessor (MP) to determine if there was significant

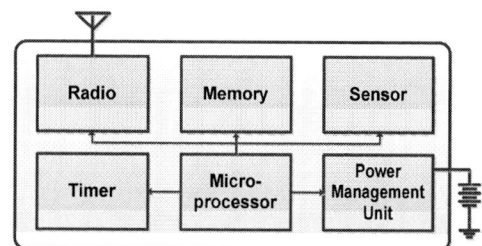

Figure 1. Simplified block diagram of a typical sensor node.

change from past images. The result is then transferred to base station. For image processing, a large (16kB) memory (compared to the other scenarios) has to be implemented in this scenario. Scenario 2 represents a temperature monitoring application where a sensor node wakes up every 10 minute and measures temperature periodically. Measured temperature is stored in memory and transferred to a base station hourly. Due to the small data size, a smaller 1kB memory can be used for this scenario. Scenario 3 is temperature monitoring sensor identical to Scenario 2, except that measured data is collected by symmetric communication among sensor nodes instead of centralized base station.

With power/energy numbers given in Table 1 and usage models in Table 2, energy usage in each scenario can be estimated as shown in Figure 2. With Scenario 1, which is relatively computationally intensive, the microprocessor dominates system energy. Therefore, optimizing the microprocessor for minimum energy operation is the key approach for minimizing overall energy consumption and maximizing lifetime of the sensor node.

With sensor nodes that wake up infrequently and spend most of their time in standby mode, memory leakage power dominates total energy as shown in Scenario 2. This is because, unlike other function blocks, memory cannot be power-gated in standby mode for minimizing leakage power since measurement data and execution code stored in the memory will be lost with power-gating. Therefore, to design a sensor node with infrequent

Table 1. Example energy/power consumption of function blocks in ULP sensor nodes.

Unit	Description	Energy /Power	Ref.
Microprocessor	Commercial ARM® Cortex-M3	29 pJ/inst.	[2]
Memory	8T SRAM (65nm)	14 pW/bit	[1]
Timer	Gate-leakage timer including LDO	8.6 nW	[1]
Radio – TX	ULP UWB transmitter energy	1.65 nJ/bit	[3]
Radio – RX	ULP UWB receiver power	1.64 mW	[3]
Image Sensor	128×128 pixel low power image sensor	140 nJ/frame	[4]
Temperature Sensor	PTAT-based	806 nJ/meas.	[1]

Table 2. Example usage model of ULP sensor nodes.

Unit	Scenario 1 (Surveillance)	Scenario 2 (Temp. monitor)	Scenario 3 (Temp. monitor /w symmetric communication)
(wake up period)	5 sec	10 min	10 min
Microprocessor	100k inst. / wake up	2k inst. / wake up	2k inst. / wake up
Memory	16kB, 229nW	1kB, 14.3nW	1kB, 14.3 nW
Timer	Not used	Not used	8.6 nW
Radio – TX	1 kb / wake up	100 b / 1 hr	100 b / 1 hr
Radio – RX	Not used	Not used	1.64mW for mismatch
Image Sensor	1 frame / wake up	Not used	Not used
Temperature Sensor	Not used	1 meas. / wake up	1 meas. / wake up

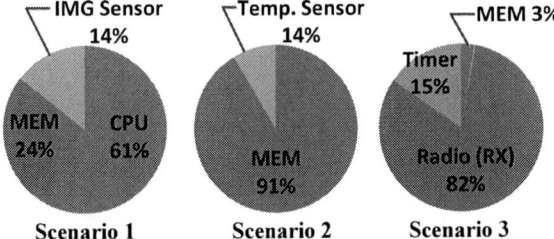

Figure 2. Energy consumption by functional unit in example sensor node scenarios.

activity, minimizing memory leakage with careful selection on memory topology is required.

Sensor nodes are required to form a wireless sensor network in some applications, which requires symmetric communication among the sensor nodes. Scenario 3 shows how energy budget could change with such sensor nodes. With an ULP sensor node, symmetric radio communication should be periodically activated since its energy budget does not allow continuous radio activation. This radio activation timing has to be precisely determined by a low power timer. If the activation time determined by one sensor node is different from the other sensor node, receiving sensor node has to activate radio receiver for mismatched amount of time until transmitter is activated. Due to the high receiver power (in the order of mW), energy wasted for receiver activation during this mismatch can dominate energy budget of this types of sensor nodes as shown in Figure 2. Therefore, accuracy and power trade-off of the timer should be carefully explored for designing sensor nodes with symmetric communication.

These three sensor node usage scenario confirms that function block which dominates the energy can vary. Once dominating function block is determined at system level, circuit level design optimization has to be done on the dominating function blocks. Therefore, in this paper, design guidelines and trade-offs for three different function blocks, namely – microprocessor, memory and timer – are discussed at the circuit level.

2. MICROPROCESSOR DESIGN CONSIDERATIONS IN ULP SENSOR NODES

For sensor nodes where microprocessor dominates system energy budget, optimizing operation conditions of microprocessor is of critical concern to reduce energy consumption. There are two key knobs that can be adjusted in

design time, namely supply voltage scaling [7] and fabrication technology selection [8].

2.1 Supply Voltage

Voltage scaling is a popular approach for microprocessor energy reduction [5][6]. Although voltage scaling incurs a performance penalty, typical sensor node performance requirements aremuch lower than that of commercial microprocessors shown in [5][6], allowing voltage scaling to the near-threshold or sub-threshold regions. However, excessive voltage scaling can actually increase energy consumption. Therefore, for sensor nodes whose microprocessor takes significant portion of system energy budget, minimum energy supply voltage (V_{min}) [7] should be carefully determined.

Microprocessor energy consumption can be represented as the sum of switching (E_{switch}) and leakage energy (E_{leak}) [EQ1]. Since switching energy is proportional to the square of supply voltage [EQ2], voltage scaling enables a quadratic E_{switch} reduction.

$$E_{Total} = E_{switch} + E_{leak} \qquad [EQ1]$$
$$E_{switch} \propto V_{DD}^{2} \qquad [EQ2]$$

E_{leak} can be represented as product of execution time (T) and leakage current (I_{leak}) of the microprocessor [EQ3]. In the super-threshold regime, E_{leak} has only a minor impact on overall energy consumption. However, as voltage scaling is extended towards the sub-threshold regime, the exponential reduction of on-current incurs an exponential increase in execution time (T), making E_{leak} non-negligible compared to the quadratically reduced E_{switch}. Therefore, with excess voltage scaling, the quadratic E_{switch} reduction can be outweighed by the exponential increase in E_{leak}.

$$E_{leak} = T \times I_{leak} \qquad [EQ3]$$

Figure 3 shows the simulated energy–supply voltage relationship for a single transition in a 51 stage inverter chain. The minimum energy operation point is approximately 220mV where the increase in E_{leak} overcomes the reduction of E_{switch}.

Meanwhile, sensor node microprocessors can have various architectures, and even with identical architecture, execution code varies across applications, making activity factor vary. Some instructions, such as multiplication/division, demand high circuit activity whereas simple instructions, such as register reads, do not. Therefore, computationally intensive applications can increase activity factor (or vice versa), andthis activity factor should be considered when determining the minimum energy

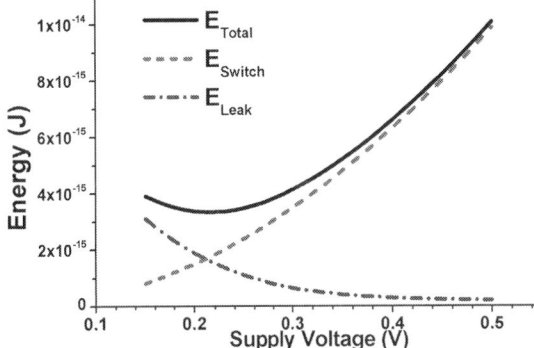

Figure 3. Energy consumption as a function of supply voltage

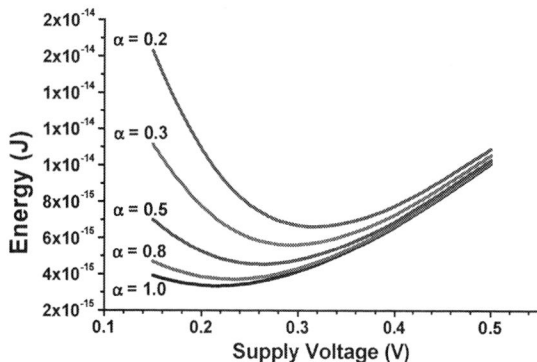

Figure 6. Energy vs Vdd for an inverter chain.

operation voltage (V_{min}). Figure 4 shows the impact of activity factor on V_{min} varies . With a higher activity factor, microprocessor energy consumption is more dominated by E_{switch}, making quadratic switching energy reduction more effective down to lower voltages. Therefore, V_{min} is lower with high activity factor.

In summary, sensor node microprocessors should leverage voltage scaling as far as performance requirements allow while comprehending the limitations posed by the minimum energy point.

2.2 Technology Selection

The previous section showed that the minimum energy operating voltage can be used for microprocessor designs. However, this discussion was limited to the time when the microprocessor is active. It also assumed that an arbitrary microprocessor frequency can be chosen for minimum energy operation. In practical sensor node design, standby time energy consumption should be considered. Moreover, there may be specific performance requirements. By carefully selecting fabrication technology, these two factors can be considered in microprocessor design.

Microprocessor energy consumption is largely affected by two key metrics – duty cycle and performance. Duty cycle is defined as the ratio of active time to total time. To first order performance is simply the clock frequency that a microprocessor runs at – we model this using 40 fan-out-of-4 (FO4) inverter delays in this discussion, which represents a typical low voltage processor microprocessor cycle time [8]. Using these two metrics, the design space for a microprocessor in a sensor node is depicted in Figure 5. Microprocessor for continuous monitoring, such as biosignal monitoring [9], have high duty cycles (bottom of Figure 5), whereas microprocessors that wake up infrequently – every 10 minutes or longer –have low duty cycles (top of Figure 5) [10]. Sensor nodes with temperature or pressure sensors would not require high microprocessor performance (left of Figure 5) while sensor nodes with imaging capabilities require higher performance (right of Figure 5).

Microprocessor energy consumption in a sensor node can be represented by [EQ4].

$$E_{total} = E_{active} + E_{standby} \qquad [EQ4]$$
$$= E_{switch} + E_{leak} + E_{standby}$$

E_{active} denotes the energy consumed when the microprocessor is active. Therefore, E_{active} consists of energy consumed by switching activity (E_{swith}) and energy consumed by leakage

current during active mode (E_{leak}). $E_{standby}$ is energy consumed in standby mode. With advanced technologies, E_{switch} can be reduced with smaller switching capacitance lower supply voltage for given performance requirement. However, due to lower threshold voltage (V_{th}) of transistors, E_{leak} and $E_{standby}$ can increase significantly. To study impact of technology selection, a chain of 40 of FO4 inverters can be simulated in SPICE for 5 different commercial CMOS technologies ranging from 250nm to 65nm. Activity ratio of microprocessor circuit is assumed to be 0.2. Supply voltage is set to meet the required performance and if the performance can be achieved at minimum energy operation voltage (V_{min}) [5], the voltage is not scaled down below V_{min}.

Figure 6 shows the technology choice for the energy optimal operation of a microprocessor. There are two clear trends in the presented technology selection analysis. First, microprocessors with small duty cycles prefer old technologies due to their high device V_{th} and low leakage current. Second, high performance microprocessors prefer new technologies, due to improved E_{switch} with lower gate/parasitic capacitance and lower supply voltage for given performance requirement. These two trends are most clear with low duty cycle/low performance (top-left of Figure 6) and high duty cycle/high performance (bottom-right) microprocessors.

Relatively old technologies are optimal in microprocessors with high duty cycle and low performance (bottom left of Figure 6). Newer technologies are expected to be preferred for high duty cycles to achieve energy optimality. However, faster speed of advanced technology can lead to task completion far ahead of required microprocessor cycle time, which results in unwanted additional idle time and waste of leakage energy. This energy penalty offsets the benefit of low E_{switch} with advanced technology and makes older technologies a better choice.

On the opposite corner of the design space in Figure 6 (low duty cycle with high performance requirements), older technologies offer reduced leakage. However, the larger supply voltage needed to meet performance requirements lead to higher active energy and overall newer technologies are preferable in such scenarios.

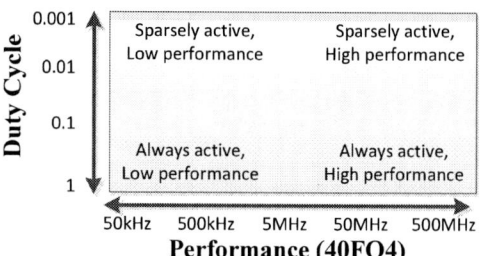

Figure 4. Sensor node design space.

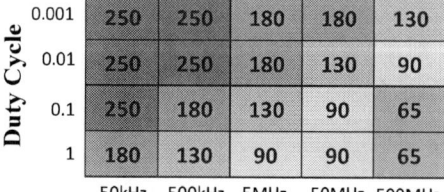

Figure 5. Optimal technology selection for microprocessor in a sensor node [8]

3. MEMORY DESIGN CONSIDERATIONS IN ULP SENSOR NODES

As wireless sensor nodes require collected data to be stored until being transmitted or processed, memories as data storage are essential elements in the system. Memories, by their nature, have a low activity factor; only a few percent of the circuit is active at a time. In addition, most wireless sensor nodes are duty-cycled and some of them spend ≥90% of the lifetime in the sleep mode. In this context, standby energy consumption dominates the memory's total energy consumption and becomes the critical factor. Non-volatile memories (e.g., Flash) have virtually zero retention (standby) power, but their active power is typically in ~mW range, which far exceeds the energy budget of smaller battery-powered sensor nodes. Thus, embedded volatile memories (e.g., SRAM, eDRAM) are preferred in such systems, however their volatility leads to significant leakage power during standby mode. Generally, smaller devices contribute to a smaller bitcell size and more leakage power. Because the bitcell array leakage power dominates the memory standby power, there exist trade-offs between the standby power and the bitcell size. These can be clearly observed in low-power memories recently reported, which we are grouping them into three as shown in Figure 7: Regular SRAMs, Sub-V_{TH} SRAMs and eDRAMs, and ULP SRAMs.

Regular SRAMs typically consume >100pW/bit of standby power, easily exceeding 10μW with just 1kB of capacity. However, their bitcell size is usually smaller than that of other types of SRAM, and the operating frequency is much faster with full supply voltages, resulting in up to >300MHz [12]. This type of SRAM is suitable for sensor nodes that require fast operations as well as a large amount of storage, with a relatively large battery size that can afford the large power consumption; both active and static powers can be more than 100μW.

Sub-V_{TH} SRAMs reduce standby power down to ~1pW/bit using low supply voltages, hence allowing smaller batteries . However, their robustness requirements, such as write margins and read margins, usually cannot be met with the conventional 6T bitcell structure due to the reduced supply voltage. A typical approach to overcome this issue involves the separation of read and write paths, which requires additional transistors. Because of these added devices, such bitcells incur 50~100% area overhead compared to regular SRAMs. 10T bitcells [14] [15] from recent literature are shown in Figure 8(a) and (b); the bitcell in (a) exploits stack effect for added leakage reduction, while the bitcell in (b) supports fully differential read and write. The area overhead could be avoided at the cost of multiple supply voltages with high power active operations [17]. For the applications requiring large amount of memory capacity, embedded DRAMs (eDRAM) [18] can be an alternative choice. They have much smaller bitcell size compared to SRAMs, as shown in Figure 8(c), resulting in much denser memory. However, the refresh power greatly depends on temperature, so the standby power may easily become orders of magnitude higher than that of SRAMs at high temperature. These sub-V_{TH} SRAMs and eDRAMs are in the middle of the spectrum (Figure 7) and can be useful for sensor nodes with a relatively small battery size (~cm³) that do not require high-speed memory access.

At the other end of the spectrum is ULP SRAM. These often use thick-oxide HVT I/O devices, achieving less than 10fW/bit of

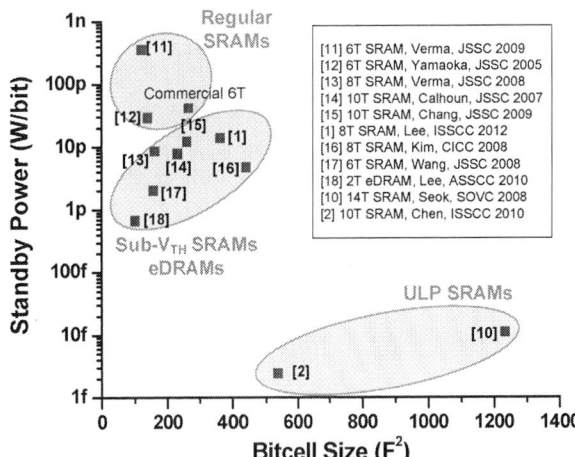

Figure 7. Low-power memories in recent publications can be grouped into three classes: Regular SRAMs, Sub-V_{TH} SRAMS and eDRAMs, and ULP SRAMs. Depending on application, designers should choose among these memory circuits.

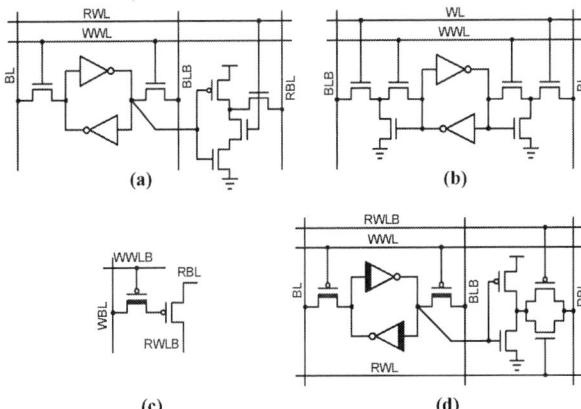

Figure 8. (a) 10T SRAM bitcell [14], (b) 10T SRAM bitcell [15], (c) 2T eDRAM Gain Cell [18], (d) 10T SRAM bitcell [2]. Thick lines indicate HVT (I/O) devices.

standby power. This can be attractive for extremely energy-constrained systems; for example, the entire 3kB ULP SRAM in [2], whose bitcell is shown in Figure 8(d), consumes only <100pW of standby power, and this enables the use of a mm³-scale battery while still providing months of lifetime. Although their operating frequency is usually limited to < 1MHz due to the combination of HVT I/O devices and the low voltage supply, they are preferred in mm³-scale sensor nodes because low-power consumption has the highest priority in such systems. However, the extremely low standby power comes with a large area penalty; the 10T HVT bitcell [2] in Figure 8(d) is ~3× larger than regular SRAMs due to the large size requirements of the HVT devices. This limits the maximum memory capacity in small form-factor systems.

4. TIMER DESIGN CONSIDERATIONS IN ULP SENSOR NODES

4.1 Role of Accurate Timers

The circuit blocks introduced in previous sections can benefit from various circuit techniques to reduce the power consumption. However, unlike those blocks, a radio transceiver block cannot operate at such low power mainly due to physical limitations of antenna efficiency and size. Since allocating substantial area for the antenna is usually not possible for ULP wireless sensor applications, the average active power of the radio of a sensor node is still in mW regime, which is 5 to 6 orders of magnitude higher than the average power of other sensor node components [20]. Therefore the operating duration of the radio transceiver block has to be limited to minimize overall system's energy consumption. To make designing the system harder, the level of instantaneous power from the radio transceiver cannot be afforded directly from power source for some applications. To cope with the problem, the system has to have a large capacitor to act as energy reservoir. Therefore the radio transceiver's operating duration not only has significant impact on total energy consumption but may also increase system volume. That is why the system has to be designed to minimize the operating duration of the transceiver. As will be shown in the next paragraph, the accuracy of a timer has substantial impact on average operating duration of the radio transceiver.

Figure 9 shows an example scenario of an ULP wireless sensor node system. A sensor node is activated every 20 minutes to take measurement and a microprocessor processes data for 100ms at 3µW. Then once in every hour, collected data is transmitted by radio module, consuming 1mW for 1ms. For any given time, timer is running in background at 1nW to track time change and initiate scheduled tasks at right timing. In total, the sensor node nominally consumes 5.5µJ per hour.

However, some applications require sensor nodes to communicate with each other. In such a scenario, each sensor is equipped with a timer to initiate communication synchronously. Unfortunately, the timers may suffer from mismatch, leading to one sensor node turning on the radio earlier than others. With 200ms hourly mismatch , the total energy consumption increases to 206µJ per hour, with 200µJ arising due to the mismatch penalty.

4.2 Design Considerations

From this example, it is clear that sensor nodes require accurate timers. However several considerations should be made in the design of such timers.

The first is the power consumption of the timer module itself. Unlike other system components, the timer is always on and cannot benefit from duty cycling. For this reason, even if the timer consumes several orders of magnitude lower power than other duty-cycled high power components such as microprocessor, it may easily dominate overall energy consumption. The second one is random error in timing. The timer will suffer from various sources of noise, such as thermal noise, which leads to random mismatch of period between different timers. Finally, process, voltage, and environmental (PVT)fluctuations will impact the timer, similar to conventional circuits. PVT sensitivity must bebe distinguished from random error. Unless the sensors are used in harsh environments, variation sources such as process or temperature are likely to act as fixed offsets between different timers, rather than manifesting as random error. With the help of simple (negligible power) digital logic, fixed offsets can be calibrated to minimize their effect. Therefore, more emphasis can be allocated to the power consumption and random error characteristics of a given timer design.

Before introducing real design examples, we introduce the concept of Allan deviation [21], the standard metric used to compare the frequency stability performance of timers. Period

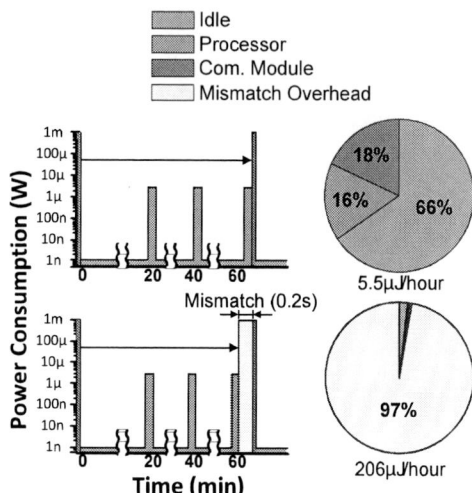

Figure 9. Example wireless sensor node operation scenario

deviation due to noise is often Gaussian and its magnitude of deviation can be characterized with RMS jitter, which is defined as the standard deviation of timer period. To compare the jitter of two different types of timers, normalized RMS jitter (expressed in ppm) is often used. However, the accuracy of a timer for measuring a multi-cycle synchronization period is not captured by these metrics since they ignore the averaging of jitter over multiple timer cycles. For example, two timers with identical ppm jitter can have different hourly accuraciessince the timer with a shorter period has more cycles to count, allowing for more averaging. In addition, long term stability estimation based on RMS jitter may give a false impression that longer averaging periods are always beneficial, which is refuted by measurements. Instead, timer uncertainty is well characterized by Allan deviation, which is defined by [EQ5], where τ is the observation period and y_n is n-th fractional frequency – that is, Δf/f for Figure 10, where f is frequency – over observation time τ. Bracket <> denotes time average of samples.

$$\sigma_y^2(\tau) = \frac{1}{2}\langle(y_{n+1} - y_n)^2\rangle, \quad \sigma_y(\tau) = \sqrt{\sigma_y^2(\tau)} \qquad [EQ5]$$

4.3 Design Examples

Here we briefly describe three recent ultra-low power timers [22-24]. Trade-off exists among three key parameters discussed in previous section, namely power consumption, random error,

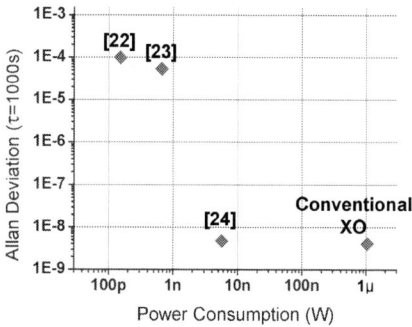

Figure 10. Allan deviation at τ=1000s and power consumption of low power timers [22-24]

and PVT variation. Figure 10 plots Allan deviation at averaging time of 1000 seconds and power consumption of the three designs. Please note that conventional XO means a 32.768kHz Pierce oscillator built with a discrete inverter running at 1V. The performance summary of the three timers is shown in Table 3. CMOS based circuits have lower power consumption at smaller area. However, the circuits need to oscillate at low frequency since its oscillation is not based on a resonant component. Inevitably, CMOS-based timers have worse performance with random error and PVT variation compared to a quartz crystal based timer. Quartz crystal based timers – commonly denoted as crystal oscillator (XO) – on the other hand, show much better frequency stability characteristics. These merits come at the expense of higher power and extra system volume due to the quartz crystal. The smallest 32.768kHz quartz crystal package available at the date of publication is 1.44mm^3 [27]. There are also MEMS-based resonators available that can be integrated on-chip while providing better accuracy than CMOS circuit-based timers [25, 26]. Power consumption of these approaches is still higher than other timers, but the numbers are expected to reduce with advances in MEMS and circuit techniques.

If the application does not require communication between sensor nodes, or synchronization period is short, CMOS based timers can save power and system volume. On the other hand, if timing has to be tightly controlled, a quartz crystal can be a good option.

Table 3. Low power timers for ULP sensor nodes

	[22]	[23]	[24]	[25]	[26]
Oscillation source	CMOS circuit	CMOS circuit	Quartz crystal	MEMS	MEMS
Frequency	≈11Hz	3.7~16Hz	32.768kHz	32.768kHz	32.768kHz
Power	150pW	660pW	5.58nW	600nW	3.2µW
σ_y	9.89×10^{-5} (τ=1400s)	5.28×10^{-5} (τ=1400s)	1.16×10^{-8} (τ=1000s)	N/A	N/A
V_{DD} sensitivity	400ppm/mV	4200ppm/mV	17.6ppm/V	N/A	<1ppm/V
Temp. stability	490ppm/°C	31ppm/°C	0.04ppm/°C^2	± 80ppm (-20°C-80°C)	± 10ppm (0°C-50°C)
Area (mm^2)	0.019	0.015	0.3 (w/o crystal)	2.25	0.156

5. CONCLUSIONS

In this paper, we have outlined guidelines for designing an ULP sensor node system. The energy budget of an ULP sensor node can vary significantly depending on application. Therefore, functional blocks that dominate system energy should be assessed early on and the design space for dominating blocks should be explored to balance and minimize the total system energy.

For microprocessors, voltage scaling down to minimum energy operation voltage (V_{min}) and technology selection for optimizing microprocessor design in consideration of performance requirement and duty cycle is discussed. Various memory topologies are presented to show trade-off between standby power and bitcell area overhead. State-of-the-art low power timers suitable for ULP sensor nodes are also discussed to illustrate the accuracy versus power trade off.

6. REFERENCES

[1] Y. Lee, et. al., "A Modular 1mm^3 Die-Stacked Sensing Platform with Optical Communication and Multi-Modal Energy Harvesting," *IEEE ISSCC*, 2012.

[2] G. Chen, et. al., "Millimeter-Scale nearly Perpetual Sensor System with Stacked Battery and Solar Cells," *IEEE ISSCC*, 2010.

[3] M. Crepaldi, et. al., "An Ultra-Low-Power Interference-Robust IR-UWB Transceiver Chipset Using Self-Synchronizing OOK Modulation," *IEEE ISSCC*, 2010.

[4] S. Hanson, et. al., "A 0.45-0.7V sub-microwatt CMOS image sensor for ultra-low power applications," *IEEE Symp. on VLSI Circuits*, 2009.

[5] Intel XScale. http://www.intel.com/design/intelxscale/

[6] IBM PowerPC. http://www.chips.ibm.com/products/powerpc/

[7] B. Zhai, et. al., "Theoretical and Practical Limits on Dynamic Voltage Scaling", *ACM/IEEE DAC*, 2004.

[8] M. Seok, et. al., "Optimal Technology Selection for Minimizing Energy and Variability in Low Voltage Applications," *ACM/IEEE ISLPED*, 2008.

[9] A. Dogan, et. al., "Multi-Core Architecture Design for Ultra-Low-Power Wearable Health Monitoring Systems," *ACM/IEEE DATE*, 2012

[10] M. Seok, et. al., "The Phoenix Processor: A 30pW Platform for Sensor Applications," *IEEE Symp. on VLSI Circuits*, 2008.

[11] N. Verma, et. al., "A High-Density 45nm SRAM Using Small-Signal Non-Strobed Regenerative Sensing," *IEEE JSSC*, vol. 44, no. 1, Jan. 2009.

[12] M. Yamaoka, et. al., "A 300-MHz 25-µA/Mb-Leakage On-Chip SRAM Module Featuring Process-Variation Immunity and Low-Leakage-Active Mode for Mobile-Phone Application Processor," *IEEE JSSC*, vol. 40, no. 1, Jan. 2005.

[13] N. Verma, et. al., "A 256kb 65nm 8T Subthreshold SRAM Employing Sense-Amplifier Redundancy," *IEEE JSSC*, vol. 43, no. 1, Jan. 2008.

[14] B. H. Calhoun, et. al., "A 256-kb 65-nm Sub-threshold SRAM Design for Ultra-Low-Voltage Operation," *IEEE JSSC*, vol. 42, no. 3, Mar. 2007.

[15] I. J. Chang, et. al., "A 32kb 10T Sub-Threshold SRAM Array with Bit-Interleaving and Differential Read Scheme in 90nm CMOS," *IEEE JSSC*, vol. 44, no. 2, Feb. 2009.

[16] T. Kim, et. al., "A Voltage Scalable 0.26V, 64kb 8T SRAM with V_{min} Lowering Techniques and Deep Sleep Mode," in *IEEE CICC*, 2008.

[17] Y. Wang, et. al., "A 1.1GHz 12µA/Mb-Leakage SRAM Design in 65nm Ultra-Low-Power CMOS Technology with Integrated Leakage Reduction for Mobile Applications," *IEEE JSSC*, vol. 43, no. 1, 2008.

[18] Y. Lee, et. al., "A 5.42nW/kB Retention Power Logic-Compatible Embedded DRAM with 2T Dual-Vt Gain Cell for Low Power Sensing Application," *IEEE ASSCC*, 2010.

[19] G. Chen, et. al., "A Cubic-Millimeter Energy-Autonomous Wireless Intraocular Pressure Monitor," *IEEE ISSCC*, 2011.

[20] Y. Lee, et. al., "Synchronization of ultra-low power wireless sensor nodes." *IEEE MWSCAS*, 2011.

[21] D.W. Allan, "Time and frequency (time-domain) characterization, estimation, and prediction of precision clocks and oscillators," *IEEE Transactions on Ultrasonics, Ferroelectrics, and Frequency Control*. Vol. UFFC-34, No. 6, pp. 647-654, Nov. 1987.

[22] Y. Lin, et. al., "A 150pW program-and-hold timer for ultra-low-power sensor platforms," *IEEE ISSCC*, 2009

[23] Y. Lee, et. al., "A 660pW multi-stage temperature compensated timer for ultra-low-power wireless sensor node synchronization," *IEEE ISSCC*, 2011.

[24] D. Yoon, et. al., "A 5.58nW 32.768kHz DLL-assisted XO for real time clocks in wireless sensing applications," *IEEE ISSCC*, 2012.

[25] J. Chang, et. al., "32kHz MEMS-based oscillator for implantable medical devices," *International Symposium on Integrated Circuits*, 2011

[26] D. Ruffieux, et. al., "Silicon resonator based 3.2µW real-time clock with ±10 ppm frequency accuracy," *IEEE JSSC*, vol. 45, no. 1, 2010

[27] Fox Electronics, 2010. FX122.
http://www.foxonline.com/pdfs/FX122.pdf

Design Exploration of Energy-Performance Trade-Offs for Wireless Sensor Networks

Ivan Beretta[1], Francisco Rincon[2], Nadia Khaled[3],
Paolo Roberto Grassi[4], Vincenzo Rana[4,1], David Atienza[1]

[1] ESL, École Polytechnique Fédérale de Lausanne, Switzerland, {ivan.beretta, david.atienza}@epfl.ch
[2] DACYA, Universidad Complutense de Madrid, Spain, francisco.rincon@fdi.ucm.es
[3] Bioanalytical Science Dept., Nestlé Research Center, Switzerland, nadia.khaled@rdls.nestle.com
[4] DEI, Politecnico di Milano, Italy, {grassi, rana}@elet.polimi.it

ABSTRACT

Wireless sensor networks (WNSs) are gradually evolving from a promising technology to a well-established reality in a large set of different domains. In order to fulfill the requirements of the specific scenario, a WSN must provide the right tradeoff between performance and lifetime, which is heavily determined by the network design. However, although the complexity of WSNs is increasing, the design space exploration is often carried out manually without the support of a general analytical methodology. In this paper, we advocate a model-based approach as an efficient and scalable way to explore the energy-performance tradeoffs during the design. In particular, we show that it is possible to define system-level models to describe wide classes of WSNs, providing a quick and accurate network evaluation. As a proof of concept, we propose a general model that describes the main characteristics of a class of WSNs for human health monitoring, and we apply it to a real case study. The results show that the energy-performance estimation error of the model never exceeds 1.74% compared to real data, while the evaluation time is reduced by up to 6 orders of magnitude with respect to an accurate network simulation.

Categories and Subject Descriptors

C.2.1 [**Computer Communication Networks**]: Network Architecture and Design; I.6.5 [**Simulation and Modeling**]: Model Development

General Terms

Design, Theory, Algorithms

Keywords

Wireless sensor networks, Cross-layer design, Model-based design, System-level modeling, Wireless body sensor networks

1. INTRODUCTION

In the last years, wireless sensor networks (WNSs) are becoming a well-established reality in many different domains, including military applications, environment control, industrial supervision and health monitoring [1]. In order to deal with the specific requirements of a given application domain, a WSN has to meet certain performance requirements as well as to guarantee a sufficient lifetime, which are often conflicting goals. The right tradeoff between these two objectives, as well as the prevention of undesired behaviors such as unbalanced performance among the different nodes of the WSN, can be guaranteed by accurately evaluating the network configurations during the design phase. In order to help the designer during the energy-performance tradeoff analysis, many design space exploration (DSE) techniques for WSNs have been proposed in the literature [3][4], and most of the classic optimization algorithms can also be adapted to WSNs with a low effort. However, providing such algorithms with an accurate system-level estimation of the WSN performance is still an open problem, and it is necessary to correctly lead the DSE algorithm to the detection of the Pareto-optimal network configurations.

The evaluation of a particular WSN includes aspects that span across multiple layers (from the network to the hardware, to the application level), and it can be performed in three ways [6]: a set of physical experiments, a network simulation or an analytical model. However, when a large number of configurations needs to be evaluated during the DSE phase, both the empirical experiments and the simulation become impractical, as the former cannot be automated, while the latter takes an unacceptable amount of time. Conversely, the analytical model enables a fast evaluation and a deep understanding of the dynamics of the network, but its definition raises several challenges related to its accuracy and reusability. In fact, a detailed characterization of a specific WSN has been shown to lead to efficient network designs [5], but such a model requires a deep knowledge about the application and the target platform, and it cannot be reused to model different classes of WSNs. On the other hand, a generic system-level model that can be easily instantiated to a specific WSN would greatly simplify the task of the designer, but no model with these characteristics has been proposed yet, as it is complex to define a characterization that can describe *all* the different classes of WSNs with a sufficient accuracy.

Although the definition of a general model is limited by

the great differences among the WSN domains, in this paper we aim at showing that it is possible to focus the scope of the model to wide classes of networks in order to capture their most relevant aspects, thus providing a model that is both detailed and reusable on many instances of WSNs. This work shows that a multi-layer characterization of the nodes and of their interactions in a well-defined class of WSNs leads to an accurate estimation with respect to both real and simulated data, and that a DSE algorithm greatly benefits from a model-based evaluation in terms of execution time. Furthermore, in order to provide a coherent system-level estimation of the network during the DSE, we propose a set of performance metrics that belong to different layers (i.e., delay, application quality, energy consumption), which lead to the determination of the optimal energy-performance tradeoffs. As a proof of concept, the model we propose targets the wide class of wearable wireless body sensor networks (WBSNs), which are a rising technology in the field of human health monitoring [2] both for medical and personal use. Experimental evaluations conducted on a real-world WBSN show that the proposed model never generates an estimation error greater than 1.74%.

2. RELATED WORK

Over the last years, model-based evaluation as a support for DSE has been extensively explored in many fields. In the WSN domain, node and network models have been traditionally proposed to characterize specific aspects of the network, and to validate new protocols [14] or energy management strategies [3]. None of them, however, guarantees a general system-level description that can be easily adapted to describe a real WSN.

At the node level, analytical models for all the most common hardware blocks were proposed even before the advent of WSNs. In particular, detailed energy characterizations are available for hardware circuits, sensors and microcontrollers [8], memory banks [7], and radio circuits [9][15]. However, these models do not consider any interdependency between the different parts of the system, hence they are not sufficient to describe the behavior of a set of networked nodes. In [3], the authors relate the energy consumption and the throughput of the node to the supply voltage of the microcontroller and the modulation level of the radio. Although the work is a good example of how different aspects of the node (i.e., sensing, processing and transmission) can be combined, the parameters that are considered are only a small subset of the ones that can be found on real nodes.

At the network level, several works propose a model of different media access control (MAC) protocols, and in particular the widely-adopted IEEE 802.15.4 [16] standard. For example, [10] characterizes the behavior of the IEEE 802.15.4 MAC layer on large-scale networks, both in terms of energy consumption and packet transmission probability. In [11], a similar analysis is proposed for WBSNs, with a particular emphasis on the radio activity of the node. The works in [17] and [18] focus on the part of the IEEE 802.15.4 standard that works in TDMA mode, and propose two separate techniques to estimate the expected packet delay. However, none of the aforementioned network models propose an in-depth analysis of the application executed by the nodes, which is crucial to have a coherent global evaluation of the WSN.

Another important aspect of the trade-off analysis is the definition of a set of metrics that capture all the relevant dy-

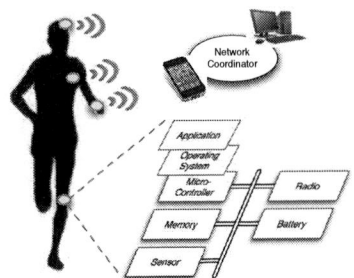

Figure 1: Overview of a typical WBSN

namics of the network. Traditionally, energy consumption is always a major concern during the network evaluation [14][3][10], but other metrics such as throughput and end-to-end delay may be considered. For example, different energy/delay tradeoffs are explored as a function of the voltage and the radio modulation in [26]. However, no application-related metric is generally proposed in order to characterize the overall behavior of the network as seen by the end user.

As a conclusion, none of the existing models provides a coherent system-level description that can be applied to real-world WSNs, mainly because they only focus on specific aspects of the system and often neglect the final application. In our work, we show that a general –and yet reliable– analytical model for the nodes and the whole network can be defined if its scope is limited to a set of WSNs sharing similar structures and application domains.

3. SYSTEM-LEVEL MODEL FOR WBSNs

In this section, we propose a general system-level model that characterizes the most important multi-layer interdependencies of the networks belonging to the same domain. As a proof of concept, we target the WBSN domain.

3.1 WBSN Domain Analysis

A typical WBSN [12] follows the structure illustrated in Figure 1. The network comprises a set of low-power nodes that can be worn by the same person or by different ones (e.g., the patients in a hospital, or a team of athletes) to monitor one or more vital signs to be sent to a central network coordinator. Once the signal has been sensed, each node performs a data pre-processing using a software application executed on a microcontroller-based hardware architecture (see Figure 1), and finally sends the output to the coordinator through the wireless channel. The coordinator is responsible for the analysis of the data, and the definition of the network activity (e.g., the enforcement of the MAC protocol). In WBSNs, a star topology network is generally employed, hence the communication between a node and the coordinator is direct [12]. Moreover, the wireless channel is shared among the nodes using a collision-free, time-division multiple access (TDMA) policy, which leads to a lower energy consumption with respect to a contention access. These assumptions are sufficient to characterize a wide set of networks in the WBSN domain, and they enable us to define an abstract model that can be easily adapted to real nodes and standards, as we show later by means of a case study.

3.2 Network Model

In the network model, we capture the interactions among

1044

the nodes and, in particular, how they share the wireless channel. For this purpose, the N nodes of the network are now considered as black boxes generating an output stream of ϕ_{out} bytes per second (B/s). The transmission is regulated by the MAC protocol, which aims at assigning a transmission interval $\Delta_{tx}^{(n)}$ (the index denotes that the quantity refers to node n) per second to each node, by acting on protocol-specific parameters that form a *configuration* χ_{mac}. Each node is then in charge of tuning the throughput $\phi_{out}^{(n)}$ in order to be able to deliver its data in the time $\Delta_{tx}^{(n)}$.

To describe the MAC layer, we introduce the following abstractions that capture its most recurring characteristics:

- a *data overhead* due to packetization and flow control, consisting of a number of extra bytes that are required to transmit ϕ_{out} (e.g., headers and tails). We indicate this overhead as $\Omega(\phi_{out}, \chi_{mac})$ (measured in B/s);

- a *control overhead*, which includes the control messages (e.g., synchronization packets and acknowledgements) that are exchanged between a node and the coordinator. As we further detail in Section 3.3, these messages generate an energy dissipation due to their transmission/reception. We identify the volume of control messages from the coordinator to the node and vice versa as $\Psi_{c \to n}(\chi_{mac})$ and $\Psi_{n \to c}(\chi_{mac})$ (measured in B/s);

- a *timing overhead* per second, i.e., time intervals where the channel is unavailable, either because of the transmission of control messages or because the network is kept idle. We call this quantity $\Delta_{control}(\chi_{mac})$;

- a *time discretization*. Since a protocol does not generally assign an arbitrary and continuous transmission time to each node. We define δ as the base time unit that is used in the selected protocol, and we express the transmission intervals as multiples of δ.

The goal of the network design is to size the transmission intervals to enable each node to deliver all its data and the corresponding control information. We model this as an assignment problem that is tailored for the typical star-topology TDMA transmission of WBSNs, but it can be also adapted to a contention access protocol (in fact, the $\Delta_{tx}^{(n)}$'s can be statistically determined as the average amount of time a node can successfully transmit per second, as shown in [19] for the CSMA/CA). In particular, the MAC protocol has to find a number $k^{(n)}$ for each node n such that:

$$\Delta_{tx}^{(n)} = k^{(n)} \cdot \delta \geq T_{tx}\left(\phi_{out}^{(n)} + \Omega\left(\phi_{out}^{(n)}, \chi_{mac}\right)\right), \quad (1)$$

where $T_{tx}(\cdot)$ denotes the transmission time required to send the specified amount of data, and depends on the physical radio. Additionally, the assignment of the transmission intervals by the MAC protocol must be constrained in order not to exceed the total of one second:

$$\sum_{i=1}^{N} \Delta_{tx}^{(i)} + \Delta_{control}(\chi_{mac}) = 1. \quad (2)$$

From the DSE perspective, allowing the network to stay silent for a long time leads to good solutions in terms of energy consumption, but in practice it increases the data delay. Hence, we define the *delay* function $d(\chi_{mac})$ to quantify the average (or the maximum) time between the generation of the data and the instant it is received by the coordinator. Such a function cannot be defined in the general case, but

it can be determined according to the specific MAC and the traffic patterns of the nodes, as we show in the case study.

3.3 Node Model

A typical WBSN node follows the microcontroller-based architecture shown in Figure 1. We hereby propose a model that captures the interdependency among the hardware components in terms of consumption and application-related metrics, as well as the influence of the network configuration on the single node. We characterize the node by means of a *configuration* χ_{node}, which includes the configurable parameters both on the hardware side (e.g., frequency, transmission power), and on the software side. All the parameters that cannot be tuned, or that are not relevant for a system-level optimization, will not be detailed in this model.

The node first samples the physiologic signal with a frequency f_s, and the samples are then quantized by an A/D converter to produce values of L_{adc} bytes, thus generating an input stream ϕ_{in} of $f_s \cdot L_{adc}$ (B/s). The sampling activity leads to an energy dissipation that can be expressed as:

$$E_{sensor} = E_{transducer} + [\alpha_{s,1} \cdot f_s + \alpha_{s,0}]. \quad (3)$$

A linear function of f_s (with coefficients $\alpha_{s,1}$ and $\alpha_{s,0}$) captures the behavior of the A/D circuit [21], whereas $E_{transducer}$ is an overhead included by the transducer.

The input stream ϕ_{in} is then processed by an application, which typically consists of filtering or data compression. The behavior of the application layer is determined by a set of parameters (e.g., approximation factors and compression ratios), which determine three key aspects:

- the *output stream* ϕ_{out}. From a quantitative perspective, the application can be modeled as a function h that processes the input stream ϕ_{in} and produces a certain amount of results to be transmitted. As a consequence, the output of the node is equal to $\phi_{out} = h(\phi_{in}, \chi_{node})$ and, if an estimation of the transmission errors is available (e.g., [9]), then the average amount of retransmitted data can be added to the original ϕ_{out};

- the *resource usage*. To represent these quantities, we define a vector $\mathbf{u} = (Duty_{app}, M_{app}, \gamma_{app}, u_4, ..., u_n)$ that contains n elements, one for each hardware resource that can be tuned on the target platform. We use a different notation to identify the duty cycle of the application on the microcontroller ($Duty_{app}$), the amount of memory required during the execution (M_{app}), and the number of memory accesses (γ_{app}), which will be used later for energy considerations. The resource usage depends on how the node is tuned (i.e., χ_{node}) and on the amount of data to be processed. Hence, we can define a function vector $\mathbf{k} = (k_1, ..., k_n)$ such that $\mathbf{u} = \mathbf{k}(\phi_{in}, \chi_{node})$, where $k_i(\phi_{in}, \chi_{node})$ computes the usage of resource i.

- the *output quality*. As the application generally introduces an approximation, we define an application-specific function $e(\phi_{in}, \chi_{node})$ that measures the loss of quality between the original and the transmitted data.

After characterizing the application, we focus on the effects of its execution. On the microcontroller side, the execution generates an energy dissipation that linearly depends on the duty cycle and on the operating frequency ($f_{\mu C}$) [21]:

$$E_{\mu C} = Duty_{app} \cdot [\alpha_{\mu C,1} \cdot f_{\mu C} + \alpha_{\mu C,0}]. \quad (4)$$

The execution also leads to an energy consumption due to memory access, which can be estimated as follows [7]:

$$E_{mem} = \gamma_{app}T_{mem} \cdot E_{acc} + (1 - \gamma_{app}T_{mem}) \, 8M_{app} \cdot E_{idle}^{bit}. \quad (5)$$

This equation includes two contributions: a dynamic consumption due to the γ_{app} memory accesses, and a residual that occurs during idle periods and is proportional to the memory size. In the equation, T_{mem} indicates the access time, E_{acc} defines the consumption of a single access, and E_{idle}^{bit} denotes the dissipation per bit due to leakage.

Finally, the output stream ϕ_{out} and the control information need to be transmitted to the coordinator by the radio unit during the assigned transmission intervals. The physical radio determines the transmission time in Equation (1) and the dissipation associated to the reception (E_{rx}) and the transmission (E_{tx}) of one bit, the latter being related to the power of the carrier signal [9], which must be chosen to achieve a low packet error rate. Thus, the energy consumption due to the radio can be expressed as:

$$E_{radio} = [8 \, (\phi_{out} + \Omega(\phi_{out}, \chi_{mac})) + 8\Psi_{n \to c}(\chi_{mac})] \cdot E_{tx}$$
$$+ 8\Psi_{c \to n}(\chi_{mac}) \cdot E_{rx} \,. \quad (6)$$

Then, after including the contribution of all the analyzed layers, the overall node consumption can be expressed as:

$$E_{node} = E_{sensor} + E_{\mu C} + E_{mem} + E_{radio} \,. \quad (7)$$

3.4 System-Level Evaluation Metrics

To complete the description of the WBSN, we combine the performance metrics of each node (i.e., E_{node} and $e(\phi_{in}, \chi_{node})$) into consistent network-level objective functions.

As mentioned in Section 1, finding balanced configurations is a major concern while combining the different metrics, in order to avoid situations where the coordinator receives data of different quality, or where heavily optimized nodes are alternated to other nodes with an insufficient lifetime. As a consequence, we define the *network-level energy consumption* (E_{net}) as a weighted combination of the average energy consumption of the nodes, and the sample standard deviation of this quantity over the WBSN:

$$E_{net} = \sum_{i=1}^{N} \frac{E_{node}^{(n)}}{N} + \vartheta \cdot \sqrt{\frac{1}{N-1} \sum_{i+1}^{N} \left[E_{node}^{(n)} - \sum_{i=1}^{N} \frac{E_{node}^{(n)}}{N} \right]^2}, \quad (8)$$

where ϑ is a positive constant that determines the importance of the balance among the nodes.

We can define a *network-level application quality* metric in a similar way, by combining all the loss-of-quality functions $e^{(n)}(\phi_{in}, \chi_{node})$ as we did in Equation (8) for $E_{node}^{(n)}$.

4. A REAL-WORLD WBSN CASE STUDY

In this section, we show that the proposed multi-layer model for WBSNs can be easily used to model a real network that uses a commercial platform and widespread standards.

4.1 Case Study Overview

We propose an illustrative case study of a WBSN for electrocardiography (ECG) monitoring. We envision a scenario that can take place in a hospital, where N patients (in this example, $N = 6$) are wearing a node that is connected to a central base station. The nodes reduce the size of the output

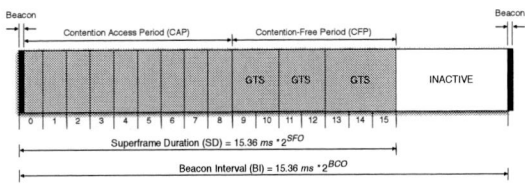

Figure 2: Structure of the IEEE 802.15.4 superframe

stream by applying one of the two available data compression techniques, i.e., digital wavelet transform (DWT) [23] and compressed sensing (CS) [13]. The two techniques have different properties in terms of complexity, signal quality and hardware requirements: for the sake of illustration, we assume that half of the nodes employ DWT, and the remaining ones execute CS.

As a node, we employ the *Shimmer* commercial platform [24], which includes an ultra low-power microcontroller, $10kB$ of RAM memory, and an IEEE 802.15.4 [16] radio module. The transmission is performed using the beacon-enabled mode of the IEEE 802.15.4 MAC layer [16]. Considering the set of parameters on the node and the MAC protocol, the number of possible network configurations of this case study exceeds the tens of millions, thus making a deep DSE impractical by using network simulation or by collecting experimental data. The proposed model, on the other hand, contains all the structures that are needed to fully describe the target network.

4.2 IEEE 802.15.4 Network Model

We hereby show how the proposed system-level WBSN model can capture the relevant dynamics of the beacon-enabled mode of the IEEE 802.15.4 [16] MAC protocol. In this MAC, a beacon is periodically sent by the coordinator to define the structure of the next *superframe*, a time interval whose structure is shown in Figure 2. The superframe is divided into an inactive and an active part, the latter being divided into 16 slots, 7 of which (known as *guaranteed time slots*, GTSs) are granted using a TDMA-like protocol.

The IEEE 802.15.4 MAC configuration is defined as $\chi_{mac} = \{L_{payload}, SFO, BCO, \Delta_{tx}^{(1)}, ..., \Delta_{tx}^{(N)}\}$, where $L_{payload}$ is the payload in a data packet, and SFO and BCO denote the superframe and the beacon orders, which in turn determine the interval between two beacons (BI) and the duration of the active part (SD) (see Figure 2) [16]. Finally, the $\Delta_{tx}^{(n)}$'s indicate the transmission time allocated to each node.

The IEEE 802.15.4 MAC protocol can be easily mapped on the structures we identified in Section 3.2. For example, the data overhead introduced by the MAC is equal to 13 bytes (11 for the header, 2 for the checksum) for each packet, hence $\Omega(\phi_{out}^{(n)}, \chi_{mac}) = 13 \cdot \phi_{out}^{(n)}/L_{payload}$. In terms of control overhead, the protocol does not require any control message from node (thus $\Psi_{n \to c}(\chi_{mac}) = 0$), whereas the coordinator sends a number of beacons (of variable length, which we denote as L_{beacon}) that depends on the number of superframes per second (i.e., $1/BI$), and an acknowledgment (4 bytes) for each transmitted packet, thus $\Psi_{c \to n}(\chi_{mac}) = 4 \cdot \phi_{out}^{(n)}/L_{payload} + L_{beacon}/BI$. Furthermore, $\Delta_{control}(\chi_{mac})$ is the time required by the coordinator to transmit $1/BI$ beacons per second, plus at least 9 slots reserved for contention access (which are not exploited in this case study), and the inactive period of the superframes.

As we mentioned in Section 3.2, the model can handle additional protocol-specific constrains on the assignment of the $\Delta_{tx}^{(n)}$'s. Firstly, the $\Delta_{tx}^{(n)}$'s cannot be arbitrarily assigned because of the time discretization imposed by the slots. Hence, we define the base transmission time δ as the slot length, i.e., $SD/16$, and we express all the $\Delta_{tx}^{(n)}$'s as multiples of δ. Then, as the protocol specifies that at most 7 slots can be used as GTSs, we formulate a constraint on the overall transmission time that can be allocated for the nodes, i.e., $\sum_{i=1}^{N}\Delta_{tx}^{(i)} \leq 7/16 \cdot SD/BI$.

Finally, thanks to the nature of data compression that leads to a uniform output rate, a simple delay model (based on the one in [17]) can be formulated. In particular, the worst-case delay for a node n occurs when the remaining nodes use all their slots (and the control overhead for all the corresponding frames) before node n is enabled to transmit:

$$d^{(n)}(\chi_{mac}) \leq \sum_{i=1,\,i\neq n}^{N}\Delta_{tx}^{(i)} + \left\lceil \frac{1}{7}\sum_{i=1,\,i\neq n}^{N}\Delta_{tx}^{(i)} \right\rceil \Delta_{control}. \quad (9)$$

4.3 Shimmer Node Model

In this section, we apply the node model described in Section 3.3 to the *Shimmer* platform [24]. As the node platform is already implemented, some parameters are fixed. In particular, the sampling frequency is determined by the nature of the ECG signal and is fixed to $f_s=250Hz$, and the resolution L_{ADC} of the A/D converter is set to 12 bits, thus generating a constant input stream $\phi_{in} = 375$ B/s. The contribution of the 10kB memory block is also constant, as the memory accesses are determined by the *Shimmer*-specific implementations of the DWT and CS algorithms [13]. At the radio level, the power of the carrier signal has been set to a sufficient level in order to minimize the probability of a packet error, thus avoiding an increment of ϕ_{out} due to retransmission. Hence, the configuration of a node is characterized as $\chi_{node} = \{CR, f_{\mu C}\}$, where CR is the compression ratio, and $f_{\mu C}$ is the frequency of the microcontroller.

The output stream ϕ_{out} can be easily expressed as a function of CR, i.e., $\phi_{out} = h(\phi_{in}, \chi_{node}) = \phi_{in} \cdot CR$, which holds for both the DWT and the CS applications. However, the two compressions show different duty cycles and loss-of-quality functions. The duty cycle of the Shimmer implementations of DWT and CS show a marginally dependency on CR, but there is a relation with respect to $f_{\mu C} \in \chi_{node}$. By analyzing the execution, we can define the resource usage function as $\mathbf{k}(\phi_{in}, \chi_{node}) = (k_{DWT}, k_{CS}) = (2265.6/f_{\mu C}, 388.8/f_{\mu C})$. To estimate the quality of the application, we select the *percentage root-mean-square difference* (PRD) [13], which quantifies the difference between the original ECG and the one reconstructed by the coordinator. Although the actual PRD value can only be determined by measuring or simulating the actual reconstructed signal, we computed an analytical estimation using two fifth-order polynomial functions $P_5^{(DWT)}(CR)$ and $P_5^{(CS)}(CR)$ that fit the experimental data provided in [13].

5. EXPERIMENTAL RESULTS

In this section, we evaluate the accuracy of the proposed model-based estimation with respect to experimental data. The results refer to the case study discussed in Section 4, but tests on different networks show a similar accuracy.

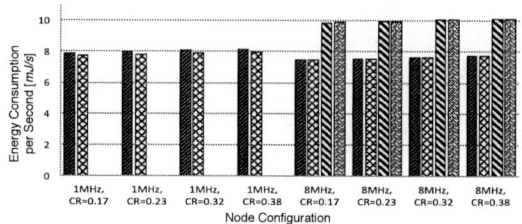

Figure 3: Estimation of the node consumption with different configurations

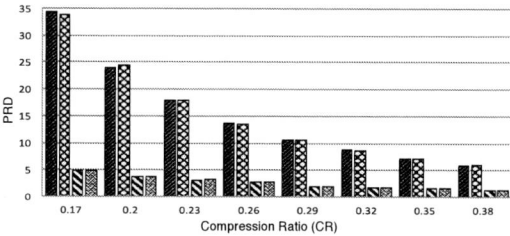

Figure 4: Estimation of the application behavior by means of the PRD metric

5.1 Estimation Accuracy

The first set of experiments aims at validating the estimation provided by the model. We first validated the model equations with respect to real experimental data obtained under different operating conditions. Figure 3 shows the estimation of the overall energy consumption of the nodes with set of realistic configurations χ_{node}. The energy estimation proves to be very accurate, as the average error on all the $f_{\mu C}$'s and CR's is equal to 0.88% for the CS, and to 0.13% for DWT, and the maximum error does not exceed 1.74%. The model also predicts that the DWT cannot complete its execution with $f_{\mu C} = 1$ MHz because its duty cycle exceeds 100%. Figure 4 shows the estimation error for the PRD's, which proves to be very low (0.92% for the CS, 0.46% for the DWT), thus showing that the model accurately estimates a crucial metric that can be exactly determined only by analyzing or simulating the actual compressed ECG.

In order to validate the network model, we compared the estimated delay to the results of a network simulation performed using the popular *Castalia* framework [25]. The choice of a network simulator over experimental data is justified by the possibility of deeply monitoring the packet flow. In spite of being a worst-case estimation, the delay function in Equation (9) provides an average overestimation lower than 100 ms over a set of 130 simulations with realistic ϕ_{out}'s and χ_{mac}'s, which is acceptable in this application.

5.2 Design Space Exploration Performance

In this paragraph, we aim at validating the proposed model-based evaluation within the context of the DSE. We employed the proposed WBSN model in a set of multi-objective optimization techniques, including genetic algorithms (which have been already used in the WSN domain [3]) and simulated annealing [27], without experiencing any relevant difference in terms of quality of the solutions. In terms of execution time, the proposed evaluation clearly outperforms

1047

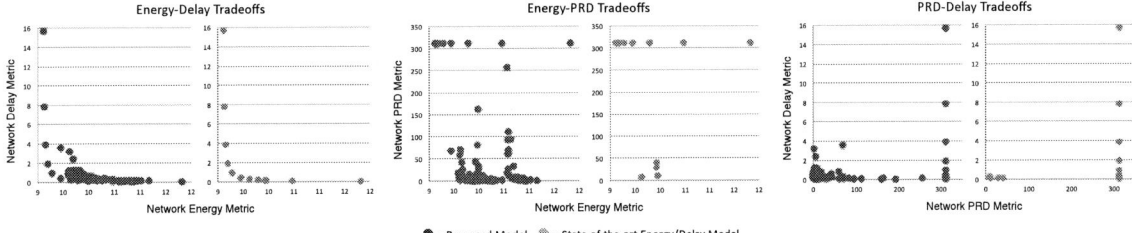

Figure 5: Tradeoffs detected using the proposed model and a state-of-the-art energy/delay model [26]

a complete network simulation, in fact, a network simulation takes 5 to 10 minutes in our case study, while the model can be evaluated approximately 4800 times per second.

Figure 5 shows the optimal tradeoffs between the three metrics we included in our model and, in order to underline the importance of considering all these metrics, the solutions are compared to the ones found by using a state-of-the-art energy/delay model [26]. It can be observed that the Pareto set generated according to the energy/delay model only contains a subset (i.e., approximately 7%) of the tradeoffs that are found using the proposed model: this is due to the fact that the energy/delay model does not include an additional application-aware metric. As a consequence, it only approximates the energy/delay curve, but it does not allow the DSE algorithm to recognize the solutions that are optimal in terms of PRD. In order to detect the large number of Pareto tradeoffs characterized by acceptable mid-range PRD's, the proposed multi-layer model must be employed.

6. CONCLUSION

In this work, we have shown the benefits of a quick and accurate analytical evaluation in the context of model-based design of WSNs. Although WSNs show a wide range of different characteristics in different fields, it is possible to formulate abstract system-level models for broad classes of networks, which share common architectural and network structures, or more generally belong to the same domain. As a proof of concept, we have considered the class of WB-SNs, and we have proved that a general and comprehensive model can be defined, and it can be applied to real networks with a low effort and a high accuracy. The results on a real case study show that the estimation error for energy and performance never exceeds 1.74% with respect to real data, while the estimation time is up to six orders of magnitude lower than an evaluation performed by a network simulator.

7. ACKNOWLEDGMENTS

This research was partially supported by the Swiss National Science Foundation (SNF), under grant 200021-127282, and by the Spanish Government under grant TIN2008-00508.

8. REFERENCES

[1] J. Yick et al., "Wireless sensor network survey," Computer Networks, no. 52, 2008, pp. 2292–2330.

[2] M. Patel and J. Wang, "Applications, challenges, and prospective in emerging body area networking technologies," Wireless Communications, vol. 17, no. 1, pp. 80–88, 2010.

[3] C.-t. Yeh et al., "Energy-Aware Data Acquisition in Wireless Sensor Networks," in Proc. of IMTC, 2007, pp. 1–6.

[4] S. Nabar et al. , "Minimizing Energy Consumption in Body Sensor Networks via Convex Optimization," in Proc. of BSN, 2010, pp. 62–67.

[5] Z. He et al., "Energy-aware portable video communication system design for wildlife activity monitoring," IEEE Circuits and Systems Magazine, vol. 8, no. 2, pp. 25–37, 2008.

[6] L. S. Bai et al., "Automated construction of fast and accurate system-level models for wireless sensor networks," in Proc. of DATE, 2011, pp. 1-6.

[7] H. Koc et al., "Minimizing Energy Consumption of Banked Memories Using Data Recomputation," in Proc. of ISLPED, 2006, pp. 358–361.

[8] V. Gutnik and A. P. Chandrakasan, "Embedded power supply for low-power DSP," IEEE Trans. on Very Large Scale Integration (VLSI) Systems, vol. 5, no. 4, pp. 425–435, 1997.

[9] C. Schurgers et al., "Power management for energy-aware communication systems," ACM Trans. on Embedded Computing Systems, vol. 2, no. 3, pp. 431–447, 2003.

[10] M. Kohvakka et al., "Performance analysis of IEEE 802.15.4 and ZigBee for large-scale wireless sensor network applications," in Proc. of PE-WASUN, 2006, pp. 48–57.

[11] N. F. Timmons and W. G. Scanlon, "Analysis of the performance of IEEE 802.15.4 for medical sensor body area networking," in Proc. of SECON, 2004, pp. 16–24.

[12] S. Ullah et al., "A Comprehensive Survey of Wireless Body Area Networks," Journal of Medical Systems, pp. 1–30, 2010.

[13] H. Mamaghanian et al., "Compressed Sensing for Real-Time Energy-Efficient ECG Compression on Wireless Body Sensor Nodes," Trans. on Biomedical Engineering, no. 99, p. 1, 2011.

[14] P. Suarez et al., "Increasing ZigBee network lifetime with X-MAC," in Proc. of REALWSN, 2008, pp. 26–30.

[15] Casilari et al., "Modeling of Current Consumption in 802.15.4/ZigBee Sensor Motes," Sensors, vol. 10, no. 6, pp. 5443–5468, 2010.

[16] IEEE, IEEE Std 802.15.4-2006, 2006.

[17] A. Koubaa et al., "GTS allocation analysis in IEEE 802.15.4 for real-time wireless sensor networks," in Proc. of IPDPS, 2006, pp. 8.

[18] P. Park et al., "Performance Analysis of GTS Allocation in Beacon Enabled IEEE 802.15.4," in Proc. of SECON, 2009, pp. 1–9.

[19] C. Buratti, "Performance Analysis of IEEE 802.15.4 Beacon-Enabled Mode," IEEE Trans. on Vehicular Technology, vol. 59, no. 4, pp. 2031–2045, 2010.

[20] R. Rieger and J. T. Taylor, "An Adaptive Sampling System for Sensor Nodes in Body Area Networks," IEEE Trans. on Neural Systems and Rehabilitation Engineering, vol. 17, no. 2, pp. 183–189, 2009.

[21] T. Burd and R. Brodersen, "Energy efficient CMOS microprocessor design," in Proc. of HICSS, 1995, pp. 288–297.

[22] F. Rincón et al., "Development and Evaluation of Multilead Wavelet-Based ECG Delineation Algorithms for Embedded Wireless Sensor Nodes," IEEE Trans. on Information Technology in Biomedicine, vol. 15, no. 6, pp. 854–863, 2011.

[23] R. Benzid et al., "Fixed percentage of wavelet coefficients to be zeroed for ECG compression," Electronics Letters, vol. 39, no. 11, pp. 830–831, 2003.

[24] A. Burns et al., "SHIMMER - A Wireless Sensor Platform for Noninvasive Biomedical Research," IEEE Sensors Journal, vol. 10, no. 9, pp. 1527–1534, 2010.

[25] Castalia Simulator, http://castalia.npc.nicta.com.au/.

[26] G. Kumar et al., "End-to-End Energy Management in Networked Real-Time Embedded Systems," Trans. on Parallel and Distributed Systems, vol. 19, no. 11, pp. 1498–1510, 2008.

[27] D. Nam and C. H. Park, "Multiobjective Simulated Annealing: A Comparative Study to Evolutionary Algorithms," Intl. Journal of Fuzzy Systems, vol. 2, no. 2, pp. 87–97, 2000.

Energy Harvesting and Power Management for Autonomous Sensor Nodes

J.F. Christmann, E. Beigné,
C. Condemine, J. Willemin
CEA-Leti, Minatec, Grenoble, France
jean-frederic.christmann@cea.fr

C. Piguet

CSEM, Neuchâtel, Switzerland
christian.piguet@csem.ch

ABSTRACT

Wireless sensor nodes that are self-powered by extracting their energy from their environment are a new opportunity for monitoring purpose. Since the available energy is not constant over time and due to very low harvested power levels, efficient energy and power management strategies are mandatory for improving their autonomy. At system level, scheduling algorithms are proposed to efficiently use multi power path architectures and avoid as much as possible the use of batteries. A data- and energy-driven architecture and its associated algorithm are presented achieving high efficiency due to fully adaptive scheme.

Categories and Subject Descriptors

B.7.1 [Integrated Circuits]: Types and Design Styles – *Advanced technologies*

General Terms

Algorithms, Design, Management

Keywords

Architecture, asynchronous, energy harvesting, power management, wireless sensor node

1. INTRODUCTION

Past ten years, new innovative microsystems appeared consisting of sensors and capable of communicating information measured in the environment through a wireless link [1]: it's the arrival of Wireless Sensor Nodes (WSN) and networks. Today's nodes may contain sensors of temperature, humidity, pressure, light intensity, light tilt and magnetic field as well as a communication component and a microcontroller. The success of those first systems has paved the way for ultra-low power communicating sensor systems. Supplying energy to these systems remains a central focus and using a battery is an appealing solution because there is no deployment limit and it allows several years of energy autonomy. Nevertheless, to increase the durability of the network, such autonomy requires an increased maintenance to replace life-ending batteries and there are networks which are impossible to access, such as sensors inserted into the structure of buildings or in the highways asphalt. To solve these battery maintenance problems, autonomous energy is one of the main research objectives in the development of wireless sensor networks.

Systems based on energy harvesting are thus an emerging solution for supplying power to wireless sensor networks nodes. Nevertheless, harvesting efficiency is low and so is the quantity of harvested energy, compared to the power consumption of an operational sensor node. Because improving the quantity and quality of service supplied by a sensor node or a sensor network is crucial in their development, the main objective of energy management systems based on energy harvesting is to optimize the energy efficiency between harvested energy and energy supplied to the sensor nodes themselves. Moreover, it is mandatory for these systems to propose an efficient storage of harvested energy to ensure a minimal quality of service when energy harvested from their environment decreases or disappears. Reconfigurable architectures allow optimizing energy efficiency according to the quantity of energy to be supplied at the node and the quantity of harvested energy. The associated control algorithm for such systems is also a key factor in improving energy efficiency. The use of innovative task scheduling allows the architecture monitoring and configuration such that the node can be functional for a longer period of time for the same initial quantity of harvested energy.

Most architectures and algorithms are based on the following power path architecture: the harvester is directly used to charge the battery which itself supplies power to the circuit. In order to improve the global power efficiency a dual power path is proposed [2] where harvesters can directly supply energy to the circuit before charging the battery. In order to decrease the total microsystem power consumption, we also propose to use a fully asynchronous harvesting platform and its dedicated power management. The idea is to wake-up the system only on energy or data events depending on environmental conditions or applicative constraints. In the case of energy-driven behaviors, the system reacts on available energy and performs internal computation according to its energy level. The system can also be woken up on-demand according to specific data-events. In both cases, the system has to be aware of its internal energy state and specific algorithms are taking full advantage of this dual-path asynchronous architecture to propose higher energy efficiency whatever the environmental conditions are.

A complete overview of energy harvesters' main principles and components is given in Section 2. Wireless Sensor network nodes power management units and their main requirements are presented in Section 3. Finally, Section 4 proposes system algorithms applied to the most efficient architectures. All along the paper, our asynchronous data- and energy-driven scheme is detailed and applied for high energy efficiency and autonomy purpose.

2. ENERGY HARVESTING

2.1 Paradox of Energy Harvesting

What are the main limitations of today proposed energy harvesting solutions? No killer applications appear, while a large scientific community is working on this topic. On one side, people designing sensor interfaces RF or digital circuits highlight an inconsistent level of harvested energy, but on another side, people working on energy harvesters blame a too large power consumption of the node. To converge to an autonomous node, the system must be redesign around the energy harvester, with an energy-driven strategy, the "only replace the battery" solution showing a dead end. As an example, an RF transfer of 100 bits at 10 meters costs several µJ, a measurement through a MEMS interface costs tens of µJ. The energy per bit per computation decreases according to the technology trend by a factor of 1.6 per year and in the same time the energy storage density increases only by 1.5/decade.

But this shrinkage has a limit, and, to reach autonomy, harvested energy must be consistent with tens of microwatts. The smaller is the harvester, the smaller is the harvested energy. First trends were to integrate in microelectronics process the harvesters, in order to reach collective fabrication. But micrometer- to millimeter-scale harvesters lead from several nanowatts to microwatts available electric energy. To be able to supply power to a complete Wireless Sensor Nodes (WSN), even with ultra low power functions, more energy is needed and multiple energy sources are required [3].

2.2 Energy Harvesters

Mostly used energy sources are mechanical, radiant and thermal gradient energies. Three kinds of sources using mechanical movement can be observed in the environment: strains, stresses and shocks, rotations and vibrations. Although physical phenomena are different, these movements have the same characteristics: low frequency (in the range from Hertz to several hundred of Hertz) and available energy depending on the weight in movement. For 1cm^3 volume harvester, maximum available energy is in the range of milliwatts.

To convert mechanical energy into electrical energy, three kinds of harvesters can be used. The first, based on piezoelectric materials [4], generates charges under constraints or deformations. Piezo materials have the advantage of generating high output voltage while presenting high output capacitance. A second kind of harvesters based on electromagnetic principle generates current through the movement between a magnet and a coil. Drawbacks are very low output voltage, quite expensive system due to magnet and difficulties to reduce coil dimensions. The last category is electrostatic [5] and based on charge pump with a variable capacitance. The major drawback is the need of an initial charge (use of battery), and high voltage generated (in range of tens of Volts to few hundred of Volts). To overcome this drawback, electrets can be included, achieving the initial polarization [6].

Another important energy source in the environment is the light, whether radiation from the sun or indoor light. The photovoltaic conversion principle is based on a photon absorption creating an electron/hole pair in a semi-conductor material separated by the internal field of a PN junction. A first key point is the different sensitivity of the PV cell technology to the light spectrum and intensity depending on the origin of the sources (e.g. sun, fluorescent light). PV cell technology (amorphous, mono- or multi-crystalline, simple or multiple junctions) [7], must be chosen according to the application constraints (irradiance and cost). In a PV cell, the V/I relations is similar to a diode translated by the PV Cell generated current. To extract the maximum of energy, the VI product must be maximal and tracked (Maximum Power Point Tracking). The MPP is function of irradiance and temperature. To realize the MPP Tracking, large number of solutions has been implemented based on approximate tracking methods, where the MPP is calculated from I-V measurements with a mathematical model, data look-up table, empirical model [8] or based on direct tracking methods where the objective is to cancel power derivative.

Thermal gradient harvesters are based on the Seebeck effect: a gradient of temperature generates an electrical voltage. Available micro-Thermo-Generators (µTEG) are very thin and efficient, but the application integration is complex due to thermal matching (thermal resistances) and area consuming (cm^3) with regard to the TEG itself (mm^3). These thermal resistances are due to the package between µTEG and hot point, the µTEG itself and passive coolers. The µTEG is electrically equivalent to a voltage source with an internal resistance. In order to extract the maximum of energy, internal resistance and output resistance must be matching. In that case only the half of the power will be harvested. Another issue is the dependence of this equivalent resistance on the thermal gradient. So, to always extract the maximum energy, output resistance has to be adapted with an MPPT system.

2.3 Energy Converters

PV cells and µTEG harvesters generate DC variable output voltage. To convert energy to an adapted power level, DC-DC converters are to be used. Mechanical vibrations harvesters generate AC signal. To transfer energy to the rest of the circuit, AC-DC converter will rectify the signal or will transfer the energy by pulse. AC-DC converters can be implemented using rectifiers based on diode, MOS used as diode [9] and zero-Vt diode [10] or using inductive converters. DC-DC converters are based on capacitive converters (Dickson or cross-coupled charge pump [11]) or inductive converters (step-down, step-up [12] or flyback). In all cases, key points are the static power consumption of the control circuitry, and the losses in the energy transfer (gate capacitances and switches R_{ON}). After power conversion, the energy can directly supply power to sensor or RF functions or can be stored for later use.

3. POWER MANAGEMENT ARCHITECTURES

Considering the low quantity of harvested energy, efficient power management is mandatory to ensure a minimal quality of service while running embedded sensors and communicating devices. Main power management objective is to efficiently provide an optimal power supply to the connected modules while reducing the global node power consumption in order to optimize the WSN's lifespan. In the next section, power management unit architecture choice is tackled with respect to existing related works.

3.1 Common Architectures and Requirements

Optimizing the global power efficiency can be done by using an adapted architecture that efficiently controls relevant operating points within the node. As illustrated in Figure 1, a wireless sensor node leveraging energy harvesting consists in three major architectural components. Energy harvesters can extract energy from different energy sources such as light, thermal gradients or vibrations. Power loads, representing the processing, sensing and communicating parts of the node, have to be power supplied to ensure the node applicative functions. In between, a Power Management Unit is designed to control and optimize the global power efficiency.

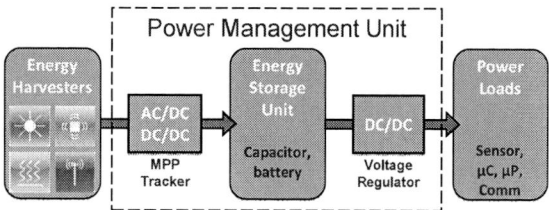

Figure 1 Usual serial architecture for WSN

Because harvested energy levels are low and not constant over time, it is relevant to use an energy storage unit ensuring the node functionality even if no energy is extracted from its environment. Voltage converters are used to provide a wide range of voltage power supplies and modularity into the system architecture. On the harvesters' side, as mentioned in section 2, AC/DC or DC/DC voltage converters are used to optimally transfer the extracted energy into the storage unit. The storage voltage value can thus be dissociated from the harvesters operating points that could thus be optimal for each type of harvested energies. Due to this architecture, an optimum energy level can still be extracted from the environment regardless of the storage unit voltage value. In the end, DC/DC converters are required to supply power to the loads with an optimal and regulated power supply voltage level. Again, in that case, the loads operating points are controlled whatever is the storage unit voltage value.

Most of related works on wireless sensor nodes power supplied by extracting their energy from the environment are fitting this serial architecture (Figure 1). Among them, the PicoCube [13] consists of a wireless sensor node powered by a vibration-based energy source. The harvester transfers energy to charge a NiMH battery through a diode bridge rectifier. Three power supply voltage levels are taken into account: the micro-controller and the sensor are power supplied under 2-2.8V thanks to a charge pump connected to the battery; a shunt regulator provides a 1V power supply for the radio digital control and a linear regulator generates a regulated 0.65V power supply for the radio analog part.

More recently, Lee et al. [14] proposed a cubic millimeter WSN realized using a five-layer three dimensional heterogeneous integration scheme. Optical communication is used to wake up, synchronize and re-program the node. Integrating a 3.2 - 4.1V thin-film Li 0.6μAh battery, the power management unit leverages 0.54mm² integrated solar cells or a 125mm² thermoelectric generator to charge the battery which provides from 0.6V to 1.2V power supply voltage levels to power the different components inside the node.

Wireless sensor nodes are embedding long-term energy storage units such as super-capacitors or batteries. Those units are mandatory to ensure the node functionality whenever no energy is harvested for a while, for example, during the night, for solar powered systems. Due to non-ideal storage technologies, those components suffer from energy losses while storing and retrieving energy from them. To overcome this leakage issue and bypass

those storage elements, Matsuo and Kurokawa [15] proposed to add an architectural direct power path into the power management unit. Bi-directional DC/DC converters were used to implement multiple power paths architecture. A generalized version of this concept is illustrated in Figure 2. Multiple energy storing stages are proposed to adapt the power supply source to the harvested energy levels. When power loads are requesting energy and if this energy can be harvested in the environment, this highest efficient power path can be used to directly supply power to the loads. In this scheme, the long-term energy storage unit is either charged when too much energy is harvested or discharged when power loads need more energy than that stored in the short-term energy storage unit.

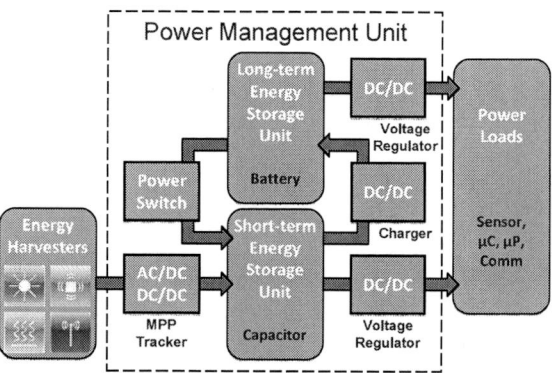

Figure 2 Multiple power paths advanced architecture

In 2005, Jiang et al. [16] presented a two-stage power supply system based on supercapacitors and on a battery. The main idea is to use the supercapacitors to handle the daily solar harvesting cycles. The battery is only used for extreme cases (e.g. day without enough sun), charge/discharge cycles are thus much wider and so is the battery lifespan. Park et al. [17] propose to combine solar and wind energy harvesting in order to power the Eco wireless sensor node [18] using this multi-stage energy storage principle.

As far as power efficiency is concerned and hardware complexity is increasing in recent wireless sensor nodes based on energy harvesting, main innovations are focusing on reducing the number of voltage converters. Multi-input and multi-output voltage converters are then considered based on single inductor. In [19], Lam et al proposed the use of a continuous voltage converter with two inputs and two outputs based on a unique inductance to carry out a battery management system. Moreover, several research groups carried out equivalent systems to hybridize a micro-fuel cell combined with a thin-film lithium battery [20] and to recharge and efficiently use a battery with the help of a photovoltaic solar module [21].

3.2 Event-Driven Multiple-Input Dual-Output Architecture

Our work aims to develop a data-driven energy harvesting platform which targets multi-sensor multi-source wireless sensor node. The platform implements different kinds of energy harvesters. Solar, thermal gradients, vibration and electromagnetic radiations energy sources provide harvested energy to the whole system through the use of a multiple power path power management unit. An overview of this architecture is illustrated in Figure 3 with an emphasis on the power management unit [2].

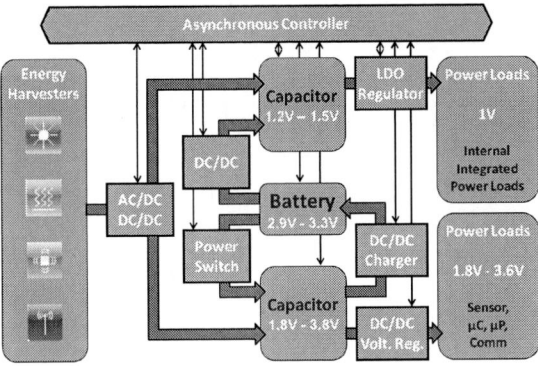

Figure 3 Multiple-input dual-output architecture overview

Two different power paths are defined in this architecture. Direct power paths go directly from the harvesters to the power loads through voltage converters and short-term storage capacitors. Indirect power paths are used to charge the battery which, subsequently, delivers its energy to the loads. On the one hand, direct power paths allow high efficient loads power supply but can only be used when both power supply is requested by the loads and energy is harvested from the environment. On the other hand, the indirect power paths provide reduced power efficiency due to additional voltage conversions and storage yield but can be used to power loads when energy harvested from the surroundings decreases or disappears. At architectural level, for energy efficiency purpose, priority is given to the 1.2V-1.5V capacitor for incoming harvested energy storage until the capacitor's voltage reaches 1.5V; energy is used afterwards to fulfill the other short-term storage capacitor and consequently charge the battery. Power path re-configuration is performed by a dedicated fully asynchronous digital controller which is implemented using Quasi Delay Insensitive (QDI) logic and a specific handshake protocol [22]. Main benefits provided by this type of asynchronous logic include robustness to power supply voltage variations, robustness to low power supply voltage levels, intrinsic sleep mode and reduced ElectroMagnetic Interferences (EMI) due to a smooth current profile. Asynchronous implementation of the digital controller is relevant in an energy- and data-driven context because these advantages are obviously correlated to energy harvesting based systems conditions [2]. The idea is to wake-up the system only on energy or data events depending on environmental conditions or applicative constraints. In the case of energy-driven behaviors, the system reacts on available energy and performs internal computation according to its energy level. The system can also be woken-up on-demand according to specific data-events.

4. POWER MANAGEMENT SCHEDULING ALGORITHMS
4.1 Common Scheduling Techniques
Although the power management unit architecture allows optimizing the power loads supply efficiency, scheduling the power loads activity is another key research issue towards wireless sensor nodes energy autonomy. To illustrate traditional power consumption reduction techniques using scheduling techniques, let us consider a load power supplied by a storage unit

of capacity $C_{storage}$ which is initially full. Equation (1) gives the autonomy of the system for a mean current $<I>$ drawn by the power load.

$$Autonomy = \frac{C_{storage}}{<I>} \qquad (1)$$

First scheduling improvement should offer the possibility to turn on and off the power loads with respect to their activity. The aim is to decrease the mean current drawn from the storage unit using an advanced sleeping mode. In the specific case of a wireless sensor node, a minimal current is necessary in sleep mode which is dedicated to wake-up oscillators and data memorization. Equation (2) gives the autonomy of the considered system where α is the duty cycle and I_{on} and I_{sleep} are respectively the active and sleeping mode currents.

$$Autonomy = \frac{C_{storage}}{\alpha <I_{on}> +(1-\alpha) <I_{sleep}>} \qquad (2)$$

The Duty Cycling concept can be extended to more than 2 modes and allows a refined control of the current drawn from the storage unit. Nevertheless, the power load has to include extra hardware to allow relevant part of its architecture to be turned on and off, thus defining varied power modes. The autonomy is then given by the generalized equation (3) where β_i and I_i are the proportion of time and the current drawn in mode i respectively.

$$Autonomy = \frac{C_{storage}}{\sum_i \beta_i I_i} \qquad (3)$$

Low-power modes usage and scheduling have been described in [23] for different applications such as processing circuits, real-time systems and sensor network. Applying this technique to energy harvesting based wireless sensor nodes is extremely difficult and can be power consuming by requiring both extra hardware and software to determine the optimum power mode and to efficiently control the platform reconfiguration.

Lots of works have been proposed using duty cycling. Considering wireless sensor nodes, it is relevant to adapt the duty cycle of the node to the energy stored into the system in order to increase the battery lifespan [13]. More generally, it is also possible to consider a wide set of Duty Cycle values to easily adapt the WSN power consumption to environmental constraints (e.g. day and night modes) but also to applicative constraints (e.g. week-end mode for nodes involved in company buildings monitoring applications). Once adaptive duty cycling tackles energy harvesting based WSN, inexhaustibility and uncertainty of harvested energy become predominant in the development of efficient algorithms which aim to precisely control the duty cycle according to the energy stored. Although harvested energy could be infinite and would theoretically allow the node to last forever, its availability, due to environmental randomness, is unpredictable and leads to critical issues. In [24], an algorithm that targets energy neutral mode of operation is presented. Based on a WSN power consumption and on harvested energy predictions, time slots are defined to dynamically adjust the duty cycle of the sensor node. Moser et al. [25] aim to optimize scheduling management by using predicted harvested energy values in order to decide whether a consuming task, among a set of tasks which have to be performed by the WSN, has to be delayed or not. The so-called "lazy algorithm" delays tasks and starts them as soon as there is enough energy into the storage unit to completely supply power to

1052

the system. A rewarding scheme is also propose ensuring that no task is outrageously delayed.

Those works are based on prediction models and complex algorithms requiring software implementation into microcontrollers. Considering sub-milliamps scaled systems, a compromise has to be found between the energy gain and the additional power consumption due to those algorithms implementation.

4.2 Energy-Driven Proposed Algorithm

As detailed in section 3.2, the proposed data-driven asynchronous implementation is a natural enabler for a fully energy-driven algorithm at system level. With respect to environmental constraints, rather than periodically and synchronously sampling energy levels for future task scheduling, the proposed algorithm scheme is based on asynchronous voltage threshold detections which are subsequently used to start the application tasks.

The main point of the proposed algorithm is the introduction of energy-driven power supply scheme. For energy-driven purpose, the system is harvesting energy until a specific voltage threshold is reached. This voltage threshold, corresponding to a capacitance energy level, is specified so that a dedicated task can be entirely performed on the energy stored in this short-term storage capacitor. The key idea is that the energy used to supply power to the load could entirely come through the direct power path with optimal energy efficiency. The algorithm consists in emitting a new energy-driven request as soon as the previous task has been performed bringing back to handshake mechanisms principles [22]. As a consequence, theoretically, energy is neither used to charge the battery nor drawn from it. To avoid useless power consumption due to an excessive duty cycle in high energy environment, a maximum value is set so that extra energy is rather used to charge the battery. In order to illustrate our proposal, a power management architecture subpart model is considered and shown in Figure 4.

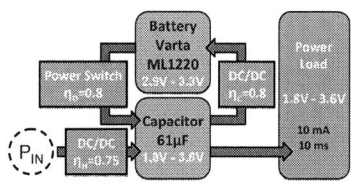

Figure 4 Energy paths and architecture considered for algorithms comparison

We aim to compare a simple *fixed duty cycle* algorithm fitting a mean input power specified by the system and our *energy-driven* duty cycle algorithm fully adaptive to available harvested energy. This fixed duty cycle value is used to define the maximum duty cycle for our algorithm.

A power load consuming 10mA in average during 10ms and functional between 1.8V and 3.6V power supply voltage levels is considered. P_{IN} input power is specified whose mean is 10mW and slowly oscillates between 1mW and 19mW. The power management architecture is defined as previously described such as the short-term storage capacitor is still first fulfilled. Once its voltage reaches 3.8V, the battery charger is turned on to charge the battery as long as the capacitor voltage does not decrease under 3.6V. If the capacitor's voltage decreases under 1.8V while delivering its energy to an active power load, the battery is discharged to compensate for the load power supply.

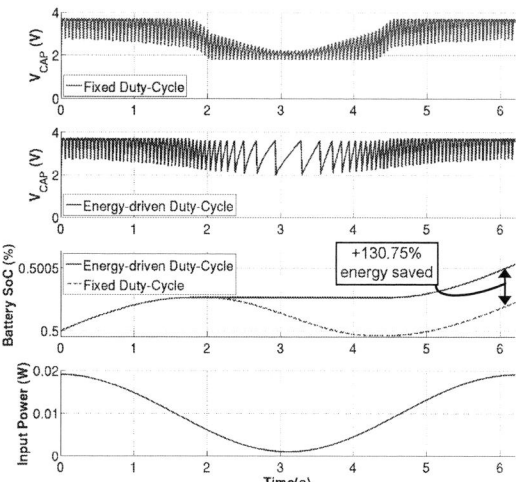

Figure 5 Algorithms comparison results

On Figure 5, from 0 to 1.5s, input power is equal or above the value used to define the fixed duty cycle and both algorithms similarly behave such as they correctly supply power to the load and use the extra energy to charge the battery. Afterwards, when input power decreases under this prior said value, the *fixed duty cycle* algorithm has to drawn current from the battery in order to supply power to the load. On the contrary, the *energy-driven* algorithm intrinsically adapts the duty cycle so that the only energy used to supply power to the load is harvested at highest architectural power efficiency. In this particular case, more than twice more energy is stored in the battery while using our energy-driven algorithm rather than the fixed duty cycle one.

The corresponding hardware implementation of such an adaptive system simply uses a continuous comparator whose output level rises when the input voltage crosses the specified voltage threshold. Due to considered capacitors value and current levels, voltage variations are slow enough to reduce the design constraints on the comparator. Consequently, a good compromise can be found between power consumption and response speed. The energy-event represented by the output rising edge is then detected by an asynchronous voltage detector based on Muller C-element and implemented in asynchronous QDI logic which fits asynchronous communicating 4-phase protocol [2][22]. The 500nA architecture of this edge detector is illustrated in Figure 6, fabricated in CMOS UMC 180nm.

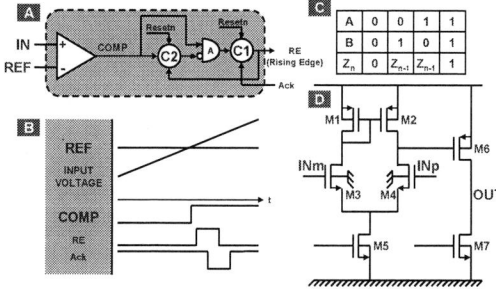

Figure 6 Energy event detection scheme (A) and behavior (B), C-element truth table (C) and comparator architecture (D)

1053

5. CONCLUSION

Harvesting energy today in wireless sensor nodes appears to be an essential solution to bring autonomy to those systems. Unfortunately, harvesting efficiency is low and so is the quantity of harvested energy, compared to the power consumption of an operational sensor node. Energy harvesters, power management architectures and scheduling algorithms have to propose together the most energy efficient solutions to supply power to the nodes during measurements or communications while respecting applicative constraints. Reconfigurable multi-inputs multi-outputs architectures proposing direct and indirect power paths are opening the way to highly efficient and autonomous systems. Multiple energy sources can be considered and efficient energy storage components are required to ensure a minimal quality of service when energy harvested from the environment decreases or disappears. Moreover, at architectural level, implementing data- and energy-driven schemes is a natural enabler to consider and use energy only when and where available with the maximum power efficiency. At system level, thanks to adaptive algorithms, the load power supply could entirely come through the harvesters and, at best, extra energy is even used to charge the battery.

6. REFERENCES

[1]   B. Warneke, M. Last, B. Liebowitz, and K. S. J. Pister, "Smart Dust: communicating with a cubic-millimeter computer," *Computer*, vol. 34, no. 1, pp. 44–51, Jan. 2001.

[2]   J.-F. Christmann, E. Beigne, C. Condemine, P. Vivet, J. Willemin, N. Leblond, and C. Piguet, "Bringing Robustness and Power Efficiency to Autonomous Energy-Harvesting Microsystems," *IEEE Design & Test of Computers*, vol. 28, no. 5, pp. 84–94, Oct. 2011.

[3]   E. O. Torres, Min Chen, H. P. Forghani-zadeh, V. Gupta, N. Keskar, L. A. Milner, H.-I. Pan, and G. A. Rincon-Mora, "SiP Integration of Intelligent, Adaptive, Self-Sustaining Power Management Solutions for Portable Applications," in *2006 IEEE International Symposium on Circuits and Systems, 2006.*, 2006, pp. 5311–5314.

[4]   M. Defosseux, M. Allain, P. Ivaldi, E. Defay, and S. Basrour, "Highly efficient piezoelectric micro harvester for low level of acceleration fabricated with a CMOS compatible process," in *2011 16th International Solid-State Sensors, Actuators and Microsystems Conference*, 2011, pp. 1859–1862.

[5]   G. Despesse, T. Jager, C. Condemine, and P.-D. Berger, "Mechanical vibrations energy harvesting and power management," in *IEEE Sensors*, 2008, pp. 29–32.

[6]   S. Boisseau, G. Despesse, and A. Sylvestre, "Optimization of an electret-based energy harvester," *Smart Materials and Structures*, vol. 19, no. 7, p. 075015, Jul. 2010.

[7]   M. A. Green, K. Emery, Y. Hishikawa, W. Warta, and E. D. Dunlop, "Solar cell efficiency tables (Version 38)," *Progress in Photovoltaics: Research and Applications*, vol. 19, no. 5, pp. 565–572, Aug. 2011.

[8]   A. Pandey, N. Dasgupta, and A. K. Mukerjee, "A Simple Single-Sensor MPPT Solution," *Power Electronics, IEEE Transactions on*, vol. 22, no. 2, pp. 698–700, Mar. 2007.

[9]   N. J. Guilar, R. Amirtharajah, and P. J. Hurst, "A Full-Wave Rectifier With Integrated Peak Selection for Multiple Electrode Piezoelectric Energy Harvesters," *IEEE Journal of Solid-State Circuits*, vol. 44, no. 1, pp. 240–246, 2009.

[10] T. T. Le, J. Han, A. V. Jouanne, K. Mayaram, and T. S. Fiez, "Piezoelectric Micro-Power Generation Interface Circuits," *IEEE Journal of Solid-State Circuits*, vol. 41, no. 6, pp. 1411–1420, May 2006.

[11] H. Shao, C.-Y. Tsui, and W.-H. Ki, "An Inductor-less Micro Solar Power Management System Design for Energy Harvesting Applications," in *IEEE International Symposium on Circuits and Systems, 2007. ISCAS 2007*, 2007, pp. 1353–1356.

[12] Y. Qiu, C. Van Liempd, B. O. het Veld, P. G. Blanken, and C. Van Hoof, "5µW-to-10mW input power range inductive boost converter for indoor photovoltaic energy harvesting with integrated maximum power point tracking algorithm," 2011, pp. 118–120.

[13] Yuen Hui Chee, M. Koplow, M. Mark, N. Pletcher, M. Seeman, F. Burghardt, D. Steingart, J. Rabaey, P. Wright, and S. Sanders, "PicoCube: A 1cm3 sensor node powered by harvested energy," in *45th ACM/IEEE Design Automation Conference, 2008. DAC 2008*, 2008, pp. 114–119.

[14] Y. Lee, G. Kim, S. Bang, Y. Kim, I. Lee, P. Dutta, D. Sylvester, and D. Blaauw, "A modular 1mm^3 die-stacked sensing platform with optical communication and multi-modal energy harvesting," in *Solid-State Circuits Conference Digest of Technical Papers (ISSCC), 2012 IEEE International*, 2012, pp. 402–404.

[15] H. Matsuo and F. Kurokawa, "New Solar Cell Power Supply System Using a Boost Type Bidirectinal DC-DC Converter," *IEEE Transactions on Industrial Electronics*, vol. IE-31, no. 1, pp. 51–55, Feb. 1984.

[16] X. Jiang, J. Polastre, and D. Culler, "Perpetual environmentally powered sensor networks," in *Fourth International Symposium on Information Processing in Sensor Networks, 2005. IPSN 2005*, 2005, pp. 463–468.

[17] Chulsung Park and P. H. Chou, "AmbiMax: Autonomous Energy Harvesting Platform for Multi-Supply Wireless Sensor Nodes," in *2006 3rd Annual IEEE Communications Society on Sensor and Ad Hoc Communications and Networks, 2006. SECON '06*, 2006, vol. 1, pp. 168–177.

[18] C. Park, J. Liu, and P. H. Chou, "Eco: an ultra-compact low-power wireless sensor node for real-time motion monitoring," in *Fourth International Symposium on Information Processing in Sensor Networks, 2005. IPSN 2005*, 2005, pp. 398–403.

[19] Yat-Hei Lam, Wing-Hung Ki, Chi-Fing Tsui, and P. K. . Mok, "Single-inductor dual-input dual-output switching converter for integrated battery charging and power regulation," in *Proceedings of the 2003 International Symposium on Circuits and Systems, 2003. ISCAS '03*, 2003, vol. 3, pp. 447–450.

[20] Suhwan Kim and G. A. Rincon-Mora, "Single-inductor dual-input dual-output buck-boost fuel-cell-li-ion charging DC-DC converter supply," in *Solid-State Circuits Conference - Digest of Technical Papers, 2009. ISSCC 2009. IEEE International*, 2009, pp. 444–445,445a.

[21] Hui Shao, Chi-Ying Tsui, and Wing-Hung Ki, "A single inductor DIDO DC-DC converter for solar energy harvesting applications using band-band control," in *VLSI System on Chip Conference (VLSI-SoC), 2010 18th IEEE/IFIP*, 2010, pp. 167–172.

[22] A. J. Martin and M. Nystrom, "Asynchronous Techniques for System-on-Chip Design," *Proceedings of the IEEE*, vol. 94, no. 6, pp. 1089–1120, Jun. 2006.

[23] A. Bogliolo, L. Benini, E. Lattanzi, and G. De Micheli, "Specification and analysis of power-managed systems," *Proceedings of the IEEE*, vol. 92, no. 8, pp. 1308–1346, Aug. 2004.

[24] J. Hsu, S. Zahedi, A. Kansal, M. Srivastava, and V. Raghunathan, "Adaptive Duty Cycling for Energy Harvesting Systems," in *Proceedings of the 2006 International Symposium on Low Power Electronics and Design, 2006. ISLPED'06*, 2006, pp. 180–185.

[25] C. Moser, L. Thiele, D. Brunelli, and L. Benini, "Adaptive Power Management for Environmentally Powered Systems," *IEEE Transactions on Computers*, vol. 59, no. 4, pp. 478–491, Apr. 2010.

Functional Timing Analysis Made Fast and General

Yi-Ting Chung[1] and Jie-Hong Roland Jiang[1,2]

[1]Graduate Institute of Electronics Engineering; [2]Department of Electrical Engineering

National Taiwan University, Taipei 10617, Taiwan

{r99943080@ntu.edu.tw, jhjiang@cc.ee.ntu.edu.tw}

ABSTRACT

Functional, in contrast to structural, timing analysis is accurate, but computationally expensive in refuting false critical paths. Although satisfiability-based analysis using timed characteristic functions has been proposed, its efficiency and generality remain room for improvement. This paper shows functional timing analysis on industrial designs can be made up to several orders of magnitude faster and more generally applicable than prior methods.

Categories and Subject Descriptors

B.8.2 [**Performance and Reliability**]: Performance Analysis and Design Aids

General Terms

algorithms, design, verification

Keywords

false path, satisfiability solving, timed characteristic function, timing analysis

1. INTRODUCTION

In modern synthesis flow of very large scale integration (VLSI) design, timing analysis is essential in identifying timing critical regions for re-synthesis, determining operable clock frequencies, and avoiding wasteful over-optimization and thus accelerating design closure in meeting stringent timing constraints. As timing analysis often has to be repeatedly performed, how to make the computation efficient and accurate becomes a crucial task.

There are two main approaches to timing analysis. *Static timing analysis* (STA), based on pure structural (or topological) analysis, though fast with linear-time complexity, can be too pessimistic in estimating circuit delay due to the ignorance of false or nonsensitizable paths [2]. *Functional timing analysis* (FTA), on the other hand, provides accurate delay calculation, but is computationally intractable, i.e., NP-hard, in identifying false critical paths [7].

Many FTA algorithms, e.g., [7, 11, 2, 4, 10, 1, 14, 13, 6, 3], have been proposed. When delay-dependency is concerned, an FTA algorithm can be delay-independent [3] or delay-dependent [2]. The former (latter) identifies true and false paths without (with) respect to some timing library. Whereas the former is incomplete in that not every delay path can be concluded true or false regardless of arbitrary delay assignments, this paper focuses on the latter analysis.

When the underlying computation engine is concerned, an FTA algorithm can be powered by an automatic test pattern generator, e.g., [4, 1], or by a satisfiability (SAT) solver, e.g., [10, 14, 6]. Since ATPG-based computation involves sophisticated circuit transformation and multi-fault testing, it is difficult to implement and scale. In contrast, SAT-based computation allows simple implementation due to its clean separation between timed characteristic function (TCF) construction [7] and SAT solving. Although recent advances in SAT solving techniques [9, 8, 5] make SAT-based FTA a viable approach, FTA for large industrial designs remains challenging due to the massive numbers of variables and clauses when translating a complex TCF into a conjunctive normal form (CNF) formula for SAT solving. Moreover, modern SAT-based FTA algorithms [14, 6] cannot handle arbitrary gate types. Although formulation for general gate types has been proposed in [10], its complex formulas make SAT solving inefficient.

This work aims to develop a scalable and general FTA framework. The main results include 1) a generalized TCF framework supporting arbitrary complex gate types for both combined and separate rise/fall-time analysis, 2) an implication-based TCF construction and its linear-time translation to CNF without extra variables being introduced, 3) a TCF reduction technique with an improved equivalence relation based on table look-up, 4) a model generation mechanism, which produces a true critical path along with its sensitization condition if the target delay is sensitizable, and 5) an algorithm to identify timing critical regions of a circuit for potential timing optimization. Experimental results show substantial speedup over prior SAT-based delay computation methods and show effective critical region identification.

The rest of this paper is organized as follows. We give a brief description of our sensitization criteria and satisfiability model in Section 2. Our general TCF formulation is introduced and compared with prior formulations in Section 3. Section 4 presents efficient algorithms for timing delay computation and critical region identification. Section 5 shows experimental evaluation. Finally, conclusion and future work are given in Section 6.

2. PRELIMINARIES

A *literal* is a Boolean variable or its negation. A *clause* (*cube*) is a disjunction (conjunction) of literals. A propositional formula is in *conjunctive normal form* (CNF) if it is written as a conjunction of clauses. The satisfiability (SAT) problem asks whether there exists a satisfying assignment to the set of variables that makes a CNF formula true. The reader is referred to [9, 8, 5] for modern SAT solving techniques, and to [15, 12] for circuit-to-CNF conversion.

2.1 Circuit Model

A (combinational) circuit $C(N, E)$ consists of nodes (or gates) N, (directed) edges $E \subseteq N \times N$. Two disjoint subsets of N are distinguished as primary inputs (PIs) and primary outputs (POs). Each node is associated with two attributes: function and delay. We assume the function can be arbitrary, from simple gate types, such as buffer, inverter, NAND, NOR, etc., to complex function units, such as XOR, multiplexer, AOI, etc. In the sequel, we sometimes do not distinguish a node from its function and its output variable when it is clear from the context. We assume the gate delay can vary from pin to pin and vary between rise and fall time. Without loss of generality, interconnect delays are assumed to be integrated into the gate delays under this timing model.

For a node f in a circuit, we let $FI(f)$ and $FO(f)$ denote the fanin and fanout nodes of f, respectively. For $g \in FI(f)$, we say g is of *controlling value*, denoted $v_c \in \mathbb{B} = \{0, 1\}$ (respectively, *non-controlling value*, denoted $v_n \in \mathbb{B}$) of f if the output value of f can (respectively, cannot) be completely determined by g with v_c (respectively, v_n) regardless of the truth assignments to other inputs. For example, any input of an AND gate is of controlling value 0.

For a complex gate, such as XOR, its inputs may likely have no controlling values at all. Nevertheless the notion of controlling values can be generalized to *controlling cubes*. For a complex gate f, a truth assignment to a minimal (strict) subset $S \subset FI(f)$ that determines the output value of f independent of other fanins forms a controlling cube. A literal in a controlling cube c is called a *controlling literal* of c. For example, the controlling cubes of the gate f with function $ab \vee c$ are $\{ab, c, \neg a \neg c, \neg b \neg c\}$, where cubes ab and c make $f = 1$ and cubes $\neg a \neg c$ and $\neg b \neg c$ make $f = 0$. In addition, $\neg a$ is a controlling literal of cube $\neg a \neg c$.

2.2 Sensitization Criteria

Among the various modes of circuit operation when functional timing analysis is concerned, *floating-mode operation* [7], which we adopt, is the most popular due to its simplicity and robustness. Under this mode of operation, the signals of a circuit are of unknown initial values and stablize to their final values induced by a set of truth assignments on the PIs.

Under the floating-mode operation, various path sensitization criteria can be defined. The *exact criterion* [2] and *viable criterion* [7] are two commonly studied criteria. When the truth and falsity of a single path is concerned, the analysis of the former is exact whereas that of the latter is conservative [2]. Nevertheless, when the timing analysis is performed for all paths of a circuit without tracing a particular path, the viable criterion becomes exact as was shown in [11]. This paper is mainly concerned with computing the longest true delay among all paths.

2.3 Satisfiability of Timing Requirement

To perform satisfiability testing on whether there exists a PI assignment that exercises a target circuit delay through some unknown true path, the condition can be translated into the so-called *timed characteristic function* (TCF) [7]. Specifically, the set of PI assignments that makes the output value of f stablize *no earlier* than time $t \geq 0$ is characterized by a (no-early) TCF, denoted $\chi^{f,t}$. In other words, a PI assignment satisfying $\chi^{f,t}$ makes the output value of f remains unknown (under the floating-mode assumption) until time t. When the stablization value of f is specific to value 0 (respec-

tively 1), the corresponding 0/1-specified TCF is denoted as $\chi^{f=0,t}$ (respectively $\chi^{f=1,t}$). Likewise one can define an early TCF, denoted $\chi^{f,t-}$, characterizing the set of PI assignments that make the output value of f stablize *earlier* than time $t \geq 0$. Note that $\chi^{f,t_1} \rightarrow \chi^{f,t_2}$ for $t_1 \geq t_2$, and $\chi^{f,t} = \neg \chi^{f,t-}$.

The circuit delay computation can therefore be formulated as searching the maximum D such that the formula

$$\bigvee_{p \in PO} \chi^{p,D} \tag{1}$$

$$= \bigvee_{p \in PO} (\chi^{p=1,D} \vee \chi^{p=0,D}) \tag{2}$$

is satisfiable. (If Formula (1) is satisfiable, the circuit delay must be equal to or larger than D because there exists some PO whose value remains unknown before time D. Otherwise, the circuit delay is strictly smaller than D.) As to be discussed in Section 3, these TCFs of Formula (1) can be constructed recursively from POs to PIs of the circuit, and Formula (1) can be converted to CNF for SAT solving.

3. TCF CONSTRUCTION

In this section we consider TCF formulations without and with 0/1-specificity. Our formulations are then compared with prior methods [14], [6], and [10]. Finally, TCF equivalence reduction techniques are proposed.

3.1 TCF without 0/1-Specificity

3.1.1 Prior Formulation

Prior work [14] reformulated the exact [2] and viable [7] sensitization criteria (with path tracing) for circuit delay computation (without path tracing) with the following TCFs

$$\chi^{f,t} = \bigvee_{g_i \in I(f)} \chi^{g_i, t-d_i} \wedge \{ \bigwedge_{g_j \in I(f)} (g_j = v_{n_j}) \vee$$
$$(g_i = v_{c_i}) \wedge \bigwedge_{g_j \in I(f)} (\chi^{g_j, t-d_j} \vee (g_j = v_{n_j})) \}, \tag{3}$$

$$\chi^{f,t} = \bigvee_{g_i \in I(f)} \chi^{g_i, t-d_i} \wedge$$
$$\bigwedge_{g_i \in I(f)} (\chi^{g_i, t-d_i} \vee (g_i = v_{n_i})) \tag{4}$$

respectively, where d_i is the pin-to-pin delay from g_i to f and v_{c_i} and v_{n_i} are the controlling and non-controlling values of g_i.[1] Equations (3) and (4) were considered in [14] as exact and approximative circuit delay computation, respectively.

The recursive definition of $\chi^{f,t}$ naturally translates to a combinational circuit. For a k-input simple gate f, Equations (3) and (4) result in $(k^2 + 13k + 2)$ and $(5k + 3)$ clauses with $(4k+1)$ and $(k+1)$ extra variables being introduced, respectively, by Tseitin's circuit-to-CNF conversion [15]. The satisfiability of such a TCF can be difficult to solve especially when the corresponding circuit is large. (Note that the number of nodes in the circuit is bounded from above by the number of possible arrival times of all nodes.)

3.1.2 Our Formulation

A close examination of Equations (3) and (4) reveals that they are essentially equivalent in circuit delay computation. In fact, as has been shown earlier in [11], Equation (4) yields exact (rather than approximative, as interpreted in [14]) analysis when path tracing is not performed.

---

[1]Equation (3) looks different from the one in [14] as it was previously expressed by both exact and viable TCFs.

Building upon Equation (4), we propose a general and compact TCF formula for arbitrary complex gates as follows.

PROPOSITION 1. *For a node f with a set C of controlling cubes, its TCF can be expressed as*

$$\chi^{f,t} = \bigvee_{g_i \in\ I(f)} \chi^{g_i, t-d_i} \wedge \bigwedge_{c \in C} \bigvee_{lit(g_i) \in c} (\chi^{g_i, t-d_i} \vee \neg lit(g_i)), \quad (5)$$

where d_i is the pin-to-pin delay from g_i to f and $lit(g_i)$ denotes the literal of g_i.

PROOF. There are exactly two possible cases for the value of f being determined before time t. First, the value of every $g_i \in FI(f)$ is determined before time $(t - d_i)$. Second, every constituent input g_i of some controlling cube c is determined to its corresponding value $lit(g_i) \in c$ before time $(t - d_i)$. Since any of the above cases makes $\chi^{f,t}$ false, the condition can be formally translated to

$$\neg\chi^{f,t} = \bigwedge_{g_i \in\ I(f)} \neg\chi^{g_i, t-d_i} \vee \bigvee_{c \in C} \bigwedge_{lit(g_i) \in c} (\neg\chi^{g_i, t-d_i} \wedge lit(g_i))),$$

whose negation equals Equation (5). ∎

Note that, for simple gates (with controlling values, in other words, with one-literal controlling cubes), Equation (5) reduces to Equation (4).

With the key observation that $\chi^{f,t}$ is recursively defined in Equation (4) with the appearance only in the positive phase without any negation, implication suffices to express the TCF constraints. The advantage of using implication, instead of equation, is that we can apply Plaisted-Greenbaum encoding [12], instead of Tseitin encoding, in converting TCFs to CNF formulas. Specifically, Equation (5) with the equality sign "$=$" being replaced by the implication sign "$\rightarrow$" can be directly translated into the CNF formula

$$(\neg\chi^{f,t} \vee \bigvee_{g_i \in\ I(f)} \chi^{g_i, t-d_i}) \bigwedge_{c \in C} (\neg\chi^{f,t} \vee \bigvee_{lit(g_i) \in c} (\chi^{g_i, t-d_i} \vee \neg lit(g_i))), \quad (6)$$

which consists of $|C| + 1$ clauses without introducing any extra variable. Hence, unlike prior methods, building TCF circuits is unnecessitated.

Note that, in converting the entire recursive definition of $\chi^{f,t}$, Tseitin encoding is still needed for parts of the original circuit that are relevant to the literals $lit(g_i)$ in individual TCFs (since these literals may appear in both positive and negative phases). Nevertheless the conversion with Tseitin encoding is applied once on the original circuit and is shared by all individual TCFs.

3.2 TCF with 0/1-Specificity

3.2.1 Prior Formulation

Prior work [6] intended to improve [14] by exploiting early TCF to simplify TCF circuits. The following equations were proposed.

$$\chi^{f,t} = \chi^{f=1,t} \vee \chi^{f=0,t}$$
$$= (f \wedge \neg\chi^{f=1,t-}) \vee (\neg f \wedge \neg\chi^{f=0,t-}) \quad (7)$$

$$\chi^{f=1,t-} = \begin{cases} \bigwedge_{g_i \in\ I(f)} \chi^{g_i=1,(t-d_{r_i})-}, & \text{for AND-gate } f \\ \bigvee_{g_i \in\ I(f)} \chi^{g_i=1,(t-d_{r_i})-}, & \text{for OR-gate } f \end{cases}$$

$$\chi^{f=0,t-} = \begin{cases} \bigvee_{g_i \in\ I(f)} \chi^{g_i=0,(t-d_{f_i})-}, & \text{for AND-gate } f \\ \bigwedge_{g_i \in\ I(f)} \chi^{g_i=0,(t-d_{f_i})-}, & \text{for OR-gate } f \end{cases} \quad (8)$$

where d_{r_i} and d_{f_i} are the corresponding rising and falling pin-to-pin delays from g_i to f, respectively. In the above expressions, TCF $\chi^{f,t}$ is obtained from two subcases $\chi^{f=1,t}$

and $\chi^{f=0,t}$, where $\chi^{f=v,t}$ is satisfiable if f stablizes to value v no earlier than time t. Note that $\chi^{f=v,t} \neq \neg\chi^{f=v,t-}$, but rather $\chi^{f=v,t} = (f \oplus \neg v) \wedge \neg\chi^{f=v,t-}$.

The advantages of separating $\chi^{f=1,t}$ and $\chi^{f=0,t}$ from $\chi^{f,t}$ are two-fold: First, it allows distinction between rising and falling delays and thus permits more accurate timing analysis. Second, since Equation (8) in circuit representation consists of a single gate, no internal variable needs to be introduced in conversion to CNF. The resultant CNF formula is easier to solve.

The disadvantages, on the other hand, are also two-fold: First, such separation doubles the TCF formula size. Second, since the formulation works for simple gates only, timing analysis of circuits with complex gates is approximative. In fact, Equation (8) can be generalized for complex gates [10] with

$$\chi^{f=1,t-} = \bigvee_{c \in C_1} \bigwedge_{lit(g_i) \in c} \chi^{g_i=v,(t-d_{r_i})-} \text{ and}$$

$$\chi^{f=0,t-} = \bigvee_{c \in C_0} \bigwedge_{lit(g_i) \in c} \chi^{g_i=v,(t-d_{f_i})-}, \quad (9)$$

where C_1 and C_0 are the sets of all prime implicants of f and $\neg f$, respectively, and $v = 0$ if $lit(g_i) = \neg g_i$ and $v = 1$ if $lit(g_i) = g_i$. When translated to CNF, Equation (9) is more complicated than Equation (8) however. Note that Plaisted-Greenbaum encoding is not applicable here due to the negations in Equation (7).

3.2.2 Our Formulation

The aforementioned disadvantages can be overcome as follows.

PROPOSITION 2. *Given a circuit, let f be a node with the set C_1 and C_0 of all prime implicants of f and $\neg f$, respectively. Then 0/1-specified TCF can be expressed as*

$$\chi^{f=1,t} = f \wedge \bigwedge_{c \in C_1} \bigvee_{lit(g_i) \in c} (\chi^{g_i=v,t-d_{r_i}} \vee \neg lit(g_i)) \text{ and}$$

$$\chi^{f=0,t} = \neg f \wedge \bigwedge_{c \in C_0} \bigvee_{lit(g_i) \in c} (\chi^{g_i=v,t-d_{f_i}} \vee \neg lit(g_i)), \quad (10)$$

where $v = 0$ if $lit(g_i) = \neg g_i$ and $v = 1$ if $lit(g_i) = g_i$.

PROOF. If $\chi^{f=1,t}$ is satisfied, it means that f valuates to true no earlier than time t. That is, for every cube in C_1, it is either not satisfied, or satisfied with at least one controlling literal valuates to true no earlier than time $t - d$. Similarly, one can prove the case of $\chi^{f=0,t}$. ∎

Since all the TCFs appear in $\chi^{f,t} = \chi^{f=1,t} \vee \chi^{f=0,t}$ and in Equation (10) without any negation, again Plaisted-Greenbaum encoding applies for CNF conversion.

3.3 Comparison on TCF Formulas

Table 1: TCF Comparison

		TCF		PO			Generality	
	Eq	#Vr	#Cl	Eq	#Vr	#Cl	CG	RF
[14]	(3)	$k+1$	$5k+3$	(1)	0	1	No	No
	(4)	$4k+1$	$k^2+13k+2$	(1)	0	1	No	No
[6]	(8)	0	$2k+2$	(7)	$2m$	$9m$	No	Yes
[10]	(9)	$k+1$	$4k+4$	(7)	$2m$	$9m$	Yes	Yes
Our	(5)	0	$k+1$	(1)	0	1	Yes	No
	(10)	0	$k+3$	(2)	0	1	Yes	Yes

Table 1 compares our formulations with those of [14], [6], and [10]. For a k-input simple gate, the number of extra variables and the number of clauses corresponding to the TCF

1057

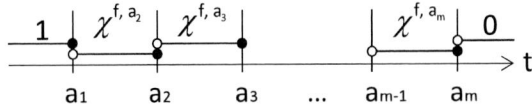

Figure 1: Equivalence intervals of $\chi^{f,t}$.

equations in Column 2 are shown in Columns 3 and 4, respectively. For a circuit with m POs, the number of extra variables and the number of clauses corresponding to the PO equations in Column 5 are shown in Columns 6 and 7, respectively. The generality for each formulation in supporting complex gate types and supporting rise/fall delays are summarized in Columns 8 and 9, respectively.

3.4 TCF Equivalence Reduction

Given a circuit with a node f, its TCFs $\chi^{f,t}$ for all t can be partitioned into equivalence classes. This equivalence relation can be exploited to simplify the recursive TCF construction. In [10], TCF equivalence based on arrival-time information is introduced. Assume that the set A of all possible arrival times of node f are sorted in an ascending order as $\{a_1, a_2, \ldots, a_m\}$ for $a_{i-1} < a_i$. Then $\chi^{f=v,t-} = \chi^{f=v,a_i-}$ if $a_{i-1} < t \leq a_i$. That is, two temporal conditions t_1 and t_2 of f are equivalent if they have the same next larger or equal arrival time in A.

For practical implementation, we propose a table lookup approach to TCF equivalence reduction with three improvements over prior works [10, 6]. First, for TCFs with 0/1-specificity, the set of arrival times of a node f is further distinguished into two sets A_1 and A_0 for those resulting in $f = 1$ and $f = 0$, respectively. This distinction reduces the number of arrival times and thus TCF equivalence classes.

Second, under boundary conditions, a TCF is substituted with a constant 0 or 1 for further reduction (constant 1 is not applicable for prior works). Specifically, Figure 1 depicts the equivalence intervals of the TCFs of node f with extended boundary conditions. If t is larger than the maximum arrival time a_m of a node f, then $\chi^{f,t}$ is unsatisfiable since f always stabilizes before t. In this case, $\chi^{f,t}$, $\chi^{f=1,t}$ and $\chi^{f=0,t}$ all equal Boolean constant 0. On the contrary, if t is no larger than the minimum arrival time a_1, then $\chi^{f,t}$ is a tautology (but $\chi^{f=1,t}$ and $\chi^{f=0,t}$ are not necessarily tautologies). That is, $\chi^{f,t}$ equals constant 1, and furthermore $\chi^{f=v,t}$ can be simplified to $f \oplus \neg v$ by $\chi^{f=v,t} = (f \oplus \neg v) \wedge \chi^{f,t}$. Observe that, in Equation (10), $\chi^{g_i=v,t-d_i}$ and $\neg lit(g_i)$ are always present together in a clause with $\neg lit(g_i) = g_i \oplus v$. When $\chi^{g_i,t-d_i} = 1$, since $\chi^{g_i=v,t-d_i} = g_i \oplus \neg v$, this clause must be satisfied due to $(\chi^{g_i=v,t-d_i} \vee \neg lit(g_i)) = ((g_i \oplus \neg v) \vee (g_i \oplus v)) = 1$. Therefore, whenever $\chi^{g_i,t-d_i} = 1$, substituting constant 1 for $\chi^{g_i=1,t-d_i}$ and $\chi^{g_i=0,t-d_i}$ is safe without altering the satisfiability of $\chi^{f,t}$. As a result, our TCFs without and with 0/1-specificity can be simplified with such constant substitution.

Third, our TCF equivalence reduction is applied to all nodes including PIs and POs. Because of the aforementioned first improvement, any TCF of a PI is either constant 1 or constant 0 because any PI has only one arrival time. On the other hand, since the arrival times at POs are the only candidate circuit delays, this information is exploited to save unnecessary checking. More precisely, only PO arrival times are checked for circuit delay by Formula (1); once some candidate delay is falsified, this delay and other larger delays are removed from the arrival-time lists of all POs. For example, assume two POs p_1 and p_2 have arrival-time lists $\{4, 5, 7\}$ and $\{6, 7\}$, respectively. If $(\chi^{p_1,7} \vee \chi^{p_2,7})$ is un-

satisfiable, we remove 7 from the two lists. Then we check $(\chi^{p_1,6} \vee \chi^{p_2,6}) = (0 \vee \chi^{p_2,6})$. Note that this removal is crucial. If 7 were not removed from the list of p_1, then $\chi^{p_1,6}$ would equal $\chi^{p_1,7}$ instead of 0 and $\chi^{p_1,7}$ would be built again.

4. ALGORITHMS

The overall algorithms of circuit delay computation and critical region identification are presented in this section.

4.1 Delay Computation

Figure 2 sketches a procedure for delay computation without rise/fall time separation. It can be easily extended under a similar framework to the computation with rise/fall time separation, which is omitted for brevity. To avoid confusion between a TCF and its output variable, in the pseudo code $x^{f,t}$ represents the output variable of TCF $\chi^{f,t}$.

While the code is self-explanatory, it should be noted that different delay search strategies can be applied depending on how functions *GetDelayList*, *GetNextDelay*, and *UpdateDelayList* are implemented. For instance, linear or binary search can be deployed with or without adaptive step-size adjustment. Counterintuitively empirical experience suggests that linear search in general works much better than binary search. Investigation reveals that, although linear search requires more SAT solving iterations than binary search, it allows the second improvement technique of Section 3.4 more applicable and thus making the CNF formula at each iteration easier to solve.

Upon termination (line 14 of *ComputeDelay*), Formula (1) must be satisfiable for $D = lowerDelay$. That is, there exists a PI assignment to sensitize some true path achieving this delay value. By applying the assignment values to PIs, we can simulate and trace one true critical path based on the exact sensitization criterion [2].

4.2 Critical Region Identification

Our delay computation algorithm can be applied to identify true timing critical regions for delay optimization. Given a target required time of a circuit, topological timing critical regions (with small slacks) can be identified by conventional STA analysis. Topological timing critical regions overapproximate functional true critical regions. The approximation can be very crude, and in this case many false critical gates and paths can be trimmed away. The true critical regions can be pinpointed by removing false arrival times with the third improvement technique of Section 3.4. Note that the TCFs of non-critical gates equal constant 0 due to the boundary condition ($t > a_m$) of TCF equivalence reduction. Effectively the computation considers only the timing critical sub-circuit, which can be much smaller than the entire circuit.

5. EXPERIMENTAL RESULTS

Our methods, named "SWIFT" for Equation (5) and "SWIFT-0/1" for Equation (10), were implemented in the C++ language using MiniSat version 2.20 [5] as the underlying SAT solver. All experiments were conducted on a Linux machine with a Xeon 3.4 GHz CPU and 32 GB RAM. Large ISCAS, ITC, and other industrial benchmark circuits were selected for experiments. For the sake of comparison with prior work [6], which handles only simple gate types, all circuits are technology mapped using only buffers, inverters, AND-gates, OR-gates, NAND-gates, and NOR-gates. It should be noted, however, that our computation is not restricted to these simple gate types and can be generally applicable to general complex gates.

1058

```
ComputeDelay(C) //compute maximum true-path delay of circuit C
begin
01    L := GetDelayList(C);
02    (lowerDelay, upperDelay) := MinMaxTopologicalDelay(C);
03    do
04        D := GetNextDelay(L);
05        Φ := (⋁_{p∈PO} x^{p,D});
06        for every PO p
07            Φ := Φ ∧ BuildTcf(p, D);
08        if IsSat(Φ)
09            lowerDelay := D;
10        else
11            upperDelay := D;
12        UpdateDelayList(C, L, lowerDelay, upperDelay);
13    while L non-empty;
14    return lowerDelay and its corresponding true path;
end

BuildTcf(f,t) //derive χ^{f,t} in CNF
begin
01    t := GetNextLargerOrEqualArrivalTime(f);
02    if χ^{f,t} has been built
03        return 1;
04    if t > f.a_m //largest arrival time of f
05        return (¬x^{f,t});
06    if t ≤ f.a_l //smallest arrival time of f
07        return (x^{f,t});
08    if f has only one fanin g_i
09        return BuildTcf(g_i, t − d_i) with x^{g_i,t−d_i} replaced by x^{f,t};
10    Φ := (¬x^{f,t} ∨ ⋁_{g_i ∈ I(f)} x^{g_i,t−d_i});
11    for each controlling cube c of f
12        Φ := Φ ∧ (¬x^{f,t} ∨ ⋁_{lit(g_i)∈c}(x^{g_i,t−d_i} ∨ ¬lit(g_i)));
13    for each g_i ∈ I(f)
14        Φ := Φ ∧ BuildTcf(g_i, t − d_i);
15        if g_i's circuit CNF has not been built
16            Φ := Φ ∧ BuildCktCnf(g_i);
17    return Φ;
end
```

Figure 2: Algorithm: Delay Computation

5.1 Delay Computation

For circuit delay computation, prior method [6], using Equations (7) and (8), was re-implemented under the same setting (including the same linear delay search strategy in a descending order) as ours for fair comparison. (We did not compare with [14] and [10] as they are not as efficient as [6].) The comparison was performed under four delay models: the unit gate delay model, fanout delay model (by calculating a gate delay as $1 + 0.2 \times$ fanout number), TSMC $0.18\mu m$ library model with combined rise/fall time (by calculating a gate delay as max{rise delay, fall delay}, and TSMC $0.18\mu m$ library model with separate rise/fall time.

Table 2 shows the experimental results under the four delay models. Column 2 shows the gate count; Column 3 shows the longest topological delay and actual true-path delay; Column 4 shows the number of SAT solving iterations needed to identify the true-path delay; Columns 5 and 8 (respectively Columns 6 and 9) show the total number of variables excluding those in original circuits (respectively clauses) involved in the CNF formulas of all SAT solving iterations; Columns 7 and 10 show the total SAT solving time in seconds. (The reported runtime excludes preprocessing time as both prior and our methods were preprocessed in a similar way. The prior method may take slightly longer time because of converting circuits to CNF formulas.) Note that SWIFT is only applicable to the first three timing models (without separating rise and fall delays) because its TCF formulation has no 0/1-specificity, and thus SWIFT-0/1 is applied in the fourth timing model with separate rise and fall delays.

The results suggest that SWIFT performs robustly and efficiently (with all runtimes within 3.06 seconds) under various delay models while the performance [6] is unpredictable (as

Table 3: Critical Region Identification

Circuit	#G	Topological		Functional		Time
		#G	#Path	#G	#Path	(s)
b05	1022	322	7435427	186	8669	0.04
b17	33741	1637	5585965	79	232	0.61
b18	117941	1101	77585298	465	20839024	18.67
c3540	1741	270	1054	91	100	0.02
c5315	25585	213	832	76	60	0.02
c7552	3827	304	97	61	8	0.01
i10	2724	452	127483	338	3071	0.08
s15850	11067	408	73984	389	22016	0.16
s38417	2608	230	476	112	10	0.06

exemplified by circuit `leon3mp`, which is solved in 12229.60 seconds under the unit delay model and 2.66 seconds under the fanout delay model) and is not as efficient. The efficiency of SWIFT stems from several factors. First, the numbers of variables and clauses encountered in SWIFT are about half of those in [6]. Second, replacing equivalence-based with implication-based TCF construction makes SAT solving easier. Third, the TCF without 0/1-specificity is more compact than that with 0/1-specificity. Fourth, perhaps most importantly, SWIFT yields more constant propagations due to equivalent TCF reduction.

On the other hand, the results also suggest that SWIFT-0/1 outperforms [6] (by a factor of 3.09 measured by geometric mean). It is interesting to note that circuit `netcard` took SWIFT-0/1 long time to solve comparable to that of [6]. (Although the timing improvement is not remarkable in this case, SWIFT-0/1 offers the generality to handle complex gates, which is not available in [6].) Compared to SWIFT, SWIFT-0/1 does not enjoy as much variable and clause reductions, and constant propagations. The formulations of SWIFT-0/1 and [6] have their own strengths. For SWIFT-0/1, there are fewer variables and clauses, and constant propagation in equivalent TCF reduction is possible. For [6], because $\chi^{f=v,t}$ in Equation (8) depends only on its fanin TCFs but not on other variables, it makes CNF formulas simple. However the formulation is only applicable to simple gates.

Table 2 also reveals that topological delay may be far pessimistic compared to true circuit delay, e.g., circuits `b05` and `b19` under the unit and fanout delay models. It suggests the importance of accurate functional timing analysis and its application on identifying true critical region for timing optimization.

5.2 Critical Region Identification

Table 3 evaluates the applicability of SWIFT on identifying timing critical regions under the unit delay model. For a circuit, its true delay is set to be the required time at its POs, and the gates and paths with non-positive slack values are declared critical. Column 2 shows the total number of gates of a circuit; Columns 3 and 4 (Columns 5 and 6) show the numbers of critical gates and paths, respectively, with respect to topological arrival times (functional true arrival times); Column 7 shows the runtime in identifying true critical regions.

The results suggest that SWIFT effectively removed spurious critical gates and paths. As a matter of fact, true critical regions can be much smaller than topological critical regions. By taking circuit `b17` as an example, SWIFT detected, in 0.61 seconds (the time spent in SAT solving), that only 79 out of its 1637 topological critical gates are true critical gates, and at least 5585733 out of its 5585965 topological critical paths are false critical paths. Pinpointing true critical regions efficiently can be beneficial to timing optimization.

6. CONCLUSIONS AND FUTURE WORK

Table 2: Circuit Delay Computation

				Unit Delay					
				[6]			SWIFT		
Circuit	#Gate	Delay	#SAT	#Var	#Clause	Time (s)	#Var	#Clause	Time (s)
b05	1022	54→42	8	15672	51104	0.03	7786	25030	0.01
b18	117941	164→159	5	18746	58335	1.72	9221	28300	0.29
b19	237959	168→158	7	84174	262549	11.50	41597	128850	0.76
c6288	2480	124→123	3	4248	12747	0.19	2118	6324	0.08
leon2	1119384	42→42	1	514	1703	0.21	250	829	< 0.01
leon3	1272597	44→44	1	354	1171	0.06	173	560	< 0.01
leon3mp	824294	40→38	3	427522	1387725	12229.60	155786	519905	0.79
netcard	983683	29→29	1	144	503	0.09	70	226	< 0.01
ray	235526	178→178	1	10338	35173	2.01	5051	17205	0.08
s35932	19876	29→26	4	100608	301828	6.95	49152	138244	0.06
uoft_raytracer	218671	178→178	1	11476	39015	1.08	5618	19109	0.02

				Fanout Delay					
				[6]			SWIFT		
Circuit	#Gate	Delay	#SAT	#Var	#Clause	Time (s)	#Var	#Clause	Time (s)
b05	1022	80.6→64.0	44	291364	945194	0.54	145404	469541	0.10
b18	117941	242.8→238.0	12	20140	62692	1.23	9861	30677	0.15
b19	237959	244.4→234.4	25	528358	1652131	77.60	262406	820125	3.06
c6288	2480	176.4→174.8	3	3222	9669	0.12	1606	4798	0.03
leon2	1119384	2070.0→2070.0	1	41544	149437	241.61	14628	55764	0.03
leon3	1272597	6854.4→6854.4	1	198674	599655	1172.05	66569	201522	0.11
leon3mp	824294	3093.2→3093.2	1	2224	7419	2.66	744	2604	< 0.01
netcard	983683	16390.2→16390.2	1	393228	1179685	0.76	131078	393231	0.27
ray	235526	383.6→383.6	1	796	2561	0.13	334	1087	0.01
s35932	19876	42.8→39.0	4	86784	260356	8.92	42240	122692	0.12
uoft_raytracer	218671	383.6→383.6	1	796	2561	0.06	334	1087	0.01

				TSMC 0.18μm Cell Library with Combined Rise/Fall Time					
				[6]			SWIFT		
Circuit	#Gate	Delay	#SAT	#Var	#Clause	Time (s)	#Var	#Clause	Time (s)
b05	1022	5.67→4.45	62	484771	1571927	0.89	241959	782062	0.13
b18	117941	13.49→13.43	3	2326	7415	0.16	1083	3457	0.02
b19	237959	14.09→13.80	15	114446	352033	1.53	56743	174462	0.25
c6288	2480	13.95→13.79	5	6916	20753	2.12	3450	10315	0.05
leon2	1119384	13.16→13.16	1	9356	28285	38.36	3142	9532	0.10
leon3	1272597	14.28→14.16	8	62946	190674	169.67	24924	75932	0.12
leon3mp	824294	14.58→14.27	17	169690	511361	876.46	57133	172517	0.19
netcard	983683	8.61→8.42	3	13180	39583	88.08	4404	13227	0.10
ray	235526	31.84→31.75	5	10310	34683	0.82	4999	16866	0.02
s35932	19876	2.83→2.64	5	85888	257669	6.06	41760	121125	0.05
uoft_raytracer	218671	31.98→31.98	1	804	2711	0.11	386	1306	0.02

				TSMC 0.18μm Cell Library with Separate Rise/Fall Time					
				[6]			SWIFT		
Circuit	#Gate	Delay	#SAT	#Var	#Clause	Time (s)	#Var	#Clause	Time (s)
b05	1022	4.67→3.52	59	433885	1407065	0.83	433148	1132193	0.72
b18	117941	11.08→10.98	7	9278	28949	0.58	9134	23404	0.48
b19	237959	11.67→11.42	14	135078	415139	1.24	134374	340148	0.92
c6288	2480	10.13→10.02	5	4068	12206	0.59	4060	10126	0.79
leon2	1119384	11.07→11.07	1	4678	14143	35.55	3142	7987	0.97
leon3	1272597	12.81→12.68	9	26375	79963	130.69	18137	46295	3.07
leon3mp	824294	12.39→12.03	4	95668	324471	603.09	64020	197810	50.22
netcard	983683	7.67→6.87	27	2915866	9076031	92964.40	2334866	6173237	90456.2
ray	235526	26.77→26.43	18	55045	183570	0.58	54186	140414	0.37
s35932	19876	2.45→2.35	5	40768	122308	2.91	39616	98148	1.67
uoft_raytracer	218671	26.40→25.82	31	213908	743738	7.02	213179	567656	2.11

This paper has shown that functional timing analysis can be made fast and general compared with sate-of-the-art methods. Based on implication relation and other technical improvements, compact CNF encoding for TCFs without and with 0/1-specificity has been devised. Thereby the power of modern SAT solvers can be fully utilized. Experiments on large designs have demonstrated promising results on delay computation and critical region identification.

Acknowledgments

The authors acknowledge Yuji Kukimoto for reference [10]. This work was supported in part by the National Science Council under grants NSC 99-2221-E-002-214-MY3, 99-2923-E-002-005-MY3, and 100-2923-E-002-008.

7. REFERENCES

[1] P. Ashar and S. Malik. Functional timing analysis using ATPG. *IEEE Trans. Comput.-Aided Design Integr. Circuits Syst.*, 14(8): 1025-1030, Aug. 1995.

[2] H.-C. Chen and D. Du. Path sensitization in critical path problem. *IEEE Trans. Comput.-Aided Design Integr. Circuits Syst.*, 12(2): 196-207, Feb. 1993.

[3] O. Coudert. An efficient algorithm to verify generalized false paths. In *Proc. Design Automation Conf.*, 2010.

[4] S. Devadas, K. Keutzer, and S. Malik. Computation of floating mode delay in combinational circuits: Theory and algorithms. *IEEE Trans. Comput.-Aided Design Integr. Circuits Syst.*, 12(12): 1913-1923, Dec. 1993.

[5] N. Eén and N. Sörensson. An extensible SAT-solver. In *Proc. SAT*, pp. 502-518, 2003.

[6] Y.-M. Kuo, Y.-L. Chang, and S.-C. Chang. Efficient Boolean characteristic function for timied automatic test pattern generation. *IEEE Trans. on Computer-Aided Design of Integrated Circuits and Systems*, 28(3): 417-425, March 2009.

[7] P. C. McGeer and R. K. Brayton. *Integrating unctional and Temporal Domains in Logic Design.* Kluwer Academic Publishers, 1991.

[8] M. Moskewicz, C. Madigan, L. Zhang, and S. Malik. Chaff: Engineering an efficient SAT solver. In *Proc. DAC*, pp. 530-535, 2001.

[9] J. Marques-Silva and K. Sakallah. GRASP: A search algorithm for propositional satisfiability. *IEEE Trans. on Computers*, vol. 48, no. 5, pp. 506-521, May 1999.

[10] P. McGeer, A. Saldanha, R. Brayton, and A. Sangiovanni-Vincentelli. Delay models and exact timing analysis. In *Logic Synthesis and Optimization*, Kluwer Academic Publishers, pp. 167-189, 1993.

[11] P. C. McGeer, A. Saldanha, P. R. Stephan, R. K. Brayton, and A. L. Sangiovanni-Vicentelli. Timing analysis and delay-fault test generation using path-recursive functions. In *Proc. Int. Conf. on Computer-Aided Design*, pages 180-183, 1991.

[12] D. Plaisted and S. Greenbaum. A structure-preserving clause form translation. . *Symbolic Computation*, 2:293-304, 1986.

[13] S. Roy, P. P. Chakrabarti, and P. Dasgupta. Event propagation for accurate circuit delay calculation using SAT. *ACM Trans. Design Autom. Electron. Syst.*, 12(3), Aug. 2007.

[14] L. Silva, J. Marques-Silva, L. Silveira, and K. Sakallah. Satisfiability models and algorithms for circuit delay computation. *ACM Trans. on Design Automation of Electronic Systems*, 7(1): 137-158, Jan. 2002.

[15] G. Tseitin. On the complexity of derivation in propositional calculus. *Studies in Constructive Mathematics and Mathematical Logic*, pp. 466-483, 1970.

Timing Analysis with Nonseparable Statistical and Deterministic Variations

Vladimir Zolotov[1], Debjit Sinha[2], Jeffrey Hemmett[3], Eric Foreman[3], Chandu Visweswariah[2],
Jinjun Xiong[1], Jeremy Leitzen[4], Natesan Venkateswaran[2]

[1] IBM Thomas J. Watson Research Center, Yorktown Heights, New York
[2] IBM Systems and Technology Group, Hopewell Junction, New York
[3] IBM Systems and Technology Group, Essex Junction, Vermont
[4] IBM Systems and Technology Group, Rochester, Minnesota

{zolotov, debjit.sinha, hemmett, eforeman, chandu, jinjun, jleitzen, natesan}@us.ibm.com

ABSTRACT

Statistical static timing analysis (SSTA) is ideal for random variations but is not suitable for environmental variations like Vdd and temperature. SSTA uses statistical approximation, according to which circuit timing is predicted accurately only for highly probable combinations of variational parameters. SSTA is not able to handle accurately deterministic sources of variation like supply voltage. This paper presents a novel technique for modeling nonseparable deterministic and statistical variations in single timing run.

Categories and Subject Descriptors

B.8.2 [**Integrated Circuits**]: Performance Analysis and Design Aids

General Terms

Algorithms, Design, Verification

Keywords

Static Timing, Statistical Timing, Variability

1. INTRODUCTION

Variability becomes one of the most limiting factors for achieving required performance and power characteristics of VLSI circuits. The variability is getting not only higher but also more complex. The number of process and environmental parameters impacting VLSI circuits is increasing with every technology node. Statistical static timing analysis (SSTA) [1, 2] proved to be a valuable alternative to multi-corner deterministic timing. SSTA predicts circuit timing for all possible combinations of process parameters.

Unfortunately, conventional SSTA models only statistical parameters. The approximations used by SSTA provide good accuracy only for highly probable combinations of variational parameters. High errors for low probability situations are not considered important because the number of chips with extreme variations is relatively small, and their impact on the total yield is insignificant. However, signal propagation is affected not only by statistical but by deterministic variations as well. Supply voltage (Vdd) is the most important deterministic parameter. Many VLSI chips are designed to operate in wide range of supply voltages. For those chips Vdd variations cannot be modeled statistically. It is wrong to say that it is enough if chips function correctly for highly probable Vdd values and may fail for low probability values. The manufactured chips must work correctly in the whole Vdd range for which they are designed.

Usually, the problem of Vdd variability is solved by running SSTA for low and high Vdd separately. However, two runs are very inconvenient both for timing closure and for circuit optimization. After fixing timing violations at low Vdd, timing at high Vdd discovers more violations, some of which were possibly introduced by fixing the circuit at low Vdd. Therefore, we have to iterate between low and high Vdd several times. The problem is even more difficult for chips designed for selective voltage binning (SVB) [3], adaptive voltage supply (AVS) [4, 5] or dynamic voltage scaling (DVS) [6]. SVB methodology assigns the manufactured chips to different voltage bins according to their performance and leakage characteristics. AVS senses characteristics of chip transistors and adjusts Vdd appropriately. DVS adjusts Vdd according to the workload. Timing analysis of such chips requires modeling of not only wide Vdd variations but also the policy of Vdd adjustment.

In spite of the high practical importance of modeling deterministic and statistical variations in a single timing run, this problem has not received enough attention in research publications. Computation of low and high timing bounds is proposed in [7]. However, that technique is expensive computationally and produces rather pessimistic bounds because it does not exploit the statistical approximation.

Delay variability due to process variations strongly depends on Vdd. Therefore, Vdd and process variations are not separable and cannot be modeled accurately with the linear canonical form [1]. Non-Gaussian and nonlinear variability attracted lots of research efforts in numerous publications. For example, [8] uses a numerical technique for computing tightness probabilities, [9] exploits chaos expansions, [10] applies moment matching of probability density functions (PDF). [11] uses polynomial fitting, [12] models skewness of PDF. However, all of them use the idea of statistical approximation: reducing error for highly probable variations at the expense of high error in other regions of variations. Therefore, these and similar methods are not suitable for modeling deterministic variations. Another issue is models that are too complex and correspondingly low computational efficiency.

This work presents a novel technique for modeling both statistical and deterministic variations in a single timing run. The deterministic parameters are modeled for the whole range of their variations, while statistical parameters are modeled using conventional statistical approximation. This approach allows computation of arrival times (ATs), required arrival times (RATs) and timing slacks in linear canonical form for all Vdd values. The computation achieves the highest accuracy at low and high Vdd corners. We take into account nonseparable interaction of Vdd variation and process variations by modeling sensitivity of delays and slews to cross-terms of deterministic and statistical sources of variations. The proposed technique is implemented in an industrial statistical timer [1] targeted to circuits with many millions of gates. The algorithms guarantee results equivalent to two separate SSTA runs at low and high Vdd.

2. BACKGROUND

2.1 Statistical Timing

Static timing analysis models the circuit with a timing graph. The graph nodes represent gate pins. The graph edges model propagation of signals through gates and interconnects. Arrival times (ATs) are propagated using two operations: propagation through an edge and computation of the latest AT at a node. The first operation sums the propagated AT with the edge delay. The second operation computes maximum of the ATs arriving at the node.

SSTA models all timing quantities (delays, slews, ATs, etc.) with a linear canonical form (LCF)

$$t = t_0 + \sum_{i=1}^{n} a_i \Delta X_i + a_R \Delta R \qquad (1)$$

where t_0 is a mean value, ΔX_i is a random variable modeling chip to chip variations, ΔR is a random variable modeling an uncorrelated variation, a_i and a_R are sensitivities to those variations respectively. Random variables ΔX_i and ΔR have zero mean values and unity variances. In most industrial applications, variability is approximated with Gaussian distributions.

SSTA uses statistical addition and maximum operations for LCFs. The statistical addition is performed by summation of the corresponding sensitivities to global variations and root square summation of the sensitivities to the uncorrelated variation. SSTA uses Clark's linear Gaussian approximation [13] for the statistical maximum. The approximation is statistical because it minimizes the error weighted by the PDF of the process parameters. The error is reduced in the central region of the distribution at the expense of large error on the distribution tails. Fig. 1 illustrates the statistical maximum for LCF with one variable. There are numerous publications on nonlinear and non-Gaussian approximations of statistical maximum [8, 9, 10, 11, 12]. All of them exploit the same idea of statistical approximation.

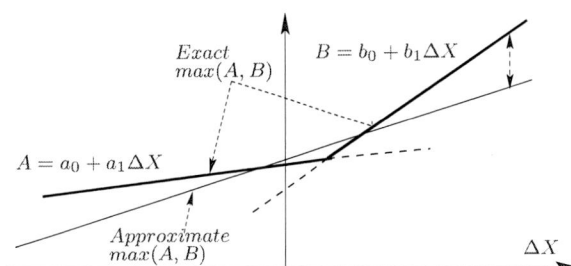

Figure 1: Approximation of statistical max function

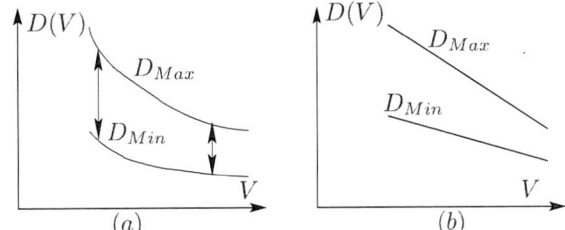

Figure 2: Delays as functions of Vdd

2.2 Non-statistical Sources of Variations

The nature of deterministic variability differs from the statistical one. Regarding statistical parameters we assume that some values of them do not require accurate modeling because the probability of those values is negligible. This is not valid for deterministic variations as we need guarantee correct chip operation for all deterministic variations. Supply voltage (Vdd) is the most important deterministic parameter. If a circuit is designed to work in a required range of Vdd, the manufactured chips must work at any Vdd value in this range.

Deterministic parameters with small impact on delays can be modeled using statistical approximation but it requires additional guard-banding similar to the one used in deterministic timing. However, parameters with large variability like Vdd cannot be modeled that way as it requires too large guard bands.

It is well known that CMOS transistors at different Vdd have different sensitivity to Leff and Vth variations. Vdd strongly affects delay sensitivities to process variations. This fact is illustrated in Fig. 2 (a) showing how maximum and minimum delays depend on Vdd. We see that the range of delay variation is different for different Vdd values. It means that the Vdd and process variations are not separable, and the separable model is not suitable for modeling combined Vdd and process variability.

2.3 Problem Formulation

The goal of this work is a technique for modeling nonseparable deterministic and statistical variations in a single timing run. Assuming a linear model for statistical variations at a fixed Vdd, and considering Vdd as a deterministic parameter, timing should compute all timing quantities in LCF at all valid Vdd values. Moreover, we require that the results for low and high Vdd corners are the same as the ones produced by two runs of the conventional SSTA at those corners.

1062

3. NONSEPARABLE MODEL

3.1 Extended Canonical Form

Modeling statistical and deterministic variability, we would like to preserve the benefits of the conventional SSTA. Therefore, we extend the canonical form (1) by introducing dependence of the mean value and sensitivities on Vdd:

$$t = t_0(\Delta V) + \sum_{i=1}^{n} a_i(\Delta V)\Delta X_i + a_R(\Delta V)\Delta R \quad (2)$$

where ΔV denotes deterministic Vdd variation, $t_0(\Delta V)$ is a mean value as a function of Vdd variations, ΔX_i and ΔR are statistical global and independent variations, and $a_i(\Delta V)$, $a_R(\Delta V)$ are their sensitivities as functions of Vdd variations. Unlike statistical variables defined by means and variances, the deterministic variable ΔV is defined with its corner values: v_h and v_l.

We define a projection of a parameterized form.

DEFINITION 1. *The operation of substitution of a constant value v for variable ΔV into parameterized form t (2) is called projection to that value v and is denoted as $t(v)$.*

$$t(v) = t_0(v) + \sum_{i=1}^{n} a_i(v)\Delta X_i + a_R(v)\Delta R \quad (3)$$

The result of projection does not depend on Vdd variation. It describes variability due to statistical parameters at fixed Vdd v.

PROPERTY 1. *Extended form (2) depends on statistical parameters ΔX_i, ΔR linearly. Therefore, any of its projections is an LCF (1).*

This property allows modeling statistical variability linearly at any value of Vdd. Form (2) is general enough to model complex interactions of Vdd and statistical parameters. However, it is too general to be computationally efficient. Being strictly limited with performance and memory requirements we approximate Vdd dependent terms $t_0(\Delta V)$, $a_i(\Delta V)$, $a_R(\Delta V)$ linearly. This transforms formula (2) into a simpler equation:

$$t_0 + a_V\Delta V + \sum_{i=1}^{n}(a_i + a_{V,i}\Delta V)\Delta X_i + (a_R + a_{R,V}\Delta V)\Delta R \quad (4)$$

Removing brackets we transform it into a bilinear form:

$$t_0 + a_V\Delta V + \sum_{i=1}^{n} a_i\Delta X_i + \sum_{i=1}^{n} a_{V,i}\Delta V\Delta X_i + $$
$$a_R\Delta R + a_{R,V}\Delta V\Delta R \quad (5)$$

where t_0, a_i and a_R have the same meaning as in formula (1); a_V is a sensitivity to deterministic variable ΔV; $a_{V,i}$ and $a_{R,V}$ are sensitivities to cross-terms $\Delta V \cdot \Delta X_i$ and $\Delta V \cdot \Delta R$ respectively. This expression can be considered as an LCF extended with cross-terms $\Delta V \cdot \Delta X_i$, $\Delta V \cdot \Delta R$ to model dependence of statistical sensitivities on Vdd variations. We call this expression an extended canonical form (ECF). The ECF is a special case of form (2). Therefore according to property 1, its projection to Vdd is a linear canonical form.

ECF is also linear with respect to Vdd variations. It means the variability of minimum and maximum delays due to Vdd is modeled linearly, as is illustrated in Fig. 2 (b).

For ECF, we have an interpolation property.

PROPERTY 2. *LCF $t(v)$ obtained by projection of ECF t to value v is a linear interpolation of LCFs $t(v_h)$ and $t(v_l)$ that are projections of $t(v)$ to high and low Vdd corners v_h, v_l.*

Algebraically this property is expressed as follows:

$$t(v) = t(v_l) + (v - v_l)\frac{t(v_h) - t(v_l)}{v_h - v_l} \quad (6)$$

Here the interpolation of the canonical forms is understood in a functional sense, i.e., point-wise. The coefficients of the interpolated canonical form are computed by the following formulas:

$$t_0(v) = t_0(v_l) + (v - v_l)\frac{t_0(v_h) - t_0(v_l)}{v_h - v_l} \quad (7)$$

$$a(v) = a(v_l) + (v - v_l)\frac{a(v_h) - a(v_l)}{v_h - v_l} \quad (8)$$

where $t_0(v)$, $t_0(v_l)$, $t_0(v_h)$ are means of LCFs $t(v)$, $t(v_l)$, $t(v_h)$; $a(v)$, $a(v_l)$, $a(v_h)$ are sensitivities to statistical variations of LCFs $t(v)$, $t(v_l)$, $t(v_h)$.

Interpolation of canonical forms is illustrated in Fig. 3 on an example of an ECF depending on deterministic variable V and statistical parameter X. We see the surface defined by the ECF is swept with the lines corresponding to the projected LCFs. This also can be derived from the fact that bilinear function (5) defines a hyperboloid, which, as it is known, can be swept with lines.

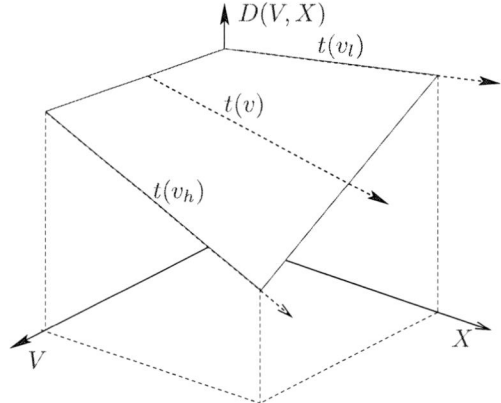

Figure 3: Linear interpolation of canonical forms

The following property states exactness of ECF for modeling LCFs at Vdd corners.

PROPERTY 3. *For any two LCFs $t(v_h)$ and $t(v_l)$ describing statistical variability at high and low Vdd corners v_h, v_l there exists a unique ECF t such that its projections to high and low Vdd corners are exactly $t(v_h)$ and $t(v_l)$.*

This property is illustrated in Fig. 4 (b). The hyperboloid defined by an ECF exactly fits four corners of the delay, which is a function of Vdd V and parameter X. On the other hand, the plane defined by an LCF cannot fit arbitrary

1063

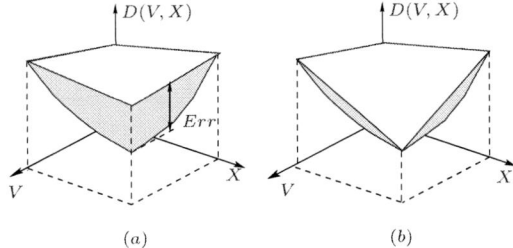

Figure 4: Linear and bilinear approximation

corners as shown in Fig. 4 (a). The LCF cannot accurately model nonseparable variations. Property 3 is important for construction of the ECF from LCFs describing statistical variability at low and high Vdd corners. This construction is described with the following formulas:

$$
\begin{aligned}
a_V &= \frac{t_0(v_h) - t_0(v_l)}{\Delta v_{hl}} & t_0 &= t_0(v_l) - a_V \cdot v_l \\
a_{i,V} &= \frac{a_i(v_h) - a_i(v_l)}{\Delta v_{hl}} & a_i &= a_i(v_l) - a_{i,V} \cdot v_l \quad (9) \\
a_{R,V} &= \frac{a_R(v_h) - a_R(v_l)}{\Delta v_{hl}} & a_R &= a_R(v_l) - a_{R,V} \cdot v_l
\end{aligned}
$$

where $t_0(v_l)$, $t_0(v_h)$, $a_i(v_l)$, $a_i(v_h)$, $a_R(v_l)$, $a_R(v_h)$ are the mean values and sensitivities of LCFs modeling statistical variations at low and high Vdd corners; $\Delta v_{hl} = v_h - v_l$ is the difference between high and low Vdd values. This property ensures that if we can construct LSFs at low and high Vdd, we can easily construct a unique ECF.

3.2 Generalizations and Extensions

It is easy to define an ECF for multiple deterministic parameters:

$$
\begin{aligned}
t_0 + \sum_{i=1}^{m} a_i^D \Delta X_i^D + \sum_{i=j}^{n} a_j^S \Delta X_j^S + \sum_{i=1}^{m} \sum_{j=1}^{n} a_{i,j} \Delta X_i^D \Delta X_j^S + \\
a_R \Delta R + \sum_{i=1}^{m} a_{R,i} \Delta X_i^D \Delta R \quad (10)
\end{aligned}
$$

where t_0 is a mean value; a_i^D, a_j^S and a_R are sensitivities to deterministic, statistical correlated and uncorrelated variations ΔX_i^D, ΔX_j^S and ΔR respectively; $a_{i,j}$ and $a_{R,i}$ are sensitivities to cross-terms of deterministic and statistical variations.

The ECF (5) models only linear Vdd variations. By adding quadratic term $a_{V,V} \Delta V^2$ we can model nonlinear dependence of delay and slew on supply voltage.

4. NONSEPARABLE TIMING

Static timing needs four operations for signal propagation: addition, subtraction, and computing minimum and maximum. Defining them for extended canonical forms we would like to provide exactness of low and high Vdd projections.

PROPERTY 4. *Let t_x, t_y, t_z be ECFs and $t_z = f(t_x, t_y)$ where function f is statistical addition, subtraction, minimum, or maximum. Then $Pr(t_z) = f[Pr(t_x), Pr(t_y)]$, where Pr is an operation of projecting to the high or low Vdd.*

This property if it were true, would guarantee that timing with ECF produces the same results as two separate runs of statistical timing at the low and high Vdd corners.

4.1 Addition and Subtraction

Let t_x, t_y, t_z be ECFs and $t_z = t_x + t_y$. All terms of t_z, except uncorrelated random term and uncorrelated cross-term, are computed by summing the corresponding terms of t_x and t_y. For example, $t_{0,z} = t_{0,x} + t_{0,y}$, $a_{V,z} = a_{V,x} + a_{V,y}$, $a_{i,z} = a_{i,x} + a_{i,y}$, $a_{V,i,z} = a_{V,i,x} + a_{V,i,y}$.

In order to achieve compatibility with property 4, the uncorrelated variations are first computed for the high and low Vdd corners. Then the uncorrelated term and uncorrelated cross-term are calculated to match these uncorrelated variations. The following algorithm gives details of this operation.

ALGORITHM 1. *Summation of uncorrelated variations.*
Input: *Uncorrelated random terms and cross-terms $a_{R,x}$, $a_{R,y}$, $a_{R,V,x}$, $a_{R,V,y}$ of ECFs t_x and t_y.*
Output: *Uncorrelated random term and cross-term $a_{R,z}$, $a_{R,V,z}$ of ECF $t_z = t_x + t_y$.*

1. *Project uncorrelated terms of t_x, t_y to low Vdd v_l:*
   $a_{R,x}(v_l) = a_{R,x} + v_l a_{R,V,x}$; $a_{R,y}(v_l) = a_{R,y} + v_l a_{R,V,y}$

2. *Compute low Vdd projection of resulting uncorrelated term $a_{R,z}(v_l) = \sqrt{a_{R,x}(v_l)^2 + a_{R,y}(v_l)^2}$*

3. *Project uncorrelated terms of t_x, t_y to high Vdd v_h:*
   $a_{R,x}(v_h) = a_{R,x} + v_h a_{R,V,x}$; $a_{R,y}(v_h) = a_{R,y} + v_h a_{R,V,y}$

4. *Compute high Vdd projection of resulting uncorrelated term $a_{R,z}(v_h) = \sqrt{a_{R,x}(v_h)^2 + a_{R,y}(v_h)^2}$*

5. *Compute resulting uncorrelated term and cross-term*

$$
\begin{aligned}
a_{R,V,z} &= \frac{a_{R,z}(v_h) - a_{R,z}(v_l)}{v_h - v_l}, \\
a_{R,z} &= a_{R,z}(v_l) - a_{R,V,z} \cdot v_l
\end{aligned}
$$

4.2 Maximum and Minimum

For ECFs maximum is computed by projecting operands to the high and low Vdd; computing statistical maximum at those corners; and then merging the results. The following algorithm describes this idea in detail:

ALGORITHM 2. *Maximum of ECF.*
Input: *ECFs t_x, t_y.*
Output: *ECF $t_z = \max(t_x, t_y)$.*

1. *Project t_x, t_y to low Vdd v_l obtaining linear canonical forms $t_x(v_l)$, $t_y(v_l)$.*

2. *Compute LCF $t_z(v_l) = max[t_x(v_l), t_y(v_l)]$ by conventional statistical maximum.*

3. *Project t_x, t_y to high Vdd v_h obtaining LCF $t_x(v_h)$, $t_y(v_h)$.*

4. *Compute LCF $t_z(v_h) = max[t_x(v_h), t_y(v_h)]$ by conventional statistical maximum.*

5. *Using formulas (9) compute ECF t_z such that its projections to low and high Vdd are LCFs $t_z(v_l)$, $t_z(v_h)$. Property 3 guarantees existence and uniqueness of the result.*

4.3 Implementation of Timing Analysis

Having addition, subtraction, minimum and maximum defined, timing is performed in the conventional manner by computing and propagating arrival and required arrival times expressed in ECF. A delay (slew) of logic gate or interconnects in ECF is computed similarly to the conventional statistical timing. First, we compute the delay (slew) LCFs $t(v_l)$, $t(v_h)$ for low and high Vdd. Then using formulas (9) we compute the ECF t, the projections of which are $t(v_l)$, $t(v_h)$. Property 3 guarantees existence and uniqueness of the result.

SSTA with nonseparable deterministic and statistical variations was implemented in an industrial tools framework [1] We implemented two techniques for computing delays and slews. The assertion technique computes sensitivities to variations and cross-terms as fractions of the mean values of signal delays and slews. The finite-differencing technique computes sensitivities to variations from delay and slew values at different Vdd and process corners using multi-corner libraries.

Our implementation handles statistically correlated variations by principal component analysis (PCA).

The run time and memory requirements are significantly reduced by sparsification of ECFs, i.e., filtering out small sensitivities. The sparsified ECF is represented with a sparse vector of its sensitivities, occupying only a fraction of memory required for dense representation.

Our implementation takes into account that some of statistical variations are perfectly separable with Vdd. Therefore, we need not compute cross-terms for them, which also saves run time and memory.

According to property 4, SSTA with ECFs is equivalent to two runs of SSTA for low and high Vdd corners. From that analysis we can derive timing results for any Vdd value by interpolating between the high and low Vdd corners. We also can compute the worst timing slack across all combinations of process parameters to guide optimization in a consistent and incremental manner.

5. EXPERIMENTAL RESULTS

The first set of results corresponds to 32nm Application Specific Integrated Circuit (ASIC) designs. Traditionally, two statistical timing runs are performed for these designs corresponding to the high- and low-voltage corners. This applies to both timing sign-off and timing optimization. High timing yield is a critical requirement for ASIC designs to ensure low cost per chip, and necessitates accurate timing analysis at the two voltage corners. At each voltage corner, statistical timing is performed with 21 statistical parameters (including Temperature, transistor strength for different Vt families, NP-skew, and Back-End-Of-Line metal characteristics). Table 1 presents comparison of the run-time and memory usage of a traditional SSTA run (SSTA) to run with ECF (SSTA ECF).

It is observed that, on average, SSTA ECF (having one deterministic variable, Vdd, and 21 statistical parameters that are non-separable with Vdd) is **1.37X** and **1.29X** more expensive in terms of run-time and memory usage, respectively, compared to a traditional SSTA run. However, when compared to two SSTA runs corresponding to two Vdd corners, the benefits of SSTA ECF are obvious; assuming the two SSTA runs take the same time and memory, depending

Table 1: Run-time and memory for ASICs

Ckt	Num. arcs	Run-times (sec)		Memory use (Mb)	
		SSTA	SSTA ECF	SSTA	SSTA ECF
A1	51.6K	10.1	12.7	9	10
A2	126.9K	286	386	96	130
A3	272.6K	711	1072	210	295
Avg.		**1.0**	**1.37x**	**1.0**	**1.29x**

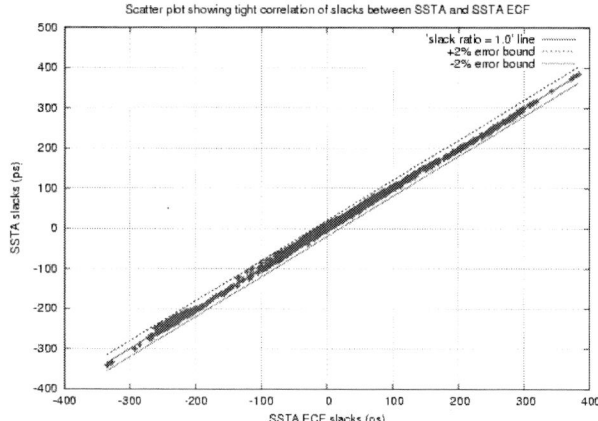

Figure 5: Plot of SSTA and SSTA ECF test slacks

on whether the jobs are run in series or parallel on the same machine, the run-time or memory usage of the SSTA ECF run is **0.685X** or **0.645X**, respectively, of the traditional flow. Even if the two SSTA runs are performed on separate machines, benefits from optimization perspective and consolidated timing-reporting are key motivations for use of SSTA ECF. Timing accuracy comparisons indicate negligible errors with SSTA ECF: *absolute* average slack differences across the designs being less then 10ps (**1%** for typical 1ns clock cycle times in 32nm technology). Fig. 5 shows a scatter plot of design slacks corresponding to multiple timing-tests in design A2 where the x-axis represents slacks (in ps) from SSTA ECF run (tool automatically reports the worst slack from the two corners of Vdd), and the y-axis represents the manually obtained worst slack from two SSTA runs at two Vdd corners (in ps). The three lines in the plot denote the 0%, and $\pm 2\%$ (relative to a 1ns clock period) error bound lines. It is observed that the slack differences are within the $\pm 2\%$ lines for the tests.

The next set of results corresponds to next-generation 22nm microprocessor designs. Variability analysis for microprocessor designs is significantly different compared to ASIC designs. The technology and the design are co-developed at the same time. As a result, accurate parameter sensitivity information is often not available. In addition, to achieve high-frequency targets, certain macros of the design are custom designed and timed at the transistor level using circuit simulation. Circuit simulation is run-time expensive; however the primary goal at design time is high frequency, and line tailoring may be used to ensure good yield. Consequently, statistical timing for microprocessor designs is limited to far fewer parameters (3 in our case, including dopant random variations) than ASICs, and additionally uses asserted sensitivities for the parameters. While the use of asserted sensitivities and fewer parameters make statistical

Table 2: Run-time and memory for MP circuits

Ckt	Num. arcs	Run-times (h:mm)		Memory use (Gb)		Slack impr. (ps)	
		SSTA	SSTA ECF	SSTA	SSTA ECF	Max	Avg
M1	5K	0:07	0:08	1.66	1.66	4.8	2.5
M2	11K	0:13	0:16	1.77	1.77	1.6	1.3
M3	24K	0:26	0:26	1.88	1.89	10.0	3.1
M4	97K	0:59	1:02	2.78	2.84	10.8	3.9
M5	108K	1:36	1:23	2.87	2.92	17.8	10.9
M6	344K	3:46	3:46	6.06	6.11	12.6	4.6
U1	4.7M	3:48	4:33	10.70	10.90	68.3	3.5
Avg		**1.0**	**1.07x**	**1.0**	**1.01x**		

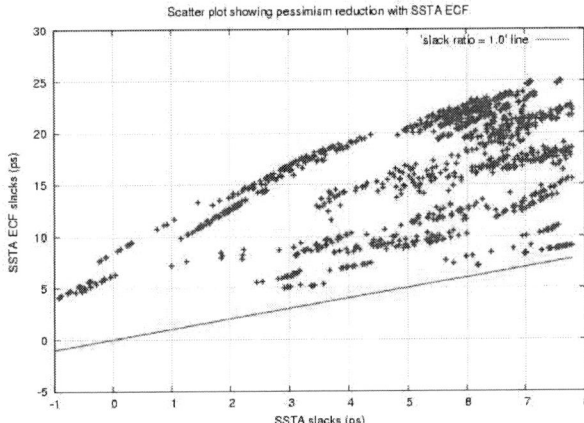

Figure 6: Plot of SSTA and SSTA ECF test slacks

analysis run-time a minimal overhead over base circuit simulation run-time, protection against variability is achieved by larger design margins.

Given the over-head of circuit simulation and custom circuit design-and-optimization, traditionally, SSTA for microprocessor designs performs only a single run with Vdd being considered a parameter and other parameters asserted with worst sensitivities across the two Vdd corners. For example, the impact of the random dopant variation is more significant at low-Vdd than high-Vdd. However, the traditional SSTA flow penalizes the timing at high-Vdd with the random variation impact from low-Vdd.

SSTA with ECF facilities pessimism relief for microprocessor design timing by modeling parameter sensitivities as a function of voltage instead of a worst-case constant value. For the example above, the timing at high-Vdd is no longer penalized more than required. Table 2 presents results comparing the run-time and memory usage of a traditional SSTA for microprocessor macro and unit (hierarchical design component containing multiple macros) designs to a SSTA ECF run. The table also highlights the slack improvement due to pessimism relief. Given that circuit simulation contributes to a major part of the run-time for these custom designs, it is natural that the run-time overhead of SSTA ECF flow is small. The negligible memory overhead is due to the small number of parameters in these runs. From the table, we see that SSTA ECF significantly reduces pessimism, and aids optimization by avoiding unnecessary buffering to fix hold violations, and helping meet higher-frequency targets, with

small run-time and memory usage overheads. Fig. 6 shows the scatter plot for timing test slacks (in ps) at high-voltage indicating pessimism relief for design M5.

6. CONCLUSIONS

We developed a novel SSTA technique for modeling nonseparable deterministic and statistical variations. The proposed technique is implemented in industrial timing tool to model Vdd and statistical variations. This allows solving several important problems of VLSI design flow. Two runs of SSTA are replaced with single timing with ECFs. Vdd variations are modeled deterministically guaranteeing accuracy for corner values of Vdd. Timing reports cover the full spectrum of process and environmental variations, including all possible Vdd values. Optimization is guided with the worst timing slack across all Vdd and process variations. Timing can analyze circuits for voltage binning [3], adaptive voltage supply [4, 5] and dynamic voltage scaling [6]. Timing can analyze circuits with multiple voltage domains.

We are planning to extend the current implementation for modeling nonlinear dependence of delays on Vdd and deterministic variations other than Vdd.

7. REFERENCES

[1] C. Visweswariah, K. Ravindran, K. Kalafala, S. G. Walker, and S. Narayan. First-order incremental block-based statistical timing analysis. *DAC*, pages 331–336, June 2004. San Diego, CA.

[2] H. Chang and S. S. Sapatnekar. Statistical timing analysis considering spatial correlations using a single PERT-like traversal. *ICCAD*, pages 621–625, November 2003. San Jose, CA.

[3] V. Zolotov, C. Visweswariah, and J. Xiong. Voltage binning under process variation. *ICCAD*, pages 425–432, 2009.

[4] J. W. Tschanz, S. Narendra, R. Nair, and V. De. Effectiveness of adaptive supply voltage and body bias for reducing impact of prameter variations in low power and high performance microprocessors. *IEEE Journal of Solid State Circuits. Vol. 38, No. 5*, pages 826–829, May 2003.

[5] T. Chen and S. Naffziger. Comparison of adaptive body bias and adaptive supply voltage for improving delay and leakage under the presence of process variation. *IEEE Trans. on VLSI. Vol. 11(5)*, pages 888–899, October 2003.

[6] B. Zhai, D. Blaauw, D. Sylvester, and K. Flautner. The limit of dynamic voltage scaling and insomniac dynamic voltage scaling. *IEEE Trans. on VLSI*, 13(11):1239–1252, November 2005.

[7] S. Onaissi and F. N. Najm. A linear-time approach for static timing analysis covering all process corners. *IEEE Trans. on CAD*, 27(7):1291–1304, July 2008.

[8] H. Chang, V. Zolotov, C. Visweswariah, and S. Narayan. Parameterized block-based statistical timing analysis with non-Gaussian and nonlinear parameters. *Proc. 2005 Design Automation Conference*, pages 71–76, June 2005. Anaheim, CA.

[9] S. Bhardwaj, P. Ghanta, and S. Vrudhula. A framework for statistical timing analysis using non-linear delay and slew model. *ICCAD*, pages 225–230, November 2005.

[10] L. Zhang, Y. Hu, and C. C. Chen. Block based statistical timing analysis with extended canonical timing model. *Proc. Asia South Pacific Design Automation Conference (ASPDAC)*, pages 250–253, January 2005. Shanghai, China.

[11] L. Chen, J. Xiong, and L. He. Non-gaussian statistical timing anmalysis using second-order polynomial fitting. *IEEE Trans. on CAD*, 28(1):130–140, January 2009.

[12] K. Chopra, B. Zhai, D. Blaauw, and D. Sylvester. A new statistical max operation for propagating skewness in statistical timing analysis. *ICCAD*, pages 237–243, November 2006.

[13] C. E. Clark. The greatest of a finite set of random variables. *Operations Research*, pages 145–162, March-April 1961.

Reversible Statistical *max/min* Operation: Concept and Applications to Timing

Debjit Sinha, Chandu Visweswariah,
Natesan Venkateswaran
IBM Systems and Technology Group
Hopewell Jn., New York, USA
{dsinha, chandu, natesan}@us.ibm.com

Jinjun Xiong, Vladimir Zolotov
IBM Thomas J. Watson Research Center
Yorktown Heights, New York, USA
{jinjun, zolotov}@us.ibm.com

ABSTRACT

The increasing significance of variability in modern sub-micron manufacturing process has led to the development and use of statistical techniques for chip timing analysis and optimization. Statistical timing involves fundamental operations like statistical-*add, sub, max* and *min* to propagate timing information (modeled as random variables with known probability distributions) through a timing graph model of a chip design. Although incremental timing during optimization updates timing information of only certain parts of the timing-graph, lack of established *reversible* statistical *max* or *min* techniques forces more-than-required computations.

This paper describes the concept of *reversible* statistical *max* and *min* for correlated Gaussian random variables, and suggests potential applications to statistical timing. A formal proof is presented to establish the uniqueness of *reversible* statistical *max*. Experimental results show run-time savings when using the presented technique in the context of chip-slack computation during incremental timing optimization.

Categories and Subject Descriptors

B.8.2 [**Hardware**]: Performance and Reliability—*Performance Analysis and Design Aids*

General Terms

Algorithms, Performance

Keywords

Statistical timing, variability

1. INTRODUCTION

With the advent of deep sub-micron technologies, a robust chip design and timing verification methodology must consider timing uncertainty due to variability in the chip manufacturing process, environmental variability (for example, variations in chip operating voltage and temperature), and device fatigue. While protection against the uncertainty was traditionally provided via timing-margins, it has been well es-

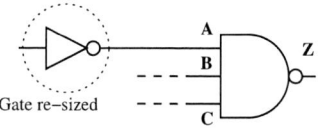

Figure 1: Gate with incremental timing update on 1 input

tablished that this approach is often pessimistic and may lead to *over-design*.

The application of statistical techniques in static timing analysis and optimization has subsequently emerged as a refinement to deterministic timing techniques. It attempts to capture the timing uncertainty due to variability by modeling timing information as random variables with known sensitivities to the *sources of variation*. These random variables with known probability distributions are propagated through a timing graph model of the design during statistical static timing analysis (SSTA) and optimization. Statistical timing propagation involves fundamental operations like *add, sub, max* and *min* on these random variables. Researchers have proposed several analytical approaches to SSTA [1–8]. Some of these works represent statistical timing information using a parameterized linear model that captures first-order timing sensitivity to Gaussian sources of variation [2–4]. While statistical timing techniques with non-linear and non-Gaussian models [5–8] have been extensively researched and claim higher accuracy, they suffer from higher complexity and thus, larger run-times. Given large modern chip designs and the need for fast optimization *turn-around-time* (TAT), it is natural that linear first-order Gaussian models have found more favorable adoption in the industry.

Chip timing optimization is an iterative and incremental process. Design updates (for example, buffering, gate-sizing, and wire re-routing) are performed incrementally on timing critical paths, and the results of the updates are analyzed periodically in the context of the entire design. The idea of incremental timing is to update timing for only sections of the chip that are affected by a given design update, thereby avoiding unnecessary re-computations through the entire design. However, statistical *max* and *min* operations have traditionally been performed as non-incremental operations. As an example, consider an incremental update (gate re-sizing) as shown in Figure 1. Incremental timing propagates updated timing information downstream to input A of the logic NAND gate. Assuming inputs B and C do not lie in the fanout cone of the re-sized gate, their arrival times (ATs) are unchanged. Computing updated timing at output Z involves a statistical *max* or *min* operation on the ATs of all three inputs (assuming no delay through the NAND gate), although the AT for only a single input changed. While this may not seem too inefficient when the number of statistical timing quantities for a *max* or *min* operation is small, the overhead could be significant for a larger number of timing quantities especially if only a small subset of the timing quantities is updated. Chip slack (or chip yield) computation involves estimating the statistical worst slack among all (often as many as hundreds of thousands) timing test slacks in a design, and involves statistical *min* operations on the test slacks. During optimization, incremental timing propagation and chip slack computation may be performed thousands of times, wherein in each iteration, only a small fraction of the total test slacks are updated. Computing the chip slack using statistical *min* operations across all the test slacks (irrespective of which or how many slacks are updated) in the inner loop of iterative timing optimization is a perfect example where non-incremental *min* operations cause inefficiency. Statistical *max* and *min* operations are non-trivial and run-time expensive (relative to other fundamental timing operations). For linear statistical timing models with Gaussian random variables [2,3], pair-wise statistical *max* operations are performed (using Clark's [9] approach for matching moments of the *max* of the distributions). The run-times for these operations could be further significant when techniques that attempt to improve the accuracy of the statistical *max* or *min* operations are employed [10]. This clearly motivates exploring techniques for incremental statistical *max* or *min* operations that would use the information about updated operands to improve efficiency. Without any loss of generality, only the statistical *max* operation is considered in the following sections of this paper. For the remainder of this paper, *max* and *min* refer to statistical *max* and *min* operations, respectively.

This paper describes the concept of *reversible* statistical *max* for Gaussian random variables, an operation that computes an operand of a *max* operation, given the other operand and the result of the *max*. A *reversible* deterministic *max* operation is trivial. As an example, given an operand 5 and the *max* 7, it is immediate that the other operand (or reversible deterministic *max*) is 7. On the other hand, if the operand and the *max* are both the same deterministic value (for example 7), the reverse *max* cannot be estimated. A statistical reversible *max* is non-trivial. However, a statistical *max* is typically a composite of its operands, which makes the reversible *max* computation feasible in most cases. A formal proof is presented to establish the uniqueness of the computed *reversible* statistical *max* barring pathological conditions where the reverse *max* cannot be computed. In computing criticality probabilities for a set of slacks, Xiong *et al.* [11] have proposed a numerical method for reversible tightness probability (extensible to reversible *max*) computation. However, their derivation is complicated, fails to answer whether or not a reversible construction is always feasible, and if yes, whether or not the solution is unique. [11] does not consider other applications of reversible *max*. In contrast, this paper provides a more elegant way of reversible *max* computation and a rigorous proof for the uniqueness of the solution. Novel applications of reversible statistical *max* in incremental timing and incremental chip slack/yield computation are introduced An extensive experimental analysis establishing the accuracy and computational efficiency of the reversible *min* operation and its dependence on the numbers of the operands and incremental changes, is presented.

2. PRELIMINARIES

Similar to prior work in [2–4], a linear statistical timing model with Gaussian random variables having a joint normal distribution is considered. A Gaussian variable X_i is formally expressed as $N(\mu_i, \sigma_i^2)$, with mean μ_i and variance σ_i^2.

The statistical *max* operation is non-trivial, and for a given set of Gaussians, performed a pair at a time. A moment matching approach [9] is commonly used to compute the *max* of any two arbitrary Gaussians X_1 and X_2 [3]. ρ_{12} represents the correlation coefficient between X_1 and X_2. Some definitions used in the approach are as follows:

$$\phi(x) \triangleq \frac{1}{\sqrt{2\pi}} exp(-\frac{x^2}{2}) \tag{1}$$

$$\Phi(y) \triangleq \int_{-\infty}^{y} \phi(x)dx \tag{2}$$

$$\theta \triangleq (\sigma_1^2 + \sigma_2^2 - 2\rho_{12}\sigma_1\sigma_2)^{1/2} \tag{3}$$

$$\alpha \triangleq \frac{\mu_1 - \mu_2}{\theta}. \tag{4}$$

The mean μ_m and variance σ_m^2 of $max(X_1, X_2)$ are expressed analytically as follows:

$$\mu_m = \mu_1\Phi(\alpha) + \mu_2[1 - \Phi(\alpha)] + \theta\phi(\alpha) \tag{5}$$

$$\sigma_m^2 = (\sigma_1^2 + \mu_1^2)\Phi(\alpha) + (\sigma_2^2 + \mu_2^2)[1 - \Phi(\alpha)] +$$
$$(\mu_1 + \mu_2)\theta\phi(\alpha) - \mu_m^2. \tag{6}$$

$\Phi(\alpha)$ denotes the tightness probability of X_1 over X_2 [3]. $max(X_1, X_2)$ is traditionally approximated to a Gaussian variable $X_m \sim N(\mu_m, \sigma_m^2)$ for timing propagation. The first and second order moments of the true *max* are matched to obtain X_m, while the higher order moments are ignored. Given X_1 and X_m, it is non-trivial to derive X_2. The following section presents a mathematical approach for this **reversible** *max* operation, that is, an approach to derive X_2.

3. REVERSIBLE *MAX* OPERATION

Given Gaussians X_1 and X_m, the reversible *max* operation attains to compute (re-construct) X_2. Performing some fundamental operations on the *max* operation yields the following.

$$max(X_1, X_2) = X_m$$
$$\Rightarrow \quad max(X_1 - X_1, X_2 - X_1) = X_m - X_1$$
$$\Rightarrow \quad max(0, X_2 - X_1) = X_m - X_1 \tag{7}$$

LEMMA 1. *If X_1 strictly dominates X_2, that is, if $Pr(X_1 \geq X_2) = 1$, then $max(0, X_2 - X_1)$ is 0, and X_2 cannot be reconstructed.*

Under the condition stated in Lemma 1, X_2 does not play any role in computation of X_m ($= X_1$), and therefore, a reversible *max* operation may have infinitely many solutions for X_2, each satisfying the property that $Pr(X_1 \geq X_2) = 1$.

When the condition stated above is not true, X_2 may be re-constructed using the following approach. For ease of notation, $(X_2 - X_1)$ is denoted as a Gaussian $X \sim N(\mu, \sigma^2)$, and $(X_m - X_1)$ is denoted as another Gaussian $X_0 \sim N(\mu_0, \sigma_0^2)$. The primary goal is to re-construct X from X_0. It should be noted that μ_0 and σ_0 are known. Once X is re-constructed, it is trivial to obtain X_2 ($= X + X_1$).

The above notations are applied to (7) to obtain

$$max(0, X) = X_0. \tag{8}$$

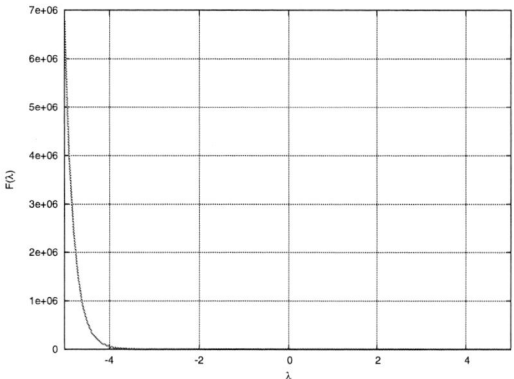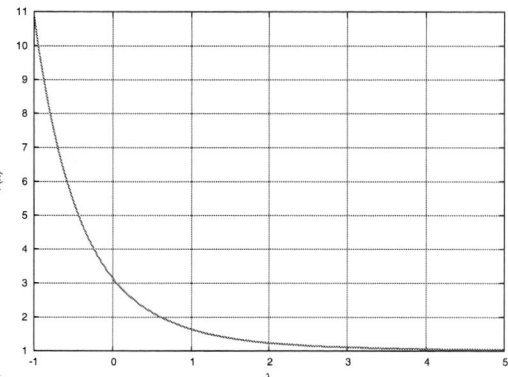

Figure 2: Plots of $F(\lambda)$ for different λ ranges

Defining $\lambda \triangleq \frac{\mu}{\sigma}$, and applying techniques for statistical *max* computation, as described in Section 2, yields the following.

$$
\begin{aligned}
\theta &\triangleq (0 + \sigma^2 - 0)^{1/2} \\
&= \sigma \quad (9) \\
\alpha &\triangleq \frac{0 - \mu}{\theta} \\
&= \frac{-\mu}{\sigma} = -\lambda \quad (10) \\
\mu_0 &= 0 + \mu\left[1 - \Phi(-\frac{\mu}{\sigma})\right] + \theta\phi(-\frac{\mu}{\sigma}) \\
&= \mu\Phi(\frac{\mu}{\sigma}) + \sigma\phi(\frac{\mu}{\sigma})
\end{aligned}
$$

Note: $\Phi(-y) = \left[1 - \Phi(y)\right]$, $\phi(-y) = \phi(y)$, so

$$
\begin{aligned}
\mu_0 &= \sigma\left[\frac{\mu}{\sigma}\Phi(\frac{\mu}{\sigma}) + \phi(\frac{\mu}{\sigma})\right] \\
&= \sigma\left[\lambda\Phi(\lambda) + \phi(\lambda)\right] \quad (11) \\
\mu_0^2 + \sigma_0^2 &= 0 + (\mu^2 + \sigma^2)\left[1 - \Phi(-\frac{\mu}{\sigma})\right] + \mu\theta\phi(-\frac{\mu}{\sigma}) \\
&= (\mu^2 + \sigma^2)\Phi(\frac{\mu}{\sigma}) + \mu\sigma\phi(\frac{\mu}{\sigma}) \\
&= \sigma^2\left[(\frac{\mu^2}{\sigma^2} + 1)\Phi(\frac{\mu}{\sigma}) + \frac{\mu}{\sigma}\phi(\frac{\mu}{\sigma})\right] \\
&= \sigma^2\left[(\lambda^2 + 1)\Phi(\lambda) + \lambda\phi(\lambda)\right] \quad (12)
\end{aligned}
$$

Equating σ^2 from (11) and (12), the following is obtained:

$$
\begin{aligned}
\frac{\mu_0^2}{\left[\lambda\Phi(\lambda) + \phi(\lambda)\right]^2} &= \frac{\mu_0^2 + \sigma_0^2}{(\lambda^2 + 1)\Phi(\lambda) + \lambda\phi(\lambda)} \\
\Rightarrow \frac{(\lambda^2 + 1)\Phi(\lambda) + \lambda\phi(\lambda)}{\left[\lambda\Phi(\lambda) + \phi(\lambda)\right]^2} &= \frac{\mu_0^2 + \sigma_0^2}{\mu_0^2} \quad (13)
\end{aligned}
$$

The right-hand side (RHS) of equation (13) is known. A function $F(\lambda)$ is defined as the left-hand side (LHS) of this equation. Mathematically,

$$
\begin{aligned}
F(\lambda) &\triangleq \frac{(\lambda^2 + 1)\Phi(\lambda) + \lambda\phi(\lambda)}{\left[\lambda\Phi(\lambda) + \phi(\lambda)\right]^2} \quad (14) \\
&= \frac{\mu_0^2 + \sigma_0^2}{\mu_0^2}. \quad (15)
\end{aligned}
$$

It is immediate from (15) that $F(\lambda) \geq 1$, for any λ. Figure 2

Algorithm: Reversible statistical *max*
Input: Gaussians $X_1 \sim N(\mu_1, \sigma_1^2)$, $X_m \sim N(\mu_m, \sigma_m^2)$
Output: Gaussian $X_2 : max(X_1, X_2) = X_m$
1   If $X_1 = X_m$, return *failure*
2   Compute $X_0 \sim N(\mu_0, \sigma_0^2) = X_m - X_1$
3   Evaluate $\frac{\mu_0^2 + \sigma_0^2}{\mu_0^2}$
4   Find λ such that $F(\lambda) = \frac{\mu_0^2 + \sigma_0^2}{\mu_0^2}$
5   Re-construct $X \sim N(\mu, \sigma^2)$ using
$\sigma = \frac{\mu_0}{\lambda\Phi(\lambda) + \phi(\lambda)}$ and $\mu = \lambda\sigma$
6   Re-construct $X_2 = X + X_1$

Figure 3: Reversible *max* operation

plots $F(\lambda)$ against λ. The plots highlight the following:

$$
\begin{aligned}
F(+\infty) &= 1.0 \\
F(-\infty) &= +\infty
\end{aligned}
$$

A formal proof is presented in the Appendix that establishes $F(\lambda)$ as a monotonically decreasing function (with slope strictly < 0). The function is thus a one-to-one function, and implies that for each value of $F(\lambda)$, there exists a unique value of λ.

λ can be computed using a combination of table lookup and binary search, or numerical methods (for example, Newton's method). Using this value of λ in (11) and (10), σ, and subsequently μ, can be computed. From (10), it is immediate that $\Pr(X \geq 0) = \Phi(\lambda)$. This (tightness probability) information is used to compute sensitivities of X which subsequently allows construction of a canonical form[1] for X (details omitted for brevity). Having computed $X \sim N(\mu, \sigma^2)$, re-construction of X_2 is trivial ($= X + X_1$). A pseudo-code for the reversible *max* operation is presented in Figure 3.

4. APPLICATIONS OF REVERSIBLE *MAX* IN STATISTICAL TIMING

Statistical timing analysis involves computation of statistical *max* and *min* operations on a set of random variables denoting statistical timing information. Given a logic gate with multiple source pins and one sink pin, latest and earli-

---

[1]Independently random terms in canonical forms [3] require some special handling, but the concept and proof of uniqueness are unaltered.

1069

est arrival time calculations at the sink pin require *max* and *min* operations, respectively. These operations are also required during required arrival time (or estimated time of arrival) propagation through the timing graph model of the design. Chip slack computation involves estimating the statistical worst slack among all timing test slacks in a design, and involves statistical *min* operations on the test slacks.

During incremental chip timing optimization, timing propagation and chip slack computation are likely performed thousands of times. Design updates (like buffer insertion, pin swapping, layer assignment, cell sizing) are performed incrementally, and the impact of these updates to overall chip timing is analyzed periodically. As described in Section 1 with reference to Figure 1, statistical *max* operations during incremental timing propagation may be needed where only a subset of operands have changed from previously computed values. A similar case applies for incremental chip slack computation. Given a set of N Gaussians, whose *max* has already been computed, and a much smaller subset of M Gaussians which have changed from their prior values (due to some incremental update during optimization), the reversible *max* technique can be employed to compute the updated *max* of N Gaussians.

Without any loss of generality, a set of N Gaussians $\{X_1, X_2, \ldots, X_{N-1}, X_N\}$ is considered with known statistical *max* X_m. Mathematically,

$$X_m = max(X_1, X_2, \ldots, X_{k-1}, X_k, X_{k+1}, \ldots, X_N).$$

In the simplest case, it is assumed that one of the N Gaussians, X_k changes to X_k^* due to an incremental update. The update requires the computation of the new *max* X_m^*, where

$$X_m^* = max(X_1, X_2, \ldots, X_{k-1}, X_k^*, X_{k+1}, \ldots, X_N).$$

Traditionally, X_m^* is computed by performing the *max* operation on the entire set of N operands, although only one operand changed. Analytical $(N-1)$ pair-wise statistical *max* operations could be performed to obtain the result [3]. For higher accuracy, greedy heuristics [10] that choose the sequence of pair-wise *max* operations based on pre-computed error metrics could be used, but are run-time inefficient (complexity as large as $N^2 logN$). In the context of chip slack computation[2], N could easily be larger than 10000. For such cases of large N, *max* operations using Monte Carlo simulations may be preferred to analytical methods to avoid large inaccuracies accumulating during pair-wise *max* operations, each of which involves approximations.

X_m^* can be computed efficiently without performing *max* on the set of N Gaussians using reversible statistical *max*. Given values prior to the update, reversible *max* yields X_m^{-k} (using algorithm in Figure 3), where, theoretically:

$$X_m^{-k} = max(X_1, X_2, \ldots, X_{k-1}, X_{k+1}, \ldots, X_N).$$

The updated *max* of N Gaussians is next computed as:

$$X_m^* = max(X_m^{-k}, X_k^*).$$

It is evident that the proposed method for calculating X_m^* is independent of N. From a theoretical perspective, it is assumed that each pair-wise *max* operation takes 1 unit of time, and a reversible *max* operation takes $(1 + \delta)$ unit of time. δ denotes a fixed overhead in the reversible *max* operation (specifically, computation of λ in step 4 of Figure 3). The presented approach has a run-time gain ratio of $\frac{N-1}{2+\delta}$, which is

---
[2] a *min* operation is performed instead of the *max* in this case

---

Algorithm: Incremental *max* for M of N Gaussians
Input: Gaussians $\{X_1, \ldots, X_N\}$, X_m, $\{X_k^*, \ldots, X_{k+M-1}^*\}$
Output: Gaussian X_m^*
1   Compute $X_a = max(X_k, X_{k+1}, \ldots, X_{k+M-1})$
2   Re-construct $X_m^{-k,M} = $ **reversible max** (X_a, X_m)
3   If *failure* in re-constructing $X_m^{-k,M}$, revert to traditional *max* for X_m^*; exit
4   Compute $X_m^* = max(X_m^{-k,M}, X_k^*, X_{k+1}^*, \ldots, X_{k+M-1}^*)$

Figure 4: Incremental *max* operation

attractive for cases of large N. Incremental chip slack computation during optimization is naturally a potential application of the reversible *min* technique.

The presented technique is extensible to cases wherein a subset containing M Gaussians is updated during an incremental design update. For simplicity, it is assumed that Gaussians $\{X_k, X_{k+1}, \ldots, X_{k+M-1}\}$ are updated and the new values are denoted as $\{X_k^*, X_{k+1}^*, \ldots, X_{k+M-1}^*\}$. The new *max* X_m^* to be computed is expressed mathematically as:

$$X_m^* = max(X_1, \ldots, X_k^*, \ldots, X_{k+M-1}^*, X_{k+M}, \ldots, X_N).$$

The simplest way to use the reversible *max* technique for computing X_m^* involves iteratively replacing Gaussians $\{X_k, \ldots, X_{k+M-1}\}$ with their new values, one at a time, and computing the new *max* in each iteration. This approach requires M iterations of using the algorithm in Figure 3.

A better technique is to compute the *max* of the subset of Gaussians that is updated, and to use the result for computing X_m^* via a single reversible *max* operation. Figure 4 outlines the steps involved.

The time required for steps 1, 2, and 4 in Figure 4 are $(M-1)$, $(1+\delta)$, and (M), respectively. The run-time gain for incremental computation of the *max* of a set of N Gaussians where a subset M has been updated, by switching from pair-wise *max* operations to the new technique is as follows.

$$\text{Run-time gain} = \frac{N-1}{(M-1)+(1+\delta)+M} = \frac{N-1}{2M+\delta}.$$

It evidently follows that the proposed technique is attractive in cases of large N and small M. However, under conditions stated in Lemma 1, it is possible that one of the updated Gaussians (for example, X_{k+1}) is the previously dominant Gaussian ($= X_m$). In this case, reversible *max* technique cannot compute X_m^*, and reverts to the traditional technique as outlined in step 3 of Figure 4.

4.1 Experimental results

Run-time and accuracy comparisons between a traditional technique and the proposed reversible technique for incremental statistical *min* of N arbitrary Gaussians are presented in this section. Notations introduced in preceding sections are used consistently. In the experiments, M of N Gaussians are updated, and the prior *min* X_m as well as the prior values of the Gaussians are known. The two techniques are used to compute the new *min* X_m^*. Each Gaussian is represented in a linear canonical form [3], with 10 sources of variation. The run-times and accuracy of the results obtained are compared for different values of N and M, using the data from the traditional technique as reference.

Table 1: Run-time gain and accuracy comparisons for reversible *min* operations

N	M	Run-times (secs) Old	New	Run-time gain (X)	Avg Mean Err (%)	Max Mean Err (%)	Avg Sigma Err (%)	Max Sigma Err (%)
50	1	0.58	0.34	1.71	0.02	0.62	0.11	2.69
100	1	1.08	0.41	2.63	0.01	0.37	0.07	1.86
500	1	4.54	0.46	9.87	0.00	0.23	0.02	0.78
1000	1	11.16	0.32	34.87	0.00	0.33	0.01	0.78
2000	1	21.58	0.33	65.39	0.00	0.19	0.01	0.51
5000	1	48.30	0.39	123.80	0.00	0.16	0.00	0.24
8000	1	87.34	0.45	194.10	0.00	0.21	0.00	0.20
10000	1	108.40	0.37	292.90	0.00	0.20	0.00	0.12
25000	1	270.20	0.26	1039.00	0.00	0.07	0.00	0.08
50	10	0.55	0.29	1.90	0.43	2.71	0.83	5.05
100	10	0.79	0.51	1.55	0.48	2.56	0.66	4.64
500	10	4.73	0.63	7.51	0.45	2.64	0.28	2.52
2000	10	20.78	0.63	32.98	0.32	2.54	0.12	1.24
5000	10	48.89	0.64	76.39	0.30	4.13	0.07	1.28
8000	10	86.34	0.61	141.50	0.37	4.97	0.05	0.74
10000	10	104.30	0.64	163.00	0.52	8.27	0.04	0.80
25000	10	257.60	0.57	451.90	0.13	2.18	0.02	0.38
500	100	4.77	1.08	4.42	1.67	4.97	0.77	3.78
1000	100	10.69	1.17	9.14	1.95	5.33	0.72	2.87
2000	100	20.83	1.60	13.02	2.09	5.73	0.61	2.93
5000	100	48.78	1.24	39.34	2.39	8.71	0.44	1.97
8000	100	84.96	1.42	59.83	3.21	11.18	0.35	1.25
25000	100	259.90	1.48	175.60	1.19	5.02	0.16	0.88
2000	1000	21.06	10.82	1.95	4.98	10.71	0.58	2.43
25000	1000	259.90	12.75	20.39	6.47	12.01	0.75	1.62

The results obtained are presented in Table 1. Run-times **"Old"** and **"New"** denote the run-times (in secs) of a traditional technique and the reversible *min* technique for computing the new *min*, respectively, normalized to some constant (T) set of calculations. Run-time gains by using the proposed technique and accuracy comparisons are additionally presented. The following set of key observations are made:

- Run-times "Old" scale linearly with N
- Run-times "Old" are independent of M
- Run-times "New" scale with M
- Run-times "New" are fairly independent of N

In general, errors in computation of the updated *min* using the reversible *min* technique are found to be small for both the mean and sigma of X_m^*. A few outliers indicate possibility of Gaussians with widely different sigmas and similar means, in which cases, the final result is significantly dependent on the order of pair-wise *min* operations. In these cases, the result obtained from traditional techniques may itself be inaccurate, and do not indicate accuracy problems in the proposed technique. To avoid such circumstances, industry statistical timers use Monte Carlo simulations for computing *max* or *min* of a large set of Gaussians. The proposed technique can be used in those cases to help achieve even better run-time savings, wherein, following an initial Monte Carlo simulation to compute the original result X_m, reversible techniques are used for incremental updates, and Monte Carlo simulations are performed periodically to preserve accuracy.

Figure 5 plots run-time gain of the proposed technique as a function of N. Multiple plots corresponding to different values of M are shown. It is intuitive that gains are largest for $M = 1$, and gradually diminish when M approaches N, the largest value of M shown in the plot being 1000. The absolute value of gain should not be used to judge the potential of the proposed approach since the gain is strongly dependent on the function implementation for computing λ given $F(\lambda)$ (step 4 of Figure 3). Accuracy-performance trade-offs for that function implementation could be performed to

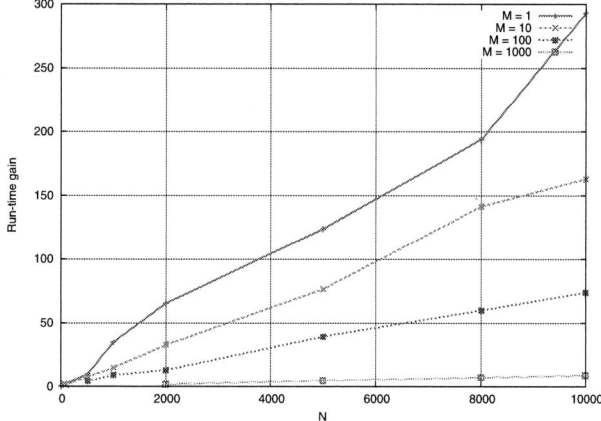

Figure 5: Run-time gain as functions of N and M

tune that function for higher overall gains. A naive binary search based implementation is used for the results presented in this paper, and has scope for significant refinement via use of Newton's method.

4.2 Practical applications in statistical timing

The concept of reversible *max* or *min* can be applied to multiple aspects of statistical timing. Based on Table 1, incremental arrival-time (required-arrival-time) calculations may not seem to be the most attractive application of the concept given that N in this case is typically less than 10 (50).

In contrast, **incremental chip slack** calculation involves a *min* operation over *tens-of-thousands* of Gaussian test-slacks, wherein only a small subset of test-slacks are updated during each incremental timing optimization step. The proposed concept is valuable in this context based on the gains observed for large N in Table 1.

A primary step in the reversible max technique is the computation of λ, as defined in Section 3. From (10), it follows that computing λ implicitly implies computation of tightness probability which is useful for **incremental timing criticality** [11] computation.

The presented technique can be used for **incremental chip timing yield** computation as well. Given a timing graph model of a chip design, and assuming virtual primary-input (PI) and primary-output (PO) nodes which are the source and sink of all original inputs and outputs, respectively, of the design, the worst statistical arrival time at the graph PO can be used to estimate design timing yield (considering late paths only for simplicity). The arrival-time (AT) and required-arrival-time (RAT) are set to 0 at both the PI and PO. Figure 6 represents this graphically.

Considering edge e_i as shown in Figure 6, the following is straightforward.

$$P \triangleq \text{Max. of path delays through } e_i$$
$$= \text{AT}_A + \text{edge-delay}_{e_i} - \text{RAT}_B \quad (16)$$
$$P' \triangleq \text{Max. of path delays not through } e_i \quad (17)$$
$$Y \triangleq \text{Max. of all path delays} = \text{AT}_{PO} \quad (18)$$

Given known values of P and Y, and assuming an incremental timing update on edge e_i which updates the delay through that edge (and subsequently P to P^*), an updated value of Y^* can be estimated without **any timing propagation**, as described in Figure 7. This technique does not propagate the impact of updated signal waveform at node B downstream. The incremental yield obtained should thus be considered an estimate, but nevertheless, has potential for guiding optimization.

The proposed concept can even be used for **incremental yield gradient** computation (with application in circuit tuning). Statistical transistor tuning involves computing the partial derivative of chip yield with respect to a given transistor width. This is done via chain ruling: the partial derivative of chip yield to delay of all paths through some timing graph edge e_i, partial derivative of the delay of all paths through e_i to delay of e_i, and partial derivative of delay of e_i to the transistor width [12]. The incremental chip yield for the first term in this chain rule can be computed using reversible max as presented in Figure 7.

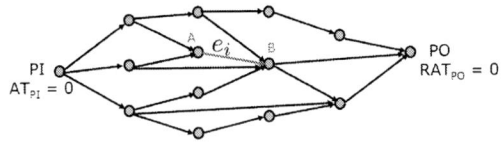

Figure 6: Timing graph model for yield estimation

Algorithm: Incremental yield estimation
Input: Gaussians P, Y, new edge-delay$^*_{e_i}$
Output: Gaussian Y^* (updated timing yield estimate)
1 Re-construct $P' = $ **reversible max** (P, Y)
2 Compute $P^* = \text{AT}_A + \text{new edge-delay}^*_{e_i} - \text{RAT}_B$
3 Estimate $Y^* = max(P^*, P')$

Figure 7: Incremental timing yield estimation

5. CONCLUSIONS

This paper introduces the concept of *reversible* statistical max and min for Gaussian random variables, and suggests potential applications in statistical timing including incremental timing propagation, incremental chip slack computation, and incremental criticality, chip yield and yield gradient estimations. A formal proof is presented to establish the uniqueness of the *reversible max*. Experimental results show significant run-time savings when using the presented technique in the context of incremental max or min of a large number of Gaussians, which is typically the case during incremental chip-slack/yield computation given that modern chip designs have thousands of timing tests (often $N > 10000$). Future work includes a more efficient implementation using Newton's method for the reversible max calculations (analytical form for $F'(\lambda)$ in Appendix).

6. REFERENCES

[1] A. Agarwal, D. Blaauw, and V. Zolotov, "Statistical timing analysis for intra-die process variations with spatial correlations," in *ICCAD*, 2003, pp. 900–907.

[2] H. Chang and S. S. Sapatnekar, "Statistical timing analysis considering spatial correlations using a single PERT-like traversal," in *ICCAD*, 2003, pp. 621–625.

[3] C. Visweswariah, K. Ravindran, K. Kalafala, S. G. Walker, and S. Narayan, "First-order incremental block-based statistical timing analysis," in *DAC*, 2004, pp. 331–336.

[4] J. Le, X. Li, and L. Pileggi, "STAC: Statistical timing with correlation," in *DAC*, 2004, pp. 343–348.

[5] H. Chang, V. Zolotov, S. Narayan, and C. Visweswariah, "Parameterized block-based statistical timing analysis with non-Gaussian parameters, non-linear delay functions," in *DAC*, 2005, pp. 71–76.

[6] L. Zhang, W. Chen, Y. Hu, and C. C. Chen, "Statistical static timing analysis with conditional linear MAX/MIN approximation and extended canonical timing model," in *IEEE Transactions on Computer-Aided Design, 25(6), pp. 1183 - 1191*, June 2006.

[7] S. Bhardwaj, P. Ghanta, and S. Vrudhula, "A framework for statistical timing analysis using non-linear delay and slew model," in *ICCAD*, 2006, pp. 225–230.

[8] L. Chen, J. Xiong, and L. He, "Non-Gaussian statistical timing analysis using second-order polynomial fitting," in *IEEE Transactions on Computer-Aided Design, vol. 28, no. 1*, 2009, pp. 130–140.

[9] C. E. Clark, "The greatest of a finite set of random variables," in *Operations Research, Vol. 9, No. 2 (Mar - Apr)*, 1961, pp. 145–162.

[10] D. Sinha, H. Zhou, and N. V. Shenoy, "Advances in computation of the maximum of a set of Gaussian random variables," in *IEEE Transactions on Computer-Aided Design, 26(8) August 2007*, pp. 1522–1533.

[11] J. Xiong, V. Zolotov, and C. Visweswariah, "Incremental criticality and yield gradients," in *DATE*, 2008, pp. 1130–1135.

[12] D. K. Beece, J. Xiong, C. Visweswariah, V. Zolotov, and Y. Liu, "Transistor sizing of custom high-performance digital circuits with parametric yield considerations," in *DAC*, 2010, pp. 781–786.

APPENDIX

This section proves that $F(\lambda)$ is a monotonically decreasing function. The following definitions are introduced:

$$A(\lambda) \triangleq (\lambda^2 + 1)\Phi(\lambda) + \lambda\phi(\lambda) \tag{19}$$

$$B(\lambda) \triangleq \lambda\Phi(\lambda) + \phi(\lambda) \tag{20}$$

$$F(\lambda) = \frac{A(\lambda)}{\left[B(\lambda)\right]^2} \tag{21}$$

Performing some mathematical operations on the above:

$$B(\lambda) = \lambda\Phi(\lambda) + \phi(\lambda)$$

$$\Rightarrow B'(\lambda) = \lambda\phi(\lambda) + \Phi(\lambda) - \lambda\phi(\lambda) \text{ since, } \phi'(\lambda) = -\lambda\phi(\lambda)$$

$$\Rightarrow B'(\lambda) = \Phi(\lambda) \quad \text{which is } > 0, \ \forall \lambda > -\infty \tag{22}$$

$$\Rightarrow B(\lambda) > 0 \quad \text{since, } B(-\infty) = 0 \tag{23}$$

$$A(\lambda) = (\lambda^2 + 1)\Phi(\lambda) + \lambda\phi(\lambda)$$

$$\Rightarrow A'(\lambda) = (\lambda^2 + 1)\phi(\lambda) + 2\lambda\Phi(\lambda) + \phi(\lambda) - \lambda^2\phi(\lambda)$$

$$\Rightarrow A'(\lambda) = 2\left[\lambda\Phi(\lambda) + \phi(\lambda)\right] = 2B(\lambda) \tag{24}$$

$$\Rightarrow A'(\lambda) > 0$$

$$\Rightarrow A(\lambda) > 0 \quad \text{since, } A(-\infty) = 0 \tag{25}$$

$$F(\lambda) = \frac{A(\lambda)}{\left[B(\lambda)\right]^2}$$

$$\Rightarrow F'(\lambda) = \frac{\left[B(\lambda)\right]^2 A'(\lambda) - 2B(\lambda)B'(\lambda)A(\lambda)}{\left[B(\lambda)\right]^4}$$

$$= \frac{B(\lambda)A'(\lambda) - 2B'(\lambda)A(\lambda)}{\left[B(\lambda)\right]^3}$$

$$= \frac{\left[B(\lambda)\right]^2 - B'(\lambda)A(\lambda)}{0.5\left[B(\lambda)\right]^3}$$

$$= \frac{\lambda^2\Phi^2(\lambda) + \phi^2(\lambda) + 2\lambda\phi(\lambda)\Phi(\lambda)}{0.5\left[B(\lambda)\right]^3} -$$

$$\quad \frac{\Phi(\lambda)\left[(\lambda^2 + 1)\Phi(\lambda) + \lambda\phi(\lambda)\right]}{0.5\left[B(\lambda)\right]^3}$$

$$= \frac{\phi^2(\lambda) + \lambda\phi(\lambda)\Phi(\lambda) - \Phi^2(\lambda)}{0.5\left[B(\lambda)\right]^3}$$

$$= \frac{G(\lambda)}{0.5\left[B(\lambda)\right]^3}, \quad \text{where} \tag{26}$$

$$G(\lambda) \triangleq \phi^2(\lambda) + \lambda\phi(\lambda)\Phi(\lambda) - \Phi^2(\lambda) \tag{27}$$

$$\Rightarrow G'(\lambda) = -\phi(\lambda)A(\lambda) < 0 \tag{28}$$

$$\Rightarrow G(\lambda) < 0 \quad \text{since } G(-\infty) = 0 \tag{29}$$

$$\Rightarrow F'(\lambda) < 0 \tag{30}$$

Predicting Timing Violations Through Instruction-Level Path Sensitization Analysis

Sanghamitra Roy   Koushik Chakraborty

USU BRIDGE LAB
Electrical and Computer Engineering, Utah State University
{sanghamitra.roy, koushik.chakraborty}@usu.edu

ABSTRACT

In this paper, we present a novel technique for early prediction of timing violations in high-performance pipelined microprocessors. We show that a static instruction in a microprocessor, identified by its *Program Counter (PC)*, is an excellent predictor of an upcoming timing violation. Our analysis combines architectural data collected from real program execution with gate level logic analysis. Exploiting this *PC* based timing violation predictability, we propose a robust system design that predicts and tolerates timing violations seamlessly in a pipelined microprocessor. Under two different faulty environments, we show 20.9–89.8% and 14.6–80.6% average performance improvements in real programs over other state-of-the-art techniques, respectively.

Categories and Subject Descriptors

B.8.1 [**Hardware**]: Reliability, Testing and Fault Tolerance

General Terms

Reliability

Keywords

Timing Faults, Path Sensitization

1. INTRODUCTION

Current and future technology nodes are increasingly susceptible to timing violations, an artifact of rapid technology scaling. Guided by a combined effect of static (process variation and wearout) and temporal (thermal, voltage or utilization) variation, timing violations can occur sporadically [14, 12]. Consequently, there is a growing need for runtime timing error detection and correction techniques in robust microprocessor designs.

A key factor behind these timing violations is the specific critical path sensitized by an instruction in a circuit block. In this paper, we demonstrate that an instruction in a microprocessor, identified by its *Program Counter (PC)*, is an excellent predictor of an upcoming timing violation. This intriguing

property—a novel contribution of our work—is due to the extensive commonality in critical paths sensitized by a given instruction during its repeated execution. Using this property, it is possible to get an indication of a timing violation in a later pipeline stage early: several clock cycles in advance.

This early indication of an upcoming timing violation can enable a host of robust system design mechanisms, otherwise impossible with existing state-of-the-art techniques. Popular error detecting techniques like RAZOR and the widely proposed embedded timing sensors are unable to predict timing violations several clock cycles in advance [6, 7]. *Our work is the first of its kind that demonstrates the instruction level predictability of timing violations in pipelined microarchitectures.*

Exploiting this timing violation predictability, we design a robust system that predicts and tolerates timing violations seamlessly in a pipelined microprocessor. We use a low overhead instruction level *Timing Violation Predictor (TVP)* that predicts upcoming timing violations in as early as the decode stage. Using the early prediction capability, we can then enable a pipeline stall signal at the appropriate cycle. Our proposed design can effectively handle timing violations in complex microprocessor designs with forwarding logic and back-to-back dependent instruction scheduling.

Existing works on timing violation detection and correction techniques fall under two broad categories: *reactive* and *proactive*. Reactive techniques detect a fault after its occurrence [6, 2]. Subsequently, the fault can be corrected in a pipelined processor by using an instruction replay, which results in a large performance overhead [5]. Proactive techniques, on the other hand, detect an upcoming timing violation before the clock edge using various sensors embedded in the pipe stages [3, 7]. Once detected, many such techniques use *time borrowing* to tolerate the violation, avoiding any performance impact. In this paper, we show that time borrowing fails to suffice as a general robust technique in high-performance microprocessors. This is because time borrowing cannot be used in forwarding bypass networks that are an integral part of microprocessors for speedy execution of back-to-back dependent instructions.

We make the following contributions in this paper:

- We show that a given instruction sensitizes similar logic paths in various circuit blocks during its repeated execution (Section 2). Our rigorous analysis integrates architectural simulation data with a gate level logic analyzer to determine critical paths sensitized during a program execution.
- Exploiting this unique property, we propose a timing violation predictor (TVP) design and its usage in a mod-

ern microprocessor (Section 3).

- Using a circuit-architectural analysis, we illustrate limitations of time borrowing techniques in high-performance microprocessors (Section 4).
- Using a full-system architectural simulation, we demonstrate average performance improvements of 20.9–89.8% and 14.6–80.6% over other state-of-the-art techniques under two diverse faulty environments, respectively (Section 6).

2. PATH SENSITIZATION: AN INSTRUCTION LEVEL PERSPECTIVE

In this section, we demonstrate the strong correlation between a given static instruction and the critical paths sensitized by many of its dynamic instances in a pipe stage (Section 2.1). This strong correlation results in the predictability of timing violations from the static instruction. This fundamental observation is derived through a rigorous circuit-architectural analysis (Section 2.4). We combine data collected using architectural simulation (e.g., instruction type and input) with a logic analyzer to identify specific sensitized paths, primary inputs, and existing logic state of the gates in the micro-architecture component.

2.1 Instruction and its Sensitized Paths: Correlations

During the program execution, each instruction has many dynamic instances. These dynamic instances show a striking commonality in their sensitized paths through a given pipeline stage. Such high commonality stems from several key factors: (a) identical changes in the micro-architecture state; (b) stability in instruction inputs; (c) identical nearby instructions during its repeated execution ensuring similar internal logic state prior to each execution. Some examples of this commonality can be traced to a few well known characteristics of instruction execution (details in Section S1) :

- For a certain PC, the decode stage interprets identical instruction bits in the same way each time.
- An instruction causes identical wakeup select activity in the Scheduler in all its dynamic instances.
- A small set of load instructions cause repeated cache misses sensitizing similar logic paths.
- Stable input values of ALU instructions [17].

2.2 Case Study with ALU

We now present a case study with ALU instructions. The ALU presents an intriguing case study for this analysis as it contains some of the deepest logic paths among all the structural units in a processor. Consequently, the likelihood of diversity in the sensitized paths through an ALU is higher than most structures in a microprocessor. We discuss other sources of commonality in details in Section S1.

For an ALU instruction, the existing logic state depends on the immediately preceding instruction that executed on that ALU. Since instructions are drawn from a static code, it is expected that the neighboring instructions are fairly stable (barring occasional wrong path executions where instructions enter the pipeline due to a mis-predicted branch). On other hand, the ALU inputs supplied by the instruction are determined by the specific task accomplished by the instruction. For example, several instructions have one of the operands

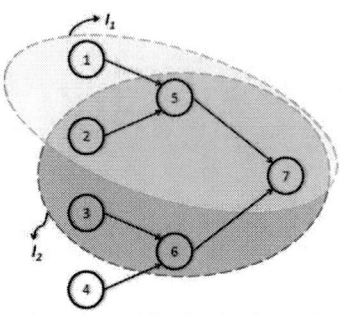

Figure 1: Dynamic path sensitization by I_1 and I_2, two dynamic instances of a static *PC I*. Each instance sensitizes the gates shown by the corresponding bounding ovals.

encoded with the instruction, thereby stabilizing one of the input values in all their dynamic instances.

Branch History: During the program execution, the code path followed before the execution of a particular instruction plays a critical role in determining the specific inputs for that instruction, as well as, the preceding instruction scheduled on the ALU. Hence, we also inspect the commonality seen for a combination of the recently followed code path and the static *PC*. In our analysis, we use outcomes of the last 32 branches to encode the recently exercised code path.

2.3 Dynamic Path Sensitization of an Instruction

Let us consider an instruction, *"add R1, R2, R3"*, where the sum of register contents in *R1* and *R2* is stored in *R3*. This instruction may execute twice (dynamic instances), but with different input values in *R1* and *R2*. These dynamic instances may then sensitize specific paths in a circuit component. Our goal is to find the commonality among all such sensitized paths from a given instruction.

Figure 1 illustrates this concept of dynamic path sensitization of a particular instruction denoted by *I* in a circuit with seven nodes. Two dynamic instances of *I*, I_1 and I_2, sensitize the paths shown by the shaded areas. $S_c\{I_i\}$ indicates the set of nodes that change state while executing instruction I_i. Hence, we see that $S_c\{I_1\} = \{1,2,5,7\}$ and $S_c\{I_2\} = \{2,3,5,6,7\}$.

The set of common gates sensitized by I_1 and I_2 is given by the intersection $(S_c\{I_1\} \cap S_c\{I_2\}) = \{2,5,7\}$. The union of the two sets is given by $(S_c\{I_1\} \cup S_c\{I_2\}) = \{1,2,3,5,6,7\}$. This analysis shows that the two dynamic instances of *I* have a 50% commonality in sensitized paths, calculated using the ratio $\frac{|(S_c\{I_1\} \cap S_c\{I_2\})|}{|(S_c\{I_1\} \cup S_c\{I_2\})|}$.

2.4 Methodology

Performing logic analysis using architectural data presents both computational and methodological challenges. A full program run consists of trillions of instructions [8]: analyzing the input and the corresponding sensitized paths from each instruction is computationally prohibitive. Even representative phases of the program consists of hundreds of million instructions, posing a massive computational challenge. To tackle this challenge, we adopted several important steps.

First, we use representative phases of 100 million instructions of several SPEC CPU2006 benchmarks using the SimPoint toolset [15]. Second, we use a profiling run of architectural simulation to identify the top 100 instructions that most

1075

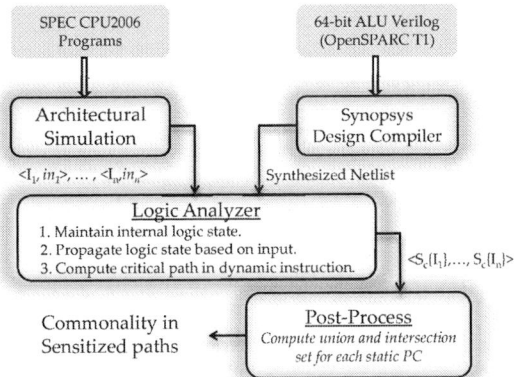

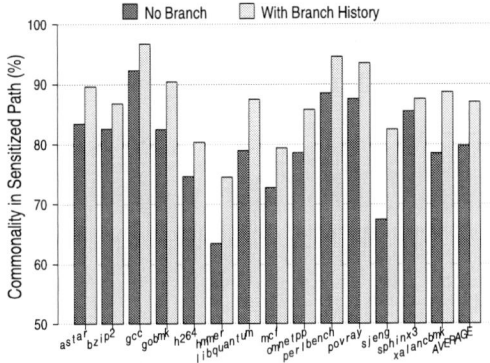

Figure 2: Estimating the commonality in sensitized paths. The dynamic instances of an instruction I and the corresponding inputs $< I_i, in_i >$ are generated from the architectural simulation. Our in-house logic analyzer runs each instruction on the synthesized netlist and generates the set $S_c\{I_i\}$ for each dynamic instance of I. The post-process step estimates the commonality in the sensitized paths of I as shown in Section 2.3.

Figure 3: Commonality in sensitized paths (64-bit integer ALU from OpenSPARC T1).

frequently exercise the 64-bit integer ALU from these phases. Many of these benchmarks have over a million static instructions: keeping track of all of them is impractical. Third, we repeat the architectural simulation to collect inputs from all the dynamic instances of these top instructions. We also collect inputs and instruction type scheduled on the ALU right before a dynamic instance of these instructions. This is a necessary step, as we use preceding instructions to set the existing internal logic state of the ALU. Subsequently, when an instance of a top PC is analyzed, we can identify the sensitized path. Figure 2 shows an overview of our methodology.

2.5 Results

Figure 3 shows the commonality in sensitized paths on a 64-bit integer ALU from the OpenSPARC T1 processor. The result shows the weighted average, based on frequencies of each instruction, of all dynamic instances from the top 100 static instructions exercising the ALU.

We notice a substantial commonality in the sensitized paths across a wide range of applications. On an average, we observe 79.7% commonality with just the static PC. When the branch history is combined, the commonality increases sub-

stantially (average of 87%). Certain benchmarks show tremendous commonality. For example, *gcc* and *perlbench* show 96.7% and 94.6% commonality, respectively. Instructions from these benchmarks operate on a fairly small range of input values, leading to this result. In contrast, *hmmer* shows relatively poor commonality as it has substantial data diversity.

2.6 Significance

The large commonality in sensitized paths from an instruction indicates that if one of its instances cause a timing violation, subsequent instances are highly likely to cause violations under identical operating conditions (voltage, temperature). Thus, we can predict these timing violations ahead in time, using a combination of instruction PC and the recent branch history.

Such early prediction of timing violations presents a tremendous opportunity for designing robust pipelined systems. Unlike Razor, where timing violations are detected only after they have actually taken place [6], precise information about an upcoming timing violation is available several clock cycles before. Consequently, it is possible to set up stall signals in appropriate pipe stages through this technique. Razor is unable to do so in aggressively clocked microprocessors due to insufficient time in propagating the stall signal throughout the requisite pipe stages [6]. Based on this unique opportunity, we now propose a novel system design.

3. SYSTEM DESIGN USING TIMING VIO-LATION PREDICTOR

In this section, we discuss our proposed techniques for designing a robust pipelined architecture that exploits the prediction of upcoming timing violations (Section 3.1). Using a *Timing Violation Predictor (TVP)* (Section 3.2), we can enable the pipeline stall signals to tolerate timing violations and preserve correct execution.

3.1 System Overview

We need a single TVP for the entire pipelined microprocessor. This TVP predicts the occurrence of timing violations in the various pipeline stages for different instruction PCs. During the decode stage, each instruction is checked for a possible timing violation using the TVP.

There are two possible outcomes from the TVP for any instruction:

Predicted No-Violation: No timing violation is predicted by the TVP. Instructions proceed normally.

Predicted Violation: A timing violation is predicted to occur at a specific pipe stage. A stall signal is then initiated at the appropriate stage of the pipeline. For example, if an instruction is decoded with a predicted violation at stage 5, then the stall signal is enabled in the pipeline when that instruction enters stage 5. This stall signal allows the pipe stage 5 to complete in two clock cycles, while the input to all other stages are recirculated to avoid forward flow of instructions during that cycle.

We use already existing circuitry in high performance microprocessors to implement pipeline stalls, and require minimal modification to simply enable the stall signal when necessary. The stalls are necessary in a pipelined architecture for a variety of reasons. One of the main reasons behind these stalls is correctly tackling data dependency, where an instruction cannot proceed due to the unavailability of input com-

1076

puted by an older instruction (e.g, an instruction depends on a load instruction, but the load may miss in the cache causing several cycles delay) [13]..

The TVP cannot rectify timing violations in the fetch and decode stages, so these are mitigated through our recovery mechanism (discussed later). However, violations in these early pipe stages are rare, as temporal variations like thermal and voltage fluctuations are predominant in the back end of the pipeline [10].

Handling Mis-prediction: The predictor can mis-predict in two ways:

- *False Positive*: If the TVP predicts a timing violation, whereas one does not actually occur, we incur the loss of one clock cycle. However, we do not face any correctness problem in the pipeline.
- *False Negative*: If the TVP does not predict a timing violation, whereas one actually occurs, we may encounter an error. In this case, an error detection and recovery mechanism is fired that detects and corrects the timing violation using a pipeline flush.

Error Recovery and Correction: Each pipeline stage in our system is equipped with error detection and recovery circuitry. The error recovery is fired only when an error is detected and there is no scheduled stall signal (during a false negative TVP prediction). This situation essentially implies a timing violation where no corrective measures has been taken. To recover from this error, we initiate a pipeline flush, thereby avoiding the propagation of incorrect value in the system. We use the voltage glitch detector circuit proposed in [4] as our error detector. This circuit requires substantially lower power and area overhead compared to techniques using duplicated shadow flip-flops [6].

3.2 Timing Violation Predictor (TVP)

Fundamentally, timing violations in a pipe stage depend on: (a) the operating conditions (e.g., local temperature, voltage fluctuation), (b) combination of process variation and aging degradation, and (c) the sensitized path. We combine the history of timing violations from an instruction (Section 2.6), and on-chip thermal and voltage sensors to design the TVP.

The TVP consists of three major components: (a) combination of *PC* and branch history; (b) *Violation History Table (VHT)*; and (c) thermal and voltage sensor. Section 2 demonstrated the strong correlation between timing violations in a pipe stage with a combination of an instruction and recent branch history. We exploit this program behavior by creating an index using both the *PC* and the *Branch History Register (BHR)*, which records the last 32 branch outcomes from the program execution.

Each entry in the VHT maintains a combination of PC and BHR that caused a recent violation, associated tag from the PC, and the pipe stage where the violation occurred. The thermal and voltage sensors provide an indication if the current condition is likely to cause a timing violation. During decode, a timing violation is predicted when there is a successful tag match in the VHT, and the sensors indicate possible conditions for timing violations.

The VHT is updated after all mis-predictions. During a false positive, an existing entry is invalidated, making it available for new faulty instructions. During a false negative, a new entry is inserted in the table, which may lead to the eviction of an existing entry.

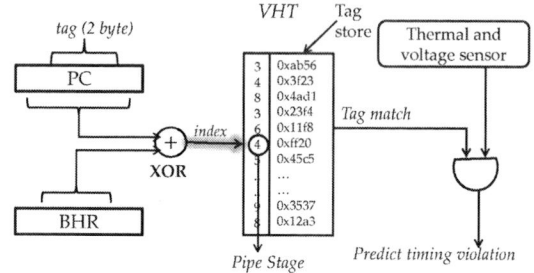

Figure 4: Timing Violation Predictor (TVP) Design

3.2.1 Predictor Performance

Superficially, it may appear that effective prediction requires a large VHT to avoid address collisions in the table. In reality, a small to moderately sized VHT is sufficient as it only tracks instructions causing timing violations.

For example, thermal emergencies in a chip may last a few milliseconds, depending upon its thermal constants [16]. During these intervals, a program may execute a few million dynamic instructions, but timing violations may be restricted to a small percentage of static instructions. We find that using a 4K sized VHT and 10% faulty dynamic instructions in such intervals, leads to less than 2% false negatives from address collisions.

4. RELATION WITH POPULAR TIMING VIOLATION TECHNIQUES

In this section, we discuss other state-of-the-art techniques to mitigate timing violations in pipelined architectures. After briefly describing *Razor* in Section 4.1, we focus on time borrowing techniques that were proposed to mitigate the large performance overhead from Razor's error correction mechanism. Combining a circuit-architectural analysis, we demonstrate how time borrowing techniques can be ineffective in masking timing violations in pipelined high-performance microprocessors. This important design issue stems from the prevalence of back-to-back dependent instructions (BDI) during program execution in these chips.

4.1 Razor

Razor, proposed by Ernst et al. [6], consists of detection and correction mechanisms for timing faults. The timing faults are detected by comparing combinational logic outputs captured at the regular clock and a delayed clock using shadow flip-flops. The correction mechanism, in a high-performance pipelined architecture, involves recovery using counter-flow pipeline. A subsequent work from the same research group have used *instruction replay* [5]. Even with an in-order machine, they report a large performance overhead: about 1% error rate results in nearly 10% performance overhead [5]. To mitigate this large overhead from Razor, several time borrowing techniques were subsequently proposed.

4.2 Time Borrowing Techniques

Time borrowing techniques constitute a class of techniques that can mask a timing violation by borrowing slack time from the adjacent pipe stages [3, 7, 4]. The promise of time borrowing techniques lies in the fact that if successful, they cause zero performance overhead. However, such techniques strongly depend on the availability of slack in some of the pipe stages to mask the timing violations in other pipe stages.

1077

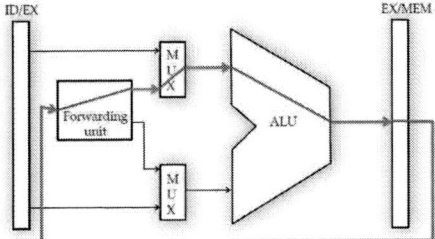

Figure 5: A typical bypass network for executing ALU back-to-back dependent instructions. This circuit constitutes a cyclic pipeline. The highlighted arrows show a critical path, where a flip-flop in the EX/MEM register is the source as well as the sink.

4.3 Problem with Time Borrowing Techniques

Time borrowing can fail when two connected critical paths in adjacent pipe stages are sensitized. Connected critical paths have a single flip-flop acting as the sink node of the first stage as well as the source node of the second stage. Using a static analysis of critical paths, Choudhury et al. show that such connections are rarely sensitized, thereby estimating a high success rate of time borrowing techniques [3].

Using only static analysis, however, **grossly underestimates the actual sensitization of connected critical paths during program execution**. Static analysis can account for all possible critical path connections, but it cannot determine the utilization rate for various connections at runtime. The frequent occurrences of the BDI issue in high-performance microprocessors during real program execution, massively increases the sensitization of connected critical paths. We now present data showing the prevalence of BDIs, and a detailed analysis of the design complexity in handling them.

4.3.1 Back-to-back Dependent Instructions (BDI)

Modern high-performance microprocessors heavily depend on the ability to issue BDIs to reap performance benefits. Using full system simulation (details in Section S3), we observe that 9–36% of instructions constitute BDIs in a high-performance microprocessor—a substantial portion. Similar results were also seen by other works [11].

4.3.2 Complexity of Handling BDIs

Figure 5 shows a bypass path in a typical pipelined microprocessor. The output of the arithmetic and logic unit (ALU) is latched in the EX/MEM register and forwards back to the ALU via the combinational forwarding unit and a combinational multiplexer (MUX) as shown in the figure. This bypass network helps in forwarding the ALU output to a back-to-back dependent ALU instruction in the immediate next clock cycle. Such forwarding logic, commonplace in high-performance microprocessors creates a cyclic pipeline with a single sequential register. Figure 5 also highlights a critical path in this circuit, where a flip-flop in the EX/MEM register constitutes both the source and the sink.

Now let us assume that the ALU has a timing violation in clock cycle t, and is followed by a BDI in clock cycle $(t + 1)$. Since the BDI uses the cyclic pipeline, cycle t needs to borrow slack time from the next cycle of its own block. At cycle $(t + 1)$, there can be another timing violation in the dependent instruction. This is because the primary forces behind a timing violation are the degree of wearout of the circuit combined with the localized temporal variation (tempera-

ture and voltage conditions). Such localized temporal variation implies repeated timing violations in a specific pipeline stage/component for several cycles [1]. In such a case, cycle t is unable to borrow time from its own stage in the next clock cycle $(t + 1)$, making it impossible for time borrowing techniques to rectify the timing violation.

5. ARCHITECTURAL METHODOLOGY

In this section, we describe our architectural simulation methodology for performance tradeoff analysis between various techniques to mitigate timing violations.

Core Microarchitecture: We use full-system simulation built on top of WindRiver SIMICS [9]. SIMICS provides the functional model of several popular ISAs, in sufficient detail to boot an unmodified operating system. For our experiments, we use the SPARC V9 ISA, and use our own detailed timing model to enforce the timing characteristics of a 4-wide out-of-order microprocessor. The core microarchitecture has 12 stages, 64 entry instruction window, 96 entry physical register file, and uses the YAGS predictor with 16K predictor table. Our TVP uses 4KB VHT, and shares the BHR with the YAGS. The core uses a two-level cache hierarchy where L1 (32KB 4-way split Instruction and Data) has a single cycle latency, while the 16-way 8MB L2 and the main memory are accessed in 25 and 240 cycles, respectively.

Workloads: We use several SPEC CPU2006 benchmarks, and focus our architectural simulation on representative phases extracted using the SimPoint toolset [15].

5.1 Fault Simulation Methodology

The goal of our fault simulation methodology is to analyze the performance *during faulty execution*. Among temporal variation, temperature induced faults may last for several milliseconds, while voltage fluctuation lasts for much smaller time frames. Consequently, we randomly select several $10ms$ time windows from the SimPoint phases in our benchmarks, and inject faults during these time windows. To cover a range of faulty executions, we show results when 1% of instructions experience faults, as well as, when 10% of instructions experience faults. All simulations are run with warmed up caches, so that cold cache effects do no mask other important performance artifacts.

Modeling Sensitized Path Commonality: When an instruction incurs a fault in a specific pipe stage, subsequent dynamic instances of the same instruction have a high probability of incurring faults (Section 2) in the same pipe stage. To model this behavior, we assign a fault probability to every faulty static instruction. For each benchmark, we set this probability to the percentage commonality from Figure 3.

6. EXPERIMENTAL RESULTS

In this section, we present the performance analysis of our technique compared to state-of-the-art techniques.

6.1 Comparative Schemes

- Razor: We model the Razor scheme for detection and correction of timing violations. The correction mechanism uses *instruction replay*, based on their most recent work [5].
- Time Borrowing: We have shown that time borrowing can fail to rectify timing violation errors in high-

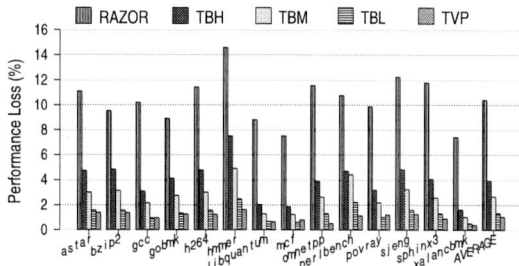

Figure 6: Performance loss during faulty execution (1% faulty instructions). *Lower is better.*

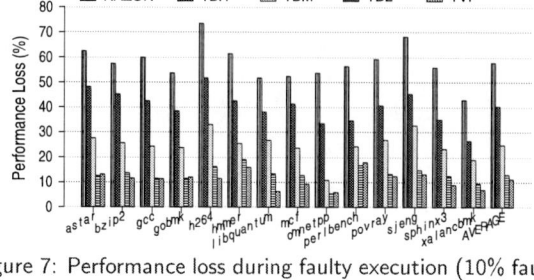

Figure 7: Performance loss during faulty execution (10% faulty instructions). *Lower is better.*

performance microprocessors (Section 4). Time borrowing if successful incurs no penalty but an unsuccessful attempt leads to a pipeline flush. Although time borrowing can fail even without BDI issue (due to non-availability of slack), in our analysis we optimistically assume that time borrowing fails only in case of BDIs. Due to the localized temporal variation (e.g., high temperature), time borrowing can fail for even up to 100% of all BDIs that incur timing violations. However, we model three different time borrowing failure rates (α) in *faulty BDIs*: 20% ($\alpha = 0.2$), 40% ($\alpha = 0.4$), and 60% ($\alpha = 0.6$). These schemes are denoted as *TBL*, *TBM*, and *TBH*, respectively. Faults are randomly distributed between BDIs and non-BDIs.

- Our technique (TVP) introduces a pipeline stall at the appropriate cycle if a timing violation is predicted. A false positive (unnecessary stall) or false negative (required pipeline flush) causes performance loss, and are fully modeled in our evaluation.

6.2 Performance Impact

Performance loss from a pipeline flush or a stall greatly depends on the intrinsic characteristics of a benchmark. Typically, benchmarks with high instructions per cycle (IPC) are more affected by these corrective measures. Timing of these events are also critical. For example, when the instruction window is full due to a pending load miss, a pipeline stall and flush have marginal and large overheads, respectively.

Figures 6 and 7 show the performance impact of various schemes we compare when 1% and 10% of all instructions incur timing violations, respectively. The performance losses are shown relative to a *fault-free* baseline. We notice that *hmmer* suffers a large performance loss, especially for TBH. In *hmmer*, the out-of-order microarchitecture is able extract substantial instruction level parallelism due to the low percentage of dependent instructions (Figure 8). Thus, even with the lower recurrence of BDIs, time borrowing suffers substantial performance loss in *hmmer*, as an instruction replay has large overhead in *hmmer* than in lower IPC *mcf* or *xalancbmk*.

Despite using optimistic models for time borrowing techniques in an out-of-order core, our proposed scheme shows substantially better performance in a faulty environment. For 1% faults, our TVP outperforms time borrowing techniques by 74.1%, 60.6%, and 20.9%, respectively. For 10% faults, we notice average performance improvements of 72.2%, 54.9%, 14.6%, respectively. Razor performs worse than time borrowing as it requires instruction replay for all faulty instructions.

7. CONCLUSION

We present a novel technique for early prediction of upcoming timing violations in a pipelined microprocessor using a combination of the instruction *PC* and temporal sensors. We exploit this phenomenon to design a robust system that predicts and tolerates timing violations seamlessly in a pipelined microprocessor.

Acknowledgments

This work was supported in part by National Science Foundation grant CNS-1117425.

8. REFERENCES

[1] ADOLFSSON, D. AND OTHERS On Scan Chain Diagnosis for Intermittent Faults. In *Asian Test Symposium* (2009), pp. 47–54.

[2] AGARWAL, M. AND OTHERS Circuit Failure Prediction and Its Application to Transistor Aging. In *VTS* (2007), pp. 277–286.

[3] CHOUDHURY, M. R. AND OTHERS TIMBER: Time borrowing and error relaying for online timing error resilience. In *Proc. of DATE* (2010), pp. 1554–1559.

[4] DADGOUR, H., AND BANERJEE, K. Aging-resilient design of pipelined architectures using novel detection and correction circuits. In *Proc. of DATE* (2010), pp. 244–249.

[5] DAS, S. AND OTHERS RazorII: In Situ Error Detection and Correction for PVT and SER Tolerance. *Solid-State Circuits, IEEE Journal of 44*, 1 (jan. 2009), 32 –48.

[6] ERNST, D. AND OTHERS Razor: A Low-Power Pipeline Based on Circuit-Level Timing Speculation. In *Proc. of MICRO* (2003), pp. 7–18.

[7] GHASEMAZAR, M., AND PEDRAM, M. Minimizing the energy cost of throughput in a linear pipeline by opportunistic time borrowing. In *Proc. of ICCAD* (2008), pp. 155–160.

[8] HENNING, J. L. SPEC CPU2006 benchmark descriptions. *SIGARCH Comput. Archit. News 34*, 4 (2006), 1–17.

[9] MAGNUSSON, P. S. AND OTHERS Simics: A Full System Simulation Platform. *IEEE Computer 35*, 2 (Feb 2002), 50–58.

[10] MESA-MARTINEZ, F. J. AND OTHERS Power model validation through thermal measurements. In *Proc. of ISCA* (2007), pp. 302–311.

[11] PALACHARLA, S. AND OTHERS Complexity-Effective Superscalar Processors. In *Proc. of ISCA* (1997), pp. 206–218.

[12] PAN, S. AND OTHERS IVF: Characterizing the vulnerability of microprocessor structures to intermittent faults. In *Proc. of DATE* (2010), pp. 238–243.

[13] PATTERSON, D. A., AND HENNESSY, J. L. *Computer Organization and Design*, 4 ed. Morgan Kaufmann, 2009.

[14] SARANGI, S. AND OTHERS VARIUS: A Model of Process Variation and Resulting Timing Errors for Microarchitects. *IEEE Transactions on Semiconductor Manufacturing 21*, 1 (2008), 3 –13.

[15] SHERWOOD, T. AND OTHERS Basic Block Distribution Analysis to Find Periodic Behavior and Simulation Points in Applications. In *PACT* (2001), pp. 3–14.

[16] SKADRON, K. AND OTHERS Temperature-Aware Microarchitecture. In *Proc. of ISCA* (2003), pp. 2–13.

[17] SODANI, A., AND SOHI, G. S. Dynamic Instruction Reuse. In *Proc. of ISCA* (1997), pp. 194–205.

Supplemental Materials

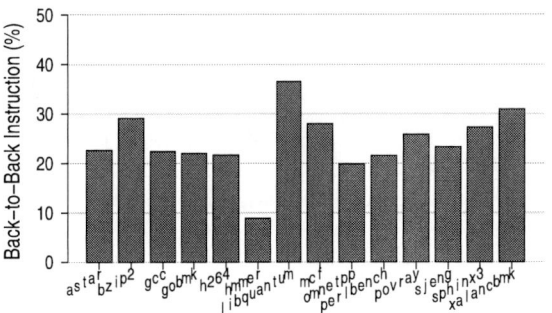

Figure 8: Percentage of back-to-back dependent instructions in SPEC CPU2006 benchmarks. These results are obtained from full system simulation of a typical 4-wide out-of-order microprocessor core (see Section 5).

S1. ARCHITECTURAL EVENTS LEADING TO COMMONALITY

Across many dynamic instances, a given instruction causes similar state changes in various pipe stages. In this section, we present a detailed discussion of many such changes in different stages of a pipeline. All of the following factors influence the high commonality in the sensitized paths of a instruction PC in its many dynamic instances.

- *Decode:* The bits encoding a given instruction A remains unchanged every time that instruction is decoded. Consequently, these bits are interpreted in an identical fashion every time, producing the same logic output from the decode stage. Moreover, the internal logic state of the combinational logic in the decode stage is set by the instruction immediately preceding the instruction A. Except for branch instruction and its target, the order of instructions remains fixed. Even for branch and target instructions, programs tend to show good predictability: leading to the low mis-prediction in modern branch predictors. These factors collectively combine to sensitize near identical logic paths in the decode stage for the instruction A, in all of its dynamic instances.

- *Rename:* In the rename stage, decoded instructions are allocated free physical registers to store their respective results. In addition, for any instruction, the rename logic determines the previously renamed instructions that the given instruction depends on. Since this dependency of an instruction remains unchanged during all of its dynamic instances (a fundamental property of the static code), the computation to accomplish this follows a similar path, sensitizing identical logic gates.

- *Schedule:* A substantial portion of the state change triggered during the issue of an instruction is directly derived from the instruction itself. For example, the type of instruction (e.g., load/store or ALU operation) determines the functional unit where the instruction is scheduled. Thus, reserving the identical functional unit in all of its dynamic instances sensitize the same path for a given instruction. Other important activities like wakeup and select are also heavily correlated to the instruction PC. The wakeup logic is invoked when an instruction produces a new value in its destination register (through a load completion or ALU), marking its dependent instructions operand ready field [11]. Since the number of dependent instructions behind a given instruction depends on the static code, and thus remains stable, every dynamic instance of the given instruction causes highly similar state changes in the wakeup/select logic of the scheduler.

- *Memory Access:* Instructions show a markedly predictable pattern in their execution characteristics during the memory access stage. For example, a large number of cache misses result from a few select instruction PCs [SR2, SR3]. A cache miss and a cache hit sensitize orthogonal set of logic paths in the circuit responsible for memory access. Therefore, we expect a very high commonality in the logic paths in this stage as well. Other key components like load-store queues in a processor also observe predictable and repeated state changes from different load and store instructions. For example, past research has demonstrated that only a few select store instructions conflict with younger loads, thereby preventing them to be issued to the cache memory early [SR1]. Thus, actions necessary to enforce this dependence at runtime requires distinct state changes (but predictable through store PCs) from the default case. These factors again combine to produce a striking commonality in the sensitized logic paths in the memory access phase.

S2. PREDICTOR OVERHEAD

To estimate the power overhead of our Timing Violation Predictor (TVP), we design a TVP in Verilog using Figure 4. Our TVP uses a 4KB Violation History Table (VHT). The instruction PC and BHR which are accessed by the TVP are already a part of the microprocessor, and hence not included inside our TVP module. We next synthesize our TVP using the Synopsys Design Compiler and a 45 nm TSMC standard cell library and measure the dynamic and leakage power. To get a relative estimate of power overhead, we also synthesize a 96 entry Physical Register File using identical methodology and measure its dynamic and leakage power. We find that the TVP consumes 15.42% power relative to the register file. Most of the power in small SRAM structures (e.g., the predictor or the register file) with low associativity (direct mapped) goes towards driving the address and data bits. Compared to the register file that drives a 64-bit data, the predictor only drives 5-bit data (pipe stage encoding). Consequently, the predictor has substantially lower power overhead compared to the register file. Assuming that the register file consumes 6.6% of the core power, a reasonable estimate in a high-performance out-of-order core [SR4], the TVP consumes only 1.02% of the core power.

S3. PROBLEM WITH TIME BORROWING

To elaborate on Section 4.3.1, we perform a detailed analysis using full system simulation of a typical 4-wide out-of-order microprocessor core to estimate the percentage of back-to-back dependent instructions (BDIs). Figure 8 shows the percentage of instructions issued that depend on the result computed in the immediately preceding cycle. We observe that 9–36% of instructions constitute BDIs in a high-performance microprocessor—a substantial portion.

S4. REFERENCES

[SR1] CHRYSOS, G. Z., AND EMER, J. S. Memory Dependence Prediction Using Store Sets. In *Proc. of ISCA* (1998), pp. 142–153.

[SR2] SHERWOOD, T. AND OTHERS Predictor-directed stream buffers. In *MICRO* (2000), pp. 42–53.

[SR3] ANNAVARAM, M. AND OTHERS Data prefetching by dependence graph precomputation. In *Proc. of ISCA* (2001), pp. 52 –61.

[SR4] MESA-MARTINEZ, F. J. AND OTHERS Power model validation through thermal measurements. In *Proc. of ISCA* (2007), pp. 302–311.

A Chip-Package-Board Co-design Methodology *

Hsu-Chieh Lee[1] and Yao-Wen Chang[1,2],

[1]Graduate Institute of Electronics Engineering, National Taiwan University, Taipei 106, Taiwan
[2]Department of Electrical Engineering, National Taiwan University, Taipei 106, Taiwan
pg30123@eda.ee.ntu.edu.tw; ywchang@cc.ee.ntu.edu.tw

ABSTRACT

In today's IC production, the design processes of chips, packages, and boards are typically separate from each other. The lack of information from other domains causes signicant design convergence problems and greatly reduces design quality. In this paper, we propose the first chip-package-board codesign methodology that provides true bi-directional information interactions among the three design domains. The codesign adopts a two-pass flow of board-package-chip followed by chip-package-board routing interactions to facilitate the overall design integration. Experimental results show that our codesign flow succeeds in the routing for all test cases, while a traditional flow and two board-driven flows fail all cases.

Categories and Subject Descriptors

B.7.2 [**Integrated Circuits**]: Design Aids—*Placement and Routing*; J.6 [**Computer-Aided Engineering**]: Computer-Aided Design

General Terms

Algorithms, Design

Keywords

Physical design, flip-chip routing, PCB routing, co-design

1. INTRODUCTION

In current chip-package-board design flows, the designers often start by designing chips, then packages, and the printed circuit boards (PCB's) are the last. In addition, the PCB congestion and routability issues are seldom considered during the planning of chips and packages. Consequently, the PCB design team often receives chip and package designs with PCB-unfriendly pin assignments, which leads to high net congestion on the board. Such PCB routing is difficult to complete on a satisfying number of layers, and the signal performance is usually degraded. If the PCB routing result cannot meet the requirements, the design flow has to go back and restart over from chip and package pin assignments. This usually incurs many iterations, which causes the bottleneck of time to market and also reduces the profit margin of a product [7, 10].

Figure 1 illustrates the importance of cross-domain codesign. On the upper-left corner of the figure, a PCB routing instance with chips and

---

*This work was partially supported by SpringSoft, TSMC, and NSC of Taiwan under Grant No's. NSC 100-2221-E-002-088-MY3, NSC 99-2221-E-002-207-MY3, NSC 99-2221-E-002-210-MY3, and NSC 98-2221-E-002-119-MY3.

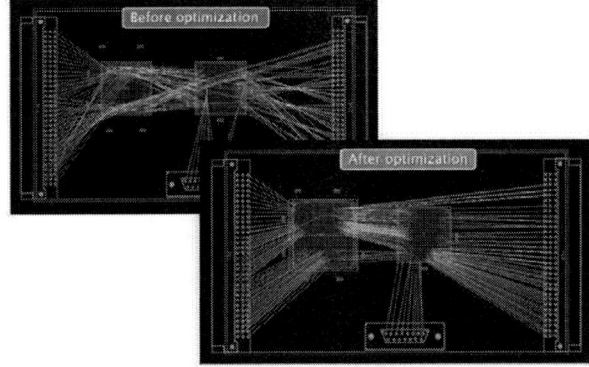

Figure 1: Without cross-domain considerations, chip and package designs create a huge number of flightline crossings in the PCB; with codesign considerations, very few flightline crossings remain on the PCB [10].

packages designed without codesign considerations is shown. Since the chip and package designers are unaware of the PCB layout, the chip and package designs might make PCB highly congested. We can tell the heavy congestion by the huge number of net flightline crossings. In addition, to remedy the huge routing congestion in the package, the package needs to be customized specifically for the design, which further increases the cost. In contrast, the "after optimization" figure shows another PCB routing instance with chips and packages designed with cross-domain knowledge. It is clear that the flightline number is dramatically decreased. Fewer net flightline crossings on a PCB lead to fewer layers, fewer vias, and improved signal quality. Also, as the congestion in packages decreases, the routing could be performed with simple standard packages.

To improve the quality of the layout, many previous works have addressed the cross-domain codesign problem in recent years. However, no previous work has attempted to bring the routing information of all chips, packages, and boards together, but only two of the three domains are considered. Peng et al. [11] proposed a chip floorplan algorithm considering the effects of interconnections between chips and packages using B*-tree based simulated annealing. Some works focused on chip-package interaction and implemented codesign of these two domains [3, 4, 6, 8, 9, 12]. While they optimized some objectives using the information of both chips and packages, their methods are still unaware of the PCB routing congestion. As a result, the package produced by their methods could still be PCB-unfriendly. Moreover, many previous works assumed fixed locations of I/O buffers, and then decided the assignment between I/O buffers and signals. This implies that the assignment is not considered during the chip placement phase, and thus much flexibility is lost. Some other works instead focused on package-board codesign [5, 7]. These works focus on the interactions between packages and boards, without considering the chip design. As a result, they may still lead to design difficulties combining chips and packages.

In this paper, we propose the first chip-package-board codesign methodology that provides information exchanges among boards, packages, and chips. With all the information combined, we can produce PCB-friendly chip and package designs with minimized net congestion on a PCB, so the PCB routing can be completed more easily in minimal layers and with better signal quality. Our board-driven codesign flow brings the information of chips, packages, and boards together and finds the balance among them, resulting in reduced wirelength and increased routability in both package and board routings. Our codesign flow contains the following key procedures:

- A PCB inverse escape routing that uses the PCB layout information to determine ball assignments.

- A package-aware I/O placement technique that optimizes package routability.

- A package bump reassignment algorithm that balance the congestion between packages and PCBs.

- A package and PCB router that completes the routing based on previous assignments.

We compare our codesign flow with traditional chip-driven design flow and other two board-driven design flows. Compared with the traditional flow, our flow completed the PCB routing with correct escape order for all test cases while the traditional flow achieves only 27.7% routability. Compared with two board-driven flows, our flow completed package routing for all test cases while other two flows achieve only 20.4% and 98.8% routability. Only can our proposed codesign flow successfully complete both package routing and PCB routing for all cases.

The rest of the paper is organized as follows. Section 2 introduces the preliminaries and formulates the problem. Section 3 describes the proposed design flow and detailed algorithms. Then, Section 4 presents the experimental results. Finally, Section 5 concludes this paper.

2. PRELIMINARIES

In this section, we first study the chip-package-board interaction and identify the key ingredients in the whole routing process in Section 2.1. We then define some terminologies in Section 2.2. Then, we formulate the chip-package-board codesign problem and the boundary signal order problem in Section 2.3 and 2.4, respectively.

2.1 Ingredients

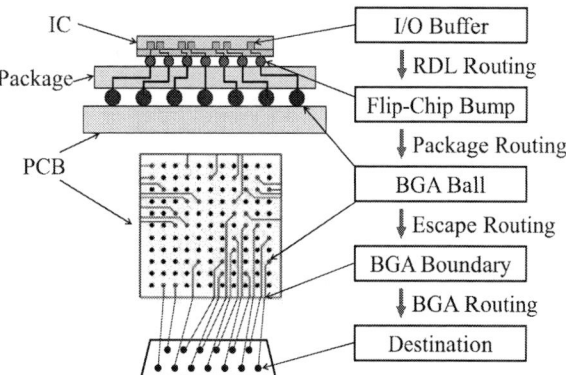

Figure 2: The upper part is a cross view of a typical single chip package using flip-chip packaging mounted on a PCB through a BGA. The lower part shows the aerial view of the BGA on the PCB. Each component on the path of a signal is identified in the right-hand side, with the corresponding routing problems being shown.

To minimize the net congestion in boards, we must first identify the essential ingredients of the design flow that has high influence over the net congestion on a PCB. Figure 2 shows the cross view of a typical single chip package mounting on a PCB in the upper-left corner. The chip is packaged using the flip-chip technique, and the package is then mounted on the PCB through a Ball Grid Array (BGA). In the rest of this paper, the terms chips, packages, and boards refer to the structures in Figure 2 without further explanation. Figure 2 also shows an aerial view of a BGA on a PCB on the lower-left part, which illustrates the path of signals after the signals reach the board. Also, each component on the path of a signal is identified in the right-hand side.

Let us explain how the signals travel from the chip to the board, according to Figure 2. A signal leaves the chip through an I/O buffer inside the chip. The physical locations of the I/O buffers are determined during the chip placement stage. To transfer the signal to the package

substrate, each I/O buffer is attached with an I/O pad, and each I/O pad must be connected to one bump pad in the Re-Distribution Layer (RDL). This process is called the *flip-chip RDL routing*. The signal enters the flip-chip solder bump through the bump pad in RDL. After that, each signal is passed from a solder bump on the top of the package to a solder ball on the bottom of the package through package routing. Once the signal reaches a solder ball, it is then passed to the PCB layer through the solder ball. Now that the signal has arrived at the PCB layer, it would be connected to other components on the PCB, such as I/O interfaces to external systems. Routing signals from solder balls to other components is usually divided into two steps: each signal inside the BGA is first routed to the boundary of the grids by a special kind of routing called *escape routing*, and then routed to the destination through general PCB routing. Figure 2 summarizes the signal path and corresponding routing problems on the right-hand side.

Our main objective is minimizing the flightline crossings and the net congestion of PCB routing. Therefore, our codesign flow must contain all steps in Figure 2 before the PCB routing. In other words, in the last step of our codesign flow, we should have performed the escape routing and route all nets to the BGA boundary.

As shown in Figure 2, the PCB routing would have minimum flightline crossings when the signal orders of BGA boundary and the destination are the same. We now analyze the flow in Figure 2 and identify the key components that could affect the signal order. First, the position of I/O buffers in chip should be carefully decided during chip placement. Then, the RDL routing should route each signal to a proper flip-chip bump. Next, the package routing routes the signals from flip-chip bumps to BGA balls, which mainly acts as a space transformer to change the pitch from the chip pitch to the board pitch. Finally, the signals should be routed to the BGA boundary through escape routing, forming the desired signal order. As shown in Figure 2, package routing from flip-chip bumps to BGA balls mainly acts as a pitch transformer between chips and boards and is often trivial. Therefore, in this paper we shall focus on the following nontrivial procedures: (1) I/O buffer placement, (2) flip-chip RDL routing, and (3) PCB escape routing.

2.2 Terminology

We first define the terms and symbols used in the rest of this paper. For a single chip package, we have the following notations:

- $O = \{o_1, o_2, \ldots\}$: the I/O buffers (pads). Each I/O buffer is assigned to a specific signal.

- $M = \{m_1, m_2, \ldots\}$: the flip-chip bump pads on top of the package.

- $B = \{b_1, b_2, \ldots\}$: the solder balls on the bottom of the package.

- $P = \{p_1, p_2, \ldots\}$: the pins in other components on a PCB.

For any two consecutive domains $(O/M, M/B, B/P)$, an assignment between them needs to be determined in order to obtain the final routing results. We use A_{OM}, A_{MB}, and A_{BP} to represent the mappings between them. Moreover, for each p_i connecting to this package, its destination must be a specific signal of the chip, which is bounded to a specific I/O buffer. Therefore, we also introduce a notation for assignment between pins (signals) and I/O buffers: A_{OP}.

A_{OM} and A_{MB} are the mappings between adjacent layers for chip-package and package-board assignments, respectively. Combined with A_{BP}, these three assignments form the signal path starting from an I/O buffer, to a flip-chip bump, then to a BGA ball, and finally to the designated pin. On the other hand, the mapping A_{OP} indicates the two ends of this whole signal path—the I/O buffer and the corresponding exterior pin on a PCB. Please note that each assignment in A_{OP} must be exactly the same as the assignment derived from A_{OM}, A_{MB}, and A_{BP} combined. As a result, if we have already decided any three of these four assignments, we can uniquely derive the last assignment. This fact would be used in Section 3.

The notations introduced previously are for a single package. For multiple packages on a PCB, there would be multiple sets of O, M, B, and P.

2.3 Codesign Problem Formulation

In this section, we introduce our problem formulation.

Chip-Package-Board Codesign Problem (CPBCP): Given M, B, P, A_{MB} and A_{OP} for each package on a PCB, find O, A_{OM}, A_{BP}, RDL routing, and PCB escape routing for each package such that the congestion in the PCB is minimized.

Note that B and P form the PCB layout, as shown in Figure 3(a). To solve the board routability issue, we let PCB designers first provide the PCB layout (B and P) and the signal assignment of each pin (A_{OP}). With the given PCB layout, the packages and chips can be designed to

1083

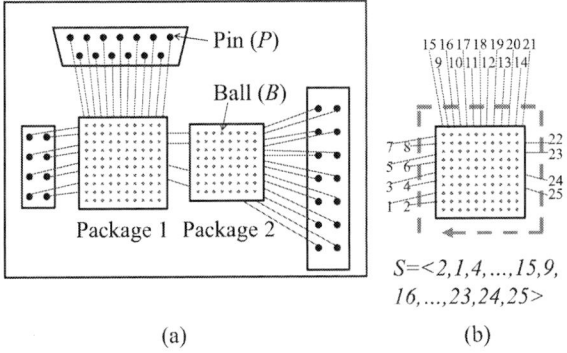

$S = <2,1,4,...,15,9,$
$16,...,23,24,25>$

(a)   (b)

Figure 3: (a) A simple PCB layout with two packages and three I/O interfaces. The red lines indicate ideal flightlines. (b) The boundary signal order of package 1.

suit this layout and minimize wire congestion on the board. The bump to ball assignment in package routing (A_{MB}) is also given as an input, as discussed in Section 2.1.

2.4 Boundary Signal Order Problem Formulation

To reduce the PCB routing congestion, we have to minimize the net flightline crossings on a PCB. With the given PCB layout and the signal-pin assignment, we can easily generate ideal PCB flightlines, as shown in Figure 3(a). As discussed in Section 2.1, the PCB congestion can be minimized if the signal net order on the boundaries of packages match the signal net order of surrounding pins and other components. An ideal wiring configuration like Figure 3(a) can be achieved by requiring each individual package to match the boundary signal order.

We introduce another terminology for the boundary signal net order:

- $S = <s_1, s_2, ..., s_{|O|}>$, where $s_i \in [1, |O|]$ represents a specific signal. S is an ordered sequence of signals formed on the boundary of a BGA.

As shown in Figure 3(b), each s_i in S represents a signal that should be routed to a certain position on the boundary of the BGA, which is the signal from I/O buffer o_{s_i}. For example, $S = <1, 3, 2>$ means that the signal from I/O buffers $O = \{o_1, o_2, o_3\}$ should be routed to the BGA boundary in the order o_1, o_3, o_2.

We then define the signal order optimization problem for a single package as follows:

Boundary Signal Order Problem (BSOP): Given M, B, A_{MB}, and S of a certain package on a PCB, find O, A_{OM}, and complete flip-chip RDL routing and PCB escape routing so that the resulting signal order on the BGA boundary is S.

By solving the BSOP for every package on the PCB, we can solve the CPBCP and minimize the PCB routing congestion.

3. ALGORITHM

In this section, we solve the BSOP from Section 2.2. We first propose a design flow that suits the spirit of chip-package-board codesign in Section 3.1. After that, we give the detailed algorithm of each step in the flow in Sections 3.2–3.5.

3.1 Overall Flow

Let us first examine the traditional chip-package-board flow. Traditional design flows start from planning chips, then packages, then PCBs. During chip and package designing, the designers might make decisions that have negative effects on board routability, due to the lack of information of boards. The congestion caused by these decisions often cannot be fixed in the board design phase. Then the design flow would have to roll back to chip and package design and try to fix those inferior decisions. However, without direct feedback of board congestion, the designers have to adjust their designs for many iterations. This usually results in the bottleneck of time to market and also reduces the profit margin, as discussed in Section 1.

To bring the information of a PCB into the design of chips and packages, a straightforward idea is to simply reverse the design order to be boards, packages, and then chips. However, the same problem still exists in this flow: the information still only propagates in one direction. The only thing changed is that now chips have to remedy all the wrong

decisions made by board and package designs, and the bottleneck will then be the chip design.

To be able to remedy the impact of previous poor decisions, the information propagation must be two-way. The chip must have the information from the board, and the board must have the information from the chip. With this in mind, we propose our codesign flow as a board-driven Λ-shaped flow, as shown in Figure 4.

The proposed flow can be viewed as a two-pass flow, while the first pass is from boards to chips, and the second pass is from chips to boards. On the first pass, the input information about the PCB layout is processed and passed to the package, and then the chip. However, unlike the simple board-package-chip flow we discussed previously, here the chip does not need to follow all the decisions made in this pass, since the decisions made by boards and packages might be chip-unfriendly due to the lack of chip information. Instead, the chip design treats the received information as "advices" provided by boards and packages. The chip design would consider this information, but would not follow it to the the degree that causes too much trouble in the chip design itself. On the second pass, the flow goes from chips to boards. With the information from chips, now the package and board design can fix the wrong decisions previously made during the first pass, while the partial result of the first pass can still be reused. By exchanging the information between different domains, the loading of different domains could be balanced with each other. Moreover, the decisions we made in earlier stages can be fixed in later stages. This feature is very important because less information is known in earlier stages, and thus we have greater chances to make inferior decisions.

The above is the general idea of the flow. The actual codesign flow can be roughly divided into four main steps: (1) inverse escape routing, (2) package-aware I/O placement, (3) flip-chip bump reassignment, and (4) RDL and escape rerouting. (As mentioned earlier, the ball-bump mapping is straightforward; so we shall omit its details in this paper.) These four steps are introduced in each of following sections. In this paper, we focus on the codesign framework and the information exchanges between different domains. Therefore, we do not address too much details of individual algorithms.

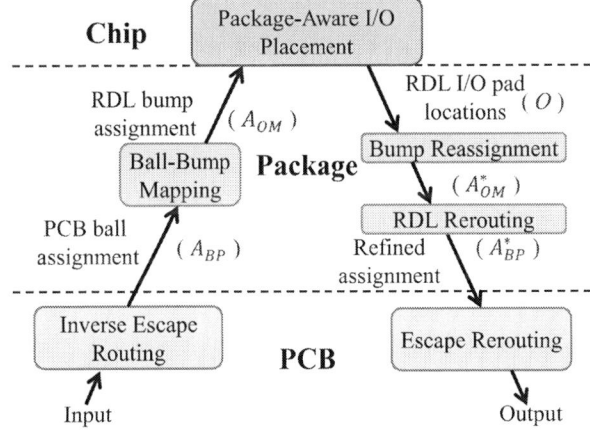

Figure 4: Proposed board-driven Λ-shaped codesign flow for the BSOP.

3.2 Inverse Escape Routing

To make the concepts easier to understand, an example test case (cd2 in Section 4) is used to demonstrate the effects of each step in the flow. Figure 5 shows the layout of the input BGA with 246 I/O signals. The blue balls in the BGA indicate signal balls. The signal balls in this example are assigned with one of the pattern proposed in Lee and Chen's work [7].

In this first step, we find the assignment between signals and balls (A_{BP}) to minimize the routing congestion in escape routing. Instead of routing from the balls to the boundary of the array, we route signal nets from the boundary to the balls. We use Yan and Wang's [13] flow tile model and network-flow construction to construct the flow network, then use the minimum-cost network-flow algorithm to solve this problem. After the global routing is done, a detailed routing based on track assignment is applied to acquire the exact location of each net. As aforementioned, we do not address the details of the algorithm here

1084

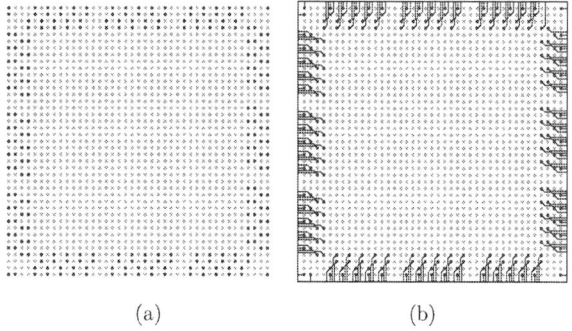

(a)   (b)

Figure 5: (a) The test case ibm01. The blue balls indicate the signal balls. (b) The red wires give the inverse escape routing results.

since it is similar to previous works.

The resulting inverse escape routing is shown in Figure 5. According to this routing result, we determine the signal assignment of each signal ball (A_{BP}), which is then passed to the next step of the flow.

3.3 Package-Aware I/O Placement

After acquiring the signal assignment of BGA balls (A_{BP}) from the inverse escape routing step, we map them into the signal assignment of flip-chip bumps, denoted as A_{MP}, through the given ball-bump assignment (A_{MB}). In this simple example, the flip-chip bump array size is exactly the same as the BGA on the PCB, and each bump is routed to the corresponding BGA ball with the same row and column indexes. The flip-chip bump array looks exactly the same as those on the BGA in Figure 5, except that the pitch is much smaller. Note that any A_{MB} would work fine; here, we just show a simple assignment example.

Now we want to determine the I/O buffer placement (O) on a chip. We use NTUplace3 [1], a leading academic analytical placer, to complete the placement of the chip. If the information from the previous step (A_{MP}) is not well-utilized, the situation would become that chips and boards both optimize themselves, which might produce significant congestion on flip-chip RDL. To demonstrate this, we run the original NTUplace3 to complete chip placement without considering A_{MP}. The resulting RDL routing instance is shown in the left-hand side of Figure 6, where the I/O buffers are denoted as green squares, and the red lines denote the flightlines between the I/O buffers and assigned flip-chip bumps. As illustrated, the whole RDL is a mess with flightline crossings, and it is apparent that this kind of RDL routing is virtually impossible to complete. Specifically, the RDL routability in this case is only 21.1%. To reduce the routing congestion in RDL, we should place the I/O buffers close to their assigned bumps during the placement procedure.

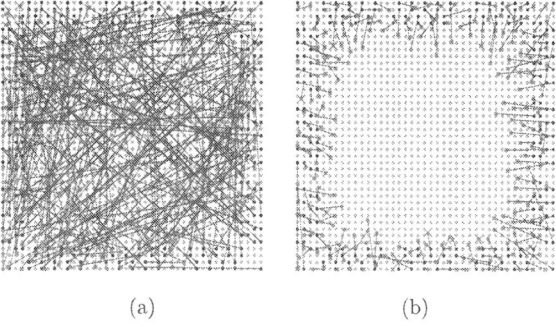

(a)   (b)

Figure 6: (a) The RDL routing flightlines produced by the original non-codesign placement scheme. The routability is only 52/246 = 21.1%. (b) The result of using package-aware I/O placement with $\alpha = 15$, the routability reaches 241/246 = 98.0%.

We modify NTUplace3 based on the force-directed concept of analytical placement to reduce the flightlines. Every net is modeled to be elastic and has a "force" that pulls the connecting blocks together, thus reducing the total wirelength. Based on this concept, we construct a

"pseudo net" for each I/O buffer, which connects the I/O buffer to an "anchor node" fixed at the location of the bump pad assigned to the I/O signal (**inspired by the work**]). During the placement procedure, the pseudo net would pull the I/O buffer close to the anchor node which is at the target location. These pseudo nets should have stronger elasticity than normal nets, so that the provided forces would be strong enough to pull the I/O buffers close to the bump locations.

Another possible approach would be forcing the I/O buffers to be fixed at the bump locations. However, the chip design itself is very complicated and has its own objectives and costs. Forcing I/O buffers at fixed locations would greatly sacrifice the chip quality.

By using the pseudo net approach, we can control the magnitude of compromise between chips and packages by adjusting the elasticity of pseudo nets, and thus balancing the costs of different domains. To achieve this goal, we define *the pseudo net weight* α as follows: each pseudo net has α times of force provided by a normal net. The bigger the α is, the I/O buffers would be placed closer to the assigned bump pads. The effect of α over RDL routing is illustrated in the right-hand side of Figure 6, which shows the RDL flightlines under $\alpha = 15$ with the routability reaching 98.0%. To complete the routing in a single layer, we need to achieve 100% routability. However, the 100% routability cannot be achieved by simply increasing α because some poor decisions made in previous steps would cause unresolvable congestion. To achieve 100% routability, we must fix previous decisions in our next step.

3.4 Flip-Chip Bump Reassignment

Although most I/O buffers are placed near the assigned bump pads, some regions in an RDL might still be heavily congested. This is because our previous inverse escape routing and bump signal assignment do not have the information of chip placement, so we might have made inferior assignments. Since we have the placement result now, we can reassign the bumps to improve previous decisions.

3.4.1 RDL Routability Improvement

In traditional pre-assignment flip-chip routing, we are not allowed to adjust the signal assignment provided. However, in our codesign flow, we use the information of flip-chip RDL routing and PCB escape routing together and provide much more flexibility than the traditional flow.

Since the RDL is still congested, we want to change the RDL bump signal assignment so that the routability is improved and the total wirelength is shortened. However, once we adjust the bump assignment, the PCB boundary signal order would also be changed unless we redo the escape routing. An example is shown in Figure 7. Figure 7(a) is the original RDL bump assignment and PCB escape routes. The two blue lines denote the flightlines of two nets in RDL. The flightline crossing indicates routing congestion. If we switch the assigned bumps of the two signals, as shown in Figure 7(b), the two nets in RDL would have shorter wirelength and lower congestion, since the flightlines are shorter and the flightlne crossing is resolved. The routability in RDL routing would be increased as a result. However, with the package routing and escape routing remain the same, the boundary signal order on the PCB would also be switched and becomes incorrect. To correct the boundary signal order, we must alter the PCB escape routes. As shown in Figure 7(c), after we adjust the escape routing on the PCB, the signal boundary order is correct again.

This situation once shows that the codesign flow is all about information communication and compromises chips, packages, and boards. As the example shown in Figure 7, the escape routing could be easily performed after adjusting the bump assignment. However, in practical cases, if we just reassign the bumps so that the RDL routing congestion is optimized without considering the PCB congestion, the PCB escape routing might suffer from severely reduced routability. Therefore, while we reassign the bumps for better routability and shorter wirelength in an RDL layer, we must also consider the escape routing. To maintain the balance between them, we propose an algorithm based on bipartite matching [2] to reassign signals to bumps.

3.4.2 Reassignment by Bipartite Matching

We model the bump reassignment problem as follows: Given a set of I/O pads (signals) O and a set of bump pads M, and the cost function $cost(o_i, m_j)$ for any $o_i \in O, m_j \in M$, find the minimum cost bipartite matching between the two sets.

To balance the congestion between an RDL and a PCB, the cost function must be defined based on the congestion of both:

$$cost(o_i, m_j) = cost_{RDL}(o_i, m_j) + \beta \cdot cost_{PCB}(o_i, m_j), \quad (1)$$

where β is a user-defined parameter to balance the weight of the two terms, and $cost_{RDL}$ and $cost_{PCB}$ should be defined to reflect the routing congestion in the RDL and the PCB, respectively.

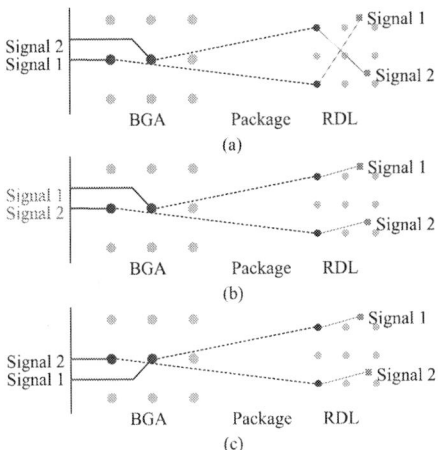

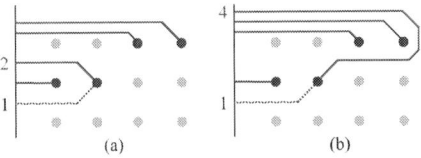

Figure 7: If the bump assignment in RDL is changed, the PCB escape routing must be adjusted accordingly. (a) The original bump assignment with RDL routing congestion. (b) The bump assignment is changed to achieve better routability in the RDL, but the boundary signal order of the BGA is now incorrect. (c) The escape routing must be adjusted to correct the signal order.

Here we define:

- $cost_{RDL} = distance(o_i, m_j)$, the flightline length between the I/O pad and the bump pad in RDL.

By minimizing the total flightline length, we implicitly assign the I/O pads to nearby bump pads, thus reducing the routing congestion in an RDL.

For $cost_{PCB}$, we must consider the impact of ball assignment on the escape routing. An example is shown in Figure 8. In Figure 8, the dotted net represents the original net routed in the inverse escape routing step, which conducts signal 1. In Figure 8(a), the ball is assigned to signal 2 during the bump reassignment, and only a minor effort is needed to make the escape order correct again. In contrast, in Figure 8(b), the ball is assigned to signal 4, and the adjustment on escape routing requires much more routing resources. In addition to the rerouting of the net itself, assigning a ball with a signal that has much different order from the original order often requires the rerouting of other nets to make room for the reassigned net.

Figure 8: The PCB rerouting effort is proportional to the difference in signal order. (a) The ball is reassigned a new signal with the order difference $(2 - 1) = 1$, and the rerouting is simpler. (b) The rerouting requires much more routing resources when the signal order difference is $(4 - 1) = 3$.

Consequently, we define the PCB cost as follows:

$$cost_{PCB}(o_i, m_j) = |i - order(A_{MB}(m_j))|^2, \qquad (2)$$

square of the difference between the original signal order of the ball $A_{MB}(m_j)$ (the ball assigned to m_j) and the new signal order i. The reason of using the square of the order difference instead of using the absolute value is as follows: during our experiment, we found that assigning balls to signals with high order difference often dramatically reduces routability. Therefore, we impose quadratic increasing penalties on this kind of assignments.

Using the bipartite matching algorithm, we match the I/O buffers (signals) with bumps using the proposed cost function. With the new signal-bump assignment (A_{OM}^*), 18 out of 246 signals are assigned to different bumps, and the routability is improved to 100%.

3.5 RDL and Escape Routing

After the reassigning of flip-chip bumps, the RDL routing and escape routing are performed using an A* maze router with rip-up and rerouting. Since the assignments are polished by previous steps, it is easier to complete the routing. The heuristic cost function of a node in the A* search process is defined as the straight-line distance between the node and the routing destination. So the estimated cost is always less than or equal to the actual cost, thus satisfying the admissible property that the A* search requires.

The flip-chip RDL routing result and the PCB escape routing result of the example is shown in Figure 9.

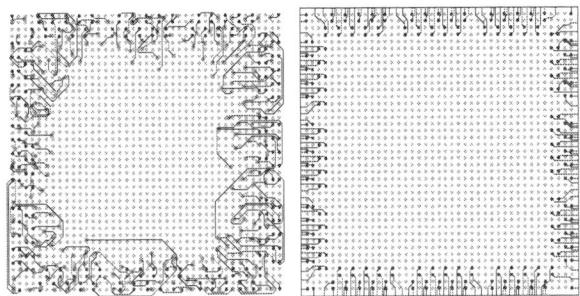

Figure 9: Results of RDL routing and PCB escape routing for ibm01.

4. EXPERIMENTAL RESULTS

In this section, we evaluate our codesign flow. In Section 4.1, information of the test cases is introduced. In Section 4.2, we compare our codesign flow with a traditional chip-package-board design flow. In Section 4.3, we compare our flow with two board-driven design flows. All experiments are conducted on a linux workstation with 16-core Intel Xeon E5620 2.4GHz CPU and 16GB RAM.

4.1 Test Cases

We generated our own test cases since no existing work has addressed the chip-package-board codesign problem. We used the ISPD'04 ibm placement benchmarks, which are modern large-scale mixed-size placement cases, as chips in our test cases. We also slightly modified the original ISPD'04 benchmarks to suit the area-I/O flip-chip structures. The PCB BGA structure was generated using Lee and Chen's pin-out assignment method [7], and the flip-chip bump array was set to have the same dimensions with the BGA. Both flip-chip RDL routing and PCB escape routing were assumed to be single-layered routing.

4.2 PCB Escape Routability

To demonstrate the effectiveness of our codesign methodology, we implemented a design flow similar to traditional chip-driven methodology called Flow TRAD. Flow TRAD is implemented as follows: the chip placement is optimized first, then in the package each signal is assigned to the closest bump with bipartite matching, and in the PCB we route the signals to the boundary while trying to maintain the required signal order.

Table 1 lists the PCB escape routing routability of the two flows. The routability using Flow TRAD is only 27.7% on average while our flow achieves 100% routability for all test cases. The results are as expected since the chip and package designs in Flow TRAD do not consider the PCB signal order. The results show the great effectiveness of our codesign flow. Note that both flows run in seconds, and thus we focus on the routability comparison.

4.3 Flip-Chip RDL Routability

We also compare our flow with other board-driven flows. As described in Section 3.1, our codesign flow is a two-pass codesign flow that is capable of fixing poor decisions from earlier stages. Here, we compare the results of our flow with two one-pass board-driven flows. Flow A optimizes chips and PCBs first, and leave the coordination to the package. As we can imagine, since the designing of chips and PCBs has no information of each other, it would be very hard to connect all the nets in flip-chip RDL routing. Flow B is similar to the first pass of our two-pass flow, the chip design uses package-aware I/O placement since it has the information of previous steps. However, unlike our two-pass flow, Flow B does not have the flip-chip bump reassignment and rerouting steps to

1086

	Chip-Driven		Our Flow		BGA size
case	routed nets	rout.	routed nets	rout.	
ibm01	71/246	30.1%	246/246	100%	41x41
ibm02	75/259	29.0%	259/259	100%	43x43
ibm03	72/283	25.4%	283/283	100%	46x46
ibm04	62/287	21.6%	287/287	100%	46x46
ibm06	61/166	36.7%	166/166	100%	29x29
ibm08	71/286	24.8%	286/286	100%	46x46
avg.		27.7%		100%	

Table 1: PCB escape routing comparison between the traditional flow and our codesign flow. Both flows run in 5 seconds. ("rout.": routability)

remedy the poor decisions in earlier stages. Therefore, Flow B might still not be able to achieve 100% routability in RDL routing.

The RDL routability of all test cases using different flows are shown in Table 2. As expected, the RDL routability of Flow A is very poor, only 20.4% on average. Flow B has a significantly higher routability due to the package-aware I/O placement algorithm. However, it still fails to achieve 100% routability for any test case since it does not has the ability to fix previous decisions. Although routability of Flow B is very close to 100%, those few unrouted nets still render the routing results unsuccessful. Only our two-pass codesign flow can remedy previous poor decisions through flip-chip bump reassignment.

	Flow A		Flow B		Our Flow	
case	# routed nets	rout.	# routed nets	rout.	# routed nets	rout.
ibm01	55/246	22.4%	241/246	98.0%	246/246	100%
ibm02	54/259	20.8%	257/259	99.2%	259/259	100%
ibm03	53/283	18.7%	282/283	99.6%	283/283	100%
ibm04	59/287	20.6%	285/287	99.3%	287/287	100%
ibm06	37/166	22.3%	161/166	97.0%	166/166	100%
ibm08	50/286	17.5%	285/286	99.7%	286/286	100%
avg.		20.4%		98.8%		100%

Table 2: The RDL routability of the three different design flows. All flows run in 5 seconds. ("rout.": routability)

The above results show that our codesign flow is superior to these one-pass flows. The main advantage of our Λ-shaped codesign flow is that the information exchange is bi-directional. For one-pass flows, the information is just transferred in one direction. In our two-pass flow, the information flow is board-package-chip in the first pass, and is chip-package-board in the second pass. In this manner, the information of different domains blend together and thus can achieve much better results.

5. CONCLUSION

In this paper, we have presented a chip-package-board codesign flow that solves the PCB routability issue in the traditional design flow. Our proposed board-driven Λ-shaped codesign flow is the first work that addresses the board routability issue in all chips, packages, and boards. It consists of the following key steps: (1) inverse escape routing, (2) package-aware I/O placement, (3) flip-chip bump reassignment, and (4) RDL routing and PCB escape routing. The experimental results have showed that our proposed flow achieves much better routability than the traditional flow and two other board-driven design flows.

References

[1] T.-C. Chen, Z.-W. Jiang, T.-C. Hsu, H.-C. Chen, and Y.-W. Chang, "NTUplace3: An analytical placer for large-scale mixed-size designs with preplaced blocks and density constraints," *TCAD*, 27(7):1228–1240, 2008.

[2] T. H. Cormen, C. E. Leiserson, R. L. Rivest, and C. Stein. *Introduction to Algorithms*. The MIT Press, 2009.

[3] J.-W. Fang and Y.-W. Chang, "Area-I/O flip-chip routing for chip-package co-design," in *Proc. of ICCAD*, pp. 518–522, 2008.

[4] J.-W. Fang and Y.-W. Chang, "Area-I/O flip-chip routing for chip-package co-design considering signal skews," *TCAD*, 29(5):711–721, 2010.

[5] J.-W. Fang, K.-H. Ho, and Y.-W. Chang, "Routing for chip-package-board co-design considering differential pairs," in *Proc. of ICCAD*, pp. 512–517, 2008.

[6] M.-F. Lai and H.-M. Chen, "An implementation of performance-driven block and I/O placement for chip-package codesign," in *Proc. of ISQED*, pp. 604–607, 2008.

[7] R.-J. Lee and H.-M. Chen, "Fast flip-chip pin-out designation respin for package-board codesign," *IEEE TVLSI*, 17(8):1087–1098, 2009.

[8] R.-J. Lee and H.-M. Chen, "Row-based area-array I/O design planning in concurrent chip-package design flow," in *Proc. of ASPDAC*, pp. 837–842, 2011.

[9] K.-S. Lin, H.-W. Hsu, R.-J. Lee, and H.-M. Chen, "Area-I/O RDL routing for chip-package codesign considering regional assignment," in *Proc. of EDAPSS*, pp. 1–4, 2010.

[10] John F. Park, "Board driven I/O planning & optimization," in *Proc. of ICCAD*, pp. 395–397, 2010.

[11] C.-Y. Peng, W.-C. Chao, Y.-W. Chang, and J.-H. Wang, "Simultaneous block and I/O buffer floorplanning for flip-chip design," in *Proc. of ASPDAC*, pp. 213–218, 2006.

[12] J. Xiong, Y.-C. Wong, E. Sarto, and L. He, "Constraint driven I/O planning and placement for chip-package co-design," in *Proc. of ASPDAC*, pp. 207–212, 2006.

[13] T. Yan and M.D.F. Wong, "A correct network flow model for escape routing," in *Proc. of DAC*, pp. 332–335, 2009.

Obstacle-Avoiding Free-assignment Routing for Flip-Chip Designs [*]

Po-Wei Lee[1], Hsu-Chieh Lee[1], Yuan-Kai Ho[1], Yao-Wen Chang[1,2],
Chen-Feng Chang[3], I-Jye Lin[3], and Chin-Fang Shen[3]

[1]Graduate Institute of Electronics Engineering, National Taiwan University, Taipei 106, Taiwan
[2]Department of Electrical Engineering, National Taiwan University, Taipei 106, Taiwan
[3]Synopsys Inc., Taipei 110, Taiwan

{webber, pg30123, yuankai}@eda.ee.ntu.edu.tw; ywchang@cc.ee.ntu.edu.tw; {cfchang, ijlin, cfshen}@synopsys.com

ABSTRACT

The flip-chip packaging is introduced for modern IC designs with higher integration density and larger I/O counts. It is necessary to consider routing obstacles for modern flip-chip designs, where the obstacles could be regions blocked for signal integrity protection (especially for analog/mixed-signal modules), pre-routed or power/ground nets, and even for through-silicon vias for 3D IC designs. However, no existing published works consider obstacles. To remedy this insufficiency, this paper presents the first work to solve the free-assignment flip-chip routing problem considering obstacles. For the free-assignment routing problem, most existing works apply the network-flow formulation. Nevertheless, we observe that no existing network-flow model can exactly capture the routability of a local routing region (tile) in presence of obstacles. This paper presents the first work that can precisely model the routability of a tile, even with obstacles. Based on this new model, a two-stage approach of global routing followed by detailed routing is proposed. The global routing computes a routing topology by the minimum-cost maximum-flow algorithm, and the detailed routing determines the precise wire positions. Dynamic programming is applied to further merge tiles to reduce the problem size. Compared to a state-of-the-art flow model with obstacle handling extensions, experimental results show that our algorithm can achieve 100% routability for all circuits while the extensions of the previous work cannot complete routing for any benchmark circuit with obstacles.

Categories and Subject Descriptors

B.7.2 [**Integrated Circuits**]: Design Aids—*Placement and routing*; J.6 [**Computer-Aided Engineering**]: Computer-aided design

General Terms

Algorithms, Design, Performance

Keywords

Physical design, Flip-chip routing, Free-assignment, Obstacle-avoiding

1. INTRODUCTION

The flip-chip packaging is introduced for modern IC designs with higher integration density and larger I/O counts. The flip-chip packaging is a technique for connecting a die to external circuitry such as package carriers or printed circuit boards, where a die is flipped over and mounted on a package carrier, with bump balls as connectors in between. The flip-chip packaging offers many advantages, such as much more area for I/Os, higher performance with shorter interconnections, and better signal integrity. In modern IC designs, I/O pads are placed along the boundaries of a chip, but bump balls and bump pads are placed in the center. To avoid changing the I/O pad locations of a design, an extra metal layer, called a *Redistribution Layer (RDL)*, is added to redistribute the I/O pads to the bump pads. Consequently, a flip-chip router is needed to connect the I/O pads to the bump pads.

Flip-chip routing problems can be classified into two categories: the *free-assignment* routing problem [2, 3, 5, 9, 11, 12, 13, 14] and the *pre-assignment* routing problem [4, 6]. For the free-assignment problem, the net assignments between I/O pads and bump pads are not predefined before routing, so a router has the freedom to assign each I/O pad to any bump pad. In contrast, for the pre-assignment one, it has predefined connections between I/O pads and bump pads before routing, and the assignments cannot be changed during routing. Both routing problems are important for real-world designs. In particular, the free-assignment problem with obstacle constraints has attracted increasing attention in real-world practice recently for two major reasons: (1) Designers on a placement team require an effective and efficient free-assignment router to help improve/evaluate the placement quality to facilitate routing. (2) Obstacles are unavoidable, where they could be regions blocked for signal integrity protection (especially for analog/mixed-signal modules), power/ground or manually pre-routed nets, even through-the-silicon vias for 3D IC designs, etc. As a result, it is desirable to develop an obstacle-avoiding free-assignment routing algorithm for flip-chip designs.

There is a series of related works on free-assignment routing. Some of them focus on flip-chip routing, and some on escape routing, a special type of PCB routing. Since the two types of routing share significant similarities, we survey previous works for both types. For the free-assignment routing problem, the network-flow formulation was extensively used in the literature [2, 3, 5, 11, 12, 13, 14]. Although there are distinguished features for the respective network-flow formulations, the network-flow based algorithms, in general, have some common steps. Given a routing instance, the routing plane is partitioned into a number of local regions called *tiles* which are typically rectangles or triangles. After that, each tile is represented by a network-flow model. Then all models are connected together, producing a global flow network which represents the original routing structure. In the flow network, each edge is associated with a flow capacity and a cost. A minimum-cost maximum-flow algorithm is then applied to the network. Finally, the network-flow result is transformed into global routing topology. Based on the routing topology, detailed routing determines the specific wiring locations and completes the routing procedure.

Yan and Wong [13] recently pointed out that the network-flow tile models used in [2, 3, 5, 11, 12, 14] are *non-exact*. That is, those models cannot precisely determine whether a routing instance is feasible or not. They might over claim a routing instance being feasible while it actually violates design rules, and/or under claim it being infeasible while it can be routed. Yan and Wong proposed an *exact* network-flow model that the global routing result correctly represents actual routability. However, we observe that in presence of obstacles, their model also becomes *non-exact*. To our best knowledge, no existing model is exact when handling obstacles. As a result, in this paper, we propose the first *exact* obstacle-aware network-flow model (OA-model) that can handle obstacles with guaranteed optimality.

Table 1 shows an optimality comparison among different network-flow models. In the table, "Yes" and "No" indicate whether the model is exact under a certain condition. There are three models in the table: "Traditional model:" traditional tile-based models (both rectangular-tile and triangular-tile based models) in [2, 3, 5, 11, 12, 14]; "YW:" Yan and Wong's model in [13]; and "Ours:" our proposed OA-model. "Tile" means optimality within a certain type tile, while "Global" concerns a whole routing instance with multiple tiles.

---

[*]This work was partially supported by IBM, SpringSoft, TSMC, and NSC of Taiwan under Grant No's. NSC 100-2221-E-002-088-MY3, NSC 99-2221-E-002-207-MY3, NSC 99-2221-E-002-210-MY3, and NSC 98-2221-E-002-119-MY3.

As shown in Table 1, our model possesses single-tile optimality, even in presence of obstacles. (A formal definition of the tile structure is given in Section 2.1.) This advancement has significant impacts for at least two reasons: (1) to our best knowledge, this is the first model that can precisely capture the routability of a tile, even with obstacles, and (2) the exact tile model provides higher success rates in routing. Note that no polynomial-time algorithm that optimally solves the obstacle-avoiding free-assignment routing problem is likely to exist since the problem is NP-complete [15].

Table 1: Optimality comparison among various network-flow models ("Traditional model:" traditional tile-based models, "YW:" Yan and Wong's model, and "Ours:" our OA-model).

Model	Obstacle-free		Obstacle-avoiding	
	Tile	Global	Tile	Global
Traditional model	No	No	No	No
YW	Yes	Yes	No	No
Ours	Yes	Yes	Yes	No

We summarize our major contributions as follows.

- We observe that no previous model can guarantee the optimality with obstacle constraints. Consequently, previous work might claim an infeasible routing instance to be feasible, and vice versa. To resolve the deficiency, we propose an obstacle-aware network-flow model that can guarantee the routing feasibility for each tile.

- Based on the network-flow model, we develop an effective two-stage technique of global routing followed by detailed routing for the flip-chip routing.

- We observe that partitioning a routing plane into triangular tiles might incur intrinsic difficulties for obstacle-avoiding routing. Therefore, we resort to rectangular tiles based on a nonuniform-grid model and obstacle configuration, and incorporate the tile model into our network-flow formulation. Based on dynamic programming, we further merge tiles to reduce the problem size and thus the running time.

- Experimental results show that our algorithm can achieve 100% routability for all circuits, while latest work cannot complete routing for any benchmark circuit, even with reasonable obstacle handling extensions.

The rest of this paper is organized as follows. Section 2 focuses on an obstacle-aware local (tile) routing problem. Section 3 considers the overall flip-chip routing problem. Section 4 reports the experimental results. Section 5 concludes this paper.

2. OBSTACLE-AWARE TILE ROUTING PROBLEM

This section considers an obstacle-aware tile routing problem (OA-TRP), a subproblem of the original global routing problem. First, Section 2.1 formulates the OATRP. Then, Section 2.2 briefly reviews Yan and Wong's model and shows its nonoptimality in obstacle handling. Finally, our obstacle-aware model is proposed to solve the OATRP optimally in Section 2.3.

2.1 Problem Formulation

We first introduce some definitions to formulate the OATRP. We define a *tile* to be a rectangular region of a routing plane without any obstacle inside the tile. As shown in Figure 1(a), a routing plane can always be partitioned into a number of tiles. However, both running time and solution quality would be poor if the tiles are too small. So it is necessary to merge adjacent tiles into bigger tiles, as shown in Figure 1(b). The tile merging mechanism and disadvantages of small tiles are detailed in Section 3.2.2.

A merged tile might have multiple obstacles on one boundary. We define an *opening* to be a segment of a tile boundary that is *not* blocked by obstacles, where wires enter/exit a tile. If there are two openings on the same side of a tile, a wire passing through these openings would form a detour inside the tile. It will be clear in Section 3.2.2 that our partitioning and merging processes result in tiles with at most one opening on each side. In other words, we prevent any detour inside a tile for better resource utilization and higher routability for practical applications.

DEFINITION 1. *An obstacle-surrounded tile is a tile of which each side has at most one opening.*

Figure 1(c) shows a single obstacle-surrounded tile extracted from Figure 1(b), of which each boundary has exactly one opening. Note that it

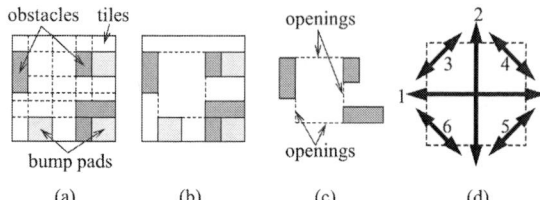

(a)   (b)   (c)   (d)

Figure 1: (a) A routing plane, partitioned into rectangular tiles, and (b) its merging result. (c) An obstacle-surrounded tile. Note that it suffices to consider such one-opening obstacle-surrounded tiles for tile routing, with our partitioning and merging processes. (d) Six types of connections.

suffices to consider such one-opening obstacle-surrounded tiles for tile routing in our work since our partitioning and merging processes will always result in tiles of this kind.

There are six possible directions for a wire to pass through a tile. Each type is characterized by two sides of a tile that intersect with the wire. Figure 1(d) shows the classification. The connection labeled i, $1 \le i \le 6$, in Figure 1(d), is said to be of *Type-i*. A *wiring configuration* inside a tile can be described by a six-tuple $(c_1, c_2, c_3, c_4, c_5, c_6)$, where c_i denotes the number of Type-i connections. We say a wiring configuration is *feasible* for a tile if the specified wires can be routed on the tile without violating design rules.

Consequently, the obstacle-aware tile routing problem (OATRP) can be formulated as follows: given an obstacle-surrounded tile, a wiring configuration, and design rule constraints including wire width, wire spacing, and routing angle, we are to determine whether the given wiring configuration is feasible. The OATRP can be viewed as a subproblem of the global routing problem, since we partition a routing plane into multiple obstacle-surrounded tiles during the routing process. Figure 1(b) shows an example of partitioning a routing plane into obstacle-surrounded tiles. The partitioning process will be detailed in Section 3.2.1.

2.2 Yan and Wong's Model

In this section, we briefly review Yan and Wong's model [13] (called YW-model, for short), shown in Figure 2, and point out its limitation on the OATRP. The model is designed for *square* tiles *without* obstacles. There are five nodes in the model. The *W-*, *N-*, *E-*, and *S-nodes* are on the respective west, north, east, and south sides of the tile, while the *C-node* is in the center. The first four nodes are called the *peripheral nodes*, and the last is called the *center node*. They define the *orthogonal capacity* (*O-cap*) as the capacity of four boundaries of a tile, and the *diagonal capacity* (*D-cap*) as the capacity of two diagonals of a tile.

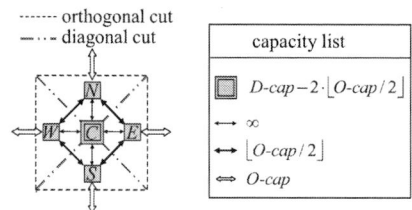

Figure 2: Yan and Wong's model (YW-model).

As shown in Figure 2, each edge connecting a peripheral node to the outside has the capacity of *O-cap*, each edge connecting adjacent peripheral nodes has the capacity of $\lfloor O\text{-}cap/2 \rfloor$, and each edge connecting the center node with a peripheral node has an infinite capacity. In addition, the center node is a *capacity node* with the capacity of $D\text{-}cap - 2 \cdot \lfloor O\text{-}cap/2 \rfloor$, which is realized by splitting the node into two and connect them with an edge of the specified capacity.

The YW-model improves over traditional models in that traditional models only use one single value to represent the capacity of a tile. They observed that both orthogonal and diagonal capacities are needed to model a tile exactly. In other words, the YW-model uses 2-dimensional routing capacity vectors, i.e., (*O-cap*, *D-cap*), to describe the capacities of tiles. As a result, their model can guarantee optimality in obstacle-free routing instances.

With obstacle-surrounded tiles, however, their model is no longer exact. Consider a counterexample shown in Figure 3(a), where (*O-cap*, *D-cap*)

1089

equals to $(2, 2)$. The YW-model is constructed accordingly as shown in Figure 3(b), on which every capacity is labeled. The model is *non-exact*, since it may cause an infeasible wiring configuration like the one shown in Figure 3(g).

Further, let us try some extensions to see whether this model can be extended to resolve the feasibility problem. One extended model is called the YW-model with adjustable orthogonal/diagnoal capacities model (YW-AODC-model). It reduces *O-cap* or *D-cap*. For the example of Figure 3(a), the $(O\text{-}cap, D\text{-}cap)$ is reduced to $(2, 1)$ (see Figure 3(c)), $(1, 2)$ (Figure 3(d)), $(1, 1)$ (Figure 3(e)), or even smaller. However, both $(2, 1)$ and $(1, 2)$ may cause the same feasibility problem shown in Figure 3(g), and $(1, 1)$ and smaller capacities may lead to another problem as shown in Figure 3(h), missing a feasible wiring configuration.

Another extended model is called the YW-model with adjustable boundary capacities model (YW-ABC-model). It adjusts each boundary capacity according to the opening. For instance, in this example, we set the north, east, south, and west boundary capacities to 1, 1, 2, and 1, respectively, as shown in Figure 3(f). Unfortunately, this model may also result in the feasibility problem shown in Figure 3(g).

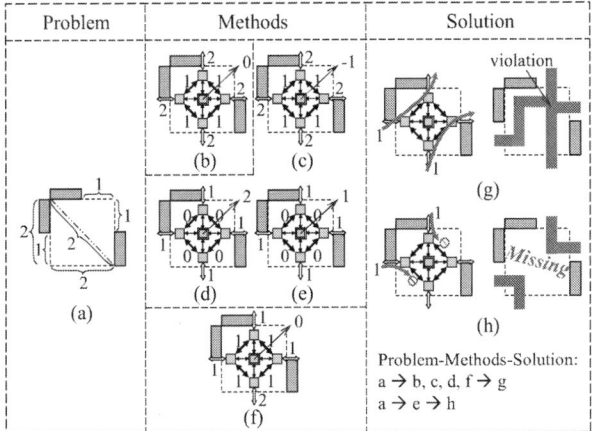

Figure 3: **Examples of how the YW-model, the YW-AODC-model, and the YW-ABC-model fail for the OATRP: (a) A given obstacle-surrounded tile and (b) its corresponding YW-model ($(O\text{-}cap, D\text{-}cap) = (2, 2)$), (c)–(e) YW-AODC-models ((c) $(2, 1)$, (d) $(1, 2)$, and (e) $(1, 1)$), and (f) the YW-ABC-model. (g) An infeasible solution that the models in (b), (c), (d), and (f) may get. (h) A feasible solution that the model in (e) may miss.**

By the analysis above, we can conclude that no matter how we choose to reduce *O-cap* or *D-cap* or to adjust each boundary capacity, we can never guarantee to exactly represent the characteristics of a given obstacle-surrounded tile with the aforementioned models. This fact provides us with a key insight: a more sophisticated model is required in order to precisely characterize the routability of an obstacle-surrounded tile.

2.3 Our Obstacle-Aware Network-Flow Model

In this section, our obstacle-aware network-flow model (OA-model) is proposed. By using the OA-model, we can solve the OATRP optimally, implying that we can *exactly* determine if a given wiring configuration is routable on a tile.

We observe that the YW-model assumes the four boundary capacities of a tile to be the same, and both the diagonal capacities also to be the same. This is a reasonable assumption for square tiles without obstacles. For an obstacle-surrounded tile, however, the tile might not be square, and some boundaries might be (partially) blocked by obstacles. Therefore, it is no longer sufficient to represent a tile with a 2-dimensional vector. To correctly model a obstacle-surrounded tile, we must use separate variables to represent the capacities of the four boundaries and the two diagonals.

DEFINITION 2. *We define a routing capacity vector (r-vector, for short) of an obstacle-surrounded tile, denoted by $\vec{r}$, as a 6-tuple $(r_1, r_2, r_3, r_4, r_5, r_6)$, where r_1 is the maximum number of connections that pass through the left boundary of the tile, i.e., the maximum number of Type-1, -3, and -6 connections combined; r_2, the top side and of Type-2, -3, and -4 connections combined; r_3, the right side and of Type-1, -4, and -5 ones combined; r_4, the bottom side and of Type-2, -5, and -6 ones; r_5,*

the descending diagonal and of Type-1, -2, -3, and -5 ones; and r_6, the ascending diagonal and of Type-1, -2, -4, and -6 ones.

Figure 4(a) illustrates the definition of the r-vector. Here, $r_1, \ldots, r_4$ are defined as the maximum numbers of wires that can go through the respective four boundaries of a tile, so they are *real orthogonal capacities*. On the other hand, r_5 and r_6 are defined as such numbers through the two diagonals, so they are *real diagonal capacities*. Moreover, Yan and Wong's 2-dimensional description of a tile can be viewed as a special case of our r-vector, where $r_1, \ldots, r_4$ all equal *O-cap*, and r_5 and r_6 both equal *D-cap*.

Given an obstacle-surrounded tile, we can compute its r-vector in constant time by first arranging Type-3, -4, -5, and -6 connections and then Type-1 and -2 connections, all compacting to the corners of the tile. Note that the r-vector computation is adaptive to various design rules. For example, an r-vector can be computed under 90-degree (Figure 4(b)) or 135-degree (Figure 4(c)) routing angles according to the design rule.

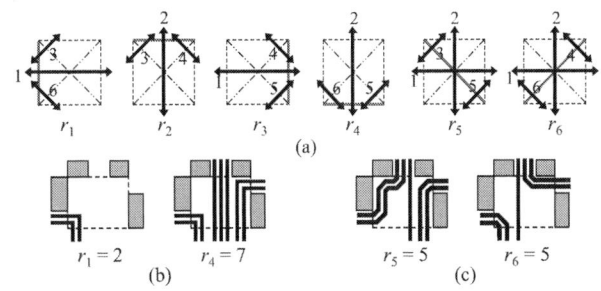

Figure 4: **(a) Illustration of the r-vector. Examples of the r-vector computation under (b) 90-degree and (c) 135-degree routing angles.**

We intend to build a flow model that correctly represents the r-vector in order that a wiring configuration is feasible *if and only if* the corresponding flow can actually flow on our model. Our proposed OA-model, shown in Figure 5, is designed for obstacle-surrounded tiles. The nodes of our OA-model are the same as those in the YW-model; however, the capacity configuration is much more sophisticated. The four edges connecting the *C-node* are each still with an infinite capacity, but all the other edges are each now with an independent variable capacity that is computed from the r-vector. Therefore, each edge is associated with a variable name. For instance, C_{WN} is the capacity of the edge between the W- and N-nodes, and C_C is that of the *C-node*. There are totally nine capacity variables in our flow model.

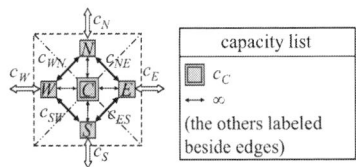

Figure 5: **Our obstacle-aware network-flow model (OA-model) with nine capacity variables.**

We say an r-vector to be *valid* if we can find an actual obstacle-surrounded tile corresponding to the r-vector. Given an r-vector, we derive the following necessary conditions for an r-vector to be valid.

THEOREM 1. *An r-vector is valid only if the following inequalities holds.*

$$r_1 \leq r_2 + r_3 + r_4, \tag{1}$$
$$r_2 \leq r_1 + r_3 + r_4, \tag{2}$$
$$r_3 \leq r_1 + r_2 + r_4, \tag{3}$$
$$r_4 \leq r_1 + r_2 + r_3, \tag{4}$$
$$r_5 \leq \min\{r_1 + r_4, r_2 + r_3\}, \tag{5}$$
$$r_5 \geq \max\{\min\{r_1, r_2\}, \min\{r_3, r_4\}\}, \tag{6}$$
$$r_6 \leq \min\{r_1 + r_2, r_3 + r_4\}, \tag{7}$$
$$r_6 \geq \max\{\min\{r_1, r_4\}, \min\{r_2, r_3\}\}, \tag{8}$$
$$r_5 + r_6 \geq \max\{r_1 + r_3, r_2 + r_4\}. \tag{9}$$

Obviously, $\vec{r} = (1, 0, 0, 0, 0, 0)$ is invalid since it violates Inequalities (1) and (9) and no obstacle-surrounded tile can have a routing configuration corresponding to the r-vector, $(1, 0, 0, 0, 0, 0)$. Here, $r_1 = 1$ means that exactly one wire can pass through the west boundary of the tile, and this wire must be of Type-1, -3, or -6. However, if this wire is of Type-1, then r_3, r_5, and r_6 should also be at least one since they all contain Type-1 wires, instead of zero as in this r-vector.

Given a valid r-vector, we intend to find the nine capacities for the flow model to correctly represent the r-vector. We have the following theorem:

THEOREM 2. *Our OA-model exactly represents the routing capacity of a given obstacle-surrounded tile if the following equations and inequalities hold:*

$$c_W = r_1, c_N = r_2, c_E = r_3, \qquad c_S = r_4, \qquad (10)$$
$$c_{SW} + c_{WN} + c_C \geq r_1, \qquad (11)$$
$$c_{WN} + c_{NE} + c_C \geq r_2, \qquad (12)$$
$$c_{NE} + c_{ES} + c_C \geq r_3, \qquad (13)$$
$$c_{ES} + c_{SW} + c_C \geq r_4, \qquad (14)$$
$$c_{WN} + c_{ES} + c_C = r_5, \qquad (15)$$
$$c_{NE} + c_{SW} + c_C = r_6. \qquad (16)$$

The above equations and inequalities do not always have a solution. For example, for $\vec{r} = (1, 0, 0, 0, 0, 0)$, we cannot find any set of capacities that satisfy the above inequalities. This is because our inequalities are derived from the characteristics of an obstacle-surrounded tile, and no obstacle-surrounded tile can have an r-vector of $(1, 0, 0, 0, 0, 0)$.

Our algorithm for solving the equations and inequalities in Theorem 2 to construct the OA-model is summarized in Algorithm 1.

Algorithm 1 OA-Model Construction

Input: $\vec{r} = (r_1, r_2, r_3, r_4, r_5, r_6)$
  ▷ $\vec{r}$ is a given valid r-vector
Output: $C = \{c_W, c_N, c_E, c_S, c_{WN}, c_{NE}, c_{ES}, c_{SW}\}$
  ▷ C is the set of the edge capacities.
1: $c_W \leftarrow r_1, c_N \leftarrow r_2, c_E \leftarrow r_3, c_S \leftarrow r_4$
2: $m_0 \leftarrow \min\{r_1, r_2, r_3, r_4\}$
3: $m_{WN} \leftarrow \min\{r_5 - m_0, \lfloor \frac{1}{2}(2r_5 + r_6 - m_0 - r_3 - r_4)\rfloor\}$
4: $m_{NE} \leftarrow \min\{r_6 - m_0, \lfloor \frac{1}{2}(2r_6 + r_5 - m_0 - r_4 - r_1)\rfloor\}$
5: $m_{ES} \leftarrow \min\{r_5 - m_0, \lfloor \frac{1}{2}(2r_5 + r_6 - m_0 - r_1 - r_2)\rfloor\}$
6: $m_{SW} \leftarrow \min\{r_6 - m_0, \lfloor \frac{1}{2}(2r_6 + r_5 - m_0 - r_2 - r_3)\rfloor\}$
  ▷ Lines 3–6 compute the upper-bounds of edge capacities.
7: $A_1 \leftarrow (m_0 + m_{SW} + m_{WN}) - r_1$
8: $A_2 \leftarrow (m_0 + m_{WN} + m_{NE}) - r_2$
9: $A_3 \leftarrow (m_0 + m_{NE} + m_{ES}) - r_3$
10: $A_4 \leftarrow (m_0 + m_{ES} + m_{SW}) - r_4$
11: $A_5 \leftarrow (m_0 + m_{WN} + m_{ES}) - r_5$
12: $A_6 \leftarrow (m_0 + m_{NE} + m_{SW}) - r_6$
  ▷ Lines 7–12 set A_i to the excess value of an edge constraint.
13: Subtract $\min\{m_{WN}, A_1, A_2, A_5\}$ from m_{WN}, A_1, A_2, A_5
14: Subtract $\min\{m_{NE}, A_2, A_3, A_6\}$ from m_{NE}, A_2, A_3, A_6
15: Subtract $\min\{m_{ES}, A_3, A_4, A_5\}$ from m_{ES}, A_3, A_4, A_5
16: Subtract $\min\{m_{SW}, A_4, A_1, A_6\}$ from m_{SW}, A_4, A_1, A_6
  ▷ Lines 13–16 eliminate excess values by decreasing m_i.
17: $c_{WN} \leftarrow \lfloor \frac{1}{2}m_0 \rfloor + m_{WN}$
18: $c_{NE} \leftarrow \lfloor \frac{1}{2}m_0 \rfloor + m_{NE}$
19: $c_{ES} \leftarrow \lfloor \frac{1}{2}m_0 \rfloor + m_{ES}$
20: $c_{SW} \leftarrow \lfloor \frac{1}{2}m_0 \rfloor + m_{SW}$
21: $c_C = m_0 \bmod 2$
  ▷ Lines 17–21 set the return values.
22: **return** C

THEOREM 3. *Given a valid r-vector, Algorithm OA-Model Construction computes the capacities of the corresponding OA-model in constant time.*

Note that, given a valid r-vector, the algorithm finds only one feasible solution, and there could be other sets of capacities satisfying Theorem 2. According to Theorem 2, however, they are all equivalent in that they all represent the same characteristics of the corresponding tile and thus share the same optimality. Consider the r-vector $(2, 2, 2, 2, 4, 4)$, for example. Algorithm OA-Model Construction will find an OA-model with the respective nine capacities $C_W, C_N, C_E, C_S, C_{WN}, C_{NE}, C_{ES}, C_{SW}$,

and C_C being 2, 2, 2, 2, 1, 3, 3, 1, and 0. The capacity configurations 2, 2, 2, 2, 2, 2, 2, 2, and 0 (or 2, 2, 2, 2, 0, 0, 0, 0, and 4) is also feasible.

Once we have an exact model to characterize an obstacle-surrounded tile, the OATRP can be solved optimally. Let us go back to the example in Figure 3(a) (redrawn in Figure 6(a)), and see how the OA-model can solve the feasibility problem not solvable by the previous models. We first compute the r-vector of the tile in Figure 6(a) as $(1, 1, 1, 2, 1, 2)$, then solve the equations and inequalities of Theorem 2 by Algorithm OA-Model Construction, resulting in the OA-model shown in Figure 6(b). Note that both the r-vector computation and Algorithm OA-Model Construction run in constant time, and so does the whole procedure. As shown in Figure 6(c), the OA-model prevents itself from getting the infeasible wiring configuration shown in Figure 3(g), because only one-unit flow is allowed to pass through the C-node. In Figure 6(d), the OA-model prevents itself from missing the feasible wiring configuration shown in Figure 6(h).

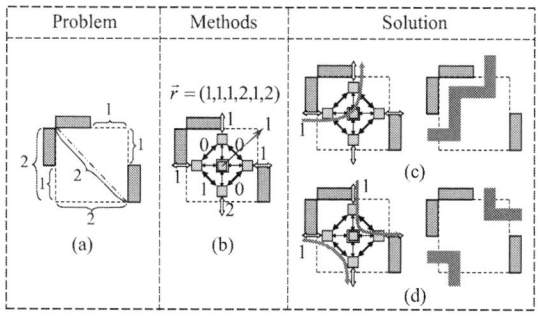

Figure 6: Examples of how the OA-model exactly characterizes the feasibility.

3. OBSTACLE-AVOIDING FREE-ASSIGNMENT FLIP-CHIP ROUTING PROBLEM

In this section, we consider the obstacle-avoiding free-assignment flip-chip routing problem (OFFRP), which is the ultimate problem we aim to solve. Section 3.1 formulates the problem. Section 3.2 presents our proposed algorithm.

3.1 Problem Formulation

To formulate the problem, we first give some notations:

- $D = \{d_1, d_2, \ldots, d_{|D|}\}$ is a set of rectangular I/O pads.
- $B = \{b_1, b_2, \ldots, b_{|B|}\}$ is a set of rectangular bump pads.
- $O = \{o_1, o_2, \ldots, o_{|O|}\}$ is a set of rectangular obstacles.
- $N = \{n_{d_1}, \ldots, n_{d_{|D|}}, n_{b_1}, \ldots, n_{b_{|B|}}\}$ is a set of numbers, where n_x is the number of nets assigned to an I/O pad or bump pad x.

The OFFRP can be formulated as follows: given D, B, O, N and design rule constraints including wire width, wire spacing, and desired routing angle, where $\forall d \in D$, $n_d = 1$, $\forall b \in B$, $n_b \geq 0$, the objective is to connect D and B according to N such that the total wirelength is minimized under maximum routability.

Note that each bump pad can be assigned with zero, one, or more I/O pads, so $n_b \geq 0$. Figure 7(a) gives an example of the problem. The instance has eight I/O pads, four bump pads, and five obstacles. Each I/O pad and bump pad is assigned with the number of nets to be routed onto the pad.

3.2 Algorithm

In this section, we present an algorithm to solve the OFFRP. The overall flow is shown in Figure 7. We adopt a two-stage technique of *global routing* followed by *detailed routing*. The global routing consists of four stages: (1) tile partitioning, (2) tile merging, (3) flow-network construction, and (4) minimum-cost-flow solving. In the tile partitioning, we partition the routing plane into a number of tiles based on the nonuniform-grid model, an obstacle-dependent model. Based on a dynamic programming algorithm, in the tile merging stage, we merge tiles to improve solution quality and reduce the problem size (and thus running time). In the flow-network construction, we create the OA-model introduced in Section 2 for each tile and combine all models to form a global flow network. After that, we solve the minimum-cost maximum-flow problem to obtain a routing topology. Finally, we carry out detailed routing based on the previous routing topology and obtain the routing result.

1091

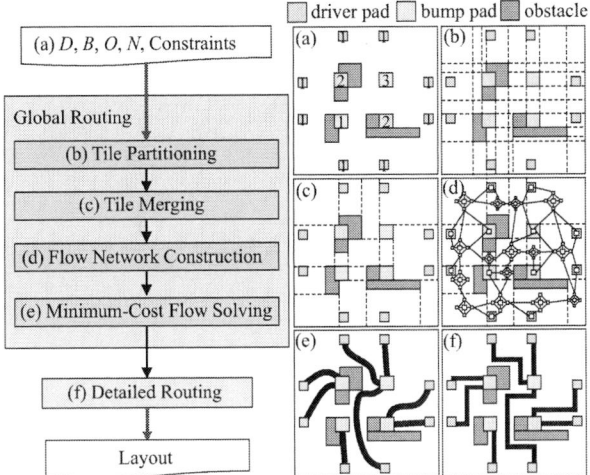

Figure 7: The overall flow of our algorithm.

3.2.1 Tile Partitioning

This stage partitions the routing plane into a number of tiles based on the nonuniform-grid model. The model was first proposed by Ohtsuki in [8], which was constructed by extending lines through the boundaries of all obstacles until they intersect with other obstacles or the boundary of the routing plane. This model is suitable for our problem since the construction of tiles utilizes the information of obstacles. In addition, all tiles obtained from this model are obstacle-surrounded tiles. Figure 7(b) shows the partitioned result for the routing instance illustrated in Figure 7(a).

3.2.2 Tile Merging

This stage merges smaller tiles into bigger ones according to predefined criterions. Tile merging improves the overall algorithm in two aspects. First, tile merging eliminates tiles that are very small or very thin. By doing so, we can reduce the loss of optimality during global routing. Second, by reducing the number of tiles, we also reduce the number of nodes and edges in the global flow network, and thus the running time of the maximum-flow algorithm.

Let us illustrate two examples of optimality loss caused by small/thin tiles. The first example is shown in Figures 8(a) and 8(b). In Figure 8(a), eight thin tiles stand side by side, each with a very small height that no wire can pass through horizontally. Since no tile allows horizontal flow, the whole combined region disallows any flow to pass. However, if we merge all the eight tiles into one larger tile, a large amount of flow, say seven, could pass through the merged tile as shown in Figure 8(b). In this case, a feasible solution might be missed if the tiles are not merged properly.

The second example is shown in Figures 8(c) and 8(d). In Figure 8(c), there are two adjacent tiles, each with a horizontal capacity of three and a vertical capacity of seven. An eight-unit flow passes through the flow model, while the corresponding routing result is actually infeasible. As shown in Figure 8(d), the sandwiched wires have insufficient space to pass through the common side. In this case, it would lead to an infeasible solution. By the observation, we shall focus on eliminating small/thin tiles during tile merging.

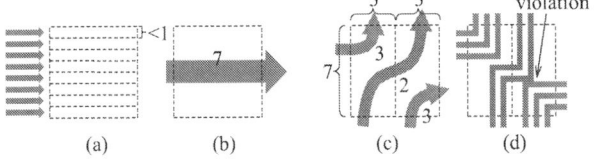

Figure 8: Examples of loss in optimality: (a) (b) missing a feasible solution and (c) (d) leading to an infeasible solution.

We now define the aspect ratio of a tile as follows:

DEFINITION 3. The aspect ratio of a tile t of width w and height h, denoted by $\Gamma(t)$, is defined as

$$\Gamma(t) = \max\{h/w, w/h\}.$$

The tile merging process is accomplished through a dynamic programming algorithm. We use a simple example shown in Figure 9 to illustrate the procedure. With an initial partitioning result, we first construct a corresponding graph, as shown in Figure 9(a), where each vertex denotes a tile, and each pair of adjacent tiles are connected with an edge. Finding a good merging result on this graph is complicated, so we reduce the graph into a tree structure by finding a minimum-cost spanning tree (MST) of the graph. Since we are going to merge tiles along edges in the tree, we should avoid edges that will produce high ratio tiles. Therefore, the cost of each edge is defined as the ratio of the merged tile.

DEFINITION 4. The cost of an edge $e = (v_1, v_2)$, denoted by $\Theta(e)$, is defined as

$$\Theta(e) = \Gamma(t),$$

where t is the merged tile of the two tiles denoted by v_1 and v_2.

After that, an arbitrary vertex is chosen to be the tree root. For example, the MST of this example is shown in Figure 9(c). The colored vertex (in orange) is chosen as the root since it is in the middle of the chip. Our dynamic programming approach then merges tiles in a bottom-up manner, from leaf vertices towards the root. For each vertex, we try all possible merging schemes for this vertex and its subtree, and the best merging result of some vertices in its subtree is stored in a table and is reused. With all merging candidates being considered, the one with the minimum cost is recorded. The cost of a merging result is defined as follows.

DEFINITION 5. Assume that the set of merged tiles in a subtree rooted at v is $T_v = \{t_1, t_2, \cdots, t_{|T_v|}\}$, the cost of a merging result rooted at v, denoted by $\Psi(v)$, is defined as

$$\Psi(v) = \beta \sum_{i=1}^{|T_v|} \Gamma(t_i) + \gamma |T_v|,$$

where β and γ are user-specified parameters, $\beta + \gamma = 1$.

The first term is the sum of the aspect ratios of merged tiles, and the second term is the number of merged tiles. For example, as shown in Figure 9(b), the arrow directions represent the order of the dynamic programming. Consider v_{11}. One possible merging candidate $\{v_3, v_4, v_8, v_{11}\}$ is shown in Figure 9(c). In this case, the best merging result of subtrees rooted at v_2, v_7, and v_{10} can be reused, which are $\{v_1, v_2\}$, $\{v_5, v_6, v_7\}$, and $\{v_9, v_{10}\}$, respectively. Alternatively, another possible merging candidate is $\{v_2, v_3, v_4, v_7, v_8, v_{11}\}$, as shown in Figure 9(d), where the best merging result of subtrees rooted at v_1, v_6, and v_{10} are $\{v_1\}$, $\{v_5, v_6\}$, and $\{v_9, v_{10}\}$. Of course, there are other merging candidates. Among all merging candidates, v_{11} will keep the best one with the minimum merging cost.

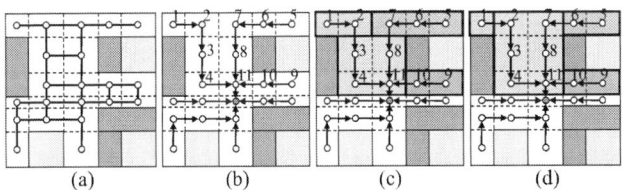

Figure 9: An example of the tile merging by dynamic programming: (a) A partitioning result and its corresponding graph. (b) The MST. (c) A merging candidate of v_{11}. (d) Another merging candidate of v_{11}.

The tile merging algorithm proceeds from leaves to the root. After the merging process, some merged tiles might not be obstacle-surrounded tiles. These tiles are further partitioned and merged by a fine-tune post-process, so that every tile will be an obstacle-surrounded tile after the tile merging stage. Figure 7(c) shows the tile merging result of the routing instance in Figure 7. One can clearly see that the number of tiles is significantly reduced after tile merging, and the tiles with high aspect ratios are also eliminated.

For practical applications, the possible ways to merge a certain number of adjacent tiles are typically a small constant, since any merging is limited only to a rectangular region. As the example in Figure 9, v_{11} has only 1, 3, 3, 2, 0, and 1 merging candidates for one to six tiles, respectively, and no candidate for more tiles. As a result, the time needed for each vertex is linear to the size of its subtree; based on this practical assumption, we have the following theorem:

THEOREM 4. The respective average- and worst-case running times of the tile merging are $\Theta(n \lg n)$ and $\Theta(n^2)$, where n is the number of vertices in the tree.

1092

Table 2: Benchmark circuits and experimental results (Wlen: count only routed nets; "†:" man-made runtime; NA: not available).

Circuit	#Nets	$|D|$	$|B|$	$|O|$	YW-ABC-method			YW-ABC-method-RR			Ours-no-TM			Ours		
					Rout. (%)	Wlen (μm)	Time (sec)	Rout. (%)	Wlen (μm)	Time (sec)	Rout. (%)	Wlen (μm)	Time (sec)	Rout. (%)	Wlen (μm)	Time (sec)
test19	19	19	19	3	94.7	> 432	0.03	94.7	> 432	300†	100.0	466	0.09	100.0	461	0.04
test143	143	143	143	3	96.5	> 10,214	0.21	96.5	> 10,214	300†	100.0	10,628	0.26	100.0	10,611	0.27
test143_2	143	143	143	5	92.3	> 10,110	0.28	97.2	> 10.110	300†	97.9	> 10,779	0.48	100.0	10,967	0.35
test150	150	150	150	5	95.3	> 20,351	0.45	95.3	> 20,351	300†	96.0	> 20,482	0.55	100.0	21,334	0.42
test200	200	200	200	10	87.0	> 23,196	0.53	90.5	> 23,196	300†	90.5	> 24,131	0.92	100.0	26,645	0.58
fc1189	513	513	784	20	95.1	> 547,536	4.53	97.7	> 549,818	600†	97.7	> 562,731	9.11	100.0	575,980	6.75
fc1458	646	646	812	15	97.4	> 777,615	7.15	97.4	> 777,615	600†	93.0	> 743,685	13.2	100.0	799,210	9.33
fc1795	639	639	1,156	15	97.3	> 606,464	6.22	97.5	> 607,493	600†	99.4	> 620,122	8.44	100.0	623,695	6.35
fc1813	657	657	1,156	15	96.5	> 672,849	6.85	97.4	> 672,849	600†	98.6	> 688,201	13.2	100.0	697,512	7.77
fc2624	1,024	1,024	1,600	15	97.0	> 1,562,845	11.53	98.7	> 1,569,372	600†	91.0	> 1.468,602	20.1	100.0	1,612,845	13.05
Average					0.95	NA	0.85	0.96	NA	1107†	0.96	NA	1.48	1.00	1.00	1.00

3.2.3 Flow-Network Construction

After the tile partitioning and merging, all remaining tiles are obstacle-surrounded. We then create the OA-model for each tile, and introduce a node for each I/O pad and bump pad. Then, all the models and nodes are connected to form a global flow network. Figure 7(d) shows an example of the constructed global flow network.

3.2.4 Minimum-Cost-Flow Solving

To convert the routing problem into the minimum-cost maximum-flow problem, we connect each bump pad node to the source and connect each I/O pad node to the sink. Each connection is assigned with a certain flow capacity according to the number of nets assigned to the corresponding pad. In addition, each edge is assigned a flow cost according to its length. After that, we can apply an existing minimum-cost maximum-flow algorithm to solve the problem in $O(|V|^3)$ time [1], where $|V|$ is the number of nodes. Figure 7(e) demonstrates the routing topology constructed from the flow network in Figure 7(d).

3.2.5 Detailed Routing

In detailed routing, we can convert the resulting routing topology to 90-degree routes by existing routing algorithms such as [8, 10]. Figure 7(f) gives the final routing for the routing topology shown in Figure 7(e).

4. EXPERIMENTAL RESULTS

We conducted four sets of experiments: (1) the method with our OA-model and algorithm ("Our"), (2) the method with our OA-model and algorithm excluding the tile merging ("Ours-no-TM"), (3) the method with the YW-ABC-model and our algorithm ("YW-ABC-method"), and (4) the YW-ABC-method with rip-up and rerouting ("YW-ABC-method-RR"). Note that although the last two methods use an extended Yan and Wong's model (i.e., the YW-ABC-model), they still adopt our algorithm, because Yan and Wong's algorithm [13] is designed for obstacle-free cases and is not applicable here.

We implemented all the methods in C++ programming language and conducted experiments on a Linux workstation with a 2.6GHz CPU and 6GB memory. Two sets of benchmark circuits were used in the experiments, both from the industry. Table 2 shows the information on the benchmark circuits, where "#Nets" denotes the number of nets; "$|D|$", "$|O|$", and "$|B|$" denote the numbers of I/O pads, bump pads, and obstacles, respectively. The table also shows the experimental results, where routability ("Rout."), wirelength ("Wlen"), and runtime ("Time") are reported. As shown in the table, our method achieved 100% routability for all the circuits, while the YW-ABC-method only achieved 87–97%. The routability loss clearly shows that Yan and Wong's model (and all previous models) cannot handle routing instances with obstacles, even with proper extensions. Note that the wirelength counts only those routed nets; therefore, only can those circuits with 100% routability report correct wirelengths.

For the YW-ABC-method-RR, each circuit is manually set a runtime long enough for rip-up and rerouting. The experimental results show that it had only 1% improvement on average routability, compared to the YW-ABC-method (and none of the circuits was complete either). The results also reveal two phenomena of flip-chip routing: (1) Generally, concurrent routing approaches (such as network-flow formulation) are better than sequential routing approaches (such as rip-up and rerouting). (2) Once preliminary routes have been determined, the room for improvement has been limited along.

The results of the Ours-no-TM is an evidence of why our tile merging matters. Without the tile merging, eight circuits became incomplete while two remained complete. We observe that it depends on whether the distribution of obstacles would generate too many small/thin tiles.

The experimental results show that our obstacle-avoiding free-assignment flip-chip routing algorithm is effective and efficient. In particular, our router completed the routing for all the benchmark circuits and consumed only 13 seconds for the largest circuit fc2624.

5. CONCLUSIONS

In this paper, we have considered the obstacle-avoiding free-assignment flip-chip routing problem, and observed that none of previous network-flow models can handle obstacles correctly. To remedy this deficiency, we have proposed an exact obstacle-aware network-flow model that can significantly improve the routability and feasibility of global routing results. A maximum-flow based global-routing algorithm utilizing our OA-model has also been proposed. The experimental results have shown that our obstacle-avoiding free-assignment routing algorithm is effective and efficient.

6. REFERENCES

[1] R. K. Ahuja, T. L. Magnati, and J. B. Orlin, *Network Flows: Theory, Algorithms, and Applications*. Englewood Cliffs, NJ: Prentice-Hall, 1993.

[2] W.-T. Chan, F. Y. L. Chin, and H.-F. Ting. "A faster algorithm for finding disjoint paths in grids," *Proc. of Int. Symp. on Algorithms and Computation*, pp. 393–402, 1999.

[3] J.-W. Fang and Y.-W. Chang, "Area-I/O flip-chip routing for chip-package co-design considering signal skews," *IEEE TCAD*, vol. 29, no. 5, pp 711–721, 2010.

[4] J.-W. Fang, C.-H. Hsu, and Y.-W. Chang, "An integer-linear-programming-based routing algorithm for flip-chip designs," *IEEE TCAD*, vol. 28, no. 1, pp. 98–110, 2009.

[5] J.-W. Fang, I.-J. Lin, Y.-W. Chang, and J.-H. Wang, "A network-flow based RDL routing algorithm for flip-chip design," *IEEE TCAD*, vol. 26, pp. no. 8, pp. 1417–1429, 2007.

[6] X. Liu, Y. Zhang, G. K. Yeap, C. Chu, J. Sun, and X. Zeng, "Global routing and track assignment for flip-chip designs," *Proc. of DAC*, pp. 90–93, 2010.

[7] F. M. Maley, *Single-layer wire routing and compaction*. Cambridge, MA: MIT Press, 1990.

[8] T. Ohtsuki, "Gridless routers-new wire routing algorithms based on computational geometry," *Proc. of ICCS*, pp. 802–809. 1985.

[9] Ma Qiang, E.F.Y Young, and M.D.F. Wong, "An optimal algorithm for layer assignment of bus escape routing on PCBs," *Proc. of DAC*, pp. 176–181, 2011.

[10] D. Staepelaere, J. Jue, T. Dayan, and W. W.-M. Dai, "SURF: Rubber-band routing system for multichip modules," *IEEE Design & Test of Computers*, vol. 10, issue 4, pp. 18–26, 1993.

[11] D. Wang, P. Zhang, C.-K. Cheng, and A. Sen. "A performance-driven I/O pin routing algorithm," *Proc. of ASP-DAC*, pp. 129–132, 1999.

[12] R. Wang, R. Shi, and C.-K. Cheng, "Layer minimization of escape routing in area array packaging," *Proc. of ICCAD*, pp. 815–819, 2006.

[13] T. Yan and M. D.-F. Wong, "A correct network flow model for escape routing," *Proc. of DAC.*, pp. 332–335, 2009.

[14] M.-F. Yu, J. Darnauer, and W. W.-M. Dai, "Interchangeable pin routing with application to package layout," *Proc. of ICCAD*, pp. 668–673, 1996.

[15] M.-F. Yu, J. Darnauer, and W. W.-M. Dai, "Interchangeable pin routing with application to package layout," technical report, UCSC-CRL-96-10, UC, Santa Cruz, CA, 1996.

[16] UMC. "0.13μm flip-chip layout guideline," pp. 6, 2004.

Clock Tree Synthesis with Methodology of Re-use in 3D IC

Fu-Wei Chen, TingTing Hwang
Department of Computer Science
National Tsing Hua University, Hsinchu, Taiwan 30013, R.O.C.
{fwchen, tingting}@cs.nthu.edu.tw

Abstract

IP reuse methodology has been used extensively in SoC (System on Chip) design. In this reuse methodology, while design and implementation cost is saved, manufacturing cost is not. To further reduce the cost, this reuse concept has been proposed at mask and die level in three-dimension integrated circuit (3D IC). In order to achieve manufacturing reuse, in this paper, we propose a new methodology to design a global clock tree in 3D IC. The objective is to extend an existing clock tree in 2D IC to 3D IC taking into consideration the wirelength, clock skew and the number of TSVs. Compared with NNG-based method, our proposed method reduces the wirelength of the new die and skew of the global 3D clock tree, on an average, 47.16% and 5.85%, respectively.

Categories and Subject Descriptors

B.7.2 **[Integrated Circuits]**: Design Aids—*Placement and routing*;
J.6 **[Computer-Aided Engineering]**: Computer-aided design (CAD)

General Terms

Design, Algorithms

Keywords

Clock tree synthesis, Through-silicon-via, 3D IC, Clock network

I. Introduction

Global interconnection has become a major performance and power consumption bottleneck for advanced VLSI technology as more functional devices are accommodated in one single chip. Three dimensional integrated circuit (3D-IC) technology solves this problem using through-silicon-via (TSV) by connecting signals among multiple stacked dies. Moreover, 3D-IC technology gives more design flexibility by heterogeneous integration [1].

Intellectual Property (IP) based design methodology has been extensively used to reduce design cost in SoC design. In this reuse methodology, while design and implementation cost is saved, manufacturing cost is not. To further reduce the cost, this re-use concept has been proposed at mask and die level in 3D-IC. For example, Alam et al. [2] proposed the reciprocal design symmetry (RDS) to reuse one mask set for 3D-ICs. Besides RDS, Hung and Lin proposed *Chipsburg* architecture in manufacturing reuse [3], [4]. Figure 1 shows an example of *Chipsburg* architecture. Figure 1(a) shows a traditional 2D SoC design applications, where IP_2 and IP_4 are used by all applications. Designers have to implement and verify them once for each application. Figure 1(b) shows an implementation in 3D-IC, where common IP, IP_2 and IP_4, are placed in the platform

---

This work is supported in part by the National Science Council of Taiwan under Grant NSC-100-2221-E-007-043-MY3.

Permission to make digital or hard copies of all or part of this work for personal or classroom use is granted without fee provided that copies are not made or distributed for profit or commercial advantage and that copies bear this notice and the full citation on the first page. To copy otherwise, to republish, to post on servers or to redistribute to lists, requires prior specific permission and/or a fee.
DAC 2012, June 3-7, 2012, San Francisco, California, USA.
Copyright 2012 ACM 978-1-4503-1199-1/12/06...$10.00

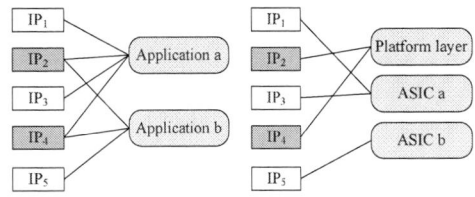

(a)   (b)
Fig. 1. Comparison between (a) traditional SoC design and (b) Chipsburg [5]

layer, and remaining IPs of each application are placed in its own ASIC layer. *Platform layer* is designed and manufactured once for all. Then, each application can perform its intended function by integrating the platform layer with its own ASIC layer in 3D-IC. In *Chipsburg* methodology [3], designers can change functionality of application by changing new ASIC layer. Due to this manufacturing reuse technique, the non-recurring engineering (NRE) costs of the platform layer and the ASIC layer are shared and reduced, respectively. Therefore, the overall cost can be reduced.

With 3D-IC integrated technique, clock tree in 3D-IC has drawn much attention recently. Jacob Minz et al. [6] proposed a thermal-aware buffered clock tree for two stacked dies. Zhao et al. [7] and Kim et al. [8] generated 3D clock tree taken into consideration the number of TSVs. Zhao et al. [9] and Kim et al. [10] proposed 3D-IC clock tree design taking pre-bond testability into account. Although a lot of research has been studied on 3D-IC clock tree construction, it can not be applied directly for manufacturing reuse in 3D-ICs because platform layer is fabricated and its clock tree is already constructed.

An intuitive methodology is to construct clock tree for an ASIC layer optimally and then to find the shortest distance between internal nodes of platform layer and ASIC layer for connection. Connection is kept added until the whole sub-trees in ASIC layer are connected to the platform layer. The following example demonstrates that using this intuitive method without considering the clock tree of platform layer when constructing the clock tree in ASIC layer can not produce good results. In Figure 2, there are six clock sinks in both ASIC and platform layers and these clock sinks do not distribute the same way in the two layers. Initially, the clock tree on platform layer has been synthesized and then the clock tree on ASIC layer starts to be designed. Figure 2 (a) is the results based on *NNG* (nearest neighbor graph) -based method [12] to generate the topology of clock tree in ASIC layer without taking the existing clock tree on platform layer into account. Note that, NNG uses nearest neighbor clock sinks to do clustering. Then, DME method is used to construct clock tree in bottom-up fashion [11]. After the clock tree is synthesized in ASIC layer, connections between platform layer and ASIC layer are established by finding shortest distance between internal nodes in platform layer and ASIC layer greedily. Figure 2 (b) is the results considering the sinks in the ASIC layer and possible connecting nodes in the existing clock tree on platform layer simultaneously when tree topology in ASIC layer is generated. Then, the same DME method is used to construct clock tree. The clock skews in Figure 2 (a) and (b) are $69.7ps$ and $35.4ps$, respectively. Figure 2 (b) is better than Figure 2 (a) because the second method is able to balance downstream capacitance in platform layer and ASIC layer when generating the

1094

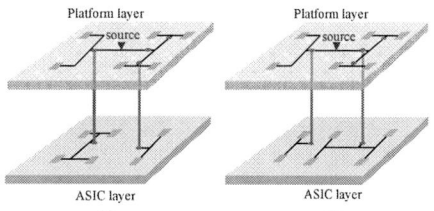

Fig. 2. Two examples of (a) *NNG*-based and (b) our proposed methods

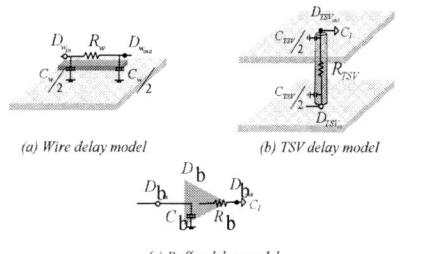

(a) Wire delay model   (b) TSV delay model

(c) Buffer delay model

Fig. 3. Delay model for 3D-IC clock tree

topology of clock tree in ASIC layer.

Based on the *Chipsburg* design methodology, we will study how to construct a clock tree. Given a platform layer where a clock tree is already constructed, we will synthesize clock tree in ASIC layer so that the skew in the global clock tree is minimized.

The rest of this paper is organized as follows. In Section II, we introduce our delay model and architecture. Section III presents our proposed 3D-IC clock tree construction algorithm for manufacturing reuse. Section IV shows our experimental results. Finally, we conclude this paper in Section V.

II. PROBLEM DEFINITION

A. Delay model

We first present the delay model used in our clock tree construction. The delay model is shown in Fig 3, and is defined as follows,

$$D_{w_{out}} = D_{w_{in}} + R_w \times (\frac{C_w}{2} + C_l)$$

$$D_{TSV_{out}} = D_{TSV_{in}} + R_{TSV} \times (\frac{C_{TSV}}{2} + C_l)$$

$$D_{b_{out}} = D_{b_{in}} + D_{buf} + R_b \times C_l$$

where $D_{w_{out}}$, $D_{TSV_{out}}$ and $D_{b_{out}}$ represent Elmore delay of wire, TSV and buffer, respectively, from source. R_w, R_{TSV} and R_b (C_w, C_{TSV} and C_b) represent the resistance (capacitance) of wire, TSV, and buffer, respectively. Finally, D_{buf} and D_l denote the delay of a buffer and downstream capacitance, respectively.

B. Architecture

Although many researchers [6]–[10] addressed the problem of generation of clock tree in 3D-IC, no method is designed for manufacturing reuse. In this paper, we propose a new methodology to design clock tree in 3D-IC. First, given a pre-fabricated die where the clock tree is already synthesized and a new die to be synthesized (no clock tree yet), we want to construct a global clock tree for both dies extending from the existing clock tree, *existing tree*, in platform layer. Figure 4(a) shows an example. To construct a global clock tree based on an existing one, a sophisticated method is to create many connecting points in *existing tree* and create clock sub-trees in ASIC layer from these connecting points as shown in Figure 4(b).

Pre-fabricated TSV-buffer:

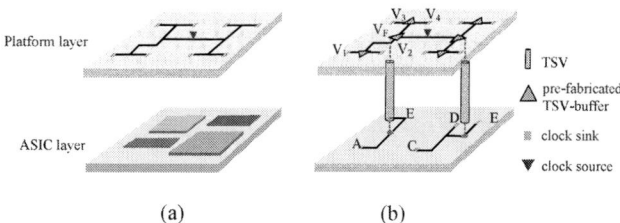

Fig. 4. 3D-IC clock tree architecture

In order to synthesize a zero-skew clock tree in platform layer **independent** of future sinks in any kind of ASIC layer, *pre-fabricated TSV buffers* are added. In platform layer, *pre-fabricated TSV-buffers* are added at all internal nodes of clock tree for future global 3D-IC clock tree construction. The *pre-fabricated TSV buffers* inserted at internal nodes of 2D clock tree in platform layer are used to drive TSVs. Since *pre-fabricated TSV-buffer* can block downstream capacitance, we can use any existing clock generation algorithm to synthesize a zero-skew clock in platform layer. In Figure 4 (b), many *pre-fabricated TSV-buffers* are built in platform layer shown as triangles.

C. Problem formulation

Given a constructed 2D clock tree with *pre-fabricated TSV-buffers* in platform layer. Our objective is to extend the existing clock tree in platform layer to ASIC layer so that wirelength and clock skew are minimized, and the number of TSVs is controlled in the global clock tree. In the following, these cost functions are further explained.

Wirelength:
In order to achieve manufacturing reuse, designer can not change the layout of existing 2D clock tree in platform layer. Therefore, the wirelength of 2D clock tree in platform layer, WL_P, is fixed. Although wirelength of clock tree on existing 2D design is unchanged, designer still has to reduce the wirelength in the new die to form a global 3D clock tree. Moreover, there may be more than one sub-trees on ASIC layer. For pre-bond testing, designer needs to route redundant clock tree to construct a complete tree for this die. Our wirelength cost is calculated as,

$$WL_{cost} = WL_A + WL_{red}$$

where WL_A is the wirelength of the clock tree in ASIC layer and WL_{red} is the wirelength for pre-bond testing. In order to save routing area, our proposed algorithm is to reduce WL_{cost} as much as possible.

Clock skew:
The clock latency of sink i, CL_i, is the delay from source to sink i. The maximum clock skew is defined as,

$$CS_{S_k,max} = \max\{CL_i - CL_j\}, \text{ for each sinks pair}(i,j) \text{ in } S_k$$

where S_k is the set of all sinks. Suppose we have two dies and the clock sinks in platform layer and in ASIC layer are in S_1 and S_2, respectively. Then, to increase maximum clock frequency, maximum clock skew $CS_{S_1 \cup S_2,max}$ of 3D-IC clock tree needs to be reduced.

The number of TSVs:
While the length of TSV is smaller than that of wire, capacitance of TSV is much larger than that of wire. Hence, we will control the number of TSVs used in the design process.

1095

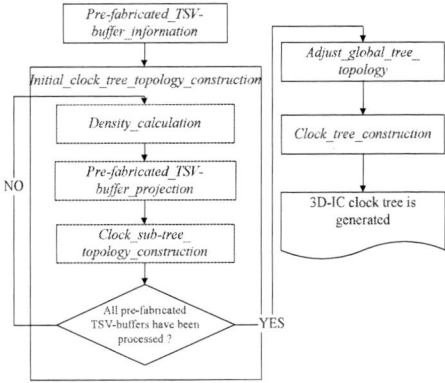

Fig. 5. Design flow

III. ALGORITHM

We propose a global clock tree construction algorithm in 3D-IC given a pre-designed clock tree in one die. Our main objective is to construct the clock tree with minimization of wirelength and clock skew.

A. Overview of Our Proposed Method

Figure 5 is the overview of our proposed 3D-IC clock tree construction algorithm. The first step, *pre-fabricated TSV-buffer_information* is to collect locations, downstream capacitance and delay data of *pre-fabricated TSV-buffers* placed on platform layer.

Then, the next step, *initial_clock_tree_topology_construction*, is performed to generate an initial clock tree topology in ASIC layer by projecting location of *pre-fabricated TSV-buffer* in platform layer into ASIC layer, creating clusters for all clock sinks on ASIC layer, and assigning a *pre-fabricated TSV-buffer* in platform layer to each cluster. Note that, each cluster represents a sub-tree topology. After this step, we use *adjust_global_tree_topology* to obtain a complete global clock tree topology in 3D-IC.

In the last step, we perform *clock_tree_construction* to route our 3D-IC clock tree network based on the tree topology generated in the previous steps. The details of *pre-fabricated_TSV-buffer_information*, *clock_sub-tree_topology_construction*, *adjust_global_tree_topology*, and *clock_tree_construction* are described in the following sections.

B. Pre-fabricated TSV-buffer Information

In this step, besides location of *pre-fabricated TSV-buffer*, downstream capacitance and the maximum delay of the parent node of *pre-fabricated TSV-buffer* in platform layer are collected. To compute the maximum delay, we take Figure 4 (b) as an example. Let node V_F represent the parent node of a *pre-fabricated TSV-buffer*, and D_{V_F,V_1}, D_{V_F,V_2}, D_{V_F,V_3}, and D_{V_F,V_4} represent downstream delay from V_F to V_1, V_2, V_3, and V_4 in platform layer, respectively. Then, for node V_F, the downstream delay of parent node of the *pre-fabricated TSV-buffer* is the maximum value among D_{V_F,V_1}, D_{V_F,V_2}, D_{V_F,V_3}, and D_{V_F,V_4}. This maximum delay is used to reduce the clock skew in ASIC layer in the later step.

C. Initial Clock Tree Topology Construction

Clock sinks are merged to generate clock tree topology taking into *pre-fabricated TSV buffer* account in bottom-up fashion in ASIC layer. *Initial_Tree_topology_construction* procedure is shown in Figure 6. Initially, each sink in ASIC layer forms a tree and S_{tree} = {all sinks in ASIC layers}, *current_h* is set to the minimum height of all *pre-fabricated TSV-buffers* (lines 1 and 2). While *current_h* is less than and equal to the tree height in platform layer,

Algorithm: *Initial_tree_topology_construction*

h_{tree} : the tree height of the clock tree in platform layer;
$h(b_i)$: the height of *pre-fabricated TSV-buffer* b_i;

Begin
1   S_{tree} = set of roots of sub-trees in ASIC layer;
2   *current_h* = $min\{h(b_i)\}$, $\forall b_i \in$ *pre-fabricated TSV-buffer*;
3   **while** *current_h* $\leq h_{tree}$
4   S_B = set of *pre-fabricated TSV-buffers* with *current_h*;
5   **density_calculation**(S_{tree});
6   **pre-fabricated_TSV-buffer_projection**(S_B);
7   S_{tree} = **clock_sub-tree_topology_construction**(*current_h*, S_{tree});
8   *current_h* + +;
End

Fig. 6. *Initial_Tree_Topology_Construction* procedure

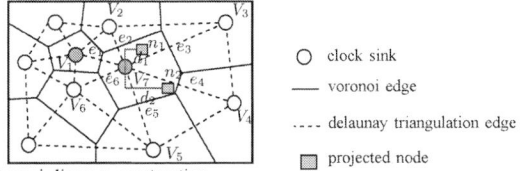

Fig. 7. Voronoi diagram construction

iterations continue (line 3). In each iteration, *pre-fabricated TSV-buffers* with the same *current_h* form a set S_B. The density among the roots of current sub-trees is computed in *density_calculation* (line 4). Next, the locations of buffers in S_B will be projected to ASIC layer in *pre-fabricated TSV-buffers projection* step (line 5). Then, based on the projected location and density function, the next higher level sub-trees are searched in *clock_sub-tree_topology_construction* (lines 6 and 7). After one loop is completed, the iteration continues with new *current_h*, with increasement of one and new S_{tree}. The iterations continue until *current_h* is larger than the tree height, h_{tree}. The details of three sub-steps are described in the following subsections.

1) Density Calculation: In this step, *voronoi diagram* is to analyze the distribution and density of clock sinks in ASIC layer. We observe that if a clock sink i in a region where clock sinks density is high, the area of clock sink i, $area(i)$ surrounded by voronoi edges is small. For example, in Figure 7, clock sink V_1 and V_7 represent clock sinks in high and low clock sinks density area, respectively. Besides the distribution information, we consider the average distance from a sink to its adjacent sinks. For a sink i, we say that sink j is adjacent to sink i if there exists a delaunay triangulation edge between them. For example, in Figure 7, adjacent sinks of sink V_7 are sinks, V_1, V_2, V_3, V_4, V_5 and V_6, and the length of edges, e_1, e_2, e_3, e_4, e_5 and e_6, are collected as the distance information for sink V_7. After *voronoi diagram* is constructed, we can obtain the area and distance information for each sink. The sparse function $sparse(i)$ for sink i is defined as follows,

$$sparse(i) = \alpha \times \sqrt{area(i)} + \beta \times \frac{\sum_{j=1}^{m} dis(i,j)}{m}$$

where m is the number of delaunay triangulation edges incident to clock sink i, and $dis(i,j)$ represents distance between sinks i and j. α and β are weight factor. Figure 7 shows that sparse function, $sparse(V_7)$, for sink V_7 is $\alpha \times \sqrt{area(V_7)} + \beta \times \frac{\sum_{k=1}^{6} dis(V_7,V_k)}{6}$. Note that, the lower the value of $sparse(i)$, the denser area the node is located.

1096

2) Pre-fabricated TSV-buffer Projection: In this step, the pre-fabricated TSV-buffers placed on platform layer are projected to ASIC layer. With these projected locations, merging of clock sinks in ASIC layer can take into account the location of the *pre-fabricated TSV-buffers* on platform layer.

In each iteration, *pre-fabricated TSV-buffers* with the same height *current_h* are projected to ASIC layer. Their coordinates on ASIC layer decide the potential sub-tree node to be connected. If there are more than one *pre-fabricated TSV-buffers* projected to the same voronoi region for a sub-tree in ASIC layer, it is to select the *pre-fabricated TSV-buffer* which has the shortest distance to the root of the sub-tree. This heuristic is to minimize total wirelength. For example, in Figure 7, two nodes, n_1 and n_2, are projected to ASIC layer and inside the area surrounded by voronoi edges of sub-tree node, V_7. Then, because the length of Hamilton distance, d_1, is shorter than that of distance, d_2, projected node n_1 is selected to be considered in *clock_sub-tree_topology_construction* with V_7.

3) Clock Sub-tree Topology Construction: In this step, we will cluster sinks to form sub-trees taking into consideration the projected nodes of *pre-fabricated TSV-buffers* account. Before we present our algorithm, we first present our observations. For a sub-tree T with height *current_h* and downstream capacitance C_T in ASIC layer, it is better to assign it to a *pre-fabricated TSV-buffer* with the same height *current_h* and similar downstream capacitance. If this assignment cannot be done, we will search a *pre-fabricated TSV-buffer* with height *current_h+1* (the next higher level internal node) in the next *while* iteration. By this assignment, it is more likely for us to obtain a balanced tree structure.

Next, for sub-trees with the same height, we will process the sub-tree with low sparse function (i.e., high density sinks) first. It means that sinks spread sparely may be assigned an internal node at farther location or an internal node at higher level. This heuristic allows that the skew of sinks that spread in a wider area is reduced through routing. If we do not have this priority, it will be difficult to tune skew of sinks in a small area (dense sinks) by routing wire.

Based on the observations, we develop our procedure, *clock_sub-tree_topology_construction*, shown in Figure 8. Initially, each sub-tree node is a clock sink or the means of clock sinks of a sub-tree. Based on the sparse function, we sort the sub-tree nodes in a list L in increasing order. Then, we find all sub-trees that is paired with a projected node and put them in a list, L_p (line 1). The rest of sub-trees in L are put into L_{up} which is not paired with any projected node (line 2). Then, we process sub-trees in L_p one by one. In each iteration, let t, p be the first element in L_p and p is its paired projected node, respectively. Then, we find a node a in L_{up} which has the shortest distance to t for merging t and a rooted at p (lines 4-6). Figure 9 shows an example. In Figure 9 (a), there are five sub-tree nodes, t, a_1, a_2, a_3, and a_4 and a projected node, p, in ASIC layer. Suppose that $L_{up}=\{a_1, a_2, a_3, a_4\}$ and a_2 has the shortest distance to t. Then, merging a_2 and t is a potential sub-tree to be formed and connected to p as shown in Figure 9 (b).

To determine whether merging these two sub-trees is beneficial. We define a cost function, $cluster_{cap}(t, a, p)$. The $cluster_{cap}(t, a, p)$ is defined as follow,

$$cluster_{cap}(t, a, p) = \frac{dist_{cap}(merge(t, a))}{dist_{cap}(platform_{sinbling}(p))}$$

$$dist_{cap}(m) = HPWL(m) \times \sqrt{N_{sinks}(m)} + \frac{C_{sinks}(m)}{C_{w_{unit}}}$$

where $cluster_{cap}(t, a, p)$ represents cost function of merging sub-tree nodes t and a to generate new sub-tree rooted at node m to be connected to the projected node p; $HPWL(m)$, $N_{sinks}(m)$ and $C_{sinks}(m)$ represent HPWL (half-perimeter wirelength) of bounding box of sub-tree m, the number of clock sinks of sub-tree m and total sinks capacitance of sub-tree m, respectively. $C_{w_{unit}}$ represents the capacitance of wire unit; and $platform_{sinbling}(p)$ represents the sinbling sub-tree node of *pre-fabricated TSV-buffer* of p.

If $cluster_{cap}(t, a, p)$ is less than a user defined ratio, γ, this merge is performed and a new sub-tree, $m(t, a)$, is formed. A large value of γ means that clustering a sub-tree nodes, a and t, and a projected node p may increase large wiring overhead (lines 7-9).

Note that, the projected node p from platform layer is only a candidate point for $m(t, a)$ to connect. If there is a more suitable connecting point at level higher than *current_h*, p will not be used. Therefore, the new sub-tree $m(t, a)$ is put to $L_{success}$ list to be processed in the next iteration (line 10). Because clock tree on ASIC layer is not actually constructed yet, we give the virtual coordinate, $(x_{m(p)}, y_{m(p)})$, to sub-tree node $m(t, a)$ as follow,

$$(x_{m(p)}, y_{m(p)}) = \left(\frac{\sum_{\forall s_i \in m(t,a)}^{N_{sinks}(p)} x_{s_i}}{N_{sinks}(p)}, \frac{\sum_{\forall s_i \in m(t,a)}^{N_{sinks}(p)} y_{s_i}}{N_{sinks}(p)} \right)$$

where s_i is the clock sink of sub-tree $m(t, a)$.

In each iteration, except the first iteration, some nodes in L_p and L_{up} may be left without merging with any other node to form a tree with higher height. These nodes then will use their connecting points formed in the previous iteration and their connecting points to the clock tree in platform layer are determined now.

In the first iteration, if there are sink nodes in ASIC layer that are not merged with other node to form a tree of (height == 2), we collect all the nodes into L_{fail} list to be processed in step *adjust_global_tree_topology* (lines 15 and 16).

The user defined ratio γ is to control the number of TSVs, if a larger γ is given, it means that we allow more wiring overhead to merge two sub-trees. The results is to have less number of sub-trees and thus less number of TSVs to connect two dies. In Section IV, we will study the effect of different γ value.

D. Adjust Global Tree Topology

In *adjust_global_tree_topology*, we are to assign internal node (root of sub-tree) to merge with each sub-tree node in S_{fail} produced from *clock_tree_topology_construction* step. Then, a complete 3D-IC clock tree topology is to be generated.

Our procedure, *adjust_clock_tree_topology*, incrementally processes sink nodes in L_{fail} from root to sink in top-down fashion as shown in Figure 10. First, we compute h_{max} and h_{min} (lines 1 and 2). For all sub-trees constructed (collected in S_{tree}) so far in the ASIC layer, we compute their heights. h_{max} and h_{min} are the maximum and minimum height of all sub-trees. Next, *current_h* is set to h_{max} (line 3). Then, we are to merge sink node in L_{fail} to some sub-trees in S_{tree}. We select the sub-trees with *current_h* and put them to a list, L_H (line 5). The sub-trees in L_H are sorted by sparse function in increasing order. Let st be the first sub-tree in L_H (line 7). Then, $cluster_{cap}(st, f, p)$ is computed. Note that, p is the *pre-fabricated TSV-buffer* that st is going to connect. If $cluster_{cap}(st, t, p)$ is less than γ, then, f is merged with the sub-tree rooted at st (lines 8 and 9). Note that, f is merged with a nearest sink in the sub-tree rooted at st. Hence, part sub-tree of st needs to be re-built. Then, f is deleted from L_{fail} and the next sink node in L_{fail} is processed (line 10). If f cannot be merged with any trees with height of *current_h*, f is put in a list, $L_{newfail}$ (lines 13 and 14).

After all sink nodes in L_{fail} list are processed, we checked if there is any sink that is not able to be merged at the *current_h* (i.e., $L_{newfail} \neq \emptyset$). If it is, we start a new iteration by decreasing

Algorithm: *Clock_sub-tree_topology_constrcution*($current\_h$, L)
$L_{success}$: list of sub-tree nodes succeed in merging with an
  unpaired sub-tree node;

Begin
1 L_p = list of sub-tree nodes with height $current\_h$ paired with
    projected nodes in increasing order;
2 L_{up} = list of sub-trees with height $current\_h$ which is not
    paired with any projected nodes, which is $L - L_p$;
3 **while** L_p is not empty
4   t = the first element in L_p;
5   p = the projected node paired with t;
6      $a = min\{dist(t, up_i)\}$, $up_i \in L_{up}$;
7   **if** $cluster_{cap}(t, a, p) \leq \gamma$ **then**
8   merge t and a using p;
9   generate new sub-tree node $m(t, a)$;
10   assign $m(t, a)$ to L_{sucess};
11   delete a from L_{up};
12   **end if**
13   delete t from L_p;
14 **end while**

15 **if** $current\_h == 1$ **then**
16   L_{fail} = { all single sink that has height = 1};
17 **endif**
18 **return** $L_{success}$;
End

Fig. 8.   *clock_sub-tree_topology_construction* procedure

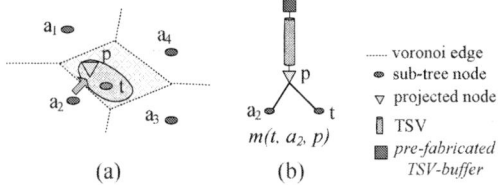

(a)   (b)

— voronoi edge
⊙ sub-tree node
▽ projected node
▯ TSV
▪ pre-fabricated
    TSV-buffer

$m(t, a_2, p)$

Fig. 9.   An example of *clock_sub-tree_topology_construction*

$current\_h$ by one and setting $L_{fail} = L_{newfail}$ (lines 17-20). If the $current\_h$ is equal to h_{min}, there are still some sinks that are not able to be merged with a sub-tree. Then, we will relax the clustering constraint by a factor of γ (lines 21-22). The new iteration starts with $current\_h = h_{max}$ (line 23).

E. Clock Tree Construction

After *global_tree_topology_construction*, the next step is to construct the actual clock tree in step *clock_tree_construction*. In this step, we adopt *DME*-based zero-skew tree construction [6], [11] and buffer insertion [14] to construct our clock tree. Finally, we can obtain a complete 3D buffered clock tree for mask reuse.

IV. EXPERIMENT

We implemented the proposed method using C++/STL programming language on Linux environment with 2.4GHz processor and 4GB memory. Our experimental results were performed on ISPD'09 clock tree contest benchmark. In our experiments, we use technology parameters based on the $45nm$ Predictive Technology Model [10], [16]. We use $10um \times 10um$ via-last TSVs with thinned die height of $20um$. The parasitic resistance and capacitance of unit wire length (TSV) are $0.1\Omega/um$ (0.035Ω) and $0.2fF/um$ ($15.48fF$), respectively. The intrinsic delay, input capacitance and output driving resistance of a buffer are $18ps$, $14fF$ and 51Ω, respectively. Our

Algorithm: *Adjust_global_tree_topology*(S_{tree}, L_{fail})
$h(t_i)$: the tree height of sub-tree t_i;

Begin
1 $h_{max} = max\{h(t_i)\}$, $t_i \in S_{tree}$;
2 $h_{min} = min\{h(t_i)\}$, $t_i \in S_{tree}$;
3 $current\_h = h_{max}$;
4 **for each** sub-tree node $f \in L_{fail}$ **begin**
5   L_H = list of the sub-tree nodes with height $current_h$
        in increasing order;
6   **while** L_H is not empty
7   st = the first element in L_H;
8   **if** $cluster_{cap}(st, f, p) \leq \gamma$ **then**
9              $merge(st, f)$;
10   delete f from L_{fail};
11   **end if**
12   **end while**
13   **if** f is not merged with any sub-tree with height $current\_h$ **then**
14   put f in $L_{newfail}$;
15   **endif**
16 **end for**
17 **if** $S_{newfail} \neq \emptyset$ **and** $current\_h \neq h_{min}$ **then**
18     $current\_h - -$;
19     $L_{fail}=L_{newfail}$;
20   **goto** line 4;
21 **else if** $L_{fail} \neq \emptyset$ **and** $current\_h == h_{min}$ **then**
22     $\gamma = \gamma \times \delta$;
23   **goto** line 3;
24 **end if**
End

Fig. 10.   *Adjust_global_tree_topology* procedure

clock frequency is set to $1GHz$ with supply voltage $1.2V$. Our weight factors, α, β, γ, and δ, are set 0.5, 0.5, 1.4, and 1.05, respectively. For pre-bond testing, the redundant tree is constructed to connect each sub-tree on ASIC layer and TG (transmission gate) is used to connect and disconnect redundant tree [9]. In order to minimize routing area of control signals, in this paper, we use the RMST-pack [13]. Finally, clock skew is reported based on SPICE simulation [17].

First, we compare two global 3D-IC clock trees constructed by *NNG*-based and *ours* methods. Note that, *NNG*-based method uses nearest neighbor graph to do clustering without *pre-fabricated_TSV-buffer_projection* step, and then *clock_sub-tree_construction* and *adjust_global_tree_topology* steps are performed to generate a whole 3D-IC clock tree topology. After the topologies are constructed by two methods, the same *DME* method [11] is used to build 3D-IC clock tree design.

Table I shows the clock skew of these two methods before the clock sub-trees in ASIC layer are inserted with buffer. Finally, the next experiment is to compare results after buffers are inserted. The same modified buffer insertion method [14] is applied. From the table I, the average improvement of clock skew of *ours* method as compared with *NNG*-based is 49%. Table II shows the results, where WL_P, WL_A, WL_{3D}, $Skew$, $\#BUF$ and $\#TSV$ represent the wirelength of the clock tree in platform layer, the wirelength of the new clock tree in ASIC layer, total wirelength of the global 3D-IC clock tree, the clock skew of the global 3D-IC clock tree, the number of buffers and the number of TSVs, respectively. In Table II, the average improvement of total wirelength of the global 3D-IC clock tree, the wirelength of ASIC layer, clock skew, and the number of buffers of our method as compared with *NNG*-based method are 2.22%, 5.85%, 47.16%, and 5.72%, respectively.

1098

TABLE II
COMPARISON OF *NNG*-BASED AND OUR METHODS METHODS

circuit name			*NNG*-based					ours				
Platform layer	ASIC layer	WL_P	WL_A	WL_{3D}	$Skew$	$\#BUF$	$\#TSV$	WL_A	WL_{3D}	$Skew$	$\#BUF$	$\#TSV$
ispd09f11	*ispd09f12*	175785	110850	286635	19	553	47	112320	288105	9	485	43
	ispd09f21		141381	317166	22	567	41	140291	316076	15	512	42
	ispd09f22		79735	255520	24	441	38	78500	254285	12	422	32
	ispd09f31		335881	511666	65	871	65	323956	499741	24	851	57
	ispd09f32		226115	401900	37	854	46	217681	393466	21	819	41
ispd09f12	*ispd09f11*	178652	116372	295024	17	576	35	107529	286181	12	571	35
	ispd09f21		140975	319627	21	583	38	132715	311367	7	552	34
	ispd09f22		65991	24643	24	433	33	67729	246381	13	421	31
	ispd09f31		327694	506349	52	937	62	322476	501128	21	874	45
	ispd09f32		243437	422089	31	1216	53	215814	394466	19	1171	51
ispd09f31	*ispd09f11*	409612	132765	542377	19	901	41	126422	536034	18	887	38
	ispd09f12		118377	527989	28	891	30	105606	515218	21	902	47
	ispd09f21		168736	578348	24	945	24	151526	561138	19	908	41
	ispd09f22		85643	495255	20	852	26	81732	491344	13	819	26
	ispd09f32		188346	597958	25	1227	61	176652	586264	21	1051	52
ispd09f32	*ispd09f11*	289892	135892	425784	18	812	32	129938	419830	12	765	36
	ispd09f12		138167	428059	23	694	45	125749	415641	21	651	44
	ispd09f21		141208	431100	28	733	42	132801	422693	23	725	39
	ispd09f22		85329	375221	20	642	31	76346	366238	14	633	28
	ispd09f31		215964	505856	25	1328	57	196145	486037	13	1129	52
ratio			1.00	1.00	1.00	1.00	1.00	0.94	0.98	0.53	0.94	0.96

TABLE I
COMPARISON OF *NNG*-BASED AND OUR METHODS METHODS BEFORE
BUFFERED

circuit name		*NNG*-based	*ours*
Platform layer	ASIC layer	skew (*ps*)	
ispd09f11	*ispd09f12*	58	27
	ispd09f21	66	43
	ispd09f22	72	31
	ispd09f31	112	57
	ispd09f32	85	38
ispd09f12	*ispd09f11*	43	21
	ispd09f21	41	26
	ispd09f22	75	35
	ispd09f31	104	54
	ispd09f32	95	43
ispd09f31	*ispd09f11*	57	24
	ispd09f12	84	29
	ispd09f21	72	41
	ispd09f22	68	33
	ispd09f32	75	51
ispd09f32	*ispd09f11*	56	30
	ispd09f12	69	33
	ispd09f21	83	41
	ispd09f22	71	42
	ispd09f31	65	41
skew (ratio)		72.56 (1.00)	37.00 (0.51)

V. CONCLUSION

In this paper, we have proposed a clock tree algorithm with methodology of reuse in 3D-IC. The wirelength and clock skew improvements of a 3D-IC clock tree of our method as compared with *NNG*-based method are 2.22% and 47.16%, respectively. Furthermore, compared with *NNG*-based, our method reduces the wirelength of ASIC layer, on an average, 5.85%.

REFERENCES

[1] C. S. Tan, Ronald J. Gutmann, and L. Rafael Reif, "Wafer Level 3-D ICs Process Technology," *Springer* , 2008

[2] S. M. Alam, R. E. Jones, S. Pozder and A. Jain, "Die/wafer stacking with reciprocal design symmetry (RDS) for mask reuse in three-dimensional (3D) integration technology," *Quality of Electronic Design, 2009 (ISQED 2009)*, pp.569-575, 2009

[3] C.-M. Hung and Y.-L. Lin, "Three-dimensional Integrated Circuits Implementation of Multiple Applications Emphasising Manufacture Reuse", to be published in *IET Computers and Digital Techniques*,vol.5 ,Iss.3 , pp.179-185

[4] Y.-L. Lin, "Chipsburger: From IP/Design Reuse for SOCs to Manufacture Reuse for 3D ICs, "D43D: System Design for 3D Silicon Integration Workshop," June 17-18, 2009, LETI, Grenoble, France

[5] D. K. Huang, "A TSV-Number-Constrained Bus System Synthesizer for Platform-based 3D IC Design", Thesis, 2010

[6] J. Minz, Xin Zhao and Sung Kyu Lim, "Buffered Clock Tree Synthesis for 3D ICs Under Thermal Variations," *in Proceedings of Asia and South Pacific Design Automation Conference*, pp.504-509, 2008

[7] Xin Zhao, Sung Kyu Lim, "Power and slew-aware clock network design for through-silicon-via (TSV) based 3D ICs," Design Automation Conference (ASP-DAC), 2010 15th Asia and South Pacific, pp.175-180, 18-21 Jan. 2010

[8] Tak-Yung Kim and Taewhan Kim, "Clock tree embedding for 3D ICs," *in Proceedings of Asia and South Pacific Design Automation Conference*, , pp.486-491, 2010

[9] Xin Zhao, D.L. Lewis, H.-H.S. Lee and Sung Kyu Lim, "Pre-bond testable low-power clock tree design for 3D stacked ICs," *in Proceedings of International Conference on Computer-Aided Design*, pp.184-190, 2-5 Nov. 2009

[10] Tak-Yung Kim and Taewhan Kim, "Clock tree synthesis with pre-bond testability for 3D stacked IC Designs," *in Proceedings of Design Automation Conference (DAC)*, pp.723-728, 2010

[11] J. Cong, A. B. Kahng, C. K. Koh and C. W. Albert Tsao, "Bounded-Skew Clock and Steiner Routing," *"ACM Transactions on Design Automation of Electronic Systems (TODAES)"*, vol.3 Issue 3, July 1998

[12] Masato Edahiro, "A clustering-based optimization algorithm in zero-skew routin", *Proceedings of the Design Automation Conference*, pp. 612-616, 1993.

[13] RMST-Pack, http://vlsicad.ucsd.edu/GSRC/bookshelf/ Slots/RSMT/

[14] Ashok Vittal and Malgorzata Marek-Sadowska, "Power Optimal Buffered Clock Tree Design," *"Proceedings of the Design Automation Conference (DAC)"*, pp. 497-502, 1995.

[15] ISPD 2009 Clock Network Synthesis Contest benchmark. http://ispd.cc/contests/09/ispd09cts.html

[16] Predictive Technology Model. http://ptm.asu.edu/

[17] NGSPICE http://ngspice.sourceforge.net/

Can Pin Access Limit the Footprint Scaling?

Xiang Qiu and Malgorzata Marek-Sadowska

University of California, Santa Barbara

Santa Barbara, CA, 93106

{xqiu, mms}@ece.ucsb.edu

Abstract

If pin density exceeds a certain threshold, pin access becomes a challenge for inter-cell signal routing and increasing the number of metal layers cannot improve routability. CMOS and FinFET layouts may never reach this threshold, but Vertical Slit Field Effect Transistor (VeSFET) ICs may exceed it. We demonstrate that VeSFET layouts are still routable within footprint using two-sided routing which achieves better wire length and via usage than one-sided routing with or without white space inserted.

Categories and Subject Descriptors

B.7.2 [**Integrated Circuits**]: Design Aids–*Placement and Routing*

General Terms

Design

Keywords

Detailed routing, pin density, VeSFET, two-sided routing, net partitioning.

1. INTRODUCTION

VLSI routing is usually performed in two steps: global and detailed routing. A global router models the routing region as a coarse grain grid with proper boundary capacities and seeks to connect all nets without overflow. Based on global routing solution, detailed router determines the exact geometry of wires connecting to cell pins. Because of limited local routing resources and densely packed local pins, pin access may be a serious problem for detailed routing [1, 2]. Taghavi, *et. al.* showed that for some 32nm industrial designs, a commercial detailed router reports many violations (opens and shorts) even if no-overflow global routing exists [2]. Routing resources can be increased in two ways: (1) by inflating cells or inserting white space between them [2]; or (2) by adding more metal layers. The first approach diminishes the benefits of scaling as transistors cannot be packed as densely as technology permits. Also performance may be reduced and power may be increased due to longer wires. Applying the second approach, difficult-to-access pins can be pulled up to upper layers, which may ease routing. This approach keeps the design footprint small and results in shorter wire lengths than is the case of a footprint expanded in 2-D plane. Reference [3] shows that adding more metal layers always achieves shorter wire lengths than adding white space. But can adding more metal layers always guarantee that a design with continuously scaled footprint can be routable?

Permission to make digital or hard copies of all or part of this work for personal or classroom use is granted without fee provided that copies are not made or distributed for profit or commercial advantage and that copies bear this notice and the full citation on the first page. To copy otherwise, to republish, to post on servers or to redistribute to lists, requires prior specific permission and/or a fee.
DAC 2012, June 3-7, 2012, San Francisco, California, USA.
Copyright 2012 ACM 978-1-4503-1199-1/12/06...$10.00

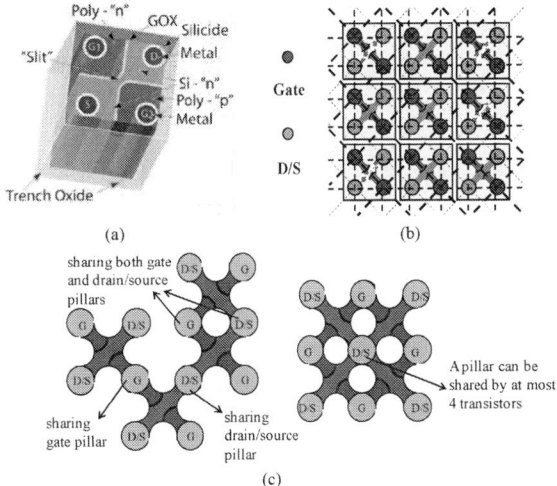

Figure 1. (a) An N-type VeSFET transistor [8]; (b) An array of isolated transistors with horizontal, vertical and diagonal routing tracks; (c) Contacted transistors with pillar sharing.

In this paper, we prove that given an unlimited number of routing layers, a placement is always routable if at least one free pin position exists or two pins to be connected are neighbors on a 2-D grid. However, when pin density becomes very high, the number of required routing layers increases rapidly. We conduct experiments on various placements with different pin densities. Routing results obtained from a commercial router suggest that there is an empirical threshold pin density beyond which the router fails. This threshold is not sensitive to Rent's exponent [4] or design size and is so high that conventional CMOS layout might never reach it, even with aggressive device footprint scaling [5]. FinFET designs are also safe because their pin density is even lower than CMOS [6]. However, future layout styles aiming at super high transistor density, such as **V**ertical **S**lit **F**ield **E**ffect **T**ransistor (VeSFET) [7-9], may exceed the threshold pin density and suffer from serious pin access problem.

VeSFET is a twin-gate junction-less transistor with gate, drain, and source terminals implemented as metal pillars [7-9]. VeSFETs are packed densely into arrays, and circuit functions are implemented by configuring interconnects and vias. Transistors can be either isolated or contacted, as illustrated in Figure 1(b) and (c). Contacted transistors share pillars and achieve denser packing. VeSFET gates can be tied or independently connected for threshold voltage control or additional logic function [10]. For example, a 2-input NAND gate can be built using two VeSFETs with independent gate inputs: P-type transistor realizing OR function and N-type VeSFET realizing AND. OPC-free inter-connect manufacturing can be applied, resulting in strictly parallel wires on each metal layer [3, 11, 12]. Wires can be diagonal, vertical or horizontal.

The unique characteristic of VeSFET transistor is that its pins can be accessed on both sides of the device layer [7-9]. Therefore, *two-sided routing* with wires routed on both sides of a 2-D chip is possible [3]. Two-sided routing reduces the equivalent pin density and can maintain footprint scaling for VeSFET ICs even when pin density on one side exceeds the threshold. In this paper, we show that designs with pin density close to the threshold are unroutable on one side for the commercial router, but routable on two sides. To make better use of two-sided routing, we propose a greedy net partitioning algorithm. We show that two-sided routing outperforms white space insertion in terms of footprint area, total wire length and via usage. For designs with medium pin density that are routable on one side without added white space, two-sided routing also achieves shorter wire lengths and fewer vias than one-sided routing.

The paper is organized as follows. In Section 2, we study the benefits and limitations of adding routing layers, and analyze the scaling effects on CMOS and VeSFET layouts. In Section 3, we explore two-sided routing for VeSFET ICs and in Section 4, we conclude the paper.

2. BENEFITS AND LIMITATIONS OF ROUTING UPWARDS

In this Section, we investigate the effectiveness of added metal layers on routability. We assume wires and vias are on grids that are uniform on all metal layers. (In real designs metal widths and spacing usually vary on different layers [1]). In addition, we assume that there are no blockages in the routing space. Thus, the routing model is a uniform 3D grid structure.

2.1 Benefits of More Routing Layers

First, we show that allowing more metal layers indeed helps routability. We use Gnl [13] to generate a netlist of 600 2-input NAND gates with 0.7 Rent's exponent. We assume that each cell occupies a 3x3 grid with pin distribution as shown in Figure 2(a). The netlist is then placed by a commercial placer with about 5% white space. We define *pin density* as:

$$\text{pin density} = \frac{\text{\# pins to be connected}}{\text{\# 2-D grid points}}$$

Thus, pin density of the cell in Figure 2(a) is 0.33. Note that if two pins on the grid belong to the same net and are already connected by intra cell wires, only one unconnected pin is counted.

After placement, a commercial router is used for full chip routing. To evaluate the benefits of more routing layers, we gradually increase the number of layers from 3 and record the number of routing violations reported by the commercial router. Similar to [2], we extract *#initial violations*—the number of violations after initial detailed routing, and *#final violations*—the number of violations after optimization. We also determine the overflow from MaizeRouter [14][1]. The routing results are listed in Table 1. Similar to the observations in [2], with 4 metal layers, the commercial router cannot find pin access paths and terminates with many violations, although the inter-bin passing nets can find their ways without overflow. By adding more layers, the same placement can be successfully routed, clearly illustrating that adding routing layers is helpful for local pin access.

---

[1] Global router usually neglects the local nets within a global bin. To rule out the effect of local nets on capacity, we assign global bin size the same as the cell size such that all nets are considered. Because the commercial router does not allow manual global bin size adjustment, we use MaizeRouter for global routing [14].

Table 1. Routability Vs. #Metal layers with pin density of 0.33

#Metal layers	Global overflow	#Initial violations	#Final violations
3L	476	5540	2898
4L	0	2249	1193
5L	0	228	0
6L	0	53	0
7L	0	0	0

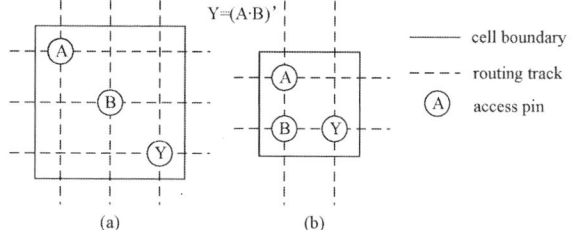

Figure 2. (a) A 2-input NAND cell occupying a 3x3 routing grid; (b) A shrunk 2-input NAND cell on a 2x2 routing grid.

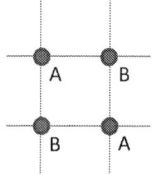

Figure 3. An unroutable net configuration.

2.2 Routability Analysis

Although more metal layers provide more routing tracks, unlimited number of layers does not guarantee a routable layout. A simple example is shown in Figure 3: nets *A* and *B* block each other regardless of how many metal layers can be used. However, we can always find a solution if at least one *spare* point exists on the 2-D grid. We have the following theorem:

Theorem 1: Any configuration of nets is routable on one side of a 2-D grid if and only if at least one grid point is spare or if 2 pins to be connected are placed at the neighboring grid positions.

Proof: Please see Section S.1 of the supplemental document.

2.3 The Limitations of Routing Upwards

A layout with extremely high pin density may need many layers for routing completion. A design theoretically routable but requiring an excessive number of metal layers may also be labeled unroutable. Thus, it is desirable to know when routing upwards becomes inefficient.

First, let's consider an example which breaks the commercial router we use. Using the same netlist as in Section 2.1, we shrink the 2-input NAND cell size to 2x2, as shown in Figure 2(b). Now the cell pin density rises to 0.75. We repeat the routing experiments and the results are listed in Table 2. In this case, the global router needs two more layers for passing nets because the global bin edge capacity is reduced by 1/3. However, detail routing fails even with 15 layers, the maximum number allowed by the commercial router. Interestingly, the number of violations saturates when the number of layers exceeds 10. The last column of Table 2 reports the top layer with violations. When sufficient number of routing layers is available (in this experiment 9), violations on upper layers begin to disappear, and violations only exist on bottom layers (in this experiment bottom 5 layers) after the violation number saturates, implying that the router cannot find paths down to the pins.

Table 2. Routability Vs. #Metal layers with pin density of 0.75

# Metal layers	Global overflow	#Initial violations	#Final violations	Top layer w/ violations
4L	289	5563	3618	L4
5L	92	4938	2972	L5
6L	0	3808	2310	L6
7L	0	2651	1651	L7
8L	0	1748	1052	L8
9L	0	1485	679	L7
10L	0	1431	439	L6
12L	0	1408	370	L5
15L	0	1402	372	L5

Table 3. # Routable instances Vs. Pin density

pin dens.	0.2	0.3	0.4	0.5	0.6	0.7	0.8
# routable inst.	1000	1000	1000	436	0	0	0

Table 4. # Routable instances Vs. Rent's exponent

Rent's exp. / pin dens.	0.3	0.4	0.5	0.6	0.7
0.4	1000	1000	1000	1000	1000
0.5	807	662	436	272	126
0.6	0	0	0	0	0

This example shows that the router may fail when pin density is extremely high. Obviously, the failure is not caused by insufficient number of routing tracks for nets crossing global bin boundaries. Detailed routing algorithms proposed recently [15, 16] are not applicable for such high pin density routing problems. Therefore, for the remainder of this paper, we will use the commercial router result as a golden solution. We want to see in what situations adding metal layers cannot effectively help the router. We study the effects of pin density, design size and Rent's exponent. For each case, we obtain 1000 netlists using Gnl [13]; place those netlists by a commercial placer; and run the commercial router to see how many instances are routable.

We analyze the effect of pin density first. Since the router allows at most 15 metal layers, we intentionally keep the designs small to ensure that the router does not fail because of too few tracks for passing inter-cell nets. We fix the design size to 100 cells and set Rent's exponent to 0.5. For simplicity, only one cell type is used in each netlist. For different pin densities, we use cells with different number of pins but the same footprint size of 5x2. For example, a netlist of 5-pin cells will have pin density of 0.5. The pins are scattered within the cell, as shown in Figure 4. We assume that only inter-cell nets are present. We gradually increase the pin density from 0.2 to 0.8. Table 3 lists the number of routable instances out of 1000. The results indicate that 0.5 is the threshold density because the router fails on more than half instances.

Rent's exponent affects the wire length distribution [17]. We experiment with Rent's exponent values from 0.3 to 0.7, and pin densities from 0.4 to 0.6. The experimental results in Table 4 show the breaking point is still for pin density of 0.5, although the number of routable cases decreases as Rent's exponent increases at this pin density. Rent's exponent does not determine routability when pin density is higher or lower.

Design size affects routing resources of passing inter-cell nets. To see whether it also correlates with pin accessibility, we gradually increase the design size from 100 to 9000 cells with Rent's exponent set to 0.6 and pin densities of 0.4 and 0.5. Experimental results in Table 5 show that 0.5 pin density examples are always

Table 5. # Routable instances Vs. Design size

# cells / pin dens.	100	500	1000	4000	9000
0.4	1000	1000	1000	1000	679
0.5	272	28	0	0	0

Table 6. Average number of violations of netlists with 9000 cells and 0.4 pin density

Rent's exp. / #layers	0.5	0.6	0.7
12L	3782	28732	109847
13L	1346	12464	58765
14L	0	3548	13894
15L	0	783	4671

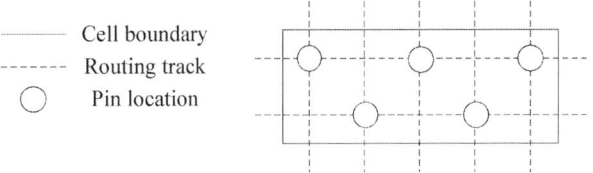

Figure 4. Footprint of a 5-pin cell for 0.5 pin density netlist.

difficult to route no matter what design size is. However, the router begins to fail for 0.4 density designs with 9000 cells. To see whether this failure is caused by too few passing tracks or by pin access, we vary the Rent's exponent and the number of routing layers. The average violation numbers for these cases are summarized in Table 6. As more routing layers are available, the number of violations reduces rapidly. Thus, the failure is more likely caused by too few tracks for passing nets rather than by pin access. According to this trend, we expect that with one or two more metal layers, designs of such a size with pin density of 0.4 may always be routable. In other words, the threshold pin density causing routing layer inefficiency is still 0.5.

So far, the statistical study indicates that 0.5 pin density is an empirical threshold for routing upwards. This threshold is not sensitive to Rent's exponent, or design size. Beyond this threshold, adding metal layers is generally not efficient for pin access. In our study, one commercial router was used. It is possible that a better routing algorithm may further increase the threshold density.

2.4 Is Scaling Limited?

The threshold pin density sets a limit to cell footprint scaling. Figure 5(a) shows a highly compacted inverter layout based on λ-rules. Transistor width is 7λ with aggressive footprint scaling [5]. Other design rules are adopted from 45nm Nangate Open Cell Library [18]. Half metal pitch is 2λ. Because relatively large space is required to separate NMOS and PMOS, at least 4 horizontal tracks are available over the cell. For row based standard cell design, at least one more horizontal track is allowed between adjacent rows due to the space required to separate transistors belonging to different cells. Such tracks are typically reserved for power and ground nets. Thus, the 2-pin inverter occupies an equivalent 2x4 routing grid such that pin density is about 0.25. Even better, the pin density will further reduce for cells with more pins and larger sizes, thus for CMOS standard cell designs it may never reach the 0.5 threshold. Hence for technology generations in the near future, CMOS designs will not be limited by pin-access routing if the number of metal layers is not a stringent constraint. FinFET layout is even less dense than bulk CMOS [6], so it is also safe. On the other hand, VeSFET layout may have much higher density than CMOS [11, 12], especially for contacted transistors as shown in Figure 1(c). For example, an inverter with contacted

1102

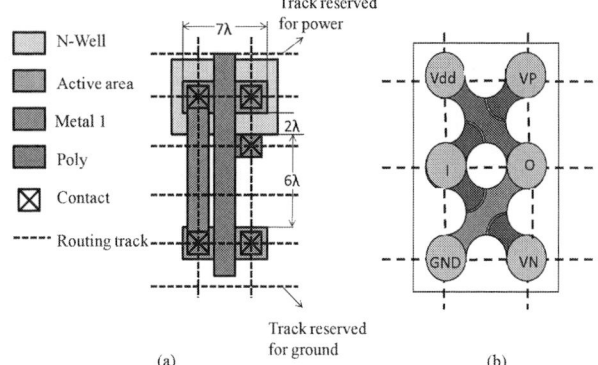

Figure 5. (a) A CMOS inverter layout with aggressive footprint scaling; (b) A VeSFET inverter layout with contacted transistors and independent-gate connection for threshold voltage adjustment. *VN* and *VP* are threshold controlling signals for N-type and P-type transistors.

transistors and independent-gates for threshold voltage adjustment has pin density of 1. Thus, circuits built from such high pin density cells may not be routable without white space, when the number of routing layers is flexible. Fortunately, the unique geometry of VeSFET transistor provides another solution — two-sided routing, that allows for routable solutions within the footprint.

3. TWO-SIDED ROUTING OF VeSFET ICs

The metal pillars of VeSFET contacts penetrate the device layer and provide accessibility on both sides of the chip. With this unique characteristic, a netlist can be partitioned such that on each side, the router is presented with a conventional one-sided routing problem of reduced pin density. To make full use of two-sided routing, net partitioning is critical.

3.1 Problem Analysis

The main purpose of net partitioning is to reduce the equivalent pin density on both sides such that pin access does not prevent routability. A secondary objective is to balance the global routing congestion to provide sufficient number of tracks for passing nets.

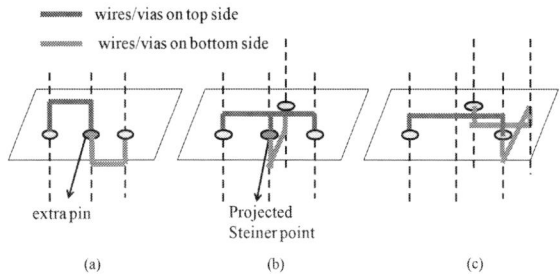

Figure 6. (a) Two-sided routing of a 2-pin net; (b) A partition of 3-pin net with extra pin for Steiner point; (c) A partition of 3-pin net without extra pin but with longer wires.

Net partitioning can be applied for different routing granularities: net (coarse) or wire segment (fine). For coarse granularity, a set of nets is divided into two disjoint subsets, while for fine granularity; a single net may be routed on both sides of the chip. Partitioning at fine granularity provides more flexibility, but may introduce extra costs and may not be efficient due to increased complexity. For example, if a 2-pin net is broken into several segments and routed on both sides of the chip, a pin of an unused transistor must serve as transition point, as shown in Figure 6(a). Not to mention that it could be difficult to find such transition pins in highly

compacted layout; each such a pin is an extra point that must be connected from both sides, potentially aggravating the pin access problem. A similar situation occurs if in a multi-pin net two wire segments connected by a Steiner point are to be separated, as shown in Figure 6(b). Constraining the transition points to existing pins, like in Figure 6(c), avoids extra access points, but may result in longer wires. Thus, we partition nets at coarse granularity: each net can only be routed on one side of the chip.

3.2 Net Partitioning Algorithm

Partitioning the nets randomly may not be helpful, as shown in Figure 7. Thus, an effective cost function is necessary to guide net partitioning. The cost function should consider both pin density and congestion of passing nets on both sides.

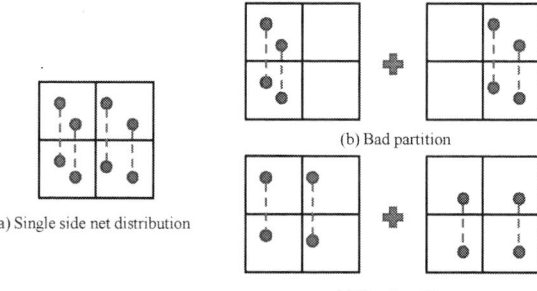

Figure 7. A net partitioning example.

As it is done by global router, we divide the placed design into small bins. In our experiments, each bin occupies a 10x10 routing grid. For each bin, the pin density cost is defined as follows:

$$cost_{density} = \sum_{top,bottom} \begin{cases} 1/(0.5 - pd) \ (pd < 0.49) \\ 100 \ (pd \geq 0.49) \end{cases}$$

where pd denotes the top or bottom side pin density of a bin, and 0.5 is the threshold pin density. When pin density is approaching the threshold, the cost increases rapidly. When pin density reaches and exceeds the threshold, the cost function remains a large constant. In practice, pin density rarely reaches the threshold with net partitioning.

The congestion cost of each bin is defined as:

$$cost_{congestion} = \sum_{top,bottom} \sum_{H,V} \begin{cases} 1/slack, \ (slack \geq 0.01) \\ 100, \ (0 \leq slack < 0.01) \\ 100 \cdot (1 - slack), \ (slack < 0) \end{cases}$$

where

$$slack = (capacity - demand)/capacity$$

We calculate horizontal and vertical congestion costs separately and add them up. The routing *capacity* of a bin is defined as the total length of horizontal/vertical routing tracks within the bin, and the routing *demand* is the estimated track usage. For example, assume a bin has 10 horizontal tracks of length 10, thus the horizontal capacity of this bin is 100. If a net has one pin in the bin and it occupies only a fraction of a track, for example, of length 6, the routing demand contributed by this net is 6. The routing demand is estimated probabilistically using the method in [19]: we find all the L-shape and Z-shape paths within the bounding box of a 2-pin net and assume equal probability for each. Multi-pin nets are decomposed into 2-pin nets based on minimum spanning tree.

The total cost of a partition is the combination of pin density and congestion costs:

$$cost_{partitioning} = \sum_{all\ bins} cost_{density} + \beta \cdot cost_{congestion}$$

where β is a user controlled weight. In our experiments, β is 1.

1103

```
Algorithm: Net Partitioning ()
Input: Circuit placement, net set N, the number of metal
layers on both sides, the number of runs k.
Output: Top side net set N_T, and bottom side net set N_B.
1.  Generate a random initial partition N_T and N_B
2.  Randomly pick an unprocessed net n_i in N_T, if moving n_i
    to N_B reduces total cost, take the change and update the
    partition.
3.  Repeat step 2 until all nets in N_T are processed.
4.  Repeat step 2 and 3 for net set N_B.
5.  Repeat step 2 to 4 until no net can be moved.
6.  Repeat step 1 to 5 k times and output the partition with
    lowest cost.
```

Figure 8. Greedy net partitioning algorithm.

Based on this cost function, we propose a greedy net partitioning algorithm, as shown in Figure 8. Nets are moved to the other side if the total cost can be reduced. The initial partition and the sequence in which the nets are processed affect the result. To help avoiding local optima, in each iteration we generate random sequences and pick the best among several initial partitions. The complexity of this algorithm is $O(m \cdot n)$, where m is the number of bins and n is the number of nets. The partitioning runtime is negligible comparing to the runtime of the routing process.

3.3 Experiments and Discussion

The net partitioning algorithm is implemented in C++ and runs on a Linux workstation with Intel 3GHz Core2 Duo CPU and 2GB of memory. We conducted experiments with LGSynth91 benchmarks mapped with a 6-cell VeSFET library using ABC synthesis tool [20]. We studied two cell libraries with medium and high pin densities. Medium pin density cells use tied-gate, isolated transistors, while high pin density cells use independent-gate, contacted transistors. In our experiments, we consider only inter-cell nets. Table 7 summarizes cell pin densities of both libraries. After technology mapping, the netlists are placed by a commercial placer. We assign 10% white space to each benchmark. The characteristics of the placed benchmarks are listed in Table 8. Comparing to circuits implemented with the medium pin density library (*Lib A*), the footprint areas of circuits implemented with the high pin density library (*Lib B*) are reduced by about 40% on average. However, the overall pin densities for *Lib B* based circuits are very close to threshold pin density even with 10% white space.

Table 7. Cell pin densities of two libraries

	INV	BUF	NAND2	NOR2	AOI21	OAI21
Lib A	0.5	0.25	0.31	0.31	0.25	0.25
Lib B	0.67	0.50	0.55	0.55	0.40	0.40

First, we route the placed benchmarks on one side. To meet VeSFET design rules, we force the commercial router to route strictly parallel wires on each layer. For each circuit, the number of available metal layers is 15. The experimental results are shown in tables 9 and 10. As expected, circuits built from high pin density cells in *Lib B* are not routable on one side without extra white space. For circuits built from medium pin density cells in *Lib A*, routing is completed in 5 to 8 metal layers.

Next, we verify the effectiveness of two-sided routing. Nets are partitioned by the algorithm described in Figure 8. After that, the commercial router is run twice to complete routing on each side. We start with 2 metal layers on each side. If routing cannot be completed, we add layers alternatively on both sides. Then we re-partition the nets based on new layer configuration and rerun the

Table 8. Benchmark characteristics

Benchmarks	#Cells	#I/O	#Nets	Pin density		Area(μm²)	
				Lib A	Lib B	Lib A	Lib B
alu4	679	22	693	0.27	0.46	505.4	307.1
C3540	1025	72	1075	0.28	0.47	742.6	454.7
C5315	1524	301	1702	0.29	0.48	1090.6	672.1
C6288	3450	64	3482	0.28	0.49	2430.2	1442.1
C7552	1958	313	2164	0.29	0.49	1359.8	832.3
des	3943	501	4199	0.28	0.46	2964.1	1799.7
frg2	819	282	962	0.30	0.50	574.2	354.3
i10	2120	481	2377	0.29	0.49	1484.2	913.5
pair	1484	310	1657	0.29	0.47	1064	659.0
pdc	4434	56	4450	0.28	0.46	3175.2	1966.7

Table 9. One and two-sided routing comparison for *Lib A* based circuits

Bench-marks	One sided routing			Two sided routing		
	#Layers	WL(μm)	#vias	#Layers	WL(μm)	#vias
alu4	6	3954	5911	4+2	3671	4598
C3540	6	5128	8560	4+2	4728	6681
C5315	6	7598	12394	4+2	7098	9438
C6288	5	11130	24113	2+2	10401	21014
C7552	6	8561	14683	4+2	7985	11361
des	6	27938	35828	4+2	26275	27134
frg2	6	3949	6653	4+2	3657	5004
i10	6	13748	17908	4+2	13128	13685
pair	6	7827	12333	4+2	7377	9507
pdc	8	48362	47147	4+4	44318	34129
AVG		1	1		0.93	0.77

router. 2 layers are added at a time in order to balance the number of horizontal and vertical routing tracks on each side. Experimental results in Table 9 indicate that with equal or fewer metal layers, two-sided routing needs about 23% fewer vias than one-sided routing for *Lib A* based circuits. It is so because one-sided routing needs to pull up pins for wires routed on upper layers. Also, with two-sided routing fewer detours are needed for pin access, because equivalent pin density is greatly reduced. This also explains why two-sided routing results in about 7% shorter total wire length than one-sided routing.

Pin access problem can also be solved by white space insertion. For benchmark circuits implemented with *Lib B*, we gradually increase the white space until the placement is routable on one side. The experimental results are listed in Table 10. About 25% extra white space is inserted to route the benchmarks on one side. In other words, more than 60% of the footprint shrinking from *Lib A* to *Lib B* is reclaimed for routing. Furthermore, the total wire length only improves about 10% but 21% more vias are used. Thus, the benefits of scaling are greatly compromised by inserting white space. On the other hand, two-sided routing achieves 13% shorter wire length and 28% fewer vias than a layout with white space insertion. It maintains the benefits of scaling: comparing to *Lib A* based circuits with one-sided routing, *Lib B* based circuits with two-sided routing achieve about 40% smaller footprint; the total wire length is 21% reduced and 13% fewer vias are used.

4. CONCLUSIONS

A highly compacted placement may be unroutable because of high congestion of crossing nets and difficulties in accessing local pins. Adding white space can solve both problems but loses the benefits of scaling: a design built from shrunk cells may need so much extra routing space that the expanded footprint compensates scaling. In this paper, we show that adding routing layers is also helpful and does not counteract scaling. However, pin access may

Table 10. One and two-sided routing comparison for *Lib B* based circuits

Benchmarks	One sided routing w/o white space	One sided routing w/ white space				Two sided routing w/o white space			
		# Layers	WL(µm)	# vias	Area(µm²)	# Layers	WL(µm)	# vias	Area(µm²)
alu4	FAIL	6	3604	7192	409.9	4+4	3113	5115	307.1
C3540	FAIL	6	4727	10326	605.7	4+4	3978	7379	454.7
C5315	FAIL	6	7074	15044	949.8	4+4	6327	10960	672.1
C6288	FAIL	6	10061	30898	2036.0	4+4	8447	21785	1442.1
C7552	FAIL	6	7545	18088	1108.4	4+4	6766	13029	832.3
des	FAIL	7	26106	42138	2400.8	4+4	22054	30312	1799.7
frg2	FAIL	6	3408	8038	471.9	4+4	3064	5751	354.3
i10	FAIL	7	11481	21766	1217.5	4+4	10581	15685	913.5
pair	FAIL	6	7312	15054	879.1	4+4	6488	10850	659.0
pdc	FAIL	11	44516	54323	2624.0	4+6	37593	40728	1966.7
AVG		1	1	1	1		0.87	0.72	0.74

still be very difficult for extremely high pin density designs even with many routing layers. Our statistical study using a commercial router indicates that an empirical threshold pin density of 0.5 exists beyond which adding more metal layers is not effective. The good news is that pin densities of CMOS and FinFET designs are far below that threshold, thus it is safe to continue footprint scaling provided more routing layers are available.

However, much denser VeSFET layouts may exceed the pin density threshold and suffer from serious pin access problem. The metal pillars of VeSFET transistors allow nets to be routed on both sides of the chip. This greatly helps pin access: designs with greater than threshold pin density are routable on two sides without inserting white space. Two-sided routing using our greedy net partitioning algorithm achieves 26% smaller footprint area, 13% shorter wire length, and 28% fewer vias comparing to benchmarks with inserted white space for one-sided routing. Benchmarks with medium pin density routable on one side can be routed on two sides with 7% less wire length and 23% fewer vias using equal or fewer metal layers.

5. ACKNOWLEDGEMENTS
This work was supported by the NSF grant #CCF 0904124. We thank Professor Wojciech Maly from Carnegie Mellon University for numerous inspiring discussions.

6. REFERENCES
[1] C. Alpert, Z. Li, M. Moffitt, G.-J. Nam, J. Roy, and G. Tellez, "What makes a design difficult to route," in *Proceedings of ISPD'10*, 2010, pp. 7-12.

[2] T. Taghavi, C. Alpert, A. Huber, Z. Li, G.-J. Nam, and S. Ramji, "New placement prediction and mitigation techniques for local routing congestion," in *Proceedings of ICCAD'10*, 2010, pp. 621-624.

[3] M. Marek-Sadowska, and X. Qiu, "A study on cell-level routing for VeSFET circuits," in *Proceedings of MIXDES'11*, 2011, pp. 127-132.

[4] B. Landman and R. Russo, "On a pin versus block relationship for partitions of logic graphs," *IEEE Trans. on Computers*, vol. C-20, no. 12, pp. 1469-1479, Dec. 1971.

[5] J. Deng, K. Kim, C.-T. Chuang, and H.-S. Wong, "The impact of device footprint scaling on high-performance CMOS logic technology," *IEEE Trans. on Electron Devices*, vol. 54, no. 5, pp. 1148-1155, May 2007.

[6] M. Alioto, "Comparative evaluation of layout density in 3T, 4T, and MT FinFET standard cells," *IEEE Trans. on VLSI*, vol. 19, no. 5, pp. 751-762, May 2011.

[7] W. Maly, and A. Pfitzner, "Complementary vertical transistors," Carnegie Mellon University, CSSI Tech. Rep., No. 08-02, 01/2008.

[8] W. Maly, "Integrated circuit device, system, and method of fabrication," Patent US 0 321 830, December 31, 2009.

[9] W. Maly, N. Singh, Z. Chen, N. Shen, X. Li, A. Pfitzner, D. Kasprowicz, W. Kuzmicz, Y.-W. Lin, and M. Marek-Sadowska, "Twin gate, vertical slit FET (VeSFET) for highly periodic layout and 3D integration," in *Proceedings of MIXDES'11*, 2011, pp. 145-150.

[10] M. Weis, "A circuit design perspective for the Vertical Slit Field Effect Transistor (VeSFET)," PhD thesis, Technical University Munich, 2009.

[11] Y.-W. Lin, M. Marek-Sadowska, and W. Maly, "Layout generator for transistor-level high-density regular circuits," *IEEE Trans. on CAD*, vol. 29, no. 2, pp. 197-210, Feb. 2010.

[12] Y-W. Lin, M. Marek-Sadowska, and W. Maly, "On cell layout-performance relationships in VeSFET-based, high-density regular circuits," *IEEE Trans. on CAD*, vol. 30, no. 2, pp. 229 - 241, Feb. 2011.

[13] D. Stroobandt, P. Verplaetse, and J. Campenhout, "Generating synthetic benchmark circuits for evaluating CAD tools," *IEEE Trans. on CAD*, vol. 19, no. 9, pp. 1011-1022, Sep. 2000.

[14] M. Moffitt, "MaizeRouter: engineering an effective global router," *IEEE Trans. on CAD*, vol. 27, no. 11, pp. 2017-2026, Nov. 2008.

[15] M. Ozdal, "Detailed-routing algorithms for dense pin clusters in integrated circuits," *IEEE Trans. on CAD*, vol. 28, no. 3, pp. 340-349, Mar. 2009.

[16] Y. Zhang and C. Chu, "RegularRoute: an efficient detailed router with regular routing patterns," in *Proceedings of ISPD'11*, 2011, pp. 45-52.

[17] J. Davis, V. De, and J. Meindl, "A stochastic wire-length distribution for gigascale integration (GSI). I. Derivation and validation," *IEEE Trans. on Electron Devices*, vol. 45, no. 3, pp. 580-589, Mar. 1998.

[18] http://www.si2.org/openeda.si2.org/projects/nangatelib/

[19] J. Lou, S. Thakur, S. Krishnamoorthy, and H.S. Sheng, "Estimating routing congestion using probabilistic analysis," *IEEE Trans. on CAD*, vol. 21, no. 1, pp. 32-41, Jan. 2002.

[20] Berkeley Logic Synthesis and Verification Group, ABC: A System for Sequential Synthesis and Verification, Release 70930. http://www.eecs.berkeley.edu/~alanmi/abc/

S1. Proof of Theorem 1

Theorem 1: Any configuration of nets is routable on one side of a 2-D grid if and only if at least one grid point is spare or 2 pins to be connected are placed at the neighboring grid positions.

Proof: We define *active* point as a point of the grid occupied by an unconnected yet pin of a net. Without loss of generality, we will route the nets one at a time.

Necessity: If all grid points are active and no two neighboring grid points belong to the same net, making any connection will block at least one active pin, and the routing will fail. Therefore, the necessity is proved.

Sufficiency: If there is one spare (not active) point on the grid, we will show a sequence of steps to create a routable solution. The idea is to maintain accessibility of all unconnected pins and access points to all incomplete sub-nets. First, we explain the *shifting* process shown in Figure 1-S of moving a spare point on the grid. In Figure 1-S(a), we want to move the spare point to the position occupied by pin *A*. On the first metal layer, we place a wire which connects to pin *B* and extends to the spare point (no pin will be blocked by this wire); then a via is inserted at the original spare point, as shown in Figure 1-S(b). A via connecting to pin *A* is also inserted. Thus, the middle point becomes spare and the rightmost point becomes the access point of pin *B* on the second metal layer. In this way, we can shift the spare point one grid point at a time using one metal layer. Similarly, on the second metal layer, we can connect pin *A* to the middle point and pull it up to the third metal layer, as depicted in Figure 1-S(c). Therefore, by using two more metal layers, the shifting process moves the spare point to the leftmost point which was originally occupied by pin *A* while both pins *A* and *B* are still accessible.

With the shifting process, we can move a spare point to any position on the grid, although such movements may introduce many metal layers because in the worst case, each shift by one grid length needs one metal layer. For any 2-pin net, we can find a path and step by step, move the spare point along the path such that the route will not block other unconnected pins, until the destination is reached. We show an example in Figure 2-S. Once a 2-pin net is routed, at least one more spare point is added, and the more spare points, the fewer shifting operations are needed to complete a route. Thus, given a netlist, we can route the nets sequentially (multi-pin nets can be decomposed into 2-pin nets). Since the number of metal layers is unlimited, and we only consider routability, the routing sequence is not important—we can start from any net. Because the number of spare points increases as more nets are routed, this process guarantees finding solution for all nets.

If no spare point exists but two neighboring pins on the grid belong to the same net, we can connect them on the first metal layer. In this way, at least one of the two grid points can be considered as spare on the second metal layer (If the net is a 2-pin net but not an I/O net, both grid points can be considered spare).

So, we have proved the sufficiency. □

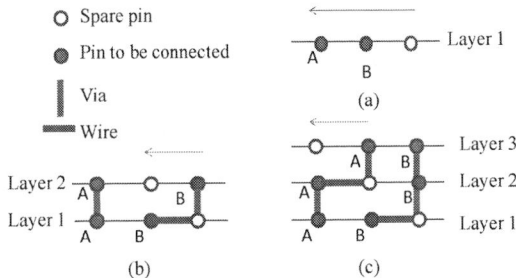

Figure 1-S. A *shifting* process moves a spare point on the grid while maintaining active pin accessibility.

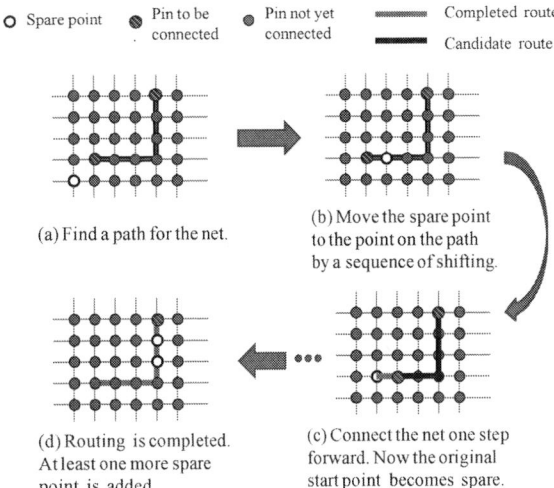

Figure 2-S. 2-pin net routing example with one spare point.

Yield Estimation via Multi-Cones

[1,3]Rouwaida Kanj, [2]Rajiv Joshi, [1]Zhuo Li, [1]Jerry Hayes, [1]Sani Nassif

[1]IBM Austin Research Labs
[2]IBM TJ Watson Labs
[3]American University of Beirut

Abstract- We propose a new yield estimation algorithm which estimates the acceptability region as the union of spherical cones. The algorithm works by dividing the input parameter space into approximately equi-probable cones, efficiently estimating the refined weight contributions for each cone, then combining the results to get the total yield. The algorithm is broadly similar to the worst-case-distances method, but is more generally applicable for cases with -for example- multiple failure regions. The algorithm is quite accurate, and offers several orders (>100x) of magnitude of speedup compared to traditional Monte Carlo. The paper includes example applications to difficult high-yield circuits like SRAM.

Categories and Subject Descriptors
B.7.2 [**Hardware**]: Integrated Circuits – *Design Aids*

General Terms
Algorithms, Performance, Design, Reliability.

Keywords
Statistical Performance Analysis, Yield Prediction, SRAM

I. INTRODUCTION

As technology scales, within-die process variations continue to increasingly impact design yield and performance gains [1, 2]. The effects are deeply felt in memory designs and there are rising concerns about their increasing impact on the functionality of logic designs [3, 4]. With millions of components on a chip, several statistical design methodologies have been developed to address the design yield degradation problem. Examples range from statistical timing analysis methods [5, 6] to the more recently developed memory yield analysis tools [7, 8, 9]. There is a rich heritage in this area and interested reader should look at reference [10]. To speed up the analysis, compared to traditional Monte Carlo, statistical sampling-based methods often rely on variance reduction methods [11]; their efficiency, however, is highly dependent on identifying the optimal importance sampling function. Integration-based methods rely on modeling the fail boundary: examples include a family of hyperplanes, a convex hull, or an ellipsoid [5, 6, 11]. With the exception of the inscribed ellipsoid method, the failure probability computation cannot be derived in closed form and requires integration by sampling of the modeled region; if the fails are rare, importance sampling may be employed. Finally, methods like the convex hull are exponential in the number of dimensions, and quickly suffer from the curse of dimensionality.

Sensitivity-based methods like the worst-case direction method [12] and other reliability estimation methods [13] are also popular techniques for design yield estimation. For the first-order reliability method (FORM), the failure region is determined from the closest failing point along a maximal sensitivity direction (Fig 1). A closed form equation is then used to estimate the failure probability. The

method is suitable for rare fail event estimation when the assumed failure boundary is a good representation of the failure region. The method fails, however, for the case of multiple failure regions.

In this paper, we propose a new weighted multi-cone based yield estimation methodology. We refer to it as the *broken spheres method* based on [14]. *The technique distinguishes itself in the ability to model the fail region by a set of multi-directional, approximately equi-probable, non-overlapping fail regions with weight refinement.* The fail boundary is modeled as the union of a set of partial (broken) hyperspherical shells; the acceptability region is modeled as the union of the corresponding multi-cones. Most importantly, multiple critical failure regions can be covered and a closed form of the yield estimate can be easily derived. The methodology is fast and accurate and enables significant speedup compared to traditional Monte Carlo. Other methods that rely on spherical sampling [15] were presented in [16]. This paper is organized as follows: Section II introduces the proposed methodology; Section III studies its convergence behavior; Sections IV and V present practical applications to an 8-D SRAM and a 10-D latch designs respectively.

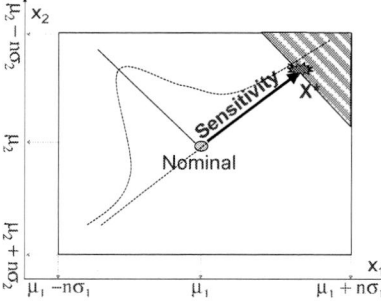

Figure 1. x* is the most probable failing point to the nominal; the fail probability in [13] is estimated as a function of ||x*||.

II. PROPOSED METHODOLOGY

A. Overview

We motivate our method by the case illustrated in Figure 2, where the failing boundary is not well represented by a single direction. This can happen for the case of multiple failure regions, or multiple competing design metrics. The overlap between the two regions in Figure 2 means that the overall yield is not simply the combination of the two yields for the two failure boundaries.

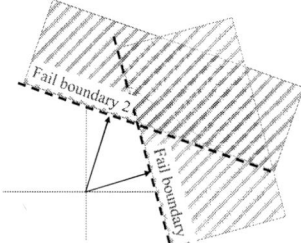

Figure 2. Example of overlapping fail regions in 2-D space.

Monte Carlo-based integration of the fail region is needed to avoid double counting the overlap. To eliminate this problem and enable multi-directional non-overlapping multiple fail regions, we propose to rely on multi-cone/broken spherical shell representations of the acceptability/fail regions respectively; see Fig. 3 for a 2-D space example. The methodology relies on the following concept.

Given a set of vector directions, $\{X_{dir}\}$ in d-D space, each failing at a distance R_{fi} from the origin, $i=1,...,n_{dir}$, and contributing to the fail probability by a weight, w_i, the fail probability, P_f, is:

$$P_f = \sum_{i=1}^{n_{dir}} w_i * p_{f_i}, where \quad p_{f_i} = prob(R > R_{fi}) \qquad (1)$$

Key Features and assumptions
1) To obtain good parameter space coverage, search directions X_{dir} are obtained by randomly sampling the directional space.
2) To explore the space properly, one would ideally rely on *equiangular directions* to obtain a uniform coverage of the directional space. While this is easy in 2-D space, it is non-trivial for multi-dimensional space, so we rely instead on uniformly distributed *equi-probable directions* as described in section B.
3) For equi-probable directions, the expected value of weight contributions w_i is: $E[w_i]=1/n_{dir}$. We propose a methodology to further refine the weight contributions, w_i, of a directional vector as will be explained in section C.
4) The spherical nature of the methodology allows for closed form representation for probability($R>R_{fi}$). No special integration method is needed, and the corresponding analytical equations will be presented in Sec. D.

For the rest of the paper, we assume that all variable distributions are standardized to a Gaussian distribution $\sim N(0,1)$.

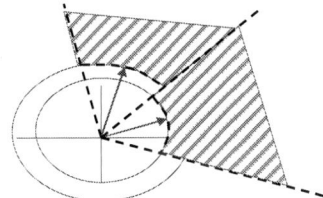

Figure 3. Breaking the space into non-overlapping cones (dashed) along critical fail directions enables non-overlapping fail regions.

Figure 4 presents the methodology flow diagram. Binary search is performed along each direction to find the corresponding fail radius, R_f. The equivalent probability is then computed from the weighted broken spherical shell probabilities. The estimate is then checked for convergence. Note that the methodology is an open sampling-based methodology. Additional directions can be added to satisfy desired convergence criteria; the weights are then recomputed. Similar to other forms of open sampling-based methods like some Quasi-Monte Carlo methods [17], one can adopt two possible convergence schemes.

- *Yield estimate convergence:* we monitor the estimate convergence as the number of directions increases.
- *Standard deviation of estimate, σ_{est} convergence:* For this, we rely on *bootstrapping*, to compute σ_{est}. E.g., for a direction sample set of size n, select k subsets each of size $m<n$. compute σ_{est} for the k subsets. σ_{est} is then evaluated for larger subset sizes, m, to monitor its convergence.

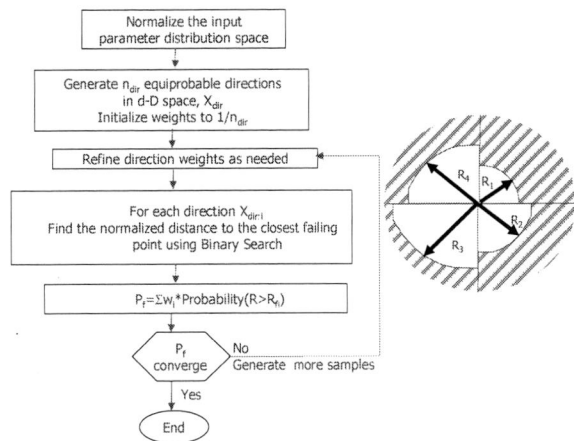

Figure 4. Methodology Flow Diagram.

B. Equiprobable Directions: Uniform sampling on a sphere

A well known property of d-dimensional (d-D) normal distribution functions $N(0, 1)$ is that they have constant probability values on the surfaces of the d-D spheres. This property can be exploited [15] to obtain points that are uniformly distributed on the surface of the d-D unit sphere as explained in the steps below and illustrated in the diagram of Fig 5.

- Start with a Gaussian random sample $X \sim N(0, 1)$.
- The normalized vectors $X_{dir:i}=X_i/\|X_i\|$ represent the direction cosines of the sample points X_i on the d-D unit sphere.
- X_{dir} vectors are uniformly distributed on the unit sphere and represent equiprobable directions that are expected to contribute equally likely to P_f.

C. Weight Refinement

Due to sampling uncertainty, the weight contributions of the different sampled directions deviate from the expected value, $E[w_i]=1/n_{dir}$. We propose the following methodology to enable an enhanced estimate of the true weight contributions.

- Given a set X_{dir} with size n_{dir}
- Sample a larger set of equiprobable directions, Y_{dir} on the unit sphere (see Fig. 6); e.g., its size can be $10 \times n_{dir}$.
- Assign every direction $Y_{dir:j}$ to the closest $X_{dir:i}$ vector. This can be accomplished with a simple brute force search, or by using any of many existing clustering algorithms. Ideally each $X_{dir:i}$ vector will have roughly the same number of $Y_{dir:j}$, and that number is a measure of the weight for each particular X_{dir}. The ratio of the ith cluster size (corresponding to $X_{dir:i}$) to card$\{Y_{dir}\}$ determines the correction factor for w_i; see Fig. 6 for details of implementation.

D. Probability Integration in Spherical Dimensions: $P(R>R_f)$

In d-D spherical space, $P(R>R_f)$ can be computed according to (2), where the spherical representation of the Gaussian probability density function (pdf) is $e^{-\frac{\rho^2}{2}}$, ρ being the radial dimension. Those integrals can be readily available and can be solved analytically according to the chi-square distribution. Table I. summarizes the analytical closed form solutions in d-D space [18, 19].

$$P(R>R_0) \sim \iint_{\Phi} \cdot \int_{R_0}^{\infty} e^{-\frac{\rho^2}{2}} \rho^{d-1} (\sin\phi_1)^{d-2} (\sin\phi_2)^{d-3} ... (\sin\phi_{d-2}) d\rho d\phi_1 ... d\phi_d \quad (2)$$

$$\propto \int_{R_0}^{\infty} \rho^{d-1} e^{-\frac{\rho^2}{2}} d\rho$$

where $\quad (\rho,\phi_1,\phi_2,...,\phi_{d-1}) \in (0,\infty) \times (0,\pi) \times (0,\pi) \times ... \times (0,2\pi)$

Table I. $P(R>R_0)$. This can be generalized for higher dimensions.

$d=1$	Erf($R_0/\sqrt{2}$)	$d=2$	$\exp(-R_0^2/2)$
$d=3$	$2\,\Gamma(3/2, 0, R_0^2/2)$	$d=4$	$\exp(-R_0^2/2)*(1+R_0^2/2)$
$d=5$	$4\,\Gamma(5/2, 0, R_0^2/2)$	$d=6$	$\exp(-R_0^2/2)*(1+R_0^2/2+R_0^4/8)$

- $X_i \sim N(0,1), i=1,...,n_{dir}$
- $X_{dir:i} = X_i / \|X_i\|$
  - $X_{dir:i} \sim$ uniformly distributed on the unit sphere

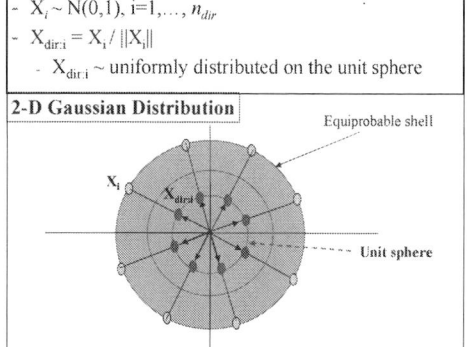

2-D Gaussian Distribution

Figure 5. Generating uniform samples on the unit sphere surface.

Weight Refinement

- Let X_{dir} be the set of equi-probable directions
  * To be used for true system simulation
  * Card $\{X_{dir}\} = n_{dir}$
- Obtain Y_{dir} additional arbitrary equiprobable random directions
  * To be used for adjusting weights
  * Card $\{Y_{dir}\} = n_{nonSimDir}$
- For j=1 to $n_{nonSimDir}$
  For i=1 to n_{dir}
  $Dist_{ji} = \|Y_{dir j} - X_{dir:i}\|$
  $[D, I] = sort(Dist_{ji})$
  $w(X_{dir}(I(1))) = w(X_{dir}(I(1)))+1;$
  $w(Xdir(.)) = w(Xdir(.))/n_{nonSimDir};$

- - - - - **Arbitrary Uniform Random Direction (Y_{dir})**
⟶ **Initial Sampling Direction Set (X_{dir})**

Figure 6. Procedure to refine direction weights.

III. THEORETICAL VALIDATION

We compare the efficiency of the methodology against that of traditional Monte Carlo. For this, we study several theoretical examples with random fail boundaries. To interpret the results properly, we list the set of definitions below.

Assumptions and Definitions

1. *Random fail regions:* For multiple fails regions, the hyperplane directions and fail distances are random.
2. *Monte Carlo experiments (MC):* Given N random samples the failure probability, P_f, and its variance σ_{Pf}^2 can be estimated from the number of failing samples N_{fails}. For low P_f, N needs to be $\sim 1/P_f$ for convergence.

$$P_f = \frac{1}{N} * N_{fails} \quad \text{, and} \quad \sigma_{Pf}^2 = \frac{P_f(1-P_f)}{N} \quad (3)$$

The number of simulations is equal to the number of samples. We set $N=$1e5 MC samples.

3. *Proposed Multi-cone Broken-sphere method (spherical):*
   i. *Binary search:*
      a. The range for the binary search distance R along a direction in the normalized space is [0, 10].
      b. The number of search iterations along a direction is 8 (binary_length).
   ii. *Number of simulations (n_{sim}) vs. number of directions (n_{dir}):* If a direction hits a failing region $n_{sim}/n_{dir}=$binary_length$=8$ for that direction, else $n_{sim}/n_{dir}=1$. Hence n_{sim} can be derived from the number of directions (n_{dir}) as follows.

   $$n_{sim} = n_{dir}*hit\_rate*binary\_length + n_{dir}*miss\_rate \quad (4)$$

   hit(miss)_rate corresponds to the number of direction vectors that hit(miss) a fail boundary.

4. *Replications:* to test for convergence, in this section, we rely on independent replications as opposed to bootstrapping.
   i. When replicating experiments with fixed n_{dir}, n_{sim} varies due to the random directions; hence we report its mean value: avg(n_{sim}).

5. *Limited Monte Carlo experiments (limitedMC):* those experiments are similar to Monte Carlo, but use a limited number of samples, avg(n_{sim}), equal to that required by the proposed methodology.

6. σ_{yield}: Instead of reporting the results in terms of P_f values, we report the equivalent sigma yield numbers according to (4);

   $$\sigma_{yield} = \varphi^{-1}(1-P_f) \quad (4)$$

   φ is the normal cumulative distribution function (cdf).

A. Accuracy and Efficiency: 2-D space

A.1 Ideal case: 360 equiangular directions

We test the accuracy for the ideal spherical coverage case of 360 equiangular directions in 2-D space. Fig. 7a illustrates excellent match between the proposed method and the computed analytical yield for the case of a single hyperplane fail bound; MC, on the other hand, fails to catch up beyond 4.1 sigma yield. We repeat the study for the case of multiple hyperplane boundaries; the number of the different hyperplanes is varied randomly (between 5 and 13), so is the distance of fails for each hyperplane. We notice excellent matching between the proposed spherical methodology and MC when the latter converges as illustrated in Fig. 7b.

A.2 Weight Refinement Effects (50 equiprobable directions)

We repeat the multiple hyperplane experiment estimation for the case of 50 equiprobable directions. Fig. 8 illustrates the estimated yield for different experiments using the 50 equiprobable directions against those of the ideal 360 equiangular directions. We notice improved estimate due to weight refinements as described in section II C compared to the strict equi-probable contribution assumption. Note that the distribution function for the conal weights illustrates a spread upto 3x the expected value for 250 replications of the experiment; the weights in Figure 8 are normalized to $1/n_{dir}$, and E[normalized weight]=1.

B. Multi-dimensional Space

In this section, we analyze the convergence trends for 2-D, 4-D, 6-D and 8-D spaces. To enforce the findings, we first present the following set of 8-D experiments. We study the 1-hyperplane fail

1109

boundary case for analytical yield values of 2, 3, 4 and 5 sigma. Figure 9 illustrates rapid convergence of the spherical estimate in 8-D space. When utilizing the same number of simulations, the limitedMC simulation weakly converges for the 4 sigma yield estimate and does not converge for 5 sigma yield case. To understand the convergence rate, we perform 100 replications of the previous example (see Fig. 10). We then compute the standard deviation of the yield estimate,

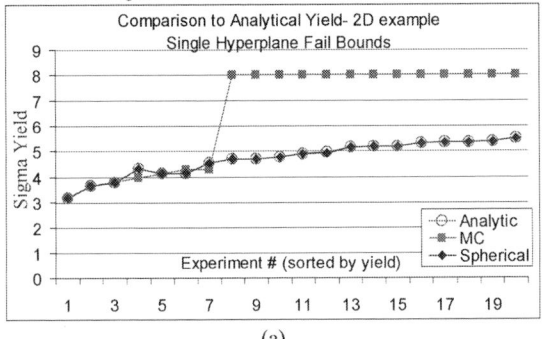

(a)

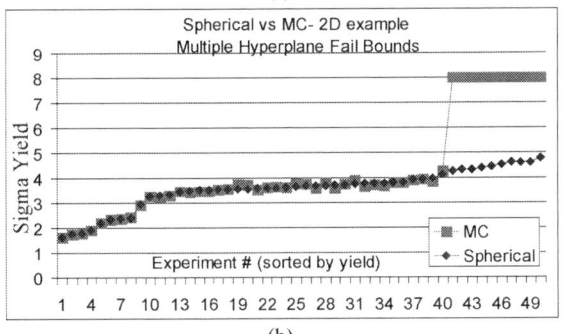

(b)

Figure 7. Excellent match between (a) Spherical and Analytic solution and (a, b) Spherical and MC when the latter converges.

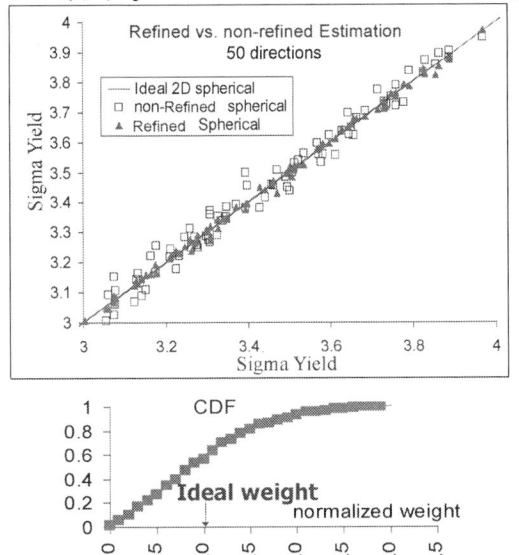

Figure 8. Weight refinement leads to improved estimate. The direction weights exhibit a wide spread around the expected value.

σ_{est}, from the different replications. S_{2_8d}, S_{3_8d}, S_{4_8d}, and S_{5_8d} in Fig. 11, represent the number of directions needed for $\sigma_{est} < 0.1$ sigma for each of the yield estimates of 2, 3, 4 and 5 sigma (in 8-D). We then study an extreme case of $2^{d/2}$ multiple random fail hyperplanes; the distance of each plane is varied randomly while maintaining multiple critical hyperplanes. Fig. 12 illustrates the case of a 5 sigma yield estimate. We note a reduction in the number of directions needed for convergence compared to 1-hyperplane bound. However, the number of required simulations is almost the same due to an increased hit rate (directions hit fail bounds more often). Fig. 13 plots the number of simulations needed for convergence for the cases of multiple and single fail hyperplanes in 2-D, 4-D, 6-D and 8-D. Fitting the data, we find a quadratic dependency between n_{sim} and the number of dimensions and a weaker quadratic relation between n_{sim} and the yield sigma for a given dimension. The large binary search radius enables good convergence due to good hit-rate even for high-yields; most importantly *the convergence rate is within few thousand simulations.*

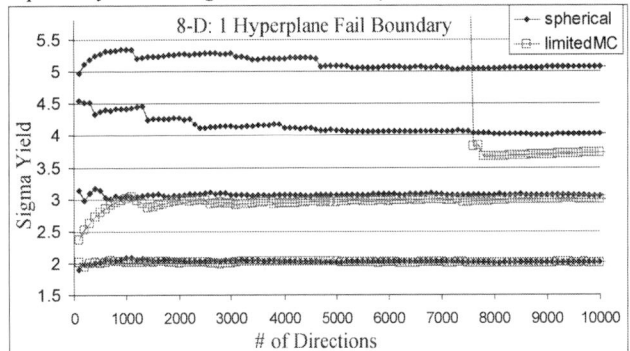

Figure 9. Spherical Method converges well for 2, 3, 4, and 5 sigma yield. LimitedMC fails to converge for 5 sigma yield.

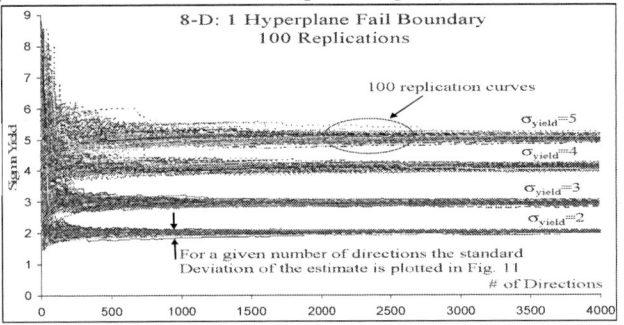

Figure 10. Replications of the yield estimate for 8-D space.

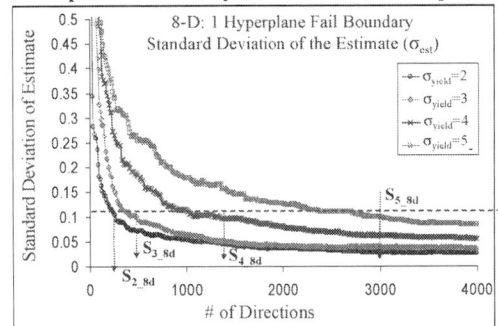

Figure 11. Standard Deviation of the estimate as function of n_{dir}.

1110

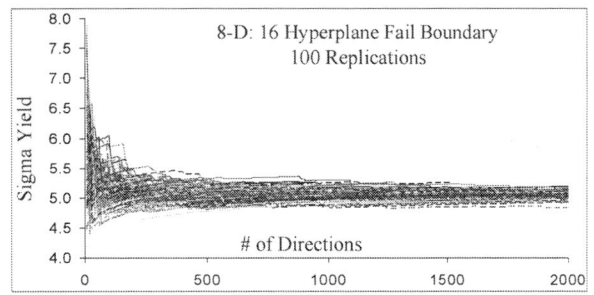

Figure 12. 5 sigma yield convergence for multiple critical bounds

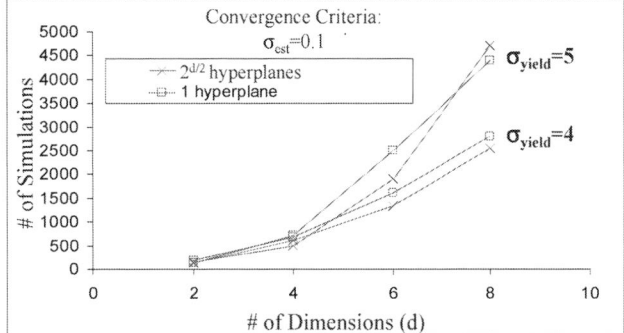

Figure 13. Number of simulations- spherical method convergence.

IV. APPLICATIONS TO 8-D SRAM DESIGN

In this section, we extend the methodology to study the read after write yield of a sub-65nm 6T SRAM design subject to random dopant (RDF) and PBTI/NBTI fluctuations [20]. All six cell transistors are subject to RDF-based V_T variation. Transistors P_1 and N_0, shown in circles in Fig 14, are also subject to systematic and random NBTI/PBTI variations based on worst-case asymmetric BTI effects with node L(R) holding '0'(1) for a long time; hence the 8-D problem.

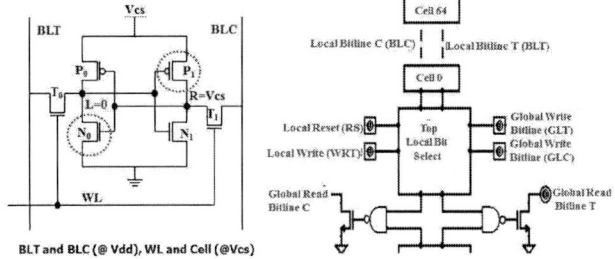

Figure 14. SRAM cell schematic view and local bit select circuitry.

For the spherical method, we relied on 10,000 directions to study convergence; we also used 100k samples for MC method. We relied on bootstrapping to construct subset replications and study σ_{est} for both MC and the proposed methodology. Table II. summarizes the results for different combinations of the cell supply (Vcs) and logic supply (Vdd). It also summarizes the number of simulations needed for convergence. With 100k simulations, the MC method did not converge beyond 4 sigma yield estimates. Whereas the spherical method required around 3100 directions (6000 simulations) for the even higher 5.25 sigma estimate which is consistent with the theoretical analysis. For the same convergence rate, the 5.25 yield estimate would require at least 10M simulations for the MC method; hence the proposed methodology enables 2-3 orders of magnitude in

savings (1600X). Fig. 15 illustrates MC bootstrap replications for the case of 4.01 sigma estimate; we also plot σ_{est} convergence on the right axis. Fig. 16 illustrates the 5.25 sigma yield estimate replications for the proposed spherical methodology; the bottom x-axis is the ndir and the upper is n_{sim}. The replications do converge early within few thousand simulations indicating good confidence in the estimate.

Table II. Spherical vs. MC techniques. Dashed implies simulation did not converge with 100k samples, and the corresponding number of MC samples needed is computed according to desired criteria and according to equation (4).

supply	Spherical			MC	
	Yield	# Directions	# simulations	yield	# simulations
(Vdd₁, Vcs₁)	3.5	600	700	3.48	11500
(Vdd₂, Vcs₂)	3.97	1180	1500	4.01	67000
(Vdd₃, Vcs₃)	4.45	2130	3100	4.11	~100k
(Vdd₄, Vcs₄)	4.88	3100	5500	--	> 1M
(Vdd₅, Vcs₅)	5.25	3120	6100	--	>10M

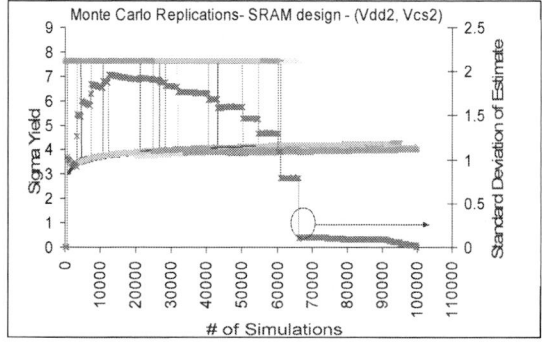

Figure 15. Bootstrap replications of the 4 sigma MC yield estimate and the corresponding σ_{est}.

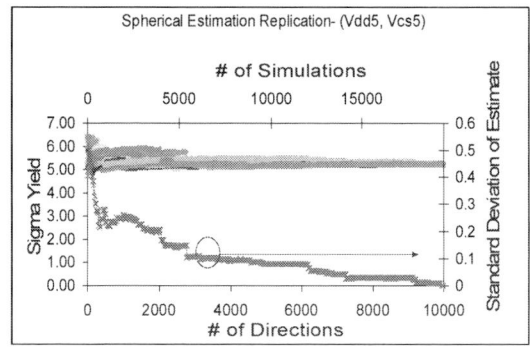

Figure 16. Bootstrap replications of the 5.25 sigma spherical yield estimate and the corresponding σ_{est}: n_{dir} and n_{sim} are illustrated on bottom and top axis.

V. 10-D LATCH DESIGN APPLICATIONS

We also study the sub-65nm master-slave latch design shown in Figure 17. The slave clock (B) is held high in flush mode, and the design yield of four setup and hold tests involving the clock-to-data master stage edges is estimated. The hold time relative to clock falling edge is intended to avoid false capture. The setup time with respect to the clock rising edge is intended to avoid the data-edge gating the clock-to-Q delay. Table III. lists the latch yield for different setup (t_s) and hold time (t_h) criteria. $t_{h(s)}$ fail criteria are normalized as function of t_d, the sum of clock and data path delays in the dotted rectangular region (Fig. 17). Ten devices corresponding to the three inverters and the tri-state inverter (dotted region) are subject to RDF V_T variations. Again, 100k Monte Carlo simulations do not converge beyond 4 sigma, whereas the spherical methodology converges within few thousand simulations and matches Monte Carlo when the latter converges. Fig. 18 illustrates the convergence for the different tests.

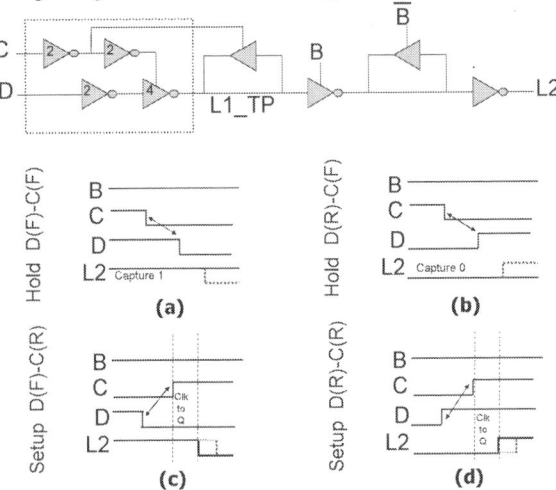

Figure 17. Schematic of Latch design and four different tests.

Table III. Monte Carlo (MC) versus Spherical yield estimate. Dashed entries did not converge in MC within 100k simulations.

test	bound	MC	Spherical		
	$t_{h(s)} <$	yield	yield	#sims	#directions
HOLD D(F) C(F)	0.2*td	2.43	2.44	1080	400
	0.4*td	3.35	3.30	2420	1100
	0.6*td	4.20	4.15	4500	2500
	0.8*td	--	4.97	3500	2500
	1.0*td	--	5.75	3000	2500
HOLD D(F) C(F)	0.2*td	--	4.84	3150	2100
	0.4*td	--	5.61	2646	2100
	0.6*td	--	6.33	2970	2700
	0.8*td	--	7.04	2700	2700
	1.0*td	--	7.67	2800	2800
SETUP D(F) C(R)	0.6*td	2.92	2.85	2030	700
	0.8*td	3.77	3.68	1800	750
	1.0*td	--	4.50	4875	2500
	1.2*td	--	5.29	4100	2500
	1.4*td	--	6.05	3600	2500
SETUP D(R) C(R)	0.6*td	1.88	1.84	800	200
	0.8*td	2.70	2.63	2040	600
	1.0*td	3.35	3.38	2100	750
	1.2*td	4.30	4.08	1725	750
	1.4*td	--	4.70	6900	4000

VI. CONCLUSIONS

We propose the multi-cone broken hyperspheres method as a novel yield estimation methodology. The technique enables yield estimation with multiple critical fail directions/regions. Intermediate steps of the algorithm involve equiprobable directions generation, weight refinements, partial spherical representation of fail bounds and closed

form derivations of the probabilities of spherical shells. The methodology is comprehensive and computationally efficient in multi-dimensional space. We verified the methodology theoretically and in application to standard memory and latch designs. It was found to achieve more than 100x savings compared to traditional Monte Carlo.

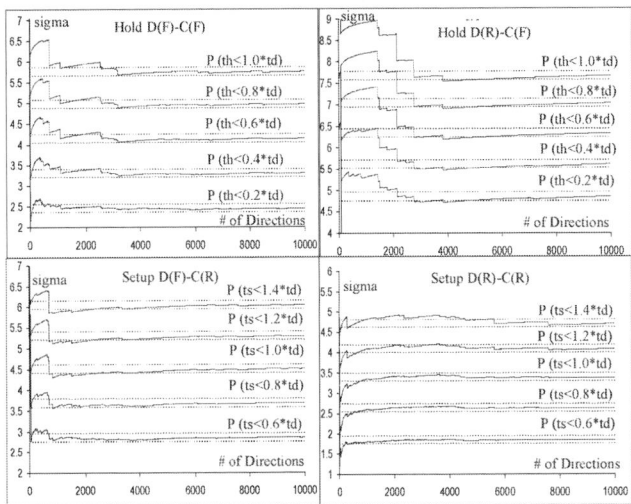

Figure 18. Yield estimate for the different setup and hold tests converges within couple of thousand directions. Dashed lines represent +/-0.1 sigma difference compared to final estimate.

Acknowledgement. Authors would like to thank Dr. Ashish Singh, and Prof. Ralf Korn for discussions and contributions.

VII. REFERENCES

[1] V. De, and S. Borkar, ISLPED '99, pp.163– 168.
[2] S. R. Nassif, ISQED, 2000, pp. 451 – 454
[3] R. V. Joshi et al., SOI Conf. 2006, pp. 211 – 214.
[4] A. Pelella et al., SOI Conference 2008 , pp. 41 – 42.
[5] J.A.G. Jess et al, DAC 2003, pp. 932 – 937.
[6] C. Visweswariah, ISQED 2008, pp. 568.
[7] R. Kanj et al., DAC '06, pp. 69-72.
[8] A. Singhee et al., *DATE*, 2007.
[9] C. Dong, X. Li, DAC 2011, pp. 200-205.
[10] S. W. Director and W. Maly, editors. *Statistical Approach to VLSI*, vol 8 of *Advances in CAD for VLSI*. North-Holland, '94
[11] D. E. Hocevar et al., TCAD, pp. 180-192, July 1983.
[12] K. Antreich, H. Graeb, C. Wieser, IEEE TCAD, 1994, pp. 57-71.
[13] G. Schueller et al., 16th ASCE Engineering mechanics conf, 2003.
[14] R. Joshi, R. Kanj, Z. Li, S. Nassif, "Broken-spheres method for improved failure probability", US patent application, 20100313070, Filed June '09.

[15] M. E. Muller, Comm. ACM, Issue 4 (April 1959) pp. 19 - 20
[16] R.A. Fonseca et al., "Statistical Simulation Method for Reliability Analysis of SRAM Core-Cells", DAC 2010.
[17] W.H. Press et al., Numerical Recipes in C, 2nd edition. New York: Cambridge University Press, 1997.
[18] A.H. Stroud. Approximate Calculation of Multiple Integrals. Prentice Hall, New Jersey, 1971.
[19] M. Abramowitz and I. A. Stegun. Handbook of Mathematical Functions. Dover Publications, New York, 1965.
[20] G. La Rosa et al., 44th IEEE IRPS 2006.

Efficient Trimmed-sample Monte Carlo Methodology and Yield-aware Design Flow for Analog Circuits

Chin-Cheng Kuo[1], Wei-Yi Hu[1,2], Yi-Hung Chen[1], Jui-Feng Kuan[1], Yi-Kan Cheng[1]

[1]Taiwan Semiconductor Manufacturing Company Ltd., Hsinchu, Taiwan
[2]Graduate Institute of Electronics Engineering, National Taiwan University, Taipei, Taiwan

{cckuoz, wyhu, simon_chen, jfkuan, yk_cheng}@tsmc.com

ABSTRACT

This paper proposes efficient trimmed-sample Monte Carlo (TSMC) methodology and novel yield-aware design flow for analog circuits. This approach focuses on "trimming simulation samples" to speedup MC analysis. The best possible yield and the worst performance are provided "before" MC simulations such that designers can stop MC analysis and start improving circuits earlier. Moreover, this work can combine with variance reduction techniques or low discrepancy sequences to reduce the MC simulation cost further. Using Latin Hypercube Sampling as an example, this approach gives 29x to 54x speedup over traditional MC analysis and the yield estimation errors are all smaller than 1%. For analog system designs, the proposed flow is still efficient for high-level MC analysis, as demonstrated by a PLL system.

Categories and Subject Descriptors

B.7.2 [Integrated Circuits]: Design Aids

General Terms

Algorithms, Design, Verification.

Keywords

Monte Carlo simulation, analog circuits, yield-aware design flow, trimmed-sample.

1. INTRODUCTION

Scaling device size significantly increases the impact of process variations on circuit performances especially sensitive analog designs. Using Monte Carlo (MC) simulations is a conventional methodology to statistically analyze the parametric yield under process variations. However, traditional MC analysis often requires several thousand of SPICE simulations [1] such that iterative design for yield at circuit level becomes almost infeasible for large analog circuits.

There are many literatures about improving the efficiency of MC analysis. Using response surface models [2-5] is popular to substitute expensive SPICE simulations. However, the equation-based approach would sacrifice accuracy for speedup such that the accuracy loss may be unacceptable, as mentioned in [1].

Another speedup strategy is to improve the capability of sample

generator for MC analysis. In traditional methodology, pseudo-random sampling (PRS) is used to generate random samples for simulations. Unfortunately, the traditional approach often requires a large number of simulations for accurate yield estimation. Therefore, variance reduction techniques are proposed to improve the accuracy or reduce the simulation runs in MC analysis. Latin Hypercube Sampling (LHS) [6-8] is one of widely used techniques, which has better sampling capability than PRS. LHS could be nearly five times more effective in yield estimation than traditional PRS-based MC analysis [8-10].

In contrast to random sampling, low discrepancy sequences (LDS) are deterministic sequences for MC simulations. LDS approach is even more frequently distributed compared with LHS [11] and the sampling error is lower [1], although LDS may have issues in high variable dimensions [12]. LHS-based, LDS-based, or mixed approach [13] indeed improves efficiency and accuracy of PRS-based MC, as demonstrated by literatures. However, these approaches still require many simulations such that the time cost of iterative yield-aware design flow is still expensive.

In this paper, efficient TSMC methodology and new yield-aware design flow are proposed to speed up analog designs for yield. This work is a simulation-based methodology for accurate yield estimation. The variation effects would be predicted quickly such that designers can fine-tune circuits earlier if the estimated yield is unsatisfied. Using this work, sign-off MC analysis with complete simulations is only performed once for detailed statistic information when the design yield was satisfied. Therefore, the proposed yield-aware design flow can be very efficient.

This methodology aims at "trimming samples" no matter which sampling technique is used in MC analysis. In other words, we do not improve the algorithms of LHS and LDS, but we keep their benefits and reduce their simulation time cost further. Figure 1 shows the proposed flow for fast MC yield analysis. First, we generate all samples according to the user-given parametric variations and simulation runs of MC. The proposed algorithm would trim these MC samples and pick out the possibly failed samples for simulations only. This algorithm includes the following steps:

- **Importance analysis:** First, this step approximates the relationships between input variables and the concerned performance by linear models. These models are used to pick out the samples whose performances are far from the design specification. This work treats such far-from-specification samples as pass/fail samples directly without simulations. According to these fail samples, the best possible yield (Y_{best}) and the worst performance case can be obtained "before" MC simulations. If Y_{best} is unsatisfied, it implies that the real yield after expensive MC simulations is certainly worse and more unacceptable due to only considering the far-from-specification samples. Therefore, this work can provide a quick yield reference for designers without running all MC simulations. This concept is useful to speed up the iterative yield checks during analog designs for yield.

- **Output partition & probability filter**: In contrast to far-from-specification samples, the around-specification samples may be numerous and critical for yield estimation. Therefore, circuit simulations are required to verify whether they pass the performance specification or not. However, simulate all these samples may be still inefficient. The steps of "Output partition" and "probability filter" would only select critical samples for simulations in order to reduce the time cost.

- **Sample grouping**: After above steps, the picked out samples are quite fewer than the original MC samples. "Sample grouping" would trim simples further to avoid running nearly duplicate simulations.

Finally, only the samples from "sample grouping" are required simulations to verify their circuit performances under process variations. Because these picked out samples have high probability to violate the specification, the analyzed yield from simulation results is always worse than the value of Y_{best}. This is why we say that Y_{best} is a reliable reference for designers to improve their circuits earlier before MC simulations. As demonstrated in experimental results, this work indeed estimate accurate results (yield value and the worst performance) compared with the sign-off MC analysis. Moreover, this methodology can combine with variance reduction techniques or low discrepancy sequences to speed up MC analysis further.

The remainder of this paper is organized as follows. Section 2 introduces the inspired ideas from the traditional MC analysis. The proposed trimmed-sample methodology is explained in Section 3. The novel yield-aware design flow is proposed in Section 4. Various experiments are provided in Section 5 for demonstration. Finally, some conclusions are given in Section 6.

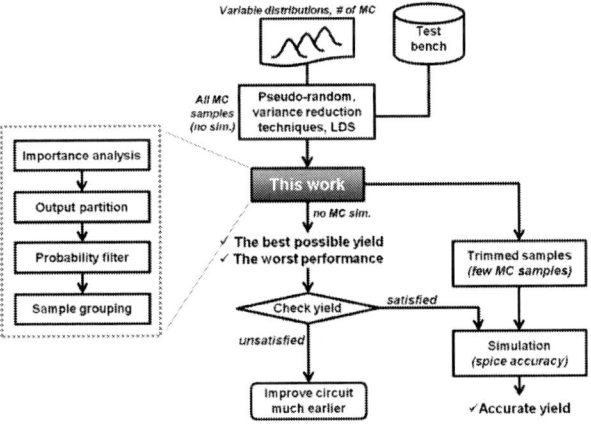

Figure 1: The proposed fast yield analysis flow

2. BACKGROUND

In this section, we use a single variable and a single output to explain the trimmed-sample concepts. This work can handle numerous variables, but single variable case is easily understood.

2.1 Monte Carlo Analysis

The variation of process parameter is often described as a Gaussian distribution in literatures and foundries' model cards. Figure 2 shows that Gaussian variable tends to cluster around its mean value (μ) such that the mean's probability is the highest. Far

from cluster, the probability of variable value is much low. The parameter σ in Figure 2 represents the standard deviation.

Monte Carlo methods for design yield analysis rely on repeated random sampling to estimate their effects of random process variations. Pseudo-random sampling (PRS) algorithm is used in traditional MC analysis to generate variable values from a specified Gaussian distribution. Then, circuit simulators use the sampled value to run a simulation and find the relative circuit performance. Therefore, the probability of output performance should be relative (not equal) to the probability of input variable no matter what the I/O relationship is. In order words, the distribution of circuit performance should also have a cluster and the probability of far-from-cluster is much lower than the cluster's probability, as illustrated in Figure 3. Actually, this probability feature had demonstrated by MC simulation results of various circuits [12-16].

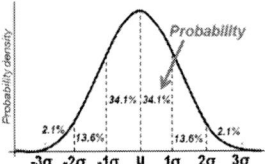

Figure 2: Probability density function of Gaussian distribution

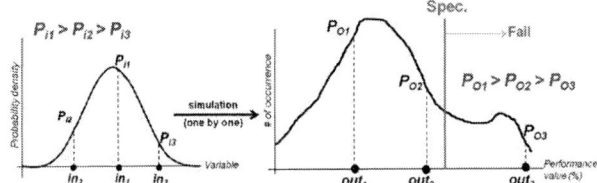

Figure 3: Probability relationship of MC analysis

Repeating the steps of sampling and simulation, an output distribution can be constructed after hundred or thousand of iterations. According to the design specification, the yield value can be calculated by $Yield \approx 1 - (N_{fail} / N_{total})$. N_{total} represents the number of total MC samples and N_{fail} is the number of failed samples which violate the specification. N_{total} is known because we must specify the simulation runs of MC for circuit simulators. Therefore, finding out the failed samples can obtain the yield value intuitively. It implies that simulating all MC samples for yield estimation may be not necessary because we can only simulate the critical samples, which have high probability to violate specification, to find the value of N_{fail}.

2.2 Trimmed-sample Concept

From above observations, the trimmed-sample concept in this paper is to pick out the possibly failed samples and simulate these samples only. Then, the yield value can be obtained because the number of failed samples is found from simulation results.

Figure 4 illustrates that the performance distribution can be classified as two regions if the specification value is known. The output performances far from specification imply that these samples certainly pass or fail the design requirement. Therefore, we can find the worst performance from these samples and estimate a rough yield value. Without considering the around-specification region, this rough yield is always better than the real value such that we called it "the best possible yield (Y_{best})" in this paper. If Y_{best} is worse than the targeted yield, designers can stop MC analysis because the real yield after MC is certainly worse and more unsatisfied.

If Y_{best} fits the yield requirement, the around-specification samples become critical and dominate the real yield value. However, simulating all samples in this region is still inefficiency. Therefore, this paper proposes an efficient TSMC methodology to reduce the simulation time cost. We utilize the probability features of MC analysis to pick out critical samples and group these samples to avoid the (nearly) duplicate simulations. This approach is independent on sampling algorithms such that users can combine with current variance reduction techniques or low discrepancy sequences to improve the efficiency of traditional MC approach further.

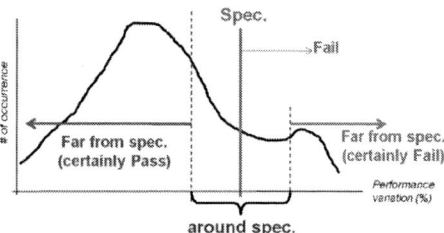

Figure 4: Categories of output distribution

3. TRIMMED-SAMPLE METHODOLOGY

This work aims at "trimming MC samples" to reduce simulation cost. Figure 5 illustrates the proposed TSMC methodology. The detailed steps will be explained in the following sections.

3.1 Importance analysis

The first step is to model the relationships between input variables and the concerned performance. This step would pick out the far-from-specification samples to provide Y_{best} value and the worst performance before running MC simulations. We change the value of process parameter by a small variation individually to find the variation effect on the concerned performance by simulation, which is sensitivity analysis. This small variation must be chosen carefully, such as 1σ of each variable's distribution, such that we can sort their importance by sensitivity values directly. Then, the process parameters are classified as three levels, which are *very important*, *important*, and *unimportant*. These levels would be considered as the weights to decide which samples would be gathered as a group in the step of "sample grouping".

For numerous variables, principle component analysis can be performed first to reduce the high-dimension MC samples by only using the first few principal components (PCs). Therefore, linear modeling cost in this work is negligible compared with the large number of MC simulations. Moreover, these independent PCs can also convert the correlated variables into uncorrelated parameters.

Therefore, correlations between input variables can be also handled in this work.

After importance analysis, the variable sensitivities (SE_{var}) are used for all MC samples to calculate the relative performance variations by (1). n is the number of variables, m is the simulation runs of MC, and ΔVar are the variable variations. Therefore, we can roughly know the output location of each sample at horizontal axis illustrated in Figure 4. Please note that we do not predict the output distribution because linear modeling errors in (1) may induce seriously erroneous judgment, especially for the around-specification samples. Therefore, we principally use (1) to find out the certainly pass/fail samples, whose output performances are far from specification. Then, Y_{best} and the worst performance sample can be obtained from these samples before running all MC simulations.

As to the around-specification samples, simulating all these samples is still inefficiency. Therefore, "output partition" and "probability filter" are proposed to pick out the critical samples and simulate them only in order to reduce the simulation time cost.

$$\begin{bmatrix} SE_{var_1} & SE_{var_2} & \cdots & SE_{var_n} \end{bmatrix} \begin{bmatrix} \Delta Var_{11} & \Delta Var_{12} & \cdots & \Delta Var_{1m} \\ \Delta Var_{21} & \Delta Var_{22} & \cdots & \Delta Var_{2m} \\ & & \cdots & \\ \Delta Var_{n1} & \Delta Var_{n2} & \cdots & \Delta Var_{nm} \end{bmatrix} = \begin{bmatrix} \Delta P_1 & \Delta P_2 & \cdots & \Delta P_m \end{bmatrix} \quad (1)$$

Sensitivity values   *All MC samples*   *Performance variations (only for partition)*

3.2 Output Partition

This section would partition the around-specification region into several parts based on the distance to performance specification (d) shown in Figure 6. The d value is relative to the distance from the nominal value to specification, which is called *margin* in this paper. The nominal value is the performance without any process variation effect. If the specification is closer to the nominal value, d should be smaller because checking the left side of nominal value in Figure 6 is not necessary to find failed samples. This work divides the *margin* into k segments such that d can be calculated by $d \approx margin / k$, where k is a user-given positive integer and *margin* is obtained by the specification value.

Region I in Figure 6 represents the next-to-specification region. Any erroneous judgment in this region would induce significant errors in yield estimation. In order words, the trimmed-sample risk is high such that we simulate all samples in Region I. Region II is the near-specification region, which has less risk of trimming simulation samples. We propose a filter function to pick out critical samples for partial simulations in this region. As to the far-from-specification region, we assume these samples certainly pass/fail design specification without simulations. Such assumption is sensible and explained in previous sections.

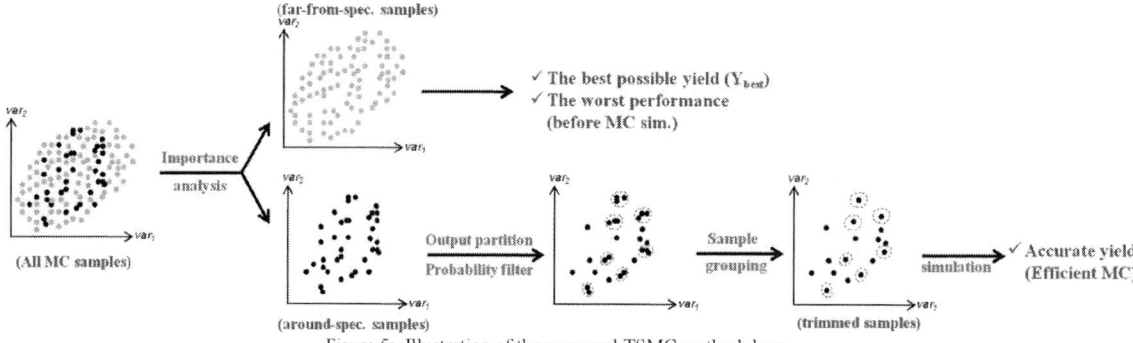

Figure 5: Illustration of the proposed TSMC methodology

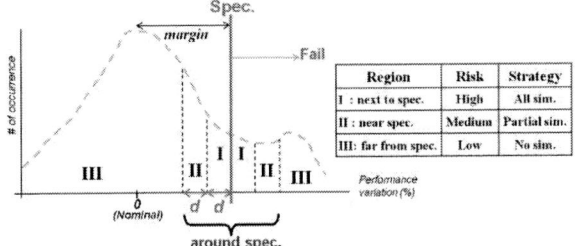

Figure 6: Output distribution partition

3.3 Probability Filter

Section 2 had pointed out the features of MC analysis: the probability of output performance is relative to the probability of input variables; the performances distributions have clusters and the probability of far-from-cluster is much low. We utilize these features and propose the "probability filter" to trim the samples in Region II. The normal probability samples would pass the filter criteria and be treated as pass/fail samples without simulations.

The first step is to find the probability of each MC sample. This work divides the probability of one input variable into several bins, as illustrated in Figure 7. These bins have the same interval at horizontal axis and the bin number must be odd for symmetry. All samples located in the same bin have the same probability. The probability value of arbitrary xy interval for a certain Gaussian variable (Var_z) is defined as P_{xy,Var_z}, as shown in Figure 7. The cumulative distribution function of Var_z describes the probability for a random variable to fall in the intervals of the form $(-\infty, x]$, where erf is Gaussian error function. Therefore, P_{xy,Var_z} can be calculated by (3) for a certain variable Var_z. Finally, the combination probability of all variables is expressed as the product of each variable's probability. Equation (4) shows the combination probability of m variables if all variables locate in the xy interval.

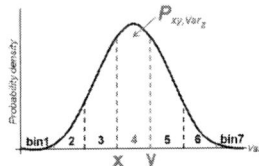

Figure 7: Interval's probability of certain variable (7 bins)

$$\Phi(x) = \frac{1}{\sqrt{2\pi}} \int_{-\infty}^{x} e^{-t^2/2} dt = \frac{1}{2}\left[1 + erf\left(\frac{x}{\sqrt{2}}\right)\right] \quad x \in R \quad (2)$$

$$P_{xy,Var_z} = \Phi(y) - \Phi(x) \quad (3)$$

$$P_{combine} = \prod_{z=1}^{m} P_{xy,Var_z} \quad (4)$$

We calculate the combination probabilities for all MC samples and normalize them by the highest probability. Then, we focus on the probabilities of samples in Region II and pick out abnormal cases for simulations. The probability far from the nominal value should be much low [12-16] such that high probability samples in Region II^F may be abnormal, as show in Figure 8. These abnormal cases may result from the error of location estimations in the step of "importance analysis". Therefore, high-probability samples in Region II^F are selected for simulations to verify whether they are

really failed. Just like a low (probability) pass filter, low-probability samples in Region II^F would pass the filter criteria to be treated as failed samples directly without simulations. In the same way, the low-probability samples in Region II^P can not pass the high (probability) pass function such that they need simulations for verification. Therefore, only partial samples in Region II require simulations in this work. As to the filter criteria, tighter criteria pick out more simulation samples and make the yield estimation more accurate with more simulation cost. These parameters are flexible for different users' requirement.

Finally, all MC samples are trimmed as a small subset, which is composed of all samples in Region I and partial samples in Region II. "Sample grouping" would pick out similar simples from this subset to avoid running (nearly) duplicate simulations.

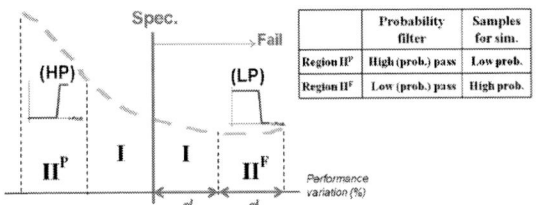

Figure 8: Probability filter for Region II

3.4 Sample Grouping

Due to large number of simulations, MC analysis may generate similar input samples and simulate them repeatedly. Especially for traditional MC analysis, pseudo-random generator has worse sampling capability such that there may be many similar input samples with similar output performances. The "sample grouping" step would gather such samples and record the sample count. Then, these samples are replaced by a single sample such that only one simulation is required for one group of similar samples. [12, 16] demonstrated that sample grouping is indeed efficient in reducing simulation runs. However, grouping "all" MC samples to predict the shape of output distribution may have significant errors in yield estimation under a certain specification value.

This work only focuses on the picked out samples, which are the small part of all MC samples. Therefore, the grouping error should affect the yield estimation slightly. Moreover, we do not predict the performance distribution such that this work has well accuracy under different specifications. Most importantly, this work can handle numerous variables, not only a single variable [12, 16].

4. YIELD-AWARE DESIGN FLOW

Traditional MC analysis performs random sampling and run circuit simulation repeatedly. Such repeating action is not stop until finish running all MC samples. After hundred or thousand of simulations, designers can obtain the yield value and the worst performance under process variations. If the analyzed results are not satisfied, they need to improve the design and run the time-consuming MC simulations again. Such yield-aware design flow becomes almost infeasible for large analog circuits because iterative MC simulations are very luxurious.

Therefore, designers may need a fast analyzer to know the parametric variation effects quickly. This paper proposes a novel design flow to speed up the process of iterative yield checks. The proposed trimmed-sample methodology can estimate the best possible yield (Y_{best}) and the worst performance value before running MC simulations. If the analyzed results are not satisfied, designers can improve their circuit immediately because the real

1116

yield after MC is certainly worse. Therefore, this work can save the MC simulation cost if Y_{best} is worse than the targeted yield. The worst performance value is also a good reference for fine-tuning the design. As to the exact yield value, only few simulations are required in this work to estimate accurate yield compared with traditional MC analysis. After several design iterations, sign-off MC analysis is only performed for more detailed analysis if necessary. Therefore, the proposed yield-aware design flow should be very useful for time to market.

5. EXPERIMENTAL RESULTS

We compare the proposed TSMC methodology with traditional MC analysis by using different analog designs. Some equation-based experiments can demonstrate that this work can handle numerous variables. We also use the LHS approach as an example to show that this work indeed reserve the benefit of variance reduction techniques and reduce the simulation time cost further.

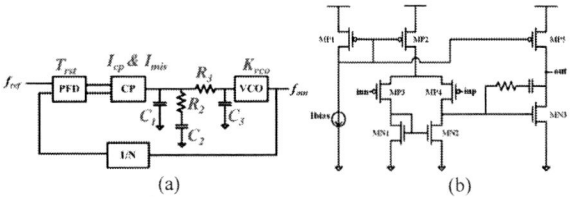

Figure 9: (a) 4th-order PLL behavior model (b) two-stage opamp

5.1 PLL System Design

For phase-locked loop (PLL) circuits, behavior-level MC analysis is a conventional approach to quickly estimate the PLL yield under process variations [12, 15-16]. We build a 4th-order PLL behavior model and let all behavior parameters have ±15% variations compared with their nominal values. There are total 9 independent variables shown in Figure 9 (a). These variables are assumed as Gaussian distributions with ±3σ variations. Locking time (T_{lock}) is the concerned performance and the nominal value (T_{lock0}) is 5.25μs. This PLL model is described by Verilog-A and Spectre is the simulator.

First, we perform 1k-run traditional MC simulations by using pseudo-random sampling (PRS) methodology and consider the analysis result as golden accuracy. The targeted design yield is > 95%. In this experiment, we sweep the T_{lock} specification to demonstrate that this work estimate yield accurately no matter what specification is. The worst T_{lock} value is 6.74μs obtained from 1k-run simulations.

From these 1k samples, we pick out the far-from-specification samples to estimate the Y_{best} value. For the worst case analysis, we select the top ten worst samples from the far-from-specification region by (1) and run simulations to obtain their performances. Table 1 shows the number of simulation for the worst and 0% error means that this work exactly finds the worst case without all MC simulations. One experiment has fewer simulations than ten because there are only seven failed samples far from specification. The last experiment has no any failed sample in the far-from-specification region such that the Y_{best} value is 100%. The worst performance in this case is obtained from the simulation results of TSMC analysis. The worst T_{lock} error is only +1.3% means that the trimmed-sample methodology is indeed accurate.

In Table 1, some Y_{best} values are already worse than the yield requirement (>95%) such that designers can stop MC analysis to improve their circuit immediately. We finish the following MC simulations to show that TSMC methodology can obtain accurate

yield with less simulation time cost. Table 1 shows that the maximum yield error of this work is only 0.6% compared with the golden values. Most importantly, this work indeed speeds up traditional MC analysis (12x to 31x), especially in the high yield cases.

Table 1. Results of "PRS+ this work" under different specifications

T_{lock} Specification	Y_{best}	# of sim. for worst case	Error of worst case	This work Yield	Golden Yield	Speedup ratio
< T_{lock0} (1+5%)	75.1%	10	0%	68.7%	68.7%	*18.5x
<T_{lock0}(1+10%)	90.8%	10	0%	82.8%	83.4%	*11.8x
<T_{lock0}(1+20%)	99.3%	7	0%	97.5%	97.7%	12.7x
<T_{lock0}(1+30%)	100.0%	-	+1.3%	99.9%	100.0%	31.3x

* MC Simulations may be not necessary because Y_{best} is already unsatisfied

5.2 Two-Stage Opamp Circuit

The second experiment is using a two-stage opamp shown in Figure 9 (b). The width (±30%), channel length (±30%), oxide thickness (±3%), threshold voltage (±30%) and RC values (±15%) are considered as design variables. We also consider the mismatch of oxide thickness and threshold voltage for differential pairs such that there are total 18 variables. The gain bandwidth product (GBW) is the concerned performance and the nominal value (GBW_0) is 34.64MHz. The performance specification is GBW > GBW_0(1–10%). Spectre is the circuit-level simulator and the targeted yield is > 97.5%.

We perform 2k-run PRS-based MC simulations and consider the analyzed results as golden accuracy. Then, this work trims these 2k samples such that only 213-run simulations are required to estimate the yield. Figure 10 shows the GBW histograms of 2k-run PRS-based MC simulations and 213-run TSMC simulations (PRS + this work). Before MC simulations, the trimmed-sample methodology picks out twelve failed samples in the far-from-specification such that Y_{best} value is 99.40% and the worst GBW is 29.68MHz. After 213-run TSMC simulations, thirty-one failed samples are found such that the total failed samples are forty-three and the estimated yield of this work is 97.85%. Table 2 shows that this work gives 9x speedup over traditional MC analysis and the yield error is only 0.1%.

The next experiment is using Latin Hypercube Sampling (LHS), which has better sampling capability, to generate MC samples. LHS can get a comparable accuracy in yield analysis with just 20% samples compared with using PRS method [7, 9-10]. Therefore, we perform 400-run LHS-based MC simulations in this experiment. Table 2 shows that the LHS approach indeed estimate similar results compared with PRS-based MC approach.

Finally, we combine this work with the LHS approach to demonstrate that the proposed methodology can reserve the benefit of variance reduction techniques and reduce the simulation time cost further. Such experiment is labeled as "LHS+ this work" in Table 2. The 400 LHS samples are trimmed as 59 samples to run simulations for yield estimation. The worst GWB and yield value are all similar to the results of LHS-based MC. Most importantly, this work has forty times speedup (40x) over traditional MC analysis with similar accuracy.

Table 2. Compare this work with traditional MC analysis

Specification: GBW > GBW_0(1–10%)	Y_{best}	Worst GBW	# of MC sim.	Real Yield
PRS-based MC (golden)	-	29.68MHz	2000	97.75%
PRS + this work	99.40%	29.68MHz	213	97.85%
LHS-based MC	-	29.25MHz	400	97.50%
LHS + this work	99.50%	29.25MHz	50	97.50%

1117

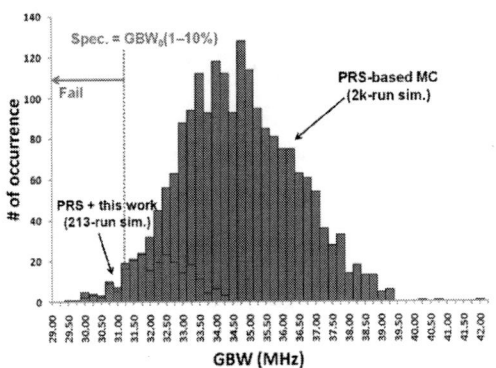

Figure 10: *GBW* histograms: PRS-based MC & PRS + this work(sim. part)

5.3 Mathematic Experiments

The 2nd-order response surface model (RSM) in (5) is often used to model the relationships between process parameters and circuit responses for analog circuits [2-5]. In this experiment, there are total 14 different RSM equations with different number of variables and coefficient values. The parameter x are the input variables, β represent the coefficients and y is the output of RSM. The mean value of each variable is 10 and the variation range is ±30%. The coefficients are randomly generated from −100 to +100 for each RSM equation. Therefore, some variables have strong effects on the output response and some are weak if coefficient values are near 0. For each equation, 10k-run PRS-based MC analysis are used to find as golden yield values.

$$y = \beta_0 + \sum_{i=1}^{k} \beta_i x_i + \sum_{i=1}^{k} \beta_{ii} x_i^2 + \sum_{i<j} \beta_{ij} x_i x_j \qquad (5)$$

Then, we trim these 10k samples for fast yield analysis, which is marked as "PRS+ this work" in the following figures. This work gives 6x to 18x speedup and the yield estimation errors are all smaller than 0.5% even if the number of variables is 1000.

The next experiment is to perform 2k-run LHS-based MC simulations for comparison. Figure 11 and 12 show that the LHS approach estimates similar yield values with 5x speedup. Then, we combine this work with LHS to reduce the MC cost further. The speedup ratio becomes 29x to 54x compared with traditional MC analysis and the analyzed yields are all closed to the results of LHS-based MC analysis. These 14 different RSM equations can also demonstrate that the trimmed-sample methodology is flexible for various cases with large number of variables.

6. CONCLUSION

This paper proposes efficient trimmed-sample MC (TSMC) methodology and novel yield-aware design flow for analog circuits. Before MC simulations, the entire MC samples are trimmed as a small part to reduce the simulation runs. Moreover, the best possible yield and the worst performance are provided as design references such that users can improve circuits earlier without MC simulations. For exact yield estimation, only few simulations are required in the proposed TSMC methodology. Combining with LHS as an example, this work gives 29x to 54x speedup over traditional MC analysis and the yield estimation errors are all smaller than 1%, as demonstrated in various experiments. For analog system designs, the proposed yield analysis flow is still efficient and suitable for high-level analysis. In future work, we will extend the trimmed-sample concepts to handle the yield analysis of multi-specification.

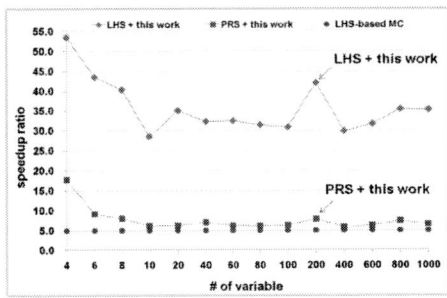

Figure 11: Speedup ratio compared with traditional MC approach

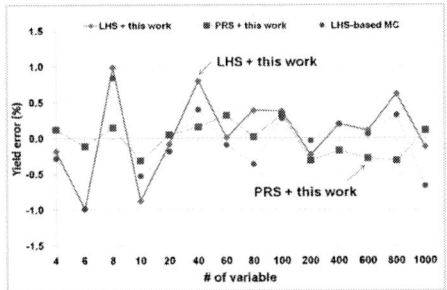

Figure 12: Yield error compared with traditional MC approach

7. REFERENCES

[1] A. Singhee and R. A. Rutenbar, "Why Quasi-Monte Carlo is Better than Monte Carlo or Latin Hypercube Sampling for Statistical Circuit Analysis," *IEEE Trans. Comput.-Aided Des. Integr. Circuits Syst.*, vol. 29, no. 11, pp. 1763-1776, 2010.

[2] G. Yu and P. Li, "Efficient look-up-table-based modeling for robust design of ΣΔ ADCs," *IEEE Trans. Circuits Syst. I, Reg. Papers*, vol. 54, no. 7, pp. 1513–1528, 2007.

[3] Z. Feng, P. Li, "Performance-oriented statistical parameter reduction of parameterized systems via reduced rank regression", in *Proc. Int. Conf. Computer-Aided Design*, 2006, pp. 868–875.

[4] X. Li, J. Le, L. T. Pileggi, and A. Stojwas, "Projection-based performance modeling for inter/intra-die variations," in *Proc. IEEE/ACM Int.Conf. CAD*, 2005, pp. 721–727.

[5] T. Fujita, K. Okada, H. Fujita, H. Onodera, and K. Tamaru, "A method for linking process-level variability to system performance," in *Proc. Asia South Pacific Design Automation Conf.*, 2000, pp. 547–551.

[6] S. Tezuka, *Uniform Random Numbers: Theory and Practice*. Boston, MA: Kluwer, 1995.

[7] J. Jaffari and M. Anis, "On Efficient LHS-Based Yield Analysis of Analog Circuits," *IEEE Trans. Comput.-Aided Des. Integr. Circuits Syst.*, vol. 30, no. 1, pp. 159–163, 2011.

[8] A. Matala, "Sample Size Requirement for Monte Carlo – simulations using Latin Hypercube Sampling," *Helsinki University of Technology, Syst. Analysis Laboratory*, 2008.

[9] J. F. Swidzinski, M. Keramat, and K. Chang, "A novel approach to efficient yield estimation for microwave integrated circuits," in *Proc. IEEE Midwest Symp. Circuit Syst.*, 1999, pp. 367–370.

[10] B. Liu, F. V. Fernández, and G. E. Gielen, "Efficient and Accurate Statistical Analog Yield Optimization and Variation-Aware Circuit Sizing Based on Computational Intelligence Techniques," *IEEE Trans. Comput.-Aided Des. Integr. Circuits Syst.*, vol. 30, no. 6, pp. 793–805, 2011.

[11] HSPICE® User Guide: Simulation and Analysis, version E-2010.12, Dec. 2010.

[12] R.O. Topaloglu, "Early, accurate and fast yield estimation through Monte Carlo-alternative probabilistic behavioral analog system simulations," in *Proc. IEEE VLSI Test Symp.*, 2006, pp. 137–142.

[13] V. Veetil, K. Chopra, D. Blaauw, and D. Sylvesterm, "Fast Statistical Static Timing Analysis Using Smart Monte Carlo Techniques," *IEEE Trans. Comput.-Aided Des. Integr. Circuits Syst.*, vol. 30, no. 6, pp. 852–865, 2011.

[14] F. Centurelli, P. Monsurrò, and A.Trifiletti, "Behavioral Modeling for Calibration of Pipeline Analog-To-Digital Converters," *IEEE Trans. Circuits Syst. I, Reg. Papers*, vol. 57, no. 6, pp. 1255-1264, 2010.

[15] C.-C. Kuo, M.-J. Lee, C.-N. Liu, and C.-J. Huang, "Fast statistical analysis of process variation effects using accurate PLL behavioral models," *IEEE Trans. Circuits Syst. I, Reg. Papers*, vol. 56, no. 6, pp. 1160–1172, 2009.

[16] R. O. Topaloglu, "Monte carlo-alternative probabilistic simulations for analog systems," In *Int. Symp. Quality Elect. Des.*, 2006, pp. 249–253.

1118

Towards Efficient SPICE-Accurate Nonlinear Circuit Simulation with On-the-Fly Support-Circuit Preconditioners

Xueqian Zhao
Department of ECE
Michigan Technological University
Houghton, MI, 49931
xueqianz@mtu.edu

Zhuo Feng
Department of ECE
Michigan Technological University
Houghton, MI, 49931
zhuofeng@mtu.edu

ABSTRACT

SPICE-accurate simulation of present-day large-scale nonlinear integrated circuit (IC) systems with millions of linear/nonlinear components can be prohibitively expensive, and thus extremely challenging. In this paper, we present a novel support-circuit preconditioning (SCP) technique for tackling large-scale nonlinear circuit simulations by exploiting sparsified graphs of a given circuit network. By extracting support graphs (SGs) from the original linear circuit networks, and combining them with nonlinear devices, support-circuit preconditioner can be efficiently computed using existing matrix solvers, allowing for on-the-fly updates during transient simulations when adopted in Krylov-subspace iterative solvers. Experimental results for a variety of large-scale circuit designs show that the proposed method achieves up to $22X$ speedups in solving the matrices involved in DC and transient (TR) simulations, and up to $8X$ reduction in memory usage, when compared with the simulator powered by the state-of-the-art direct solver KLU.

Categories and Subject Descriptors

B.7.2 [**Design Aids**]: simulation—*Integrated Circuits*

General Terms

Algorithms

Keywords

Transient Simulation, Iterative Methods, Preconditioner

1. INTRODUCTION

As the relentless technology scaling reaches into the nanoscale regime, integrated circuit (IC) designs have been hindered by the every-increasing design complexity brought by billions of transistors and interconnect components that need to be accurately modeled and analyzed. SPICE-accurate simulation of nowadays large-scale nonlinear integrated circuit (IC) systems with millions of linear/nonlinear circuit components can still be prohibitively computationally expensive, and thus extremely challenging. Traditional SPICE-accurate simulators are typically built upon reliable direct

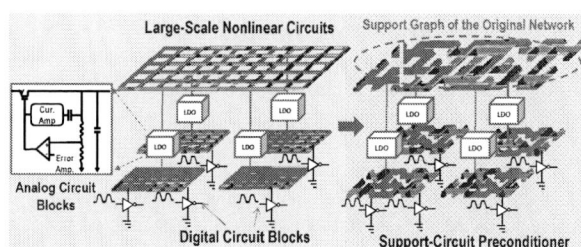

Figure 1: Support-circuit preconditioner for SPICE-accurate nonlinear circuit simulation.

sparse matrix solvers. Unfortunately, solving such circuit problems usually requires factorizing extremely large yet less sparse matrices derived from the circuit networks, which may not be runtime and memory efficient. For instance, a state-of-the-art LU solver, KLU [1], can be very efficient for solving moderate-sized circuit matrices (over $10X$ faster compared to Sparse1.3) but still exhibit poor scalability with large-scale problems due to the exponentially increased memory and runtime cost [1].

As alternatives, iterative matrix solvers have been proposed for large-scale circuit simulations based on Krylov-subspace iterative methods, such as GMRES algorithm [2], to achieve better memory efficiency as well as parallelism [3–6], but their performance usually depends on the effectiveness and efficiency of the preconditioners. As shown in [4], a parallelizable preconditioning algorithm for transient analysis is proposed by obtaining the block-triangular form (BTF) according to the sparse matrix structure at DC solution point, and reusing the sparse matrix pattern for preconditioner constructions throughout the following transient simulation steps. However, during nonlinear circuit transient simulations, the nonlinear dynamic systems can vary dramatically, which may make the BTF preconditioner less effective and sometimes lead to divergence, as reported in [4]. As demonstrated in [7], incomplete matrix factor preconditioners are usually less robust, since the matrix fill-ins of the preconditioner are merely determined based on their magnitudes instead of the more important circuit design information, such as topologies and connectivities.

In this work, we present a novel circuit-oriented *support-circuit* preconditioning technique that is inspired by recent state-of-the-art support graph and graph sparsification research [8–10]. Compared with previous matrix-oriented preconditioners, the proposed support-circuit preconditioner (SCP) can be more efficiently constructed through the following steps, as shown in Fig. 1: (1) extract ultra-sparsifier support graphs (SGs) from the original linear circuit networks, which is one-time task and thus induces negligible cost; (2) combine the support-graph circuits with nonlinear devices to form the complete *support circuit*; (3) use existing sparse matrix solvers [1] to compute the full LU matrix factors for the sparse matrices of the support circuit, and use them as *support-circuit preconditioners* in the preconditioned Krylov-subspace iterative solvers, such as GMRES solvers [2]. It should be emphasized that this circuit-oriented preconditioner will be particularly useful when solving *parasitics-dominant circuits*, since the circuit graphs can be significantly sparsified and their preconditioners can be efficiently updated on the fly during transient simulations. Our experimental results show that the proposed technique achieves up to $22X$ speedups for large-scale DC and transient (TR) nonlinear circuit analysis, and can significantly reduce the memory cost by a factor of eight when compared with the SPICE simulator powered by the state-of-the-art direct matrix solver KLU [1].

2. BACKGROUND ON NONLINEAR CIRCUIT SIMULATIONS

In this section, the background of nonlinear circuit simulation using direct and iterative methods will be reviewed. In the last, the proposed support-circuit preconditioner will be briefly introduced.

2.1 Nonlinear Circuit Simulation Methods

General nonlinear electronic circuit simulation techniques rely on Newton-Raphson method to solve the following nonlinear differential equations:

$$f(x(t)) + \frac{d}{dt}q(x(t)) + u(t) = 0, \qquad (1)$$

where $f(\cdot)$ and $q(\cdot)$ denote the static and dynamic nonlinearities, $x(t)$ is a vector including nodal voltages as well as branch currents, and $u(t)$ is the input excitation vector. Sophisticated numerical methods can be used to solve the above nonlinear differential equations by first linearizing the nonlinear circuit system at a given solution point, and subsequently solving the corresponding linear matrix problems. For instance, after linearizing the system, conductance matrix $G\left(x^k\right) = \frac{\delta f}{\delta x}\big|_{x^k}$ and capacitance matrix $C\left(x^k\right) = \frac{\delta q}{\delta x}\big|_{x^k}$ can be easily obtained which are typically asymmetric matrices. The major computational cost for solving small circuit problems is mainly due to the nonlinear device evaluations, while for much larger circuits solving the asymmetric Jacobian matrices using direct solution method can be much more expensive due to the exponentially increased runtime and memory cost [1].

Iterative methods [3, 4, 11] exhibit much higher memory efficiency since only the sparse matrix and a few vectors are stored during the computation, but can suffer from slow convergence or divergence issues. It is critical to construct effective preconditioners on the fly when using Krylov-subspace iterative solvers, which can effectively control the number of preconditioned GMRES iterations during transient simulations [4]. Unfortunately, generating effective matrix factorization-based preconditioners on the fly can be prohibitively expensive and not practical for large-scale problems.

2.2 Support-Circuit Preconditioner

In this work, unlike existing matrix-oriented preconditioning methods, we present a circuit-oriented preconditioning method that is inspired by recent state-of-the-art support graph and graph sparsification research [8–10]. Compared with previous matrix-oriented preconditioners, the proposed support-circuit preconditioner (SCP) can be very effective for both DC and transient simulations, while the involved computational cost of this preconditioner is very low, and therefore can be frequently updated on the fly during transient simulations.

The basic idea behind our approach is to sparsify the linear networks by applying advanced support graph theory [9], and subse-
quently combine the sparsified linear networks (also called ultra-sparsifiers [8]) with nonlinear devices to form the support circuit that is very close to the original circuit system. Finally, matrix factorization for the support circuit can be quickly generated in near-linear time, and used for the preconditioning procedures of GMRES iterations [2].

It is worth noting that the proposed preconditioned iterative simulation method is very robust and targeting at SPICE-accurate simulations: for every linearized system during Newton-Raphson iterations, we generate the latest support-circuit preconditioner and solve the corresponding linearized system iteratively, so there is no approximation made throughout the computations.

3. SUPPORT GRAPHS FOR LINEAR CIRCUIT NETWORKS

3.1 Support Graph

The support graph is usually a subgraph extracted from the original graph. Support-graph preconditioning is to first construct a graph according to a given Laplacian matrix A, and then extract the support graph and subsequently build a preconditioner P based on the support graph. For instance, maximum spanning tree or low-stretch spanning tree can be used as a support graph, which have been proposed for solving linear symmetric diagonally-dominant (SDD) matrices in [12, 13].

Support-graph preconditioning seeks to compute a preconditioner P such that the generalized eigenvalues and the condition number of the matrix pencil (A, P) are bounded [9]. If both A and P are symmetric positive definite (SPD) matrices, the convergence depends on the condition number $\kappa(A, P)$ computed by

$$\kappa(A, P) = \lambda_{max}(A, P)/\lambda_{min}(A, P), \qquad (2)$$

where $\lambda(A, P)$ denotes the generalized eigenvalues. A stronger theoretical result on convergence can be derived as follows. Define the support of (A, P), denoted by $\sigma(A, P)$, as follows [9]:

$$\sigma(A, P) = min\{\tau \in \mathbb{R} | x^T(\tau P - A)x \geq 0 \ for \ all \ x \in \mathbb{R}^n\}. \qquad (3)$$

Subsequently, if one can split A and P into $A = A_1 + A_2 + ... + A_m$ and $P = P_1 + P_2 + ... + P_m$ such that all $\tau P_i - A_i$ are positive semidefinite matrices, one can show that the largest generalized eigenvalue of (A, P) is bounded by τ [9].

3.2 Ultra-Sparsifier Support Graph

A weighted graph $\tilde{P}_{graph}$ α-approximates a weighted graph A_{graph} if

$$\mathcal{L}(\tilde{P}_{graph}) \preceq \mathcal{L}(A_{graph}) \preceq \alpha\mathcal{L}(\tilde{P}_{graph}), \qquad (4)$$

where $\mathcal{L}(A_{graph})$ is the Laplacian matrix A of A_{graph} [8], and $\tilde{P} \preceq A$ means that for all the $x \in \mathbb{R}^n$,

$$x^T \tilde{P} x \leq x^T A x, \qquad (5)$$

where the Laplacian quadratic form of A_{graph} is

$$x^T A x = \sum_{\text{edge } i} w_i (\Delta x_i)^2. \qquad (6)$$

In linear circuit systems, if A_{graph} denotes a linear circuit network, $x^T A x$ means the total power dissipated in the resistive network. Note that, compared to the original graph, the support graph has fewer edges. For instance, the spanning tree of a graph that includes n vertices and m ($m \geq n$) edges retains only $n - 1$ edges, thus the power dissipated on the support graph is much smaller than the power dissipated in the original system. If a preconditioner can approximate not only the eigenvalues of the original system, but also the power dissipation, the preconditioner can be more effective [8].

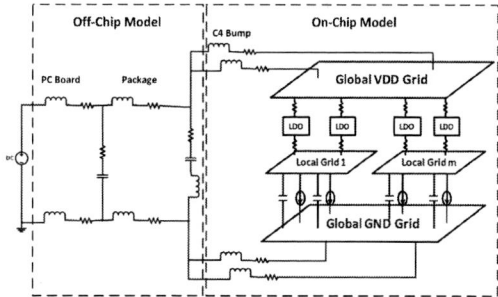

Figure 2: Die-package PDN model with on-chip LDO packages.

4. SUPPORT-CIRCUIT PRECONDITIONER FOR NONLINEAR CIRCUIT ANALYSIS

In this section, the support-circuit preconditioned GMRES method for general nonlinear circuit analysis will be described in details.

4.1 Support Circuit

Preconditioned iterative solvers have been proposed by using the Jacobian matrix obtained at the initial Newton-Raphson step of DC analysis as the preconditioner for DC and transient analysis [3]. Due to the extremely high computational cost, it is not practical to create and factorize the preconditioner on the fly. However, during transient simulations, the DC Jacobian matrix can be quite different from the realistic Jacobian matrices, resulting in a less robust/effective preconditioner for iterative solvers. It has been shown that using the constant preconditioner or the sparse matrix pattern obtained at DC solution point in transient simulations may result in slow convergence, and sometimes even divergence [3,4].

4.1.1 A Motivating Example for A Simple Support Circuit

First, let us review the model of a power delivery network (PDN) with on-chip low-dropout regulators (LDOs) and package parasitics [11], as shown in Fig. 2. The global VDD grid and local grids are coupled with on-chip analog LDO circuit modules which consist of tens of thousands of transistors operating in saturation region, while local grids and global GND grid are coupled with active digital functional gates. All the on-chip PDNs are connected with C4 bumps which further get connected with PC package-level power supply wires. To build the proposed support-circuit preconditioner for the above large nonlinear circuit system, we can first extract the support graphs using maximum spanning trees of the linear resistive networks (power grid meshes), and subsequently combine them with other nonlinear devices (LDO modules and digital functional blocks) and package parasitics to form a new nonlinear circuit system, *support circuit*, that has the same number of nodes but a much sparser system matrix. It therefore becomes much easier to factorize and solve the support-circuit system than the original circuit during the preconditioned GMRES iterations of each Newton-Raphson iteration, as illustrated in Fig. 1.

4.1.2 Ultra-Sparsifier for Building A Much Better Support Circuit

Generating the spanning trees requires removing noncritical interconnects (wires) from the original linear networks, which will reduce the overall conductivities of these networks. As described in Section 3.2, the conductance matrix corresponding to the spanning tree, P_{graph} of A_{graph} is 0-ultra-sparse, and α-approximates the A_{graph}, where α can be very large. To further improve the approximation of the spanning-tree support graph, the energy-conservation

principles and weighted degree lemmas [10] have been studied and deployed in this work. The power (Joule heat) dissipated by the linear circuit network can be calculated through

$$P_{Joule} = \sum A_{ij}(x_i - x_j)^2 = x^T A x, \qquad (7)$$

where x is the nodal voltage vector and A is the Laplacian matrix.

It is obvious that by adding extra edges to the initial spanning-tree support graph, the new conductance matrix $\tilde{P}_{graph}$ can better approximate the power dissipated by the original network graph A_{graph}. For example, (4) indicates that if the power consumption of the original network can be exactly matched by the support graph, α equals to 1. Consequently, a much better support circuit can be built by adopting new support graphs by adding extra edges to the original spanning-tree support graphs. Note that if all the removed edges are restored for building the ultra-sparsifier support graph, the support circuit can exactly approximate the original circuit. However, factorizing the resultant less sparse matrix is usually much more time consuming. Therefore, there is a need to tradeoff between the effectiveness of the preconditioner and the runtime efficiency of the matrix factorization algorithms.

4.1.3 Support Circuit Construction

Algorithm 1 Ultra-sparsifier support circuit construction algorithm.

Input: Linear circuit netlists and nonlinear devices netlist.
Output: Support circuit netlist.

1: Initialization phase:
　a) Load the netlist.
　b) Build the conductance matrices of the linear networks.
2: **for** linear network k: **do**
3: 　Arbitrarily pick one starting node and compute the maximum spanning tree.
4: 　**for** node $i \in$ network k: **do**
5: 　　Compute the weighted degree $wd(i)$ with (8).
6: 　　Set $mark(i)$ to 0.
7: 　**end for**
8: 　Compute the average weighted degree $awd(k)$ with (9) for network k.
9: 　**for** node $i \in$ grid k: **do**
10: 　　**if** $wd(i) > awd(k)$ & $mark(i) < uplink$ **then**
11: 　　　Find the unpicked most weighted edge and add it to the support circuit.
12: 　　　$mark(i)$++.
13: 　　**end if**
14: 　**end for**
15: **end for**
16: Combine the support graph netlists with nonlinear devices netlist.
17: Return the final support circuit netlist.

When building maximum spanning trees, the most critical conductive edges of each node will be selected. However, this method does not consider the original conductivities and degrees of the each node, which in turn may lead to very coarse approximation of the original system. To address the above issue, in the process of finding the ultra-sparsifier support graph, we use a *weighted degree* metric to select the most important edges (wires) to be added to the prior spanning trees. The weighted degree of a vertex v in a graph A_{graph} is defined as the ratio [10]

$$wd(v) = \frac{vol(v)}{\max_{u \in N(v)} w(u, v)}, \qquad (8)$$

where $vol(v)$ denotes the total weight incident to the node v, and $w(u, v)$ denotes the weight of the edge connected to node v and node u. The average weighted degree of the graph is defined as [10]

$$awd(A_{graph}) = \frac{1}{n} \sum_{v \in V} wd(v), \qquad (9)$$

where V denotes the set of all the vertices in graph A_{graph}, and n is the total number of vertices. In a 2D mesh grid, each node has four connected edges. Thus the weighted degree of a node k should be within the range that $1 \leq wd(k) \leq 4$. If the weighted degree of node k approaches one, it indicates that node k has only one dominating edge that has to be chosen in the maximum spanning tree, and as a result, node k is not a critical node. However, if the weighted degree of node k approaches four, it indicates that node k has four evenly critical edges, and node k is therefore a critical node. Consequently, for the maximum spanning tree of a 2D mesh network that includes only one or two of the four edges, restoring some of the unpicked edges of the critical nodes can effectively improve the support circuit approximation.

We conclude the proposed ultra-sparsifier support circuit construction procedures in Algorithm 1. In our algorithm, $mark(i)$ of node i is the number of extra edges that have already been added to the critical node i. $uplink$ is a user-defined parameter denoting the upper bound of $mark(i)$. For instance, $uplink$ can be set to 1 to achieve the best runtime performance for 2D mesh circuits. It is worth noting that for networks (3D meshes) with higher node degrees, $uplink$ should be set with greater values accordingly.

4.2 The Support-Circuit Preconditioned GMRES Algorithm

Algorithm 2 Support-circuit preconditioned GMRES method.

Input: Original and support circuit netlist.
Output: Circuit simulation solution.

1: Initialization phase:
   a) Load original and support circuit netlists.
   b) Build the device lists for original and support circuits.
   c) Allocate the sparse matrices for the original circuit and the support circuit.
2: **while** for a Newton-Raphson iteration during DC or Transient analysis: **do**
3:    Evaluate devices and compute linearized circuit systems.
4:    Create the original circuit matrix A and the support-circuit matrix $\tilde{A}$.
5:    Preconditioned GMRES iterations start:
      a) Factorize $\tilde{A}$ using full LU decomposition algorithm to get the preconditioner.
      b) Perform preconditioned GMRES iterations by resolving $\tilde{A}$.
6:    Update the solution vector.
7:    Check convergence.
8: **end while**
9: Return the solution.

The complete algorithm flow of the proposed support-circuit preconditioned GMRES method is concluded in Algorithm 2. We note that since the support circuit is usually much sparser than the original circuit, solving the support circuit is usually much cheaper than solving the original system, leading to more desirable near-linear runtime and memory efficiency.

4.3 Adaptive Preconditioner Updating

To further improve the simulation efficiency, we propose an adaptive preconditioner update technique. Since between two Newton-Raphson steps, the system Jacobian matrices may not change too much. Consequently, the previous preconditioner can be reused at a cost of N_{gmres_extra} more GMRES iterations, and save the runtime (t_{factor}) for preconditioner re-factorization. Note that the runtime (t_{factor}) for preconditioner factorization is much greater than the runtime (t_{gmres}) for one GMRES iteration. Thus, the allowed maximum extra number of GMRES iterations can be calculated by

$$N_{extra\_allowed} = \frac{t_{factor}}{t_{gmres}}. \qquad (10)$$

Consequently, if the extra number (N_{gmres_extra}) of GMRES iterations required by using the previous preconditioner exceeds the threshold number (N_{bound}), then the preconditioner needs to be re-evaluated and re-factorized.

The key ideas of the proposed adaptive support circuit preconditioned GMRES method are summarized as follows, where N_{cur} denotes the number of GMRES iterations in current Newton-Raphson step, and N_{prev} denotes the number of GMRES iterations in previous Newton-Raphson step.

1. The parameter N_{bound} is determined by runtime for re-factorizing the preconditioner and the runtime for performing one GMRES iteration. N_{bound} should be no greater than $N_{extra\_allowed}$.

2. Using the previous preconditioners may result in more GMRES iterations, but can also save preconditioner factorization time. If $N_{gmres\_extra} \leq \frac{t_{factor}}{t_{gmres}}$, where $N_{gmres\_extra} = N_{cur} - N_{prev}$, the adaptive method can effectively reduce the runtime.

3. In a Newton-Raphson step, if using the previous preconditioner costs more than N_{gmres_extra} iterations, the preconditioner needs to be re-evaluated and re-factorized.

5. EXPERIMENTAL RESULTS

In this section, extensive experiments have been conducted to evaluate our proposed support-circuit preconditioned GMRES method for the following three large-scale nonlinear circuits with different problem sizes and nonlinearities. For all the following test cases, transistors are evaluated using BSIM4 models while the power gating circuits, LDO circuits and clock buffers are implemented in a standard $90nm$ CMOS technology.

1. Power-Gated PDN. The modeling of die-package power-gated PDN is similar to the one shown in Fig. 2 by replacing the on-chip LDO packages with sleep transistors, where the package parasitics parameters are obtained from [14]. The power-gated PDN contains package power delivery interconnects and on-chip power delivery network. The package PDN models the parasitics of the PC board power supply interconnects that connect on-chip PDN through multiple VDD/GND C4 bumps. The on-chip PDN contains global VDD/GND grids, gated local grids, current loadings, decoupling and intrinsic capacitors, and sleep transistors for power gating. All the global and local networks are composed of resistive mesh grids. The global VDD grid supplies the power to local grids through multiple sleep transistors. The dynamic current loadings between local grids and global GND grid approximate the currents drawn by active digital functional blocks using triangular waveforms [15].

2. PDN with On-Chip Voltage Regulators. The modeling of a die-package PDN with low-dropout voltage regulators (LDOs) illustrated in Fig. 2 is also derived from [14]. It contains package power delivery interconnects and on-chip power delivery network integrated with on-chip LDO packages. The package PDN uses the same lumped package model as the die-package power gated PDN model, while the on-chip PDN contains global VDD/GND grids, LDO-driven local grids, current loadings, decoupling and intrinsic capacitors, and LDO packages. All the global and local networks are composed of resistive mesh grids. The dynamic current loadings between local grids and global GND grid approximate the currents of active digital functional blocks using triangular waveforms [11].

3. Hybrid Clock Distribution Network. The hybrid clock distribution networks are built based on the design in [16], which consist of the resistive mesh grid and the buffer-driven H-tree networks. Random capacitor values are assigned to the sink capacitors connected to the mesh grid, modeling the loading capacitance of sequential circuit elements.

Table 3: Experimental results comparison of runtime and memory between direct solver [1] and our proposed adaptive ultra-sparsifier support-circuit preconditioned method. Runtime includes the total time for matrix factorizations and GMRES iterations, but does not include device evaluation time.

CKT		Direct Solver		Non-Adapt.		Adapt.		
	#NR.	Runtime(s)	Mem(MB)	#GMRES	Runtime(s)	#GMRES	Runtime(s)	Mem(MB)
ldo small	421	1,111.05	168.8	3,889	123.45(8.9X)	4,312	112.23(9.8X)	38.7(4.4X)
ldo medium	441	7,440.10	733.5	4,908	534.73(13.9X)	5,297	495.12(15.0X)	127.4(5.8X)
ldo large1	443	17,941.50	1,580.0	3,265	954.83(18.8X)	4,437	945.38(18.9X)	260.0(6.1X)
ldo large2	443	18,591.42	1,542.6	5,156	1,176.71(15.7X)	6,092	1,170.83(15.8X)	260.7(5.9X)
pg small	318	737.72	166.3	1,620	73.77(10.0X)	2,231	70.07(10.5X)	37.6(4.4X)
pg medium	320	5,022.58	722.9	2,144	344.18(14.6X)	2,404	328.27(15.3X)	100.8(7.2X)
pg large1	321	20,074.09	1,742.0	5,119	1,075.85(18.6X)	5,550	1,044.52(19.2X)	267.3(6.5X)
pg large2	321	16,027.20	1,683.0	5,007	995.52(16.1X)	5,567	965.57(16.6X)	254.7(6.6X)
clk small	157	368.32	183.2	1,103	36.11(10.2X)	1,574	31.62(11.6X)	34.3(5.3X)
clk medium	157	2,653.30	633.0	1,366	173.04(15.3X)	1,730	164.80(16.1X)	76.3(8.3X)
clk large	159	9,333.23	1,304.0	2,105	437.75(21.3X)	2,276	425.15(22.0X)	232.7(5.6X)

Table 1: Experimental setup of large-scale nonlinear circuits. "#unk" denotes the number of unknowns, "R" denotes the number of resistors, "L" denotes the number of inductors, "C" denotes the number of capacitors, "M" denotes the number of transistors, and "i" denotes the number of current sources.

CKT	#unk	R	L	C	M	i
ldo small	201,204	384,602	5,005	63,590	924	63,504
ldo medium	577,027	1,106,008	14,455	182,471	1,540	182,329
ldo large1	1,068,095	2,048,994	26,917	337,759	2,156	337,561
ldo large2	1,074,367	2,049,778	26,917	338,543	10,780	337,561
pg small	200,544	384,518	5,005	63,506	168	63,504
pg medium	575,921	1,105,868	14,455	182,331	294	182,329
pg large1	1,066,541	2,048,748	26,917	337,563	406	337,561
pg large2	1,066,541	2,048,798	26,917	337,563	2,030	337,561
clk small	190,976	381,066	261	190,969	130	-
clk medium	546,128	1,090,766	519	546,121	259	-
clk large	1,014,056	2,026,086	771	1,014,049	384	-

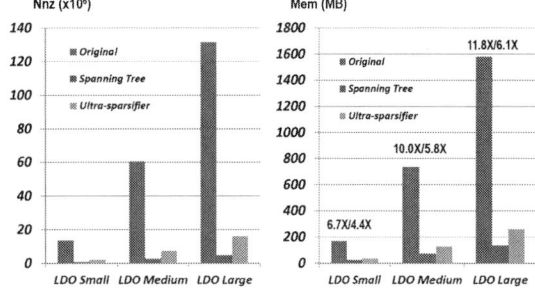

Figure 3: Memory cost and #non-zeros of LU matrix factorizations.

Table 2: Runtime of matrix factorization and solving for LDO test cases.

LDO	Factor			Solving		
	Orig.	Tree	Ultra	Orig.	Tree	Ultra
small	2.15	0.07(30.7X)	0.14(15.4X)	0.04	0.005(8.0X)	0.01(4.0X)
medium	14.96	0.26(58.2X)	0.55(27.1X)	0.17	0.02(8.5X)	0.03(5.6X)
large1	40.20	0.48(84.5X)	1.17(34.1X)	0.35	0.04(8.8X)	0.08(4.3X)

5.1 Experimental Setup

We compare our proposed method with the state-of-the-art direct solver KLU [1]. The same KLU solver is adopted for complete LU factorization of support circuits. A set of large-scale nonlinear circuits (power gated PDNs, PDNs with on-chip LDO, and hybrid clock distribution networks) has been tested. Our algorithms have been implemented in C++. All experiments are performed using a single CPU core of a computing platform running 64-bit Red-Hat 6.0 with 2.67GHz 12-core CPU and 48GB DRAM memory. All runtime results are measured in seconds. The details of these nonlinear circuits have been concluded in Table 1, in which "ldo" denotes the PDNs with on-chip LDOs, "pg" denotes power gated PDNs, and "clk" denotes hybrid clock distribution networks.

5.2 Preconditioner Construction Cost

The results of matrix factorizations for PDNs with on-chip LDO models has been concluded in Fig. 3 and Table 2, in which "Orig." denotes the original circuit, "Tree" denotes the support circuit using spanning trees, and "Ultra" denotes the support circuit using ultra-sparsifier graphs. All results are reported in seconds. Compared with direct solver, our proposed support-circuit preconditioned iterative method can reduce up to 85% for the total number of non-zero elements in matrix LU factors and up to 80% for the peak memory usage during complete LU matrix factorizations, achieving up to 34.1X and 5.6X speedups for matrix factorization and solving, respectively.

5.3 Experimental Performance

First, we demonstrate the benefits of the proposed method by comparing it to the state-of-the-art direct solver KLU [1]. The simulation results for 11 test cases with different circuit models and problem sizes are summarized in Table 3. "Direct Solver" denotes the results of state-of-the-art KLU solver [1], and "Adapt." denotes the results of our proposed adaptive ultra-sparsifier support-circuit preconditioned method. For "Direct Solver", "Mem." denotes the memory required by LU factorization for system matrix, and for "Adapt.", "Mem." denotes the memory required by LU factorization for support circuit. "#GMRES" denotes the total number of GMRES iterations for DC and TR analysis, in which the time step is constant though the proposed support-circuit preconditioning technique is also applicable to other adaptive time step control schemes. "Runtime" denotes the total time spent for matrix factorization and preconditioned iterative GMRES iterations for 100-time-step transient analysis including DC analysis, and ratio that is obtained by comparing to "Direct Solver". We want to emphasize that the proposed preconditioned iterative algorithms share the same numbers of Newton-Raphson iterations as the conventional SPICE simulation algorithm that uses direction solution method,

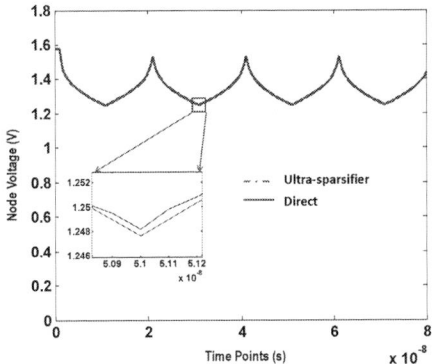

Figure 4: Transient waveforms comparison between the state-of-the-art direct solver and proposed ultra-sparsifier support-circuit preconditioned method.

since throughout the analysis we always accurately solve every linearized system.

The proposed ultra-sparsifier preconditioned method uses the same direct solver [1] to create preconditioner's matrix factors. Due to the *tree-like* structures of the ultra-sparsifier support graphs, the preconditioner matrix $\tilde{A}$ is much sparser and thus easier to factorize than the original conductance matrix A. As a result, the factorization time of $\tilde{A}$ is much smaller than that of matrix A. As illustrated in Table 2, for problem sizes varying from 200k to 1M, the reductions in matrix factorization time for the ultra-sparsifier support-circuit preconditioner compared to the original system are $15X$, $27X$, and $34X$, respectively.

Moreover, we observe from Table 3 that for all the 11 test cases, our proposed non-adaptive method can achieve up to $21X$ speedups in solving for the unknowns in DC and transient analysis, compared to the results using the direct solver. While by using the adaptive preconditioned method, it can achieve up to $22X$ speedups, which is slightly faster than the non-adaptive method. Since both non-adaptive method and adaptive method use the similar support-circuit preconditioners, they require the similar memory consumption during the matrix factorizations, achieving up to $8X$ memory reduction when compared to the direct solution method. We also compare the runtime and number of GMRES iterations for test cases with similar problem sizes, but different numbers of transistors, such as $ldolarge1$ and $ldolarge2$ test cases. As observed from Table 3, both $ldolarge1$ and $ldolarge2$ require the same number of Newton-Raphson iterations. However, $ldolarge2$ that has $5X$ more LDO blocks (transistors) requires a bit more GMRES iterations in each Newton-Raphson step and thus longer runtime than the $ldolarge1$ test case. In the last, the transient simulation solution accuracy of the proposed adaptive ultra-sparsifier support-circuit preconditioned GMRES algorithm is also examined, as shown in Fig. 4. For test cases with a $1.8V$ power supply, the solution differences in DC and transient analysis using our solver and the direct solver are indistinguishable, and solution errors are always less than $0.5mV$.

6. CONCLUSION

We propose a novel support-circuit preconditioning method for large-scale nonlinear circuit simulations. By extracting ultra-sparsifier support graphs from the linear networks, and combining them with the nonlinear analog or digital circuit blocks, the support-circuit preconditioner can be efficiently computed for accelerating the GMRES-based iterative methods. Since support circuit can effectively re-

duce the condition numbers of the preconditioned system, much less GMRES iterations are required for reaching the desired accuracy level. Our experimental results on several large-scale nonlinear circuits show that when compared with the state-of-the-art direct solver KLU, the proposed support-circuit preconditioned GMRES solver is much more runtime and memory efficient, achieving up to $22X$ speedups in solving matrices in DC and transient analysis, and up to $8X$ memory cost reduction.

7. REFERENCES

[1] T. Davis and E. Palamadai Natarajan. Algorithm 907: KLU, a direct sparse solver for circuit simulation problems. *ACM Trans. Math. Softw.*, 37:36:1–36:17, 2010.

[2] Y. Saad and M. Schultz. GMRES: a generalized minimal residual algorithm for solving nonsymmetric linear systems. *SIAM J. Sci. Stat. Comput.*, pages 856–869, 1986.

[3] Z. Li and C.-J.R. Shi. An efficiently preconditioned GMRES method for fast parasitic-sensitive deep-submicron VLSI circuit Simulation. In *Proc. IEEE/ACM DATE*, pages 752 – 757, 2005.

[4] H. Thornquist, E.R. Keiter, R.J. Hoekstra, D.M. Day, and E.G. Boman. A parallel preconditioning strategy for efficient transistor-level circuit simulation. In *Proc. IEEE/ACM ICCAD*, pages 410–417, 2009.

[5] Z. Feng and Z. Zeng. Parallel multigrid preconditioning on graphics processing units (GPUs) for robust power grid analysis. In *Proc. IEEE/ACM DAC*, pages 661–666, 2010.

[6] X. Zhao, J. Wang, Z. Feng, and S. Hu. Power grid analysis with hierachical support graphs. In *Proc. IEEE/ACM ICCAD*, pages 543–547, 2011.

[7] H. Peng and C. Cheng. Parallel transistor level circuit simulation using domain decomposition methods. In *Proc. ACM/IEEE ASP-DAC*, pages 397–402, 2009.

[8] D. A. Spielman and S. Teng. Nearly-linear time algorithms for graph partitioning, graph sparsification, and solving linear systems. In *Proc. ACM STOC*, pages 81–90, 2004.

[9] M. Bern, J. R. Gilbert, B. Hendrickson, N. Nguyen, and S. Toledo. Support-graph preconditioners. *SIAM J. Matrix Anal. Appl.*, 27:930–951, 2006.

[10] I. Koutis, G.L. Miller, A. Sinop, and David Tolliver. Combinatorial preconditioners and multilevel solvers for problems in computer vision and image processing. Technical report, CMU, 2009.

[11] Z. Zeng, X. Ye, Z. Feng, and P. Li. Tradeoff analysis and optimization of power delivery networks with on-chip voltage regulation. In *Proc. ACM/IEEE DAC*, pages 831 –836, 2010.

[12] E. Boman, D. Chen, B. Hendrickson, and S. Toledo. Maximum-weight-basis preconditioners. *Numerical Linear Algebra and Applications*, 11:695–721, 2004.

[13] E. Boman, B. Hendrickson, and S. Vavasis. Solving elliptic finite element systems in near-linear time with support preconditioners. *SIAM J. Numer. Anal.*, 46:3264–3284, 2004.

[14] M.S. Gupta, J.L. Oatley, R. Joseph, G. Wei, and D.M. Brooks. Understanding voltage variations in chip multiprocessors using a distributed power-delivery network. In *Proc. IEEE/ACM DATE*, pages 1 –6, 2007.

[15] Z. Zeng, Z. Feng, and P. Li. Efficient checking of power delivery integrity for power gating. In *Proc. ISQED*, pages 1 –8, 2011.

[16] P.J. Restle, T.G. McNamara, D.A. Webber, P.J. Camporese, K.F. Eng, K.A. Jenkins, D.H. Allen, M.J. Rohn, M.P. Quaranta, D.W. Boerstler, C.J. Alpert, C.A. Carter, R.N. Bailey, J.G. Petrovick, B.L. Krauter, and B.D. McCredie. A clock distribution network for microprocessors. *IEEE JSSC*, 36(5):792 –799, 2001.

Sparse LU Factorization for Parallel Circuit Simulation on GPU

Ling Ren, Xiaoming Chen, Yu Wang, Chenxi Zhang, Huazhong Yang
Department of Electronic Engineering
Tsinghua National Laboratory for Information Science and Technology
Tsinghua University, Beijing, China
{rl08@mails, chenxm05@mails, yu-wang@mail}.tsinghua.edu.cn

ABSTRACT

Sparse solver has become the bottleneck of SPICE simulators. There has been few work on GPU-based sparse solver because of the high data-dependency. The strong data-dependency determines that parallel sparse LU factorization runs efficiently on shared-memory computing devices. But the number of CPU cores sharing the same memory is often limited. The state of the art Graphic Processing Units (GPU) naturally have numerous cores sharing the device memory, and provide a possible solution to the problem. In this paper, we propose a GPU-based sparse LU solver for circuit simulation. We optimize the work partitioning, the number of active thread groups, and the memory access pattern, based on GPU architecture. On matrices whose factorization involves many floating-point operations, our GPU-based sparse LU factorization achieves 7.90× speedup over 1-core CPU and 1.49× speedup over 8-core CPU. We also analyze the scalability of parallel sparse LU factorization and investigate the specifications on CPUs and GPUs that most influence the performance.

Categories and Subject Descriptors

J.6 [**Computer-Aided Engineering**]: Computer-aided design (CAD)

General Terms

Performance

Keywords

GPU, Parallel Sparse LU Factorization, Circuit Simulation

1. INTRODUCTION

SPICE (Simulation Program with Integrated Circuit Emphasis) [1] is the most frequently used circuit simulator today. However, the rapid development of VLSI (Very Large Scale Integration) presents great challenges to SPICE simulators' performance. In modern VLSI, the dimension of circuit matrices after post-layout extraction can easily reach several million. It may take SPICE simulators days or even weeks to perform post-layout simulation on modern CPUs. The two most time-consuming steps in SPICE simulation are the sparse matrix solver by LU factorization and the model evaluation. These two steps have to be performed iteratively for many times

In the last decade, Graphic Processing Units (GPU) prove useful in many fields [2]. There have been works on GPU-based model evaluation [3, 4]. The parallelization of model evaluation in SPICE simulation is straightforward since it consists of a large amount of independent tasks, i.e. model evaluation for each component in the circuit. However, it is difficult to parallelize the sparse solver because of the high data-dependency during the numeric LU factorization and the irregular structure of circuit matrices. To our knowledge, there has been no work on GPU-based sparse solver for circuit simulation.

Previous works on GPU LU factorization mainly focus on dense matrices [5–7]. Though the performance of GPU-based dense LU factorization is very promising, up to 388 Gflop/s (Giga FLoating point OPerations per second) on GTX 280 [6], a simple calculation shows that sparse matrices should not be factorized as dense matrices. Take *onetone2* (36k by 36k) as an example. Even if we calculate the performance of dense factorization as 1000 Gflop/s, it would still cost 15.5 second to factorize *onetone2* as a dense matrix, while a straightforward sequential sparse factorization on a single core usually costs less than 1 second.

Up till now, sparse LU solvers are mainly implemented on CPUs. SuperLU [8–10] incorporates *Supernode* in Gilbert / Peierls (G/P) left-looking algorithm [11], enhancing the computing capability with dense blocks. PARDISO [12] also adopts *Supernode*. Christen et al. mapped PARDISO to GPU [13]. Their work still follows the idea of GPU-based dense LU factorization. They compute dense blocks on GPU and the rest of the work is still done on CPU.

However, it is hard to form supernodes in extremely sparse matrices such as circuit matrices. This feature makes *Supernode*-based algorithms less efficient than column-based algorithms for circuit simulation. So in KLU [14], which is specially optimized for circuit simulation, G/P left-looking algorithm [11] is adopted directly without *Supernode*.

KLU only has the sequential version. Due to the high data-dependency during the numerical factorization, parallel G/P left-looking algorithm is efficient only on shared-memory computing platforms, such as FPGA (Field Programmable Gate Array), multi-core CPU and GPU. Several studies developed G/P left-looking algorithm on FPGA [15–17], but the scalability to large-scale circuits is limited by FPGA on-chip resources. Chen et al. parallelized the algorithm on multi-core CPU [18]. Their implementation scales well with the number of CPU cores, but the number of CPU cores sharing the same memory is often limited. Most commodity CPUs (e.g., Intel Xeon, AMD Phenom) have no more than 6 cores. The state of the art GPUs provide a possible solution to the above problems. Compared with CPUs, CPUs have far more cores sharing the same memory (e.g. GTX580 GPU has 512 CUDA cores), and higher global memory bandwidth. Therefore, in this work, we for the first time present a parallel sparse LU solver (without pivoting) for circuit simulation on GPU. Our contributions are

- **exposing more parallelism for many-core architecture.** We must expose enough parallelism to make the sparse LU solver efficient on GPU (many-core architecture). Two kinds of parallelism are proposed in [18] to describe the

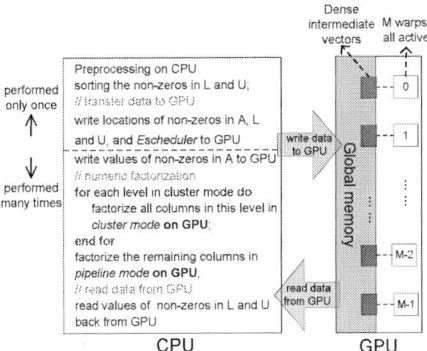

Figure 1: The workflow of sparse LU factorization on GPU

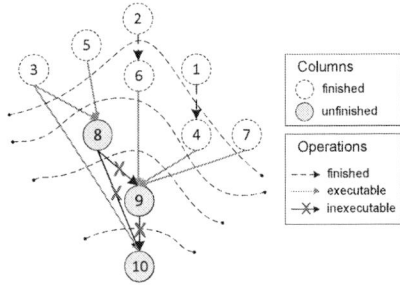

Figure 2: Parallelism and timing order in parallel G/P algorithm

2.2 Exposing More Parallelism

Algorithm 1 Sequential G/P left-looking algorithm

$\mathbf{L} = \mathbf{I}$
for $k = 1 : n$ **do**
    // solving $Lx = b$, $b = A(:, k)$, the kth column of $\mathbf{A}$
    $x = b;$
    for $j = 1 : k - 1$ where $U(j, k) \neq 0$ **do**
        // Vector MAD
        $x(j + 1 : n) = x(j + 1 : n) - L(j + 1 : n, j) \cdot x(j);$
    end for
    $U(1 : k, k) = x(1 : k);$
    $L(k : n, k) = x(k : n)/U(k, k);$
end for

Algorithm 1 is the sequential G/P left-looking algorithm [11]. The core operation in the algorithm is vector multiple-and-add (MAD). Two parallel modes are proposed in [18] to describe the parallelism between vector MAD operations, i.e. the *cluster mode* and the *pipeline mode*. We take Fig. 2 as an example to explain this level of parallelism. Suppose the columns in dashed circles are already finished, and column 8, 9 and 10 are being processed. The MAD operations represented by green solid arrows are executable. Parallelism exists in these operations, though the operations to the same column must be executed in a strict order.

The parallelism between MAD operations alone cannot take full advantage of GPUs' memory bandwidth. We utilize another intrinsic level of parallelism in sparse LU factorization: the parallelism within vector operations. Now we consider how to partition the workload to fully utilize GPU resources. In the process, several factors should be considered.

For convenience, we refer to the threads that process the same column as a virtue group. Threads in a virtue group operate on nonzeros in the same column and must synchronize. Smaller virtue groups can reduce idle threads. But virtue groups should not be too small for two reasons. First, too small virtue groups result in too few threads in total (the number of columns factorized simultaneously, i.e. the number of virtue groups, is limited by the storage space and cannot be very large), which is undesirable in GPU computing. Second, GPUs schedule threads in a SIMD (Single-Instruction-Multiple-Data) manner. A group of SIMD threads are called a *warp* on NVIDIA GPUs or a *wavefront* on AMD GPUs. If threads within a warp diverge, all necessary paths are executed serially. Different virtue groups process different columns and hence often diverge. Thus, too small virtue groups increase divergence between SIMD threads.

Taking all the above factors into consideration, we propose the following work partitioning strategy. In *cluster mode*, columns are very sparse, so while ensuring enough threads in total, we make virtue groups as small as possible to minimize idle threads. In *pipeline mode*, columns usually contain enough nonzeros for a warp or several warps. So the size of virtue groups matters little in the sense of reducing idle threads. We use one warp as one virtue group. This strategy not only reduces divergence between SIMD threads, but also saves the cost of synchronization, since synchronization between threads within the same warp is automatically guaranteed by GPU's SIMD architecture.

---

parallelism between vector operations. This level of parallelism alone is not enough for the thousands of threads running concurrently on GPU. We also utilize the parallelism within the vector operations. To efficiently deal with the two levels of parallelism, we partition the workload based on the features of the two modes and GPU architecture. Our strategy minimizes idle threads, saves synchronization costs, and ensures enough threads in total.

- **ensuring timing order on GPU.** In parallel left-looking algorithm, appropriate timing order must be guaranteed. Ensuring timing order on GPU involves carefully controlling the number of thread groups.

- **optimizing memory access pattern.** (1) We design the suitable data format of the intermediate vectors on GPU; (2) We propose sorting the nonzeros to improve the data locality for more coalesced accesses to global memory.

Experimental results on 36 matrices show that the GPU-based LU solver is efficient with matrices whose factorization involves many (more than 200M with our platforms) flops. On these matrices, the GPU achieves 7.90× speedup over 1-core CPU and 1.49× speedup over 8-core CPU. For matrices with denormal numbers [19], the speedup is even greater. We further analyze the scalability on different CPUs and GPUs and investigate which specifications of the devices have the greatest influence on the performance of sparse LU solver.

The rest of this paper is organized as follows. In Section 2, we introduce our GPU-based sparse LU factorization in detail. Experimental results and discussion are presented in Section 3. Section 4 concludes the paper.

2. SPARSE LU FACTORIZATION ON GPU

In this section, we present our GPU-based sparse LU factorization in detail. Fig. 1 is the workflow. The preprocessing is performed only once on CPU (Section 2.1). Numeric factorization is done on GPU in two parallel modes, where we optimize the workload partitioning based on the different features of the two modes and the GPU architecture (Section 2.2). Then we discuss several important points in our GPU-based sparse LU solver, including timing order between GPU thread groups (Section 2.3) and the optimizations to the memory access pattern (Section 2.4).

2.1 Preprocessing

The preprocessing consists of three operations: (1) HSL_MC64 algorithm [20] to improve numeric stability; (2) AMD (Approximate Minimum Degree) algorithm [21] to reduce fill-ins; (3) G/P algorithm based pre-factorization (a complete numeric factorization with partial pivoting) [11] to calculate the symbolic structure of the LU factors. We denote the matrix after preprocessing as $\mathbf{A}$. In circuit simulation, the nonzero structure of $\mathbf{A}$, $\mathbf{L}$ and $\mathbf{U}$ remains unchanged through the iterations of numeric factorization.

1126

2.3 Ensuring Timing Order on GPU

Algorithm 2 Pipeline mode parallel left-looking algorithm

---
for each work-group in parallel **do**
  while there are still unfinished columns **do**
    Get a new column, say column k;
    Put the nonzeros of $\mathbf{A}$ into the intermediate vector x;
    for all column j that $U(j,k) \neq 0$ **do**
      Wait until column j is finished;
      $x(j+1:n) = x(j+1:n) - L(j+1:n,j) \cdot x(j)$;
    end for
    $U(1:k,k) = x(1:k)$;
    $L(k:n,k) = x(k:n)/U(k,k)$;
    Mark column k finished;
  end while
end for

Algorithm 2 is the *pipeline mode* parallel left-looking algorithm. In this parallel mode, appropriate timing order between columns must be guaranteed. If column k depends on column t, only after column t is finished, can column k be updated by column t. We still use the example in Fig. 2 to explain the required timing order. Suppose column 8, 9 and 10 are being processed, and other columns are finished. Column 9 can be first updated with column 4, 6, 7, corresponding to the solid green arrows. But currently column 9 can not be updated with column 8. It must wait for column 8 to finish. Similar situation for column 10.

Ensuring timing order on GPU deserves special attention. The number of warps in the GPU kernel must be carefully controlled. It has to do with the concept of resident warps on GPU [22]. Resident warps refer to warps that reside on Streaming Multiprocessor (SMs) and are active for execution. A GPU kernel can have many warps. Often, due to the limited resources, some warps are not resident at the beginning. Rather, they have to wait for other resident warps to finish execution and then become resident.

However, in *pipeline mode* of sparse LU factorization, we have to ensure all the warps to be resident from the beginning. If a column is allocated to an non-resident warp, columns depending on it have to wait for this column to finish. But in turn, the non-resident warp would have no chance to become resident because no resident warp can ever finish. This results in a deadlock. Fig. 3 is an illustration of the situation. Suppose we have issued 3 warps on a GPU that supports only 2 resident warps. There is no problem in *cluster mode*, since warp 1 and 2 will eventually finish execution so that warp 3 can start. But in *pipeline mode*, column 9 and column 10 depend on column 8, which is allocated to the non-resident warp 3, so the resident warps (warp 1 and 2) fall in dead loops, waiting for column 8 forever. This in turn leaves no chance for warp 3 to become resident.

Therefore the maximum number of columns that can be factorized simultaneously in *pipeline mode* is exactly the number of resident warps in this kernel. This number depends on factors such as the resource usage and the number of branches or loops [22],

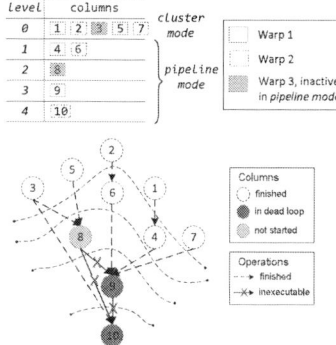

Figure 3: An illustration of deadlocks resulting from non-resident warps

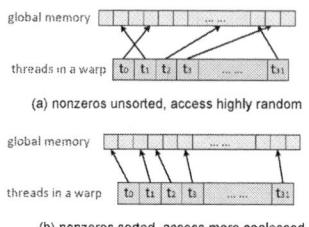

Figure 4: More coalesced memory accesses after sorting

and greatly influences the performance of our GPU-based sparse LU solver. But processing too many columns simultaneously is also undesirable. It is likely that for a certain column, its corresponding warp has nothing to do but to wait, because columns it depends on are unfinished. (In this case, this warp still consumes GPU cores and memory bandwidth.) Our experiments in Section 3.3 will confirm this point.

2.4 Optimization of Memory Access Pattern

Optimization for GPU-based sparse LU factorization is mainly about memory optimization. In this subsection, we discuss the data format for intermediate vectors, and the sorting process for more coalesced accesses to global memory.

Intermediate Vectors' Format. We have two alternative data formats for the intermediate vectors (x in Algorithm 2): CSC (Compressed Sparse Column) sparse vectors and dense arrays. CSC sparse vectors save space and can be placed in shared memory, while dense arrays have to reside in global memory. Dense arrays are preferred in this problem for two reasons. First, CSC format is inconvenient for indexed accesses. We have to use Binary Search, which is very time-consuming even within shared memory. Moreover, using too much shared memory would reduce the number of resident warps per SM:

$$\text{resident warps per SM} \leq \frac{\text{size of shared memory per SM}}{\text{size of a CSC sparse vector}}$$

which results in severe performance degradation.

Improving Data Locality. Higher global memory bandwidth is achieved on GPU if memory accesses are coalesced [22]. But the nonzeros in $\mathbf{L}$ and $\mathbf{U}$ are out of order after preprocessing, which affects the coalesced accesses. We sort the nonzeros in $\mathbf{L}$ and $\mathbf{U}$ by their row indices to improve the data locality. As shown in Fig. 4, after sorting, neighboring nonzeros in each column are more likely to be processed by consecutive threads.

In Fig. 5, we use the 21 matrices in the Group B in Table 2 to show the effectiveness of our sorting process. On average, GPU bandwidth is significantly increased from 37.69 GB/s to 91.17 GB/s (2.4× higher). It's worth mentioning that CPU sparse LU factorization also benefits from sorted nonzeros, but the performance increase is only 1.15×. The sorting overheads are negligible, since sorting is performed only once and the time for sorting is usually less than one factorization. We incorporate the sorting procedure in the preprocessing stage.

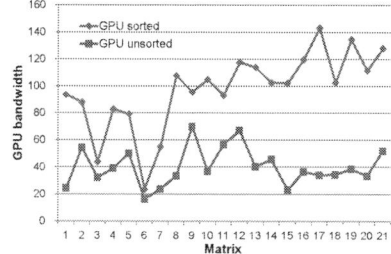

Figure 5: Performance increases by sorting the nonzeros

1127

Table 1: Related specifications of different devices

Devices	Xeon E5405	Xeon X5680	Radeon 5870	GTX580
Peak Bandwidth	——	——	153.6 GB/s	192.4 GB/s
Number of Cores	2×4 = 8 cores	2×6 = 12 cores	20 CUs 320 cores[1]	16 SMs[2] 512 cores
Active groups Active threads	—— 8	—— 8	160 10240	512 16384
L1 cache L2 cache L3 cache	32KB/core 12MB/4 cores ——	32KB/core 256KB/core 12MB/6 cores	8KB/CU 512KB/all ——	16KB/SM 768KB/all ——
Clock rate	2.0 GHz	3.2 GHz	850 MHz	772 MHz

[1] CU = Compute Unit. In Radeon 5870, each core contains 5 processing elements (PEs). But PEs are combined for double precision floating point operations [25], so they can be regarded as a single core in our problem.
[2] SM = Stream Multiprocessor

3. EXPERIMENTAL RESULTS AND DISCUSSION

3.1 Experiment Setup

We test the performance of parallel sparse LU factorization on the following four computing platforms: 2 Xeon E5405 CPUs, 2 Xeon X5680 CPUs, AMD Radeon 5870 GPU, and NVIDIA GTX 580 GPU. The experiments on CPU are implemented with C (SSE used) on a 64-bit Linux server. Radeon 5870 is programmed using OpenCL v1.1 [23]. GTX580 is programmed with CUDA 4.0 [22]. The related specifications of all the four devices are listed in Table 1. 36 matrices from University of Florida Sparse Matrix Collection [24] are used to evaluate our GPU sparse LU factorization. Though our intention is for circuit matrices, we also include some matrices from other applications to show that our GPU-based sparse solver is not confined to circuit simulation.

3.2 Performance and Speedup

In Table 2, we present the performance of our GPU-based sparse LU factorization on GTX 580 and compare with KLU and our CPU implementation on Xeon X5680 with different number of cores. We also intended to compare with SuperLU. But the parallel version of SuperLU fails on more than 1/3 of the matrices, and for those successfully factorized matrices, it is also 4.4× slower than our CPU implementation on average. Thus the detailed results of SuperLU are not presented.

The listed time is only for numeric factorization, excluding preprocessing and right-hand solving. Some data have to be transferred between CPU and GPU in every factorization (see Fig. 1). Time for these transfers are included in GPU runtime. We find that GPU bandwidth is strongly related to the Mflops (Mega FLoating-point OPeration) in factorization. So the average bandwidth and speedup in the last row of the table do not convey much useful information.

We categorize our test matrices into three groups. The first two groups are according to the Mflops in their factorization, less than 200M flops in Group A and more than 200M flops in Group B. We show the relation between Mflops and GPU bandwidth for these two groups of matrices in Fig. 6. From the figure, we can see the GPU bandwidth is positively related to Mflops, which indicates that in sparse LU factorization, the high memory bandwidth of GPU can be exploited only when the problem scale is large enough. The low bandwidth for Group A indicates that some overheads (e.g. data transfer, launching kernels) account for most of the runtime. For matrices in Group B, GPU achieves 7.90× speedup over 1-core and 1.49× speedup over 8-core CPUs.

Group C is in some sense special. Many denormal floating point numbers occur when factorizing these matrices. Denormal numbers are used to represent extremely small real numbers. CPUs deal with denormal numbers much slower than with normal represented numbers [19]. This is the major reason why CPU achieves very poor bandwidth on these matrices. In contrast, the state of the art GPUs can handle denormal numbers at the same speed as normal numbers. So GPU speedups for these matrices are very high. Full speed support for denormal numbers is an advantage of GPU in sparse LU factorization and other general purposed computing.

3.3 Scalability Analysis

Table 3: Bandwidth achieved on different devices in GB/s

Devices	Xeon E5405	Xeon X5680	Radeon 5870	GTX 580
Bandwidth Achieved	20.76	61.25	38.18	91.17

The average performance on the four devices are listed in Table 3. The detailed performance on the 21 matrices in Group B are presented in Fig. 7. On different platforms, the factors that restrict the performance are different. The cache size and speed has great influence on the performance of parallel sparse LU factorization on CPU. But cache has little influence on GPU performance. We have tried to declare all the variables on GPU as 'volatile' so that no data are cached. This only results in less than 10% performance loss. The reason is possibly that the cache on GPU is very small so that the cache hit rate is low even in the original kernel. The dominant factors on GPU performance are the peak global memory bandwidth, and whether there are enough active threads to bring out the high bandwidth.

We have mentioned in Section 2.3 that processing too many columns simultaneously may decrease the performance. This phenomenon is not seen on CPUs, but is indeed the case with GTX 580. Presented in Fig. 8 is the achieved bandwidth on 4 matrices on GTX 580 with different number of resident warps. The best performance is attained with about 24 resident warps per SM, rather than with maximum resident warps. This suggests we have fully utilized the *pipeline mode* parallelism on GTX 580 with 24 resident warps per SM. On GTX 580, we achieve 74% peak bandwidth at most (on *twotone*). Considering that the memory

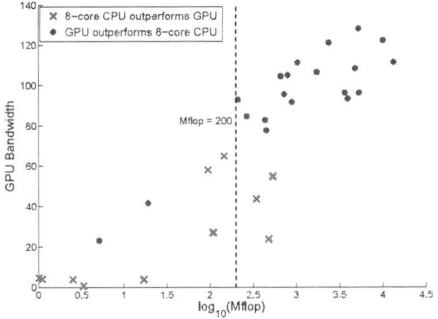

Figure 6: Relation between Mflops and GPU speedups

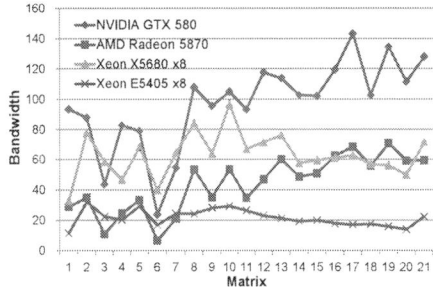

Figure 7: Performance of different devices on the second group of matrices

Table 2: Performance of sparse LU factorization on GTX 580 and Xeon X5680

	Matrix	[1]N (K)	[2]nonzeros (K)	[3]Mflops	1-core CPU time (s)	1-core CPU Bandwidth (GB/s)	GPU time (s)	GPU Bandwidth (GB/s)	GPU speedup over CPU over 1-core	over 4-core	over 8-core	speedup over KLU
A1	hcircuit	103.2	513.1	1.0	0.016	1.76	0.006	4.57	2.60	1.22	0.99	2.45
A2	lung2	109.5	492.6	1.1	0.007	4.44	0.007	4.37	0.99	0.80	0.70	1.80
A3	circuit_4	80.2	307.6	2.5	0.016	4.24	0.018	3.79	0.89	0.38	0.30	0.74
A4	rajat21	402.0	1893.4	3.3	0.051	1.77	0.146	0.63	0.35	0.34	0.28	0.56
A5	bcircuit	67.3	375.6	5.1	0.019	7.27	0.006	23.07	3.17	1.72	1.13	4.24
A6	dc1	116.8	766.4	16.9	0.053	8.74	0.123	3.77	0.43	0.17	0.13	0.51
A7	trans4	116.8	766.4	16.9	0.054	8.56	0.123	3.77	0.43	0.20	0.13	0.54
A8	hvdc2	189.9	1347.3	19.2	0.069	7.63	0.013	41.56	5.45	3.42	2.21	23.35
A9	onetone2	36.1	227.6	94.1	0.217	11.84	0.043	59.71	5.04	1.56	0.86	10.47
A10	transient	178.9	961.8	107.8	0.306	9.64	0.109	26.92	2.79	0.92	0.53	2.63
A11	ckt11752_dc_1	49.7	333.0	144.6	0.305	12.94	0.059	67.30	5.20	1.71	0.98	0.75
A	**Average**					**6.00**		**9.68**	**1.61**	**0.77**	**0.54**	**1.88**
B1	TSOPF_RS_b300_c3	42.1	4413.5	211.1	0.480	12.04	0.062	93.33	7.75	4.99	2.85	1.86
B2	epb3	84.6	463.6	267.2	0.567	12.88	0.083	87.68	6.81	2.12	1.13	6.69
B3	raj1	263.7	1302.5	340.7	0.820	11.36	0.214	43.60	3.84	1.24	0.74	452.71
B4	ASIC_680ks	682.7	2329.2	436.5	1.446	8.26	0.144	82.74	10.02	3.27	1.77	9.70
B5	thermomech_TC	102.2	711.6	449.4	0.950	12.93	0.155	79.03	6.11	2.05	1.16	6.13
B6	ASIC_680k	682.9	3871.8	474.8	1.622	8.00	0.547	23.73	2.96	0.97	0.60	4.23
B7	ASIC_100k	99.3	954.2	529.6	1.253	11.55	0.265	54.68	4.73	1.51	0.85	7.47
B8	ASIC_100ks	99.2	578.9	663.0	1.465	12.37	0.168	107.81	8.71	2.34	1.28	15.22
B9	rma10	48.6	2374.0	730.6	1.513	13.21	0.209	95.52	7.23	2.84	1.50	6.89
B10	onetone1	36.1	341.1	799.8	1.370	15.96	0.208	104.94	6.58	1.94	1.09	59.74
B11	thermomech_dM	204.3	1423.1	898.8	1.914	12.84	0.264	93.14	7.25	2.40	1.39	7.24
B12	venkat50	62.4	1717.8	1043.5	2.220	12.85	0.243	117.52	9.14	3.09	1.65	9.03
B13	Zhao1	33.9	166.5	1737.1	4.088	11.62	0.417	113.80	9.79	2.78	1.50	10.00
B14	thermomech_dK	204.3	2846.2	3637.9	8.348	11.92	0.969	102.69	8.62	3.11	1.78	8.75
B15	crashbasis	160.0	1750.4	3933.0	10.388	10.35	1.052	102.18	9.87	3.00	1.72	9.24
B16	G2_circuit	150.1	726.7	4780.0	12.101	10.80	1.094	119.45	11.06	3.17	1.95	10.74
B17	twotone	120.8	1222.4	5245.0	13.229	10.84	1.002	143.08	13.20	3.70	2.29	66.27
B18	sme3Dc	42.9	3148.7	5291.9	12.821	11.29	1.411	102.56	9.09	2.75	1.81	8.80
B19	xenon1	48.6	1181.1	10066.3	24.802	11.10	2.047	134.47	12.12	3.77	2.40	11.83
B20	helm2d03	392.3	2741.9	13331.8	32.357	11.27	3.273	111.37	9.89	3.47	2.23	9.88
B21	denormal	89.4	1156.2	2387.1	5.676	11.50	0.510	127.94	11.13	3.27	1.79	11.14
B	**Average**					**11.54**		**91.17**	**7.90**	**2.58**	**1.49**	**11.73**
C1	torso2	1033.5	116.0	651.4	4.243	4.20	0.184	97.02	23.11	6.09	3.20	20.13
C2	majorbasis	160.0	1750.4	3933.0	25.490	4.22	1.050	102.37	24.26	7.10	3.81	23.65
C3	ASIC_320k	321.8	2635.4	584.1	25.060	0.64	0.384	41.58	65.25	15.61	8.14	64.12
C4	ASIC_320ks	321.7	1827.8	651.6	28.470	0.63	0.170	104.52	167.02	41.07	21.25	163.28
C	**Average**					**1.63**		**81.06**	**49.72**	**12.90**	**6.77**	**45.78**

[1] the matrix dimension [2] the number of nonzeros in Matrix **A**
[3] the number of Mega floating point operates [4] all the average values are geometric average

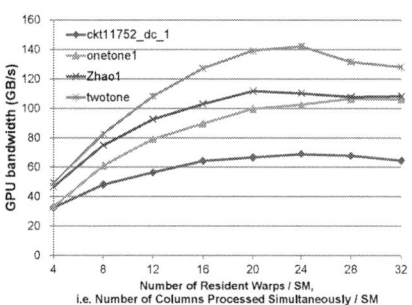

Figure 8: GPU Bandwidth on vs. number of resident warps on GTX 580

accesses are not fully coalesced and that warps sometimes have to wait, this is already close to the peak 192 GB/s. The performance can be improved if GPU peak bandwidth increases in the future.

On Radeon 5870, we achieve 45% peak bandwidth at most (on *xenon1*). A primary reason is that there are too few active wavefronts on Radeon 5870 to fully utilize the global memory bandwidth. On the two CPUs and Radeon 5870 GPU, the bandwidth keeps increasing with the issued threads (wavefronts), as shown in Table 4, which means the performance of our sparse LU solver

Table 4: Performance on matrices Group B on CPUs and Radeon 5870

Devices	Bandwidth achieved (GB/s)		
	1-core	4-core	8-core
Xeon E5405	4.99	14.41	20.38
Xeon X5680	11.54	35.38	61.22
	2 wavefronts / CU	4 wavefronts / CU	8 wavefronts / CU
Radeon 5870	26.96	39.20	52.59

can be improved if there are more CPU cores sharing the same memory, or Radeon 5870 supports more active wavefronts.

3.4 Hybrid Sparse LU solver

We have observed that matrices with few flops in factorization are not suitable for GPU acceleration, so we propose a CPU/GPU hybrid sparse solver for circuit simulation. The entire workflow is shown in Fig. 9.

For an input matrix, we first factorize it on CPU with partial pivoting [26] (this is part of the preprocessing). From the preprocessing, we obtain the FLOPs in factorizing the input matrix, and based on this information choose the appropriate platform for numeric factorization. If GPU is chosen, auto-tuning is performed to find the optimal number of warps to be issued. After each iteration, we check the residual between the solutions in two consecutive iterations. If the residual is large, we perform the preprocessing again; otherwise, we enter the next iteration directly. In most cases, the nonzero values do not change rapidly during

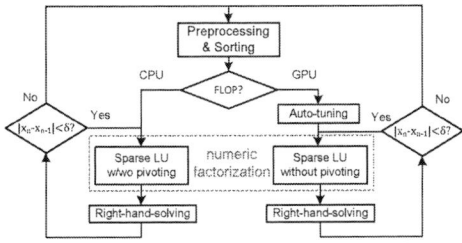

Figure 9: The proposed workflow of CPU/GPU hybrid sparse solver for circuit simulation.

several round of iterations. Thus the pivoting results from the preprocessing can be used in subsequent factorizations to ensure the numerical stability. When nonzeros values vary greatly from previous iterations, we may consider factorizing with partial pivoting on CPU again. In this way the hybrid solver also solves the problem of numerical accuracy.

4. CONCLUSIONS

This is the first work on GPU-based sparse LU factorization intended for circuit simulation. We have presented our GPU sparse solver in detail and analyzed the performance. Our experiments demonstrate that GPU outperforms CPU on matrices with many floating point operations in factorization.

One limitation of our GPU-based sparse LU solver is the inability to handle matrices with too many nonzeros in $L + U - I$, because of the relatively small global memory (1GB) of GTX 580. Yet, current state of the art GPUs such as NVIDIA GTX590 already have 3GB global memory, which is sufficient for factorizing most matrices. With the development of GPU, the scalability to matrices with more nonzeros and fill-ins can be improved.

5. ACKNOWLEDGMENTS

This work was supported by National Science and Technology Major Project (2011ZX01035-001-001-002, 2010ZX01030-001, 2011ZX03003-003-01), National Natural Science Foundation of China (No.60870001, 61028006), Microsoft/AMD and Tsinghua University Initiative Scientific Research Program.

6. REFERENCES

[1] L. W. Nagel, "SPICE 2: A computer program to stimulate semiconductor circuits." *Ph.D. dissertation, University of California, Berkeley*, 1975.

[2] *GPGPU'10: Proceedings of the 3rd Workshop on General-Purpose Computation on Graphics Processing Units.* New York, NY, USA: ACM, 2010.

[3] N. Kapre and A. DeHon, "Performance comparison of single-precision spice model-evaluation on FPGA, GPU, Cell, and multi-core processors," in *Field Programmable Logic and Applications. International Conference on*, 2009, pp. 65 –72.

[4] K. Gulati, J. F. Croix, S. P. Khatr, and R. Shastry, "Fast circuit simulation on graphics processing units," in *Proceedings of the 2009 Asia and South Pacific Design Automation Conference.* IEEE Press, 2009, pp. 403–408.

[5] V. Volkov and J. Demmel, "LU, QR and cholesky factorizations using vector capabilities of GPUs," EECS Department, University of California, Berkeley, Tech. Rep. UCB/EECS-2008-49, May 2008.

[6] S. Tomov, J. Dongarra, and M. Baboulin, "Towards dense linear algebra for hybrid gpu accelerated manycore systems," *Parallel Comput.*, vol. 36, pp. 232–240, June 2010.

[7] S. Tomov, R. Nath, H. Ltaief, and J. Dongarra, "Dense linear algebra solvers for multicore with gpu accelerators," *IEEE International Symposium on Parallel Distributed Processing Workshops and Phd Forum IPDPSW*, pp. 1–8, 2010.

[8] J. W. Demmel, S. C. Eisenstat, J. R. Gilbert, X. S. Li, and J. W. H. Liu, "A supernodal approach to sparse partial pivoting," *SIAM J. Matrix Analysis and Applications*, vol. 20, no. 3, pp. 720–755, 1999.

[9] J. W. Demmel, J. R. Gilbert, and X. S. Li, "An asynchronous parallel supernodal algorithm for sparse gaussian elimination," *SIAM J. Matrix Analysis and Applications*, vol. 20, no. 4, pp. 915–952, 1999.

[10] X. S. Li and J. W. Demmel, "SuperLU_DIST: A scalable distributed-memory sparse direct solver for unsymmetric linear systems," *ACM Trans. Mathematical Software*, vol. 29, no. 2, pp. 110–140, June 2003.

[11] J. R. Gilbert and T. Peierls, "Sparse partial pivoting in time proportional to arithmetic operations," *SIAM J. Sci. Statist. Comput.*, vol. 9, pp. 862–874, 1988.

[12] O. Schenk and K. Gartner, "Solving unsymmetric sparse systems of linear equations with pardiso," *Computational Science - ICCS 2002*, vol. 2330, pp. 355–363, 2002.

[13] M. Christen, O. Schenk, and H. Burkhart, "General-purpose sparse matrix building blocks using the NVIDIA CUDA technology platform," 2007.

[14] T. A. Davis and E. Palamadai Natarajan, "Algorithm 907: KLU, a direct sparse solver for circuit simulation problems," *ACM Trans. Math. Softw.*, vol. 37, pp. 36:1–36:17, September 2010.

[15] J. Johnson, T. Chagnon, P. Vachranukunkiet, P. Nagvajara, and C. Nwankpa, "Sparse LU decomposition using FPGA," *International Workshop on State-of-the-Art in Scientific and Parallel Computing (PARA)*, 2008.

[16] N. Kapre, "SPICE2 - a spatial parallel architecture for accelerating the spice circuit simulator," Ph.D. dissertation, California Institute of Technology, 2010.

[17] T. Nechma, M. Zwolinski, and J. Reeve, "Parallel sparse matrix solver for direct circuit simulations on FPGAs," *Circuits and Systems (ISCAS), Proceedings of IEEE International Symposium on*, pp. 2358–2361, 2010.

[18] X. Chen, W. Wu, Y. Wang, H. Yu, and H. Yang, "An escheduler-based data dependence analysis and task scheduling for parallel circuit simulation," *Circuits and Systems II: Express Briefs, IEEE Transactions on*, vol. 58, no. 10, pp. 702–706, oct. 2011.

[19] L. de Soras, "Denormal numbers in floating point signal processing applications," 2002.

[20] I. S. Duff and J. Koster, "The design and use of algorithms for permuting large entries to the diagonal of sparse matrices," *SIAM J. Matrix Anal. and Applics*, no. 4, pp. 889–901, 1997.

[21] P. R. Amestoy, Enseeiht-Irit, T. A. Davis, and I. S. Duff, "Algorithm 837: AMD, an approximate minimum degree ordering algorithm." *ACM Trans. Math. Softw.*, vol. 30, pp. 381–388, September 2004.

[22] NVIDIA Corporation, "NVIDIA CUDA C programming guide v3.2," 2010.

[23] Khronos OpenCL Working Group, "The opencl specification v1.1," 2010.

[24] T. A. Davis and Y. Hu, "The University of Florida sparse matrix collection," *to appear in ACM Transactions on Mathematical Software*.

[25] Advanced Micro Devices, Inc, "AMD accelerated parallel processing OpenCL programming guide v1.3," 2011.

[26] X. Chen, Y. Wang, and H. Yang, "An adaptive LU factorization algorithm for parallel circuit simulation," *17th Asia and South Pacific Design Automation Conference*, pp. 359–364, 2012.

Is Dark Silicon Useful?

Harnessing the Four Horsemen of the Coming Dark Silicon Apocalypse

Michael B. Taylor
Computer Science & Engineering Department
University of California, San Diego

ABSTRACT

Due to the breakdown of Dennardian scaling, the percentage of a silicon chip that can switch at full frequency is dropping exponentially with each process generation. This utilization wall forces designers to ensure that, at any point in time, large fractions of their chips are effectively dark or dim silicon, i.e., either idle or significantly underclocked.

As exponentially larger fractions of a chip's transistors become dark, silicon area becomes an exponentially cheaper resource relative to power and energy consumption. This shift is driving a new class of architectural techniques that "spend" area to "buy" energy efficiency. All of these techniques seek to introduce new forms of heterogeneity into the computational stack. We envision that ultimately we will see widespread use of specialized architectures that leverage these techniques in order to attain orders-of-magnitude improvements in energy efficiency.

However, many of these approaches also suffer from massive increases in complexity. As a result, we will need to look towards developing pervasively specialized architectures that insulate the hardware designer and the programmer from the underlying complexity of such systems. In this paper, I discuss four key approaches – the four horsemen – that have emerged as top contenders for thriving in the dark silicon age. Each class carries with its virtues deep-seated restrictions that requires a careful understanding of the underlying tradeoffs and benefits.

Categories and Subject Descriptors B.7.1 [*Integrated Circuits*]: Types and Design Styles

General Terms Design, Performance, Economics

Keywords Dark Silicon, Multicore, Dim Silicon, Utilization Wall, Dennardian Scaling, Near Threshold, Specialization

1. INTRODUCTION

Recent trends in VLSI technology have led to a new disruptive regime for digital chip designers, where Moore's Law continues but CMOS scaling ceases to provide the fruits that it once did. As in prior years, the computational capabilities of chips are still increasing by 2.8× per process generation; but a *utilization wall* [27] limits us to only 1.4× of this benefit – resulting in large swaths of our silicon area remaining underclocked, or dark – hence the term *dark silicon* [20, 9].

These numbers are easy to derive from simple scaling theory, which is a good thing, because it allows us to think intuitively about the problem. Transistor density continues to improve by 2× every two years, and native transistor speeds improve by 1.4×. But energy efficiency of transistors is improving only by 1.4×, which, under constant power-budgets, results in a 2× shortfall in energy budget to power a chip at its native frequency. Therefore, our rate of utilization of a chip's potential is dropping *exponentially* by a jaw-dropping 2× per generation. Thus, if we are just bumping up against the dark silicon problem in last generation's product line, then in eight years, we will be faced with designs that are 93.75% dark!

In the title of this paper, we refer to this widespread disruptive factor informally as the *dark silicon apocalypse*, because it officially marks the end of one reality ("Dennardian Scaling") – where progress could be measured by improvements in transistor speed and count – and the beginning of a new reality ("post-Dennardian Scaling") – where progress is measured by improvements in transistor energy efficiency. In the past, we tweaked our circuits to reduce transistor delays and turbo-charged them with dual-rail domino to reduce FO4 delays. In the new regime, we will tweak our circuits to reduce transistor toggles per function, and we will strip them down and starve them of voltage to squeeze out every femtojoule. Where once we would spend exponentially increasing amounts of silicon area to buy performance, **now, we will spend exponentially increasing amounts of silicon area to buy energy efficiency.**

A direct consequence of this breakdown in CMOS scaling is the industrial transition to multicore in 2005. Because filling chips with cores does not circumvent utilization wall limits, multicore is not the final solution to dark silicon [9] – it is merely industry's initial, liminal response to the shocking onset of the dark silicon age. Wikipedia defines liminality as "an in-between situation characterized by the reversal of hierarchies, and uncertainty regarding the continuity of tradition and future outcomes". With multicore, industry as a whole was uncertain as to the ramifications and scale of the power problems it was going to have, but it knew it needed to do something to address the problem. Over time, in this liminal phase, we are realizing more and more the ramifications and the semiconductor community as a whole is coming to a realization of what the new regime holds.

Due to the breakdown of Dennardian scaling, multicore chips will not be able to scale with die area; the fraction of a chip that can be filled with cores running at full frequency is dropping exponentially with each process generation [9, 27]. This reality will force designers to ensure that, at any point in time, large fractions of their chips are effectively *dark* or *dim* – either idle or significantly underclocked. As exponentially larger fractions of a chip's transistors become dark transistors, silicon area becomes an exponentially cheaper resource relative to power and energy consumption. This shift calls for new architectural techniques that "spend" area to "buy" energy efficiency.

In this paper, we examine some of the potential approaches that are coming to light about the dark silicon regime. We start by recapping the utilization wall that is the cause of dark silicon in Section 2, and by examining why the multicore response to the utilization wall is inherently limited [9]. We will look at recently proposed responses that are emerging as solutions as we transition beyond the transitional multicore stop-gap solution. Looking back, all of these responses appeared to be unlikely candidates from the beginning, carrying unwelcome burdens in design, manufacturing, and programming. None would appear ideal from an aesthetic engineering point of view (hence the analogy to the "four horsemen"). But the success of complex multi-regime devices like MOSFETs has taught us that engineering as a field has an enormous tolerance for complexity if the end result is better. As a result, we believe that future chips will apply not just one of these alternatives, but all of them.

In Section 3 we examine perhaps the most grim of the four candidates, which we refer to as *shrinking silicon*: simply scaling down the size of chips to reduce the amount of dark silicon. Section 4 examines the promise of underclocked, or dim silicon. Section 5 discusses the promise of specialized co-processors in dark silicon dominated technology. Finally, we examine the promise of new classes of circuits in Section 6, before concluding.

2. THE UTILIZATION WALL THAT CAUSES DARK SILICON

In this section, we show that a *utilization wall* [27] is the cause of dark silicon [20, 9]. Table 1 shows how this utilization wall is derived. We employ a scaling factor, S, which is the ratio between the feature sizes of two processes (e.g., $S = 32/22 = 1.4x$ between 32 and 22 nm process generations.) In both Dennardian and Post-Dennardian (Leakage-Limited Scaling), transistor count will scale by S^2, and transistor switching frequency will scale by S. Thus our net increase in compute performance from scaling is S^3, or $2.8x$.

However, to maintain a constant power envelope, these gains must be offset by a corresponding reduction in transistor switching energy. In both cases, scaling reduces transistor capacitance by S, improving energy efficiency by S. In Dennardian Scaling, we are able to scale the threshold voltage and thus the operating voltage, which gives us another S^2 improvement in energy efficiency. However, in today's Post-Dennardian, leakage-limited regime, we cannot scale threshold voltage without exponentially increasing leakage, and as a result, we must hold operating voltage roughly constant. The end result is that today, we have a shortfall of

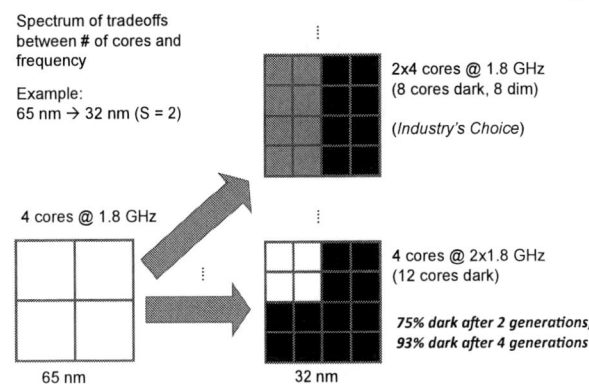

Figure 1: Multicore scaling leads to large amounts of Dark Silicon. (From [9].)

S^2, or $2\times$ per process generation. *This is an exponentially worsening problem that accumulates with each process generation.*

2.1 Silicon's New Potential: 40% energy savings per generation, or 1.4x performance

It is this shortfall that causes problems with multicore as the solution to scaling [9, 27]. Although we have enough transistors to increase the number of cores by $2\times$, and they would run $1.4\times$ faster, we only have the energy budget to receive a $1.4\times$ improvement. As shown in Figure 1, across two process generations ($S = 2$), we can either increase core count by $2\times$, or frequency by $2\times$, or some middle ground between the two. The remaining $4\times$ potential goes unused. This $4\times$ reflects itself in either dark or dim silicon, depending on our preference for frequency versus core count. A quick survey of recent designs such as Tilera TileGx, Intel Gulftown and Nvidia Firmi shows that industry has pursued various combinations of core count increase, and frequency increase/decrease that correlate very closely with the utilization wall. (For subsequent work to [9, 27] on dark silicon and multicore scaling that explores more sophisticated models that incorporate factors such as application space and cache size, see [6, 13, 16].)

3. THE SHRINKING HORSEMAN (#1)

When confronted with the possibility of dark silicon, an immediate response of many chip designers is "Area is expensive. Chip designers will just build smaller chips instead of having dark silicon in their designs!" Of all of the four dark horses, we believe that *shrinking chips* are the most pessimistic outcome, and although all chips may eventually experience "shrinkage", the ones that shrink the most will be those for which dark silicon cannot be applied fruitfully to actually result in a better product, and will rapidly turn into low-margin businesses for which further generations of

Transistor Property	Dennardian	Post-Dennardian
Δ Quantity	S^2	S^2
Δ Frequency	S	S
Δ Capacitance	$1/S$	$1/S$
Δ V_{dd}^2	$1/S^2$	1
$\implies$ Δ Power $= \Delta\, QFCV^2$	1	S^2
$\implies$ Δ Utilization $= 1/$Power	1	$1/S^2$

Table 1: Dennardian vs. Post-Dennardian (leakage-limited) scaling In contrast to the Classical regime proposed by Dennard [4], under the Post-Dennardian regime, the total chip utilization for a fixed power budget drops by a factor of S^2 with each process generation. The result is an exponential increase in the quantity of dark silicon for a fixed-sized chip under a fixed area budget. From [27].

Moore's Law provide small benefit. To examine this question in further detail, we look at a spectrum of second-order effects associated with shrinking chips.

Misconceptions of Dark Silicon. First, it is worth saying that dark silicon does not mean blank, useless or unused silicon – it is just silicon that is not used all the time, or at its full frequency. Even during the best days of CMOS scaling, microprocessor and other circuits were chock full of "dark logic" that is used only for some applications – for example, SIMD SSE units on x86 processors are not used for irregular applications and a doubling of last-level cache conveys benefits only for a narrow band of applications for which 1) cache misses comprise a major percent of program execution time and 2) a large fraction of the working set suddenly fits. For instance, many streaming applications experience experience no benefit from today's last level caches. And for SSE functional units, and especially last-level caches, this logic is often not used every cycle even in programs that do make use of them, which makes them "dark-silicon friendly".

Going into the future, the exponential growth of dark silicon area will push us beyond logic targeted for direct performance benefits towards swaths of low-duty cycle logic that exists not for direct performance benefit, but for the purpose of improving energy efficiency, which causes an indirect improvement in performance because it frees up more of the fixed power budget.

Cost Side of Shrinking Silicon. Understanding shrinking chips calls for us to consider semiconductor economics. There is a ring of truth to the "build smaller chips" argument – after all, designers spend much of their time trying to meet area budgets for existing chip designs. Smaller chips are generally cheaper, their leakage should be lower depending on power-gate efficiency, and in the small-signal regime of design optimization they are cheaper linearly (or better) with area. But *exponentially smaller chips are not exponentially cheaper* – even if they start out at being 50% of the cost of the system, after a few process generations, the cost of the silicon will be a tiny fraction of packaging and test costs, let alone system, marketing, sales, support and other costs. (For instance, for a typical $100mm^2$ desktop processor die, the silicon cost itself is only $10 or so, but the chip sells at $100-$300.) I/O pad area, design costs, masks costs will fail to be amortized, leading to a rising cost per mm^2 of silicon that ultimately will result in the lack of incentive to move the design to the next process generation. These designs will be "left behind" on older generations. (If this happens at a large scale across too many designs, then fab construction costs would be amortized more slowly as wafer quantities plummeted, and fab investments would become less attractive relative to alternative investments, signifying an unhappy economic ending to Moore's Law ...)

Revenue Side of Shrinking Silicon. On the other side of the shrinking silicon is the selling price of the chip. In a competitive market, if there is a way to use the next process generation's bounty of dark silicon to attain a benefit to the end product, then competition will force companies to do it. Otherwise, they will generally be forced into the low-end, low-margin, high-competition part of the market and their competitor will take the high end and enjoy high margins and achieve market superiority, much as happened with AMD and Intel in recent years [21]. Thus, in scenarios where dark silicon can be used profitably, decreasing area in lieu of exploiting it would certainly decrease system costs, *but these decreases in area would have much more catastrophic effects on sale price*, if it results in compromised performance or functionality relative to the competition. Thus, the shrinking chips scenario is likely to happen only if we can find no practical use for dark silicon.

Power and Packaging Issues with Shrinking Chips. A major consequence of exponentially shrinking chips is a corresponding exponential rise in power density. Recent work in analyzing the thermal characteristics of manycore chips [15] has shown that peak hotspot temperature rise can be modeled as $T_{max} = TDP \times (R_{conv} + k/A)$, where T_{max} is the rise in temperature, TDP is the target thermal design power of the chip, R_{conv} is the heatsink thermal convection resistance (lower is a better heatsink), k incorporates manycore design properties, and A is the area of the chip. If area drops exponentially, then the second term dominates and chip temperatures will rise exponentially. This in turn will force a lowering of the TDP so that temperature limits are met and reduce scaling below even the nominal 1.4× gain expected from energy efficiency gains. Thus, if thermals drive your shrinking chip strategy, it is much better to hold your frequency constant and increase cores by 1.4× with a net area decrease of 1.4× than it is to increase your frequency by 1.4× and shrink your chip by 2×. On the other hand, there is a concern that even without shrinking chips, the power-density of hotspots is still increasing exponentially, and could be a concern. A recent paper [16] suggests that this is not a significant concern, because as the hotspots shrink, the heat transfer to neighboring non-hotspots becomes proportionally more efficient.

Shrinking chips also present a host of practical engineering issues. Barring scalable innovations in 3-D integration along the lines of through-silicon vias (TSVs), designs would be increasingly pin-limited and would have trouble shrinking even though transistor area is shrinking, since I/O pads have not scaled well with Moore's Law.

4. THE DIM HORSEMAN (#2)

If we move beyond the prospect of shrinking silicon and consider populating dark silicon area with logic that we only use part of the time, then we are faced with two choices: do we try to make the logic in question general-purpose, or special purpose? In this section, we look at low-duty cycle alternatives that try to retain general applicability across many applications. We employ the term *dim silicon* [24, 16] to refer to general-purpose logic that is typically underclocked or used infrequently to meet the power budget.

Dim silicon techniques include scaling up the amount of cache logic, employing near-threshold voltage (NTV) processor designs, using Coarse-Grained Reconfigurable Array (CGRA)-based architectures that attempt to reduce energy by reducing the multiplexing of processor datapaths, and employing temporal dimming techniques.

Near-Threshold Voltage Processors. One recently emerging approach is the use of near-threshold voltage (NTV) logic [5], which operates in the near-threshold regime, providing more less-extreme tradeoffs between energy and delay than conventional subthreshold circuits.

Recently, researchers have looked at wide-SIMD implementations of NTV processors [19, 14] which seek to exploit data-parallelism, the most energy-efficient form of parallelism, and also a NTV many-core implementation [2] and an NTV x86 (IA32) implementation [18].

Although per-processor performance of NTV processors drops faster than the corresponding savings in energy-per-instruction (say a 5× energy improvement for a 8× performance cost), the performance loss can be offset by using 8× more processors in parallel if the workload allows it.

So assuming perfect parallelization, NTV could offer 5× the throughput improvement while absorbing 40× the area – approximately eleven generations of dark silicon. If 40× more free parallelism exists in the workload relative to the parallelism "consumed" by an equivalent energy-limited super-threshold manycore processor, it is a net win to employ NTV in deep-dark silicon limited technology. As we will see with specialization, the more energy-limited the domain (i.e. runs off a small solar panel or battery), the less total parallelism in the workload needed to break even, and thus the broader the applicability across workloads.

NTV presents a variety of circuit-related challenges that have seen active investigation, especially because technology scaling is likely to exacerbate rather than ameliorate these factors. A significant challenge with NTV has been susceptibility to process variability. As the operating voltage is dropped, variation in transistor threshold due to random dopant fluctuation (RDF) is proportionally higher, and the variation in operating frequency can vary greatly. Since NT designs expand the area consumption of designs by $\sim 8\times$ or more, variation issues are exacerbated, especially in SIMD machines which typically have tightly synchronized lanes. Recent efforts have looked at making SIMD designs more robust to these variations [25, 19]. Other challenges include the penalties involved in designing SRAMs that can operate at lower voltages and the increased energy consumption due to longer interconnect caused by the spreading of computation across a large number of slower processors.

Bigger Caches. An often proposed dim-silicon alternative is to simply use dark silicon area for caches. We can imagine, for instance, expanding per-core cache at a rate that soaks up the remaining dark silicon area; at a rate of $1.4 - 2\times$ more cache per core per generation. Increased cache sizes can carry both performance and energy benefits for miss-intensive applications, since off-chip accesses are power hungry. The miss-rate of the workload is a key parameter in determining the optimality of increasing cache size.

Going into the future with lower power off-chip interfaces and 3-D integrated memories, the benefits of larger on-chip caches are likely to be reduced; according to a recent study on dark-silicon limited server workloads, one crossover point for server workloads is when caches become large enough that the system ceases to be bandwidth-limited [13] and becomes power-limited.

Coarse-Grained Reconfigurable Arrays. One recurring "dim alternative" is the use of reconfigurable logic. Since the bit-level granularity and long wires of conventional bit-level FPGAs usually incurs high energy overheads, the most promising option is coarse-grained reconfigurable arrays (CGRAs) which have optimized paths for word-level operations. The idea is to naturally lay out the datapaths of the computation in space to avoid the multiplexing costs that are inherent to processor pipelines. The duty cycle of CGRA elements is very low, making it a potential fit exploiting dark silicon. Research in CGRAs has been ongoing prior to the days of dark silicon [7, 8] and continues into the dark silicon era [12]. Commercial success has been limited, but new constraints often make us look at old designs with fresh eyes.

Computational Sprinting and Turbo Boost. Other techniques work through the use of "temporal dimness" as opposed to "spatial dimness", temporarily exceeding the nominal thermal budget but relying on thermal capacitance to buffer against temperature increases, and then ramping back to a comparatively dark state. Intel's Turbo Boost 2.0 [23] uses this approach to boost performance up until the processor reaches, nominal temperature, relying upon the innate capacitance of the heatsink. Computational Sprinting [22] takes this a step further, by proposing the use of phase-change materials to allow chips to exceed their sustainable thermal budget by an order of magnitude or more for sub-second durations, providing a short but substantial computational boost.

5. THE SPECIALIZED HORSEMAN (#3)

As exponentially larger fractions of a chip's transistors become dark transistors, silicon area becomes an exponentially cheaper resource relative to power and energy consumption. This shift calls for new architectural techniques that "spend" area to "buy" energy efficiency. One approach is to use this dark silicon to implement a host of specialized co-processors, each of which is either much faster or much more energy-efficient (100–1000×) than a general purpose processor [27]. Execution hops among coprocessors and general purpose cores, executing where it is most efficient. At the same time, the unused cores are power- and clock- gated to keep them from consuming precious energy.

The promise for a future of widespread specialization is already being realized: we are seeing a proliferation of specialized *accelerators* that span diverse areas such as baseband processing, graphics, computer vision, and media coding. These accelerators enable orders-of-magnitude improvements in energy-efficiency and performance, especially for computations that are highly parallel. Recent proposals [27, 13] have extrapolated this trend and anticipate that in the near future we will see systems that are comprised of more coprocessors than general-purpose processors. In this paper, we term these systems Coprocessor Dominated Architectures, or *CoDAs*.

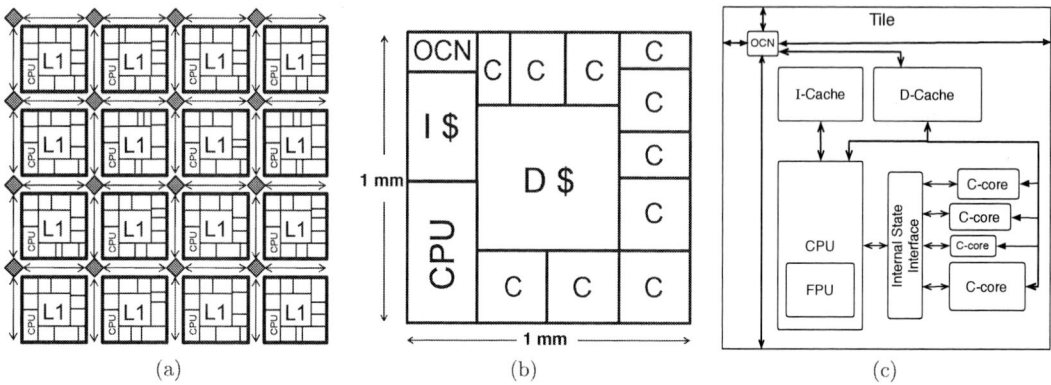

Figure 2: The GreenDroid architecture, an example of a Coprocessor-Dominated Architecture (CoDA). The GreenDroid Mobile Application Processor (a) is made up of 16 non-identical tiles. Each tile (b) holds components common to every tile—the CPU, on-chip network (OCN), and shared L1 data cache—and provides space for multiple c-cores of various sizes. (c) shows connections among these components and the c-cores.

As the use of specialization grows to combat the problem of dark silicon, we are faced with the reality of a modern-day specialization "tower-of-babel" crisis that fragments our notion of general purpose computation and eliminates the traditional clear lines of communication that we have between programmers and software and the underlying hardware. Already, we see the deployment of specialized languages such as CUDA that are not usable between similar architectures (e.g. AMD and Nvidia), and we see over-specialization problems between accelerators that causes them to become inapplicable to closely related classes of computations (e.g. double-precision scientific codes running incorrectly on GPU floating-point hardware that has been specialized for graphics.) We also see problems with adoption due to the excessive costs of programming heterogeneous hardware (e.g., the slow uptake of Sony Playstation 3 due to the difficulty of porting games to exploit the Cell Processor.) Specialized hardware also runs the risk of being obsolete as standards are revised (e.g., an update of the JPEG standard.)

Insulating Humans from Complexity. All of these factors speak to potential exponential increases in the human effort required to both design and program these CoDAs. Combating the tower-of-babel problem requires that we define a new paradigm for how specialization is expressed and exploited in future processing systems. We need new scalable architectural schemas that employ pervasively specialized hardware to minimize energy and maximize performance while at the same time insulating the hardware designer and the programmer from the underlying complexity of such systems.

Overcoming Amdahl-Imposed Limits on Specialization. Amdahl's Law provides an additional roadblock for specialization. The issue is that we need to find broad-based specialization approaches that save energy across the majority of the computation in question, including not only regular, parallel code, but also irregular code. One such CoDA-based system that targets both irregular and regular code is the UCSD GreenDroid processor [9, 10, 26, 11], which is a mobile application processor that targets the hotspots of the Android mobile environment using hundreds of specialized cores called *conservation cores*, or *c-cores* [27, 28, 24], that are automatically generated from C/C++ source code. The c-cores support a patching mechanism that allows

them to track software changes. They attain an estimated $\sim 8 - 10\times$ improvement in energy efficiency, at no loss in serial performance, even on non-parallel code, and without any user intervention required.

In contrast to Near-Threshold Voltage Processors, there is no need to find additional parallelism in the workload in order to cover a serial performance loss. As a result, conservation cores are likely to work across a wider range of workloads including collections of serial programs. However, for highly-parallel workloads where execution time is loosely concentrated, Near-Threshold Voltage Processors may hold an area advantage due to their reconfigurability.

6. THE DEUS EX MACHINA HORSEMAN (#4)

Of the four horsemen, this is by far the most unpredictable. *Deus Ex Machina* refers to a plot device in literature or theatre in which the protagonists seem utterly doomed, and then something completely unexpected and unforeshadowed comes out of nowhere to save the day. In the case of dark silicon, one Deus Ex Machina would be a breakthrough in semiconductor devices. However as we shall see, the breakthroughs that would be required would have to be quite fundamental – in fact most likely would require us to build transistors out of devices other than MOSFETs. The reason is that leakage is set by fundamental principles of device physics, and is limited to a sub-threshold slope of 60 mV/decade at room temperature; that is, in the typical case, a reduction of $10\times$ for every 60 mV that the threshold voltage is above the V_{ss}, which is determined by properties of thermionic emission of carriers across a potential well. Thus, although innovations like Intel's FinFET/Tri-Gate transistor, high-K dielectrics, etc, represent significant achievements maintaining sub-threshold slope close to their historical values, they still remain within the scope of the MOSFET-imposed limits and are one-time improvements rather than scalable changes.

Two VLSI candidates that bypass these limits because they are not based on thermal injection, are Tunnel Field Effect Transistors (TFETS)(e.g., [17]), which are based on tunneling effects, and Nano-Electro-Mechanical switches (e.g, [3, 1]), which are based on physical switches. Both of

1135

them hint at orders-of-magnitude improvements in leakage, but remain to be tamed from the wild.

Perhaps one source of our optimism at finding new devices is the efficiency and density of the human brain. The brain integrates 100 trillion synapses that operate at < 100 mv and embody an existence proof of highly parallel, mostly dark operation.

7. CONCLUSION

In this paper, we have examine four possibilities for our dark silicon dominated future. Although silicon is getting darker, for researchers the future is bright and exciting; dark silicon will cause a transformation of the computational stack, and from that transformation will come many opportunities for investigation.

8. ACKNOWLEDGEMENTS

We thank Brucek Khailany (NVidia), Chris Batten (Cornell), Dreslinski (U Mich), Steven Swanson, Jack Sampson and the GreenDroid Team (UC San Diego), Mattan Erez (UT Austin), Dean Tullsen (UC San Diego), Arun Raghavan (U Penn), Milo Martin (U Penn), Doug Burger (Microsoft), and Thomas Wenisch (U Mich) for productive (and challenging) discussions.

9. REFERENCES

[1] Chen et al. "Demonstration of integrated micro-electro-mechanical switch circuits for vlsi applications." In *ISSCC*, Feb. 2010.

[2] D. Fick et al. "Centip3de: A 3930 dmips/w configurable near-threshold 3d stacked system with 64 arm cortex-m3 cores." In *ISSCC*, Feb. 2012.

[3] H. Dadgour, and K. Banerjee. "Design and analysis of hybrid nems-cmos circuits for ultra low-power applications." In *DAC*, june 2007.

[4] R. Dennard, F. H. Gaensslen, V. L. Rideout, E. Bassous, and A. R. LeBlanc. "Design of Ion-Implanted MOSFET's with Very Small Physical Dimensions." In *JSSC*, October 1974.

[5] R. Dreslinski, M. Wieckowski, D. Blaauw, D. Sylvester, and T. Mudge. "Near-threshold computing: Reclaiming moore's law through energy efficient integrated circuits." *Proceedings of the IEEE*, Feb. 2010.

[6] H. Esmaeilzadeh, E. Blem, R. St. Amant, K. Sankaralingam, and D. Burger. "Dark silicon and the end of multicore scaling." *SIGARCH Comput. Archit. News*, June 2011.

[7] W. et al. "Baring it all to software: Raw machines." In *IEEE Computer*, September 1997.

[8] Goldstein et al. "Piperench: A reconfigurable architecture and compiler." *Computer*, Apr. 2000.

[9] N. Goulding, J. Sampson, G. Venkatesh, S. Garcia, J. Auricchio, J. Babb, M. Taylor, and S. Swanson. "GreenDroid: A mobile application processor for a future of dark silicon." In *HOTCHIPS*, 2010.

[10] N. Goulding-Hotta, J. Sampson, G. Venkatesh, S. Garcia, J. Auricchio, P.-C. Huang, M. Arora, S. Nath, V. Bhatt, J. Babb, S. Swanson, and M. Taylor. "The GreenDroid mobile application processor: An architecture for silicon's dark future." *Micro, IEEE*, March 2011.

[11] N. Goulding-Hotta, J. Sampson, Q. Zheng, V. Bhatt, S. Swanson, and M. Taylor. "Greendroid: An architecture for the dark silicon age." In *ASPDAC*, 2012.

[12] V. Govindaraju, C.-H. Ho, and K. Sankaralingam. "Dynamically specialized datapaths for energy efficient computing." In *HPCA*, 2011.

[13] N. Hardavellas, M. Ferdman, B. Falsafi, and A. Ailamaki. "Toward dark silicon in servers." *IEEE Micro*, 2011.

[14] Hsu, Agarwal, Anders et al. "A 280mv-to-1.2v wide-operating-range ia-32 processor in 32nm cmos." In *ISSCC*, Feb. 2012.

[15] W. Huang, M. R. Stant, K. Sankaranarayanan, R. J. Ribando, and K. Skadron. "Many-core design from a thermal perspective." In *DAC*, 2008.

[16] W. Huang, K. Rajamani, M. Stan, and K. Skadron. "Scaling with design constraints: Predicting the future of big chips." *IEEE Micro*, july-aug. 2011.

[17] A. Ionescu, and H. Riel. "Tunnel field-effect transistors as energy-efficient electronic switches." In *Nature*, November 2011.

[18] Jain, Khare, Yada et al. "A 280mv-to-1.2v wide-operating-range ia-32 processor in 32nm cmos." In *ISSCC*, Feb. 2012.

[19] E. Krimer, R. Pawlowski, M. Erez, and P. Chiang. "Synctium: a near-threshold stream processor for energy-constrained parallel applications." *IEEE Computer Architecture Letters*, Jan. 2010.

[20] R. Merrit. "ARM CTO: power surge could create 'dark silicon'." *EE Times*, October 2009.

[21] C. Nosko. "Competition and quality choice in the cpu market." 2010.

[22] Raghavan et al. "Computational sprinting." In *HPCA*, Feb. 2012.

[23] E. Rotem. "Power management architecture of the 2nd generation intel core microarchitecture, formerly codenamed sandy bridge." In *Proceedings of Hotchips*, 2011.

[24] J. Sampson, G. Venkatesh, N. Goulding-Hotta, S. Garcia, S. Swanson, and M. B. Taylor. "Efficient complex operators for irregular codes." In *HPCA*, 2011.

[25] Seo, Dreslinski, Woh et al. "Process variation in near-threshold wide simd architecture." In *DAC*, June 2012.

[26] S. Swanson, and M. Taylor. "GreenDroid: Exploring the next evolution for smartphone application processors." In *IEEE Communications Magazine*, March 2011.

[27] Venkatesh, Sampson, Goulding, Garcia, Bryksin, Lugo-Martinez, S. Swanson, and M. B. Taylor. "Conservation cores: Reducing the energy of mature computations." In *ASPLOS*, 2010.

[28] G. Venkatesh, J. Sampson, N. Goulding, S. K. Venkata, M. B. Taylor, and S. Swanson. "QsCores: trading dark silicon for scalable energy efficiency with quasi-specific cores." In *MICRO*, 2011.

Platform 2012, a Many-Core Computing Accelerator for Embedded SoCs: Performance Evaluation of Visual Analytics Applications

Diego Melpignano[1], Luca Benini[1,2], Eric Flamand[1],
Bruno Jego[1], Thierry Lepley[1], Germain Haugou[1],
Fabien Clermidy[3], Denis Dutoit[3]

[1]STMicroelectronics - AST,
Grenoble, France

{Diego.Melpignano,
Eric.Flamand,
Bruno.Jego,Thierry.Lepley,
Germain.Haugou}@st.com

[2]University of Bologna – DEIS,
Bologna, Italy

Luca.Benini@unibo.it

[3]CEA-LETI,
Grenoble, France

{Fabien.Clermidy,
Denis.Dutoit}@cea.fr

ABSTRACT

P2012 is an area- and power-efficient many-core computing accelerator based on multiple globally asynchronous, locally synchronous processor clusters. Each cluster features up to 16 processors with independent instruction streams sharing a multi-banked one-cycle access L1 data memory, a multi-channel DMA engine and specialized hardware for synchronization and aggressive power management. P2012 is 3D stacking ready and can be customized to achieve extreme area and energy efficiency by adding domain-specific HW IPs to the cluster. The first P2012 SoC prototype in 28nm CMOS will sample in Q3, featuring four 16-processor clusters, a 1MB L2 memory and delivering 80GOPS (with 32 bit single precision floating point support) in 18mm^2 with 2W power consumption (worst-case). P2012 can run standard OpenCL^{TM} and proprietary Native Programming Model SW components to achieve the highest level of control on application-to-resource mapping. A dedicated version of the OpenCV vision library is provided in the P2012 SW Development Kit to enable visual analytics acceleration. This paper will discuss preliminary performance measurements of common feature extraction and tracking algorithms, parallelized on P2012, versus sequential execution on ARM CPUs.

Categories and Subject Descriptors

C.1.4 [**Processor Architectures**]: Parallel Architectures - *Distributed Architectures*

General Terms

Algorithms, Measurement, Performance, Design.

Keywords

Low-power, many-core, computer vision, feature extraction, SoC,

DAC 2012, June 3-7, 2012, San Francisco, California, USA.
Copyright 2012 ACM 978-1-4503-1199-1/12/06 ...$10.00.

1. INTRODUCTION

The Platform 2012 (P2012) project aims at moving a significant step forward in programmable accelerator architectures for next-generation data-intensive embedded applications such as multi-modal sensor fusion, image understanding and mobile augmented reality. P2012 is an area-,power-efficient and process aware many-core computing fabric, and it provides an architectural harness that eases integration of hardwired IPs.

The P2012 computing fabric is highly modular and scalable, as it is based on multiple processor clusters implemented with independent power and clock domains, enabling aggressive fine-grained power, reliability and variability management.

Clusters are connected via a high-performance fully-asynchronous network-on-chip (NoC) and feature up to 16 processors with independent instruction streams (Multiple Program Multiple Data - MPMD) sharing single-cycle access multi-banked/multi ported level-1 data memories, a multi-channel DMA engine, and specialized hardware for synchronization and advanced variability-aware temperature and power management. P2012 achieves extreme area and energy efficiency by exploiting domain-specific acceleration at the processor and cluster level. Hardware-software interaction is facilitated by the local and global interconnect which efficiently supports point-to-point stream communication. The ultimate goal of P2012 is to fill the area and power efficiency gap between general-purpose embedded CPUs and fully hardwired application accelerators.

From the software viewpoint, P2012 supports different programming models for a wide range of performance and platform portability objectives. Programming Models are based on industry standards that can be implemented effectively on the P2012 platform. OpenCL^1 1.1 is supported since 2011. The Native Programming Model (NPM) is closely coupled to the platform and provides the highest level of control on application-to-resource mapping, at the expense of a lower-level abstraction.

---

^{1} OpenCL is a trademark of Apple Inc., used by permission by Khronos [5]

Advanced Programming Models (typically data flow variations) tuned to exploit combinations of HW and SW processing elements within a cluster are also available. P2012 programming tools assist the developer from high-level application capture and simulation, to the analysis, debug and visualization of the performance- and power-optimized version of the application mapped onto the fabric, including its interaction with a host processor.

The P2012 ecosystem is rapidly shaping up. A complete SDK for OpenCL and NPM is available today for a community with more than thirty R&D partners. The full runtime source is available to promote independent development of innovative software environments. OpenCV bindings have also been prototyped to facilitate acceleration of computer vision and image understanding applications. A variety of virtual platform options are supported for software development, performance analysis and optimization, HW-SW design exploration. An FPGA emulation board will be available by Q3 2012, and the first P2012 SoC will sample in STM 28nm CMOS technology [1] by Q4 with 4 fully populated clusters, each capable of 20GOPS (with full floating point support) in 3.7mm^2 of silicon with a 0.4W peak power consumption.

2. P2012 POSITIONING

The digital IC industry shift toward multi-core platforms is now well under way. There are two main architectures that are driving this evolution: general purpose shared memory multi-cores (GP-SMPs) and programmable graphic units (GP-GPUs). Both families have to strike a reasonable balance between power budget and achievable performance. GP-GPUs are the forefront for what concerns the number of cores, with up to thousands of them, strongly factorizing hardware resources, and heavily relying on a data-parallel workload assumption, whereby the same program/instruction is executed on many different data items. Maximum performance is achieved assuming that thousands of small threads, mostly identical, can be exposed to exploit a massive hardware multi-threading that allows hiding the memory latency. On the other hand, general purpose SMP, are driven by coarse grain parallelism and strong memory consistency models, which limit the core count to less than 10 today. Maximum performance is achieved assuming that a few large independent threads can be exposed.

At the same time, we see a rapidly growing demand for a new type of interactions between the user and the device based on the understanding of the environment sensed in multiple manner (image, motion, sound, …) striving to create more friendly user interfaces (augmented reality, virtual reality, haptic, …). The good news is that this class of applications is showing a relatively high degree of data parallelism; the bad news is that, differently from traditional graphics and image processing, parallel threads usually expose a behavior heavily dependent on the local data content, resulting into many truly independent parallel computations. In such a situation, general purpose coarse grain parallelism does not meet performance requirements while GP-GPU loses efficiency due to large branch divergence between threads ([7], [8], [11]). In this gap P2012 is positioned. In addition, the following design requirements have been taken:

• Easily scalable architecture even from the point of view of design effort, in practice this means a "step and repeat" based approach, where the building blocks are highly decoupled.

• Deep sub-micron process friendly design; for this purpose, sophisticated adaptive schemes need to be used to reduce the

safety margins part of the sign-off process. These adaptive schemes will be mixed with extremely power efficient solutions.

• Easily programmable and open architecture, sticking as much as possible to existing parallel languages such as OpenCL, to enable a rich software ecosystem.

• Possibility to quickly derive heterogeneous implementations combining software with hardware accelerated blocks using the same programming model. This requirement allows addressing different market needs. Combining all these requirements, P2012 was designed as a flexible, parametric and scalable architecture that covers a spectrum going from a pure SMP cluster-based solution down to a software-controlled hardware pipeline with all the intermediate points. Figure 1 summarizes the key characteristics of the proposed architecture, with respect to GP-SMP, GP-GPU and hardwired special-function pipeline.

The radar chart combines several normalized figures of merit, namely: peak performance (GOPS), peak performance vs. average performance, performance per power unit (GOPS/W), performance per cost unit (GOPS/mm^2), flexibility. In the next section, we will see that P2012 can be configured as a hardware pipeline with software control (P2012-HWSW), as a homogeneous multi-core cluster (P2012-SW) as well as a full spectrum of HW/SW intermediate solutions.

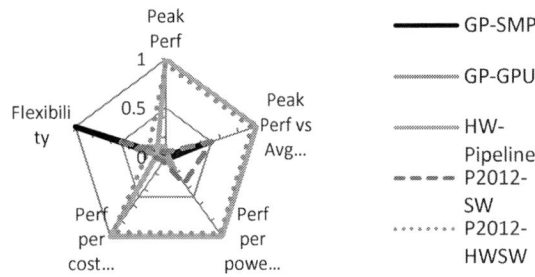

Figure 1: P2012 positioning.

The bottom line of Figure 1 is that P2012-SW is a competitive architecture for embedded computing, with more architectural flexibility than current GP-GPUs as well as significantly higher power and area efficiency than GP-SMP (and moderately higher efficiency than GP-GPUs). On the other hand, it also offers a viable design platform to build customized HW pipelines with high design productivity. In the remainder of the paper we focus primarily on P2012-SW, as this is the target of our first silicon implementation; we analyze performance results of visual analytics applications on P2012-SW.

3. ARCHITECTURE

P2012 can be described as a Globally Asynchronous Locally Synchronous (GALS) fabric of tiles, called clusters, connected through an asynchronous global NoC [2] (Global ANOC). The P2012 cluster, depicted in Figure 2, aggregates a multi-core computing engine, called ENCore (red box) and a cluster controller (yellow box). The ENCore cluster can host a number of processors (processing elements – PEs) varying from 1 to 16. The STxP70-V4 processor is the PE for ENCore. It is a cost effective and extensible 32-bit RISC core supported by a comprehensive state-of-the-art development toolset. The STxP70-V4 has a 32-bit load/store architecture with a variable-length instruction-set encoding (16, 32 or 48-bit) for minimizing the code footprint. Instructions can manipulate 32-bit, 16-bit or 8-bit data words, as well as small vectors of 8 bits and 16 bits elements. The instruction set is fully predicated to minimize the branch penalty impact. The STxP70-V4 core is implemented with a 7-stage pipeline for reaching 600MHz and it can execute up to two instructions per clock cycle (dual issue).

A STxP70-4 version with a floating-point unit extension (FPx) is used for the first silicon implementation of P2012. In this configuration, one single ENCore16 cluster can execute up to 16 floating point operations per cycle. The dual-issue architecture of STxP70-V4 ensures that this FLOP rate is sustainable as loads, stores and integer operations can be issued in parallel with the FLOPs. In addition, since all cores have independent instruction issue pipelines, there is no single-instruction, multiple-data restriction on execution, which is a common restriction of GPUs. This greatly simplifies application development and optimization.

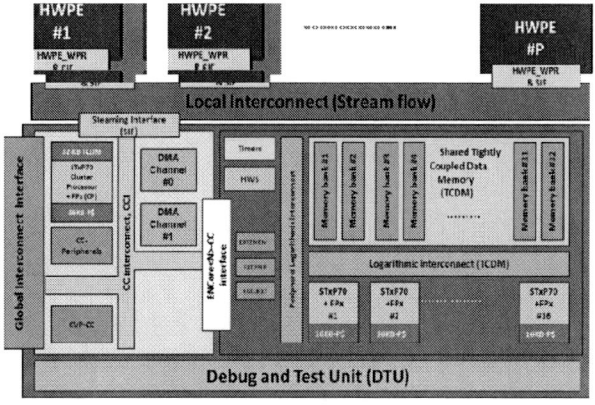

Figure 2: The P2012 configurable cluster architecture.

The Cluster Controller (CC) consists of a cluster processor sub-system, a DMA sub-system, a CC interconnect, and three interfaces: one to ENCore, one to global fabric-level ANOC and one to the local asynchronous network for plugging in hardwired accelerators. The CC contains a cluster processor based on a STxP70-V4 dual-issue core with FPX, 16-KB of program cache and 32-KB of local data memory. The cluster processor, in conjunction with its cluster controller peripherals, is in charge of booting and initializing the ENCore16 PE. It also performs application deployment on the ENCore16 PEs, some of the error handling, as well as energy management through the Clock Variability and Power (CVP) module. The CVP, whose registers are accessible in P2012 address map, controls process, variability and temperature sensors information and generates clocks (with ultra-fast adaption capability) and voltage. The DMA sub-system is made of 2 independent DMA channels. It performs the data block transfers from the external memory to the internal memory and vice versa while the various PE are operating. The CC interconnect supports intra and inter-cluster communication.

As depicted in Figure 2, the P2012 cluster can link the ENcore, described above, with a set of hardware accelerators (Hardware Processing Elements, or HWPEs) that provide cost-optimized implementation for those functions for which a software implementation would be too expensive in terms of area and energy (e.g. advanced motion compensation in video encoding). Further details about the P2012 architecture can be found in [3].

P2012 provides a modular architectural template to build programmable accelerators: the CC and the Local Interconnect create a bridge between the maturing high-level synthesis tools and scalable multi-core architecture. Moreover, the design flow ensures continuity in the cost optimization trajectory. Starting from a software-centric implementation for best time to market, computationally intensive kernels can be progressively implemented as dedicated hardware blocks. These hardware blocks can either be exposed as OpenCL custom devices or as nodes of a dataflow graph with the automatic generation of control code for orchestrating dataflow between hardware and software components.

Power, thermal and variability management are essential features in computing architectures targeting deep-submicron CMOS implementation. P2012 makes use of several hardware-assisted control loops to reduce design-time margin and to improve energy efficiency. Each cluster has a local clock, generated with a small-size and highly reactive Frequency-Locked-Loop (FLL). Clock speed can be adjusted in a few cycles on a per-cluster basis with no inter-cluster constraints thanks to the GALS architecture. The fabric interconnect is fully asynchronous, hence no global chip-wide clock distribution is required. Static and dynamic variability are managed though a number of distributed sensors, both direct (critical path monitors, both embedded and replica-based) and indirect (thermal sensors, both absolute and relative). Sensors are accessible through memory-mapped registers clustered in the CVP module. Hence feedback-based software policies can be implemented for operating point selection. In addition, hardware-based programmable triggers are also available to provide an ultra-fast reaction mechanism to emergency signals coming from the sensors.

4. SOFTWARE AND TOOLS

As depicted in Figure 3, the P2012 software stack provides at the top the programming models used to develop parallel applications, namely OpenCL and the Native Programming Model (NPM). The MIND [4] SW component tool-chain helps compiling applicative binaries (to be dynamically downloaded on the cluster by the host), bindings for cluster-local or remote communication as well as parts of the cluster runtime. SW components lifetime is controlled by the Comete Component Manager (CM) in the host and operated by the Comete Component Operator (CO) in the cluster controller.

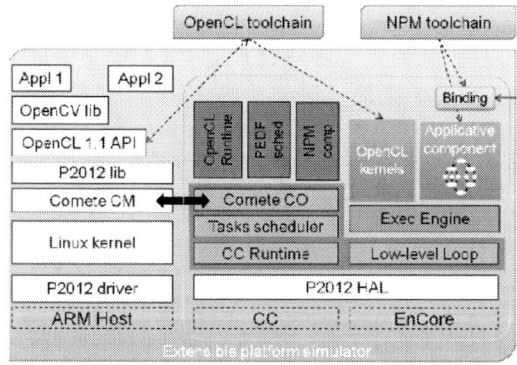

Figure 3: The P2012 Software Stack.

The other functions of the so-called resident runtime (orange boxes) include the task scheduler and dispatcher, memory allocators, resource and power monitoring, the communication driver with the host and the low-level execution loop in the Encore processors. Different execution engines can be used in the

1139

cluster, one example being the Reactive Task Manager that allows easy fork/join and duplication of worker jobs on the available ENCore processors. P2012 HW resources (e.g. DMA for memory transfers) are accessed by the runtime and by application kernels or components through a HW Abstraction Layer (HAL).

Depending on the programming model, the CC can run either NPM application components or a data flow scheduler (to use the HW blocks in the cluster) or an OpenCL runtime. A host OpenCL 1.1 API library can be used to execute computation kernels on P2012. A host Linux driver allows complete control of a P2012 fabric.

The flexible SW architecture described above makes it easy to integrate new programming models into P2012.

The P2012 programming tools provide solutions for multiple levels of the development cycle: from high-level application capture and simulation, using the supported programming models, to the analysis, debug and visualization of the performance- and power-optimized version of the application mapped onto the fabric, leveraging the P2012 runtime. This range of solutions is embodied in a Software Development Kit (SDK), which also includes the plugin-extensible virtual platform models for simulation and execution trace analysis and an Eclipse-based IDE.

From the software viewpoint, the P2012 architecture is PGAS (partitioned global address space). All processors have full visibility on all the memories with no aliasing: hence, it is possible for processors in one cluster to load and store directly in remote L1 memories in other clusters. The same holds for L2 and main memory. Loads are blocking, while stores are posted. A relaxed memory consistency model is hardware-supported through memory barrier instructions. Synchronous and asynchronous DMA-assisted memory copy functions are the preferred way to hide the access latency to remote memories.

From an application viewpoint, P2012 can efficiently accelerate computationally intensive machine vision algorithms. Therefore a dedicated version of the well-known OpenCV library is provided in the SDK.

In order to provide an open platform for visual analytics acceleration, standard APIs need to be supported; in the future, P2012 is going to offer an implementation of the new Vision HAL [13] API, currently being standardized by Khronos.

5. VISUAL ANALYTICS ACCELERATION

Feature extraction of image corners and their tracking are computationally intensive tasks whose acceleration on GPGPU architectures does not achieve high speedups compared to non-accelerated CPU execution [8]. Such class of algorithms is computation intensive and data adaptive and can leverage on the parallel and flexible many-core P2012 architecture. All applications presented in this section have been programmed for P2012 with OpenCL 1.1 [5].

Figure 4 shows the execution times of a baseline FAST corner detection algorithm on single STxp70 core, on a P2012 cluster with 16 processors at 600MHz, on one NVidia GTX 280 GPU (240 cores at 1.5 GHz in a high end PC) and on a single core ARM CA9 at 1GHz, all processing the same VGA input image (Shangai) with 777 corners detected.

The characteristics of the FAST algorithm and the way data transfers are managed on P2012 results in quasi optimal speedup on a 16 processors cluster, compared to the sequential version running on a single STxp70 core. This version of the FAST

algorithm has been used to derive performance metrics other than execution time, as further discussed in section 5.1, which take into account area and power efficiency.

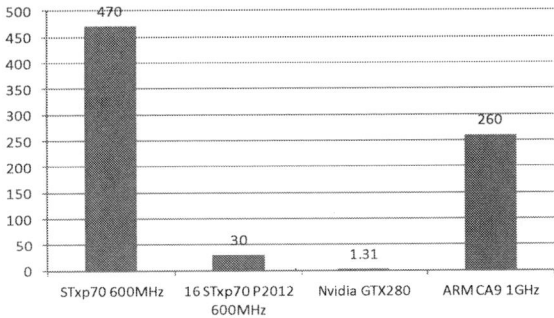

Figure 4: FAST corner detection execution times [ms].

If we consider an implementation of the same algorithm [5], where loop unrolling and other speed optimization techniques were applied (at the price of much larger code footprint), we find the results plotted in Figure 5.

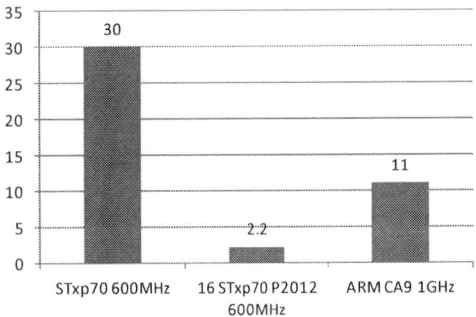

Figure 5: Optimized FAST execution times [ms].

The speed-up factor of the P2012 parallel OpenCL version vs. the sequential C version is reduced compared to Figure 4 (instruction cache misses effect due to large code footprint), but it is still satisfactory, also with respect to sequential execution on ARM.

Once corners are identified in an image, they can be tracked on successive pictures using the PKLT algorithm [15].

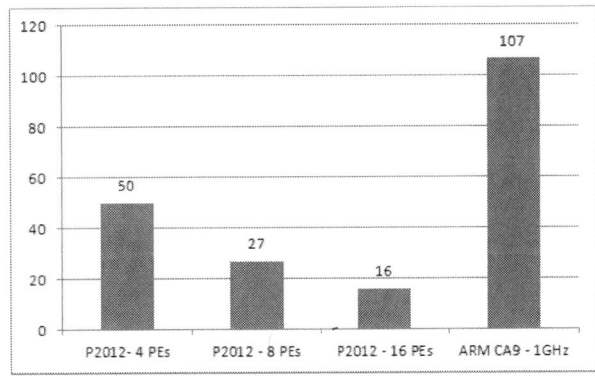

Figure 6: PKLT execution times [ms].

Figure 6 shows execution times of the PKLT algorithm running on a single P2012 cluster when 4, 8 and 16 PEs (600 MHz) are

1140

used respectively to process our reference VGA image with 800 corners. It shows a good scalability since doubling the number of cores activated in a P2012 cluster provides an average speedup of 1.75. For this application, a P2012 full cluster gives a speedup of 6.7 compared to a single 1Ghz A9 ARM core (with NEON floating point unit) executing the PKLT kernel sequentially on the same dataset.

In feature extraction algorithms, SIFT [8] is considered as the most robust, at the cost of high computational requirements. STMicroelectronics implemented a parallel OpenCL version of the SIFT algorithm, based on the vlfeat library [9]. Figure 7 shows the execution of SIFT on a P2012 cluster and compare it to the execution of the reference vlfeat C code running on a single core ARM Cortex A9 with the NEON floating point unit at 1Ghz. These execution times are calculated for a DoG (Difference of Gaussian) pyramid of 4 octaves, starting at octave 0, with 3 search levels in each octave, on the 'box' image (320x226), giving as output 250 keypoint descriptors.

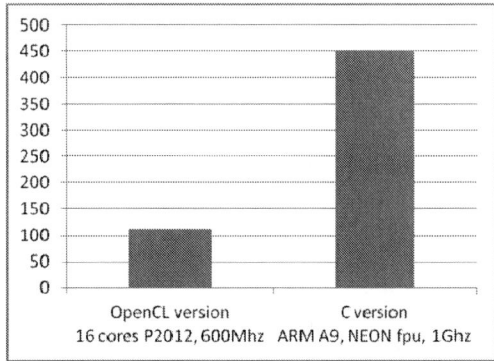

Figure 7: SIFT execution times [ms].

This version, not yet deeply optimized for P2012, gives a 4x speedup compared to the ARM CPU. In this example, 65 OpenCL kernels are successively executed on the P2012 cluster with intermediate data being stored in the slow L3 memory. Most of the L3 memory latency is hidden by combining asynchronous copies (exploiting the P2012 cluster DMA), double buffering and software pipeline in OpenCL kernels.

5.1 Comparing Architectural Efficiency

Metrics used to compare P2012 against other architectures (see Figure 1) include GOPS/mm^2 and GOPS/mm^2/W.. In Table 1, the efficiency of the baseline FAST algorithm running on P2012 is compared with a general purpose CPU, a GPGPU and a dedicated HW implementation, all translated into a 32nm technology node.

Table 1: P2012 performance comparison.

	ARM CA9 single core	GPGPU	P2012 16PEs	HW[2]
GOPS/mm^2	1	3	6	>100
GOPS/mm^2/W	1	2.3	25	>100

The advantages of the P2012 architecture are noticeable even for a single cluster homogeneous fabric with no dedicated HW IPs;

---

[2] Estimated figures.

moreover the cluster can be easily scaled to provide more computing, depending on the application being targeted.

5.2 Other Performance Factors

Overall performance of visual analytics tasks on a heterogeneous system is strongly influenced by the memory hierarchy, especially by the way data buffers are transferred between the host and the accelerator internal memory. In P2012 this data transfer must be explicitly managed by the programmer with use of DMA and synchronization primitives. Performance analysis shows that proper use of the DMA can hide transfer times of buffers to be processed in the P2012 accelerator, but special care should be taken in synchronizing the processing and I/O stages, in order to achieve good load balancing.

Figure 8 shows an example of execution traces for the OpenCL version of the baseline FAST algorithm: at the top, DMA transfers are reported, the bottom part of the figure depicting activity of each processor. It is noticeable that overall performance is limited primarily by unbalanced load (two processors still working, while all others are idling on a synchronization barrier); code rework is necessary in these cases.

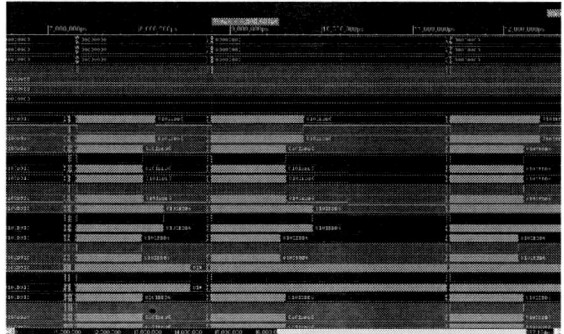

Figure 8: Cluster execution traces example.

Results shown in previous sections refer to execution times of individual visual analytics functions; in order to complete our performance analysis, the complete SW stack depicted in Figure 3 needs to be considered, e.g. to avoid useless memory copies. In particular, attention must be paid to high level CV applications where several visual analysis algorithms are combined, a challenge in terms of memory management, especially for the large buffers that store intermediate results between kernels. This study is part of future work.

6. THE P2012 FLEXIBLE SOC

The first silicon embodiment of P2012 is the flexible SoC depicted in Figure 9. The main objective of the SoC is to demonstrate software productivity and computational efficiency of a P2012-based homogeneous computing fabric, rather than minimizing power or area through HW specialization. Hence, the four clusters in the SoC are featuring the maximal configuration ENCore16 with FP support. The SoC features an additional 1MB L2 memory for code and data buffering and a Fabric Controller (FC) for fabric-level runtime and interaction with the off-chip host. The chip is GALS, with 5 clock and power domains (one per cluster, plus one for FC and L2).

One key innovation in the physical implementation of the SoC is its flexibility in off-chip connectivity. The die can be configured as an "accelerator chiplet" for three-dimensional die-stacking by appropriately setting the static MUXes shown on the right hand side of Figure 9. In this 3D mode (denoted 3D ANOC) the fabric interface to host and main memory goes through three 32 bits data-wide asynchronous IO ports (two initiators, one target) driven by micro-buffers and tied to micro-pads for die stacking. The bandwidth supported in this configuration is 3.2GBps. In addition (not shown in the figure), power and ground are also delivered through a "vertical plug". In this configuration the die will be flipped and stacked on top on a host SoC with CPU, peripherals, standard IOs and DRAM interfaces.

A second 2D configuration is supported by the static MUXes. In this mode traditional board-level high-speed interface (denoted NOC IF) links the fabric with the external host and main memory. This interface is physically driven through a smaller number of standard IO pads (two 81-pin ports). The 2D configuration is featuring a lower throughput (only 1 GBps) and energy efficiency than the 3D one but allows simple interfacing with on-board FPGA-based hosts (e.g. Xilinx Zynq [14] or Altera ArriaV SX [15] devices).

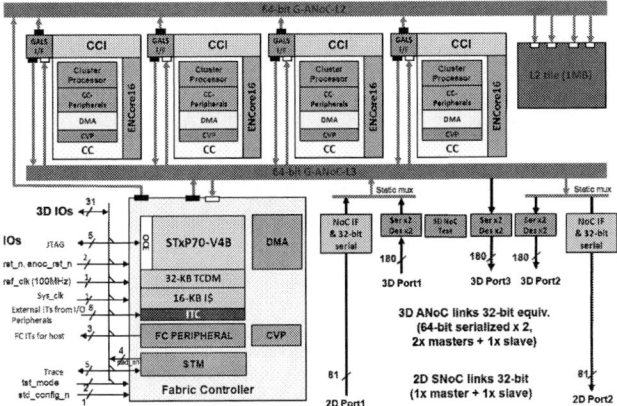

Figure 9: Block diagram of the flexible SoC.

The SoC is being implemented in STMicroelectronics' low-power 28nm CMOS process [1]. Target chip area is below 26mm2, with 15.2mm2 dedicated to the 4 clusters. On-going backend trials confirm that clock frequency is between 500 and 600 MHz (depending on PVT corner). The flexible clocking scheme with closed-loop control enables fast adjustment to the operating conditions, thereby greatly reducing conservative guard-banding on clock frequency at run-time. Each cluster features 16 (ENCore 16) +1 (CC) dual issue processors, hence peak GOPS per cluster is between 17 and 20. The power distribution grid of the SoC is designed to handle power delivery in both 3D and 2D configurations. Expected power consumption per cluster under heavy workload is upper-bounded at 0.5W. The chip power grid is designed to sustain up to 4W of power dissipation (at 1.1V, 125C), but its aggressive power management features enables energy-proportional operation up to a few hundreds mW average power.

7. CONCLUSION

The P2012 architecture provides a flexible acceleration fabric to span the efficiency spectrum between fully programmable homogeneous many-cores and application specific accelerators. It is designed from the ground up for variability tolerance, energy management, reliability and it comes with a complete set of programming models and tools to accelerate ecosystem buildup and productive software development. The P2012 SDK and virtual platforms are available today. This paper has analyzed P2012 performance in accelerating image corner extraction and tracking. Preliminary results show that a P2012 solution is competitive in terms of efficiency compared to general purpose CPUs and to GPGPUs, even more when considering power consumption. Further work is needed to characterize the acceleration factor when multiple visual analytics functions are invoked in sequence, a typical pattern found in OpenCV applications.

8. REFERENCES

[1]   F. Arnaud, S. Colquhoun, A.L. Mareau, S. Kohler, S. Jeannot, F. Hasbani, R. Paulin, S. Cremer, C. Charbuillet, G. Druais, P. Scheer, 2011. Technology-Circuit Convergence for Full-SOC Platform in 28 nm and Beyond. International Electron Devices Meeting.

[2]   Y Thonnart, P. Vivet, F. Clermidy, DATE 2010. A fully-asynchronous low-power framework for GALS NoC integration.

[3]   L. Benini, E. Flamand, D. Fuin, D. Melpignano, DATE 2012. P2012: Building an Ecosystem for a Scalable, Modular and high-efficiency Embedded Computing Accelerator.

[4]   MIND Component framework project - Online: mind.ow2.org

[5]   Khronos OpenCL - Online: http://www.khronos.org/opencl/

[6]   FAST corner detection – Online: http://www.edwardrosten.com/work/fast.html

[7]   Jason Clemons, Haishan Zhu, Silvio Savarese, and Todd Austin, 2011. MEVBench, An Embedded Vision Benchmarking Suite IEEE Interational Symposium on Workload Characterization. Online: http://www.eecs.umich.edu/mevbench/

[8]   David G. Lowe, 2004. Distinctive image features from scale-invariant keypoints. *International Journal of Computer Vision*, 60, 2 (2004), pp. 91-110.

[9]   The VLFeat open source library – Online: www.vlfeat.org

[10]   In Kyu Park et al., 2011. Design and Performance Evaluation of Image Processing Algorithms. *IEEE Transactions on Parallel and Distributed Systems*, vol.22, no.1 (Jan 2011)

[11]   A. Ensor, S. Hall, 2011. GPU-based Image Analysis on Mobile Devices. *Twenty-sixth International Conference Image and Vision Computing New*

[12]   *Zealand* (IVCNZ 2011)

[13]   Khronos Vision – Online: http://www.khronos.org/vision

[14]   Zynq-7000 Extensible Processing Platform - Online: www.xilinx.com

[15]   Arria V FPGA SX SoC – Online: www.altera.com

[16]   Jean-Yves Bouguet , Pyramidal Implementation of the Lucas Kanade Feature Tracker. – Online: http://robots.stanford.edu/cs223b04/algo_tracking.pdf .

Assessing the Performance Limits of Parallelized Near-Threshold Computing

Nathaniel Pinckney, Korey Sewell, Ronald G. Dreslinski, David Fick, Trevor Mudge, Dennis Sylvester, and David Blaauw

EECS Department University of Michigan, Ann Arbor, MI.

npfet@umich.edu

ABSTRACT

Supply voltage scaling has stagnated in recent technology nodes, leading to so-called "dark silicon." In this paper, we investigate the limit of voltage scaling together with task parallelization to maintain task completion latency. When accounting for parallelization overheads, minimum task energy is obtained at "near threshold" supply-voltages across 6 commercial technology nodes and provides 4X improvement in overall CMP performance.

Categories and Subject Descriptors
B.8.0 [**Hardware**]: Performance and Reliability – *general*.

General Terms
Performance, Design.

Keywords
Near-Threshold Computing, low-power design, parallelization.

1. INTRODUCTION

Moore's law [1] has historically enabled increased microprocessor performance with each technology node while maintaining constant power density. However, conventional voltage scaling has slowed in recent years. As transistors have become leakier, it has become more difficult reduce the threshold voltage and hence supply voltage scaling has stagnated as well in order to maintain sufficient overdrive and performance [2]. This deviation from constant-field scaling theory [3], combined with continuing scaling of transistor density has resulted in an increase in power density (W/mm^2) beyond the 130nm node, as shown in Figure 1.

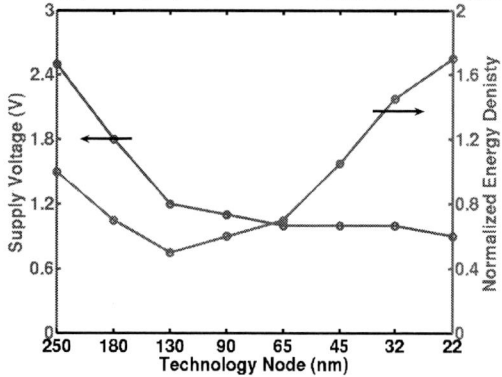

Figure 1: Nominal supply voltage and energy density from 250 nm to 22 nm.

The first consequence of this supply voltage stagnation has been the inability to increase processor frequency while still meeting power density constraints. Instead, processor designs have added more cores without significant increase in their frequency, leading to a prevalence of chip multiprocessor (CMP) [4] in contemporary commercial architec-

tures. However, since the die area of a server class chip has remained approximately constant at ~300–600mm^2, and since the number of cores has been increasing geometrically with each process step, the total chip power has again started to increase, despite relatively flat core frequencies. In practice the maximum allowable power dissipation of a single die is constrained by thermal cooling limits and is roughly 150W without advanced cooling technologies [1]. Hence, the second consequence of supply voltage stagnation is a limit on the number of cores that can be active simultaneously on a die and thus the maximum attainable performance of a modern CMP.

For instance, a 600mm^2 CMP could accommodate 23 Intel Westmere cores [1] in 22nm CMOS, which would dissipate 211W when all simultaneously executing, far exceeding the practical thermal dissipation limit. This would result in 40% of the cores (9 of 23) being idle. The problem of power-constrained core under-utilization has been recently observed in the literature and is sometimes referred to as *dark silicon* [5]. If scaling trends continue to 16nm, a similar CMP would consist of 46 cores, consume a max of 300W, and 50% of the cores (23 of 46) would be idle. As a result the most recent server-class CMPs have incorporated extensive power gating methods to turn off idle cores to free thermal budget for active cores [1].

Because modern CMP performance is now limited by power and not die-area, it is necessary for a paradigm shift in CMP design: cores are plentiful but powering them is not. The overall CMP performance can be best measured as *task throughput*: the number of completed tasks per second. In a power-constrained CMP, the task throughput is limited by the number of tasks that can be simultaneously active on the CMP within the thermal constraint. Thus, if we are able to lower task energy, the number of simultaneous tasks on the CMP (and hence activated cores) can be increased, improving task throughput.

The most effective knob for reducing energy consumption of a task running on a microprocessor is lowering the operating voltage. In tandem, processor frequency is reduced and task completion latency is increased. This type of voltage reduction has been widely used in DVFS, but since it impedes task completion latency, it is not generally applicable for high-performance applications. To address this, parallelization can be used to counteract lower clock frequencies and maintain latency. In this approach, the execution code of the task is parallelized so that the task executes on multiple cores in the CMPs, each operating at a lower frequency and voltage. In this way, the completion time remains the same as when the same task was executed serially on a single core at full voltage, while significant savings in total energy expended for completing the task is obtained. This reduction in energy consumption in turn allows more tasks to be executed on the CMP, thereby increasing overall CMP task throughput.

This combined voltage / parallelization approach is similar to the simpler *circuit* based parallelization approach proposed earlier in [6] which trades-off energy for latency. The method envisioned here instead parallelized the *algorithm* and thereby maintains task latency while still obtaining energy improvement. In fact, with the emergence of CMPs, many key applications are currently being parallelized by software developers. However, parallelization entails a number of overheads, which tend to increase as the task is parallelized into smaller subtasks. These overheads limit the obtainable energy improvement from the proposed approach, as the overhead eventually dominates over the quadratic energy gains from voltage reduction. Hence, there exists a minimum energy point, at which a task is optimally parallelized and voltage scaling reaches its efficiency limit.

To our knowledge, no systematic analysis has been performed to determine where this energy minimum lies. Hence, in this paper, we study this energy minimum, its associated energy gains, core operating voltage and task parallelization. We model three key factors limit energy-efficient parallelization in modern CMOS technologies: The leakage of a transistor — *Leakage Overheads*; the inability to achieve ideal code parallelization — *Amdahl Overheads*; the impact of coherence, interconnect, and memory system design — *Architectural Overheads*. All these overheads are interrelated and limit the obtainable energy efficiency gains from voltage scaling, the optimal energy voltage (V_{opt}) and the number of parallel subtasks required for frequency drop compensation (N_{opt}). In addition, we study the behavior of V_{opt} and N_{opt} across process nodes from 180nm to 32nm technology, using commercial process models.

Our key finding is that when realistic application-dependent overhead is included the optimal operating voltage is near threshold, roughly 200-400mV above the threshold voltage, and that this voltage range is valid across the six generations of industrial technologies as well as across transistor Vt selection. When accounting for all three overheads, operation at V_{opt} yields an energy efficiency gain of ~4× compared to operation at nominal voltage in 32nm and therefore allows a 4× increase in CMP task throughput under thermal constraints. Additionally, we find the maximum amount of energy-efficient parallelism, N_{opt}, across SPLASH2 benchmarks has a median value of approximately 12. Because running at lower supply voltage increases sensitivity to variation, we also explore the impact of variation on V_{opt} and include this in our analysis.

2. SCALING LIMITERS

There are three key limiters to energy-efficient scaling when a task is parallelized to maintain constant latency: leakage, Amdahl, and architectural. Each of these contributes to increased minimum energy and raises the energy-efficient operating point V_{opt}. The three key limiters are analyzed in the following subsections.

2.1 Leakage

First, we will assume a task can be perfectly parallelized across cores to compensate for frequency loss at a lower voltage, and the only nonideality from running at a slower clock frequency is transistor leakage. It is well known that reducing the supply voltage initially increases energy efficiency of a computation quadratically, yielding dramatic energy efficiency gains [7]. In the last 7 years, it has also been shown that leakage energy poses a *fundamental limiting factor* to energy efficiency gains through voltage reduction [8,9]. The required energy to complete a task can be divided into two categories, dynamic and static, and the classic relationship between energy and operating voltage is:

$$E_{total} = E_{dynamic} + E_{static} = CV_{dd}^2 + I_{leak}V_{dd}T_{task}$$

Dynamic or active energy is the energy consumed in charging and discharging the transistor and interconnect capacitances associated with the task being executed. Static or leakage energy is due to the always present subthreshold and gate oxide currents integrated over the time T_{task} to complete a task. While dynamic energy represents the energy needed to complete a task, static energy is parasitic and only poses an overhead on the computation. Although leakage can be mitigated in standby mode using techniques such as power gating and body biasing, it is more difficult to do so in active mode. Hence, leakage forms an unavoidable and fundamental limit on energy-efficiency.

To understand how leakage energy scales with V_{dd}, clock frequency scaling must be considered since as clock frequency is reduced the time to complete a task increases. For illustration purposes, the relationship between operating voltage and clock frequency is approximately:

$$\frac{1}{T_{task}} \propto f \propto \frac{(V_{dd} - V_t)^\alpha}{V_{dd}}$$

where Vt is the threshold voltage and α is process dependent but close to 2. For our results we simulated industrial transistor models in Cadence Spectre to obtain energy and performance. The canonical circuit topology was a chain of 31 fanout-of-4 inverters along with dummy devices for realistic input and output slew rates. The logic activity factor was chosen as 15% to emulate a core where 15% of the logic gates switch on average per clock cycle [10]. In addition chains of other types of logic gates were simulated to confirm that the result obtained for an inverter chain were representative of other logic structures as well.

Initially, when V_{dd} is large relative to Vt, frequency scales proportionately to V_{dd} as shown in Figure 2. As V_{dd} is further reduced and nears Vt, frequency scales exponentially with V_{dd} because the transistor is no longer fully activated. Instead the transistor drive current comes from subthreshold leakage current which scales exponentially with the gate-to-source voltage, and thus exponentially with V_{dd}.

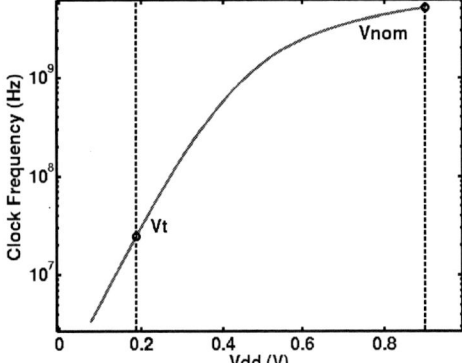

Figure 2: Clock frequency of a logic chain versus operating voltage.

As operating voltage is lowered, the static energy increases since the time to complete a task scales inversely with clock frequency. Eventually, at very low voltages, static energy dominates over dynamic energy, Figure 3. The operating voltage where total energy is minimized is called V_{opt} and occurs when the derivatives with respect to V_{dd} of the two energies are equal, $\frac{dE_{static}}{dVdd} = \frac{dE_{dynamic}}{dVdd}$ [9]. Beyond this point static energy increases more rapidly than dynamic energy decreases, and the total energy increases away from the energy minimum. For a 32nm node, V_{opt} when considering leakage overheads is ~300 mV.

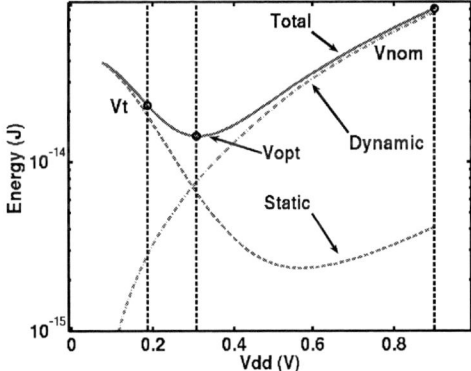

Figure 3: Total, static, and dynamic energy across V_{dd} for a 32nm process.

In recent years, several sensor processors that operate at this V_{opt}, which typically lies below the device threshold voltage, have been designed and demonstrated as much as 10× energy efficiency gains over operation at nominal supply voltage [11,12]. However, these sensor processors also incur phenomenal frequency loss, often operating at clock frequencies of 100s of kHz.

To fully compensate for a frequency loss of X because reduced voltage operation, a task with k instructions must be parallelized across X cores. If frequency is not compensated the total execution time would increase proportionally to X*k. But, since the task is parallelized, each of the X cores runs k/X instructions so the total execution time is X*(k/X) = k. Thus, no performance is lost from parallelizing. While most scientific and high-performance applications have been parallelized to operate on CMPs, it is not practical to recover a factor of 100's or 1000's in frequency loss without enormous parallelization overheads. Therefore, V_{opt}, considering optimization overheads will be at a higher voltage level, which will be discussed in the next subsection.

1144

2.2 Amdahl

As discussed earlier, scaling voltage is essential in the CMPs to achieve maximum computational performance for a fixed thermal budget, since the number of simultaneous tasks that can fit in a TDP is directly proportionate to the energy efficiency of the task. When scaling supply voltage for a latency-sensitive task, slower clock frequency can be compensated by executing the task in parallel across more cores. For real applications the process of subdividing a task includes non-idealities, such as serial portions of code, and thus incurs parallelization overhead. To compensate a task of k instructions for a frequency loss of X requires X cores (each running k/X instructions) plus m additional instructions of parallelization overhead. These extra m instructions consume additional energy, penalizing lower voltage operation, and therefore increase V_{opt}. Hence, parallelization overheads compound the impact of leakage overheads which limits the voltage scalability of a latency-sensitive task. Compensating below V_{opt} by further subdividing the task results in a net energy increase due to leakage and parallelization overheads.

The well-known Amdahl's law [13] shows that speedup of algorithms as they are parallelized over an increasing number of cores is limited by the parallelizable portion of the code and by new code introduced to initialize and decompose the program. Speedups are therefore bounded asymptotically as parallelization increases because the serial portion eventually dominates. These overheads will be referred to as A*mdahl overheads* and include only the impacts of algorithmic parallelization.

The gem5 [14] system simulator is used to evaluate the impact of Amdahl overheads on a Network-on-Chip (NoC) system. We evaluated the SPLASH-2 benchmark suite which is a set of highly parallelized scientific algorithms applicable to CMPs. Each core is an Alpha architecture with one instruction-per-cycle running at 1GHz. To separate Amdahl overheads from additional architectural non-idealities, we simulated the system with infinite interconnect bandwidth and an ideal memory with 1 cycle latency. Architectural non-idealities are addressed in the next subsection.

The effective speedup of parallelizing by running on 1 to 64 cores is shown in Figure 4. For illustration only three representative benchmarks are labeled, but the entire suite is plotted in the figure. Some benchmarks, such as Barnes, have nearly ideal speedup indicating very little Amdahl overheads and perfect parallelization. Other benchmarks, such as LUNC, reach a speedup of only 10 with 64 cores indicating a high percentage of serial code. These benchmarks represent a range of parallelized scientific applicable to CMPs and, as the number of CMP cores continues to increase, more high-performance applications will be similarly parallelized.

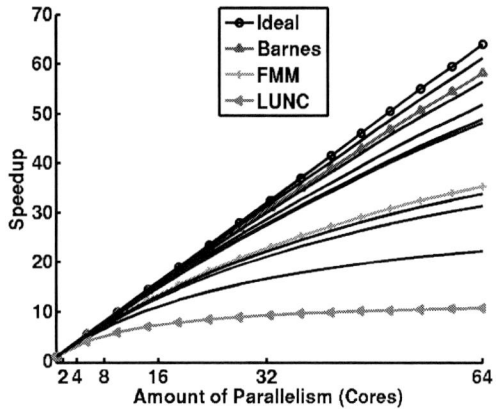

Figure 4: Speedup versus amount of parallelism demonstrating application-dependent Amdahl Overheads.

Amdahl's Law [15] gives $Speedup = \frac{n}{1-P_s+P_s n}$, where n is the number of cores parallelized over and P_s is the Amdahl serial coefficient. We fitted the SPLASH-2 benchmark speedups to Amdahl's law and applied it to the voltage scaling calculations to obtain V_{opt} when considering

non-ideal parallelization. The benchmarks were parallelized to fully compensate for frequency loss from lower voltage operation. Figure 5 shows V_{opt} increasing in 32nm due to Amdahl overheads. When Amdahl overheads are added, the V_{opt} operating range for most overheads is ~25-150 mV above the leakage overheads only case. Although the serial coefficient is highly application-dependent, the range of V_{opt} for the benchmarks is small, varying by only ~150mV. If the serial coefficient is 100% (e.g., none of the code is parallelizable) then nominal voltage would be optimal.

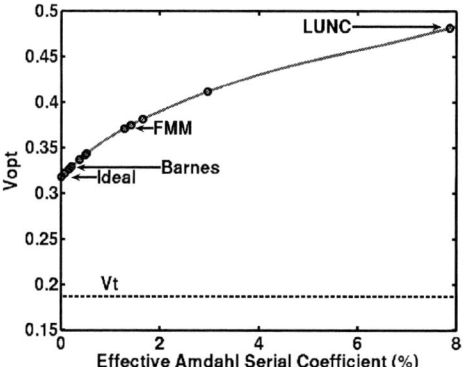

Figure 5: V_{opt} vs. Amdahl coefficient for all SPLASH-2 benchmarks (three labeled) in 32nm.

2.3 Architectural

Architectural features, such as coherency, inter-core communications, and cache pollution, further add overhead to a CMP system as voltage is reduced and a task is parallelized. Furthermore, application memory access patterns can affect overhead. For example, a subtask competes for L2 cache resources and may evict another subtask's data. Coherence overhead is added when a multiple subtasks share a single block of data. Communication overhead is increased when there is heavy communication between distant cores on an NoC because data must transverse multiple hops.

To quantify architectural overheads, the SPLASH-2 benchmarks were simulated with gem5 as in Section 2.2 but the configuration was changed to add non-ideal memories, caches, and interconnect. The NoC simulations were run using a tiled Mesh topology where each tile contains a core, private L1 caches, and a slice of a shared L2 cache. A MOESI directory protocol is used to maintain coherence. Table 1 lists the detailed simulation parameters.

Feature	Description
Cores	1 to 64 one-IPC Alpha cores @ 1GHz
L1 Caches	32 kB, 1 cycle latency, 4-way associative, 64-byte line size
L2 Caches	Shared 1MB divided evenly between cores, 10 cycle latency, 8-way associative, 64-byte line size
Interconnect	2-GHz Routers, 128-bit, 2-stage routers, 50 cycle-access to main memory

Table 1: A list of simulation parameters to measure architectural overheads.

Architectural overheads from memory and interconnect non-idealities reduce the obtainable speedup when parallelizing. These non-idealities were added in the V_{opt} calculation when parallelizing and the benchmarks were again parallelized to fully compensate for frequency loss. Like leakage and Amdahl overheads, architectural overheads further increase the minimum energy consumption and V_{opt} as shown in Figure 6. Certain benchmarks are highly parallelizable before caches and coherency is introduced, while others have negligible architectural overheads. For example, ocn has almost no Amdahl overheads but significant architectural overheads. To contrast, lun has little architectural but significant Amdahl overheads. Across the benchmarks shown V_{opt} increases by no more than 200mV. Thus, architectural overheads are another key

1145

limiter to voltage scaling, increasing V_{opt} and the minimum obtainable energy consumption when a task is parallelized to compensate for frequency loss. Though not included in this paper for brevity, we also simulated SPLASH-2 running on a bus-based architecture and the corresponding increase in V_{opt} is similar.

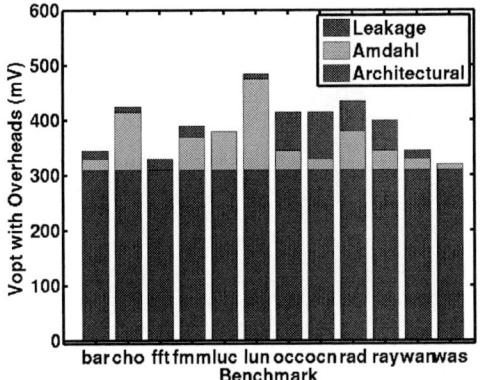

Figure 6: Architectural and Amdahl overheads increase V_{opt} by no more than 200 mV.

3. IMPACT OF TECHNOLOGY AND CIRCUIT FEATURES ON NTC

The previous section discussed the three key limiters of energy-efficient scaling. However, V_{opt} is also impacted by additional technology and circuit factors, including technology node, transistor Vt, and process variation, which are discussed below.

3.1 Technology

In the previous section V_{opt} was analyzed at single 32nm technology node. To identify if there is a voltage scaling and parallelization guideline consistent across many technologies, we calculated V_{opt} for SPLASH-2 across 6 industrial technologies when accounting for all three voltage scaling overheads, as shown in Figure 7. Circuit simulations of energy and performance were done in Cadence Spectre using industrial foundry technology kits from 32nm to 180nm.

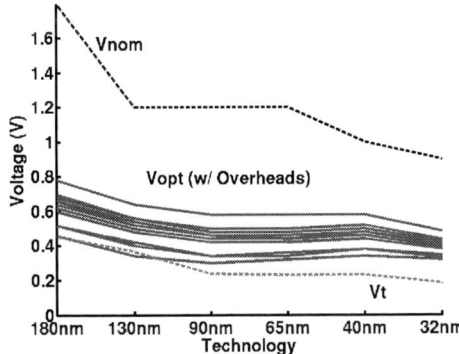

Figure 7: V_{opt} across technologies when including all three overheads. V_{opt} has been trending downward with each generation and is ~200-400mV above Vt for most benchmarks.

The process node affects V_{opt} primarily because technologies have become more leaky generation-to-generation due to reduced threshold voltage. Higher leakage increases V_{opt}, however, the lower threshold voltage will also improve the frequency degradation with voltage scaling which will reduce V_{opt}. A key finding of this work is that V_{opt} consistently tracks ~200-400mV above the threshold voltage for most benchmarks across the six technology nodes. We define this region above the threshold voltage as the near-threshold (NTC) region. Three benchmarks, Barnes, FFT, and Water Spatial, were close to ideally parallelizable and are not contained in the NTC region. However, most general-purpose,

high-performance CMP applications will have some degree of parallelization overhead and thus lie in the NTC region.

The median energy gains at V_{opt} operation and optimal number of cores to parallelize across to compensate for clock frequency loss, N_{opt}, for SPLASH-2 across technology nodes is shown in Figure 8. Table 2 includes a breakdown of energy gains and optimal number of cores for each benchmark in the SPLASH-2 suite. Energy gains have diminished by ~1.8× from 180nm to 32nm as leakage has increased and the dynamic range available for voltage scaling has narrowed from 180nm to 32nm. This difference is less dramatic with less scalable benchmarks, since the parallelism overheads are higher and thus the amount of voltage scaling in older technologies is limited. The energy gains in newer technology from operating at V_{opt} instead of at nominal voltage are ~4× and N_{opt} has a median of ~12, and no more than 25, cores for SPLASH-2 in 32nm. This increased energy efficiency directly increases CMP performance when limited by a thermal budget. Thus, to maximize thermally-limited CMP performance, tasks should operate in the near-threshold region and parallelize on no more than 25 cores in 32nm. The energy gains and optimal amount of parallelism has decreased with each generation.

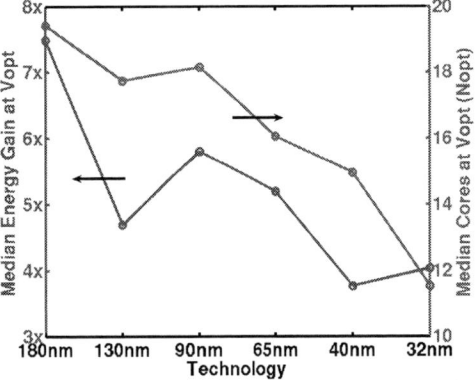

Figure 8: Median energy gains and optimal number of cores, N_{opt}, when operating at V_{opt} as compared to nominal voltage for SPLASH-2 benchmarks.

	180nm	130nm	90nm	65nm	40nm	32nm
bar	12.7× (71)	7.9× (55)	10.0× (69)	7.8× (43)	5.7× (44)	5.2× (19)
cho	6.1× (13)	3.9× (10)	4.6× (14)	4.3× (12)	3.2× (11)	3.5× (9)
fft	13.2× (68)	8.3× (82)	10.4× (67)	8.0× (42)	5.8× (43)	5.2× (23)
fmm	7.7× (21)	4.8× (20)	6.0× (19)	5.3× (16)	3.9× (16)	4.1× (12)
luc	8.6× (26)	5.3× (25)	6.7× (24)	5.9× (18)	4.2× (21)	4.4× (13)
lun	4.1× (8)	2.7× (6)	3.1× (7)	3.0× (6)	2.3× (6)	2.6× (6)
occ	6.9× (14)	4.3× (16)	5.3× (17)	4.8× (14)	3.5× (13)	3.8× (9)
ocn	6.8× (15)	4.2× (12)	5.2× (17)	4.7× (14)	3.5× (14)	3.8× (9)
rad	5.7× (11)	3.6× (12)	4.3× (12)	4.0× (10)	3.0× (9)	3.4× (8)
ray	7.3× (17)	4.6× (15)	5.6× (16)	5.0× (17)	3.7× (13)	4.0× (11)
wan	12.6× (71)	7.8× (56)	9.9× (70)	7.8× (32)	5.7× (45)	5.2× (19)
was	18× (186)	11.3× (250)	13.0× (121)	9.0× (51)	6.8× (79)	5.5× (25)

Table 2: Energy gain and optimal number of cores N_{opt} (in parenthesis) across SPLASH-2 benchmarks and technologies when including the three voltage scaling overheads.

1146

If Amdahl and architectural overheads can be neglected because an application is latency insensitive (for instance, sensor applications) then only the fundamental leakage overhead needs to be considered. To provide a comparison with the trend of V_{opt} for latency-sensitive applications, we show in Figure 9 the fundamental lower bound on V_{opt} across technologies, where leakage is the only voltage scaling overhead. In 180nm and 130nm V_{opt} for a perfectly parallelizable task is below threshold. Because technologies are becoming leakier with process scaling, V_{opt} has been trending upward with each generation and becomes super-threshold in 90nm. For a perfectly parallelizable task the energy gain has decreased from 52× in 180nm to 6× in 32nm. Likewise, the optimal number of cores N_{opt} has decreased from ~21,000 (clearly unachievable) in 180nm to 29 in 32nm. Though parallelizing across thousands of cores in older technologies is not achievable, the gains and N_{opt} in recent technology nodes have dramatically decreased even when neglecting Amdahl and architectural overheads.

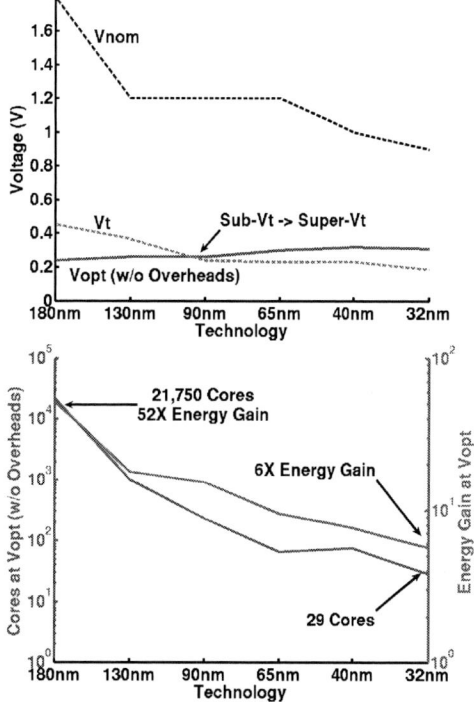

Figure 9: Theoretical maximum energy-efficient parallelism, showing V_{opt}, N_{opt}, and gains across six technology nodes with leakage overheads only.

3.2 Process Variation

A challenge of operating at a reduced voltage is increased sensitivity to process, temperature, and supply voltage variations that causes variability in circuit delay and energy consumption. A slower critical path and leakier devices decrease energy-efficiency thus increasing V_{opt}. Figure 10 (top) shows the 3-sigma delay variation relative to mean for a single gate and a chain of logic in 40nm technology using industrial variation models. Process variation can be global, affecting all transistors uniformly across a die, or local which causes delay mismatch between different devices and paths on a chip.

In the NTC region, local variation accounts for 30% of 3-sigma delay of a single gate. Since a CMP's maximum clock frequency is limited by the worst-case critical path, mismatch between different critical paths raises V_{opt} as the leakiest path runs at clock frequency set by the slowest path. However, local variation is reduced for deeper logic depths since local variation is usually uncorrelated and hence averages out along a path. Thus, for a chain 31 gates, local variation is only 10% of total variation, so its impact is minor.

Global variation raises V_{opt}, Figure 10 (bottom, for three technology nodes) but all paths will either: (1) slow and have less leakage or (2) have more leakage but run fast, so the total leakage overhead is relatively constant. This is unlike local variation where the leakiest path is run at the slowest clock frequency, thus global variation's contribution to increasing V_{opt} is less than local variation. Total delay variation at V_{opt} is significant, but high-performance CMPs are usually binned for speed so that each die can run at its optimal frequency. Thus, the range of bins will increase but, since each die is tuned to its optimum speed, global variation does not significantly increase V_{opt}.

The increase in V_{opt} when considering 3-sigma delay variation and the parallelization overheads described above is 30mV-60mV for an average SPLASH-2 benchmark. The delay variation also depends on the number of critical paths in a design, since local variation reduces by taking a maximum across multiple paths. As the number of paths increases the mean shifts up, but the variation is reduced, shown in Figure 10 (bottom). Thus, variation does impact delay but its impact on V_{opt} and minimum energy are small.

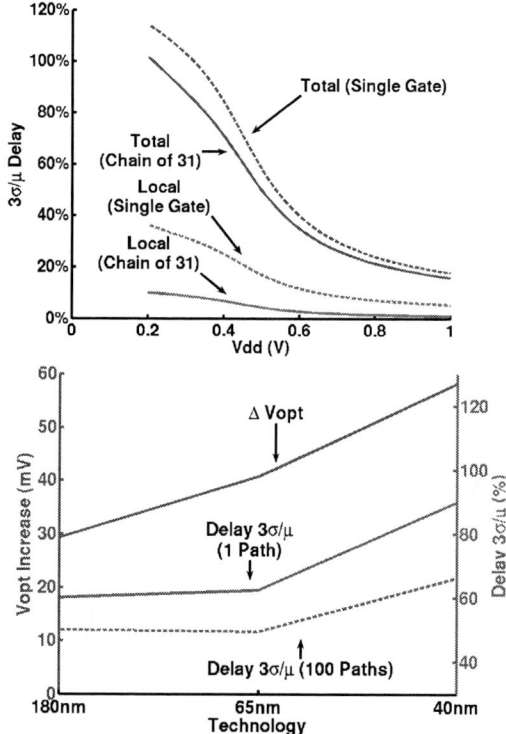

Figure 10: Change of delay for total and local process variation of a single gate and a logic chain of 31 gates (top). Increase in V_{opt} because of 3-sigma variation across generations (bottom).

3.3 Transistor Threshold Voltage

The energy-efficient operating voltage V_{opt} also depends on transistor threshold voltage selection. Conventionally regular threshold voltage transistors are used for high-performance applications, since they have the best drive strength, whereas the higher threshold voltage transistors are used where low static-power is a concern, such as in mobile applications.

When considering a parallelized task, Amdahl and architectural overheads limit the energy-efficiency and voltage scalability, thus setting V_{opt}, Figure 11 (top). As threshold voltage is reduced, leakage begins to dominate until the voltage scalability is limited by leakage overheads and not Amdahl or architectural overheads. The energy drops initially as threshold is reduced, since the task can run faster, and V_{opt} correspondingly tracks. Once leakage dominates the energy stays relatively constant. As a rule-of-thumb, the optimal threshold voltage is at the inflection point (~250mV in figure) between the parallelism-dominant and

1147

leakage-dominant region, since above this point energy increases and below this point the process becomes unnecessarily leaky.

For comparison, Figure 11 (bottom) shows V_{opt} when a task only includes leakage overheads. Since there is no Amdahl or architecture overheads, V_{opt} lowers as threshold voltage increases, since the integrated leakage current is reduced, until V_{opt} enters the subthreshold regime. Once V_{opt} is subthreshold, raising the threshold voltage does not change the energy-efficient operating voltage or energy consumption to first-order [8].

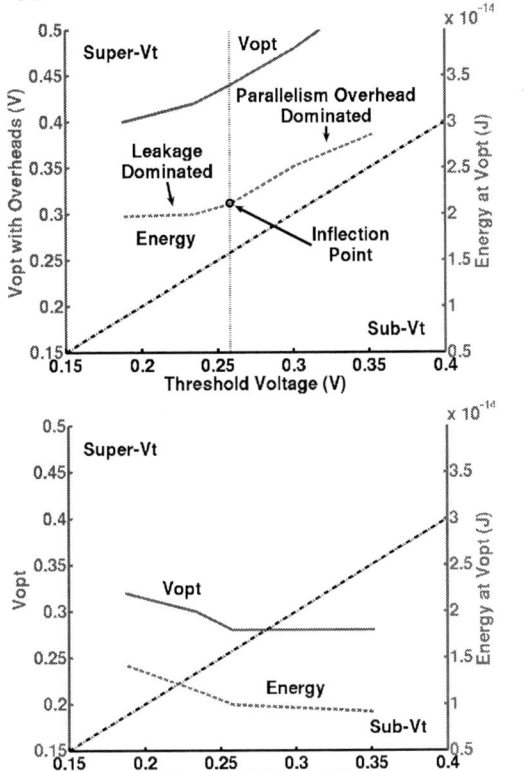

Figure 11: V_{opt} vs. Vt in 32nm for with Amdahl and architecture overheads (top) and leakage only (bottom).

Tasks that are not latency sensitive can operate in subthreshold with high Vt transistors, but this is not optimal for latency-sensitive applications. To achieve maximum performance in latency sensitive applications, even when limited by a thermal budget, the threshold voltage should be reduced until leakage starts to dominate voltage scalability.

4. CONCLUSIONS

We have detailed the limits of voltage scaling for latency-sensitive applications, when slower clock frequency is compensated by parallelization across multiple cores. As CMPs become limited by thermal cooling constraints, near-threshold operation is needed to maximize computations for a fixed thermal design power. The three voltage scaling limiters, leakage, Amdahl, and architectural, contribute to increasing the minimum energy and optimal supply voltage V_{opt} to maximize total CMP performance. As a guideline, the near-threshold region for maximum energy-efficiency is roughly 200mV-400mV above threshold voltage for most applications and this trend held for the six technology nodes we examined.

NTC operation increases energy-efficiency of a core by approximately $4\times$ in 32nm for the SPLASH-2 benchmarks we investigated, roughly translating to a $4\times$ improvement in performance for a thermally-limited CMP. Additionally, the maximum amount of energy-efficient parallelism is no more than 25 cores in 32nm. Delay variation increases in the NTC region, but has little impact on V_{opt}. For latency sensitive applications threshold voltage should be minimized until leakage dominates voltage scalability, whereas latency insensitive applications benefit from subthreshold operation.

5. ACKNOWLEDGEMENTS

The authors gratefully acknowledge the National Science Foundation and ARM Holdings for their support of this project.

6. REFERENCES

[1] G. Moore, "No exponential is forever: But 'forever' can be delayed!" *IEEE International Solid-State Circuits Conference* Keynote address, 2003.

[2] N. Weste, D. Harris. *CMOS VLSI design: a circuits and systems perspective. (3rd edition)*, Location: Boston, MA, 2005.

[3] R. Dennard, F. Gaensslen, V. Rideout, E. Bassous, and A. LeBlanc. "Design of ion-implanted mosfet's with very small physical dimensions," Solid-State Circuits, IEEE Journal of, 9(5):256 – 268, Oct. 1974.

[4] S. Sawant, et. al., "A 32nm Westmere-EX Xeon® enterprise processor," *Solid-State Circuits Conference Digest of Technical Papers (ISSCC), 2011 IEEE International* , vol., no., pp.74-75, 20-24 Feb. 2011

[5] H. Esmaeilzadeh, E. Blem, R. St. Amant, K. Sankaralingam, and D. Burger, "Dark Silicon and the End of Multicore Scaling," *The 38th Annual International Symposium on Computer Architecture ISCA*, June 2011.

[6] A. Chandrakasan, S. Sheng, and R. Brodersen, "Low-power CMOS digital design," JSSC, vol. 27, no. 4, Apr. 1992, pp. 473-484.

[7] A. Wang and A. Chandrakasan, "A 180mV FFT processor using subthreshold circuit techniques," *IEEE International Solid-State Circuits Conference*, pp. 292-529, 2004.

[8] B. Zhai, D. Blaauw, D. Sylvester, and K. Flautner, "Theoretical and practical limits of dynamic voltage scaling," *Design Automation Conference, 2004. Proceedings. 41st,* pp. 868 - 873, Jan 1 2004.

[9] L. Nazhandali, B. Zhai, R. Helfand, M. Minuth, J. Olson, S. Pant, A. Reeves, T. Austin, and D. Blaauw, "Energy Optimization of Subthreshold-Voltage Sensor Processors," *The 32nd Annual International Symposium on Computer Architecture ISCA*, Madison, Wisconsin USA, June 4-8, 2005.

[10] J.M. Rabaey, A.P. Chandrakasan, and B. Nikolic, *Digital Integrated Circuits: A Design Perspective*, Prentice Hall, 2003.

[11] S. Hanson *et al.*, "Performance and variability optimization strategies in a sub-200mV, 3.5pJ/inst, 11nW subthreshold processor," *Symposium on VLSI Circuits*, pp. 152-153, 2007.

[12] M. Seok, S. Hanson, Y. Lin, Z. Foo, D. Kim, Y. Lee, N. Liu, D. Sylvester, D. Blaauw, "The Phoenix Processor: A 30pW platform for sensor applications," *IEEE Symposium on VLSI Circuits*, pp. 188-189, 2008.

[13] Amdahl, Gene (1967). "Validity of the Single Processor Approach to Achieving Large-Scale Computing Capabilities". *AFIPS Conference Proceedings* (30): 483–485.

[14] N. Binkert, B. Beckmann, G. Black, S. K. Reinhardt, A. Saidi, A. Basu, J. Hestness, D. R. Hower, T. Krishna, S. Sardashti, R. Sen, K. Sewell, M. Shoaib, N. Vaish, M. D. Hill, and D. A. Wood, "The gem5 simulator," *Computer Architecture News (CAN)*, June 2011.

[15] G. M. Amdahl. "Validity of the single processor approach to achieving large scale computing capabilities," *AFIPS '67 Proceedings*, April 1967

Near-Threshold Voltage (NTV) Design—Opportunities and Challenges

Himanshu Kaul, Mark Anders,
Steven Hsu, Amit Agarwal,
Ram Krishnamurthy,
Shekhar Borkar
Intel Corp.
2111 NE 25th Avenue, JF2-04
Hillsboro, OR 97124

Shekhar.Y.Borkar@intel.com

ABSTRACT

Moore's Law will continue providing abundance of transistors for integration, only to be limited by the energy consumption. Near threshold voltage (NTV) operation has potential to improve energy efficiency by an order of magnitude. We discuss design techniques necessary for reliable operation over a wide range of supply voltage—from nominal down to subthreshold region. The system designed for NTV can dynamically select modes of operation, from high performance, to high energy efficiency, to the lowest power.

Categories and Subject Descriptors

B 7.1 [Microprocessors and microcomputers, VLSI]

General Terms

Performance, Design, Reliability, Experimentation, Verification.

Keywords

NTV, subthreshold, power, energy, performance.

1. INTRODUCTION

VLSI technology scaling has continued over the last several decades enabling affordable, efficient gadgets enriching lives, which we now take for granted. There were several challenges on the way threatening progress: design productivity in the 80's, power consumption in the 90's, and leakage issues in the last decade. Advances in design automation for productivity, clock gating, and power management came to the rescue. Although the technology scaling treadmill has continued, doubling transistors every generation, reduction in supply voltage scaling does not reduce energy per operation to utilize all the transistors. Therefore, the next challenge we face is of energy efficiency—not just low power—but to continue to deliver logic throughput with much less energy consumption. An order of magnitude reduction in energy per operation will be required.

Subthreshold operation of circuits, where supply voltage is reduced below the threshold voltage of the transistor, was believed to be the most efficient operating point. Although this mode of operation consumes much lower power, it is not necessarily the most energy efficient, as we will show later. Rather, near threshold voltage (NTV) operation, where supply voltage is reduced close to the threshold, provides higher energy efficiency. We will describe the benefits of the NTV operation, issues and design challenges, and opportunities for design automation to enable this new design paradigm.

2. BENEFITS OF NTV OPERATION

At nominal operating point, the frequency of operation reduces almost linearly with the supply voltage, reducing performance linearly, and reducing active energy per operation quadratically. Leakage power too reduces exponentially, and therefore reducing supply voltage should not only reduce power but also improve energy efficiency. Our test chip experiment was conducted to prove this benefit [1], and we expected energy efficiency benefit to extend even in the subthreshold region of operation, with even greater efficiency.

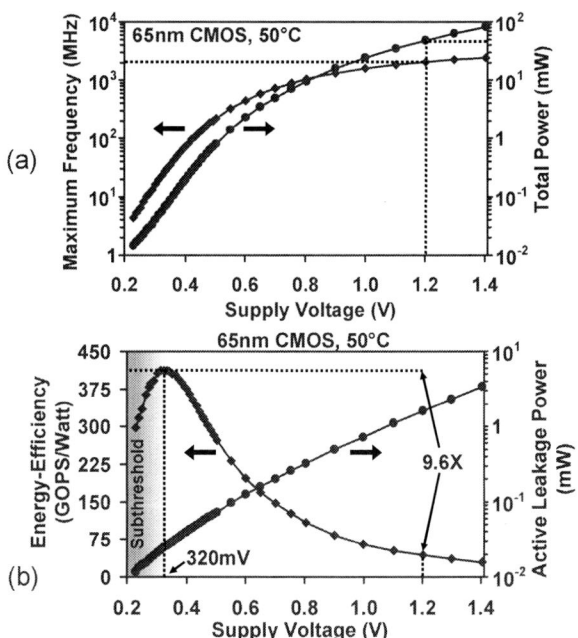

Figure 1: Energy efficiency of NTV operation

The results were, however, a little surprising as shown in Figure 1. As the supply voltage is reduced the frequency reduces (a), and the energy efficiency increases (b) as expected; however, it peaks near the threshold voltage of the transistor and then starts reducing in the subthreshold region. This unexpected reduction in the subthreshold region is explained by noticing the following. In the subthreshold region leakage power dominates, and it reduces with voltage but the reduction in frequency is larger than reduction in the leakage power, reducing energy efficiency. Therefore, it is desirable to operate close to the threshold voltage of the transistor for maximum energy efficiency, almost 10X compared to at nominal supply voltage. Subthreshold operation will yield even lower power, but at the expense of reduced energy efficiency, and may be desired in some applications.

The NTV operation is shown to scale well with technology with measurements confirming benefits on 45 nm, 32 nm, and 22 nm technologies [2][3][4]. Notice that it includes even the new tri-gate transistor technology (22 nm), clearly showing benefits across technology generations.

3. DESIGN CHALLENGES & SOLUTIONS

A conventional design will scale in voltage improving energy efficiency; however, the voltage scaling will be limited because voltage sensitive circuits will start failing much before the threshold voltage. The circuits have to be designed to operate in the NTV mode comprehending the side-effects of lowering the voltage. In this section we discuss some of the major design challenges and solutions. Detailed discussion on this topic may be found in [5].

3.1 Variations in circuit performance

As the supply voltage approaches threshold voltage a small change in the supply voltage results in a large change in the logic delay or frequency of operation. Figure 2 shows modeling of frequency of a logic block with voltage scaling. The frequency reduces almost linearly as expected. However, even a 5% change in the supply voltage or in the threshold voltage has increasingly larger spread in the frequency as the voltage is reduced. As much as 50% variation in the frequency may be expected near the threshold voltage.

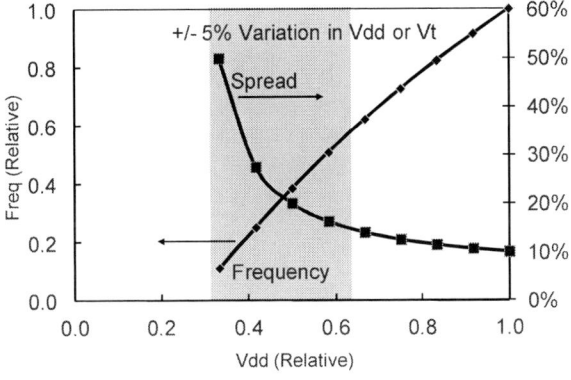

Figure 2: Modeling frequency variation

Detailed variation analysis and measurements are described in [1] and shown in Figure 3.

Figure (a) shows Monte-Carlo simulations showing spread of frequency at nominal voltage as well as NTV. At nominal voltage the spread is +/- 18%, and it increases to +/- 2X at NTV. Figure

(b) shows impact of temperature increasing the spread from +/- 5% at nominal voltage to +/- 2X at NTV across the temperature range.

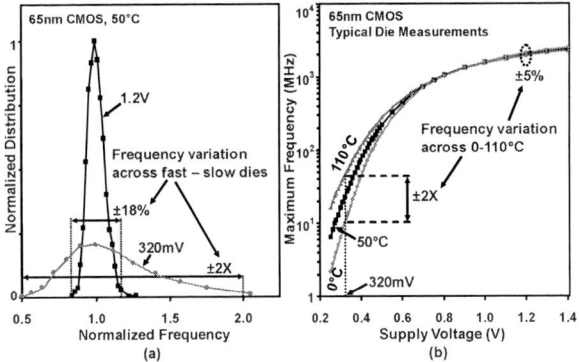

Figure 3: Modeling and measurements of variation

To compensate for logic performance variations several techniques have been proposed, including applying body bias. These conventional techniques will have limited scope because deeply scaled technologies have either no body, or little body effect left, and the energy cost of fine grain variation control could reduce the energy benefit.

We propose to tolerate the effect of variations using system level techniques. For example, in a many core system where the number of cores is very large, the cores will exhibit different frequency of operation due to variations. Assign the nearest frequency of operation to these cores, and due to the law of large numbers, the overall logic throughput of the chip will not be affected, as shown in Figure 4.

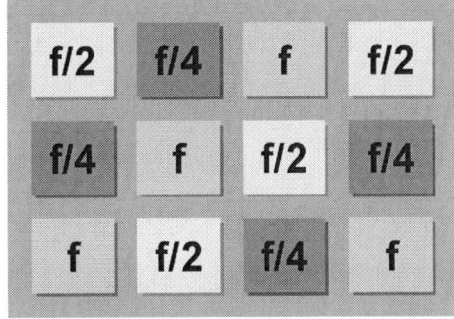

Figure 4: Frequency assignment in many-core system

3.2 Subthreshold Leakage

The subthreshold leakage power will have two adverse effects: (1) disproportionately large leakage power, and (2) higher variability in the leakage power. Figure 1 shows that across the entire supply voltage range the total power reduces by four orders of magnitude, but the leakage power reduces by only three orders of magnitude. The active power reduces cubically, but leakage power does not, and that is why, expect disproportionately larger percentage of subthreshold leakage power with NTV operation.

Figure 5 shows modeling of subthreshold leakage power in successive generations of technologies. Assuming 20% of the total power is in leakage in each generation, it shows percentage of leakage power increasing with NTV. As much as 50% of the total power could be in leakage and that too with much increased

1150

variability. The total power consumption of the system is much lower, but substantial portion of the power will be in leakage.

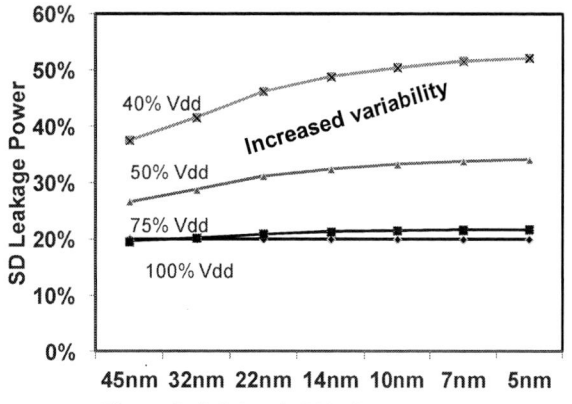

Figure 5: Subthreshold leakage power

At low logic activity the active power is low and the leakage power dominates, reducing the effectiveness of NTV for energy efficiency. Therefore, fine grain leakage power management, with sleep transistors or power gating techniques will be even more important.

3.3 SRAM and Register File

Small signal arrays, such as static memory, are designed to operate in a narrow voltage range and need significant design considerations. 6T static memory cells are typically designed with small transistors for higher density, and thus have stability and yield issues at lower voltages. There are two potential solutions for static memory: (1) employ larger 6T memory cells, or 8T, 10T cells which can operate at lower voltages, all compromising area, and (2) do not operate static memory blocks at NTV. Since static memory energy consumption is relatively low in a system it may be a good compromise.

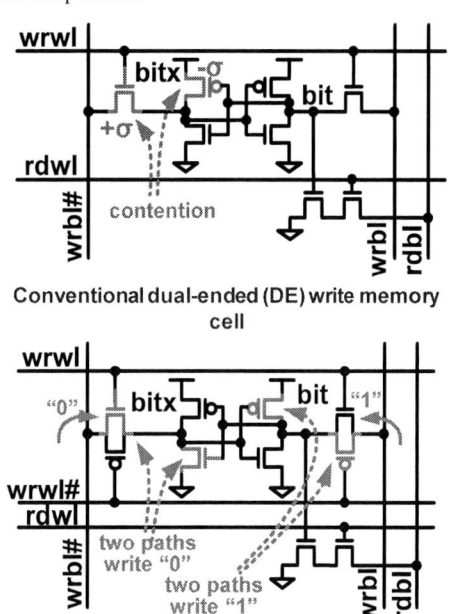

Figure 6: NTV Tolerant Register File

Register file circuits at NTV are limited by contention in read/write circuits due to parameter variation which becomes worse with technology scaling; minimum sized devices are worse in this respect. Also at lower voltages, increased write contention between strong PMOS pull-up and weak NMOS transfer devices across parameter variations could result in faulty behavior.

The register file circuit can be made NTV friendly by replacing the conventional dual-ended write cell equipped with a transmission gate [6], as shown in Figure 6. Upsizing the NMOS transfer devices in a conventional dual-ended write cell improves write contention; however, higher threshold voltage in cross-coupled inverter devices caused by parameter variation still increases write completion delay, limiting voltage scaling. By replacing NMOS transfer devices with full transmission gates improves both contention and voltage scaling because: (a) it provides two paths to write "1" or "0" to both node bit lines, averaging random variation across two transistors, (b) strong "1" and "0" writes on both sides, and (c) cell symmetry (NMOS and PMOS) reduces the effect of systematic variation.

3.4 Latches, Flip-Flops, Multiplexers, Gates

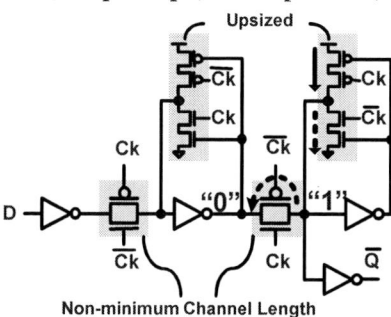

Figure 7: NTV friendly flip-flop design

The storage nodes in latches and flip-flops have weak keepers and large transmission gates. When the transmission gate for the slave stage of a conventional master-slave flip-flop circuit is turned off, the weak on-current from the slave-keeper contends with the large off-current through the transmission gate. This causes the node voltage to drop, affecting the stability of the storage node. Low voltage reliability of the flip-flops can be improved by the use of non-minimum channel length devices in the transmission gates to reduce off-currents, and with upsized keepers to improve on-currents to restore charge lost due to leakage. The write operation remains unaffected since the keepers are interruptible. The circuit modifications shown in Figure 7 reduce the worst-case droop by 4X in the ultra-low voltage optimized design.

To tolerate effects of variations at low voltages, averaging technique can be employed, as shown in Figure 8, described in [4]. Vector flip-flops across two adjacent cells with shared local minimum sized clock inverters to average variation, reducing low voltage hold time violations and improving minimum supply voltage by 175mV. The stacked min-delay buffers also limit variation-induced transistor speed up, improving hold time margin at low voltage by 7%-30%.

1151

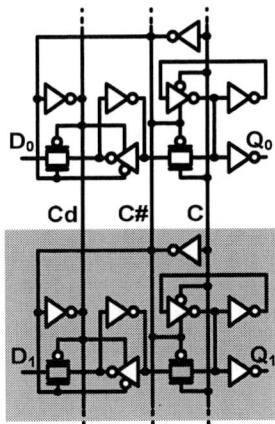

Figure 8: Vector flip-flop

Wide multiplexers are also prone to static droops on nodes shared by transmission gates at low voltages. Such structures are typical for one-hot multiplexers, where the on-current of one of the selected inputs contends with the off-current of the remaining unselected inputs. To avoid this effect, wide multiplexers should be remapped using 2:1 multiplexers as shown in Figure 9, thereby reducing the worst-case off-current contention. Remapping a one-hot 4:1 multiplexer to an encoded 4:1 multiplexer composed of 2:1 multiplexers results in up to 3X reduction in the worst-case static droop.

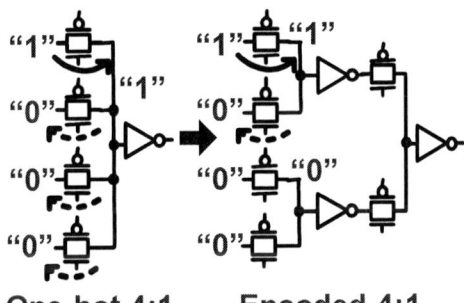

One-hot 4:1   Encoded 4:1

Figure 9: Multiplexers redesigned for NTV

Other optimizations in the logic include remapping deep stacked combinational logic and series connected transmission gates to a maximum of 3 transistor stack to reduce body effect, and to minimize effects of large off-current paths and weak on-current paths on circuit nodes.

3.5 Level Converters

The use of multiple supply voltage domains results in the need for level shifter circuits at the low-to-high voltage domain boundaries. A conventional level shifter uses a CVSL stage to provide the up-conversion functionality, with the associated contention currents contributing to a significant portion of the level shifter power. Driving the output load directly with the CVSL stage increases its size, while use of additional gain stages after the level shifter to reduce CVSL stage loading results in increased delay.

Figure 10 shows a 2-stage cascaded split-output level shifter. An intermediate supply voltage for up-conversion over such a large voltage range limits the maximum current ratio between the higher-supply PMOS pull-up and lower-supply NMOS pull-down devices for correct CVSL stage functionality. Energy-efficient up-

conversion from subthreshold voltage levels to nominal supply outputs is achieved by decoupling the CVSL stage of this level shifter from the output, enabling a downsized CVSL stage for the same load without extra gates in the critical path. Reduced contention currents in a downsized CVSL stage enable the split-output design to achieve up to 20% energy reduction for equal fan-out and delay.

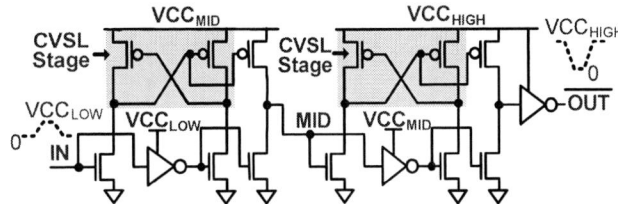

Figure 10: Two stage cascaded split-output level shifter

Ultra-low voltage split-output level shifters are described in [4] and shown in Figure 11.

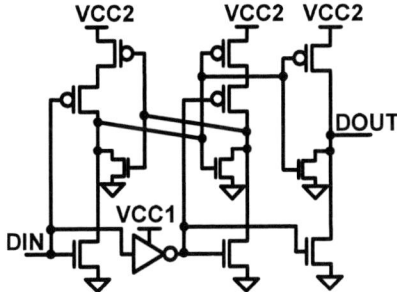

Figure 11: Ultra-low voltage split-output level shifter

This level shifter decouples the CVSL stage from the output driver stage and interrupts contention devices, improving minimum supply voltage by 125mV. For equal fan-in/out, the level shifter weakens contention devices, thereby reducing power by 25%-32%.

3.6 Soft Errors & Reliability

Single event upsets (soft-errors) is of concern, and especially with NTV operation [7] [8]. These errors are caused by alpha particles and more importantly cosmic rays (neutrons), hitting silicon chips, creating charge on the nodes to flip a memory cell or a logic latch. These errors are transient and random. It is relatively easy to detect these errors in memories by protecting them with parity, and correcting these errors in memory is also relatively straight forward by employing error correcting codes. However, if such a single event upset occurs in random logic state then it is difficult to detect and correct.

Soft error rate per bit has been decreasing with technology scaling; however the number of bits is almost doubling each generation, with the net effect of increased soft errors as shown in Figure 12 (a). Notice that figures (b) and (c) show soft errors increasing with reduced supply voltage, making NTV designs more vulnerable to soft errors, and needs further investigation.

NTV operation will have some positive impacts on reliability. Due to reduced supply voltage electric fields are reduced, and lower power consumption will yield lower junction temperature. Therefore, device aging effects, such as NBTI will be less of a concern. Lower temperature and lower currents will also reduce electromigration related defects.

1152

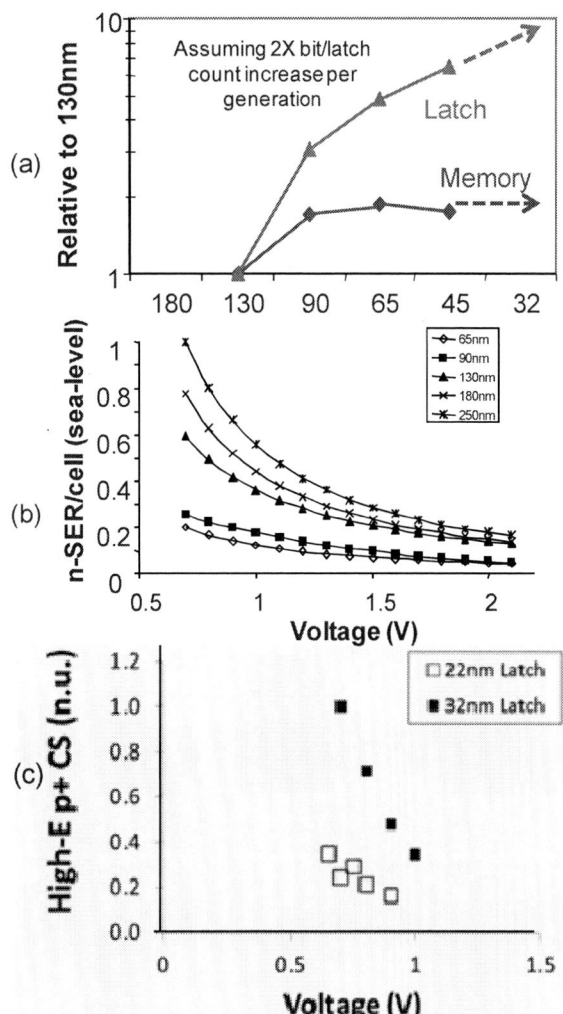

Figure 12: Soft error projections

4. EXPERIMENTAL NTV PROCESSORS

Following the NTV design guidelines several experimental designs have been reported [9][10] with encouraging results. We highlight the experimental Pentium® processor designed to operate from nominal to NTV, as well as in the subthreshold region, with varying performance, power, and energy efficiency.

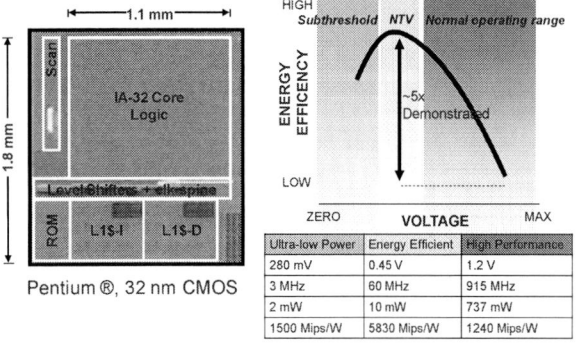

Figure 13: NTV Pentium® processor

The experimental processor was designed on 32 nm bulk CMOS process, following all of the design guidelines mentioned above, with the goal to operate over the full voltage range—from nominal to subthreshold. At nominal supply voltage it provides the highest performance with modest power and energy efficiency. In the subthreshold region it provides the lowest power, with reduced performance and modest energy efficiency. At NTV it provides the highest energy efficiency, with three orders of magnitude lower power than the original design two decades ago on 0.7μ technology, yet delivering the same performance. This example shows that an NTV design can provide a wide dynamic range—from high performance to low power to high energy efficiency as shown in Figure 13. Notice that this experiment reports only 5X improvement in energy efficiency at NTV due to two reasons: (1) the original design on 0.7μ technology did not comprehend NTV guidelines, and (2) the SRAM (caches) in this design do not operate at NTV.

5. SYSTEM LEVEL OPTIMIZATION

Although NTV has a potential to improve energy efficiency of logic throughput by an order of magnitude a careful system level optimization is required to determine the most efficient NTV operating point.

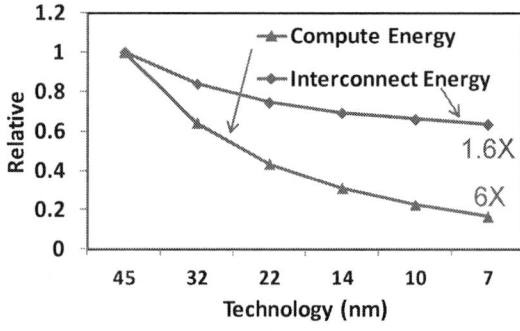

Figure 14: Compute & global interconnect energy scaling

Logic energy (with its own local interconnect) will scale disproportionately with respect to global interconnect energy as shown in Figure 14. That is, energy per operation will reduce faster than energy to move data over a fixed distance. Since NTV reduces frequency of operation, thus reducing throughput of the logic block, more logic will be needed for constant throughput (for example parallelism). This may incur more data movement, adding data movement energy to the system. As the supply voltage comes closer to the threshold with NTV, system's logic energy reduces but the data movement energy increases. Hence a global optimization at the system level is required to determine the optimal NTV operating point.

6. PROSPECTIVE OF NTV

The great old days of Moore's law scaling and performance; typified by dramatic improvements in transistor density, speed, and energy, delivered 1000-fold performance improvement. The progress continues, but will be more difficult, with technology scaling producing continuing improvements in transistor density, but comparatively little improvement in transistor speed and energy. As a result, in the future, the frequency of operation will increase slowly; and energy will be the key limiter of performance. That is why there is a fear of Dark Silicon—unused silicon, or idle transistors—due simply because of energy [11]. With business as usual, and without continued innovations, this would be a likely scenario, but far from reality.

Future designs will use large-scale parallelism, with heterogeneous cores, a few large cores and a large number of small cores, operating at low frequency and low voltage near threshold—NTV for extreme energy efficiency [12]. Aggressive use of various types of customized accelerators will yield the highest performance and greatest energy efficiency on many applications. The objective will be the purest form of energy proportional computing, and at the minimum levels of energy possible. Heterogeneity in compute and communication hardware will be essential to optimize for performance for energy proportional computing and to cope with variability—all made possible by NTV.

7. DESIGN AUTOMATION

Design automation will play a critical role in enabling NTV operation, by devising tools and methodologies, hiding the details of the NTV design constraints from the designer. We list a few such desired design technologies:

Device modeling	Transistors models are not very accurate near threshold. They need to be modeled accurately.
Variation modeling	Model variations and impact on circuit timings and leakage power.
Identification of NTV circuit blocks (fine grain)	Identify circuits that are not in critical path that can take advantage of NTV.
Identification of NTV system blocks (coarse grain)	Identify large system blocks that can use NTV without sacrificing system performance.
Insertion of level converters	Automatically insert level converters and interface circuits.
Synthesis	Synthesize circuit blocks comprehending variations and timing criticalities.
Leakage power management	Automatically insert sleep transistors (power gates) in the design.
System level modeling and optimization	Model the system, synthesize, and determine the optimum NTV operating point.
System level synthesis	Synthesize the entire system with selective operation of system blocks at NTV.
Testing	Testing strategies for wide supply voltage range.

8. CONCLUSION

Moore's Law will continue providing abundance of transistors for integration, only to be limited by the energy consumption. Near threshold voltage (NTV) operation of logic can improve energy efficiency by an order of magnitude. We have discussed several NTV design techniques for such future designs, allowing them to operate over a wide range of supply voltage, to dynamically select modes of operation, from high performance, to high energy efficiency, to the lowest power.

9. ACKNOWLEDGMENTS

This research was, in part, funded by the U.S. Government under contract number HR0011-10-3-0007. The views and conclusions contained in this document are those of the authors and should not be interpreted as representing the official policies, either expressed or implied, of the U.S. Government.

10. REFERENCES

[1] H. Kaul et al, "A 320 mV 56 µW 411 GOPS/Watt Ultra-Low Voltage Motion Estimation Accelerator in 65 nm CMOS", JSSC, Volume: 44 , Issue: 1 , 2009.

[2] H. Kaul et al, "A 300 mV 494GOPS/W Reconfigurable Dual-Supply 4-Way SIMD Vector Processing Accelerator in 45 nm CMOS", JSSC, Volume: 45 , Issue: 1, 2010.

[3] H. Kaul et al, "A 1.45GHz 52-to-162GFLOPS/W Variable-Precision Floating-Point Fused Multiply-Add Unit With Certainty Tracking in 32nm CMOS", ISSCC, 2012.

[4] S. Hsu et al, "A 280mV-1.1V 256b Reconfigurable SIMD Vector Permutation Engine with Vertical Shuffle in 22nm CMOS", ISSCC, 2012.

[5] R. Dreslinski et al, "Near-Threshold Computing: Reclaiming Moore's Law Through Energy Efficient Integrated Circuits", Proceedings of the IEEE, Volume: 98 , Issue: 2, 2010.

[6] A. Agarwal et al, "A 32nm 8.3GHz 64-entry × 32b variation tolerant near-threshold voltage register file", VLSI Circuits Symposium, 2010.

[7] N. Seifert et al, "Radiation-Induced Soft Error Rates of Advanced CMOS Bulk Devices", 44th Reliability Physics Symposium, 2006.

[8] N. Seifert et al, "Soft Error Susceptibilities of 22nm Tri-Gate Devices", NSREC, 2012

[9] M. Seok et al, "The Phoenix Processor: A 30pW platform for sensor applications", VLSI Circuits Symposium, 2008.

[10] S. Jain et al, "A 280mV-to-1.2V Wide-Operating-Range IA-32 Processor in 32nm CMOS", ISSCC 2012.

[11] H. Esmaeilzadehy et al, "Dark Silicon and the End of Multicore Scaling", ISCA '11.

[12] S. Borkar et al, "The Future of Microprocessors", Communications of the ACM, May 2011.

Near-Threshold Operation for Power-Efficient Computing? It Depends...

Leland Chang and Wilfried Haensch
IBM T. J. Watson Research Center
Yorktown Heights, NY 10598

{lelandc, whaensch}@us.ibm.com

ABSTRACT

While it has long been argued that near-threshold (~0.5V) operation of CMOS technologies can dramatically improve power efficiency, widespread application of such low voltage operation to VLSI systems has yet to materialize. This is due in part to practical system workload demands, in which single-thread performance needs can limit strategies to improve parallelizeable throughput performance, but also due to barriers in the ability of supporting hardware to counter variability and reliability concerns while maintaining power efficiency throughout the system. This paper describes the issues on which the realization of near-threshold computing depends to explain why this strategy is not yet pervasive today. However, recent advancements across the spectrum of system design – including heterogeneous architectures, transistor and memory technologies, power delivery, packaging, and I/O – suggest that as the market for throughput performance grows, hardware technologies may soon become available to practically harness the promise of near-threshold operation.

Categories and Subject Descriptors

B.7.0 [**Integrated Circuits**]: General.

General Terms

Performance, Design, Reliability

Keywords

VLSI circuits, Digital circuits, Power efficiency, Low voltage, Near-threshold computing, Parallelism, Throughput performance, Single-thread performance, Power delivery, Power management, Technology optimization, SRAM, Soft errors, Variability.

1. INTRODUCTION

In recent years, fundamental physical limitations have forced CMOS scaling to deviate from historical trends. In particular, non-scalability of the transistor threshold voltage and underlying limits on the subthreshold slope have limited supply voltage reduction due to the need to balance leakage power and device performance. Challenges in gate dielectric scaling and manufacturing variability have further contributed to this trend. Because raw speed has historically been of paramount importance, the supply

voltage in modern technologies is now significantly higher than originally suggested by scaling theory [9], which has led directly to dramatic increases in power consumption. As cooling and battery life constraints today place severe restrictions on product performance, power efficiency will be key to sustaining continued performance enhancement in future VLSI systems.

Near-threshold operation of CMOS technologies has long been touted as a strategy to improve power efficiency for logic computation [2,3,5,11,16] by depending upon parallelism to compensate for frequency loss. Practical application of such techniques, however, has remained elusive. Fundamentally, the viability of near-threshold computing depends upon the extent to which parallelism can be used in practical systems. Despite growing acceptance of parallel algorithms, single-thread performance needs have thus far limited the ability of systems to utilize low-voltage operation. In addition, the requisite system hardware needed to support parallel systems operating at near-threshold voltages has not been available, which diminishes gains in power efficiency and, at worst, precludes low-voltage functionality altogether. Recent progress, however, suggests that solutions to these issues may be within reach, thereby enabling practical realization of the power efficiency benefits of near-threshold computing.

2. A PROMISE OF POWER EFFICIENCY

Power dissipation in the active mode, which is comprised of dynamic switching and static leakage, has a roughly cubic dependence on voltage. Circuit speed, however, is empirically observed to have an approximately linear dependence on voltage [5]. As such, a reduction in voltage can dramatically reduce power dissipation at a modest performance loss, which can be compensated

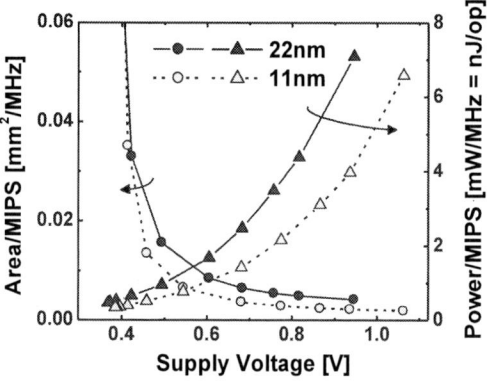

Figure 1. Near-threshold (~0.5V) operation balances parallelizeable performance (area/MIPS) and power efficiency (power/MIPS). From Chang, et al. [5]

for in parallelizeable workloads by linearly increasing the number of processors (and thus chip area). Figure 1 plots the results of a detailed technology optimization program [10], in which each point represents a technology with all parameters tuned for optimal performance at a given supply voltage. Power/MIPS, which is related to energy per operation, and area/MIPS, which is inversely related to performance in a parallel system, achieve a balance at ~0.5V across different technology generations. As compared with the ~1V supplies widely used today, operation at 0.5V can provide an 8x improvement in power efficiency with a moderate 4x frequency loss that could potentially be compensated for by parallelism. While the specific optimum voltage will vary somewhat between different applications, such a near-threshold voltage regime is likely to provide a practical compromise for power efficiency to improve parallelizeable throughput performance.

3. IT DEPENDS...

Realization of the promise of power efficiency of near-threshold computing depends strongly on the ability to utilize parallel algorithms in systems of interest as well as to engineer appropriate hardware to support such systems. To date, both have proved to be significant restrictions. However, with increasing acceptance of parallelism in computing architectures and algorithms as well as recent advancements in hardware technology, there is hope that these limitations may be overcome.

While the pervasiveness of near-threshold operation will ultimately be limited by single-thread performance needs, the invocation of heterogeneity in system design may be key to enabling its practical use. With such a strategy in place, challenges in near-threshold hardware design are exposed, including variability tolerance, reliability, power delivery, packaging, and I/O – each of which could offset power efficiency gains in computation. Solutions to these issues are within reach, however, and will all need to be developed to enable near-threshold computing. Market forces and future product needs will thus ultimately determine whether these techniques can be developed for widespread application.

3.1 ...On the Advent of Heterogeneity

The performance of computing systems today is characterized by both single-thread, which relies on the raw speed of a single processor core, and throughput, which can be improved linearly with parallelism, metrics. Their relative importance depends on the workloads represented in performance benchmarks, which may vary greatly across the application space. Some systems, such as for HPC or exascale computing [13], may be sufficiently specialized to focus on throughput performance, but many mainstream systems must instead consider both throughput and single-thread

performance. Despite growing acceptance of parallel algorithms, single-thread performance remains important for applications or portions thereof that cannot be parallelized. Housekeeping overheads for parallelization as well as fundamental limits in the fraction of computation that is sequential in nature ultimately limit the benefits that can be achieved through parallelism [1] and drive the need for continued single-thread performance improvements. In addition, some key applications depend on legacy software code for which parallelization may simply be impractical. Since near-threshold operation necessarily degrades operating frequency, practical systems will require a separate solution to meet single-thread performance requirements.

To balance throughput and single-thread performance needs, heterogeneous systems combining near-threshold cores with traditional ~1V cores may provide an optimal overall solution [2]. As shown in Figure 2, such a system could either be implemented as a parallel system that is heterogeneous by design with a few fast, high-voltage cores and many efficient, near-threshold cores or as a dynamically adjustable parallel system in which the voltage of a few cores can be raised to improve single-thread performance and lowered to improve throughput performance. Heterogeneity by design allows for separate optimization of the two types of processing cores, but requires distinct core designs and fixes heterogeneity within the overall chip at design time. Dynamic heterogeneity allows more chip-level flexibility and utilizes only a single core design, but this one entity must balance optimization across a wide voltage range, which is inevitably less optimal for either single-thread or throughput performance. Either strategy, however, can use near-threshold operation to improve throughput performance while maintaining the ability to provide single-thread performance at high voltage. Associated advancements will also be needed to effectively manage the overhead of such heterogeneity, including efficient and granular regulation of on-chip voltages and synchronization of I/O channels due to non-uniform core operation frequencies. As such, to address single-thread performance needs, near-threshold computing may require significant work in the development of heterogeneous systems.

3.2 ...On Supporting Hardware Technology

Practical realization of near-threshold systems has been challenging due not just to applicability in the marketplace, but also due to technical issues in building reliable and efficient systems. Variability and soft error reliability have often been heralded as key concerns for logic and memory at low voltages, but recent techniques have been developed to both mitigate root causes at the device level and augment operating margins at the circuit level. Other challenges include efficient delivery and management of voltage rails as well as packaging and I/O to support increased levels of parallelism – issues for which new solutions have also recently become available. Without commensurate advances in each of these areas, gains in the efficiency of computation from near-threshold operation may be negated by penalties in operating margins or inefficiencies in power delivery and communication.

3.2.1 New and Re-optimized Device Technologies

Existing CMOS technologies are traditionally optimized for ~1V operation. As a result, direct usage at near-threshold voltages would lead to a significant degradation of operating margins due to the enhanced impact of variability. However, to mitigate such concerns, technologies can instead be optimized to achieve low-voltage robustness by trading off performance in the ~1V regime. In and of itself, near-threshold operation does not drive fundamental changes in device technology. Rather, near-threshold operation

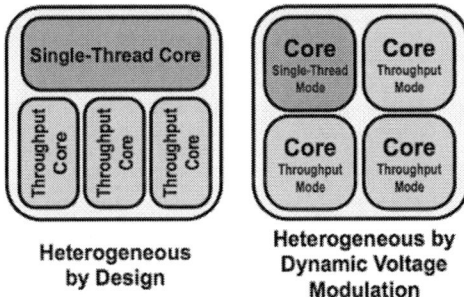

Figure 2. Examples of heterogeneity in microprocessors, which may be essential to utilize near-threshold operation while balancing single-thread performance needs

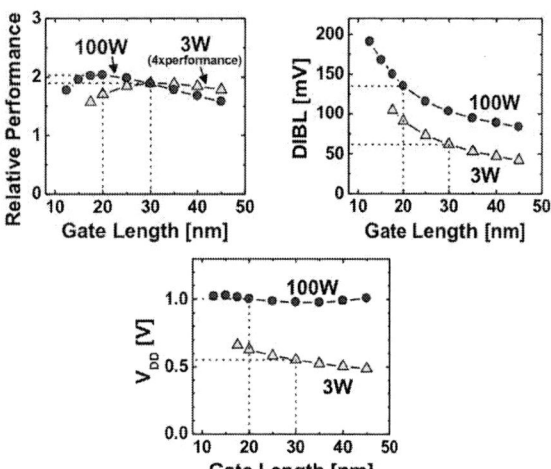

Figure 3. Devices optimized for near-threshold voltages (3W processor case) prefer long gate lengths and improved electrostatics to mitigate variability. From Chang, et al. [5]

drives to a different optimization point – one that focuses on variability tolerance rather than raw performance.

In traditional ~1V technologies, transistor gate lengths are scaled aggressively to achieve maximum device performance with only moderate short-channel effect control; these devices can be very sensitive to process-induced variation, especially when operated at low voltage. Devices optimized for near-threshold operation instead exhibit significantly improved control of short-channel effects, which can be achieved by longer gate lengths and better device electrostatics [5]. This is demonstrated in the optimization results in Figure 3, which optimize all technology parameters other than gate length for two different chip power levels. It can be seen that in the low-power case, optimal parallelizeable performance is achieved when drain-induced barrier lowering (DIBL), as characterized by the difference between the linear and saturated threshold voltages, is significantly reduced. While longer gate lengths linearly worsen area and capacitance, improved short-channel effects reduce variability to enable more aggressive voltage scaling, which quadratically improves power consumption. Thus, by emphasizing transistor electrostatics instead of raw device speed, variability can be controlled to achieve a significant reduction in voltage and power.

The introduction of tri-gate [8] or other thin-body transistor structures [4], may be especially advantageous for near-threshold operation. Instead of scaling gate length, the improved electrostatic integrity of such devices could be used to better control short-channel effects and further lower voltages to improve power efficiency. Additionally, since the thickness of the silicon channel ultimately controls short-channel effects, low channel doping could potentially be used, which may eliminate discrete dopant fluctuation concerns. While such structures can alleviate traditional variability mechanisms, it should be noted that new sources of variability might also be introduced. In particular, the body or fin thickness dimension must be sufficiently thin and strictly controlled to achieve a net improvement in variability. In addition, as these devices are fully-depleted and thus require gate work function engineering to set device threshold voltages, the availability of appropriate gate materials to achieve target threshold voltages may limit the realization of low channel doping, which may forfeit benefits in reduced discrete dopant fluctuation.

For heterogeneous systems that need to achieve both single-thread and throughput performance targets, a key question is whether devices optimized for near-threshold voltages can be efficiently integrated onto the same chip with devices optimized for ~1V. Without such integration, any single technology inevitably causes degradation in single-thread and/or throughput performance. Utilizing a technology optimized for ~1V degrades near-threshold throughput performance due to variability limits on voltage scaling. Employing instead a technology optimized for near-threshold operation degrades single-thread performance due to long gate lengths. Alternatively, a technology that balances requirements across a wide operating voltage range is necessarily non-optimal for both single-thread and throughput performance.

3.2.2 Low-Voltage SRAM and Latch Topologies

Low-voltage memory today is already severely limited by variability. Even at ~1V operation, SRAM operating margins are degraded by random variation as caused by discrete dopant fluctuation and gate line edge roughness. The use of aggressive peripheral assist circuits and dedicated higher voltage supplies is now commonplace in a variety of VLSI products. In order to realize the full benefit of near-threshold computing, SRAM voltages will need to scale accordingly to ensure that memory does not dominate total power; this greatly exacerbates what is already a challenging issue. Techniques to minimize variability at the device level, such as re-optimization to longer gate lengths or migrating to transistor structures with low channel doping, can enable SRAM voltage scaling, but it is likely that additional techniques will be necessary due to the extreme voltage reduction required. Transitioning from the traditional 6T cell to an 8T cell [7] as shown in Figure 4 can enhance variability tolerance at the circuit level by eliminating read and write contention within the cell while still preserving competitive bit cell density. The 8T cell concept is already seeing use in high-performance caches, but may need to be extended to high-capacity cache levels to, in conjunction with device level variability reduction and aggressive assist circuits, enable near-threshold computing.

In addition to variability, soft errors as induced by alpha particles and cosmic rays can be a significant reliability concern at near-threshold voltages. This is due both to reduced critical charge (Q_{crit}) at low voltage as well as increased device count via parallelism. Since memory cells are usually protected by parity or error correction codes, latches are generally of primary concern. To improve soft error resilience, radiation-hardened latch circuit topologies, including DICE [14] and stacked SOI [20] latches, can be used at the cost of some amount of increased latch area. In addition, introduction of FinFET and thin-body transistor structures can significantly decrease charge collection volume, which

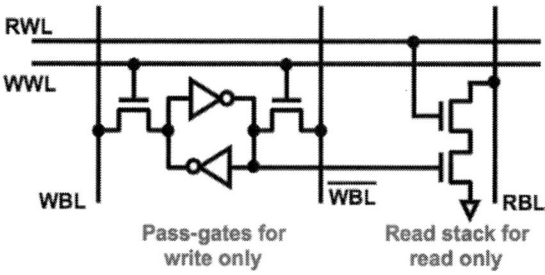

Figure 4. Migration to an 8T cell could help to enable near-threshold SRAM functionality while preserving competitive bit cell area. From Chang, et al. [7]

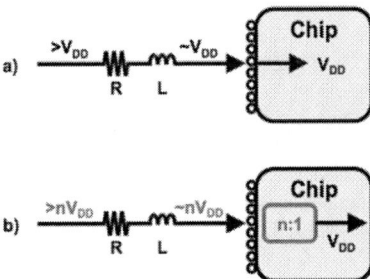

Figure 5. Power delivery in a) a traditional system, and b) a system with on-chip voltage down-conversion, which may be needed to enable near-threshold computing

may improve device-level soft error immunity. In particular, Fin-FET technologies are inherently comprised of parallel devices, which may offer further Q_{crit} enhancement [17]. At near-threshold voltages, it is likely that moderate tradeoffs in latch area and power can mitigate soft error resilience concerns.

3.2.3 Integrated Power Delivery and Management
Power delivery and management circuits in the form of board-level voltage regulator modules already play a critical role in VLSI systems; however, near-threshold computing may drive the integration of such functionality into the chip itself. Point-of-load voltage conversion may be essential to ensure accurate and efficient power delivery at near-threshold voltages while fine-grain on-chip voltage generation could be necessary to power heterogeneous cores. Integrated voltage regulators may thus be an important enabling technology for near-threshold systems.

Systems today are often designed to provide maximum performance within a given power budget constraint. At this constant power level, a reduction in operating voltage to the near-threshold regime leads to a corresponding increase in the current that must be delivered to the chip. This inevitably increases IR drops in the power delivery network and $L\partial I/\partial t$ supply noise – both of which require expanded voltage margins and can significantly increase power dissipation. Efficient and accurate delivery of near-threshold supplies may thus require dramatic improvements in chip packaging technology to reduce R and L and increase decoupling capacitance. A more effective strategy could be to instead utilize point-of-load power conversion as depicted in Figure 5b, in which power is delivered to the chip at a voltage an integer ratio, n, higher than the desired chip operating voltage and locally down-converted by an on-chip conversion circuit. After considering voltage conversion, both I^2R losses and $L\partial I/\partial t$ can be im-

proved by a net factor of up to n^2 [19] depending on the efficiency of conversion. While traditionally available on-chip passive components place severe limitations on the efficiency of switching regulator circuits, recent advancements such as the integration of deep trench capacitors into high-performance CMOS processes can enable efficiencies in upwards of 90% [6] – a level of loss that may be significantly outweighed by the IR and $L\partial I/\partial t$ improvement of high-voltage delivery.

Efficient power management may also be needed to provide voltage regulation capability in a heterogeneous system. An initial step in this direction is power gating, in which unused logic blocks or microprocessor cores can be turned off to conserve power. This technique, which is already widely used, establishes the infrastructure for voltage regulation by incorporating header switch devices and partitioned voltage rails [18]. As shown in Figure 6, by controlling the gate voltage of the header device to adjust impedance, linear regulation of the output voltage can be achieved. Combined with frequency control, such dynamic voltage and frequency scaling (DVFS) enables heterogeneity to balance throughput performance at near-threshold voltages and single-thread performance at ~1V. In such a scheme, power reduction is roughly a cubic function of voltage, which more than compensates for the conversion loss associated with linear regulation, which is a linear function of voltage.

3.2.4 Parallelism Overheads: Packaging and I/O
Fundamentally, voltage reduction to achieve near-threshold computing depends upon the utilization of increased parallelism to improve throughput performance. Inevitably, there are overheads associated with the resources needed to support and control the many parallel processing cores. In particular, increased parallelism necessarily increases silicon area usage for computation and memory, which must be addressed by increasing chip dimensions or the number of chips in a system – options that both result in tradeoffs in packaging cost and form factor. I/O congestion in the form of bandwidth and memory capacity limitations will also restrict the efficacy of parallelization and could result in increased I/O power. The severity of these constraints will vary significantly between applications, which may lead to a variety of solutions in practice. In particular, recent advances in 3D-integration [12] and integrated optical communication [15] may provide new opportunities to improve package form factor and memory bandwidth. To handle the area and communications overheads associated with parallelism, near-threshold systems will no doubt require aggressive techniques in packaging and I/O.

4. DISCUSSION
The practical utility of near-threshold operation fundamentally depends on the extent to which parallelizeable throughput performance is demanded by market trends. While some market segments can already leverage such power-efficient computing, they do not yet represent a critical mass to drive technology development. A key step may lie in the advent of heterogeneous systems integrating varied processor or customized accelerator cores to address the fundamental inability of near-threshold computing to provide single-thread performance. Requisite hardware support, from technology re-optimization and memory redesign to integrated voltage regulation and 3D-integration, depends more on the development and execution of known techniques rather than a search for the invention of new solutions. Thus, alignment with market trends is critical in justifying the investment needed to realize near-threshold systems.

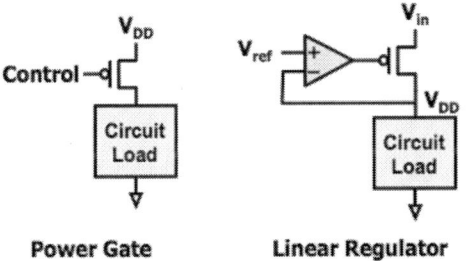

Power Gate   **Linear Regulator**

Figure 6. For on-chip power management in heterogeneous systems, linear regulators to achieve dynamic voltage modulation can mimic power gates as already used today.

The design automation community will play a key role in the enablement of near-threshold computing. Developing heterogeneous systems will require the ability to efficiently complete multiple core designs and to analyze interaction between cores to project system performance. In particular, communication and synchronization between cores running at disparate clock frequencies may present new modeling challenges. Due to a strong sensitivity to variability at low voltages, transistors models and standard cell libraries will require recalibration and revision to ensure accuracy and functionality. Latches, clock buffers, and register file cells will generally need to be redesigned for low voltage robustness, which likely leads to small tradeoffs in area and performance. In addition, statistical methods for timing and yield analysis may need continued improvement and refinement to achieve sufficient accuracy at low voltages. Whether as new developments or extensions of existing techniques, it is likely that the entire suite of design tools will need to be updated to ensure that efficient and robust circuits can be designed at near-threshold voltages.

5. CONCLUSIONS

Near-threshold operation offers the promise of significant improvements in the power efficiency of VLSI systems for parallelizeable workloads. A reduction to practice, however, requires an architecture that can also meet single-thread performance needs as well as hardware that can support near-threshold voltages by mitigating variability and reliability concerns and ensuring efficiency in the power delivery and I/O subsystems. The extent to which these developments are needed may vary between applications, but it is likely that all issues will need to be simultaneously addressed. While these concerns have thus far limited widespread adoption of near-threshold computing in mainstream applications, recent advances across the spectrum of technology, circuits, and systems – from transistor, memory, packaging, and I/O technologies to power delivery and management to heterogeneous computer architectures – hold hope that the promise of near-threshold computing can be realized.

6. ACKNOWLEDGMENTS

The authors acknowledge technical discussions and support from many IBM colleagues, in particular D. Frank and G. Shahidi.

7. REFERENCES

[1] Amdahl, G. M. 1967. Validity of the single-processor approach to achieving large-scale computing capabilities. In *Proc. Am. Federation of Information Processing Societies Conf.*, AFIPS Press, pp. 483-485.

[2] Borkar S. and Chien, A. A. 2011. The future of microprocessors. *Comm. ACM.* 54, 5 (May 2011), 67-77.

[3] Chandrasakan, A. P., Sheng, S., and Broderson, R. W. 1992. Low-power CMOS digital design. *IEEE J. Solid-State Circuits.* 27, 4 (Apr. 1992), 473-483.

[4] Chang, L., Choi, Y.-K., Ha, D., Ranade, P., Xiong, S., Bokor, J., Hu, C., and King, T.-J. 2003. Extremely scaled silicon nano-CMOS devices. *Proc. IEEE*, 91, 11 (Nov. 2003), 1860-1873.

[5] Chang, L., Frank, D. J., Montoye, R. K., Koester, S. J., Ji, B. L., Coteus, P. W., Dennard, R. H, and Haensch, W. 2010. Practical strategies for power-efficient computing technologies. *Proc. IEEE.* 98, 2 (Feb. 2010), 215-236.

[6] Chang, L., Montoye, R. K., Ji, B. L., Weger, A. J., Stawiasz, K. G., and Dennard, R. H. 2010. A fully-integrated switched-capacitor 2:1 voltage converter with regulation capability and 90% efficiency at 2.3A/mm^2. In *Symp. VLSI Circuits.* 55-56.

[7] Chang, L., Montoye, R. K., Nakamura, Y., Batson, K. A., Eickemeyer, R. J., Dennard, R. H., Haensch, W. and Jamsek, D. 2008. An 8T-SRAM for variability tolerance and low-voltage operation in high-performance caches. *IEEE J. Solid-State Circuits.* 43, 4 (Apr. 2008), 956-963.

[8] Damaraju, S., George, V., Jahagirdar, S., Khondker, T., Milstrey, R., Sarkar, S., Siers, S., Stolero, I., and Subbiah, A. 2012. In *Int'l Solid-State Circuits Conf.* 56-57.

[9] Dennard, R. H., Gaensslen, F. H., Yu, H. N., Rideout, V. L., Bassous, E., and LeBlanc, A. R. 1974. Design of ion-implanted MOSFETs with very small physical dimensions. *IEEE J. Solid-State Circuits.* SC-9, 5 (Oct. 1974), 256–268.

[10] Frank, D. J., Haensch, W., Shahidi, G., and Dokumaci, O. 2006. Optimizing CMOS technology for maximum performance. *IBM J. Res. Dev.* 50, 4/5 (Jul./Sept. 2006), 419-431.

[11] Horowitz, M., Indermauer, T., and Gonzalez, R. 1994. Low-power digital design. In *IEEE Symp. Low Power Electronics.* 8-11.

[12] Knickerbocker, J. U., Andry, P. S., Dang, B., Horton, R. R., Interrante, M. J., Patel, C. S., Polastre, R. J., Sakuma, K., Sirdeshmukh, R., Sprogis, E. J., Sri-Jyantha, S. M., Stephens, A. M., Topol, A. W., Tsang, C. K., Webb, B. C., and Wright, S. L. 2008. Three-dimensional silicon integration. *IBM J. Res. Dev.* 52, 6 (Nov. 2008), 553-569.

[13] Kogge, P. et al. 2008. Exascale Computing Study: Technology Challenges in Achieving an Exascale System; http://users.ece.gatech.edu/mrichard/ExascaleComputingStudyReports/exascale_final_report_100208.pdf.

[14] Krueger, D., Francom, E., and Langsdorf, J. 2008. Circuit design for voltage scaling and SER immunity on a quad-core Itanium® Processor. In *Int'l Solid-State circuits Conf.* 94-95.

[15] Miller, D. A. B. 2000. Rationale and challenges for optical interconnects to electronic chips. *Proc. IEEE.* 88, 6 (Jun. 2000), 728-749.

[16] Meindl, J. 1995. Low power microelectronics: Retrospect and prospect. *Proc. IEEE.* 83, 4 (Apr. 1995), 619-635.

[17] Oldiges, P., Dennard, R., Heidel, D., Ning, T., Rodbell, K., Tang, H., Gordon, M., and Wissel, L. 2009. Technologies to further reduce soft error susceptibility in SOI. In *Int'l Electron Devices Meeting.* 405-408.

[18] Preston, R. 2011. Design and process optimization for "Green" SoCs. In *Int'l Solid-State Circuits Conf.* Forum: Design of "Green" High-Performance Processor Circuits.

[19] Schrom, G., Hazucha, P., Hahn, J.-H., Kursun, V., Gardner, D., Narendra, S., Karnik, T., and De, V. 2004. Feasibility of monolithic and 3D-stacked DC-DC converters for microprocessors in the 90nm technology generation. In *IEEE Symp. Low Power Electronics and Design.* 263-268.

[20] Warnock, J., Sigal, L., Wendel, D., Muller, K. P., Friedrich, J., Zyuban, V., Cannon, E., and KleinOsowski, A. J. 2010. POWER7™ local clocking and clocked storage elements. In *Int'l Solid-State Circuits Conf.* 178-179.

Not so Fast my Friend: Is Near-Threshold Computing the Answer for Power Reduction of Wireless Devices?

Matt Severson
Qualcomm CDMA Tech.
9600 N. Mopac
Austin, TX 78759
severson@qualcomm.com

Kendrick Yuen
Qualcomm CR&D
5775 Morehouse Dr.
San Diego, CA 92121-1714
kyuen@qualcomm.com

Yang Du
Qualcomm CR&D
5775 Morehouse Dr.
San Diego, CA 92121-1714
ydu@qualcomm.com

ABSTRACT

In addition to battery life, power is limiting the performance, feature set, and form factor of most mobile communication devices. Many advanced low power techniques have been developed to deal with the problem, however, the cost of these techniques in terms of area, time to market, and quality has to be traded off against the power savings of each. Near threshold Computing (NTC) is a good example of this. NTC is showing great promise as a technique to extend battery life through optimizing energy efficiency. However, the end-user's overall experience is still the most important metric and that experience is influenced by performance, response time and battery life; not to mention price. Therefore NTC is limited in its impact to certain mobile products and applications after all the tradeoffs have been considered.

Categories and Subject Descriptors

B.0 [**Hardware**]: General. The ACM Computing Classification Scheme: http://www.acm.org/class/1998/

General Terms

Algorithms, Performance, Design, Reliability.

Keywords

Near Threshold Computing, NTC, wireless, mobile, low power.

1. INTRODUCTION

Power consumption is a key differentiator across all tiers of electronic products. This is especially true in the highly competitive mobile communications market. The integration of new features and demand for higher performance have driven down time between charges and customer satisfaction [23]. On top of this, classic voltage scaling has slowed giving rise to a power density crisis. Thus, in addition to battery life, power is now limiting the performance, feature set, and form factor of wireless devices. Two types of solutions have emerged. Those that employ process advances to extend Moore's law and classic transistor scaling are collectively referred to as "More Moore".

Those that use circuit, architecture or methodology improvements to compensate for limits to process scaling are referred to as "More than Moore." [22]

"More Moore" advancements to process like Strained-Silicon, Hi-K Metal Gate, FD-SOI and FinFET have allowed traditional scaling to continue [22]. These techniques provide increased transistor densities, higher performance and lower power. But how long can or will it continue? "More than Moore" techniques improve the power but usually come with an increased cost and or decreased performance.

Near-Threshold Computing is one of the many advanced low power techniques that have been developed to deal with the power problem. By lowering the supply voltage and operating "near threshold", energy efficiency can be improved by an order of magnitude compared to operating in super-threshold regions [15, 22, 9] However, as with most "more than Moore" techniques it comes at a cost to variability, decreased performance, higher latency and increased cost [24]. In this paper we explore what things should be considered when applying NTC to mobile SoCs.

2. PERFORMANCE

Mobile phone and mobile computing chips compete on benchmark performance related to the quality of user experience on the end product. At near-threshold supply voltages, performance is reduced by approximately 10x or more compared to super-threshold [26, 27]. This makes near-threshold operation a more attractive technique for eliminating wasted energy in mid/low performance circuits or modes of operation that don't require super-threshold performance all the time.

Near-threshold operations could also be applied to high-performance cores in mobile SOCs by adding parallelism [7]. However, adding a high level of parallelism to offset the performance degradations is not practical for most products in the mobile phone or tablet SoCs due to cost. Where cost is not an issue, parallelism can add control complexity and add pressure to the memory hierarchy. So for cores with high performance or low latency requirements, near-threshold operation is more likely to be used to extend the bottom end of the AVS/DVFS range for light workload modes. But is it worth it?

Architecture and circuit design optimization for near- to super-threshold scalable circuits is a tradeoff between the high-performance, high voltage and low/mid-performance, low-voltage operating points. Designing for robust ultra-low voltage operation can make meeting peak performance requirements more difficult. Simultaneous optimization requires advanced design methodologies and can extend time to market.

3. DELAY VARIATION

Delay variation at near-threshold can be 20x that of nominal Vdd over corners [2, 6]. Worst-case Vdd guard bands to account for the increased variation can severely limit performance or negate energy benefits of a near-threshold design. Some of the variation can be addressed with improvements to corner-based design methodology like statistical analysis, however this may impact design cycle time. Adaptive techniques & fault-tolerant circuits (e.g., ABB, Razor, etc.) can be used to address still more of the variation [8]. Although increased global variation at low-voltages can be countered with global body biasing [26], the amount of overhead required to counter the increased local variation (e.g. RDF) makes practical implementation of adaptive compensation difficult for cost-sensitive commercial chips. Fault-Tolerant circuits show promise for dealing with local variation at a cost to area [8]. The increased affects of supply voltage noise on near-threshold circuits also puts more stringent requirements on PDN design. All of this adds back energy overhead for the additional compensation circuitry.

4. COST & AREA

While NTC can achieve >10x power reduction, the additional area and therefore cost penalty needs to be carefully considered. This is especially pertinent to consumer electronics where ownership costs must remain flat or decrease. NTC cost overheads are primarily due to area increases associated with mitigating performance variations [25, 6].

To combat variation, many sensitive cells are trimmed out of the standard cell library. In addition, minimum width transistors have to be sized up to widths >2x the technology minimum allowed [12]. This can result in 5-10% additional chip area for near-threshold operation.

In SRAM, 8T to 10T bit cells are used to achieve stability and regain yield loss [12, 4, 3], leading to more than 20-40% area penalty in on-die memory. Near-threshold register files, flip-flops and latch circuits also experience a similar area impact.

Performance tracking and adaptive power management circuits also add significant area overhead. A recent design using the bubble RAZOR technique [8] reported a 21% area increase for retiming and up to 87% area hit when all latch cells incorporated timing error checking.

Voltage regulation to support NTC, like the distributed on-chip DC-DC converter of an A9 multi-core design was shown to consume 10% and 30% of the main chip area in 45nm and 130nm, respectively [17]. Such underlining cost penalties have been a major hurdle for the widespread adoption of NTC in mobile devices. Addressing the problem requires a systematic approach from fundamental technology improvement, circuit optimization and selective application.

5. PROCESS TECHNOLOGY

Certain process technologies have advantages when it comes to near-threshold operation. Near-threshold design requires special attention to sub-threshold swing (SS), short channel effects (SCE) and random doping fluctuation (RDF). Process optimization of FDSOI has been shown to achieve 16x faster transistor and over 50x lower energy-delay product with little area penalty compared to bulk CMOS operating at 0.3V [29]. The undoped body of FDSOI or FinFET offers better sub-threshold swing and lower RDF [11, 29, 28]. ETSOI and FinFET independent double gate

may also provide effective back gate bias [18, 20] for dealing with intrinsic transistor performance variation. Ultimately, proper selection of process technology can minimize the variation and area overhead of NTC.

Conventional NTC circuit designs focus on optimizing supply voltage and device size from a discrete set of transistor Vts. Making Vt a tunable design parameter [19] could allow more degrees of freedom to achieve the optimal power-performance-cost point. Dynamic Vt tuning using adaptive back bias in ETSOI together with the dual Vdd technique is a good example [14]. As Vt tuning has a direct impact on the transistor, it can provide additional performance through forward biasing and control leakage during sleep through reverse biasing and avoid the extra cost of multi-Vt and length modulation. Circuit techniques that exploit the latest process technology (e.g. FinFET, ETSOI) go hand in hand with NTC.

6. APPLICATION OF NTC

End user experience is the most important factor in the success of a mobile product. That experience is comprised of HW and SW performance, response time and battery life at a given price point. Users are not willing to give up performance and response time to extend the battery life. They want both. Many end users are discontented or annoyed by low power features that slow response times or performance [21]. They may even disable features or delay entry to low power modes when given the choice. However, battery performance is an even less satisfying aspect of the end user experience [23]. Thus it is important to consider all aspects of the end user experience when applying NTC and only apply it to those applications that can tolerate lower HW performance without affecting the end-user experience.

Today's mobile phones are complex multi-mode devices which support a wide variety of applications like talk, text, e-mail, web browsing, music playback, video streaming, gaming, position location, and camera. The result is a highly integrated heterogeneous compute platform. NTC can effectively be used to lower power of digital cores and is especially beneficial when digital components dominate the use-case power profile. NTC should be used sparingly with applications that are dominated by the power of other system components or non-digital power domains when using NTC means those components must be ON for longer. This can apply to the Display, Radio, and crystal oscillator but also the SRAM [24] or SDRAM memory. Overall, NTC should only be used when total energy from the battery can be saved when considering all system components.

The emergence of multi-core is a direct response to the need to improve performance without increasing power density. By extending the use of pipelining and parallelism to enable NTC we can maintain performance while improving energy efficiency. However, this is only possible for applications where more parallelism can be exploited when considering all the control overhead [24]. NTC will not benefit applications that are dominated by single-threaded performance or latency where the Timeline would be extended by its use [24, 21]. Furthermore, designing a core for near-threshold operation may increase the power when operating in the super-threshold region.

Finally, designing hardware to support NTC increases costs and therefore should be used for applications where low power is as important as or more important than price. Small price increases of mobile phone components that raise customer satisfaction can be justified, but should be carefully considered.

7. CONCLUSION

We need "more Moore" and "more than Moore" techniques to get passed the power wall. Near-threshold computing is an effective "more than Moore" technique for increasing energy efficiency, but its impact on other metrics like performance, latency, and cost must be considered. Therefore it is limited in its application. If employed correctly on mobile products, it can be a differentiator, but if not careful, it may not provide any benefit at all.

8. REFERENCES

[1] Borkar, S. 2011. 3D Integration for Energy Efficient System Design. DAC 2011, 214.

[2] Calhoun, B.H., Brooks, D. 2010. Can Subthreshold and Near-Threshold Circuits Go Mainstream? Micro, IEEE Vol. 30, Iss. 4. 80-85.

[3] Calhoun, B.H., Chandrakasan, A.P. 2007. A 256-kb 65-nm Subthreshold SRAM Design for Ultra-Low-Voltage Operation, Solid-State Circuits, IEEE Journal of, vol. 42, 680-688.

[4] Chang, L. et al. 2007. 5.3 GHz 8 T-SRAM with operation down to 0.41 V in 65 nm CMOS, VLSI Circuits, 252.

[5] Damaraju, S. et al. 2012. A 22nm IA Multi-CPU and GPU System-on-Chip. ISSCC Dig. Tech. Papers, 55-6.

[6] Dreslinski, R. *et al.*, 2010. Near-threshold computing: Reclaiming Moore's law through energy efficient integrated circuits, *Proc. IEEE*, Vol. 98, no. 2, 253–266, Feb. 2010.

[7] Fick, D. et al. 2012. Centip3De: A 2920DMIPS/W configurable Near-Threshold 3D Stacked System with 64 ARM Cortex-M3 Cores ISSCC 2012, 190.

[8] Fojtik, M. et al. 2012. Bubble Razor: An Architecture-Independent Approach to Timing-Error Detection and Correction. ISSCC Dig. Tech. Papers, 488-89.

[9] Hsu, S. et al., 2012 A 280mV-to-1.1V 256b Reconfigurable SIMD Vector Permutation Engine with 2-Dimensional Shuffle in 22nm CMOS. ISSCC Dig. Tech. Papers, 177-9.

[10] Hu, C., et al. 2010. Prospect of Tunneling Green Transistor for 0.1V CMOS, IEDM, 387.

[11] Islam, A., et al. 2010. Energy Efficient and Process Tolerant Full Adder Design in Near Threshold Region using FinFET ISED, no. 19, 56-60.

[12] Jain, S. et al. 2012. A 280mV-to-1.2V Wide-Operating-Range IA-32 Processor in 32nm CMOS. ISSCC Dig. Tech. Papers, 65-67.

[13] Jan, C.H., et al., 2009 A 32nm SoC Platform Technology with 2nd Generation High-k/Metal Gate Transistors Optimized for Ultra Low Power, High Performance, and High Density Product Applications, IEDM Dig. Tech. Papers, 1-4.

[14] Kakoee, M., and Benini, L. 2011. Fine-Grain Power and Body-Bias Control for Near-Threshold Deep Sub-Micron CMOS Circuits IEEE Transactions on Emerging and Selected Topics in Circuits and Systems, Vol. 1, No. 2 June 2011, 131.

[15] Kaul, H., et al., 2008 A 320mV 56μW 411GOPS/Watt Ultra-Low Voltage Motion Estimation Accelerator in 65nm CMOS, ISSCC Dig. Tech. Papers, 316-616.

[16] Kaul, H. et al. 2012. A 1.45GHz 52-to-162GFLOPS/W Variable-Precision Floating-Point Fused Multiply-Add Unit with Certainty Tracking in 32nm CMOS ISSCC Dig. Tech. Papers, 182-84.

[17] Kim, W. et al. 2011 A fully-integrated 3-level DC/DC converter for nanosecond-scale DVS with fast shunt regulation ISSCC Dig. Tech. Papers, 268-270.

[18] Liu, Q. et al. 2011 Impact of Back Bias on Ultra-Thin Body and BOX (UTBB) SOI Symposium VLSI, 160.

[19] Markovic, D., et al. 2010. Ultralow Power Design in Near-Threshold Region Proceedings of the IEEE, Vol. 98, No. 2, Feb. 2010, 237.

[20] Mathew, L., et al. 2004. CMOS Vertical Multiple Independent Gate Field Effect Transistor (MIGFET) IEEE International SOI Conference 2004, 187.

[21] Mudge, T. and Holzle, U. 2010. Challenges and Opportunities for Extremely Energy-Efficient Processors. Micro, IEEE vol. 30, Iss, 4, 20-24.

[22] Perlmutter, D. 2012 Sustainability in Silicon and Systems Development. ISSCC Dig. Tech Papers, 30-34.

[23] J.D. Power and Associates 2012 Smartphone Battery Life has become a Significant Drain on Customer Satisfaction and Loyalty. 2012 Wireless Smartphone and Traditional Mobile Phone Satisfaction Studies – Vol. 1.

[24] Pu, Y., et al., 2010. Misleading Energy and Performance Claims in Sub/Near Threshold Digital Systems, ICCAD 625-631, 2010.

[25] Seok, M. et al. 2011. CAS-FEST 2010: Mitigating Variability in Near-Threshold Computing, IEEE Transactions on Emerging and Selected Topics in Circuits and Systems, Vol. 1, no. 1, 42-49, March.

[26] Sylvester, D., et. al. 2011 CAS-FEST 2010: Mitigating Variability in Near-Threshold Computing, IEEE Transactions on Emerging and Selected Topics in Circuits and Systems, Vol. 1, no. 1, March.

[27] Sylvester, D. et. al. Ultralow-voltage, minimum-energy CMOS, IBM Journal of Research and Development Vol. 50, Iss. 4.5.

[28] Vandooren, A. et al.2003. Mixed-signal performance of Sub-100nm fully-depleted SOI devices with metal gate, high K (HfO2) dielectric and elevated Source/Drain extension IEDM, 975-977.

[29] Vitale, S. et al. 2010. FDSOI Process Technology for Subthreshold-Operation Ultralow-Power Electronics Proceedings of the IEEE, Vol. 98, No.2, Feb. 2010.

Accurate Process-Hotspot Detection Using Critical Design Rule Extraction[*]

Yen-Ting Yu[1], Ya-Chung Chan[2], Subarna Sinha[3], Iris Hui-Ru Jiang[1], and Charles Chiang[4]

[1]Dept. of Electronics Engineering and Inst. of Electronics, National Chiao Tung University, Hsinchu, Taiwan
[2]MStar Semiconductor, Inc., Chupei, Taiwan
[3]Computer Science Department, Stanford University, Stanford, CA, USA
[4]Synopsys, Inc., Mountain View, CA, USA

ABSTRACT

In advanced fabrication technology, the sub-wavelength lithography gap causes unwanted layout distortions. Even if a layout passes design rule checking (DRC), it still might contain process hotspots, which are sensitive to the lithographic process. Hence, process-hotspot detection has become a crucial issue. In this paper, we propose an accurate process-hotspot detection framework. Unlike existing DRC-based works, we extract only critical design rules to express the topological features of hotspot patterns. We adopt a two-stage filtering process to locate all hotspots accurately and efficiently. Compared with state-of-the-art DRC-based works, our results show that our approach can reach 100% success rate with significant speedups.

Categories and Subject Descriptors

B.7.2 [**Integrated Circuits**]: Design Aids

General Terms

Algorithms, Design.

Keywords

Design for manufacturability; process hotspot; pattern matching, design rule checking; lithography.

1. INTRODUCTION

As the printed feature size becomes smaller than the actual lithographic wavelength in advanced fabrication technology, the sub-wavelength lithography gap leads to unwanted shape distortions of the printed patterns. Even in a DRC-clean layout, some layout patterns are still sensitive to the lithographic process. These potentially problematic layout patterns, referred to as *process hotspots*, should be replaced with yield-friendly configurations. Therefore, detecting process hotspots in a layout is of particular importance to enable this correction process.

Recently, many research endeavors have been devoted to process-hotspot detection (also known as *pattern matching*) [1–8]. These approaches can be classified into four categories:

---

* This work was partially supported by Synopsys and NSC of Taiwan under Grant No. NSC 100-2220-E-009-047.

Permission to make digital or hard copies of all or part of this work for personal or classroom use is granted without fee provided that copies are not made or distributed for profit or commercial advantage and that copies bear this notice and the full citation on the first page. To copy otherwise, or republish, to post on servers or to redistribute to lists, requires prior specific permission and/or a fee.
DAC 2012, June 3-7, 2012, San Francisco, California, USA.
Copyright 2012 ACM 978-1-4503-1199-1/12/06...$10.00.

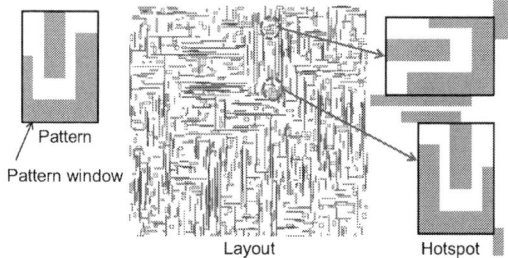

Fig. 1. Hotspots are accurately identified by our approach. We consider eight possible orientations and can handle the hotspots surrounded by arbitrarily-shaped polygons.

1) Graph-based hotspot detection: Kahng *et al.* present a pioneering work on hotspot detection in [1]. They create a dual graph to represent a given layout. Then, they filter out over-weighted edges and faces according to a user-specified threshold value. However, this method may generate false alarms due to the simplified error model.

2) Machine-learning-based hotspot detection: Ding *et al.* propose a machine learning kernel in [2], where they extract hotspot features to train their artificial neural network. Recently, Ding *et al.* extend it to a hierarchical learning framework in [3]. Wuu *et al.* devise another machine learning engine in [4]. This approach suffers from long training time and false alarms.

3) String-matching-based hotspot detection: In [5] and [6], Yao *et al.* and Xu *et al.* use worm-like movement to investigate all possible windows within a layout. Each window is converted to a layout matrix, and the matrix and pattern are encoded by strings. String matching is then applied to identify hotspots. This approach is accurate, but the layout matrix conversion is time consuming because the grid size is very small for advanced technology nodes.

The above three approaches may suffer from inaccuracy or long running time. Different from them, the fourth approach is *DRC-based hotspot detection*, which leverages on DRC to improve the accuracy of the detection process. Typically, DRC-based hotspot detection first converts the topological features of process hotspots to design rules and then analyzes the DRC report to identify hotspots [7][8].

In [7], Pikus and Collins extract all lengths of polygon edges and distances between adjacent polygons inside a given pattern as the topological features. At the analysis step, they construct a search graph to record the locations reported by DRC. They traverse the search graph to identify hotspots. In [8], Gennari *et al.* exploit two techniques, 2D image-based DRC and hashing, to improve DRC-based hotspot detection. They use a hash table to store the location and the configuration around each edge or corner in a layout. They then compute match factors between the pattern and layout to determine hotspots. However, a sophisticated hash function is required to prevent hash collisions.

To avoid inaccuracy and long running time, in this paper, we propose an accurate process-hotspot detection framework based on the DRC-based approach. We consider eight possible orientations (including combinations of four rotations (0°, 90°, 180°, 270°) and two mirrors (horizontal and vertical mirrors)) and allow that arbitrarily-shaped polygons surround around the pattern window as shown in Fig. 1.

To achieve this goal, unlike existing DRC-based works, we target to find the topological features that can express a given hotspot pattern sufficiently and efficiently. Consequently, first of all, we adopt an efficient representation to model the pattern, and then we extract only critical topological features from the representation and convert them to design rules. Second, considering eight orientations, we apply DRC to find all locations that match any of these rules within a layout. Finally, we propose a two-stage filtering process to identify all hotspot locations accurately and efficiently. The key features of our approach include:

- We extract only critical rules: The classic way is to interpret all topological relations of edges within a pattern to design rules. However, doing so may generate numerous design rules and induce tremendous locations reported by DRC, thus making subsequent analysis difficult. On the other hand, too few extracted rules may also result in tremendous locations reported and complicated analysis. Hence, we extract only critical design rules to facilitate DRC and subsequent analysis.

- Our hotspot detection is accurate: We propose a two-stage filtering process; pre-filtering indicates potential locations, while finalization verifies exact locations. Compared with other analysis techniques adopted by state-of-the-art DRC-based approaches, our results show that our approach can detect hotspots 100% accurately with significant speedups.

The remainder of this paper is organized as follows. Section 2 briefly introduces design rule checking and gives the problem formulation. Section 3 details our process-hotspot detection framework. Section 4 shows our experimental results. Finally, Section 5 concludes this paper.

2. PRILIMINARIES

In this section, we briefly introduce design rule checking (DRC) and give the problem formulation.

2.1 Design Rule Checking

Design rules are a set of parameters to guarantee the manufacturability of a layout. For a specific manufacturing process, foundries provide the corresponding set of rules to ensure sufficient margins to compensate the variability during manufacturing. If these rules are violated, the design may not operate correctly. Fig. 2 indicates the most fundamental design rules. For a single layer, a width rule specifies the minimum width of any shape in the layout, while a spacing rule specifies the

minimum distance between two neighboring objects. For two layers, an enclosure rule specifies an object should be covered with some additional margin by some object on the other layer.

In addition to the fundamental rules, modern DRC tools can perform general dimensional checks within a single polygon (including length, width, area, overlap, ratio, and density calculations) or between polygon edges (including intersecting polygon spacings, enclosure spacings and external polygon spacings). Given a *runset* file (design rules for a specific process) and a layout, a DRC tool reports design rule violations (indicating locations and violated rules). Basically, design rules can be expressed by equations and/or inequalities. For example, the minimum spacing rule can be described as the spacing between any two adjacent polygon edges is smaller than the specified value as shown in Fig. 2. The DRC tool then indicates the locations where there exist some edges violate the minimum spacing value.

2.2 Problem Formulation

As mentioned in Section 1, detecting process hotspots in a layout is a crucial issue. The hotspot detection problem is formulated as follows.

The Hotspot Detection Problem:
Given a hotspot pattern and a layout, our goal is to report all hotspot locations with eight possible orientations in the layout.

Hotspot patterns are patterns with exact dimensions which are provided by foundries. A hotspot location means the layout configuration at this location exactly matches that of the pattern inside the pattern window.

3. OUR HOTSPOT DETECTION FRAMWORK

In this section, we detail our DRC-based hotspot detection framework as shown in Fig. 3. First of all, we extract only the critical topological features of a given hotspot pattern and convert them to design rules. We devise an efficient representation–Modified TCG–to facilitate the extraction. Second, we apply these rules to DRC. DRC reports all locations that fit any generated rule considering eight possible orientations. Finally, we adopt a two-stage filtering process to analyze the DRC results: Pre-filtering indicates all potential locations that match the given pattern, while finalization identifies all true locations. We can easily extend our approach to consider multiple patterns simultaneously.

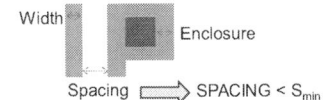

Fig. 2. Basic design rules: width, spacing, and enclosure.

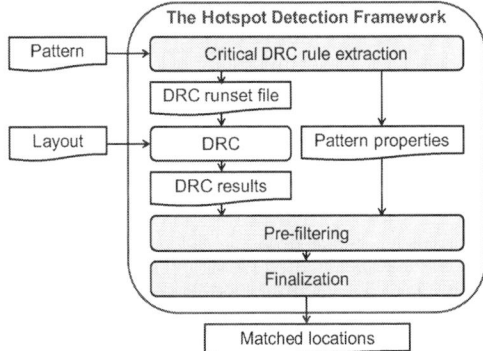

Fig. 3. The overview of our hotspot detection framework.

3.1 Modified TCG and Critical DRC Rule Extraction

To use the aid of DRC to realize hotspot detection, we shall extract design rules from the given pattern. The classic way is to interpret all topological relations of edges within a pattern to design rules. However, doing so may generate numerous design rules and induce tremendous locations reported by DRC thus making subsequent analysis difficult. On the other hand, too few extracted rules may also result in tremendous locations reported and complicated analysis. Hence, our goal is to extract *only* critical design rules to facilitate DRC and subsequent analysis.

There are two tasks: 1) to model the given pattern by a good representation that can reflect topological features, and 2) to select critical features from the representation and translate them to design rules.

To accomplish the first task, we extend *transitive closure graph* (TCG) representation proposed by Lin and Chang in [9]. TCG is widely used to represent a compact placement; it uses a pair of constraint graphs, C_h and C_v, to record geometric relations among modules. However, hotspots may not be in a compact form because of spacing among polygons (see Fig. 4(a)). The spacing among polygons (i.e., white spaces) is essential for hotspot detection, which contains topological features of a pattern. In order to consider spacing by TCGs, we tile the pattern. We stretch the horizontal edges of each polygon until they reach other polygons or the window boundaries (see Fig. 4(a)). After horizontal tiling, a pattern is composed of block tiles (contributed by polygons) and space tiles (contributed by white spaces), and the pattern becomes compact.

We convert the horizontally tiled pattern to a horizontal modified TCG (MTCG). It is known that *each compact placement can be represented by a unique TCG*; a modified TCG inherits this property, i.e., MTCG can represent a unique tiled pattern. In an MTCG, each vertex represents a block tile (dot) or a space tile (circle), while each edge represents some topological relation among tiles.

MTCGs can be constructed by the sweep line algorithm. In the vertical constraint graph C_v, a directed edge is added between any two adjacent tiles if their projections on *x*-axis overlap. Similarly, in the horizontal constraint graph C_h, a directed edge is added between any two adjacent tiles if their projections on *y*-axis overlap. Moreover, the diagonal relations among block tiles can be extracted when a space vertex with one incoming and one outgoing edges connected to the same pair of block vertices in C_v and C_h. The diagonal relation between two corner-touched block tiles is directly checked from constraint graphs. Since spacing is considered, and the tiled pattern is compact, the transitive edges (which are redundant) can be simplified during MTCG construction. Without these transitive edges, our MTCGs are thus sparse.

To fully represent a given pattern, we adopt not only a horizontal MTCG but also a vertical MTCG. Consider the example shown in Fig. 4(b). If we use only the horizontal MTCG, we cannot extract

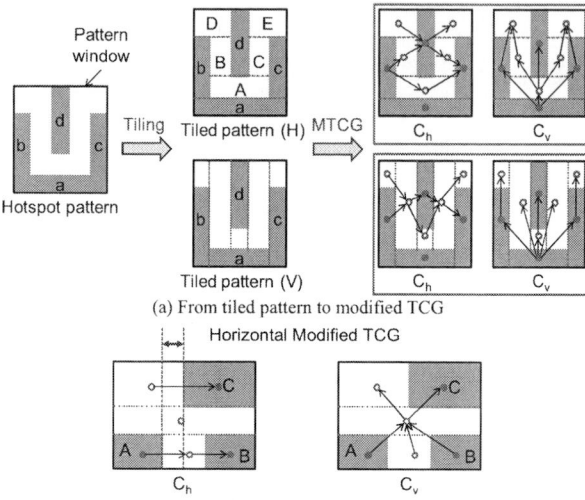

(a) From tiled pattern to modified TCG

(b) One-way tiling

Fig. 4. Modified TCG construction.

the horizontal distance between block tiles *A* and *C.* The properties of MTCGs are summarized as follows.

Lemma 1: MTCG is compact for a given tiled pattern.

Theorem 1: MTCG is unique for a given tiled pattern.

Moreover, recall that in the aforementioned classic way, a complete (dense) graph is needed to record all topological relations. In contrast, since redundant relations are not included, our MTCGs are sparse. The good representation–MTCGs–facilitates the second task, critical feature selection.

To accomplish the second task, we extract the following critical topological features. One of our goals is to handle patterns that are surrounded by arbitrarily-shaped polygons. Hence, we first focus on the internal topological relations (see Fig. 5). These primary rules can be expressed by equations.

1) Rule one–the width and height of a block tile: As shown in Fig. 5(a), we find the dimension of each block tile that does not touch the window boundary. Given an MTCG, we extract all block vertices whose incoming and outgoing edges are connected to space vertices.

2) Rule two–the distance between two adjacent block tiles: As shown in Fig. 5(b), we find the dimensions of all space tiles that do not touch the window boundary and are located in between block tiles. Given an MTCG, we extract any space vertex which lies in between exactly two block vertices.

3) Rule three–the diagonal relations between two convex corners of block tiles: As shown in Fig. 5(c), we find the diagonal relations between any two convex corners of block tiles. Given an MTCG, we extract space vertices whose in and out degrees are larger than two and also check their diagonal relations and distance.

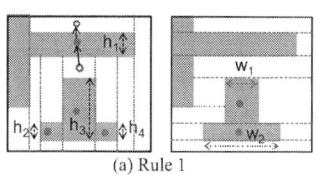 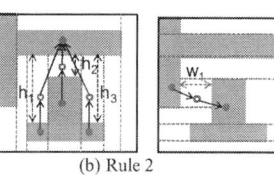 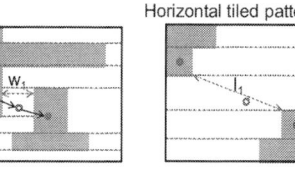 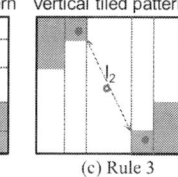 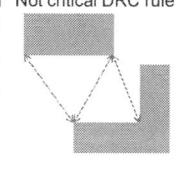

(a) Rule 1   (b) Rule 2   (c) Rule 3

Fig. 5. Primary critical rules.

1165

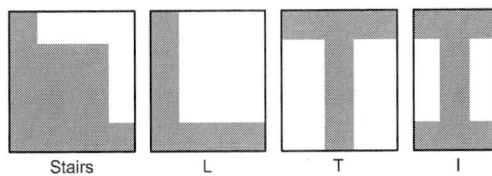

Fig. 6. Special cases.

The primary rules can handle most patterns. However, the primary rules may be insufficient for some special cases. As shown in Fig. 6, we cannot extract any primary rules for patterns "Stairs", "L", and "T", and we have only one rule for pattern "I". Too few rules imply that too many redundant locations could be reported. To overcome this difficulty, we add two secondary rules for tiles that touch the window boundary. The secondary rules can be expressed by inequalities.

4) Rule four–the space or block tile with one edge touching the window boundary: As shown in Fig. 7(a), we identify boundary tiles. Given an MTCG, we extract a space vertex that has zero indegree or outdegree in C_v (C_h) and has nonzero incoming and outgoing edges connected to block vertices in C_h (C_v).

5) Rule five–the space tile with two edges touching the window boundary or space tiles: As shown in Fig. 7(b), we extract the dimensions of space boundary tiles.

The secondary rules can handle the cases that the primary rules cannot, e.g., rule 4 can handle "T" and "I", while rule 5 can handle "Stairs" and "L". However, rule 5 is too general and may induce too many design rules. Hence, if we can extract critical rules based on the first four types of rules, we do not generate rules for rule 5 to speed up the subsequent process.

So far, we convert a pattern to MTCG and extract critical rules. It can be seen that MTCG is an efficient representation and can capture critical topological features well. Later, our results will show that we can achieve a 100% success rate to detect all hotspots.

A pattern may have eight possible orientations as shown in Fig. 8(a). Based on MTCG, the extracted critical rules express vertical and horizontal geometric relationships. Rules extracted from C_h are always perpendicular to rules extracted from C_v (see Fig. 8(b)). Consequently, we divide these eight orientations into two sets (see Fig. 8(a)). We generate a runset file for each set and run DRC twice to obtain the locations that hit any generated rule.

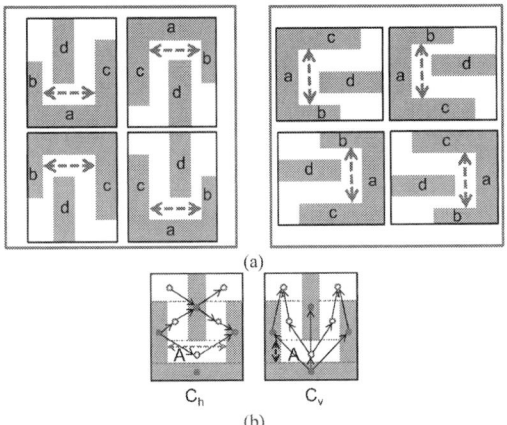

(a)

(b)

Fig. 8. (a) Eight orientations. (b) Rules extracted from C_h and C_v are mutually perpendicular, e.g., space tile A.

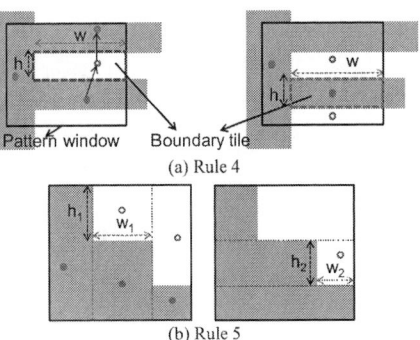

(a) Rule 4

(b) Rule 5

Fig. 7. Secondary critical rules.

3.2 Pre-filtering

Based on the DRC results and pattern properties, pre-filtering is applied to find the potential hotspot locations. Based on the way we set primary and secondary rules, we have the following property:

Theorem 2: Each extracted critical rule corresponds to a rectangle.

DRC reports the locations and dimensions of all polygons or spaces in the layout that match some specified critical rule. Based on Theorem 2, these reported block/spacing polygons are *rectangles*.

Hence, we create *rule rectangles* to record the pattern properties as follows. Given a pattern, a *reference point* is set to the bottom-left corner of its pattern window. Each extracted rule is modeled as a rule rectangle: A rule rectangle is associated with a width, a height, the relative distance (d_x, d_y) between the reference point and the bottom-left corner of this rectangle. As shown in Fig. 9(a), the spacing between polygons b and d is extracted based on rule 2 and is recorded by a rule rectangle. Totally five rule rectangles are recorded for this pattern.

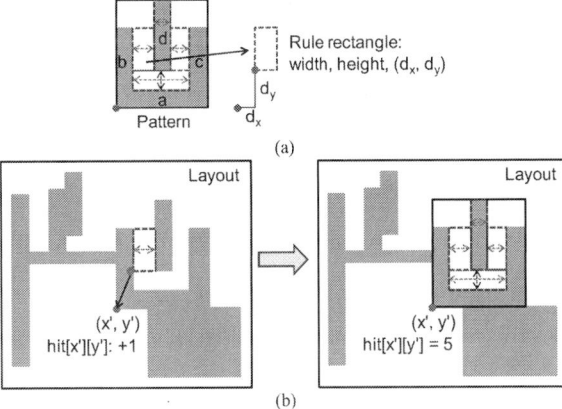

(b)

Fig. 9. Finding the potential locations. (a) Pattern properties. (b) Pre-filtering.

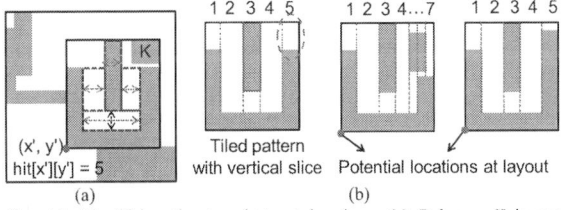

(a)

(b)

Fig. 10. Identifying the true hotspot locations. (a) Polygon K is not identified during pre-filtering. (b) Finalization. These two potential locations are not hotspots.

1166

Pre-filtering indicates the potential hotspot locations by analyzing the rectangles reported by DRC and the rule rectangles (pattern properties) of the given pattern. To facilitate the analysis, we use a variable $hit[x][y]$ to record the total number of rules matched at coordinate (x, y) and use a queue Q to store all $hit[x][y]$ values. When parsing the DRC results, for each reported rectangle, we calculate the corresponding reference point (x', y') in the layout according to (d_x, d_y) set by the rule rectangle and increment its $hit[x'][y']$ value by 1. Finally, once the hit value in Q is equal to or greater than the number of rule rectangles, we find a potential hotspot location. For example, as shown in Fig. 9(b), we collect 5 matched rules at (x', y'), and thus (x', y') is a potential hotspot location.

3.3 Finalization

Pre-filtering indicates all locations in layout that match the extracted critical rules. However, some non-hotspot locations might pass pre-filtering. For example, in Fig. 10(a), (x', y') is reported by pre-filtering; nevertheless, there exists an extra polygon K, and (x', y') should be excluded. Hence, finalization is indeed necessary to identify true hotspot locations.

The finalization stage verifies the potential locations reported by pre-filtering. Since arbitrarily-shaped polygons may surround hotspots, first of all, we frame a checking window based on each potential location (which is the bottom-left corner of the window). Second, we vertically slice the layout inside the window. If the number of generated slices or the area of each tile within each slice is different from the given pattern, it is not a hotspot (see Fig. 10(b)).

4. EXPERIMENTAL RESULTS

Our algorithm was implemented in the C++ programming language on a Linux platform with a 2.4 GHz CPU with 16 GB RAM. The experiments were conducted with two layouts and with seven hotspot patterns. Table I summarizes the statistics of these layouts. Fig. 11 lists 7 patterns: "Stair 1", "I", and "S" are killer cases used to test the capability of our approach, "Mountain" and "Stair 2" are from [5], and "Ind1" and "Ind2" are extracted from real designs. These layouts are from real designs, and we randomly inject 6,400 and 9,600 exact hotspots for each pattern into Layout1 and Layout2, respectively. In addition, we also inject similar-shaped hotspots for patterns "Stair 1", "Stair 2", "I" to test the robustness of our framework. We adopt a state-of-the-art industrial DRC engine into our framework. The pattern and layout are described in GDS format. We have two experiments as follows.

4.1 The Impact of Critical Rule Extraction

Since rule five may induce too many design rules, *hierarchical* rule extraction used in our experiments means that rule 5 is applied only if no rules are extracted for rules 1 to 4. In the first experiment, we show the impact of the hierarchical rule extraction on the running time of the detection flow.

Table II compares hierarchical critical rule extraction with complete critical rule extraction in terms of running time, the number of matched rules reported by DRC, the number of finalized hotspot locations, and the success rate. First of all, the success rates of these two rule extraction strategies are both 100%. Among these patterns, "Stair 1" contains only rule 5. As expected, the rules matched by DRC are tremendous, and thus pre-filtering and finalization are slow. It is reasonable that hierarchical critical rule extraction performs equally well as complete critical rule

extraction for "Stair 1". Patterns "S" and "Ind1" have primary and secondary DRC rules. In these cases, the hierarchical extraction achieves significant speedups.

4.2 Comparison with Other DRC-based Approaches

The second experiment compares our flow with other DRC-based approaches. For fair comparison, we incorporate the ideas of edge-based rule extraction and corner-based hotspot analysis proposed in [8] into our detection framework.

Table III lists the results of using edge-based rule extraction. Edge-based rule extraction models the lengths of all polygon edges within the pattern window as design rules. Redundant and duplicated rules are removed in our experiments. Although the edge-based method extracts fewer rules, it incurs tremendous locations reported by DRC. In contrast, our critical rule extraction identifies critical topological features well (fewer locations reported by DRC) and thus leads to faster analysis (shorter running times of pre-filtering and finalization).

Table IV lists the results of using the corner-based hotspot analysis instead of using our finalization. The corner-based hotspot analysis uses a hash table to store the feature of each polygon corner. The feature used here is the polygon area inside a small window centered at the investigated corner, and the window size is set to 10% of the height and 10% of the width of the given pattern window. It can be seen that both analysis methods are correct, but our method outperforms corner-based hotspot analysis.

5. CONCLUSION

In this paper, we propose an accurate process-hotspot detection framework. Unlike existing DRC-based approaches, we extract only critical design rules to express the topological features of hotspot patterns. We adopt a two-stage filtering process to locate all hotspots accurately and efficiently. Our results show that our approach not only reaches 100% success rate but also results in a short DRC report and superior efficiency. Future work includes handling multi-layer, range, and incompletely-specified patterns.

6. REFERENCES

[1] A. B. Kahng *et al.* Fast dual graph based hotspot detection. In *Proc. SPIE*, vol. 6349, pp. 628–635, 2006.

[2] D. Ding *et al.* Machine learning based lithographic hotspot detection with critical feature extraction and classification. In *Proc. ICICDT*, pp. 219–222, 2009.

[3] D. Ding *et al.* High performance lithographic hotspot detection using hierarchically refined machine learning. In *Proc. ASP-DAC*, pp. 775–780, 2011.

[4] J.-Y. Wuu, *et al.* Rapid layout pattern classification. In *Proc. ASP-DAC*, pp. 781–786, 2011.

[5] H. Yao *et al.* Efficient process-hotspot detection using range pattern matching. In *Proc. ICCAD*, pp. 625–632, 2006.

[6] J. Xu *et al.* Accurate detection for process-hotspots with vias and incomplete specification. In *Proc. ICCAD*, pp. 839–846, 2007.

[7] F. G. Pikus and T. W. Collins, Jr. Topological pattern matching. *US Patent Application* 2010/018594 A1, Jul. 2010.

[8] F. E. Gennari *et al.* Fast pattern matching. *US Patent* 7818707, Oct. 2010.

[9] J.-M. Lin and Y.-W. Chang. TCG: A Transitive closure graph based representation for non-slicing floorplans. In *Proc. DAC*, pp. 764–769, 2001.

TABLE I. LAYOUTS

	Layout1	Layout2
Area (mm²)	1.2×1.2	1.2×1.2
#Polygon*	661,056	1,028,622

*#Polygon: the number of polygons.
The layouts are based on 32nm process.

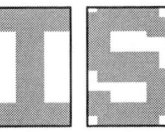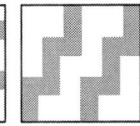

Stair 1   I   S   Mountain   Stair 2   Ind1   Ind2

Fig. 11. Hotspot patterns.

TABLE II. HIERARCHICAL VS. COMPLETE CRITICAL RULE EXTRACTION

	Pattern	Layout	Critical rule extraction time (s)	#ruleC* (R)	DRC time (s)	#ruleD** (A)	Pre-filtering time (s)	#hotspot (B)	Finalization time (s)	#hotspot*** (C)	Total time (s)	Ratio (A/RB)	Ratio (B/C)	Success rate
Hierarchical	Stair 1	Layout1	<0.01	(0,0,0,0,3)	41	2,396,343	106.68	38,400	57.86	6,400	205.54	20.80	6.00	100%
		Layout2	<0.01		63	3,710,439	164.20	57,592	125.42	9,600	352.62	21.48	6.00	100%
	I	Layout1	<0.01	(1,0,0,1,0)	9	45,004	1.51	25,600	24.79	6,400	35.30	0.88	4.00	100%
		Layout2	<0.01		9	67,640	2.27	38,400	56.25	9,600	67.52	0.88	4.00	100%
	S	Layout1	<0.01	(1,3,0,7,0)	17	609,013	39.81	6,400	1.60	6,400	58.41	8.65	1.00	100%
		Layout2	<0.01		35	914,217	61.16	9,600	3.55	9,600	99.71	8.66	1.00	100%
	Mountain	Layout1	<0.01	(3,0,0,1,0)	8	64,698	1.92	6,400	1.60	6,400	11.52	2.53	1.00	100%
		Layout2	<0.01		11	97,418	2.98	9,600	3.54	9,600	17.52	2.54	1.00	100%
	Stair 2	Layout1	<0.01	(2,1,0,0,0)	6	77,159	2.49	12,800	6.26	6,400	14.75	2.01	2.00	100%
		Layout2	<0.01		9	115,950	3.99	19,200	14.18	9,600	27.17	2.01	2.00	100%
	Ind1	Layout1	<0.01	(3,3,0,6,0)	23	596,472	31.91	6,400	1.59	6,400	56.50	7.77	1.00	100%
		Layout2	<0.01		35	895,377	50.44	9,600	3.61	9,600	89.05	7.77	1.00	100%
	Ind2	Layout1	<0.01	(2,4,0,4,0)	24	512,941	25.53	6,400	1.58	6,400	51.11	8.01	1.00	100%
		Layout2	<0.01		37	769,949	40.10	9,600	3.53	9,600	80.63	8.02	1.00	100%
Ratio					1.00	1.00	1.00	1.00	1.00	1.00	1.00			
Complete	Stair 1	Layout1	<0.01	(0,0,0,0,3)	41	2,396,343	106.68	38,400	57.86	6,400	205.54	20.80	6.00	100%
		Layout2	<0.01		63	3,710,439	164.20	57,592	125.42	9,600	352.62	21.48	6.00	100%
	I	Layout1	<0.01	(1,0,0,1,0)	9	45,004	1.51	25,600	24.79	6,400	35.30	0.88	4.00	100%
		Layout2	<0.01		9	67,640	2.27	38,400	56.25	9,600	67.52	0.88	4.00	100%
	S	Layout1	<0.01	(1,3,0,7,2)	39	2,876,939	420.66	6,400	1.59	6,400	461.25	34.58	1.00	100%
		Layout2	<0.01		89	4,431,688	652.42	9,600	3.56	9,600	744.98	35.51	1.00	100%
	Mountain	Layout1	<0.01	(3,0,0,1,0)	8	64,698	1.92	6,400	1.60	6,400	11.52	2.53	1.00	100%
		Layout2	<0.01		11	97,418	2.98	9,600	3.54	9,600	17.52	2.54	1.00	100%
	Stair 2	Layout1	<0.01	(2,1,0,0,0)	6	77,159	2.49	12,800	6.26	6,400	14.75	2.01	2.00	100%
		Layout2	<0.01		9	115,950	3.99	19,200	14.18	9,600	27.17	2.01	2.00	100%
	Ind1	Layout1	<0.01	(3,3,0,6,1)	56	2,479,937	105.02	6,400	1.67	6,400	162.69	29.81	1.00	100%
		Layout2	<0.01		84	3,835,998	167.63	9,600	3.58	9,600	255.21	30.74	1.00	100%
	Ind2	Layout1	<0.01	(2,4,0,4,0)	24	512,941	25.53	6,400	1.58	6,400	51.11	8.01	1.00	100%
		Layout2	<0.01		37	769,949	40.10	9,600	3.53	9,600	80.63	8.02	1.00	100%
Ratio					1.48	1.98	3.17	1.00	1.00	1.00	2.13			

TABLE III. EDGE-BASED RULE EXTRACTION (VS. OUR CRITICAL RULE EXTRACTION)

	Pattern	Layout	Critical rule extraction time (s)	#ruleC (R)	DRC time (s)	#ruleD (A)	Pre-filtering time (s)	#hotspot (B)	Finalization time (s)	#hotspot (C)	Total time (s)	Ratio (A/RB)	Ratio (B/C)	Success rate
Edge-based	Stair 1	Layout1	<0.01	(0,0,0,0,3)	41	2,396,343	106.68	38,400	57.86	6,400	205.54	20.80	6.00	100%
		Layout2	<0.01		63	3,710,439	164.20	57,592	125.42	9,600	352.62	21.48	6.00	100%
	I	Layout1	<0.01	(0,0,0,0,2)	29	1,533,876	139.97	25,600	25.08	6,400	194.05	29.96	4.00	100%
		Layout2	<0.01		43	2,409,841	221.92	38,400	55.64	9,600	320.56	31.38	4.00	100%
	S	Layout1	<0.01	(0,0,0,0,5)	48	2,742,895	503.23	6,400	1.59	6,400	552.82	85.72	1.00	100%
		Layout2	<0.01		91	4,431,688	650.17	9,600	3.51	9,600	744.68	92.33	1.00	100%
	Mountain	Layout1	<0.01	(0,0,0,0,5)	45	2,538,307	151.36	6,400	1.59	6,400	197.95	79.32	1.00	100%
		Layout2	<0.01		67	3,923,568	241.98	9,600	3.53	9,600	312.51	81.74	1.00	100%
	Stair 2	Layout1	<0.01	(0,0,0,0,3)	41	2,319,777	150.35	12,800	6.25	6,400	197.60	60.41	2.00	100%
		Layout2	<0.01		60	3,607,091	239.57	19,200	14.10	9,600	313.67	62.62	2.00	100%
	Ind1	Layout1	<0.01	(0,0,0,0,4)	46	2,614,473	397.87	6,400	1.62	6,400	445.49	102.13	1.00	100%
		Layout2	<0.01		67	4,037,635	643.34	9,600	3.55	9,600	713.89	105.15	1.00	100%
	Ind2	Layout1	<0.01	(0,0,0,0,3)	25	1,307,058	174.33	6,400	1.61	6,400	200.94	68.08	1.00	100%
		Layout2	<0.01		36	2,025,831	277.28	9,600	3.56	9,600	316.84	70.34	1.00	100%
Ratio					2.15	3.64	7.59	1.00	1.00	1.00	4.34			

TABLE IV. CORNER-BASED ANALYSIS (VS. OUR FINALIZATION)

	Pattern	Layout	Critical rule extraction time (s)	#ruleC (R)	DRC time (s)	#ruleD (A)	Pre-filtering time (s)	#hotspot (B)	Finalization time (s)	#hotspot (C)	Total time (s)	Ratio (A/RB)	Ratio (B/C)	Success rate
Corner-based	Stair 1	Layout1	<0.01	(0,0,0,0,3)	41	2,396,343	106.68	38,400	72.61	6,400	220.29	20.80	6.00	100%
		Layout2	<0.01		63	3,710,439	164.20	57,592	158.03	9,600	385.23	21.48	6.00	100%
	I	Layout1	<0.01	(1,0,0,1,0)	9	45,004	1.51	25,600	30.49	6,400	41.00	0.88	4.00	100%
		Layout2	<0.01		9	67,640	2.27	38,400	70.03	9,600	81.30	0.88	4.00	100%
	S	Layout1	<0.01	(1,3,0,7,0)	17	609,013	39.81	6,400	1.92	6,400	58.73	8.65	1.00	100%
		Layout2	<0.01		35	914,217	61.16	9,600	4.37	9,600	100.53	8.66	1.00	100%
	Mountain	Layout1	<0.01	(3,0,0,1,0)	8	64,698	1.92	6,400	2.00	6,400	11.92	2.53	1.00	100%
		Layout2	<0.01		11	97,418	2.98	9,600	4.49	9,600	18.47	2.54	1.00	100%
	Stair 2	Layout1	<0.01	(2,1,0,0,0)	6	77,159	2.49	12,800	7.85	6,400	16.34	2.01	2.00	100%
		Layout2	<0.01		9	115,950	3.99	19,200	17.58	9,600	30.57	2.01	2.00	100%
	Ind1	Layout1	<0.01	(3,3,0,6,0)	23	596,472	31.91	6,400	1.98	6,400	56.89	7.77	1.00	100%
		Layout2	<0.01		35	895,377	50.44	9,600	4.53	9,600	89.97	7.77	1.00	100%
	Ind2	Layout1	<0.01	(2,4,0,4,0)	24	512,941	25.53	6,400	1.94	6,400	51.47	8.01	1.00	100%
		Layout2	<0.01		37	769,949	40.10	9,600	4.52	9,600	81.62	8.02	1.00	100%
Ratio					1.00	1.00	1.00	1.00	1.25	1.00	1.07			

*#ruleC: the number of critical rules extracted (#rule1, #rule2, #rule3, #rule4, #rule5).   **#ruleD: the number of locations reported by DRC
***#hotspot: the number of hotspots

Improved Tangent Space Based Distance Metric for Accurate Lithographic Hotspot Classification

Jing Guo[1], Fan Yang[1*], Subarna Sinha[3], Charles Chiang[2] and Xuan Zeng[1*]

[1]State Key Lab of ASIC & System, Microelectronics Dept., Fudan University, China
[2]Synopsys Inc., U.S.A.
[3]Stanford University, U.S.A.

ABSTRACT

A distance metric of patterns is crucial to hotspot cluster analysis and classification. In this paper, we propose an improved tangent space based metric for pattern matching based hotspot cluster analysis and classification. The proposed distance metric is an important extension of the well-developed tangent space method in computer vision. It can handle patterns containing multiple polygons, while the traditional tangent space method can only deal with patterns with a single polygon. It inherits most of the advantages of the traditional tangent space method, e.g., it is easy to compute and is tolerant with small variations or shifts of the shapes. Compared with the existing distance metric based on XOR of hotspot patterns, the improved tangent space based distance metric can achieve up to 37.5% accuracy improvement with at most 4.3x computational cost in the context of cluster analysis. The improved tangent space based distance metric is a more reliable and accurate metric for hotspot cluster analysis and classification. It is more suitable for industry applications.

Categories and Subject Descriptors:
J.6 [Computer-Aided Engineering]: Computer-Aided Design
General Terms: Algorithm, Design
Keywords: Lithographic, Hotspot, Classification, Distance Metric

1. INTRODUCTION

The lithographic printability issue becomes more and more critical as the technology node continues to shrink. Although various resolution enhancement techniques (RET) have been proposed to improve the lithographic printability, there still exist lithographic hotspots which will cause manufacturability problems hence deterioration of the yield. Therefore, it is essential to detect and clear those problematic patterns at early design stages.

Traditionally, design rules are used to model the hotspots [1] [2]. However, the more and more complex lithographic effects cannot be efficiently described by simple geometric rules. As a golden verification approach, lithographic simulation can also be used to detect the hotspots [3] [4]. Its high computational cost makes it impractical for hotspot detection at early design stages.

---

*Corresponding authors. Email: {yangfan, xzeng}@fudan.edu.cn.

In recent years, modern machine learning and pattern matching based methods have been proposed for hotspot analysis and detection. In the machine learning based methods [5] [6], neural network or support vector machine based regression model is built from a set of training hotspot patterns. The regression models are then used to predict or detect the hotspots. A good set of training patterns is very important for the successful application of these regression models for hotspot detection.

Pattern matching based methods employ explicit models rather than regression models to depict the hotspot patterns. The hotspot detection is actually a matching process based on these explicit models. These methods are believed to be faster and more accurate than the machine learning based methods for hotspot detection, if an accurate model for the hotspot patterns is defined. In [7], the concept of range pattern is proposed by incorporating tolerant variations into the traditional design rules.

In [8], a pattern matching based hotspot classification and detection scheme is proposed. Firstly, the extracted hotspots are classified into clusters by data mining methods. The representative hotspot in each cluster is then identified and stored in a hotspot library for future hotspot detection. Besides hotspot library generation, the classification approach also has tremendous application value for automatic hotspot correction and diagnosis. For hotspots of a cluster, engineers no longer need to manually analyze the failure reason and correct the hotspots one by one. Instead, they only need to focus on the representative hotspot in each cluster and the remaining hotspots in the same cluster can be corrected automatically according to the same correction template for the representative hotspot.

The hotspot classification approach in [8] heavily relies on a distance metric of different pattern samples. The distance metric is a quantitative measure of the difference of a pair of pattern samples. With the distance metric, the pattern samples which are close to each other are clustered into a group by cluster analysis. This is the process of hotspot classification. For an ideal distance metric, it should have the ability to capture the sketch of the hotspot pattern and also be tolerant with small variations or shifts of the shapes. The distance metric proposed in [8] is defined as a weighted integral over the area where a pair of hotspot patterns differs (XOR of patterns). This metric can exactly characterize the shapes of the hotspot patterns. If there are differences between two patterns, it will be reflected by the XOR operation directly. However, this distance metric is quite sensitive to the small variations or shifts of the shapes. With such a sensitive distance metric, the classification accuracy will be remarkably low and the hotspot detection process will also be error-prone.

In the computer vision community, a well-developed tangent space method [9] [10] has been successfully applied to polygon matching. The tangent space method defines a distance metric of a pair of polygons, which is the L_2 norm of the difference of the corresponding turning functions of the polygons. The turning function of polygon measures the angle of the counterclockwise tangent as a function of the normalized arc length, measured from some reference point of the polygon. Please refer to section 3 for the details of the definition. The tangent space method has many superb advantages for shape matching. It is invariant under translation, rotation and change-of-scale. More importantly, it is easy to compute and can deal with noise. This metric is a good candidate for hotspot classification, if the following difficulties can be overcome. Firstly, this metric is invariant under change-of-scale. But for lithography, the distance metric should be sensitive to the change-of-scale. Secondly, this metric can only deal with a single polygon. But for lithography, the distance metric should deal with hotspot patterns containing multiple polygons.

In this paper, we propose an Improved Tangent Space (ITS) based explicit metric for pattern matching based hotspot classification. The proposed distance metric is an important extension of the well-developed tangent space method. It can handle patterns containing multiple polygons. At the same time, it inherits some important advantages of the traditional tangent space method, e.g., it is easy to compute and is tolerant with small variations or shifts of the shapes. Compared with the existing distance metric based on XOR of hotspot patterns, the ITS based distance metric can achieve up to 37.5% accuracy improvement with at most 4.3x computational cost in the context of cluster analysis. Although the computational cost of the ITS based metric is a bit higher than that of XOR based metric, we remark that the ITS based distance metric is more reliable and accurate for hotspot cluster analysis and classification. It is more suitable for industry applications.

The rest of this paper is organized as follows. In section 2, a brief overview of cluster analysis and XOR based distance metric [8] is presented. In section 3, we propose the Improved Tangent Space (ITS) method. In section 4, we will show some experimental results in the context of hotspot clustering. Conclusions and future work are presented in section 5.

2. BACKGROUND REVIEW

Since we verify the distance metric in the context of cluster analysis of hotspots, we will give a brief introduction to cluster analysis in this section. We will also review the XOR based distance metric [8] in this section.

2.1 Introduction to Cluster Analysis

Cluster analysis is defined as classification of objects into groups so that the objects within the same group are closer to each other than those from different clusters, according to the predefined distance metric. We will briefly review state-of-the-art clustering methods which are applied to hotspot classification in [8].

2.1.1 Hierarchical Clustering Algorithm

Hierarchical clustering algorithms [11] produce a hierarchical representation of the data objects in which the clusters at each level of the hierarchy are created by merging clusters at the next lower level. The traditional representation of the hierarchy is a tree. At the lowest level of the tree, each cluster contains a single object. At the highest level, there is only one cluster containing all the objects. The tree can be built in two ways: bottom-up or top-down. Bottom-up strategies start at the bottom and at each level recursively merge a selected pair with the smallest inter-distance into a single cluster. Top-down methods start at the top and at

each level recursively split one of the existing clusters with the largest intra-distance at that level into two clusters.

After the tree is built, it is still necessary to choose the number of clusters, because in practical it is unknown what number of clusters is most suitable before clustering. The number of clusters is chosen by quantitatively examining the quality of the clustering [12] [13]. The final clustering result is derived such that the objects within the same group are more similar to each other than those from different clusters.

Hierarchical clustering can produce nearly ideal clustering result. However, it requires a pre-computed pair-wise distance matrix of all the objects, which is computationally intensive.

2.1.2 Incremental Clustering Algorithm

The incremental clustering method [14] is a heuristic clustering algorithm, in which data objects are sequentially inserted into incrementally evolving clusters. In the incremental clustering algorithm, the cluster is represented using cluster feature (CF). CF consists of the main feature of the cluster including: the number of objects in the cluster, the central object of the cluster and the radius of the cluster. A CF-tree is used to organize all the existing clusters. The CF-tree can be viewed as a search tree for incremental clustering to guide the new object to insert into the most appropriate cluster. The tree is updated every time a new object has been inserted.

The incremental clustering does not require computation of a complete pair-wise distance matrix, it is therefore time and space economic in clustering large datasets. However, since the data objects are inserted into the CF-tree sequentially, the clustering result is significantly affected by the insertion sequence of the data objects.

2.2 XOR Based Distance Metric

In this subsection, we briefly review the XOR based distance metric [8]. The limitations of XOR based distance metric will also be discussed.

2.2.1 Definition of the Distance Metric

From the viewpoint of lithography, the hotspots result from their local surrounding context. A hotspot and its local surrounding context are defined as a clip in [8]. Every pixel of the clip is either light or dark. If we overlay two clips, every pixel of the two clips either match (both are light or both are dark) or differ (if one is light and one is dark). In [8], the distance metric ρ is defined as the square root of the weighted integral over the regions where two clips Γ_1 and Γ_2 differ (XOR of two clips),

$$\rho(\Gamma_1, \Gamma_2) = \left[\int \int_{\Gamma_1 \neq \Gamma_2} \omega(x, y) dA \right]^{\frac{1}{2}}. \qquad (1)$$

The weighting function $\omega(x, y)$ is derived from lithographic system to describe the magnitude of the effect of each point of the clip on the hotspot at the center [8].Patterns that are within certain distance have remarkable important effect on the hotspot, while the patterns that are several wavelengths or more away from the hotspot center have negligible contributions to the hotspot.

In order to address the rotation and reflection of the clips, the distance metric is redefined as

$$\rho'(\Gamma_1, \Gamma_2) = \min_{\tau \in D_8} \rho(\Gamma_1, \tau(\Gamma_2)), \qquad (2)$$

where D_8 represents the set of eight transformations. The eight transformations are combinations of four rotations ($0°$, $90°$, $180°$, $270°$) and two mirrors (x mirror, y mirror).

In order to accelerate the calculation of the weighted integral over the regions of the XOR results of two clips, a

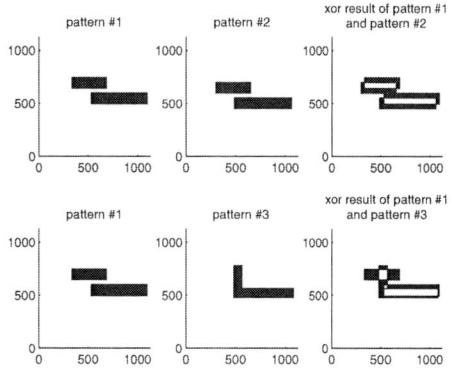

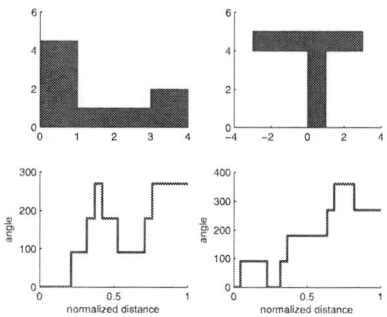

Figure 2: Turning function representations of polygons.

Figure 1: XOR based distance metric is sensitive to geometric shifts and bumps.

rectangle based look-up-table algorithm is employed in [8]. Due to limited space of the paper, please refer to [8] for the details of the algorithm.

2.2.2 Limitations

The XOR based distance metric is quite sensitive to the small variations or shifts of the patterns. With such a sensitive distance metric, the classification accuracy will be remarkably low and the hotspot detection process will also be error-prone. We use some simple patterns as shown in Figure 1 to illustrate the limitations. In Figure 1, pattern #2 is derived by shifting pattern #1 to the right-bottom direction for less than 3.98% of the longest edge length in pattern #1. From the viewpoint of lithography, pattern #2 is similar to pattern #1. The distance between pattern #2 and pattern #1 should be quite small. However, according to the definition in [8], the distance between pattern #1 and pattern #2 is 2.2468 (The average of intra-cluster distance is 0.341 in this testcase). For comparison, we show the XOR result of pattern #1 and a totally different pattern #3. According to the definition in [8], the distance between pattern #1 and pattern #3 is 2.1047, which is even smaller than the distance between pattern #1 and pattern #2. The XOR based distance metric does not match our intuition. The classification and detection procedure based on this distance metric will be error-prone.

3. IMPROVED TANGENT SPACE BASED DISTANCE METRIC

In this section, we will firstly introduce the tangent space method in the community of computer vision. The Improved Tangent Space (ITS) based distance metric for hotspot classification will be proposed afterwards.

3.1 Tangent Space Method for Shape Matching

In the computer vision community, a well-developed tangent space method [9] has been successfully applied to shape matching.

Tangent space method uses a turning function $\Theta_A(s)$ to represent a polygon. The turning function $\Theta_A(s)$ of a polygon A gives the angle of the counterclockwise tangent as a function of the normalized arc length s, measured from some reference point O on the boundary of A. $\Theta_A(s)$ keeps track of the turning that takes place, increasing with left-hand turns and decreasing with right-hand turns. Samples of turning functions are shown in Figure 2.

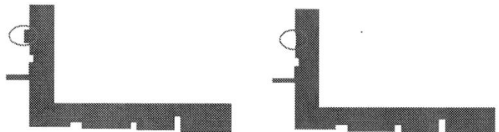

Figure 3: Polygon with noise on the boundary.

The distance metric based on tangent space method is defined as the L_2 norm of the difference of the corresponding turning functions of the polygons,

$$d(A, B) = \left(\int_0^1 |\Theta_A(s) - \Theta_B(s)|^2 ds \right)^{\frac{1}{2}}. \qquad (3)$$

The distance metric should be insensitive to the rotation of the polygons and the choice of the referenced point O. Therefore, the distance metric is revised as the minimal distance over all the choices of rotations and referenced point,

$$d(A, B) = \left(\min_{\Theta \in R, t \in [0,1]} \int_0^1 |\Theta_A(s+t) - \Theta_B(s) + \theta|^2 ds \right)^{\frac{1}{2}}, \qquad (4)$$

where t represents the shift amount from the reference point O along the boundary of polygon A. θ represents the rotation angle of polygon A. $\Theta_A(s+t)+\theta$ represents the turning function of polygon A with the shift of referenced point O by t and rotation by angle of θ. An algorithm with complexity $O(mn \log (mn))$ is proposed in [9] to calculate the minimal distance in (4), where m represents the number of vertices in one polygon and n represents the number in the other polygon. The piecewise-constant characteristic enables easy and fast calculation of the integral in (4).

As the edge should be normalized to 1, noise on the boundary of polygon A may affect the shape matching, as shown in the left polygon in Figure 3. In [10], the authors introduce the following definition to recognize noise on the boundary of polygon,

$$K(s_1, s_2) = \frac{\beta(s_1, s_2) \, l(s_1) \, l(s_2)}{l(s_1) + l(s_2)}, \qquad (5)$$

where s_1 and s_2 represent two adjacent edges, $l(s_1)$ and $l(s_2)$ represent the length of edge s_1 and s_2, respectively. $\beta(s_1, s_2)$ represent the angle between s_1 and s_2. If the value of $K(s_1, s_2)$ is smaller than a threshold, the edges s_1 and s_2 are regarded as noise and should be eliminated. This method can be applied to eliminate the bumps in hotspot pattern matching. As shown in Figure 3, the small bump found with (5) can be eliminated to improve the accuracy of shape matching.

1171

In conclusion, the distance metric defined by tangent space method is invariant under translation, rotation and change-of-scale. It can deal with noise, e.g., bumps in hotspots. It is also fast and easy to compute. These advantages make it a potential method for hotspot pattern clustering.

However, the tangent space based distance metric encounters the following difficulties for hotspot pattern clustering. Firstly, this metric is invariant under change-of-scale. But for hotspot pattern clustering, the distance metric should be sensitive to the change-of-scale. More importantly, this metric can only deal with a single shape. But for hotspot pattern matching, the distance metric should deal with hotspot patterns containing multiple shapes. In the following subsection, we will propose an improved tangent space based distance metric to overcome these difficulties.

3.2 Improved Tangent Space Based Distance Metric

The limitations of traditional tangent space based metric restrict its application in hotspot clustering. In this subsection, we make an important improvement to the basic tangent space based distance metric, which is referred as Improved Tangent Space (ITS) based distance metric.

Similar to [8], we define a hotspot and its local surrounding context as a clip. In ITS based distance metric, we define a polar coordinate system for each clip, as shown in Figure 4. The origin of the polar coordinate system is defined as the center of the clip. For each polygon in the clip, we use two functions to describe the polygon, i.e., the radial function and angular function. Radial function gives the radial coordinates of the turning vertices as a function $R_A(s)$ of the normalized polygon edge length s, measured from the bottom left point of the polygon. Angular function gives the angular coordinates of the turning vertices as a function $\Theta_A(s)$ of the normalized polygon edge length. For illustration, the radial and angular function representations of polygons shown in Figure 4 are illustrated in Figure 5.

Based on the definitions of the radial and angular functions, we firstly define the ITS based distance metric of a pair of polygons A and B,

$$d(A, B) = d_R(A, B) * d_\Theta(A, B), \qquad (6)$$

where

$$d_R(A, B) = \left[\int_0^1 \omega(x, y) |R_A(s) - R_B(s)|^2 ds \right]^{\frac{1}{2}},$$
$$d_\Theta(A, B) = \left[\int_0^1 |\Theta_A(s) - \Theta_B(s)|^2 ds \right]^{\frac{1}{2}}.$$

Here $\omega(x, y)$ is the weighting function the same as that defined in XOR based distance metric [8]. It is also used to describe the magnitude of the effect of each point of the clip on the hotspot at the center. $d_R(A, B)$ is the square root of the weighted integral of the difference of the radial functions of polygons A and B over the normalized edge length s. $d_\Theta(A, B)$ is the square root of the integral of the difference of the angular functions of polygons A and B. One should note that both the radial function and angular function are piecewise constant functions. The integrals of radial function and angular function are easy to compute. We use the definition in [10] as shown in (5) to recognize and remove the noise on the boundary of the polygon.

In order to deal with clips with multiple polygons, a distance metric should be able to capture not only the shapes of the polygons, but also the spacing between the polygons. In the ITS based distance metric, both the absolute position and the shape information of the polygons are encoded into the distance by integrating the differences of radial and angular functions of polygons A and B. The difference in positions hence spacing or the difference in shapes of poly-

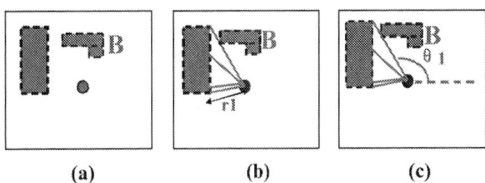

Figure 4: (a) hotspot center (b) radius of polygon edge (c) angle of polygon edge

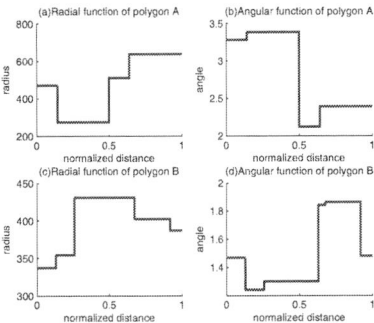

Figure 5: Improved tangent space representation of radius and angle.

gons will both be reflected by this distance metric directly. It is possible to deal with clips with multiple polygons based on the ITS based distance metric.

Now, we consider two clips containing multiple polygons. For illustration, we show two clips in Figure 6. From Figure 6, we can find that the difference of Clip Γ_1 and Clip Γ_2 consists of three parts, i.e., the difference of polygon A in Γ_1 and polygon A' in Γ_2, the difference of polygon B in Γ_1 and polygon B' in Γ_2, the difference of polygon C in Γ_1 and a null polygon in Γ_2. This is because polygon A'/B' in Γ_2 is most similar to polygon A/B in Γ_1, while polygon C in Γ_1 cannot find a matched polygon in Γ_2. We can use the distance metric defined in (6) to judge whether two polygons are matched. If the distance is within a predefined threshold, then the two polygons are regarded as matched polygons. Otherwise, they are regarded as unmatched polygons. The distance of matched polygon pair A and A' can be obtained by

$$\begin{aligned} \text{distance(matched)} &= d(A, A') \\ &= d_R(A, A') * d_\Theta(A, A'). \end{aligned} \qquad (7)$$

The distance of polygon C in Γ_1 and a null polygon in Γ_2 can be expressed as

$$\begin{aligned} \text{distance(unmatched)} &= d(C, 0) \\ &= d_R(C, 0) * d_\Theta(C, 0) \\ &= \left(\int_0^1 |R_C(s)|^2 ds \right)^{\frac{1}{2}} \\ &\quad * \left(\int_0^1 |\Theta_C(s)|^2 ds \right)^{\frac{1}{2}}. \end{aligned} \qquad (8)$$

We summarize the process of computing ITS distance metric of Clip Γ_1 and Clip Γ_2 in Algorithm 1.

To address the rotation and reflection of the clips, the ITS based distance metric is redefined as

$$\rho'(\Gamma_1, \Gamma_2) = \min_{\tau \in D_8} \rho(\Gamma_1, \tau(\Gamma_2)), \qquad (9)$$

Algorithm 1 Compute ITS distance metric of Clip Γ_1 and Clip Γ_2

Input: Clip Γ_1 and Clip Γ_2
Output: ITS distance metric $\rho(\Gamma_1, \Gamma_2)$

1: Let $\{p_1, p_2, \cdots, p_m\}$ be the polygons in Clip Γ_1
2: Let $\{p_1', p_2', \cdots, p_n'\}$ be the polygons in Clip Γ_2
3: Initialize the states of all the polygons in Clip Γ_1 and Clip Γ_2 as unmatched.
4: **for** $i = 1 : m$ **do**
5:    For polygon p_i in Clip Γ_1, find a unmatched polygon p_j' in Clip Γ_2 such that the distance of p_i and p_j', i.e., $d(p_i, p_j')$, is minimized.
6:    **if** $d(p_i, p_j')$ is within a predefined threshold ϵ **then**
7:       Set polygons p_i and p_j' as matched pair.
8:       Set the states of polygons p_i and p_j' as matched.
9:    **end if**
10: **end for**
11: Calculate the distances $\text{distance}_k(\text{matched})$ of the matched pairs of polygons, according to equation (7).
12: Calculate the distances $\text{distance}_k(\text{unmatched})$ of the polygons with unmatched states in Clips Γ_1 and Γ_2, according to equation (8).
13: $\rho(\Gamma_1, \Gamma_2) = \sum_k \text{distance}_k(\text{matched}) + \sum_k \text{distance}_k(\text{unmatched})$

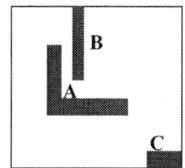 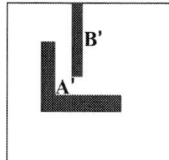

Figure 6: Clip Γ_1 and clip Γ_2.

where D_8 represents the set of eight transformations. The eight transformations are combinations of four rotations ($0°$, $90°$, $180°$, $270°$) and two mirrors (x mirror, y mirror).

The ITS based distance metric is sensitive to the change-of-scale because the distance metric definition includes the absolute position information of the polygons. It can handle patterns containing multiple polygons. At the same time, it inherits most of the advantages of the traditional tangent space method, i.e., it is easy to compute and can deal with noise. The ITS based distance metric is a superb metric for hotspot classification.

4. EXPERIMENTAL RESULTS

In this section, we will present experimental results to demonstrate the advantages of ITS based distance metric in the context of hotspot cluster analysis.

4.1 Sensitivity to Shifts and Bumps

In this subsection, we will test the sensitivities of XOR and ITS based metric to noise (e.g., shifts and little bumps). For simplification, we use some simple testing patterns here.

In the first experiment, we will test the sensitivities of XOR and ITS based metric to little shifts. We show three testing patterns in Figure 7. Clip #1 has only one "L" shape. Clip #2 and Clip#3 both have two rectangles. Clip #3 is obtained by shifting the two polygons in Clip #2 toward right direction. We show the distance calculated by XOR and ITS metric in Table 1 and Table 2, respectively. From Table 1 and Table 2, we can find that the distance between Clip #2 and Clip #3 is approximately the same as the distance between Clip #1 and Clip #2 for XOR based metric. However, the distance between Clip #2 and Clip #3 is only one seventh of the distance between Clip #1 and Clip #2

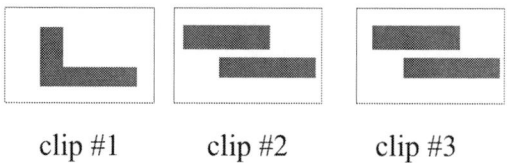

Figure 7: Clips with geometric shifts.

for ITS based metric, which matches the intuition of lithography.

Table 1: Distance matrix by XOR metric for clips with geometric shifts.

	clip #1	clip #2	clip #3
clip #1	0	2.1942	2.2066
clip #2	2.1942	0	2.2943
clip #3	2.2066	2.2943	0

Table 2: Distance matrix by ITS metric for clips with geometric shifts.

	clip #1	clip #2	clip #3
clip #1	0	2.581	1.932
clip #2	2.581	0	0.322
clip #3	1.932	0.322	0

In the second example, we will test the sensitivities of XOR and ITS based metric to bumps. We show three clips in Figure 8. Clip #1 has only one "L" shape. Clip #2 has a small bump compared to Clip #1. Clip #3 has two rectangles. From the viewpoint of lithography, Clip #1 and Clip #2 should be clustered into the same group, while Clip #3 should be in a different cluster. We show the distance calculated by XOR and ITS measure in Table 3 and Table 4, respectively. From Table 3 and Table 4, we can find that the distance between Clip #1 and Clips #2 is remarkably large by XOR based metric, which is almost approximately equals to inter-cluster distance (distance between Clip #3 and Clip #1 / Clip #2). However, with the ITS based metric, the inter-cluster distance is about 20x of the intra-cluster distance, which provides great gap between intra-cluster distance and inter-cluster distance, making it accurate to distinguish clips of different clusters.

4.2 Clustering Accuracy and Speed Comparison

In this subsection, we will provide the clustering speed and accuracy comparison with different clustering algorithms (hierarchical clustering and incremental clustering algorithms) and different distance metrics (XOR and ITS based distance metrics). The test cases are derived from industry. We show the clustering time and accuracy comparison in Table 5. In Table 5, "xor" means XOR distance metric, "its" means ITS distance metric, "hier" and "incr" mean the hierarchical and incremental clustering methods. We check each pattern and assign a classification label to them from the viewpoint of lithography manually. These patterns are then classified by the aforementioned cluster analysis methods. For a pattern, if the classification label obtained by cluster analysis matches the manually assigned one, we say the clustering result is correct for this pattern. The accuracy is defined as the number of the correctly classified patterns over the total number of patterns.

From Table 5, we can find that ITS metric based clustering methods can achieve nearly ideal accuracy for all the test cases except the C15_640 and C16_1280 cases. The degeneracy of the accuracy of the cases C15_640 and C16_1280 for ITS metric based incremental clustering method is mainly

Table 5: Comparison of accuracy and speed with different clustering algorithms (hierarchical clustering, incremental clustering algorithms) and different distance metrics (XOR based and ITS based distance metrics).

cases	# patterns	xor+hier		xor+incr		its+hier		its+incr	
		accu	time(s)	accu	time(s)	accu	time(s)	accu	time(s)
C1_40	40	1	3.54	1	2.99	1	10.11	1	5.53
C2_40	40	1	3.79	1	2.40	1	10.01	1	4.07
C3_40	40	1	2.89	1	2.07	1	7.52	1	4.52
C4_80	80	1	11.78	0.75	5.39	1	33.32	1	12.34
C5_70	70	0.8571	10.39	1	5.74	1	24.66	1	11.33
C6_80	80	0.875	12.14	0.9875	5.73	1	38.08	1	14.63
C7_80	80	1	12.37	1	6.68	1	36.16	1	14.76
C8_80	80	1	13.06	1	5.68	1	39.23	1	14.60
C9_80	80	1	11.72	1	5.62	1	31.98	1	11.40
C10_80	80	0.625	10.86	0.625	6.38	1	32.25	1	13.12
C11_80	80	0.75	13.24	0.9125	5.99	1	37.53	1	13.32
C12_80	80	0.8125	13.12	0.625	5.30	1	37.12	1	13.63
C13_160	160	0.75	47.00	1	15.16	1	145.66	1	49.12
C14_320	320	1	189.57	1	41.09	1	591.37	1	153.05
C15_640	640	0.625	630.57	0.6625	115.54	1	2028.50	0.875	495.30
C16_1280	1280	0.625	2572.22	0.5625	536.86	1	7911.77	0.8125	2249.74
average	/	0.8699	/	0.8828	/	1	/	0.9847	/
worst	/	0.625	/	0.5625	/	1	/	0.8125	/

clip #1   clip #2   clip #3

Figure 8: Clips with bumps.

Table 3: Distance matrix by XOR method for clips with bumps.

	clip #1	clip #2	clip #3
clip #1	0	2.241	2.585
clip #2	2.241	0	2.622
clip #3	2.585	2.622	0

due to the limitation of incremental clustering method. Because the incremental clustering method is sensitive to the sequence of data input. Nevertheless, the accuracy of the cases C15_640 and C16_1280 for ITS metric based incremental clustering method is still acceptable. On the other hand, the accuracy of XOR metric based clustering methods is remarkably lower for some test cases.

Statistically, the CPU time for computing an ITS based distance metric is about $2 \sim 4$ times of that for computing an XOR distance metric. However, we remark that ITS based metric is more reliable and accurate for hotspot classification. It is more suitable for industry applications.

5. CONCLUSION

In this paper, we propose an Improved Tangent Space (ITS) based distance metric for hotspot classification and detection. Compared with the existing XOR based distance metric, the ITS method is less sensitive to the noises, e.g., geometric shifts and little bumps. It can achieve nearly ideal accuracy in the context of cluster analysis. ITS metric is a more reliable and accurate metric for hotspot classification. The ITS based distance metric can also be applied to hotspot detection. In the future work, we will reduce the computational cost of computing ITS based distance metric and apply this metric to hotspot detection.

Acknowledgment

This research is supported partially by National Natural Science Foundation of China (NSFC) research projects 61125401, 61006030, 60976034 and 61076033, National Basic Research Program of China under the grant 2011CB309701, National Major Science and Technology Special Project 2011ZX01035-001-001-003 of China during the 12-th five-year plan period.

Table 4: Distance matrix by ITS method for clips with bumps.

	clip #1	clip #2	clip#3
clip #1	0	0.242	5.759
clip #2	0.242	0	6.491
clip #3	5.759	6.491	0

6. REFERENCES

[1] Chul-Hong Park, Yoo-Hyon Kim, Ji-Soong Park, Kwan-Do Kim, Moon-Hyun Yoo, and Jeong-Taek Kong. A systematic approach to correct critical patterns induced by the lithograhpy process at the full-chip level. In *Proc.SPIE*, 1999.

[2] Gerard T. Luk-Pat, Alexander Miloslavsky, Frank Tseng, and Linni Wen et. al. Correcting lithography hotspots during physical-design implementation. In *Proc.SPIE*, 2006.

[3] Li-Da Huang and Wong M.D.F. Optical proximity correction (OPC)-friendly maze routing. In *DAC*, 2004.

[4] J.Mitra. P.Yu, and D.Z.Pan. RADAR: Ret-aware detailed routing using fast lithography simulations. In *DAC*, 2005.

[5] Duo Ding and David Z.Pan. Machine learning based lithographic hotspot detection with critical feature extraction and classifications. In *International conference on integrated circuit design technology*, 2009.

[6] Duo Ding, Andres J. Torres, Fedor G. Pikus, and David Z. Pan. High performance lithographic hotspot detection using hierarchically refined machine learning. In *ASPDAC*, 2011.

[7] H.Yao, S.Sinha, J.Xu, C.Chiang, Y.Cai, and X.Hong. Efficient range pattern matching algorithm for process-hotspot detection. In *ICCAD*, 2006.

[8] Ning Ma. *Automatic IC Hotspot Classification and Detection using Pattern-Based Clustering*. PhD thesis. Engineering IC Mechanical Engineering, University of California, Berkeley, 2008.

[9] Daniel P.Huttenlocher Klara Kedem and Joseph S.B.Mitchell Esther M.Arkin. L.Paul Chew. An efficiently computable metric for comparing polygonal shapes. (3):209–216, 1991.

[10] Longin Jan Latecki and Rolf Lakamper. Shape similarity measure based on correspondence of visual parts. (10):1–6, 2000.

[11] S. C. Johnson. Hierarchical clustering schemes. *Psychometrika*, pages 241–254, 1967.

[12] L. Hubert and J. Schultz. Quadratic assignment as a general data-analysis strategy. *Br. J. Math. Stat. Psychol*, pages 190–241, 1976.

[13] G. W. Milligan. A monte carlo study of thirty internal criterion measures for cluster analysis. *Psychometrika*, (2):187–199, 1981.

[14] Venkatesh Ganti, Raghu Ramakrishnan, and Johannes Gehrke. Clustering large datasets in arbitrary metric spaces. *Proceedings of the 15th International Conference on Data Engineering*, pages 502–511, 1999.

Simultaneous Flare Level and Flare Variation Minimization with Dummification in EUVL[*]

Shao-Yun Fang[1] and Yao-Wen Chang[1,2]

[1]Graduate Institute of Electronics Engineering, National Taiwan University, Taipei 106, Taiwan
[2]Department of Electrical Engineering, National Taiwan University, Taipei 106, Taiwan
yuko703@eda.ee.ntu.edu.tw; ywchang@cc.ee.ntu.edu.tw

ABSTRACT

Extreme Ultraviolet Lithography (EUVL) is one of the most promising Next Generation Lithography (NGL) technologies. Due to the surface roughness of the optical system used in EUVL, the rather high level of flare (i.e., scattered light) becomes one of the most critical issues in EUVL. In addition, the layout density non-uniformity and the flare periphery effect (the flare distribution at the periphery is much different from that in the center of a chip) also induce a large flare variation within a layout. Both of the high flare level and the large flare variation could worsen the control of critical dimension (CD) uniformity. Dummification (i.e., tiling or dummy fill) is one of the flare compensation strategies to reduce the flare level and the flare variation for the process with a clear-field mask in EUVL. However, existing dummy fill algorithms for Chemical-Mechanical Polishing (CMP) are not adequate for the flare mitigation problem in EUVL due to the flare periphery effect. This paper presents the first work that solves the flare mitigation problem in EUVL with a specific dummification algorithm flow considering global flare distribution. The dummification process is guided by dummy demand maps, which are generated by using a quasi-inverse lithography technique. In addition, an error-controlled fast flare map computation technique is proposed and integrated into our algorithm to further improve the efficiency without loss of computation accuracy. Experimental results show that our flow can effectively and efficiently reduce the flare level and the flare variation, which may contribute to the better control of CD uniformity.

Categories and Subject Descriptors

B.7.2 [**Integrated Circuits**]: Design Aids

General Terms

Algorithms, Design, Performance

Keywords

Extreme Ultraviolet Lithography, Flare, Dummification, Manufacturability

---

[*]This work was partially supported by IBM, SpringSoft, TSMC, and NSC of Taiwan under Grant No's. NSC 100-2221-E-002-088-MY3, NSC 99-2221-E-002-207-MY3, NSC 99-2221-E-002-210-MY3, and NSC 98-2221-E-002-119-MY3.

1. INTRODUCTION

Extreme Ultraviolet Lithography (EUVL) is one of the most promising Next Generation Lithography (NGL) technologies since the ten times reduction in wavelength in EUVL offers the capability of a continuation of Moore's law beyond the 22 nm technology node [1, 17]. However, the used light of 13.5 nm wavelength is not transmitted, but absorbed by most of materials, and thus only reflective optical components and masks can be used. Due to the surface roughness of the optical system, flare, which is undesired scattered light contribute to wafer exposure, is one of the most critical issues in EUVL, as illustrated in Figure 1(a).

For the process using a clear-field mask that is also made of reflective materials in EUVL, the layout patterns are formed by absorbers on the mask, as illustrated in Figure 1(b). Thus, during an exposure process on the wafer, the vacant regions not covered by layout patterns will be exposed by the light, and vice versa. However, the scattered flare reduces the contrast between bright regions (vacant regions) and dark regions (layout patterns), and may result in critical dimension (CD) distortion.

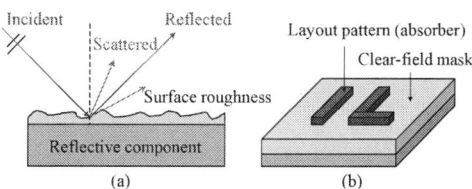

Figure 1: (a) Flare is undesired scattered light due to the surface roughness of reflective optical components and masks used in EUVL. (b) A clear-field mask on which layout patterns are formed by light absorbing materials.

Since the flare is proportional to the surface roughness of the optical system and inversely proportional to squared wavelength [12, 19], EUVL suffers from rather high level of flare compared to traditional lithography technologies. It is reported by the alpha demo tool (ADT) at IMEC that the intrinsic flare is about 16% [12]. On the other hand, the regions at the periphery of a chip receive much less flare compared to the regions in the center of a chip (assuming the regions outside the chip boundaries are dark-fields), causing a large flare variation [9]. We refer to the phenomenon as *the flare periphery effect*. In addition, the non-uniformity of layout patterns may contribute to the flare variation within a chip as well. For the process with a clear-field mask, regions with lower pattern density contribute to more flare distribution than those with higher pattern density [9, 10]. Since the high flare level causes CD distortion and the flare variation damages CD uniformity, flare compensation strategies are required.

There are two strategies for flare compensation. One is applying global CD resizing similar to optical proximity correction (OPC) on pattern features according to the flare value received by each feature [13, 19]. However, previous work has reported that one percent change in flare level may cause 10 nm change in CD (CD sensitivity on flare is 10 nm/% Δflare level) at the 22 nm technology node and may be considerably larger for more advanced technology nodes [10]. As a result, a large flare variation may not be fully compensated by applying a global CD resizing and may cause the difficulty of controlling CD uniformity accurately. Another flare compensation strategy is dummification (i.e., tiling or dummy fill) [2, 9, 13]. By adding dummy patterns according to global flare distribution, intra-chip flare variation could be reduced. Although dummification may be limited to design rules and layout constraints, it has been shown that dummification can simultaneously reduce intra-chip flare level and flare variation in EUVL for the process with a clear-field mask, and thus may greatly simplify the flare compensation methodology with global CD resizing [10].

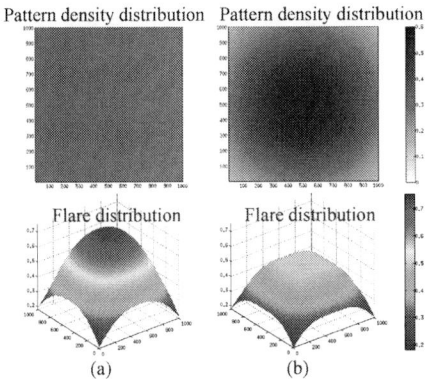

Figure 2: Flare comparison between layouts with different density distributions. (a) A layout with uniform density distribution may have large flare variation due to the flare periphery effect. (b) A layout with density distribution conforming to the global flare distribution has smaller flare variation.

There are many existing dummy fill algorithms for improving Chemical-Mechanical Polishing (CMP) quality [3, 4, 7, 18, 6]. The objectives of these algorithms mainly focus on density variation minimization or density gradient minimization for a layout. However, the flare variation of a layout with minimized layout density variation could be far from optimal; that is, achieving maximum pattern density uniformity is not equivalent to achieving maximum flare uniformity in EUVL. As shown in Figure 2(a), a layout with merely completely uniform layout density distribution may still suffer from large flare variation due to the flare periphery effect. Another layout with smaller flare variation is shown in Figure 2(b), in which the layout density distribution is less uniform but conforms to the global flare distribution (trend) of the layout. Thus, a more sophisticated dummification algorithm for flare variation minimization in EUVL is required.

In this paper, we propose the first work that solves the flare mitigation problem in EUVL for the process with a clear-field mask with a dummification algorithm considering global flare distribution. Since it is computationally expensive for deriving the global flare distribution of a layout, we first propose an error-controlled fast flare map computation approach to improve the efficiency of the algorithm without loss of computational accuracy. Then, we present a dummification process consisting of two stages: (1) the global dummification stage followed by (2) the local refinement stage. Experimental results based on the MCNC and the Faraday benchmarks show that our flow can effectively and efficiently reduce the flare level and the flare variation, which can contribute to the better control of the CD uniformity within a layout.

The rest of this paper is organized as follows: Section 2 gives some preliminaries and the problem formulation of this paper. In Section 3, an error-controlled fast flare map computation method is introduced. Section 4 details the proposed dummification algorithm for simultaneous flare level and flare variation minimization. Experimental results are reported in Section 5. Finally, we conclude our work in Section 6.

2. PRELIMINARIES

In this section, the preliminaries and the problem formulation of flare optimization with dummification are given.

2.1 Flare Map Computation

Practically, flare in EUVL can be modelled as a scattering point spread function (PSF), and the flare distribution can be obtained by convolving the PSF with the original image intensity I_0. Since flare in EUVL could result in significant change in the image intensity, accurate flare map computation with high resolution is required for implementing flare compensation. However, the flare PSF has a very large coverage. It has been reported that an area described by a radial distance of about $1000\ \mu m$ accounts for only about 95% of the flare seen at a point on a wafer [15]. Due to the long-range effects of the flare PSF and the high complexity of I_0, directly convolving the PSF with I_0 could be very computationally expensive. To tackle this problem, previous work has shown that dividing a layout into suitably sized grids (e.g., $1\mu m \times 1\mu m$) and calculating the layout density for each grid can achieve a good approximation of I_0 [13, 19]. Then, by convolving the generated density map $I_D(x, y)$ at the coordinate (x, y) with the discrete $PSF(x, y)$, we can derive the flare map $I_F(x, y)$ of a layout as follows:

$$I_F(x, y) = I_D(x, y) \otimes PSF(x, y). \tag{1}$$

Note that for a clear-field mask, flares are distributed from vacant regions without the coverage of patterns. Thus, in our work, density maps are referred to as *vacancy density maps*.

However, for full chip flare map generation, the computation process may still be too time-consuming. Therefore, other speed-up techniques without lose of computational accuracy are required for flare map computation.

2.2 Flare Reduction with Dummification

Dummification (tiling or dummy pattern insertion) is a simple method to reduce the flare effects in EUVL. Although dummification may be constrained by layout patterns, the technique can significantly mitigate the flare level and the flare variation for the process with a clear-field mask [13]. In addition, dummification may also greatly simplify the flare compensation methodology with global CD resizing [10].

Some previous work has proposed ideas to perform dummification for flare mitigation in EUVL. Singh et al. developed an iterative methodology that adds auxiliary patterns by utilizing commercial tools which are mainly driven by polishing requirements (e.g., CMP) [15]. However, as pointed out in Section 1, achieving maximum pattern density uniformity is not equivalent to achieving maximum flare uniformity in a dummification process due to the flare periphery effect in EUVL. Thus, the dummification algorithms for CMP are not suitable for flare mitigation in EUVL.

Another idea is to vary the size of dummies according to the location within a field [16]. This inspires an intuitive dummification method that varies dummy densities as a linear function of distance from the center of a layout. For regions far from the center of a layout, the flare values are expected to be smaller, and thus fewer dummy patterns are required. In contrast, more dummy patterns are required for those regions near the center of a layout. Although dummification with a linear function considers the flare periphery effect in EUVL, the solution space is limited and global flare distribution is not considered as well. Thus, a dummification method considering global flare distribution is desired.

2.3 Problem Formulation

Given an input layout, we first divide the layout into fine grids and analyze the vacancy density and the maximum available dummy density (the maximum areas dummy patterns can be inserted without conflicting with the original design) for each grid. A flare map is then computed by convolving the discrete flare PSF with the vacancy density map. After that, the dummification process can be formulated as a dummy value assignment problem considering global flare distribution. The problem formulation of simultaneous flare level and flare variation mitigation with dummification for the process with a clear-field mask in EUVL can be described as follows:

PROBLEM 1. *Given a grid-based layout, assign a dummy density value to each grid such that the flare level and the flare variation in the flare map of the layout with dummification are simultaneously minimized.*

After deriving a dummy density value d_i for each grid g_i, we simply insert dummies in g_i with their total area equal to $d_i \times a_i$, where a_i is the area of a grid. In addition, dummies are inserted in the dummy available regions of a grid to ensure that the inserted dummies would not conflict with the original design.

3. ERROR-CONTROLLED FAST FLARE MAP COMPUTATION

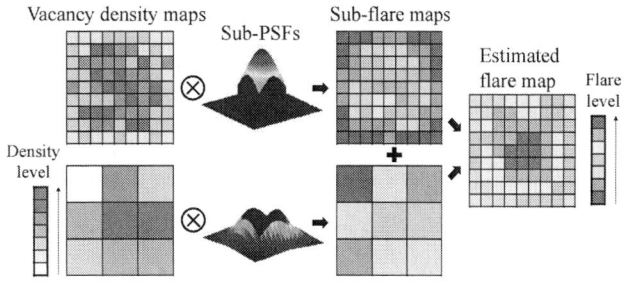

Figure 3: Estimated flare map computation with multiple convolutions between density maps and sub-PSFs.

As mentioned in Section 2.1, the full chip flare map generation is computationally expensive. Some previous work proposed a technique applying multiple convolutions with coarsened grids of different sizes instead of performing one convolution with very fine grids to speed up the computation process [10, 12, 15]. Figure 3 illustrates the method. The PSF is divided into several sub-PSFs with different, yet uniform, grid sizes. For a sub-PSF with larger variation, its grid size is chosen to be smaller; otherwise, the grid size is chosen to be larger. By convolving each sub-PSF and a density map with the same grid size and by summing the generated sub-flare maps, the estimated flare map is generated.

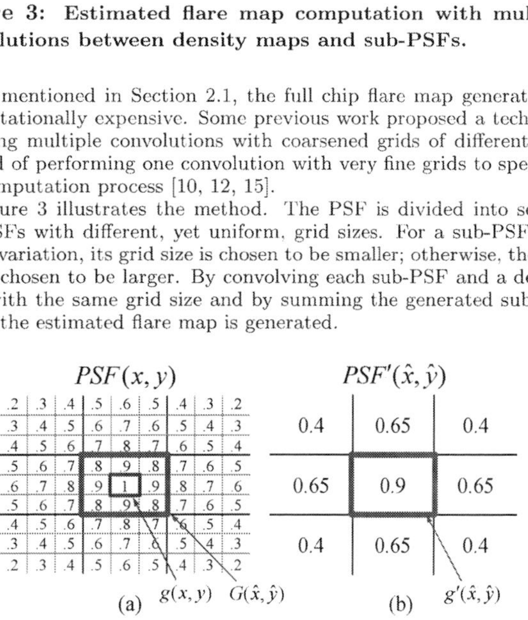

Figure 4: Illustration of deriving a sub-PSF from the original PSF. The values indicate the function values of the original PSF and the sub-PSF. (a) The original PSF with finest grids. (b) A sub-PSF with its grid size equal to 3.

However, the accuracy of an estimated flare map strongly depends on the division of the PSF and the resolution of each sub-PSF. No previous work mentioned how to control the accuracy of an estimated flare map. Therefore, we define a criterion for judging whether a sub-PSF is a good approximation of the original PSF or not. In addition, we propose an error-controlled sub-PSFs generation method such that the error between an original flare map and its estimated flare map can be controlled better. We first give some notations used as follows:

- PSF': a sub-PSF.
- $g(x, y)/g'(\hat{x}, \hat{y})$: a grid in the PSF/$PSF'$.
- $PSF(x, y)/PSF'(\hat{x}, \hat{y})$: a function value of the PSF/PSF' at $g(x, y)/g'(\hat{x}, \hat{y})$.
- $G(\hat{x}, \hat{y})$: a set of grids $g(x, y)$ contained in $g'(\hat{x}, \hat{y})$.

Figure 4(a) illustrates a PSF and Figure 4(b) shows a sub-PSF PSF' with its grid size equal to 3. The values shown in Figure 4 are the function values of the PSF and the sub-PSF. As illustrated in Figure 4, the grids belonging to $G(\hat{x}, \hat{y})$ (enclosed by the red/thicker window) in Figure 4(a) are contained in $g'(\hat{x}, \hat{y})$ in Figure 4(b). Based on the notations, we give the following definitions as a criterion for judging how a sub-PSF is approximated to the original PSF.

DEFINITION 1. *A function value $PSF'(\hat{x}, \hat{y})$ in a PSF' is said to be ϵ-controlled if for each $g(x, y) \in G(\hat{x}, \hat{y})$, the error between $PSF(x, y)$ and $PSF'(\hat{x}, \hat{y})$ is less than ϵ%.*

DEFINITION 2. *A PSF' is an ϵ-controlled sub-PSF if all function values $PSF'(\hat{x}, \hat{y})$ are ϵ-controlled.*

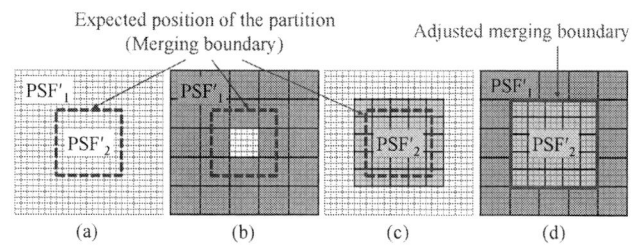

Figure 5: Illustration of the error-controlled sub-PSFs generation. (a) The original PSF is to be partitioned into two sub-PSFs according to the expected position of the partition. (b) PSF'_1 with its maximized grid size is error-controlled. (c) PSF'_2 with its maximized grid size is error-controlled. (d) The merging boundary is adjusted such that the two sub-PSFs are disjoint.

Thus, based on Definition 2, our objective is to find a set of ϵ-controlled sub-PSFs such that the union of the sub-PSFs is a good approximation of the original PSF. The set of sub-PSFs should be chosen such that the union of the sub-PSFs covers the original PSF and the PSFs are disjoint with each other. The ϵ-controlled sub-PSFs generation process is shown in Figure 5. Figure 5(a) shows an original PSF with finest grids. Suppose we want to partition the PSF into two sub-PSFs, PSF'_1 and PSF'_2. The blue (dashed) window in Figure 5(a) indicates the expected position of the partition. For each sub-PSF, we maximize its grid size and keep the sub-PSF to be ϵ-controlled to get better efficiency and accuracy. The maximized grid sizes of sub-PSFs may be different since the gradients of PSF values vary from the center to the periphery of the PSF. In addition, each sub-PSF needs to cover all function values in its corresponding region according to the partition positions. As illustrated in Figures 5(b) and 5(c), PSF'_1 and PSF'_2 have different maximized grid sizes, and they both cover function values in their corresponding regions. After separately deriving each sub-PSF, the sub-PSFs might be overlapped with each other on the merging boundaries. Therefore, as we merge the sub-PSFs, the merging boundaries are adjusted such that the sub-PSFs are disjoint with each other. As illustrated in Figure 5(d), the adjusted merging boundary is indicated by the red (solid) window, and the two sub-PSFs are disjoint. After getting the set of ϵ-controlled sub-PSFs, an estimated flare map can be efficiently computed with better error control.

4. FLARE LEVEL AND FLARE VARIATION MINIMIZATION WITH DUMMIFICATION

In this section, we present our dummification algorithm for simultaneous flare level and flare variation minimization. We first give the algorithm flow in Section 4.1, and two major stages in the flow, the global dummification stage and the local refinement stage, are then detailed in Section 4.2 and 4.3.

4.1 Algorithm Flow

Figure 6 shows our flare optimization algorithm flow with dummification. Given a grid-based mask layout as the input, we first generate its vacancy density map and compute the corresponding flare map. After that, dummy value assignment is performed with the guidance generated by using a quasi-inverse lithography technique to evaluate the dummy demands of different regions. The dummification process is composed of two major stages: the global dummification stage followed by the local refinement stage.

In the global dummification stage, we simultaneously optimize flare level and flare variation with a top-down dummy value assignment method. Since flare in EUVL has a long-range effect, the top-down

1177

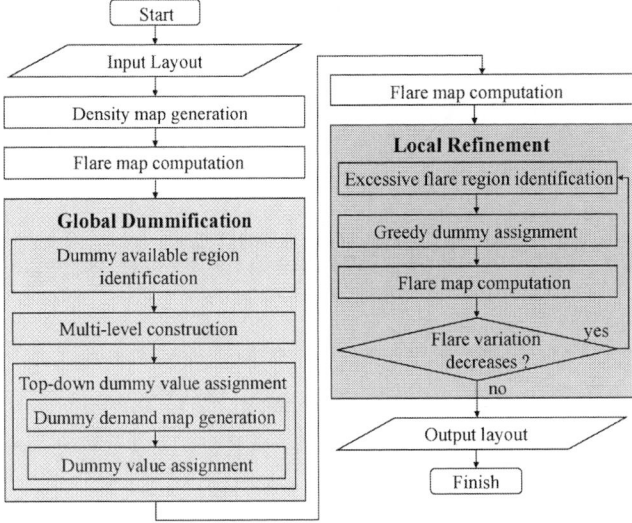

Figure 6: Overall flow of the proposed simultaneous flare level and flare variation minimization algorithm with dummification.

approach can capture the information of global flare distribution better than a bottom-up counterpart since the top-down one has a more global view. After identifying the available dummy region for each grid, we construct the multilevel structure by accumulating the flare map and the available dummy map. Then we repeatedly assign dummy values to subregions from the top level to sub-levels by applying a quasi-inverse lithography technique, which evaluates the demand of dummies for each subregion.

In the local refinement stage, the objective is to further minimize the flare variation of a chip such that the CD uniformity can be controlled better. We iteratively identify grids with excessive flare and greedily assign maximum available dummy values to grids with higher refinement demands by applying a similar quasi-inverse lithography technique. The optimization process terminates when no improvement in flare variation reduction can be made through this local refinement approach.

Note that flare map computation for a whole chip is needed for each stage in our algorithm flow. Especially in the local refinement stage, flare map computation is performed for each iteration for judging whether the flare variation is reduced or not. Consequently, efficient and accurate flare map computation is necessary. Thus, we accommodate the error-controlled fast flare map computation introduced in Section 3 to improve the program efficiency without loss of computational accuracy.

We detail the two stages in Section 4.2 and Section 4.3 respectively.

4.2 Global Dummification

As mentioned in Section 1, for the process in EUVL with a clear-field mask, it is possible to simultaneously reduce the flare level and the flare variation of a layout with dummification. In addition, even for a layout with uniform pattern density, the flare variation may be still large due to the flare periphery effect; that is, achieving maximum pattern density uniformity is not equivalent to achieving maximum flare uniformity in a dummification process. Furthermore, although the flare level of a chip can be minimized by inserting as many dummy patterns as possible, the flare variation should also be minimized to get better CD uniformity control. Therefore, a dummification algorithm considering global flare distribution for flare optimization is desirable.

From Equation (1), a flare map is computed by convolving the PSF and a vacancy density map $I_D(x, y)$. Thus, a flare map of a layout with dummification can be computed as follows:

$$I'_F(x, y) = (I_D(x, y) - I_{dummy}(x, y)) \otimes PSF(x, y), \qquad (2)$$

where $I_{dummy}(x, y)$ is a dummy density map. Thus, the flare reduction

due to dummification can be computed with the following equation:

$$I_F(x, y) - I'_F(x, y) = I_{dummy}(x, y) \otimes PSF(x, y). \qquad (3)$$

To minimize flare variation, the regions with higher flare levels require more dummies than those with lower flare levels. Therefore, by Equation (3), we propose a quasi-inverse lithography technique to guide the dummy value assignment. After computing the original flare map of an input layout, we use a quasi-inverse PSF to propagate the flare reduction demand of a region to the dummy demands of neighboring regions, inspired by the work [5] for OPC optimization. The quasi-inverse PSF function is defined as follows:

$$Q(x, y) = \int_{-\infty}^{\infty} \int_{-\infty}^{\infty} PSF(x - f, y - g) PSF(f, g) \, df \, dg, \qquad (4)$$

This quasi-inverse PSF models the relation of multiple points on a wafer on one point of the mask. Since the flare of a region can be compensated by inserting dummies into neighboring regions, this quasi-inverse kernel function propagates the flare reduction demand of a region to the dummy demands of neighboring regions. Therefore, the dummy demand of a region is the sum of propagated dummy demands, and thus a dummy demand map can be obtained by convolving the quasi-inverse PSF with a flare map. A dummy demand map can be computed as follows:

$$D(x, y) = \int_{-\infty}^{\infty} \int_{-\infty}^{\infty} I_F(x - f, y - g) Q(f, g) \, df \, dg. \qquad (5)$$

The larger value of a region in a dummy demand map indicates that the region requires more dummy patterns than those regions with smaller dummy demand values for flare compensation.

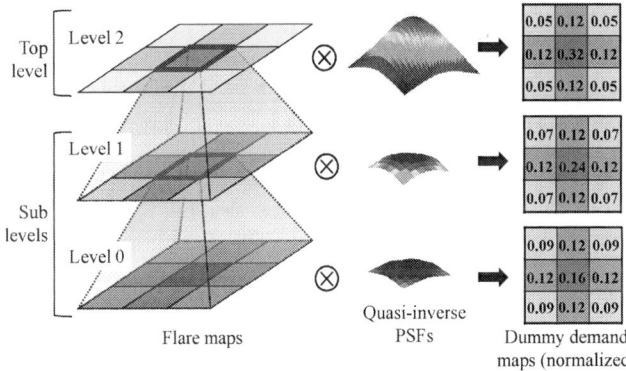

Figure 7: Illustration of the top-down dummy demand maps generation. The larger value of a grid in a dummy demand map indicates that the grid requires larger dummy value than those regions with smaller dummy demand values for flare compensation.

After identifying the available dummy area of each grid, we need to assign a dummy value not exceeding the available dummy value for each grid. However, performing dummy value assignment with a dummy demand map generated by directly using Equation (5) not only requires large computational effort, but also get an undesired dummy assignment solution, due to the lack of global information. Therefore, a top-down approach is applied which provides a more global view as solving the dummy value assignment problem. We propose a top-down framework using the quasi-inverse lithography technique with Equation (5) to derive dummy demand maps for each level. Figure 7 shows an example of generating multilevel dummy demand maps. First, the multilevel structure is constructed and $W \times W$ fine grids in level i are merged into one coarsened grid in level $i + 1$. The flare values of level i are accumulated for constructing the flare map of level $i + 1$, and the process is repeated until the top-level flare map is constructed. The multi-level quasi-inverse kernel functions are also constructed for each

1178

level according to the range covered by $W \times W$ grids. Then, the dummy demand maps are computed by convolving a flare map and a quasi-inverse kernel function from the top level to the bottom level. Note that each dummy demand map is normalized such that the summation of all demand values is one.

After deriving a dummy demand map for each level, the dummy assignment processes are also performed in the top-down manner. We solve the dummy assignment problems of the top-level and sub-levels by linear programming (LP), which are respectively detailed in Section 4.2.1 and Section 4.2.2.

4.2.1 Top-Level Dummy Value Assignment

For a top-level dummy value assignment problem, the inputs are a top-level dummy demand map and a top-level available dummy map, and the output is a top-level dummy value assignment map. For the dummy value assignment problem with a top-level dummy demand map, the objective is to maximize the total amount of assigned dummies and make the relative dummy values of any two grids conform to their relative dummy demand values. As illustrated in Figure 8(a), for the given top-level dummy demand map and the available dummy map, the assigned dummy values are maximized and conform to the dummy demand map. For example, the relative dummy values of grids i and j (10 and 1.6) conform to their relative dummy demand values (0.32 and 0.05).

We use an LP formulation to solve this top-level dummy value assignment problem. The notations used in the LP formulation are listed as follows:

- d_i: assigned dummy value of grid i.
- t_i: target dummy value of grid i.
- r_i: dummy demand value of grid i in the dummy demand map.
- a_i: available dummy value of grid i.
- m: the index of a grid with the maximum dummy demand value.
- α: user-defined parameter.

Based on the notations, the top-level dummy assignment problem can be formulated as follows:

$$maximize \quad \sum_i d_i - \alpha \sum_i |d_i - t_i|, \quad (6)$$

$$subject\ to \quad t_i = a_m \cdot (r_i/r_m), \forall i, \quad (7)$$

$$0 \le d_i \le a_i, \forall i. \quad (8)$$

In this formulation, the objective is to maximize each d_i constrained by a_i and to minimize the deviation between d_i and t_i for each grid i, where t_i is set to be a fraction of the a_m corresponding to the ratio r_i/r_m. Although the objective function is not linear due to the absolute values, the above formulation can be transformed into a linear model. We have the following theorem:

THEOREM 1. *The top-level dummy assignment problem can be solved by liner programming with linear numbers of constraints and variables.*

4.2.2 Sub-Level Dummy Value Assignment

Different from the top-level dummy value assignment problem, a sub-level dummy value assignment problem has one more input, the target total dummy value T of all grids in a subproblem. For a dummy value assignment problem in sub-level i, T is derived from an assigned dummy value in the previous level $i+1$. Thus, the objective of a sub-level dummy value assignment problem is to minimize the deviation between T and the sum of all dummy values and let the relative dummy values of all grids conform to a sub-level dummy demand map. As illustrated in Figure 8(b), a sub-level dummy value assignment map is generated with a sub-level dummy demand map, a sub-level available dummy map, and a target total dummy value T. In the dummy value assignment result, the sum of all dummy values equals T, and the dummy values conform to the dummy demand map. For example, the relative dummy values of grid i and grid j conform to their relative dummy demand values.

The sub-level dummy assignment problem can be formulated as follows:

$$minimize \quad |T - \sum_i d_i| + \alpha \sum_i |d_i - t_i|, \quad (9)$$

$$subject\ to \quad t_i = r_i \cdot T, \forall i, \quad (10)$$

$$0 \le d_i \le a_i, \forall i. \quad (11)$$

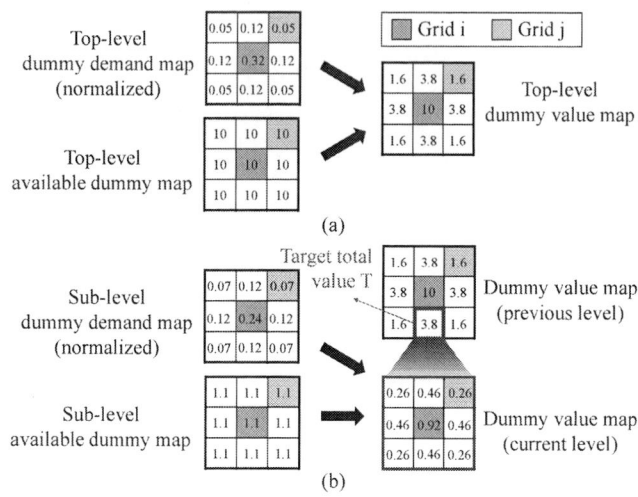

(a)

(b)

Figure 8: (a) Illustration of a top-level dummy value assignment problem. The assigned dummy values are maximized and conform to the dummy demand map. (b) Illustration of a sub-level dummy value assignment problem. The sum of all dummy values equals the target total value, and the dummy values conform to the dummy demand map.

In this formulation, the objective is to minimize the deviation between $\sum_i d_i$ and T, and the deviation between d_i and t_i for each grid i, where t_i is set to be a fraction of T according to its dummy demand value r_i. Note that a dummy demand map is normalized such that $\sum_i r_i = 1$, and thus $\sum_i t_i = T$. Similar to the top-level dummy assignment formulation, we have the following theorem.

THEOREM 2. *The sub-level dummy assignment problem can be solved by linear programming with linear numbers of constraints and variables.*

4.3 Local Refinement

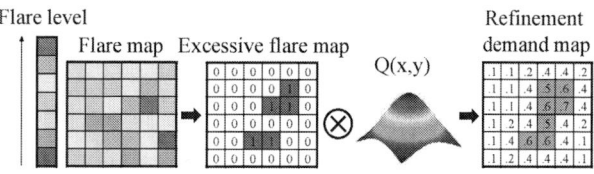

Figure 9: Local refinement. A refinement demand map is computed by convolving an excessive flare map and the quasi-inverse PSF.

After deriving a dummy value assignment solution in which the flare level and the flare variation are simultaneously optimized with our top-down framework, we try to further minimize the flare variation of a chip such that the CD uniformity can be controlled better in the local refinement stage. Figure 9 shows the local refinement process. Given the flare map of a current layout with dummification, we first identify grids with excessively large flare values in the map and construct the corresponding excessive flare map with 0/1 values. As shown in Figure 9, the red (shaded) grids marked as '1' have larger flare values than other grids. Then, we again utilize the quasi-inverse PSF (Equation (4)) to propagate the flare reduction demand of grids with value '1' in the excessive flare map to the refinement demands of neighboring regions. Then, the refinement demand of a grid is the sum of propagated refinement demands, and thus a refinement demand map can be obtained by convolving the quasi-inverse PSF with the excessive flare map (Equation (5)). For grids with refinement demand values exceeding a threshold value, we greedily assign dummy values to their

1179

maximum available dummy values and check if the flare variation of the refined layout is reduced or not. As illustrated in Figure 9, the larger refinement demand values of the green (shaded) grids indicate that the grids are desirable to be assigned with more dummies. The local refinement process is performed in an iterative manner, and the process terminates as no improvement in flare variation can be made by using this refinement approach.

5. EXPERIMENTAL RESULTS

Our algorithm was implemented in the C++ programming language on a 2.40 GHz Linux workstation with 16 GB memory. We used the lp_solve package as the LP solver [11] and used the FFTW library [8], which is a C subroutine library for computing the discrete Fourier transform (DFT), to further improve the efficiency of convolution operations. The experiments were based on two suites of benchmarks, the MCNC and the industrial Faraday benchmarks. We used the metal-1 of each circuit as the input layout.

We first pre-processed an input layout by dividing the layout into grids of size 1 $\mu m \times$ 1 μm, analyzing the pattern density of each grid, and identifying the available dummy region for each grid. The PSF was set to be a discrete Gaussian function. Since previous work has been reported that an area described by a radial distance of about 1000 μm accounts for about 95% of the flare seen at a point on a wafer [15], we set the PSF to be 2000 $\mu m \times$ 2000 μm with resolution identical to the grid-based layout, and set its standard deviation σ to be 500 μm to capture the flare distribution within 2σ. In addition, we assumed 1 unit of flare may be generated from a vacancy region of size 1 $\mu m \times$ 1 μm, and the density upper bound of each grid was set to be 0.6, which is the default value set in most commercial tools [6].

We compared the flare optimization results between two dummification approaches: (1) dummification according to a linear function of distance from the center of a layout and (2) our algorithm flow. For the first approach, the target dummy value t_i of each grid i was set according to a linear function as follows:

$$t_i = \left(1 - \frac{dist(i, center)}{dist_{max}}\right) \cdot a_{max}, \qquad (12)$$

where $dist(i, center)$ and $dist_{max}$ are the respective distances from the center of a chip to the grid i and to the farthest grid, and a_{max} is the maximum available dummy value of an empty grid (without any patterns), which is the density upper bound we set. The function tries to assign the maximum available dummy value to the central grid and make the dummy values of the farthest grids be 0. Then, we assigned a dummy value $d_i = \min(t_i, a_i)$ to each grid i, where a_i is the available dummy value of grid i.

Table 1 shows the comparison results. In the table, "Original", "Linear," and "Ours" respectively list the flare information of the original layout before dummification, the layout with dummification according to a linear function, and the layout with dummification by using our algorithm flow. "Avg." gives the average flare level, and "Var." gives the flare level variation of each circuit. Observing from the table, our algorithm achieves 36% average flare level reduction and 37% flare level variation reduction over the linear dummification approach. Furthermore, for the circuits with larger sizes (e.g., Struct, Primary1, and Primary2), the reductions in the flare level variation are about 50%. The significant improvements may result from the consideration of global flare distribution (trend) during the dummification process in our algorithm. These results show that our algorithm can effectively reduce the average flare level and the flare level variation, and thus the CD uniformity within a layout can be controlled better.

6. CONCLUSIONS

This paper has presented a simultaneous flare level and flare level variation optimization flow with dummification. Unlike the previous dummy fill algorithms for CMP and other heuristic methods, our algorithm performs dummification by considering global flare distribution (trend) in EUVL. In addition, the error-controlled fast flare map computation was integrated into the flow to further improve the algorithm efficiency without loss of computational accuracy. Experimental results based on two suites of benchmarks have shown that our algorithm can effectively and efficiently reduce the average flare level and the flare level variation, and thus the CD uniformity within a layout can be controlled better.

Table 1: Comparison of dummification results between a linear approach and our algorithm.

Circuit	Original		Linear			Ours		
	Avg.	Var.	Avg.	Var.	CPU (sec)	Avg.	Var.	CPU (sec)
Struct	0.850	0.743	0.606	0.535	66	0.382	0.288	155
Primary1	0.877	0.748	0.626	0.588	93	0.381	0.282	221
Primary2	0.898	0.745	0.645	0.627	157	0.388	0.283	417
S5378	0.061	0.007	0.045	0.005	16	0.029	0.003	23
S9234	0.054	0.005	0.040	0.004	17	0.025	0.003	22
S13207	0.129	0.031	0.094	0.022	18	0.061	0.015	37
S15850	0.143	0.038	0.104	0.028	17	0.068	0.018	39
S38417	0.289	0.161	0.207	0.110	19	0.139	0.077	78
S38584	0.322	0.205	0.230	0.138	21	0.159	0.099	95
Dma	0.101	0.016	0.073	0.011	19	0.043	0.007	31
Dsp1	0.258	0.102	0.184	0.071	20	0.115	0.045	59
Dsp2	0.224	0.076	0.160	0.054	19	0.098	0.033	52
Risc1	0.407	0.262	0.288	0.160	22	0.190	0.122	24
Risc2	0.386	0.235	0.274	0.161	22	0.183	0.109	100
Comp.	—	—	1.00	1.00	—	0.64	0.63	—

7. ACKNOWLEDGMENT

We would like to thank Mr. H.-C. Wang, Mr. Y.-C. Cheng and Mr. R.-G. Liu in TSMC for providing precious suggestions related to this work.

8. REFERENCES

[1] H. Aoyama et al., "Applicability of extreme ultraviolet lithography to fabrication of half pitch 35nm interconnects," *Proc. SPIE 7636*, Feb. 2010.

[2] M. Chandhok et al., "Determination of the flare specification and methods to meet the CD control requirements for the 32 nm node using EUVL," *Proc. SPIE 5374*, Feb. 2004.

[3] Y. Chen, A. B. Kahng, G. Robins, and A. Zelikovsky, "Practical iterated fill synthesis for CMP uniformity," *Proc. DAC*, pp. 671–674, Jun. 2000.

[4] Y. Chen, A. B. Kahng, G. Robins, and A. Zelikovsky, "Closing the smoothness and uniformity gap in area fill synthesis," *Proc. ISPD*, pp. 137–142, Apr. 2002.

[5] T.-C. Chen, G.-W. Liao, and Y.-W. Chang, "Predictive formulae for OPC with applications to lithography-friendly routing," *IEEE TCAD*, Vol. 29, No. 1, pp. 40–50, Jan. 2010.

[6] H.-Y. Chen, S.-J. Chou, and Y.-W. Chang, "Density gradient minimization with coupling-constrained dummy fill for CMP control," *Proc. ISPD*, pp. 105–111, Mar. 2010.

[7] L. Deng, M. D. F. Wang, K.-Y. Chao, and H. Xiang, "Coupling-aware dummy metal insertion for lithography," *Proc. ASP-DAC*, pp. 13–18, Jan. 2007.

[8] FFTW3 (Fastest Fourier Transform in the West). http://www.fftw.org/.

[9] C. Krautschik, M. Ito, I. Nishiyama, and S. Okazaki, "Impact of EUV light scatter on CD control as a result of mask density changes," *Proc. SPIE 4688*, Mar. 2002.

[10] J. Lee et al., "A study of flare variation in extreme ultraviolet lithography for sub-22nm line and space pattern," *Jpn. J. Appl. Phys.*, pp. 06GD09, Jan. 2010.

[11] lp_solve 5.5.2.0. http://lpsolve.sourceforge.net/5.5/.

[12] A. Myers et al., "Experimental validation of full-field extreme ultraviolet lithography flare and shadowing corrections," J. Vac. Sci. Technol. B, Vol. 26, No. 6, pp. 2215–2219, Nov. 2008.

[13] F. Schellenberg et al., "Layout compensation for EUV flare," *Proc. SPIE 5751*, Mar. 2005.

[14] M. Shiraishi, T. Oshino, K. Murakami and H. Chiba, "Flare modeling and calculation for EUV optics," *Proc. SPIE*, Feb. 2010.

[15] V. K. Singh et al., "US patent 6,625,802: Method for modifying a chip layout to minimize within-die CD variations caused by flare variations in EUV lithography," 2003.

[16] V. Singh, "The importance of layout density control in semiconductor manufacturing," *Proc. EDPW*, pp. 70-74, 2003.

[17] O. Wood et al., "Integration of EUV Lithography in the fabrication of 22 nm node devices," *Proc. SPIE 7271*, Feb. 2009.

[18] H. Xiang, L. Deng, R. Puri, K.-Y. Chao, and M. D. F. Wang, "Dummy fill density analysis with coupling constraints," *Proc. ISPD*, pp. 3–9, Mar. 2007.

[19] C. Zuniga et al., "EUV flare and proximity modeling and model-based correction," *Proc. SPIE7969*, Feb. 2011.

A Novel Layout Decomposition Algorithm for Triple Patterning Lithography*

Shao-Yun Fang[1], Yao-Wen Chang[1,2], and Wei-Yu Chen[1]

[1]Graduate Institute of Electronics Engineering, National Taiwan University, Taipei 106, Taiwan
[2]Department of Electrical Engineering, National Taiwan University, Taipei 106, Taiwan
yuko703@eda.ee.ntu.edu.tw; ywchang@cc.ee.ntu.edu.tw; cweiyu@eda.ee.ntu.edu.tw

ABSTRACT

While double patterning lithography (DPL) has been widely recognized as one of the most promising solutions for the sub-22nm technology node to enhance pattern printability, triple patterning lithography (TPL) will be required for gate, contact, and metal-1 layers which are too complex and dense to be split into only two masks, for the 15nm technology node and beyond. Nevertheless, there is very little research focusing on the layout decomposition for TPL. The recent work [16] proposed the first systematic study on the layout decomposition for TPL. However, the proposed algorithm extending a stitch-finding method used in DPL may miss legal stitch locations and generate conflicts that can be resolved by inserting stitches for TPL. In this paper, we point out two main differences between DPL and TPL layout decompositions. Based on the two differences, we propose a novel TPL layout decomposition algorithm. We first present two new graph reduction techniques to reduce the problem size without degrading overall solution quality. We then propose a stitch-aware mask assignment algorithm, based on a heuristic that finds a mask assignment such that the conflicts among the features in the same mask are more likely to be resolved by inserting stitches. Finally, stitches are inserted to resolve as many conflicts as possible. Experimental results show that the proposed layout decomposition algorithm can achieve around 56% reduction of conflicts and more than 40X speed-up compared to the previous work.

Categories and Subject Descriptors

B.7.2 [**Integrated Circuits**]: Design Aids

General Terms

Algorithms, Design, Performance

Keywords

Triple Patterning Lithography, Layout Decomposition, Manufacturability

---

*This work was partially supported by IBM, SpringSoft, TSMC, and NSC of Taiwan under Grant No's. NSC 100-2221-E-002-088-MY3, NSC 99-2221-E-002-207-MY3, NSC 99-2221-E-002-210-MY3, and NSC 98-2221-E-002-119-MY3.

1. INTRODUCTION

In order to overcome the resolution limit of conventional optical lithography, various next-generation lithography systems have been proposed, such as extreme ultraviolet (EUV) lithography and electron beam (e-beam) lithography. However, these next-generation lithography systems are still not available for mass production due to several barriers. Consequently, multiple patterning lithography, which extends current 193nm immersion lithography to patterning the sub-32nm half pitch node, corresponding to the sub-22nm technology node, has been regarded as one of most promising solutions to overcome the resolution limit of conventional optical lithography [1, 11].

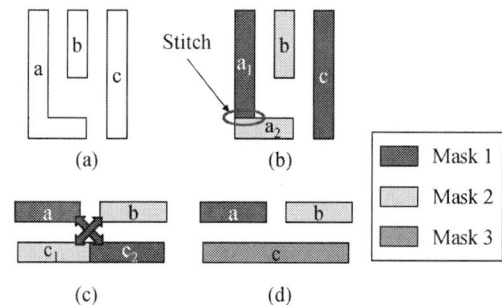

Figure 1: Multiple patterning lithography. (a) An example layout. (b) Layout decomposition for DPL with an inserted stitch. (c) An example layout which cannot be decomposed into two masks with stitch insertion. (d) The layout can be decomposed into three masks with TPL.

The simplest case of multiple patterning lithography is double patterning lithography (DPL). During the layout decomposition process in DPL, if the distance between two features is less than the minimum coloring spacing min_{cs}, they should be assigned to different masks. Otherwise, there will be a conflict between these two features. In some cases, to resolve a conflict, a feature may be split into two touching parts and assigned to different masks. As shown in Figures 1(a) and (b), the layout can be decomposed into two masks by inserting a *stitch*. However, even with stitch insertion, some conflicts still cannot be resolved in DPL. Figure 1(c) shows an example of an irresolvable conflict in DPL. Since there exists a region in feature c which is within min_{cs} from features a and b, even though we split feature c into two parts, each of the two parts still forms a conflict with a and b. As a result, it is impossible to produce a conflict-free solution for this layout with DPL. However, this problem can be easily resolved if three masks are available; that is, the layout can be decomposed without conflict with triple patterning lithography (TPL), as shown in Figure 1(d).

For the 15nm technology node and beyond, TPL will be required for gate, contact, and metal-1 layers, which are too complex and dense to be split into two masks [2, 3, 9, 11, 12, 16]. With the same principle of DPL, the original layout is decomposed into three masks and manufactured through three exposure/etching steps in TPL. Although TPL can resolve some conflicts that cannot be resolved in DPL, the

TPL layout decomposition problem is not easier than the DPL layout decomposition problem. Actually, since the minimum coloring spacing min_{cs} set in TPL is larger than that in DPL, more pairs of features are within min_{cs} of each other; that is, more pairs of features need to be assigned to different masks in TPL. Thus, the problem turns out to be more complicated and difficult.

There is very little research focusing on the layout decomposition for TPL. As the DPL layout decomposition problem is generally regarded as a two-coloring problem, the TPL layout decomposition problem can be modeled as a three-coloring problem. Cork et al. [3] investigated the challenges involved in the TPL layout decomposition problem for the contact layer. They proposed a SAT-based three-coloring algorithm and then compared the scalability of different coloring algorithms using a variety of contact patterns based on Penrose tiles, which have been proven to be three-colorable. However, this work only deals with contact arrays, not general layouts, and thus the stitch issue is not considered. Yu et al. [16] proposed the first systematic study on the general layout decomposition for TPL. The algorithm first generates a set of stitch candidates by using the projection method proposed by Kahng et al. [7] and widely used in DPL. The TPL layout decomposition is formulated as an integer linear programming (ILP) to simultaneously minimize the numbers of conflicts and stitches. Since solving ILP is time-consuming, they also proposed three acceleration techniques without loss of solution quality. To improve the scalability, they further proposed semidefinite programming (SDP) based approximation algorithm. However, the projection method they used to find a set of stitch candidates may miss legal stitches, which we will prove in Section 2.2.2. Thus, the algorithm may generate some conflicts that can be resolved by inserting stitches for TPL.

In this paper, we point out two main differences between DPL and TPL layout decompositions. Based on the two differences, we propose a novel TPL layout decomposition algorithm. Experimental results show that our layout decomposition algorithm can achieve around 56% reduction of conflicts and more than 40X speed-up, compared to the previous work proposed by Yu et al. [16]. In addition, we compute the cost of our layout decomposition results by using the objective function proposed in [16], and the resulting cost is also reduced by more than 34% compared to their algorithm. This phenomenon reveals that their method may fail to resolve conflicts which can be resolved by inserting stitches.

The rest of this paper is organized as follows: Section 2 gives the problem formulation of this paper and points out the two differences between DPL and TPL. In Section 3, our layout decomposition approach for TPL is presented. Experimental results are reported in Section 4. Finally, we conclude our work in Section 5.

2. PRELIMINARIES

In this section, we first formally define the layout decomposition problem for TPL in Section 2.1. Then, we discuss two main differences between DPL and TPL layout decompositions in Section 2.2, which show that the TPL layout decomposition problem is more difficult than the DPL one.

2.1 Problem Formulation

In this work, we solve the layout decomposition problem for TPL. There are two critical issues in the layout decomposition problem: conflicts and stitches. If the distance between two features in the layout is less than the minimum coloring spacing min_{cs}, they should be assigned to different masks. Otherwise, a conflict exists between these two features. To resolve a conflict, a feature may be split into two parts and assigned to different masks. However, this introduces stitches, which may cause yield loss due to overlay error and increase manufacturing cost. Therefore, conflict and stitch minimizations are the main challenges in the layout decomposition for TPL. Thus, the layout decomposition problem for TPL can be formulated as follows:

PROBLEM 1. *Given a layout represented by a set of polygonal features, the minimum wire width, the minimum spacing, the overlay margin, and the minimum coloring spacing for TPL, assign all features in the layout to three masks while the numbers of conflicts and stitches are minimized.*

2.2 Differences between DPL and TPL

In this section, we will discuss two main differences between DPL and TPL layout decompositions, which show that the TPL layout de-

composition problem is more complicated and more difficult than the DPL layout decomposition problem.

2.2.1 Problem Complexity

The DPL layout decomposition problem is generally modeled as a two-coloring problem on a conflict graph. Then, determining whether a graph is two-colorable can be done in linear time by checking if there exists any odd cycle. Moreover, two-coloring a two-colorable graph can also be done in linear time with a breadth-first search (BFS) algorithm.

However, for the TPL layout decomposition, the problem becomes much more complicated. As the DPL layout decomposition problem is modeled as a two-coloring problem, the TPL layout decomposition problem can be modeled as a three-coloring problem. It can be shown that a three-coloring problem is NP-complete. First, determining whether a general graph is three-colorable is NP-complete, even for a planar graph [5]. Furthermore, even though the graph is three-colorable, coloring a three-colorable graph with four colors is NP-complete [8], and thus three-coloring a three-colorable graph is also NP-complete. Therefore, the TPL layout decomposition problem is NP-complete even if the conflict graph used to model TPL conflicts in a layout is three-colorable and planar [16].

2.2.2 Stitch Finding

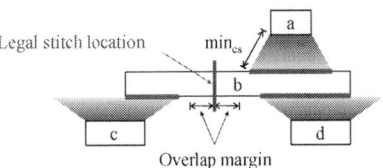

Figure 2: The projection method for stitch location finding.

Most previous works for DPL and TPL [16] utilize a projection method proposed by Kahng et al. [7] to find legal stitch locations before mask assignment. The reason is that this projection method has been proven to be able to find all possible stitch locations in DPL [7]. As illustrated in Figure 2, the method computes the projections on each feature from its neighboring features within the minimum coloring spacing min_{cs}. Then, a stitch can be inserted at a location where no projection covers and no overlap margin violation occurs.

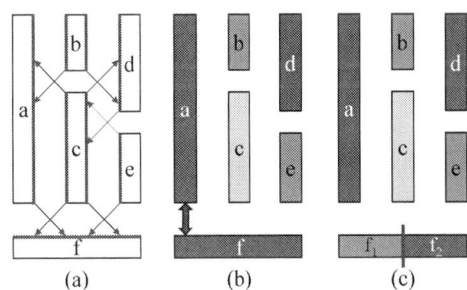

Figure 3: An example showing that not all legal stitch locations can be found by the projection method in TPL. method.

However, the property does not hold in TPL. That is, we cannot find all legal stitch locations by using the projection method in TPL. Figure 3 shows an example. In the layout depicted in Figure 3(a), no legal stitch position can be found by using the projection method since all features are covered by projections. Since the layout is not three-colorable, at least one conflict exists after a mask assignment process, as shown in Figure 3(b). In fact, as illustrated in Figure 3(c), we can find a legal stitch location and resolve the conflict by inserting a stitch. With the above example, we can conclude that not all legal stitch locations can be found by the projection method in TPL. Therefore, if we only insert stitches at the locations found by the projection method, we may miss legal stitches and generate conflicts that can be resolved with stitch insertion.

1182

3. LAYOUT DECOMPOSITION FOR TPL

In this section, we first present the overall flow of our layout decomposition approach for TPL in Section 3.1. Then, we detail each step in the flow in Sections 3.2–3.5.

3.1 Algorithm Flow

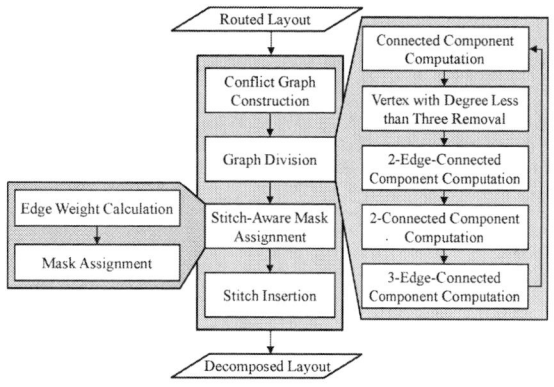

Figure 4: The flow of our layout decomposition for TPL.

The overall flow of our layout decomposition for TPL is shown in Figure 4. First, we construct a conflict graph to transform the original geometric problem into a graph problem, and thus the TPL layout decomposition problem can be modeled as a three-coloring problem. Since the three-coloring problem is NP-complete, the time required to exactly solve it increases dramatically with the numbers of vertices and edges in the graph. Therefore, to reduce the problem size, we shall divide the conflict graph into a set of coloring-independent groups, and then the color assignment problem can be solved independently for each group without affecting overall solution quality. The graph division method incorporates three graph simplification methods proposed by Yu et al. [16]: *connected component computation, vertex with degree less than three removal*, and *2-edge-connected component computation*. In addition, we propose two new graph reduction techniques, *2-connected component computation* and *3-edge-connected component computation*, to further enhance the program efficiency. Next, to solve the color (mask) assignment problem, we propose a stitch-aware mask assignment algorithm, which is based on a heuristic that finds a mask assignment such that the conflicts among the features in the same mask are more likely to be resolved by inserting stitches. Finally, we resolve as many conflicts as possible with stitch insertion for each mask. The four steps in the flow are detailed in the following sections.

3.2 Conflict Graph Construction

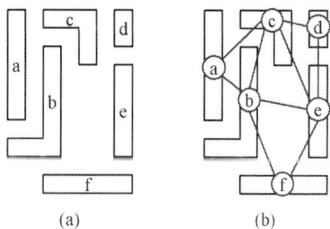

Figure 5: Conflict graph construction. (a) An input layout represented by a set of polygonal features. (b) The corresponding conflict graph.

First, for the purpose of simplification, we construct a conflict graph [7] to transform the original geometric problem into a graph problem. As shown in Figure 5, given a layout composed of a set of polygonal features, the corresponding conflict graph $G = (V, E)$ is constructed in which each vertex $v_i \in V$ represents a feature i and an edge $e_{i,j} \in E$

exists between two vertices if the distance between the two corresponding features i and j is less than the minimum coloring spacing min_{cs}. An edge in the conflict graph is a conflict candidate and is defined as a *conflict edge* (*CE*). Note that in most previous works for the layout decomposition in DPL and TPL, stitch edges (SEs) indicating legal stitch locations are generated with the projection method and inserted in a conflict graph as well. However, in our work, we do not insert SEs in a conflict graph because we have proved that not all legal stitch locations can be found by the projection method in TPL. After the construction of a conflict graph, the layout decomposition problem can be modeled as a three-coloring problem, and a conflict exists if two vertices connected by a CE are assigned the same color.

3.3 Graph Division

Since the three-coloring problem is NP-complete, the time required to exactly solve it increases dramatically with the numbers of vertices and edges in a conflict graph. Therefore, to reduce the problem size, we first incorporate three graph simplification methods proposed by Yu et al. [16]: *connected component computation, vertex with degree less than three removal*, and *2-edge-connected component computation*. In addition, we propose two new graph reduction techniques, *2-connected component computation* and *3-edge-connected component computation*, to further reduce the problem size and enhance the efficiency. Note that all of the five techniques can be performed in linear time $O(|V| + |E|)$. We detail the graph reduction techniques in the following subsections.

3.3.1 Connected Component Computation

A connected component of a graph is a subgraph in which any two vertices are connected to each other with paths. Since no edge exists between any two components, the color assignment problem can be solved independently for each component, and the final solution can be obtained by taking the union of sub-solutions.

3.3.2 Vertex with Degree Less than Three Removal

A component in a conflict graph can be simplified by temporarily removing all vertices with degree less than three. As illustrated in Figure 6(a), vertex v_a is a vertex with degree less than three, and thus we temporarily remove it from the component. After solving the three-coloring problem on the remaining graph, as shown in Figure 6(b), the removed vertex v_a is then added and colored according to the colors of its adjacent vertices. The final coloring solution is shown in Figure 6(c).

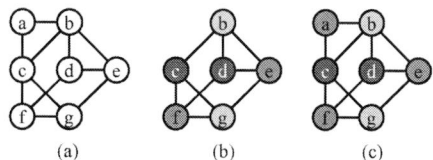

Figure 6: An illustration of the removal of vertices with degree less than three.

3.3.3 2-Edge-Connected Component Computation

Two-edge-connected-component computation is also defined as bridges computation [16]. A bridge of a graph is an edge such that the graph will be decomposed into two components if the edge is removed. If a bridge exists in a conflict graph, the three-coloring problem can be solved independently on the two components connected by the bridge, and then the two sub-solutions can be merged with a simple color remapping process. An example is shown in Figure 7. The conflict graph in Figure 7(a) can be decomposed into two components by removing the edge e_{cd}. After solving the three coloring problem on each component as shown in Figures 7(b) and (c), the two components can be merged and recolored such that no conflict occurs between the vertices v_c and v_d.

3.3.4 2-Connected Component Computation

In addition to the above three methods proposed by Yu et al. [16], we propose two new graph reduction techniques to further reduce the problem size and enhance program efficiency. The first technique is 2-connected component computation. A 2-connected component of a

1183

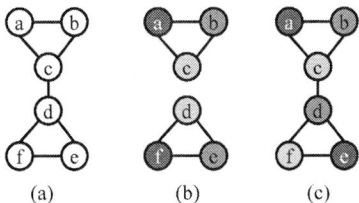

Figure 7: An illustration of 2-edge-connected component computation.

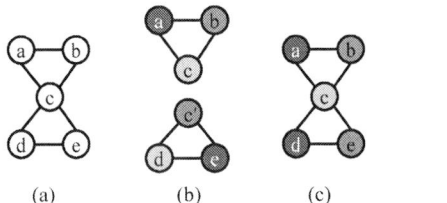

Figure 8: An illustration of 2-connected component computation.

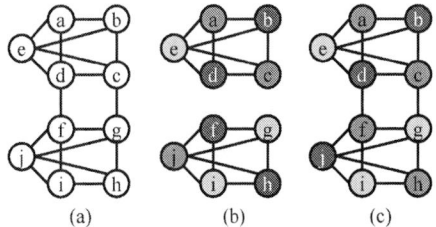

Figure 9: An illustration of 3-edge-connected-component computation.

connected graph is a maximal connected subgraph that remains connected while removing any single vertex. In addition, a vertex of a connected graph is called a *cut vertex* if its removal decomposes the graph into two or more connected components. Thus, we can identify 2-connected components of a connected graph by finding cut vertices.

Two 2-connected components derived by duplicating a cut vertex can be colored independently and merged with a similar color remapping method used in 2-edge-connected-component computation. As shown in Figure 8(a), vertex v_c is a cut vertex. By creating a pseudo-vertex c', the conflict graph can be divided into two 2-connected components, as illustrated in Figure 8(b). After solving the three-coloring problem on the two components independently, the two sub-solutions can be merged into one three-coloring solution with a simple recoloring process on one component, as shown in Figure 8(c).

For the 2-connected component computation, we have the following theorem:

THEOREM 1. *Solving the three-coloring problem on each 2-connected subgraph and then merging the sub-solutions at the cut vertex into one three-coloring solution with color remapping does not degrade the overall solution quality; that is, the number of conflicts does not increase with this graph division method.*

We apply a depth-first search (DFS) algorithm presented by Tarjan [13] to find all the 2-connected components of a conflict graph. The algorithm computes all the 2-connected components and identifies all the cut vertices simultaneously during only one DFS in linear time $O(|V|+|E|)$, where $|V|$ and $|E|$ are respectively the number of vertices and the number of edges in a conflict graph $G = (V, E)$.

3.3.5 3-Edge-Connected Component Computation

The second graph reduction technique we propose is 3-edge-connected component computation. A 3-edge-connected component of a graph G is a maximal connected subgraph of G such that for every two distinct vertices in the subgraph, there are at least three edge-disjoint paths in G connecting them. Three-edge-connected components of a conflict graph can be identified by finding 2-cuts in the graph, where a 2-cut is a pair of edges whose removal would disconnect the graph [15].

Similar to 2-edge-connected component computation, each 3-edge-connected component can be colored independently, and then components can be merged with a color remapping process. As shown in Figure 9(a), the pair of edges e_{df} and e_{cg} is a 2-cut, and the conflict graph can be decomposed into two connected components by removing the 2-cut, as illustrated in Figure 9(b). After solving the three-coloring problem on the two components independently, the two sub-solutions can be merged into one three-coloring solution with a simple recoloring process on one component, as shown in Figure 9(c).

For 3-edge-connected component computation, we have the following theorem:

THEOREM 2. *Solving the three-coloring problem on each 3-edge-connected subgraph and then merging the sub-solutions at the 2-cut into one three-coloring solution with color remapping does not degrade the overall solution quality; that is, the number of conflicts does not increase with this graph division method.*

We adopt an algorithm presented by Tsin [14] to find all the three-edge-connected components of a conflict graph in linear time $O(|V|+|E|)$, where $|V|$ and $|E|$ are respectively the number of vertices and the number of edges in a conflict graph $G = (V, E)$. The algorithm computes all the three-edge-connected components and identifies all the 2-cuts by performing only one depth-first search (DFS) over the given graph.

3.4 Stitch-Aware Mask Assignment

After the graph division operations, we solve the three-coloring problem on each sub-conflict graph. We propose a stitch-aware mask (color) assignment algorithm, which is based on a heuristic that finds a mask (color) assignment such that the conflicts among the features in the same mask are more likely to be resolved by inserting stitches.

3.4.1 Edge Weight Calculation

We observe that some conflicts are more likely to be resolved by inserting stitches, whereas some conflicts are difficult to be resolved. Therefore, while performing mask assignment, we intend to find a mask assignment such that the conflicts among the features in the same mask are more likely to be resolved by inserting stitches. To achieve this goal, we first assign a weight to each conflict edge to reflect how hard the corresponding conflict can be resolved by inserting stitches. For an edge of a vertex, we calculate the weight w as follows: an edge between the target vertex and an adjacent vertex indicates that the two corresponding features are within min_{cs}. Thus, the adjacent feature creates a projection p on the target feature. If there are more other projections created by other adjacent features overlapped with p, the target conflict is more difficult to be resolved by inserting stitch. Thus, we set the weight of the target edge to be the maximum density (number of projections overlapped with each other) of other projections which are overlapped with p.

See Figure 10 for an example. In the layout depicted in Figure 10(a), feature b is the target feature. For edge e_{ab}, the projection from feature a to b is p_a, and projections of edges e_{cb} and e_{db} are overlapped with p_a. Since the projections of e_{cb} and e_{db} are not overlapped with each other, the maximum density of projections overlapped with p_a is one. Thus, $w_{ab} = 1$ (similarly, $w_{cb} = w_{db} = 1$). During the mask assignment procedure, if the three adjacent features a, c, and d are assigned different colors, a conflict must exist at e_{ab}, e_{cb}, or e_{db}. In this case, as shown in Figure 10(b), we can insert a stitch at feature b to resolve the conflict. In contrast, in the layout depicted in Figure 10(c), the maximum density of projections overlapped with p_a is two, and thus $w_{ab} = 2$ (similarly, $w_{cb} = w_{db} = 2$). In this case, if the three adjacent features a, c, and d are assigned different colors, the conflict at e_{ab}, e_{cb}, or e_{db} cannot be resolved by inserting a stitch at b, since there exists a section b_2 that cannot be assigned any color without conflict, as shown Figure 10(d).

3.4.2 Mask Assignment

After calculating the weight for each edge, we present a stitch-aware mask assignment algorithm to find a mask assignment such that the

1184

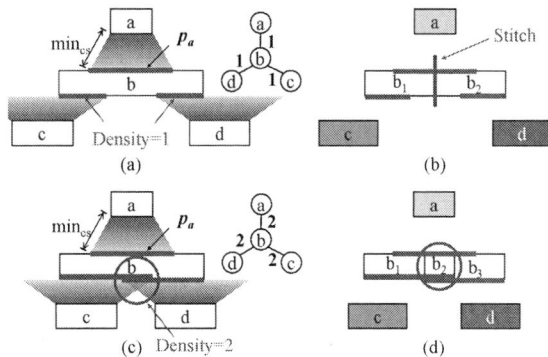

Figure 10: Edge weight calculation. (a) The projections of e_{cb} and e_{db} are overlapped with the projection of e_{ab}, but are not overlapped with each other. Thus $w_{ab} = 1$. (b) A stitch can be inserted to resolve the conflict. (c) The projections of e_{cb} and e_{db} are overlapped with the projection of e_{ab} and are overlapped with each other. Thus $w_{ab} = 2$. (d) In this case, stich cannot be inserted to resolve the conflict.

Algorithm: Stitch-Aware Mask Assignment Algorithm

Input:	$G(V, E)$,	/* a conflict graph */
	$CNUM$,	/* # candidate independent sets */
	$TNUM$,	/* a threshold number of vertices */
	$CSIZE$,	/* size of a candidate vertex list */
Output:	$C_1, C_2,$ and C_3	/* three color classes */

```
1   i ← 0
2   V' ← V
3   while V' ≠ ∅ and i < 3
4       i ← i + 1
5       mincost ← ∞
6       for j = 1 to CNUM
7           U ← V', C ← ∅, X ← ∅
8           while |U| > TNUM
9               CL ← {v_a| ∑_{e_ax ∈ E, v_x ∈ X} w_ax is max}, where CL
                    is a candidate list of size CSIZE
10              Select v_a ∈ CL at random
11              N(v_a) ← {v_b|e_ab ∈ E}
12              U ← U - {v_a} - N(v_a)
13              C ← C ∪ {v_a}
14              X ← X ∪ N(v_a)
15          end while
16          C' ← Use exhaustive search to find an independent set
                that maximizes ∑_{e_ab ∈ F} w_ab, where
                F = {e_ab|v_a ∈ C', v_b ∈ V' - C ∪ C'}
17          C ← C ∪ C'
18          cost ← ∑_{e_ab ∈ E, v_a, v_b ∈ V' - C} w_ab
19          if cost < mincost
20              C_i ← C
21              mincost ← cost
22          end if
23      end for
24      V' ← V' - C_i
25  end while
26  if V' ≠ ∅
27      Assign each remaining vertex v_a ∈ V' to a color class C_i
            such that the conflict occurs at an edge with the smallest
            weight
28  end if
```

Figure 11: Stitch-aware mask assignment algorihtm

conflicts among the features in the same mask are more likely to be resolved by inserting stitches. The proposed mask assignment algo-

rithm is based on a modified recursive largest first (RLF) algorithm. The RLF algorithm was proposed by Leighton [10] and improved as the XRLF algorithm by Johnson et al. [6], which is a very successful heuristic for solving the graph coloring problem. Therefore, we modify the XRLF algorithm to solve the mask (color) assignment problem on an edge-weighted conflict graph. With an input conflict graph $G(V, E)$, the algorithm generates three output subgraphs $G_1(V_1, E_1)$, $G_2(V_2, E_2)$ and $G_3(V_3, E_3)$, where $V_1 \cup V_2 \cup V_3 = V$, $V_i \cap V_j = \emptyset$ for $i \neq j$, and $E_i = \{e_{ab}|v_a \in V_i$ and $v_b \in V_i\}$. Each subgraph represents a color class, and thus vertices in a subgraph will be colored with the same color. Since the weight of an edge indicates how hard the corresponding conflict can be resolved with stitch insertion, an edge with a larger weight is expected not to appear in any of the three subgraphs.

The stitch-aware mask assignment algorithm is summarized in Figure 11. When constructing a color class, the objective is to maximize the size of the color class and minimize the total edge weight of the residual graph. Our algorithm constructs the three color classes by first constructing three independent sets sequentially. For constructing each independent set, a number of candidate independent sets are first generated, and the one minimizing the total edge weight of the residual graph is chosen. After three independent sets are generated, each remaining vertex not included in the independent sets is assigned to one of the three independent sets such that the corresponding conflict occurs at an edge with the smallest weight. Finally, the three color classes are generated and no vertex is left.

An example of a mask assignment process is shown in Figure 12. For an edge-weighted conflict graph shown in Figure 12(a), vertices v_c, v_e are chosen in the first independent set, as shown in Figure 12(b). After the second and the third independent sets are constructed as illustrated in Figures 12(c) and (d), v_g is left as a remaining vertex. Since the weight of the edge e_{fg} is smaller than those of the other adjacent edges of v_g, v_g is assigned to the third color class such that the corresponding conflict is more likely to be solved, as shown in Figure 12(e).

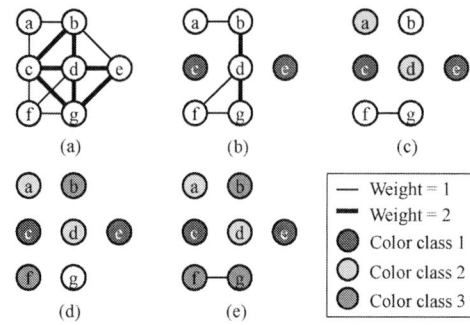

Figure 12: A stitch-aware mask assignment process. (a) An edge-weighted conflict graph. (b)(c)(d) The constructions of three independent sets. (e) The remaining vertex v_g is assigned to C_3 since the edge weight of e_{fg} is smaller than those of the other adjacent edges of v_g.

3.5 Stitch Insertion

After mask assignment, we resolve the conflicts between features by inserting stitches. We first compute the projections from neighboring features on the conflicting feature. See Figure 13(a) for an example. Then, the conflicting feature can be partitioned into several segments according to the ends of these projections, and these segments have different available colors (masks), as illustrated in Figure 13(b). Thus, if we can find at least one available color for each segment, then we can solve the conflict by inserting stitches. Otherwise, the conflict cannot be resolved with stitch insertion. Since there may be several stitch insertion combinations to resolve a conflict, we apply a plane sweep method to solve a conflict with a smaller number of stitches. As shown in Figure 13(c), the plane sweep method scans the conflicting feature from the left to the right (or from the right to the left) to find the longest continuous segments. After that, as illustrated in Figure 13(d), we insert a stitch at a suitable location on the feature and assign each segment an appropriate color.

1185

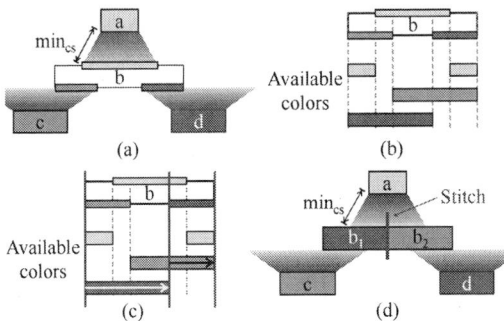

Figure 13: An example of stitch insertion. (a) Projection computation. (b) Feature partition with the ends of the projections (c) Feature scanning with the plane sweep method. (d) Conflict removal with stitch insertion.

Table 1: Comparison of the decomposition results.

Design	SDP Based [16]				Ours			
	#C	#S	Cost	cpu (s)	#C	#S	Cost	cpu (s)
C432	3	1	31	0.14	0	6	6	0.01
C499	0	0	0	0.19	0	0	0	0.01
C880	1	6	16	0.27	1	15	25	0.01
C1355	1	6	16	0.21	1	7	17	0.02
C1908	0	1	1	0.29	1	0	10	0.04
C2670	2	4	24	0.53	2	14	34	0.06
C3540	5	6	56	0.72	2	15	35	0.08
C5315	7	7	77	1.01	3	11	41	0.11
C6288	82	131	951	4.49	19	341	531	0.13
C7552	12	15	135	1.72	3	46	76	0.17
S1488	1	1	11	0.33	0	4	4	0.03
S38417	44	55	495	21.67	20	122	322	0.62
S35932	93	18	948	96.45	46	103	563	2.13
S38584	63	122	752	99.80	36	280	640	2.26
S15850	73	91	821	87.22	36	201	561	2.14
Avg.	2.28	0.40	1.51	40.29	1.00	1.00	1.00	1.00

4. EXPERIMENTAL RESULTS

The proposed layout decomposition algorithm for TPL was implemented in the C++ programming language on a 2.93 GHz Linux machine with 48 GB memory. We used the ISCAS-85 & 89 benchmarks provided by the authors of [16] to evaluate our algorithm. In addition, the metal-1 layer was used because it is one of the most complex layers. The minimum coloring spacing min_{cs} was set as 120nm for the first ten cases and as 100nm for the last five cases, which are the same to the updated setting provided by the authors of [16].

We compare the decomposition results between our algorithm and the SDP-based algorithm proposed by Yu et al. [16]. In Table 1, '#C' denotes the number of conflicts, "#S" the number of stitches, and "cpu" the computation time for the decomposition process. In addition, "Cost" is set as $\#C+0.1 \times \#S$, which is the same as the objective function in [16] (the cost for a conflict is typically much larger than that for a stitch since the conflict incurs much higher manufacturing cost). The updated experimental results of the SDP-based algorithm are also provided by the authors of [16]. As shown in Table 1, compared to the SDP-based algorithm, our layout decomposition approach can averagely reduce the number of conflicts by 56% and achieve 40X speedup. In addition, although more stitches are used to resolve conflicts, the cost is reduced by more than 34%. The significant improvement in cost reduction reveals the fact that their proposed algorithm only inserts stitches at the locations found by the projection method, and thus some legal stitches are missed and some conflicts are generated, which can actually be resolved by inserting stitches in TPL. Figure 14 shows the layout decomposition result of the circuit C432.

5. CONCLUSIONS

In this paper, we have presented a novel layout decomposition approach for TPL. We have proven that the widely used projection method

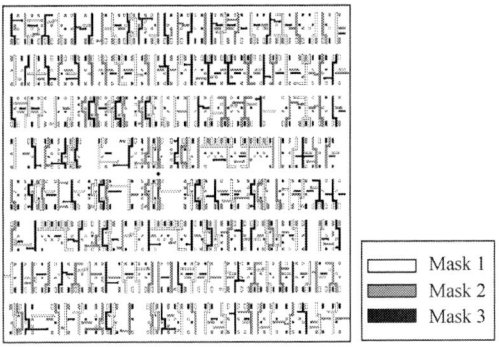

Figure 14: Layout decomposition result of C432.

cannot find all legal stitch candidates for TPL. As a result, we have proposed a stitch-aware mask assignment algorithm which is based on a heuristic to find a mask assignment such that the conflicts among the features in the same mask are more likely to be resolved by inserting stitches. To further reduce the problem size and enhance program efficiency, we have also proposed two new conflict graph reduction techniques. Experimental results have shown that our layout decomposition algorithm can efficiently and effectively generate a good layout decomposition result with fewer conflicts and smaller total costs, compared to the previous work.

6. ACKNOWLEDGMENT

We would like to thank Mr. Bei Yu and the authors of [16] for providing benchmarks and updated experimental results.

7. REFERENCES

[1] G. E. Bailey, A. Tritchkov, J.-W. Park, L. Hong, V. Wiaux, E. Hendrickx, S. Verhaegen, P. Xie, and J. Versluijs. "Double pattern EDA solutions for 32nm HP and beyond," *Proc. SPIE*, vol. 6521, pp. 65211K, 2007.

[2] Y. Borodovsky, "Lithography 2009 overview of opportunities," in Semicon West, 2009.

[3] C. Cork, J.-C. Madre, and L. Barnes. "Comparison of triple-patterning decom- position algorithms using aperiodic tiling patterns," *Proc. SPIE*, vol. 7028, pp. 702839, 2008.

[4] T. H. Cormen, C. E. Leiserson, R. L. Rivest, and C. Stein "Introduction to Algorithms." The MIT Press, 2009.

[5] M. R. Garey, D. S. Johnson, and L. Stockmeyer, "Some simplifed NP-complete problems," *Proc. STOC*, pp. 47–63, 1974.

[6] D. S. Johnson, C. R. Aragon, L. A. McGeoch, and C. Schevon, "Optimization by simulated annealing: an experimental evaluation; part II, graph coloring and number partitioning." Operations Research, 39:378–406, 1991.

[7] A. B. Kahng, C.-H. Park, X. Xu, and H. Yao, "Layout decomposition approaches for double patterning lithography," *IEEE TCAD*, vol. 29, no. 6, pp. 939–952, 2010.

[8] S. Khanna, N. Linial, and S. Safra, "On the hardness of approximating the chromatic number," *Proc. ISTCS*, pp. 250–260, 1993.

[9] M. LaPedus. "SPIE: Intel to extend immersion to 11-nm." EE Times, Feb. 22, 2010.

[10] F. T. Leighton, "A graph coloring algorithm for large scheduling problems," Journal of Research of the National Bureau of Standards, 84(6):489–506, 1979.

[11] L. Liebmann and A. Torres, "A designer's guide to sub-resolution lithography: enabling the impossible to get to the 15nm node," *Proc. DAC*, 2011.

[12] R. Merritt, "Otellini: Intel to ship more SoCs than PC CPUs–someday." EE Times, Sep. 22, 2009.

[13] R. Tarjan, "Depth-first search and linear graph algorithms," *Proc. SWAT*, pp. 114–121, 1971.

[14] Y. H. Tsin, "A simple 3-edge-connected component algorithm," Theory of Computing Systems, 40:125–142, 2007.

[15] D. B. West, "Introduction to Graph Theory," Prentice Hall, 2001.

[16] B. Yu, K. Yuan, B. Zhang, D. Ding, and D. Z. Pan, "Layout decomposition for triple patterning lithography," *Proc. ICCAD*, 2011.

PS3-RAM: A Fast Portable and Scalable Statistical STT-RAM Reliability Analysis Method

Wujie Wen, Yaojun Zhang, Yiran Chen
University of Pittsburgh
Pittsburgh, PA 15261, USA
{wuw2,yaz24,yic52}@pitt.edu

Yu Wang
Tsinghua University
Beijing 100084, CHINA
yu-wang@mail.tsinghua.edu.cn

Yuan Xie
Pennsylvania State University
University Park, PA 16802, USA
yuanxie@cse.psu.edu

ABSTRACT

Process variations and thermal fluctuations significantly affect the write reliability of spin-transfer torque random access memory (STT-RAM). Traditionally, modeling the impacts of these variations on STT-RAM designs requires expensive Monte-Carlo runs with hybrid magnetic-CMOS simulation steps. In this paper, we propose a fast and scalable semi-analytical simulation method – PS3-RAM, for STT-RAM write reliability analysis. Simulation results show that PS3-RAM offers excellent agreement with the conventional simulation method without running the costly macro-magnetic and SPICE simulations. Our method can accurately estimate the STT-RAM write error rate at both MTJ switching directions under different temperatures while receiving a speedup of multiple orders of magnitude (five order or more). PS3-RAM shows great potentials in the STT-RAM reliability analysis at the early design stage of memory or micro-architecture.

Categories and Subject Descriptors

B.8.2 [**Performance and Reliability**]: Performance Analysis and Design Aids

General Terms

Design, Reliability

Keywords

STT-RAM, process variation, thermal fluctuation, reliability

1. INTRODUCTION

Conventional memory technologies, i.e., SRAM, DRAM, and Flash, have achieved remarkable successes in modern electronic designs. Following technology scaling, the shrunk feature size and the increased process variations impose serious power and reliability concerns on these technologies. Many new memory technologies, including spin-transfer torque random access memory (STT-RAM), have emerged above the horizon. By leveraging a good

---

*This work was supported in part by NSF 1116171 and University of Pittsburgh - Central Research Development Fund.

combination of the non-volatility of Flash, the comparable cell density to DRAM, and the nanosecond programming time like SRAM, STT-RAM has shown great potentials in embedded memory and on-chip cache designs [10, 13, 16].

In STT-RAM, the data is represented as the resistance state of a magnetic tunneling junction (MTJ) device. The MTJ resistance state can be switched by applying a switching current with different polarizations. Compared to the charge-based storage mechanism of conventional memories, the magnetic storage mechanism of STT-RAM shows less dependency on the device volume and better scalability. However, process variation continues to be an issue in STT-RAM designs. Also, the thermal fluctuations in MTJ resistance switching process may generate the intermittent failures of STT-RAM write operations. Many researches were conducted to evaluate the impacts of process variations and thermal fluctuations on STT-RAM reliability [5, 9]. The general evaluation method is the follows: First, Monte-Carlo SPICE simulations are run extensively to characterize the distribution of the MTJ switching current I during the STT-RAM write operations, by considering the MTJ and MOS transistor device variations. I samples are then sent into the macro-magnetic model to obtain the MTJ switching time (τ_{th}) distributions under the thermal fluctuations. Finally, the τ_{th} distributions of all I samples are merged to generate the overall MTJ switching performance distribution. A write failure happens when the applied write pulse width is shorter than the needed τ_{th}. Nonetheless, the costly Monte Carlo runs and the dependency on the macro-magnetic and SPICE simulations incur huge computation complexity of such a method and limits its application at the early stage STT-RAM design and optimization.

In this work, we propose a fast, portable and scalable statistical STT-RAM reliability analysis method – "PS3-RAM", which includes three integrated steps: 1) characterizing the MTJ switching current distribution under both MTJ and CMOS device variations; 2) recovering MTJ switching current samples from the characterized distributions in MTJ switching performance evaluation; and 3) performing the simulation on the thermal-induced MTJ switching variations based on the recovered MTJ switching current samples. Our major contributions in developing the above method are:

1. We developed a sensitivity analysis technique to capture the statistical characteristics of the MTJ switching at the scaled technology nodes. It achieves multi-order($> 10^5$) run time cost reduction and marginal accuracy degradation, compared to SPICE-based Monte-Carlo simulations;

2. We proposed using dual-exponential model for the fast and accurate recovery of MTJ switching current samples in statistical STT-RAM thermal analysis;

3. We decoupled PS3-RAM with SPICE and macro-magnetic modeling and simulations, and extended it into STT-RAM array level analysis and design space exploration.

2. PRELIMINARY

2.1 STT-RAM Basics

Figure 1(c) shows the "one-transistor-one-MTJ (1T1J)" STT-RAM cell structure, including a MTJ and a NMOS transistor. In the MTJ, an oxide barrier layer (e.g., MgO) is sandwiched between two ferromagnetic layers. '0' and '1' are stored as the different resistances of the MTJ, respectively. When the magnetization directions of two ferromagnetic layers are parallel (anti-parallel), the MTJ is in its low (high) resistance state. Figure 1(a) and (b) shows the high and the low MTJ resistance states, which are denoted by R_H and R_L, respectively. The MTJ switches from 0 to 1 when the current drives from reference layer to free layer, or from 1 to 0 when the current drives in the opposite direction.

2.2 Operation Errors of MTJ

In general, the MTJ switching time decreases when the switching current increases. A write failure happens when the MTJ switching does not complete before the switching current is removed. There are two reasons can cause this failure:

2.2.1 Persistent errors

The current through the MTJ is affected by the process variations of both transistor and MTJ. For example, the driving ability of the NMOS transistor is subject to the variations of transistor channel length (L), width (W), and threshold voltage (V_{th}). The MTJ resistance variation also affects the NMOS transistor driving ability by changing its bias condition. The degraded MTJ switching current leads to a longer MTJ switching time and consequently, results in an incomplete MTJ switching before the write pulse ends. This kind of errors is categorized as "persistent" errors, which are incurred by only device parametric variations. Persistent errors can be measured and repeated after the chip fabrication is done.

2.2.2 Non-persistent errors

Another kind of errors is called "non-persistent" errors, which happen intermittently and may not be repeated. The non-persistent errors of STT-RAM are mostly caused by the intrinsic thermal fluctuations during MTJ switching [11]. Due to thermal fluctuations, the MTJ switching time will not be a constant value but rather a distribution even under a constant switching current.

3. PS3-RAM METHOD

Figure 2 depicts the overview of our proposed PS3-RAM methods, including the sensitivity analysis for MTJ switching current (I) characterization, the I sample recovery, and the statistical thermal analysis of STT-RAM. Array-level analysis and design optimizations can be also conducted by using PS3-RAM.

3.1 Sensitivity Analysis on MTJ Switching

In this section, we presents our sensitivity model used for the characterization of the MTJ switching current distribution. We then

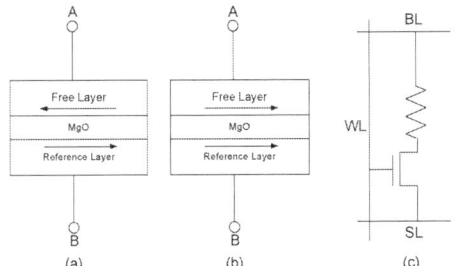

Figure 1: STT-RAM basics. (a) Anti-parallel (high resistance). (b) Parallel (low resistance). (c) 1T1J cell structure.

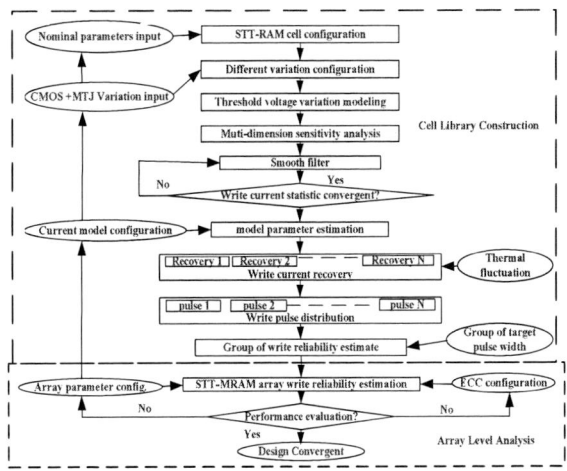

Figure 2: Overview of PS3-RAM.

analyze the contributions of different variation sources to the distribution of the MTJ switching current in details. The definitions of the variables used in our analysis are summarized in Table 1.

3.1.1 Threshold voltage variation

The variations of channel length, width and threshold voltage are three major factors incurring the variations of transistor driving ability. V_{th} variation mainly comes from random dopant fluctuation (RDF) and line-edge roughness (LER), which is also the source of some geometry variations (i.e., L and W) [8, 14]. It is known that the V_{th} variation is also correlated with L and W and its variance decreases when the transistor size increases. The deviation of the V_{th} from the nominal value following the change of L (ΔL) can be modeled by [14]:

$$\Delta V_{th} = \Delta V_{th0} + V_{ds} exp(-\frac{L}{l'}) \cdot \frac{\Delta L}{l'}. \qquad (1)$$

The the standard deviation of V_{th} can be calculated as:

$$\sigma_{V_{th}}^2 = \frac{C_1}{WL} + \frac{C_2}{exp\left(L/l'\right)} \cdot \frac{W_c}{W} \cdot \sigma_L^2. \qquad (2)$$

Here W_c is the correlation length of non-rectangular gate (NRG) effect, which is caused by the randomness in sub-wavelength lithography. C_1, C_2 and l' are technology dependent coefficients. The first term in Eq. (2) describes the RDF's contribution to $\sigma_{V_{th}}$. The second term in Eq. (2) represents the contribution from NRG, which is heavily dependent on L and W. Following technology scaling, the contribution of this term becomes prominent due to the reduction of L and W.

3.1.2 Sensitivity analysis on variations

Although the contributions of MTJ and CMOS parameters to the MTJ switching current distribution cannot be explicitly expressed, it is still possible for us to conduct a sensitivity analysis to obtain the critical characteristics of the distribution. Without loss of generality, the MTJ switching current I can be modeled by a function of W, L, V_{th}, A, and τ. A and τ are the MTJ surface area and MgO layer thickness, respectively. The 1st-order Taylor expansion of I around the mean values of every parameter is:

$$\begin{aligned} I\left(W, L, v_{th}, A, \tau\right) &\approx I\left(\overline{W}, \overline{L}, \overline{V}_{th}, \overline{A}, \overline{\tau}\right) + \frac{\partial I}{\partial W}\left(W - \overline{W}\right) \\ &+ \frac{\partial I}{\partial L}\left(L - \overline{L}\right) + \frac{\partial I}{\partial V_{th}}\left(V_{th} - \overline{V}_{th}\right) \\ &+ \frac{\partial I}{\partial A}\left(A - \overline{A}\right) + \frac{\partial I}{\partial \tau}\left(\tau - \overline{\tau}\right). \end{aligned} \qquad (3)$$

1188

W, L and τ generally follow Gaussian distribution [5], A is the product of two independent Gaussian distributions, V_{th} is correlated with W, L, as shown in Eq. (1) and (2). Because the MTJ resistance $R \propto \frac{e^\tau}{A}$ [5], we have:

$$\frac{\partial I}{\partial A}\Delta A + \frac{\partial I}{\partial \tau}\Delta \tau = \frac{\partial I}{\partial R}\left(\frac{\partial R}{\partial A}\Delta A + \frac{\partial R}{\partial \tau}\Delta \tau\right)$$
$$= \frac{\partial I}{\partial R}\Delta R. \qquad (4)$$

It indicates that the combined contribution of A and τ is the same as the impact of MTJ resistance. The difference between the actual I and its mathematical expectation μ_I can be calculated by:

$$I\left(W, L, V_{th}, R\right) - E\left(I\left(\overline{W}, \overline{L}, \overline{V}_{th}, \overline{R}\right)\right) \approx \qquad (5)$$
$$\frac{\partial I}{\partial W}\Delta W + \frac{\partial I}{\partial L}\Delta L + \frac{\partial I}{\partial V_{th}}\Delta V_{th} + \frac{\partial I}{\partial R}\Delta R.$$

Here we assume $\mu_I \approx E\left(I\left(\overline{W}, \overline{L}, \overline{V}_{th}, \overline{R}\right)\right) = I\left(\overline{W}, \overline{L}, \overline{V}_{th}, \overline{R}\right)$ and the mean of MTJ resistance $\overline{R} \approx R\left(\overline{A}, \overline{\tau}\right)$. Combining Eq. (1), (2), and (5), the standard deviation of I (σ_I) can be calculated as:

$$\begin{aligned}
\sigma_I^2 &= \left(\frac{\partial I}{\partial W}\right)^2 \sigma_W^2 + \left(\frac{\partial I}{\partial L}\right)^2 \sigma_L^2 + \left(\frac{\partial I}{\partial R}\right)^2 \sigma_R^2 \\
&+ \left(\frac{\partial I}{\partial V_{th}}\right)^2 \left(\frac{C_1}{WL} + \frac{C_2}{exp\left(L/l'\right)} \cdot \frac{W_c}{W} \cdot \sigma_L^2\right) \\
&+ 2\frac{\partial I}{\partial L}\frac{\partial I}{\partial V_{th}}\rho_1\sqrt{\frac{C_1}{WL}}\sigma_L + 2\frac{\partial I}{\partial W}\frac{\partial I}{\partial V_{th}}\rho_2\sqrt{\frac{C_1}{WL}}\sigma_W \\
&+ 2\frac{\partial I}{\partial L}\frac{\partial I}{\partial V_{th}}V_{ds}exp(-\frac{L}{l'})\frac{\sigma_L^2}{l'}.
\end{aligned} \qquad (6)$$

Here $\rho_1 = \frac{cov(V_{th0}, L)}{\sqrt{\sigma_{V_{th0}}^2 \sigma_L^2}}$ and $\rho_2 = \frac{cov(V_{th0}, W)}{\sqrt{\sigma_{V_{th0}}^2 \sigma_W^2}}$ are the correlation coefficients between V_{th0} and L or W, respectively [14]. $\sigma_{V_{th0}}^2 = \frac{C_1}{WL}$. Our further analysis shows that the last three terms at the right side of Eq. (6) are significantly smaller than other terms, and can be safely ignored in the simulations of STT-RAM normal operations.

The accuracy of the coefficient in front of the variances of every parameter at the right side of Eq. (6) can be improved by applying window based smooth filtering. Take W as an example, we have:

$$\left(\frac{\partial I}{\partial W}\right)_i = \frac{I\left(\overline{W} + i\Delta W, L, V_{th}, R\right) - I\left(\overline{W} - i\Delta W, L, V_{th}, R\right)}{2i\Delta W}. \qquad (7)$$

where $i = 1, 2, ...K$. Different $\frac{\partial I}{\partial W}$ can be obtained at the different step i. K samples can be filtered out by a windows based smooth filter to balance the accuracy and the computation complexity as:

$$\overline{\frac{\partial I}{\partial W}} = \sum_{i=1}^{K}\omega_i\left(\frac{\partial I}{\partial W}\right)_i. \qquad (8)$$

Here ω_i is the weight of sample i, which is determined by the window type, i.e., Hamming window or Rectangular window [4].

3.1.3 Variation contribution analysis

The variations' contributions to I are mainly represented by the first four terms at the right side of Eq (6) as:

$$S_1 = \left(\frac{\partial I}{\partial W}\right)^2 \sigma_W^2, S_2 = \left(\frac{\partial I}{\partial L}\right)^2 \sigma_L^2, S_3 = \left(\frac{\partial I}{\partial R}\right)^2 \sigma_R^2$$
$$S_4 = \left(\frac{\partial I}{\partial V_{th}}\right)^2 \left(\frac{C_1}{WL} + \frac{C_2}{exp\left(L/l'\right)} \cdot \frac{W_c}{W} \cdot \sigma_L^2\right). \qquad (9)$$

As pointed out by many prior-arts [15], an asymmetry exists in STT-RAM write operations: the switching time of '0'→'1' is longer than that of '1'→'0', and suffers from a larger variance. Also, the switching time variance of '0'→'1' is more sensitive to the transistor size changes than '1'→'0'. As we shall show later, this phenomena can be well explained by using our sensitivity analysis. To

Table 1: Simulation parameters and environment setting

Parameters	Mean	Standard Deviation
Channel length	$\overline{L} = 45$nm	$\sigma_L = 0.05\overline{L}$
Channel width	$\overline{W} = 90 \sim 1800$nm	$\sigma_W = 0.05\overline{L}$
Threshold voltage	$\overline{V}_{th} = 0.466$V	by calucaltion
Mgo thickness	$\overline{\tau} = 2.2$nm	$\sigma_\tau = 0.02\overline{\tau}$
MTJ surface area	$\overline{A} = 45 \times 90$nm^2	by calculation
Resistance low	$R_L = 1000\Omega$	by calculation
Resistance high	$R_H = 2000\Omega$	by calculation

the best knowledge of authors, this is the first time the asymmetric variations of STT-RAM write performance and their dependencies on the transistor size are explained and quantitatively analyzed.

As shown in Figure 1, when writing '0', the word-line(WL) and bit-line(BL) are connected to V_{dd} while the source-line(SL) is connected to ground. $V_{gs} = V_{dd}$ and $V_{ds} = V_{dd} - IR$. The NMOS transistor is mainly working in triode region. Based on short-channel BSIM model., the MTJ switching current supplied by a NMOS transistor working in saturation region can be calculated by:

$$I = \frac{\beta \cdot \left[(V_{dd} - V_{th})(V_{dd} - IR) - \frac{a}{2}(V_{dd} - IR)^2\right]}{1 + \frac{1}{v_{sat}L}(V_{dd} - IR)}. \qquad (10)$$

Here $\beta = \frac{\mu_0 C_{ox}}{1 + U_0(V_{dd} - V_{th})}\frac{W}{L}$. U_0 is the vertical field mobility reduction coefficient, μ_0 is electron mobility, C_{ox} is gate oxide capacitance per unit area, a is body-effect coefficient and v_{sat} is carrier velocity saturation. Based on short-channel PTM model [6] and BSIM model [1, 7], we derive $\left(\frac{\partial I}{\partial W}\right)^2$, $\left(\frac{\partial I}{\partial L}\right)^2$, $\left(\frac{\partial I}{\partial R}\right)^2$, and $\left(\frac{\partial I}{\partial V_{th}}\right)^2$ as:

$$\left(\frac{\partial I}{\partial W}\right)_0^2 \approx \frac{1}{(A_1 W + B_1)^4}, \quad \left(\frac{\partial I}{\partial L}\right)_0^2 \approx \frac{1}{\left(\frac{A_2}{W} + B_2 W + C\right)^2}$$

$$\left(\frac{\partial I}{\partial R}\right)_0^2 \approx \frac{1}{\left(\frac{A_3}{W} + B_3\right)^4}, \quad \left(\frac{\partial I}{\partial V_{th}}\right)_0^2 \approx \frac{1}{\left(\frac{A_4}{\sqrt{W}} + B_4\sqrt{W}\right)^4}$$

Our analytical deduction shows that the coefficients A_{1-4}, B_{1-4} and C are solely determined by $W/L/V_{th}/R$. Here R is the high resistance state of the MTJ, or R_H. For a NMOS transistor working in triode region at '0'→'1' switching, the MTJ switching current is:

$$I = \frac{\beta}{2a}\left[(V_{dd} - IR - V_{th}) - \frac{I}{WC_{ox}v_{sat}^2}\right]^2. \qquad (11)$$

Here R is the low resistance state of the MTJ, or R_L. We have:

$$\left(\frac{\partial I}{\partial W}\right)_1^2 \approx \frac{1}{(A_5 W + B_5)^4}, \quad \left(\frac{\partial I}{\partial L}\right)_1^2 \approx \frac{1}{\left(\frac{A_6}{W} + B_6\right)^2}$$

$$\left(\frac{\partial I}{\partial R}\right)_1^2 \approx \frac{1}{\left(\frac{A_7}{W} + B_7\right)^4}, \quad \left(\frac{\partial I}{\partial V_{th}}\right)_1^2 \approx \frac{1}{\left(\frac{A_8}{W} + B_8\right)^2}$$

Again, A_{5-8} and B_{5-8} can be expressed as the function of $W/L/V_{th}/R$.

In general, a large S_i corresponds to a large contribution to I variation. When W is approaching infinity, only S_3 is nonzero at '1'→'0' switching while both S_2 and S_3 are nonzero at '0'→'1' switching. It indicates that the residual values of S_1–S_4 at '0'→'1' switching is larger than that at '1'→'0' switching when $W \rightarrow \infty$. In other words, '0'→'1' switching suffers from a larger MTJ switching current variation than '1'→'0' switching when NMOS transistor size is large.

3.1.4 Simulation results of sensitivity analysis

Sensitivity analysis [3] can be used to obtain the statistical parameters of MTJ switching current, i.e., the mean and the standard deviation, without running the costly SPICE and Monte-Carlo simulations. It can be also used to analyze the contributions of different

1189

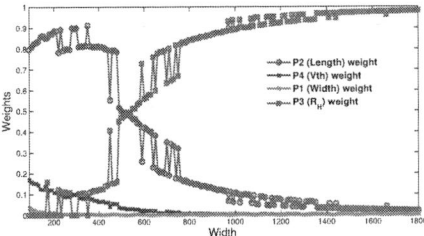

Figure 3: The normalized contributions under different W at '1'→'0' switching.

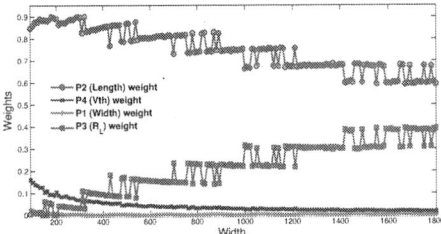

Figure 4: The normalized contributions under different W at '0'→'1' switching.

variation sources to I variation in details. The normalized contributions (P_i) of variation resources $W/L/V_{th}/R$ are defined as:

$$P_i = \frac{S_i}{\sum\limits_{i=1}^{4} S_i}, i = 1, 2, 3, 4 \qquad (12)$$

Figure 3 and Figure 4 show the normalized contributions of every variation source at '0'→'1' and '1'→'0' switchings, respectively, at different transistor sizes. We can see that L and V_{th} are the first two major contributors to I variation at both switching directions when W is small. At '1'→'0' switching, the contribution of L ramps up until reaching its maximum value when W increases, and then quickly decreases when W further increases. At '0'→'1' switching, however, the contribution of L monotonically decreases, but keep being the dominant factor over the simulated W range. At both switching directions, the contributions of R rises up when W increases. At '1'→'0' switching, the normalized contribution of R becomes almost 100% when W is really large.

3.2 Write Current Distribution Recovery

After the I distribution is characterized by the sensitivity analysis, the next question becomes how to recover the distribution of I from the characterized information in the statistical analysis of STT-RAM reliability. We investigate the typical distributions of I in various STT-RAM cell designs and found that dual-exponential function can provide an excellent accuracy in modeling and recovering these distributions. The dual-exponential function we used to recover the I distributions is shown as the below:

$$f(I) = \begin{cases} a_1 e^{b_1(I-u)} & I \le u \\ a_2 e^{b_2(u-I)} & I > u. \end{cases} \qquad (13)$$

Here a_1, b_1, a_2, b_2 and u are the fitting parameters, which can be calculated by matching the the first and the second order momentums of the actual I distribution and the dual-exponential function as:

$$\begin{aligned} \int f(I)dI &= 1 \\ \int If(I)dI &= E(I) \\ \int I^2 f(I)dI &= E(I)^2 + \sigma_I^2. \end{aligned} \qquad (14)$$

Here $E(I)$ and σ_I^2 can be obtained from the sensitivity analysis. The recovered I distribution can be used to generate the MTJ switching current samples, as shown in Fig 5. At the beginning of

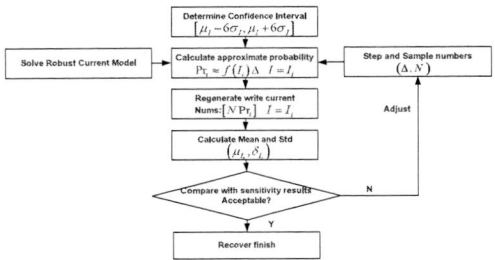

Figure 5: Basic flow for MTJ switching current recovery.

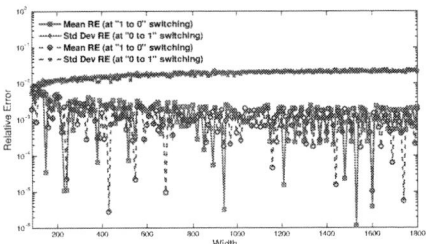

Figure 6: Relative Error of the recovered I w.r.t. the result from sensitivity analysis.

the sample generation flow, the confidence interval for STT-RAM design is determined, e.g., $[\mu_I - 6\sigma_I, \mu_I + 6\sigma_I]$ for a six-sigma confidence interval. Assuming we need to generate N samples within the confidence interval, at the point of $I = I_i$, a switching current sequence of $[N \Pr_i]$ samples must be generated. Here $\Pr_i \approx f(I_i) \Delta$. Δ equals $\frac{12\sigma_I}{N}$, or the step of sampling generation. $f(I_i)$ is the dual-exponential function.

Fig. 6 shows the relative error of the mean and the standard deviation of the recovered I distribution w.r.t. the results directly from the sensitivity analysis (see Eq. (5) and (6)). The maximum relative error $< 10^{-2}$, which proves the accuracy of our dual-exponential model.

Fig. 7 and Fig. 8 compare the probability distribution functions (PDF's) of I from SPICE Monte-Carlo simulations and from the recovery process based on our sensitivity analysis at two switching directions. Our method achieves good accuracy at both simulated transistor channel widths ($W = 90$nm or $= 720$nm).

3.3 Statistical Thermal Analysis

The variation of the MTJ switching time (τ_{th}) incurred by the thermal fluctuations follows Gaussian distribution when τ_{th} is below 10~20ns [15]. In this range, the distribution of τ_{th} can be easily constructed after the I is determined. The distribution of MTJ switching performance can be obtained by combining the τ_{th} distributions of all I samples.

4. WRITE RELIABILITY ANALYSIS

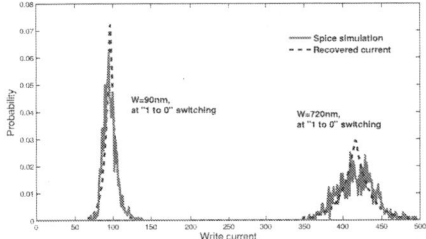

Figure 7: Recovered I vs. Monte-Carlo result at '1'→'0'.

1190

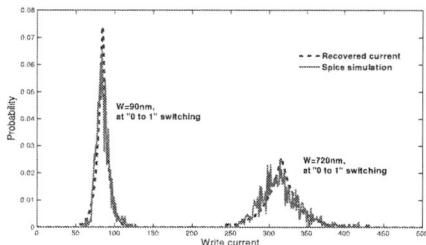

Figure 8: Recovered I vs. Monte-Carlo result at '0'→'1'.

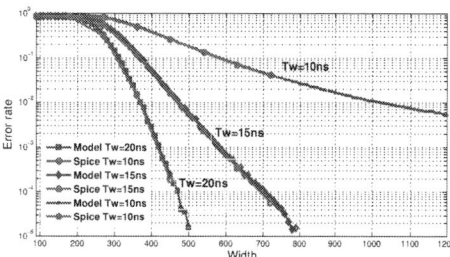

Figure 9: Write failure rate at '0'→'1' when T=300K.

In this section, we conduct the statistical analysis on the write reliability of STT-RAM cells by leveraging our PS3-RAM method. Both device variations and thermal fluctuations are considered. We also extend our method into array level analysis and demonstrate its effectiveness in STT-RAM design optimizations.

4.1 Reliability Analysis of STT-RAM Cells

The write failure rate P_{WF} of a STT-RAM cell can be defined as the probability that the actual MTJ switching time τ_{th} is longer than the write pulse width T_w, or $P_{WF} = P(\tau_{th} > T_w)$. τ_{th} is impacted by the MTJ switching current, MTJ and MOS device variations, MTJ switching direction, and thermal fluctuations. The conventional simulation of P_{WF} requires the costly Monte-Carlo runs with hybrid SPICE and macro-magnetic modeling steps. Instead, we can also use PS3-RAM to analyze the statistical STT-RAM write performance. The simulation environment is summarized in Table 1.

Figure 9 and 10 show the P_{WF}'s simulated by PS3-RAM for both switching directions at 300K. For comparison purpose, the Monte-Carlo simulation results are also presented. Different T_w's are selected at either switching directions due to the asymmetric MTJ switching performances [15]: ($T_w = 10, 15, 20ns$ at '0'→'1' and $T_w = 6, 8, 10, 12ns$ at '1'→'0'). Our PS3-RAM results are in excellent agreement with the ones from Monte-Carlo simulations.

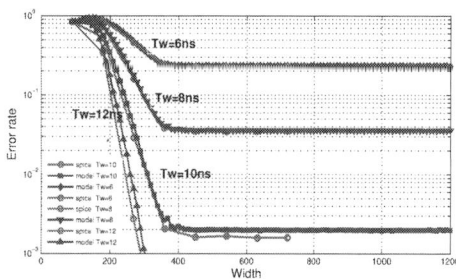

Figure 10: Write failure rate at '1'→'0' when T=300K.

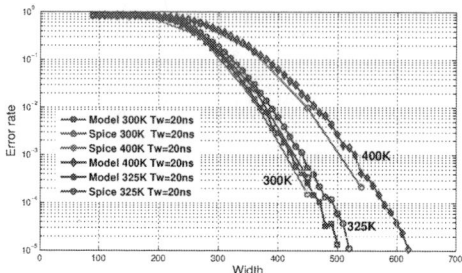

Figure 11: P_{WF} under different temperatures at '0'→'1'.

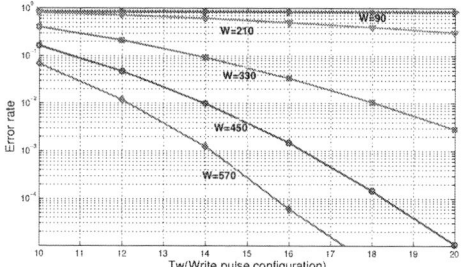

Figure 12: STT-RAM design space exploration at '0'→'1'.

Since '0'→'1' is the limiting switching direction for STT-RAM reliability, we also compare the P_{WF}'s of different STT-RAM cell designs under different temperatures at this switching direction in Figure 11. The results show that PS3-RAM can provide very close but pessimistic results compared to that of the conventional simulations. PS3-RAM is also capable to precisely capture the small error rate change due to a little temperature shift (from T=300K to T=325K).

It is known that either prolonging the write pulse width or increasing the MTJ switching current (by sizing up the NMOS transistor) can reduce the P_{WF}. In Figure 12, we demonstrate an example of using PS3-RAM to explore the STT-RAM design space: the tradeoff curves between P_{WF} and T_w are simulated at different W's. For a given P_{WF}, for example, the corresponding tradeoff between W and T_w can be easily identified on Figure 12.

4.2 Computation Complexity Evaluation

We compared the computation complexity of our proposed PS3-RAM with the conventional simulation method. Suppose the number of variation sources is M, For a statistical analysis of a STT-RAM design, the numbers of SPICE simulations required by conventional flow and PS3-RAM are $N_{std} = N_s{}^M$ and $N_{PS3-RAM} = 2KM + 1$, respectively. Here K denotes the sample numbers for window based smooth filter in sensitivity analysis, N_s is average sample numbers of every variation in the Monte Carlo simulations in conventional method, $K \ll N_s$. The speedup $X_{speedup} \approx \frac{N_s{}^M}{2KM}$ can be up to multiple orders of magnitude: for example, if we set $N_s = 100$, $M = 4$, (Note: V_{th} is not an independent variable) and $K = 50$, the speed up is around 2.5×10^5.

4.3 Array Level Analysis and Design Optimization

We use a 45nm 256Mb STT-RAM design [12] as the example to demonstrate how to extend our PS3-RAM into array level analysis and design optimizations. The number of bits per memory block $N_{bit} = 256$ and the number of memory blocks $N_{word} = 1M$. ECC (error correction code) must be applied to correct the write failures of memory cells. Two types of ECC's with differ-

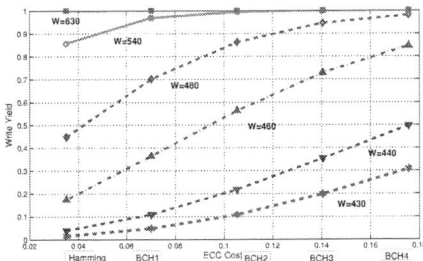

Figure 13: Write yield with ECC's at '0'→'1', T_w=15ns.

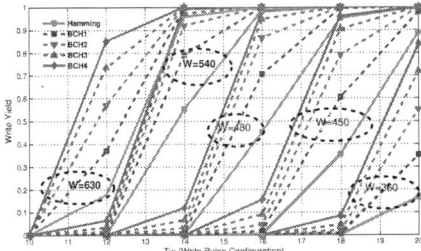

Figure 14: Design space exploration at '0'→'1'.

ent implementation costs are under considerations , i.e., single-bit-correcting Hamming code and a set of multi-bits-correcting BCH codes. We use (n, k, t) to denote an ECC with n codeword length, k bit user bits being protected (256 bit here) and t bits being corrected. The ECC codes corresponding to the error correcttion capability t from 1 to 5 are Hamming code $(265, 256, 1)$ and four BCH codes – BCH1 $(274, 256, 2)$, BCH2 $(283, 256, 3)$, BCH3 $(292, 256, 4)$ and BCH4 $(301, 256, 5)$, respectively. The write yield of the memory array Y_{wr} can be defined as:

$$Y_{wr} = P(n_e \le t) = \sum_{i=0}^{t} C_n^i P_{WF}^i (1 - P_{WF})^{n-i}. \quad (15)$$

Here, n_e denotes the total number of error bits in a write access. Y_{wr} indeed denotes the probability that the number of error bits in a write access is smaller than the error correction capability.

Figure 13 depicts the Y_{wr}'s under different combinations of ECC scheme and W when $T_w = 15$ns at '0'→'1' switching. The ECC schmes required to satisfy $\sim 100\%$ Y_{wr} for different W are: (1) Hamming code for $W = 630$nm; (2) BCH2 for $W = 540$nm; and (3) BCH4 for $W = 480$nm. The total memory array area can be estimated by using the STT-RAM cell size equation $\text{Area}_{\text{cell}} = 3(W/L + 1)(F^2)$ [2]. Calculation shows that (3) offers us the best STT-RAM array area, which is only 88% and 95% of the ones of (1) and (2), respectively. We note that PS3-RAM can be seamlessly embedded into the existing deterministic memory macro models [2] for the extended capability on the statistical reliability analysis and the multi-dimensional design optimizations on area, yield, performance and energy.

Figure 14 illustrates the STT-RAM design space in terms of the combinations of Y_{wr}, W, T_{sw} and ECC. After the pair of (Y_{wr}, T_w) is determined, the tradeoff between W and ECC can be found in the corresponding region on the figure. It shows that PS3-RAM provides a fast and efficient method to perform the device/circuit/architecture co-optimization for STT-RAM designs.

5. CONCLUSION

We developed a fast and scalable statistical STT-RAM reliability analysis method called PS3-RAM. PS3-RAM can be used to simulate the impact of process variations and thermal fluctuations on the statistical STT-RAM write performance, without running the costly Monte-Carlo simulations on SPICE and macro-magnetic models. Simulation results show that PS3-RAM can achieve very high accuracy compared to the conventional simulation method, while achieving a speedup of multiple orders of magnitude. The potentials of PS3-RAM on the device/circui/achitecture co-optimization of STT-RAM designs are also demonstrated.

6. REFERENCES

[1] BSIM. http://www-device.eecs.berkeley.edu/bsim3/. *UC Berkeley*.

[2] X. Cong, N. Dimin, Z. Xiaochun, K. H. Seung, N. Matt, and Y. Xie. "Device Architecture Co-Optimization of STT-RAM Based Memory for Low Power Embedded Systems". In *ICCAD*, pages 463–470, Nov 2011.

[3] P. Doubilet, C. Begg, M. Weinstein, P. Braun, and B. McNeil. "Probabilistic Sensitivity Analysis Using Monte Carlo Simulation. A Practical Approach". 1985.

[4] F. Harris. "On the Use of Windows for Harmonic Analysis with the Discrete Fourier Transform". *Proceedings of the IEEE*, 66(1):51 – 83, Jan. 1978.

[5] J. Li, H. Liu, S. Salahuddin, and K. Roy. "Variation-Tolerant Spin-Torque Transfer (STT) MRAM Array for Yield Enhancement". In *CICC*, pages 193 –196, Sep. 2008.

[6] P. T. M. (PTM). http://www.eas.asu.edu/ ptm/. *ASU*.

[7] B. Sheu, D. Scharfetter, P.-K. Ko, and M.-C. Jeng. "BSIM: Berkeley short-channel IGFET model for MOS transistors". *JSSC*, 22(4):558 – 566, Aug 1987.

[8] R. Singha, A. Balijepalli, A. Subramaniam, F. Liu, and S. Nassif. "Modeling and Analysis of Non-Rectangular Gate for Post-Lithography Circuit Simulation". In *44th DAC*, pages 823 –828, June 2007.

[9] C. W. Smullen, A. Nigam, S. Gurumurthi, and M. R. Stan. "The STeTSiMS STT-RAM Simulation and Modeling System". In *ICCAD*, pages 318–325, Nov 2011.

[10] G. Sun, X. Dong, Y. Xie, J. Li, and Y. Chen. "A Novel Architecture of the 3D Stacked MRAM L2 Cache for CMPs". In *15th HPCA*, pages 239–249. IEEE, 2009.

[11] X. Wang, Y. Zheng, H. Xi, and D. Dimitrov. "Thermal Fluctuation Effects on Spin Torque Induced Switching: Mean and Variations". *JAP*, 103(3):034507–034507–4, Feb. 2008.

[12] W. Xu, Y. Chen, X. Wang, and T. Zhang. "Improving STT MRAM storage density through smaller-than-worst-case transistor sizing". In *46th DAC*, pages 87 –90, July 2009.

[13] W. Xu, H. Sun, Y. Chen, and T. Zhang. "Design of Last-Level On-Chip Cache Using Spin-Torque Transfer RAM (STT-RAM)". In *IEEE Trans. on VLSI System*, pages 483–493. IEEE, 2011.

[14] Y. Ye, F. Liu, S. Nassif, and Y. Cao. "Statistical Modeling and Simulation of Threshold Variation under Dopant Fluctuations and Line-Edge Roughness". In *45th DAC*, pages 900 –905, June 2008.

[15] Y. Zhang, X. Wang, and Y. Chen. "STT-RAM Cell Design Optimization for Persistent and Non-Persistent Error rate Reduction: A statistcal Design View". In *ICCAD*, pages 471–477, Nov. 2011.

[16] P. Zhou, B. Zhao, J. Yang, and Y. Zhang. "Energy Reduction for STT-RAM Using Early Write Termination". In *ICCAD*, pages 264–268. ACM, 2009.

Exploiting Narrow-Width Values for Process Variation-Tolerant 3-D Microprocessors

Joonho Kong   Sung Woo Chung

Dept. of Computer and Radio Communication Engineering, Korea University
Anam-dong, Seongbuk-Gu, Seoul, 136-713 Korea
{luisfigo77, swchung}@korea.ac.kr

ABSTRACT

Process variation is a challenging problem in 3D microprocessors, since it adversely affects performance, power, and reliability of 3D microprocessors, which in turn results in yield losses. In this paper, we propose a novel architectural scheme that exploits the narrow-width value for yield improvement of last-level caches in 3D microprocessors. In a energy-/performance-efficient manner, our proposed scheme improves cache yield by 58.7% and 17.3% compared to the baseline and the naïve way-reduction scheme (that simply discards faulty cache lines), respectively.

Categories and Subject Descriptors

B.3.2 [**Memory Structures**]: Design Styles - *cache memories*

General Terms

Performance, Design, Reliability

Keywords

3D microprocessor, Last-level cache, Narrow-width value, Process variation, Yield

1. INTRODUCTION

A typical way to improve performance of chips (microprocessors) is to employ advanced process technologies. More advanced process technologies bring performance improvement due to their smaller feature size and also have several advantages such as power and area reduction. However, process variation is a challenging problem, since it adversely affects performance (mainly clock frequency) as well as power consumption, which in turn results in yield losses. Without any preventive technique, we inevitably face severe yield losses.

3D integration technology is one of the promising technologies. It vertically stacks several dies, enabling power and chip footprint (area) reduction, and performance improvement due to wire length reduction. However, 3D microprocessors are not free from process variation since die manufacturing process is same as 2D chip manufacturing process. Moreover, 3D chips are even more vulnerable than 2D chips since the typical 3D manufacturing process bonds dies that are generated from different wafers. It means that 3D chips may suffer severe wafer-to-wafer (W2W) variation. Since different wafers are likely to be manufactured from different environment such as temperature, W2W variation may occupy a huge portion of variation source in 3D chips. Though die-to-die (or die-to-wafer) bonding (which bonds already tested dies) can be used to relieve W2W variation, it is not attractive due to its low manufacturing throughput [19] Moreover, its testing cost is high since each die has to be tested separately.

Among many structures in microprocessors, the components that are composed of SRAM cells are known to be most vulnerable to process variation. For example, caches, register files,

and buffer structures (e.g., issue queues) are known to be most vulnerable. In typical 2D microprocessors, among these SRAM-based components, L1 caches are known to be most vulnerable to process variation [21]. On the other hand, in 3D microprocessors, last-level caches (LLC: L2 or L3 caches) are known to be the most vulnerable components. As presented in [19], over half of 3D microprocessors have their critical path in the L2 cache (LLC). Even worse, since LLCs are typically composed of several layers (dies) due to their huge capacity, LLCs are much more vulnerable to W2W variation compared to the other components that can be implemented within only one die (layer).

To efficiently relieve process variation in LLCs of 3D microprocessors, in this paper, we propose a novel architectural scheme leveraging on an architectural insight called as narrow-width values. By exploiting this feature, our proposed scheme can save lots of faulty cache lines under severe W2W variation. In other words, we can correctly store the data values under lots of faulty SRAM cells in LLCs. Our proposed scheme significantly improves yield with negligible performance loss and area overhead. Moreover, by adopting Gated-Vdd [22] to faulty cache subblocks, our proposed scheme also reduces energy consumption of LLCs.

The rest of this paper is organized as follows. In Section 2, we provide essential backgrounds for process variation in 3D microprocessors and narrow-width values. In Section 3, our novel yield improvement scheme is presented. In Section 4, we explain our evaluation methodology. In Section 5, we provide evaluation results in terms of yield, energy, performance, and area. In Section 6, we briefly introduce recent literatures relevant with our proposed scheme. Lastly, in Section 7, we conclude this paper.

2. BACKGROUND

2.1 Process Variation in 3D Microprocessors

Process variation is a major hurdle in deep-submicron process technologies. However, many studies have been focusing on mitigating process variation in 2D microprocessors [2][6][11][14][16][21][24]. In 3D microprocessors, process variation is much more severe than in 2D microprocessors. The main reason is that different dies (layers) in 3D chips are manufactured from different wafers. It may cause severe W2W variation, which can be another major source of parametric variation in 3D chips, while it is not a severe problem in 2D chips. Moreover, as we stack more layers in a 3D chip, much more severe parametric variation is expected due to W2W variation.

Among the major components in the microprocessor, caches are known to be most vulnerable to parametric variation. 6T SRAM cells are exposed to many failure mechanisms such as delay, read, write, and leakage failures. Among various levels of caches, L1 caches are known to be the most vulnerable component in 2D microprocessors [21]. However, a recent study has revealed that over 52% of 3D microprocessors have their critical paths in L2 caches (LLC) [19]. LLCs are also major leakage sources in microprocessors, which makes them vulnerable to leakage failures. Moreover, 6T SRAM cells are not free from read and write failures. Though recent studies have introduced the implementation of LLCs with DRAM cells [15][29], they are targeted at data-intensive server microprocessors, which need massive on-chip storages for performance improvement. On the other hand, in commercial embedded or desktop microprocessors, SRAM cells are still widely used to constitute LLCs. Another major problem of LLCs in 3D microprocessors is that LLCs are typically constructed by using several different dies, which means LLCs are most vulnerable to W2W variation; LLCs are composed of several layers in 3D microprocessors due to their large capacity (over 1MB). Thus, employing preventive techniques is necessary for LLCs to prevent severe yield losses.

2.2 Narrow-width Values

The narrow-width value contains meaningful data values in the LSB side and the remaining bit portion in the MSB side is filled with all '0's. Thus, only storing the meaningful portion of data values is enough and the remaining part of the data values can be filled by using zero-extension logics. In fact, narrow-width values have been widely exploited for soft-error protection or power/performance efficiency in register files or L1 caches [4][7][8][10][12]. However, the narrow-width value feature can also be used in LLCs. According to our simulation results, the ratio of narrow-width values is significantly high (over 75% of accessed data in the L2 cache, on average), encouraging the exploitation of the narrow-width value feature in LLCs. In this paper, we divide data values into four groups: 16-bit narrow-width, 32-bit narrow-width, 48-bit narrow-width, and 64-bit full-width. For example, 16-bit narrow-width value has its meaningful bit portion in 16 bits of LSB side and the remaining 48 bits (MSB side) of data are all '0's. The main benefit by exploiting the narrow-width value is that it makes storage utilization more efficient. In other words, with same storage capacity, we can store more data by exploiting the narrow-width value feature.

3. A NOVEL 3D LAST-LEVEL CACHE ARCHITECTURE FOR YIELD IMPROVEMENT

3.1 Preliminaries

3.1.1 Base architecture and terminology

The base architecture is shown in Figure 1 (a). In this paper, we assume that four layers (layer 1~4) are forming LLCs. In layer 0 (the lowest layer), there is a microprocessor core (or cores). There are also through silicon vias (TSVs) for interconnection between each layer. It is a widely used 3D architecture that is already introduced in [15][29]. Though we restrict base architecture shown in Figure 1 (a), our scheme can be applied to any other 3D architecture where the LLC is composed of several layers (dies). In this paper, we assume that the L2 cache (LLC) is 8-way set-associative with 64-byte line size and the total capacity is 1MB (each layer has 256KB data array size). The total number of cache sets is 2048.

Figure 1 (b) describes a block hierarchy constructing one cache line. One 'cache line' is 64-byte size and it can be divided into 8 'cache word subblocks' (64-bit size). In a cache word subblock, one word is stored and it can be further divided into 4 'cache bit subblocks' (16-bit size). In this paper, we use the terminology introduced in this subsection to avoid misunderstanding.

3.1.2 Layer-partition schemes

Another important design decision is which layer-partition scheme is used for constructing 3D LLCs. There can be several possible ways to divide cache structures into each layer (i.e., which layer-partition scheme is used) in 3D microprocessors. In this work, we introduce three layer-partition schemes: set-partition, way-partition, and bit-partition. The set-partition scheme divides cache sets into four layers. Thus, each of 512 sets is mapped to each layer. The way-partition scheme divides cache ways into four layers, mapping each of two ways to each layer. The bit-partition scheme divides each 64-bit word into four layers, mapping 16-bit in each word to each layer. The bit-partition scheme is similar to the 3D architecture introduced in [23] (the microprocessor has four layers and each layer has 16-bit data paths).

3.1.3 Naïve way-reduction schemes

Due to negative impacts of process variation, there can be lots of faulty SRAM cells in LLCs. Without any preventive technique (baseline), only a single faulty SRAM cell in a chip may lead to a yield loss. In order to reduce yield losses, the simplest method is

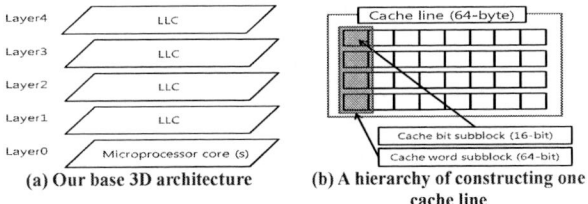

(a) Our base 3D architecture   (b) A hierarchy of constructing one cache line

Figure 1. Base 3D architecture and cache block/line hierarchy

Table I. Possible types of data words that can be allocated according to the # of '0's in the fault bits (4-bit)

The # of '0's in fault bits of the cache word subblock	Possible types of the data word that can be allocated to the cache word subblock
0	X (the entire cache line cannot be used)
1	16-bit narrow-width value
2	16-bit, 32-bit narrow-width value
3	16-bit, 32-bit, 48-bit narrow-width value
4	16-bit, 32-bit, 48-bit narrow-width value, 64-bit full-width value

to discard (i.e., do not use) the cache line that has faulty SRAM cells (named 'naïve way-reduction scheme' in this paper). In this case, the available number of cache lines (ways) in a cache set is decreased. If there is no available cache line in any cache set, this chip is regarded as a yield loss. We can also discard faulty cache sets and access the main memory instead of LLCs when accessing the specific memory addresses (which correspond to faulty cache sets). However, in this case, we may suffer severe performance loss due to frequent main memory accesses. It hurts performance yield of the microprocessor. For example, if an application frequently accesses the specific memory address that corresponds to faulty cache sets, we may suffer approximately 10X performance overhead in the worst-case. Typically, the main memory access latency is much higher than the LLC access latency (more than 10X longer). It can be more deteriorated if there is severe bus/interconnect contention. Moreover, to support such kind of LLC bypassing schemes, we need additional logic which may be another burden. Thus, there should be at least one non-faulty cache line in each cache set, so that the chip works well without severe performance loss.

To reduce energy consumption, unused (faulty) cache lines are permanently power gated. Power gating can be simply implemented by using Gated-Vdd [22] (PMOS power gating has negligible area overhead with leakage energy reduction of 86%). The naïve way-reduction scheme needs the fault bitmap (where fault bits reside) to record whether the cache lines are faulty or not. We need one fault bit for one cache line. Note that '1' in the fault bit means that the corresponding cache line is faulty, and vice versa. Though the naïve way-reduction scheme can reduce yield losses compared to the baseline, it hurts performance due to the reduced number of available cache lines.

3.2 Main Idea

For energy/performance efficiency as well as further yield loss recovery, in this work, we exploit the narrow-width value feature. Our proposed scheme operates in a *finer-grain* manner compared to the naïve way-reduction scheme. As we explained in Section 3.1.3, in the naïve way-reduction scheme, discarding decision is made in a cache line (64-byte) granularity. On the other hand, in our proposed scheme, discarding decision is made in a cache bit subblock (16-bit) granularity to exploit the narrow-width value feature. Obviously, our proposed scheme needs a larger fault bitmap compared to the naïve way-reduction scheme. However, overall area and energy overhead is insignificant. As in the naïve way-reduction scheme, we also adopt Gated-Vdd [22] to faulty cache bit subblocks in our proposed scheme for energy-efficiency. Among the layer-partition schemes introduced in Section 3.1.2, our proposed scheme adopts *the bit-partition scheme* to implement the LLC. Four cache bit subblocks that form a cache word subblock are allocated to four separate layers. Due to W2W variation, different layers are likely to have quite different device characteristics. By using the bit-partition scheme, though some layers suffer severe process variation, *the other layers can save the entire chip with lots of available cache lines* by exploiting the narrow-width value feature.

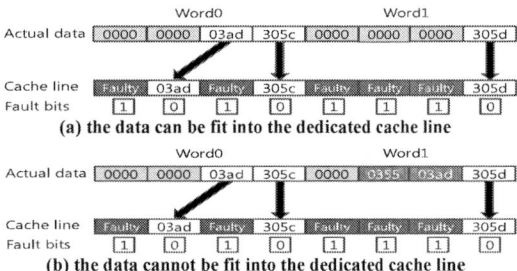

(a) the data can be fit into the dedicated cache line

(b) the data cannot be fit into the dedicated cache line

Figure 2. Examples of the case when the data can be fit into the dedicated cache line and the opposite case

In our proposed scheme, when the data is allocated to the cache, the data words are classified into one of the four types of data (16-bit, 32-bit, 48-bit narrow-width, and 64-bit full-width value). After the data word classification is performed, the fault bitmap is checked if the data words can be fit into the dedicated location of the cache. Our proposed scheme counts the number of '0's in the fault bits (4-bit) of the corresponding cache word subblock (64-bit) to determine which types of the data word can be fit into that cache word subblock. Table I shows the possible types of data words that can be allocated according to the number of '0's in the fault bits (4-bit). We then determine whether each word can be fit into the dedicated cache word subblock in the cache line or not. If *all* of the cache word subblock can contain each data word, the whole data (cache line size) can be allocated in the cache line. Compared to the naïve way-reduction scheme, our proposed scheme can save much more cache lines. In case of the naïve way-reduction scheme, only one faulty SRAM cell in the cache line leads to the failure of the entire cache line. However, in case of our proposed scheme, in spite of faulty SRAM cells, we can still use the cache line unless we stuck in case of the first row in Table I (the number of '0's in fault bits of the cache word subblock=0). It brings better performance due to the increased number of available cache lines as well as more yield loss recovery. In case of our proposed scheme, as the case of the naïve way-reduction scheme, if there is no available cache line in at least one cache set, this chip is regarded as a yield loss.

To better understand our data allocation mechanism, we provide simple examples of the case when the data can be fit into the cache line and the opposite case in Figure 2. To simplify the examples, we assume that one cache line is 16-byte size (originally, 64-byte size in this paper) in these examples. Thus, there are two cache word subblocks in one cache line. In Figure 2 (a), the word0 is 32-bit narrow-width value and the number of '0's in the fault bits is '2'. Thus, the word0 can be fit into the cache word subblock. Likely, the word1 is 16-bit narrow-width value and the number of '0's in the fault bits is '1', fitting well into the dedicated cache word subblock. Since both of two words fit well in two cache word subblocks, the entire data values (16-byte: cache line size in this example) can be allocated to the cache line. On the other hand, in Figure 2 (b), though word0 can be fit into the dedicated cache word subblock, word1 cannot be fit (it is a 48-bit narrow-width value while the number of '0's in the fault bits is '1'). Thus, the data cannot be allocated to the dedicated cache line.

3.3 Algorithms

In this subsection, we explain the algorithm of our proposed

scheme in detail. In a typical cache access, there can be three possible consequences: cache miss, cache read hit, and cache write hit. In this paper, we assume our cache hierarchy uses the inclusive/write-back cache. Note that our proposed scheme can also be applied to exclusive or write-through cache with a little modification. Figure 3 depicts the flow charts of our proposed scheme.

In case of cache misses (Figure 3 (a)), new data is fetched from the main memory and this data should be stored to the LLC. In this case, we should select a victim cache line that will be evicted from the LLC. The conventional way to select the victim cache line is to choose the least recently used (LRU) line in a cache set. However, in our proposed scheme, we should additionally check if data delivered from the main memory can be fit into the cache line that is selected by the LRU policy. In case that the data can be fit into the selected cache line, the data can be stored without any problem. However, there is a case where the data cannot be fit into the selected (LRU) cache line. In order to deal with this case, we adopt *'a devised LRU policy'*. First, in a cache set, our devised LRU policy searches the cache lines that the data value can be fit into. Among the searched cache lines (that the data can be fit into), the least recently used cache block is chosen as a victim cache line. There can be a case where the data is not fit into any cache line in the dedicated cache set. In this case, our proposed scheme does not allocate the data in the LLC. Since this case rarely occurs, performance loss is negligible. Note that the latency for all of these operations during cache misses can be overlapped with the memory access latency. Thus, our proposed scheme does not incur any additional latency for cache misses.

In case of read hits (Figure 3 (b)), we should check the fault bits that correspond to the data which will be read from the cache. In our proposed scheme, when a cache line (64-byte) is accessed, 32-bit of the fault bits should be accessed. Fault bit information is used in zero-extension logic. In the zero-extension logic, by referring to the fault bit information, '0's are filled in the empty MSB side (by using MUXs). The recovered data words are aligned *in order*, forming the complete data (cache line size). Finally, the data is delivered to the microprocessor cores. In order to save more cache lines, we can adopt a rotation scheme that can change the word sequence in a cache line. In this case, data can be stored more compactly. However, it needs quite complex additional logic and may hurt performance due to the additional logic overhead. Thus, in this paper, we do not adopt the complex rotation scheme. Though the zero-extension logic has negligible additional latency, we conservatively assume that accessing the zero-extension logic incurs 1-cycle delay in addition to the cache access latency, since it may be the critical path in manufactured chip due to process variation. Note that accessing the fault bitmap does not incur any additional latency. Since recent L2 cache design employs a serial (sequential) access of tag and data arrays [20], we can sufficiently overlap the fault bitmap access latency with the tag array access latency.

In case of write hits (Figure 3 (c)), the dirty (modified) data from the upper-level caches should be written to the LLC. In this case, like the case of the cache misses, we should check the fault bits of the dedicated cache line to confirm the modified (dirty) data can be fit into that cache line. If the data can be fit into the cache line, the dirty data is written to the cache line. Otherwise, the dirty data is written to the main memory and the cache line is

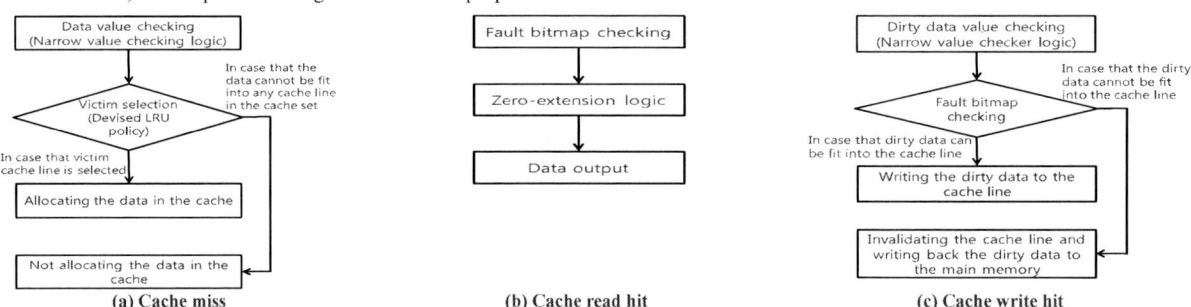

(a) Cache miss **(b) Cache read hit** **(c) Cache write hit**

Figure 3. Flow charts of our proposed scheme according to three cases: cache miss, cache read hit, and cache write hit

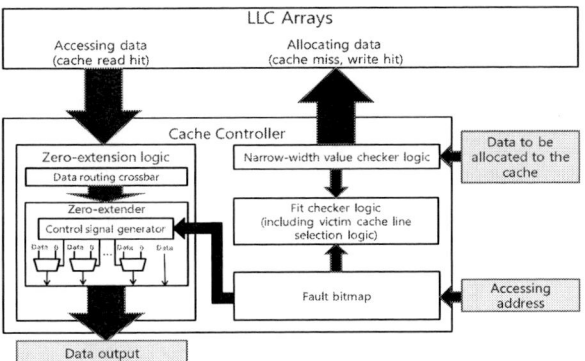

Figure 4. A high-level implementation of the cache controller

invalidated to guarantee the operation correctness. Due to the write buffers, the write access to the LLC does not cause any stall of the processor cores. Thus, there is no additional latency in case of the write hit.

3.4 Implementation

Figure 4 depicts the high-level implementation of the cache controller to support our proposed scheme. Note that we depict only newly added logics in the cache controller in Figure 4. There are four newly added logics to support our proposed scheme: narrow-width value checker logic, fault bitmap, fit checker logic, and zero-extension logic. The narrow-width value checker logic takes a data value as an input and classifies the data value into four types (16-bit, 32-bit, 48-bit narrow-width, and 64-bit full-width). The fault bitmap contains information on whether each cache bit subblock is faulty or not. The values in the fault bitmap can be determined during chip testing procedure. In our proposed scheme, we need one fault bit per one cache bit subblock (16-bit). Since we use 1MB L2 cache, 64KB fault bitmap is required. The fault bitmap takes the address as an input to access the fault bits which correspond to the accessing cache line. The fit checker logic is needed for the case when either a cache miss or a write hit occurs. The fit checker logic checks if new data or dirty data can be fit into the dedicated cache line. It also includes the victim selection logic for our devised LRU policy. The zero-extension logic restores the narrow-width data to the original data. It includes a data routing crossbar, a control signal generator, and 2-to-1 MUXs. Referring to the fault bit information, the data routing crossbar aligns the meaningful data from the LSB side (i.e., the meaningful data is routed to the input of proper MUXs or wires). The control signal generator provides proper control signals (by looking at the fault bit information) to the 2-to-1 MUXs where either data values from LLC arrays or '0's are selected. By counting the '0's in the fault bits that correspond to each cache word subblock, the data is selected in the LSB side of data words while the '0's are selected in the MSB side of data words. Note that we do not need MUXs which correspond to 16 bits in the rightmost side (bit [0-15] in a 64-bit word) of each word since there should be always meaningful data. Thus, we need only 3 MUXs for each word. The additional logics are simple and easy to implement. In our evaluation, we also analyze energy and area overhead of additional logics in more details.

4. EVALUATION METHODOLOGY

For yield evaluation of our proposed scheme, we model process variation of SRAM cells in LLCs. Furthermore, we build an architectural simulation framework for energy and performance evaluation. In the following subsections, we explain our simulation framework in detail.

4.1 Process Variation Modeling for 3D Integrated Chips

In this work, to precisely model process variation, we assign different effective gate length (Leff) and threshold voltage (Vth) to each device of SRAM cells in the L2 cache. In order to reflect W2W variation effects in our yield simulation, we generate process variation maps of dies from different wafers so that they

have quite dissimilar device characteristics (Vth and Leff). Conversely, among the variation maps from the same wafer, device characteristics are more similar than those in the variation maps from the different wafers. To model the systematic within-die variation, we use a method introduced in [16], which is originally based on [1]. In our work, one grid is mapped to 1-byte (8-bit SRAM cells) and each layer is formed by 512X512 grids. After generating the variation maps, we gain the mean Vth (denoting M_{grid_Vth}) and mean Leff (denoting M_{grid_Leff}) for each grid. These values are mapped to proper SRAM cells according to the layer-partition scheme. Thus, with same variation map, the yield results are different according to the layer-partition scheme. We also model random variation effects such as random dopant fluctuation (RDF). Within 8-bit cells that are mapped to the same grid, we map Vth and Leff of each device by generating Gaussian random variables with $N(M_{grid\_Vth}, \sigma_{Vth}^2)$ and $N(M_{grid\_Leff}, \sigma_{Leff}^2)$, respectively.

We perform Monte Carlo simulation to evaluate yield across five variation severities with 240 chips for each variation severity level (denoting 'PV level'). Thus, total 1200 chips are simulated (the total number of variation maps: 1200 * 4 layers = 4800). By using different standard deviation values such as σ_{Vth} and σ_{Leff}, we can adjust PV levels. PV level 0 is the case of the lowest process variation severity while PV level 4 is the case of the highest. We assume that wafer-to-wafer bonding technology is used for 3D integration [19]. Note that we use 45nm technology node and the nominal device parameters for the MOSFETs are based on the device parameters used for ITRS-lstp (low standby power) SRAM cells [18]. Note that this type of SRAM cells is typically used in L2 caches for leakage reduction.

4.2 SRAM Failure Models

For yield evaluation, we adopt four SRAM failure models: delay [25], read [17], write [3], and leakage (BSIM leakage model [5]) failures. Note that the delay, read, and write failure models we use in this paper are originally derived from Mukhopadhyay et al.'s SRAM failure model [17]. While the delay, read, and write failures occur at the cell-level, the leakage failure occurs at the chip-level. The calculated leakage power values of each cell are added to calculate leakage power consumed by the chip. If consumed leakage power of the chip is higher than 3 * *leakage consumption of the chips without process variation*, this chip is regarded as a yield loss. Most of the previous works only modeled delay and/or leakage failure in SRAM cells [6][11][14][16][19][21][24]. However, in this work, we additionally model the read and write failures that also have negative impacts on cache yield. Consequently, it leads to more accurate cache yield simulation.

4.3 Energy Parameters

For energy evaluation, we derived leakage power and per access dynamic energy values of the L2 cache from CACTI 6.5 [18] and 3DCACTI [28]. Note that we use ITRS-lstp cells for L2 cache arrays and assume that tag and data arrays are sequentially accessed [20]. We derived energy values for three different layer-partition schemes which have different physical layouts of L2 caches. Table II shows the derived dynamic energy and leakage power for three different layer-partition schemes. We obtain the dynamic energy values from CACTI and properly scale them for 3D configurations to reflect reduced routing energy in 3D chips. Note that in the set-partition and way-partition schemes, per-access dynamic energy consumption is assumed as same since their main difference of dynamic energy consumption comes from vertical routing between different layers (i.e., which layer among four layers is accessed). We also model per-layer vertical routing access energy (including TSV accessing energy) by properly scaling the energy values of routing energy obtained from CACTI, in order to reflect a reduced wire length effect in 3D chips. Regarding the leakage power, we assume that same leakage power is consumed across three layer-partition schemes because data and tag arrays have same area and capacity regardless of which layer-partition scheme is used. However, actual leakage energy consumption can be different from chip to chip by selectively adopting the Gated-Vdd scheme [22] to faulty cache bit subblocks (lines).

Table II. Derived energy/power parameters

Energy parameters for cache arrays		
	Set-partition & Way-partition	Bit-partition (including our proposed scheme)
Dynamic energy per access (J)	0.133012e-9	0.184780e-9
Vertical routing energy (J)	0.000875e-9	0.001110e-9
Leakage power (W)	0.013785	0.013785
Energy parameters for additional logics		
	Naïve way-reduction schemes	Our proposed scheme
Dynamic energy (J)	0.001349e-9	0.018703e-9
Leakage power (W)	0.0000984	0.0013810

As shown in Table II, we also model energy overhead of the additional logics for naïve way-reduction schemes and our proposed scheme. In our proposed scheme, the energy overhead mainly comes from the fault bitmaps. Since the fault bitmap structure is similar to the cache data array structure, we derived energy consumption of fault bitmaps from CACTI. Note that the other logics (the narrow-width value checker logic, zero-extension logic, and fit checker logic) have negligible energy overhead.

4.4 Architectural Simulation Framework

For performance evaluation, we use M-Sim 3.0 simulator [26], which is derived from SimpleScalar toolset [27]. We model the processor architecture of our simulator similar to the commercial embedded microprocessors such as ARM Cortex-A9. From the architectural simulator, we collect cache access traces of the L2 cache required for calculating energy consumption and evaluate IPC (Instruction Per Clock cycles). In our evaluation, the access latency of the L2 cache is assumed to be 10 cycles. We also assume that there is one cycle additional delay for accessing the zero-extension logic in our proposed scheme, as we explained in Section 3.3. The clock frequency of the simulated microprocessor is set to be 1.0 GHz. We use eight L2 cache sensitive benchmarks [13] from SPEC CPU 2006 INT benchmark suite.

5. EVALUATION RESULTS

In this section, we show the effectiveness of our proposed scheme. We compare our proposed scheme with six different schemes. The baseline scheme does not mitigate process variation. The naïve way-reduction scheme simply discards the faulty cache lines to improve yield, as we explained in Section 3.1.3. Since different layer-partition schemes can be applied to the baseline and naïve way-reduction schemes, the total number of schemes including our proposed scheme is seven. In this section, we use abbreviations for those schemes as shown in Table III.

5.1 Yield

In this subsection, we present yield results of seven different schemes. Table IV shows yield results of seven different schemes. The numbers in Table IV represent the number of passed chips. Note that the criteria of determining passed/failed chips according to three different schemes (baseline, naïve way-reduction, and our proposed scheme) were explained in Section 3.1.3 and 3.2.

Table IV. Yield results of 1200 chips

PV level	base-sp	base-wp	base-bp	nw-sp	nw-wp	nw-bp	prop
level0	240	240	240	240	240	240	240
level1	13	13	11	240	240	240	240
level2	0	0	0	193	231	208	240
level3	0	0	0	11	29	13	110
level4	0	0	0	9	9	4	127
Total	253 (21.1%)	253 (21.1%)	251 (21.0%)	693 (57.8%)	749 (62.4%)	705 (58.8%)	957 (79.8%)

Table III. Classification of schemes and their abbreviations

Categories	Set-partition	Way-partition	Bit-partition
Baseline	base-sp	base-wp	base-bp
Naïve way-reduction	nw-sp	nw-wp	nw-bp
Our proposed scheme	X	X	prop

Our proposed scheme shows the best yield (79.8%) across the seven schemes. The baseline schemes (base-sp, base-wp, and base-bp) show the lowest yield (21.0~21.1%) since none of the preventive schemes are adopted. Even worse, when PV level is higher than 2, the baseline scheme shows 0% yield. Due to severe W2W variation, the baseline yield in 3D chips is significantly low. When adopting naïve way-reduction schemes (nw-sp, nw-wp, and nw-bp), yield is improved compared to the baseline schemes by 36.7~41.3%. However, our proposed scheme shows still higher yield than the naïve way-reduction schemes by 17.4~22.0%.

The advantage of our proposed scheme is robustness to severe process variation. In PV level 3 and 4, yield of our proposed scheme is 49.4% (237/480) while that of the nw-wp scheme (it shows the highest yield among three naïve way-reduction schemes) is only 7.9% (38/480). As process variation becomes more severe (particularly, due to W2W variation), our proposed scheme can save lots of chips that cannot be saved by the naïve way-reduction scheme. Moreover, our proposed scheme achieves high yield *without any redundant cells*. Actually, we can achieve much higher yield by adopting our proposed scheme together with the redundancy schemes [6][24].

5.2 Energy

For energy evaluation, we assume that three kinds of failures (delay, read, and write) are randomly distributed across the SRAM cells in the L2 cache. Figure 5 presents normalized energy consumption based on two *cache line-level fault rates*: 20% and 30%. For instance, if the cache line-level fault rate is 20%, 20% (on average) of cache lines in the LLC are faulty. The fault rate of 20~30% is approximately corresponds to the PV level 2~3 (under a little severe process variation). We only consider the energy consumed by the L2 cache and all of the results are normalized to the energy results of the base-wp scheme.

In case of the fault rate of 20%, our proposed scheme shows energy reduction of 10.5% and 0.2% (on average), compared to the baseline schemes (averaged across base-wp, base-sp, and base-bp) and the naïve way-reduction schemes (averaged across nw-wp, nw-sp, and nw-bp), respectively. Due to the Gated-Vdd, the naïve way-reduction schemes and our proposed scheme saves more energy compared to the baseline schemes. Since the naïve way-reduction schemes adopt Gated-Vdd in cache line-level while our proposed scheme adopts Gated-Vdd in cache bit subblock-level, more cache bit subblocks can be turned off in the naïve way-reduction schemes, under the same fault rate. However, the naïve way-reduction schemes suffer more performance degradation (the detailed performance results are provided in the next subsection) due to the reduced number of available cache lines. This performance loss brings more leakage energy consumption due to longer execution time. Thus, the performance overhead reduction of our proposed scheme also brings energy reduction. In case of the fault rate of 30%, our proposed scheme shows much lower energy consumption (2.6% lower than the naïve way-reduction schemes, on average). As the fault rate becomes higher, our proposed scheme can save more energy compared to the baseline and naïve way-reduction schemes. Note that leakage reduction from our proposed scheme potentially recovers leakage yield losses.

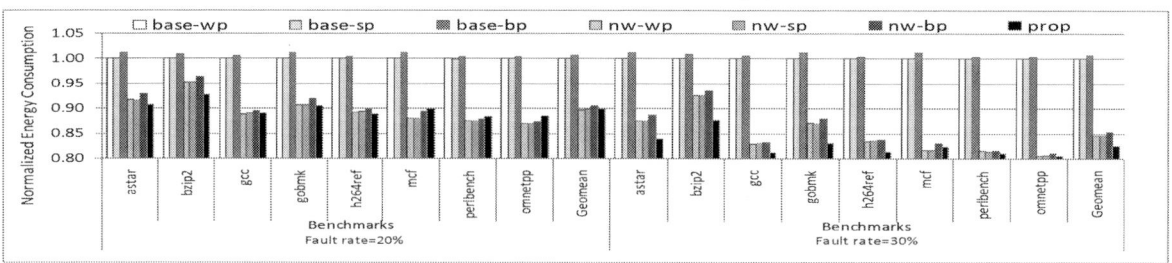

Figure 5. Normalized energy results across seven different schemes in case of the fault rate of 20% and 30%

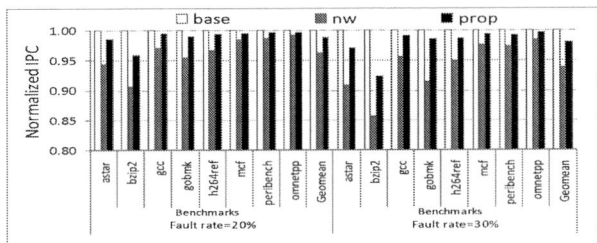

Figure 6. Normalized IPC results across three different schemes in case of the fault rate of 20% and 30%

5.3 Performance

Figure 6 shows the IPC (normalized to the *base*) for two cache line-level fault rates: 20% and 30%. Since we assume that the cache access cycle is same across the layer-partition scheme, we show performance results across three different schemes: baseline (*base*), naïve way-reduction (*nw*), and our proposed scheme (*prop*). As shown in Figure 6, our proposed scheme shows negligible performance loss compared to the baseline scheme (1.2%, on average) in case of the fault rate of 20%. However, the *nw* scheme suffers higher performance degradation compared to our proposed scheme (2.6%, on average). Furthermore, when the fault rate is 30%, performance degradation in case of the *nw* scheme becomes 6.1%, on average. In contrast, our proposed scheme shows almost consistent performance results (of 0.9% difference) across two fault rate cases. Accordingly, our proposed scheme is robust to severe process variation (particularly, under severe W2W variation).

5.4 Area

Our proposed scheme has small area overhead. The fault bitmap occupies the largest area overhead (6.04% of the entire L2 cache area obtained from CACTI) among our additional logics. The zero-extension logic needs 2-to-1 MUXs, comparators, and some logic gates (e.g., XOR gates for fault bit addition), which are negligible area overhead compared to the entire L2 cache area. Narrow-width value checker logic needs comparators and AND gates, which are also trivial area overhead. The fit checker logic can be implemented by small modification from the conventional victim selection logic. Thus, we conservatively estimate our additional logic overhead as 7% of the entire L2 cache area. Since our proposed scheme has small area overhead, the negative impact on yield due to the area increase is also trivial. The area overhead of our proposed scheme can be alleviated by adjusting mapping granularity of the fault bits, which has been already introduced in [14]. In case that two cache bit subblocks are mapped to one fault bit, the area overhead is reduced to approximately 4%.

6. RELATED WORK

Most studies for mitigating process variation are focused on 2D chips [2][11][14][16][21] rather than 3D chips. There have been a few works that are aimed at 3D microprocessors. In [29], Zhao et al. proposed DRAM-based LLC architecture to mitigate process variation in 3D microprocessors. Their target is DRAM-based cache where data-intensive server application is preferred for their design. Ferri et al. [9] also proposed several 3D integration techniques to improve performance as well as sales profit of chips. Ozdemir et al. [19] proposed a CLAPS (Cross LAyer Path Splitting) technique for variation-aware 3D microprocessor design. Their technique is focused on critical path splitting for yield improvement in 3D microprocessors. Though those techniques we introduce above are aimed at mitigation of process variation in 3D microprocessors, our proposed scheme can be differentiated with those techniques due to the exploitation of the narrow-width value feature for process variation-tolerant LLCs.

The narrow-width value feature has been extensively explored for power-performance efficiency or soft-error protection. Brooks and Martonosi [4] exploited the narrow-width value feature for power reduction and performance improvement of the microprocessor data path. A register packing technique [7] of which main goal is performance improvement also utilizes the narrow-width value feature. Hu et al. [10] also proposed an in-register duplication technique for soft-error tolerant register file design. There have been also several works that apply the narrow-width value feature to cache memories. Ergin et al. [8] explored soft-error resilient L1 data cache. Islam and Stenstrom [12] proposed a separated narrow-width cache structure for energy reduction and performance improvement. However, those techniques are applied to register files or L1 data caches. Moreover, the main goal of those techniques is not process variation (hard-error) resilience, but performance improvement, energy reduction, or soft-error resilience.

7. CONCLUSION

In this paper, we presented a new novel architecture-level scheme to mitigate process variation in 3D microprocessors. As shown in evaluation results, our proposed scheme significantly improves yield in an energy-/performance-efficient manner with small area overhead. We believe that our proposed scheme is a good alternative for future 3D microprocessor design under severe process variation.

8. ACKNOWLEDGEMENT

This work was supported by the Smart IT Convergence System Research Center funded by the Ministry of Education, Science and Technology as Global Frontier Project (SIRC 2011-0031863).

9. REFERENCES

[1] A. Agarwal, et al., "Path-based statistical timing analysis considering inter and intra-die correlations", TAU 2002.

[2] A. Agarwal, et al., "A process-tolerant cache architecture for improved yield in nanoscale technologies", IEEE Trans. on VLSI Systems, 2005.

[3] K. Agarwal and S. Nassif, "Statistical Analysis of SRAM Cell Stability", DAC 2006.

[4] D. Brooks and M. Martonosi, "Dynamically exploiting narrow width operands to improve processor power and performance", HPCA 1999.

[5] BSIM MOSFET Model. http://www-device.eecs.berkeley.edu/~bsim3/bsim4.html

[6] A. Das, et al., "Microarchitectures for managing chip revenues under process variations", IEEE Computer Architecture Letters, 2007.

[7] O. Ergin, et al., "Register packing: Exploiting narrow-width operands for reducing register file pressure", MICRO 2004.

[8] O. Ergin, et al., "Exploiting narrow values for soft error tolerance", IEEE Computer Architecture Letters, 2006.

[9] C. Ferri, et al., "Parametric yield management for 3D ICs: Models and strategies for improvement", ACM Journal on Emerging Technologies in Computing Systems, 2008.

[10] J. S. Hu, et al., "In-register duplication: Exploiting narrow-width value for improving register file reliability", DSN 2006.

[11] E. Humenay, et al., "Impact of parameter variations on multi-core chips", ASGI held in conjunction with ISCA 2006.

[12] M. M. Islam and P. Stenstrom, "Characterization and exploitation of narrow-width loads: The narrow-width cache approach", ACM CASES 2010.

[13] X. Jiang, et al., "ACCESS: Smart scheduling for asymmetric cache CMPs", HPCA 2011.

[14] J. Kong, et al., "Fine-grain voltage tuned cache architecture for yield management under process variations", IEEE Trans. on VLSI Systems, published online.

[15] G. H. Loh, "Extending the effectiveness of 3D-stacked DRAM caches with an adaptive multi-queue policy", MICRO 2009.

[16] K. Meng and R. Joseph, "Process variation aware cache leakage management", ISLPED 2006.

[17] S. Mukhopadhyay, et al., "Modeling of failure probability and statistical design of SRAM array for yield enhancement in nanoscaled CMOS", IEEE Trans. on CAD of Integrated Circuits and Systems, 2005.

[18] N. Muralimanohar, et al., "Cacti 6.0: A tool to model large caches", Technical Report HPL-2009-85, 2009.

[19] S. Ozdemir, et al., "Quantifying and coping with parametric variations in 3D-stacked microarchitectures", DAC 2010.

[20] H. Park, et al., "A novel tag access scheme for low power L2 cache", DATE 2011.

[21] Y. Pan, et al., "Selective wordline voltage boosting for caches to manage yield under process variations", DAC 2009.

[22] M. Powell, et al., "Gated-Vdd: A circuit technique to reduce leakage in deep-submicron cache memories", ISLPED 2000.

[23] K. Puttaswamy and G. H. Loh, "Thermal herding: Microarchitecture techniques for controlling hotspots in high-performance 3D-integrated processors", HPCA 2007.

[24] B. F. Romanescu, et al., "Reducing the impact of intra-core process variability with criticality-based resource allocation and prefetching", ACM Int'l Conference on Computing Frontiers, 2008.

[25] S. R. Sarangi, et al., "A model for timing errors in processors with parameter variation", ISQED 2007.

[26] J. J. Sharkey, et al., "M-Sim: A flexible, multithreaded architectural simulation environment", Technical Report CS-TR-05-DP01, Department of Computer Science, State University of New York at Binghamton, 2005.

[27] SimpleScalar toolset. http://www.simplescalar.com

[28] Y.-F. Tsai, et al., "Three-dimensional cache design exploration using 3DCacti", ICCD 2005.

[29] B. Zhao, et al., "Variation-tolerant non-uniform 3D cache management in die stacked multicore processor", MICRO 2009.

A Supplemental Material
Exploiting Narrow-Width Values for Process Variation-Tolerant 3-D Microprocessors

In this supplemental material, we provide detailed examples of our proposed scheme, a detailed description of evaluation methodology, and further discussion of our evaluation results. Note that the reference numbers in this supplemental material correspond to the reference numbers in the 6-page main manuscript.

S1. THE RATIO OF NARROW-WIDTH VALUES

In this section, we provide the detailed results regarding the ratio of narrow-width values. The simulation framework is presented in Section 4.4. To examine the ratio of the narrow-width value, whenever the L2 cache is accessed, we check the type (among four types) of accessed data and count their appearance along the execution time of benchmarks. Thus, if a specific data is accessed many times, the appearance of this data is also counted repeatedly. As shown in Figure S1, the ratio of the accessed narrow-width value is over 75%, on average. Particularly, in five benchmarks (gcc, gobmk, h264ref, mcf, and omnetpp) the ratio of accessed

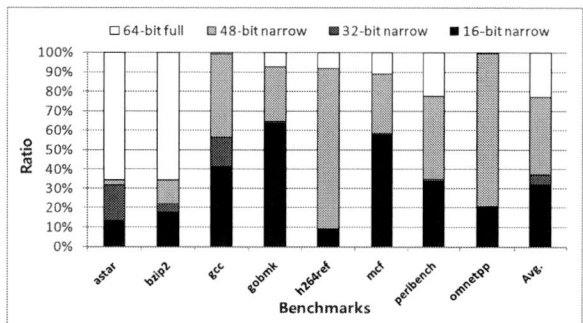

Figure S1. The ratio of three types of narrow-width values and full-width values that are actually accessed in L2 caches

narrow-width value is over 90%. It means that we can sufficiently gain the storage advantage by using the narrow-width value feature.

S2. A DETAILED EXAMPLE OF OUR PROPOSED SCHEME

In this section, we provide more detailed descriptions and examples of our proposed scheme for better understanding of readers.

S2.1 Operations in Zero-extension Logics

In case of cache read hit, the zero extension logic should fill the empty bit portion in the MSB side with '0's. The zero-extension is performed by MUXs and the recovered data words form the complete data (cache line-size). In this subsection, we provide an example of the operation in the zero-extension logic with more detailed descriptions.

Figure S2 shows an example of the operation in the zero-extension logic. In this example, though we show only a data word (word0) which corresponds to the cache word subblock0, there are eight data words to form the complete data (64 bits * 8 = 64 bytes). As we explained in Section 3.1, our base 3D microprocessor is composed of five layers. The microprocessor cores are in the lowest layer (layer 0) and the LLC arrays are in the remaining four layers (layer 1~4). In Figure S2, the data (16-bit) in each cache bit subblock (forming a cache word subblock) comes from each layer (layer 1~4). By looking at the fault bit information, the data routing crossbar routes the data from non-faulty cache bit subblocks to the MUXs/wires which are connected to the LSB side of the data word (i.e., the data routing crossbar aligns the meaningful data from the LSB side). In the example shown in Figure S2, the fault bits of the cache word subblock0 are '1010'. Thus, the data (a cache bit subblock size– 16-bit) in layer 1 and layer 3 are routed to the wire (which are directly connected to the data output) and MUX3, respectively. Note that we do not need MUXs which correspond to 16 bits in the rightmost side (bit [0-15] in a 64-bit word) of each word since

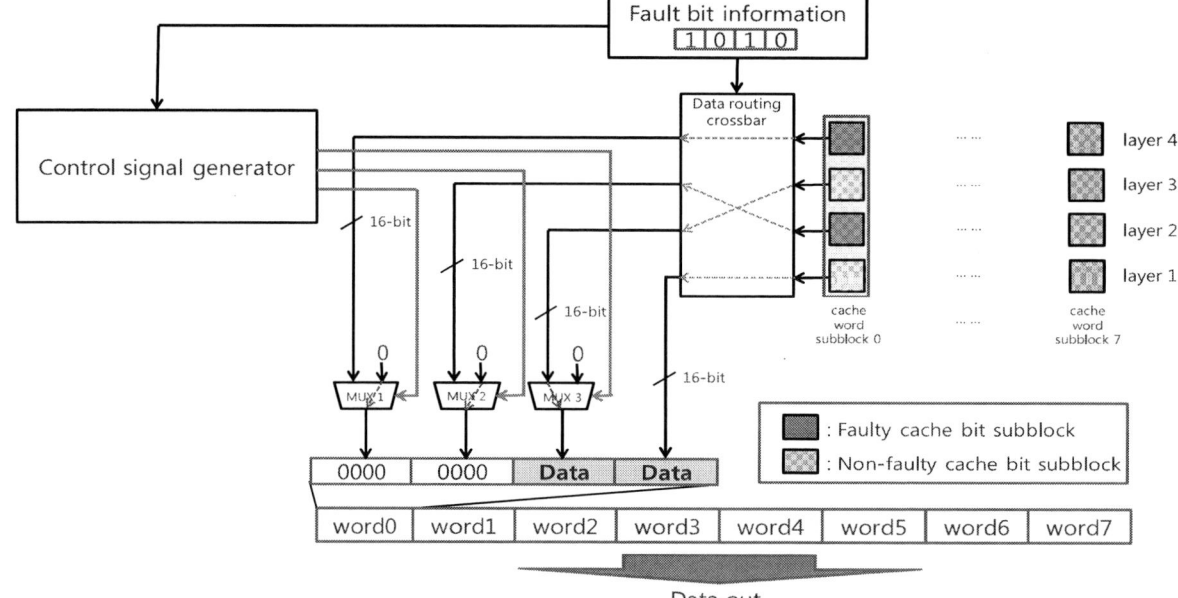

Figure S2. An example of the operation in zero-extension logic

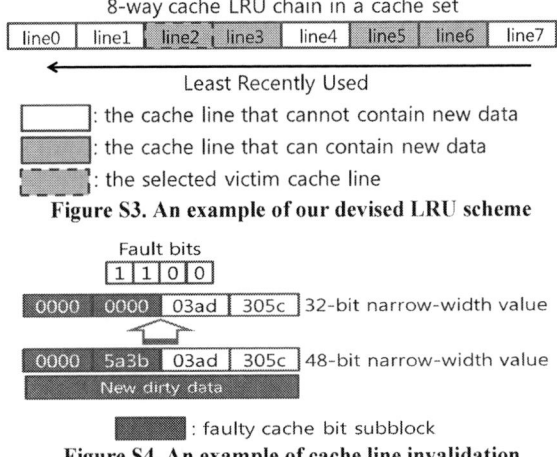

Figure S3. An example of our devised LRU scheme

Figure S4. An example of cache line invalidation

there should be always meaningful data. The fault bit information is also transferred to the control signal generator, which generates control signals for MUXs to correctly recover the data (by zero-extension). After going through the MUXs, the meaningful data is filled in the LSB side and '0's are filled in the remaining MSB side. In this example, the data from the data routing crossbar is selected in MUX3 while '0's are selected in MUX1 and MUX2. The operation for the remaining 7 words is entirely same and performed in parallel.

S2.2 Our Devised LRU Scheme

Figure S3 depicts the example of our devised LRU scheme. As we already explained in Section 3.3, our devised LRU scheme operates in two steps. First, our proposed scheme searches the cache lines that can contain the new data in the dedicated cache set. In the example shown in Figure S3, four cache lines that can contain the data are found (line2, line3, line5, and line6). Among the searched four lines, line2 is finally selected as a victim cache line since it is the least recently used line among the four searched cache lines.

S2.3 Cache Line Invalidation

As we explained in Section 3.3, a cache line may be invalidated due to the modified (dirty) data in case of the write hit. For better understanding, we provide an example of the case when the cache line invalidation occurs. Figure S4 describes an example of the cache line invalidation. Originally, the 32-bit narrow-width value (0x0000000003ad305d) is in a cache word subblock of a cache line. This cache word subblock can only store 16-bit and 32-bit narrow-width values. However, the cache write hit occurs and this data is modified to a 48-bit narrow-width value. Since the cache word subblock cannot store the 48-bit narrow-width value (0x00005a3b03ad305d), the entire cache line (that includes this cache word subblock) should be invalidated. In order to guarantee the operation correctness, the modified data is written to the main memory.

S3. A DETAILED DESCRIPTION OF OUR EVALUATION METHODOLOGY
S3.1 Process Variation Modeling for 3D Integrated Chips

For process variation modeling for 3D microprocessors, we assume that we use the wafer-to-wafer bonding technique to manufacture 3D chips [19] due to its high productivity. To model

wafer-to-wafer variation, the variation maps in different layers have different mean values (denoted as M_{wafer}). These mean values are Gaussian random variables. Assuming $X_{wafer} \sim N(0, 1)$, M_{wafer} values are obtained by using the following equation:

$$M_{wafer} = 0 + \sigma_{W2W} * X_{wafer} \qquad (1)$$

Regarding within-die variation, there are two types of variation, systematic and random variation. To model the systematic variation, we build variation maps by using the model described in [1]. In a die (there are total four dies in LLCs), cache data arrays are divided into 512X512 grids with 10 hierarchical correlation layers (it is totally different from four 3D LLC 'layers'). The hierarchical correlation layers to build a process variation map are shown in Figure S5. Within one hierarchical correlation layer, the values that are assigned to the grids represent Gaussian distributions with $N(M_{wafer}, \sigma_{within-layer}^2)$. To generate the spatially correlated variation map, the random variables in the same 2D location across hierarchical correlation layers are added up (this added value is denoted by M_{grid}). In our work, since there are 10 hierarchical correlation layers, 10 independent Gaussian random variables are added to generate a value for one grid. Assuming

$$X_{layer-n} \sim N(M_{wafer}, \sigma_{within-layer}^2) \text{ in hierarchical correlation layer number} = n \qquad (2)$$

M_{grid} can be calculated as follows:

$$M_{grid} = X_{layer-0} + X_{layer-1} \dots X_{layer-8} + X_{layer-9} \qquad (3)$$

With this variation map, *mean Vth* ($M_{grid\ Vth}$) and *mean Leff* ($M_{grid\ leff}$) value of each grid are calculated as follows:

$$M_{grid\ Vth} = Nominal\ Vth + \sigma_{Vth} * M_{grid} \qquad (4)$$
$$M_{grid\ leff} = Nominal\ Leff + \sigma_{Leff} * M_{grid} \qquad (5)$$

Note that Vth and Leff are generated from the same variation maps in our work, since as Vth increases, Leff is also increased, and vice versa [11]. In our work, one grid is mapped to 1-byte (8-bit SRAM cells), composing a 256K byte data array in a layer (die). Since the L2 cache in our evaluation is composed of four layers, capacity of the entire cache is 1MB.

To model random variation effects such as random dopant fluctuation (RDF), within 8 bit cells (48 devices: 6T SRAM * 8) that are mapped to the same grid, we generate Gaussian random variables by using the value generated above as a mean value. The Vth and Leff of devices that are mapped to the same grid represent Gaussian distribution with $N(M_{grid\ Vth}, \sigma_{Vth}^2)$ and $N(M_{grid\ Leff}, \sigma_{Leff}^2)$, respectively.

We perform Monte Carlo simulation to evaluate yield across five variation severities (PV level 0~4) with 240 chips for each variation severity level. Thus, total 1200 chips are generated. By using different σ_{W2W}, $\sigma_{within-layer}$, σ_{Vth}, and σ_{Leff}, we can adjust process variation severity levels. As shown in Table S1, PV level 0 is the case of the lowest process variation severity while PV

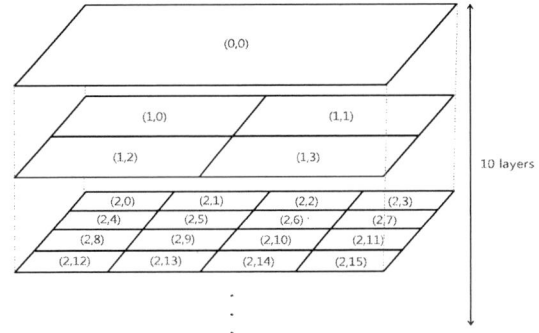

Figure S5. 10 hierarchical correlation layers to model spatial correlation of process variation

Table S1. Five levels of process variation severity

	Level0	Level1	Level2	Level3	Level4
σ_{W2W}	0.01	0.02	0.03	0.04	0.05
$\sigma_{within-layer}$	0.00	0.03	0.06	0.09	0.12
σ_{Vth}	0.02*Nominal Vth	0.04*Nominal Vth	0.06*Nominal Vth	0.08*Nominal Vth	0.1*Nominal Vth
σ_{Leff}	0.0111111*Nominal Leff	0.0166666*Nominal Leff	0.0222222*Nominal Leff	0.0266666*Nominal Leff	0.0333333*Nominal Leff

Table S2. SRAM failure models used in this paper

Categories	Descriptions
Delay failure model	Simplified delay model proposed by Sarangi et al. [25]
Read failure model	Long channel transistor read failure model proposed by Mukhopadhyay et al. [17]
Write failure model	Simplified write failure model proposed by Agarwal and Nassif [3]
Leakage failure model	BSIM leakage model [5]

level 4 is the case of the highest. Note that we use 45nm technology node and the nominal device parameters for the MOSFETs are based on the device parameters used for ITRS-lstp SRAM cells.

S3.2 SRAM Failure Models

For evaluation of yield, we adopt several SRAM failure models. We model four types of SRAM failures; delay, read, write, and leakage failures. We summarize each failure model we use in this paper in Table S2.

For delay model, we use Sarangi et al.'s SRAM delay model [25]. This model is a simplified version of [17], proposed by Mukhopadhyay et al. The delay cutoff boundary is set to be the mean delay without process variation + 1.5 * σ_{delay}. Assuming the delay of SRAM cells without process variation is 1.0, the delay cutoff boundary used in this paper is 1.308. In order to find delay failing cells, we calculate the delay of each SRAM cell in the L2 cache and compare it with the delay cutoff boundary. If the calculated delay of the cell is higher than the delay cutoff, this cell is regarded as delay failing cells.

When the read failure occurs, the cell data is destroyed during the read operation (a destructive read). Since it adversely affects functionality of the microprocessor, consideration on the read failure is also important. For a read failure model, we use the long channel transistor read failure model proposed by Mukhopadhyay et al. [17]. We calculate V_{read} and V_{triprd} of each SRAM cell by using the model in [17] and find the cells of which V_{read} is higher than V_{triprd} (the condition where the read failure occurs).

The write failure occurs when the write delay (T_{write}) is slower than the wordline activation delay (T_{word}). The write failures also negatively affect the functionality of the microprocessor since wrong values may be stored in the microprocessor caches. For a write failure model, we use a simplified write failure model proposed by Agarwal and Nassif [3]. If the calculated write delay of the SRAM cell is higher than the wordline activation delay, this cell is regarded as a write failing cell. Assuming the mean write delay is 1.0, we use 1.320 as wordline activation delay in our work.

The above three SRAM failures occur at the cell-level. However, leakage failure occurs at the chip-level. In order to calculate chip-level leakage power consumption, we calculate the leakage power of each SRAM cell by using the BSIM leakage model [5]. The calculated leakage power values of each cell are added to calculate leakage power consumed by the chip. If consumed leakage power of the chip is higher than *3 * leakage consumption of the chips without process variation*, this chip is regarded as a yield loss.

S4. A DETAILED DESCRIPTION OF EVALUATION RESULTS

In this section, we provide our evaluation results with more detailed evaluation data. The main purpose of this section is to provide more precise evaluation results and more discussion that is not mentioned in our main manuscript.

S4.1 Yield

Figure S6 represents the yield results across the seven schemes. The yield results in Figure S6 are entirely same as Table IV, which are already shown in our 6-page main manuscript. As we explained in Section 5.1, in case of the PV level 0, all seven schemes show 100% yield results. However, in case of the PV level 1, *base* schemes show significantly low yield results since even only one faulty SRAM cell in the L2 cache leads to the yield loss. On the other hand, the *nw* schemes and our proposed scheme (*prop*) can save lots of chips. This trend is almost same in case of the PV level 2. However, in case of the PV level 3 and 4, only our proposed scheme can save lots of chips.

Regarding the yield results, there are two important points that should be addressed. 1) In case of our proposed scheme, the yield result of PV level 4 is a little higher than that of PV level 3. It implies that our proposed scheme is considerably robust to severe process variation. While *nw* schemes are vulnerable to severe process variation, our proposed scheme shows relatively consistent yield results across various PV levels. 2) In case of *base* and *nw* schemes, among the three partition schemes, the way-partition (*wp*) scheme shows the best yield results. The *wp* scheme is relatively robust to W2W variation since the cache lines in the same cache set are distributed across four different layers. In contrast, in case of the set-partition (*sp*) and bit-partition (*bp*) schemes, the cache lines in the same cache set are located in the same layer. In this case, even when only one layer suffers severe process variation, the entire chip may be endangered since all of the cache lines in the cache sets located in that layer (which suffers severe process variation) are likely to be faulty.

Our huge yield improvement comes from a synergistic effect between our bit-partitioned architecture and narrow-width value feature. Our bit-partitioned architecture (applied to 3D microprocessors) spreads each of four bit partitions in a data word among different layers. In this case, though some layers suffer severe process variation, the other layers can *save an entire chip with a lot of available cache lines* by exploiting the narrow-width value feature. It also means our proposed scheme is more suitable for 3D microprocessors (than 2D microprocessors) where severe W2W variation is expected.

S4.2 Energy

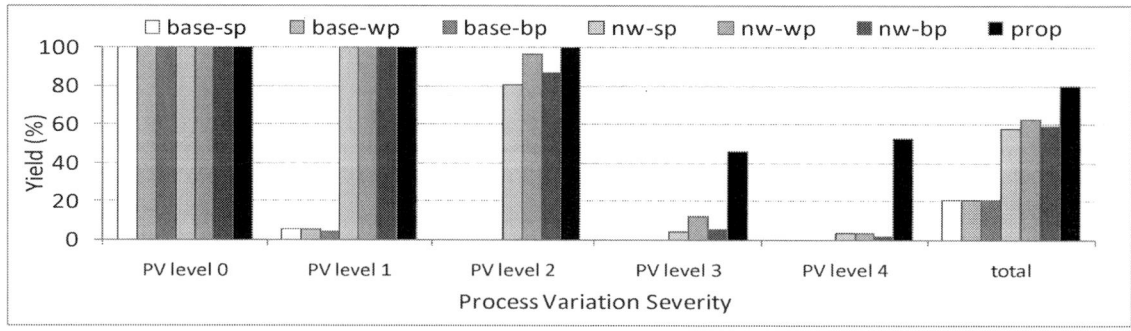

Figure S6. Yield results across five PV levels (0~4)

1201

Table S3. Cache line-level (64-byte) fault rate and cache bit subblock-level fault rate (16-bit)

Line-level fault rate	Bit subblock-level fault rate
20%	0.6949%
30%	1.1084%

In Section S4.2 and S4.3, we show two cases of the cache line-level fault rates. For fair comparison, we additionally calculate the cache bit subblock-level fault rate according to the cache line-level fault rate (Table S3). For example, in case that cache line-level fault rate is 20%, the probability where a cache bit subblock is faulty is 0.6949%. Depending on these probabilities, we randomly distribute the faulty cache lines or cache bit subblocks in the L2 cache. According to the distribution of faulty cache lines or faulty cache bit subblocks in the L2 cache, energy and performance results may also vary, though its impact is not so significant. Thus, we run each simulation for five times with different distributions of faulty cache lines or faulty cache bit subblocks. We provide the average results of these five runs. We run selected eight benchmarks (astar, bzip2, gcc, gobmk, h264ref, mcf, perlbench, and omnetpp) from SPEC CPU 2006. For precise evaluation, we fast-forward 20 billion instructions and actually run 500 million instructions.

Table S4 shows a table version of the energy results, already shown in Figure 5. In case of the fault rate of 20%, our proposed scheme consumes slightly higher energy compared to the *nw-wp* and *nw-sp* scheme by 0.06% and 0.03% (on average), respectively. However, in overall, our proposed scheme shows energy reduction of 0.2% (on average), compared to the naïve way-reduction schemes (averaged across *nw-wp*, *nw-sp*, and *nw-bp*). In some benchmarks (mcf, perlbench, and omnetpp), all of the *nw* schemes (*nw-wp*, *nw-sp*, and *nw-bp*) consume less energy compared to our proposed scheme (*prop*) by 1.3%, on average. The main reason is that these benchmarks have relatively less performance overhead reduction from our proposed scheme than the other benchmarks (detailed performance results are provided in the next subsection). Note that longer execution time results in more leakage energy consumption. In case of the fault rate of 30%, our proposed scheme consumes less energy compared to the naïve way-reduction schemes by 2.6% (much higher reduction

compared to the case of the fault rate of 20%), on average. Only when executing mcf, our proposed scheme consumes higher energy compared to the *nw-wp* and *nw-sp* scheme by 0.8% and 0.9%, respectively. As the fault rate becomes higher, our proposed scheme can save more cache lines (which also brings better performance) compared to the *nw* schemes. Consequently, our proposed scheme can save more energy under severe process variation compared to the *nw* schemes.

S4.3 Performance

Table S5 shows a table version of the performance results, already shown in Figure 6. In all benchmarks, our proposed scheme reduces performance overhead (due to the reduced number of available cache lines) more than the *nw* schemes. Under the same fault rate, in our proposed scheme, the LLC has more available cache lines compared to the *nw* schemes. Because off-chip main memory access accompanies huge latency overhead, cache miss rate reduction in the LLC results in better performance. However, when executing mcf, perlbench, and omnetpp, performance gap between our proposed scheme and the *nw* schemes is less than when executing the other five benchmarks. The main reason is that these benchmarks are relatively insensitive to the reduced number of available cache lines. When adopting the *nw* schemes, even compared to the baseline scheme, these benchmarks (mcf, perlbench, and omnetpp) show relatively small performance overhead (less than 1.6% and 2.7% in case of fault rate=20% and 30%, respectively). On the other hand, the other benchmarks show relatively large performance gap between the *nw* schemes and our proposed scheme. In case of bzip2, performance gap between the *nw* schemes and our proposed scheme is highest among eight benchmarks (5.5% and 7.3% in case of fault rate=20% and 30%, respectively). As we explained in Section S4.2, performance significantly affects energy consumption since the leakage energy consumption is closely related to the execution time (in our evaluation, over 99% of energy is consumed by leakage). Performance overhead reduction also brings energy saving of LLCs. As the fault rate becomes higher (more severe process variation), our proposed scheme shows more performance overhead reduction and more energy reduction compared to the naïve way-reduction (*nw*) schemes.

Table S4. Normalized energy results

Fault rate=20%									
	astar	bzip2	gcc	gobmk	h264ref	mcf	perlbench	omnetpp	Geomean
base-wp	1.000000	1.000000	1.000000	1.000000	1.000000	1.000000	1.000000	1.000000	1.000000
base-sp	1.000005	1.000002	0.999989	1.000051	1.000001	1.000002	0.999588	1.000000	0.999955
base-bp	1.011918	1.008844	1.00504	1.012768	1.004314	1.013099	1.003262	1.00436	1.007943
nw-wp	0.916491	0.95358	0.888789	0.906549	0.892164	0.880374	0.874734	0.870025	0.897479
nw-sp	0.915672	0.953396	0.890752	0.906319	0.893518	0.880345	0.874738	0.870255	0.897773
nw-bp	0.928546	0.962673	0.894592	0.918547	0.898685	0.892939	0.878878	0.873323	0.905613
prop	0.906333	0.92721	0.88929	0.905192	0.888928	0.899006	0.883688	0.885646	0.898059

Fault rate=30%									
	astar	bzip2	gcc	gobmk	h264ref	mcf	perlbench	omnetpp	Geomean
base-wp	1.000000	1.000000	1.000000	1.000000	1.000000	1.000000	1.000000	1.000000	1.000000
base-sp	1.000005	1.000002	0.999989	1.000051	1.000001	1.000002	0.999998	1.000000	1.000006
base-bp	1.011918	1.008844	1.00504	1.012768	1.004314	1.013099	1.003674	1.00436	1.007995
nw-wp	0.87446	0.926234	0.828282	0.871729	0.833314	0.817377	0.814942	0.804723	0.845528
nw-sp	0.874064	0.925314	0.830238	0.869289	0.834797	0.816319	0.813759	0.805952	0.845387
nw-bp	0.886321	0.935527	0.833033	0.879245	0.837732	0.829897	0.815863	0.809714	0.852486
prop	0.83911	0.876557	0.811033	0.829987	0.813656	0.823763	0.809549	0.804542	0.825739

Table S5. Normalized performance (IPC) results

Fault rate=20%									
	astar	bzip2	gcc	gobmk	h264ref	mcf	perlbench	omnetpp	Geomean
base	1.000000	1.000000	1.000000	1.000000	1.000000	1.000000	1.000000	1.000000	1.000000
nw	0.94379	0.905786	0.970661	0.954234	0.965918	0.983831	0.986113	0.991769	0.962394
prop	0.98533	0.958456	0.994735	0.989759	0.993622	0.994515	0.995809	0.996267	0.988489

Fault rate=30%									
	astar	bzip2	gcc	gobmk	h264ref	mcf	perlbench	omnetpp	Geomean
base	1.000000	1.000000	1.000000	1.000000	1.000000	1.000000	1.000000	1.000000	1.000000
nw	0.908599	0.856408	0.955443	0.914715	0.94972	0.97583	0.973484	0.984472	0.938925
prop	0.970995	0.924064	0.990606	0.984824	0.986366	0.993025	0.992556	0.996165	0.979564

Hardware Synthesis of Recursive Functions through Partial Stream Rewriting

Lars Middendorf
University of Rostock
Richard Wagner Str. 31
18119 Rostock-Warnemünde
lars.middendorf@uni-rostock.de

Christophe Bobda
University of Arkansas
504 J. B. Hunt Building
Fayetteville, AR 72701
cbobda@uark.edu

Christian Haubelt
University of Rostock
Richard Wagner Str. 31
18119 Rostock-Warnemünde
christian.haubelt@uni-rostock.de

ABSTRACT

Current high-level synthesis tools based on C/C++ offer only limited support for recursion and functions pointers. We present a novel approach for high-level synthesis that represents the program as a term rewriting system. Based on this concept, dynamic creation of threads, parallel recursive tasks and data-dependent branching can be supported in hardware. Complex examples are used to show the effectiveness of our method.

Categories and Subject Descriptors

B.5.2 [**Hardware**]: REGISTER-TRANSFER-LEVEL IM-PLEMENTATION —*Design Aids, Automatic synthesis*

General Terms

Algorithms, Design, Languages, Theory

Keywords

High-Level Synthesis, Term Rewriting, Recursion, Function Pointer, Stream Processing

1. INTRODUCTION

The synthesis of hardware components from a high-level software specification improves design productivity. In addition, resource consumption and performance can be optimized by automated scheduling and design space exploration. Further, when compared to a manual implementation, the automated process is less error prone and can be verified, so that the quality of the resulting design is greatly improved [8]. However, most tools only support a subset of the C/C++ language and focus on either data flow or control flow dominant applications [7]. While data flow graphs can be mapped very efficiently into hardware, the synthesis of dynamic control flow and especially recursive function calls is often not possible. Usually, pointers are also not supported in general, because the tools rely on static analysis to

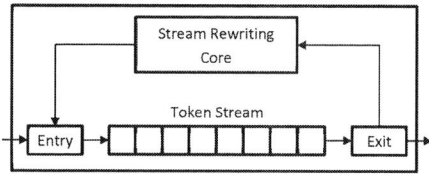

Figure 1: Stream Rewriting Machine

constrain the set of possible locations [11] [10]. If a program does not match the expectations of the synthesis software the compilation either fails or produces inappropriate large structures. Consequently, a program must be adapted according to the restrictions of the synthesis tools before it can be implemented in hardware [3].

Although high-level synthesis can be considered as a great success, we can identify three open problems: First, often a fixed mapping between high-level functionality and hardware components is constructed, so that the generated design cannot react to varying workloads at runtime. As a second consequence, arbitrary communication between two modules requires at least one logical channel implemented via a network or direct connections. Third, when recursion is supported, the local state is often stored on a stack which enforces sequential access and hinders concurrent execution of multiple branches.

In this paper, we propose a novel method for high-level synthesis of RTL code from C programs that overcomes some of these drawbacks. Recursive functions, indirect calls via pointers and the dynamic creation of new threads are supported natively by stream rewriting. However, unlike a software-based processor, which uses a stack for recursion, the synthesized hardware is able to execute multiple branches in parallel. It uses a distributed light-weight scheduling mechanism that makes it possible to spawn and synchronize multiple threads per clock cycle. In fact, all branch operations are implemented as asynchronous func-

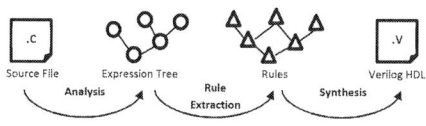

Figure 2: Synthesis tool chain

tion calls by the synthesis front end when it splits the source program into basic blocks. Hence, even normal control flow can take advantage of the hardware scheduler to hide the latency of long pipelines.

Figure 1 shows the concept of our computation model called the *Stream Rewriting Machine*. A stream of tokens is circulating in a ring while the synthesized hardware performs modifications on the stream. The program code is compiled into a term rewriting system implemented in hardware. Similarly, the current state of the system is represented by a sequence of tokens. Threads and local data are encoded by different types of tokens in this stream. By replacing certain patterns, the program is executed and a new state is written on the stream. New threads are created by inserting additional tokens and terminated threads are removed from the stream. Input data is read from the entry point at the left side and results are sent to the output port at the right side. The continuous creation and removal of threads causes the size of the token stream to vary frequently and a computation is finished when the ring is empty.

We propose a tool chain consisting of parser, rule extraction and synthesis software that generates RTL code for this model (Fig. 2). First, the parser analyzes the source file and converts the program into an expression tree similar to a functional language. This tree is then used to extract rewriting rules which describe functional units of the hardware. In the last step, the hardware implementation of these rules is generated and written to a Verilog source file.

The rest of the paper is organized as follows. In section 2, we will discuss related work. Section 3 contains a detailed specification of our computational model. Section 4 describes the hardware architecture generation during the compilation of a C program into RTL code. The results of several examples are presented in section 5 and the last section 6 is reserved for conclusion and discussion of further improvements.

2. RELATED WORK

There exist two different types of recursion. Structural recursion allows us to create hierarchical designs and, similar to macros, templates or inline functions, it is always expanded during synthesis. As a result, most synthesis tools, including VHDL and Verilog compilers, provide support for structural recursion. Contrary, this paper presents a method for generating hardware from a recursive behavior specification. There already have been several attempts of hardware synthesis of recursive functions[12]:

The stack-based approach presented in [13] offers a general solution to this problem. Two stacks for storing local variables and execution position are used to remember the previous state. The control flow is implemented as a hierarchical state machine controlling these stacks which are implemented in fast on-chip memory. Although the computations within each function can be parallelized, it is not possible to execute several recursive branches in concurrently. Similarly, the recent work on synthesis of affine recursion [5] uses a single stack but does not permit multiple invocations.

A stack-less solution has been proposed in [4] to enable parallel processing of recursive calls. Similar to our architecture, a chain of processing nodes is connected in a pipeline. However, if the recursion level exceeds the maximum supported depth, an expensive dynamic reconfiguration of the chip is required. Contrary, our solution is fully functional

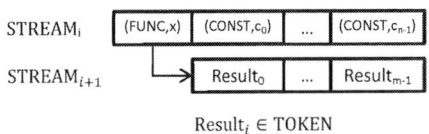

Result$_i$ ∈ TOKEN

Figure 3: Basic rewriting rule

with only one node and the depth is only limited by the available memory.

Term rewriting systems have been already used for both verification [1] and synthesis [6] of hardware components. Also the *Bluespec* language [2], which is based on Haskell, transforms the program into a set of rules for hardware implementation. Though, their concept is fundamental different from our approach. The rules in *Bluespec* are guards monitoring the state and signals of each module. A rule executes by performing an atomic modification of the module's state. Concurrency is employed by allowing different rules to fire simultaneously and partitioning the functionality into separate modules. Contrary, our rules work on a stream of tokens which allows us to substitute several non-overlapping parts of the stream in parallel.

The most similar approach to ours is the *queue machine* [9] which also performs iterative replacement operations on a stream. Is is also optimized for pipelined operations, but does not support flow control nor recursion or indirect calls. Most important, it always operates on neighboring tokens and does not provide random access. As a result, often half of the execution time is spent on moving arguments into the right position by swapping and duplicating tokens.

The concept of computation by term rewriting has been presented for the software development language *Joy* [14] which is similar to *FORTH*. Our model is related to the approach of *Joy*, but supports random access of arguments to avoid the costly reordering like the *queue machine*.

3. CONCEPT

In this section, we present the computational model of our high-level synthesis.

3.1 Definitions

The system illustrated in Figure 1 can be described by the contents of the ring and a function *execute* representing the hardware module. Formally, we will describe the ring as a stream of tokens:

$$\text{STREAM} := \langle t_i \rangle \text{ with } t_i \in \text{TOKEN} \qquad (1)$$

Each token is a tupel storing a type and an integer:

$$\text{TOKEN} := \text{TYPE} \times \mathbb{Z} \qquad (2)$$

$$\text{TYPE} := \{\text{CONST}, \text{FUNC}, \text{OUT}, \text{END}\} \qquad (3)$$

Although, we are using integer valued tokens, arbitrary data-types, including floating-point numbers, can be stored by choosing an appropriate encoding. The different types of tokens act as markers for the replacement rules defined in the hardware module. The CONST token represents a literal

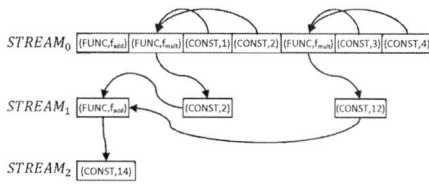

Figure 4: Evaluation of a nested rewriting rule.

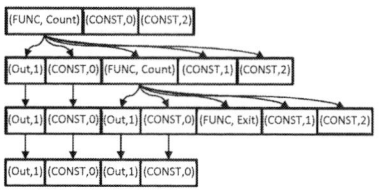

Figure 5: Recursive loop counting from 0 to 1

value on the stream and has no further semantics associated. Function calls are marked using the FUNC token and may refer to multiple subsequent literals. The OUT token extracts a fixed number of literals at the exit point. Finally, the END token is always the last token of the sequence and remembers the end of the stream in the ring. As the tokens are cycling, the hardware module modifies the contents of the stream. An iteration is defined as the process of testing and applying each matching rewriting rule at most once. It corresponds to one invocation of *execute*.

$$execute : \text{STREAM} \to \text{STREAM} \qquad (4)$$

Multiple iterations are defined recursively as an infinite sequence of state vectors.

$$\text{STREAM}_n := execute(\text{STREAM}_{n-1}) \qquad (5)$$

The final result of the computation is defined as the limiting value of this sequence:

$$\text{RESULT} := \lim_{n \to \infty} \text{STREAM}_n \qquad (6)$$

Therefore, we can conclude that results are fixed points of *execute*. However, if a computation does not terminate, there might be always a different subsequent state, so that the sequence does not converge to a single value. In the next section, we will specify *execute* in more detail.

3.2 Arithmetic Operations

Beside from the token OUT, our model of computation defines only the single type of rule displayed in Fig. 3. It looks for a FUNC token followed by n literals and then replaces the corresponding segment by the result of a function f. The function f takes n inputs and produces m output tokens. Since all inputs are literals, the domain can be set to a product of $\mathbb{Z}$ instead of TOKEN:

$$f : \mathbb{Z}^n \to \text{TOKEN}^m \text{ with } n, m \in \mathbb{N}_0$$

Both n and m can be zero, so that there are also constant rules, which take no parameters and rules that produce no outputs. We expect the synthesis tool to build macro rules like f automatically from commonly used functionality in the C-program. Therefore, the program becomes a finite set of rewriting functions $\{f_0, ..., f_{n-1}\}$. The following example defines a set containing two rules for basic addition and multiplication:

$$\text{RULES} := \{f_{add}, f_{mult}\}$$

$$f_{add} : \mathbb{Z} \times \mathbb{Z} \to \text{TOKEN}$$
$$(x, y) \mapsto (\text{CONST}, x + y)$$

$$f_{mult} : \mathbb{Z} \times \mathbb{Z} \to \text{TOKEN}$$
$$(x, y) \mapsto (\text{CONST}, x \cdot y)$$

An example using these rules in three iterations is shown in Figure 4. Instead of an integer, the FUNC tokens are annotated using the name of the function. In the first invocation of *execute*, only the inner multiplication can be resolved and the addition has to wait for the temporary results. In the next iteration, the arguments for the addition are ready and the final result can be produced.

This example shows two important properties of the term rewriting system: First, data dependencies are handled in general and automatically. Rules can be nested with arbitrary depth and execute as soon as their input data becomes available. Otherwise the corresponding tokens are kept and checked again in the next cycle. Second, multiple rules belonging to different branches of the call-tree are evaluated concurrently. As a result, also recursive traversal will be accelerated by visiting multiple paths in parallel.

3.3 Functions and Recursion

Function calls are the most basic operation in our system and can be represented directly using the FUNC token. Due to the pattern matching, multiple nested frames persists on the stream until their arguments are computed, so that all types of recursion are supported. Similar to branching, function pointers are implemented by choosing the integer value of the FUNC token dynamically. The example in Fig. 5 implements a recursive counter, which outputs the numbers from i to n using two rules:

$$\text{RULES} := \{\text{COUNT}, \text{EXIT}\}$$

The function COUNT prints the current iteration and invokes itself recursively. But, if the upper bound of the loop is reached, it calls EXIT instead:

$$\text{COUNT} : \mathbb{Z}^2 \to \text{TOKEN}^4 \ (i, c) \mapsto$$
$$((\text{OUT}, 1), (\text{CONST}, i), x, (\text{CONST}, i), (\text{CONST}, c))$$

with x defined as:

$$x = \begin{cases} (\text{FUNC}, \text{COUNT}) & x < c \text{ goto next iteration} \\ (\text{FUNC}, \text{EXIT}) & x \geq c \text{ exit loop} \end{cases}$$

The rule EXIT consumes both arguments and removes them from the stream. Hence, the current thread is terminated.

$$\text{EXIT} : \mathbb{Z}^2 \to \text{TOKEN}^0$$
$$(i, c) \mapsto$$

1205

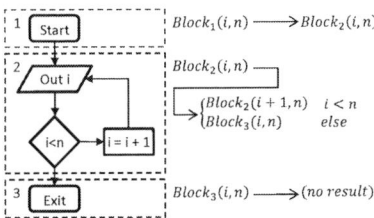

Figure 6: Mapping of Flow Control into Rules

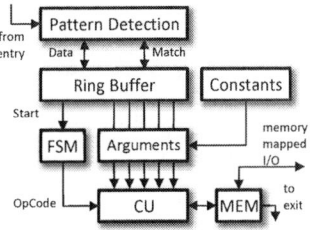

Figure 7: Structure of the Rewriting Core

Since the rewriting system does not define an explicit order of evaluation, all calls are asynchronous by default. The actual order is determined by data dependencies, so that inner calls are evaluated first. Synchronous calls and barriers are special cases that can be implemented by introducing artificial dependencies.

3.4 Branching and Loops

Branches and loops are translated into indirect calls and recursive functions similar to the example in the previous section. First, all high-level flow-control statements like IF, WHILE and DO are converted into GOTO instructions and labels. As a result, we can build a flow-control graph and split the source program into basic blocks. Each block is then mapped into a separate rule and control transfer between blocks is converted into calls. This general approach allows us to support loops both with a fixed and a variable number of iterations. Since the architecture does not provide registers, local variables must be passed to the next block via additional arguments. As a result, the number of local variables also determines the performance of the system because less variables produce shorter rules.

Figure 6 shows the control flow graph for the previous counter if implemented in a high-level language. It contains three blocks called $Block_1$ to $Block_3$ and two local variables which are appended to each state transition. The function starts in $Block_1$ which directly calls the body of the loop in $Block_2$. Similar to the previous example, the current value of the loop counter i is printed. Likewise, the dependent branch at the end is implemented by choosing between the two different successor blocks $Block_2$ or $Block_3$. The last block $Block_3$ exits the program and therefore does not call any further functions.

3.5 Outputs

Similar to the function rule, the *Stream Rewriting Machine* can return the results of a computation using the $OUT(n)$ token where n is a fixed number of literal arguments. However, the sequence is not replaced by another term but instead written to the exit point. Due to the automatic scheduling via pattern matching, the OUT token circulates in the ring until the results are ready.

4. STREAM REWRITING MACHINE

In this section, we will describe the hardware architecture of the *Stream Rewriting Machine* and validate the decisions made in section 3.

4.1 Rewriting Core

Our tool chain generates a structure similar to the abstract computation model shown in Fig. 1. The actual stream processing takes place inside the rewriting core which is illustrated in Fig. 7. The token stream coming from the entry point is arriving at the top-left corner of this scheme. First, it goes through the pattern detector which marks all FUNC tokens followed by a sufficient number of arguments. Each of these tokens can be executed immediately and is tagged with an extra bit indicating the positive match. The ring buffer is used as a temporary storage for the arguments allowing them to be later accessed in any order. The buffer can hold a pending rule while another one is evaluated, so that argument fetch and execution can be overlapped. As a result, even the last argument can be read in the first cycle of the term rewriting process. Depending on the state of the match bit, the FSM either issues the corresponding sequence of tokens or initiates a pass-through operation. The pass-through operation is effectively a NOP, so that unmatched parts of the stream are preserved for further iterations. The computational unit (CU) is responsible for calculating types and values of the new tokens by arithmetic or logic operations. Arguments are received either from the ring buffer or from a special constant RAM. The CU itself consists of several functional units (FU), each one implementing a special function in hardware. During synthesis, the operations required by all rules determine the concrete set of FUs built into the computational unit. In addition to the formal specification, the last stage provides a general I/O interface, allowing rules to access external memory or periphery. Though, the execution order of rules is based purely on data-dependencies and does not account for side-effects. Hence, memory accesses must be serialized similar to synchronous calls if necessary.

4.2 Analysis

The definition of the rewriting system allows for an sufficient hardware implementation due to the following reasons:

All rules using the FUNC token belong to a single class of prefix rules, so that a single parametrized pattern detector component can detect all of them. As a result, the complexity of this module does not depend on the number of rules. Further, all rules are prefix rules, so that no backtracking is required either.

Each rule corresponds to an acyclic data flow graph containing a fixed number of inputs and outputs. Therefore, the substitution of packets can be pipelined to increase the maximum clock rate of the synthesized circuit. However, deep pipelines also increase the latency of the module, but it can be partly compensated if the ring contains an appropriate number of matching function calls. As a result, complex functions containing both compute-intensive arithmetic and irregular flow control can be implemented without branch prediction.

1206

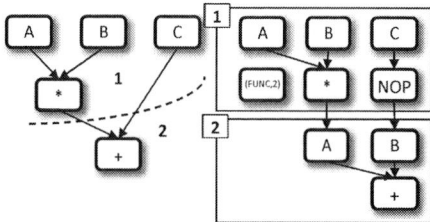

Figure 8: Expression tree split into multiple rules

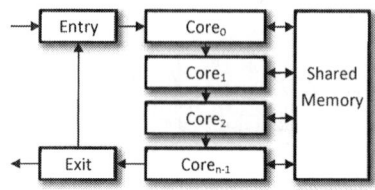

Figure 9: Stream Rewriting Machine

4.3 Rule Extraction and Synthesis

The tree generated by the analysis phase (Fig. 2) does not yet represent hardware operations. First, the abstract operations like addition or multiplication are mapped into functional units (FU) from a library. In addition, multiple simple operation can be combined, but this merge may hinder the reusablity of a FU. Since every core must be able to execute all functions, we are also able to share hardware resources across of the whole module. As a result, the hardware costs of the CU (Fig. 7) does not increase if an operation is used multiple times. It will be composed of all functional units selected in this phase.

Almost the complete execution is pipelined to support complex arithmetic like floating-point at high clock rates. Hence, there is no possibility to reuse temporary results. All inputs must be either arguments of the rule or can be fetched from the constant memory. As a consequence, rules containing multiple FUs in a chain must be split and are executed in multiple steps. However, the proposed architecture allows us to pipeline these calculations, because the next step is then evaluated on the next core.

The expression tree shown on the left side of Fig. 8 implements a multiply-add operation. For this example, we assume that there exist only separate functional units for addition and multiplication. Hence, we insert a break after the multiplication and move the addition into another function. Both rules are linked by inserting a (FUNC,2) token from the first rule to pass the temporary results to the second rule. The argument C is used in the sum and is therefore copied using a NOP operation.

4.4 Scalability

All rules perform local rewrite operations by removing n tokens and inserting m results. Further, the abstract model does not define an explicit execution order, so that multiple matching rules can be replaced simultaneously without affecting the final result. Hence, we can accelerate the computation by instantiating several identical *Rewriting Cores* which modify the stream of tokens at multiple positions.

Figure 9 shows a *Stream Rewriting Machine* consisting of n chained *Rewriting Cores* named $core_0$ to $core_{n-1}$ and a shared memory block that may be later replaced by a more sophisticated approach. The results of one core are passed to the next stage, so that n subsequent iterations can be overlapped and executed in parallel. Formally, the function $execute$ is composed n times to create a pipeline of n stages.

$$pipe_n := execute^n \qquad (7)$$

Hence, the number of iterations required to calculate the result is reduced by the factor n.

Table 1: Execution Time in Cycles

Test \ Cores	1	2	4	8	16
sum(100)	32148	16589	9229	6479	6255
rsum(100)	3460	1989	1262	1223	1215
fib(9)	1530	959	709	734	735
fib(12)	5827	3226	1894	1335	1463
fib(16)	38698	20443	10502	5871	4930
ack(2,5)	6821	6253	5981	5845	5777
ack(3,2)	53789	37945	35729	34917	34649

Table 2: Resources (LUT/FF/BRAM)

Test	1 Core	4 Cores	16 Cores
sum	1311/1085/7	4443/3809/27	17109/14378/105
rsum	1727/1402/8	6170/5074/29	23829/19770/113
fib	1335/1089/7	4539/3825/27	17493/14442/105
ack	1318/1096/7	4471/3853/27	17221/14554/105
all	2060/1669/10	7580/6142/34	29344/24046/130

5. RESULTS

The proposed concept has been implemented into a synthesis tool that generates RTL Verilog source code from a C description. First results using this tool are presented next.

5.1 Generic Tests

Table 1 shows the execution time in cycles for several recursive functions and multiple cores. All examples have been generated by our tool chain from C-like sources and are simulated as Verilog RTL code. The source code of these functions can be found in the supplementary material. The number of cores is modified using a generic parameter and does not require a new synthesis run.

The *sum* function is a simple recursive sum using integer addition with a latency of one cycle. It runs slowly because every instance creates only one additional thread, but due to pipelining it achieves a moderate speed-up. The *rsum* function is almost 10 times faster, because it uses a balanced binary call-tree which better utilizes the architecture of the ring. The Fibonacci (fib) examples show a similar behavior and also computes multiple branches in parallel. The ackermann function (ack) also contains two recursive invocations, but they have data dependencies. Hence, a concurrent evaluation is not possible and only a minor speed-up is achieved.

All functions have been synthesized for the XC6VLX240T-FF1156-1 FPGA from Xilinx. Each entry in Table 2 shows the number of 6-input look-up tables(LUT), flip-flops(FF) and embedded RAMs (BRAM) consumed by this particular implementation. The synthesis time of our tool chain always remained below one second using an Intel i7-2720

1207

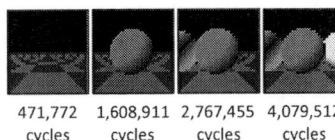

| 471,772 cycles | 1,608,911 cycles | 2,767,455 cycles | 4,079,512 cycles |

Figure 10: Results of the ray tracing simulation

CPU and 8GB RAM. In addition, a *Stream Rewriting Machine* implementing all four functions have been synthesized for comparison. Since every functional unit is shared globally across all rules, the combination requires only 30% more resources.

We do not achieve a true linear speed-up, because the function of multiple iterations *pipe* also consumes more cycles than the single-core variant (*execute*). However, according to the definition of the term rewriting system, a full parallel evaluation is possible, so that these results are unrelated to the computational model and only reflect a drawback of the current implementation.

5.2 Ray Tracing

In order to provide a more complete example we have implemented a ray tracing application. It shows the usage of function pointers, recursive functions and the parallel evaluation of branches. Hence, all elements of our computational model are implemented in this example.

Ray tracing generates images by sending rays from the camera through every pixel into a three-dimensional scene. The color of the pixel is then determined by the object at the nearest intersection point. In our example, the objects are stored as a set of parameters and a pointer to an intersection function. It takes origin and direction of a ray and returns the color of the corresponding object at the intersection point. Similar to virtual functions, the usage of the function pointer allows us to handle the three types of objects (Plane, Sphere, Node) equally and without switch statements. Also the scene itself is stored as an object and represents a tree that is visited recursively.

The ray tracing simulation has been run to produce 32x32 pixel images of different scenes. We assume 12 cycles latency for floating-point addition, 8 cycles for multiplication and 28 cycles for division and square root. These values nearly correspond to the parameters of the floating-point cores from Xilinx when optimized for maximum frequency. The resulting images and the number of cycles are shown in Fig. 10. As a result, we can see that the computational model is able to evaluate complex C programs containing recursion and function pointers. The source code of this example is located in the supplementary section.

6. CONCLUSIONS

Our achievements can be summarized as follows: We have specified an abstract term rewriting system which is able to express recursion and indirect calls with a single type of rule. Due to its simplicity, the formal model leads to a direct hardware implementation and shows a novel approach for high-level synthesis. Several examples of varying complexity have been evaluated regarding speed and scalability. Since the general feasibility of this approach has been shown, our next task is to increase the performance. Further, we

plan to place the state machine into a RAM, so that already synthesized hardware could be patched by updating the microcode.

7. REFERENCES

[1] M. S. Chandrasekhar, J. P. Privitera, and K. W. Conradt. Application of term rewriting techniques to hardware design verification. In *Proceedings of the 24th ACM/IEEE Design Automation Conference*, DAC '87, pages 277–282, New York, NY, USA, 1987. ACM.

[2] N. Dave, A. Pellauer, and M. Pellauer. Scheduling as rule composition. In *Formal Methods and Models for Codesign, 2007. MEMOCODE 2007. 5th IEEE/ACM International Conference on*, pages 51 –60, 30 2007-june 2 2007.

[3] S. Edwards. The challenges of hardware synthesis from c-like languages. In *Design, Automation and Test in Europe, 2005. Proceedings*, pages 66 – 67 Vol. 1, march 2005.

[4] G. Ferizis and H. A. ElGindy. Mapping recursive functions to reconfigurable hardware. In *FPL*, pages 1–6, 2006.

[5] D. R. Ghica, A. Smith, and S. Singh. Geometry of synthesis iv: compiling affine recursion into static hardware. In *ICFP*, pages 221–233, 2011.

[6] J. C. Hoe and Arvind. Hardware synthesis from term rewriting systems. In *VLSI*, pages 595–619, 1999.

[7] G. Martin and G. Smith. High-level synthesis: Past, present, and future. *Design Test of Computers, IEEE*, 26(4):18 –25, july-aug. 2009.

[8] M. McFarland, A. Parker, and R. Camposano. The high-level synthesis of digital systems. *Proceedings of the IEEE*, 78(2):301 –318, feb 1990.

[9] H. Schmit, B. Levine, and B. Ylvisaker. Queue machines: Hardware compilation in hardware. In *Proceedings of the 10th Annual IEEE Symposium on Field-Programmable Custom Computing Machines*, pages 152–, Washington, DC, USA, 2002. IEEE Computer Society.

[10] L. Séméria and G. D. Micheli. Resolution, optimization, and encoding of pointer variables for thebehavioral synthesis from c. *IEEE Trans. on CAD of Integrated Circuits and Systems*, 20(2):213–233, 2001.

[11] L. Séméria, K. Sato, and G. D. Micheli. Synthesis of hardware models in c with pointers and complex data structures. *IEEE Trans. VLSI Syst.*, 9(6):743–756, 2001.

[12] I. Skliarova and V. Sklyarov. Recursion in reconfigurable computing: A survey of implementation approaches. In *Field Programmable Logic and Applications, 2009. FPL 2009. International Conference on*, pages 224 –229, 31 2009-sept. 2 2009.

[13] V. Skylarov, I. Skilarova, and B. Pimentel. Fpga-based implementation and comparison of recursive and iterative algorithms. In *Field Programmable Logic and Applications, 2005. International Conference on*, pages 235 – 240, aug. 2005.

[14] M. von Thun. A rewriting system for joy. Available from the author, 1996.

S. SUPPLEMENTARY MATERIAL

S.1 Example: sum

```
// writes x to the exit point
// generates OUT token
void print(int x);

// Calculates the sum of [0..x] recursively
int sum(int x)
{
    if (x == 0)
        return 0;
    return x + sum(x-1);
}

void main(int x)
{
    print(sum(x));
}
```

S.2 Example: rsum

```
// writes x to the exit point
// generates OUT token
void print(int x);

// Calculates the sum of [a..b]
// by dividing the range recursively
int rsum(int a, int b)
{
    if (b - a == 1)
        return a;

    int m = (a+b) >> 1;
    return rsum(a,m) + rsum(m,b);
}

void main(int x)
{
    print(rsum(0,x+1));
}
```

S.3 Example: fib

```
// writes x to the exit point
// generates OUT token
void print(int x);

// Calculates the fibonacci number x
int fib(int x)
{
    if (x < 2)
        return x;
    return fib(x-1) + fib(x-2);
}

void main(int x)
{
    print(fib(x));
}
```

S.4 Ray Tracing

```
// Print data at the exit point
void print(int x);
void print2(int arg0, int arg1);
// Used to simulate set_pixel x,y,color
void print3(int arg0, int arg1, int arg2);

// Similar to OpenCL, there are
// built-in vector types float1,...,float4
// and functions like dot3f for dot-product

struct Node;
struct Intersection;

// Generic intersection function
typedef Intersection (*IntersectFunc)
    (Node *node, float3 start, float3 dir);

// Stores one object of the scene
struct Node
{
    // position
    float4 pos;

    // intersection function
    IntersectFunc intersect;

    // left child node
    int left;

    // right child node
    int right;

    // color rgb
    float3 color;
};

// Represents an intersection between a
// ray and the scene
struct Intersection
{
    float dist;    // distance from the start
    float3 color;  // color at this position
};

// intersection functions for all three types
Intersection plane_intersect
    (Node *node, float3 start, float3 dir);
Intersection sphere_intersect
    (Node *node, float3 start, float3 dir);
Intersection node_intersect
    (Node *node, float3 start, float3 dir);

// raytracing function
float3 raytrace_color(float3 start, float3 dir);
```

```c
// Array containing scene objects
Node nodes[7] =
{
{ float4(0,1,0,0), plane_intersect,
    0, 0, float3(1,0,0) },
{ float4(0,1,2,1), sphere_intersect,
    0, 0, float3(0,0,1) },
{ float4(1,0,0,1), node_intersect,
    0, 1, float3(1,1,1) },
{ float4(-2,1,2,1), sphere_intersect,
    0, 0, float3(0,1,0) },
{ float4(1,0,0,1), node_intersect,
    3, 2, float3(1,1,1) },
{ float4(2,1,2,1), sphere_intersect,
    0, 0, float3(0,1,1) },
{ float4(1,0,0,1), node_intersect,
    4, 5, float3(1,1,1) },
};

// Helper function to construct an
// intersection structure
Intersection intersection
    (float dist, float3 color)
{
    Intersection result;
    result.dist = dist;
    result.color = color;
    return result;
}

// Generates a checkerboard pattern mapped
// onto the XY plane
float3 get_color(float3 pos)
{
    int ix = fabs(pos.x);
    int iy = fabs(pos.y);
    int iz = fabs(pos.z);
    return ((ix ^ iz) & 1) ?
        float3(1,0,0) :
        float3(0.25,0,0);
}

// Calculates intersection between a
// plane and the ray start + t * dir
Intersection plane_intersect
    (Node *node, float3 start, float3 dir)
{
    // Solves the following system:
    // ray(t) = start + t * dir
    // plane: dot(plane.xyz, x) + plane.w = 0

    float4 plane = node->pos.xyzw;
    float n_dot_dir = dot3f(plane.xyz, dir);
    float n_dot_start = dot3f(plane.xyz, start);

    // Solve for parameter t of the ray
    // We ignore the case n_dot_dir==0
    float t = (-plane.w - n_dot_start)
            / n_dot_dir;

    // p is the intersection point
    float3 p = start + t * dir;

    // Calculate color at position p
    float3 c = get_color(p);

    return intersection(t, c);
}

// Calculates intersection between a
// sphere and the ray start + t * dir
Intersection sphere_intersect
    (Node *node, float3 start, float3 dir)
{
    float4 pos = node->pos;

    // The position vector contains the center
    // the sphere and the squared radius in w
    float3 center = pos.xyz;
    float squared_radius = pos.w;

    // Solves the following system
    // ray(t) = start + t * dir
    // sphere: (p-center)^2 == squared_radius

    float3 v = start - center;

    float a = dot3f(dir, dir);
    float b = 2 * dot3f(v, dir);
    float c = dot3f(v,v) - squared_radius;

    // Determinant of the quadratic equation
    float det = b * b - 4 * a * c;

    // There is not intersection point
    if (det < 0)
    {
        return intersection(det, float3(0,0,0));
    }

    // One or two intersection points exists
    float s = sqrt(det);

    // Calculate parameter t of both points
    float t0 = (-b + s) / (2 * a);
    float t1 = (-b - s) / (2 * a);

    // Choose nearest intersection point
    float dist = t0 < t1 ? t0 : t1;

    // Calculate actual position
    float3 n = v + dist * dir;

    // Perform simple lighting
    float diffuse = fabs(n.x - n.y + n.z);

    // Clamp diffuse to [0..1]
    diffuse = diffuse < 0 ? 0 : diffuse;
    diffuse = diffuse > 1.0 ? 1.0 : diffuse;

    // Multiply diffuse light with sphere color
    return intersection(dist,
        diffuse * node->color);
}
```

```
// Intersection function for scene node
// Invokes intersection function of it childs
// and returns the nearest intersection point
Intersection node_intersect
    (Node *node, float3 start, float3 dir)
{
    Intersection a, b;

    int left = node->left;
    int right = node->right;

    // Call intersect on child nodes
    a = nodes[left].
        intersect(&nodes[left], start, dir);
    b = nodes[right].
        intersect(&nodes[right], start, dir);

    if (a.dist < 0)
        return b;

    if (b.dist < 0)
        return a;

    return (a.dist < b.dist) ? a : b;
}


// Calculates the direction of the ray
// at screen position (x,y)
float3 view_vec(int x, int y)
{
    float vx = (2.0/31.0) * x - 1.0;
    float vy = (-2.0/31.0) * y + 1.0;

    return float3(vx, vy, 1);
}


// Packs floating point RGB color
// into integer value
int rgb(float r, float g, float b)
{
    return (int)(r * 255)
        + ((int)(g * 255) << 8)
        + ((int)(b * 255) << 16);
}



// Sends a ray into the scene and returns
// the color of the nearest intersection
// or black if no intersection was found
float3 raytrace_color
    (float3 start, float3 dir)
{
    Intersection i, i0, i1;
    i = nodes[6].intersect
        (&nodes[6], start, dir);
    return i.dist < 0.0
        ? float3(0,0,0)
        : i.color;
}
```

```
// entry function invoked by the test bench
// Calculates the color for pixel (x,y)
void raytrace(int x, int y)
{
    // Get ray direction for pixel (x,y)
    float3 v = view_vec(x,y);

    // Camera position is (0,1,0)
    float3 start = float3(0,1,0);

    float3 color = raytrace_color(start, v);

    // Convert color into packed integer
    int c = rgb(color.x, color.y, color.z);

    // Set color of pixel (x,y) to c
    print3(x,y, c);
}
```

Chisel: Constructing Hardware in a Scala Embedded Language

Jonathan Bachrach, Huy Vo, Brian Richards, Yunsup Lee,
Andrew Waterman, Rimas Avižienis, John Wawrzynek, Krste Asanović
EECS Department, UC Berkeley *
{jrb|huytbvo|richards|yunsup|waterman|rimas|johnw|krste}@eecs.berkeley.edu

ABSTRACT

In this paper we introduce *Chisel*, a new hardware construction language that supports advanced hardware design using highly parameterized generators and layered domain-specific hardware languages. By embedding Chisel in the Scala programming language, we raise the level of hardware design abstraction by providing concepts including object orientation, functional programming, parameterized types, and type inference. Chisel can generate a high-speed C++-based cycle-accurate software simulator, or low-level Verilog designed to map to either FPGAs or to a standard ASIC flow for synthesis. This paper presents Chisel, its embedding in Scala, hardware examples, and results for C++ simulation, Verilog emulation and ASIC synthesis.

Categories and Subject Descriptors

B.6.3 [**Logic Design**]: [Design Aids – automatic synthesis, hardware description languages]

General Terms

Design, Languages, Performance

Keywords

CAD

1. INTRODUCTION

The dominant traditional hardware-description languages (HDLs), Verilog and VHDL, were originally developed as hardware *simulation* languages, and were only later adopted as a basis for hardware *synthesis*. Because the semantics of these languages are based around simulation, synthesizable

---

*Research supported by DoE Award DE-SC0003624, and by Microsoft (Award #024263) and Intel (Award #024894) funding and by matching funding by U.C. Discovery (Award #DIG07-10227).

designs must be inferred from a subset of the language, complicating tool development and designer education. These languages also lack the powerful abstraction facilities that are common in modern software languages, which leads to low designer productivity by making it difficult to reuse components. Constructing efficient hardware designs requires extensive design-space exploration of alternative system microarchitectures [9] but these traditional HDLs have limited module generation facilities and are ill-suited to producing and composing the highly parameterized module generators required to support thorough design-space exploration. Recent extensions such as SystemVerilog improve the type system and parameterized generate facilities but still lack many powerful programming language features.

To work around these limitations, one common approach is to use another language as a macro processing language for an underlying HDL. For example, Genesis2 uses Perl to provide more flexible parameterization and elaboration of hardware blocks written in SystemVerilog [9]. The language called Verischemelog [6] provides a Scheme syntax for specifying modules in a similar format to Verilog. JHDL [1] equates Java classes with modules. HML [7] uses standard ML functions to wire together a circuit. These approaches allow familiar and powerful languages to be macro languages for hardware netlists, but effectively require leaf components of the design to be described in the underlying HDL. This combined approach is cumbersome, combining the poor abstraction facilities of the underlying HDL with a completely different high-level programming model that does not understand hardware types and semantics.

An alternative approach is to begin from a domain-specific application programming language from which a hardware block is generated. Esterel [2] uses event-based statements to program hardware for reactive systems. DIL [4] is an intermediate language targeted at stream processing and hardware virtualization. Bluespec [3] supports a general concurrent computation model, based on guarded atomic actions. While these can provide great designer productivity when the task in hand matches the pattern encoded in the application programming model, they are a poor match for tasks outside their domain. For example, the design of a programmable microprocessor is not well described in a stream programming model, and guarded atomic actions are not a natural way to express a high-level DSP algorithm. Furthermore, in general it is difficult to derive an efficient microarchitecture from a higher-level computation model, especially if the goal is a programmable engine to run many applications, where the human designer would prefer to write a generator to explore this design space in detail.

In this paper, we introduce Chisel (Constructing Hardware In a Scala Embedded Language), a new hardware design language we have developed based on the Scala programming language [8]. Chisel is intended to be a simple platform that provides modern programming language features for accurately specifying low-level hardware blocks, but which can be readily extended to capture many useful high-level hardware design patterns. By using a flexible platform, each module in a project can employ whichever design pattern best fits that design, and designers can freely combine multiple modules regardless of their programming model. Chisel can generate fast cycle-accurate C++ simulators for a design, or generate low-level Verilog suitable for either FPGA emulation or ASIC synthesis with standard tools. We present several design examples and results from emulation and synthesis experiments.

2. CHISEL OVERVIEW

Instead of building a new hardware design language from scratch, we chose to embed hardware construction primitives within the Scala programming language. We chose Scala for a number of reasons: Scala 1) is a very powerful language with features we feel are important for building circuit generators, 2) is specifically developed as a base for domain-specific languages, 3) compiles to the JVM, 4) has a large set of development tools and IDEs, and 5) has a fairly large and growing user community. Chisel comprises a set of Scala libraries that define new hardware datatypes and a set of routines to convert a hardware data structure into either a fast C++ simulator or low-level Verilog for emulation or synthesis. This section describes the features of the base Chisel system, whereas the next two sections describe how Chisel supports abstraction and powerful generators.

2.1 Chisel Datatypes

The basic Chisel datatypes are used to specify the type of values held in state elements or flowing on wires. In Chisel, a raw collection of bits is represented by the `Bits` type. Signed and unsigned integers are considered subsets of fixed-point numbers and are represented by types `Fix` and `UFix` respectively. Boolean values are represented as type `Bool`. Note that these types are distinct from Scala's builtin types such as `Int`. Constant or literal values are expressed using Scala integers or strings passed to constructors for the types.

Chisel provides a *Bundle* class, which the user extends to make collections of values with named fields (similar to *structs* in other languages):

```
class MyFloat extends Bundle {
  val sign       = Bool()
  val exponent   = UFix(width = 8)
  val significand = UFix(width = 23)
}
val x  = new MyFloat()
val xs = x.sign
```

The keyword `val` is part of Scala, and is used to name variables that have values that won't change. The `width` named parameter to the `UFix` constructor specifies the number of bits in the type. Chisel also provides *Vecs* for indexable collections of values:

```
// Vector of five 23-bit signed integers.
val myVec = Vec(5) { Fix(width = 23) }
val reg3 = myVec(3) // Connect to one element of vector.
```

Bundles and Vecs can be arbitrarily nested to build complex data structures. The set of primitive classes (`Bits`, `Fix`, `UFix`, `Bool`) plus the aggregate classes (`Bundles` and `Vecs`) all inherit from a common superclass, `Data`. Every object that ultimately inherits from `Data` can be represented as a bit vector in a hardware design.

2.2 Combinational Circuits

A circuit is represented as a graph of nodes in Chisel. Each node is a hardware operator that has zero or more inputs and that drives one output. A literal is a degenerate node that has no inputs and drives a constant value on its output. One way to create and wire together nodes is using textual expressions:

```
(a & b) | (~c & d)
```

where `&` and `|` represent bitwise-AND and -OR respectively, and `~` represents bitwise-NOT. The names `a` through `d` represent named wires of some (unspecified) width. Any simple expression can be converted directly into a circuit tree, with named wires at the leaves and operators forming the internal nodes. The final circuit output of the expression is taken from the operator at the root of the tree, in this example, the bitwise-OR.

Simple expressions can build circuits in the shape of trees, but to construct circuits in the shape of arbitrary directed acyclic graphs (DAGs), we must describe fan-out. In Chisel, we can name a wire holding a subexpression by declaring a variable, then referencing it multiple times in subsequent expressions:

```
val sel = a | b
val out = (sel & in1) | (~sel & in0)
```

The named Chisel wire `sel` holds the output of the first bitwise-OR operator so that the output can be used multiple times in the second expression.

Bit widths are automatically inferred unless set manually by the user. The bit-width inference engine starts from the graph's input ports and calculates node output bit widths from their respective input bit widths, always preserving exact results unless an explicit truncation is requested.

2.3 Functions

We can define functions to factor out a repeated piece of logic that we later reuse multiple times in a design:

```
def clb(a: Bits, b: Bits, c: Bits, d: Bits) = (a & b) | (~c & d)
val out = clb(a,b,c,d)
```

The `def` keyword is part of Scala and introduces a function definition, with each argument followed by a semicolon then its type, and the function return type given after the semicolon following the argument list. The equals (=) sign indicates the start of the function definition.

2.4 Ports and Components

Ports are used as interfaces to hardware components. A port is simply any `Data` object that has directions assigned to its members. An example port declaration is as follows:

```
class FIFOInput extends Bundle {
  val ready = Bool(OUTPUT)
  val valid = Bool(INPUT)
  val data  = Bits(32, INPUT)
}
```

`FIFOInput` becomes a new type that can be used in component interfaces or for named collections of wires.

The direction of an object can also be assigned at instantiation time:

1213

```
class ScaleIO extends Bundle {
  val in = new MyFloat().asInput
  val scale = new MyFloat().asInput
  val out = new MyFloat().asOutput
}
```

By folding directions into the object declarations, Chisel is able to provide powerful wiring constructs described later.

In Chisel, *components* are very similar to *modules* in Verilog, defining a hierarchical structure in the generated circuit. The hierarchical component namespace is accessible in downstream tools to aid in debugging and physical layout. A user-defined component is defined as a *class* which: (1) inherits from `Component`, (2) contains an interface stored in a port field named `io`, and (3) wires together subcircuits in its constructor. As an example, consider defining a two-input multiplexer as a component:

```
class Mux2 extends Component {
  val io = new Bundle {
    val sel = Bits(1, INPUT)
    val in0 = Bits(1, INPUT)
    val in1 = Bits(1, INPUT)
    val out = Bits(1, OUTPUT)
  }
  io.out := (io.sel & io.in1) | (~io.sel & io.in0)
}
```

The wiring interface to a component is a collection of ports in the form of a `Bundle`, held in a field named `io`. The `:=` assignment operator, used here in the body of the definition, is a special operator in Chisel that wires the input of left-hand side to the output of the right-hand side.

Port classes represent the interface to a component, and users can organize interfaces into hierarchies using standard Scala facilities. For example, a user could define a simple link for handshaking data as follows:

```
class SimpleLink extends Bundle {
  val data = Bits(16, OUTPUT)
  val rdy  = Bool(OUTPUT)
}
```

We can then extend `SimpleLink` by adding parity bits using bundle inheritance:

```
class PLink extends SimpleLink {
  val parity = Bits(5, OUTPUT)
}
```

From there we can define a filter interface by nesting two PLinks into a new `FilterIO` bundle:

```
class FilterIO extends Bundle {
  val x = new PLink().flip
  val y = new PLink()
}
```

where `flip` recursively changes the "gender" of a bundle, changing input to output and output to input.

We can now define a filter by defining a filter class extending component:

```
class Filter extends Component {
  val io = new FilterIO()
  ...
}
```

where the `io` field contains `FilterIO`.

Beyond single elements, vectors of elements form richer hierarchical interfaces. For example, in order to create a crossbar with a vector of inputs, producing a vector of outputs, and selected by a UFix input, we utilize the `Vec` constructor:

```
class CrossbarIo(n: Int) extends Bundle {
  val in  = Vec(n){ new PLink().flip() }
  val sel = UFix(ceilLog2(n), INPUT)
  val out = Vec(n){ new PLink() }
}
```

where `Vec` takes a size as the first argument and a block returning a port as the second argument.

We can now compose two filters into a filter block as follows:

```
class Block extends Component {
  val io = new FilterIO()
  val f1 = new Filter()
  val f2 = new Filter()
  f1.io.x <> io.x
  f1.io.y <> f2.io.x
  f2.io.y <> io.y
}
```

where `<>` bulk connects interfaces of opposite gender. Bulk connections connect leaf ports of the same name to each other. After all connections are made and the circuit is being elaborated, Chisel warns users if ports have other than exactly one connection.

2.5 State Elements

The simplest form of state element supported by Chisel is a positive-edge-triggered register, which can be instantiated functionally as:

```
Reg(in)
```

This circuit has an output that is a copy of the input signal `in` delayed by one clock cycle. Note that we do not have to specify the type of `Reg` as it will be automatically inferred from its input when instantiated in this way. In the current version of Chisel, clock and reset are global signals that are implicitly included where needed.

Using registers, we can quickly define a number of useful circuit constructs. For example, a rising-edge detector that takes a boolean signal in and outputs true when the current value is true and the previous value is false is given by:

```
def risingedge(x: Bool) = x && !Reg(x)
```

2.6 Conditional Updates

In the previous examples with registers, we simply wired their inputs to combinational logic blocks. When describing the operation of state elements, it is often useful to instead specify when updates to the registers will occur and to specify these updates spread across several separate statements. Chisel provides conditional update rules in the form of the `when` construct to support this style of sequential logic description. For example,

```
val r = Reg() { UFix(width = 16) }
when (c === UFix(0)) {
  r := r + UFix(1)
}
```

where register `r` is updated at the end of the current clock cycle only if `c` is zero. The argument to `when` is a predicate circuit expression that returns a `Bool`. The update block following `when` can only contain update statements using the update operator `:=`, simple expressions, and named wires defined with `val`.

In a sequence of conditional updates, the last conditional update whose condition is true takes priority. For example,

```
when (c1) { r := Bits(1) }
when (c2) { r := Bits(2) }
```

leads to `r` being updated according to the following truth table:

c1	c2	r	
0	0	r	r unchanged
0	1	2	
1	0	1	
1	1	2	c2 takes precedence over c1

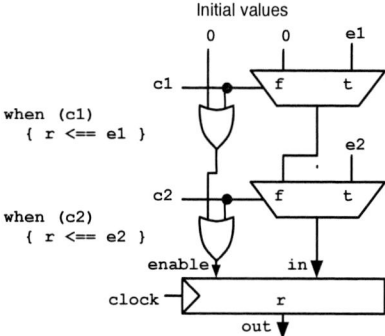

Figure 1: Equivalent hardware constructed for conditional updates.

Figure 1 shows how each conditional update can be viewed as inserting a mux before the input of a register to select either the update expression or the previous input according to the **when** predicate. In addition, the predicate is OR-ed into a firing signal that drives the load enable of the register. The compiler places initialization values at the beginning of the chain so that if no conditional updates fire in a clock cycle, the load enable of the register will be deasserted and the register value will not change.

3. ABSTRACTION

In this section we discuss abstraction within Chisel. Abstraction is an important aspect of Chisel as it 1) allows users to conveniently create reusable objects and functions, 2) allows users to define their own data types, and 3) allows users to better capture particular design patterns by writing their own domain-specific languages on top of Chisel.

3.1 Polymorphism and Parameterized Types

Scala is a strongly typed language and uses parameterized types to specify generic functions and classes. Chisel users can define their own reusable functions and classes using parameterized classes. For instance we can write a parameterized function for defining an inner-product FIR digital filter generically over Chisel Num's. The inner product FIR filter can be mathematically defined as:

$$y[t] = \sum_j w_j * x_j[t-j] \tag{1}$$

where x is the input and w is a vector of weights. In Chisel this can be defined as:

```
def innerProductFIR[T <: Num] (w: Array[Int], x: T) =
  foldR(Range(0, w.length).map(i => Num(w(i))
                                   * delay(x, i)), _ + _)

def delay[T <: Bits](x: T, n: Int): T =
  if (n == 0) x else Reg(delay(x, n - 1))

def foldR[T <: Bits] (x: Seq[T], f: (T, T) => T): T =
  if (x.length == 1) x(0) else f(x(0),
                          foldR(x.slice(1, x.length), f))
```

where **delay** creates an n-cycle delayed copy of its input and **foldR** (for "fold right") constructs a reduction circuit given a binary combiner function **f**. In this case, **foldR** creates a summation circuit. Finally, the **innerProductFIR** function is constrained to work on inputs of type **Num** for which Chisel multiplication and addition are defined.

Like parameterized functions, we can also parameterize classes to make them more reusable. For instance, we can generalize the Filter class, defined in section 2.4, to use any kind of link. We do so by parameterizing the **FilterIO** class and defining the constructor to take a zero-argument type constructor function as follows:

```
class FilterIO[T <: Data]()(type: => T) extends Bundle {
  val x = type.asInput.flip
  val y = type.asOutput
}
```

We can now define **Filter** by defining a component class that also takes a link type constructor argument and passes it through to the **FilterIO** interface constructor:

```
class Filter[T <: Data]()(type: => T) extends Component {
  val io = (new FilterIO()){ type }
  ...
}
```

3.2 Abstract Data Types

Through support for abstract data types, Chisel permits much simpler code than would otherwise be possible. For example, consider constructing a block, such as the FFT, requiring arithmetic on complex numbers. In Chisel, complex numbers can be defined as follows:

```
class Complex(val real: Fix, val imag: Fix) extends Bundle {
  def +(b: Complex): Complex =
    new Complex(real + b.real, imag + b.imag)
  ...
}
```

where we overload infix operators to provide an intuitive algebraic interface. Complex numbers can now be used in both the interface and in arithmetic:

```
class Example extends Component {
  val io = new Bundle {
    val a = new Complex(Fix(2, INPUT), Fix(2, INPUT))
    val b = new Complex(Fix(2, INPUT), Fix(2, INPUT))
    val out = new Complex(Fix(2, OUTPUT), Fix(2, OUTPUT))
  }
  val c = io.a + io.b
  io.out.r := c.r
  io.out.i := c.i
}
```

4. GENERATORS

A key motivation for embedding Chisel in Scala is to support highly parameterized circuit generators, a weakness of traditional HDLs.

4.1 Cache Generator

One example of a highly parameterized subsystem is a memory cache generator. In Chisel, the basic configuration options can first be defined:

```
object CacheParams {
  val DIR_MAPPED = 0
  val SET_ASSOC  = 1
  val WRITE_THRU  = 0
  val WRITE_BACK  = 1
}
```

The main body of the cache generator component can then be declared with desired generator parameters and optional default values. The **io** bundle then references two IO interface bundles, one specifying a connection to a CPU and the other defining the memory interface. Computed parameters are then defined, followed by the main body of the generator:

```
class Cache(cache_type: Int    = DIR_MAPPED,
            associativity: Int = 1,
            line_size: Int     = 128,
            cache_depth: Int   = 16,
            write_policy: Int  = WRITE_THRU
            ) extends Component {
  val io = new Bundle() {
    val cpu = new IoCacheToCPU()
    val mem = new IoCacheToMem().flip()
  }
  val addr_idx_width = ( log(cache_depth) / log(2) ).toInt
  val addr_off_width = ( log(line_size/32) / log(2) ).toInt
  val addr_tag_width = 32 - addr_idx_width - addr_off_width - 2
  val log2_assoc     = ( log(associativity) / log(2) ).toInt
  ...
  if (cache_type == DIR_MAPPED)
    ...
}
```

The resulting `Cache` generator can then be used in a larger system:

```
...
val data_cache = new Cache(cache_type = SET_ASSOC, line_size = 64)
connection_to_cpu <> data_cache.io.cpu
connection_to_mem <> data_cache.io.mem
...
```

4.2 Sorting Network

In addition to offering flexible parameterization, Chisel supports recursive creation of hardware subsystems. In the example below a simple sorting network is specified using a two-input `SortBlock` defined with handshaking ports. First, a simple queue IO interface data type is defined by extending the `Bundle` class. This data type will be used to define connections between the sorting primitives:

```
class IoSortBlockOut extends Bundle() {
  val output     = Bits(sort_data_size, OUTPUT)
  val output_rdy = Bool(OUTPUT)
  val has_output = Bool(OUTPUT)
  val pop        = Bool(INPUT)
}
```

The `SortBlock` primitive is then defined to output the minimum of the two inputs, subject to handshaking:

```
class SortBlock extends Component() {
  override val io = new Bundle() {
    val in1 = new IoSortBlockOut()
    val in2 = new IoSortBlockOut()
    val out = new IoSortBlockOut()
  }
  ...
}
```

Using this sorting primitive, it is then possible to define a recursive architecture to find the minimum of a vector of numbers. `SortVector` below recursively finds the minimum of the first and second halves of the input vector, and returns the minimum of the two results. This example also demonstrates the power of using `Bundle` to combine inputs and outputs along with arrays of `Bundle` using `Vec`.

```
class SortVector(in_width: Int) extends Component() {
  val io = new Bundle() {
    val in_vec = Vec(in_width) { new IoSortBlockOut().flip }
    val out    = new IoSortBlockOut()
  }
  val min1 = new SortPair()
  min1.io.out <> io.out
  val midpoint = in_width / 2
  if (in_width < 4) {
    // Connect first input directly to min1
    min1.io.in1 <> io.in_vec(0)
  } else {
    val min_first_half = new SortVector(midpoint)
    for (i <- 0 until midpoint)
      min_first_half.io.in_vec(i) <> io.in_vec(i)
```

```
    min1.io.in1 <> min_first_half.io.out
  }
  if (in_width < 3) {
    min1.io.in2 <> io.in_vec(1)
  } else {
    val min_second_half = new SortVector(in_width - midpoint)
    for (i <- midpoint until in_width)
      min_second_half.io.in_vec(i - midpoint) <> io.in_vec(i)
    min1.io.in2 <> min_second_half.io.out
  }
}
```

Note that Verilog is not able to describe this type of recursion, and a designer would need to use a different language, such as `Python`, to generate Verilog from a recursive routine.

4.3 Memory

Memories are given special treatment in Chisel since hardware implementations of memory have many variations, e.g., FPGA memories are instantiated quite differently from ASIC memories. Chisel defines a memory abstraction that can map to either simple Verilog behavioral descriptions, or to instances of memory modules that are available from external memory generators provided by foundry or IP vendors. In the simplest form, Chisel allows memory to be defined with a single write port and multiple read ports as follows:

```
Mem(depth: Int,
    target: Symbol = 'default, readLatency: Int = 0)
```

where `depth` is the number of memory locations, `target` is the type of memory used, `readLatency` is the latency of read ports to be defined on the memory. A memory object can then be read from using the `read(rdAddress)` method. For example, an audio recorder could be defined as follows:

```
def audioRecorder(n: Int) = {
  val addr = counter(UFix(n));
  val ram = Mem(n).write(button(), addr)
  ram.read(Mux(button(), UFix(0), addr))
}
```

where a counter is used as an address generator into a memory. The device records while `button` is `true`, or plays back when `false`.

We can use simple memory to create register files. For example we can make a one write port, two read port register file with 32 registers as follows:

```
val regs = Mem(32)
regs.write(wr_en, wr_addr, wr_data)
val idat = regs.read(iaddr)
val mdat = regs.read(maddr)
```

where a new read port is created for each call to read.

Additional parameters are available to mimic common memory behaviors, to aid with the process of mapping to real-world hardware. The following is an example of a memory that is first defined with no memory ports, after which read, write, or read/write ports are added:

```
val regfile =
  Mem(64, readLatency = 1,
      hexInitFile = "hex_init_values.txt");
regfile.write(addr_in, data_in1, wen, w_mask = bit_mask);
val read_data = regfile.read(addr_in);
```

By default, this memory will be compiled to Verilog RTL. To produce a reference to a Verilog instance of a memory module, one adds `target = 'inst` to the constructor call. When Chisel compiles to Verilog, a second file will be generated, e.g., `design.conf`, which can be used by the synthesis design flow to construct the requested memory objects.

5. FAST C++ SIMULATOR

Fast simulation is crucial to reduce hardware development time. Custom logic simulation engines can provide fast cycle-accurate simulation, but are generally too expensive to be used by individual designers. FPGA emulation approaches are valuable but the FPGA tool flow can take hours to map a design iteration. Conventional software Verilog RTL simulators are popular, as they can be run by individual designers on workstations or shared server farms, but are slow.

For Chisel, we have developed a fast C++ simulator for RTL debugging. The Chisel compiler produces a C++ class for each Chisel design, with a C++ interface including clock-low and clock-high methods. We rely on two techniques to accelerate execution. First, the simulator code generation strategy is based on a templated C++ multi-word bit-vector runtime library that executes all the basic Chisel operators. The C++ templates specialize operations for bit vectors using a two-level template scheme that is first parameterized on bits and then on words. In particular, all overhead is removed for the case where the RTL bit vector fits into the host machine's native word size. Second, we remove as much branching as possible in the code so that we can best utilize the ILP available in a modern processor and minimize the number of stalls.

6. RESULTS

In this section, we present preliminary results on using Chisel for various hardware designs. To measure designer productivity, we took a simple 3-stage 32-bit RISC processor that was originally hand-written in Verilog, and converted it to equivalent Chisel code. The original Verilog code was 3020 lines of code whereas the resulting Chisel code was only 1046 lines, yielding a nearly 3× reduction.

To compare quality of results, we used a set of floating-point primitive components we have designed in Chisel, including multiplication, addition, and several data conversion operators. A 64-bit Fused-Multiply-Add (FMA) unit has been mapped to both Verilog and C++ emulation code, and both results have been simulated in testbenches using SoftFloat and TestFloat [5] to verify IEEE-754-2008 compliance. The generated Verilog was mapped to a commercial 65 nm process and compared to the same design described using hand-coded Verilog, and as expected there was no significant difference in results:

Source	Clock Period	Total Area	Logic Area
Chisel	7ns	62197 um^2	60801 um^2
Verilog	7ns	62881 um^2	61485 um^2
Chisel	2.5ns, Retimed	66472 um^2	61279 um^2
Verilog	2.5ns, Retimed	67034 um^2	62227 um^2

To compare the speed of simulation using the Chisel C++ simulator, we used a more sophisticated 64-bit five-stage in-order RISC pipeline with a floating-point unit, MMU, and caches. We compared the speed of Chisel C++ simulation and Synopsys VCS Verilog simulation when booting a research OS on this processor ($88,291,350$ cycles total) with results as follows:

Simulator	Time (s)	Speedup
VCS RTL simulator	5390	1.00
Chisel C++ RTL simulator	694	7.77

The Chisel-generated C++ simulator is approximately 8× faster than VCS.

Finally, we have developed a complete 64-bit vector processor including FPUs, MMUs, and 32 K 4-way set-associative instruction and data caches. The Chisel code was used to generate an LVS and DRC-clean GDSII layout in an IBM 45 nm SOI 10-metal layer process using memory-compiler-generated 6T and 8T SRAM blocks. Total area was $1.76\,mm^2$, with a critical path of 1 ns.

7. CONCLUSION

Chisel makes the power of a modern software programming language available for hardware design, supporting high-level abstractions and parameterized generators without mandating a particular computational model, while also providing high-quality Verilog RTL output and a fast C++ simulator.

The Chisel system and hardware libraries are being made available as an open-source project available at:

http://chisel.eecs.berkeley.edu

to encourage wide adoption.

8. ACKNOWLEDGEMENTS

We'd like to thank Christopher Batten for sharing his fast multiword C++ template library that inspired our fast emulation library. We'd also like to thank all the Berkeley EECS graduate students who participated in the Chisel bootcamp and have given feedback on Chisel after using it in various classes and research projects.

9. REFERENCES

[1] BELLOWS, P., AND HUTCHINGS, B. JHDL – an HDL for reconfigurable systems. *IEEE Symposium on FPGAs for Custom Computing Machines* (1998).

[2] BERRY, G., AND GONTHIER, G. The Esterel synchronous programming language: Design, semantics, implementation. *Science of Computer Programming 10*, 2 (1992).

[3] BLUESPEC INC. *Bluespec(tm) SystemVerilog Reference Guide: Description of the Bluespec SystemVerilog Language and Libraries*. Waltham, MA, 2004.

[4] GOLDSTEIN, S., AND BUDIU, M. Fast compilation for pipelined reconfigurable fabrics. *ACM/FPGA Symposium on Field Programmable Gate Arrays* (1999).

[5] HAUSER, J. The softfloat and testfloat packages. *http://www.jhauser.us/arithmetic/index.html*.

[6] JENNING, J., AND BEUSCHER, E. Verischemelog: Verilog embedded in scheme. *Proceedings of DSL '99: The 2nd conference on Domain Specific Languages* (Oct 1999).

[7] LI, Y., AND LEESER, M. HML – a novel hardware description language and its translation to VHDL. *IEEE Transactions on Very Large Scale Integration (VLSI) Systems 8*, 1 (Oct 2000).

[8] ODERSKY, M. E. A. Scala programming language. *http://www.scala-lang.org/*.

[9] SHACHAM, O., AZIZI, O., WACHS, M., QADEER, W., ASGAR, Z., KELLEY, K., STEVENSON, J., SOLOMATNIKOV, A., FIROOZSHAHIAN, A., LEE, B., RICHARDSON, S., AND M., H. Rethinking digital design: Why design must change. *IEEE Micro* (Nov/Dec 2010).

10. SUPPLEMENTAL

In this section we give more detailed examples, results, and discussion of Chisel.

10.1 Builtin Operators

Chisel defines a set of hardware operators for the builtin types which can be found in Table 1.

10.2 Layers of Languages

Scala was designed to support the creation of embedded domain-specific languages. In fact, it is easy to create a series of languages, one layered on top of another, resulting in improved clarity and efficiency in specification. As a small example, we can easily build a `switch` statement involving a series of comparisons against a common key, based on the Chisel conditional updates introduced earlier.

As a small example, we can easily build a `switch` statement involving a series of comparisons against a common key, based on the Chisel conditional updates introduced earlier.

switch construct	translates into

`switch(idx) {`
`  is(v1) { u1 }`
`  is(v2) { u2 }`
`}`
	`when (idx === v2) { u2 }`
	`when (idx === v1) { u1 }`

The `switch` construct supports simple specification of FSMs:

```
val s_even :: s_odd :: Nil = Enum(2){ UFix() }
val state = Reg(resetVal = s_even)
switch (s.in) {
  is (s_even) { state <== s_odd  }
  is (s_odd)  { state <== s_even }
}
```

We are exploring embedding new domain-specific languages in Chisel to provide high-level behavioral synthesis.

10.3 Scala Embedding Discussion

Embedding Chisel in Scala gave a number of advantages but also presented a number of challenges.

In Scala, we are able to cleanly integrate Chisel components, bundles and interfaces with Scala classes. Using introspection, we can find all relevant fields and their names in Scala objects. Scala also provides a number of facilities for writing domain-specific languages including operator overloading.

Unfortunately, there are other areas where it is still challenging to customize the language seamlessly. The first one is providing a succinct literal format. Unfortunately, unlike Common Lisp, in Scala it is impossible to define new tokens. The second one is that, at least in standard Scala, it is impossible to overload existing syntax, such as `if` statements. In general, there is no way to extend the Scala syntax in arbitrary ways. Higher-order functions and lightweight thunks help, but the result is that the Chisel syntax is slightly more awkward than we'd ideally like.

Yet another challenge is providing informative error messages. When errors occur, it is possible to provide stack backtraces to report to users on what line number an error occurred. Unfortunately, it is challenging to filter the stack trace to give the user the exact line the error occurred.

Although Scala has a large number of data types, we are not able to completely layer our hardware data types on to these Scala ones. We instead built a parallel type hierarchy. Scala has a very powerful parameterized type system that allows us to create generic functions and classes that can

be precisely type checked. Unfortunately, the type system is not able to infer bit widths automatically, so we have to add a separate bit-width inference pass, as described below. The advantage is that our Chisel design is more portable to other host languages.

10.4 Bitwidth Inference

Users are required to set bitwidths of ports and registers, but otherwise, bit widths on wires are automatically inferred unless set manually by the user. The bit-width inference engine starts from the graph's input ports and calculates node output bit widths from their respective input bit widths according to the following set of rules:

operation	bit width
`z = x + y`	`wz = max(wx, wy) + 1`
`z = x - y`	`wz = max(wx, wy) + 1`
`z = x & y`	`wz = max(wx, wy)`
`z = Mux(c, x, y)`	`wz = max(wx, wy)`
`z = w * y`	`wz = wx + wy`
`z = x << n`	`wz = wx + maxNum(n)`
`z = x >> n`	`wz = wx - minNum(n)`
`z = Cat(x, y)`	`wz = wx + wy`
`z = Fill(n, x)`	`wz = wx * maxNum(n)`

where for instance wz is the bit width of wire z, and the & rule applies to all bitwise logical operations.

The bit-width inference process continues until no bit width changes. Except for right shifts by known constant amounts, the bit-width inference rules specify output bit widths that are never smaller than the input bit widths, and thus, output bit widths either grow or stay the same. Furthermore, the width of a register must be specified by the user either explicitly or from the bitwidth of the reset value. From these two requirements, we can show that the bit-width inference process will converge to a fixpoint.

10.5 BlackBox's

Users can create wrappers for existing opaque IP components using `BlackBoxes` which are Components with only IO and no body. For example, a Verilog-based memory controller module can be linked in by defining it as a subclass of `BlackBox`:

```
class MemoryController extends BlackBox {
  val io = new MemoryIo();
}
```

and then by instantiating it and connecting to it as done with any other Chisel component. The emitted Verilog will then contain code to create and wire in the module.

10.6 Vending Machine FSM Example

Here is an example of a vending machine FSM defined with a `switch` statement:

```
class VendingMachine extends Component {
  val io = new Bundle {
    val nickel = Bool(INPUT)
    val dime   = Bool(INPUT)
    val rdy    = Bool(OUTPUT) }
  val s_idle :: s_5 :: s_10 :: s_15 :: s_ok :: Nil = Enum(5){UFIx()}
  val state = Reg(resetVal = s_idle)
  switch (state) {
    is (s_idle) {
      when (io.nickel) { state <== s_5 }
      when (io.dime)   { state <== s_10 }
    } is (s_5) {
      when (io.nickel) { state <== s_10 }
      when (io.dime)   { state <== s_15 }
    } is (s_10) {
      when (io.nickel) { state <== s_15 }
```

1218

Example	Explanation
Bitwise operators. Valid on Bits, Fix, UFix, Bool.	
`val invertedX = ~x`	Bitwise-NOT
`val hiBits = x & Bits("h_ffff_0000")`	Bitwise-AND
`val flagsOut = flagsIn	overflow`	Bitwise-OR
`val flagsOut = flagsIn ^ toggle`	Bitwise-XOR
Bitwise reductions. Valid on Bits, Fix, and UFix. Returns Bool.	
`val allSet = andR(x)`	AND-reduction
`val anySet = orR(x)`	OR-reduction
`val parity = xorR(x)`	XOR-reduction
Equality comparison. Valid on Bits, Fix, UFix, and Bool. Returns Bool.	
`val equ = x === y`	Equality
`val neq = x != y`	Inequality
Shifts. Valid on Bits, Fix, and UFix.	
`val twoToTheX = Fix(1) << x`	Logical left shift.
`val hiBits = x >> UFix(16)`	Right shift (logical on Bits & UFix, arithmetic on Fix).
Bitfield manipulation. Valid on Bits, Fix, UFix, and Bool.	
`val xLSB = x(0)`	Extract single bit, LSB has index 0.
`val xTopNibble = x(15,12)`	Extract bit field from end to start bit position.
`val usDebt = Fill(3, Bits("hA"))`	Replicate a bit string multiple times.
`val float = Cat(sign,exponent,mantissa)`	Concatenates bit fields, with first argument on left.
Logical operations. Valid on Bools.	
`val sleep = !busy`	Logical NOT.
`val hit = tagMatch && valid`	Logical AND.
`val stall = src1busy || src2busy`	Logical OR.
`val out = Mux(sel, inTrue, inFalse)`	Two-input mux where sel is a Bool.
Arithmetic operations. Valid on Nums: Fix and UFix.	
`val sum = a + b`	Addition.
`val diff = a - b`	Subtraction.
`val prod = a * b`	Multiplication.
`val div = a / b`	Division.
`val mod = a % b`	Modulus
Arithmetic comparisons. Valid on Nums: Fix and UFix. Returns Bool.	
`val gt = a > b`	Greater than.
`val gte = a >= b`	Greater than or equal.
`val lt = a < b`	Less than.
`val lte = a <= b`	Less than or equal.

Table 1: Chisel operators on builtin data types.

```
    when (io.dime)   { state <== s_ok }
  } is (s_15) {
    when (io.nickel) { state <== s_ok }
    when (io.dime)   { state <== s_ok }
  } is (s_ok) {
    state <== s_idle
  }
}
io.rdy := (state === s_ok)
}
```

10.7 Simulation Performance

In Section 6 we compared simulation speed for a 64-bit five-stage RISC processor design using a Chisel-generated C++ simulator and Synopsys VCS Verilog simulation. Table 2 is a more complete breakdown of the results in terms of compile time, run time, and total time. We also include results for a Chisel-generated FPGA emulation, which provides the fastest per-cycle emulation performance but with a large compile time.

Because of compilation time, the fastest backend for sim-

ulation performance depends on the number of target cycles to be simulated. While the Chisel C++ emulator runs approximately 10× faster than VCS, as shown in Figure 2, this advantage is only realized when simulating millions of cycles or more. FPGA emulation is only fastest for simulations exceeding billions of target cycles. We are planning to experiment with techniques to improve the compile-time performance of the Chisel-generated C++ code, possibly with switches to optimize for compile-time or run-time.

10.8 FIFO

A generic FIFO could be defined as shown in Figure 3 and used as follows:

```
class DataBundle() extends Bundle {
  val A = UFix(width = 32);
  val B = UFix(width = 32);
}

object FifoDemo {
  def apply () = (new Fifo(32)){ new DataBundle() };
```

Simulator	Compile Time (s)	Compile Speedup	Run Time (s)	Run Speedup	Total Time (s)	Total Speedup
VCS RTL simulator	22	1.000	5368	1.00	5390	1.00
Chisel C++ RTL simulator	119	0.184	575	9.33	694	7.77
Virtex-6 FPGA	3660	0.006	76	70.60	3736	1.44

Table 2: Comparison of simulation time between Chisel C++ simulator, Synopsys VCS Verilog simulation, and FPGA emulation, on a 64-bit five-stage RISC processor running an OS boot test.

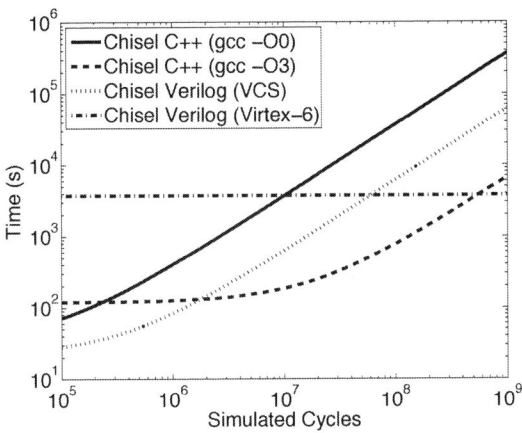

Figure 2: A comparison of total time required to compile and simulate a system using various backends from Chisel.

```
}
```

It is also possible to define a generic decoupled interface:

```
class ioDecoupled[T <: Data]()(data: => T) extends Bundle() {
  val ready = Bool(INPUT)
  val valid = Bool(OUTPUT)
  val bits  = data.asOutput
}
```

This template can then be used to add a handshaking protocol to any set of signals:

```
class decoupledDemo extends ioDecoupled()( new DataBundle() )
```

The FIFO interface in Figure 3 can be now be simplified as follows:

```
class FifoIO[T <: Data]()(gen: => T) extends Bundle() {
  val enq = new ioDecoupled()( gen ).flip()
  val deq = new ioDecoupled()( gen )
}
```

10.9 Generated Verilog

Running the Chisel compiler on the FIFO example generates the Verilog code shown in Figure 4 .

The Verilog output from Chisel might need to be simulated together with other existing Verilog IP blocks. We compared the Verilog simulation speed of the Chisel-generated Verilog versus hand-written behavioral Verilog for a 64-bit data-parallel processor design, including pipelined single and double-precision FMA units, and a pipelined 64-bit integer multiplier. We ran 92 test assembly programs on both VCS-generated simulators. The Chisel-generated Verilog simulator was $1.65\times$ slower in total than the behavioral Verilog

```
class FifoIO[T <: Data]()(gen: => T) extends Bundle() {
  val enq_val = Bool(INPUT)
  val enq_rdy = Bool(OUTPUT)
  val deq_val = Bool(OUTPUT)
  val deq_rdy = Bool(INPUT)
  val enq_dat = gen.asInput
  val deq_dat = gen.asOutput
}

class Fifo[T <: Data] (n: Int)(gen: => T) extends Component {
  val io        = new FifoIO()( gen )
  val enq_ptr   = Reg(resetVal = UFix(0, sizeof(n)))
  val deq_ptr   = Reg(resetVal = UFix(0, sizeof(n)))
  val is_full   = Reg(resetVal = Bool(false))
  val do_enq    = io.enq_rdy && io.enq_val
  val do_deq    = io.deq_rdy && io.deq_val
  val is_empty  = !is_full && (enq_ptr === deq_ptr)
  val deq_ptr_inc = deq_ptr + UFix(1)
  val enq_ptr_inc = enq_ptr + UFix(1)
  val is_full_next =
    Mux(do_enq && ~do_deq && (enq_ptr_inc === deq_ptr), Bool(true),
    Mux(do_deq && is_full,                              Bool(false),
        is_full))
  enq_ptr <== Mux(do_enq, enq_ptr_inc, enq_ptr)
  deq_ptr <== Mux(do_deq, deq_ptr_inc, deq_ptr)
  is_full <== is_full_next
  val ram = Mem(n, do_enq, enq_ptr, io.enq_dat)
  io.enq_rdy := !is_full
  io.deq_val := !is_empty
  ram.read(deq_ptr) <> io.deq_dat
}
```

Figure 3: Parameterized FIFO example.

simulator due to the low-level structural nature of the Verilog code generated by Chisel. However, we have not yet tuned the Verilog output for Verilog simulation performance, and we believe even the current slowdown is acceptable to enable co-simulation.

10.10 Chisel Components

Chisel has been in use for over a year and a number of components have been written in it. We developed the following components as part of our research infrastructure, many of which are used in the vector processor described in Section 10.11:

- clock dividers
- queues
- decoders, encoders, popcount
- scoreboards
- integer ALUs
- LFSR
- Booth multiplier, iterative divider
- ROMs, RAMs, CAMs
- TLB
- direct-mapped caches, set-associative blocking caches

1220

```
module Fifo(input clk, input reset,
    input  io_enq_val,
    output io_enq_rdy,
    output io_deq_val,
    input  io_deq_rdy,
    input [31:0] io_enq_dat_A,
    input [31:0] io_enq_dat_B,
    output[31:0] io_deq_dat_A,
    output[31:0] io_deq_dat_B);

  wire T0;
  wire is_empty;
  wire T1;
  reg[4:0] deq_ptr;
  wire[4:0] T2;
  wire[4:0] deq_ptr_inc;
  wire do_deq;
  reg[4:0] enq_ptr;
  wire[4:0] T3;
  wire[4:0] enq_ptr_inc;
  wire do_enq;
  wire T4;
  reg[0:0] is_full;
  wire is_full_next;
  wire T5;
  wire T6;
  wire T7;
  wire T8;
  wire T9;
  wire T10;
  wire T11;

  assign io_deq_val = T0;
  assign T0 = ! is_empty;
  assign is_empty = T11 && T1;
  assign T1 = enq_ptr == deq_ptr;
  assign T2 = do_deq ? deq_ptr_inc : deq_ptr;
  assign deq_ptr_inc = deq_ptr + 1'h1/* 1*/;
  assign do_deq = io_deq_rdy && io_deq_val;
  assign T3 = do_enq ? enq_ptr_inc : enq_ptr;
  assign enq_ptr_inc = enq_ptr + 1'h1/* 1*/;
  assign do_enq = io_enq_rdy && io_enq_val;
  assign io_enq_rdy = T4;
  assign T4 = ! is_full;
  assign is_full_next = T7 ? 1'h1/* 1*/ : T5;
  assign T5 = T6 ? 1'h0/* 0*/ : is_full;
  assign T6 = do_deq && is_full;
  assign T7 = T9 && T8;
  assign T8 = enq_ptr_inc == deq_ptr;
  assign T9 = do_enq && T10;
  assign T10 = ~ do_deq;
  assign T11 = ! is_full;

  always @(posedge clk) begin
    if(reset) begin
      deq_ptr <= 5'h0/* 0*/;
    end else if(1'h1/* 1*/) begin
      deq_ptr <= T2;
    end
    if(reset) begin
      enq_ptr <= 5'h0/* 0*/;
    end else if(1'h1/* 1*/) begin
      enq_ptr <= T3;
    end
    if(reset) begin
      is_full <= 1'h0/* 0*/;
    end else if(1'h1/* 1*/) begin
      is_full <= is_full_next;
    end
  end
endmodule
```

Figure 4: Verilog Generated from Chisel for the FIFO example

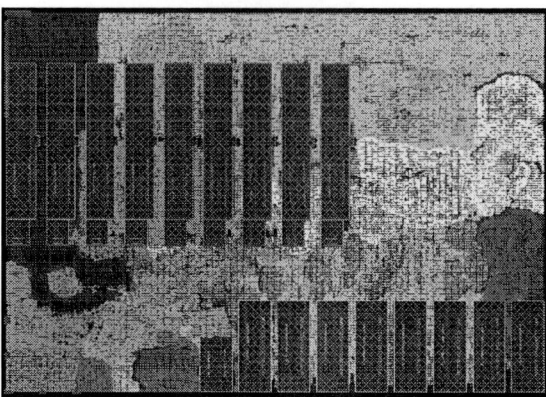

Figure 5: Data-parallel processor layout results

- direct-mapped caches, set-associative non-blocking caches
- prefetcher
- fixed-priority arbiters, round-robin arbiters
- single-precision, double-precision floating-point units
- 64-bit decoupled in-order single-issue 5-stage processor
- 64-bit vector unit (data-parallel processor)

We are working to factor these components into a standard library from which developers can more readily build large-scale designs.

We have taught a class in advanced computer architecture design where all students produced projects in Chisel. Example projects included accelerators for security, FFT, and spatial computing.

Additionally, Berkeley EECS graduate student Chris Celio is developing a number of educational processor microarchitectures with associated labs to help undergraduates learn computer architecture. These included a microcoded processor, one-stage, two-stage, and five-stage pipelines, and an out-of-order processor, all with accompanying visualizations.

10.11 Data-Parallel Processor Layout Results

The data-parallel processor layout results using IBM 45nm SOI 10-metal layer process using memory compiler generated 6T and 8T SRAM blocks are shown in Figure 5.

Specification and Synthesis of Hardware Checkpointing and Rollback Mechanisms

Carven Chan, Daniel Schwartz-Narbonne, Divjyot Sethi, Sharad Malik
Princeton University, New Jersey
{carvenc, dstwo, dsethi, sharad}@princeton.edu

ABSTRACT

The increasing pressure to make hardware resilient to runtime failures has prompted development of design techniques for specific classes of systems, e.g. processors and routers. However, these techniques come at increased design and verification costs, thus limiting their broader application. In this work we describe a methodology for general RTL designs based on the widely usable checkpointing and rollback resiliency mechanism. We take a modeling and language approach that provides an appropriate set of abstractions for the resiliency logic. This cleanly separates the main design behavior from the resiliency behavior, leading to ease of design. Further, as the language abstractions can be automatically synthesized into resiliency logic, our methodology can merge with existing design flows. The concerns of verifying this additional resiliency logic can be addressed by synthesizing behavioral assertions capturing correct behavior. We demonstrate the use of this methodology on four examples, with synthesis for performance and area to estimate the overhead of the additional synthesis logic.

Categories and Subject Descriptors
B.5.2 [**RTL Implementation**]: Design Aids—*automatic synthesis, hardware description languages, verification*

General Terms
Design, Languages, Reliability, Verification

Keywords
CpR-Verilog, backward error recovery

1. INTRODUCTION

In modern and future highly integrated hardware, faults at runtime are likely. Integration density, high error-rate post-silicon fabrics, or hazardous environments will contribute to transient errors. Process variation, device aging, and thermal breakdown contribute to permanent failures. The subtle, concurrent interactions in large designs may even result in functional bug escapes. To cite the International Technology Roadmap for Semiconductors [2]: *In general, automatic insertion of robustness into the design will become a priority as systems become too large to be functionally tested at manufacturing exit.*

Of the potential solutions, Backwards Error Recovery (BER) is widely recognized as a general method for improving system robustness. The idea of BER is to take snapshots of system states at appropriate intervals, then roll back and resume execution after an error, possibly including some recomputation. Because the error diagnosis is decoupled from the recovery, coverage of a wide variety of runtime faults is possible. BER is also lightweight, in comparison to approaches based on physical redundancy. BER implementations for designs, such as processors, have been proposed in the literature (e.g., [6]).

While BER techniques can potentially benefit many designs, there is a usability barrier that has yet to be overcome. Proposed BER techniques are customized towards classes of designs, such as processors and routers. This manual effort is time consuming and prone to additional design errors, such as interference with the intended error-free operation. Incorporating BER capability into a module involves complex interplay of behaviors in the RTL. Even for small modules, these issues become nontrivial, and to scale the manual approach to many modules becomes a daunting effort. Given increasing design complexity, this ad-hoc design approach may become prohibitively expensive.

To address these challenges, we propose a design methodology for BER that uses higher level specification and automation. Our solution has the following advantages:

Applicability. A framework (model, language, tools) to specify BER for many RTL designs.

Separation of concerns. Design code and BER code are kept separate, with the BER behavior requiring minimal if any modifications to the original design code.

Abstractions. The designer uses the abstractions to focus on the specification of BER, the error checking and functional requirements for recovery and re-computation relating to their design, and not on its implementation details.

Synthesizability. The BER specification is automatically synthesized, thus improving design productivity.

Improved verification. Correct by construction synthesis and automatically generated assertions significantly lower the verification burden.

We make the following contributions in this paper:

- Checkpoint/Recovery-Verilog (CpR-Verilog), a superset of synthesizable Verilog to specify BER for any design (Sec. 3)
- Automatic synthesis of BER logic to RTL (Sec. 4)
- A series of case studies with evaluation of effort and overhead (Sec. 5)
- Techniques for verification of the additional BER logic (Sec. 6)

First we cover the conceptual groundwork of the methodology in the next section.

2. DESIGNING FOR BACKWARD ERROR RECOVERY

The basic recomputation in BER is described as follows. Suppose that in a system we can store a known error-free state S_{Valid}, as well as the inputs fed to the system after storing this state, $I_0 \ldots I_{t-1}$. Then, we can recover from an error at time k, $k < t$ by restoring the state S_{Valid} and then re-executing the system by replaying the stored inputs $I_0 \ldots I_{t-1}$. To implement this there are five interrelated processes:

- Checkpoint: store one or more previously valid states.

- Restore: bring the system back to a previously valid state.

- Log: store a set of signal (input) values needed for recomputation.

- Replay: apply previous signal (input) values to execute recomputation.

- Checking: detect online errors within a defined latency.

Figure 1 shows a generic Verilog module with BER implemented. The state of the module is stored in variable *state* and the state update function is f. The output b is a function g of input a and the system state S. BER is implemented by maintaining a single checkpoint, *state_shadow*. A new checkpoint is stored into *state_shadow* every $CP\_LATENCY$ cycles, based on the *cp_counter* register. The error checker e asserts the *err* signal on error; when this occurs the system state is restored by copying from *state_shadow* within one clock cycle. This simple example is representative of the checkpointing of key state bits of a small module, such as the SHA-1 algorithm in the case studies (the full design utilizes logging as well).

```
module generic (input clk, input a, output b);
    reg[4:0] state, state_shadow;
    assign b = g(state, a);
    always @(posedge clk) begin
        if (!err)
            state <= f(state, a);
        else
            state <= state_shadow;
        if (cp_counter == CP_LATENCY)
            state_shadow <= state;
    end
    parameter CP_LATENCY;
    reg[4:0] cp_counter;
    always @(posedge clk)
        cp_counter <= cp_counter + 1;
    wire err = e(state, a);
endmodule;
```

Figure 1: A generic Verilog module with basic BER mechanism

BER implementations for realistic designs have a number of challenges. State and signal values may be numerous so only the relevant bits should be checkpointed or stored. The control logic, as in the example, interweaves the main design code; this reduced code legibility, in combination with timing details of the BER, allows for subtle timing errors to be easily introduced.

Meanwhile, checkers (error detection hardware) are diverse, extremely application dependent, and may cut across levels of abstraction (above or below RTL); some examples are RTL (registers), error detecting codes, functional checks (e.g. reduced precision checker). When and how faults are detected depends on the specific checker. Error checker design is already a well studied area ([13], [11]). Thus, in our current methodology, the error detection aspect of BER is managed by the designer, who can exploit understanding of both microarchitecture and application context.

In the next section, we show how a language based approach can be used to address these challenges in implementing BER.

3. LANGUAGE DESIGN

3.1 CpR-Verilog Language Abstractions

BER techniques can be implemented at a number of different abstraction levels. The specification and implementation of BER varies across different designs due to considerations such as: area and performance trade offs; concurrency i.e. I/O constraints or interaction with other BER-capable modules; error coverage and detection latency. Language formalisms help constrain this design problem, and increased analyzability and automation as a consequence can improve productivity and verification.

Our approach is to add some necessary abstractions to reduce the manual effort of writing BER hardware. To this end, CpR-Verilog consists of language primitives for specifying what state and data storage is needed for BER how they are to be used the context of a synchronous hardware design. Together with the error detection mechanisms, this should cover common cases of hardware-based BER protocols and implementations. Although this paper focuses on language extensions for the Verilog hardware description language, the basic primitives are similarly applicable to other hardware design language such as VHDL or SystemC.

CpR-Verilog augments Verilog with a small number of syntactic additions which represent the essential keywords and constructs. One of the design goals of CpR-Verilog is to maintain a separation of concerns: designers should be able to add BER to their designs with only minimal changes to the underlying design code. We accomplished this by describing CpR-Verilog actions as *guarded operations* ([7]) on the underlying RTL module.

The CpR-Verilog language has two basic primitives:

3.2 Statesets

A stateset is a named object associated with a set of bits, as well as rules defining when those bits should be checkpointed and restored.

3.2.1 Declaration

Syntactically, stateset declarations are very similar to Verilog module instantiations.

stateset #(.depth (D)) S (clk, {s0, s1, s2});

Since the error detection latency in a BER system may be greater than the checkpointing latency, it may be necessary to store multiple checkpoints [22]. The .depth (D) parameter allows the user to specify the minimum number of checkpoints that should be recorded. CpR-Verilog is synchronous, so clk defines the clock that should be used to control the checkpointing system. S is the name of the stateset, and s0, s1, s2 are the RTL variables that are tracked by this particular stateset.

An RTL design may have multiple statesets, representing different logical sets of state bits that have different checkpointing and recovery conditions. For example, in a hyper-threaded processor, there might be a stateset and checkpoint associated with each active instruction. Some bits may not be tracked by any stateset. In a processor, for example, cache lines may not need to be checkpointed since the cache can be rebuilt from main memory.

3.2.2 Checkpointing

Checkpointing is accomplished using guarded actions. A checkpoint consists of the values of the stateset named in the operation argument, at the time that the operation was triggered. For each stateset, up to "depth" checkpoints are stored, with new checkpoints overwriting the least recent checkpoint.

checkpoint (S, p);

means to checkpoint the stateset S iff boolean condition p is true.

Since a checkpointing operation may require multiple clock cycles, we provide a signal

checkpoint_done (S)

which can be queried to check if the checkpointing operation has completed.

3.2.3 Restore

Restoring values is also accomplished using guarded actions. Rollback recovery schemes can require multiple checkpoints ([22]); therefore, the restore operation takes a third argument, an index i specifying how many checkpoints ago to restore to. All saved checkpoints more recent than the restored point are discarded.

restore (S, q, i)

restores variables tracked by S to the values they had when the checkpoint i was taken iff boolean condition q is true. restore_done (S) is set to true when the operation has completed. If i is out of range, the unreachable_history (S) signal will be raised.

3.3 Buffersets

1223

A bufferset is a named object associated with a set of signals, which logs those signals for future replay.

3.3.1 Declaration
Buffersets are declared using a syntax similar to that of Verilog modules.

    bufferset #(.depth (D)) Y (clk,{y0, y1, y2});

.depth(D) records the minimum number of cycles of signals the buffer must store. clk specifies which clock the buffer should be synchronized to. Y is the name of the bufferset. y0, y1, y2 are the signals tracked by this particular bufferset.

3.3.2 Logging
Logging is enabled using the guarded action:

    buffer (Y, p);

which logs the variables associated with Y if and only if boolean condition p is true. Depending on the I/O protocols, it may not be necessary to log signals every cycle. For example, a memory controller might hold the address constant until the action had completed. The boolean guard p allows for selective logging in these cases, reducing overhead.

Buffersets are finite, so whenever a bufferset is full, old entries are discarded to make room for new values.

3.3.3 Replay
Unlike for statesets, where checkpoints are not closely related to each other, buffersets store sequences of values that may require multiple operations to effect the necessary recomputation. To this end we provide two distinct recovery operations for buffersets.

Rollback specifies how far back to replay from.

    rollback (Y, q, i);

If the guard q is true, bufferset Y will set to replay from i entries ago. The language provides a useful index, cprv_max_index, for rolling back to the furthest possible, i.e., least recent, entry.

If i exceeds the number of valid entries, the

    unreachable_history (Y)

signal will be set to true.

Unbuffer. Values recorded in the bufferset are replayed using the unbuffer operation:

    unbuffer (Y, q);

A single entry from the bufferset is restored to the set of signals in Y.

The buffer operation may be concurrent with both of these replay operations, so that inputs arriving in the event of an error recovery are still logged. Naturally, if new inputs are being buffered faster than the buffer is being cleared, the buffer might overflow, thus:

    buffer_stall (Y);

is a signal that can be used to check for or block further buffering. This does not occur dring regular execution, when older entries are always overwritten by new values.

The replay portion of recovery is typically completed once the relevant logs are empty. Thus the language provides a control signal replay_done (Y), to indicate that the bufferset Y has no further entries to unbuffer.

3.4 Semantics
More precisely, we can define the semantics of all the above primitives in terms of pointer-based indexed memories, each being utilized in a way similar to a finite circular queue (i.e., the head and tail pointers h, r in Fig. 2). The following sections describe the state changes that occur.

3.4.1 Statesets
Each stateset S with checkpointing condition p and restore condition q is associated with some set of registers $\mathcal{R}$. A stateset can be represented as a queue as in Fig. 2a, with width equal to the total bitwidth of $\mathcal{R}$, and depth as specified by the designer in the stateset declaration. Whenever p is true, $\mathcal{R}$ is shifted left onto the stateset, potentially shifting off the oldest existing checkpoint. The signal checkpoint_done (S) will be set to false while this is occurring, and true otherwise.

During regular computation, q is false. Once q is true in the event of a restore operation, the registers of $\mathcal{R}$ are assigned the

values in the ith entry of the stateset register. The stateset register is shifted right i entries.

3.4.2 Buffersets
Each bufferset B with buffer condition p, rollback condition q and unbuffer condition u is associated with some set of signals $\mathcal{S}$.

A bufferset can be in one of two modes. In *normal* mode, the bufferset will enqueue $\mathcal{S}$ whenever p is true, shifting off the oldest values if necessary. Normal mode continues until a rollback command is received.

When a rollback(index) command is received, the replay pointer, r, is set to the index, and *recovery mode* is entered. A bufferset in recovery mode will continue to enqueue signals whenever p is true, but will now raise an error if its shift register becomes full, since this represents potential data loss.

Every signal in $\mathcal{S}$ is associated with a multiplexer. When the bufferset is in normal mode, the signal takes its normal values. When the bufferset is in recovery mode, the signal takes the value given by $Q[r]$.

Each unbuffer command increments r, updating all relevant signals in the design. Once r points to the end of the bufferset, all data has been replayed, and the bufferset returns to normal mode.

3.5 Running example

```
module generic (input clk, input a, output b);
    reg[4:0] state;
    assign b = g(state, a);
    always @(posedge clk)
        state <= f(state, a);

    stateset (.depth (1)) S (clk, {state});
    parameter CP_LATENCY;
    checkpoint (S, cp_counter == CP_LATENCY);
    reg[4:0] cp_counter;
    always @(posedge clk)
        cp_counter <= cp_counter + 1;
    wire err = e(state, a);
    restore (S, err, 1);
endmodule
```

Figure 3: Example with BER specified using CpR-Verilog.

Figure 3 shows our previous example implemented using CpR-Verilog. The manual checkpointing in Figure 1 is replaced with the stateset S and operation primitives. As a result the code is better organized, with the code for the main design separate from that for error detection, checkpointing, and recovery.

During normal operation the main design's state update (*state*) proceeds in parallel with periodic checkpoint operations. Here we assume the checkpoint_done signal is not needed because the checkpoint is small and requires one clock cycle. When an error is detected the hardware state is restored from the most recent checkpoint entry, again taking place in one clock cycle. In the SHA-1 case study, this BER implementation pattern is used to verify and correct key state registers, such as the SHA digest value, over the course of computation.

4. SYNTHESIS
In this section we show how Verilog can automatically be synthesized from CpR-Verilog. This automated synthesis enables integration of CpR-Verilog into the standard design flow. Moreover, it is feasible because we chose the set of abstractions in CpR-Verilog to be close to that of Verilog.

The synthesis algorithm needs to perform the following tasks:

- Synthesize Statesets: Each stateset S is synthesized by instantiating a Verilog module, Q_S, which represents a queue with size equal to depth. The I/O to these modules then represents the operations on this stateset.

1224

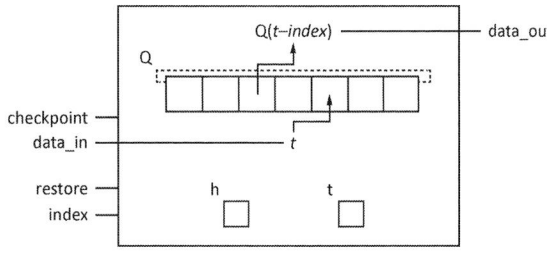

(a) Stateset template

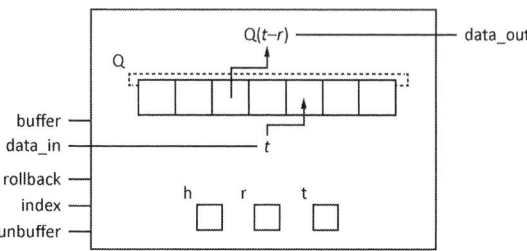

(b) Bufferset template

Figure 2: RTL-level reference models for semantics of the stateset and bufferset language primitives, consisting of external interface signals and internal indexed memory state using up to three pointers.

- Synthesize Buffersets: The synthesis of a Bufferset is similar to that of the stateset, with the key difference being in the logic to enable the replay operation.

- Interface with Design: Finally, the control logic for enabling data flow between the stateset modules and the system state needs to be synthesized. This logic enables checkpointing, buffering, restoration and replay.

4.1 Synthesis Algorithm

This algorithm rewrites a CpR-Verilog design into standard Verilog. As with conventional synthesis tools, the Verilog portions of the input design are assumed to follow the IEEE standard for RTL synthesis [1]. The notation is as follows:

- Q denotes an instance of the RTL module for a stateset or bufferset, with the module implementing the queue-based semantics of these two primitives.

- M.id refers to the port named id in module M.

- := refers to a net connection; the algorithm adds or replaces connections by generating either behavioral or structural Verilog.

- *concatenation* and *partselect* refer to generating the AST nodes for the corresponding Verilog operations.

- op(S, g, i) denotes an operation op on some stateset or bufferset S, guarded by g, optionally specifying an index i

First, the syntax tree is analyzed to compute instantiation parameters, bit-width requirements for new variables, and stateset/bufferset symbol information. Then, the translation to RTL Verlog is as in the following pseudocode:

let D denote the design represented as a CpR-Verilog AST
for each stateset or bufferset S in D **do**
  instantiate a stateset or bufferset module Q_S
  Q_S.data_in := concatenation of S.variables
  for each variable v in S.variables **do**
    declare v_{shadow}
    instantiate multiplexer m_v
    v_{shadow} := partselect of Q_S.data_out // retrieve v's value
    m_v.in1 := v
    m_v.in2 := v_{shadow}
    for each statement that updates v (v := expr) **do**
      rewrite as v := m_v.out
  for op(S, g, i) in D **do**
    Q_S.op := g
    if op $\in \{$checkpoint, restore$\}$ **then**
      m_v.sel := g
    if op $\in \{$restore, rollback$\}$ **then**
      Q_S.index := i
  output D

The first top-level loop generates circular queues for each stateset and bufferset. The nested loops generate the datapath needed

for each variable. Although Verilog contains so-called "behavioral" code, the translation can be done because the synthesizable subset of Verilog itself has well-defined semantics separating state and variant representations of combinational logic. The second top-level loop generates code for the operations.

This algorithm gives a reference implementation based on the language semantics; in particular, the queue semantics for statesets and buffersets can be directly implemented into RTL modules. Thus, in our synthesis algorithm we use Verilog module definitions for statesets and buffersets (as per Fig. 2). This reference algorithm is useful for implementing BER in small designs, or for functional verification of BER in more complex designs. On the other hand, optimizations for performance or area are still possible, such as pipelining or transferring checkpoints to main memory instead of dedicated registers.

4.2 Synthesis Correctness

A validated synthesis algorithm enables correct-by-construction design. This entails formally proving that the synthesized Verilog is a refinement of CpR-Verilog specification. We outline the proof as follows; the full proof is future work:

- The synthesized stateset and bufferset modules obey the queue semantics, with corresponding state transitions and output behavior. Statesets and buffersets are implemented using Verilog RTL templates, which can be proven to implement the CpR-Verilog semantics using formal techniques such as model checking or theorem proving. Since this is a one-time check, the amortized cost is acceptable.

- The synthesized control logic corresponds to the guarded operations. This can be verified by having the synthesis algorithm generate PSL (Property Specification Language) assertions which can be checked using existing formal or semi-formal techniques.

- The synthesized datapath correctly implements the dynamic changes to the states and signals in the main design. This can be checked using similar PSL assertions.

5. EVALUATION

In this paper the goal of the experiments is to use a series of small modules to illustrate the common case issues in implementing BER. We implemented BER using CpR-Verilog for four designs with openly available IP cores with source code: a multicycle FPU module, a SHA-1 module, and a pipelined FPU multiplier (all from [3]), and an SDRAM controller from [4]. Details on the design and BER implementation for each case study are given in Appendix A. Generally, deriving a BER specification requires design decisions. The main computation requires considerations such as choosing the variables to represent system state, whether and how to replay the computation, microarchitectural issues (pipelining), and environmental issues (input/output latency or stalling, or consistency and coordination with other modules). The latency from error to checker influences the precise timing of the checkpointing and recovery mechanisms, and ultimately defines the fault coverage. For each example module we have implemented one particular BER behavior, but other BER protocols are possible depending on application context. To evaluate our approach we use code metrics and gate-level synthesis results.

5.1 Code Comparison

The checking, checkpointing, and recovery was specified in a single section of code within the total design, which helps separate the intended error-free mode and recovery mode of operation. From Table 1, the programming of the checkpointing and recovery mechanism presents up to 8% lines of code overhead. The overhead presented by lines of error detection code by the checker modules is at most 9%. The total lines of code overhead is the sum of the two ratios in each column (= BER/Base). However, such metrics present a partial picture, as the bigger productivity benefit is due to CpR-Verilog providing better code organization and clarity.

Design	Base	BER	%CpR	%Checker
fpu_double	1500	21	.006	.008
fpu_mul	240	41	.08	.09
sha	500	25	.04	.01
sdrm	900	29	.02	.01

Table 1: Lines of code comparison.

In Table 2 shows a measure of productivity benefit. A programmer can add *CpR* lines of specification code, which are automatically synthesized into *BER* lines of implementation code. In a manual design, all *BER* lines would have to be added manually, so a CpR-Verilog programmer saves *Ratio* design effort. This combination of code reuse, standardization, and code generation allows the designer to focus on higher level issues of design correctness. (The Base code in this table is synthesized Verilog from the source Verilog in the previous table).

Design	Base	CpR	Synthesized	BER	Ratio
fpu_double	1755	21	1998	210	10
fpu_mul	574	41	855	178	4.3
sha	430	25	711	250	10
sdrm	1710	29	1990	239	8.2

Table 2: Code expansion by synthesis.

5.2 BER Cost

Our initial results, showing the cost of implementing BER in our reference designs, are shown in Table 3. We expect that a direct comparison against hand-implemented BER designs would be close to identical, because the synthesis algorithm for these cases produces BER control and datapaths similar to that of hand-designed versions.

Design	Verilog	CpR-Verilog	Ratio
fpu_double	92.0	94.0	1.02
fpu_mul	54.4	62.4	1.15
sha	13.3	17.9	1.35
sdrm	0.19	0.25	1.29

Table 3: Synthesized areas in 1k tech library (lsi_10k) units, of original versus BER-enabled designs.

Although we restricted our evaluation to small modules, this set was sufficient to exercise and test the language constructs. It is conceivable to use such small modules enhanced with BER, integrating them within larger designs. More generally, further

design experiments will help us better understand the scalability of this approach.

6. APPLICATIONS IN VERIFICATION

Verifying highly integrated fault tolerant systems is a hard problem ([2], [15]), given the intricacies in specifying the state variables, checker latencies, rollback indices, or any part of the re-computation. In this section we propose how the information available in the CpR-Verilog model can potentially be used to simplify verification of such systems. Specifically, we describe assertions which would previously require manual annotation which can now be generated automatically from a CpR-Verilog model. These assertions can then be checked using existing debugging and verification tools.

We classify three main categories of BER design errors which our methodology can help detect:

- *Design errors mapping to semantics violations:* Is the user design semantically valid CpR-Verilog?

- *Design errors mapping to incorrect usage:* Does the design represent a valid BER protocol?

- *Functional errors:* Does the BER protocol ensure the required high-level robustness properties?

Semantics Violations Given the language semantics of CpR-Verilog, it is possible to automatically generate assertions that check basic operation behavior, such as:

- Using Unavailable Checkpoints: The restore/replay operations might erroneously try to recover to a nonexistent entry of state/signal values. These errors map to an out of bound index value for statesets/buffersets, and are easily detected.

- Buffer Overflow/Underflow: A bufferset will continue to log input values in recovery mode. If the bufferset depth is too small, and the recovery computation too slow, it is possible that the buffer might overflow before the old data can be processed. This property can be detected with an assertion on the replay (r) pointers underlying each bufferset.

Incorrect Usage A design can be semantically correct, and yet not implement a valid BER protocol. BER protocols tend to share common features that can be checked for automatically, because the various operations are explicit, analyzable primitives in the language.

For statesets, it is possible that checkpointing and recovery operations may conflict in different ways. A simultaneous checkpoint and recovery is not a semantics violation but a potential conflict (an ambiguity in design intent) that affects how the BER protocol indexes into future checkpoints. With the language such a case becomes easy to detect: a) checkpoints occurring after an error, up to and including the next restore, are potentially invalid, and b) any subsequent restore operation that uses these should raise a warning. Other examples include the semantics involving multiple (nondisjoint) statesets; the guarded operation semantics may help with reasoning about the consistency between different pieces of checkpoint data.

More generally, it becomes possible to analyze the BER protocol in terms of sequences and interleavings of the different stateset and bufferset operations.

Functional Errors Finally, it is possible that a design has a valid BER protocol that fails to guarantee the required system resiliency properties, perhaps because the designer neglected to protect important state. This can be tested using functional verification techniques. With CpR-Verilog, functional verification can be broken down into two cases. The error-free case must show that all the BER processes do not interfere with the main computation. In the erroneous case the obligation is to show that the system is guided back by the BER operations to a consistent state.

7. RELATED WORK

Two prior efforts are closely related to our work in attempting to develop a methodology for adding BER to hardware modules. Micro-rollback [21] is aimed at the problem of the latency of online checkers. It proposes new register and register file designs for reverting a module's state by a small number of clock cycles. In [10] a gate-level (netlist) point of entry is used for specifying checkpointed registers, and three FPGA-specific automated synthesis implementations are proposed with trade-offs in performance overhead versus FPGA resource usage. Our work differs from the above in four key ways: the CpR-Verilog synthesis is targeted at general designs (e.g. ASICs); we use a language approach to formally describe checkpointing and logging behaviors; the point of design entry is RTL rather than gate level, because it allows for better human reasoning about details of concurrency and also enables functional simulation of BER behavior at the source level; the language allows us to provide support for verification through the use of assertions.

The area of microprocessor design has explored a great variety of BER schemes, with differing checking and recovery mechanisms and goals [6]. Recent examples include an end-to-end checker with pipeline flush and restart [5]; timing skew checks at the gate level with recovery of individual pipeline stages [9], and online circuit testing with rollback-capable registers and caches [19]. In the schemes for computer architectures, the computation and checkpoint states are often shared, and in some designs [9] the error checking and recomputation are closely intertwined. One perspective is to view these diverse schemes as a collection of optimizations, some of which could be described or synthesized using the CpR-Verilog primitives; further investigation is needed. Meanwhile our proposed methodology is suitable for typical cases of designs using ASIC methodology.

There is a large body of research on the challenges of implementing error recovery at the system level ([18], [20], [8]), where multiple processors, network nodes, and both hardware and software are involved. In relation to our work, multiple levels of recovery can coexist, each level providing redundancy or covering different error classes ([18]). At even higher levels, backward error recovery has also been utilized in operating systems and software, particularly in scientific applications [17]. Our approach is complementary with the above methods. As chips continue to increase in density and complexity, it behooves designers to include principles and benefits of dynamic verification in hardware, with the aid of models and automation of synthesis and verification.

Hardware checkpointing, logging, and rollback have applications outside the domain of resiliency. The work in [14] uses these as part of speculative execution in high-performance computing; in [16], design-specific hardware logging is manually added for post-silicon fault diagnosis. There may be utility in a methodology that unifies these separate problem areas with shared modeling and language.

8. CONCLUSION

In this paper we take a domain-specific modeling approach, extending the Verilog hardware description language with a small, simple set of abstractions for describing backward error recovery mechanisms. We implemented a reference synthesis and evaluated four designs, demonstrating the cost of using BER. The advantages of a disciplined methodology using language primitives are threefold: improved programming productivity, compatibility with existing RTL flows, and easier design verification of the BER logic via assertions.

This initial framework opens up avenues for future work, including: scaling to larger examples, synthesis optimizations, evaluation using fault injection simulation, automatic generation of the verification assertions, and incorporating techniques beyond BER into the language. Meanwhile, the flow that described here is of immediate practical application.

9. ACKNOWLEDGMENTS

This work is fully supported by a grant from the Semiconductor Research Corporation (SRC Task: 1813.001).

10. REFERENCES

[1] IEEE standard for verilog register transfer level synthesis. *IEEE Std 1364.1-2002*, 2002.

[2] International Technology Roadmap for Semiconductors. http://www.itrs.net/, Dec. 2011.

[3] OpenCores. http://www.opencores.org/, Dec. 2011.

[4] Xilinx. http://www.xilinx.com/, Dec. 2011.

[5] T. M. Austin. DIVA: a reliable substrate for deep submicron microarchitecture design. In *MICRO-32.*, pages 196–207. IEEE, 1999.

[6] N. S. Bowen and D. K. Pradham. Processor- and memory-based checkpoint and rollback recovery. *Computer*, 26(2):22–31, Feb. 1993.

[7] E. W. Dijkstra. Guarded commands, nondeterminacy and formal derivation of programs. *Commun. ACM*, Aug. 1975.

[8] E. N. M. Elnozahy, L. Alvisi, Y. Wang, and D. B. Johnson. A survey of rollback-recovery protocols in message-passing systems. *ACM Computing Surveys*, 34:375–408, Sept. 2002.

[9] D. Ernst, N. S. Kim, S. Das, S. Pant, R. Rao, T. Pham, C. Ziesler, D. Blaauw, T. Austin, K. Flautner, and T. Mudge. Razor: a low-power pipeline based on circuit-level timing speculation. In *MICRO-36.*, Dec. 2003.

[10] D. Koch, C. Haubelt, and J. Teich. Efficient hardware checkpointing. page 188. ACM Press, 2007.

[11] I. Lee, M. Basoglu, M. Sullivan, D. H. Yoon, L. Kaplan, and M. Erez. Survey of error and fault detection mechanisms. Technical Report TR-LPH-2011-002, The University of Texas at Austin, April 2011.

[12] J. Lo. Reliable floating-point arithmetic algorithms for error-coded operands. *IEEE Transactions on Computers*, 43(4):400–412, Apr. 1994.

[13] A. Mahmood and E. J. McCluskey. Concurrent error detection using watchdog processors-a survey. *IEEE Trans. Comput.*, 37:160–174, February 1988.

[14] J. Martinez, J. Renau, M. Huang, and M. Prvulovic. Cherry: Checkpointed early resource recycling in out-of-order microprocessors. In *MICRO-35*, 2002.

[15] V. P. Nelson. Fault-tolerant computing: fundamental concepts. *Computer*, 23(7):19–25, July 1990.

[16] S.-B. Park and S. Mitra. IFRA: Instruction footprint recording and analysis for post-silicon bug localization in processors. In *DAC 2008.*, 2008.

[17] J. S. Plank, M. Beck, G. Kingsley, and K. Li. Libckpt: transparent checkpointing under unix. In *Proceedings of the USENIX 1995 Technical Conference Proceedings*, TCON'95, Berkeley, CA, USA, 1995. USENIX Association.

[18] M. Prvulovic, Z. Zhang, and J. Torrellas. ReVive: cost-effective architectural support for rollback recovery in shared-memory multiprocessors. In *ISCA, 2002*. IEEE.

[19] S. Shyam, K. Constantinides, S. Phadke, V. Bertacco, and T. Austin. Ultra low-cost defect protection for microprocessor pipelines. *SIGARCH Comput. Archit. News*, 34(5):73–82, Oct. 2006.

[20] D. J. Sorin, M. M. Martin, M. D. Hill, and D. A. Wood. SafetyNet: improving the availability of shared memory multiprocessors with global checkpoint/recovery. In *29th Annual International Symposium on Computer Architecture, 2002*, pages 123–134. IEEE, 2002.

[21] Y. Tamir, M. Tremblay, and D. A. Rennels. The implementation and application of micro rollback in fault-tolerant VLSI systems. In , *Eighteenth International Symposium on Fault-Tolerant Computing, 1988. FTCS-18, Digest of Papers*, pages 234–239. IEEE, June 1988.

[22] R. Teodorescu, J. Nakano, and J. Torrellas. SWICH: a prototype for efficient Cache-Level checkpointing and rollback. *IEEE Micro*, 26(5):28–40, Oct. 2006.

APPENDIX

A. DESIGN EXAMPLE DETAILS

In each of the sections below we describe the original functionality and the rationale for modifications to provide BER.

A.1 FPU module

A.1.1 Description

This is a 64-bit, nonpipelined floating point unit from [3], supporting four basic arithmetic operations. An *enable* input signal increments the computation by a single step, and a *ready* output signal goes high after a fixed number of steps, e.g., 20 steps for addition and 71 steps for division.

A.1.2 BER Design

Error Checking Direct checking of floating point arithmetic is an open problem [12], but we here demonstrate the use of two simple checks: standard IEEE FPU exceptions, and a timeout counter to check that the arithmetic computation finishes within 20–72 steps.

Recovery Specification In this example we suppose that the intended BER behavior is to simply re-do the current floating-point operation. As mentioned previously this is one of several possible BER mechanisms; e.g. alternative checks and external requirements could warrant recomputing multiple FPU operations.

Recovery Implementation The language primitives immediately suggest two ways to specify the implementation: either buffer the input operations, or take a snapshot of the registers during the initial steps of execution.

The abstractions help the designer to focus on correctness issues, including:

1. Avoid infinitely looping into recomputation for a single set of FPU operands.

2. Do not take checkpoints or perform buffering during recovery.

3. After recomputation completes, the module is either idle or starting a new input operation.

4. The output ready signal should be low until possibly the last clock cycle of recovery.

5. In the original design, it is possible to override/interrupt the current computation with a new set of inputs. The modified version should preserve this behavior.

A.2 SHA-1 module

A.2.1 Description

This Secure Hash Algorithm module [3] accepts a starting command followed by 16 message words. The output is ready after 81 clock cycles, at which time the design's busy signal goes low. To implement the algorithm, the module maintains a digest value, held in a register that is updated and persists across computations.

A.2.2 BER Design

Error Checking Again we demonstrate only some simple concurrent checking mechanisms. We added a state transition check i.e, that certain counters (`round` and `read_counter`) increment every cycle during computation. We also added a checker that replicated and compared some parts of the hash function logic.

Recovery Specification The desired BER behavior is to revert the SHA module's state, including the digest register, to that at the beginning of the current computation, and then recompute using the previously logged input data.

Recovery Implementation Implementing BER requires both checkpointing and input logging. Because the module interface at the signal level does not distinguish between data words and null input, the buffering logic introduces a 16-cycle counter to hold the buffer operation guard high during the 16 consecutive cycles of input. This is a simple example of the kind of design-dependent control needed to implement BER.

A.3 SDRAM controller

A.3.1 Description

This module translates memory operations to low-level signals interfacing with physical memory. Internally it consists of several counters and a finite state machine, to generate properly timed signals.

A.3.2 BER Design

Error Checking For demonstration our checker is only a simple timeout check on the controller state, checking the behavior of a single memory access. It is still possible to implement BER across multiple memory accesses, but then the coordination with client modules initiating the memory requests becomes more complex.

Recovery Specification As with the previous designs, the overall strategy is to provide checkpointing and logging for each incoming memory access so as to enable recomputation after error detection.

Recovery Implementation The SDRAM state machine is checkpointed once at the beginning of every memory access. Meanwhile, inputs are buffered, consisting of a command word and any data words accompanying a write command. This requires a counter to determine the guard for the buffer operation. Recovery consists of restoring the controller state machine and reading the inputs from the buffer contents. In this design the `replay_done` language primitive is used to check that the last value has been read from the bufferset.

Again we can list several of the behavioral properties to be verified in terms of BER operation:

1. Buffer operations may be concurrent with unbuffer

2. The correct (dynamic) index was used for buffer rollback

3. Preservation of the SDRAM autorefresh behavior

4. The SDRAM output status signal is consistent with recovery phases

A.4 Pipelined FPU multiplier

A.4.1 Description

This is a pipelined FPU multiplier implementation supporting all four rounding modes, from [3]. The computation latency is 21 clock cycles. An `enable` signal single-steps the pipeline flow.

A.4.2 BER Design

Error Checking Here we use a simple checker with limited coverage: replicates 11 of the combinational logic functions used in the pipeline. Thus the checker latency is 0 cycles but only manifests at a particular pipeline stage.

Recovery Specification This example illustrates application of BER for pipelined microarchitectures. The difficulty with using checkpoints is that registers change frequently due to pipeline flow. Thus a more economical solution is to use the bufferset primitive to save as much of the input sequence as required for recomputation.

Recovery Implementation Upon error, the pipeline must flush and redo some of the ongoing computations. With respect to the environment, all new inputs are stalled (although buffering them is an alternative), and the output valid signal is held low until the recomputed values reach the end of the pipeline. These behaviors are carefully coordinated, for example using counters.

Optimizing Memory Hierarchy Allocation with Loop Transformations for High-Level Synthesis

Jason Cong, Peng Zhang, Yi Zou
Computer Science Department
University of California, Los Angeles
Los Angeles, CA 90095, USA

{cong, pengzh, zouyi}@cs.ucla.edu

ABSTRACT

For the majority of computation-intensive application systems, off-chip memory bandwidth is a critical bottleneck for both performance and power consumption. The efficient utilization of limited on-chip memory resources plays a vital role in reducing the off-chip memory accesses. This paper presents an efficient approach for optimizing the on-chip memory allocation by loop transformations in the imperfectly nested loops. We analytically model the on-chip buffer size and off-chip bandwidth after affine loop transformation, loop fusion/distribution and code motion. Branch-and-bound and knapsack reuse techniques are proposed to reduce the computation complexity in finding optimal solutions. Experimental results show that our scheme can save 40% of on-chip memory size with the same bandwidth consumption compared to the previous approaches.

Categories and Subject Descriptors: B.5.2 [**Hardware**]: Design Aids – *optimization*

General Terms: Algorithms, Design, Experimentation

Keywords: High-Level Synthesis, Loop Transformation, Memory Hierarchy Optimization, Data Reuse

1. INTRODUCTION

Off-chip memory bandwidth is a dominant bottleneck for performance and power consumption in digital hardware systems. On-chip memories have sufficient bandwidth but limited sizes due to implementation cost [1]. Allocating a portion of large arrays in on-chip buffers has proven to be an efficient technique to reduce the off-chip memory accesses in digital system designs. High-level synthesis tools enable these optimizations to be performed at the C-code level by traditional compiler techniques. A great deal of attention has been paid over the past two decades to optimizing the off-chip memory bandwidth by improving data reuse and locality [2-25]. The research can be classified into two categories.

The work in the first category focuses on improving data locality and date reuse by code transformation, especially loop transformation [2-11]. By changing the accessing order of array references in the loop nests, the co-located references become

This work was supported in part by the Semiconductor Research Corporation under Contract 2009-TJ-1879, the National Science Foundation under the Expeditions in Computing Program CCF-0926127, and Gigascale Systems Research Center (GSRC).

temporally "closer", which means a smaller data reuse buffer. Specific loop transformations, such as loop interchange, loop skewing, loop merging and loop tiling, were studied one by one in [2, 3], including the feasibility and profitability of the transformations, and the sequential combination of them to form complex transformations. But in practice, it is much harder to analytically model the sequential combinations of these loop transformations. Polyhedral-based loop transformation is widely used to unify the combination of a sequence of specific loop transformations into one single affine transformation matrix [4-9]. The pioneering work [4, 5] used unimodular transformation matrices for a unified representation of loop interchange, loop reversal and loop skewing transformations. To support more general transformations and objectives, affine transformation frameworks were established based on parametric integer linear programming [6, 7]. Data dependence and transformation legality constraints are expressed with a polyhedral model in a linear form. To improve data locality, iteration distances between dependent array instances are formulated in the objective function. Beside affine transformation, loop fusion/distribution, code motion and tiling for imperfectly nests loops have also been studied in recent work [8-11]. However, these models [2-11] use simple platform-independent objective functions, which cannot accurately model the impact of memory hierarchy allocation in hardware synthesis. For example, memory accesses to on-chip and off-chip memories will have significantly different cost models.

The work in the second category optimizes the allocation of the reuse buffers in the memory hierarchy for a fixed loop order. The data transfer and storage exploration (DTSE) methodology [1, 12] established an integrated design flow for the memory hierarchy optimization for customized memory systems. The optimization flow first analyzes the data reuse graph which presents all the possible data reuse buffer candidates of the array references at each loop level in the source program [13, 14]. Then, heuristics based on reuse buffer size and bandwidth reduction are applied to decide the allocation of the reuse candidates and their memory hierarchy [15-18]. In contrast to the heuristic approaches, an optimal allocation was proposed by formulating the problem into a mixed linear programming optimization problem [19]. For all these hierarchy allocation approaches, an independent loop transformation preprocessing is assumed to optimize data locality. The final result of memory hierarchy optimization may be greatly affected by this preprocessing.

Recently, researchers noticed the importance of considering platform-dependent cost modeling in optimizing the loop transformation [20]. Loop transformation and memory hierarchy allocation are loosely coupled by introducing fast hierarchical memory size estimators [21, 22] to evaluate the promising transformations. But the search process lacks analytic model for guidance which makes it inefficient to search a large transformation space. Other researchers use analytic optimization formulations to optimize the loop tiling parameters and memory hierarchy allocation simultaneously [23, 24]. Their formulations are solved using non-linear optimization, such as sequential quadratic programming [23] and geometric programming [24]. However, these schemes [23, 24] still need affine transformations as a preprocessing procedure to improve data locality and enable tiling. Different from previous approaches, the recent study in [25] combines affine transformation and hierarchy allocation in an analytic and systematic way. A model-guided searching-based approach is used to find the optimal affine loop transformation and hierarchy allocation. But the work is limited to perfectly nested loops, so it is not applicable to real applications.

In this paper we propose an efficient approach for optimizing the on-chip memory allocation with loop transformations for imperfectly nested loops. The contributions of this work are:

- We propose an analytical modeling of hierarchy allocation problem with loop transformations for imperfectly nested loops. Buffer size is calculated after loop transformations such as affine loop transformations, loop fusing/distribution and code motion.
- We develop an efficient and optimal solution to the combined problem, which uses the branch-and-bound approach to prune the sub-optimal transformation space and the knapsack reuse technique to reduce the complexity of each transformation.

The remainder of this paper is organized as follows. Section 2 demonstrates a motivation example to show the benefits of combined optimization. Section 3 describes some preliminaries and the formulation of our combined optimization problem. Section 4 proposes an efficient solution to the formulated optimization problem. Section 5 gives the experimental results, and is followed by conclusions in Section 6.

2. MOTIVATION EXAMPLE

The work in [25] made a good case for combining loop transformation and memory hierarchy, but it was limited to perfectly nested loops. Here we use an example with two loop nests in Fig. 1(a) to show the necessity for supporting imperfectly nested loop and loop fusion/distribution. Data reuse between array references can be exploited by allocating on-chip buffers. For example, Fig. 1(b) shows the data reuse from reference *A[i,j]* to *A[i-2,j]*. The on-chip reuse buffer size is $2N$, because the data fetched by reference *A[i,j]* will be used by reference *A[i-2,j]* after two loop *i* iterations, and the $2N$ data elements accessed during this period (two loop *i* iterations) need to be stored in the reuse buffer for continuous data reuse. After the data in the buffer is reused, it can be replaced to store new reusable data. The modulo operation in the reuse buffer addressing indicates that the buffer is accessed and updated in a cyclic way. By allocating the reuse buffer, off-chip memory accesses by reference *A[i-2,j]* are saved*.

Loop transformation can be used to reduce the buffer size by improving the data locality of array accesses. T0 in Fig. 1(c) is the result of the traditional loop optimizers, which minimizes the size of the largest reuse buffer. The largest reuse buffer in the original code is from *S1:B[i,j]* to *S2:B[i,j-3]*, whose size is N^2 because S1 and S2 are in different loop nests. In T0 the largest buffer size (from *S1:A[i,j]* to *S2:A[i-3,j]*) is reduced to $3N$ because reuse buffer can be cyclically reused every three loop *i* iterations.

However, traditional loop transformation do not consider the impact of memory hierarchy allocation (the selection of the reuse

* In Fig.1(b), bandwidth of *A[i-1,j]* can also be saved using the same reuse buffer, but we do not show it in the figure for the clarification of data reuse from *A[i,j]* to *A[i-2,j]*.

```
// original
for i=0 to N
   for j=0 to N
S1:   B[i,j]=f0(A[i, j], A[i-1, j], A[i-2, j]
                A[i-3, j]);
for i=0 to N
   for j=0 to N
S2:   C[i,j]=f1(B[i, j],B[i, j-1],B[i, j-2],
                B[i,j-3]);
```

(a)

```
data_type buf[2, N];
for i=0 to N
   for j=0 to N {
S1:   B[i,j]=f0(A[i, j], A[i-1, j],
                buf[i%2, j], A[i-3, j]);
      buf[i%2, j] = A[i, j];
   }
...
S2: ...
```

(b)

```
// T0 transformation
for i=0 to N
   for j=0 to N {
S1:  B[i,j]=f0(A[i, j], A[i-1, j], A[i-2, j]
               A[i-3, j]);
S2:  C[i,j]=f1(B[i, j],B[i, j-1],B[i, j-2],
               B[i,j-3]);
   }
```

(c)

```
// T1 transformation
for j=0 to N
   for i=0 to N
S1:   B[i,j]=f0(A[i, j], A[i-1, j], A[i-2, j]
                A[i-3, j]);
for i=0 to N
   for j=0 to N
S2:   C[i,j]=f1(B[i, j],B[i, j-1],B[i, j-2],
                B[i,j-3]);
```

(d)

Figure 1. (a) Original code. (b) Buffer allocation. (c) Transformation T0 with a fused loop. (d) Transforamtion T1 with separately optimized loops.

TABLE I. BUFFER SIZES FOR LOOP TRANSFORMATIONS UNDER BANDWIDTH REQUIREMENTS FOR THE EXAMPLE IN FIGURE 1

Given AC	Original	T0	T1
$2N^2$	N^2	**3*N***	N^2
$4N^2$	$3N$	N	**6**

buffers to be allocated on-chip). T0 is the optimal transformation when all the data reuse buffers are allocated. But in practical cases, on-chip memory may not be sufficient for all these buffers. Trying to optimize the locality of off-chip accesses does not have much benefit because off-chip memories always have high density, but will generate over-constraints for the locality optimization of on-chip accesses. The results of traditional loop transformation may not be optimal when a certain amount of bandwidth needs to be traded for on-chip memory requirement. In our example, T1 shown in Fig1. (d) has smaller on-chip buffer size compared to T0 if we allow two reusable references to stay in off-chip memory.

Table I shows the comparison of total on-chip buffer size (BS) for two loop transformations (T0 and T1 in Fig. 1) under different off-chip memory bandwidth requirements which are expressed as off-chip access counts (AC). When the given AC is $2N^2$, references *S1:A[i,j]* and *S2:C[i,j]* have to access off-chip memory, so all the reuse buffers are allocated on-chip. T0 has minimal buffer size in this case. When the given AC increases to $4N^2$, we can select two reusable references to be allocated off-chip. For T1, after two largest buffers (*S1:B[i,j]* and *S2:B[i,j]*) are removed, all the data reuse occurs in the inner loops. But for T0, at least one reuse buffer is carried on the outer loop. So T1 has smaller total buffer size than T0 in this case. The detailed explanation of Table I is given in [26].

We can see from Table I that loop transformation is important to improve the final result of memory hierarchy allocation, and inversely the trade-off between buffer size and bandwidth in hierarchy allocation impacts the optimality of the loop transformation. This paper investigates the interactions between loop transformation and hierarchy allocation in imperfectly nested loops, and optimizes these two steps simultaneously to achieve the optimal result.

3. PROBLEM FORMULATION

The challenge of combined loop transformation and memory allocation is the modeling of links between the design space and the overall physical design metrics, such as off-chip memory bandwidth and on-chip memory utilization. Our problem is specified as follow: Given the high-level program with affine loop bounds and memory accesses, find the optimal loop transformations and two-level memory hierarchy allocation to minimize the on-chip buffer size under a specified off-chip bandwidth constraint. The dual problem, minimizing bandwidth with given buffer size, can be optimized by solving a sequence of the primal problems using binary searching.

3.1 Loop Transformation

We use the polyhedral model [6] to represent the program in the linear form. A program consists of a set of statements $P = \{S_n \mid n = 0..N-1\}$. Each statement describes a set of array references and the computation between these references. A statement in the loops has multiple instances which are indexed by the iteration vector. The iteration vector specifies the iterations of the loops surrounding the statement from the outermost level to the innermost level: $\vec{i}^S = (i_0, i_1, ..., i_{L_S-1})^T$, where L_S is the number of loops surrounding S. For example, in Fig. 1(d) iteration vectors for S1 and S2 are $\vec{i}^{S1} = (j,i)^T$ and $\vec{i}^{S2} = (i,j)^T$ respectively. The access instance in a statement can be indexed by the iteration vector of the statement as well.

For a program with affine array accesses, standard data flow analysis [2] can derive the dependence between array references. The set of all the read-after-write dependence is defined as D^{wr}, each element in D^{wr} is a pair of iteration vectors of the dependent access instances. For example, in Fig. 1(a) references $S1:B[i,j]$ and $S2:B[i,j-1]$ have true data dependence, so all the dependent iteration vector pairs $\{(\vec{i}^{S1}, \vec{i}^{S2}) \mid \vec{i}^{S1} = \vec{i}^{S2} - (0,1)^T\}$ are in D^{wr}. We can similarly define data reuse set D^{rr} by considering the read-after-read relations. Other kinds of dependence are not considered because they can be eliminated by renaming techniques [2].

The $2d+1$-dimensional representation $\vec{\phi}$ for affine schedules (d being the maximal loop depth in the program) is widely used to model loop fusion/distribution, code motion and affine loop transformations [27, 28, 11]. The even components of schedule are constants representing positions of the statement at different loop levels, and the odd components are the linear combinations of loop iterators which models the affine transformations. By labeling the position constants as $\vec{c}^S = (c_0^S, c_1^S, ..., c_d^S)^T$, and the affine transformation matrix as $\Theta^S = (\vec{\varphi}_0^S, \vec{\varphi}_1^S, ..., \vec{\varphi}_{d-1}^S)^T$, we have $\phi_{2l}^S = c_l^S$ and $\phi_{2l+1}^S = \vec{\varphi}_l^S \cdot \vec{i}^S$.

Fig. 2 shows the schedule representation of the loop transformations in Fig. 1 (c) and Fig. 1(d). For transformation T0, S1 and S2 has two levels of outer common loops (loop i and j), the first two components of the position vectors are the same ($c_0^{S1} = c_0^{S2} = c_1^{S1} = c_1^{S2} = 0$); within loop j, the relative order of the two statements are determined ($c_2^{S1} = 0, c_2^{S2} = 1$). In T1, S1 and S2 are in the distributed loops at top level, so $c_0^{S1} = 0, c_0^{S2} = 1$. For the odd levels, T0 has the same loop iteration scanning order with the original code, which is iterator i in the outer level and then iterator j in the inner level for both statements. So matrices Θ for both statements are identity matrices. But in T1, the loops surrounding S1 are permutated, so Θ^{S1} is a permutation matrix and $\phi_1^{S1} = j, \phi_3^{S1} = i$.

This schedule representation can model all the combinations of affine loop transformations, fusion/distribution, and code motion.

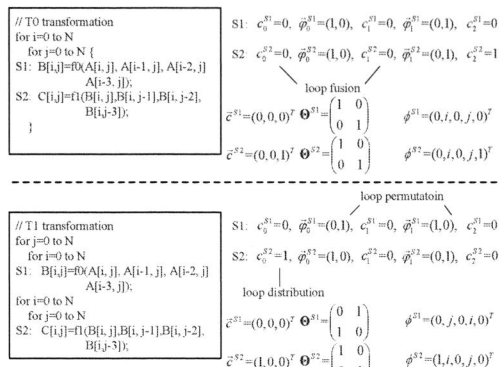

Figure 2. Loop transformations and interleaved schedule vectors.

And legality condition of a loop transformation can be expressed as $\forall (\vec{i}^x, \vec{i}^y) \in D^{wr}, \vec{\phi}^y(\vec{i}^y) \succ_{lex} \vec{\phi}^x(\vec{i}^x)$, which preserves the order of the dependent statement instances after transformations. The proof of the expressiveness and legality condition of loop transformations are provided in [26].

3.2 Reuse Graph and Hierarchy Allocation

Data reuse graph (DRG) is widely used to represent data reuse candidates [14, 17, 18] as Fig. 3(a). Nodes of the graph are array references, and edges are the data reuse between the nodes. Nodes are weighted by the access count (AC) of the reference, and edges are weighted by the reuse buffer size (BS).

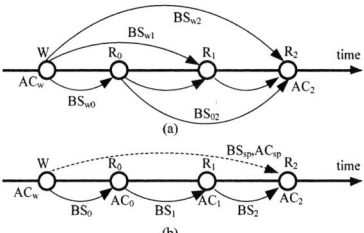

Figure 3. (a) Full data reuse graph. (b) Simplified data reuse graph.

We extend the standard DRG in the following two aspects. First, while traditional DRG only considers read nodes, we also model write nodes in DRG. Thus, it is possible to save the bandwidth of the first read node (in access order), because it can reuse data from the write node. And bandwidth of the write node can even be saved if all the read nodes of the same array are reused and the data is not the primary output of the design. Second, we simplify the DRG by pruning sub-optimal buffer allocations. In most affine programs, the reuse distance [16] for each array reference is a constant, or can be converted to a constant by array partitioning. By ignoring the boundary data elements, we can assume that all the data of each node are reused as a whole. For each node, same AC saving can be obtained by data reuse from different nodes. So we only consider the reuse from the nearest neighboring node as Fig. 3(b), which has the minimal buffer size. The hierarchy allocation is modeled as binary variables $\{b_y\}$ where b_y indicates whether node y is reused from its nearest neighbor. To model the case where write accesses are saved, we introduce a special edge as dashed line in Fig. 3(b) weighted by the total AC and BS of the array. If the special edge b_w^γ of array γ is allocated, no other edges (b_x) of the same

1231

array γ need to be allocated because those buffers are already allocated by the special edge. Using the simplified DRG and binary variables, we can model and evaluate each hierarchy allocation candidate:

$$BS = \sum_y b_y BS_y, \quad AC = AC_{total} - \sum_y b_y AC_y$$

$$\forall x \in R_\gamma, b_w^\gamma \cdot b_x = 0$$

where constant AC_{total} is the total AC without data reuse, and R_γ is the set of read references to array γ.

3.3 Reuse Buffer Size Calculation

As shown in the motivation example, cyclic reuse buffers are used to reduce the buffer size. But seldom previous work has addressed the analytic calculation of cyclic reuse buffer size in imperfectly nested loops. Furthermore, various loop transformations make it harder to calculate the accurate buffer size.

We analytically calculate the size of transformed reuse buffer for node y in three steps. First, determine the access order of the nodes, and find the nearest neighbor reference x which has the same array with y and accesses before node y. Second, calculate the reuse distance between nodes x and y using standard data flow analysis [2], and get the loop level carrying the reuse (l_r) and the distance carried on the loop level (srd_{xy}). Third, calculate the number of array elements accessed by y in one iteration of level l_r (notated as $Q(l_r)$), and then the buffer size is $srd_{xy} \times Q(l_r)$.

For example we calculate the buffer size for reference S1:A[i-2,j] in Fig 1(c). The nearest neighbor node is S1:A[i-1,j]. By data flow analysis, we can know the reuse between the two nodes is carried at the outer loop i, and the distance at this level is 1. Within each iteration of i, there are N data accessed by both references, so the total buffer size is $1 \times N = N$. The analytic calculation of $Q(l_r)$ is performed by building linear constraints for the accessed data elements and counting the integer points in the polytope by the Barvinok library [29]. The detailed derivation is given in [26].

3.4 Formulated Optimization Problem

From the discussion above, we can summarize our formulation as Problem 1 stated below. Eqn. 1 and Eqn. 2 sum up the total BS and AC where $D = D^{wr} \bigcup D^{rr}$. Eqn. 3 and Eqn. 4 model legal loop transformations. Eqn. 5 defines the constraints of allocation variables for the special edges in Section 3.2. Finally, Eqn. 6 calculates the buffer sizes where analytic form of $Q_{l_r}(\vec{\phi}^x, \vec{\phi}^y)$ is given in [26].

PROBLEM 1. *Given an affine program P with array accesses, and a bound of off-chip access count, find the optimal loop transformation $(\vec{c}, \Theta)$ for each statement and memory hierarchy allocation b_y for each reuse edge to*

$Minimize\ BS = \sum_{y \in D} b_y BS_y(\vec{\phi}^x, \vec{\phi}^y)$	(1)
$Subject\ to\ AC = AC_{total} - \sum_{y \in D} b_y AC_y \leq AC_{required}$	(2)
$\phi_{2l}^y = c_l^y, \phi_{2l+1}^y = \vec{\varphi}_l^y \cdot \vec{i}^y, l = 0..L-1$	(3)
$\forall (\vec{i}^x, \vec{i}^y) \in D^{wr}, \vec{\phi}^y \succ_{lex} \vec{\phi}^x$	(4)
$\forall x \in R_\gamma, b_w^\gamma \cdot b_x = 0$	(5)
$BS_y(\vec{\phi}^x, \vec{\phi}^y) = Q_{l_r}(\vec{\phi}^x, \vec{\phi}^y) \cdot srd_y(\vec{\phi}^x, \vec{\phi}^y)$	(6)

4. EFFICIENT SOLUTION

The buffer size calculation is non-convex, which means a enumeration-based approach as used in [25] is needed. But the

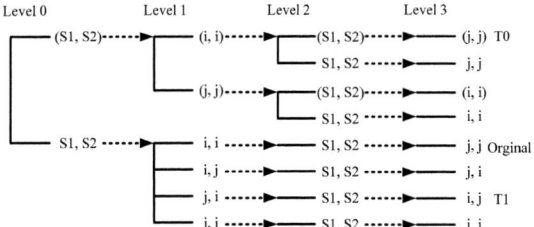

Figure 4. Enumeration tree of loop transformation.

design space pruning in [25] is no longer applicable because of the much larger design space for imperfectly nested loops.

4.1 Enumeration with Branch-and-Bound

Instead of enumerating all legal transformations in a brute-force way, the features of buffer allocation are used to prune sub-optimal partial transformations to greatly reduce the computation complexity of finding optimal solutions for large designs.

Fig. 4 shows the enumeration tree of the motivation example in Fig. 1. For each even level l, the statements are partitioned into ordered groups where c_l indicates the group index of each statement. For the loop levels, we limit the matrix coefficients $\vec{\varphi}_l^S$ as -1, 0 or 1. These matrices can model the most useful affine transformations such as permutation, reversal and skewing as in [25]. Illegal branches are pruned according to Eqn. 4.

Branch-and-bound approach was adopted in [27] to find good loop transformations for parallelism, but they did not consider memory hierarchy allocation. We propose an efficient branch-and-bound scheme for our problem considering interaction between partial schedule and minimal buffer size. For each loop transformation candidates $(\vec{c}, \Theta)$, we can calculate $\vec{\phi}^x$ for each statement to eliminate Eqn. 3, Eqn. 4 and Eqn. 6. The remaining problem is optimized the hierarchy allocation to minimize total BS subjects to AC constraints. Because a reuse candidate carried on an outer loop will generate a much larger buffer than that carried on an inner loop. We travel the enumeration tree branch by branch from outer levels to inner levels. For a branch B, we calculate the lower and upper bounds of the minimal total buffer size for the branch according to the partial transformation of the outer loop levels. The determined buffer set D_B^d contains all the edges carried on the outer levels that are determined by the branch, and non-determined buffer set D_B^n contains the remaining edges. The lower bound (LB_B) of the buffer size at branch B is calculated by optimizing the allocation of buffers in D_B^d and assuming the bandwidth of D_B^n are saved without BS cost. The upper bound (UB_B) of buffer size at branch B is calculated by adding the maximal possible sizes of all the buffers in D_B^n into LB_B. We can prove that the BS range of a child branch is always covered by that of its parent branch: $LB_{Parent} \leq LB_{Child} \leq UB_{Child} \leq UB_{Parent}$. If two branches have non-overlapped BS ranges, we can prune the branch with larger BS and all its sub-branches without losing optimality.

4.2 Knapsack Reuse Technique

In addition to pruning the searching branches, we also reduce the complexity of hierarchy allocation for each branch. In general, the hierarchy allocation problem can be solved by integer linear programming with an exponential computation complexity. But we observe that hierarchy allocation problem consisting of Eqn. 1, Eqn. 2 and Eqn. 5 can be converted into an extended knapsack problem. Each reuse buffer is an item to be put into the knapsack. BS_y and AC_y are the value and weight of item y respectively, and $t_y = 1 - b_y$ indicates whether item y is taken into the knapsack.

If we first ignore the constraint of Eqn. 5, the standard knapsack problem can be solved by dynamic programming. Let $BS[i, AC]$ be the maximum value that can be attained with weight no more than AC using up to first i items, and we have

$$BS[i, AC] = max(BS[i-1, AC], BS[i-1, AC-AC_i] + BS_i).$$

The complexity of the dynamic programming is O(mn), where m is number of reuse buffers, and n is total AC which can be greatly reduced by normalizing the AC of reuse buffers to the loop iteration count.

The extended knapsack problem cannot be directly solved in polynomial time because the solution of the sub-problem is dependent on the decisions of its super-problem. Fortunately, the original dynamic programming does not specify the order of the items, and we can make the sub-problem independent by reordering the items. [26] shows the details of the reordering approach and proves that the complexity remains O(mn) for the extended knapsack problem.

Another feature of the hierarchy allocation problem is that the determined reuse buffer set of the child branch will always contain that of the parent branch, which means the intermediate knapsack results of the parent branch can be reused by its child branches. By reusing the knapsack results, the average computation complexity can be reduced to O(Δmn), where Δm is number of newly determined reuse buffers in the current branch, and n is the total AC. The extended knapsack problem can also reuse intermediate data from parent branches [26]. The reordering process will invalidate the intermediate results because the first i items are changed after reordering. But this kind of computation complexity overhead is small because only one reordering process is needed for each array, and only a part of the intermediate results are invalidated.

5. EXPERIMENTAL RESULTS

Our memory hierarchy optimization algorithm is performed as a source-to-source preprocessing step to a high-level synthesis tool. Our design flow takes loop kernels in high-level specifications like C/C++ as input, and analyzes the polyhedral intermediate representation (IR) with dependence and reuse distances using the ROSE compiler infrastructure [30]. The core optimizer finds the optimal loop transformation and on-chip buffer allocation. In the code generator, loop transformation is performed by CLooG [31], and the on-chip buffer is generated by the ROSE infrastructure. The optimized loop kernels are synthesized into VHDLs and then

circuit netlists by the high-level synthesis tool AutoPilot [32, 33] and the FPGA implementation tool Xilinx ISE [34]. Our test designs include a set of real-life data-intensive loop kernels: FDTD and JACOBI are stencil codes chosen from polybench 3.0 [35]; DENOISE smooths a 3D image by averaging 13 neighboring pixels [36]; REG is one of the major parts of a 3D medical image registration algorithm [36], and SEG is a two-phase image segmentation algorithm [36]. We include the whole programs of the benchmarks with multiple loop nests, instead of only one main loop nest in [25]. The proposed combined loop transformation (LT) and memory hierarchy allocation (HA) scheme for imperfectly nested loops (LTHA-INL) is compared with two reference points in our experiments. The first reference point is the combined LT and HA scheme for perfectly nested loops (LTHA-PNL) in [25], in which the loop nests are optimized independently. And the second point is the separate LT and HA scheme (LT+HA) that was done in [11, 19].

Experimental results of the three approaches are reported in Table II. The second column (AC) shows the normalized access count per loop iteration for each design, which is calculated from the given bandwidth and performance requirement. We set the clock frequency as 10ns, and all design implementations satisfy the timing constraint. The FPGA implementation results of the three approaches in the Xilinx Virtex-6 xc6vlx365t platform are compared, such as the utilization of logic slice and on-chip Block RAM (BRAM), the execution latency in cycles, and the power consumption in mW. We also list the runtime (in seconds) of our proposed algorithm in the last column. We normalize the four metrics to the values of the LTHA-PNL scheme, and calculate the geometric mean of the normalized data in the last row of Table II.

From the results, it is clear that loop fusion/distribution and code motion are important to the results of hierarchy allocation. Compared to the only LTHA-PNL scheme, the separated LT+HA scheme can save the on-chip memory size by 60%. And our LTHA-INL scheme gains an additional 40% memory reduction and 19% power saving compared to the separated LT+HA scheme. In some cases, for example JACOBI_3D, LT+HA get worse results than LTHA-PNL because the original distributed loops are good enough for relative sufficient bandwidth. And in other cases such as FDTD_2D, LTHA-PNL has a much large buffer size because of the data reuse between different distributed loop nests. To isolate the impact of performance, we keep the initiation interval (II) unchanged for the loop pipelining in high-level synthesis. The II of a fused loop is set as the sum of the II of the original loops. The logic slice saving mainly comes from the better resource and data sharing after loop fusion. By our efficient pruning techniques, over 100x speed-up is achieved, and the overall execution time is within several seconds for the real-life benchmarks with four levels of loops and tens of array references.

TABLE II. EXPERIMENT RESULTS

Design	AC	LTHA-PNL [25]				Separated LT+HA [11]+[19]				Proposed LTHA-INL				
		Slice (#)	BRAM (#)	Latency (cycle)	Power (mW)	Slice (#)	BRAM (#)	Latency (cycle)	Power (mW)	Slice (#)	BRAM (#)	Latency (cycle)	Power (mW)	Runtime (s)
JACOBI_3D	3	487	203	4.86E+8	133	504	347	4.86E+8	163	378	203	4.86E+8	112	5.52
JACOBI_4D	3	-	6054*	1.22E+11	-	548	757	1.22E+11	244	520	427	1.22E+11	179	0.28
FDTD_2D	4	398	422	6.08E+6	148	281	18	6.08E+6	57	271	10	6.08E+6	56	0.31
FDTD_3D	6	-	1218*	3.64E+8	-	390	406	3.64E+8	150	378	204	3.64E+8	112	1.34
DENOISE	4	1074	609	2.62E+8	325	836	407	2.62E+8	240	836	306	2.62E+8	221	0.32
REG	12	1569	610	1.22E+9	645	1209	636	1.22E+9	582	1170	306	1.22E+9	393	0.14
SEG	10	-	882*	8.54E+8	-	932	714	8.54E+8	315	932	586	8.54E+8	291	2.50
Geomean		1.00	1.00	1.00	1.00	0.44	0.40	1.00	0.74	0.41	0.24	1.00	0.60	

*The maximum number of BRAMs in the experimental FPGA is 832, so there are no final implementation results for these cases.

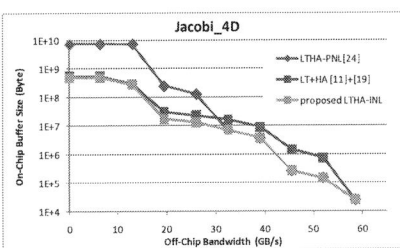

Figure 5. Design space exploration.

Fig. 5 investigates the trade-off between bandwidth and buffer size. When the bandwidth is low, LTHA-PNL is not good because loop fusion is needed to optimize large buffers. When the bandwidth is high, separate LT+HA performs poorly because it has over-constraints for small buffers. However, Our LTHA-INL can consistently attain optimal solutions in all cases.

6. CONCLUSION

This paper presents the first in-depth study that combines loop transformation and hierarchy allocation for imperfectly nested loops. It gains 40% reduction of the on-chip buffer usage under the same off-chip bandwidth constraint with no performance overhead. The proposed space pruning techniques are shown to be highly effective to speed up the execution of our algorithm. Our future work will integrate loop tiling transformation and tiling size selection into our memory hierarchy optimization.

7. REFERENCES

[1] F. Catthoor, E. d. Greef, and S. Suytack, *Custom Memory Management Methodology: Exploration of Memory Organisation for Embedded Multimedia System Design*. Norwell, MA, USA: Kluwer Academic Publishers, 1998.

[2] K. Kennedy and J. R. Allen, *Optimizing compilers for modern architectures: a dependence-based approach*. San Francisco, CA, USA: Morgan Kaufmann Publishers Inc., 2002.

[3] K. S. McKinley, S. Carr, and C.-W. Tseng, "Improving data locality with loop transformations," *ACM Trans. Program. Lang. Syst.*, vol. 18, pp. 424–453, July 1996.

[4] M. E. Wolf and M. S. Lam, "A data locality optimizing algorithm," in PLDI '91. New York, NY, USA.

[5] M. E. Wolf and M. S. Lam, "A loop transformation theory and an algorithm to maximize parallelism," *IEEE Trans. Parallel Distrib. Syst.*, vol. 2, pp. 452–471, Oct. 1991.

[6] P. Feautrier, "Some efficient solutions to the affine scheduling problem: Part II. multidimensional time," *International Journal of Parallel Programming*, vol. 21, pp. 389–420, 1992.

[7] A. W. Lim, G. I. Cheong, and M. S. Lam, "An affine partitioning algorithm to maximize parallelism and minimize communication," in ICS '99. New York, Aug. 1999.

[8] N. Ahmed, N. Mateev, and K. Pingali, "Synthesizing transformations for locality enhancement of imperfectly-nested loop nests," *IJPP*, vol. 29, pp. 493–544, Oct. 2001.

[9] U. Bondhugula, A. Hartono, J. Ramanujam, and P. Sadayappan, "A practical automatic polyhedral parallelizer and locality optimizer," in *PLDI '08*, pp. 101–113, New York, NY, USA,.

[10] U. Bondhugula, O. Gunluk, S. Dash, and L. Renganarayanan, "A model for fusion and code motion in an automatic parallelizing compiler," in *PACT '10*, pp. 343–352, Sept. 2010,.

[11] L.-N. Pouchet, U. Bondhugula, C. Bastoul, A. Cohen, J. Ramanujam, P. Sadayappan, and N. Vasilache. "Loop transformations: convexity, pruning and optimization," in *POPL '11*, pp. 549-562, Jan. 2011.

[12] F. Catthoor, K. Danckaert, K. Kulkarni, E. Brockmeyer, P. Kjeldsberg, T. v. Achteren, and T. Omnes, *Data access and storage management for embedded programmable processors*. Norwell, MA, USA: Kluwer Academic Publishers, 2002.

[13] T. Van Achteren, G. Deconinck, F. Catthoor, and R. Lauwereins, "Data reuse exploration techniques for loop-dominated applications," in *DATE '02*, pp. 428–435, Mar. 2002.

[14] I. Issenin, E. Brockmeyer, M. Miranda, and N. Dutt, "DRDU: A data reuse analysis technique for efficient scratch-pad memory management," *ACM Trans. Des. Autom. Electron. Syst.*, April 2007.

[15] S. Wuytack, J.-P. Diguet, F. V. M. Catthoor, and H. J. De Man, "Formalized methodology for data reuse: exploration for low-power hierarchical memory mappings," *IEEE Trans. on VLSI*, 1998.

[16] M. Kandemir and A. Choudhary, "Compiler-directed scratch pad memory hierarchy design and management," in *DAC '02*, pp. 628—633, June 2002.

[17] E. Brockmeyer, M. Miranda, and F. Catthoor, "Layer assignment techniques for low energy in multi-layered memory organisations," in *DATE '03*, pp. 1070–1075, Mar. 2003.

[18] J. Cong, H. Huang, C. Liu, and Y. Zou, "A reuse-aware prefetching scheme for scratchpad memory," in *DAC '11*, pp. 960–965, 2011.

[19] I. Issenin, E. Brockmeyer, B. Durinck, and N. D. Dutt, "Data-reuse-driven energy-aware cosynthesis of scratch pad memory and hierarchical bus-based communication architecture for multiprocessor streaming applications," *IEEE trans. on CAD*, vol. 27, no. 8, pp. 1439–1452, 2008.

[20] M. Palkovic, F. Catthoor, and H. Corporaal, "Trade-offs in loop transformations," *ACM Trans. Des. Autom. Electron. Syst.*, vol. 14, pp. 22:1–22:30, April 2009.

[21] P. R. Panda, N. D. Dutt, and A. Nicolau, "Local memory exploration and optimization in embedded systems," *IEEE Trans. on CAD*, vol. 18, no. 1, pp. 3–13, 1999.

[22] Q. Hu, P. G. Kjeldsberg, A. Vandecappelle, M. Palkovic, and F. Catthoor, "Incremental hierarchical memory size estimation for steering of loop transformations," *ACM Trans. Des. Autom. Electron. Syst.*, vol. 12, September 2007.

[23] M. M. Baskaran, U. Bondhugula, S. Krishnamoorthy, J. Ramanujam, A. Rountev, and P. Sadayappan, "Automatic data movement and computation mapping for multi-level parallel architectures with explicitly managed memories," in PPoPP'08, pp. 1–10, Feb. 2008.

[24] Q. Liu, G. A. Constantinides, K. Masselos, and P. Cheung, "Combining data reuse with data-level parallelization for FPGA-targeted hardware compilation: A geometric programming framework," *IEEE Trans. on CAD*, vol. 28, no. 3, 2009.

[25] J. Cong, P. Zhang, and Y. Zou, "Combined loop transformation and hierarchy allocation in data reuse optimization," in *ICCAD'11*, pp. 185–192, Nov. 2011.

[26] J. Cong, P. Zhang, and Y. Zou, "Optimizing Memory Hierarchy Allocation with Loop Transformations for High-Level Synthesis," *Technical Report, Computer Science Department, UCLA*, TR200019, 2012.

[27] Kelly, W. A. *Optimization within a Unified Transformation Framework*, Ph.D. dissertation, University of Maryland, 1996.

[28] S. Girbal, N. Vasilache, C. Bastoul, A. Cohen, D. Parello, M. Sigler, and O. Temam, "Semi-Automatic Composition of Loop Transformations for Deep Parallelism and Memory Hierarchies," *IJPP*, vol. 34, no. 3, June 2006.

[29] Barvinok library. http://freshmeat.net/projects/barvinok

[30] ROSE compiler infrastructure. http://rosecompiler.org/

[31] The CLooG code generator. http://www.cloog.org/

[32] J. Cong, B. Liu, S. Neuendorffer, J. Noguera, K. Vissers, and Z. Zhang, "High-level synthesis for FPGA: From prototyping to deployment," *IEEE Trans. on CAD*, vol. 30, no. 4, 2011.

[33] Z. Zhang, Y. Fan, W. Jiang, G. Han, C. Yang, and J. Cong, "AutoPilot: A Platform-Based ESL Synthesis System," *High-Level Synthesis: From Algorithm to Digital Circuit*, ed. P. Coussy and A. Morawiec, Springer Publishers, 2008.

[34] Xilinx ISE Design Suite. http://www.xilinx.com/products/design-tools/ise-design-suite/.

[35] Polyhedral benchmark suite v3.1. http://www.cse.ohio-state.edu/~pouchet/software/polybench/ .

[36] J. Cong, V. Sarkar, G. Reinman and A. Bui, "Customizable Domain-Specific Computing," *IEEE Design and Test of Computers*, vol. 28, no. 2, pp. 5-15, March/April 2011.

A Metric for Layout-Friendly Microarchitecture Optimization in High-Level Synthesis

Jason Cong   Bin Liu

Computer Science Department, University of California, Los Angeles

{cong,bliu}@cs.ucla.edu

ABSTRACT

In this work we address the problem of managing interconnect timing in high-level synthesis by generating a layout-friendly microarchitecture. A metric called *spreading score* is proposed to evaluate the layout-friendliness of microarchitectural netlist structures. For a piece of connected netlist, spreading score measures how far the components can be spread from each other with bounded length for every wire. The intuition is that components in a layout-friendly netlist (e.g., a mesh) can spread over the layout region without introducing long interconnects. We propose a semidefinite programming relaxation to allow efficient estimation of spreading score, and use it in a high-level synthesis tool. On a number of test cases, a normalized spreading score shows a stronger bias in favor of interconnect structures that have better timing after layout, compared to the widely used metric of total multiplexer inputs. We also justify our metric and motivate further study by relating spreading score to other metrics and problems for layout-friendly synthesis.

Categories and Subject Descriptors

B.6.3 [**Logic Design**]: Design Aids—*automatic synthesis, optimization*

General Terms

Algorithms, Design, Experimentation

Keywords

High-Level Synthesis, Interconnect, Layout

1. INTRODUCTION

High-level synthesis (HLS) is the process of automatically generating RTL models from behavioral specifications. Compared to the traditional RTL-based design flow, the potential advantages of HLS include better management of design complexity, code reuse across platforms and performance targets, and easy design space exploration. HLS is getting wide adoption. In our experience, a primary challenge to HLS is the generation of results that consistently meet timing (i.e., the frequency target). We consider timing to be a vital factor to the success of HLS, not just because an RTL implementation that cannot meet timing is often unacceptable, but because actions the designer could take to circumvent timing failures tend to seriously undermine the advantages of HLS.

A straightforward approach to fixing timing failures is to manipulate the RTL code directly, but this requires an understanding of the generated code, and is time-consuming and error-prone. Some tools allow specification of explicit clock boundaries and/or sharing decisions. For example, the Handel-C language requires cycle-accurate input [1]. While this approach is useful when fine-tuning is needed, it is questionable whether the level of abstraction is really raised with this method. In the extreme case, input to the HLS tool is just an RTL model specified in a high-level language. Furthermore, code reuse and design space exploration can become more difficult when decisions on scheduling/sharing are included in the specification. Another common practice is to add an extra timing margin to tolerate excessive interconnect delays. This partially solves the problem, but not always, because delay of a long interconnect can exceed the target clock period, especially when the synthesis engine is not sophisticated enough concerning interconnect complexity. This approach often leads to unnecessarily long latency and low throughput, and is thus undesirable in many target applications of HLS such as signal processing. In addition, a large timing margin can cause overhead in area and power because the synthesis tool needs to use faster components and insert more registers.

Therefore, an HLS tool needs to manage interconnect delay intelligently in order to fully realize its advantages. This is challenging due to the absence of information about the gate-level netlist and the layout. Most existing solutions to the problem can be categorized as follows.

1. Use a regular architecture for global interconnects. The chip area is typically divided into regular islands (or clusters), where inter-island data transfers are performed on regular multicycle interconnects, such as a mesh [6, 19, 22]. The approach is effective for managing global interconnects; yet the regularity is a strong unnecessary constraint and can lead to suboptimal performance and resource usage.

2. Incorporate a rough layout. There are numerous efforts to combine HLS with floorplanning to help interconnect estimation and optimization [9, 10, 13, 26, 28]. This is quite a natural approach and can potentially

work well. However, since layout itself is nontrivial, implementation of a stable and fast layout engine in the inner iteration of microarchitecture optimization is a challenge.

3. Use structural metrics to evaluate netlist structures. It is recognized that interconnect complexity (and thus timing) depends largely on netlist structure. An HLS tool can easily explore many different microarchitectures, guided by structural metrics. Such metrics are usually derived from a graph representation of the netlist without performing layout. Widely used structural metrics include total multiplexer inputs [4, 14, 15, 20], number of global interconnects [7, 21], adhesion [17], etc. These metrics generally lead to efficient heuristics, but in general the interconnect delay after layout cannot be guaranteed.

In this paper we propose a new structural metric called *spreading score*. The intuition is that components in a layout-friendly netlist can often be spread apart from each other without introducing many long interconnects, and that such long interconnects should have larger allocated slacks. Spreading score captures these properties in a mathematical programming formulation and can be estimated efficiently using semidefinite programming (SDP). Compared to the approach of performing a layout, our metric is stable and fast, because the globally optimal solution to the SDP problem can be obtained in polynomial time. Compared to previous structural metrics, our metric is more layout-oriented. Experimental evaluation using a normalized spreading score to guide HLS optimization shows encouraging timing improvement on a series of test cases without large area overhead, when compared to the previous metrics, such as total multiplexer inputs.

The remainder of this paper is organized as follows. In Section 2 we describe spreading score as the optimal value of an optimization problem. An SDP relaxation is presented in Section 3 to allow efficient estimation. Experimental results are reported in Section 4. A few interesting connections with other problems and metrics are discussed in Section 5, followed by a conclusion in Section 6.

2. THE SPREADING METRIC

An HLS tool typically represents an optimized input specification as a control-data flow graph (CDFG). The synthesis engine performs module selection, operation scheduling and resource sharing, and then generates a microarchitecture-level netlist.[1] The netlist often consists of components (including functional units, registers, memories, I/O ports, multiplexers, pre-synthesized modules, etc.) and wires which connect the components. To simplify the discussion, we first consider the simple case where each component has only one output port, and delays between all input ports and the output port are the same.

We construct a directed graph $G = (V, E)$ to model the component-level connectivity, where $V = \{1, 2, \ldots, n\}$ is the set of vertices with each representing a component, and $E \subseteq V \times V$ is the set of directed edges with each representing a wire from the source component to the sink component. Note that an edge is present only when there are data transfers between the two components; if two components are connected in the netlist only because they are both

---

[1]Unless otherwise noted, the term netlist refers to a micro-architecture-level netlist in this paper.

sinks of a net, no edge is created between the corresponding vertex pair. In addition, connections from a component to itself are discarded to avoid self-loop in the graph; this is reasonable because such connections can be regarded as local interconnects within the component.

A layout of the netlist is regarded as an embedding of G in the 2-dimensional Euclidean space $\mathbb{R}^2$. Each vertex i is associated with a column vector $p_i = (x_i, y_i)^T$ to represent its position in the embedding. The length of the connection $(i, j) \in E$ can be measured as the Euclidean distance in $\mathbb{R}^2$, i.e., $\|p_i - p_j\| = \sqrt{(x_i - x_j)^2 + (y_i - y_j)^2}$.

Consider the following optimization problem.

$$
\begin{aligned}
\text{maximize} \quad & \sum_{i=1}^n w_i \|p_i\|^2 \\
\text{subject to} \quad & \sum_{i=1}^n w_i p_i = 0 \\
& \|p_i - p_j\| \le l_{ij} \quad \forall (i, j) \in E
\end{aligned}
\tag{1}
$$

Here $w = (w_1, w_2, \ldots, w_n)^T$ is the nonnegative weight vector with w_i being the area of component i; l_{ij} is the maximum allowed length for the wire connecting i and j. The objective function measures how far components are spread from their weighted center of gravity, using a weighted 2-norm of the distance vector. Thus the problem in Eqn. 1 asks to maximize component spreading, under the constraint that the length of every connection $(i, j) \in E$ does not exceed l_{ij}.

With the proper selection of l_{ij}, we claim that the optimal value of the above problem can be used to evaluate the layout-friendliness of a netlist. This is based on the following observation: if components in a netlist can be spread over the chip area without introducing long wires, it will be easy for the layout tool to remove overlaps between components without significant increase in interconnect delay.

This argument can be supported by examining well-known hand-designed interconnect topologies. For example, mesh [6, 19], ring [16] and couterflow pipeline [24] can all spread without long interconnects, and they are regarded as scalable and layout-friendly topologies; on the other hand, spreading the full crossbar or hypercube on the 2D plane inevitably introduces long interconnects, and these topologies are generally much more expensive in interconnect cost.

Note that increasing the allowed wire length l_{ij} can often lead to better spreading in Eqn. 1, and this explains why an extra timing margin can help layout. However, l_{ij} is limited by timing constraints in practice. To capture first-order timing information, we use d_{ij} to denote the delay of the wire $(i, j) \in E$, and consider it to be a monotone single-variate function of wire length

$$
d_{ij} = \mathcal{D}(l_{ij}).
\tag{2}
$$

Two additional variables t_i and τ_i are attached to each $i \in V$, where t_i denotes the arrival time (after clock edge) at input ports of i, and τ_i denotes the arrival time at the output port. We then have

$$
\tau_i + d_{ij} \le t_j, \forall (i, j) \in E.
\tag{3}
$$

If the corresponding component is combinational, we use d_i to denote its delay, and then

$$
\tau_i = t_i + d_i.
\tag{4}
$$

Otherwise, if the component is sequential, τ_i will be a constant, and t_i should be bounded by the required time at the input of i, T_i, that is,

$$
t_i \le T_i.
\tag{5}
$$

1236

For a register i, T_i can be regarded as equal to the clock period subtracted by the setup time. A primary output also has a required time depending on the interface timing specification.

We can then treat l_{ij} as variables and optimize spreading under timing constraints.

$$
\begin{aligned}
\text{maximize} \quad & \sum_{i=1}^n w_i \|p_i\|^2 \\
\text{subject to} \quad & \sum_{i=1}^n w_i p_i = 0 \\
& \|p_i - p_j\| = l_{ij} && \forall (i,j) \in E \\
& d_{ij} = \mathcal{D}(l_{ij}) && \forall (i,j) \in E \\
& \tau_i + d_{ij} \le t_j && \forall (i,j) \in E \\
& \tau_i = t_i + d_i && \forall \text{ combinational } i \\
& t_i \le T_i && \forall i \in V
\end{aligned}
\tag{6}
$$

The formulation in Eqn. 6 effectively combines interconnect slack allocation with node spreading, and captures both structural property and timing property of the netlist. We refer to the optimal value of the above problem as the spreading score of the netlist.

The graph construction and labeling procedure can be extended easily to handle more complex cases. For example, for a component with multiple input ports and multiple output ports, if the delay varies significantly between different inputs and outputs, we can create a vertex for each port, so that the delay between each pair of ports can be characterized individually as done in [18]; constraints on the distances between ports can be enforced to keep the geometry of the component. Similar treatment on a very large component can make the estimation of interconnect length aware of port positions, instead of regarding all ports as being located at the center of the component. The required time at a port can be manipulated easily to capture nontrivial situations like multicycle paths, multiple clock domains, or other complex I/O timing requirements.

3. EFFICIENT EVALUATION

It is difficult to solve the problem in Eqn. 6 directly, because maximizing a convex function (like the objective function in Eqn. 6) is generally NP-hard (note that minimizing a convex function is easy). We hereby propose a tractable relaxation and use the solution of the relaxed problem to estimate the spreading score.

Consider the graph G with n vertices, we use a $2 \times n$ matrix $P = (p_1, p_2, \ldots, p_n)$ to represent its embedding in $\mathcal{R}^2$, i.e.,

$$
P = \begin{pmatrix} x_1 & x_2 & \cdots & x_n \\ y_1 & y_2 & \cdots & y_n \end{pmatrix}.
\tag{7}
$$

Let $Q = P^T P$. Then Q is a symmetric semidefinite matrix with rank at most 2, and

$$
Q_{ij} = p_i^T p_j = x_i x_j + y_i y_j.
\tag{8}
$$

We can use Q as variables in the formulation in Eqn. 6 without losing any useful information, as indicated by the following theorem.

THEOREM 1. *Given a semidefinite matrix Q, we can always reconstruct the embedding of the graph, in the sense that the distance between any pair of vertices is preserved.*

PROOF. Since Q is semidefinite, we can perform a Cholesky decomposition and get matrix $U = (u_1, u_2, \ldots, u_n)$, so that $Q = U^T U$. Let $P = (p_1, p_2, \ldots, p_n)$ be another matrix such that $Q = P^T P$. We have $\|p_i - p_j\|^2 = (p_i - p_j)^T (p_i - p_j) = Q_{ii} + Q_{jj} - 2Q_{ij}$, and $\|u_i - u_j\|^2 = (u_i - u_j)^T (u_i - u_j) =$

$Q_{ii} + Q_{jj} - 2Q_{ij}$. Thus $\|p_i - p_j\| = \|u_i - u_j\|$; this means that pairwise distances between vertices are decided given Q. □

Using Eqn. 8, we can rewrite objective and constraint functions in Eqn. 6 as follows.

$$
\sum_{i=1}^n w_i \|p_i\|^2 = \sum_{i=1}^n w_i Q_{ii} = \langle \text{diag}(w), Q \rangle
\tag{9}
$$

$$
\left\| \sum_{i=1}^n w_i p_i \right\|^2 = \sum_{i=1}^n \sum_{j=1}^n w_i w_j Q_{ij} = \langle w w^T, Q \rangle
\tag{10}
$$

$$
\begin{aligned}
\|p_i - p_j\|^2 &= Q_{ii} + Q_{jj} - 2Q_{ij} \\
&= \left\langle (e_i - e_j)(e_i - e_j)^T, Q \right\rangle \\
&= \left\langle K^{ij}, Q \right\rangle
\end{aligned}
\tag{11}
$$

Here $\text{diag}(w)$ is the $n \times n$ diagonal matrix with w on its diagonal. e_i is the ith standard basis vector in $\mathcal{R}^n$ and we define matrix $K^{ij} = (e_i - e_j)(e_i - e_j)^T$ to simplify the equations. $\langle X, Y \rangle$ is the element-wise inner product (Frobenius inner product) of matrices X and Y, i.e., $\langle X, Y \rangle = \sum_i \sum_j X_{ij} Y_{ij}$. Then we can rewrite the problem in Eqn. 6 to use Q as variables.

$$
\begin{aligned}
\text{maximize} \quad & \langle \text{diag}(w), Q \rangle \\
\text{subject to} \quad & \langle w w^T, Q \rangle = 0 \\
& \langle K^{ij}, Q \rangle = l_{ij}^2 && \forall (i,j) \in E \\
& d_{ij} = \mathcal{D}(l_{ij}) && \forall (i,j) \in E \\
& \tau_i + d_{ij} \le t_j && \forall (i,j) \in E \\
& \tau_i = t_i + d_i && \forall \text{ combinational } i \\
& t_i \le T_i && \forall i \in V \\
& Q \succeq 0 \\
& \text{rank}(Q) \le 2
\end{aligned}
\tag{12}
$$

The above problem is equivalent to that in Eqn. 6, and is thus equally hard. Yet after relaxation of the rank constraint, the resulting problem is easy to solve when a quadratic delay model is used. That is, $\mathcal{D}(l_{ij}) = \alpha l_{ij}^2$, where α is a constant that depends on technology. Then we get the following relaxed problem.

$$
\begin{aligned}
\text{maximize} \quad & \langle \text{diag}(w), Q \rangle \\
\text{subject to} \quad & \langle w w^T, Q \rangle = 0 \\
& \tau_i + \alpha \left\langle K^{ij}, Q \right\rangle \le t_j && \forall (i,j) \in E \\
& \tau_i = t_i + d_i && \forall \text{ combinational } i \\
& t_i \le T_i && \forall i \in V \\
& Q \succeq 0
\end{aligned}
\tag{13}
$$

This problem is convex. In fact, it can be solved as an SDP problem. Like linear programs, SDP problems can be solved optimally in polynomial time, and efficient solvers have been developed in recent years [2]. Due to page limitations, we will not discuss background on convex programming and SDP here. Interested readers may refer to [3, 25] on these topics.

The problem in Eqn. 13 essentially asks for an embedding in $\mathcal{R}^n$ instead of $\mathcal{R}^2$, and its optimal value is a lower bound of the spreading score. It would be interesting to see how good the bound is. For this, we refer to the following result.

THEOREM 2 (GÖRING, HELMBERG AND WAPPLER [12]). *For a relaxed version of the problem in Eqn. 1 with $p_i \in \mathcal{R}^n$, an optimal embedding always exists in $\mathcal{R}^{tw(G)+1}$, where $tw(G)$ is the tree-width [23] of G.*

Although a rigorous proof is yet to be derived, we conjecture that the same result holds for the problem in Eqn. 13; this is based on the intuition that additional variables for slack allocation do not interfere with variables for graph embedding. The result implies low distortion when the optimal solution in $\mathcal{R}^n$ is embedded back in $\mathcal{R}^2$. In the extreme case where $tw(G) = 1$ (i.e., the netlist is a tree), an optimal solution always exists in $\mathcal{R}^2$, and our relaxation is exact. This can also be empirically explained as follows: for a vertex i at position p_i, the direction from origin (weighted center of gravity) to p_i is the direction of steepest ascent for the objective function, and this direction is within the vector subspace spanned by existing direction vectors, because $p_i = -\frac{1}{w_i} \sum_{j \neq i} w_j p_j$; thus the objective function intrinsically prefers moves that do not increase the dimension of the embedding.

The quadratic delay model is used in Eqn. 13 to simplify the relaxation. It is also possible to use the linear delay model, but that leads to more variables and less sparse matrices in the formulation.

4. EXPERIMENTAL EVALUATION

4.1 Normalization

While spreading score characterizes the layout-friendliness of a given netlist, using it directly to compare different netlists tends to favor netlists with larger area, because more components and larger weights can increase the spreading score naturally. To avoid this problem, we can normalize the spreading score of a netlist against that of a uniform mesh with the same area, and use the resulting value in comparison.

Consider a 2-dimensional $m \times m$ uniform mesh with unit component weight and unit interconnect length between neighboring components, the total weight $W = m^2$. Assuming m is odd and $r = \frac{m-1}{2}$, we can calculate its spreading score as

$$\sum_{i=-r}^{r} \sum_{j=-r}^{r} \left(i^2 + j^2 \right) = \frac{1}{6}(W^2 - W) = O(W^2). \quad (14)$$

This means that the spreading score of a mesh grows quadratically with regard to the total area. Thus, we can divide the spreading score by $\left(\sum_{i=1}^{n} w_i \right)^2$ to get a normalized value, which can be used to compare different netlists without a bias on area.

4.2 Experiment Setup

We have implemented a simulated annealing algorithm to perform microarchitecture exploration in the xPilot HLS tool [5]. Based on an initial solution, perturbations are performed to generate alternative microarchitectures. Feasibility and cost are evaluated to decide whether the new solution is accepted. We compare two cost functions: (A) a weighted sum of total area and normalized spreading score (with a negative weight, i.e., larger spreading score will lead to smaller cost), and (B) a weighted sum of total area and total multiplexer inputs. In both cases, feasibility check includes legality check (dependency, resource hazard, combinational loop, and performance constraint) and timing check (without considering interconnect delay). Area and timing information about components are obtained from a precharacterized library. Random perturbations are performed in the following ways.

- Move an operation from one functional unit to another.
- Move a variable from one register to another.
- Merge two functinal units or registers.

- Insert (or delete) an additional register before the input port of a component, if the input data have been buffered at least two cycles before its use. This creates a multicycle interconnect and doesn't require changes in operation scheduling.
- Reschedule an operation one cycle earlier or later, and update the schedules of related operations if necessary.

When estimating spreading score, we construct sparse matrices to describe the problem in Eqn. 13 and use CSDP 6.1.1 [2] to solve the SDP problem. To speed up the solver, we use the solution of the SDP problem for the previous netlist as the starting point when solving the problem for the perturbed netlist. In addition to the primal solution, CSDP is able to give solutions to the dual problem as well. The dual solution can potentially provide information for sensitivity analysis and can be used to guide the perturbation; however, this is not yet implemented.

We perform HLS on several designs and implement the results on a 65nm ASIC technology, using Synopsys Design Compiler and IC Compiler. In all cases, the layout region is a square and the target density is 80%. Since the designs are blocks to be integrated in a larger chip, we only use lower-level metal layers (M1 to M5) to route wires in the design; upper metal layers are reserved for system-level connections and power/clock networks.

4.3 Results and Analysis

We take a few snapshots of the simulated annealing process and plot the area, total multiplexer inputs and spreading score for each snapshot in Figure 1 for the design "QAM." For both cases, the area and spreading score are normalized so that either metric in the initial solution (with no sharing) is one, and the total multiplexer inputs metric is normalized so that it is one in the final solution of the case with optimization for cost (A).

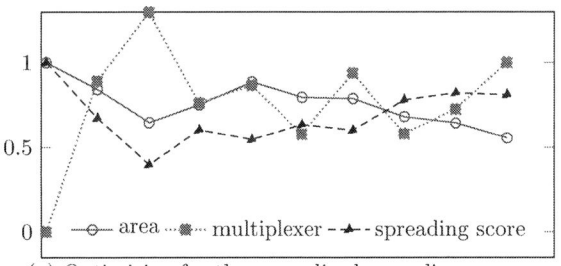

(a) Optimizing for the normalized spreading score.

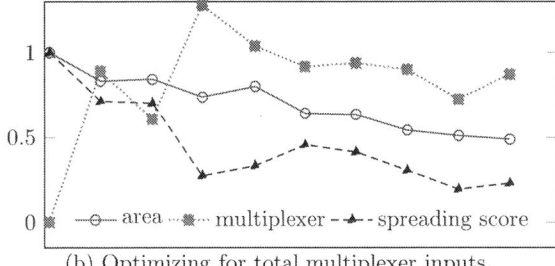

(b) Optimizing for total multiplexer inputs.

Figure 1: Evolution of the solution quality.

From Figure 1, we observe that total area is generally reduced in the optimization process. This is probably because the design has a lot of compatible operations that can share functional units, and sharing them does reduce total area

1238

despite potential area overheads caused by the added multiplexers. The initial netlist without any sharing has the best normalized spreading score. This is not due to the bias toward larger area because normalization has been performed; it can be explained by the fact that sharing often creates connections between components that were not directly connected, and thus makes the netlist harder to spread. While the optimization of the normalized spreading score and total area tends to limit total multiplexer inputs, optimizing the total multiplexer inputs directly does not necessarily lead to better result on the normalized spreading score. With similar amount of total multiplexer inputs in the netlist, the normalized spreading score can still vary significantly; this indicates that normalized spreading score and total multiplexer inputs point to different optimization directions.

We report the worst-case slack and area after implementation for each benchmark in Table 1. Modern RTL synthesis tools typically have a large amount of freedom in making trade-offs between timing and area/power through logic refactoring, cell selection, buffer insertion, etc. Therefore, the achieved clock period after layout tends to be very close to the clock target, making advantages in timing less obvious. Despite this effect, the approach that optimizes for the normalized spreading score does lead to consistently better timing. More significant advantages are expected with less powerful downstream tools and simpler libraries.

Table 1: Timing and area results after layout.

name	clock	slack_A	slack_B	area_A	area_B
DCT	3.00	0.17	0.08	23105	21733
QAM	3.50	0.00	-0.20	63068	58100
Sphere	2.50	-0.12	-0.33	125091	131484
Industry1	3.08	0.00	0.00	348817	347092
Industry2	3.08	0.03	-0.18	506692	464060

Clock target and slacks is in ns; area is in nm^2.

Due to the trade-off between area and timing, a better structured netlist tends to have more slack, which in turn gives the RTL synthesis tool more freedom in reducing area. On the other hand, less aggressive sharing may be needed to obtain a more layout-friendly netlist in HLS, and this can increase the overall area. These effects can affect timing in different directions; we are unable to draw a conclusion as to which cost function leads to superior area.

In our experiments, evaluation of the normalized spreading score leads to a significantly longer runtime of the tool. Theoretically, the worst-case complexity of SDP is $O(n^3)$, while feasibility checks and evaluation of total multiplexer inputs can all be finished in $O(n)$. For the design "Industry2" with more than 500 operations, the tool with cost (A) takes about 30 minutes on a workstation with dual 2.6GHz CPUs and 4GB memory, while the tool with cost (B) takes about 5 minutes on the same workstation. Fortunately, hierarchical design style is often used in engineering practices, and this helps to control the number of components in a given level of hierarchy to make our approach feasible for very large designs.

5. FURTHER DISCUSSION

In this section we relate spreading score to other problems and metrics, to further study its properties and to motivate future research.

One may want to measure how far the vertices are spread by looking at pairwise distances, instead of distances between vertices and their center. Consider the embedding of $(p_1, p_2, \ldots, p_n)$ centering at $c = \frac{1}{n}\sum_{i=1}^{n} p_i$, we have

$$\sum_{i=1}^{n} \|p_i - c\|^2 = \sum_{i=1}^{n} \left(p_i^T p_i + c^T c - 2c^T p_i \right)$$
$$= \sum_{i=1}^{n} p_i^T p_i - \frac{1}{n}\sum_{i=1}^{n}\sum_{j=1}^{n} p_i^T p_j,$$
$$\sum_{i=1}^{n-1}\sum_{j=i+1}^{n} \|p_i - p_j\|^2 = \frac{1}{2}\sum_{i=1}^{n}\sum_{j=1}^{n}(p_i - p_j)^T(p_i - p_j)$$
$$= n\left(\sum_{i=1}^{n} p_i^T p_i - \frac{1}{n}\sum_{i=1}^{n}\sum_{j=1}^{n} p_i^T p_j \right).$$

Thus the two metrics differ only by a constant factor n in the unweighted version. However, the weighted sum of square for pairwise distances, i.e., $\sum_i \sum_j w_{ij}\|p_i - p_j\|^2$, does offer more flexibility, because of the larger number of weights (n^2, compared to n in the formulation for spreading score). The reformulation and relaxation techniques in Section 3 can still be applied to handle the revised formulation with pairwise distances; the only difference is that the coefficient matrix in the objective function will be dense (as opposed to diagonal in Eqn. 13). The added flexibility is useful for certain purposes. For example, when two components are connected by a path with many registers and plenty of slack, their distance in the embedding is probably long. In such a case, further increasing their distance does not offer a clear advantage, and then we can reduce the corresponding weight in the objective.

We now discuss the relation between the proposed embedding and placement. We conjecture that our embedding problem is related to the dual of the placement problem in some sense. Roughly speaking, the placement problem asks to minimize wire length, with lower bounds on pairwise distances so that components do not overlap; the embedding problem asks to maximize pairwise distances, with upper bounds on wire length. Such "duality" indicates connections between spreading score and wire length after layout. This further justifies the use of spreading score as an estimator of layout-friendliness.

Kudva, Sullivan and Dougherty propose to use *sum of all-pairs min-cut* (in contrast to *sum of all-pair distance* in an embedding) to evaluate the *adhesion* of a gate-level netlist, and use it in logic synthesis to improve routability [17]. Their idea of using a structural metric to guide the generation of a layout-friendly netlist influences our work. However, we use different approaches: their metric is related to techniques used in the analysis of social networks [27], which has roots in classic graph theory; our technique is influenced by the geometric embeddings of graphs and the associated algebraic structures. We consider our metric advantageous in two aspects: (1) it is more layout-oriented; (2) thus it has the ability to capture timing information by relating interconnect delay to distance. On the other hand, the adhesion metric is probably advantageous in another two aspects: (1) it can be evaluated more efficiently;[2] (2) since

---

[2] An approximation of adhesion can be obtained in $O(n^2)$. The complexity of SDP with a fixed error bound is $O(n^3)$ for the dense case; sparsity and incremental solving can improve scalability for our problem.

cut size is used, it may be more closely related to average wire density in layout, and thus help to reduce congestion.[3]

Spreading score is indirectly related to cut size as well, as suggested by the following result from [12]. For a simplified version of the problem with $w_i = 1$ and $l_{ij} = 1$, the dual problem of the relaxation in Eqn. 13 can be transformed to an SDP formulation for calculating $\frac{n}{\hat{a}(G)}$. Here $\hat{a}(G)$ is the *absolute algebraic connectivity* of graph G, and it is shown to be related to the node connectivity as well as edge connectivity of G [11]. According to the SDP duality theory, the estimated spreading score we get is equal to $\frac{n}{\hat{a}(G)}$.

One limitation of spreading score is that it does not capture timing related to control signals (from the FSM controller) very well. This is because RTL synthesis tools often change the FSM controller drastically in optimization. Capturing controller timing is intrinsically difficult without logic synthesis. In our implementation, we exclude the FSM when constructing the graph; instead, we generate a Moore-style one-hot FSM to alleviate the problem.

6. CONCLUSION

A new metric of a netlist, spreading score, is proposed to measure layout-friendliness. It captures both structural properties and timing information, and can be estimated efficiently and stably. The usefulness of spreading score has been justified both theoretically and experimentally. New techniques introduced in this paper can potentially lead to interesting observations and solutions for other related metrics and problems. We consider this an interesting direction for future work.

Acknowledgements

This work was supported in part by the Semiconductor Research Corporation under Contract 2009-TJ-1879. The authors thank Janice Martin-Wheeler for her help in editing this paper.

REFERENCES

[1] Agility Design Solutions Inc. *Handel-C Language Reference Manual*, 2007.
[2] B. Borchers. CSDP, a C library for semidefinite programming. *Optimization Methods and Software*, 11(1):613–623, 1999.
[3] S. Boyd and L. Vandenberghe. *Convex Optimization*. Cambridge University Press, New York, NY, 2004.
[4] D. Chen and J. Cong. Register binding and port assignment for multiplexer optimization. In *Proc. Asia and South Pacific Design Automation Conf.*, pages 68–73, 2004.
[5] J. Cong, Y. Fan, G. Han, W. Jiang, and Z. Zhang. Platform-based behavior-level and system-level synthesis. In *Proc. IEEE Int. SOC Conf.*, pages 199–202, 2006.
[6] J. Cong, Y. Fan, G. Han, X. Yang, and Z. Zhang. Architecture and synthesis for on-chip multicycle communication. *IEEE Trans. on Computer-Aided Design of Integrated Circuits and Systems*, 23(4):550–564, April 2004.
[7] J. Cong, Y. Fan, and W. Jiang. Platform-based resource binding using a distributed register-file microarchitecture. In *Proc. Int. Conf. on Computer-Aided Design*, pages 709–715, 2006.

[8] J. Cong, B. Liu, G. Luo, and R. Prabhakar. Towards layout-friendly high-level synthesis. In *Proc. Int. Symp. on Physical Design*, pages 165–172, 2012.
[9] W. Dougherty and D. Thomas. Unifying behavioral synthesis and physical design. In *Proc. Design Automation Conf.*, pages 756–761, 2000.
[10] Y.-M. Fang and D. F. Wong. Simultaneous functional-unit binding and floorplanning. In *Proc. Int. Conf. on Computer-Aided Design*, pages 317–321, 1994.
[11] M. Fiedler. Laplacian of graphs and algebraic connectivity. *Combinatorics and Graph Theory*, 25:57–70, 1989.
[12] F. Göring, C. Helmberg, and M. Wappler. The rotational dimension of a graph. *Journal of Graph Theory*, 66(4):283–302, 2011.
[13] Z. Gu, J. Wang, R. Dick, and H. Zhou. Unified incremental physical-level and high-level synthesis. *IEEE Trans. on Computer-Aided Design of Integrated Circuits and Systems*, 26(9):1576–1588, 2007.
[14] C.-Y. Huang, Y.-S. Chen, Y.-L. Lin, and Y.-C. Hsu. Data path allocation based on bipartite weighted matching. In *Proc. Design Automation Conf.*, pages 499–504, 1990.
[15] T. Kim and X. Liu. Compatibility path based binding algorithm for interconnect reduction in high level synthesis. In *Proc. Int. Conf. on Computer-Aided Design*, pages 435–441, 2007.
[16] M. Kistler, M. Perrone, and F. Petrini. Cell multiprocessor communication network: Built for speed. *IEEE Micro*, 26(3):10–23, May 2006.
[17] P. Kudva, A. Sullivan, and W. Dougherty. Measurements for structural logic synthesis optimizations. *IEEE Trans. on Computer-Aided Design of Integrated Circuits and Systems*, 22(6):665–674, June 2003.
[18] A. Kuehlmann and R. A. Bergamaschi. Timing analysis in high-level synthesis. In *Proc. Int. Conf. on Computer-Aided Design*, pages 349–354, 1992.
[19] H. Kung. Why systolic architectures? *Computer*, 15(1):37–46, Jan. 1982.
[20] M. C. McFarland. Reevaluating the design space for register transfer hardware synthesis. In *Proc. Int. Conf. on Computer-Aided Design*, pages 262–265, 1987.
[21] B. M. Pangre. Splicer: a heuristic approach to connectivity binding. In *Proc. Design Automation Conf.*, pages 536–541, 1988.
[22] H. Park, K. Fan, S. A. Mahlke, T. Oh, H. Kim, and H.-s. Kim. Edge-centric modulo scheduling for coarse-grained reconfigurable architectures. In *Proc. Int. Conf. on Parallel Architecture and Compilation Techniques*, pages 166–176, 2008.
[23] N. Robertson and P. D. Seymour. Graph minors. II. Algorithmic aspects of tree-width. *J. Algorithms*, 7(3):309–322, 1986.
[24] I. E. Sutherland, C. E. Molnar, and R. F. Sproull. Counterflow pipeline processor architecture. Technical Report 94-25, Sun Microsystems Laboratories, 1994.
[25] L. Vandenberghe and S. Boyd. Semidefinite programming. *SIAM Review*, 38(1):49–95, 1996.
[26] J.-P. Weng and A. Parker. 3D scheduling: high-level synthesis with floorplanning. In *Proc. Design Automation Conf.*, pages 668–673, 1991.
[27] D. R. White and F. Harary. The cohesiveness of blocks in social networks: Node connectivity and conditional density. *Sociological Methodology*, 31(1):305–359, 2001.
[28] L. Zhong and N. K. Jha. Interconnect-aware low-power high-level synthesis. *IEEE Trans. on Computer-Aided Design of Integrated Circuits and Systems*, 24(3):336–351, 2005.

---

[3]Our metric is also useful for the optimization of wire density. In [8], we report some results on the correlation between a simpler version of spreading score and channel width in FPGA designs.

Computer Generation of Streaming Sorting Networks

Marcela Zuluaga
ETH Zurich
zuluaga@inf.ethz.ch

Peter Milder
Carnegie Mellon University
pam@ece.cmu.edu

Markus Püschel
ETH Zurich
pueschel@inf.ethz.ch

ABSTRACT

Sorting networks offer great performance but become prohibitively expensive for large data sets. We present a domain-specific language and compiler to automatically generate hardware implementations of sorting networks with reduced area and optimized for latency or throughput. Our results show that the generator produces a wide range of Pareto-optimal solutions that both compete with and outperform prior sorting hardware.

Categories and Subject Descriptors

B.5.2 [**Register Transfer Level Implementation**]: Design Aids—*automatic synthesis*; F.2.2 [**Theory of Computation**]: Nonnumerical Algorithms and Problems—*sorting and searching*

General Terms

Design

Keywords

Hardware Sorting, Design Space Exploration, HDL Generation

1. INTRODUCTION

Sorting is a fundamental operation that is required in a wide range of applications with different requirements in performance and cost. For example, continuous data processing applications may have throughput requirements that cannot be matched by sequential sorting algorithms running on a processor. On the other hand, applications with more relaxed performance requirements can benefit from offloading of time consuming operations such as sorting, by running them on dedicated hardware that can be implemented using simple components.

Two main categories of hardware sorters can be distinguished: sorting networks and linear sorters. Sorting networks operate in parallel over the input elements, processing them through a network of comparison-exchange elements. This solution offers great performance but becomes prohibitively expensive for large data sets because of the area cost or since elements can no longer be provided at the same time. Instead, linear sorters input one element at

a time, sorting them into a register array as they are received. This approach can offer small designs but fails to scale in performance.

In this paper, we fill the gap between these two approaches by introducing sorting networks that can offer a variable streaming width, while maintaining high throughput capabilities. We show that this approach dramatically reduces the footprint of sorting networks by effectively reusing hardware components to operate sequentially over the input elements. To the best of our knowledge, this type of reuse had not been done for sorting networks. The likely reason is in the challenge of streaming the permutation stages. We solve this problem using the method in [14].

Furthermore, we built a hardware generator that enables the automatic and systematic exploration of a large design space for hardware sorters. This way, designers can identify Pareto-optimal solutions that match their specific requirements. The generator makes use of a small Domain-Specific Language (DSL) to represent at a high level different known sorting networks. These DSL expressions are then annotated with implementation directives that capture different types and degree of reuse, which gives rise to a large and complex design space. Finally, a special DSL compiler generates Register-Transfer Level (RTL) descriptions for each desired design.

In summary, the contributions of this paper are

- to our knowledge the first high throughput streaming sorting networks with arbitrary streaming width;
- a DSL-based framework to describe sorting networks and architectural parameters;
- a hardware description generator that automatically translates DSL expressions into a hardware description language, thus enabling the automatic exploration of a large design space to identify Pareto-optimal solutions; and
- a wide range of hardware solutions for sorting that are highly optimized for area, throughput, latency and memory usage.

1.1 Related Work

Many sorting methods have been invented and studied in detail in the past decades [5, 8]. Among those, sorting networks are attractive due to their inherent parallelism and input-independent structure. Bitonic sorting networks, first presented in [2], have a regular structure and can sort $n = 2^t$ elements in $\frac{1}{2} \log_2 n (\log_2 n + 1)$ stages of $\frac{n}{2}$ parallel compare operations. At every stage, elements are permuted such that a sorted sequence is obtained at the final stage. Although other approaches have shown to require $O(\log_2 n)$ stages [1], the lack of regularity and the large constants hidden in this bound make them in practice less efficient than bitonic networks [12].

Several techniques have been proposed to reduce the area of sorting networks. In [15], it is demonstrated that a single physical stage, composed of an array of $\frac{n}{2}$ compare modules followed by a perfect shuffle permutation, can be used to sort n elements by recirculating them $(\log_2 n)^2$ times. A version of this implementation that can support several input sizes (up to some maximum) is presented in [6] but does not gain performance. Reconfigurable logic is used to change the required shuffles depending on the input width. The work in [7] explores area-throughput trade-offs with a pipelined architecture composed of $\log_2 n$ stages that recirculates the data within each stage. Our generator can produce each of these designs as special case. Further, we will show that through our novel approach to implementing reuse, we can considerably further improve the area/performance tradeoff and at the same time offer a much larger space of Pareto-optimal designs.

Not based on sorting networks is [11], which uses linear sorters [8] with the goal to improve throughput. The authors show how a stack of linear sorters can be used to parallelize the sorting of a joint list by interleaving the inputs of the system and the outputs of the individual linear sorters. Results show that interleaving creates significant delays due to conflicts and increases the complexity of the design, such that input widths greater than eight no longer represent a performance gain due to the low execution frequency of the resulting datapath. We include this work in our comparison and again show considerable and systematic gains in area and performance.

Finally, as one application example of hardware sorters, we cite [10], which evaluates various sorting networks on FPGAs, focusing on accelerating database applications. The authors demonstrate that sorting networks on FPGAs represent a competitive solution to sorting data sets of small sizes; our work may help to remove this restriction.

2. DSL FOR SORTING NETWORKS

In this section we introduce a small domain-specific language (DSL) to represent sorting networks, borrowing concepts from [3] and [9]. The elements of the language are structured operators (viewed as data flow graphs) that map vectors to vectors of the same length. For instance, the operator S_2 transforms a vector of two elements into a vector of two ordered elements. Each operator is either a first order operator or a composition using higher order operators. In the following, vectors are written as $x = (x_i)_{0 \le i < n} = (x_0, \ldots, x_n)$, and operators A_n are functions on these vectors of length n. Thus, $y = A_n x$ indicates that an operator A_n is applied to an input vector x, generating an output vector y. We formally introduce first order operators and higher order operators next.

First order operators. We define the following first order operators:

$$S_2 : (x_0, x_1) \mapsto (\min(x_0, x_1), \max(x_0, x_1))$$
$$I_n : x \mapsto x$$
$$J_n : (x_i)_{0 \le i < n} \mapsto (x_{n-1-i})_{0 \le i < n}$$
$$X_2^c : \begin{cases} I_2, & c = 0 \\ J_2 \circ S_2, & c = 1 \\ S_2', & c = 2 \end{cases}$$
$$L_m^n : (x_{ik+j})_{\substack{0 \le i < m \\ 0 \le j < k}} \mapsto (x_{jm+i})_{\substack{0 \le i < m \\ 0 \le j < k}}; \quad n = km$$

I_n is the identity operator, J_n flips the input vector, and L_m^n performs a stride-by-m permutation on n elements (also called corner turn, transposition, or shuffle). $L_{n/2}^n$ is called perfect shuffle. S_2 and X_2^c are the basic kernels for sorting networks. S_2 sorts 2 elements into ascending order, whereas X_2^c can be configured to

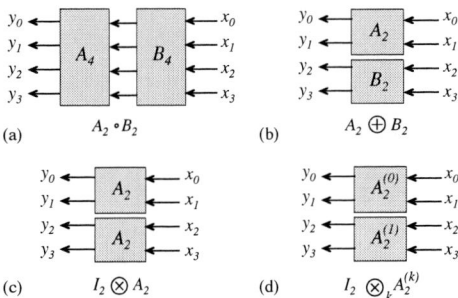

Figure 1: Higher order operators and associated data-flow graph structures: (a) composition, (b) direct sum, (c) tensor product, and (d) parameterized tensor product.

perform ascending sorting, descending sorting, or to preserve the original order of the input elements.

Higher order operators. The purpose of higher order operators is to recursively compose operators into more complex data-flow structures. We define the following.

- Composition ($\circ$): $A_n \circ B_n$ is the composition of operators, as shown in Fig. 1(a): the input vector is first mapped by B and then by A. The symbol $\circ$ may be omitted to simplify expressions. For an iterative composition we use the product sign: $A_n^{(0)} \circ \cdots \circ A_n^{(t-1)} = \prod_{i=1}^t A_n^{(i)}$. Since composition is applied from right to left, we draw dataflow diagrams accordingly.

- Direct sum ($\oplus$): $A_n \oplus B_m$ signifies that A_n maps the upper n elements and B_m the bottom m elements of the input vector (see Fig. 1(b)).

- Tensor product ($\otimes$): The expression $I_m \otimes A_n = A_n \oplus \cdots \oplus A_n$ replicates m times the operator A_n to operate in parallel on the input vector (see Fig. 1(c)).

- Indexed tensor product ($\otimes_k$): The expression $I_m \otimes_k A_n^{(k)} = A_n^{(0)} \oplus \cdots \oplus A_n^{(m-1)}$ allows for a change in the replicated operator through the parameter k (see Fig. 1(d)).

Bitonic sorting networks: Example. Our DSL is designed to represent sorting networks with a regular structure, specifically those based on bitonic sorting [5, pp. 230]. We define S_n as the sorting operator that transforms an input vector of size n into a sorted ascending sequence of the same size. We assume $n = 2^t$ is a two-power. Further, we define M_n as a bitonic merge operator that transforms a regular-bitonic sequence of size n into a sorted sequence of the same size. A regular-bitonic sequence is a concatenation of an ascending sequence and a descending sequence of size $n/2$. In the case of $n = 2$, $M_2 = S_2$.

The classical bitonic sorting network [2] consists of a sequence of t merging stages. In our DSL it becomes

$$S_{2^t} = \prod_{i=t}^1 [(I_{2^{t-i}} \otimes M_{2^i})(I_{2^{t-i}} \otimes (I_{2^{i-1}} \otimes J_{2^{i-1}}))]. \quad (1)$$

Specifically for $n = 8$,

$$S_8 = M_8 (I_2 \otimes J_4)(I_2 \otimes M_4)(I_2 \otimes (I_2 \oplus J_2))(I_4 \otimes M_2), \quad (2)$$

and the associated dataflow graph is shown in Fig. 2(a).

Each merger in (1) is recursively decomposed into smaller mergers. In our DSL,

$$M_{2^t} = (I_2 \otimes M_{2^{t-1}})L_2^{2^t}(I_{2^{t-1}} \otimes S_2)L_{2^{t-1}}^{2^t}. \quad (3)$$

1242

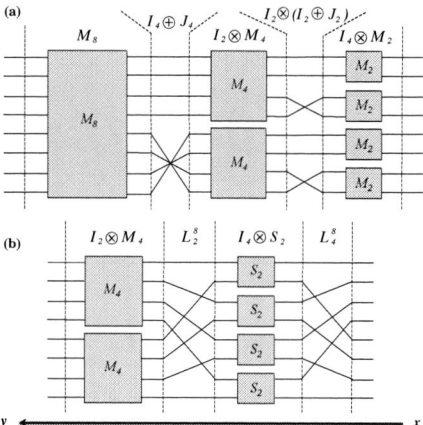

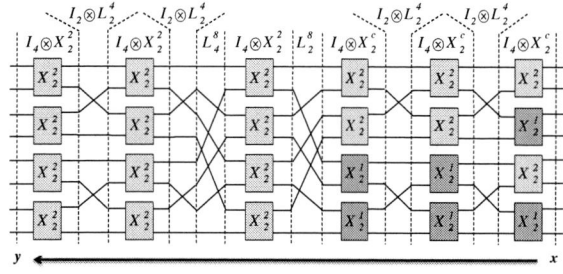

Figure 2: A bitonic sorter for $n = 8$: (a) S_8 is based on (1) and sorts by a sequence of mergers; each merger is again decomposed. For example, (b) M_8 is based on (3).

Figure 3: Data flow graph representation of a bitonic sorter of size 8 based on SN3. X_2^1 sorts the inputs in descending order and X_2^2 sorts the inputs in ascending order.

Fig. 2(b) shows the example M_8.

Recursive expansion of (3) yields a complete decomposition into basic blocks S_2:

$$M_{2^t} = \prod_{j=1}^{t} \left[(I_{2^{t-j}} \otimes L_2^{2^j})(I_{2^{t-1}} \otimes S_2)(I_{2^{t-j}} \otimes L_{2^{j-1}}^{2^j}) \right]. \quad (4)$$

Note that each of the t stages has a different permutation that connects the parallel S_2's. Similar to the Pease FFT [13], this expression can be manipulated using tensor product identities [4] into the "constant geometry" form

$$M_{2^t} = \prod_{j=1}^{t} \left[(I_{2^{t-1}} \otimes S_2) L_{2^{t-1}}^{2^t} \right]. \quad (5)$$

The permutation is now the same in each iteration. This manipulation is also applied to sorting networks in [6].

A complete sorting network is obtained by inserting either (4) or (5) into (1). Besides these two, several other variants with slightly different structure and number of stages have been reported in the literature. In our generator, we consider the following five sorting networks; the corresponding DSL expressions are in Table 1.

- SN1 is obtained by inserting (4) into (1) [2].
- SN2 is obtained by inserting (5) into (1) [6]. By using (5) as a

breakdown rule for M_n, the regularity of the design increases, but it also increases the complexity of the occurring permutations.

- SN3 is obtained by rewriting SN1 to use the operator X_2^c instead of S_2 to eliminate the J_n permutations [2]. SN3 represents the trade-off of eliminating a permutation at the cost of the additional control logic for X_2^c. SN3 for $n = 8$ is shown in Fig. 3.
- SN4 is obtained by rewriting SN2 to use the operator X_2^c, eliminating the permutations J_n [6].
- Each of the prior networks has $\frac{t}{2}(t + 1)$ stages with different permutations in each stage. SN5 has t^2 stages but perfect regularity [15]. Each stage consists of parallel pairwise comparisons, followed by the same perfect shuffle. Thus, SN5 increases the latency of S_n in exchange of the regularity in the permutation stages.

3. HARDWARE GENERATOR

The DSL expressions in Table 1 represent algorithms that can directly be translated into combinatorial logic. With proper pipelining, these designs offer maximal performance, but the area cost of these implementations quickly becomes prohibitive as n increases. To solve this problem our generator considers a much richer design space that arises from exploiting the regularity of the sorting networks to "fold" them by reusing sorting elements S_2 and X_2^c. We explain the procedure and the overall generator in the following.

Datapath reuse. The sorting networks in Table 1 exhibit regularity expressed through the iterative composition ($\prod$) and the tensor product ($\otimes$). These types of regularity can be mapped to a variety of sequential datapaths [9] as explained next.

First, $\prod$ indicates that an operator will be used in sequence more than once. Fig. 4 (top-left) illustrates a direct interpretation: each iteration gives an independent module in series. However, by employing *iterative reuse*, this same computation can be performed by building a single A_n module and recirculating the data as many times as required, as shown in Fig. 4 (top-right). In our DSL, we express this freedom by annotating the formula with a *depth* parameter d that indicates the number of modules to be implemented. A depth d smaller than the maximum t in $\prod_{j=1}^{t} A_n$ is possible if d evenly divides t and the permutations in the block A_n are not dependent on the iteration index j. This is the case in the inner composition in SN2 and SN4 and in both compositions in SN5.

Second, $I_k \otimes A_m$ indicates that operator A_m will be used k times in parallel. This is illustrated in Fig. 4 (bottom-left) for $k = \frac{n}{2}$, $m = 2$, where n data words flow into the system at the same time. Alternatively, we could perform the same computation by instead *folding* our data stream and datapath *vertically* as shown in Fig. 4 (bottom-right). Now the data words stream in and out of the system at a rate of 2 words per cycle, and flow through a single instance of A_2. We refer to this type of reuse as *streaming reuse*. We express it in our DSL by annotating expressions as shown in Fig. 4 with their *width* w, which indicates the number of inputs taken in each cycle. Our generator also supports $w = 1$, which requires that even the basic modules S_2 and X_2 are folded. In summary, any $w \geq 1$ that evenly divides n is a legal width for the expression $I_{n/2} \otimes A_2$.

By setting values of d and w, we specify a particular hardware implementation of an expression from our DSL, each with its own area and throughput. Larger values of d and w correspond to higher costs and throughput. Streaming and iterative reuse can also be combined. Table 2 shows for each network the number of stages in the datapath, the legal choices of d and w, and the associated number of sorters used.

1243

$$\text{SN1:} \quad \prod_{i=1}^{t-1}\left[(I_{2^{t-1}}\otimes S_2)\prod_{j=2}^{t-i+1}\left[\left(I_{2^{t-j}}\otimes(I_2\otimes L_{2^{j-1}}^{2^{j-2}})L_2^{2^j}\right)(I_{2^{t-1}}\otimes S_2)\right]\left(I_{2^i-1}\otimes\left(L_{2^{t-i+1}}^{2^{t-i}}(L_{2^{t-i}}\otimes J_{2^{t-i}})\right)\right)\right](I_{2^{t-1}}\otimes S_2)$$

$$\text{SN2:} \quad \prod_{i=1}^{t-1}\left[\prod_{j=2}^{t-i+1}\left[(I_{2^{t-1}}\otimes S_2)(I_{2^i-1}\otimes L_{2^{t-i}}^{2^{t-i+1}})\right]\left(I_{2^i-1}\otimes(I_{2^{t-i}}\otimes J_{2^{t-i}})\right)\right](I_{2^{t-1}}\otimes S_2)$$

$$\text{SN3:} \quad \prod_{i=1}^{t-1}\left[(I_{2^{t-1}}\otimes_m X_2^{g(i,m)})\prod_{j=2}^{t-i+1}\left[\left(I_{2^{t-j}}\otimes(I_2\otimes L_{2^{j-1}}^{2^{j-2}})L_2^{2^j}\right)(I_{2^{t-1}}\otimes_m X_2^{g(i,m)})\right]\left(I_{2^i-1}\otimes L_{2^{t-i+1}}^{2^{t-i}}\right)\right](I_{2^{t-1}}\otimes_m X_2^{g(i,m)})$$

$$\text{SN4:} \quad \prod_{i=1}^{t-1}\left[\prod_{j=2}^{t-i+1}\left[(I_{2^{t-1}}\otimes_m X_2^{g(i,m)})(I_{2^i-1}\otimes L_{2^{t-i+1}}^{2^{t-i}})\right]\right](I_{2^{t-1}}\otimes_m X_2^{g(t,m)}); \quad g(i,m)=\begin{cases}1, & m[t-i]=1\text{ and }i\neq 1\\ 2, & (m[t-i]=0\text{ or }i=1)\end{cases}$$

$$\text{SN5:} \quad \prod_{i=0}^{t-1}\prod_{j=0}^{t-1}\left[(I_{2^{t-1}}\otimes_m X_2^{f(i,j,m)})L_{2^t}^{2^{t-1}}\right]; \quad f(i,j,m)=\begin{cases}0, & t-1<j+i\\ 1, & m[t-1-j-i]=1\text{ and }i\neq 0\\ 2, & m[t-1-j-i]=0\text{ or }i=0\end{cases}$$

Table 1: DSL representation of five bitonic sorting networks S_{2^t}. $m[x]$ represents the value of the bit in position x of the binary representation of m.

Network	Sorting Stages	Implementation directives			Cost	
		Reuse	Ranges	Constraints	Number of 2-input sorters ($w\geq 2$)	2-input sorters used
SN1	$\frac{t}{2}(t+1)$	w	$1\leq w\leq n$	$w\mid n$	$\frac{w}{4}t(t+1)$	S_2
SN2	$\frac{t}{2}(t+1)$	w,d_i	$1\leq d_i\leq j$ $1\leq w\leq n$	$d_i\mid j$ $w\mid n$	$\frac{w}{2}\left(1+\sum_{i=1}^{t-1}d_i\right)$	S_2
SN3	$\frac{t}{2}(t+1)$	w	$1\leq w\leq n$	$w\mid n$	$\frac{w}{4}t(t+1)$	X_2 (2 states)
SN4	$\frac{t}{2}(t+1)$	w,d_i	$1\leq d_i\leq j$ $1\leq w\leq n$	$d_i\mid j$ $w\mid n$	$\frac{w}{2}\left(1+\sum_{i=1}^{t-1}d_i\right)$	X_2 (2 states)
SN5	t^2	w,d,d'	$1\leq d,d'\leq t$ $1\leq w\leq n$	$d\mid t,d'\mid t$ $w\mid n$	$\frac{w}{2}dd'$	X_2 (3 states)

Table 2: Implementation characteristics of the different break down rules in Fig. 1 when applying streaming and iterative reuse. The constraint $a\mid b$ means that a must divide b evenly. Last column: one state means that S_2 is used; otherwise X_2 is used.

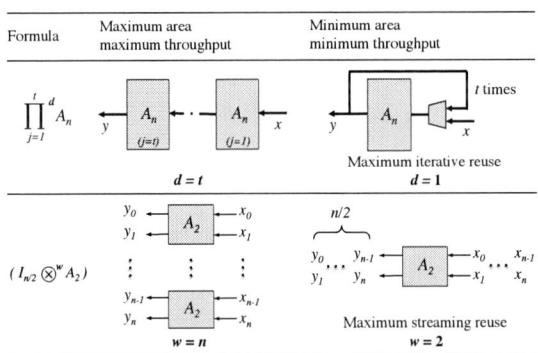

Figure 4: Iterative reuse and streaming reuse applied to expressions.

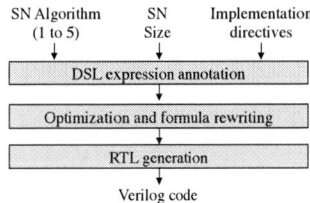

Figure 5: Flow diagram of the generator.

To the best of our knowledge, sorting networks with streaming reuse ($w<n$) have not been implemented before. The likely reason is in the challenge of streaming the required permutations. In our generator we use the method from [14], which uses simple two-ported memories and two-input switches to solve the problem for all permutations in Table 2.

RTL generation. A sorting network expressed in our DSL along-side its implementation directives w and d completely specifies a set of algorithmic and hardware implementation choices. We have created an automated hardware generator (an extension of Spiral [9]) that produces annotated DSL expressions and compiles them to synthesizable register-transfer level Verilog. Fig. 5 illustrates our system's flow.

First, one of the five algorithms in Table 1 is selected and adapted to reflect the user's choices for sorting network size n and implementation directives d and w. The operations for basic blocks S_2 and X_2^c are specified by an intermediate code representation. Next, a set of optimization and rewriting rules are applied with the goal of simplifying the expressions and improving the quality of the generated design. Then, the result is translated into synthesizable register-transfer level Verilog. The design is automatically

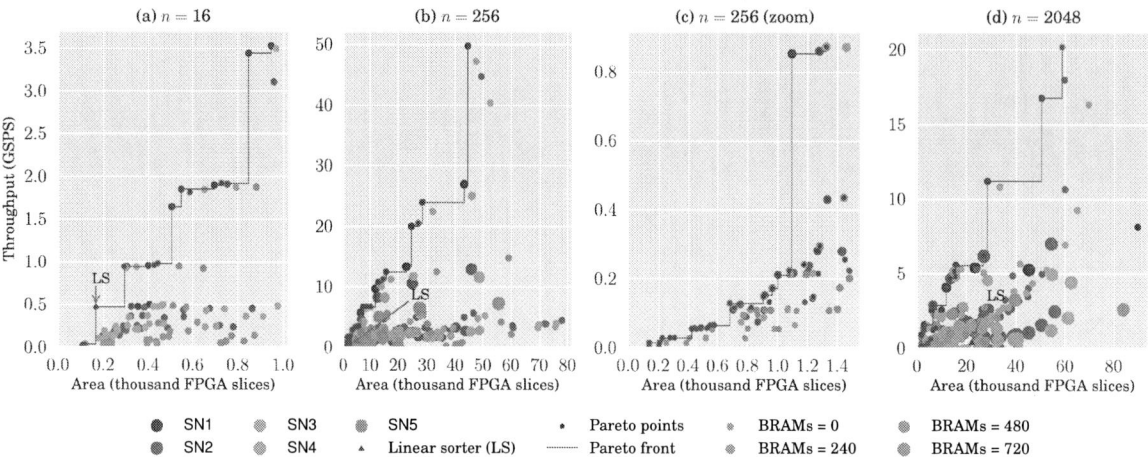

Figure 6: Throughput-area trade-offs for sorting networks of sizes: 16, 256 and 2048.

pipelined to maximize its achievable clock frequency, and timing analysis is performed to ensure that all signals route through the system and any feedback paths with correct timing. Other implementation options such as data type are additionally fed to the RTL generation stage.

Streaming permutations (where w is less than the number of points permuted) are implemented by the system as explained in [14]. Permutations that operate on non-streaming data (i.e., when all points to be permuted are available in parallel at the same time) are simply implemented with wires.

The generator additionally calculates the system's memory requirements and the latency and throughput of each block. It outputs the final design alongside Verilog testbenches for the verification of the created modules.

4. EVALUATION

With our generator we can systematically explore the large design space spanned by the five different sorting networks and the various reuse options. Each design has a different trade-off between area cost, performance, and memory requirements. In this section we present various experiments on an FPGA for small, medium, and large sizes. We identify Pareto-optimal designs and compare to prior work.

Experimental setup. Verilog descriptions are synthesized and place-and-routed for the Xilinx Virtex-6 XC6VLX760 FPGA using Xilinx ISE 13.1. All designs are generated to process 16-bit fixed-point data. We characterize each design by its cost: hard onchip memory units called Block RAM (BRAM) and the FPGA's reconfigurable slice usage, and by its performance: latency or throughput. The target FPGA includes 720 BRAMs of 36Kbits and 118,000 slices. Latency is measured in microseconds and throughput is measured in giga samples per second (GSPS); both are calculated with the system's maximum execution frequency, given by Xilinx ISE after place-and-route. BRAM and FPGA slice usage are taken from place-and-route reports. An additional parameter to enable or disable the usage of BRAMs was added to our generator, which allows us to further explore platform dependent trade-offs.

Exploring the design space. Fig. 6 shows an evaluation of the design space for the sizes $n = 16, 256, 2048$. All designs that fit onto the target FPGA are shown. 150 designs are shown for $n = 16$, 412 for $n = 256$, and 349 for $n = 2048$. Generating

the RTL for each design took a few milliseconds, while obtaining precise resource usage and timing information for each design took from minutes to hours, depending on the complexity of the implementation. In each plot, the x-axis is the number of slices used, the y-axis the throughput, and the size of the marker is proportional to the number of BRAMs used. Fig. 6(c) zooms into a small region of Fig. 6(b).

Among the designs, only the Pareto-optimal ones have to be considered for an application. The (throughput, area)-optimal designs are connected by a line, and all Pareto-optimal (considering area, throughput and BRAM-usage the target objectives) are marked by a black dot. As expected, the smallest design in all cases is obtained with SN5, $w = 2$ and $d = d' = 1$.

Fig. 6(a) shows the solutions found for $n = 16$. All possible designs fit in the target FPGA. The highest throughput is obtained with $w = 16$. Plot (b) shows all the solutions obtained for $n = 256$. The design that achieves best performance is SN1 with $w = 128$. Close to it we find SN2, SN3, SN4, all with $w = 128$ and no iterative reuse, which require more area and achieve lower execution frequencies. Plot (c) shows designs for $n = 256$ that fit in 2000 slices. The smallest designs are SN5, followed by SN2 and SN4. The smallest fully streaming designs, i.e., without iterative reuse, achieve best performance in the range: SN1, SN2, SN3, SN4 all with $w = 2$. Plot (d) shows all the solutions obtained for $n = 2048$. The design that achieves best performance is SN3 with $w = 64$.

For comparison purposes, we added to our design space an implementation (by hand) of a linear sorter, as described in [8, pp. 5]. It is part of the Pareto optimal set only for $n = 16$.

In summary, the set of Pareto-optimal points is non-obvious and composed of different networks and implementation directives. Our generator enables a systematic exploration and identification of these designs. The area requirements of a design could be estimated from the number of 2-input sorters given in Table 2 to guide designers on selecting only a subset of designs for hardware synthesis.

Comparison to prior work. A key contribution is not only the ability to generate designs but, to the best of our knowledge, the first designs for sorting networks with streaming reuse to dramatically reduce area cost, even for high throughput designs. We illustrate this contribution in Fig. 7, which shows the area cost (y-axis) of various sorters for increasing n (x-axis) that fit onto the

1245

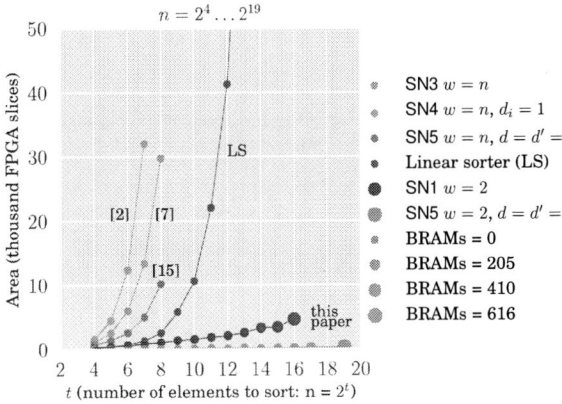

Figure 7: Area usage growth with the number of elements to sort $n = 2^t$.

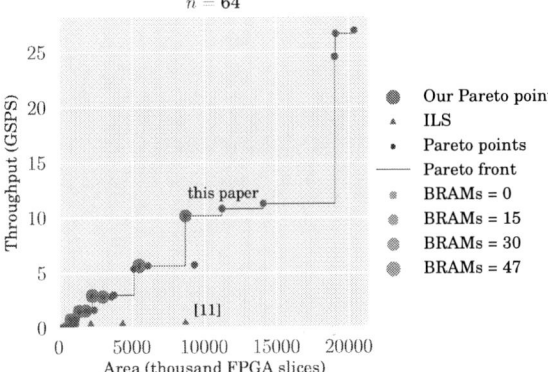

Figure 8: Streaming sorting networks generated in this paper in comparison with the interleaved linear sorters (ILS) from [11].

target FPGA. With prior work it was possible to build the designs of the four uppermost lines: networks without reuse (here: SN3 with $w = n$), iterative reuse in networks with constant geometry ([7] implements SN4 with $d_i = 1$ and [15] implements SN5 with $d = d' = 1$), and the linear sorter (which is not based on a network) from [8, pp. 5].

Our work enables, for example, the two bottom lines: fully streaming designs (here SN1 which processes $w = 2$ elements per cycle) and a design with maximal reuse (SN5 with $w = 2$ and $d = d' = 1$), i.e., it uses only one sorter X_2^c. Both designs require more BRAM but considerably less slices. Even a sorter for 2^{19} elements can be fit onto the target FPGA.

We also compared our designs with the interleaved linear sorters (ILS) presented in [11]. For $n = 64$, Fig. 8 shows throughput and area of our Pareto-optimal designs connected by a line and the ILS designs with $w = 1, 2, 4, 8$ from [11]. Again, the improvement is considerable.

5. CONCLUSIONS

We believe that for well-understood kernel functions IP core generators based on small domain-specific languages offer a practical solution that is hard to match by human designs. We have presented

such a generator for sorting. Using the generator, we could generate a large set of candidate designs that capture the existing algorithm knowledge of bitonic sorting networks as well as the various implementation options. The Pareto-optimal designs are non-obvious. Equally important, we used prior work on streaming permutations to present the first sorters with streaming reuse to dramatically reduce area and hence to enable much larger sorting problems to be processed at high throughput on FPGAs.

6. REFERENCES

[1] M. Ajtai, J. Komlós, and E. Szemerédi. An O(n log n) Sorting Network. In *Symposium on Theory of computing (STOC)*, pages 1–9. ACM, 1983.

[2] K. E. Batcher. Sorting Networks and their Applications. In *Spring Joint Computer Conference (AFIPS)*, pages 307–314. ACM, 1968.

[3] F. Franchetti, F. de Mesmay, D. McFarlin, and M. Püschel. Operator Language: A Program Generation Framework for Fast Kernels. In *IFIP Working Conference on Domain Specific Languages (DSL WC)*, volume 5658 of *Lecture Notes in Computer Science*, pages 385–410. Springer, 2009.

[4] J. R. Johnson, R. W. Johnson, D. Rodriguez, and R. Tolimieri. A Methodology for Designing, Modifying, and Implementing Fourier Transform Algorithms on Various Architectures. *Circuits, Systems, and Signal Processing*, 9:449–500, 1990.

[5] D. Knuth. *The Art of Computer Programming: Sorting and searching*. The Art of Computer Programming. Addison-Wesley Pub. Co., 1968.

[6] C. Layer and H.-J. Pfleiderer. A Reconfigurable Recurrent Bitonic Sorting Network for Concurrently accessible data. In *Field Programmable Logic and Application*, volume 3203 of *Lecture Notes in Computer Science*, pages 648–657. Springer, 2004.

[7] C. Layer, D. Schaupp, and H.-J. Pfleiderer. Area and Throughput Aware Comparator Networks Optimization for Parallel Data Processing on FPGA. In *Symposium Circuits and Systems (ISCAS)*, pages 405–408, 2007.

[8] F. Leighton. *Introduction to Parallel Algorithms and Architectures: Arrays, Trees, Hypercubes*, volume 1. M. Kaufmann Publishers, 1992.

[9] P. A. Milder, F. Franchetti, J. C. Hoe, and M. Püschel. Formal Datapath Representation and Manipulation for Implementing DSP Transforms. In *Design Automation Conference (DAC)*, pages 385–390, 2008.

[10] R. Mueller, J. Teubner, and G. Alonso. Data Processing on FPGAs. In *VLDB Endowment*, volume 2, pages 910–921, 2009.

[11] J. Ortiz and D. Andrews. A Configurable High-throughput Linear Sorter System. In *International Symposium on Parallel Distributed Processing (IPDPSW)*, pages 1–8, 2010.

[12] M. S. Paterson. *Improved Sorting Networks with O (log N) Depth*, volume 5. Springer, 1990.

[13] M. C. Pease. An Adaptation of the Fast Fourier Transform for Parallel Processing. *Journal of the ACM*, 15(2):252–264, 1968.

[14] M. Püschel, P. A. Milder, and J. C. Hoe. Permuting Streaming Data Using RAMs. *Journal of the ACM*, 56(2):10:1–10:34, 2009.

[15] H. Stone. Parallel Processing with the Perfect Shuffle. *IEEE Transactions on Computers*, C-20(2):153 – 161, 1971.

SUPPLEMENTAL MATERIAL

This section includes material that clarifies concepts from Section 2 or results from Section 4.

Fig. 9 shows the data flow graph representation of a bitonic sorter of size $2^3 = 8$ based on SN5. SN5 is a sorting network introduced in Section 2. It is composed of $3^2 = 9$ stages. Each stage performs $8/2 = 4$ parallel compare operations followed by the perfect shuffle permutation.

Table 3 shows FPGA slices and BRAMs used by some of the sorting networks that we generate. The table also includes a linear sorter that was separetaly implemented for comparison. The last design in the table (last 2 rows) are the smallest sorting networks that we generate for a given n. This data was used for the plot in Fig. 7.

Fig. 10 shows the area and latency of designs for size $n = 2048$ that fit into 10,000 slices. The smallest Pareto-optimal designs are provided by SN5 with various degrees of reuse. Most of the low-latency Pareto-optimal designs in the range are found with SN1 and SN3.

Fig. 11 and Fig. 12 zoom into a small region of the plots in Fig. 6(a) and (d) respectively. Both figures display the smallest solutions found in each design space.

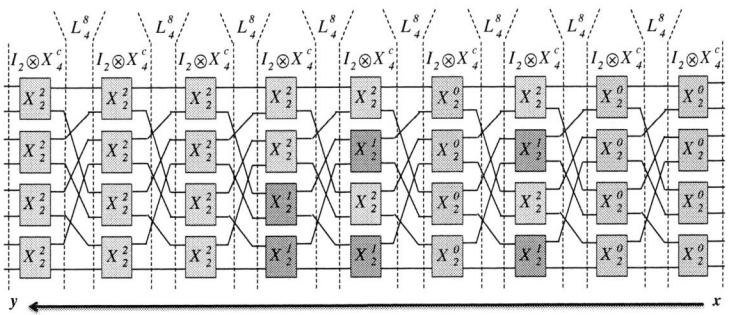

Figure 9: Data flow graph representation of a bitonic sorter of size 8 based on SN5. X_2^0 preserves the original order of the input elements, X_2^1 sorts the inputs in descending order, and X_2^2 sorts the inputs in ascending order.

n	16	32	64	128	256	512	1,024	2,048	4,096	8,192	16,384
t	4	5	6	7	8	9	10	11	12	13	14
SN3 $w = n$ FPGA slices	1,579	4,476	12,325	32,008							
BRAMs	0	0	0	0							
SN4 $w = n, d_i = 1$											
FPGA slices	1,067	2,546	6,009	13,337	29,776						
BRAMs	0	0	0	0	0						
SN5 $w = n, d = d' = 1$											
FPGA slices	727	1,372	2,591	5,007	10,154						
BRAMs	0	0	0	0	0						
Linear sorter (LS)											
FPGA slices	167	434	749	1,332	2,572	5,840	10,641	22,076	41,327	82,566	
BRAMs	0	0	0	0	0	0	0	0	0	0	
SN1 $w = 2$											
FPGA slices	303	483	582	890	1,092	1,391	1,567	1,876	2,134		
BRAMs	9	14	20	27	35	44	54	67	86	117	172
SN5 $w = 2, d = d' = 1$											
FPGA slices	120	122	124	136	136	161	180	231	159	249	195
BRAMs	1	1	1	1	1	1	1	2	5	10	19

Table 3: Area and BRAM usage for different network sizes n. This data corresponds to the trend lines displayed in Fig. 7.

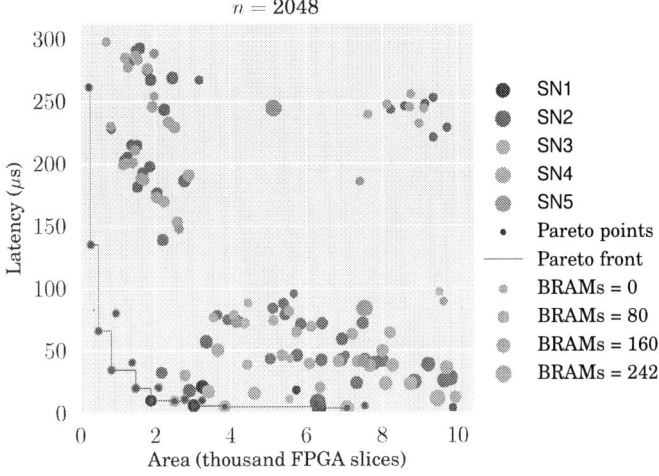

Figure 10: Latency-area trade-offs for sorting networks of size 2048.

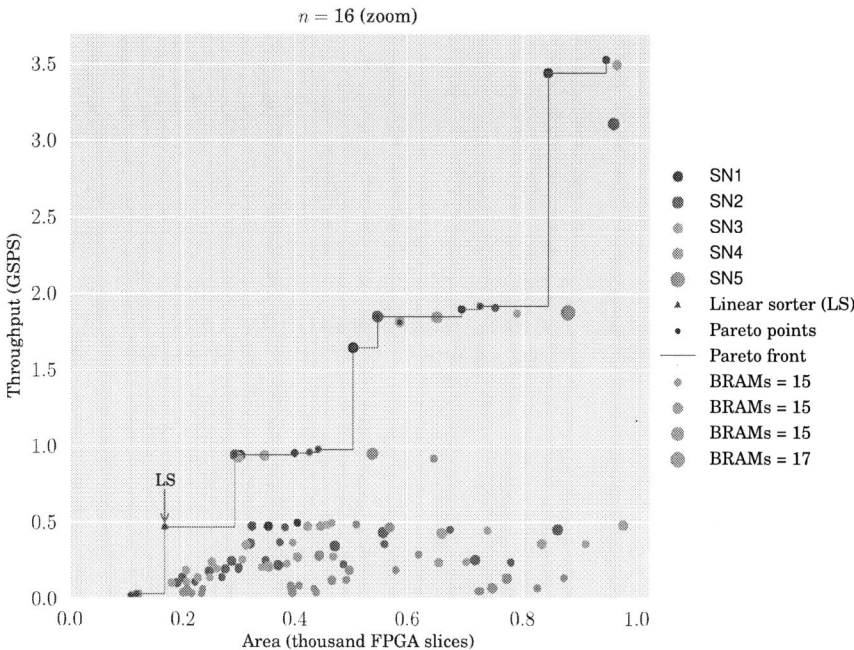

Figure 11: Throughput-area trade-offs for sorting networks of size 16 that fit in 1000 FPGA slices. All the design points are displayed in Fig. 6(a).

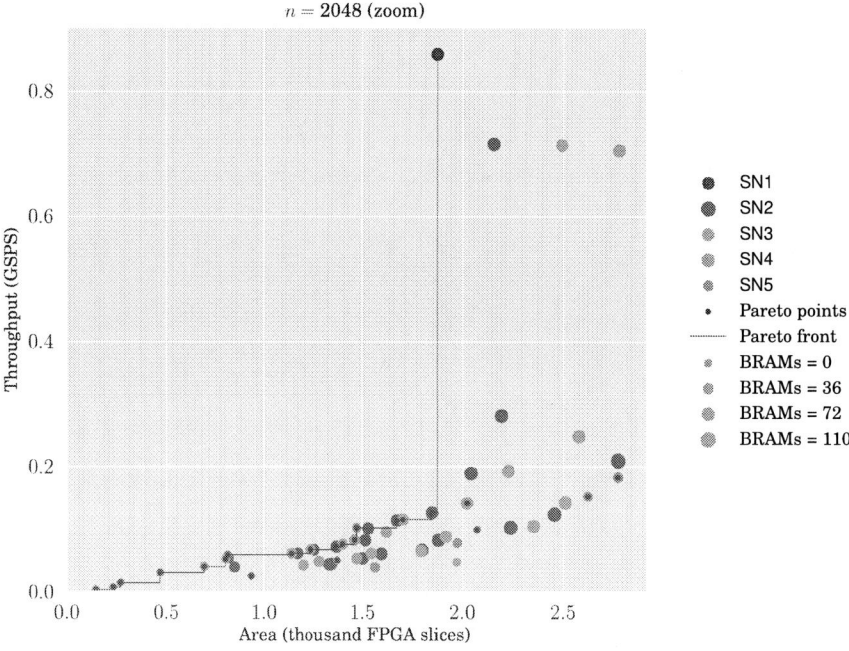

Figure 12: Throughput-area trade-offs for sorting networks of size 2048 that fit in 3000 FPGA slices. All the design points are displayed in Fig. 6(d).

CrowdMine: Towards Crowdsourced Human-Assisted Verification

Wenchao Li
UC Berkeley
wenchaol@eecs.berkeley.edu

Sanjit A. Seshia
UC Berkeley
sseshia@eecs.berkeley.edu

Somesh Jha
UW Madison
jha@cs.wisc.edu

ABSTRACT

We propose the use of crowdsourcing and human computation to help solve difficult problems in verification and debugging that can benefit from human insight. As a specific scenario, we explain how non-expert humans can assist in the verification process by finding patterns in portions of simulation or execution traces which are represented as images. Such patterns can be used in a variety of ways, including assertion-based verification, improving coverage, bug localization, and error explanation. Several related issues are discussed, including privacy and incentive mechanisms.

Categories and Subject Descriptors

B.7.2 [**Design Aids**]: Verification; H.1.2 [**User/Machine Systems**]: Human factors

General Terms

Algorithms, Verification, Human Factors

Keywords

Specification, verification, crowdsourcing, human computation

1. INTRODUCTION

The field of electronic design automation (EDA), in general, and formal verification, in particular, has relentlessly pushed for automation. For several problems, this is indeed the right strategy. But for many problems human insight and involvement remain invaluable. Consider, for example, the process of verifying a design. First of all, one needs to write a specification, typically in the form of properties (assertions) or a reference model. Second, one must create an environment model, typically in the form of constraints on the inputs or a state machine description. Next, one runs the verifier, such as a model checker, which is usually thought of as a "push-button" technique. While this is largely true, human insight is not entirely absent; e.g., one might need to supply hints to the verifier in the form of suitable abstraction techniques or (templates for) inductive invariants. If the verifier returns with a counterexample trace, one must debug the design by localizing the cause of error in time (relevant part of the trace) and space (relevant part of the design). Finally, the process of repairing the design to eliminate the bug is also one that needs human input. To summarize, even after decades of work on automating the verification process, we continue to need human insight in a variety of tasks, including writing specifications, creating models, guiding the verification engine, debugging and error localization, and repair.

This paper takes the position that while we cannot completely remove human insight from the verification process, we can change the way humans provide insight to the verifier. Today, such input typically comes from expert verification engineers, trained in the tools of their field. But such experts are few and expensive. And even experts have a hard time answering questions such as: When are we done verifying? Have we written enough properties? Where is the bug? And so on. We contend that the experts and automated

tools can be assisted in the verification process by a large crowd of non-expert humans performing simple, repetitive tasks. Each task involves a pattern recognition or other cognitive operations that humans are typically good at. The main technical challenges are to identify steps in the verification process where human insight is critical, find ways to transform these steps into tasks that non-expert humans can perform, and combine the results to resolve those steps in the verification process. As preliminary evidence to show that these challenges can be met, we present a system called CrowdMine for finding specifications from traces based on pattern recognition by humans.

The idea of tapping into a crowd of humans to assist in a computational task is not new. *Crowdsourcing* is the act of taking a job traditionally performed by a designated agent (usually an employee) and outsourcing it to an undefined, generally large group of people in the form of an open call [4]. *Human computation* is a paradigm for utilizing human processing power to solve problems that computers cannot yet solve [10]. (See Quinn and Bederson [7] for a more detailed description of these and related terms.) Our proposal is to use a combination of crowdsourcing and human computation to improve the state-of-the-art in verification. The availability of tools like Amazon's Mechanical Turk [1] and TurKit [6] make such a combination easier to deploy today.

In recent years, others have also advocated the use of crowdsourcing and human computation in design and verification, both for hardware and software. DeOrio and Bertacco [2] propose having humans assist in solving NP-complete problems arising in EDA, such as Boolean satisfiability (SAT) solving. Schiller and Ernst [8] propose the use of crowdsourcing and human computation for solving problems in software engineering, including software verification. The important difference between our proposal and these works is that we target steps in the verification process that *already* require human input, and which we think are unlikely to be automated entirely (similar to hard AI problems in the class of passing the Turing test, but unlike many NP-hard problems). We seek to leverage crowdsourcing and human computation to scale up the productivity in these steps manyfold.

To summarize, this paper makes the following contributions:

- Advocate the use of crowdsourcing and human computation for sub-tasks in verification that require human insight;
- Demonstrate the idea through CrowdMine, a novel game devised for finding patterns from system traces that can suggest likely specifications (Section 2), and
- Sketch out the landscape of similar applications (Section 3).

2. CROWDMINE

Many existing behavioral or specification mining techniques rely on the use of templates [3, 5]. Hence, it is the user's responsibiltiy to come up with a good set of templates. This process requires expert insight and is often incomplete. The main idea of CrowdMine is to tap into the human ability to recognize patterns in images to assist the process of mining specifications. For example, a trace can be visualized as a 2D image, where the rows are signals and the columns are cycles. CrowdMine first transforms segments of a trace into images and then queries a non-expert crowd to identify common patterns in those images. We have designed a game, described below, that incorporates these ideas.

2.1 Game Design

The purpose of our game is to discover other interesting patterns that do not match any of the pre-defined templates.

Game Design:

1. The player is presented with a set of two images along with

$m(1 \leq m \leq 3)$ patterns, as shown in Figure 1. Each pattern is a collection of squares which are not necessarily adjacent.

2. The player is asked to identify a pattern that is common in the two images but does not match any of the m patterns given. Additional constraints for the pattern can also be specified.

3. If the identified pattern is indeed common in the two images, points will be awarded. The number of points awarded is equal to the number of squares in the identified pattern. This scoring function directs the player's attention to more complex patterns.

4. A time limit of T (e.g. 15) seconds is enforced for each play.

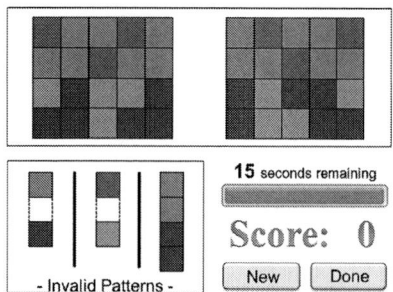

Figure 1: Example game interface: the task is to find a pattern that is common in the top two images but is not one of the invalid patterns.

Backend:

1. First, we sample a small set of n (e.g. 4) signals of interest and assign them an arbitrary order. Next, we randomly select two segments of k (e.g. 5) cycles in the trace and project them onto the n signals. This generates the two sub-traces for the game.

2. A template-based pattern mining algorithm is run to find common patterns that exist in these two sub-traces. Three of these mined patterns are selected in random as invalid patterns.

3. When each play finishes, the pattern selected by the player is checked if it is indeed a common pattern in both images.

4. A correct pattern is added to a database for further refinement.

GUI design:

- Different color ranges are used to encode different types of signals, e.g. input signals vs. output signals.
- Different shades of color are used to encode different signal values, e.g. dark for 0 and light for 1 for binary signals.
- The player can click on a square to select it. The selected square will be highlighted.
- When the player clicks "Done", the collection of selected squares in each image is considered as the identified pattern.

The example above is a 2-input and 2-output round-robin arbiter. We use this simple example to illustrate our workflow. The bottom two rows are the two request signals req_0 and req_1. The top two rows are the two response signals $resp_0$ and $resp_1$. The meaning of the patterns given are described below.

- (A): "When req_0 is low, $resp_0$ is low."
- (B): "When req_1 is high, $resp_1$ is high."
- (C): "All signals are low in one cycle."

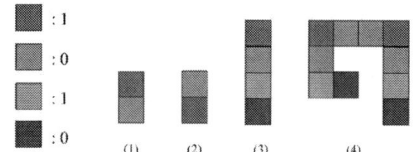

Figure 2: Color Code and Example Player-Identified Patterns

Some player-identified patterns are shown in Figure 2.

- (1) and (2): "Only one response is high at any cycle."
- (3): "When req_1 is high and there is no competing req_0, $resp_1$ is high at the same cycle."
- (4): This is the largest common pattern in the two images. Note that this pattern is an artifact of the two sub-traces chosen rather than a universal behavior that characterizes the arbiter. We can

rank identified patterns by frequency of occurrence, thus filtering artifacts that appear rarely across many pairs of sub-traces.

2.2 Discussion

Privacy. Our design is particularly attractive for companies that value confidentiality because the internals of the circuit are not revealed. For IP protection, the mapping of sub-traces to images should be kept confidential. This mapping include the correspondence of signals in the circuit, the color code, and any additional transformation on the subtraces. Randomization can also be used in selecting sub-traces and the mapping to images. Finally, *secret sharing* methods such as the *threshold schemes* developed by Shamir [9] are particularly relevant in this context.

Incentives. Three mechanisms are possible. (1) *Necessity:* Authentication systems such as reCAPTCHA [10] embed queries into a human challenge with partially known answers. Our game design can be augmented for this purpose. For example, two plays are presented to the user in series in which the answer is known for one of the plays. (2) *Enjoyment:* Our game design can be viewed as a puzzle game and the player derives enjoyment by solving it. In addition, the scoring-based system invites human competition and can attract a larger crowd. (3) *Profit:* Platforms such as Amazon's Mechanical Turk [1] provide a *for-profit* medium for crowdsourcing any human intelligence task. We plan to deploy our game on the Mechanical Turk to evaluate the proposed approach.

Human-Computer Collaboration. It is possible to combine algorithmic techniques with inputs from humans to achieve something better than what can be accomplished by either solely humans or a completely automated approach. In our setting, the human-identified patterns can be further refined (e.g. ranked) to produce the most relevant ones based on feedback from the back-end verification and debugging processes. They can also be used in automated tasks such as bug localization [5].

3. LOOKING AHEAD

We believe several games similar to CrowdMine can be created and applied to a range of applications in verification, debugging, and related areas. For example, one can improve coverage of a design by properties (or tests) by highlighting parts of a trace corresponding to variables not covered by (enough) properties, and users can be provided incentives to find patterns involving those parts. Properties generated by a system like CrowdMine can be hypothesized as auxiliary inductive invariants to speed up verification. Human-observed patterns in spurious counterexamples could potentially enable better abstraction-refinement in model checking. Finally, the process of debugging has similarities to investigating a crime scene (!) — the "crime" is the manifestation of the error (the failure), and one seeks to find a cause-and-effect chain that explains how the failure happened; this analogy suggests a natural game that could be formulated for non-expert humans to assist in debugging.

Acknowledgements. This research was supported in part by NSF grant CNS-0644436, an Alfred P. Sloan Research Fellowship, and the Gigascale Systems Research Center, one of six research centers funded under the Focus Center Research Program (FCRP), a Semiconductor Research Corporation entity.

4. REFERENCES

[1] Amazon's mechanical turk. www.mturk.com/mturk/welcome.

[2] A. DeOrio and V. Bertacco. Human computing for EDA. In *Design Automation Conference (DAC)*, pages 621–622, 2009.

[3] M. D. Ernst, J. H. Perkins, P. J. Guo, S. McCamant, C. Pacheco, M. S. Tschantz, and C. Xiao. The daikon system for dynamic detection of likely invariants. *Sci. Comput. Program.*, 69(1-3):35–45, 2007.

[4] J. Howe. Crowdsourcing: A definition. http://crowdsourcing.typepad.com.

[5] W. Li, A. Forin, and S. A. Seshia. Scalable specification mining for verification and diagnosis. In *Proceedings of the Design Automation Conference (DAC)*, pages 755–760, June 2010.

[6] G. Little, L. B. Chilton, M. Goldman, and R. C. Miller. TurKit: Human computation algorithms on mechanical turk. In *Proc. 23nd ACM symposium on User interface software and technology (UIST)*, pages 57–66, 2010.

[7] A. J. Quinn and B. B. Bederson. Human computation: A survey and taxonomy of a growing field. In *ACM Conference on Human Factors in Computer Systems (CHI)*, 2011.

[8] T. W. Schiller and M. D. Ernst. Rethinking the economics of software engineering. In *In Workshop on the Future of Software Engineering Research*, pages 325–330, 2010.

[9] A. Shamir. How to share a secret. *Commun. ACM*, 22:612–613, November 1979.

[10] L. von Ahn. *Human Computation*. PhD thesis, Carnegie Mellon University, December 2005.

1251

Extracting Design Information from Natural Language Specifications

Ian G. Harris
Center for Embedded Computer Systems
University of California Irvine,
harris@ics.uci.edu

ABSTRACT

Natural language specifications are the first concrete behavioral description which is the basis for any manually generated formal behavioral model. Natural language is preferred as the initial description method mainly because it is much simpler for a designer to use than existing hardware description languages. The focus of this project is the extraction of behavioral information from a natural language specification to generate a formal behavioral description with clear and unambiguous semantics. In the initial effort presented here, we employ semantic parsing to identify key information describing bus transactions in the natural language specification. The identified information is used to generate Verilog tasks which embody bus transactions. To our knowledge, the work presented here is the first attempt to generate simulatable Verilog from natural language descriptions.

Categories and Subject Descriptors

B.1.2 [**Control Structure Performance Analysis and Design Aids**]: Automatic Synthesis, Formal Models

General Terms

Design

Keywords

Natural Language Processing, Synthesis, Behavioral Modeling

1. INTRODUCTION

The integrated circuit (IC) design process has evolved greatly, from the manual layout of a small number of components to the automated design of ICs containing billions of transistors. To accommodate the dramatic increases in design complexity, the field of electronic design automation (EDA) was born, starting with simple schematic capture tools and culminating in the complex automation tools available today. EDA tools have proven effective in supporting synthesis and verification tasks, but the initial behavioral model must be generated manually by human experts. Manually generating a behavioral description is ex-

pensive, requiring significant time and a large number of well-trained design and verification engineers. A large part of the verification process is devoted to detecting and fixing design errors created during the processes of creating a behavioral description. Natural language specifications are the first concrete behavioral description which is the basis for the manually generated formal behavioral model. The task of interpreting natural language specifications has been exclusively manual because, generally speaking, only humans with expert design knowledge have had the ability to properly interpret specification documents.

The focus of this research is to **automatically formalize natural language specifications**. We investigate the extraction of behavioral information from a natural language specification to generate a formal behavioral description with clear and unambiguous semantics. This line of research has the following benefits.

1. *Reduced time-to-market:* The time-consuming process of manually generating a formal behavioral description will be reduced or eliminated.

2. *Reduced number of design errors:* The number of design errors introduced while generating a formal behavioral description will be reduced as the process becomes less dependent on manual interaction.

3. *Early identification of incomplete and inconsistent specifications:* Flaws in the original specification can be identified through their impact on the formal behavioral model.

The expressiveness of natural language guarantees that there are a multitude of different, valid descriptions for any single behavior, which vary morphologically and syntactically. Our information extraction technique must be robust in the face of syntactic variations commonly used by designers in hardware specifications. **Semantic parsing** is a technique for semantic analysis which has been applied to extract document information in many different applications including world geography [5], air travel [2], and more recently biomedical data [4]. Semantic parsing uses a context-free grammar (CFG) and includes domain-specific symbols with explicit semantic meaning. We apply semantic parsing to the domain of hardware specifications, enabling syntactic constructs to be mapped to behavioral templates which will be used to structure behavioral descriptions. By adding appropriate productions to the context-free grammar, we can easily extend the range of syntactic variations which can be recognized using our approach.

2. SYSTEM OVERVIEW

Figure 1 depicts the structure of the proposed system for transactor generation from a natural language description.

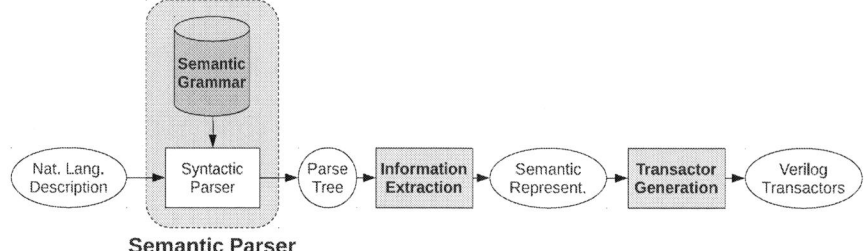

Figure 1: Structure of the Natural Language Transactor Generation System

1. Transmitting an address is performed by sending 7 address bits.
2. Sending an address bit is performed by setting SDA to a value while SCL is low and generating a pulse on SCL.

Figure 2: I^2C Write Transaction, selected sentences

The natural language description is processed using a **Semantic Parser**, which we have developed, to generate a parse tree. The semantic parser is built using an off-the-shelf syntactic parser, the Natural Language Toolkit [1], and a semantic grammar which we define for this application. Information Extraction is applied to the resulting parse tree to generate a semantic representation which contains all behavioral information about each transaction. The semantic representation is used to perform Transactor Generation and generate a set of Verilog transactors which accurately model the behavior of the specified transactions.

3. EXPERIMENTAL RESULTS

The system was implemented in Python using the API provided in the Natural Language Tool Kit [1] to create a recursive descent syntactic parser. All results were generated on a 2 GHz Intel Core 2 Duo processor with 1 GB RAM. To evaluate our system we have generated a bus transactor for the I^2C serial protocol developed by Philips [3] to support on-board communication. The protocol uses two wires, the data line SDA and the clock line SCL whose rising edges synchronize data transmission. We use a 10 sentence natural language specification of the write transaction which we have developed manually.

An unique **Transaction** object is created to capture the information in each sentence of the natural language description, forming a transaction graph, with the *Write transaction* as the top-level node. The bus transactor generation process was performed in 61.5 seconds of CPU time. The resulting transactor, named **i2c_write**, is composed of 32 lines of Verilog code. We have validated the correctness of the resulting **i2c_write** code manually and by simulating several write transactions.

We examine in detail the part of the I^2C write transaction which defines the act of *Transmitting an address* as defined in the two sentences, in Figure 2. Figure 3a shows a graph representation of the *Transmitting an address* transaction. The transaction in Figure 3a includes the *Sending an address bit* transaction from sentence 2 which is annotated with the number 7 to represent the number of iterations. The **q_extract** task call removes a bit from the address input queue queue and assigns SDA to the bit value. The resulting Verilog code for *Transmitting an address* is shown in Figure 3b.

4. CONCLUSIONS

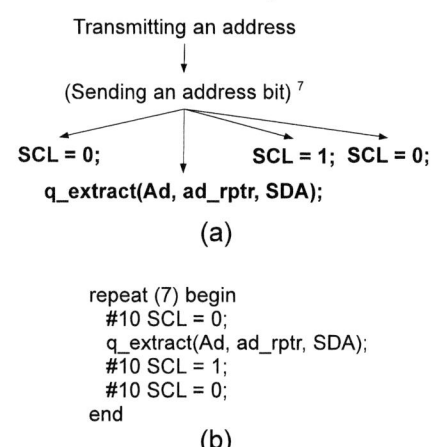

```
repeat (7) begin
    #10 SCL = 0;
    q_extract(Ad, ad_rptr, SDA);
    #10 SCL = 1;
    #10 SCL = 0;
end
```

(b)

Figure 3: "Transmitting an address" transaction, (a) transaction graph, (b) Verilog code

We have presented an approach to processing natural language protocol specifications to generate simulatable Verilog transactors. We define a semantic grammar which is used to identify key information in the natural language description during the parsing process. The range of English grammatical constructs which can be parsed with our current tool is broad enough to capture a transaction in a well-used bus protocol. To our knowledge, this work represents the first effort to generate a simulatable model from a natural language description.

5. REFERENCES

[1] Steven Bird, Ewan Klien, and Edward Loper. *Natural Language Processing with Python — Analyzing Text with the Natural Language Toolkit*. O'Reilly Media, 2009.

[2] S. Issar and W. Ward. Cmu's robust spoken language understanding system. In *EUROSPEECH'93*, pages 2147–2150, 1993.

[3] NXP Semiconductors. *I2C-bus specification and user manual*, rev. 03 edition, June 2007. UM10204.

[4] Pedro D. Poon H. Unsupervised semantic parsing. In *Proceedings of EMNLP*, 2009.

[5] D. H. D. Warren and C. N. Pereira. An efficient easily adaptable system for interpreting natural language queries. *American Journal of Computational Linguistics*, 8(3-4):110–122, 1982.

Material Implication in CMOS: A New Kind of Logic

Elkim Roa, Wu-Hsin Chen and Byunghoo Jung
Electrical and Computer Engineering Department
Purdue University, West Lafayette, IN USA
{elkim, chen279, jungb@purdue.edu} *

ABSTRACT

For more than seventy years, all the development in digital electronics have been founded on Shannon's work based on the fact that Boolean logic operators, OR, AND and NOT, can form a computationally complete logic framework. We propose a new paradigm in logic circuit design using material implication logic operators, different from the traditional logic gates in implementation and operation. In this paper we present early evidences, with experimental silicon results, showing that this new logic framework significantly improves performance, power and speed, over an equivalent conventional-logic framework in CMOS. This new computing paradigm would enable the continuance of increasing computing functionality and performance with decreasing cost in silicon technologies.

Categories and Subject Descriptors

B.6 [**Logic Design**]: Design Styles - Combinatorial Logic

General Terms

Performance, Design, Measurement, Theory

Keywords

Material Implication, High-Speed Logic

1. INTRODUCTION

A hundred years ago, the first volume of Whitehead and Russell's monumental work Principia Mathematica was published [1]. The book presented how the truths of math could be derived from logic operations. They described four fundamental logic operations, three of them used 27 years later by Shannon [2], and a fourth logic operation called material implication, A IMP B (also denoted as A → B), which reads A implies B, and is equivalent to (NOT A) OR B. Russell and Whitehead emphasized the relevance of this logic

---

*E. Roa is also with U. Industrial de Santander in Colombia.

operation by showing that only the IMP and NOT operations are enough to form a computationally complete logic basis. Because of the fact that Boolean logic operations, OR, AND, and NOT, form a computationally complete logic basis and they can be easily implemented using switching devices, modern digital electronics have been founded on Boolean algebra, often referred to as switching algebra. In addition, the great achievement in modern electronics seems to indicate no need for extra logic functions. Consequently, the fourth fundamental logic operator, material implication (A IMP B), has been ignored during last a few decades and disappeared in many electronics textbooks (still plays an important role in logic theory textbooks). A year ago, a letter in the journal Nature, presented that memristors can naturally execute the material implication operation, and inherently provides logic-in-memory [3]. However, these devices were fabricated using two layers of platinum-wire separated by an active layer of TiO_2, requiring a different process platform than current silicon process. Furthermore, reported speed performance was considerably low.

In this research, we will show that CMOS circuits can also be used to perform material implications, IMP and NIMP (negated IMP). They are unit gates, which are not based on traditional Boolean logic operation, (NOT A) OR B. After presenting their effectiveness for high-speed and low power operation in a frequency divider design, we will discuss about their potential extension to a complete logic basis, which would affect modern computing paradigm if successful.

2. IMPLICATION LOGIC IN CMOS

A set of logic functions is called complete if and only if any possible logic function can be composed by a combination of functions of the set. The operation logic NAND by itself, or NOR by itself, can be used to construct any Boolean logic, making the NAND and the NOR complete operations. It has been proved that the following set of pairs are complete: {OR, NOT}, {AND, NOT}, {IMP, NOT}, {IMP, FALSE}, {NIMP, NOT}, {NIMP, TRUE}, {IMP, XOR}, {NIMP, XNOR} and {NIMP, IMP}. As a consequence, we can take any pair and develop a complete logic system with their respective algebra. Interestingly, the operation cells, IMP and NIMP, are common in most of the pairs. The truth table for the basic material implication operation is indicated in Figure 1a, meaning that if A is true, then the output after the implication operation will be B. The stateful nature of the implication operation, automatically provides a stateful logic operation if the operation can incorporate a memory. For example, implementing the IMP and NOT op-

1254

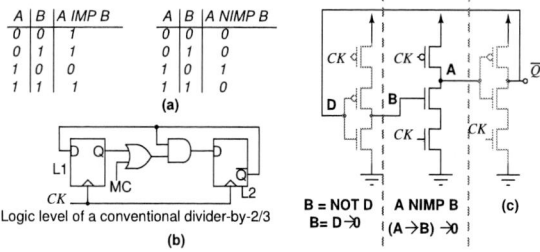

(a)

Logic level of a conventional divider-by-2/3

(b)

B = NOT D
B = D → 0

A NIMP B
(A → B) → 0

(c)

Figure 1: (a) Truth table of IMP and NIMP. (b) Conventional frequency divider-by-2/3. (c) Frequency divider-by-2/3 using NIMP operation.

erations in such a way that they behave as memory devices at the same time, enables them to recursively synthesize all the logic operations, as shown in Table 1.

Table 1: Boolean operations using material implication and false operations.

NOT A	$A \to 0$
A OR B	$(A \to 0) \to B$
A NOR B	$((A \to 0) \to B) \to 0$
A AND B	$(A \to (B \to 0)) \to 0$
A NAND B	$A \to (B \to 0)$
A XOR B	$(A \to B) \to ((B \to A) \to 0)$
A XNOR B	$((A \to B) \to ((B \to A) \to 0)) \to 0$

3. DESIGN EXAMPLE

The operations NAND, NOR, OR, AND and NOT, have been used so far as the key gates to synthesize most of the digital logic in CMOS. However IMP and NIMP operators, can provide a shortcut function to optimize certain logic operations at circuit level. We present a dual-modulus divide-by-2/3 shown in gate level at Figure 1b as an example. This is an appropriate example considering it involves combinatorial and sequential operators. This circuit is commonly used in frequency synthesizers where speed performance and power consumption are key features. The conventional wisdom is to synthesize the circuit using the traditional switching logic concepts proposed by Shannon. For instance, recently published prescalers have proposed to add the AND and OR logic gates within the flip-flops to reduce the number of stages in cascade, reducing the path delay [4]. However, in this approach, parasitic capacitances at internal nodes increase, requiring greater current driving capability and consequently more power consumption and slower speed.

Alternatively, the logic gate that includes AND and OR can by replaced by a dynamic NIMP to provide the same functionality. The NIMP operation is achieved by shifting the DC level of the clock signal connected to the flip-flop shown in Figure 1c. By doing this, the operation performed at node A is $(A \to B) \to 0$, which is the same as A NIMP B. The complete performed operation within the flip-flop is explained in Table 2, where 0+ and 1+ indicate the new logic values with the clock level shifted up. As a result, the final operation is $(A \to (D \to 0)) \to 0$, where B is D → 0. This operation is equivalent to A AND D, which is the

Table 2: NIMP operation explained.

A	B	$A_{CK=0}$	$A_{CK=1}$	$A_{CK=1+}$	$A_{CK=0+}$
0	0	1	0	0	0
0	1	1	0	0	0
1	0	1	1	1	1
1	1	1	0	0	0

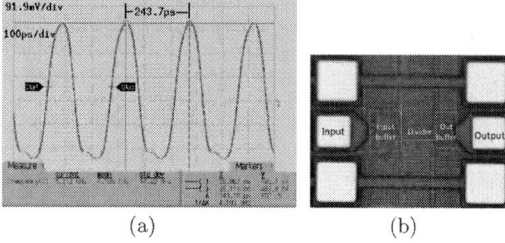

(a)   (b)

Figure 2: (a) Measured output in divide-by-3 operation mode. (b) Microphotograph of divider-by-2/3.

required function to have the circuit operating in divide-by-3 mode. The divide-by-2 operation mode is performed by the using the normal clock DC level, where the circuit operates as a flip-flop connected in feedback mode. Consequently, a divider-by-2/3 is designed using just one flip-flop by considering the NIMP and NOT operations without requiring extra logic gates. Experimental results indicate a 2.5 folds of speed improvement and a 4 folds of power reduction compared to the traditional AND and OR based frequency dividers presented in [4]. The measured output of the frequency prescaler in divide-by-3 operation mode, is presented in Figure 2(a) given an input frequency of 12.3GHz. The prescaler is implemented in $0.13\mu m$ 1.2V CMOS technology. Figure 2(b) shows a chip microphotograph.

4. DISCUSSION

We have introduced new implication logic cells significantly different from those conventionally used in Boolean logic cells. The initial evaluation based on a reconfigurable 2/3 bit counter shows that, these new logic cells are effective in achieving high-speed and low-power operation and have a great potential to form a highly efficient complete functional logic library based on {IMP, NOT}, {IMP, FALSE}, {NIMP, NOT}, {NIMP, FALSE}, or {NIMP, IMP} pairs. If successful, it will open a plethora of new applications and approaches for logic synthesis for future high-speed and low-power computing systems.

5. REFERENCES
[1] A. Whitehead and B. Russell, *Principia Mathematica.* Cambridge at the University Press, 1910.

[2] C. Shannon, "A Symbolic Analysis of Relay and Switching Circuits," Master's thesis, MIT, 1937.

[3] J. Borguetti et al., "'Memristive' Switches Enable 'Stateful' Logic Operations Via Material Implication," *Nature*, vol. 464, pp. 873–876, April 2010.

[4] M. Khrisna et al., "Design and Analysis of Ultra Low Power True Single Phase Clock CMOS 2/3 Prescaler," *IEEE Transaction on Circuits and Systems I*, vol. 57, no. 1, pp. 72–82, Jan. 2010.

Boolean Satisfiability using Noise Based Logic

Pey-Chang Kent Lin, Ayan Mandal, and Sunil P. Khatri
Texas A&M University, College Station, TX 77843, USA

ABSTRACT

Noise-based Logic (NBL) is a probabilistic logic system which can be used to simultaneously apply a superposition of arbitrarily many input vectors to a SAT instance. Using this property, we can determine whether an instance is SAT in a single operation. A satisfying solution can be found by iteratively performing SAT checks up to n times, where n is the number of variables in the SAT instance. In this paper, we formulate NBL-based SAT, and discuss its scalability. The NBL-based SAT engine has been simulated in software for validation purposes, although the focus of the paper is on the theory of NBL-based SAT.

Categories and Subject Descriptors

B.6 [**Hardware**]: Logic Design; I.1 [**Computing Methodologies**]: Symbolic and Algebraic Manipulation

General Terms

Algorithms, Theory

Keywords

Boolean Satisfiability, noise based logic

1. INTRODUCTION

Recently, it was shown that noise can be used to realize logic circuits [1, 2, 3]. We refer to this logic scheme as Noise-based Logic (NBL) in the sequel. In NBL, a plurality of uncorrelated noise sources (referred to as *noise bits*) are utilized where each noise source has zero mean. NBL is a *probabilistic logic scheme* with a precisely quantifiable threshold of error. NBL can be utilized to realize multi-valued logic as well [3, 4].

The orthogonality of the noise bits yields some powerful properties. One of these is the ability to apply all possible inputs to an n input NBL circuit *simultaneously* using $2n$ uncorrelated noise sources. Consider a combinational circuit with n inputs $x_1, x_2, \cdots x_n$. For each input x_i, let us assume we have a noise source (noise bit) $N_{\overline{x_i}}$ to represent the $\overline{x_i}$ literal, and a noise source N_{x_i} to represent the x_i literal. Hence, for $1 \leq i \leq n$, we may apply the input $(N_{x_i} + N_{\overline{x_i}})$ to the ith input of the circuit. This in effect means that we applied *all* 2^n inputs to the circuit *simultaneously*. We will see how a variation of this idea ends up being important in Section 2.

In this paper, we present an approach to solve the SAT problem utilizing NBL. The resulting approach can provide a SAT/UNSAT decision in a single operation, and can provide a satisfying input vector in a number of such operations which is linear in the number of inputs. This is possible because NBL allows us to apply all 2^n minterms to the SAT instance *simultaneously*.

2. OUR APPROACH

We begin with some definitions pertaining to Noise-based Logic (NBL).

DEFINITION 1. **Independent Noise Processes:** *Consider two noise processes* $V_i(t)$ *and* $V_j(t)$. *These noise processes are independent iff the correlation operator* $\langle \rangle$ *applied to* $V_i(t)$ *and* $V_j(t)$ *yields* $\langle V_i(t)V_j(t)\rangle = \delta_{i,j}$, *where* $\delta_{i,j}$ *is the Kronecker symbol* ($\delta_{i,j} = 1$ *when* $i = j$, *and* $\delta_{i,j} = 0$ *otherwise*).

Permission to make digital or hard copies of all or part of this work for personal or classroom use is granted without fee provided that copies are not made or distributed for profit or commercial advantage and that copies bear this notice and the full citation on the first page. To copy otherwise, to republish, to post on servers or to redistribute to lists, requires prior specific permission and/or a fee.
DAC 2012 June 3-7, 2012, San Francisco, California, USA.
Copyright 2012 ACM 978-1-4503-1199-1/12/06 ...$10.00.

DEFINITION 2. **Basis (Reference) Noise Processes (Bits):** *Consider* M *noise processes* $V_1(t), V_2(t), \cdots, V_M(t)$. *If these processes are pairwise independent, then* $V_1(t), V_2(t), \cdots, V_M(t)$ *are referred to as basis (reference) noise processes (bits).*

For convenience, we assume that all the noise processes in the sequel have a zero mean value.

Consider two orthogonal basis noise bits $V_i(t)$ and $V_j(t)$ ($i \neq j$). The product $Z_{i,j}(t) = V_i(t) \cdot V_j(t)$ of two orthogonal basis noise bits is orthogonal to $V_k(t)$ ($k = 1, 2, \cdots, M$). This property was used [3] to realize a **logic hyperspace**. In other words, $\langle Z_{i,j}(t), V_k(t)\rangle = 0$

DEFINITION 3. **Noise-based Logic Hyperspace:** *Using* $2m$ *basis noise bits* $V_1^0(t), V_1^1(t), \cdots V_m^0(t), V_m^1(t)$, *we can compute a noise hyperspace* $\mathcal{H}$ *with dimensionality* 2^m, *by multiplying these noise bits appropriately, and performing their additive superposition [3].*

2.1 SAT to NBL-SAT Transformation

In this subsection, we describe the process of transforming a SAT decision problem S into an equivalent NBL SAT instance S_N. Consider a decision problem expressed as a CNF S with m clauses ($S = c_1 \cdot c_2 \cdots c_m$) on a set of binary variables $X = (x_1, x_2, \cdots, x_n)$. We would like to determine if S is satisfiable, and if so, find a satisfying assignment. S_N is comprised of the product of 2 sets of clauses τ_N and Σ_N, where τ_N contains all 2^n valid minterms for the instance S, while Σ_N includes all satisfying minterms for S.

For each clause c_j, we create $2n$ basis noise sources which are used to represent both literals of each variable $x_1, x_2, \cdots, x_n$. Let $N_{x_i}^j$ be the noise source corresponding to literal x_i in clause c_j, and $N_{\overline{x_i}}^j$ be the noise source corresponding to literal $\overline{x_i}$ in clause c_j. In total, we create $2mn$ independent basis noise sources as there are m clauses, each requiring $2n$ noise sources. Note that the noise sources are independent across clauses, such that the product of any noise, for any literals x_p and x_q from clauses c_j and c_k respectively, (where $j \neq k$), has a zero mean ($\langle N_{x_p}^j \cdot N_{x_q}^k \rangle = 0$).

Construction of τ_N: First we construct the noise hyperspace τ_N which contains all 2^n valid minterms to be applied to the SAT instance Σ_N. The hyperspace τ_N is constructed as the product of n terms. The ith term consists of the sum of two noise products, $N_{x_i}^1 N_{x_i}^2 \cdots N_{x_i}^m$ and $N_{\overline{x_i}}^1 N_{\overline{x_i}}^2 \cdots N_{\overline{x_i}}^m$. These products correspond to the product of noise sources for literals x_i and $\overline{x_i}$ respectively, used in all clauses for Σ_N.

$$
\begin{aligned}
\tau_N = \ & ((N_{x_1}^1 N_{x_1}^2 \cdots N_{x_1}^m) + (N_{\overline{x_1}}^1 N_{\overline{x_1}}^2 \cdots N_{\overline{x_1}}^m)) \\
& \cdot ((N_{x_2}^1 N_{x_2}^2 \cdots N_{x_2}^m) + (N_{\overline{x_2}}^1 N_{\overline{x_2}}^2 \cdots N_{\overline{x_2}}^m)) \\
& \cdots ((N_{x_n}^1 N_{x_n}^2 \cdots N_{x_n}^m) + (N_{\overline{x_n}}^1 N_{\overline{x_n}}^2 \cdots N_{\overline{x_n}}^m))
\end{aligned}
\tag{1}
$$

Construction of Σ_N: Now we construct the NBL-based SAT instance Σ_N from the SAT instance S by replacing the positive literal of variable x_i in clause c_j by cube subspace $T_{x_i}^j$, and the negative literal of variable x_i in clause c_j by cube subspace $T_{\overline{x_i}}^j$, where $T_{x_i}^j = (N_{x_1}^j + N_{\overline{x_1}}^j) \cdot (N_{x_2}^j + N_{\overline{x_2}}^j) \cdots (0 + N_{x_i}^j) \cdots (N_{x_n}^j + N_{\overline{x_n}}^j)$. Similarly $T_{\overline{x_i}}^j = (N_{x_1}^j + N_{\overline{x_1}}^j) \cdot (N_{x_2}^j + N_{\overline{x_2}}^j) \cdots (N_{\overline{x_i}}^j + 0) \cdots (N_{x_n}^j + N_{\overline{x_n}}^j)$. By binding the literal v ($\overline{v}$) to the cube subspace T_v^j ($T_{\overline{v}}^j$) in clause j, we obtain additive superposition of noise minterms satisfying clause c_j.

Example 1: Consider the CNF formula $S = c_1 \cdot c_2 \cdot c_3 \cdot c_4 = (\overline{x_1}) \cdot (x_2 + x_3) \cdot (x_1 + \overline{x_3}) \cdot (\overline{x_1} + \overline{x_2} + x_3)$. The NBL-SAT instance Σ_N is as follows:
$\Sigma_N = (T_{\overline{x_1}}^1) \cdot (T_{x_2}^2 + T_{x_3}^2) \cdot (T_{x_1}^3 + T_{\overline{x_3}}^3) \cdot (T_{\overline{x_1}}^4 + T_{\overline{x_2}}^4 + T_{x_3}^4)$

When Σ_N is expanded out, the noise vectors for minterms from each clause form products with noise vectors of minterms from all other clauses. A valid satisfying minterm for Σ_N would be such that its final noise product contains a product of noise vectors from all clauses that represent the same minterm. All other combination of noise vectors are logically invalid (in the sense that they are not present in τ_N). Note that expansion is performed only for purposes of illustrating the approach. In practice, the SAT check is done by simply observing the average value of $\tau_N \cdot \Sigma_N$, without explicit expansion.

Consider a SAT formula S where the number of variables and clauses are $n = 2$ and $m = 3$ respectively. An example of a valid noise-based minterm is $N_{x_1}^1 N_{x_1}^2 N_{x_1}^3 N_{\overline{x_2}}^1 N_{\overline{x_2}}^2 N_{\overline{x_2}}^3$, which corresponds to the minterm $x_1 \overline{x_2}$ of S. An example of an invalid noise-based minterm is $N_{x_1}^1 N_{\overline{x_1}}^2 N_{\overline{x_1}}^3 N_{x_2}^1 N_{x_2}^2 N_{x_2}^3$, which corresponds to the (invalid) term $x_1 \overline{x_1} x_2$ of S.

Thus Σ_N is the additive superposition of all valid (satisfying) and invalid minterms of the SAT instance. Since τ_N only contains all valid minterms as shown in Equation 1, the product of $\tau_N \cdot \Sigma_N$ is the additive superposition of the self-correlation of each of the valid minterms. The average value of $\tau_N \cdot \Sigma_N$ is zero if the instance S is unsatisfiable, and positive if the instance S is satisfiable.

2.2 Satisfiability Check using NBL-SAT

Algorithm 1 describes the procedure for a single operation satisfiability check using NBL-SAT. The check for satisfiability is done with an observation of the average value of $S_N = \tau_N \cdot \Sigma_N$. The key to the single operation

---

Algorithm 1 Pseudocode of NBL-SAT checker

$NBL - SAT\_check(S_N)$
$S_N \leftarrow (\tau_N \cdot \Sigma_N)$
if S_N output has a zero average **then**
  return(S is unsatisfiable)
else
  return(S is satisfiable)
end if

SAT check achieved by this algorithm are the superposition and correlation properties of the noise basis sources. In Σ_N, each clause c_j contains any number of cube subspaces T_v^j. The disjunction of all the T_v^j in clause c_j results in a new noise vector Z^j. Thus Z^j is the additive superposition of all noise-based minterms that satisfy clause c_j. The output of Σ_N is the conjunction (product) of all Z^j noise vectors from the clauses. Hence Σ_N includes the additive superposition of all noise-based minterms that satisfy S. Multiplying Σ_N with τ_N simply yields the additive superposition of the self correlation of the noise-based minterms of S. We recall that the average value of the product of two independent noise sources is 0. The product $\tau_N \cdot \Sigma_N$ has a zero average only in the case where Σ_N and τ_N do not share any noise minterms. Hence no minterm exists in Σ_N that correlates to any of the valid minterms in τ_N. But τ_N is the additive superposition of *all* valid minterms for S. Hence, if S_N has a zero average, then there is no valid minterm that exists across all clauses and we conclude S is unsatisfiable (line 4).

However, if Σ_N and τ_N contain common minterm(s), the product of the noise vectors results in a positive average for S_N. Then if $\tau_N \cdot \Sigma_N$ has a positive DC offset, we can conclude that at least one common satisfying minterm exists across all clauses (line 6).

THEOREM 2.1. *The product of the NBL-SAT instance Σ_N and hyperspace τ_N produces a zero average iff S unsatisfiable.*

PROOF. *Omitted due to lack of space.* □

The reason why we are able to perform the SAT check with a single operation is that we are able to simultaneously apply all minterms to the NBL-SAT instance, since each of the minterms in NBL are orthogonal basis noise vectors. This is not possible in traditional SAT solvers.

It is easy to see that determining the satisfying assignment for any SAT problem can be done by at most n applications of Algorithm 1. This is done as follows. Assuming that the instance is shown to be satisfiable by Algorithm 1, we bind variables iteratively to their (WLOG) positive literal. In iteration i, if the new instance is SAT, we append literal x_i to the satisfying solution, otherwise we append literal $\overline{x_i}$ to the solution. In the subsequent iterations, we bind variable x_i to its positive value (negative value) if the satisfying assignment contains x_i in its positive (negative) polarity.

2.3 Scaling Issues

In order to discuss how NBL-SAT scales with the number of variables and clauses, consider 3-SAT instances (in which each clause has 3 literals) with n variables and m clauses. We assume that each basis noise source ($N_{x_i}^j$) is a uniform random variable between [-0.5, 0.5]. Recall that the average value of $\tau_N \cdot \Sigma_N$ is proportional to the number of satisfying minterms,

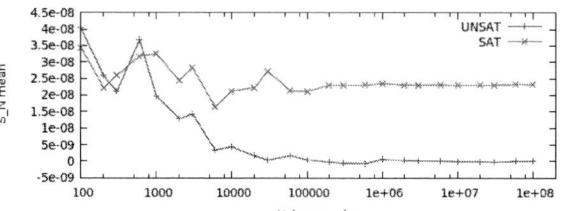

Figure 1: S_N mean for UNSAT and SAT instances

since such minterms are present in both τ_N and Σ_N. Hence the ability of NBL-SAT to discriminate between an instance with one satisfying minterm and another instance which is unsatisfiable needs to be considered. We define the SNR of NBL-SAT as SNR $= \frac{\hat{\mu}_1 - 3\hat{\sigma}_1}{\hat{\mu}_0 + 3\hat{\sigma}_0}$, where $\hat{\mu}_i$ is the expectation of the mean of the average value of $\tau_N \cdot \Sigma_N$ when there are i satisfying minterms, and $\hat{\sigma}_i$ is the expectation of the standard deviation of the average value of $\tau_N \cdot \Sigma_N$, when there are i satisfying minterms. Note that $\hat{\mu}_0 = 0$. Assuming that there are N samples in each noise source, we have $\hat{\mu}_1 = E(\frac{1}{N}\Sigma_{i=1}^{N}\{\Pi_{j=1}^{nm} x_j^2\})$, where x_j is uniformly distributed within [-0.5, 0.5]. The product is over nm since there are nm noise products in any satisfying minterm in NBL-SAT. Simplifying, we have $\hat{\mu}_1 = (\frac{1}{12})^{nm}$.

Similarly, the unbiased estimate of the variance of the mean of the product of nm independent uniform distributions [5] (over N samples) is given by $\hat{\sigma}^2 = \frac{1}{N-1}(\frac{1}{12})^{2nm}$.

Now the total number of products in a NBL-based 3-SAT instance with n variables and m clauses is $(2^n) \cdot (2^n - 2^{n-3})^m \sim O(2^{nm})$. The first term refers to the number of products of τ_N, while the second term is the number of products in Σ_N. Since these $O(2^{nm})$ products are independent, their variances will add up, and so we have $\hat{\sigma}_1 = \hat{\sigma}_0 = \frac{1}{\sqrt{N-1}}(\frac{1}{12})^{nm} \cdot 2^{nm}$.

For SNR $\gg 1$, we can ignore $\hat{\sigma}_1$ in the SNR expression, yielding: SNR $= \frac{\hat{\mu}_1}{3\hat{\sigma}_0} = \frac{\sqrt{N-1}}{3 \cdot 2^{nm}}$.

Note that if it is known that the instance has K satisfying minterms, then the SNR expression above is multiplied by K.

3. EXPERIMENTAL RESULTS

To validate our NBL-SAT algorithm, we simulated several small NBL-SAT instances in the C programming language. In our simulations, each basis noise source ($N_{x_i}^j$) is a uniform random variable between [-0.5, 0.5]. Each instance is simulated until the mean value of S_N has converged to the third significant digit or until 10^8 noise samples have been reached. Our experiments focus on the SAT checker from Algorithm 1, as the satisfying assignment determination for the SAT instance simply consists of iterative applications of the SAT checker.

We use the following two examples, one unsatisfiable and one satisfiable, to validate the correctness of our scheme.
$$S_{UNSAT} = (x_1 + x_2) \cdot (x_1 + \overline{x_2}) \cdot (\overline{x_1} + x_2) \cdot (\overline{x_1} + \overline{x_2})$$
$$S_{SAT} = (x_1 + \overline{x_2}) \cdot (\overline{x_1} + \overline{x_2}) \cdot (x_1 + \overline{x_2}) \cdot (\overline{x_1} + \overline{x_2})$$
The first two clauses in our satisfiable example are redundant, but they bring the number of clauses m to 4, which makes the S_N values comparable with our unsatisfiable example which also has $m = 4$. In Figure 1, the average values of S_N of both examples are plotted as a function of number of noise samples.

Although no NBL circuits exist today, realizing the NBL-SAT solution approach of our paper would require widely studied, and ubiquitously available circuit components such as wideband amplifiers, analog adders, analog multipliers and low-pass filters. NBL-SAT may be implemented on FPGAs or ASICs as well. We hope that the result of this paper will encourage development of NBL circuits.

4. REFERENCES

[1] L. B. Kish, "Thermal noise driven computing," *Applied Physics Letters*, vol. 89, no. 144104, 2006.

[2] L. B. Kish, "Noise-based logic: Binary, multi-valued, or fuzzy, with optional superposition of logic states," *Physics Letters A*, vol. 373, pp. 911–918, 2009.

[3] L. B. Kish, S. Khatri, and S. Sethuraman, "Noise-based logic hyperspace with the superposition of 2^N states in a single wire," *Physics Letters A*, pp. 1928–1934, 2009.

[4] K. C. Bollapalli, S. P. Khatri, and L. B. Kish, "Implementing digital logic with sinusoidal supplies," in *Proceedings of the Conference on Design, Automation, and Test in Europe*, DATE '10, (3001 Leuven, Belgium, Belgium), pp. 315–318, European Design and Automation Association, 2010.

[5] C. P. Dettmann and O. Georgiou, "Product of n independent uniform random variables," *Statistics & Probability Letters*, vol. 79, no. 24, pp. 2501 – 2503.

Cognitive Computing with Spin-Based Neural Networks

Mrigank Sharad[1], Charles Augustine[2], Georgios Panagopoulos[1], and Kaushik Roy[1]

[1]School of Electrical and Computer Engineering, Purdue University, West Lafayette, Indiana 47907, USA

[2]Circuit Research Lab, Intel labs, Intel Corporation, Hillsboro, OR, US

msharad@purdue.edu, charles.augustine@intel.com, gpanagop@purdue.edu, kaushik@purdue.edu

ABSTRACT

We model a step transfer function neuron with lateral spin valve (LSV) and propose its application in low power neural network hardware. The computational task in such a network is performed by nano-magnets, metal channels and programmable conductive elements, that constitute the neuron-synapse units and operate at a terminal voltage of ~20 mV. CMOS transistors provide peripheral support in the form of clocking, power gating and inter-neuron signaling. Simulations for cognitive as well as Boolean computation applications show more than 94% improvement in power consumption as compared to a conventional CMOS design at the same technology node.

Categories and Subject Descriptors

B.7.1 [Integrated Circuits] Types & Design Styles - Advanced

Technologies

General Terms

Performance, Design

Keywords

Neural network, spin valve, low power design, magnets

1. INTRODUCTION

Recent experiments on lateral spin valves (LSV) have demonstrated switching of *nano-magnets* with non local spin transfer torque (STT) [1, 2]. It involves generation of spin potential difference across a magnet-metal interface using spin polarized current flow in the metal, and, results in flipping of the magnet with pure spin current, without direct charge current injection into it [3, 4]. The metallic device structure of an LSV allows application of very small terminal voltages, resulting in low switching energy for the magnets.

Owing to analog nature of spin current, an LSV can perform non-Boolean computation. Application of LSV as a majority gate has been proposed earlier [5]. In this work we propose a device model for neuron based on LSV. We show that, with appropriate clocking scheme, an LSV with complementary polarizer inputs acts as a neuron. The proposed neuron device can be integrated in a large-scale crossbar-network architecture, with programmable or fixed conductive elements as synapses. We also propose a model for cellular neuron with domain wall magnets (DWM) [6] as programmable synapses. It can be suitable for low power, cellular neural network (CNN) hardware which employs neighborhood connectivity.

The compact, low resistance, magneto-metallic neuron units allow synapse current flow across a small terminal voltage

of ~20 mV, resulting in low computation power. In this paper we briefly describe the system level integration scheme for the proposed neuron model.

Rest of the paper is organized as follows. We introduce the proposed device structures for spintronic neuron in section 2 System level information is briefly discussed in section 3. Section 4 presents the simulation framework, synthesis and simulation results for the proposed design scheme. Finally, section 5 concludes the paper.

2. SPIN BASED NEURON-SYNAPSE UNITS

Device structure for the neuron model is shown in fig. 1a. The input magnets m_1 and m_2 act as complementary spin polarizers, and, receive input current from positive and negative weight synapses, that constitute of via's or programmable conductive elements like memristor or phase change memory (PCM) [7]. The neuron magnet m_4 forms the free layer of a magnetic tunneling junction (MTJ). The third input magnet m_3 is used to realize current-mode Bennett clocking for the neuron magnet m_4

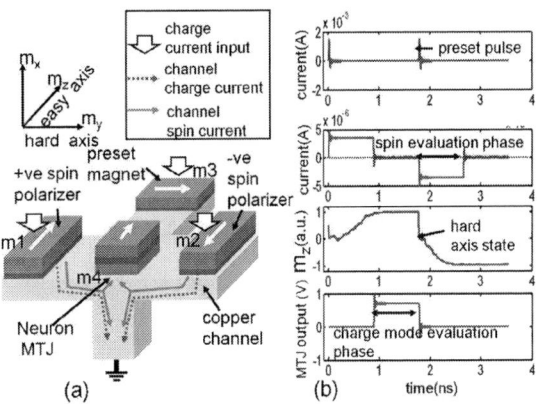

Fig. 1 (a) Neuron with via-synapse (b) neuron switching waveform

Current injected through m_3 into the channel, pushes the neuron magnet to the unstable hard axis state (fig. 1b). After removal of the hard axis bias, the neuron magnet makes a fast transition to one of the stable states, governed by the net spin polarity of the charge current injected into the channel. Thus m_4 in effect compares the amount of current received from the positive and the negative weight synapses, connected to m_1 and m_2 respectively. Owing to low stability of the hard axis state, even a small difference in the currents received by m_1 and m_2 effects the switching. Hence, in effect, a zero threshold, bipolar step function of a neuron is realized. The 'near' zero spin current threshold for the neuron also reduces the required current injection through input synapses to few micro amperes.

The proposed spintronic neuron can be integrated with large number of programmable or non-programmable 'conductive' synapses to arrive at low power computational networks. Programmable 'spintronic' synapses can be realized using domain wall magnets (DWM). For the lateral DWM-channel interface shown in fig. 2b, the spin polarity of the current injected into the channel through the DWM depends upon the DW location (fig. 2a). The DW can be programmed by a lateral injection of charge current through the DWM [6]. Hence, a DWM can acts as a low resistance, compact and programmable synapse, leading to homogenous and modular spin-mode neuron-synapse unit (fig. 2c). The limited spin diffusion length of metal channel however restricts the number of DWM synapses per neuron [3], and hence this structure is more suitable for CNN architecture with neighborhood connectivity. Both the neuron models, with conductive and spintronic synapses respectively, can compute with analog as well as binary inputs. This is because, the device structure involves current mode operation, that is analog in nature.

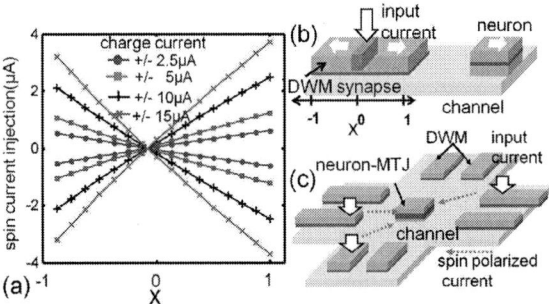

Fig. 2(a) spin polarization as a function of DW location (b)DWM synapse with channel interface (c) cellular neuron with programmable domain wall synapse

3. SYSTEM INTEGRATION

Since spin signals can not be transmitted over a distance longer than a few spin diffusion length of the metal channel [3, 4], we employ charge mode signaling to interconnect the spintronic neurons. A dynamic CMOS latch (fig. 3a) senses the state of the neuron MTJ and drives the current source transistor, which transmits synapse current to all the fan-out neurons. Note that, the synapse input currents, involved in computation, flow across a small terminal voltage ΔV (fig. 3b), thereby reducing the static power consumption resulting from large number of synaptic communications.

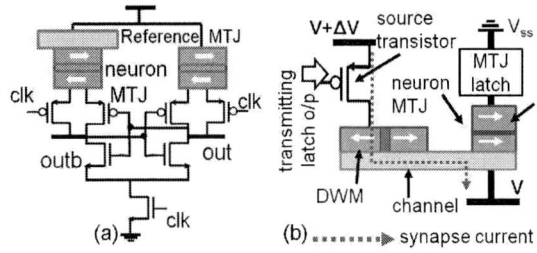

Fig. 3(a) CMOS detection unit senses the state of the neuron magnet and (b) transmits synpse current to the fanout neurons across a small terminal voltage ΔV.

The aforemensioned detection scheme can be applied to both the neuron strcutures described in the previous section. As mensioned before, the DWM based neuron is suitable for a CNN design (fig. 4a) whereas the four terminal neuron in fig. 1a is a suitable candidate for a network with programmable or fixed conductive crossbar elements (fig.4b)

4. SIMULATION AND PERFORMANCE

Simulation of the spin device structures presented in this work, involves self consistent solution of the spin transport using Valet Fert's spin diffusion model and the magnet dynamics using Landau-Lifshitz-Gilbert equation LLG [3]. *This simulation*

framework has been benchmarked with experimental results for lateral spin valves.

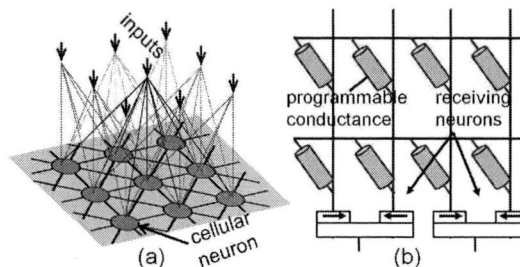

Fig. 4(a) CNN architecture with neighborhood connectivity (b) Fully connected crossbar network architecture

Neural networks can perform both Boolean as well as cognitive computations (fig. 5). The network weights are obtained by offline training. For arithmetic computation blocks, like multipliers and adders, the required network size grows exponentially beyond an input dimension of 4x4, because of large training set. Hence, for larger number of input bits, the overall computation is decomposed into 3x3 or 4x4 units in order to obtain maximum benefits. Fig. 5c compares the performance of the spin-CMOS (45nm) hybrid network with 45nm-CMOS design for some benchmark cognitive and Boolean applications.

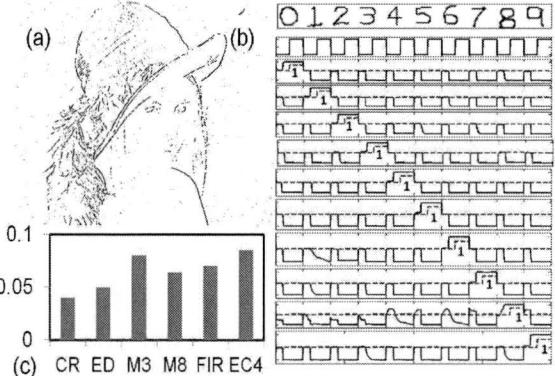

CR : character recognition M8: 8-bit mult. EC4: 4 bit encoder
ED : edge detection M3: 3bit mult. FIR : 8-b,8-stage filter

Fig. 5 (a) Edge detection from grey scale image using DWM based CNN (b) Charater recognition with DWM based network (c) Power consumtion of proposed desgin scheme relative to 45nm CMOS at 500 MHz frequecy.

5. CONCLUSION

We employed spin device phenomena like, majority evaluation and hard axis switching to model compact and low energy neuron-synapse units. The proposed circuit/architecture can lead to ultra low power spin-CMOS hybrid neural networks for cognitive as well as Boolean computations that can achieve more than 94% lower power as compared to a CMOS design.

ACKNOWLEDGEMENT: This research was funded in part by Nano Research Initiative and by the INDEX center

REFERENCES [1] Kimura et. al, Phys. Rev. Lett. 2006 [2] Sun et. al., Appl. Phys. Lett. 2009. [3] Behin-Ain et. al., Nature Nano. 2010 [4] Behin-Ain et. al, Appl.Phys.Lett. 2011 [5]. Augustine et al., Nanoarch, 2011 [6] M. Sharad et al., IJCNN, 2012 [7]Kuzum et. al., Nano Lett., 2011

Capacitance of TSVs in 3-D Stacked Chips a Problem?
Not for Neuromorphic Systems!

Antoine Joubert
CEA-LETI, Minatec Campus
joubert.antoine@cea.fr

Marc Duranton
CEA-LIST, Nano-INNOV
marc.duranton@cea.fr

Bilel Belhadj
CEA-LETI, Minatec Campus
bilel.belhadj@cea.fr

Olivier Temam
Inria Saclay Ile-de-France
olivier.temam@inria.fr

Rodolphe Héliot
CEA-LETI, Minatec Campus
rodolphe.heliot@cea.fr

ABSTRACT

In order to cope with increasingly stringent power and variability constraints, architects need to investigate alternative paradigms. Neuromorphic architectures are increasingly considered (especially spike-based neurons) because of their inherent robustness and their energy efficiency. Yet, they have two limitations: the massive parallelism among neurons is hampered by 2D planar circuits, and the most cost-effective hardware neurons are analog implementations that require large capacitors, We show that 3D stacking with Through-Silicon-Vias applied to neuromorphic architectures can solve both issues: not only by providing massive parallelism between layers, but also by turning the parasitic capacitances of TSVs into useful capacitive storage.

Categories and Subject Descriptors

B.7.1 [**Integrated Circuits**]: Types & Design Styles - Advanced technologies

General Terms

Design

Keywords

3D architectures, Neuromorphic systems, analog circuits

1. INTRODUCTION

Due to increasingly stringent energy constraints (i.e., Dark Silicon [4]), heterogeneous multi-cores composed of a mix of cores and energy-efficient accelerators are becoming mainstream. However, on top of energy constraints, defects are becoming prominent constraints as well, so that the research focus is progressively shifting to designing energy-efficient *and* defect-tolerant accelerators. Interestingly, these technology constraints come at a time where high-performance applications experience a dramatic shift of their own, from scientific computing applications to *Recognition, Mining, Synthesis* applications, as coined by Intel [2]. And few approaches are better positioned than *neuromorphic architectures* to tackle Recognition and Mining applications while providing inherent defect-tolerant and energy-efficient properties, as recently illustrated by the IBM Cognitive Chip [1].

Thanks to very efficient coding of information, spiking neurons are usually considered to be the neuron model with the greatest potential for applications. Analog neurons have a significantly smaller area footprint than digital neurons because they can leverage physics laws for implementing several elementary functions (see Figure 1): temporal integration is realized through capacitive integration, weight summation through Kirchhoff's law, and leakage is an intrinsic behavior of microelectronic devices.

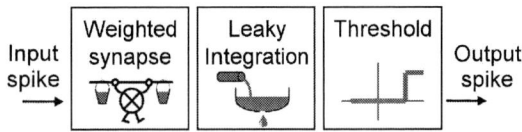

Figure 1: Block diagram of a leaky-integrate-and-fire neuron spiking neuron: weighted summation, temporal integration with leakage, and threshold.

However, analog spiking neuron implementations suffer from two limitations. First, they require large capacitances to store the internal membrane voltage of the neuron. Typical capacitances values are in the order of $0.5 - 1\ pF$, which corresponds to $50 - 200\ \mu m^2$ capacitors in $32 - 65\ nm$ technologies, and up to 50% of the total neuron area. Second, neuromorphic architectures at large are fundamentally 3D structures with massive parallelism, but the 2D planar circuits on which they are implemented considerably limit the bandwidth between layers (or significantly increase the area and energy cost required to achieve sufficient bandwidth).

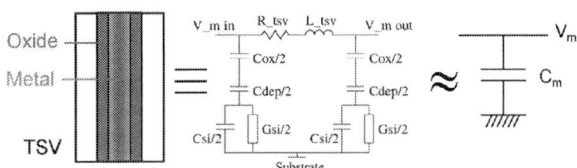

Figure 2: TSV structure and electrical model.

Both issues can be addressed with 3D stacking. It provides massive parallelism between layers by stacking and directly connecting them via a large number of Through-Silicon Vias (TSVs). Normally, these TSVs are a significant limitation of 3D stacking: they consume on-chip area and suffer from parasitic capacitances [3]. However, we can actually take advantage of these capacitances to implement the capacitors of spiking neurons, turning a weakness into a useful feature.

2. TSV MODEL

Through Silicon Vias are used to create vertical interconnections in 3-D stacked chips. As shown in Figure 2, they are composed of a metal wire isolated from the substrate. They are not perfect wires however, and act as MOS capacitors. We use an RLC π-shaped electrical model of TSVs as described in [3] (see Figure 2, middle). The TSV model features a resistance R_{tsv}, and an inductance L_{tsv}. It is isolated from the substrate by three capacitors C_{ox}, C_{dep} and C_{si}, while parasitic silicon substrate losses are represented by a conductance G_{si}. All of these values are process dependent and can change with TSV density, height, diameter, operating frequency and oxide thickness.

3. NEURON DESIGN AND SIMULATION

We designed an analog leaky integrate-and-fire neuron using the TSV model presented in Section 2 (TSV-neuron), and compared it through Spice simulations against a neuron using a standard capacitor. Figure 3 shows the behavior of the neuron internal potential V_m for standard (top) and TSV (bottom) neurons (for different densities). When V_m reaches a threshold voltage V_{th}, V_m is reset to a resting potential. With a medium density TSV ($400/mm^2$), it can be seen that the TSV-neuron exhibits the exact same behavior as the standard neuron. TSVs with other densities exhibit similar behavior, albeit with different time constants.

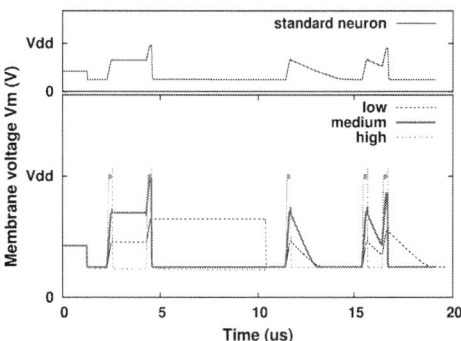

Figure 3: Membrane potential of standard- and TSV-based (several TSV densities) neurons.

4. ARCHITECTURAL OPPORTUNITIES

In Figure 4, we give examples of neuron connectivity and silicon area. Each neuron is composed of three blocks: weight injection, capacitance, and threshold detection (see Figure 1), respectively of areas $A/4$, $A/2$ and $A/4$, considering a standard neuron with area A. These blocks are respectively represented in orange, red and blue in Figure 4. TSVs can either be classically used as wires to transmit information (b), or be used as capacitive elements (c,d). Topology (c) splits neurons into two halves on each side of the TSV; topology (d) embeds both weight injection and threshold detection on

Table 1: Expected gains

Topology	(a)	(b)	(c)	(d)
Silicon footprint	A	A	A/2	A
Mask set footprint	A	A	A/4	A/2
Connectivity	4 IOs	6 IOs	4 Is & 4 Os	8 IOs

both layers around the TSVs, allowing the neuron to operate on each layer, thus dramatically increasing connectivity. In Table 1, we show the area and connectivity gains that can be expected. Two areas are considered: one is the total silicon footprint of a neuron, the other is its required mask area. For topologies (c) and (d), the same mask set can be re-used for each layer in the 3D stack, thus reducing the neuron footprint on mask layers costs.

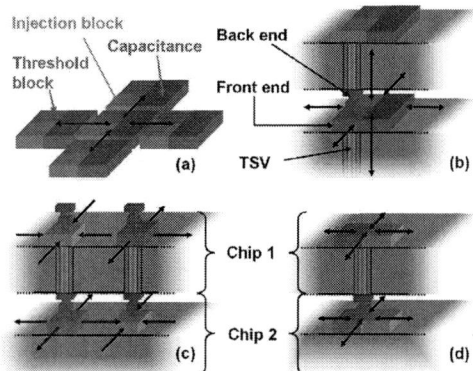

Figure 4: Illustrations of (a) a standard 2D neuromorphic architecture (b) a 3D-IC with standard 2D neurons, and (c) (d) a 3D-IC with TSV-based neurons.

5. DISCUSSION

We showed that it is possible to take advantage of TSVs for implementing the capacitive functionality of spiking neurons in neuromorphic architectures, dramatically increasing density and connectivity. TSVs with different characteristics (length, width, and pitch) could serve various needs: compact TSVs to quickly transmit information from one die to another; large TSVs to act as capacitors while still transmitting information vertically. While TSV-based capacitors are subject to substrate noise and might thus be influenced by neighboring TSVs [5], this phenomenon could again be leveraged to transmit information among neighboring neurons, in the spirit of local field potentials in biological systems.

6. REFERENCES

[1] G. Brumfiel. Inside ibm's cognitive chip. *Nature News*, Aug 2011.

[2] P. Dubey. Recognition, mining and synthesis moves computers to the era of tera. *Technology at Intel Magazine*, 09, Feb 2005.

[3] C. Fuchs, J. Charbonnier, and S. Cheramy. Process and RF modelling of TSV last approach for 3D RF interposer. *IITC/MAM 2011*, pages 9–11, 2011.

[4] M. Muller. Dark Silicon and the Internet. In *EE Times "Designing with ARM" virtual conference*, 2010.

[5] R. Weerasekera et al. Compact modelling of through-silicon vias (TSVs) in three-dimensional (3-D) integrated circuits. In *Proc. IEEE Int. Conf. on 3D System Integration (3D IC)*, pages 1–8, 2009.

Communication-Aware Mapping of KPN Applications onto Heterogeneous MPSoCs

Jeronimo Castrillon, Andreas Tretter, Rainer Leupers, Gerd Ascheid

Institute for Communication Technologies and Embedded Systems (ICE)

RWTH Aachen University, Germany

maps@ice.rwth-aachen.de

ABSTRACT

Kahn Process Networks (KPNs) are a widely accepted programming model for MPSoCs. Existing KPN mapping techniques mainly focus on assigning processes to processors. However, with embedded interconnect becoming more complex, communication has started to play an equally important role to that of computation. This paper presents a new KPN mapping algorithm that addresses communication and computation jointly. The algorithm is tested on two platforms with real applications and with randomly generated KPNs. We show that the algorithm finds solutions in situations where bare process mapping fails. It also reduced the average application makespan considerably when compared to previous heuristics.

Categories and Subject Descriptors

D.1.2 [**Software**]: Automatic Programming – Program synthesis

General Terms

Algorithms, Design, Performance

Keywords

mapping, real time, dataflow graphs, heterogeneous MPSoC, embedded systems

1. INTRODUCTION

In the last ten years we have seen how Multi-Processor Systems on Chip (MPSoCs) have permeated several segments of the embedded systems market [1]. MPSoCs, and especially heterogeneous ones, have emerged to provide the performance required by new demanding applications with the limited energy budget of portable devices, e.g., in mobile phones [2]. In order to fully leverage the computing power of MPSoCs while attaining a competitive productivity, diverse programming models are being investigated. In the

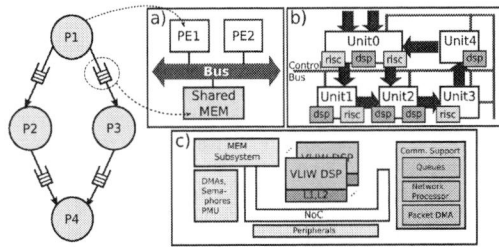

Figure 1: MPSoC architectures. (a) Simple MPSoC with communication over shared memory (b) Sketch of a data stream architecture similar to the BWC200. (c) Simplified view of TI's Keystone.

baseband processing and multimedia domains. different flavors of process networks are widely used. Initially, the focus was set on analyzable models, e.g., Synchronous Data Flow (SDF) graphs [3]. Lately, more expressive models like Kahn Process Networks (KPN) [4] started to be addressed.

From the early times, dataflow programming models were conceived within a *software synthesis* environment. A myriad of techniques have been developed to automatically map these applications onto parallel architectures. While only a couple of these works focused on heterogeneous architectures, very few considered the impact of data communication. Application mapping is still regarded as the assignment of tasks to processing elements. Nowadays, however, the ever increasing application demands for data throughput under energy constraints have made interconnect architectures become more complex (see Figure 1). That is the case of so-called pipelined MPSoCs [5, 6] and data streaming architectures, e.g., BlueWonder's BWC200 [7], where communication paths are restricted, or new architectures like Texas Instruments (TI) Keystone [8] where many different communication primitives including HW support are available. Therefore, the role of communication during application synthesis can no longer be ignored. Moreover, for some applications, carefully mapping communication may be even more important than doing it for computational tasks.

In this paper we propose a new algorithm that directly addresses process and communication mapping. The algorithm was integrated in the MAPS framework [9], extending the algorithms presented in [10]. Within MAPS, the algorithm can be used iteratively in order to compute mappings for real time applications or to explore memory-performance trade-offs. As benchmark, we present a case study on two different platforms with real applications and with randomly generated KPN graphs. For the JPEG application, an average speedup of 2.4X was observed over mappings obtained with state-of-the-art heuristics. We also show that the algorithm finds solutions where bare process mapping fails.

The rest of the paper is organized as follows. In Section 2 we discuss our contribution against previous work. Section 3 introduces the nomenclature used to formalize the mapping problem. Communication-aware KPN mapping is addressed in Section 4 and results are presented in Section 5. Finally, conclusions are drawn in Section 6.

2. RELATED WORK

Several manual attempts to map KPN applications onto modern architectures have shown how important the issue of communication is (on Intel IXP [11] or Cell BE [12]). However, few automatic mapping approaches take communication explicitly into account. Initial works ignored communication and focused on process scheduling [13, 14]. Later, communication was modeled by annotating latencies on the edges of the application graph (for example in [15], with fundamental underlying techniques in [16]). This approach can only distinguish between local communication, usually with zero latency (*edge zeroing* algorithms), and inter-processor communication (IPC). The latter supposes a single IPC option, which does not correspond to the situation in today's MPSoCs. Other authors treat communication mapping as an afterthought, i.e., as a consequence of the process mapping [17, 18]. This approach can lead to suboptimal or even impossible solutions. This is sometimes circumvented by actually synthesizing HW FIFO channels on reconfigurable architectures [19, 20]. However, with increasing SoC production costs, application specific MPSoCs will be rare.

In this paper we present a communication-aware mapping heuristic for fixed HW platforms (i.e., no HW synthesis). Processes and communication channels are mapped in no predefined order, but depending on the application and the target platform. We abstract from interconnect and storage and rather map to *communication primitives*. This reflects the fact that communication can be implemented differently over the same resources, e.g., lock-less FIFOs or FIFOs with packaging over the same memory. In this way, we capture the variety of IPC SW interfaces in today's MPSoCs and add flexibility to the mapping process.

We address the mapping problem by devising heuristics. Authors in [21, 22], instead, employ evolutionary algorithms. We believe that with clever heuristics, similar results to that of evolutionary algorithms can be achieved in less time. Furthermore, heuristics can be more robust to inaccurate performance estimates, common in early SW development phases. For instance, heuristic decisions based on the KPN topology would hold even with wrong time estimates.

3. PROBLEM DEFINITION

This section formalizes the KPN mapping problem and introduces the notation used in the rest of this paper. The main concepts are illustrated in Figure 2.

3.1 Target MPSoC

A heterogeneous MPSoC is composed of Processing Elements (PE) of different types, storage elements, peripherals and interconnect. Let $PT = \{pt_1, pt_2, \ldots, pt_{N_{PT}}\}$ be the set of all Processor Types (PT) and pe_j^v represent the j-th

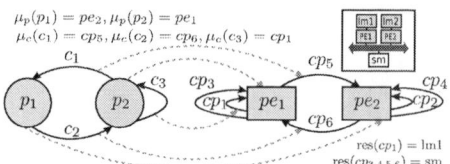

Figure 2: Illustration: Problem definition. Sample MPSoC with local and shared memories (lm, sm). Note that every PE can also communicate with itself over shared memory.

instance of type $v \in PT$. The set of all PEs can be then defined as the disjoint union $PE = \sqcup_{v \in PT} PE^v$, with $PE^v = \{pe_1^v, \ldots, pe_{N_v}^v\}$. Similarly, the set of all Communication Resources (CR) can be defined as $CR = \{cr_1, \ldots, cr_{N_{CR}}\}$. We distinguish two constraints on a CR: the number of logical channels that can be mapped to it (chanmax(cr)) and the available storage in bytes (memmax(cr)).

We model the target platform as a directed multigraph $MPSoC = (PE, CP)$, with CP a multiset of ordered pairs $(pe_i^u, pe_j^v) \in PE \times PE$ representing a Communication Primitive (CP) that can be used to send data from pe_i^u to pe_j^v. As mentioned in Section 2, CPs abstract from actual HW resources. However, in order to apply resource constraints (e.g., chanmax(), memmax()), a function is needed that associates a CR for every CP. Let res : $CP \rightarrow CR$ be such a function. Note that this function need not be injective.

Finally, let $\zeta^{cp} : CP \times \mathbb{N} \rightarrow \mathbb{N}$ be a function, so that $\zeta^{cp}(cp, b)$ returns the cycle count of sending b bytes over cp.

3.2 KPN Application and Buffer Sizes

An application is represented as a directed multigraph $KPN = (P, C)$. Since processes in a KPN can have an arbitrary control flow, their timing characterization is more involved than for static dataflow models. For this purpose, we use the MAPS trace generation flow presented in [10]. Every process $p \in P$ is represented as an infinite sequence of *segments* $T_p = (S_1^p, S_2^p, \ldots) = (S_i^p), i \in \mathbb{N}$ called trace[1]. A segment is a portion of the execution of the process that finishes with an access (read or write) to any of its channels.

Let $\mathcal{T}$ and $\mathcal{S}$ be the set of all traces and all segments of the application respectively. With the timing information obtained from tracing, it is possible to define $\zeta^{seg} : \mathcal{S} \times PT \rightarrow \mathbb{N}$ and $\Xi : \mathcal{T} \times PT \rightarrow \mathbb{N}$. These functions return the cycle count of a segment and of an entire trace for every processor type. (Note that $\Xi(T_p, PT) = \sum_{S \in T_p} \zeta^{seg}(S, PT)$.)

Buffer sizing is an important procedure when analyzing KPNs. Since it is already included in MAPS, we only focus on mapping and consider buffer sizes as given. Let $\sigma : C \rightarrow \mathbb{N}$ represent the result of the buffer sizing procedure, i.e., bytes assigned to every KPN channel. Let $\hat{\sigma}(c)$ be the buffer size in tokens assigned for channel $c \in C$.

3.3 KPN Mapping

Given a $KPN = (P, C)$ and an $MPSoC = (PE, CP)$, a **correct** KPN mapping is a pair of unique assignments $(\mu_p : P \rightarrow PE, \mu_c : C \rightarrow CP)$ such that:

- It is consistent, i.e., $\forall c \in C, \text{src}(\mu_c(c)) = \mu_p(\text{src}(c)) \land \text{dst}(\mu_c(c)) = \mu_p(\text{dst}(c))$, and

---

[1]The sequence is conceptually infinite. In reality, traces are finite, bounded by the length of the profiling run.

1263

- $\forall cr \in CR, |A^{cr}| \leq \text{chanmax}(cr)$ and $\sum_{c \in A^{cr}} \sigma(c) \leq \text{memmax}(cr)$, with A^{cr} the set of KPN channels mapped to the CR cr, i.e., $A^{cr} = \{c \in C, \text{res}(\mu_c(c)) = cr\}$.

4. COMMUNICATION-AWARE MAPPING

This section presents our proposed algorithm. However, before introducing it, a simple algorithm is presented that adds a channel mapping to an existing process mapping. This allows us to compare the joint mapping algorithm against existing process mapping heuristics in Section 5.

4.1 Independent Channel Mapping (ICM)

ICM refers to the process of finding a channel mapping $\mu_c : C \rightarrow CP$, given a pre-computed process mapping $\mu_p : P \rightarrow PE$, so that (1) The KPN mapping (μ_p, μ_c) is correct and (2) the application makespan is minimized.

In order to solve this NP complete problem, a simple greedy heuristic was devised. It starts by creating a set with all possible CPs for every KPN channel $A^c = \{cp = (pe_i, pe_j) \in CP, pe_i = \mu_p(\text{src}(c)) \wedge pe_j = \mu_p(\text{dst}(c)) \wedge \sigma(c) \leq \text{memmax}(\text{res}(cp))\}$. Then the channels are sorted, in decreasing order, according to the amount of traffic observed during tracing. While traversing the sorted list, for every channel c' with total traffic X in bytes, the algorithm assigns $\mu_c(c') = cp^*$ over a resource $cr^* = \text{res}(cp^*)$, with

$$cp^* = \underset{cp \in A^{c'}}{\operatorname{argmin}} \zeta^{cp}(cp, X)$$

subject to $\sigma(c') + \sum_{c \in A^{cr^*}} \sigma(c) \leq \text{memmax}(cr^*) \wedge |A^{cr^*}| < \text{chanmax}(cr^*)$. Additionally, a simple backtracking was implemented for CPs over HW FIFOs. This improved the success rate of ICM on platforms with restricted communication links. If at any point it is not possible to assign a CP, the algorithm fails.

4.2 Joint Process and Channel Mapping

In a joint mapping, both μ_p and μ_c must be defined, so that the KPN mapping (μ_p, μ_c) is correct and the makespan is minimized. To solve this problem, we developed the *Group-Based Mapping* (GBM) algorithm with two main underlying goals: (1) Narrow the mapping space while avoiding early selection of specific HW resources, (2) analyze jointly processes and channels in no specific order.

To achieve the first goal the algorithm was split into two phases. In the first one, KPN elements (channel and processes) are iteratively mapped to *groups* of HW resources. This reflects the fact that different resources may display the same timing characteristic for a given KPN element. In the second phase, a sort of homogeneous mapping is performed in which the actual HW resources are selected.

To achieve the second goal, the algorithm was designed so that during mapping, processes and channels are selected according to an improvement measure. This measure is not biased to processes or channels, thus, they are selected in no prescribed order (see Section 4.2.1).

Figure 3 shows an overview of the algorithm The first phase of the algorithm (while loop) works on so-called assignment sets (A^p and A^c), i.e., the sets of HW resources to which KPN elements can be mapped. These sets are reduced iteratively (call to MakeProposal) until no more reductions are required (call to CanMakeProposals). The function MakeProposal selects a KPN element e^* and reduces its set from A^{e^*} to A^*. In the first call to Assess,

```
Algorithm: GroupBasedMapping
Input: KPN = (P, C), MPSoC = (PE, CP), σ : C → ℕ
Output: (μ_p, μ_c)
  E = P ∪ C;
  ∀p ∈ P  A^p := PE;
  ∀c ∈ C  A^c := CP;
  while CanMakeProposals ({(e ∈ E, A^e)}) do
      (e*, A*) := MakeProposal ({(e ∈ E, A^e)});
      (s, {(e, A'^e)}) := Assess ({(e ∈ E, A^e)}, (e*, A*));
      if s = false then
          (s, {(e, A'^e)}) := Assess ({(e ∈ E, A^e)}, (e*, A^{e*} \ A*));
          if s = false then
              fatal error ;
      ∀e ∈ E  A^e := A'^e;
  return DoHomogeneousMapping ({(e ∈ E, A^e)});

  function Assess({(e, A^e)}, (e*, A*))
      {(e, A'^e)} := Propagate ({(e ∈ E, A^e)}, (e*, A*));
      if ∃e ∈ E, A'^e = {} then return (false, {});
      if LoadControl ({(e ∈ E, A'^e)}) = Fail then return (false, {});
      if e* ∈ C then
          {(e, A'^e)} := ConsistCheck ({(e ∈ E, A'^e)}, e*);
          if ∃e ∈ E, A'^e = {} then return (false, {});
      return (true, {(e, A'^e)});
  end
```

Figure 3: GBM Algorithm

it is checked whether the reduction is feasible. If it is not, the proposed A^* is removed from A^{e^*} and the feasibility is checked anew (second call to Assess). After the first phase has succeeded, there is a group of resources (A^e) for every KPN element. It is these groups where the second phase performs homogeneous mapping on. Further details of the GBM algorithm are described in the following.

4.2.1 Making a Proposal

Proposals are generated by analyzing a DAG representation of the process traces. In particular, the algorithm analyzes the *Dominant Sequence* [23], i.e., the critical path of a partially mapped graph. The trace DAG is a graph $TG = (V = \mathcal{S} \cup RE \cup \{v_s, v_e\}, E \subseteq V \times V)$, where $\mathcal{S}$ is the set of all segments, RE contains a node for every read event and $\{v_s, v_e\}$ are help nodes. Edges represent dependencies.

Let $RE = \{r_i^c\}, c \in C, i \in \mathbb{N}$, with r_i^c the i-th read access to channel c. For convenience, let $S_{r_i^c} \in \mathcal{S}$ be the segment that finishes with read event r_i^c (note that $S_{r_i^c} \in T_{\text{dst}(c)}$). Similarly, let $WE = \{w_i^c\}$ be the set of all write events and $S_{w_i^c} \in \mathcal{S}$ the segment that finishes with write event w_i^c ($S_{w_i^c} \in T_{\text{src}(c)}$). Finally, let $\forall S \in \mathcal{S}, \text{nextseg}(S)$ denote the segment coming directly after S, i.e., $S_{i+1}^p = \text{nextseg}(S_i^p)$. For the sake of brevity, suppose that every channel access corresponds to a single token[2]. With this simplification, we distinguish following dependencies in the trace graph:

1. Sequential order: $\forall S_i^p \in \mathcal{S}, (S_i^p, S_{i+1}^p) \in E$.

2. Read after compute: $\forall r_k^c \in RE, (S_{r_k^c}, r_k^c) \in E$, i.e., the read event happens only after the segment has finished.

3. Block-reads: $\forall w_k^c \in WE, (S_{w_k^c}, r_k^c) \in E$, i.e., the k-th token can be read only after it has been written.

4. Unblock-read: $\forall r_k^c \in RE, (r_k^c, \text{nextseg}(S_{r_k^c})) \in E$, i.e., after the read event issued by a segment is done, the next segment in the trace is unblocked.

---

[2]This assumption only serves to simplify the presentation in this paper and is not built in the algorithm.

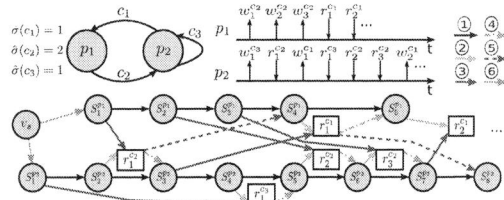

Figure 4: Example: trace DAG.

5. **Block-writes:** $\forall w_k^c \in WE, (r_{k-\hat{\sigma}(c)}^c, \text{nextseg}(S_{w_k^c})) \in E$, i.e., due to buffer sizing, the k-th token can only be written if at least $(k - \hat{\sigma}(c))$ tokens have been read.

6. **Single root, single leaf:** There is an edge from v_s to every root node and from every leaf node to v_e.

Figure 4 illustrates the construction of a trace DAG for a simple KPN. In the upper part, a trace of events is displayed for each process. In the lower part, the corresponding DAG is shown. The colors of the arrows correspond to the afore-mentioned dependencies (see keys in the upper right corner). As an example, consider the *block-write* edge $(r_1^{c_2}, S_4^{p_1})$. In this case, since two tokens fit in channel c_2, the third write would block if no token has been read.

At every call to MakeProposal, the critical path of the trace DAG is determined. This is done by using the timing provided by the slowest resource in the assignment set $A^e, e \in P \cup C$. The algorithm compares the impact of alternative assignments for the nodes in the critical path (better mapping than the slowest resource). The KPN element corresponding to the nodes for which this impact is the highest is selected. The function then returns this element together with its assignment set reduced to the resource group that produced the highest improvement. Once all the elements in the critical path have been assigned to a single group, no more proposals are done.

4.2.2 Mapping Propagation

After a proposal has been made, the function Propagate updates all the assignment sets accordingly. If the assignment set of a process is reduced, then the assignment sets of all incoming and outgoing edges have to be updated. Obviously, the same holds for channels. For this reason, the mapping propagation proceeds recursively. Consider as an example a new assignment set consisting of a single $cp' \in CP$ over a HW FIFO for a KPN channel $c' \in C$. In such a case, the assignment sets of the source and destination processes become $A^{\text{src}(c')} = \{\text{src}(cp')\}$ and $A^{\text{dst}(c')} = \{\text{dst}(cp')\}$.

4.2.3 Load Control

The function LoadControl tries to prevent proposals selecting always the same group of resources. It is therefore required to define a measure of the occupation of a group. This measure has to take into account that the *bigger* the KPN elements are, the more difficult it is to distribute them within a resource group. Consider for example an accumulated total memory requirement of 100 kB to be distributed on 3 memories of 40 kB each. While it is easy to distribute 20 *small* channels of 5 kB among the group, it is impossible to do it for 4 *bigger* channels of 25 kB. A similar observation holds for processes.

Consider a group of resources $G \subset PE \cup CR$ which has been assigned a group of KPN elements $A^G \subset P \cup C$. Generally speaking, if $|A^G| > |G|$, the load control ensures that the utilization of the group U stays below a variable threshold:

$$U < \frac{\beta_x \cdot |A^G|}{|A^G| + |G| \cdot (\beta_x/\beta_y - 1)}$$

The threshold is controlled by the amount of KPN elements and the amount of resources. The parameters $\beta_x \in (0, 1), \beta_y \in (0, \beta_x)$ provide further control. In the extreme cases $|A^G| = |G|$ and $|A^G| >> |G|$ the utilization will be compared against β_y and β_x respectively. For memories we use $\beta_x = 1, \beta_y = 0.5$ and for PEs $\beta_x = 0.95, \beta_y = 0.75$.

The utilization U for a group of memories is determined by the ratio of the required and the available memory. For PEs it is more involved, since it is difficult to define an *available time*. We do it by computing a pessimistic estimation of the application makespan T, based on a list scheduling of the trace DAG. With this, the utilization of a group of PEs is defined as $U = t/(|G| \cdot T)$, with $t = \sum_{p \in A^G} \Xi(T_p, u)$ the accumulated time of the assigned processes.

4.2.4 Consistency Check

When a KPN channel is assigned to a group of CRs, inconsistencies may appear. Consider a group of HW FIFOs G_{fifo}^{CP} that sparsely connect processors within a group G^{PE}. In this case, it is not enough to perform local checks, i.e. whether all producers and consumers of the KPN channels mapped to G_{fifo}^{CP} are mapped to G^{PE}. It has to be further checked that there actually exist at least one fixed feasible mapping. This is performed within the function ConsistCheck, with a worst case execution time of $O(|C|^2 \cdot |CP| \cdot |G^{PE}|)$. Further details of this function are not presented in this paper.

4.2.5 Homogeneous Mapping

After the heterogeneous phase, the assignment sets of the nodes in the critical path of the trace DAG consist only of resources of the same type. For these nodes, the problem is reduced to multiple smaller homogeneous mapping problems, where the main concern is to decide how to share HW resources. Since sharing PEs has a higher impact on the runtime than sharing CRs, we focus on process mapping. Channel mapping can be done, in this case, as an afterthought, if the assignment sets are correct. The same holds for the KPN elements outside the critical path.

Any homogeneous mapping algorithm can be used in this phase, like those derived from the *first fit bin packing* algorithm. We use the information available from the previous phase to better guide this final step. First, note that due to constraints imposed by channel assignment sets, mapping a process $p \in P$ will sometimes imply mapping a group of processes $G_p^P \subseteq P$. Processes are sorted primarily by the size of these induced sets ($|G_p^P|$) and then by the size of the assignment sets ($|A^p|$). The bigger the induced group and the smaller the assignment set, the earlier the process has to be mapped. Finally, the selected process p is mapped to the $pe' \in A^{PE}$ on which a local list scheduler returns the best timing. For speed reasons, the local list scheduler considers only processes that are already mapped to pe' and ignore other dependencies. Formally, let $\text{ALAP}(S_i^p)$ be the ALAP time of segment S_i^p, obtained during the critical path computation on the trace DAG. Similarly, let $t_{S_i^p}^{pe}$ be the time

1265

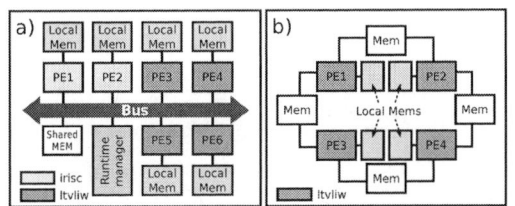

Figure 5: Test platforms. (a) Densely Connected Platform (DCP). (b) Sparsely Connected Platform (SCP).

computed by the local list scheduler for the same segment on a given $pe \in PE$. The final mapping is $\mu_p(p) = pe'$, with

$$pe' = \operatorname*{argmin}_{pe \in A^p} \sum_{S_i^p \in T_p} \max\left(0, t_{S_i^p}^{pe} - \text{ALAP}(S_i^p)\right).$$

During homogeneous mapping, similar operations to those described in Sections 4.2.2–4.2.4 are performed.

5. CASE STUDY

In this section we compare the results of the GBM algorithm against previous mapping heuristics, extended with the ICM algorithm described in Section 4.1.

5.1 Experimental Setup

5.1.1 Target Platforms

Two target platforms were created that expose some of the characteristics of today's MPSoCs as discussed in Section 1. A simplified view of the platforms is shown in Figure 5. The *Densely Connected Platform* (DCP) in Figure 5a stands for MPSoCs where PEs can communicate by various means (like TI's Keystone). The *Sparsely Connected Platform* (SCP) in Figure 5b, instead, represents platforms with a restricted interconnect (like BlueWonder's BWC200). The PEs *irisc* and *ltvliw* are in-house cycle accurate models developed with Synopsys Processor Designer [24]. The bus in the DCP is a transaction accurate model of an AMBA AHB bus. For the bus and the memories, models from the Synopsys IP library [25] are used. The runtime manager in Figure 5a controls the execution of processes on the different PEs.

5.1.2 Test KPNs

We selected two real applications: a low-pass audio filter (LP-AF) and a JPEG encoder/decoder. The former is an implementation in the frequency domain for stereo audio streams. The latter is a similar JPEG implementation to the one in [10]. Additionally, we test the algorithm with 800 randomly generated KPNs. This serves to support the observations performed for the two real applications.

To create random KPN graphs we used the graph generation facility of the SDF-for-free tool (SDF3) [26]. We keep the topology of the generated SDF and replace some of the actors by either CSDF actors or by KPN processes. After doing this, we generate traces that correspond to each newly created node (SDF, CSDF or KPN). For the tests, we generated two sets of random graphs: low-communication (LC-RKPN) and high-communication (HC-RKPN). Each set contains 2 groups of 200 graphs, one with high and one with low variance (in terms computation and communication times). The graphs in LC-RKPN and HC-RKPN have

an average communication-to-computation ratio of 0.1% and 10% respectively.

5.1.3 Performance Estimation

The MAPS framework contains a so-called *Trace Replay Module* (TRM) with which the execution time of different mappings can be estimated provided with the KPN traces. For the LP-AF and JPEG applications we obtained the KPN traces (T_p) using the actual cycle accurate processor simulators. We further compared the results of the TRM against a full system simulator of the DCP MPSoC built with Synopsys Platform Architect [25]. For over 10 different application mappings, the average absolute error in the application makespan was of only 2.82%. For this reason, we base further analysis on the cycle counts reported by MAPS TRM.

5.2 Results

We compare the GBM algorithm against six heuristics from [10], which we will further refer to as *ICM heuristics*: *Random mapping* (RAN), *Random Walk* (RW), *Load balancing* (LB), *Simulated Annealing* (SIM), *Output-rate Balancing* (ORB) and *Affinity* (AFF). RAN serves as reference point, since, in average, any heuristic should outperform it. RW is commonly used to solve problems with complex interactions, which is the case of KPN mapping. LB and SIM are very often used for mapping KPNs. ORB and AFF are implementations of the heuristics presented in [10].

The results for platform DCP are summarized in Figure 6. The figure shows the achieved makespan for all the test KPNs relative to the makespan obtained by GBM. The makespan of LP-AF and JPEG obtained by GBM was of 78.5 and 159.1 Mcycles respectively. As can be seen in Figure 6a, GBM matches the performance of the ICM heuristics for the simple LP-AF application. It only reports a slight average improvement of 2.2% (10.8% including the outlier – AFF). For a more demanding application such as JPEG, the average makespan obtained by the ICM heuristics is 142% higher than GBM's result corresponding to a speedup of 2.42X speedup (2.9X with outlier – ORB). The speedup over the best of the ICM heuristics, namely RW, was of 1.16X. For the JPEG application, with traces of around 25 MB, the runtime of the whole mapping flow including GBM was of 2.1 s (On a AMD Phenom host processor running at 3.2 GHz with 8 GB of RAM).

The results for the 4 different groups of random KPNs are shown in Figure 6b. For the low communication case (LC-RKPN1-2), GBM reported a speedup of 6%, whereas for cases HC-RKPN1-2 the speedup was of 16%. Also here, the benefit of GBM increases with the complexity of the application. Notice also, that GBM was almost always the best algorithm, while the second best varied along the experiments. GBM was outperformed once by ORB in the LC-RKPN2 case, but only by 1.8%. Finally, note that the results of heuristics like LB, ORB, SIM and AFF were worse than those of RW. This has not been the case in previous works which neglect or oversimplify the interconnect, which shows the impact of data communication.

GBM performed well on platform DCP, which is not a very challenging platform for channel mapping. The benefits of a joint mapping algorithm becomes more evident when targeting platform SCP. For LP-AF and JPEG, it was impossible for the ICM heuristics to find a correct mapping provided the pre-computed process assignment. GBM, instead, found correct mappings with a makespan of 109.6 and 175.6 Mcy-

1266

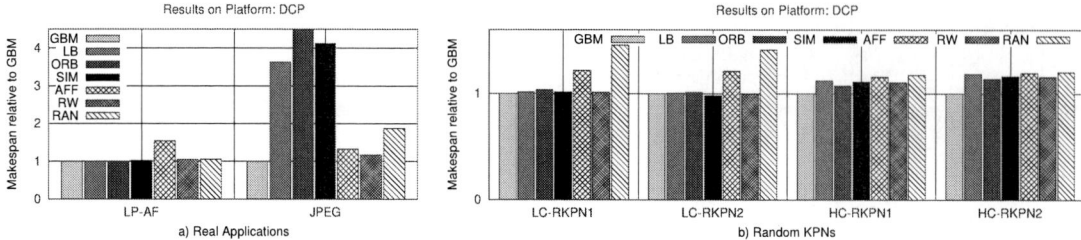

Figure 6: Results for the platform DCP

cles respectively. A similar situation was observed in the case of random graphs. GBM managed to map all random graphs, whereas the ICM heuristics failed in over 96% of the cases (see Table 1).

6. CONCLUSIONS

In this paper we presented our formulation of the KPN mapping problem. By defining communication primitives, we abstracted from communication resources, which allowed us to better capture new complex MPSoC interconnects. We then presented the Group-Based Mapping (GBM) algorithm for jointly assigning processes and channels of KPN applications onto heterogeneous MPSoCs. In a densely connected platform, GBM outperformed state-of-the-art heuristics. The margin for random KPNs was of 11%, for an audio filter application of 2.2% and 142% for JPEG. In a sparsely connected platform, only GBM found a correct mapping for the real applications. Finally, with a moderate runtime under three seconds for 25 MB of KPN traces, the GBM algorithm proved to be efficient.

Further benchmarking of GBM on commercial platforms is in our future work. We also plan extensions to more complex runtime systems and low power modes.

Acknowledgment

This work has been supported by the UMIC Research Centre, RWTH Aachen University (www.umic.rwth-aachen. de). Special thanks to the other members of the MAPS team: W. Sheng, A. Stulova, S. Schürmans, M. Odendahl.

7. REFERENCES
[1] W. Wolf et al., "Multiprocessor System-on-Chip (MPSoC) Technology," IEEE Trans. Computer-Aided Design Integr. Circuits Syst., vol. 27, no. 10, pp. 1701 –1713, oct. 2008.
[2] C. H. K. van Berkel, "Multi-core for Mobile Phones," in in Proc. of DATE '09, pp. 1260–1265.
[3] E. Lee et al., "Synchronous Data Flow," in Proc. of the IEEE, vol. 75, no. 9, pp. 1235–1245, Sept. 1987.
[4] G. Kahn, "The Semantics of a Simple Language for Parallel Programming," in IFIP Congress'74, J. L. Rosenfeld, Ed.

Table 1: Test results for 500 random KPNs on platform SCP

	Successes	Failures	Success Ratio (%)
GBM	500	0	100
LB	19	481	3.8
ORB	13	487	2.6
SIM	19	481	3.8
AFF	24	476	4.8
RW	16	484	3.2
RAN	6	494	1.2

[5] S. Carta et al., "A Control Theoretic Approach to Energy-efficient Pipelined Computation in MPSoCs," ACM Trans. Embed. Comput. Syst., vol. 6, September 2007.
[6] H. Javaid et al., "Low-power Adaptive Pipelined MPSoCs for Multimedia: An H.264 Video Encoder Case Study," in in Proc. of DAC '11, june, pp. 1032 –1037.
[7] "Blue Wonder Communications' BWC200 Passed First IOT Tests against ZTE Environment," D & R, April 2010. [Online]: http://www.designreuse.com/news/23192/lte-ip-iot-tests.html
[8] "Keystone Device Architecture," Texas Instruments. [Online]: http://processors.wiki.ti.com/index.php/Keystone
[9] R. Leupers et al., "MPSoC Programming using the MAPS Compiler," in in Proc. ASP-DAC'10, pp. 897–902.
[10] J. Castrillon et al., "MAPS: Mapping Concurrent Dataflow Applications to Heterogeneous MPSoCs," IEEE Trans Ind. Informat., vol. PP, no. 99, p. 19, Oct 2011.
[11] S. Meijer et al., "Automatic Partitioning and Mapping of Stream-based Applications onto the Intel IXP Network Processor," in in Proc. the SCOPES'07. ACM, pp. 23–30.
[12] D. Nadezhkin et al., "Realizing FIFO Communication When Mapping Kahn Process Networks onto the Cell," in in Proc. of SAMOS'09, pp. 308–317.
[13] T. M. Parks, "Bounded scheduling of process networks," Ph.D. dissertation, EECS Department, UCB, CA, USA, 1995.
[14] M. Geilen et al., "Requirements on the Execution of Kahn Process Networks," in in Proc. of ESOP'03. Springer Verlag, pp. 319–334.
[15] S. Sriram et al., Embedded Multiprocessors: Scheduling and Synchronization. Marcel Dekker, Inc., 2000.
[16] Y.-K. Kwok et al., "Static scheduling algorithms for allocating directed task graphs to multiprocessors," ACM Comput. Surv., vol. 31, no. 4, pp. 406–471, 1999.
[17] E. Cheung et al., "Automatic Buffer Sizing for Rate-constrained KPN Applications on Multiprocessor System-on-Chip," in in Proc. IEEE Int. High Level Design Validation Workshop. IEEE Computer Society, 2007, pp. 37–44.
[18] S. Stuijk et al., "Multiprocessor Resource Allocation for Throughput-Constrained Synchronous Dataflow Graphs," in in Proc. DAC '07. ACM, pp. 777–782.
[19] H. Nikolov, "System-Level Design Methodology for Streaming Multi-processor Embedded Systems," Ph.D. dissertation, Universiteit Leiden, 2009.
[20] S. Meijer et al., ""throughput modeling to evaluate process merging transformations in polyhedral process networks"," in Proc. DATE'10, pp. 747–752.
[21] C. Erbas et al., "Multiobjective Optimization and Evolutionary Algorithms for the Application Mapping Problem in Multiprocessor System-on-Chip Design," IEEE Trans. Evol. Comput., vol. 10, no. 3, pp. 358 – 374, june 2006.
[22] L. Thiele et al., "Mapping Applications to Tiled Multiprocessor Embedded Systems," in in Proc. of ACSD '07. Washington, DC, USA: IEEE Computer Society, pp. 29–40.
[23] G. C. Sih et al., "A Compile-Time Scheduling Heuristic for Interconnection-Constrained Heterogeneous Processor Architectures," IEEE Trans. Parallel Distrib. Syst., vol. 4, no. 2, pp. 175–187, 1993.
[24] Synopsys, "Processor Designer." [Online]: http://www. synopsys.com/Tools/SLD/ProcessorDev/Pages/default.aspx
[25] ——, "Platform Architect." [Online]: http://www.synopsys.com/Tools/SLD/VirtualPrototyping/ Pages/PlatformArchitect.aspx
[26] S. Stuijk et al., "SDF³: SDF For Free," in in Proc. of ACSD'06. IEEE Comp. Soc., June, pp. 276–278.

Unrolling and Retiming of Stream Applications onto Embedded Multicore Processors

Weijia Che and Karam S. Chatha,
Faculty of Computer Science and Engineering,
Arizona State University, Tempe, AZ 85287.
{weijia.che, kchatha}@asu.edu

ABSTRACT

In recent years, we have observed the prevalence of stream applications in many embedded domains. Stream applications distinguish themselves from traditional sequential programming languages through well defined independent actors, explicit data communication, and stable code/data access patterns. In order to achieve high performance and low power, scratch pad memory (SPM) has been introduced in today's embedded multicore processors. Programing on SPM based architecture is both challenging and time consuming. In this paper we address the problem of automatic compilation of stream applications onto SPM based embedded multicore processors through unrolling and retiming. In our technique, code overlay and data overlay are implemented to overcome the limited SPM capacity. Smart double buffering and code prefetching are introduced to amortize memory access delays. We evaluated the efficiency of our technique through compiling several stream applications onto the IBM Cell processor and compared their performance with existing approaches.

Categories and Subject Descriptors

D.3.4 [**Software-Processors**]: Compilers

General Terms

Algorithms, Design, Experimentation

Keywords

Stream, multicore, SPM, unrolling, retiming, overlay

1. INTRODUCTION

In the past few years, stream formats have been recognized as an important model of computation in many embedded system domains. Examples include mobile computing, image/video processing, and network processing. Stream applications share common characteristics such as well defined independent filters, explicit exposed data communication, and stable code/data access patterns. Due to these characteristics, several languages have been developed to model stream applications. Examples of such languages (formally

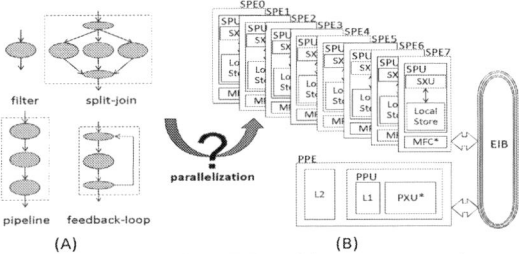

Figure 1: IBM Cell BE architecture overview.

referenced as stream languages) include StreamIt [21] from MIT, CUDA [1] from Nvidia, Brook [2] from Stanford, and CAL [7] from Berkeley. We discuss StreamIt language in more detail here since it serves as the input specification to our experiments. In Figure 1 (A) we provide the four basic structures in StreamIt, namely *filter*, *split-join*, *pipeline*, and *feedback-loop*. Any stream program is constructed out of these four basic structures. StreamIt language models a compute intensive unit as a filter/actor and exposes the data communication among distinct filters as FIFOs. As a result, a stream application is naturally represented as an synchronous data flow (SDF) graph and can be analyzed by a compiler.

Increasing demand for high performance and low power processors in many embedded systems has led to the advent of SPM based architectures. Compared to caches, SPM provides us with simpler logic, smaller chip area, faster access time and lower power consumption, all of which are key requirements for embedded multicore architecture design. Examples of multicore processors that incorporate SPMs include IBM Cell Broadband Engine (BE) [11], Nvidia GeForce series [12], Ageia's PhysX [23], TI TMS320C6472 [22] and many DSPs. In an SPM based architecture, the workload of dynamic management of the limited on-chip memory is shifted from the hardware to the programmer. Code and data transfers among various memory elements are completely software managed and realized through direct memory access (DMA) engine. In Figure 1 (B) we provide an example of such an embedded multicore processor that also serves as the target architecture in our experiments. As shown in the figure, IBM Cell BE has one PPE and eight SPEs. The PPE works as a control plane and eight SPEs serve as high performance data processing plane. Each SPE hosts a 256 KB SPM that is formally referenced as SPE local store. Each SPE or PPE also hosts a DMA engine.

Current parallelizing compilers for embedded multicore processors are still in their infancy, consequently designs are

Table 1: SDF and Architecture Specification

	Constant	Description
SDF	$d(v)$	delay/runtime of actor v
	$C(v)$	code size of actor v
	$w(e)$	initial delays on edge e
	$C(e)$	steady-state token size of edge e
	N_{user}	user specified software pipeline stages
Arch.	$|P|$	number of processing engines
	$C(p)$	processor memory of processor p
	T_{slope}	increasing rate of a DMA transfer
	D_{init}	largest code/data size that can be transferred with T_{init}
	T_{init}	initial cost of a DMA transfer

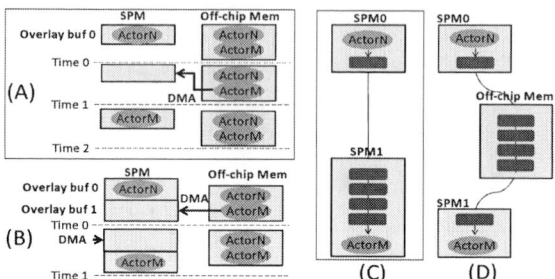

Figure 2: Code overlay and data overlay.

implemented manually. The programmer bears the burden of scheduling actors to processing engines (PEs) and managing the on-chip SPMs throughout the program life time. In order to cope with the limited size of each SPM, code overlay and data overlay have to be implemented. To amortize DMA delay, double buffering that overlaps computation with communication has to be exploited. Double buffering requires storing an extra copy of data and thus results in memory overhead. Finally, actor executions, data/code communication, double buffering, and code/data overlays, they all come together to determine the final performance of the schedule. In this paper, we propose a heuristic approach that automatically compiles a stream application onto embedded multicore processors with the objective of throughput optimization. Our heuristic approach is able to:

- **U**nroll and **R**etime **S**tream formats onto **E**mbedded **M**ulticore processors (URSEM) with the objective of throughput maximization.
- Exploit trade-offs among code overlay, data overlay and double buffering, thus efficiently address the limited on-chip SPMs and DMA delay.
- Schedule stream applications with loop structures and also accept an upper bound on the resulting software pipeline stages.

2. PROBLEM DESCRIPTION

The input to our problem is composed of an SDF representation of the stream application and a hardware description of the target architecture. The SDF specification is given by $G < V, E >$ where V in G represents actors/filters and E represents edges. Prior to invoking our optimization technique, we transform the given SDF into a single appearance SDF (discussed in Section 4.1). The resulting SDF and the target architecture is described in Table 1. DMA latency is approximated by $T_c(x) = T_{init}$, if $x \leq D_{init}$ and $T_c(x) = T_{init} + (x - D_{init}) * T_{slope}$, otherwise. In the equation, $T_c(x)$ denotes the latency of transferring x bytes of code/data. The output of our technique is an actor to PE mapping and a software pipelined schedule together with double buffering, and code/data overlays that maximize the throughput.

The classical retiming techniques [14] [6] alters delays among various function units of a circuit and retains its original logic. The retiming approaches are intriguing to our problem in that they handles loop structures inherently. By properly constraining the retiming delays, an upper bound on the resulting number of software pipeline stages can be imposed. However, the existing approaches that employ retiming for throughput optimization are not directly applicable to our problem due to memory access delays, dou-

ble buffering, limited on-chip SPMs, and code/data overlays. Code/data overlays reduce memory usage by sharing the same physical memory with different code segments and data sets over time. In our basic code overlay scheme, we allocate one overlay buffer for all actors that are (i) scheduled on processor p and (ii) mapped to the off-chip main memory. Figure 2 (A) depicts the program behavior of this code overlay scheme. In the example, ActorM is scheduled to execute next and ActorN is currently present in the overlay buffer. At Time 1, the DMA engine brings ActorM from the off-chip memory and evicts ActorN. As soon as the DMA transfer is completed, we invoke the execution of ActorM (at Time 2). Since the DMA engine and the execution unit in each PE operate independently, we can pre-fetch ActorM while executing ActorN, as shown in Figure 2 (B). Code pre-fetching improves performance at the expense of one extra buffer allocation. In our data overlay scheme (if triggered), we allocate one buffer for each edge that belongs to processor p and push the rest of buffers back to the off-chip main memory. Figure 2 (C) depicts the data memory without data overlay and Figure 2 (D), with data overlay. Data overlay reduces the data memory usage at the expense of circling through the off-chip memory.

In the rest of this paper, Section 3 discusses related work. Section 4 presents our URSEM approach. Section 5 provides experimental results and Section 6 concludes the paper.

3. RELATED WORK

Several previous approaches have addressed the problem of implementing stream workload on embedded multicore processors. A hierarchical framework for scheduling SDF onto multicore processors was discussed by Pino et al. [19]. More recently Ostler et al. [18] proposed techniques for mapping stream based applications onto network processing processors. Liao et al. [16] investigated parallelizing Brook language onto general purpose multicore processors through data and code transformations. Stratton et al. [20] developed a framework MCUDA that executes CUDA language on shared memory multicore processors. In contrast to the above approaches, our technique focuses on embedded multicore processors that incorporate SPMs. In addition to actor to PE mapping and double buffering, we also face the challenge of dynamic management of the limited on-chip SPM for program code and data.

There have been approaches that concentrate on automatic compilation of stream applications onto multicore processors. Gordon et al. [8] explored the trade-offs between data and task level parallelisms and developed a heuristic to generate multi-threaded code for the RAW architecture. Hormati et al. [9] [10] proposed compiler frameworks for mapping stream languages onto GPUs and heterogeneous architectures. Kudlur et al. [13] came up with an ILP that unfolds and partitions a stream application onto multicore processors. An improved version of this work that addresses memory constraint was later presented by Choi et al. [5]. Our approach is distinguished from the above approaches in that we explore the trade-offs between double buffering, code overlay and data overlay, thus efficiently address the SPM constraint.

The previous work that comes closest to us is the CSMP [4] and RETM [3] approaches proposed by Che et al.. The CSMP approach utilizes batch fusion and fission operations to map an SDF model onto embedded multicore processors. In this work, a loop structure is treated as a high-level actor and the resulting software pipeline stages of the schedule is uncontrollable. Our URSEM approach on the contrary handles loop structures inherently and can accept an upper bound on the resulting software pipeline stages. The proposed RTEM approach compiles a stream program onto SPM based multicore processors through retiming. Compared with this work, our URSEM performs unrolling and retiming simultaneously. As a result, we can achieve better performance and scalability. Further, we implemented our code overlay with pre-fetching, which reduces the overall overhead. In the case when the SPM capacity is extremely restricted, we also introduce data overlay.

4. URSEM HEURISTIC APPROACH

Prior to entering our URSEM heuristic approach, we first discuss the pre-processing steps that are performed to construct a single appearance SDF.

4.1 Pre-processing for Single Appearance SDF

In our URSEM heuristic, we require a single appearance SDF. Given a regular SDF without loop structures, we can simply combine all executions of the same actor in a PASS into a high-level execution to derive a single appearance SDF. In the case when loop structures are present, we block process an actor as many times as permitted by the delays on its feedback loop edge. The resulting delay in the single appearance SDF is given by $\lfloor w(e)/N_v \rfloor$, where $w(e)$ denotes the delays/tokens on feedback loop edge e and N_v denotes the tokens consumed by actor v (consumer of edge e) in the steady-state execution. If $w(e) < N_v$ the entire loop is treated as a one high-level actor. The resulting actor is stateful and the delays on e becomes a state variable.

4.2 URSEM Heuristic Algorithm

In this section, we discuss the high level routine of our algorithm as illustrated in $AlgorithmURSEM$. It iteratively performs unrolling and retiming to schedule a stream format G onto an embedded multicore processor P. It terminates when no further performance improvement can be achieved or the unroll factor exceeds $|P|$, Line 19. II_f in the condition indicates the minimum II achieved by scheduling G with an unroll factor f. We divide II_f by f to derive the corresponding II of the original program G. At each iteration, we first unroll the stream program by the given factor and store the unrolled graph in G_f. We utilize the graph unrolling algorithm by Chao et al. [15]. Then the minimum II achieved at unroll factor f (II_f) is calculated through a binary search. Binary search is conducted within the range of $\{\sum_{v \in V_f} C(v)/|P|, \sum_{v \in V_f} C(v)\}$, and $AlgorithmRDL$ is invoked to check whether a given II is achievable. The parameters G_f, P, N, l, and db passed to the binary search capture

---

```
1   II_f ← +∞, f ← 0;
2   repeat
3       II_{f-1} ← II_f, f ← f + 1, G_f ← Unroll(G, f);
4       /* retime with min{|P|, N_user} pipeline stages */
5       l ← 0, db ← 0, N ← min{|P|, N_user} ;
6       II_f ← BinarySearch(G_f, P, N, l, db);
7       |RGs| ← r_max(v) − r_min(v) + 1;
8       if |RGs| < |P| then
9           /* retime with list scheduling */
10          l ← 1, db ← 0, N ← |RGs| ;
11          II_f ← BinarySearch(G_f, P, N, l, db);
12      else
13          /* retime with double buffering */
14          l ← 0, db ← 1, N ← N_user, stages ← 0;
15          while stages < min{|P|, N_user − |P|} do
16              Identify RG (r) that results in min II_f by
                    calling II_f ← BinarySearch(G_f, P, N, l, db);
17              Update II_f and set_db(r, 1);
18              stages ← stages + 1;
19  until II_f/f ≥ II_{f-1}/(f − 1) or f > |P|;
20  return II_{f-1}/(f − 1);
```

$AlgorithmURSEM(G, P)$

the unrolled graph, the multicore architecture, the maximum number of retiming groups (RGs) to be generated, and whether list scheduling and double buffering are enabled, respectively. RG is defined as a group of actors that have the same retiming delay ($r(v)$). In the remainder of this paragraph, we focus on the high-level overview of our URSEM heuristic[1]. The actual number of RGs generated by the retiming procedure is given by $|RGs| \leftarrow r_{max}(v) - r_{min}(v) + 1$, where $r_{max}(v)$ and $r_{min}(v)$ are the maximum and the minimum retiming delays. $|RGs|$ could be less than P due to inter-iteration dependencies or user specified limitation on software pipeline stages. In $AlgorithmURSEM$, a list scheduling is implemented to improve on the initial solution if $|RGs| < |P|$, Line 11. Otherwise if $|RGs| = |P|$, a smart double buffering scheme is implemented, Line 13-18. We greedily introduce double buffering to each RG. The RG that provides us with the most significant performance improvement is selected at each iteration. The process terminates when all RGs are double buffered or there is no extra pipeline stage left, Line 15.

4.3 AlgorithmRDL

$AlgorithmRDL$ determines whether a given II is achievable for scheduling graph G on P through retiming. The retiming delay of each actor is set to zero in the initialization. Then the algorithm enters an iterative procedure where we construct G_0 by preserving all actors from G_r and exactly those edges with $w_r(e) = 0$, Line 3. A scheduling order is generated from G_0 such that if there is an edge directing from actor u to v in G_0, then u must be scheduled before v. We calculate the completion time of each actor $\Delta(v)$ by applying $AlgorithmDeltaCD$. $AlgorithmDeltaCD$ schedules a retimed graph G_r onto a multicore processor P with code and data overlays. We discuss it in Section 4.4. Starting from the 5th line of $AlgorithmRDL$, we compare the completion time of each actor with a given II. If its completion time is larger than II, we increase the retiming delay of v. In the algorithm, $get\_db(r(v))$ returns one if $r(v)$ is double buffered (zero otherwise). If double buffering is enabled for $r(v)$ then $r(v) \leftarrow r(v) + 2$, indicating that one

---

[1]The discussion of $AlgorithmRDL$ is provided in Section 4.3.

```
1  ∀v ∈ V_f set r(v) ← 0;
2  for i = 0 to |V_f| − 1 do
3      Construct G_0 and a scheduling order S;
4      ∀v ∈ V_f, apply AlgorithmDeltaCD to calculate Δ(v);
5      forall the v ∈ S do
6          if Δ(v) > II then
7              if r(v) − r_min(v) < N_user − (get_db(r(v)) + 1)
                 then
8                  r(v) ← r(v) + (get_db(r(v)) + 1);
9              else
10                 return −1;
11     Compute G_r based on the retiming r of each actor v;
12     ∀e ∈ E_f, if w_r(e) < 0 then return −1;
13 II_min ← AlgorithmDeltaCD(G_r, P);
14 if II_min ≤ II then return II_min else return −1;
```

AlgorithmRDL(G_f, P, II, N, l, db)

addition delay is allocated for DMA transfers. Otherwise, we increase $r(v)$ by 1, Line 8. We only alter the retiming delay of an actor when there are enough delays left to be scheduled, Line 7. If an actor's completion time is larger than II and its retiming delay cannot be altered due to lack of pipeline stages (captured by $r(v) − r_{min}(v) \geq N_{user}$, where $r_{min}(v) \leftarrow \min_{v \in V} r(v)$) then we immediately return -1 (failure), Line 10. At each iteration, after the retiming process, we compute the new retimed graph G_r by setting $w_r(e) \leftarrow w(e) + r(v) − r(u)$ for each e in the unrolled graph. For the purpose of double buffering, we occasionally increase the retiming delay of an actor by two instead of one. In this case, the validity of the retiming needs to be verified. If an invalid retiming is found, we return -1, indicating that double buffering cannot be introduced. This scenario could happen when we have loop structures with limited delays on their feedback edges. Upon termination of the iterative retiming procedure, we apply $AlgorithmDeltaCD$ to calculate the resulting II and store it to II_{min}. Finally, we return II_{min} if $II_{min} \leq II$ and return -1 (failure), otherwise.

4.4 AlgorithmDeltaCD

Construction of RG to PE mapping.

$AlgorithmDeltaCD$ schedules a given retimed graph onto a multicore processor with code and data overlays. $P(r(v))$ in the algorithm denotes the set of processors that an actor with retiming $r(v)$ could be scheduled on. If $l = 0$, then $|P(r(v))| = 1$. In this case, each RG is mapped to exactly one processor. Otherwise ($l = 1$), the RG with the maximum parallelism[2] is scheduled on $|P(r(v))| = |P| − |RGs| − 1$ processors with list scheduling [17] and the remaining RGs are scheduled on one processor each. We schedule each actor v following S and update the completion time of actor v ($\Delta(v)$) and the workload of processor p ($\Delta(p)$) accordingly. The calculations of $\Delta(v)$ and $\Delta(p)$ requires the knowledge of code/data memory usage and the memory state of processor p. Their calculations are provided below.

Calculation of code, data memory usage.

The calculation of code memory of processor p after scheduling v is given by,

$$C_{code}(p) \leftarrow C_{code}(p) + C(v) \qquad (1)$$

The processor data memory after scheduling v on p is given

---

[2]RG that has the maximum width following a BFS search.

```
1  ∀v ∈ V, set Δ(v) ← d(v);
2  Calculate RG to PE mapping (P(r(v)));
3  forall the v ∈ S do
4      Schedule v on p ∈ P(r(v)) (list scheduling if l = 1);
5      Update code/data memory, memory state of p;
6      Calculate Δ(v) and set Δ(p) ← Δ(v);
7  return Max_{v∈V} Δ(v);
```

AlgorithmDeltaCD(G_r, P, S, l, db)

as follows. For every edge e that has v as a consumer

if $get\_state(p) \neq$ DATA_OVERLAY
$$C_{data}(p) \leftarrow C_{data}(p) + w_r(e) * C(e) \qquad (2)$$
else $\quad C_{data}(p) \leftarrow C_{data}(p) + C(e)|e : u \to v, u \notin p, v \in p$

For every edge e that has v as a producer

$$C_{data}(p) \leftarrow C_{data}(p) + (1 + get\_db(r(v))) * C(e) \qquad (3)$$

In Equation (2), $get\_state(p)$ returns the memory stage of processor p. The memory state of a processor p could be SF (sufficient), CO (code overlay), DO (data overlay), and IF (infeasible). The transitions of memory states and their conditions are illustrated in Figure 3. The memory state of each processor is first initialized to be SF. Then as we keep on scheduling actors on p, the processor memory state changes to CO when the code and data size becomes larger than $C(p)$. In memory state CO, $\tau_o(v)$ is recalculated for every v that has been scheduled since an actor that has been mapped to the on-chip SPM may be relocated to the off-chip memory when we try to schedule another actor on p. If the SPM is only able to accommodate the program internal data and two overlay buffers ($2 * C_{max}(v)$), the memory state changes to DO. $C_{max}(v)$ denotes the largest actor code size. We conservatively allocate the overlay buffer size to be $C_{max}(v)$ such that every actor can be placed in it. The data memory and $\tau_o(v)$ for each actor being scheduled is recalculated in this case with data overlay enabled. Finally when the data memory (with data overlay) plus an overlay buffer is larger than the SPM, the memory enters IF state, indicating that this actor cannot be scheduled on the p.

Calculation of processor workload.

The calculation of the workload after scheduling actor v on p is given by

$$\Delta(p) \leftarrow d(v) + Max\{\tau_c(v), \Delta(p) + \tau_o(v)\} \qquad (4)$$

In Equation(4), $d(v)$ is the computation delay of actor v. $\tau_c(v)$ models the earliest start time of v due to data dependencies. $\Delta(p)$ indicates the workload of p before scheduling v on it. $\tau_o(v)$ indicates the code overlay overhead of scheduling v on p. $\Delta(p) + \tau_o(v)$ models the earliest start time of v due to limited PEs and code overlay. The calculation of $\tau_c(v)$ and $\tau_o(v)$ are discussed in the following,

Calculation of $\tau_c(v)$.

There could be two categories of data dependencies, namely intra-pipeline dependencies and inter-pipeline dependencies in our schedule. For an intra-pipeline dependency, the edge that connects the producer and the consumer has no delay on it. Therefore the consumer can only start after its producer finishes execution and transfers its data to the consumer side. For an inter-pipeline dependency, the producer and consumer have at least one delay between them. The consumer can execute with the producer simultaneously in a pipelined manner. Given the above discussion, the calcu-

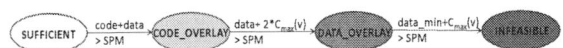

Figure 3: Processor memory state transitions.

lation of $\tau_{intra}(v)$ is given by,

$$\tau_{intra}(v) \leftarrow \max_{e:u \rightarrow v, v \in p, u \notin p, w_r(e)=0} \Delta(u) + T_c(C(e)) \quad (5)$$

In Equation (5), $v \in p$, $u \notin p$, and $w_r(e) = 0$ indicate the condition for edge e to have an intra-pipeline communication. $\Delta(u)$ indicates the completion time of producer v. Since we schedule actors following the scheduling order S, by the time we schedule v on p, $\Delta(u)$ is known to us. $T_c(C(e))$ computes the cost of transferring data from the producer to the consumer. Function $T_c(x)$ is a hardware feature and is defined in Section 2. The calculation of $\tau_{inter}(v)$ is given by,

$$\tau_{inter}(v) \leftarrow \begin{cases} T_c(C(e)), \text{if } get\_db(r(v)) = 0, \\ \max\{0, T_c(C(e)) - d(u)\}, \textbf{otherwise.} \end{cases} \quad (6)$$

where $e : u \rightarrow v, v \in p, u \notin p, w_r(e) \geq 1$.

In Equation (6), the condition for an edge to have inter-pipeline communication overhead is given by $v \in p, u \notin p$, and $w_r(e) \geq 1$. When double buffering is disabled, the communication overhead equals the DMA transfer cost. Otherwise, DMA transfer is overlapped with actor computation and the effective cost is given by $\max\{0, T_c(C(e)) - d(u)\}$.

Calculation of $\tau_o(v)$.

When the on-chip SPM of p is not able to accommodate all its code and data, code overlay overhead is encountered. The calculation of $\tau_o(v)$ is given by,

$$\tau_o(v) \leftarrow \begin{cases} 0, & \text{if } get\_state(p) = \text{SF}, \\ \max\{0, T_c(C(v)) - d(u)\}, & \text{if } get\_state(p) = \text{CO}, \\ T_c(C(v)), & \text{if } get\_state(p) = \text{DO}, \\ +\infty, & \text{if } get\_state(p) = \text{IF}. \end{cases} \quad (7)$$

In Equation (7), $T_c(C(v))$ captures the code overlay overhead without code pre-fetching and $\max\{0, d(u) - T_c(C(v))\}$ captures the code overlay overhead with code pre-fetching.

4.5 Algorithm Complexity

Without unrolling, *AlgorithmDeltaCD* runs in $O(|E| + |V|)$. *AlgorithmFDL* wraps *AlgorithmDeltaCD* within a loop of $|V|$. Therefore *AlgorithmFDL* runs in $O(|V|(|E| + |V|))$. The binary search adds another complexity of $O(log_2 U)$ ($U = \sum_{v \in V} d(v)$). As a result, URSEM without unrolling runs in $O(|V|(|E| + |V|)log_2 U)$. Since the unroll factor is bounded by $|P|$ in our technique, the overall algorithm complexity is given by $O(|V||P|^2(|E| + |V|)(log_2 U + log_2 |P|))$.

5. EXPERIMENTAL RESULTS

We adopted StreamIt language as our input specification. URSEM was implemented as an optimization pass in the StreamIt compiler 2.1.1. The hardware platform we experimented with is a PlayStation3 system running Fedora 9 at 3.2 GHz. In this platform, 1 PPE and 6 SPEs are available to the programmer. Table 2 presents the benchmark details.

Overall Performance Comparison.

We first compared the performance of our URSEM with two existing approaches, namely CSMP [4] and RTEM [3]. CSMP has no control over the number of software pipelines being generated. For a fair comparison, we set N_{user} to be infinity for RTEM and URSEM. Figure 4 presents their performance results. The x-axis and y-axis provide the benchmark names and their steady-state execution time normal-

Table 2: Benchmark Specifications

Benchmarks	$|V|$	$|E|$	Benchmarks	$|V|$	$|E|$
BeamFormer	40	72	FilterBank	51	65
BitonicSort	26	31	FMRadio	29	39
ChannelVocoder	55	70	MPEG2	17	20
DCT	15	22	SerpentFull	37	36
DES	8	7	TDE	29	28
FFT	17	16	Vocoder	97	128

ized to the lower bound. The lower bound is given by $L = \max\{\max_{v \in V} d(v)/f, \sum_{v \in V} d(v)/|p|\}$. Our URSEM outperforms RTEM due to the fact that we perform unrolling and retiming simultaneously. A better load balancing is expected as the unroll factor increases. Compared with RTEM, we also implemented code prefetching and data overlay that further improves the overall performance. Our URSEM also performs better than CSMP in most cases. This is due to the fact that in CSMP, as the fusion operation proceeds, the granularity that the algorithm can operate on becomes larger and larger. Whereas, our URSEM always works on the granularity of a single actor. Overall our URSEM outperforms CSMP by 21% and RTEM by 6%. The algorithm run time of our URSEM, based on the benchmark size, is hundreds of seconds. Whereas CSMP and RTEM finishes in less than ten seconds. The increased algorithm run time is due to the *Unrolling* operation. Nevertheless, a user can provide an upper bound on the unroll factor for a shorter algorithm run time, or terminate the *Unrolling* as soon as an acceptable solution is achieved.

Impact of optimizations.

We examine the impact of each optimization in URSEM in Figure 5. N_{user}, $|P|$ and SPM size are set to infinity, 28, and 4K respectively. We applied Retiming, *Unrolling*, *Double Buffering*, *Code Overlay*, *Code Pre-fetching*, and *Data Overlay* incrementally. We didn't show the results for *FilterBank*, *FMRadio*, and *Serpentfull* because they reduce to the mapping of scheduling every actor on PPE. As observed from Figure 5, *Unrolling* delivers the most significant performance gain due to improved parallelism in an unrolled graph. Double buffering is another optimization that has a significant impact. It takes effect when the code and data memory is still tolerable compared to the SPM size. *Code Overlay*, *Code Prefetching*, and *Data Overlay* further reduce the steady-state execution time whenever they can be applied. Overall, the performance improvement by applying all six optimizations is over 47%.

Performance scaling with PEs.

In this section, we examine the scalability of URSEM with different number of PEs. N_{user} and SPM are set to infinity and 256K respectively so that they do not become the limiting factors. The number of processors are set to 7, 14, 21, and 28 respectively. The experimental results are shown in Figure 6. The y axis in the figure presents the normalized steady-state execution time scaled to 7 PEs. Overall our approach scales with the number of processors as observed from Figure 6. This property results from the fact that we perform iterative unrolling in our algorithm. When the number of processors increases, the algorithm will search for a larger unroll factor. Overall the performance improvement is around 70% when we increased the number of PEs from 7 to 28. The results validate the scalability of our approach.

Performance scaling with SPMs.

In this experimental setup, we examine the performance

1272

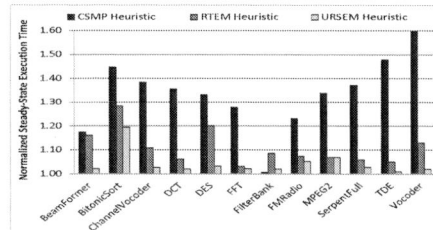

Figure 4: Overall performance comparison.

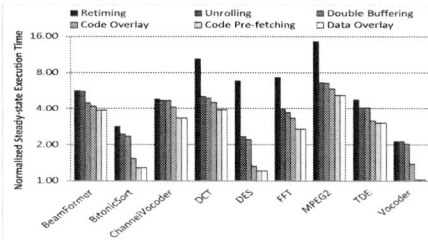

Figure 5: Impact of optimizations.

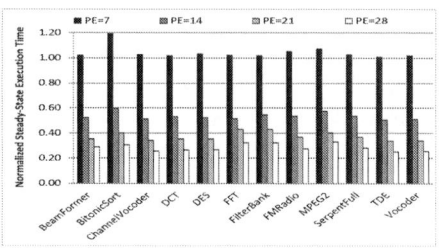

Figure 6: Performance scaling with PEs.

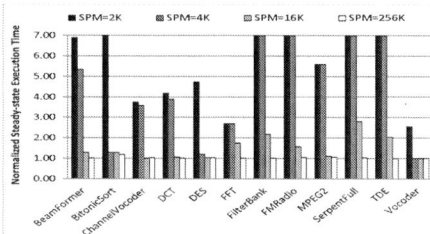

Figure 7: Performance scaling with SPMs.

of our URSEM algorithm under tight SPM constraints. The sizes of each SPM is set to be 2K, 4K, 16K, and 256K respectively. N_{user} and $|P|$ are set to infinity and 7 respectively. Figure 7 provides the experimental results under this setup. As observed from Figure 7, our URSEM always generates a valid solution. When the SPM size is extremely limited, many solutions reduce to mapping all actors to PPE. From Figure 7, when the SPM size is set to 2K, 6 out of 12 benchmarks map everything to PPE. When the memory increases from 2K to 16K, the steady-state execution time of each benchmark drops down dramatically. This behavior suggests that the on-chip SPM is very precious when the code and data memory are comparatively large. The results validate the rationale for introducing code overlay and data overlay in our technique.

6. CONCLUSION

In this paper, we propose an unrolling and retiming approach for scheduling stream applications onto embedded multicore processors. In our technique, a user specified number of software pipeline stages can be imposed. Compared to the the existing approaches, our URSEM algorithm efficiently unrolls and schedules a stream application with loop structures. Our URSEM scales well over a wide range of PEs, delays, and SPMs. Further, our heuristic performs code pre-fetching and data overlay under tight SPM constraints, thus is able to handle extreme cases with tolerable performance results. Our future work will address stream applications with dynamic behavior and execution time.

7. ACKNOWLEDGMENT

The research presented in this article was supported in part by grants from Science Foundation Arizona (SFAz), Stardust Foundation, National Science Foundation (CCF-0903513), and Semiconductor Research Corporation (SRC).

8. REFERENCES

[1] *Compute Unified Device Architecture Programming Guide.* NVIDIA: Santa Clara, CA, 2007.

[2] I. Buck, T. Foley, and D. Horn et al. Brook for GPUs: stream computing on graphics hardware. ACM, 2004.

[3] W. Che and K. Chatha. Compilation of stream programs onto scratchpad memory based embedded multicore processors through retiming. In *DAC11*, pages 122 –127, june 2011.

[4] W. Che, A. Panda, and K. Chatha. Compilation of stream programs for multicore processors that incorporate scratchpad memories. *DATE, 2010*.

[5] Y. Choi and Y. Lin et al. Stream compilation for real-time embedded multicore systems. In *CGO '09*, pages 210–220, 2009.

[6] A. Darte and G. Huard. Loop shifting for loop compaction. *International Journal of Parallel Programming*, 28, 2000.

[7] J. Eker and J. W. Janneck. Cal language report. 2003.

[8] M. I. Gordon and W. Thies et al. Exploiting coarse-grained task, data, and pipeline parallelism in stream programs. *SIGOPS Oper. Syst. Rev.*, 40:151–162, October 2006.

[9] A. Hormati, Y. Choi, and M. Kudlur et al. Flextream: Adaptive compilation of streaming applications for heterogeneous architectures. In *PACT '09*, pages 214 –223, sept. 2009.

[10] A. Hormati, M. Samadi, and M. Woh et al. Sponge: portable stream programming on graphics engines. *SIGPLAN Not.*, 46(3):381–392, Mar. 2011.

[11] J. A. Kahle, M. N. Day, and H. P. Hofstee et al. Introduction to the Cell multiprocessor. *IBM Journal of Research and Development*, 49:589 –604, 2005.

[12] E. Kilgariff and R. Fernando. The GeForce 6 series GPU architecture. In *SIGGRAPH*. ACM, 2005.

[13] M. Kudlur and S. Mahlke. Orchestrating the execution of stream programs on multicore platforms. In *Proceedings of ACM SIGPLAN*, 2008.

[14] C. Leiserson and J. Saxe. Retiming synchronous circuitry. *Algorithmica*, 6:5–35, 1991.

[15] C. Liang-Fang. *Scheduling And Behavioral Transformations For Parallel Systems*. PhD thesis, Princeton University, 1993.

[16] S.-w. Liao and Z. Du et al. Data and computation transformations for Brook streaming applications on multiprocessors. In *Proceedings of CGO*, 2006.

[17] G. D. Micheli. *Synthesis and Optimization of Digital Circuits*. McGraw-Hill Higher Education, 1st edition, 1994.

[18] C. Ostler et al. Ilp and heuristic techniques for system-level design on network processor architectures. *TODAES*, 2007.

[19] J. Pino and E. Lee. Hierarchical static scheduling of dataflow graphs onto multiple processors. In *ASSP*, volume 4, 1995.

[20] J. Stratton et al. MCUDA: An efficient implementation of CUDA kernels for multi-core CPUs. In *LCPC*, 2008.

[21] W. Thies, M. Karczmarek, and S. Amarasinghe. Streamit: A language for streaming applications. 2304:49–84, 2002.

[22] L. Truong. White paper: Low power consumption and a competitive price tag make the six-core TMS320C6472 ideal for high performance applications. *Processing Business*, oct. 2009.

[23] S. Wasson. Ageia's physx physics processing unit. *The tech report, PC hardware explored*, Last accessed July 2008.

Exploiting Spatiotemporal and Device Contexts for Energy-Efficient Mobile Embedded Systems

Brad Donohoo[†], Chris Ohlsen[†], Sudeep Pasricha[†‡], Charles Anderson[‡]

† Department of Electrical and Computer Engineering
‡ Department of Computer Science
Colorado State University, Fort Collins, CO
bdonohoo@rams.colostate.edu, ohlsensc@rams.colostate.edu, sudeep@colostate.edu, anderson@cs.colostate.edu

ABSTRACT

Within the past decade, mobile computing has morphed into a principal form of human communication, business, and social interaction. Unfortunately, the energy demands of newer ambient intelligence and collaborative technologies on mobile devices have greatly overwhelmed modern energy storage abilities. This paper proposes several novel techniques that exploit spatiotemporal and device context to predict device interface configurations that can optimize energy consumption in mobile embedded systems. These techniques, which include variants of linear discriminant analysis, linear logistic regression, non-linear logistic regression with neural networks, and k-nearest neighbor are explored and compared on synthetic and user traces from real-world usage studies. The experimental results show that up to 90% successful prediction is possible with neural networks and k-nearest neighbor algorithms, improving upon prediction strategies in prior work by approximately 50%. Further, an average improvement of 24% energy savings is achieved compared to state-of-the-art prior work on energy-efficient location-sensing.

Categories and Subject Descriptors: H.1.2 [**Models and Principles**]: User/Machine Systems – *human factors*

General Terms – Algorithms, Performance, Human Factors

Keywords: Smartphone, Energy Optimization, Machine Learning

1. INTRODUCTION

Mobile phones and other portable devices (tablets, PDA's, and e-readers) are fundamental everyday tools used in business, communication, and social interactions. As newer technologies (e.g. 4G networking, multicore/GPUs) and applications (e.g. 3D gaming, Apple's FaceTime™) gain popularity, the gap between device usage capabilities and battery lifetime continues to increase, much to the annoyance of users who are now becoming more and more reliant on their mobile devices. The growing disparity between functionality and mobile energy storage has been a strong catalyst in recent years to develop software-centric algorithms and strategies for energy optimization [1]-[17]. These software techniques work in tandem with well-known energy optimizations implemented in hardware including CPU DVFS, power/clock gating, and low power mode configurations for device interfaces and chipsets [18]-[21].

The notion of "smart" mobile devices has recently spawned a number of research efforts on developing "smart" energy optimization strategies. Some of these efforts employ strategies that are *context-aware* including utilization of device, user, spatial, temporal, and application awareness that attempt to dynamically modify or learn optimal device configurations to maximize energy savings with little or negligible impact on user perception and quality of service (QoS) [10][13][17]. This general theme of a *smart* and *context-aware* energy optimization strategy is further explored in this paper, in which a select number of machine learning algorithms are proposed and evaluated for their effectiveness in learning a user's mobile device usage pattern pertaining to spatiotemporal and device contexts, to predict data and location interface configurations. These resulting predictions manage the network interface states allowing for dynamic adaptation to optimal energy configurations when respective interfaces are not required while maintaining an acceptable level of user satisfaction. This idea is further motivated by considering the power distributions of the Google Nexus One smartphone illustrated in Figure 1 [22]. Even when 3G, WiFi, and GPS interfaces are all enabled and idle, they account for more than 25% of total system power dissipation. Furthermore, when only one of the interfaces is active, the other two idle interfaces still consume a non-negligible amount of power. Our work exploits this fact to save energy by dynamically managing data and location interfaces, e.g., turning off unnecessary interfaces at runtime.

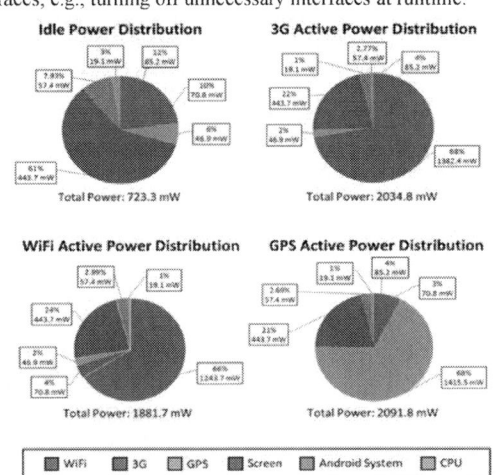

Figure 1. Google Nexus One smartphone power distributions

In this paper, we propose and demonstrate the use of four different classes of machine learning algorithms *(i) linear discriminant analysis, (ii) linear logistic regression, (iii) k-nearest neighbor, and (iv) non-linear logistic regression with neural networks*, on predicting user data/location usage requirements using spatiotemporal and device contexts. These strategies are tested on both synthetic and real-world user usage patterns, which demonstrate that high and consistent prediction rates are possible. The proposed techniques are also compared with prior work on device configuration prediction using self-organizing maps [13] and energy-aware location sensing [9], showing an improvement upon these state-of-the-art techniques.

The remainder of this paper is organized as follows. Section 2 reviews several key related works. Section 3 provides a brief overview of the machine learning concepts used in the study. Section 4 discusses the acquisition and creation of real and synthetic user profiles that are used in our analysis studies. Section 5 describes our device power modeling effort. Section 6 presents the results of our experimental studies. Finally, Section 7 presents our conclusions.

2. RELATED WORK

A large amount of work has been done in the area of energy optimization for mobile devices in recent years. Much of this work focuses on optimizing energy consumed by the device's wireless interfaces by intelligently selecting the most energy-efficient data interface (e.g. 3G/EDGE, WiFi) [1], [2]. Other work [3]-[8] focuses on energy-efficient location-sensing schemes aiming to reduce high battery drain caused by location interfaces (e.g. WiFi, GPS) by deciding when to enable/disable location interfaces or modify location acquisition frequency. Lee et al. [9] in particular propose a Variable Rate Logging (VRL) mechanism that disables location logging or reduces the GPS logging rate by detecting if the user is standing still or indoors. The authors in [10] propose a context-aware method to determine the minimum set of resources (processors and peripherals) that results in meeting a given level of performance, much like our work. They determine if a user is moving/stationary and indoors/outdoors and control resources using a static lookup table. In contrast, *our work controls resources dynamically by using machine learning algorithms that are trained on real user activity traces.* Zhuang et al. [11] propose an adaptive location-sensing framework that involves substitution, suppression, piggybacking, and adaptation of applications' location-sensing requests to conserve energy. Their work is directed towards LBAs (location-based applications) and only focuses on location interfaces, while *ours is a system-wide optimization strategy that is capable of saving energy regardless of the foreground application type.*

A substantial amount of research has been dedicated to utilizing machine learning algorithms for the purpose of mobile user context determination. Batyuk et al. [13] extend a traditional self-organizing map to provide a means of handling missing values, then use it to predict mobile phone settings such as screen lock pattern and WiFi enable/disable. Other works attempt to predict the location of mobile users using machine learning algorithms. In [14] the authors propose a model that predicts spatial context through supervised learning, and the authors in [15] take advantage of signal strength and signal quality history data and model user locations using an extreme learning machine algorithm. These works are focused on using user context for device self-configuration and location prediction, *whereas our work is focused on using user context for optimizing energy consumed by both data transfer and location interfaces.*

One of the key motivations for applying pattern recognition and classification algorithms to mobile device usage is the observation that user usage patterns are often mutually-independent, in that each user generally has a unique device usage pattern. The use of pattern recognition then allows for energy optimization algorithms to be fine-tuned for each user, achieving energy savings without perturbing user satisfaction levels. This idea is further confirmed in mobile usage studies [16], which additionally focused on smartphone usage pattern analysis and its implications on mobile network management and device power management. Although their work had a slightly different focus than our work, the key relevant take-away is that the authors demonstrated from a two month real smartphone usage study that *all users have unique device usage patterns.* Our previous work [17] also found that usage patterns are unique in the way users interact with different apps on their mobile devices. In Section 4 of this study, the real user usage patterns further confirm this claim – all five users had unique usage patterns in the amount of interaction as well as when and where the interactions most often took place.

3. ALGORITHM OVERVIEW

The notion of searching for patterns and regularities in data is the fundamental concept in the field of pattern recognition and data classification. *Machine learning* is often focused on the development

and application of computer algorithms in this field [23]. In this work we focus on exploiting learning algorithms to discover opportunities for energy saving in mobile embedded systems through the dynamic adaptation of data transfer and location network interface configurations without any explicit user input. Consequently, we enable energy-performance tradeoffs that are unique to each user. This is accomplished by learning both the spatiotemporal and device contexts of a user through the application of a given algorithm and using these to control device network configuration to save energy. In other words, given a set of input contextual cues, the algorithms will exploit learned user context to dynamically classify the cues into a system state that precisely governs how data and location interfaces are utilized. The goal is to achieve a state classification that saves energy while maintaining user satisfaction. An overview of the basic underlying concepts and application of the machine learning algorithms used in this study is briefly discussed below.

3.1 Linear Discriminant Analysis

Linear discriminant analysis (LDA) makes use of a Bayesian approach to classification in which parameters are considered as random variables of a prior distribution. This concept is fundamentally different from data-driven linear and non-linear discriminant analyses in which what is learned is a function that maps or separates samples to a class. Bayesian estimation and the application of LDA is also known as *generative modeling,* in that what is learned is a probabilistic model of the samples from each class. By considering parameters as random variables of a prior distribution one can make use of prior known information. For example knowing that a mean μ is very likely to be between a and b, the probability can be determined in such a way that the bulk of the density lies between a and b [24]. Given a prior probability distribution and a class likelihood, Bayes' theorem (equation 1) can be invoked to get an inferred posterior probability to derive a class prediction (C_k) for a new observed sample x_n using a *maximum a posteriori* (*MAP*; equation 2) [24]:

$$p(C_k|x_n) = \frac{p(C_k)p(x_n|C_k)}{p(x)} \qquad (1)$$

$$\underset{C}{\operatorname{argmax}}\, p(C_k|x_n) \qquad (2)$$

LDA is applicable to a wide range of classification problems of both univariate or multivariate input spaces and binary- or multi-class classification. A number of statistical distribution functions can be applied, but the most common is the *Gaussian* or *Normal* distribution (which we use in our study) as shown in equation 3 below.

$$N(x|\mu,\sigma^2) = \frac{1}{\sqrt{2\pi\sigma^2}} exp\left\{-\frac{1}{2\sigma^2}(x-\mu)^2\right\} \qquad (3)$$

There are a number of reasons the Gaussian distribution is a prominent model used in several fields of study, including natural and social sciences, but a simplistic view is that the shape of the Gaussian distribution, the well-known bell shape, is often an acceptable model for complex and partly random phenomena. This notion is brought about by the fact that the shape of the Gaussian is controlled by the mean μ and variance σ^2. The location of the mean (or the peak of the curve) gives an expectation of where random variables tend to cluster and the variance (or width of the curve) gives an indication of the variability of the data.

3.2 Linear Logistic Regression

Similar to LDA, linear logistic regression (LLR) is a technique used to derive a linear model that directly predicts $p(C_k|x_n)$, however it does this by determining linear boundaries that maximize the likelihood of the data from a set of class samples instead of invoking Bayes' theorem and generating probabilistic models from priori information. LLR expresses $p(C_k|x_n)$ directly by requiring all linear function values to be between 0 and 1 and that they all sum to 1 for any value of x_n, as shown in equation 4:

$$p(C_k|x_n) = \frac{f(x_n, \beta_k)}{\sum_{m=1}^{K} f(x_n, B_m)} \qquad (4)$$

With LDA, Bayes' theorem, class priors, and class probability models were used to infer the class posterior probabilities, which were then used to discriminate between the different classes for a given sample x_n. In contrast, LLR solves for the linear weight parameters, β_k, directly using gradients to maximize the data likelihood. This is done by enumerating the likelihood function, $L(\beta)$, using a 1-of-K coding scheme for the target variables, as shown in equations 5, 6 [23], in which every value of $t_{n,k} \in \{0,1\}$ and each row only contains a single '1'. The class variable transformations are known as indicator variables and are used in the exponents of the likelihood function to select the correct terms for each sample x_n.

$$C = \{k_1, k_2, k_3, \dots, k_n\}$$
$$\downarrow$$
$$\begin{pmatrix} t_{1,1} & t_{1,2} & \cdots & t_{1,k} \\ t_{2,1} & t_{2,2} & \cdots & t_{2,k} \\ \vdots & & & \\ t_{n,1} & t_{n,2} & \cdots & t_{n,k} \end{pmatrix} \qquad (5)$$

$$L(\beta) = \prod_{n=1}^{N} \prod_{k=1}^{K} p(C_k|x_n)^{t_{n,k}} \qquad (6)$$

In order to find the β that maximizes the data likelihood, the product of products is transformed to a sum of sums using the natural logarithm to simplify the gradient calculation with respect to β. Since equation 7 is non-linear, iterative methods like gradient ascent or scaled conjugate gradient [24] can be used to solve for the gradient of the log likelihood, $\nabla_\beta LL(\beta)$, , and obtain the respective β weights.

$$LL(\beta) = \sum_{n=1}^{N} \sum_{k=1}^{K} t_{n,k} \log p(C_k|x_n) \qquad (7)$$

3.3 Non-linear Logistic Regression with Neural Networks

Neural network models, also known as *Artificial Neural Networks*, are inspired by the way the human brain is believed to function. Many of the normal basic everyday information processing requirements handled by the brain, for example sensory processing, cognition, and learning, surpass any capable computing system out there today. Although a human brain is quite different than today's computing hardware, it is believed that the basic concepts still apply in that there is a computational unit, known as a *neuron*, and connections to memory stored in *synapses*. The main difference being that the human brain consists of billions of these simple parallel processing units, neurons, which are interconnected in a massive multi-layered distributive network of synapses and neurons [24].

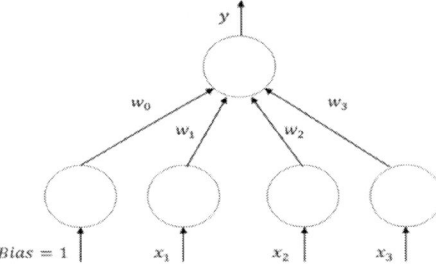

Figure 2. Neural network perceptron model

In machine learning, these concepts are modeled as what is called a *perceptron*, the basic processing element, connected by other *perceptrons* through weighted connections, as illustrated in Figure 2. The output of a perceptron is simply a weighted sum of its inputs including a weighted bias, as shown in equation 8.

$$y = \sum_{i=1}^{n} w_i x_i + w_0 \qquad (8)$$

To compute the output y given a sample x_i, backpropagation using the gradient with respect to the weights is performed using a training dataset to find the weight parameters, w_i, that minimize the mean squared error between the neural network outputs, y_i, and the target outputs, t_i. By default, the neural network consists of a hyperplane (for multiple perceptrons) that can be used as a linear discriminant to linearly separate the classes. To improve prediction accuracy, we make it non-linear, by applying a sigmoidal or hyperbolic tangent to hidden unit layer perceptrons, as denoted in equation 9. This allows for non-linear boundaries with the output of the neural network being linear in the weights, but non-linear in the inputs.

$$y_i' = sigmoid(y_i) = \frac{1}{1 + \exp(\boldsymbol{w}^T \boldsymbol{x})} \qquad (9)$$

For classification with a neural network (non-linear logistic regression), the number of parallel output perceptrons is equal to the number of classes. The output from each perceptron, y_i, is then sent to post processing as in equation 10 to determine the respective class by taking the maximum of the post-processed outputs:

$$C_i \; if \; y_i = \max_i \frac{\exp(y_i)}{\sum_i \exp(y_i)} \qquad (10)$$

One of the biggest criticisms about the use of neural networks is the time required for training. Although this can be a major issue if using a simple gradient descent approach, newer training techniques, such as the scaled conjugate gradient (SCG) [25], can greatly minimize the time required for training. SCG, a method for efficiently training feed-forward neural networks, was used for training the neural networks in this study.

3.4 K-Nearest Neighbor

The k-nearest neighbor (KNN) algorithm is a fairly simple non-parametric unsupervised approach for the data classification problem. A key assumption of non-parametric estimation is that similar inputs have similar outputs [23]. In KNN, new samples are classified by assigning them the class that is the most common among the k closest samples in the attribute space. This method requires some form of distance measure for which Euclidean distance is typically used. The Euclidean distance between two points a and b, each containing i attributes, is defined in equation 11.

$$d(a,b) = \sqrt{\sum_{i=1}^{n} (b_i - a_i)^2} \qquad (11)$$

4. USER INTERACTION STUDIES

In order to compare the relative effectiveness of the different algorithms at predicting a user's data/location usage requirements, five real user usage profiles for four different Android smartphones (HTC myTouch 3G, Google Nexus One, Motorola Droid X, HTC G2, and Samsung Intercept) were collected over a one week period with a custom Context Logger application. The application logged user context data on external storage, which was acquired at the end of the one week session and used in our algorithm analysis.

4.1 Context Logger

We created a custom *Context Logger* application that ran in the background as an Android service and gathered both spatiotemporal and device usage attributes at a one minute interval. Table 1 lists the attributes recorded by the logger and used for algorithm analysis and indicates whether the attribute was a continuous variable (floating point), discrete variable (integer), or logical variable (true/false). The *GPS Satellites* attribute is used as an indirect correlation to GPS signal

1276

strength and *WiFi RSSI* is a measure of WiFi signal strength. In addition to more common device attributes such as *Battery Level* and *CPU Utilization*, we gathered several uncommon OS attributes: *Context Switches*, *Processes Created*, *Processes Running*, and *Processes Blocked*. We hoped to aid prediction by using these as inputs to the machine learning algorithms. The three target variables (*Data Needed*, *Coarse Location Needed*, and *Fine Location Needed*) were obtained by examining the requested Android permissions of all of the device's current running foreground applications and services. The *Device Moving* attribute was determined by using the accelerometer sensor and a metric for movement that is the sum of the unbiased variance of X, Y, and Z acceleration [7], given as:

$$Var(m_1 \dots m_n) = \frac{\sum_{i=1}^{N} m_i^2 - \frac{1}{N}\left(\sum_{i-1}^{N} m_i\right)^2}{N-1} \quad (12)$$

$$Metric = Var(x_1 \dots x_n) + Var(y_1 \dots y_n) \\ + Var(z_1 \dots z_n) \quad (13)$$

Table 1. Recorded data attributes

Context	Attribute	Type
Temporal	Day of week	Discrete
	Time of day	Discrete
Spatial	Latitude	Continuous
	Longitude	Continuous
	GPS Satellite Count	Discrete
	WiFi RSSI	Discrete
	Number of WiFi APs Available	Discrete
	3G Network Signal Strength	Discrete
	Device Moving	Logical
	Ambient Light	Discrete
Device	Call State	Discrete
	Battery Level	Discrete
	Battery Status	Discrete
	CPU Utilization	Continuous
	Context Switches	Discrete
	Processes Created	Discrete
	Processes Running	Discrete
	Processes Blocked	Discrete
	Screen On	Logical
Targets	Data Needed	Logical
	Coarse Location Needed	Logical
	Fine Location Needed	Logical

4.2 Data Preparation

As mentioned earlier, GPS location coordinates were recorded along with the other attributes at one minute intervals. The Android SDK GPS location data returns the user's longitude and latitude coordinates in decimal degrees as reported by the onboard GPS chipset [26]. Although exact accuracy is dependent on the actual GPS hardware, the returned values were truncated to a given precision and each longitude and latitude coordinate pair was mapped to a unique location identifier. The truncated location resolution generalized the number of unique locations in which a user spends his/her time, given that for example, a user's home may consist of several different samples of different longitude and latitude pairs. In addition, temporal conditions can be applied to further reduce the number of unique locations (e.g. disregard locations where user spent less than x minutes). Figure 3 shows the effect of truncation and application of temporal conditions for a real user and how the primary locations where the user spent most of his/her time are revealed. For our study we used a location precision of 4 decimal places.

The desired data/location interface configurations were partitioned into eight different states based on the desired target variables (*Data Required*, *Coarse Location Required*, and *Fine Location Required*). Table 2 maps the logical values of the three target variables to a state. The states define the device's current required resources. *Efficiently predicting one of these 8 states using temporal, spatial, and device*

context input variables in Table 1 may ultimately allow opportunistic shutdown of location/wireless radios. If all interfaces are enabled, they would consume a significant amount of energy in their idle states without this dynamic control.

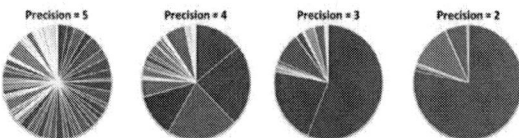

Figure 3. Unique locations identified for varying GPS precisions

Table 2. Interface configuration states

State	Data Required	Coarse Location Required	Fine Location Required
1	No	No	No
2	No	No	Yes
3	No	Yes	No
4	No	Yes	Yes
5	Yes	No	No
6	Yes	No	Yes
7	Yes	Yes	No
8	Yes	Yes	Yes

4.3 Real User Profiles

We distributed the Context Logger application to five different mobile device users, and logged their context data over the course of one week. Figure 4 demonstrates the relative state distributions as described in Table 2 for the five different real users. As can be seen, states 2 – 4 are never realized as data was always required when location was required. In addition, the distributions highlight that each user had a considerably different usage pattern. User 1 can be categorized as a minimal user, where the user rarely utilized their phone with only brief periods of interaction, while users 2 and 5 used their phones rather frequently and for longer periods of time. Users 3 and 4 can be categorized as moderate users, primarily utilizing their devices for only certain times of the day.

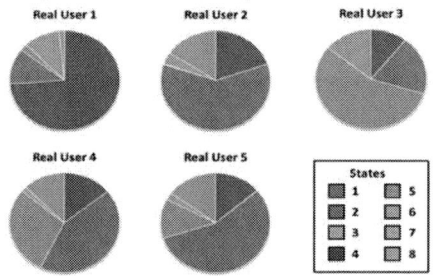

Figure 4. Real user state distributions

4.4 Synthetic User Profiles

Prior to the algorithm evaluation studies, a subset of synthetic user profiles were created for five different idealized and generalized models of average user usage patterns including the following: *(i) 8 – 5 Business Worker, (ii) College Student, (iii) Social Teenager, (iv) Stay At Home Parent,* and *(v) Busy Traveler*. Both an indoor/outdoor location timeline and an interface state profile were created for each synthetic user. Given the difficulty of generating realistic device system data, such as context switches, CPU utilization, and processes created, only a subset of the attribute space was considered. The remaining attributes were based on both the desired state and/or location. For example, if a user was at an outdoor location, larger GPS satellite values and weak WiFi RSSI values were used as opposed to when the user was indoors.

5. DEVICE POWER MODELING

In order to quantify the energy-effectiveness of using machine learning algorithms to predict energy-optimal device states, power

1277

analysis was performed on the real Android based smartphones used, with the goal of creating power models for the data and location interfaces. We use a variant of Android OS 2.3.3, (Gingerbread) and the Android SDK Revision 11 as our baseline OS. We built our power estimation models using real power measurements, by instrumenting the contact between the smartphone and the battery, and measuring current using the Monsoon Solutions power monitor [27]. The monitor connects to a PC running the Monsoon Solutions power tool software that allows real-time current and power measurements over time. We manually enabled the data/location interfaces one by one and gathered power traces for each interface in their active and idle states. The power traces from the Monsoon Power Tool were then used to obtain average power consumption measurements for each interface.

6. EXPERIMENTAL RESULTS

6.1 Prediction Accuracy Analysis

Recall that the input attributes for the learning algorithms come from the gathered spatiotemporal and device context data (Table 1), and the predicted output is one of the 8 interface configuration states (Table 2). To evaluate the prediction accuracy of the different algorithms, the data for each real user was randomly partitioned into training and test sets using a common 80/20 partitioning scheme. The algorithms were then trained on the training data and evaluated on the test data. This was repeated five times for each implementation and the net prediction accuracy is presented in Figure 5.

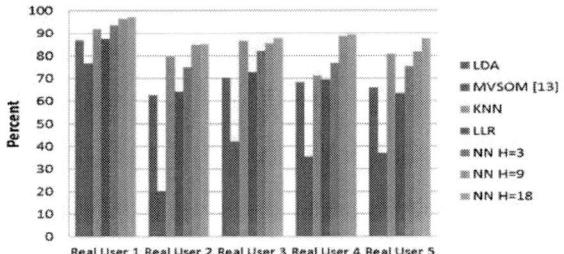

Figure 5. Real user algorithm prediction accuracy

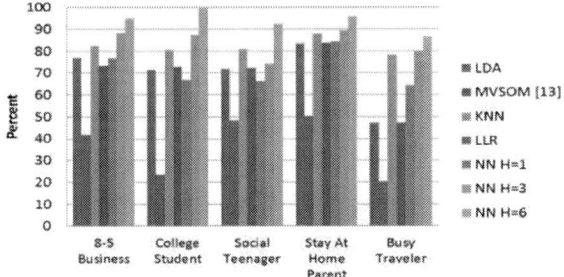

Figure 6. Synthetic user algorithm prediction accuracy

Three different neural network (NN) implementations with a varying number of hidden units, equal to the total (H=18), half (H=9), and one-sixth (H=3) the size of the attribute space were evaluated. We compare the prediction accuracy of our algorithms with the configuration prediction strategy presented in [13] (*MVSOM – Missing Values Self-Organizing Map*). As illustrated in Figure 5, the application of neural networks with a number of hidden units of at least half the size of the attribute space resulted in the highest prediction rates. K-nearest neighbor (KNN), linear logistic regression (LLR), and linear discriminant analysis (LDA) also performed fairly well, with prediction accuracies in the range of 60 – 90 %. However these approaches were much more sensitive to the usage pattern. MVSOM performed the worst and had a high degree of variance in both the usage pattern and random training data selection.

The same algorithms were applied to the synthetic user profiles; however, the attribute space was reduced to only *Day, Time, Location, GPS Satellites, WiFi RSSI, Network Signal Strength, Data Needed,*

Coarse Location Needed, and *Fine Location Needed.* The same strategy for selecting numbers of hidden units for the neural network implementations was applied for the reduced attribute space. Figure 6 illustrates the algorithm prediction rates for the synthetic users showing similar trends as in the real user data.

6.2 Energy Savings

It is important to note that despite high prediction accuracy, the amount of potential energy savings is still highly dependent on the user usage pattern and if the algorithms are positively or negatively predicting states where energy can be conserved. Figures 7 and 8 illustrate the energy savings achieved by the individual algorithms when the algorithm's prediction target states are applied to the real and synthetic user profiles, and compare them against the *VRL* technique (*Variable Rate Logging* [9]). Note that as VRL does not predict system state, results for its prediction accuracy are not shown in Figure 5. Although simpler linear models can achieve high energy savings, it is important to note that energy savings themselves are not good discriminants of an algorithm's *goodness* because user satisfaction must also be considered. For example, if a user spends a significant amount of time in the energy consuming state 8 and the algorithms are predicting a less energy-consuming state during these instances, then more energy can be conserved at the cost of user-satisfaction. Directly correlating the prediction accuracy of the algorithms is especially important for highly interactive users such as users 2 and 5, as opposed to minimally interactive users, e.g., user 1.

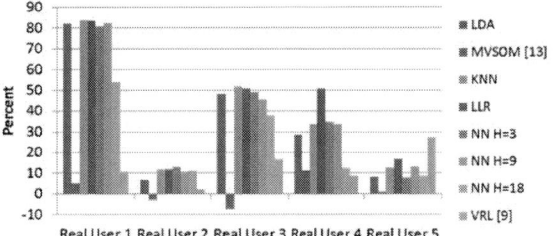

Figure 7. Percent energy saved for real users

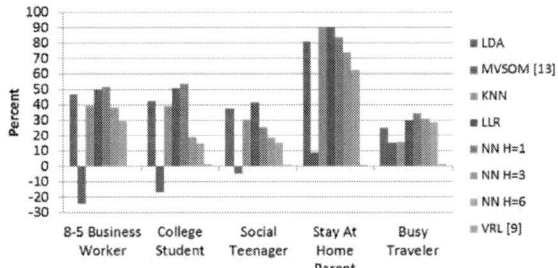

Figure 8. Percent energy saved for synthetic users

All energy savings are relative to the baseline case for Android systems without proactive multi-network interface management. Lower prediction and more generalized models result in the highest energy savings as in the case of LDA and LLR. However, again, these higher savings come at the cost of degraded user satisfaction. *KNN overall performs fairly well in terms of both prediction accuracy and energy savings potential, as does the nonlinear logistic regression with NN approach.* With the latter, an important point to note is that prediction accuracy is proportional to the complexity of the neural network and indirectly proportional to the net energy savings. This outcome is expected as less complex neural networks will result in more generalized models relaxing the constraint for inaccurate predictions that result in higher energy savings. MVSOM, with its low prediction rates, also led to instances of negative energy savings, as it often predicted higher energy states when the true target state was one of less energy consumption. Thus we believe that the MVSOM approach is not very viable for use in mobile embedded systems.

1278

VRL's energy saving capability is constrained because it does not disable device interfaces (only deactivates location logging or reduces logging rate), ignoring idle energy consumption. *Overall, compared to VRL, the average energy savings of our KNN algorithm is 25.6%, and that of the NN approaches is 11.7% (H=18), 24.1% (H=9), and 24% (H=3) for real user patterns.*

6.3 Implementation Overhead

The prediction and energy savings results presented in the previous sections were obtained using a Python implementation of the algorithms on a 2.6 GHz Intel® Core i5™ processor. When considering real-world implementation, it is important to consider the implementation overhead of the individual algorithms. Current hardware in mobile devices on the market today is quickly catching up to the abilities of modern stationary workstations (e.g. Google's Galaxy Nexus – 1.2 GHz duo-core processor). We determined the maximum implementation overhead for the Google Nexus One 1 GHz Qualcomm QSD 8250 Snapdragon ARM processor [22], as shown in Table 3 below. The values shown in the table are average prediction times for each algorithm at runtime, and do not include initial training time. KNN's run time is several orders of magnitude larger than any of the other algorithms, because all computations are deferred until classification. Therefore, although KNN is as good as or better than the neural network (NN) based approach in terms of energy savings and prediction, *the non-linear logistic regression with NN approach is preferable because of its fast execution time.*

Table 3. Average algorithm run times

Algorithm	Average Run time (seconds)
LDA	0.00361
MVSOM [13]	2.15023
KNN	254.131
LLR	0.00307
NN (all 3 variants)	0.02501
VRL [9]	0.05140

Learning algorithm training is also an important consideration. There are several approaches to when and how the algorithms can be trained quickly and in an energy-efficient manner without affecting user QoS on a real mobile platform. One possibility is to train only while the mobile device is plugged in and charging. Our plans for future work include implementation of the abovementioned energy-saving training techniques on a real mobile device.

In summary, our LDA and LLR approaches have the lowest implementation overhead and can result in high energy savings, but often at the cost of user satisfaction. Although our KNN approach is very effective in terms of prediction accuracy and energy savings, its unreasonable implementation overhead renders it unacceptable for real-world applications. The prior work with MVSOM [13] provides low energy savings as a result of its poor prediction accuracy, and takes a long time to run; whereas VRL [9] has low run time but also very low energy savings. Our *non-linear logistic regression with neural network approach that uses the fast scaled conjugate gradient training method and with the number of hidden unites equal to half the attribute space* provides good accuracy, good energy savings, and demonstrates the best adaptation to various unique user usage patterns, while maintaining a low implementation overhead.

7. CONCLUSIONS

In this work we demonstrated the effectiveness of using various machine learning algorithms on user spatiotemporal and device contexts in order to dynamically predict energy-efficient device interface configurations. We demonstrated up to a 90% successful prediction using neural networks and k-nearest neighbor algorithms, showing improvements over the self-organizing map prediction approach proposed in [13] by approximately 50%. In addition, approximately 85% energy savings was achieved for minimally active users with an average improvement of 24% energy savings compared the variable rate logging algorithm proposed in [9] for our best

approach involving non-linear logistic regression with neural networks that also has high prediction accuracy and low overhead.

8. REFERENCES

[1] H. Petander, "Energy-aware network selection using traffic estimation," in *MICNET '09*, pp. 55-60, Sept. 2009.

[2] M. Ra, et al., "Energy-delay tradeoffs in smartphone applications," In *MOBISYS '10*, pp. 255-270, Jun. 2010.

[3] I. Constandache, et al., "EnLoc: energy-efficient localization for mobile phones," in *INFOCOM '09*, pp. 19-25, Jun. 2009.

[4] K. Lin, et al., "Energy-accuracy trade-off for continuous mobile device Location," in *MOBISYS*, pp. 285-298. Jun. 2010.

[5] F. B. Abdesslem, et al., "Less is more: energy-efficient mobile sensing with SenseLess," in *MOBIHELD '09*, pp. 61-62, Aug. 2009.

[6] J. Paek, et al., "Energy-efficient rate-adaptive GPS-based positioning for smartphones," in *MOBISYS '10*, pp. 299-314, Jun. 2010.

[7] I. Shafer, M. L. Chang, "Movement detection for power-efficient smartphone WLAN localization," in *MSWIM '10*, pp. 81-90, Oct. 2010.

[8] M. Youssef, et al., "GAC: energy-efficient hybrid GPS-accelerometer-compass GSM localization," in *GLOBECOM '10*, pp. 1-5, Dec. 2010.

[9] C. Lee, M. Lee, D. Han, "Energy efficient location logging for mobile device," in *SAINT '11*, pp. 84, Oct. 2010.

[10] K. Nishihara, K. Ishizaka, J. Sakai, "Power saving in mobile devices using context-aware resource control," in *ICNC '10*, pp. 220-226, 2010.

[11] Z. Zhuang, et al., "Improving energy efficiency of location sensing on smartphones," in *MOBISYS '10*, pp. 315-330, Jun. 2010.

[12] Y. Wang, et al., "A framework of energy efficient mobile sensing for automatic user state recognition," in *MOBISYS '09*, pp. 179-192, 2009.

[13] L. Batyuk, et al., "Context-aware device self-configuration using self-organizing maps" in *OC '11*, pp. 13-22, June 2011.

[14] T. Anagnostopoulos, C. Anagnostopoulos, S. Hadjiefthymiades, M. Kyriakakos, A. Kalousis, "Predicting the location of mobile users: a machine learning approach," in *ICPS '09*, pp. 65-72, July 2009.

[15] T. Mantoro, et al., "Mobile user location determination using extreme learning machine," in *ICT4M*, pp. D25-D30, 2011.

[16] J. Kang, S. Seo, J. W. Hong, "Usage pattern analysis of smartphones," in *APNOMS '11*, pp. 1-8, Nov. 2011

[17] B. K. Donohoo, C. Ohlsen, S. Pasricha, "AURA: An Application and User Interaction Aware Middleware Framework for Energy Optimization in Mobile Devices," in *ICCD '11*, pp. 168-174, Oct. 2011.

[18] S. Choi, et al., "A selective DVS technique based on battery residual microprocessors and microsystems," Elsevier Sc., 30(1):33–42, 2006.

[19] F. Qian, et al., "TOP: Tail Optimization Protocol for Cellular Radio Resource Allocation," In *ICNP*, 2010.

[20] N. Balasubramanian, A. Balasubramanian, and A. Venkataramani. "Energy Consumption in Mobile Phones: a Measurement Study and Implications for Network Applications," In *IMC*, 2009.

[21] S. Swanson. M.B. Taylor. "Greendroid: Exploring the next evolution of smartphone application processors," Communications Magazine, IEEE. Vol 49. Issue 4. April 2011.

[22] HTC, "Google Nexus One Tech Specs," http://www.htc.com/us/support/nexus-one-google/tech-specs

[23] C. M. Bishop, "Pattern Recognition and Machine Learning," 1st ed. New York: Springer Science+Business Media, 2006

[24] E. Alpaydin, "Introduction to Machine Learning," 2nd ed. Massachusetts: The MIT Press, 2010.

[25] M. Moller, "Efficient Training of Feed-Forward Neural Networks," Ph.D. dissertation, CS Dept., Aarhus Univ., Arhus, Denmark, 1997.

[26] Android Developers, official website, http://developer.android.com/index.html.

[27] Monsoon Solutions Inc., official website, http://www.msoon.com/ LabEquipment/PowerMonitor, 2008.

EPIMap: Using Epimorphism to Map Applications on CGRAs

Mahdi Hamzeh, Aviral Shrivastava, and Sarma Vrudhula
School of Computing, Informatics, and Decision Systems Engineering
Arizona State University, Tempe, AZ
{mahdi, aviral.shrivastava, vrudhula}@asu.edu

ABSTRACT

Coarse-Grained Reconfigurable Architectures (CGRAs) are an attractive platform that promise simultaneous high-performance and high power-efficiency. One of the primary challenges in using CGRAs is to develop efficient compilers that can automatically and efficiently map applications to the CGRA. To this end, this paper makes several contributions: i) *Using Re-computation for Resource Limitations:* For the first time in CGRA compilers, we propose the use of re-computation as a solution for resource limitation problem. This extends the solutions space, and enables better mappings, ii) *General Problem Formulation:* A precise and general formulation of the application mapping problem on a CGRA is presented, and its computational complexity is established. iii) *Extracting an Efficient Heuristic:* Using the insights from the problem formulation, we design an effective global heuristic called EPIMap. EPIMap transforms the input specification (a directed graph) to an Epimorphic equivalent graph that satisfies the necessary conditions for mapping on to a CGRA, reducing the search space. Experimental results on 14 important kernels extracted from well known benchmark programs show that using EPIMap can improve the performance of the kernels on CGRA by more than 2.8X on average, as compared to one of the best existing mapping algorithm, EMS. EPIMap was able to achieve the theoretical best performance for 9 out of 14 benchmarks, while EMS could not achieve the theoretical best performance for any of the benchmarks. EPIMap achieves better mappings at acceptable increase in the compilation time.

Categories and Subject Descriptors

C.3 [**SPECIAL-PURPOSE AND APPLICATION-BASED SYSTEMS**]: [Real-time and embedded systems]; D.3.4 [**Processors**]: [Code generation, Compilers, Optimization]

General Terms

Algorithms, Design, Performance

Keywords

Coarse-Grained Reconfigurable Architectures, Compilation, Modulo Scheduling

1 Introduction

The fundamental challenge faced by all segments of the microelectronics industry is the simultaneous demand for high-performance and higher power-efficiency. The next generation ultra high definition TVs and software defined radios require performance and power efficiencies of tens of Giga (10^9) operations per second per watt (Gops/W) [25]. On the other hand, experts believe that for Exascale computing (10^{18} ops/s $= 10^9$ Gops) to be really practical, power efficiencies of at least hundreds of Gops/W are necessary. Even under the most aggressive scaling scenarios and the most optimistic assumptions on the impact of high clock frequencies on power and thermal characteristics, fundamental innovations at the architecture level are also needed [5].

At the architecture level, accelerators are a promising approach to improve both the performance and power-efficiency of execution. Although special purpose or function specific hardware accelerators (e.g. for FFT) can be very power efficient, they are expensive, not programmable, and therefore limited in usage. Graphics Processing Units (GPUs) are becoming very popular; although programmable, they are limited to accelerating only "parallel loops." Field Programmable Gate Arrays are general-purpose, but they lose a lot of power-efficiency in managing the fine-grain reconfigurability they provide. Coarse-Grained Reconfigurable Architectures or CGRAs have been shown to be an excellent alternative as they not only have power efficiencies close to hardware accelerators, but can be utilized for a wide range of applications because they are *programmable*. For instance, the ADRES CGRA has been shown to achieve performance and power efficiency of 60 GOPS/W in 32 nm CMOS technology [6]. A catalog of existing CGRA designs and architectures is given in [13].

A CGRA is simply an array of processing elements (PEs) interconnected by a 2-D grid. A PE typically consists of an arithmetic logic unit (ALU) and a few registers. It is referred to as *coarse grained reconfigurable* because each PE can be programmed to execute different instructions at the cycle level granularity. CGRAs are completely statically scheduled. Computation is laid out on CGRA, with PEs operating on the output of their neighboring PEs. Very little power, other than the PE power is expended in performing an ALU operation; therefore CGRAs are very power-efficient.

Even though the possible performance and power-efficiency of a CGRA is very high, what can be achieved is critically limited by the compiler technology. A CGRA compiler is much more complex than a regular compiler, since in addition to the regular task of "expressing application in terms of machine instructions," a CGRA compiler must: i) perform explicit pipelining of operations (or software pipelining), ii) map operations to PEs, and iii) route data between PEs so as not to violate data dependencies. Since all of these are hard problems, existing CGRA compilers use "search

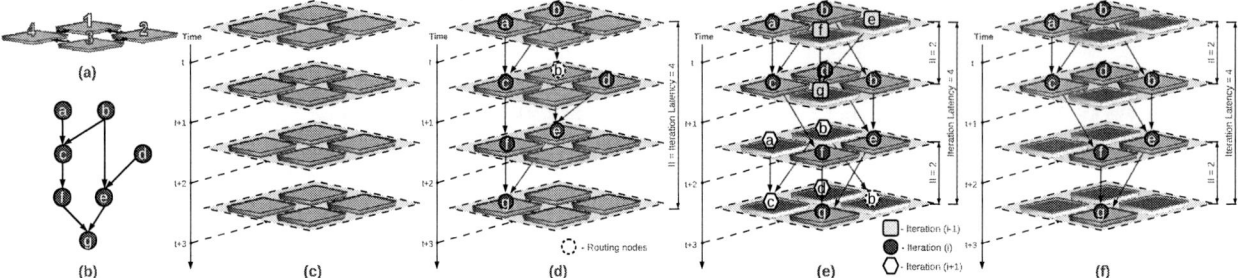

Figure 1: (a) a 2 × 2 CGRA, (b) an input DFG, (c) a time extended CGRA, (d) a valid mapping of the given DFG *(b)* on CGRA *(a)* with iteration latency =II= 4, (e) another mapping for the given DFG with iteration latency= 4 and II = 2, lower II is achieved because two iterations of the loops are executed simultaneously, Dark color PEs in *(e)* execute an operation of another iteration. (f) Only nodes from one iteration of the loop is shown for *(e)*. II is a key performance metric, with the lesser II the better throughput.

based" heuristics to map applications to the CGRA, but the quality of mapping is not good. Towards improving compilers for CGRA, this paper makes three important contributions:

1. Re-computation for limited resource problem: A major challenge to effective application mapping on CGRA is resource limitation. Traditional approach is to use routing to find a solution within the given resource limitation. However our formulation and method utilizes both routing and re-computation to find a solution within the resource limitation which often leads to better mappings.

2. General problem formulation and complexity analysis: We show that the mapping problem can be described as that of finding a subgraph in a minimally Time-Extended CGRA (TEC) graph that is *Epimorphic* to the input graph. This is important, because even though several application mapping heuristics exist for CGRA, the problem has not been formulated before, and we believe that this is the reason for the poor performance of existing heuristics.

3. Distilling an effective heuristic: Insights from the problem formulation enable us to identify the necessary conditions for a feasible solution, and distill an effective and theoretically justified heuristic, which we name EPIMap. EPIMap allows for a systematic search of the solution space which results in high quality mapping.

We compare the quality of mappings generated by EPIMap with the EMS [23] algorithm, which is one the best existing mapping algorithm both in terms of quality and compilation time [1]. Experimental results on 14 important kernels extracted from standard benchmarks, including SPEC2006 show that i) EPIMap generates mappings that have on average 2.85X better performance than EMS; ii) In 9 out of 14 benchmarks, EPIMap finds the optimum mapping while EMS could not find the optimum mapping for any of these benchmarks, and iii) the improvement in mapping quality comes at acceptable (one time) cost of increase in compile time.

2 Background and Related Works

A CGRA is a 2-D mesh of PEs, with each PE having an ALU and a register file (Figure 2). Each PE is connected to its neighbors, and the output of a PE at cycle t is accessible to its neighboring PEs in the next cycle. In addition, a common data bus from the data memory provides data to all the PEs in a row. The earlier work on CGRAs include XPP [4], PADDI [7], PipeRench [11], KressArray [14], Morphosys [18], MATRIX [21], and REMARC [22].

Since many applications spend most of the time on loops [24], this paper focuses on the problem of mapping the innermost loops on a CGRA. A single iteration of a loop is represented as a Data Flow Graph (DFG), which is a directed graph $D = (V_d, E_d)$ where nodes represent operations and arc $(a, b) \in E_d$ iff the output of operation a is an input of operation b. $(a, b) \in E_d$ implies that operation b can only be executed after operation a has been completed.

The inputs to the problem are a DFG of a loop (extracted by the compiler) and a $M \times N$ CGRA. The output is a valid mapping of the nodes in the DFG to the PEs in the CGRA. The goal is to minimize

the total execution time of the entire loop. Figure 1 illustrates all the aspects of the problem. Figure 1(a) shows a 2 × 2 CGRA, and Figure 1(b) shows a DFG of a loop. To map the loop on the CGRA, first the CGRA must be extended in time. Figure 1(c) shows the CGRA extended 4 steps in time. The loop DFG must be mapped on TEC. Figure 1(d) shows a valid mapping of the loop DFG onto the TEC. The mapping is valid, because data dependencies between nodes are preserved. For example, node a at time t and c at time $t + 1$ are mapped onto the PE_4. Since the result of node a is stored in PE_4, the output of a (PE_4 at time t) will be available for c (PE_4 at $t + 1$). Nodes b and e are mapped on PE_1 at times t and $t + 2$ respectively. The output of b must be retained in PE_1 until e is executed at $t + 2$. Shown by a dashed node, this represents **routing** from PE_1 at time t to itself at time $t + 2$.

It is important to note that the execution on the CGRA is pipelined. Pipelining essentially means that we can execute instructions of different iterations of the loop at the same time. This explicit pipelining of schedule is referred to as software pipelining and modulo scheduling [24] is one of the most popular techniques for the same. The performance metric here is *initiation interval* (II), rather than the schedule length. II is the number of cycles between the start of two consecutive iterations of a loop. Figure 1(e) shows another way to execute the same loop DFG with instructions from other iterations also executing simultaneously. The dark circle nodes in the diagram represent the operations of the ith iteration, while the light square nodes represent the operations of the $(i - 1)^{th}$ iteration, and the light hexagon nodes represent the operations of the $(i + 1)^{th}$ iteration. Note that the schedule length is 4 cycles, but the II is only 2 cycles. This is because the dark circle nodes representing the ith iteration span from time t to $t + 3$, but the operations of the next iteration, i.e., the light hexagon nodes start from time $t + 2$. This mapping of one iteration is then shown in Figure 1(f), where we have removed the nodes from other iterations for clarity and easier comparison with Figure 1(d). Note that PEs that are shaded

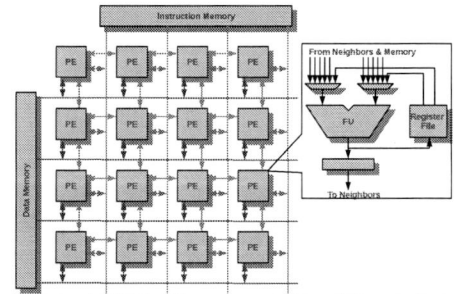

Figure 2: A 4 × 4 CGRA. PEs are connected in a 2-D mesh. Each PE is an ALU plus a local register file.

1281

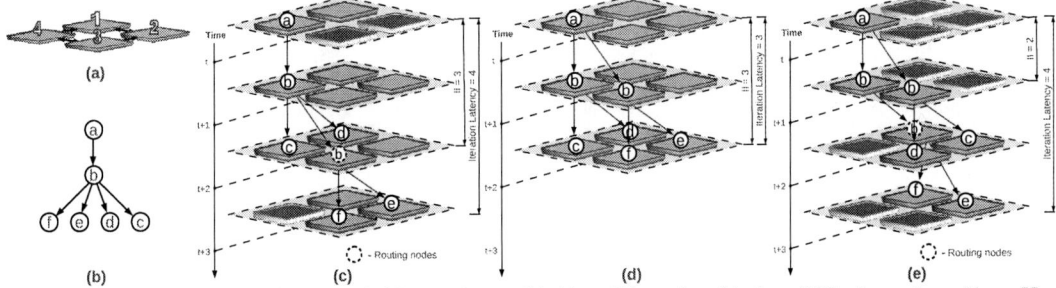

Figure 3: (a) a 2 × 2 CGRA, (b) input DFG where node b has out-degree of 4. (c) a valid mapping of the input DFG using routing, achieves $II = 3$. (d) shows the mapping of the input DFG using re-computation, achieves $II = 3$. (d) another mapping using both routing and re-computation, achieves $II = 2$.

black indicate their use in other iterations. As compared to Figure 1(d), since a new iteration starts every 2 cycles in Figure 1(f), the throughput and performance improve by a factor of 2.

This mapping and scheduling of the loop DFG on the CGRAs is done by the compiler. Recognizing the significant influence of the compiler on the achieved performance, much recent research has focused on efficient application mapping schemes [2, 3, 8, 9, 12, 15, 17, 20, 23, 26]. Each of these schemes impose their own restrictions and employ different, albeit intuitively justifiable, heuristic local decisions. For instance, many of these earlier works partition the problem in to three subproblems, namely, scheduling (when to perform an operation), placement (on which PE to perform an operation) and routing (how to route data between PEs), and solve these independently using a generic search method such as simulated annealing [15, 20], or use techniques developed in high-level hardware synthesis [9]. Since the key metric used in all the methods is II, the existing methods select an II, attempt to solve the three problems (possibly in different order), and if a feasible solution is not found, increase the II and repeat.

One of the major drawbacks of compiler research in this area is the lack of a precise and general formulation of the problem. Proper problem formulation allows us to systematically solve the mapping problem. Next, we first explain the concept of re-computation and why it is needed, then we show our problem formulation which allows for both *routing* and *re-computation*, and finally present our EPIMap heuristic before showing the experimental results.

3 What is Re-computation and Why?

Resource limitation is a major problem when trying to map a loop DFG onto a CGRA. One kind of resource limitation that is seen very often is the *out-degree problem*. Figure 3 illustrates the out-degree problem, and its solutions. The node b in the DFG in Figure 3(b) has an out-degree of 4 whereas the maximum out-degree of PEs in the TEC is 3 (if b is mapped on PE_4 at time $t + 1$, then a dependent operations can only be mapped on PE_4, PE_1, or PE_3 at time $t + 2$). Normally, if a node has out-degree greater than the maximum out-degree of the TEC, then it cannot be mapped onto the CGRA. In this example, only three of the dependent operations (among c, d, e, and f) can be mapped at time $t + 2$, and the fourth will have to be mapped at time $t + 3$; but if we do that, the fourth operation cannot receive the value of b.

Figure 3(c) shows how routing can be used to solve this out-degree problem. Operation b is performed on PE_4 at time $t + 1$. At time $t + 2$, PE_3, being adjacent to PE_4, copies the result of b to its output register. At time $t + 3$, PE_3 uses its own output (which has b) to perform operation f, and PE_2, which is adjacent to PE_3, uses the result in PE_3's output register (result of b) to perform operation e. Thus PE_3 just routes the output of operation b at time $t + 2$, to make it available at time $t + 3$. Another, slightly non-intuitive way to solve the out-degree problem is **re-computation**. Figure 3(d) shows that operation b is performed on both PE_4 and

PE_3 at time $t + 1$. Then their results are used by dependent nodes c, d, e and f in time $t + 2$.

Note that this is not routing, but re-computation because there is no path between two PEs on which b is mapped. In this example, routing, and re-computation both enable an II of 3. Another example of re-computation, explained in detail in Appendix A illustrates a case, where re-computation leads to better II than possible by routing. But, even in this example, Figure 3(e) shows that performing both routing and re-computation, the input graph can be mapped with an II of 2. Using only one of them, the best achievable II is 3. Consequently, the problem formulation must uniformly account for routing and re-computation.

4 Problem and Complexity

Let $D = (V_d, E_d)$ be the input DFG, and $C = (V_c, E_c)$ the TEC, extended to some number k time steps[1]. Let $C^* = (V_{c^*}, E_{c^*})$ be a subset of C where $V_{c^*} \subseteq V_c$ and $E_{c^*} \subseteq E_c$. In Appendix B, we give a formal definition of a *valid* mapping of operations in D to PEs in C, but intuitively, a valid mapping is one that allows correct execution of the DFG by satisfying all the data dependencies. Of course, valid mappings must allow overlapping operations in different time steps, and resolve out-degree and resource conflict problems by routing and/or re-computation. We define valid mappings in terms of **Epimorphisms**, which forms the basis of our algorithm and heuristic.

DEFINITION 1. *Let G and H be two digraphs. A mapping f : $V(G) \to V(H)$ is a Homomorphism if $(f(u), f(v)) \in E(H) \Rightarrow (u, v) \in E(G)$. It is called an Epimorphism if it is node surjective, i.e., every node in H is an image of some node in G. Note that node surjective implies arc surjective [16].*

Simply stated, for every valid mapping there exists a smallest k for which there is an Epimorphic map $M : C^* \to D$ that satisfies certain conditions. Conversely, an Epimorphic map from $M : C^* \to D$, for some k, that satisfies the same conditions corresponds to a valid mapping. Thus the optimization problem is to construct an Epimorphism $M : C^* \to D$, where C^* is a subgraph of a minimally extended TEC C. The problem formulation and the NP-completeness proof is explained in detail in Appendix C.

Now we see how the formulation works. Consider a node (i.e. a PE) $i \in V_{c^*}$. Let $i' = M(i) \in V_d$. i' be the operation that is mapped to PE i. For example, if the node i is PE_4 at time t in Figure 1(d), then, in the mapping shown, $i' = M(PE_{4,t}) = a$. Similarly, let $j \in V_{c^*} : \exists (j, i) \in E_{c^*}$, and let j' be the operation that is mapped to PE j. For example, j is the PE_4 at time $t + 1$, and $j' = M(PE_{4,t+1}) = c$. Then epimorphism requires that if there is an arc between a and c, then there must be an arc between PE_4 at time t, and PE_4 at time $t + 1$. This example illustrates how epimorphism ensures that data dependencies are preserved.

---

[1]To avoid cumbersome notation, we don't show the explicit dependence on k, which is to be determined and minimized.

Our problem formulation seamlessly captures routing and re-computation. Whenever we use routing/re-computation, a set of PEs in the TEC map to one operation in the DFG. For example, in Figure 3(c), in which the out-degree problem is resolved using routing, PE_4 of time $t+1$, and PE_3 of time $t+2$ are mapped to operation b. Since operation a has an arc to operation b, epimorphism requires that there be at least one arc between the set of PEs that are mapped to a, and the set of PEs that are mapped to b. This is true, since there is an arc from PE_4 at time t (where operation a is mapped), to PE_4 at time $t+1$ (where operation b is mapped). Similarly the data dependencies with operations c, d, e, and f are satisfied. In Figure 3(d), in which the out-degree problem is resolved using re-computation, a set of two PEs, PE_4 of time $t+1$, and PE_3 of time $t+1$ are mapped to operation b. In this case also, the data dependencies with the other operations are satisfied.

Again, to use routing or re-computation, a set of PEs in the TEC map to one operation in the DFG. The only difference is the presence/absence of arc between the PEs in the set. In Figure 3(c), PE_4 of time $t+1$, and PE_3 of time $t+2$ have an arc between them, so they transfer data through routing. On the other hands, in Figure 3(d), PE_4 of time $t+1$, and PE_3 of time $t+1$ do not have an arc between them, therefore the operation b has to be recomputed.

We now explain the conditions. Essentially, the condition is just to ensure that the PE at which computation, or re-computation happens, receives all the input operands. Thus, if we consider a node (i.e. a PE) $i \in V_{c^*}$. Let $i' = M(i) \in V_d$. i' is the operation that is mapped to PE i. Let $j \in V_{c^*} : \exists(j, i) \in E_{c^*}$. Then if there is no PE that is connected to i onto which operation i' is mapped, (i.e. if $(j, i) \in V_{c^*}$ and $M(j) \neq i'$), then for any $k' : \exists(k', i') \in V_d$, there must exist a $k \in V_{c^*} : \exists(k, i) \in E_{c^*}$ and $M(k) = k'$.

5 EPIMap

In this section, we describe our heuristic algorithm called EPIMap. EPIMap reduces search space because of three main reasons: i) it changes the input DFG to ensure that the graph meets necessary mapping conditions; ii) it determines a more accurate lower bound on II, and iii) when a placement is impossible, it changes nodes (re-computation or routing) that are left unmapped and attempts a new placement. A combination of routing and re-computation can be achieved when a node cannot be placed multiple times.

5.1 Overview

The EPIMap algorithm is presented in Algorithm 1. EPIMap initially changes the input DFG to hold the necessary mapping conditions (line 1-4). In the prepared DFG, it determines the minimum II (line 5). Then it attempts to find a valid mapping for this II. EPIMap achieves this in the While loop shown in line 6. If such a mapping cannot be found, it collects the set of nodes left unmapped and changes their input through routing or re-computation (line 24). In addition, it stores the number of unmapped nodes in the previous attempt (line 25). In the next attempt, if the number of unmapped nodes increases, it avoids further attempt for the prepared DFG, restores the original DFG, and decreases the number of operations at each cycle (line 21, 22) and retries again. If at any point of time MII increases and becomes greater than II (line 12), EPIMap restores the original DFG, increases II, and attempts for a new mapping (line 12-16). EPIMap repeats these steps until a valid mapping can be achieved.

5.2 Necessary Mapping Conditions

We first characterize the properties of a DFG so that it is mappable.
1. Out-degree of each operation must be less than the out-degree of PEs in the TEC. If there is a node u with out-degree

larger than that of PEs in the TEC, then a feasible mapping cannot be found. This can be fixed by either routing or re-computation. To perform routing, EPIMap adds a new node v and moves some outgoing arcs to v. Then it adds the arc (u, v). EPIMap performs re-computation by creating a new node v and connecting all nodes that have an outgoing arc to node u, to v, i.e. $\forall r \in V_d : (r, u) \in E_d$, add a new arc (r, v). Function *Constraint_Outdegree* in EPIMap is called to perform this step. From definition 1, we can immediately conclude that the modified graph is epimorphic to the input graph.

2. DFG must be balanced. A linear ordering of the nodes can be obtained by topological sorting. EPIMap schedules operations using this order. If there is an arc (i, j) between nodes i (scheduled at time t_i) and j (scheduled at time t_j) where $t_i - t_j > 1$, then DFG is not balanced. The formal definition of a balanced graph is given in the Appendix D. When such an arc is found, EPIMap adds extra nodes and balances the graph. From Definition 1, we can conclude that the balanced graph is epimorphic to the input graph. For example, in Figure 1(b), there is a distance of 2 between orders of node b and e. To overcome this problem, another b node is inserted between those nodes as illustrated in Figure 1(d). It should be noted that EPIMap removes every cycle in the DFG because topological sorting is impossible for cyclic graphs. This step partially schedules the operations while maintaining the routing. EPIMap calls *Balance* function for this phase.

3. The number of nodes at each level must be less than the number of PEs in the CGRA (non-time extended). If a level violates this constraint, EPIMap finds a set of nodes that can be moved to the other levels (previous or next) with the minimum cost. Here, the cost is the number of extra nodes that should be added to the DFG to preserve the data dependency of nodes. Again, from definition 1, this graph is also epimorphic to the input DFG. This step completes scheduling while maintaining routing. At the end of this step, cycles in the DFG are restored.

5.3 Determining Minimum II

While result of previous steps schedules the operations based on the necessary conditions on the DFG, this step finds a modulo schedule for the operations. Let MII denote the minimum II. This can be expressed as $MII = Max(ResMII, RecMII)$, where $ResMII = \lceil \frac{n}{M \times N} \rceil$ is the resource constrained minimum II and n is the number of operations in the DFG, and $RecMII$ is recurrence-constrained MII. Recurrence-constrained indicates inter-iteration dependency of operations in a loop. When such a dependency exists, the next iteration cannot start until the results from the previous iteration becomes available [24]. When MII is found, EPIMap folds (modulo schedules) the DFG (after preprocessing). To create such a graph, called MDFG, EPIMap assigns a level to each node. The level of each node in MDFG is the level of that node in the preprocessed DFG modulo MII. In MDFG, if the number of nodes at each level is less than the number of PEs in the CGRA, MII is feasible, otherwise, EPIMap moves operations to next or previous levels to satisfy this constraint. If satisfying this constraint is impossible, EPIMap increases MII until such a graph can be achieved. This phase is completed by *DetermineMII* and *UpdateMII* functions. It should be noted that EPIMap determines II accurately without any attempt to actually place the operations onto the CGRA.

5.4 Placement

EPIMap places operations (assigns nodes in the DFG to the nodes in the TEC) by finding the maximum common subgraph (MCS) (line 17) between the TEC and MDFG using Levi's algorithm [19]. If MCS is isomorphic to DFG, the mapping is completed. Otherwise, DFG must be changed. Since MCS is an NP-Complete

Algorithm 1 EPIMap(Input D, Input C)

1:   $D_p \leftarrow$ Constraint_Outdegree(D);
2:   $M \leftarrow |V_c|$;
3:   $D_p \leftarrow$ Balance(D_p);
4:   $D_p \leftarrow$ Constraint_Levels(D_p, M);
5:   $II \leftarrow$ DetermineMII(D_p); $D_i \leftarrow D_p$;
6:   **while** Mapping is not found **do**
7:     N=∞;
8:     **while** true **do**
9:       $D_i \leftarrow$ Balance(D_i);
10:      $D_i \leftarrow$ Constraint_Levels(D_i, M);
11:      $D_{mp}, MII \leftarrow$ UpdateMII(D_i, M);
12:      **if** $MII > II$ **then**
13:        $II \leftarrow II + 1$;
14:        $M \leftarrow |V_c|$;
15:        $D_i \leftarrow D_p$; break;
16:      **end if**
17:      $CS \leftarrow$ MCS (D_{mp}, II, C);
18:      **if** $V_{CS} = V_{D_{mp}}$ **then**
19:        **return** CS;
20:      **else**
21:        **if** $V_{D_i} - V_{CS} > N$ **then**
22:          $M \leftarrow M - 1$; $D_i \leftarrow D_p$; break;
23:        **else**
24:          $D_i \leftarrow$ ChangeInput($V_{D_i} - V_{CS}, D_i$);
25:          $N \leftarrow |V_{D_i} - V_{CS}|$;
26:        **end if**
27:      **end if**
28:     **end while**
29: **end while**

problem [10], we have modified Levi's algorithm to keep track of the number of attempts that did not increase the number of nodes in the common subgraph. When the number of unsuccessful attempts reaches a threshold value, EPIMap avoids further attempts.

6 Experimental Results

We have modified GCC and defined a new C pragma directive to specify the loops in the source code. We have taken loops from different applications including SPEC2006 benchmarks programs.

To evaluate the effectiveness of EPIMap, we compare the average II achieved using EPIMap with that achieved by EMS [23]. EMS uses a node selection scheme to map operations. To show that EPIMap generates better mappings than EMS even if EMS uses a better node selection scheme, we created 500 random node selection orders and selected the mapping that EMS generates with the best II. The mapping results for this case are labeled **BCEMS**. The original EMS results are labeled **EMS**.

We assume that a 4×4 homogeneous CGRA, and PEs are capable of performing fixed-point and logical operations. Access to the memory as well as other operations have latency of 1 cycle. We assume that there is enough memory to hold the instructions and variables of loops where PEs have 2 local registers. In addition, we assume that all PEs have access to the data memory but the data bus is shared among PEs in a row and is mutually exclusive. For load and store operations, two instructions are executed, one for address generation and one that generates/loads data.

6.1 EPIMap Generates Better Mapping

Figure 4(a) shows the average achieved II of different benchmarks using different mapping techniques. The first observation is that on average, EPIMap achieves a lower II than both EMS and BCEMS

in all applications. We also note that even with better node selection scheme, EMS results is worse than EPIMap.

In the best case, EPIMap-II $= 0.19 \times$ EMS-II (*Swim_calc2*), and EPIMap-II $= 0.28 \times$ BCEMS-II (*LowPass*). This means that in the best case, EPIMap accelerates *Swim_calc2* by a factor of 5.25X more than EMS and by 3.5X more than BCEMS, as the performance is inversely proportional to II. On average, EPIMap finds mappings with II of 0.35X and 0.44X of that EMS and BCEMS, which translates to a 2.26X and 2.84X better performance, respectively.

6.2 EPIMap's II is close to Minimum II

In Figure 4(b), the relative II (in percentage) is calculated by dividing MII by resulting II, i.e. $\frac{MII}{II}$; thus, the higher value implies a mapping that is closer to an optimal II.

We observe that EPIMap achieves the lower bound (MII) in 9 out of 14 benchmarks. In the five cases where EPIMap could not achieve the MII, it was either because there was no mapping that can achieve MII or the threshold value was reached before a valid mapping was found.

The average relative II gets worse for the applications with complex memory access patterns including *Bzip2* which has many memory related store instructions (bus is shared and mutually exclusive). In addition, there are many loops in *Bzip2* with the number of nodes nearly equal to the number of nodes in the TEC. When EPIMap reduces the number of operations at each level of the preprocessed DFG, some nodes will be rescheduled. To maintain data dependency for those nodes, extra nodes need to be added to the DFG. Therefore, $ResMII$ increases.

We also note that EMS and BCEMS could not achieve MII for any benchmark. The average relative II of EMS is 45% of the average relative II of BCEMS, and never exceeds 60% for average relative II of BCEMS. However, on the average, $\frac{MII}{EPIMap-II} \approx 0.92$, whereas $\frac{MII}{EMS-II} \approx 0.30$ and $\frac{MII}{BCEMS-II} \approx 0.47$. EPIMap produces mapping that achieve II which is much closer to MII than either EMS or BCEMS.

6.3 DFG Modification Reduces Search Space Significantly

Experimental results show that EPIMap generates valid mappings in 70% of loops in 6 benchmarks including: *Sor, Sobel, Lowpass, Laplace, wavelet, and Sjeng* in the first try before any attempt is made to reduce the number of operations at each level (second heuristic). However, EMS tries on average 7.87 unsuccessful IIs to find a valid mapping for these benchmarks. For each successful II, EMS searches for different node mappings and routing. This shows that changing input graph to ensure that it satisfies the necessary mapping conditions reduces search space substantially.

6.4 Counter-Intuitive Minimum IIs after Limiting Number of Nodes at Each Cycle

In the first glance, it seems that reducing number of nodes at each cycle increases II. However, for 50% of loops in *Swim_call, Swim_cal2, H.264, Jpeg, Libquantum and Sjecg*, this constraint is used on average 3 times but it did not increase the achieved II.

This is because the number of nodes in the modified DFG of these loops is less than the number of nodes in the TEC. This difference allows EPIMap to add more nodes to the DFG without increasing $ResMII$. Hence, we conclude that level constraint does not reduce mapping quality unless the number of nodes in the extended CGRA and modified DFG are close.

6.5 Reasonable Compilation Time

We measured the running time of EPIMap and EMS on an Intel Core2 machine with CPU frequency of 2.66 GHz. Figure 4(c)

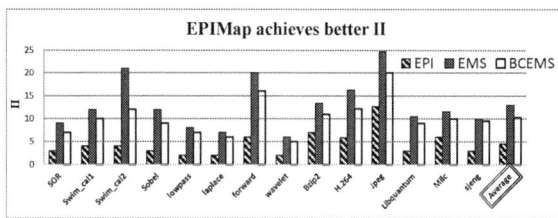

(a) EPIMap achieves mapping with average II of 0.35X lower than EMS.

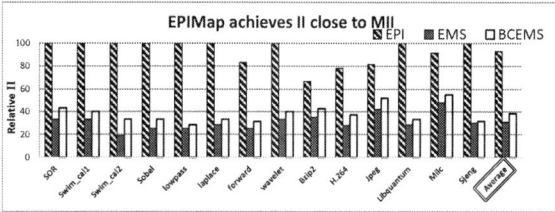

(b) In 9 out of 14 applications, EPIMap generates mapping with $II = MII$.

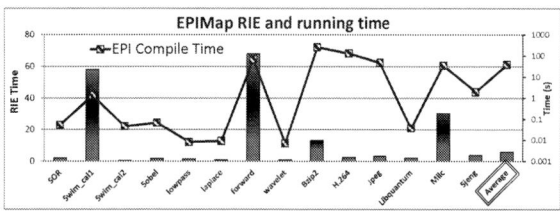

(c) Better mapping is achieved at the cost of increase in compilation time. However, the actual compilation time is usually negligible.

Figure 4: The comparison of achieved II of different applications using EPIMap, EMS, and BCEMS. On average, EPIMap achieves lesser II than other techniques. This better mapping quality comes at the cost of more compilation time of those applications.

shows the relative increase in execution (RIE) time of EPIMap and its actual execution time for different applications, where the threshold of EPIMap is 10^7. The bars show the RIE time of EPIMap to EMS, i.e. $\frac{T_{EPI} - T_{EMS}}{T_{EMS}}$ for different benchmarks. The actual EPIMap execution time is depicted by the solid line.

It can be observed that EPIMap achieves better average II at the cost of more execution time, the time it takes for EPIMap to generate a valid mapping or compilation time of application. In fact, the average RIE time of EPIMap to EMS is around 5.8. However, we can observe that the actual compilation time is fairly low, mostly less than 30 seconds on average.

7 Summary

CGRAs are promising structures that provide high-performance and high power-efficiency. However, achieving high power-efficiency is challenging primarily because compilation for CGRA is difficult. In this paper, we formulate the problem of mapping application onto a CGRA and establish its complexity. We also characterize the necessary conditions for application specification to find a feasible mapping. To tackle the mapping problem, we proposed a heuristic algorithm called EPIMap. EPIMap is different from the existing methods in the sense that it systematically searches the solution space to find a valid mapping. Our experimental results show that EPIMap generates mapping which leads to significant performance improvement compared to EMS.

8 Acknowledgments

This work was supported in part by NSF IUCRC Center for Embedded Systems under Grant DWS-0086, by the Science Foundation of Arizona (Grant SRG 0211-07), and by the Stardust Foundation.

9 References

[1] AA, T. V., RAGHAVAN, P., MAHLKE, S. A., SUTTER, B. D., SHRIVASTAVA, A., AND HANNIG, F. Compilation techniques for cgras: exploring all parallelization approaches. In *CODES+ISSS* (2010), pp. 185–186.

[2] AHN, M., YOON, J., PAEK, Y., KIM, Y., KIEMB, M., AND CHOI, K. A spatial mapping algorithm for heterogeneous coarse-grained reconfigurable architectures. In *Proc. DATE* (2006), pp. 363–368.

[3] BANSAL, N., GUPTA, S., DUTT, N., NICOLAU, A., AND GUPTA, R. Network topology exploration of mesh-based coarse-grain reconfigurable architectures. In *Proc. DATE* (2004), pp. 474–479.

[4] BECKER, J., AND VORBACH, M. Architecture, memory and interface technology integration of an industrial/ academic configurable system-on-chip (csoc). In *Proc. ISVLSI* (2003), pp. 107–112.

[5] BORKAR, S. Design challenges of technology scaling. *IEEE Micro 19*, 4 (1999), 23–29.

[6] BOUWENS, F., BEREKOVIC, M., SUTTER, B. D., AND GAYDADJIEV, G. Architecture enhancements for the adres coarse-grained reconfigurable array. In *Proc. HiPEAC* (2008), pp. 66–81.

[7] CHEN, D. C. *Programmable arithmetic devices for high speed digital signal processing.* PhD thesis, University of California, Berkeley, 1992.

[8] DIMITROULAKOS, G., GALANIS, M., AND GOUTIS, C. Exploring the design space of an optimized compiler approach for mesh-like coarse-grained reconfigurable architectures. In *Proc. IPDPS* (2006), pp. 113–122.

[9] FRIEDMAN, S., CARROLL, A., VAN ESSEN, B., YLVISAKER, B., EBELING, C., AND HAUCK, S. Spr: an architecture-adaptive cgra mapping tool. In *Proc. FPGA* (2009), pp. 191–200.

[10] GAREY, M. R., AND JOHNSON, D. S. *Computers and Intractability: A Guide to the Theory of NP-Completeness.* W. H. Freeman & Co., New York, NY, USA, 1979.

[11] GOLDSTEIN, S., SCHMIT, H., MOE, M., BUDIU, M., CADAMBI, S., TAYLOR, R., AND LAUFER, R. Piperench: a coprocessor for streaming multimedia acceleration. In *Proc. ISCA* (1999), pp. 28 –39.

[12] GUO, Y., HOEDE, C., AND SMIT, G. A pattern selection algorithm for multi-pattern scheduling. In *Proc. PDPTA* (2006), pp. 198–205.

[13] HARTENSTEIN, R. A decade of reconfigurable computing: a visionary retrospective. In *Proc. DATE* (2001), pp. 642–649.

[14] HARTENSTEIN, R., AND KRESS, R. A datapath synthesis system for the reconfigurable datapath architecture. In *Proc. ASP-DAC* (1995), pp. 479 –484.

[15] HATANAKA, A., AND BAGHERZADEH, N. A modulo scheduling algorithm for a coarse-grain reconfigurable array template. In *Proc. IPDPS* (2007), pp. 1–8.

[16] HELL, P., AND NESETRIL, J. *Graphs and Homomorphisms.* Oxford University Press, New York, NY, USA, 2004.

[17] LEE, J.-E., CHOI, K., AND DUTT, N. D. Compilation approach for coarse-grained reconfigurable architectures. *IEEE Design and Test of Computers 20*, 1 (2003), 26–33.

[18] LEE, M.-H., SINGH, H., LU, G., BAGHERZADEH, N., KURDAHI, F. J., FILHO, E. M. C., AND ALVES, V. C. Design and implementation of the morphosys reconfigurable computingprocessor. *J. VLSI Signal Process. Syst. 24* (2000), 147–164.

[19] LEVI, G. A note on the derivation of maximal common subgraphs of two directed or undirected graphs. *Calcolo 9* (Dec. 1973), 341–352.

[20] MEI, B., VERNALDE, S., VERKEST, D., AND LAUWEREINS, R. Design methodology for a tightly coupled vliw/reconfigurable matrix architecture: a case study. In *Proc. DATE* (2004), pp. 1224–1229.

[21] MIRSKY, E., AND DEHON, A. Matrix: a reconfigurable computing architecture with configurable instruction distribution and deployable resources. In *Proc. FPGAs for Custom Computing Machines* (1996), pp. 157 –166.

[22] MIYAMORI, T., AND OLUKOTUN, K. Remarc: Reconfigurable multimedia array coprocessor. *IEICE Trans. on Information and Systems* (1998), 389–397.

[23] PARK, H., FAN, K., MAHLKE, S. A., OH, T., KIM, H., AND KIM, H.-s. Edge-centric modulo scheduling for coarse-grained reconfigurable architectures. In *Proc. PACT* (2008), pp. 166–176.

[24] RAU, B. R. Iterative modulo scheduling: an algorithm for software pipelining loops. In *Proc. MICRO* (1994), pp. 63–74.

[25] WOH, M., MAHLKE, S., MUDGE, T., AND CHAKRABARTI, C. Mobile supercomputers for the next-generation cell phone. *Computer 43*, 1 (2010), 81 –85.

[26] YOON, J., SHRIVASTAVA, A., PARK, S., AHN, M., AND PAEK, Y. A graph drawing based spatial mapping algorithm for coarse-grained reconfigurable architectures. *IEEE Trans. on VLSI Systems 17*, 11 (2009), 1565–1578.

APPENDIX

A Example of the Need for Re-computation

One form of resource limitation problem is resource conflict problem. If due to the modulo scheduling, the number of nodes using the same resource exceeds the number of available resources, we call it resource conflict problem. There is a resource conflict between nodes b, e and f in Figure 3(b) when it is mapped onto a 2×2 CGRA (Figure 3(a)). For such a mapping $MII = \lceil \frac{6}{4} \rceil = 2$. To meet data dependencies, node b has to be executed on a PE that is adjacent to PEs executing nodes c and d. In addition, nodes e and f should be mapped on PEs which are also adjacent to PEs executing nodes c and d. In mapping shown in Figure 3(c), nodes c and d are mapped onto PE_4 and PE_2 respectively. Therefore, there are two PEs adjacent to these PEs, PE_1 and PE_3. Due to the modulo scheduling, nodes b from iteration i, e, and f from iteration $i-1$ are executed at the same time. At this time, there are three nodes but two available resources. In Figure 3(c), node b is executed on two different PEs to overcome this problem. Without re-computation, it is impossible to find a valid mapping for $II = 2$.

B Formal Problem Definition

DEFINITION 2. *Valid Mapping: Let n be the number of nodes in V_d, i.e. $n = |V_d|$, and $C = (V_c, E_c)$ be TEC. Let $C^* = (V_{c^*}, E_{c^*})$ be a subset of C, i.e., $C^* \subseteq C$. Let $S = \{s_1, s_2, s_3, ..., s_n\}$ be a set of n disjoint subsets of V_{c^*} such that $1 \leq \forall i \leq n, |s_i| \geq 1$. A mapping function $f : V_d \to S$ is a valid mapping iff $\forall u, v \in V_d$: $(u, v) \in E_d$,*

> *$\forall v' \in f(v)$, there must be a path from a node $u' \in f(u)$ to v'. The nodes in this path must only be nodes in $f(v)$.*

In this definition, the set of PEs in the TEC are partitioned into some sets s_i. The number of sets in this partition must be n to ensures that every node in the DFG will be mapped into a set. $s_i > 0$ ensures that every operations is executed at least once. If the number of PEs in a set s_i is more than one, it implies that the nodes mapped into s_i will be executed, and/or routed, and/or re-computed.

A mapping is valid if data dependencies between nodes in the DFG are preserved after mapping. When there is an arc between two nodes $(u, v) \in E_d$ in the DFG, every PE in the set s_v that executes node v, can receive the result of execution of u as input. Thus, for every PE in the set s_v, there must be a path from PEs in the set s_u where u is mapped into. More precisely, every PE in the set s_u either executes u or routes the result of execution of u.

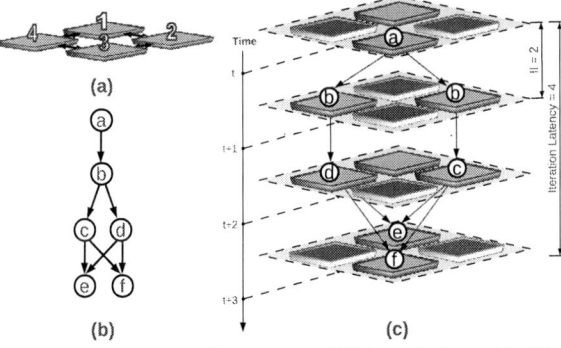

(a) **(b)** **(c)**

Figure 5: (a) An input DFG, (b) a 2×2 CGRA, (c) the best achievable mapping. There is resource conflict between nodes b and nodes e or f ($II = 2$). Re-computing node b allows us to avoid this conflict.

Therefore, the result of u can be routed from PEs in this set to the PEs executing node v that need u as an input.

C Problem Formulation

In this section, we formulate the problem of mapping a DFG onto a CGRA as finding an Epimorphic subgraph of TEC onto the input DFG. For simplicity, we assume that input DFG is acyclic.

To do this, Theorem 1 states that when a node a is mapped to a set of nodes in TEC, the nodes in this set are only connected to each other (routing) or they are connected to the nodes on which the adjacent nodes of a are mapped (re-computation is there if more than one node in that set). Then Theorem 2 states that every valid mapping implies an Epimorphic function from a subgraph of TEC to the input DFG. Using Theorem 1 and 2, we prove that every node a in the DFG whose in-degree is greater than 0 would be mapped onto a node in TEC which has input either from the nodes that a is mapped onto (routing) or from all the nodes that the input subgraph of a is mapped onto in Theorem 3. Then, Theorem 4, also proves the converse, that every Epimorphic function from a subgraph of TEC to the input DFG that holds the aforementioned properties implies a valid mapping. Consequently, they are equivalent.

Finally, since we prove the equivalence of problem of mapping application onto a CGRA with graph Epimorphism which is an NP-Complete problem [16], therefore CGRA application mapping is NP-Complete.

THEOREM 1. *In a valid mapping, every node $u \in V_d$ that is mapped onto a set $s_i \subset V_{c^*}$ where $|s_i| > 1$, $\forall g, h \in s_i$, if there is a path P between g and h then P only passes through some nodes in s_i.*

PROOF. Let's assume that there is a path P' between g and h which passes through some nodes not in s_i. Without loss of generality, let's assume that P' starts from g passes through at least one node $k \notin s_i$ and ends in h. Therefore, there is an arc between $f^{-1}(g)$ and $f^{-1}(k)$ and also an arc between $f^{-1}(k)$ and $f^{-1}(h)$ where $f^{-1}(h) = f^{-1}(g)$ in the DFG. This implies that there is a cycle in the input DFG which contradicts with our assumption that the DFG is acyclic. $\square$

THEOREM 2. *Let $D(V_d, E_d)$ be the input DFG and $C_k(V_c, E_c)$ be the CGRA graph extended in time k. Every valid mapping implies an Epimorphic function $M : C^* \longrightarrow D$ where C^* is a subgraph of C.*

PROOF. Let V_{c^*} be the set of nodes in C^* and E_{c^*} be the set of arcs.

- m is function. A mapping is valid if each PE executes the maximum of one operation per cycle. Therefore, $\forall i \in V_{c^*}, M(i)$ is exactly one element in V_d.

- m is surjective. A valid mapping maps implies that every nodes in graph D must be mapped onto some PEs in TEC. Therefore, it is surjective.

- S is homomorphic to D. In a valid mapping, every node $i \in V_d$ is mapped to a set of nodes $s_i \subset V_{c^*}$. Therefore, $\forall a, b \in V_{c^*}$, if there is an arc $(a, b) \in E_{c^*}$, then either $a, b \in s_i$ or $a \in s_i, b \in s_j$ where $s_i \neq s_j$. The first case simply implies homomorphism according to definition. We claim that the second case also implies homomorphism. Let's assume node $j \in V_d$ is mapped to s_j. If there is an arc between nodes i and j then it implies a homomorphism. Otherwise, it is trivial to see that if there is no arc between i and j but between a and b, then we can remove (a, b) from E_s and mapping will be still valid.

Both graphs in Figure 3(c) and Figure 3(d) are Epimorphic to Figure 3(b). All arcs in the mapping correspond to an arc in the input DFG. However, there is an arc between PE_4 at time t and PE_3 at time $t+2$. The second one is used for routing. Figure 3(d) shows another mapping. In this mapping there is an arc between PE_4 time t and PE_4 at time $t+1$. This arc is mapped into the arc between nodes a and b in the input DFG. Similarly, there is an PE_4 time t and PE_3 at time $t+1$ which is also mapped into the same arc. This case, node b is executed two times (re-compuation).

$\square$

DEFINITION 3. *Input subgraph: for every node i in a digraph $D(V_d, E_d)$, there exists a subgraph $G(V_g, E_g)$ such that V_g is the set of all nodes $j : (j, i) \in E_d$ in addition to node i; also E_g is the set of all arcs $(j, i) : (j, i) \in E_d$.*

DEFINITION 4. *An isomorphism from $G(V_g, E_g)$ onto $H(V_h, E_h)$ is defined as $f : V_g \longrightarrow V_h$ such that:*

1. $|E_g| = |E_h|$

2. $|V_g| = |V_h|$

3. $\forall u, v \in V_g : (u, v) \in E_g$ iff $(f(u), f(v)) \in E_h$ [10].

THEOREM 3. *Every valid mapping implies an Epimorphic function $M : C^* \longrightarrow D$ such that $\forall i \in V_{c^*}$, **input subgraph** of $M(i)$ $K(V_k, E_k)$, **input subgraph** of i $L(V_l, E_l)$: if $Indegree(M(i)) > 0 \Rightarrow$*

1. $\forall j \in V_l : M(j) = M(i)$ or

2. K and L are isomorphic.

PROOF. • Theorem 2 proves that every valid mapping implies an Epimorphic function $M : C^* \longrightarrow D$.

• Let's assume $a \in V_d$ is mapped onto set $s_a \subset V_{c^*}$. $\forall i \in V_{c^*} : Indegree(i) > 0$ either:

– all incoming arcs of i are from nodes $j \in s_a$ which implies $\forall j \in V_l : M(j) = M(i)$.

– input arcs of i are from a combination of nodes $j \in s_a$ and nodes $k \notin s_a$. This case cannot happen according to Theorem 1.

– all incoming arcs are from nodes $j \notin s_a$. In this case, we show that in order to have a valid mapping, K and L should be isomorphic. Let's assume that K and L are not isomorphic. Then either $V_k \neq V_l$ or $E_k \neq E_l$. If the number of nodes or arcs are different then the mapping cannot be valid. It should be noted that the only node in K with incoming arcs is $M(i)$ and i is the node in L with incoming arcs. Therefore, the connection topology cannot violate isomorphism constraint if the mapping is valid.

$\square$

THEOREM 4. *Every Epimorphic function $M : C^* \longrightarrow D$ with following constraint implies a valid mapping. Constraint: $\forall i \in V_{c^*}$, **input subgraph** of $M(i)$ $K(V_k, E_k)$, **input subgraph** of i $L(V_l, E_l)$: if $Indegree(M(i)) > 0 \Rightarrow$*

1. $\forall j \in V_l : M(j) = M(i)$ or

2. K and L are isomorphic.

PROOF. • Because the function is surjective, all nodes in D must be covered by at least one node in C^*.

• Since epimorphism preserves node adjacency [16] and the function is surjective, then all arcs in E_d are covered by an arc in E_{c^*} which implies, for all $u, v \in V_d : (u, v) \in E_d$ there is an arc between $(m^{-1}(u), m^{-1}(v))$.

• $\forall u \in V_d : indegree(u) > 0$ where $K(V_k, E_k)$ is input subgraph of u and $L(V_l, E_l)$ is input subgraph of $m^{-1}(u)$:

– if $\forall j \in V_l : M(j) = u$, there must be a node $v \in V_{c^*}$ such that input subgraph of v is isomorphic to K. Because function is surjective, all arcs of E_k must be mapped. Therefore, according to the constraint, since the first case cannot happen, there exists a node whose input subgraph is isomorphic with K which implies u is mapped properly.

– if K and L are isomorphic, then mapping of u is valid.

$\square$

THEOREM 5. *Every valid mapping from an input DFG $D(V_d, E_d)$ onto CGRA graph extended in time k $C_k(V_c, E_c)$ is equivalent to an Epimorphic function $M : C^* \longrightarrow D$ such that $\forall i \in V_s : K(V_k, E_k)$ **input subgraph** of $M(i)$, $L(V_l, E_l)$ **input subgraph** of i : if $Indegree(M(i)) > 0 \Rightarrow$*

1. $\forall j \in V_l : M(j) = M(i)$ or

2. K and L are isomorphic.

where $C^*(V_{c^*}, E_{c^*}) : C^* \subseteq C$.

PROOF. • From Theorem 3, every valid mapping implies an Epimorphic function $M : C^* \longrightarrow D$ with above-mentioned constraints.

• From Theorem 4, every Epimorphic function $M : C^* \longrightarrow D$ with above-mentioned constraints implies a valid mapping.

$\square$

THEOREM 6. *The problem of mapping an input DFG onto a CGRA is NP-Complete.*

PROOF. Mapping an input DFG onto a CGRA is in NP. The decision problem of the above problem is: Given an input DFG D, a certificate which is a T, a given graph G' that is a subgraph of the extended in time graph of CGRA and a mapping from subsets of G' onto nodes in D, is mapping to G' a valid mapping. The yes-instance of this problem can be verified in polynomial time. For every arc in the G' we can verify the valid mapping conditions by checking the corresponding arc in the input DFG. Therefore, mapping is in NP. Finding minimum Epimorphism has been proved to be NP-Complete [16]. We have shown the equivalence of valid mapping and Epimorphism, therefore, we conclude that the problem of finding a valid mapping is NP-Complete. $\square$

D Formal Definitions

DEFINITION 5. *Balanced graph $G(V_g, E_g)$: a graph is balanced if $\forall i, j \in V_g$ such that there are two paths $P_1(i, j)$ and $P_2(i, j)$ between them then the length of paths are equal. In addition, $\forall i, j \in V_g : if \exists k, l \in V_g$ such that $P_1(i, k), P_2(j, k), P_3(i, l),$ and $P_4(j, l)$ then $L(P_1) - L(P_2) = L(P_3) - L(P_4)$ where $L(P_i)$ is the length of path P_i.*

Instruction Scheduling for Reliability-Aware Compilation

Semeen Rehman, Muhammad Shafique, Jörg Henkel

Karlsruhe Institute of Technology (KIT), Chair for Embedded Systems, Karlsruhe, Germany

semeen.rehman@student.kit.edu; {muhammad.shafique, henkel}@kit.edu

Abstract

An instruction scheduling technique is presented that targets at improving the reliability of a software program given a user-provided tolerable performance overhead. A look-ahead-based heuristic schedules instructions by evaluating the reliability of dependent instructions while reducing the impact of spatial and temporal vulnerabilities of various processor components. Our reliability-driven instruction scheduler (implemented into the GCC compiler) provides on average a 22% reduction of program failures compared to state-of-the-art.

Categories and Subject Descriptors: D.3.4 [Processors]: *Code Generation, Compilers;* B.8 [**Performance and Reliability**]: *Reliability, Testing, and Fault-Tolerance*

General Terms: Algorithms, Design, Reliability, Performance

Keywords: Reliability, dependability, reliability estimation, instruction vulnerability estimation, reliable software, code generation, instruction scheduling, embedded systems, technology scaling

1. Introduction and Related Work

Soft errors have emerged as a non-negligible design challenge in hardware/software systems and their importance is likely to increase further in upcoming technology generations [1][2][5] because the transistor dimensions and operating/threshold voltages keep on shrinking [2] eventually leading to further reduction of the critical charge. This results in a high susceptibility to soft errors [1] manifesting as bit flips in the underlying circuits which may influence the correct software program execution. To address this issue, several soft error mitigation techniques have been proposed at both hardware and software levels. Because of the prevailing fact that hardware level techniques [3][4][5][12] are both area-/power-wise costly and the verification/validation of a reliable hardware is expensive and time consuming [17], software-level soft error mitigation techniques have evolved. The prominent software-level soft error mitigation techniques are: instruction or register duplication [17][9][10], control flow checking [6][7][17] and register vulnerability reduction [11]. However, these state-of-the-art techniques incur significant performance degradation along with large memory overhead. In [25][26], both hardware and software level error-mitigation techniques are jointly employed along with application-specific knowledge to achieve power-efficient reliability. The work in [23] introduced several reliability-aware software transformations at the compiler front-end. This work does not consider the algorithms in the compiler back-end. Especially, the instruction scheduling impacts the instruction vulnerability by affecting the vulnerable periods of different instructions and their operands in pipeline resources (see terminology in Section 1.1). This paper aims at developing and evaluating a soft error-driven instruction scheduling technique as a low-level optimization in the back-end of a reliability-aware compiler.

Several reliability-aware instruction scheduling approaches have been proposed [11][12][19][20] that reorder the instruction profile of a program while incurring relatively limited performance degradation and almost no memory overhead compared to instruction redundancy techniques. The work of [12] minimizes the residency cycles (see definition in Section 1.1) of vulnerable bits inside the issue queue of a superscalar processor by performing instruction scheduling at runtime. However, this technique requires architecture modification of the hardware scheduler and introduces a significant hardware overhead. In contrast, ISSE [19] reschedules a program's assembly code to minimize the operands' vulnerable periods via exploiting the slack time. The slacks are identified after a performance-driven instruction scheduling, which already tries to minimize the slacks as much as possible to avoid pipeline stalls for improving the performance. Since the available slack after the performance-optimized scheduling is limited, state-of-the-art instruction scheduling techniques [19] and [11] provide limited reliability improvements of 2% and 9%, respectively. Moreover, state-of-the-art techniques like [19][11] provide limited reliability improvements as they primarily improve the reliability of the register file, which typically covers a small portion of the processor layout compared to the pipeline and instruction execution unit (IEU) [13].

Since both performance and reliability are key design parameters, an instruction scheduler is required that optimizes for software program's reliability ('goal') under a tolerable performance overhead ('constraint').

1.1 Terminologies and Basics

Before proceeding to our core idea, we present terminologies and basics that are used in this paper.

Vulnerable Period and Pipeline Stage Residency: the *vulnerable period* of a variable operand used in an instruction is defined as the time (in cycles) between a value written into it and its usage, i.e. write-read interval. A variable vulnerable period is different for different instructions. The *pipeline stage residency* is given as the vulnerable periods of an instruction in a certain pipeline stage. The 'pipeline residency' of an instruction is the sum of residency cycles in all pipeline stages. Pipeline stage residency may vary depending upon the number and duration of pipeline stalls, instruction's latency in that particular pipeline stage, and the latency of instructions in the succeeding pipeline stages.

Software Program Error Types: depending upon the nature of the software program's output, program error types are classified into three categories, (i) correct output value, (ii) incorrect output value, and (iii) software program failure (like 'abort', 'crash', 'exception', etc.). Bit flips in the data values of arithmetic instructions (except address generation) may lead to incorrect output values that may appear as a glitch and do not appear in the next application iteration or bears no impact because of control flow etc. In contrast, bit flips in the variable containing the address value may lead to a program failure due to a load from or store to a wrong location of data memory, or a wrong branch/call address. Since a program failure requires a software restart, it appears to be more 'crucial' for the user and may be intolerable. Therefore, a program failure is categorized as a more severe error compared to the incorrect output value.

'Crucial' and 'Non-Crucial' Instructions: with respect to program failures, an instruction is categorized into 'crucial' and 'non-crucial' instruction. If a fault during an instruction's execution leads to an incorrect output value but does not terminate or halt the software program, it is classified as a *'non-crucial' instructions*, e.g. arithmetic and logical (add, sub, and, or, etc.) data computing instructions (except address generation). In contrast, an instruction is categorized as a *'crucial' instruction* if a fault during its execution leads to (or may possible lead to) a program failure, e.g. load, store, address computation, jump/branch, etc. (as discussed above). A large number of 'crucial' instruction executions denotes high susceptibility towards program failures.

Spatial and Temporal Vulnerability: Different processor components (like instruction decoder, execution unit, cache controller, etc.)

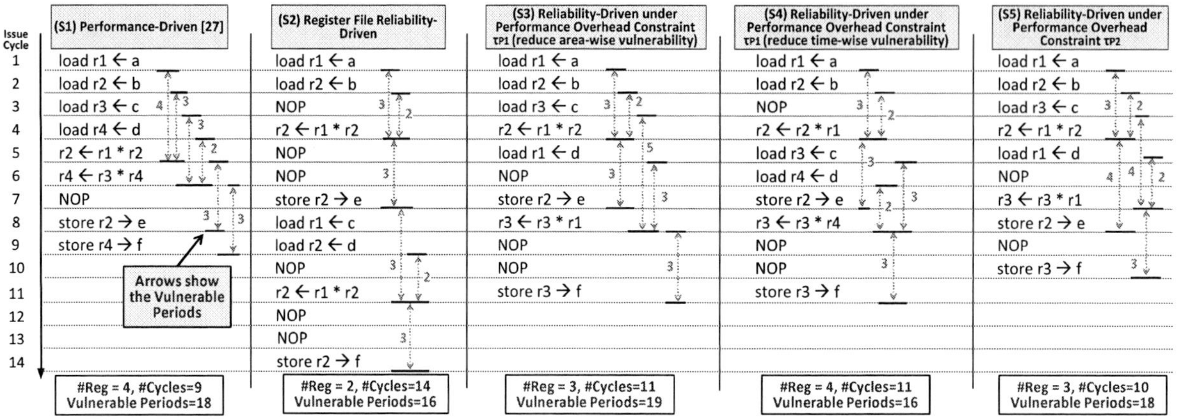

Issue Cycle	(S1) Performance-Driven [27]	(S2) Register File Reliability-Driven	(S3) Reliability-Driven under Performance Overhead Constraint τP1 (reduce area-wise vulnerability)	(S4) Reliability-Driven under Performance Overhead Constraint τP1 (reduce time-wise vulnerability)	(S5) Reliability-Driven under Performance Overhead Constraint τP2
1	load r1 ← a	load r1 ← a	load r1 ← a	load r1 ← a	load r1 ← a
2	load r2 ← b	load r2 ← b	load r2 ← b	load r2 ← b	load r2 ← b
3	load r3 ← c	NOP	load r3 ← c	NOP	load r3 ← c
4	load r4 ← d	r2 ← r1 * r2	r2 ← r1 * r2	r2 ← r2 * r1	r2 ← r1 * r2
5	r2 ← r1 * r2	NOP	load r1 ← d	load r3 ← c	load r1 ← d
6	r4 ← r3 * r4	NOP	NOP	load r4 ← d	NOP
7	NOP	store r2 → e	store r2 → e	store r2 → e	r3 ← r3 * r1
8	store r2 → e	load r1 ← c	r3 ← r3 * r1	r3 ← r3 * r4	store r2 → e
9	store r4 → f	load r2 ← d	NOP	NOP	NOP
10		NOP	NOP	NOP	store r3 → f
11		r2 ← r1 * r2	store r3 → f	store r3 → f	
12		NOP			
13		NOP			
14		store r2 → f			
	#Reg = 4, #Cycles=9 Vulnerable Periods=18	#Reg = 2, #Cycles=14 Vulnerable Periods=16	#Reg = 3, #Cycles=11 Vulnerable Periods=19	#Reg = 4, #Cycles=11 Vulnerable Periods=16	#Reg = 3, #Cycles=10 Vulnerable Periods=18

Arrows show the Vulnerable Periods

Fig. 1 Comparing the performance and reliability of different scheduling heuristics

have a dissimilar chip area. For an even fault distribution, the probability of a fault in a certain processor component depends upon its area, transistor size/type/density, etc. [1][13]. *Spatial vulnerability* is the probability of fault occurrence during the execution of an instruction depending upon the area and type (like sequential or combinatorial logic) of the specific processor components that it uses. Furthermore, different instructions typically exhibit distinct residency in different pipeline stages, e.g., the *execute stage residency* of an *add* instruction is 1 cycle and of a *multiply* instruction is 3 or more cycles (depending upon the microarchitecture). *Temporal vulnerability* is defined as the probability of the occurrence of fault during the execution of an instruction depending upon its pipeline residency. Note, the reliability of a program is a complex function of spatial and temporal vulnerabilities of its instructions (w.r.t. the processor components they use in different pipeline stages) and different error types. For instance, long vulnerable periods of variables (stored in the register file) result in an increased temporal vulnerability, while using more live variables results in an increased spatial vulnerability. Moreover, even for the same overall performance, longer pipeline residency of a 'crucial' instruction leads to higher susceptibility towards program failures compared to a longer pipeline residency of a 'non-crucial' instruction. A performance-driven instruction scheduler ignores this fact of instruction cruciality and may improve the performance by scheduling the instructions with the longest delay or reducing the pipeline residency irrespective of an instruction type/cruciality and potential program errors.

1.2 Motivational Instruction Scheduling Scenarios and Problem Statement

An instruction schedule determines the instruction execution sequence which can highly affect the vulnerable periods of an instruction in different processors components like variable values in the register file or pipeline stage residency. An instruction scheduler optimizing only for performance may result in degraded reliability due to (i) increased program's susceptibility to failures because it scheduled a 'crucial' instruction after a pipeline stalling instruction in order to increase the performance; (ii) increased spatial vulnerability because it used more registers or instruction execution units to achieve a higher performance. Alternatively, an instruction scheduler optimizing only for reliability may lead to degraded performance due to data hazards (occur due to the data dependencies between different instructions, e.g., read-after-write, write-after-write, write-after-read).

We investigate these issues and motivate the need of considering both reliability and performance in instruction scheduling with the following example scenarios (Fig. 1). Our assumptions are: execution cycles (execute stage residency) of load/store = 2 (two load-store units are available), multiply = 3 (considering two dedicated multiply hardware units), NOP = 1.

S1). *Performance-driven instruction scheduling (GCC List Scheduler [27][22]):* This schedule provides the best performance by prioritizing the instructions on a program's critical path. However, it does not consider the number of registers and their vulnerable periods in the scheduling decision function, thus it results in increased spatial and temporal vulnerabilities. This schedule uses 4 registers and results in a total vulnerable period of 18. Note, if a potential pipeline stalling instruction (like a multi-cycle *multiply* instruction or a blocking *load* instruction) precedes a 'crucial' instruction, it may result in an increased pipeline residency of the 'crucial' instruction (e.g., due to a cache miss), which may lead to an increased susceptibility towards program failure. In this schedule, four consecutive load instructions result in an increased susceptibility towards failure. It might be (reliability-wise) beneficial to *avoid* scheduling a 'crucial' instruction right after a pipeline stalling instruction.

S2). *Register file reliability-driven instruction scheduling:* This schedule aims at reducing the vulnerable periods between different registers. Assuming the register allocation has been determined before instruction scheduling, it will also incur a reduced spatial redundancy (only 2 registers are used). However, to avoid *data hazards*, several NOPs are inserted that introduce a performance overhead of 5 cycles (i.e. 55%) compared to S1 (14 cycles of S2 vs. 9 cycles of S1). As a side effect, this schedule also avoids the 'crucial' instructions to enter in the pipeline during the execution of a potential pipeline-stalling instruction.

S3). *Reliability-driven instruction scheduling under tolerable performance overhead τP1 reducing only spatial vulnerability:* Unlike S2, this schedule reduces the spatial vulnerability of instructions for all processor components (and not only register file), while keeping the performance overhead under the tolerable limit (τP1=20%). However, this schedule results in significantly high temporal vulnerability, i.e. a vulnerable period of 19.

S4). *Reliability-driven instruction scheduling under tolerable performance overhead τP1 reducing only temporal vulnerability:* Unlike S3, this schedule optimizes for temporal vulnerabilities under the tolerable performance overhead of τP1=20%. It reduces the temporal vulnerability of variable 'c' (stored in register *r3*) by moving the *load* instruction right after the first *multiply* instruction. The S4 schedule results in a total vulnerable period of 16 with an execution time of 11 cycles (similar performance as of S3). The reduction in temporal vulnerability comes at the cost of an increased spatial vulnerability (4 registers are used in S4 instead of 3 registers used in S3). However, as a side effect it increases the program's susceptibility towards failures by scheduling 'crucial' instructions like *load* and *store* right after the multi-cycle *multiply* instruction.

S5). *Reliability-driven instruction scheduling under tolerable performance overhead τP2:* Unlike S3 and S4, this schedule optimizes for both spatial and temporal vulnerabilities under the tolerable performance overhead of τP2=10% (i.e. even a tighter performance overhead constraint compared to S3 and S4). This schedule

1289

reduces both spatial and temporal vulnerabilities compared to S4 and S3 schedules, respectively, while achieving a performance close to the performance-driven S1 schedule. Unlike S4, in order to reduce the susceptibility towards failures, it avoids scheduling *load* and *store* instructions after the *multiply* instruction. Among the five schedules, S5 provides a good compromise between performance overhead and reliability improvement.

The soft-error-driven instruction scheduling turns even more challenging in case of *structural hazards*[1] that may cause frequent pipeline stalls.

Summarizing the above scenarios: Since an instruction scheduling algorithm determines the execution sequence of instructions, it also affects the spatial and temporal vulnerabilities of instructions in different processors components. A performance-driven instruction scheduler may provide reduced temporal vulnerability for instructions on the critical path, but it may lead to a significantly higher spatial vulnerability (by using more resources, see Fig. 1) or even longer vulnerable periods of the operands of the 'crucial' instructions if they do not lie on the critical path. Therefore, it may lead to an overall higher vulnerability. A reliability-driven instruction scheduler on the other side would not only *balance the spatial and temporal vulnerabilities*, but also *account for 'crucial' instructions* to improve the software reliability, though it may incur a performance loss.

Therefore, given a user-provided performance overhead, a reliability-driven instruction scheduler needs to optimize the software reliability which is a complex function of spatial and temporal vulnerabilities of different instructions w.r.t. different processor components/resources used during their execution. Finally, a soft-error-driven scheduler needs to incorporate the knowledge of 'crucial' and 'non-crucial' instructions in order to reduce the program's susceptibility towards failures.

1.3 Novel Contributions in a Nutshell
We present a *soft-error-driven instruction scheduler for reliability-aware compilation* that schedules instructions with the objective to enhance a software program's reliability under a user-provided tolerable performance overhead. We classify instructions as 'crucial' and 'non-crucial' based on the severity of potential program errors. As a comprehensive reliability cost function, our soft-error-driven instruction scheduler incorporates the idea of *instruction reliability weight*, which is a joint cost function of statically-estimated instruction vulnerability (spatial and temporal vulnerabilities of various processor components used during the execution of a certain instruction) along with an instruction's cruciality, probabilities of different error types, and dependent instructions. The proposed scheduler employs a *lookahead*-based heuristic for evaluating the reliability weights of various candidate instructions while taking into account the reliability weights of the successor instructions. The *lookahead* property reduces the risk of scheduling a 'crucial' instruction after a pipeline stalling instruction.

2. Estimation Model for Software Reliability
Several software program reliability estimation techniques have been proposed PVF [8], RVF [11]. However, these reliability estimation models have in common that they do not consider the knowledge of hardware-specific details, like chip footprint with processor details (area of different components, number of physical registers, etc.), for fault distribution and fault injection under different fault rates. Moreover, these techniques do not consider the temporal vulnerability within a pipeline. Therefore, these techniques are based on too abstract fault models and they will lead to over- or under-estimation of the reliability requirement. The program reliability estimation technique in [23][24] bridges the gap between hardware and software by quantifying the effects of hardware-level faults at the instruction level while taking into account the knowledge of the processor architecture and layout. Therefore,

in this paper, we employ the reliability estimation model "Instruction Vulnerability Index (IVI)" [23] that quantifies the program vulnerability as a function of spatial vulnerability and temporal vulnerability. The IVI is statically estimated during compilation using the methodology of [24]. Note: the reliability model is not a contribution of our paper. In order to also give readers the possibility to reproduce our results, we have listed the deployed reliability model in the Section S2.

3. Soft-Error-Driven Instruction Scheduling
Fig. 2 illustrates the compiler flow with our novel soft-error-driven instruction scheduler. It operates at the basic block level and the input is an instruction dependency graph with the estimated IVI of each instruction (ξ, Section 2), processor model, and the predicted probability of each basic block. The estimated IVI may vary depending upon the activated compiler optimization options and upon the processor's microarchitecture. Different compiler optimization options lead to different target code and thus lead to different potentials for code reorganization. Our scheduler is flexible and outputs a reliability-enhanced assembly code given a performance constraint. The scheduler employs a *lookahead*-based heuristic that maximizes the reliability weight of a basic block. In the following, we discuss the input, output, and optimization goal on a formal basis followed by our *lookahead*-based heuristic. The following formalism/formalization is necessary for the reproducibility of our implemented scheduler.

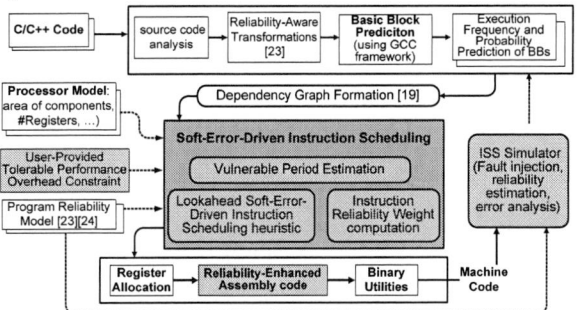

Fig. 2 Reliability-Aware Compiler Flow with Soft-Error-Driven Instruction Scheduler

3.1 Formal Problem Modeling
Input: A Basic Block (BB) is given as a directed acyclic graph $G=(N_G, E_G)$, where N_G is a set of nodes representing instructions, such that $N_G=\{n_1,..., n_N\}$ and N is the number of instructions in the BB. Each node is given as a tuple $n_i=\{SN_i, PN_i, \xi_i, T_i, eT_i, d_i, o_i\}$. T_i, eT_i, d_i, $o_i=\{o_1,..., o_m\}$ are instruction execution time, earliest time an instruction can be scheduled, and destination/source operands, respectively. For a node i, $SN_i=\{s_1,..., s_x\}$ and $PN_i=\{p_1,..., p_y\}$ are the sets with successor and predecessors nodes, respectively. E_G is a set of edges representing the instruction dependencies, such that $E_G=\{e_{ni \to nj} \mid n_i, n_j \in N_G\}$, where $e_{ni \to nj}$ denotes the latency of moving from instruction n_i to n_j. The execution frequencies and probabilities of basic blocks are predicted using the GCC framework [18].

Output: A reliability-optimized instruction schedule G_S.

Constraint: A user-provided tolerable performance overhead $P\tau$, i.e. tolerable performance loss compared to the performance-driven instruction schedule. In this work, we use the instruction scheduler of the GCC framework as a basis. It schedules the instruction with the maximum delay (δ_{MAX}) first.

Optimization Goal: An *Instruction Reliability Weight* (ψ) is employed as the objective function, such that the *instruction with the highest reliability weight* (ψ_{MAX}) *is scheduled first* in case the performance loss introduced by scheduling that instruction is below $P\tau$. The *Instruction Reliability Weight* (ψ, Eqs. 1–4) of the ith instruction constitutes the following parameters:

- *Estimated Instruction Vulnerability Index* (ξ, Section 2, S2) as the aggregated vulnerability of the ith instruction in different processor components ($r \in R=\{$pipeline, register file, ...$\}$).

---

[1] Occur due to resource conflicts in case the hardware cannot support simultaneous overlapped execution of instructions. For instance, only one register-file write port is available and the pipeline wants to perform two writes in the same cycle.

1290

- *Probabilities of failure and incorrect output (P_F, P_I)* considering the knowledge of instruction type, i.e. 'crucial' or 'non-crucial' instruction (*CIi* or *nCIi*). These probabilities are obtained using fault injection experiments at a certain fault rate '*f*'.
- *Dependent instructions (I_D)* in the same basic block (BB, with *N* instructions); an instruction with several dependents/successors exhibits a higher potential for fault propagation (see Eq. 3). Some dependent instructions may also exist in other basic blocks. It may result in fault propagation to those dependent basic blocks (DB). In this case, the goal is to keep such an instruction near to the end of the basic block. This incurs a negative cost (see Eq. 4). Note, an instruction might become reliability-wise important if it has dependent 'crucial' instructions. This observation motivated our *lookahead*-based instruction scheduling heuristic.

$$\psi_{nCIi} = \prod_{\forall r \in R} P_I \times \xi_i \qquad (1)$$

$$\psi_{CIi} = (1 + \prod_{\forall r \in R} P_F + \prod_{\forall r \in R} P_I) \times \xi_i \qquad (2)$$

$$\psi_i{}' = \sum_{\forall j \in \{CI, nCI\}} \psi_{ji} \times I_{Dj} / N \qquad (3)$$

$$\psi_i = \psi_i{}' - \sum_{\forall j \in \{CI, nCI\}} \psi_{ji} \times \left(\sum_{\forall k \in DB} I_{Djk} / N_k\right) \qquad (4)$$

The reliability weights of the complete basic block and software function are quantified as *basic block reliability weight* (ψB, Eq. 5) and *function reliability weight* (ψF, Eq. 6), respectively.

$$\psi B = \sum_{i=0 to N} \psi_i / N \qquad (5)$$

$$\psi F = \sum_{\forall j \in \{BB\}} \left(\sum_{k=0}^{nExec_j} \psi B_{jk} / T_F\right) \qquad (6)$$

$nExec_j$ is the number of predicted executions of BB_j and T_F is the execution time of the function 'F'.

3.2 Soft-Error-Driven Instruction Scheduling Heuristic

Due to its lookahead property, the soft-error-driven instruction scheduling heuristic evaluates the reliability weight of an instruction in conjunction with the reliability weights of its dependent instructions in order to determine a scheduling decision. Moreover, it keeps track of the performance loss compared to a performance-driven instruction scheduler. Fig. 3 shows an abstract example of the lookahead scheduler. Let an instruction node 'a' be scheduled at level 'L-1'. Next choices at level 'L' are instruction nodes 'b', 'c', and 'd'. However, it might be reliability-wise beneficial, if another level 'L+1' is explored, as the dependent instructions of (for instance) 'b' are 'crucial' instructions and 'c' and 'd' are 'non-crucial' instructions. An example scenario is shown in Fig. 3, where a soft-error-driven schedule without *lookahead* leads to a schedule of "+, +, +, *, *, store". However, scheduling the left-side *add* instruction after the right-side *add* and scheduling *store* after the *multiply* instruction result into two reliability issues: (i) the vulnerable period of the *store* input operand is increased, (ii) the vulnerable period of the address value of the *store* instruction is significantly longer. Therefore, when jointly considering the reliability of two levels 'L' and 'L+1', the following schedule is obtained "+, +, store +, *, *", where the left-side *add* and *store* instructions are scheduled before the right-side *add*. The second schedule exhibits a higher reliability weight (ψB) compared to the first case.

In the above example, only two levels are explored by the *lookahead* heuristic. Similarly, further levels could also be explored, but that would exponentially increase the scheduling complexity. Although the proposed *lookahead* is generic in its concept, we adopted exploring two levels in the *lookahead* as a compromise between complexity and reliability improvement. Considering the example of Fig. 3 and *lookahead* for two levels, the optimization goal is:

Maximize $[(\psi_b + max_{i=1,...,l} \psi_{bi}), (\psi_c + max_{j=1,...,m} \psi_{cj}), (\psi_d + max_{k=1,...,n} \psi_{dk})]$ (7)

Due to the *lookahead* into the subsequent scheduling levels, the instruction scheduler avoids selecting locally best solutions at a certain scheduling level. It is due to the fact that the reliability weights of the dependent instructions on the subsequent scheduling levels are also incorporated in the optimization goal. Fig. 4 presents the pseudo-code for the *lookahead* heuristic representing the implemented scheduler.

1. **Lookahead ( ): Input:** Instruction Graph **G=(N_G,E_G),** Tolerable Performance Overhead **Pτ, Output:** Instruction Schedule **G_S**
2. $G_S := \emptyset$; $G_{SC} := \emptyset$; // *Set of scheduled and candidate instructions*
3. $\forall n \in N_G$ $SN[n] := n.getSucc( )$; $PN[v] := n.getPred( )$;
4. $\forall n \in N_G$ $if(ready(i))$ $G_{SC} := G_{SC} \cup i$; // *Add to the ready list*
5. $\forall n \in G_{SC}$ $eT[n] := 0$;
6. $T_{Curr} := 0$;
7. $i_{PSel} := \emptyset$; $i_{RSel} := \emptyset$;   // *Perf.- and Reliability-maximizing instr.*
8.   **while** $(G_{SC} \neq \emptyset)$ {
9. $\forall i \in G_{SC}$ $\psi[i] := estimateReliabilityWeight(i)$;   // *Eq. 1–4*
10. $\psi_{MAX} := -\infty$; $\delta_{MAX} := -\infty$;
11. $\forall i \in G_{SC}$
12. $if(\delta[i] > \delta_{MAX}$ & $eT[i] \leq T_{Curr})$ $\delta_{MAX} := \delta[i]$; $i_{PSel} := i$;
13. $if(P\tau > 0)$ {
14. $\forall i \in G_{SC}$ {
15. $\forall j \in (G_{SC} - \{i\} + j.getSchedulableSN( ))$ {
16. $\psi_{ij} := \psi[i] + \psi[j]$; $\delta_{Loss} := (\delta_{MAX} - \delta[i])$
17. $if(\psi_{ij} > \psi_{MAX}$ & $\delta_{Loss} < P\tau)$
18. $\psi_{MAX} := \psi_{ij}$; $i_{RSel} := i$;
19.   **else if**$((\psi_{ij} = \psi_{MAX} \& \delta_{Loss} < P\tau) \& (\delta[i] > \delta[i_{RSel}] \& eT[i] \leq T_{Curr}))$
20. $\psi_{MAX} := \psi_{ij}$; $i_{RSel} := i$;
21.   }
22. $if(i_{RSel} \neq \emptyset)$ $i_{Sel} := i_{RSel}$; $P\tau := P\tau - (\delta_{MAX} - \delta[i_{Sel}])$;
23.   **else** $i_{Sel} := i_{PSel}$; }
24.   **else** $i_{Sel} := i_{PSel}$;
25. $T_{Curr} := T_{Curr} + T[i_{Sel}]$;
26. $G_{SC} := G_{SC} - i_{Sel}$; $G_S := G_S \cup i_{Sel}$;
27. $\forall s \in SN[i_{Sel}]$
28. $if(\forall m \in PN[s]$ $\exists t | N_{GS}[t] = m)$
29. $G_{SC} := G_{SC} \cup s$; $eT[s] := T_{Curr} + e_{i_{Sel} \to s}$;
30. } **return** G_S;

Fig. 4 Our soft-error-driven instruction scheduler

Instruction Scheduling Algorithm: First, the sets of scheduled and scheduling-candidate instructions are initialized (line 2). The successor and predecessor instructions for each instruction are obtained (line 3) and the ready list is initialized (line 4). Afterwards, initialization of earliest time, current time (T_{Curr}), performance-maximizing and reliability-maximizing selected instruction (i_{PSel}, i_{RSel}) is performed (lines 5-7). The loop continues executing until all scheduling candidates are processed. As discussed in Section 1.2 the vulnerable periods of instructions depend upon the scheduled instruction. Therefore, in every loop iteration, reliability weights of all instructions are re-estimated using Eqs. 1–4 (line 9). First, an instruction using a performance-driven scheduler is obtained for comparing it with the delay of the instruction selected by a soft-error-driven scheduler (line 12). If the tolerable performance overhead is greater than zero, the instructions in the candidate set are evaluated for reliability using a *lookahead* approach (lines 13-23). Otherwise, the performance-maximizing instruction is scheduled (line 24). The *lookahead* heuris-

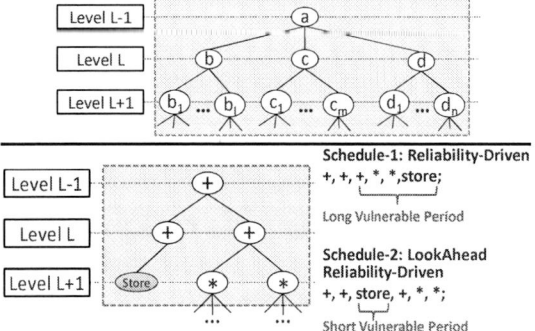

Fig. 3 An example of our scheduler with Lookahead heuristics

1291

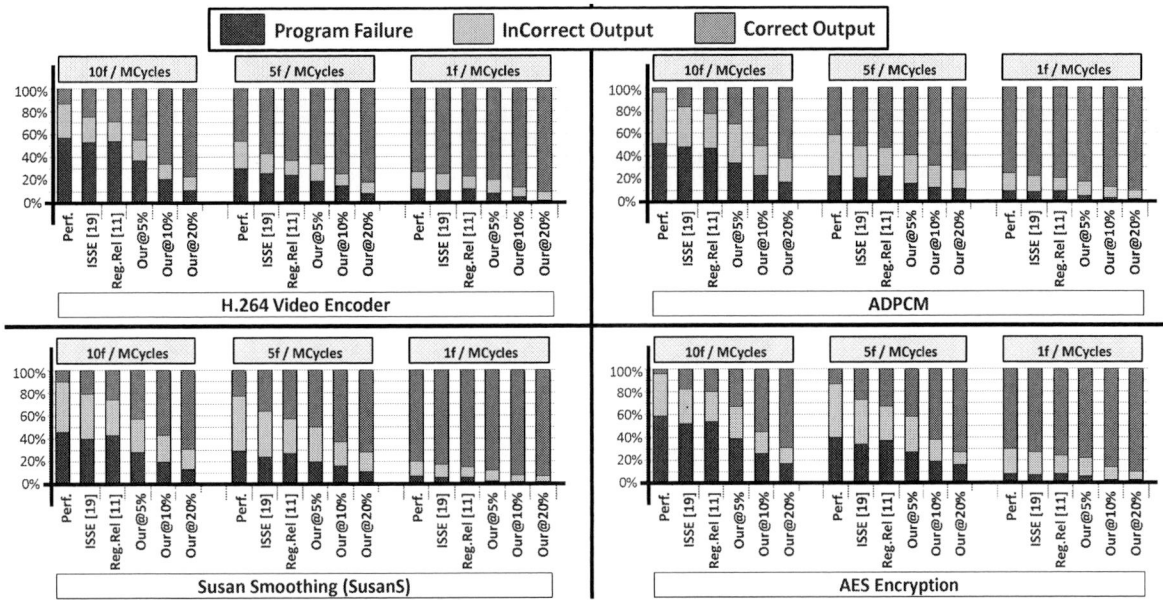

Fig. 5 Comparing the error distribution of our soft-error-driven instruction scheduler (at three different tolerable performance overhead constraints) to the performance-driven and state-of-the-art ISSE [19] and register file reliability improving [11] instruction schedulers for three different faults rates

tic accounts for the reliability weights of the candidate instructions and their schedulable successor instructions (line 15). It selects an instruction 'i' if the sum of the reliability weights of the instruction 'i' and its successor instruction 'j' with the highest reliability weight (ψ_{ij}) is greater than the currently highest reliability weight (ψ_{MAX}), and the performance loss of the instruction (δ_{Loss}) is less than the tolerable performance overhead (Pτ); see lines 16-18. Note, there may be several combinations with the same highest accumulated reliability weights that fulfill the tolerable performance overhead. In that case the one with the minimum performance loss is selected (lines 19-20). If there is no such solution that fulfils the tolerable performance overhead, a performance-maximizing instruction is scheduled. Otherwise, the reliability-optimizing instruction is scheduled and the corresponding δ_{Loss} is subtracted from the tolerable performance overhead (lines 22-23). Afterwards, the sets of scheduled instructions and scheduling candidates are updated accordingly (line 26). If the dependencies are resolved, the instructions are moved to the set of scheduling candidates (lines 27-29).

4. Experimental Setup and Results

Our Fault model: single bit flip transient faults that are spatially and temporally evenly distributed, i.e. evenly distributed over the processor footprint and randomly injected in different processor components w.r.t. their respective area. This is done during the execution of a software program for a given fault rate. The fault rate is determined using the neutron flux calculator [14] and fault probability (see detailed parameter list in Section S3). The manifesting effects of these transient faults are observed at the program layer as diverse error types (see error classification in Section 1.1). Like in prominent industrial and research projects (AMD [16], IBM [15], [13]), the caches are assumed to be protected by parity or ECC.

We have integrated our soft-error-driven instruction scheduler and reliability estimation technique in the GCC framework. For reliability evaluation and soft error analysis, we have developed a reliability-aware instruction-set simulator (see details in Section S3) that has an integrated fault injection, reliability analysis and reliability estimation methods [23]. The parameters for the reliability evaluation are: (i) fault rate in #faults/10MCycles which is obtained from the neutron flux [14] for a given city coordinates and altitude, (ii) fault probability, (iii) processor layout and frequency (in this work a Leon-II processor at 100 MHz is deployed [13]). Three different fault rates 1, 5, and 10 faults/MCycles are used to cover the terrestrial and

aerial use cases (also considered by well-known related work like [9]). Single-bit transient faults are randomly injected in processor components with respect to their area. [see Section S3 for details of the fault injection and hardware-level fault modeling at ISS-level].

For evaluation, we have used an H.264 video encoder [21], and ADPCM, AES, and SusanS applications from MiBench. Additional detailed experimental evaluation for various compute-intensive functions of the H.264 video encoder ('DCT', 'HT', 'MC-FIR', 'IPRED', 'SAD', 'SATD') are provided in Section S1. The compiler option O3 is enabled.

4.1 Comparison to State-of-the-Art

Fig. 5 presents the comparison of error distribution of our instruction scheduler (under three different tolerable performance overhead constraints, for three different fault rates) to the performance-driven scheduler [22] and state-of-the-art reliability-aware schedulers ISSE [19] and register-file reliability improving scheduler [11]. Fig. 5 shows that compared to state-of-the-art [19] and [11], our scheduler provides reduction in program failures of up to 15% at 5% performance loss and up to 45% at 20% performance loss, as it incorporates the knowledge of 'crucial' and 'non-crucial' instructions in the reliability weight. Specifically, the reduction in program failures is due to a lesser number of corrupted address operands of the 'crucial' instructions (load/store instructions) as a result of reduced temporal vulnerability of these 'crucial' instructions that are not on the critical path. Compared to the scheduler of [11], our scheduler achieves a reduced number of program failures due to reducing the pipeline residency of 'crucial' instructions. The primary reason of performance loss is because of reducing the spatial vulnerability at the cost of temporal vulnerability, such that the overall vulnerability is reduced. It is visible by the reduced number of program failures when moving from 5% to 10% to 20% performance loss cases. For ADPCM and AES applications, when moving from 10% to 20% performance loss, the reliability improvement is small. It is because of the balancing of spatial and temporal vulnerabilities. Moreover, with an increasing tolerable performance, the percentage of program failures decreases due to the lookahead nature, that prefers scheduling a 'crucial' instruction of the upcoming scheduling level first to reduce its vulnerable periods. However, this comes at the cost of a slight increase in the percentage of incorrect output, which is less severe compared to the program failures (as discussed in Section 1.1). On overall, our scheduler provides an increase in the correct output cases by average 22% (averaged over various tolerable performance overheads), when compared to state-of-the-art instruction schedulers [19][11].

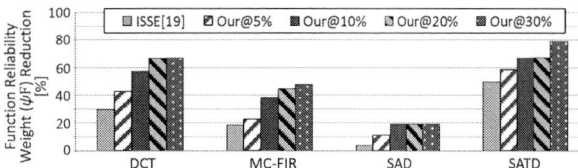

Fig. 6 Comparing function reliability weight reductions

For fair comparison to state-of-the-art, we have evaluated all (ours and state-of-the-art techniques) with the same reliability model (see Section S2). Fig. 6 presents the reduction in the function reliability weight (ψF, Eq. 6) of our instruction scheduler (under different tolerable performance overheads) and ISSE [19] compared to a performance-driven instruction scheduler [22] for different kernels of the H.264 video encoder application. Fig. 6 illustrates that our scheduler reduces the ψF value by up to 79% (on average 53%) at the cost of a 30% performance loss. For a 5% tolerable performance loss, our soft-error-driven scheduler reduces the ψF value by up to 58% (on average 33%). Note, when moving from 20% to 30% tolerable performance loss, in some cases (DCT, MC-FIR) there is limited reliability improvement, because the potential for reliability improvement at the software level for these applications has already been exploited. In case of SAD, due to excessive load instructions, the reliability improvement potential is limited. Therefore, it is fully exploited for a 5% tolerable performance, and there is no apparent reliability improvement by the instruction scheduler after the 10% case. Still in these cases, the reliability improvement of our scheduler is better compared to that of ISSE [19]. For large-sized programs (like SATD), our scheduler provides consistent reliability improvements.

Compilation Time Overhead: The reliability improvement of our instruction scheduler comes at the cost of an increase of approximately 3x (depending upon the number of successor instructions) in the compilation time for instruction scheduling due to the re-estimation of vulnerable periods and compile-time reliability estimation of all instructions in each iteration of the scheduler and prediction of basic block probabilities.

Summary: state-of-the-art by principle cannot achieve the level of reliability that our approach can, because of the following conceptual differences:

- our scheduler considers temporal *and* spatial vulnerabilities of the underlying fault injection model (state-of-the-art does not)
- our scheduler considers program failures based on soft errors observed when the software program is executed in a processor pipeline stage by stage (state-of-the-art does not provide these details and therefore cannot optimize for it)
- state-of-the-art tries to incur no performance overhead whereas we leave it up to the user to specify a tolerable performance overhead that may be associated with a gain in reliability. Often, a small performance loss may enable a high reliability. We can exploit those scenarios, state-of-the-art cannot.

5. Conclusion

We demonstrate that our soft-error-driven instruction scheduling provides a significant improvement towards reliable hardware/software systems. This can be attributed to the fact that we incorporate the knowledge of spatial and temporal vulnerabilities, distinction between 'crucial' and 'non-crucial' instructions, and probabilities of different error types under a user-provided tolerable performance overhead. Our instruction scheduler may be applied in conjunction with other hardware-level and software-level methods as (in a highly dependable system) all abstraction layers should contribute to increase the reliability of a system. Finally, state-of-the-art by principle cannot achieve the level of reliability that our approach can, because of conceptual advantages of our approach as summarized in Section 4.1.

Acknowledgement

This work is supported in parts by the German Research Foundation (DFG) as part of the priority program "Dependable Embedded Systems" (SPP 1500 - spp1500.itec.kit.edu).

6. References

[1] R.Baumann, "Radiation-induced soft errors in advanced semiconductor technologies," IEEE TDMR, vol. 5, no. 3, pp. 305-316, 2005.

[2] P.Shivakumar, M.Kistler, "Modeling the effect of technology trends on the soft error rate of combinational logic". IEEE DSN, vol.47, no. 6, pp. 2586-2594, 2002.

[3] R. Vadlamani, J. Zhao, W.Burleson, R.Tessier,"Multicore soft error rate stabilization using adaptive dual modular redundancy", DATE, pp. 27-32, 2010.

[4] D. Ernst, S. Das, S. Lee, D. Blaauw, T. Austin, T. Mudge, N. S. Kim, K.Flautner, "Razor: circuit-level correction of timing errors for low-power operation," IEEE MICRO, vol. 24, no. 3, pp. 10-20, 2004.

[5] S. S. Mukherjee, C. Weaver, J. Emer, S.K. Reinhardt, T.Austin, "A systematic methodology to compute the architectural vulnerability factors for a high-performance microprocessor", MICRO, pp. 29-40, 2003.

[6] R. Venkatasubramanianw, J. P. Hayes, B. T. Murray, "Low cost on-line fault detection using control flow assertions". IEEE IOLTS, pp.137–143, 2003.

[7] P. P. Shirvani, N. R. Saxena, E. J. McCluskey, "Software implemented EDAC protection against SEUs". IEEE Transactions on Reliability, vol 49, pp. 273–284, 2000.

[8] V. Sridharan, "Introducing Abstraction to Vulnerability Analysis", Ph.D. Thesis, March 2010.

[9] J. Hu, S. Wang, S. G. Ziavras, "In-Register Duplication: Exploiting Narrow-Width Value for Improving Register File Reliability," DSN, pp. 281-290, 2006.

[10] J. S. Hu, F. Li, V. Degalahal, M. Kandemir, N. Vijaykrishnan, M. J. Irwin, "Compiler-Directed Instruction Duplication for Soft Error Detection," DATE, vol.2, pp. 1056-1057, 2005

[11] J. Yan, W. Zhang, "Compiler guided register reliability improvement against soft errors," IEEE EMSOFT, pp. 203-209, 2005.

[12] X. Fu, W. Zhang, T. Li, J. Fortes, "Optimizing Issue Queue Reliability to Soft Errors on Simultaneous Multithreaded Architectures", International Conference on Parallel Processing, pp. 190-197, 2008.

[13] J. Gaisler, "A portable and fault-tolerant microprocessor based on the SPARC v8 architecture", DSN, pp. 409-415, 2002.

[14] Flux calculator: www.seutest.com/cgi-bin/FluxCalculator.cgi.

[15] IBM® XIV® Storage System cache: http://publib.boulder.ibm.com/infocenter/ibmxiv/r2/index.jsp.

[16] AMD PhenomTM II Processor Product Data Sheet 2010.

[17] G. A. Reis, J. Chang , N. Vachharajani, R. Rangan, D. I. August, S. S. Mukherjee, "Software controlled fault tolerance," ACM TACO, vol. 2, pp. 366-396, 2005.

[18] T. Ball, J. R. Larus, "Branch Prediction for Free", ACM SIGPLAN, vol. 28, pp. 300-313, 1993.

[19] J. Xu, Q. Tan, R. Shen, "The Instruction Scheduling for Soft Errors based on Data Flow Analysis", IEEE Pacific Rim International Symposium on Dependable Computing, pp. 372-378, 2009.

[20] A. Benso, S. Chiusano, P. Prinetto, L. Tagliaferri, "A C/C++ Source-to-Source Compiler for Dependable Applications", DSN, pp. 71-78, 2000.

[21] M. Shafique, L. Bauer, J. Henkel, "Optimizing the H.264/AVC Video Encoder Application Structure for Reconfigurable and Application-Specific Platforms", JSPS, vol. 60, no. 2, pp. 183-210, 2010.

[22] A. Parikh, S. Kim, M. Kandemir, N. Vijaykrishnan, M. J. Irwin, "Instruction scheduling for low-power", Journal of VLSI Signal Processing systems, vol 37, no. 1, pp. 129-149, 2004.

[23] S. Rehman, M. Shafique, F. Kriebel, J. Henkel, "Reliable software for unreliable hardware: Embedded code generation aiming at reliability", Codess+ISSS, pp. 237-246, 2011.

[24] S. Rehman, M. Shafique, F. Kriebel, J. Henkel, "RAISE: Reliability-Aware Instruction SchEduling for Unreliable Hardware", ASP-DAC, pp. 671-676, 2012.

[25] M. Shafique, B. Zatt, S. Rehman, F. Kriebel, J. Henkel, "Power-Efficient Error-Resiliency for H.264/AVC Context-Adaptive Variable Length Coding", IEEE DATE, pp. 697-702, 2012.

[26] S. Rehman, M. Shafique, J. Henkel, "ReVC: Computationally Reliable Video Coding on Unreliable Hardware Platforms: A Case Study on Error-Tolerant H.264/AVC CAVLC Entropy Coding", IEEE ICIP, pp. 405-408, 2011.

[27] Haifa Scheduler: http://gcc.gnu.org/, http://opensource.apple.com/source/gcc_os/gcc_os-1660/gcc/haifa-sched.c

Supplementary Sections

S1. Additional Experimental Findings and Results

Fig. 7 shows the error distribution for various computational kernels of the H.264 video encoder application to identify which kernels contribute more towards the failures. Compared to state-of-the-art [19] and [11], our scheduler provides reduction in program failures of up to 17%-26% at 5% performance loss (25%-40% at 20% performance loss). In general, the reduction in program failures is due to a lesser number of corrupted address operands of the 'crucial' instructions as a result of a reduced temporal vulnerability.

It can be seen that kernels with more 'crucial' instructions like 'DCT' and 'IPRED' are reliability-wise more critical. In case of SAD, the reliability improvement diminished when moving from 10% to 20% performance constraint, because the reduction in spatial vulnerability is canceled out by the increase in the temporal vulnerability. This can also be said for the 'SAD' application, the performance constraint should be set to 10% as a good tradeoff between performance loss and reliability improvement.

On overall, our scheduler provides an increase in the correct output cases by average 16.3%, 22.1%, and 28.5% for 5%, 10%, 20% tolerable performance overhead, respectively, when compared to state-of-the-art instruction schedulers [19][11].

S2. Software Program Reliability Estimation Model

As discussed in Section 2, we employ the reliability model that is in essence the same as in [23] and quantifies the program vulnerability as function of spatial vulnerability (area of different processor components like register file, pipeline, obtained from layout, processor architectural features) and temporal vulnerability (instruction vulnerable periods in processor components). It estimates the *Instruction Vulnerability Index* (ξ) of an instruction '*i*' as its collective or combined vulnerability in different processor components ($R=\{pipeline, register\ file, address\ generation\ unit, cache/memory\ controller\}$) according to the respective area (A_r) and error probability $P_E(r)$ of a component '*r*' depending upon its microarchitecture [1].

$$\xi_i = \left(\sum\nolimits_{\forall r \in R} \xi_{ir} \times A_r \times P_E(r) \right) / \sum\nolimits_{\forall r \in R} A_r \qquad (8)$$

The ξ_i of the instruction '*i*' at a certain processor component '*r*' of an architecturally-defined size in bits (β_{rTotal}) is given as the product of its vulnerable periods in that processor component (υ_{ir}) and vulnerable bits (β_r vulnerable bits required for Correct Execution), as shown in Eq. 9. Note, A_r and β_r capture the spatial vulnerability, while υ_{ir} captures the temporal vulnerability.

$$\xi_{ir} = (\upsilon_{ir} \times \beta_r) / \sum\nolimits_{\forall r \in R} \beta_{rTotal} \qquad (9)$$

The reliability of a software program can only be estimated after obtaining its binary and executing it for a given input data set. Because of changing control flow and pipeline stalls, the challenges are the compile-time prediction of vulnerable periods and vulnerable bits of different processor components used by an instruction.

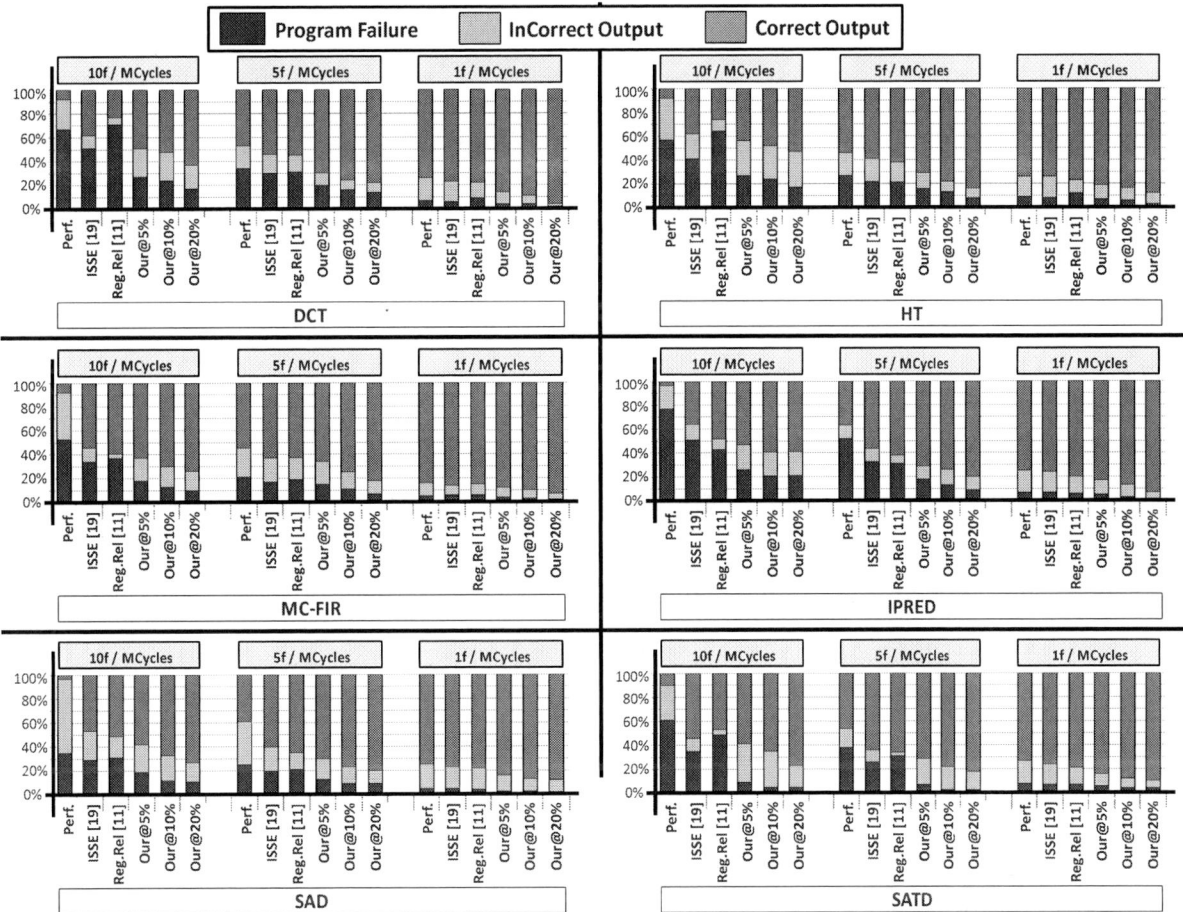

Fig. 7 Comparing the error distribution of our soft-error-driven instruction scheduler (at three different tolerable performance overhead constraints) to the performance-driven and state-of-the-art ISSE [19] and register file reliability improving [11] instruction schedulers for three different faults rates

The compile-time reliability estimation technique incorporates (a) spatial and temporal vulnerabilities, and (b) knowledge of predicted conditional probabilities of basic block executions.

For a pipeline resource of different pipeline stages (P), the estimated vulnerability is given as ξ_{iPipe} (Eq. 10, where υ_{ip} denotes the pipeline residency of an instruction 'i' in a certain pipeline stage 'p' and depends upon the execution time of the instruction in the succeeding pipeline stage. Similarly, the vulnerability of an instruction in other processor components like the address generation unit, and cache/memory controller are computed.

$$\xi_{iPipe} = \left(\sum\nolimits_{\forall p \in P} \upsilon_{ip} \times \beta_p\right) / \sum\nolimits_{r \in R} \beta_{rTotal} \qquad (10)$$

For the register file resource, the estimated vulnerability of an instruction is given as Eq. 11.

$$\xi_{iReg} = \frac{\left(\sum\nolimits_{\forall r \in operands} \left(\sum\nolimits_{\forall i \in prevBBs} P_i \times \upsilon_{ir}\right) \times \beta_{op}\right)}{\sum\nolimits_{r \in R} \beta_{rTotal}} \qquad (11)$$

In case the operands of an instruction stem from the previous basic blocks (*prevBBs*), the estimation of vulnerable period requires the execution probability of basic blocks. We employ the branch prediction feature of the GCC framework (activated using option -*fguess-branch-prob*) for gathering the predicted execution probability of each basic block. The branch probability is estimated using heuristics of [18] available inside the GCC framework

Vulnerable Bits: A fault in the hardware does not necessarily lead to a user-visible error e.g., due to control flow or due to subsequent instructions. This is mainly because of the masking effect from the microarchitecture-state to the ISA-visible state [8]. To demonstrate this observation, let us consider the following example scenario. A fault can occur in the value of either R1 or R2. In case of R1, a fault in the upper 16 bits will not have an effect on the final value stored in R3. However, in case of R2, a fault in any of the 32-bits will affect the final value stored in R3. Therefore, bits 0-15 of R1 are vulnerable-bits, whereas the bits ranging from 16 to 31 of R1 are not vulnerable. Contrarily, all 32 bits of R2 are vulnerable. Therefore, vulnerable bit analysis is important to be considered in program reliability estimation.

*R1=0x**ABCD**1234; R2=0x0000FFFF; R3=R1&R2=0x0000FFFF*

From this observation, we classify the bits into two categories i.e. vulnerable and non vulnerable bits.

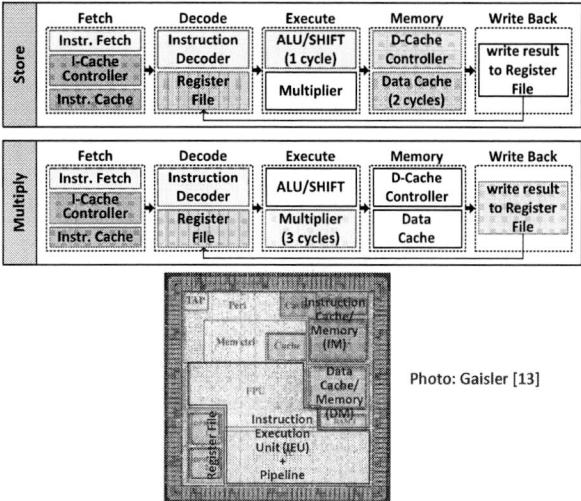

Fig. 8: Showing Spatial and Temporal Vulnerabilities during the execution of Multiply and Store Instructions [23]

Spatial and temporal vulnerability: Fig. 8 illustrates the temporal and spatial vulnerability comparing a *store* and a *multiply* instruction executing on the same processor. Regarding the spatial vulner-

ability in the *Execute* pipeline stage the *multiply* instruction is more vulnerable, as it uses a multiplier which is area-wise larger than the ALU used by the *store* instruction. Comparing the mentioned instructions for the *Memory* pipeline stage, however, the *store* instruction uses the Data Cache Controller and the Data Cache while the *multiply* instruction passes this stage passively. Therefore, the former is more vulnerable in this stage. Regarding the temporal vulnerability, it can be observed that the multiplication needs more time (Multiplier, 3 cycles) in the Execute pipeline stage than the addition (ALU, 1 cycle) which is performed for the *store* instruction. This results in a higher temporal vulnerability for the *multiply* instruction.

S3. Details of Fault Injection and Simulation Setup

In the following, details about the fault injection and simulation setup are presented. Fig. 9 shows a broader overview of our program reliability analysis methodology. This methodology works in two main phases (see Fig. 9) that are automated using scripts (due to >10^4 experiments per program):

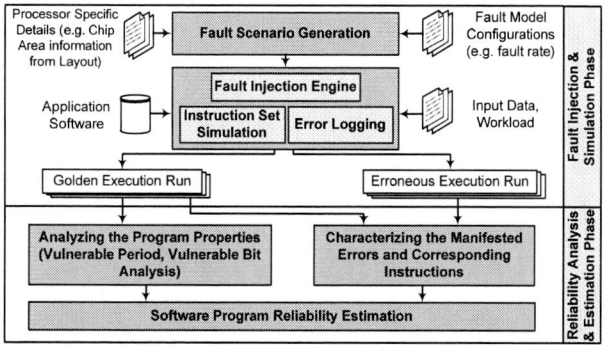

Fig. 9: Overview of our program reliability analysis and estimation setup with processor-aware configurable fault injection

Fault Injection and Simulation Phase: An Instruction Set Simulator (ISS) based fault injection technique is employed which is equipped with a configurable fault generation engine that generates different fault scenarios considering different fault models (e.g., number of bit flips and distribution), fault rates and faults at different processor components (e.g., register file, PC, Instruction Word (IW)). The processor-specific details (chip layout, component area, number of registers, etc.) and fault model configurations are passed as input. Afterwards, the fault injection engine injects the faults according to the fault scenarios (e.g., bit flips in the operands, opcode, and/or register values) during the application program execution (using an ISS) to obtain a potential 'erroneous run'. The application program is additionally simulated without fault injection to obtain a 'golden run' (i.e., correct execution). It is later on used for comparison with the 'erroneous run' to identify the potential errors in the program output. The details of generating fault scenarios with the knowledge of hardware and the procedure of modeling processor-level faults at the ISS-level is shown in Fig. 9.

Fault Models and Configurable Fault Generation Engine: The configurable fault generation engine considers different parameters (see Fig. 10) for generating various fault scenarios at different fault rates. A fault rate (in faults/10MCycles) is computed from the neutron flux (determined using the place and altitude where the device will be under use), fault probability, and the processor frequency.

We obtain the fault distribution in different processor components by considering the chip layout (i.e., area of different components, see Fig. 10) from Gaisler [13] for a Leon-II embedded processor. Fig. 12 elaborates the procedure of how faults and their impacts in different processor components are modeled at the ISS level. For example, it is shown that a fault in the instruction decoder or instruction word (resulting in a wrong operation or source/destination addresses) of processor is modeled as corrupting

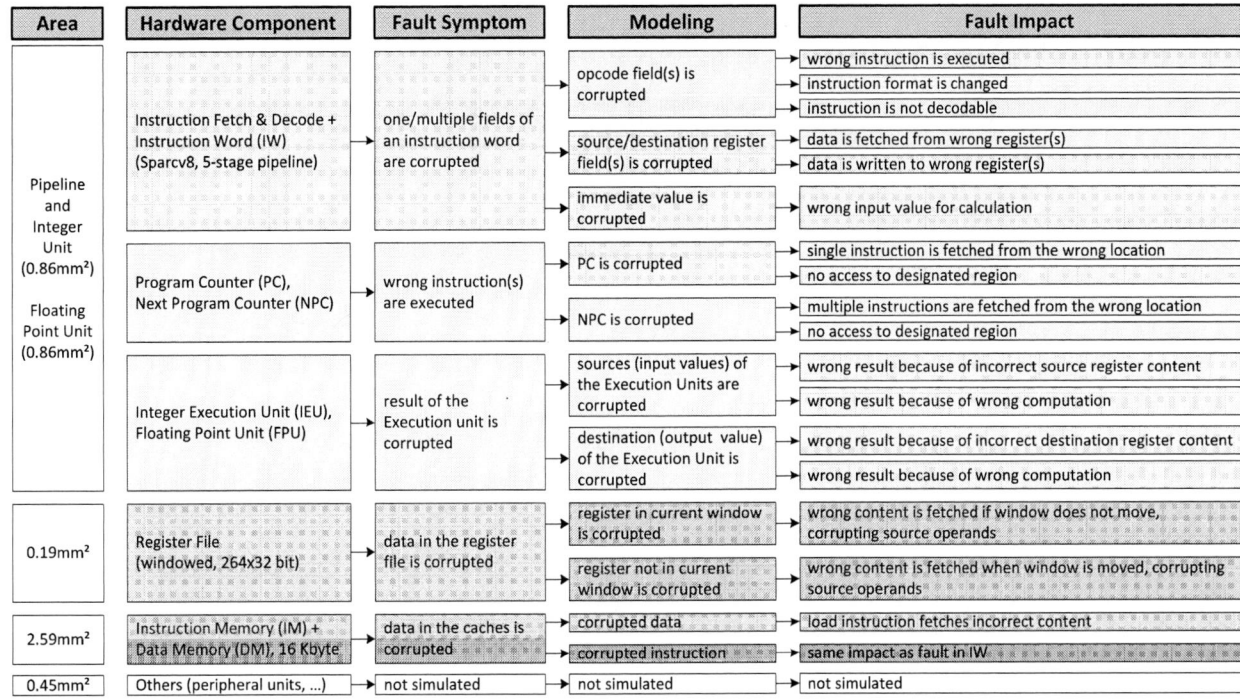

Area	Hardware Component	Fault Symptom	Modeling	Fault Impact
Pipeline and Integer Unit (0.86mm²) Floating Point Unit (0.86mm²)	Instruction Fetch & Decode + Instruction Word (IW) (Sparcv8, 5-stage pipeline)	one/multiple fields of an instruction word are corrupted	opcode field(s) is corrupted	wrong instruction is executed / instruction format is changed / instruction is not decodable
			source/destination register field(s) is corrupted	data is fetched from wrong register(s) / data is written to wrong register(s)
			immediate value is corrupted	wrong input value for calculation
	Program Counter (PC), Next Program Counter (NPC)	wrong instruction(s) are executed	PC is corrupted	single instruction is fetched from the wrong location / no access to designated region
			NPC is corrupted	multiple instructions are fetched from the wrong location / no access to designated region
	Integer Execution Unit (IEU), Floating Point Unit (FPU)	result of the Execution unit is corrupted	sources (input values) of the Execution Units are corrupted	wrong result because of incorrect source register content / wrong result because of wrong computation
			destination (output value) of the Execution Unit is corrupted	wrong result because of incorrect destination register content / wrong result because of wrong computation
0.19mm²	Register File (windowed, 264x32 bit)	data in the register file is corrupted	register in current window is corrupted	wrong content is fetched if window does not move, corrupting source operands
			register not in current window is corrupted	wrong content is fetched when window is moved, corrupting source operands
2.59mm²	Instruction Memory (IM) + Data Memory (DM), 16 Kbyte	data in the caches is corrupted	corrupted data	load instruction fetches incorrect content
			corrupted instruction	same impact as fault in IW
0.45mm²	Others (peripheral units, …)	not simulated	not simulated	not simulated

Fig. 12: Modeling hardware-level faults in different processor components at the ISS-level (an example for the case of Leon 2) [23]

one/multiple fields of the instruction word in ISS that results in a wrong opcode or wrong operand. The faults in different components can be single- or multiple bit flips (due to particles with different energy). The fault generation engine creates a set of numerous fault files which are provided as an input to the fault injection engine that injects faults during the program execution.

Parameter	Description	Properties/Values
Distribution	Distribution models for fault generation	random, uniform
Bit Flips	Min/Max number of bits flipped	1/1, 1/2, 1/3…
Fault Probability	Probability that strike becomes a fault	10%-100%
Fault Location	List of target processor components	Register file, PC, IW, IM, DM, etc.
Processor Layout/Area	Size of the complete target device	in mm²
Component Area	Area of different processor components given as percentage of processor area	0%-100%
Place and Altitude	City and altitude at which the device is used to determine the flux rate[4]	59° 55' N, 10° 45' O; 1- 20km
Frequency	Operating frequency of the processor	100 MHz

Fig. 10: Different Parameter for Fault Injection [23]

The fault generation engine creates a set of numerous fault files (see Fig. 11 for an excerpt), which are provided as an input to the fault injection engine that injects faults during the program execution.

		[Cycle]:	[Location], [Vector], [Address]
226668:	6, 131074, 2,;	Cycle:	In which cycle the fault should be injected
4458402:	4,32768,0,;	Location:	In which component the fault should be injected
5271986:	3,65602,227,;	Vector:	# of bits and their positions for bit flips
94276206:	1,71680,1,;	Address:	sub-address of fault location (e.g. register number in case of the register file)

Fig. 11 Format and an excerpt of a Fault Protocol

Fault Injection Engine: The step-by-step operational flow of the fault injection engine is shown in Fig. 13.

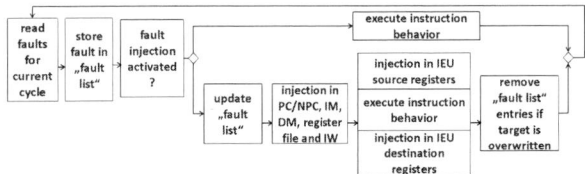

Fig. 13 Flow of Fault Simulation Process

The following steps are taken:
a) The faults for the current cycle are read from the fault file and are stored in a *fault list* which contains all faults that are currently injected.
b) If the fault injection is activated (a functionality implemented to be able to inject faults only in specific functions/parts of a program), the *fault list* is updated and bit flips are injected in the respective components
c) Afterwards the current instruction is executed.
d) After that, the *fault list* is inspected for entries whose target is overwritten. Those entries are removed from the *fault list*.
e) If the fault injection is deactivated, only the instruction is executed.

(Note that the fault injection does not introduce any unwanted side effects like changing performance counters)

Reliability Analysis and Estimation Phase: In this second phase of the methodology (Fig. 9), an error characterization is performed for program reliability analysis considering the program properties (e.g., histograms of the executed instructions). The error characterization and the properties of programs are used to obtain reliability metric at the instruction, function level and application program-level i.e. (IVI, FVI and AVI respectively). These metrics are then used to quantify the susceptibility of a program towards failures.

Compiling for Energy Efficiency on Timing Speculative Processors

John Sartori and Rakesh Kumar
University of Illinois at Urbana-Champaign

ABSTRACT

Timing speculation is a promising technique for improving microprocessor yield, in field reliability, and energy efficiency. Previous evaluations of the energy efficiency benefits of timing speculation have either been based on code compiled for a traditional target [2] – a processor that produces no errors, or code that relies on additional hardware support [6]. In this paper, we advocate that binaries for timing speculative processors should be optimized differently than those for conventional processors to maximize the energy benefits of timing speculation. Since the program binary determines the utilization pattern of the processor, which in turn influences the error rate of the processor and the energy efficiency of timing speculation, binary optimizations for timing speculative processors should attempt to manipulate the utilization of different microarchitectural units based on their likelihood of causing errors. An exploration of targeted and standard compiler optimizations demonstrates that significant energy benefits are possible from TS-aware binary optimization.

Categories and Subject Descriptors
D.3.4 [Processors]: Compilers, **General Terms:** Design
Keywords: error resilience, binary optimization, computer architecture, energy efficiency, timing speculation

1. INTRODUCTION

Timing speculation [2] is a promising technique for improving microprocessor yield, in field reliability, and energy efficiency. In the most common usage model, timing speculation involves relaxing voltage or frequency guardbands to improve energy efficiency at the expense of timing errors. Errors are corrected or tolerated by a hardware or software error resilience mechanism [2] to maintain acceptable output quality.

Previous evaluations of the energy efficiency benefits of timing speculation have either been based on code compiled for a traditional target [2] – a processor that produces no errors – or code that relies on instruction set extensions and additional hardware support [6]. For example, [6] advocates the use of instruction set extensions whose circuit implementations have shorter critical paths. Unfortunately, physical design tools render most pipeline stages critical in power-optimized processors [8,11], reducing the effectiveness of such approaches. Also, instruction set extensions may not be feasible in many settings.

In this paper, we make a case for compiling differently for timing speculative processors in a way that increases energy efficiency without additional hardware support or instruction set extensions. To motivate our approach, we first explain the nature of benefits afforded by timing speculation (TS). The magnitude of energy efficiency benefits available from exploiting TS depends on two factors – (a) *where* and (b) *how often* the processor produces errors when operating at an overscaled voltage or frequency. (For more details, see supplemental Section S1.) The path slack distribution of a timing speculative processor determines which paths do not meet timing constraints (negative slack paths) and thus cause errors when they are toggled.

Likewise, the activity distribution of the processor describes how often paths are toggled, and thus determines the frequency of errors caused by a path when it has negative slack. Together, the slack and activity distributions dictate the error distribution of a processor, i.e., the locations and frequencies of errors produced in an overscaled processor – i.e., a processor operating below nominal voltage or above nominal frequency.

Altering the error distribution of a timing speculative processor has the potential to increase the energy benefits from exploiting error resilience. For example, previous works have demonstrated that modifying the slack distribution (*where* errors are produced) can increase the energy efficiency of a timing speculative design [4, 9, 10, 12]. In this paper, we focus on the activity distributions (*how often* errors are produced) of timing speculative processors and make a case for *timing speculation-aware binary optimization*. Since the program binary, in conjunction with the processor architecture, determines a processor's activity distribution, optimizing a program binary for timing speculative processors can manipulate the utilization of different microarchitectural units based on their timing slack distribution to deliver energy efficiency benefits. For example, binary optimizations can be used to change the set of frequently exercised paths in a processor to avoid activating the longest paths. Since these paths are the first to have negative slack when the processor is overscaled, throttling their activity reduces early onset timing violations. Similarly, binary optimizations can be used to reduce error rate by throttling activity in structures of the processor that cause the most errors. Other possibilities include optimizations to overlaps errors in a single cycle to reduce the effective errors per cycle and optimizations to redistribute errors in the processors to reduce the effective error recovery overhead.

This paper on timing speculation-aware binary optimization makes the following contributions.

- We show that the activity distribution of a processor, and by extension, the error distribution, can be altered through binary optimizations.

- We demonstrate that the energy efficiency of timing speculative processors can be improved by altering their activity distributions through binary optimizations, without any additional hardware support.

- Through careful analysis of the main factors that influence processor error rate, we show that several optimizations that are already supported by existing compilers can improve the energy efficiency of TS.

- We quantify the energy savings from targeted and standard binary optimizations for a family of timing speculative processor architectures. We observe up to 39% additional energy savings from TS-aware binary optimization for a Razor-based processor.

2. BASELINE ARCHITECTURE

Which optimizations are most effective for a processor depend on which processor modules cause the most errors. In this section, we describe the family of processor architectures we study to develop binary optimization strategies and identify their error-critical modules. We also discuss how the error criticality of modules may depend on program characteristics.

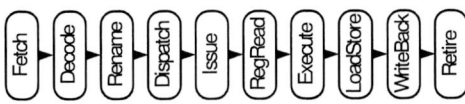

Figure 1: The FabScalar Pipeline. [1]

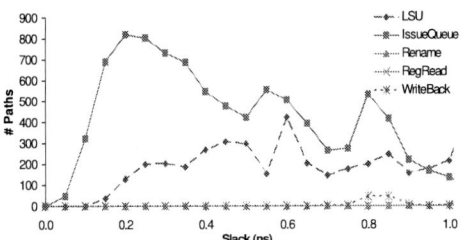

Figure 2: The static slack distributions for the pipeline stages show how many critical paths they have but do not provide information about how often the paths toggle, which is essential in characterizing error rate.

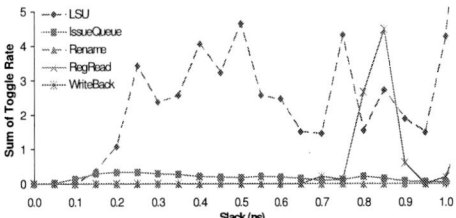

Figure 3: The activity-weighted (dynamic) slack distributions for different pipeline stages indicate how much timing critical activity they have, and by extension, how frequently they will produce errors for a given level of overscaling – i.e., a given $(voltage, frequency)$ pair.

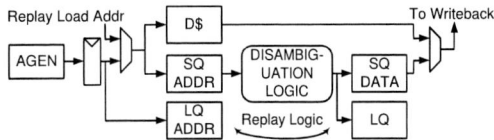

Figure 4: Memory disambiguation is on the critical path of the LSU [1]. The path delay is longest when store-to-load forwarding is required, since this necessitates an access to the SQ data RAM, in addition to the other disambiguation operations.

2.1 FabScalar Architecture

We use the FabScalar [1] framework for our architectural evaluations. FabScalar is a parameterizable, synthesizable processor specification that allows for the generation and simulation of RTL descriptions for arbitrarily configured scalar and superscalar processor architectures. FabScalar allows for the configuration of many microarchitectural parameters, including superscalar width (ss), fetch width and depth (fw,fd), numbers and types of functional units, issue width and depth (iw,id), issue queue size (iq), select logic depth (sel), register file depth (rrd), re-order buffer entries (rob), physical registers (reg), and load and store queue sizes (lsq). In this paper, we study a family of superscalar processors by selecting interesting candidates from the available configurations space of FabScalar. Figure 1 shows the FabScalar pipeline.

2.2 Error Criticality Analysis

Different pipeline stages cause errors at different rates, depending on their slack and activity distributions. Figure 2 shows the static slack distributions for the pipeline stages that cause the most errors. While our highly-optimized design flow removes excess slack in all stages, two stages in particular – the issue queue (IQ) and the load store unit (LSU) – have the highest number of critical paths. Based on Figure 2, one might expect that the IQ, having many more critical paths than all other modules combined, would produce the most errors in the processor. However, the static slack distribution only shows the *potential* for paths to cause errors. Not all stages exercise their critical paths often. Stages with frequently exercised critical paths cause the most errors. In Figure 3, we create activity-weighted, *dynamic* slack distributions by showing the sum of toggle rates for all the paths at each value of timing slack (activity from SPEC benchmarks). The more timing critical *activity* a module has, the more errors it is likely to cause. From Figure 3, it is clear that the LSU dominates the error distribution of the processor.

2.3 Program Dependence of Error Criticality

As demonstrated in the previous section, the LSU, and secondarily, the IQ are the primary sources of timing violations for the family of processor architectures that we studied. Below, we describe the implementation of the LSU and the IQ in the FabScalar processor to understand the dependence of error rate on program characteristics.

The LSU (Figure 4) performs memory disambiguation for the processor. This involves checking for dependencies between loads and stores. After address resolution, a store must search the address CAM of the LQ and process all entries with matching addresses to determine if any load issued out of order and broke a RAW dependence. Load disambiguation is more com-

plicated because it may include store-to-load forwarding. In addition to a search through the SQ address CAM, a load must generate a mask vector indicating all preceding stores in program order. Matching entries from the CAM search are filtered by the mask vector, and the latest resulting entry, if any, forwards data from the SQ data RAM to the load.

LSU delay depends on program characteristics for several reasons. The primary reason is that the store-to-load forwarding path is on the static critical path of the LSU. Since many RAW dependencies in a code lead to more forwarding, the timing error rate will be higher for code with a relatively large number of RAW dependencies. Program characteristics also determine the utilization of the LQ and the SQ, which, in turn, dictates access delays for the structures. For example, when the LQ or SQ are nearly full, as may be common for memory-centric codes, more entries must be accessed in a single cycle to generate mask vectors. This increases the length of the propagation path, and consequently, increases delay. Additionally, when there are many dependencies between memory operations, address CAM searches generate many hits, increasing load capacitance and delay for the CAM access. Finally, propagation delay increases when many hits are signaled in parallel (due to many potential dependencies), since the average length of the propagation path from the CAM entries to the port increases. Hence, the average delay is higher for memory-centric codes with a large number of dependencies.

We confirmed that the forwarding paths are timing critical in the LSU, and that more dependencies result in activation of longer paths, by observing the activity-weighted (dynamic) slack distribution of the LSU for two different instruction streams (Figure 5). The first contains a stream of memory operations that access the same address. Each load depends on the previous store and activates the forwarding paths in the LSU. In the second stream, the dependencies are removed. Figure 5 demonstrates that activity on the critical paths of the LSU is greatly reduced when the dependencies are removed and forwarding is not required.

The wakeup-select logic used in the IQ is similar in nature to the memory disambiguation logic in the LSU. For example, wakeup consists of finding all instructions that depend on the destination register of another instruction. This CAM-based dependence check in the IQ is performed in much the same way as the dependence checks in the LSQ. Likewise, select logic, which selects a ready, waiting instruction to execute is somewhat akin to the masking logic that identifies valid, conflicting stores for forwarding. Because of their similarities, the LSU and IQ have similar timing considerations.

1298

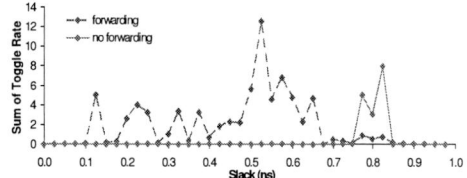

Figure 5: Since forwarding paths are critical in the LSU, eliminating dependencies and the need for store-to-load forwarding reduces activity on the critical paths of the LSU.

Table 1: Average processor-wide Razor overheads.

Hold buffering	Razor FF	Counterflow	Error Recovery
2% energy	23% energy	<1% energy	P cycles

3. METHODOLOGY

To understand the impact of different binary optimizations on the error behavior and energy efficiency of different processor architectures, we used a detailed methodology that carefully models the relationships between execution behavior, power, performance, and reliability. Designs are implemented with the *TSMC* 65GP library (65nm), using *Synopsys Design Compiler* for synthesis and *Cadence SoC Encounter* for layout. In order to evaluate the power and performance of designs at different voltages and to provide V_{th} sizing options for synthesis, *Cadence Library Characterizer* was used to generate low, nominal, and high V_{th} libraries at each voltage (V_{dd}) between $1.0V$ and $0.5V$ at $0.01V$ intervals. Designs are implemented at 500 MHz. Power, area, and timing analyses are performed in *Synopsys PrimeTime*. Gate-level simulation is performed with *Cadence NC-Verilog* to gather activity information for the design, which is subsequently used for dynamic power estimation and error rate measurement. (For more details on error rate measurement, see supplemental Section S2.)

In our evaluations, we compile and run several microbenchmarks to demonstrate architecture-specific TS-aware optimizations. We also run instruction traces from the SPEC benchmark suite (bzip,gap,mcf,parser,vortex). after fast-forwarding the benchmarks to their Simpoints [5]. All benchmarks are compiled with gcc-2.7.2.3 (SPEC benchmarks and gcc version correspond to those supported by FabScalar).

We model Razor-based error resilience [2] in this paper (though our proposed techniques are generally applicable to any timing error resilient processor). Table 1 summarizes the processor-wide static and dynamic overheads of Razor-based error detection and correction. In our design flow, we measure the percentage of die area devoted to sequential elements, as well as the timing slack (with respect to the shadow latch clock skew of 1/2 cycle) of any short paths that need hold buffering. When evaluating energy at the processor level, we account for the increased area and power of Razor flip-flops, hold buffering on short paths, and implementation of the recovery mechanism. Most of the static overhead is due to Razor FFs. Buffering overhead is small, and the availability of cells with high and low V_{th} provides more control over path delay, eliminating the need for buffering on most paths. We also add energy and throughput overheads proportional to the error rate to account for the dynamic cost of correcting errors over multiple cycles. We use a counterflow pipeline Razor implementation [2] with correction overhead proportional to the number of processor pipeline stages (P). We conservatively replace all sequential cells with Razor FFs.

4. RESULTS AND DISCUSSION

We now discuss different architecture-specific binary optimizations that may increase the efficiency of timing speculative processors. The proposed optimizations are primarily geared toward error avoidance in the LSU and IQ. We first discuss targeted loop-based optimizations and quantify their benefits through the use of microbenchmarks. Then, we evaluate the

```
for(i=0; i<N; i++)        for(i=0; i<N; i+=4){
  sum += A[i];              sum1 += A[i];
                            sum2 += A[i+1];
                            sum3 += A[i+2];
                            sum4 += A[i+3];
                          }
                          sum=sum1+sum2+sum3+sum4;
```

Figure 6: Original loop (left) and unrolled loop (right).

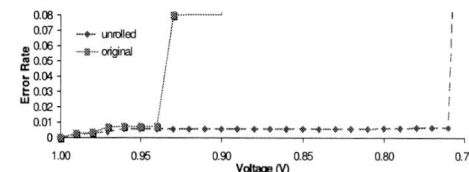

Figure 7: Loop unrolling reduces activity on LSU forwarding paths, resulting in a significant error rate reduction.

benefits of combining standard *gcc* optimizations using *O* levels for SPEC benchmarks.

4.1 Targeted Optimizations for TS Processors

4.1.1 Loop Unrolling

As described above (Section 2), activity on the static critical paths of the LSU can be reduced by avoiding dependent memory operations and scenarios that cause the LSQ to fill up. This can enable significantly deeper voltage overscaling, since the LSU is often the source of many timing violations.

Loop unrolling is a classic compiler optimization that can eliminate and spread out loop carried dependencies, and thus has the potential to reduce LSU delay. Normally, unrolling would only be used when spin up and spin down costs are overcome by reducing the number of executed instructions. However, TS-aware compilation provides a new use for unrolling – avoiding errors to increase the efficiency of TS by grouping often independent instructions (like vector math) and eliminating often dependent instructions (like branches and loop index updates). Unrolling allows optimization of register allocation over multiple loop iterations that can eliminate load and store disambiguation, thus reducing pressure on the LSU. Unrolling can also reduce pressure on the branch resolution unit and arithmetic unit, since the number and frequency of branch instructions and loop index updates are reduced. Thus, in addition to fostering critical path avoidance by reducing dependencies, loop unrolling can also be an agent for activity throttling.

Unrolling can cause binary size to increase, which may reduce instruction cache efficiency and may be undesirable in some embedded processors. Unrolling may also cause an increase in dynamic power. When exploiting TS-aware binary optimization, it is important to consider the impact on performance and power, as well as energy efficiency.

Figure 6 shows an example of loop unrolling by a factor of 4. Figure 7 shows the error rate of the processor when executing the two code sequences of Figure 6. Unrolling significantly reduces the error rate by reducing activity on the forwarding paths in the LSU. This error rate reduction enables additional overscaling and results in a substantial energy reduction for a Razor based TS processor, as shown in Table 2. Microarchitectural parameters not specified in Table 2 are $iq = 16, rob = 64, reg = 64, lsq = 8 + 8 = 16$.

In the error-free case, the same unrolled loop causes dynamic power to increase significantly, even as it increases throughput. Thus, unrolling has the potential to reduce error rate but may also increase power for a conventional processor where TS is not allowed. So, most energy-efficient binary optimization depends

Table 2: Razor-based TS and error-free energy savings (%) for loop unrolling. (*ss* = superscalar width)

CORE	original	unrolled	unrolled error-free
ss1	11.8	43.1	1.6
ss2	6.4	20.8	2.0
ss4	4.0	42.9	3.2

1299

```
for(i=0; i<N; i++)        for(i=0,j=N/2;i<N/2;i++,j++){
  sum += A[i];              sum1 += A[i];
                            sum2 += A[j];
                          }
                          sum = sum1+sum2;
```

Figure 8: Original code (left – ILP 1) and code with more ILP exposed (right – ILP 2).

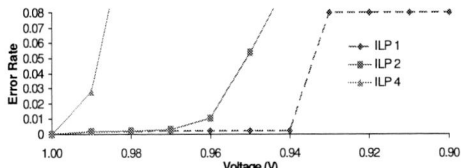

Figure 9: When hardware parallelism is not available (*ss1*), exposing parallelism floods backend queue structures and increases the error rate.

on whether the target uses TS. This demonstrates the need for TS-aware compiler analysis and optimization.

4.1.2 Balancing Instruction-Level Parallelism

In an out-of-order processor, instructions are dispatched to the processor backend as long as there is available space in the appropriate backend structures, namely, the reorder buffer (ROB), IQ, and LSQ. However, when there are not enough execution units to handle ready, waiting instructions, backend structures fill up and remain full. As discussed above, this leads to longer propagation delays for these structures – especially for queues.

Thus, we observe that when hardware parallelism is limited, optimizing the binary to promote software parallelism can actually increase energy in a timing speculative processor by increasing logic delay and limiting overscaling. Consequently, when hardware parallelism is limited, a TS-aware compiler should actually throttle parallelism to prevent instructions from reaching the backend. This kind of compiler optimization is contrary to conventional wisdom, which promotes ILP whenever possible for potential performance gains.

On the other hand, when hardware parallelism is available, the scenario is reversed. Dependencies that hinder ILP keep queues full and increase the delay of dependence-checking logic. Thus, when adequate hardware resources are available, enhancing parallelism can eliminate dependencies and lead to better TS efficiency.

To illustrate the above points, we have run the codes in Figure 8 on TS processors with different superscalar widths. Figure 9 compares the error rates of the code sequences for the *ss1* case. In this case, hardware parallelism is not available, and exposing more instructions to the processor backend causes queue structures to fill, increasing propagation delays. Thus, the error rate increases as more parallelism is exposed (e.g., ILP4).

For a processor with more hardware parallelism (e.g., *ss2*), the backend can handle increased software parallelism without putting excessive fill pressure on queue structures. In this case, the reduced dependencies of the more parallel code reduce activity in the timing critical disambiguation logic and enable more overscaling. Figure 10 compares error rates for the codes on a *ss2* processor. The error rate for the code without exposed parallelism (ILP1) increases abruptly and surpasses the error rates for the more parallel codes. Table 3 shows energy results for Razor-based TS, demonstrating that TS efficiency increases *when hardware and software parallelism are balanced*. The table also demonstrates that enhancing parallelism does not provide any significant energy savings in the error-free case, motivating the need for TS-specific compiler analysis and optimization.

Table 3: Razor-based TS and error-free energy savings (%) for balancing parallelism.

CORE	ILP 1	ILP 2	ILP 4	ILP 2 error-free
ss1	13.1	5.5	0.0	1.4
ss2	5.4	9.8	9.7	0.8

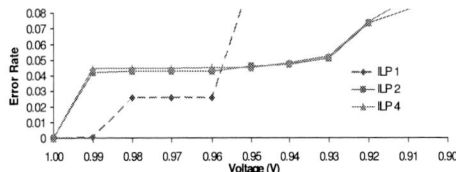

Figure 10: When hardware parallelism is available (*ss2*), exposing parallelism eliminates dependencies and reduces error rate.

```
j = N-1;                  sum = A[0] + A[N-1];
for(i=0; i<N; i++){       for(i = 1; i < N; i++){
  sum += A[i] + A[j];       sum += A[i] + A[i-1];
  j = i;                  }
}
```

Figure 11: Original code (left) and code with a dependence peeled from the loop (right).

4.1.3 Loop Splitting

Loop splitting or peeling can also be used to break dependencies in code by peeling dependent instructions out of the loop body. The original code in Figure 11 contains two dependencies – a loop carried dependence for the accumulator variable (sum), and a dependence between the array indices (i, j). By peeling one of the iterations from the loop, we can eliminate one of the dependencies. This reduces the load on the CAM structure that performs dependence checking, and eliminates occurrences of forwarding. Figure 12 shows how peeling a dependence from the loop reduces the error rate for *ss1*, *ss2*, and *ss4* processors. Table 4 compares the energy savings achieved by Razor-based TS and error-free operation before and after loop splitting is performed. In all cases, the additional overscaling enabled by loop splitting results in energy savings for Razor-based TS. For error-free operation, loop splitting actually increases energy slightly, because it causes a small reduction in performance (IPC). This divergence between the best decision for TS and error-free cases motivates the need for TS-specific compiler analysis and optimization.

4.1.4 Loop Fusion

Another technique for manipulating dependence patterns in code is loop fusion. Loop fusion merges independent instructions in separate loops into the same loop. Grouping independent instructions can help to break up long chains of dependent instructions by spreading them further apart in the binary. This can reduce the need for forwarding, since conflicting instructions are able to clear the LSQ before their dependent instructions are dispatched to the processor backend. As a side effect, loop fusion may decrease locality of access, which can degrade cache performance. In general, it is important to consider the potential performance impacts of TS-aware binary optimization along with the energy savings it enables.

Figure 13 compares code sequences with (right) and without (left) loop fusion. Note that loop fusion and loop splitting are inverse operations. I.e., the original code can be produced by performing loop splitting on the fused code. In the *ss1* case (Figure 14), grouping independent instructions does not provide benefits, since there are not adequate hardware resources to handle the exposed ILP. In this case, the unfused (split) code has a lower error rate, because the activity of the LSU (the module that causes the most errors) is throttled by the interleaving of branches and loop index updates with the loads and stores. This activity throttling leads to increased TS energy efficiency, as shown in Table 5.

In the *ss4* case (Figure 15), the clustering of independent instructions in the fused code spaces out dependent instructions

Table 4: Razor-based TS and error-free energy savings (%) for loop splitting.

CORE	original	split	split error-free
ss1	5.8	13.8	-0.2
ss2	0.0	9.0	-0.4
ss4	3.6	13.4	-0.1

1300

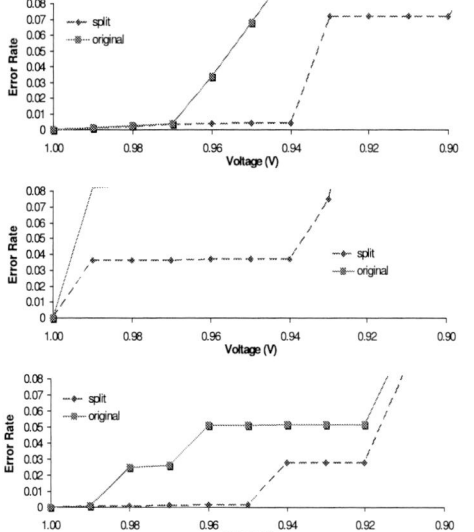

Figure 12: By removing a dependence from the loop, loop splitting reduces the error rates of the *ss1* (top), *ss2* (middle), and *ss4* (bottom) processors.

in the pipeline, thus eliminating many occurrences of forwarding and reducing activity on timing critical paths in the LSU. This critical path avoidance reduces error rate and enhances TS efficiency, as shown in Table 5. Again, energy savings from loop fusion in the error-free case are only meager ($< 1\%$), motivating the need for TS-aware compiler analysis and optimization.

```
for(i=0; i<N; i++)          for(i = 0; i < N; i++){
  sum1 += A[i];               sum1 += A[i];
for(i=0; i<N; i++)            sum2 += B[i];
  sum2 += B[i];               sum3 += C[i];
for(i=0; i<N; i++)            sum4 += D[i];
  sum3 += C[i];             }
for(i=0; i<N; i++)
  sum4 += D[i];
```

Figure 13: Original code (left) and code with fused loops (right).

Several other TS-aware binary optimizations are possible. The goal of this paper is to demonstrate that significant energy benefits may be possible from TS-aware binary optimization. An exhaustive exploration of all possible binary optimizations is beyond the scope of this work.

4.2 Standard gcc Optimizations for Timing Speculative Processors

Fortunately, many standard gcc optimizations have similar goals as the targeted optimizations discussed above. For example, optimizing for a higher O level has the potential to reduce dependencies and bolster ILP. Similarly, optimizing for a lower O level may effectively restrict ILP. Below, we evaluate the TS efficiency of SPEC binaries that have been optimized at different O levels.

For architectures without available hardware parallelism (e.g., *ss1*), highly optimizing compute-limited applications can cause pipeline backend structures to fill, resulting in longer delays and higher error rates. On the other hand, for memory-bound ap-

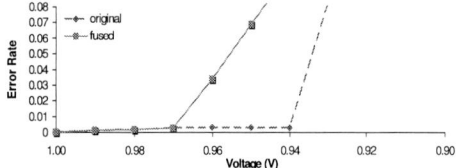

Figure 14: When hardware parallelism is limited (*ss1*), the unfused (split) code has a lower error rate, since LSU activity is throttled.

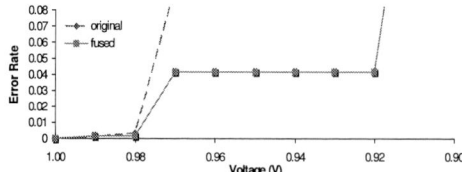

Figure 15: When hardware parallelism is available (*ss4*), the fused code spaces out dependent instructions, reducing forwarding, and consequently, error rate.

Table 5: Razor-based TS and error-free energy savings (%) for loop fusion.

CORE	original	fused	fused error-free
ss1	12.2	5.9	0.2
ss4	4.2	12.3	0.5

plications with many indirect memory references, critical LSU paths are not frequently exercised. Instead, IQ contributes most substantially to the error rate, so optimizing at a higher O level, which reduces average IQ entries, and consequently, IQ delay, reduces the error rate. Thus, when hardware parallelism is limited, compute-limited applications should be optimized for a lower O level ($O0$), while memory-bound, pointer-chasing codes can be optimized for a higher O level.

For architectures with available hardware parallelism (e.g., *ss2*), highly optimizing compute-limited applications can reduce dependencies, activity on critical LSU paths, and error rate. Optimizations do not have much effect on memory-bound, pointer-chasing codes, since available hardware parallelism allows average IQ entries to remain low, and critical LSU paths are not frequently exercised. Below, we test these intuitions for SPEC benchmarks with standard gcc O levels.

Figure 16 shows the error rates of SPEC benchmarks we evaluated at available O levels, running on the *ss1* core. Although higher optimizations (e.g., $O2$) generally improve performance (IPC), they increase error rate and degrade TS energy efficiency for compute-limited codes (Table 6). This is because optimizing at the higher O level enhances software parallelism, but there is not sufficient hardware parallelism to handle the dispatched instructions. Thus, backend structures (LSQ and IQ) fill, and propagation delay increases, limiting overscaling. Consequently, performing no optimizations ($O0$) is preferable for compute-limited applications on the *ss1* core when TS is used. Note that this is an interesting result, as the choice of O level would be different when compiling for the error-free case, since increasing the O level improves performance.

For pointer-chasing codes like *vortex*, which performs object-oriented database lookups, and thus contains many indirect memory references, critical LSU forwarding paths are not frequently exercised. Rather than the LSU, the IQ dominates the processor error rate for the $O0$ binary on this core. Optimizing for a higher O level results in fewer average IQ entries, reducing delay and error rate, and significantly increasing energy savings (Table 6).

For the *ss2* core, the backend queue structures are not overly stressed. Optimizing at a higher O level reduces dependencies for compute-limited codes, and by extension, activity on the critical paths of the LSU. This reduces error rate (Figure 17) and allows more overscaling and reduced energy (Table 7). Thus, higher optimization (O) levels are beneficial, in general, for Razor-based TS when hardware parallelism is not restricted. Choosing the correct optimization level that balances hardware and software parallelism maximizes energy savings. Note that results in this section demonstrate that the best optimization level is different for TS and non-TS cases. For example, $O1$ achieves the most energy benefits for TS on the *ss2* core, even though $O2$ has higher performance in the error-free case.

As expected, memory-bound, pointer-chasing codes see little impact from optimizations on the *ss2* core. The many indirect memory references in *vortex* cannot be optimized at compile time, and thus, optimizations do not significantly impact LSU activity. Also, since HW parallelism is available to relieve IQ

1301

fill pressure, optimizations do not significantly reduce the IQ error rate either.

In the error-free case, optimizing at a higher level ($O2$) can increase performance, but this performance comes with a significant increase in power consumption. Thus, energy is not significantly improved with $O2$ in the error-free case (Tables 6, 7). Distinctions between the best strategy in TS and non-TS cases further demonstrates the need for TS-aware compiler analysis and optimization.

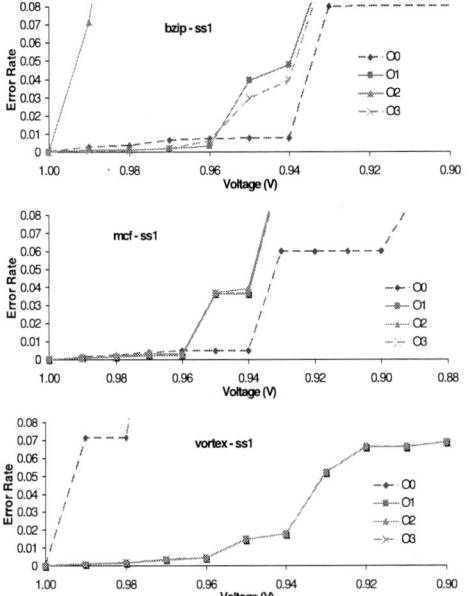

Figure 16: For the *ss1* core, highly optimizing compute-bound code (e.g., bzip) can increase the error rate, because fill pressure increases the delays of highly utilized pipeline backend structures and limits overscaling. Optimizing memory-bound code (e.g., vortex) can reduce error rate, because critical LSU paths are not exercised, and optimizations reduce IQ fill pressure.

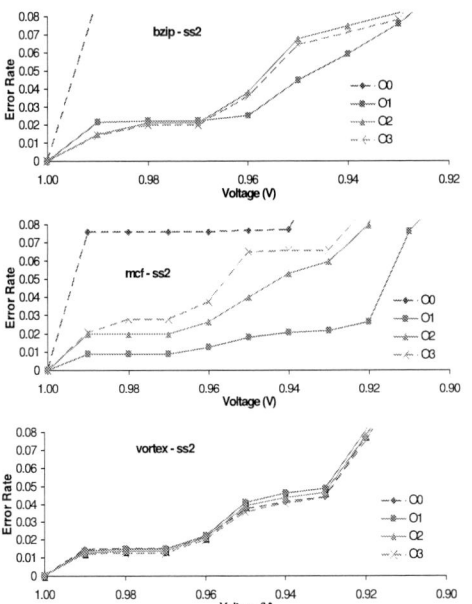

Figure 17: Optimizing compute-bound code (e.g., bzip) can reduce dependencies and activity on the critical paths of the LSU for the *ss2* core. Choosing the right optimization level that balances HW and SW parallelism can be important. This results in lower processor error rates. The effect of optimizations is limited for memory-bound code (e.g., vortex).

Table 6: Razor-based TS and error-free energy savings (%E), performance (IPC), and binary size (MB) for SPEC benchmarks at different O levels (*ss1*).

ss1	bzip			mcf			vortex		
OPT	%E	IPC	MB	%E	IPC	MB	%E	IPC	MB
O0	11.8	0.45	0.32	14.7	0.56	0.31	0.0	0.55	1.70
O1	7.5	0.79	0.29	9.2	0.67	0.29	14.0	0.49	1.48
O2	0.0	0.77	0.29	9.0	0.54	0.29	14.0	0.51	1.47
O3	7.2	0.75	0.31	9.2	0.59	0.30	14.0	0.51	1.49
O2 no-error	1.2	0.77	0.29	0.1	0.56	0.29	0.0	0.55	1.47

Table 7: Razor-based TS and error-free energy savings (%E), performance (IPC), and binary size (MB) for SPEC benchmarks at different O levels (*ss2*).

ss2	bzip			mcf			vortex		
OPT	%E	IPC	MB	%E	IPC	MB	%E	IPC	MB
O0	0.0	0.65	0.32	0.0	0.69	0.31	10.4	0.61	1.70
O1	7.7	1.39	0.29	13.4	1.45	0.29	10.0	0.74	1.48
O2	5.7	1.32	0.29	9.1	1.37	0.29	10.2	0.75	1.47
O3	7.5	1.34	0.31	8.5	1.26	0.30	10.4	0.76	1.49
O2 no-error	1.2	1.5	0.29	1.0	1.49	0.29	0.4	0.78	1.48

5. SUMMARY AND CONCLUSIONS

Previous work on improving energy efficiency of timing speculative processors relied on code targeting conventional processors or assumed additional hardware support and instruction set extensions. In this paper, we have demonstrated that careful binary optimization can increase the energy efficiency of error resilient processors without additional hardware support. Since the program binary determines the utilization pattern of the processor, which in turn influences the error rate of the processor and the energy efficiency of timing speculation, optimizing a binary specifically for timing speculative processors can manipulate the utilization of different microarchitectural units based on their timing slack distribution to deliver energy efficiency benefits. We have demonstrated up to 39% additional energy savings with timing speculation-aware binary optimization for Razor-based processors. We expect the energy benefits to grow as more sophisticated compiler techniques are developed.

6. REFERENCES

[1] N. Choudhary, S. Wadhavkar, T. Shah, S. Navada, H. Najaf-abadi, and E. Rotenberg. Fabscalar. In *WARP*, 2009.

[2] D. Ernst, N. S. Kim, S. Das, S. Pant, R. Rao, T. Pham, C. Ziesler, D. Blaauw, T. Austin, K. Flautner, and T. Mudge. Razor: A low-power pipeline based on circuit-level timing speculation. In *MICRO*, page 7, 2003.

[3] Y. Fujimura, O. Hirabayashi, T. Sasaki, A. Suzuki, A. Kawasumi, Y. Takeyama, K. Kushida, G. Fukano, A. Katayama, Y. Niki, and T. Yabe. A configurable sram with constant-negative-level write buffer for low voltage operation with $0.149\mu m^2$ cell in 32nm high-k/metal gate cmos. In *ISSCC*, 2010.

[4] B. Greskamp, L. Wan, W. Karpuzcu, J. Cook, J. Torrellas, D. Chen, and C. Zilles. Blueshift: Designing processors for timing speculation from the ground up. *HPCA*, 2009.

[5] G. Hamerly, E. Perelman, J. Lau, and B. Calder. Simpoint 3.0: Faster and more flexible program analysis. In *JILP*, 2005.

[6] G. Hoang, R. Findler, and R. Joseph. Exploring circuit timing-aware language and compilation. In *ASPLOS*, pages 345–356, 2011.

[7] Intel Corporation. Intel atom processor z5xx series, 2008.

[8] A. Kahng, S. Kang, R. Kumar, and J. Sartori. Designing processors from the ground up to allow voltage/reliability tradeoffs. In *HPCA*, 2010.

[9] A. Kahng, S. Kang, R. Kumar, and J. Sartori. Recovery-driven design: A methodology for power minimization for error tolerant processor modules. In *DAC*, 2010.

[10] A. Kahng, S. Kang, R. Kumar, and J. Sartori. Slack redistribution for graceful degradation under voltage overscaling. In *ASPDAC*, 2010.

[11] J. Patel. CMOS process variations: A critical operation point hypothesis, 2008.

[12] S. Sarangi, B. Greskamp, A. Tiwari, and J. Torrellas. Eval: Utilizing processors with variation-induced timing errors. *MICRO*, pages 423–434, 2008.

[13] J. Sartori and R. Kumar. Architecting processors to allow voltage/reliability tradeoffs. *CASES*, 2011.

S1. Understanding How Slack and Activity Distributions Determine Error Rate

In this supplemental section, we explain in greater detail how slack and activity distributions determine the error rate of a processor. The extent of energy benefits gained from exploiting timing error resilience depends on the error rate of a processor. In the context of voltage overscaling-based timing speculation, for example, benefits depend on how the error rate changes as voltage decreases. Likewise, in the context of frequency overscaling, benefits depend on how the error rate changes as frequency increases. If the error rate increases steeply, only meager benefits are possible [8]. If the error rate increases gradually, greater benefits are possible. In this paper, we have considered voltage overscaling-based timing speculation, though our conclusions should also be applicable to other forms of timing speculation.

The timing error rate of a processor in the context of voltage overscaling depends on the timing slack and activity distributions of the paths of the processor. Figure 18 shows an example slack distribution. The slack distribution is a histogram that shows the number of paths in a design at each value of timing slack. As voltage scales down, path delay increases, and path slack decreases. The slack distribution shows how many paths can potentially cause errors because they have negative slack (shaded region). Negative slack means that path delay is longer than the clock period.

From the slack distribution, it is clear which paths can cause errors (timing violations) at a given voltage and frequency. In order to determine the error rate of a processor, however, the activity of the negative slack paths must be known. A negative slack path causes a timing error when it toggles. Therefore, knowing the cardinality of the set of cycles in which any negative slack path toggles reveals the number of cycles in which a timing error occurs.

For example, consider the circuit in Figure 19 consisting of two timing paths. P_1 toggles in cycles 2 and 4, and P_2 toggles in cycles 4 and 7. At voltage V_1, P_1 is at critical slack, and P_2 has 3ns of timing slack. Scaling down the voltage to V_2 causes P_1 to have negative slack. Since P_1 toggles in 2 of 10 cycles, the error rate of the circuit is 20%. At V_3, the negative slack paths (now P_1 and P_2) toggle in 3 of 10 cycles (cycles 2,4,7), and the error rate is 30%.

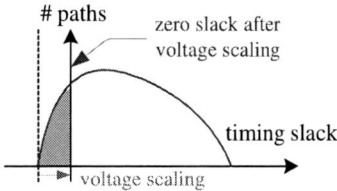

Figure 18: Voltage scaling shifts the point of critical slack. Paths in the shaded region have negative slack and cause errors when toggled.

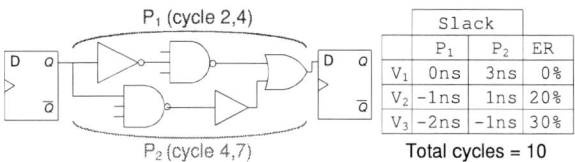

Figure 19: Slack and activity distributions determine the error rate.

S2. Details of Activity-based Error Rate Calculation

This supplemental section provides additional details on how we calculate error rate in our design flow. To calculate the error rate produced by a binary running on a processor implementation, we run a gate-level simulation of the binary on the synthesized, placed, and routed processor RTL, and we capture switching information for the nets in the design in the form of a value change dump (VCD) file. To calculate the error rate of a design at a particular voltage, toggled nets from the VCD file are traced to find the paths that have toggled in each cycle. The delays of toggled paths are measured using PrimeTime, and any cycle in which a negative slack path toggles is counted as an error cycle. The error rate (ER) of the design is equivalent to the cardinality of the set of error cycles, divided by the total number of simulation cycles (X_{tot}), as shown in Equation 1,

$$ER = \frac{|\bigcup_{p \in P_n} \chi_{toggle}(p)|}{X_{tot}} \qquad (1)$$

where P_n is the set of negative slack paths and $\chi_{toggle}(p)$ is the set of cycles in which path p toggles.

Figure 20 shows an example VCD file and illustrates the path extraction method. The VCD file contains a list of toggled nets in each cycle, as well as their new values. Toggled nets in each cycle are marked, and these nets are traversed to find toggled paths. A toggled path is identified when toggled nets compose a connected path of toggled cells from a primary input or flip-flop to a primary output or flip-flop. In Figure 20, nets a, b, and c have toggled in the first and fourth cycles (#1, #4), and nets d and c have toggled in the second and fourth cycles (#2, #4). Two toggled paths are extracted: $a-b-c$ and $d-c$. Paths $a-b-c$ and $d-c$ both have toggle rates of 40% ($|\chi_{toggle}(p)| = 2$ and $X_{tot} = 5$). Therefore, if only one of the paths has negative slack, the error rate is 40% in this example. If both paths have negative slack, then timing errors will occur in cycles #1, #2, and #4, for an error rate of 60%.

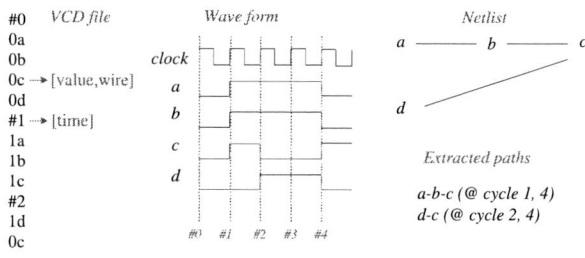

Figure 20: VCD file format and path extraction from the VCD file.

In addition to inducing timing errors by increasing logic delays, voltage scaling may prompt reliability concerns for SRAM structures, such as insufficient Static Noise Margin (SNM). Fortunately, the minimum energy voltage for our processors is around 750mV, while production-grade SRAMs have been reported to operate reliably at voltages as low as 700mV [3]. Research prototypes have been reported to work for even lower voltages. In any case, modern processors typically employ a "split rail" design approach, with SRAMs operating at the lowest safe voltage for a given frequency [7].

S3. Related Work

Previous work on TS-aware design has focused on optimizing *hardware* to improve the efficiency of TS. Work has been done primarily at the design level [4, 8–10, 12] and the architecture level [13] to reshape the slack distribution of a processor to enhance the energy efficiency benefits of TS. These optimizations primarily focus on making the static slack distribution of a processor more amenable to overscaling.

This work, however, focuses on optimizations at the *software* level that influence the activity and *dynamic* slack distributions of a processor (see Section 2). Since the error rate of a timing speculative processor depends on both slack and activity (see Section S1), TS-aware compilation has just as much potential to optimize processor error rate as hardware-based techniques. A promising direction of work involves co-optimization of software and hardware to reshape the dynamic slack distribution and maximize the energy efficiency benefits of exploiting TS.

The closest related work [6] focuses on extending the instruction set to include instructions for which the circuit implementation has a shorter critical path. Replacing instructions with these new instructions increases timing slack and enables more overscaling. The instruction set extensions proposed by [6] primarily focus on reduced-complexity arithmetic operations. We optimize program binaries to improve energy efficiency for TS processors without requiring hardware support.

In a typical ASIC design flow, all paths with excess timing slack are optimized to remove the timing slack, thus reducing power consumption and area. This design style produces a design with a critical slack wall [11], so that the vast majority of timing paths have near-critical slack. Since all circuit modules in our designs have many critical paths, as we would expect in a processor implemented by a typical CAD flow, we are unable to utilize optimizations that redirect instructions to units with more timing slack [6]. Instead, our optimizations focus on avoiding activation of the critical paths in a hardware unit and throttling the activity of units that cause the most errors. Additionally, we focus on binary optimizations that do not require instruction set extensions, and thus, may be more generally applicable. Finally, since the architectures that we evaluate are different than the architecture studied in [6], the modules that cause the most errors are different. Therefore, our architecture-specific optimizations focus on different regions of the processor.

AUTHOR INDEX

Aadithya, Karthik . V. 311
Abelein, Ulrich 205
Agarwal, Amit 1149
Agarwal, Anant 265
Agosta, Giovanni 77
Ahmadyan, Seyed Nematollah 1018
Akesson, Benny 988
Al Maashri, Ahmed 579
Aliakbarian, Hadi 542
Alpert, Charles J. 465, 768, 774
Anders, Mark 1149
Anderson, Charles 1274
Andrade, Hugo 656
Ascheid, Gerd 121, 1262
Ashouei, Maryam 962
Atasu, Kubilay 350
Athikulwongse, Krit 741
Atienza, David 636
Auerbach, Joshua 271
Augustine, Charles 1258
Bachrach, Jonathan 1212
Bacon, David F. 271
Baj-Rossi, Camilla 6
Bamis, Athanasios 163
Bampi, Sergio 866
Bao, Min 197
Barenghi, Alessandro 77
Bartolini, Davide B. 856
Bathen, Luis Angel D. 214, 447
Beckmann, Nathan 265
Beigne, E. 1049
Belhadj, Bilel 1260
Benini, Luca 1137
Benkeser, Christian 510
Beretta, Ivan 1043
Bernstein, Gary 476
Bertacco, Valeria 115, 729, 955
Bhardwaj, Kshitij 382
Bhuyan, Laxmi 1006
Blaauw, David 406, 980, 1037
Bobba, Shashikanth 42
Bobda, Christophe 1203
Boero, Cristina 6
Borkar, Shekhar 1149
Brisk, Philip 26
Bromberg, David 486
Brooks, David 277
Browy, Chris 936
Brunhaver, John 623
Bryson, William 133
Buckl, Christian 188, 430
Burcea, Ioana 271
Burg, Andreas 510
Burleson, Wayne 12
Burns, Steven 603

Butler, Kenneth M. 808
Campanoni, Simone 277, 856
Cancare, Fabio 856
Cao, Yu 139, 283
Carrara, Sandro 6
Castrillon, Jeronimo 1262
Chaji, G. Reza 182
Chakrabarti, Chaitali 579
Chakrabarty, Krishnendu 18, 1024
Chakraborty, Koushik 382, 1074
Chakraborty, Samarjit 688
Chan, Carven 1222
Chan, Ya-Chung 1163
Chandra, Vikas 283
Chandramoorthy, Nandhini 579
Chandrasekar, Karthik 988
Chang, Chen-Feng 1088
Chang, Hua-Yu 802
Chang, Kai-Hui 936
Chang, Leland 1155
Chang, Naehyuck 516, 522
Chang, Yao-Wen 549, 762, 802, 1082,
        1088, 1175, 1181
Chang, Yuan-Hao 882
Chang, Yu-Chi 376
Chao, Mango C.-T. 1012
Charkrabari, Chaitali 980
Chatha, Karam S. 672, 1268
Chatterjee, Debapriya 115, 955
Che, Weijia 672, 1268
Chebira, Amina 636
Chen, Chia-Hsin 398
Chen, Chi-Hao 453
Chen, Fu-Wei 1094
Chen, Gang 430
Chen, Haibo 106
Chen, I-Che 613
Chen, Wei-Yu 1181
Chen, Wu-Hsin 1254
Chen, Xiang 1000
Chen, Xiaoming 1125
Chen, Xinke 876
Chen, Yi-Hung 1113
Chen, Yiran 585, 1000, 1187
Cheng, Perry 271
Cheng, Yi-Kan 1113
Cher, Chen-Yong 642
Chiang, Charles 1163, 1169
Chiang, Patrick 974
Chien, Hsing-Chih Chang 549
Chiou, Derek 790
Choi, Junchul 664
Choi, Yunju 536
Chou, Chun-Nan 327
Chou, Sheng 762

AUTHOR INDEX

Christmann, J. F. 1049
Chu, Chris 597
Chung, Jaewoong 888
Chung, Sung Woo 1193
Chung, Yi-Ting 1055
Clark, Shane S. 12
Cline, Brian 283
Condemine, C. 1049
Condley, Walter 145
Cong, Jason 843, 1229, 1235
Coskun, Ayse K. 648
Cotter, Matthew 579
Crop, Joseph 974
Csaba, Gyorgy 476
Danowitz, Andrew 623
Das, Reetuparna 406
Davoodi, Azadeh 709
Daya, Bhavya 398
De Gyvez, Jose Pineda 962
De Marchi, Michele 42
De Micheli, Giovanni 6, 42
Debole, Michael 579
Derrien, Steven 48
Dey, Sujit 826
Ding, Duo 756
Ding, Huping 412
Ding, Wei 834
Dingler, Aaron 476
Dong, Xiangyu 253
Donkoh, Eric 62
Donohoo, Brad 1274
Dreslinski, Ronald G. 406, 980, 1143
Du, Yang 1160
Duranton, Marc 1260
Dutt, Nikil D. 214, 447
Eberl, Michael 205
Edith, Beigne 994
Eles, Petru 197
El-Shambakey, Mohammed 437
Erez, Mattan 850
Eusse, Juan 121
Fallah, Farzan 561
Fang, Shao-Yun 1175, 1181
Fang, Zhenman 106
Fazzari, Saverio 133
Fedder, Gary K. 176
Feng, Zhuo 1119
Fey, Gorschwin 941
Fick, David 1143
Finder, Alexander 941
Fink, Stephen J. 271
Flamand, Eric 1137
Fong, Neric 289
Foreman, Eric 1061
Foroutan, Sahar 366

Forte, Domenic 96
Fu, Kevin 12
Garg, Siddharth 697
Gester, Michael 459
Ghasemi, Hamid Reza 56
Ghodrat, Mohammad Ali 843
Ghosal, Arkadeb 656
Gill, Michael 843
Glaß, Michael 205
Goossens, Kees 988
Goswami1, Dip 688
Grassi, Paolo Roberto 1043
Grigorian, Beayna 843
Grissom, Daniel 26
Gruenwald, Charles 265
Guo, Jing 1169
Guo, Xiaofei 573
Gupta, Priyank 529
Gupta, Sumeet 492
Guthaus, Matthew R. 145
Ha, Soonhoi 664
Haensch, Wilfried 1155
Hagleitner, Christoph 350
Hakim, Nagib 561
Hamzeh, Mahdi 1280
Hao, Kecheng 344
Harris, Ian G. 1252
Haubelt, Christian 1203
Haugou, Germain 1137
Hayes, Jerry 1107
Heliot, Rodolphe 1260
Hemmett, Jeffrey 1061
Henkel, Jorg 866, 1288
Ho, Yen-Sheng 327
Ho, Yuan-Kai 1088
Hodges, Ben R. 723
Hoffmann, Henry 259, 856
Holloway, Glenn 277
Holt, Jim 259
Hong, Ted 561
Horowitz, Mark 623, 783
Hsieh, Chiao 327
Hsiu, Pi-Cheng 453
Hsu, Meng-Kai 762
Hsu, Steven 1149
Hu, Miao 498, 585
Hu, Wei-Yi 1113
Hu, Xiaobo Sharon 476
Hu, Xuchu 145
Hu, Yibin 106
Huang, Chung-Yang 327
Huang, He 169
Huang, Jia 188
Huang, Jiawei 504
Huang, Kai 188, 430

AUTHOR INDEX

Huang, Po-Chun.....................882
Huang, Rei-Fu.....................1012
Huang, Shi-Yu.....................1031
Huang, Shuai.....................876
Huang, Yoshi Shih-Chieh.....................376
Huang, Yu-Hung.....................127
Huisken, Jos.....................962
Hwang, Tingting.....................1094
Ienne, Paolo.....................229
Irwin, Mary Jane.....................678
Jaffari, Javid.....................182
Jane, Mary.....................834
Jang, Ohyoung.....................834
Jego, Bruno.....................1137
Jeong, Min Kyu.....................850
Jha, Somesh.....................1250
Jiang, Iris Hui-Ru.....................802, 1163
Jiang, Jie-Hong Roland.....................1055
Jiang, Lei.....................907
Jimenez, Xavier.....................229
Jog, Adwait.....................243
Jones, Timothy.....................277
Joshi, Rajiv.....................1107
Joubert, Antoine.....................1260
Jovic, Jovana.....................121
Jung, Byunghoo.....................1254
Jung, Deokwoo.....................163
Jung, Moongon.....................317
Jung, Seobin.....................536
Kahng, Andrew B......................392, 820
Kandemir, Mahmut.....................678, 834
Kang, Seokhyeong.....................820
Kanj, Rouwaida.....................1107
Karakonstantis, Georgios.....................510
Karri, Ramesh.....................83, 573
Kasture, Harshad.....................265
Kaul, Himanshu.....................1149
Kawakami, Katsutoshi.....................648
Kelley, Kyle.....................783
Keng, Brian.....................947
Keskin, Gokce.....................176
Khaled, Nadia.....................1043
Khatri, Sunil P......................734, 1256
Kim, Dae Hyun.....................888
Kim, Dongki.....................888
Kim, Heesoo.....................808
Kim, Jaeha.....................536
Kim, Lee-Sup.....................630
Kim, Myung-Chul.....................747
Kim, Nam Sung.....................56
Kim, Sungchan.....................664
Kim, Yejoong.....................1037
Kim, Younghyun.....................522
Kim, Youngsik.....................897
Kirsch, Christoph M......................913

Klefstad, Raymond.....................1006
Knoll, Alois.....................188, 430
Kong, Joonho.....................1193
Koushanfar, Farinaz.....................68, 90, 133, 220
Koyfman, Anatoly.....................955
Kozhikkottu, Vivek.....................796, 826
Krishna, Tushar.....................398
Krishnaswamy, Smita.....................814
Kuan, Jui-Feng.....................1113
Kuang, Jilong.....................1006
Kuehlmann, Andreas.....................814
Kumar, Jayanand Asok.....................808, 1018
Kumar, Pratyush.....................688
Kumar, Rakesh.....................918, 1297
Kunz, Wolfgang.....................334
Kuo, Chin-Cheng.....................1113
Kuo, Tei-Wei.....................453, 882
Kurian, George.....................259
Kursun, Eren.....................642
Kurtz, Steve.....................476
Lach, John.....................504
Lau, Eric.....................259
Lauwereins, Rudy.....................1
Leary, Glenn.....................672
Leblebici, Yusuf.....................42
Lee, D. T......................613
Lee, Hsu-Chieh.....................1082, 1088
Lee, Po-Wei.....................1088
Lee, Sunggu.....................897
Lee, Sungkwang.....................888
Lee, Yoonmyung.....................1037
Lee, Yunsup.....................1212
Lepley, Thierry.....................1137
Leupers, Rainer.....................121, 1262
Li, Hai.....................498, 585
Li, Jian.....................106
Li, Kai.....................90
Li, Peng.....................476, 876
Li, Wenchao.....................1250
Li, Xin.....................176, 642
Li, Zhuo.....................465, 768, 774, 1107
Liang, Yun.....................412
Lim, Sung Kyu.....................157, 317, 741
Limaye, Rhishikesh.....................656
Lin, Bill.....................392
Lin, David.....................561
Lin, I-Jye.....................1088
Lin, Pey-Chang Kent.....................734, 1256
Lin, Shih-Chin.....................1012
Lin, Xue.....................516
Lin, Yu-Hsiang.....................1031
Lingamneni, Avinash.....................924
Lionel, Vincent.....................994
Liu, Beiye.....................585
Liu, Bin.....................1235

AUTHOR INDEX

Liu, Bo...542, 962
Liu, Chih-Hung 613
Liu, Frank..723
Liu, Haotian...289
Liu, Sizhe...567
Liu, Xiao..555
Lo, Daniel..421
Lowery, Alicia 62
Lu, Yi...106
Lu, Yi-Shan..127
Luo, Yan..18
Ma, Qiang..591
Maggio, Martina 259
Mahlke, Scott...980
Malburg, Jan...941
Malik, Sharad...1222
Mandal, Ayan..1256
Marek-Sadowska, Malgorzata.......................1100
Markov, Igor L...747
McCants, Carl..133
Mei, Arie...301
Melpignano, Diego 1137
Meng, Jie...648
Middendorf, Lars...1203
Milder, Peter 1241
Miller, D. Michael 36
Min, Qinghao..106
Mirhoseini, Azalia 68
Mishra, Asit K...243
Mitra, Subhasish...561
Mitra, Tulika..412
Mojumder, Niladri 492
Moon, Seok-Hwan 630
Morad, Ronny 955
Morris, Daniel 486
Mudge, Trevor 406, 1143
Mukherjee, Tamal...176
Muller, Dirk...459
Murillo, Luis Gabriel.......................................121
Nahas, Joseph..476
Najm, Farid N...151
Nam, Gi-Joon...465
Narayanan, Vijaykrishnan.................................579
Nassif, Sani..1107
Nath, Siddhartha..392
Neuman, Sabrina M...259
Nieberg, Tim..459
Niemier, Michael..476
Novo, David...229
Oh, Hyunok..664
Ohlsen, Chris..1274
Ou, Hung-Chih..549
Paek, Seungwook 630
Palem, Krishna...924
Pan, David Z...317, 756

Panagopoulos, Georgios.................................1258
Panten, Christian 459
Park, Sang Phill..492
Park, Sangyoung 398, 522
Park, Yongjun..980
Pasricha, Sudeep 1274
Patel, Hiren D..697
Pathak, Mohit..741
Paver, Nigel...850
Pawlowski, Robert..974
Payer, Hannes..913
Pedram, Massoud 516
Pelosi, Gerardo 77
Peng, Zebo..197
Petrot, Frederic 366
Phelps, Andrew..176
Philippe, Maurine 994
Piguet, C...1049
Pileggi, Larry...486
Pileggi, Lawrence T...176
Pinckney, Nathaniel 1143
Pino, Robinson E...585
Pino, Yougok...83
Poku, Osei...567
Porod, Wolfgang 476
Potkonjak, Miodrag 68, 90, 133, 133
Prasad, Ankita..656
Purandare, Mitra 350
Püschel, Markus..1241
Qian, Haifeng...642
Qiu, Xiang...1100
Raabe, Andreas 188
Rabbah, Rodric 271
Radiom, Soheil...542
Raghunathan, Anand 492, 796, 826
Rajendiran, Aravindkumar................................697
Rajendran, Jeyavijayan 83
Ramanathan, Parmeswaran...............................709
Rana, Vincenzo...1043
Ranieri, Juri...636
Ransford, Benjamin 12
Ravindran, Binoy 437
Ravindran, Kaushik 656
Ray, Sandip...344
Reddy, Lakshmi...768
Rehman, Semeen 1288
Reinman, Glenn 843
Ren, Ling...1125
Richards, Brian...1212
Rinard, Martin..930
Rincon, Francisco..1043
Roa, Elkim...1254
Robins, Gabriel 504
Rose, Garrett S...498
Roth, Christoph 510

AUTHOR INDEX

Rotner, Jonathan 176
Roveda, Janet 529
Roy, Kaushik 492, 796, 1258
Roy, Sanghamitra 382, 1074
Roychowdhury, Jaijeet 301, 311
Ryzhenko, Nikolai 603
Sabne, Amit 796
Sadeghi, Ahmad-Reza 220
Sankar, Vjiay Karthik 476
Sartori, John 918, 1297
Sasanian, Zahra 36
Sato, Takashi 139
Satpathy, Sudhir 406
Savvides, Andreas 163
Scheffer, Louis K. 717
Scheuermann, Michael 157
Schulte, Christian 459
Schulte, Michael J. 56
Schwartz-Narbonne, Daniel 1222
Sciuto, Donatella 856
Sentieys, Olivier 48
Seo, Sangwon 980
Seok, Mingoo 968
Seshia, Sanjit A. 356, 1250
Sethi, Divjyot 1222
Seudie, Herve 220
Severson, Matt 1160
Sewell, Korey 1143
Shacham, Ofer 623
Shafique, Muhammad 866, 1288
Shao, Zili 214
Sharad, Mrigank 1258
Sharifi, Akbar 678
Sheibanyrad, Abbas 366
Shen, Chin-Fang 1088
Shin, Donghwa 516
Shin, Wongyu 630
Shojaei, Hamid 709
Shrivastava, Aviral 1280
Shriver, Emily 62
Shukla, Sunil 271
Sim, Jaehyeong 630
Sinangil, Mahmut 259
Sinangil, Yildiz 259
Sinanoglu, Ozgur 83
Sinha, Debjit 1061, 1067
Sinha, Saurabh 283
Sinha, Subarna 1163, 1169
Sinkar, Abhishek A. 56
Sironi, Filippo 856
Sloan, Joseph 918
Srikantaiah, Shekhar 678
Srivastava, Ankur 96
Stevenson, John 783
Stoffel, Dominik 334

Su, Yangfeng 295
Sudanthi, Chander 850
Suh, G. Edward 421
Sun, Jin 529
Sutaria, Ketul 139
Suto, Gyuszi 471
Suzanne, Lesecq 994
Sylvester, Dennis 406, 1037, 1143
Sze, Cliff 465, 768, 774
Taurino, Irene 6
Taylor, Michael B. 1131
Tehranipoor, Mohammad 703
Teich, Jurgen 205
Temam, Olivier 1260
Tovinakere, Vivek D. 48
Tran, Trung N 656
Tretter, Andreas 1262
Tripakis, Stavros 656
Tripunitara, Mahesh V. 697
Tsai, Tsung-Chan 376
Tsay, Ren-Song 127
Tuzzio, Nicholas 703
Ukhov, Ivan 197
Urdahl, Joakim 334
Vassiliev, Artem 623
Vasudevan, Shobha 808, 1018
Velamala, Jyothi Bhaskarr 139
Veneris, Andreas 947
Venkataramani, Swagath 796
Venkateswaran, Natesan 1067
Vetterli, Martin 636
Vincenzi, Alessandro 636
Vinco, Sara 115
Viswanathan, Natarajan 465, 768, 774
Visweswariah, Chandu 1061, 1067
Vo, Huy 1212
Vrudhula, Sarma 1280
Vygen, Jens 459
Wachs, Megan 623, 783
Walter, Fabio Leandro 866
Wang, Cheng-Yuan Michael 453
Wang, Chundong 235
Wang, Fa 176
Wang, Guoqiang 656
Wang, Hongfei 567
Wang, Huandong 876
Wang, Jue 253
Wang, Qing 289
Wang, Yanzhi 516, 522
Wang, Yi 214
Wang, Yu 1125, 1187
Ward, Samuel 756
Wedler, Markus 334
Wei, Gu-Yeon 277
Wei, Sheng 90

AUTHOR INDEX

Wei, Yaoguang 768, 774
Welp, Tobias....................................814
Wen, Wujie 1187
Wentzlaff, David 265
Wille, Robert 36
Willemin, J.....................................1049
Woh, Mark 980
Wong, Martin D. F...........................591
Wong, Ngai 289
Wong, Weng-Fai 235
Woo, Dong Hyuk 888
Wu, Hsin-I 127
Wu, Qing...498
Xiao, Yang 579
Xie, Fei 344
Xie, Qing 522
Xie, Yuan 243, 253, 1187
Xiong, Jinjun 1067
Xu, Cong 243
Xu, Qiang..555
Xue, Chun Jason 1000
Yakoushkin, Sergey 121
Yang, Chia-Lin 453
Yang, Fan..295, 1169
Yang, Hao-Yu 1012
Yang, Huazhong 1125
Yang, Jun..907
Yang, Qing 169
Yao, Shi-Chune 642
Ye, Fangming 1024
Yeric, Greg 283
Yoo, Sungjoo 888, 897
Yoon, Dongmin................................1037
York, Johnathan 790
Youseff, Lamia 265
Yu, Xiaochun..................................567
Yu, Yen-Ting...................................1163
Yuan, Feng......................................555
Yue, Siyu 516
Yuen, Kendrick 1160
Zang, Binyu 106
Zatt, Bruno 866
Zeng, Xuan 295, 1169
Zhang, Chenxi 1125
Zhang, Guangfei..............................876
Zhang, Hongbo 591
Zhang, Peng....................................1229
Zhang, Weihua 106
Zhang, Xiaorong 169
Zhang, Xuehui 703
Zhang, Yang 289
Zhang, Yanheng 597
Zhang, Yaojun.................................1187
Zhang, Youtao.................................907
Zhang, Yuanrui 834

Zhao, Bo 907
Zhao, Hui 834
Zhao, Mengying...............................1000
Zhao, Xin..157
Zhao, Xueqian 1119
Zheng, Jian 1000
Zhou, Huapeng 642
Zhou, Keyong..................................106
Zhou, Nancy Y................................465
Zhu, Jian-Gang................................486
Ziv, Avi 955
Zolotov, Vladimir 1061, 1067
Zou, Yi 1229
Zuluaga, Marcela 1241

9781450311991